Fire
Protection
Handbook

Contents

Contents

Contents

Contents

Fire
Protection
Handbook
Eighteenth Edition

Arthur E. Cote, P.E.
Editor-in-Chief

Jim L. Linville
Managing Editor

Meri-K. Appy
Robert P. Benedetti
Ron M. Coté
Martha H. Curtis
Casey C. Grant
John R. Hall, Jr.

Wayne D. Moore
Pamela A. Powell
Robert E. Solomon
Gary O. Tokle
Robert J. Vondrasek

Section Editors

National Fire Protection Association
Quincy, Massachusetts

Editor-in-Chief: Arthur E. Cote
Managing Editor: Jim L. Linville
Project Editor: Susan Merrifield
Indexing: Hagerty and Holloway
Copy Editors: Hilary Davis,
           Joan Fitzgerald,
           Angela Morrison,
           Deborah Liehs
Art Editor: Todd Bowman
Artwork: George Nichols,
           Todd Bowman,
           Gayle Croissant-Madden,
           Nancy J. Anderson,
           Charles Liversedge,
           José Diaz
Text Processing: Louise Grant,
           Kathy Barber,
           Muriel McArdle,
           Claire Ross,
           Nancy Maria,
           Beth Turner,
           Laurie Ruszcyk,
           Carole Cowing,
           Nancy Walker
Composition: Argosy
Cover Design: Greenwell Associates, Inc.
Production Coordinators: Debra Rose,
           Donald McGonagle,
           Ellen Glisker,
           Cathy Ray
Printed by: R.R. Donnelly & Sons

The following are other registered trademarks of the
National Fire Protection Association, Inc.

Fire Command
Fire Journal
NFPA Journal
Learn Not to Burn
Life Safety Code and *101*
National Electrical Code and *NEC*
National Fire Codes
NFPA and Design (logo)
Sparky and Sparky the Fire Dog

First Printing, February 1997

Previous Editions

First	1896
Second	1901
Third	1904
Fourth	1909
Fifth	1914
Sixth	1921
Seventh	1924
Eighth	1935
Ninth	1941
Tenth	1948
Eleventh	1954
Twelfth	1962
Thirteenth	1969
Fourteenth	1976
Fifteenth	1981
Sixteenth	1986
Seventeenth	1991
Eighteenth	1997

Library of Congress Card Number: 62-12655
NFPA Number FPH1897
ISBN: 0-87765-377-1
National Fire Protection Association
One Batterymarch Park, Quincy, MA 02269-9101

Dedication

To George D. Miller, president of NFPA, whose leadership has enabled NFPA to begin its second century of service in pursuit of a fire safe world.

Contents

Contents

Contents

Introduction

Since its first edition a century ago, the *Fire Protection Handbook* has endeavored to fulfill the needs of the fire protection community for a single-source handbook on the state of the art in fire protection and fire prevention practices.

It was originally known as the *Handbook of the Underwriter's Bureau of New England* and was first published in 1896, the same year that the National Fire Protection Association was founded. The original author, Everett U. Crosby, was manager of the Underwriter's Bureau of New England. He was also the first secretary of NFPA, serving from 1896 to 1903, and chairman of the NFPA executive committee from 1903 to 1907. His father, Umberto C. Crosby, was the first chairman of the NFPA executive committee, serving in 1896 and 1897, and the second president of NFPA, serving from 1897 to 1900. Henry A. Fiske joined Crosby as coeditor of the 2nd edition in 1901, and the *Handbook* later became known as the *Crosby-Fiske Handbook of Fire Protection*. H. Walter Forster joined the editorial team in 1918, and in 1935, Crosby, Fiske, and Forster donated all rights to their handbook to the NFPA. NFPA has published all successive editions since that 8th edition.

The prologue to the 100th anniversary edition, "What Time Has Crystallized into Good Practice," was prepared by Gordon P. (Mac) McKinnon and details the history of the *Handbook* from 1896 to 1996.

The *Fire Protection Handbook* has changed significantly in the last 100 years. While the body of knowledge in the field of fire protection has proliferated, the *Handbook* has kept pace, expanding from fewer than 200 pages in the first edition to over 2,400 pages in this 18th edition. As the most pressing concerns of fire protection have evolved, from citywide conflagrations in the late 1800s, through group fires and single building fires and more recently to room fires, the number of subjects covered by the *Handbook* has increased greatly. This is evidenced by the expansion of the text from the short, running commentary that made up the first edition to material organized into 190 chapters in this volume. Today, there are more chapters than there were pages in 1896!

This handbook is organized around the six major strategies that are the building blocks of a systems approach to fire safety through balanced fire protection:

—Prevention of ignition

—Design to slow early fire growth

—Detection and alarm

—Suppression

—Confinement of fire

—Evacuation of occupants

Production of the *Fire Protection Handbook* through 18 editions and 100 years has involved literally thousands of fire protection experts from within and outside NFPA. However, in addition to its founders, over the last half-century five individuals have been especially responsible for establishing the *Handbook* as the "bible" of fire protection practitioners.

—Robert S. Moulton, late NFPA technical secretary, who, during his 40 years of service, edited the 9th, 10th, and 11th editions;

Introduction

—George H. (Hitch) Tryon, late NFPA assistant vice president, who edited the 12th and 13th editions;

—Richard E. Stevens, retired NFPA vice president and chief engineer, who, during his 35 years with the Association, contributed to five editions of the *Handbook* and served as chief technical consultant for the 14th and 15th editions;

—Gordon P. (Mac) McKinnon, retired NFPA editor-in-chief, who was directly involved in the preparation of five editions of the *Handbook* over a quarter century. He served as editorial coordinator of the 12th edition, managing editor of the 13th edition, editor of the 14th and 15th editions, and consulting editor for the 16th edition.

—Jim L. Linville, managing editor of the 16th and 17th editions, and this the 18th and 100th anniversary edition.

I am especially proud to have had the privilege of acting as editor-in-chief for the 16th, 17th, and 18th editions.

In offering this edition of the *Fire Protection Handbook*, the editors solicit suggestions for improvements in the interest of making future editions increasingly useful to all concerned. Every effort has been made to ensure that the text is consistent with the best available information on current fire protection practices. However, the National Fire Protection Association, as a body, is not responsible for the contents, as there has been no opportunity for the membership to review the *Handbook* before its publication. If readers discover errors or omissions, the editors would appreciate those shortcomings being called to their attention.

Arthur E. Cote, P.E.
Editor-in-Chief

Preface

Without a doubt, supervising the editing of the oldest, the largest, and the most comprehensive reference book on fire protection in the world is one of the most challenging and complicated tasks in the field of technical book publishing. When compared to the 17th edition, the 18th edition has grown from 2,000 pages with 2,100 illustrations to more than 2,400 pages with more than 2,000 illustrations. The number of authors involved—all of whom are experts in their specialties—has increased from 204 to 232.

Each existing chapter has been thoroughly updated by the addition of new code and standards provisions, new techniques, and the latest fire experience statistics. But that's only part of the story.

The changes made in this edition are too numerous to list individually, but in general:

1. Every chapter has been thoroughly updated, and there are 29 totally new chapters.
2. The more than 2,000 line drawings were recreated from scratch in electronic format.
3. In order to reduce redundancy and further refine the *Handbook's* organizational structure, 38 chapters with similar and/or overlapping content were combined.
4. More than 60 "new" authors contributed to the *Handbook* for the first time.
5. The chapter Bibliographies were expanded and updated.

The result is a professional reference book with a grand total of more than 2,400 pages, all of which have been revised and updated.

Acknowledgments

It may be that only another editor can fully understand the complexity of a project like the *Fire Protection Handbook*, but the statistics are mind-boggling: 190 chapters by 232 authors; hundreds of reviewers; more than 2,000 illustrations to be rendered, captioned, cropped, and proofed; 4 or 5 million words to edit, keyboard, proof, typeset, and proof again; hundreds of tables and equations; 2,400+ pages to lay out, review, correct, and lay out again; a readership that ranges from volunteer fire fighters and fire protection engineers to university professors and research scientists; a breakneck schedule with a small but dauntless staff augmented by freelancers; and an end-product with more pages than most major magazines publish in an entire year. But most important of all is the haunting knowledge that this book is different because, to tens of thousands of fire protection professionals—professionals on whom hundreds of millions of people depend for life safety, and protection of property from fire—it's THE HANDBOOK. There are times when one simply has to strive for excellence, and editing the *Fire Protection Handbook* is one of those times.

Each edition of any book as massive and as important as the *Fire Protection Handbook* is an enormous undertaking requiring the cooperation of hundreds of people, and this 18th edition is no exception. In one way or another, virtually every NFPA staff member has been involved—many made generous personal sacrifices of their time and talents. One member of our team, however, deserves special mention. Dr. John R. Hall, Jr., not only served as section editor of Section 11, he authored or co-authored three of the chapters, and reviewed the vast majority of the statistical information.

Acknowledgments

Obviously, no single person can know everything about a subject as enormous and complex as fire protection. While every word in this centennial edition was reviewed by the Editor-in-Chief, Section Editors, and other appropriate NFPA staff members, selected chapters were also reviewed by qualified outside experts from all over the world. Therefore, the editors wish to acknowledge and thank all of those experts who contributed their time and energy to improving the *Handbook*. Special thanks are due to NFPA staffers Rita F. Fahy, Michael J. Karter, Jr., Carl E. Peterson, L. Charles "Chuck" Smeby, Jr., and Gregory Harrington who gave selflessly of their time; and to James Arvin, of the Aga Gas Company, who helped out with the Gases chapters; Vic Suski of the American Trucking Association, who reviewed the Motor Vehicle chapter; and Morgan Hurley, of the U.S. Coast Guard who reviewed the Marine chapter. Recognition is also due Leonard Belliveau, Jr., of Worcester Polytechnic Institute, who spent most of the summer of 1996 revising and updating most of the expanded bibliographies.

Our thanks are also extended to Dr. Richard N. Wright, Director, National Institute of Standards and Technology, Building and Fire Research Laboratory; Mr. Paul M. Fitzgerald, Chief Executive Officer, Factory Mutual System; Mr. Stanley J. Couvillon, Vice-President Engineering, Industrial Risk Insurers; and Mr. G. Thomas Castino, President and Chief Operating Office, Underwriters Laboratories Inc. Their willingness to allow their respective staffs to devote the time required to participate as both reviewers and authors of many of the chapters in this *Handbook* is gratefully acknowledged. Without their help the book would have suffered.

This editor would like to personally thank Susan Merrifield and Hilary Davis, who spent many sleepless nights to ensure the success of the 18th edition, Joan Fitzgerald, Angela Morrison, Deborah Liehs, Peggy Holloway, Todd Bowman, and George Nichols who were there when they were needed. Finally, thanks again to Louise Grant: This is the fourth time Louise has composed the *Fire Protection Handbook*—I don't know if that's a record, but I guarantee you that it's a lot of hard, dedicated work.

Over the six-year period between this and the previous edition, many conscientious readers have called errors and inconsistencies to our attention. We wish we had time and space to mention them all. Nothing is perfect, and so we continue to solicit your comments. We do try, we do care, and we will respond to each and every letter.

Jim L. Linville
Managing Editor

What time has crystallized into good practice

The *Fire Protection Handbook* from 1896 to 1996

The late Robert S. Moulton, for 40 years technical secretary for the National Fire Protection and for 20 years editor of the NFPA *Handbook of Fire Protection* before retiring in 1961, was not a man of great patience. He was, in fact, a feisty leader who did not suffer fools or inattention. His knowledge of fire protection was unparalleled, and he could be erudite on the complexities of the discipline. But giving detailed explanations of why things were as they were would sometimes raise his temperature, particularly if he spotted glazed eyes or befuddled looks. He would then bring the discussion (with Bob Moulton it was more a lecture than a conversation!) to a close with the pronouncement that the subject at hand was "what time had crystallized into good practice."

Time does have a way of sorting out what is important from the mundane—a truism to all sorts of endeavors including producing 18 editions of the *Handbook* over precisely 100 years. That is why George H. Tryon III, who succeeded Bob Moulton as NFPA technical secretary and *Handbook* editor, and who had learned his lessons well as one of Moulton's more patient acolytes, wrote in the preface of the 12th edition: "Like all previous editions, beginning in 1896, this *Handbook* aims to provide, in compact form, the essential information on fire prevention and fire protection that time has crystallized into good practice." And that is what all 18 editions of the *Handbook* have done.

The question then is: "What is it that makes a practice good?" Aside from divine revelation, it comes down to human experience. It includes the intellectual effort to sort out and evaluate what is learned through honest inquiry into the nature of things and experience gained in applying the lessons learned.

Fire protection as an art and technology is not really new. The Romans with their aqueducts and Corps of Vigilantes and the burghers of *Nieuw Amesterdam* (Manhattan Island) with their fire prevention and building code, had a pretty good idea of what they wanted to do to prevent fire and to control it when it got out of hand. They had their "good" practices, as rudimentary as they may have been.

But what is new (we're talking centuries now) is that fire protection and fire prevention have evolved from mostly parochial problem-solving to mature and universal disciplines that systematically address all facets of fire problems and present to the world solutions that are accepted as "good practice."

But good practices are almost worthless if no one knows about them or how they came to be. I'm sure Everett U. Crosby had no idea of pontificating on "good practices" in putting together the handbook he published in 1896, the first predecessor of today's

Prologue to the Centennial Edition

Fire Protection Handbook. He had a problem to solve. He needed something quickly that would summarize for inspectors employed by the insurance company members of the Underwriters' Bureau of New England (UBNE) the fire protection rules and regulations the companies were observing in their underwriting practices. Thus came the first edition of the *Handbook of the Underwriters' Bureau of New England*. Here "good practices" were being gathered together, albeit for a small parochial audience. Nevertheless, it was a beginning.

But to back up a bit. The last quarter of the 19th century was a time of post-Civil War technical ferment. New technologies were emerging at a phenomenal rate. Among them were refinements in methods of controlling fire and electricity and the hazards of electricity and its exciting applications. And, too, names began to emerge of men who would have a profound effect on the future of fire protection technology. There was John R. Freeman whose classic studies of the flow of water through pipes and orifices resulted in reference tables that were a mainstay of fire protection hydraulic calculations until the advent of computers. (His tables appeared in all editions of the *Handbook* until the 16th.) Freeman also espoused a systematic approach to implementing fire protection, including consistent sprinkler rules, that he practiced in his work with the Factory Mutual Insurance Companies. Then there was C. J. H. Woodbury, chief engineer of the Factory Mutual Companies, who as early as 1884 established test procedures to evaluate the response time of 15 different sprinklers, a first in reliability testing. Woodbury was also much concerned with the hazards of electricity as it became more frequently used at insured properties.

Another pioneer was William H. Merrill, a young Boston electrician who conceived the idea of examining and testing electrical equipment in a laboratory that later became Underwriters' Laboratories Inc. Another of Merrill's concerns was the evolution of too many separate and independent electrical codes, the electrical equivalent at that time to the plumber's nightmare of too many existing sprinkler installation standards. These were two gnawing problems that eventually had a lot to do with the *Handbook*.

It was more than coincidence that 1896 saw Crosby's first edition of the *Handbook* and the birth of the NFPA. Crosby and Freeman, along with Frederick Grinnell of the Providence Steam and Gas Pipe Company were among the handful of stock fire insurance company executives who met in Crosby's Boston office early in the previous year as a committee to look into the problem of lack of consistent sprinkler rules and the need for an organization to oversee them. There was agreement, and early the next year consistent sprinkler rules were developed. The committee met again on November 6, 1897 to finalize the governing organization–the National Fire Protection Association.

One of Crosby's biggest editorial problems was solved. He finally had consistent sprinkler rules for his forthcoming handbook. In all, 37 of the 183 pages of the first edition were devoted to the sprinkler rules and 49 pages to water supplies for sprinkler systems. Curiously, alarm valves for wet-pipe sprinkler systems were given a separate 4-page treatment.

As to addressing electrical hazards, the *Handbook* printed verbatim the "Rules and Requirements of the National Board of Fire Underwriters," one of the many competing proprietary codes of the time. It wasn't until the following year that electrical interests came together to promulgate a single agreed-upon electrical code–the *National Electrical Code* of 1897.

Prologue to the Centennial Edition

Other broad subjects covered in the first edition included building construction, fire doors, shutters, and exposure protection. They were essentially the rules promulgated by the Factory Improvement Committee of the New England Insurance Exchange. Text on 'Thermostats' (alarm systems) appears to have relied heavily on data sheets from manufacturers of the systems.

From this brief description of contents one could surmise that Crosby just cobbled together a text of rules and regulations fleshed out with contributions from manufacturers and let it go at that. Not so. Crosby took it on himself to insert here and there cogent comments on some of the rules to give inspectors a bit of "background" on the rules and some of the pitfalls to avoid in implementing them.

A few outside authors were also involved. John Freeman held forth on the Underwriters' steam pump, citing its advantages and warning inspectors to be wary of "ordinary commercial pumps" dressed up with Underwriter fittings. W. H. Stratton of the Factory Insurance Association, who later would become chairman of the NFPA executive committee (1897-1904), wrote on the organization of private fire departments and construction of hydrant and hose houses. C. E. Worthington described "fire-proof" oil and paint storage rooms. Beyond that, no authors are identified, although it is known that Henry A. Fiske, then an inspector with UBNE, contributed many of the sketches and drawings.

Fiske joined Crosby as co-editor for the 2nd edition in 1901. The book's audience was still mostly technical staff of the stock fire insurance companies, which, co-incidentally, were the original organization members of the NFPA.

Two unrelated events of the first decade of the 20th century must have had considerable influence on the direction the *Handbook* would take in the ensuing years. At the 4th annual NFPA meeting in 1900, Everett Crosby, by then NFPA secretary, wondered aloud that perhaps industrial plants might employ "competent" men to carry on regular fire inspections and not leave it all to insurance inspectors. Obviously he felt the need for the fire protection effort to go beyond the parochial interests of the stock fire insurance companies. It probably crossed his mind that if "competent" men were needed, the *Handbook* would have a role in training them.

Thus he wrote in the preface to the 3rd edition of what now was the *Handbook of Fire Protection for Improved Risks* that it was the intention of the authors to make the "hand-book a text for the novice as well as a valuable aid to the experienced fire protection engineer . . . to architects and property owners . . ." The *Handbook* was indeed enlarged and was in sync with what Crosby foresaw at the 1900 meeting.

The second was in 1903 when the first "Quarterly Bulletin" of the NFPA Committee on Special Hazards was published. Henry Fiske, by then manager of UBNE, was chairman of the committee and editor of the "Bulletin," a post he held through 17 issues. (In 1907 the "Bulletin" was replaced by the "NFPA Quarterly." Fiske continued as editor.) He was now wearing two editorial hats—that of the "NFPA Bulletin" and the *Handbook*. The NFPA publication focused on, among other things, current fire protection practices in various industries. It was no coincidence then, that the 3rd edition contained an extensive checklist of "chief hazards" to look for in a variety of industries.

The early editions continued to grow in scope of subject matter and number of pages. More of the text was instructive commentary by the editors and less verbatim recitation of insurance rules and regulations. By the 4th edition in 1909, now titled the *Crosby-Fiske Handbook of Fire Protection*, the editors allowed as how they needed help in expanding coverage. They enlisted Major John Stephen Sewell, who had been special representative of the federal government in investigations of the Baltimore (1904) and San Francisco (1906) conflagrations, to write extensively on protecting exposed steel and cast iron columns in "fireproof" buildings. "The term 'fireproof' is best of all," he wrote in describing his recommendations. He corrected himself in the 6th edition observing the "modern tendency" to use fire resistive " . . . for no material available to man is proof against damage by fire . . ."

Others joining the editors were Captain Greely S. Curtis, engineer for the Boston Fire Department, for text on the organization of public fire departments, the first mention of that subject in the handbook series; E.V. French, chairman of the NFPA Committee on Private Fire Supplies from Public Mains, who wrote on meters and "private fire pipes"; and William Morrill of Underwriters' Laboratories who described and illustrated (with colored plates no less!) the new labeling service offered by the laboratory.

If there was a landmark edition in the first quarter of the 20th century it was the 5th in 1914. The impetus for its importance was the 1911 Triangle Shirtwaist fire that took the lives of 146 garment workers in New York City. That tragic fire broadened the mission of fire protection to protecting lives and not just property. The NFPA quickly appointed the Committee on Safety to Life to study fire safety in schools, hotels, and other public occupancies. The chairman was H. Walter Forster, and under his aggressive leadership it produced in 1912 the report "Suggestions for Organization and Execution of Fire Drills."

Forster, using the 1912 report as his source, authored a chapter in the 5th edition focusing on safety to life. He made no bones about his intent. In the first sentence he wrote, "The conservation of life is far more important than that of property." Here was the sea change in the direction for all subsequent editions of the *Handbook*. Life safety would be paramount where prior to 1914 property conservation was the focus.

Not immodestly the editors in their preface to the 5th edition observed that the *Handbook* "is now regarded as a standard publication, and we feel a responsibility in maintaining and advancing its helpfulness, as it now goes to nearly all countries, whether or not English speaking." The book by now had 563 pages compared to 183 in the 1st edition.

Forster obviously was an important contributor to the *Handbook*. By the 6th edition he was invited to join Crosby and Fiske as an editor to share in the responsibilities of producing the book and the royalties. The *Handbook*, until 1935, was published and distributed through the resources of commercial publishers. It must have been a fairly profitable operation for both the editors and publishers as the relationships continued for more than a quarter century. Unfortunately, figures on press runs for the early editions are not available.

For the *Crosby-Fiske-Forster Handbook of Fire Protection*, as it was to be known through to the 10th edition in 1948, the triumvirate of editors could exult "that the increasing attention being given fire protection by individual plants and public fire departments is most encouraging and in the preparation of this edition the needs of these organizations have been in the minds of the authors." By now the *Handbook* had ex-

Prologue to the Centennial Edition

panded to 758 pages with 41 chapters in nine parts. The NFPA and its technical committees were increasingly the source of the good practices expounded upon in the text.

Forster, much involved with the NFPA Safety to Life Committee, with great confidence based his revision of the chapter on "Egress Facilities and Drill" in the sixth edition on the forthcoming Building Exits Code even though it had not been officially adopted by the Association. An important provision he reported was that the "unit width of stair measure in new buildings was to be 22 inches," a measure that has remained constant to this day.

The editors were also realizing that the *Handbook* wasn't the only kid on the block as a general source of fire protection information. The 6th edition, in a separate chapter, listed availability of other sources, a *Handbook* feature that has continued to this day. Among the publications mentioned were the "NFPA Quarterly," standards and pamphlets on specialized subjects; and *Field Practices*, an inspection manual first published by the NFPA in 1914. The last named was the forerunner of today's NFPA *Inspection Manual*.

Among books listed was Gorham Dana's *Automatic Sprinkler Protection*, published in 1914 with later supplements. It was the authoritative text on sprinklers at the time. Dana was for many years a member of the NFPA Sprinkler Committee and supplied the photos of sprinklers for the first edition edited by Crosby and Fiske.

The year 1935 was another benchmark for the *Handbook*. In the prior year Fiske and Forster stopped at the NFPA's Boston office with a generous gift—rights to the *Handbook* which had been published commercially since its inception. Henceforth the *Handbook* would become an NFPA publication. To ensure that *Handbook* text would be compatible with NFPA standards, the Association's board of directors appointed a special committee whose job it was to review the text. Bob Moulton was appointed editor.

Under Moulton's leadership extensive revisions and additions were undertaken for the 8th edition. In all, more than 100 fire protectionists were involved either as authors, text reviewers or technical sources. At the risk of omitting some important names, the following are mentioned for their major contributions to the 8th edition:

—Franklin Wentworth, NFPA's first managing director;

—Percy "Perk" Bugbee, then assistant managing director and later general manager—"Mr. Fire Protection" to generations of protectionists;

—Horatio Bond, then an NFPA field engineer and later chief engineer, who made significant contributions in the areas of public water supply and municipal fire protection.

—Warren Y. Kimball, then a youngster in the NFPA Fire Records Department and later director of NFPA's Fire Service Department.

—S. H. Ingberg, chief, Fire Resistance Department, National Bureau of Standards. His tests on the fire resistance of materials were landmark criteria in assessing fire behavior in buildings.

—W. D. Milne, assistant manager, Eastern Underwriters Bureau, who had a principal interest in industrial fire protection and was editor of *Industrial Fire Hazards* first published in 1927 and the predecessor to today's NFPA handbook of the same name.

—A. J. Steiner of Underwriters' Laboratories and a chairman of the NFPA Fire Tests Committee. He was instrumental in developing what is known as the Steiner Tunnel Test for assessing the hazard of interior finish materials as they are used in buildings.

—Norman J. Thompson, engineer, Associated Factory Mutual Insurance Companies. Thompson was later to be instrumental in tests that led to the standard, or spray sprinkler, in the early 50s that replaced "old-style" sprinklers with much improved distribution patterns. He was author of the NFPA book *Fire Behaviors and Sprinklers* published in 1964.

—And, of course, Fiske and Forster who continued their interest in the volume they had created. Crosby was deceased by this time.

An important chapter in the 8th edition was "Measurement of Fire Severity and Resistance." Here was reported in the *Handbook* for the first time the scientific basis for predicting fire behavior in buildings—the time-temperature curve. The curve is the standard scale of measuring fire severity that has formed the basis for almost all fire resistance testing. The curve was a long time in creation. Professor Ira H. Woolson of Columbia University, chairman of the NFPA Committee on Fire Resistive Construction from 1902 to 1912, conducted extensive tests that supplied data for the time and temperature relationship in developing fires. Ingberg at NBS later confirmed Woolson's findings in a series of room burnout tests. A conference of interested parties in 1918 led to the adoption of the curve in the same year.

While the time-temperature test procedure had been a reality for some years, it wasn't until the 8th edition that text was set aside, at Moulton's insistence, to explain in some detail the history of the curve and how it could be applied in practice. The "cookbook" role of the *Handbook* was becoming less pervasive to the more important task of educating users to the rationales for the burgeoning fire technology.

The 8th edition also marked the start of reporting accumulated data on fire losses. In a chapter, "Loss of Life by Fire," the NFPA Fire Record Department covered ages and sex of fire fatalities, causes of loss of life, and a listing of large loss of life fires. At the time loss of life annually from fire was estimated at approximately 10,000 annually compared with an estimate of 15,000 lives for 1920.

The 8th edition for the first time also covered extinguishing systems other than sprinklers. In a chapter, "Smothering Systems," there was text on foam, carbon dioxide, water spray, carbon tetrachloride and steam smothering system; and inert gas for fire and explosion prevention.

By the 11th edition in 1954, the book, now known as the NFPA *Handbook of Fire Protection (Crosby-Fiske-Forster)* had 1,560 pages in 73 chapters with more than 130 acknowledgments listed. Keeping the *Handbook* to a reasonable size was becoming a problem. The ever increasing body of technical data made the selection process a difficult one that in retrospect was not always successful. Giving up the idea that the book could be exhaustingly complete was hard to do.

Prologue to the Centennial Edition

A case in point. Water spray protection came into widespread use in the era of World War II. It behooved the handbook staff to cover the subject as completely as possible from fixed systems to portable nozzles. Consequently, the 11th edition's text included copious tables on the discharge patterns, operating pressure and capacities of every fixed and adjustable nozzle manufactured at the time. The tables, illustrations and text on water spray ran on for pages. There was a bit of overkill here. By the 12th edition text had been reduced to representative examples of nozzles available and brief descriptions of their operating principles—a decision the editors felt was adequate as an explanation of a technology that had become a good practice. That exercise was to be repeated time and again as the editors struggled to keep things to budget size without sacrificing breadth and accuracy in text. It was a tough act.

Fashioning the *Handbook* as a textbook as well as a general resource covering fire protection and prevention from the simple rules of good housekeeping to the sophistication of extinguishing systems became a goal of the editors in planning the 12th edition. George Tryon and Richard E. Stevens, then technical secretary and assistant technical secretary respectively on the NFPA staff, along with A.L. Brown retired chief engineer of the Factory Mutual Engineering Division, were instrumental in completely revising the outline for the *Handbook*, which would be entitled, *Fire Protection Handbook* as it is known today.

(The author of this Prologue started his handbook career in 1959 as a "gofer" for Hitch Tryon and Dick Stevens as they were cranking up for the 12th edition. He finally got his turn at the editor's chair for the 14th and 15th editions.)

It was an exhaustive review of content. The text was reorganized into sections and chapters that would make it more practical to keep the text current with technical improvements in the future. The same basic format would hold for the 13th edition.

But the explosion of fire related technology as the last quarter of the 20th century approached dictated some extensive changes in content and format and the actual size of the 14th edition in 1976. (It weighed in at a hefty 6 pounds packed into an 8-$\frac{1}{2}$ by 11-inch trim size and containing more than 1,400 pages.) New fire problems and solutions to them that in the previous decade were only beginning to make themselves known were now deserving of extensive attention. High-rise buildings and their hazards to life are a good example.

The previous 13 editions gave only passing mention to that type of structure and no direct reference to its threats to life. But a series of tragic high-rise fires worldwide in the years between the 13th and 14th editions prompted considerable research on the movement of smoke in tall structures and a more systematic approach to designing life safety into structures. Thus were included two new important chapters: "Smoke Movement in Buildings" and "Systems Concept of Firesafe Building Design," the latter focusing on a "decision tree" that was the brainchild of the NFPA Committee on Systems Concept for Fire Protection in Structures. It identified a system of fire safety for all buildings, not just high rises. For the first time in the handbook series a thrust of technology was introduced that would have tremendous impact on how fire protection and prevention would be approached in years to come.

Emphasis on life safety continued to expand in coverage. The 13th edition had a single chapter on exits; the 14th edition devoted a complete section of seven chapters to life safety hazards of occupancies paralleling the ever expanding concerns of NFPA's Safety to Life Committee. The positive dedication of text to life safety continued in all subsequent editions through to this 18th edition, which contains 37 chapters on life safety.

Analyses of data on loss of life and property from fire in earlier editions relied heavily on reports of fire incidents collected by the NFPA Fire Record Department. They were not always consistent in the information they contained. It was a Herculean task, and was directed for many years by the late Chester I. Babcock who later became editor of the NFPA "Fire Journal" and for years was a principal contributor to handbook text. Quantifying and analyzing the myriad bits of information taxed the department's resources. Even newspaper clippings got into the act.

All that has changed now. In the 70s NFPA and federal fire authorities, working jointly and independently, put together a National Fire Incident Reporting System based on a uniform coding for fire protection prepared by the NFPA Fire Reporting Committee. NFPA's growing analytical expertise continued with federal support and has produced a fire reporting and analysis system unequaled in the world.

The later editions of the *Handbook* reflect this growing sophistication in data collection and analysis and the role of computers in the collection process and analysis. Gone are the detailed tabulations on losses of life and injuries, losses by occupancies, building losses by cause and a host of other statistical categories (still available annually in the *NFPA Journal*). Replacing them were chapters on how the information is gathered, processed and analyzed. And, more importantly, there were chapters that give good guidance on what you can do with the information once you have it—a point not really addressed in any detail in earlier editions.

When Dick Stevens retired, Art Cote, as NFPA's chief engineer took over as editor-in-chief for the 16th, 17th, and 18th editions with Jim Linville serving as managing editor.

As system analysis became more prevalent, not only in data collection but in building and extinguishing systems design as well, the handbook editors once again took a look at the organization of contents and did major surgery. Dr. John R. Hall, Jr., head of NFPA's Fire Analysis Department, conceived the idea of reorganizing the text to underscore the systems approach to fire protection. He settled on six major strategies to systematically develop a logical path from prevention of fire through to fire control, and getting people out of harm's way. It took 183 chapters, more than 2,000 pages and the efforts of 203 leading fire protectionists to get the job done in the 17th edition. The same intensive effort went into preparation of this, the 18th edition.

So, there you have it—an almost microscopic view of the first 100 years of a continuing book. If there is any one document series that can summarize the sturdy growth of a noble endeavor it is each edition of the *Fire Protection Handbook* and its predecessors. This view is minuscule, but the intent is to show how the *Handbook* grew from a simple hip-pocket-sized book for limited "nuts and bolts" use to a prestigious volume with a worldwide audience (there have been Spanish and Chinese editions.)

For this author it has been a privilege to have spent most of his professional life on the handbook staff. In the course of more than a quarter century he had the honor of associating with many of the leaders in fire protection. Names have been mentioned in this Prologue, but they are looked upon as representative of the thousands (that's no exaggeration) who have made contributions to the book, directly as authors and reviewers or through membership on NFPA technical committees and the NFPA staff. It is the documents produced by the committees and processed by a dedicated staff that underlay the good practices found in the *Handbook*.

If it could all be summarized in one sentence it would be this: Good practices come from good people. That's what it's all about.

Gordon P. "Mac" McKinnon
West Springfield, NH
December 1996.

BASICS OF FIRE
AND FIRE SCIENCE

Section 1 of the *Fire Protection Handbook* contains the chapters you should read before you read anything else in the book. As the name says, it presents the basics—about the fire problem, fire protection, fire safety, fire science, and everything else that makes a systematic approach to fire possible.

Chapter 1 says, "This edition of the *Fire Protection Handbook* is more than an order of magnitude larger than the original edition." Chapter 2 says, "A little over one hundred years ago, structural steel was unknown, reinforced concrete had not been used in structural framing applications, and the first 'high-rise' building had just been built in the United States." It can be hard to realize just how much the fire problem in the United States has changed since the first Handbook appeared and how much our knowledge of fire has grown.

Section 1 provides statistical information on the long-term trends in fire and on short-term trends and patterns that help to show the relative status and importance of various strategies to reduce that fire problem. Section 1 also provides a number of basic organizing principles to help provide unity and understandability to the large, sprawling subject of fire safety and fire protection. These principles and themes are then repeated and used throughout the Handbook to help show the connections between subjects. Finally, Section 1 presents the most basic and general systems concepts and physical-science concepts related to fire protection engineering.

As Chapter 1 explains, Sections 2 through 8 of the Handbook have been defined in terms of major types of strategies to prevent or mitigate fire. The section sequence roughly corresponds to the order in which these strategies become relevant in the timeline of fire development. Section 9 continues this strategy-type approach but organizes its chapters by specific property classes. Each chapter addresses the different strategies of fire safety with respect to its property class. Section 10 then addresses the organizations that provide fire protection, set up by statistical information on the U.S. fire service in Chapter 1.

Chapter 1 says of the post-ignition strategies, which are collectively known as fire protection, "It is important to remember that fire protection requires the development of an integrated system of balanced protection which uses many different design features and systems to reinforce one another and to cover for one another in case of the failure of any one. . . . Success is measured by the extent of usage of effectively designed integrated fire protection systems."

The need to emphasize systems design and systems thinking is developed further in Chapters 2 and 3. Chapter 2 presents a systems approach to fire prevention and fire protection within the context of the critical *building design* phase of a property's life. Chapter 3 presents a brief overview of systems thinking and systems analysis in

general, then shows how these general concepts can be applied to fire protection through the *fire safety concepts tree*. The fire safety concepts tree is a very general model and is particularly appropriate for Section 1 because it can be used simply as a conceptual or logical model or as a framework for detailed quantitative analysis, such as hazard analysis. As Chapter 3 says, "In the broadest sense, systems analysis is simply the methodical study of an entity as a whole."

Chapter 4 provides an overview of building and fire codes and standards, from their origins in the time of Hammurabi of Babylon, nearly 4,000 years ago. Those of us in the fire protection field are so immersed in the details of specific codes and standards that it can be hard to remember how much of an innovation it was to have codes at all and how much the focus of codes and principal fire hazards have changed.

Chapters 5 through 9 address the physical-science knowledge that underpins modern fire protection engineering. An understanding of these basic definitions and laws is essential to fire protection in the 1990s, when it is increasingly important to know not only what the codes, standards, and other rules of good practice say but why they say it.

Chapter 5 addresses "basic definitions of some of the physical properties and chemical terms applicable to the chemistry and physics of fire" to the extent of "basic background reference material applicable to this and other sections of this handbook." Chapter 6 extends this discussion to the particular subject of explosions.

Chapter 7 provides an overview of the physics of fire growth in a compartment, the principal phenomenon underlying fire protection. Chapter 8 provides an overview of the mechanics of fire extinguishment, and Chapter 9 examines the rapidly growing field of environmental issues in fire protection.

Also look for these: The early chapters of Sections 4 and 8 also present basic scientific material but material specific to one type of strategy and hence to those sections only. Chapter 2 of Section 4 addresses the biochemistry of combustion-product effects on people, and Chapter 1 of Section 8 addresses the human behavior patterns that affect evacuation. The whole of Section 11 may be seen as an elaboration of the brief introduction to modeling provided by the discussion of the fire safety concepts tree, as Section 11 addresses other modeling approaches and associated data sources and measurement methods. Finally, those interested in the science of fire should read *The SFPE Handbook of Fire Protection Engineering*, also published by NFPA.

—Arthur E. Cote
December 1996

AMERICA'S FIRE PROBLEM AND FIRE PROTECTION

John R. Hall, Jr. and Arthur E. Cote

When the first *Fire Protection Handbook* was published in 1896, it contained no information on the size, trends, or patterns of the U.S. fire problem. Like most scientific and engineering handbooks, especially in years gone by, the first *Fire Protection Handbook* merely compiled a series of rules, tables, and formulas, which the knowledgeable fire safety professional of that time would find useful. One had to read between the lines to realize that the information provided in that handbook said nothing about hazards to life and limb and little about fire prevention as a strategy to reduce fire loss. (One also saw very little about hazards related to fires and explosions, such as electrical shock, hazardous materials, and environmental risks, topics which later editions have steadily expanded.)

In the century since that first edition appeared, the field of fire protection has grown tremendously. This edition of the *Fire Protection Handbook* is more than an order of magnitude larger than the original edition. With that growth of information has come an increasing need for context, perspective, and systematic approaches to organizing knowledge.

This chapter has two goals. One is to provide a perspective on the size, trends, and patterns of America's fire problem. Special attention has been paid to the incidents that led to changes in codes and standards—and the changes in fire experience that followed those code and standard changes. The other goal is to provide a basic structure for the elements of fire protection—a structure that can be tied to particular measures of the fire problem and that is carried forward in the first-level organization of this handbook. There are eleven major sections in this handbook. The first eight correspond to the eight principal types of strategies for fire protection. The ninth contains information on systems approaches to fire protection for particular property classes. The tenth addresses organizational considerations for fire protection organizations, such as fire departments. And the last major section addresses general sources of information and methods of analysis.

To understand the sequencing of the first six major sections, think of fire protection as a series of opportunities to intervene against a hostile fire, arrayed along a timeline of potential growth in fire severity. First, there are the opportunities to prevent the fire entirely. Second, there are opportunities to slow the initial growth and spread of fire or to reduce the severity of fires through the design, selection, and handling of materials and products. Third, there are

opportunities to detect fire early, permitting effective intervention before damage becomes too severe. Fourth, there are opportunities for automatic or manual suppression. Fifth, there are opportunities to confine the fire in space through compartmentation and other passive fire protection methods. Sixth, there are opportunities to move occupants to safe locations, taking advantage of the extra time provided by the earlier steps to move people from hazardous locations to safety, or to defend them where they are.

Before examining each of these fire protection strategies, it is useful to assess the size of the fire problem that faces us and the success achieved in reducing it over the past decades.

U.S. FIRE LOSS TRENDS

Table 1-1A and Figures 1-1A through 1-1D show that the number of fires has declined over the past decade and a half, although it has been stuck around 2.0 million for the last five years. Civilian fire

TABLE 1-1A. *The U.S. Fire Problem (1977–1994) Fires Reported to U.S. Fire Departments*

Year	Fires	Civilian Deaths	Civilian Injuries	Direct Property Damage*
1977	3,264,000	7,395	31,190	$4,709,000,000
1978	2,817,500	7,710	29,825	$4,498,000,000
1979	2,845,500	7,575	31,325	$5,750,000,000
1980	2,988,000	6,505	30,200	$6,254,000,000
1981	2,893,500	6,700	30,450	$6,676,000,000
1982	2,538,000	6,020	30,525	$6,432,000,000
1983	2,326,500	5,920	31,275	$6,598,000,000
1984	2,343,000	5,240	28,125	$6,707,000,000
1985	2,371,000	6,185	28,425	$7,324,000,000
1986	2,271,500	5,850	26,825	$6,700,000,000
1987	2,330,000	5,810	28,215	$7,159,000,000
1988	2,436,500	6,215	30,800	$8,352,000,000
1989	2,115,000	5,410	28,250	$8,655,000,000
1990	2,019,000	5,195	28,600	$7,818,000,000
1991	2,041,500	4,465	29,375	$9,467,000,000
1992	1,964,500	4,730	28,700	$8,295,000,000
1993	1,952,500	4,635	30,475	$8,546,000,000
1994	2,054,500	4,275	27,250	$8,151,000,000

* Direct property damage figures do not include indirect losses, such as business interruption, and have not been adjusted for inflation.
Source: NFPA National Fire Experience Survey.

Dr. John R. Hall, Jr., is NFPA's assistant vice president for Fire Analysis and Research. Arthur E. Cote, P.E., is NFPA's senior vice president, operations.

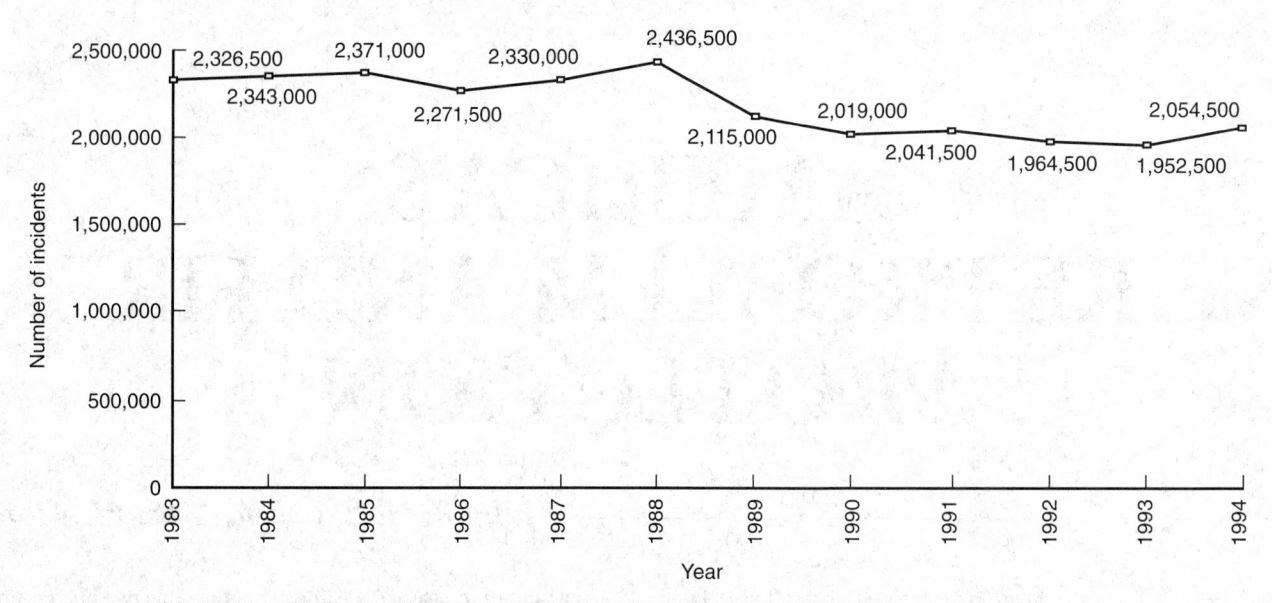

FIG. 1-1A. U.S. fire incident trend (1983-1994). (Source: NFPA National Fire Experience Survey)

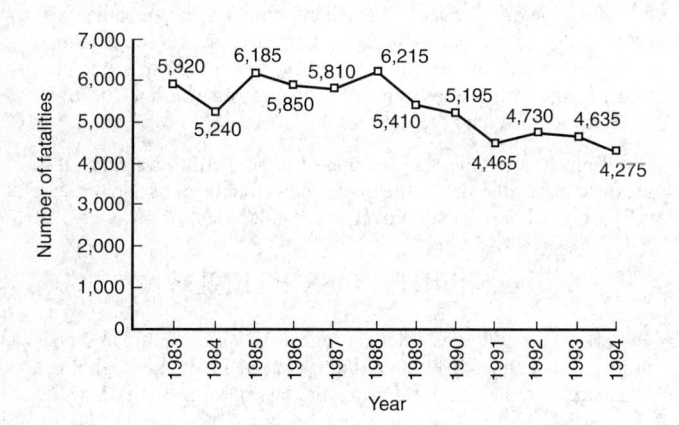

FIG. 1-1B. U.S. civilian fire death trend (1983-1994). (Source: NFPA National Fire Experience Survey)

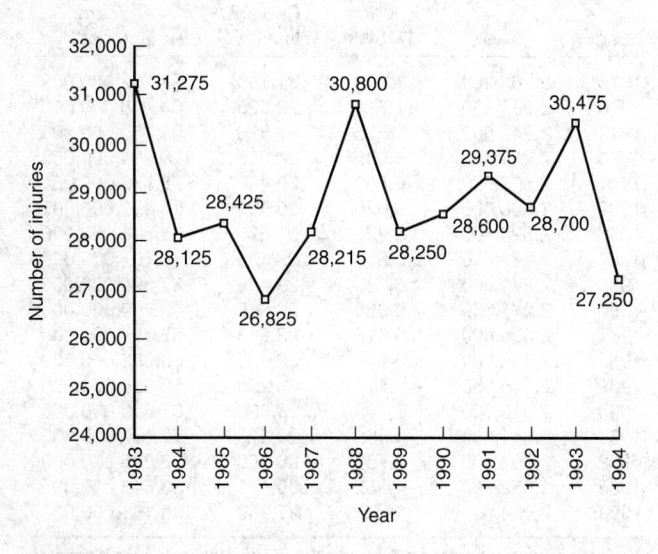

FIG. 1-1C. U.S. civilian fire injury trend (1983-1994). (Source: NFPA National Fire Experience Survey)

deaths were stuck around 6,000 during 1982 to 1988, but have declined by nearly a third since then. Civilian fire injuries have shown no consistent trend up or down, but the 1995 total was the lowest recorded since NFPA's current statistics began. Direct property damage has risen dramatically, but most of that is due to inflation, although there have been years when individual fires of historic size have driven the total, adjusted for inflation, to new heights.

Table 1-1B and Figure 1-1E provide a longer perspective on U.S. fire deaths by switching to trends in the death-certificate data base. (NFPA uses this data base only for long-term trend analysis, where there is no alternative data base with as much consistency in methodology. Fires involving post-crash vehicle fires or arson are often excluded or missed from the fire totals in the death-certificate data base.)

Fire deaths have fallen nearly two-thirds in the 76 years since their peak levels around World War I. Fire death *rates* have fallen by 85 percent. The decline in fire deaths was fairly irregular until the mid-1960s, in part because individual incidents with death tolls in the hundreds used to occur with some regularity. A few such incidents could significantly affect the total U.S. fire death toll in a

TABLE 1-1B. *History of U.S. Fire and Burn Fatalities*

Year	Fire Deaths	Fire Deaths per 100,000 Population
1913	8,900	9.1
1923	9,100	8.1
1933	6,800	5.4
1943	8,700	6.5
1953	6,600	4.2
1963	8,200	4.3
1973	6,500	3.1
1983	5,000	2.2
1993	4,000	1.6

Fire deaths are calculated on the basis of death certificates and are given to the nearest hundred. Fire deaths involving arson or post-collision vehicle fires may be omitted. Classification changes in 1968, 1958, and especially 1948 limit comparability of figures.

Source: National Safety Council, Accident Facts, National Safety Council, Chicago, 1994, pp. 31-33.

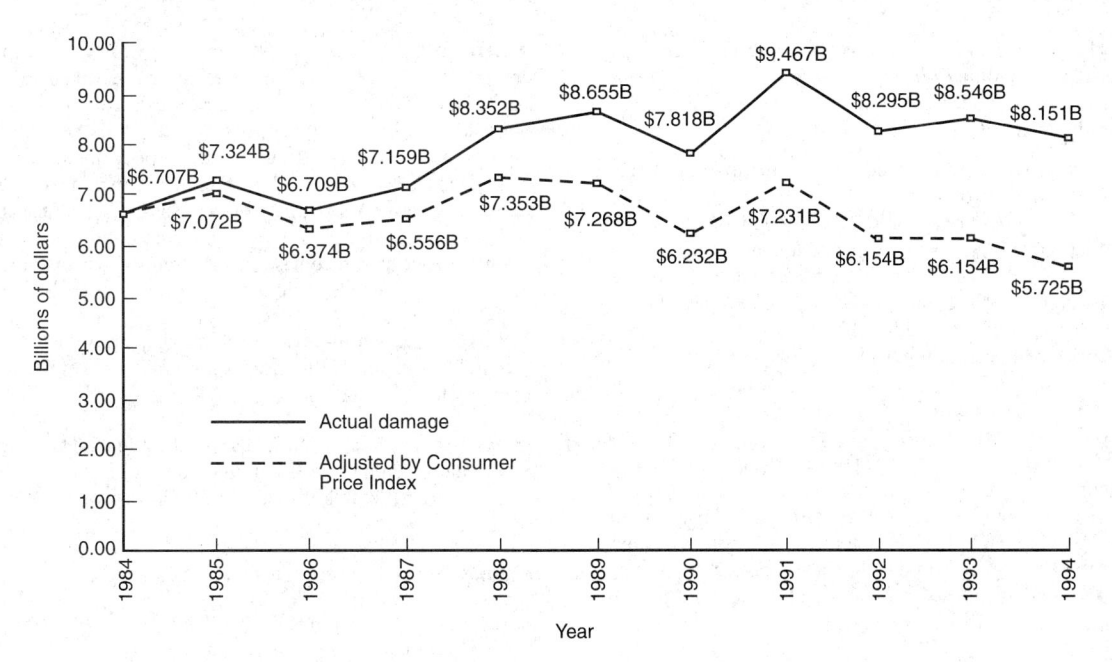

FIG. 1-1D.   *U.S. direct property damage trend (1984-1994). (Source: NFPA National Fire Experience Survey)*

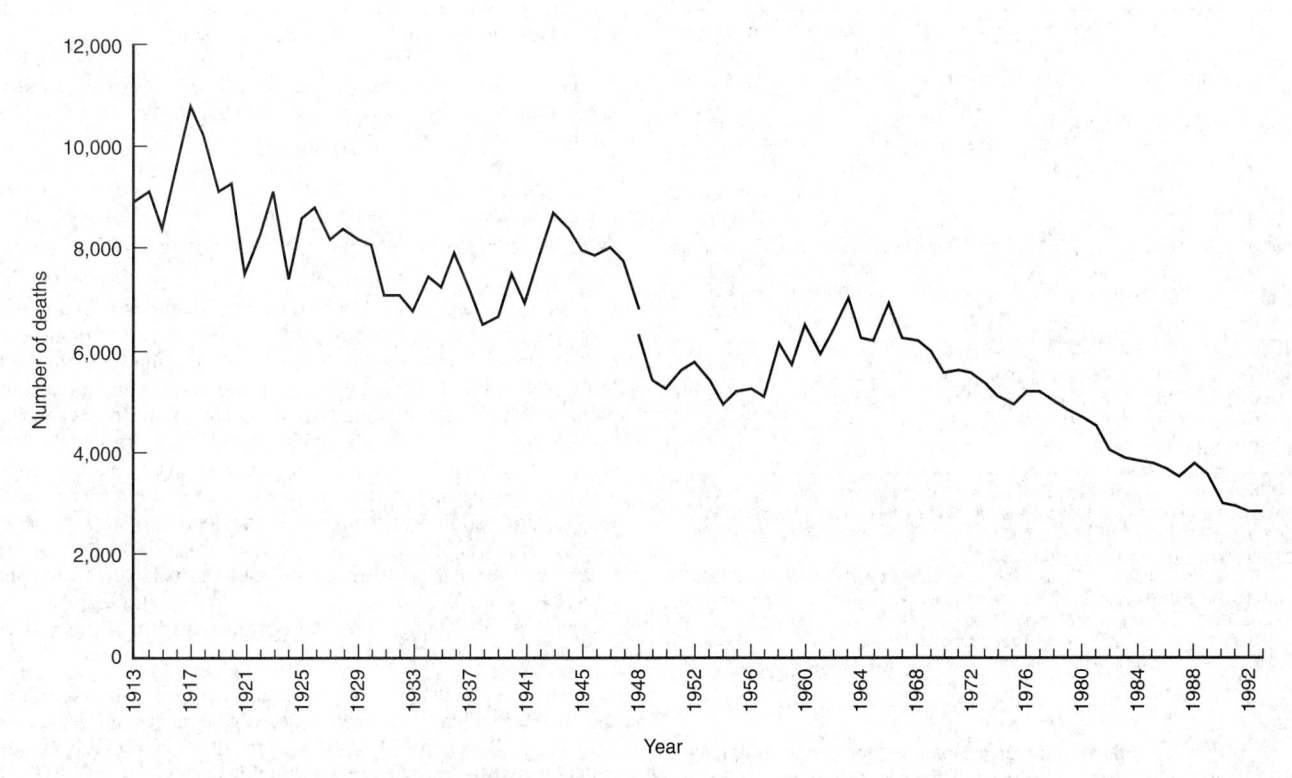

FIG. 1-1E.   *U.S. fire and burn deaths (1913-1993). (Source: National Center for Health Statistics, National Safety Council)*

given year. From 1955 to the present, there have been only two U.S. incidents with a death toll of 100 or more—the Beverly Hills Supper Club fire in 1977, and the Oklahoma City, OK, bombing of 1995. By contrast, the previous 55 years (1900 through 1954) produced 44 incidents involving death tolls of 100 or more. That's a decline from nearly one fire per year of that severity to one every two decades.

"America Burning,"[1] the report of the National Commission on Fire Prevention and Control (NCFPC), set a goal in 1973 of reducing America's fire death toll by one-half in a generation, which is usually

understood to mean 25 years. With that generation nearly gone, fire deaths measured by death certificates have fallen by nearly one-half. If current trends continue, this objective will be met.

Fire is second only to falls as a cause of accidental death in the home, and it is the number one cause for children and young adults.

Fire death rates in the U.S. and Canada were twice as high as those in western Europe and Japan 20 years ago. They were 50 percent higher 10 years ago and only about 25 percent higher in the early 1990s. The gap is closing because the U.S. and Canada have been improving fire safety faster than other countries. Throughout, analysis has shown that the U.S. does well in holding down the average severity of fire (i.e., deaths per fire) but fares poorly in holding down the fire incident rate, i.e., preventing fires.

Table 1-1C provides a longer perspective on property damage in fire, using estimates from the Insurance Services Office (ISO). (NFPA uses this data base only for long-term trend analysis, where there is no alternative data base with as much consistency in methodology. The data base has considerable uncertainty due to its adjustments for uninsured and unreported losses.) Fire losses have grown more than a hundred-fold in the past 110 years, but this reflects a 13- to 14-fold decline in the purchasing power of the dollar and a more than 40-fold increase in the nation's economy [measured by gross national product (GNP)] after adjustment for those changes in purchasing power. As a fraction of GNP, fire loss has declined by more than 80 percent since the turn of the century, a decline very similar to the decline in the fire death rate relative to population. America's fire losses, relative to the size of its economy, are quite comparable to the loss rates in western Europe and Japan. As a nation, the U.S. appears to do a better job protecting property than protecting its citizens, at least compared to the other fully industrialized democracies.

TABLE 1-1C. *History of U.S. Property Damage in Fire*

Year	Property Damage (in Millions)*	Property Damage (in Millions of 1990 Dollars)†	Property Damage per Capita)‡	Property Damage per Capita (in 1990 Dollars)	Property Damage as Percentage of GNP§
1880	$ 75	$1,006	$ 1.49	$20.01	0.67%
1890	$ 109	$1,577	$ 1.73	$25.01	0.83%
1900	$ 161	$2,515	$ 2.11	$33.05	0.86%
1910	$ 214	$2,986	$ 2.32	$32.32	0.61%
1920	$ 448	$2,917	$ 4.21	$27.40	0.49%
1930	$ 502	$3,923	$ 4.08	$31.87	0.56%
1940	$ 286	$2,665	$ 2.17	$20.20	0.29%
1950	$ 649	$3,517	$ 4.29	$23.25	0.23%
1960	$1,108	$4,880	$ 6.18	$27.21	0.22%
1970	$2,238	$7,519	$11.01	$36.98	0.23%
1980	$5,579	$8,849	$24.63	$39.06	0.20%
1990	$8,609	$8,609	$34.61	$34.61	0.15%

* Insurance Services Office Estimates, including adjustments for estimated uninsured and unreported losses, as reported by Insurance Information Institute, *The Fact Book: 1992 Property/Casualty Insurance Facts*, Insurance Information Institute, New York, 1992, p. 6.
† Based on Consumer Price Index, as calculated by U.S. Bureau of Labor Statistics, including estimates for historical times by U.S. Bureau of the Census.
‡ Based on U.S. residential population.
§ Based on gross national product estimates by U.S. Bureau of Economic Analysis, including estimates for historical times by U.S. Bureau of the Census.

To put the total in another perspective, the property damage caused by fire in 1993 was more than the cost of building 185 new homes of average cost (which was $126,500 in 1993) *each day* for one year.

FIRE PATTERNS BY PROPERTY CLASS

The patterns of fire by property class can be examined using a property classification procedure first used in NFPA's long-range planning exercises, as follows:

Homes and Garages: This includes dwellings, duplexes, mobile homes, apartments, townhouses, condominiums, and detached dwelling garages.

American Community: This includes all public assembly properties; all educational, health care, or correctional facilities; and stores and offices. It also includes any residential properties that do not qualify as homes, such as hotels and motels, rooming or boarding houses, dormitories, fraternity or sorority houses, and barracks. This revised definition of the American Community also incorporates the American Heritage property class, which consists of historically important places and objects. Its direct fire loss is quite small, but it deserves and receives special attention because losses in this property class may be quite literally irreplaceable.

Industrial Environment: This includes all industrial, manufacturing, defense, and utility properties, and all storage facilities other than dwelling garages.

Other Structures: This includes all buildings or structures that are vacant or under construction, demolition, or renovation. It also includes all structures that are not buildings, such as bridges and tunnels.

Mobile Environment: This includes all vehicles.

Outdoor: This includes all wildlands, forests, brush, grass, timber, and crops. It also includes outdoor trash, such as dumpsters or litter. Note that the statistics used here capture only fires reported to local fire departments; therefore national, state, and private forests and parks may be significantly underrepresented.

Other: This includes all other and unclassified properties. It is needed for completeness but is a factor only for fire incidents, not for human or property losses.

Figures 1-1F through 1-1I summarize the patterns of fire loss in these seven property categories in recent years. The Outdoor category accounts for the largest share of fire reported incidents.

Fire deaths, and to a lesser degree fire injuries, are overwhelmingly concentrated in Homes and Garages. The three other building and structure categories combined (i.e., American Community, Industrial Environment, and Other Structures) account for a fire death toll less than half the size of the death toll in the Mobile Environment, which is itself only one-sixth the death toll in Homes and Garages. The majority of these vehicle fire deaths are due to fires following survivable crashes in people's personal cars or trucks. The U.S. fire death problem therefore is not a matter of large death tolls in large buildings, although these incidents dominate the news. Most people who die in fire die in ones or twos in the very places where they feel safest—their own homes and vehicles. Home and garage fires also lead the dollar loss due to fire, although here the share for other buildings is much larger than it is for deaths and injuries.

Most of the visibility of America's fire problem and a large share of the public's fear and concern centers around very large fires. Here, the relative importance of the different property classes can be very different, as is shown in Tables 1-1D through 1-1G. Tables 1-1D and 1-1E list the ten deadliest U.S. fires and explosions of all time and from 1985 to 1994, respectively, while Tables 1-1F and 1-1G list the ten costliest U.S. fires and explosions of all time and from 1985 to 1994, respectively.

The ten deadliest fires and explosions of all time include five from the Mobile Environment, two that began in the Outdoor Environment, and one from the Industrial Environment. Only two, the Iroquois Theater and the Cocoanut Grove nightclub, involved the American Community, which people tend to think of as the area of greatest risk of multiple-death fires. In the 100 fires and explosions from 1900 to the present that killed at least 50 people, similar results, which may surprise the fire community, were found.[2] The

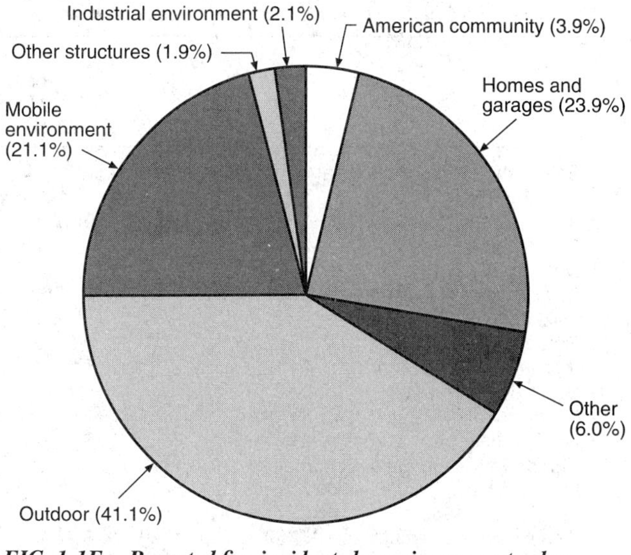

FIG. 1-1F. *Reported fire incidents by major property class (1989-1993). (Source: NFIRS and NFPA National Fire Experience Survey)*

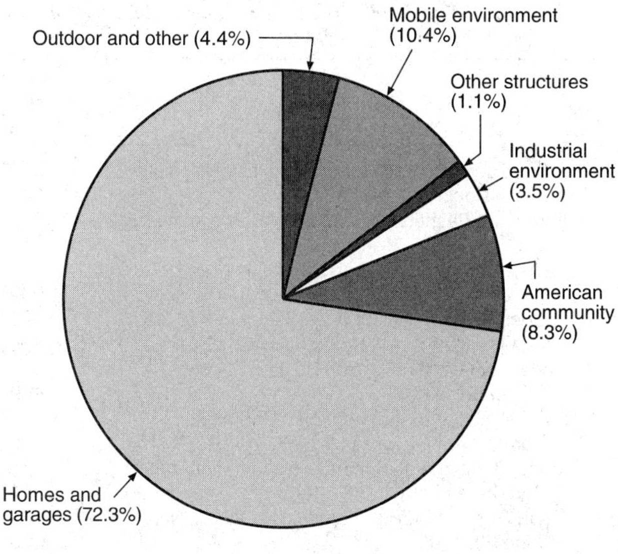

FIG. 1-1H. *Civilian fire injuries by property use (1989-1993). (Source: NFIRS and NFPA National Fire Experience Survey)*

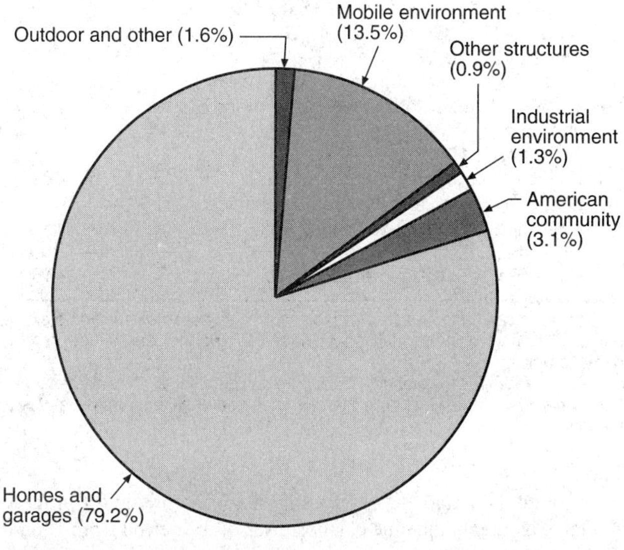

FIG. 1-1G. *Civilian fire deaths by property use (1989-1993). (Source: NFIRS and NFPA National Fire Experience Survey)*

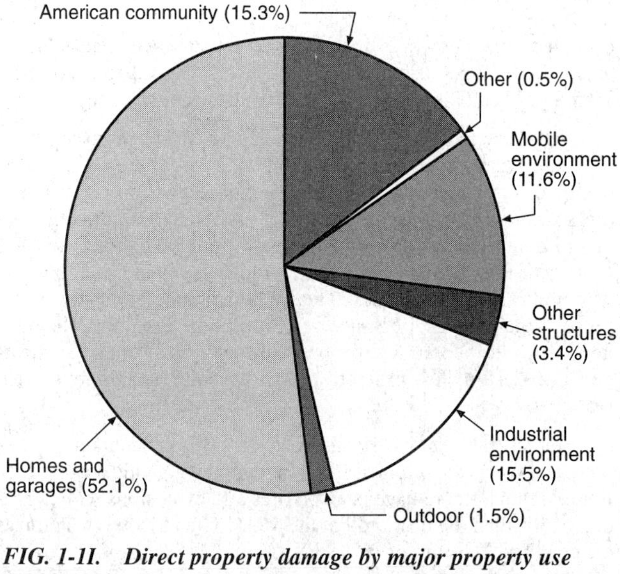

FIG. 1-1I. *Direct property damage by major property use (1989-1993). (Source: NFIRS and NFPA National Fire Experience Survey)*

leading category was the Industrial Environment with 68 incidents, 59 of them being mine fires or explosions. The American Community had 22, the Mobile Environment had seven, the Outdoor Environment had two, and the San Francisco earthquake and fire is hard to tie to any single environment. Even in the period from 1985 to 1994, the ten deadliest fires and explosions include five from the Mobile Environment, two from the Industrial Environment, and one that started in the Outdoor Environment, in contrast with two from the American Community.

From 1948 through 1995, only six building fires in the U.S. resulted in 80 or more deaths—the Winecoff Hotel fire (119), the Our Lady of the Angels School fire (95), the Beverly Hills Supper Club fire (165), the MGM Grand Hotel fire (85), the Happy Land Social Club fire (87), and the Oklahoma City bombing of an office building (168). That's six building fires in 48 years (1948-1995), compared to 14 building fires and explosions of comparable severity in the

previous 48 years (1900-1947). As previous comparisons in this chapter also have shown, any threshold of severity chosen demonstrates the tremendous progress made in the first half-century in reducing the rate of loss of life in fire, both in general and especially in major life-loss fires. Later in this chapter, there is a more extended discussion of the "century of accomplishment" in fire safety that has been the legacy of NFPA's first century.

Another statistic may help to further explain this pattern. During the period from 1989 to 1993 in structure fires outside of homes, 61 percent of all persons killed had been familiar with the building they died in for more than a year. More than 84 percent had been familiar with the building for more than a week. This suggests a fire fatality problem that affects employees (and possibly some long-term patients) far more than customers. (Long-term guests or tenants are not the answer because the percentage of fatal victims with long-term familiarity goes up if all residential properties are

TABLE 1-1D. *The Ten Deadliest U.S. Fires and Explosions in History*

	Number of Deaths
1. S.S. Sultana steamship boiler explosion and fire Mississippi River, April 27, 1865	1,547
2. Forest fire Peshtigo, WI, and environs, October 8, 1871	1,152
3. General Slocum excursion steamship fire New York, NY, June 15, 1904	1,030
4. Iroquois Theater Chicago, IL, December 30, 1903	602
5. Forest fire Northern MN, October 12, 1918	559
6. Cocoanut Grove night club Boston, MA, November 28, 1942	492
7. S.S. Grandcamp and Monsanto Chemical Company plant Texas City, TX, April 16, 1947	468
8. Monongah Mine coal mine explosion Monongah, WV, December 6, 1907	361
9. North German Lloyd steamship Hoboken, NJ, June 30, 1900	326
10. Explosion of two ammunition ships at depot Port Chicago, CA, July 18, 1944	322

Source: *1984 Fire Almanac* and National Safety Council's *Accident Facts.*

excluded.) In other words, the number one fear of Americans—a fatal fire in strange surroundings—has been reduced to a minority share of a shrinking problem, even if homes are excluded.

All these patterns underline the fact that life safety is a major issue everywhere—in homes and vehicles where most deaths occur, in the American Community where public concerns are highest, and in America's workplaces where the significant risk of fire death still exists. Conversely, Figure 1-1I shows that the Industrial Environment, which is typically considered the focus of potential property loss, actually contributes about the same property damage as the American Community. In short, the dangers of fire to people and property are everywhere, and so strategies of design, fire protection, and other programs must also reach everywhere.

Tables 1-1D and 1-1E also deliver a very encouraging message, and that is the progress made in reducing gigantic deadly fires. In this century, there have been 55 fires and explosions with higher death tolls than the death toll at the MGM Grand Hotel, which was the deadliest U.S. fire of the 1980s.

TABLE 1-1E. *The Ten Deadliest U.S. Fires and Explosions* (1985-1994)*

1. Happy Land social club New York, NY, March 25, 1990	87
2. Branch Davidian complex Waco, TX, April 19, 1993	47
3. Galaxy Airlines #203 Reno, NV, January 21, 1985	42
4. United Air Lines #232 Sioux City, IA, July 19, 1989	38
5. Continental #1713 Denver, CO, November 15, 1987	28
6. Church school bus Carrollton, KY, May 14, 1988	27
7. Oakland Hills (forest) fire storm Oakland, CA, October 20, 1991	26
8. Imperial Food Products chicken plant Hamlet, NC, September 3, 1991	25
9A. Phillips Polyolefin plant Pasadena, TX, October 23, 1989	23
9B. F-16 and C-130 collision Fort Bragg, NC, March 23, 1994	

* Excludes non-fire deaths in incidents with both.
Source: NFPA's Fire Incident Data Organization.

TABLE 1-1F. *The Ten Largest U.S. Fire Losses in History**

	Loss (in Millions of 1994 Dollars)
1. San Francisco earthquake and fire April 18, 1906 (Loss in year of occurrence—$350 million)	$5,748
2. Great Chicago Fire October 8, 1871 (Loss in year of occurrence—$168 million)	$2,069
3. Oakland (CA) Fire Storm October 20, 1991 (Loss in year of occurrence—$1,580 million)	$1,631
4. Great Boston Fire November 9, 1872 (Loss in year of occurrence—$75 million)	$924
5. Phillips Petroleum plastics manufacturing and storage plant, Pasadena, TX October 23, 1989 (Loss in year of occurrence—$750 million)	$897
6. Baltimore conflagration February 7, 1904 (Loss in year of occurrence—$50 million)	$821
7. Los Angeles civil disturbance April 29—May 1, 1992 (Loss in year of occurrence—$567 million)	$599
8. "Laguna Fire," Orange County, CA October 27, 1993 (Loss in year of occurrence—$541 million)	$541
9. U.S.S. Lafeyette (formerly S.S. Normandie ocean liner) New York, NY February 9, 1942 (Loss in year of occurrence—$53 million)	$482
10. S.S. Grandcamp and Monsanto Chemical Company plant, Texas City, TX April 16, 1987 (Loss in year of occurrence—$67 million)	$445

* The list is limited to fires for which some reliable dollar-loss estimate exists, is limited to fires occurring in or over the U.S., and includes direct property damage only.
Source: *1984 Fire Almanac*, NFPA Fire Incident Data Organization, and Consumer Price Index, including the U.S. Bureau of the Census's estimates of the index for historical times.

Tables 1-1F and 1-1G tell a slightly different tale for property damage. Note that the nine costliest fires of the period from 1985 to 1994 were also nine of the 20 costliest fires of all time. In fact, these nine incidents all occurred from 1987 on, a period of eight years, while only two incidents large enough to make the top 20 had occurred in the previous 20 years. Despite the dramatic progress in reducing fire loss relative to the size of the U.S. economy, recent years have seen a resurgence of individual fires with historic-size losses, particularly California forest fires, which account for four of the nine recent incidents.

Tables 1-1F and 1-1G also show how much the character of large costly fires has changed. Five of the ten costliest fires of all time were fires that engulfed huge parts of cities, while only one of the ten costliest fires of the period from 1985 to 1994—the Los Angeles civil disturbance fires—could be similarly described, and it was not a conflagration like the others. Except for the wildland/urban interface, the risk of conflagration has declined significantly from the pattern of a century ago.

"Conflagration," "group fire," and "city fire" (the last term used in Japan) lack fully standardized definitions. The definitions used to code death certificates use the term conflagration to mean any building fire. More often, conflagration is used colloquially to mean any

TABLE 1-1G. *The Ten Largest U.S. Fire Losses (1985-1994)**

	Loss (in Millions of 1994 Dollars)
1. Oakland (CA) Fire Storm October 20, 1991 (Loss in year of occurrence—$1,580 million)	$1,631
2. Phillips Petroleum plastics manufacturing and storage plant, Pasadena, TX October 23, 1989 (Loss in year of occurrence—$750 million)	$ 897
3. Los Angeles civil disturbance April 29—May 1, 1992 (Loss in year of occurrence—$567 million)	$ 599
4. "Laguna Fire," Orange County, CA October 27, 1993 (Loss in year of occurrence—$541 million)	$ 541
5. Shell Oil Company petroleum refinery Norco, LA, May 5, 1988 (Loss in year of occurrence—$330 million)	$ 413
6. One Meridian Plaza office building, Philadelphia, PA February 23, 1991 (Loss in year of occurrence—$325 million)	$ 354
7. Hoechst Celanese chemical manufacturing plant Pampa, TX November 14, 1987 (Loss in year of occurrence—$154 million)	$ 281
8. "Paint Fire" Goleta forest fire, Santa Barbara County, CA June 27, 1990 (Loss in year of occurrence—$237 million)	$ 269
9. "Cleveland Fire" forest fire, Placerville, CA October 1, 1992 (Loss in year of occurrence—$246 million)	$ 260
10. World Trade Center office building bombing, New York, NY February 26, 1993 (Loss in year of occurrence—$230 million)	$ 236

* Reported losses are taken from best sources known to NFPA and cover only direct property damage.

Source: NFPA Fire Incident Data Organization and Consumer Price Index.

large fire with significant flame showing outside. NFPA generally uses the term conflagration to describe a fire with major building-to-building flame spread over some distance. Significant building-to-building fire spread within a complex or among adjacent buildings is typically called a group fire.

Table 1-1H describes four different types of conflagrations, or group fires, and gives recent examples of each. The first type listed is also the most common conflagration scenario in recent years, which is the forest fire or brush fire that spreads to nearby buildings in what is called the urban/wildland interface. The classic conflagration scenario of the last century, involving narrow streets and closely packed buildings, is now comparatively rare. The 1973 Chelsea, MA, conflagration is the only major example in recent decades. The construction and density conditions for such fires are largely confined to the old manufacturing areas of some old cities, particularly in the Northeast. A more common problem, particularly in the Southwest, is the phenomenon of conflagrations driven by strong, hot, dry winds and untreated wood shingles. This problem has been around for as long as NFPA has been collecting information on major fires. Finally, the group fire scenario, in which fire extends to most or all of the buildings of the complex, also continues to occur. The conditions for these fires also are most common in the old manufacturing areas of old cities.

TABLE 1-1H. *Illustrative Recent Conflagrations and Group Fires*

1. Urban/Wildland Interface, Malibu, CA, 1987.
   A fire of incendiary origin with multiple points of origin was driven by Santa Ana winds gusting over 60 mph. The fire moved 12 mi. in 10 hrs, entering Malibu, where it damaged or destroyed 74 dwellings and mobile homes and 54,000 acres of property. Losses were estimated at nearly $9 million.
   Four of the 20 costliest U.S. fires and explosions of all time are recent California conflagrations of this type, all shown in Table 1-1G.
2. Congested, Combustible District, Chelsea, MA, 1973.
   The fire began in a yard area of a congested district of combustible buildings, including multi-family housing and storage properties. The heavy fuel load inside and outside the buildings, and the narrow separations between buildings created an ideal environment for rapid multi-building fire growth. The wind was strong, gusting up to 48 mph. The water supply was inadequate to support master streams, and some sprinklered properties had broken piping that depleted the water supply without fighting the fire. Eventually, hundreds of buildings over 17 blocks were destroyed. Damage was estimated at $4 million.
3. Untreated Wood Shingles, Anaheim, CA, 1982.
   A fire initially reported as a tree fire spread over several blocks. Untreated wood shingles provided a ready source of secondary ignitions and firebrands, which were transported by high winds, gusting up to 60 mph. These hot, dry Santa Ana winds spread damage to 53 structures, for total estimated damages of $50 million. More than any other fire in recent history, this illustrates the devastating potential hazard of untreated wood shingles.
   Other recent major fires involving untreated wood shingles as significant factors include a 1983 apartment complex fire in Dallas ($5.4 million damage), a 1985 brush fire in and near Los Angeles ($11.2 million damage), a 1985 office building fire in San Antonio ($4.2 million damage), and a 1988 home hotel fire in Lihue, HI ($5.5 million damage).
4. Group Fire—Commercial or Industrial Complex, Fall River, MA, 1987.
   A boiler flue pipe ignited wood components in a ceiling/floor space of a manufacturing plant. Multiple penetrations of the fire wall helped the fire spread into the main area of the plant, where it overwhelmed the plant's complete-coverage sprinkler system. Once the first building was fully involved, high winds (35 mph) helped spread fire to the other five buildings in the complex. Total damages were estimated at $50 million.
   Other recent group fires include a 1987 Lowell, MA, fire ($30 million damage), exacerbated by a shut-off sprinkler system and oil-soaked wood floors, and a $70 million fire that damaged or destroyed 20 buildings in Lynn, MA, in 1981.

Source: NFPA Fire Incident Data Organization.

FIRE PREVENTION

Every hostile fire requires an initial heat source, an initial fuel source, and something to bring them together. That something nearly always has a human component, usually an immediate act or omission that brings heat and fuel together or sometimes the delayed effects of an error in design or installation. Fire also requires oxygen and sustainable chemical chain reactions, both of which can be useful points of attack for some suppression systems. Nevertheless, the three components of heat, fuel, and human error are central to nearly all fires and can be used as a framework for thinking about fire prevention, without much fear of oversimplification.

It is important to remember that prevention can occur through successful action on the heat source, the fuel source, or the behavior that brings them together. No one line of attack is clearly preferable to the other, and success is most likely to come if all three components are treated as real options at all times.

Product redesign is the means by which a heat source or a fuel source is changed. Such a change can come about in several ways: (1) regulation, (2) design to a set of voluntary, consensus codes and standards, (3) a voluntary industry-wide program to change product designs, not driven by laws or codes, and (4) product redesign in response to consumer demand.

Changing behavior means education, which also may be done by targeting a variety of audiences. School children are a group that can be reached when they are ready and able to learn substantial quantities of information. NFPA's *Learn Not to Burn*®[3] curriculum is designed for this audience. Manufacturers' labels and instructions reach consumers when they may be most disposed to learn rules for the firesafe use of products. Product advertisements and public service announcements (PSAs), such as the NFPA *Learn Not to Burn*[3] PSAs, have the potential to reach large audiences. Brochures, educational kits, community meetings, and the like can reach small, motivated audiences that may act as change agents. And Fire Prevention Week, of which NFPA is the originator and a sponsor, provides heightened attention to fire safety each October.

"America Burning,"[1] the report of the U.S. National Commission on Fire Prevention and Control, stated in 1973:

Indifferent to fire as a national problem, Americans are similarly careless about fire as a personal threat. It takes the careless or unwise action of a human being, in most cases, to begin a destructive fire. In their home environments, Americans live their daily lives amid flammable materials, close to potential sources of ignition. Though Americans are aroused to issues of safety in consumer products, fire safety is not one of their prime concerns.

This is an understandable perception for people who work for fire safety full-time and give it the very highest priority. But more than two decades later, it can be seen that this view is at least overstated. Many people do tend to believe that fire "only happens to the other guy," and that view ironically is strongest in populations with above-average fire risk. But many people probably come to these views out of a lack of understanding of the risks of what they do and, even more, a lack of perspective on how they can achieve greater safety without unacceptably high costs or other sacrifices of things that also matter to them.

People can be educated and motivated if approached in the right way. Addressing people effectively may involve nothing more than emphasizing what to do rather than what not to do. This is why an important milestone of the late 1970s was the introduction into the nation's schoolrooms of fire safety education based upon sound behavioral principles and good learning techniques. NFPA's *Learn Not to Burn*[3] campaign, which includes both a curriculum and a public information program, is a good example of a thoughtfully prepared educational campaign with specific, far-reaching goals. The *Learn Not to Burn*[3] curriculum, in particular, seeks to teach school children key fire safety behaviors that can be retained for a lifetime.

Product design strategies need to be carefully examined and not treated as simple technological "quick fixes." Many design improvements can be undone by product users if the users have not also been educated to the importance of firesafe product usage. Some product design proposals have high costs relative to the problem they address, and some involve the loss of other significant product performance features, even non–fire safety features. And with any product design strategy, except one driven by consumer demand, there should be concern over the loss of freedom of choice.

Educational programs have concerns, too. NFPA's *Learn Not to Burn*[3] curriculum is in use in roughly 5 percent of U.S. schools. While many more schools have some fire safety education, there is a danger that brief exposure to fire safety slogans will be considered an adequate means of teaching firesafe behavior. This is not the case.

Also, in any fire prevention program, the highest-risk groups tend to be the hardest to reach. The poor may not be able to afford safer products. Small rural communities are too scattered to be reached efficiently by targeted means of communications. Preschool children are harder to reach with curriculum programs. Elderly people may resist changes in products and practices. This means that a program with less-than-national coverage is likely to miss a disproportionate number of those most in need. The challenge to fire prevention is to recognize this danger and design programs that will reach everyone. It can be done. For example, universal fire safety education, particularly in the schools but also in neighborhoods, seems to be a crucial factor in the still lower fire death rates in western Europe and Japan.

Leading Causes of Fire

As noted, hostile fire requires an initial heat source, an initial fuel source, and usually a behavioral error. Therefore, causes can be defined and ranked in terms of any of these three components, as is done in Table 1-1I for structure fires. The cause categories shown are those that accounted for the largest number of civilian deaths per year in structure fires from 1989 through 1993. Many of these categories overlap. For example, "abandoning or discarding something"

TABLE 1-1I. *Major Causes of U.S. Structure Fires Reported to Fire Departments (1989-1993 Annual Average)*

Cause	Civilian Deaths	Civilian Injuries	Direct Property Damage (in Millions)	Fires
Defined by Heat Source				
Smoking material (i.e., lighted tobacco product)	1,100	3,280	$ 417	42,800
Heating equipment	570	2,570	$ 868	102,300
Match	380	2,130	$ 672	51,400
Electrical distribution equipment	360	1,900	$ 923	60,000
Cooking equipment	330	5,620	$ 496	122,400
Defined by First Ignited Item				
Upholstered furniture	770	2,060	$ 271	18,800
Mattress or bedding	670	3,730	$ 373	37,700
Cotton or rayon fabric or finished goods	630	14,770	$ 947	130,600
Sawn wood	500	7,550	$3,657	252,300
Manmade fiber fabric or finished goods	500	9,920	$ 802	82,200
Structural member or framing	310	1,070	$1,012	53,300
Combustible or flammable liquid	310	11,720	$1,851	95,700
Defined by Behavior				
Incendiary or suspicious causes	780	3,210	$1,954	112,700
Abandoning or discarding something	720	2,580	$ 371	42,200
Child playing with fire	410	2,690	$ 259	28,200

Source: NFIRS and NFPA Survey.

usually means smoking materials, and most match fires involve incendiary or suspicious causes or a child playing.

Homes and Garages: Table 1-1J provides a ranking of major fire causes for Homes and Garages.

Smoking materials are the number one cause of civilian fire deaths, accounting for nearly one-fourth of deaths, and most begin with ignition of *upholstered furniture, mattresses,* or *bedding.* The ignitability of these items by lighted tobacco products has been declining for some time, due to a federal law restricting mattress construction and a voluntary program by the upholstered furniture industry.

Incendiary or suspicious causes (i.e., arson and suspected arson) are the number one cause of property damage for home and garage fires, accounting for one of every five dollars lost. Nearly half of all the people arrested for arson are juveniles.

Cooking equipment is the leading cause of home fire and home fire injuries, and is involved in the majority of unreported home fires. Unattended cooking is the principal behavioral factor.

Heating equipment is the second leading cause of home fire incidents. Most involve portable or space heaters; however, the current size of the heating fire problem is much smaller than the peak it reached shortly after the rapid growth in usage of portable and other space heaters in the 1970s and early 1980s.

Child fire play, typically involving matches or lighters, accounts for only one of every 10 fire deaths, but it is the leading cause of preschooler fire deaths, accounting for more than one in three.

Electrical distribution system equipment accounts for a much smaller share of the home fire problem than many people realize, ranking no higher than third among the twelve major cause categories for any measure of loss. The widespread use and enforcement of NFPA 70, *National Electrical Code®,* is probably the primary reason why electrical systems are not among the leading causes of fire. Moreover, even a fire cause such as this, which seems so totally an equipment problem, usually involves human error. One benchmark study found that 61 percent of home electrical fires involved code violations, particularly the general workmanship provisions.[4] Exposed elements, such as cords, are even more subject to abuse by occupants.

American Community: Arson and smoking dominate the ignition scenarios of fatal fires in the American Community, as Table 1-1K shows. Together, they accounted for more than half of all fire deaths in these property classes from 1989 to 1993. Incendiary and suspicious fires (i.e., arson and suspected arson) are the leading cause for all four measures of fire loss specified in the table. Electrical distribution system components are also significant contributors to fire problems in these properties. Cooking fires are a major problem in all properties where food or drink is served. They account for two out of every five fires in eating and drinking establishments, but even there, incendiary and suspicious fires account for the largest share of property loss.

Industrial Environment: The many types of specialized equipment used in industry collectively accounted for the largest share of 1989 through 1993 structure fires and associated losses in the Industrial Environment, as Table 1-1L indicates. Incendiary and suspicious fires ranked second in fires and direct property damage. It is important to note, however, that these properties also had significant problems with open-flame fires (e.g., torches), electrical distribution systems, and heating equipment. Exclusive concentration on the more exotic hazards of any one industrial setting would be a mistake. Effective prevention requires a broad view encompassing all potential ignition scenarios.

Other Properties: Post-collision fires and arson or suspected arson are the leading causes of fires and associated losses in vehicles, particularly for deaths, where they collectively account for more than half the total. Electrical system fires also are significant, as are overheated objects, such as tires, brakes, or exhaust pipes.

Arson and suspected arson also is far and away the leading cause of outdoor fires in general and wildland fires in particular. Debris burning, careless handling of smoking materials, lightning, and children playing are other leading causes of wildland fires.

FIRE PROTECTION

Recognizing that prevention will never be 100 percent successful, it is necessary to plan and design so as to mitigate damages when fire occurs. The various strategies to do this constitute what is usually called fire protection. It is important to remember that fire protection

TABLE 1-1J. *Major Fire Causes for Homes and Garages (1989-1993 Annual Average)**

Major Cause	Fires		Civilian Deaths		Civilian Injuries		Direct Property Damage (in Millions)	
Cooking equipment	101,800	(21.1%)	320	(8.2%)	5,060	(24.1%)	$ 382.9	(8.6%)
Heating equipment	86,200	(17.9%)	540	(13.9%)	2,220	(10.5%)	$ 599.0	(13.6%)
Incendiary or suspicious causes	63,300	(13.1%)	680	(17.6%)	2,350	(11.2%)	$ 868.7	(20.1%)
Other equipment	45,500	(9.4%)	240	(6.3%)	1,730	(8.2%)	$ 406.6	(9.4%)
Electrical distribution system	40,400	(8.4%)	370	(9.5%)	1,520	(7.2%)	$ 564.8	(12.8%)
Appliance, tool, or air conditioning	31,400	(6.5%)	110	(2.8%)	1,090	(5.2%)	$ 233.8	(5.3%)
Smoking materials	27,100	(5.6%)	900	(23.4%)	2,580	(12.3%)	$ 259.2	(5.9%)
Child playing	23,700	(4.9%)	390	(10.1%)	2,550	(12.1%)	$ 232.0	(5.3%)
Open flame	23,500	(4.9%)	130	(3.4%)	800	(3.8%)	$ 206.7	(4.8%)
Exposure (to other hostile fire)	20,200	(4.2%)	40	(1.1%)	190	(0.9%)	$ 384.2	(8.8%)
Other heat source	10,400	(2.2%)	130	(3.3%)	780	(3.7%)	$ 104.8	(2.4%)
Natural causes	9,000	(1.9%)	20	(0.4%)	170	(0.8%)	$ 136.0	(3.2%)
Total	482,800	(100.0%)	3,870	(100.0%)	21,020	(100.0%)	$4,462.3	(100.0%)

*This ranking uses a hierarchical sorting procedure developed by analysts at the U.S. Fire Administration, which focuses on heat sources and behavioral causes. Results will be affected by the sorting rules used and the sequence in which unknown-cause fires are proportionally allocated to known causes and results by the property-use class are aggregated. Fires are expressed to the nearest hundred, and deaths and injuries to the nearest ten.
Source: NFIRS and NFPA National Fire Experience Survey.

TABLE 1-1K. *Major Fire Causes in the American Community (1989-1993)**

Major Cause	Fires		Civilian Deaths		Civilian Injuries		Direct Property Damage (in Millions)	
	Average Loss per Year							
Incendiary or suspicious causes	19,800	(24.9%)	49	(32.4%)	650	(27.1%)	$ 619.2	(47.4%)
Cooking equipment	10,800	(13.7%)	7	(4.9%)	310	(12.9%)	$ 59.0	(4.5%)
Electrical distribution system	10,600	(13.3%)	8	(5.5%)	250	(10.3%)	$ 164.5	(12.6%)
Other equipment	7,600	(9.5%)	11	(7.1%)	220	(8.9%)	$ 107.0	(8.2%)
Appliance, tool, or air conditioning	7,500	(9.5%)	5	(3.3%)	170	(7.0%)	$ 53.3	(4.1%)
Smoking materials	6,400	(8.1%)	34	(22.9%)	320	(13.1%)	$ 44.2	(3.4%)
Heating equipment	5,800	(7.3%)	9	(6.0%)	150	(6.1%)	$ 80.6	(6.2%)
Open flame	4,600	(5.9%)	9	(6.2%)	190	(7.9%)	$ 55.5	(4.2%)
Exposure (to other hostile fire)	2,600	(3.2%)	1	(0.5%)	10	(0.4%)	$ 66.3	(5.1%)
Natural causes	1,400	(1.8%)	1	(0.5%)	40	(1.6%)	$ 36.0	(2.8%)
Other heat source	1,200	(1.5%)	9	(5.9%)	70	(2.8%)	$ 15.5	(1.2%)
Child playing	900	(1.1%)	7	(4.8%)	50	(2.0%)	$ 5.8	(0.4%)
Total	79,200	(100.0%)	150	(100.0%)	2,430	(100.0%)	$1,307.0	(100.0%)

* This ranking uses a hierarchical sorting procedure developed by analysts at the U.S. Fire Administration, which focuses on heat sources and behavioral causes. Results will be affected by the sorting rules used and the sequence in which unknown-cause fires are proportionally allocated to known causes and results by property-use class are aggregated. Fires are expressed to the nearest hundred, and deaths and injuries to the nearest ten.
Source: NFIRS and NFPA National Fire Experience Survey.

TABLE 1-1L. *Major Fire Causes in the Industrial Environment (1989-1993)**

Major Cause	Fires		Civilian Deaths		Civilian Injuries		Direct Property Damage (in Millions)	
	Average Loss per Year							
Other equipment	9,400	(22.2%)	18	(29.2%)	380	(37.1%)	$ 568.7	(42.8%)
Incendiary or suspicious causes	7,300	(17.1%)	9	(14.6%)	70	(6.5%)	$ 222.0	(16.7%)
Open flame	6,400	(15.1%)	6	(9.6%)	130	(12.5%)	$ 91.3	(6.9%)
Electrical distribution system	4,300	(10.0%)	4	(5.8%)	100	(9.5%)	$ 134.9	(10.2%)
Natural causes	3,300	(7.7%)	13	(20.4%)	100	(9.6%)	$ 77.1	(5.8%)
Exposure (to other hostile fire)	3,100	(7.3%)	0	(0.4%)	10	(1.5%)	$ 36.4	(2.7%)
Heating equipment	2,900	(6.9%)	2	(2.9%)	80	(7.4%)	$ 105.2	(7.9%)
Appliance, tool, or air conditioning	1,500	(3.5%)	3	(4.6%)	50	(4.6%)	$ 21.4	(1.6%)
Child playing	1,200	(2.9%)	0	(0.4%)	20	(1.8%)	$ 6.9	(0.5%)
Other heat source	1,200	(2.8%)	3	(4.6%)	30	(2.8%)	$ 22.6	(1.7%)
Smoking materials	1,100	(2.6%)	4	(6.7%)	30	(2.9%)	$ 31.4	(2.4%)
Cooking equipment	800	(1.9%)	1	(0.8%)	40	(3.8%)	$ 10.5	(0.8%)
Total	42,500	(100.0%)	63	(100.0%)	1,040	(100.0%)	$1,328.4	(100.0%)

*This ranking uses a hierarchical sorting procedure developed by analysts at the U.S. Fire Administration, which focuses on heat sources and behavioral causes. Results will be affected by the sorting rules used and the sequence in which unknown-cause fires are proportionally allocated to known causes and results by property-use class are aggregated. Fires are expressed to the nearest hundred, and deaths and injuries to the nearest ten.
Source: NFIRS and NFPA National Fire Experience Survey.

requires the development of an integrated system of balanced protection that uses many different design features and systems to reinforce one another and to cover for one another in case of the failure of any one. Defense in depth and engineered redundancy are concepts that also are relevant here. The process of achieving that integration, balance, and redundancy to attain fire safety objectives is the essence of fire protection engineering, including codes and standards.

This means that success is not measured by the extent of use of any one technology or system or code. Success is measured by the extent of usage of effectively designed, integrated fire protection systems. No one system should be considered disposable, and no one system should be considered a panacea.

Once fire has started, the first opportunity to reduce its impact comes in the design of burnable items, i.e., the choice of materials and products and their environments. Both the growth of the fire from small to large and its spread along vertical or horizontal surfaces may be slowed through such design.

Active fire protection systems provide the next opportunity. Automatic detection systems will tend to activate first, followed by automatic sprinklers or other automatic suppression systems, although this will vary depending on the design of the detection and suppression systems.

Passive fire protection provides the final opportunity to stop the fire and smoke but also plays an essential role in providing automatic systems with a manageable fire to act on. Passive protection is designed to confine fire and smoke in zones, a concept called compartmentation. Special attention is given to protection of the building's structural integrity and the spaces through which occupants will move to safety.

Occupant evacuation depends on effective detection and a system to alert occupants, along with a total fire safety design that will defend the occupants where they are or provide protected routes to safe refuges, inside or outside of the building. Evacuation also depends on the knowledge of the occupants.

MATERIALS, PRODUCTS, AND ENVIRONMENTS

If prevention fails, the first opportunity to reduce fire damage comes in the design of materials, products (including assemblies), and environments so as to slow down the growth and spread of fire. Tables 1-1M through 1-1O show that most structure fires never grow beyond the first involved room, but most deaths and property damage occur in the one-fourth of fires that do. This pattern holds for homes and for structures other than homes. Clearly, then, a fire that can be slowed in its growth so that it can be discovered and controlled before fire leaves the first room is likely to result in far less damage to people and property. (Injuries are something of an exception, as they are almost as likely to occur in small fires as in large ones.)

TABLE 1-1M. *Sizes of Structure Fires (1989-1993)*

Extent of Flame Damage	Fires	Civilian Deaths	Civilian Injuries	Property Damage
Confined to room	71.4%	20.6%	58.5%	19.4%
Beyond room, confined to floor	4.6%	12.1%	10.8%	8.6%
Beyond floor	24.0%	67.2%	30.7%	72.0%
Total	100.0%	100.0%	100.0%	100.0%

Source: NFIRS and NFPA National Fire Experience Survey.

TABLE 1-1N. *Sizes of Home Fires (1989-1993)*

Extent of Flame Damage	Fires	Civilian Deaths	Civilian Injuries	Property Damage
Confined to room	73.6%	20.0%	56.4%	20.3%
Beyond room, confined to floor	5.2%	12.4%	11.8%	11.1%
Beyond floor	21.2%	67.5%	31.9%	68.5%
Total	100.0%	100.0%	100.0%	100.0%

Source: NFIRS and NFPA National Fire Experience Survey.

TABLE 1-1O. *Sizes of Fire in Structures Other Than Homes (1989-1993)*

Extent of Flame Damage	Fires	Civilian Deaths	Civilian Injuries	Property Damage
Confined to room	65.5%	28.5%	69.6%	18.0%
Beyond room, confined to floor	3.0%	8.0%	5.8%	4.9%
Beyond floor	31.5%	63.5%	24.6%	77.1%
Total	100.0%	100.0%	100.0%	100.0%

Source: NFIRS and NFPA National Fire Experience Survey.

The following are some of the fire protection approaches that may be taken under this overall strategy:

1. *Restrict materials used in contents and furnishings*
   - Reduce heat release rate
   - Reduce smoke generation rate
   - Prevent unusual toxic hazard relative to quantity of smoke generated
2. *Add fire retardant to materials*
   - Slow growth of heat release rate

3. *Use fire-resistive barriers*
   - Slow spread of fire to large secondary items
4. *Restrict total fuel load*
   - Limit contents based on total fuel potential
5. *Restrict linings of rooms to prevent rapid flame spread*
   - Restrict wall coverings
   - Restrict ceiling coverings
   - Restrict floor coverings
6. *Restrict materials in concealed spaces*
   - Restrict concealed combustibles
   - Restrict concealed space linings
7. *Require safe handling of large quantities of potential fuel*

With regard to the last point, large quantities of combustible or flammable liquids or gases are the most obvious examples of materials that could make a small fire huge very quickly.

It is important to handle and store these materials safely, following procedures such as those in NFPA 54, *National Fuel Gas Code*; NFPA 58, *Standard for the Storage and Handling of Liquefied Petroleum Gases*; and NFPA 30, *Flammable and Combustible Liquids Code*. Unusually hazardous environments, such as oxygen-enriched atmospheres, also receive special attention.

DETECTION AND ALARM

The impact of automatic detection and alarm systems has been particularly great in the Homes and Garages environment. From less than 5 percent of homes having smoke detectors in 1972, the United States has risen to a position where 93 percent of homes had at least one smoke detector in 1994.[5] (See Figure 1-1J.) Detectors cut the risk of dying in a home fire nearly in half, and they do this despite the fact that one-fifth of detectors—and one-third of detectors in homes that have fires—are non-operational (primarily because of dead or missing batteries) and many homes with detectors do not have all the detectors they need for code-compliant, every-level protection. Nevertheless, thousands of lives have been saved through the widespread use of home smoke detectors.

FIG. 1-1J. *Growth in detector coverage (1970-1994). (Sources for homes with detectors: 1977, 1980, and 1982 estimates from sample surveys by the U.S. Fire Administration; 1983-1994 estimates from "The Prevention Index," a Louis Harris survey for Prevention Magazine)*

This success has been achieved in several stages. The initial surge in detector usage was driven primarily by manufacturer advertising of an attractive, affordable product and secondarily by public service announcements promoting home smoke detectors. Since then, state and local laws have come along to help complete the process of equipping America's homes with smoke detectors, which is important, because more than 40 percent of reported home fires occur in the tiny fraction of homes without smoke detectors. In 1977, most states had no home smoke detector requirements of any kind. By 1983, most states had a home smoke detector law, but most states still did not cover existing single-family homes, by far the largest segment of the population. By 1988, most states had laws that extended to all homes, new or existing, but most still did not mandate code-compliant, every-level coverage. The trend in state laws is clear—more complete and more thorough laws—but the process is still far from complete.

As noted, non-operational detectors typically are that way because of dead or missing batteries. Frustration over nuisance alarms, which typically outnumber real fire alarms by an order of magnitude, may be a factor in battery removal. Sensitivity drift, which is a gradual shift in the range of fire effects that will activate the detector, is an example of the problems that can affect older detectors. The consensus of expert opinion now is that detectors more than ten years old should be replaced, and this affects tens of millions of detectors now in use.

The success of smoke detectors in the home—and the importance of completing the job of providing complete coverage by operational detectors in all homes—should not obscure the fact that automatic detection and alarm systems are an important part of an integrated system of fire protection in *any* property class. Major life-loss fires frequently cite problems with detection or alarm as significant contributing factors. Among the frequently cited problems in major fires with large loss of life are: (1) the absence of needed systems, (2) misapplication, i.e., using equipment or systems that were not appropriate for the property, (3) lack of maintenance, and (4) improper response by occupants to notification of the fire.

In too many property classes, the majority of fires occur in facilities with no detectors. This is true, for example, for two-thirds of fires in public assembly properties, for two-thirds of store or office properties, and for two-thirds of industrial properties.

SUPPRESSION

Automatic sprinklers are highly effective elements of fire protection systems design for buildings. When sprinklers are present, the chances of dying in a fire and property loss per fire are cut by one- to two-thirds, compared to fires reported to fire departments where sprinklers are not present.[6] Furthermore, this simple comparison understates the potential value of sprinklers, because it lumps together all sprinklers, regardless of type, coverage, or operational status. If unreported fires could be included and complete, well-maintained, properly installed and designed systems could be isolated, sprinkler effectiveness would be seen as even more impressive.

When sprinklers do not produce satisfactory results, the reasons usually involve one or more of the following: (1) partial, antiquated, poorly maintained, or inappropriate systems; (2) explosions or flash fires that overpower the system before it can react; or (3) fires very close to people who can be killed before a system can react.

Except for health care facilities, hotels and motels (especially high rise), department stores, high-rise office buildings, and industrial sites, sprinkler usage is rare in properties with large potential for life loss. Sprinkler usage is growing in these properties, but most

fires still occur in properties without sprinklers. Most reported fires in storage properties—even in general-purpose warehouses—also occur in properties where sprinklers were reported as absent. There is considerable potential for expanded use of sprinklers to reduce the loss of life and property to fire.

In recent years, particular attention has gone to quick-response residential sprinklers for the home. The evidence of their potential for substantial reductions in loss of life and property is clear, but usage is still quite limited. Several communities have adopted requirements for these new sprinklers in new housing developments, and their experience provides the principal real-world proof to date of the tremendous fire protection value such systems deliver.

The suppression strategy for fire protection involves much more than home sprinklers or even sprinklers in general. Portable fire extinguishers are in use in roughly half of all homes and in many other properties, as well. Specialized, non-water-based suppression systems, using alternative suppression agents or other design characteristics adapted to the special needs of a particular property class, also exist. In the second half of the 1980s, the world became aware of and concerned about the potentially adverse environmental impacts of suppression systems using halogenated agents (the so-called halon/ozone problem). This development has led to research on alternative fire protection technologies and new agents for critical electronic facilities. It also serves to underline the importance of considering major non-fire factors in any fire protection design.

CONFINING FIRES

The design of building features to effectively contain fires and their effects is one of the most technically complex aspects of fire protection and is certainly the most difficult to evaluate statistically. In all the other phases—prevention, detection, suppression, evacuation—systems thinking also is essential, but it is at least possible to focus attention on one set of products or systems or occupants. But in fire confinement, systems thinking is both essential and unavoidable, because one must deal with every facet of building design and operation. One must consider not only the direct effects of wall assemblies, ceiling/floor assemblies, door assemblies, and the like on fire confinement but also their effectiveness in creating the design conditions assumed by every other part of the overall fire protection system.

The effectiveness of detectors and sprinklers can be demonstrated to some degree by comparisons of fire experience in buildings that have these systems to buildings that do not. Unfortunately, however, every building has walls, floors, ceilings, and doors, and the various types of assemblies have proved too diverse to be routinely documented in the national fire data bases. The best analysis possible with existing statistics shows that, for many property classes, the proportion of fires confined to the room of origin rises as the type of construction becomes more fire-protective, from unprotected wood frame to fire resistive.

Problems with building in effective fire confinement tend to be more subtle, too. For detectors and sprinklers, the leading problem is turning off or disabling the equipment—a clear-cut, yes/no change that is easy to document. For fire confinement features, the problems tend to be more a matter of degree, i.e., barriers partially breached for utility networks, some but not all doors blocked open, etc.

Setting the context for this important strategy therefore must be approached in other ways. One is to point out the strong correlation between fire damage, i.e., to life and property, and fire spread, as was done in Tables 1-1M through 1-1O. Table 1-1P provides more evidence on this point, as it shows that the majority of fatal fire victims, including roughly 40 percent of the victims of fatal fires in structures other than homes, are located away from the room of fire origin. More than 25 percent of the victims are on another floor from the point of fire origin. What these figures indicate is that there are still many fatal victims of fire who might be saved by strategies that block or delay the passage of fire and smoke between rooms and floors. That's where fire protection design for fire confinement comes in.

TABLE 1-1P. *Locations of Fatal Victims of Structure Fires (1989-1993)*

Victim Location	All Structures	Homes	Structures Other Than Homes
Intimate with ignition	17.4%	16.7%	26.4%
Not intimate but in same room	24.1%	23.7%	28.8%
Not in room, but on same floor	30.2%	31.3%	16.4%
Not on same floor, but in building	26.2%	26.4%	23.5%
Outside building of origin	1.0%	0.8%	2.7%
Unclassified and other known	1.1%	1.0%	2.2%
Total	100.0%	100.0%	100.0%

Source: NFIRS and NFPA National Fire Experience Survey.

Some of the options available for fire confinement are the following:

1. *Use construction barriers to block fire spread between zones*
   - Wall assemblies
   - Ceiling/floor assemblies
   - Barriers between occupied and concealed spaces
   - Firestopping in concealed spaces
   - Exterior barriers to vertical spread between floors
2. *Design doors and windows to block fire spread between zones*
   - Fire door assemblies
   - Window restrictions to block spread between buildings
3. *Separate buildings enough to prevent fire spread between buildings*
4. *Regulate design and operations of ducts to permit shutoff of air movement in fires*
5. *Regulate design and operation of systems for heating, venting, and air conditioning to prevent their serving as mechanisms to transfer smoke and gases into uncontaminated, occupied areas.*

Barriers around concealed spaces are of importance for two reasons. First, fires in concealed spaces cannot hurt people until smoke or flames break out into occupied spaces. Second, fires in occupied spaces may spread faster if they break into concealed spaces, because many interior barriers do not extend into concealed spaces. For example, fire may spread vertically through concealed wall spaces if they are not firestopped at each floor or horizontally through concealed floor/ceiling spaces if the walls do not extend past the suspended ceilings that many buildings have below their main ceilings.

Exterior flame spread can occur in several ways, all of which must be addressed. Flying brands or heat from other fires can ignite exterior materials. Flames from one floor can "lap" up the exterior walls to ignite combustible materials on other floors by radiation from flame fronts.

Large losses of life or property are virtually unknown in buildings that comply with the fire protection requirements of modern codes. This is clear evidence of their effectiveness. Nevertheless, large losses continue to occur because so many buildings are not in compliance with modern codes. A major reason is the frequent exclusion of existing buildings from many requirements by "grandfathering" or a lack of retroactive application of codes. Even where good codes are on the books, enforcement may be hurt by a scarcity of resources in overburdened building and fire departments. And some communities do not routinely update their codes, which can create areas of vulnerability.

Sustained, consistent attention to fire safety is key to the long-term success of an integrated system design for fire protection. Furthermore, universal application is essential to make a substantial impact on the fire problem.

Balanced fire protection really works, but there are no simple, one-time "quick fixes." The installed system must be maintained every day. If this is done, however, such systems will virtually eliminate large losses of life or property. In fact, very large fires usually involve multiple major failures of fire protection design, but sometimes one or two deficiencies can be enough to produce disaster in a building with many excellent features. Remember that fire protection succeeds or fails as a system and is therefore only as sound as its weakest link.

Universality means, in simple terms, that any successful program must go where the fires are, including poor neighborhoods, rural communities, big-city neighborhoods, preschoolers, the elderly, and existing buildings. These are among the hardest groups to reach and the most frequently overlooked in the design of programs. But the fight for American fire safety will largely be won or lost in these areas.

EVACUATION OF OCCUPANTS

Most fire protection strategies are designed to slow or divert the movement of smoke and fire, not stop it, so key questions are whether, where, and how to move the occupants. The occupant evacuation strategy, which includes defending in place and safe interior refuges as possibilities, involves both building design principles and behavioral/educational elements. The building design principles include the following:

1. *Two ways out from any location*
2. *Adequate numbers, sizes, and spacing of exits*
3. *Adequate capacity of all escape route parts*
4. *Protection for escape paths*
   - Mark exit paths clearly and light them
   - Restrict fuel loads and finishes in exit paths
   - Enclose stairways
   - Use construction barriers to keep fire out
   - Use smoke control methods to protect atmosphere in exit paths
5. *Escape to outside or to protected places, or adequate defense of places where occupants should remain*
6. *Avoidance of makeshift security systems*

Protection of the exit paths begins with construction barriers to separate hallways from other rooms where fire may begin. Stairways need to be enclosed to separate them from all other spaces.

Exits from any room or area must be adequate for the number of people those exits will serve. This dictates minimum numbers and sizes of exits. Multiple exits from a room must be sufficiently separate to provide two distinct paths from the room to the outside so that escape will still be possible if one escape route is blocked.

The paths themselves must not involve long distances, which dictates maximum distances from any occupied room to a stairway access or to the access to some other protected area. Restrictions on the fuel loads and finishes in areas that serve as parts of exit paths also are needed to prevent rapid spread of fire into and through these paths.

Once the building is designed for safe evacuation, the occupants have to be educated to the principles of escape behavior:

1. *Know whether to escape and where to go (e.g., stay in place, go to safe refuge, go outside)*
2. *Know two ways out*
3. *Get out fast*
4. *Practice escape*
5. *Check paths for safety before proceeding*
   • *Feel the door*
6. *Crawl low under smoke*

It is worth emphasizing the need for practice. Learning the rules of safe escape as slogans is not enough. (That is why it is a matter of concern that the majority of American households have not developed and rehearsed an escape plan.) If fire occurs, one doesn't have time to make the mistakes that typically occur during learning. To use an analogy, suppose one has read about driving a car and even passed the state's written test. One doesn't want the first time behind the wheel to be a drive to the hospital emergency room. The lack of learned confidence in the key behavior could cost time when it can be least afforded.

Most people have not seen a real hostile fire. They have no idea how fast fire can grow or how bad it can get. They are not familiar with the phenomenon of flashover. Therefore, they spend time they can't afford confirming that there is a fire or gathering up valuables. They also tend to try to leave by their customary route, even in the face of serious fire hazard. Rehearsing safe behaviors is the only way to be sure that they will be used when necessary.

Special mention should be made of the problems posed by unusually vulnerable groups. Small children will need help to escape and are likely, on their own, to think hiding makes them safe. People who are elderly, handicapped, or impaired by substance abuse are likely to have physical or mental limitations that may present a need for help in escaping. Patients in health care facilities and inmates of confinement facilities present the largest problem, as they are large groups that cannot evacuate on their own. Specific plans need to be developed and rehearsed to address these situations.

SYSTEMS APPROACHES FOR PROPERTY CLASSES

In applying the general principles of fire protection, it usually becomes clear that particular property classes have special considerations that must be understood for an effective systems approach. Some of these special considerations are briefly summarized here.

Homes and Garages

Fire death rates in the biggest cities are roughly 50 percent higher than in the small cities, but fire death rates in rural communities of less than 5,000 population are more than 100 percent higher. The unusually high fire risk in rural communities is one of the hidden parts of America's fire problem. The peak rates in very small and very large communities are explained by the fact that poverty rates are highest in these two sizes of communities.

The American South has the highest fire death rates in the country (23 percent above the national average in the period from 1989 through 1993). This is in large part because that region has proportionally more rural poverty than any other U.S. region. However, the South has been closing that gap and was not the leading region in two of the last five years. Several of the states in the South have been pursuing state-wide fire safety initiatives as ambitious as anything undertaken anywhere in the U.S., and the results are only just starting to appear in lower fire death rates.

Fire death rates are highest for preschool children and older citizens. Preschool children have twice the fire death rate of the population as a whole. At the other end, fire death rates begin climbing at age 50 and climb higher and higher as people age. People age 65 and over have fire death rates twice the national average. People age 75 and over have fire death rates three times the national average. People age 85 and over have fire death rates four times the national average.

Since fire deaths occur primarily in Homes and Garages, patterns like these are true for those fire deaths and for all fire deaths, as well.

American Community

Community activities often concentrate large numbers of people, creating the risk of large loss of life should a fire occur in the properties that make up the American Community. For this reason, properties in the American Community are generally subject to legally binding fire codes, such as NFPA *101*®, *Life Safety Code*®.

Historically, fires resulting in a major loss of life have resulted in important changes in building and fire codes and in standard fire protection or prevention practices. In the early 1900s, four building fires—the Iroquois Theatre in Chicago (1903), the Rhoades Opera House in Boyertown, PA (1908), the Lakeview Grammar School in Collinwood, OH (1908), and the Triangle Shirtwaist Factory in New York City (1911)—were largely responsible for the appointment in 1913 of the NFPA Committee on Safety to Life. The opening summary of the "Origin and Development of *101*" in the current NFPA *101*, *Life Safety Code*, states:

For the first few years of its existence, the Committee devoted its attention to a study of the notable fires involving loss of life and in analyzing the causes of this loss of life. This work led to the preparation of standards for the construction of stairways, fire escapes, etc., for fire drills in various occupancies, and for the construction and arrangement of exit facilities for factories, schools, etc., which form the basis of the present *Code*.

The 1937 fire at the Consolidated School in New London, TX, tragically pointed out the need for state laws to protect public buildings not subject to municipal ordinance and inspection. Then, in the 1940s, a series of multiple-death fires—including those at The Rhythm Club, The Cocoanut Grove, and the La Salle, Canfield, and Winecoff Hotels—focused national attention upon the need for adequate exits and other fire safety features in hotels and public buildings. These fires resulted in major changes to the "Building Exits Code" (as NFPA *101*, *Life Safety Code*, was then known) over a period of almost two decades. NFPA 102, *Standard for Grandstands, Folding and Telescopic Seating, Tents, and Membrane Structures*, was the result of still another multiple-death fire of the 1940s—the 1944 Hartford, CT, circus tent fire in which 168 people were killed.

Three hospital fires—St. Anthony's in Effingham, IL, in 1949 (74 killed); Mercy Hospital in Davenport, IA, in 1950 (41 killed); and Hartford Hospital, Hartford, CT, in 1961 (16 killed)—moved hospital administrators and fire prevention officials across the nation to assess the quality of construction and fire protection systems in hospitals.

The Our Lady of the Angels School fire in Chicago on Dec. 1, 1958, probably resulted in the swiftest action in the wake of any major fire since World War II. Within days of the fire, state and local officials throughout the nation ordered fire inspections of schools, and within one year, it was reported that major improvements in life safety had been made in 16,500 schools across the country. Improvements in the frequency and quality of exit drills and inspections, in the storage of combustible supplies, and in the disposal of waste materials were also reported in almost every community where schools were surveyed. This fire and the 1961 Hartford Hospital fire, which killed 16 people, also focused attention on the hazards of combustible ceiling finishes.

During a four-year period from 1970 to 1973, there were eight care-of-aged facility fires, each killing at least ten people and collectively killing 112 people. This rapid succession of catastrophic fatal fires stimulated action in a way that the deadliest nursing-home fire of all time, which occurred in isolation at the Golden Age Nursing Home of Fitchville, OH, on November 23, 1963, and killed 63 people the day after President John F. Kennedy was assassinated, could not. A pattern was found of insufficient fire protection and sprawling, undivided construction based on adding extensions to aging, converted farmhouses. Codes were reexamined, and the new Medicare program provided clout by tying reimbursement eligibility to code-compliance with NFPA *101, Life Safety Code.*

During the period from 1976 through 1985, attention shifted to boarding homes, spurred by a rash of major fatal fires in those facilities in 1979 through 1981. Six severe fires, causing 114 deaths in all, with the first three fires all occurring in the same month, commanded attention. During the period from 1979 to 1984, boarding home residents had a risk of dying in a multiple-death fire that was five times the risk for residents of other residential properties. Analysis of these incidents and other smaller fatal fires showed that many of the residents had characteristics of age or chronic mental illness that should have received ongoing medical attention that the facilities were not licensed or equipped to provide. This gap between the fire protection and other features of the facilities and the needs of their occupants was a more difficult problem than the one posed by the nursing home fires because there were fewer mechanisms to enforce codes.

The factors in what came to be called the "board-and-care" home fire problem were all too familiar and can be summarized as an overall lack of basic fire protection provisions. Contributory factors include inadequate means of egress, combustible interior finishes, unenclosed stairways, lack of automatic detection or sprinkler systems, and lack of emergency training for the staff and residents. Many of the facilities were either licensed as something other than a boarding home (such as a hotel) or were unlicensed, "underground" boarding homes. None of the facilities was provided with automatic sprinkler protection.

There is no particular mystery about how to reduce these risks, either. Enclosed stairs, providing two ways out, avoiding the use of combustible interior finishes, compartmentation, providing automatic detection and sprinkler systems, and training staff and residents in emergency procedures all would greatly reduce the risk of multiple deaths if fire occurs.

These analyses and others (such as an NFPA study of two decades of hotel fires causing ten or more deaths, done for Congressional testimony) consistently point to the difficulty of delivering fire protection to and enforcing codes and standards on the facilities that cater to the poorest and most vulnerable populations. Simply creating and adopting good codes and standards is not enough. State and local authorities must adopt them and enforce them, being sure to pay attention to existing buildings, or this knowledge will lack practical impact where it is most needed.

But while fire safety professionals stay aware of the difficulties and challenges still to be met, it is useful to look back at what has been accomplished, which is considerable. The next major section of this chapter, entitled "A Century of Accomplishment," does just that.

Our American Heritage: Special mention also should be made of the small part of the American Community referred to as our American Heritage. Our American Heritage includes historic buildings, museums, art galleries, and memorial structures. Such structures often attract large numbers of visitors. Since many of these structures are owned and operated by nonprofit charitable organizations, funds for fire protection are often limited. In addition, many historic structures are located in remote areas where public protection is minimal at best. Museums and art galleries incorporate the special hazards of workrooms for restoration and storage. Also, many properties that are not wholly devoted to historical preservation contain major areas of historical importance. The large-loss fire in the Los Angeles Central Library in 1986 was a recent reminder of this point, as a number of unique collections were destroyed, damaged, or endangered.

There is a tendency to forget the vulnerability to fire of these vestiges of the American heritage that were so important to our ancestors and will be important to future generations. The small share of total dollar loss they typically represent does not begin to capture the real loss fires in such facilities can mean to American culture. The actual loss sustained when artifacts and ancient structures burn cannot be measured in dollars alone.

During the 1980s, fires damaged or destroyed a long list of historic buildings. Included are the Paul Revere house in Boston, MA; the Wayside Inn in Sudbury, MA; Sutter's Mill in Sacramento, CA; the Daniel Boone dwelling in Defiance, MO; the Franklin Roosevelt home in Hyde Park, NY; the Legislative Office Building in Concord, NH; the John C. Calhoun mansion in Clemson, SC; Hundred Oaks Castle in Winchester, TN; the Confederate Memorial Building in Greenwood, MS; the Benbow Hall dormitory of Oak Ridge Military Academy in Greensboro, NC; and a barn on President Dwight D. Eisenhower's farm in Gettysburg, PA.

Mobile Environment

The principal point of interest about the Mobile Environment fire problem is how large it is, relative to every other environment but Homes and Garages. As Figure 1-1G shows, the Mobile Environment's fire death toll dwarfs all other properties combined, excluding the roughly 80 percent in Homes and Garages. Figure 1-1H shows the Mobile Environment fire injury toll is second only to Homes and Garages, and Figure 1-1I shows the Mobile Environment fire damage total is not far behind the structure environments other than Homes and Garages.

Fire safety programs and requirements for the Mobile Environment have been almost entirely delegated, by NFPA and others, to the federal government. The majority of vehicle fire deaths occur in post-crash fires, and the death toll is far less than in the crashes themselves, so the difference may be understandable. Nevertheless, the mission of fire safety must encompass vehicles, which are a large, and increasingly large, share of the total.

Some special issues and trends pertain to the Mobile Environment. One development of the past two decades has been an increased use of recreational vehicles, such as campers and trailers for travel, and tents for lodging. This increased usage creates a potential for increased fire hazard in these recreational-type properties, but so far the actual fire experience has been small by comparison to other road vehicles. For example, of the 331,500 passenger road vehicle fires per year that occurred during the period from 1988 to 1992,

3,600 involved all types of all-terrain vehicles (including motorcycles, golf carts, dune buggies, and snowmobiles), 3,100 involved motor homes (not mobile homes, which typically stay in one place), and 1,500 involved travel or camping trailers. Collectively, they represented roughly one out of every 40 passenger road vehicle fires. (Tents and other similar outdoor sleeping quarters contributed less than 50 fires a year to the structure fire total.)

Another concern that is so far based more on changing activity patterns than on confirmed fire problems is the increasing use of automobile fuels other than gasoline and diesel (e.g., LP-Gas, LNG, and CNG). These new fuels introduce hazards with which the public and authorities are not totally familiar.

Changes in transportation preference, such as the growing use of public transportation, have opened up new problems for fire protection. The fire potential problems of public transportation center around equipment design and materials, as well as growing dependency on automation.

Transportation of raw materials and finished goods throughout the nation and in U.S. coastal waters by common carriers represents the potential for substantial economic loss if fire destroys the materials transported or the vehicles in which they are being moved. This type of transportation at the interstate level is subject to federal regulation.

Another key area is hazardous materials. This problem not only poses the potential for costly loss, but, more important, carries with it substantial risk to public safety. Hazardous materials often are transported over land, even through highly congested urban communities. The fire record shows an abundance of examples of disastrous accidents involving hazardous materials transportation.

Air transportation is also regulated by the federal government, with regard to the level of fire safety in aircraft design and fire potential following ground impact. Other fire protection concerns associated with air transportation include fuel servicing, aircraft maintenance, design of airport facilities, and aircraft rescue and fire fighting techniques.

Outdoor Environment

A significant part of the great American heritage is that created by nature. Forests and wildlands represent resources that have been recognized nationally in legislation as a part of the American heritage only since the days of Theodore Roosevelt's presidency (1901-1909). Forests and open lands provide raw materials, recreational facilities, and scenic breaks from the urban sprawl.

Great portions of the American landscape, including wildlands, forests, and deserts, are under the control of the U.S. Department of Agriculture. Departments of this federal agency have their own regulations, which include fire prevention and suppression programs.

Wildland fires are once again much in the news. Environmental trends have produced a succession of years that are among the hottest and driest on record in the United States. These weather conditions have combined with the expanded scale of activity in the wildland/urban interface to produce a number of very large property-loss fires, as seen in Tables 1-1F and 1-1G. These major wildfires have also taken a toll in fire fighters' lives, most notably in 1994, when more than 30 fire fighters died on-duty at or while responding to, or returning from, wildland-related incidents, roughly a third of the total on-duty fire fighter death toll for that year. These figures point to a devastating toll of life and property that deserves the increased level of attention it has recently received through federal/private cooperative efforts, such as the Wildland/Urban Interface Program.

A CENTURY OF ACCOMPLISHMENT

NFPA has been a force for fire protection and fire safety for the past 100 years. In that time, most of NFPA's influence has come through its codes and standards, which have changed the rules that guide building specifications and operations and, in that way, have changed the environments in which people live. As in past editions of the *Fire Protection Handbook*, this edition cites examples of major NFPA code changes that responded to needs demonstrated in major fire incidents.

This section describes several of the most dramatic examples of fire experience that responded favorably to changes in codes or to changes in the extent of compliance with codes. The literally hundreds of thousands of people who have been part of the NFPA "family" since 1896 have helped to save untold lives from the scourge of unwanted fire. What follows is some of the clearest evidence available of that good work.

Because most codes and standards focus on the risk of large fire losses outside the home, this section will focus on major incidents outside the home involving multiple loss of life. In particular, NFPA's fire incident record-keeping is best for incidents in which 10 or more people died, and these are used here as the principal tracking mechanism for the impact of codes, standards, and other fire protection requirements. In keeping with the normal approach used by NFPA's statisticians, most incidents are cited anonymously, to keep the emphasis on statistical patterns. In general, only incidents involving 50 or more deaths are given specific identifiers.

Schools and Hospitals

Since NFPA was founded in 1896, NFPA's fire incident records show a total of nine school fires in which at least 10 people died. Two of these are among the ten deadliest single-building fires in U.S. history.

The first of the nine was also the only one not to involve grades K-12. It was a 1903 university building fire in which 11 people died. NFPA files indicate the building had no fire exits and escape was cut off. The second incident, in 1908, was also the second deadliest school fire in history. Inadequate fire escapes, inadequate fire drills, and a building layout that created bottleneck exit paths, all contributed to the death toll of 176 at the Lakeview Grammar School in Collinwood, OH. A parochial school fire in 1915 was the third of the nine fires and cost 21 people their lives. After the fact, the exits were deemed inadequate, even though they complied with existing laws. Again, the failure to include all exits in fire drills made those drills inadequate and proved critical in this fire, as many exits were not used when they could have made a difference.

The first of three incidents in the 1920s was a 1923 fire that killed 77 people. Fleeing occupants jammed the only stairway serving the Cleveland School of Beulah, SC, and it collapsed during the evacuation. A 1924 fire in a one-room schoolhouse spread rapidly from its origin in a Christmas tree. The fire department's report of the fire said "the struggling crowd became wedged in the doorway, which was only 3 ft (0.9 m) wide." A total of 33 people died. Finally, a 1927 incident led to 46 deaths when a "crazed farmer" used dynamite to bomb a school.

The deadliest school incident in history stemmed from a gas explosion; 294 people died in the Consolidated School of New London, TX, in 1937. The next school incident killing at least 10 people came in 1954, when a faulty heating system led to a gas buildup in the attic of a school. Ignition by an undetermined heat source produced a flash fire in an unoccupied area. Delayed discovery allowed the fire to spread into occupied areas, and 15 people died.

The last three incidents took place over two-and-a-half decades, from 1927 to 1954, and all three involved rapidly developing situations—two explosions and a flash fire—that were unlike the earlier incidents. Officials might have been forgiven for believing that the older school fire problem, consisting of too many ordinary combustibles and insufficient or inadequate exits, was essentially under control. If so, that belief proved illusory when the 1958 Our Lady of the Angels School fire occurred in Chicago, IL, killing 95 people, 92 of them children.

The reaction to that fire was more rapid, more sweeping, and arguably more effective than the reaction to any other single fire in U.S. history. NFPA codes and standards were quickly revisited, and stricter requirements for interior finish and exiting were established. An April 1959 issue of the *Journal of American Insurance* caught the mood of the country at the enforcement end, noting in an article subhead: "U.S.A., aroused by Chicago lesson, is overhauling its school buildings as never before."

In the nearly four decades since Our Lady of the Angels, there has *never* been another school fire killing 10 or more people. In recent years, there has not even been an incident to come close. In the period from 1989 through 1993, grades K-12 averaged one civilian fire death per year—a typical annual death toll for schools since at least 1980. Detailed examination of fatal school fires by NFPA analysts indicates that the few deaths that do occur are all or nearly all adults (e.g., maintenance workers) or juvenile firesetters fatally injured by the fires they set. Innocent children are not the victims in school fires any more. Few lessons have been learned as thoroughly as the ones from the Our Lady of the Angels incident appear to have been learned.

A similar story applies to U.S. hospitals. There were a similar number of incidents in which 10 or more people died, including a 1929 fire that killed 125 people in the Cleveland Clinic Hospital of Cleveland, OH, and a 1949 fire that killed 74 people in St. Anthony's Hospital of Effingham, IL. The last hospital fire to kill 10 or more people was the Hartford Hospital fire of 1961, in which 16 people died in Hartford, CT. Reports on the fire focused on combustible ceiling tile that led to rapid flame spread down a corridor, delayed reporting to the fire department, and the fact that all the deaths involved doors left open to the corridor. Closed doors meant survival in every case.

The NFPA *Building Exits Code* of 1961 already mandated complete sprinkler systems for hospitals for many combinations of height and construction. The next edition after the Hartford Hospital fire was the 1963 edition, which extended this requirement to even more combinations of height and construction. The 1966 edition changed the name of the document to a wordier version of today's NFPA *101, Life Safety Code*, signaling right from the cover that this code was not just about exits, as it had not been for some time.

By 1980, when NFPA first developed its nationally representative statistics on the use of sprinklers in buildings that had fires, nearly half (43.5 percent) of all fires in facilities that care for the sick were in properties with sprinklers. This undoubtedly meant that most hospitals were sprinklered by then, because it is known from occasional special studies on sprinkler usage that reported fires are more likely to occur in unsprinklered properties. By 1993, the latest year for which statistics are available, 66.7 percent of fires in facilities that care for the sick were in properties with sprinklers, and 89.1 percent were in properties with automatic fire detectors. And this equipment was making a measurable difference. For example, statistics from 1984 through 1993 indicated that sprinklers cut the chances of dying in a fire by 58 percent for facilities that care for the sick.

Since hospitals are required to comply with NFPA *101, Life Safety Code,* to receive reimbursement under Medicare and Medic-

aid, other provisions of NFPA *101* are undoubtedly also in nearly universal use. This helps explain why, in the latest available statistics, hospitals averaged fewer than seven civilian fire deaths per year between 1989 and 1993.

As in the case of schools, the characteristics of the people who still die in hospital fires underline the extent to which NFPA codes have solved the type of hospital fire problem they targeted decades ago. About four of every five hospital fire deaths involve a victim who is so close to the start of the fire as to be described as "intimate with ignition." Smoking and incendiaries dominate the causes of fatal fires in facilities that care for the sick. These circumstances involve very rapid fatal injuries, often too rapid to be reliably prevented by even the best fire protection provisions. Fire prevention strategies are needed and, indeed, the past decade has seen widespread adoption of more aggressive controls and prohibitions on smoking.

Hotels and Motels

From 1900 to the present, four of the 100 U.S. fires and explosions that killed at least 50 people involved hotels or motels. This compares to eight in industrial settings, five that involved ships, two that involved industrial settings and nearby ships, eight that involved places of assembly, four in schools, and 59 in mines. (The other 10 were two forest fires, two hospital fires, two nursing home fires, one prison fire, one missile silo incident, one office building bombing, and one post-earthquake fire.) Three of the four hotel fires occurred in a four-year period, 1943 through 1946—the Gulf Motel fire in Houston, TX, in 1943, which killed 54 people; the LaSalle Hotel fire in Chicago, IL, in 1946, which killed 61 people; and the Winecoff Hotel fire in Atlanta, GA, in 1946, which killed 119 people. National conferences on hotel fire safety were convened in early 1947, one by NFPA and another by President Harry Truman; the result included significant change to the hotel section of NFPA's *Building Exits Code* and model legislation for hotel fire safety, prepared by the Fire Marshals Association of North America, NFPA's member section for fire marshals.

A third of a century later, the 1980 MGM Grand Hotel fire, in Las Vegas, NV, inspired an industry response that combined unprecedented widespread code compliance with fire safety provisions that often ran ahead of code requirements. The result has been a dramatic change both in the fire death toll in hotels and motels and in the use of proven fire protection systems in that industry.

In 1981, NFPA published an overview of hotel and motel fire experience leading up to the 85 deaths in the MGM Grand Hotel fire. In the two decades from 1961 through 1980, NFPA documented 53 hotel or motel fires that each killed at least five people. (In those days, the threshold for a multiple-death fire was three deaths in any property; today, it is five deaths in residential properties.) With more than two fires a year causing at least five deaths apiece, chronic public concern over fire safety when traveling was understandable.

The MGM Grand Hotel fire was not the only fire that shaped public concerns at this time. Certainly, the MGM Grand Hotel fire ran counter to the conventional industry wisdom that fire problems were primarily a matter of older or poorer facilities. MGM Grand was neither. But the MGM Grand fire also was not an isolated incident. Two months after the MGM Grand fire, a second Las Vegas, NV, hotel, this one part of a national chain, had a fire that killed eight people. Then, the year after that, 12 more people died in a fire in another hotel of the same chain. A second chain lost 10 people in each of two fires during each of the two years preceding the MGM Grand Hotel fire. All of these events combined to create a picture of an industry in need of improvement.

Led by strong industry associations and fire-safety-conscious professionals at the major chains, the industry began to respond. In 1980, the year of the MGM Grand Hotel fire, sprinklers were reported present in only one of every nine hotel or motel fires reported to U.S. fire departments. Detectors were reported present in just over one-fourth of reported hotel or motel fires.

By 1993, sprinklers were reported present in one-third of hotel and motel fires and in three-fourths of high-rise hotel fires. An industry-sponsored study of sprinkler usage in 1988 found sprinklers present in roughly half of all properties, suggesting the percentage today is much higher still. By 1993, detectors were reported present in three-fourths of all hotel or motel fires. And for both detectors and sprinklers, it is reasonable to assume that the new level of built-in fire protection had much to do with the dramatic drop in the number of hotel and motel fires since 1980. NFPA statistics from 1984 through 1993 indicated that sprinklers cut the chances of dying in a given fire by 81 percent and also reduced the average property loss per fire by 57 percent.

In terms of the deadliest fires, beginning in 1983, there have only been two hotel or motel fires that killed 10 or more people, and each of them was on the outer fringes of the industry. A 1984 fire killed 15 people in a facility that called itself a rooming house but was classified as a hotel by NFPA, because it had too many occupants to qualify as a rooming or boarding house. This facility also included a number of deinstitutionalized former mental patients among its occupants, raising questions about the health-care needs of the occupants *vs.* the fact that no such care was provided by the facility. A 1993 fire killed 20 people in a facility licensed for, and principally run for, long-term residents; yet the fire began in, and most deaths occurred in, a section housing transient guests.

In terms of the more familiar names in the lodging industry and the bulk of the facilities in operation, the changes of the past 15 years have moved the fire experience picture in hotels and motels to roughly the same position as nursing homes and hospitals. That is, a steadily shrinking number of fatal victims tend more and more to be people who die in fires they caused or that began very close to them. In the period from 1988 through 1992, for example, nearly half of all hotel and motel fire deaths came in fires started by smoking materials or associated lighting implements (i.e., matches, lighters).

As with schools and hospitals, so with hotels and motels, the past century has seen dramatic accomplishments—a clear sequence from major fires to major code changes to major changes in the practices in the concerned facilities to major declines in the loss of life due to fire.

Places of Assembly

Seven of the 11 deadliest single-building fires and explosions in U.S. history have involved places of assembly. (The other four were two school fires discussed above, one prison fire, and the 1995 bombing of an Oklahoma City, OK, office building.) They include the Brooklyn Theater fire of 1876, which killed 285 people and preceded NFPA's founding. The Iroquois Theater fire of 1903, which killed 602 people in Chicago, IL, and the Rhoades Opera House fire of 1908, which killed 170 people in Boyertown, PA, occurred early in NFPA's life and led to an early focus on the dangers posed by inadequate exiting provisions and the kinds of combustible loads that make rapid fire development likely.

From 1909 through 1939—a period of more than three decades—there were three incidents that killed 10 or more people in places of assembly, one of which involved an explosion. A 1919 Louisiana restaurant and dance hall fire led to 25 deaths when relatives rushing into the building to try to save occupants collided with occupants fleeing onto the only exit stairway, leading to the collapse of that stairway. A 1928 dance hall incident started by ignition of gasoline fumes, and killed 38 people. A 1929 night club fire spread rapidly via the decorations, and many of the 22 who died took refuge in rooms with no way out.

In five years, from 1940 to 1944, three of the worst fires in U.S. history occurred, all involving places of assembly. More than one-fourth of the 746 occupants of the Rhythm Club in Natchez, MS, died in a 1940 fire. Overcrowding, inadequate exits, and highly combustible decorations were all cited as factors in the 207 deaths. Nearly half of the more than 1,000 occupants of the Cocoanut Grove night club in Boston, MA, died in a 1942 fire. Overcrowding, inadequate exits, and highly combustible decorations again were cited as key factors in the 492 deaths, with roughly 200 trapped behind one revolving door alone. And 168 of the roughly 7,000 patrons at the Ringling Brothers Barnum & Bailey Circus in Hartford, CT, died in a 1944 fire, many of them children.

These three incidents, particularly Cocoanut Grove, the sixth deadliest fire or explosion in U.S. history, are credited with inspiring a rush of tougher codes, focusing on exiting provisions and the use of combustible materials, but also pushing a more comprehensive and appropriate definition of places of assembly. Incredibly, many jurisdictions excluded eating and drinking places from the classification "place of assembly" prior to Cocoanut Grove.

In the half-century since 1944, no incident in a place of assembly has matched the death toll in any of these three incidents, and only one incident came close—the 1977 Beverly Hills Supper Club fire in Kentucky, which killed 165 people. This fire, the deadliest U.S. fire of the past 50 years (and the second deadliest fire or explosion incident, after the 1995 Oklahoma City, OK, bombing), echoed many of the problems of the 1940s incidents. The 165 victims represented less than 10 percent of the 2,400 to 2,800 occupants reportedly on site, but those occupants represented two to three times the facility's capacity, indicating severe overcrowding. All types of exiting problems were also recorded.

In terms of accomplishments, the post-Cocoanut-Grove code changes worked. Large-loss-of-life fires have not occurred where code compliance has been observed, except for rapid-onset situations, such as explosions. But large-life-loss incidents do continue to occur—less severe individually but still too large and too frequent to be acceptable—because of spotty enforcement. One type of incident, most dramatically embodied in the Happy Land social club fire of 1990, in New York, NY, has occurred in each of the last five decades and is worthy of special attention. That is an arson fire using accelerants started in a location that interferes with normal exiting.

The first of these arson incidents is also the least well documented. In NFPA's sixth decade, 1946 through 1955, a 1947 gambling hall fire began when a gallon of gasoline was thrown on the floor and ignited. The exiting provisions are not known, but 15 people died in the fire. In NFPA's seventh decade, 1956 through 1965, a 1965 restaurant fire began when an arsonist poured gasoline at the main entrance. The 40 to 50 occupants may have encountered a security delay at the rear exit, which also was too narrow; 13 people died, or, roughly one-fourth to one-third of the number present.

The succeeding fires involved even worse exiting problems and death tolls that accounted for even larger fractions of the total occupants. In NFPA's eighth decade, 1966 through 1975, a 1973 lounge fire was set in the stairway of the lounge's normal exit. The second exit was poorly marked, and the windows were blocked or barred. The death toll of 32 people represented roughly half of the occupants present. In NFPA's ninth decade, 1976 through 1985, a 1976 fire was set in the only exit of a New York City social club, leaving only windows for escape. The death toll of 25 represented roughly half of the occupants present.

In NFPA's tenth decade, 1986 through 1995, a 1990 arson fire was set in the main exit of New York City's Happy Land social club. There were no other marked exits, nor were the windows useable for escape. Only six of the 93 occupants survived, five by going out an obscure, normally locked door that one of them had a key to, and the sixth by literally running through the fire at the entrance, sustaining critical burns.

None of these properties had complete sprinkler systems or sprinklers in the area of the fire, so they emphasize the value of such systems. But taken together, they also illustrate how the consequences become worse as the deviation from code-compliant exiting provisions becomes worse. If there is literally nowhere to go, then everyone will die.

Finally, consider what it will take to accomplish even more in terms of life safety from fire, in any of the types of properties discussed thus far. People who die in fires, die in either the *kinds of fires* that codes do not reach or the *properties* that codes do not reach due to lack of adoption or lack of enforcement.

For schools and hospitals, codes reach nearly everywhere. These properties are tightly controlled. Nursing homes and the lodging industry are not quite so tightly controlled, but are very broadly compliant. Both have active industry associations that have broad membership and are sensitive to fire safety. Both industries have difficulty primarily in exerting control over properties on the fringes of the industry, such as board-and-care homes. Notwithstanding the emergence in recent years of legitimate board-and-care homes that provide only personal care to residents who need nothing more, there is a sizable group of facilities, long referred to as board-and-care homes, that provide only lodging to residents who legitimately need health care as well.

Places of assembly have the problem of widespread non-compliance far more than do hotels or nursing homes. Proportionally fewer properties belong to national chains, which in other industries often lead the move to greater fire safety. Industry associations have less of a track record as leaders in fire safety and, more importantly, capture a smaller proportion of their industry's operating facilities. There is more turnover in these facilities, which also hampers enforcement efforts, because educating owners and managers about fire safety is a gradual, incremental process, which has to start all over whenever an existing facility "goes under" and a new facility takes its place.

Despite all this, the deadliest fires (e.g., 50 or more deaths) in places of assembly are less frequent and less deadly than they were a half-century ago, which suggests that enforcement and sensitivity to fire safety is better among at least the largest facilities. This is an achievement worth building upon.

Manufacturing Facilities

Manufacturing properties are worthy of note because they also include a significant single incident that led to dramatic changes in NFPA codes. The 1911 Triangle Shirtwaist Company fire in New York City still ranks as the deadliest manufacturing facility fire (excluding explosions) in U.S. history. NFPA code requirements for the workplace, particularly exiting provisions, were substantially changed as a result of that fire.

Since NFPA's founding, according to NFPA records, there have been 38 fires or explosions in manufacturing facilities that caused 10 or more deaths. Of these, 30 involved explosions, two involved flash fires, and one involved a large spill of burning fuel from a crashed aircraft; therefore, only four are truly comparable to the Triangle Shirtwaist Company fire.

One such fire took place only two years after the Triangle Shirtwaist Company fire and also involved a New York state clothing manufacturer. The death toll in that fire reached 35, as an external fire escape was engulfed in flames early in the incident.

More than 40 years later, two more garment factory fires, one like the Triangle Shirtwaist Company located in New York City, occurred in 1957 and 1958, killing 15 and 24 people, respectively. In the former, one stairway was open and filled early with smoke, and one fire escape was enveloped in flames and collapsed. In the latter, fire filled a key stairway early when the door to the stairway was wedged open.

In 1991, a North Carolina food processing plant fire killed 25 people. NFPA investigators and the U.S. Occupational Safety and Health Administration (OSHA) both pointed to the similarities between this incident and the Triangle Shirtwaist Company fire, in terms of grossly inadequate exiting provisions.

With the North Carolina fire still so recent in memory, it is particularly difficult to declare any kind of century of accomplishment for manufacturing facility fires. Major fires of the Triangle Shirtwaist Company variety were never particularly common, so there is no particular evidence of a statistical trend. NFPA codes for fire safety in the workplace are certainly stricter today, but there is no clear way with existing fire incident data to demonstrate that actual industrial practices are markedly better than they were a century ago.

One exception is in the use of sprinklers, where manufacturing properties have shown steady improvement (up from 45 percent of reported fires being in sprinklered properties in 1980 to 50 percent in 1993) and significant impact (civilian deaths per fire were one-third lower from 1984 through 1993 when sprinklers were present in manufacturing facilities). Within the manufacturing group, food product manufacturing facilities—the type of facility involved in the 1991 North Carolina fire—had the lowest percentage of reported fires with sprinklers present.

However, a battle for life safety in manufacturing facilities is still being fought with respect to explosions, which involve a number of different NFPA codes. During the five decades from 1916 through 1965, the U.S. averaged just over four manufacturing facility explosions per decade that each killed 10 or more people. During the three decades of 1966 through 1995, the average was just under three such incidents per decade. It is on the thin reed of one fewer incident per decade, on average, that a claim can be made for "a century of accomplishment" in manufacturing facility fire safety.

As with places of assembly but even more so, an undeniable "century of accomplishment" in writing safer codes has been undercut by far less clear-cut progress in achieving code-compliance. OSHA and the American labor movement, both strong champions of a safer industrial workplace in years gone by, have been substantially weakened in recent years. Perhaps they can still be part of a revitalized coalition, with NFPA, concerned state and local authorities, and the thousands of responsible fire-safety-conscious owners and managers of the better manufacturing sites, to improve a situation that still has much room for improvement.

Mines

Mining is cited here in order to end this section on a true high note, even though NFPA had little input into what has been one of the most remarkable centuries of accomplishment in any class of properties. Most of the 100 U.S. fires and explosions that have killed 50 or more people since 1900 have been in mining, specifically coal mining. (Only three of the 59 mining incidents were not coal mines; they were metal mines.)

In the first decade that NFPA tracked, from 1900 through 1909, 15 of the 59 mining incidents occurred, killing a total of 2,286 people. At the start of the second decade, from 1910 through 1919, the U.S. Bureau of Mines was formed (in 1910). In that second decade, there were 18 incidents, killing a total of 1,882 people. Sometime in the second and third decades, the U.S. Bureau of Mines began issuing rules and guidelines, because by the end of the third decade (1929), they were publishing summaries of what they issued. In the third decade, from 1920 through 1929, there were a total of 14 incidents, killing a total of 1,407 people.

In the fourth decade, from 1930 through 1939, which includes the Great Depression, there were only two incidents, killing a total of 136 people. NFPA 493, *Spontaneous Heating and Ignition of Coal and Other Mining Products*, was first issued in 1936, but, by then, the big decline in the frequency of serious incidents had already occurred.

In the fifth decade, from 1940 through 1949, which includes the wartime demands of World War II, there were seven incidents, killing a total of 533 people. The next three decades had one incident each, killing a total of 288 people. The decades from 1980 through 1989 and 1990 through 1999 (as of this writing) have had no such incidents.

A natural question, which also applies to the fire experience of garment manufacturing facilities discussed earlier, is how much the reduction in fire deaths reflects a reduction in economic activity and associated exposure. Employment in coal mining peaked around the end of the second decade of the 20th century and declined dramatically until around 1970, when it rose sharply for about a decade, then began declining again to its present levels, the lowest in this century. The decline in mining incidents killing 50 or more people and in the number of people killed in those incidents has been even steeper than the decline in employment in the industry, but both have been so steep that one has to consider both the reduction in exposure and increased safety among those exposed as major factors in the trends.

What makes these two property groups different, then, is that their century of accomplishment in fire safety, which has been dramatic and has involved changes in both the adequacy of the rules and the breadth of enforcement, has been overshadowed by nearly a century of decline in activity.

Implications for the Next Century

Of the many property classes that had tremendous risk of death in fire when NFPA was born, several have nearly eliminated life loss from fire and have achieved nearly all that can be achieved by fire protection after ignition occurs. Others have moved a long way in that direction but still have pockets where code compliance remains spotty. Still others have accomplished the development of adequate codes for fire safety but have major gaps in enforcement and compliance that still leave thousands, even millions, of people at risk. For the first group of properties, the next century's agenda will be principally fire prevention and maintaining the gains in fire safety already won. For the other properties, to varying degrees, the next century's agenda will be finding ways to extend effective fire safety practices throughout.

ORGANIZING FOR FIRE PROTECTION

Fire department organization is a critically important element of fire protection. The effectiveness of the organization and management of U.S. fire departments determines whether the more than $14 billion (1992 dollars) in annual expenditures for local fire protection are spent well or badly. More important, it also determines whether the more than one million U.S. fire fighters will be used effectively and safely. Protecting the community and protecting themselves are

dual responsibilities of a fire department, and both are more likely when planning and preparation are fully used.

Table 1-1Q gives a brief overview of the resource base of local fire departments. Only 10 percent of local fire departments are composed entirely or mostly of career fire fighters, but they protect 58 percent of the population, and 25 percent of all local fire fighters are career fire fighters. The majority of volunteer fire fighters serve in small, rural communities of less than 2,500 population.

TABLE 1-1Q. *Selected U.S. Local Fire Department Resources*

- 259,650 career fire fighters
- 795,400 volunteer fire fighters
- $14.4 billion in total public expenditures
- 30,500 fire departments
- 44,400 fire stations
- 61,450 pumper trucks (at least 500 gpm)
- 7,200 aerial apparatus

Source: 1992-1993 NFPA National Fire Experience Survey, and Fire Service Inventory, U.S. Bureau of the Census.

Figure 1-1K shows that the workload for fire departments has been climbing, up 42 percent from 1980 to 1993, led by a 73 percent increase in emergency medical calls, which constitute more than half of all fire department calls. (See Figure 1-1L.) Since many communities still do not offer emergency medical service, those that do may find medical calls running at 80 to 90 percent of their calls.

Fires now constitute only one-eighth of all fire department calls, depending on how mutual-aid calls are categorized. (See Figure 1-1L.) But fire calls demand the majority of fire department resource utilization, including stress on people and equipment. The potential to overload a department's total resources still comes primarily from fires.

Hazardous materials calls are a small fraction of the total but a growing one, and they also can involve a lot of resources per incident. False alarms today are more often non-emergency activation's of automatic detection systems and less often the malicious box-alarm activations that were of so much concern in the 1960s and 1970s.

The increasingly diverse responsibilities of the modern fire department are both an opportunity and a challenge. On the one hand, more duties mean more value delivered to the community. On the other hand, more duties mean a greater chance of overloading the system with simultaneous major calls. Since the rapid rise in calls for service has occurred during a period of generally declining or level staffing, the risk of overload is real.

A more subtle challenge comes from the need to design an emergency response system to cope with different types of emergencies having very different patterns. Emergency medical calls require swift response with a minimum of equipment and personnel. Fire calls require swift response with a large complement of equipment and personnel and may tend to occur in different parts of the community than those where emergency medical calls are concentrated. These competing bases for decision-making require sophisticated planning. In many major cities, fire fighters are now required to also be trained as paramedics or emergency medical technicians (EMTs).

Organizing Before Fire Occurs

Preparing for fire begins with the promotion of fire prevention and built-in fire protection, which fire departments can do in a variety of ways.

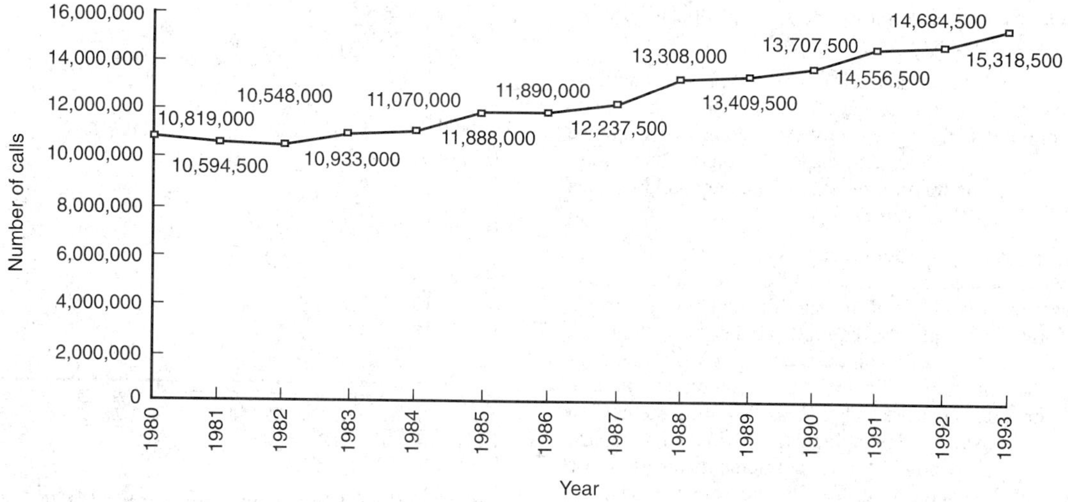

FIG. 1-1K.   Fire department emergency calls (1980-1993). (Source: NFPA National Fire Experience Survey)

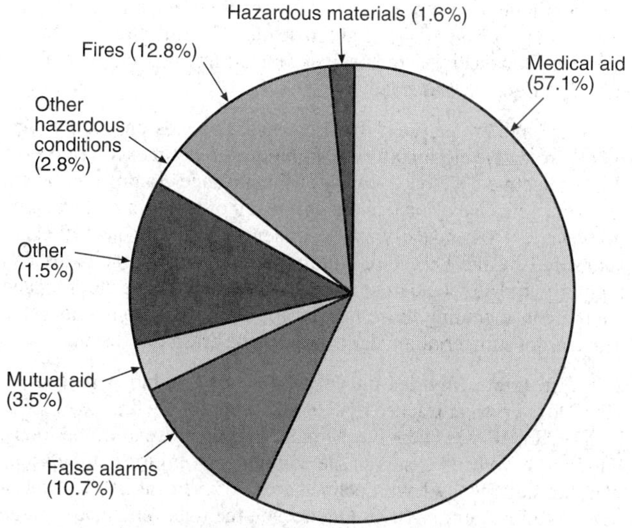

FIG. 1-1L.   Fire department calls (1993). (Source NFPA National Fire Experience Survey)

First, fire departments typically are involved as inspectors and sometimes as certifiers for a wide range of fire-related regulations. Sometimes fire departments support building inspectors in their reviews of building code requirements for new and remodeled buildings. More often, fire departments handle ongoing fire code requirements for commercial buildings. Various permit requirements, from the general ones, such as occupancy permits and business licenses, to specific permits for hazardous materials and processes, also may be handled by fire departments. Smoke detector laws are among the few examples where existing homes may be covered by these activities. These code and permit inspections not only empower fire departments to control fixed hazards, they also provide an excellent opportunity to educate and motivate occupants on general rules of fire safety. Research has shown this educational effect can reduce the frequency of fire significantly.

Second, fire departments investigate the causes of fires and pass on what they learn to a community whose attention is focused by the immediacy of a recent tragedy. Support for prosecution of ar-

son cases is the most obvious example of the effect investigations can have. Changes in codes and standards can result from these findings, too.

The fire department also is the one constant in local programs to educate the public about fire safety, even though other groups may have primary authority for some programs. Fire departments can serve as catalysts for, and participants in, programs to counsel juvenile firesetters or to teach fire safety to school children. Fire departments often arrange for extensive programs of fire safety speeches to community groups, increasing community attention to the special programs of Fire Prevention Week in October. Fire departments can circulate fire safety messages by promoting local media use of fire safety public service announcements and by pointing out lessons from news coverage of local fires. Some fire departments are able to pursue even more ambitious programs, such as home fire inspections, using their own resources or acting as change agents to enlist a group of volunteers to do the job.

In summary, the fire department is central to local fire prevention and protection, and fire prevention and protection is central to the duties of the modern fire department.

With regard to preparing for non-fire emergencies, fire departments are increasingly called upon to serve as the local agents for enforcement of federal and state laws to achieve "cradle-to-grave" control of hazardous materials. This may involve regulations on storage, use, and transportation. In many communities, the fire department resources for gathering, processing, and acting on information in this new area are severely lacking. This is one of the areas where local fire departments are most in need of help to do the new jobs that are now expected of them.

Organizing Effectively at Fires

Effective fire suppression requires clear policies and objectives, with tactics that follow logically from those policies. A suppression policy or objective is a concise description of priorities for the use of resources available at a fire. Particularly in properties with large numbers of people, containment of the fire and protection of the large numbers of people outside the fire-involved zone may need first priority, ahead of any efforts to save people and property in areas already involved in fire. Some losses, such as injury or even death of occupants in the room of fire origin, occur so early in a fire that they cannot be consistently prevented by any suppression strategy of the fire department alone.

Objectives imply tasks, such as forcible entry, rescue, application of water, ventilation, and salvage. And each task implies needs for staffing, equipment, the water delivery system, and response time. These needs are also affected by the level of on-site prevention and protection the community or the particular building has adopted.

For example, suppose a building has a certain size and a set of on-site hazards associated with the kind of business or use at the property. These imply a potential fire size for planning purposes. The potential fire size implies a required number of gallons per minute (L/min) of water that would be needed to control and suppress that fire. The gallons-per-minute (L/min) requirement, in turn, implies needs for water pressure and pipe sizes serving the nearest hydrants and for hose sizes and pumping capacity in the fire department apparatus that respond. All these requirements also imply a certain staffing level to handle all the needed fireground activity, including: (1) forcing entry if necessary, (2) finding and evacuating or protecting any occupants in certain zones of the building, (3) ventilating as necessary, (4) carrying the needed hoses to the potentially distant fire location, and (5) applying water there until the fire is controlled and extinguished. This process is a task analysis of suppression resources and their use. Implicitly or explicitly, such an analysis should underlie all resource decisions.

It should be clear from this example that on-site sprinkler protection or limitations on quantities of burnable materials on-site can change the potential fire size, which might permit longer response times or smaller personnel complements. However, if anything and everything is permitted on-site, the fire department must plan for the worst, and if they lack the resources to do so, then disaster is there, just waiting to happen.

Pre-incident planning can range from the review and rehearsal of strategy and tactics for a single facility to the elaborate planning required to prepare for natural disasters or civil-defense emergencies. Large-scale planning of the latter type may involve the development and maintenance of multi-agency coordination systems and large-scale incident command and control systems.

Protecting Fire Fighters

Fire fighter deaths on duty have generally been declining, though the trend has been far from steady. Even the large jump from 1993 to 1994 may be seen as bringing the total back closer to what has been the long-term trend line. (See Figure 1-1M.) The high death tolls up to the early 1980s were led by career fire fighters, but the death tolls since then have been dominated by volunteers. These totals reflect both immediate and delayed effects of acute injuries or illnesses but do not reflect deaths due to long-term chronic illnesses, such as cancer.

In a typical year, nearly half of all fire fighter deaths in the line of duty involve heart attacks. These deaths are nearly always limited to older fire fighters, at least 38 years of age, and are particularly prevalent in volunteer fire departments, where older fire fighters are more likely to serve. A large fraction of these heart attack deaths involve fire fighters with serious pre-existing health conditions, including arteriosclerosis, hypertension, and previous heart attacks. Since nearly half of all fire fighter deaths involve heart attacks, physical fitness is one of the most important elements of an occupational health and safety program.

In a typical year, one-fourth of on-duty fire fighter deaths occur during response to, or return from, an emergency call. Vehicle accidents and heart attacks while operating vehicles are both major causes. This scenario affects volunteers more than career fire fight-

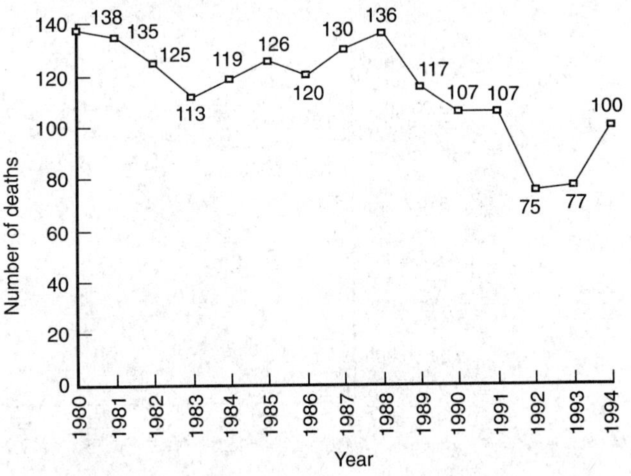

FIG. 1-1M.   U.S. fire fighter death trend (1980-1994). (Source: NFPA FIDO)

ers. Driver training is needed to prevent vehicle accidents. Safe passenger practices must be emphasized because falls from vehicles are also a major cause of fire fighter deaths. Careful attention to the safety of the vehicles themselves, including inspection, maintenance, and repair, is needed, too.

Fire fighter safety on the fireground depends on good equipment. Protective clothing for fire fighters is sometimes referred to as the "protective envelope." Many of the advances in this equipment in recent years have come about as a result of a program of research in the late 1970s and early 1980s, under the name "Project FIRES," sponsored by the U.S. Fire Administration. NFPA maintains and regularly updates standards for each type of protective clothing and equipment, including those needed for special environments, such as hazardous material incidents or airplane crash fire rescue.

Fire fighter injuries on duty have been within 4 percent of 100,000 every year tracked by NFPA, except for the first year, 1977, i.e., 1978 to 1993. There has been no particular trend in the totals, but there have been trends in the components. Injuries at the scene of a fire emergency have declined from 65 percent of the total in 1981 to 52 percent in 1993. One reason the total remained so high was that injuries at the scene of a non-fire emergency have been increasing, from 9 percent of the total in 1981 to 16 percent in 1993.

INFORMATION AND ANALYSIS

Fire protection is becoming increasingly scientific. The advent of sophisticated new data bases, measurement techniques, and computer-based models has produced a pace of change unlike anything seen before. The *Fire Protection Handbook* has reflected this acceleration of knowledge, as dozens of whole chapters simply did not exist as recently as two editions ago.

There is an increasing demand for hard evidence of the effectiveness and the cost-effectiveness of fire protection features, systems, codes, and standards. New technologies judged by old standards may be admitted too slowly or too quickly to the marketplace. The old "build and burn" approach to gathering information on the fire performance of materials, products, assemblies, buildings, systems, and features is very expensive and is increasingly seen as wasteful when valid alternatives exist.

Validity is the key, however. Fire protection engineers will find themselves having to deal with and control change in the content of their discipline at an ever-increasing rate. Use of unvalidated tests or models can bring dire consequences, but so can a failure to adapt to new techniques. Much of the material within this section, as well as the scientific fundamentals that underline new methods, receives more elaborate treatment in *The SFPE Handbook of Fire Protection Engineering*, published jointly by NFPA and the Society of Fire Protection Engineers.[7]

BIBLIOGRAPHY

References Cited

1. NCFPC, "America Burning," 1973, report of the U.S. National Commission on Fire Prevention and Control, U.S. Government Printing Office, Washington, DC.
2. Fire Analysis Division, "The Deadliest Fires and Explosions of the 1900s," *Fire Journal*, Vol. 82, No. 3, May/June 1988, pp. 48-54.
3. *Learn Not to Burn* (curriculum and public service announcements), National Fire Protection Association, Quincy, MA, 1987.
4. Hall, J. R., Jr., Bukowski, R., and Gombeg, A., "Analysis of Electrical Fire Investigations in Ten Cities," NBSIR 83-2803, Dec. 1983, National Bureau of Standards, Center for Fire Research, Gaithersburg, MD, p. 56.
5. Hall, J. R., Jr., "U.S. Experience with Smoke Detectors," National Fire Protection Association Fire Analysis and Research Division, Quincy, MA, June 1995.
6. Hall, J. R., Jr., "U.S. Experience with Sprinklers," National Fire Protection Association Fire Analysis and Research Division, Quincy, MA, June 1995.
7. Society of Fire Protection Engineers and National Fire Protection Association, *SFPE Handbook of Fire Protection Engineering*, 2nd ed., DiNenno, P.J., *et al.*, eds., SFPE and NFPA, Quincy, MA, 1995.

NFPA Codes, Standards, and Recommended Practices

Reference to the following NFPA codes, standards, and recommended practices will provide further information on the elements of fire protection discussed in this chapter. (See the latest version of *The NFPA Catalog* for availability of current editions of the following documents.)

NFPA 30, *Flammable and Combustible Liquids Code*
NFPA 54, *National Fuel Gas Code*
NFPA 58, *Standard for the Storage and Handling of Liquefied Petroleum Gases*
NFPA 70, *National Electrical Code®*
NFPA 101®, *Life Safety Code®*
NFPA 102, *Standard for Grandstands, Folding and Telescopic Seating, Tents, and Membrane Structures*

Additional Reading

The NFPA Fire Analysis and Research Division, through its "One-Stop Data Shop" program, offers more than a hundred reports and packages that elaborate on the points in this chapter. Most are updated annually.
"1994 Learn Not to Burn Champions," *NFPA Journal*, Vol. 88, No. 3, May/June 1994, pp. 67–69.
"Arson Fire Kills 87 in Worst Mass Slaying in U.S. History," *Fire Control Digest*, Vol. 16, No. 4, April 1990, pp. 1–4.
Brushlinsky, N. N., *et al.*, " Russia's 1993 Fire Statistics," *Fire Technology*, Vol. 30, No. 4, 1994, pp. 458–467.
Catchpole, L., "Fires in the Food Industry," *Fire Prevention*, No. 285, December 1995, pp. 21–25.
Damant, G. H., and Nurbakhsh, S., "Christmas Trees—What Happens When They Ignite?," *Fire and Materials: An International Journal*, Vol. 18, No. 1, January—February 1994, pp. 9–16.

"Fire Statistics United Kingdom, 1993," Home Office, London, September 1995, 191 pages.
"Fire Statistics: Summary of Major Fires, 1992," *Fire Protection*, Vol. 20, No. 4, December 1993, pp. 24–25.
Hall, J. R., Jr., "Total Cost of Fire in the United States Through 1993," National Fire Protection Association, Quincy, MA, October 1995, 21 pages.
Hall, J. R., Jr., "U.S. Arson Trends and Patterns, 1990," National Fire Protection Association, Quincy, MA, October 1991, 56 pages.
Hall, J. R., Jr., "U.S. Fire Problem Overview Report Through 1990. Leading Causes and Other Patterns and Trends," National Fire Protection Association, Quincy, MA, February 1992, 122 pages.
Hall, J. R., Jr., "U.S. Fire Problem Overview Report Through 1994. Leading Causes and Other Patterns and Trends," National Fire Protection Association, Quincy, MA, April 1996, 131 pages.
Hall, J. R., Jr., Taylor, K. T., and Sullivan, M. J., "Large-loss Fires Top $2.6 Billion in Damage in 1991," *NFPA Journal*, Vol. 86, No.6, November/December 1992, pp. 40–47, 74, 80.
Jerome, I., "Update on Arson Fires, 1990," *Fire Prevention*, No. 242, September 1991, p. 16.
Karter, M. J., Jr., "Fire Loss in the United States During 1990," *Fire Journal*, Vol. 85, No. 5, September/October 1991, pp. 36-38, 40–42, 44–46, 48.
Karter, M. J., Jr., "U.S. Fire Experience by Region: 1986-1990," National Fire Protection Association, Quincy, MA, January 1992, 40 pages.
Karter, M. J., Jr., "U.S. Fire Experience by Region: 1989-1993," National Fire Protection Association, Quincy, MA, March 1995, 47 pages.
Louderback, J., "Arverne Conflagration of 1992: 141 Structures Destroyed in the Rockaways Section of Queens, New York," *Firehouse*, Vol. 18, No.1, January 1993, pp. 67, 69.
Miller, A. L. and Tremblay, K. J., "342 Die in Catastrophic Fires in 1991," *NFPA Journal*, Vol. 86, No. 4, July/August 1992, pp. 62-73.
"National Fire Loss 1994," *Fire Protection*, Vol. 22, No. 2, June 1995, pp. 12–22.
Nolan, D. P., "Statistical Review of Fires and Explosion Incidents in the Gulf of Mexico 1980-1990," *Journal of Fire Protection Engineering*, Vol. 7, No. 3, 1995, pp. 99-105.
Pagni, P. J., "Causes of the 20 October 1991 Oakland Hills Conflagration," *Fire Safety Journal*, Vol. 21, No. 4, 1993, pp. 331–340.
Queen, P. R., "Conflagration in Oakland!," *American Fire Journal*, Vol. 43, No. 12, December 1991, pp. 12–15.
Randall, J., and Jones, R.T., "Teaching Children Fire Safety Skills," *Fire Technology*, Vol. 29, No. 4, 1993, pp. 268-280.
Runyan, C. W., *et al.*, " Risk Factors for Fatal Residential Fires," *Fire Technology*, Vol. 29, No. 2, 1993, pp. 183–193.
Scoones, K., "Fires in Hotels During 1990," *Fire Prevention*, No. 249, May 1992, pp. 11–12.
Scoones, K., "FPA Large Loss Analysis for 1993," *Fire Prevention*, No. 286, January/February 1996, pp. 42–50.
Scoones, K., "Serious Fires in Educational Establishments During 1993," *Fire Prevention*, No. 283, October 1995, pp. 35–37.
Scoones, K., "Serious Fires in Manufacturing Industries in 1993," *Fire Prevention*, No. 285, December 1995, pp. 31–33.
Scoones, K., "Serious Fires in the Leisure/Retail Industry During 1992," *Fire Prevention*, No. 272, September 1994, pp. 17–18.
Sekizawa, A., "Statistical Analysis on Fatalities Characteristics of Residential Fires," *Proceedings of the Third International Symposium on Fire Safety Science*, Elsevier Applied Science, London, 1991, pp. 475–484.
Taylor, K. T., and Sullivan, M. J., "Large Loss Fires in the United States in 1990," *NFPA Journal*, Vol. 85, No. 6., November/December 1991, pp. 28–32, 34, 70, 92–98.
Tremblay, K. J., "Catastrophic Fires and Deaths Drop in 1992," *NFPA Journal*, Vol. 87, No. 5, September/October 1993, pp. 56–62, 64, 66–69.
Tremblay, K. J., "Catastrophic Fires of 1994," *NFPA Journal*, Vol. 89, No. 5, September/October 1995, pp. 48–54, 56–58, 60, 62, 64, 66–68, 70.

FUNDAMENTALS OF FIRE SAFE BUILDING DESIGN

Revised by Robert W. Fitzgerald

Building design and construction practices have changed significantly during the past century. A little over one hundred years ago, structural steel was unknown, reinforced concrete had not been used in structural framing applications, and the first "high-rise" building had just been built in the United States.

The design professions have also advanced significantly during the past century. The practice of architecture has changed markedly, and techniques of analysis and design that were unknown a century or even a generation ago are available to engineers today. Building design has become a very complex process, with many skills, products, and technologies integrated into its system.

Fire protection has made developmental strides in the building industry similar to those of other professional disciplines. At the turn of the century, conflagrations were a common occurrence in cities. In later years, increased knowledge of fire behavior and building design enabled buildings to be constructed in such a manner that a hostile fire could be confined to the building of origin rather than to the block or larger areas. Progress has continued in the field of fire protection so that, at the present time, knowledge is available that enables a hostile fire to be confined to the room of origin or even to smaller spatial subdivisions in a structure.

DESIGN AND FIRE SAFETY

Much activity is taking place today regarding fire safe building design. The general thrust of some developments appears to be directed toward quantification procedures and identification of a rational design methodology to parallel or supplement the traditional "go or no go" specifications approach. Knowledge in the field of fire protection is undergoing development and reorganization that will enable buildings to be designed for fire safety more rationally and efficiently. This section of the Handbook identifies the components of a field that is changing dynamically in its analysis and design capabilities.

"America Burning," the report of the National Commission on Fire Prevention and Control,[1] identifies several areas in which building designers create unnecessary hazards, often unwittingly, for the building occupants. In some cases, these unnecessary hazards are the result of oversight or insufficient understanding of the interpretations of test results. In other cases, they are due to a lack of knowledge of fire safety standards or failure to synthesize an integrated fire safety program.

The Commission's report cites that conscious incorporation of fire safety into buildings too frequently is given minimal attention by the designer and, further, that building designers and their clients are often content only to meet the minimum safety standards of the local building code. Frequently, both assume incorrectly that the codes provide completely adequate measures rather than minimal ones, as is the case. In other instances, building owners and occupants see fire as something that will never happen to them, as a risk that they will tolerate because fire safety measures can be costly, or as a risk adequately balanced by the provisions of fire insurance or availability of public fire protection.

Conditions arising from these attitudes need not exist, much less continue. Information is available for design professionals to incorporate a greater measure of fire protection into their designs. Use of fire protection information requires that the various members of the building design team recognize that fire conditions are a legitimate element of their design responsibilities. This requires a greater understanding of the special loadings that fire causes on building features and of the countermeasures that can be incorporated into designs.

The material in this chapter identifies the components of a complete fire safety system. The organizational structure may be used as a basis with which to evaluate the relative safety both of new designs and also of existing buildings.

Evaluating a design for building fire safety represents a systematic approach to the six principal types of fire safety strategies identified in Section 1, Chapter 1, "America's Fire Problem and Fire Protection," i.e., prevention, slowing of initial growth and spread, detection, suppression, compartmentation, and evacuation, in the context of the choices that can be built into or installed in a building. This chapter describes in general terms the processes required to create such a design. More specific guidance requires joining the general processes described here with the more detailed guidance in later chapters on the specific fire safety strategies. Also, this chapter addresses an additional building strategy: designing the building to facilitate fire department operations should these become necessary.

Objectives of Fire Safe Design

The conscious, integrated process of design for building fire safety, if it is to be effective and economical, must be integrated into the complete architectural process. All members of the traditional

Robert W. Fitzgerald, Ph.D., P.E., is professor of fire protection engineering and civil engineering at Worcester Polytechnic Institute, Worcester, MA.

building design team should incorporate, as an integral part of their work, design for emergency fire conditions. The earlier in the design process that fire safety objectives are established, alternative methods of accomplishing those objectives are identified, and engineering design decisions are made, the more effective and economical the final results.

As the first step in the process, setting objectives is part of clearly identifying the specific needs of the client with regard to the function of the building. After the building functions and client needs are understood, the designer must consciously ascertain both the general and the unique conditions that influence the level of fire risk that can be tolerated in the building. The acceptable levels of risk and the focus of the fire safety analysis and design process are concentrated in the following three areas:

1. Life safety.
2. Property protection.
3. Continuity of building operations.

It is difficult to ascertain the level of risk that will be tolerated by the owner, occupants, and community. Often it is necessary to put a conscious effort into recognizing the sensitivity of the occupants, contents, and mission of the building as to the products of combustion. Consequently, fire safety criteria often are not identified in a clear, concise manner that enables the designer to provide appropriate protection for the realization of the design objectives. Unfortunately, it is impossible to provide more than general guidelines that must be considered in building design to assist in the identification of the fire safety objectives in this handbook. Specific objectives must be developed for each individual building.

Life safety: Adequate life safety design for a building is often related only to compliance with the requirements of local building regulations. This may or may not provide sufficient occupant protection, depending upon the particular building function and occupant activities.

The first step of life safety design is to identify the occupant characteristics of the building. What are the physical and mental capabilities of the occupants? What are the range of their activities and locations during the 24-hour, seven-day-a-week periods? Are special considerations needed for certain periods of the day or week? In short, the designer must anticipate the special life safety needs of occupants during the entire period in which they inhabit the building.

The identification of life safety objectives is usually not difficult, but it does require a conscious effort. In addition, it requires an appreciation of the time and extent to which the products of combustion can move through the building. The interaction of the building response to the fire and the actions of its occupants during the fire emergency determines the level of risk that the building design poses.

Property protection: Specific items of property that have a high monetary or other value must be identified in order to protect them adequately in case of fire. In some cases, specially protected areas are needed. In other cases, a duplicate set of vital records in another location may be adequate. The establishment of the fire safety objectives should ascertain if the user of the building has property that requires special fire protection.

Continuity of building operations: The maintenance of operational continuity after a fire is the third major design concern. The amount of "downtime" that can be tolerated before revenues begin to be seriously affected must be identified. Frequently, certain functions or locations are more essential to the continued operation of the building than others. It is important to recognize those areas particularly sensitive to building operations, so that adequate protec-

tion is provided for the vital business operations conducted in them. Often, these areas need special attention that is not required throughout the building.

In modern buildings, the value of the contents of a single room may be extremely high. This value may be due to the cost of equipment or records, or to the high cost of business interruption. The sensitivity of equipment and data to the effects of heat, smoke, gases, or water must be addressed. In any event, the designer should protect the specially sensitive rooms from products of a fire originating either inside or outside of the room.

In addition to objectives relating to the people, property, and mission that is provided by the building under consideration, two additional types of objectives are useful for some occupancy functions. The first is design for the protection of neighbors. A recognition of the potential effect of exposure fires may influence the design sufficiently to mitigate potential problems. The second considers the impact on the environment from problems such as runoff of chemicals housed in the building that dissolved in fire department water applications. Waterborne or airborne products of combustion produced in buildings that house certain chemicals can affect the environment significantly. Also consideration for the safety of firefighting personnel responding to a building fire should be taken into account.

ELEMENTS OF BUILDING FIRE SAFETY

Fire Prevention

As noted earlier, the first opportunity to achieve fire safety in a building is through fire (ignition) prevention, which involves separating potential heat sources from potential fuels. Table 1-2A lists

TABLE 1-2A.   *Fire Prevention Factors*

1. Heat Sources
   a. Fixed equipment
   b. Portable equipment
   c. Torches and other tools
   d. Smoking materials and associated lighting implements
   e. Explosives
   f. Natural causes
   g. Exposure to other fires
2. Forms and Types of Ignitable Materials
   a. Building materials
   b. Interior and exterior finishes
   c. Contents and furnishings
   d. Stored materials and supplies
   e. Trash, lint, and dust
   f. Combustible or flammable gases or liquids
   g. Volatile solids
3. Factors that Bring Heat and Ignitable Material Together
   a. Arson
   b. Misuse of heat source
   c. Misuse of ignitable material
   d. Mechanical or electrical failure
   e. Design, construction, or installation deficiency
   f. Error in operating equipment
   g. Natural causes
   h. Exposure
4. Practices that Can Affect Prevention Success
   a. Housekeeping
   b. Security
   c. Education of occupants
   d. Control of fuel type, quantity, and distribution
   e. Control of heat energy sources

common factors in fire prevention and identifies major candidate heat sources and ignitable materials, common factors that bring them together, and practices that can affect the success of prevention.

Most building fires are started by heat sources and ignitable materials that are brought into the building, not built into it. This means the design of the building, from the architect's and builder's standpoint, provides limited potential leverage on the building's future fire experience. The building's owners, managers, and occupants, however, will have numerous opportunities to reduce fire risks through prevention, and they should be urged to do so.

For design purposes, fire prevention will be enhanced by careful observance of codes and standards in the design and installation of the electrical and lighting system, the heating system, and any other major built-in equipment, such as cooking, refrigeration, air conditioning, and clothes washing and drying. Venting systems need to be designed carefully to carry carbon monoxide and potential fuels along protected paths. These venting systems will need to be inspected and cleaned regularly.

Protection from lightning and exposure fires will affect the external design of the building, particularly in certain parts of the country, such as areas near wildlands. A fire in one building creates an external fire hazard to neighboring structures by exposing them to heat by radiation, and possibly by convective currents, as well as to the danger of flying brands of the fire. Any or all of these sources of heat transfer may be sufficient to ignite the exposed structure or its contents.

When considering protection from exposure fires, there are two basic types of conditions: (1) exposure to horizontal radiation, and (2) exposure to flames issuing from the roof or top of a burning building in cases where the exposed building is higher than the burning building. Radiation exposure can result from an interior fire where the radiation passes through windows and other openings of the exterior wall. It can also result from the flames issuing from the windows of the burning building or from flames of the burning facade itself. A source for guidelines and data on exposure protection is given in NFPA 80A, *Recommended Practice for Protection of Buildings from Exterior Fire Exposures.*

Inside the building, design features may make incendiarism, arson, or other human-caused fires more or less likely by making security and housekeeping easier or harder to perform. The interaction of the design with these critical support activities should be thought through and planned into the design from the outset.

Spread of Fire and Products of Combustion

The concern here is with slowing the fire to provide other fire safety measures with sufficient time to be effective. A systematic design for this purpose should address the possible ways that hazard can grow rapidly, e.g., flame spread, rapid growth in rate of heat release or rate of mass release, unusually toxic gases, unusual corrosivity, quantity of fuel available to feed the fire, and so forth. Each of these can be evaluated separately in terms of the threat to exposed people, property, and mission of the building. The building design should provide effective countermeasures.

In a building fire, the most common hazard to humans is from smoke and toxic gases. Most building-related fire deaths are directly related to these products of combustion. Death often results from oxygen deprivation in the bloodstream, caused by the replacement of oxygen in the blood hemoglobin by carbon monoxide. In addition to the danger of carbon monoxide, many other toxic gases that are present in building fires cause a wide range of symptoms, such as headaches, nausea, fatigue, difficult respiration, confusion, and impaired mental functioning.

Smoke, in addition to accompanying toxic and irritant gases, contributes indirectly to a number of deaths. Dense smoke obscures visibility and irritates the eyes and can cause anxiety and emotional shock to building occupants. Consequently, the occupant may not be able to identify escape routes and utilize them. (For more information, see Section 4, Chapter 2, "Combustion Products and Their Effects on Life Safety.")

Although heat injuries do not compare in quantity to those caused by inhalation of smoke and toxic gases, they are painful, serious, and cause shock to victims. In addition to deaths from thermal products of combustion, the pain and disfigurement caused by nonfatal burns can result in serious, long-term complications.

Property also is affected by the thermal and nonthermal products of combustion, as well as by extinguishing agents. Smoke may damage goods located long distances from the effects of the heat and flames. Fires that are not extinguished quickly often result in considerable water damage to the contents and the structure, unless special measures are incorporated to prevent that damage. It should be noted, however, that the water damage caused in extinguishing a fire rarely exceeds the fire damage resulting from a fire that is not suppressed.

Fast flame spread over finish materials or building contents and vertical propagation of fire are serious concerns. The ability of the fire service to contain or extinguish a fire is diminished significantly if the fire spreads vertically to two or more floors. With a given potential for fire growth, the prevention of vertical fire spread is influenced principally by architectural and structural decisions involving details of compartmentation, which are discussed later.

Designing Countermeasures to Fire Growth

The building fire safety system can be organized around the fire growth and its resulting products of combustion, i.e., flame/heat and smoke/gas. The ease of generation and movement of these products is influenced by the countermeasures provided by the building. The effectiveness of the building fire safety systems determines the speed, quantity, and paths of movement of these products of combustion.

The speed and certainty of fire growth and development in rooms can vary greatly. The contents and interior finish in some rooms are quite safe, and, for this type of situation, it is unlikely that, once ignited, a fire can grow to full involvement of the room. On the other hand, the interior design of other rooms poses a high hazard which, if an ignition were to occur, could lead to an almost certain full room involvement.

The traditional method of describing the fire growth hazard has been through fire loads reflected in use and occupancy classifications. Building types, rather than rooms within buildings, have been grouped with regard to their relative hazard. For example, residential and educational occupancies are considered low hazard, because they normally contain relatively low fuel loads in the rooms. Mercantile buildings are normally a moderate hazard, while certain industrial and storage buildings may be considered a high hazard because they contain a high fuel load.

This type of classification is a basis for building and fire code requirements, and, historically, it has been quite useful. However, a more detailed look at the fire growth potential within the rooms of a building can be a valuable part of a detailed fire safety design. The fire growth hazard potential, which identifies the speed and relative likelihood of a fire reaching full room involvement, is a useful base from which to design suppression interventions and to evaluate life safety problems. For example, situations where fast, severe fires will occur may call for automatic sprinkler protection, even though it may not be required by a building or fire code.

The basis for a fire growth hazard analysis is the combustion characteristics in a room. The main factors that influence the likelihood and speed with which full room involvement occurs are: (1) fuel load (i.e., the quantity, type of materials, and their distribution); (2) interior finish of the room; (3) air supply; and (4) size, shape, and construction of the room.

Fire development in a room is neither uniform nor a guaranteed progression from ignition to full room involvement. Fires develop through several stages, called realms. Table 1-2B provides guidance on descriptions of the realms. Within any realm a fire may continue to grow or it may be unable to sustain continued development and die down. Table 1-2B includes a rough guide to the approximate flame sizes that may be used to describe the fire size of the realms. It also describes the major factors that influence growth within a realm. Absence of a significant number of the factors would indicate that the fire would self-terminate, rather than continue to develop.

TABLE 1-2B. *Major Factors Influencing Fire Growth*

Realm	Approximate Ranges of Fire Sizes	Major Factors that Influence Growth
1 (Preburning)	Overheat to ignition.	Amount and duration of heat flux. Surface area receiving heat material ignitability.
2 (Initial Burning)	Ignition to radiation point. [10 in. (254 mm) high flame].	Fuel continuity. Material ignitability. Thickness. Surface roughness. Thermal inertia of the fuel.
3 (Vigorous Burning)	Radiation point to enclosure point [10 in. to 5 ft high flame (254 mm to 1.5 m)].	Interior finish. Fuel continuity. Feedback. Material ignitability. Thermal inertia of the fuel. Proximity of flames to walls.
4 (Interactive Burning)	Enclosure point to ceiling point [5 ft. (1.5 m) high flame to flame touching ceiling].	Interior finish. Fuel arrangement. Feedback. Height of fuels. Proximity of flames to walls. Ceiling height. Room insulation. Size and location of openings. HVAC operation.
5 (Remote Burning)	Ceiling point to full room involvement.	Fuel arrangement. Ceiling height. Length/width ratio. Room insulation. Size and location of openings. HVAC operations.

Different rooms pose different levels of risk regarding the likelihood of reaching full room involvement and the time in which fire development takes place. The factors in Table 1-2B provide a general guide to the important types of factors.

If one were to focus on a single event that might be used to represent the relative level of risk posed by the contents and interior finish in a room, it would be the ability of flames to reach the ceiling. The arrangement of contents and types of fuels where it would be difficult for a fire to grow to touch the ceiling pose a relatively low fire growth hazard potential. On the other hand, where furniture combustibility and density will allow a fire to develop to ceiling height, or when combustible interior finish is present, the fire growth hazard potential usually is comparatively high.

Detection and Alarm

Fire detection provisions are needed so that automatic or manual fire suppression will be initiated, any other active fire protection systems will be activated (e.g., automatic fire doors for compartmentation and protection of escape routes), and occupants will have time to move to safe locations, typically outside the building.

One reason for concern over any rapid initial fire growth is that it can shrink the time available after detection for these life- and property-saving responses. Therefore, detection provisions must be designed systematically to reflect the building's other features, its occupants, and its other fire safety features.

For example, smoke is often the first indicator of fire, so a system of automatic detection based on smoke detectors often makes sense. In certain properties or areas, however, detectors based on heat or rate of increase in heat may be more appropriate because of the types of fires likely to occur in those areas or because of the potential for non-fire activations in those areas. Whatever type of detection system is chosen, it is important that, for each area of the building, a realistic assessment be made of the implications for response time after the fire is detected and before a lethal or other high-hazard condition develops.

Alarm provisions need not be linked to the detection sensor locations, but should be designed systematically to tell occupants what they need to do, based on where they are and their ability to respond. This would include the possible use of central annunciator panels and monitors to inform responsible staff, voice messages to provide instructions on occupant movements, and direct remote alarms to supervised stations or fire departments. All of these options will have an impact on the time available for some type of response and, possibly, on the efficiency of that response. A timeline can be constructed to provide a quantitative base for analysis and design of this and related building fire safety features.

Automatic Suppression

For nearly a century and a half, automatic sprinklers have been the most important single system for automatic control of hostile fires in buildings. Many desirable aesthetic and functional features of buildings that might offer some concern for fire safety because of the fire growth hazard potential can be protected by the installation of a properly designed sprinkler system.

Among the advantages of automatic sprinklers is the fact that they operate directly over a fire and are not affected by smoke, toxic gases, and reduced visibility. In addition, much less water is used because only those sprinklers fused by the heat of the fire operate, particularly if the building is compartmented.

Other automatic extinguishing systems, e.g., carbon dioxide, dry chemical, clean (halon replacement) agents, and high-expansion foam, may be used to provide protection for certain portions of buildings or types of occupancies for which they are particularly suited.

An automatic sprinkler system has been the most widely used method of automatically controlling a fire. The major elements for determining the effectiveness of an automatic sprinkler system are: (1) its presence or absence; (2) if present, its reliability; and (3) if reliable, its design and extinguishing effectiveness.

Although automatic sprinkler systems have a remarkable record of success, it is possible for the system to fail. Often failure is due to a feature that could have been avoided if appropriate attention had been given at the time of the system's design, installation, or maintenance. Table 1-2C describes common failure modes and their causes. During the design stages, these factors should be addressed to increase the probability of successful extinguishment by the sprinkler system.

TABLE 1-2C. *Common Automatic Sprinkler Failure Modes*

Failure Mode	Potential Causes
Water supply valves are closed when sprinkler fuses.	Inadequate valve supervision. Owner attitude. Maintenance policies.
Water does not reach sprinkler.	Dry pipe accelerator or exhauster malfunctions. Pre-action system malfunctions. Maintenance and inspection inadequate.
Nozzle fails to open when expected.	Fire rate of growth too fast. Response time and/or temperature of link inappropriate for the area protected. Sprinkler link protected from heat. Sprinkler link painted, taped, bagged, or corroded. Sprinkler skipping.
Water cannot contact fuel. (Note: The intent of this failure mode is to ensure that discharge is not interrupted in a manner that will prevent fire control by a sprinkler.)	Fuel is protected. High-piled storage is present. New construction (walls, ductwork, ceilings) obstructs water spray.
Water discharge density is not sufficient.	Discharge needs are insufficient for the type of fire and the rate of heat release. Change in combustible contents occurred. Number of sprinklers open is too great for the water supply. Water pressure is too low. Water droplet size is inappropriate for the fire size.
Enough water does not continue to flow.	Water supply is inadequate because of original deficiencies, changes in water supply, or changes in the combustible contents. Pumps are inadequate or unreliable. Power supply malfunctions. System is disrupted.

Compartmentation

Barriers, such as walls, partitions, and floors, separate building spaces. These barriers also delay or prevent fire from propagating from one space to another. In addition, barriers are important features in any fire-fighting operation.

The effectiveness of a barrier is dependent upon its inherent fire resistance, the details of construction, and the penetrations, such as doors, windows, ducts, pipe chases, electrical raceways, and grilles. Although the hourly ratings of fire endurance do not always represent the actual time that the barrier can withstand a building fire, unpenetrated rated barriers seem to perform rather well. This may be due to the rather large factor of safety inherent in the codes. On the other hand, it is quite common for rated barriers to fail because of inattention to penetrations. For example, the fire resistance of a rated floor-ceiling assembly can be voided because of large or

numerous poke throughs. The fire resistance of a rated partition is lost when a door is left open.

Fire resistance requirements imposed by the regulatory system often have comparatively little meaning because of inattention to the functional and construction details. To predict field performance of barriers, the penetrations and details of construction must be considered, in addition to the fire endurance of the base construction.

The major function of barriers is to prevent heat and flame spread from causing an ignition in an adjacent room or floor. It is useful to classify barrier failure in two categories. One is a massive barrier failure, which would occur when a part of the barrier collapses or when a large penetration, such as a door or a large window, is open. When a massive failure occurs, the adjacent room can become fully involved in a short period of time. The second type of failure is a localized penetration failure. This occurs when flames or heat penetrates small poke throughs or small windows. A localized penetration failure causes a hot spot to occur. If fuel is present and ignition occurs, this could lead to a full room involvement by the normal fire development progression.

Smoke and gases move through a building much faster and more easily than flames and heat. The time duration from ignition until a building space is untenable is an important aspect of fire safety, and the loss of tenability may be due to smoke and gases more often than flames and heat. Therefore, barriers need to be designed and considered as barriers to the spread of smoke and fire gases, too.

In addition to its value as means of containing the fire, compartmentation also addresses specific needs for protection, such as structural integrity of the building and escape routes.

The collapse of structural building elements can be a serious life safety hazard. Although statistically it has not resulted in many deaths or injuries to building occupants, structural collapse is a particular hazard to fire fighters. A number of deaths and serious injuries to fire fighters occur each year because of structural failure. While some of these failures result from inherent structural weaknesses, many are the result of renovations to existing buildings that materially, though not obviously, affect the structural integrity of the support elements. A building should not contain surprises of this type for fire fighters.

The potential for structural collapse must be determined. Building codes address this aspect through construction classification requirements. The relationship between fire severity and fire resistance to collapse are the principal factors.

Collapse can occur when the fire severity exceeds the fire endurance of the structural frame. However, this is comparatively rare. Structural collapse is more commonly associated with deficiencies in construction. These deficiencies are not evident under normal, everyday use of the building. They become a problem when the fire weakens supporting members, triggering a progressive collapse.

Design for Evacuation and Occupant Movement

The design for life safety may involve one or a combination of the following three alternatives: (1) evacuation of the occupants, (2) defending the occupants in place, or (3) providing an effective area of refuge. These alternatives can be evaluated by the likelihood that the building spaces will be tenable for the period of time necessary to achieve the expected level of safety. The criteria for tenability, therefore, becomes an important part of the design.

Evacuation: The design for building evacuation involves two major components: (1) the availability of an acceptable path or paths for escape, and (2) the effective alerting of the occupants in sufficient time to allow egress before segments of the path of egress become untenable.

Alerting occupants to the existence of a fire is a vital part of the life safety design. A useful performance objective could be to identify that occupants should have at least *x* minutes to escape from the time they *know* of a fire until the escape route is blocked. To accomplish this, the designer either must ensure that the fire and the movement of its products of combustion will be slow enough to provide that time, or incorporate special provisions into the building to achieve that objective.

Defending in place: The second life safety design alternative is to defend the individual in place. This may be appropriate for occupancies such as hospitals, nursing homes, prisons, and other institutions. It may be an appropriate alternative for other buildings when the size or design may show that evacuation has an unacceptably low likelihood of success. Defend-in-place design also uses a performance criteria of time and tenability levels.

The performance criteria relating to time might state that the building space should be tenable for *y* minutes after the start of the fire. The duration for *y* could be identified as a period much longer than the duration of any possible fire. The definition of tenability may be quite different from that acceptable for evacuation because of the influence of both time and the products of combustion.

Refuge: The third alternative is to design for an area of refuge. This involves occupant movement through the building to specially designed refuge spaces. This type of design is more difficult than either of the other two alternatives because it involves the major design aspects of each. In certain types of buildings this may be a reasonable alternative. However, an evaluation of the effectiveness of the area of refuge design and its likelihood of success are extremely important.

Life safety design for a building is difficult. It involves more than a provision for emergency egress. It must also address the population who will be using the building and what they will be doing most of the time. Consideration must then be given to communication, the protection of escape routes, and temporary or permanent areas of refuge for a reasonable period of time for the building occupants to achieve safety.

Even occupants familiar with their surroundings often experience difficulty in locating means of egress. The problem is compounded for transients and occasional visitors to the building. Architectural layout and normal circulation patterns are important elements in emergency evacuation. For example, many large office buildings are a maze of offices, storage areas, and meeting rooms. Clearly marked emergency travel routes can enhance life safety features in all buildings.

Design for Fire Department Operations

The protection offered by a community fire department has an important influence on building fire design. Some buildings are designed in a manner that helps the fire department extinguish fires while they are small; others are designed in a manner that hinders a fire department. Rarely does the designer consciously design the building for emergency operations. The following discussion provides some guidelines for building design to enhance the building's ability to allow the fire department to extinguish a fire with minimal threat to life and property.

Ideally, a building is designed so that should a fire occur, it can be attacked before it extends beyond the room of origin. If that is not possible, the building design and construction features should retard fire spread so that the fire department will encounter a relatively small, easily controllable fire. The major aspects of this part of building design include: (1) fire department notification, (2) initial agent application, (3) fire extinguishment, (4) ventilation, (5) water supply and use, (6) water removal, and (7) barrier effectiveness. These aspects are discussed briefly to provide guidance for incorporating features into the building that enable departments to be more effective and less harmful to the building.

Fire department notification: The complete chain of events, i.e., (1) detection of the fire, (2) decision to inform the fire department, (3) sending of the message, and (4) correct receipt of the information by the fire department, should be a part of every building fire safety design. It should be consciously designed, rather than left to chance. The time durations for completing the events through agent application are very dependent upon the speed of the fire spread. Buildings have been lost because of insufficient attention to the method of notifying the local fire department.

Agent application: The next critical event is fire department application of agent to the fire. This involves three distinct events for its success: (1) arrival at the site, (2) nozzle entrance into the room, and (3) water discharge from the nozzle. Each of these events can be affected by site or building access considerations in the design.

Ideal exterior accessibility occurs where a building can be approached from all sides by fire department apparatus. This is not always possible. In congested areas, only the sides of buildings facing streets may be accessible. In other areas, topography or constructed obstacles can prevent effective use of apparatus in combating the fire.

Some buildings located some distance from the street make the approach of apparatus difficult. If obstructions or topography prevent apparatus from being located close enough to the building for effective use, fire-fighting equipment, e.g., aerial ladders, elevating platforms, and water tower apparatus, are rendered useless. Valuable labor must be expended to hand carry hose lines or ground ladders long distances.

The matter of access to buildings has become far more complicated in recent years especially in light of the movement to secure buildings against possible terrorist attacks. The building designer must consider this important aspect in the planning stage. Inadequate attention to site details can place the building in an unnecessarily vulnerable position. If its fire defenses are compromised by preventing adequate fire department access, the building must compensate with more complete internal building protection.

The arrival at the site is only a part of the agent application evaluation. The fire fighters then must be able to enter the building, reach the floor of the fire, and find the involved room or rooms. This is often a time-consuming, difficult task. Considerable attention must be given to the problem of finding the fire and getting fire fighters and equipment to the fire.

Access to the interior of a building can be greatly hampered where large areas exist and where buildings have blank walls, false facades, solar screens, or signs covering a high percentage of exterior walls. Obstacles that prevent ventilation allow smoke to accumulate and obscure fire fighters' vision. Lack of adequate interior access also can delay or prevent fire department rescue of trapped occupants.

Windowless buildings and basement areas present unique fire-fighting problems. The lack of natural ventilation facilities, such as windows, contributes to the buildup of dense smoke and intense heat, which hamper fire-fighting operations. Fire fighters must attack fires in these spaces despite heat and smoke. This can result in lengthy times for fire extinguishment and greater damage by the products of combustion.

Fire extinguishment: After the time-consuming and sequential events of notification and initial agent application have transpired, the fire department is ready to fight the fire. The size of fire that is present at the time of initial agent application determines the firefighting strategy and likelihood of success of the operation.

Broadly speaking, three categories of fire conditions may be expected: (1) comparatively small fires may be extinguished by direct application of water; (2) when the fire is larger than can be directly extinguished, the building is opened (ventilated), and the hose streams drive the fire, heat, and smoke out of the building; and (3) fires that are too large for this operation must be surrounded. All available techniques of ventilation and heat absorption by water evaporation are used; however, the fire area is lost, and the main purpose of this strategy is to protect exposures, both external and internal.

Ventilation: Ventilation is an important fire-fighting operation. It involves the removal of smoke, gases, and heat from building spaces. Ventilation of building spaces performs the following important functions:

1. Protection of life by removing or diverting toxic gases and smoke from locations where building occupants must find temporary refuge.
2. Improvement of the environment in the vicinity of the fire by removal of smoke and heat. This enables fire fighters to advance close to the fire to extinguish it.
3. Control of the spread or direction of fire by setting up air currents that cause the fire to move in a desired direction. In this way, occupants or valuable property can be more readily protected.
4. Provision of release for unburned, combustible gases before they develop a flammable mixture, thus avoiding a backdraft or smoke explosion.

The building designer should be conscious of these important functions of fire ventilation and provide effective means of facilitating venting practices whenever possible. This may involve access panels, movable windows, skylights, or other means of readily opened spaces in case of a fire emergency. Emergency controls on the mechanical equipment or inclusion of an engineered smoke-control system may also be an effective means of accomplishing the functions of fire ventilation. Each building has unique features, and, consequently, a unique solution should be incorporated into the design.

Water supply and use: Water is the principal agent used to extinguish building fires. Although other agents may be employed occasionally (e.g., carbon dioxide, dry chemical, foams and surfactants, and clean halon replacement agents), water remains the primary extinguishing agent of the fire service. Consequently, the building designer should anticipate the needs of both the fire department and automatic extinguishing systems and provide an adequate supply of water at adequate residual pressure.

Water normally is supplied to the building site by mains that are part of the water distribution system. Few cities can supply a sufficient amount of water at required pressures to every part of the city. Consequently, water supplied to hydrants, standpipes, or sprinklers must be boosted by pumps located on fire department apparatus or in the buildings themselves. Buildings that do not have an adequate, reliable water source for fire fighting must either provide supplemental water or incorporate other fire defense measures to compensate for this deficiency.

Careful attention must be given to water supply, distribution, and pressure for emergency fire conditions. High-rise buildings are particularly sensitive in this respect because the water pressures that are required depend upon building height. The water supply needs of large buildings must also be given careful attention.

Fire conditions that require operation of a large number of sprinklers or use of a large number of hose streams can reduce pressure in standpipe and sprinkler systems to the point where residual pressures in the distribution system are adversely affected. Fire department connections for sprinkler and standpipe systems are important components of building fire defenses. The building designer must carefully consider installation details of fire department connections to make sure they will be easily located, readily accessible, and properly marked. Locations should be approved by the local fire department.

Water removal: Watertight floors are important in this respect. Salvage efforts can be greatly affected by the integrity of the floors. Of greater importance is the number and location of floor drains. If interior drains and scuppers are available, salvage teams can effectively remove water with a minimum of damage to the structure.

BIBLIOGRAPHY

Reference Cited

1. NCFPC, "America Burning," the report of the National Commission on Fire Prevention and Control, 1973, Superintendent of Documents, U.S. Government Printing Office, Washington, DC.

NFPA Codes, Standards, and Recommended Practices

Reference to the following NFPA codes, standards, and recommended practices will provide further information on the fundamentals of fire safe building design discussed in this chapter. (See the latest version of _The NFPA Catalog_ for availability of current editions of the following documents.)

NFPA 13, _Standard for the Installation of Sprinkler Systems_
NFPA 14, _Standard for the Installation of Standpipe and Hose Systems_
NFPA 22, _Standard for Water Tanks for Private Fire Protection_
NFPA 24, _Standard for the Installation of Private Fire Service Mains and Their Appurtenances_
NFPA 70, _National Electrical Code®_
NFPA 72, _National Fire Alarm Code_
NFPA 80A, _Recommended Practice for Protection of Buildings from Exterior Fire Exposures_
NFPA 92A, _Recommended Practice for Smoke-Control Systems_
NFPA 92B, _Guide for Smoke Management Systems in Malls, Atria, and Large Areas_
NFPA _101®_, _Life Safety Code®_
NFPA 232, _Standard for the Protection of Records_
NFPA 241, _Standard for Safeguarding Construction, Alteration, and Demolition Operations_
NFPA 263, _Standard Method of Test for Heat and Visible Smoke Release Rates for Materials and Products_
NFPA 264A, _Standard Method of Test for Heat Release Rates for Upholstered Furniture Components or Composites and Mattresses Using an Oxygen Consumption Calorimeter_
NFPA 1231, _Standard on Water Supplies for Suburban and Rural Fire Fighting_

Additional Reading

Allen, W., "Fire and the Architect—the Communication Problem," _Fire Safety Journal_, Vol. 14, No. 4, 1989, pp. 205–220.
Anchor, R. D., Malhotra, H. L., and Purkiss, J. A., eds., _Proceedings_ of the International Conference on Design of Structures Against Fire, Elsevier, New York, 1986.
Barker, J. A., "Training: The Key to Fire Safety," _Fire Prevention_, No. 224, Nov. 1989, pp. 28–30.

Beck, V. R., and Poon, S. L., "Results from a Cost-Effective Decision-Making Model for Building Fire Safety and Protection," *Fire Safety Journal*, Vol. 13, Nos. 2 & 3, 1988, pp. 197–210.

"Design Guide: Structural Fire Safety—CIBW14 Workshop Report," *Fire Safety Journal*, Vol. 10, No. 2, 1986, pp. 75–138.

"Designing Effective Fire Protection," *Civil Engineering*, July/August 1988, p. 2.

Dunning, G., *Fireproof, Fire-Resistant or Semi-Combustible Designing for Fire Safety in High-Rise Buildings: A Partially Annotated Bibliography*, Vance Bibliographies, Monticello, IL, 1989.

Egan, M. D., *Concepts in Building Firesafety*, R. Krieger, Malabar, FL, 1986.

Fire Test Standards, 2nd ed., American Society for Testing and Materials, W. Conshohocken, PA, 1988.

Gilardi, S. A., "Planning for Fire Safety," *Construction Specifier*, Vol. 41, No. 4, April 1988, pp. 33–34.

Glon, P. L., "Building Safety—A New Educational Approach," *Building Standards*, Vol. 56, No. 2, 1987, pp. 20–21.

Harmathy, T. Z., "Properties of Building Materials," *SFPE Handbook of Fire Protection Engineering*, National Fire Protection Association, Quincy, MA, 1995.

The Institution of Engineers, *Fire Safety Engineering for Building Structure and Safety*, E. A. Books, Crows Nest, Australia, 1989.

Iwankiw, N., "Design Aids for Fire Protection," *Proceedings*—Solutions in Steel, The National Engineering Conference, AISC, 1986, p. 23.

Koyamatsu, T. J., "Automatic Fire-Extinguishing Systems," *Building Standards*, Vol. 56, No. 3, 1987, pp. 5–8.

Ley, M., *Grading Methods of Building Fire Safety*, M. S. Thesis, Worcester Polytechnic Institute, Worcester, MA 1987.

Luck, H., *et al.*, "New Technologies in Automatic Detection and Suppression," *Proceedings* of the Conference on New Technologies to Reduce Fire Losses and Costs, Elsevier, 1986, pp. 211–218.

Malhotra, H. L., *Fire Safety in Buildings*, Building Research Establishment, Borehamwood, England, 1987.

Morikawa, T., Yanai, E., and Nishina, T., "Evaluation of Toxic Gases from Experimental Fires in an Existing Building," 9th Joint Panel Meeting of the UJNR Panel on Fire Research and Safety, NBSIR 88-3753, April 1988, National Bureau of Standards, Gaithersburg, MD, pp. 443–454.

Morris, J., "Life Safety and Sprinkler Protection," *Fire Prevention*, No. 209, May 1988, pp. 18–21.

Nakamura, H., "Investment Model of Fire Protection Equipment for Office Buildings," *Proceedings* of the First International Symposium on Fire Safety Science, Hemisphere, Washington, DC, 1986, pp. 1019–1028.

Nelson, H. E., "An Engineering Fire Protection Design Assessment System," 9th Joint Panel Meeting of the UJNR Panel on Fire Research and Safety, NBSIR 88-3753, April 1988, National Bureau of Standards, Gaithersburg, MD, pp. 157–171.

Nelson, H. E., "Room Fires as a Design Determinate—Revisited," *Fire Technology*, Vol. 26, No. 2, May 1990, pp. 99–105.

Nelson, H. E., and Walton, W. D., "Basic Structure of the Fire Protection Design Assessment System," NBSIR 85-3298, Feb. 1986, National Bureau of Standards, Gaithersburg, MD.

Ow, C. S., and Thin, C. C., "Upgrading Fire Protection in Existing Buildings," *Fire International*, Vol. 12, No. 112, August/September 1988, pp. 39–40.

Quintiere, J., "Analytical Methods for Firesafety Design," *Fire Technology*, Vol. 24, No. 4, 1988, pp. 333–352.

Rogowski, B. F. W., "Investigating the Contribution to Fire Growth of Combustible Materials Used in Building Components," *Proceedings* of the Conference on New Technology to Reduce Fire Losses and Costs, Elsevier, New York, 1986, pp. 88–105.

Shields, T. J., *Buildings and Fire*, Wiley, New York, 1987.

Singh, J., and Thomas, P. M., "Calculating the Overall Fire Risk in New Building Designs," *Fire Prevention*, No. 224, Nov. 1989, pp. 32–36.

Twilt, L., and Witteveen, J., "Trends in Fire Safety Design of Buildings," *Heron*, Vol. 32, No. 4, 1987, pp. 95–114.

Vance, M., *Fireproof Building: A Bibliography*, Vance Bibliographies, Monticello, IL, 1988.

Volkmann, P., "Three Leading Factors to Save Life and Property," *Fire International*, Vol. 12, No. 109, February/March 1988, pp. 39–40.

White, A. G., *Fire Safety Workplace Security: A Selected Bibliography*, Vance Bibliographies, Monticello, IL, 1986.

Yurkonis, P. R., "Evaluating the Fire Safety of an Existing Building," *Facilities Management, Operation, and Engineering*, Vol. 16, No. 2, 1989, p. 16.

SYSTEMS CONCEPTS FOR BUILDING FIRE SAFETY

Revised by John M. Watts, Jr.

The problem of fire safety in buildings is overwhelming both in the number of variables and in the subsequent difficulty of obtaining detailed data. The established approach of specification codes and standards has limitations for many modern structures and for older buildings of historic significance. An alternative approach to dealing with this kind of situation is systems analysis. In the broadest sense, systems analysis is simply the methodical study of an entity as a whole. The objective is to define a credible process for making the best decision from among the alternatives.

This chapter introduces the concepts of systems and systems thinking. It then briefly describes principles of, and a generalized procedure for, systems analysis. Systems approaches to fire problems are discussed, and the structure and uses of the NFPA fire safety concepts tree are described.

Systems analysis is a way of identifying and using interrelationships of the factors and principles presented in Section 1, Chapter 2, "Fundamentals of Fire Safe Building Design." It also encompasses many of the ideas and procedures discussed in Section 11, Chapter 7, "Fire Hazard Analysis," and Section 11, Chapter 8, "Fire Risk Analysis."

SYSTEMS

Russell Ackoff, in his book *Redesigning the Future*,[1] says we are living in the "systems age," an emerging intellectual era that is producing the "post-industrial revolution." Attention has expanded from elements to wholes with interrelated parts, i.e., to systems. Phenomena are now thought of as parts of a larger system rather than in terms of their components. It is recognized that optimizing the components does not optimize the system, in the same way that an all-star team does not play as well as the league champion.

The idea of a system has many common examples, such as automobile exhaust system, digestive system, democratic system of government, sewer system, ecosystem, fire suppression system, etc. By definition a system is a set of components that work together for an overall objective. Sprinklers, piping, alarm valve, and water supply are components of an automatic sprinkler system, with an overall objective of fire control. Similarly, fire resistance, fire detection, and automatic sprinklers may be thought of as components of a sys-

tem, the objective of which is to provide an acceptable level of fire safety in a building.

One of the fundamental characteristics of a systems analysis is that the system being studied is rigorously defined, i.e., in systematic fashion. This enables one to apply the concept of a system to many activities that are not ordinarily thought of as systems, e.g., a ship at sea, supermarket checkout, a warehouse, bidding for a contract, a family, the world,* etc.

Characteristics of Systems

A system is often highly complex. This makes a rigorous definition difficult. All systems, however, have certain common characteristics that, if properly identified, will usually define the system adequately. These characteristics are:

1. Boundary,
2. Input and output,
3. Variables, and
4. Structure.

The interrelationships among these characteristics are illustrated in Figure 1-3A.

Boundary: The difference between systems analysis and a more generalized systems approach is the difference between "closed" and "open" systems.[2] In systems analysis, a system is considered to be bounded in such a manner that the system behavior of interest is generated entirely within the boundary; thus, it is a closed system. The system may interact with its environment, but outside events do

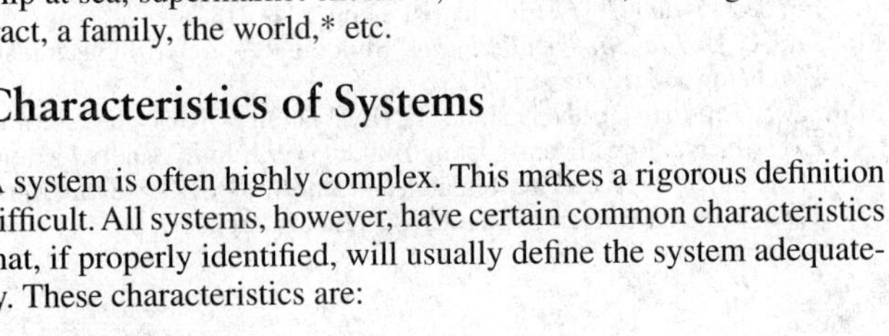

FIG. 1-3A.   *Characteristics of systems.*

John M. Watts, Jr., Ph.D., is director of the Fire Safety Institute, a not-for-profit information, research, and educational corporation, located in Middlebury, VT. He also serves as editor of NFPA's quarterly technical journal, *Fire Technology.*

*Indeed, the world has been systematically analyzed. See *World Dynamics*, by J. W. Forrester, Wright-Allen, Cambridge, MA, 1973.

not themselves govern the behavior of the system; i.e., there is no feedback mechanism between the system and external elements. For example, power failure is an event that may affect a fire safety system, but electrical generation and distribution may not be included within the system boundary. In building fire safety, the boundary consists of the structure and its contents. When fire service personnel enter the building, they become part of the system.

Input and output: A system that has input and output is a dynamic system. The system must act in some manner to convert the input to output. The conversion mechanism is the essence of the system. Building fire safety is a dynamic system, with time varying interactions between the fire and structure. The input is an ignition scenario and the output is some measure of success of the system, e.g., limitation of fire spread.

Variables: Certain components of the system may be under control of the designer or decisionmaker. These are the variables of the system. A system usually consists of a very large number of variables, many of which defy quantification. And the complexity of the system is a rapidly increasing function of the number of variables. There is a very large number of decision variables in building fire safety design, e.g., physical and chemical properties of the materials of the building and its contents, geometrical features, attributes of detection and suppression systems, etc. The purpose of a systems analysis is to arrive at the most desirable values of these variables.

Structure: System structure is the overall framework relating the variables within the system boundary. There are many basic structural forms that systems may take. Networks are a common structure of many systems. Among the various types of network structures are series-parallel, source-sink, decision, hierarchical, time sequence, logic, information flow, open-loop-closed-loop, and signal flow.

This list is not exhaustive, nor are the items mutually exclusive. Most systems are a combination of several such basic structures. Later in this chapter a hierarchical logic network approach to building fire safety—the fire safety concepts tree—is discussed.

SYSTEMS ANALYSIS

Systems analysis evolved in the 1950s as a popular spinoff of space and defense research. It is not a well-defined discipline in that it combines elements of many principles of scientific study with less formalized craftsmanship. The term "systems analysis" is used to describe a careful, methodical, comprehensive, and well-documented step-by-step approach to solving a problem about a system.

Systems analysis is both systemic and systematic. These two important terms are derived from the concept of a system. Systemic refers to the wholeness of things in an inclusive sense. For example, a systemic disease is one that afflicts the whole body. A systemic approach to fire safety is one that considers every part of the system. The alternative to a systemic approach is to look at only one piece

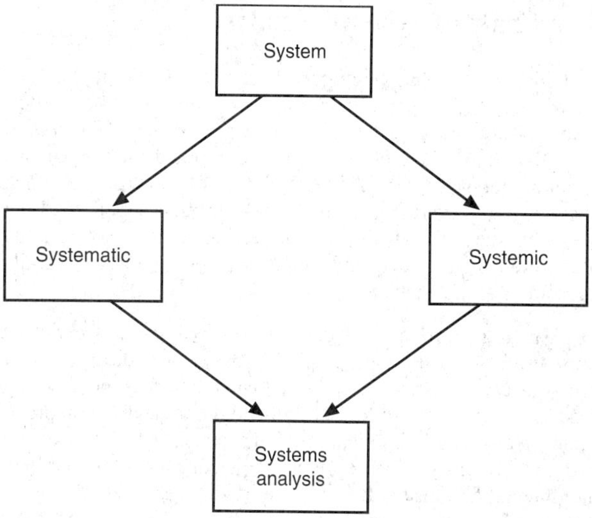

FIG. 1-3B. Systems analysis.

of a problem at a time, ignoring the mechanisms by which the pieces are interrelated. This is sometimes referred to as disjoint incrementalism and is an all-too-frequent approach to fire safety.

Systematic refers to being methodical or acting according to a plan and not casually or at random. Repeatable methodology is the key characteristic of science, and science is the most acceptable way of increasing knowledge. A systematic approach establishes a basis for widespread acceptance, as opposed to a visceral approach. Systems analysis is therefore methodical or systematic, and inclusive or systemic. Thus, a systems analysis can be defined as a methodology applied to the whole system. (See Figure 1-3B.) This is also referred to as a systems approach.

Systems Analysis and Scientific Method

Scientific method may be thought of as a systems concept. However, it is generally applied to "hard" rather than "soft" systems. Hard systems have components whose interrelationships are well defined and easily quantified, while soft systems contain inexact expressions. It is the need to resolve problems relating to soft systems that has given prominence to systems analysis.

Intellectual activities, such as science and systems analysis, may be represented on a scale of reproducibility. This scale is defined by its endpoints. At one end of the scale is pure science, which is characterized by its 100 percent reproducibility. At the other end of the scale is pure art, which has a reproducibility of zero (e.g., the Mona Lisa). In between these points lie the problem-solving activities of engineering and systems analysis, with engineering being closer to science and systems analysis closer to art. (See Figure 1-3C.)

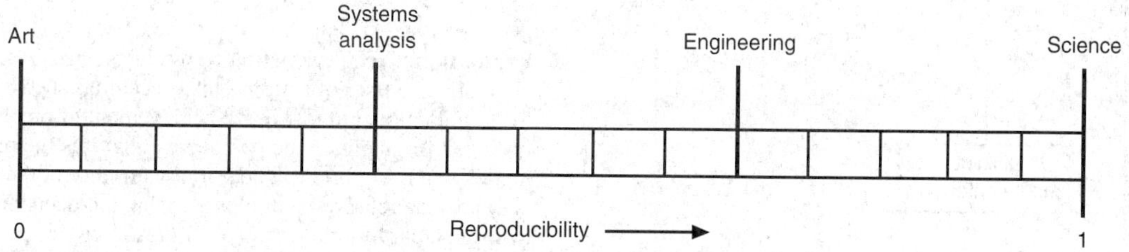

FIG. 1-3C. Problem-solving methodologies.

The Craft of Systems Analysis

Systems analysis may be defined as the systematic application of knowledge, skills, logic, and intuition to solve a problem about a system. Knowledge of the system undergoing analysis is required, as well as knowledge of techniques appropriate to the analysis. A systems analysis is often a multidisciplinary team project so that the knowledge of several individuals can be brought together. Skills in the techniques of analysis are as necessary as in any other craft or profession. Among such skills, logic is necessary for the overall organization of the analysis.

The logical framework of a systems analysis is essentially the same used by most people implicitly or unconsciously in making everyday decisions. As the decision system becomes more complex, it becomes necessary to formalize the logic of the decision process. Intuition also has a significant role in systems analysis. A complex system has many gaps in verifiable knowledge that must be filled by judgment and intuition. It is this use of judgment and intuition that distinguishes systems analysis from more structured techniques, and it is their systematic application that distinguishes systems analysis from visceral problem-solving.

Choosing Problem-Solving Methodologies

Approaches to problem-solving can be categorized in a hierarchical fashion according to the size of the system they address. (See Figure 1-3D.) The extremes of this hierarchy are the total systems approach, which takes the most global view of the problem, and disjoint incrementalism, which addresses only the smallest pieces, individually, without consideration of their interrelationships. Most present-day fire safety analyses fall between these end points.

The systemic approach yields the most efficient solution, but requires a great deal of time-consuming analysis. It is desirable to move as far up this scale as is feasible. Where deterministic physical modeling can take one only so far toward a total systems approach, it is necessary to consider alternative mathematical models, such as probability theory, decision analysis, utility theory, etc., and, beyond their limits, to consider more qualitative models, such as heuristics, delphi, expert systems, etc.

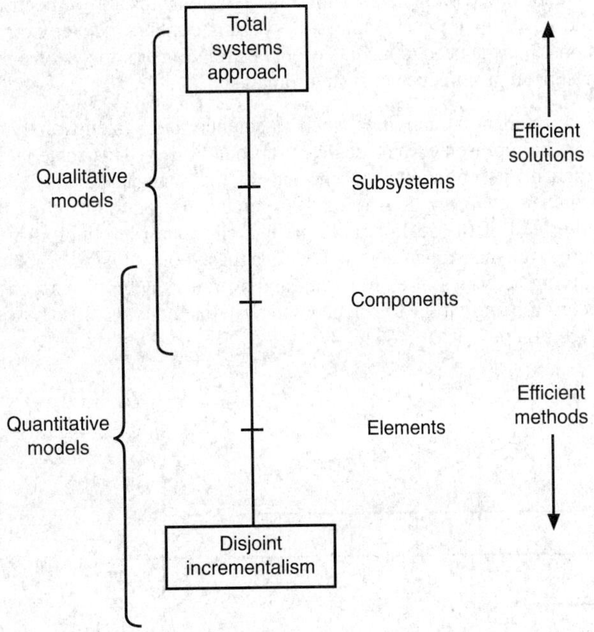

FIG. 1-3D. Problem-solving hierarchy.

Systems Analysis—A Procedure

Most descriptions of systems analysis outline a sequence of tasks to be performed in a more or less iterative fashion. Many of these schemes are quite varied and complex, yet may include similar components. Systems analyses tend to take their character from the analyst and from the problem addressed; hence, they may bear little resemblance to each other in detail. Specific techniques differ from study to study, and there may be only a thread of similarity that ties these studies together. The most elementary representation of this thread comprises three basic steps: (1) definition, (2) modeling, and (3) evaluation. (See Figure 1-3E.) This generalized three-part procedure begins with an analysis not of the problem but of the system within which the problem occurs. The second step is the synthesis of system activity into a conceptual model, and the third step is a comparison and questioning of the responsiveness of the model.

Definition: Problem definition is the first and most important step in a systems analysis. It is difficult to get a right answer from the wrong problem. Concise definition of the problem is the output of this initial step.

The problem definition process should provide a basis for understanding, communication, and verification. Real-world problems (as opposed to textbook theory) evolve in an amorphous form. It is necessary to reformulate the problem into a form amenable to analysis, and this requires a detailed understanding of the system. Thus, problem definition has two parts: (1) rigorous study of the system and (2) development of a well-defined statement of the problem.

Problem definition usually includes identification of the scope, objectives, measures of effectiveness, variables, and interrelationships. The scope limits the commitment to the problem within the system and includes assumptions and constraints. Objectives are the desirable output of the system, while measures of effectiveness identify the degree to which objectives are achieved. The variables are those factors that can be manipulated to achieve objectives, and the interrelationships identify the known interactions of the variables within the system structure. Thorough documentation of the problem definition is essential to communication and validation of the analysis.

Modeling: Modeling is the formal representation of the understanding gained of the system in the problem definition step. This representation takes the form of a symbolic model of the system. It may be diagrammatic, mathematical, computerized, or, often, some combination of all three. The behavior of the system may be conveniently studied by manipulating the model rather than manipulating the system itself. For such a study to have validity, the model must closely represent the system's behavior of interest. However, the more the model is like the system, the more difficult it is to manipulate (like the system itself); therefore, approximations and simplifying assumptions are required to make the model tractable. Thus, an acceptable model must be sufficiently analogous to the real problem to evaluate alternatives with the accuracy to permit sound decisions, yet simple enough to be amenable to analysis.

Evaluation: The evaluation step selects, analyzes, and compares alternative courses of action. This is accomplished by manipulating the system model to maximize achievement of objectives. The analyst needs to maintain the perspective that it is the problem, not the model, that is important. The primary purpose of the evaluation step is to provide relatively simple rules that the decisionmaker can use to eliminate inferior alternatives. In a comprehensive study, evaluation also includes implementation of the best alternative and monitoring to ensure that expected results are actually realized.

Any scientific investigation is essentially an iterative process, and these steps are not always followed sequentially but, more often, cyclically. The modeling process may suggest refinements to the problem definition, while evaluation may suggest revisions to the model or additions to the problem definition. (See Figure 1-3E.)

Documentation

A professional systems analysis requires that, as the study proceeds, a record be made of assumptions, data, and parameter estimates, and why they were chosen; model structure and details; steps in the analysis; relevant constraints; results; sensitivity tests; validation; etc. This documentation is invaluable to the analyst, as well as to those who may be considering use of the results. One example of the type of documentation necessary is found in ASTM E1472, *Standard Guide for Documenting Computer Software for Fire Models.*

Clear and complete documentation enhances confidence in the analysis; its absence inevitably carries with it the opposite effect. No approach to building fire safety should be taken seriously, unless its development has left an adequate "paper trail."

SYSTEMS APPROACHES APPLIED TO THE FIRE PROBLEM

Fire safety can be incorporated into building design by three different methods:

1. Mandate that design and construction conform to prescriptive requirements in specification-oriented building codes and standards. Such requirements are based on fire experiences and are generally strict.
2. Use performance codes to overcome the inflexibility of specification codes. A present limitation of performance criteria is that building components are considered on an *ad hoc* basis. The definition of fire safety performance of all the separate components and the measurement of that performance are difficult to obtain. Also, when considering the entire building as a system, the satisfactory physical performance of individual components does not ensure the desired level of safety for the building as a whole, as some components have a higher probability of successful performance than others.
3. Use a systems analysis that incorporates an integrated method that addresses the aesthetic, functional, structural, electrical, mechanical, and other relevant subsystems. Buildings can be designed for fire safety through the use of systems analysis rather than by strict compliance with codes. This approach to fire safety requires a higher degree of professional technology; however, it can achieve a greater level of cost effectiveness than strict enforcement of codes, and most often allows greater flexibility.

Evaluation of current levels of building fire safety, and the recommendations for upgrading, can only be accomplished properly if the impact of a change in a structure can be evaluated in context. For example, if a building is to be rehabilitated, which requires bringing it up to code level, does this mean strict compliance, or are there alternative means to achieve the same level of fire safety? It is also desirable to consider all other factors bearing on the "fire safety system" when evaluating the impact of a code or standard change on the level of fire safety. In addition, before catastrophic fires occur, it is highly desirable to determine the impact of new processes and materials on the level of fire safety provided through compliance with a given code or standard.

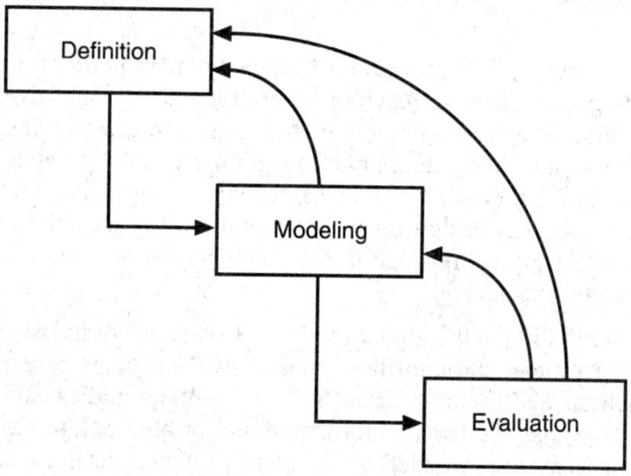

FIRE SAFETY CONCEPTS TREE

During the 1960s there was a growing awareness that modern high-rise buildings designed to building codes and standards contained potential fire safety weaknesses. Various researchers and organizations, principally the U.S. General Services Administration (GSA) and the National Research Council of Canada (NRCC), identified aspects of weakness. Studies identifying the time of evacuation of high-rise buildings were published, and the problems relating to air movement in tall buildings were identified.

About the same time, NFPA formed a committee on high-rise structures. After initial meetings, this committee determined that its contribution should not be in developing standards for high-rise construction, but rather in identifying a system of fire safety for all buildings. It became the NFPA Technical Committee on Systems Concepts for Fire Protection in Structures. After many modifications, the fire safety concepts tree, shown in Figures 1-3F through 1-3J, evolved. The tree was first published in 1974 and was incorporated into NFPA 550, *Guide to the Fire Safety Concepts Tree* (hereinafter referred to as NFPA 550) in 1986. The Committee on Systems Concepts was discharged in 1990, and the NFPA Standards Council assumed responsibility for reconfirmation of NFPA 550 in 1995.

Structure of the Tree

The fire safety concepts tree, as presented in Figures 1-3F through 1-3J, shows the elements that must be considered in building fire safety and the interrelationship among those elements. It enables a

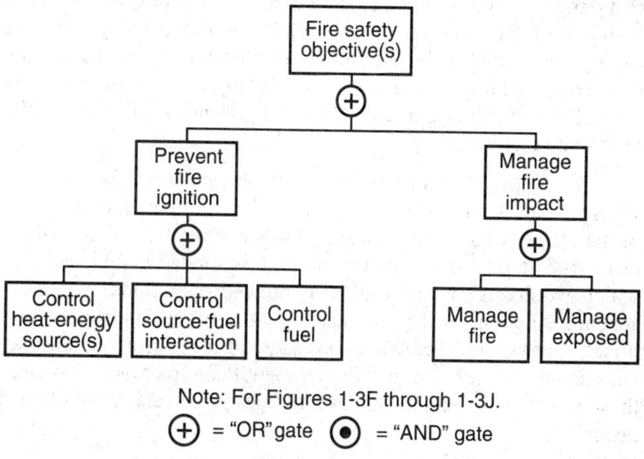

FIG. 1-3E.   Basic steps and cycles of a systems analysis.

Note: For Figures 1-3F through 1-3J.
⊕ = "OR" gate   ⦿ = "AND" gate

FIG. 1-3F.   The principal branches of the fire safety concepts tree.

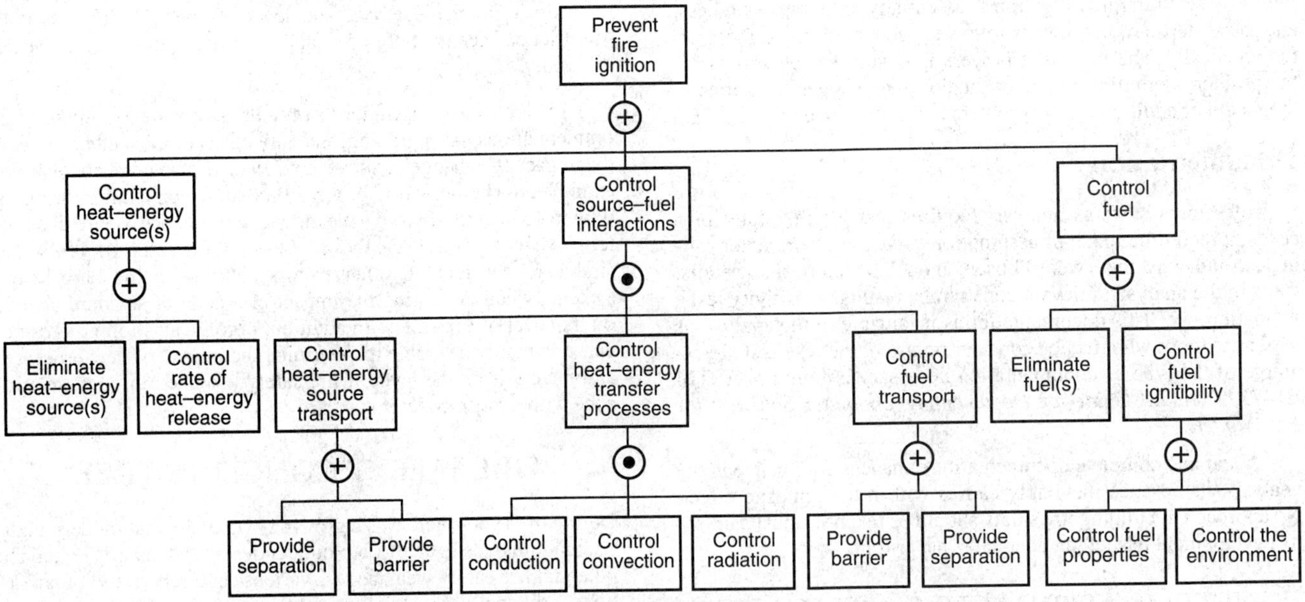

FIG. 1-3G.   *Components of the "prevent fire ignition" branch of the fire safety concepts tree.*

building to be analyzed or designed by progressively moving through the various concepts in a logical manner. Its degree of success depends upon how completely each level is satisfied. Lower levels on the tree, however, do not represent a lower level of importance or performance; they represent a means for achieving the next higher level. In addition to the material in this chapter, see NFPA 550 for detailed guidance in the use of the tree.

Fire Safety Objectives

Use of the fire safety concepts tree requires that the fire safety objectives (goals) be clearly identified. These objectives describe the degree to which the building should protect its occupants, property and contents, continuity of operations, and neighbors. The objectives should be quantified wherever possible, rather than stated in broad or general terms.

The life safety objective, for example, might state that all occupants be safeguarded against the intolerable or untenable effects of the fire. It may be further stated that emergency personnel, such as fire fighters, who may be expected to stay in areas considered too dangerous for the occupants, should be protected against possible collapse of the building or entrapment. A range of specific life safety objectives may be appropriate for different types of occupancies. Nursing home requirements, for example, are vastly different from those for offices, and both differ from industrial occupancies or storage facilities.

To identify the property protection objectives, the building designer must ask a number of questions. For example, does a portion of the property have a significantly higher value than the rest of the property? If destroyed, which property cannot be replaced and which would significantly affect continued operations? Are there specific functions at the site that are vital to continuity of operations? Further, are there functions that can be performed at other sites on an emergency basis? Proper identification of factors such as these enable the designer to tailor the design to reflect the client's needs.

The fire safety concepts tree is also especially useful in situations that are not addressed by other approaches, such as traditional codes and standards. An example is historic structures, which are irreplaceable artifacts of history and culture with immeasurable intangible value. Here, special consideration is necessary to design fire and life safety into such structures without unduly sacrificing historic integrity.

"Prevent Fire Ignition" or "Manage Fire Impact"

The fire safety concepts tree provides the logic required to achieve fire safety; i.e., it provides conditions whereby the fire safety objectives can be satisfied, but it does not provide the minimum condition required to achieve those objectives. Thus, according to the tree, the fire safety objectives can be met if fire ignition can be prevented or if, given ignition, the fire can be managed. This logical "OR" function is represented by the symbol (+) under fire safety objectives in Figure 1-3F.

The "prevent fire ignition" branch of Figure 1-3G essentially represents a fire prevention code. Most of the concepts described in this branch require continuous monitoring for success. Consequently, the responsibility for satisfactorily achieving the goal of fire prevention is essentially an owner/occupant responsibility. The designer, however, may be able to incorporate certain features into the building that may assist the owner/occupant in preventing fires.

It is impossible to prevent completely the ignition of fires in a building. Therefore, to reach the overall fire safety objective, from a building design viewpoint, a high degree of success in the "manage fire impact" branch assumes a significant role. Essentially, this branch of the fire safety concepts tree may be considered as a building code by the design team. After ignition occurs, all considerations shift to the "manage fire impact" branch to achieve the fire safety objectives.

According to the logic of the tree, the impact of the fire can be managed either through the "manage fire" or "manage exposed" branches. (See Figure 1-3F.) The "OR" (+) gate indicates that the objectives may be reached through either or both design branches, as long as the avenue selected completely satisfies the fire safety objective. Naturally, it is acceptable to do both, which will increase the likelihood of success over using one branch only.

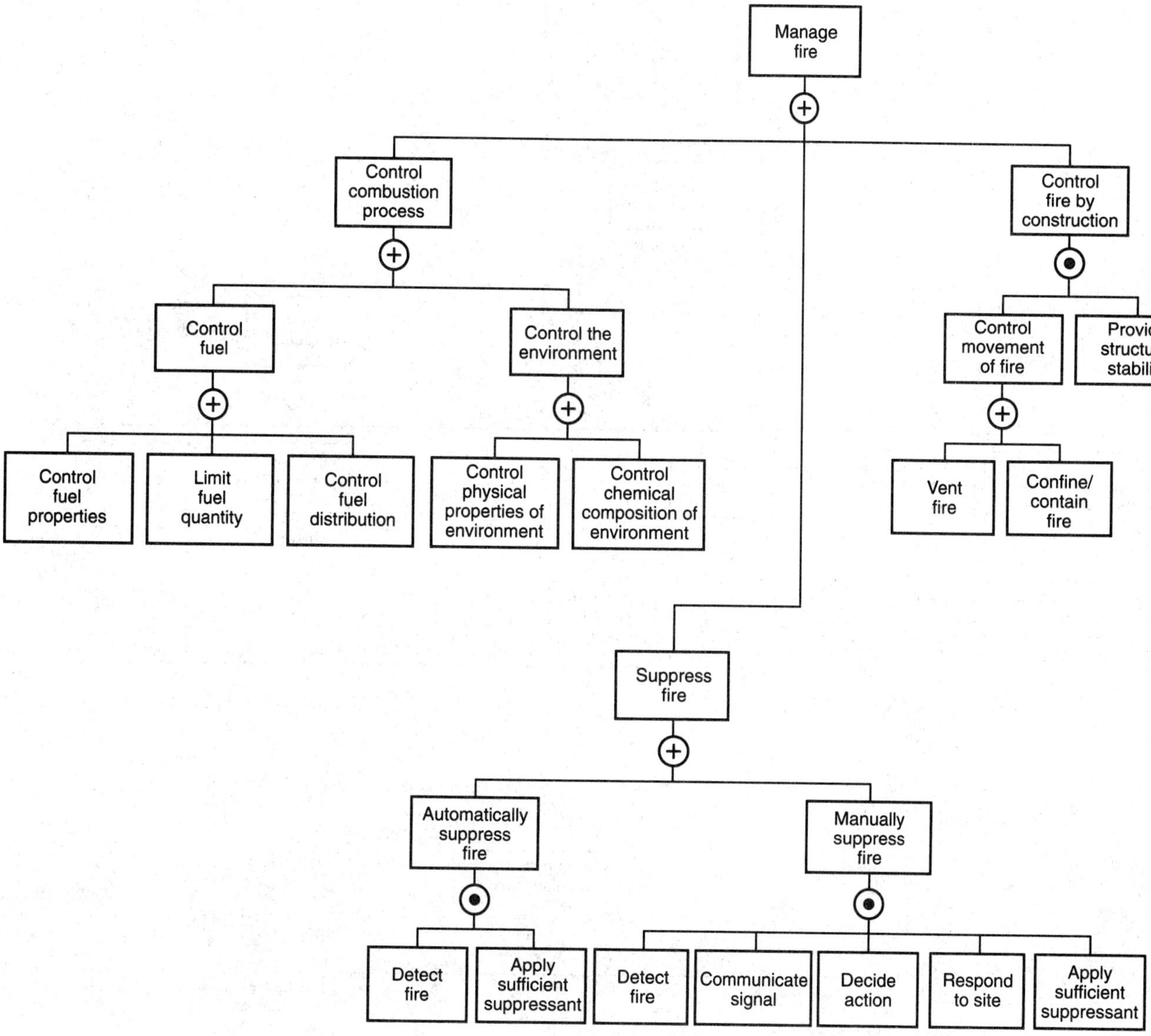

FIG. 1-3H.   *Components of the "manage fire" branch of the fire safety concepts tree.*

Through the "manage fire" branch, the fire safety objectives can be achieved by managing the fire itself. Figure 1-3H shows that this can be accomplished by: (1) controlling the combustion process, (2) suppressing the fire, or (3) controlling the fire by construction. Here, again, any one of these branches of the tree will satisfy the "manage fire" concept. For example, in some fires, success is achieved where the building construction controlled the fire. In other fires, success is achieved by controlling the combustion process, either through control of the fuel or the environment.

The "suppress fire" event and its branches are also shown in Figure 1-3H. In this figure, the symbol (•) represents a logical "AND" gate, and signifies that all of the elements in the level immediately below the gate are necessary to achieve the concept above the gate. To accomplish automatic suppression, for example, both concepts, i.e., detecting the fire and applying sufficient suppressant, are necessary. Similarly, to manually suppress the fire, five concepts must occur. The omission of any single concept is sufficient to break the chain and cause the failure of suppression to manage the fire.

In considering the "control fire by construction" concept, structural integrity must be provided, and the movement of the fire itself must be controlled. As shown in Figure 1-3H, this can be accomplished either by venting, confining, or containing the fire.

Manage Exposed

As shown in Figure 1-3I, the fire impact can be lessened by managing the "exposed," i.e., people, property, or functions, depending upon the design aspects being considered. In considering the "manage exposed" branch, it can be successful either by limiting the amount exposed or by safeguarding the exposed. For example, the number of people as well as the amount or type of property in a space may be restricted. Often, this is impractical. If this is the case, the objectives can still be met by incorporating design features to safeguard the exposed.

The exposed people or property may be safeguarded either by moving them to a safe area of refuge or by defending them in place. For example, people in institutionalized occupancies, such as hospitals, nursing homes, or prisons, must generally be defended in

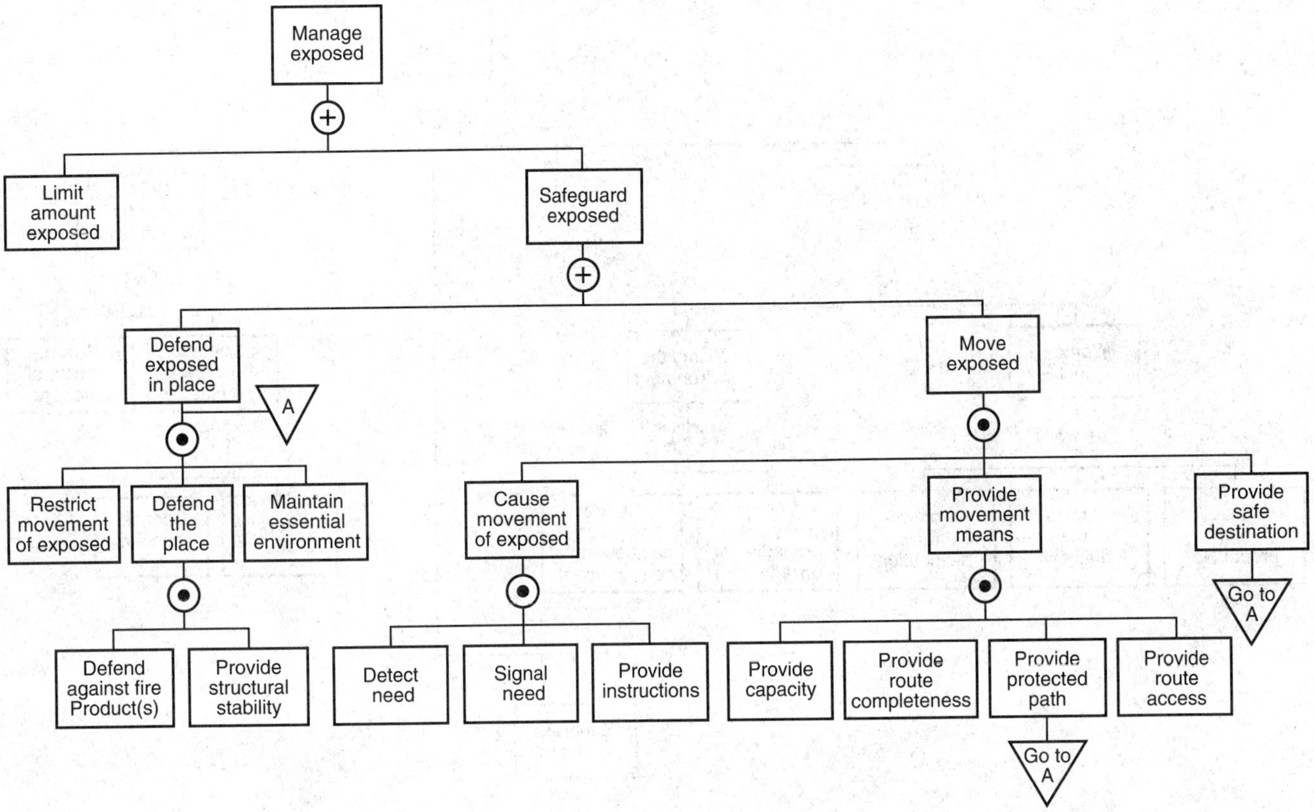

FIG. 1-3I.   Components of the "manage exposed" branch of the fire safety concepts tree. (∇ = transfer/entry point)

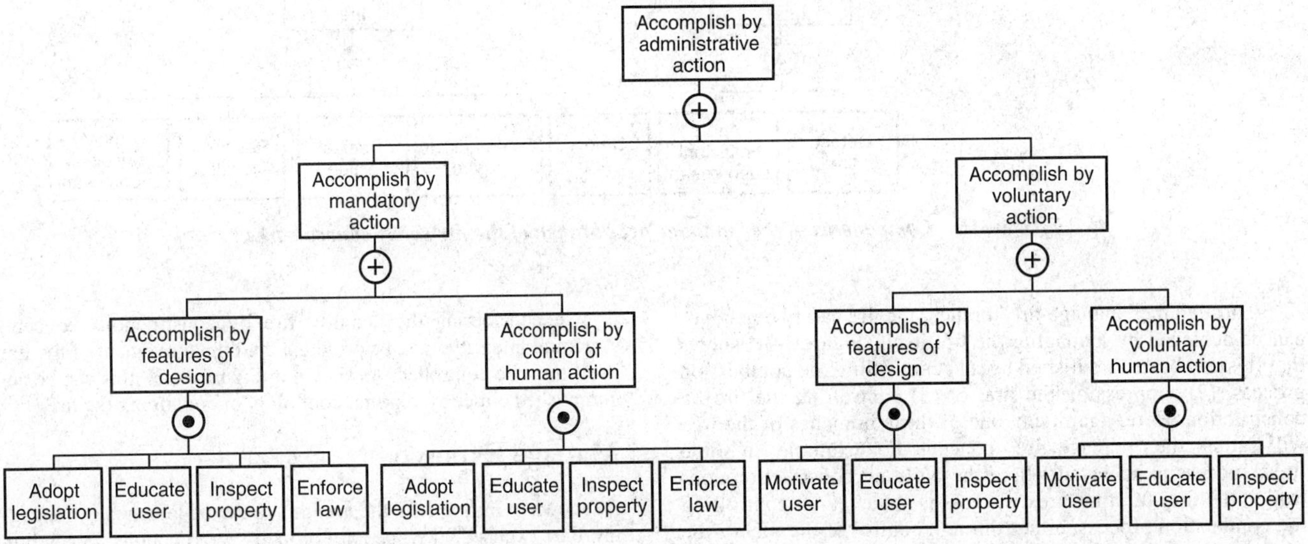

FIG. 1-3J.   Components of the "administrative action guide"—a third dimension to the fire safety concepts tree.

place. To do this, the "defend exposed in place" branch shown in Figure 1-3I would be considered. On the other hand, alert, mobile individuals, such as those expected in offices or schools, could be moved to safeguard them from fire exposure on either a short-term or long-range basis, depending upon other key design elements. Figure 1-3I illustrates the concepts that must be achieved if fire safety objectives are to be met when managing the exposed property and people.

Administration Action Guide

The administration action guide organizes into a logical structure those actions that might be taken to carry out the elements of the basic fire safety concepts tree. (See Figure 1-3J.) It can be thought of as the third dimension of the tree and, as such, a checklist of actions that might be taken. For example, "provide safe destination" (see Figure 1-3I) could be accomplished through code enforcement ("mandatory action") or public fire safety education ("voluntary action").

USE OF THE FIRE SAFETY CONCEPTS TREE

The fire safety concepts tree can be adapted for a number of different functions. A particularly useful aspect of the tree is its relationship to traditional code requirements.

Fire Prevention/Building Codes

The "prevent fire ignition" branch is essentially a fire prevention code. (See Figure 1-3G.) It describes and shows the interrelationships among the essential features of such a code.

The "manage fire impact" branch is essentially a building code. (See Figures 1-3F, 1-3H, and 1-3I.) It enables the code writer to organize the important elements of the building code and to identify their interrelationships. An important feature for building codes, for example, is the subject of alternatives, or "tradeoffs." The only legitimate areas in which alternatives can be established is among those factors below an "OR" (+) gate in the decision tree. The factors below an "AND" (•) gate are necessary and thus cannot be considered as alternatives. It is up to the Authority Having Jurisdiction to decide what levels of protection are adequate to be allowed as tradeoffs for code or tree-mandated items.

Building Analysis

The fire safety concepts tree provides a framework to support fire safety analyses. It is a fundamental necessity that objectives be identified and agreed upon before they can be addressed. Once the fundamental fire safety objectives for a particular building are identified, the designer/architect can analyze the building's design by progressing successively through the various levels of concepts. Redundancies and deficiencies can be identified specifically and evaluated. Often, the weaknesses become so apparent that specific and effective solutions can be devised economically.

The building analysis, supported by the framework of the tree, has among its advantages the capacity to consider separately fire prevention considerations and the requirements involved in managing the fire impact in the event an ignition occurs. This definite separation of events can enable the owner/occupant to take immediate, corrective action for fire prevention, using thoughtful and effective methods.

Building Design

The fire safety concepts tree can be employed effectively in building design. If the architect incorporates the tree during the preliminary planning phase of design, many important decisions and alternatives can be defined more effectively. For example, decisions can be made regarding evacuation *vs.* temporary refuge, including their implications on the functions of the building. Specific needs with regard to the decision are then recognized.

The tree also provides for the separation of the functions of fire prevention and building design. In this way, the responsibilities of the owner/occupant can be differentiated from those of the building design team. Those concepts that are eventually incorporated into the design can be identified with a specific member of the building design team.

Building codes and NFPA standards are important factors in building design, and the fire safety concepts tree should not supersede them. Rather, the tree enables those documents to be interrelated and, consequently, used more effectively. It also aids in design decision-making with regard to fire safety.

Specific minimum design guidelines and standards are still the appropriate domain of the building codes and the standards writers, because the concepts tree does not, in its present form, provide a fire safety score for a building. Other techniques have been developed to provide such scores. The major systems, some of which provide static measures, while others provide dynamic measures, are described elsewhere in this handbook.

BIBLIOGRAPHY

References Cited

1. Ackoff, R., *Redesigning the Future*, John Wiley, New York, 1974.
2. Checkland, P. B., "Toward a Systems-Based Methodology for Real-World Problem Solving," *Journal of Systems Engineering*, Vol. 3, No. 2, 1972, pp. 87-116.

NFPA Codes, Standards, and Recommended Practices

Reference to the following NFPA codes, standards, and recommended practices will provide further information on systems concepts for building fire safety discussed in this chapter. (See the latest version of *The NFPA Catalog* for availability of the current edition of the following document.)

NFPA 550, *Guide to the Fire Safety Concepts Tree*

Additional Reading

ASTM E1472, *Standard Guide for Documenting Computer Software for Fire Models*, American Society for Testing and Materials, Philadelphia, PA, 1992.

Beard, A., "Some Ideas on a Systemic Approach," *Fire Safety Journal*, Vol. 14, 1986, pp. 943–952.

Budnick, E., "Systematic Approach for Fire Risk Assessment," VHS Video, Federal Fire Forum, Current Technical Fire Issues, Gaithersburg, MD, 1993.

Hinks, A.J., "Systemic Modeling of Life Safety in Building Fires: LISA," *Proceedings* of the One-Day Conference to Review the Potential and Limitations of Fire Safety Models for Building Design, University of Salford and Association of Building Engineers, 1994, pp. 63–72.

House, J.M., and Smith, T.F., "System Approach to Optimal Control for HVAC and Building Systems," *ASHRAE Transactions*, Vol. 101, No. 2, 1995.

Marrion, C.E., "Design and Analysis of Fire Detection Systems Using Goal Oriented Systems Approach," *Proceedings* for the Society of Fire Protection Engineers Engineering Seminars on Performance-Based Fire Safety Engineering, Society of Fire Protection Engineers, Boston, MA, 1993, pp. 63–68.

Sekizawa, A., Sagae, K., and Sasaki, H., "A Systemic Approach for Optimum Fire Fighting Operation Against Multiple Fires Following a Big Earthquake," *Proceedings* of the Second International Symposium on Fire Safety Science, Hemisphere Publishing Corporation, New York, 1989, pp. 423–432.

BUILDING AND FIRE CODES AND STANDARDS

Revised by Arthur E. Cote and Casey C. Grant

Throughout history there have been building regulations for preventing fire and restricting its spread. Over the years these regulations have evolved into the codes and standards developed by committees concerned with fire protection. In many cases, a particular code dealing with a hazard of paramount importance may be enacted into law.

THE HISTORY AND DEVELOPMENT OF FIRE PROTECTION REGULATIONS

King Hammurabi, the famous law-making Babylonian ruler who reigned from approximately 1955 to 1913 B.C., is probably best remembered for the *Code of Hammurabi*, a statute primarily based on retaliation. The following decree is from the *Code of Hammurabi*:

> In the case of collapse of a defective building, the architect is to be put to death if the owner is killed by accident; and the architect's son if the son of the owner loses his life.

Today, society no longer endorses Hammurabi's ancient law of retaliation but seeks, rather, to prevent accidents and loss of life and property. From these objectives have evolved the rules and regulations that represent today's building codes and standards for fire prevention, fire protection, and fire suppression.

Early Building and Fire Laws

The earliest recorded building laws apparently were concerned with the prevention of collapse. During the rapid growth of the Roman Empire under the reigns of Julius and Augustus Caesar, the city of Rome became the site of a large number of hastily constructed apartment buildings—many of which were erected to considerable heights. Because building collapse due to structural failure was frequent, laws were passed that limited the heights of buildings—first to 70 ft (21.3 m) and then to 60 ft (18.2 m).

Later in history there evolved many building regulations for preventing fire and restricting its spread. In London, during the 14th century, an ordinance was issued requiring that chimneys be built of tile, stone, or plaster; the ordinance prohibited the use of wood for this purpose. Among the first building ordinances of New York City was a similar provision, and among the first legislative acts of Boston was one requiring that dwellings be constructed of brick or stone

and roofed with slate or tile (rather than being built of wood and having thatched roofs with wood chimneys covered with mud and clay similar to those to which the early settlers had been accustomed in Europe). Obviously, the intention of these building ordinances was to restrict the spread of fire from building to building in order to prevent conflagrations. As an inducement for helping to prevent fires, a fine of ten shillings was imposed on any householders who had chimney fires. This encouraged the citizenry to keep its chimneys free from soot and creosote. Thus was the first fire code in America established and enforced.

In colonial America, the need for laws that offered protection from the ravages of fire developed simultaneously with the growth of the colonies. The laws outlined the fire protection responsibilities of both homeowners and authorities. Some of these new laws were planned to punish people who put themselves and others at risk of fire. For example, in Boston no person was allowed to build a fire within "three rods" of any building, or in ships that were docked in Boston Harbor. It was illegal to carry "burning brands" for lighting fires except in covered containers, and arson was punishable by death. Regardless of such precautions, in Boston and in other emerging communities, fires were everyday occurrences. Therefore, it became necessary to enact more laws with which to govern building construction and to make further provisions for public fire protection. There emerged a growing body of rules and regulations concerning fire prevention, protection, and control. From these small beginnings, various codes and types of codes have evolved in this country, ranging from the most meager of ordinances to comprehensive handbooks and volumes of codes and standards on building construction and fire safety.

Development of Building and Fire Codes

The rapid growth of early American cities inspired much speculative building, and the structures usually were built close to one another. Construction often was started before adequate building codes had been enacted. For example, the year before the great Chicago fire of 1871, Lloyd's of London stopped writing policies in Chicago because of the haphazard manner in which construction was proceeding. Other insurance companies had difficulty selling policies at the high rates they had to charge. Despite these excessively high rates, many insurance companies suffered great losses when fire spread out of control.

The National Board of Fire Underwriters [NBFU; later the American Insurance Association (AIA) and now the American Insurance Services Group (AISG)], organized in 1866, realized that

---

Arthur E. Cote, P.E., is senior vice president, operations at NFPA. Casey C. Grant, P.E., is NFPA's assistant vice president, codes and standards administration.

the adjustment and standardization of rates were merely temporary solutions to a serious technical problem. This group began to emphasize safe building construction, control of fire hazards, and improvements in both water supplies and fire departments. As a result, the new tall buildings constructed of concrete and steel conformed to specifications that helped limit the risk of fire. These buildings were called Class A buildings. In 1905 the National Board of Fire Underwriters published the first edition of its *Recommended Building Code* [later the *National Building Code* (NBC)]. This was a first and very useful attempt to show the way to uniformity.

In San Francisco in early 1906, although there were some new Class A concrete and steel buildings in the downtown section, most of the city consisted of fire-prone wood shacks. Concerned with such conditions, the National Board of Fire Underwriters wrote that "San Francisco has violated all underwriting traditions and precedents by not burning up."

On April 18 of that same year, the city of San Francisco experienced a conflagration—started by an earthquake—that killed 452 people and destroyed some 28,000 buildings. Although the contents of many of the new Class A buildings were destroyed in the San Francisco fire, most of the walls, frames, and floors remained intact and could be renovated. (See Figure 1-4A.)

Following analysis of the fire damage caused by the San Francisco disaster and other major fires, the National Board of Fire Underwriters became convinced of the need for more comprehensive standards and codes relating to the design, construction, and maintenance of buildings. With this increasing recognition of the importance of fire protection came more knowledge about the subject. Engineers started to accumulate information about fire hazards in building construction and in manufacturing processes, and much of this information became the basis for the early codes and standards.

Several chapters in this handbook have a bearing on the provisions of building codes and their enforcement. Of particular interest is Section 7, Chapter 2, "Building Construction," which contains information on the various types of construction and how they are classified in building codes as a basis for fire protection requirements.

BUILDING CODE DEFINED

A building code is a law or regulation that sets forth minimum requirements for design and construction of buildings and structures. These minimum requirements, established to protect the health and safety of society, generally represent a compromise between optimum safety and economic feasibility. Although builders and building owners often establish their own requirements, the minimum code requirements of a jurisdiction must be met. Features covered include structural design, fire protection, means of egress, light, sanitation, and interior finish.

There are two general types of building codes. Specification or prescriptive codes spell out in detail what materials can be used, the building size, and how components should be assembled. Performance codes detail the objective to be met and establish criteria for determining if the objective has been reached; thus, the designer and builder are free to select construction methods and materials as long as it can be shown that the performance criteria can be met. Performance-oriented building codes still embody a fair amount of specification-type requirements, but provisions exist for substitution of alternate methods and materials ("trade offs"), if they can be proved adequate.

BUILDING CODES AND FIRE PROTECTION

The requirements contained in building codes are generally based upon the known properties of materials, the hazards presented by various occupancies, and the lessons learned from previous experiences, such as fire and natural disasters. The promulgation of modern building codes in the United States began with the disastrous conflagrations and earthquakes that occurred at the turn of the century.

For a number of years, building codes dealt mainly with structural safety under fire or earthquake conditions. Since then, codes have grown into documents prescribing minimum requirements for structural stability, fire resistance, means of egress, sanitation, lighting, ventilation, and built-in safety equipment. More than 50 percent

FIG. 1-4A.   *A great earthquake and ensuing conflagration devastated San Francisco in 1906. As many as 452 people were killed, and 28,000 buildings were damaged or destroyed. Total financial loss was $350 million, which is over $6 billion in estimated 1996 dollars.*

of a modern building code usually refers in some way or another to fire protection.

Building codes usually establish fire limits or fire districts in certain areas of a municipality. Only specific types of construction are allowed within the fire limits. Such a restriction is said to reduce the conflagration potential of the more densely populated areas. Use of a given type of building construction alone, however, is not necessarily a deterrent to conflagration. Outside the fire limits, the restriction of certain construction types is relaxed, due to such factors as decreased building density (i.e., increased spacing between buildings). Unfortunately, as areas outside the fire limits are developed, building density increases and the fire limits frequently must be extended. In addition, without construction restrictions, areas outside the fire limits invite the erection of large buildings despite public protection that is weak or lacking.

Another example of the impact of building codes on fire protection and prevention is the establishment of height and area criteria. The criteria establish the maximum height and area of a particular building, based on its intended use. These requirements have varied considerably from one area to the next; there is no nationally recognized standard for setting height and area limitations. However, the Council of American Building Officials (CABO) developed standardized height and area limitations in the 1980s for use in each of the model building codes. The types of building construction are important factors in establishing height and area limitations.

Other requirements found in building codes that directly relate to fire protection include: (1) enclosure of vertical openings such as stair shafts, elevator shafts, and pipe chases; (2) provision of exits for evacuation of occupants; (3) requirements for flame spread of interior finish; and (4) provisions for automatic fire suppression systems. Exit requirements found in most building codes are based on requirements in NFPA 101®, Life Safety Code® (hereafter referred to as Life Safety Code.)

Inasmuch as a building code is actually a law, various state and local jurisdictions write their own codes. Because of the complexities of modern building code development, several organizations develop model building codes for use by jurisdictions which can then adopt the model codes into law.

CODES FOR THE BUILT ENVIRONMENT

Although much focus is provided by building codes, a variety of other related codes also readily serve the built environment. Specifically, these codes address distinct interrelated topics that are essential components in structures of all kinds.

Topics that are typically addressed include electrical, plumbing, mechanical, fuel gas, energy, and fire prevention. Yet this is not an all-inclusive list, and any particular subject that lends itself to specific and detailed criteria is eligible, and thus the evolution of "electrical codes," "plumbing codes," "mechanical codes," etc. Often the reference to "building codes" is intended to include, in a general sense, a reference to all of these related codes for the built environment.

Of these different related topics, fire prevention codes are somewhat unique. It often is difficult to differentiate between items that should go into a fire prevention code and those best included in a building or other related code. Generally, those requirements that deal specifically with construction of a building are part of a building or similar code administered by the building department. A fire prevention code, on the other hand, includes information on fire hazards in a building, and usually is regulated by the fire official.

Requirements for exits and fire extinguishing equipment generally are found in building codes, while the maintenance of such items is covered in fire prevention codes. More simply stated, building and other related codes generally address the original design or major renovation of a building, while a fire prevention code addresses the building during its useful life after the initial construction or renovation is complete.

MODEL CODE ORGANIZATIONS

With the exception of the independent operations of some of the largest cities, the business of code development for the built community in the United States is mainly in the hands of the model code organizations. The most prominent organizations are: National Fire Protection Association (NFPA), International Conference of Building Officials (ICBO), Southern Building Code Congress International (SBCCI), Building Officials and Code Administrators (BOCA), and the International Code Council (ICC). The primary objectives of these organizations is to provide standardization of construction regulations and/or support of the enforcement of these regulations. Each organization and its model codes are discussed in the following paragraphs.

NFPA (National Fire Protection Association)

NFPA does not publish a model building code per se; however, NFPA 101, Life Safety Code, establishes minimum requirements necessary for providing a reasonable degree of life safety from fire. The Life Safety Code addresses those construction, protection, and occupancy features necessary to minimize danger to life from fire, smoke, fumes, or panic. It also identifies the minimum criteria for design of egress facilities to permit prompt escape of occupants from buildings or, where preferable, evacuation into safe areas within buildings. The Life Safety Code does not attempt to address those general fire prevention or building construction features that normally are functions of fire prevention and building codes. The NFPA Committee on Safety to Life was founded in 1913, following the tragic Triangle Shirtwaist Factory Fire in New York City that killed 146 workers, and initially developed a number of pamphlets on exits from various occupancies. The first edition of the NFPA Building Exits Code (later the Life Safety Code), developed by the Committee on Safety to Life, was published in 1927.

The Life Safety Code is one of more than 300 codes, standards, recommended practices, and guides developed and published by NFPA. The complete compendium of these documents, which encompass the entire scope of fire protection and prevention, practices, and safeguards, comprises the NFPA National Fire Codes® (NFC). Included within the NFC are a number of individual codes related to fire safety, e.g., NFPA 70, National Electrical Code®, one of the most widely adopted codes in the world; NFPA 30, Flammable and Combustible Liquids Code; NFPA 1, Fire Prevention Code; and NFPA 54, National Fuel Gas Code—just to name a few.

NFPA codes and standards are developed by more than 225 NFPA technical committees, each representing a full balance of affected interests. All interests are afforded an equal opportunity to participate in a system that is not dominated by any particular interest, thus evolving full and complete consensus on technical issues. More than 5,000 individuals serve on the NFPA committees, all on a voluntary, unpaid basis. Committees operate according to a detailed Regulations Governing Committee Projects and are administered by a Standards Council that reports to the NFPA's Board of Directors.

Built into the codes and standards development and adoption process is the publication of calls for proposals to amend existing documents, or to shape the content of new documents. Only after a public review and comment cycle has been completed is the final committee report brought before the NFPA membership for action. This democratic legislative procedure allows proponents and opponents to be heard freely. Once adopted by the NFPA membership at either the NFPA Annual or Fall meeting, and issued by the NFPA Standards Council, the documents are published and made available for voluntary adoption. Many of these documents are referenced by the model building codes.

The NFPA codes and standards process is accredited by the American National Standards Institute (ANSI) and is considerably more accessible to the general public than the processes used by other code organizations. Other model code organizations, e.g., BOCA, ICBO, and SBCCI, have been traditionally by and for building officials, which significantly restricts involvement and voting to the building official community. The newly formed ICC uses a code development process that also restricts final voting to the code official only. Conversely, NFPA documents are developed and maintained in an open full-consensus process that allows widespread involvement, and thus provides documents that are more technically balanced and economically fair.

ICBO (International Conference of Building Officials)

ICBO first published the *Uniform Building Code* (UBC) in 1927. The UBC is principally used in the western United States, but has been adopted in municipalities as far east as Michigan.

In addition to publishing the building code, the Conference has also published a *Mechanical and Plumbing Code* and a *Fire Prevention Code*. The *Fire Prevention Code* was published in conjunction with the International Fire Code Institute (IFCI), an organization formed in 1995, by ICBO, and the Western Fire Chiefs Association, a division of the International Association of Fire Chiefs (IAFC).

The Conference maintains a staff with headquarters in Whittier, CA, which provides governmental bodies with technical assistance in the administration and enforcement of the ICBO codes.

SBCCI (Southern Building Code Congress International)

The SBCCI was organized in 1940 and first published the *Southern Standard Building Code* in 1945. Now called the *Standard Building Code* (SBC), this code is used principally throughout the southern portion of the United States.

Like ICBO, SBCCI also has published mechanical, plumbing, and fire prevention codes. They also have published a gas code to be used in conjunction with the SBC. SBCCI maintains a technical staff headquartered in Birmingham, AL.

BOCA (Building Officials and Code Administrators)

Originally known as the Building Officials Conference of America, BOCA published the first edition of the *Basic/Building Code* (BBC) in 1950. Since that time, the organization has changed its name to Building Officials and Code Administrators, International. For a period of time, the BBC was called the *Basic/National Building Code*, but is now published as the *National Building Code*. It is principally used in the Midwest and northeast portions of the United States.

BOCA also has published mechanical, plumbing, and fire prevention codes. BOCA maintains a technical staff headquartered in Country Club Hills, IL.

ICC (International Code Council)

In 1995 the International Code Council (ICC) was established. This organization is comprised of representatives from BOCA, ICBO, and SBCCI. The purpose of the ICC is to combine the codes of the three model building code organizations into single national models. The ICC has developed the *International Mechanical Code* and the *International Plumbing Code*. They plan to develop an International Fire code, and an International Building code by the year 2000. These international codes are designed to form a "single model building code" for adoption and use primarily in the United States.

Other Code Producers

There are several other notable organizations that have been directly involved in the development of model codes for the built environment.

IAPMO (International Association of Plumbing and Mechanical Officials) also develops the *Uniform Mechanical Code*. IAPMO had been a joint sponsor with ICBO on the development of the *Uniform Mechanical Code* but is now proposing to develop this code independently. The future of this activity is unclear at this time.

UPC (Uniform Plumbing Code): This plumbing code is administered by the ANSI A40 Committee, which is an ANSI-accredited committee. Although the ANSI A40 Committee has existed for some years, this initiative is now jointly supported by a coalition involving IAPMO, MCA (Mechanical Contractors Association), and NAPHCC (National Association of Plumbing-Heating-Cooling Contractors). IAPMO had previously been a joint sponsor with ICBO on another plumbing code that was also entitled the *Uniform Plumbing Code*. The first edition of the *Uniform Plumbing Code* generated by the ANSI A40 Committee as an ANSI-accredited document is expected by 1997. This will essentially be a full consensus alternative to the *International Plumbing Code* developed by ICC.

AISG (American Insurance Services Group): As previously noted, the National Board of Fire Underwriters (NBFU), renamed the American Insurance Association (AIA), and now known as the American Insurance Services Group (AISG), first published the *National Building Code* in 1905. The code was used as a model for adoption by cities, as well as a basis to evaluate the building regulations of towns and cities for town grading purposes. The code was periodically reviewed by the NBFU staff, revised as necessary, and republished. The last code revision was the 1976 edition. Since then, the AISG has discontinued updating and publishing the *National Building Code*, and Building Officials and Code Administrators (BOCA) has acquired the right to use the name "*National Building Code*." The AISG also developed a fire prevention code, most recently published in 1976, but has also discontinued the updating and publishing of this document.

Code Adoption and Use

It must be recognized that model codes are only representations of possible regulations, and they do not actually become law until enacted by state and municipal legislatures. The general areas of model code adoption and use can be seen in Figures 1-4B through 1-4E.

Other codes beside the building codes are adopted independently and not necessarily with the related building code from the respective organization. The world's most widely adopted code, NFPA 70, *National Electrical Code*, is adopted in virtually every state in the United States, Mexico, and numerous other countries.

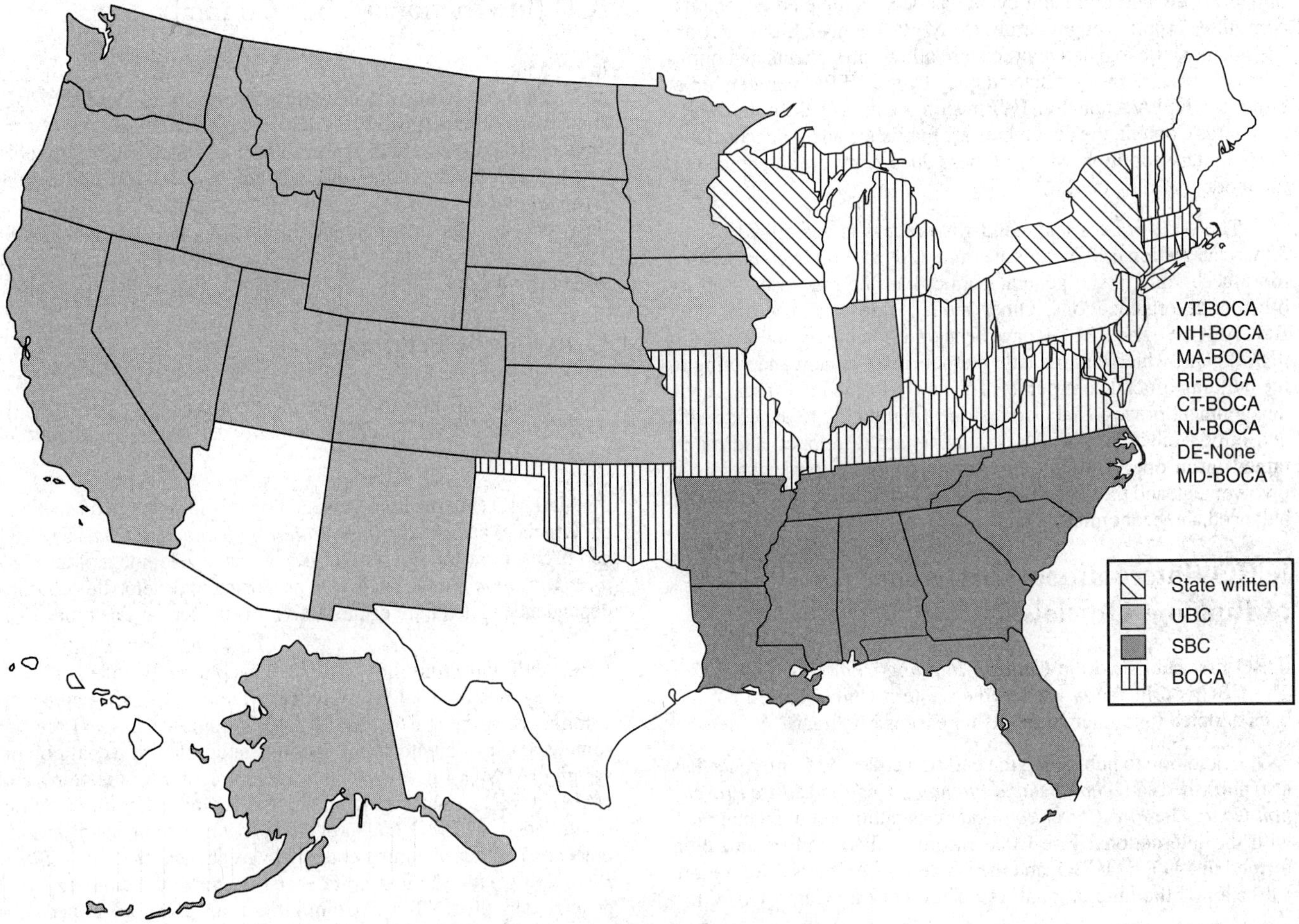

VT-BOCA
NH-BOCA
MA-BOCA
RI-BOCA
CT-BOCA
NJ-BOCA
DE-None
MD-BOCA

State written
UBC
SBC
BOCA

FIG. 1-4B. *Adoption of a regional model building code.*

BUILDING CODE DEVELOPMENT

The operations of model code organizations are based on the input of affected interest groups. Any individual or interested organization can provide input to the model building code organizations (i.e., ICBO, SBCCI, BOCA, or ICC), but only building officials get a final vote on the acceptance of code changes. In contrast, the NFPA and other ANSI-accredited processes are not dominated by any one particular interest group that can solely adjudicate overall code changes based on their final vote.

Model building codes are revised annually. Revised editions are typically published at three-year intervals, with annual supplements in the interim years.

TOWARD UNIFORMITY OF BUILDING CODES

Although model codes originally were known as regional codes, boundary lines for their adoption are much less significant today. In fact, much effort has been expended toward the elimination of differences among the model building codes.

The Joint Committee on Building Codes, composed of representatives of the organizations sponsoring the model codes and others concerned with the development of codes and standards, was formed in 1949 to work toward uniformity among the codes. The Joint Committee later became the Model Code Standardization Council and then the Board for the Coordination of Model Codes

(BCMC). Representatives of BOCA, ICBO, SBCCI, and NFPA participated in BCMC. The purpose of this committee was to develop uniform code language to be included in each of the model building codes and NFPA *101, Life Safety Code.* Proposals resulting from BCMC were subsequently processed in accordance with the code change procedures of each code organization. The formation of the ICC has resulted in the discontinuance of BCMC. However, a new entity now exists known as the Board for the Development of Model Codes (BDMC) involving the same four organizations (NFPA, BOCA, ICBO, and SBCCI), addressing many of the issues previously addressed by BCMC.

OTHER CODE-RELATED ORGANIZATIONS

Other organizations involved in codes- and standards-related issues include the following.

National Conference of States on Building Codes and Standards (NCSBCS)

NCSBCS is a nonprofit corporation founded in 1967 as a result of Congressional interest in reform of building codes. It attempts to foster increased interstate cooperation in the area of building codes and standards and coordinates intergovernmental code administration reforms. NCSBCS is an executive-branch organization of the National Governors Association, and includes as members, gover-

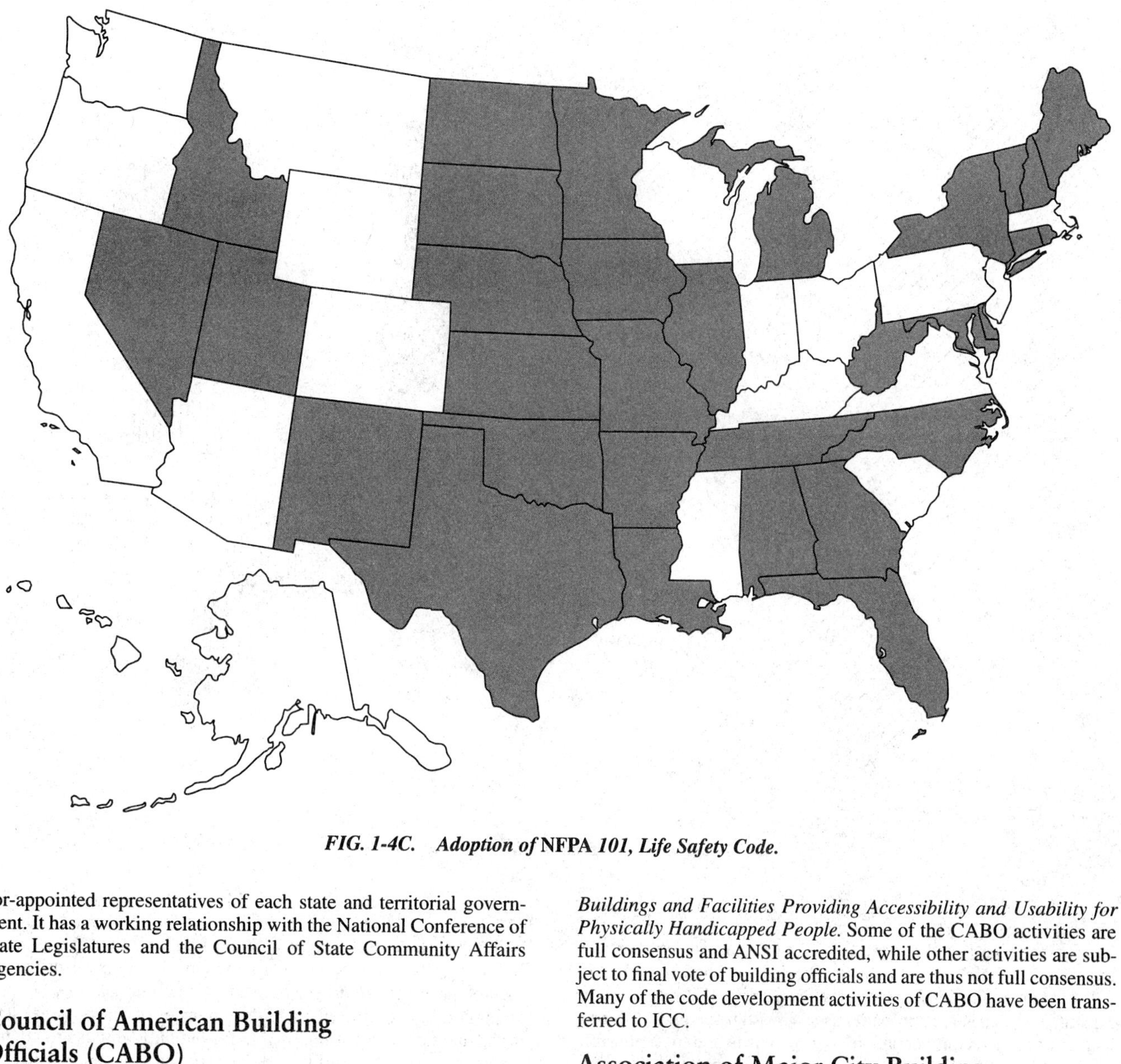

FIG. 1-4C.   Adoption of NFPA 101, Life Safety Code.

nor-appointed representatives of each state and territorial government. It has a working relationship with the National Conference of State Legislatures and the Council of State Community Affairs Agencies.

Council of American Building Officials (CABO)

The Council of American Building Officials (CABO) is an organization formed by the three major model building code groups, i.e., ICBO, SBCCI, and BOCA, in 1972. CABO was formed to represent the model building code process on a national level and to promote the model code process.

Two of the major accomplishments of CABO have been: (1) to organize the National Evaluation Services, and (2) publish the *One- and Two-Family Dwelling Code*. Previously referred to as the National Research Board, the National Evaluation Services coordinates the research and evaluation programs of the three model code groups (i.e., ICBO, SBCCI, and BOCA) to eliminate the need for a manufacturer working with three different organizations. The *One- and Two-Family Dwelling Code* contains regulations for detached housing that meet the requirements of the three model building codes organizations. NFPA is responsible for the establishment and revision of the electrical section of the *One- and Two-Family Dwelling Code*, under special agreement formerly with CABO and now with ICC. CABO also develops the *Model Energy Code* and is the secretariat to ANSI A117, the *American National Standard for Buildings and Facilities Providing Accessibility and Usability for Physically Handicapped People*. Some of the CABO activities are full consensus and ANSI accredited, while other activities are subject to final vote of building officials and are thus not full consensus. Many of the code development activities of CABO have been transferred to ICC.

Association of Major City Building Officials (AMCBO)

The Association of Major City Building Officials (AMCBO) was formed in 1974. This group focuses on issues of building codes, administrative techniques, and public safety in buildings. The association has 36 members and provides a national forum of city and county building officials united to discuss topics of mutual interest.

AMCBO is affiliated with the National Conference of States on Building Codes and Standards (NCSBCS). The activities of AMCBO include encouraging the development of comprehensive training and educational programs for building code enforcement personnel, providing scientific and technical resources for the improvement of building codes, and enhancing building technology and products to reduce the cost of construction and maintain safety levels.

National Institute of Building Sciences (NIBS)

NIBS was authorized by Congress in 1974, under Public Law 93-383, as a nongovernmental, nonprofit organization governed by a

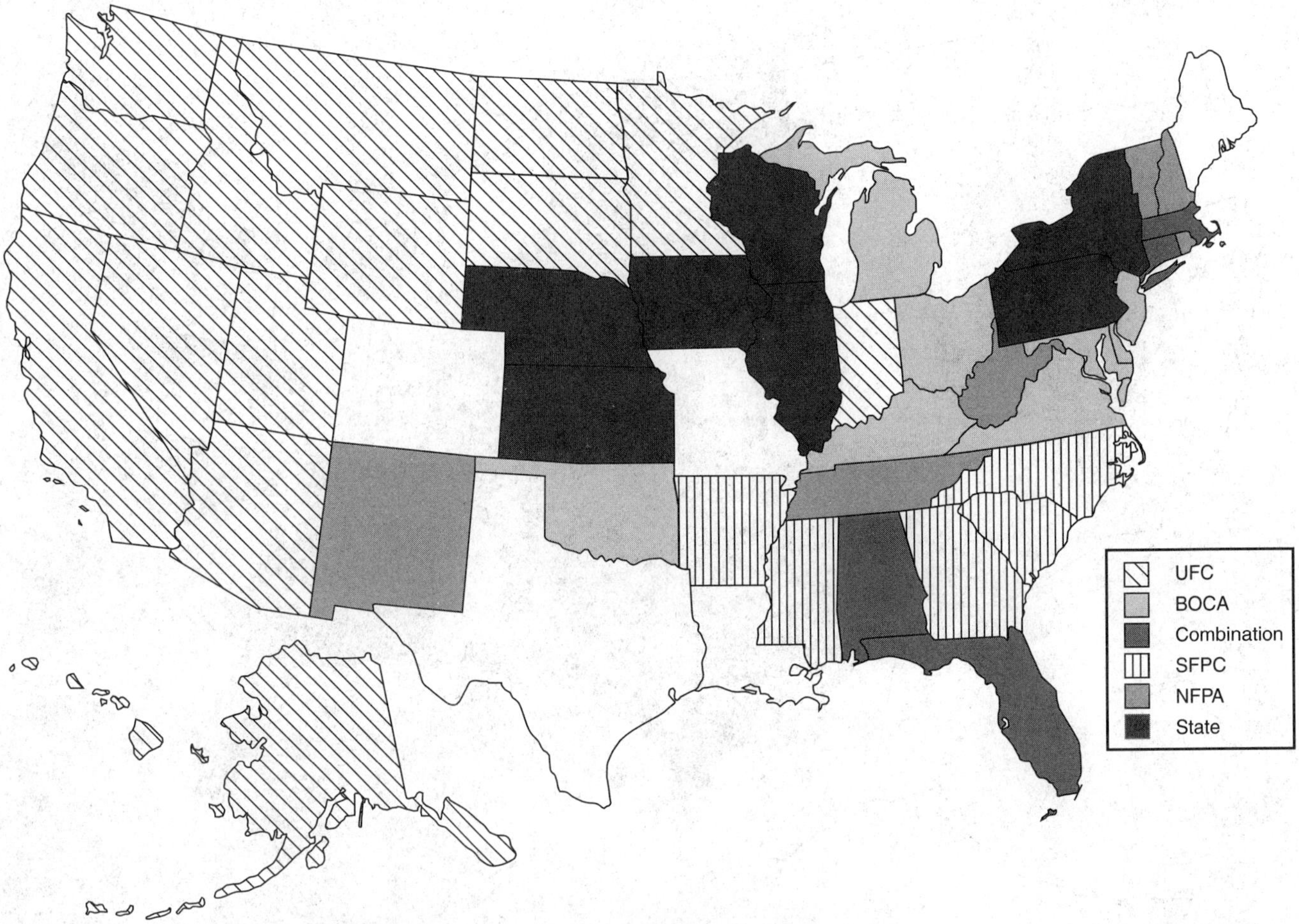

FIG. 1-4D. *Adoption of fire prevention code.*

21-member board of directors. Fifteen of the board members are elected and six are appointed by the President of the United States, with the advice and consent of the U.S. Senate. The Institute is a core organization that serves primarily as an investigative body, offering its findings and recommendations to government and to responsible private-sector organizations for voluntary implementation. It carries out its mandated mission essentially by identifying and investigating national problems confronting the building community, and proposing courses of action to bring about solutions to the problems. NIBS activities are board-based and center around regulatory concerns, technology for the built environment, and distribution of technical and other useful information.

Working under its very broad mandate, NIBS has established a Consultative Council, with membership available to representatives of all appropriate private trade, professional, and labor organizations; private and public standards, codes, and testing bodies; public regulatory agencies; and consumer groups. The Council's purpose is to ensure a direct line of communication between such groups and the Institute and to serve as a vehicle for representative hearings on matters before the Institute.

World Organization of Building Officials (WOBO)

WOBO was founded in 1984, with the primary objective of advancing education through worldwide dissemination of knowledge in building science, technology, and construction. WOBO was established because of increased participation of nations in the global marketplace; the rapid development of new international building technologies and products; and development of international standards that now make it impossible for building officials to confine their concern to activities within their own national boundaries.

ENFORCEMENT OF CODES AND STANDARDS

The types of government and the characteristics of governing authorities around the world vary considerably, yet despite the differences, there are some aspects that are common with relation to legislative adoption of codes and standards. For the sake of illustration, the following discussion focuses on this topic, based on a form of government similar to that used in the United States.

Today the life and property of every citizen is safeguarded to at least some extent by fire legislation enacted by the Congress of the United States, state legislatures, city councils, town meetings, and many other jurisdictions and levels of government. The implementation and enforcement of this legislation is in the hands of administrative agencies of government, such as federal departments and agencies, state fire marshal offices and other appropriate state agencies, and local fire departments, building departments, electrical inspectors, etc.

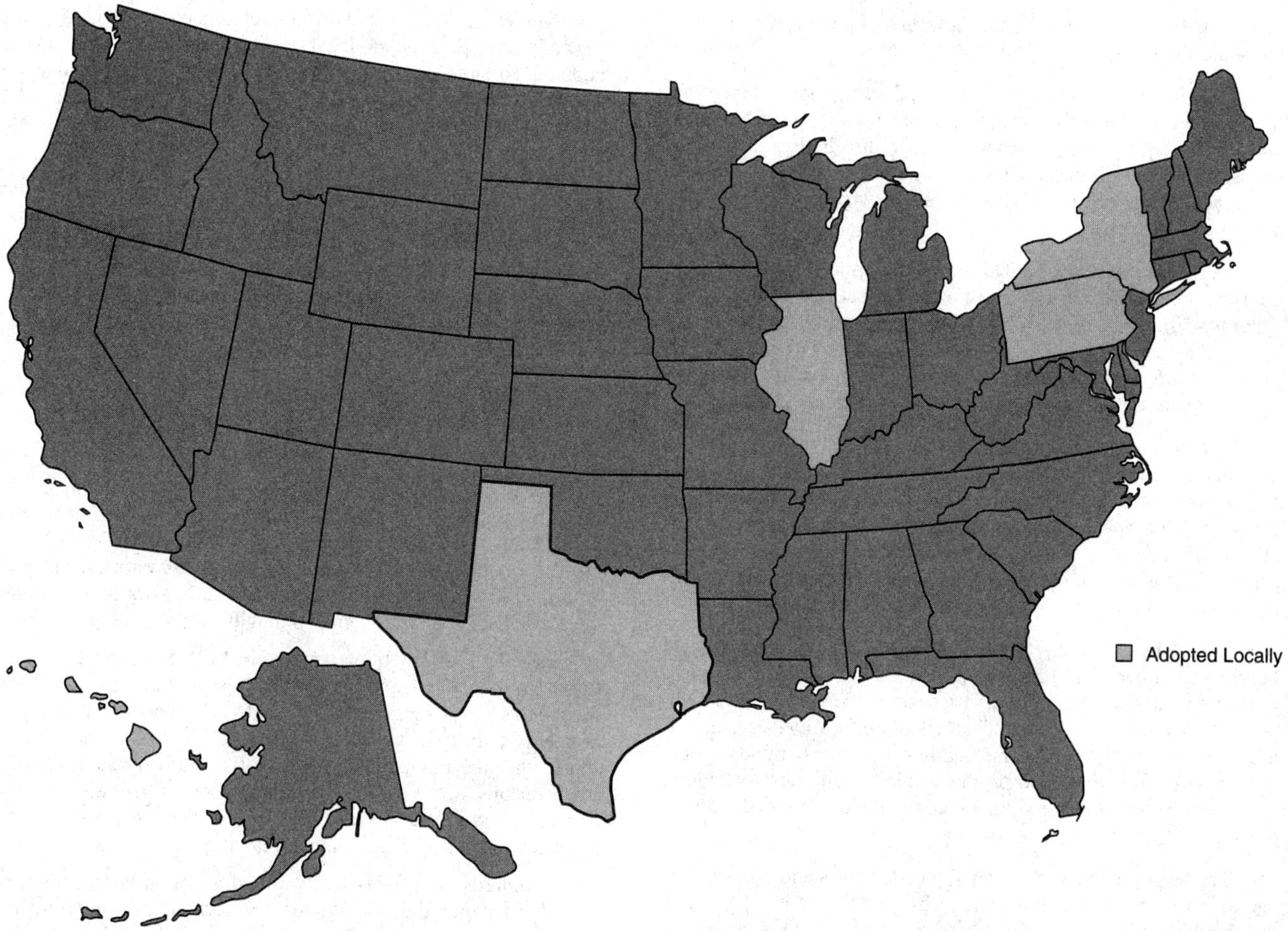

FIG. 1-4E.   State adoption of an electrical code.

☐ Adopted Locally

In the earlier days of the United States, the protection of citizens from fire was solely the concern of the local community. Present-day fire fighting is carried on by local fire departments. While most communities have had some type of building code since the beginning of the 20th century, they have not had fire prevention codes until more recently.

With the need for more detailed, comprehensive standards and codes relating to the construction, design, and maintenance of buildings came the knowledge that regulations based on such codes certainly could prevent most fires and reduce losses in the fires that did occur.

Regulations relating to fire safety are determined and enforced by different levels of government. While some functions overlap, federal and state laws generally govern those areas that cannot be regulated at the local level.

Federal Fire Safety Regulations

There is a substantial amount of federal regulation pertaining to fire safety. Under the Constitution in the United States, Congress has the power to regulate interstate commerce. This power has been interpreted to permit Congress to pass laws authorizing various federal departments and agencies to adopt and enforce regulations to protect the public from fire hazards.

Any federal department or agency in the United States can promulgate fire safety regulations only if authority to do so is granted by a specific act of Congress. These regulations have the force of law, and violations can result in legal action. In general, such federal laws can be enacted to provide: (1) that all state laws on the same subject are superseded by the federal law, (2) that state laws not conflicting with the federal law remain valid, or (3) that any state law will prevail if it is more stringent than the federal law. Among the federal agencies that have the authority to promulgate fire safety regulations are the Occupational Safety and Health Administration (OSHA), the Department of Health and Human Services (HHS), and the Consumer Product Safety Commission (CPSC).

Fire Safety Regulations of State and Local Government

Within the scope of the police power of state government in the United States is the regulation of building construction for the health and safety of the public—a power usually delegated to local governments of the state.

Building code requirements usually apply to new construction or to major alterations to buildings. Retroactive application of code requirements is very rare. Building code applicability usually ends with the issuance of an occupancy permit or certificate of occupancy. The basic premise that fire legislation should regulate for the safety of current occupants and for current risk is not generally the province of building codes once a structure is occupied. Then, after-occupancy fire codes or fire prevention codes apply. Also, this usually is the point at which the authority of the building official ends and the fire official begins.

This division of authority, however, does not preclude interaction between the two officials during both a building's development and its subsequent use. In practice, many jurisdictions assign responsibilities to building and/or fire officials in both building and fire code application. The division of authority varies considerably among communities.

In most states in the United States, the principal fire official is the state fire marshal. For the most part, the state fire marshal is the statutory official charged by law with responsibility for the administration and enforcement of state laws relating to safety to life and property from fire. Usually the state fire marshal also has the power to investigate fires and to suppress arson.

The manner in which each state handles the promulgation of building and fire regulations varies widely. In some states, each local government may have its own code, while in others the local authority has the option of adopting the state building code. In still others, the state code establishes the minimum below which the local regulations cannot go. Finally, in some states the local government has no choice and must adopt the state code.

These situations have resulted in a plethora of different local building and fire codes. Some of the local governments adopt one or more of the model codes or codes based upon the model codes. Others draft their own local codes. This lack of uniformity has been criticized by materials producers, building designers, builders, and others, and some years ago prompted the appointment of federal commissions to study the situation and make recommendations to the administration.[1,2,3]

The legal procedure for adopting codes and standards into law can also vary from one enforcing jurisdiction to another. Usually, the simplest and best way is to adopt by reference. This method, applicable to public authorities as well as to private entities, requires that the text of the law or rule cite the code or standard by its title and give adequate publishing information to permit its exact identification. The code or standard itself is not reprinted in the law. All deletions, additions, or changes made by the adopting authority are noted separately in the text of the law. Adoption of a current edition of a code or standard obviates outdated editions maintained as law until a new law referencing a new edition is adopted.

Where local laws do not permit adoption by reference, a code or standard can be adopted by transcription. This requires that the text of the adopted code or standard be transcribed into the law. Existing material can be deleted and new material added only if such material does not change the meaning or intent of the existing or remaining material. Under adoption by transcription, the code or standard cannot be rewritten, although changes can be made for administrative provisions. Because the text of the code or standard is transcribed into the law, due notice of the copyright of the document's developer is required. As a result, most code groups copyright their codes or standards to prevent misuse and unlawful use.

THE ROLE OF STANDARDS IN THE BUILT ENVIRONMENT

Many requirements found in building codes are excerpts from, or based on, the standards published by nationally recognized organizations. The most extensive use of the standards is their adoption into building codes by reference, thus keeping the building codes to a workable size and eliminating much duplication of effort. Such standards are also used by specification writers in the design stage of a building to provide guidelines for the bidders and contractors.

Numerous NFPA standards are referenced by the model building codes and, thus, obtain legal status where these model codes are adopted. Notable examples of such referenced NFPA standards are those that deal with extinguishing systems, flammable liquids, hazardous processes, combustible dusts, liquefied petroleum gas, electrical systems, and fire tests.

The model building codes contain appendices that list standards published by many organizations, including standards-making organizations, professional engineering societies, building materials trade associations, federal agencies, and testing agencies. The appendices are prefaced with a statement indicating that the standards are to be used where required by the provisions of the code or where referenced by the code.

NATIONAL STANDARDS

The voluntary standards development system in the United States is unmatched in the world: it is efficient, cost-effective, highly productive, and results in the promulgation of thousands of quality standards each year. U.S. voluntary standards are developed by a diverse, decentralized network of private-sector entities. Many different organizations are involved in standards work in this country, a feature that is one of the great strengths of the system.

Based on information compiled in 1991, the U.S. standardization community currently maintains approximately 94,000 standards in active status.[4] The number of U.S. standards at any given moment in time, however, is difficult to identify. Today, it is assumed that the number 94,000 is still a relatively valid estimate, since various newly created standards tend to offset a trend of the largest U.S. producer of standards, the U.S. Department of Defense, to retire more standards each year than it generates.

Standards exist for virtually all industries and product sectors. The oldest standard developing organization in the United States is the U.S. Pharmacopoeial Convention, which published standards for 219 drugs in 1820.[5]

For a variety of reasons, data on the number of standards must be treated with respect and caution. These reasons include: (1) uncertainty on whether to consider as a standard a product description, specification, definition of a term, or description of a procedure; (2) the distinction between a single standard with many sections and a series of separate but related standards may be arbitrary; (3) the influence and impact of various standards on the economy can vary dramatically; (4) many documents become technologically obsolete, but remain in a technically active status; (5) information on the number of state and local government standards is extremely limited and fragmented; and (6) de facto standards are typically not included in statistical information (i.e., unsponsored and unwritten yet usually widely accepted standards, such as the configuration of typewriter and computer keyboards).

The 94,000 standards in the United States generally comprise 41,500 private-sector standards and approximately 52,500 federal government standards.[5] Furthermore, private-sector standards can be further subdivided based on the type of sponsoring organization: standards-developing organizations; scientific and professional societies; and industry associations. Table 1-4A provides a summary of this information.

In comparison to most foreign systems, the institutional structure of the U.S. voluntary consensus standards system is highly decentralized. More than 400 private-sector standards developers exist in the United States, with approximately 275 engaged in ongoing standards-setting activities that are mostly organized around an academic discipline, profession, or given industry.[5] The remainder typically have a small number of standards that were developed in the past, that may or may not be occasionally updated.

TABLE 1-4A.   *U.S. Standards and Their Developers*[5]

Private Sector	Number of Standards	Percentage
Standards-Developing Organizations	14,000	16%
Scientific and Professional Societies	13,000	14%
Industry Associations	14,500	15%
Total of Private Sector	41,500	45%
Federal Government		
Department of Defense (DOD)	38,000	40%
General Services Administration (GSA)	6,000	6%
Other	8,500	9%
Total of Federal Government	52,500	55%
Overall Total	94,000	100%

TABLE 1-4B.   *Top 20 Private Sector Standards-Developing Organizations*[5] *(active standards as of 1991)*

Organization	Numbers of Standards
ASTM (formerly American Society of Testing of Materials)	8,500
Society of Automotive Engineers (SAE)	5,100
U.S. Pharmacopoeia	4,450
Aerospace Industries Association	3,000
Association of Official Analytical Chemists	1,900
American National Standards Institute (ANSI)*	1,400
Association of American Railroads	1,350
American Association of State Highway and Transportation Officials (AASHTO)	1,100
American Petroleum Institute (API)	880
Cosmetic, Toiletry, and Fragrance Association	800
American Society of Mechanical Engineers (ASME)	745
American Conference of Governmental Industrial Hygienists	700
Underwriters Laboratories Inc. (UL)	630
Electronic Industries Association (EIA)	600
Institute of Electrical and Electronics Engineers (IEEE)	575
American Association of Cereal Chemists	370
American Oil Chemists Society	365
American Railway Engineers Association	300
National Fire Protection Association (NFPA)	275
Technical Association of the Pulp and Paper Industry	270
Total	33,310

* Published and copyrighted by ANSI, but sponsored and developed by numerous member organizations.

It is interesting to note that, of the more than 400 private-sector standards developers in the U.S., the 20 largest developers account for 80 percent of all private-sector development.[6] Table 1-4B lists the number of standards handled by each of these top 20 developers, which account for more than 33,000 standards. This summary includes the 1,400 separate standards that are published and copyrighted by American National Standards Institute (ANSI), but are sponsored and developed by numerous other organizations. Even if these ANSI standards are excluded from the top 20 developers, the remaining standards still account for more than three-fourths of all private-sector standards.

American National Standards Institute (ANSI)

The significant private-sector standards development system in the United States is largely self-regulated, with oversight and coordination provided by ANSI, a federation of U.S. codes and standards developers, and company and government users of those standards.

Originally known as the American Engineering Standards Committee, ANSI's first meeting was held on January 17, 1917, by the following founding organizations: American Institute of Electrical Engineers, American Institute of Mining Engineers, American Society of Civil Engineers, American Society of Mechanical Engineers, and the American Society for Testing and Materials. The Government Departments of War, Navy, and Commerce were soon involved, along with NFPA and other organizations. One of the early documents that was accepted and registered under the established rules as an "American Standard" was the 1920 edition of NFPA 70, *National Electrical Code*.[6]

In 1928 the name of the American Engineering Standards Committee was changed to the American Standards Association. This organizational title was used until 1968 when the organization became known briefly as the United States of America Standards Institute (USASI) before adopting the current title of American National Standards Institute.

Organizational membership in ANSI comprises 250 U.S. professional and technical societies, and trade associations, along with 1,100 U.S. companies.[6] ANSI is able to fulfill its coordinating role for the voluntary standards system in the United States because of the support it receives from those actively involved in standards work. NFPA is an ANSI-accredited codes and standards organization, resulting in ANSI accreditation for virtually all NFPA codes and standards. In 1991 approximately 9,500 standards approved by ANSI were designated as "American National Standards."[5]

ANSI coordinates and harmonizes private-sector standards activity in the United States. In order for a document to be designated an American National Standard, the principles of openness and due process must have been followed in its development, and consensus among those directly and materially affected by the standard must have been achieved. ANSI also represents the U.S. interests in the international standardization activities of the International Organization for Standardization (ISO) and the International Electrotechnical Commission (IEC).

The ANSI arrangement is unique in the international arena, since most countries are represented by a single organization that is either fully or partially funded by that country's federal government. The United States, however, is represented by a single private organization (ANSI) that further represents the interests of numerous private standards-development organizations (e.g., ASTM, IEEE, NFPA, etc.). This results in a complex legal and business environment involving international copyright. Further complicating this situation is that U.S. standards developers do not limit their activities to only U.S. constituents and often have members involved from other countries.

Under ANSI procedures, all American National Standards must be reviewed and reaffirmed, modified, or withdrawn no longer than every five years—a requirement that ensures that voluntary standards in the United States keep pace with developing technology and innovations. Thus, the voluntary system produces quality standards that do not become outdated.

Standards-Developing Organizations (SDOs)

Authority and technical expertise in the U.S. standards-developing system is highly decentralized and linked to specific industry sectors. This has evolved based on the development of a wide range of consensus standards processes in many different standards-developing organizations (SDOs). The basic common principles of consensus standards development are thus applied in different ways, with procedures and objectives specific to the needs of a particular industry or professional community.

Standards developed by the private sector are generally developed by three types of organizations.[4,5]

1. *Standards-Developing Organizations.* These organizations typically have standards development as one of their central activities or missions. Membership-oriented standards-developing organizations are the most prominent of these organizations, and they tend to have the most diverse membership among all SDOs, since they are not limited to a particular industry or profession. These membership organizations have a notable number of international members, which is a feature of many U.S. SDOs in general, and makes the U.S. standards-developing system distinct among the rest of the world. Standards-developing membership organizations, because of their diverse membership, tend to have the strictest due-process requirements. Aside from membership organizations, standards development is also a key activity of certain testing and certification organizations, such as Underwriters Laboratories Inc. or the American Gas Association.

Two examples of standards-developing organizations are ASTM and NFPA, both of which are membership oriented. ASTM, originally referred to as the American Society for Testing and Materials, has a membership of 35,000. The 132 ASTM technical committees are responsible for more than 8,500 standards, and approximately 33 percent of ASTM's sales of standards are to international users. NFPA, for sake of comparison, has about 67,000 members. The 235 consensus technical committees of NFPA are responsible for about 300 safety-oriented documents, which is dramatically fewer than ASTM. This difference in committee structure provides some indication of the distinction of product standards handled by ASTM, and safety codes and standards handled by NFPA.

Furthermore, despite NFPA having substantially fewer documents than ASTM and some other standards developers, the total number of pages generated by NFPA (because they are mostly safety-oriented documents rather than product oriented) is often comparable and, in some cases, clearly more.

2. *Scientific and Professional Societies.* These societies are a refined form of membership organizations that support the practice and advancement of a particular profession. The most recognized of these societies involve the engineering disciplines. A unique characteristic of these societies is that the participants, as part of their standards-development processes, typically function as individual professionals, and not as specific representatives of their sponsoring organization or industry.

Prominent examples of scientific and professional societies include the Institute of Electrical and Electronics Engineers (IEEE) and the American Society of Mechanical Engineers (ASME). IEEE has a worldwide membership of more than 300,000 engineering professionals. The approximately 600 standards published by IEEE focus specifically on areas of electrotechnology. ASME has an international membership of more than 100,000. The ASME standards process has more than 700 committees responsible for 745 codes and standards. ASME has responsibility for the *Boiler and Pressure Vessel Code*, which comprises 11,000 pages and is one of the most prominent single documents in the U.S. standards-development arena and in the world.

3. *Industry Associations.* Industry or trade associations are organizations of manufacturers, service providers, customers, sup-

pliers, and others that are active in a given industry. The development of technical standards is specifically intended to further the interests of their particular industry sector.

The Association for the Advancement of Medical Instrumentation (AAMI) is a good example of a trade organization that develops standards. Approximately 5,000 health-care professionals support their activities and include representatives from industry, health care facilities, academia, research centers, and government agencies, such as the Food and Drug Administration (FDA). Industry association SDOs are likely to be more openly responsive to commercial market concerns than other types of SDOs. Some other examples of industry associations include the American Petroleum Institute (API) and the National Electrical Manufacturers Association (NEMA).

SAFETY *VS.* RISK IN CODES AND STANDARDS DEVELOPMENT

There are two broad categories of voluntary codes and standards: (1) safety codes and standards and (2) product standards. These documents are not strictly a matter of science, especially safety codes and standards. Codes and standards oriented toward safety tend to be more complicated and extensive than product standards. Furthermore, safety codes and standards are often adopted with the power of law, and thus require more extensive technical advisory support.

Safety is dependent upon risk, and the degree of safety achieved depends on how much willingness there is to "pay" to eliminate the risk. Total elimination of some risks would be exorbitant, and elimination of all risk is not feasible, even apart from cost. In addition, part of the cost of risk elimination is the reduction of freedom. Many aspects of safety systems or materials standards have this effect, as they come to bear on the establishment of an "acceptable level" of risk.

Assessments of levels of risk also are needed with respect to calculations of onerousness and cost of use; for, if tolerance limits are exceeded, codes and standards will be modified in practice, or ignored. Also, the more onerous and costly compliance becomes, the more carefully critics will examine the "degree of contribution to a safe environment" that the code or standard will bring about.

Although code- or standards-making does not deal with the "science" of safety, amenable to full laboratory control, an aggregation of subjective factors enters in: social, economic, political, legal, and competitive.

One of the strengths of the voluntary consensus codes- and standards-development system in the United States is that the deliberative committee structure, which comprises a balanced representation of all affected interests, including users, consumers, manufacturers, suppliers, distributors, labor, testing laboratories, enforcers, and federal, state, and local government officials, can consider all of the diverse factors at hand and develop a consensus on an acceptable level of standardization. It has been observed that "this may be one of the greatest strengths of the present private standards-writing system, insofar as it truly represents variety, and one of the greatest insufficiencies of a governmental system."[7]

BIBLIOGRAPHY

References Cited

1. "Building the American City," Report of the National Commission on Urban Problems, Superintendent of Documents, U.S. Government Printing Office, Washington, DC, 1968.

2. *Building Codes: A Program for Intergovernmental Reform*, Advisory Commission on Intergovernmental Relations, Superintendent of Documents, U.S. Government Printing Office, Washington, DC, 1966.
3. "Report of the President's Commission on Housing," Superintendent of Documents, U.S. Government Printing Office, Washington, DC, 1982.
4. National Research Council, "Standards, Conformity Assessment, and Trade into the 21st Century," National Academy Press, Washington, DC, 1995.
5. Toth, R. B., "Standards Activities of Organizations in the United States," NIST Special Publication 806, National Institute of Standards and Technology, Gaithersburg, MD, 1991.
6. Hemenway, D., "Industrywide Voluntary Product Standards," Ballinger Publishing Company, Cambridge, MA.
7. Dixon, R. G., Jr., *Standards Development in the Private Sector: Thoughts on Interest Representation and Procedural Fairness*, National Fire Protection Association, Quincy, MA, 1978.

NFPA Codes, Standards, and Recommended Practices

Reference to the following NFPA codes, standards, and recommended practices will provide further information on building and fire codes and standards discussed in this chapter. (See the latest version of *The NFPA Catalog* for availability of current editions of the following documents.)

NFPA 1, *Fire Prevention Code*
NFPA 30, *Flammable and Combustible Liquids Code*
NFPA 54, *National Fuel Gas Code*
NFPA 70, *National Electrical Code®*
NFPA 70A, *Electrical Code for One- and Two-Family Dwellings and Mobile Homes*
NFPA 80, *Standard for Fire Doors and Fire Windows*
NFPA 80A, *Recommended Practice for Protection of Buildings from Exterior Fire Exposures*
NFPA 88A, *Standard for Parking Structures*
NFPA 88B, *Standard for Repair Garages*
NFPA 90A, *Standard for the Installation of Air Conditioning and Ventilating Systems*
NFPA 90B, *Standard for the Installation of Warm Air Heating and Air Conditioning Systems*
NFPA 92A, *Recommended Practice for Smoke-Control Systems*
NFPA 92B, *Guide for Smoke Management Systems in Malls, Atria, and Large Areas*
NFPA 99, *Standard for Health Care Facilities*
NFPA 101®, *Life Safety Code®*
NFPA 105, *Recommended Practice for the Installation of Smoke-Control Door Assemblies*
NFPA 203, *Guide for Roof Coverings and Roof Deck Constructions*
NFPA 204M, *Guide for Smoke and Heat Venting*
NFPA 220, *Standard on Types of Building Construction*
NFPA 241, *Standard for Safeguarding Construction, Alteration, and Demolition Operations*
NFPA 703, *Standard for Fire Retardant Impregnated Wood and Fire Retardant Coatings for Building Materials*

Model Codes (Building and Fire Codes)

BOCA, *National Building Code*, Building Officials and Code Administrators, International, Country Club Hills, IL, 1993.
BOCA, *National Fire Prevention Code*, Building Officials and Code Administrators, International, Country Club Hills, IL, 1993.
ICBO, *Uniform Building Code*, Vols. 1, 2, and 3, International Conference of Building Officials, Whittier, CA, 1994.
International Plumbing Code—ICC.
International Mechanical Code—ICC.
NFPA 1, *Fire Prevention Code*.
NFPA 54, *National Fuel Gas Code*.
NFPA 70, *National Electrical Code®*.
NFPA 70A, *Electrical Code for One- and Two-Family Dwellings and Mobile Homes*.
NFPA 101®, *Life Safety Code®*.
One- and Two-Family Dwelling Code, Council of American Building Officials, Falls Church, VA, 1995.

SBCCI, *Standard Building Code*, Southern Building Code Congress International, Inc., Birmingham, AL, 1994.
Standard Fire Prevention Code, Southern Building Code Congress, International, Inc., Birmingham, AL, 1994.
Uniform Fire Code, Vols. 1 and 2, International Fire Code Institute, 1994.
Uniform Plumbing Code—IAPMO (NCA/NAPHCC).
Uniform Mechanical Code—IAPMO.

Additional Reading

ASTM Standards in Building Codes, 27th ed., American Society for Testing and Materials, W. Conshohocken, PA, 1990.
Babrauskas, V., "Designing Products for Fire Performance: the State of the Art of Test Methods and Fire Models," *Fire Safety Journal*, Vol. 24, No. 3, 1995, pp. 299–312.
Batik, A. L., "A Layman's View of the Relationship of Standards to Product Liability," *Standards Engineering*, Dayton, OH, Jan./Feb. 1990.
Baker, D. R., "Meeting High-Rise Requirements for Fire Detection/Alarm/Supression," *Consulting—Specifying Engineer*, Vol. 3, No. 2, Feb. 1988, pp. 56–59.
Baker, D. R., "Performance by Computer Modeling or Prescription by Model Code," *TR 86-5*, Society of Fire Protection Engineers, Boston, MA, 1986.
Breitenberg, M. A., *The ABC's of Standards-Related Activities in the United States*, U.S. Department of Commerce, National Bureau of Standards, Gaithersburg, MD, May 1987.
Bukowski, R. W., and Babrauskas, V., "Developing Rational, Performance-based Fire Safety Requirements in Model Building Codes," *Fire and Materials: An International Journal*, Vol. 18, No. 3, May–June 1994, pp. 173–192.
Cooke, P. W., *A Review of U.S. Participation in International Standards Activities*, U.S. Department of Commerce, National Bureau of Standards, Gaithersburg, MD, Jan. 1988.
Cooke, P. W., *A Summary of the New European Community Approach to Standards Development*, U.S. Department of Commerce, National Bureau of Standards, Gaithersburg, MD, Aug. 1988.
Cooke, P. W., *An Update of U.S. Participation in International Standards Activities*, U.S. Department of Commerce, National Institute of Standards and Technology, Gaithersburg, MD, Jan. 1988.
Corcoran, D., "Fire Prevention and Building Restoration Activities," *Fire Engineering*, Vol. 146, No. 12, December 1993, pp. 94–98, 100.
Cote, R., *Life Safety Code Handbook*, 6th ed, National Fire Protection Association, Quincy, MA, 1994.
Deakin, A. G., "Fire Safety in Buildings: Standards for 1992's Europe," *Fire International*, No. 121, February/March 1990, pp. 15–16.
Finnimore, B., "Need for Atria Fire Codes," *Fire Prevention*, No. 205, Dec. 1987, pp. 30–33.
Gross, J. G., "Harmonization of Standards and Regulations: Problems and Opportunities for the United States," *Building Standards*, National Institute of Standards and Technologies, Gaithersburg, MD, March/April 1990, pp. 32–35.
Harvey, C. S., "Flexible Approach to Fire-Code Compliance," *Architectural Record*, No. 10, Oct. 1988, pp. 130–135.
Hosker, H., and Waters, C., "Building Regulations Determined," *Fire Prevention*, No. 224, Nov. 1989, pp. 37–38.
Johnson, P. F., "International Implications of Performance Based Fire Engineering Design Codes," *Journal of Fire Protection Engineering*, Vol. 5, No. 4, 1993, pp. 141–146.
Kaufman, S., "1990 National Electric Code—Its Impact on the Communication Industry," 38th International Wire and Cable Symposium, U.S. Army Communication Electronics Command, Atlanta, GA, 1989, pp. 301–305.
Korman, R., and Post, N. M., "The Code System, It Ain't Pretty . . . But It Works, Codes, ENR," *Construction Weekly*, June 22, 1989.
Lathrop, J. K., "Life Safety Code Key to Industrial Fire Safety," *NFPA Journal*, Vol. 88, No. 4, July/August 1994, pp. 36–46.
"Legal Aspects of Code Enforcement: A Report on the 1993 Annual Conference Education Program," *Building Standards*, National Institute of Standards and Technologies, Gaithersburg, MD, Vol. 63, No. 1, January/February 1994, pp. 27–30.
Lucht, D. A., Kime, C. H., and Traw, J. S., "International Developments in Building Code Concepts," *Journal of Fire Protection Engineering*, Vol. 5, No. 4, 1993, pp. 125–133.
"Major Changes to the 1995 Codes," *Consensus*, Spring 1995, p. 25.

Mawhinney, J. R., "Development of Regulations in the 1990 National Fire Code of Canada on Storage of Dangerous Goods," *Fire Technology*, Vol. 26, No. 3, August 1990, pp. 266–280.

Meacham, B. J., and Custer, R. L. P., "Performance-Based Fire Safety Engineering: An Introduction of Basic Concepts," *Journal of Fire Protection Engineering*, Vol. 7, No. 2, 1995, pp. 35–54.

Moss, D., "Fire Safety and Compliance of 1992," *FM Journal*, July/August 1995, pp. 15–19.

Murphy, J. J., Jr., "Fire Safety Operations and Code Compliance Concerns. Part 2," *Fire Engineering*, Vol. 148, No. 1, January 1995, pp. 7882, 84, 86.

Neale, R. A., "When Code Equivalencies Don't Work," *American Fire Journal*, Vol. 48, No. 1, January 1996, pp. 2023.

Oey, K. H., and Passchier, E., "Complying with Practice Codes," *Batiment International/Building Research and Practice*, Vol. 21, No. 1, 1988, pp. 30–36.

Peralta, M., "Statement of the American National Standards Institute Concerning International Voluntary Standardization," American National Standards Institute, New York, July 25, 1989.

Richardson, L. R., "Determining Degrees of Combustibility of Building Materials–National Building Code of Canada," *Fire and Materials: An International Journal*, Vol. 18, No. 2, March–April 1994, pp. 99–106.

Robertson, J. C., "Development and Enactment of Fire Safety Codes," *Introduction to Fire Prevention*, 3rd ed., Macmillan, New York, 1989, pp. 112–132.

Sabatini, J., "Ensuring Code Compliance in High-Hazard Buildings," *Plant Engineering*, Vol. 44, No. 9, May 10, 1990, pp. 57–59.

Schirmer, C., "Helping Develop the Codes and Standards," *Fire Journal*, Vol. 84, No. 3, 1990, p. 44.

Standards Activities of Organizations in the United States, National Institute of Standards and Technology, U.S. Dept. of Commerce, Washington, DC, 1991.

Steiner, V. M., "Building Codes—Bane or Blessing?" *Plant Engineering*, July 21, 1988.

Strength, R. S., "Status Report Model Building Codes 1992, NEC-93 and IEC-89," Fire Retardant Chemicals Association Fall Conference: Industry Speaks Out on Flame Retardancy: Coatings; Polymers and Compounding; Test Method Development; New Products, Technomic Publishing Co., Lancaster, PA, 1992, pp. 41–46.

Stubbs, M. S., "The Widening Web of Codes and Standards," *AEBO Section Newsletter*, Fall 1988, National Fire Protection Association, Quincy, MA, 1988. (Reprinted from *Doors and Hardware*.)

Swankin, D. A., *How Due Process in the Development of Voluntary Standards Can Reduce the Risk of Anti-Trust Liability*, U.S. Department of Commerce, National Institute of Standards and Technology, Washington, DC, Feb. 1990.

Terio, C., *Introduction to Building Codes and Standards*, The American Institute of Architects, State Government Affairs, Washington, DC, Apr. 1987.

Todd, N. W., and Ryan, J. D., "Improving Codes by Predicting Product Performance in Real Fires," *Fire Journal*, Vol. 84, No. 2, 1990, p. 64.

Traw, J. S., "ICBO Code Interpretation Policy," *Building Standards*, Jan.-Feb. 1990.

Turner, M., "New Code Governing the Means of Escape for Disabled People," *Fire Prevention*, No. 215, Dec. 1988, pp. 36–37.

Use of Building Codes in Federal Agency Construction, Building Research Board, National Research Council, Commission on Engineering and Technical Systems, Washington, DC, 1989.

"World Trade Center Bombing May Bring Code Reviews," *Consulting-Specifying Engineer*, Vol. 13, No. 5, April 1993, p. 13.

"1991 Updated: Legislation and Codes Affecting the Fire Sprinkler Industry," *Sprinkler Age*, Vol. 10, No. 11, November 1991, pp. 12–15, 17.

CHEMISTRY AND PHYSICS OF FIRE

Revised by D. D. Drysdale

This chapter presents basic definitions of some of the physical properties and chemical terms applicable to the chemistry and physics of fire; it also discusses combustion, the principles of fire, heat measurement, heat transfer, and heat energy sources (i.e., sources of ignition).

The material contained in this chapter does not attempt to offer a comprehensive course of instruction on the subject, but is intended to present basic background reference material applicable to this and other sections of this handbook.

BASIC DEFINITIONS AND PROPERTIES

Atoms: The basic particles of chemical composition are called atoms. Atoms are extremely minute. Substances composed of only one type of atom are called elements. An atom has a compact core or nucleus around which electrons (negatively charged units of matter) travel in orbit. The nucleus comprises protons (positively charged) and neutrons (no charge). Substances that have loosely bound electrons generally are good electrical and thermal conductors. Those with rigidly bound electrons (so that no easy transfer of electrons is possible) are good insulators.

Molecules: Groups of atoms combined in fixed proportions are called molecules. Substances composed of molecules that contain two or more different kinds of atoms are called compounds.

Chemical formula: The chemical formula shows the number of atoms of the various elements in the molecule. For example, the chemical formula for propane is C_3H_8, where C stands for carbon and H for hydrogen. (See Table 1-5A for symbols of other elements.) A formula may be written to indicate the arrangement of the atoms in the molecule (e.g., propane is $CH_3CH_2CH_3$).

Atomic number of an element: The number of protons in the nucleus of a particular atom determines the atomic number of that element and its place in the periodic table.

Atomic weight of an element: The atomic weight of an element is proportional to the weight of its atom. The atomic weight of carbon (isotope C-12) is arbitrarily assigned the value of 12.00. Table 1-5A gives the atomic weights of elements.

Molecular weight of a compound: The molecular weight of a compound is the sum of the atomic weights of all atoms in its molecule. [In a chemical formula, the subscripts following an element's symbol indicate the number of atoms of that element present in the compound. Thus, the molecular weight of propane (C_3H_8) is $(3 \times 12) + (8 \times 1) = 44$.]

Gram molecular weight: The gram molecular weight of a substance (in grams) is equal to its molecular weight.

Specific gravity: Specific gravity is the ratio of the weight of a solid or liquid substance to the weight of an equal volume of water. The scales of the most commonly used hydrometers are based on a specific gravity of 1 for water at 4°C (1 cc of water at 4°C weighs 1 g).

Gas specific gravity: Gas specific gravity is the ratio of the weight of a gas to the weight of an equal volume of dry air at the same temperature and pressure. It may be evaluated by the following: gas specific gravity equals its molecular weight divided by 29, where 29 is the composite molecular weight of dry air.

Buoyancy: Buoyancy is the upward force exerted on a body or volume of fluid by the ambient fluid surrounding it. If a volume of a gas has positive buoyancy, then it is lighter than the surrounding gas and will tend to rise. If it has negative buoyancy, it is heavier and will tend to sink. The buoyancy of a gas depends upon both its molecular weight (e.g., gas specific gravity) and its temperature.

If a flammable gas with a gas specific gravity greater than 1 escapes from its container, it will typically sink to a low level and can travel considerable distances to potential sources of ignition. Propane (C_3H_8), which has a molecular weight of 44, is heavier than air and, if leaking from a cylinder, will accumulate near the ground.

The density of a gas will decrease as its temperature is increased. Thus, hot products of combustion will tend to rise. On the other hand, if there is a spillage of liquefied natural gas (LNG), the vapor is heavier than air because of its very low temperature, despite the fact that the gas specific gravity of methane (CH_4), the principal component, is less than 1 (its molecular weight is 16). As with propane at ambient temperature, LNG spills can be very dangerous, as the vapor can spread over a wide area with little dilution.

Vapor Pressure and Boiling Point

Because molecules of a liquid are always in motion (the amount of motion depends upon the temperature of the liquid), molecules are

---

D. D. Drysdale, Ph.D., is a reader in the Unit of Fire Safety Engineering, Department of Civil and Environmental Engineering, University of Edinburgh, Scotland.

TABLE 1-5A. *The chemical elements based on the assigned relative atomic mass of the Carbon-12 isotope equal to 12.00. Most elements consist of isotope mixtures. Elements with atomic weights in parentheses are unstable isotopes.*

Element	Symbol	Atomic No.	Atomic Weight
Actinium	Ac	89	(227)
Aluminum	Al	13	226.9815
Americium	Am	95	(243)
Antimony, stibium	Sb	51	121.75
Argon	Ar	18	39.948
Arsenic	As	33	74.9216
Astatine	At	85	(210)
Barium	Ba	56	137.34
Berkelium	Bk	97	(247)
Beryllium	Be	4	9.0122
Bismuth	Bi	83	208.980
Boron	B	5	10.81
Bromine	Br	35	79.904
Cadmium	Cd	48	112.40
Calcium	Ca	20	40.08
Californium	Cf	98	(251)
Carbon	C	6	12.011
Cerium	Ce	58	140.13
Cesium	Cs	55	132.9055
Chlorine	Cl	17	35.453
Chromium	Cr	24	51.996
Cobalt	Co	27	58.9332
Columbium, see niobium			
Copper	Cu	29	63.546
Curium	Cm	96	(247)
Dysprosium	Dy	66	162.50
Einsteinium	E	99	(254)
Erbium	Er	68	167.26
Europium	Eu	63	151.96
Fermium	Fm	100	(257)
Fluorine	F	9	18.9984
Francium	Fr	87	(223)
Gadolinium	Gd	64	157.2
Gallium	Ga	31	69.72
Germanium	Ge	32	72.59
Gold, aurum	Au	79	196.9665

Element	Symbol	Atomic No.	Atomic Weight
Hafnium	Hf	72	178.49
Helium	He	2	4.0026
Holmium	Ho	67	164.9303
Hydrogen	H	1	1.0080
Indium	In	49	114.82
Iodine	I	53	126.9045
Ilridium	Ir	77	192.2
Iron, ferrum	Fe	26	55.85
Krypton	Kr	36	83.8
Lanthanum	La	57	38.905
Lawrencium	Lr	103	(257)
Lead, plumbum	Pb	82	207.2
Lithium	Li	3	6.
Lutetium	Lu	71	174.97
Magnesium	Mg	12	24.305
Manganese	Mn	25	54.9380
Mendelevium	Md	101	(256)
Mercury, hydrargyrum	Hg	80	200.59
Molybdenum	Mo	42	95.94
Neodymium	Nd	60	144.2
Neon	Ne	10	20.179
Neptunium	Np	93	237.0482
Nickel	Ni	28	58.71
Niobium, columbium	Nb	41	92.9064
Nitrogen	N	7	14.0067
Nobelium	No	102	(254)
Osmium	Os	76	190.2
Oxygen	O	8	15.9994
Palladium	Pd	46	106.4
Phosphorus	P	15	30.9738
Platinum	Pt	78	195.09
Plutonium	Pu	94	(244)
Polonium	Po	84	(210)
Potassium, kalium	K	19	39.10

Element	Symbol	Atomic No.	Atomic Weight
Praseodymium	Pr	59	140.9077
Promethium	Pm	61	(145)
Protoactinium	Pa	91	231.0359
Radium	Ra	88	226.0254
Radon	Rn	86	(222)
Rhenium	Re	75	186.2
Rhodium	Rh	45	102.9055
Rubidium	Rb	37	85.4678
Ruthenium	Ru	44	101.07
Samarium	Sm	62	150.4
Scandium	Sc	21	44.9559
Selenium	Se	34	78.96
Silicon	Si	14	28.086
Silver, argentum	Ag	47	107.868
Sodium, natrium	Na	11	22.9898
Strontium	Sr	38	87.62
Sulfur	S	16	32.06
Tantalum	Ta	73	180.9479
Technetium	Tc	43	98.9062
Tellurium	Te	52	127.60
Terbium	Tb	65	158.9254
Thallium	Tl	81	204.37
Thorium	Th	90	232.0381
Thulium	Tm	69	168.9342
Tin, stannum	Sn	50	118.69
Titanium	Ti	22	47.90
Tungsten	W	74	183.8
Uranium	U	92	238.029
Vanadium	V	23	50.9414
Xenon	Xe	54	131.30
Ytterbium	Yb	70	173.04
Yttrium	Y	39	88.9059
Zinc	Zn	30	65.38
Zirconium	Zr	40	91.22

continually escaping from the free surface of the liquid to the space above. Some molecules remain in this space while others, due to random motion, collide with the liquid and are "recaptured."

If the liquid is in an open container, molecules (collectively called vapor) continuously escape from the surface, and the liquid evaporates. If, on the other hand, the liquid is in a closed container, the escaping molecules are confined to the vapor space, and a point of equilibrium eventually is reached where the rate of escape of molecules from the liquid equals the rate of their return to the liquid. The pressure exerted by the vapor at the point of equilibrium is called vapor pressure. It is measured in kilopascals (kPa) or pounds per square inch absolute (psia).*

As the temperature of a liquid increases, its vapor pressure approaches and ultimately exceeds atmospheric pressure. In the open air, the liquid will boil when its vapor pressure equals atmospheric pressure; the corresponding liquid temperature is known as the boiling point.

Although the vapor pressure of a liquid varies with its temperature, a great many variables affect the rate at which it actually evap-

---

*Absolute pressure equals the total force exerted against a unit of area. It is measured in Pascals (Newtons per square meter) or in pounds per square inch (psi). It often is expressed in fractions or multiples of atmospheric pressure, or in terms of the height of a column of liquid (usually mercury) that will balance the absolute pressure. In situations where pressure gages are used, absolute pressure is determined by adding gage pressure to atmospheric pressure. Atmospheric, or ambient, pressure equals 101.3 kPa, 14.7 psia, or 30 in. of mercury. A kiloPascal equals 1,000 Pascals.

orates when exposed to air. These variables include atmospheric temperature and pressure, and air movement.

Vapor-air specific gravity (vasg): Vapor-air specific gravity is the ratio of the weight of a vapor-air mixture (resulting from the vaporization of a liquid at equilibrium temperature and pressure) to the weight of an equal volume of air under the same conditions. The specific gravity (density) of a vapor-air mixture thus depends upon the ambient temperature, the vapor pressure of the liquid at the temperature, and the molecular weight of the liquid. At temperatures well below the boiling point of a liquid, the vapor pressure of the liquid may be so low that the vapor-air mixture, consisting mostly of air, has a density that approximates that of pure air, i.e., a vapor-air specific gravity near 1. As the temperature of the liquid increases to the boiling point, the rate of vaporization increases, the vapor displaces the surrounding air, and the vapor-air mixture specific gravity approaches that of the pure vapor specific gravity.

A vapor-air mixture with a density significantly above that of air at the ambient temperature will seek lower levels. On the other hand, diffusion and mixing by convection will limit the distance of travel of mixtures having densities near or less than 1. The vapor-air specific gravity of a substance at ambient temperature may be calculated as follows:

Let P equal the ambient pressure, p the vapor pressure of the liquid at ambient temperature, and s its pure vapor specific gravity. Then,

$$vsag = \frac{ps}{P} + \frac{P-p}{P}$$

The first term, $\frac{Ps}{P}$, is the contribution of the vapor to the specific gravity of the mixture; the second term, $\frac{P-p}{P}$, is the contribution of air.

EXAMPLE: Find the vapor-air specific gravity at 38°C and atmospheric pressure for a flammable liquid whose vapor specific gravity is 2, and whose vapor pressure at 38°C is 10.1 kPa, or one-tenth of atmospheric pressure.

$$vsag = \frac{(10.1)(2)}{101} + \frac{101-10.1}{101} = 0.2 + 0.9 = 1.1$$

Endothermic and Exothermic Chemical Reactions

Heat of reaction is the energy that is absorbed or released when a given reaction takes place. In endothermic reactions, the new substances formed contain more energy than the reacting materials, so energy must be absorbed by the reaction. Exothermic reactions produce substances with less energy than that of the reacting materials, so energy is released by the reaction. While the energy may be in different forms, it is usually in the form of heat, added to or released from a chemical reaction.

COMBUSTION

Combustion is an exothermic, self-sustaining reaction involving a solid, liquid, and/or gas-phase fuel. The process is usually (but not necessarily) associated with the oxidation of a fuel by atmospheric oxygen. Some solids can burn directly by glowing combustion or smoldering, but in flaming combustion of solids and liquid fuels, vaporization takes place before burning. It is necessary to distinguish between two types of flaming: (1) premixed, in which gaseous fuel is mixed intimately with air before ignition, and (2) diffusive, in

which combustion takes place in the regions where the fuel and oxygen are mixing. If premixed burning takes place in a confined space, a rapid pressure rise will occur, giving rise to an explosion.

Oxidation Reactions

Fire involves oxidation reactions that are exothermic, i.e., heat is generated. The reactions are complex and are not understood in their entirety, although certain generalizations can be made.

For an oxidation reaction to take place, a combustible material (fuel) and an oxidizing agent must both be present. Fuels include innumerable materials which, due to their chemistry, can be oxidized to yield relatively stable species, such as carbon dioxide and water; thus, for example:

$$C_3H_8 + 5O_2 = 3CO_2 + 4H_2O$$

Hydrocarbons, such as propane (C_3H_8), consist entirely of carbon and hydrogen and may be regarded as "prototype fuels." Virtually all common fuels, whether solid, liquid, or gaseous, contain significant proportions of carbon and hydrogen.

In the present context, the most common oxidizing agent is molecular oxygen (O_2) in the air, which consists approximately of one-fifth oxygen and four-fifths nitrogen. However, there are certain chemicals that are powerful oxidizing agents, such as sodium nitrate ($NaNO_3$) and potassium chlorate ($KClO_3$), which if intimately mixed with a solid or liquid fuel, produce highly reactive mixtures. Thus, gunpowder is a physical mixture of carbon and sulfur (the fuel) with sodium nitrate as the oxidizer. If reactive groups, such as the nitrate group, are incorporated chemically into a fuel, such as in cellulose nitrate or trinitrotoluene (TNT), the resulting species can be highly unstable and will decompose violently under appropriate conditions.

There are circumstances involving reactive species in which combustion may take place without oxygen being involved. Thus, hydrocarbons may "burn" in an atmosphere of chlorine, while zirconium dust can be ignited in pure carbon dioxide.

Ignition (piloted ignition and autoignition): Ignition is the process by which self-sustaining combustion is initiated. Considering first a flammable gas- or vapor-air mixture (see below), piloted ignition can be achieved by the introduction of an ignition source, such as a flame or spark. However, if the temperature is raised sufficiently, the mixture will undergo autoignition, in which the onset of combustion is spontaneous.

In general, for the combustion process to become self-sustaining, the molecules of fuel and oxygen must be excited to an activated state, which results in the formation of highly reactive intermediate species (free radicals). These initiate rapid, branched chain reactions that convert fuel and oxygen into products of combustion, with the release of heat (energy). The chain reaction will be self-sustaining for as long as the rate of production of the radicals equals (or exceeds) their natural rate of removal (decay). Once ignition has occurred, combustion will continue until all the available fuel or oxidant has been consumed, or until the flame is extinguished. In general, a self-sustained ignition can occur only in those situations that are capable of supporting self-sustained combustion. For example, if the ambient pressure (or ambient oxidant concentration) is insufficient for sustaining combustion, it also will be insufficient for ignition.

For combustible liquids and solids, the initiation of flame occurs in the gas phase. Thermal energy (heat) must first be supplied to convert a sufficient part of the fuel to vapor, thus creating a flammable vapor-air mixture in the vicinity of the surface. For liquid fuels, this is principally a process of evaporation, but almost all solid fuels must undergo chemical decomposition before vapor is released. One can usually identify a minimum liquid or solid temperature that is capable of generating a flammable mixture close to the fuel surface. For liquid fuels, this is defined in terms of the bulk liquid temperature, and is called the flash point. The same phenomenon can be recorded for combustible solids, but must be defined as a surface temperature. Note that these are piloted ignition temperatures, because an external pilot is needed to ignite the mixture and only the flammable vapor-air mixture will burn. A slightly higher temperature—the fire point—must be achieved if the liquid (or solid) fuel is to continue burning after the flammable mixture has been consumed.

In practice, the piloted ignition temperatures of solids and liquids can be influenced by the rate of air flow (oxidant), the rate of heating, and the size and shape of the fuel bed. As a result, reported piloted ignition temperatures, particularly for solids, depend somewhat upon the specific test methods.

In general, piloted ignition of a gas- or vapor-air mixture is affected by composition, ambient pressure, and the dimensions of the containing vessel, as well as the nature and energy of the pilot. For a given fuel-air mixture, there is a minimum pressure below which ignition does not occur. As the temperature increases, less and less pilot energy is required to ignite the mixture until, at a sufficiently high temperature, the mixture will ignite spontaneously. This temperature is referred to as the autoignition (or spontaneous ignition) temperature (AIT).

The autoignition temperature of a gaseous fuel also depends on composition and pressure, but it is particularly sensitive to the size and shape of the vessel in which the measurement is made. Differences in test apparatus can lead to significant differences in the results; for example, different values of AIT have been reported in the literature for the same vapor. (See Tables 1-5B and 1-5C.) (The same comments are relevant to the measurement of AIT for solid fuels.)

TABLE 1-5B. Variation of Autoignition Temperature with Mixture Composition

% Propane in Air	Autoignition Temperature °F	°C
1.5	1018	548
3.75	936	502
7.65	889	476

TABLE 1-5C. Variation of Autoignition Temperature for Carbon Disulphide (CS₂) with Vessel Size

Volume cm³	in.³	Autoignition Temperature °F	°C
200	12	248	120
1000	61	230	110
10000	610	205	96

Explosions

Explosions generally occur in situations where the fuel and oxidant have been allowed to mix intimately before ignition. As a result, the combustion reaction proceeds very rapidly without being delayed by the need for first mixing fuel and oxidant. If premixed gases are confined, their tendency to expand on burning can cause a rapid

pressure rise or explosion. This is in contrast to fires in which fuel and oxidant are initially separate and the combustion rate is controlled by the rate at which they are able to mix. As a result, the burning rate per unit volume is much lower for fires, and the very rapid increase in pressure characteristic of explosions is not encountered.

For ignition of a fuel that has been premixed with air, the fuel concentration must lie within the limits of flammability. Once ignition has occurred, flame will propagate through the unburnt mixture until all the flammable mixture has been consumed. The fuel may be in the form of gas or vapor, or a suspension of droplets (e.g., a mist) of a combustible liquid, or solid particles of an explosible dust.

The Limits of Flammability

The limits of flammability are the extreme concentration limits of a combustible in an oxidant through which a flame, once initiated, will continue to propagate at the specified temperature and pressure. For example, hydrogen-air mixtures will propagate flame for concentrations between 4 and 74 percent by volume of hydrogen at 21°C and atmospheric pressure. The smaller value is the lower (lean) limit, and the larger value is the upper (rich) limit of flammability. When the mixture temperature is increased, the flammability range widens; when the temperature is decreased, the range narrows. (See Figure 1-5A.) A decrease in temperature can cause a flammable mixture to become nonflammable by placing it either above or below the limits of flammability for the specific environmental conditions.

Note in Figure 1-5A that, for liquid fuels in equilibrium with their vapors in air, a maximum temperature exists for each fuel below which there is insufficient vapor to form a flammable vapor-air mixture: This is the flash point. There is also a maximum temperature above which the fuel vapor concentration is too high to propagate flame; gasoline in a partially full gasoline tank is an example of this. These may be referred to as the lower and upper flash points in air, respectively, although the term "flash point" on its own invariably signifies the former. The flash point temperatures for a combustible liquid increase with environmental pressure.

FIG. 1-5A. Saturated vapor-oxidant mixtures should be associated with the sloping vapor pressure line rather than just the upper flash point vertical line.

Fire Point

Fire point may be defined as the lowest temperature of a liquid in an open container at which vapors evolve fast enough to support continuous combustion. The fire point is usually a few degrees above the flash point. For typical fuels, the minimum rate of vaporization required to support combustion is of the order of 2 g/m².

It should be emphasized that, if a source of ignition has been established, fires can spread over liquids whose temperatures are considerably below their lower flash points. In such situations, the ignition source or fire itself heats the liquid surface locally so that its temperature rises above the fire point.

Catalysts and Inhibitors

Catalyst: A catalyst is a substance that greatly affects the rate of a chemical reaction but is not changed by the reaction itself. For example, platinum in a car's catalytic converter causes residual fuel to burn without permanently consuming the platinum catalyst.

Inhibitors: Inhibitors, also called stabilizers, are chemicals that may be added in small quantities to an unstable material to prevent a vigorous reaction. For example, premature polymerization of styrene monomer is inhibited by the addition of at least 10 ppm (parts per million) of tertiary-butyl-catechol (TBC). Fire-retardant chemicals usually act as inhibitors. For example, small quantities of chlorine or bromine containing compounds, when added to plastics, can reduce plastics' ignitability and ability to support the spread of small flames. Fire-retardant chemicals are, in general, not effective in large fires.

Stable and Unstable Materials

Stable materials: Stable materials are materials that have the capacity to resist changes in normal environmental exposure to air, water, heat, shock, or pressure. Most combustible materials can be classified as stable, although, of course, they can be made to burn.

Unstable materials: Unstable materials polymerize, decompose, condense, or become self-reactive when exposed to air, water, heat, shock, or pressure. For example, decomposition of pure gaseous acetylene, hydrazine, or ethylene oxide can result in violent explosions.

PRINCIPLES OF FIRE

Considerable technical knowledge exists concerning the ignition, burning, and fire spread characteristics of combustible materials (solids, liquids, and gases). However, most of this knowledge is for very simple geometric arrangements and thus is inadequate for predicting ignition and resultant burning in realistic situations. Nevertheless, insight can be gained from an understanding of these simplified situations.

Perhaps the simplest combustion situation is the burning of premixed, fuel-air gas mixtures involved in explosions. Much is known about premixed flames, as they have been the subject of much experimental research. The flammability limits and associated burning rates have been catalogued for most common vapor and gas mixtures.[1] In addition, scientists now can accurately calculate the burning rates for simple hydrocarbon fuel-air mixtures in terms of their multiple individual chemical reactions.[2]

Other simple situations involve the steady burning of fuel droplets or small samples of simple plastics. In such situations, the burning takes place in gas-phase laminar diffusion flames where the rate of combustion is controlled by the rate of fuel and air supply.

In contrast to explosions, fires occur in situations where fuel and oxidant are initially unmixed. Their burning rates are restricted primarily by the supply of fuel and oxidant (air) to the flames rather than the basic chemical reaction rates within the flames. In fires, the basic gas-phase combustion process usually occurs along thin flame sheets, called "diffusion flames," which separate regions rich in fuel vapor from regions rich in oxidant. Fuel vapor and oxidant diffuse toward each other from opposite sides of the flame sheet where they combine to produce combustion products and heat, which in turn diffuse away from the flame sheet.

When diffusion flames are small—e.g., from a burning match or candle—they typically appear quite smooth and steady. They are called "laminar" diffusion flames. If the fire is allowed to grow (e.g., by spreading over an extended surface), the flames will become unstable, evidenced as a characteristic flickering. Eventually, they will become fully turbulent when the fire diameter exceeds 30 to 50 mm.

Scientists have obtained a relatively clear understanding of small fires involving laminar diffusion flames. For example, they can calculate the rate of flame spread over solid surfaces,[3] and steady burning rates[4] in terms of basic combustion properties for a variety of simple geometries (e.g., smooth flat surfaces, cylinders, etc.). In these situations, the burning rates are controlled by the convective heat transfer from the flame to the solid fuel which, in response, gasifies and supplies fuel vapors to the flames. The oxidant (air) is supplied to the flames by the upward flow induced by the buoyant hot combustion products. The buoyant flow also may enhance the convective heat transfer from the flames to the solid fuel. In the case of a spreading flame, the spread rate is governed by the forward heat transfer from the flames to the as yet uninvolved fuel, which must be preheated before it can provide fuel vapors to the flames. (An analysis of many different fire spread situations that shows how one might predict their spread rates is contained in Williams, 1977.)[3]

Larger fires involving turbulent diffusion flames are less well understood because of difficulties in describing the turbulent gas motion, and the flame radiation, which is usually the dominant form of heat transfer for larger fires (fuel bed diameters greater than 30 to 50 mm). Experience and measurements have shown that this enhanced role of flame radiation in larger fires often alters the relative flammability ranking of fuels in comparison to their small-scale flammability rankings.

The study of these larger (hazardous) scale fires is at the forefront of current fire research. In recent years, scientists have developed new laboratory tests that are capable of providing data relevant to full-scale fire behavior.[5] Furthermore, they have prepared complex computer models capable of predicting fire development from ignition through various stages of growth, to full room involvement (flashover), propagation to adjacent rooms, and possibly even to other buildings.

Ignition and Combustion

To illustrate the many physical and chemical processes involved in fires, it is necessary to first discuss the ignition, burning, and eventual extinction of a wood slab in a typical situation, such as a fireplace.[6]

1. Suppose the wood slab is initially heated by thermal radiation. As its surface temperature approaches the boiling point of water, gases (principally steam) slowly evolve from the wood. These initial gases have little, if any, combustible content. As the slab temperature increases above the boiling point of water, the "drying" process penetrates deeper into the wood.

2. With continued heating, the wood surface begins to discolor when the surface temperature approaches 300°C. This discoloring is visible evidence of pyrolysis, the chemical decomposition

of matter through the action of heat. When wood pyrolyzes, it releases combustible gases while leaving behind a black, carbonaceous residue called char. This pyrolysis process penetrates deeper into the wood slab as the heating continues.

3. Soon after active pyrolysis begins, combustible gases typically evolve rapidly enough to support gas-phase combustion. However, combustion will occur only if there is a pilot flame or some other source of energy sufficient to cause piloted ignition. If no such pilot is present, the wood surface must be heated to a much higher temperature before autoignition occurs.

4. Once ignition occurs, a diffusion flame rapidly covers the pyrolyzing surface. One the diffusion flame is established, little, if any, oxygen will reach the pyrolyzing surface. Meanwhile, the flame heats the fuel surface and causes an increase in the rate of pyrolysis. If the original radiant heat source is withdrawn at ignition, the burning will continue provided the wood slab is thin enough (less than 19 mm). Otherwise, the flames will go out, because the slab surface loses too much heat by thermal radiation to the surroundings and thermal conduction into its interior. If there is an adjacent, parallel wood surface (or insulating material) facing the ignited slab, some of the surface radiation loss is returned as the adjacent surface heats up and begins to radiate back. Under these circumstances, the ignited slab can continue burning even after the withdrawal of the initial heat source. This explains why one cannot burn a single large log in a fireplace, but must use several logs to capture the radiant heat losses.

5. As the burning continues, a char layer builds up. This char layer, which is a good thermal insulator, restricts the flow of heat to the wood interior, and consequently the rate of pyrolysis tends to reduce. The pyrolysis rate also will decrease when the supply of unpyrolyzed wood runs out. When the pyrolysis rate decreases to the point of not being able to sustain gas-phase combustion, oxygen will diffuse in sufficient amounts to the char surface, permitting it to undergo direct glowing combustion (provided that the radiant heat losses are not too large).

6. This scenario presumes an ample (but not excessive) supply of air (oxidant) for combustion. If there were insufficient oxidant to burn the available fuel vapor, the excess vapors would travel with the flow and possibly burn where they eventually would find sufficient oxidant. For example, this happens when fuel vapors emerge and burn outside a window of a fully involved but under-ventilated room fire. Generally, under-ventilated fires produce large amounts of smoke and toxic products (e.g., carbon monoxide).

If, on the other hand, one was to impose a forced draft onto the pyrolyzing surface, the oxidant supply may exceed that required for complete combustion of the fuel vapors. In this case, the excess oxidant can cool the flames sufficiently to suppress their chemical reaction and extinguish them, as happens, for example, when one blows out a match. In the case of larger fires with ample supply of fuel vapors, imposing a forced draft on them simply increases their rate of burning by increasing the flame-to-fuel-surface heat transfer, which in turn enhances the fuel supply rate.

7. Following the ignition of a certain portion of our wood slab, the flames are likely to spread over the entire fuel array. Flame spread can be thought of as a continuous succession of piloted ignitions where the flames themselves provide the heat source. One commonly observes that upward flame spread on a vertical surface is much more rapid than downward or horizontal flame spread. This is because hot flames generally travel upward and contribute their heat over a greater area in an upward direction. Thus, each successive "upward ignition" adds a much greater burning area to the fire than a corresponding "downward" or "horizontal" ignition.

Generally, materials that ignite easily (rapidly) also propagate flames rapidly. The ignitability of a material is controlled by its resistance to heating (thermal inertia) and by the temperature rise required for it to begin to pyrolyze. Materials with low thermal inertia, such as foamed plastics or balsa wood, heat rapidly when subjected to a given heat flux. These materials are often easy to ignite and can cause very rapid flame spread. On the other hand, dense materials, such as ebony wood, tend to have relatively high thermal inertias and are difficult to ignite.

8. The burning rates of larger, more hazardous fires are principally governed by the radiant heat transfer from the flames to the pyrolyzing fuel surface.[7] This flame radiation comes primarily from the luminous soot particles in the flames. Combustibles that tend to produce copious amounts of soot or smoke, such as polystyrene, also tend to support more intense fires, despite the fact that their fuel vapors burn less completely, as evidenced by their higher smoke output.

Well-ventilated fires generally release less smoke than poorly ventilated ones. In well-ventilated fires, the surrounding air can gain speedy access to the unburned fuel vapors and soot before the fuel vapors cool down by radiation. Poorly ventilated fires can release copious amounts of smoke and products of incomplete combustion, such as carbon monoxide. In poorly ventilated fires, fuel vapors have insufficient air to burn completely before cooling off and leaving the fire area.

Fires occurring in oxygen-enriched atmospheres have higher flame temperatures, increased fractions of heat release by radiation, and increased burning rates per unit fuel area. These higher flame temperatures generally cause a much greater conversion of fuel vapors into soot, resulting in significantly increased smoke release rates. For example, a well-ventilated methanol fire typically burns with a blue (i.e., soot-free) flame in normal air. However, a similarly well-ventilated methanol fire can burn with a brightly luminous smoky flame in an oxygen-enriched atmosphere. This sensitivity to ambient oxygen concentration significantly increases the flame radiation, burning rates, and resultant fire hazard.

Flammability Properties of Solid and High Fire Point Liquid Combustibles

The above discussion can be summarized by describing the combustible properties that contribute to fire hazards in typical fire situations.

Heat of combustion: This is a measure of the maximum amount of heat that can be released by the complete combustion of unit mass of combustible material.

Stoichiometric oxidant: The mass of oxidant required for complete combustion of a unit mass of combustible is called the stoichiometric oxidant requirement. Combustibles with large stoichiometric oxidant requirements often produce large flame heights, which in turn present greater fire spread hazards. The stoichiometric oxidant requirements for typical (organic) combustibles is approximately proportional to their heat of combustion, so that organic combustibles all release approximately the same amount of heat per unit mass of consumed oxidant. Thus, it is found that 13LJ of energy is released for every gram of oxygen consumed in the combustion of most common materials. This is the basis for the measurement of heat release in the cone calorimeter.

Heat of gasification: The heat required to vaporize a unit mass of combustible material that is initially at ambient temperature is called the heat of gasification. This quantity is very important because it determines the amount of combustible vapor supplied to a fire in response to a given supply of heat to the pyrolyzing surface. The fire hazard of some plastics can be reduced by adding inert fillers, such as alumina trihydrate, which increase their effective heats of gasification.

Ignitability (piloted): Ignitability is inversely proportional to the time it takes for a given applied heat flux to raise the surface temperature of a material to its piloted ignition (fire point) temperature. This property is important both for ignition and fire spread.

Char formation: Char is a black, carbonaceous residue, formed during the pyrolysis of some materials, e.g., wood and some thermosetting plastics. The insulating properties of the resulting char layer can be very effective in reducing burning rates by restricting the flow of heat to the unpyrolyzed material. Intumescent paints simulate formation of char by swelling and forming an insulating protective layer when heated. Note that thermoplastics, such as polymethylmethacrylate, tend not to char, but soften, melt, and flow instead.

Soot formation: Soot consists of minute solid carbonaceous particles formed as a result of incomplete combustion and pyrolysis, particularly in the fuel-rich regions of diffusion flames. Combustibles whose flames produce significant amounts of soot are generally more hazardous because the soot increases flame radiation, which in turn increases the burning rate. The soot is also the source of smoke that is released from the flame and can result in significant smoke spread to points remote from the fire.

Inhibitors: The addition of small quantities of chemical inhibitors to the fuel or oxidant can impede gas-phase reactions. Such flame inhibitors can be quite effective in retarding ignition and flame spread involved in small fires. Flames also can be inhibited by introducing additives to solid combustibles, which either promote char formation or produce higher concentrations of water vapor in the pyrolysis gases (or both).

Melting: Combustibles that tend to melt are often more hazardous, because the molten combustible can flow to form a pool, thus increasing the area of pyrolyzing surface. The molten material itself also can be a hazard.

Toxicity: Carbon monoxide usually is the principal toxicant produced by a fire. It is typically present in pyrolysis gases and combustion products that have not been completely oxidized. Some materials, especially those containing elements other than C, H, and O (e.g., polyvinylchloride or polyurethane), can produce additional toxicants (such as hydrogen chloride or hydrogen cyanide, respectively). (See Section 4, Chapter 2, "Combustion Products and Their Effects on Life Safety.")

Geometry: Last, but not least, the geometry of a material strongly influences its flammability. In general, thin materials ignite more readily and spread flame more rapidly than thick. Upward flame spread is more rapid than downward or horizontal flame spread. Finally, geometric arrangements that provide ample air (oxidant) access while providing shielding to prevent radiant losses are usually the most hazardous.

Flammability Principles

The science of fire protection rests upon the following principles:

1. An oxidizing agent, a combustible material, and an ignition source are essential for combustion.
2. The combustible material must be heated to its piloted ignition temperature before it can be ignited or support flame spread.

3. Subsequent burning of a combustible is governed by the heat feedback from the flames to the pyrolyzing or vaporizing combustible.
4. The burning will continue until:
   a. The combustible material is consumed; or
   b. The oxidizing agent concentration is lowered to below the concentration necessary to support combustion; or
   c. Sufficient heat is removed or prevented from reaching the combustible material to prevent further fuel pyrolysis; or
   d. The flames are chemically inhibited or sufficiently cooled to prevent further reaction.

All the material presented in this handbook for the prevention, control, or extinguishment of fire is based upon these principles.

HEAT MEASUREMENT

The temperature of a material is the condition that determines whether it will transfer heat to or from other materials. Heat always flows from higher to lower temperatures. Temperature is measured in degrees.

Temperature Units

Celsius: A Celsius (or centigrade) degree ($^\circ$C) is 1/100 the difference between the temperature of melting ice and boiling water at 1 atmosphere pressure. On the Celsius scale, zero is the melting point of ice; 100 is the boiling point of water. This unit is approved by the International System (SI) of units.

Kelvin: A Kelvin degree or Kelvin (K) is the same size as the Celsius degree. However, the zero on the Kelvin scale (sometimes called Celsius Absolute) is -459.67°F (-273.15°C). Zero on the Kelvin scale is the absolute lowest achievable temperature; hence, the Kelvin scale provides us with so-called absolute temperatures. The Kelvin also is an approved SI unit.

Fahrenheit: A Fahrenheit degree ($^\circ$F) is 1/180 the difference between the temperature of melting ice and boiling water at 1 atmosphere pressure. On the Fahrenheit scale, 32 is the melting point of ice (0°C); 212 is the boiling point of water (100°C).

Rankine: A Rankine degree ($^\circ$R) is the same size as the Fahrenheit degree. On the Rankine scale, zero is -459.67°F (-273.15°C), so that the Rankine scale also provides an absolute temperature.

Fahrenheit and Rankine degrees are not approved SI units, and their use is greatly discouraged.

Heat Units

Joule (J): Conventionally, the Joule is defined as the energy (or work) expended when unit force (1 Newton) moves a body through unit distance (1 m). It is the most convenient unit of energy to use and can be related to the calorie (q.v.), which is defined in terms of the heat energy required to raise the temperature of unit mass of water by 1°. The Joule is an approved SI unit.

Watt (W): This is a measure of power, or the rate of energy release. One Watt is equal to one Joule per second (1 W = 1 J/s). The rate of heat release from a fire can be expressed in kilowatts (kW) or megawatts (mW)—units that are familiar to the electrical engineer.

Calorie: The amount of heat required to raise the temperature of one gram of water one degree Celsius [measured at 59°F (15°C)] is called a calorie. One calorie equals 4.183 Joules.

British thermal unit (Btu):   The amount of heat required to raise the temperature of 1 pound of water one degree Fahrenheit [measured at 60°F (15.5°C)] is called the British thermal unit. One Btu equals 1,054 Joules (252 calories); therefore, it can be inferred that 1.054 kilowatts would heat 1 pound of water one degree Fahrenheit in 1 second.

Btu and calories are not approved SI units.

Heat energy has quantity, as well as potential (intensity). For example, consider the following analogy:

Two water tanks stand side by side. If the first tank holds twice as many gallons as the second, then the first tank can hold twice the quantity of water as the second. But if the level of water in the two tanks is equal, then their pressures or potentials are equal. If the bottoms of the two tanks are connected by a pipe, water will not flow from one to the other because both tanks have the same equilibrium pressure. In a similar manner, one body may hold twice the quantity of heat energy (measured in Joules or Btu) as a second body. But if the potentials or temperatures of the energies of the bodies are equal, no heat energy will flow from one body to the other when they are brought into contact, since the bodies are at equilibrium. If a third body at a lower temperature were brought into contact with the first body, heat would flow from the first to the third until both body temperatures became equal. The amount or quantity of heat flowing until this equilibrium is reached depends upon the heat retention capacities of each body involved.

[Note that, essentially, ignition involves the addition of sufficient heat (by heat transfer, q.v.) to raise the temperature to the appropriate value. On the other hand, extinction may be accomplished by the removal of heat. "Chemical" extinguishment works by another mechanism, i.e., by interrupting chemical reactions that are important in the combustion process.][2]

Temperature Measurement

Devices that measure temperature depend upon physical change (expansion of a solid, liquid, or gas), change of state (solid to liquid), energy change (changes in electrical potential energy, i.e., voltage), or changes in thermal radiant emission and/or spectral distribution. The principles of operation of the more common temperature measuring devices are discussed here.

Liquid expansion thermometers: These thermometers consist of a tube partially filled with a liquid, which measures expansion and contraction of the liquid with changes in temperature. The tube is calibrated to permit reading of the level of the liquid in degrees of a temperature scale.

Bimetallic thermometers: Bimetallic thermometers contain strips of two metals, which are laminated, with different coefficients of expansion. As the temperature changes, the two metals expand or contract to different extents, causing the strip to deflect. The amount of deflection is measured on a scale that is calibrated in degrees of temperature.

Solid fusion: Solid fusion makes use of the melting or fusion point of a solid (metal, chemical, or chemical mixture) to indicate whether or not a hot object is above or below the melting point of the solid. For example, the eutectic metal in a sprinkler link melts in a fire environment, which activates the sprinkler.

Thermocouples: Thermocouples consist of a pair of wires of different metals or alloys welded together at a point to form a junction. A voltage is generated across this junction, the magnitude of which depends on the nature of the metals and the temperature. This is compared against a compensating junction at 0°C, and calibrated to give readings in degrees.

Pyrometers: Pyrometers measure the intensity of radiation from a hot object. Since intensity of radiation depends upon temperature, pyrometers can be calibrated to give readings in degrees of temperature. Optical pyrometers measure the intensity of a particular wavelength of radiation.

Specific Heat

The specific heat of a substance defines the amount of heat it absorbs as its temperature increases. It is expressed as the amount of thermal energy required to raise the unit mass of a substance one temperature degree and is measured in J/g(°C) or cal/g(°C) or Btu/lb(°F). Water has a specific heat of 4.2 J/g(°C), or 1 Btu/lb°F.

The specific heats of various substances vary over a considerable range; for most common substances, except water, they are less than 4.2 J/g(°C). Specific heat figures are significant in fire protection because they indicate the relative quantity of heat needed to raise the temperature to a point of danger, or the quantity of heat that must be removed to cool a hot substance to a safe temperature.

One reason for the effectiveness of water as an extinguishing agent is that its specific heat is higher than that of most other substances.

Latent Heat

Heat is absorbed by a substance when it is converted from a solid to a liquid, or from a liquid to a gas. This thermal energy is called latent heat. Conversely, heat is released during conversion of a gas to a liquid or a liquid to a solid.

Latent heat is the quantity of heat absorbed by a substance in passing between liquid and gaseous phases (latent heat of vaporization), or between solid and liquid phases (latent heat of fusion). It is measured in Joules per unit mass. The latent heat of fusion of water (normal atmospheric pressure) at the freezing or melting point of ice (0°C) is 333.4 J/g; the latent heat of vaporization of water at its boiling point (100°C) is 2,257 J/g (or 970.3 Btu/lb or 539.5 cal/g). The larger heat of vaporization of water is another reason for the effectiveness of water as an extinguishing agent. It requires 3 million Joules to convert 1 kilogram of ice at 0°C to steam at 100°C. The latent heats of most other common substances are substantially less than that of water. Thus, the heat absorbed by water is subtracted from the burning system so it cannot be used to propagate the flame further by vaporizing more liquid, or by pyrolyzing more solid.

HEAT TRANSFER

Transfer of heat governs the ignition, burning, and extinguishment of most fires. Heat is transferred by one or more of three methods: (1) conduction, (2) convection, or (3) radiation.

Conduction

Heat transferred by direct contact from one body to another is a process called conduction. Thus, a steam pipe in contact with wood transfers its heat to the wood by actual contact; in this example, the pipe is the conductor.

The quantity of heat energy transferred by conduction through a body or in a given time is a function of the temperature difference and the conductance of the path involved. Conductance depends on thermal conductivities, cross-sectional areas normal to the flow path, and length of flow path. The rate of heat transfer, then, is simply the quantity of heat per unit time, while the heat "flux" is the quantity of heat per unit cross-sectional area per unit time. The thermal conductivity of a material is the heat energy flux resulting from a unit temperature gradient (falloff of one degree per unit of distance). A typical unit of thermal conductivity is J/(m•s•°C), i.e., W/m °C.

The conduction of heat through air or other gases is independent of pressure in the usual range of pressures. It approaches zero only at very low pressures. No heat is conducted in a perfect vacuum. Solids are better heat conductors than gases; the best commercial insulators consist of fine particles or fibers, with the spaces between the particles filled with air.

Heat conduction cannot be completely stopped by any "heat-insulating" material. Thus, the flow of heat is unlike the flow of water, which can be stopped by a solid barrier. Heat-insulating materials have a low heat conductivity. No matter how thick the insulation, solidly insulating the space between the source of heat and the combustible material may be insufficient to prevent ignition. If the rate of heat conduction through the insulating material is greater than the rate of dissipation from the combustible material, the temperature of the combustible material may increase to the point of ignition. For this reason, there should always be an air space or some manner of carrying the heat away by convection, rather than relying solely upon the heat-insulating material for protection. By the transmission of heat over a long period, fires have occurred through as much as a 0.6-m thickness of solid concrete.

For heat conduction, the most important physical properties of a material are thermal conductivity (k), density (ρ), and specific heat (c). Unfortunately, the last two quantities are usually listed separately, although it is their product (ρc) that is of interest in the field of heat conduction. The product ρc is a measure of the amount of heat necessary to raise the unit volume of the material by unit temperature. A typical unit might be Joules per cubic meter per degree Celsius [J/(m^3•°C)], and a possible name for the quantity is thermal capacity per unit volume.

Thermal conductivity and thermal capacity per unit volume are rarely of much importance individually. In fact, the solution to heat conduction problems is so complex that it cannot be adequately presented here. However, one or two interesting features can be appropriately mentioned. Probably the most useful feature is concerned with the time constant of a thickness (x) of a material. Thus, if the surface of a material is suddenly exposed to a temperature rise, then the temperature at a depth (x) within the material will begin to change substantially at a certain time.

$$t = \frac{x^2 \rho c}{k}$$

The dimensions of the time expression are given as:

$$\frac{cm^2 \cdot \left(J/cm^3 \cdot °C \right)}{\left(J/cm \cdot sec \cdot °C \right)} = sec$$

Meters, Joules, and degrees Celsius cancel out, leaving the result of the dimension in seconds, i.e., time. This is a time constant; thus, the higher the number, the slower the transfer. It can be seen that the time required for a thermal wave to penetrate a body increases with the square of its thickness.

Convection

Convection involves the transfer of heat by a circulating medium—either a gas or a liquid. Thus, heat generated in a stove is distributed throughout a room by initially heating the air in contact with the stove by conduction; the circulation of heated air through the room to distant objects is heat transfer by convection. Finally, heat is transferred from the air to the objects by conduction. Heated air expands and rises, and, for this reason, heat transfer by convection often occurs in an upward direction, although air currents can be made to carry heat by convection in any direction, e.g., by use of a fan or blower.

Note that the term "convective heat transfer" is also used to describe the mode by which heat is transferred between a fluid and a solid surface. Indeed, the convective heat transfer coefficient (h) is defined by the expression

$$q = h\Delta T$$

where q is the rate of heat transfer per unit surface area (W/m^2), and ΔT is the temperature difference (°C) between the fluid and the surface.

Radiation

Radiation is a form of energy traveling across a space or through materials as electromagnetic waves, such as light, radio waves, or X-rays. In a vacuum, all the radiant energy waves travel at the same speed as light. On arrival at a body, they are absorbed, reflected, or transmitted. Visible light consists of wavelengths between 0.4×10^{-6} to 0.7×10^{-6} m (violet to red). Emission from combustion processes occurs principally in the infrared region (wavelengths longer than the red wavelength). Our eyes see only the tiny fraction that is emitted within the visible region.

A candle flame gives a common example of radiation. Air heated by the flame rises upward while cooler air moves in toward the candle to supply the flame with more oxygen, thus sustaining the burning process. If a hand is held to the side of the flame, it would experience a sensation of warmth. This energy is called radiant heat or radiation.

In the case of small fires, such as one involving a candle, most of the heat leaves the combustion zone by vertical convection, e.g., a hand held over the flame instead of beside it will sense more heat. However, larger, more hazardous fires may release 30 percent to 50 percent of the total amount of energy as radiation, exposing nearby surfaces to high levels of radiant heat transfer.

Radiated energy travels in straight lines. Obviously, the heat received from a small source would be less than that received from a large radiating surface, provided the sources emit comparable energies per unit area. (See Figure 1-5B.)

Radiant heat passes freely through gases consisting of symmetrical diatomic molecules, such as hydrogen (H_2), oxygen (O_2), and nitrogen (N_2), but is absorbed by suspended particles, such as smoke, and in narrow wavelength bands by unsymmetrical molecules, such

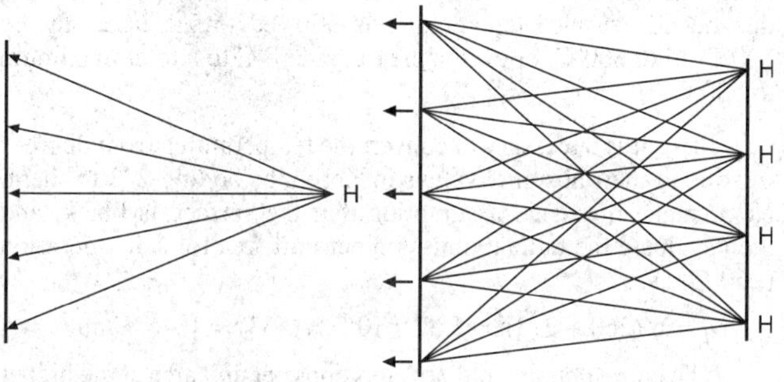

FIG. 1-5B. *A comparison of heat absorption by surfaces of similar area from a pinpoint source (left) and a large radiating surface (right).*

as water vapor (H_2O), carbon monoxide (CO), carbon dioxide (CO_2), and sulfur dioxide (SO_2). Although the concentration of these gases in the atmosphere is low, the presence of carbon dioxide and water vapor prevents radiation in the vicinity of 2.8 microns and 4.4 microns (in particular) from reaching the surface of the earth. Advantage is taken of this fact in the development of infrared flame detectors, some of which are designed to be blind to solar radiation and respond only to 2.8 microns or 4.4 microns. These wavelengths are emitted strongly by hot water vapor and carbon dioxide in flames. Note also that water vapor and carbon dioxide in the atmosphere can absorb appreciable amounts of radiation at distances from large fires. This helps to explain why forest fires or large LNG fires are less hazardous on days when humidity is high. Also, since water droplets absorb almost all of the incident infrared radiation, mists or water sprays are effective attenuators of radiation.

When two bodies face each other and one body is hotter than the other, radiant energy will flow from the hotter body to the cooler body until thermal equilibrium is achieved. The ability to absorb this radiated heat is a function of the kind of surface of the cooler body and the area of the radiating surface of the hotter body. If the receiving surface is shiny or polished, it will reflect most of the radiant heat away; if it is black or dark in color, it will absorb most of the heat. Most nonmetallic materials are effectively "black" to infrared radiation, despite the fact that they may appear light or colored to visible radiation. Some substances, such as water or glass, are transparent to visible radiation and will allow it to pass through them with minimal absorption; however, both liquid water and glass are opaque to most infrared wavelengths. Glass greenhouses and solar panels operate on the principle of being transparent to the sun's visible radiation, while at the same time being opaque to the infrared radiation attempting to escape from the greenhouse or solar panel.

Shiny metallic materials are excellent reflectors of radiant energy. For example, aluminum foil often is used together with fiberglass in building insulation. Sheet metal is often used beneath stoves or on heat-exposed walls.

The Stefan-Boltzmann law states that the radiant emission per unit area from a black surface is proportional to the fourth power of its absolute temperature. The law can be expressed by the formula:

$$q = \varepsilon \sigma T^4$$

where q is the radiant emission per unit surface area; ε is the surface emissivity correction factor, which is 1.0 for an ideal black surface and typically close to 1 for most nonmetallic materials in the infrared wavelength region; σ is the Stefan-Boltzmann proportionality constant [equal to 5.67×10^{-12} W/(cm$^2 \cdot$K^4)]; and T is the absolute temperature expressed in Kelvin. To appreciate the importance of this fourth power, consider the following situation.

EXAMPLE: A heater is designed to operate safely with an outside surface temperature of 260°C. What is the increase in radiation if this outside surface temperature is allowed to increase, say by 100°C up to 360°C, or even increase by 258°C to a total maximum to 500°C?

First, it is necessary to convert the temperatures from degrees Celsius to their absolute values in Kelvin by adding 273 to them. Next, make the usual assumption that the surface is black, and then evaluate the radiant emission per unit area for safe operation (260°C) as:

$$q = \sigma (260 + 273)^4 = 5.67 \times 10^{-12} \times (533)^4 = 0.46 \text{ W/cm}^2$$

The corresponding radiant emissions per unit area at the higher temperatures are:

$$q = \sigma (360 + 273)^4 = 0.91 \text{ W/cm}^2$$

$$q = \sigma (500 + 273)^4 = 2.0 \text{ W/cm}^2$$

Thus, it can be seen that, by increasing the stove temperature by only 100°C, one approximately doubles its emission from 0.46 to 0.91 W/cm^2. This is the dramatic effect of the fourth power. If heat radiation was merely a function of the first power of the absolute temperature, then it would increase by only about 20 percent as the absolute temperature increased from 533 to 633 K. Finally, if one were so careless as to allow the stove to reach 500°C, it would emit 2.0 W/cm^2, which is sufficiently high to ignite many typical home furnishings.

Because residential coal or wood stoves can sometimes undergo a temperature "runaway" if given too much air, it is important to limit the maximum received radiant heat transfer by keeping all nearby furnishings away from the stove. The radiant energy transmitted from a point-like source to a receiving surface will vary inversely, as the square of their separation distance. If a stove is small relative to its distance from nearby objects, then it behaves like a point source; doubling its separation distance will decrease the incident radiant heat (per unit area) by a factor of four. However, if the nearby object is close to the stove, the stove appears like a large surface, and small changes in separation will have little or no effect on the severity of the radiative heat transfer (e.g., a large oven within a few centimeters from a combustible wall). In this case, one must protect the wall by some other means, e.g., a shiny, metal-coated, noncombustible board.

Generally, scientists and engineers have a good understanding of radiation between solid surfaces. They also can reliably estimate radiant heat transfer absorbed or emitted by gases of a known composition and temperature. However, they have difficulty in estimating the radiation from flames, because most of this radiation comes from the soot in the flames and there have been very few measurements of soot in flames. This lack of soot data often prevents the estimate of fire burning rates. There is, however, one useful rule of thumb: The total radiant output from flames (higher than about 20 cm) is usually about 30 to 40 percent of the maximum heat output, assuming complete combustion.[7] A comparable amount of energy leaves by convection, with the remaining fraction leaving in the form of incomplete combustion.

HEAT ENERGY SOURCES OR SOURCES OF IGNITION

Since fire prevention and extinguishment depend upon the control of heat, it is important to be familiar with the more common ways in which heat energy can be produced. Four sources of heat energy are: (1) chemical, (2) electrical, (3) mechanical, and (4) nuclear.

Chemical Heat Energy

Oxidation reactions usually produce heat. They are the source of the heat that is of primary concern to fire protection engineers.

Heat of combustion: The heat of combustion is the amount of heat released during a substance's complete oxidation (combustion, i.e., conversion to carbon dioxide and water). Heat of combustion, commonly referred to as calorific or fuel value, depends upon the kinds and numbers of atoms in the molecule, as well as their arrangement in the molecule. Calorific values are commonly expressed in Joules per gram, but are sometimes reported in Btu/lb or calories/g (1 Btu/lb = 2.32 J/g; and 1 cal/g = 4.18 J/g). In the case of fuel gases, calorific values are commonly reported in Btu/cu ft. Calorific values are used in calculating fire loading, but do not necessarily indicate relative fire hazard, since the fire hazard depends upon rate of burning, which is determined by the distribution of fuel, as well as upon the total amount of heat produced.

Also, heat is produced in incomplete or partial oxidations, which occur at some stage in almost all accidental fires and in spontaneous heating by oxidation. For almost all compounds of carbon and hydrogen, or of carbon, hydrogen, and oxygen (which include substances of vegetable or petroleum origin), the heat of oxidation, whether complete or partial, depends upon the amount of oxygen consumed. For these common substances (e.g., coal, natural gas, common plastics, oils, wood, cotton, sugar, and vegetable and mineral oils), the heat of oxidation is approximately 3 kJ/g (3700 kJ/m^3, or 100 Btu/cu ft) of air consumed, regardless of the completeness of combustion. For this reason, the heat produced in a fire or in a spontaneous heating oxidation often is limited by the air (oxygen) supply.

Spontaneous heating: Spontaneous heating is the process whereby a material increases in temperature without drawing heat from its surroundings. If this results in a material reaching its ignition temperature, spontaneous ignition or spontaneous combustion can be said to have taken place. The fundamental causes of spontaneous heating are few, but the conditions are many and varied under which these fundamental factors may operate to create a dangerous condition. Three conditions that have much to do with whether an oxidation reaction will cause dangerous heating are: (1) rate of heat generation, (2) air supply, and (3) insulation properties of the immediate surroundings.

If exposed to the atmosphere, most all-organic substances capable of combination with oxygen will oxidize, with evolution of heat. The rate of oxidation at normal temperatures is usually so slow that it is usually imperceptible (e.g., the perishing of rubber); under these circumstances, the released heat is transferred to surroundings as rapidly as it is formed, with the result that there is no appreciable temperature increase in the combustible material being oxidized. This, however, is not true of all combustible materials, since certain oxidation reactions at normal temperatures (e.g., oxidation of powdered zirconium in air) generate heat more rapidly than it can be dissipated, with spontaneous ignition being the result.

In order for spontaneous ignition to occur, sufficient air must be available to permit oxidation, yet not so much air that the heat is carried away by convection as rapidly as it is formed. An oil-soaked (e.g., vegetable oil) rag that might heat spontaneously in the bottom of a wastebasket would not be expected to do so if hung on a clothesline where air movement would remove heat; nor would it be likely to heat up if it were sealed in a tightly packed bale of rags. On the other hand, because of more air and the insulating effect of the bale, a loosely packed bale might provide ideal conditions for heating. Because of the many possible combinations of the interrelated factors of air supply and insulation, it is impossible to predict with certainty whether a material will heat spontaneously.

In the presence of air, substances subject to oxidation will first form products of partial oxidation, which may act as catalysts to further oxidation. For example, olive oil that has turned rancid from exposure to the air will oxidize at a higher rate than fresh, pure oil.

Additional heat can initiate spontaneous heating in some combustible materials not subject to this phenomenon at ordinary temperatures. In these instances, preheating increases the rate of oxidation sufficiently so that more heat is produced than can be lost. Many fires have been caused by the spontaneous heating of foam rubber following preheating in a dryer.

A common cause of heating of agricultural produce is due to microbiological activity that generates heat. Since most bacteria cannot live at temperatures much above the range of 70 to 80°C, continued heating of such materials to their ignition temperatures is thought to be due to oxidation processes that become significant at these elevated temperatures.

The moisture content of agricultural products has a definite influence on the spontaneous heating by bacteria. Wet or improperly cured hay is very likely to heat in barn lofts. Experience has indicated that such heating may result in ignition within a period of two to six weeks after storage. Alfalfa meal that has been exposed to rain and then stored in bins or piles is very susceptible to spontaneous heating. Soybeans stored in bins have been known to sustain what is called "bin burn," i.e., the beans next to the bin walls are charred due to the moisture condensation on the inside surfaces of the wall and the self-heating of the beans. Other agricultural products susceptible to spontaneous heating are those with a high content of oxidizable oils, such as cornmeal feed, linseed, rice bran, and pecan meal.

Heat of decomposition: The heat of decomposition is the heat released by the decomposition of compounds that have been formed from their elements by endothermic reactions. Such compounds are intrinsically unstable, and when decomposition is started, such as by heating the substance above a critical temperature, decomposition continues with the liberation of heat. Acetylene and cellulose nitrate are well known for their tendency to decompose, with the liberation of dangerous quantities of heat. The chemical action responsible for this effect in many commercial and military explosives is the rapid decomposition of an unstable compound.

Heat of solution: The heat of solution is the heat released when a substance is dissolved in a liquid. Most materials release heat when dissolved, although the amount of heat is usually not sufficient to have any significant effect on fire protection. In the case of some chemicals, e.g., concentrated sulfuric acid, the heat evolved may be sufficient to be dangerous. The chemicals that react with water in this manner are not themselves combustible, but the liberated heat may be sufficient to ignite nearby combustible materials.

In contrast to most materials, ammonium nitrate absorbs heat when dissolved in water. (It is said to have a negative heat of solution.) Some first-aid products, for use where cold is recommended, consist of dry ammonium nitrate in watertight packages. These packages become cold when water is added.

Electrical Heat Energy

When current flows through a conductor, electrons pass from atom to atom within the conductor. The better conductors, such as copper and silver, have the most easily removed outer electrons, so that the force or voltage required to establish or maintain any unit electric current (or electron flow) through the conductor is less than for substances composed of more tightly bound electrons. Thus, the electrical resistance of any substance depends upon atomic and molecular characteristics; the electrical resistance is proportional to the energy required to move a unit quantity of electrons through the substance against the forces of electron capture and collision. This energy expenditure appears in the form of heat.

Resistance heating: Resistance heating is characterized by a rate of heat generation proportional to the resistance and the square of the current. Since the temperature of the conductor resulting from resistance heating depends upon dissipation of heat to the surroundings, bare wires can carry more current than insulated wires, without heating dangerously, and single wires can carry more current than closely grouped wires, or wires bundled into a cable of equivalent cross-sectional area.

The heat generated by incandescent and infrared bulbs is due to the resistance of the filaments in the bulbs. Material of very high melting point is used for the "white-hot" filaments of incandescent lamps, and destruction of the filament by oxidation is prevented by partial evacuation of the bulb and by removal of oxygen.

The filaments of infrared lamps operate at a much lower temperature (a "red" heat); the most efficient infrared lamp reflectors are gold, because gold is one of the best reflectors of infrared radiation.

Dielectric heating: Whenever atoms are subjected to electric potential gradients from external sources, the arrangement of the atom (or of a molecule of several atoms) is distorted. A tendency exists for electrons to move in the direction of the positive potential and for protons to move in the opposite direction. This is true whether the externally applied potential is due to a battery or a generator, or is the result of a magnetic field. Even though the external potential is insufficient to break away any electrons, the distortion of the normal atomic or molecular arrangement represents an energy expenditure. This is of practically no consequence if the external force is unidirectional, but can be substantial if the potential is pulsating or alternating. For example, the heating of a dielectric (a good insulator) may be considerable if the frequency of alternation of external potential becomes high.

Induction heating: Whenever a conductor is subject to the influence of a fluctuating or alternating magnetic field, or whenever a conductor is in motion across the lines of force of a magnetic field, potential differences appear in the conductor. These potential differences result in the flow of fire, with attendant resistance heating in the conductor. For rapidly changing or alternating potentials, additional energy is expended, and, as the polarity changes, this energy appears as heat energy due to the mechanical and electrical distortion of the molecular structure. This latter type of heating increases with the frequency of alternation. Food in a microwave oven, for example, is heated by the molecular friction induced by absorbed microwave energy.

A useful form of induction heating is created by passing a high-frequency alternating current through a coil surrounding the material to be heated.

An alternating current passing through a wire can induce a current in another wire parallel to it. If the wire in which a current is induced does not have adequate current-carrying capacity for the size of the induced current, resistance heating will occur. In this example, the heating is due primarily to resistance to flow, and only in a small degree to molecular friction.

Leakage current heating: Since all available insulating materials are imperfect insulators, there is always some current flow when the insulators are subjected to substantial voltages. This flow is commonly referred to as a leakage current and is usually not important from the standpoint of heat generation. However, if the insulating material is not suited for the service, or the material is too thin (for reasons of economy, space saving, or attempts to attain the maximum capacity in a condenser), leakage currents may exceed safe limits, resulting in heating of the insulator with consequent deterioration of the material and ultimate breakdown.

Heat from arcing: Arcing occurs when an electric circuit that is carrying current is interrupted either intentionally, as by a knife switch, or accidentally, as when a contact or terminal becomes loosened. Arcing is especially severe when motor or other inductive circuits are involved. The temperatures of arcs are very high, and the heat released may be sufficient to ignite combustible or flammable material in the vicinity. In some instances, the arc may melt the conductor and scatter molten metal. One requirement of an intrinsically safe electrical circuit is that arcing, due to accidental current interruption, will not release sufficient energy to ignite the hazardous atmosphere in which the circuit is located.

Static electricity heating: Static electricity (sometimes called frictional electricity) is an electrical charge that accumulates on the surfaces of two materials that have been brought together and then separated. One surface becomes charged positively, the other negatively. If the substances are not bonded or grounded, they will eventually accumulate sufficient electrical charge so that a spark may occur. Static arcs are ordinarily of very short duration and do not produce sufficient heat to ignite ordinary combustibles, such as paper. Static sparks, however, are capable of igniting flammable vapors and gases, and clouds of explosible dust. Hydrocarbon fuel (e.g., gasoline) flowing in a pipe can generate enough static electricity of sufficient energy to ignite a flammable vapor.

Heat generated by lightning: Lightning is the discharge of an electrical charge from a cloud to an opposite charge on another cloud or on the ground. Lightning passing between a cloud and the ground can develop very high temperatures in any material of high resistance in its path, such as wood or masonry.

Mechanical Heat Energy

Mechanical heat energy is responsible for a significant number of fires each year. Frictional heat is responsible for most of these fires, although there are a few notable examples of ignition by the mechanical heat energy released by compression.

Frictional heat: The mechanical energy used in overcoming the resistance to motion when two solids are rubbed together is known as frictional heat. Any friction generates heat. The danger depends upon the available mechanical energy, the rate at which the heat is generated, and its rate of dissipation. An example of frictional heat is the heat caused by friction of a slipping belt against a pulley.

Friction sparks: Friction sparks include the sparks that result from the impact of two hard surfaces, at least one of which is usually metal. Some examples of friction sparks often reported as responsible for fires are sparks from dropping steel tools on a concrete floor, from falling tools striking machinery or piping, from tramp metal in grinding mills, and from shoe nails on concrete floors.

Friction sparks are formed in the following manner: heat, generated by impact or friction, initially heats the particle. The maximum temperature is usually determined by the lowest melting point of the materials involved, but with some metals, the freshly exposed surface of the particle may oxidize at the elevated temperature, with the heat of oxidation increasing the temperature until the particle is incandescent.

Although the temperatures necessary for incandescence vary with different metals, in most cases they are well above the ignition temperatures of flammable materials, e.g., the temperature of a spark from a steel tool approaches 1400°C; sparks from copper-nickel alloys with small amounts of iron may be well above 300°C. However, the ignition potential of a spark depends upon its total heat content; thus, the particle size has a pronounced effect on spark ignition. The practical danger from mechanical sparks is limited by the fact that usually they are very small and have a low total heat content, even though each spark may have a temperature of 1100°C or higher. Mechanical sparks cool quickly and start fires only under favorable conditions, such as when they fall into loose dry cotton, combustible dust, or explosive materials. Larger particles of metal, able to retain their heat longer, usually are not heated to dangerous temperatures. Although the hazard of ignition of flammable vapors or gases by friction sparks is often overemphasized, it is best to avoid the use of grinding wheels and other sources of mechanical sparks in areas where any flammable liquids, gases, or vapors are, or may be, present. The possibility of ignition due to some unusual condition should not be overlooked.

Nickel, Monel metal, and bronze have a very slight spark hazard; stainless steel has a much lower spark hazard than ordinary tool steel. Special tools of copper-beryllium and other alloys are designed to minimize the danger of sparks in hazardous locations. Such tools cannot, however, completely eliminate the danger of sparks because a spark may be produced under several conditions. Little or no benefit is gained by using nonsparking hand tools in place of steel to prevent explosions of hydrocarbons.[8] Leather, plastic, and wood tools are, however, free from the friction spark hazard.

Heat of compression: The heat of compression is the heat released when a gas is compressed. This is also known as the diesel effect. The temperature of a gas increases when compressed, and this has found practical application in diesel engines in which heat of compression eliminates the need for a spark ignition system. Air is first compressed in a diesel engine cylinder, after which an oil spray is injected into the compressed air. The heat released when the air is compressed is sufficient to ignite the oil spray.

Nuclear Heat Energy

Nuclear heat energy is energy released from the nucleus of an atom. The nucleus is composed of matter held together by tremendous forces, which can be released when the nucleus is bombarded by energized particles. Nuclear energy is released in the form of heat, pressure, and nuclear radiation. In nuclear fission, energy is released by splitting the nucleus; in nuclear fusion, energy is released by the joining of two nuclei.

The energy released by bombardment of the nucleus is commonly a million times greater than that released by ordinary chemical reaction. Instantaneous release of large quantities of nuclear heat energy results in an atomic explosion. Controlled release of nuclear energy is a source of heat for everyday use (i.e., steam generation for electric generating stations).

BIBLIOGRAPHY

References Cited

1. Zabetakis, M. G., "Flammability Characteristics of Combustible Gases and Vapors," Bulletin 627, 1965, Bureau of Mines, U.S. Dept. of Interior, Washington, DC.
2. Westbrook, C. K., and Dryer, F. L., "Chemical Kinetics and Modeling of Combustion Processes," Eighteenth Symposium (International) on Combustion, The Combustion Institute, Pittsburgh, PA, 1981.
3. Williams, F. A., "Mechanisms of Fire Spread," Sixteenth Symposium (International) on Combustion, The Combustion Institute, Pittsburgh, PA, 1977, pp. 1281-1294.
4. Kim, J. S., de Ris, J., and Kroesser, F. W., "Laminar Free-Convective Burning of Fuel Surfaces," Thirteenth Symposium (International) on Combustion, The Combustion Institute, Pittsburgh, PA, 1971, pp. 949-961.
5. Babrauskas, V., *New Technology to Reduce Fire Losses and Costs*, S. J. Grayson and D. A. Smith, eds., Elsevier, London, 1986, pp. 78-87.
6. Browne, F. L., "Theories of the Combustion of Wood and Its Control," Report No. 2136, 1958, Forest Products Laboratory, U.S. Dept. of Agriculture, Madison, WI.
7. de Ris, J., "Fire Radiation—A Review," Seventeenth Symposium (International) on Combustion, The Combustion Institute, Pittsburgh, PA, 1979, pp. 1003-1016.
8. NFPA, "Friction Spark Ignition of Flammable Vapors," *NFPA Quarterly*, Vol. 53, No. 2, Oct. 1959, pp. 155-157.

Additional Reading

Alpert, R. L., and de Ris, J., "Prediction of Fire Dynamics," Final and Fourth Quarterly Report, National Institute of Standards and Technology, Report: NIST-GCR-94-642, June 1994, 36 pages.
Atreya, A., and Abu-Zaid, M., "Effect of Environmental Variables on Piloted Ignition," *Proceedings* of the Third International Symposium on Fire Safety Science, Elsevier Applied Science, London, 1991, pp. 177–186.

Bertelli, G., *et al.*, "Structure—Char Forming Relationship in Intumescent Fire Retardant Systems," *Proceedings* of the Third International Symposium on Fire Safety Science, Elsevier Applied Science, London, 1991, pp. 537–546.
Brehob, E. G., and Kulkarni, A. K., "Time-dependent Mass Loss Rate Behavior of Wall Materials Under External Radiation," *Fire and Materials: An International Journal*, Vol. 17, No. 5, September-October 1993, pp. 249–254.
Butcher, E. G., "Nature of Fire Size, Fire Spread and Fire Growth," *Fire Engineers Journal*, Vol. 47, No. 144, 1987, pp. 11–14.
Carty, P., and White, S., "Smoke/Char Relationships in PVC Formulations," *Journal of Fire Sciences*, Vol. 13, No. 4, July/August 1995, pp. 289–299.
Chen, Y., *et al.*, "Effects of Fire Retardant Addition on the Combustion Properties of a Charring Fuel," *Proceedings* of the Third International Symposium on Fire Safety Science, Elsevier Applied Science, London, 1991, pp. 527–536.
Combustion Fundamentals of Fire, Cox, G., ed., Academic Press Limited, London, 1995.
Damant, G. H., "Cigarette Ignition of Upholstered Furniture," *Journal of Fire Sciences*, Vol. 13, No.5, September/October 1995, pp. 337–349.
de Ris, J., "A Scientific Approach to Flame Radiation and Material Flammability," *Proceedings* of the Second International Symposium on Fire Safety Science, Hemisphere Publishing Corporation, New York, 1989, pp. 29–48.
Delichatsios, M. A., "Basic Polymer Material Properties for Flame Spread," *Journal of Fire Sciences*, Vol. 11, No. 4, July/August 1993, pp. 287–295.
Delichatsios, M. A., *et al.*, "Flame Radiation Distribution from Fires," *Proceedings* of the Fourth International Symposium on Fire Safety Science, International Association for Fire Safety Science, 1994, pp. 149–160.
Di Blasi, C., "On The Influence of Physical Processes on the Transient Pyrolysis of Cellulosic Samples," *Proceedings* of the Fourth International Symposium on Fire Safety Science, International Association for Fire Safety Science, 1994, pp. 229–240.
Drysdale, D. D., and Thomson, H. E., "The Ignitability of Flame Retarded Plastics," *Proceedings* of the Fourth International Symposium on Fire Safety Science, International Association for Fire Safety Science, 1994, pp. 195–204.
Emmons, H. W., "The Ceiling Jet in Fires," *Proceedings* of the Third International Symposium on Fire Safety Science, Elsevier Applied Science, London, 1991, pp. 249–260.
Fan, W. C., and Wang, J., "Predictions of Unsteady Burning of a Fuel Bed," *Proceedings* of the Third International Symposium on Fire Safety Science, Elsevier Applied Science, London, 1991, pp. 325–334.
Fredlund, B., "Modeling of Heat and Mass Transfer in Wood Structures During Fire," *Fire Safety Journal*, Vol. 20, No. 1, 1993, pp. 39–70.
Friedman, R., *Principles of Fire Protection Chemistry*, 2nd ed., National Fire Protection Association, Quincy, MA, 1989.
Friedman, R., "Some Unresolved Fire Chemistry Problems," *Proceedings* of the First International Symposium on Fire Safety Science, Hemisphere, 1986, pp. 349–359.
Gann, R. G., *et al.*, "Cigarette Ignition of Soft Furnishings," *Proceedings* of the Second International Symposium on Fire Safety Science, Hemisphere Publishing Corporation, New York, 1989, pp. 77–86.
Glassman, I., *Combustion*, 2nd ed., Academic Press, Orlando, FL, 1987.
Griffiths, J. F., and Barnard, J. A., *Flame and Combustion*, Blackie Academic & Professional, London, 1995.
Hasemi, Y., and Nishihata, M., "Fuel Shape Effect on the Deterministic Properties of Turbulent Diffusion Flames," *Proceedings* of the Second International Symposium on Fire Safety Science, Hemisphere Publishing Corporation, New York, 1989, pp. 275–284.
Heat Release in Fires, Babrauskas, V. and Grayson, S. J., eds., Elsevier Science Publishers, Ltd., London, 1992.
Hirano, T., "Physical Aspects of Combustion in Fires," *Proceedings* of the Third International Symposium on Fire Safety Science, Elsevier Applied Science, London, 1991, pp. 27–44.
Hwang, C. C., and Litton, C. D., "Ignition of Combustible Dust Layers on a Hot Surface," *Proceedings* of the Third International Symposium on Fire Safety Science, Elsevier Applied Science, London, 1991, pp. 187–196.
Iqbal, N., and Quintiere, J., "Flame Heat Fluxes in PMMA Pool Fires," *Journal of Fire Protection Engineering*, Vol. 6, No. 4, 1994, pp. 153–163.

Janssens, M., "A Thermal Model for Piloted Ignition of Wood Including Variable Thermophysical Properties," *Proceedings* of the Third International Symposium on Fire Safety Science, Elsevier Applied Science, London, 1991, pp. 167–176.

Karpov, A. I., and Bulgakov, V. K., "Prediction of the Steady Rate of Flame Spread Over Combustible Materials," *Proceedings* of the Fourth International Symposium on Fire Safety Science, International Association for Fire Safety Science, 1994, pp. 373–384.

Kashiwagi, T., Omori, A., and Brown, J. E., "Effects of Material Characteristics on Flame Spreading," *Proceedings* of the Second International Symposium on Fire Safety Science, Hemisphere Publishing Corporation, New York, 1989, pp. 107–118.

Kennedy, L. A., and Cooper, L. Y., "Before the Smoke Clears—Heat and Mass Transfer in Fires and Controlled Combustion," *Mechanical Engineering*, Vol. 109, No. 4, 1987, pp. 62–67.

Kokkala, M. A., "Experimental Study of Heat Transfer to Ceiling from an Impinging Diffusion Flame," *Proceedings* of the Third International Symposium on Fire Safety Science, Elsevier Applied Science, London, 1991, pp. 261–270.

Kumar, S., and Cox, G., "Radiation and Surface Roughness Effects in the Numerical Modeling of Enclosure Fires," *Proceedings* of the Second International Symposium on Fire Safety Science, Hemisphere Publishing Corporation, New York, 1989, pp. 851–860.

Kumar, S., Gupta, A. K., and Cox, G., "Effects of Thermal Radiation on the Fluid Dynamics of Compartment Fires," *Proceedings* of the Third International Symposium on Fire Safety Science, Elsevier Applied Science, London, 1991, pp. 345–354.

Law, M., "Behavior of a Fire," *Proceedings* of the International Conference on the Design of Structures Against Fires, Elsevier, New York, 1986, pp. 15–20.

Levine, R. S., and Pagni, P. G., eds., *Fire Science for Fire Safety*, Gordon and Breach, New York.

Lyons, J. W., *Fire*, Scientific American Books, New York, 1986.

Mikkola, E., "Charring of Wood Based Materials," *Proceedings* of the Third International Symposium on Fire Safety Science, Elsevier Applied Science, London, 1991, pp. 547–556.

Most, J. M., Bellin, B., and Sztal, B., "Interaction between Two Burning Vertical Walls," *Proceedings* of the Second International Symposium on Fire Safety Science, Hemisphere Publishing Corporation, New York, 1989, pp. 285–294.

Ohlemiller, T., et al., "Assessing the Flammability of Composite Materials," *Journal of Fire Sciences*, Vol. 11, No. 4, July/August 1993, pp. 308–319.

Ohlemiller, T. J., "Smoldering Combustion Propagation on Solid Wood," *Proceedings* of the Third International Symposium on Fire Safety Science, Elsevier Applied Science, London, 1991, pp. 565–574.

Ohtani, H., Maejima, T., and Uehara, Y., "In-Situ Heat Release Measurement of Smoldering Combustion of Wood Sawdust," *Proceedings* of the Third International Symposium on Fire Safety Science, Elsevier Applied Science, London, 1991, pp. 557–564.

Pagni, P. J., "Fire Physics—Promises, Problems, and Progress," *Proceedings* of the Second International Symposium on Fire Safety Science, Hemisphere Publishing Corporation, New York, 1989, pp. 49–66.

Parker, W. J., "Prediction of the Heat Release Rate of Douglas Fir," *Proceedings* of the Second International Symposium on Fire Safety Science, Hemisphere Publishing Corporation, New York, 1989, pp. 337–343.

Quintiere, J. G., *Fire Growth and Development*, Center for Fire Research, Gaithersburg, MD, 1989.

Quintiere, J. G., and Cleary, T. G., "Heat Flux from Flames to Vertical Surfaces," *Fire Technology*, Vol. 30, No. 2, 1994, pp. 209–231.

SFPE Handbook of Fire Protection Engineering, "Section 1," 2nd ed., DiNenno, P.J., *et al.*, eds. National Fire Protection Association, Quincy, MA, 1995.

Suzuki, T., *et al.*, "Polyurethane Foam Smoldering Supported by External Heating," *Proceedings* of the Fourth International Symposium on Fire Safety Science, International Association for Fire Safety Science, 1994, pp. 397–408.

Tewarson, A., and Macaione, D. P., "Polymers and Composites—An Examination of Fire Spread and Generation of Heat and Fire Products," *Journal of Fire Sciences*, Vol. 11, No. 5, September/October 1993, pp. 421–441.

Tewarson, A., "Flammability Parameters of Materials: Ignition, Combustion, and Fire Propagation," *Journal of Fire Sciences*, Vol. 12, No. 4, July/August 1994, pp. 329–356.

Tewarson, A., Chu, F., and Jiang, F. H., "Combustion of Halogenated Polymers," *Proceedings* of the Fourth International Symposium on Fire Safety Science, International Association for Fire Safety Science, 1994, pp. 563–574.

Torero, J. L., Fernandez-Pello, A. C., and Kitano, M., "Downward Smolder of Polyurethane Foam," *Proceedings* of the Fourth International Symposium on Fire Safety Science, International Association for Fire Safety Science, 1994, pp. 409–420.

Tuovinen, H., "Modeling of Laminar Diffusion Flames in Vitiated Environments," *Proceedings* of the Fourth International Symposium on Fire Safety Science, International Association for Fire Safety Science, 1994, pp. 113–124.

Weast, R. C., ed., *Handbook of Chemistry and Physics,* 69th ed., CRC Press, Boca Raton, FL, 1988.

Zdanowski, M., Teadorczyk, A., and Wojcicki, S., "A Simple Mathematical Model of Flashover in Compartment Fires," *Fire and Materials*, Vol. 10, 1986, p. 145.

Zhou, L., and Fernandez-Pello, A. C., "Turbulent Burning of a Flat Fuel Surface," *Proceedings* of the Third International Symposium on Fire Safety Science, Elsevier Applied Science, London, 1991, pp. 415–424.

Zukoski, E. E., "Mass Flux in Fire Plumes," *Proceedings* of the Fourth International Symposium on Fire Safety Science, International Association for Fire Safety Science, 1994, pp. 137–148.

Zukoski, E. E., *et al.*, "Combustion Processes in Two-Layered Configurations," *Proceedings* of the Second International Symposium on Fire Safety Science, Wakamatsu, Hemisphere Publishing Corporation, New York, 1989, pp. 295–304.

EXPLOSIONS

William J. Cruice

The term "explosion," in its most widely accepted sense, means a bursting associated with a loud, sharp noise and an expanding pressure front, varying from a supersonic shock wave to a relatively mild wind. Unfortunately, the term also has been extended to signify chemical or physical/chemical events that produce explosions. In addition, terms denoting specific chemical reactions are used as synonyms for "explosion." This blurring of meanings has caused confusion. Nevertheless, certain terms incorporating the word "explosion" that actually signify causative events must be used, since many such terms have relatively wide acceptance and reasonably specific meanings.

This chapter discusses explosions by detailing fundamental concepts, classifications by sources, blast effects and damage potential, physical explosions, gases and vapors, dusts and mists, thermal explosions and runaway reactions, deflagration/detonation, assessment of explosion potential, and countermeasures.

For related topics, see the following: Section 1, Chapter 5, "Chemistry and Physics of Fire"; Section 1, Chapter 8, "Theory of Fire Extinguishment"; Section 4, Chapter 7, "Gases"; and Section 4, Chapter 15, "Dusts."

FUNDAMENTAL CONCEPTS OF AN EXPLOSION

An explosion is defined as a rapid release of high-pressure gas into the environment. The primary key word is "rapid"; the release must be sufficiently fast so that energy contained in the high-pressure gas is dissipated in a shock wave. The second key term is "high pressure," which signifies that, at the instant of release, the gas pressure is above the pressure of the surroundings.

Note that the basic definition is independent of the source or mechanism by which the high-pressure gas is produced. An explosion may result from the overpressurization of a containing vessel or structure by physical means (e.g., bursting of a balloon), physical/chemical means (e.g., boiler explosion), or a chemical reaction (e.g., combustion of a gas mixture). Certain explosions result from chemical reactions that proceed so quickly that the high-pressure gas is generated instantaneously, even though no confining vessel or structure exists (e.g., detonation of high explosives). The classification of explosions by their source of high-pressure gas is extremely

useful. The fundamental concept of an explosion, then, is that of a sudden release of high-pressure gas and its dissipation of energy in the form of a shock wave.

The released high-pressure gas comes to equilibrium with the surroundings, creating certain effects while doing so. The characteristics of the effects of the gas on the surroundings depend upon: (1) the rate of release, (2) the pressure at release, (3) the quantity of gas released, (4) directional factors governing the release, (5) mechanical effects coincident with release, and (6) the temperature of the released gas. The latter two considerations are relatively straightforward in basic concepts. The striking of surrounding elements by high-temperature gases can cause severe heat damage, including direct damage to surfaces, thermal distortion, and fires in combustibles. Projectiles launched in the release process can impinge and cause crushing, piercing, and/or other structural failures. However, some explosions do not involve high-temperature gases and projectiles; the characteristics of the pressure wave are determined by the first four considerations.

Pressure equilibrates at the velocity of sound; in ambient air, the velocity of sound is approximately 1,100 ft per sec (300 m/s). For energy to be dissipated as a shock wave, the velocity of release must be sonic or supersonic. The initial shock wave spreads radially from the point of release. Its form near the point of release is of very short duration (milliseconds or microseconds) and includes a high-intensity impulse. As the distance from the point of release increases, the intensity (amplitude) decreases and the duration (period) increases until, far in the field, the impulse has a form similar to a mild breeze. (See Tables 1-6A and 1-6B, and Figure 1-6A.)

The initial intensity (amplitude) of the shock wave is determined by the pressure of the gas at the instant of release. The overall

TABLE 1-6A. *Effect of Shock Waves (Physiological)*

Physiological Effect	Peak Overpressure	
	(psi)	(kPa)
Knock personnel down	1	7
Eardrum rupture	5	34
Lung damage	15	100
Threshold for fatalities	35	240
50% fatalities	50	345
99% fatalities	65	450

William J. Cruice is vice president of Hazards Research Corporation, Mt. Arlington, NJ.

TABLE 1-6B.   *Effect of Shock Waves (Structural)*

Structural Element	Failure	Peak Overpressure (psi)	(kPa)
Glass windows, large and small	Shattering usually, occasional frame failure	0.5–1	3.5–7
Corrugated asbestos siding	Shattering	1–2	7–14
Corrugated steel or aluminum paneling	Connection failure followed by buckling	1–2	7–14
Wood siding panels, standard house construction	Usually failure occurs in main connections allowing a whole panel to be blown in	1–2	7–14
Concrete or cinder block wall panels 8 or 12 in. (200 or 300 mm) thick (not reinforced)	Shattering of the wall	2–3	14–21
Self-framing steel panel building	Collapse	3–4	21–28
Oil storage tanks	Rupture	3–4	21–28
Wooden utility poles	Snapping failure	5	34
Loaded rail cars	Overturning	7	48
Brick wall panel, 8 or 12 in. (200 or 300 mm) thick (not reinforced)	Shearing and flexure failures	7–8	48–55

energy of the shock wave is related to the total quantity of gas released and the pressure and temperature at the instant of release. An estimate of the overall energy available in a compressed gas is obtained by multiplying the pressure by the volume. In the event that high-pressure gas is generated following the initial release, the total energy released increases, but the shock wave intensity does not; the duration (period) of the impulse is longer.

If the gas is uniformly released simultaneously in all directions, the shock wave travels in an expanding sphere. However, the velocity of release must be at least sonic, and only certain types of events can bring about a uniform release at sonic or supersonic velocities. The vast majority of explosions involve some form of mechanical confinement (vessel or other structure) in which gas accumulates to high pressure. When its pressure exceeds the strength of the confining vessel, the vessel fails at its weakest point, and the shock wave travels in the direction of the failure. Thus, vessel failure explosions create a pressure wave that is generally not uniform in all directions.

Since the form of the initial shock wave impulse is dependent upon the total energy of release and is usually highly directional, the terms "near field" and "far field" are not numerically definable. "Near field" refers to the radial distance in which shock effects (shattering) predominate, whereas "far field" refers to the area beyond that, where observed damage corresponds to wind loading effects.

CLASSIFICATION OF EXPLOSIONS BY SOURCE

Physical *vs.* Chemical

The basic division among explosions caused by a high-pressure gas is between physical and/or physical/chemical sources and chemical reaction sources. In certain cases, all of the high-pressure gas is produced by mechanical means or by events that do not involve a basic change of the chemical substance. Gas may be raised to high pressure mechanically; by external heating of gases, liquids, or solids; or a superheated liquid may be suddenly released by mechanical means and create high pressure by flash vaporization. None of these events involve changes in the basic chemical nature of the substances involved; the entire process of generating high pressure, and releasing and creating the effects of explosions, can be understood in terms of basic physics. These events are commonly called "physical explosions." (See Figure 1-6B.)

In other cases, generation of the high-pressure gas is the result of chemical reactions where the fundamental nature of the product(s) differs substantially from the material(s) previously present (reactant). The most common chemical reaction involved in explosions is combustion, in which a fuel (such as methane) mixed with air is ignited and burns to produce carbon dioxide, water vapor, and other products. Many other chemical reactions also give rise to high-pressure gases. Explosions result from decomposition of pure substances, detonation, combustion, hydration, corrosion, and various interactions of two or more chemical substances in greater or

FIG. 1-6A.   *Decay of shock wave with distance. (Wave form at various times)*

FIG. 1-6B.   *Classification of hazardous reactions.*

lesser degrees of mixture. Any chemical reaction can cause an explosion if gaseous products are produced by the reaction, if substances not otherwise involved are vaporized by heat released by the reaction, or if gases already present are raised to higher temperatures by heat released by the reaction.

Chemical reactions are classified thermodynamically as either exothermic (heat releasing) or endothermic (heat absorbing). Whether heat is released or absorbed depends upon the conditions under which the substance(s) reacts. However, since a chemical reaction proceeds faster at higher temperatures and slower at lower temperatures, a chemical reaction that absorbs heat will cool and self-extinguish if external heat is not continually supplied. Thus, endothermic reactions can result in explosions only in special circumstances, where the reactant is brought into contact with a copious supply of external heat, and the reaction itself produces gaseous products. Examples of such reactions include certain decarboxylations, pyrolyses, and other forced decompositions.

Exothermic reactions, on the contrary, produce a temperature increase in the reaction mass, causing the reaction to proceed more rapidly. If heat is produced in the reaction mass faster than it can be lost to the surroundings, exothermic reactions easily become self-sustaining and self-accelerating. Such reactions can quickly become uncontrollable, even under carefully designed conditions, as well as under accidental conditions where few or no control mechanisms exist. Since heat is released, high-pressure gas can be created by vaporizing reactants, products, or surroundings, or by heating gases already present, even if no gases are directly produced by the reaction. Accordingly, nearly all chemical explosions result from exothermic reactions.

Uniform *vs.* Propagating Reactions

Chemical reactions can be classified as *uniform* reactions, in which the chemical transformations occur essentially throughout the entire reaction mass, and *propagating* reactions, in which there is a clearly defined reaction zone separating the unreacted materials from the products of reaction, which moves through the reaction mass.

Uniform reactions: The rate of a uniform reaction depends only upon the temperature and the concentration(s) of the reacting agent(s) and is the same throughout the reaction mass.

As the temperature of the mass is raised, the energy-releasing reaction proceeds more rapidly and eventually leads to observable self-heating, in which the heat generated within the reaction mass exceeds that lost to the surroundings (in most cases, the walls of a vessel). Since heat is generated throughout the reaction mass, but lost more slowly from the center than from the outside surface, the center becomes hotter and its reaction rate increases. This rapidly increasing reaction rate in the center continues until the reactant(s) in the hot zone are either entirely consumed or the center erupts to dissipate the high temperature. (See Figure 1-6C.)

Uniform reactions occur in solids, liquids, and gases. Gases undergo convection and diffusion quite readily; hot material can be transferred quickly from the center to the periphery, where heat can be lost more easily to the surroundings. Liquids also undergo convection and diffusion, although the effective rate of transfer is lower than with gases. Solids transfer heat exclusively by conduction and have the same higher heat production per unit volume as liquids. In general, solids present the greatest heat transfer problem. However, since mixtures of solids can react only at contact points between the reacting agents, most uniform reactions of solids occur with compounds that decompose, rather than with mixtures.

Pure liquids and true solutions can react rapidly and continuously. Liquid mixtures that form two separate phases, and liquids

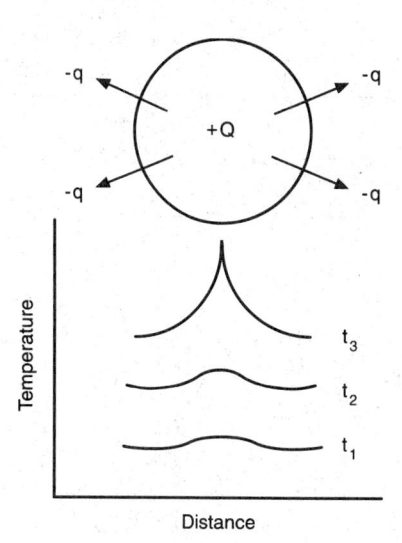

FIG. 1-6C. *Growth of thermal explosion (+Q = heat generated, -q = heat lost, t = time).*

that react with undissolved solids, react only at the interface; the bulk rate of the reaction depends upon the surface area of globules or particles of the reacting phases where contact occurs. Stirring improves heat transfer from the center, which reduces the reaction rate of the system. However, stirring can also increase the reaction rate by bringing fresh, unreacted materials together (especially in multiphase systems, such as slurries and immiscible liquids). The exact effect of agitation depends upon the specific system and the conditions existing within it.

Usually, uniform reactions generate gas too slowly to produce high pressure in the absence of a confining structure. A confined reaction can generate high-pressure gas by producing gaseous products, by vaporizing reactants or other materials, or by raising the temperature of gases present in the vessel. If a reaction produces a sufficiently high pressure, the confining vessel will rupture and an explosion will occur.

In practice, most chemical systems capable of uniform reaction can undergo alternative reaction paths, whether to the same or different products. The rates of reaction are quite different—largely determined by temperature, as shown in Figure 1-6D. Most uniform reactions, if not quenched by early release and/or dissipation of the high-temperature center, undergo a transformation to propagating reactions.

Propagating reactions: A mixture of hydrogen and oxygen can be stored at normal room temperatures for extensive periods with no sign of chemical reaction. However, most mixtures of hydrogen with oxygen react violently if an igniter is applied. The reaction begins at the igniter and travels through the mixture. Three distinct zones can be observed: (1) the reaction zone (flame), (2) the product zone (behind the flame), and (3) the unreacted zone (in front of the flame). This is a classic illustration of a propagating reaction. (See Figure 1-6E.)

A propagating reaction is always exothermic. The reaction is started by the creation of a relatively small high-temperature zone, whether by an external igniter (e.g., match, spark, or shock) or by heat buildup in the core of a uniform reaction system. For the reaction to propagate, the reaction core, which is activated by the igniter, must sufficiently raise the temperature of the surrounding material so that it reacts in like manner. The higher the initial temperature of the system, the easier it is to ignite, and the more likely

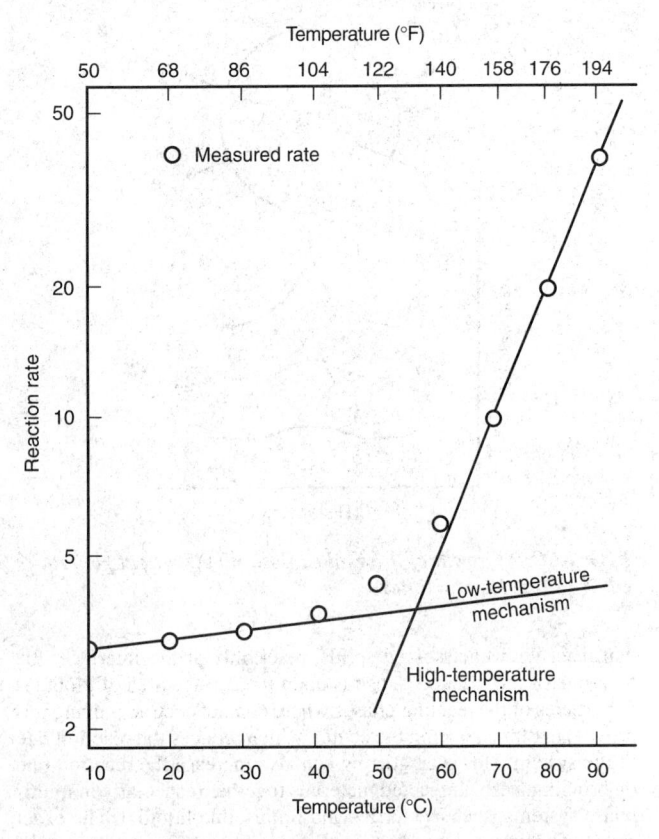

FIG. 1-6D. *Change in dominant reaction path with temperature.*

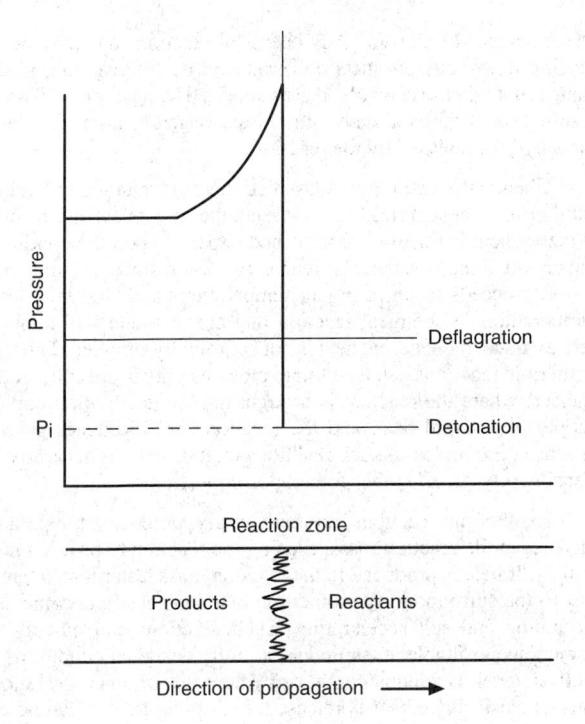

FIG. 1-6E. *Instantaneous pressure profiles of propagating reactions.*

it is to support a propagating reaction, since relatively less energy transfer is required to initiate the surrounding unreacted material.

Heat not transferred to the unreacted material is retained in the products of reaction or lost to the surroundings. The temperature of the products is determined by dividing the total heat retained by the heat capacity. Any pressure increase is determined by the vapor pressure of the system and conventional pressure-temperature-volume (PTV) relationships for any gases present or produced (directly or indirectly) by the reaction.

Since a propagating reaction begins at some specific point and travels through the reaction mass, the rate of energy release is related to the propagation velocity or rate of travel of the reaction zone. Propagation velocities vary from nearly zero to several times the speed of sound, depending upon composition, temperature, pressure, degree of confinement, and other considerations. Since pressure equilibrates at the speed of sound in the unreacted materials, a critical difference exists between subsonic reactions (deflagrations) and supersonic reactions (detonations).

Deflagrations and detonations propagate in gases, liquids, solids, pure compounds, and single-phase and multi-phase mixtures. In many cases, a propagating reaction occurs only under some form of external confinement, as the reaction otherwise dissipates energy too rapidly to be self-sustaining. Confinement required for propagation varies widely with the rate of reaction and the physical state of the materials. It is necessary only that the reactants be kept in place long enough to react, and that the reaction zone retain energy for sufficient time to ignite unreacted material. Relatively weak confinement can be adequate for these purposes when the reaction time is only a few milliseconds or microseconds. Detonations and deflagrations can propagate in paper, plastic, glass and metal containers,

pipes, reactors, tanks, drums, boreholes, wells, open pits, buildings, caves, tunnels, etc. Many materials sustain violent propagating reactions with no external confinement.

BLAST EFFECTS AND DAMAGE POTENTIAL

The primary procedures for estimating the damage potential of an explosion relate to the intensity of the shock wave produced by the release of the high-pressure gas. The most common means of estimating involve comparing explosions with TNT detonations, as much information exists on TNT-produced shock waves. However, the specific means of relating a potential explosion to TNT detonation differs with the type of explosion assumed.

Unconfined Materials

Unconfined materials may be analyzed by conventional thermodynamics to ascertain the anticipated heat release from a total reaction to most stable products. Once the available energy is determined, an assumption is made that the material can detonate, and the energy potential is considered to be released from an energy-equivalent mass of TNT. The intensity of the assumed shock wave then is obtained as a function of distance, using standard references.

It is essential to realize that this only provides an *estimate* of the possible damage potential *if the material can* detonate. Many materials will not support a detonation when unconfined. Whether or not a material can support a detonation must be determined by experimentation, using appropriately scaled samples. If the material cannot detonate, a very conservative estimate of deflagration pressure can be obtained by assuming that the thermal energy is released, and that the gases are formed at thermodynamic pressures within the geometry of the subject charge. A comparison then can be made using an energy equivalent mass of TNT.

Confined Materials

Confined materials must be analyzed in three ways. First, the charge should be evaluated in the same manner as for unconfined materials. The effective pressure front must be sufficient to rupture the confinement, or external damage will not occur due to an expanding shock front. Damage, however, may result from physical displacement of the confining structure due to unequal loading of internal shock waves.

Second, an analysis should be performed based upon thermodynamic pressure. The thermodynamic pressure must be sufficient to rupture the confining structure, or external damage from an expanding shock front will not be experienced. Conventional constructions (buildings, rooms, etc.) generally fail at quite low pressures; vessels generally fail at approximately four times the allowed working pressure. In all cases, the ultimate failure pressure of the confinement should be obtained from a competent structural engineer, with appropriate allowances for temperature, corrosion, and other weakening factors.

The third method of estimation is based upon equating the failure of the confinement to a shock front from a TNT charge. Vessel failure should be assumed and the amount of TNT required to produce a shock wave of intensity equal to the ultimate failure pressure of the vessel at the vessel wall should be determined, assuming the charge to be at the geometric center of the vapor space of the vessel. The shock wave then can be projected radially.

In all cases of confined materials, the pressure/shock wave estimates must be followed by an intelligent evaluation of the projectile effects, as ruptured vessels or building elements invariably produce some projectiles.

The initial estimation of the damage potential is frequently based upon the worst possible conditions of energy release, reaction velocity, etc. Using experimental data, the estimates can be refined. However, the confidence that can be placed in any estimate must be inversely proportional to the cost of being incorrect; conservative estimates of damage potentials are preferred, unless very strong evidence is cited to demonstrate that the worst case cannot occur.

Most combustions and explosions generate and release high-temperature gases and/or condensed-phase materials. These hot products can cause severe heat damage to personnel, structures, and adjacent equipment. Secondary fires and process upsets also are common. In assessing the damage potential associated with combustions and explosions, it is essential to consider the heat effects associated with the high-temperature products of reaction.

PHYSICAL EXPLOSIONS

Physical explosions are releases of high-pressure gas that do not involve chemical reactions, although most physical explosions involve vaporization.

Most physical explosions involve a confining vessel, such as a boiler, gas cylinder, or other tank. High pressure in the vessel is created by a mechanical compression of the gas, heating of the contents, or introduction of high-pressure gas from another vessel. When the pressure reaches the ultimate strength of the weakest portion of the vessel, failure occurs. In some cases, relatively small elements of the overall vessel assembly may be the weakest portions, and these elements are expelled as projectiles. In other cases, the actual vessel walls or seams fail, and the vessel is flung open with great violence.

Damage to surroundings depends primarily upon the failure mode. If small elements fail but the vessel remains essentially intact, the expelled projectiles are as dangerous as bullets or cannon-balls; gas release, however, is highly directional and is controlled by the diameter(s) of the hole(s) created when the small elements are expelled. Under these conditions, the damage to the surroundings may be limited to projectile penetration, scorching, or other harmful effects of hot gases, and/or minor displacement of weak structural elements. If the expulsion of small parts and the release of gases is unbalanced, the vessel will be thrust in the direction opposite to the gas expulsion, and may be toppled, hurled, or otherwise moved. Such an event can produce damage by crushing objects in the vessel's path, and/or can lead to the collapse of buildings or other structures if load-bearing elements are displaced or destroyed.

The failure of vessel walls or seams generally releases major projectiles and causes severe thrust of the vessel body in the direction opposite to the gas release. However, the gas release is extremely rapid under such circumstances, and a generally severe and broad shock wave is produced. The damage potential may be crudely estimated by multiplying the volume of gas in the vessel by the actual pressure at the moment of failure, although the accuracy of the estimate may vary considerably with the compressibility and the condensibility of the gas (which cools on expansion from the original vessel volume). The pressure wave from the rupture of the vessel is highly directional, traveling largely in the direction of release and creating equivalent pressure effects at much greater distances in that direction than in others, but substantial pressure effects are created in essentially all directions.

If the vessel contains a superheated liquid (i.e., a liquid at a temperature above its normal boiling point, or a liquefied gas, such as ammonia or carbon dioxide), flash vaporization of the liquid occurs when the vessel ruptures. A sufficient amount of liquid vaporizes to cool the discharged material to its normal boiling temperature, and the vaporized material adds significantly to severe pressure effects, depending upon the total amount of vaporizing per unit of time. This phenomenon is called a boiling liquid expanding vapor explosion (BLEVE).

An analogous phenomenon is the flash vaporization of a liquid (or, rarely, a solid) suddenly brought into contact with another substance at a temperature far above the ordinary boiling point of the liquid. Such events include spillage of cryogenic liquids or cold liquefied gases into the normal environment, or flash vaporization of liquids by hot metals or minerals. In such cases, the liquid receives heat from the hotter surface at a sufficient rate that a high-pressure gas is created instantly at the hot surface. If the interface between the cold liquid and the hot surface is large, vaporization can occur quickly enough to create a shock wave, spreading in all directions. Actually, the vaporization rate is usually significantly lower, but the rapid generation of gas frequently creates pressures high enough to destroy buildings or vessels. Sudden introduction of water into high-temperature boiler tubes, heat exchangers, or heat-transfer fluid tanks can produce severe explosions.

In the event that the discharged material can burn in air or otherwise chemically react with the environment into which it is discharged, a severe danger of secondary explosion exists.

GASES AND VAPORS

The most common form of chemical reaction that produces high-pressure gases from other gases or vapors is the combustion of gaseous fuels in air. However, other oxidizing gases (such as oxygen, chlorine, fluorine, and a variety of gaseous compounds and mixtures) may be substituted for normal air, frequently producing more severe combustion processes. Certain gases (such as acetylene, ethylene, ethylene oxide, butadiene, nitrous oxide, and others) can propagate a decomposition flame under proper conditions of temperature and pressure in the absence of any other gas. This latter phenomenon is generally called disproportionation.

For the general case of combustion involving a fuel gas and an oxidizing gas (such as air), mixtures are flammable only within a certain range of compositions. A minimum fuel proportion is required to sustain combustion; there is also a maximum fuel proportion above which combustion cannot be self-sustaining. Near these limiting fuel proportions, flames propagate through the mixture rather slowly, but in the middle of the flammable range, combustion velocities can be faster than the speed of sound. (See Figures 1-6F and 1-6G.)

In general, increasing the temperature and/or pressure of the mixture widens the flammable range and increases the combustion velocity throughout the flammable range. Generally, limits must be determined experimentally for the specific mixture, temperature, and pressure of interest.

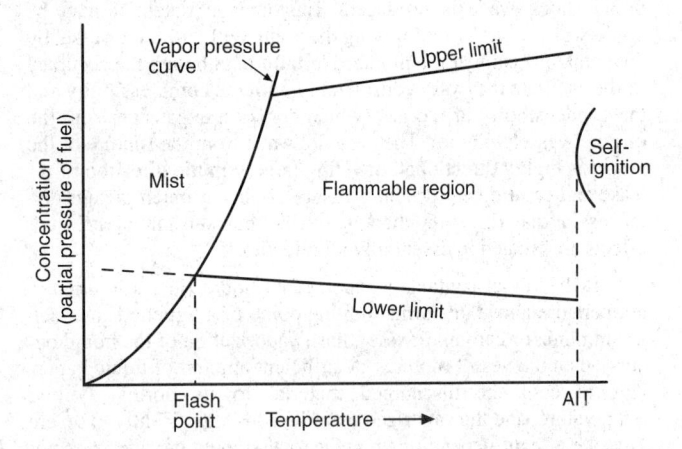

FIG. 1-6F.   Flammability limits as a function of temperature.

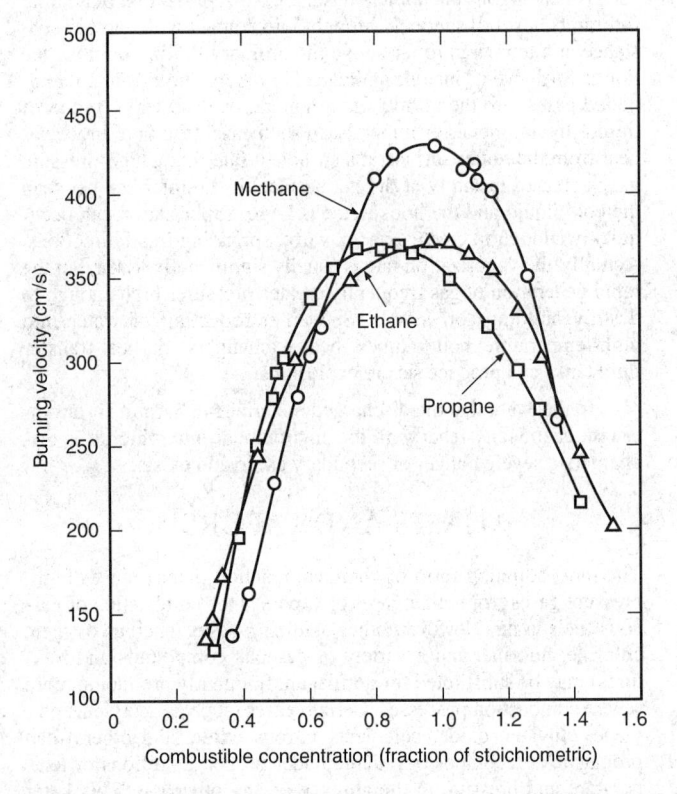

FIG. 1-6G.   Combustion velocity vs. concentration.

The great majority of combustible gas mixtures are stable at ordinary temperatures and pressures. The combustion reaction must be initiated by some means. Once ignition occurs, the combustion reaction is self-sustaining by transferring heat and activated agents from the burned gas (products) to the unburned gas (reactants) at the flame front (reaction zone), and the reaction propagates from the ignition point to the physical limits of the combustible mixture.

Gas mixtures are generally more difficult to ignite near the flammable limits and most easily ignited somewhat above the precisely balanced (stoichiometric) mixture. In general, however, gas mixtures are very easily ignited anywhere in the flammable range. The concept of minimum energy requirement for ignition is very difficult to apply in practical situations, particularly since the minimum energy input required varies with the type of ignition source. The minimum electrical discharge energy required to initiate combustion is very different from the minimum shock energy, for example, and neither relates well to the energy requirements for hot surface ignition. Some mixtures (e.g., chlorine/fuel) can be ignited by radiant energy (e.g., visible light, ultraviolet light, etc.) of the correct frequency. (See Figure 1-6H.)

Most reactive gas mixtures are capable of self-ignition under specific conditions of temperature and pressure. At any given temperature, certain molecules are energetic enough to react when they meet; ordinarily such reactions and the heat liberated by them are not detectable. At some temperature, the intermolecular reactions are frequent enough so the gas mixture begins to self-heat and eventually reaches its kindling temperature. Propagating combustion then is initiated if the overall mixture is flammable at the temperature and pressure conditions attained.

Autoignition depends upon the specific mixtures of gases, the volume and geometry of the vessel, the materials of construction, and the initial temperature and pressure of the mixture and the surroundings. Published autoignition temperatures are specific to the method of determination and cannot be used indiscriminately. The existence of thermal and catalytic wall effects makes application of published autoignition temperatures to practical situations a delicate and potentially very dangerous activity. (See Figure 1-6I.)

Pressure generated by the combustion of gas mixtures results primarily from the heat liberated and the consequent high temperature of the product gases. In most gas combustions, heat is lost to the surroundings by radiation from the flame, convection currents, etc.

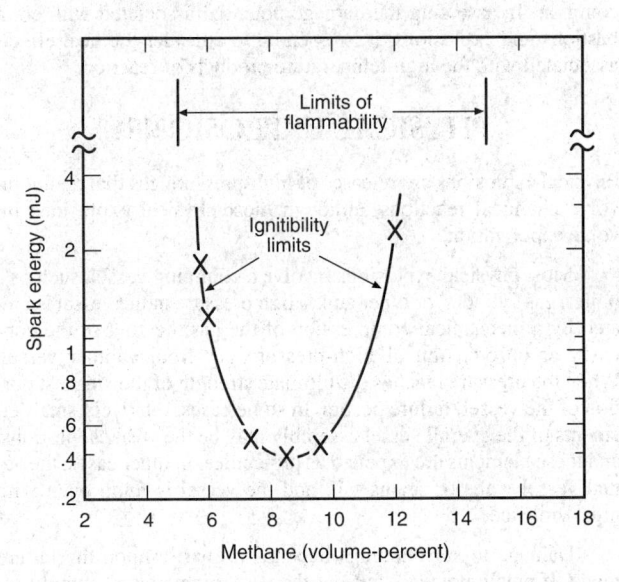

FIG. 1-6H.   Ignition energy vs. concentration.

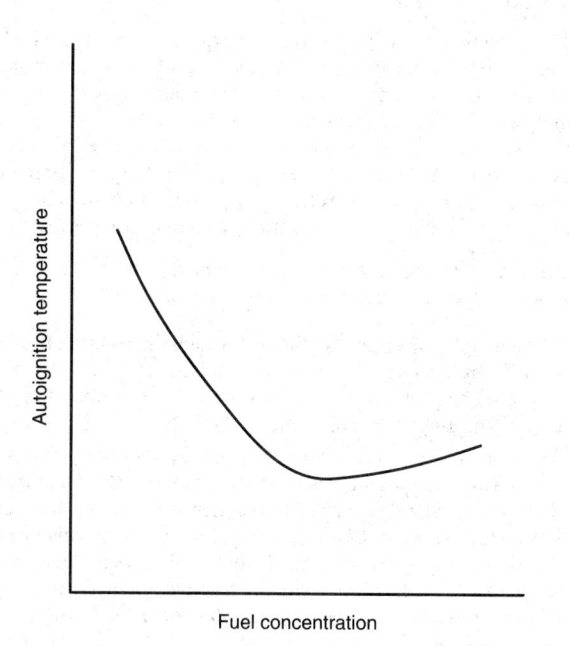

FIG. 1-6I. *Autoignition temperature vs. concentration.*

Generally, the highest pressures are attained when combustion occurs rapidly. For most confined subsonic combustions, the maximum pressure produced is approximately ten times the initial pressure, since the flame temperature is limited by dissociation reactions. This ratio may differ if the starting mixture is composed of compressible gases or if the fuel, oxidizer, or an otherwise inert component of the mixture can decompose and/or yield a significant change in the number of moles of gas in the system.

However, if the gas mixture is unconfined or if confinement is breached, the burning gas expands as a fireball at normal atmospheric pressure, and the maximum fireball volume is approximately ten times the initial volume of the mixture. As in the case of pressure generation, expansion of the mixture is least near the limiting concentrations and greatest near the middle of the range.

Gases that propagate decomposition flames generally exhibit minimum absolute concentrations (i.e., minimum reactive gas pressure and minimum ratio of gas to other gases in a mixture) below which the flame does not propagate. However, most such gases do not have upper limits, but are flammable from the lean limit to 100 percent. The characteristics of pressure generation, combustion velocity, fireball size, etc., are specific to each gas and do not relate readily to principles applicable to conventional fuel/oxidizer mixtures.

Certain gases and mixtures are inherently capable of supersonic, shock wave driven reaction (detonation) under appropriate conditions of temperature and pressure. For flammable gas/oxidizer mixtures, the detonable range is largely dependent upon the source of ignition, but always includes the stoichiometric and the fastest-burning ratios for the normal (subsonic) combustion reaction. Gases capable of sustaining a detonation ordinarily do not reach actual supersonic conditions unless initiated by a high-intensity shock wave. However, in certain confinement geometries, such as pipes, where the length exceeds the diameter by a factor of ten or more, a conventional combustion reaction can self-accelerate to the point where a transition from deflagration to detonation occurs if the mixture is in the detonable range of compositions. The final pressure generated in detonation is the same as the final pressure from deflagration, but the transient pressures created by the shock wave in the detonation

process can be more than double the final pressure, and the effects on objects in the path of the shock wave can characteristically be more than four times the final pressure. Accordingly, vessels capable of containing deflagration maximum pressures can sustain partial or total destruction if the gas mixture detonates rather than deflagrates.

DUSTS AND MISTS

Dusts and mists (i.e., finely divided liquids) can generate high-pressure gas by combustion in air or another reactive gaseous environment. However, since the necessary chemical reactions can occur only at the interface of the dispersed particles or droplets with the surrounding gas, the rates of pressure generation are limited by the available surface area of the dispersed material. For any given mass of dust or mist, the surface area increases as the particle or droplet diameter decreases, so the violence of combustion for any given substance increases as the particles or droplets become finer. These dispersed condensed phases are quite similar in combustion behavior.

Combustion can occur with virtually any particle or droplet, but, in practice, most explosion hazards with dispersed materials are encountered with 20-mesh (840 micron) sizes or smaller. As the particle or drop size decreases, dispersion is more easily created and is relatively more stable and enduring. Accordingly, finer particles or droplets constitute graver hazards by making dispersions more likely, by remaining dispersed for a longer time, and by burning more rapidly than coarser particles. For most combustible agents, however, any ignition of a dispersion in air or other reactive gas can create pressures sufficient to destroy process equipment and structures.

There is a minimum uniform concentration for a propagating combustion to occur. There is no reliable upper concentration limit for dusts or mists, however, since the reaction is surface-area controlled. Most upper limits actually represent that concentration of dust or mist that renders some experimental igniter ineffective, rather than a true thermodynamic limiting concentration.

As a first step, the combustion process involves vaporization of volatile combustibles from the surface of the particle. Accordingly, coal with less than approximately 20 percent by weight volatiles (excluding moisture) does not propagate a dust combustion, but clay particles coated with volatile organics can produce dust explosions. In this context, "volatiles" include materials (e.g., soaps, flowing agents, silicones, etc.) that can be pyrolyzed or otherwise vaporized at high temperatures, as well as low-boiling solvents, etc.

The rate of pressure rise and the maximum pressure attained in dust combustions and mist combustions increase as the concentration of dispersed particles increases, generally passing through a maximum at a concentration that depends upon particle size, then slowly decrease as the concentration increases further. A substantial amount of data has been published for a wide variety of dusts, based on specific apparatus configurations. However, while useful for comparison with other dusts under the same conditions, the numerical values can be very misleading if improperly used. In particular, combustions of dispersed particulates are highly subject to turbulence, which greatly increases the rate of pressure rise actually experienced. Maximum pressures are about ten times the starting pressure, but are seldom attained. (See Figure 1-6J.)

Since dusts and mists frequently occur as dispersions in lightly constructed equipment, and since, in practice, there is usually a large supply of quiescent material that can be hurled into the air by the initial combustion, dust combustions and many mist combustions can self-propagate for large distances into areas where no flammable mixture previously existed. Dust explosions are frequently described

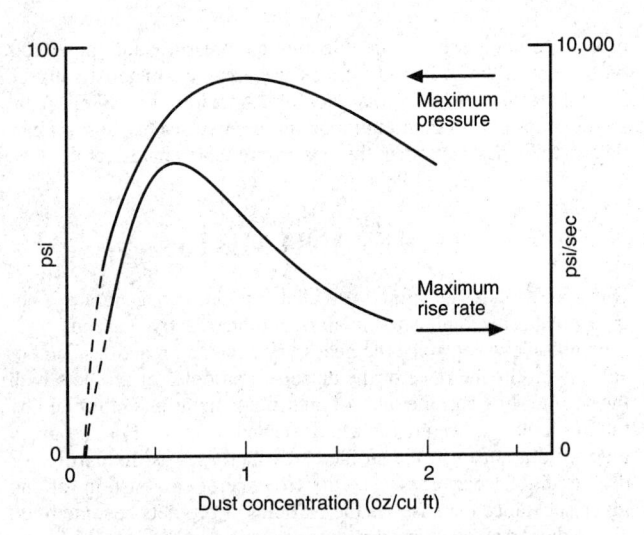

FIG. 1-6J. *Severity of dust explosion vs. concentration.*

as rolling explosions, characterized as sounding like distant thunder. Rolling explosions generally are initiated by a primary dust explosion that creates and ignites secondary dust dispersions, which in turn create and ignite further dust dispersions.

Ignition of dust and/or mist dispersions in air commonly is achieved by an electric discharge, open flame, or hot surface. The energy required for ignition is generally higher than that needed to ignite vapor/air mixtures, but is low compared to the energy sources available in the environment. Dusts and mists frequently cause abrasion and binding, with consequent generation of high-temperature surfaces or flames in mechanical equipment, which can, in turn, ignite the dispersion. Moving dust or mist particles frequently create electrostatic accumulations that eventually discharge and ignite the dispersion. Most insidious in dusty environments is the layering of particles on surfaces that become hot and initiate a combustion in the layer. A small puff of wind from the combustion in the layer creates adjacent airborne dispersions that are ignited by hot products from the original combustion. The resulting cloud formation can travel throughout an entire dusty area, creating widespread and massive destruction.

The presence of flammable vapors in dust dispersions, even far below the normal lower limit for the vapors, recently has been found to greatly enhance the ignitibility and velocity of dust combustions. This synergistic effect emphasizes the uniqueness of many chemical systems and operating environments. Assumptions of behavior not experimentally or experientially confirmed or supported can be disastrous.

Effects of dust and mist explosions on the surroundings are analyzed in essentially the same manner as gas explosions. The same principles of vessel failure, structural failure, pressure generation, fireball generation, and secondary effects apply. However, unlike gaseous combustions, dust and mist combustions tend to create substantial amounts of condensed-phase products at very high temperatures, e.g., hot solids, tars, gums, oils, etc. These hot products tend to adhere to surfaces they contact, creating severe heat damage and frequently causing additional fires.

THERMAL EXPLOSIONS AND RUNAWAY REACTIONS IN CONDENSED PHASES

Thermal explosions and runaway reactions begin as uniform reactions, whether in reaction vessels, storage vessels, pools, or piles of material. All follow essentially the same pattern of heat accumula-

tion and reaction rate acceleration. At some environmental and system temperatures, depending upon the size of the reaction mass, nonuniformities in temperature and the reaction rate lead to a change in the favored reaction mechanism so that it becomes a propagating reaction. If high-pressure gases are released, explosive effects are created in the environment, their form being determined by the rate at which the gas is released. In some cases, the change in mechanism may result in a detonation, with disastrous effects.

The general thermal runaway is frequently accelerated by either, or both, of two other chemical phenomena:

1. Autocatalytic reactions are those in which the reaction products chemically accelerate the initial reaction in which they were formed. The chemical acceleration often is more important than the self-heating effect. (See Figure 1-6K.)
2. The critical concentration effect can take two basic forms. In both cases, the initial reaction creates a metastable condition in the mass by slowly generating reactive products. In one form, these reactive products are capable of rapid exothermic decomposition, with the reaction rate being highly dependent upon the temperature. In the other form, the reactive products are capable of further exothermic reaction with the unreacted portion of the original chemical(s) present, and this form also is highly dependent upon temperature. In either event, the reactive products may accumulate in the overall mass until a slight increase in overall temperature produces an immense surge of reaction, rapid liberation of enormous quantities of heat and/or gaseous products, and usually a disastrous explosion.

Thermal explosions and runaway reactions differ from combustions of gases, mists, and dusts in several ways; these differences create significantly greater explosion potentials. First, the amount of material physically present in a unit volume is much greater in a condensed-phase system, and the total heat available within the reacting mass is many times greater than that for dispersions in air. Second, the attainable pressure is generally much

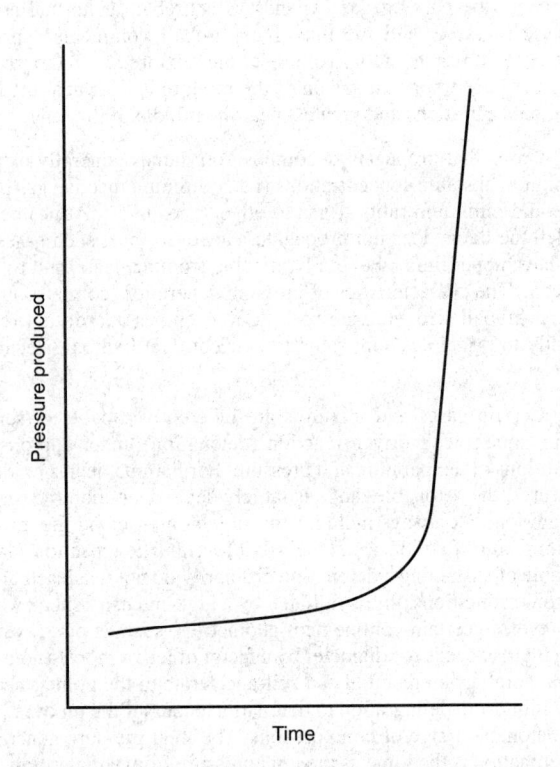

FIG. 1-6K. *Autoacceleration of thermal decomposition.*

higher for condensed phases, because the higher temperature reaction mechanisms usually produce large amounts of gaseous products in a limited vapor space. Third, the reaction mass generally swells and forms gas bubbles, with thermal expansion of the condensed phase. In some cases the thermal expansion alone can hydraulically fail a vessel that theoretically could contain the gases produced. This produces a BLEVE effect when the confinement is ruptured. Fourth, whether because of vessel rupture or boilover, large quantities of hot chemicals often are discharged into the environment. Under such circumstances, as the chemicals are usually combustible in air, catastrophic secondary explosions and fires occur.

Estimating the potential effects of thermal explosions and runaway reactions may be approached through the criteria just described; however, this method has limitations. It often is assumed that the ultimate pressure of a confining vessel is that at which it ruptures under any circumstances. In fact, the physical rupturing of a vessel requires a certain amount of time. If the reaction taking place in the vessel creates high pressure faster than the rate at which the vessel expands, the actual gas pressure upon release may be much higher than the pressure at which the vessel will fail under sustained load. Specific instances have occurred in which the effects of environmental damage clearly indicate that the actual gas pressure at release was two to ten times the maximum pressure the vessel would be expected to hold. Accordingly, if thermodynamic estimations of the damage potential predict significantly worse effects than would be anticipated from a rupturing vessel, it may be foolhardy to rely on the simple rupture model.

It is unwise to predict reaction paths in the absence of complete experimental information. Various evaluation methods are available to assess the actual severity and sensitivity of chemical systems under thermal explosion conditions; all the methods have limitations. Nevertheless, experimental assessments are essential for properly evaluating potential hazards of runaways. Furthermore, it is unrealistic to assess only the theoretical or desired compositions; process upsets in composition also must be anticipated and evaluated.

Thermal explosions and runaway reactions usually are associated with nitration, polymerization, and diazotization processes, and with chemical structures involving unsaturation or N-N, N-O, Cl-O, O-O, N-Cl bonds, among others. Generally, any reaction that is self-sustaining under existing environmental conditions theoretically can give rise to a thermal explosion. In chemical processing, this includes any reaction that does not require a continuous application of heat to proceed to completion. All such reactions must be assessed for thermal explosion/runaway reaction potential, with due regard for the fact that process equipment, controls, and personnel are subject to various failures and errors.

DEFLAGRATION/DETONATION IN CONDENSED PHASES

Theoretically, any exothermic reaction can be the basis for a deflagration or a detonation, provided the heat of reaction is directed in sufficient quantity into the unreacted materials. Accordingly, these features of the chemical system and its environment that promote such energy transfer favor propagation, whereas those that promote loss of the energy do not.

Propagating reactions deliver energy to the unreacted materials or to the surroundings. The amount of energy delivered is determined by the distance that the energy must travel and the ability of the energy to be transferred to the next reacting element. Consider a column of material, such as a full pipe, initiated at one end. The reaction zone may be represented as a cross-sectional element, with the radius being the distance the energy must travel to reach the surroundings. When the energy reaches the outer edge of the reaction zone, its loss to the surroundings is determined by the confinement characteristics (e.g., strength, heat capacity, etc.) of the containment vessel. Thus, propagation is generally favored when charges with larger diameters and stronger confinement vessels are used. There is a minimum diameter of charge below which a reaction will not propagate. This characteristic—the critical diameter—is specific to each chemical system and the effective characteristics of confinement. The principle of critical diameter applies equally to unconfined materials (e.g., solids, gels, etc.), although the critical diameter for an unconfined material is generally greater.

Propagating reactions that initiate in condensed phases are commonly associated with moving equipment, such as pumps and fast-acting valves, but they also begin from thermal decomposition in localized elements. Localized high-temperature conditions, whether arising from cavitation, friction, adiabatic compression, self-heating, external sources such as welding, electrical phenomena, mechanical impingement, fire, etc., can give rise to a propagating reaction if the chemical system and surroundings favor it. The effects of the reaction depend upon the velocity of propagation.

Deflagration of condensed phases produces higher pressures than deflagration of gases, mists, or dusts in air, as there is more energy released per unit of volume. The propagation proceeds by mass transfer (i.e., movement of heat and activated agents into the unreacted materials) at velocities ranging from millimeters per hour to hundreds of meters per second. The reaction zone is very hot and composed of gassified matter, much of which loses heat to confining pipes and vessels. In high-energy, low-velocity systems, the confinement frequently weakens due to excessive heating, and causes localized failures. Deflagrations are extremely pressure sensitive, since loss of confinement permits products and reactants to vaporize, and thus permits energy from the reaction zone to be lost as heat of vaporization. Accordingly, many deflagrations can be extinguished by releasing the pressure through venting mechanisms or by failure of confinement. The concept of critical diameter applied to deflagration relates more to heat capacity and thermal conductivity of the confinement vessel than to strength or rigidity. Since the reaction proceeds at mass transfer rates, heat losses to the walls are important, but the heat may travel through the walls to the unreacted material, sensitizing the system. Use of critical diameter principles in precluding deflagration is not widespread, since other, more reliable methods are available. (See Figure 1-6L.)

The detonation of condensed phases invariably produces extremely high pressures in the confinement vessel. The reaction proceeds in the unreacted material by shock compression at velocities ranging from slightly supersonic to several times the velocity of sound. There is a large impulse in the direction of propagation. The classic testing method for detonability of liquids involves placing a mild steel plate 0.375 in. (9.5 mm) thick above the 1.7-oz. (50-mL) sample and using a high-explosive charge to initiate the sample. If the sample propagates a detonation, a hole of the same diameter as the sample charge is produced in the unrestrained plate. (The test is usually done in a bunker, because in the open the plate may stay airborne for over 10 sec.)

Lateral pressures (at right angles to the direction of propagation) frequently are millions of pounds per square inch (psi) (1 million psi equals 6,895 MPa), enduring at such levels for millionths of a second at any given point. Such tremendous impulses cause shattering of the confinement vessel, which flies apart, propelled by the still-high pressure of the products. Usually, the shock wave in the chemical system passes the fragmenting confinement vessel faster than the parts of the vessel can physically move. Detonations cannot be quenched by conventional venting methods. Propagation can be interrupted only by reducing the diameter of the unreacted material below the critical diameter in the specific confinement vessel, or by breaking the continuity of the chemical system creating the shock

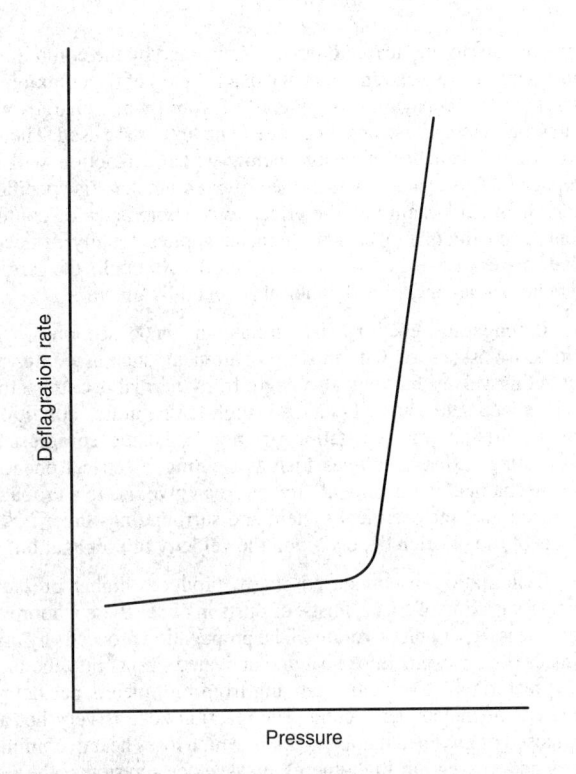

FIG. 1-6L.   *Effect of initial pressure on deflagration rate.*

wave. Since the shock wave is traveling thousands of meters per second, the latter option is rarely feasible. The use of critical diameter traps with certain chemicals is well established, however.

Detonations obviously are more dangerous than deflagrations. Theoretically, however, any material that is exothermic can sustain a detonation under some set of circumstances. In keeping with theory, it is frequently found that condensed-phase deflagrations can self-accelerate under confinement to a point where a deflagration becomes a detonation. The transition is attributed to the preheating of unreacted materials by energy transmissions through confinement vessels and by pressure-piling and precursor wave effects in the unreacted material itself, as well as to physical changes, such as bubble formation, etc.

Propagating reactions in condensed-phase systems are particularly hazardous, in that large masses of liquids, solids, and slurries frequently are connected to operating equipment by pipelines, conveyors, etc. Accordingly, it is possible for a minor malfunction (e.g., pump cavitation, auger friction, valve compression) to initiate a reaction that can deflagrate or detonate through the mass transfer system to storage vessels or transportation tanks, with catastrophic results. The most insidious condition occurs when a deflagration reaction propagates in piping that is small enough to prevent transition to detonation. If the deflagration enters a vessel, it may easily become a detonation, because the vessel diameter is no longer the controlling factor. Under this condition, the entire volume of material in the vessel will probably detonate, because the shock wave travels through the material faster than the material can be moved by the pressure.

ASSESSMENT OF EXPLOSION POTENTIAL

All explosions involve materials that release energy to the surroundings. All accidental explosions result from abnormal conditions or "upsets" in a system at equilibrium. Assessing the explosion poten-

tial involves two distinct sets of information. The first involves the fundamental nature of the substances being processed, stored, handled, used, or transported. The second includes the characteristics of the specific equipment in which the substances are present. While the data sets are distinct, it is necessary to consider the effects of each on the other, so the actual assessment process involves a high degree of iteration. Consideration must also be given to the environment in which the system is located, including both abnormal energy inputs and the effects of system abnormalities.

Properties of Materials

This information may best be understood in terms of two basic concepts corresponding to two basic questions: (1) *Severity*—type and power of reactions, i.e., "What can it do?" and (2) *Sensitivity*—initiation modes, i.e., "What can cause it?" The distinction appears obvious, but confusion and error in explosion assessment results if the distinction is not made.

Severity: The various types of reactions have been discussed with particular characteristics highlighted. For assessment, however, it is inadequate to qualitatively describe reactions. It becomes essential to determine the maximum pressures and temperatures of reactions, rates of increase of pressure and temperature, heat of reaction, etc., and the conditions under which such reactions can occur. These determinations should be independent of the conditions under which handling, etc., is expected to occur, but maximum attention is often given to those conditions that are expected. Such practice can be risky, because all such characteristics and conditions are not predictable. Preferred practice includes a determination of the material properties under extreme conditions, at least initially.

Severity is a more critical element in assessing subject materials for several reasons. First, if the materials cannot undergo an explosion-generating reaction, questions regarding sensitivity do not apply. Second, if the material can react to produce explosive forces, it is reasonable to anticipate that input energy can somehow concentrate to initiate the reaction. Third, since severity experiments employ strong rather than marginal initiators, severity experiments are more reliable than sensitivity experiments.

Usually the first step in assessing severity involves comparing the material structures with those of materials having known properties. The thermodynamic potential is evaluated by manual or computerized calculations. [The manual process is arduous; use of the ASTM Chemical Thermodynamic and Energy Release Evaluation Program (CHETAH) is far superior.] In the final analysis, however, the results of such efforts must be tested by experimentation.

Sensitivity: All initiation mechanisms involve an input of energy into the subject materials. If a specific form of potential initiation energy is chosen, the minimum amount of such energy required to produce a reaction can be determined. However, it is essential to realize that the minimum energy requirement varies widely with the type of initiation energy chosen. There is no reliable quantitative relationship between sensitivity to electrical sparks and sensitivity to mechanical impact, for example. Relationships between different types of energy inputs are complicated in that any source delivers energy in multiple forms, and only those forms that can be absorbed by molecular bonds are effective in initiating chemical reactions. Even under carefully controlled conditions, sensitivity determinations are apparatus-dependent. Applying minimum energy values to real situations presumes that the same general phenomenon (e.g., electric arc) will exhibit the same distribution of energy forms in the laboratory and in the field. Such an assumption is never 100 percent reliable, and can be disastrous.

In practice, energy inputs that create localized high temperatures are those most likely to initiate a chemical reaction. The most commonly evaluated forms of energy input are flames, electrical discharges, hot surfaces, mechanical compression, and shock-wave compression. Other commonly recognized sources of initiation energy are special cases of the above, e.g., friction that creates hot surfaces and/or mechanical compression. Such initiation sources often are evaluated separately, but the measurements thus obtained are so highly apparatus-dependent that applying them to real situations is possible only by analogy.

Assessing thermal sensitivity is actually a severity procedure (i.e., responds to "What can it do?") rather than a sensitivity procedure. Although materials are ranked as more or less thermally sensitive according to the temperature at which self-accelerating decomposition appears, any determined threshold temperature for self-acceleration is highly apparatus-dependent. Further, thermal decomposition is dependent only upon the mass temperature and how it is attained. Thus, there is no way of measuring a discrete energy input that initiates uniform thermal decomposition. Essentially all sensitivity evaluations refer to propagating reactions of one form or another.

Thermal stability evaluations are, however, the most comprehensive and reliable assessments that can be performed, primarily since the reactivity is independent of the mode of energy input. The thermal stability of a chemical system, properly evaluated, provides more widely applicable information than any other evaluation procedure. Generally, if the subject materials can release energy, the phenomenon will appear in a proper thermal stability evaluation.

Assessing severity and sensitivity characteristics depends heavily upon selecting materials for evaluation. The potentials of the system must be probed, or the exercise can yield a false sense of security. Consideration must be given not only to those materials and ratios that are planned, but also to those that may arise from malfunctions, human errors, and normal variations in the operating system.

Properties of the System

The "system" used here refers to the operating equipment in which the materials are or can be found. Any system includes *active* or operating elements, and *passive* or nonoperating elements.

Active elements: This term refers to units that use moving parts, such as pumps, blowers, grinders, valves, stirrers, classifiers, etc. These represent paths by which external energy is supplied to the materials under normal operating conditions. Further, most such equipment has a built-in capacity that is greater than required for normal use, and, in the event of malfunction, can usually input energy at levels far above normal and can input energy in form(s) quite different from normal. (A blower or fan, for example, normally supplies energy for transporting gas; the energy is moderate and supplied discretely through the motion of the blades. If a blade is distorted, however, energy is concentrated in the form of molten particles from the fan blade or housing, which are highly efficient ignition sources.)

Assessing the explosion potential requires determining the type and intensity of energy transmitted to the materials and comparing it with the sensitivity characteristics of the materials. First, the analysis is performed with respect to the normal operation of active elements over the range of materials that could be affected. All active elements should be found safe in the normal mode by substantial margins. The types and intensities of energy input possible under various failure modes are then assessed, and the results compared again with the sensitivity of the materials. This analysis must indicate further safety margins, or action must be taken to prevent any failure mode from supplying excessive energy inputs, or, ultimately, action must be taken to protect personnel from the effects of the reaction(s) that may be produced.

Assessment of active element failure must be succeeded by the opposite consideration, i.e., the results expected when the active elements fail to function. Pump or valve failure may prevent the transfer of hot materials and consequently result in excessive high-temperature exposure. Filter failure may produce dangerous material concentrations downstream. Fan failure may permit accumulation of flammable vapors. Assessing this phase requires a good working knowledge of the system and all its elements, so that an elemental failure can be understood in terms of effects on other parts of the system. The ultimate objective is to ensure that the system is so designed that all such failures cannot lead to an explosion, or to ensure that proper design features are incorporated to control explosive effects and protect personnel and surroundings.

Passive elements: The nonoperating elements of the system do not supply energy to the materials. Instead, these elements confine materials and the energy that the materials carry or release. Passive elements thereby affect the type of reaction that can occur as well as the magnitude and form of pressure that is released. Passive elements also can create conditions under which energy in the materials can be concentrated and/or transformed, creating potential ignition conditions. Examples of such phenomena include generation of electrostatic charges in flowing materials, adiabatic compression of vapors in dead-ended pipes, and catalytic effects of container surfaces. Passive elements frequently can be corroded and/or dissolved by the contents, altering the composition of the materials that subsequently reach other points in the system. The geometry of passive elements can produce collection points for highly sensitive materials that tend to separate from the normal flow, thereby creating localized concentration gradients that may cause a disaster. The passive elements of the system cannot be ignored while assessing the explosion potential. Examining the passive elements of the system must be guided by a consideration of the characteristics of the materials to be contained and of the active elements. The passive elements must be appropriate in strength, geometry, and chemical properties for the contents under the conditions of energy inputs that can prevail during normal and malfunctioning conditions.

The Environment

Here "environment" refers to the immediate surroundings of the system that can interact with materials in the system. The immediate environment often is treated as an extension of the operating system, because environmental features also can be active or passive. Since all interactions are by definition abnormal, the evaluation procedures are often judgmental.

The environment can contribute energy to the system by lightning, welding or tapping, external fire or explosion, impingement of moving machinery or vehicles, and many other means. Some such inputs can be precluded by design, others by protective structures, and still others by administrative controls. For all practical purposes, any environmental energy input to the system that can be credibly anticipated must be considered, as much harm can be produced.

Environmental effects, such as loss of water, power, gas, etc., must be considered, and frequently present more severe problems than environmental energy sources. While utility losses frequently will be anticipated when considering the functional failure of active elements, it is desirable to separately assess the effects of losing utilities. This separate assessment requires evaluating the simultaneous failure of multiple functioning units, a condition often overlooked or considered unlikely, but an obvious result of utility losses.

Finally, the environment, e.g., a building, should be considered in regard to the escape of materials from the system, both by leakage and by explosive breaches of containment. In assessing the effects of leaks, concern should be given to the risk of fire/explosion within the escaped materials alone or as they are mixed with air. Explosive containment breaches create severe potentials for secondary explosions of the expelled materials in air; however, they also create failures in adjacent structures and systems, including the loss of utilities in otherwise unrelated areas. Assessing the explosion potential must address such secondary effects that can change a minor system upset into a disaster.

COUNTERMEASURES

Despite the best efforts of design and operating personnel, human errors occur, equipment breaks, and instruments fail, frequently producing potentially explosive conditions. However, by properly assessing explosion potential, the character and severity of such upsets and resultant explosive reactions can be determined, and countermeasures can be designed into the operating system. In this context, countermeasures mean system elements or operations designed to cope with explosive or potentially explosive reactions, rather than preventive measures.

Countermeasures can be grouped into five general types: (1) containment, (2) quenching, (3) dumping, (4) venting, and (5) isolation.

Containment

In many cases it is feasible to design the operating system to contain the maximum credible pressures and/or other forces that can be produced by the anticipated explosive reaction. Because of the design, there is no sudden release of pressure to the environment, there is no explosion, properly speaking.

The prime benefits of containment design are that it is *passive* (i.e., no operating function is involved), and it is clean (i.e., no dissipation of materials occurs). The primary disadvantage is that, since the energy released in the shock wave is closely related to the failure pressure, containment places a very high premium on accuracy of forecasted intensity. If the reaction should be intense enough to breach containment design, the resulting explosion will be far worse than would result from a normal operating design.

Containment is most easily practiced in cases involving gasphase combustions, where maximum pressures of up to ten or twenty times the initial pressure are the rule. Containment can also be practiced in many cases involving dust and mist combustions, and in certain low-intensity condensed-phase runaway or thermal decomposition reactions.

Containment of normally intense thermal runaway/decomposition reactions is extremely difficult because of the overall volume of the reactants and the maximum pressures that can be attained to render containment design impractical from both engineering and economic viewpoints. Practical containment for normally intense thermal reactions is limited to relatively low-volume systems in bench-scale, pilot-scale, or custom processing operations. The same considerations apply to deflagration in condensed phases.

Containment of condensed-phase detonations is practical only in very limited cases, as the pressure and impulse characteristics are such that extremely heavy equipment is needed, and the lifetime of such equipment is usually very brief.

Quenching

Quenching of potentially or actually explosive conditions involves either heat removal or chemical inhibition. Heat removal may be performed by external means (e.g., emergency additional heat-sinking capacity of equipment) or by introduction of a new material to absorb heat of reaction. Chemical inhibition procedures involve introduction of new material(s) into the chemical system to abate reaction by dilution or by removing active chemical species. Quenching is attractive in many cases because no release to the environment is involved, but is practical only when the function (i.e., heat removal, dilution, etc.) requires a short time relative to the time to explosion.

The most common quenching application is the use of flame arrestors to prevent propagation of vapor/air combustions, wherein heat from the flame front is lost to the flame arrestor, the reaction zone is cooled, and the reaction rate drops sharply. To be effective, these passive units must have a high heat-sinking capacity relative to the heat output of the flame, must have a high surface area in contact with the flame, and must be large enough to cool the gas mixture below the temperature at which reignition can occur. Flame arrestors are generally designed to cope with a specific range of flames. It is essential in selection of flame arrestors that the designer be confident that the chosen unit will, in fact, cope with the potential flame.

Gas, dust, and mist combustions can also frequently be combated by dilution with carbon dioxide, water, mist, steam, certain dry powders, etc., to create enough internal heat-sinking capacity to abate or extinguish the flame front. Use of halogenated hydrocarbon suppressants for extinguishment of gas, dust, and mist combustions involves chemical interference with the flame propagation mechanism, as well as heat-sinking and dilution effects. In all cases, operation of an injection system is needed to introduce the quenchant. This system must operate fast enough to catch the flame front, an operation involving a detection subsystem and a discharge subsystem. Reliability of these subsystems must be considered in the assessment of explosion potential. Further, it must be considered in design whether the introduction of the quenchant will independently create sufficient pressure and/or other forces to cause damage, possibly equivalent in severity to the potential damage from the "prevented explosion." This latter consideration applies particularly to many applications in powder-handling equipment, storage vessels, and building-type structures.

Quenching by introduction of heat-absorbing, diluting, or chemically interfering reagents into condensed-phase systems is invariably a more time-consuming process, generally applicable only to uniform reactions caught early in the self-accelerating stage. In most condensed-phase systems, the mass of reagents is large, and a correspondingly large mass of quenchant must be introduced and thoroughly mixed to be effective. In certain cases, chemical inhibitors can be effective in relatively small concentrations, but the required extent of mixing is also very high. Selection of quenchants is a highly technical matter, but certain generally desirable characteristics may be mentioned. The quenchant should have a high effective heat capacity, fluidity, and miscibility with the reagent mixture, but should also have a low vapor pressure even at high temperatures. It is worth restating that, in all cases, the quenching process requires operation of subsystems for detection and injection; the reliability of these subsystems must be considered in a reassessment of explosion potential after the quenchant system has been designed.

Quenching is generally inapplicable in coping with propagating reactions in condensed phases, with the exception of critical-diameter traps for reaction in pipelines. Introduction and proper mixing of quenchants are generally too slow to be valuable for deflagration or detonation processes, especially in pipelines or other confinement of similar geometry.

Dumping

Dumping essentially means the release of the reaction mixture itself (as opposed to venting, which usually means release of gas). Dumping is ordinarily a relatively slow process applicable to the early stages of a self-accelerating uniform reaction, and, as ordinarily understood, implies the transfer of the reaction mixture to an alternate containment of some kind. The term is more commonly applied to condensed phases, but, in actuality, the procedure is often used with gas-phase systems that may be "blown down" in the event of an upset condition.

Dumping does not stop the reaction, of course, but merely transfers the problem to a (presumably) more favorable location for other treatment. Most commonly, condensed phases are dumped to a cold quenchant-filled vessel; this vessel must itself be capable of withstanding or coping with a potentially explosive condition in the event that the quench process does not proceed properly. Gas mixtures may be dumped through scrubbers, into catch tanks, and/or to the environment, depending on the gases involved. Again, receivers must be examined relative to the consequences of continued reaction after dumping. Further, dumping is again an active process requiring operation of subsystems whose reliability must be factored into the overall assessment of explosion potential.

Venting

Venting refers specifically to the release of gas from a vessel or confining structure in a controlled manner. The venting may be performed by manually operated elements; automated elements; preloaded elements; or fully passive elements, such as burst discs, rupture panels, or blowout walls. This countermeasure is useful for coping with gas, mist, and dust combustions, uniform or propagating condensed-phase reactions, and most conditions leading to physical explosions. The essential requirements are that the vent system be sufficiently fast acting to become fully operational in a timely manner, and that the vent be capable of releasing gas from the confining structure at the maximum rate at which gas is being formed by the potentially explosive situation.

Venting of all gas systems is, in general, fairly well understood in practical terms. The rate of gas generation is reasonably easy to establish, and the equations describing the fluid dynamics are common currency in engineering. The situation changes markedly when the confining structure contains a liquid near or above its normal boiling point or a liquid with substantial amounts of gas dissolved under pressure. The sudden release of pressure almost invariably results in sudden boiling, swelling, and foaming of the liquid, and the vented matter is commonly a mixture of gas and liquid that obeys a radically different fluid dynamic.* In many cases, essentially all of the liquid is expelled from the tank in the venting process. Accordingly, even when the gas phase can safely be vented to atmosphere, provision must be made for effluent liquid as part of the overall design of the vent system.

In addition to the engineering design of the venting process elements, there are two other critical aspects of vent system design that must be considered:

1. The venting process releases material from the original location to another location. Venting frequently includes discharge of hot gases, liquids, and/or solids, which may burn or otherwise react with materials in the receiving location. Venting frequently includes discharge of flammables, and improper selection of a receiving location can produce a hazardous condition, possibly of greater severity than the original. Venting of "harmless"

gases into occupied areas can reduce oxygen levels and cause personnel injuries. In general, venting should discharge to the outside or to locations designed specifically to cope with the vented materials.

2. The venting process produces dynamic effects on the equipment from which material is discharged and on equipment elements in the receiving area. Since the discharge is generally not uniform in all directions, the discharging element must be capable of withstanding the thrust effect, which is equal and opposite in direction to the discharge. The discharging element must also be capable of withstanding other dynamic effects of venting, such as vibration, internal materials flows, and sustained flowing pressure, during the venting process. In the receiving location, equipment and structural elements are subjected to shock and/or pressure effects from the venting process, and load-bearing elements must bear the mass of discharged material and (in cases of rupture or blowout elements) potential projectile effects. Accordingly, the overall design of the venting system extends beyond speed and capacity of the vent element to inherently include consideration of the effects of venting on the relieved location and on the receiving location, although different aspects of the overall design are frequently handled by different specialists.

Isolation

Isolation encompasses separation of an element from surroundings that may be adversely affected by an explosion. Separation may be achieved by remote location of the potentially hazardous element, or by blast-resistant structures designed to deflect, abate, or contain blast waves, missiles, and/or expelled materials.

Isolation simply by remote location is practical for facilities that routinely perform potentially hazardous work, such as explosives facilities, etc. Site selection must consider the magnitude of the potential explosion, blast range, and missile range relative to presently and foreseeably occupied areas, as well as those other considerations generally applicable to site selection activities. The same considerations apply to isolation by distance of a unit process from the balance of a plant. It is important to realize that distance is of primary benefit for the primary explosion only. Expelled materials may travel well beyond the pertinent blast range, and consideration of possible secondary effects is critical. Evaluation of secondary explosion potentials is especially important where expelled materials include, for example, flammable gas or atomized liquids, but assessment of effects on the overall environment is important even when no secondary explosion potential exists.

Isolation by blast-resistant structures requires a complex design process. The magnitude of the explosion must be assessed, as well as the form of the blast wave and the missiles and product materials produced. The structure must then be designed to withstand the shock or impulse load of the explosion, the static pressure created by gas released within the structure, the range and depth of penetration of missiles, and any secondary effects of expelled materials on the blast-resistant structure itself (combustion of expelled flammable vapors, for example).

Blast-resistant structures are seldom designed to fully contain effects of an explosion, but are employed to deflect or abate effects of the blast wave and missiles. Full containment of all explosion effects is practical only when the potentially explosive element is quite small. Certain bench-scale operations can be performed using full containment by blast-resistant structures, and certain flow reactors are small enough that full containment is feasible, but, generally, potentially explosive elements are too large to permit practical full-containment design. Accordingly, secondary effects of deflected blast waves, missiles, and expelled materials must be

---

*Venting of process vessels involving such two-phase flow has been extensively investigated by a consortium of commercial firms acting as the Design Institute for Emergency Relief Systems (DIERS), under the auspices of the American Institute of Chemical Engineers (AIChE). A number of papers based on this activity have been published in *Chemical Engineering Progress*. A DIERS project manual was published in the late 1980s.

anticipated in the design of facilities incorporating blast-resistant structures.

Blast-resistant structures may be constructed of reinforced concrete, steel, timber, earth, sand, or a wide variety of other materials. Many designs utilize multiple elements of different materials to achieve specific effects under specific circumstances. Timber, earth, and sand are frequently used in construction of simple deflecting walls or revetments, often employed for relatively short-duration conditions and/or for separation of elements in remotely sited facilities. Construction of steel or reinforced concrete blast-resistant structures is more common in industrial process operations. A substantial body of information on blast-resistant structures has been compiled for military purposes[1] and is available to non-military designers for general application. An excellent short treatment has been published in *Chemical Engineering Progress.*[2]

In certain cases, where the potentially explosive condition is large and the environmental elements requiring protection are small, blast-resistant structures may be used to enclose key environmental elements rather than to confine the effects of explosion. Typical of such strategy is the use of blast-resistant control rooms in refineries and other large process plants. In such cases, the overall design must also consider personnel safety as regards oxygen supply, thermal loads, and toxic effects of expelled materials.

It is essential to understand that a blast-resistant structure is ordinarily rated in terms of a specific charge loading at a specific location. Use of a blast-resistant structure for purposes other than that for which it was originally designed requires reassessment of blast effects to be sustained. Any alteration of the system within the blast-resistant structure that involves movement of potentially explosive elements may render the structural design inadequate for continued protection. In particular, alterations that move elements closer to walls and/or corners increase resultant stresses of explosion on the structure, and failure of the structure may occur in the event of an explosion.

CONCLUSION

The foregoing is a necessarily brief summation of countermeasures technology. Generally speaking, countermeasures design incorporates more than one form of countermeasure in a single operating system, to provide intermediate levels of response. A reaction vessel may well be equipped with a pressure-relief valve, a dump valve, and a rupture disc, and yet may also be placed in a blast-resistant structure if the severity of the maximum credible reaction so warrants.

The most important single criterion in the selection of countermeasures is the ability of the system design to reliably perform the necessary functions in the available time. In general, it is preferred to buttress use of active countermeasures with passive countermeasures so that, in the event of failure of the active countermeasure, a second line of protection exists. Obviously, any system of countermeasures must be properly inspected, exercised, and maintained on a regular basis. Countermeasures can be frustrated by inadvertence. Circumstances change in the environment, and new technology continually becomes available. Finally, installation of countermeasures does not mean the system is "safe." Safety is maintained only by constant vigilance.

BIBLIOGRAPHY

References Cited

1. U.S. Army, "Structures to Resist the Effects of Accidental Explosions," U.S. Army Technical Manual, No. 5-1300, 1969, Department of Defense, Washington, DC.

2. Tunkel, Steven J., "Barricade Design Criteria," *Chemical Engineering Progress*, Vol. 79, No. 9, Sept. 1983, pp. 50-55.

NFPA Codes, Standards, and Recommended Practices

Reference to the following NFPA codes, standards, and recommended practices will provide further information on explosions discussed in this chapter. (See the latest version of *The NFPA Catalog* for availability of current editions of the following documents.)

NFPA 61, *Standard for the Prevention of Fires and Dust Explosions in Agricultural and Food Products Facilities*
NFPA 65, *Standard for the Processing and Finishing of Aluminum*
NFPA 68, *Guide for Venting of Deflagrations*
NFPA 69, *Standard on Explosion Prevention Systems*

Additional Reading

Cashdollar, K. L., and Hertzberg, M., *Industrial Dust Explosions*, American Society for Testing and Materials, W. Conshohocken, PA, 1987.
Chubb, M., "Explosion in Michigan Air-Bag Assembly Plant: Lessons Learned," *Industrial Fire Safety*, Vol. 2, No. 5, September/October 1993, pp. 30–34.
Cohen, A., *et al.*, "Heats of Explosion at Low Pressures. Part 2," Final Report, October 1989-January 1990, Ballistic Research Lab., Aberdeen Proving Ground, MD, BRL-TR-3188, December 1990, 35 pages.
"Confined Vented Explosions," Work Package No. BL2, British Gas Research and Technology, UK, February 1991, 49 pages.
Cooper, M. G., Fairweather, M., and Tite, J. P., "On the Mechanisms of Pressure Generation in Vented Explosions," *Combustion and Flame*, Vol. 65, 1986, pp. 1–14.
Eckoff, R. K., *Dust Explosions in the Process Industries*, Butterworth--Heinemann, Ltd., Oxford, England, 1991.
"Effects of Simplification of the Explosion Pressure-Time History," Work Package No. BR1, British Gas Research and Technology, UK, February 1991, 115 pages.
Eisner, H., "Explosion and Fire Disrupt World Trade Center," *NFPA Journal*, Vol. 87, No. 6, November/December 1993, pp. 91–104.
"Explosions in Highly Congested Volumes," Work Package No. BL3, British Gas Research and Technology, UK, February 1991, 61 pages.
Fire and Explosion Index: Hazard Classification Guide, 6th ed., American Institute of Chemical Engineers, New York, NY, 1987.
"Fire/Blast Performance of Explosion/Fire Damaged Structural and Containment Steelwork," Work Package No. FR6, Steel Construction Inst., UK, June 1991, 49 pages.
"Fireworks Factory Explosion: Near Kuala Lumpur, Malaysia, May 7, 1991," *Hazardous Cargo Bulletin*, Vol. 12, No. 7, July 1991, pp. 71–72.
Garrison, W. G., "Major Fires and Explosions Analyzed for 30-Year Period," *Hydrocarbon Processing*, Sept. 1988.
Goldfarb, Z., and Kuhr, S., "EMS Response to the Explosion," *Fire Engineering*, Vol. 146, No. 12, December 1993, pp. 74–83.
Harris, R. J., *Gas Explosions in Buildings and Heating Plant*, British Gas Corporation, E & F N Spon, Ltd., London, England, 1983.
Hartman, J. M., "Explosion in New Jersey Chemical Plant Kills Five," *Fire Engineering*, Vol. 148, No. 12, December 1995, pp. 33–42.
Hashagen, P. "First Report: World Trade Center Explosion and Fire," *Firehouse*, Vol. 18, No. 4, April 1993, pp. 32-33.
Hinman, E. E., "Explosion and Collapse: Disaster in Milliseconds," *Fire Engineering*, Vol. 148, No. 10, October 1995, pp. 14–15.
Isner, M. S., "Natural Gas Explosion in Motel, Hagerstown, Maryland, February 18, 1990," Fire Investigation Report, National Fire Protection Association, Quincy, MA, 28 p. 1990.
Isner, M. S., and Klem, T. J., "World Trade Center Explosion and Fire, New York, New York, February 26, 1993," National Fire Protection Association, Quincy, MA, Fire Investigation Report, 1993, 72 pages.
Isner, M. S., "Explosion and Fire Kill Four in Maryland Motel," *Fire Journal*, Vol. 84, No. 6, November/December 1990, pp. 74-78,80–82.
Johannsson, O., "The Disaster of San Juanico," *Fire Journal*, Jan. 1986, p. 32.
Jihu, Q., Kejun, X., and Changhao, H., "Retrospective Discussion on the Shenzhen Warehouse Explosion," *Proceeding* of the First International Conference on Fire Science and Engineering, ASIAFLAM '95, Interscience Communications Limited, China, 1995, pp. 69–75.

Kaufman, C. W., Srinath, S. R., and Tai, C. S., *Proceedings* of the International Symposium on the Explosion Hazard Classification of Vapors, Gases and Dusts, Nat. Acad. Press, Washington, DC, 1987, pp. 41–46.

Keefer, R., "Explosion at SRI," *American Fire Journal*, Vol. 44, No. 6, June 1992, pp. 28–31.

Kirby, R. E., "Major Propane Gas Explosion and Fire, Perryville, Maryland, July 6, 1991," USFA Fire Investigation Technical Report Series, Federal Emergency Management Agency, Washington, DC, Report 053, 1992, 65 pages.

Lamkie, A. J. and Davis, D., "Night Into Day: The Edison, New Jersey, Gas Pipeline Explosion," *Fire Engineering*, Vol. 148, No. 5, May 1995, pp. 34–42.

Leiker, D., and Mesch, T., "Natural Gas Explosions Level Westminster Homes. On the Job: Colorado," *Firehouse*, Vol. 20, No. 7, July 1995, pp. 50–52,54.

Lewis, B., and Von Elbe, G., *Combustion, Flame, and Explosions of Gases*, 3rd ed., Academic Press, Orlando, FL, 1987.

Massa, R. J., "Vulnerability of Buildings to Blast Damage and Blast-Induced Fire Damage," *Fire Engineering*, Vol. 146, No. 12, December 1993, pp. 103–105.

McDaniel, J. T., "Explosion of a Benfield Solution Storage Tank," *Plant/Operations Progress*, Vol. 5, No. 1, 1986, pp. 45–48.

Nolan, D. P., "Statistical Review of Fires and Explosion Incidents in the Gulf of Mexico 1980-1990," *Journal of Fire Protection Engineering*, Vol. 7, No. 3, 1995, pp. 99–105.

Norman, J., "Street Management at Explosions and Collapses," *Firehouse*, Vol. 20, No. 9, September 1995, pp. 116,118.

Patrick, K. A., "Balboa Gas Line Rupture, 'Up Close and Personal'," *Fire Engineering*, Vol. 147, No. 8, August 1994, pp. 62-64,66–68.

"Prediction of the Pressure Loading on Structures Resulting From an Explosion," Work Package No. BL4, British Gas Research and Technology, UK, February 1991, 43 pages.

Proceedings of the Symposium on Industrial Dust Explosions, STP 958, American Society for Testing and Materials, W. Conshohocken, PA, 1987.

Rasbash, D. J., "Quantification of Explosion Parameters for Combustible Fuel-Air Mixtures," *Fire Safety Journal*, Vol. 11, Nos. 1 & 2, 1986, pp. 113–125.

"Recipe for a Dust Explosion," *Record*, Vol. 72, No. 4, Fourth Quarter 1995, pp. 3–12.

Shope, R. L., and Keenan, W. A., "Safety Window Shield to Protect Against External Explosions," Final Report, October 1989-September 1990, Naval Explosive Ordnance Disposal Technology Center, Indian Head, MD, TN-1834, Final Report, July 1991, 41 pages.

"World Trade Center Blast. BOCA/NFiPA Investigation," *Building Official and Code Administrator*, Vol. 47, No. 6, November/December 1993, pp. 18-18,40–45.

Zalosh, R.G., "Explosion Protection," *The SFPE Handbook of Fire Protection Engineering*, 2nd ed., DiNenno, P.J., *et al.* eds. National Fire Protection Association, Quincy, MA, 1995.

DYNAMICS OF COMPARTMENT FIRE GROWTH

Richard L.P. Custer

Over the past twenty years, research scientists and engineers have worked to develop an understanding of the factors and physical processes that enter into and control the growth and spread of fire and its products. Much of the work has focused on two general types of fire scenarios: (1) the pool fire and (2) the compartment fire. Research on pool fires has developed an understanding of energy production from a burning surface and the dynamics of the plume of hot gases and other products of combustion that rise from the surface. Compartment fire understanding, in part, is built on the work involving pool fires in order to define the characteristics of a simplified fire environment without the effects that compartment boundaries (such as walls and ceilings) and compartment openings or vents (such as doors and windows) have on the development of the growth rate of the fire. Other work has produced literally hundreds of test fires in compartments with varying fire sources, physical dimensions and materials of construction, and venting arrangements. As a result of this work, it is now possible to quantify many aspects of pool and compartment fires in order to predict their effect for use in hazard analysis, analysis and design of fire protection systems, and fire reconstruction. Fire spread and growth in the context of this chapter is limited to the compartment of origin and the fuel packages within it.

The purpose of this chapter is to provide the reader with a basic understanding of the concepts involved in modern-day applied fire dynamics as the basis for using the calculation methods described elsewhere in this handbook. See Section 11, "Information and Analysis for Fire Protection." This chapter is intended to introduce the general concepts of fire growth in a compartment and not to focus on the detailed mathematics involved. References to other chapters of this handbook and to the published literature will be provided, as needed, to guide the reader to sources of information for additional study.

For the purposes of this chapter, the discussion will deal with fire growth from the time of established burning to the time when the fire involves the entire compartment and is controlled by airflow out of and into the compartment vents.

Established burning is defined as the point in fire development when the size of the flame is sufficiently large so that flaming combustion will continue without an independent external ignition source and the fire will grow to the extent permitted by the fuel. The

flame height at established burning is frequently considered to be approximately 254 mm (10 in.) on a horizontal fuel surface. It is suggested that, at this point, there is sufficient energy feedback from the flame to the fuel so that the flame will not go out without external influences.

FIRE GROWTH

The following discussion assumes that ignition has taken place and the fire has reached the point of established burning. Beginning with the first materials ignited, the early stages of a fire provide the driving force for growth and spread, both within the compartment and to other portions of the building. The fire serves not only as a source of energy providing flame and heated gases for the spread of fire, it is the source of the smoke particulate and the toxic and corrosive gases that form the products of combustion. The rate and amount of energy produced by the initial fire in a compartment will frequently determine whether or not the fire will spread beyond that compartment.

The fuel available for fire growth and spread can be characterized in two ways: (1) the rate at which it burns and releases energy into the compartment environment and (2) the total energy available that could be released from the fuel. Each of these characteristics is used to describe potential fire severity.

Rate of burning is commonly described using the term heat release rate. Heat release rate is quantified in terms of the kilowatts (Btu per second) released instantaneously at a given point in time during the fire. This describes how fast the energy is being released. The concept of fuel loading severity is based on the amount of energy that would be available if all the fuel were to be consumed regardless of how long it would take. Fuel load is generally expressed in terms of kg of fuel per square m (BTUs per sq ft) of floor area of the space being evaluated. Fuel loading does not consider the speed at which the fuel burns but rather addresses the issue of how long a fire might burn until the fuel is consumed. These concepts will be discussed in detail below.

HEAT RELEASE RATE

The amount of heat released by a fire per unit of time depends on its heat of combustion, which is the amount of energy produced for each unit of mass burned, the mass of fuel consumed per unit of time, and the efficiency of the combustion process. The heat release rate then is determined by multiplying the mass loss rate (mass/

Richard L.P. Custer, M. Sc., is a principal of Custer Powell, Inc., a fire safety consulting firm located in Wrentham, MA. Mr. Custer is a Fellow, Society of Fire Protection Engineers; and Adjunct Associate Professor of Fire Protection Engineering, Worcester Polytechnic Institute, Worcester, MA.

time) by the heat of combustion (energy/mass) and the combustion efficiency (fraction of the mass converted to energy) to yield the heat release rate, Q, in units of energy per unit time. Various units are used, such as kilowatts (kW), Btu/sec, or joules/sec. The kilowatt (1,055 Btu/sec) is the most common unit. The heat release rate is important during the growth phase of the fire when air for combustion is abundant and the characteristics of the fuel control the burning rate. During this phase, the instantaneous heat release rate will increase over time.

The burning rate of a given fuel is controlled both by its chemistry and by its form. Fuel chemistry refers to its composition, e.g. cellulosic vs. petrochemical. Cellulosic materials would include wood, paper, cotton, fabric, etc. Petrochemical materials, in general, refer to plastics that are largely composed of formulations derived from petroleum. The form of the material also has an effect on the burning rate.

One way of looking at the form of fuel is in terms of the surface area available to burn compared to the mass of the material; this is called the surface area to mass ratio. In the case of wood, for example, a solid block of wood weighing 2.2 lb (1 kg) will burn more slowly than that same weight if converted into thin sheets of paper, and may burn explosively if converted into very fine wood dust and dispersed throughout a compartment volume.

Another example of form that is, in part, related to chemistry deals with the differences between foam plastic and rigid plastic. Foam plastic, in general, burns more rapidly than a similar formulation in rigid form. Two examples are: (1) flexible urethane vs. rigid urethane and (2) Styrofoam™ vs. rigid styrene. Another difference in physical characteristics related to burning of plastics is whether or not the plastic will melt when it burns. Those plastics that will change shape or form when heated are referred to as thermoplastics, and may melt and release their energy more rapidly than those plastics that remain rigid when heated. The latter are referred to as thermosetting plastic materials. Thermosetting materials generally tend to form a char layer and burn more slowly.

While this is not an exhaustive discussion of the fuel burning characteristics, it does represent some factors that the reader might consider in assessing the potential heat release rate or growth rate of a fire, given certain types of fuels. More detailed discussions of individual fuels are found in the appropriate chapters of this handbook.

FUEL LOADING

The concept of fuel loading is a way of characterizing the hazard of a compartment fire or building fire in terms of the length of time the building would be expected to burn, based on the total amount of fuel available and the total energy.

Fuel loading is determined by adding up all fuel present and dividing it by the area of the compartment or fire space. The fuel load is expressed as a mass of fuel equivalent to wood. When plastics or other materials are present, a conversion is made by multiplying the number of lbs (kg) of plastic or other materials by the heat of combustion for those materials and dividing by the heat of combustion for wood. This produces an equivalent number of lbs (kg) of wood.

Fuel loading is related to the expected length of time a fire will burn once it is controlled by the amount of air available for the fuel to burn. The amount of heat produced by a fire during this time is controlled by the air that can be supplied through openings, such as doors and windows. For fire duration analysis, all doors and windows are generally assumed to be open.

A fire burning at a constant heat release rate burns fuel mass at a constant rate, as well. Given the mass of material being burned per

minute and the amount of material available to be burned, it is possible to estimate the total burning time.

CLASSIFICATIONS OF FIRE

It is frequently useful to classify fires in order to simplify communication regarding certain common characteristics. Fires have been characterized in four general ways: (1) type of combustion process, (2) growth rate, (3) ventilation, and (4) fire stage.

Classification by Type of Combustion Process

Perhaps the simplest description of a fire classification would be to divide the fire into three regimes: (1) pre-combustion, (2) smoldering combustion, and (3) flaming combustion. No sequence is necessarily implied by this classification. Pre-combustion is considered the process of heating fuels to their ignition point, during which time vapors and particulates are released from the fuel.

Smoldering is defined as glowing combustion on the fuel surface and may or may not be related in any way to the oxygen content in the vicinity of the smoldering process. What is implied here is that the fuel vapor production rate and temperatures involved may not be sufficient to support flaming combustion.

Flaming combustion is almost self-explanatory, in that the production of sufficient energy and fuel vapors in the combustible range is the condition that underlies and supports the presence of flame.

These conditions of burning may exist simultaneously within a given fire. As flames spread from one point to another on a given item of fuel or within the building, the pre-combustion or pre-ignition situation will exist at the perimeter of the fire. The presence of both smoldering and flaming, even in the same compartment, is quite common as the fire spreads through different types of fuels by different mechanisms.

Classification by Rate of Growth

Fires may also be classified on the basis of growth rate. A fire that increases its instantaneous energy output or heat release rate over time is said to be a growing fire. Typically, growing fires have more air available than is needed for combustion of the fuel gases being generated and will continue to grow until limited either by the amount of fuel available or the amount of air for combustion.

A second category based on growth rate is the steady-state fire. Under steady-state conditions, the fire's heat output or heat release rate remains relatively constant over time. This is not to say that there will not be variations, but there is no rapid continuing increase or continuing decrease in energy release rate. An example of this phenomenon might be a flammable liquid pool fire of fixed diameter where, once the entire surface is involved in flame, the amount of energy produced is controlled by the surface area and will be essentially constant until the fuel is exhausted. Another example would be the production of energy from a fire becoming limited due to air supply.

A third category is the burnout or decay condition, where there is plenty of air for combustion but the heat release rate is decreasing, due to fuel consumption.

(See Figure 1-7A for a graphical representation of these categories.)

Fires Classified on the Basis of Ventilation

Fires may also be classified based on whether the fire is dominated by the fuel available to burn or by the oxygen or air available for the

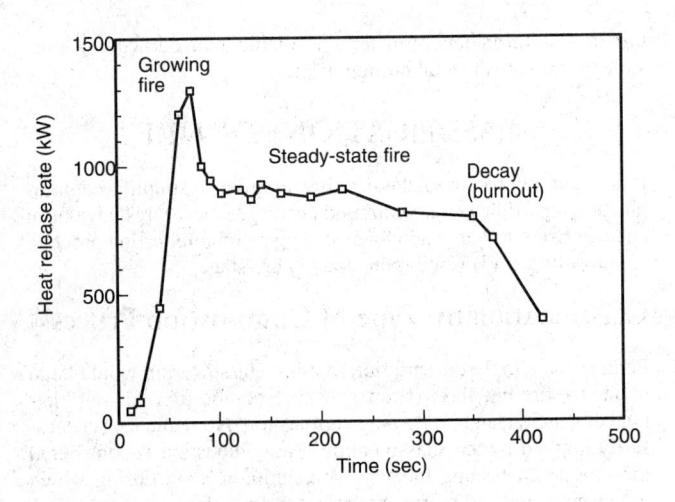

FIG. 1-7A.   Categories of fire growth, based on growth rate.

combustion process to continue. When a fire is burning in the open, or is in the early stages of development within a compartment where there is excess air for combustion, this fire is said to be a fuel-controlled fire. In a compartment fire with sufficient fuel available, the window or door openings may ultimately serve to control the amount of air available for combustion within the compartment. Once the fire develops to a point where it is producing more fuel vapors than can be consumed in the compartment with the available air, it is considered to be a ventilation-controlled fire.

Classification by Fire Stage

The fire service has typically classified the course of a fire in three stages: (1) incipient, (2) free burning, and (3) smoldering.[1] The first stage or phase, called the incipient, is related to the start of the fire during which time there is no active flaming. The fire may be smoldering for several hours.

The second phase or stage, called the free burning or flame production period, is accompanied by increased fuel consumption and heat generation.

The third phase or stage, called smoldering, is characterized by reduced oxygen in the compartment and rapidly decreasing heat production. In discussions of the three phases or stages of fire, it is generally pointed out that, while these are typically the steps in which a fire progresses, fires may, because of changes in ventilation, return to phase two and continue to free burn in the flame production stage.

This classification of fire types by stages has been useful in the past in describing general conditions of burning, but should not be relied upon as a rigorous description of the sequence of events involved in ignition and in fire growth and spread.

EFFECTS OF COMPARTMENT BOUNDARIES ON FIRE

The presence of compartment boundaries, such as walls and ceilings, can have a significant influence on the manner in which a fire grows and spreads by affecting and controlling heat losses and heat release rate.

Fire in the Open

The simplest arrangement of a fire is in the open where it is unaffected by either walls or ceilings. A fire in the open is considered a

free-burning fire and is generally fuel controlled. Figure 1-7B shows a steadily burning fire in the open, with no confining ceiling or walls. This situation can either represent a fire outdoors or a small compartment fire that has not grown to the size where the compartment boundaries may have an influence.

Directly above the fire, a column of hot gases and combustion products rises into the air. This column is referred to as the plume and forms a narrow inverted cone-shaped column of rising heated combustion products and smoke. Under stable conditions, the plume will be symmetrical, about the center of the fire. The action of the hot gases rising due to buoyancy causes air to flow in from the surrounding air at the base of the fire and along the boundaries between the plume and the surrounding air. This process is called entrainment, and is due to the cooling effects of the entrained air. When the temperature in the plume reaches the temperature of the surrounding air, the gases and smoke stop rising. This phenomenon is frequently observed, particularly on a calm day when smoke from a chimney rises to a certain level and suddenly stops and begins to spread out horizontally. This is the result of the equalization of temperature of the smoke in the plume with the surrounding air.

Fire Under a Ceiling (Far from Walls)

When a ceiling is located above a fire plume, the rising hot gases and combustion products impinge on the ceiling and begin to flow outward away from the plume centerline. On a smooth, flat (nonsloping) ceiling, this flow ideally would be equal in all directions. Figure 1-7C shows the general effects of a ceiling. When a ceiling interposes the plume, the flow can be considered in two regions: (1) the plume region and (2) the ceiling jet region. The temperature and

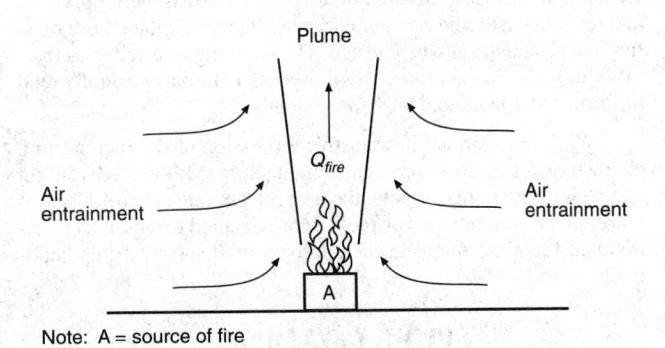

Note: A = source of fire

FIG. 1-7B.   Fire in the open.

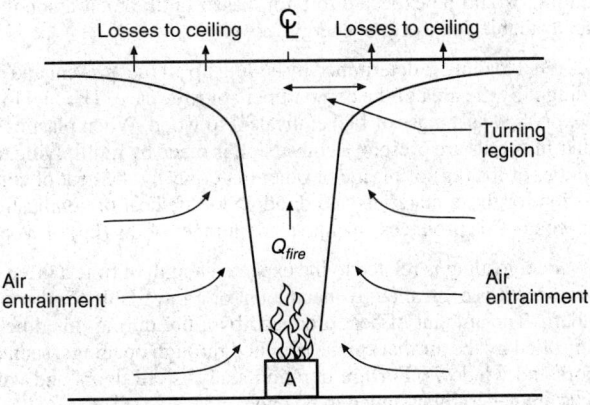

Note: A = source of fire

FIG. 1-7C.   Fire under ceiling, far from walls.

velocities that can be estimated within the plume and the ceiling jet are strongly dependent on location, with respect to radial distance from the centerline.

Research has shown that one set of relationships holds for the immediate area of the impingement point where the plume turns to flow out horizontally underneath the ceiling. This area is known as the turning region. In this region, the flow gas is buoyancy-dominated, and the temperatures and velocities are a function primarily of the height of the ceiling above the base of the fire, since the height affects the amount of entrainment. While the actual temperature above the plume is related to the heat release rate of the fire (Q), for any given fire it can be shown that the temperature along the centerline of the plume decreases with increasing height. (See Figure 1-7D.)

The velocity and temperature of the ceiling jet outside of the turning region is a function of the distance or radius from the plume. Temperatures in the plume will decrease as the radius increases, due to losses to the ceiling and to the entrainment of cooler air from the surroundings. Flow rate outside the turning radius is dominated more by momentum than by buoyancy.[2]

Fire in a Compartment Away from Walls

Figure 1-7E represents the early stages of a compartment fire with a single vent where there is no effect of the compartment. There are two items in the compartment: A and B. A is the source of the fire, and B is an initially unignited target sufficiently far from A that it cannot be ignited directly.[3] As the fire in object A increases in intensity, the gases accumulating at the ceiling spread out and become trapped at the soffit of the door. (See Figure 1-7F.) As the fire continues to produce smoke and hot gases, the layer at the ceiling will thicken and ultimately begin to flow out under the soffit into the next compartment, which may be either another room or a hallway, for example. (See Figure 1-7G.) If the fire were to stop growing or become steady state at this time, the thickness of the upper layer in the fire compartment would become essentially constant, with the excess combustion products leaving the compartment at a constant rate.

However, if the fire continues to increase in heat release rate and the opening is too small to carry away the combustion products at the rate at which they are generated, the upper layer will continue to increase in thickness and descend to the floor, even though there is a vent. (See Figure 1-7H.)

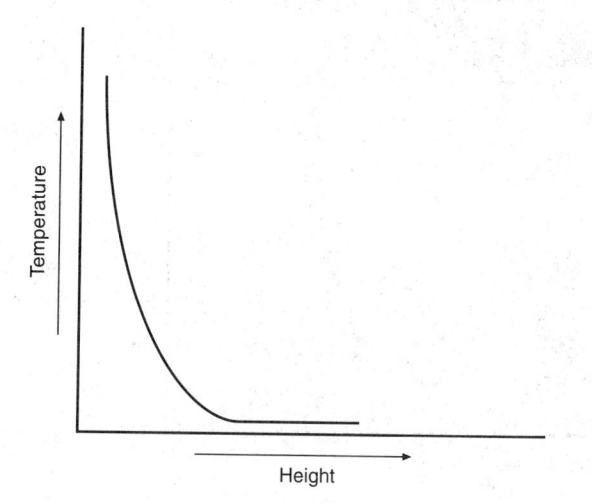

FIG. 1-7D. *Effect of ceiling height on jet temperatures in the turning region.*

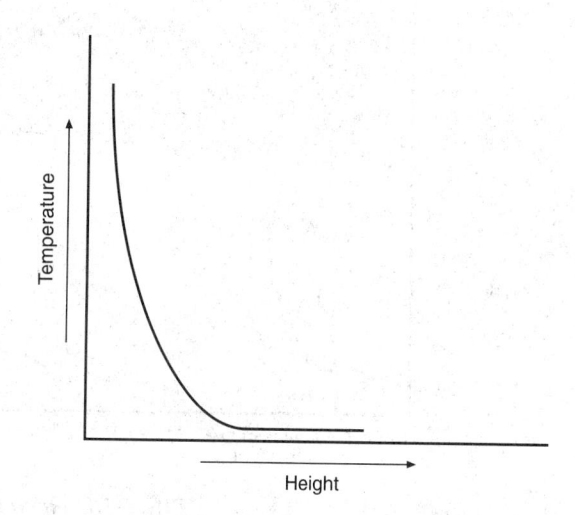

FIG. 1-7E. *Early fire with no compartment effect.*

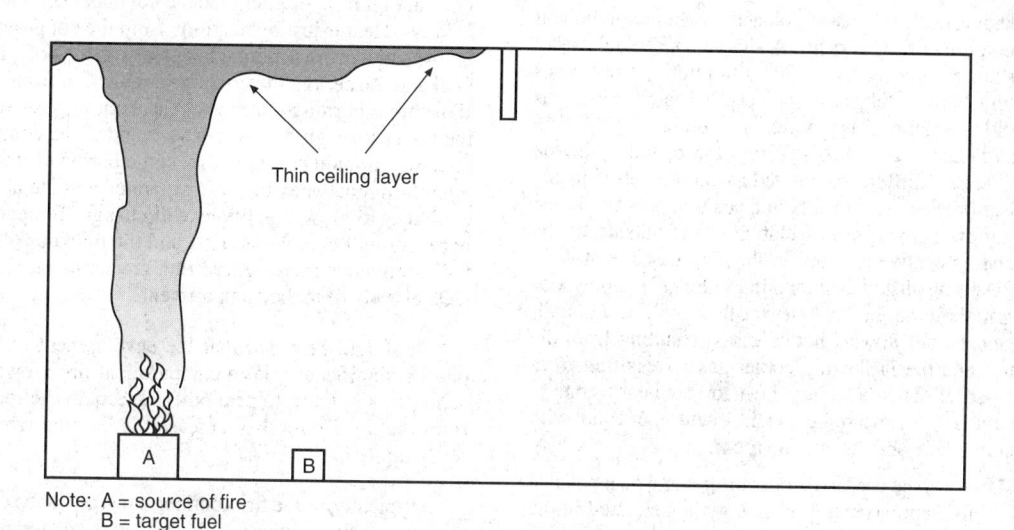

Note: A = source of fire
       B = target fuel

FIG. 1-7F. *Initial ceiling effect.*

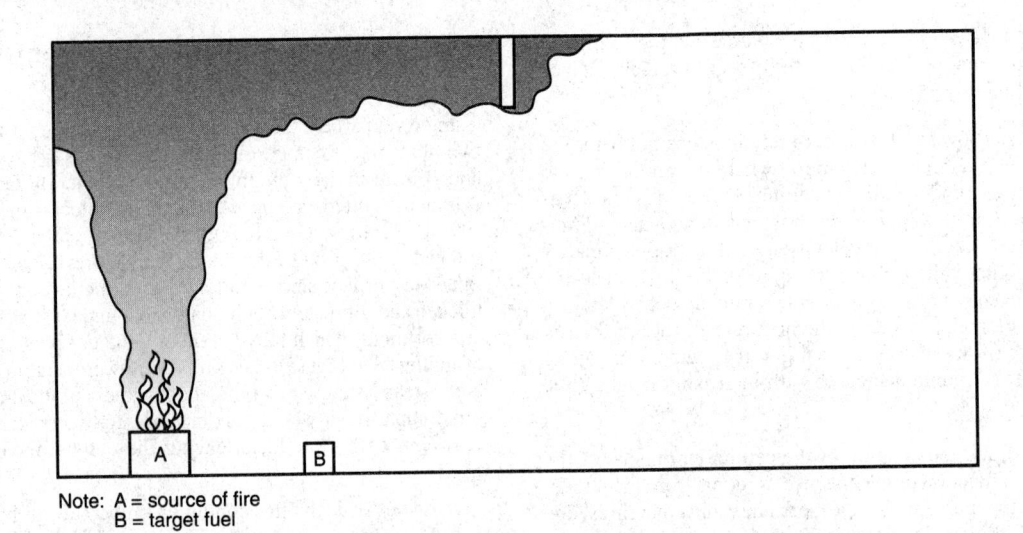

Note: A = source of fire
B = target fuel

FIG. 1-7G. *Initial smoke discharge from compartment of origin.*

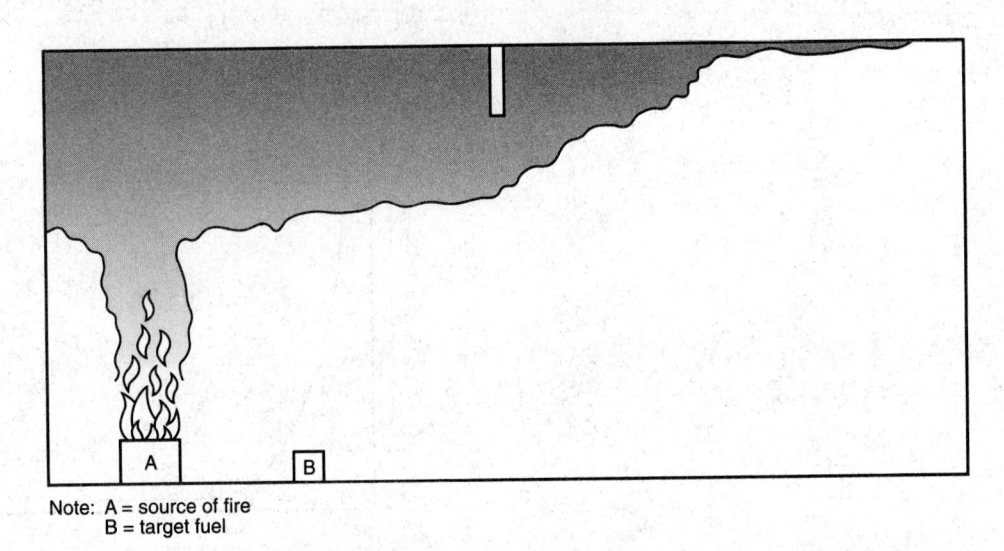

Note: A = source of fire
B = target fuel

FIG. 1-7H. *Increasing fire size and layer depth.*

At this time, it is useful to take a look at the various component parts of the compartment fire "system." A simple conceptual model for this compartment fire system consists of a plume of hot gases above the initially burning object or objects, a heated upper gas layer, and a cool layer below. This model is commonly referred to as a zone model.[4] (See Figure 1-7I.) Across the boundary of the door, there is a pressure difference created by the hot gases inside and the cool gases outside. This results in a positive pressure in the hot gases leaving the compartment, relative to the outside of the compartment, and a negative pressure in the cool gases relative to the inside. The resulting airflow is shown in Figure 1-7J, along with the ceiling layer and plume. In the doorway, there will be a boundary layer between the out-flowing hotter gases (resulting from the positive pressure) and the in-flowing cooler gases (resulting from the negative pressure). This boundary layer or boundary zone is commonly referred to as the neutral plane, i.e., neutral or equal with respect to pressure inside and outside the room.

Figure 1-7I can also be used to discuss the general heat balance within the zones of a compartment fire. Energy is generated in the combustion region of the fire. The plume acts as a pump to provide combustion products and hot gases to the upper layer. This is the

Q_{fire} in Figure 1-7I. Energy in the hot upper layer is lost in a number of ways. Heat is lost by radiation from the hot gases to the cool area below and by convection of hot gases out the door. Heat is lost to the wall and the ceiling materials by conduction from the hot gas layer. If the fire is burning at a constant heat output, the layer thickness and the losses through the door or vent and to the compartment boundaries will remain constant after some period of time to establish an equilibrium between the heat generated and the heat lost. If the fire continues to grow, conditions will change. Temperatures in the upper layer will continue to rise, and the thickness of the upper layer will continue to increase, creating greater radiation to unignited objects elsewhere in the compartment.

It should be noted that the development of the ceiling layer plays a significant role in compartment fire growth. In addition to acting as a radiator to heat other objects in the room, the radiation from the layer will also increase the burning rate of the items already ignited.

Typically, as the fire continues to grow in the compartment with a corresponding increase in the thickness and temperature of the upper gas layer, a transition will occur from a fire that is dominated by

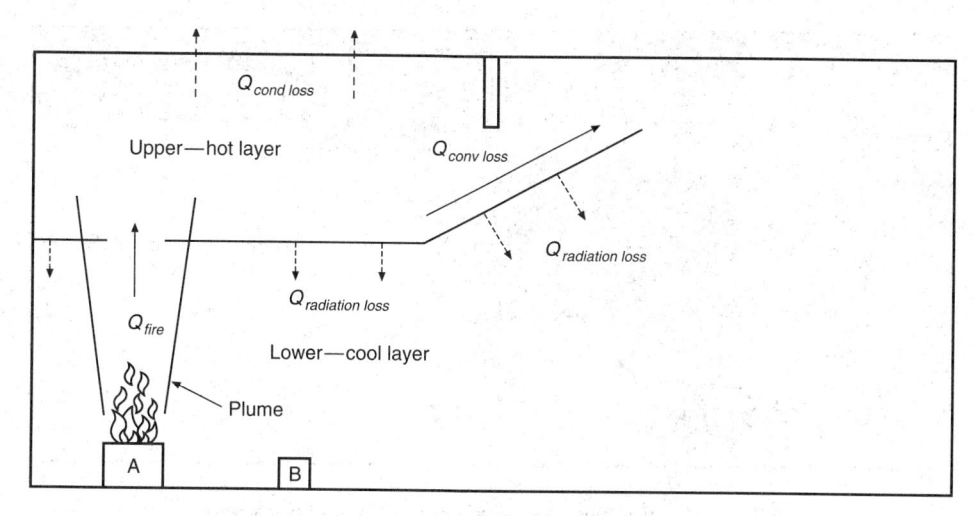

FIG. 1-7I.   Compartment fire zones and heat transfer.

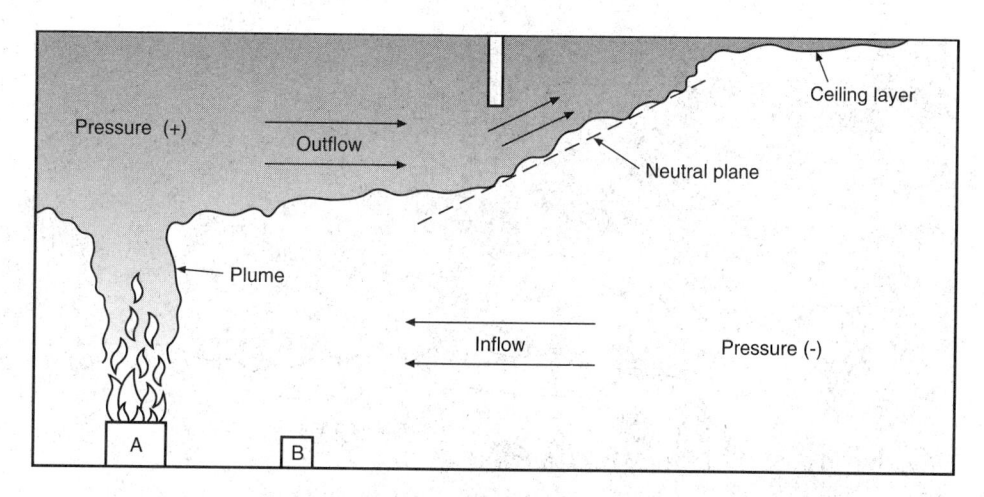

FIG. 1-7J.   Compartment fire pressures and airflow.

the first materials ignited to a fire that is dominated by the burning materials throughout all of the room. This transition is called flashover. Ventilation at flashover becomes controlled by the size of the room openings and the position of the layer in the opening. As the layer descends, the effective ventilation area of the opening is decreased.

The triggering conditions[5] for the flashover transition are reached when: (1) the upper gas layer is approximately 600°C, and (2) the radiant flux on unignited materials in the room is approximately 20 kW/m². This triggering condition is known as flashover. Figure 1-7K represents flashover, the transition stage between preflashover and the fully involved compartment fire (called full-room involvement).

Full-room involvement is shown in Figure 1-7L. Full-room involvement is characterized by the production of excess fuel vapors that cannot be consumed within the compartment with the combustion air available. This results in flame extension through vent openings into adjacent compartments or out of windows, should they fail. Window failure generally occurs shortly before or after flashover conditions are reached and can provide additional vent area.

Flashover is not an inevitable result of a compartment fire. In the event that the fuel is limited or that there is a sufficiently large ventilation opening, the ceiling layer may not develop adequately to

make the transition through flashover to full room involvement. Application of suppression agents, either automatically or manually, can also interrupt the process at or prior to flashover. It should be noted that some research indicates that the heat release rate of burning objects, such as mattresses, can be increased by a factor of two (2) in a post-flashover room fire.[6]

Once the transition from flashover to full-room involvement begins, the fire approaches ventilation control. Smoke from under the neutral plane is frequently recirculated back toward the fire, along with smoke that may be accumulating in adjacent compartments in the hallway. This process reduces the oxygen available for combustion, causing reduced heat release rate or the fire to approach steady-state burning.

EFFECTS OF FIRE LOCATION

Under some circumstances, the location of the fire in a room can have an effect on the rate of growth of the fire, in terms of ceiling jet temperature velocity.[7] When a fire is burning in a room far from walls, air is free to be entrained into the plume from all directions. (See Figure 1-7M.) If the fire is close to a wall or in a corner, the amount of air entrainment into the plume is decreased, and adjustments can be made for heat release rate in the correlations used to calculate temperature and velocity. For fires adjacent to a wall, $2Q$

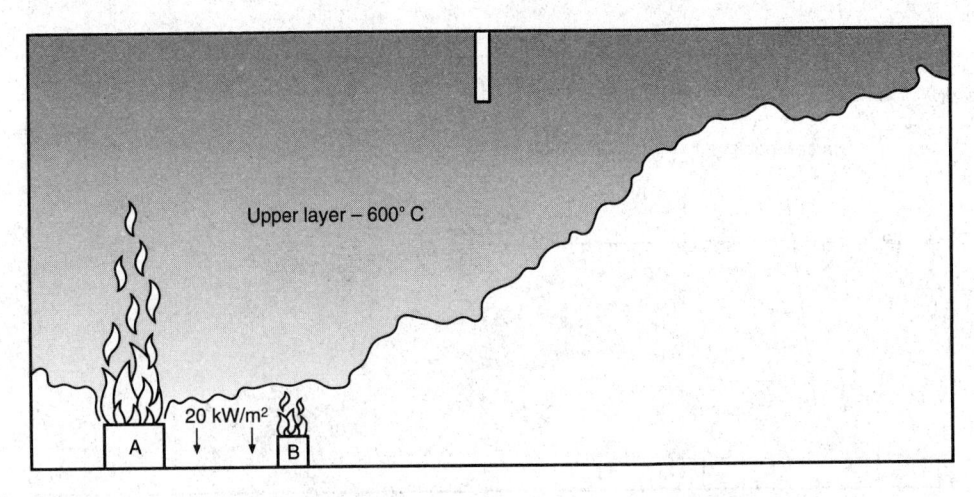

Upper layer – 600° C

20 kW/m²

A   B

FIG. 1-7K.   Flashover—transition to full-room involvement.

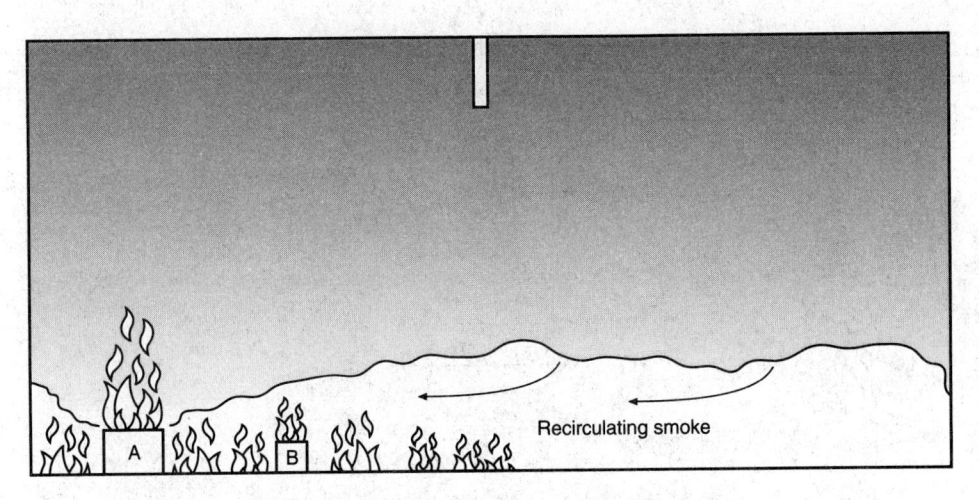

Recirculating smoke

A   B

FIG. 1-7L.   Full-room involvement (post-flashover).

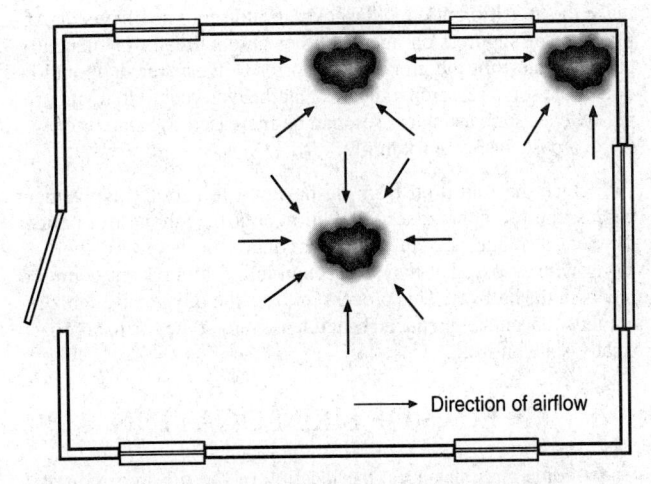

→ Direction of airflow

FIG. 1-7M.   Effect of fire location on air entrainment.

is substituted for Q; for a fire in a 90-degree corner, Q is multiplied by four (4) in the correlations. It should be noted, however, that, experiments have shown that if a circular burner is placed so that only one point is in contact with the wall, the fire behaves almost identically to a fire away from the wall.[8,9]

SUMMARY

The above discussion was presented to provide the reader with an overview of the processes involved in fire growth and spread in and beyond the compartment. The reader is encouraged to study the equations and relationships discussed elsewhere in this handbook and is directed to the additional reading section of this bibliography.

BIBLIOGRAPHY

References Cited

1. Layman, L., *Attacking and Extinguishing Interior Fires*, National Fire Protection Association, Quincy, MA, 1995, pp. 12–15.
2. Evans, D., "Ceiling Jet Flows," *SFPE Handbook of Fire Protection Engineering*, 2nd ed., DiNenno, P.J., *et al.* eds., National Fire Protection Association, Quincy, MA, 1995.
3. Babrauskas, V., "Will the Second Item Ignite," *Fire Safety Journal*, Vol. 4, No. 4, 1982, pp. 281–292.
4. Quintiere, J., "Fundamentals of Enclosure Fire 'Zone' Models," *Journal of Fire Protection Engineering*, Vol. 1, No. 3, 1989, pp. 99–119.
5. Drysdale, D. D., *Introduction to Fire Dynamics*, John Wiley and Sons, New York, 1985.
6. Babrauskas, V., "Burning Rates," *SFPE Handbook of Fire Protection Engineering*, 2nd ed., DiNenno, P.J., *et al.* eds., National Fire Protection Association, Quincy, MA, 1995.

7. Alpert R., and Ward, E., "Evaluation of Unsprinklered Fire Hazards," *Fire Safety Journal*, Vol. 7, No. 2, 1984.

8. Alpert, R., "Calculation of Ceiling-Mounted Fire Detectors," *Fire Technology*, Vol. 8, No. 3, 1972, p. 181.

9. Zukowski, E. E., Kubuta, T., and Cetegen, B., "Entrainment in Fire Plumes," *Fire Safety Journal*, Vol. 3, No. 2., 1981, p. 107.

Additional Reading

Bishop, S. R., *et al.,* "Nonlinear Dynamics of Flashover in Compartment Fires," *Fire Safety Journal*, Vol. 21, No. 1, 1993, pp. 11–45.

Chow, W. K., and Ng, Y. S., "Experimental Studies of Compartment Fire," *Journal of Applied Fire Science*, Vol. 4, No. 1, 1994-1995, pp. 17–30.

Cooper, L. Y., "Compartment Fire-Generated Environment and Smoke Filling," *SFPE Handbook of Fire Protection Engineering*, 2nd ed., DiNenno, P. J., *et al.* eds., National Fire Protection Association, Quincy, MA, 1995.

Luochan, S., Jianping, Y., and Jianren, F., "Numerical Study of the Compartment Fire With Transient Developing Source," First Asian Conference on Fire Science and Technology (ACFST), October 9-13, 1992, Hefei, China, International Academic Publishers, China, 1992, pp. 330–334.

Thomas, P. H., "Two-Dimensional Smoke Flows From Fires in Compartments: Some Engineering Relationships," *Fire Safety Journal*, Vol. 18, No. 2, 1992, pp. 125–137.

THEORY OF FIRE EXTINGUISHMENT

Raymond Friedman

Fire can be extinguished by:

1. Physically separating the combustible substance from the flame.
2. Removing or diluting the oxygen supply.
3. Reducing the temperature of the combustible or the flame.
4. Introducing chemicals that modify the combustion chemistry.

Any specific extinguishment technique may involve one of these mechanisms or, more often, several of them simultaneously.

For example, when water is applied to a fire of a solid combustible burning in air, several extinguishing mechanisms are involved simultaneously. The solid is cooled by contact with water, causing its rate of pyrolysis, or gasification, to decrease. The gaseous flame is cooled, causing a reduction in heat feedback to the combustible solid, and a corresponding reduction in the endothermic pyrolysis rate. Steam is generated, which, under some confined conditions, may prevent oxygen from reaching the fire. Water in the form of fog may block radiative heat transfer.

As another example, consider the application of a blanket of aqueous foam to a burning pool of flammable liquid. Several mechanisms may be operative. The foam prevents the fire's radiant heat from reaching the surface and supplying the needed heat of vaporization. If the fire point of the flammable liquid is higher than the temperature of the foam, the liquid is cooled and its vapor pressure decreases. If the flammable liquid is water-soluble, such as alcohol, then, by a third mechanism, it will become diluted by water from the foam, and the vapor pressure of the combustible will be reduced.

As yet another example, when dry chemical is applied to a fire, the following extinguishing mechanisms may be involved: (1) chemical interaction with the flame, (2) coating of the combustible surface, (3) cooling of the flame, and (4) blockage of radiative energy transfer.

Ideally, any fully successful theory of fire extinguishment should be able to predict the quantity and rate of application of extinguishing agent needed for a given fire. Such a theory would be better than empirical measurements that yield the same information, because the empirical data would be fully reliable only in circumstances identical to those employed in the empirical testing. Fur-

thermore, the theory would provide guidance toward improvement of extinguishment performance.

Unfortunately, the agents mentioned above—water, foam, dry chemical—each work by a combination of several mechanisms, and the relative importance of the various contributions will vary with the circumstances. The degree of complexity resulting from this situation, as well as other problems, has up to now prevented completion of a quantitative fundamental theory of extinguishment action. However, much is known about the various extinguishment modes, and this knowledge will be outlined herein. A more detailed treatment can be found in Friedman.[1]

THE COMBUSTION PROCESS

Much scientific information has been developed about the combustion process, the understanding of which is central to the understanding of fire extinguishment. For details, one may refer to the texts by Friedman,[1] which is rather elementary; Drysdale[2] and Glassman,[3] which are more advanced; or Strehlow,[4] which is the most advanced. Only some simple, basic concepts of combustion especially relevant to fire extinguishment will be presented here.

The term *combustion* usually refers to an exothermic, or heat-producing, chemical reaction between some substance and oxygen. Chemical analysis of combustion products shows the presence of certain molecules involving combinations of oxygen atoms with other types of atoms, such as CO_2, H_2O, SO_2, NO_2, Al_2O_3, or SiO_2.

A slow reaction is a reaction between some substance and oxygen that requires weeks or months to complete. Such a reaction, which is *not* combustion, releases heat so slowly that the temperature never increases more than a degree or so above the temperature of the surroundings. One example of this process is the rusting of metal.

The difference between a slow oxidative reaction and a combustion reaction is that the latter occurs so rapidly that heat is generated faster than it is dissipated, causing a substantial temperature rise of at least hundreds of degrees, and often several thousand degrees. Very often, the temperature is so high that visible light is emitted from the combustion reaction zone.

The concern in fire protection is generally with combustion reactions between various materials and the oxygen of the air.

A flame is a gaseous oxidation reaction that: (1) occurs in a region of space much hotter than its surroundings and (2) generally

Dr. Raymond Friedman was vice president in charge of fire research for Factory Mutual Research Corporation from 1969 to 1987. Now semi-retired, he is a consultant, and an adjunct professor at Worcester Polytechnic Institute.

emits light. Familiar examples include the yellow flame of a candle and the blue flame of a gas burner.

The flame has been referred to as gaseous. When a solid such as a match or candle burns, a portion of the heat of the gaseous flame is transferred to the solid, causing the solid to vaporize. (See Figure 1-8A.) This vaporization can occur with or without chemical decomposition of the molecules. If chemical decomposition occurs, it is called pyrolysis.

There is another mode of combustion that does not involve any flame. It is called smoldering, glowing, or nonflaming combustion. A cigarette burns in this way. Upholstered furniture containing cotton batting or polyurethane foam can also smolder. A large pile of wood chips, sawdust, or coal can smolder for weeks or even months.

Smoldering generally is limited to porous materials that can form a carbonaceous char when heated. The oxygen in the air slowly diffuses into the pores of the material, and there is a glowing reaction zone within the material, even though the glow might not be visible from the outside. These porous materials are poor conductors of heat, so even though the combustion reaction occurs slowly, enough heat is retained in the reaction zone to maintain the elevated temperature needed to sustain the reaction.

It is not uncommon for a piece of upholstered furniture, once ignited, to smolder for several hours. During this time, the reaction zone spreads only 2 or 4 in. (50 or 100 mm) from the ignition point, and then, suddenly, the furniture can burst into flames. The rate of burning during flaming combustion is many times greater than the rate of burning during smoldering combustion.

Combustion requires a high temperature, and the reactions must proceed fast enough at this high temperature to generate heat as fast as it is dissipated, so that the reaction zone will not cool down. If anything is done to upset this heat balance, such as introducing a coolant, it is possible that the combustion will be extinguished. It is not necessary for the coolant to remove heat as fast as it is being generated, because the combustion zone in a fire is already losing some heat to the cooler surroundings. In some cases,

only a modest additional loss of heat is needed to tip the balance toward extinguishment.

Extinguishment can be accomplished by cooling either: (1) the gaseous combustion zone or (2) the solid or liquid combustible. In the latter case, the cooling prevents the production of combustible vapors. This is probably the primary mode of action when a wood fire is extinguished by applying water.

As an alternative to removing heat from the combustion zone to slow the reactions, it is also possible to reduce the temperature of the flame by modifying the air that supplies the oxygen. Air is 21 percent by volume oxygen, the remainder being almost entirely the inert gas nitrogen. The nitrogen, which is drawn into the flame along with the oxygen, absorbs heat, with the result that the flame temperature is much lower than it would be in a fire burning in pure oxygen. If additional nitrogen or some other chemically unreactive gas, such as steam, carbon dioxide, or a mixture of combustion products, were to be added to the air entering the flame, the heat absorbed by these inert molecules would cause the flame temperature to be even lower.

The flame temperature is so important, because the rate of a key combustion reaction ($H + O_2 \rightarrow OH + O$) is very sensitive to temperature. A small decrease in temperature causes a disproportionately large decrease in the rate of this reaction, according to Arrhenius's law.

The symbol H denotes a free hydrogen atom, as contrasted with the ordinary stable form of hydrogen, H_2. Combustion consists of rapid *chain reactions* involving these hydrogen atoms and other active species; hydroxyl free radicals, OH; and free oxygen atoms, O. Figure 1-8B shows a sequence of reactions occurring in the hydrogen-oxygen flame. It is seen that a single H atom, when introduced into an H_2—O_2 mixture at an elevated temperature, will be transformed by a sequence of rapid reactions, requiring a fraction of a millisecond, to form two molecules of H_2O and three new H atoms. Each of these new H atoms can immediately initiate the same sequence, and a branching chain reaction is produced, which continues until the reactants are consumed. Then, the remaining H, O, and OH species recombine according to the reactions

$$H + O \rightarrow OH$$

and

$$H + OH \rightarrow H_2O$$

Similar chain reactions occur in flames of any hydrogen-containing species. Hydrogen is present in the vast majority of combustibles, except for metals and pure carbon.

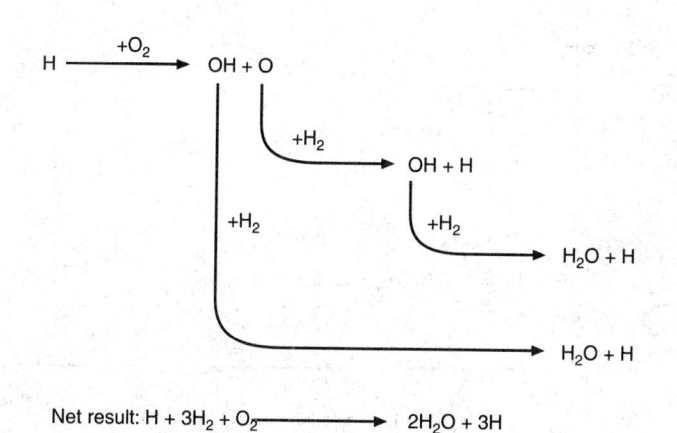

Inverted beaker is trapping carbon dioxide combustion products from flame.

Inverted beaker is lowered, bringing carbon dioxide down on burning wax flame.

Freely burning candle   Self-extinguished candle

FIG. 1-8A.   *Flaming combustion (left) and extinguishment (right) by its own combustion products of a wax candle.*

H $\xrightarrow{+O_2}$ OH + O

+H_2

OH + H

+H_2

+H_2

H_2O + H

+H_2

H_2O + H

Net result: $H + 3H_2 + O_2 \longrightarrow 2H_2O + 3H$

FIG. 1-8B.   *Chain reaction mechanism in the hydrogen-oxygen flame.*

The ability of the hydrogen atoms to multiply rapidly in a flame depends, then, on the prevailing temperature in the flame, which is modified by heat loss or by inert gases, thus leading to extinguishment.

It is also possible to remove hydrogen atoms or other active species from the flame by purely chemical means, i.e., by introducing a species capable of chemical inhibition. This process will be discussed later.

Accordingly, there are two fundamental ways of reducing combustion intensity in a flame and ultimately causing extinguishment: (1) reducing the flame temperature or (2) adding a chemical inhibitor to interfere with the chain reaction.

Consider the effect of adding additional nitrogen to a fuel vapor-air mixture. Suppose the fuel vapor is methane, CH_4. Figure 1-8C shows that, if more than about 35 percent additional nitrogen is added to a 9.5 percent methane-air mixture at 25°C, the resulting mixture is nonflammable. That is, the maximum flame temperature is reduced from about 1900°C to about 1200°C by the added nitrogen, and the rate of heat generation is then too low to sustain the flame against the heat losses from the flame to the surroundings.

Notice also from Figure 1-8C that a 9.5 percent methane-air mixture can be made nonflammable not only by adding nitrogen but, alternately, by adding additional air or additional methane.

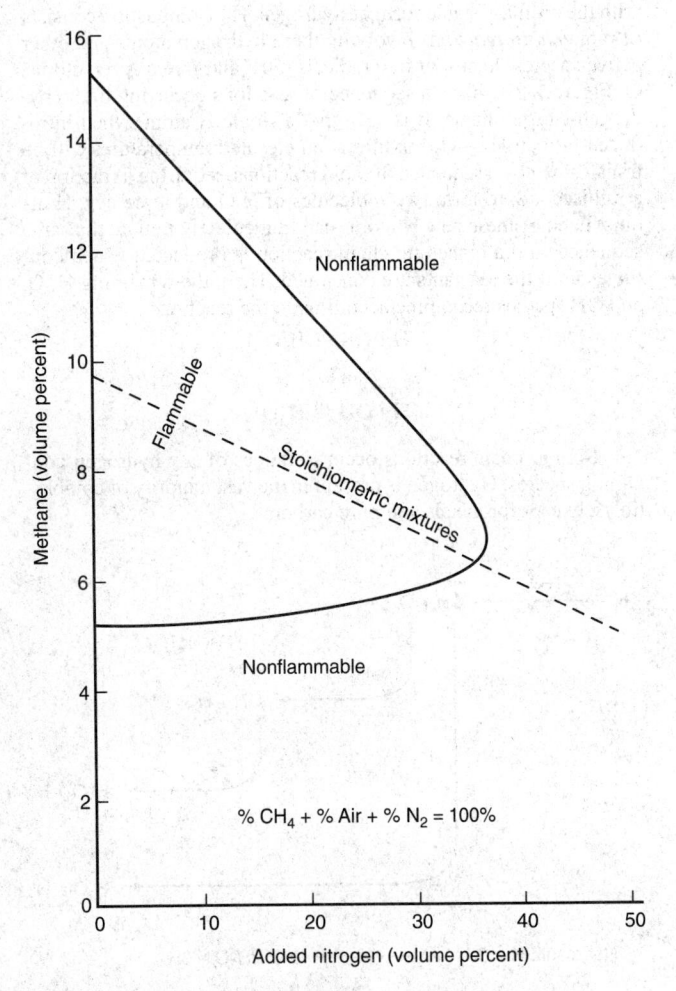

FIG. 1-8C. *Limits of flammability of various methane-air-nitrogen mixtures at 25°C and 1 atm. (From Zabetakis.[5])*

Such additions would produce an excess of one of the reactants and, therefore, a dilution and reduction of flame temperature.

The foregoing discussion of flammability limits applies to premixed combustion, or combustion of a uniform mixture of fuel and air and possibly a third component. This is often the case for explosions, but fires are generally diffusion flames rather than premixed flames. That is, a solid or liquid combustible is vaporizing, and air approaches the vapor cloud from the sides. The flame burns at the interface of the interdiffusing combustible vapors and air. The hot combustion products then rise because of buoyancy.

Such a flame is clearly more complex than a premixed flame, but much the same principles apply to its extinguishment. If more inert gas is added to the air feeding a diffusion flame, extinguishment will occur when the flame temperature is reduced to about 1200 or 1300°C. However, another important way of extinguishing a diffusion flame over a solid or liquid is to cool the solid or liquid enough to interrupt the gasification process. If the gasification rate can be reduced to less than a few g/m^2s, the flame becomes unstable and can no longer sustain itself.

EXTINGUISHMENT WITH WATER

One might suppose that water is the most widely used extinguishing agent because of its low cost and ready availability, relative to other conceivable liquids. However, it turns out that, quite aside from cost and availability, water is superior to any other known liquid for fighting the majority of fires.

Water has a very high heat of vaporization per unit mass, at least four times as high as that of any other nonflammable liquid. It is also outstandingly nontoxic; even a chemically inert liquid, such as liquid nitrogen, can cause asphyxiation. Water can be stored at atmospheric pressure and normal temperatures. Its boiling point (100°C) is well below the 250 to 450°C range of pyrolysis temperatures for most solid combustibles, so evaporative cooling of the pyrolyzing surface is efficient. No other liquid, regardless of cost, can match these properties.

However, water is not an absolutely perfect extinguishing agent. It does freeze below 0°C. It does conduct electricity. It can irreversibly damage some items, although, in many cases, it is practical to salvage water-damaged items. Water may not be effective for flammable liquid fires, especially flammable liquids that are insoluble in water and float on water, such as hydrocarbons. Water is not compatible with certain hot metals or certain chemicals. With fires in these materials, other agents, e.g., aqueous foam, inert gases, halons (with specific limitations due to atmospheric concerns), and dry chemical, will be preferred.

Water may extinguish a fire by a combination of mechanisms: cooling the solid or liquid combustible; cooling the flame itself; generating steam that prevents oxygen access; and, as fog, blocking radiative transfer. While all these mechanisms may contribute to extinguishment, probably the most important is cooling a gasifying combustible.

In order for a solid to burn, a portion of it must be at a high enough temperature so that pyrolysis occurs at a sufficient rate to maintain the flame. For most solids, this temperature is 300 to 400°C, and the pyrolysis rate must be a few g/m^2s. If even a small amount of liquid water, with its high heat of vaporization, can reach this region, the solid can be cooled sufficiently to reduce or stop the pyrolysis, and the flame will be extinguished. Even deep-seated fires can be suppressed in this way. Accordingly, water is the obvious agent of choice for burning solids.

The two most common means of applying water are by: (1) a solid stream or spray from a hose and (2) spray from automatic sprinklers. The practical aspects of manual fire fighting and the use of sprinklers are discussed elsewhere in this handbook. From a scientific viewpoint, studies reviewed by Heskestad[6] and Rasbash[7] have investigated the minimum rate of water application to a burning solid surface that will cause extinguishment.

In an important paper by Magee and Reitz,[8] burning slabs of various plastics, horizontal and vertical, were simultaneously heated with radiant heaters and cooled with controlled water sprays. Extinguishment conditions were then determined. Figure 1-8D shows a linear relationship between the radiative heating rate and the water application rate required for extinguishment. The reciprocal of the slope of the line is found to be approximately the heat of vaporization of water, as theory would predict.

To extinguish burning polymethyl methacrylate (Lucite®, Plexiglas®, Perspex®), enough water must be applied to reduce the burning rate to less than about 4 g/m²s. Depending on the intensity of the externally imposed radiative flux—up to 18 kW/m²—a water application rate of from 1.5 to 8 g/m²s was required. This is a very small application rate of water. For extinguishment with no external radiative flux, it was only necessary to spray enough water so that the heat absorbed by its vaporization was 3 percent of the heat of combustion.

Experiments with other plastics and with wood cribs have given similar results; only a few g/m²s of water must be applied to the burning surface to cause extinguishment, and the rate of heat absorption by the water is only a few percent of the rate of heat generation by the fire before water application.

The reason for this high efficiency is now well understood. Consider a horizontal slab of polymethyl methacrylate, 0.3 × 0.3 m², which is burning steadily on its top surface. Measurements[1] have shown that only about 12 percent of the energy released by combustion is transferred back to the surface. Of this energy arriving at the surface, primarily by radiation from the flame above, about 40 percent is reradiated from the hot surface to the surroundings, and only 60 percent of 12 percent, or 7 percent of the available

combustion energy, is used to decompose and gasify the plastic. It is only necessary to apply enough water to the surface to drain off a substantial portion of this 7 percent of the combustion energy, and the rate of burning is then reduced to a point at which the flame can no longer sustain itself.

Of course, the evaporation of the water produces steam that dilutes the flame and reduces the flame temperature, causing some reduction in the rate of burning, but this effect is generally small and need not be considered in a first-approximation model of the extinguishment process.

It is interesting to note that, in ignition experiments, a solid surface is progressively heated, with a gradual increase in the rate of pyrolysis, or gaseous decomposition, but it is not possible to ignite the vapors to obtain a self-sustaining flame until the pyrolysis rate reaches a certain minimum value. This is roughly the same value to which the pyrolysis rate must be reduced when water is applied to a burning surface to accomplish extinguishment.

In practical fire fighting, water must be applied at 10 to 100 times the rates used in the research described above because of the difficulty of delivering the water directly to the burning surface.

In the case of fire suppression by use of sprinklers at the ceiling, it is possible to calculate the fraction of the water droplets able to penetrate the fire plume and arrive at the burning surface beneath. Such a calculation is complex and requires a powerful computer.[9] It is necessary to know the heat release rate of the fire, the location of the fire relative to the nearby ceiling sprinklers, and the distribution of drop sizes of the water. The calculation takes into account the aerodynamic drag on the downward-moving droplets encountering the upward-moving fire gases, and calculates the change in motion of the fire gases because of the downward momentum of the water spray. Droplet evaporation and cooling of the fire gases are included in the calculation.

The drop-size distribution of the water depends not only on the sprinkler design, but also on the pressure drop across the sprinkler. The mean drop size varies inversely with the cube root of the pressure drop. It also depends on the surface tension of the water, which could be modified with additives.

In some cases—e.g., a purely gaseous fire—water may extinguish the fire by cooling the flame rather than the source of the fuel vapor. The theory of this action was discussed previously.

For a pool fire of a flammable liquid with a high flash point (e.g., diesel oil), water can be effective by reducing the temperature of the liquid below its flash point. However, water impinging on the flammable liquid with high velocity can cause burning liquid to be scattered, increasing the fire intensity. Water applied as a foam or as a fine mist avoids this situation.

EXTINGUISHMENT WITH AQUEOUS FOAMS

The principal application of aqueous foam agents is for fighting flammable liquid fires. If the flammable liquid is lighter than water and is insoluble in water, then application of water would simply result in the liquid floating on it and continuing to burn. If the flammable liquid is an oil or fat, the temperature of which is substantially above the boiling point of water, then the water will penetrate the hot oil, turn into steam below the surface, and cause an eruption of oil that will accelerate the burning rate and that might spread the fire.

Foams are the primary tools for fighting fires that involve substantial quantities of petroleum products, such as those found at refineries, tankers, and storage areas.

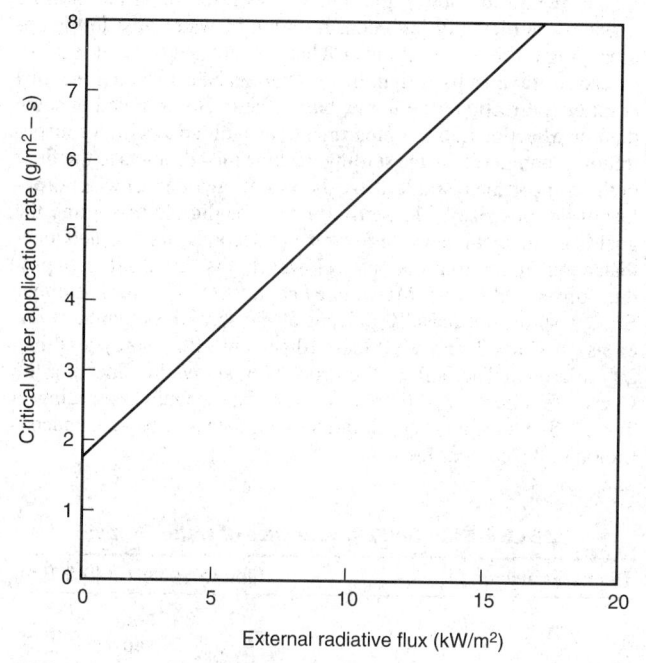

FIG. 1-8D.   Water application rate needed to extinguish a fire on a vertical polymethyl methacrylate sheet.

If the flammable liquid is water-soluble, like alcohols, then the addition of sufficient water will dilute the liquid to the point where it is no longer flammable. If there is a deep pool of alcohol rather than a shallow spill, however, the time required to obtain sufficient dilution might be so great that an aqueous foam would be a better extinguishing agent. If the nature of a liquid is unknown, an aqueous foam might be chosen instead of direct application of water.

Another important application of aqueous foam agents is on liquids or solids that are burning in difficult-to-access spaces, such as a room in a basement or the hold of a ship. The foam is used to flood the compartment completely.

Fire-fighting foam is a mass of bubbles formed by various methods from aqueous solutions of specially formulated foaming agents. Because foam is much lighter than any flammable liquid, it floats on the liquid, producing an air-excluding, cooling, continuous layer of vapor-sealing, water-bearing material that can halt or prevent combustion.

Fire-fighting foams are formulated in several ways for fire extinguishing action. Some foams are thick and viscous, forming tough heat-resistant blankets over burning liquid surfaces and vertical areas. Other foams are thinner and spread more rapidly. Some are capable of producing a vapor-sealing film of surface-active water solution on a liquid surface, and some are meant to be used as large volumes of wet gas cells for inundating surfaces and filling cavities. There are various methods for applying foams.

The use of foam for fire protection requires attention to its general characteristics. Foam breaks down and vaporizes its water content under attack by heat and flame. Therefore, it must be applied to a burning surface in sufficient volume and rate to compensate for this loss and to provide an additional amount to guarantee a residual foam layer over the extinguished portion of the burning liquid. Before starting to apply foam to a large fire, one must accumulate a sufficient amount of foam concentrate to do the job. Nothing will be accomplished by putting out only part of a fire and then running out of foam, because the fire will build back to its original intensity.

Foam is an unstable air-water emulsion and can be broken down easily by physical or mechanical forces. Certain chemical vapors or fluids can destroy foam quickly. When certain other extinguishing agents are used in conjunction with foam, severe breakdown of the foam can occur. Turbulent air or violently uprising combustion gases can divert light foam from the burning area.

There is no theoretical basis available to serve as a guide to the needed rates of application of the various types of foams in different situations. The guidelines come from experience or from empirical tests.

One useful laboratory measure of foam effectiveness is to fill a graduated cylinder with the foam and observe the time required for a certain fraction of the water in it to drain to the bottom. The more stable the foam, the slower it will drain. Clearly, the time for collapse of a foam layer should be greater than the time required to coat the entire surface of a large spill with foam. Thus, a basis is available for calculating the necessary rate of application. Of course, fire causes foam to break down at a greater rate than indicated by the drain test in a cylinder, so empirical information is still needed.

Another laboratory tool, useful for formulating film-forming foams, is the measurement of surface tension of the foam solution, F, and of the flammable liquid, L, and of the interfacial tension between the two liquids, FL. The film will spread over the surface only if F plus L is greater than FL.

EXTINGUISHMENT WITH WATER MIST

There has been recent interest in developing equipment for applying a fine mist of water to a fire as an alternate to halogenated agents. Three methods are as follows:

1. Fixed installation in which a fine mist is used to inert a compartment in which a fire may occur, perhaps in a concealed and unpredictable location.
2. Fixed spray nozzles positioned around the site of an anticipated fire.
3. A portable extinguisher using a fine spray or mist.

Three mechanisms by which a fine water mist might extinguish a flame are as follows:

1. The mist droplets, while evaporating, remove heat, either at the surface of the combustible or within the gaseous flame. This cooling can cause extinguishment, as discussed previously.
2. The fine droplets evaporate in the hot environment even before reaching the flame, generating steam that dilutes the oxygen percentage in the air approaching the flame, thus causing extinguishment by a mechanism similar to that of an inert gas, e.g., carbon dioxide.
3. The mist blocks radiative heat transfer between the fire and the combustible.

In those tests in which mists have successfully extinguished fires, some combination of the above effects appears to have occurred in each case.

In regard to mechanism 1, it may be said that it is much easier for a large drop to reach a burning surface than a very fine drop (mist), which would tend to be blown away from the surface by the pyrolysis gases, if indeed the fine drop could get through the flame to the vicinity of the underlying surface in the first place. This difficulty disappears when the fine mist is directed at the burning surface with high momentum.

In regard to cooling the flame gases rather than the burning surface, a sufficiently high concentration of water mist in the approaching air must be achieved. This is estimated to be at least 15 percent water mist by weight in air. This can be achieved if the mist is sprayed directly at the flame; but, if the mist is sprayed in some random direction into a compartment, the buildup to a high concentration is hampered by the settling of some mist droplets to the floor of the compartment. Table 1-8A shows settling times of water droplets of various sizes. The settling rate is negligible (assuming the goal is extinguishment in a few tens of seconds) for droplets finer than about 10 microns diameter. However, it is very difficult to produce droplets this fine. Mawhinney et al.[10] classify mists as either Class 1 sprays (at least 10 percent of the spray less than 100 microns) or Class 2 sprays (at least 10 percent of the spray less than 200 microns). The bulk of the droplets in sprays just meeting the Class 1 requirement settle out in a very few seconds, according to Table 1-8A; thus, it is very difficult to build up an adequate concentration in the compartment.

TABLE 1-8A. *Settling Velocities of Water Drops*[12]

Diameter (microns)	Time to Settle 1 ft (0.305 m)
5	391 sec
10	99 sec
20	25 sec
50	4 sec

The blockage of radiation by a mist will often be effective in reducing the intensity or spread rate of a fire, but will rarely be sufficient in itself to extinguish a fire.

In summary, the effectiveness of a fine mist depends on: (1) momentum and direction of the spray relative to the fire and (2) the compartment geometry. Further discussion of water mists can be found in the material referenced in 10 and 11.

EXTINGUISHMENT WITH INERT GASES

Water acts to extinguish fires primarily by cooling, although the formation of steam helps to dilute the concentration of oxygen. On the other hand, inert gases act to extinguish a fire primarily by dilution. Carbon dioxide is the most commonly used inert gas, although nitrogen or steam could be used. Theoretically, helium, neon, or argon could be used, but they are expensive, and there is no reason to use them except in certain special cases, such as magnesium fires.

Table 1-8B presents the minimum proportions of carbon dioxide or nitrogen gas which, if added to air, will form an atmosphere in which various vapors will not burn. On a volume basis, carbon dioxide is substantially more effective than nitrogen. Note, however, that a given volume of carbon dioxide is 1.57 times as heavy as nitrogen (44 to 28 molecular weight ratio), so the two gases have nearly equal effectiveness on a weight basis. Either gas, in sufficient quantity, will prevent the combustion of anything except certain metals or unstable chemicals such as pyrotechnics, solid rocket propellants, hydrazine, etc.

TABLE 1-8B. *Minimum Required Volume Ratios of Carbon Dioxide or Nitrogen to Air that Will Prevent Burning of Various Vapors at 25°C*

Vapor	Carbon Dioxide		Nitrogen	
	(CO_2/Air)	(% O_2)	(Extra N_2/Air)	(% O_2)
Carbon disulfide	1.59	8.1	3.0	5.2
Hydrogen	1.54	8.2	3.1	5.1
Ethylene	0.68	12.5	1.00	10.5
Ethyl ether	0.51	13.9	0.97	10.6
Ethanol	0.48	14.2	0.86	11.3
Propane	0.41	14.9	0.78	11.8
Acetone	0.41	14.9	0.75	12.0
n-Hexane	0.40	15.0	0.72	12.2
Benzene	0.40	15.0	0.82	11.5
Methane	0.33	15.7	0.63	12.9

See Friedman.[1]

If available, steam also can be used as an inert extinguishing agent. The percentage by volume required is intermediate between that required for carbon dioxide and for nitrogen.

Table 1-8B shows that the required addition of either carbon dioxide or nitrogen reduces the oxygen level to a point at which exposed humans will suffer undesirable effects. In the case of carbon dioxide, an additional serious physiological effect will occur at the concentrations required to extinguish a fire.

Table 1-8B refers only to vapors, but the data are relevant to liquids or solids because they burn only by vaporizing or pyrolyzing. Accordingly, application of an inert gas can extinguish the flame over a liquid or solid. However, if the inert gas dissipates after several minutes, because, for example, the enclosure is not airtight, it is possible that a glowing ember or hot metal could reignite the fire. Reignition is common for a deep-seated fire, such as might occur in upholstered furniture or in a carton of documents.

Some explanation of the physical forms of carbon dioxide is appropriate. Carbon dioxide is unusual in that it can exist only as a gas or solid at normal atmospheric pressure, but not as a liquid. Figure 1-8E shows the phase diagram of carbon dioxide.

The solid form of carbon dioxide, commonly known as "dry ice," at atmospheric pressure, exists only below –79°C, at which temperature it undergoes sublimation directly to vapor, without melting. However, carbon dioxide liquid can exist at elevated pressures, as long as the temperature is above –57°C and the pressure is above 5.2 atm. This temperature and pressure condition is known as the triple point of carbon dioxide because it is the only condition at which solid, liquid, and vapor can coexist.

Liquid carbon dioxide can be kept in a pressure vessel at any temperature between –57°C and +31°C (the critical temperature). Above the critical temperature, there will no longer be a liquid-gas interface in the pressure vessel, so the fluid in the vessel would be a gas. A pressure vessel at 21°C containing liquid carbon dioxide would be at a pressure of 58 atm, which is the vapor pressure of carbon dioxide at that temperature. This pressure is used to expel liquid carbon dioxide from a cylinder in fire fighting. The cylinder normally would contain an internal dip tube reaching to the bottom so that liquid rather than vapor would be discharged. As the liquid droplets emerge from a nozzle into the lower-pressure environment, instantaneous evaporation occurs, with evaporative cooling of the residual liquid in each drop. This process causes solidification of the residual portion into "dry ice" particles at –79°C. If the liquid was originally at 21°C, about 75 percent of the discharged liquid would have evaporated and about 25 percent would have been converted to "dry ice" particles.

Some of the "dry ice" particles might impinge on a combustible surface and have a cooling effect. However, because the heat of sublimation of carbon dioxide is only about one-fourth the heat of

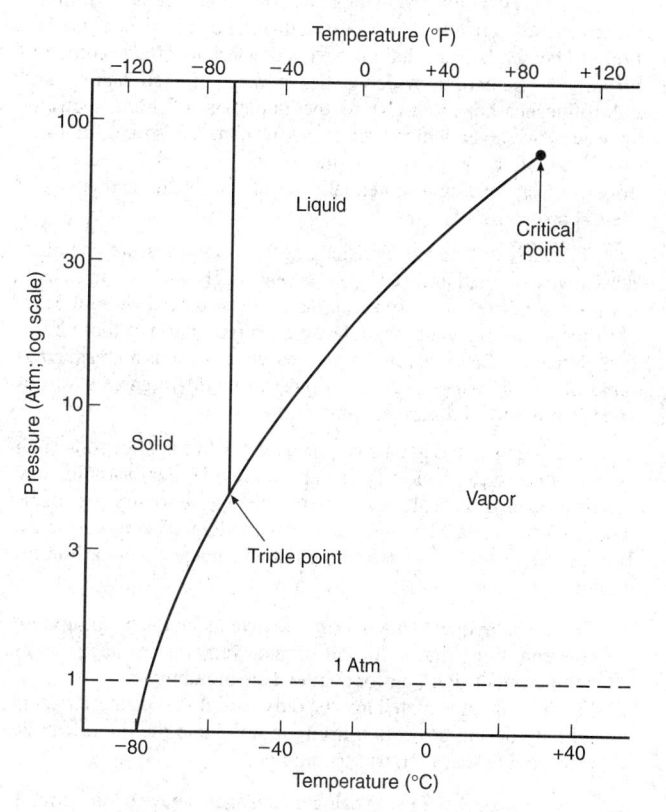

FIG. 1-8E. *Phase diagram of carbon dioxide.*

vaporization of water and because only about one-fourth of the carbon dioxide discharged is converted to "dry ice," the cooling effect on a hot surface is only about one-sixteenth that produced by water discharged at an equal rate (on a mass basis).

In comparing carbon dioxide and nitrogen, carbon dioxide has the advantage that it can be stored in a cylinder as a liquid at a relatively moderate pressure of 58 atm (21°C), while nitrogen at the same temperature must be stored as a gas, usually at about 140 atm. A given size of cylinder at 21°C and these pressures could hold about three times as large a volume of carbon dioxide as nitrogen (measured after expansion, at atmospheric conditions). Nitrogen also could be stored for short periods more compactly as a cryogenic liquid at –196°C and 1 atm. However, long-term storage results in continuous loss of nitrogen. As a result of these factors, carbon dioxide is used more commonly than nitrogen as an inerting gas.

Semi-permeable membranes are being developed that can inexpensively separate the oxygen and nitrogen of the air. When and if such systems become cost-effective, it might be practical to provide permanent nitrogen-inerting of hazardous spaces that do not require human presence. A reduction of the oxygen percentage in the air from 21 percent to 10 percent by volume would make fires and explosions impossible, except for a few special gases, e.g., hydrogen, acetylene, or carbon disulfide, which would require greater dilution.

EXTINGUISHMENT WITH HALOGENATED AGENTS

Halogenated agent extinguishing systems are a relatively recent innovation in fire protection, but, despite this, they already face extinction. As of January 1, 1994, the global production of fire protection halons in many countries ceased.

The reason halon production has come to a halt has nothing to do with its effectiveness as an extinguishing agent. Halon production has ceased because halons have a deleterious effect on the environment. Scientific evidence has strongly linked halons and chlorofluorocarbons (CFCs) to the depletion of Earth's stratospheric ozone layer, which protects us from the sun's harmful ultraviolet radiation. Depletion of the ozone layer may reduce its effectiveness, leading to potentially significant health and environmental problems.

The halogenated extinguishing agents, or halons, are chemical derivatives of methane (CH_4) or ethane (CH_3—CH_3), in which some or all of the hydrogen atoms have been replaced with fluorine, chlorine, or bromine atoms, or by some combination of these halogen elements. These agents are liquids when stored in pressurized tanks at normal temperatures, but most of them are gases at atmospheric pressure and normal temperatures.

The halogenated agents can be used for fire applications such as those discussed previously for carbon dioxide. For example, they can be used on electrical fires, in cases where water or dry chemicals would cause damage, or for inert-gas flooding of compartments. Halogenated agents have two principal advantages over carbon dioxide:

1. Certain halogenated agents are effective in such low volumetric concentrations that sufficient oxygen remains in the air after compartment-flooding for comfortable breathing.
2. For several halogenated agents, only partial vaporization occurs initially during projection from a nozzle, and the liquid can be projected farther than carbon dioxide.

The drawbacks of using halogenated agents have to do with the toxicity and corrosivity of their decomposition products, and with

the detrimental effect halogenated compounds have on Earth's ozone layer.

Of the various halons, Halon 1301 (bromotrifluoromethane) is by far the most commonly used in fire protection because it has the lowest toxicity, as well as the highest effectiveness on a weight basis. Among the highly effective halons, it has the highest volatility, which is desirable for flooding applications. If a halon liquid is needed for direct application to a burning surface to accomplish cooling as well as inerting of the nearby region, however, a less volatile halon, such as Halon 1211 (bromochlorodifluoromethane) or Halon 2402 (dibromotetrafluoroethane), would be preferred.

Table 1-8C gives the physical properties of these three halons. They are all liquids at normal temperatures when stored in pressurized tanks. They can be stored under high-pressure nitrogen if the liquid must be expelled from the tank more rapidly than under the vapor pressure of the halon alone. The use of nitrogen for pressurization is especially important for outdoor storage in the winter.

TABLE 1-8C. *Physical Properties and Chemical Formulas of Three Halon Extinguishing Agents*

	Halon 1301 CF_3Br	Halon 1211 CF_2ClBr	Halon 2402 $C_2F_4Br_2$
Boiling point (°C)	–58	–4	+47
Liquid density at 20°C (g/cc)	1.57	1.83	2.17
Latent heat of vaporization (J/g)	117	134	105
Vapor pressure at 20°C (atm)	14.5	2.5	0.46

The inerting capabilities of Halon 1301 and Halon 1211 are shown in Table 1-8D. If methane in any proportion is combined with a mixture containing 5.4 volumes of Halon 1301 and 100 volumes of air, at 25°C, then no combustion can result. By contrast, a mixture containing 33 volumes of carbon dioxide and 100 volumes of air is required to obtain the same result. This suggests that a molecule of Halon 1301 is 33/5.46 = 6.1 times as effective as a molecule of carbon dioxide.

Note, however, that the molecular weight of Halon 1301 is 149, while that of carbon dioxide is 44, so the ratio of molecular weights is 149/44 = 3.39. Accordingly, on a weight basis, Halon 1301 is only 6.1/3.39 = 1.8 times as effective as carbon dioxide for methane fires.

Table 1-8D shows that the inerting proportion of halon needed varies somewhat depending on the nature of the combustible, and substantially more halon is needed for hydrogen, carbon disulfide, or ethylene fires than for most other combustibles. Table 1-8D also shows that Halon 1301 and Halon 1211 have similar effectiveness on a volume basis for most combustibles. On the basis of molecular weights, a molecule of Halon 1211 is 165.5/149 = 1.11 times as heavy as a molecule of Halon 1301.

Table 1-8D is based on experiments in which a strong ignition source is applied to a uniform combustible-air-halon mixture, and the occurrence or nonoccurrence of a propagating flame is noted. A somewhat smaller quantity of halon would be needed to cause an existing flame on a burner to become unstable and then extinguish itself; the quantity of halon needed will depend on the details of the burner. A flame burning on a solid is even more easily extinguished. Grant[13] found that the flames on most solids could be extinguished with 4 to 6 percent by volume of Halon 1301 in the surrounding atmosphere.

TABLE 1-8D. *Minimum Required Volume Ratios of Halons to Air at 25°C that Will Prevent Burning of Various Vapors*

Vapor	Halon 1301		Halon 1211	
	(1301/Air)	(% O_2)	(1211/Air)	(% O_2)
Hydrogen	0.29	16.2	0.43	14.7
Carbon disulfide	0.15	18.2	—	—
Ethylene	0.13	18.5	0.114	18.8
Propane	0.073	19.5	0.065	19.7
n-Hexane	—	—	0.064	19.7
Ethyl ether	0.070	19.6	—	—
Acetone	0.059	19.8	0.054	19.9
Methane	0.054	19.9	0.062	19.7
Benzene	0.046	20.0	0.052	19.9
Ethanol	0.045	20.0	—	—

Calculated from tabulations by Kuchta[14].

Still, the discussion in the previous section about problems with extinguishment of deep-seated fires by carbon dioxide or nitrogen is equally valid when a halon agent is used. Unless a sufficient quantity of the halon in liquid form can reach the seat of the fire and cool all the solid sufficiently, reignition can occur after the agent has dissipated. If the halon reaches the combustible as a gas via compartment-flooding, then no such cooling can occur, and the effect of the halon is to extinguish the gaseous flame without affecting the pyrolysis or smoldering.

Table 1-8D also shows that the addition of Halon 1301 or 1211 needed in the air will only reduce the oxygen percentage from 21 percent to about 19 percent for most combustibles, while the required amount of carbon dioxide would have reduced the oxygen level to 14 percent or 15 percent. Furthermore, the physiological effects of carbon dioxide on humans at the concentrations needed for inerting are greater than those of Halon 1301.

In recent years, it has been found that halons and other chemicals work their way into Earth's upper atmosphere, where they appear to act as catalysts for the conversion of ozone, O_3, to normal oxygen, O_2. The ozone in the upper atmosphere plays a valuable role in filtering out the far-ultraviolet radiation of the sun, which would otherwise damage plant and animal life on Earth's surface. Destruction of the ozone layer also might affect the world's weather. Accordingly, there has been international activity directed toward eliminating the production of halons and/or the release of halons into the atmosphere.

The stratospheric ozone layer depletion issue is a problem confronting the global community unlike any other. Late in 1987, the United States and 24 other countries (including the European Economic Community) signed the Montreal Protocol to protect stratospheric ozone. Originally, the protocol restricted the consumption of ozone-depleting chlorofluorocarbons (CFCs) to 50 percent of the 1986 use levels by 1998, and halon production was to be frozen in 1993 at 1986 production levels. But the November 1992, Copenhagen revision to the Montreal Protocol accelerated this, such that all production of the chemicals ceased worldwide as of January 1, 1994.

The Montreal Protocol is based on unprecedented trade restrictions and is the first time nations of the world have joined forces to address an environmental threat in advance of fully established effects. The trade restrictions concern nations not participating in the agreement (the nonsignatories).

The halons used in fire protection make up only a small fraction of the total current halogenated hydrocarbon use, which includes refrigerants, blowing agents for foamed plastics, solvents, and propellants for aerosol products in cans, such as hair sprays or deodorants. For these nonfire uses, substitute fluids are available, and a changeover is occurring in various countries.

The regulation of Halon 1301 under the Montreal Protocol has generated tremendous research and development efforts across the world in a search for replacements and alternatives. Over the past several years, several total flooding, clean agent alternatives to Halon 1301 have been commercialized, and development continues on others. In addition to clean total flooding gaseous alternatives, new technologies, such as water mist and fine solid particulate, are being introduced.

It would, of course, be highly desirable if chemists discovered a substitute for Halon 1301 or Halon 1211 that provided both fire protection and breathability qualities and did not attack the ozone layer. In considering this possibility, a review follows of what is known about why the CF_3Br and CF_2ClBr molecules are so effective.

Figure 1-8F shows a methane-air flammability limit diagram as modified by various volumetric proportions of bromotrifluoromethane or carbon dioxide. The enormous difference is obvious.

It has been established that carbon dioxide acts by absorbing heat and reducing the flame temperature from about 1900°C for a stoichiometric fuel-air mixture to about 1200 to 1300°C. Below these temperatures, most flames can no longer burn.

If nitrogen were added to a stoichiometric fuel-air mixture instead of carbon dioxide, a somewhat larger volume of inert agent would be needed because the heat capacity of the nitrogen molecule is less than that of the carbon dioxide molecule. Similarly, if argon were added—argon has an even lower heat capacity per molecule—an even larger volume of argon would be needed for inerting. In each case, the flame would go out when the temperature dropped below 1200 to 1300°C.

However, if a small volume of bromotrifluoromethane were added to a flame, so that the temperature dropped to only about 1500°C, the flame would be extinguished. Clearly, the mechanism is different.

The important chemical reactions in flames involve the free atoms H and O and the free radical OH, which undergo chain reactions with the fuel and oxygen. In particular, the branching chain reaction $H + O_2 \rightarrow OH + O$ is very important. It is believed that the CF_3Br molecule decomposes in the flame to form HBr, and the HBr then acts to remove H atoms and OH radicals, by the following two combustion reactions:

$$HBr + H \rightarrow H_2 + Br$$

and

$$HBr + OH \rightarrow H_2O + Br$$

It happens that the molecules HF and HCl cannot react as rapidly with H or OH as can HBr, so bromine appears to be essential to the inerting molecule. It has been found that hydrogen iodide, HI, is about as effective as HBr, but iodine is more expensive and heavier than bromine, as well as quite toxic, so iodine offers no advantage over bromine for flame extinguishment.

In addition to the destruction of chain carriers, it has been speculated that a secondary contribution of halons to flame extinguishment comes from the extreme sootiness of halogen-containing flames. The more sooty or luminous the flame, the greater the radiative heat loss and the lower the temperature.

The role of the fluorine in halogenated agents is twofold. First, the fluorine atoms replace hydrogen atoms in the methane or ethane, thereby reducing the flammability of the inerting agent itself. Second, the toxicity of the agent is reduced. For example, CH_3Br is

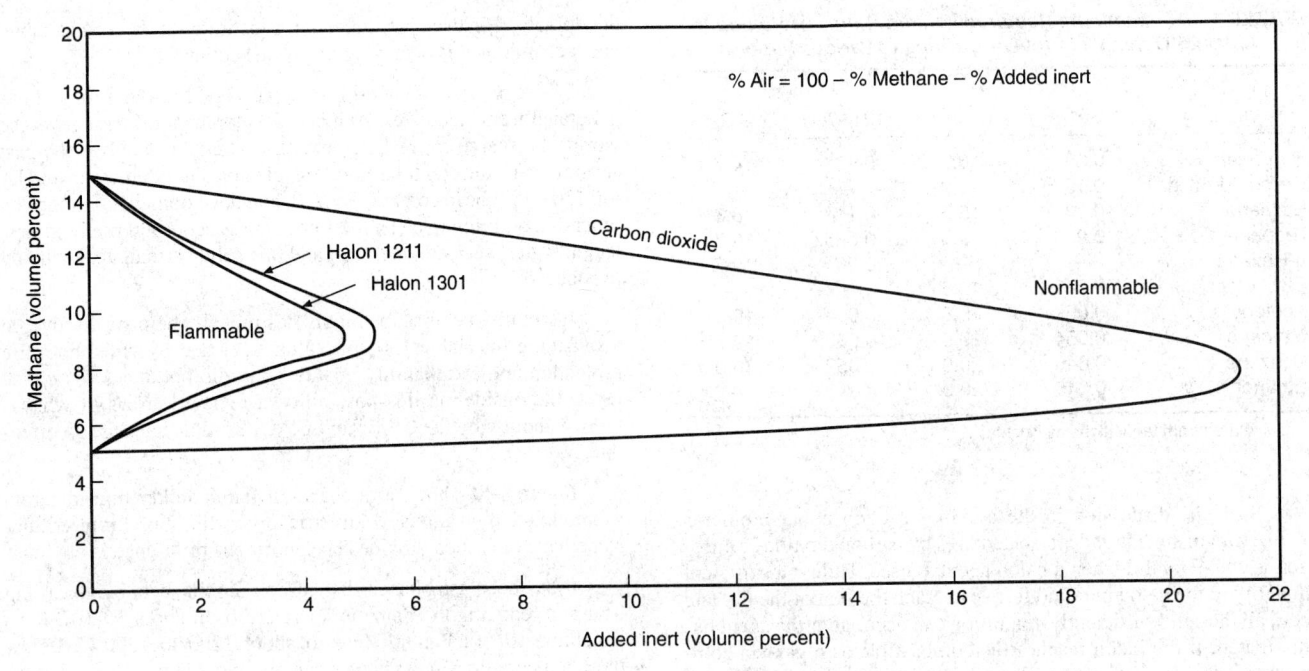

FIG. 1-8F. *Flammability limits for methane-air mixtures with added inerting agents.*[14]

much more toxic than CF_3Br, and, again, CH_2ClBr is much more toxic than CF_2ClBr.

This current degree of understanding about why CF_3Br is such a good inerting agent for flames does not provide clear guidance as to how other equally effective gaseous agents might be found. The vast majority of known molecules are liquids or solids, not gases, at room temperature and 1 atm. The known gaseous molecules have all been considered for inerting effectiveness, but no practical substitute as effective as CF_3Br has yet emerged.

EXTINGUISHMENT WITH DRY CHEMICAL AGENTS

Dry chemicals provide an alternative to carbon dioxide or the halons for extinguishing a fire without the use of water. These powders, which are 10 to 75 microns in size, are projected by an inert gas. Of the five types of dry chemicals in use, only one, monoammonium phosphate, is effective against deep-seated fires because of a glassy phosphoric acid coating that forms over the combustible surface. All forms of dry chemical act to suppress the flame of a fire.

One reason that dry chemical agents other than monoammonium phosphate are popular has to do with corrosion. Any chemical powder can produce some degree of corrosion or other damage, but monoammonium phosphate is acidic and corrodes more readily than other dry chemicals, which are neutral or mildly alkaline. Furthermore, corrosion by the other dry chemicals is stopped by a moderately dry atmosphere, while phosphoric acid has such a strong affinity for water that an exceedingly dry atmosphere would be needed to stop corrosion.

Application of any dry chemical agent on electrical fires is safe, from the viewpoint of electric shock, for fire fighters. However, these agents, especially monoammonium phosphate, can damage delicate electrical equipment.

For the special case of kitchen fires involving hot cooking oil, monoammonium phosphate is not recommended because it does not create a foam layer (saponification) on the surface of the oil. An alkaline dry chemical, such as sodium bicarbonate, is preferred.

Table 1-8E lists the chemical names, formulas, and popular or commercial names of the various dry chemical agents. In each case, the particles of powder are coated with an agent, such as zinc stearate or a silicone, to prevent caking and to promote free flowing. The effectiveness of any of these agents depends on the particle size. The smaller the particles, the less agent is needed, as long as particles are larger than a critical size.[15] The reason for this is believed to be that the agent must vaporize rapidly in the flame to be effective.[16] However, if extremely fine agent were used, it would be difficult to disperse and apply to the fire.

It is difficult to compare precisely the effectiveness of one dry chemical with another because a comparison to reveal chemical differences would require that each agent have identical particle size, which is difficult to achieve. Furthermore, gaseous agents can be compared by studying the flammability limits of uniform mixtures at rest. If particles were present, however, they would settle out unless the mixture were agitated, thus modifying the combustion behavior.

It seems clear that the effective powders act on a flame by some chemical mechanism, presumably forming volatile species that re-

TABLE 1-8E. *Dry Chemical Agents*

Chemical Name	Formula	Popular Name(s)
Sodium bicarbonate	$NaHCO_3$	Baking soda
Sodium chloride	$NaCl$	Common salt
Potassium bicarbonate	$NHCO_3$	"Purple K"
Potassium chloride	KCl	"Super K"
Potassium sulfide	K_2SO_4	"Karate Massiv"
Monoammonium phosphate	$(NH_4)H_2PO_4$	"ABC" or Multipurpose
Urea + potassium bicarbonate	$NH_2CONH_2 + KHCO_3$	"Monnex"

act with hydrogen atoms or hydroxyl radicals. However, the precise reactions have not been established firmly. Although the primary action is probably removal of active species, the powders also discourage combustion by absorbing heat; by blocking radiative energy transfer; and, in the case of monoammonium phosphate, by forming a surface coating.

DEEP-SEATED FIRES

As was previously mentioned, combustion may occur in a smoldering, rather than a flaming, mode. Extinguishment of such fires is usually quite difficult.

Application of water or foam to the surface of a smoldering fire is not always effective, because the water cannot penetrate the hot interior where the combustion is occurring. Surrounding the smoldering material with an inert gas or a halon gas will only be effective if such an atmosphere can be maintained for the extensive time required for the interior to cool.

If the smoldering object can be sealed for a long time to prevent access of oxygen, the combustion will eventually cease. However, the pyrolysis and combustion product gases being generated in the interior of the porous material will generate pressure, which will tend to break through any sealant applied.

One practical approach is to remove the smoldering object from the building and either let it burn outside or submerge it in water for a long time. If the smoldering is occurring in a large outdoor pile, extinguishment will generally require digging into the pile and extinguishing the hot material with water as it is exposed.

Much research has been done on the theory of the smoldering process, but this has not yet led to new practical techniques for extinguishment.

SPECIAL CASES OF EXTINGUISHMENT

Three-dimensional Gas Fires

Extinguishment of a fire involving a *continuously flowing* combustible gas often is very difficult. The best tactic is to shut off the flow of gas. If extinguishment is accomplished while the gas is still flowing inside a building, then the danger of filling the building with an explosive gas mixture is introduced. In some cases, it might be preferable to let the flame continue to burn if the flow of combustible gas cannot be stopped.

If it is not possible to shut off the gas supply, fire fighting by any of several techniques is possible. One approach is to attack the *base* of the flame with a dry chemical nozzle, carbon dioxide, steam, or a halon. Whichever agent is used, it should be projected in the same general direction as the burning jet or plume.

When this tactic is used, it is advisable to cool any hot metal in the vicinity and to remove or de-energize any other ignition sources *before* attacking the fire itself. Otherwise, re-ignition is likely to occur after temporary extinguishment, and the supply of agent could be depleted by that time.

Metal Fires

Water is usually the wrong agent for fires involving metals because a number of metals can react exothermically with water to form hydrogen, which, of course, burns rapidly. Furthermore, violent steam explosions can result if water enters molten metal.[17] As an exception, extinguishment has been accomplished when large quantities of water were applied to small quantities of burning magnesium, in the absence of pools of molten magnesium.

Table 1-8F is a list of extinguishing agents used for various metal fires. In general, metal fires are difficult to extinguish because of the very high temperatures involved and the correspondingly long cooling times required. Note that certain metals react exothermically with nitrogen, so the only acceptable inert gases for these metals are helium and argon. Halons should not be used on metal fires.

TABLE 1-8F.   *Extinguishing Agents for Metal Fires*

Agent	Main Ingredients	Used on
Powders		
"Pyrene" G–1 or "MetalGuard"	Graphitized coke + organic phosphate	Mg, Al, U, Na, K
"Met–L–X"	NaCl + Ca$_3$(PO$_4$)$_2$	Na
Foundry flux	Mixed chlorides + fluorides	Mg
"Lith–X"	Graphite + additives	Li, Mg, Zr, Na
"Pyromet"	(NH$_4$)$_2$H(PO$_4$) + NaCl	Na, Ca, Zr, Ti, Mg, Al
"T.E.C."	KCl + NaCl + BaCl$_2$	Mg, Na, K
Dry sand	SiO$_2$	Various
Sodium chloride	NaCl	Na, K
Soda ash	Na$_2$CO$_3$	Na, K
Lithium chloride	LiCl	Li
Zirconium silicate	ZrSiO$_4$	Li
Liquids		
"TMB"	Trimethoxyboroxine	Mg, Zr, Ti
Gases		
Boron trifluoride	BF$_3$	Mg
Boron trichloride	BCl$_3$	Mg
Helium	He	Any metal
Argon	Ar	Any metal
Nitrogen	N$_2$	Na, K

From Prokopovitsh.[17]

Chemical Fires

In addition to metals, certain inorganic chemicals are not compatible with water. For example, alkali and alkaline earth carbides, of which the best known is calcium carbide, react with water to form acetylene, which is highly flammable. Lithium hydride, sodium hydride, or lithium aluminum hydride react with water to produce hydrogen. The peroxides of sodium, potassium, barium, and strontium react exothermically with water. Cyanide salts react with acidified water to form a highly toxic gas, hydrogen cyanide. Even if these chemicals are not combustible themselves, they could be packed in combustible cartons and thus become involved in a fire. Or they could be stored in racks above combustible items.

Certain organic peroxides, used as polymerization catalysts in plastics manufacturing, are so unstable that they are stored under refrigeration to avoid exothermic heating. If water at normal room temperature were to be applied, it would provide heat to the peroxide and promote its exothermic decomposition.

A problem in applying water to fires involving toxic chemicals, such as pesticides, is associated with the runoff of contaminated water, which could cause groundwater pollution. If no other agent but water is available or practical, in such cases the only alternatives might be to use the minimum quantity of water possible or to allow a building to burn, thus producing downwind air pollution if the fire does not completely destroy the toxic chemicals.

BIBLIOGRAPHY

References Cited

1. Friedman, R., *Principles of Fire Protection Chemistry*, 2nd ed., National Fire Protection Association, Quincy, MA, 1989.
2. Drysdale, D., *An Introduction to Fire Dynamics*, Wiley, New York, 1985.
3. Glassman, I., *Combustion*, 2nd ed., Academic Press, New York, 1987.
4. Strehlow, R. A., *Combustion Fundamentals*, McGraw-Hill, New York, 1984.
5. Zabetakis, M. G., *Flammability Characteristics of Combustible Gases and Vapors*, Bulletin 627, U.S. Bureau of Mines, Washington, DC, 1965.
6. Heskestad, G., "The Role of Water in Suppression of Fire," *Fire and Flammability*, Vol. 11, 1980, pp. 254–259.
7. Rasbash, D. J., "The Extinction of Fire with Plain Water: A Review," *Fire Safety Science: Proceedings of the First International Symposium*, Hemisphere, New York, 1986, pp. 1145–1163.
8. Magee, R. S., and Reitz, R. D., "Extinguishment of Radiation-Augmented Plastic Fires by Water Sprays," Fifteenth Symposium (International) on Combustion, The Combustion Institute, Pittsburgh, PA, 1975, pp. 337–347.
9. Alpert, R. L., "Numerical Modeling of the Interaction Between Automatic Sprinklers Sprays and Fire Plumes," *Fire Safety Journal*, Vol. 9, Nos. 1–2, 1985, pp. 157–163.
10. Mawhinney, J. R., Dlugogorski, B. Z., and Kim, A. K., "A Closer Look at the Fire Extinguishing Properties of Water Mist," Fire Safety Science—*Proceedings of the Fourth International Symposium*, National Institute for Standards and Technology, Gaithersburg, MD, 1994, pp. 47–60.
11. International Conference on Water Mist Fire Suppression Systems, Swedish National Testing and Research Institute, Boras, Sweden, Nov. 1993.
12. Friedlander, S., *Smoke, Dust, and Haze*, Wiley, New York, 1977.
13. Grant, C., "Halon Design Calculations," *SFPE Handbook of Fire Protection Engineering*, 2nd ed., DiNenno, P.J., *et al.*, eds. National Fire Protection Association, Quincy, MA, 1995, Section 4, pp. 123–144.
14. Kuchta, J. M., *Investigation of Fire and Explosion Accidents in the Chemical, Mining, and Fuel-Related Industries—A Manual*, Bulletin 680, U.S. Bureau of Mines, Washington, DC, 1985.
15. Ewing, C. T., Faith, F. R., Hughes, J. T., and Carhart, H. W., "Flame Extinguishment Properties of Dry Chemicals," *Fire Technology*, Vol. 25, 1989, pp. 134–149.
16. Iya, K. S., Wollowitz, S., and Kaskan, W. E., "The Mechanism of Flame Inhibition by Sodium Salts," *Fifteenth Symposium (International) on Combustion*, The Combustion Institute, Pittsburgh, PA, 1975, pp. 329–336.
17. Prokopovitsh, A. S., "Combustible Metal Agents and Application Techniques," *Fire Protection Handbook*, 16th ed., National Fire Protection Association, Quincy, MA, 1986, pp. 49–54.

Additional Reading

Almgren, L. E., "Designing for Fire and Explosion Safety," *Proceedings* of the National Meeting of the American Institute of Chemical Engineers, American Institute of Chemical Engineers, New York, 1988, p. 10.
Application Guide for Explosion Suppression Systems, Fenwal Inc., Ashland, MA, Nov. 1979.
"Availability and Properties of Passive and Active Fire Protection Systems," Work Package No. FR4, Steel Construction Inst., UK, June 1991, 90 pages.
Bodurtha, F. T., *Industrial Explosion Prevention and Protection*, McGraw-Hill, New York, 1980.
Bruderer, R. E., "Design Example: Explosion Protection Selected for a Spray Drying Installation," *Plant/Operation Progress*, Vol. 8, No. 3, July 1988, pp. 141–146.

Caprano, M. A., and Strickland, J. H., "Preventing Fires and Explosions in Pilot Plants," *Plant/Operation Progress*, Vol. 8, No. 4, 1989, pp. 189–194.
diMarzo, M., *et al.*, "Dropwise Evaporative Cooling of a Low Thermal Conductivity Solid," *Proceedings* of the Third International Symposium for Fire Safety Science, Elsevier Applied Science, New York, 1991, pp. 987–996.
Dow's Fire and Explosion Index: Hazard Classification Guide, 6th ed., American Institute for Chemical Engineers, New York, 1987.
"Fire Prevention and Control Master Planning: Introductory Summary," National Fire Prevention Control Administration, Fire Safety and Research Office, Washington, DC, 1978.
Ewing, C. T., *et al.*, "Flame Extinguishment Properties of Dry Chemicals: Extinction Weights for Small Diffusion Pan Fires and Additional Evidence for Flame Extinguishment by Thermal Mechanisms," *Journal of Fire Protection Engineering*, Vol. 4, No. 2, 1992, pp. 35–52.
Gmurczyk, G. W., Grosshandler, W. L., and Lowe, D. L., "Suppression Effectiveness of Extinguishing Agents Under Highly Dynamic Conditions," *Proceedings* of the Fourth International Symposium for Fire Safety Science, Intl. Assoc. for Fire Safety Science, Boston, MA, 1994, pp. 925–936.
Groothuizen, T. M., and Pasman, H. J., "Explosions in Liquids and Solids," *Loss Prevention*, No. 9, 1975, pp. 91–96.
Hirst, R., Savage, N., and Booth, K., "Measurement of Inerting Concentrations," *Fire Safety Journal*, Vol. 4, No. 3, 1981/82, p. 147.
International Progress in Fire Safety, Fire Retardant Chemicals Association, Technomic, Lancaster, PA., 1987.
Klueg, E. P., "Liquid Nitrogen as a Powerplant Fire Extinguishant," *Fire Technology*, Vol. 5, No. 3, Aug. 1969, pp. 197–202.
Najario, F. N., "Preventing or Surviving Explosions," *Chemical Engineering*, Aug. 15, 1988.
Notarianni, K. A., "Water Mist Fire Suppression Systems," *Proceedings* for the Technical Symposium on Halon Alternatives, Society of Fire Protection Engineers and PLC Education Foundation., 1994, pp. 57–64.
Nyden, M. R., *et al.*, "Flame Inhibition Chemistry and the Search for Additional Fire Fighting Chemicals," National Institute of Standards and Technology, Gaithersburg, MD, NIST SP 861, April 1994.
Persson, B. and Dahlberg, M., "Simple Model of Foam Spreading on Liquid Surfaces," Swedish National Testing and Research Institute, Boras, Sweden, SP REPORT 1994:27, 1994, 28 pages.
Stull, D. R., *Fundamentals of Fire and Explosion, Monograph Series*, No. 10, Vol. 73, American Institute of Chemical Engineers, New York, 1977.
Swift, I., "Design of Deflagration Protection Systems," *Journal of Loss Prevention in the Process Industries*, Vol. 1, 1988, pp. 5–15.
Swift, I., "Developments in Explosion Protection," *Plant/Operation Progress*, Vol. 7, No. 3, July 1988, pp. 159–167.
Tucker, D. M., Drysdale, D. D., and Rasbash, D. J., "The Extinction of Diffusion Flames Burning in Various Oxygen Concentrations by Inert Gases and Bromotrifluoromethane," *Combustion and Flame*, 1981.
Tuhtar, D., *Fire and Explosion Protection: A Systems Approach*, Halsted Press, New York, 1989.
Tewarson, A., and Khan, M. M., "Role of Active and Passive Fire Protection Techniques in Fire Control, Suppression and Extinguishment," *Proceedings* of the Third International Symposium for Fire Safety Science, Elsevier Applied Science, New York, 1991, pp. 1007–1017.
Wagner, H. G., "Gas Explosions, Their Origin and Effects" (translated from *Chemie Ing. Tech.*, 1975, Vol. 47, No. 6, pp. 236–241), United Kingdom Atomic Energy Authority, Risley Translation 2861, Risley, England, 1975.
Weldon, G. E., "Damage-Limiting Construction," *Fire Technology*, Vol. 9, No. 4, Nov. 1973, pp. 263–270.
Wrenn, C., "Inerting for Fire Safety," *Plant/Operations Progress*, Vol. 5, No. 4, Oct. 1986, pp. 225–227.
Zalosh, R. G., "Explosion Protection," *SFPE Handbook of Fire Protection Engineering*, 2nd ed., DiNenno, P.J., *et al.*, eds. National Fire Protection Association, Quincy, MA, 1995, Section 3, pp. 312–329.

ENVIRONMENTAL ISSUES IN FIRE PROTECTION

Jane I. Lataille

Almost all human activity can harm the environment. Manufacturing products, storing and using chemicals, burning fuel, or discarding wastes can cause pollution. More often than not, efforts made to protect the environment from these activities have raised new fire protection concerns. Likewise, many fire protection measures have added to environmental problems. Therefore, whenever attempts are made to reduce pollution or to protect a structure or a process from fire, the interrelationship between fire protection and the environment must be considered.

This chapter presents an overview of the environmental issues related to fire protection. It is not an exhaustive review of environmental incidents that have affected fire protection engineering. For more specific information, see the bibliography at the end of this chapter.

The first part of this chapter—Pollution Control—describes fire protection concerns introduced by pollution control measures. The second part of this chapter—Fire Protection Systems—describes environmental problems that are related to fire protection systems and methods. The last part of this chapter explores the impact these interrelationships have on fire protection engineers.

POLLUTION CONTROL

Pollution may be classified as air, water, land, thermal, or noise pollution. Air, water, and land pollution generally imply pollution by substances. Thermal pollution is most often associated with water, and noise pollution with air. Efforts to control any type of pollution may raise fire protection concerns.

Air Pollution

Particles: Process-generated particles, such as dust, ash, or fibers, must be removed from the air to keep it clean. Such particles are removed with dry or wet dust collectors, or with electrostatic precipitators. Particle collection equipment can involve combustible dusts, heat- and friction-producing mechanical equipment, combustible filters, combustible collection bags, transformers, and other electrical equipment—all of which require protection.

General building ventilation systems use filters to remove particles from both fresh and recirculated air. If the filters burn, a ven-

tilation system can spread smoke throughout a building. Since ventilation systems are enclosed, fires in them are difficult to locate and extinguish. Therefore, ventilation system filters should always be noncombustible, with the lowest possible smoke-generation ratings. Ventilation systems also can be arranged to exhaust those portions of a building in which smoke has been detected and to pressurize adjacent areas simultaneously to prevent smoke leakage into them.

An extreme case of controlling air quality is seen in today's cleanrooms. The massive amounts of filters used in cleanroom air-handling systems make fire protection crucial. Although the filters may be noncombustible, the materials the filters collect can burn, causing damage to expensive cleanroom equipment and materials. Products of combustion from coated wiring and other plastics materials also can do extensive damage when circulated through the cleanroom air-handling system. Carefully designed air-handling systems with fast smoke detection and flexible air control are necessary parts of the fire protection design.

Vapors: One way to control air pollution is to collect vapors from vapor-producing processes. Collected vapors are then converted, incinerated, or recycled (recovered). Each of these three options for treating vapors needs protection.

Catalytic conversion can involve enclosed vessels, high temperatures, and fluidized beds. The hazards needing protection include heated combustible materials, fuel-fired equipment, and the presence of ignitable vapors in enclosed spaces. Process monitoring and control are also important to safe operation.

Incineration is done in flare stacks, in fuel-fired equipment designed primarily for incineration, or in fuel-fired equipment designed for some other purpose and modified to burn waste liquids or vapors. Incineration by flare often is used with emergency venting systems to burn off emergency discharges. It also is used when the non-emergency venting of vapors is not frequent enough to warrant their recovery. A flare stack is just a tall stack into which vapors are piped to be burned in a pilot flame at the top of the stack. The flame normally is monitored with a closed-circuit television camera or with thermal sensors. Stacks also must be monitored for safe air flows and oxygen levels so the stack gas mixture does not become explosive.

Besides the protective measures normally taken for fuel-fired equipment, incineration requires proper design of burners and combustion safeguard logic to account for the different physical and chemical properties of the materials to be burned. Sometimes, this

Jane I. Lataille, P.E., ARM, is a research consultant with Industrial Risk Insurers, Hartford, CT.

design includes the ability to switch from fuel to fuel or to burn different fuels simultaneously.

Solvent recovery systems reduce pollution, as well as recycle solvents. Recycling is now more common because laws require collection of vapors that were previously released to the atmosphere. Most solvent recovery systems have vapor collection ductwork, condensers, and liquid storage tanks. Activated carbon recovery systems absorb solvent that is then collected by steaming it out of the carbon. In some cases, collected solvent must be filtered, scrubbed, or distilled before it can be reused. Often, both "clean" and "dirty" solvent storage tanks are needed in a solvent recovery system. Each tank, piece of equipment, and process involving a solvent requires proper fire protection.

Plastic ductwork often is used to collect vapors that corrode metal. The interior of this ductwork is a combustible concealed space that requires sprinkler protection. In addition, burning plastics emit dense, acidic smoke, so fire response must include actions that minimize corrosion of exposed equipment.

Fluorocarbon gases: It is now believed that the ozone level in Earth's upper atmosphere is depleted when normally stable fluorocarbon molecules are broken down by ultraviolet radiation. Fluorine atoms thus freed from the fluorocarbon molecules then combine with single oxygen atoms from oxygen molecules that also have been broken down by ultraviolet radiation. These single oxygen atoms would normally combine with oxygen molecules to make ozone if the fluorine atoms were not present. Bromine and chlorine have the same effect.

As early as the 1970s, it was suspected that fluorocarbons had detrimental effects on the environment. International action to study these effects took place during the early 1980s and led to the Vienna Convention of 1985, when 27 countries signed an agreement to protect the ozone layer. Research continued, and in 1987, the Montreal Protocol was signed. This protocol froze the production of most fluorocarbons to 1986 levels by July 1989. It also called for further cuts in 1993 and 1998.

After the protocol was enacted, scientists determined that the ozone-depletion problem was even more serious than previously thought. So, the protocol was revised in June 1990 to mandate *no* fluorocarbon production by the year 2000. It was further revised in November 1992 to prohibit halon production as of January 1, 1994. (Halon is any bromine-containing fluorocarbon.) Therefore, the countries following the protocol no longer produce halons.

Fluorocarbons are used as refrigerants and propellants. To safeguard the environment, these fluorocarbons are being replaced by much more hazardous materials, such as hydrocarbons, alcohol, or ammonia. Use of the replacement substances creates new fire protection problems. Fluorocarbons are also used as extinguishing agents.

Other gases: Any gas that is not normally present in Earth's atmosphere may present a pollution problem if it gets into the atmosphere. The same is true of gases normally found in the atmosphere that are present in concentrations higher or lower than usual. Changes in the composition of the atmosphere affect its complex chemical balance and cause many other changes, the most well known of which is global warming.

Gases believed to contribute to global warming are called greenhouse gases. These gases warm Earth by reflecting back to Earth's surface sunlight that would otherwise have escaped the atmosphere. CO_2 is the greenhouse gas that has received the most publicity. The greatest source of excess CO_2 in the atmosphere is combustion of fossil fuel. The cleanest burning fossil-fuel-fired equipment, which may release no conventional pollutants, releases water and CO_2 as products of complete combustion. This puts more CO_2 into the atmosphere than is present naturally.

Fossil-fuel-burning equipment includes industrial boilers and furnaces; motor vehicles; and household appliances, such as furnaces, lawn mowers, leaf blowers, and snowblowers. It also includes the very large utility boilers used to generate the massive amounts of electricity used today. It is unlikely that there will be a reduction in the amount of excess CO_2 entering the atmosphere, but attempts can be made to control the increase.

Two ways of controlling greenhouse gas emissions that may have fire protection concerns are: (1) changing chemical processes, which may add an unknown hazard while removing a known one, or (2) running fired equipment intermittently that once ran continuously, which increases the number of startups and shutdowns.

Water Pollution

Building floors were once commonly sloped to holes called scuppers, to allow water and other liquids to flow out of the building. Scuppers are no longer allowed in buildings. Now, all liquids must be contained and treated before they are released. Even large fire department training grounds are now built with thick, underground clay liners to contain all water discharged from practice fire fighting. This keeps the earth's water from becoming (more) polluted. It also introduces an additional hazard if any pooled liquids are combustible or flammable.

Polluted water is now treated with various types of chemicals. These chemicals can neutralize the water, react with harmful substances to inert them, or cause contaminants to settle out of the water. Fire protection must be considered when storing and using treatment chemicals that are hazardous.

Sometimes, pollutants removed from the water are hazardous. A good example of this is oil spilled into the ocean by a tanker. Appropriate fire protection measures must be taken during collection and storage of such materials. Mixed pollutants may have to be separated further after removal because they may have different disposal requirements. Separation and other treatment equipment must be designed and arranged to handle the materials safely.

The need to clean up Earth's water affects more than the water being treated. Chemicals that were once allowed to be dumped into rivers and streams must now be collected and stored. Fire protection measures appropriate for storage and handling of hazardous materials must then be taken for these chemicals.

Land Pollution

Waste disposal is increasing due to the rising world population and industrial growth. Dumps and landfills are ravaging the land and contaminating groundwater that serves as the source of public water supplies. Recycling helps restore the land, but recycling systems often entail more fire protection concerns.

Recycling is an industry of its own, with many plants that do nothing else. These plants need basic fire protection and risk management programs, just as any other industry does. They must also deal with the hazards unique to their recycling processes.

Plastics recycling involves grinding, melting, and extruding processes, as well as warehousing. Due to the burning characteristics of plastics, fire protection systems for their storage and handling must be designed carefully. Process equipment must be designed with due regard for the capacity of plastics to hold static charges.

A fast-emerging recycling industry is the so-called "trash to cash" business. In this industry, municipal waste is collected, separated, shredded, and burned to generate electric power. The hazards inherent in the generation of electrical power must be protected, as well as those unique to the storage of large amounts of combustible waste inside buildings and to the processing and burning of this waste. Unique protection problems include the large distance between the combustible waste and roof sprinklers, due to the need for large cranes or other materials-handling systems, and the special design of systems that safely burn waste with varying chemical content.

Other materials now commonly recycled are glass, metal, paper, and wood. In each case, protection must be provided for the buildings, materials, and processes associated with recycling of materials that used to be buried in the land and forgotten.

Another strategy to eliminate sources of land pollution is to replace underground liquids storage tanks, especially those for gasoline and oil, with aboveground tanks. This strategy reverses the 1960s policy of burying tanks to reduce their fire protection problems, except that now, there are even more problems.

To contain leaks from aboveground tanks, they are provided with dikes sized to hold the contents of the largest tank within the dike. To keep spills from seeping into the ground, the surfaces of diked areas must be nonporous, or, if there is a spill, the ground must be dug up and tested. Contaminated ground must then be removed and treated as hazardous waste.

Rainwater that collects in the dike must be tested before it can be released. If there are any contaminants, the water must be treated as hazardous waste. While this treatment greatly reduces the chance of pollution, it increases the amount of fire protection needed.

If underground tanks are necessary, they must be designed to minimize the potential for leakage. Design features include proper preparation of the tank bed, anchoring the tank, careful backfilling, using double walls on the tank, leak detection, groundwater monitoring, and liquid level alarms. As an option to double walls, fiberglass tanks or tanks of other nonmetallic materials may be used. These tanks require especially careful installation.

Thermal and Noise Pollution

Concern for the environment rests mainly with the addition of physical pollutants to the air, water, and land. However, the need to control heat and sound also has had an effect on the fire protection engineer.

Even pure water should not be dumped into a stream if it will raise the temperature of the stream enough to affect its ecology. Elevated water temperatures can cause excessive algae growth and can be harmful to fish and other water life.

Cooling towers reprocess cooling water to help alleviate thermal pollution of waterways and to help preserve Earth's water. Because cooling towers are often of combustible construction or have combustible fill, they need fire protection systems.

Heat preservation is good for global ecology. It is also economical because fuel and fuel-burning equipment are expensive to buy and maintain. Insulating is a common way to preserve heat, but it introduces fire protection concerns, in that many of the materials with good insulating properties are combustible and can emit toxic and acidic products of combustion when burning. The problem increases when such insulation is installed inside inaccessible spaces, such as closed-off attics and crawl spaces.

Noise control requires careful design and placement of process equipment, as well as the use of soundproofing. Soundproofing materials introduce the same concerns that insulating materials do. They may be combustible, they may release toxic and acidic products of combustion when burning, and they may be installed in inaccessible places, such as inside air-handling ducts or process equipment.

FIRE PROTECTION SYSTEMS

All fire protection system extinguishing agents can affect the environment. Of primary concern are halons. Like refrigerants and propellants, fire protection halons are fluorocarbons. The relationship between the fire protection halons and depletion of the ozone layer has drawn much attention. The most damaging halon is considered to be Halon 1301, which is the most common fluorocarbon used for fire protection. While much attention is focused on halons, however, fire protection engineers must be aware of other sources of potential harm to the environment.

Halons and the Ozone Layer

Research has found that Halon 1301 has the highest ozone-depletion potential of all the fluorocarbons, at least ten times higher than the four most common fluorocarbon refrigerants. As a result, the fire protection community has had to move quickly to work out ways of acceptably controlling loss potential while minimizing the use of this previously well-accepted extinguishing agent.

Full-flooding halon acceptance tests are now often replaced with pressure and puff tests to prove the integrity of the piping, and with door fan tests to prove room tightness. Pressure tests of piping can find leaks, which are usually caused by incorrectly fastened fittings. Puff tests confirm that there are no serious piping blockages. If properly done, a door fan test gives an accurate measure of expected extinguishing agent leakage. Procedures and calculations for door fan testing are now well developed.

These alternative methods eliminate the dumping of halon into the environment from extinguishing systems testing. Halon release can be reduced further by carefully designing detection systems to suit the occupancies they protect, and by periodically inspecting, testing, and maintaining halon systems so that inadvertent discharge does not occur. Recovery of halon discharged from cylinders during hydrostatic testing is another way to reduce the amount of halon released into the environment. (See Figure 1-9A.)

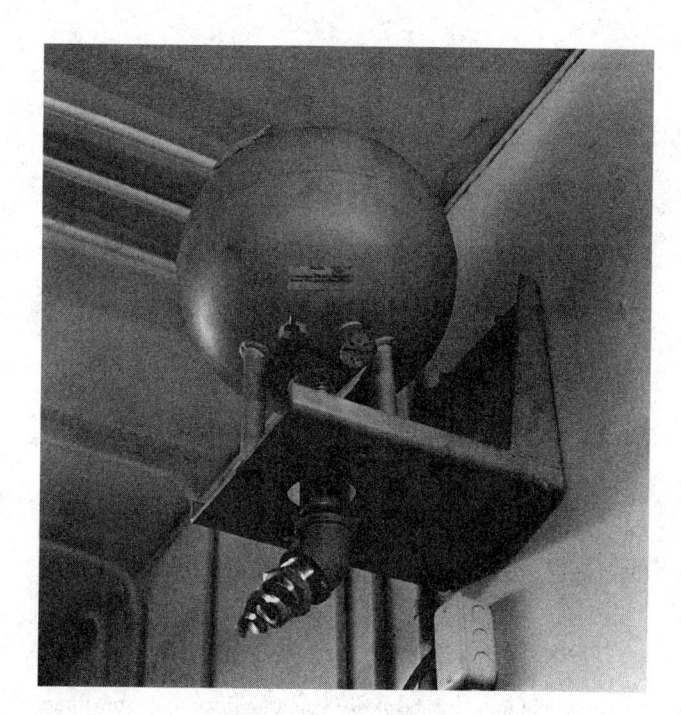

FIG. 1-9A. A Halon 1301 storage container. (Source: IRI and Fenwal)

According to current Environmental Protection Agency (EPA) regulations, existing Halon 1301 systems may be kept in service, existing Halon 1301 reserves may be maintained, and Halon 1301 reserves may even be used in new systems when handled by qualified contractors. However, new Halon 1301 systems should be discouraged. One of many alternative protection methods should be appropriate for the hazard being protected. Two methods to immediately consider are: (1) conventional sprinkler and (2) CO_2 protection.

Alternative gaseous agents to Halon 1301, sometimes called clean agents, have also been developed. NFPA 2001, *Standard on Clean Agent Fire Extinguishing Systems*, addresses the use of these agents in extinguishing systems. The agents shown in Table 1-9A are now acceptable to the EPA for use in total flooding systems in occupied areas.

TABLE 1-9A. *Clean Agent Alternatives to Halon 1301*

Agent	Chemical Name	Trade Name	Primary Manufacturer
FC-218	Octafluoro-propane	FE-38	3M
HFC-23	Trifluoro-methane	FE-13	E. I. duPont
HFC-227ea	Heptafluoro-propane	FM-200	Great Lakes Chemical Company
IG-541	N, Ar, and CO_2	Inergen	Ansul
FC-3-1-10	Perfluoro-butane	PFC-410	3M
Blend A	Blend of 3 HCFCs and a scavenger	NAF S-III	North American Fire Guardian
HFC-125	Pentafluoro-ethane	FE-25	E. I. duPont
*HBFC-22B1	Bromodifluoro-methane	FM-100	Great Lakes Chemical Company

* Production ceased.

Due to their bromine content, which contributes to ozone depletion, Blend A and HBFC-22B1 will have to be phased out by 1997. EPA accepts additional alternate agents for use in local application systems or in portable extinguishers. These are called streaming agents.

Another possible alternative to Halon 1301 systems is specially designed water mist systems. These are water-based systems that produce very fine mist and use much less water than conventional water-based systems. These systems have long been used aboard ships, but have only recently been developed for industrial use. At this time, only a few pre-engineered-type water mist systems have been listed for specific applications, such as enclosed machine rooms and turbine enclosures. The technology for these systems is still in its infancy. NFPA has just published NFPA 750, *Standard on Water Mist Fire Protection Systems*.

Other Special Extinguishing Agents

Substances that get into the ground can contaminate groundwater. These substances include agents used in special extinguishing systems, such as dry and wet chemicals, fire protection foams, and the antifreezes sometimes added to wet-pipe sprinkler systems. It is becoming more common to enclose or dike areas protected with these agents so that they may be cleaned up after discharge. Sometimes, areas protected with these agents are drained to a suitably sized holding tank for later disposal. (See Figure 1-9B.)

FIG. 1-9B. Foam storage tank. (Source: IRI and Rockwood)

One of the early fire protection agents was carbon tetrachloride. This liquid, a precursor of the halogens, was known to be an effective extinguishing agent. It was either put into pressurized containers with sprinkler-like operating mechanisms at the bottom, or it was sealed into glass bulbs for throwing at a fire manually. (See Figure 1-9C.) These containers or bulbs were once commonly placed in unsprinklered dead records storage rooms and in other small rooms. Carbon tetrachloride is no longer used because of its high toxicity. It has been replaced with alternate extinguishing agents, all of which must now be reviewed for their effects on the environment.

Carbon dioxide is a nontoxic, relatively inert gas that is found naturally in the atmosphere. It extinguishes fire by displacing the

FIG. 1-9C. Old-style carbon tetrachloride extinguisher. (Source: IRI)

oxygen required for combustion and, therefore, poses a suffocation hazard. Careful system design with appropriate safety interlocks is used to mitigate this danger to people. Otherwise, carbon dioxide has been an ideal extinguishing agent for protecting many hazards. However, the warming of the Earth through the greenhouse effect is thought to be aggravated by excessive carbon dioxide. So now, even the use of this agent must be considered carefully.

Fire Protection Water

With most extinguishing agents posing some sort of environmental threat, an obvious solution is to use water-based extinguishing systems, such as sprinklers, water spray, deluge, and so on, wherever water is a suitable extinguishing agent. However, even this presents environmental concerns. Indeed, it is likely that discharge from water-based systems will be one of the biggest fire protection concerns of the coming century.

When a sprinkler system operates, the water flows over whatever is burning and over anything else that is in the area of the sprinkler discharge. The water then collects and flows even farther, carrying potential contaminants, which seep into the ground unless the protected area has been designed to contain the sprinkler discharge.

There have been many incidents in which sprinkler discharge has carried hazardous materials and posed a contamination hazard to drinking water. In a May 1987 paint plant fire in Dayton, OH, sprinkler discharge spread flammable liquids near potable water sources.

There have also been incidents in which fire protection discharge contributed to air pollution. In these cases, chlorine gas was released from stored swimming pool chemicals when they were wetted by sprinklers. Two well-known incidents of this type occurred in June 1988 in Springfield, MA, and in August 1988 in Glendale, AZ. The effects of fire protection discharge must therefore be considered in the design of a protected facility and its site.

Even if a facility existed that had no possible contaminants, the sprinkler system fed by a public water supply could still pose an environmental threat—that is, contamination of the public water supply through backflow. Pollutants can pass from the sprinkler system to the water supply through the normal backflow permitted by check valves. The pollutants come from the water pumped into the sprinkler system through the fire department connection. This water could be contaminated if a fire department used a nonpotable auxiliary source, such as a pond, or if the water in the fire department tankers was treated with anticorrosion chemicals.

To keep sprinkler system water from contaminating public water supplies, many states and municipalities have passed laws pertaining to the arrangement of sprinkler systems. Some do not permit connections to the public supply if the fire protection system includes a suction or gravity tank. Others do not permit booster pumps. Antifreeze systems may not be allowed. In some cases, systems with tanks or pumps may be connected to the public water supply if double check valves are used.

A requirement that is becoming more common for sprinkler systems is the installation of a reduced-pressure-principle backflow preventer wherever the potential for contamination of the public water supply exists. (See Figure 1-9D.) Interpretation of where a potential for contamination exists is often left to a state or municipality. Examples of the potential for contamination are the presence of a fire department connection on a sprinkler system, the proximity of a nonpotable suction source, or the addition of anticorrosion chemicals to fire department tanker water.

The reduced-pressure-principle backflow preventer uses two check valves, two taps for pressure sensing, and a diaphragm-oper-

FIG. 1-9D. *Reduced-pressure-principle backflow preventer. (Source: IRI and Cla Val)*

ated dump valve. The dump valve releases pressure in the backflow preventer's zone of reduced pressure, which prevents backflow whenever the first check valve is closing. Although reduced-pressure-principle backflow preventers prevent backflow better than double check valves, they have many disadvantages, including high cost and excessive friction loss. Backflow preventers also can rob water from the sprinkler system, if the dump valve fails to close when it should. (See Figure 1-9E.)

Manual fire-fighting efforts introduce the same environment-contaminating potential that automatic fire protection systems do. Any water used to extinguish a fire can carry hazardous materials that had been in the area of the fire. These materials can seep into the groundwater. Therefore, even unprotected facilities need careful site layout.

Fire Protection Measures

Many measures have been taken for protection from fire beyond installing automatic extinguishing systems. Not surprisingly, these, too, are a source of environmental concern.

PCBs (polychlorinated biphenyl) were used to replace most transformer dielectric fluids due to their low fire and explosion hazard compared with the formerly used combustible oils. Unfortunately, it was later learned that the stability that made PCBs safe from a fire protection standpoint also caused serious pollution problems—that is, PCBs are poisons that do not break down. PCB and PCB-contaminated transformers now must be retrofitted or replaced. PCBs also are found in capacitors, large circuit breakers (usually over 1,000 V), and isolating switches.

Structural steel members of buildings or process structures often are fireproofed to withstand the heat of possible fires. The materials used in fireproofing formulations can be hazardous to the environment. Asbestos is a primary example of such a material. Application of spray-on fireproofing can result in overspray. After application, any fireproofing can crumble when fire protection hose streams are applied. Therefore, fireproofing materials must be reviewed for their contamination potential.

Fire-retardant chemicals are used to reduce the flamespread of plastics, wood, and other materials. When a treated product is used outside, the fire-retardant chemicals can leach due to harsh weather and get into the soil. Both inside and out, these chemicals escape into the air. Like all chemicals, the fire retardants must be reviewed

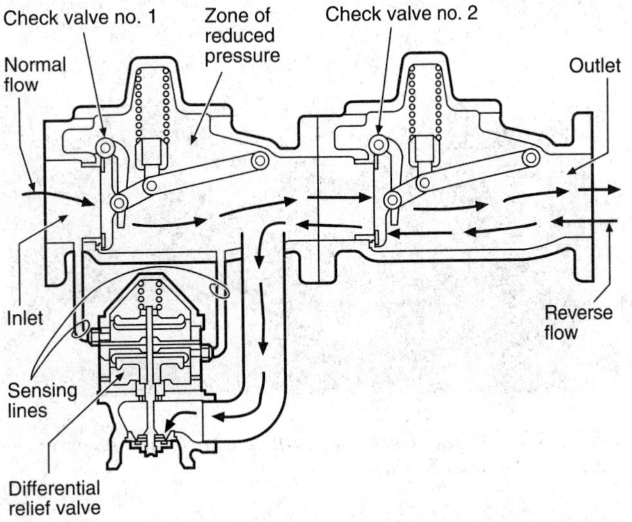

FIG. 1-9E. Reduced-pressure-principle backflow preventer system diagram. (Source: Cla Val)

for their potential to contaminate the environment, both before their use and after they have become part of the finished product.

To keep the bottom of a steel suction tank from rusting, a specified thickness of the sand under the tank is saturated with No. 2 fuel oil. This practice, required by NFPA 22, *Standard for Water Tanks for Private Fire Protection*, must now be reviewed and environmentally acceptable alternatives must be developed.

CONCLUSIONS

Recently, the PCB, landfill, ozone, and oil spill problems have focused more attention on the environment. So, too, have incidents of groundwater contamination by fire protection discharge. Many new laws have been passed to protect and preserve the environment, and there are more to come. Every fire protection engineer must be aware of these laws and must apply them properly when designing and installing fire protection systems.

Among the fire protection engineer's present and future concerns are the following:

1. Backflow preventers may have to be installed. Sprinkler system hydraulics must then take backflow preventer friction loss into account. Testing and maintenance of backflow preventers will have to be added to fire protection equipment inspection programs. Where the preventers are not used, careful analysis of the potential for contamination will have to be made.

2. The use of halon systems must be reviewed very carefully. All other options for extinguishing agents and types of protective systems must be considered. Emergency operating procedures and disaster recovery plans must be developed and tested. A thorough business interruption analysis is necessary to develop emergency plans and to specify the appropriate protection.

3. Underground tanks must be designed and installed carefully. Proper tank bed preparation, tank anchoring, and backfilling are critical. Double-walled tanks with leak detection between the walls often are required. The quality of groundwater around the tank must be monitored. Aboveground tanks will be more common, and fire protection engineers will have to decide where to install such tanks on crowded sites and how to protect the tanks and their exposures.

4. For every fire protection system installed, consideration must be given to how the discharge from that fire protection system will be contained, handled, and disposed of. If fire protection discharge is found to have contaminants, the discharge must be handled as hazardous waste.

5. Any places on a site where rainwater collects, such as sumps, catch tanks, and dikes, will have to be monitored for the presence of contaminants, and, if any are found, this water must be treated as hazardous waste.

6. Increased storage of hazardous chemicals will be more common. Every environmentally hazardous substance used in manufacturing facilities will have to be recovered to comply with the new laws. Fire protection engineers will have to decide where to place that storage and how to protect it.

Fire protection engineers have begun to plan fire protection systems with concern for the environment, but there is still much to be done. The fire protection discipline must be coordinated with the civil, structural, mechanical, and electrical disciplines to ensure a total facility design that preserves the environment while allowing the most economical and effective fire protection possible. High cost and much effort will be involved, but this is necessary because an environment safe from fire and explosion is of little value if one cannot live in it in good health.

BIBLIOGRAPHY

References

"Controlling Fire Protection Halon Emissions," *Fire Technology*, Feb. 1988, p. 70.

"Final Report of the Halon-Technical Options Committee of the United Nations Environmental Programme," *Montreal Protocol Assessment Technology Review*, Aug 11, 1989. (Available from ICF Inc., 9300 Lee Highway, Room 1171, Fairfax, VA, 22031-1207.)

Halon Alternatives Research Corporation, "EPA Releases List of Acceptable Alternatives," *HARC NEWS*, Vol. 3, No. 1, Apr. 1993, pp. 1–2.

Hazardous Waste: A Guide to RCRA Requirements, National Fire Protection Association, Quincy, MA, 1986.

Industrial Risk Insurers, "Clean Agent Halon Replacements," *IRInformation*, IM.13.6.0, Dec. 1995.

Industrial Risk Insurers, "Door Fan Testing for Enclosure Integrity," *IRInformation*, IM.13.0.5.2, Sept. 1992.

Isman, K. E., P.E., "New Standard for Backflow Prevention," *Sprinkler Quarterly*, Winter 1990, p. 64.

Mitchell, Jr., J. M., "Carbon Dioxide and Future Climate," *Weatherwise*, Aug./Sept. 1991, pp. 17–23.

"Significant New Alternatives Policy," *The Federal Register*, May 12, 1993.

Su, J. Z., Kim, A. K., and Mawhinney, J. R., "Review of Total Flooding Gaseous Agents as Halon 1301 Substitutes," *Journal of Fire Protection Engineering*, Vol. 8, No. 2, 1996, pp. 45–64.

Additional Reading

General

Dungan, K. W., "Fire Protection and the Environment," *The Locomotive*, Fall 1991, Vol. 67, No. 7, pp. 163–167.

HSB Professional Loss Control, "Fire Protection and Chemical Recycling," *Fire and Current Technologies*, Vol. 2, No. 1, Spring 1993, pp. 1–2.

Hall, Jr., J. R., "Fire Protection and the Future," *NFPA Journal*, May/June 1991, pp. 38–53.

Holmes, W. H., "Converting Waste to Energy," *Plant Engineering*, Mar. 21, 1991, pp. 120–122.

Teague, P. E., "Fire Protection and the Environment," *NFPA Journal*, Jan./Feb. 1991, pp. 34–43.

Air quality

ASME Research Committee on Industrial and Municipal Waste, "Rotary Kiln Incinerators: The Right Regime," *Mechanical Engineering*, Sept. 1989, pp. 78–81.

"Controlling Particulate Emissions from Utility and Industrial Boilers," *Power Special Report*, June 1980, Vol. 124, No. 6, pp. S-1–S-20.

Corcoran, E., "Cleaning up Coal," *Scientific American*, May 1991, pp. 106–116.

Elliott, T. C., "Coal Handling and Preparation," *Power*, Jan. 1992, pp. 17–32.

Gordon, V. R., and Holness, P. E., "Human Comfort and Indoor Air Quality," *Heating/Piping/Air Conditioning*, Feb. 1990, pp. 43–52.

Mawhinney, J. R., P.E., "Tire Fire Pollutes the Environment," *NFPA Journal*, Jan./Feb. 1991, pp. 50–58.

McKee, B. A., "Yearning to Breathe Free," *Nation's Business*, Feb. 1990, pp. 46–47.

Backflow prevention

AWWA, "Recommended Practice for Backflow Prevention and Cross-Connection Control," AWWA M14, American Water Works Association, Denver, CO, 1980.

Backflow Protection for Fire Sprinkler Systems, National Fire Sprinkler Association, Inc., Patterson, NY, 1988.

Ballanco, J. E., P.E., "Automatic Fire Sprinkler Backflow Protection," *Heating/Piping/Air Conditioning*, Mar. 1992, pp. 87–88.

"Protecting Drinking Water Systems from Backflow," *Consulting-Specifying Engineer*, Nov. 1989, p. 64.

"Recommended Practice for Backflow Prevention and Cross-Connection Control," AWWA M14, American Water Works Association, Denver, CO, 1980.

Fire retardants

"Fire-Resistant Plywood—Hot Topic for Builders," *Engineering News Record*, Nov. 30, 1989, p. 17.

Fluorocarbons and their substitutes

"An Inside Look at How CFCs Affect Your Life," *ASHRAE Journal*, Nov. 1989, p. 42.

Anderson, S. O., "Halons and the Stratospheric Ozone Issue," *Fire Journal*, Vol. 81, No. 3, May/June 1987, p. 56.

Blatt, M. H., "Electric Chillers: Cost-Effective Choice for the Future," *Heating/Piping/Air Conditioning*, Mar. 1993, pp. 75–82.

Chines, S. A., "Halon, Ozone, and the Environment," *The Sentinel*, Third Quarter 1989, p. 3.

Cohn, B. M., "Laboratory Halon Systems: Rehab, Replace or Remove?," *Journal of Applied Fire Science*, Vol. 5, No. 1, 1995/1996, pp. 45–54.

DiNenno, P. J., Hanauska, C. P., and Forssell, E. W., "Design and Engineering Aspects of Halon Replacements," *Process Safety Progress*, Jan. 1995, pp. 57–62.

Gallup, J. G., P.E., "Ozone Fears Limit Halon Use," *Plant Engineering*, Sept. 6, 1990, pp. 92–93.

Gilkey, H. T., P.E., "The Coming Refrigerant Shortage," *Heating/Piping/Air Conditioning*, Apr. 1991, pp. 41–63.

Grant, C. C., "Controlling Fire Protection Halon Emissions," *Fire Technology*, Feb. 1988, pp. 70–78.

Grant, C. C., "Fire Protection Halons and the Environment," *Fire Technology*, Feb. 1988, pp. 63–64.

Grant, C. C., "Halons and the Ozone Layer: An Overview," *Fire Journal*, Vol. 83, No. 5, Sept./Oct. 1989, p. 59.

Grant, C. C., "Life Beyond Halon," *Fire Journal*, Vol., No. 3, May/June 1990, pp. 52–59.

Halon Alternatives Research Corporation, "President Announces Accelerated Phaseout of Ozone-Depleting Substances," *HARC NEWS*, Vol. 2, No. 1, Apr. 1992, p. 1.

Harrington, J. L., "The Halon Phaseout Speeds Up," *NFPA Journal*, Mar./Apr. 1993, pp. 38–42.

Industrial Risk Insurers, "Clean Agent Systems—NFPA 2001-1994," *IRInformation*, IM.13.6.1, Dec. 1995.

Makansi, J., "Ammonia: It's Coming to a Plant Near You," *Power*, May 1992, pp. 16–22.

Mauzerall, D. L., "Protecting the Ozone Layer: Phasing out Halon by 2000," *Fire Journal*, Vol. 84, No. 5, Sept./Oct. 1990, pp. 22–31.

McCon, P., "Replacements for Halogenated Fire Extinguishing Agents in Fixed Systems—Where We Stand," *Professional Safety*, Apr. 1996, pp. 22–26.

Moore, T., "Refrigerants for an Ozone-Safe World," *EPRI Journal*, Jul./Aug. 1992.

Nadel, B., "Ozone-Friendly Cooling," *Popular Science*, Jul. 1990, pp. 57–59.

Stamm, R. H., "The CFC Problem: Bigger than You Think," *Heating/Piping/Air Conditioning*, Apr. 1989, p. 51.

Stoecker, W. F., "Opportunities for Ammonia Refrigeration," *Heating/Piping/Air Conditioning*, Sept. 1989, pp. 93–108.

Stouppe, D. E., P.E., "CFCs, The Law and You!" *The Locomotive*, Fall 1992, Vol. 68, No. 3, pp. 51–56.

Stouppe, D. E., P.E., "Taxes Help the Ozone Layer," *The Locomotive*, Vol. 67, No. 1, Spring 1990, pp. 11–13.

Willey, A. E., "Conference on Halons and the Environment Held in Switzerland," *Fire Journal*, Vol. 82, No. 6, Nov./Dec. 1988, p. 47.

Wojdon, W., and George, M., "How to Replace CFC Refrigerants," *Hydrocarbon Processing*, Aug. 1994, pp. 107–112.

Ziemba, J., and Waters, S., "Halon: The Search for Alternatives," *Firewatch!*, Vol. 30, No. 3-4, 1993, pp. 6–7.

Ziffer, F. E., "Managing Refrigerants in a CFC-Free Era," *Plant Engineering*, Sept. 3, 1992, pp. 53–56.

Zimmer, C., "Unintended Consequences," *Discover*, Mar. 1995, pp. 32–33.

Zurer, P., "Fate of CFC Alternatives Remains Up in the Air," *C&EN*, Jul. 16, 1990, pp. 5–6.

PCBs

McPartland, B. J., "PCBs . . . Time Is Running Out," *Electrical Construction and Maintenance*, Sept. 1988, p. 62.

"PCBs and PCB-Containing Equipment—Disposal and Treatment Options," *Environmental Contractor Magazine*, Nov. 1989, p. 73.

Spills and fire protection discharge

Andrews, Jr., R. C., P.E., "The Environmental Impact of Firefighting Foam," *Industrial Fire Safety*, Nov./Dec. 1992, pp. 26–31.

Constantine, P., "Lining up Against Oil," *Civil Engineering*, Jan. 1990, p. 70.

"Foam-Water Sprinkler Protection for Flammable Liquids," *Plant Operations Progress*, Oct. 1989, p. 218.

"Guidelines for Safe Handling and Storage of Calcium Hypochlorite and Chlorinated Isocyanurate Pool Chemicals," Monsanto-Olin-PPG (St. Louis, MO; Cheshire, CT; Pittsburgh, PA), 1989.

Industrial Risk Insurers, "Dry Chemical Extinguishing Agents—Corrosive and Contaminating Effects," *IRInformation*, IM.13.1.3, Feb. 1990.

Isner, M. S., "$49 Million Loss in Sherwin-Williams Warehouse Fire," *Fire Journal*, Vol. 83, No. 2, Mar./Apr. 1989, p. 65.

Isner, M. S., *Flammable Liquids Warehouse Fire, Dayton, Ohio, May 27, 1987*, Investigation Report, 1987, National Fire Protection Association, Quincy, MA.

PVI Industries, "HVAC Technology Speeds Alaskan Oil Cleanup," *Engineered Systems*, Jan./Feb. 1990, pp. 26–32.

Underground tanks

Donovan, B. C., and O'Connor, P., "Underground Storage Tank Update," *Plant Engineering*, June 22, 1989, p. 60.

EPA—LUST, *Locating Underground Storage Tanks*, U.S. Environmental Protection Agency, Office of Underground Storage Tanks, Washington, DC, 1985. (Video)

Gellop, J. G., "Ozone Fears Limit Halon Use," *Plant Engineering*, Sept. 6, 1990, pp. 92–93.

"Government Regs Force Loss Prevention Changes," *Sentinel*, Vol. 47, No. 4, pp. 3–7.

Gross, R. A., "Preventing Contamination of the Earth," *The Sentinel*, Third Quarter 1989, p. 8.

Hazardous Waste: From Cradle to Grave (Video, 40 min), distributed by National Fire Protection Association, Quincy, MA, 1986.

Hazardous Waste: Small Quantities, Big Challenges (Video, 27 min), distributed by National Fire Protection Association, Quincy, MA, 1985.

Henry, M. F., "Update on the Underground Leakage Problem," *Fire Journal*, Vol. 80, No. 1, Jan. 1986, pp. 26–27 and 86.
Murarka, I. P., "MOSES: Mineral Oil Spill Evaluation System," *EPRI Journal*, Jul./Aug. 1991, pp. 46–48.
Silveria, V., "Basic UST Leak Detection Systems," *Plant Engineering*, Aug. 13, 1992, pp. 74–76.
Smalley, J. C., "Underground Storage Tanks: Rest in Peace," *Fire Command*, Mar. 1986, pp. 40–42.
Underground Storage Tanks: A Guide to Installation, National Fire Protection Association, Quincy, MA, 1988.
Underground Storage Tanks: A Question of When (Video, 37 min), distributed by National Fire Protection Association, Quincy, MA, 1988.
Underground Storage Tanks: Close to Home (Video, 9 min), distributed by National Fire Protection Association, Quincy, MA, 1988.
Underground Storage Tanks: In Your Own Backyard (Video, 26 min), distributed by National Fire Protection Association, Quincy, MA, 1988.
Underground Storage Tanks: Rest in Peace (Video, 30 min), distributed by National Fire Protection Association, Quincy, MA, 1986.

NFPA Codes, Standards, and Recommended Practices

Reference to the following NFPA codes, standards, and recommended practices will provide further information on environmental issues in fire protection discussed in this chapter. (See the latest version of *The NFPA Catalog* for availability of current editions of the following documents.)

NFPA 11, *Standard for Low-Expansion Foam*
NFPA 11A, *Standard for Medium- and High-Expansion Foam Systems*
NFPA 12, *Standard on Carbon Dioxide Extinguishing Systems*
NFPA 12A, *Standard on Halon 1301 Fire Extinguishing Systems*
NFPA 13, *Standard for the Installation of Sprinkler Systems*
NFPA 15, *Standard for Water Spray Fixed Systems for Fire Protection*
NFPA 16, *Standard for the Installation of Deluge Foam-Water Sprinkler and Foam-Water Spray Systems*
NFPA 17, *Standard for Dry Chemical Extinguishing Systems*
NFPA 17A, *Standard for Wet Chemical Extinguishing Systems*
NFPA 20, *Standard for the Installation of Centrifugal Fire Pumps*
NFPA 22, *Standard for Water Tanks for Private Fire Protection*
NFPA 24, *Standard for the Installation of Private Fire Service Mains and Their Appurtenances*
NFPA 30, *Flammable and Combustible Liquids Code*
NFPA 33, *Standard for Spray Application Using Flammable or Combustible Materials*
NFPA 34, *Standard for Dipping and Coating Processes Using Flammable or Combustible Liquids*
NFPA 43B, *Code for the Storage of Organic Peroxide Formulations*
NFPA 43D, *Code for the Storage of Pesticides*
NFPA 45, *Standard on Fire Protection for Laboratories Using Chemicals*
NFPA 54, *National Fuel Gas Code*
NFPA 58, *Standard for the Storage and Handling of Liquefied Petroleum Gases*
NFPA 75, *Standard for the Protection of Electronic Computer/Data Processing Equipment*
NFPA 231, *Standard for General Storage*
NFPA 231C, *Standard for Rack Storage of Materials*
NFPA 231D, *Standard for Storage of Rubber Tires*
NFPA 231E, *Recommended Practice for the Storage of Baled Cotton*
NFPA 231F, *Standard for the Storage of Roll Paper*
NFPA 232, *Standard for the Protection of Records*
NFPA 329, *Recommended Practice for Handling Underground Releases of Flammable and Combustible Liquids*
NFPA 471, *Recommended Practice for Responding to Hazardous Materials Incidents*
NFPA 703, *Standard for Fire Retardant Impregnated Wood and Fire-Retardant Coatings for Building Materials*
NFPA 750, *Standard on Water Mist Fire Protection Systems*
NFPA 850, *Recommended Practice for Fire Protection for Electric Generating Plants*
NFPA 2001, *Standard on Clean Agent Fire Extinguishing Systems*

FIRE AND LIFE SAFETY EDUCATION

For public fire protection organizations and for business and industry, fire and life safety education is a key strategy for preventing incidents and for minimizing loss due to property damage, indirect losses (such as business interruption), injuries, and death. Such a broad-based strategy requires a variety of approaches and techniques.

As fire and life safety education matures, approaches and techniques borrowed from other disciplines blend to create a new and distinct discipline. This section provides an overview of the discipline of fire and life safety education and describes some of the approaches and techniques that comprise it.

Chapter 1 describes the evolving state of the art of fire and life safety education, including a discussion of the benchmarks of a new profession. The next two chapters examine the planning that fire and life safety educators must undertake before appearing in front of an audience. This planning includes using data to target resources effectively, discussed in Chapter 2, and designing disaster response education programs, discussed in Chapter 3. (Chapter 3 exemplifies how fire and life safety educators adapt the techniques of allied professionals to the communication of their own information.)

The fourth chapter, "Fire and Life Safety Education: Theory and Techniques," applies the knowledge base of teaching children and adults to the specialized topics of fire and life safety education.

The following two chapters focus more narrowly on specific audiences and tools. Chapter 5, "Reaching High Risk Groups," provides critical information on identifying and designing programs for groups with a high risk of fire, burn injury, or related hazard. Note that the analysis described in Chapter 2 will be key input to Chapter 5. Chapter 6 provides guidance on working with the media to convey educational messages to a broad-based audience.

Chapter 7 closes this section with a review of evaluation techniques for fire and life safety education. This review assists the fire and life safety educators perform the basic program evaluation functions that are necessary components for planning future efforts.

Also look for these: Much of the raw material of fire and life safety education is considered elsewhere in the *Fire Protection Handbook*. Chapters on human behavior in fire, on life safety concerns, and on fire protection devices commonly found in the home (such as smoke detectors) are particularly germane to fire and life safety educators.

—Meri-K Appy
—Pamela A. Powell
December 1996

FIRE AND LIFE SAFETY EDUCATION: THE STATE OF THE ART

Pamela A. Powell and Meri-K Appy

Contemporary fire departments provide a rich variety of public services: fire suppression, emergency medical services, plans review, and inspection services, among others. In many career and volunteer fire departments, this list of services includes fire and life safety education. Similarly, many corporate safety offices now provide fire and life safety education, as well as fire prevention services, fire brigade training, etc. Schools and other organizations teach fire and life safety skills, often as part of comprehensive health education or injury prevention programs.

Whoever the sponsor, *public fire and life safety education* may be defined as "comprehensive community fire and injury prevention programs designed to eliminate or mitigate situations that endanger lives, health, property, or the environment."[1] This definition from NFPA 1035, *Standard for Professional Qualifications for Public Fire and Life Safety Educator,* illustrates that fire and life safety education is multi-hazard education. In other words, fire and life safety educators often teach bicycle safety, pedestrian safety, poisoning prevention, cardiopulmonary resuscitation (CPR) and other emergency medical procedures, safe baby-sitting techniques, electrical safety, and water safety, as well as the more traditional subjects of fire and burn prevention.

This chapter provides an overview of fire and life safety education, including a review of current trends [as exemplified by the National Fire Protection Association's (NFPA) Champion *Learn Not to Burn*® Award Program], a discussion of fire and life safety education as a profession, and a discussion of "all-injury" prevention as a future role for fire and life safety education.

OVERVIEW OF FIRE AND LIFE SAFETY EDUCATION

Role in Injury Prevention and Loss Reduction

As multi-hazard educators, fire and life safety educators are very much part of a comprehensive or holistic approach to injury control. Injury control includes three phases: (1) prevention, (2) treatment, and (3) rehabilitation.

Three general strategies are available to prevent injuries:

Persuade persons at risk of injury to alter their behavior for increased self-protection—for example, to use seat belts or install smoke detectors.

Require individual behavior change by law or administrative rule—for example, by laws requiring seat belt use or requiring the installation of smoke detectors in all new buildings.

Provide automatic protection by product and environmental design—for example, by the installation of seat belts that automatically encompass occupants of motor vehicles, or built-in sprinkler systems that automatically extinguish fires.

Fire and life safety educators sometimes call these three injury prevention strategies "the three Es," i.e., (1) behavior change through public *education,* (2) legislative changes and *enforcement* of mandates, and (3) *environmental change* and technological development.[2] (See Figure 2-1A.)

The three injury prevention strategies, i.e., education, enforcement, and environmental change, are closely linked and most effective when used together. For example, in the area of preventing injuries caused by children playing with lighters and matches:

1. *Education* includes efforts to teach parents and other caregivers to use child-resistant lighters and to keep all lighters and matches out of the sight and reach of children.

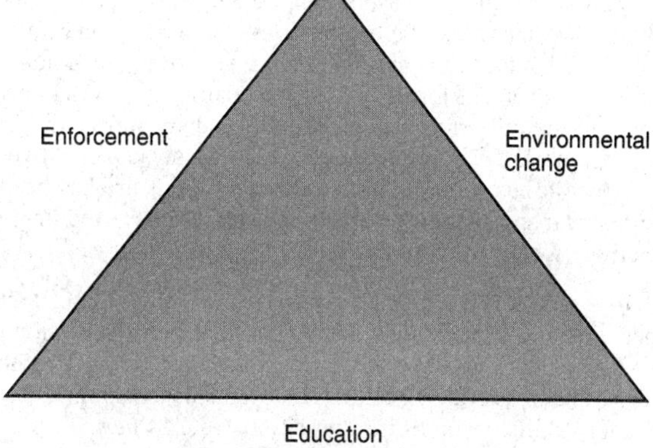

Pamela Powell is Director of Training and Education with Custer Powell, Inc., a consulting firm located in Wrentham, MA. She is nationally recognized in the fields of fire safety education, training, and meeting facilitation. Meri-K Appy is vice president of public education at the National Fire Protection Association. She directs the education component of NFPA's worldwide mission to reduce the burden of fire on society.

FIG. 2-1A. *Injury prevention strategies.*

(Enforcement / Environmental change / Education)

2. *Enforcement* includes a U.S. Consumer Product Safety Commission (CPSC) standard requiring that most lighters sold in the United States be child-resistant.
3. *Environmental change* includes the manufacture of child-resistant lighters or the storage of all lighters and matches away from children's reach.

The educational strategy often strengthens enforcement or environmental change strategies. For example, educational messages on the need to keep all lighters and matches out of the reach of children help compensate for gaps in enforcement: about 30 million new lighters each year and all matches are excluded from the CPSC standard. Until an environmental change produces a maintenance-free residential smoke detector, educational messages on how to test and maintain smoke detectors will be needed.

Clearly, each strategy contributes to a comprehensive injury prevention program.

However, a basic finding from research is that the second strategy—requiring behavior change—will generally be more effective than the first, and that the third—providing automatic protection—will be the most effective. A fundamental reason for this is that members of high-risk groups tend to be the hardest to influence with approaches that involve either voluntary or mandated changes in individual behavior. Teenagers, for example, are much less likely than adults to wear seat belts, whether or not a law requires them to do so.[3]

History of Fire and Life Safety Education

In one form or another, fire and life safety education has existed since at least the early 1900s. (See Figure 2-1B.)

Three factors have been especially significant in shaping fire and life safety education as it is today:

1. The publication of *America Burning*.[4]
2. The evolution of positive educational messages.
3. The availability of incident data for program planning.

The impact of *America Burning*:[4]   With rare exceptions, fire and life safety education activities were occasional and fragmented until the 1970s. However, in 1968, President Lyndon B. Johnson signed the Fire Research and Safety Act, which created the National Commission on Fire Prevention and Control. By 1973 the Commission had completed a nationwide series of fact-finding hearings and delivered its final report to President Richard M. Nixon. The Commission's report, *America Burning*,[4] included strong recommendations for stronger public education activities. According to *America Burning*,

Among the many measures that can be taken to reduce fire losses, perhaps none is more important than educating people about fire. Americans must be made aware of the magnitude of fire's toll and its threat to them personally. They must know how to minimize the risk of fire in their daily surroundings. They must know how to cope with fire, quickly and effectively, once it has started. Public education about fire has been cited by many Commission witnesses and others as the single activity with the greatest potential for reducing losses.[4]

Fire service advocates cited statements such as the one above to persuade fire departments (and others) to support education programs. Fire departments, in turn, cited *America Burning*[4] in budget requests and in convincing schools to extend the teaching of fire and life safety skills beyond Fire Prevention Week. *America Burning*[4] became a rallying point for those who advocated continuous and comprehensive fire and life safety education programs.

Positive educational messages: Shortly after the publication of *America Burning*,[4] several organizations began research on how to make fire and life safety education messages more effective. For example, the National Fire Protection Association commissioned opinion research on how the public felt about fire and education. The unpublished research by Strother Associates of Cambridge, MA, included these findings:

1. People are aware of the threat of fire, but fear (unless continually maintained) has little long-term, positive effect on behavior. Although concerned about fire safety, the public often did not know what to *do* about fire safety.
2. The public wanted clear, concise, and positive actions to take for fire safety.[5]

A 1991 study by the American Red Cross (on whether seeing slides of natural disasters encouraged people to take more or less action to prepare for disasters) confirmed the earlier research. According to the Red Cross study, seeing graphic images of disasters did not encourage people to take action to prepare for a disaster. In fact, seeing images of disasters confused people and actually discouraged them from taking action.[6]

As a result of research such as this, fire and life safety education materials acquired a new tone and a new look. Messages became less preachy and far less prone to discuss what people *should* do. Instead, fire and life safety focused more on teaching behaviors. At the same time, images of burned buildings and injured, perhaps scarred, children were replaced with, for example, diagrams of how and where to install home smoke detectors.

Data for program planning: *America Burning*[4] provided motivation for fire and life safety education, and the use of positive images became a fundamental technique. Improved access to fire incident data was the third factor that shaped the state-of-the-art of fire and life safety education.

The development of the National Fire Incident Reporting System (NFIRS) by the U.S. Fire Administration and user-friendly reports from the National Fire Protection Association influenced the day-to-day practice of fire and life safety education. Using NFIRS data, educators could identify the causes of, for example, fatal home fires and target educational resources at those causes. Also using NFIRS data, educators could compare their community's loss experience with that of similarly sized communities statewide and nationwide.

Fire incident data became much more user-friendly. Articles in *Fire Journal* (and its successor, the *NFPA Journal*) focused on the statistical background of issues of interest to fire and life safety educators, such as children playing with lighters and matches. A column—in lay language—on "Firesafety Education by the Numbers" became a regular feature of the *NFPA Education Section Newsletter*.

The impact of widely available and understandable data had a profound impact on fire and life safety education. Data became a planning and evaluation tool for the educator.

CURRENT TRENDS

The United States and Canada, still at the top of the list of Western industrialized nations with the highest fire death rates *per capita*, have seen a long-term downward trend in fire fatalities in the past two decades. The dramatic increase in home smoke detector usage, supported by widespread public education efforts, has clearly played a major role in reducing the fire problem in North America.

Yet the percentage of full-time staff and annual budgets devoted by U.S. fire departments to public education is a fraction of what the suppression forces receive. And too often even the most

1909	NFPA's Franklin Wentworth begins sending fire prevention bulletins to correspondents in 70 cities, with the hope that local newspapers will publish the bulletins as news articles.
1911	Fire Marshals Association of North America proposes the October 9 anniversary of the Great Chicago Fire as a day to observe fire prevention.
1912	NFPA publishes *Syllabus for Public Instruction in Fire Prevention*—fire safety topics for teachers to use in the classroom.
1916	NFPA and the National Safety Council establish a Committee on Fire and Accident Prevention. Communities nationwide organize Fire Prevention Day activities.
1920	President Woodrow Wilson signs first presidential proclamation for Fire Prevention Day.
1922	President Warren G. Harding signs first Fire Prevention Week proclamation.
1923	23 states have legislation requiring fire safety education in schools.
1927	NFPA begins sponsoring national Fire Prevention Contest.
1942	New York University publishes *Fire Prevention Education*.
1946	U.S. government publishes *Curriculum Guide for Fire Safety*.
1947	Hartford Insurance Group begins the Junior Fire Marshal Program, perhaps the first nationally distributed fire safety program for children.
1948	American Mutual Insurance Alliance publishes first edition of *Tested Activities for Fire Prevention Committees*, based on Fire Prevention Contest entries.
1950	In October, 7,000 newspapers receive the ad "Don't Gamble with Fire—The Odds Are Against You," developed by the Advertising Council and NFPA.
1954	Ted Royal of the Advertising Council creates Sparky®, the Fire Dog as NFPA fire safety mascot.
1965	*Fire Journal* begins a regular column on "Reaching the Public."
1966	"Wingspread Conference" highlights the need for public education.
1973	• The National Commission on Fire Prevention and Control publishes its report, *America Burning*. • The Fire Department Instructors' Conference offers its first presentation on fire and life safety education, delivered by Cathy Lohr of North Carolina.
1974	• NFPA and the Public Service Council release the first television *Learn Not to Burn®*, public service announcements, starring the actor Dick Van Dyke. • The Fire Prevention and Control Act establishes the National Fire Prevention and Control Administration.
1975	The National Fire Prevention and Control Administration holds its first national fire safety education conference.
1977	• NFPA 1031, *Standard for Professional Qualifications for Fire Inspector, Fire Investigator, and Fire Prevention Education Officer*, is published. • National Fire Prevention and Control Administration releases *Public Fire Education Planning: A Five-Step Process*. • National Fire Prevention and Control Administration launches national smoke detector campaign.
1979	• J. C. Robertson's *Introduction to Fire Prevention* is published by Glencoe Press. • *Learn Not to Burn Curriculum* is published by NFPA. • International Fire Service Training Association (IFSTA) releases IFSTA 606, *Public Fire Education*.
1981	NFPA establishes its Education Section.
1985	• The National Education Association recommends the *Learn Not to Burn Curriculum*. • NFPA publishes *Firesafety Educator's Handbook*.
1986	Learn Not to Burn Foundation is incorporated.
1987	• The first edition of NFPA 1035, *Standard for Professional Qualifications for Public Fire Educator*, encourages civilians to become public fire educators in the fire department. • TriData Corporation publishes *Overcoming Barriers to Public Fire Education*.
1990	Oklahoma State University publishes the first issue of the *Public Fire Education Digest*.
1994	TriData Corporation releases *Proving Public Fire Education Works*. • NFPA launches *Learn Not to Burn* Champion Award Program.
1996	• *Learn Not to Burn* Foundation incorporated into NFPA Public Education Division as the NFPA Center for High-Risk Outreach.

This timeline was originally prepared for IFSTA's *Fire and Life Safety Educator*, based on information from Pam Powell's "Firesafety Education: It's Older than You Think" (*Fire Journal*, May 1986) and information provided by Nancy Trench, Fire Service Training, Oklahoma State University. See Bibliography for further information.

FIG. 2-1B. Timeline.

highly regarded prevention education programs are reduced or eliminated during budget cuts, because they lacked tangible data to prove their positive impact in reducing the risk of fire in the community.

The scenario is changing, however. Having struggled for years in isolation, practitioners in the field are realizing that fire is not the sole responsibility of the public educator or even the fire department, but rather a *community* problem requiring support and involvement from the whole community. Programs to educate the public are now being designed with a more complete knowledge of the target audience, taking into account such factors as age, culture, learning styles, and setting. Efforts to create positive values about fire safety and to motivate people to change their risky behavior is increasingly viewed as a long-term commitment, not something that can be achieved through an occasional, one-size-fits-all approach.

Further, the fire and life safety education community is growing more sophisticated in using sound evaluation techniques to target and document program successes, leading community decision-makers—both within the fire service and without—to view public education as an essential, cost-effective, high-impact component of an integrated fire protection strategy.

NFPA CHAMPION PROGRAM—A MODEL FOR THE FUTURE

The National Fire Protection Association is at the forefront of the movement to provide a national focus for measurable fire safety education efforts aimed at young children. In 1979, the Association, in consultation with educators, fire and burn prevention experts, and curriculum specialists, developed and field-tested the *Learn Not to*

Burn Curriculum in selected schools throughout the United States. Published in 1981 and revised in 1983 and 1987, the *Curriculum* has been used to teach fire and burn prevention lessons to hundreds of thousands of children worldwide. In 1991 NFPA released the *Learn Not to Burn Resource Books*,[7] featuring lessons and hands-on learning activities for children in kindergarten through Grade 3. Also in 1991, NFPA published the *Learn Not to Burn Preschool Program*,[8] created by the *Learn Not to Burn* Foundation to reach the high-risk population of children ages 3 to 5 years old.

The availability of solid educational tools, such as the *Learn Not to Burn* line of products, did not guarantee consistent, universal usage, however. More support was needed to encourage communities to establish and maintain a dedicated focus on comprehensive fire and burn prevention education. To meet this need, NFPA launched the *Learn Not to Burn* (LNTB) Champion Award Program in 1994, with the goal of enhancing effective usage and documentation of the *Learn Not to Burn* program.

The Champion model promotes development of a community-based strategy involving the fire department, schools, other safety organizations, and the private sector, working in coalition to deliver positive fire and burn prevention lessons to children and families. Inspired by the 1990 TriData Corporation Study, *Proving Public Fire Education Works*,[9] that cited the importance of a Champion (i.e., a motivated spark plug to carry the program through to its success), the LNTB Champion Award Program is creating an active network of trained public education leaders capable of organizing, evaluating, and maintaining a comprehensive *Learn Not to Burn* effort at the local, state, and national levels.

Since 1994, approximately 60 communities throughout North America have been selected each year as *Learn Not to Burn* Champion sites. Under the direction of the designated Champion, NFPA provides, at no cost to the community, *Learn Not to Burn* program materials and technical support to establish or enhance the implementation of the LNTB program in local elementary schools and day-care centers. A similar program initiated in 1995, "*Learn Not to Burn*: Safe Cities," offers tools and specialized training to promote comprehensive use of LNTB program materials in major urban centers.

Evaluation results from the first group of 1994 Champions were encouraging. Pre- and post-test scores indicated an average knowledge gain of 12 percentage points among the 14,528 students who participated in the pilot test. (See Table 2-1A.)

TABLE 2-1A.   *Educational Gain in LNTB Champion Sites*

Grade level	Kinder-garten	1st grade	2nd grade	3rd grade	Average
Educational gain	14.0	11.0	14.0	9.1	12.0

A further indication of program success is the fact that, within the first year of implementation, there were 10 *Learn Not to Burn* life saving incidents involving children exposed to LNTB lessons through the Champion program. More research will be conducted to compare results of each subsequent LNTB Champion and Safe Cities group and to track program performance longitudinally.

NFPA has identified several fundamental qualities that appear to correlate with long-term success in implementing the *Learn Not to Burn* program. These "Cs to Success" are used as the criteria by which applications are judged to determine which communities will receive *Learn Not to Burn* Champion and Safe Cities awards. Listed below, in order of importance, they may also be used by fire departments and other safety advocates to assess a community's readiness

to maintain any substantive safety education partnership targeting children in the school environment.

A Committed Chief

This is perhaps the single most important predictor for successful implementation of any fire and life safety program. Increasingly, fire chiefs see prevention and education as critical elements in their overall operational strategy to serve the safety needs of their community. While resource decisions about fire department activities still clearly favor traditional suppression efforts, a growing number of innovative fire chiefs are highlighting fire prevention and public education as the most efficient, cost-effective approach for both improved community relations and public fire safety.

Learn Not to Burn "Champion Chiefs" understand the need for comprehensive, ongoing implementation by supporting a true partnership with their school community, and are willing to commit the time and talent of one Champion, operating with sufficient freedom and organizational support, to oversee the project locally and share the results with others in the field.

A Dedicated Champion

The Champion, dedicated to *Learn Not to Burn* both personally and by professional function, is responsible for planning, conducting, evaluating, and expanding the LNTB program from the initial pilot to full-scale implementation. Often served by the fire department's fire and life safety educator, this is more of a managerial role than a teaching role. Long-term success depends on the Champion's ability to think strategically, to motivate and coach others toward common goals, and to document and communicate program results to key decision makers and participants.

Champions are advised to begin their LNTB implementation project with a small pilot test of the materials in only 10 classrooms—two at each level from preschool through Grade 3. This enables a Champion to become thoroughly familiar with the LNTB materials and process, and to practice critical planning, management, and evaluation skills on a small scale. Through pre- and post-testing of student knowledge gain and surveys of pilot teachers, caregivers, and other program participants, Champions document initial program impact and use concrete data to make informed decisions about future expansion.

Collaboration

In addition to solid fire department support, successful *Learn Not to Burn* Champion sites enjoy a high level of commitment from the community's board of education. At the outset, Champions secure a letter of endorsement, preferably from the community's superintendent of schools, committing the full participation of the teachers selected for the pilot phase. Assuming the pilot is successful and that outside funding is secured to cover needed program materials, the school system agrees to continue to work in partnership with the fire department to achieve full implementation of the program in every classroom from preschool through Grade 3.

In the *Learn Not to Burn* Champion model, classroom teachers are responsible for delivering the fire safety lessons included in the program manuals. The role of the fire department is to support and reinforce the classroom lessons by maintaining close ties with participating teachers and by visiting the schools as often as possible to answer questions and assess students' understanding and performance.

A Compelling Case

It makes sense to focus the fire safety effort in areas that need it most. By analyzing the incidence of fires and burns in the community and focusing first on those neighborhoods at highest risk, public educators can achieve dramatic results early in the program and use that momentum to gain support for further expansion. The extent to which program managers can articulate an urgent need and define the tangible benefits of comprehensive fire safety education will largely determine the level of media attention, funding, and full community involvement, thereby having a direct effect on the program's documented impact.

Continuity

Because children learn best through repetition and practice, the most effective way to teach fire safety is through repeated, age-appropriate learning opportunities extended over a period of years. The *Learn Not to Burn* Champion model targets the high-risk preschool population with simple messages they can understand and apply to their own experiences, and helps educate their caregivers about how to create a safer home environment for their children. The learning process is extended into grade school, as well, at least through Grade 3; this allows children to acquire the basic information they need to be safer from fires and burns, much of which is too complex for preschoolers to grasp.

Continuous, consistent fire and burn prevention education prepares children to make informed decisions about fire safety for themselves and to influence the behavior of the adults around them. In many of the *Learn Not to Burn* life saving incidents documented by NFPA, children who had received *Learn Not to Burn* instruction in school helped their parents respond properly to a fire or burn emergency, and many more encouraged their families to install and maintain smoke detectors and adopt other potentially life-saving measures.

Coalition

Maintaining an effective fire and burn prevention program requires sufficient human and financial resources to plan, conduct, and track the effort, year after year. Typically, no one organization has the wherewithal to do this alone. By organizing a team of community members who either have—or can be inspired to embrace—a commitment to fire safety, fire and life safety educators can establish programs that will deliver long-term results.

The key to success when working through a coalition is to develop a written strategic plan that outlines a clear set of goals, measurements, and time frames. Each participating organization will have its own individual priorities; however, successful coalitions are able to come to consensus on an achievable number of objectives toward which everyone in the group can make a concrete contribution. Progress in meeting the stated objectives must be carefully documented; and as the coalition develops, the plan must be updated and the leadership structure redefined.

Creativity

The *Learn Not to Burn* Champion model prescribes a consistent approach to delivering LNTB lessons to children in preschool through Grade 3. The strategy is advantageous for two reasons. First, it enables fire and life safety educators implementing the Champion model in communities throughout North America to learn from each other how to improve and enhance their own program. And by collecting comparable program impact data from individual Champion communities, states, and regions, NFPA hopes in time to be able to document a national public education success story. For

these reasons, adherence to the basic Champion model is highly recommended.

However, communities are encouraged to supplement the Champion model with program elements designed to address local needs, e.g., as Los Angeles, CA, did by including an esteem-building component in the design of the Los Angeles Fire Department School Outreach Program. The creative use of meaningful enhancements to reinforce the *Learn Not to Burn* classroom lessons, such as safety trailers, clown and puppet characterization techniques, coloring and comic books, etc., helps customize the program and provides an important sense of ownership. Creative reinforcement techniques can also enhance student knowledge gain and retention, so long as the information presented does not conflict with the core LNTB content and behaviors.

National Trends

The critical success factors incorporated in NFPA's *Learn Not to Burn* Champion Award Program are reflected in other national initiatives, as well. A strong coalition effort between the United States Fire Administration and the American Red Cross in 1994 used market research and demographic analysis to develop media materials, a door hanger, brochure, and an activity poster in Spanish and English. The "Home Fire Safety—Act on It" campaign enables fire and life safety advocates to address an underserved population by delivering needed fire safety messages to Spanish-speaking audiences. In another example of collaborative program development, Allstate Insurance Company launched the "For a Safer America Campaign" to develop a multifaceted advertising campaign in partnership with the Ad Council and several consumer, education, and fire safety organizations.

Greater emphasis is being placed on the need for strategic planning at the national level. In 1995, a volunteer group of fire and life safety leaders met in Portland, OR, to begin an ongoing process to set strategic goals and objectives for the field. Having identified some key research needs and new program ideas, the group formed the North American Coalition for Fire and Life Safety Education which hopes to spur coordinated action among a growing number of groups and individuals who share a commitment to fire and life safety. This kind of participatory decision-making and a more sensible allocation of available resources will help avoid duplication of effort and ensure delivery of consistent, behavior-changing public education messages and programs.

FIRE AND LIFE SAFETY EDUCATION AS A PROFESSION

Benchmarks of a Profession

Over the last generation, fire and life safety education appears to have become more "professional." But what constitutes a "profession" or "discipline"?

In describing the emerging profession of fire protection engineering in 1951, Illinois Institute of Technology President H. T. Heald cited six key components of a profession.[11] They are:

1. Body of knowledge
2. Strong motive of service over profit
3. Qualifications in individual competency and character
4. Education
5. Recognition of status (including licensing)
6. Assuring the public of member competence

Three of these components—body of knowledge, education, and recognition of status—are especially germane to fire and life safety education and will be discussed here. (See Table 2-1B.)

Body of knowledge: Publications and research illustrate the presence of a professional body of knowledge. Although the quantity and quality of publications devoted to fire and life safety education appears to be growing, the profession still lacks a foundation in research.

Professional literature for fire and life safety educators dates from at least 1912, the year in which NFPA published the *Syllabus for Public Instruction in Fire Prevention.* (See Figure 2-1B.) Thirty years passed before the release of the next significant text, New York University's *Fire Prevention Education.* Soon after World War II, the *Curriculum Guide for Fire Safety and Tested Activities for Fire Prevention Committees* was published. Texts in fire and life safety education essentially ceased until the 1970s, when the National Fire Prevention and Control Administration (now the U.S. Fire Administration) released a number of original publications, including *Public Fire Education Planning: A Five-Step Process,*[12] *The Media Ideas Workbook,*[13] *Teaching Fire Safety Through Exhibits,*[14] *Fire Safety on Television for Preschoolers,*[15] and a series of manuals on firesetter intervention programs.[16,17,18]

Publication activities then shifted from the public to the private sector, as seen by Glencoe Press's *Introduction to Fire Prevention;*[19] the International Fire Service Training Association's IFSTA 606, *Public Fire Education;*[20] and NFPA's *Learn Not to Burn Curriculum.*[21] More recently, TriData Corporation has released three significant publications: *Overcoming Barriers to Public Fire Education,*[22] *Proving Public Fire Education Works,*[7] and *Reaching the Hard-to-Reach: Techniques from Fire Prevention Programs and Other Disciplines.*[23]

There has been less activity in the area of periodicals on fire and life safety education. Two notable exceptions are: (1) the *Education Section Newsletter,*[24] published by NFPA since 1981 for members of its Education Section and (2) the *Public Fire Education Digest,* available by subscription from Oklahoma State University.[25]

Different editions of two publications illustrate the growth and change of fire safety education as a profession. Fire and life safety has grown from an occasional chapter in the *Fire Protection Handbook,* to a small section in the 16th edition, to a much expanded section in this, the 18th edition. Similarly, the 1979 edition of IFSTA

606, *Public Fire Education,*[20] was largely a collection of descriptions of local fire and life safety education programs. The second edition was retitled *Fire and Life Safety Education*[26] and contains minimal program narrative. Instead, the second edition applies knowledge from other disciplines (such as education, program evaluation, public speaking, and media relations) to fire and life safety education.

The fire and life safety educator's body of knowledge is progressing from oral history and program descriptions to the stage of adapting relevant knowledge from other disciplines. While the professional literature reflects an increasing body of knowledge in fire and life safety education, there is still room for growth. For example, there is very little credible research in fire and life safety education. Peer-reviewed articles on fire and life safety education research are extremely rare and appear in publications, such as NFPA's *Fire Technology,* that fire and life safety educators typically do not read.

Education: Professional educational activities may be viewed on a continuum of academic rigor, ranging from conferences (at the least rigorous end of the scale), to courses at nonaccredited institutions, to courses at accredited institutions of higher learning (at the most academically rigorous end of the scale).

Fire and life safety education conferences have been an integral part of the profession for many years. While a number of statewide agencies or membership organizations provide conferences, three conference series are especially noteworthy.

1. The North Carolina Department of Fire and Rescue Services has been conducting fire and life safety education conferences since February 1976. The 1995 conference attracted 132 paid participants from several states.
2. Fire Service Training at Oklahoma State University has offered fire and life safety education conferences since 1976. More than 200 people from 28 states, Puerto Rico, and Canada participated in the 1995 conference.
3. The Illinois Fire Inspectors Association held its 20th Annual Public Education Conference in April 1996.

Like publications, fire and life safety education conferences are evolving from descriptions of local programs and materials to include more sessions on techniques, hands-on workshops, etc.

The National Fire Academy (NFA) is a nonaccredited institution within the U.S. Fire Administration/Federal Emergency Man-

TABLE 2-1B. *Benchmarks of a Profession*

Benchmark	Fire and Life Safety Education Indicator
Body of knowledge	
Publications	Increasing number of "stand-alone" publications
	Two active periodicals
	No peer-reviewed periodicals
Research	Extremely limited
Education	
Conferences	Many state-level fire and life safety education conferences
Courses (non-accredited institutions)	Two courses available at the National Fire Academy
Courses (accredited institutions)	NA
Recognition of professional status	
Professional organizations	Many state or regional fire and life safety organizations
	Limited interest groups within national/international organizations
	No national/international organization for fire and life safety education
Certification	Georgia, Illinois, Maryland, and North Carolina are the only states that currently certify to NFPA 1035, *Standard for Professional Qualifications for Public Fire and Life Safety Educator.* Other states (e.g., California) conduct training, leading to in-state certification.

agement Agency (USFA/FEMA) and located in Emmitsburg, MD. In 1995, two of the Academy's 31 on-campus courses and programs were on fire and life safety education.

"Presenting Effective Public Education Programs" is a five-day course for "new public educators whose primary responsibilities are presenting educational programs to public audiences" and "designed for participants who present fire, burn, injury prevention, or general community safety education programs," according to the NFA USFA/FEMA *On-Campus Course Schedule.*[27] Potential students must be currently responsible for delivering public education programs and have a maximum of two years of experience in public education.

The more advanced two-week "Developing Fire and Life Safety Strategies" course "presents effective strategies for developing and promoting fire and life safety education programs, including identification of community needs, development of implementation strategies, and evaluation of effective programs," according to the *On-Campus Course Schedule.*[27] Course themes include fire prevention as an injury control issue, prevention education as a way to form community-based coalitions, and the need for fire department leadership. Potential students need at least two years of experience in fire and life safety education, burn prevention education, or community safety. The National Fire Academy's "Fire Service Instructional Methodology" course (or equivalent) is a prerequisite.

These and other National Fire Academy courses "are equal in difficulty to those at the college or university level."[28] Academy courses are reviewed by the American Council on Education (ACE), which recommends equivalent credit, curricula, and level. "Presenting Effective Public Education Programs" has not yet been reviewed by ACE. ACE recommends that "Developing Fire and Life Safety Strategies" is equivalent to three semester hours in adult education, fire science, general education, public health, or safety studies in the category of upper division of baccalaureate (bachelor's degree).

The most academically rigorous professional education programs are formal courses offered by accredited institutions of higher learning. Fire and life safety education courses included in community college fire science curricula would be one example. However, a request for information about such courses over CompuServe did not yield any responses.

Recognition of professional status: Two well-known ways of recognizing professional status are: (1) membership in professional organizations and (2) licensing or certification. (See Table 2-1B.)

Many state or regional fire and life safety education organizations exist. In addition, there are a few special-interest groups of national or international organizations, e.g., the Prevention Committee of the American Burn Association and the Education Section of the National Fire Protection Association. In 1996, however, there is no known national or international organization devoted solely to fire and life safety education.

Licensing or certification is the remaining measure of professional status. As of 1995, Georgia, Illinois, Maryland, and North Carolina certified that public fire and life safety educators met NFPA 1035, *Standard for Professional Qualifications for Public Fire and Life Safety Educator,* criteria.[1] Other jurisdictions are developing accreditation programs to certify fire and life safety educators.

Career Issues

In recent years, fire and life safety education has broadened from its fire service base to include burn prevention personnel, community health educators, and others. Note, however, the fire service is still the largest employer of fire and life safety educators. Therefore, NFPA professional qualifications standards have a great impact on career fire and life safety educators.

NFPA 1031, *Standard for Professional Qualifications for Fire Inspector, Fire Investigator, and Fire Prevention Education Officer,* was first published in 1977. Under NFPA 1031, the prerequisite for qualifying as an inspector, investigator, or fire prevention education officer was satisfying the requirements of NFPA 1001, *Standard for Fire Fighter Professional Qualifications.* In other words, NFPA 1031 required that fire prevention education officers first be fire fighters. Civilians—even those with credentials as classroom teachers or health educators—could not fulfill the requirements in NFPA 1031. Previous experience as a fire fighter was the only door into fire and life safety education.

That situation changed in 1986, due to a two-part decision by the Professional Qualifications Board/Joint Council of National Fire Service Organizations. First, the Joint Council separated fire inspectors, investigators, and fire prevention education officers, calling for separate professional standards for each of these three specialties. (In effect, this decision "promoted" fire prevention education by putting it on an equal footing with the more traditional fire service tasks of inspection and investigation.) At the same time, the Joint Council decided that the career ladder for public fire educators need not start with fire fighting. Civilians could now qualify as educators.

Subsequently, the NFPA Technical Committee on Public Fire Educator Professional Qualifications developed NFPA 1035, *Standard for Professional Qualifications for Public Fire Educator,* which the Association adopted in 1987. The document was revised and renamed *Standard for Professional Qualifications for Public Fire and Life Safety Educator* in 1993.

The 1993 edition defines and sets out minimum qualifications for three levels of public fire and life safety educator. Those definitions are:

Public Fire and Life Safety Educator I. The individual who has demonstrated the ability to coordinate and deliver existing educational programs and information as specified in this standard for the Public Fire and Life Safety Educator I level.

Public Fire and Life Safety Educator II. The individual who has demonstrated the ability to prepare educational programs and information to meet identified needs as specified in this standard for the Public Fire and Life Safety Educator II level.

Public Fire and Life Safety Educator III. The individual who has demonstrated the ability to create, administer, and evaluate educational programs and information as specified in this standard for Public Fire and Life Safety Educator III level.[1]

Note that the National Fire Academy course "Presenting Effective Public Education Programs" closely corresponds to Public Fire and Life Safety Educator I. "Developing Fire and Life Safety Strategies" corresponds to Public Fire and Life Safety Educator II. There is currently no known course of study (at the National Fire Academy or elsewhere) that corresponds to Public Fire and Life Safety Educator III.

The 1993 edition of NFPA 1035, *Standard for Professional Qualifications for Public Fire and Life Safety Educator,* has already been an influential document in the field of fire and life safety education. In addition to state certification programs and National Fire Academy courses, NFPA 1035 guided the preparation of the new

International Fire Service Training Association manual, *Public Fire and Life Safety Educator*.[26]

ALL-INJURY PREVENTION—THE WAVE OF THE FUTURE

Injury is probably the most underrecognized major public health problem facing the nation today, and the study of injury represents unparalleled opportunities for reducing morbidity and mortality and for realizing significant savings in both financial and human terms—all in return for a relatively modest investment.[29]

For America's children under the age of 14, the number-one health risk isn't violence, drugs, or disease. It's injuries. A landmark 1985 study by the Committee on Trauma Research, *Injury in America: A Continuing Public Health Problem*,[3] documented the magnitude of the injury problem and set forth a research agenda, calling upon Congress to establish a center for injury control within the federal government. Reviewing the progress made since 1985, the National Academy of Sciences' Committee to Review the Status and Progress of the Injury Control Program at the Centers for Disease Control concluded in 1988 that "the message is clear: injury remains a major public health problem. Prevention and control efforts must expand to reflect the magnitude of the challenge."[29]

The National Committee for Injury Prevention and Control recommends a multifaceted, coordinated process at the community level to address local needs. Advocates must first use data to define the local injury problem, collaborate with others to develop a program based on these findings, select a mix of interventions that reflects the state of the art in injury control, and evaluate the program's achievement of process and outcome objectives.

Fire departments can serve a vital role in the concerted effort to reduce the national incidence of preventable injury and death. The public safety needs of communities increasingly place fire department personnel in the role as emergency responders—first on the scene not only in a fire emergency, but in medical emergencies as well. With a proven record in educating the public to prevent and respond effectively to fire emergencies, fire safety advocates often have the credibility and expertise to organize their communities around broader safety issues. Increasingly, fire and life safety advocates view such incidents as traffic injuries, drownings, firearm injuries, falls, and poisonings; that is, not random accidents, but predictable events that, with proper education, are largely preventable.

Until recently, injury prevention practitioners have lacked the educational tools and support to target the major risk areas that affect school-age children. The Phoenix Fire Department was the first to adopt a more comprehensive prevention focus by developing the *Urban Survival Fire and Life Safety Program*[30] in 1991. Designed to encourage community involvement from Phoenix teachers, fire fighters, PTA members, local medical and health establishments, and other partners, the program provides information, classroom lessons, and reinforcement activities on a wide range of life safety topics. The goal is "to teach school children, adults, elderly, and the disabled the skills necessary to protect themselves and their families by responding promptly and effectively when confronted with a fire or life safety hazard."[30]

In 1995 the National Fire Protection Association took the issue of injury prevention education to the national level. With partial funding from the Lowe's Home Safety Council, a nonprofit organization founded in 1993 to enhance the quality of American home life through better knowledge and practice of home safety, NFPA is developing a new curriculum entitled *RiskWatch®*. Scheduled for release in 1998, *RiskWatch* is a school-based educational curriculum for children in preschool through Grade 8, carefully designed and tested to provide information to promote good safety choices for children and their families.

Working with NFPA and Lowe's Home Safety Council is the National SAFE KIDS® Campaign®, the first nationwide movement to prevent childhood injury. The goal of the National SAFE KIDS Campaign is to make communities safer for America's children through legislation, education, product modification, and public awareness. Rounding out the development team is a select group of injury prevention professionals with nationally recognized expertise in the risk areas of traffic injuries, drowning, fires and burns, unintentional firearm injuries, falls, and poisoning. Serving as the project's technical advisory group, they will provide research data, technical content, and appropriate prevention messages for each of the identified risk areas, and review the curriculum lessons developed to address their safety specialty.

CONCLUSION

Over the last generation, fire and life safety education has evolved from teaching fire safety exclusively to a more comprehensive educational effort that incorporates fire safety and other subjects as part of an overall injury prevention strategy. Over the same period of time, fire and life safety education practice has grown from a "Fire Prevention Week only" event to ongoing programs, as seen in the activities of the *Learn Not to Burn* Champions and other programs. Similarly, fire and life safety education has become more sophisticated as a profession. Improvements still need to be made, however, in three benchmarks of a profession: body of knowledge, education, and recognition of status (including licensing).

BIBLIOGRAPHY

References Cited

1. NFPA 1035, *Standard for Professional Qualifications for Public Fire and Life Safety Educator*, National Fire Protection Association, Quincy, MA, 1993.
2. The National Fire Academy, for example, uses these terms in its course "Developing Fire and Life Safety Strategies."
3. *Injury in America: A Continuing Public Health Problem*, Committee on Trauma Research, Commission on Life Sciences, National Research Council, and the Institute of Medicine, National Academy Press, Washington, DC, 1985.
4. *America Burning*, Report of the National Commission on Fire Prevention and Control, May 1973, U.S. Government Printing Office, Washington, DC.
5. Powell, P. A., "Learn Not to Burn: A Decade of Progress," *Fire Journal*, Vol. 79, No. 2, March 1985, p. 8.
6. Lopes, R., Ph.D., "Public Perception of Disaster Preparedness Presentation Using Disaster Damage Images," *NFPA Education Section Newsletter*, Fall/Winter 1992, National Fire Protection Association, Quincy, MA.
7. Appy, M.K., and Comoletti, J.L., *Learn not to Burn® Resource Book*, National Fire Protection Association, Quincy, MA, 1991.
8. Gamache, S., Powell, P., *et al*, *Learn Not to Burn Preschool Program*, *Learn Not to Burn* Foundation, National Fire Protection Association, Quincy, MA, 1991.
9. Schaenman, P. et al., *Proving Public Fire Education Works*, TriData Corporation, Arlington, VA, 1990, p.113.
10. Appy, M. K., "The Cs to Success: Critical Success Factors in Implementing NFPA's Learn Not to Burn Program," *NFPA Education Section Newsletter*, Fall/Winter 1995, National Fire Protection Association, Quincy, MA.
11. Lucht, D. A., "Coming of Age," *Journal of Fire Protection Engineering*, Vol. 1, No. 2, 1989, p. 38.
12. Strother, R. A., Powell, P. A., and Buchbinder, L., eds., *Public Fire Education Planning: A Five-Step Process*, GPO Stock Number 003-000-00544,1977.

13. Hollister, G. C., ed., *The Media Ideas Workbook*, U.S. Fire Administration, FA-30, Emmitsburg, MD, Oct. 1980.
14. *Teaching Fire Safety Through Exhibits*, U.S. Fire Administration, FA-36, Emmitsburg, MD, Oct. 1980.
15. Peel, B., Schauble, L., and the Sesame Street® Research Staff (Children's Television Workshop) for the U.S. Fire Administration, *Fire Safety on Television for Preschoolers*, FA-2A, July 1980.
16. *Child (Preadolescent) Firesetter Handbook: Age 7 and Under*, U.S. Fire Administration and the International Association of Fire Chiefs, Washington, DC, Dec. 1988.
17. *Child (Preadolescent) Firesetter Handbook: Age 7-13*, U.S. Fire Administration and the International Association of Fire Chiefs, Washington, DC, Feb. 1988.
18. *Adolescent Firesetter Handbook: Ages 14-18*, U.S. Fire Administration and the International Association of Fire Chiefs, Washington, DC, 1988.
19. Robertson, J. C., *Introduction to Fire Prevention*, Glencoe Publishing Company, Inc., Encino, CA, 1979.
20. IFSTA 606, *Public Fire Education*, Oklahoma State University, Stillwater, OK, 1979.
21. *Learn Not to Burn Curriculum*, National Fire Protection Association, Quincy, MA, 1979.
22. Schaenman, P., et al., *Overcoming Barriers to Public Fire Education*, TriData Corporation, Arlington, VA, 1987.
23. Kulenkamp, A., Lundquist, B., and Schaenman, P., *Reaching the Hard-to-Reach: Techniques from Fire Prevention Programs and Other Disciplines*, TriData Corporation, Arlington, VA, Oct. 1994.
24. *Education Section Newsletter*. Available to members of NFPA's Education Section. For Section membership information, contact the Public Education Division, National Fire Protection Association, Quincy, MA.
25. *Public Fire Education Digest*. For additional information, contact Fire Service Training, Oklahoma State University, Stillwater, OK 74074.
26. *Fire and Life Safety Educator*. The International Fire Service Training Association validated publication in July 1995 and expects to publish in January 1997.
27. NFA USFA/FEMA, *On-Campus Course Schedule*, National Fire Academy, Emmitsburg, MD.
28. *Catalog of Training Activities*, 1993-1995, National Fire Academy, Emmitsburg, MD.
29. IFSTA, *Public Fire and Life Safety Educator*, International Fire Service Training Association, Oklahoma State University, Stillwater, OK, 1996.
30. *Injury: Meeting the Challenge*, National Committee for Injury Prevention and Control. Education Development Center, Inc., Newton, MA, 1988.
31. Phoenix Fire Department, *Urban Survival Fire and Life Safety Program*, City of Phoenix, AZ, 1991.

NFPA Codes, Standards, and Recommended Practices

Reference to the following NFPA codes, standards, and recommended practices will provide further information on fire and life safety education discussed in this chapter. (See the latest version of *The NFPA Catalog* for availability of current editions of the following documents.)

NFPA 1001, *Standard for Fire Fighter Professional Qualifications*
NFPA 1031, *Standard for Professional Qualifications for Fire Inspector*
NFPA 1035, *Standard for Professional Qualifications for Public Fire and Life Safety Educator*

Additional Reading

"'Alarming' Day," *Fire Fighting in Canada*, Vol. 39, No. 8, October 1995, pp. 14–15.
"CPSC Advance Notice of Proposed Rulemaking on Children's Sleepwear Standards [58 FR 4111, January 13,1993]," *Product Safety and Liability Reporter*, January 27, 1993, pp. 87–91.
"Curious Kids Set Fires. Fire Safety Program," Federal Emergency Management Agency, Washington, DC, 1991.
Damant, G.H. and Nurbakhsh, S., "Christmas Trees—What Happens When They Ignite?," *Fire and Materials: An International Journal*, Vol. 18, No. 1, January-February 1994, pp. 9–16.

"Directory of National Community Volunteer Fire Prevention Program. Community-Based Fire Prevention Education Initiatives, 1984-1992.," National Criminal Justice Assoc., Washington, DC, Federal Emergency Management Agency, Emmitsburg, MD, FA-92, April 1993, 130 pages.
"Fire Departments and Communities: Partners in Prevention," *Fire Engineering*, Vol. 148, No. 6, June 1995, pp. 101–110.
Hall, J. R., Jr., "Children Playing With Fire: U. S. Experience, 1980-1993," National Fire Protection Assoc., Quincy, MA, August 1995, 35 pages.
Hayashi, T., "Disaster Protection for the Aged," Firesafety Frontier '94, International Fire Conference and Exhibition in Tokyo, Creating a Safe Tomorrow, October 18-22, 1107-112, Tokyo, Japan, 1994, pp. 277–280.
Henricsson, L., "Measures for the Safety of the Lives of the Elderly and the Physically Handicapped," Firesafety Frontier '94, International Fire Conference and Exhibition in Tokyo, Creating a Safe Tomorrow, October 18-22, 1107-112, Tokyo, Japan, 1994, pp. 281–286.
Jones, R.T. and Randall, J., "Rehearsal–Plus: Coping With Fire Emergencies and Reducing Fire-Related Fears," *Fire Technology*, Vol. 31, No. 4, 1994, pp. 432–444.
Kakegawa, S., *et. al.*, "Evaluation of Fire Safety Measures in Care Facilities for the Elderly by Simulating Evacuation Behavior," *Proceedings* of the Fourth International Symposium for Fire Safety Science, Intl. Assoc. for Fire Safety Science, Boston, MA, 1994, pp. 645–656.
Larson, R. D., "From Ashes to Education," *Firehouse*, Vol. 18, No. 1, January 1993, pp. 44, 46.
Lerner, N. D. and Huey, R. W., "Residential Fire Safety Needs of Older Adults," Thirty-fifth Annual Meeting of the Human Factors Society, Part 1 of 2, September 2-6, 1991, San Francisco, CA, Human Factors Soc., Inc., Santa Monica, CA, 1991, pp. 172–176.
"Let's Retire Fire. A Fire Safety Program for Older Americans," Federal Emergency Management Agency, Washington, DC, 1990, 15 pages.
Millsap, S., "Planning for Disasters in Your Community," *Fire Engineering*, Vol. 147, No. 12, December 1994, pp. 28–33.
O'Rourke, J. J., "Woodstock '94: Fire Planning for Large Public Events. Part 1: Fire Safety Preparations," *Fire Engineering*, Vol. 148, No. 1, January 1995, pp. 74–78
Perrault, M. E., "Home Security and Fire Safety Meeting Report, August 14-15, 1994, Quincy, Massachusetts," National Fire Protection Assoc., Quincy, MA, Meeting Report, December 1994, 22 pages.
Perroni, C., "'Get Alarmed, South Carolina' Lessons Learned From Its Success," USFA Fire Investigation Technical Report Series, Special Report, Federal Emergency Management Agency, Washington, DC, 1991, 61 pages.
Randall, J. and Jones, R.T., "Teaching Children Fire Safety Skills," *Fire Technology*, Vol. 29, No. 4, 1993, pp. 268–280.
Runyan, C.W., *et. al.*, "Risk Factors for Fatal Residential Fires," *Fire Technology*, Vol. 29, No. 2, 1993, pp. 183–193.
Teague, P. E., "1992 FPW Offers 'Sound' Advice," *NFPA Journal*, Vol. 86, No. 5, September/October 1992, pp. 68–71.
Thomas, J., "California Still Has Lessons To Learn," *Fire International*, No. 143, May 1994, pp. 37,40,42.
Walker, B. L., *et. al.*, "Short-Term Effects of a Fire Safety Education Program for the Elderly," *Fire Technology*, Vol. 28, No. 2, May 1992, pp. 134–162.
Walker, B.L., "The Effects of a Burn Prevention Program on Child-Care Providers," *Fire Technology*, Vol. 31, No. 3, August 1995, pp. 244–264.
Wood, B., "Childrens Fire Setting Behavior," *Fire Engineers Journal*, Vol. 55, No. 179, November1995, pp. 31–35.
"1994 Learn Not to Burn Champions," *NFPA Journal*, Vol. 88, No. 3, May/June 1994, pp. 67–69.

USING DATA FOR PUBLIC EDUCATION DECISION MAKING

John R. Hall, Jr.

Successful fire and life safety education begins with good decisions by the managers of fire and life safety education—decisions that identify the major fire problems and target audiences, design a program best able to address those problems with that audience, and deliver the right program to the right people. At every stage of this process, good decisions depend on the proper use of good data.

As a fire and life safety education program is developed and applied, questions needing data arise and may be grouped into the following six sections, which will be used to organize this chapter.

- What is the problem to be addressed?

- What is the strategy to address the problem?

- Who is the target audience?

- Was the target audience reached by the strategy?

- Did the strategy change the target audience as intended?

- Did the fire problem decline?

The first three question groups give the design for the program. The second three question groups indicate what was accomplished. Because most programs are not one-time affairs, the lessons learned regarding accomplishments will then feed into the next cycle of program redesign decisions.

This chapter addresses matching the right data with the questions posed, interpreting that data correctly, and eventually using data and all other available resources to save lives and property. Many of the issues and themes of this chapter have counterpoints elsewhere in the *Fire Protection Handbook*. The reader is particularly encouraged to read Section 11, Chapter 3, "Use of Fire Incident Data and Statistics," which provides additional guidance on techniques of data interpretation for fire safety not limited to public education; Section 2, Chapter 7, "Evaluation Techniques for Public Education," which provides more detailed guidance on questions 4 through 6 than will be provided here; and Section 2, Chapter 5, "Reaching High-Risk Groups," which provides more information for those (common) occasions when the target audience includes a number of high-risk groups.

This chapter—and this author—do not propose that analysis of readily available data is necessary or sufficient to answer all questions in the design and evaluation of public fire safety education programs (or any other kind of program). It is claimed that answers are sounder when they reflect the best data available and are ill-advised if they contradict, without engaging, the best data analysis.

Use data to challenge assumptions, even the ones that seem most obvious. If the data isn't strong enough or detailed enough to dictate the specific answer needed, then use the data to reduce the range of possibilities before applying judgment to make the final choices. Don't be afraid to use data to support choices and persuade third parties; however, don't confuse that exercise with the use of data to shape and assess choices. Decision-making and advocacy of decisions are both essential, and both can be done better with data analysis.

For advocacy purposes, representative statistics may be less effective than more specific examples that allow people to put a human face on the fire losses and the program activities. Match appropriate data to the target audience. Supervisors, managers, local legislative officials, and budget officials tend to want hard numbers, although even they will have an easier time grasping the need and the plan with examples. The general public, the press, and organizations outside the fire safety arena may relate primarily to the examples. The resources cited in the last part of this chapter, particularly the NFPA (National Fire Protection Association) "One-Stop Data Shop," can help locate accurate statistics and relevant stories.

WHAT IS THE PROBLEM?

This first stage requires the fire safety education team to define the objectives explicitly. What aspects of the fire problem should be addressed? What scales measuring fire loss are the team willing to use to measure success and hold the team accountable for results? The answer may appear obvious—save lives and prevent injuries, for example—but here, and at every other stage in the data use process, the team will learn a great deal about real objectives as it wrestles with the problem of setting performance criteria that are measurable, meaningful, and realistically achievable.

Start with the basics. Is the team comfortable being judged by changes in the size of the fire problem? If so, they will be taking responsibility for outcomes that are not totally under their control. Fire safety or risk is affected by both the behaviors people learn and the products and other physical aspects of their environment. Education

---

Dr. John R. Hall, Jr., is assistant vice president for fire analysis and research at the National Fire Protection Association. He has been involved in studies of fire experience patterns and trends, models of fire risk, and studies of fire department management since 1974 at NFPA, the National Bureau of Standards, the U.S. Fire Administration, and the Urban Institute.

directly affects only the former. On the other hand, if success is measured only by the number of children taught or the fraction of all children reached, then the team can guarantee success, given sufficient resources. But reaching children does not necessarily mean teaching children, and learning does not necessarily mean safety.

The following three examples are different ways to fail, relative to the only criterion that ultimately matters—are people safer? And yet each of these examples of unequivocal failure would be judged a success by some, even many, methods of evaluation.

1. To take an extreme example, an educator can stand in front of a class full of sight- or hearing-impaired or non-English-speaking children and deliver a standard fire safety education program. Although this may add 20 to 25 children to the tally of people reached, one can say with confidence, based on the mismatch between the standard medium of delivery and the special needs of these audiences, that the behaviors of this target audience will not change at all.

2. Or, an educator can deliver a program to a classroom of children who pass the test at the end of the day with "flying colors" but who forget everything they learned by the start of the next day's classes.

3. Or, an educator can teach a program that stays with the children for the rest of their lives, changing their behaviors as intended, but that somehow misses the behaviors that will prove to be most important in coping with the fire hazards they will face as adults.

Why address evaluation/success at this early stage? Because the design of a program—and the data appropriate to help design the program—should anticipate the way it will be evaluated. That is, at the very end, when evaluating whether the "problem" was reduced, it should be the same "problem" identified at the outset that inspired the design of the program.

Type of Loss

There are four principal measures of fire loss: (1) fire incidents (count each fire once), (2) deaths, (3) injuries, and (4) monetary loss. In some settings, two other measures may be of value: (1) environmental impact and (2) continuity of operations. The latter two measures are increasingly important in commercial settings. If a particular education effort leads to, for example, presenting a training session at the local chemical plant that employs a large number of the community's people and provides a large part of the community's tax base, then the educator may be less concerned about preventing a "typical" $10,000 fire at the plant than at preventing a catastrophic fire that might put thousands out of work and release a vapor cloud that could force evacuation of thousands more.

Which is the right measure to use? Most fire and life safety educators have a principal concern with saving lives, so deaths should be a measure of fire loss of concern to them. But should it be the only measure used?

Injuries are another measure of loss often targeted by fire safety educators. Injuries are several times more common than deaths, and some injuries are extraordinarily expensive, painful, and tragic. Most people value reducing the risk of death much more highly than reducing the risk of injury. However, injuries are more common, which means that injury statistics can provide significant, hard evidence of a program's positive effects much sooner. For that reason, measuring success by fire incidents, regardless of severity, also has advantages.

In the end, decisions must be made on which measure(s) of loss will be used. For each measure of loss, there needs to be an explicit rationale for the measure, which will provide guidance on how to interpret it and how much importance to attach to it. Different measures of loss will yield different priorities and multiple measures of loss can pull in different directions. For example, child-playing and smoking fires are major contributors to deaths and injuries, but only minor contributors to fire incidents. (See Table 2-2A.)

TABLE 2-2A. *Choice of Loss Measure Affects Cause Rankings for Home Fires, 1989-1993*

Major Cause	Fires	Civilian Deaths	Civilian Injuries	Direct Property Damage (in Millions)
Cooking equipment	101,200 (21.7%) #1	320 (8.3%) #6	5,060 (24.3%) #1	$382.0 (8.7%) #5
Heating equipment	85,400 (18.3%) #2	540 (14.0%) #3	2,180 (10.5%) #5	$601.8 (13.7%) #2
Incendiary or suspicious causes	57,900 (12.4%) #3	680 (17.6%) #2	2,310 (11.1%) #4	$866.6 (19.8%) #1
Other equipment	44,000 (9.4%) #4	240 (6.3%) #7	1,700 (8.2%) #6	$407.4 (9.3%) #4
Electrical distribution system	39,600 (8.5%) #5	370 (9.5%) #5	1,490 (7.2%) #7	$566.4 (12.9%) #3
Appliance, tool or air conditioning	31,200 (6.7%) #6	110 (2.8%) #10	1,080 (5.2%) #8	$234.3 (5.4%) #8
Smoking material (e.g., cigarette)	26,300 (5.6%) #7	900 (23.3%) #1	2,570 (12.4%) #2	$258.2 (5.9%) #7
Child playing	22,400 (4.8%) #8	390 (10.1%) #4	2,530 (12.1%) #3	$231.0 (5.3%) #9
Open flame	22,200 (4.8%) #9	130 (3.3%) #8	770 (3.7%) #9	$206.0 (4.7%) #10
Exposure (to other hostile fire)	17,500 (3.8%) #10	40 (1.1%) #11	180 (0.9%) #11	$383.7 (8.8%) #6
Other heat source	10,000 (2.2%) #11	130 (3.2%) #9	770 (3.7%) #10	$104.5 (2.4%) #12
Natural causes	8,700 (1.9%) #12	20 (0.5%) #12	160 (0.8%) #12	$136.3 (3.1%) #11
Total	466,300 (100.0%)	3,860 (100.0%)	20,810 (100.0%)	$4,378.2 (100.0%)

Cooking fires are major contributors to fire incidents and injuries, but much less so to deaths. Smoke detectors have major impact on the risk of death but very minor impact on the risk of injury. And so on. Design the education program to produce results on the measures that matter most, not the measures that will show results the soonest. For example, suppose the team chooses to measure fire incidents as early evidence of program success and fire deaths the measure of the real objective. Then they should design the program not for its leverage on the rate of fire incidents but only for its leverage on fire deaths.

Even after a measurement scale is chosen, more decisions are needed to select a specific measure. For example, suppose fire deaths are chosen as a measure of fire loss. This can be translated into four very different specific measures: (1) deaths in fires, (2) fatal fires, (3) multiple-death fires, or (4) deaths in multiple-death fires. Counting fatal fires rather than total deaths reduces the emphasis on deaths occurring in multiple-death fires. Under what circumstances might this make sense? Imagine a fire and life safety education manager in Las Vegas in the early 1980s. The MGM Grand Hotel fire caused more deaths in one fire than the city experienced in all other fires combined in many years. In that kind of situation, counting deaths individually—and targeting programs accordingly—will mean devoting all educational resources to ensuring that the city never has another fire like the MGM Grand Hotel fire. This is a worthy goal, but it is not the only worthy goal. How many lives would be worth losing in ordinary home fires in order to ensure, say, that the odds against another MGM Grand-sized fire were a billion to one instead of only a hundred million to one?

Alternatively, the team could decide up front to concentrate exclusively on multiple-death fires. The public often seems to be far more upset by one fire that kills five people than by five fires that kill one person each. If the intention is not so much to increase safety as to increase feelings of safety or to reduce public distress over gaps in safety, then the team might devote more attention to boosting the safety factors of programs that prevent large fires rather than programs more likely to save people in the circumstances where deaths are actually occurring.

Type of Property

A similar set of choices arises in choosing the properties of interest. Suppose the concern is with fire deaths, and the community, like most, suffers 80 percent of its fire deaths in home fires. This clearly argues for special attention to homes. But should the team target all homes and should homes be the only target? If not all homes, which homes should be targeted? If not exclusively homes, what other properties should be targeted?

Data on the fire problem will indicate where to target programs if targeting is done, but the same data will indicate the high price paid for targeting of any kind. Targeting based on fire problem data will mean targeting high-risk places and people. However, targeting the highest-risk places and people may mean giving up in advance any chance of reducing the large share of the fire problem cumulatively accounted for by places and people having less than the highest risk. It may be that 1 percent of the community accounts for 20 percent of fire deaths, for example, but if that group is targeted exclusively, then 80 percent of fire deaths will be left untouched, even if the program works perfectly on its targeted group.

In addition, there are other legitimate concerns, many of which also have data, that can be used in choosing the types of properties to target. For example, there are costs involved in delivering the program. If there is need to travel to selected sites (e.g., schools, homes) to deliver the program (e.g., school-based fire safety education, home inspections), then those costs are likely to be sensitive to the scatter of locations of targeted properties. It will cost much less,

for example, to go to the highest-risk block in the city than to go to the 25 highest-risk households in the city. (See Table 2-2B for an example of a data analysis technique for prioritizing and targeting parts of a community.)

TABLE 2-2B. *Targeting by Part of Community Using Data— An Illustration from a Dallas Anti-Arson Program*

District Number	Number of Arson Incidents	Percent of Total Arson Incidents	Cumulative Percentage
5	94	16.7	16.7
6	79	14.0	30.7
4	70	12.4	43.1
3	69	12.2	55.3
1	66	11.7	67.0

Setting priorities is a matter of drawing a line on this table at a point where resources available for the program will be exhausted or where the value of adding one more district drops sharply. The city drew that line here.

2	42	7.4	74.4
10	40	7.1	81.5
9	38	6.7	88.2
8	33	5.9	94.1
7	33	5.9	100.0
Total	564	100.0	100.0

Also, the program itself needs to have multiple parts if it is to address several very different property types. After homes, the property class that accounts for, by far, the largest share of fire deaths is road vehicles. But the fire problems of cars and trucks are far different from those of homes. An educational program designed for homes would be much easier to adapt to other residential properties—or to any group of buildings—than to vehicles, where the educator might as well start over from scratch.

In fact, it is difficult to find any educational program designed for vehicles, and this brings up another problem. Data analysis may do more than help choose among alternative educational programs. It may indicate a need to go beyond education in designing an effective prevention or mitigation program, for example, through codes and standards or other means for changing the physical characteristics of an engineered environment.

Public priorities do not always follow patterns of risk, either. The public wants most of all to be protected from fire risks associated with strangers, even though they are much more at risk from themselves and their families and friends. Risks are less acceptable when they are unfamiliar or involuntary, hence, the tendency to focus on strangers.[1] Homes and private vehicles account for more than 90 percent of all fire deaths, but they account for much less than 90 percent of people's exposure, as measured by, say, time.[2] The public worries about fire risks to children, even though school-age children are the lowest-risk age group in the population.

Much of the public worries about fire risks to people like themselves more than they worry about fire risks to people in other social groups.[3] Hence, programs targeted on the comparatively small high-risk groups may draw less public support than programs that make people with average-to-low risk even safer. This may seem like a purely philosophical or ethical question, but most fire safety educators are able to operate only because other people decide to give them funding and other resources. A program designed with an eye to the special concerns of the people who control the resources is more likely to obtain the resources it needs to succeed.

All of this needs to be considered, and all of it can be reduced to data. Should the educator target properties with a history of fire deaths? Or places with unusually vulnerable populations (e.g., schools, nursing homes, hospitals), even if their actual fire history has been extremely low? Or places with high occupancy (e.g., arenas, high-rise buildings), just because of the number of people potentially exposed to fire? Or places with unusual hazards, because the complex knowledge and behaviors required for their fire safety are the furthest from simple common sense? Should only buildings be targeted? Are other agencies or organizations being implicitly expected to address the fire safety needs of places not targeted?

Trend *vs.* Current Size

Another point to examine in data analysis to define the fire problem is the question of trends. From a national perspective, for example, total fire deaths have been declining, and the risk of death from fire relative to the size of the population has been declining even faster. Trends for fire deaths involving particular fire causes, however, have shown different trends.

In 1980 the top two major fire causes, in terms of home fire deaths, were: (1) smoking and (2) heating. Since then, these two causes have been among the fastest declining, so that, in 1992, both accounted for smaller shares of fire deaths than they had in 1980. (See Figure 2-2A.) Arson and suspected arson declined in terms of fire deaths, but not so steadily and not so steeply; and it moved past heating as the second leading cause of home fire deaths. Child-playing fire deaths rose in terms of share of fire deaths, ranking fourth in 1993 behind smoking, arson and suspected arson, and heating. In 1993, child-playing fire deaths increased so much that the total represented an absolute increase in deaths, not just percentage share of deaths, over 1980.

Which cause should a new educational program target first? Should it be smoking, which is still, as far as the most current statistics indicate, the leading cause of fire deaths? Should it be arson and suspected arson, which are likely to be the leading cause in a few years if trends continue? Should it be heating, which of all these four causes involves behaviors that are easiest to change through education and so might still produce the most life-saving value for a given level of effort? Or should it be child-playing fires, the only

well-defined major cause of fire for which the fire death problem was actually increasing as of 1993?

There is no one best answer to these questions. Any declining problem that is given priority in a new program may turn out to be a case of overkill. Any increasing problem that is given priority may turn out to have been varying around a constant or even declining trend line; that is, the "increase" may turn out to have been an artifact of when and how the baseline statistics were analyzed. Then again, any problem *not* given priority may suddenly "take off" and, once increasing, may lead to considerable loss and suffering before it can be controlled. For example, heating fire losses increased from the late 1970s to the early 1980s before intensified educational efforts, combined with other factors, brought that fire problem back under some degree of control.

Putting It All Together

The previous paragraphs were heavier on questions and issues than on answers. For those who prefer answers and guidance, the following tips may provide a reasonable way forward.

Design the program to reduce the risk of fire death among the people whose safety is the responsibility of the educator's team, which may be a community, an age group, a company's employees, an organization's members, etc. Treat every death as equally important and potentially preventable. Although fire deaths should not be the only measure, because they take too long to show statistically significant effects, deaths should be regarded as the primary measure of success.

When extending the scope to include injuries, property damage, or some other objective, treat all such objectives as secondary to the risk of fire death. That is, design the program to achieve a reduction in the fire death toll, and only then, if some latitude in the design of the program exists (i.e., room for a few more behaviors to be taught), look for program design features that will address the other objectives. The reason for this is that, even though other kinds of losses are more frequent, deaths are preemptively more important.

Use at least a five-year baseline of the community's or organization's fire experience for analysis. More years will be needed if the community or organization has fewer than 100,000 people. If the target population is closer to 10,000, or even less, the community or

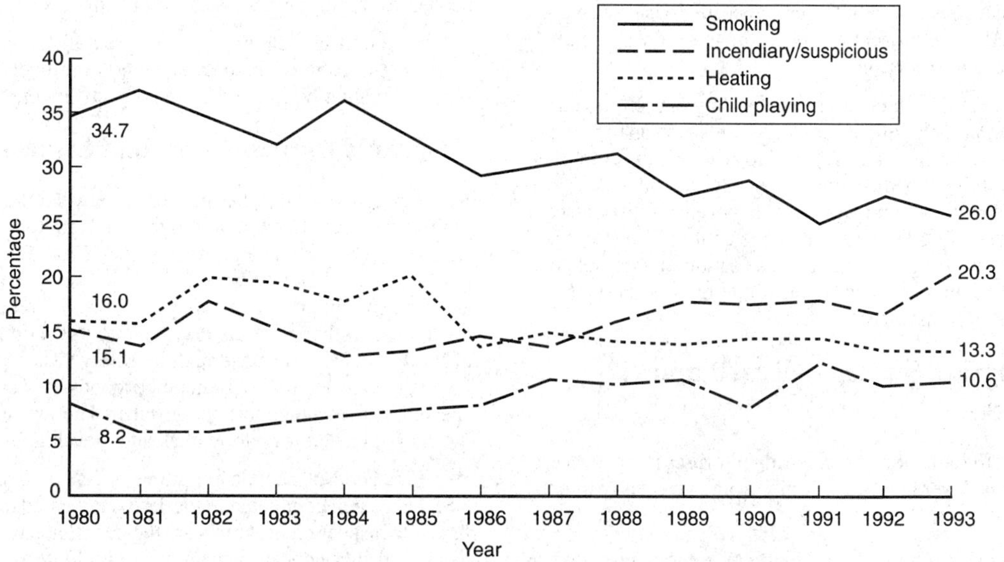

FIG. 2-2A.   Trends in four leading causes of fire deaths.

organization's own data will comprise too small a base to analyze or track success by the fire death toll in any reasonable period of time. In that event, use historical national fire death data—or, if obtainable, state data or data on a group of similar communities—for analysis to design the program.

Analyze the historical data base for major patterns that will be useful in selecting or refining programs or in targeting or reaching selected audiences. (See the following two sections.)

Do a "snapshot" analysis of the baseline, i.e., one that ignores trends. Then, only after the initial conclusions have been set, look at major trends and consider making a few adjustments based on the trends. Minimizing use of any apparent trend information other than major trends will: (1) avoid the dangers of over-reacting to statistical fluctuations that aren't real trends and (2) avoid the temptation to change the program before it has finished "doing its job" (which is analogous to stopping the antibiotics as soon as symptoms improve).

WHAT IS THE STRATEGY?

The purpose of data analysis is to set the stage for design or selection of a program that will reduce the fire problem as defined. However, many programs already exist; so don't "reinvent the wheel"— or any other parts of the car. This means the analysis will be most useful if it identifies patterns in the fire problem that favor some strategies and programs or help to set specifications for the selected strategies and programs.

For example, if the education team has data on fire deaths by home smoke detector status (e.g., how many in homes with no detectors, how many in homes where detectors were not working, how many in homes with an insufficient number of detectors, how many in homes with detectors in the wrong place) or related variables (how many in homes where there was no escape plan), then they will have a much better idea whether to build the program around home smoke detectors, and if so, what specific points the education should address. Some of the data cited in this example will be retrievable from a typical state or local fire department data base on fire incidents. Other data might require a survey. The analysis done will depend on the data available. If data specific to the target audience is not available, analysis based on national data may be available and used.

Fire cause data is similarly of interest because each cause corresponds to specific fire prevention behaviors that may be emphasized if the cause is a major concern.

A key danger at this stage lies in committing oneself too early to a particular approach. For example, if fires are analyzed by cause—and only by cause—this may lead to an education emphasis on prevention, to the exclusion of the kind of education that can save lives even when fire occurs. It is far better to develop at the outset a comprehensive view of all the programs one could pursue, then let the specifics of those programs guide the selection of analyses that will help decide the best programs and refine them to best fit the team's particular circumstances.

Keys to Program Success—Reach and Effectiveness

The central point to remember when setting priorities is that success requires substantial leverage, which comes from two equally important sources:

1. The educational program has to be effective enough to make a major difference in the fire risks faced by everyone the program reaches.

2. The program has to reach most of the people, representing most of the existing fire risk, or else even a highly effective program will have little effect on the size of the total fire problem.

People cannot be saved if they are not reached or if they are reached by a program that makes little or no difference. Frequently, the need to reach a large audience and the need to be effective with those reached will compete with one another. One can affordably reach many people with billboards and radio public service announcements, but it is not clear that such media, by themselves, make much difference. The educator can dramatically reduce the fire risk to people by using a comprehensive fire safety curriculum, e.g., NFPA's *Learn Not to Burn® Curriculum*, but it takes significant resources to do so, and even then, such a program directly reaches only school-age children.

Effectiveness, in turn, can only be assessed against a clearly stated objective, and a single program may be used to achieve many distinct objectives. What are the objectives with a school-based program, for example? (1) Is the program objective to teach the school children how to be more fire-safe during their school-age years? (2) Is the program objective to teach them fire safety for the rest of their lives? (3) Is the program objective to have them deliver fire safety to their families?

The first program objective involves a narrow reach (only the students and only for a limited time), but a high degree of effectiveness that can be readily measured. The second objective involves a wider reach (only the students but for their entire lives), but a lesser degree of effectiveness (many will forget what they have learned without practice) or a need to expand the program (how can fire safety refresher training be given to adults who learned fire safety in schools as children?). The third objective involves the widest reach (potentially everyone who knows a school-age child), but the least effectiveness (a lot will be lost in translation before the students can pass on their knowledge to their families).

To evaluate the impact of a program, a baseline is needed that addresses all the objectives. If the educator intends only to teach the school children while they are present, only a pre-test and a post-test are needed to check gains in knowledge, what fraction of the class learned all behaviors, etc. However, if the goal is to teach children for life, methods must be found to test adults five, ten, etc., years out of school, to determine their retention levels. And if the goal is to teach families through their children, then pre- and post-tests are needed for family members as well as for school children.

Selection and refinement of the program using data analysis requires application of the data collection protocols that support the kinds of evaluation implied by the objectives set.

Programs Implement Broad Strategies

At this stage, consider the difference between the strategy and the program. For an educational program, the strategy is to impart specific knowledge and change specific behaviors for a designated target audience. The program includes the materials used to accomplish that end, the people taught, and the people they teach, together with the materials they have to use. For example, the city of Portland, OR, won a national fire safety award for a program that involved targeting high-risk neighborhoods by working through respected, trusted neighborhood groups.[4] The fire department taught the church leaders who took the lead in teaching the neighborhood.

Fire code inspection programs are also education programs. NFPA research conducted in the 1970s indicated that the successful fire code inspection programs are the ones that have nearly universal reach and that educate and motivate building occupants to make changes, even when the inspector isn't there.[5] Most fires and losses involve behaviors and transitory hazards that may not show up in an annual inspection. Unless the occupant has been convinced to act as a year-round inspector, which means education, the value of the inspection wears off very fast, even if the inspector is very thorough and enforcement is very effective.

Six General Approaches to Prevent Fire Loss

During development of the initial list of candidate educational programs, for which objectives will be assigned and reach and effectiveness assessed, it may be useful to use the following set of categories, introduced in Section 1, Chapter 1, of this handbook: (1) prevention, (2) slowing the rate of growth of initial fire, (3) early detection, (4) early suppression, (5) preventing unusually rapid spread of fire or its effects, and (6) evacuation. Each of these categories has engineering and educational approaches, which ideally should work together. And each can be pinpointed as a possible priority through data analysis.

Analysis of fire deaths by major causes, defined in terms of behaviors and heat sources, can identify a number of possible priorities for prevention education. Most educational programs include a number of prevention-oriented behaviors and so can benefit from an analysis by major causes. Bear in mind, however, that conventional educational techniques are of limited or unproven effectiveness in dealing with most of the leading causes of fatal fires, i.e., smoking, arson, and children playing with fire. Make sure the program is designed to succeed with these causes before assigning a high effectiveness to the prevention education part.

Analysis of fire deaths by the first burnable item ignited may shed some light on the problem of rapid early fire growth. Flammable and combustible liquids and gases can promote such rapid growth, as can large pieces that are ordinary combustibles, like upholstered furniture. Most educational programs contain some guidance on storage of such materials, on limiting quantities, or on the use of potential heat sources near them, but the educator might also consider guidance on selecting items to reduce risk (e.g., choosing upholstered furniture made to the industry's voluntary standard on resistance to cigarette ignition) and on positioning items (e.g., how close are furniture pieces to walls and to each other?)

Analysis of fire deaths by detector performance will always show the value of home smoke detectors and of keeping them operational. National statistics show smoke detectors cut the chances of dying when fire occurs in a home nearly in half.[6] So will analysis by sprinkler performance, which should be included as an option in any fire safety program. Sprinklers reduce the chances of dying and the average property loss per fire by one-half to two-thirds.[7]

Analysis of fire deaths by extent of flame damage—and, in particular, whether or not fire is confined to the room of fire origin—will always show the importance of blocking fire spread. For an educational program, the analysis will be enough to justify including guidance on the need to avoid creating avenues of rapid flame spread (e.g., untreated wood paneling on walls, especially on stairway or corridor walls).

Analysis of fire deaths by the victim's condition before fire began and activity at time of fatal injury will give some insight into the importance of evacuation issues. Typically, most victims never get as far as trying to escape. They were asleep, impaired by drugs or alcohol or disability, or they were too young or too old to act effectively. Every such finding translates into a point for education, but be realistic about what can be achieved. The educator probably can't convince people not to get drunk—and certainly can't change disabilities or the limitations associated with various ages—but he/she may be able to convince people to develop and practice escape plans that are tailored to their circumstances.

Note that in each of these cases, NFPA has conducted analyses of national fire data that can be used both to make the general points that help sell the importance of particular behaviors to key third parties and to demonstrate formats for analyses that can be done on the community's or organization's fire experience data.[8]

Putting It All Together

Assemble and brainstorm a list of promising educational programs. Include any available evidence on their track record of effectiveness. Lacking such evidence, try to develop some estimates, for example, by asking for quantitative estimates from a variety of national experts and others familiar with the programs.

Set up a flow chart showing how specific actions will lead to changes in and actions by other parties, and from them to changes in the target audience, and from those changes to reductions in fire deaths and other forms of fire loss and risk. Fill in anticipated effects of changes at each stage and how effective the changes will be, based on the predicted effects.

Based on a review of available resources (and other considerations in the following section), rough out how large a version of each program can be launched. Translate that into an estimate of reach, i.e., the fraction of the target audience that can probably be reached.

Despite the emphasis on data analysis throughout this exercise, it is important to think qualitatively more than quantitatively. Look for big differences in likely impact, and don't be too wrapped up in small percentages. It should be obvious by now that much of the information will involve best guesses. Pay particular attention to any points in the flow chart where there is a real risk that the whole program will go "off track" if some key group does not do their part.

For programs that still look attractive after this analysis, use data analysis of fire experience to specify program details (e.g., which prevention behaviors will be emphasized?). Based on these details, estimate what fraction of the total fire problem the program will target.

WHO IS THE TARGET AUDIENCE?

Most public education programs target one of the following four groups: (1) the general population, (2) high-risk populations, (3) school children, or (4) people where they work. The first two are targeted because they represent all or a large share of the total fire problem, which gives the program a large reach. The second two are targeted because they find people under circumstances where they can and will spend significant time in an ideal learning environment, able to listen to lessons, practice behaviors, and be tested for what they learn. Conversely, the first group is hard to reach in ways that are highly effective, and the second group is just hard to reach. The last two groups have below-average fire risk to begin with, which may mean that high impact requires extending the reach of the program.

For programs aimed at the general population, serious questions have to be asked at the outset as to how effective the program will be. It is almost impossible to devise a program for the general population in which the participants practice fire-safe behaviors. Can the program even ensure that everyone will read or hear the lessons? Mailed materials and materials dropped off by people on rounds will often be discarded without being read. Billboards and public service announcements are within viewing range of only a fraction of the population and will be ignored by many of those.

Cost alone is usually enough to force a program to target only a fraction of the general population, in which case the program should be targeted at high-risk populations. If forced instead to target a partial program based on greatest ease or lowest cost, the education will almost inevitably, albeit inadvertently, end up targeting low-risk populations, the people who need the program least. The safest neighborhoods for home inspections, the most receptive neighborhoods for community meetings, the participants in fraternal organizations and other social groups, and the best and most receptive schools, all tend to favor the more affluent, better educated, and lower fire-risk people. It takes a great deal of work to hit the target of high-risk populations, so much so that Section 2, Chapter 5, of this handbook is devoted to that topic alone.

Once the decision to target high-risk populations is made, they must be identified and creative ways must be determined to reach them. If high-risk populations are defined by where people live, use actual fire history to find parts of the community where fire rates and fire death rates, relative to population, are unusually high. This will lead naturally to programs that deliver fire safety within those neighborhoods. (High-risk areas can easily have rates several times the community-wide average. Again, do not focus too much on small differences, such as slightly above-average fire rates.)

At this stage, effectiveness considerations, already discussed, will become as important as extent of coverage, or reach. Door-to-door programs offer more potential for substantial learning but are very resource-intensive. How many can the program afford to reach in this fashion? Assembling people through the guidance of some community group is more cost-effective but permits less individualized teaching and therefore may not result in as much practice of behaviors by the entire target audience. Such programs also tend to draw disproportionately from the lower-risk occupants of even a high-risk neighborhood. Mailing or dropping off materials is even more affordable and even less likely to change behavior. Use available data on effectiveness with estimates of projected numbers of people reached to compare different programs.

High-risk populations need not be defined by neighborhood or community, however. Older adults are a high-risk population, for example. Use data analysis to check every assumption for these programs.

For example, in the early 1990s, there were just under 2 million resident patients in nursing homes and related facilities, compared to a total of more than 32 million United States residents who were age 65 or older.[9] That is, there were more than 15 older adults living at home for every one living in a nursing home; therefore, any program intending to target older adults as a high-risk population would only "scratch the surface" by targeting the nursing home population.

Further, older adults living in nursing homes already have much lower fire risk than older adults living at home, because of the much greater built-in fire protection and supervision afforded in nursing homes. The population ratio may be roughly 15:1, but the ratio of fire deaths in the same period has been more like 80:1. Therefore, within the high-risk older adult population, the people living in nursing homes need additional fire safety least.

With creativity, an educator might try to target a program for older adults in homes but without ignoring nursing homes altogether. For example, nursing homes or like facilities could be used as magnets to draw in older adults from homes throughout the community?

The need for data analysis lies in the need to check assumptions, because assumptions are an inevitable part of crafting a fire and life safety education program to fit within limited resources. The example just discussed started with a (presumed) fact: It is not affordable to target everyone in the population, so limited program resources should be applied where they will do the most good. That means large-reach and/or high effectiveness. Targeting high-risk populations gives the most reach (in terms of share of the fire prob-

lem) per population reached. First assumption: Older adults are a high-risk population. Data confirms this; they have more than twice the fire death rate of the general population. Second assumption: Nursing homes are a compact, efficient way to reach older adults. Data disproves this, as shown. Third assumption: A program can be devised with enough effectiveness to produce high impact when targeted to older adults. Data can best address this only after the program is specified (targeting nursing homes didn't work) and alternative programs are specified, against which to compare the older-adult program.

If targeting nursing homes had proved to be a good idea, it would still be necessary to be creative, challenge assumptions, and analyze data to specify details of a program. For example, should the educational program target the patients or the staff? Both are possibilities, and both types of programs may have data on effectiveness. Or would a mixed program, targeting both groups, be best (where "best" takes account of affordability, too)?

In fact, every fire-related characteristic can be examined as a basis for identifying high- vs. low-risk populations, and, hence, for targeting fire and life safety education programs. Smoking material fires lead the list of causes of fatal fires. How about an educational program that works through points of sale of cigarettes? Heating equipment fires still rank fairly high. How about educational materials distributed where portable or space heaters are sold, serviced, or maintained? These are examples of what can be an iterative process, wherein data helps in design and selection of existing educational programs but also suggests unmet needs that may spark ideas for entirely new kinds of educational programs. Data analysis can be a stimulus to that kind of creativity as well as a reality check on assumptions about what will work and who most needs attention.

Targeting schools or workplaces presents its own set of questions, similar in form but different in specifics. Does the program provide that participants will practice fire-safe behaviors, or (less effective) take a written or oral test after being taught about fire-safe behaviors, or (even less effective) be lectured on fire safety with no testing or interaction, or (least effective) be notified of the availability of materials on fire safety?

Note the variations implied by this question. After a target audience and a set of information to be used in an educational program are specified, success (effectiveness) in getting the audience to learn, know, and be able (and likely) to act on the information may depend on the manner and intensity of the interaction. The easier it is for the audience to "drop out" of the exchange, by ignoring the materials or even by not examining them at all, the more likely it will be that many people will not be educated.

And reach takes many forms. Do all classrooms in the school, all schools in the system, all areas in the workplace, and all facilities in the industry participate? Or do only the most motivated, self-selected ones, who probably have the lowest fire risk already, participate?

WAS THE TARGET AUDIENCE REACHED?

At this stage, data is used not to design or select the program but to evaluate its impact. That evaluation may lead to some redesign, in which case it is necessary to move back to an earlier point, as appropriate. This first step in evaluation corresponds to the first effects the educator can hope to measure. It relies on *activity* measures—measures of the work already done.

The third section called for estimating the likely reach of programs and strategies as a basis for choosing those to be implemented. After program implementation, it is necessary to measure the reach that was actually achieved.

To do this, begin with data on people and groups to whom the program was delivered. For example, which schools were sent materials? Usually, the materials will not have been sent directly to the target audience, so it will be necessary to survey the addressees to determine how many of them passed on the materials, and to what fraction of the target audience members within their organization. Then it may be necessary to survey those individuals to determine how many of them used the materials.

Take the school materials example again. The program data will indicate which schools were sent the materials, but the educator will need to survey the principals to determine which classrooms took the materials, then survey the teachers from those classrooms to determine which ones used the materials and to what extent, and finally survey the teachers or the students to determine how many students in the classrooms participated in the exercise (e.g., omitting students who were sick or excused for other reasons).

Whatever method is used to build up the data base, the goal is to estimate the fraction of the original target audience that was actually reached. Therefore, count people, not some more convenient distribution unit, such as classrooms, schools, workplaces, etc.

This evaluation needs to be this thorough so that a realistic view of what was achieved can be obtained. There are many steps in a program where targeted people may be missed, and so any such step not evaluated is a place where one might inadvertently overestimate the effective reach of the program.

DID THE STRATEGY WORK?

This is the evaluation step that corresponds to the design step that specified the strategy. It usually involves pre- and post-tests of knowledge or learned behaviors for the target audience. This step calls for *change* measures—the most direct measures possible of the difference the program has made.

This comes later in the time sequence because, for most programs, the program objectives will imply a goal of long-term learning. The educator needs to ensure that the program has effects that last past the end of the program. This means, ideally, a sequence of post-tests, e.g., immediately after the program, one year out, five years out, etc.

In time, enough may be known about patterns of retention of fire safety knowledge that it will no longer be necessary to conduct elaborate post-testing regimens for specific programs; but, at present, such information is quite useful.

Even at this stage, look primarily for changes that are substantive and closely linked to how lives are saved. Measure changes in knowledge (pre- and post-tests) and behavior (by surveys, interviews, and other sample observation techniques). It may be interesting to note, for example, that school officials and fire officials are working together more now than before or that they enjoyed participating in the program or that they believe it made a difference; but these kinds of changes are too far removed from the hard evidence of program impact to deserve much emphasis in the evaluation. They may even reflect a mistaken notion of how success occurs. A recent study of workplace teams found that successful teams rarely sought consensus and that consensus was not a pre-condition for team success. An evaluation that assumed consensus was valuable in itself or that consensus was a good indicator of task success and program impact would be off the mark.

WERE LIVES SAVED OR LOSSES REDUCED?

This is the ultimate measure of benefit and impact—and the only one that truly matters. It is the evaluation counterpart to the design step that asked what the problem was. It involves comparing fire death rates and other fire loss rates after the program to those before the program. This step calls for *outcome* measures.

These are change measures, too, but they are measures of the intrinsically valuable changes to which all the work was meant to lead. If people learn more or know more, that is valuable only if they have fewer fires and fewer of them die [or are injured or whatever the chosen loss measure(s) for the objectives may have been]. At the same time, these change measures may be affected by more than just the program. Time has passed since the baseline period. People may have gotten poorer or richer, smarter or less knowledgeable, more mature or more impaired by advanced age. And other groups will have been conducting programs, some directly and deliberately focusing on fire safety, others not intending to address fire safety but having some impact on it nevertheless. Therefore, while these measures are more directly indicative of the program objectives, their changes or lack of changes may reflect more than the program under scrutiny.

Therefore, it is useful to apply experimental techniques that can isolate the fire safety program from other influences. Some common phenomena should be checked, as described below.

Fire loss rates, including death rates, will vary randomly over time, often significantly. If the educator introduces a program at a time when those variations have pushed loss rates up to historically high levels, then significant reduction relative to that baseline can be expected, even if the program itself has no effect. Loss rates will return to "normal" levels or trends, and that return will look like an improvement. Since a random variation pushing up loss rates looks the same as a problem "spinning out of control," in need of immediate special attention, it is not unusual for programs to be launched at the top of one of these artificial peaks. Consider that possibility when interpreting data.

Consider, too, the Hawthorne effect. First noted in studies of industrial productivity, the "Hawthorne effect" refers to the tendency of populations to do better just because someone is paying more attention to them. Just being in a fire and life safety program may inspire participants to be more fire-conscious, and so more fire-safe, even if they really haven't learned anything more. These motivational effects rarely outlast the program, or even the experimental phase of the program when publicity is most intense, so this possibility is another reason to check for effects over the long term. (It also can be checked by running a "placebo" program that has no educational content but delivers to a control group all the attention of being part of a program.)

Analyze data not just for how much impact a program produces but for how the program achieves its impact. It is not unusual for a successful program to produce effects differently than expected. This is important for design choices when moving from pilot programs to full-scale programs or from experimental programs to institutionalized programs, because it is important to emphasize those features that actually produced success.

One classic example was a fire department home inspection program whose introduction coincided with a large decline in home fire rates.[5] Detailed analysis, however, showed the decline often preceded the arrival of inspectors in a particular neighborhood, suggesting that the fire safety effects were produced more by the motivational effects of the program's introduction and the desire to "look good" for the inspectors when they arrived. This had implications for the program, because it suggested a need to emphasize mass-media publicity in support of the program on a continuing basis and to look for ways to encourage people to "clean up for the inspectors" rather than relying on the inspectors to find hazards on the spot and have them corrected.

See Section 2, Chapter 7, "Evaluation Techniques for Public Education," for more detailed guidance on measurement techniques and guidelines for interpretation of these last three sections. Also, there is extensive guidance on evaluation in NFPA's *Learn Not to Burn Curriculum* materials.

AVAILABLE DATA AND ANALYSIS RESOURCES

Much of the design and all of the evaluation work needs to be based on local data; this includes data specific to the team's program, community or organization, and target audiences. But a fair amount of the design work can be done on the basis of existing national data analysis, and even the evaluation can take advantage of existing data analysis already done on similar programs.

Guidance on analysis methods is given in Section 11, Chapter 3, "Use of Fire Incident Data and Statistics." For more details on sophisticated statistical analysis and modeling techniques, consult *The SFPE Handbook of Fire Protection Engineering*, published by NFPA.[10]

The most important caution to keep in mind regarding methods is this: The key weakness in most data bases is not the quality of the reporting but the representativeness of the coverage. Data bases do not have to be complete to be comprehensive; if they are properly representative, statistical methods will allow valid projection of totals from samples. Data bases do not have to be totally accurate within each record to be very accurate for statistical analysis; errors usually balance each other out, as one person's miscoding is offset by an opposite miscoding by someone else. (See Section 11, Chapter 3, "Use of Fire Incident Data and Statistics.")

The best national data for public education program design consists of what are called national estimates based on two national fire-department-based data bases—the U.S. Fire Administration's National Fire Incident Reporting System (NFIRS) and NFPA's annual fire experience survey. Through the "One-Stop Data Shop," NFPA maintains a listing of dozens of reports covering all aspects of the home fire problem, including patterns, trends, comparative risk factors, major causes, victim patterns, and detection and suppression equipment roles in fire safety. NFPA encourages educators to seek this support.

The U.S. Fire Administration's National Fire Data Center (NFDC) is another valuable data analysis resource. In addition to its role in maintaining and distributing NFIRS to all fire data users, the NFDC has done more than any other organization to develop analyses at the state and local levels. NFIRS is not a complete data base, having data on about half of all reported U.S. fires each year, so analyses below the national level are not possible for all areas; however, the NFDC is an excellent resource.

Another major source of fire data analysis at the national level is the U.S. Consumer Product Safety Commission (CPSC). In addition to a wide range of national estimate analyses focused on home fires—and a key role working with NFPA analysts in standardizing the methods used for making national estimates—CPSC is the nation's leading organization for special data collection and analysis projects applicable to home fires. Watershed analyses of home electrical fires and smoke detector usage and problems are only two of the best known examples. Their Hazard Analysis Group conducts these analyses and can describe what is available.

The best compilation of data-based evaluations of fire safety education programs is in a report by TriData Corporation. Anyone interested in evaluation of fire safety education programs should read *Proving Public Fire Education Works*, which also served as the basis for Section 2, Chapter 7.

Finally, the roughly 30,000 communities in the U.S. are resources for each other. Remember that, when creating programs or analyzing data for programs, it is likely that some other community has done so already and learned some useful lessons. NFPA's Champions program is building a network of leading-edge communities that have not only successfully demonstrated fire safety education programs but also committed themselves to sharing their knowledge and methods with others. Contact NFPA Public Education for more on the Champions network.

BIBLIOGRAPHY

References Cited

1. Slovic, P., Fischoff, B., and Lichtenstein, S., "Rating the Risks," *Readings in Risk*, ed. by Theodore S. Glickman and Michael Gough, Resources for the Future, Inc., Washington, DC, 1990.
2. Karter, M. J., Jr., *Fire Loss in the United States During 1994*, Fire Analysis & Research Division, National Fire Protection Association, Quincy, MA, 1995; and interval estimates of average percentages of time spent in various locations, done by John R. Hall, Jr., National Fire Protection Association, 1996.
3. This is an example of people's tendency to show concern with situations where benefits go to others and costs or risks go to "us." See M.W. Merkhofer, *A Complete Evaluation of Quantitative Decision-Making Approaches*, Final Report, SRI Project 2102, National Science Foundation, April 1983, Washington, DC.
4. Teague, P. E., "Member Profile: Jim Crawford," *NFPA Journal*, September/October 1993, p. 23.
5. Hall, J. R., Jr., et al., *Fire Code Inspections and Fire Prevention: What Methods Lead to Success?*, National Fire Protection Association and the Urban Institute, Boston, MA, 1979.
6. Hall, J. R., Jr., *U.S. Experience with Smoke Detectors*, Fire Analysis & Research Division, National Fire Protection Association, Quincy, MA, June 1995.
7. Hall, J. R., Jr., *U.S. Experience with Sprinklers*, Fire Analysis & Research Division, National Fire Protection Association, Quincy, MA, June 1995.
8. A listing of reports and a description of analyses that can be done may be obtained from the "One-Stop Data Shop," National Fire Protection Association, 1 Batterymarch Park, Quincy, MA 02296-9101. Phone: (617) 984-7450; fax (617) 770-0700. Guidance on how to analyze state or local data bases in the same way can be obtained from NFPA's analysts at the same address.
9. *Statistical Abstract of the United States 1994*, Bureau of the Census, Washington, DC, 1995, Tables 192 and 13.
10. *The SFPE Handbook of Fire Protection Engineering*, 2nd ed., DiNenno, P.J., *et al.*, eds, National Fire Protection Association, Quincy, MA, 1995.

Recommended Reading

NFPA, *Learn Not to Burn® Curriculum*, National Fire Protection Association, Quincy, MA.

Additional Resources

NFPA Champions Network, NFPA Public Education, National Fire Protection Association, Quincy, MA. Phone: (617) 984-7285.
NFPA, "One-Stop Data Shop," National Fire Protection Association, Quincy, MA. Phone: (617) 984-7450.
Proving Public Fire Education Works, TriData Corporation, 1500 Wilson Blvd., Arlington, VA 02209. Phone: (703) 351-8300.
U.S. Consumer Product Safety Commission (CPSC), Hazard Analysis Group. Phone: (301) 504-0470.
U.S. Fire Administration's National Fire Data Center (NFDC), 16825 S. Seton Ave., Emmitsburg, MD 21727. Phone: (301) 447-1272.

DESIGNING DISASTER EDUCATION PROGRAMS

Rocky Lopes

Research shows that educational activities play a vital role in saving lives, minimizing injuries, and enabling people to cope more effectively with disasters. Two challenges that face many organizations in the community that share a role in protecting life and property are: (1) getting families and individuals involved in preparing for disasters that may happen to them and (2) finding ways to prevent or reduce the effects of disaster.

This chapter provides information on how to select, design, and enhance disaster education programs, including presentations and materials. Also addressed are the broad aspects of disaster education activities, including many natural and technological hazards that can happen and ways to focus educational efforts on high-priority issues.

When consistent, clear, action-oriented information is provided to the public, the public is more likely to take action to prevent, prepare for, and respond appropriately to disaster. This chapter provides guidelines to effect this.

FUNDAMENTALS

Priorities for Disaster Education

Community disaster education (CDE) guides the community on how to prepare for, prevent, and cope with disasters. Various organizations do so with a variety of materials and activities. Disaster education can address a variety of community needs in a dynamic and flexible way.

Disaster education can influence people to prepare for disasters. Disaster education activities and materials can help reach:

1. Groups not previously involved with a specific unit.
2. Segments of the population that are vulnerable to injuries or losses from hazards.
3. The largest number of people, reflecting a community's diversity.

Each community is unique in the challenges it faces and the resources it can muster to meet them. Therefore, a disaster education program must identify those hazards, demographics, and resources specific to that community. Two resource examples are: (1) financial

---

Rocky Lopes, Ph.D., is director, community disaster education, in the Disaster Services Department at the National Headquarters of the American Red Cross, Washington, DC.

support and human resources and (2) investment in building relationships with other organizations that share common goals.

Distributing brochures and making presentations can be useful, but a disaster education program requires a more multi-layer approach. The real measure of the effectiveness of a disaster education program is the degree to which people:

1. Learn from it.
2. Use it to *do* something to prepare ahead of time.
3. Know what to do when faced with disaster.
4. Respond appropriately.

Disaster education is an ongoing process that involves progressive steps to be successful. It is essential that each participating organization choose appropriate target audiences and design an overall approach that will be effective.

Disaster education is more manageable when efforts are concentrated on specific segments of a community. That is, when a targeted group(s) is identified, it can enhance:

1. Development of achievable goals.
2. Identification of key leaders and community representatives who can participate in CDE.
3. Development of a strategy to meet the needs of targeted populations.
4. Identification of the best information sources and ways to reach them.
5. Development of an ongoing relationship with targeted groups.

A systematic and realistic method for designing a disaster education plan specific to a community is needed in order to reach the maximum audience and ensure that real learning takes place. Hazard and community profiles, as described in the following subsections, provide elements necessary to effect these goals.

A Hazard Profile

A hazard profile must address those hazards that a specific community faces. It should focus on natural or technological events that can cause community disruption and/or displace people from their homes temporarily or permanently, e.g., a flood or a residential fire. Public health authorities often conduct education programs on community health hazards, such as substance abuse or AIDS; however, these subjects are usually not part of community disaster education programs.

A list of probabilistic hazards specific to a community should be prepared. Many of these hazards cause frequent emergencies and are well known. Other hazards may happen rarely and may be forgotten in planning. Inclusion of all pertinent opinions can help ensure all hazards are included on the list. Sources of hazard information include:

1. City or county's disaster plan, which is developed by the local emergency management or civil defense office;
2. State office of emergency management (or equivalent);
3. Fire department(s);
4. National Weather Service; and
5. State offices of geology or U.S. Geological Survey (USGS).

Hazards to consider are:

Blizzard	Lightning
Chemical emergency	Nor'easter (extratropical storm)
Earthquake	Radiological emergency
Explosion	Tornado
Fire (residential)	Transportation accident
Flash flood	Tropical storm
Flood	Tsunami
Frost and freeze	Volcanic eruption
Heat wave	Wildfire
Hurricane	Winter storm

After a list is created, rank each hazard, as follows:

1. *Likelihood:*   Place a rank of 1, 2, or 3 next to the hazard, with those ranked 3 being most likely, and those ranked 1 possible, but least likely to occur. For example, heat waves and flooding are both very likely events for Memphis, TN, and would be ranked a 3. Winter storms, while possible, do not happen very often in Memphis, so they would be ranked 1.
2. *Frequency:*   Rank from 1 to 3 the frequency with which hazards occur. Again, 3 indicates rather frequent occurrence. Use 1 to indicate low occurrence.
3. *Community Disruption:*   Finally, rank the hazards by the amount of community disruption the event would have. For example, an earthquake in Memphis is quite possible, and would cause widespread community disruption. A residential fire is also very possible, but does not cause widespread disruption. Rank the hazards that can cause the most community disruption with a 3, and the least community disruption with 1.

It is important to understand specific terminology and data applied to ranking. For example, a "100-year flood" does not mean a flood can happen every 100 years. It means there is a 1 percent chance of flooding to a certain depth every year. Further, it should be realized that if a community just had an event, such as a hurricane or a tornado, it does not alter the likelihood or timing of the same event happening again.

After the hazards are ranked according to likelihood, frequency, and community disruption, multiply the numbers together for each hazard. Hazards with the highest multiplied totals indicate those on which disaster education activities should be focused first.

Community Profile

A community profile should be developed to determine: (1) a general picture of the population makeup, (2) the organizations that serve a community, and (3) those geographic areas most at risk. This profile must be refined for specific target audiences determined later.

The public comprises many audiences with diverse needs and interests. Understanding the makeup of a community helps: (1) prioritize at-risk populations and (2) identify those audiences to target first, next, etc.

Information to develop a demographic profile of a community can come from many sources, such as:

Chamber of Commerce	Local library
City/county manager/executive	Mayor's office
Congressional representative(s)	Planning and zoning board(s)
Emergency management	Social service agencies
agency	United Way or other federated
Local board(s) of education,	fundraising organizations
college, or university	U.S. Bureau of Census

Many factors can be useful when compiling community profile information: population clusters; socioeconomic status; building types, sizes, and locations; age; level of education; and ethnic diversity. (Such information provides details about the target audience.)

A profile should also include information about the organizations and groups that serve the community, such as:

1. Location of, and populations served by, schools, religious organizations, and community-based service organizations, e.g., community centers and health clinics;
2. Number and population served by organized labor groups;
3. Location and nature of the business community and business organizations; and
4. All media that serve the community, e.g., radio, newspapers, magazines, television (direct, satellite broadcast, and cable), and computer networks.

Determination of those geographic areas that are at highest risk for certain hazards, and if there are particular groups or populations at risk in these areas, is the most critical aspect of a community profile. For example, is the area subject to windstorms, tornadoes, or hurricanes? If so, mobile home parks and other vulnerable structure sites should be a major focus of disaster education. Is deteriorating/poorly constructed housing concentrated in certain areas of the community? Residential fire risk is a high priority in these areas.

The community profile should be merged with the hazard profile to determine: (1) the areas where disaster education efforts should be concentrated and (2) which disasters should be addressed.

TARGET AUDIENCES

Disaster research shows that certain segments of the U.S. population face a greater risk from hazards because of age, ethnic background, income, physical capabilities, limited English proficiency, and geographic location. For example, some buildings or neighborhoods may be more vulnerable to the effects of disaster, and people with limited English proficiency may not receive important warning or preparedness information. Further, some people may have more difficulty in recovering from a disaster because of a lack of available resources or a lack of knowledge about available assistance.

A community disaster plan developed by an emergency management agency should contain an assessment of the potential impact of the hazards on the community (e.g., families affected; demographics, with attention to "special" populations; impact on infrastructure, such as roads and bridges; and utilities).

It is important to target initial community disaster education efforts toward those groups who are most vulnerable to disaster. When reviewing the hazards in a community, determine which groups would be most affected by those hazards or were affected by past disasters.

Use the following questions about factors in the community that may put people at greater risk to disaster as a guide to identifying target audiences.

1. How is the land used? This includes location patterns, use types, and public/private access.
2. What is the population distribution in specific areas?
3. How does this population distribution change at times during the day or on weekends? This includes rush-hour and work-day patterns.
4. What is the economic development forecast for the community?
5. What is the building stock like? This includes the age of buildings in the community, type of construction, elevation, and nature of occupancy.
6. What are the high-occupancy structures and how are they used? This includes large apartment buildings, offices, major employment or shopping centers, theaters, auditoriums, stadiums, and schools.
7. What are the involuntary occupancy structures? This includes prisons, mental institutions, hospitals, and convalescent and nursing homes. It is important to not only know the location, but their occupancy, frequency of use, and time and duration of use.
8. What and where are the "lifelines" in the community? Lifelines include the transportation network, communications, water, sewer, gas, and electrical systems. Also, remember to include airports, train and bus stations, subway systems, cargo terminals, docks, water or gas storage facilities, power plants or stations, water or sewage treatment facilities, bridges, interchanges, etc.
9. Where are facilities that manufacture, store, or transfer hazardous materials? Include power plants, dams, and manufacturing or storage facilities for toxic materials.
10. What methods are available to get information to the community? This includes radio, direct and satellite broadcast, cable television, newspapers, magazines, emergency broadcast system, sirens, public address systems, etc.

This is not a scientific process, but it should yield information that helps determine where to focus community disaster education efforts.

Forced-Choice Priorities

Forced-choice priorities for disaster education efforts follow: (1) assessing and ranking hazards that can happen in the community and (2) determining those parts of the community, and its residents, that are at highest risk for these hazards.

No one organization can provide disaster education to everyone in the community. Limited resources, both human and financial, prohibit this. The following factors can assist in making choices for disaster education priorities:

1. Who else in the community is conducting educational efforts on a specific hazard? (See the following section entitled "Identify Current Disaster Education Activities.") If another organization is already conducting such efforts, it would be prudent to concentrate educational efforts on other hazards.
2. What is the severity of the problem(s)? For example, does an identified hazard cause many injuries or deaths when it happens, or does it cause more property damage? Be careful not to let media reports influence this decision. On occasion, the media will focus on two or three deaths, but treat property damage lightly, thus skewing the severity of the problem. Media reports on Hurricane Andrew, which occurred in September 1992, is an example. Although 15 people died from the direct impact of the

hurricane in Florida, more than 10,000 people were displaced from their homes due to wind damage, because they did not adequately protect their property. It can be inferred that home protection activities, called "mitigation" by government authorities, could have reduced the impact of the hurricane. Logically, this focus quite possibly could have reduced the number of deaths.
3. What is the community's perception of the risk? For example, earthquakes can be a hazard in areas where people do not perceive them as such (e.g., the Central U.S.; Charleston, SC; or New England). In this case, the public will be less likely to prepare for earthquakes because they deny that earthquakes can happen in their community. This denial blocks their ability to perceive that they are at risk, and therefore, no matter how good the educational effort, little public preparedness activity follows. It is often easier to get people to prepare for something they think can happen to them (e.g., flooding in the Central U.S.). If people are already prepared for one event, they possibly will be better prepared to deal with another.
4. How can the most vulnerable populations be reached? These populations in the community include those who speak a language that no one in the disaster education program can speak; and those with cultural differences and norms that prevent program attendance. These populations must be addressed beyond the belief that a conventional notice will lead to program attendance and, thus, disaster education. The section entitled "Working with Community Coalitions" provides information that can increase participation of all community members.

Prioritize the selection of disaster education activities by first addressing those that can be met with the most success. Choose target groups that are initially easier to reach, and then later try to reach those that are more difficult. Doing disaster education for groups that other organizations are already reaching can be made a lower priority.

IDENTIFY CURRENT DISASTER EDUCATION ACTIVITIES

There are many organizations already providing disaster education to the public. This is not a new activity. Disaster education goes by several names, such as family protection, population protection, human services, emergency education, and other titles. The title is not important; the educational effort is.

What other disaster education activities are being done in the community already? There may be several groups, organizations, and individuals actively doing some form of disaster education. Some of these groups are the following:

American Red Cross chapters	National Weather Service office
Other nearby American Red Cross units	Neighborhood organizations
County extension service	Other voluntary agencies active in disaster response and recovery
Disaster supply vendors	
Emergency Management or Civil Defense Office	PTAs, PTSAs, PTOs, and the like
Fire department	
Insurance companies	Religious groups
Law enforcement agency	Schools
Local media	Telephone and utility companies
Local organizers/activists	

It is unlikely that one can contact each of these organizations to find out what they are doing in disaster education. However, it is likely that, in every community, one or more of the above organizations has been involved for some time in public education efforts.

Contact two or three of the most active organizations in the area, and ask the following questions:

1. Who are the key players and contacts?
2. What hazards are they addressing?
3. Who are they targeting?
4. With whom are they working?
5. What messages are they delivering?
6. What is the focus of their information?
7. What materials and media are they using?
8. Is their effort working?
9. Who are they unable to reach?
10. How can you work together?

Identify the needs being met by these efforts. Are they successful? Do these groups need help with their programs? What needs are not being met by these efforts? Is there a gap that needs to be filled? Are any potential collaborators? If possible, try to unite efforts.

Educational Materials

Many materials have already been developed by others and are available for free or at minimal cost. Because development is expensive and time consuming, it is worthwhile to conduct a thorough search of available materials before deciding to self-produce products. Carefully review all materials to ensure that the information is correct.

A note about using other organization's materials: Much work goes into developing quality materials. The reasons why a product has been developed may not be readily apparent. Before deciding to create a new product that may be modeled after an existing product, try to contact the creator of the item to determine how it was developed, its uses, and how evaluation information was obtained. *A copyrighted product cannot be modified or duplicated without written permission.* For noncopyrighted materials, courtesy and ethics *require* that one obtain permission to use another individual's or organization's product, especially if the intent is to put one's own organizational name or local contact information on the item. In summary, plagiarizing other's materials or modifying them without permission can either be illegal or can create ill-will, both of which are detrimental to the organization's reputation and/or future.

General Resources

The following is a list of organizations that have produced materials for use in public education activities. Contact each organization for further information about available materials. Some organizations publish catalogs, product listings, or both; be specific when asking for materials.

American Red Cross: For further information on available materials and their costs, contact the local American Red Cross chapter that serves your area. (The national headquarters cannot fulfill requests for materials; only local chapters can place orders for Red Cross materials). The Red Cross also provides Community Disaster Education information on the World Wide Web at the following URL (uniform resource locator): http://www.redcross.org

Federal Emergency Management Agency (FEMA): The FEMA Family Disaster Preparedness Program and Mitigation Directorate and the U.S. Fire Administration publish many materials for use in public education. The *Emergency Preparedness Materials Catalog* is available free by writing: FEMA, P.O. Box 2012, Jessup, MD 20794-2012. Note that three divisions of FEMA produce materials: (1) Family Disaster Preparedness Program, (2) Mitigation Directorate, and (3) USFA. Telephone requests for FEMA materials are accepted at (800) 480-2520.

National Fire Protection Association (NFPA):   NFPA develops many printed and video-based educational materials. Call 1-800-344-3555 to request a catalog or to place an order for materials.

National Weather Service: The National Weather Service has educational materials available in print, video, and film formats. For more information on available materials, contact the local office of the National Weather Service, and ask for the warning coordination meteorologist.

U.S. Fire Administration: *Resources on Fire*, a catalog of fire education materials published by the U.S. Fire Administration, is available by writing: U.S. Fire Administration, 16825 S. Seton Ave., Emmitsburg, MD 21727.

U.S. Geological Survey (USGS):   The U.S. Geological Survey has produced abundant information about earthquakes, volcanic activity, and hydrology. Two good initial contacts to consider are: (1) the state office of geology or mineralogy and (2) a university geology department. The USGS headquarters can also be contacted for information: U.S. Geological Survey Information Services, Box 25286, Denver Federal Center, Lakewood, CO 80225. Phone: 1-800-USA-MAPS.

State Resources: Many state agencies also publish educational materials, often providing information unique to the state. Contact:

1. State fire marshal's office,
2. State office of emergency management,
3. State department of public health, and
4. State director of fire service training.

Each state has different names for its lead agencies that deal with disasters. A state contact can provide information on those agencies that may have disaster education materials available and can make further resource suggestions.

Another state resource is the agricultural extension service (sometimes called cooperative extension service). Extension service offices can be found in every county of every state, often in the county seat. An alternative is the state agricultural extension service, which is usually housed on a large university campus (often at land-grant institutions).

Quality/Accuracy Assessment

Any educational materials handed out or shown to the public reflect on the distributing organization's reputation. It is particularly important to review materials produced by other organizations before using them in a community education program.

The following guidelines can help determine the quality and accuracy of materials:

1. What year was the item published? Materials more than three years old may contain outdated information.
2. Does the information in one organization's materials differ from that of another? For example, does one recommend taping windows before a hurricane makes landfall, and another discourage this practice? Taping windows used to be a recommendation in the 1970s, but has been proved to be ineffective. If an organization's materials recommend outdated practices, even if recently published, it is an indication that little attention has been given to recent information updates and state-of-the-art practices.
3. What evaluation criteria are available? That is, have the materials been proved to be effective for the intended audience?
4. What does the material cost? Higher cost does not necessarily mean greater accuracy or suitability for a target audience.

Accuracy of materials can be confirmed by professional review. Contact practicing geologists, meteorologists, hazardous materials coordinators, and other appropriate professionals to conduct reviews.

Budgetary Considerations

Many materials exist for disaster education efforts, but some of them are quite costly. Therefore, direct costs for acquiring materials must be considered, as well as the following cost issues:

1. Are there shipping charges?
2. Can the item be reproduced locally?
3. Can one educational video be purchased for duplication? (Note copyright laws.)
4. Has an item been produced using color (such as a four-color brochure) for a reason? If so, it is not appropriate to photocopy the item using a black-and-white process, because marketing research has probably shown that doing this results in an ineffective way to reach the target audience.

Avoid compromising a presentation, e.g., photocopying four-color brochures, which can only result in unsatisfactory, illegible results. Rather, consider combining resources/forces with a local business, hospital, insurance company, school system, utility company, or other organizations that share a common goal in reaching people with important information, but may lack the resources to do so. They may appreciate new educational efforts and subsidize costs of acquiring or printing quality materials.

DISASTER EDUCATION PLANNING

Identify Community's Information Needs

Different communities have different needs for information. For example, Californians are very aware of earthquake risk and are accustomed to receiving earthquake education information. However, the Central U.S. is also at high risk for great earthquake damage, yet fewer people are aware of this risk and simple ways to prepare.

Determine Information Sources

It is necessary to identify those organizations, specific media, and opinion leaders that can disseminate disaster education information for each target audience.

1. What media reaches the target population? These can include:
   (a) Radio and television stations, newspapers, and periodicals.
   (b) Minority-oriented or ethnic community newspapers, periodicals, and radio and television stations.
   (c) Special newsletters or bulletins distributed by community organizations, agencies, hospitals, etc.
   (d) Billboards, posters on buses, kiosks, etc.
2. Which groups serve the target audience? They can include:
   (a) State and local agencies.
   (b) Religious groups and centers.
   (c) Community centers.
   (d) Local merchants.
   (e) Neighborhood associations.
3. Who are the community's opinion shapers and leaders that influence and represent the targeted population? They can include:
   (a) Religious leaders.
   (b) Business leaders.
   (c) Radio and television personalities.
   (d) Local politicians.
   (e) Local influential persons, e.g., the head of the neighborhood association, the Urban League, the National Association for the Advancement of Colored Peoples (NAACP),

La Raza, Ser, League of United Latin American Citizens (LULAC), and Puerto Rican Forum.

It is important to reach high-risk audiences in ways that will be effective. For example, if a high-risk group has a great fear or distrust of anyone in a uniform, then a uniformed fire department representative may not be the best choice as a presenter of this information.

Cultural influences and changes must also be addressed, since they can have an impact on the way information is absorbed. For example, people who speak Asian languages in the United States may not be able to read the language in its written form. They can understand it when they hear it, but may have lost the ability to read the language. Also, generally speaking, it is not part of the culture of Asian language speakers to "read for information." Contrary to traditional educational methods employed in the United States, people who speak Asian languages learn differently. In a recent survey ("How FEMA's Community Outreach Worked After Northridge," *Los Angeles Times,* June 18, 1994), Asian-language speakers were asked, "Where would you go for information on how to prepare for a disaster?" None mentioned written materials (e.g., brochures); instead, they suggested that their best sources were television and a respected elder.

When reaching out to hard-to-reach audiences, it is very important to include same-culture representatives from hard-to-reach target audiences to help ensure that methods, materials, and approaches will be understood and successful.

Determine Activities and Approaches

Three factors need to be addressed when determining those activities and approaches that will produce the best results for the target audience. They are: (1) media selection, (2) message selection, and (3) cost/resources.

Media Selection: Different target audiences have different ways of receiving information and must be approached appropriately. For example, a print brochure designed for adults but given to a child is probably worthless. Or, showing a video in English to a non-English-speaking group will not result in much retention of information, and thus few results.

People can be reached successfully when using appropriate materials. Again, involve representatives from the target audience; they can help determine the best ways to conduct outreach activities for their community.

Message Selection: What messages are appropriate for what audiences? Do not make similar recommendations to all audiences. For example, it is appropriate to address how burglar bars on windows prevent escape from fire in areas where these devices are common, but it is obviously inappropriate to address this issue where they are not.

The following information is useful when creating educational materials. These seven points, derived from research, can significantly add to the success of local community disaster education efforts.

1. *Choose and Limit Messages.*

   The average person has a limited potential for retaining information received. Studies have shown that the average adult remembers less than half of what he or she hears after one hour, less than one-fourth of what he or she hears after one day, and less than one-tenth of what he or she hears after one week. Therefore, be careful to: (1) choose the most important messages and (2) limit the number to no more than seven messages per setting.

In a presentation, reinforce the messages and theme. Repeat important messages several times in different ways, and supplement the presentation with a handout, given at the end, not at the beginning or during the presentation. Emphasize the messages by using a variety of tools, e.g., newsprints, brochures, posters, films, videos, and slides. Also, learning activities can increase information retention. The more images people see, the more they will remember.

Public presentations should be limited to 30 minutes. Beyond that time, it can be difficult to manage and hold audience attention.

2. *Use Positive Messages.*

Use messages that are short, specific, and positive; i.e., those that tell people what to do. Negative messages like "Don't Panic" do not tell people what to do. In fact, messages like this can actually cause people to think about panic, because the message plants a negative thought in their minds.

Avoid potentially mixed messages. For example, "in case of fire, don't use the elevator." This is an accurate message, but doesn't tell people to use the stairs (the right thing to do). Instead, it tells people what not to do. Research has shown that people remember the last thing they heard better than the first. So, in this case, they are more likely to remember "elevator" instead of "don't."

3. *Avoid Mixed or Value-Laden Messages.*

The reason for undertaking community disaster education is to teach people what to do. Therefore, when mixed messages or value-laden statements are given, the educator's own credibility is compromised, and some of the audience may be offended. A mixed message is when one thing is said and then another is communicated. For example, "in case of a tornado, avoid chasing it." The intended message is that people should take cover and protect themselves. But what this message does is plant a thought about a potentially harmful activity—chasing tornadoes.

Avoid value-laden statements, i.e., those statements that express personal beliefs or opinions without identifying them as such, and that sometimes may imply that the educator's values are more important or better than those of the audience. An example of a value-laden statement is, "I think it's stupid to try to outdrive a tornado in a car." This statement implies that the educator thinks people who try to "outdrive" tornadoes are stupid. But they aren't stupid—they are just engaging in inappropriate behavior based on an incorrect belief.

4. *Use Awareness and Educational Messages Appropriately.*

Messages can be categorized in various ways. Disaster education activities usually include both awareness messages and educational messages. It is important to recognize what each type of message is designed to do, and when to use each one.

An *awareness message* informs people about the nature of hazards around them, tries to motivate action, and may give simple safety tips. An awareness message demonstrates that there is more to know. To be effective, however, it should be part of an overall strategy that first heightens people's awareness and then educates them.

An *educational message* also gives information about hazards, warnings, and how to stay safe, i.e., prevent personal injury or loss of life. But educational messages give more in-depth information than awareness messages and are designed to have the audience learn this information.

It is important to keep these distinctions in mind when preparing community disaster education materials. Present awareness messages first. People need to be aware of the hazard and the nature of the threat and be motivated to learn more. Awareness is a crucial link in the chain of how people receive information. Also remember that one message is not enough—i.e., repeat messages to get people to retain information. It is therefore necessary to include both awareness messages and educational messages in the disaster education plan.

5. *Correct Myths and Misinformation.*

Another purpose of the messages given in disaster education activities is to correct myths and misinformation. The best method is to identify a myth or misconception, provide the correct information (i.e., expose the myth), and follow up by telling people what to do.

6. *Overcome the Desire to "Tell Everything."*

The term "information overload" applies to community education activities. Don't overwhelm the audience. Determine the most important messages to deliver, i.e., don't try to tell everything known about the subject all at once. This is a difficult challenge. There is so much information to give, so much to learn, and so much misinformation to correct.

The following points can help avoid the risk of "information overload":

(a) Give out precise information.
(b) Give out only information that will heighten awareness and encourage positive behaviors among the public.
(c) Provide information in stages. Start with awareness-raising messages, which will focus the attention of the audience, then provide education messages and go into more detail.
(d) Focus on a few key points—and repeat them. Remember that one time is not enough.

For example, is the community as aware of its hazards as Californians are of earthquakes, or does it have misconceptions about the hazards? In California, the emphasis may be on making residents aware of the importance of home improvements to reduce the risks and on reinforcing safety behaviors. In contrast, in the New Madrid Seismic Zone in the Central U.S., it may be more appropriate to raise awareness about the threat of earthquakes before providing information on home improvements or safety behaviors.

7. *Time the Messages.*

Consider the timing of both awareness and educational activities. For example, design activities around a local historic disaster event or at the beginning of a severe weather season. Be prepared to provide information to reach a community during a "teachable moment" when external events heighten interest, such as during media coverage of a disaster in another area. After interest subsides, it is still important to continue community disaster education activities. It is also important to have information available and "on the shelf" to use when the opportunity presents itself.

Cost and Resource Requirements: Every educational activity has some cost attached to it, even if materials for distribution are available at no cost. It is therefore necessary to determine direct and indirect costs, and those resources required to carry out effective disaster education.

First, identify the direct costs. They are easier to assess. Some direct costs are:

1. Salary of staff assigned to the project;
2. Cost of materials, including shipping and handling charges;
3. Photocopying/duplicating charges;
4. Office supplies;
5. Equipment (or equipment rental);
6. Advertising or promotional fees;

7. Space (for presentations); and
8. Postage and telephone charges.

Indirect costs are harder to assess. Indirect costs include expenses absorbed in other budgets, such as salaries or compensatory time for presenters, postage and telephone charges, office supplies taken from a general source, and those that are usually not "charged" per activity, such as photocopying in small amounts, utility charges (e.g., electricity), etc.

A financial analyst or specialist, comptroller, or accountant can help determine the direct and indirect costs of the activities, and can provide information on what expenses must be paid directly and what expenses can be covered by other resources.

Implement the Activity

The following recommendations can help guide a successful disaster education activity.

1. *Be Prepared.*
   (a) Order materials to support the program. Allow sufficient time for orders to be received, sorted, and distributed to presenters.
   (b) Schedule activities. Involve others in the community who share the same disaster education goals, as identified earlier in this chapter. It is important to contact them well in advance, state the request, and allow time for arrangements to be made.
   (c) Outline an agenda for the program. Get approval from everyone involved. Allocate time to allow appropriate participation from all presenters and for questions from the audience. Limit the presentation to 30 minutes, as recommended earlier in this section. If a longer presentation is essential, allow time for a brief break(s) as necessary.
   (d) Check the location where the activity will take place. Is there sufficient room? Adequate utilities, e.g., lighting, heating, etc.? Chairs? Tables? Audiovisual equipment?
   (e) Check equipment. If audiovisual equipment is required to support the educational effort, learn how it works beforehand. Provide extra bulbs, screen, extension cord, etc., if needed.
   (f) Prepare newsprints, overhead transparencies, or other audiovisuals and graphic projections well in advance. Set them up in the presentation room and then sit in the back of the room. Can the text be read? Are the graphics clear? Is the sound audible?
   (g) If using a videocassette recorder (VCR) and monitor, cue the first video segment in advance, and check the monitor angle for visibility and glare.
   (h) Address safety concerns. Is there safe access and egress? Are electrical cords taped down?

2. *Make a Good First Impression.*
   (a) Arrive at the designated site early. Make sure everything is set up so that the audience perceives a professional presentation.
   (b) Ensure co-presenters and other coalition members arrive early, and briefly review with them planned activities.
   (c) Welcome each audience member. Be interested in them as individuals and as participants. Nametags help—for both presenters and participants. Using people's names shows interest.
   (d) Ask each participant to register and provide his or her name, address, telephone number, and those organizations represented, if appropriate. This information can be important later. Funding agencies, supporters, and sponsors often ask how many people were reached and who they were. This information can also be used to verify if the target audience identified in the disaster education plan was reached.
   (e) Start on time. People who arrive on time do not appreciate having to wait for latecomers and getting a message that their time is unimportant.
   (f) Introduce speakers, sponsors, hosts, community representatives/liaisons, etc., before starting the activity. Thank the host for assistance provided.

3. *Conduct an Efficient Program.*
   (a) Explain the agenda and activity objectives upfront. Ensure that all participants understand the purpose of the presentation. Clarify audience expectations.
   (b) Adhere to the plan. Establish a private "signal system" in advance so that presenters can be informed unobtrusively that they are running over time, straying from the plan, or that they need to stop to answer a question.
   (c) Allow sufficient time for audience questions. Any immediately unanswerable question should be referred to fellow presenters or someone in the audience who may know the answer. Recognizing someone from the audience to help answer questions can be beneficial in two ways: (1) it increases target audience participation in the educational process and (2) decreases the perception that the presenters "know all."
   (d) If a question cannot be answered, record it and include the name and telephone number of the person asking it. "I don't know" is an acceptable answer, when followed by, "I'll find out for you." Find the answer later and call the person back.
   (e) Before the activity is finished, review the intended objectives with the audience. Ask them if: (1) they think the objectives of the activity were met and (2) whether their expectations were met.
   (f) Conclude at the designated time. People value their time and appreciate it when this is recognized.

4. *Don't Forget Follow-up Activities.*
   (a) Offer a way for people to make contact later if they have questions.
   (b) Clean up the facility after the presentation, or, better yet, try to leave it in better condition than it was found. Pick up any disaster education materials that the audience may have left behind.
   (c) Arrange to meet with presenters, sponsors, and the disaster education planning team to discuss the activity and ways to improve it, while it is fresh in everyone's memory.

Evaluate and Build the Program

What worked? What didn't? Improve the program, based on an objective evaluation. Several methods exist to gather information useful for evaluation. See Section 2, Chapter 7, "Evaluation Techniques for Public Education," for further information on how to evaluate results.

WORKING WITH COMMUNITY COALITIONS

A coalition is a formal effort that results from a network. A network is an informal arrangement of information and resource sharing. A coalition draws on resources discovered through a network to form a "project team" for development and implementation of an educational program or project.

When a coalition comes together, the members share a common goal. Before work is undertaken by coalition members, the group should record the common goal and obtain agreement, both from group members and from organization leaders that the group members may represent.

The public often perceives information from community coalition efforts as more credible. This is due to the active participation of different segments of the community in planning and implementing projects, e.g., disaster education programs. Building coalitions with other organizations takes time, but is well worth the effort. Since other organizations in the community may share this common goal, i.e., disaster education, and no one organization can reach everyone at once, it is desirable to work together from the beginning in these efforts.

Organizational Accountability

Each participant in a collaborative effort brings to it his or her own organization's goals, values, and culture. It is also necessary to be thoroughly familiar with one's own organization's commitment to this collaborative work, especially when working with others is considered valuable or when something must be done for political reasons.

An organization's management will more likely support this work if it understands what is being done, the value of the work, the potential impact on the community, and the expected results, including recognition and positive consequences. Be prepared to explain these factors/results. Management often views time as an investment and may measure it in terms of dollar value. The educator is, therefore, accountable to the organization for the time put into the activity, as well as for the financial resources invested in the activity.

Promises and Their Limits

When people meet and plan coalition activities, resources are needed to accomplish the planned work. Often, particularly with new members, people promise organizational resources without checking with management first. Avoid unsupported promises of time, resources, etc., to a coalition when working for an organization; management usually prefers to assign both work and the time devoted to it. Remember, one's own organization comes first.

The following sequence should help ensure that the educator's organization and management become co-participants and supporters of the coalition effort:

1. Attempt to determine just what resources are needed.
2. Return to the organization's management and discuss plans and expected results.
3. Describe the resources that are needed, and suggest those resources the organization could provide.
4. Get management's commitment *before* making promises for delivery; this will ensure personal success, earn respect both from management and from coalition group members, and turn promises into realities.

Available Resources

Resources are necessary for any education program. The educator should consult with his or her organization's management to determine those resources currently available for a collaborative program and those that may be available at other times of the year, since organizational resources can change with time.

If organizational funding is a problem, perhaps services, e.g., printing, advertising, or educational materials, may be preferred, as well as input of time, which is a valid, measurable resource.

Further note that management might be more willing to provide additional resources once positive evaluations come back from initial work efforts.

Coalition-Building

Working with a coalition can take much time and effort, and perhaps it may seem easier to do an educational project alone. However, the benefits of investing this time often outweigh the disadvantages.

The following steps are recommended when building and maintaining coalitions for disaster education activities:

1. *Assess the Needs.*

   A needs assessment sets the stage for a planning process, by identifying needs in the community, available services to meet those needs, and any gaps between needs and services. See Section 2, Chapter 2, "Using Data for Public Education Decision Making," for more information on needs assessment.

2. *Create a Mission Statement and Goals.*

   A mission statement is a clearly written document of purpose that outlines the reasons why the coalition exists. It must be narrow enough to provide direction and broad enough to encompass all relevant, potential activities. Goals are general statements about what the coalition intends to accomplish. They must be consistent with the mission statement. Together, the mission statement and goals define: (1) the general intentions of the coalition's activities and (2) the expected results.

3. *Develop Objectives.*

   Objectives are specific, measurable statements of intentions that refine the goal statements. They describe what will be done, who will do it, and a time frame of expected completion.

4. *Develop a Budget.*

   After the goals and objectives have been determined, a budget for accomplishing each goal and objective can be developed. Available resources can be compared with the needed resources. If a gap exists between the budget needs and the available resources, one of two actions is required:

   (a) Additional funds must be raised to support the implementation of the goals and objectives, or

   (b) Goals and objectives should be revised to conform with available funds.

5. *Create a Coalition Action Plan.*

   Create an action plan that includes:

   (a) Specific tasks needed to accomplish an objective,

   (b) Assignments of responsibilities, and

   (c) A timetable with deadlines for each task.

   An action plan is the actual day-to-day operating plan. It denotes each task clearly, so that deadlines for completion can be clearly marked. It identifies specific individuals responsible for accomplishing each task and clarifies the tasks to be done and their appropriate sequence.

6. *Monitor and Adjust the Plan.*

   Periodic monitoring of the coalition action plan is necessary to assess how the plan is working and allows the coalition to make any required adjustments. Monitoring should be built into the action plan at regular intervals.

7. *Evaluate the Results.*

   On a regular basis, and especially near the end of the planning cycle, the coalition should evaluate the results of its operation. Evaluation results and an updated needs assessment are useful tools that assist the coalition in the next round of planning. More information on evaluation of educational programs can be found in Section 2, Chapter 7, "Evaluation Techniques for Public Education."

BIBLIOGRAPHY

References

American National Red Cross, *Community Disaster Education Resource,* The American National Red Cross, Washington, DC, 1996.

American National Red Cross, *Conducting Disaster Education Activities in Your Community,* The American National Red Cross, Washington, DC, 1991.

American National Red Cross, *Instructor Candidate Training Participant's Manual,* The American National Red Cross, Washington, DC, 1990.

"Directory of National Community Volunteer Fire Prevention Program. Community-Based Fire Prevention Education Initiatives, 1984-1992," National Criminal Justice Association, Washington, DC, Federal Emergency Management Agency, Emmitsburg, MD, FA-92, April 1993, 130 pages.

Drabek, T. E., *Human System Responses to Disaster: An Inventory of Sociological Findings,* Springer-Verlag, Inc., New York, 1986.

Dynes, R. R., and Quarantelli, E. L., "The Family and Community Context of Individual Reactions to Disaster," *Emergency and Disaster Management: A Mental Health Sourcebook,* Parad, H., Resnik, H. L. P., and Prad, L., eds., The Charles Press, Bowie, MD, 1976, pp. 231–245.

"Fire Departments and Communities: Partners in Prevention," *Fire Engineering,* Vol. 148, No. 6, June 1995, pp. 101–110.

Haas, J. E., and Drabek, T. E., "Community Disaster and System Stress: A Sociological Perspective," *Social and Psychological Factors in Stress,* McGrath J. E., ed., Holt, Rinehart and Winston, Inc., New York, 1970, pp. 264–286.

Hall, John R., Jr., "Children Playing With Fire: U.S. Experience, 1980-1993," National Fire Protection Association, Quincy, MA, August 1995.

Hall, John R., Jr., "U.S. Fire Death Patterns, By State 1980-1992," National Fire Protection Association, Quincy, MA, August 1996.

Hall, John R., Jr., "U.S. Experience With Smoke Detectors and Other Fire Detectors: Who Has Them? How Well Do They Work? When Don't They Work?" National Fire Protection Association, Quincy, MA, August 1996.

Ho, J., "How FEMA's Community Outreach Worked after Northridge," *Los Angeles Times,* June 18, 1994.

Jones, E. V., and Cooper, C. M., "Adult Education Programming and Memory Research," *Lifelong Learning: The Adult Years,* 6, No. 3, Nov. 1988, pp. 22–23, 31.

Jones, R.T., and Randall, J., "Rehearsal–Plus: Coping With Fire Emergencies and Reducing Fire-Related Fears," *Fire Technology,* Vol. 31, No. 4, 1994, pp. 432–444.

Larson, R.D., "From Ashes to Education," *Firehouse,* Vol. 18, No. 1, January 1993, pp. 44, 46.

Lerner, N.D., and Huey, R.W., "Residential Fire Safety Needs of Older Adults," Thirty-fifth Annual Meeting of the Human Factors Society, Part 1 of 2, September 2-6, 1991, San Francisco, CA, Human Factors Soc., Inc., Santa Monica, CA, 1991, pp. 172–176.

Lopes, R., "Public Perception of Disaster Preparedness Presentations Using Disaster Damage Images," Working Paper 72, Institute of Behavioral Science, University of Colorado, Boulder, CO, 1992.

Lopes, R., *Transfer of Technical Concepts to Actual Practice: American Red Cross First-Aid and CPR Training Programs,* Department of Education, University of Southern California, Los Angeles, CA, 1988.

Mileti, D. S., Hutton, J. R., and Sorensen, J. H., *Earthquake Prediction Response and Options for Public Policy,* Institute of Behavioral Science, University of Colorado, Boulder, CO, 1981.

Mileti, D. S., Drabek, T. E., and Haas, J. E., *Human Systems in Extreme Environments,* Institute of Behavioral Science, University of Colorado, Boulder, CO, 1975.

Millsap, S., "Planning for Disasters in Your Community," *Fire Engineering,* Vol. 147, No. 12, December 1994, pp. 28–32.

Perroni, C., "'Get Alarmed, South Carolina' Lessons Learned From Its Success," USFA Fire Investigation Technical Report Series, Special Report, Federal Emergency Management Agency, Washington, DC, 1991.

Perry, R. W., and Lindell, M. K., "The Psychological Consequences of Natural Disaster: A Review of Research on American Communities," *Mass Emergencies,* 3, 1978, pp. 105–115.

Perry, R. W., Lindell, M. K., and Greene, M. R., "Crisis Communications: Ethnic Differentials in Interpreting and Acting on Disaster Warnings," *Social Behavior and Personality,* 10, No. 1, 1982, pp. 97–104.

Quarantelli, E. L., *Disasters: Theory and Research,* Sage Publications, Beverly Hills, CA, 1978.

Scanlon, T. J., Luukko, R., and Morton, G., "Media Coverage of Crisis: Better than Reported, Worse than Necessary," *Journalism Quarterly,* 55, Spring 1978, pp. 68–72.

Slayton, Donna M., and Miller, Alison L., "Patterns of Fire Casualties in Home Fires By Age and Sex, 1989-1993," National Fire Protection Association, Quincy, MA, September 1995.

Smith, Linda E., "Fire Incident Study," *National Smoke Detector Project,* U.S. Consumer Product Safety Commission, Bethesda, MD, January 1995.

Stewart, Laurence J., "The U.S. Home Product Report, 1989-1993," *Forms and Types of Materials First Ignited in Fires,* National Fire Protection Association, Quincy, MA, February 1996.

Stewart, Laurence J., "The U.S. Smoking-Materials Fire Problem through 1993 and The Role of Lighted Tobacco Products in Fire," National Fire Protection Association, Quincy, MA, February 1996.

Thomas, J., "California Still Has Lessons to Learn," *Fire International,* No. 143, May 1994, pp. 37, 40, 42.

Tremblay, K.J., "Catastrophic Fires of 1994," *NFPA Journal,* Vol. 89, No. 5, September/October 1995, pp. 48–69.

Turner, R. H., Nigg, J. M., and Paz, D. H., *Waiting for Disaster: Earthquake Watch in California,* University of California Press, Los Angeles, CA, 1986.

Walker, B.L., *et al.,* "Short-Term Effects of a Fire Safety Education Program for the Elderly," *Fire Technology,* Vol. 28, No. 2, May 1992, pp. 134–162.

Walker, B.L., "The Effects of a Burn Prevention Program on Child-Care Providers," *Fire Technology,* Vol. 31, No. 3, August 1995, pp. 244–264.

USFA, *Proving Public Fire Education Works,* United States Fire Administration, Emmitsburg, MD, 1990.

Zemke, R. and Zemke, S., "30 Things We Know for Sure About Adult Learning," *Training,* July 1988, pp. 57–61.

"1994 Learn Not to Burn Champions," *NFPA Journal,* Vol. 88, No. 3, May/June 1994, pp. 67–69.

FIRE AND LIFE SAFETY EDUCATION: THEORY AND TECHNIQUES

Pamela A. Powell

Public fire and life safety education may be defined as "comprehensive community fire and injury prevention programs designed to eliminate or mitigate situations that endanger lives, health, property, or the environment."[1] Organizations whose personnel conduct fire and life safety education programs include fire departments, state or provincial fire agencies, schools, burn prevention and treatment organizations, city health departments, and corporate safety offices. Fire and life safety educators blend and apply seven distinct areas of expertise in their professional practice: (1) education theory, (2) education techniques, (3) methods of coalition building, (4) presentation skills, (5) working with the media, (6) education program planning, and (7) education program evaluation.

This chapter will focus on education theory and techniques. For further information, see the following chapters in Section 2: Chapter 2, "Using Data for Public Education Decision Making"; Chapter 3, "Designing Disaster Education Programs"; Chapter 5, "Reaching High-Risk Groups"; Chapter 6, "Using the Media for Educational Outreach"; and Chapter 7, "Evaluation Techniques for Public Education."

EDUCATION THEORY FOR FIRE AND LIFE SAFETY EDUCATORS

What Is Learning?

The terms *learning*, *education*, and *information* are related, but there are important differences between these three concepts.

Learning is knowledge gained through observation and study, resulting in a change of attitude or behavior.[2] Stated another way, learning cannot take place without change.

Education is the process of training, teaching, or instructing students in new skills. Training, teaching, and instruction all deal with preparing students for some kind of action or activity, such as driving a car, operating a machine, or testing a smoke detector. "How to do" something is at the core of education.

Information is facts, knowledge, or data. Information is the raw material of the new skills of education and the change of learning.

For example, a presentation for the chamber of commerce on how to develop an occupant emergency plan for the workplace is an educational activity. It teaches a skill (i.e., how to develop an occupant emergency plan). Later, when the participants actually do their plans, a change has happened, i.e., learning has taken place.

A presentation to the same group about how the community's Insurance Services Office (ISO) rating is set is a public information activity. Interesting facts have been shared. But the participants do not have any new skills, such as how to use an ISO grading schedule.

Characteristics of Learning

For the practicing fire and life safety educator, the complex field of learning theory can be summarized in a few points.[2]

1. Learning is a life-long activity. Learning does not stop when school ends.
2. Because learning causes a change in attitude or behavior, learning can be stressful. For example, it is uncomfortable to change a fire and life safety attitude from "My kids would never play with fire" to "I need to be more careful about where I put my cigarette lighter."
3. People learn at different rates and in different ways. How people learn is not related to intelligence.
4. Learning must be reinforced to be effective. This characteristic argues against once-a-year fire and life safety education programs, but is a strong argument for ongoing activity.
5. Effective learning requires support from the people who influence or control the students. In the workplace, supervisors should support an employee's new interest in keeping exits clear. Parents need to support a child's pressure to install more smoke detectors.
6. Learning is enhanced when multiple senses are stimulated. Students need to *hear* words and *see* illustrations. Through props (such as burned items from a house fire) students *touch* and even *smell* the effects of fire. This characteristic of learning applies to adults as well as children, although great care should be taken in the use of potentially threatening props with children.
7. Learning is most effective when it is focused. In focused learning, the fire and life safety educator begins with an overview (e.g., "This morning, we will learn what to do in case your clothes catch fire") and then explains the individual actions (e.g., how to "stop, drop, and roll"; how to cool a burn; and when to get medical attention).

Pamela Powell is director of training and education with Custer Powell, Inc., a consulting firm located in Wrentham, MA. She is nationally recognized in the fields of fire safety education, training, and meeting facilitation.

Motivation and Learning[3]

Fire and life safety educators often describe the need to motivate their audiences to learn. Since 1954, motivation has often been described according to a "hierarchy of needs," as identified by researcher Abraham H. Maslow.

According to Maslow, people have five basic needs that motivate all actions: (1) physiological, (2) safety and security, (3) belonging and social activity, (4) esteem and social status, and (5) self-actualization or fulfillment. The needs can be drawn as a pyramid, with the more urgent needs—physiological needs—at the bottom. People must first satisfy the needs at the bottom of the pyramid before they can move on to the needs higher up. *But once a need is satisfied, it no longer motivates action.* Table 2-4A applies Maslow's hierarchy of needs to fire and life safety education.

The Domains of Learning

Education is not based solely on fundamental human needs, such as those described by Maslow. Education is also based on audience needs. Audience needs are often described according to three domains of learning:

1. The *affective domain* involves how people feel about a situation and is the realm of values and opinions (e.g., whether or not they worry that their cigarette may have started a grass fire; whether they care about damage to wildlands).
2. The *cognitive domain* involves what people understand (e.g., how fast a grass fire is likely to grow).
3. The *psychomotor domain* deals with what people can do (e.g., how to dispose of smoking materials properly).

Each domain can be related to specific kinds of educational needs. (See Table 2-4B.) As discussed later in this chapter, the domains of learning are very important to the process of writing educational objectives.

Lifelong Learning

People learn from cradle to grave. *How* people learn, on the other hand, changes throughout life. The twin factors of "development stage" and age determine how people learn throughout their lives. Generally, people pass through six development stages, as outlined in Table 2-4C.[4]

Aging happens automatically, whereas psychological development happens only when the person is ready and when environmental factors (such as family, friends, school, and work situation) are right.

Age and Learning

The second factor that influences learning is age. Preschoolers, elementary school children, adolescents, adults, and seniors all learn very differently. In learning, age combines the influences of psychological/social/personal development with the effects of motor ability and communication skills. Being aware of how people learn at different ages is the first step in making sure that education programs and materials are "age appropriate."

Preschool children: Children ages 3 to 5 grow and change more rapidly than any other fire safety audience. Some basic facts, though, will be true throughout the preschool years.

1. Children ages 3 to 5 will learn fire and burn safety the same way they learn everything else—by seeing and doing.[5] What young children *do* will have a more lasting effect than what they *see*. For example, an adult will remember how to "stop, drop, and roll" after seeing a demonstration. A preschool child must practice before learning happens.
2. What young children *see* is more influential than what they *hear*. A child will watch an adult find a lighter, but will not remember the adult saying: "Always tell me when you find a lighter. Never touch a lighter."
3. Young children love repetition. Fire and life safety educators will need far fewer programs and activities for this group than for other audiences.
4. Young children have a very limited view of the world. Using the right approaches with this audience is critical, because the wrong approach may be dangerous.

Children's Television Workshop (the producers of *Sesame Street*®) studied how preschoolers would interpret fire and burn safety messages on television. Their findings also apply to non-televised programs. Their findings include:[6]

1. Showing a dangerous activity, even while warning against it, is not recommended for preschoolers. *Children respond more to what they see than to what they hear.*
2. The extremely limited vocabulary of preschool children limits the fire and burn safety concepts that can be presented. "Hot" and "fire" will be understood, but "scald," "avoid," "prevent," "appliance," and "boil" are beyond preschoolers.
3. Young children have difficulty relating one event to another. "If/then" thinking or making choices (e.g., "If the door is hot, don't open it" or "Use your second way out") is simply more than many preschool children can do.
4. Young children view the world unpredictably. They may focus on one part of a program or story, but ignore the rest. It is difficult for them to link the beginning and end of even the simplest story. Preschoolers do not necessarily link a problem and its solution.

TABLE 2-4A.   *Maslow's Hierarchy for Fire and Life Safety Educators*

Type of Need	Example	Application to Fire and Life Safety Education
Physiological	Food, water, shelter, immediate safety from life-threatening situations	The need to escape from an actual fire overrides all other activities, including gathering possessions
Safety and security	Dangers that are not immediately life-threatening	Escape planning
Sense of belonging and social activity	Identifying with and spending time with a group	For children, completing a home hazard inspection with a group of classmates; for older teenagers and adults, participating in a fire department "ride-along" program
Esteem and social status	Recognition of worth or achievement within the group	For children, a badge for completing a fire safety lesson; for adults, a "certificate of fire safety achievement" from their child after testing a smoke detector
Self-actualization or fulfillment	The U.S. Army recruiting slogan "Be all you can be"	"I want to be the best parent I can. So I'll start testing the smoke detectors."

TABLE 2-4B. *Education Needs and Learning Domains*

Target Domain	Audience Needs	Audience Situation
Affective	Motivation	Does not value something
Cognitive	Information	Does not know something
Psychomotor	Skills	Cannot do something

Findings such as these have very practical implications for fire and life safety educators. For example, "all fire safety messages should involve the two basic techniques by which preschoolers actually gain knowledge. By seeing an image, it is implanted; by doing, the action is learned. These two principles should guide fire safety messages for the preschool audience."[6]

Clearly, special care is needed in teaching safety to preschool children. Rather than develop their own, fire and life safety educators may prefer to rely on programs and materials such as the *Learn Not to Burn: Preschool Program*,[7] those available from Oklahoma State University, and the *Sesame Street Fire Safety Resource Book*.[8]

Elementary school children: The elementary school years are marked by steady physical, mental, and emotional growth. Considering physical growth, for example, "there are some marvelous hap-

TABLE 2-4C. *Developmental Stages and Implications for Education*

Stage	Characteristics	Educational Implications
Stage 0, Blind trust	Develops only biologically; does not interact with surroundings	Too early for education
Stage 1, Focused on self	Responds out of fear; recognizes and begins to interact with surroundings; bases decisions on authority, obedience, and punishment	Wants to avoid the pain of burns; may fear punishment for fireplay
Stage 2, "Me and Thee"	Responds out of need for personal gain; will satisfy needs of others for personal gain; bases decisions on "What's in it for me?"	Wants approval of others: for children, approval of fire fighters, teachers, and parents is important. Adults will seek approval of individuals who matter to them.
Stage 3, Group stage	Responds from a need for belonging and loyalty; accepts group rules as price for being part of the group; tries to meet group expectations	Responds to pressure from peers (whether children, adolescents, or adults) for fire safety; may respond to group pressure to be fearless
Stage 4, Group focus stage	Makes groups more abstract, does things "for the good of the order"; values social stability; accepts group rules for the good of other people	Responds to fire safety rules (such as bans on outdoor burning) and ordinances (such as smoke detector codes)
Stage 5, Interactive and independent stage	Regulates own behavior; negotiates for what the individual thinks is best	Will be fire safe only if it personally matters; may ignore rules that "don't apply to me"

penings going on in the body, such as the shedding and erupting of teeth, the body's changing proportions. Along with the increase in height and weight, the internal organs—heart, kidneys, and liver—are increasing in their ability to function."[9]

Children ages 5 to 11 are probably the most frequent audience for fire and life safety education programs. Two strong trends in elementary education relate directly to fire and burn prevention: (1) *developmental learning* and (2) *comprehensive education*. Fire and life safety educators may need to tailor their approach and techniques to conform to these educational trends in elementary schools.

Developmental learning.[10]   University of Chicago educator Robert Havighurst identifies nine major groups of "developmental tasks" for elementary school children. Several of these tasks relate directly to fire and burn safety.

1. Physical skills.   Children become much more coordinated so that tasks that were impossible or difficult a short time ago—such as lighting a match—become possible or easy.
2. Self-attitudes.   Children develop strong attitudes about themselves: their adequacy, personal hygiene and appearance, vulnerability, or their safety.
3. Social skills.   By elementary school age, children learn to get along with others, to make friends. Courtesy and other rules (including safety behavior) become important to them.
4. Social roles.   Boys and girls become more interested in gender differences. Children this age are curious about—and want to follow—what society perceives as masculine and feminine roles.
5. Academic skills.   Elementary school children learn a fantastic array of academic skills. In the mid-elementary years, abstract learning begins; speed reading, spelling, cursive writing, and higher math skills follow.
6. Living concepts.   Concepts of everyday living (such as social, work, and civic issues) are included here.
7. Values.   With the development of a conscience, children learn to distinguish between good and bad, right and wrong.
8. Independence.   Between the ages of 5 and 11, children realize that adults are imperfect. As children spend more time away from their home, peers become more important.
9. Social attitudes.   Feelings about home, religion, government, and other groups develop.

Fire and life safety programs that actually teach bigger areas, such as Havighurst's developmental tasks, have a far better chance of being taught in the classroom than a stand-alone fire safety effort. For example:

1. Fire and burn safety helps children develop positive *self-attitudes*.
2. A fire station tour teaches *social skills* (how the fire fighters live and work together), *social roles* (both men and women can be fire fighters), and *living concepts* (the fire service as a career), in addition to *self-attitudes* (learning about safety).

Comprehensive education.   The education trend of the 1990s is toward comprehensive education and away from categorical education. Rather than being taught as a stand-alone subject, fire and burn safety can be taught as part of a larger safety program or as part of a comprehensive health education program.

In many ways, fire and life safety education has been comprehensive education for many years. For example, for years fire departments have conducted babysitter programs that teach first aid and child care along with fire safety. Many fire safety educators include electrical safety, water safety, and citizen cardiopulmonary resuscitation (CPR) in their presentations. Many fire departments teach emergency response to hurricanes, earthquakes, or other natural disasters.

The Phoenix Fire Department adopted the idea of comprehensive education when it developed the *Urban Survival Program.*[11] In addition to fire and burn safety, the program's 32 modules cover how to:

- Be a safe pedestrian.
- Understand firearm safety.
- Practice transportation safety.
- Practice basic electrical safety.
- Practice water safety.
- Identify poisons and harmful substances.
- Protect your pet.
- Understand desert survival.
- Be a happy latchkey child.
- Practice outdoor recreation safety.
- Avoid confined spaces and dangerous workplaces.
- Identify the need for CPR training.
- Understand basic first aid.

Adolescents:[12] Knowing how adolescents develop and learn helps make the job of teaching them easier.

[Adolescents] live in the present and seem reluctant, even fearful, of looking too far beyond next weekend. Taking precautions in advance or being too concerned about the possible dangers that could harm them are not considered very important by most young people. This may be in part because teenagers characteristically have an it-can't-happen-to-me attitude, but the underlying reason may have more to do with appearing "babyish" or "nerdlike" to their friends than to any real sense of invulnerability.[12]

Of all the characteristics of adolescence, fire and life safety educators especially need to understand their developmental tasks and how adolescents learn.

Developmental tasks. Adolescence is a bridge between childhood and adulthood. Becoming an adult is much more than going through puberty, learning how to drive, and deciding what to do after high school. Adolescence involves several developmental tasks that H. D. Thornburg listed in his 1975 book, *Development in Adolescence.* The tasks are:

1. Develop conceptual thinking and problem-solving skills.
2. Form more mature relationships with peers of both sexes.
3. Achieve a masculine or feminine social role.
4. Prepare for marriage and family life.
5. Prepare for a career.
6. Acquire a set of ethics to guide behavior.
7. Develop civic competence and responsible community behavior.

A major challenge for fire safety educators is to find ways to fit their message into the big picture of adolescent development. For example, a discussion about knowing who is pulling false alarms can be presented as ethics. The role of smoking in residential fire death can be taught in terms of responsibility toward others.

How adolescents learn. Adolescents have their own distinct learning style, especially when it comes to learning behaviors and attitudes (rather than facts). Ways that adolescents learn include the following.

1. Direct experience. Although direct experience with a fire or burn must be avoided, adolescents can have indirect experience with a fire or burn through fire department "ride-along" programs, serving coffee and donuts at the fire scene, or by volunteering at a burn unit or burn camp.
2. Hypothetical projection. Adolescents might think about how a fire would destroy their clothes or stereo or about how a burn injury would change their appearance.
3. Role model emulation. This involves following the attitudes and actions of role models. For example, young adolescents will observe whether their heroes smoke and how they dispose of their smoking materials.
4. Instruction/demonstration. This technique is particularly effective in teaching skills, such as changing a smoke detector battery or CPR.
5. Rehearsal. Through simulation, adolescents practice their response to emergencies.
6. Teaching "best practice" to others. For example, showing younger adolescents the safe use of flammable liquids is a good test of how well the "teacher" can perform and gives a sense of competence.

Adults:[3] Adults as students reflect the many roles they have in life.

They are parent, spouse, employer, employee, homeowner, renter, community worker, volunteer, child to their parents, hero to their children, and emotional and financial provider. . . . They are expected to face the trials of the very young, the confusion of the teenager, the responsibility of adjusting to a changing world, and, somehow, balance their own desires and needs for relaxation and happiness.

Adults appreciate the need for training and education for professional purposes. They also invest untold amounts of money and time in leisure activities. The task of the educator is to make fire safety just as worthy of their time, interest, and effort.[13]

When it comes to learning, adults are not simply "grown-up children." Instead, adults learn in specific ways. General principles of how adults learn include the following.

1. Adults can and do learn. The old saying "You can't teach an old dog new tricks" does not apply to people. In fact, the ability to learn through comprehension and synthesis is far greater in adults than in children.

   Adults are especially skilled at informal learning (e.g., figuring out how to use a new VCR) and they also learn through "learning projects" (e.g., researching consumer-oriented publications to find top-rated compact disk players, or mastering a new computer program).
2. Adults need to know why they must learn something. Establishing a need to know is the cornerstone of teaching adults. For the most part, adults feel that they need to know practical, nuts-and-bolts-type information. Being able to perform some task better is usually a strong motivator for the adult learner.
3. Adults need to be self-directing. Adults like to control their lives and make their own decisions. Similarly, adults like to control what and how they learn. In other words, adults are participatory—not passive—learners.

   Instructional methods that encourage participation include role play, debate, problem-solving discussion, demonstration, case discussion, coaching others, and skills-practice exercises.
4. Adult learners bring their total life experiences into the learning environment. Instructional methods that encourage participation take advantage of the experiences and knowledge of adult learners. However, adults may have already made up their minds

about how fire safe they are—or want to be. Brainstorming exercises or group discussion may help keep minds open.

5. Adults are task-centered learners. Adults tend to center their learning around completing tasks, solving problems, or handling the way they live. Adults are, for example, much more interested in "writing better business letters" than in diagramming sentences, or in "doing my time sheet right" than in addition.

6. Adults have external and internal motivators to learn. External motivators to learning include objectives such as getting promoted, getting a raise, or finding a job with better working conditions. For the most part, external motivators are visible or measurable. Internal motivators are related to the hierarchy of needs. They are usually invisible and impossible to measure.

Older adults:[14]   Older adults, senior citizens, seniors, the elderly, elders—perhaps no other group has so many names. Older adults are an important audience for fire and life safety education for two reasons: (1) older adults are the fastest growing segment of the population, and (2) older adults have an unusually high risk of fire and burns.

When planning fire and life safety programs for older adults, remember to:

1. Allow plenty of time. Like anyone else, older adults may get anxious if time appears short.
2. Count on a talkative audience, and include a question-and-answer session.
3. Watch out for too much sunlight or glare in a room.
4. Darken the room, as possible, when using audiovisuals.
5. Keep the room at a steady, warm temperature.
6. Limit programs to 30 min.
7. Keep a low-pitched voice; speak slowly and clearly.
8. Use or design simple handouts. Avoid "busy" materials and fancy typography. Look for double-spaced materials.

For more guidance on working with older adults, see Section 2, Chapter 5, "Reaching High-Risk Groups."

Different Audiences Require Different Messages

Fire safety messages *even on a single subject* will have very different forms and be taught quite differently to different age groups. For example, education messages on smoking and fire safety will change with the age of the audience. (See Table 2-4D.)

TABLE 2-4D.   *Age and Message*

Audience Age	Message
Preschoolers	"Hot things hurt." "Tell a grownup when you find matches or lighters."
Elementary school children	"A match is a tool." "Smokers need watchers."
Adolescents, adults, and older adults	"Keep matches and lighters where children can't reach them."

Positive Messages Are Effective

There has been surprisingly little research about what kinds of messages motivate people to be firesafe. The research that does exist, though, gives clear direction to fire and life safety educators.

A Study of Motivational Psychology Related to Fire Preventive Behavior in Children and Adults,[15] commissioned by the NFPA and completed in 1974, states:

There is ample evidence that fear and threats of violence, unless continually maintained, have little long-term positive effect on behavior. People tend to tune out or suppress information which is disturbing or contrary to their belief or inclinations.

People under threat or tension will seek to reduce that tension. Some will seek new information in order to act effectively. Others will block out that which is creating the tension. Therefore, fire prevention information should be directed toward positive motives which already exist within the recipient of the message.[15]

The *Motivational Psychology*[15] study shifted the tone of fire and life safety education from negative to positive. Messages such as "stop, drop, and roll" began replacing negative messages, such as "don't run if your clothes catch fire." The study also contributed to the development of the National Fire Protection Association's *Learn Not to Burn* programs. That study challenged the conventionally held view of the public's apathy about fire safety by concluding that people *do* care about fire safety, even though they might not know how to prevent or respond to fires. Based on interviews with people who had and had not experienced a fire, the study noted that "awareness of a fire as a powerful and potentially tragic force is not deeply buried in most people's consciousness." For both those who had and had not experienced a fire, "fire prevention awareness is increased when a person is responsible for others."

The Association's research also included questions about how the public wanted fire safety information presented. The messages from the public were clear: Teach, don't preach. Don't scare us to death, but make us think about fire safety. Give us information and if you can't make it interesting, at least make it short.[16]

A 1991 study by the American Red Cross about whether people took action to prepare for natural disasters after seeing slides of disasters reached conclusions that were similar to the Strother study.[15] Highlights from the American Red Cross report are as follows:

1. People remember more when they see disaster damage images in presentations.
2. Seeing disaster images does not encourage people to take action to get ready for a natural disaster.
3. Seeing disaster images discourages people from taking action.
4. After seeing disaster images, people were confused about what to do.
5. Avoiding or denying something unpleasant caused people not to take action, among those who saw disaster images.
6. "Just not getting around to it" caused people not to take action, among those who did not see the disaster images.[17]

EDUCATION TECHNIQUES FOR FIRE AND LIFE SAFETY EDUCATORS

This part of the chapter will cover curriculum development, as well as two special topics: (1) questioning techniques and (2) managing the learning environment.

Curriculum Development

There are seven basic steps to curriculum development: (1) write educational objectives, (2) develop a course outline, (3) develop a lesson plan, (4) select instructional methods, (5) choose instructional materials, (6) develop testing tools, and (7) allocate time.

Write educational objectives: Educational objectives are sometimes called instructional objectives, behavioral objectives, or learning objectives. Regardless of what they are called, objectives answer the question: "What will happen as a result of the education program?"

Increasingly, objectives focus on what people can *do* after education, not on what they will *know* after education. For example, "the student will *show how to* change the battery in a home smoke detector" rather than "the student will *describe how to* change the battery in a home smoke detector," or "the student will appreciate the importance of changing the battery in a home smoke detector."

Objectives need to be realistic and measurable. Making objectives realistic is the responsibility of the educator. Making objectives measurable—so that students and programs can be evaluated—is also necessary. A standardized language and format for educational objectives help make them measurable.

Objectives are often written in a standardized format, originally developed by Robert F. Mager in his landmark book, *Preparing Instructional Objectives*.[18] This standardized format has three components.

1. The *performance* or behavior component of the objective describes what learners must be able to do to demonstrate what they have learned.
2. The *condition* component of the objective describes the conditions and resources available to the learner.
3. The *standard* or *criterion* component describes how well the action must be performed.[19]

Table 2-4E presents sample objectives for the affective, cognitive, and psychomotor domains of learning. Note that the affective objective deals with values (i.e., what is important). The cognitive objective covers understanding or knowledge. The psychomotor objective describes what the participant will be able to do. These three sample objectives are measurable.

TABLE 2-4E. *Educational Objectives and Domains of Learning*

Domain	Sample Objective
Affective (values, feelings, or opinions)	At the conclusion of the 30-min. presentation on residential fire safety (condition), the adult participant will be able to state two factors (standard or criterion) that illustrate the importance of testing and maintaining smoke detectors (performance or behavior).
Cognitive (knowledge or understanding)	At the conclusion of the 30-min. presentation on residential fire safety (condition), the adult participant will be able to describe three key steps (standard or criterion) in testing and maintaining smoke detectors (performance or behavior).
Psychomotor (skills)	At the conclusion of the 30-min. presentation on residential fire safety and given a smoke detector for demonstration purposes (condition), the adult participant will be able to change the smoke detector's battery (performance or behavior) so that the detector alarms when tested (standard or criterion).

Develop a course outline: The course outline is developed to meet the educational objectives and may be as straightforward as determining the sequence for teaching several lessons. For these instances, fire safety educators teach a single audience a single lesson, and developing a course outline is not necessary.

Develop a lesson plan: A series of steps, known as the "five-step method of instruction," comprises the act of "teaching." The steps are: (1) pretest, (2) preparation, (3) presentation, (4) application, and (5) evaluation. The steps of preparation, presentation, application, and evaluation will become part of the written lesson plan.

A lesson plan is a step-by-step guide for presenting a lesson. It outlines the material to be taught and the procedures to be followed in teaching the fire safety lesson. The lesson plan helps ensure efficient use of time and accurate coverage of the subject matter. When several fire and life safety educators are involved, the lesson plan helps ensure uniform teaching and helps avoid overlapping instruction.

The lesson plan guides how the students and educator spend their class time, but does not dictate what will happen in class minute-by-minute. A lesson plan is *not* a script. A good lesson plan has some built-in flexibility—which is very valuable if, for example, the pretest reveals that the class already knows what the educator planned to teach, but does not understand what the educator expected the class to already know.

Pretest. Pretests determine what the student/participant already knows (and doesn't know) before the fire safety education program. Pretests allow the public fire educator to customize the presentation to meet the needs of the individual participants.

In addition, pretests are useful tools for audience assessment. Some public fire educators use pretests as motivational tools. Pretests can also provide students with an overview of what they will learn. Finally, pretests can serve as learning aids by providing questions for practice quizzes and by guiding individual study and review.

The most common type of pretest is the pencil-and-paper written test, although oral or performance tests are also used. Written or oral pretests are somewhat more convenient, simply because the entire class can take part in a written or oral pretest simultaneously. However, written or oral pretests do not assess psychomotor skills; e.g., a score of 100 percent on a written pretest about home smoke detectors is no guarantee that the student can actually change the battery.

Preparation. In everyday speech, the term "preparation" describes the necessary "homework" that educators do prior to a class or presentation. In educational use and this chapter, the term *preparation* describes how the educator prepares the student to learn.

Preparation involves getting the students' attention and letting them know why the material is important to them. Arousing curiosity and developing interest and a sense of personal involvement on the part of the student are all part of preparation. Preparation and motivation are closely linked.

Presentation. Presentation is the actual delivery of instructional materials and ideas. This activity involves explaining information, using supplemental training aids, and demonstrating methods and techniques.

Application. During the application step, students use or apply what the instructor has taught. Through application, students practice new techniques and skills. Whenever possible, each student should apply new knowledge by performing the task or solving problems. For example, students could demonstrate how to report a fire, perform the "stop, drop, and roll" technique, or test a smoke detector. The fire and life safety educator should supervise the application step closely, checking key points and correcting errors.

Evaluation. During evaluation or testing, the fire and life safety educator finds out whether the educational objectives have been met. Seen another way, evaluation shows whether students can perform a task independently.

Select instructional methods: Instructional methods and educational materials are the vehicles that carry the fire and life safety education message to the audience.

Fire and life safety educators have many types of instructional methods available to them. The challenge is to select the most appropriate instructional method for the educational objective. Types of instructional methods include:

1. Lecture. In a lecture, the fire and life safety educator tells, talks, and explains. Today's audiences are accustomed to the fast pace, as well as the visual and audio stimulation, of television, computers, and video games. Lectures can bore the audience. For this reason, fire and life safety educators should use the lecture format sparingly—and only for short periods.
2. Illustration. Through the illustration method of instruction, the fire and life safety educator shows the audience something, e.g., how an overloaded electrical outlet looks. Unlike a *demonstration*, illustration does not show the audience how to perform a task.

    The illustration method relies heavily upon educational materials (such as drawings, posters, photos, slides, overhead transparencies, videotapes or films, models, and diagrams) and props. Illustration is especially helpful in combination with the lecture format.
3. Demonstration. The demonstration method teaches new skills. The educator actually does the task, usually explaining the task step-by-step. The audience may then practice the skill through drills. *Safety is always a concern with demonstrations and drills,* especially true when live fire is involved. Whenever the safety of a live demonstration is in question, use another method or material.
4. Discussion. Group discussion can be valuable, especially when the audience already has basic knowledge. Types of discussion are: *guided discussion* (an exchange of ideas directed toward a goal), the *conference method* of discussion (in which the group directs its thinking toward solving a common problem), the *case study* (the group reviews real or hypothetical events), *role-playing* (in which the group acts out various scenarios), and *brainstorming* (where the objective is to identify as many ideas or approaches as possible).

Matching instructional method and educational objective. Selecting the most appropriate instructional method to address a specific educational objective is a critical skill for fire and life safety

educators; it is a matter of using the right tool for a particular task. Table 2-4F presents examples of matched instructional methods and educational objectives.

Choose instructional materials:[20] Educational materials are any material—printed matter, audiovisual materials, or props—that the educator uses to teach new fire and life safety skills to the audience. Through materials, the audience will actually *see* the effects of fire and burns, as well as the specific steps to take to avoid them. The materials literally give shape and color to the information and skills that the educators are teaching. Effective materials are important tools in getting and keeping the attention of the audience.

As important as they are, materials alone do not make an education program. An education program has specific measurable objectives and evaluation instruments, uses instructional methods that match the objectives, and uses materials to reach the objectives and reinforce the methods.

Materials are also important from the standpoint of resources. Next to the educator's time, materials are the most expensive item in the budget.

For these reasons, fire and life safety educators must use special care in selecting or creating their materials. Recognizing that materials are tools (rather than a program) and matching the material to the educational objective and to the audience are especially key to selecting educational materials.

Kinds of educational materials. The three kinds of educational materials are: (1) print matter, (2) audiovisual materials, and (3) props.

1. Print materials include brochures (also called flyers or folders), posters, fact sheets, coloring books, activity sheets, display or bulletin boards, educational card or board games, and pretests/posttests.
2. The variety of audiovisual materials is expanding and now includes films and filmstrips, videotapes and audiotapes, 35-mm slides and slide-tape combinations, overhead transparencies (sometimes called "vu-graphs"), and computer simulators. Flip charts and blackboards are also examples of audiovisual materials.
3. Props are artifacts or samples that the audience can see, touch, smell, or hear. A burned object from a home fire, a piece of melted glass, a manual fire alarm pull station, or a smoke detector could each be an effective prop during a fire and life safety presentation. Props are effective teaching aids because they make

TABLE 2-4F. *Matching Instructional Methods and Educational Objectives*

Educational Domain	Sample Educational Objective	Effective Instructional Method
Affective (values, feelings, or opinions)	"At the conclusion of the 30-min. presentation on residential fire safety, the adult participant will be able to state two factors that illustrate the importance of testing and maintaining smoke detectors."	Group discussion, role play, and case study
Cognitive (knowledge or understanding)	"At the conclusion of the 30-min. presentation on residential fire safety, the adult participant will be able to describe three key steps in testing and maintaining smoke detectors."	Lecture, illustration, guided discussion, conference discussion, and case study
Psychomotor (skills)	"At the conclusion of the 30-min. presentation on residential fire safety and given a smoke detector for demonstration purposes, the adult participant will be able to change the smoke detector's battery so that the detector alarms when tested."	Illustration, demonstration and drills, and role play

the subject real. Since they often involve several of the senses, props are an excellent memory aid.

Evaluating educational materials. There are vast differences in the quality of fire and life safety educational materials. Whether purchased commercially, borrowed or adapted from another educator, or created in-house, all educational materials should be evaluated.

At a minimum, educational materials should be evaluated before their first use and should be re-evaluated every year or so. The purpose of the follow-up evaluation is to make sure that what was acceptable initially still is.

The techniques for evaluating materials range from asking a few simple questions to sophisticated testing. The amount of evaluation needed depends on: (1) how much money the fire and life safety educator is investing, (2) how many people the material is expected to reach, and (3) how long the material will be used.

The kinds of evaluation techniques for educational materials can also vary greatly. For example, qualitative approaches are fairly subjective and rely on the fire and life safety educator's experience, judgment, and interpretation. Quantitative approaches are likely to be more sophisticated and to involve formal testing.

In evaluating educational materials, the key point is to evaluate all materials being considered in the same way.

Qualitative approaches. Evaluating educational materials can be as simple as reviewing the material against a standard checklist. Questions to ask during a review include:

1. How well does the material match the specific educational objectives of the fire and life safety program?
2. Is the material technically accurate? Is the material produced by a reputable organization?
3. Is the information clearly understood by members of the potential audience? Answering this question can range from asking several members of the audience to describe or demonstrate skills that the material is trying to teach, to more elaborate testing.
4. Is the material age appropriate? With 3- to 4-year-old children, for example, are stories or activities limited to the 5-minute typical attention span for that age group? For guidance in determining how age-appropriate material is, see *Developmentally Appropriate Practice in Early Childhood Programs: Serving Children from Birth Through Age Eight.*[21]
5. Is the material free of bias? Does the material reflect how people in the audience really live and work? The only way to make sure that material is bias-free is to invite (and *accept*) comments from representative members of the potential audience.

These questions essentially call for "yes" or "no" responses. The Pan-Educational Institute has developed a more sophisticated approach to evaluating materials as part of its *Practical Criteria and Instruments for Selecting and Implementing Sound Fire and Burn Education Programs in School and Community.*[22] The package includes worksheets for assessing posters and stand-alone print visuals, written informational materials (brochures, stand-alone handbooks, storybooks, coloring books, and workbooks/worksheets), and audiovisual materials (media campaigns, public service announcements, and audiovisual presentations). When the scoring is done, a material is found to fit one of five categories: (1) limited quality, (2) below average quality, (3) medium quality, (4) high quality, or (5) comprehensive high quality.

Quantitative approaches. Reading ease is a major consideration in selecting written fire and life safety education materials. A seventh to eighth grade reading level is typical for adults in the United States. Daily newspapers are often written at about that reading level.

Several readability indexes measure how easy a passage is to read. Commonly used indexes are a passive sentence index (a simple percentage of passive sentences within a passage), Flesch Reading Ease/Flesch Grade-Level Index, and the Gunning Fog Index.

The Flesch Index is based on a 100-word passage of the written material and counts the average number of words per sentence and the average number of syllables. With the Flesch Index, 17 words per sentence and 147 syllables per 100 words is "standard" reading ease. The Gunning Fog Index, on the other hand, combines the average sentence length and the number of difficult words (three or more syllables). Curriculum specialists with local school departments, or the education and journalism departments of universities, can help fire and life safety educators use these or other readability indexes.

These indexes can be used manually and are included in some word-processing programs. Whether the indexes are used manually or through the word processor, *the important point is to check whether sample materials are harder or easier to read than some standard that reflects the audience.* A highly precise test is not necessary, so long as the fire and life safety educator knows whether written materials are "in the ballpark" for readability.

Testing materials. In some cases, fire and life safety educators may need to test materials before going into full production. Questions to ask when deciding to test materials include:

1. Could someone get hurt if the educational message is misunderstood?
2. How widely will the material be distributed?
3. How difficult or expensive will it be to change the material once it is produced?

After considering questions such as these, the *Learn Not to Burn* Foundation decided to test a planned television public service announcement (PSA). The results are published in *Pre-Production Evaluation of the "Tell a Grown-Up to Put It Away" Public Service Announcement for 3-6 Year Olds.*[23] In that test, 50 children ages 3 to 6 saw the pre-production version of a 30-second PSA twice. The children were then interviewed, one at a time, after each viewing to determine understanding. The PSA was found to be ineffective with the target audience. The test results influenced the *Learn Not to Burn* Foundation's decision not to produce the PSA.

The *Learn Not to Burn* Foundation experience offers some lessons for all fire and life safety educators. The pre-production test had a *set* interview procedure and questions—so the results compared "apples and apples." The test evaluated how well the *target* audience responded; e.g., if only middle-income children had been in the study, the results would have been different. Perhaps most important, the material was tested when *changes were still possible.*

Create or purchase educational materials? Many fire and life safety educators like to prepare their own educational materials, rather than buy them (from a business or non-profit organization) or adopt what other educators have already created. The interest in developing materials often is related to saving money ("Those other materials are too expensive for me") or the desire to give materials a "local flavor."

Before deciding to create their own materials, fire and life safety educators can use the following as a checklist:

1. How much will it really cost to develop materials (including the cost of time spent)?
2. How important is "local flavor"? How can it be added at the lowest possible cost?

3. What is the motivation to create the materials? Are locally created materials really needed and cost-effective? Or does the educator just prefer to create materials?
4. Does the fire and life safety educator have the technical knowledge, experience, and equipment to create materials?

Selecting audiovisual materials. Audiovisual materials range from the simple—flip charts, 35-mm slides, and overhead transparencies—to the sophisticated—films and filmstrips, videotapes and audiotapes, slide-tape combinations, and computer simulators.

Flip charts, 35-mm slides, and overhead transparencies can be very effective audiovisual educational materials and can be created in-house. The key points are: (1) limit the text, (2) have effective visuals in audiovisual materials, and (3) keep the visual image as simple as possible. Large-size or projected words are simply words.

Flip charts. Flip charts are often used to record ideas from meetings and can also be used as audiovisual educational materials. Flip charts are inexpensive, easy-to-change, and effective with small groups. Although the paper is always white, marking pens come in a wide variety of colors. Flip charts can be re-used a few times but will need to be replaced when they get torn, rumpled, or soiled.

35-mm slides. 35-mm slides are a very popular medium for do-it-yourself audiovisuals. They are far superior to flip charts in illustrating concepts (e.g., the extensive damage caused by a home fire), emotions (e.g., the face of a fire fighter after a rescue), and objects (e.g., the parts of a smoke detector). Personal computers have vastly improved the quality of text slides—making slides of transfer letters obsolete.

Slides are also easy to rearrange into different presentations, are effective with large groups, and are more durable than flip charts or overhead transparencies.

There are, though, several disadvantages with slides. The lights must be down or off, making eye contact with the audience difficult or impossible. Once the presentation has started, the flexibility to change is gone. The fan on the slide projector may be noisy.

Overhead transparencies. "Overheads" are among the most popular teaching aids of all time—for good reason. They are inexpensive and easy to make, are durable, and can be used in rooms ranging in size from a classroom to an auditorium. They show text and line art to good advantage. The user can mark on them to underscore a point during a presentation. The lights can be on full or almost full. Overheads can be rearranged for use in several presentations.

Like flip charts and 35-mm slides, overhead transparencies do have some disadvantages. Photographs do not reproduce particularly well on transparencies. Like the slide projector, the overhead transparency projector fan may be noisy.

Develop testing tools: Testing tools can be pencil-and-paper tests or nonwritten tests, such as practical skills tests. Some fire and life safety materials include testing instruments. Local schools may develop testing tools or help fire and life safety educators develop them.

Section 2, Chapter 7, "Evaluation Techniques for Public Education," covers the analysis of test results.

Allocate time: A key part of developing a lesson plan is deciding how much time to spend on various subjects. A basic rule of thumb is to spend time on topics and drills that support the educational objective. Topics and drills that do not support the educational objective should receive little (if any) time in the lesson plan.

Questioning Techniques[24]

Questioning is an extremely effective educational technique. Questions encourage thinking, the exchange of ideas, and the molding of opinions, attitudes, and values. As a result, what the fire and life safety educator *asks* the audience may be as important as what the educator *tells* the audience.

Effective questions:

• Are "open-ended" or require more than a "yes" or "no" answer.

• Do not suggest the answer.

• Seek information, but do not make the audience feel uninformed.

There are four kinds of questions:

1. *Direct questions* are aimed at one person in the audience. Because direct questions can intimidate the audience, they are not often used with adults.
2. *Overhead questions* are aimed at the whole group, and anyone is free to respond. Because these questions are particularly helpful in starting thinking or in bringing out ideas and opinions, they are frequently part of a case study or brainstorming discussion.
3. *Rhetorical questions* are directed at the entire group, but an answer is not expected. A rhetorical question can be an effective "attention-getter." For example, a fire and life safety educator might open a presentation by saying, "What are the most important things that you and your family can do to protect yourselves from a home fire? By the end of this presentation, you will know how to answer that question."
4. *Relay questions* are those that the audience asks, and the educator sends back to the audience. Relay questions are a good technique for opening up discussion. Relay questions should not be used to avoid a question that the educator cannot answer or to force the audience to guess about facts or techniques.

Managing the Learning Environment

The term *learning environment* refers to the physical facilities where the learning will take place. The learning environment supports everything that the educator does in the classroom. For this reason, managing the physical learning environment is as important as selecting the best instructional method or educational material.

Individual factors that make up the learning environment include the *room size, temperature,* the level of *light, ventilation,* background *noise, acoustics,* the *comfort* of chairs and tables, and the *physical arrangement* of the room.

Practiced educators inspect the room before arriving for a presentation. It may be possible to move the program to another room or to correct what is wrong with the room. For example, removing extra chairs and rearranging chairs into a circle can make a huge room feel smaller. Videos or films *can* be shown in rooms with uncurtained windows—if the screen or monitor is placed so that the windows are behind the audience.

Rooms can be arranged in several ways, depending on the size of the group and whether or not tables are used. The choice of a room arrangement lies in satisfying three criteria:

1. Can everyone in the audience see and hear the educator and audiovisual materials?
2. Can everyone in the audience see and hear each other?
3. Does the arrangement place as many people as possible as close to the educator as possible?

When tables are used for small groups, three common arrangements are: (1) a U-shaped setup, (2) a hollow square, and (3) a conference table.

The U-shaped arrangement offers eye contact between the educator and audience, as well as among the audience. The fire and life safety educator is able to move around the open U. All participants are able to see a screen or video monitor placed behind the educator's table. The U-shape can either be flat or long. The flat U-shape lets more members of the audience be closer to the educator and audiovisual material. (See Figure 2-4A.)

The hollow square is very similar to the U-shaped room setup. Participants may feel farther away from each other, even though the distance is the same as the U-shaped arrangement. A conference table setup is essentially a solid U-shaped arrangement. This arrangement is more commonly used for meetings but may be used for a class. (See Figure 2-4B.)

For larger groups that need tables, several classroom styles are available. The traditional classroom style setup is the most restrictive and provides little eye contact among participants. The chevron (sometimes called herringbone) or fan-style classroom setups place more members of the audience closer to the educator—and let participants see each other. (See Figure 2-4C.)

The most common way to arrange chairs without tables is the amphitheater or auditorium style. Rearranging the chairs into a slight semi-circle provides for better eye contact between the educator and the audience (and among members of the audience). The semi-circle also improves the audience's line of sight to a screen or video monitor. (See Figure 2-4D.)

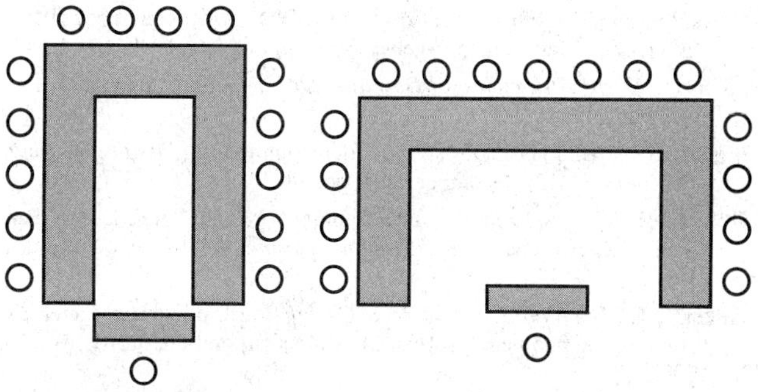

FIG. 2-4A.   Examples of U-shaped arrangements.

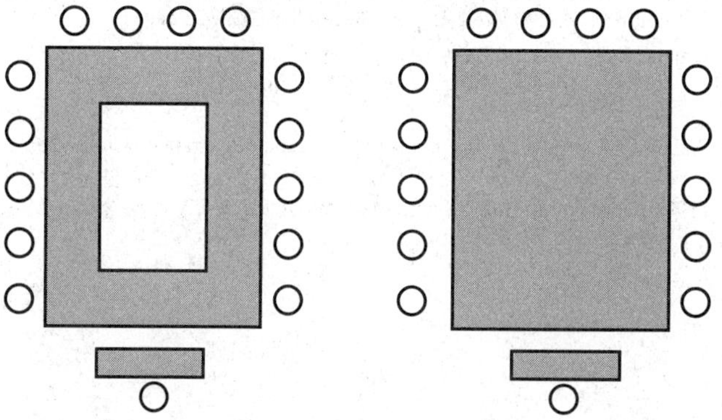

FIG. 2-4B.   Examples of the hollow square (left) and conference table (right) arrangements.

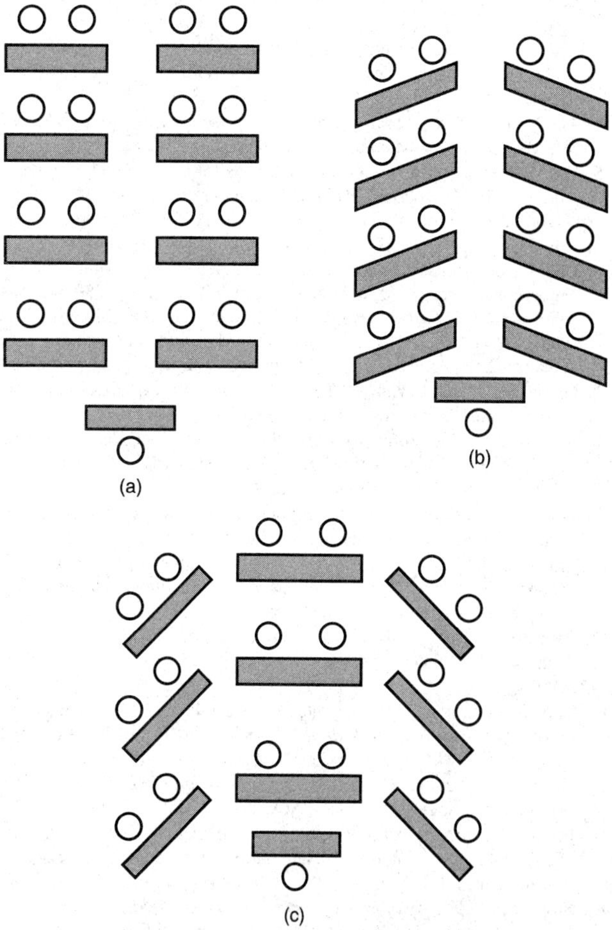

FIG. 2-4C.   (a) Traditional classroom setup; (b) chevron arrangement; (c) fan-style classroom. (Note that these same seating arrangements can be used without tables.)

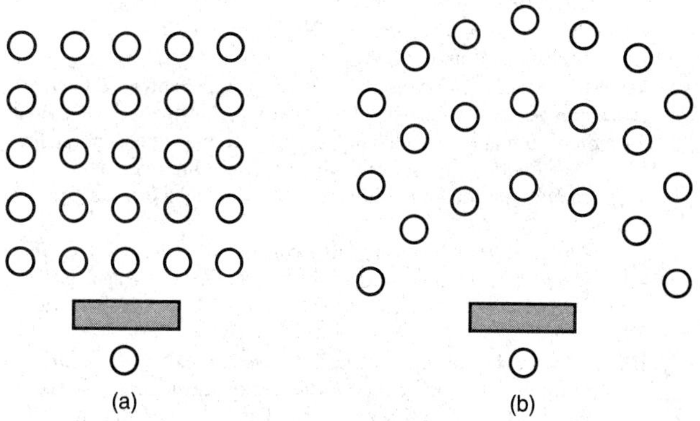

FIG. 2-4D.   (a) Amphitheater or auditorium; (b) semi-circle.

BIBLIOGRAPHY

References Cited

1. NFPA 1035, *Standard for Professional Qualifications for Public Fire and Life Safety Educator*, National Fire Protection Association, Quincy, MA, 1993.

2. Lambert, C., *Secrets of a Successful Trainer: A Simplified Guide for Survival*, Wiley, New York, 1986.

3. This material is largely based on R. Custer and P. Powell, "Teaching Fire Safety in the Workplace," in *Fire Safety at NYNEX: Your Piece of the Action*, 1991.

4. The National Fire Academy's course, *Developing Fire and Life Safety Strategies*, covers the concept of life stages in some detail.

5. For more information on teaching methods for preschool children, see Smalley, T., "Preschool Children," *Fire Safety Educator's Handbook*, National Fire Protection Association, Quincy, MA, 1983, pp. 43–51.

6. *Fire Safety on Television for Preschoolers*, prepared by the Children's Television Workshop, under a grant from the U.S. Fire Administration/ Federal Emergency Management Agency, FA-2A, July 1980; and *Sesame Street Fire Safety Resource Book*, prepared by the Children's Television Workshop, under a grant from the U.S. Fire Administration/ Federal Emergency Management Agency, Emmitsburg, MD, 1982.

7. Gamache, S., and Powell, P., *The Learn Not to Burn® Preschool Program*, National Fire Protection Association, Quincy, MA, 1991.

8. *Sesame Street Fire Safety Resource Book*, Children's Education Services, Children's Television Workshop, One Lincoln Plaza, Dept. FS, New York, NY 10023. [Telephone (212) 595–3456.]

9. Gross, C., "Elementary School Children," *Fire Safety Educator's Handbook*, National Fire Protection Association, Quincy, MA, 1983, p. 56.

10. This material is largely based on C. Gross's chapter, "Elementary School Children," in *Fire Safety Educator's Handbook*, National Fire Protection Association, Quincy, MA, 1983, p. 56.

11. Pedrotti, D., *Urban Survival Program*, Phoenix Fire Department, Community Services Division, 520 West Van Buren, Phoenix, AZ 85003, 1991.

12. This material is largely based on J. Sowers' chapter, "Adolescents," in *Fire Safety Educator's Handbook*, National Fire Protection Association, Quincy, MA, 1983.

13. Gratton, J., "Adults," *Fire Safety Educator's Handbook*, National Fire Protection Association, Quincy, MA, 1983, p. 85.

14. For more information, see R. Lightman's chapter, "Elderly," in the *Fire Safety Educator's Handbook*, National Fire Protection Association, Quincy, MA, 1983.

15. Strother, R., *A Study of Motivational Psychology Related to Fire Preventive Behavior in Children and Adults*, National Fire Protection Association, Quincy, MA, 1974.

16. "*Learn Not to Burn®:* A Decade of Progress," *Fire Journal*, March 1985, p. 8.

17. Lopes, R., "Public Perception of Disaster Preparedness Presentation Using Disaster Damage Images," *NFPA Education Section Newsletter*, Fall/Winter 1992.

18. Mager, R., *Preparing Instructional Objectives*, rev. 2nd ed., David S. Lake Publishers, Belmont, CA, 1984.

19. See the *Association Education Handbook*, p. 30, Glenn H. Tecker, ed., American Society of Association Executives, Washington, DC, 1984.

20. For more information on selecting, creating, and using educational materials, see *Public Fire and Life Safety Educator*, International Fire Service Training Association, Stillwater, OK (scheduled for January 1997 release).

21. Bredenkamp, S., *Developmentally Appropriate Practice in Early Childhood Programs: Serving Children from Birth Through Age Eight*, National Association for the Education of Young Children, Washington, DC, 1989.

22. *Practical Criteria and Instruments for Selecting and Implementing Sound Fire and Burn Education Programs in School and Community*. This document is informally known as "The Stillwater Standard." The package is available for $19.95 from the Pan-Educational Institute, 10922 Winner Road, P.O. Box 520347, Independence, MO 64052. [Telephone (816) 461–0201; FAX (816) 461–0210.]

23. Pretsfelder, G., Ed.M., *Pre-Production Evaluation of the "Tell a Grown-Up to Put It Away" Public Service Announcement for 3 to 6 Year Olds*, for the *Learn Not to Burn® Foundation*, National Fire Protection Association, Quincy, MA, Dec. 1990.

24. Adapted from M. Hunter, *Mastery Teaching*, TIP Publications, El Segundo, CA, 1982, pp. 85–90.

Additional Reading

Comer, D., *Developing Safety Skills with the Young Child*, Delmar, 1987.

"Directory of National Community Volunteer Fire Prevention Program. Community-Based Fire Prevention Education Initiatives, 1984–1992," National Criminal Justice Assoc., Washington, DC, Federal Emergency Management Agency, Emmitsburg, MD, FA-92, April 1993, 130 pages.

"Fire Departments and Communities: Partners in Prevention," *Fire Engineering*, Vol. 148, No. 6, June 1995, pp. 101–110.

Hall, J. R., Jr., "Children Playing With Fire: U. S. Experience, 1980–1993," National Fire Protection Assoc., Quincy, MA, August 1995, 35 pages.

Henricsson, L., "Measures for the Safety of the Lives of the Elderly and the Physically Handicapped," Firesafety Frontier '94, International Fire Conference and Exhibition in Tokyo, Creating a Safe Tomorrow, October 18–22, 1107–112, Tokyo, Japan, 1994, pp. 281–286.

Jones, R. T. and Randall, J., "Rehearsal—Plus: Coping With Fire Emergencies and Reducing Fire-Related Fears," *Fire Technology*, Vol. 31, No. 4, 1994, pp. 432–444.

Kakegawa, S., *et. al.*, "Evaluation of Fire Safety Measures in Care Facilities for the Elderly by Simulating Evacuation Behavior," *Proceedings* of the Fourth International Symposium for Fire Safety Science, Intl. Assoc. for Fire Safety Science, Boston, MA, 1994, pp. 645–656.

Knowles, M. S., "Adult Learning," *The Training and Development Handbook*, 3rd ed., McGraw-Hill, 1987.

Lambert, C., *Secrets of a Successful Trainer: A Simplified Guide for Survival*, Wiley, New York, NY, 1986.

"Let's Retire Fire. A Fire Safety Program for Older Americans," Federal Emergency Management Agency, Washington, DC, 1990, 15 pages.

Lichty, T., *Design Principles for Desktop Publishers*, Scott, Foresman, Glenview, IL, 1989.

Millsap, S., "Planning for Disasters in Your Community," *Fire Engineering*, Vol. 147, No. 12, December 1994, pp. 28–33.

O'Rourke, J. J., "Woodstock '94: Fire Planning for Large Public Events. Part 1: Fire Safety Preparations," *Fire Engineering*, Vol. 148, No. 1, January 1995, pp. 74–78

Parker, R. C., *The Aldus Guide to Basic Design* (the developers of PageMaker® desktop publishing software), Aldus Corporation, Seattle, WA, 1987.

Parker, R. C., *Looking Good in Print*, Ventana Press, Chapel Hill, NC, 1988.

Peoples, D., *Presentations Plus*, Wiley, New York, NY, 1988.

Perrault, M. E., "Home Security and Fire Safety Meeting Report, August 14–15, 1994, Quincy, Massachusetts," National Fire Protection Assoc., Quincy, MA, Meeting Report, December 1994, 22 pages.

Walker, B. L., *et. al.*, "Short-Term Effects of a Fire Safety Education Program for the Elderly," *Fire Technology*, Vol. 28, No. 2, May 1992, pp. 134–162.

Walker, B. L., "The Effects of a Burn Prevention Program on Child-Care Providers," *Fire Technology*, Vol. 31, No. 3, August 1995, pp. 244–264.

"1994 Learn Not to Burn Champions," *NFPA Journal*, Vol. 88, No. 3, May/ June 1994, pp. 67–69.

REACHING
HIGH-RISK GROUPS

Sharon Gamache

This chapter will focus on those groups that, determined by fire safety statistics and sometimes anecdotal information, are at greatest risk to be killed or injured by fire or burns. The studies "Patterns of Fire Casualties in Home Fires by Age and Sex, 1987-91,"[1] "Children Playing with Fire: U.S. Experience, 1990-1993,"[2] and "U.S. Fire Death Patterns, by State, 1980-1991"[3] provided the foundation for determining those most at risk from fire: (1) children age 5 and younger, (2) adults age 65 and older, and (3) people with low incomes in both urban and rural communities.

"At risk" may also refer to other circumstances or characteristics, such as groups that are more difficult to reach or at risk from other related health and safety problems. Most groups listed have received fewer resources when it comes to public fire and life safety education materials and resources. However, many organizations and fire departments have made great strides in creating effective and innovative programs to reach high-risk groups.

For information on the use of statistics to determine high-risk groups or target audiences, see Section 2, Chapter 2, "Using Data for Public Education Decision Making." For additional information on evaluating programs for high-risk groups, see Section 2, Chapter 7, "Evaluation Techniques for Public Education."

PRESCHOOL CHILDREN

Children age 5 and younger and adults age 65 and older have the highest fire death rates in the United States.* Preschool children have a fire death rate of 44 fire deaths per million, or more than twice the national average.[1]

Young children who play with matches and lighters are more vulnerable to fire than from any other cause. Children age 5 and younger are also less able than older children to control their environments, are more dependent on adults, and are less likely to have received or understand information about fire safety.

*Equivalent statistics not available for Canada.

Sharon Gamache is executive director of the NFPA Center for High-Risk Outreach, formerly the Learn Not to Burn Foundation. At the Center and previously at the National Safety Council she developed programs for people at high risk for fire and burns. Before working in safety, Ms. Gamache was a community organizer in St. Paul and Chicago.

General Characteristics of This Age Group

An instructive video, *From Theories to Play: Providing a "Creative" Developmentally Supportive Environment for Young Children,*[4] reviews characteristics of play and environment for young children. This video can help fire safety instructors with developing or using appropriate safety activities with young children.

The following are some general principles:

1. Children need a variety of play environments. They need outdoor and indoor play and a variety of play materials that support sensorimotor, dramatic, and constructive play. Construction materials can range from fluid, such as sand and paints, to more complex, such as puzzles or interlocking building blocks.
2. Activities should provide opportunities for children to interact with adults and other children in a language-rich environment. Children should be provided experiences with music, puppets, stories, and dramatic play to enhance language development.
3. Children like to create or recreate dramatic activities in the miniature, using items such as little characters, building blocks, puppets, etc., or with activities, such as acting out family scenes with costumes and props.
4. Children at this age have difficulty waiting for turns. Games with rules are inappropriate at this time, as children of this age do not understand win and lose.[4]

In a Federal Emergency Management Agency (FEMA) funded investigation on how children respond to disasters, Children's Television Workshop (CTW) found that young children ages 3 to 4 look to adults to help interpret their world, whereas children ages 5 through 7 begin to recognize cause and effect in relation to the disaster.[5]

Interdisciplinary Approaches

To reduce fire deaths among preschool children, it is important to look at several different approaches to reduce deaths and injuries.

Examples of approaches are: legislation and engineering, reaching parents and other caregivers, and teaching older siblings directly. Review these approaches, as follows, in context of the problem of children playing with matches and lighters.

Legislation and engineering: In July 1994 the U.S. Consumer Product Safety Commission (CPSC) enacted legislation making it a requirement that all disposable and some novelty cigarette lighters be child-resistant. This legislation specified that 85 percent of children age $2^1/_2$ years to $4^1/_2$ years tested must not be able to activate these lighters. The CPSC, however, allowed lighter manufacturing companies to manufacture their pre-regulated products that year until July 12. The manufacturers, in response, produced a quantity of lighters numbering as many as the greatest number they manufactured in any one of the previous three years; the result: there will still be non-child-resistant lighters on the market for possibly another 2 years. Many stores at this time still have both child-resistant lighters and non-child-resistant lighters on the shelf.

Canada enacted legislation June 14, 1995, making child-resistant lighters mandatory; this legislation went into immediate effect, thus the effects of the change will probably become evident earlier in Canada than in the United States.

A significant reduction in drug- and caustic-substance-related deaths among children resulted from legislated child-resistant packaging. If this precedence set by poison prevention packaging can be translated to lighter manufacture legislation, one can expect to see a significant reduction of fire deaths among children because of the changes.*

Although educational programs can be effective in reducing deaths among young children, no educational program can take the place of an engineering change, i.e., a change in the product. However, engineering change is not enough. Some children will be able to operate child-resistant lighters, and a significant number of fires that kill young children are begun by matches.† It is, therefore, extremely important to get the message to adults and caretakers to keep all lighters and matches hidden from children.

Reaching the caretaker: An adult component must be included in a comprehensive safety education program for preschoolers. It is, however, not documented how much behavior can be changed among adults and caretakers; and when caretakers are reached, whether or not the parents of the children who are most likely to become victims of fire are also reached.

One way to reach adults is through public service announcements (PSA). In 1987 the *Learn Not to Burn®* Foundation produced a PSA on matches and lighters, entitled *Got a Light*. The message: keep lighters and matches out of a child's sight, out of a child's hands. The announcement was: (1) clear and simple for adults and (2) conveyed the message to children without showing matches and lighters. (See Figure 2-5A, a drawing from the West Virginia Project funded by *Learn Not to Burn* Foundation, NFPA, and the Northeastern Area State and Private Forestry, USDA Forest Service.) Unfortunately, the PSA was not used by public advertising as much as was expected or needed.

In 1995 the *Learn Not to Burn* Foundation again released the PSA to fire departments which then delivered the PSA to their local television stations. These efforts were directed mostly to areas that the Foundation knew to be at greatest risk to fire, i.e., urban areas with a number of poor communities and rural areas with poor communities. The PSA rerelease was part of the strategy the Foundation used to reduce deaths in states where they were also working to reduce deaths and injuries through implementation of the *Learn Not to Burn Preschool Program* and other programs.

---

*Since 1960 the death rate per 100,000 population has more than doubled from 0.9 to 2.0 in 1990. The 25- to 44-year age group had the greatest increase in death rate, from 1.0 in 1960 to 4.0 in 1990. However, the death rate for the 0- to 4-year age group has fallen dramatically, from 2.2 in 1960 to 0.3 in 1990.[6]

†The annual average of 1980-93 structure fire civilian deaths as reported to fire departments showed 97 deaths among children age 5 and younger were due to matches, compared to 133 begun by lighters.[2]

FIG. 2-5A. Fire safety messages, such as keep matches and lighters high out of reach of children, preferably in a locked cabinet, are important to convey to parents and other caregivers of young children.

Another strategy to reach parents and caretakers of preschoolers is through the implementation of a parent component in a child-based preschool program. The *Learn Not to Burn Preschool Program* has three letters for educators to send home to parents or caretakers at different times during the course that teach eight key behaviors to young children.

1. The introductory letter tells parents what behaviors the children are learning in school but reminds parents that, although children are learning these fire safety behaviors, fire safety is still the parents' responsibility. It emphasizes four points for parents, e.g., keep matches and lighters up high out of the reach of children, preferably in a locked cabinet.

2. A second letter is sent home when the match and lighter lesson is taught. This letter emphasizes that children must be taught to stay away from matches and lighters and to tell an adult when they see matches and lighters. It then repeats the information to parents to keep matches and lighters away from children.[7]

3. The third letter describes how to design and practice a fire escape plan and the importance of smoke detectors.

Drawings or art activities from preschool programs can also be sent home to parents or caretakers to reinforce understanding of what the children are learning in school.

When training day-care providers, many teachers requested a video approach for reaching adults. The perception was that parents did not always read information sent home, and they would be more likely to respond to a video shown at parent meetings or at home.

Teaching young children: Fire safety materials must: (1) be appropriate for preschoolers and (2) not be the same materials used to teach children in elementary schools.

In 1979 the U.S. Fire Administration (USFA) asked Children's Television Workshop (CTW) if *Sesame Street*®, an educational television series created for in-home viewing by preschoolers, could be used to help teach fire safety to young children. CTW agreed to explore ways of including burn prevention and fire safety messages in its programming. Initial research resulted in a final report, *Fire Education for Sesame Street: A Research Study on Mass Media Fire Education.*[5] The research study recommended that it is prudent to be cautious and conservative when using television to communicate fire and burn messages.

Because of the one-way nature of viewing, television presents a difficult challenge especially with regard to such serious and important topics. For example, we need to be concerned that if a child viewed [the program] alone, questions could not be answered or clarified by an adult and a dangerous misunderstanding could occur.[5]*

Resulting from this initial research was the USFA-funded *Sesame Street* fire safety curriculum, one of the first available fire safety programs for preschool children.

In 1990 Anne Marie Santoro (one of the original developers of the *Sesame Street* fire safety curriculum), Children's Television Workshop, and the *Learn Not to Burn* Foundation came to the conclusion that public service advertising to young children with messages on matches and lighters would most likely be ineffective. For the Foundation report "Pre-Production Evaluation of the 'Tell a Grown-up to Put It Away' Public Service Announcement for 3-6 Year Olds,"[8] the Foundation tested 50 children in day-care settings with a video prototype and found that the youngest children (3 and 4 years old) did not understand the safety message well enough to risk putting the PSA on television; that is, mere exposure to the message and pictures of matches and lighters without comprehension can have the opposite effect on children, possibly causing them to have more of an interest in playing with or exploring matches and lighters.

The Foundation concluded that match and lighter messages would be more effective in its *Learn Not to Burn Preschool Program* wherein the teaching of eight key fire safety behaviors would include interaction with caregivers.

Reaching young children in child-care settings directly is probably the easiest way to deliver a program. There are day-care centers; preschools; and pre-Kindergarten programs, in some public school systems. It is somewhat easier to get a response from day-care providers to introduce safety materials into the day-care center, since they tend to be somewhat flexible on the material they can introduce into the classroom.

Since day-care providers are not necessarily linked under one organization, it is often necessary to contact several groups before identifying all day-care providers in a community. Many states have government organizations that license day-care and early childhood programs. These organizations often provide training and are a good resource when working at the state level. In Canada, the Ministry of Social Services and various province day-care associations help coordinate activities.

Content of Programs

The *Learn Not to Burn Preschool Program* created by NFPA's *Learn Not to Burn* Foundation gives the following basic approaches when teaching young children.

1. Good teaching of young children teaches fire safety awareness and skills in a non-threatening way. It says *"Don't Scare Children: Teach Them What to Do."*

   The *Program* teaches fire safety awareness and skills in a non-threatening way. Caregivers do not scare children by showing them burned toys or pictures of burned people.

2. Young children learn in several ways. Lessons to preschoolers should use a variety of activities to get behaviors across. Children learn best when they can use all the senses—through what they see, hear, smell, and touch, as well as how they move.

   Activities should vary and be participatory. The lessons should be short, but repeated to reinforce the concepts.

3. New adults, such as fire fighters, can be introduced into the child's environment.[7]

It is important when teaching young children that fire safety behaviors be taught in a comprehensive program. Because they do not understand "if . . . then . . . situations," young children need to act out different scenarios. For example, if preschoolers only learn the behavior "stop, drop, and roll," they will do that behavior for every fire situation (e.g., when there is fire and smoke in the room; if they burn themselves; etc.).

Match and lighter messages are important for prevention, because children playing with fire is the leading cause of fire deaths and injuries to preschool children. Children need to be taught that lighters and matches are not toys and that they should tell a grown-up when they find matches and lighters.

In *Follow the Footsteps to Fire Safety, A Prevention Program for Young Children*,[9] one lesson is dedicated to the topic of "good fires/bad fires." This lesson introduces the concept of fire and lays the groundwork for the next lesson, which teaches match and lighter safety.

The *Learn Not to Burn Preschool Program*, in both English and French versions, has reached nearly 40,000 day-care centers in the United States and Canada. Much of its success has been in the easy-to-use lessons offering a variety of activities. The songs, which were written by Jim Post for the program, are simple, repetitive, and have accurate and direct messages. Children will often remember messages that they learn in a song and remember what to do in a fire situation, such as shown in the following verse from "Crawl Low Under Smoke."

> *On your hands and your knees like a dog in a chase,*
>
> *Crawl low under smoke and leave that place;*
>
> *Crawl low under smoke where you can breathe;*
>
> *Crawl low under smoke and leave.*[7]

Another lesson important to children is understanding to go to a fire fighter if they are trapped in a fire situation and cannot escape. Messages children might learn are to yell out to the fire fighter and not hide from the fire fighter even though the fire fighter may look or sound different.

The *Play Safe! Be Safe! Children's Fire Safety Education Program, Ages 3-5,*[10] includes a video component that introduces the friendly fire fighter and shows the fire fighter acting in different teaching situations, dressing, and explaining the gear.

Presentations by fire fighters should not scare children with equipment, but rather the child should get to know the fire fighter without his or her protective clothing or equipment followed by a demonstration of the clothing and equipment donned one piece at a time. (See Figure 2-5B, an activity from the *Learn Not to Burn Preschool Program*.)

---

*Since that time, CTW has incorporated some fire safety education messages in its television programming.

Circle the fire fighter ready to fight a fire.

FIG. 2-5B. The Learn Not to Burn Preschool Program teaches children (1) to recognize the fire fighter as a helper, (2) not to be afraid of fire fighters with protective clothing and equipment, and (3) not to be afraid to yell out to the fire fighters.

Evaluation of Programs for Young Children

Preschool programs should be field tested and pre- and post-tested in their development and implementation stages. However, one cannot use written tests to evaluate young children's knowledge gain. Using and documenting one-on-one interviews, in which children demonstrate specific behaviors prior to and after the program, is one way to access children's knowledge gain.

Using illustrations in which children have to point to the correct behavior is another. The illustrator must be careful about subtle cues, such as showing the child or adult in the correct behavior illustration with a smile on his or her face.

"Good ways to assess young pupils include observation, portfolios, and interviews. We do far too much testing to the degree that our teaching looks like testing: children sitting quietly, filling in blanks." Instead, "the influence ought to flow in the opposite direction: assessment should resemble good instruction."[11]

For more information on evaluation, see Section 2, Chapter 7, "Evaluation Techniques for Public Education."

OLDER ADULTS

A well-rounded fire safety education program should include programs reaching older adults. In the early 1990s, people age 65 or older accounted for one-ninth to one-eighth of the population in the United States and Canada and this share is steadily growing.[13,14] People in this age group are twice as likely to die in a fire than the general population. And the risk goes up with age. People age 75 and older are three times as likely to die in a fire, and at age 85 and older, four times as likely.[1]

In targeting programs to reach older adults, it should be noted that older adults vary in their physical and mental abilities and prob-

ably in their risk of death from fires and burns. As health care has improved and more is known about exercise and nutrition, people are more active and in good health and involved longer in the community. Also many older people are financially secure and living in good housing.

There is a greater rate of fire death among the very old or "vulnerable elderly." From this, one may infer that this is a different population from those people age 65 to 75. Disabilities, such as immobility and hearing and vision impairments, affected 46 percent of people over age 85 in the early 1980s.[15]

Further, it is more likely that older people who are also poor will be victims of fire because they have two documented characteristics, i.e., age and low income, that put them at greater risk of fire death and injury.

About 90 percent of older adults are housed in their own homes rather than in nursing institutions; therefore, education programs should primarily target the self-housed elderly and those with whom they share housing.[16]

Fire Patterns for Older Adults

Most fire safety messages for older adults are similar to those for other adults and older children, but some variances of patterns may affect what an educator may emphasize.

Smoking materials were the leading cause of fire deaths for all age groups combined—25 percent of all deaths. Older Americans (50 to 74) were at an even greater risk from these fires, with smoking material fires accounting for more than 35 percent of their deaths. Smoking materials were also the leading cause of fire deaths for those aged 75 to 84, and 85 and older. Heating fires accounted for nearly as many deaths among the 85 and older age group as smoking materials.

Victims older than 49 were less likely to be asleep and unimpaired (e.g., by drugs or alcohol) when the fire occurred, but more likely to be bedridden or physically challenged than all age groups. Approximately 25 percent of fire victims in this age group were intimately involved in the ignition of the fire, compared with 16 percent of the victims of all age groups.[1]

As far as injuries are concerned, people 85 and older have the highest risk of non-fatal injuries in addition to the highest risk of death, with the injury risk nearly 60 percent higher than the national average. Older adults age 85 and over had a higher percentage of fire deaths involving fixed local heating units.

Combustibles placed or located too close to the heat source were the cause of 29 to 41 percent of fatalities in fires involving heating equipment of this high-risk group. In these instances, mattresses, bedding, and other soft goods were the materials most commonly ignited. Adults age 75 and over suffered high percentages of fatalities (16 to 22 percent) from fires igniting clothing on a person.[1]

In cooking-related deaths, the leading cause of death for adults younger than 65 was unattended cooking, whereas for adults age 75 and older a higher portion of fire deaths involved combustibles located too close to the heat source. Clothing was the material first ignited in 83 percent of the deaths involving "combustible too close to a heat source" for persons 75 and older.

Males are at higher risk to fire deaths than females. The largest difference in fire death rate was for males 85 and over.[1]

Presentations to Older People

Fear of crime among older people, and the reality of crime among people living in low-income communities in urban areas or poor rural areas, may cause people to do more to protect themselves in their homes from intruders.* On the high scale of home security systems, with automated lights and alarms, there is little risk of being trapped in fires. However, many older people who use bars, grates, double-cylinder dead-bolt locks, a door club, or any related devices may be unknowingly trapping themselves in their homes in case of a fire. (See subsection entitled "Home Security and Fire Safety.")

The *Learn Not to Burn* Foundation and the American Association of Retired Persons (AARP) collaborated on an instructional piece for educators in a 1995 NFPA fire prevention week kit[17] that gave the following basic principles for reaching older adults:

1. *Presentations*

    Educators should make presentations interactive. Older adults have a lot of experience and want to share that experience. Presenters should ask questions, invite input, and look to the older audience frequently for responses. If a video is used, the presenter must make sure everyone can hear and see it clearly. Demonstrations and interactive activities, such as having participants design their own escape plans or having discussions about real fire case studies, are important. Older adults have a lot of experience, and they want to have a chance to share it.

    (a) Many older people have hearing impairments but may be too embarrassed to tell you that they cannot hear, so it's important that presenters speak clearly and project their voices without shouting.

    (b) Although meeting places for all groups of people should be accessible, people are more likely to have a physical disability as they grow older; therefore top on the list for someone hosting a safety presentation is to make sure that the building room is accessible.

    (c) Older adults can be reached through a variety of organizations including senior housing, parks and recreation departments, places of worship, nutrition centers, or AARP Chapters.[17]

2. *Printed Materials*

    (a) When developing or choosing print materials for the older audience, one should use a "sans serif" type. Print size should be at least 12 point. Use a 14 point or larger font for so-called large-print books. (See Figure 2-5C.) Uncoated paper with a matte finish helps to cut glare. Black print on white or buff paper is most readable. A designed background may be interesting to the eye, but it will make the copy more difficult to read; therefore designs and pictures are better kept separate from the actual print.

    (b) Illustrators should portray older adults in positive, active roles. Stereotypes are likely to make the target audience avoid the message.[17]

3. *Special Considerations*

    (a) Some frail older people, who are probably the most vulnerable to fire, may not be able to leave their homes to attend presentations. Train home health-care workers and other home visitors to find and correct fire hazards in the older person's home. Provide these workers with printed materials so they can communicate fire safety information to homebound older persons.

    (b) Many senior citizens are also interested in the health and safety of their families and other people in their communities. They may have regular visits from grandchildren. In addition, AARP estimates that 3.4 million children currently live in a household headed by grandparents. The older caretaker needs to know the importance of storing matches and lighters up high, away from children, having working smoke detectors on every level, and practicing fire escape plans involving everyone in the household.[17]

In the article "Older Consumers Don't Believe You,"[19] it was reported that older groups of people are very skeptical groups of people, especially when it comes to consumer advertising.

Avoid talking down. Seniors aren't newly arrived on planet earth. They may look old, but they were all young once and most of them still have many of the same values and interests.[19]

Do not patronize older people. Older people don't think they are cute and they hate it when you treat them like little dolls. Never forget that they know more about this world than you do. They aren't waiting for you to give them a badge for still being alive.[19]

Although certain aspects of brain power decline with age, it is not as severe as one would expect.

*Across all localities of residence, rates of crime decreased as the age of victim or head of household increased. While rates of victimization were highest in cities for all age categories, there were no significant differences between rates of victimization from most crimes among the elderly (age 65 or older) who resided in suburban or rural locations. However, rural residents in the oldest age group were more likely than their suburban counterparts to have experienced burglary.[18]

This is an example of "sans serif" type.

This is an example of 12-point type.

This is an example of 14-point type.

FIG. 2-5C. Examples of type style and size for printed materials directed at an older audience.

Don't simplify to the point of banality. The older brain is packed with gigabytes of information and with advanced age-blessed software, so naturally retrieval time takes longer. It doesn't mean seniors are stupid, and they will resent it if you fail to appreciate this. Don't produce advertising that requires rapid informational retrieval to be understood.[19]

Sample Programs for Older Adults

One program that successfully reached older adults was "Project Fire Wise, A Fire and Burn Risk Reduction Strategy for Seniors,"[20] which began in St. Paul, MN. This project consists of an educational program and a system for the referral of older adults into smoke detector and installation programs. In 5 years, "Project Fire Wise" reached more than 10,500 senior citizens statewide, and in St. Paul distributed or installed 1,250 battery-operated smoke detectors and installed 350 hard-wired smoke detectors.

The program, begun in 1990, was a collaboration between the St. Paul Department of Fire and Safety Services and the American Red Cross, St. Paul area chapter. The Fire Department trained volunteers recruited by the Red Cross. The volunteers were trained to conduct presentations using a scripted slide show with prevention and protection messages. The program also included a "fire safety checklist for senior citizens."

The International Association of Fire Fighters, Local 21, provided money to purchase battery-operated detectors and recruited members to help install detectors in homes of older people who were unable to install them. In addition, the fire department organized a hard-wired smoke detector program in which the International Brotherhood of Electrical Workers #110 and the National Electrical Contractor's Association provided and installed the detectors at a low cost to homeowners.

In order to reach homebound older people, "Project Fire Wise" involved the Ramsey County Public Health Nursing Service and the "Meals on Wheels" program in the fire safety program delivery.[20]

The Mount Prospect, IL, Fire Department developed the "Save Our Seniors Fire and Life Safety Training."[21] This was a 4-hour outdoor event conducted for 450 people. Six behavioral topics, based on the *Learn Not to Burn* Curriculum and other programs for older adults, were presented at instructional learning stations. Those behaviors were: (1) stop, drop, and roll;* (2) fire prevention equipment; (3) kitchen and household hazards; (4) electrical heating hazards; (5) preventable fires; and (6) childproofing the home.

All the instructional methodology was written to relate to the particular restrictions of an older population. The learning stations were staffed by professional fire fighters and fire safety educators. The public education officer of the Mount Prospect Fire Department planned the event with an organized team of senior citizens in order to learn the needs of older people and to help with publicity on the event.

One enticement was the free food provided by local businesses. Other information booths included were: (1) insurance companies explaining what insurance was available to older adults, (2) a local burn foundation explaining the process to rehabilitate a burn victim, (3) a community hospital discussing what happens when a person is transported to an emergency room, and (4) the police department describing crime prevention practices.[21]

---

*The *Learn Not to Burn* Foundation recommends *demonstration only* of stop, drop, and roll for older adults because of the danger of causing hip and other fractures in practicing. In a real clothing fire situation, it's worth the risk.

PEOPLE IN LOW-INCOME COMMUNITIES

Today's urban environment offers challenges to anyone working to: (1) improve schools or (2) provide health services, city services, crime prevention, or fire and life safety measures. Traditionally programs for fire safety tended to reach those who needed it least. And without a concentrated effort to understand the urban or rural poor situations, it is unlikely that programs developed for more affluent communities will be effective. (See Table 2-5A.)

It is important to understand who are the poor today in U.S. urban, suburban, or rural settings. The article "No More Pity for the Poor"[22] identifies myths and truths about poor people today:

More than one of seven Americans lives in poverty including more than one of every five children. This means that people do not earn enough to feed, clothe, and house their families at a level which the federal government considers adequate; the minimum incomes needed to do this are $7,363 for singles to $14,763 for a family of four to $29,529 for a family of nine or more.[22]

Not just incomes set the poor apart. Stereotyped images often do too. Three stereotypes are true:

1. *The poor are less educated than other Americans.* Only 56 percent of poor household heads are high school graduates, compared with 81 percent of all household heads.
2. *Most of the poor don't work.* Only 9 percent of poor working-age adults work full-time, year round, compared with 42 percent of all working-age adults; 40 percent of the poor do work part-time or part of the year, compared with 69 percent of all adults.
3. *Most of the poor live in single-parent households headed by women.* Just over half of the poor families (52 percent) in America are headed by single females compared with only 18 percent of all U.S. families.[22]

Four stereotypes are statistically false:

1. *The poor have too many kids.* The average poor family receiving Aid to Families with Dependent Children (AFDC) consists of 2.9 people. The typical American family: 3.2 people.
2. *Most of the poor live in inner-city ghettos.* Only about one in eight poor Americans lives in census tracts with poverty rates of 40 percent or higher.
3. *Most of the poor are black.* Although blacks are overrepresented among the poor, whites make up 67 percent of America's poor, compared with 83 percent of the overall population. Blacks, who represent 13 percent of the population, make up 29 percent of the poor. The black ratio has barely budged since the mid-1960s.
4. *Poverty is a lifelong affliction.* Studies show that, while about a third of the poor are elderly or disabled and thus unlikely to climb out of poverty, one of every five poor people in a given year isn't poor the following year. Isabel V. Sawhill, a senior fellow at the Urban Institute, a nonpartisan think tank, estimates that roughly a third of the nation's poor remain impoverished only temporarily, owing to job loss, illness, or divorce.[22]

In the 1980 project report "An Assessment of Public Fire Education Programs for Low-Income Residents,"[23] produced by the National Safety Council for the U.S. Fire Administration, the Council interviewed representatives from fire departments and other organizations from large cities. They cited the two most common reasons for not having a fire safety education program targeted to low-income residents: (1) insufficient funding and (2) concern about ability to work with low-income people.

TABLE 2-5A. *Fire and Death Rates by Degree of Poverty In U.S. Cities with a Population of More Than 250,000*

Percent of Persons Below Poverty Level	Median Residential Fire Rate	Median Residential Fire Death Rate	Number of Cities
25.1% and over	179 (80-230)	2.82 (0.55-5.22)	6
20.1 to 25.0	164 (22-676)	2.30 (0.24-5.79)	14
15.1 to 20.0	168 (83-389)	1.27 (0.00-2.61)	17
10.1 to 15.0	147 (79-242)	0.92 (0.09-3.37)	14
5.1 to 10.0	122 (90-217)	0.87 (0.00-1.20)	4

We examined the most recent fire and death rates from the nation's largest cities. This analysis looks at each city's residential fire rate, or the amount of fires per 100,000 people; death rate in residential fires, or the number of deaths per 100,000 people; and percent of the population below the poverty line as of 1992, which was $14,343 in annual cash income for a family of four.

The fire and death rates were based on data reported to the NFPA by fire departments in 1993 or 1994. Of the 64 U.S. cities with populations greater than 250,000, according to the 1992 census, 55 reported their loss experience to the NFPA.

The relationship between fire rates and high levels of poverty can't be shown in as much detail as it can in other reports because they use comparisons by census tract, rather than rates citywide. However, the table shows that, overall, cities with higher levels of poverty do have higher rates of residential fires and deaths in residential fires.

Source: *NFPA Journal*, Jan/Feb. 1996.

Since that time, many efforts have been made to address the limitations that existed in programming for low-income communities. Probably one of the programs having the greatest breakthrough for public educators was the Portland, OR, project. The Portland Fire Bureau conducted a needs analysis which revealed that an area of the community containing 5 percent of the city's population was experiencing 26 percent of all the residential fires. A market research company revealed that the target audience, a low-income, mostly African-American group, would not respond to information from the fire bureau telling them to put in smoke detectors. Nor would they allow fire service personnel to come into their homes to install detectors. However, by developing a campaign with community volunteers, the bureau succeeded in distributing 700 smoke detectors in the target area.

Evaluation of Portland's campaign showed the following impact:

1. A community network was developed that provided the impetus for fire safety in the target neighborhood.
2. The number of working smoke detectors rose from 77 percent to 85 percent.
3. The number of persons able to recall proper maintenance of smoke detectors rose from 37 percent to 58 percent.
4. Fire deaths in the city dropped to nearly half of the previous year's number from 10 to 6.[24] As of August 1995, the area of the city that had 26 percent of all residential fire deaths had only one death since the program began in 1987.[25] "The program was successful because the community took ownership of the problem, and created [its] own solution."[25] Planting the notion of fire safety played a major part.[26]

In 1991 TriData Corporation launched a multi-year community-based fire safety program to test the application of Portland's methods in other low-income communities. The program's objective was to see if the approach would work in a variety of geographic areas and in communities with different demographic characteristics.[27]

In one project, the city of Cleveland, OH, Fire Department already had a smoke detector program in which 21,000 smoke detectors had been distributed. In terms of detectors installed, the program appeared to be successful, and by general consensus smoke detectors save lives. However, fire-related deaths did not go down.

It was discovered through work with the city and market research that the people most likely to die or be injured in a fire did not take advantage of Cleveland's program. The Cleveland Fire Department took another look at the program and asked, "Where do fires and fire deaths occur? Which residents are not responding to the city-wide smoke detector program?" The answer was that Ward 14, on the city's west side, had the highest fire rate and the Latino community, centered in Ward 14, did not respond in great numbers to the existing smoke detector programs.[27]

The Cleveland Fire Department then launched a targeted community-based fire safety education program with extensive involvement of Latino fire fighters, the mayor's Latino liaisons, community residents, churches, and other organizations. The fire department received 381 requests for free smoke detectors for this target population in the first 45 days of the program. That was more than 10 times the response from the Latino community than had been received from the city's nontargeted, city-wide program during its entire 2-year history.

The community-based fire safety education model was designed to address the challenges of the poor urban areas, i.e., working with tight budgets, large numbers of children unattended, non-English-speaking residents, and other issues related to social and economic changes.[27]

An important concept of the community-based approach is that no two communities are exactly alike. Fire departments must adjust approaches to fire safety education to fit the audience they are trying to reach.[27]

In another project, the New Orleans, LA, Fire Department undertook a program in the central city area where 90 percent of residents were African-American, 27 percent of households were headed by women, and most were low income. The Fire Department launched a smoke detector giveaway program with detectors installed by fire fighters. A key component of the program was the hiring of two African-American women as community liaisons. This idea came out of market research which suggested that women would better communicate with the many young mothers in the target community. These community liaisons worked closely with community groups to: (1) win support for the program and (2) recruit volunteers for a community-wide door-to-door canvass.

The result: 1,299 smoke detectors were installed, and the local cable company donated an additional 1,000 detectors. Preliminary and follow-up surveys in New Orleans showed the percentage of target area homes with at least one smoke detector jumped from 54 percent before the program to 76 percent afterward.

Target-area residents also adopted better fire safety behaviors. Following the program, 58 percent reported having fire escape plans, up from 40 percent prior to the program.[27]

The basic principles of the community-based approach are:

1. **Geographical targeting.**   Identify specific neighborhoods or areas where the fire problem is most severe and concentrate efforts on reaching residents of these high-risk areas.
2. **Market research.**   Use methods developed by the advertising industry to learn more about the target audience to understand how they get information and what will motivate them to listen and respond to fire safety messages.
3. **Grassroots community involvement.**   Invite the active participation of as many of the targeted audience as possible in each aspect of the program.

Community involvement is not just a means for disseminating information but should be a goal in itself. When people actively participate in a program, they develop a personal stake in fire safety that they bring back to family, friends, and neighbors.[28]

The following basic steps are the foundation of a successful community-based fire safety education program:*

1. Educators analyze fire data to identify the geographic target area with most fires and deaths and to determine the main causes.
2. Someone, usually a market researcher, conducts focus groups among people in the target area to:
   (a) identify the leaders in the community and
   (b) determine the best strategies for communicating fire safety messages to residents.
3. Meetings are held with the leaders of community groups identified during the market research.
4. The educators implement the program with the cooperation of the community groups. Some programs give free smoke detectors and educational materials as part of each pilot program.[27] Most of the programs included a door-to-door canvass of households in the target area. Fire fighters or trained volunteers installed the detectors.
5. The success of each program was evaluated by documenting citizens' response and analyzing fire data by comparing numbers of fires or fire deaths and injuries from before the program and after the program.[27]

The qualities necessary to be an effective fire and life safety educator in low-income and a variety of other communities are very similar to those necessary to be a community organizer trying to improve conditions in neighborhoods. Some of those qualities are:

1. An ability to find leaders in the community.
2. Finding the networks and organizations in the community that really work for people.
3. Understanding the enlightened self-interest of the person or persons to be motivated.
4. Recognizing differences without stereotyping people.
5. Imagination or empathy.
6. Readiness to give volunteers and leaders in the community recognition and to empower others to act and lead on their own.
7. A sense of humor.[29]

---

*These approaches are based on the U.S. Fire Administration's five-step planning process.

School-Based Programs

School-based programs have been very popular in reaching children and their parents. Example programs are: (1) ones in which the fire department does all the teaching directly to students in their classrooms or in auditoriums, (2) fire safety houses and trailers, and (3) ones that concentrate on training the teacher to teach fire safety as part of the ongoing curriculum.

To better reach children in low-income urban communities, one should have an understanding of: (1) problems existing in many inner-city schools and neighborhoods and (2) the children attending these schools. The book *Savage Inequalities, Children in America's Schools*[30] gives examples of various city school systems and the problems people face. A chapter on Chicago schools outlines conditions that children deal with every day:

One neighborhood, North Lawndale, "according to the *Chicago Tribune*," has one bank, one supermarket, 48 state lottery agents, and 99 licensed bars and liquor stores . . . . "With only a single supermarket, food is of poor quality and is overpriced . . . . With only one bank, there are few loans available for home repair; private housing has therefore deteriorated rapidly."

Typically in the U.S., people in very poor communities place high priority on education and they often tax themselves at higher rates than do the very affluent communities. But even though they tax themselves several times the rate of an extremely wealthy district, they are likely to end up with far less money for each child in their schools.

All of these disparities are also heightened, in the case of larger cities like Chicago, by the disproportionate number of entirely tax-free institutions—colleges and hospitals and art museums, for instance—that are sited in such cities. In some cities, according to Jonathan Wilson, former chairman of the council of Urban Boards of Education, 30 percent or more of the potential tax base is exempt from taxes compared to as little as 3 percent in the adjacent suburbs.

In the school system itself the city often relies on low-paid subs, who represent one fourth of Chicago's teaching force. Even substitute teachers are in short supply. On an average morning in Chicago, 5700 children in 190 classrooms come to school to find no teacher.[30]

The public educator needs to take into account what teachers and administrators are dealing with in schools in urban communities. There are many groups eager to provide their messages to children in these schools; therefore, in order to get fire and life safety on the priority list, one must be able to: (1) communicate the urgency of the safety issue and (2) provide new resources to the school system one is trying to reach.

NFPA's *Learn Not to Burn*: Safe Cities Program

In 1995, in an effort to help major urban centers with the fire problem, NFPA chose 10 cities from across the United States to receive a *Learn Not to Burn* Safe Cities Award. This program designates a manager, called a "*Learn Not to Burn* Safe Cities Champion," to direct a project. The city receives fire safety education materials along with technical support to establish or enhance the implementation of the *Learn Not to Burn* program in local elementary schools and day-care centers.

The program managers are fire safety educators from local fire departments, who plan, conduct, and evaluate the program. The Safe Cities model calls for a three-person team effort: (1) the "Champion," (2) a school system representative from the city school district, and (3) a business or agency committed to providing additional resources for the project.[31]

The model is a replication of one used in Los Angeles for three years. The Los Angeles program was a three-phase program funded by the U.S. Fire Administration. The first phase was a target program in three schools. The schools were chosen based on cultural and ethnic diversity. The pre- and post-tests administered in the pilot showed a 25 percent increase in knowledge. The second phase was an expansion to include all schools in the mid-city and south-central portions of Los Angeles. The third phase was the full implementation of the program. The Los Angeles Fire Department anticipates having the *Learn Not to Burn* program in 440 elementary schools by 1997.

To be chosen as a "Champion," applicants must agree to conduct and evaluate a pilot program in at least two classrooms at each of five grade levels: preschool and Kindergarten through grade 3.

The Safe Cities resources value $15,000 per city and can help boost fire safety education in cities that are always under pressure to cut budgets. Often the fire safety education budget is the first cut. NFPA is funding 10 more "Safe Cities" in 1997.

Reaching the Poor in Rural Areas

People living in large urban communities are not the only people dealing with poverty in this country. In the United States there is also rural poverty, especially in the southeastern states such as Mississippi, Louisiana, Alabama, North Carolina, South Carolina, Georgia, Kentucky, Arkansas, Tennessee, West Virginia, and the panhandle of Florida.

In southern rural areas, besides having high fire death rates, there is usually less accessibility to health care; problems with crime; and, in many cases, a lack of educational resources. A reliance on volunteer fire departments may mean that there is a longer response time to a fire and fewer fire service personnel to provide education, inspection, and other fire services.

In 1988 the South Carolina State Fire Commission appointed a committee to develop a strategy to reduce the state's fire problems. Eighty participants gathered to develop this strategy to reduce the fire fatalities in the state, which was listed among the top three states for the worst fire death rates.

The factors that made South Carolina high risk included:

1. The low economic level in the state forces citizens to seek low-cost, substandard housing, and, in turn, these families must turn to alternative means of heating and cooking.
2. The low education level of many citizens prohibits the ability to read and understand manufacturer's instructions for installation, use, and maintenance of alternative heating devices.
3. Older homes are not maintained and need electrical repairs, chimney repairs, and cleaning.
4. Children, disabled persons, and older citizens are left alone and cannot escape when fires occur.
5. Many families do not understand the importance of smoke detectors and do not consider them a necessity.
6. A large rural population living great distances from fire departments contributes to the fire-related fatality rate problem.[31]

A "Get Alarmed" program was the result of activities developed at the strategic conference. Included were smoke detector installment, development and implementation of a prevention program for school children, media coverage, and coalition building. In 1991 South Carolina reported the lowest number of fire deaths in more than 10 years—a 53 percent decrease in three years.

In 1993 the *Learn Not to Burn* Foundation was invited to Mississippi.[32] Mississippi had alternated with South Carolina from one

year to another on placing first in the highest fire death rate in the country. Preliminary program groundwork, regarding the terrible fire problem in Mississippi, took place with the Department of Insurance. A Mississippi Fire Chief's Association meeting in January 1993 resulted in the beginning of an effort to reduce fire deaths in the state with the implementation of the *Learn Not to Burn Preschool Program*.[7]

The *Learn Not to Burn Preschool Program* was picked for several reasons: (1) the cost was low, (2) it reached a high-risk group, and (3) the use of interactive materials, songs, and colorful materials made it both fun for instructors to use and accessible to children.

In a state dealing with poverty and a lack of resources, it is important that those involved in education have success on the first attempt at coalition building.

To call attention to the fire problem, a summit was convened in September 1993, bringing together 150 leaders of volunteer organizations, fire service organizations, and political leaders; this helped focus efforts on the issue and gave a boost to people who were involved. With funds from the *Learn Not to Burn* Foundation and local organizations, an initiative to reach all preschool children began and continued. The *Learn Not to Burn* Foundation provided technical assistance, such as training to Head Start and day-care teachers/providers in all regions of the state. The training sessions resulted in 25 public educators who could also provide training for day-care providers throughout the state. All training sessions included information on the fire problem and causes in Mississippi. Adult caretakers, too, were educated on the problem and on the need for smoke detectors, escape planning, and teaching children fire safety behaviors.

Coalition building was very important. The key fire service organizations involved were the Mississippi Fire Chief's Association and the Fire Fighters' Association. Local fire departments, such as Jackson, MS, and Madison, MS, also allowed their staffs to work many days to support all the efforts in the state. The *Learn Not to Burn* PSA, *Got a Light,* was aired on television in order to reach adults in the community with a match and lighter safety message.

Part of the coordination for activities came from the Mississippi Fire Safety Educators Association. Implementation of the *Learn Not to Burn Preschool Program*[7] helped build the association so that it would have the power to deal with other fire and burn issues.

It was very important that the Head Start Association and the Mississippi Department of Public Health/Early Childhood Unit be involved with the organizing efforts since training was coordinated with them. As of 1996, teachers from nearly 500 day-care centers and approximately 12,000 children have been reached in Mississippi. A reduction of deaths has started to occur, but it can only be determined over several years if this trend will continue.

These efforts were successful, due to using people that the kids already trust; the people that buy the message have to be visible in the community—the educators and the fire fighters. You need the people also in the regulatory business, day-care associations—so they could see that part of the coalition involvement was increased fire safety in the day care and among young children.[32]

This effort can best be maintained through continued communication. "There must be an effort to get to the unlicensed day cares, as well. [Many exist] and they need the fire safety message as well."[32]

Coalition building is also essential to the effort and requires contact with the "street mayor," i.e., the informal leadership in the towns who make decisions.[32]

The *Learn Not to Burn* Foundation initiated similar projects in Arkansas and West Virginia, also states with high fire death rates. In Arkansas, leadership came from the Arkansas Early Childhood Commission, Arkansas Fire Fighters' Association, the Arkansas Fire Chiefs' Association, the governor's office, and the Department of Public Health. Training on the *Learn Not to Burn Preschool Program*[7] took place in day-care centers throughout the state, reaching hundreds of day-care providers.

The importance of smoke detector use in saving lives was addressed simultaneously with the *Learn Not to Burn Preschool Program*[7] in Arkansas. The Arkansas Department of Public Health received a grant to install smoke detectors in homes of children in Head Start programs and in the homes of older adults in poor counties in the state. The people involved with the smoke detector program were also part of the state coalition formed as a result of the *Learn Not to Burn Preschool Program.*[7]

In 1994 a West Virginia project, funded by the *Learn Not to Burn* Foundation, the Northeastern Area State and Private Forestry, and USDA Forest Service, and coordinated with the West Virginia Fire Marshal's office, set out to reach every day-care provider in the state. With assistance from local fire department personnel, nine training sessions were conducted for fire service personnel, forestry personnel, and facilitators from the West Virginia University Cooperative Extension Fundamentals Program. The Fundamentals Program facilitators were people who trained in-state child-care workers in a variety of skill areas. About 1,000 day-care centers and 25,000 preschoolers were reached in the state.

Some special efforts were necessary in the most rural areas. Agencies that could establish personal contacts in those areas were able to reach day-care providers. In some cases, *Learn Not to Burn* materials were distributed through the Head Start Association home visitor program in those areas where there are no day-care providers and fire departments are remote.

The *Learn Not to Burn* Foundation also attempted to reach adults through this program by distributing printed reproducible materials that teachers could share with parents, and church and synagogue bulletins and community newsletters could publish. Local forestry and fire officials decided on four messages that could be used throughout the state: (1) have working smoke detectors, (2) keep matches and lighters out of the reach of children, (3) report forest fires, and (4) practice safe trash burning.

Each cited state presented common characteristics that affected the development of coalitions. Perhaps the most important was the willingness of fire departments with resources to share training, materials, and technical assistance with other full-time and volunteer fire departments. This process has been termed the "mutual aid" of public fire safety education.[33] Other characteristics include the following elements that were summarized in a presentation on the three state's efforts, West Virginia, Arkansas, and Mississippi, at the Oklahoma Public Fire Safety Education Conference in Oklahoma City, OK, in August 1995:

1. Identify local agencies relating to the target groups.
2. Recognize that "one size does not fit all" when working with a coalition in a particular state. Each state may have different participants. One state may have a fire marshal's office that is key to the organizing effort. In another state it could be the early childhood agency.
3. A coalition has to have multiple leaders. One person may chair the coalition meeting but the organization cannot depend on only one leader. One leader may be the "spark plug" to get things going; however, that person's life circumstances (e.g., job, location, interest in the organization) may change. The mission must continue under the leadership of someone else ready to "carry the baton."

Mean pre- and post-test scores by age

■ Mean pre-test   □ Mean post-test

FIG. 2-5D.   Evaluation is an important component of statewide implementations. A knowledge and performance test conducted among preschool children from eight preschools in West Virginia showed improvement in children's knowledge and ability. (Source: Carson and Powell, December 1995)

4. The program has to be more than a "book drop." Training and technical assistance must be provided for teachers or others who will be doing teaching or delivery of services.
5. Coalition members must share responsibility and roles. One person cannot handle all the work or, if that person does, he or she will "burn out" and the organization will be in trouble.
6. Initial efforts should be ones in which the participants can find success. If the process is too complicated, people will feel that they are not "winning" and lack the wherewithal to succeed and to make a difference.
7. There must be evaluation based on institutional changes, people reached, knowledge gained, and change in the fire problem. Sometimes the changes may be anecdotal, e.g., the case of a child in the program saving her own life and those of others because of what was learned in the program. (See Figure 2-5D, a graph from the *Learn Not to Burn* Foundation final report to U.S. Forest Service, Northeastern Area Rural Fire Protection.)
8. In states that have a lot of poverty and high fire death rates, it makes sense to do state-wide organizing because there are usually pockets of poverty throughout the state.[34]

OTHER GROUPS

Other Language Groups

In most urban and some suburban and rural communities, one must be prepared to provide instruction and information in languages other than English. According to the U.S. Census Bureau, foreign-born residents made up 8.7 percent of the U.S. population in 1994—the highest share since World War II.* That is 22.6 million people or nearly one in 11 who were foreign-born, nearly double the 1970 level.[35]

One-fifth of all immigrants, or 4.5 million people, arrived in the last five years. One-third live in California. Forty-six percent of all immigrants are Hispanic. The majority of immigrants are from Mexico, followed in descending order by the Philippines, Cuba, El Salvador, Canada, Germany, China, the Dominican Republic, South Korea, Vietnam, and India.[35]

Canada has two official languages, English and French. Of a population of about 27,300,000, approximately 6,500,000, or nearly 24 percent, have French as their mother tongue. Also in Canada, the 1991 census showed that foreign-born people made up 16 percent of Canada's population.[36]

---

*The high point for numbers of foreign-born in the United States was 14.7 percent in 1910, and the low point was 4.8 percent in 1970.

To reach people foreign born, it is important to understand language and cultural differences. When the *Learn Not to Burn* Foundation developed its *Learn Not to Burn Preschool Program*[7] for French-speaking Canadians, project staff not only translated the book into French but also produced original music that accompanied the book in the French language using a French Canadian musician. The songs had different melodies and styles and incorporated French idioms. The music and teacher's guide were then tested in "garderies," or day-care centers, in Montreal to make sure the correct meaning of the behaviors was captured in the new program. Montreal fire officials helped in the development and testing and many educators and fire officials throughout Canada reviewed the program.

In Bakersfield, CA, one of the key factors in a program to reach Latinos with a smoke detector giveaway program and an educational program was the establishment of a 24-hour Spanish-speaking answering service, through which the Latino community could ask questions and make requests. Also the fire fighter teams that installed smoke detectors in the Latino communities always included at least one Latino fire fighter.[37]

The Division of Fire Prevention of the Massachusetts Department of Public Safety created the "Fire Safety for Newcomers" booklet containing 26 fire safety tips in seven languages: English, Chinese, Creole, Khmer, Portuguese, Spanish, and Vietnamese. Illustrations, created by an immigrant in the United States, served as an "eighth language" to reach the illiterate and semiliterate. The booklet also addresses fire safety needs of Massachusetts residents who recently emigrated from other countries and how they may be at increased risk because of cultural traditions regarding fire. For example, some cultures may be accustomed to cooking with open flames in the home.[38]

It is essential to use a proficient translator to create foreign-language versions of fire safety materials. An advisory group of people who speak the language to which one is translating must review the copy. Further, a second translator should translate the copy back into English to confirm that the message holds.[39]

Be sensitive to the audience. Foreign-language populations may include some who are not literate in their own spoken language; therefore, one may have to rely on illustrations to describe the message. Also, realize that many people who emigrate to North America may have jobs at a much lower level than what they had in their homelands, e.g., physicians who now drive taxis or teachers who now work in factories.

Further, translated materials may be best given to the consumer with one side in English and one side in the foreign language, because household members may be at different levels in their reading skills in English, with some primarily reading English and others primarily reading their mother tongue.

First Nation Populations

In 1994 people living in the First Nation Reservations in Canada had a fire death rate 2.7 times higher than the Canadian national average. Children comprised half these deaths.[40]

House fires are the leading cause of death from unintentional injuries in U.S. Native American homes. Fire death rates for Native Americans substantially exceed those of all U.S. races.[41]

At particular risk are Native American communities in the upper Midwestern United States (i.e., North Dakota, South Dakota, Minnesota, and Wisconsin). This population has a death rate 6 times the U.S. national rate of all ages and races.[42] In response, the Indian Health Services and the U.S. Fire Administration formed a coalition that used a community health promotion model, with emphasis on the principles of social marketing, community development, and community organization, for a permanent fire safety program. Four

Native American communities were selected to be part of this program. Preliminary results revealed that the majority of smoke alarms in these communities were purposely disarmed because of nuisance alarms. Therefore, nuisance alarms became a prominent part of the fire safety program.

The Red Lake Indian Reservation, MN, Injury Prevention Project[43] was created to address the high rates of injury and death from fires and falls that occurred in this Northern Minnesota reservation. The reservation population addressed was 3,700, with almost half the population under the age of 18, and 51 percent below the poverty level. Like other Native communities the people of Red Lake had a disproportionately high fire death and injury rate.

After two child fire-play deaths occurred in 1990 and the problem of absence of smoke detectors was recognized, Red Lake officials contacted the State Fire Marshal's Division of the Minnesota Safety Council with the goal to work together to diminish the fire safety problem. A coalition of the State Fire Marshal's office, Indian Health Services, and Red Lake Tribal Council obtained a grant for a 2-year program that involved a variety of elements.

Two volunteer fire fighters were hired as community injury prevention coordinator and school coordinator. They performed home inspections, suggested improvements, and conducted community and school safety education programs. The inspections focused on electrical and heating systems; smoke detectors; fire extinguishers; and construction issues, such as number of exits in a home. By June 1993 more than 40 percent of the occupied homes were inspected. More than 800 safety improvements were made.

Ongoing fire safety education programs were presented to Head Start, elementary school children, and community groups. Also, *Keepers of the Fire*,[43] a culturally specific fire safety curriculum for second graders, was created to complement the *Learn Not to Burn* Curriculum. Representatives from the Minnesota State Fire Marshal Division, the Minnesota Safety Council, and the University of Minnesota also collaborated on the curriculum development.

A University of Minnesota instructor of the Ojibwa language, and a native of Red Lake, added Ojibwa words and traditional opening and closing ceremonies to the curriculum. He created the phrase "boo-da-way-we-ayaa-jig," or "fire-keepers." This is a reference to those who, by tradition, gathered wood and watched over fires on cold nights. This became the central concept of the curriculum. The program also focused on education and training of Red Lake fire fighters; car fires, auto extrication, and the use of self-contained breathing apparatus were addressed.[43]

Working with several leaders to create the program and getting them involved in data collection were important aspects of the success of the program. You can't just take a shotgun approach. You have to have a comprehensive approach which deals with the fire problem in that Native community, involves leaders in the community, and uses materials appropriate to that community. It's really the same principles you would use in any community to be successful.[44]

SPECIAL ISSUES

Home Security and Fire Safety

There are issues pertaining specifically to high-risk or poor populations that must be addressed by fire safety education. One example is entrapment caused by bars on windows and doors installed to keep out intruders. Even though fire deaths in general are going down in this country, those related to bars on windows and doors are on the rise. An average of less than one fire death per year was attributed to illegal gates or locks for the years 1980 through 1985.

This average increased dramatically to nearly 16 deaths a year for the period between 1986 and 1991.[45]

Incidents occurring in recent years show the seriousness of fires when bars trap people from escaping. In February 1993, seven family members died in a fire in Bruce, MS. Bars on windows kept a grandmother, five grandchildren, and one infant great-grandchild from escaping. A fire in Detroit, MI, that same month killed seven children left unattended for a short time. The children trying to escape were trapped by bars on windows and locked doors.

In 1991 the *Learn Not to Burn* Foundation's Technical Advisory Council put priority on this issue, and in 1993 formed the Home Security and Fire Safety Task Group to address it. This group met in August 1994 and identified the following elements:

1. People are afraid of being victims of crime.
2. It is important to recognize the crippling fear many people live with daily.
3. Loss of life can occur when exits are blocked in a fire.
4. People's behavior reflects a lack of understanding of the many components of this issue.
5. The threat of more tragedies is very real.[45]

The Task Group recognized that solutions to this issue would have to take into account that high-risk groups are at greater risk of becoming victims of murder and other crimes as well as being at greater risk of fire.

Engineering, enforcement, and education recommendations were made by the Task Group. The educational messages recommended by the Task Group and the Technical Advisory Council were included in the *Learn Not to Burn* Foundation's Community Education packet and in other educational materials. They are as follows:

1. Windows should open easily and fully enough for escape.
2. All security-barred windows needed for escape should have quick-release devices. Make sure everyone can open them.
3. Locked or barred doors should operate quickly and easily. Make sure everyone knows how to open them.
4. Practice exit drills in the home and use them to identify and correct obstructions of doors and windows needed for escape.[45] (See Figure 2-5E.)

Some recommendations that came from the Task Group were: (1) develop a standard for quick-release devices on home-security bars and a monitoring system to ensure compliance; (2) improve

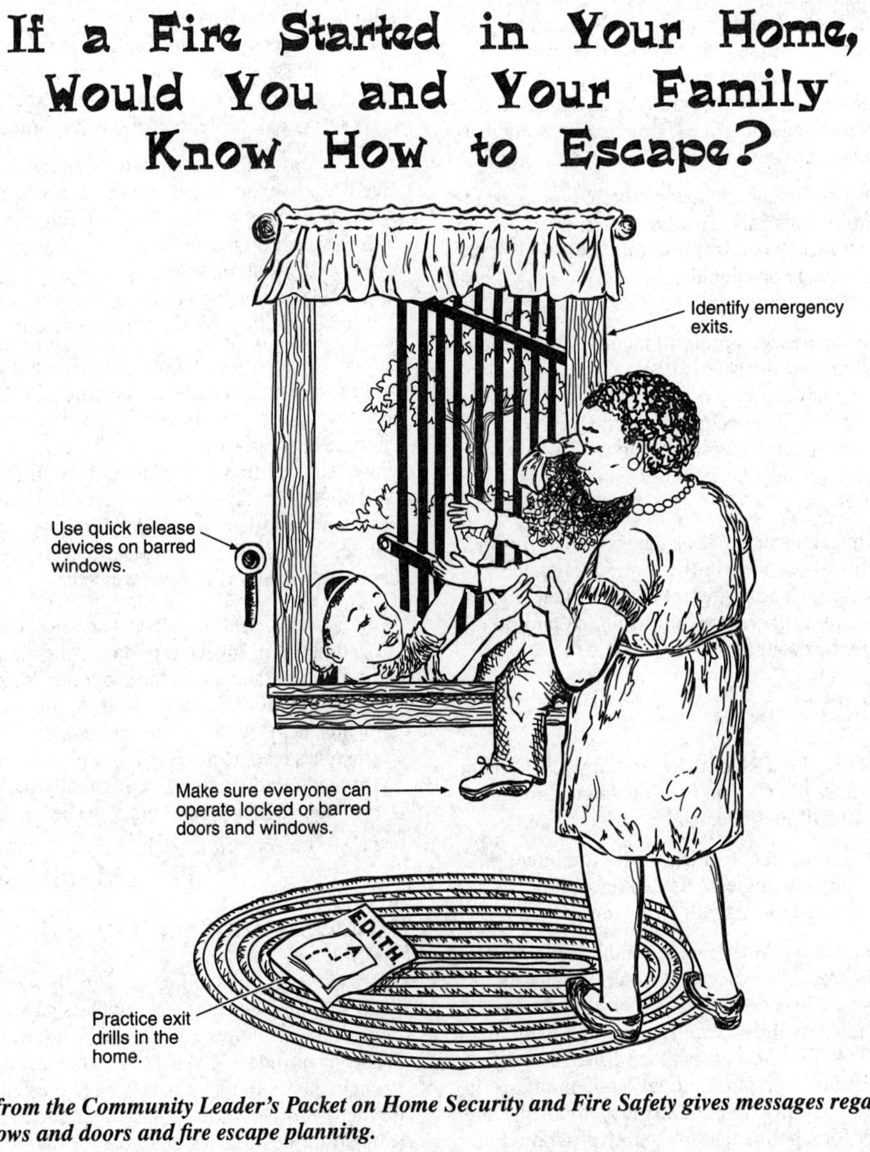

If a Fire Started in Your Home, Would You and Your Family Know How to Escape?

Identify emergency exits.

Use quick release devices on barred windows.

Make sure everyone can operate locked or barred doors and windows.

Practice exit drills in the home.

FIG. 2-5E. A flyer from the Community Leader's Packet on Home Security and Fire Safety gives messages regarding quick release devices on barred windows and doors and fire escape planning.

data collection regarding the incidents with bars and other security devices on windows and doors as well as total usage by consumers of such devices; (3) educate police, law enforcement agencies, and social service agencies to include fire safety concerns in addition to crime prevention in their educational programs; and (4) educate code enforcement officials about the standards and potential hazards of security devices.[45]

Manufactured Homes

The occupants of manufactured homes, particularly those occupants of manufactured housing made previous to the 1976 U.S. Housing and Urban Development (HUD) standard, could be classified as a group at higher risk to death from fire.

The fire incidence rate for manufactured homes is substantially lower than the rate for other dwellings; however, the fire death rate for manufactured homes is 63 percent to 89 percent higher.

Death rates may be higher due to: (1) room sizes are smaller, so conditions associated with flashover in the room of fire origin can occur more rapidly than in other housing; and (2) occupants of manufactured homes tend to be poorer than occupants of other dwellings, especially occupants of the older, pre-HUD-regulated manufactured homes, and poverty is associated with higher fire death rates.[46]

Education efforts should emphasize smoke detector use, installation, and maintenance programs. Smoke detectors can reduce death rates in both pre- and post-HUD-regulated manufactured homes. All post-regulated manufactured homes are required to have factory-installed smoke detectors, yet a lack of smoke detectors in manufactured homes is still a problem.* In 1993, 38.4 percent of post-regulated manufactured home fires were reported as having no detector(s) present, implying that detectors were removed.[46]

Another fire safety strategy is retrofitting older, pre-regulated housing with smoke detectors.

Prevention messages, such as information on heating systems, electrical systems, kitchen fires, and smoking hazards, are equally important in fire protection.

BIBLIOGRAPHY

References Cited

1. Karter, Jr., M. J., and Miller, A. L., "Patterns of Fire Casualties in Home Fires by Age and Sex, 1987–91," May 1994, National Fire Protection Association, Quincy, MA.
2. Hall, Jr., J., "Children Playing with Fire: U.S. Experience, 1990–1993," July 1995, National Fire Protection Association, Quincy, MA.
3. Hall, Jr., J., "U.S. Fire Death Patterns, by State, 1980–1991," Sept. 1994, National Fire Protection Association, Quincy, MA.
4. Phelps, P. C., Ph.D., *From Theories in Play: Providing a "Creative" Developmentally Supportive Environment for Young Children* (videotape), Diane Wilkins Productions in cooperation with the Creative Center for Childhood Research and Training, Creative Preschool, Tallahassee, FL, 1990.
5. Davis, E. P., Friedman, G. I., and Martin, L., "Community Education and Child-Care Programs," *Educational Research and Development*, Vol. 38, No. 4, 1990, pp. 46–50.
6. Hoskin, A. F., *et al.*, "Home Accidents, 1992," *Accident Facts*, 1993 ed. National Safety Council, Itasca, IL, p. 103.
7. Gamache, S., Powell, P., *et al.*, *Learn Not to Burn Preschool Program*, *Learn Not to Burn* Foundation, National Fire Protection Association, Quincy, MA, 1991.

8. Pretsfelder, Ed., M. Gary, "Pre-Production Evaluation of the 'Tell a Grown-up to Put It Away' Public Service Announcement for 3-6 Year Olds," Dec. 1990, *Learn Not to Burn* Foundation, NFPA, Quincy, MA.
9. Peterson, P., and Bergeron, S., *Follow the Footsteps to Fire Safety, A Prevention Program for Young Children*, Saint Paul Department of Fire and Safety Services, St. Paul, MN, May 1993, pp. 3–9.
10. Fire Proof Children, a division of National Fire Service Support Systems, Inc., and Rowan & Blewitt, Inc. *Play Safe! Be Safe! Children's Fire Safety Education Program, Ages 3-5*, BIC Corporation, Milford, CT, 1993.
11. Bredenkamp, S., and Rosegrant, T., *Reaching Potentials: Appropriate Curriculum and Assessment for Young Children*, Vol. I, National Association for the Education of Young Children, Washington, DC, (NAEYC), 1992.
12. *A Profile of Older Americans*, Program Resources Department, American Association of Retired Persons (AARP) and the Administration on Aging (AOA), U.S. Department of Health and Human Services, Washington, DC, 1995.
13. American Association of Retired Persons, "A Profile of Older Americans," AARP, Washington, DC, 1995.
14. *Statistics Canada, 1991 Census*, Catalogue No. 95-337, Ottawa, Canada.
15. "Aging America: Trends and Projections," prepared by U.S. Senate Special Committee on Aging with American Association of Retired Persons, Federal Council on the Aging, Administration on Aging, 1985–1986 ed., p. 87.
16. Friedan, B., "Denial and the Problem of Age," *The Fountain of Age,* Simon & Schuster, New York, 1993, p. 68.
17. Gamache, S., and Turner, M., "Reaching Older Adults: Watch What You Heat," 1995 Fire Prevention Week Kit, Oct. 1995, Department of Public Affairs, National Fire Protection Association, Quincy, MA.
18. Bureau of Justice Statistics, "Crime Victimization in City, Suburban, and Rural Areas, Annual Average 1977-89," U.S. Department of Justice, Rockville, MD.
19. Haller, T., "Older Consumers Don't Believe You," *Advertising Age,* August 14, 1995, p. 14.
20. Fuller, T. K., "*Project Fire Wise, A Fire and Burn Risk Reduction Strategy for Seniors*," Applied Research project submitted to the National Fire Academy as part of the Executive Officer Program, Jan. 1995, Department of Fire & Safety Services, St. Paul, MN.
21. Author interview with Skip Hart, formerly of Mount Prospect, IL, Fire Department, but presently with Buffalo Grove Fire Department, IL, Jan. 1995.
22. Topolnicki, D. M., "No More Pity for the Poor," *Money Magazine,* May 1995, p. 123.
23. Klones, S., *et al.*, "An Assessment of Public Fire Education Programs for Low-Income Residents," Sept. 1980, U.S. Fire Administration, Washington, DC, p. 74.
24. Crawford, J., "Firesafety Education: Not Just Kid Stuff," *Fire Command*, Vol. 56, No. 9, Sept. 1989, pp. 12–17.
25. Author interview with James Crawford, Fire Marshal, Portland, OR, August 1995.
26. Hershfield, V., and Johnson, J. T., "Planting the Seed of Fire Safety," *NFPA Journal,* May-June 1995, pp. 75–79.
27. Rossomando, C., *The Community-Based Fire Safety Education Handbook*, National Association of State Fire Marshals, 1995, pp. 1–6.
28. Author interview with Christina Rossomando, President, Rossomando & Associates, August 1995.
29. Gamache, S., "Community Organizing Skills and the Fire Safety Educator," presentation at the Oklahoma Public Fire Safety Education Conference, August 1995, Tulsa, OK.
30. Kozol, J., "Other People's Children," *Savage Inequalities, Children in America's Schools,"* Crown, New York, 1990.
31. Bosell, L. H., "South Carolina Initiatives in Public and Private Collaboration for Fire Prevention," presented at the Annual Meeting of State Coordinators of the National Community Volunteer Fire Prevention Program, March 18, 1992.
32. Author interview with Chief Tom Lariviere, Madison Mississippi Fire Department, and president of Mississippi Fire Chief's Association (1993-1995), July 1995.
33. Author interview with Mike Barrick, Deputy Fire Marshal, West Virginia, Jan. 1995.
34. Gamache, S., *et al.*, "*Building State Coalitions to Reach a High-Risk Group*," presentation on West Virginia, Mississippi, and Arkansas

---

* For manufactured homes, NFPA 501A, *Standard for Fire Safety Criteria for Manufactured Home Installations, Sites, and Communities*, has most of the HUD requirements; some manufactured housing may have the same safety features as post-HUD-regulated manufactured homes.

efforts at Oklahoma Public Fire Safety Education Conference, August 1995, Oklahoma City, OK.

35. Hansen, K. A., and Bachu, A., "The Foreign-Born Population: 1994," *Current Population Reports,* P20-486, August 1995, Bureau of the Census, Economics, and Statistics Administration, U.S. Department of Commerce, Washington, DC.

36. *Statistics Canada, 1991 Census*, Catalogue No. 95-337, Ottawa, Canada.

37. Kulenkamp, A., Lundquist, B., Schaenman, P., "Bakersfield, California, Fire Department: Smoke Detector Program for Hispanics," *Reaching the Hard-to-Reach, Techniques from Fire Prevention Programs and Other Disciplines*, TriData Corporation, Arlington, VA, 1994, p. 20.

38. Kulenkamp, A., Lundquist, B., Schaenman, P., "Massachusetts Department of Public Safety: Reaching Poor, Non-English-Speaking Groups," *Reaching the Hard-to-Reach, Techniques from Fire Prevention Programs and Other Disciplines*, TriData Corporation, Arlington, VA, 1994, pp. 38-39.

39. Author interview with Jennifer Mieth, Fire Safety Officer, Division of Fire Prevention of the Massachusetts Department of Public Safety, August 1995.

40. "Fire Loss Report—1994," Indian and Northern Affairs Canada, EA-HQ-94-04, Ottawa, Ontario, Canada 1995.

41. "National Indian Safe Home Coalition," 1993, Indian Health Services, Washington, DC.

42. Fanaselle, W. L., "Community Health Promotion, Social Marketing, and Institutionalizing Fire Safety in Native American Communities," 1995, Indian Health Services, Washington, DC.

43. Kulenkamp, A., Lundquist, B., and Schaenman, P., "Red Lake Indian Reservation, Minnesota: Injury Prevention Project," *Reaching the Hard-to-Reach, Techniques from Fire Prevention Programs and Other Disciplines*, TriData Corporation, Arlington, VA, 1994, pp. 53–55.

44. Author interview with Mary Nachbar, Deputy State Fire Marshal, Minnesota State Fire Marshal's Office, August 1995.

45. Perrault, M. E., "Home Security and Fire Safety Meeting Report," Dec. 1994, *Learn Not to Burn* Foundation, National Fire Protection Association, Quincy, MA.

46. Hall, Jr., J., "Manufactured Home Fires in the U.S. (1980-1993)," July 1995, Fire Analysis & Research Division, National Fire Protection Association, Quincy, MA.

NFPA Codes, Standards, and Recommended Practices

Reference to the following NFPA standard will provide further information on reaching high-risk groups discussed in this chapter. (See the latest version of *The NFPA Catalog* for availability of the current edition of the following document.)

NFPA 501A, *Standard for Fire Safety Criteria for Manufactured Home Installations, Sites, and Communities.*

USING THE MEDIA FOR EDUCATIONAL OUTREACH

Dena Schumacher

In disaster coverage, reporters provide the public with pictures, sounds, and words. Public information officers (PIO) at the scenes prepare press releases, give interviews on camera, and communicate with throngs of reporters.

In the midst of confusion at the bombing of the nine-story Alfred P. Murrah Federal Building in Oklahoma City, OK, on April 9, 1995, the media—via public information specialists—played a vital role in informing the world. Emergency specialists everywhere watched newscasts, listened to radiocasts, and read newspapers to receive updates. Media actions would have been incomplete without the diligent efforts of fire service and emergency rescue personnel willing and able to share the information.

Beyond the obviously critical need for fire-rescue teams, a secondary need to provide information to frightened citizens and family members became immediately apparent. Reporters provided insight into the emergency. The public information specialists provided the information to the reporters.

A satellite "city" sprouted up across from the bomb site. Media professionals informed Oklahoma City citizens of a variety of life safety topics during those days. They provided the nation with updates on citizen and emergency worker needs. Supplies arrived by the semi-trailer load.

Usually, the media works under less stressful circumstances, perhaps on a series of articles examining various departmental operations, including staffing and fireground tactics; fireworks regulation fact sheets; or news releases announcing the National Fire Protection Association (NFPA) *Learn Not to Burn*® Champion Program award winners.

How does one convey information to reporters? Who must be contacted? What does the media expect? Where do fire and life safety educators begin?

To provide information effectively, fire personnel must understand the media and communications: how they work, how reporters and editors view their jobs, and how the fire service can best utilize ever-changing communication resources.

Dena Schumacher is a fire and safety education specialist with the Champaign (IL) Fire Department. She holds a B.S. in child and human development and an M.S. in journalism. She is a member of the NFPA Technical Committee on Fire and Life Safety Education. She is also the NFPA's Fire Safety Education Representative in the Midwest.

The ability to communicate with others is essential to progress; failures and frustrations result from the *inability* to communicate. Before venturing into the realm of education and media, understand that

Communication is both simple and complex. The results can be rewarding or disappointing, exciting or frustrating, important or trivial. It is the process we use most in our daily lives and perhaps understand the least.[1]

Many books have been written on communications, media, and education. This chapter simply outlines the basics, and aims to provide practical information to aid fire personnel who communicate with the media. It discusses community media services, print as well as broadcast, noting their similarities and differences. This chapter also explores legal issues involved in working with the media and the public. In addition, it describes four situations in which fire and life safety educators engage the media:

1. Fire service personnel most frequently encounter the media during emergency situations. To bolster their news stories, reporters approach any available personnel for further information.
2. Fire service personnel depend on the media to share information with citizens—perhaps to announce a child care project, a cardiopulmonary resuscitation (CPR) course, or a smoke detector campaign—through public service announcements (PSAs) and news releases.
3. The media allow the fire service to send messages in practical ways, such as billboards, newsletters, flyers, and hot air balloons.
4. Fire service personnel work with the media to educate, providing coursework, communication technologies, and other materials. Educators find it necessary to become acquainted with and adapt to ever-changing vocabulary and technology—e.g., the Internet, World Wide Web, etc.

DEFINING CRITICAL WORDS

Public Relations *vs.* Public Education

There is a fine distinction between public relations and public education. Public educators perform both functions.

Public relations and public education are exactly the same . . . only different. There are many similarities between the two. Both require a carefully thought-out planning process, including a solid implementation strategy. Both can (and should) be evaluated for their effectiveness. The goals for each are different.

The goal of a public relations strategy is to improve (or sustain) the positive working relationship between the fire service and the community that supports it.[2]

The primary goal of a good education program is not to enhance public relations.

That's a by-product. Rather, the goal of education in its purest sense is to enlighten someone, giving them the knowledge and skill to improve their existence.[2]

The fire and life safety educator can help citizens put knowledge to work. Therefore, the educator must: (1) understand and know what people need; (2) communicate the information in a timely, meaningful way; and (3) make sure people receive and understand the message.

Information can be shared in various ways; however, this chapter only addresses information shared through media.

Media

Media is defined as the vehicles used to convey information to an audience. They include radio, television, newspapers, magazines, motion pictures, direct mail, cable television, books, records, tapes.[3]

Public educators must investigate various media outlets within the community to discover which are most useful, given the problems and messages individual communities must address and send. These outlets can reach various audiences within the community.

Communications

Communications is the process by which persons interact and, thereby, influence each other.[4]

To communicate is to impart knowledge. However,

. . . information is not necessarily communications. Information converts most effectively into communications when it is needed, useful—and sought.[4]

The educator must not only recognize community needs from the fire department's perspective, but must also address community perceptions. Respecting fellow citizens' wants and needs is a necessary block of community building. Along with knowing what problems a community faces, public educators must know and act upon what the community's citizens believe they need.

Keep two points in mind:

1. The audience is *always* in the driver's seat in any communications situation. Audience members ultimately decide whether to listen to, accept, reject, or act on any new information.
2. No one—including our audience—has much interest in operating totally within the confines of someone else's agenda.[5]

Active listening, focus groups, and community surveys can help with this process of understanding.

Educational Outreach

Educational outreach takes place on various fronts in various ways. The *Random House Unabridged Dictionary* defines *education* as "the act or process of imparting or acquiring general knowledge, and developing the powers of reasoning and judgment, and generally of preparing oneself or others intellectually for mature life."[6] Here, educational outreach is defined as the process of sharing meaningful information with fellow employees and citizens, young and old alike.

Educational outreach through media technologies grows hourly. Fire departments, large and small, must not only acknowledge and accept changes in computer technologies and satellite transmission, but also explore and use them.

Practical Publicity

Practical publicity includes handouts, magnets, fire department musters, and open houses—ways in which information may be passed but educational gain cannot be measured, except by sensing whether the community liked or disliked what the fire department educator presented.

FIRE DEPARTMENT'S COMMUNICATION GOALS AND OBJECTIVES

Communications Within the Fire Department

Communication goals within the fire department serve several purposes. Inhouse—with fellow employees—the educator must communicate, listen, and interact:

1. Employee to employee,
2. Staff to suppression, and
3. Fire fighter to fire chief.

The fire and life safety educator must have backing and support of the fire department before proceeding into the community and introducing him/herself to the media. The educator must know which areas the department wishes to highlight, which departmental goals need an explanation, and how the department wishes to address potentially embarrassing situations. Are other members willing to be available to the press? Are media representatives welcome to stop by the department?

After devising a plan to work with the media, department members must help prepare campaigns and programs. The fire department also needs to review news releases prior to dissemination to prepare for reporters' calls.

Communications with Citizens

A second, more obvious, mission includes communicating information to the citizens:

1. Before an emergency occurs (prevention), and
2. At the scene of the emergency.

Fire service members possess extensive information. Media representatives, on the other hand, are interested in the information educators can provide. While the fire service may serve as "eyes and ears" for reporters, the media generally serves as the voice to the public.

One key to successful media relations includes understanding the media's various roles. This is discussed later in the next section, "Understanding Community Media Services." Another key to successful media relations is identifying the audience. Identifying a very general audience may ease the communications planning, but complicate all other parts of the process. It results in diluted messages and weak delivery, and compromises the timing of messages. It also wastes time and materials, forcing fire chiefs, fire and life safety educators, and public information officers (PIO) to send messages to persons who have neither need for, nor interest in, them.[4] Chapters 3, 4, and 5 of this section provide useful information on defining audiences.

The audience should be specifically identified; this greatly increases any program's potential effectiveness:

> If you find yourself coming up with answers like "everyone" or "the general public," you don't have a sufficiently clear understanding of who you're trying to reach—do not proceed any further until you clarify this.[5]

Each news station, newspaper, and news outlet claims a particular audience and a particular area as its own. Media outlets within the fire department's service delivery area can provide information on their particular target audiences.

UNDERSTANDING COMMUNITY MEDIA SERVICES

Identify Local Media

To create a positive relationship with the media, fire service personnel must take time to understand each media outlet's responsibilities and target audience.

Some fire and life safety educators may be able to spend the time and effort necessary to become competent professional-level producers in one or more of the various media. Because the majority of full-time educators cannot, the essential issues become: (1) what sort of relationship can be developed with local media representatives and (2) how to work cooperatively with writers, editors, and broadcasters.[5]

Educators should make diligent effort to meet reporters, editors, and station managers; get to know *them* and become familiar with their specific job responsibilities.

Job turnover occurs regularly in the media business. Reporters and editors move frequently, so getting to know new personnel is a never-ending task. Keep a contact list and update it regularly. Media sources can be identified in several ways: (1) check media directories, (2) scan the Yellow Pages of the phone directory, and (3) talk with other public relations personnel in the community. A wide variety of books, information, and services are available to aid fire and life safety educators in identifying local resources. (See Appendix A.) Balance the created media list by including neighborhood ethnic and suburban papers. Collect and learn names, titles, phone numbers, and fax numbers of those in charge. An organizational flow chart proves useful in determining who to contact about a particular project. (See Figure 2-6A.)

Figure 2-6A shows an example of a media quick call list that can be used during an emergency. Pertinent names and phone numbers are readily accessible to the public information officer (PIO).

Medium	Contact Person	Phone #	Fax #
Print organizations			
1.			
2.			
3.			
Notes			
Radio			
1.			
2.			
3.			
Notes			
Television			
1.			
2.			
3.			
Notes			

FIG. 2-6A. Example quick call list form.

Keep copies of the completed form at home, in the office, and in the vehicle generally used in emergencies.

Establish Credibility

Building the fire department's reputation with the media is critical. Reporters, editors, station managers, and producers can quickly form opinions about the educator's operating style. Reporters often rely on *credible* communicators at nonprofit agencies. They need to know the latest trends and opportunities, activities, and resources. Assure that accurate and timely facts will be provided, promises will be kept, and deadlines will be met; then *follow through.* Deadlines are critical in broadcasting and journalism; *being late is inexcusable.*

To establish credibility: (1) understand what makes news, (2) know what the media needs to know, and (3) use imagination and creativity to find new ways to keep the media updated on the fire organization.[7]

Never omit crucial information. For example, an Illinois fire fighter's union hired an out-of-state fund-raising group to host concerts. Telephone solicitations across several counties stressed that the money went to public fire education funds. Local reporters discovered that less than 10 percent went to the union treasury and only 1 percent to public education funding. Reporters found fault, fire fighters looked foolish, and public cynicism reigned.

Identify Key Players

Educators must identify key players at local television and radio stations and newspapers. Understand what reporters, station managers, and editors do, and they, in turn, will most likely better understand the fire and life safety educator's role in the fire service.

MATCHING THE MEDIUM WITH THE MESSAGE

All communications channels have distinguishing features, each situation is different—and each requires a fresh look.[4]

Broadcast Media

Radio: Radio is the medium of the mind.[7] Radio is cost effective. Radio is efficient. Radio is usually local. Over 95 percent of the population 12 years and older listen to radio at least once a week, and two out of three listeners tune to FM stations, according to Arbitron, a radio rating service.[7] Radio appeals to audiences for many reasons. An intimate medium far less intrusive than television, it demands less concentration than newspapers. Portable, it can be listened to while doing most anything.

Due to its temporal nature, radio works best when used to notify, remind, or tell easily remembered, uncomplicated stories. Radio messages must be simple and concise. Radio reporters rely on lively interviews to add color to their stories. Broadcasters suggest stories 20, 30, and 60 seconds in length.

The station manager (general manager) oversees all operations. The program director oversees news and sports. Fire service personnel work most frequently with the news director. Note, however, at smaller stations, the station manager is often the owner, program director, and chief engineer. (See Figure 2-6B.)

Figure 2-6B shows the general organization of a radio station. Individual stations vary considerably, depending on their size and programming. The publicist should identify persons in the offices shown, however, and should maintain contact with the news director

and individual reporters, as well as the promotion director and public affairs director.

Television: For 72 percent of the U.S. population, television is the primary source of news, according to an April 1994 Roper Organization for Television Information Office report.[8]

Key contacts in the television news department include the news director, who oversees news staff and broadcast; the assignment editor, who assigns reporters to stories and ensures coverage of major stories and "beats"; the news producer, who determines which reports will air; and reporters, who cover the stories. (See Figure 2-6C.) For information on television public affairs, approach the public affairs director or the public service director.

Figure 2-6C shows the general organization of a television station. Most stations have large staffs, sometimes numbering into the hundreds. Key contacts include the vice president for news, the news director, the assignment editor, and individual reporters, as well as the vice president for public affairs and members of the public affairs staff, such as the public service director and the editorial writer.

Before approaching a television news department with a story idea, know the station's format and to whom the idea should be pitched.

Television reporters believe the best stories have strong visuals that grab and hold a viewer's interest. With a camera crew in tow, a television reporter is attracted to activities that make exciting or interesting footage and interview subjects who talk in clear, concise language.[7]

Appendix B provides guidance and information to improve professional communication with the media.

Public service announcements: One common way to inform the public is through public service announcements (PSAs). Prepared for radio and television stations, PSAs are short announcements, generally running between 10 and 60 seconds.

The Federal Communications Commission (FCC) defines a PSA as

... any announcement ... for which no charge is made and which promotes programs, activities, or services of federal, state, or local governments ... or the programs, activities, or services of nonprofit organizations ... and other announcements regarded as serving community interests.[9]

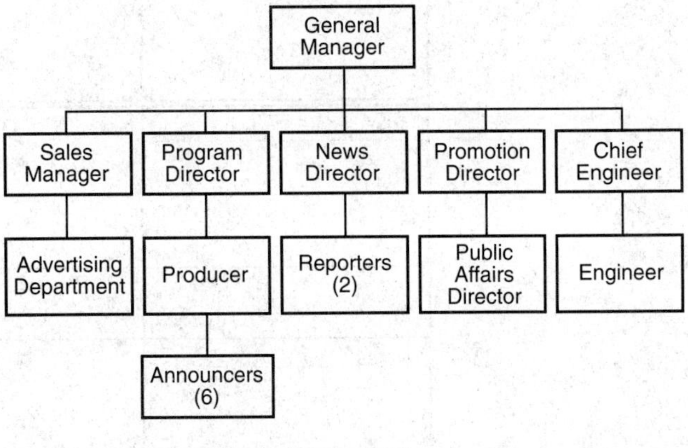

FIG. 2-6B. *Generalized organization of a radio station.* (Source: Tedone, p. 75)

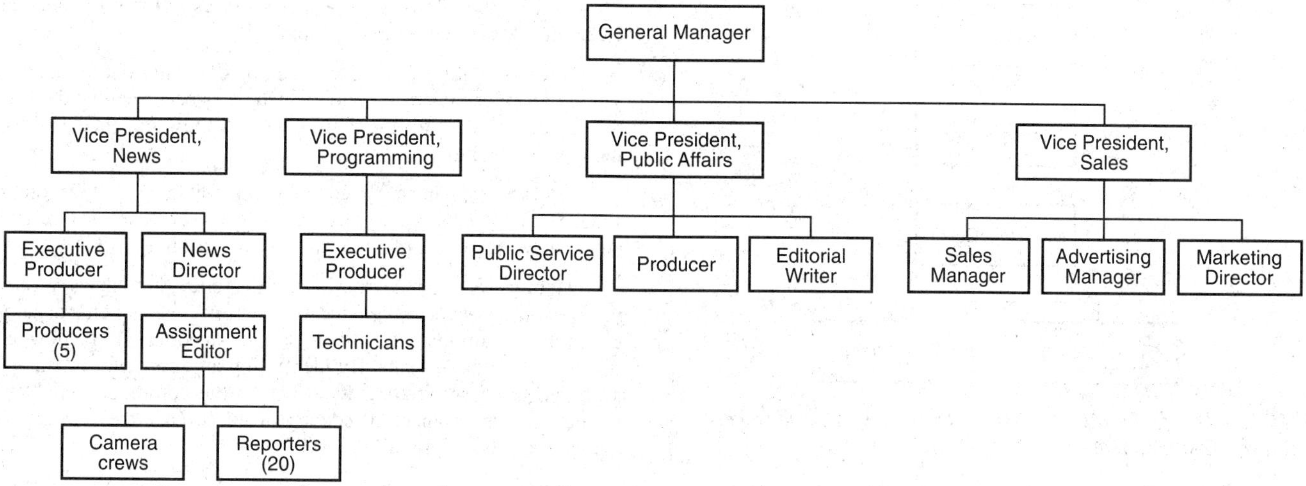

General Manager

Vice President, News — Vice President, Programming — Vice President, Public Affairs — Vice President, Sales

Executive Producer | News Director | Executive Producer | Public Service Director | Producer | Editorial Writer | Sales Manager | Advertising Manager | Marketing Director

Producers (5) | Assignment Editor | Technicians

Camera crews | Reporters (20)

FIG. 2-6C.   Generalized organization of a television station. (Source: Tedone, p. 94)

PSAs give the listener important information. Generally, the first two-thirds of the copy attracts attention and directs the listener to further information. Much can be said in 20 seconds; note, however, at this writing, stations have no obligation to provide a minimum amount of public affairs and public service programs.

In a mid-sized market the educator may deliver PSAs and reasonably expect they will be used. In major markets, running PSAs has become increasingly difficult.

Because competition is strong, many stations stylize their newscasts to stand out on the crowded dial. They simply refuse to run anything their competition might have also received. In most large markets, it's best simply to become a news source. Provide information in a regular and timely fashion.[7]

Public access television: Public access television is another excellent and inexpensive way to provide community education programs. Approximately 2,000 public educational or governmental access channels in the United States are either city-supported, member-based, or franchised from the cable company. The community cable company can explain how the system operates locally.

Local talk shows also provide an excellent forum for sharing information. Take time to view a segment of a program before agreeing to appear.

Television tends to be underutilized for educational purposes.

In the emerging information age, when the trend is to move "from institutional help to self-help," television and video tapes can be the educator's forum to speak with millions of people in the places where they live, study, work, and play. The equipment is in place. The channels are open. The medium is available and waiting for our message.[7]

Print Media

The educator cannot overestimate the importance of the press. Newspapers offer the best place for an educational news item to reach a large audience. Print journalism includes daily and weekly newspapers, magazines, and wire services.

Reporters need good sources to inform them about timely, important stories. By understanding how the press operates, how to prepare press releases, and with whom to work, the educator can be

that good source. The educator can identify the types of press available in the area by consulting newsstands, media directories, and phone books. (See Appendix A.) After creating a complete list of print media, determine who's best suited to convey information for a particular campaign. This is a good first step, but not the only step.

The second step is to identify appropriate editors, reporters, and other newspaper and magazine contacts. This increases the likelihood that the copy will be read and professionally handled.

As in broadcast media, understanding the print media hierarchy is essential. Most newspapers are staffed similarly with general titles; however, the work done under those titles varies from paper to paper, city to city. To better understand this, look at the major divisions in a large daily newspaper. There are generally three: (1) the corporate board and executive committee, (2) the business division, and (3) the editorial division.[10]

The first two divisions have little to do with the educator's publicity and media needs. The editorial and news staffs can provide more assistance. A large daily, e.g., *The Washington Post,* contains many sections. Identify: (1) the sections of the paper most likely to use safety information and (2) the person on the staff best suited to help.

When dealing with a medium-sized newspaper, include the news editor, assistant news editor, features editor, and one or two reporters on the mailing list. (See Figures 2-6D and 2-6E.)

Figures 2-6D and 2-6E show the organization of weekly newspapers. The small weekly has a circulation of about 5,000. The medium-sized weekly has a circulation of about 100,000. The editors and reporters of a paper this size can all be used as publicity contacts. Efforts should be made, however, to identify precisely the "beats" or areas for which each editor and reporter is responsible.

After identifying the target audience, the educator must determine what reporters need.

All reporters are attracted to a good story, but each has somewhat different ways of covering it. A newspaper or magazine reporter will generally interview sources in person or on the telephone, looking for photographic possibilities or statistics for an eye-catching piece of graphic art.[7]

Understand What Makes News

News might be defined as any piece of information that will affect your head, heart, or pocketbook.[7]

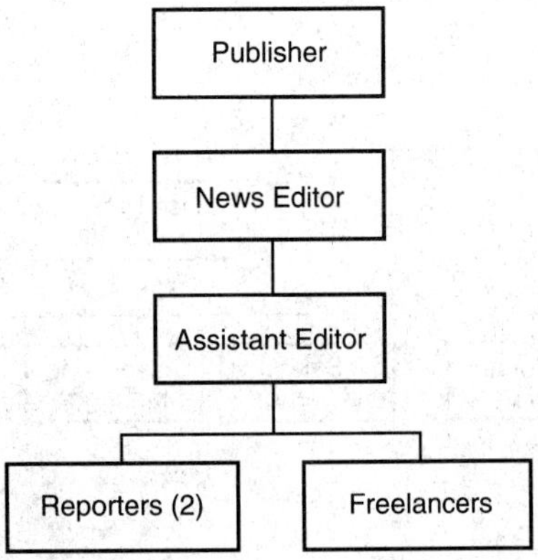

FIG . 2-6D.   Generalized editorial structure of a small weekly. (Source: Tedone, p. 43)

News reports something new—information that is a break from everyday events; it is information that people need to make sound decisions about their lives.[11] News editors decide if information they receive is indeed newsworthy by checking news values or news pegs. These news pegs are the basis for deciding what makes print and what does not.

From the media's view point, six factors (i.e., news values) determine an event's or an idea's newsworthiness:

1. Consequence: Does the event affect many people?
2. Interest: Is event an unusual, rare, odd, or interesting situation?
3. Timeliness: Are the events immediate or recent?
4. Prominence: Does the event involve well-known personalities or institutions?
5. Proximity: Does the event occur within the circulation or broadcast area?
6. Conflict: Does the event show dissent or trouble between people or institutions?

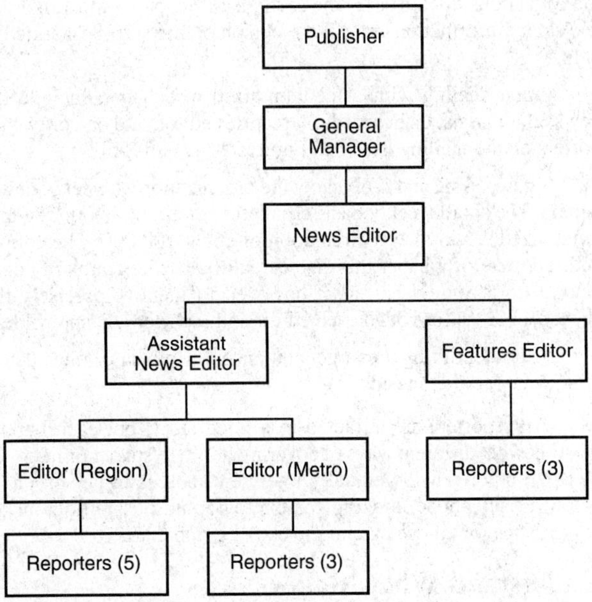

FIG. 2-6E.   Generalized editorial structure of a medium-sized weekly. (Source: Tedone, p. 43)

At least three-fourths of all news stories fall into the general categories of Consequence and Interest.[11]

News is relative. Each day, changing dynamics define "news." These dynamics include: (1) how much news occurred that day, (2) the advertising load for the days, (3) the type of audience, and (4) the type of medium.

"Hard" news must be covered immediately, or it becomes stale. Today's news is not necessarily tomorrow's news. Fire and emergency disasters are in the category of hard news.

Many news items do not fit the emergency (hard news) model yet are still newsworthy or educational. "Soft" news items usually cannot be committed to a specific day; news editors decide how or when to place these items. By identifying an event, writing an appropriate headline, relating its effect on the community, and explaining its importance, an educator can increase the chances of media using the information.

Most local fire service stories lack the inherent news value of a national event; however, ample community interest exists to warrant sharing information. Many fire service activities are considered news, and announcements of them may be locally broadcast and printed. Local media and community members may want to know of meetings, rallies, seminars, and open houses. Workshops, National SAFE KIDS Campaign® activities, bicycle safety demonstrations, and checkpoints can raise citizens' safety consciousness.

Public statements on local affairs, awards given and received, fund drives, calls for membership, and the appointment and resignation of organization officials can generate considerable interest. Announcements of the availability of free speakers, films for loan, and the findings of reports and surveys conducted by the department have community news value. There is no limit to the activities that can generate publicity—especially for a well organized, positively focused, and energetic group.

Let the media know—in their terms—how fire service information provides their audiences with useful, creative, positive information. (See Appendix C.)

Remember, messages that destroy do not reflect the intent of educational news, hard news, or practical publicity credibility. Never overstate or distort the significance of an event, inflate a serious issue, or be theatrical.

PROVIDING PRACTICAL PUBLICITY

Fire and life safety educators must carefully plan how to spend limited money and time. One approach is practical publicity. Practical publicity, e.g., distributing frisbees shaped like smoke detectors and imprinted with "Test Your Detector," helps provide departmental recognition. More complex examples might include an open house or a mass CPR training.

No one method is proven best to run a publicity campaign. However, a three-phase publicity framework from which to build exists; it can also be used as a benchmark for evaluation.[10]

1. *Kick-off phase*:   This "name recognition" phase explains who the group is, i.e., education division of the fire department, its foundation and origin, and its plans for the future. Billboards, lawn signs, placards, news releases, flyers, and rallies support this phase.
2. *Expansion/issue-oriented phase*:   This phase comes after the public knows what the education department does and believes it to be credible. At this stage, the goal is to acquaint citizens with issues important to the fire department. The key is to emphasize issues, educate, and inform. Providing accurate information maintains credibility. Use seminars and workshops,

radio and television editorials, newsletters, PSAs, and news releases. Meet people face to face.[10]

3. *Culmination phase*: After establishing an identity, informing citizens, and focusing on important issues the fire department can educate once.[10]

Call upon the already informed public if a popular issue, e.g., passing fireworks reform legislation, arises. Phone trees, newsletters, direct mailings, letters to the editor, letters to politicians, press releases, news releases, citizen editorials, and media coverage are effective after publicity efforts have made issues newsworthy.

The effect is an amplification of your group's plea—a far greater noise than your group could ever make alone.[10]

The educator might choose to include alternative outlets and campaign strategies such as:

- Exhibits and displays
- Newsletters
- Poster contests
- Attention-grabbing methods, e.g., hot-air balloons
- Instructional slide shows or videos
- Signs, billboards, and bumper stickers
- Safety trailers
- Flyers, brochures, grocery bags, and placemat announcements

EVALUATING EFFORTS

In any process, the final stage is evaluation. To assess success, compare the results of a program with its original objectives.

While it is not this chapter's purpose to discuss evaluation in depth, understand the importance of evaluation and feedback. For example, if the number of bike accidents decreases by half after the fire and life safety educator provides three public service announcements and appears on local community calendar talk shows, the value of the effort is evident.

If fireworks injure four people after the fire and life safety educator has spent a large percentage of the budget on local billboard advertising to tell citizens that a strict fireworks ordinance has just passed, the value of the message is also clear. A question might, however, be raised as to the effectiveness of the campaign.

Evaluations may be conducted either: (1) throughout the program or (2) after completion of the program. If the objectives are met, pertinent information for future planning becomes available. If the objectives are only partially met, reasons for failure can be explored.

PROVIDING PUBLIC INFORMATION AT A FIRE/EMERGENCY SCENE

The PIO is responsible for gathering, preparing, and providing information to the media. The PIO helps convey the needed story safely, quickly, and efficiently; and serves as the contact between the incident commander and the media. The PIO's goal is to help the media present an accurate, understandable, fact-filled account of the emergency, and to ensure that reporters are safe and do not interfere with those who are attempting to secure the scene.

Public information officers link the fire service and the community. The PIO, acting in a professional capacity, assembles as much timely and accurate information as possible and informs reporters so they, in turn, can provide it to their viewers, listeners, and readers. The job of PIO is a "role rather than a rank."[12]

The professional PIO does not attempt to stand in the limelight or upstage a fire officer. After gathering information, the PIO

streamlines the process of providing reporters needed information as fire fighters continue their work.

As in every other aspect of the educator's job, the PIO will be more efficient if planning and thought have been put into practice ahead of time, *before* the inevitable emergency arises.

Create and Follow Standard Operating Procedures

Creating standard operating procedures (SOPs) expedites work with news media at an emergency scene or at any other time or place reporters might be contacted. SOPs establish: (1) who in the department may speak with the media and (2) how the fire chief wants the matter handled. (See Figure 2-6F.)

The PIO provides the bridge between the fire department and the community served. Often the community does not understand the fire department's day-to-day operations and activities. It may not know how the modern fire service operates or the various tasks fire personnel perform, despite impressive efforts in evidence. In most jurisdictions, the percentage of citizens who call the fire department for emergency assistance is low. Although the fire service conducts many public education and community service activities, a high percentage of the community learns of these activities only through the media. Providing clear, concise, and timely information to the media benefits everyone.[13]

PIO Tools

When facing reporters at an emergency scene, the PIO needs essential tools:

1. A radio to communicate with fire command and other suppression personnel.
2. Two pencils, a notepad, and a small tape recorder. Addresses, names of persons injured, telephone numbers, etc., can easily be forgotten in the rush of a large emergency. Write them down. If the PIO's style is to include every detail, conversation(s) should be taped.

   Taping conversations allows self-critique later. If inaccurate reporting becomes a problem, the tape recorder may show reporters that the PIO expects them to provide correct information.
3. A media quick-call list provides up-to-date phone numbers. The quick-call list is a short version of the media address and phone list. In an escalating emergency, the PIO must be prepared to notify media sources or update them as the situation changes. The list includes essential local broadcast and print media news reporters and personnel. (See Figure 2-6A.)
4. A mobile telephone may be useful for updating the news media and communicating with others involved in the situation.
5. Basic question list—media worksheet. The Colorado Springs, CO, Fire Department offers company officers and PIOs a laminated 3-× 4- in. trifold flyer to keep as a convenient guide. It includes interview tips; reminders of questions to ask fire, EMS, and hazmat personnel; and rescue information and phone numbers. (See Appendix C and Figure 2-6G.)

Managing the Media Scene

Reporters need accurate and timely information. If a fire department experiences little contact with the press, a plan of action prepared by the PIO is essential.

The incident commander is responsible for managing public information during an emergency situation. When needed or designated in the department's SOPs, the incident commander may

```
A. News Releases (Emergency Operations)
   1. News releases to the press, radio, and television will be given by
      the public information officer or a chief officer utilizing the media
      information release format (MIRF).
   2. A close rapport shall be maintained concerning all press releases
      between the fire chief, the public information officer, the
      investigator, and the battalion chief's office.

B. Fire Causes
   1. Fire causes will be listed according to one of the following:
   a. Natural
   b. Accidental
   c. Suspicion/incendiary
      1. In the event arson is suspected or involved, the inspector shall
         make this determination. Until they do, the cause will be listed
         as under investigation.

C. Fire Damage and Estimated Dollar Loss
   1. The terms "light," "moderate," and "heavy" will be used to describe
      fire damage on the MIRF report.
   a. Light, moderate, and heavy will define the "range" of fire loss by
      percent.
      1. Light — a fire loss of 1 to 10 percent of the value of the
         vehicle/structure.
      2. Moderate — fire loss of 11 to 25 percent of the value of the
         vehicle/structure.
      3. Heavy — fire loss of more than 25 percent of the value of the
         vehicle/structure.

   Example: When a structure is estimated to be worth $100,000, and
            the fire loss is estimated to be $4,000, damage is
            considered to be light. If the estimated fire loss is $15,000,
            it is considered to be moderate, etc.

D. Damage Estimates:
   1. Damage estimates for large businesses, industrial/commercial
      structures, and major complexes will not be released until the
      insurance companies provide us with that information.
   2. We will estimate, in dollars, the fire loss for building and contents
      on the MIRF report.
```

FIG. 2-6F.   An example of a standard operating procedures memo regarding news releases to the press. (Adapted from a Champaign, IL Fire Department memo.)

Damage and Injury

Conditions found:

Actions taken:

Injury _____

Name _____ Age _____ To/By_____

Injury _____

Name _____ Age _____ To/By_____

Dollar loss: Structure _____   Contents_____

Cause of Fire

Location of ignition _____

Source of ignition _____

Contributing factors _____

Comments by investigator _____

_____

Insurance:

Agency support:

Interviewer information

Interviewer _____   Media source _____

Interviewer _____   Media source _____

FIG. 2-6G.   Sample damage and injury report form. (Adapted from Colorado Springs Fire Department materials.)

establish a sector for public information. In this area reporters receive up-to-date information about the emergency.

Colorado Springs Action Plan

The following procedure was developed by the Colorado Springs Fire Department to guide the PIO.[13]

1. Report to the incident command (IC) for a briefing on the situation. Receive instructions. Use a media worksheet to record information. This worksheet anticipates the questions a reporter will ask. (See Figure 2-6G.)

2. Establish a media area for releasing information. Choose an area fairly close to the IC, yet away from danger. Inform the IC of the location of the media sector. Identify it clearly.

3. The media should know the PIO location. Reporters should also know when to check back for updates of the breaking story; establish intervals. Adhere to these established meeting times. Recap the situation and be there for late-arriving reporters, even when no new information is available. Establish ground rules with the reporters. Where is it safe to be?

4. Prepare a news release. Gather the media and share that information. Note any questions left unanswered. Attempt to find answers.

5. Assist media photographers in safely obtaining effective images.

6. Continue to check with the IC. Solicit updated information and listen to the emergency radio for any status changes. Report facts only. Do not release personal information on victims or losses until the IC verifies and clears names. When gathering this information, learn as much as possible: the full name, including middle initial; age; and to which hospital the person was taken. Never release the identities of fire casualties to the media unless all appropriate next of kin have been notified.

Most reporters recognize that this type of information must be delayed. They will, however, want to know identities as soon as possible. Show human concern and empathy when talking about sensitive issues, such as death and loss of home.

Ask for assistance when the situation warrants. The PIO's function is important to the flow of operations. Be courteous. Be professional. Be brief.

EXPLORING LEGAL ISSUES

The fire and life safety educator must be aware of various legal terms and issues. When dealing with media, leave no room for carelessness, misunderstanding, or inaccuracy. Copyright laws, privacy acts, broadcast regulations, and the federal Freedom of Information Act (FOIA) should be understood and addressed.

Copyrights

Copyright questions frequently arise. Copying another's work without permission is stealing. The fire and life safety educator must receive permission from the author or publisher. Copyright laws dictate that copyright owners have the exclusive right to reproduce, distribute, and use original works of expression fixed in a tangible medium.

NFPA's Sparky® is an example of a registered trademark. The Association's legal office must grant permission before a fire department may use Sparky for any promotion, activity, or brochure.

Copyright statutes, revised in 1976, take into account developments in photocopiers, videotapes, motion pictures, broadcasting, cable television, and other technologies developed after the act's adoption in 1909.

To claim a copyright:

1. Place a copyright notice on the work. This information, written in the upper-right-hand corner of the work, claims the copyright.
2. Complete a copyright form and pay the designated fee.
3. Register the work with the copyright office in Washington, DC.

Further, when using copyrighted material, include:

1. The word "copyright," the abbreviation "Copr.," or the symbol ©.
2. The name of the owner of the copyright.
3. The year of publication.

Registration is not necessary for a valid copyright, but if a suit is brought, the work must be registered.

Generally copyrights last the author's lifetime, plus fifty years.[14]

For copyright information, contact: Copyright Office, Information and Publications Section, LM-455, Library of Congress, Washington, DC 20559. Publications available include: (1) Catalogue of materials published, (2) Highlights of new copyright law (#R99), (3) Copyright basics (#R1), and (4) Trademarks (#R13).

Freedom of Information Act

What rules pertain when a citizen asks for access to a private meeting? When must meeting transcripts be released? Freedom of information is defined as: "The access to government documents that do not compromise national security or violate privacy rights of individuals."[3]

"Sunshine" laws address media access to public meeting records. Varying from state to state, these laws generally cover:

1. The amount of notice required to hold a public meeting.
2. Whether or not a meeting is open to the public.
3. Requirements on meeting minutes or transcripts.
4. Executive session rules.

The federal Freedom of Information Act, a 1966 statute, requires all federal executive and administrative agencies to furnish information to the public when it is requested, unless that information is in one of nine protected categories. The scope of the act was expanded in 1974.[3] A model for many state laws, it is designed to make government information available to the public. The government has 10 days to respond to a request and 20 days to respond to an appeal if records are denied.[15]

Libel and slander laws protect the reputation of a person or institution. Libel is written defamation.

It is essentially a false or defamatory attack in written form on a person's reputation or character. Broadcast defamation is libel because there usually is a written script. Oral or spoken defamation is slander.[11]

The educator must be constantly watchful for pictures and statements that might defame or invade another's privacy.

COMMUNICATION TECHNOLOGIES

Advances in computer technology have strengthened the relationship between communications and education. Telephone lines, fax machines, and videos can be hooked into the computer.

Bruce Piringer, director of the University of Missouri's Fire and Rescue Training Institute, envisions a time in the not-so-distant future when self-education via CD ROM will change the fire and life safety educator's position to one of managing, rather than delivering, programs.

Imagine the potential for personalized educational programs: a life safety library where citizens can pull down a computer menu to create a personal home escape plan, hunt for hazards, or search for specific injury or prevention information. Piringer envisions "voyagers" searching the computer for fires of the week, perhaps finding before-and-after fire photos and information on fire causes.

Seasonal messages, tips, statistics, and medical self-help menus are presently available to the fire and life safety educator.

Jonathan Seybold, prominent computer and electronic publishing analyst, said recently that with the power of PCs doubling every eighteen months, he foresees further integration of all digital technologies. Computing, publishing, entertainment, communication, and consumer electronic devices will all use and produce digital data and our world will be in the middle of an uncharted information environment in which users will be able to copy, move around and even interchange all types of data.[9]

Educating with computer-based media offers numerous possibilities, e.g., one-to-one teaching, continuing education, and professional development for fire and life safety educators. Many North American universities use computer-based communication technologies. Among the technological possibilities are audio conferencing, satellite video conferencing, and computer-based audio graphics. Portions of these educational resources are available, at minimal cost, to fire and life safety educators through: (1) cooperative extension offices that operate through partnerships with federal and state authorities and (2) community colleges.

Audio Conferencing

This "listening-only" medium combined with previously distributed printed materials lets instructors schedule seminars and workshops. Currently, the Illinois Department of Children and Family Services uses this educational tool to train and certify child-care providers. Child-care providers receive an information packet along with a schedule of when class will air on cable television.

Satellite Video Conferencing
Remote Downlink

Participants must find a site, e.g., a community college, where they can attend a scheduled course. The course may be taught at several sites simultaneously. This educational outlet allows fire personnel to become involved with community college training. It is a form of face-to-face instruction.

Computer-Based Audio Graphics

Interactive computer-based education furthers a student's ability to learn by linking with remotely located instructors. Audio and visual media combined at selected sites improves an instructor's ability to work with students. Video clips, slides, and student-teacher feedback make this an attractive alternative. Currently a number of continuing education courses are taught using this technology.

APPENDIX A: MEDIA GUIDES AND DIRECTORIES

The following media guides and directories may be found at public libraries.

Ayer Directory of Publications, the publicist's bible, lists nearly every newspaper and magazine. It includes publishers' addresses, circulation, topics covered, and maps with coordinates. A subject index helps users identify audiences by interests. A revised edition is published annually.

Bacon's PR and Media Information Systems presents information on every radio and television station in the U.S. and includes contact names, formats, target audiences, and network affiliations.

Burrelle's Media Directories includes more than half a dozen directories listing minority media and local media contacts in Chicago, New England, New Jersey, New York, Pennsylvania, and the Washington, DC, metropolitan area. Contact names, deadlines, and editorial focus are given.

Gale Directory of Publications and Broadcast Media, an annual guide to publications and broadcast stations, includes newspapers, magazines, journals, radio stations, television stations, and cable systems.

APPENDIX B: MEDIA RELATIONS TERMS

The following definitions clarify several terms commonly used in media relations.

Advance: A news story written before an event and held for later release. Also called an "advancer."[3]

Fact sheet: An information handout about a person or event, provided to assist the media.[3] It may be more than one page. The name and telephone number of a contact person should be provided at the top of the front page. Each page ends with a completed paragraph and "more" typed across the bottom of the page if necessary. The last page ends with "-30-" or a number sign (#) typed at the bottom. Figure 2-6B1 is an example.

News release: A short, factual description of an event prepared for the media. The terms "press release" and "news release" are interchangable.

The first sentence (the "lead") in a news release summarizes what needs to be conveyed followed by who, what, where, when, why, and how. Always provide source(s) of information, headline, date(s), and who to contact for additional information.

Present the most important facts in the beginning paragraphs; follow with facts of lesser significance. An editor can then decide the article's value with only a glance. This writing style is termed "inverted pyramid." The news release should be doubled-spaced, with 2-in. margins. Use agency letterhead and print on only one side of the paper. Date the release at the top. Insert "FOR IMMEDIATE RELEASE" when prompt publishing is desired. Use "FOR RELEASE ON (DATE)" on releases that need to be held until a certain date. Number the pages, place "more" at the bottom of continued pages, and place "end" or "-30-" or # on the final page. Limit news releases to one page whenever possible.

Unless the fire and life safety educator has much broadcast experience, providing the news release information in outline form may be more efficient. This lets the news director create the story. When necessary contact media professionals for input and advice.

Press conference: A meeting between members of the press and a newsworthy figure in which statements are made and/or questions are asked.[3] Figure 2-6B2 is an example.

Press kit: A portfolio of news releases, pictures, background information, etc., distributed to the press for publicity purposes.[3]

Public service announcements (PSA):

Any announcement . . . for which no charge is made and which promotes programs, activities, or services of federal, state, or local governments . . . or programs, activities, or services of nonprofit organizations . . . and other announcements regarded as serving community interests.[9]

News stations are not as committed to public service airtime as they once were; however, many stations keep card and computer files of information provided and use it when possible.

APPENDIX C: INTERVIEWING TIPS

An interview is a formal exchange of information; know as much as possible about the reporter(s), audience(s), and issue(s) before providing one. The fire and life safety educator must be aware that the message goes to all readers, listeners, and viewers to whom the interviewing reporter provides information.

Communicate clearly and credibly, and define those "headlines" that the audience should see or hear. In addition:

1. *Be prepared.*   Review all pertinent information before the interview. Anticipate and prepare for questions the interviewer is likely to ask. Have reference notes.

2. *Be professional.*   Dress appropriately. Avoid unnecessary movements and gesture calmly. Move more slowly than you normally would. Avoid chewing gum, smoking, or eating when providing reporters with information. Be confident. Look the reporter in the eye. During an interview, resist the temptation to look straight into the camera or at yourself on the TV monitor.

3. *Be newsworthy, factual, and correct.*   Double check information before releasing it to the media.

4. *Plan points to be made.*   Present two or three main points as quotable single-sentence summaries. Begin the interview by stating the most important points or conclusions.

5. *Be understood.*   Use common language; avoid fire fighter jargon and technical terms. When technical terms must be used, explain them.

FIRE FACTS

U.S. Fire Loss Statistics for 1994

Every 15 seconds, a fire department responds to a fire somewhere in the United States.

Public fire departments responded to 2,054,500 fires in the United States, of which 614,000 occurred in structures (including 438,000 in homes), 422,000 occurred in vehicles, and 1,018,500 occurred in outside properties.

Nationwide, there was a civilian fire death every 123 minutes.

4,275 civilians (non-firefighters) died in fires, down 7.8% from the previous year, and 3,425 of those deaths occurred in home fires, down 7.9%. About 80% of all U.S. fire deaths occurred in home fires.

Nationwide, there was a civilian fire injury every 19 minutes.

There were an estimated 27,250 civilian fire injuries, of which 19,475 occurred in homes.

There were 1,073,600 firefighters in the United States, serving in 30,495 departments. Of these firefighters, 265,700 were career and 807,900 were volunteer. In 1993, there were 7,700

women in firefighting and fire prevention occupations, according to the Bureau of Labor Statistics.

A total of 104 firefighters died while on duty. Of these, 33 were career, 38 were volunteer, and 33 were non-municipal (those not employed by local, public fire departments). That same year, 95,400 firefighters were injured while on duty. Of those, 52,875 occurred on the fireground.

All career fire departments totaled 1,778 (or 5.8% of all departments) in 1994, protecting 42.8% of the population. Mostly career departments totaled 1,448 (4.8%), protecting 14.6% of the population. Mostly volunteer departments totaled 4,400 (14.4%), protecting 17.9% of the population. All volunteer departments totaled 22,869 (75.0%), protecting 24.7% of the population.

Source: NFPA Survey of Fire Departments for United States Fire Experience During 1994

1995 U.S. fire loss statistics will be available from NFPA's One-Stop Data Shop in August 1996.

LEADING CAUSES OF 1989-1993 HOME STRUCTURE FIRES, DEATHS, INJURIES AND DIRECT PROPERTY LOSS

(National Estimates — Annual Averages)

Major Cause	Fires	Civilian Deaths	Civilian Injuries	Direct Property Damage (in Millions)
Cooking equipment	101,200 - #1	320 - #6	5,060 - #1	$382.0 - #5
Heating equipment	85,400 - #2	540 - #3	2,180 - #5	$601.8 - #2
Incendiary or suspicious causes	57,900 - #3	680 - #2	2,310 - #4	$866.6 - #1
Other equipment	44,000 - #4	240 - #7	1,700 - #6	$407.4 - #4
Electrical distribution system	39,600 - #5	370 - #5	1,490 - #7	$566.4 - #3
Appliance, tool or air conditioning	31,200 - #6	110 - #10	1,080 - #8	$234.3 - #8
Smoking material (e.g., cigarette)	26,300 - #7	900 - #1	2,570 - #2	$258.2 - #7
Child playing	22,400 - #8	390 - #4	2,530 - #3	$231.0 - #9
Open flame	22,200 - #9	130 - #8	770 - #9	$206.0 - #10
Exposure (to other hostile fire)	17,500 - #10	40 - #11	180 - #11	$383.7 - #6
Other heat source	10,000 - #11	130 - #9	770 - #10	$104.5 - #12
Natural causes	8,700 - #12	20 - #12	160 - #12	$136.3 - #11
Total	**466,300**	**3,860**	**20,810**	**$4,378.2**

Source: 1989-1993 NFIRS and NFPA Survey. For the latest statistics on all aspects of fires and fire safety in the United States, contact NFPA's One-Stop Data Shop at 617-984-7450 or nschwartz@nfpa.org

FIG. 2-6B1.   *Example: Use of NFPA fact sheets for Fire Prevention Week.*

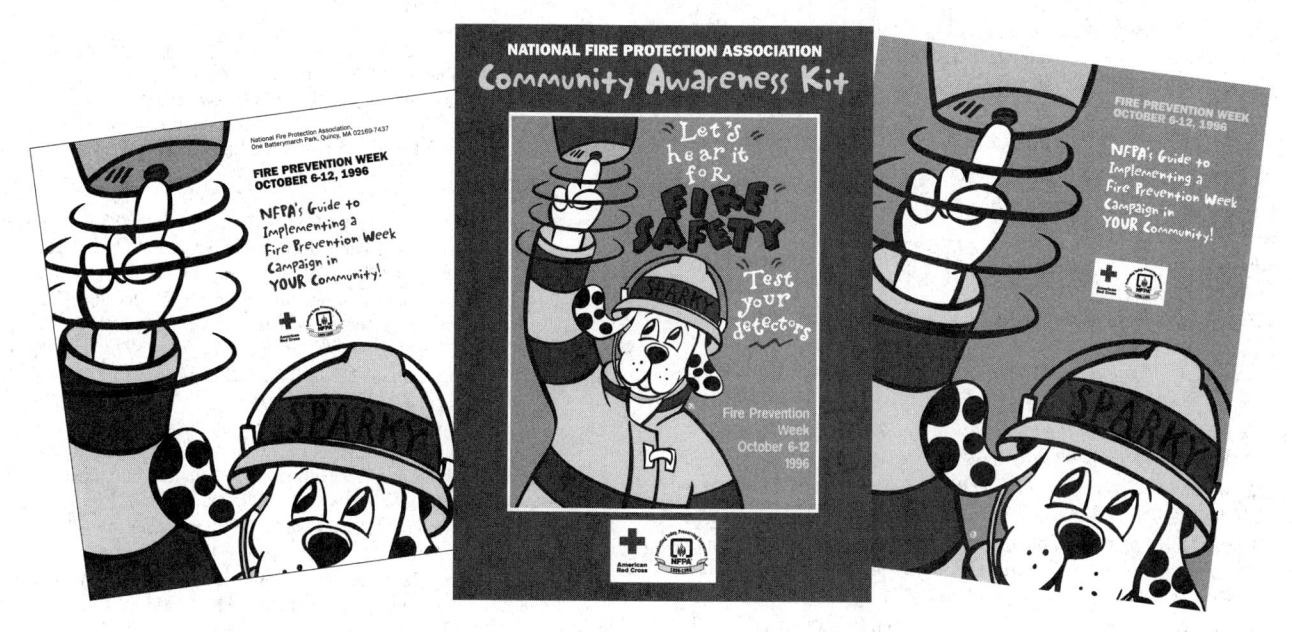

FIG. 2-6B2.   Example: NFPA Fire Prevention Week press kit.

6. *Assume that all is "on the record."*   Anything said can and likely will be used.

7. *Never say "no comment."*   It invites speculation. Rather, say "I don't have enough information to properly answer that question."

8. *Avoid "traps."*   If the reporter's question contains objectionable words, don't repeat them in any answer, even to refute them. If a reporter poses a hostile or inaccurate remark, respond by saying, "First let me correct a misperception that was part of your question."

9. *Maintain composure.*   Some interviewers try to provoke angry responses to get additional information. Always remain calm and courteous.

10. *Be prepared to restate your policy or position.* Do not attempt to answer hypothetical questions. Refocus the interview on the issue by saying, "Here's the situation we are focusing on."

11. *Don't be afraid to say, "I don't know."*   If the answer to a question is not known, say so. Follow with, "As soon as I get the information, I'll get right back to you." Make note of the question and get the answer.

12. *Stay on track.*   Don't digress from the subject. Lead a reporter who starts to wander back to the subject at hand.

13. *Be accountable.*   Never assign blame or criticize.

14. *Be available.*   Offer to answer any follow-up questions a reporter may have and help clarify problems that may arise as the story is being written or edited.

15. *Be honest.*   Never lie or "stretch the truth." Such tactics rarely work and, when discovered, can have severe consequences.

16. *Be patient and willing to teach.*   Reporters have different professional knowledge and backgrounds. They need education on some fire and life safety issues.

17. *Stand up and be counted.*   Keep records of media calls. See how and if department comments were used. If a serious misinterpretation becomes apparent, call the reporter or editor and ask for a correction. If reporters used comments in an appropriate and effective way, tell them so.

APPENDIX D: INFORMATION RESOURCES

The Associated Press Style Book and Libel Manual provides a handy reference for writing press releases. It's alphabetically arranged by subject.

The Advertising Council (261 Madison Avenue, New York, NY, 10016-2303) obtains media time and space worth $1.2 billion for nonprofit organizations and government agencies. The Council, ". . . freely counsel[s] leaders in government and private philanthropy who come to us for help, whether or not a Council campaign eventually emerges."

The Council's four criteria for projects are:

1. The project must be noncommercial, nondenominational, nonpartisan politically, and not designed to influence legislation.
2. The purpose of the project . . . [is] such that the advertising methodology can help achieve its objectives.
3. If the organization is a fund-raising one, the Council will take into consideration whether or not it currently meets the standards of public and private accreditation organizations.
4. The project is national in scope, or sufficiently national so that it is relevant to media audiences in communities throughout the nation.

Public Interest Public Relations, Inc. (PIPR 225 W. 34th Street, Suite 1500, New York, NY 10122) promotes issues and ideas of not-for-profit organizations.

Additional guidance on news releases is contained in "Improving On-Scene Releases," by Cooksey, P., Ph.D. (*Fire Chief Magazine,* Vol. 29, July 1985, pp. 30-33).

BIBLIOGRAPHY

References Cited

1. Read, H., *Communication Methods for All Media,* University of Illinois at Urbana-Champaign, College of Agriculture, 1982.
2. Crawford, J., "How to Tell the Difference—What Every Fire Service Leader Should Know about Public Education and Public Relations," *IAFC On Scene,* Vol. 6, No. 13, July 15, 1992.

3. Ellmore, T., *NTC's Mass Media Dictionary,* National Textbook Company Business Books, Lincolnwood, IL, 1991.

4. Evans, J. F., "Education Campaign Planning," course reference, University of Illinois at Urbana-Champaign, 1985.

5. Blackburn, D. J., ed., *Extension Handbook, Processes and Practices,* Thompson Educational Publishing, Inc., Toronto, 1994.

6. Berg Flexner, S., ed., *Random House Unabridged Dictionary,* 2nd ed., Random House, 1993.

7. Calvert, P., ed., *The Communicator's Handbook, Techniques and Technologies,* Maupin House, Gainesville, FL, 1993.

8. Shaul, D., news director, WCIA-CBS Channel 3, Champaign, IL, May 1995.

9. Yale, D. R., *The Publicity Handbook,* NTC Business Books, Lincolnwood, IL, 1991.

10. Tedone, D., *Practical Publicity, How to Boost Any Cause,* Harvard Common Press, Boston, MA, 1983.

11. Mencher, M., *News Reporting and Writing,* Wm. C. Brown Publishers, Dubuque, IA, 1987.

12. Harber, J. B., ed., *The Public Information Officer: A Media Relations Training Program for the Fire Service,* Action Training Systems, Inc., Vancouver, WA, 1993.

13. Kirtley, E., Colorado Springs Fire Department, PIO training materials, 1993.

14. Pinson, L., and Jinnett, J., *Out of Your Mind . . . and into the Market Place, Marketing, Researching and Reaching Your Target Market,* Fullerton, CA, 1989.

15. Rich, C., *Writing and Reporting News: A Coaching Method,* Wadsworth Publishing Co., Belmont, CA 1994.

Additional Reading

Cavagnaro, E., "Broadcast Media: The Key to Communicating with the Public in Large Scale Disasters," KCBS Newsradio 74, San Francisco, CA, CIB W14/95/1 (J); Tokyo Fire Department, Firesafety Frontier '94, International Fire Conference and Exhibition in Tokyo, on Creating a Safe Tomorrow, October 18-22, 1107-112, Tokyo, Japan, 1994, pp. 229-232.

Federal Emergency Management Agency, "Home Fire Safety: Act on It! How the Media Can Promote Public Fire Safety Education," Media Kit, Federal Emergency Management Agency, Washington, DC, 1992, 25 pages.

Federal Emergency Management Agency, "Public Fire Education Today: Fire Service Programs Across America," Federal Emergency Management Agency, Washington, DC, FA-98, September 1990, 237 pages.

Hansen, J., "Working with the Media," *Fire Engineering,* Vol. 148, No. 10, October 1995, pp. 116-119.

King, K. B., "Buying the Front Page/Media Marketing," Keller Fire-Rescue, TX, International Association of Fire Chiefs (IAFC), Fire-Rescue International Conference Proceedings, 1993 Annual Conference, August 28-September 1, 1993, Dallas, TX, 1993, pp. 27-33.

Murphy, J. J., Jr., "McDonald's and Media Teach Fire Prevention," *Fire Chief,* Vol. 35, No. 4, April 1991, pp. 70-71.

Sanders, R. E., "Long-Term Answer Is Public Education," *NFPA Journal,* Vol. 87, No. 6, November/December 1993, pp. 14, 26.

Schaenman, P., *et al., Proving Public Fire Education Works,* TriData, Arlington, VA, 1990, 143 pages.

Stittleburg, P. C., "Making the Media Work for You," *NFPA Journal,* Vol. 88, No. 6, November/December 1994, pp. 25, 102.

U.S. Fire Administration, "Home Fire Safety: Act on It! How to Use the Media to Promote Public Fire Safety Education," Media Kit, Fire Admin., Emmitsburg, MD, 1992, 20 pages.

EVALUATION TECHNIQUES FOR FIRE AND LIFE SAFETY EDUCATION

Philip S. Schaenman and Paul Gunther

The term *evaluation* means different things to different people. That leads to various concepts of how to evaluate a public fire and life safety education program. Some people evaluate programs based on how well the public education program is accepted and used by teachers, or how well it is liked by the target audience. Others evaluate a public education program from the point of view of features that are thought to constitute a good program, such as whether it has good graphics, is well targeted, and has input from various community groups.[*]

Another type of evaluation asks whether a program causes "institutional changes," such as getting additional funding for public education materials, obtaining additional slots for public educators in the prevention bureau, having public education incorporated in the school curriculum, or stimulating local businesses and service organizations to participate in public fire and life safety education.

All of these concepts of evaluation are useful. All help indicate the path to program success. But they are several steps removed from being able to prove that public fire and life safety education works in achieving its main purpose: reduction of deaths, injuries, and dollar loss from fire.

Many programs have side effects, secondary missions, or even hidden agendas. Public education, for example, can enhance the image of the fire department, demonstrate a caring city administration, and raise or lower public fears.

A sophisticated, comprehensive set of measures of effectiveness of a public program would consider such issues, which often are important in the minds of managers under their day-to-day political pressures. But the main purpose of public fire and life safety

education is to improve safety, and unless one can demonstrate that a program is achieving this goal, it will be hard to convince decision makers in the long run that the program should be supported. Sooner or later questions will arise about the bottom line.

A HIERARCHY OF EVALUATION MEASURES

The concept of evaluation, for the purposes of this chapter, is to focus on "bottom line" or end impacts: Did the program work in the sense of reducing fire deaths, injuries, dollar loss, or the number of fires? Being able to show that a program made a difference in these end measures is the most persuasive type of evidence of its success.

In order to achieve a bottom-line effect from public education, a sequence of events must take place:

1. *Outreach.* Safety information must be delivered to the target audience and reach enough of the audience to make a difference.
2. *Knowledge gain or refresh.* The audience must understand the material and must remember it. The information must be relevant and accurate for improving safety. It also must add to what the audience already knows, or remind them of what they know.
3. *Behavioral change or maintenance.* The target audience must act on the information gained (or refreshed). They have to recall it accurately and be motivated to put it in use.
4. *Environmental change.* Actions taken to improve the safety of the environment need to be done correctly, and the changes maintained.
5. *End impact.* The behavioral or environmental changes must have a significant impact on the types of problems that actually occur and not be overwhelmed by factors beyond control or not addressable by public education.

Ideally, a program should prove that it caused the desired end impacts. Where it is not possible to demonstrate such "bottom-line" effects directly, then look for the next best evidence earlier in the sequence leading to end impacts: Were there changes in behavior or environment that are likely to produce a bottom-line effect? For example, did the program change fire and life safety behaviors, e.g., get people to maintain or install detectors, practice escape plans, or identify and discuss an outside meeting place with their family? Did the program reduce fire safety hazards that lead to fires, such as defective wiring, dirty chimneys, or wood stoves installed too close to walls? Since people usually have to change their behavior in order to change the environment, such as doing more maintenance, or hiring

---

[*]See, for example, the excellent materials in the Pan-Education Institute's *Community Public Fire Education Assessment* package. These materials are helpful in shaping programs that will have bottom-line effects. Available from the Pan-Education Institute, 10922 Winner Road, P.O. Box 520347, Independence, MO 64052.

Philip S. Schaenman is president of TriData Corporation of Arlington, VA. Prior to founding TriData, Mr. Schaenman was associate administrator of the U.S. Fire Administration from 1976 to 1981, where he directed the National Fire Data Center and the Fire Administration's programs in fire protection technology.

Paul Gunther is a mathematical statistician who has served as a senior fire data analyst for the National Fire Data Center, U.S. Fire Administration and as senior consulting statistician for TriData Corporation. He also was an applied mathematician for the U.S. manned space program, the Internal Revenue Service, and aerospace programs in the United States and abroad.

someone to clean a chimney, the change in hazards can be considered another way to measure changes in behavior.

If a change in behavior or environment cannot be shown, the evaluation can retreat one step further and ask whether a program caused a change in awareness or knowledge of key fire safety information that leads to safer behavior and environment if the knowledge is applied. Examples include knowing what kind of fuel is supposed to go into a kerosene heater, the way to extinguish a grease fire, or the need to crawl low under smoke.

Even further back in the chain of proof is the extent to which people are reached by a program, especially when measured in terms of the percent of the target audience that was contacted by the program, and the frequency of contacts. Reaching 90 percent of the school children in Grade 3 twice a year would be an example of this measure.

Table 2-7A shows the hierarchy of measures. Try to use the highest level of proof possible on the list to get as close to evaluating the goals of prevention directly; avoid assumptions that can stand between what is measured and what really was effected.

People reached may not learn. People who learn may not act. And acting as instructed does not always work. Therefore, the closer one comes to measuring the end impact, the surer one can be that there really is an end impact. However, showing that only some hierarchy of measures changed is vastly better than doing no evaluation. The intermediate measures also can be excellent proof that the public education program did, indeed, change things that led to an observed change in the bottom line and that an observed change in end impact did not happen by chance.

The intermediate measures can also be diagnostic and help show where in the chain the education process is breaking down if there is not any change in the bottom line. Intermediate measures are best used in conjunction with end measures.

Sources of data for the measures range from existing, routine data collected by every department to special studies. Approaches to undertaking evaluations of these data measures are discussed in the next section.

Outreach data may be obtained by counting attendees at public education presentations, classes, and exhibits, or tallying households visited. For television, radio, and newspapers, information about circulation or audience usually is known and available. When a reasonably significant portion (10 percent or more) of the population is thought to have been reached, a citizen survey can be used, in the form of phone calls to random households or a mail survey. Total population of a target group often is available from U.S. Cen-

sus data or local planning departments; this forms the denominators for computing the percentage reached.

Knowledge changes usually are measured by before and after tests.

Behavior changes and environment changes can be measured by using telephone or mail surveys, by visiting and inspecting households, or by asking school children about their households.

End impact changes are measured from fire incident and casualty report data or special studies.

BASIC APPROACHES TO MAKING COMPARISONS

This section presents a short, condensed summary of the complex subject of data measure evaluation. A more in-depth treatment on evaluation in general is available from the Urban Institute's book, *Public Program Evaluation.*[1] Another excellent reference specifically on evaluating public fire education is the National Fire Academy's fire education evaluation guide.[2]

Perhaps the most basic concept of an evaluation is to demonstrate that some benefit can come from having a prevention program compared to not having it. There are several approaches to making the desired comparisons.

1. Compare the targeted community's experiences over time, to see if there is a difference after a program started compared to the "baseline" before it started. (Changes in the community that occurred during the period used in the comparison and that could affect the results must be taken into consideration.)
2. Compare one part of the community to another, to see if there is a difference between areas that have public education programs *vs.* other areas that do not yet have the program, or had it for a shorter time or with less intensity.
3. Compare the community to other communities, especially to similar communities that have not had a comparable public education program recently. A variation is to compare like parts of different communities, such as inner-city neighborhoods or high-rise dwellers.
4. Compare changes in a targeted type of fire (or fire safety behavior) to the trend in other types of fires (or other fire safety behaviors). For example, consider the change in cooking fires *vs.* the change in heating fires after a campaign that targeted cooking fires.

TABLE 2-7A.   *A Hierarchy of Evaluation Measures for Public Fire and Life Safety Education*

	Aspect Measured	Examples of Evaluation Measures
Strongest Proof	1. End results	• Number of deaths, injuries, dollar loss, or fires per capita • Anecdotes of saves linked to programs
	2. Behavior or the environment	• Percent of households with a working smoke detector • Percent of households sprinklered • Percent of chimneys cleaned at least annually
	3. Awareness, knowledge	• Percent of public who know how to extinguish a grease fire • Percent of public who know how to use extinguishers • Percent of public aware of need to crawl low under smoke
	4. Extent of program outreach	• Percent of population (or a subgroup) receiving public education materials • Percent of elderly receiving safety lecture • Percent of schoolchildren with x hours of safety instruction each year
	5. Likeableness and usage of programs	• Percent of teachers who think program materials are good and use them
Weakest Proof	6. Institutional change	• Introduction of safety curriculum in schools • Addition of service organization to aid dissemination

Measuring Change over Time

The most straightforward way to show that a public fire education program has had an impact is to demonstrate that things got better after the program started. (See Figure 2-7A.)[3]

Sometimes the effects are not immediate, and time must be allowed to observe results. Other times the effect is expected to be immediate and wear off with time, in which case speed of measurement is necessary. And sometimes the changes result from uncontrollable factors outside the program. All of these issues must be addressed.

While ideally a public education program produces a drop in fires or deaths, the impact sometimes may be to reduce the magnitude of an increase, or change the rate of an increase. Data from statewide fire and life safety education programs in South Carolina provide further information. (See Figure 2-7B.)[3]

Often, there are random fluctuations in fire experience; these are especially noticeable when the community is small and the number of fires few. It may be necessary to observe data that zigzags for weeks or years before being able to discern the trend. (See Figure 2-7C.) Identifying patterns with fluctuating or "noisy" data sometimes can be accelerated by using averages over several points of data or more sophisticated statistical analysis tests that help measure the likelihood that a true change has occurred.

The speed with which a prevention program has impact varies with the fire safety behavior being targeted, the frequency and timing of the message, and the receptiveness of the audience. It also may vary with the size of the audience reached. For example, if one-third of the population of a city is reached by a public service announcement (PSA) associated with a popular television show—a very large audience for television—one might expect an immediate drop in the type of fires targeted. If instructions in the PSA were on how to prevent and extinguish a cooking fire, and the city averaged nine reported cooking fires per week, it might be reasonable to expect an average of six or fewer after the broadcast, if the program was designed to reach a broad cross section of households. The behavior change can be immediate for those who saw the PSA. It does not require any accumulation of knowledge over time, because the message is simple.

Effects of a program may be delayed when the safety behavior being sought takes time and money to implement. For example, when people are instructed to test their detectors and replace the batteries, it may take some time before they get around to doing the test, then buying the batteries or detectors, and then installing them. In contrast, there is no delay in being able to implement "crawl low under smoke" or "put a lid on a grease fire."

The speed with which a prevention program has an impact also depends on the nature of the audience. If the audience comprises mostly "safe households" with a low fire rate, it may take a while to see any further impact.

In contrast, a PSA targeted to areas with disproportionately high fire rates may impact faster, because there is more opportunity to create a larger change. However, if these households tend to ignore safety information, there could be no change. One needs to consider the status of the audience in assessing program impact, especially when comparing different programs or the same program in different areas.

The effects of a program also may be delayed when the target audience is young. For example, teaching students in junior high school to put a lid on a grease fire may not have its full effect until they begin cooking more for themselves (though they may pass the information to other members of their households or intervene in a fire themselves).

Measuring a drop in reported fires as a result of a public education program can be thwarted due to the unpredictable effects publicity can have on people's likelihood to report a fire. It is known that many fires go unreported. Almost all of the unreported ones are

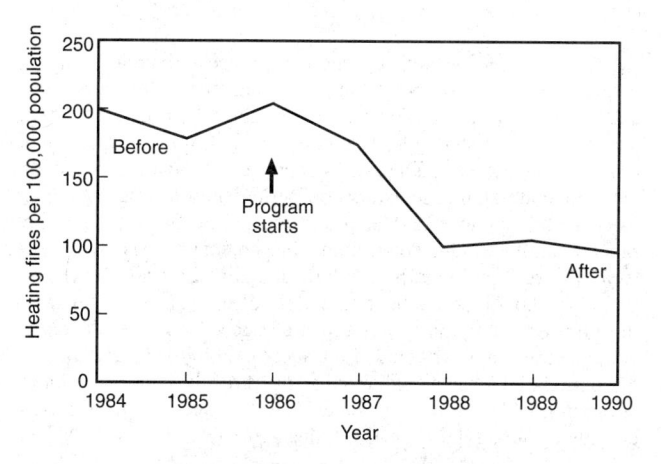

FIG. 2-7A.   *Example of a change over time that shows a successful program's impact (hypothetical example).*

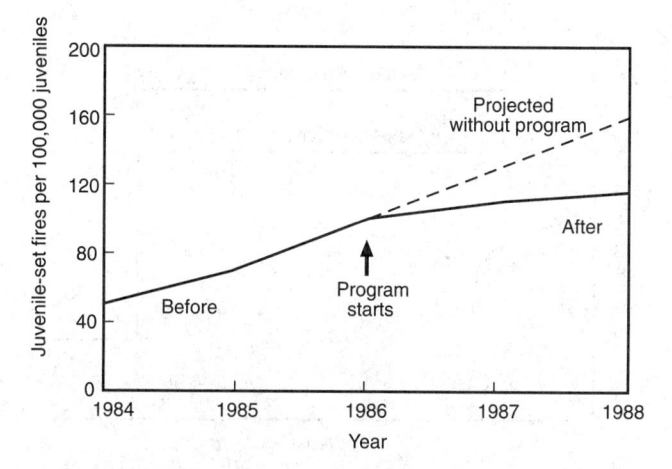

FIG. 2-7B.   *Example of a change in the rate of increase that shows a successful program's impact (hypothetical example).*

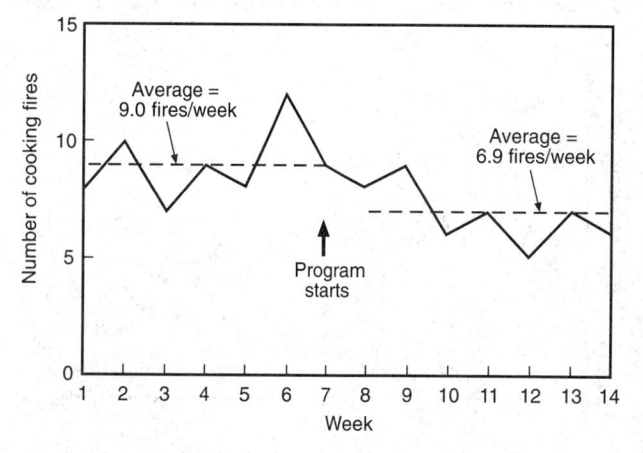

FIG. 2-7C.   *Change in trend visible even with fluctuating data (hypothetical example).*

small, though they include fires that injure and that cause hundreds or even thousands of dollars in damages.

When a fire and life safety education program brings attention to a fire problem, more of these small fires might get reported than is usual, making it appear to the fire department that the problem is worsening when it actually is not. For example, encouraging people to evacuate their home quickly may cause some who previously would have extinguished their own fire and not reported it, to leave and report the fire. The education program thus could cause an upsurge in reporting of fires. As another example, asking people to refer children who set fires to a juvenile firesetter program could encourage reporting of fires that might otherwise have gone unreported because of the perceived futility of reporting them.

The only way to know for sure whether a surge in reported fires is due to reporting of previously unreported fires is to periodically survey the population to determine the degree of their underreporting and the reasons for it. Then one can determine whether fires are being shifted from being unreported to being reported or are truly increasing in number.

Increasing Chances for Identifying Changes

Although there are a number of reasons why it may not be easy to see a change in trend as a result of public education, there are several ways to increase the likelihood of detecting a success.

One way is to use more sophisticated statistical techniques to see if there is some signal of a change amidst the noisy, erratic data coming from the outside world. Some fire departments have people on their staff or from nearby universities or other organizations who have training in statistical analysis. Modern, easy-to-use statistical computer software packages may help, too. By all means use them if available. But if it takes extremely sophisticated statistical techniques to identify a change caused by a program, the program is not likely to be having a large effect. That is, if one can't see the effect of the program looking at the data with the "naked eye," chances are that the effect is probably not very large. There are exceptions, and even if a change is small, it may be cost effective if only a correspondingly small effort were required to produce it.

An easier approach than employing sophisticated techniques is to collect more data over time. Average the data before the program started and compare that to the average after the program starts, as in Figure 2-7C. The averaging will take out some of the random fluctuations, and some of the problems inherent in having only small numbers of deaths, injuries, or fires for smaller areas. Compute the trend lines. (See Figure 2-7D.) Look at fires, not just deaths, because there are usually about 100 times more fires than deaths.

The impact of a program aimed at ignition prevention will show up sooner and more clearly for fires than deaths. If the program is directed toward education about escape or preventing bodily harm after a fire starts, rather than preventing or mitigating the fire itself, look at the number and severity of civilian injuries, and not just deaths. There tends to be about four times more injuries than deaths, and again the effects of a program may be more readily detected.

If the targeted community's population has not remained constant over the period of analysis, look at the data per 10,000 or per 100,000 population. (One also should look at the data on a per capita basis when comparing the community to others.) If the targeted community's population is increasing but the number of fires or deaths remains approximately the same, the rate per 10,000 people will be going down. That is, when more people are at risk but the community is not experiencing more fires or deaths, that can be an indication of a successful program.

The evaluation period can be accelerated by: (1) comparing different parts of the community or (2) comparing the community against others.

Comparing different parts of the community: As an evaluation tool, comparing parts of the community against each other is most useful when a program is introduced to one area of the community and not to the others. Then one can compare the "treated" population to the "untreated" population, as is done in medical research. (See Figure 2-7E.)

For example, an educator might introduce to the schools in one neighborhood a program directed at getting smoke detectors tested and properly installed. The students might be asked to count the number of detectors they have at home and to test them. This can be a homework exercise. Then the students can be taught the importance of maintaining detectors and periodically testing them. Results can be monitored for the next three months by looking at fires in that school area, and determining the percent of fires in which detectors were found working *vs.* how that area did prior to the program, and how it compares to the areas of the community without the program. If the program appeared successful immediately, it could be tried in other areas. If not successful immediately, the material might be repeated and monitoring of results continued. An attempt might also be made to determine if any outside factors were causing a counter, i.e., negative, impact.

As another example, for one year one might equip the schools in a selected neighborhood or district with a fire safety curriculum. Results would then be monitored for the trial neighborhood *vs.* the

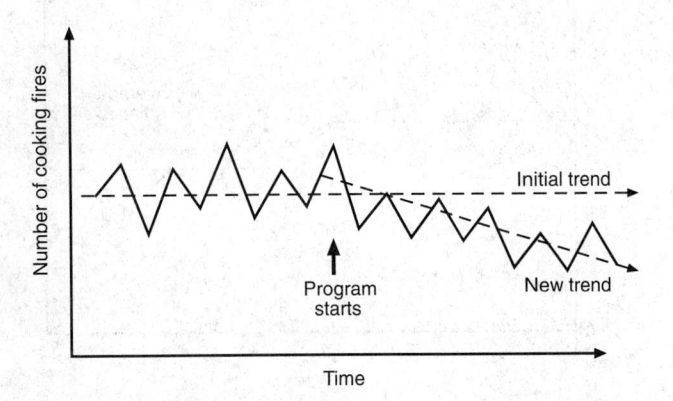

FIG. 2-7D. *Example of plotting trends in noisy data (hypothetical example).*

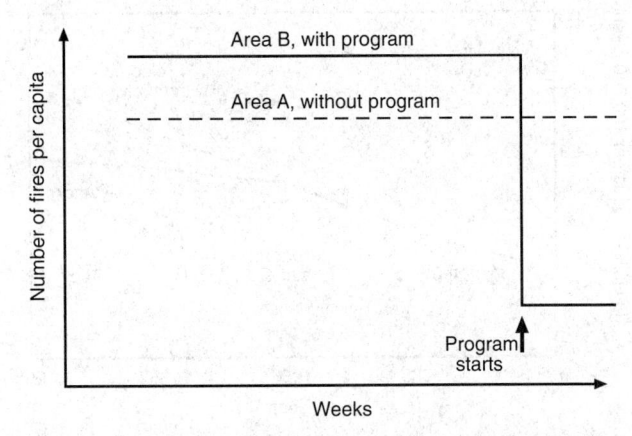

FIG. 2-7E. *Comparing area with prevention program to area without program (hypothetical example).*

rest of the community. If the program shows some success, perhaps in a few types of fires such as "children playing," then it might be implemented in the rest of the community. Chances of getting expanded implementation are much greater if one can show quantitatively that the program really made a difference in the area of the trial.

Sometimes a program applied in one area of a community inadvertently may have effects on the whole community. A notable example came from Edmonds, WA, where a door-to-door home inspection program in the mid-1970s using paid elderly women to make the inspections was found to have had about as much impact in the census tracts that had *not* been visited by the program as in the census tracts that had been visited. The suspicion was that the publicity attendant to the program caused people to "clean up their act" in anticipation of having a stranger inspecting their house. The program raised awareness of fire safety even in areas that did not receive the door-to-door inspections. If one compared the areas visited by the program to the areas not visited by the program, it could be concluded that the program was not very effective, when in fact the overall impact of the program was a drop in fire incidence by a remarkable two thirds. The before-and-after comparison for the city as a whole made clear what had most likely happened.

It is sometimes politically difficult to run an education program in one area of the community and not in others unless: (1) it is clearly labeled a pilot program, and (2) a plan exists for subsequently introducing the program everywhere. Nevertheless, whenever possible, introduce a program in one area first, and test the results in that area *vs.* the rest of the city. Even for proven programs this initial test can help the educator make adjustments in the program before introducing it everywhere.

When comparing different parts of the community, it is best to compare *like areas*, if possible. For example, compare a program in a low-income area against how other low-income areas are doing without the program, or compare a moderate-income area to other moderate-income areas. This is especially important if all similar areas have been trending in the same direction, and the impact of the program is to reduce the rate of rise or accelerate the drop in fires or casualties.

Comparisons among areas are best made over a period of time. Data is needed for each area from before and after the time when a program was first started, to know whether differences between areas result from differences in their characteristics or differences in the program.

Comparing the community with others: Similar in principle to comparing different areas within a community is to compare the community to other communities—especially to other communities that are generally similar to the target community. A variation on this is to compare the community's experience to that of its state. Comparisons among communities (or areas) should be made on a per capita basis or in terms of percentage changes or trends in the magnitude of the problem. Otherwise, differences observed in the numbers of fires or casualties may be due to differences in the numbers of people protected rather than from the results of the program.

HANDLING UNCONTROLLABLE FACTORS

It often is not enough to show that there was a change in knowledge or behavior or bottom-line measures, such as injuries or dollar loss. One also needs to demonstrate that the observed changes were caused by the education program and not by other factors, such as an unusually warm winter that caused fewer heater fires or a dramatic national fire in the news that raised awareness. Likewise, one needs

to check whether positive effects of a program may have been masked by uncontrollable factors, such as a weakening economy that leads to an increase in vandalism and arson fires during the implementation of an arson control program.

The sociology of fire is complex. There are many causes of fire, and many uncontrollable factors that may affect the environment or behaviors leading to fires. A social scientist or statistician might never be totally satisfied with the degree of rigor of most evaluations that are practical, because it is difficult to account for all of the uncontrollable variables without much effort, and perhaps not even then. Nevertheless, evaluations can be useful if undertaken with a little care, because they generally lead one in the right direction.

Table 2-7B lists some of the factors that might affect evaluation results. The first list in Table 2-7B contains variables that are largely out of the hands of the fire and life safety educator to affect. The most classic examples perhaps are the weather and the economy. The second list has factors that may not be controllable in the short run, though they may be affected by codes and other prevention efforts in the longer run. The third list has the initial conditions that need to be considered, even though they can be affected by public education.

TABLE 2-7B.   *Examples of Factors That Affect Evaluation Results*

Uncontrollable Factors
- Age profile of population
- Income distribution of population
- Education level of population
- Geographical scatter of population
- Ethnic groups in population
- Weather or climate change
- Economic changes
- Migration of people in or out of community
- Nature of local business and industry
- Changes in fire reporting procedures

Semi-Controllable Factors
- Condition of housing
- Architecture of the homes
- Hazards of new technology

Starting Conditions
- Severity of fire problem (fire and death rates)
- Previous exposures of population to fire safety information
- Current level of detector usage and condition

Evaluations can be made more meaningful by comparing results for situations in which these external variables are reasonably similar, or at least by acknowledging the possible influence of these variables.

The need to consider uncontrollable variables can cut more ways than one. A modest reduction in the number of wood-burning stove fires that would have followed the introduction of an intense education program on the need to clean chimneys might be masked or negated by an unusually severe winter. The colder temperatures might cause a sharp increase in the use of stoves and a corresponding increase in fires that overwhelms the decrease from more people cleaning their chimneys. But the increase might be less than in earlier years when no prevention effort took place, which could be a demonstrable positive impact of the program. If there were a decrease in chimney fires in spite of a particularly cold winter, that would suggest that the program was being extremely effective, not only in reducing the normal toll, but doing so in the face of increased stove use.

The reverse situation is also possible: An ineffective public education program might appear to be effective if chimney fires drop because the winter is relatively mild and the use of stoves decreases. That type of external factor has to be considered before one starts claiming success. It should be noted, however, that the chimney fire situation is a somewhat difficult example because relatively mild winter weather can cause people to reduce airflows to the firebox to reduce the intensity of the fire, which in turn can increase creosote buildup and lead to more chimney fires, instead of fewer.

To understand whether changes in fire incidence or casualties are due to the public fire and life safety education program or to climate change, one needs to look at the data on the weather in the period in question, and perhaps some nearby communities that did not change their public safety education programs but had the same weather. Looking at the data averaged across several winters may also reveal whether a program is having an impact independent of fluctuations in climate (unless the climate takes a several-year shift in one direction).

Another approach is to compare the data from years when the uncontrollable factor—in this case the climate—was the same. Nebraska's Forest Service provided a brilliant example of this approach. Fire rates in years of relatively similar dryness were compared with respect to the intensity of the prevention programs in those years. Comparing results from years with similar weather allowed the impact of the program to stand out. Otherwise the fluctuations in dryness masked the effects of the fluctuation in the intensity of the program.

Dealing with Pre-Existing Conditions

Besides considering factors that are uncontrollable by the fire department, evaluation of public education programs needs to consider the relevant pre-existing fire situation in the community. This is especially important to do when effectiveness is compared to resources expended; that is, when productivity is to be measured. Table 2-7B includes a list of several starting conditions to consider. The existing level of smoke detector usage, for example, will affect the apparent effectiveness of a new effort on smoke detectors. It is easier to get the middle 20 percent of households to install detectors than the last "hard-core" 20 percent. It usually is easier to reduce a high fire death rate than it is to further reduce a very low death rate. It is easier to maintain new detectors than old detectors, because the new ones use less expensive batteries, begin with fresh batteries, and are less worn out. While one may use the change in the percent of households with working detectors as a measure of effectiveness,

it must be realized that it gets harder and harder to reach the last ten households.

FOCUSING THE EVALUATION PROPERLY

Selecting Appropriate Measures

When a fire safety education program is targeted at a particular type of fire, such as grease fires on the stove, Christmas tree fires, or kerosene heater fires, the evaluation, too, should focus on how that type of fire changes over time, and how it compares to other types of fires for which there was no current program. Sometimes focusing on one type of fire may raise awareness and have a beneficial effect on many types of fires, and so the overall change in the fire problem needs to be monitored, too. But the primary focus should be on the type of fire that was the target of the program.

The same philosophy of targeting the evaluation properly applies to programs aimed at changing people's response to fire, such as escaping quickly or closing doors or using extinguishment methods properly.

Unfortunately, those attempting to undertake evaluations often look for changes only in the larger universe of fires rather than in the target group they were after. They measure the change in all fires when a program was targeted just to residences; or they measure the change in all residential fires when the program was aimed solely at cooking fires; or they look at all cooking fires when the prevention message was targeted solely at grease fires.

The chance of detecting results increases if one keeps the focus as narrow as the category the program message addresses. Otherwise, the results are diluted amid a larger pool of fires, and one may not be able to detect them.

Table 2-7C gives examples of some targeted messages and associated measures of their effects. In each case one also should compare the change in the targeted type of fire to the change in other types of fires, or fires in general, as a control. If cooking fires are targeted and they dropped in number, check whether heating and children-playing fires dropped, too. And check whether cooking fires dropped in nearby communities where there was no program. If there were reductions in other types of fires or in other communities, forces other than the education program may be at work. Or the program may have broader impact than what it targeted.

It may be difficult to obtain the data for exactly what one wants to measure. Therefore, it may be necessary either to: (1) retreat and find a surrogate measure or (2) measure the impact on the next

---

TABLE 2-7C. *Sample Measures for Specific Education Focuses*

Education Theme	Examples of Measures to Use
Use of smoke detectors	# households with detectors # reported fires (early detection leads to occupant extinguishment and fewer reports) # fire deaths
Getting out quickly from residential fires	# injuries while attempting fire control in residential fires # fire deaths # severe injuries
Need to clean chimneys	# chimney fires
Fire hazards of smoking	# fires or deaths involving smoking
Safe storage of flammable liquids at home	# non-arson fires where flammable liquid was material first ignited
Children playing with lighters or matches	# residential fires where heat of ignition was a match (or lighter) and ignition factor was "children playing" # children injured in above type of fire

larger category that includes the type of fire or behavior targeted. For example, one runs a PSA about putting a lid on grease fires, but cannot easily measure the number of grease fires; therefore, the number of cooking fires might be used instead. Further, if a school program is focused on the need for children to escape quickly from home fires, but one cannot measure the number of injuries from not escaping quickly, it might still be useful to track the number of injuries to children in home fires.

Further, consider the uncontrollable factors most relevant to the type of fires being targeted. For example, the number of reported careless smoking fires will be affected by changes in the use and maintenance of smoke detectors, because detectors allow fires to be detected and extinguished before they get large enough to be reported to the fire department. However, smoke detectors will have a negligible effect on the reported number of barbecue fires or outdoor dumpster fires.

See Table 2-7D for other examples of uncontrollable factors that should be considered when making evaluations of programs targeted at particular types of fires.

Measuring Change in Knowledge

One of the most common types of evaluations of fire safety education is the classic multiple-choice test. Often such tests are given before a program starts, or at the beginning of the first class, and then repeated at the end of the training using the same test instrument. Sometimes the "after" test uses a slightly different set of questions. The results of the "before" test not only establish a baseline, but also give insight into exactly where the group is weak. The "after" test shows the increase in knowledge.

TABLE 2-7D. *Examples of Uncontrolled Variables to Consider for Particular Types of Fires*

Type of Fire	Variables to Consider
Children playing	Change in number of single-headed households with children Number of youths Economic levels Ethnic makeup of the community
Heating	Climate (average degree-days; abrupt changes in weather)
Careless smoking	Upholstered furniture and mattress regulations Level of alcoholism Usage and maintenance of detectors Cigarette consumption

Pre- and post-tests do not have to be elaborate. Sometimes a few well-chosen questions can suffice.

Figure 2-7F shows profiles of test scores before and after a class that was conducted over a four-day period. The graphs show the profile of scores for different questions. It is very clear that the class dramatically improved. Although this example was drawn from a class of fire fighters taught about fire safety education, the data could have been for a classroom of children or a group of elderly.

In addition to the graphic presentation in Figure 2-7F, the average score and the range of scores for the pre-test *vs.* the post-test describe the change quantitatively. This type of evaluation can be done separately for each class or group of people, or all classes can be lumped together. The separation by class is especially useful

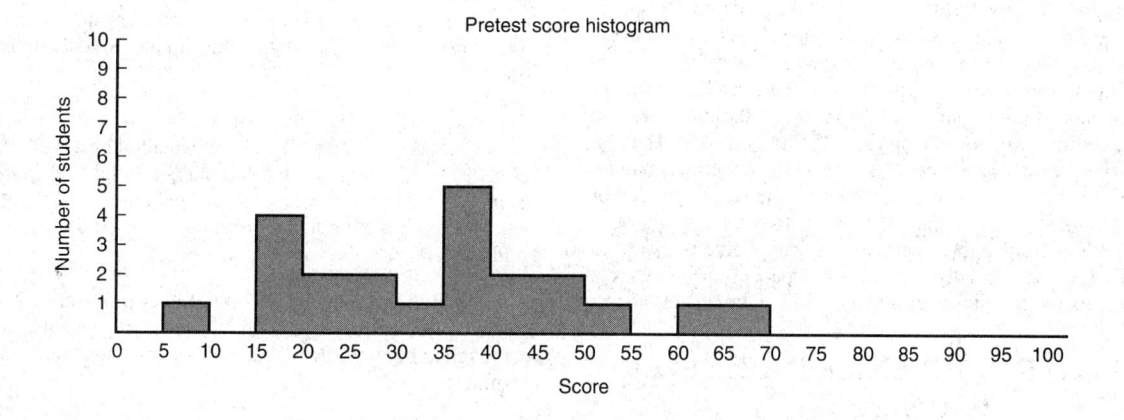

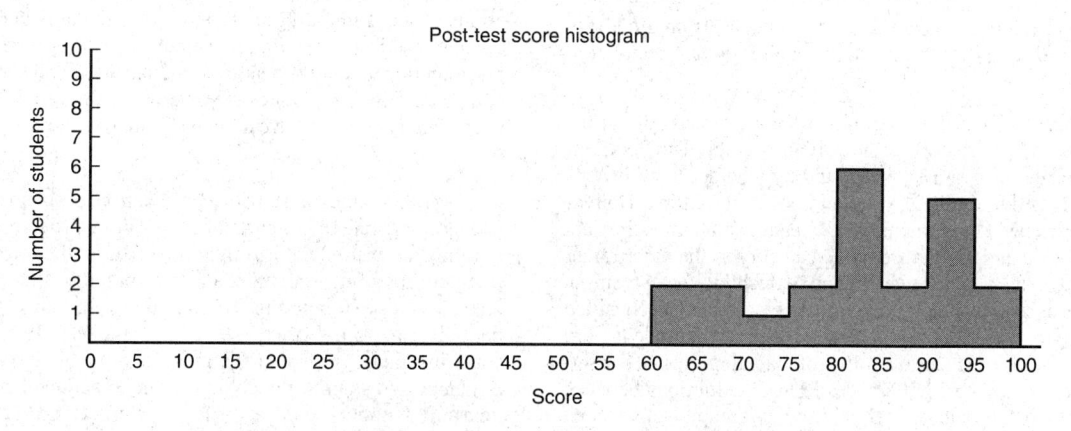

FIG. 2-7F. Profile of test scores based on Bolitin/Histogram.

when there are differences in the composition of the classes or when the classes are taught by different instructors.

Retention tests: Giving a post-test immediately at the end of training may reflect only short-term retention, though the test itself may promote retention. It is desirable, therefore, to administer another test several weeks to a year after the last training session. That test would measure retention and present an opportunity for a "booster shot" or reminder of the safety lesson to the students.

Practical tests: Another testing problem is that a paper test may not reflect what a person will remember or do in a crisis. Physical, hands-on training and demonstrations of competence are better indicators of actual performance. Few people would certify someone as competent in cardiopulmonary resuscitation (CPR) or first aid based solely on a multiple-choice test, yet this can occur in fire safety program evaluation.

Different pre- and post-tests: When the same test is used for pre- and post-testing of the same group, there might be some improvement in the score even if nothing was taught, due to increased familiarity with the questions and discussion of answers among the students. This is not necessarily detrimental—the test itself can be part of the learning experience. But there are ways to circumvent the problem:

1. Slightly different wording of questions can be used in pre- and post-tests, though some of the changes measured might result from differences in the test wording rather than changes in knowledge.
2. Test one group before they are taught the prevention material, and test another group after they are taught the material. If the groups are reasonably similar in composition, the difference in test results will show the effectiveness of the program in improving knowledge, while requiring only one test per class and allowing all students to receive the prevention material.

    This approach works particularly well when a large number of reasonably similar classes are to be taught the same fire and life safety material, for example, all Grade 5 classes. The test might be given to every other class before the material is taught, with the remainder given the post-test. The large number of classes is likely to make the difference between the two groups unimportant. To take the other extreme, one would not want to use this approach on only two classes, especially if one were a mainstream or remedial class and the other an honors class.[8]

EVALUATING JUVENILE FIRESETTER PROGRAMS

Juvenile firesetter programs are different in nature from other prevention programs, and their evaluation requires some special considerations.

First, juvenile firesetter programs often are aimed both at the curious child and the disturbed or malicious child. The "bottom-line" effect of the program is manifested by whether there is a reduction in: (1) children-playing fires and/or (2) fire setting. Both of these categories need to be monitored, because sometimes fires are simply re-labeled and shifted from one category to the other rather than truly reduced. A "get-tough" policy with children can result in labeling fires as fire setting that previously would have been called "children playing." Conversely, a more sympathetic attitude toward juvenile firesetters could cause a shift from labeling fires as fire setting to "children playing." Differences in legal definitions between jurisdictions also can muddy inter-community comparisons if both categories of fires are not considered.

Second, a common practice in evaluating juvenile firesetter programs is to consider "recidivism rates," i.e., the rates of repeat fires from children who have been treated by the program.

Recidivism rates usually are quite different for the curious firesetter than for the disturbed or malicious firesetter. It is easier to "cure" the curious child than the disturbed child. Rates therefore should be measured separately for these two categories. If not, one program can appear to be more successful than another, simply because one has a greater proportion of the easier-to-cure, curious-child cases.

Also, curious firesetters may have a low recidivism even without any program. A truly fair evaluation would consider their recidivism rates, with and without the program. A recidivism rate of 5 percent might appear good at first, but would not be so good if the rate without the program was 4 percent.

STATISTICAL CONFIDENCE—IS THE CHANGE A FACT OR A FLUKE?

The evaluator often must address factors that confound simple statistical approaches. The majority of fire departments do not have a trained statistician, though many departments have someone with some knowledge of formal statistical techniques. Unfortunately, real-world statistical situations often require considerable expertise to be able to say that "with such and such a confidence level" the impact of a program is real.

Standard statistical formulas actually may be invalid for many situations, such as when the samples collected are not truly random samples and the assumptions behind the formulas are not met. It is recommended that fire departments seek statistical expertise either from among their own personnel or from outside sources, such as local universities or statistical consultants, to assist in their evaluations.

Despite these caveats, some basic idea of statistics can give at least a rough, "ballpark" feel for whether observed changes are likely to be significant. This material is offered later in this chapter, as much to give the reader a feel for the kinds of situations that are not "statistically significant" as they are to be a guide for quoting statistics.

What many statisticians do not like to reveal is that, in many cases, common sense and good judgment are needed to say whether a change is likely to be real or not, and not just complex statistical formulas.

It is also useful practice to periodically review previous evaluations to see if initial findings about effectiveness or lack of effectiveness held up over time. Sometimes what seemed like a true program impact may appear to be a fluke after a few years of data is collated. Sometimes it goes the other way; a program with little impact initially starts to have a large cumulative impact after a year or two.

Another kind of statistical evaluation problem comes when trying to separate the impacts of two or more programs that overlap in time. For example, it is difficult to separate the impact of fire department programs on smoke detector purchases *vs.* the impact of television ads. Comparing communities with and without fire department programs but exposed to the same television ads may give some insight into whether the programs in question add anything. Sophisticated statistical techniques can be applied but are virtually unusable without expert statisticians. Suffice it to be flagged as a complex problem here.

PROVING EFFECTIVENESS AFTER A PROGRAM IS TERMINATED

Sometimes the value of a program is not evident until after it is gone. Proving that there was a real loss of benefit after a budget cut or reallocation of resources causes the demise of a program is one way to fight for its restoration. It also is a valuable type of hindsight from which the profession can learn.

Just as a drop in fires, fire casualties, or dollar loss after a program is initiated can prove it worked, so can a rise in these measures after the program is dropped. Likewise, drops in smoke detector maintenance, drops in fire safety knowledge, and increases in numbers of hazards found per home after a program terminates can all indicate the effectiveness it had.

Of special importance after a program ends (or after an initial test of a program in one area) is to see how long the good results last. One might be able to show, for instance, that half of the people exposed to a public service announcement continue to test their detectors for three months after seeing it and then begin to forget and stop testing. Information on how fast the initial effect wears off is highly valuable for identifying the minimum repetition frequency a message needs to have to produce a continuing impact on the majority of those who receive it. This information can be used to optimize the use of scarce prevention resources.

USE OF ANECDOTES

One of the favorite means of evaluating public education programs is through anecdotes of survivors who testify that they learned a safety behavior from a public education program, and it saved their own life or that of another.

Anecdotes are wonderful human interest stories and can attract the attention of the media. To most people, including decision makers, they are more interesting and more understandable than statistics. But anecdotes alone can be dismissed as weak evidence, especially if there are only one or two. Despite the frequently made statement that "this program is worth it if only one life is saved," budget managers do not always see it that way.

Nevertheless, anecdotes are a legitimate form of evaluation, especially when used in conjunction with statistics. An anecdote provides insight into a program and proves that at least in one case the program worked. *Multiple anecdotes are much stronger evidence than single anecdotes, because one event might be considered a fluke.*

Anecdotes are most credible when documented in writing by survivors or first-hand witnesses and when they describe not only an appropriate behavior but where that behavior was learned. Preferably the anecdote should be verified by at least a telephone call or second witness. The more anecdotes on a program, the better. NFPA's *Learn Not to Burn Curriculum*[4] has accumulated more than 200 documented cases where its messages saved lives or reduced injury.

Anecdotes are also more impressive when at least some of them are fresh (because programs and program delivery change) and when there are a series of anecdotes, indicating consistent, long-term results.

Since it is virtually impossible to collect anecdotes for all saves associated with a program, one needs the relevant overview statistics as well. A budget manager is much more likely to continue to fund a public education program where fire deaths decreased by ten over the previous year and anecdotes were used to show that the program was a factor in the decline, than where only the statistics or only the anecdotes are used. Together the picture portrayed is much more convincing.

Some people denigrate the use of anecdotes because they don't add up to statistical "proof." But at the local level or even nationally it does not take that many anecdotes to drastically change the fire loss picture. If public education programs reached half the people in each community and saved more persons per 100,000 people reached by public education per year than at present, the nation's fire death rate would be cut in half. (To see this, consider that one out of 100,000 is equivalent to 2,500 saves out of 250,000 people and the annual fire death toll per year has been about 5,000.)

UNINTENTIONAL IMPACTS

The evaluation of a program should include a check for unintentional impacts along with expected desirable results. Some common unexpected side effects are:

1. Greater reporting of minor fires that previously went unreported. Encouraging the public to flee a fire and call 911 instead of trying to put it out can lead to an increase in reported fires. Other positive aspects of a public education program sometimes are partly or entirely masked by this surge. Looking at the number of small fires apart from large fires can be a check on this.
2. Increase in fires set by children. Sometimes raising the subject of fire to children boomerangs and greater curiosity is generated instead of being lessened.
3. Scares, nightmares, and fear-related problems. Young children can have a new fear added unwittingly by a program that is too scary for their age group. Children may become fearful instead of careful.
4. Intrusion into parents' lives. Getting a child to harass parents to improve home safety or to purchase safety devices the parents feel they cannot afford can lead to complaints by the parents. Usually this is outweighed by positive feedback from parents whose safety consciousness is improved. But complaints should be taken seriously, and the curriculum and instructors reviewed to see if the safety message can be delivered effectively without unwanted side effects.

STATISTICAL METHODOLOGY APPLIED TO EVALUATING PUBLIC EDUCATION

This section discusses statistical confidence limits on evaluation results.[10] Perhaps the most important point in the detailed discussion below is that, in general, data from more than one year after the start of the education program is needed to be reasonably sure that the results are not a fluke.

Confidence Limits

When trying to measure the impact of a public education program, it is desirable to be able to compute the probability that the changes observed after the program is implemented were not due to chance or normal year-to-year fluctuations, but rather were true changes. With a limited amount of data it may be difficult to tell, and the numerical computation of probability can be a valuable aid.

Table 2-7E presents probability values computed for different amounts of data on fires (or injuries or deaths) before and after a public education program was implemented. The number of fires can refer to all fires or to any subcategory of fire, such as single-family dwellings only, or cooking fires, or fires in low-income areas, etc.

The following example shows how to use Table 2-7E. Brief explanations are also given of the statistical quantities computed.

TABLE 2-7E. *Confidence Limits and Statistical Significance for a Given Percent Change in the Average Number of Fires as a Result of a Prevention Program*

		Δ = 10%				Δ = 20%				Δ = 30%			
		$N_1=N_2=1$	$N_1=2, N_2=1$	$N_1=5, N_2=1$	$N_1=N_2=2$	$N_1=N_2=1$	$N_1=2, N_2=1$	$N_1=5, N_2=1$	$N_1=N_2=2$	$N_1=N_2=1$	$N_1=2, N_2=1$	$N_1=5, N_2=1$	$N_1=N_2=2$
		$\bar{n}_2=18$				$\bar{n}_2=16$				$\bar{n}_2=14$			
$\bar{n}_1=20$	Δ_L	–24.6%	–18.0%	–14.0%	–13.2%	–11.5%	–5.8%	–2.4%	–1.2%	1.5%	6.4%	9.3%	10.8%
	Δ_U	35.3%	33.5%	32.2%	28.6%	43.4%	41.8%	40.8%	37.2%	51.3%	50.1%	48.3%	45.7%
	K_{P_o}	0.324	0.371	0.412	0.459	0.667	0.756	0.830	0.943	1.03	1.15	1.26	1.45
	P_o	38%	36%	34%	32%	25%	22%	20%	17%	16%*	12.5%	10%	7%
		$\bar{n}_2=45$				$\bar{n}_2=40$				$\bar{n}_2=35$			
$\bar{n}_1=50$	Δ_L	–10.5%	–7.1%	–4.9%	–4.0%	1.3%	4.2%	6.0%	7.1%	13.0%	15.5%	17.1%	18.3%
	Δ_U	26.9%	25.3%	24.2%	22.2%	35.5%	34.2%	33.3%	31.3%	44.1%	43.1%	42.4%	40.3%
	K_{P_o}	0.513	0.587	0.651	0.725	1.05	1.19	1.31	1.49	1.63	1.83	1.99	2.30
	P_o	30%	28%	26%	23.5%	15%*	12%	9.5%	7%	5.2%	3.4%†	2.4%	1.1%
		$\bar{n}_2=90$				$\bar{n}_2=80$				$\bar{n}_2=70$			
$\bar{n}_1=100$	Δ_L	–4.0%	–1.9%	–.5%	2.8%	7.1%	9.0%	10.2%	11.1%	18.3%	19.9%	20.9%	21.9%
	Δ_U	22.2%	21.0%	20.1%	18.8%	31.3%	30.2%	29.5%	28.1%	40.3%	39.4%	38.8%	37.4%
	K_{P_o}	0.725	0.830	0.921	1.03	1.49	1.69	1.86	2.11	2.30	2.58	2.81	3.25
	P_o	23%	20%	18%	16%*	7%	4.5%†	3.1%	1.8%	1.1%	0.5%	0.25%	0.06%
		$\bar{n}_2=180$				$\bar{n}_2=160$				$\bar{n}_2=140$			
$\bar{n}_1=200$	Δ_L	0.3%	1.8%	2.7%	3.2%	11.1%	12.4%	13.3%	13.8%	21.9%	23.1%	24.0%	24.4%
	Δ_U	18.8%	17.8%	17.1%	16.5%	28.1%	27.3%	26.5%	25.8%	37.4%	36.5%	35.9%	35.3%
	K_{P_o}	1.03	1.17	1.30	1.45	2.11	2.39	2.62	2.98	3.26	3.64	3.98	4.58
	P_o	15%*	12%	10%	7.3%	1.7%†	0.84%	0.44%	0.14%	0.06%	0.01%	0	0
		$\bar{n}_2=450$				$\bar{n}_2=400$				$\bar{n}_2=350$			
$\bar{n}_1=500$	Δ_L	4.0%	4.8%	5.4%	5.8%	14.5%	15.3%	15.8%	16.1%	25.0%	25.7%	26.2%	26.5%
	Δ_U	15.7%	15.0%	14.5%	14.1%	25.2%	24.6%	24.1%	23.8%	34.7%	34.2%	33.8%	33.4%
	K_{P_o}	1.62	1.86	2.06	2.29	3.32	3.76	4.14	4.71	5.15	5.79	6.29	7.27
	P_o	5.3%*	3.1%†	2.0%	1.1%	0.05%	0.01%	0	0	0	0	0	0

NOTES:

$\bar{n}_1$ = Average number of annual fires in a given category before public education program starts; N_1 = Number of years of data available before program starts.

$\bar{n}_2 \Delta$ = percent change $\dfrac{\bar{n}_1 - \bar{n}_2}{\bar{n}_1} \times 100\%$

Δ_U and Δ_L = Upper and lower 68% confidence limits, respectively, for Δ.

K_{P_o} = Number of standard errors corresponding to percent change, Δ, computed from K_{P_o} = $\qquad K_{P_o} = (\bar{n}_1 - \bar{n}_2)/\sqrt{\dfrac{\bar{n}_1}{N_1} + \dfrac{\bar{n}_2}{N_2}}$

P_o = Degree of statistical significance = probability that percent change, Δ, could have occurred by chance (from normal distribution tables of P_o vs K_{P_o}; $K_{P_o} < 1$ ($P_o > 16\%$) implies not significant change; $K_{P_o} > 1.645$ ($P_o < 5\%$) implies significant change; and $1 < K_{P_o} < 1.645$ ($16\% > P_o > 5\%$) implies possibly significant change.

*Denotes beginning of moderate 16% significance.

†Denotes beginning of definite 5% significance.

More detailed discussion of statistical concepts are provided later in this chapter.

Example and Discussion: Suppose your fire data shows that you had an average of 20 cooking fires per year for the past two years before a public education program was implemented. (In Table 2-7E, then, $\bar{n}_1 = 20$, and $N_1 = 2$.) Suppose also that during the year after the public education program was implemented, there were 16 cooking fires. (In Table 2-7E, then, $\bar{n}_2 = 16$, and $N_2 = 1$.) The percent change, Δ, is then 20 percent $[(\bar{n}_1 - \bar{n}_2)/\bar{n}_1]$. This change can be interpreted as an estimate of the "true" effect of the program, as determined in principle from many years of before and after data,

with nothing else changing. The statistical accuracy of this estimate, which is affected by the normal year-to-year variability in the occurrence of fires and by the amount of available data, can be described in terms of *confidence limits*. These limits can be found in Table 2-7E, upon entering the data of the example: i.e., the values $\bar{n}_2 = 20$, Δ = 20 percent (or $\bar{n}_2 = 16$), $N_1 = 2$, and $N_2 = 1$.

The lower confidence limit for the change (Δ_L) is shown to be –5.8 percent, and the upper confidence limit (Δ_U) is +41.8 percent. These limits were computed assuming a confidence (probability) level of 68 percent. This means that one would be willing to give (or take) approximately 2:1 odds (68:32) that the true change as determined from having many years of data would be between –5.8 percent and 41.8 percent. That is, the true change in fires as a result of the program would be somewhere between a 5.8 percent rise and a 41.8 percent drop. This is a very wide range because of the small amount of before and after data that is available in the example.

The confidence limits can also be interpreted as the statistical "error" of the estimate associated with the observed change of 20 percent. The more data one has, the closer one gets to the true change, and the smaller is the error and the narrower the confidence limits. The negative value of –5.8 percent for the lower confidence limit does not mean that the public education program had negligible effect or did harm, but rather that more data is needed. With the available data, the observed change could well have occurred by chance due to the normal fluctuation in fires from year to year.

A more precise measure of the possibility of the change being a fluke is given by the last entry in Table 2-7E, $P_o = 22$ percent, which is the probability that the change (the 20 percent observed drop in the example) would occur by chance alone. If this probability is very low (it can never get down to exactly zero), then one can be "fairly sure" that the observed change was due to the public education program; while if it is high, the change could well have been due simply to chance alone.

To interpret a change in fire data, it is recommended using "significance levels" for P_o of 5 percent (which corresponds to a 90 percent confidence level) and 16 percent (which corresponds to a 68 percent confidence level) as decision values. If the value of P_o in Table 2-7E is less than 5 percent, then the change is considered to be "definitely significant"; if it is between 5 percent and 16 percent, it is considered to be "moderately significant"; and if it is greater than 16 percent, as in the example (where it is 22 percent), it is not significant. As noted, even if the change is not statistically significant, this is not to say that the public education program had no effect, but rather that there is just not enough data to determine the true change with sufficient accuracy.

One can verify from Table 2-7E that, whenever the 68 percent lower confidence limit is negative (meaning there could have been an increase rather than a decrease in fires), then P_o is greater than 16 percent, and the situation is "not significant."

Finally, Table 2-7E shows a value of K_{P_o} (0.756 in the example). This quantity is an alternative and useful measure of the degree of statistical significance; it represents the number of "standard deviations" or "standard errors" that the change is from zero (that is, from no change). The standard deviation is a measure of the spread in the data. If K_{P_o} is less than one (standard deviation), then P_o is greater than 16 percent and the change is not significant; if K_{P_o} is greater than 1.645, the change is definitely significant; and if K_{P_o} is between 1 and 1.645, it is moderately significant. In practice, P_o is determined from K_{P_o} using tabulations presented in many statistical textbooks. Even without such tables, K_{P_o} can be used directly, in the manner just described, to determine the statistical significance of a change. Moreover, K_{P_o} is simple to compute with a handheld calculator.

An examination of Table 2-7E shows how the degree of statistical significance varies as number of years of data increases. For example, if one has two years of data after the public education program is implemented rather than one year, along with two years of data from before the program, $P_o = 17$ percent, which is better than in the given example but still not significant. A change from before to after of 21 percent rather than 20 percent would actually be needed to reach the threshold of 16 percent significance level. But even a change of 30 percent, which leads to $P_o = 7$ percent, would still not yield a definitely significant result. When one has only a single year's data after the public education program, no matter how many years of data one uses before the program, the results will not be as good as having two years of data before and after. That is, one

needs to have more than one year of data after a program is implemented to be reasonably sure the result is not a fluke.

Table 2-7E also shows that, when one has a "before" average of 50 fires ($\bar{n}_1$), with only one year of data before and after ($N_1 = N_2 = 1$), moderate significance (15 percent) is reached with a 20 percent observed change, while even a 30 percent change is not quite enough ($P_o = 5.2$ percent) for definite significance. With a before average of 100 fires, definite significance is obtained with a 20 percent change and two years of before data.

The above discussion suggests that it would be useful to know the precise (or critical) change, Δ_o, needed to barely obtain either moderate (16 percent) or definite (5 percent) significance. The curves shown in Figure 2-7G give this information for various values of $\bar{n}_1$, and for one and two years of data. For the above example, the critical change required is 26.2 percent for moderate significance and 41.8 percent for definite significance. This set of curves can be useful for judging how detailed one can categorize the type of fire when studying the impact of the public education program. For example, is there enough data on the change in cooking fires involving oil to get significance in determining whether a "put the lid on grease fires" campaign worked?

Many excellent introductory textbooks are available if one chooses to pursue statistical analysis. (See additional reading in the bibliography.)

Special Analyses and Simplifying Assumptions

In analyzing fire data, circumstances often arise that require special analyses and simplifying assumptions. Outside professional statistical help may often be the best course if the data calls for such things as: (1) determining how many former years' data it is statistically valid to use, (2) how to adjust for pre-program trends, and (3) how to analyze results when a public education program is implemented gradually.

Quite likely the entry values appearing in Table 2-7E will not cover all applications; the table is intended to provide merely an indication of what results one can expect to obtain. A summary of specific formulas used in the computations is presented at the end of this chapter.

ADDITIONAL STATISTICAL CONCEPTS

The remainder of this chapter discusses in more detail the statistical concepts introduced in the above synopsis. The goal is to explain the statistical ideas underlying the notions of confidence limits for, and significance of, the change in fire rates following the introduction of a public education program. Significant factors include: (1) nature and measurement of fire data variability, (2) why the relative variability is relevant, (3) meaning and measurement of confidence limits for a simple average and for a percent change in the average, and (4) statistically significant change.

Why There Is Fluctuation

The essential rationale underlying statistical analysis of fire data is that fires are accidental in nature, depending only on: (1) individuals' propensities (behavior patterns) and (2) the random effects of related physical phenomena contributing to an ignition. A person may or may not toss a match into the garbage on a particular day, and if he/she does, the garbage may or may not ignite. A person may or may not decide to add one more electrical appliance to a socket, and doing so may or may not start a fire. A cigarette may or may not be dropped, and it may or may not ignite the material it falls on. And so on. Arson fires, on the other hand, are perhaps an exception—these are more willful rather than accidental in nature, but still may

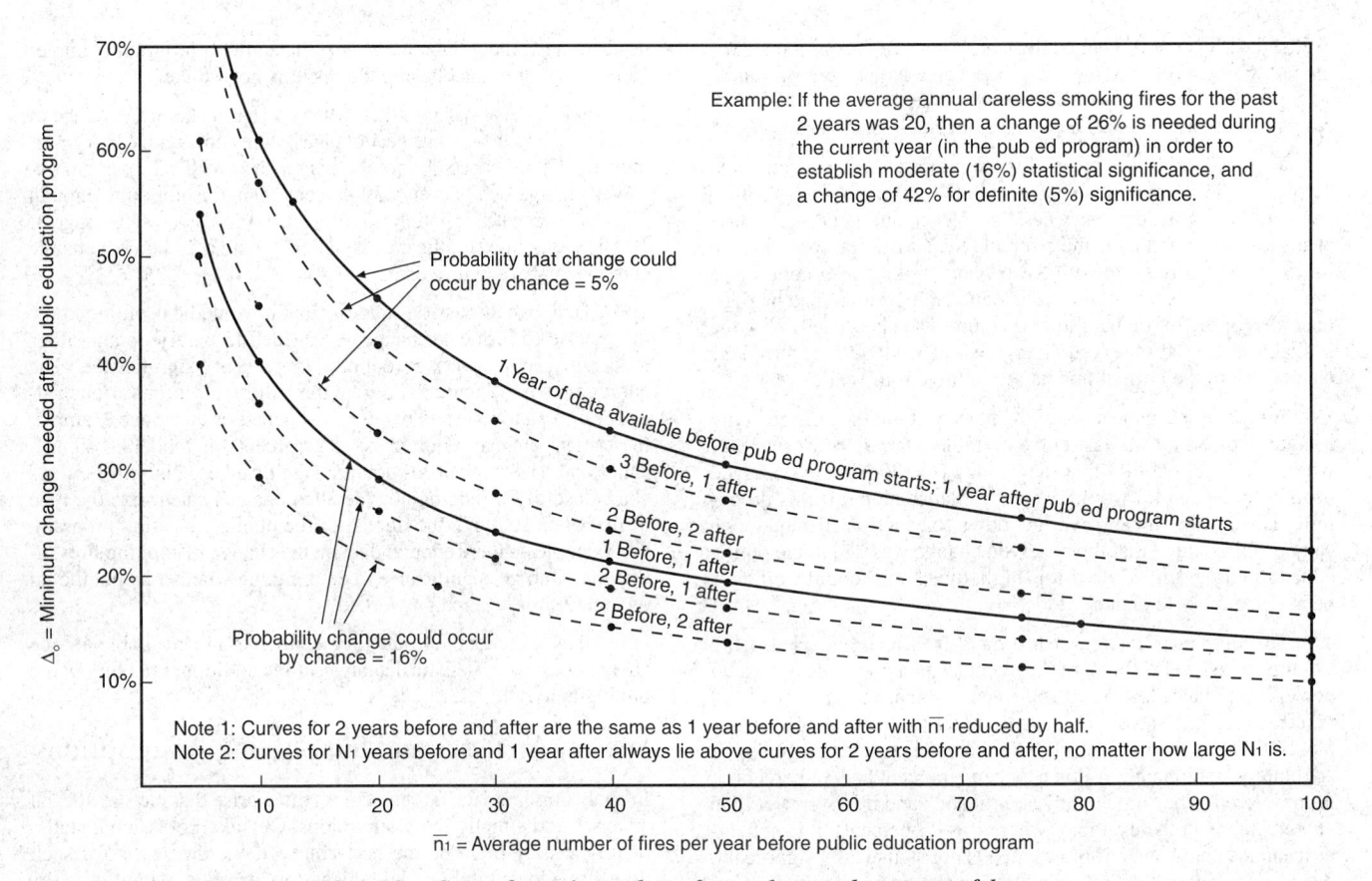

Example: If the average annual careless smoking fires for the past 2 years was 20, then a change of 26% is needed during the current year (in the pub ed program) in order to establish moderate (16%) statistical significance, and a change of 42% for definite (5%) significance.

Probability that change could occur by chance = 5%

1 Year of data available before pub ed program starts; 1 year after pub ed program starts

3 Before, 1 after

2 Before, 2 after

1 Before, 1 after

2 Before, 1 after

2 Before, 2 after

Probability change could occur by chance = 16%

Note 1: Curves for 2 years before and after are the same as 1 year before and after with $\overline{n_1}$ reduced by half.
Note 2: Curves for N_1 years before and 1 year after always lie above curves for 2 years before and after, no matter how large N_1 is.

Δ_o = Minimum change needed after public education program

$\overline{n_1}$ = Average number of fires per year before public education program

FIG. 2-7G. *Curves for various values of n_1, and one and two years of data.*

have random aspects to them, e.g., the youth may or may not decide to vandalize a property on a whim, the lovers' quarrel may or may not lead to arson for revenge, etc.

Suppose one has data on the number of annual fires of a particular type of fire, such as cooking fires, for several previous years. And assume that the nature of the fire experience, i.e., of the cooking fires, for these years is essentially the same. The year-to-year variability (or fluctuation) of accident data then will typically follow the well-known bell-shaped curve, called the "normal" frequency distribution. (See Figure 2-7H.) Variation in heights, weights, IQs, school grades, and many other characteristics of the general population all have been found to follow the pattern of this "normal" distribution. The variability (or spread) of the distribution is called the standard deviation and is usually denoted by σ. Figure 2-7H shows that this spread is a way to express how wide the bell-shaped curve

is for a particular situation. The formula for computing the magnitude of this variability depends on the type of measurement being studied. For fire and other accident data, the variability, σ, is surprisingly simple to compute: It is the square root of the average (or mean) number of fires. If, for example, there were an average of 36 cooking fires per year over the past 12 years, then the variability, σ, would be equal to 6 ($\sqrt{36}$).

This standard deviation of the "normal" distribution is important to know because it enables one to make a variety of probability statements. In particular, 68 percent (or approximately $2/3$) of the data—or of the so-called population—will probably be within one standard deviation of the mean. In the cooking fire example, approximately eight ($2/3 \times 12$) of the years would have had between 30 and 42 cooking fires (36 ± 6), two of the years would have had fewer than 30 fires, and two years more than 42 fires. Moreover, if the current year follows the pattern of the past, the odds are 2:1 that there will be between 30 and 42 fires. Similarly, for the normal distribution, 90 percent of the data will be within 1.645 standard deviation of the mean. Hence, if there were an average of 36 fires for a 10-year period, one can expect that in all but one of the years the number of fires would be between 26 and 46 ($36 \pm 1.645 \times \sqrt{36}$), and the odds are 9:1 that the current year will fall between these same limits.‡ Statistical tables of the normal distribution, which are presented in most statistical textbooks, provide the correspondence between the number of standard deviations about the mean *vs.* the amount of probability included.

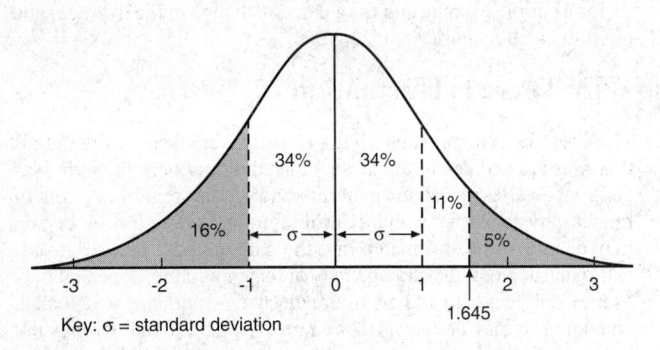

Key: σ = standard deviation

FIG. 2-7H. *"Normal" frequency distribution (hypothetical example).*

‡A value of 95 percent, corresponding approximately to two standard deviations, is frequently used in statistical studies. This, however, appears to be too stringent for fire data applications, though many researchers have used it in fire-related papers.

A more precise statement of the statistical assumptions is that the probability distribution for accidental events, known as the Poisson distribution, is well approximated by the normal distribution, with $\sigma = \sqrt{n}$, where n is the mean (or average) number of annual fires. (If n is too small, say less than 5, the approximation is not too satisfactory.) A good check is to compare this formula for σ with the more general statistical formula for standard deviation, namely,

$$\sigma = \sqrt{\sum_{i=1}^{n} \frac{(n_i - \bar{n})^2}{N-1}}$$

where n_i is the number of fires in the ith of N years; one should get $\sigma = \sqrt{n}$. One might wish to examine historical fire data (in various categories) to verify this equivalence, and also the number of years for which the data fall within various σ limits. Such an exercise may serve not only to build confidence in the formula, but also to provide an intuitive feeling for the statistical fluctuations that occur. If, substantially, the same general conditions are not present in each of the years, $\sqrt{n}$ would underestimate the degree of fluctuation.

The variability, σ, of the fire data can actually be computed from the fires in only a single year or even for only a portion of one year (after correcting for seasonal effects). In principle, the square root formula is applicable also to fire deaths, although multiple deaths occurring in a single fire create statistical complications. Multiple (that is, exposure) fires in several buildings arising from a single ignition fall into this same category, but these ordinarily represent only a small fraction of the total fires. For injuries, an additional problem relates to consistent reporting of minor injuries.

Why "Relative" Fluctuation Is a More Relevant Measure

Relative fluctuation or variability is the standard deviation expressed as a percentage of the sample mean. Compare the variabilities in the following two cases: In one case there was an average of 25 fires and hence a variability, σ, of 5. In the second case there was an average of 400 fires and hence a variability of 20. Which case appears to have the greater fluctuation? Most people would select the former despite the fact that a σ of 5 is much less than a σ of 20. In the first case the relative fluctuation (that is, relative to the mean) is 20 percent ($^5/_{25} \times 100$ percent). In the second it is 5 percent ($^{20}/_{400} \times 100$ percent). This example should show why the *relative* variability, expressed as a percentage of the mean, rather than the *absolute* variability, is the more relevant measure.

Relative variability can be computed as the *inverse* of the square root of the average number of fires—in the examples, 100 percent / $\sqrt{25}$ and 100 percent / $\sqrt{400}$, respectively. It increases as the number of fires decreases, and *vice versa,* i.e., when the number of fires increases the relative variability decreases. For a fire category that has only one-fourth as many fires as a second category, the relative fluctuation is doubled. In many statistical analyses it is necessary to combine categories or to consider biennial rather than annual data in order to reduce the relative fluctuation to an acceptable level.

Confidence Intervals

A statistical estimate of the average fire rate has by itself only limited usefulness, unless one specifies also the statistical error associated with the estimate. Confidence intervals are a precise way of specifying this error, the way now almost universally accepted. Moreover, they show clearly the positive impact of using several years of data.

A *confidence interval* is by definition a plus or minus range about a statistical estimate (that has been derived from a sample of data), which reflects the error associated with that estimate. One

could be estimating the average weight of fire fighters, or the proportion of persons who say they would vote for the Republican candidate, or the overall average number of annual fires based upon previous years' data, or the percent change in average annual fires due to a public education program.

The narrower the interval, the smaller is the statistical error and the better the estimate. But one can never guarantee 100 percent that the "true" value (that is, the value obtained from an infinite amount of data) will be within a specified interval derived from a finite sample of data. It is necessary to associate a degree of confidence (which can best be expressed in terms of a probability) with the interval. This is the reason for including the word "confidence" along with the word "interval."

By convention, confidence (or probability) levels of 68 percent, 90 percent, 95 percent, and/or 99 percent have been adopted as standard benchmarks. For fire data, where in practice one is severely limited in the ability to obtain many years of comparable data, the confidence levels of 68 percent and 90 percent are usually most appropriate. The 68 percent confidence limits may be said to provide "moderate" confidence (i.e., one is willing to give or take 2:1 odds that the true mean would turn out to be within the confidence limits if there were unlimited years of data), while the 90 percent confidence limits provide "strong" confidence (9:1 odds).

For a given data set, one gets a smaller confidence interval by reducing the associated confidence (probability), and *vice versa.* But for a given confidence level the smaller the interval the more accurate the estimate. The benefit of using several years of data will be reflected in a smaller confidence interval for the estimate.

Calculation of confidence intervals for mean annual fires: The confidence interval is calculated in terms of multiples of the so-called statistical *standard error* (SE) of the estimate. Quantitatively, the standard error of an estimate varies inversely with the square root of the number of years of data, and directly with the square root of the average annual fire rate. If one has four years of data rather than one year, the standard error is cut in half. For two years of data, the error is reduced by 30 percent ($1/\sqrt{2} \times 100$ percent = 70 percent). The multiple of the standard error that is used to obtain the actual confidence interval itself depends upon the confidence level one is using: 1.645 for 90 percent confidence and 1.0 (that is, the standard error itself) for 68 percent confidence.

For any type of statistical data, not just fire data, the standard error (SE) is proportional to the standard deviation, σ, of the data and inversely proportional to the square root of the number of years of data. Algebraically, for N years of fire data with average annual rate n

$$SE = \sigma / \sqrt{N} = \sqrt{n} / \sqrt{N} = \sqrt{n/N}$$

How to determine confidence intervals for the mean, or average, fire rate for past years, before the introduction of a public education program, is illustrated below. (See the next subsection for the somewhat more difficult case of percent change in fire rate after introducing the public education program.)

Assuming that the data show there was an average of 64 annual fires, the following are examples of the types of confidence interval statements that can be made. (Which one is most appropriate depends on the particular application at hand.)

1. With two years of data the (confidence) probability is 68 percent that the "true" average number of fires is between the confidence limits of 58.3 and 69.7 ($64 \pm \sqrt{64}/\sqrt{2}$), and

2. The probability is 90 percent that the true average number of fires is between 54.7 and 72.2 ($64 \pm \sqrt{64}/\sqrt{2}$), and

3. With four years of data the probability is 68 percent that the true average is between 60 and 68 ($64 \pm \sqrt{64} / \sqrt{4}$), and

4. The probability is 90 percent that the true average will be between 57.4 and 70.6 ($64 \pm 1.645 \times 8/2$).

Confidence intervals for percent change: The above discussion about confidence limits for the average annual fire rate of past years applies fairly well to the percent change after introducing a public education program. Although the concepts are basically the same, the precise procedure is somewhat involved and will be discussed only through rough approximations. One complication is that there is variability in both the "before" mean and the "after" mean. The effect of this is to increase the standard error in the estimate of the change in fire rate from before to after by about 40 percent ($\sqrt{2} \cong 1.4$) of the standard error of the estimate of the before rate.§

A second complication is that the number of "before" years may differ from the number of "after" years. But when the number of years is the same before and after, the situation is the same as the estimate of annual fires. That is, if there are two years each before and after, rather than one year, the standard error and resulting confidence limits are reduced by 30 percent ($1/\sqrt{2} = 0.70$).¶

Finally, since the percent change is relative to the "before" annual fires, one must consider the relative fluctuation, discussed previously, rather than the absolute fluctuation used in connection with the standard error for estimating the mean. If the average number of annual "before" fires is doubled, the relative standard error and relative confidence limits are reduced by 30 percent. Putting these three effects together yields the approximate standard error for percent change, and thus the approximate confidence interval.#

$$(\sqrt{2} / \sqrt{N}) / \sqrt{\bar{n}_1} = \sqrt{\frac{2}{N\bar{n}_1}}$$

As an example, when the average number of before fires is 20 and there are two years each before and after, the standard error is 22 percent ($\sqrt{2} / \sqrt{2} / \sqrt{20} \times 100$ percent). One can then obtain confidence limits in the usual way. If, for example, the change is 20 percent (corresponding to average annual fires of 16 for the two after years), the approximate 68 percent confidence limits are 42 percent and –2 percent (20 percent ± 22 percent). Similarly, the 90 percent confidence limits are 57 percent and –17 percent (20 percent ± 1.645 × 22 percent.)**

Is the Change Statistically Significant?

Although confidence limits in essence sum up the statistical contribution to the data, in the case of public education programs—and

for a very large variety of essentially similar investigations—one wishes to translate the result into a definitive answer (or conclusion) to the question: Was the public education program effective? Equivalently, but in more precise terms, could the observed change have occurred by chance alone if the public education program had truly been ineffective?

The answer is implicit in the confidence limits with only a change in emphasis (perhaps mostly, if not solely, in the terminology ordinarily employed). A lower confidence limit that is negative, as in the above example, actually implies a negative conclusion, i.e., the observed change is not statistically significant and could indeed have occurred by chance. If the 68 percent confidence interval was used, this implies that an observed change of 20 percent or more (using the value in the example) would occur by chance at least 16 percent of the time, even if the public education program had no effect; similarly, it would occur at least 5 percent of the time with a negative lower 90 percent confidence limit.†† Sixteen percent is considered to be a "moderate significance level," and if the lower 68 percent confidence limit is positive, then the change can be considered to be moderately significant; similarly, if the lower 90 percent interval is still positive, then the change is considered definitely significant. If the lower confidence limit is negative, then the converse is true. In the above example, the 20 percent change is definitely not statistically significant.

There is an alternative way to view the question of significance. Rather than considering predefined moderate and definite significance levels, it is simpler—and actually more precise—to reverse the procedure. One first computes the number of standard errors (say K_{P_o}) represented by the observed change, and then checks, in tables of the normal distribution, what probability (say P_o) this corresponds to.‡‡ If P_o is judged to be too high, then the change is considered to be not statistically significant.

One can reverse the process and for a specified value of P_o (that is, 16 percent or 5 percent) determine the critical change, Δ_o, needed to give exactly a zero lower confidence limit. This change thus becomes the borderline value separating a significant change from a not significant change. Inverting Equation 1a leads to Equation 2a. (See Summary of Formulas.) Equation 2 for unequal Ns was used to plot the curve in Figure 2-7H.

More often than not with fire data, "not significant" is likely to be the situation encountered. But the most appropriate practical

---

§An intuitive explanation of how the $\sqrt{2}$ arises is as follows. First, the fluctuation in the difference between two rates is the same as the fluctuation in the sum. The normal distribution does not really distinguish between positive and negative values—it treats both in the same way—and it is more convenient to consider the sum. In particular, let $\bar{n}_1 + \bar{n}_2 = 2\bar{n}$ be the two-year sum (or biennial fire rate) of each of the two previous years' fire rates. The former biennial rate is twice the latter average annual fire rate, i.e., $2n$ vs. n. The σ of the former is $\sqrt{2n}$, whereas the σ of the latter is $\sqrt{n}$; the σ of the two-year sum or difference is $\sqrt{2}$ ($\cong 1.4$) times as large as the one-year (or before) rate.

¶When the average number of "before" years differs from the number of "after" years, one cannot use a simple arithmetic mean. This is the reason that two years both before and after yield better statistical accuracy than any large number of "before" years' data when compared with only a single "after" year's data.

#The algebraic formula for the approximate SE is where N is the (same) number of before and after years. The before average, $\bar{n}_1$, should actually be replaced by a more complicated function of n1 and the after average fire, $\bar{n}_2$.

**For the numerical example used, the correction factors turn out to be 0.77 for the lower 68 percent confidence limit and 0.95 for the upper 68 percent confidence limit; and 0.72 and 1.03 for the 90 percent lower and upper confidence limits. The approximate 68 percent confidence limits of 0.424 and 0.024 agree roughly with the exact limits, shown in Table 2-7E of 0.372 and 0.012.

††Sixteen percent corresponds to the probability that is below the moderate 68 percent lower confidence limit, the probability that is above the upper confidence limit, also 16 percent, is not of interest in the present context. Similarly, 5 percent corresponds to the probability below the 90 percent lower confidence limit. Figure 2-7H may be helpful in this regard. Confidence limits exclude both "tails" of the normal distribution, whereas a significant change concerns only the one tail beyond the lower confidence limit.

‡‡For this calculation, one uses an SE formula similar to the one for the algebraic formula for the approximate SE (noted in earlier footnote), but with replaced by its corrected value corresponding to a zero lower confidence value. The result is the simple formula shown as Equation 1a in the Summary of Formulas. Using $\bar{n}_1 = 20$, $\bar{n}_2 = 16$, $N = 2$), this equation gives a value for K_{P_o} of 0.94 (shown in Table 2-7E), i.e., slightly less than one SE. The corresponding probability, P_o, from tables of the normal distribution, is 17 percent (also shown in Table 2-7E) or slightly more than the moderate 16 percent significance level (which corresponds to exactly one SE). Thus, the observed 20 percent change is not statistically significant.

conclusion in most cases is that more years of data are needed in order to really establish whether or not the public education program is effective. Although such additional data improve the accuracy (that is, shrinks the width of the confidence interval), this must be balanced against the fact that more years ordinarily leads to less homogeneity of the data.

Summary of Formulas

The following definitions are used:

N_1 = number of years of data before public education program

N_2 = number of years of data after public education program

n_1 = average number of fires before public education program

n_2 = average number of fires after public education program

K_{P_o} = normal distribution deviation computed on the assumption that the public education program had no real effect

1. Probability that the absolute change is due to chance.

$$K_{P_o} = (\bar{n}_1 - \bar{n}_2)/\sqrt{\frac{\bar{n}_1}{N_1} + \frac{\bar{n}_2}{N_2}} \qquad (1)$$

P_o is then determined from normal distribution tables. When $N_1 = N_2 = N$, Equation 1 becomes

$$(1a)$$

$$K_{P_o} = (\bar{n}_1 - \bar{n}_2)/\sqrt{\frac{\bar{n}_1 + \bar{n}_2}{N}}$$

2. Confidence limits for the absolute change can be derived from the following equation

$$\bar{n}_1 - \bar{n}_2 \pm 1.96 \sqrt{\frac{\bar{n}_1}{N_1} + \frac{\bar{n}_2}{N_2}}$$

Acknowledgment

This chapter is largely based on the book *Proving Public Fire Education Works* published by TriData Corporation in 1990 under the authorship of Philip Schaenman *et al*. Many of the tables in this chapter originally appeared in that book.

BIBLIOGRAPHY

References Cited

1. Hatry, H., *et al.*, *Public Program Evaluation* (2nd ed.), The Urban Institute, Washington, DC, 1989. Note: Many other Urban Institute publications address the art and science of evaluation.
2. Short Guide to Evaluating Local Public Fire Education Programs, U.S. Fire Administration, 1991.
3. Schaenman, P., *et al.*, *Proving Public Fire Education Works*, TriData, Arlington, VA, 1990.
4. *Learn Not to Burn® Curriculum*, National Fire Protection Association, Quincy, MA.

Additional Reading

Wallis, W. A., and Roberts, H. V., *Statistics: A New Approach*, The Free Press, 1960.
Weiss, N. A., and Hassett, M. J., *Introductory Statistics*, 2nd ed., Addison-Wesley, Reading, MA, 1987.

FIRE PREVENTION

Prevention is probably the most effective means for reducing loss due to fire. It can be achieved by changing or controlling the source of heat that leads to ignition, the source of fuel that is first ignited, or the circumstances by which the two are brought together.

Section 3 of this handbook (Section 2 in prior editions) discusses various sources of heat and fuel, some common to most occupancies and operations, others specific to certain sources of heat and fuel, and the ways in which they can be prevented from causing a fire. Methods of fire prevention that involve educating the general public are covered extensively in Section 2, "Fire and Life Safety Education."

Most fuels first involved in a fire are also important contributors to rapid growth of the fire, and these are discussed in more detail in Section 4, "Materials, Products, and Environments." The relationship of building design and construction to fire growth is covered in Section 7, "Confining Fires." Discussion of the fire hazards of particular occupancies is covered in Section 9, "Systems Approaches to Property Classes." Several major causes of fire, such as careless use of smoking materials and cooking-related fires, are not covered extensively in this handbook, but are discussed in detail in NFPA's *Learn Not to Burn®* materials. Arson investigation is discussed in Section 11, Chapter 1.

The purpose of this section is threefold: (1) to discuss particular sources of heat and fuel, including some which may be part of a specific occupancy (discussed in more detail in Section 9); (2) to identify the fire hazards associated with these; and (3) to discuss applicable means of fire prevention. In this latter context, the reader should recognize that certain techniques of fire detection, protection, suppression, and confinement can be successfully applied to several heat and fuel sources. Therefore, there will be common elements among the chapters and sections of the Handbook. Chapters

1 through 6 and Chapters 31, 32, and 33 address fire hazards that are common to most occupancies or commercial, institutional, or industrial operations. Other chapters focus on specific categories.

Chapter 1 covers the electrical system, the leading source of heat that results in property damage in building fires. Chapters 2 and 3 discuss control of static electricity and protection from lightning. Chapter 4 deals with emergency and standby electrical power supplies, and their interconnection with the electrical system.

Chapter 5 discusses the leading sources of building fire incidents in the United States, heating systems and appliances, which can cause fires because they typically operate above the ignition temperature of many common materials. Chapter 6 addresses boiler-furnaces and their fuels. Chapter 7 covers heat transfer systems using nonwater fluids, which can operate at temperatures above 350°F. Chapter 8 covers industrial equipment, such as process ovens, dryers, and furnaces. Chapter 9 covers oil quenching and molten salt baths for heat treating and tempering. Chapter 10 discusses stationary combustion engines used as prime movers for standby electrical generators and fire pumps.

Other chapters in Section 3 address specific processes or operations. Included are chapters on storage of flammable and combustible liquids, gases, chemicals, and solid fuels in small containers and in tanks and silos at processing plants. Chapter 14 discusses welding and cutting operations, Chapter 28 covers refrigeration systems, and Chapter 31 examines waste handling.

Finally, Chapter 32 discusses the regulation of hazardous waste.

—Robert P. Benedetti
December 1996

ELECTRICAL SYSTEMS AND APPLIANCES

Revised by John M. Caloggero

This chapter discusses electrical systems and appliances. Specific topics include origins of electrical fires; codes and standards; building wiring, design, and protection; electrical household appliances; industrial and commercial equipment; electrical equipment for outdoor use; locations exposed to moisture and dust; communications systems; and special occupancy electrical problems.

Other chapters in this handbook detail the related components of fire hazards of building services, most notably Section 3, Chapter 5, "Heating Systems and Appliances"; and Section 7, Chapter 14, "Air-Conditioning and Ventilating Systems."

Special considerations for the installation of electrical systems can be found elsewhere in this handbook. See Section 3, Chapter 3, "Lightning Protection Systems"; Section 3, Chapter 8, "Industrial and Commercial Heat Utilization Equipment"; Section 4, Chapter 8, "Medical Gases"; and Section 8, Chapter 2, "Concepts of Egress Design."

If properly designed, installed, and maintained, electrical systems are both convenient and safe; otherwise they may be a source of both fire and personal injury. Electricity may cause a fire if it arcs or overheats electrical equipment and can cause injury or death through shocks and burns.

When an electric circuit carrying a current is interrupted either intentionally, as by a switch, or unintentionally, as when a connection at a terminal becomes loosened, arcing at the switch contacts or heating from the high-resistance connection at the terminal is produced. The intensity of the arc and degree of heat depend largely upon the current and resistance of the contact at the terminal, and also the voltage and inductance of the circuit load. The temperature may easily be high enough to ignite any combustible material in the vicinity.

An electric arc may not only ignite combustible material in its immediate vicinity, e.g., the insulation and covering of the conductor, but it may also melt the metal of the conductor. Hot sparks from burning combustible material and hot metal may splatter or fall on combustible material, setting it on fire.

When an electrical conductor carries current, heat is generated in direct proportion to the resistance (ohms) of the conductor and to the square of the current (amperage). The resistance of conductors

used to convey current to the location in which it is used, or to convey it through the windings of a piece of apparatus (except resistance devices and heaters), should be as low as practical. Metals, such as copper and aluminum, are used for this purpose. In other instances, such as in electric heaters, electric cooking equipment, and soldering irons, the heat from the current serves a useful purpose.

The heating of electrical conductors is almost never a fire hazard under design conditions. NFPA 70, *National Electrical Code®* (hereinafter referred to as *NEC*), specifies the maximum safe current a conductor can carry without overheating. (See Tables 3-1A through 3-1D.) This depends upon the size of conductor, the temperature in which the equipment is installed, and the type of insulation. Where the specified current is exceeded, the generation of heat causes deterioration of the electrical insulation, and the excess heat generated may ignite combustibles in the area. Apparatus or appliances that use electric conductors as heating elements or use an electric arc to generate heat (e.g., arc welders) can be fire hazards if improperly installed and used.

One common method of reducing the degree of hazard is to use the proper size conductor for the load, suitable insulation for the environment, and the correct size overcurrent protection (fuse or circuit breaker). Where heat-generating equipment is utilized, ensure that clearances and sufficient air circulation is provided to prevent unsafe temperatures and premature breakdown of electrical insulation, or ignition of nearby combustibles.

All standards governing electric equipment include requirements intended to prevent fires caused by arcing and overheating, and to prevent accidental contact, which may cause an electric shock. These primary sources of electrical hazards should be kept in mind whenever any work is being done on electric equipment.

ORIGINS OF ELECTRICAL FIRES IN BUILDINGS

Table 3-1E provides an overview of 1989 through 1993 structure fires reported to U.S. fire departments that were coded as caused by electrical failure. This table will miss some relevant incidents, because it is limited to fires coded as involving a short circuit, ground fault, or other electrical failure. Other causes, some of which may include a mix of electrical failures with other fires, are not included. Also, these statistics reflect cause determinations by fire officers on the scene, so some errors of inclusion or exclusion may occur.

John M. Caloggero is principal electrical specialist on the staff of NFPA.

TABLE 3-1A. *Allowable Ampacities of Insulated Conductors Rated 0 through 2000 Volts, 60° to 90°C (140° to 194°F) Not More Than Three Current-Carrying Conductors in Raceway or Cable or Earth (Directly Buried), Based on Ambient Temperature of 30°C (86°F)*

Size	Temperature Rating of Conductor. See *NEC* Table 310-13.						Size
	60°C (140°F)	75°C (167°F)	90°C (194°F)	60°C (140°F)	75°C (167°F)	90°C (194°F)	
AWG kcmil	TYPES TW*, UF*	TYPES FEPW*, RH*, RHW*, THHW*, THW*, THWN*, XHHW* USE*, ZW*	TYPES TBS, SA SIS, FEP*, FEPB*, MI RHH*, RHW-2, THHN*, THHW*, THW-2*, THWN-2*, USE-2, XHH, XHHW* XHHW-2, ZW-2	TYPES TW*, UF*	TYPES RH*, RHW*, THHW*, THW*, THWN*, XHHW*, USE*	TYPES TBS, SA, SIS, THHN*, THHW*, THW-2, THWN-2, RHH*, RHW-2, USE-2, XHH, XHHW, XHHW-2, ZW-2	AWG kcmil
	COPPER			ALUMINUM OR COPPER-CLAD ALUMINUM			
18	....	....	14	....	....	....	....
16	....	....	18	....	....	....	....
14	20*	20*	25*	....	....	....	....
12	25*	25*	30*	20*	20*	25*	12
10	30	35*	40*	25	30*	35*	10
8	40	50	55	30	40	45	8
6	55	65	75	40	50	60	6
4	70	85	95	55	65	75	4
3	85	100	110	65	75	85	3
2	95	115	130	75	90	100	2
1	110	130	150	85	100	115	1
1/0	125	150	170	100	120	135	1/0
2/0	145	175	195	115	135	150	2/0
3/0	165	200	225	130	155	175	3/0
4/0	195	230	260	150	180	205	4/0
250	215	255	290	170	205	230	250
300	240	285	320	190	230	255	300
350	260	310	350	210	250	280	350
400	280	335	380	225	270	305	400
500	320	380	430	260	310	350	500
600	355	420	475	285	340	385	600
700	385	460	520	310	375	420	700
750	400	475	535	320	385	435	750
800	410	490	555	330	395	450	800
900	435	520	585	355	425	480	900
1000	455	545	615	375	445	500	1000
1250	495	590	665	405	485	545	1250
1500	520	625	705	435	520	585	1500
1750	545	650	735	455	545	615	1750
2000	560	665	750	470	560	630	2000

	CORRECTION FACTORS						
Ambient Temp. °C	For ambient temperatures other than 30°C (86°F), multiply the allowable ampacities shown above by the appropriate factor shown below.						Ambient Temp. °F
21-25	1.08	1.05	1.04	1.08	1.05	1.04	70-77
26-30	1.00	1.00	1.00	1.00	1.00	1.00	78-86
31-35	.91	.94	.96	.91	.94	.96	87-95
36-40	.82	.88	.91	.82	.88	.91	96-104
41-45	.71	.82	.87	.71	.82	.87	105-113
46-50	.58	.75	.82	.58	.75	.82	114-122
51-55	.41	.67	.76	.41	.67	.76	123-131
56-60	....	.58	.71	....	.58	.71	132-140
61-70	....	.33	.58	....	.33	.58	141-158
71-80	....	....	.41	....	....	.41	159-176

*Unless otherwise specifically permitted elsewhere in the *Code*, the overcurrent protection for conductor types marked with an asterisk (*) shall not exceed 15 amperes for No. 14, 20 amperes for No. 12, and 30 amperes for No. 10 copper; or 15 amperes for No. 12 and 25 amperes for No. 10 aluminum and copper-clad aluminum after any correction factors for ambient temperature and number of conductors have been applied.

Source: NFPA 70-1996, *National Electrical Code.*

TABLE 3-1B. *Allowable Ampacities of Single Insulated Conductors, Rated 0 through 2000 Volts, in Free Air Based on Ambient Air Temperature of 30°C (86°F)*

Size	Temperature Rating of Conductor. See *NEC* Table 310-13.						Size
	60°C (140°F)	75°C (167°F)	90°C (194°F)	60°C (140°F)	75°C (167°F)	90°C (194°F)	
AWG kcmil	TYPES TW*, UF*	TYPES FEPW*, RH*, RHW*, THHW*,THW*, THWN*,XHHW* ZW*	TYPES TBS, SA, SIS, FEP*, FEPB*, MI,RHH*, RHW-2, THHN*, THHW*, THW-2*, THWN-2*, USE-2, XHH, XHHW*, XHHW-2, ZW-2	TYPES TW*, UF*	TYPES RH*, RHW*, THHW*, THW*, THWN*, XHHW*	TYPES TBS, SA, SIS, THHN*, THHW*, THW-2, THWN-2, RHH*, RHW-2, USE-2, XHH, XHHW*, XHHW-2, ZW-2	AWG kcmil
	COPPER			ALUMINUM OR COPPER-CLAD ALUMINUM			
18	....	....	18	....	....	....	....
16	....	....	24	....	....	....	....
14	25*	30*	35*	....	....	....	....
12	30*	35*	40*	25*	30*	35*	12
10	40*	50*	55*	35*	40*	40*	10
8	60	70	80	45	55	60	8
6	80	95	105	60	75	80	6
4	105	125	140	80	100	110	4
3	120	145	165	95	115	130	3
2	140	170	190	110	135	150	2
1	165	195	220	130	155	175	1
1/0	195	230	260	150	180	205	1/0
2/0	225	265	300	175	210	235	2/0
3/0	260	310	350	200	240	275	3/0
4/0	300	360	405	235	280	315	4/0
250	340	405	455	265	315	355	250
300	375	445	505	290	350	395	300
350	420	505	570	330	395	445	350
400	455	545	615	355	425	480	400
500	515	620	700	405	485	545	500
600	575	690	780	455	540	615	600
700	630	755	855	500	595	675	700
750	655	785	885	515	620	700	750
800	680	815	920	535	645	725	800
900	730	870	985	580	700	785	900
1000	780	935	1055	625	750	845	1000
1250	890	1065	1200	710	855	960	1250
1500	980	1175	1325	795	950	1075	1500
1750	1070	1280	1445	875	1050	1185	1750
2000	1155	1385	1560	960	1150	1335	2000

CORRECTION FACTORS

Ambient Temp. °C	For ambient temperatures other than 30°C (86°F), multiply the allowable ampacities shown above by the appropriate factor shown below.						Ambient Temp. °F
21-25	1.08	1.05	1.04	1.08	1.05	1.04	70-77
26-30	1.00	1.00	1.00	1.00	1.00	1.00	78-86
31-35	.91	.94	.96	.91	.94	.96	87-95
36-40	.82	.88	.91	.82	.88	.91	96-104
41-45	.71	.82	.87	.71	.82	.87	105-113
46-50	.58	.75	.82	.58	.75	.82	114-122
51-55	.41	.67	.76	.41	.67	.76	123-131
56-60	....	.58	.71	....	.58	.71	132-140
61-70	....	.33	.58	....	.33	.58	141-158
71-80	....	....	.41	....	....	.41	159-176

*Unless otherwise specifically permitted elsewhere in the *Code*, the overcurrent protection for conductor types marked with an asterisk (*) shall not exceed 15 amperes for No. 14, 20 amperes for No. 12, and 30 amperes for No. 10 copper; or 15 amperes for No. 12 and 25 amperes for No. 10 aluminum and copper-clad aluminum.
Source: NFPA 70-1996, *National Electrical Code.*

TABLE 3-1C. *Allowable Ampacities of Three Single Insulated Conductors Rated 0 through 2000 Volts, 150° to 250°C (302° to 482°F), in Raceway or Cable Based on Ambient Air Temperature of 40°C (104°F)*

Size	Temperature Rating of Conductor. See *NEC* Table 310-13.				Size
	150°C (302°F)	200°C (392°F)	250°C (482°F)	150°C (302°F)	
AWG kcmil	TYPE Z	TYPES FEP, FEPB, PFA	TYPES PFAH, TFE	TYPE Z	AWG kcmil
	COPPER		NICKEL OR NICKEL-COATED COPPER	ALUMINUM OR COPPER-CLAD ALUMINUM	
14	34	36	39	....	14
12	43	45	54	30	12
10	55	60	73	44	10
8	76	83	93	57	8
6	96	110	117	75	6
4	120	125	148	94	4
3	143	152	166	109	3
2	160	171	191	124	2
1	186	197	215	145	1
1/0	215	229	244	169	1/0
2/0	251	260	273	198	2/0
3/0	288	297	308	227	3/0
4/0	332	346	361	260	4/0
250	....	....	....	....	250
300	....	....	....	....	300
350	....	....	....	....	350
400	....	....	....	....	400
500	....	....	....	....	500
600	....	....	....	....	600
700	....	....	....	....	700
750	....	....	....	....	750
800	....	....	....	....	800
1000	....	....	....	....	1000
1500	....	....	....	....	1500
2000	....	....	....	....	2000

CORRECTION FACTORS

Ambient Temp. °C	For ambient temperatures other than 40°C (104°F), multiply the allowable ampacities shown above by the appropriate factor shown below.				Ambient Temp. °F
41-50	.95	.97	.98	.95	105-122
51-60	.90	.94	.95	.90	123-140
61-70	.85	.90	.93	.85	141-158
71-80	.80	.87	.90	.80	159-176
81-90	.74	.83	.87	.74	177-194
91-100	.67	.79	.85	.67	195-212
101-120	.52	.71	.79	.52	213-248
121-140	.30	.61	.72	.30	249-284
141-160	....	.50	.65	....	285-320
161-180	....	.35	.58	....	321-356
181-200	....	....	.49	....	357-392
201-225	....	....	.35	....	393-437

Source: NFPA 70-1996, *National Electrical Code.*

Electrically powered equipment is involved in several times this number of fires each year but in non-electrical failure scenarios, such as installation or use of equipment with hot surfaces too close to combustibles. In fact, nearly all fires involving electrically powered equipment stem from foreseeable or avoidable human error, such as:

Improper installation: Failure to follow manufacturer's installation instructions or observe all provisions of the *NEC*, including the workmanship provisions, can result in installation of equipment in ways that will lead to overloading, damage to equipment, or excessive heat exposure to nearby combustibles. Mismatching of wire gage size with equipment wattage or over-fusing are examples.

Lack of maintenance: Equipment wears out in service, e.g., through the deterioration of electrical insulation. Deterioration can be accelerated through improper installation, excessive heat, environmental conditions, or physical damage due to use of the incor-

TABLE 3-1D. *Allowable Ampacities for Single Insulated Conductors Rated 0 through 2000 Volts, 150° to 250°C (302° to 482°F), in Free Air Based on Ambient Air Temperature of 40°C (104°F)*

Size	Temperature Rating of Conductor. See *NEC* Table 310-13.					Size
	150°C (302°F)	200°C (392°F)	Bare or covered conductors	250°C (482°F)	150°C (302°F)	
AWG kcmil	TYPE Z	TYPES FEP, FEPB, PFA		TYPES PFAH, TFE	TYPE Z	AWG kcmil
	COPPER			NICKEL OR NICKEL-COATED COPPER	ALUMINUM OR COPPER-CLAD ALUMINUM	
14	46	54	30	59	....	14
12	60	68	35	78	47	12
10	80	90	50	107	63	10
8	106	124	70	142	83	8
6	155	165	95	205	112	6
4	190	220	125	278	148	4
3	214	252	150	327	170	3
2	255	293	175	381	198	2
1	293	344	200	440	228	1
1/0	339	399	235	532	263	1/0
2/0	390	467	275	591	305	2/0
3/0	451	546	320	708	351	3/0
4/0	529	629	370	830	411	4/0
250	....	....	415	....	....	250
300	....	....	460	....	....	300
350	....	....	520	....	....	350
400	....	....	560	....	....	400
500	....	....	635	....	....	500
600	....	....	710	....	....	600
700	....	....	780	....	....	700
750	....	....	805	....	....	750
800	....	....	835	....	....	800
900	....	....	865	....	....	900
1000	....	....	895	....	....	1000
1500	....	....	1205	....	....	1500
2000	....	....	1420	....	....	2000

CORRECTION FACTORS

Ambient Temp. °C	For ambient temperatures other than 40°C (104°F), multiply the allowable ampacities shown above by the appropriate factor shown below.					Ambient Temp. °F
41-50	.95	.97	....	.98	.95	105-122
51-60	.90	.94	....	.95	.90	123-140
61-70	.85	.90	....	.93	.85	141-158
71-80	.80	.87	....	.90	.80	159-176
81-90	.74	.83	....	.87	.74	177-194
91-100	.67	.79	....	.85	.67	195-212
101-120	.52	.71	....	.79	.52	213-248
121-140	.30	.61	....	.72	.30	249-284
141-160	....	.50	....	.65	....	285-320
161-180	....	.35	....	.58	....	321-356
181-200	....	....	....	.49	....	357-392
201-225	....	....	....	.35	....	393-437

Source: NFPA 70-1996, *National Electrical Code.*

rect wiring methods. Even when the installation is properly installed, aging can cause deterioration of materials and equipment; therefore, periodic inspection and maintenance of electrical equipment should be performed.

Improper use: Listed equipment that is not installed in accordance with the manufacturer's instruction for its intended use and for the conditions under which it is used can cause fires or injure personnel. Use of equipment listed for use in dry locations, but installed in damp or wet locations, or general-purpose equipment installed in locations where flammable liquids are used can be sources of ignition.

Carelessness or oversight: Even momentary lapses of caution in the use of equipment can cause fires. Failure to turn off devices (e.g., space heaters) not in use, dropping combustibles into equipment,

TABLE 3-1E. *Structure Fires Due to Electrical Failure (Annual Average of 1989 Through 1993 Fires Reported to U.S. Fire Departments)*

Equipment Involved in Ignition	Average Number of Fires per Year
Heating Equipment	5,830
Central heating unit	2,010
Water heater	1,020
Fixed area heating device	1,260
Indoor fireplace	30
Portable heating device	900
Chimney or flue	40
Chimney connector	20
Heat transfer system	130
Unclassified type	300
Unknown type	120
Cooking Equipment	5,210
Stove	2,390
Oven or rotisserie	830
Fixed area cooking or warming device	250
Deep fat fryer	130
Portable cooking or warming device	130
Open-fired grill	20
Grease hood or duct	480
Unclassified type	270
Unknown type	120
Air-Conditioning or Refrigeration Equipment	3,890
Central air-conditioning or refrigeration unit	980
Water cooling device or tower	170
Fixed area refrigeration unit	1,250
Fixed area air-conditioning unit	670
Portable air conditioning or refrigeration unit	490
Unclassified type	190
Unknown type	140
Electrical Distribution Equipment	40,350
Fixed wiring	15,850
Transformer	790
Meter or meter box	740
Overcurrent protection device (e.g., fuse, circuit breaker)	2,880
Switch, receptacle, or outlet	4,420
Lighting fixture, lampholder, ballast, or sign	5,130
Cord or plug	7,040
Lamp or light bulb	720
Unclassified type	1,210
Unknown type	1,570
Appliance or Tool	11,110
Television, radio, or phonograph	1,800
Dryer	2,330
Washing machine	1,020
Floor-care equipment (e.g., vacuum cleaner)	130
Separate motor or generator	840
Hand tool (e.g., drill, soldering iron)	90
Portable appliance to produce controlled heat (e.g., iron, electric blanket)	1,600
Portable appliance not to produce controlled heat (e.g., can opener, electric razor)	630
Unclassified type	1,830
Unknown type	860
Special Equipment	1,430
Electronic Equipment (e.g., telephone, computer, x-ray)	500
Vending machine or drinking fountain	70
Office machine	120
Biomedical equipment	40
Separate pump or compressor	260
Internal combustion engine	30
Conveyer	20

TABLE 3-1E. *(Continued)*

Equipment Involved in Ignition	Average Number of Fires per Year
Printing press	10
Unclassified type	340
Unknown type	60
Processing Equipment	390
Furnace, oven, or kiln	90
Casting, molding, or forging equipment	20
Heat-treating equipment	20
Working or shaping machine	60
Coating machine	10
Painting equipment	30
Chemical process equipment	20
Waste recovery equipment	10
Unclassified type	70
Unknown type	60
Service or Maintenance Equipment	450
Incinerator	10
Bearing or brake	10
Rectifier or charger	130
Tarpot or tar kettle	0*
Arc or oil lamp	10
Elevator	170
Torch	20
Unclassified type	100
Unknown type	10
Other	12,210
Vehicle	500
No equipment involved	11,110
Unclassified type	420
Unknown type	190
Unknown Type	4,620
Total	85,510

Source: 1989-1993 NFIRS, NFPA survey.
NOTES: "Electrical failure" defined as Ignition Factor 54 (short circuit or ground fault) or 55 (other electrical failure). Numbers rounded to nearest ten. Sums may not equal totals because of rounding error.
*Not zero, but rounds to zero.

and draping combustibles over operating equipment (e.g., clothes on a lamp) are all examples.

Some special studies have been done that provide additional insights going beyond the level of detail in Table 3-1E, although no other sources of representative national fire statistics exist. These studies include the annual tabulations by the International Association of Electrical Inspectors (IAEI) of fire and shock incidents that come to their attention and the several field studies done by the U.S. Consumer Product Safety Commission. These studies appear to support a few qualitative observations:

1. Cord fires include many fires involving extension cords, as opposed to lamp or other appliance cords. Extension cords are probably overused as semi-permanent extensions of wiring systems.

2. Overcurrent protection device fires appear to involve circuit breakers more often than fuses, both overall and relative to the degree of use of each type of device. This is even more notable given that fuses are used primarily in much older homes, where deterioration due to age or damage due to system alterations would be more likely, and that fuses are more susceptible to tampering.

3. Color televisions pose a greater fire hazard than black-and-white televisions due to differences in the energy usage of the two types.

CODES AND STANDARDS

All electrical installations in the United States should be made, used, and maintained in accordance with the *NEC* and other standards that apply in special situations, e.g., NFPA 70E, *Standard for Electrical Safety Requirements for Employee Workplaces*; NFPA 79, *Electrical Standard for Industrial Machinery* (hereinafter referred to as NFPA 79); NFPA 75, *Standard for the Protection of Electronic Computer/Data Processing Equipment*; and NFPA 99, *Standard for Health Care Facilities* (hereinafter referred to as NFPA 99). There are also special standards for electrical installations on shipboard, aircraft, in hazardous (classified) locations, and other locations.

National Electrical Code (ANSI/NFPA 70)

The *NEC* provides for the practical safeguarding of persons and property from hazards arising from the use of electricity. The *NEC* was first issued under its present name in 1897. It is revised every three years by the National Electrical Code Committee.

National Electrical Safety Code (ANSI C2)

As interest increased in electrical safety in the United States, a need arose for a code to cover the practices of public utilities and others when installing and maintaining overhead and underground electric supply and communications lines. Accordingly, the first edition of the *National Electrical Safety Code* was completed in 1916. Currently this code is updated and published by the Institute of Electrical and Electronics Engineers (IEEE).

Canadian Electrical Code

The Canadian Standards Association sponsors and publishes the *Canadian Electrical Code*. This code establishes essential requirements and minimum standards for the installation and maintenance of electrical equipment. This code can be adopted and enforced by electrical inspection departments throughout Canada and was prepared with due regard for the *NEC* and the *National Electrical Safety Code*. Its several parts are the Canadian equivalent of the *NEC*, the *National Electrical Safety Code*, and the standards of Underwriters Laboratories Inc. (UL) in the U.S.

The *National Electrical Code Handbook*

While not in itself a code or standard that can have the force of law, the *National Electrical Code Handbook* is a valuable resource in implementing provisions of the *NEC*. The *NEC Handbook* contains the entire text of the *NEC* supplemented by comments, diagrams, examples, and illustrations that are intended to clarify and explain some of the intricate requirements of the *NEC*. The *NEC Handbook* is published by the NFPA; a new edition is published with each new edition of the *NEC*.

Listed and Labeled Electrical Equipment Manufacturers

Three publications are available that list the names of companies making electric appliances and materials that have been listed or labeled by UL, ETL Testing Laboratories, Inc., (ETL), and other testing laboratories. Devices and materials in the product categories listed have been tested for actual field use in accordance with the *NEC* and against other ANSI safety standards. The lists are published so that the names of manufacturers of devices that meet test criteria may be readily obtained. UL publishes the *Hazardous Location Equipment Directory*, *Electrical Appliance and Utilization Equipment Directory*, and *Electrical Construction Materials Directory*.

They are published annually by UL with six-month supplements. Certain types of electrical equipment are covered in other testing laboratory directories, such as the *Marine Products Directory*.

There are sections of the *NEC* that influence the design and construction of electric equipment. Testing labs have complete published standards for most electric fittings, materials, and equipment, including specifications for performance under tests and in service, and further details of design and construction. Other testing laboratories in the United States and Canada that test electric equipment include Factory Mutual Research Corporation (FMRC); ETL Testing Laboratories, Inc.; U.S. Testing; Applied Research Laboratories; Dash, Straus, and Goodhue; and the Canadian Standards Association (CSA). Some laboratories stipulate compliance with the *NEC* in testing equipment. (Where no published standards are available for special equipment, arrangements usually can be made with testing laboratories for testing on an individual basis.)

BUILDING WIRING, DESIGN, AND PROTECTION

This section contains information on various types of building wiring and equipment, including panelboards and overcurrent protection; types of electric conductors; identification of conductors, terminals, etc.; how to calculate loads; and similar subjects.

Service Entrance

Good practice requires that service entrance conductors comply in all respects with the detailed requirements of Article 230 of the *NEC*. Figure 3-1A provides typical arrangements of service entrance conductors.

The basic requirement is that a building or other structure can be supplied by only one service. Where more than one service is installed as permitted, a permanent plaque or directory is required to be installed at each service drop or lateral or at each service equipment location, denoting all of the other services on or in that building or structure, and the area served by each. This plaque or directory should be constructed of sufficient durability to withstand the ambient environment.

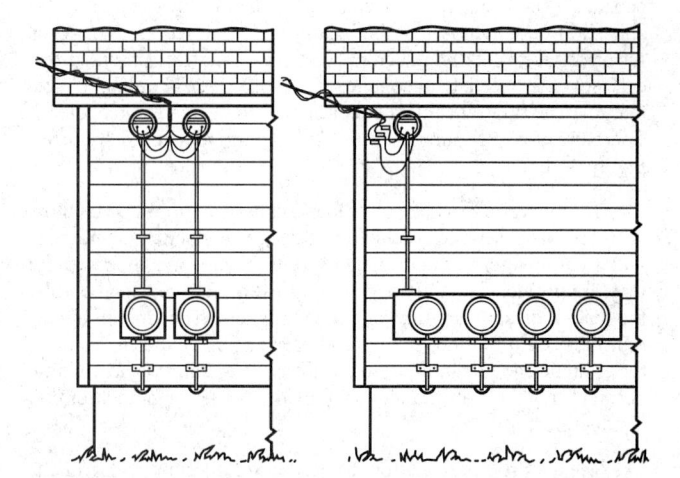

FIG. 3-1A.   The diagram at left shows an installation in which two sets of service entrance conductors are tapped from one service drop, with meters mounted on the outside of the building. In the diagram at right, a set of main service entrance conductors is connected to the service drop and is carried through a trough and then through four sets of subservice entrance conductors to each service equipment. Here again, the meters are on the exterior wall.

A new requirement has been added: where any combination of branch circuits, feeder, and services supply power to a building or other structure, a permanent plaque or directory must be installed at each service disconnect location, to indicate where the other disconnects that feed the building or structure are located.

The basic rule is that a building or other structure is to be supplied by only one service. There are, however, several exceptions.

Where separate services are permitted, such as for fire pumps, emergency or legally-required standby electrical systems, optional standby electrical systems, or parallel power production systems, six additional switches or circuit breakers are allowed, in addition to the six service disconnects normally allowed. The intent here is that a disruption of the main building service should not disconnect fire pump equipment, or emergency or legally required standby electrical systems, optional standby electrical systems, or parallel power production systems.

Special permission is required for the installation of more than one service to a sufficiently large building or other structure. Expansion of buildings, shopping centers, or industrial plants often necessitates the addition of one or more services. It might, for example, be impractical or impossible to install one service for an industrial plant with sufficient capacity to compensate for any and all future load requirements. It is also impractical to run feeders for extremely long distances, due to high costs and voltage drop problems.

More than one service is allowed for different characteristics, such as different voltages, frequencies, single-phase or 3-phase services, or different utility rate schedules. For example, different service characteristics exist between a 3-wire, 120/240-V single-phase service and a 3-phase, 4-wire, 480Y/277-V service. For different applications, such as different rate schedules, it is intended to allow a second service for supplying a second meter on a different rate for equipment, such as an electric water heater.

A maximum of six size No. 1/0 or larger underground conductors that are connected together at their supply end and not at their load end are considered to be *one* service lateral. These conductors are permitted to supply up to six service disconnecting means installed at one location in a group, as required by *NEC* Section 230-72(a). This exception is included in the *NEC* to increase the cable impedance by not paralleling the conductors, thereby lowering the available short-circuit current at the building. It is not required that each set of service lateral conductors be of the same size, provided that each set of service lateral conductors is No. 1/0 AWG or larger. Note that this lateral arrangement is considered to be *one* service lateral. Therefore, the total number of disconnects cannot exceed six. (See Figure 3-1B.)

It is necessary to keep all feeder and branch-circuit conductors separated from service conductors. Service conductors are not provided with overcurrent protection where they receive their supply; they are protected against overload conditions at their load end by the service disconnect fuses or circuit breakers. The amount of current that could be imposed upon feeder or branch circuit conductors, should they be in the same raceway and a fault occurs, would be much higher than the ampacity of the feeder or branch-circuit conductors.

Means are provided for disconnecting all conductors in the building from the service entrance conductors. The principle is to permit disconnection of all conductors of the service at that location with no more than six operations of the hand. (See Figure 3-1C.) The limitation is applicable to all of the service disconnecting means located, or grouped, in one place. It does not include the services for fire pumps, emergency electrical systems, etc., which the *NEC* recognizes as being separate services for specific purposes.

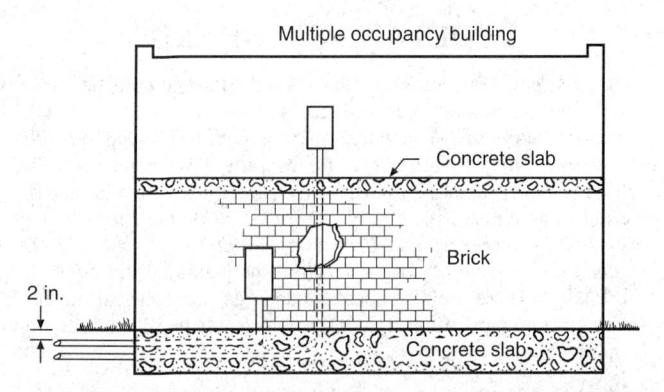

FIG. 3-1B. *Conductors are considered outside a building where installed under not less than 2 in. (51 mm) of concrete beneath a building or in a raceway encased by 2 in. (51 mm) of concrete or brick within a building. They are also considered outside the building where installed in a transformer vault.*

The disconnecting means is either an outside or inside installation, at a readily accessible location nearest the point of entrance of the service entrance conductors. Service entrance conductors must be of sufficient size to carry the calculated load.

No maximum distance is specified from the point of entrance of service conductors to a readily accessible location for the installation of a service disconnecting means. The authority enforcing the *NEC* has the responsibility for, and is charged with, making a decision as to how far inside the building the service entrance conductors are allowed to travel within the building to the main disconnecting means.

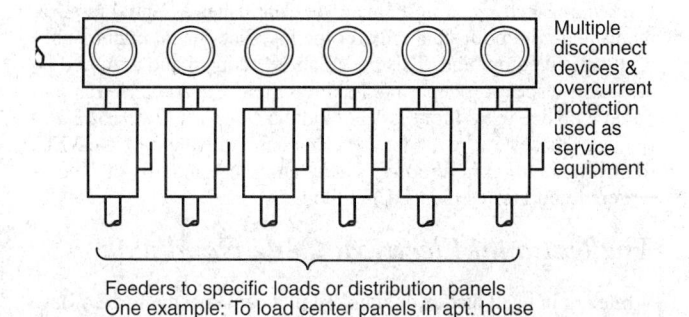

Feeders to specific loads or distribution panels
One example: To load center panels in apt. house

FIG. 3-1C. *Typical layout for multiple disconnecting means. The maximum number of switches is six at any one location, and the principle is to permit disconnecting all conductors of the service at that location with no more than six operations of the hand.*

The length of service entrance conductors should be kept to a minimum inside buildings, since power utilities provide limited overcurrent protection and, in the event of a "fault," the service conductors could ignite nearby combustible materials.

Some local jurisdictions have local ordinances that allow service entrance conductors to run within the building up to a specified length to terminate at the disconnecting means. The Authority Having Jurisdiction may permit service conductors to bypass fuel storage tanks or gas meters, etc., permitting the service disconnecting means to be located in a readily accessible location. However, if the authority judges the distance as being excessive, the disconnecting means may be required to be located on the outside of the building or near the building at a readily accessible location that is not necessarily nearest the point of entrance of the conductors.

One set of service entrance conductors, either overhead or underground, is permitted to supply two to six service disconnecting means in lieu of a single main disconnect. A single-occupancy building can have up to six disconnects for each set of service entrance conductors. Multiple-occupancy buildings (residential or other than residential) can be provided with one main service disconnect or up to six main disconnects for each set of service entrance conductors.

Multiple-occupancy buildings may have service entrance conductors run to each occupancy, and each such set of service entrance conductors may have from one to six disconnects.

Where service entrance conductors are routed outside the building, each set of service entrance conductors is permitted to supply not more than six disconnecting means at each occupancy of a multiple-occupancy building.

Service entrance conductors, overhead or underground, are the supply conductors between the point of connection to the service drop or service lateral conductors and the service equipment. Service equipment is intended to constitute the main control and means of cutoff of the electrical supply to the premises wiring system. At this point, an overcurrent device, which usually consists of circuit breakers or a set of fuses, is required to be installed in series with each ungrounded service conductor.

The service overcurrent device will not protect the service conductors under conditions of short circuit or ground fault on the line side of the disconnect. A degree of protection under these overcurrent conditions is provided by the special requirements for service conductor physical protection and location.

Figure 3-1D diagrams the wiring and components of a fire pump circuit designed to comply with requirements of the *NEC* and NFPA 20, *Standard for the Installation of Centrifugal Fire Pumps*. The pump motor represented is a 100-hp, 460-V, 3-phase, squirrel-cage induction type, and its full-load current, taken from *NEC* Table 430-150, is 124 A.

The service conductors, feeder conductors, and branch-circuit conductors are sized at 125 percent of the motor full-load current

rating (124 A at 125 percent = 155 A). According to *NEC* Table 310-16, the next closest copper wire size for 155 A is a No. 2/0 XHHW conductor with an allowable ampacity of 175 A; therefore, all of the conductors would be sized at not less than No. 2/0 XHHW copper. Where the voltage drop at the controller under starting conditions exceeds 15 percent, conductors with an ampacity greater than 155 A are necessary. The locked-rotor current of the motors in the installations illustrated in Figure 3-1D is 725 A, derived from *NEC* Table 430-151B. The power supply protection may consist of either fuses or circuit breakers sized to carry locked-rotor current. In either case, the next standard ampere rating for fuses or circuit breakers is 800 A.

Most fire pumps rated 100-hp and over would require a disconnecting means rated at 1,000 A or more. However, due to the emergency nature of their use, fire pumps are exempt from the ground-fault provisions below.

Ground-Fault Protection for Equipment

Ground-fault protection of equipment on services rated 1,000 A or more, operating at 480Y/277 V, was first required in the 1971 *NEC* because of the unusually high number of burndowns that were reported on this type of service. Alternating current, due to its sine wave characteristics, can be a dangerous source of ignition. Where the voltage is greater than 250 V to ground, even though the voltage sine wave decreases to zero, the voltage can cause a re-strike of the arc, causing continuation of burning at the ground-fault location.

Ground-fault protection of services will not protect the conductors on the supply side of the service disconnecting means, but is designed to provide protection from line-to-ground faults that occur on the load side of the service disconnecting means. An alternative to installing ground-fault protection may be to provide multiple disconnects rated less than 1,000 A. For instance, up to six 800 A disconnecting means may be used and, in this case, ground-fault protection would not be required. The second fine print note (FPN) in *NEC* Section 230-95(b) recognizes that ground-fault protection may be desirable at lesser amperages on solidly grounded systems for voltages exceeding 150 V to ground, but not exceeding 600 V phase-to-phase.

The maximum setting for ground-fault sensors is 1,200 A; however, there is no minimum, and it should be noted that setting at lower levels increases the likelihood of nuisance tripping. The *NEC* requirements place a restriction on fault currents greater than 3,000 A and limit the duration of the fault to not more than 1 sec. This will minimize the amount of damage done by an arcing ground fault, which is directly proportional to the time it is allowed to burn.

Where interconnection is made between multiple-supply systems, or multiple buildings, care should be taken to ensure that the interconnecting of different systems or buildings does not negate the proper ground-fault sensing by the ground-fault protection equipment. A careful engineering study should be made to ensure that fault currents do not take parallel paths back to the supply system, thereby bypassing the ground-fault detection device.

Physical Protection

Underground service conductors must be protected against physical damage. (See Figure 3-1E.) Tables 3-1F and 3-1G give minimum cover requirements for underground installations. ("Cover" is defined as the distance between the top surface of direct-buried cable, conduit, or other approved raceways and the finished grade.)

For circuits over 600 V, there are variations in burial depths shown in Table 3-1G, under certain circumstances. For example, the burial depth beneath areas subject to vehicular traffic is 24 in. (610 mm). Under airport runways and adjacent areas, the burial depth

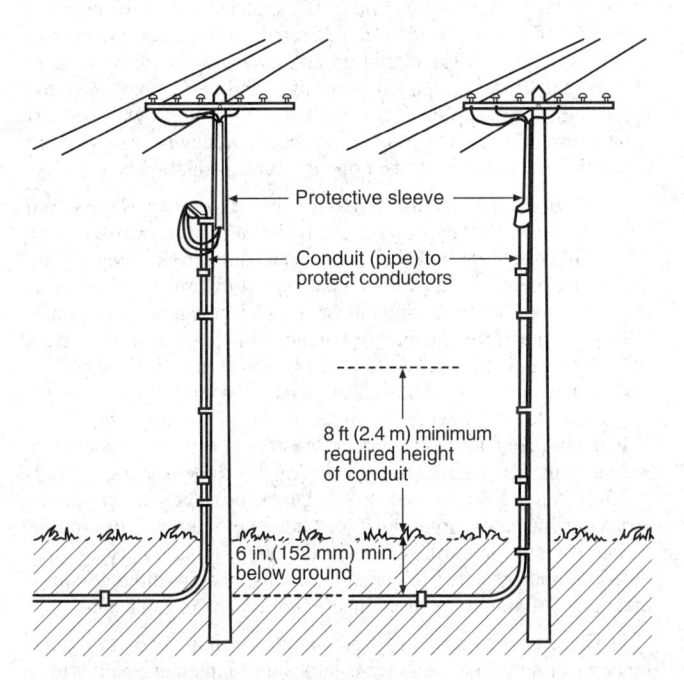

FIG. 3-1D.   Two different configurations permitted for connecting to electric-utility-owned service drops.

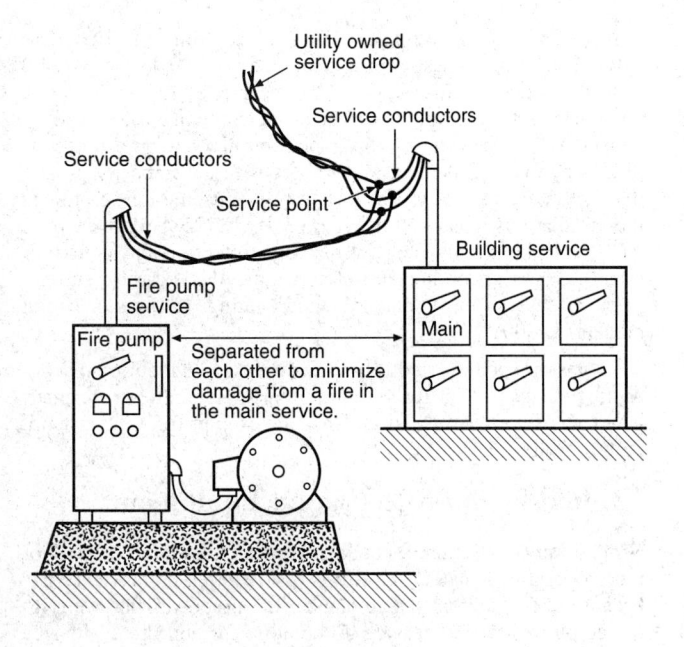

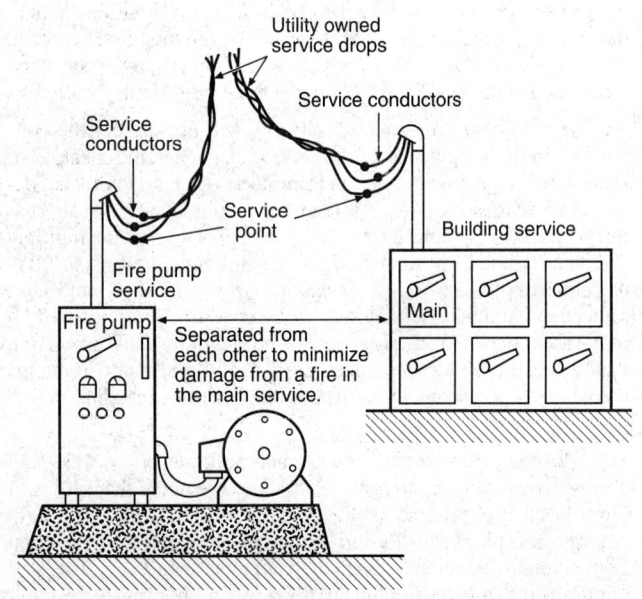

FIG. 3-1E. Service conductors from the outside pole are installed in rigid metal conduit underground from the overhead supply line. Note the service head to protect the conductors against the entrance of water and that the point of connection of the service entrance conductors is below the level of the service head.

must be a minimum of 18 in. (457 mm). Lesser burial depths are permitted where access is necessary, or where termination or splicing points are located.

Lightning (Surge) Arresters—Building Wiring

Lightning and surge arresters are not required but may be used, particularly where buildings are supplied by an overhead service and where thunderstorms are prevalent in the area. [See NFPA 780, *Lightning Protection Code* (hereinafter referred to as NFPA 780)]. Surge arresters are required to be installed on electric and telephone services, and on radio and television antenna lead-ins. Installation of such devices serves to minimize damage to electronic equipment, since equipment using solid-state electronic components is very sensitive to voltage surges.

Each lead-in from an outdoor antenna to a radio or television set or to an amateur radio transmitting station needs protection from lightning. Antenna discharge units should be installed to provide lightning protection for the equipment. Additional protection from power line surges or from connected communications equipment may also be desirable.

Grounding Requirements—Building Wiring

Dangerous voltages, which may be a fire hazard and a personal injury (shock) hazard, may be imposed on electric distribution systems and equipment. These voltages may be caused by lightning, inadvertent contact of low-voltage conductors with a high-voltage primary system, breakdown of insulation in transformers, surface leakage due to dirt or moisture, or by a wire coming loose from its connection.

Grounding one conductor of the electric system and then grounding all exposed noncurrent-carrying metal parts of an electrical installation that may accidentally come in contact with an ungrounded circuit conductor ensures that a low-resistance path is provided for the fault-current to actuate the fuse or circuit breaker. The fault current finds a path through the equipment grounding conductor and can cause the operation of an overcurrent device (fuse or circuit breaker) in the ungrounded circuit conductor, thereby eliminating the dangerous condition. The impedance of the fault path must be low enough to permit sufficient current flow to cause the overcurrent device to operate quickly.

The metal enclosures of conductors (metal armor of cables, metal raceways, boxes, cabinets, and fittings) must be grounded. (Article 250 of the *NEC* contains general requirements for grounding electrical installations.)

The *NEC* also requires grounding of exposed noncurrent-carrying metal parts of cord- and plug-connected equipment likely to become energized in hazardous (classified) locations, where operated by persons standing on the ground or metal surfaces, or where operated at more than 150 V to ground, except for guarded motors. In some isolated cases, the metal frames of electrically heated appliances should also be grounded. The *NEC* also specifies grounding of particular cord- and plug-connected appliances in residential occupancies. These appliances include refrigerators, freezers, air conditioners, clothes washers and dryers, dishwashers, sump pumps, electric aquarium equipment, portable hand-held motor-operated tools, and some appliances, e.g., hedge clippers, lawn mowers, snow blowers, and wet scrubbers. Appliances may be exempt from grounding if equivalent protection via double insulation is provided.

Where available, a metallic underground water piping system that has at least 10 ft (3.1 m) of pipe in the earth must be used as the grounding electrode. However, a metal underground water pipe grounding electrode must be supplemented by an additional electrode, in the event that the metal water pipe is replaced by a nonmetallic pipe; therefore, the integrity of the grounding electrode system will be maintained. In addition to the above, the following electrodes, if available, must be bonded together to make up the grounding electrode system: (1) the metal frame of the building where effectively grounded, (2) a concrete-encased electrode, and (3) a ground ring. An example of a grounding electrode is shown in Figure 3-1F. Piping should be inspected to ensure that there are no intervening insulating joints of nonconducting materials that might isolate it from the earth. Effective bonding jumpers must be provided around insulated joints and sections, and around any equipment that is likely to be disconnected for repairs or replacement.

The electrical system service neutral conductor is grounded by connecting it to a metal water pipe and supplementing it with a driven ground rod. The metal parts of the service equipment are allowed to be grounded by bonding to the neutral (grounded) con-

TABLE 3-1F. *Minimum Cover Requirements, 0 to 600 Volts, Nominal, Burial in Inches*
(Cover is defined as the shortest distance measured between a point on the top surface of any direct buried conductor, cable, conduit, or other raceway and the top surface of finished grade, concrete, or similar cover.)

	Type of Wiring Method or Circuit				
Location of Wiring Method or Circuit	1 Direct Burial Cables or Conductors	2 Rigid Metal Conduit or Intermediate Metal Conduit	3 Nonmetallic Raceways Listed for Direct Burial without Concrete Encasement or Other Approved Raceways	4 Residential Branch Circuits Rated 120 Volts or Less with GFCI Protection and Maximum Overcurrent Protection of 20 Amperes	5 Circuits for Control of Irrigation and Landscape Lighting Limited to Not More than 30 Volts and Installed with Type UF or in Other Identified Cable or Raceway
All Locations Not Specified Below	24	6	18	12	6
In Trench Below 2-in. Thick Concrete or Equivalent	18		12	6	6
Under a Building	0 (in raceway only)	0	0	0 (in raceway only)	0 (in raceway only)
Under Minimum of 4-in. Thick Concrete Exterior Slab with No Vehicular Traffic and the Slab Extending Not Less than 6 In. Beyond the Underground Installation	18	4	4	6 (direct burial) 4 (in raceway)	6 (direct burial) 4 (in raceway)
Under Streets, Highways, Roads, Alleys, Driveways, and Parking Lots	24	24	24	24	24
One- and Two-Family Dwelling Driveways and Outdoor Parking Areas, and Used Only for Dwelling-Related Purposes	18	18	18	12	18
In or Under Airport Runways, Including Adjacent Areas Where Trespassing Prohibited	18	18	18	18	18

Note 1. For SI units: 1 in. = 25.4 mm.
Note 2. Raceways approved for burial only where concrete encased shall require concrete envelope not less than 2 in. thick.
Note 3. Lesser depths shall be permitted where cables and conductors rise for terminations or splices or where access is otherwise required.
Note 4. Where one of the wiring method types listed in columns 1–3 is used for one of the circuit types in columns 4 and 5, the shallower depth of burial shall be permitted.
Note 5. Where solid rock is encountered, all wiring shall be installed in metal or nonmetallic raceway permitted for direct burial. The raceways shall be covered by a minimum of 2 in. of concrete extending down to rock.
Source: NFPA 70-1996, *National Electrical Code.*

TABLE 3-1G. *Minimum Cover Requirements (Cover is defined as the shortest distance in inches measured between a point on the top surface of any direct buried conductor, cable, conduit, or other raceway and the top surface of finished grade, concrete, or similar cover.)*

Circuit Voltage	Direct Buried Cables	Rigid Nonmetallic Conduit Approved for Direct Burial*	Rigid Metal Conduit and Intermediate Metal Conduit
Over 600-22 kV	30	18	6
Over 22kV-40 kV	36	24	6
Over 40 kV	42	30	6

For SI units: 1 in. = 25.4 mm.
*Listed by a qualified testing agency as suitable for direct burial without encasement. All other nonmetallic systems require 2 in. (51 mm) of concrete or equivalent above conduit in addition to above depth.
Source: NFPA 70-1996, *National Electrical Code.*

ductor. The service enclosure is bonded to the grounded neutral conductor at point "B." This connection is often a removable strap or green-colored screw. The service raceway (conduit) is grounded through the metal of the panelboard. Since service equipment is not protected by an overcurrent device, a bonding jumper is one of the methods employed to ensure electrical continuity at service equipment; this electrical connection is indicated at "A." The meter box on the line side of the service equipment is usually grounded by means of the neutral terminal being bonded directly to the meter box, and would be required where service-entrance cable is used. The upper portion of the service raceway or cable is grounded through threaded connections to the meter box where metal conduit or electrical metallic tubing is used. The conduit between the meter and panelboard may be grounded by bonding to the meter enclosure, or bonding to the panelboard. The branch circuit equipment grounding conductor is connected to ground through the equipment grounding terminal block, which is grounded at point

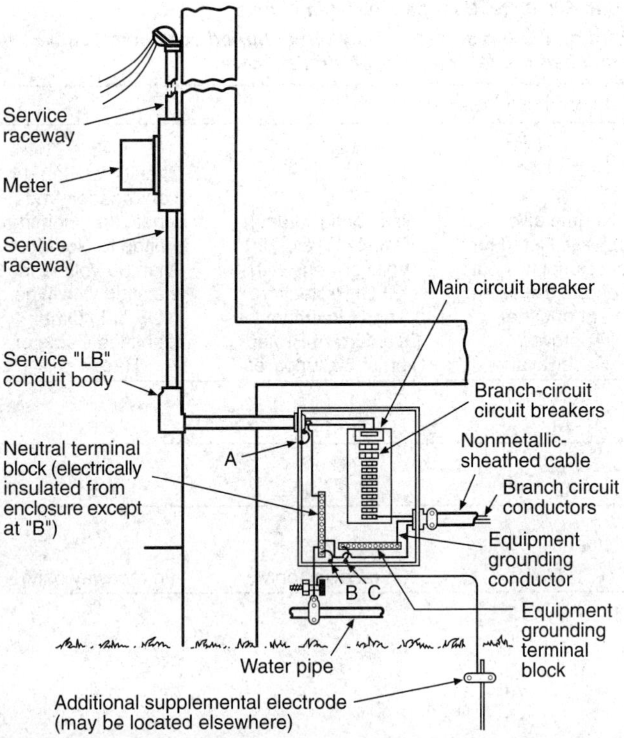

FIG. 3-1F.   Grounding at a typical small service [ac, single-phase, three-wire (120/240 V)].

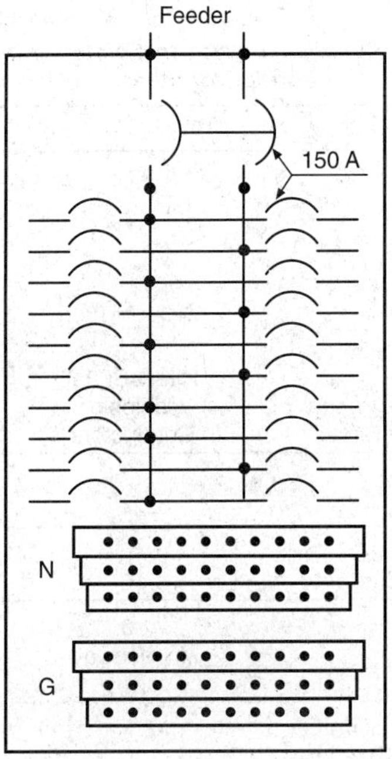

FIG. 3-1G.   One main provided (two are permitted). Grounding terminal block may not be required, depending upon wiring system used.

"C." This connection is commonly the mounting screw for this terminal block. The grounding connection shown is to be an underground cold-water piping system. The connection should be as required by the *NEC*.

Panelboards and Overcurrent Protection—Building Wiring and Equipment

Panelboards: A panelboard consists of an assembly that may include switches, fuses, or circuit breakers that are mounted on copper or aluminum bus-bars, whose function is the control and protection of light, heat, or power circuits. The buses are mounted in a cabinet or cutout box, which may be placed in or on a wall and is accessible only from the front.

The *NEC* does not allow a lighting and appliance branch circuit panelboard to have more than 42 overcurrent devices installed in any one cabinet or cutout box. Lighting and appliance panelboards are not allowed to have more than two main circuit breakers or two main sets of fuses to protect the buses in a lighting and branch-circuit panelboard. (See Figures 3-1G and 3-1H.) Figures 3-1I and 3-1J show exceptions to the general rule for individual protection of panelboards.

Panelboards installed in wet locations need weatherproof cabinets and must be mounted so that there is at least $1/4$ in. (6 mm) of air space between the cabinet and the wall or other supporting surface. Where located in corrosive atmospheres, equipment must be suitable for the conditions. Special requirements in Chapter 5 of the *NEC* govern installation of panelboards in hazardous (classified) locations.

Overcurrent protection: Conductors are provided with overcurrent protection to open the circuit, if the current is greater than the value that will cause the conductor temperature to exceed the tem-

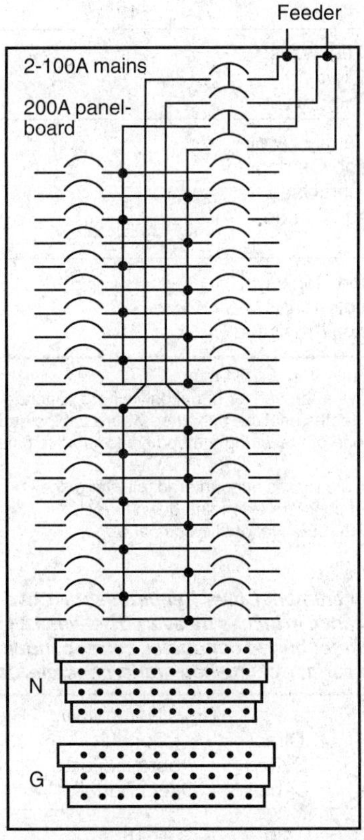

FIG. 3-1H.   Two mains required. Grounding terminal block may not be required, depending upon wiring system used.

perature rating of the conductor insulation.   Equipment also needs overcurrent protection to minimize the effects of an electrical fault within the equipment, thereby reducing the amount of arcing and

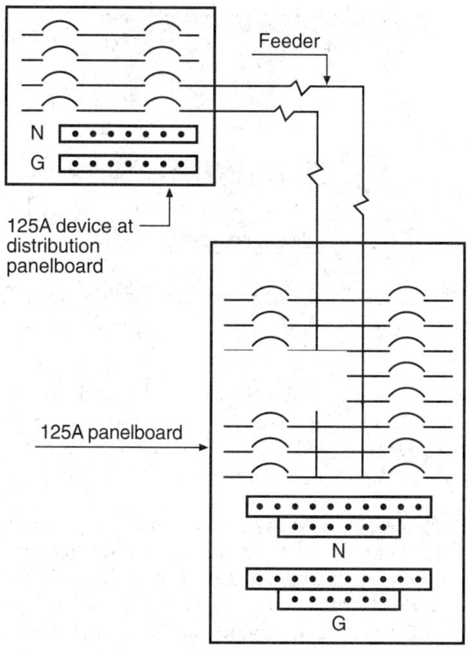

FIG. 3-1I. No mains required. Grounding terminal block may not be required, depending upon wiring system used.

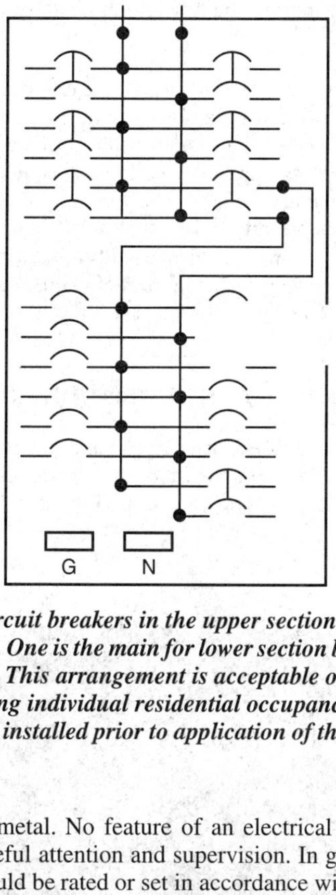

FIG. 3-1J. Circuit breakers in the upper section are the service circuit breakers. One is the main for lower section buses supplying branch circuits. This arrangement is acceptable only for the service to an existing individual residential occupancy where the panelboard was installed prior to application of the 1981 NEC requirements.

ejection of hot metal. No feature of an electrical installation deserves more careful attention and supervision. In general, overcurrent devices should be rated or set in accordance with the tables for conductor ampacity in the *NEC*. They should be installed in each ungrounded conductor of each circuit and feeder, at the point where the conductor to be protected receives its supply. (See Tables 3-1A through 3-1D.) Exceptions for fixture wires, cords, taps, motor circuits, remote-control circuits, and fire protective signaling circuits are covered in Articles 240, 430, 725, and 760 of the *NEC*.

Overcurrent Protective Devices

The most commonly used overcurrent protective devices for feeders, circuits, and equipment are fuses and circuit breakers. (Overcurrent and undervoltage relays, etc., are used on high-voltage, high-current systems). Thermal motor running overload sensing devices are not considered as overcurrent protection, but are used as overload protection.

Plug fuses: There are two basic types of plug fuses: (1) the ordinary Edison-base type and (2) Type S. Either of these may or may not be of the time-delay type. An Edison-base fuseholder will take an Edison-base fuse of any size up to a maximum 30 ampere rating. Plug fuses of the Edison-base type are allowed only for replacements in existing installations where there is no evidence of tampering.

The Type S fuse is designed to minimize tampering or bridging. Adapters are available that fit the Edison-base fuseholders. After an adapter has been properly installed, it cannot be removed without damaging the fuseholder. The adapters are designed to prevent using Edison-base fuses in the fuseholder and to prevent using larger (higher rated) Type S fuses in an adapter designed for a lower rating. They also prevent the use of pennies in the fuseholder and other common bypassing schemes.

Time-delay plug fuses: Whether Type S or Edison-base design, time-delay fuses permit short-time current surges, such as motor starting currents, without interruption of the circuit. These momentary surges are harmless unless repeated a number of times at short intervals. This makes it possible to use Type S fuses in sizes small enough to give better protection than a non-time-delay type, which must be oversized to allow for such surges. In the case of a short circuit or a high-current fault, however, the time-delay type will operate and clear the circuit as rapidly or even more rapidly than the non-time-delay type. Some representative plug fuses are shown in Figures 3-1K, 3-1L, and 3-1M.

Cartridge fuses: These are of both the time-delay and the non-time-delay types. They are also of the one-time and the renewable link types. When a one-time fuse opens, the entire fuse must be replaced. But when a renewable link fuse opens, the fuse link can be replaced inside the same cartridge unless the cartridge has also been damaged.

Renewable link fuses have two disadvantages: (1) the links can be doubled or tripled, etc., thereby defeating their purpose and usefulness, and (2) the links, upon replacement, can be left with loose connections. Several representative cartridge fuses are shown in Figures 3-1N and 3-1O.

Circuit breakers: There are two basic types of breakers: (1) adjustable trip and (2) nonadjustable trip. The adjustable-trip type may be of either the air- or oil-immersed type. The setting of the trip point

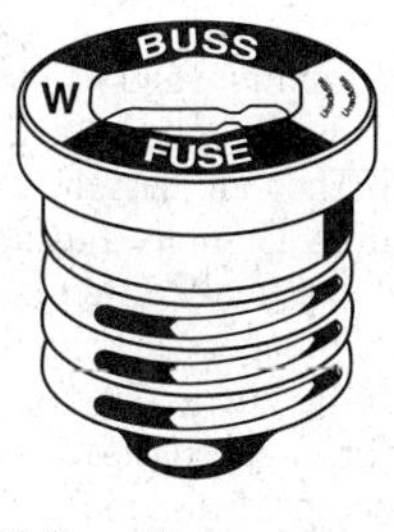

FIG. 3-1K. A typical Edison-base nonrenewable, single-element fuse. (Source: Bussman Mfg. Div., McGraw-Edison Co.)

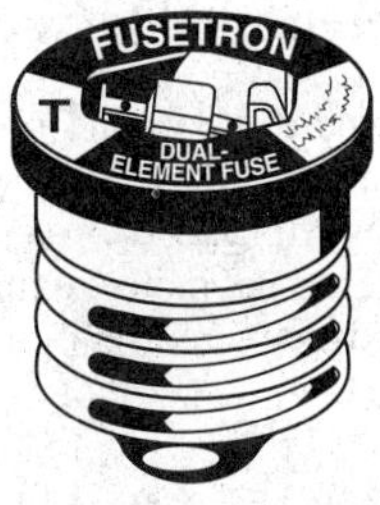

FIG. 3-1L. An Edison-base nonrenewable, dual-element fuse. (Source: Bussman Mfg. Div., McGraw-Edison Co.)

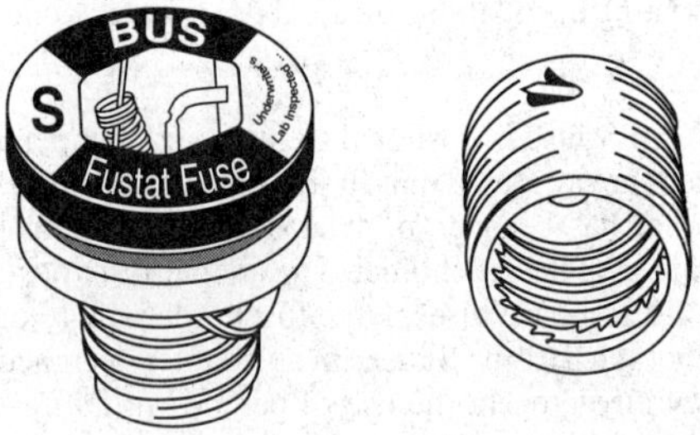

FIG. 3-1M. A Type S nonrenewable fuse. The time-lag type of fuse shown is acceptable but not required by the NEC. These fuses have been designed to minimize tampering or bridging. The NEC specifies that fuseholders for plug fuses of 30 ampere or less must not be used unless they are designed to accept Type S fuse or to accept a Type S fuse through use of an adapter. (Source: Bussman Mfg. Div., McGraw-Edison Co.)

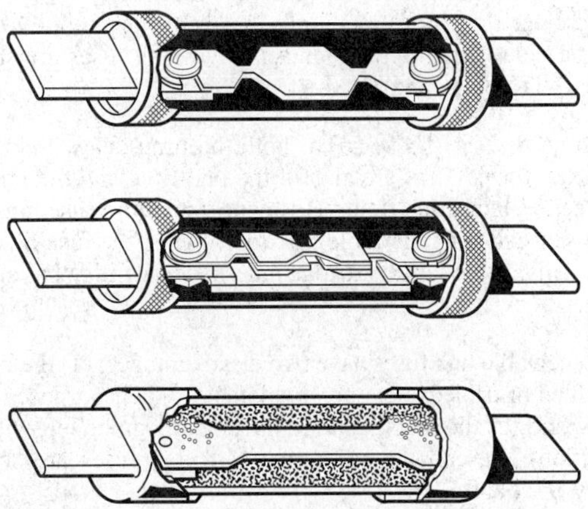

FIG. 3-1N. Three types of cartridge fuses: at top, an ordinary dropout link renewable fuse; at center, a super-lag renewable fuse; and at bottom, a one-time fuse. (Source: Bussman Mfg. Div., Cooper Industries)

is adjustable between a minimum and a maximum range and is usually used only on large installations having qualified operators and maintenance personnel. They are designed to trip when the current reaches that of the setting. The nonadjustable-trip type comes in a molded case, making it extremely difficult or impossible to change its rating. A molded-case breaker can be of the inverse-time thermal-magnetic type, or of the magnetic-only inverse-time type. Inverse-time thermal-magnetic breakers consist of two elements: (1) an electro-magnetic trip element that trips quickly on high fault current and (2) a bimetallic trip element that provides a time delay. Molded-case circuit breakers are designed so that the current has to exceed the rat-

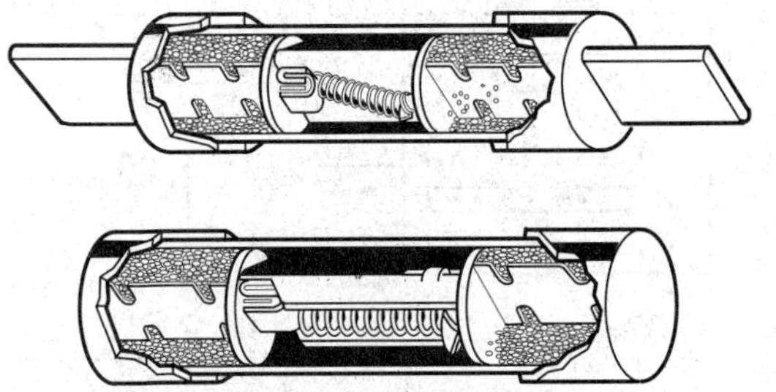

FIG. 3-1O. A dual-element cartridge fuse, blade and ferrule type. (Source: Bussman Mfg. Div., Cooper Industries)

ing (as is also true with all types of fuses) before it will trip. Unless it is the thermally compensated type, high ambient temperatures can reduce the current required to trip the circuit breaker (or fuse). A representative nonadjustable circuit breaker is shown in Figure 3-1P.

Thermal devices: These are not intended for protection against short circuits, but only for the protection of overload currents of a comparatively lower magnitude unless otherwise designed. An example is overload protection for a motor where short-circuit and ground-fault protection is provided by the branch-circuit fuses or circuit breakers, and motor running overload protection is provided by the thermal relays. Another thermal device is that used in recessed lighting fixtures, which will temporarily open the circuit when the fixture exceeds a predetermined temperature. For detailed requirements for motors and other equipment, see Articles 410 and 430 of the *NEC*.

Current-limiting overcurrent protective devices: These devices are provided in some installations where high current (more than 10,000 amperes) is available under short-circuit conditions. They are designed so that, when interrupting a specified current, the devices will consistently limit the short-circuit current in that circuit to a magnitude substantially less than that obtainable in the same circuit if the device were replaced with a solid conductor having comparable impedance. (See Figure 3-1Q.) Fuseholders for current-limiting fuses are designed to make it difficult to install a noncurrent-limiting fuse.

Ground-Fault Circuit-Interrupters

Circuit breakers and fuses open a circuit and stop the flow of electricity when the flow exceeds the ratings of the circuit breaker or fuse. Lighting and receptacle circuits are usually rated 15 or 20 amperes.

FIG. 3-1P. A 15-ampere, single-pole, 120-V branch-circuit breaker. (Source: Westinghouse Electric Corp.)

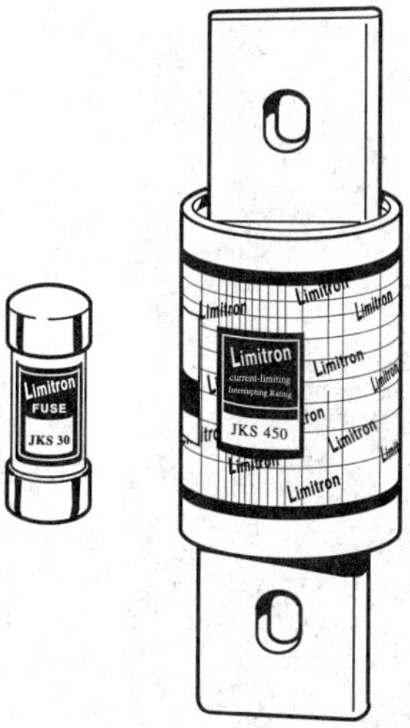

FIG. 3-1Q. A current-limiting overcurrent protection device (rated 0 to 600 amperes). (Source: Bussman Mfg. Div., McGraw-Edison Co.)

Under certain conditions, as little as 20 mA (milliamperes) can kill a normal healthy adult; thus, the rating of lighting and receptacle circuits is high, relative to the amount of current that could kill a person.

Ground-fault circuit-interrupters (GFCI) are devices that sense when even a small amount of current passes to ground through any path other than the proper conductor. When this condition exists, the GFCI trips almost instantly, stopping all current flow in the circuit and through the person receiving the ground-fault shock.

Simplicity of design is one reason for the reliability of the GFCI. Figure 3-1R shows a typical circuit arrangement of a GFCI for personnel protection. Figure 3-1S shows types of GFCI units and a GFCI tester.

Present GFCIs are set to operate when line-to-ground currents exceed 5 mA. Evaluation standards permit a differential of 4 to 6 mA. Even at trips of 5 mA, it should be clearly understood that the instantaneous current will be higher, and any shock during the time the fault is being cleared will be uncomfortable. A shock at 5 mA is unpleasant. The key to the ground-fault circuit-interrupter is the time-current characteristic. Trip-out time is about $1/40$ of a sec (25 milliseconds) when the fault reaches or exceeds 6 mA.

There are two classes of GFCIs—Class A and Class B. The Class A GFCI trips at not more than 6 mA. The Class B GFCI trips when the current exceeds 20 mA, and is for use only with swimming pool underwater lighting fixtures that were installed prior to local adoption of the 1965 edition of the *NEC*. The Class A GFCI is used where personnel protection against ground fault is required or desired.

GFCIs at construction sites: Precautions must be taken to aid the efficient operation of GFCIs on construction sites. Most laboratory-

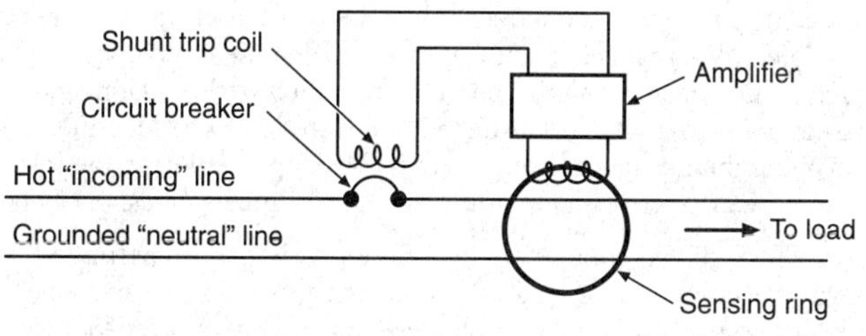

FIG. 3-1R. Circuit arrangement for a typical ground-fault circuit-interrupter for personnel protection. (Source: I-T-E Imperial)

tested 120-V appliances have 0.5 mA leakage or less under normal operating conditions. However, moisture and improper maintenance on portable handheld tools, which is common at construction sites, can create conditions under which GFCIs can be expected to trip. Flexible cords with standard attachment plug cap and connector when dropped in water, e.g., puddles, can be expected to cause leakage currents (100 to 300 mA or greater), far in excess of GFCI trip currents. Motors with dirty brushes, carbon tracking on commutators, or moisture in the windings contribute to leakage current. A commonsense approach to installing, using, and maintaining GFCI circuits goes a long way toward eliminating nuisance tripping at construction sites. (Actually, tripping under any of the conditions mentioned is not nuisance tripping, but merely a device performing its intended function.)

Moisture is the major culprit in current leakage on wiring and equipment. Panelboards, receptacles, and cord caps and connectors intended for dry locations must not be subjected to wet conditions. Construction receptacles must be centrally located to enable cords 150 ft (46 m) or less in length to be used. There should be sufficient circuits available to minimize the number of tools necessary on each circuit. Receptacles must not be on the same circuit as lighting.

GFCIs in residential and commercial occupancies: The *NEC* requires ground-fault protection for personnel (i.e., a GFCI) on all 125-V, single-phase 15- and 20-ampere receptacles installed in the following locations in dwelling units: bathrooms, garages and grade-level portions of unfinished accessory buildings used for storage or work areas, outdoors, unfinished basements, crawl spaces that are at or below grade level, receptacles that are installed to serve kitchen countertop surfaces, and receptacles that are located within 6 ft. (1.8 m) of the outside edge of a wet-bar sink.

A single or duplex receptacle does not need GFCI protection if: (1) it is located within a dedicated space for an appliance that is not easily moved from one location to another or (2) it is not readily accessible and supplied by a dedicated circuit for snow-melting or de-icing equipment. In other than dwelling units, receptacles that are installed must be GFCI protected in all bathrooms and rooftops.

All 125-V receptacles, whether of the parallel blade or twist locking type, that are located within 20 ft (6.1 m) of a swimming pool, fountain, or outdoor spa or hot tub are required to be GFCI protected. Where spas or hot tubs are installed indoors, receptacles that provide power to the spa or hot tub must be GFCI protected. Hydromassage bathtubs and their associated equipment are required to be GFCI protected, as well as any receptacles within 5 ft (1.5 m) of the tub.

In addition, in all occupancy types, non-GFCI-protected receptacles that are replaced due to damage or other causes that were not previously required to be GFCI protected under previous *NEC* editions must be GFCI protected if the present *NEC* requires it.

Types of Wiring Methods and Materials

The *NEC* recognizes a number of standard wiring methods, some of which are suitable for general use while others are suitable only for special purposes. In some localities, regulations restrict the use of some of the wiring methods recognized by the *NEC*. The most widely used wiring methods are rigid metal conduit; rigid nonmetallic conduit; intermediate metal conduit (IMC); electrical metallic tubing (EMT); metal-clad cable (MC); armored cable (AC; trade name: "BX"); nonmetallic sheathed cable (NM, NMC; trade name "Romex"); surface metal and nonmetallic raceways; wireways; busways; underfloor raceways; and cellular metal floor raceways. (See Figures 3-1T through 3-1W.) In some areas of the country, flexible metal conduit (trade name: "Greenfield") is used for complete wiring systems. None of these wiring methods should be used for

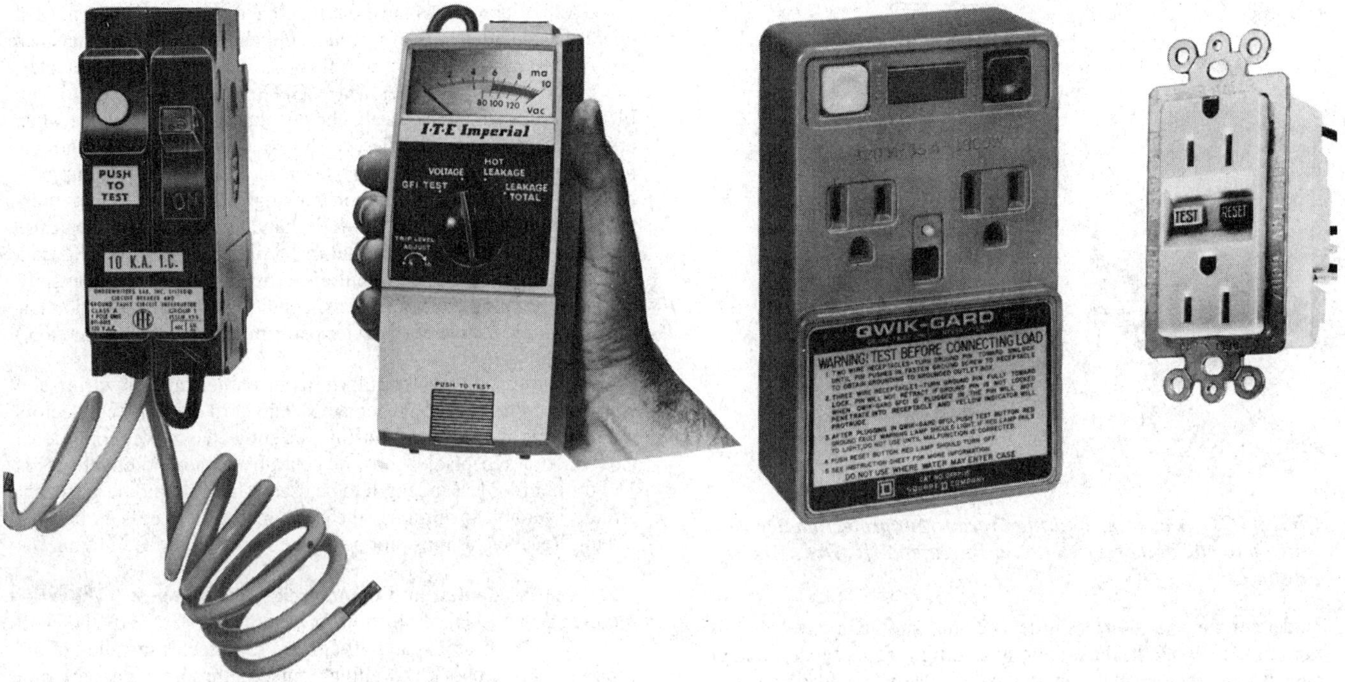

FIG. 3-1S. A variety of GFCIs are available, including plug-in-circuit-breaker types and duplex-receptacle types. These also come in portable units, and each has a test switch so that the unit can be checked periodically to ensure continuous proper operation. (Sources: I-T-E Imperial, Square D Co.)

either temporary or permanent work where they do not fully conform to *NEC* safety requirements.

Identification of Conductors, Terminals, Circuits, and Branch Circuits

With few exceptions, all interior wiring systems have a grounded circuit conductor that is continuously identified throughout. The identification for conductors of No. 6 or smaller (except for Type MI cable) should consist of an outer white or natural gray color. Insulated conductors larger than No. 6 should have similar identification or be identified by distinctive white markings at their terminations during installation. The grounded conductors of Type MI cable are also identified by distinctive markings at their terminations during installation.

FIG. 3-1T. A Type UF sheathed cable. The insulated conductors are jacketed in parallel and the overall covering must be flame retardant, moisture resistant, fungus resistant, corrosion resistant, and suitable for direct burial in the earth. (Source: Plastic Wire & Cable Corporation)

FIG. 3-1U. A typical armored cable (Type AC) that is frequently referred to as "BX." Another type (ACL) has lead-covered conductors and is often referred to as "BXL." (Both "BX" and "BXL" are trade names.) (Source: General Electric Company)

In general, the terminals of electrical devices to which a grounded conductor is to be connected are identified by being made of metal substantially white in color, have a metallic plating substantially white in color, or the word "white" located adjacent to the terminal. If the terminal is not visible, the conductor entrance hole may be colored white. In the case of screw-shell-type lampholders, the white terminal is the one that is connected to the screw shell.

Where a terminal is provided for the connection of an equipment grounding conductor, it must be designated by a green-colored hexagonal-head screw or nut; or, if the terminal is not visible, the conductor entrance hole must be marked with the word "green," or the letters "G" or "GR," or the grounding symbol.

Lighting and Appliance Branch Circuits

A branch circuit is that portion of a wiring system extending between the final overcurrent device protecting the circuit and the outlets on the circuit. Branch circuits are further defined as: (1) a small-appliance branch circuit supplying one or more outlets to which appliances are to be connected that has no permanently connected lighting fixtures other than those that are part of the appliance, (2) a general-purpose branch circuit supplying two or more outlets for lighting and appliances, and (3) an individual branch circuit supplying only one piece of equipment.

The equipment grounding conductor of a circuit can be bare or insulated. Where it is insulated, it must be identified by a continuous green color or green with one or more yellow stripes. Conductors larger than No. 6 that are insulated or covered must be identified by stripping the insulation from the exposed length, coloring the insulation green, or covering it with green tape or green adhesive labels.

Ungrounded conductors of different voltages are permitted to be of any color other than green, white, or gray. In some special applications, e.g., isolated (ungrounded) power systems used in hospital operating rooms, the circuits are identified as follows: isolated conductor No. 1—orange, and isolated conductor No. 2—brown.

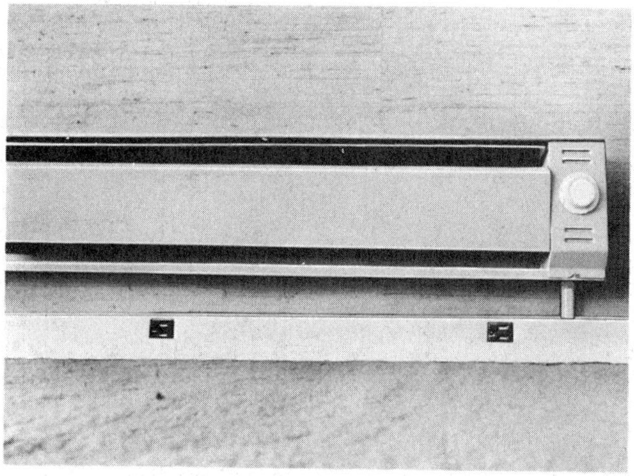

FIG. 3-1V. Installation of a typical metal raceway system showing an all-steel baseboard and a multi-outlet system. (Source: The Wiremold Company)

FIG. 3-1W. The use of a surface metal raceway in a domestic kitchen installation. Frequent spacing of outlets eliminates interference from cords and reduces risk of injury. (Source: The Wiremold Company)

Lampholders, when connected to circuits having a rating of more than 20 amperes, must be of the heavy-duty type with a rating of not less than 660 watts if of the admedium type, and not less than 750 watts if of any other type.

Receptacles, when connected to circuits having two or more receptacles or outlets, must not supply loads greater than the values given in Table 3-1H. They must not be supplied by circuits of ratings greater than the values listed in Table 3-1I. Where larger than 50 amperes, the receptacle rating is not permitted to be less than the branch circuit rating, except for motor loads.

Individual branch circuits may supply any loads. Branch circuits having two or more outlets may supply only loads as follows.

Branch circuits are for lighting units or appliances rated at 15 and 20 amperes. The rating of any one cord- and plug-connected appliance must not exceed 80 percent of the branch circuit ampere rating. The total rating of fixed appliances must not exceed 50 percent

TABLE 3-1H. *Maximum Cord- and Plug-Connected Load to Receptacle*

Circuit Rating (Amperes)	Receptacle Rating (Amperes)	Maximum Load (Amperes)
15 or 20	15	12
20	20	16
30	30	24

Source: NFPA 70-1996, *National Electrical Code.*

of the branch circuit ampere rating when lighting units or other appliances are also supplied.

Branch circuits rated at 40 and 50 amperes are permitted to supply fixed lighting units with heavy-duty lampholders, infrared heating units, or other utilization equipment in other than dwelling units. Branch circuits rated larger than 50 amperes may supply only nonlighting loads.

Receptacle outlets in dwelling occupancies, including guest rooms in hotels and motels, must be installed in every kitchen, family room, dining room, living room, parlor, library, den, sun room, bedroom, recreation room, or similar rooms. Insofar as practical, the outlets should be spaced equal distances apart. (See Figure 3-1X.)

Most appliances and portable lamps are provided with flexible cords at least 6 ft (1.8 m) in length. Countertop appliances for use in the kitchen are usually supplied with 2-ft (0.6-m) cords. Therefore, the *NEC* requires receptacles to be installed so that no point along the wall line is more than 24 in. (0.6 m), measured horizontally, from a receptacle. All receptacles that serve countertop spaces are required to be GFCI protected. (See Figure 3-1Y.)

Receptacles installed on 15- and 20-amperes branch circuits must be of the grounding type and must be effectively grounded. This does not mean that all cord- and plug-connected equipment must be of the grounded type. (See the current *NEC* for specific details on grounding.)

TABLE 3-1I. *Receptacle Ratings for Various Size Circuits*

Circuit Rating (Amperes)	Receptacle Rating (Amperes)
15	Not over 15
20	15 or 20
30	30
40	40 or 50
50	50

Source: NFPA 70-1996, *National Electrical Code.*

Calculation of Loads

The methods specified in the *NEC* provide a basis for calculating the expected branch circuit and feeder loads and for determining the number of branch circuits required. Where in normal operation the maximum load of a branch circuit will continue for 3 hrs or more, such as store or office lighting, the loads specified must be increased by 25 percent.

The minimum unit lighting loads in volt-amperes per square foot for various types of occupancies are given in Article 220 of the *NEC*. In determining the load, open porches and garages in connection with dwelling occupancies are not included. Unfinished and unused spaces in dwellings need not be included unless adaptable for future use.

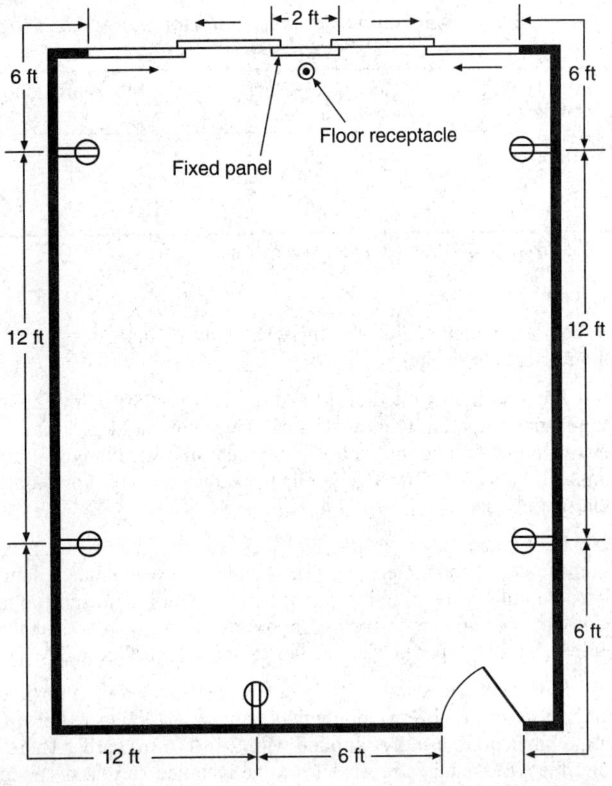

FIG. 3-1X. Illustration of the requirements in the NEC on the placement of receptacles in dwelling occupancies. Receptacles should be installed so that no point along the floor line is more than 6 ft (1.8 m), measured horizontally, from an outlet so as to permit a lamp or appliance equipped with a 6-ft (1.8-m) cord to be located anywhere along the wall, in the room. (For SI units: 1 ft = 0.305 m.)

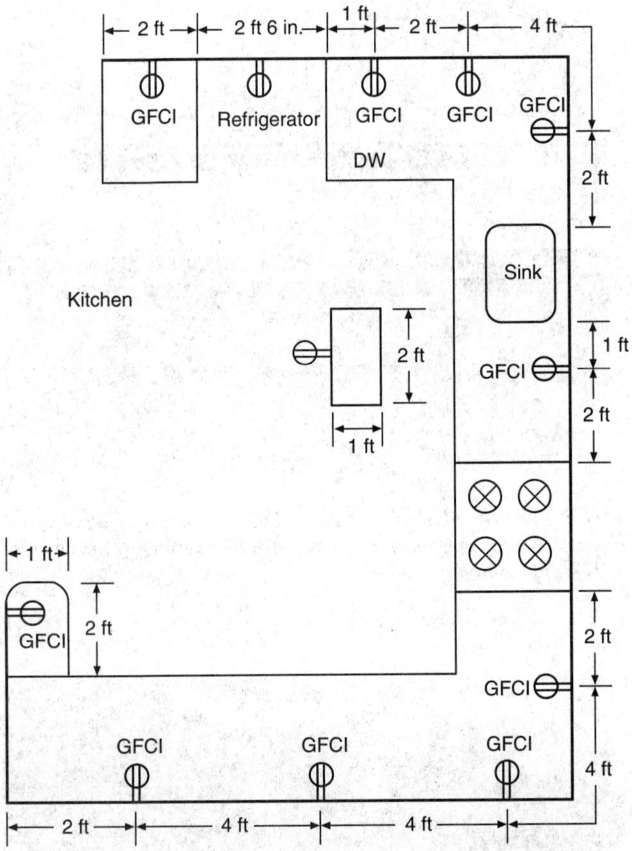

FIG. 3-1Y. Illustration of the requirements in the NEC on the placement of receptacles at countertops and islands in kitchens and dining areas of dwelling units. (For SI units: 1 in. = 25.4 mm; 1 ft = 0.305 m.)

For other than general illumination lighting, general-use receptacles, appliances, and loads other than motor loads, the following minimum unit load should be included for each outlet:

1. Outlets supplying specific appliances and other loads—ampere rating of appliance or load served.
2. Outlets supplying heavy-duty lampholders—600 VA.
3. Other outlets—180 VA. (This does not apply to the small-appliance branch circuit.)

Feeder Loads

Feeder conductors must have sufficient ampacity to supply the load served. The computed load of a feeder must not be less than the sum of all branch-circuit loads supplied by the feeder, subject to the demand factor provisions of the *NEC*. Tables of demand factors for general illumination, electric ranges and other cooking appliances, and for clothes dryers are found in the *NEC*. Table 3-1J lists the demand factors that may be applied to that portion of the total feeder load computed for general illumination.

An optional method of calculating the load for a single dwelling unit served by a 120/240V or 208/120V-system, with feeder or service conductors rated 100 amperes or greater, is to use the percentage values provided in Table 3-1K.

Examples of branch circuit and feeder calculations for other buildings and occupancies are given in the *NEC*.

These calculations are not applicable to receptacle outlets as required by the *NEC* for dwelling units. The 20-ampere small-appliance branch circuits and laundry branch circuits in dwelling units are calculated at 1,500 VA per circuit. Other receptacle outlets rated 20 amperes or less in dwelling units are considered part of the general lighting load, based on the area and calculated at 3 VA per sq ft. (For SI units: 1 sq ft = 0.0929 m².) For receptacle outlets in other than dwelling units, each single or multiple receptacle on a single yoke or strap is considered at not less than 180 VA.

An example from the *NEC* of how to calculate the load in a single-family dwelling is presented to illustrate the principles involved.

For a dwelling that has a floor area of 1,500 sq ft (139 m²) (exclusive of an unoccupied cellar, unfinished attic, and open porches), a 12-kW range, and 5-kW clothes dryer, the load may be computed as follows:

General Lighting Load: 1,500 sq ft (139 m²) at 3 volt-amperes per sq ft (0.09 m²) = 4,500 volt-amperes.

Number of Branch Circuits Required:

General Lighting Load:

4,500 ÷ 120 = 37.5 amp; 37.5 ÷ 15 = 2.5 or three 15-ampere, 2-wire circuits; or two 20-ampere 2-wire circuits.

Small-Appliance Circuit: Two 2-wire, 20-ampere circuits.

Laundry Circuit: One 2-wire, 20-ampere circuit.

TABLE 3-1J. *Lighting Load Feeder Demand Factors*

Type of Occupancy	Portion of Lighting Load to which Demand Factor Applies (volt-amperes)	Demand Factor Percent
Dwelling Units	First 3000 or less at	100
	From 3001 to 120,000 at	35
	Remainder over 120,000 at	25
Hospitals*	First 50,000 or less at	40
	Remainder over 50,000 at	20
Hotels and Motels — Including Apartment Houses without Provision for Cooking by Tenants*	First 20,000 or less at	50
	From 20,001 to 100,000 at	40
	Remainder over 100,000 at	30
Warehouses (Storage)	First 12,500 or less at	100
	Remainder over 12,500 at	50
All Others	Total volt-amperes	100

*The demand factors of this table shall not apply to the computed load of feeders to areas in hospitals, hotels, and motels where the entire lighting is likely to be used at one time, as in operating rooms, ballrooms, or dining rooms.

Source: NFPA 70-1996, *National Electrical Code.*

Minimum Size Feeders Required:

Computed Load	
General Lighting	4,500 volt-amperes
Small-Appliance Load	3,000 volt-amperes
Laundry	1,500 volt-amperes
Total (without range and dryer)	9,000 volt-amperes
3,000 volt-amperes at 100 percent	3,000 volt-amperes
9,000–3,000 = 6,000 volt-amperes at 35 percent =	2,100 volt-amperes
Net Computed (without range and dryer)	5,100 volt-amperes
Range Load	8,000 volt-amperes
Clothes Dryer	5,000 volt-amperes
Net Computed (with range and dryer)	18,100 volt-amperes

For 120/240 V system feeders 18,100 ÷ 240 = 75 amperes. Therefore, feeder size for total load may be selected on basis of 75-ampere load. Net computed load exceeds 10 kVA, so service conductors must have a rating of 100 amperes.

Flexible Cords and Cables

Flexible cords are made in many types for various kinds of service, ranging from wiring for portable lamps to elevator cables. Article 400 of the *NEC* gives a description, the ampacities, and the intended use of the various flexible cords now available. The ampacity of flexible cords is limited by type and gage of wire used. They must not be overloaded nor their size reduced where the gage is intended for mechanical strength.

Flexible cords are frequently subject to physical damage and rapid wear. Grounds or short circuits may occur if the insulation is damaged, and the resulting arc may ignite the insulation or nearby combustible material. Replacement of flexible cords as soon as they show damage or appreciable wear is of utmost importance.

Under present UL labeling practices, only replacement nondetachable power supply cords; range and dryer power supply cords; and cord sets, such as extension cords; carry the UL label on the product itself. Formerly, the original cords and power supply cords

TABLE 3-1K. *Optional Calculation for One-Family Dwelling Unit*

Largest of the following five selections.

(1) 100 percent of the nameplate rating(s) of the air conditioning and cooling, including heat pump compressors.
(2) 100 percent of the nameplate ratings of electrical thermal storage and other heating systems where the usual load is expected to be continuous at the full nameplate value. Systems qualifying under this selection must not be figured under any other selection in this table.
(3) 65 percent of the nameplate rating(s) of the central electric space heating including integral supplemental heating in heat pumps.
(4) 65 percent of the nameplate rating(s) of electric space heating if less than four separately controlled units.
(5) 40 percent of the nameplate rating(s) of electric space heating of four or more separately controlled units.

Plus: 100 percent of the first 10 kVA of all other load. 40 percent of the remainder of all other load.

Source: NFPA 70-1996, *National Electrical Code.*

were labeled, and many consumers thought a label on the cord of an electrical appliance signified the appliance itself was investigated for safety. Nondetachable cords on new appliances no longer carry the UL label, so the consumer cannot be misled into assuming the appliance itself has been UL tested, which may not be the case.

Flexible cords are not considered by the *NEC* as a wiring method. They are covered in Chapter 4 of the *NEC*, "Equipment for General Use." Flexible cords, including extension cords, are not acceptable as a substitute for the fixed wiring of a structure, but they are acceptable for use in extending the length of a cord on a portable lamp or on cord- and plug-connected appliances. They should be used in accordance with any instructions included with the listing or labeling, and the flexible cord itself should be used in accordance with Article 400 of the *NEC*. Flexible cords are not limited solely to "temporary wiring."

Switches

Switches are required for the control of lights and appliances, and as the disconnecting means for motors and their controllers. Switches may be of either the air-break or the oil-break type. In the oil-break type, the interrupting device is immersed in oil.

A chief fire hazard in switches is the arcing produced when the switch is opened. This hazard is somewhat greater with oil-break switches. If operated much beyond their rated capacity, or if the condition of the oil is poor or its level is not properly maintained, the arc may vaporize the oil, rupture the case, and cause a fire. However, the amount of oil is comparatively small (except in high-voltage equipment), and these switches present no hazard if properly used and maintained.

ELECTRICAL HOUSEHOLD APPLIANCES

Electric Heating Equipment

Electric heating equipment is used widely, and it is important that all such devices be tested and listed by qualified electrical testing laboratories. Some fixed space heating equipment covered in Article 424 of the *NEC* is acceptable for installation in direct contact with combustible material. Baseboard heaters listed by UL, for example, have been tested and found to incorporate suitable safeguards against fire hazards that might result from contact with draperies, furniture, carpeting, bedding, etc., although discoloration or scorching (but no glowing embers or flaming) may result on adjacent materials. Other electric air heaters, however, may present fire hazards if they come in contact with combustible materials or if they are covered or blocked in any manner. Space heating systems must not be installed where exposed to severe physical damage unless adequately protected, nor in damp or wet locations unless approved for such locations.

A heater installed in an air duct or plenum must be of a type suitable for that purpose. Article 424 of the *NEC* covers installation of heaters for ducts and plenums and such items as airflow reliability, problems of condensation, fan circuit interlock, limit controls, and location of disconnecting means. NFPA 90B, *Standard for the Installation of Warm Air Heating and Air Conditioning Systems*, also contains details on central systems.

Electric Ranges, Wall-Mounted Ovens, and Counter-Mounted Cooking Units

Each of these devices needs a means for disconnection from all ungrounded conductors of the supply circuit. A separable connector or a plug and receptacle may serve as the disconnecting means for free-standing household ranges. A plug and receptacle connection at the rear base of the range, if accessible from the front by removal of a drawer, is considered satisfactory. Grounding requirements for ranges are covered in Article 250 of the *NEC*.

In new installations, the grounded conductor (neutral) is not allowed as a means of grounding the frame of an electric range or counter-mounted cooking unit. Therefore, the flexible power cord must contain 4 conductors, and a 4-blade plug. Where there is an existing branch circuit containing only 3 conductors supplying a 3-wire receptacle, the grounded conductor (neutral) is allowed to be used for grounding the frame of the appliance.

The *NEC* gives the methods for calculating feeder loads for household electric ranges and other cooking appliances.

Refrigerators

Fire problems with refrigerators include deterioration in service of the fractional horsepower motors used, and the possibility of overheating. Refrigerant coils and motors are susceptible to accumulations of lint and oily deposits, so cleanliness is a primary consideration in fire prevention. New devices have sealed motors not requiring oiling or maintenance, but overheating can result from improper use or inefficient cooling of the refrigerant due to coil dirt or damage. Ordinary household or commercial refrigerators must not be used to store flammable liquids. [Refrigerators for use in Class I hazardous (classified) locations, and refrigerators for flammable materials storage in laboratories in health-related institutions in accordance with Chapter 10 of NFPA 99, are listed by UL.] Exposed noncurrent-carrying metal parts of refrigerators and freezers, which are likely to become energized, need to be grounded under provisions of the *NEC*.

Room Air-Conditioning Units

These have the same fire hazards as other motor-operated appliances. The exposed noncurrent-carrying metal parts must be grounded as required by Article 250 of the *NEC*.

Incandescent Lamps

Because incandescent lamps produce considerable heat, they inherently possess the hazard of heating and igniting combustible material in contact with them. Under normal conditions, with incandescent lamps in approved lampholders and fixtures where properly guarded, the heating hazard is negligible, but ignition of combustible material may result if lamps are surrounded by or laid on such combustible material. Table 3-1L presents information on surface and base temperatures of standard lamps in open sockets. The temperature measurements are all taken at an ambient temperature of 77°F (25°C). In most instances where lamps are incorporated in various kinds of lighting equipment, the ambient temperature at which the lamps operate is higher; thus, the surface temperatures of the bulbs are also higher. The bulb temperatures may be further increased if the lamp is in other than a vertical, base-up position. Figure 3-1Z shows surface temperatures of 100-W A-19 and 500-W PS-35 lamps in various positions; but, again, these temperatures were measured in an ambient temperature of 77°F (25°C); so where enclosed in fixtures, the temperatures will be higher. The older style conventional Christmas tree lamps of U.S. manufacture have approximate average bulb surface temperatures at the hottest spot of approximately 260°F (127°C) for the blue and green colors, and somewhat lower for the white, red, and yellow colors.*

In locations where there are flammable vapors or gases, combustible dusts, or readily ignitable fibers or flyings, etc., lamps in fixtures specially approved for the location must be used. Lighting fixtures approved for Class I, Division 1 locations must be marked to indicate the maximum wattage of the lamps for which they are approved, and they must be protected against physical damage by a suitable guard or located beyond the area of potential damage.

Fixed lighting fixtures in Class I, Division 2 locations must be protected from physical damage by a suitable guard or located out of the area of potential damage. Where there is the potential for falling sparks or hot metal from the lamps that might ignite localized concentrations of flammable vapors or gases, suitable enclosures or other means must be provided. Where lamp size might be increased so that, under normal operating conditions, the surface temperature of the fixture could exceed 80 percent of the ignition temperature in degrees Celsius of the gas or vapor involved, the fixture must be of a type that has been tested for the marked temperature operating range.

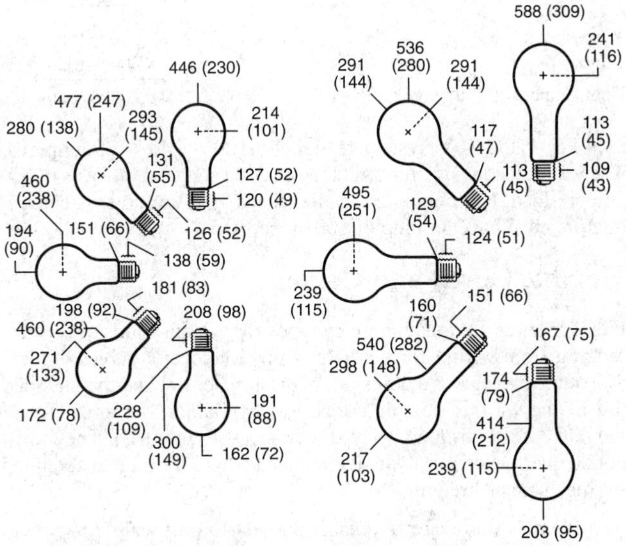

FIG. 3-1Z. *Surface temperatures of lamps in various positions. At left is a 100-W, A-19 lamp, and at right a 500-W, PS-35 lamp.*

---

*The data in Table 3-1L was extracted from the IES Handbook, 5th ed., of the Illuminating Engineering Society, and is used here with the permission of the Society. The reader is asked to keep in mind the above comments in the text regarding use of data in Table 3-1L and Figure 3-1Z.

TABLE 3-1L. *General Service Lamps for 115-, 120-, and 125-Volt Circuits (Will Operate in Any Position, but Lumen Maintenance Is Best for 40 to 1500 Watts when Burned Vertically Base-Up)* *

Watts	Bulb and Other Description	Base	Fila-ment	Rated Aver-age Life (hrs)	Maxi-mum Overall Length in In. (mm)	Average Light Cen-ter Length in in. (mm)	Approx-imate Initial Filament Temp. (K)	Max. Bare Bulb Temp. °F (°C)	Base Temp.† °F (°C)	Approxi-mate Initial Lumens	Rated Initial Lumens Per Watt‡	Lamp Lumen Deprecia-tion# (percent)
10	S-14 inside frosted or clear	Med.	C-9	1500	3½ (89)	2½ (63)	2420	106 (41)	106	80	8.0	89
15	A-15 inside frosted	Med.	C-9	2500	3½ (89)	2⅜ (60)	—	—	—	126	8.4	83
25	A-19 inside frosted	Med.	C-9	2500	3⅞ (98)	2½ (63)	2550	110 (43)	108 (42)	230	9.2	79
40	A-19 inside frosted and white	Med.	C-9	1500	4¼ (108)	2¹⁵⁄₁₆ (75)	2650	260 (127)	221 (105)	455	11.4	87.5
40	S-11 clear	Inter-med.	CC-2V or C-7A	350 500	3⁵⁄₁₆ (59)	1⅝ (41)	2800 —	570 (299)	390 (199)	477 —	11.9 —	— —
50	A-19 inside frosted	Med.	CC-6	1000	4⁷⁄₁₆ (113)	3⅛ (79)	—	—	—	680	13.6	—
60	A-19 inside frosted and white¶	Med.	CC-6	1000	4⁷⁄₁₆ (113)	3⅛ (79)	2790	255 (124)	200 (93)	860	14.3	93
75	A-19 inside frosted and white¶	Med.	CC-6	750	4⁷⁄₁₆ (113)	3⅛ (79)	2840	275 (135)	205 (96)	1180	15.7	92
100	A-19 inside frosted and white¶	Med.	CC-8	750	4⁷⁄₁₆ (113)	3⅛ (79)	2905	300 (149)	208 (98)	1740	17.4	90.5
100§	A-19 inside frosted and white	Med.	CC-8	1000	4⁷⁄₁₆ (113)	3⅛ (79)	—	—	—	1680	16.8	—
100	A-21 inside frosted	Med.	CC-6	750	5¼ (133)	3⅞ (98)	2880	260 (143)	194 (90)	1640	16.9	90
100§	A-23 inside frosted or clear	Med.	C-9	1000	5¹⁵⁄₁₆ (151)	4⁷⁄₁₆ (113)	—	—	—	1480	14.8	—
150	A-21 inside frosted	Med.	CC-8	750	5½ (140)	4 (102)	2960	—	—	2880	19.2	89
150	A-21 white	Med.	CC-8	750	5½ (140)	4 (102)	2930	—	—	2790	18.6	89
150	A-23 inside frosted or clear or white	Med.	CC-6	750	6³⁄₁₆ (157)	4⅝ (117)	2925	280 (138)	210 (99)	2780	18.5	89
150	PS-25 clear or inside frosted	Med.	C-9	750	6¹⁵⁄₁₆ (176)	5¼ (133)	2910	290 (143)	210 (99)	2660	21.2	87.5
200	A-23 inside frosted or white or clear	Med.	CC-8	750	6⁵⁄₁₆ (160)	4⅝ (117)	2980	345 (174)	225 (107)	4000	20.0	89.5
200	PS-25 clear or inside frosted	Med.	CC-6	750	6¹⁵⁄₁₆ (176)	5¼ (133)	—	—	—	3800	19.0	—
200	PS-30 clear or inside frosted	Med.	C-9	750	8¹⁄₁₆ (205)	6 (152)	2925	305 (152)	210 (99)	3700	18.5	85
300	PS-25 clear or inside frosted	Med.	CC-8	750	6¹⁵⁄₁₆ (110)	5³⁄₁₆ (132)	3015	401 (205)	234	6360	21.2	87.5
300	PS-30 clear or inside frosted	Med.	C-9	750	8¹⁄₁₆ (205)	6 (152)	3000	275 (135)	175 (79)	6100	20.3	82.5

*Lamp burning base-up in ambient temperature of 77°F (25°C).
† At junction of base and bulb.
‡ For 120-V lamps.
§ Used mainly in Canada.
¶ Lumen and lumen per watt values of white lamps are generally lower than for inside frosted.
# Percent initial light output at 70 percent of rated life.

TABLE 3-1L. *(Continued)*

Watts	Bulb and Other Description	Base	Filament	Rated Average Life (hrs)	Maximum Overall Length in in. (mm)	Average Light Center Length in in. (mm)	Approximate Initial Filament Temp. (K)	Max. Bare Bulb Temp. °F (°C)	Base Temp. °F (°C)	Approximate Initial Lumens	Rated Initial Lumens Per Watt‡	Lamp Lumen Depreciation# (percent)
300	PS-30 clear or inside frosted	Mog.	CC-8	1000	$8^5/_8$ (219)	7 (178)	—	—	—	5960	19.8	—
300	PS-35 clear or inside frosted	Mog.	C-9	1000	$9^3/_8$ (238)	7 (178)	2980	330 (166)	215 (102)	5860	19.6	86
500	PS-35 clear or inside frosted	Mog.	CC-8	1000	$9^3/_8$ (238)	7 (178)	3050	415 (213)	175 (79)	10600	21.2	89
500	PS-40 clear or inside frosted	Mog.	C-9	1000	$9^3/_4$ (248)	7 (178)	2945	390 (199)	215 (102)	10140	20.3	—
750	PS-52 clear or inside frosted	Mog.	C-7A	1000	$13^1/_{16}$ (332)	$9^1/_2$ (241)	2990	—	—	15660	20.9	—
750	PS-52 clear or inside frosted	Mog.	CC-8 or 2CC-8	1000	$13^1/_{16}$ (332)	$9^1/_2$ (241)	3090	—	—	17000	22.6	89
1000	PS-52 clear or inside frosted	Mog.	C-7A	1000	$13^1/_{16}$ (332)	$9^1/_2$ (241)	2995	480 (249)	235 (113)	21800	21.8	—
1000	PS-52 clear or inside frosted	Mog.	CC-8 or 2CC-8	1000	$13^1/_{16}$ (332)	$9^1/_2$ (241)	3110	—	—	23600	23.6	89
1500	PS-52 clear or inside frosted	Mog.	C-7A	1000	$13^1/_{16}$ (332)	$9^1/_2$ (241)	3095	510 (265)	265 (129)	34000	22.6	78

*Lamp burning base up in ambient temperature of 77°F (25°C).
† At junction of base and bulb.
‡ For 120-V lamps.
§ Used mainly in Canada.
¶ Lumen and lumen per watt values of white lamps are generally lower than for inside frosted.
# Percent initial light output at 70 percent of rated life.

Electric Discharge (Fluorescent) Lamps

The operating temperature of a fluorescent tube is lower than that of the glass envelope of the incandescent lamp, but high voltage is often used to start the lamp. This requires the use of transformers, reactors, capacitors, and switches; the heat produced by this equipment must be taken into account, and the equipment properly safeguarded.

Bulb temperatures on the surface of fluorescent lamps average between 100 and 110°F (38 and 43°C) over most of their length, since this is a requirement for efficient light production. There is a small area of higher bulb temperature directly above the cathode at each end of a fluorescent lamp that ranges from 120 to 250°F (49 to 121°C), depending upon the type involved.

Where fluorescent lamps having an open circuit voltage of more than 300 V are installed in dwelling occupancies, they *must* have no exposed live parts when lamps are being inserted, are in place, or are being removed.

Extreme care is required in mounting fluorescent lamp fixtures containing a ballast on combustible, low-density, cellulose fiberboard. Low-density refers to combustible low-density, cellulose fiberboard tiles or sheets that have a density of 20 lb per cu ft (0.32 g/cm³) or less, that are formed of bonded plant material, but do not include solid or laminated wood or fiberboard that has a density in excess of 20 lb per cu ft (0.32 g/cm³). Fixtures containing ballasts should not be mounted in contact with low-density, cellulose fiberboard, unless they are specifically listed by a qualified electrical testing laboratory for mounting in that manner. If not specifically listed for this use, they should be spaced not less than 1 1/2 in. (38 mm) from the surface of the combustible material. A principal hazard is ignition of the low-density cellulose fiberboard, which under some conditions can ignite at relatively low temperatures. Integral thermal ballast protection is now provided for fluorescent fixtures installed indoors to protect against ballast overheating as a result of failure of capacitors, lamps, ballast winding shorts, etc.

Electric (High-Intensity) Discharge Lamps

The operating temperature of a high-intensity discharge lamp, such as a mercury vapor lamp, is usually much higher than the operating temperature of either Edison-base incandescent or fluorescent lamps, exceeding 550°F (288°C) in some sizes. Extreme care is therefore necessary where such lamps are encountered to prevent their contacting combustible material. Although such lamps were at one time used primarily for street lighting and other outdoor lighting applications, their higher lighting efficiency and new designs providing good color correction and lower wattages have resulted in increased use indoors. Such lamps usually require ballasts (transformers), sometimes capacitors, and are designed to fit into Edison-base lampholders. There are, however, self-ballasted lamps that can be substituted directly for incandescent lamps. High-intensity discharge lamps should be used only in fixtures listed by a qualified electrical testing laboratory for their use, as indicated on the fixture.

Portable Hand Lamps

Metal-shell, paper-lined lampholders are not designed to be used as portable lamps. A fire hazard may result from a defective or worn cord or from the breaking of a lamp. A personnel hazard may result from personal contact with bare spots on the cord or by contact with a live metal lamp socket. Portable hand lamps ideally have a handle and, where subject to physical damage or where the lamp may come in contact with combustible material, they have a substantial guard.

Portable hand lamps used in hazardous (classified) locations must be suitable for the purpose. (See Figure 3-1AA.) Portables used on grounded surfaces or in damp or wet locations must be grounded, unless supplied through an isolating transformer with an ungrounded secondary of not more than 50 V, or be of the double insulation type.

Television and Radio Equipment

Outdoor antennas and lead-in conductors must be of corrosion-resistant material and securely supported. (See Figure 3-1BB.) They must not be attached to poles or similar structures carrying electric light or power wires or trolley wires, nor should they cross over or be near electric light or power circuits (to avoid accidental contact). Where antennas or lead-in conductors are in the proximity of outdoor electric light or power circuits operating at less than 250 V, they must have at least 2 ft (0.61 m) of clearance. Lead-in conductors also must be kept at least 6 ft (1.8 m) from any conductor forming a part of a lightning protection system. The metal sheath of the coaxial cable used for cable television systems must be grounded in accordance with Article 820 of the *NEC*, to avoid electric shock and fire hazards. The sheath must be electrically bonded to the electrical system power ground. Where a separate ground rod is installed for the cable television ground, it must be bonded to the electrical system grounding electrode system or to the electrical service equipment.

Masts and metal structures supporting antennas must be well grounded, and each conductor of a lead-in from an outdoor antenna must be protected by an approved antenna discharge unit. Metal masts on buildings are required by NFPA 780 to be bonded to the nearest lightning conductor (where available). This is normally done with standard lightning conductors.

FIG. 3-1AA.   A representative portable hand lamp of a type suitable for use in Class I, Group C and D locations, in accordance with Article 501 of the NEC. It has an aluminum guard and globe holder. (Source: Stewart R. Browne Mfg. Co., Inc.)

Radio interference eliminators and noise suppressors connected to power supply leads and devices, intended to permit an electric supply circuit to be used in lieu of an antenna, should be listed for the purpose by a qualified electrical testing laboratory.

Operating television sets, even the new solid-state sets, can develop considerable heat, so most cabinets housing such equipment are provided with ventilation openings at the rear and bottom. It is important not to install television receivers so the desired ventilation is cut off or significantly reduced (e.g., by recessing the set in a wall, bookcase, etc.), unless the receiver is designed for such use.

Clothes Washers and Dryers

These appliances are required to have a means for disconnecting all ungrounded conductors from the supply circuit. A separable cord connector or an attachment plug and receptacle may serve as the disconnecting means.

Electric clothes washers and dryers and similar appliances are usually installed within reach of a person who can make contact with a grounded surface or object. Consequently, the exposed, non-current-carrying metal parts of these machines must be grounded to remove the danger of electric shock.

Accumulations of lint in dryers and in lint traps also present a potential fire hazard if not periodically cleaned.

Smoothing Irons

These appliances intended for use in residences are required by the *NEC* to be equipped with approved means to limit the temperature. Even since the widespread use of automatic irons, some hazard still remains because satisfactory ironing of many fabrics requires a degree of heat sufficient to cause ignition if the iron is left in contact with some combustible materials for a considerable period of time.

Electrically Heated Pads and Bedding

The user must be warned by means of a marking on the appliance and by instructions packed with it against the common possible abuses that would increase the fire or shock hazard.

Because many fabrics readily absorb moisture, pads listed by UL (unless of the waterproof type) are provided with moisture-resistant envelopes. This envelope should be examined frequently for signs of deterioration. Blankets are normally designed so that they may be laundered; some cannot be dry cleaned (as with Stoddard Solvent or perchloroethylene) and are normally labeled "Do Not Dry Clean."

INDUSTRIAL AND COMMERCIAL EQUIPMENT

Furnaces

Industrial furnaces generally employ transformers, which may be either the dry type, askarel-insulated type, oil-insulated type, or less flammable fluid type, e.g., dimethyl silicone. Although new transformers are no longer made with askarel (a material containing polychlorinated biphenyl, or PCB), there are many such transformers still in use. The *NEC* requires oil-insulated transformers of a total rating exceeding 75 kVA to be located in a fire-resistant vault.

Oil-filled circuit breakers that control arc furnaces are subjected to unusually severe duty and, unless frequently inspected and properly maintained, may fail with disastrous results. Circuit breakers on

FIG. 3-1BB. A dangerous arrangement of a lightly supported television antenna in the proximity of power lines.

circuits operated at more than 600 V and which are used to control oil-filled transformers must be located outside the transformer vaults. Vents on high-voltage circuit breakers must be piped outdoors. Electric arc furnace circuits, due to the nature of the operation, are also subject to high surge voltages that can cause failure of the arc circuit breakers used. In these cases, shunt capacitors are installed in the circuit to prevent these high-voltage surges.

Inductive and dielectric heat-generating equipment employing high-frequency alternating currents is used in many heat-treating processes. To eliminate both the personnel hazards and fire hazards of such equipment, construction and installation should comply with the special requirements of the *NEC*.

Other hazards of electric furnaces are similar to those of furnaces employing other means of heating.

Motors

Ignition of the motor insulation or nearby combustible material may be caused by sparks or arcs when the motor winding short circuits or grounds, or when brushes operate improperly. Bearings may overheat because of improper lubrication, and sometimes excessive bearing wear allows the rotor to rub on the stator. The individual drives of machines of many different types sometimes make it nec-

essary to install motors in locations and under conditions that are injurious to motor insulation. Dust that can conduct electricity may be deposited on the insulation, or deposits of textile fibers, etc., may prevent the normal dissipation of heat. Motors should be cleaned and lubricated regularly. All motor installations should comply with the requirements of the *NEC*, which includes special rules for motors in hazardous (classified) locations.

Machine Tools

The electrical equipment of a modern machine tool or plastics processing machine may vary from a simple single-motor drill press or extruder to a large complicated multimotor automatic machine involving highly complex control systems and equipment. This latter type is generally custom designed and factory wired. Machine tools incorporate many devices and safeguards to provide safety to life, safety from fire, reduction of lost time due to replacement of parts, safety to the machine itself, and safety to the work in process.

NFPA 79 covers the specific electrical equipment, apparatus, and wiring furnished as a part of an industrial machine, starting at the electrical supply connection, provided that the voltage is 600 V or less. The *NEC* contains requirements for general application, particularly for protection of several motors on one branch circuit.

Electrical equipment is subject to damage by oil, metal chips, coolants, and moving parts. The fixed wiring to the machines specified in the *NEC* generally should be conductors in conduit, tubing, or Type MI cable. Exceptions are connections to continuously moving parts, which must be extra-flexible, multi-conductor cable. For flexible connections where small or infrequent movement is involved, as at motor terminals, flexible metal conduit or liquid-tight flexible metal conduit may be used.

Motor Control Center Rooms

Automation in industrial plants has led to the development of what is sometimes called "motor control center rooms." Automation of production machinery has greatly increased the use of motors and their related control circuits. This has resulted in grouping the motor control equipment in relatively large, compartmented, metal enclosures. These are specially designed, factory-built equipment, commonly referred to as "motor control centers." They are usually located in large rooms, and electrical failures or fires in the motor control center have often resulted in the destruction of the equipment in many compartments of the control center due to rapid temperature rise.

A factor in some large losses has been that multi-conductor circuits have been installed stacked one above the other in cable trays. Heat from arcing or a fire in the control center could raise the temperature in the control center room to a point where normally slow-burning cables become readily combustible. A fire in such cables can shut down a process or an entire plant for weeks or months. The *NEC* covers the use and construction of cable trays, and should be strictly followed where this type of support is used.

The number of motor control centers in one room should be limited or arranged so that exposure of a large number of circuits or equipment to any one fire is avoided. Motor control center rooms should be of noncombustible construction, with a roof or ceiling not readily weakened by a major electrical disturbance or fire. In addition to normal ventilation, the room should be provided with emergency ventilation for removal of heat and smoke in event of a severe fire.

Suggested fire protection for motor control center rooms may consist of preaction sprinkler systems, fixed water spray protection to cover exposed cables, fire detectors to initiate an alarm at a central point in the event of abnormally high temperatures or smoke conditions, total flooding carbon dioxide or other gaseous agent system, portable extinguishers for Class C fires located at each entrance to the control room, and small hose connections with suitable hose and adjustable spray nozzles readily available near the control room.

Switchboards

Ideally, switchboards are installed in clean, dry locations. They should be under competent supervision and accessible only to qualified persons. Switchboards operating at voltages over 600 V must be provided with protection, to avoid damage from condensation leaks and breaks in piping. Piping for fire suppression is allowed within the electrical room for the protection of the electrical installation. Where it is necessary to install a switchboard in a wet location or outside a building, it must be in a weatherproof enclosure. Ample space must be provided for maintenance operations, as stipulated in the *NEC*.

Insulated conductors grouped within switchboards, as well as the instrument and control wiring, should have a flame-retardant outer covering.

Circuit breakers and switches must have ample ratings for the maximum loads and sufficient interrupting capacity for the maximum short-circuit currents. Contacts of switches and circuit breakers should be kept in good condition, and the oil in oil circuit breakers renewed periodically and kept at the proper level.

Capacitors

Capacitors may be insulated with a combustible or a nonflammable liquid. Capacitors containing more than 3 gal (11 L) of flammable liquid must be enclosed in vaults or outdoor fenced enclosures. These safeguards prevent persons from coming into accidental contact or bringing conducting materials into accidental contact with exposed energized parts, terminals, or buses associated with them.

Capacitors must have a means of draining the stored charge, such as through a suitable resistance. If no means were provided for draining off the charge stored in a capacitor after it is disconnected from the line, a severe shock might be received by a person servicing the equipment, or the equipment might be damaged by a short circuit.

Resistors and Reactors

Except when installed in connection with switchboards or control panels whose locations are suitably guarded from physical damage and accidental contact with live parts, resistors should always be completely enclosed in properly ventilated metal boxes. A resistor is always a source of heat, and, when mounted on a combustible wall, a thermal barrier should be required if the space between resistors and reactors and any combustible material is less than 12 in. (305 mm).

Large reactors are commonly connected in series, with the main leads of large generators or the supply conductors from high-capacity network systems assisting in limiting the current delivered under short-circuit conditions. Small reactors are used with lightning arresters to offer a high impedance to the passage of a high-fre-quency lightning discharge and aid in directing the discharge to ground. Another type of reactor, having an iron core and closely resembling a transformer, is used as a remote-control dimmer for stage lighting in some of the older installations. Reactors are sources of heat and therefore should be mounted in the same manner as resistors.

Motion Picture Projectors and Studios

Motion picture projectors of the professional type [i.e., employing 35 or 70 mm film having 5.4 perforations per in. (25.4 mm) on each edge] must be located in approved enclosures. Motor-driven projector units consist of a listed projector and lamp with motors designed or guarded to prevent ignition of film by sparks or arcs. All professional projectors should be operated by qualified personnel.

Motor-generator sets, transformers, rectifiers, rheostats, and similar equipment for the supply or control of current to projection or spotlight equipment should, if practical, be located in a separate room. If placed in the projection room, the equipment must be so located or guarded that arcs or sparks cannot come in contact with film. Switches, overcurrent devices, or other equipment not normally required or used for projectors, sound reproduction, flood or other special-effect lamps, or other equipment (except remote-control switches for the auditorium lights, or a switch for the motor operating the curtain and masking of the motion picture screen) must not be installed in projection rooms.

In projection rooms suitable for use with cellulose acetate (safety) film only, a sign reading "SAFETY FILM ONLY PERMITTED IN THIS ROOM" should be posted on the outside of each projection room door and within the projection room itself in a conspicuous location.

Approved projectors of the nonprofessional or miniature type, when employing cellulose acetate (safety) film, may be operated without a projection room.

Special rules are also given in the *NEC* to cover electrical installations in motion picture studios, factories, laboratories, stages, or areas of buildings in which work is done on cellulose nitrate film. Provisions include wiring on stages, sets, dressing rooms, viewing, cutting, and patching tables, and film storage vaults.

Cranes and Hoists

Wiring methods and installation of electrical equipment for cranes and hoists are covered in Article 610 of the *NEC*. Where bare contact conductors are objectionable because of the presence of easily ignitable material, current may be conducted to a crane or hoist via a multiple cable on a cable reel or suitable takeup devices.

Where a crane operates over combustible material, the resistors must either be placed in (1) a well-ventilated cabinet of noncombustible material that will not emit flames or molten metal, or (2) a cage or cab constructed of noncombustible material that encloses the sides of the cage or cab from the floor to a point at least 6 in. (152 mm) above the top of the resistors.

Collectors must be designed to reduce sparking between them and the contact conductors to a minimum. When operated in rooms where easily ignitable fibers or materials producing combustible flyings are handled, manufactured, used, or stored, the installation must comply with the special requirements of Article 503 of the *NEC*.

All exposed metal parts of cranes, monorail hoists, hoists, and accessories (including pendant controls) should be metallically joined together into a continuous electrical conductor so the entire crane or hoist will be grounded.

Elevators, Dumbwaiters, Escalators, and Moving Walks

Installation of electric equipment and wiring for elevators, dumbwaiters, escalators, and moving walks are covered in Article 620 of the *NEC*.

All live parts of electrical apparatus in the hoistways, at landings, in or on cars of elevators and dumbwaiters, or in the wellways or landings of escalators or moving walks must be enclosed to prevent accidental contact. The conductors to the hoistway door interlocks from the hoistway riser must be flame-retardant and suitable for a temperature of not less than 392°F (200°C). Other wiring in raceways must have flame-retardant insulation.

Traveling cables for operating, control, and signal circuits must be suspended at the car and hoistway in a manner that reduces strain on individual copper conductors to a minimum. Supports for the traveling cables must also be arranged to prevent damage to cables coming in contact with the hoistway or equipment. Where necessary, suitable guards can protect the cables.

To reduce the danger of electric shock, the following equipment must be effectively grounded in accordance with the grounding requirements of the *NEC*: (1) metal conduit, Type MC cable, or Type AC cable attached to elevator cars, (2) the frames of all motors, elevator machines, and controllers, (3) the metal enclosures for all electric devices in or on the car or hoistway, and (4) the frames of nonelectric elevators if accessible to persons and if any electric conductors are attached to the car.

All 125-V, single-phase, 15- and 20-ampere receptacles installed in machinery spaces, pits, elevator car tops, and in escalator and moving walk wellways must be of the ground-fault circuit-interrupter type. This protection does not include a GFCI circuit breaker located in a distribution panel; the intent is to allow personnel to reset the GFCI receptacle without climbing around in the dark. Receptacles installed in the machine room are allowed to be protected by either a GFCI receptacle or a GFCI circuit breaker.

Bonding of elevator rails to lightning protection system grounding down conductors is recommended, but they must not themselves be used as down conductors. Lightning protection down conductors are not allowed within the hoistway.

Equipment mounted on members of the structural metal frame of a building is considered to be grounded. Metal car frames supported by metal hoisting cables attached to or running over sheaves or drums of elevator machines are deemed to be grounded when the machines are properly grounded.

Heating Cable

Recognized testing laboratories have listed a number of heating cables for use within buildings. These cables are of the resistance type, primarily designed to be used around water pipes to prevent freezing and facilitate the flow of viscous liquids. The cable is secured to the pipe with straps or wrapped around the pipe with the pitch recommended by the manufacturer. Electric heating cables are required to have a grounded metal outer covering; this *NEC* rule became effective July 1, 1996. The pipe and heating cable are usually encased in thermal insulation. Where the pipe passes vertically through flooring or where the pipelines run along the floor level, the installation is best protected by sheet metal not less than No. 10 gage over the insulation, with the protection extending at least 4 ft (1.22 m) above the floor.

Some units incorporate a thermostat that automatically turns on the heating cable when the temperature drops below a predetermined value. The pipe heating cables are intended to be connected to a permanent supply wiring system. Unless specifically indicated otherwise by marking on the heating cables or in the installation instructions, the heating cables are intended for use only on metallic pipes. Ground-fault protection of equipment must be provided for each branch circuit supplying electric heating equipment; this is not GFCI protection. An exception is allowed for industrial establishments where only qualified persons will service the installation, and continued circuit operation is necessary for safe operation of the equipment or process. In this case, an alarm indication of a ground fault is required.

Electrical heating cables for outdoor ice- and snow-melting systems are covered in the *NEC*.

ELECTRICAL EQUIPMENT FOR OUTDOOR USE

Electric equipment that is installed outside of buildings must be weatherproof in design, i.e., constructed so exposure to the weather will not interfere with its successful operation; otherwise it must be enclosed in weatherproof cabinets. These enclosures are designed to prevent moisture or water from entering and accumulating within them. If mounted on a wall or other supporting surface, there should be at least $^1/_4$ in. (6 mm) air space between the box or cabinet and the supporting surface.

Description of Devices

Electric equipment suitable for outdoor use is tested by recognized electrical testing laboratories. The tests encompass: (1) suitability of the materials used and the protective coatings for exposure to sunlight, rain, and snow, (2) effects of heat and cold, and (3) provisions for grounding, etc. The term "raintight," as used in the *NEC*, is applied to equipment that on exposure to a beating rain will not let water enter. The term "weatherproof," as used in the *NEC*, is applied to equipment that on exposure to the weather will still operate successfully.

Rainproof, raintight, or watertight equipment can fulfill the requirements for weatherproof equipment where varying weather conditions other than wetness, e.g., snow, ice, dust, or temperature extremes, are not a factor.

Electric Signs and Outline Lighting

Electric signs and outline lighting, fixed, mobile, or portable, are required to be listed by a recognized testing laboratory. Except for portable indoor signs, signs and outline lighting equipment are usually constructed of metal or other noncombustible material. The spacing between wood or other combustible materials and an incandescent or High Intensity Discharge (HID) lamp or lampholder must not be less than 2 in. (51 mm). Enclosures for outside use must be weatherproof. Signs and outline lighting systems must be constructed and installed so that adjacent combustible materials are not subjected to temperature in excess of 194°F (90°C). All steel parts of enclosures must be galvanized or otherwise protected from corrosion. Signs, troughs, tube terminal boxes, and other metal frames must be grounded in the manner specified in the *NEC*.

Each electric sign (other than cord-connected) and each outline lighting installation must be controlled by an external operable switch or circuit breaker (handle on outside of switch enclosure), which will open all ungrounded conductors. The switch or breaker must be within sight of the sign or outline lighting installation unless it is capable of being locked in the open position.

Portable outdoor electric signs must be equipped with factory-installed GFCI protection. The GFCI must be an integral part of the attachment plug or located in the power supply cord within 12 in. (305 mm) of the attachment plug.

Electric Fences

Wire fences with electrical connections to produce a shock when animals come in contact with them are widely used on farms. To avoid hazard to persons and to livestock, the current and the time interval during which the current is on and off must be limited to values which will not cause fatalities or injuries to persons or animals, while still causing an unpleasant sensation of shock. Fatalities and fires have resulted from homemade equipment supplied from ordinary lighting circuits. The open circuit voltage need not be limited if the current is properly limited. UL Research Bulletin No. 14 and UL 69-1993, "Standard for Safety Electric-Fence Controllers," give detailed information on these installations.

The output characteristics of some controllers are such that combustible material may be readily ignited when a grounded object occupies a position respective to the energized fence to produce an electric arc. To protect against fire from such a cause, users should determine that the controller has been designed and tested with respect to this hazard. Electric fence controllers listed in accordance with the requirements of UL 69 have been so tested.

Marina and Boatyard Wiring

Marina and boatyard wiring presents special outdoor electric equipment wiring problems that are covered in Article 555 of the *NEC*. Electric service equipment for floating docks or marinas must be located adjacent to, but not on or in, the floating structure. Metal raceways and metal boxes must not be depended upon for grounding; a continuous insulated copper conductor, not smaller than No. 12 AWG, must be provided as an equipment grounding conductor from outlet boxes and receptacles to the service ground. Electrical equipment and wiring located at gasoline dispensing stations must follow the requirements in Article 514 of the *NEC* and NFPA 30A, *Automotive and Marine Service Station Code* (hereinafter referred to as NFPA 30A). Wiring over and under navigable water should be subject to approval by the Authority Having Jurisdiction.

LOCATIONS EXPOSED TO MOISTURE AND NONCOMBUSTIBLE DUSTS

Special electric equipment is required for use where moisture and noncombustible dusts may be present. For years, the *NEC* referred to such equipment as "vaportight," but this term was dropped from the *NEC* because of confusion in the field between equipment so designated and "explosionproof" equipment. Many users assumed that "vaportight" equipment was safe to use in atmospheres containing flammable gases or vapors (Class I locations), combustible dusts (Class II locations), or easily ignitable fibers or flyings (Class III locations). To avoid this type of misunderstanding, the term "enclosed and gasketed" was developed by UL for this type of equipment. Some inspection authorities permit enclosed and gasketed lighting fixtures in Class I, Division 2; Class II, Division 2; and in Class III, Divisions 1 and 2 locations when marked to show the operating temperature and maximum wattage of permissible lamps. The *NEC* requires lighting fixtures to be marked for use in wet and damp locations when so used.

SIGNALING AND COMMUNICATIONS SYSTEMS

Communications systems, including telephone, telegraph, fire and burglar alarms, and watchman and sprinkler supervisory systems, usually operate with low voltages and currents, and if kept free from accidental contacts with higher voltage systems, they present no unusual hazards. There are, however, *NEC* requirements for the wiring of these systems if the wiring is in a duct, plenum, or other space for environmental air, because of the hazard of spreading products of combustion from one location to another. There are also *NEC* requirements for separation from other systems, such as lighting and power circuits, to minimize imposing higher voltages and current on the communication equipment and conductors that are rated for low energy levels. Where fire-rated walls, floors, or ceilings are penetrated, the fire-rating integrity must be maintained by proper sealing around cables to prevent the spread of fire or smoke.

Fire alarm systems are classified as nonpower limited or power limited. Nonpower-limited cables (NPLFA) must be identified and listed as: NPLFP, where used in other space used for environmental air; NPLFR, where used as a vertical riser from floor to floor or in shafts; and NFPLF, where used for general-purpose in other than risers, ducts, plenums, or environmental air-handling spaces. Power-limited systems (PLFA) must be identified and listed with the following markings: FPLP, FPLR, and FPL; these are similar to those of nonpower-limited cables.

Signaling systems and communications systems must be installed in accordance with Articles 725, 760, and 800 of the *NEC*, and NFPA 72, *National Fire Alarm Code*, which covers the proper electrical wiring and equipment for fire alarm services. NFPA 72 cross references to the *NEC*. Telephone communications circuits may be used for completing the fire-protective circuits between the protected premises and fire alarm headquarters.

EMERGENCY SYSTEMS

Requirements for the installation, operation, and maintenance of emergency system circuits and equipment are given in Article 700 of the *NEC*. The systems are intended to supply light and power when the normal supply fails. They are generally installed in places of assembly where artificial lighting is required, such as auditoriums, theaters, sports arenas, etc., occupied by large numbers of persons. (See Figures 3-1CC and 3-1DD.)

Legally required standby systems installed to serve loads, such as heating and refrigeration systems, communications systems, ventilation and smoke removal systems, sewage disposal, lighting systems, and industrial processes, that, when stopped during any interruption of the normal electrical supply, could create hazards or hamper rescue or fire fighting operations, are covered in Article 701 of the *NEC*. NFPA 110, *Standard for Emergency and Standby Power Systems*, covers the performance requirements for such systems.

NFPA *101*®, *Life Safety Code*®, specifies where emergency lighting is considered essential for life safety. NFPA 99, Chapter 7, gives minimum factors governing the design, operation, and maintenance of those portions of health care facility electrical systems where any degree of interruption would jeopardize the effective and safe care of hospitalized patients. The provisions in NFPA 99 do not supersede the recommendations in NFPA *101* or the *NEC*, which both limit the type of alternate source of electrical power allowable for use to ensure electric power continuity in health care facilities. NFPA 99 recognizes the progressively greater dependence being placed upon electrical apparatus for the preservation

FIG. 3-1CC. *The heavy lines represent walls required by typical local building codes to be of fire-rated construction. The light lines represent walls not required by typical local building codes to be of fire-rated construction.*

of life of hospitalized patients, and is a guide in the selection of the electrical services for emergency supply to all lighting and power equipment considered essential, and for their design and maintenance in health care facilities.

The sources of electric current that can be used for emergency lighting equipment are: (1) a storage battery of suitable capacity, (2) a generator driven by some form of prime mover, and (3) a second electric service widely separated electrically and physically from the regular service to minimize the possibility of simultaneous interruption of both services. NFPA 99 stipulates that the alternate source of power for health care facilities be prime-mover-driven generators or, where the normal source is an on-site generator, an external utility service may be the alternate source. Means must be provided for automatically energizing emergency lights upon failure of the regular lighting system supply. For hospitals, the transition time from the instant of failure of the normal power source to an emergency generator source must not exceed 10 sec.

Audible and visual signal devices are provided, where practical, to: (1) warn of derangement of the emergency source, (2) indicate that the battery or generator is carrying load, (3) indicate when a battery charger is functioning properly, and (4) indicate a ground fault on large 480Y/277-V systems.

No appliances or lamps, other than those required for emergency use, are to be supplied by the emergency lighting circuits. A good emergency lighting system is designed so that the failure of any individual lighting element, such as the burning out of a light bulb, cannot leave any space in total darkness.

The emergency lighting circuit wiring is independent of all other wiring and equipment and must not enter the same raceway, cable, box, or cabinet with other wiring except at transfer switches and exit lighting fixtures supplied from both the normal and the emergency sources.

A transfer switch that supplies emergency power is not allowed to also supply normal loads from the same transfer switch. Where other than emergency loads are to be supplied, they must also be supplied from a transfer switch other than the one supplying emergency circuits.

It is good practice to test the complete system upon installation and periodically thereafter to ensure it is maintained in proper operating condition. A written record should be kept of such tests and maintenance.

FIG. 3-1DD. *Typical installation of a diesel-powered standby generator such as might be used for emergency light and power systems. (Source: P & H Harnischfeger Corp.)*

SPECIAL OCCUPANCY ELECTRICAL PROBLEMS

This section discusses hazardous (classified) locations and other special problems requiring individual attention for which the *NEC* has specific recommendations.

Hazardous (Classified) Locations—General

The *NEC* now recognizes two methods of classification of hazardous locations. The new method uses zones that are consistent with the rules of the International Electrotechnical Commission (IEC). The new method is described in Article 505 of the 1996 edition of the *NEC* and allows area classification to be performed by a qualified professional engineer.

Electric power for lighting, motors, controls, and instrumentation, along with their associated wiring, is necessary or desirable in many hazardous locations. Due to the presence of flammable liquids, gases, vapors, combustible dusts, or readily ignitable fibers or flyings in these areas, special electrical equipment and wiring methods are necessary, to prevent ignition sources from occurring and to minimize the occurrence of fire and explosion.

A flammable liquid, as it pertains to Article 500 of the *NEC* is defined as: "A liquid having a flash point below 100°F (37.8°C) and having a vapor pressure not exceeding 40 psia at 100°F (37.8°C) shall be known as a Class I liquid."

Explosionproof apparatus is defined in the *NEC* as: "Apparatus enclosed in a case that is capable of withstanding an explosion of a specified gas or vapor that may occur within it and of preventing the ignition of a specified gas or vapor surrounding the enclosure by sparks, flashes or explosion of the gas or vapor within, and that operates at such an external temperature that a surrounding flammable atmosphere will not be ignited thereby."

In Article 500 of the *NEC*, hazardous (classified) locations are divided into three classes depending upon the kind of hazardous material involved; each class is further divided into two divisions according to the likelihood of the potential hazard existing. For the purposes of testing for approval and area classification, various air mixtures (not oxygen enriched) have been grouped on the basis of their characteristics. The groups are given in Tables 3-1M through 3-1P [statistics from Section 500-3 of the *NEC* and tables from NFPA 497M, *Manual for Classification of Gases, Vapors, and Dusts for Electrical Equipment in Hazardous (Classified) Locations* (hereinafter referred to as NFPA 497M)]. For Group A, B, C, and D materials with flash points of 100°F (38°C) and higher, see NFPA 497M. For a more complete table of Group G dusts and information on Group E dusts, see the following and NFPA 497M.

Group G includes combustible dusts, such as wood, flour, grain, corn starch, cocoa, food, chemical, and plastics dusts. Coal, coke, and carbonaceous dusts are classified in Group F. Group E includes metal dusts and other dusts of similarly hazardous characteristics.

Determining the proper group classification for flammable gases and vapors involves a determination of explosion pressures and maximum safe clearance between parts of a clamped joint under several conditions, and comparison of the values obtained with those obtained for materials that have already been evaluated.

Listed equipment is marked to show the class, group, and operating temperature, or temperature range, referenced to a 104°F (40°C) ambient. The temperature range, if provided, is indicated by identification numbers, as shown in Table 3-1Q. The identification numbers are marked on equipment nameplates.

TABLE 3-1M. *Group Classification and Autoignition Temperature (AIT) of Selected Flammable Gases and Vapors of Liquids Having Flash Points below 100°F (37.8°C)*

Material	Group	AIT °F	AIT °C
Acetaldehyde	C*	347	175
Acetone	D*	869	465
Acetonitrile	D	975	524
Acetylene	A*	581	305
Acrolein (inhibited)†	B (C)*	455	235
Acrylonitrile	D*	898	481
Allyl alcohol	C*	713	378
Allyl chloride	D	905	485
Ammonia	D*‡	928	498
n-Amyl acetate	D	680	360
sec-Amyl acetate	D	—	—
Benzene	D*	1040	560
1,3-Butadiene†	B (D)*	788	420
Butane	D*	550	288
1-Butanol	D*	650	343
2-Butanol	D*	761	405
n-Butyl acetate	D*	790	421
iso-Butyl acetate	D*	790	421
sec-Butyl acetate	D	—	—
Butylamine	D	594	312
Butylene	D	725	385
Butyl mercaptan	C	—	—
n-Butyraldehyde	C*	425	218
Carbon disulfide§	—*	194	90
Carbon monoxide	C*	1128	609
Chlorobenzene	D	1099	593
Chloroprene	D	—	—
Crotonaldehyde	C*	450	232
Cyclohexane	D	473	245
Cyclohexene	D	471	244
Cyclopropane	D*	938	503
1,1-Dichloroethane	D	820	438
1,2-Dichloroethylene	D	860	460
1,3-Dichloropropene	D	—	—
Dicyclopentadiene	C	937	503
Diethyl ether	C*	320	160
Diethylamine	C*	594	312
Di-isobutylene	D*	736	391
Di-isopropylamine	C	600	316
Dimethylamine	C	752	400
1,4-Dioxane	C	356	180
Di-n-propylamine	C	570	299
Epichlorohydrin	C*	772	411
Ethane	D*	882	472
Ethanol	D*	685	363

* Material has been classified by test.

† If equipment is isolated by sealing all conduit 1/2 in. or larger, in accordance with Section 501-5(a) of the *NEC*, equipment for the group classification shown in parentheses is permitted.

‡ For classification of areas involving Ammonia, see "Safety Code for Mechanical Refrigeration,"ANSI/ASHRAE 15, and "Safety Requirements for the Storage and Handling of Anhydrous Ammonia," ANSI/CGA G2.1.

§ Certain chemicals may have characteristics that require safeguards beyond those required for any of the above groups. Carbon disulfide is one of these chemicals because of its low autoignition temperature and the small joint clearance to arrest its flame propagation.

¶ Petroleum naphtha is a saturated hydrocarbon mixture whose boiling range is 68 to 274°F (20° to 135°C). It is also known as benzine, ligroin, petroleum ether, and naphtha.

Sources: NFPA 497M. Autoignition temperatures listed above are the lowest value for each material as listed in NFPA 325, *Guide to Fire Hazard Properties of Flammable Liquids, Gases, and Volatile Solids*, or as reported in an article by Hilado, C. J., and Clark, S. W., in *Chemical Engineering*, Sept. 4, 1972.

TABLE 3-1M. (Continued)

Material	Group	AIT °F	AIT °C
Ethyl acetate	D*	800	427
Ethyl acrylate (inhibited)	D*	702	372
Ethylamine	D*	725	385
Ethyl benzene	D	810	432
Ethyl chloride	D	966	519
Ethylene	C*	842	450
Ethylenediamine	D*	725	385
Ethylene dichloride	D*	775	413
Ethylenimine	C*	608	320
Ethylene oxide†	B (C)*	804	429
Ethyl formate	D	851	455
Ethyl mercaptan	C*	572	300
n-Ethyl morpholine	C	—	—
Formaldehyde (gas)	B	795	429
Gasoline	D*	536–880	280–47
Heptane	D*	399	204
Heptene	D	500	260
Hexane	D*	437	225
2-Hexanone	D	795	424
Hexenes	D	473	245
Hydrogen	B*	752	400
Hydrogen cyanide	C*	1000	538
Hydrogen selenide	C	—	—
Hydrogen sulfide	C*	500	260
Isoamyl acetate	D	680	360
Isoamyl alcohol	D	662	350
Isobutyl acrylate	D	800	427
Isobutyraldehyde	C	385	196
Isoprene	D*	743	395
Isopropyl acetate	D	860	460
Isopropylamine	D	756	402
Isopropyl ether	D*	830	443
Isopropyl glycidyl ether	C	—	—
Liquefied petroleum gas	D	761–842	405–450
Manufactured gas (containing more than 30% H₂ by volume)	B*	—	—
Mesityl oxide	D*	652	344
Methane	D*	999	537
Methanol	D*	725	385
Methyl acetate	D	850	454
Methylacetylene	C*	—	—
Methylacetylene-propadiene (stabilized)	C	—	—
Methyl acrylate	D	875	468
Methylamine	D	806	430
Methylcyclohexane	D	482	250
Methyl ether	C*	662	350
Methyl ethyl ketone	D*	759	404
Methyl eormal	C*	460	238
Methyl eormate	D	840	449
Methyl isobutyl ketone	D*	840	440
Methyl isocyanate	D	994	534
Methyl mercaptan	C	—	—
Methyl methacrylate	D	792	422
2-Methyl-1-propanol	D*	780	416
2-Methyl-2-propanol	D*	892	478
Monomethyl hydrazine	C	382	194
Naphtha (petroleum)¶	D*	550	288
Nitroethane	C	778	414
Nitromethane	C	785	418
Nonane	D	401	205
Nonene	D	—	—
Octane	D*	403	206
Octene	D	446	230
Pentane	D*	470	243

TABLE 3-1M. (Continued)

Material	Group	AIT °F	AIT °C
1-Pentanol	D*	572	300
2-Pentanone	D	846	452
1-Pentene	D	527	275
Propane	D*	842	450
1-Propanol	D*	775	413
2-Propanol	D*	750	399
Propionaldehyde	C	405	207
n-Propyl acetate	D	842	450
Propylene	D*	851	455
Propylene dichloride	D	1035	557
Propylene oxide†	B (C)*	840	449
n-Propyl ether	C*	419	215
Propyl nitrate	B*	347	175
Pyridine	D*	900	482
Styrene	D*	914	490
Tetrahydrofuran	C*	610	321
Toluene	D*	896	480
Triethylamine	C*	—	—
Tripropylamine	D	—	—
Turpentine	D	488	253
Unsymmetrical dimethyl hydrazine (UDMH)	C*	480	249
Valeraldehyde	C	432	222
Vinyl acetate	D*	756	402
Vinyl chloride	D*	882	472
Vinylidene chloride	D	1058	570

* Material has been classified by test.

† If equipment is isolated by sealing all conduit 1/2 in. or larger, in accordance with Section 501-5(a) of the *NEC*, equipment for the group classification shown in parentheses is permitted.

‡ For classification of areas involving Ammonia, see "Safety Code for Mechanical Refrigeration," ANSI/ASHRAE 15, and "Safety Requirements for the Storage and Handling of Anhydrous Ammonia," ANSI/CGA G2.1.

§ Certain chemicals may have characteristics that require safeguards beyond those required for any of the above groups. Carbon disulfide is one of these chemicals because of its low autoignition temperature and the small joint clearance to arrest its flame propagation.

¶ Petroleum naphtha is a saturated hydrocarbon mixture whose boiling range is 68 to 274°F (20° to 135°C). It is also known as benzine, ligroin, petroleum ether, and naphtha.

Sources: NFPA 497M. Autoignition temperatures listed above are the lowest value for each material as listed in NFPA 325, *Guide to Fire Hazard Properties of Flammable Liquids, Gases, and Volatile Solids*, or as reported in an article by Hilado, C. J., and Clark, S. W., in *Chemical Engineering*, Sept. 4, 1972.

TABLE 3-1N. *Selected Nonconductive Dusts Classified as Group G—Ignition Sensitivity Equal to or Greater than 0.2; Explosion Severity Equal to or Greater than 0.5*

	Minimum Cloud or Layer Ignition Temp.* °F	°C
AGRICULTURAL DUSTS		
Alfalfa meal	392	200
Cellulose	500	260
Cinnamon	446	230
Cocoa, natural, 19% fat	464	240
Corn	482	250
Corncob grit	464	240

*Normally, the minimum ignition temperature of a layer of a specific dust is lower than the minimum ignition temperature of a cloud of that dust. Since this is not universally true, the lower of the two minimum ignition temperatures is listed. If no symbol appears between the two temperature columns, then the layer ignition temperature is shown.

TABLE 3-1N. *(Continued)*

	Minimum Cloud or Layer Ignition Temp.*		
	°F		°C
Corn dextrine	698		370
Cornstarch, commercial	626		330
Cork	410		210
Cottonseed meal	392		200
Garlic, dehydrated	680	NL	360
Malt barley	482		250
Milk, skimmed	392		200
Potato starch, dextrinated	824	NL	440
Rice	428		220
Rice bran	914	NL	490
Rice hull	428		220
Safflower meal	410		210
Soy flour	374		190
Soy protein	500		260
Sucrose	662	CI	350
Sugar, powdered	698	CI	370
Wheat	428		220
Wheat flour	680		360
Wheat starch	716	NL	380
Wheat straw	428		220
Woodbark, ground	482		250
Wood flour	500		260
Yeast, torula	500		260
CHEMICALS			
Acetoacetanilide	824	M	440
Adipic acid	1022	M	550
Anthranilic acid	1076	M	580
Azelaic acid	1130	M	610
2,2-Azo-bis-butyronitrile	662		350
Benzoic acid	824		440
Benzotriazole	824	M	440
Bisphenol-A	1058	M	570
Chloroacetoacetanilide	1184	M	640
Diallyl Phthalate	896	M	480
Dihydroacetic acid	806	NL	430
Dimethyl isophthalate	1076	M	580
Dimethyl terephthalate	1058	M	570
3,5-Dinitrobenzoic acid	860	NL	460
Diphenyl	1166	M	630
Ethyl hydroxyethyl cellulose	734	NL	390
Fumaric acid	968	M	520
Hexamethylene tetramine	770	S	410
Hydroxyethyl cellulose	770	NL	410
Isotoic anhydride	1292	NL	700
Paraphenylene diamine	1148	M	620
Paratertiary butyl benzoic acid	1040	M	560
Pentaerythritol	752	M	400
Phthalic anhydride	1202	M	650
Salicylanilide	1130	M	610
Sorbic acid	860		460
Stearic acid, aluminum salt	572		300
Stearic acid, zinc salt	950	M	510
Sulfur	428		220
Terephthalic acid	1256	NL	680
DRUGS			
Aspirin	1220	M	660
Gulasonic acid, diacetone	788	NL	420
Mannitol	860	M	460
l-Sorbose	698	M	370
Vitamin B1, mononitrate	680	NL	360
Vitamin C (ascorbic acid)	536		280
DYES, PIGMENTS, INTERMEDIATES			
Green base harmon dye	347		175

TABLE 3-1N. *(Continued)*

	Minimum Cloud or Layer Ignition Temp.*		
	°F		°C
Red dye intermediate	347		175
Violet 200 dye	347		175
PESTICIDES			
Crag No. 974	590	CI	310
Dieldrin (20%)	1022	NL	550
Dithane	356		180
Ferban	302		150
Manganese vancide	248		120
Sevin	284		140
THERMOPLASTIC RESINS AND MOLDING COMPOUNDS			
Acetal resins			
Acetal, linear (polyformaldehyde)	824	NL	440
Acrylic resins			
Acrylamide polymer	464		240
Acrylonitrile polymer	860		460
Acrylonitrile-vinyl chloride-vinylidene chloride copolymer (70-20-10)	410		210
Methyl methacrylate polymer	824	NL	440
Methyl methacrylate-ethyl acrylate copolymer	896	NL	480
Methyl methacrylate-ethyl acrylate-styrene copolymer	824	NL	440
Methyl methacrylate-styrene-butadiene-acrylonitrile copolymer	896	NL	480
Methacrylic acid polymer	554		290
Cellulosic resins			
Cellulose acetate	644		340
Cellulose triacetate	806	NL	430
Cellulose acetate butyrate	698	NL	370
Nylon (polyamide) resins			
Nylon polymer (polyhexa-methylene adipamide)	806		430
Polycarbonate resins			
Polycarbonate	1310	NL	710
Polyethylene resins			
Polyethylene, high-pressure process	716		380
Polyethylene, low-pressure process	788	NL	420
Polyethylene wax	752	NL	400
Polymethylene resins			
Carboxypolymethylene	968	NL	520
Polypropylene resins			
Polypropylene (no antioxidant)	788	NL	420
Rayon resins			
Rayon (viscose) flock	482		250
Styrene resins			
Polystyrene molding cmpd.	1040	NL	560
Polystyrene latex	932		500
Styrene-acrylonitrile (70-30)	932	NL	500
Styrene-butadiene latex (>75% styrene; alum coagulated)	824	NL	440

*Normally, the minimum ignition temperature of a layer of a specific dust is lower than the minimum ignition temperature of a cloud of that dust. Since this is not universally true, the lower of the two minimum ignition temperatures is listed. If no symbol appears between the two temperature columns, then the layer ignition temperature is shown.

NOTES:

CI means the cloud ignition temperature is shown.

NL means that no layer ignition temperature is available and the cloud ignition temperature is shown.

M signifies that the dust layer melts before it ignites; the cloud ignition temperature is shown.

S signifies that the dust layer sublimes before it ignites; the cloud ignition temperature is shown.

TABLE 3-1N. (Continued)

	Minimum Cloud or Layer Ignition Temp.*		
	°F		°C
Vinyl resins			
Polyvinyl acetate	1022	NL	550
Polyvinyl acetate/alcohol	824		440
Vinyl chloride-acrylonitrile copolymer	878		470
Vinyl toluene-acrylonitrile butadiene copolymer	936	NL	530
THERMOSETTING RESINS AND MOLDING COMPOUNDS			
Allyl resins			
Allyl alcohol derivative (CR-39)	932	NL	500
Amino resins			
Urea formaldehyde molding compound	860	NL	460
Urea formaldehyde-phenol formaldehyde Molding compound (wood flour filler)	464		240
Epoxy resins			
Epoxy	1004	NL	540
Epoxy-bisphenol A	950	NL	510
Phenolic resins			
Phenol formaldehyde	1076	NL	580
Phenol formaldehyde molding compound (wood flour filler)	932	NL	500
Polyester resins			
Polyethylene terephthalate	932	NL	500
Styrene modified polyester-glass fiber mixture	680		360
Polyurethane resins			
Polyurethane foam, no fire retardant	824		440
SPECIAL RESINS AND MOLDING COMPOUNDS			
Ethylene oxide polymer	662	NL	350
Ethylene-maleic anhydride copolymer	1004	NL	540
Petroleum resin (blown asphalt)	932		500
Rubber, crude, hard	662	NL	350
Rubber, synthetic, hard (33% S)	608	NL	320

*Normally, the minimum ignition temperature of a layer of a specific dust is lower than the minimum ignition temperature of a cloud of that dust. Since this is not universally true, the lower of the two minimum ignition temperatures is listed. If no symbol appears between the two temperature columns, then the layer ignition temperature is shown.

NOTES:

CI means the cloud ignition temperature is shown.

NL means that no layer ignition temperature is available and the cloud ignition temperature is shown.

M signifies that the dust layer melts before it ignites; the cloud ignition temperature is shown.

S signifies that the dust layer sublimes before it ignites; the cloud ignition temperature is shown.

TABLE 3-1O. Selected Carbonaceous Dusts Classified as Group E—Ignition Sensitivity Equal to or Greater than 0.2; Explosion Severity Equal to or Greater than 0.5

	Minimum Cloud or Layer Ignition Temp.*		
Material†	°F		°C
Aluminum, atomized collector fines	1022	CI	550
Aluminum, A422 flake	608		320
Aluminum—cobalt alloy (60-40)	1058		570
Aluminum—copper alloy (50-50)	1526		830
Aluminum—lithium alloy (15% Li)	752		400

TABLE 3-1O. (Continued)

	Minimum Cloud or Layer Ignition Temp.*		
Material†	°F		°C
Aluminum—magnesium alloy (downmetal)	806	CI	430
Aluminum—nickel alloy (58-42)	1004		540
Aluminum—silicon alloy (12% Si)	1238	NL	670
Boron, commercial—amorphous (85% B)	752		400
Calcium silicide	1004		540
Chromium, (975) electrolytic, milled	752		400
Ferromanganese, medium carbon	554		290
Ferrisilicon (885, 9% Fe)	1472		800
Ferrotitanium (19% Ti, 74.1% Fe, 0.06% C)	698	CI	370
Iron 98% H_2 reduced	554		290
Iron 99% carbonyl	590		310
Magnesium, Grade B, milled	806		430
Manganese	464		240
Tantalum	572		300
Thorium, 1.2% O_2	518	CI	270
Tin, 96%, atomized (2% Pb)	806		430
Titanium, 99%	626	CI	330
Titanium hydride (95% Ti, 3.8% H_2)	896	CI	480
Vanadium, 86.4%	914		490
Zirconium hydride (93.6% Zr, 2.1% H_2)	518		270

*Normally, the minimum ignition temperature of a layer of a specific dust is lower than the minimum ignition temperature of a cloud of that dust. Since this is not universally true, the lower of the two minimum ignition temperatures is listed. If no symbol appears between the two temperature columns, then the lower ignition temperature is shown. "CI" means the cloud ignition temperature is shown. "NL" means that no layer ignition temperature is available and the cloud ignition temperature is shown.

† Certain metal dusts may have characteristics that require safeguards beyond those required for atmospheres containing the dusts of aluminum, magnesium, and their commercial alloys. For example, zirconium, thorium, and uranium dusts have extremely low ignition temperatures [as low as 68°F (20°C)] and minimum ignition energies lower than any material classified in any of the Class I or Class II groups.

Source: NFPA 497M.

TABLE 3-1P. Selected Carbonaceous Dusts Classified as Group F—Ignition Sensitivity Equal to or Greater than 0.2; Explosion Severity Equal to or Greater than 0.5

	Minimum Cloud or Layer Ignition Temp.*		
Material	°F		°C
Asphalt (blown petroleum resin)	950	CI	510
Charcoal	356		180
Coal, Kentucky bituminous	356		180
Coal, Pittsburgh experimental	338		170
Coal, Wyoming	—		—
Gilsonite	932		500
Lignite, Californi	356		180
Pitch, coal tar	1310	NL	710
Pitch, petroleum	1166	NL	630
Shale, oil	—		—

*Normally, the minimum ignition temperature of a layer of a specific dust is lower than the minimum ignition temperature of a cloud of that dust. Since this is not universally true, the lower of the two minimum ignition temperatures is listed. If no symbol appears between the two temperature columns, then the layer ignition temperature is shown. "CI" means the cloud ignition temperature is shown. "NL" means that no layer ignition temperature is available and the cloud ignition temperature is shown.

Source: NFPA 497M.

TABLE 3-1Q. *Temperature Identification Numbers*

Maximum Temperature		Identification Number
°F	°C	
842	450	T1
572	300	T2
536	280	T2A
500	260	T2B
446	230	T2C
419	215	T2D
392	200	T3
356	180	T3A
329	165	T3B
320	160	T3C
275	135	T4
248	120	T4A
212	100	T5
185	85	T6

FIG. 3-1EE. *Explanation of the principle of "explosionproof" equipment, indicating containment of hot gases within the enclosure. (Source: Crouse-Hinds Company)*

FIG. 3-1FF. *Representative lighting fixtures for use in various hazardous locations. (Left) A lighting fixture for use in Class I hazardous locations (60-300 W). (Right) A dust-ignitionproof light fixture for use in Class II hazardous locations. (Source: Crouse-Hinds Company)*

The ignition temperature of a liquid, solid, or gaseous substance is the minimum temperature required to initiate or cause self-sustained combustion, independent of the heating or heated element. The flash point and the ignition temperature are not the same and the flash point is always lower than the ignition temperature.

Since there is no consistent relationship among explosion properties, explosion pressure, maximum experimental safe gap, minimum ignition energy, and ignition temperature, testing requirements for groups and classes are independent of each other.

The operating temperature markings specified in Table 3-1Q are not to exceed the ignition temperature of the specific gas or vapor to be encountered.

For information regarding ignition temperatures of gases and vapors, see Table 3-1M, NFPA 497M, and NFPA 325, *Guide to Fire Hazard Properties of Flammable Liquids, Gases, and Volatile Solids.*

Complete definitions of the several classes and divisions of hazardous (classified) locations and the methods of wiring and types of electrical equipment to be used in each are covered in detail in the *NEC.* Rules applying specifically to commercial garages, aircraft hangars, gasoline dispensing facilities and service stations, bulk storage plants, finishing processes, and areas in health care facilities containing flammable anesthetics are also covered in the *NEC.*

Class I, Division 1: Locations in which: (1) ignitable concentrations of flammable gases or vapors can exist under normal operating conditions, or (2) ignitable concentrations of such gases or vapors may exist frequently because of repair or maintenance operations or leakage, or (3) breakdown or faulty operation of equipment or processes might release ignitable concentrations of gases or vapors, and might also cause simultaneous electrical equipment failure. Motors and other rotating electrical machinery, lighting fixtures, most switches, circuit breakers, and similar electrical equipment in these locations must be of the explosionproof type or purged type approved for Class I locations of the proper group. [See NFPA 496, *Standard for Purged and Pressurized Enclosures for Electrical Equipment* (hereinafter referred to as NFPA 496).] (See Figures 3-1EE, 3-1FF, and 3-1GG.) Approved intrinsically safe equipment and circuits may also be used. These are usually limited to low-energy control, signal, and instrumentation systems.

Class I, Division 2: Locations: (1) in which volatile flammable liquids or flammable gases are handled, processed, or used, but in

which the liquids, vapors, or gases will normally be confined within closed containers or closed systems from which they can escape only in case of accidental rupture or breakdown of such containers or systems, or in case of abnormal operation of equipment; or (2) in which ignitable concentrations of gases or vapors are normally prevented by positive mechanical ventilation, and which might become hazardous through failure or abnormal operation of the ventilating equipment; or (3) that are adjacent to Class I, Division 1 locations, and to which ignitable concentrations of gases or vapors might occasionally be communicated unless such communication is prevented by adequate, positive pressure ventilation from a clean air source, and effective safeguards against ventilation failure are provided. In general, ordinary types of motors that do not have brushes, switching mechanisms, etc., may be installed in these locations, but motors that do have sliding contacts, switches, etc., must have the switches in explosionproof or other type enclosure approved for Class I, Division 2 locations. Lamps that operate at temperatures above the ignition temperature of the gas or vapor involved must have approved explosionproof or purged enclosures. Units located where falling sparks or hot metal from broken lamps might ignite localized concentrations of flammable gases below the Division 2 location must also have suitable enclosures. Switches, circuit breakers, controllers,

FIG. 3-1GG. Representative switching and panelboard equipment for use in various hazardous locations. (Left) An example of a panelboard for use in Class I hazardous locations. (Right) Tumbler switch for use in Class I and Class II hazardous locations. (Source: Crouse-Hinds Company)

and other devices that create arcs or sparks and that are intended to interrupt ignition-capable energy in normal operation must either be explosionproof, have their contacts in hermetically sealed chambers, or have contacts immersed in oil. Enclosures for electrical equipment that does not interrupt ignition-capable energy or otherwise represent an ignition source under normal conditions may be of the general-purpose type. Portable lamps must be of the type satisfactory for use in a Class I, Division 1 location.

Class II, Division 1: Locations: (1) in which combustible dust is in the air under normal operating conditions in sufficient quantities to produce explosive or ignitable mixtures; or (2) where mechanical failure or abnormal operation of machinery or equipment might cause such explosive or ignitable mixtures to be produced, and might also provide a source of ignition through simultaneous failure of electric equipment, operation of protective devices, or from other causes; or (3) in which combustible dusts of an electrically conductive nature may be present in hazardous quantities. Motors must be of an enclosed type approved as dust-ignitionproof for the proper Class II locations. Lighting fixtures, switches, circuit breakers, controllers, and fuses that are intended to interrupt current in normal operation must be provided with dust-ignitionproof enclosures approved for the proper Class II locations. Maximum surface temperatures on equipment in Class II locations under actual operating conditions must not exceed the ignition temperature of the dust. See Tables 3-1N through 3-1P for information regarding ignition temperatures of dusts. For additional information, see NFPA 497M.

Class II, Division 2: Locations: (1) in which combustible dust is not normally in the air in quantities sufficient to produce explosive or ignitable mixtures; and (2) where dust accumulations are normally insufficient to interfere with normal operation of electrical equipment or other apparatus, but combustible dust may be in suspension in the air as a result of infrequent malfunctioning of handling or processing equipment; or (3) where combustible dust accumulations on, in, or near the electrical equipment may be sufficient to interfere with the safe dissipation of heat from the electrical equipment or may be ignitable by abnormal operation or failure of electrical equipment. In general, totally enclosed motors are suitable. Under certain conditions, standard, open-type motors without sliding contacts, switches, etc., may be used in many of these locations. Fixed lamps and lampholders must have enclosures designed to minimize the deposit of dust and prevent the escape of sparks, burning material, or hot metal. Switches, circuit breakers, controllers, and fuses

should be provided with dust-tight enclosures. Division 1 temperature limits also apply to Division 2 locations.

Class III, Division 1: Locations in which easily ignitable fibers or materials producing combustible flyings are handled, manufactured, or used. In general, motors in these locations must be of the enclosed type except that, where only moderate accumulations of lint are present and cleaning and maintenance are satisfactory, self-cleaning, squirrel-cage, textile motors or standard, open-type motors without sliding contacts, switches, etc., may be installed. Lamps and lampholders, switches, circuit breakers, controllers, and fuses must be dusttight. Maximum surface temperatures under actual operating conditions must not exceed 329°F (165°C) for equipment that is not subject to overloading, and 248°F (120°C) for equipment such as motors, power transformers, etc., which may be overloaded.

Class III, Division 2: Locations in which easily ignitable fibers are stored or handled (except in manufacture). Motors must be of the enclosed type. Lamps, lampholders, switches, circuit breakers, controllers, and fuses must have enclosures similar to those specified for Class III, Division 1 locations. Division 1 temperature limits apply to Division 2 locations.

Equipment for Use in Hazardous Locations

Equipment for use in Class I, Division 1 hazardous (classified) locations, as defined in the *NEC*, is sometimes referred to as "explosionproof" equipment. The two basic design criteria for explosionproof apparatus for Class I locations are: (1) that it withstand internal explosions of flammable gas or vapor-air mixtures and (2) that it prevent propagation of the internal explosion to the surrounding flammable atmosphere. In other words, it is recognized that surrounding flammable gas or vapor-air mixtures will enter the enclosure of this equipment and the possibility exists of their ignition within the enclosure. To prevent the propagation of flame to the outside surrounding atmosphere, which may likewise contain flammable vapor-air mixtures, the enclosures of this equipment must: (1) arrest flame at joints or other openings to the outside, (2) be of sufficient strength to resist (without rupture or serious distortion) the internal pressure, and (3) ensure that the external surface temperature of the enclosure is not high enough to ignite the surrounding gas or vapor. The various gas or vapor-air mixtures vary considerably with respect to: (1) the propagation of flames through joints of such assemblies, (2) the pressure developed within the enclosure following ignition, and (3) the ignition temperature of the gas or vapor-air mixture. Thus, equipment should be tested and listed for use, and the various groups listed in Table 3-1M should be addressed, as necessary.

Equipment for use in Class I hazardous (classified) locations must also be designed to operate under full load and, in the case of equipment such as motors likely to be overloaded, must operate under overload conditions without developing surface temperatures above the ignition temperature of the flammable gas or vapor with which it is intended to be used.

As an alternative to "explosionproof" motors, controllers, lighting fixtures, switches, etc., or equipment approved for use in Class I locations, it is permissible to use either of the following two designs:

1. Totally enclosed equipment supplied with positive pressure ventilation from a source of clean air with discharge to a safe area. Controls are arranged to prevent energizing the machine until ventilation has been established and the enclosure purged with at least four volumes of air, and to automatically de-energize the equipment when the air supply fails. Electric motors require at least 10 volumes of air while maintaining an internal pressure of

at least 25 Pa (0.1 in. water), in order to accomplish proper purging of the voids between windings.

2. Totally enclosed inert gas-filled equipment supplied with a suitably reliable source of inert gas for pressurizing the enclosure. Devices are provided to ensure a positive pressure in the enclosure as well as to automatically de-energize the equipment when the gas supply fails.

When either of these two designs is used, no external surface of the motors, generators, or other rotating electrical machinery can have an operating temperature (in degrees Celsius) that exceeds 80 percent of the ignition temperature (in degrees Celsius) of the gas or vapor involved, as determined by ASTM E659, "Autoignition Temperature of Liquid Chemicals." Appropriate devices (heat sensors) must also be provided to detect any increase in temperature of the equipment beyond its design limits and then to automatically de-energize the equipment. The detectors and control equipment used are to be suitable for the atmosphere involved without positive pressure; in other words, they must be explosionproof or intrinsically safe if the location is Class I, Division 1.

NFPA 496 provides information for the design of purged enclosures for the purpose of eliminating or reducing within the enclosure a Class I hazardous (classified) location (gases or vapors in air in quantities sufficient to produce explosive or ignitable mixtures). Protective measures include supplying an enclosure with clean air or an inert gas at sufficient flow and positive pressure to achieve and maintain an acceptable safe level of the atmosphere.

Another alternative is the use of "intrinsically safe" equipment and wiring, which is defined as equipment and wiring incapable of releasing sufficient electrical energy under normal or abnormal conditions to cause ignition of a specific hazardous atmospheric mixture.

In many cases, the amount of "explosionproof," purged, or intrinsically safe equipment required can be reduced through the exercise of ingenuity in the layout of electrical installations by locating much of the equipment in nonhazardous areas. The extent of the hazardous areas is normally defined by the codes and standards relating to the storage and handling of the specific flammable liquids, gases, or solids.

Equipment for use in Class II and Class III hazardous (classified) locations presents a somewhat different problem, because the equipment is designed to be dust-ignitionproof for Class II, Division 1 and for some Class II, Division 2 locations, and to be dust-tight or totally enclosed with telescoping covers for some Class II, Division 2 locations and for Class III locations. It is thus not intended to resist internal explosions of dust-air mixtures. Such equipment is tested in specific dust-air mixtures to determine that the enclosures are dust-ignitionproof for Class II locations and that overheating does not occur when the device is blanketed with dust or lint and flyings.

Garages, Commercial (Repair and Storage)

The requirements of Article 511 of the *NEC* apply to locations used for the servicing and repair of passenger automobiles, buses, tractors, trucks, etc., in which flammable liquids or flammable gases are used. NFPA 88A, *Standard for Parking Structures,* and NFPA 88B, *Standard for Repair Garages,* should also be consulted in connection with further classifications of, and the fire protection recommendations applying to, garages. The specific hazardous (classified) areas in these garages are defined in the *NEC.*

Aircraft Hangars

Where aircraft containing gasoline, jet fuels, or other flammable liquids are stored or serviced, the specific (classified) hazardous areas are as defined in Article 513 of the *NEC.* Reference should also be made to NFPA 409, *Standard on Aircraft Hangars,* for guidance on the construction and protection of these structures.

Service Stations

This occupancy group includes locations where gasoline or other volatile flammable liquids or liquefied petroleum gases are transferred to the fuel tanks or auxiliary fuel tanks of self-propelled vehicles. Reference should be made to NFPA 30, *Flammable and Combustible Liquids Code* (hereinafter referred to as NFPA 30), and NFPA 30A as well as Article 514 of the *NEC* for information on the hazardous (classified) areas and other guidance on fire safety in service stations.

Bulk Storage Plants (Flammable Liquid)

Where gasoline or other volatile flammable liquids are stored in tanks having an aggregate capacity of one carload or more and from which such products are distributed, the hazardous (classified) areas are defined in NFPA 30 and Article 515 of the *NEC.* NFPA 30 also contains many requirements for the safe construction and utilization of bulk storage plants.

Paint Spray Booths and Areas

The *NEC* hazardous (classified) locations requirements apply to areas where paints, lacquers, or other flammable finishes are regularly or frequently applied by spraying or dipping, or by other means and where volatile flammable solvents or thinners are used, or where readily ignitable deposits or residues from such paints, lacquers, or finishes may occur. Further information regarding safeguards for finishing processes are contained in NFPA 33, *Standard for Spray Application Using Flammable or Combustible Materials,* and NFPA 34, *Standard for Dipping and Coating Processes Using Flammable or Combustible Liquids.* Hazardous (classified) areas with respect to flammable vapors are defined in Article 516 of the *NEC.*

Chemical Plants

In chemical plants where flammable liquids or gases are processed or handled, NFPA 497A, *Recommended Practice for Classification of Class I Hazardous (Classified) Locations for Electrical Installations in Chemical Process Areas,* provides information on the extent of the hazardous (classified) location.

Flammable Anesthetics

The use of flammable anesthetics in health care facilities has been greatly reduced; however, there may be some older facilities that still use them. Special rules apply in hospital operating rooms and other locations where flammable anesthetics are or may be administered to patients. Article 517 of the *NEC* defines the anesthetizing location as: "Any area of a health care facility that has been designated to be used for the administration of any flammable or nonflammable anesthetic agent in the course of examination or treatment, including the use of such agents for relative analgesia."

In a flammable anesthetizing location, the entire area is considered to be a Class I, Division 1 location extending upward to a level 5 ft (1.5 m) above the floor. The remaining volume up to the structural ceiling is considered to be above a hazardous (classified) area.

Any room or location in which flammable anesthetics or volatile flammable disinfecting agents are stored must be considered to be a Class I, Division 1 location from floor to ceiling.

For further information on the subject of flammable anesthetics, reference should be made to Chapter 3 of NFPA 99.

Places of Assembly, Theaters, and Similar Locations

In places of assembly, theaters, and similar locations where numbers of people congregate, it is important that the electric equipment be properly designed and installed. The general rules of the *NEC*, as well as its special requirements for these occupancies, should be followed in installing the equipment. NFPA *101* should be consulted for further information on use of emergency lighting.

BIBLIOGRAPHY

NFPA Codes, Standards, and Recommended Practices

Reference to the following NFPA codes, standards, and recommended practices will provide further information on the safeguards for electrical systems and appliances discussed in this chapter. (See the latest version of *The NFPA Catalog* for availability of current editions of the following documents.)

NFPA 12, *Standard on Carbon Dioxide Extinguishing Systems*
NFPA 20, *Standard for the Installation of Centrifugal Fire Pumps*
NFPA 30, *Flammable and Combustible Liquids Code*
NFPA 30A, *Automotive and Marine Service Station Code*
NFPA 33, *Standard for Spray Application Using Flammable or Combustible Materials*
NFPA 34, *Standard for Dipping and Coating Processes Using Flammable or Combustible Liquids*
NFPA 70, *National Electrical Code®*
NFPA 70B, *Recommended Practice for Electrical Equipment Maintenance*
NFPA 70E, *Standard for Electrical Safety Requirements for Employee Workplaces*
NFPA 72, *National Fire Alarm Code*
NFPA 75, *Standard for the Protection of Electronic Computer/Data Processing Equipment*
NFPA 77, *Recommended Practice on Static Electricity*
NFPA 79, *Electrical Standard for Industrial Machinery*
NFPA 88A, *Standard for Parking Structures*
NFPA 88B, *Standard for Repair Garages*
NFPA 90B, *Standard for the Installation of Warm Air Heating and Air Conditioning Systems*
NFPA 99, *Standard for Health Care Facilities*
NFPA 101®, *Life Safety Code®*
NFPA 110, *Standard for Emergency and Standby Power Systems*
NFPA 325, *Guide to Fire Hazard Properties of Flammable Liquids, Gases, and Volatile Solids*
NFPA 409, *Standard on Aircraft Hangars*
NFPA 496, *Standard for Purged and Pressurized Enclosures for Electrical Equipment*
NFPA 497A, *Recommended Practice for Classification of Class I Hazardous (Classified) Locations for Electrical Installations in Chemical Process Areas*
NFPA 497M, *Manual for Classification of Gases, Vapors and Dusts for Electrical Equipment in Hazardous (Classified) Locations*
NFPA 501A, *Standard for Fire Safety Criteria for Manufactured Home Installations, Sites, and Communities*
NFPA 501C, *Standard on Recreational Vehicles*
NFPA 501D, *Standard for Recreational Vehicle Parks and Campgrounds*
NFPA 513, *Standard for Motor Freight Terminals*
NFPA 780, *Lightning Protection Code*

Additional Reading

An Illustrated Guide to Electrical Safety, U.S. Department of Commerce, Occupational, Safety and Health Administration, Washington, DC, 1983.
Barnes, M. A. and Matheson, A. F., Technical Evaluation of Fire Related Characteristics of Vinyl Based and Non Halogenated Cable Materials,"

EVC Compounds Ltd., Helsby, UK, Hydro Polymers Ltd., Aycliffe, UK.
Beck, P. E, "1996 NEC: An Engineers Guide," *Consulting-Specifying Engineer*, Vol. 18, No. 4, Oct. 1995, pp. 46–48.
Bernstein, T., "Investigation of Alleged Appliance Electrocutions and Fires Caused by Internally Generated Voltages," IEEE *Transactions* on Industry Applications, Vol. 25, No. 4, July-August 1989, pp. 664-668.
Braun, E., Shields, J. R., and Harris, R. H., "Flammability Characteristics of Electrical Cables Using the Cone Calorimeter," *NISTIR 88-4003*, Jan. 1989, Center for Fire Research, Gaithersburg, MD.
Earley, M. W., Caloggero J. M., and Sheehan, J. V., eds., *National Electrical Code Handbook*, 7th ed., National Fire Protection Association, Quincy, MA, 1996.
Electrical Appliance and Utilization Equipment Directory, Underwriters Laboratories Inc., Northbrook, IL.
Electrical Construction on Materials Directory, Underwriters Laboratories Inc., Northbrook, IL.
"Electricity in Emergencies," *Industrial Fire World*, Vol. 2, No. 4, August 1987, pp. 24-27.
Fanney, A. H. and Dougherty, B. P, "Thermal Performance of Residential Electric Water Heaters Subjected to Various Off-Peak Schedules," Department of Energy, Washington, DC, *Solar Engineering*, Vol. 2, 1992, pp. 1221–1230; International Solar Energy Conference, 1992, April 5–9, 1992, Maui, HI, Am. Soc. of Mechanical Engineers, New York, NY, 1992.
Farrell, G. W., "Designing for Safety," Electrical Consultant, Cary, IL, *Consulting-Specifying Engineer*, Vol. 19, No. 5, May 1996, pp. 47–48,50.
Freund, A., "Fire-Testing Cast-Coil Transformers," *EC&M*, NY, 1986.
Gibbons, J. A. M., and Stevens, G. C., "Limiting the Corrosion Hazard from Electrical Cables Involved in Fires," *Fire Safety Journal*, Vol. 15, No. 2, 1989, pp. 183-190.
Goto, K., *et al*., "Flame Retardance and Flammability Test of Insulating Components of Electrical Apparatus," *Proceedings* of the Symposium on Electrical Insulating Materials, Institute of Electrical Engineers of Japan, 1986, pp. 293-296.
Handbook of Electrical Hazards and Accidents, Geddes, L.A., Editor, CRC Press, Inc., Boca Raton, FL, 1995.
Hazardous Location Equipment Directory, Underwriters Laboratories Inc., Northbrook, IL.
Hilado, C. J., *Flammability of Electrical and Electronic Materials*, Technomic Publishing Co., Lancaster, PA, 1985.
Hoover, J., *et al*, "Full-Scale Fire Research on Concealed Space Communication Cables," *Proceedings* of Interflam '96, Seventh International Interflam Conference, Interscience Communications Ltd., London, England, 1996, pp. 295–304 pp.
Isner, M. S., "Telephone Exchange Fire, Los Angeles, California, March 15,1994," Fire Investigation Report, National Fire Protection Association, Quincy, MA, 1994, 36 pages.
Kassabian, A., "Electrical Baseboard Heaters," Fire Investigation Report, Office of the Fire Marshall, Canada, March 31, 1993, 27 pages.
Kaufman, S., "1990 National Electrical Code—Its Impact on the Communications Industry," 38th International Wire and Cable Symposium, U. S. Army Communications Electronics Command (CECOM), 1989, pp. 301–305.
Lazar, I., "System and Equipment Grounding for Industrial Plants," *Consulting-Specifying Engineer*, Vol. 3, No. 3, 1988, pp. 86-89.
Liu, S. T., Kelly, G. E., and Terlizzi, C. P., "Performance Testing of a Family of Type I Combination Appliances," Department of Energy, Washington, DC, NISTIR 5626, April 1995, 30 pages.
Magison, E. C., *Electrical Instruments in Hazardous Locations*, 3rd ed., Instrument Society of America, Pittsburgh, PA, 1978.
McGuinnes, W. J., *Mechanical and Electrical Equipment for Buildings*, 7th ed, Wiley, New York, 1986.
Menke, K., "Improving Apparatus Electrical Systems," *Fire Engineering*, Vol. 149, No. 2, February 1996, pp. 43–46.
Northrup, S. D., "Evaluation of Less Flammable Transformer Liquids," *Proceedings* of the 5th BEAMA International Electrical Insulation Conference, Cavanagh Associates, Leatherhead, England, 1986, pp. 101-106.
National Electrical Safety Code, Institute of Electrical and Electronics Engineers, NY, 1989.
Orbeck, T., "Development of Guides for Fire Hazards Assessments for Electrical Insulating Materials and Systems," *Conference Record* of the 1986 IEEE International Symposium on Electrical Insulation, IEEE, 1986, pp. 265-269.

Palko, E., "1990 National Electrical Code: What the Changes Mean to Plant Engineers," *Plant Engineering*, Vol. 44, No. 7 April 12, 1990, p. 79.

Rakosnik, R.J., "Back-Wiring Poses Fire Hazard," *Fire Engineering*, Vol. 148, No. 3, March 1995, pp. 84–87.

Schram, P. J., and Earley, M. W., *Electrical Installations in Hazardous Locations*, National Fire Protection Association, Quincy, MA, 1988.

Scoones, K., "Fires With Electrical Causes 1990," *Fire Prevention*, No. 245, December 1991, p. 18.

Scoones, K., "Serious Fires Due to Electrical Causes During 1992," *Fire Prevention*, No. 264, November 1993, pp. 14–16.

Smith, L. E., and McCoskrie, D., "What Causes Wiring Fires in Residences?" *Fire Journal*, Vol. 84, No. 1, Jan./Feb. 1990, p. 18.

Snell, J. E., "Measuring Hazards of Products of Combustion from Electrical Systems," *Fire Journal*, Vol. 81, No. 5, Sept./Oct. 1987, pp. 108-109.

Sundin, D. W., and Northrup, S. D., "Relationship Between Liquid Flammability Tests and Transformer Fire Situations," *Conference Record*—Industrial and Commercial Power Systems Technical Conference 1987, IEEE, 1987, pp. 117-121.

Swingler, S. G., Stevens, G. C., and Gibbons, J. A. M., "Small-scale Assessment of Flame Propagation and Heat Release Rate of Electric Cable Materials," *Proceedings* of the Fifth International Conference on Dielectric Materials, Measurements and Applications, IEE Conference Publication No. 289, IEE, London, 1988, pp. 45-50.

Teague, P. E., "1992 NEC Addresses Current Concerns," *NFPA Journal*, Vol. 86, No. 5, September/October 1992, pp. 50–52,54–55.

UL 94, "Standard for Safety Test for Flammability of Plastic Materials for Parts in Devices and Appliances," UL 94, Underwriters Laboratories Inc., Northbrook, IL, 1991.

UL 1410, "Standard for Safety Television Receivers and High-Voltage Video Products," Underwriters Laboratories Inc., Northbrook, IL, 1986.

"Understanding Electricity and Electrical Dangers," *Fire Engineering*, Vol. 149, No. 4, April 1996, pp. 57–92.

Whittington, B. W., "Electrical Installations in Industrial Locations," *Industrial Fire Hazards Handbook*, 3rd ed., Cote, A. E., ed., National Fire Protection Association, Quincy, MA, 1990, pp. 1103-1126.

Yereance, R. A., *Electrical Fire Analysis*, Charles C. Thomas, Springfield, IL, 1987.

CONTROL OF ELECTROSTATIC IGNITION SOURCES

Revised by Don R. Scarbrough

Static electricity as a source of ignition is a hazard common to a wide variety of industries and processes. This chapter explains the nature of static electricity and means and methods of eliminating or minimizing it as an ignition source. Information on the hazards of static electricity in specific circumstances and processes can be found in the following chapters: Section 1, Chapter 6, "Explosions"; Section 3, Chapter 1, "Electrical Systems and Appliances"; Section 3, Chapter 16, "Spray Finishing and Powder Coating"; Section 3, Chapter 21, "Storage of Flammable and Combustible Liquids"; Section 4, Chapter 7, "Gases"; Section 4, Chapter 14, "Explosion Prevention and Protection"; and Section 4, Chapter 15, "Dusts."

The term "static electricity" as used in this chapter refers to the electrification of materials through the physical processes described below, and the effects of the positive and negative charges so formed, particularly where discharges may result that constitute a fire or explosion hazard. Other electrical terms used in this chapter are defined in an Appendix at the end of the chapter.

When electricity is present on the surface of a nonconductive body, where it is trapped or prevented from escaping, it is termed *static electricity*. Electricity on a conducting body that is in contact only with nonconductors is also prevented from escaping and is therefore nonmobile or "static." In either case, the body on which this electricity is evident is said to be "charged."

Around 1890, electricity was described as a "subtle agent, without weight or form, that appears to be diffused through all nature, existing in all substances, which gives no indication of its presence in the latent state, but is capable of producing sudden and destructive effects in its active state." With the discovery of the electron, this definition now seems prophetic, and the adjective "static" more meaningful. Electricity can move freely through some substances, such as metals, that are called "conductors," but can flow with difficulty or not at all through or over the surface of a class of substances called "nonconductors" or "insulators." This latter group includes gases, glass, amber, resin, sulfur, paraffin, most synthetic plastics, and dry petroleum oils.

The development of electrical charges in itself may not be a potential fire or explosion hazard. There must be a discharge or sudden recombination of separated positive and negative charges. In order for static electricity to be a source of ignition, four conditions must be fulfilled:

1. There must be an effective means of static generation.
2. There must be a means of accumulating the separate charges and maintaining a suitable difference of electrical potential.
3. There must be a discharge of adequate energy.
4. The discharge must occur in an ignitable mixture.

Static electricity may appear as the result of motions that involve the separation or pulling apart of contacting surfaces, usually of dissimilar substances, either liquid or solid, one or both of which usually must be a poor conductor of electricity. Charge generated by this mechanism is usually referred to as the result of "separation" or "frictional" or "tribo" charging. Examples of such motion that are commonly found in industry are:

1. Motions of all sorts that involve friction between contacting surfaces, usually of dissimilar liquids or solids.
2. Steam, air, or gas flowing from any opening in a pipe or hose, when the steam is wet or the air or gas stream contains particulate matter.
3. The separation of a stream of liquid from contact with a hose, faucet, or pouring spout.
4. Pulverized materials passing through chutes or pneumatic conveyors.
5. Nonconductive power or conveyor belts in motion.
6. Moving vehicles.

Charge may appear upon the surface of an object when that object is simply brought near another object that bears a static charge. This charge mechanism is known as "induction" and results fundamentally from manipulation (through repulsion and attraction) of the inherently balanced charges within an uncharged object to produce separated regions upon its surface that bear balanced, but opposite charges.

An initially uncharged surface may become charged through "contact" with another object or surface that is charged. For example, a pail placed to collect a falling stream of liquid or pulverized material poured from a faucet or other container (the stream therefore being charged by "separation") will become charged by contact with the charged stream of material.

Yet another mechanism for placing charge upon a surface involves "bombardment" of the surface with charged molecules (ions) of air, which themselves have become charged by bombardment with electrons within a corona near a high-voltage electrode. In this

Don R. Scarbrough is a product safety staff consultant for the Nordson Corporation, Westlake, OH, and a long-standing member of NFPA's Technical Committee on Static Electricity.

process the target surface becomes charged through attachment of the impacting ions, some of which may remain attached, and some of which may surrender their charge to the surface and then float away.

Other processes through which charge may be imparted to an object or fluid involve exposure to an electron beam or flow around a high-voltage electrode. These techniques are commonly referred to as "charge injection."

The generation of static electricity cannot be prevented absolutely, because its intrinsic origins are present at every interface. The object of most static-corrective measures is to provide a means whereby charges separated by whatever cause may recombine harmlessly before sparking potentials are attained. If hazardous static conditions cannot be avoided in certain operations, means must be taken to ensure that there are no ignitable mixtures at points where sparks may occur.

STATIC GENERATION

When two bodies composed of dissimilar materials are in close physical contact, there is likely to be a transfer of free electrons between them, one giving up electrons to the other, and an attractive force is established. When the bodies are separated, work must be done in opposition to these attractive forces. The expended energy reappears as an increase in electrical tension or voltage between the two surfaces. If the bodies are insulated from their surroundings, both are said to be "charged"; the one having the excess of electrons is said to have a negative charge, while the other is said to have an equal positive charge. If a conductive path is available between them, the charges thus separated will reunite immediately. If no such path is available, as is the case with insulators, the potential increase with separation may easily reach values of several thousand volts.

In many cases, one of the objects has a deliberate or inherently conductive path to the earth, and its charge is immediately lost to the earth, which is considered to have an infinite capacity to absorb or give up electrons. The other (insulated) object now retains its "charge" (often called potential or voltage) with respect to its surroundings. It would hold this charge indefinitely, except for the fact that it must somehow be supported and no supporting insulator (even air) is a perfect nonconductor.

If one of the bodies is itself a nonconductor, the flow of electrons across its surface is inhibited, and the charge tends to remain at the points where electron transfer originally occurred. A highly charged insulating surface can be discharged with the appearance of a spark by bringing an "earthed" conductor close to it, but only a limited area will be so discharged, and sparks so produced seldom release enough energy to cause ignition. Thus, nonconductors, the bodies most directly involved in charge separation, are usually not directly responsible for fires and explosions. (An exception is gases and flammable liquids, which are discussed later in this chapter.) However, such charges can in some cases be the agency for building up or accumulating a charge on a conductive body, which can release all of its stored energy in an incendive spark.

On a conductive body the charge is free to move. Because like charges repel each other, the charge will distribute itself over the surface. If no other body is in close proximity, the concentration will favor the surface having the least radius of curvature; for a point, if the voltage is high enough, the voltage gradient may exceed the breakdown potential of air (about 30,000 V/cm) and the air can be ionized, and a brush discharge may occur.

Like charges repel each other and unlike charges attract, because of forces resident in the electrical fields that surround them. These forces have a strong influence on nearby objects. If the neigh-

boring object is a conductor it will experience a separation of charges by induction. Its repelled charge is free to give or receive electrons as the case may be; if another conductor is brought near, the transfer may occur through the agency of a spark, very often an energetic spark. When the inducing charge is moved away from the insulated conductor, there is a reversed sequence of events, and sparks may result. Thus, in many situations, induced charges are far more dangerous than the initially separated ones upon which they are dependent.

If the object close to the highly charged nonconductor is itself a nonconductor, it will be polarized, i.e., its constituent molecules will be oriented to some degree in the direction of the lines of force, since the electrons have no true migratory freedom. Because of their polarizable nature, insulators and nonconductors are often called dielectrics. Their presence as separating media enhances the accumulation of charge.

Storage

Two conductive bodies separated by an insulator constitute a capacitor or condenser, and where a potential difference is applied between these bodies, electricity can be stored. One body receives a positive and the other an equal negative charge. In many instances involving accumulation of static electricity, one of the bodies is the earth, the insulating medium is the air, and the insulated body is some object to or from which a charge (electrons) has been transferred by one of the mechanisms previously described.

When a conducting path is made available, the stored energy is released, or the condenser is "discharged," possibly producing a spark. The energy so stored and released is related to the capacitance of the condenser (C) and the voltage (V) in accordance with the following:

$$\text{Energy} = \frac{C}{2} \times V^2$$

Energy (millijoules) = C/2 (picofarads) × V^2 (volts × 10^{-9}).

Discharge Energy

The ability of a discharge to produce ignition of a flammable mixture is governed largely by the energy transferred to the mixture, which will be some fraction of the total stored energy available because some energy is expended in heating the electrodes. Experiments at atmospheric pressure with plane electrodes have shown that the spark breakdown voltage has a minimum value at a critical short gap distance—about 350 volts for the shortest measurable gap (e.g., 0.01 mm). Increased gaps require proportionally higher voltages. At close spacing the heat loss or flame quenching effect virtually precludes ignition of the electrodes.

Progressing from lowest to highest intensity, discharges through air may occur as corona, brush, propagating brush, or spark. Corona is generally thought of as not intense enough to ignite gases. Brush discharges range from corona levels to higher intensity that is capable of igniting gases, but not capable of igniting common dusts. Propagating brush and spark discharges may ignite both gases and dusts.

At most favorable electrode spacing, tests have shown that optimum mixtures of saturated hydrocarbon vapors and gases in air require about 0.25 millijoules of discharge energy to produce ignition. Examples of the capacitance necessary to store the required 0.25 millijoules at various voltages are listed in Table 3-2A.

For gaps of 1.5 mm or more, substantially longer than the quenching distance, the total energy required to produce ignition is increased somewhat in proportion to the excess of spark length over the diameter of the necessary critical flame volume. (For U.S.

TABLE 3-2A. *Examples of Conditions Required to Produce Ignition*

	Potential (volts)	Capacitance (pf)*	Gap Length	
A.	350	4,000	Minimum voltage to jump shortest measurable gap. Quenching effect precludes ignition.	
B.	1500	222	Gap about 0.5 mm, just exceeding quenching distance. Incendiary sparks.	
C.	5000	20		1.5 mm
D.	10,000	5	Object the size of a baseball.	3+ mm
E.	20,000	1¼	Object the size of a large marble.	7 mm

*Picofarads.
For U.S. units: 1 in. = 25.4 mm.

customary unit: 1 in. = 25.4 mm.) This, in turn, may require somewhat greater capacitance than indicated above. This explains why corona discharge from a sharp point at very high voltage may not be incendive.

Ignitable Mixtures

Elimination of ignitable mixtures in the areas where sparks of static electricity can occur is the surest method of preventing static-caused fires. This is practical in certain areas, and is better discussed later in connection with the specific process involved.

Summary

In summary, static electricity will be manifest only where highly insulated bodies or surfaces are found. If a body is "charged" with static electricity, there will always be an equal and opposite charge produced. If a hazard is suspected, the situation should be analyzed to determine the location of both charges and to see what conductive paths are available between them. Tests of the high-resistance paths should be made with an applied potential of 500 volts or more, in order that a minor interruption (e.g., paint or grease film or air gap) will be broken down and a correct reading of the instrument obtained. Resistances as high as 10,000 megohms will provide an adequate leakage path in many cases; when charges are generated rapidly, however, a resistance as low as 1 megohm might be required. Where bonds are applied, they should connect the bodies on which the two opposite charges are expected to be found.

DISSIPATION OF STATIC ELECTRICITY

A static charge that already exists can be removed or allowed to dissipate itself. The acts or conditions that can accomplish this are the same as those acts or conditions that permit the separation of charges to occur in the first place. Thus, dissipation of static and prevention of its generation are opposite approaches to the same objective.

Humidification

A static charge cannot persist except on a body insulated from its surroundings. Most commonly encountered materials that are not usually thought of as conductors, such as fabrics, paper, wood, concrete or masonry foundations, etc., contain a certain amount of moisture in equilibrium with that in the surrounding atmosphere. This moisture content varies, depending on weather (and to a large measure controls the conductivity of the material), and hence its

ability to prevent the escape of static electricity. In an analogous manner, under some conditions water vapor may condense on the surface of some nominally insulating materials, notably glass and porcelain, to render the surface conductive.

The conductivity of the materials under discussion—wood, paper, etc.—is not controlled by the absolute water content of the air but by its relative humidity. This figure, as ordinarily recorded in weather reports and comfort charts, is the ratio of the partial pressure of the moisture in the atmosphere to the partial pressure of water at the prevailing atmospheric temperature. Under conditions of high relative humidity (50 percent or higher) the materials in question will reach equilibrium conditions containing enough moisture to make the conductivity adequate to prevent static accumulations. The generating mechanism may be present, but the generated charge leaks away so fast that no observable accumulation results.

At the opposite extreme, with relative humidities of 30 percent or less, these same materials may dry out, become good insulators, and static manifestations become noticeable. There is no definite boundary line between these two conditions.

It should be emphasized that the conductivity of these materials is a function of relative humidity. At any constant moisture content, the relative humidity of an atmosphere decreases as the temperature is raised. In cold weather, the absolute humidity of the outdoor atmosphere may be low, even though the relative humidity may be high. When this same air is brought indoors and heated, the relative humidity becomes very low. As an example, a saturated atmosphere at an outdoor temperature of 30°F (–1°C) would have a relative humidity of only a little over 20 percent if heated up to a room temperature of 70°F (21°C). This phenomenon is responsible for the common belief that static generation is always more intense during winter months. The static problem is usually more severe during this period because static charges on a material have less ability to dissipate when relative humidities are low.

Where static electricity has introduced operational problems, such as the adhesion or repulsion of sheets of paper, layers of cloth, fibers, and the like, humidifying the atmosphere has proved to be a solution. It is usually stated that a relative humidity of about 50 percent or higher will avoid such difficulties.

Unfortunately, it is not practical to humidify all occupancies in which static might be a hazard. It is necessary to conduct some operations in an atmosphere having a low relative humidity to avoid deleterious effects on the materials handled. High humidity can also cause intolerable comfort conditions in operations where the dry bulb temperature is high. On the other hand, a high humidity may advantageously affect the handling properties of some materials, thus providing an additional advantage. In some cases, localized humidification produced by directing a steam jet onto critical areas may provide satisfactory results without the need for increasing the humidity in the whole room. However, it must be remembered that steam that contains droplets of water may itself generate static. Local static can be reduced by providing a low-velocity jet of humidified air.

It does not follow that humidification is a cure for all static problems. In particular, it must be remembered that the conductivity of the air is not appreciably increased by the presence in it of water in the form of a gas. If static electricity accumulates on some surface that is heated above normal atmospheric temperature, e.g., cloth passing over heated rollers, alterations in the relative humidity in the surrounding air may do no good whatsoever.

Another situation where the control of atmospheric humidity appears to accomplish little is in controlling the charge of static electricity that appears on the surface of oils under some circumstances. Such a surface does not absorb water vapor in the same way as paper or wood does, so it can remain an insulating surface capable of accumulating static charges even though the atmosphere above it may have a relative humidity up to 100 percent.

In summary, humidification of the atmosphere to a relative humidity of 50 percent or higher may be a cure for static problems where the surfaces on which the static electricity accumulates are those materials that reach equilibrium with the atmosphere, such as paper or wood, and that are not abnormally heated. For heated surfaces and for static on the surface of oils and some other liquid and solid insulating materials, high humidity will not provide a means for draining off static charges, and some other solution must be sought.

Bonding and Grounding

Where natural conditions, including humidity, do not ensure a conductive path to prevent static accumulation, artificial conducting paths may be necessary.

"Bonding" is the process of connecting two or more conductive objects together by means of a conductor. "Grounding" (earthing) is the process of connecting one or more conductive objects to the earth, and is a specific form of bonding. A conductive object may also be grounded by bonding it to another conductive object that is already connected to the ground. Some objects are inherently bonded or inherently grounded by their contact with the ground. Examples are underground piping or large storage tanks resting on the ground. Bonding minimizes potential differences between conductive objects. Grounding minimizes potential differences between objects and the earth.

Bond wires and ground wires should have adequate capacity to carry the largest currents that may be anticipated for any particular installation. When currents to be handled are small, the minimum size of wire is dictated by required mechanical strength rather than current carrying capacity. The currents encountered in the bond connections used in the protection against accumulations of static electricity are in the order of microamperes (one millionth of an ampere).

The acceptable resistance in a ground connection depends upon the type of hazard for which it is intended to give protection. To prevent the accumulation of static electricity, the resistance need not be less than 1 megohm and in most cases may be even higher. To protect electrical power circuits, the resistance must be low enough to ensure operation of the fuse or circuit breaker under fault conditions. Any ground that is adequate for power circuits or lightning protection is more than adequate for protection against static electricity.

A bond or ground is composed of suitable conductive materials having adequate mechanical strength, corrosion resistance, and flexibility for the service intended. Since the bond or ground does not need to have low resistance, nearly any conductor size will be satisfactory from an electrical standpoint. Solid conductors are satisfactory for fixed connections. Flexible conductors are used for bonds that are to be connected and disconnected frequently. Conductors may be insulated or uninsulated. Some prefer uninsulated conductors so that defects can be easily spotted by visual inspections. If insulated for mechanical protection, the concealed conductor is checked for continuity at regular intervals, depending on the inspector's experience. Connections may be made with pressure-type ground clamps, brazing, or welding. Battery clamps, or magnetic or other special clamps, provide metal-to-metal contact.

A special situation requiring substantial conductors may arise if there is a possibility that a ground wire may be called upon to carry current from power circuits or lightning protection systems.

Ionization

Under certain circumstances air may become sufficiently conductive to bleed off static charges.

Static comb: A static charge on a conducting body is free to flow, and on a spherical body in space it will distribute itself uniformly over the surface. If the body is not spherical, the self-repulsion of the charge will make it concentrate on the surfaces having the least radius of curvature.

If the body is surrounded by air (or other gas) and the radius of curvature is reduced to almost zero, as with a sharp needle point, the charge concentration on the point can produce ionization of the air, rendering it conductive. As a result, whereas a surface of large diameter can receive and hold a high voltage, the same surface equipped with a sharp needle point can reach only a small voltage before the leakage rate equals the rate of generation.

A "static comb" is a metal bar equipped with a series of needle points. Another variation is a metal wire surrounded with metallic tinsel. If a grounded static comb is brought close to an insulated charged body (or a charged insulating surface), ionization of the air at the points will provide enough conductivity to make the charge speedily leak away or be "neutralized." This principle is sometimes employed to remove the charge from fabrics, power belts, and paper.

Electrical neutralization: The electrical neutralizer is a line-powered high-voltage device that is an effective means for removing static charges from materials, such as cotton, wool, or silk, or paper in process, manufacturing, or printing. It produces a conducting ionized atmosphere in the vicinity of the charged surfaces. The charges thereby leak away to some adjacent grounded conducting body. Electrical neutralizers should not be used where flammable vapors, gases, or dust may be present, unless approved specifically for such locations.

Radioactive neutralizer: Another method for dissipating static electricity involves the ionization of the air by radioactive material. Such installations require no redesign of existing equipment. However, engineering problems are involved, such as elimination of health hazards, dust accumulations, determination of radiation required, and proper positioning in relation to the stock, machine parts, and personnel. These considerations are best worked out in consultation with radiation specialists.

Open flame: Ionization of the air can also be obtained by an open flame. This method is frequently used in the printing industry to remove static from paper sheets as they come off the press, thus avoiding the mechanical problems involving one sheet of paper adhering to another, but obviously not avoiding an ignition source.

CONTROL OF IGNITABLE MIXTURES

Despite efforts to prevent accumulation of static charges, which should be the primary aim of good design, there are many operations involving the handling of nonconductive materials or nonconductive equipment, which do not lend themselves to this built-in solution. It may then be desirable, or essential, depending on the hazardous nature of the materials involved, to provide other measures to supplement or supplant static dissipation facilities. For example, where a normally ignitable mixture is contained within a small enclosure, such as a processing tank, an inert gas may be used to bring the mixture well below its flammable range. When operations are normally conducted in an atmosphere above the upper flammable limit, it may be practicable to apply the inert gas only during the periods when the mixture passes through its flammable range.

Mechanical ventilation may be applied in many instances to dilute an ignitable mixture to well below its normal flammable range. Also by directing the air movement, it may be practical to prevent the flammable liquids or dusts from approaching an operation where an otherwise uncontrollable static hazard may exist. To be considered reliable, the mechanical ventilation should be interlocked with the equipment to ensure its proper operation.

Where a static-accumulating piece of equipment is unnecessarily located in a hazardous area, it is preferable to relocate the equipment to a safe location rather than to rely upon prevention of static accumulation.

FLAMMABLE LIQUIDS

Static is generated when liquids move in contact with other materials. This occurs commonly in such operations as flowing through pipes, and in mixing, pouring, pumping, filtering, or agitating. Under certain conditions, particularly with liquid hydrocarbons, with high-velocity flow, and with filtration through insulative, resin-bonded media, static may accumulate. If the accumulation is sufficient, a static spark may occur. If the spark occurs in the presence of a flammable vapor-air mixture, an ignition may result. Therefore, steps must be taken to prevent the simultaneous occurrence of the two conditions.

Standard control measures are designed to prevent incendive sparks, or the formation of ignitable vapor-air mixtures. In many cases, air that may form an ignitable mixture with the vapor can be eliminated or reduced in concentration to render the mixture nonflammable.

Before a container is filled, contact is made between the filling nozzle and the container, and the contact should be maintained throughout the filling operation. By this procedure, any difference in potential between the container and nozzle will be dissipated before the filling operation is started, and differences in potential between the nozzle and container are prevented from forming during the filling operation. See Figure 3-2A for methods of bonding containers and nozzles during container filling.

The relative static-accumulating tendencies of a number of petroleum products have been measured in laboratory tests. In general, aliphatic solvents and lower boiling hydrocarbons exhibited lower charging tendencies than higher boiling products. However, the charging tendency of any given product was found to vary widely from one sample to the next.

The conductivity of a liquid is a measure of its ability to hold a charge. The lower the conductivity, the greater the ability of the liquid to hold a charge. If the conductivity of a liquid is greater than 50 pS/m (picosiemens per meter), any charges that are generated will dissipate without accumulating to a hazardous potential.

Free Charges on Surface of Liquid

If an electrically charged liquid is poured, pumped, or otherwise transferred into a tank or container, the unit charges of similar sign (i.e., + or −) within the liquid will be repelled from each other toward the outer surfaces of the liquid, including not only the surfaces in contact with the container walls but also the top surface adjacent to the air space, if any. It is this latter charge, often called the "surface charge," that is of most concern in many situations.

In most cases the container is of metal, and hence conducting. Two situations can occur, somewhat different with respect to protective measures, depending on whether the container is in contact with the earth or is insulated from it. These two situations are: (1) an or-

dinary storage tank resting on earth or concrete or other slightly conducting foundation, and (2) a tank truck on dry rubber tires.

In the first situation, the metal container is connected to ground. The charges that reach the surfaces in contact with the vessel will reunite with charges of opposite sign that have been attracted there. During this process, the tank and its contents, considered as a unit, are electrically neutral, i.e., the total charge in the liquid and on its surface is exactly equal and opposite to the charge on the tank shell. This charge on the tank shell is "bound" there, but gradually disappears as it reunites with the charge migrating through the liquid. The time required for this to occur is called "relaxation time." The relaxation time depends primarily on the conductivity of the liquid. It may be a fraction of a second or several minutes.

During this process, the tank shell is at ground potential. Externally, as already mentioned, the container is electrically neutral. But internally, there may be differences of potential between the container wall and the fluid, lasting until charges on the fluid have gradually leaked off and reunited with the unlike charges on the tank walls.

If the potential difference between any part of the liquid surface and the metal tank shell should become high enough to cause ionization of the air, electrical breakdown may occur and a spark may jump to the shell. A spark across the liquid surface is an ignition hazard where flammable vapor-air mixtures are present. No bonding or grounding of the tank or container can remove this internal surface charge.

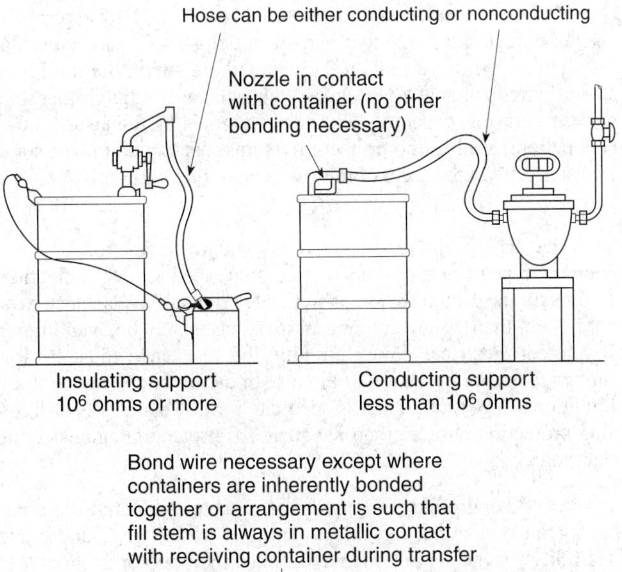

FIG. 3-2A. Recommended methods of bonding flammable liquid containers during container filling.

In the second situation, when the tank shell is highly insulated from the earth, the charge on the liquid surface attracts an equal and opposite charge to the inside of the container. This leaves a "free" charge on the outside of the tank, of the same sign as that in the liquid, and of the same magnitude. This charge can escape from the tank to the ground in the form of a spark. In filling a tank truck through an open dome, it is this source of sparking that is suspected of having caused some fires; in this case, the spark jumps from the edge of the fill opening to the fill pipe, which is at ground potential. This hazard can be controlled by grounding the container before filling starts or by bonding the fill pipe to the tank. If grounding the tank is used, the fill pipe must also be grounded.

The foregoing discusses the distribution of charges delivered into a container with a flowing stream. Further generation or separation may occur within the container in several ways to produce a surface charge: (1) flow with splashing or spraying of the incoming stream, (2) disturbance of water bottom by the incoming stream, (3) bubbling of air or gas through a liquid, or (4) jet or propeller blending within the tank.

These charges on the surface of a liquid cannot be prevented by bonding or grounding, but can be rendered harmless by inerting the vapor space, by displacing part of the oxygen with a suitable inert gas, or by increasing the concentration of flammable gas in the vapor space to above the upper flammable limit with a gas, such as natural gas. Use of conductivity additives will rapidly relax the surface charge and prevent the buildup of a hazardous potential.

It must be recognized that mists and foams of flammable and combustible liquids can be ignited by static sparks in much the same way dusts can be ignited. Ignition is possible even though the liquid in the mist is below its flash point.

GASES

Gases not contaminated with solid or liquid particles have been found to generate little, if any, electrification in their flow. When the flowing gas is contaminated with dust, metallic oxides, scale particles, etc., or with liquid particles or spray, electrification may result. A stream of such particle-containing gas directed against a conductive body will charge the latter unless it is grounded or bonded to the discharge pipe.

When any gas is in a closed system of piping and equipment, the system need not be electrically conductive or electrically bonded. Compressed air or steam containing particles of condensed water vapor often manifests strong electrification when escaping.

Carbon dioxide, discharged as a liquid from orifices under high pressure (where it immediately changes to a gas and "snow"), can result in static accumulations on the discharge device and the receiving container. This condition is not unlike the effect from contaminated compressed air or from steam flow where the contact effects at the orifice play a part in the static accumulation. High-pressure carbon dioxide should not be discharged into flammable atmospheres because it presents a high risk of ignition due to static spark.

Hydrogen-air and acetylene-air mixtures may be ignited by a spark energy of as little as 0.017 millijoule. In the pure state, no static charges are generated by the flow of hydrogen. However, as gaseous hydrogen is commercially handled in industry, such as flowing through pipelines, discharging through valves at filling racks into pressure containers, or flowing out of containers through nozzles, the hydrogen may be found to contain particles of oxide carried off from the inside of pipes or containers. In this contaminated state, hydrogen may generate static.

The liquefied petroleum gases (LP-Gases) behave in a manner similar to uncontaminated gases in the gas phase and to contaminated gases in the mixed phase. Bonding is not required when LP-Gas vehicles are loaded or unloaded through closed connections so that there is no release of vapor at a point where a spark could occur, irrespective of whether the hose or pipe used is conducting or nonconducting. (A closed connection is one where contact is made before flow starts and is broken after flow has ended.)

DUSTS AND FIBERS

As previously pointed out, the flow of a stream of gas containing small particles can result in a separation of electrons and the accumulation of a static charge on any insulated conductive body with which it comes in contact. Also, dust displaced from a surface on which it rests may develop a considerable charge. The ultimate charge depends on the inherent properties of the substance, size of particle, amount of surface contact, surface conductivity, gaseous breakdown, external field, and leakage resistance in a system. Greater charges develop from smooth rather than from rough surfaces, probably because of greater initial surface contact. Electrification develops during the first phase of separation. Subsequent impact of airborne particles on obstructions may affect their charge slightly, but if the impact surface becomes coated with the dust, this effect is slight.

Charge generation seldom occurs if both materials are good electrical conductors, but it is likely to occur with a conductor and a nonconductor or two nonconductors. When like materials are separated, as in dispersing quartz dust from a quartz surface, positive and negative charges are developed in the dispersed dust in about equal amounts to give a net zero charge. With materials differing in composition, a charge of one polarity may predominate in the dust. Each of the materials becomes equally charged but with opposite polarity. With a metallic and an insulating material, the former usually assumes a positive and the latter a negative polarity.

Electrostatic charge generation in moving dust normally cannot be prevented. High humidity or grounding of the surface from which dust is dispersed will not eliminate the charge generation. The method of dispersion of the dust, the amount of energy expended in dispersal, the degree of turbulence, and the composition of the atmosphere usually affect the magnitude or distribution of the charges.

Not only can dust participate in the generation of static, it may also be the material ignited by static sparks. A suspension of finely divided combustible particles in air has much the same properties as a flammable gas-air mixture, except for a generally higher ignition energy. (With some metallic dust, ignition energy is comparable to gas-air mixtures.) It can burn to produce explosive effects. It has a lower flammable limit, but no strictly definable upper limit.

Ignition of Dust by Static Discharge

Dust clouds and layers of many combustible materials (with or without a volatile constituent) have been ignited experimentally by static discharge. In some instances, the charge was generated by movement of dust, in others by a static generator (Wimshurst machine), or by electronic equipment. With dust clouds, it has been shown that a minimum dust concentration exists below which ignition cannot take place regardless of the energy of the spark. At the minimum dust concentration a relatively high energy is required for ignition. At higher dust concentrations (5 to 10 times the minimum), the energy required for ignition is at a minimum.

The energy stored in a capacitive circuit has been previously discussed, where it was tacitly assumed that all of this energy was released in the spark (zero circuit resistance). For the ignition of dusts it has been shown that a circuit resistance of 10,000 to 100,000 ohms may be required for optimum igniting power, indicating that some of the energy is dissipated in the circuit instead of being released in the spark, with a corresponding increase in the total energy required.

A layer of combustible dust can be ignited by static discharge and will burn with a bright flash, glow, or, for some metallic dusts, with flame. Apparently, there is little correlation in the minimum energy required for ignition of dust layers and clouds. Layers of some metallic dusts, such as aluminum, magnesium, titanium, and zirconium, require less energy for ignition than carbonaceous materials.

Primary explosives, e.g., mercury fulminate and tetryl, are readily detonated by static spark discharge. Steps necessary to prevent accidents caused by static electricity in explosives manufacturing operations and storage areas vary considerably with the static sensitiveness of the material being handled. In all instances in which static electricity was authentically established as the cause of ignition, the spark occurred between an insulated conductor and ground. It has not been verified experimentally that a dust cloud can be ignited by static discharge within itself.

STATIC DETECTORS

The following devices have an application in the measurement and determination of static electricity within the limitations of each device as described.

Electroscopes

The leaf electroscope is a simple but sensitive device that demonstrates the presence or absence of electrical charges by the repulsion of its leaves when the device is charged. Only units intended as portable dosimeters for ionizing, radiation, and one or two classroom demonstration models are available.

Neon Lamps

A small neon lamp or fluorescent tube will light up feebly when one terminal is grounded (or held in the hand) and the other makes contact with any sizeable conductor that carries a charge potential of 70 volts or more. Like the electroscope, it gives but little quantitative information; however, when it passes current, it may give a rough idea of the rate at which charges are being produced in certain operations. Adjustable series-parallel groupings of such lamps and small capacitors can be arranged to give a semblance of quantitative information.

Electrostatic Voltmeters

These meters operate by electrostatic attraction between movable and stationary metal vanes. No current is passed to maintain deflection because one set of vanes (usually the stationary one) is very highly insulated. Small portable, accurately calibrated instruments are available in several ranges from 100 to 10,000 volts. This type of meter may be used for quantitative electrostatic analysis.

Electrometers

Electrometers are frequently used for laboratory and field investigations of static electricity. These instruments employ special input stages designed for high input resistance and low grid current and may be used in several ways. With a small antenna mounted on the grid terminal, they are very suitable for detecting transient electrostatic charge effects; in the presence of a constant field, the charge induced on the grid will leak off and the meter pointer will return to zero. However, as the rate of leakage is not high, the electrometer

finds considerable use for "on-the-spot" static checks. Often electrometers are equipped with high-resistance terminal shunts to convert them to current meters in the nanoampere range. They can thus be used in many applications to indicate the rate of charge development. A simple adaptation converts the instrument into a megohm meter, as explained in the manufacturer's instruction manual.

Field Mills

A "field mill" is a device that overcomes the serious limitation of the electrometer in that it provides a continuous indication of charge by providing its own transients in the form of continuous grid modulation. It can therefore "look" at a distant charge and determine its potential. The electrical field extending from the charge is chopped by a motor-driven variable condenser or window. The resulting pulsations are transformed, amplified, and rectified to give a dc meter deflection proportional to the strength of the field. A similar instrument of suitable design can be immersed in a liquid, as in a pipeline, to measure charge density.

APPENDIX

Some electrical terms used in this chapter are defined as follows:

Brush discharge: A form of corona in which brighter "fingers" or "streamers" can be seen.

Capacitance: Electrons received by an electrically neutral body of material, such as a person, a car, or an aircraft, raise the voltage at a rate determined by the surface area and shape of the body. The voltage is determined by the surface characteristics (capacitance) of the body and the number of electrons on this surface. The larger the body, the more electrons are needed to raise the voltage a specific amount; hence, the higher the capacitance of this body.

Capacitance is measured in terms of "farads" or millionths of a farad, i.e., "microfarads." Millionths of one millionth of a farad are called "picofarads."

Charge: Measured in coulombs or fractions thereof, the static charge on a body is the number of separated electrons on the body (negative charge) or the number of separated electrons not on the body (positive charge). Electrons cannot be destroyed; so when an electron is removed from one body, it must go to another body. Thus there are always equal and opposite charges produced. Since it would be awkward to say there are 6,240,000,000,000,000,000 electrons on a body, instead it is said that the body has a charge of one "coulomb." A coulomb is simply a name for this specific quantity of electrons. In electrostatics an even more practical unit is a microcoulomb, representing a charge of 6.24×10^{12} electrons.

Corona: A type of air discharge that may occur between flat electrodes or as a single-point discharge at an electrode that has a very small radius or a point. The discharge has a faint blue glow and makes a hissing sound.

Current: Just as water flow is measured in terms of the amount of water that passes a certain point in a specific period of time (gallons or liters per minute), so too is the flow of electrons past a certain point measured by time. The flow of electrons is called current. Current is measured in terms of electrons per second, coulombs per second, or amperes. When dealing with static electricity, current is commonly limited to the microampere range.

Energy: Energy is measured in joules or fractions thereof. A spark is energy being expended. Energy is required to do work. The measure of energy takes several forms. Often it is physical energy, which is measured in foot-pounds or joules or watts. If it is heat energy, it is measured in British thermal units (Btu) or calories (cal); and if it is electrical energy it is measured in watt-seconds or in joules. A joule is the energy expended in 1 sec by an electric current of one ampere in a resistance of one ohm (approximately 0.738 foot-pound). Static discharge energy is usually measured in thousandths of a joule (millijoules or mJ). A static spark needs a minimum amount of energy to cause problems.

Incendive: A discharge that has enough energy to ignite an ignitible mixture is said to be incendive. Thus an incendive spark can ignite an ignitible mixture and cause a fire or explosion. A nonincendive discharge does not possess the energy required to cause ignition even if it occurs within an ignitible mixture.

Incendivity: The ability of a discharge to ignite an ignitible mixture is called incendivity. The energy level required for incendivity varies as described in the text and can be calculated.

Potential: Stored energy is able to do work. In electricity, this ability is expressed in terms of the potential of doing work. Potential in electricity is measured in terms of "volt," "kilovolts," or "millivolts." Potential, or voltage, is measured from a base point. This point can be any voltage but is usually "ground," which is theoretically zero voltage. When one point with a potential of "x" volts to ground is compared with another point with a potential of "y" volts to ground, then it is said that a potential difference of "x-y" volts exists between the two. Then, when a point with a potential of 2500 (+) volts to ground is compared with a point with a potential of 1500 (–) volts to ground, the potential difference is 4,000 volts.

Propagating brush discharge: An energetic discharge from an insulative surface upon which an intense charge has been stored. To store adequate charge, the insulator must be thin [less than $^1/_4$ in. (6 mm)] and be backed by a conductive surface that is either grounded or at opposite polarity. The discharge is branched, occurs over an area of several to many square inches, and produces a loud popping sound. (For SI conversion: 1 sq. in. = 645.16 mm^2.)

Resistance: Electrical current encounters difficulty in passing through an electrical circuit or conductor. This difficulty can be measured and is called resistance. Resistance can be measured in terms of voltage drop over a part of the circuit but is usually measured in terms of "ohms" or "megohms." The resistance of a circuit in ohms is equal to the ratio of voltage in volts to current in amperes, i.e.,

$$\frac{10 \text{ V}}{1 \text{ amp}} = \text{ohms}, \frac{1 \text{ V}}{1\mu \text{ amp}} = \text{megohm}.$$

Spark: An energetic discharge, usually accompanied by a popping sound, that occurs when air or gas between two charged conductors becomes highly ionized, enough to break down, conducting current through a distinct, luminous channel.

BIBLIOGRAPHY

NFPA Codes, Standards, and Recommended Practices

Reference to the following NFPA codes, standards, and recommended practices will provide further information on electrostatic ignition sources discussed in this chapter. (See the latest version of *The NFPA Catalog* for availability of current editions of the following documents.)

NFPA 30, *Flammable and Combustible Liquids Code*

NFPA 33, *Standard for Spray Application Using Flammable or Combustible Materials*

NFPA 45, *Standard on Fire Protection for Laboratories Using Chemicals*

NFPA 61, *Standard for the Prevention of Fires and Dust Explosions in Agricultural and Food Product Facilities*

NFPA 65, *Standard for the Processing and Finishing of Aluminum*

NFPA 70, *National Electrical Code®*

NFPA 77, *Recommended Practice on Static Electricity*

NFPA 385, *Standard for Tank Vehicles for Flammable and Combustible Liquids*

NFPA 407, *Standard for Aircraft Fuel Servicing*

NFPA 480, *Standard for the Storage, Handling, and Processing of Magnesium Solids and Powders*

NFPA 481, *Standard for the Production, Processing, Handling, and Storage of Titanium*

NFPA 482, *Standard for the Production, Processing, Handling, and Storage of Zirconium*

NFPA 650, *Standard for Pneumatic Conveying Systems for Handling Combustible Materials*

NFPA 651, *Standard for the Manufacture of Aluminum Powder*

NFPA 654, *Standard for the Prevention of Fire and Dust Explosions in the Chemical, Dye, Pharmaceutical, and Plastics Industries*

NFPA 655, *Standard for Prevention of Sulfur Fires and Explosions*

NFPA 664, *Standard for the Prevention of Fires and Explosions in Wood Processing and Woodworking Facilities*

NFPA 850, *Recommended Practice for Fire Protection for Electric Generating Plants*

NFPA 1124, *Code for the Manufacture, Transportation, and Storage of Fireworks*

NFPA 1125, *Code for the Manufacture of Model Rocket and High Power Rocket Motors*

Additional Reading

American Petroleum Institute, *Recommended Practice for Protection Against Ignitions Arising Out of Static, Lightning, and Stray Currents* (RP 20003), American Petroleum Institute, Washington, DC, 1991.

Berkey, B. D., Pratt, T. H., and Williams, G. M., "Review of Literature Related to Human Spark Scenarios," *Plant/Operations Progress*, Vol. 7, No. 1, Jan. 1988, pp. 32–36.

Britton, L. G., "Static Hazards Using Flexible Intermediate Bulk Containers for Powder Handling," *Process Safety Progress*, Vol. 12, No. 4, Oct. 1993, pp. 240–250.

Britton, L. G., "Systems for Electrostatic Evaluation in Industrial Silos," *Plant/Operation Progress*, Vol. 7, No. 1, Jan. 1988, pp. 45–50.

Britton, L. G., "Using Material Data in Static Hazard Assessment," *Plant/Operations Progress (Plant Safety Progress)*, Vol. 11., No. 2, April 1992, pp. 56–70.

Britton, L.G., and Smith, J. A., "Static Hazards of Drum Filling, Parts 1 & 2," *Plant/Operations Technology*, Vol. 7, No. 1, Jan. 1988, p. 53.

Bustin, W. M., and Dudek, W. G., *Electrostatic Hazards in the Petroleum Industry*, Research Studies Press, Ltd., England, 1983.

Crow, R. M., "Static Electricity: A Review of Literature," Defence Research Establishment Ottawa, Ontario, Canada, DREO-TN-91-28, November 1991, 21 pages.

Dahn, C. J., Kashani, A., and Reyes, B., "Static Electricity Hazards of Flexible Intermediate Bulk Containers," *Process Safety Progress*, Vol. 13, No. 3, July 1994, pp. 123–127.

Davis, K., "Electrostatic Discharges—Stopping the Sparks Flying," *Fire Prevention*, No. 228, April 1990, pp. 21–24.

Expert Commission for Safety in the Swiss Chemical Industry, "Static Electricity: Rules for Plant Safety," *Plant Operations Progress*, Vol. 7, No. 1, Jan. 1988, pp. 1–22.

"Fire and Explosion Due to Electrostatic Charges in the Plastics Industry," Bulletin No. 6, Society of the Plastics Industry, NY.

Generation and Control of Static Electricity, National Paint and Coatings Association, Washington, DC. 1988.

Glor, M., *Electrostatic Hazards in Powder Handling*, Research Studies Press/Wiley, Letchworth, UK.

Gomez, A. and Chen, G., "Charge-Induced Secondary Atomization in Diffusion Flames of Electrostatic Sprays," *Combustion and Flame*, Vol. 96, No. 1-3, 1994, pp. 47–59.

Handbook of Electrical Hazards and Accidents, Geddes, L.A., Editor, CRC Press, Inc., Boca Raton, FL, 1995.

Hansheng, L., "Experimental Investigation of Electrostatic Fire Accidents After Aircraft Landing and Preventive Measures," *Journal of Aircraft*, Vol. 26, No. 5, May 1989, pp. 405–419.

Howells, P., "Electrostatic Hazards of Foam Blanketing Operations," *Industrial Fire Safety*, Vol. 2, No. 4, July/August 1993, pp. 18,21–24.

"Instantaneous Fire Detection/Suppression System for Jaguar Paint Spray Electrostatic Plant," *Product Finishing*, Vol. 40, No. 7, 1987, pp. 14–15.

Jones, T. B., and King, J. L., *Powder Handling and Electrostatics: Understanding and Preventing Hazards*, Lewis Publishers, Michigan, 1991.

Kinney, P. D., *et al.*, "Use of Electrostatic Classification Method to Size 0.1 mum SRM Particles. A Feasibility Study," *Journal of Research of the National Institute of Standards and Technology*, Vol. 96, No. 2, March/April 1991, pp. 147–176.

Louver, J. F., Mauren, B., and Biocourt, G. W., "Tame Static Electricity," *Chemical Engineering Progress*, Vol. 90, No. 11, Nov. 1994, pp. 75–79.

Luo, H., "Experimental Investigation of Electrostatic Fire Accidents After Aircraft Landing and Preventive Measures," *Journal of Aircraft*, Vol. 26, No. 5, 1989, pp. 405–409.

Mancini, R. A., "The Use (and Misuse) of Bonding for Control of Static Ignition Hazards," *Proceedings* of the 21st Annual Loss Prevention Symposium, AICHE, August 1987.

Meal, L., "Static Electricity," *Fire Engineering*, Vol. 142, No. 5, 1989, p. 61.

Owens, J. E., "Spark Ignition Hazards Caused by Charge Induction," *Plant/Operations Progress*, Vol. 7, No. 1, Jan. 1988, pp. 37–39.

Pratt, T., "Electrostatic Ignitions in Enriched Oxygen Atmosphere: A Case History," *Process Safety Progress*, Vol. 12, No. 4, Oct. 1993, pp. 203–205.

Pratt, T., "Static Electricity in Pneumatic Transport Systems: Three Case Histories," *Process Safety Progress*, Vol. 13, No. 3, July 1994, pp. 109–113.

Rosenthal, L. A., "Static Electricity and Plastic Drums*," Plant Operations/Progress*, Vol. 7, No. 1, Jan. 1988, pp. 51–52.

Schram, P. J., and Earley, M. W.*, Electrical Installations in Hazardous Locations*, National Fire Protection Association, Quincy, MA, 1988.

"Understanding Electricity and Electrical Dangers," *Fire Engineering*, Vol. 149, No. 4, April 1996, pp. 57–92.

Walmsley, H. L., "The Avoidance of Electrostatic Hazards in the Petroleum Industry," *Journal of Electrostatics*, Vol. 27, Nos. 1 and 2, 1992.

LIGHTNING PROTECTION SYSTEMS

Revised by John M. Caloggero

Lightning causes 25,000–30,000 fires per year. Roughly 100 people per year are killed by lightning or, less often, by the fires it causes. Lightning fires cost about $150 million a year.

Unlike many other causes of death and injury that one can run away from, lightning strikes before the warning thunder. The principles of protection and personal safety are well known and are spelled out in NFPA 780, *Lightning Protection Code* (hereinafter referred to as NFPA 780). It must be remembered, nevertheless, that lightning involves many uncertainties and that, while a given pattern of lightning behavior may be probable, there is no guarantee that a lightning discharge will not deviate from that pattern.

Frequency and severity of thunderstorms: The frequency of thunderstorms varies throughout the world. The severity of thunderstorms, as distinguished from their frequency of occurrence, is much greater in some locations than in others. Hence, the need for protection varies geographically, although not necessarily in direct proportion to thunderstorm frequency. A few severe thunderstorms a season may make the need for protection greater than a relatively large number of storms of lighter intensity. (See Figures 3-3A and 3-3B for statistical data on the frequency of thunderstorms in the United States and Canada, respectively, and Figure 3-3C for world data.)

Earth resistivity: Figure 3-3D illustrates the estimated average earth resistivity in the United States in ohm-meters. The highest earth resistivity is located in the northeast segment of the country, with 2000 ohm-meters in Long Island, NY, northern New Hampshire, and Vermont. Structures in these areas are susceptible to more frequent lightning strikes.

Value and nature of building and contents: Buildings and structures that are essential to life safety, such as hospitals, fire stations, police stations, and military and civil defense facilities, should have a lightning protection system installed if they are located in areas where lightning commonly occurs. Some buildings have an uninsurable historical or sentimental value. The type of building construction also influences the degree of protection to be considered. Some buildings require no supplemental lightning protection due to their construction. The real or intangible value of the building's contents must also be considered, and, in some cases, attention must be given to the nature of and the susceptibility of the contents to damage by induced lightning currents, e.g., storage of explosives.

Appendix H of NFPA 780 contains a risk assessment guide, which provides guidelines for assessing the need for lightning protection. Not part of the requirements of NFPA 780, the appendix is included for informational purposes only.

Personnel hazards: The lightning hazard to personnel in buildings or structures is a major concern. Because a stroke of lightning may lead to a considerable discomfort, if not injury or death, lightning protection may be necessary to eliminate possible personnel hazards in buildings of any type (other than those constructed of structural steel framing). In places of public assembly, the potential of panic should be considered, because a lightning discharge above

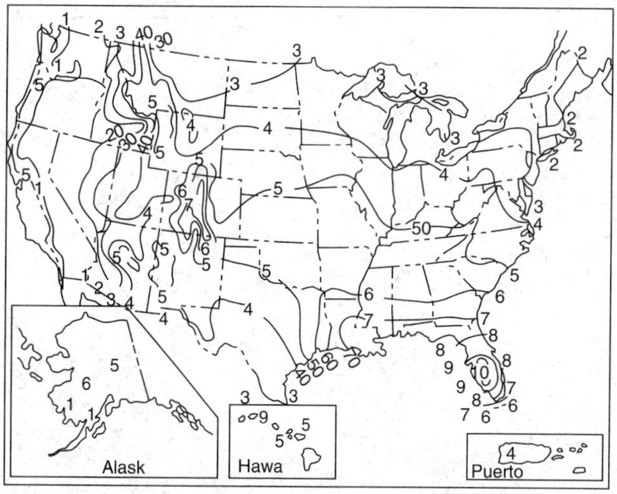

FIG. 3-3A. *Statistics for continental United States showing mean annual number of days with thunderstorms. The highest frequency is encountered in south-central Florida. Since 1894, the recording of thunderstorms has been defined as the local calendar day during which thunder was heard. A day with thunderstorms is so recorded, regardless of the number occurring on that day. The occurrence of lightning without thunder is not recorded as a thunderstorm. (Data supplied by Environmental Science Service Administration, U.S. Department of Commerce)*

John M. Caloggero is principal electrical specialist at NFPA and NFPA staff liaison to NFPA's Technical Committee on Lightning Protection.

or adjacent to a structure, even if not accompanied by a fire, can have visual and audible effects that may induce panic.

Relative exposure: In closely built-up towns and cities, the hazard is not as great as in open country. In the latter, farm barns are most frequently hit. In hilly or mountainous areas, a building located on high ground is usually subject to greater hazard than one in a valley or otherwise sheltered area.

Indirect losses: In addition to direct losses, e.g., damage of buildings or their contents by lightning, fire resulting from lightning,

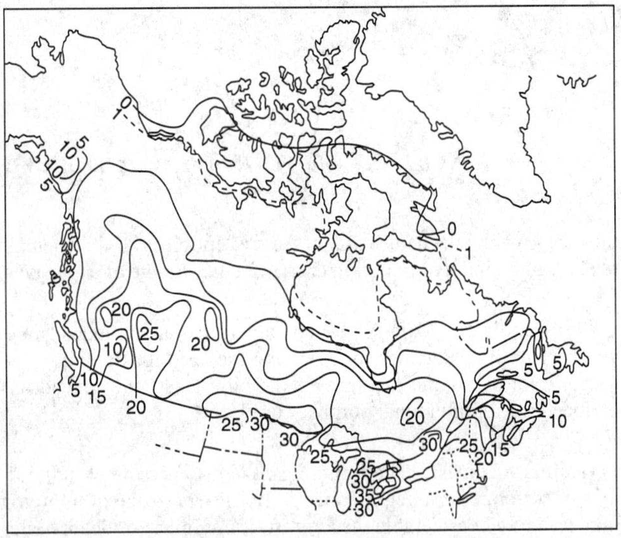

FIG. 3-3B. *Canadian statistics showing annual average of days with thunderstorms. Data based on the period 1957-1972. (Source: Meteorological Division, Department of Transportation, Canada)*

death of livestock, etc., there may be indirect losses. Interruption of business, computer, or farming operations, especially at certain times of the year, may involve losses quite distinct from and in addition to those arising from direct property damage. In some cases, whole communities depend on the integrity of a single structure for a measure of safety and comfort, for example, the brick chimney of a water pumping plant. A stroke of lightning to the unprotected chimney of a plant of that sort might have serious consequences, resulting in lack of sanitary drinking water, irrigating water, water for fire protection, or some similar effect.

NATURE OF LIGHTNING

The planet and its atmosphere can be viewed as a large spherical capacitor, with the earth at about 300,000 V negative, with respect to the upper ionosphere. This potential difference powers the flow of electricity from the earth to the atmosphere. The rate of current flow, calculated at approximately 1400 to 1800 amp, is spread out over the earth's entire surface. Therefore, this amounts to about 0.00009 amp per sq mi. Under clear normal conditions, the electrostatic field intensity at the earth's surface is 150 to 200 V/m of elevation. As the thunderstorm approaches, the potential difference rises to 10,000 to 30,000 V/m of elevation.

Air currents pick up moisture as they pass over large bodies of water, such as lakes, rivers, and the oceans. This air mass is heated where it passes over warmer land masses, causing it to rise. As the moist air mass passes through the 14,000-ft elevation, the moisture condenses and forms ice crystals. The ice crystals in an active cloud become positively charged, while the water droplets usually have negative charges. Due to these conditions, the thundercloud normally develops a positive charge at the upper level and negative charge at the lower level. An electric field now exists within the cloud as well as above and below it. This negative charge in the lower level of the cloud induces a positive charge in grounded objects below it. Sharp pointed objects, such as flagpoles, steeples, corners of buildings, and trees, develop concentrated fields at these

FIG. 3-3C. *Annual frequency of thunderstorm days as compiled by the World Meteorological Organization, 1956. (Source: World Meteorological Organization)*

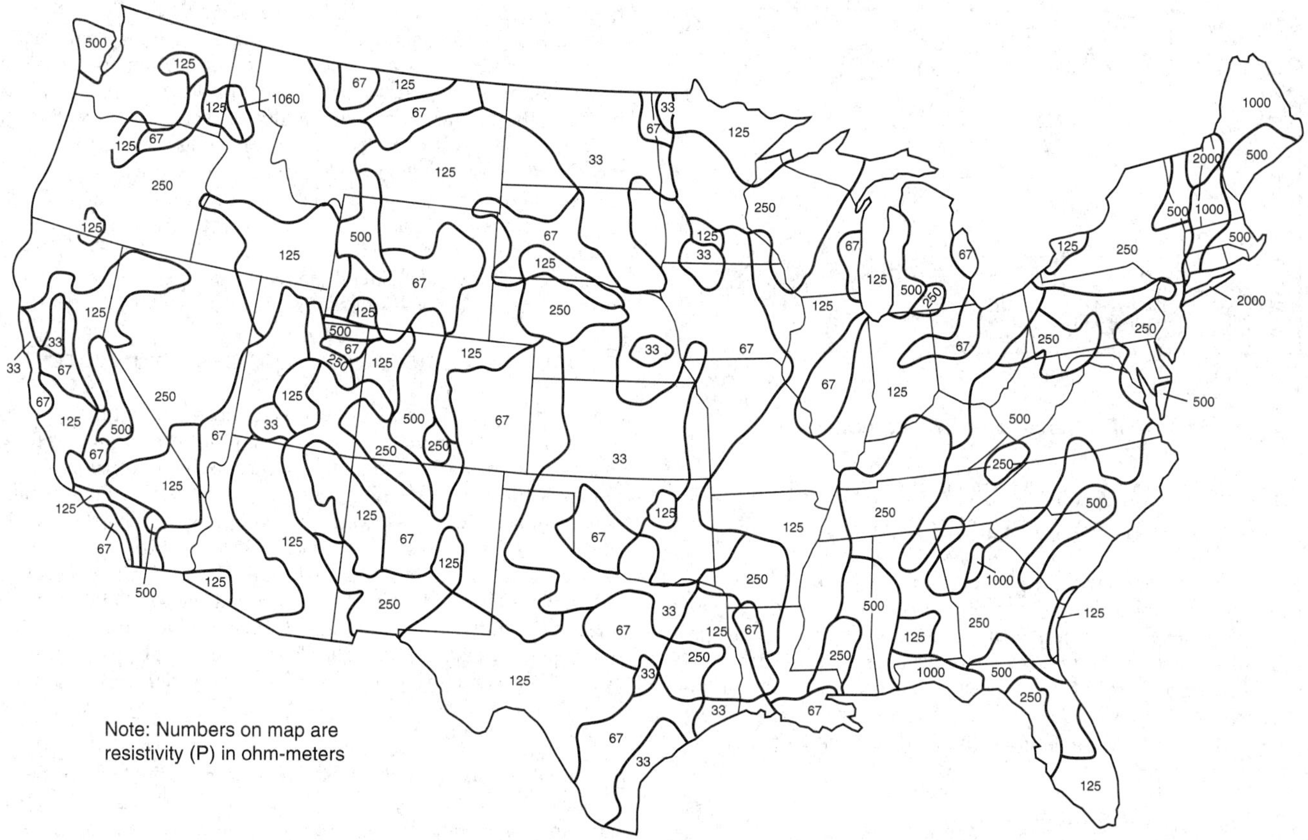

FIG. 3-3D.   *Estimated average earth resistivity in the continental United States.*

Note: Numbers on map are resistivity (P) in ohm-meters

locations, resulting in a point-discharge current of a few amp that moves upward in the electric field.

As the voltage between the earth and cloud increases, the lightning discharge to earth starts at the cloud, progressing in incremental steps with a faintly luminous discharge, called a leader stroke in lengths of about 20 m. The current in these leaders is estimated at about 200 amp. When the faint lightning leader reaches an upward leader, an intense flash can be seen traveling upward to the cloud. This is called the return stroke. This is a direct short circuit between the negative portion of the cloud and the electrostatically induced positive charge at the earth.

The development of a stepped leader is diagrammed in Figure 3-3E. When sufficient charge has built up on a cloud, a "pilot streamer" will develop in a generally downward direction. This streamer ionizes a path in the air. The current associated with it is small, on the order of only a few amp. The streamer initially extends on the order of 150 ft (45.7 m) below the cloud. After a pause of about 50 microseconds, the streamer will proceed for a second step, again generally downward but usually in a somewhat different direction from the initial step. Subsequent steps occur at intervals of about 50 microseconds. The path of each step is essentially straight, but each new step generally takes a different direction. The change in direction at each junction results in the zigzag path characteristic of lightning. Branches may (and usually do) occur but, for simplification, these are not shown in Figure 3-3E.

As the leader progresses downward, it creates an ionized path of high conductivity in the air. Thus, the tip of the leader essentially remains at cloud potential, and the voltage gradient between the tip and the earth increases as the tip progresses downward. At some critical point (point 6 in Figure 3-3E), the voltage gradient becomes high enough to break down the remaining air gap, and the initial

stroke to ground is completed. The point from which the final breakdown occurs is called the point of discrimination; and the distance over which this breakdown occurs is defined as the striking distance.

As many as 40 component strokes have been observed in a single flash. Speeds range from 100 mi per sec (161 km/s) for the first pilot leader stroke to 20,000 mi per sec (32 190 km/s) for the main stroke. Currents average about 24,000 amp but range up to 270,000 amp in extreme cases, lasting for a few millionths of a second, but lesser currents are present for a longer period. Potentials have been estimated as high as 15 million V.

Because of the high voltage and rapid changes in current flow, induced charges are important. Thus, NFPA 780 requires the interconnection of metallic masses as a part of any lightning protection system.

Lightning causes fire when sufficient heat is produced to ignite combustible materials in the vicinity. Substantial damage, however, can be produced without resulting in fire. Dry wood beams in houses struck by lightning may be severely splintered, and windows blown outward. Such damage primarily results from pressure generated by the expanding lightning channel. Because some effects of lightning can be indirect, it is not necessary for lightning to strike a building to cause damage. Lightning striking an overhead wire can be conducted over the wire to a building. Therefore, lightning arresters (surge arresters) should be provided to minimize such damage. Although there are several types of surge arresters, all permit the free flow of lightning charges to ground, while preventing the flow of ordinary electric current over the same path. The surge arrester should be backed up with a transient voltage surge suppressor on branch circuits that serve computers and electronic appliances.

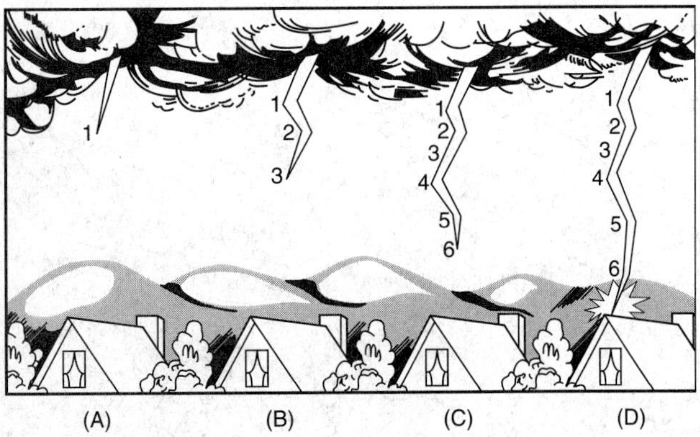

FIG. 3-3E. *Development of a stepped leader.*

TRADITIONAL THEORY OF LIGHTNING PROTECTION

The theory of lightning protection is simple—to provide means by which a lightning discharge may enter or leave the earth without damaging the property protected. There is no evidence that any form of protection can prevent the occurrence of a lightning discharge. A lightning protection system has two functions: (1) to intercept a lightning discharge before it can strike the object protected, and (2) to discharge the lightning current harmlessly to earth.

Conventional Concept

The zone protected by a grounded rod or mast is conventionally taken as the space enclosed by a cone that has its apex at the top of the mast. Similarly, the zone protected by a grounded horizontal overhead wire is conventionally taken as a triangular prism, with the upper edge along the wire. In either case, the degree of protection is generally assumed to be a function of the shielding angle, i.e., the angle between an element of the cone and a vertical line through the apex, or between the side of the prism and the vertical plane through the horizontal wire. This is sometimes expressed as the ratio of the horizontal distance protected to the height of the mast or wire.

Figure 3-3F illustrates this concept and tabulates some generally accepted zones of protection. NFPA 780 states that, for structures not greater than 50 ft in height, a horizontal distance equal to the height of the rod or mast has been found to be substantially immune to direct strokes. This horizontal distance extends to twice the height of the rod or mast, if the structure does not exceed 25 ft in height. The *British Standard Code of Practice*, "The Protection of Structures Against Lightning,"[1] states that a shielding angle of 45 degrees provides an acceptable level of protection for ordinary structures, but that for structures with explosives or highly flammable contents, the shielding angle should not exceed 30 degrees. Both codes indicated that complete protection cannot be guaranteed in any case. It is assumed, however, that the criteria presented reduces the probability of lightning strikes to a practical minimum. While these criteria provide essentially complete protection in most cases, theoretical consideration indicates that even the 30-degree shielding angle is not always adequate. Experience with electric power transmission lines corroborates this.

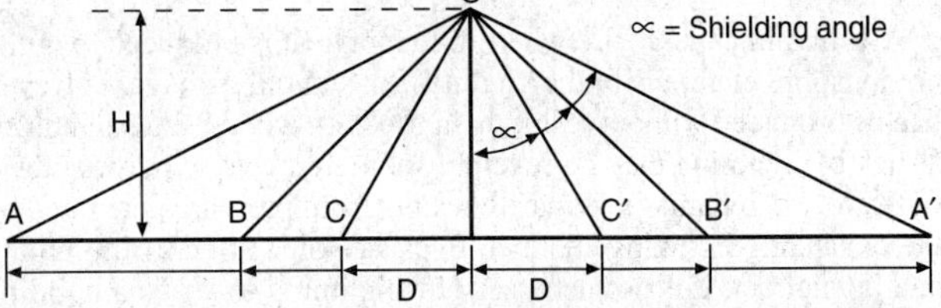

Geometric Concept of Lightning Protection

The geometry of lightning protection is illustrated in Figure 3-3G, which shows a hypothetical cross section of an area protected by two vertical masts (or horizontal overhead wires). Consider first a downward leader descending to the right of Mast A. Final breakdown will occur when the leader tip approaches within a distance X (the striking distance) of any grounded object or surface. If the point of discrimination, P (the point at which final breakdown will occur), is equidistant from the tip of the mast and the ground, either the mast or the ground may be struck. If P is close to the mast, it will intercept the stroke; if P is farther from the mast, the ground will be struck. If any structure to the right of the mast extends above ground to within a distance less than X of point P, the structure will be struck.

It can be seen that the zone protected by a single mast (or by an outside mast of a multi-mast system, in a direction away from the other masts) is bounded by the arc of a circle of radius X, passing through the tip of the mast and tangent to the ground. Geometrically, it can be shown that the protected distance D at an elevation B above ground is given by the expression shown, where H is the height of the mast above ground and X is the striking distance.

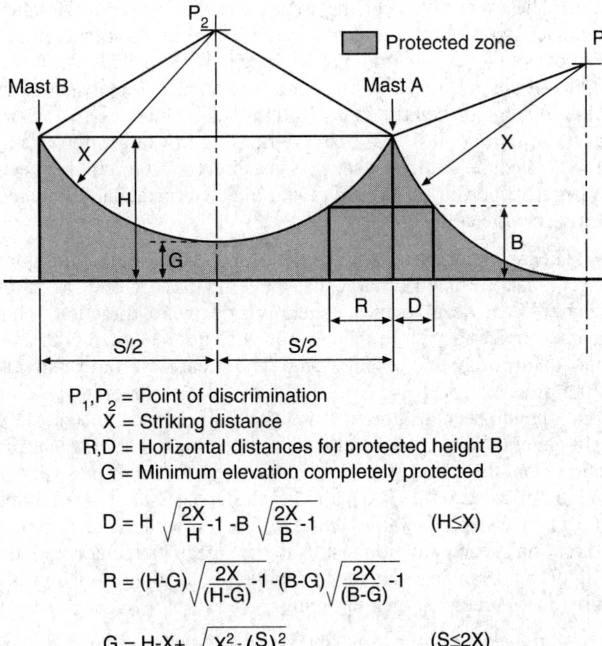

P₁,P₂ = Point of discrimination
X = Striking distance
R,D = Horizontal distances for protected height B
G = Minimum elevation completely protected

$$D = H\sqrt{\frac{2X}{H}-1} - B\sqrt{\frac{2X}{B}-1} \qquad (H \le X)$$

$$R = (H-G)\sqrt{\frac{2X}{H-G}-1} - (B-G)\sqrt{\frac{2X}{B-G}-1}$$

$$G = H-X+\sqrt{X^2-\left(\frac{S}{2}\right)^2} \qquad (S \le 2X)$$

For H>X,
$$D = X-B\sqrt{\frac{2X}{B}-1}$$

FIG. 3-3G. *Geometric concept of lightning protection.*

Zone	D/H	∝	Authority	Recommended for
AOA'	2/1	63°	NFPA	Ordinary cases
BOB'	1/1	45°	NFPA	Important cases
COC'	0.58/1	30°	British code	Ordinary structures
			British code	Danger structures

FIG. 3-3F. *Conventional concept of protected zones.*

Similarly, for a leader extending downward between the two masts, the protected zone is bounded by the arc of a circle of radius X passing through the tips of the two masts. If the masts are close enough together, the space between them is completely protected for an elevation above ground, which is a function of mast height H and spacing S. In this case, the protected distance R at elevation B and the minimum elevation G, which is completely in the protected zone, are given by the two expressions shown.

These equations hold only if H is less than or equal to X, and S is less than or equal to 2X. Further, G must be positive. If G is negative, complete protection does not exist between the two masts, and the equation for D holds.

The geometric concept shows that the protected zone is not a cone or prism, but is bounded by surfaces generated by concave arcs. This fact is not generally appreciated, although it is at least partially recognized in transmission line shielding. Further, the protected distance is seen to be a function of:

1. Striking distance.
2. Height of mast or overhead wire above ground.
3. The distance between two or more masts or overhead wires.

Figure 3-3G shows why the protected zone between two masts is greater than the total of the zones of the two masts considered individually. Previous NFPA lightning protection codes recognized this qualitatively but not quantitatively. The figure also shows why the shielding angle required on transmission lines decreases as tower (and overhead ground wire) height increases.

Striking Distance

The striking distance is related to the potential of the lightning stroke, which is directly related to the charge on the cloud. Since the peak stroke current is also related to the charge on the cloud, it can be shown that the striking distance is directly related to the peak stroke current. For a given mast or ground wire height, the protected distance increases with increasing striking distance. Thus, an adequately designed lightning protection system provides maximum protection against what could be the most damaging strokes. Since minimum protection exists for a minimum striking distance (and minimum stroke current), the lightning protection system should be designed to protect against minimum anticipated striking distances.

A review of lightning stroke current data indicates that the peak stroke current is usually at least 10,000 amp. This corresponds to a striking distance of slightly over 100 ft (30.5 m). It is therefore believed that a protective system based on a minimum striking distance of 100 ft (30.5 m) will provide essentially complete protection. This is considered to be a reasonable design basis.

Design Curves

Using the derived equations and an assumed striking distance, the protection provided by a single mast (or horizontal overhead wire) may be readily calculated. Similarly, the configuration of masts or overhead wires required to protect a given structure may be investigated. However, the calculations are tedious, and it is not convenient to determine directly by simple algebra the mast or overhead wire height required to protect a structure of given dimensions. Hence, curves showing the relationship between protected zones and various mast or horizontal overhead wire configurations are useful. Such curves can be readily constructed graphically. (See Figures 3-3H and 3-3I.) These curves are based on a striking distance of 100 ft (30.5 m).

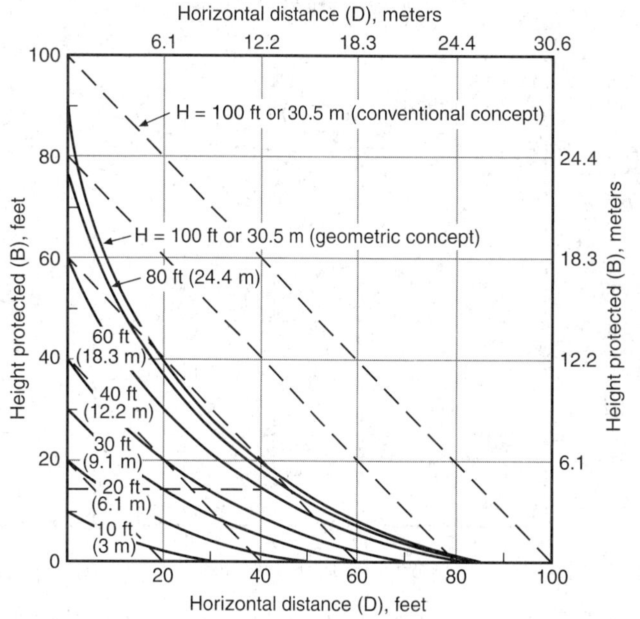

FIG. 3-3H. Zone protected by a single mast, horizontal wire, or air terminal. Solid curves are the geometric concept, with a striking distance of 100 ft (30.5 m). Dashed lines are the conventional concept, with a shielding angle of 45 degrees. "H" equals height of mast or horizontal wire. (For SI units: 1 ft = 0.305 m.)

PROPERTY PROTECTION

Components

Air terminals: According to current United States practice, as indicated in NFPA 780, conditions required for protection of ordinary buildings are met by placing metal air terminals on the uppermost parts of the building or its projections, with conductors connecting the air terminals with each other and to the ground. By this means, a relatively small amount of metal, properly proportioned and distributed, affords a satisfactory degree of protection, and it can be installed in a way that interferes only minimally with the contour and appearance of the building. In the current *British Standard Code of Practice*,[1] the use of pointed air terminals (vertical finials) is not considered essential, except where dictated by practical considerations.

One theory states that a protective system based on a minimum striking distance of 100 ft (30.5 m) for structures containing flammable liquids and gases will provide 99.9 percent protection, and a minimum striking distance of 150 ft (45.7 m) will provide 99.5 percent protection for ordinary structures that are generally capable of withstanding minor direct strokes.[2,3,4] This theory supports the use of 10-in. (254-mm) minimum height air terminals above the object to be protected at 20-ft (6.1-m) maximum intervals. A convenient geometric equation:

$$Y = \sqrt{R^2 - (R - h)^2}$$

where

Y = radius of horizontal protected zone,

R = striking distance, and

h = height of air terminal

illustrates that a striking distance of 100 ft (30.5 m) will provide a 12.9-ft (3.9-m) radius of horizontal protected zone or 25.8-ft

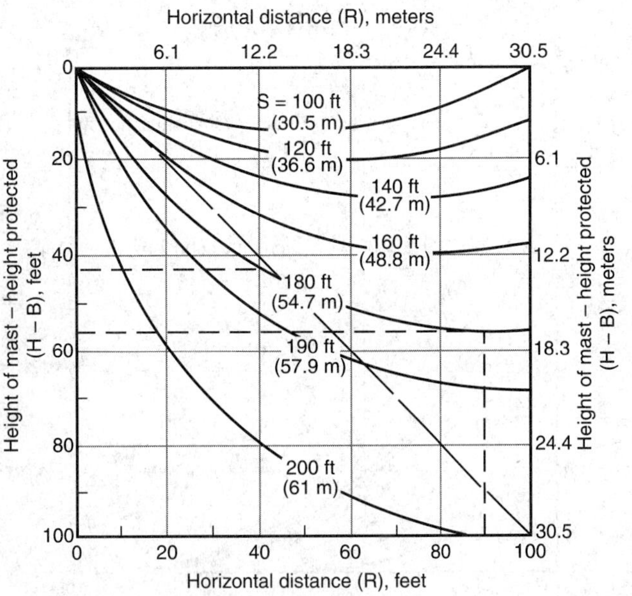

FIG. 3-3I. *A zone protected by two masts, two horizontal wires, or two air terminals; the striking distance is 100 ft (30.5 m). "H" equals the height of the horizontal wire, "S" is the spacing between masts or wires, and "B" is the height protected above ground. (For SI units: 1 ft = 0.305 m.)*

(7.9-m) horizontal protected zone between air terminals, and a striking distance of 150 ft (45.7 m) will provide a 15.8-ft (4.8-m) radius of horizontal protected zone or 31.6-ft (9.6-m) horizontal protected zone between air terminals.

The subject of sharply pointed *vs.* blunt air terminals is under study by the NFPA Lightning Protection Committee. According to Moore *et al.*,[5] pointed air terminals have a built-in defense against lightning strikes. The strength of the electric field around the tip of an air terminal is limited by a phenomenon called "point discharge." When the electric field reaches a certain strength, but before lightning strikes, the stepped leader current flows from the air through the air terminal to the ground. The effect is called point discharge because the point is letting the surrounding air lose charge, or discharge electricity. A pointed air terminal will presumably discharge sooner than a blunt one. Accordingly, Moore *et al.*[5] concluded that a blunt air terminal is more likely to intercept a lightning strike and complete a circuit.

Conductors: NFPA 780 specifies the use of copper or aluminum metals for lightning conductors. Where copper and aluminum are used together, special precautions must be taken against electrolytic corrosion. Conductors are coursed along ridges of sloping roofs, around the edges of flat roofs, and vertically from these and from the air terminals to ground. Vertical runs are designated as "down conductors," and there are never fewer than two on any kind of structure, except for flagpoles, masts, spires, and similar structures. The down conductors are normally placed at diagonally opposite corners of square or rectangular structures, but other factors, such as direct coursing, security against displacement, location of metallic elements in the structure, location of water pipes, and favorable ground conditions, need to be considered in determining optimum placement. For detailed instructions about the installation of conductors, their form and size, and about the installation of air terminals, refer to the latest edition of NFPA 780.

The down conductors to ground must be substantial and must provide a reasonably direct path. The blocking effect (choke) of electrical induction must be avoided. For example, a lightning conductor that is run through a piece of iron pipe for protection, unless bonded to the pipe at the top and bottom, largely cancels the value of the conductor.[6]

Grounding: Ground connections are essential to the effectiveness of a lightning protection system, and every effort should be made to provide ample contact with the earth. (See Figure 3-3J.) Resistance of the protective system to ground is not critical. However, resistance and surge impedance must be low enough to prevent side flashes between the down conductor and other grounded objects. Also, the resistance must be low enough to prevent excessive voltage gradients in the soil surrounding the ground electrode. Excessive voltage gradients can be dangerous or fatal to people or animals that may be standing or walking in the vicinity when a lightning stroke occurs. If the down conductor is remote from other grounded objects, and people or animals are not likely to be near the ground electrode during a lightning storm, a relatively high ground resistance can be tolerated. Otherwise, an extensive wire network will need to be laid on the surface.

NFPA 780 does not specify any maximum recommended ground resistance. The *British Standard Code of Practice*[1] recommends a maximum resistance of 10 ohms for a lightning protection system that will normally include two or more ground electrodes. Also 10 ohms is generally considered a reasonably good ground resistance for transmission towers. In this case, a high ground resistance may result in flashover to the line conductors, even though the overhead ground wire intercepts the direct lightning stroke. The U.S. Department of Agriculture recommends a ground resistance not greater than 50 ohms as acceptable for protection of farm structures.

The ground resistance of a ground electrode is directly proportional to soil resistivity. Soil resistivity varies over wide limits (see Figure 3-3D); nevertheless, 10,000 ohm/cm is a reasonable order of magnitude average for many areas. Figure 3-3K, which is based on derived equations,[7] has been prepared to show the variation in resistance to earth of vertical rods or pipes as a function of length in 10,000 ohm/cm soil. For other soil resistivities, the resistance to ground may be obtained by multiplying the indicated resistance by the known soil resistivity divided by 10,000. Vertical rods driven into the ground are commonly used as grounding electrodes. In addition, where buried metallic water pipes are nearby, it is required to connect the lightning protection ground system to the piping system.

If bedrock is on or near the surface, it would be impossible to make a ground connection in the ordinary sense of the term because most kinds of rock are insulating, or at least of high resistivity, and in order to obtain effective grounding, other and more elaborate means are necessary. One effective means is to use an extensive wire network (grid) laid on the surface of the rock surrounding the building, to which the down conductors could be connected. Another is to use ground ring conductors. One ring should be installed at the bottom of the foundation to which the down conductors or structural steelwork are connected, with a second ring conductor provided by blasting a circular trench around the periphery and placing in it a strip conductor. The blasted area should be thoroughly compacted with backfill, and the two ring conductors electrically interconnected. Using one of these methods would make the distribution of the electrical potential around the protected building substantially the same as if it were resting on conducting soil, and the resulting protective effect would be equal.

For sandy soil, or any dirt covering the foundation, a concrete-encased electrode (commonly called a "Ufer" ground, after the in-

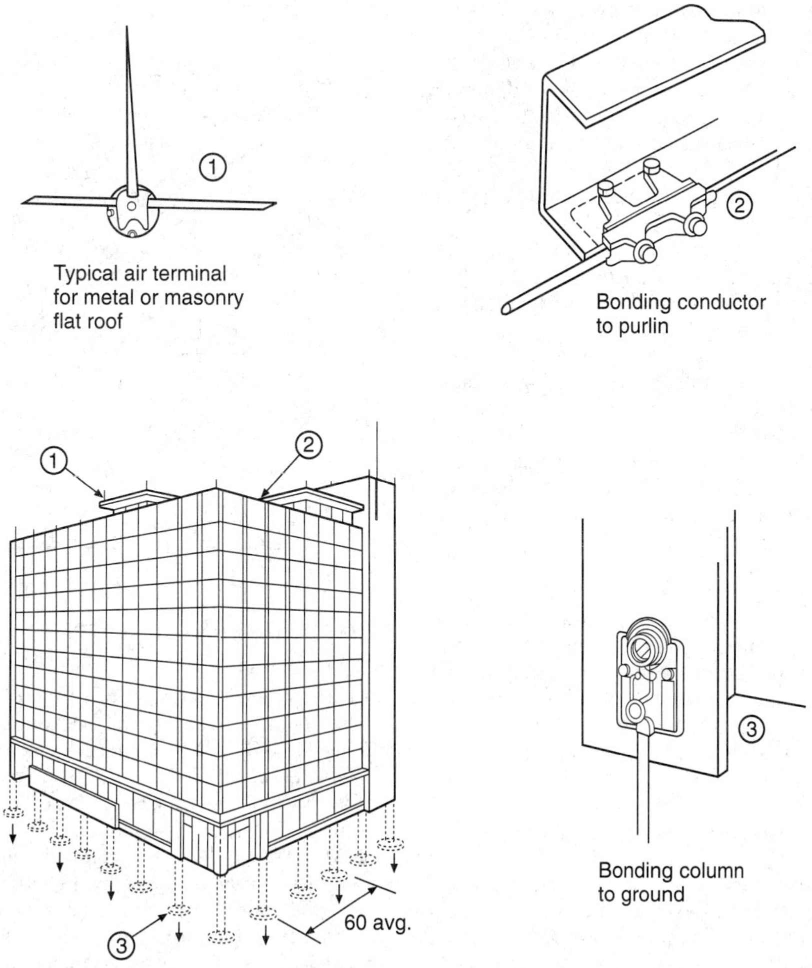

Typical air terminal
for metal or masonry
flat roof

Bonding conductor
to purlin

Bonding column
to ground

60 avg.

FIG. 3-3J.   Grounding and bonding of lightning down conductors. Water pipe grounds (if pipes are metallic) can be made at 1, 2, or 3.

venter Herb Ufer) should be installed for new construction. This electrode consists of tying in the steel reinforcing bars or rods in the footing of the foundation that is in direct contact with the earth.

In general, the extent of the grounding arrangements will depend upon the character of the soil, ranging from simple extension of the conductor into the ground where the soil is deep and of high conductivity, to an elaborate buried network where the soil is very dry or of very poor conductivity. Where a network is required, it should be buried if there is soil enough to permit it, as this adds to its effectiveness. Its extent will be determined largely by the judgment of the person planning the installation, following the minimum requirements of NFPA 780, and keeping in mind that, as a rule, the more extensive the underground metal available, the more effective the protection.

Protection of Structures (Traditional Systems)

Underwriters Laboratories Inc. "Master Label Service" for lightning protection systems: UL has had a "Master Label Service" for lightning protection systems since 1923. It provides for both factory inspection and labeling of lightning protection materials (components), as well as field inspections of a substantial number of installations for which Master Labels have been issued. The service covers the installation of labeled lightning protection materials (components) on all types of structures, with the exception of those used for the production, handling, or storage of ammunition,

explosives, flammable liquids or gases, and explosive ingredients. Protection of electrical transmission lines and equipment is not within the scope of the "Master Label Service."

Structural steel buildings: Buildings constructed of structural steel framing may be protected by the installation of air terminals at the high parts of the building, connecting such air terminals to the metal framing, and grounding the framing at the bottom end. This assumes that the structural steel framework is electrically continuous, or is made electrically continuous by bonding. Grounding is required from approximately every other steel column around the perimeter and is arranged so that grounds average not more than 100 ft (30.5 m) apart, except as noted below.

Air terminals are not required if the steel columns extend at least 10 in. (254 mm) above the object to be protected. Air terminals are not required on metal roofs where the metal is not less than $3/16$ in. (4.8 mm) thick.

Additional grounding is not required if the steel column extends 10 ft (3.05 m) or more into the earth or is connected to a concrete-encased electrode.

Reinforced concrete structures: Reinforced concrete buildings, in which the reinforcing rods are electrically bonded together and grounded, are similar to structural steel buildings in regard to lightning protection. In usual building practice, as successive reinforcing bars are fitted, they are made to overlap lengthwise with bars already installed and are then tied together with metal binding wire. With thousands of such connections in a completed framework, the electrical resistance is fully acceptable for the purpose of lightning protection. However, if the reinforcing rods are electrically discontinuous, the building should be treated the same as a building of nonconducting materials. Lightning strokes to reinforced concrete buildings where there are insulating gaps between reinforcing rods can cause cracks at places where beams and floor slabs are connected to their supports. Prestressed concrete buildings are a particular problem. If the wires in precast units are not interconnected (as they frequently are not), individual units are isolated and when a lightning charge is induced, serious damage can result.

Metal-roofed and metal-clad buildings: Metal roofing and siding must not be substituted for the main conductors and air terminals of a lightning protection system, unless constructed of $^3/_{16}$-in. (4.8-mm) minimum sheet metal that has been made electrically continuous by bonding or an approved interlocking contact. This is because there is a likelihood that lightning will puncture thinner sheet metals and ignite wood framing, etc.

Tanks containing flammable and combustible liquids and gases: Tanks containing flammable and combustible liquids or flammable gases stored at atmospheric pressures have been ignited by lightning. Fires may be started by direct hits, which ignite vapors escaping from the tank, or may be started on the roofs of wood-roofed tanks. A lightning stroke in the vicinity may, by induction, produce sparks that may ignite such vapors. If there are openings in the roof, either intentional or accidental, externally ignited vapors may carry flame inside the tank, resulting in an explosion or fire if it contains a flammable or combustible vapor-air mixture.

Aboveground tanks storing flammable or combustible liquids or flammable gases at atmospheric pressures are considered to be reasonably well protected against lightning if constructed entirely of steel, and if: (1) all joints between steel plates are riveted, bolted, or welded; (2) all pipes entering the tanks are metallically con-

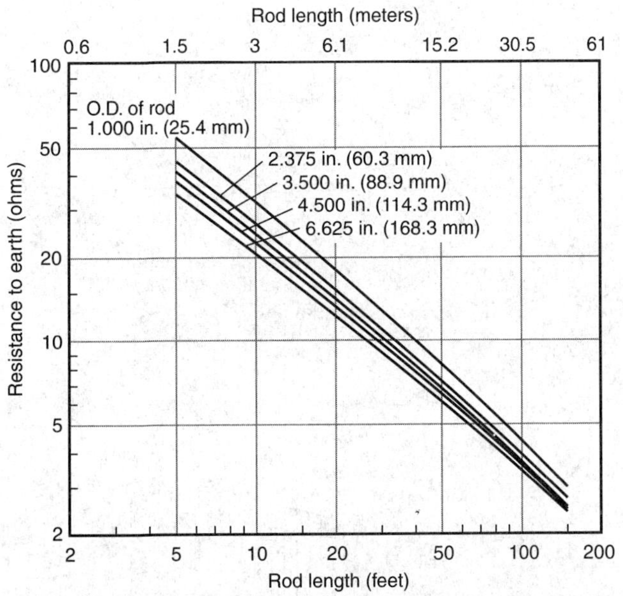

FIG. 3-3K.   *Resistance of vertical ground rods in 10,000 ohm/cm soil.*

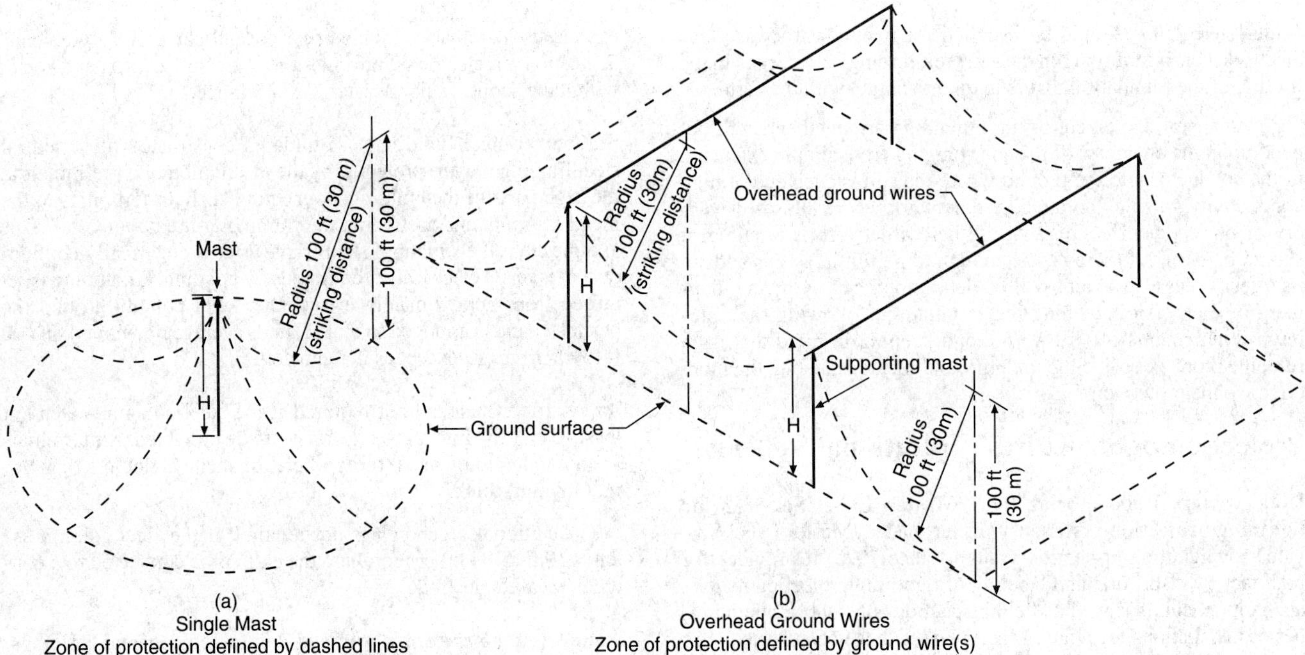

nected to the tank at point of entrance; (3) all vapor openings are closed or are provided with flame arresters (Class I or II liquids, as defined in NFPA 321, *Standard on Basic Classification of Flammable and Combustible Liquids*); (4) the tank and roof are constructed of $^3/_{16}$-in. (4.8-mm) minimum thick sheet metal so holes will not be burned through by lightning strokes; and (5) the roof is continuously welded, bolted, or riveted and caulked to the shell to provide a vaportight seam and electrical continuity. Where additional protection is desired, the internal supporting members of the roof may be bonded to the roof at 10-ft (3.05-m) intervals, or an overhead ground-wire system or mast protection may be installed in accordance with NFPA 780. (See Figure 3-3L.)

FIG. 3-3L.   *Zone of protection for mast height "H."*

Steel tanks in direct contact with the ground or aboveground steel tanks connected to extensive properly grounded metallic piping systems are considered to be inherently grounded.

Steel tanks with wood or other nonmetallic roofs are not considered self-protecting, even if the roofs are essentially vaportight or sheathed with thin metal. Such tanks should be protected by an overhead ground-wire system or mast protection. (See Figure 3-3L.)

Floating roof tanks with hangers located within a vapor space may be protected by bonding the roof to the shoes of the seal at 10-ft (3.05-m) intervals around the circumference of the tank and by providing insulated joints or installing jumper bonds around each pinned joint of the hanger mechanism. Based upon experience, floating roof tanks without vapor spaces would not appear to need lightning protective measures.

Aboveground storage tanks or containers of flammable liquids or liquefied petroleum gas under pressure are considered to be safe from lightning-caused explosions since the vapor-air mixture is "too rich" to burn, and the vapor is contained within the tank.

Tall structures: Spires, masts, and flagpoles of materials other than metal and other slender structures require one air terminal extending 10 in. (254 mm) or more above the uppermost point, a down conductor, and a ground terminal in compliance with NFPA 780. Church steeples will require one air terminal with a two-way path to ground at the extremities and additional air terminals on the lower-level portion that is outside of the steeple zone of protection.

In Great Britain, brick chimneys, church spires, etc., are protected by using two down conductors connected at the top to an existing metal cap, or to a circular conductor with no air terminals. (See Figure 3-3M.) Metal smokestacks need no protection against lightning other than that provided by their construction if they are grounded.

Trees, other specialized structures, and facilities: Trees are sometimes provided with lightning protection where they are especially valuable, are of historical significance, overhang buildings, or provide a shelter for livestock. Guidance on the installation of lightning protection is given in Appendix F of NFPA 780. Other appendices of NFPA 780 cover: (1) parked aircraft; (2) livestock in fields; and (3) picnic grounds, playgrounds, ballparks, and other open places. (See Figure 3-3N.)

Grounding metal masses (bonding): Extensive masses of metal that form part of a building or its appurtenances, such as metal ventilators and television or radio antennas, must be bonded to the lightning protection system. Nevertheless, the grounding of metal windowsills could be a hazard to human life in the event of a direct lightning strike. Details are given in NFPA 780, under "Metal Bodies," "Potential Equalization," and "Bonding of Metal Bodies," which determine the need for bonding and require loop conductors around a structure at various levels to be bonded to all down conductors. This provides an equipotential ground grid that causes the different metal components in the structure to be at the same voltage, thereby minimizing the shock and arcing hazard and also decreasing the amount of current that could affect computers and other electronic equipment.

Maintenance

Proper maintenance of lightning protection systems is essential to effective protection. Particular attention should be given to air terminals and connections in the vicinity of water gutters, and ground rod connections, as rods may be broken or corroded at ground level or just below, where the damage is not apparent. Materials (components) may be missing because of storm damage or stolen because of their value.

Surge Arresters and Suppressors on Electrical Apparatus and Circuits

The installation of surge arresters on power and communications lines where they enter structures and at power utility plants is covered in Article 280 of NFPA 70, *National Electrical Code®*. More information on the protection of electrical apparatus and circuits

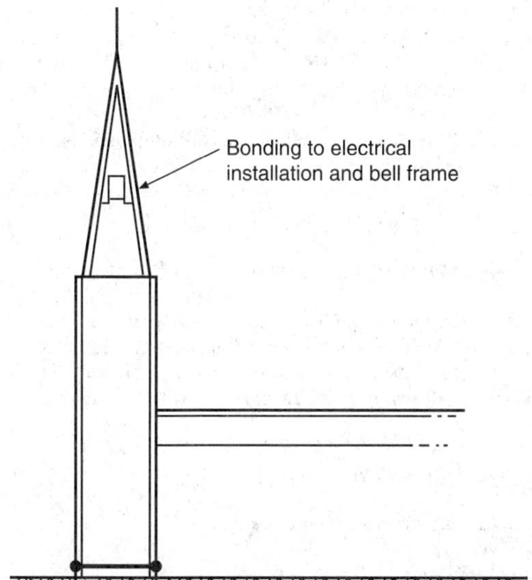

FIG. 3-3M.   *An installation of a lightning protection system in a church steeple according to British Standard Code of Practice[1] requirements. (Note absence of air terminals on lower-level portion; protection is provided by an elevated conductor above the ridge.)*

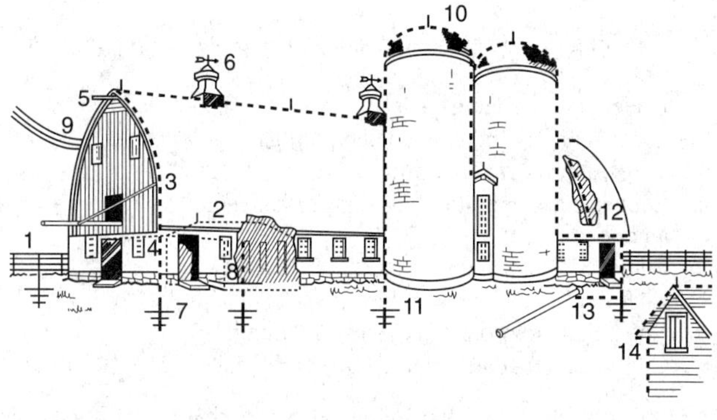

FIG. 3-3N.   *Lightning protection system on a typical large barn. The numbers indicate the following features: (1) ground-attached wire fence; (2) extend protection to any addition; (3), (4), and (5) show bonding conductors to litter, metal door, and hay tracks; (6) shows air terminal on cupolas, ventilators, etc.; (7) indicates ground connection (there should be at least two grounds); (8) illustrates tie-in of metal stanchions; (9) power lines to building need lightning arresters; (10) at least one air terminal should be on each domed silo (at least two on each flat roof or unroofed silo); (11) shows special ground for silo, if required; (12) illustrates connections to vents and (13) connections to water pipes; and (14) illustrates desirability of protecting adjacent buildings.*

against damage due to lightning can be found in the "Additional Reading" section of the bibliography, under "Surge Arresters." The use of secondary service arresters on electric services is required for buildings equipped with a UL Master-Labeled Lightning Protection System. Such arresters may be installed at the yard pole, at the outside electric service entrance, or at the interior service entrance equipment, depending on local regulations. Home lightning protectors are available that are designed for installation at the dwelling weatherhead or indoors at the service entrance equipment. These protectors (arresters) drain lightning surge-induced charges harmlessly to ground, and then open to restore electrical service to normal. Before installing a secondary service arrester, it should be determined that the neutral wire is grounded.

The surge arrester should be coupled with a transient voltage surge suppressor on branch circuits that serve computers and electronic appliances. These devices are constructed as portable (plug-in) or permanently wired.

In the United States, except on the Pacific Slope where lightning storms are infrequent, most electric utilities install lightning arresters on the primaries of important transformers on systems of 44,000 V or less.

PROTECTION OF PERSONS

The risk of a direct lightning strike is greatest among persons whose occupations keep them outdoors. About one hundred people a year are non-fatally injured in fires due to lightning, most of them in house fires. Prevention of these injuries involves a combination of protection of the structure from fire due to lightning and the usual steps to protect occupants of a building from injury due to fire.

Guide for Personal Safety During Thunderstorms

Do not go out of doors or remain outside during thunderstorms, unless it is necessary. Seek shelter inside buildings, vehicles, or other structures or locations that offer protection from lightning.

Seek shelter in the following places that may protect persons from lightning:

1. Dwellings or other buildings that are protected against lightning.
2. Underground shelters, such as subways, tunnels, and caves.
3. Large metal or metal-framed buildings.
4. Large unprotected buildings.
5. Enclosed automobiles, buses, and other vehicles with metal tops and bodies.
6. Enclosed trains and streetcars.
7. Enclosed metal boats or ships.
8. Boats that are protected against lightning.
9. City streets that are shielded by nearby buildings.

If possible, avoid the following places that offer little or no protection from lightning:

1. Small unprotected buildings, barns, sheds, etc.
2. Tents and temporary shelters (not lightning protected).
3. Automobiles (nonmetal top or open).
4. Recreational vehicles (nonmetal or open).
5. Electrical appliances, telephones, and plumbing fixtures.

Certain locations are extremely hazardous during thunderstorms and should be avoided if at all possible. Approaching thunderstorms should be anticipated and the following locations avoided when storms are in the immediate vicinity:

1. Open fields, athletic fields, and golf courses.
2. Parking lots and tennis courts.
3. Swimming pools, lakes, and seashores.
4. Near wire fences, clotheslines, overhead wires, and railroad tracks.
5. Under isolated trees.

In the above locations, it is especially hazardous to be riding in or on any of the following during lightning storms:

1. Open tractors and other farm machinery operated in open fields.
2. Golf carts, scooters, bicycles, and motorcycles.
3. Open boats (without masts) and hovercraft.
4. Automobiles (nonmetal top or open).

It may not always be possible to choose an outdoor location that offers good protection from lightning. Follow these rules when there is a choice in selecting locations:

1. Seek depressed areas—avoid hilltops and high places.
2. Seek dense woods—avoid isolated trees.
3. Seek buildings, tents, and shelters in low areas—avoid unprotected buildings and shelters in high areas.
4. If you are hopelessly isolated in an exposed area and you feel your hair stand on end, indicating that lightning is about to strike, crouch down with both feet close together. Do not lie flat on the ground or place your hands on the ground.

BIBLIOGRAPHY

References Cited

1. "The Protection of Structures Against Lightning," *British Standard Code of Practice*, BS 6651, 1985, British Standards Institute, Milton Keynes.
2. Offerman, P. F., "Lightning Protection of Structures," IEEE Conference Record of 1969 Fourth Annual Meeting of the IEEE Industry and General Applications Group, 69 C5-IGA, 1969.
3. Lee, R. H., "Protection Zone for Buildings Against Lightning Strokes Using Transmission Line Protection Practice," IEEE *Transactions* on Industry Applications, Vol. 1A-14, No. 6, Nov./Dec. 1978.
4. Davis, N. H., "The Rolling Sphere Interception Concept," *Proceedings* of the 19th International Conference on Lightning Protection, Graz, Austria, Vol. I, Apr. 25-29, 1988.
5. Moore, Brook, and Krider, "A Study of Lightning Protection Systems," The Atmospheric Science Program of the Office of Naval Research (under Contract No. 00014-78M-0090).
6. Kaufmann, R. H., "Some Fundamentals of Equipment—Grounding Circuit Design," AIEE *Transactions*, 54-244, Nov. 1954.
7. Sunde, E. D., *Earth Conduction Effects in Transmission Systems*, D. Van Nostrand Co., Inc., 1949, pp. 75-80.

NFPA Codes, Standards, and Recommended Practices

Reference to the following NFPA codes, standards, and recommended practices will provide further information on lightning protection systems discussed in this chapter. (See the latest version of *The NFPA Catalog* for availability of current editions of the following documents.)

NFPA 70, *National Electrical Code®*
NFPA 302, *Fire Protection Standard for Pleasure and Commercial Motor Craft*
NFPA 780, *Lightning Protection Code*

Additional Reading

Adams, R., "When Lightning Strikes," *Firehouse*, Vol. 16, No. 7, July 1991, p. 18.
Electrical Construction Materials Directory, published annually in May with November Supplements, Underwriters Laboratories Inc., Northbrook, IL.
Factory Mutual Research Corp., "Lightning Arresters and Grounds," *Loss Prevention Data Sheets*, 5-11/14-19, FMRC, Norwood, MA.

Frydenlund, M. G., "Understanding Lightning Protection," *Plant Engineering*, Vol. 44, No. 9, May 10, 1990, pp. 62-66.

Frydenlund, M. M., "Protecting Industrial Plants Against Lightning's Dangers," *Consulting/Specifying Engineer*, Vol. 8, No. 3, September 1990, pp. 60–62,64,66,68.

Golde, R. H., "Protection of Structures Against Lightning," *Proceedings*, The Institution of Electrical Engineers, Vol. 115, No. 10, Oct. 1968.

Golde, R. H., *Lightning Protection*, Academic Press, New York, 1977.

Hart, C. W., and Edgar, W. M., *Lightning and Lightning Protection*, Interference Control Technologies, Gainesville, VA, 1988.

Hedlund, C. F., "Lightning Protection for Buildings," IEEE *Transactions* on Industry and General Applications, Vol. IGA-3, No. 1, Jan./Feb. 1967, Institute of Electrical and Electronics Engineers, New York.

IEEE Standard 142-1982 (ANSI/IEEE), Chapter 3, "Static and Lightning Protection Grounding."

Jackson, A. L. and Vallas, P. S., "Struck by Lightning: An Investigation," *Fire Engineering*, Vol. 149, No. 3, March 1996, pp. 61–62,64–65.

Journal of the Franklin Institute, Philadelphia, Vol. 283, No. 6, June 1967 (special issue on lightning research).

"The Lightning Flash," published bi-monthly by Lightning Technologies, Inc., 10 Downing Parkway, Pittsfield, MA.

"Lightning Protection System, General with Regard to Installation," German Standard DIN 57185, Part 1/VDE 0185, English Translation, 1983.

Marinoff, E., "Fire Causes by Lightning, Sandwich South Township, September 7, 1990," Fire Investigation Report, Office of the Fire Marshall, Canada, Fire Investigation Report, June 22, 1993, 8 pages.

Plumer, J. A., "We Need Better Lightning Protection," *Fire Journal*, Vol. 81, No. 1, 1987, p. 41.

Proceedings of the International Conference on Lightning Protection, published bi-annually by different publishers, 20th ed. by Swiss Electromechanical Assn., Interlocken, Switzerland.

Queen, P. L., "Thunder and Lightning," *American Fire Journal*, Vol. 45, No. 10, October 1993, p. 7.

Rosenthal, J., "Concern Over Claims for Lightning Conductors," *Fire Prevention*, No. 248, April 1992, p. 10.

Towne, H. M., *Lightning—Its Behavior and What to Do About It*, United Lightning Protection Association, Inc., Webster, NY, 1956.

Uman, M.A., *Lightning*, Dover Publication, Inc., Mineola, NY, 1983.

Uman, M.A., *All About Lightning*, Dover Publication, Inc., Mineola, NY, 1986.

Underwriters Laboratories Inc., UL 96, *Standard for Safety Lightning Protection Components*, 4th ed., 1995, Northbrook, IL.

Underwriters Laboratories Inc., UL 96A, *Standard for Safety Installation Requirements for Lightning Protection Systems*, 10th ed., 1995, Northbrook, IL.

Viemeister, P.E., *The Lightning Book*, MIT Press, Cambridge, MA, 1975.

Surge Arresters, ANSI Standards

ANSI/IEEE C62.1-1989, *Gapped Silicon-Carbide Surge Arresters for Alternating-Current Power Circuits.*

ANSI/IEEE C62.11-1993, *Standard for Metal-Oxide Surge Arrestor for AC Power Circuits.*

ANSI/IEEE C62.31-1987, *Specifications for Gas Tube Surge-Protective Devices, Test.*

ANSI/IEEE C62.32-81, *Standard Test Specifications for Low-Voltage Air Gap Surge-Protective Devices (Excluding Valve and Expulsion Devices)* (R 1994).

EMERGENCY AND STANDBY POWER SUPPLIES

Revised by George W. Flach

The Northeast blackout of 1965 dramatically demonstrated the need for a source of electric power independent of electric public utilities. The Northridge, California, earthquake of January 1994 is the most recent example of the need for localized electric power in an emergency. Power interruptions occur due to natural or man-made causes. Natural causes are storms, floods, earthquakes, and the like; man-made causes are many, such as vehicles striking transformers or utility poles, equipment failure from one or more causes, and human operational error.

NFPA 70, *National Electrical Code®* (hereinafter referred to as NFPA 70 or *NEC*), recognizes two types of private power supply systems: (1) emergency and (2) standby, which is further subdivided into (a) legally required standby systems and (b) optional standby systems. Requirements for the proper installation of any of these systems are found in Articles 700, 701, and 702 of the *NEC*.

Emergency power systems are systems supplying power for illumination; elevators, critical heating, ventilation, and air conditioning (HVAC); fire pumps and alarms; public address systems; and other loads essential to the safety of life and property when public utility power is unavailable and where required by local, state, or federal codes. (See Figure 3-4A.)

Legally required standby systems are provided as an alternate source to the normal power source (e.g., a public utility), where an electric power interruption to a continuous industrial process could endanger lives or hamper rescue or fire fighting operations. These systems are also installed for heating, refrigeration, and ventilation systems, or to serve other electrical loads that are specified by municipal, state, federal, or other codes or by any governmental agency having jurisdiction.

Optional standby systems are intended to protect private business or property where life safety does not depend on the performance of the system. Some examples of loads that are connected to these systems are heating and cooling equipment, data processing, and communications systems. Optional standby systems, selected by the owner, are also used on farms and in residences to supply loads during a power failure.

The loads to be served by the emergency or standby power system can be classified as critical and non-critical. Critical loads are loads that must be served within a relatively short time period after the power is needed in order to preserve life or property. Non-critical loads are all other loads that are permitted to be connected to the system.

In hospitals, alternate power source wiring and equipment are termed an essential electrical system. One or more on-site generators must be used to supply this system. The essential electrical system comprises an emergency system and an equipment system. The emergency system is further divided into (1) the life safety branch and (2) the critical branch.

Several terms overlap each other, e.g., alternate power systems, standby power systems, legally required standby systems, alternate power sources, etc. NFPA 110, *Standard for Emergency and Standby Power Systems*, places all such systems into categories defined in terms of type, class, and level, rather than using such aforementioned terms. Type defines the maximum allowable time that the load is without acceptable electrical power. Class defines the minimum allowable time that the alternate source has the capability of providing its rated load without being refueled. Level defines the equipment performance stringency requirements.

This chapter provides information on the components that, alone or in combination, form emergency and standby power systems.

ENERGY SOURCES

NFPA 70 recognizes five electric sources for emergency power supplies: (1) storage batteries, (2) generator set(s), (3) uninterruptible power supplies, (4) a separate service, and (5) unit equipment. Item 4 is not recognized as an acceptable emergency power source for those occupancies covered by NFPA *101®*, *Life Safety Code®*. To cope with emergencies, a plaque or directory is required at each electric power source. Each sign must indicate where all other sources are located.

Energy sources for emergency and standby power supplies are generally one of the following: liquid petroleum products, liquefied petroleum gas, natural or synthetic combustible gas, or batteries. Other sources, e.g., steam, rotating mass (flywheel), and compressed noncombustible gas, have limitations that restrict their use for these purposes.

On-site storage of the energy source to supply power for the duration of the need for emergency and standby power is an important consideration, especially in areas subject to severe environmental disturbances, such as storms, earthquakes, flooding, etc. Section

George W. Flach is president of Flach Consultants in New Orleans, LA. He is a member of the NFPA Technical Committee on the National Electrical Code—Panel 15.

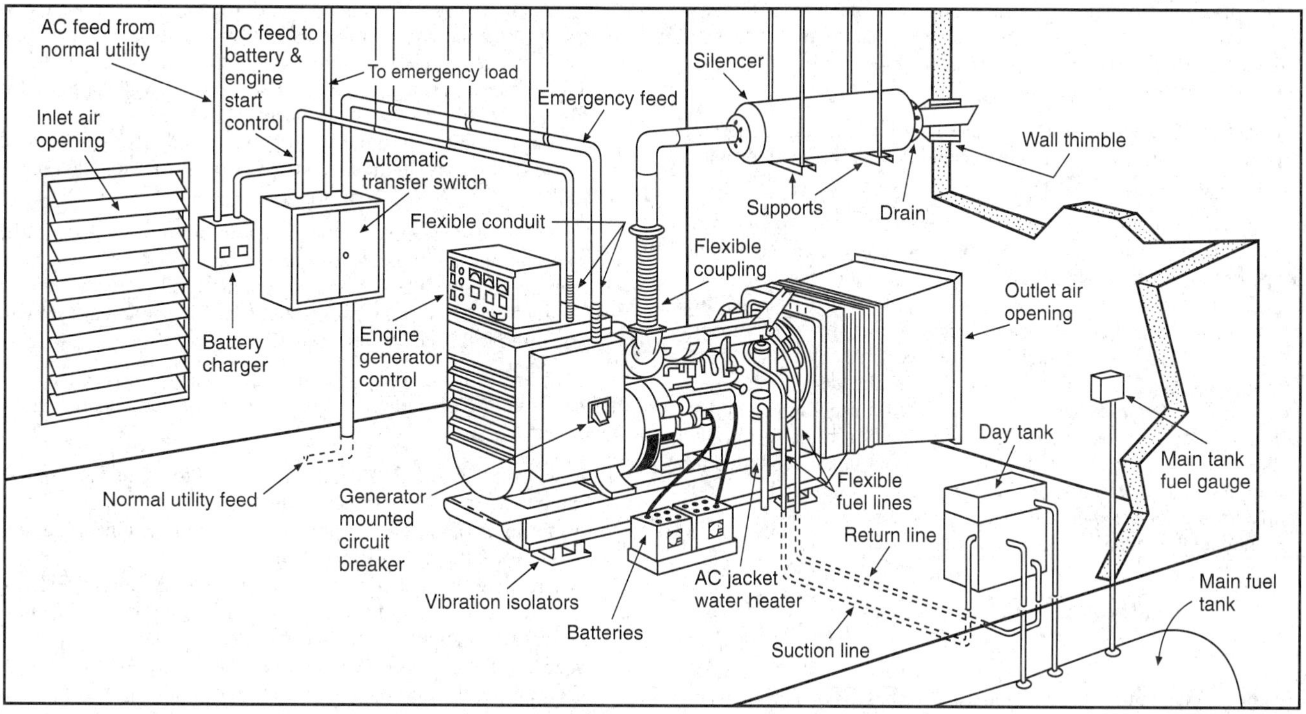

FIG. 3-4A. A typical generator installation used to supply emergency power. (Source: Caterpillar)

3, Chapter 10, "Stationary Combustion Engines," provides details on safeguards for on-site fuels.

Batteries

Batteries, in general, are the most dependable energy source for emergency and standby power and are primarily used for emergency lighting, either as a central battery system or as a self-contained lighting assembly defined as "unit equipment" in NFPA 70. They are also used with uninterruptible power supplies (UPS), a system that automatically provides power, without delay or transients, during a period when the normal power source is incapable of performing acceptably. A single battery or a group of batteries is limited in the amount of usable energy it is capable of delivering before becoming discharged; hence, backup to the battery source is often supplied in the form of other on-site emergency or standby power supplies.

Batteries used for this service must have sufficient capacity to supply the total emergency load for at least 1 1/2 hrs. During this time the voltage cannot drop below 87.5 percent of normal. They are usually either lead-acid or nickel-cadmium type. The lead-acid battery is less costly and requires less space for the same capacity; but the nickel-cadmium type has a longer life, more rugged construction, and requires less maintenance. Sealed batteries, i.e., those not requiring periodic replacement of water, are favored for use with self-contained lighting assemblies.

After use, batteries must be recharged to have power available again when needed. A self-contained lighting assembly normally is provided with an automatic recharger and a sensing relay for determining normal power supply loss to the normal illuminating means. Consideration should be given to having the battery recharger connected to the emergency or standby power source (when such a source is installed) upon loss of normal power where evacuation of the installation is not expected during a normal power outage or where a large number of people could be stranded in a high-rise building for a duration longer than the expected usual discharge

time of the battery without recharging. Central battery systems and UPS systems are normally provided with more complex battery rechargers, such as dual-rate, regulated, etc., to maximize battery life and minimize gassing.

Proper location of batteries with adequate ventilation is an important environmental consideration. The gaseous products of electrolysis (hydrogen and oxygen) can be a fire hazard.

UNINTERRUPTIBLE POWER SUPPLIES AND OTHER STORED ENERGY POWER SUPPLIES

A true UPS system providing continuous and transient-free single-phase or polyphase ac power is used to supply such critical load applications as sensitive electronic and computer equipment for data processing, life support, communications, and control equipment. Such a system, upon loss of normal power, continues to supply power without waveform distortion, and the load is totally unaware of normal power loss.

There are applications where such purity is not required, and either waveform distortion (transient or phase displacement of waveform) or a brief lapse of power to the load, or both, is acceptable. Economic considerations will bear upon the type of stored energy power supply used.

Batteries, combined with rectifiers and solid-state inverters (converters of dc power to ac power), provide power to the majority of UPS systems in use. Such systems are known as static UPS systems. This system rectifies the incoming ac power to dc power, then inverts the dc power back to ac power for delivery to the load. A battery is permanently connected between the rectifier and inverter (on the dc bus) and takes over as the power source immediately upon loss of the incoming ac power. Upon return of the incoming ac power, this power serves to power the load as well as to recharge the battery. The inverter serves to determine the characteristics of the ac power to the load, so that any voltage or frequency variations or transients on the incoming power are absent from the load. The rectifier is continuously sized to the capacity of the load plus that required for battery recharging, while the inverter is continuously sized to the capacity of the load.

Other UPS systems are variations of a rotating generator driven by various drivers (engines, ac and dc motors), with or without a flywheel.

A stored energy power supply system similar to the UPS system, but economically more attractive, is one in which the normal ac power is fed through to the load with a static transfer switch transferring the load to an inverter supplied by a battery upon loss of the normal ac power. The inverter receives a continuous synchronizing signal from the normal ac power, so that, upon loss of this power, the inverter is still approximately in phase with the power just lost. This, coupled with a transfer time of a few milliseconds, permits the load to be without power for only a brief period and without serious transients occurring upon application of inverter power. Note, however, that the inverter need only be sized intermittently for the load power and that the rectifier be sized only to accommodate the recharging power of the battery.

ENGINE-DRIVEN GENERATOR POWER SYSTEMS

An engine generator power system usually consists of a prime mover, a generator coupled thereto (either direct or geared), transfer equipment, and distribution and protective equipment. (See Figure 3-4B.) Such a system is supplied for either standby or emergency service, and is most common where such power is required for a considerable period of time.

Prime Mover

The prime mover may be a reciprocating engine fueled by gasoline, diesel fuel, or natural or synthetic gas. Gas turbines are also used, particularly in larger power outputs. Gasoline and natural and syn-

thetic gas engines are less costly than diesel engines, but have disadvantages that are considerable. For example, the gasoline engine and fuel storage therein are more susceptible to fire and explosion. Also, the natural gas engine usually obtains its energy supply from an off-site gas utility, a supply that may not be available at the same time of unavailability of the normal power supply (for this reason, an on-site fuel supply might be required by the Authority Having Jurisdiction).

Gas turbines are much smaller than reciprocating engines of the same output, and operate at much higher speeds. They require no cooling-water connection and burn natural gas, kerosene, diesel fuel, or fuel oil. They are, however, quite inefficient and are slow in coming up to full speed (more than 10 s).

The prime mover and its associated auxiliary equipment are required by the *NEC* to have the following characteristics for emergency service:

1. It must start automatically and come up to full speed and power capability within 10 s of demand.
2. A minimum on-site fuel supply permitting full engine-generator output for 2 hrs must be available.
3. Where water-cooled engines are used (generally above 50 kW), the municipal water supply cannot be used for cooling (i.e., municipal water supply input, waste to drain output).
4. Where a battery is used for engine starting, the battery must have an automatic charging means independent of the prime mover.
5. Where the prime mover requires more than 10 s to come up to full speed and power capability, an auxiliary power supply may supply the required power (within 10 s) to the load, provided it can continue to do so until the load is assumed by the prime mover.

A prime mover driving a legally required standby power system must meet conditions 2, 3, and 4 above, and condition 1 within 60 s.

A governor maintains the speed of the prime mover approximately constant throughout the full power output range by varying

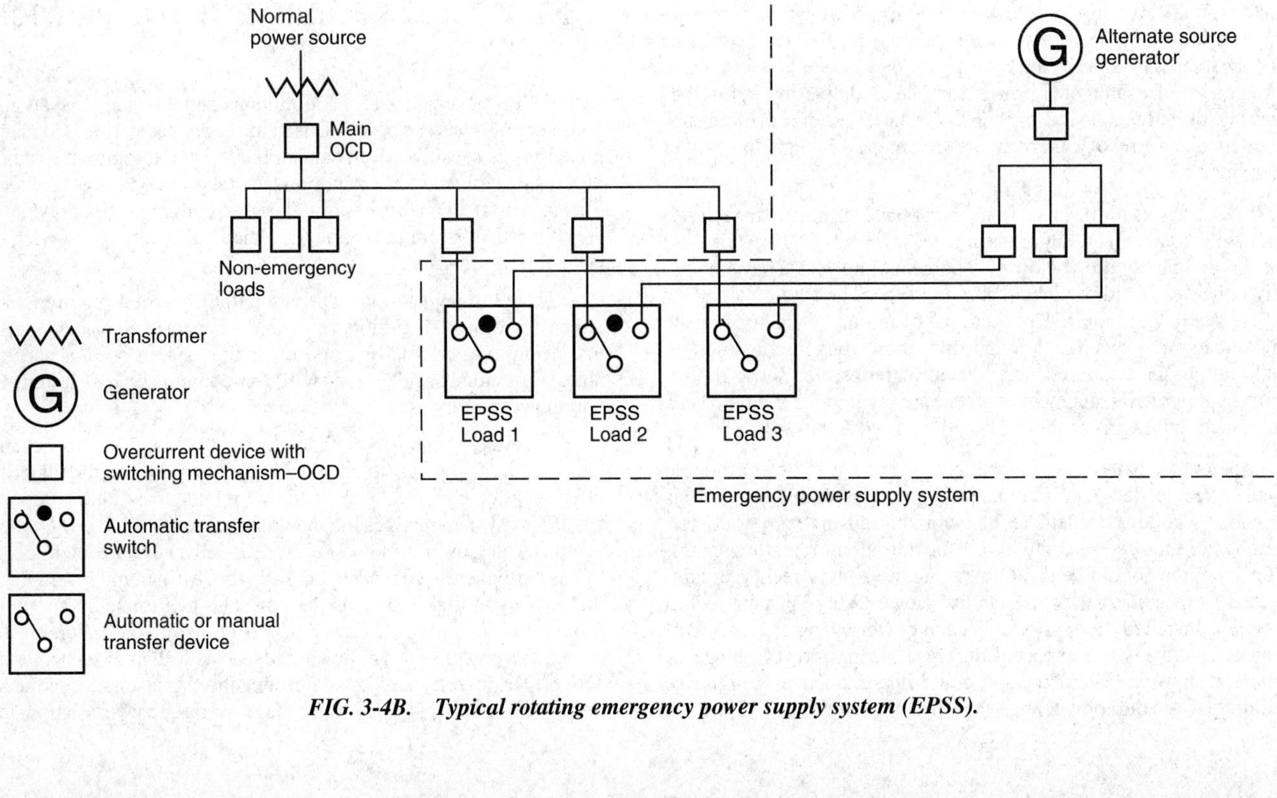

FIG. 3-4B.   Typical rotating emergency power supply system (EPSS).

the fuel input to the prime mover. Governors may be mechanical, electrical, hydraulic, or a combination thereof.

The Generator

Generators convert the mechanical shaft output of the prime mover into the electrical power for use by the loads connected thereto. The performance of the generator and its voltage regulator under varying load conditions must be considered in determining its suitability. The output voltage of a generator drops suddenly when a load is applied, and rises suddenly when a load is removed. At present, generators larger than a few kilowatts in size have automatic voltage regulators that quickly bring the voltage back to the specified value. The initial voltage drop is important because if the drop is excessive, even for a few cycles, electromagnetic components, e.g., relays, motor starters, etc., may drop out and may or may not, depending upon circuitry, reclose when the voltage is restored, or they may cycle open and close repeatedly, causing equipment damage. Voltage rise (overshoot) upon load removal may cause damage to equipment remaining connected. The initial voltage drop is not usually of concern when the loads being connected are within the ratings of the generator. However, it is customary to exceed the rated current of the generator by two to three times during motor starting only. The initial voltage dip during these current overloads is, therefore, one of the limiting factors in sizing the generator. The performance of the generator when subjected to a short-circuit or heavy overload condition is also a matter of consideration. Under this condition, the generator current output decays with time, a characteristic inherent in the generator, so that, in a relatively short time, the current lessens to a magnitude of only several times its rating. If the decay rate is too rapid, it may not be possible to have selective tripping of overcurrent devices in the current loop.

Transfer Equipment

Electromechanical automatic transfer switches are normally used to transfer selected loads to the engine generator power supply when the normal source fails. The transfer switch senses normal power, usually all phases, and upon loss of normal power automatically signals the prime mover to start. A brief time delay is generally built into the transfer switch logic to delay the signal so that a momentary normal power interruption or flicker does not activate the prime mover. The transfer switch also senses the voltage and frequency of the generator output and, when these reach acceptable levels, either immediately thereafter or after a time delay (for selective load pickup), the switch transfers the load to the engine generator output. Upon restoration of the normal power supply to acceptable levels, the transfer switch may transfer the load automatically (usually after a time delay to allow for permanent normal power restoration) back to the normal source, or manual transfer of the load may be required, the choice having been originally made by the characteristics of the load. Transfer switches, in general, immediately retransfer the load back to the normal source (if available and at an acceptable level), should the emergency source fail. Maintenance of a transfer switch may require deactivation of loads connected to the switch while the maintenance is being performed. Some loads are so critical that such deactivation is unacceptable. In such cases, the transfer switch is furnished with an isolation by-pass switch, making it possible for the load to be served by the normal or emergency source while the automatic transfer switch is being serviced.

Distribution

Distribution of the available auxiliary power (emergency or standby) concerns itself with the loads served, the order of priority, and the total loading of the engine generator set. Protection concerns itself primarily with equipment protection and with continued availability of power to unfaulted loads upon a system electrical fault.

Engine generator set loading prioritizes the need for power and the magnitude of the various loads. Critical loads must be served first, followed by non-critical large motor loads. A motor load produces a heavy power and kVA load upon first starting, which influences momentary voltage and frequency output of the generator, causing a decrease in each. Step loading of the engine generator set is accomplished by using multiple transfer switches, timed to operate in sequence in the order of priority. Other more complex arrangements with programmable controllers or computers are sometimes used.

Permissible loading of the engine generator set during periods of loss of normal power is governed by codes and standards. Generally, during such periods, it is permissible to use available power not being used for critical loads to power non-critical loads and to resort to load-shedding techniques to pick up the critical loads when needed.

Protection, while generally in accordance with established procedures for the normal power supply, must take into account the usual situation of limited power availability of the engine generator set vs. the larger power availability of the normal source. As indicated earlier, generators, under short-circuit or heavy overload conditions, have collapsing outputs such that the sustained total current output of the generator drops to a magnitude comparable to its full-load value. Rapid clearing of the fault is essential (prior to the drop in current to its sustained value); otherwise the protective device will not see the short-circuit condition. Rapid fault clearing is also important to enable the generator to reestablish itself (voltage restoration) and continue to serve the loads still connected to it.

MISCELLANEOUS CONSIDERATIONS

Multiple-engine generator sets are sometimes used to supply emergency and standby power. Such sets are usually operated in parallel, with load-shedding provisions incorporated should one of the sets fail. Because of the sizable monetary investment in an emergency or standby power system, some installations use the engine generator set in conjunction with the normal power supply for peak-shaving purposes. Peak shaving is the transfer of some loads from the normal power source to the auxiliary power source during periods of maximum loading of the normal power source. The economic benefit of peak shaving is a result of the utility rate structure. If a user can keep the ratio of maximum (or peak) power usage to average power usage low, then the utility charge is lessened. By transferring selected loads to the engine generator set during periods of peak power usage, the aforementioned ratio is lowered.

Motor-starting methods for large motors is another significant consideration. Two matters of concern are: (1) the power required from the prime mover, and (2) the kVA required from the generator. Selection of the proper type of reduced voltage starting is equally important. Open-transition reduced voltage starting is not a recommended method, nor is primary resistance reduced voltage starting. Auto-transformer starting and closed transition wye-delta starting are generally preferred to other methods.

BIBLIOGRAPHY

References

Battery Service Manual, 11th ed., Battery Council International, Chicago, IL 1995.

"IEEE Recommended Practice for Emergency and Standby Power Systems for Industrial and Commercial Applications," ANSI/IEEE 446-1987, Institute of Electrical and Electronics Engineers, Inc., Piscataway, NJ.
"Kohler Engineer's Guidebook to Power Systems," *Consulting Engineer*, Vol. 64, No. 2, Part 2 of 2, Feb. 1985.

NFPA Codes, Standards, and Recommended Practices

Reference to the following NFPA codes, standards, and recommended practices will provide further information on emergency and standby power supplies discussed in this chapter. (See the latest version of *The NFPA Catalog* for availability of current editions of the following documents.)

NFPA 37, *Standard for the Installation and Use of Stationary Combustion Engines and Gas Turbines*
NFPA 70, *National Electrical Code®*
NFPA 99, *Standard for Health Care Facilities*
NFPA *101®*, *Life Safety Code®*
NFPA 110, *Standard for Emergency and Standby Power Systems*

HEATING SYSTEMS AND APPLIANCES

Revised by David S. Johnston

Heat-producing appliances and associated equipment are among the most prevalent causes of fire because they operate at temperatures above the ignition temperature of many common materials. In addition, combustion-type appliances may involve the hazards of an accumulated combustible mixture, the discharge of unburned fuel, and possible exposure of fuel to ignition sources.

Additional information about heating systems and appliances can be found in: Section 1, Chapter 5, "Chemistry and Physics of Fire"; Section 3, Chapter 1, "Electrical Systems and Appliances"; Section 3, Chapter 6, "Boiler-Furnaces"; Section 3, Chapter 8, "Industrial and Commercial Heat Utilization Equipment"; Section 3, Chapter 21, "Storage of Flammable and Combustible Liquids"; Section 3, Chapter 22, "Storage of Gases"; Section 4, Chapter 7, "Gases"; and Section 7, Chapter 14, "Air-Conditioning and Ventilating Systems."

FUELS AND METHODS OF FIRING

Combustion may be defined as a chemical reaction between a fuel and oxygen, with evolution of heat and light. In a fuel-burning, heat-producing device, this reaction needs to be continuous so a balance will be established between the rate at which oxygen, or air, and fuel are supplied and the rate at which heat and the products of combustion are removed. Complete combustion takes place when all of the fuel is oxidized by the air supplied to it. Air left over after complete oxidation is called excess air. If only enough air is supplied for complete combustion, perfect combustion can theoretically result under suitable conditions. Because this is impractical in most heat-producing appliances, the appliances operate with some excess air.

Air for combustion is supplied in two ways: (1) primary air is introduced through or with the fuels, and (2) secondary air is supplied to the combustion zone. In some cases, all of the air supplied is primary air, while in others only part is primary air and the balance is secondary air.

Incomplete combustion in a fuel-burning device can produce hazardous carbon monoxide. This condition indicates poor efficiency, since the fuel is not completely oxidized, and its total heating value of the fuel is not obtained. Incomplete combustion usually results from inadequate air supply, insufficient mixing of air and gases, or a temperature too low to produce or sustain combustion.

David S. Johnston is president of DSJ Technical Services in Wooster, OH. He is chairman of the NFPA Technical Committee on Chimneys, Fireplace, and Venting Systems for Heat-Producing Appliances.

Solid Fuel—Coal

Coal is one of the principal solid fuels used in heat-producing appliances. There are a number of types of coal, each having widely different characteristics. In many cases, there is no distinct line of demarcation between them, and the qualities of one type can overlap those of another. Pulverized coal systems, however, require the use of coals having characteristics within the specific range of the coal-handling and coal-burning equipment.

The seven principal types of coal used in heat-producing appliances are:

Anthracite: A clean, dense, hard coal. It burns with a minimum of smoke and has a minimum dust hazard during handling.

Semianthracite: A coal having a higher volatile content than anthracite but not as hard.

Bituminous: This includes many types of coal with different properties, depending upon where they are mined. It gives off considerable smoke and soot if improperly fired. Bituminous coal may be subject to spontaneous heating under some storage conditions and has a dust hazard during handling.

Semibituminous: A dusty soft coal that tends to break up. It may be subject to spontaneous heating under some storage conditions. It normally produces less smoke than bituminous coal.

Subbituminous: A coal subject to spontaneous heating under some storage conditions. It burns with very little smoke and soot.

Lignite: A coal having a woody structure and normally a high moisture content. It is subject to spontaneous heating under some storage conditions. It burns with little smoke and soot.

Coke: A product of the destructive distillation of coal. The type of coke produced is determined by the coal used and the temperatures and time of distillation. Petroleum coke is produced from the destructive distillation of oil.

Combustion of Coal

When coal is burned in a firebox, oxygen in the air passing through the grate (primary air) unites with the carbon in the lower portion of the fuel bed, called the oxidation zone, to form carbon dioxide. Some of this carbon dioxide is then reduced to carbon monoxide in the upper part of the fuel bed, called the reduction zone. Gases liberated from the fuel bed are carbon monoxide, carbon dioxide, nitrogen, and some oxygen. Oxygen from the air admitted over the fuel bed (secondary air) combines with some of the carbon monoxide to form carbon dioxide. When fresh coal is applied to the fire, moisture is driven off as steam and the hydrocarbon gases are distilled, combined with oxygen, and burned above the grate in what is called the distillation zone.

Methods of Firing Coal

Coal is fired either manually or automatically by stokers or pulverized coal burners.

Hand firing: There are many acceptable hand-firing methods. With highly volatile coals, the objective is to leave a suitable bed of glowing fuel to ignite the gases as they are driven off from the fresh coal charge; otherwise the gas may accumulate and explode. Draft regulation is the only automatic feature of a hand-fired furnace. Draft can be controlled by a room thermostat, steam pressure, or water temperature. But, as with any hand-fired solid fuel, the fire is continuous and overtemperature conditions are difficult to control unless the fire is continually attended.

Mechanical stokers: These are classified according to their coal burning capacities, and range from those that handle 10 lb (4.5 kg) per hr to those handling over 1200 lb (540 kg) per hr. A stoker feeds fuel to the combustion chamber and usually supplies air for combustion under automatic control. There are four basic types of stokers: (1) underfed, (2) overfed, (3) traveling and chain grate, and (4) spreader, with fired grate or continuous ash discharge.

Underfed stokers move the coal by screw conveyor or rams. (See Figure 3-5A.) Overfed stokers feed the coal pneumatically or by rotors. Spreader stokers similarly feed coal by rotors or paddles and discharge onto stationary or traveling grates. Stokers may feed from hoppers or directly from bins. They are equipped with fueling controls that either change the firing rate or stop the fuel altogether. Some stokers are equipped with a control that will stop the feeding of fuel if the fire goes out.

Pulverized coal systems: These are of two general types: (1) the direct-fired, or unit, system, and (2) the storage, or bin, system. In the direct-fired system, raw coal is fed directly to a pulverizer where it is pulverized, mixed with air, and blown to the burners. In the storage system, coal is delivered to a raw fuel bin and then passed to a pulverizer where it is reduced to a powder and dried for storage in a pulverized fuel bin before delivery to the furnace.

Solid Fuels—Miscellaneous

Miscellaneous solid fuels include: wood, logs, scrap lumber, and wood waste from lumber and paper mills (sawdust, shavings, bark); hogged fuel (sawmill refuse run through a disintegrator or "hog" to form uniform chips or shreds); charcoal; briquets; peat; bagasse;

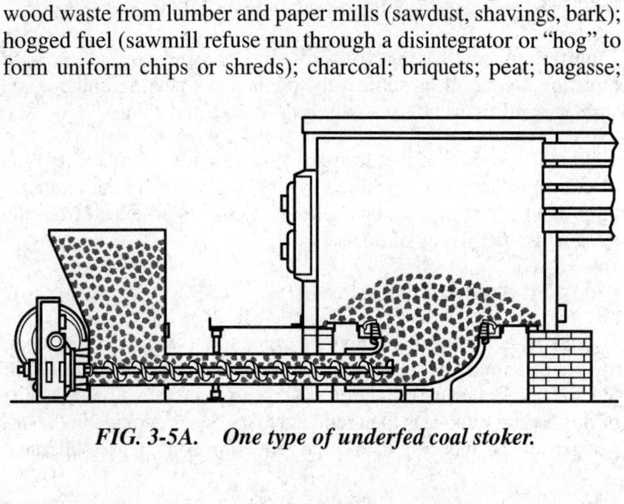

FIG. 3-5A. One type of underfed coal stoker.

sugar cane; and a host of other combustibles such as tanbark, wet bark, straw, city refuse, paper, etc. The methods of firing these fuels vary considerably with the type of device and the nature of the particular fuel.

Since the mid-1970s, the most prevalent form of solid fuel usage has been in residential cordwood stoves, furnaces, and fireplaces. Concerns about price and availability of conventional residential energy sources sparked a surge in installation and use of such devices. While the "crisis" attitude toward conventional energy has eased, solid fuel appliances remain a popular alternative source of supplemental heating.

The rise in installation of wood-burning appliances was accompanied by a dramatic increase in heating-related home fires, especially in the early 1980s. The severity of this problem has also decreased since then, due to decreased usage of installed appliances, and better installation, use, and maintenance practices. Nevertheless, solid-fuel-related fires remain a significant part of the residential fire picture. The hazards associated with wood-burning devices are discussed in detail in a later section of this chapter.

Another important development in the use of solid fuels is the introduction of pellet-fuel-burning appliances. Processed solid fuels of various types have long been used in commercial and industrial operations. Pellet-fuel-burning appliances are relatively small-capacity, residential appliances. Pellet fuel is produced by milling wood waste to fine particles, then extruding under high pressure to form pellets about $1/4$ in. in diameter by $1/2$ in. to 1 in. long. (For SI units: 1 in. = 25.4 mm.) The fuel is placed in the hopper of a pellet-fuel-burning appliance, from where it is metered to a burn pot, depending on heat demand. Pellet-fuel-burning appliances thus bring a degree of automatic operation to residential solid fuel usage.

Combustion of wood takes place in three stages. Unless the wood has been purposely kiln-dried before use, the first stage is the driving-off of moisture. The second stage is destructive distillation such as that which occurs in a charcoal kiln, where the heat drives off combustible gases from the wood and leaves charcoal. The third stage involves burning of the gases in the air above the wood and burning of the charcoal combined with oxygen, forming an intensely hot and luminous bed of coals. The heat from both burning gases and charcoal distills more volatiles from the wood and increases the temperature and rate of combustion. The acceleration of combustion is limited only by the rate at which air can be brought into contact with the burning wood.

Liquid Fuel—Fuel Oils

This section discusses fuel oil used in heat-producing appliances and equipment.

Fuel oil, like other petroleum products, is composed of varying compounds of hydrocarbons. Even the same grade of fuel oil will vary as to its chemical composition, depending upon such factors as the type of crude oil used and the refining process employed.

Types of fuel oil: Fuel oils obtained from present-day refining processes are known as "cracked" oils, and practically all petroleum products on the market today are obtained by the cracking process. No. 1 and No. 2 fuel oils, as well as those fuels commonly known as kerosene, range oil, furnace oil, star oil, and diesel oil, may still be broadly classed as distillates, and No. 4, No. 5, and No. 6 oils (as well as Bunker C) as residuals.

Oil burners are constructed (and subsequently tested and listed by testing laboratories) for a specific grade or grades of fuel oil. For example, burners listed by Underwriters Laboratories Inc. (UL) have the grade or grades of fuel oil that may be used in each inscribed on the UL listing mark applied to the unit. (See Figures

3-5B and 3-5C.) It is important that only the proper grade or grades of oil be used in each burner for safety as well as efficiency. Specifications for fuel oil are outlined by the American Society for Testing and Materials (ASTM) in ASTM D396, *Standard Specification for Fuel Oils*, and in Canadian Government Specification 3-GP-28. (See Table 3-5A.)

Methods of Firing Fuel Oil

Two methods, i.e., vaporization and atomization, are used to prepare fuel oil for combustion. Air for combustion is supplied by natural or mechanical draft. Ignition is by an electric ignition system, a gas pilot, an oil pilot, or manual means. Operation may be continuous, modulating with high-low flame, or intermittent. While most burners operate from automatic temperature- or pressure-sensing controls, some simpler types are operated manually.

Oil burners may be classified in several different ways, i.e., by application, by type of vaporizer or atomizer, by firing rates, etc. They are divided into two major groups: (1) residential and (2) commercial-industrial. Vaporizing burners and atomizing burners having capacities of not more than 7 gal/hr (25 L/hr) are considered residential types and are intended to be used with fuel oils not heavier than No. 2. The major portion of residential-type burners produced is the pressure atomizing type, commonly referred to as the gun type.

Vaporizing burners: These burners include the sleeve type, the pot type, and the vertical rotary wall flame type. Sleeve-type burners (see Figure 3-5D) are used in residential heating and cooking stoves. Natural draft pot burners (see Figure 3-5E) have applications similar to sleeve-type burners, while forced-draft pot burners may be used in central heating furnaces. Vertical rotary wall flame vaporizing burners are used in residential boilers and furnaces for central heating. While the gun-type burner is now the most common, many of the burners listed above are still in service.

FIG. 3-5B.   *Listing mark used on an oil burner listed by Underwriters Laboratories Inc. Note that the grade of oil satisfactory for use in the device appears at the bottom of the marker.*

FIG. 3-5C.   *Listing mark used on an oil-fired boiler assembly listed by Underwriters Laboratories Inc. Note the grade of oil satisfactory for use in the device appears at the bottom of the marker.*

Atomizing: These burners include high- and low-pressure types, horizontal rotary cup types, and air and steam atomizing types, the latter used primarily for commercial and industrial applications. High-pressure, gun-type burners consist of a motor, oil pump (with integral or separate pressure-regulating and shutoff valve), strainer, fan, ignition transformer, nozzle, and electrode assembly. (See Figure 3-5F.) The motor-driven pump draws oil from the supply tank and delivers it to the nozzle at pressures of from 100 to 300 psi (690 to 2,068 kPa). The nozzle atomizes the oil into fine particles and swirls it into

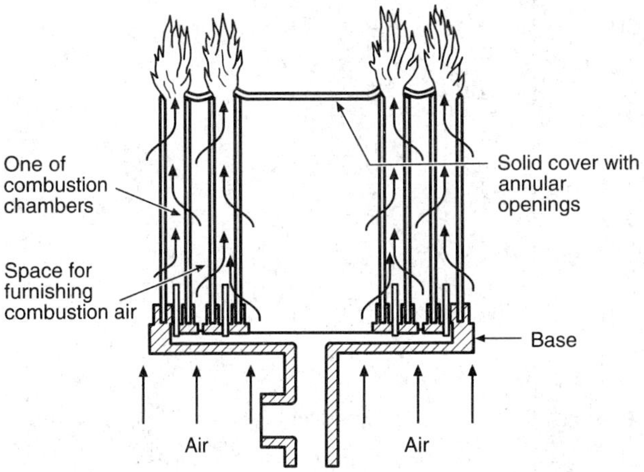

FIG. 3-5D.   *A sleeve-type burner that consists essentially of a flat, cast-iron or pressed steel base having two or more interconnecting grooves. Perforated metal cylindrical sleeves are placed in the grooves so that the space between the inner two sleeves and the outer two sleeves becomes a combustion chamber. The space between the second and third sleeves is for the purpose of furnishing combustion air. Covers with annular openings are usually placed on top of the sleeves. Sleeve burners may or may not employ wicks in the annular spaces in the base, depending upon the make and type of burner.*

FIG. 3-5E.   *A vaporizing pot-type burner. The oil is introduced into the bottom of the burner and is continually vaporized by the heat of the fire. The holes on the side permit primary and secondary air to enter the burner where they mix with the oil vapors for proper combustion. The primary air is mixed with oil vapors before combustion takes place, and the secondary air is supplied to complete the combustion process. At full fire, the velocity of the vapors and the amount of air required result in the flame burning only at the top of the burner.*

TABLE 3-5A. Detailed Requirements for Fuel Oils* ††

Grade of Fuel Oil	Flash Point, °F (°C) Min	Pour Point, °F (°C) Max	Water and Sediment, Percent by Volume Max	Carbon Residue on 10 Percent Bottoms, Percent Max	Ash, Percent by Weight Max	Distillation Temperatures, °F (°C) 10 Percent Point Max	90 Percent Point Min	90 Percent Point Max	Saybolt Viscosity, sec Universal at 100°F (38°C) Min	Universal Max	Furol at 122°F (50°C) Min	Furol Max	Kinematic Viscosity, centistokes At 100°F (38°C) Min	At 100°F Max	At 122°F (50°C) Min	At 122°F Max	Gravity, deg. API Min	Copper Strip Corrosion Max	Sulfur, Percent Max
No. 1: A distillate oil intended for vaporizing pot-type burners and other burners requiring this grade of fuel	100 or legal (38)	0§	trace	0.15	—	420 (215)	—	550 (288)	—	—	—	—	1.4	2.2	—	—	35	No. 3	0.5 or legal
No. 2: A distillate oil for general-purpose domestic heating for use in burners not requiring No. 1 fuel oil	100 or legal (38)	20§ (−7)	0.05	0.35	—	—	540§ (282)	640 (338)	(32.6)#	(37.93)	—	—	2.0§	3.6	—	—	30	—	0.5 §† or legal
No. 4: Preheating not usually required for handling or burning	130 or legal (55)	30 (−7)	0.50	—	0.10	—	—	—	45	125	—	—	(5.8)	(26.4)	—	—	—	—	¶
No. 5 (Heavy): Preheating may be required for burning and, in cold climates, may be required for handling	130 or legal (55)	—	1.00	—	0.10	—	—	—	150	300	—	—	(32)	(65)	—	—	—	—	¶
No. 5 (Light): Preheating may be required for burning and, in cold climates, may be required for handling	130 or legal (55)	—	1.00	—	0.10	—	—	—	350	750	(23)	(40)	(75)	(162)	(42)	(81)	—	—	¶
No. 6: Preheating required for burning and handling	150 (65)	—	2.00**	—	—	—	—	—	(900)	(9000)	45	300	—	—	(92)	(638)	—	—	¶

*It is the intent of these classifications that failure to meet any requirement of a given grade does not automatically place an oil in the next lower grade unless, in fact, it meets all requirements of the lower grade.

†Outside U.S., the sulfur limit for No. 2 must be 1.0 percent.

‡Legal requirements to be met.

§Lower or higher pour points may be specified whenever required by conditions of storage or use. When pour point less than 0°F is specified, the minimum viscosity must be 1.8cSt (32.0 sec, Saybolt Universal) and the minimum 90 percent point must be waived.

¶The 10 percent distillation temperature point may be specified at 440°F (226°C) maximum for use in other-than-atomizing burners.

#Viscosity values in parentheses are for information only and not necessarily limiting.

**The amount of water by distillation plus the sediment by extraction must not exceed 2.00 percent. The amount of sediment by extraction must not exceed 0.50 percent. A deduction in quantity shall be made for all water and sediment in excess of 1.0 percent.

††This table reprinted from ASTM D396. See complete specification ASTM D396.

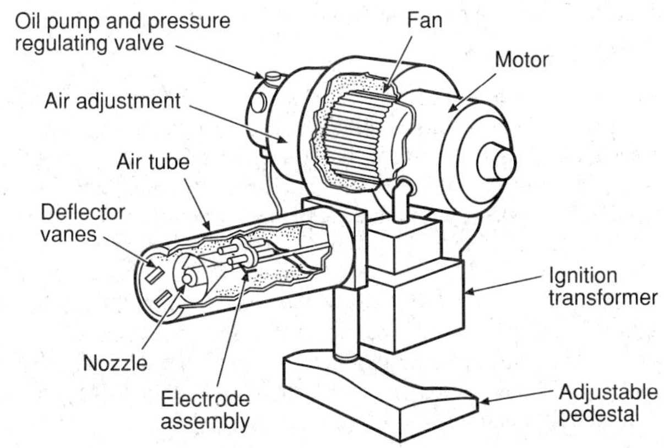

FIG. 3-5F. A high-pressure atomizing oil burner. (Source: National Oil Fuel Institute, Inc.)

the combustion chamber as a cone-shaped spray where it mixes with air and is ignited. These burners are generally designed to burn No. 2 oil, although some of the larger sizes are made for No. 4 oil.

The industrial version of the high-pressure burner, known as the mechanical atomizing burner, is a high-capacity burner for use with large boilers and industrial furnaces. Figure 3-5G shows a typical piping arrangement for one or more burners supplied by separate pump sets.

The low-pressure gun burner is similar to, but differs from, the high-pressure gun burner in two ways. First, the burner includes an air pump to supply compressed air for atomization, and second, the oil and air are delivered to the nozzle at pressures of 15 psi (100 kPa) or less. The size ranges, ignition, and fuel used are as described for the high-pressure gun burner.

The horizontal rotary-cup oil burner atomizes the oil by spinning it in a thin film from a horizontal rotating cup and injecting high-velocity primary air into the oil film through an annular nozzle that surrounds the rim of the cup. Secondary air for combustion is supplied from a separate fan that forces air through the burner wind box. The introduction of secondary air by means of natural draft is not recommended. These burners are used for firing boilers and furnaces and may be used singly or as multiple units on a single appliance. No. 2 fuel may be used with some of the smaller sizes, but Nos. 4, 5, and 6 are generally used with the larger sizes. (See Figure 3-5H.)

In the steam atomizing burner, atomization is accomplished by the impact and expansion of steam. Oil and steam flow in separate channels through the burner gun to the nozzle where the steam and oil mix before being discharged through an orifice into the combustion space. This type of burner is used mainly on large boilers generating steam at 100 psi (690 kPa) and higher and having capacities above 12,000,000 Btu per hr (3.5 MW) input.

Air-atomizing burners are designed to use compressed air at either high or low pressure for atomization. The high-pressure burner is nearly identical with the steam atomizing burner in design and application. Some burners operate well with either medium.

The low-pressure air atomizing burner uses comparatively large volumes of air at low pressure of 5 psi (35 kPa) or less. Air from the blower passes through the burner body and is discharged through annular slots between the body and the nozzle tip, where it meets the film of oil at an angle as it issues from the tip. The impact of the air stream upon the oil film produces a fine mist that is projected into the combustion space. These burners may be fired with either light or heavy oil and are most often used to fire industrial heating and processing furnaces. (See Figure 3-5I.)

Combination oil and gas burners, which combine the features of certain oil burners with provisions for burning gas, have been developed. These permit the operator to choose either fuel as circumstances may dictate.

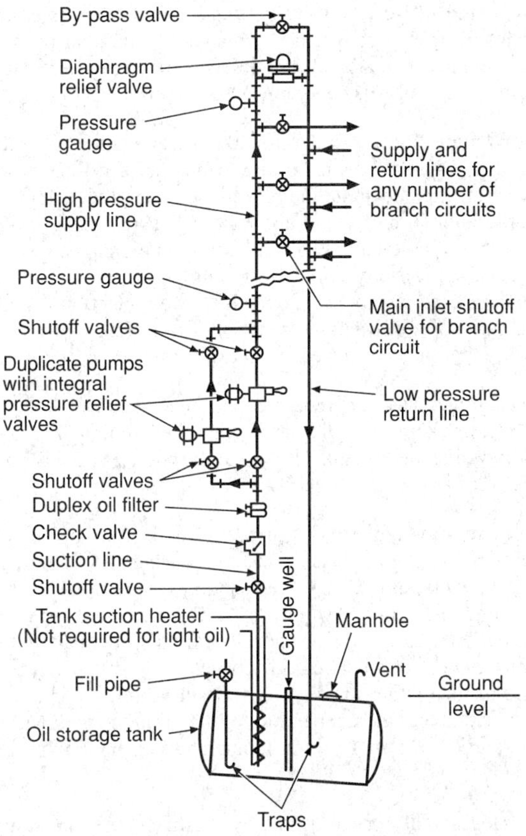

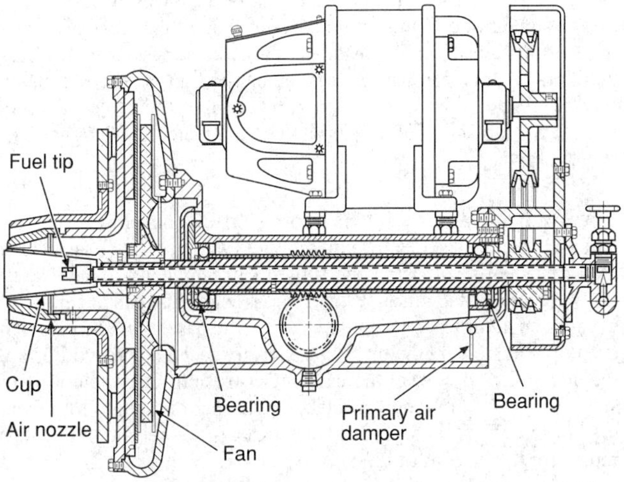

FIG. 3-5G. Typical piping arrangement for burners supplied by separate pump sets. (Source: Combustion Handbook, North American Mfg. Co.)

FIG. 3-5H. Horizontal rotary-cup oil burner. (Source: National Oil Fuel Institute, Inc.)

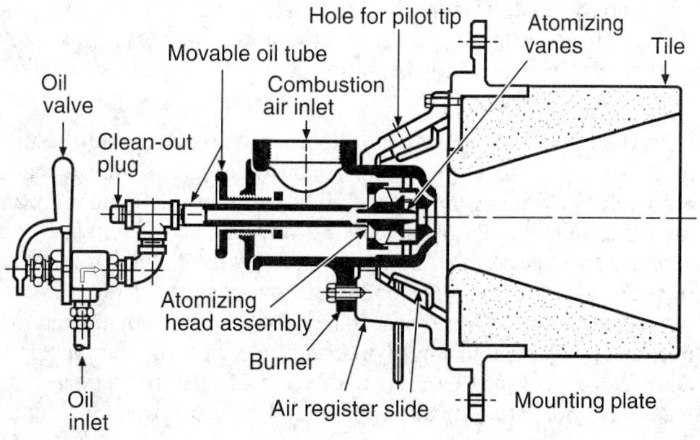

FIG. 3-5I. Low-pressure air atomizing oil burner. (Source: Combustion Handbook, North American Mfg. Co.)

Fuel Oil Storage

Details on the installation of fuel oil tanks and related piping are described in NFPA 31, *Standard for the Installation of Oil-Burning Equipment* (hereinafter referred to as NFPA 31). NFPA 31 covers the installation of tanks underground, unenclosed and enclosed inside buildings, and outside aboveground. The installation requirements in NFPA 31 differ slightly from the requirements for tanks in NFPA 30, *Flammable and Combustible Liquids Code* (hereinafter referred to as NFPA 30), which covers liquids with a wider range of flash points, which are used in a variety of locations.

Unenclosed tanks in buildings: NFPA 31 currently permits the installation of individual unenclosed tanks of capacities up to 660 gal (2,500 L) in the lowest story or basement of buildings. Two such tanks may be installed in any one area, but the aggregate capacity of tankage connected to any one burner may not exceed 660 gal (2,500 L). If separation is provided for each tank with capacities of less than 660 gal (2,500 L) each, the aggregate capacity may be greater. The separation must be either an unpierced masonry wall or a partition with a fire-resistance rating of not less than 2 hr and should extend from the lowest floor area to the ceiling above the tanks. This type of installation is frequently found in multiple-unit housing where each unit has its own heating system.

If an unenclosed tank is located above the lowest story or basement of a building, it must not be larger than 60-gal (230-L) capacity. All unenclosed tanks inside buildings must be located not less than 5 ft (1.5 m) from any fire or flame, either in or external to any fuel-burning appliance. Furthermore, such tanks must not obstruct quick and safe access to any utility service meter, switch panel, or shutoff valve.

Corrosion of tanks: Internal corrosion of unenclosed fuel oil tanks, caused by the electrolytic action of water on steel, has been a major problem. Corrosion and subsequent tank leakage have actually been more of a nuisance than a fire hazard. Fire can occur, however, when a burner is located in a pit in the basement and oil from a leaking tank enters the pit. Two generally accepted procedures are recommended to correct the cause of corrosion. One method is to slope the tank to one end and take the burner oil supply line from the bottom of that end. In this way, water is not permitted to accumulate in the bottom of the tank. The quantity of water that comes from condensation inside the tank is so small that it does not present a problem in the safe operation of the burner. The second method is to add a small amount of an alkaline solution to the oil itself or to add a small amount of the solution directly into the tank at periodic intervals.

Enclosed tanks in buildings: Tanks larger than 660-gal (2,500-L) capacity may be installed in buildings, provided the tanks are within an enclosure constructed of walls, floor, and ceiling that has a fire-resistance rating of not less than 3 hr. The walls must be bonded to the floor. Any opening into the enclosure is required to have a non-combustible sill or ramp at least 6 in. (152 mm) high and, since it is a Class A (horizontal) opening, it must be provided with a self-closing fire door. The top and walls of the enclosure must be independent of building construction, except that an exterior building wall having a fire resistance rating of not less than 3 hr may also serve as a wall for the tank enclosure. This method of installation is generally found only in commercial and industrial buildings.

Outside aboveground tanks: Tankage not in excess of 1,320 gal (5,000 L) may be installed outside aboveground in built-up areas. The tanks may be adjacent to buildings, but must be kept away from the line of adjacent property. Not more than one or two tanks having an aggregate capacity of not more than 660 gal (2,500 L) are permitted to be connected to one oil-burning appliance. Tankage in excess of 1,320 gal (5,000 L) may be installed. However, such installations must meet all of the relevant requirements of NFPA 30.

Underground tanks: These are tanks built to conform to NFPA 31, which also gives special precautions regarding building foundations, property line setbacks, protection against traffic (e.g., vehicle movement over the buried tanks), and areas subject to flooding. (See NFPA 30 for additional information on protecting tanks containing flammable liquids in locations that may be flooded.)

Centralized oil distribution systems: A system of piping that supplies oil from a central tank or tanks to a number of buildings, mobile homes, recreational vehicles, etc., may be used under certain conditions. For details covering such systems see NFPA 31.

Fill and vent pipes: Underground tanks and tanks larger than 10-gal (38-L) capacity inside buildings must have fill and vent pipes terminating outside of buildings. Vent openings and vent pipes must be large enough to prevent abnormal pressure during filling, and must not be smaller than 1 1/4 in. (30 mm) nominal pipe size for 550-gal (2,082-L) capacity, and correspondingly larger for tanks having greater capacity. In addition, outside aboveground tanks must have some form of construction or device to relieve excessive internal pressure that may be caused by an exposure fire.

Gaging devices are required, but on inside tanks they must not allow oil or vapor to be discharged into the building. Test wells are not allowed on inside tanks. Piping to the burner(s) and the required accessories are covered in NFPA 31.

Small-capacity tanks: There are several types of small tanks used with cooking appliances and room heaters. An auxiliary tank of not over 60-gal (227-L) capacity may be provided between the burner and the main fuel supply tank. An unenclosed supply tank of not more than 10-gal (38-L) capacity for an individual appliance is permitted, provided it is placed not less than 2 ft (0.6 m) from a source of heat either inside or outside the appliance being served, and provided that the temperature of the oil in the tank will not exceed 25°F (–4°C) above room temperature at the maximum firing rate.

Integral tanks are furnished by manufacturers as a component part of some oil-burning appliances, e.g., kerosene and oil stoves. Stoves with integral tanks [not over 5-gal (20-L) capacity] operate on the gravity or barometric feed principle. The tank reservoir oil and the burner oil are at the same level. As the oil level drops due to burning, oil runs into the reservoir from the tank, either because a float drops and opens a valve (gravity feed) or because air enters the tank through the cap and allows oil to enter the reservoir until it covers the air opening (barometric feed).

NFPA 31 contains specific provisions on supply tanks for kerosene and oil stoves, portable kerosene heaters, and conversion range oil burners.

Liquid Fuels—Miscellaneous

Alcohol and gasoline are used for some heat-producing appliances and equipment to a limited extent. Alcohol stoves and torches, and gasoline torches are examples. NFPA 30 gives additional basic data on the safe storage practices for these liquids.

Gas Fuels

Generally, gas-fired, heat-producing devices and equipment use natural gas, LP-Gas (liquefied petroleum gas), an LP-Gas/air mixture, or mixtures of these gases. Natural gas is the most commonly used gas fuel in the United States, because gas transmission pipeline systems supply most areas of the country. LP-Gas-fired heating devices also are popular, particularly in sparsely settled areas and in recreational vehicles. A number of other flammable gases are used as fuel in many industrial processes for special purposes.

Natural gas consists principally of methane and some ethane, propane, butane, and small amounts of carbon dioxide and nitrogen. Fuel gases can be made in a variety of ways from coal, or by the cracking of oils. Liquefied petroleum gases are largely propane, propylene, butane, and butylene, or mixtures of these gases.

Typical calorific values of various gases as fuel are given in Section 4, Chapter 7, "Gases." Various procedures for sampling, analyzing, measuring, and testing gaseous fuels are outlined in 13 ASTM standards, which have been compiled in the publication ASTM *Standards on Gaseous Fuels.*

Methods of Firing Fuel Gases

Air is either mixed with gas at the burner of a gas-fired appliance or premixed with gas. A wide variety of types and sizes of burners are used to mix the gases effectively under different conditions. Every burner is designed to transform the potential energy of the fuel gas into useful heat, which can be absorbed in the most effective manner.

Injection (bunsen) burners: Practically all residential and commercial gas appliances and some industrial gas-fired appliances use injection-type gas burners. A jet of gas injects primary air for combustion into the burner and mixes it with the gas. This mixing occurs before the gas reaches the burner ports or point of ignition. Figure 3-5J shows a typical injection-type gas burner.

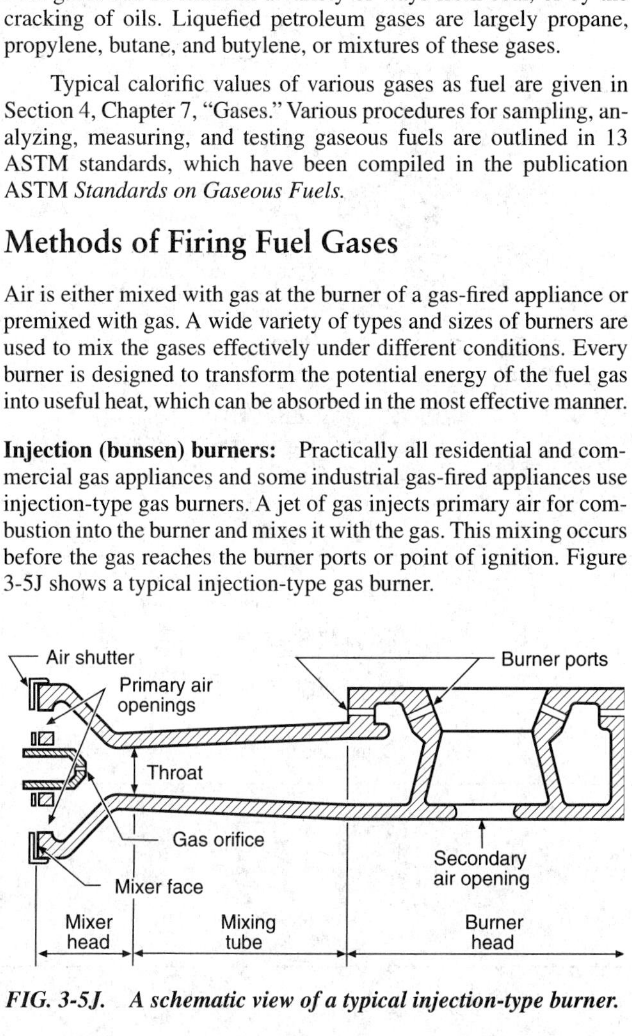

FIG. 3-5J. A schematic view of a typical injection-type burner.

Luminous or yellow-flame burner: In the luminous or yellow-flame burner, only air externally supplied at the point of combustion is used for burning the gas. The flame is produced without premixing air with the gas.

Catalytic burners: Another type of burner is the catalytic burner that permits combustion of the gas at temperatures well below the normal ignition temperature of fuel gas/air mixtures.

Power burners: In a power burner, either gas or air or both are supplied at pressures exceeding the line pressure of the gas and atmospheric pressure of the air. The added pressure is applied at the burner. When air for combustion is supplied by a fan ahead of the appliance, the appliance is known as a forced-draft burner. A premixing burner is a power burner in which all or nearly all of the air for combustion is mixed with the gas as primary air (air that mixes with the gas before it reaches the burner port or ports). A pressure burner is supplied with an air/gas mixture under pressure [usually from 0.5 to 14 in. of water (0.1 to 3.5 kPa) and occasionally higher].

Figure 3-5K illustrates some of the types of industrial appliance gas burners.

Interchanging gases: Changing from one kind of gas to another may cause serious problems unless the appliance is designed to accept the change. Appliance designs certified by the American Gas Association (AGA) laboratories are tested for satisfactory performance with one or more gases as requested by the manufacturer. These include natural and mixed gases, as well as LP-Gas and LP-Gas/air mixtures. The specific gas or gases that may be used in an appliance are marked on the appliance. Although such appliances may be converted in the field to any one of the marked gases, the conversion may require some interchange of parts and may involve some element of hazard unless carefully performed by experienced, qualified personnel.

Appliance and Piping Installation

Installation, alteration, and repair of gas appliances and gas piping should be done only by qualified agencies fully experienced in this work. Provisions outlined in NFPA 54, *National Fuel Gas Code* (hereinafter referred to as NFPA 54), provide the necessary guidance. Gas must be turned on *only after* the system has been thoroughly tested to be certain there is no leakage. All air in piping must be purged, as a remaining slug of air will extinguish the flame. When gas flows again, unburned gas will escape, resulting in a potentially hazardous condition.

The uncontrolled flow of unburned gas into any gas appliance could form an explosive gas/air mixture. Safeguards to prevent this are extremely important, particularly in industrial and other large gas-burning devices. Without such safeguards, unburned gas may accumulate in combustion chambers or ovens at the time of lighting or when the gas pressure fluctuates. If the pressure is reduced below a certain point, the flame may be extinguished. With a subsequent return of normal pressure, unburned gas might continue to flow. An increase in pressure may increase the velocity of flow through the burner in excess of the flame propagation rate. This will force the flame away from the burner. A number of automatic devices are available to safeguard against this hazard and are commonly found on the larger installations.

Electricity

Most electrical heating appliances for heating small rooms and for other small heating jobs involve resistor heating elements, which are usually one or more metal alloy wires, nonmetallic carbon rods, or printed circuits. Resistor heating is used in radiators, unit heaters, convectors, central hot water systems, central warm air heating systems, and panel-type radiant heat installations for walls, floors, and ceilings. Resistor heating is also used for household electrical appliances, such as stoves, irons, toasters, etc.

Units of electrical heat: Heating calculations for electrical heating are expressed in kilowatt hours (kWh). One kWh equals 3,412 Btu.

Controls: Automatic regulation of the electrical input (usually through proportional-type step controllers) allows heaters to be energized and deenergized in increments small enough to prevent excessive total operation of all heaters during temperature drops.

Method of installation: Electric heating systems should be installed in accordance with NFPA 70, *National Electrical Code®* (hereinafter referred to as NFPA 70).

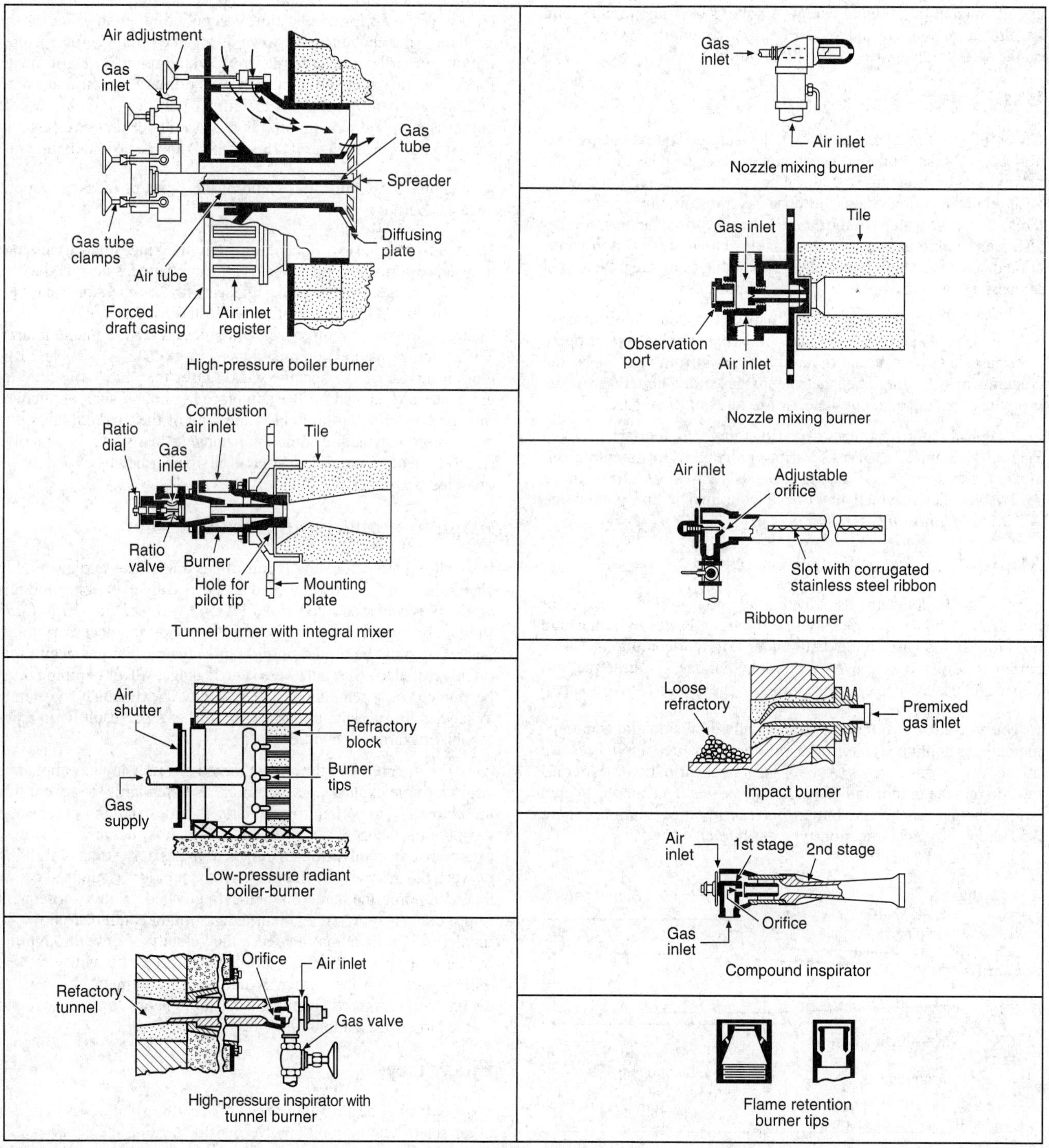

FIG. 3-5K. Types of industrial appliance gas burners.

CONTROLS FOR FUEL BURNERS

The uncontrolled flow of unburned fuel into appliances and burners can lead to serious consequences. When fired with pulverized coal, gas, or oil, a combustible mixture could accumulate within the confines of the appliance or equipment. If ignited, it could result in an explosion. To guard against the discharge of unburned fuel or improper mixtures of fuel and air, certain controls are required for all fuel burners.

Primary Safety Controls

These required controls cause the fuel to shut off in the event of ignition failure or flame failure. Except for pot-and-sleeve-type vaporizing burners, shutoff is usually accomplished via a primary safety-control. For gas appliances, an automatic gas ignition system is used. Both types of controls sense the presence or absence of flame. Upon ignition or flame failure, they cause the fuel to be shut off in the prescribed period of time. All fuel to the burner should be shut off. Primary safety controls usually provide for starting and stopping the burner in response to changes in demand (room thermostat, process controller, etc.). Controls for commercial and industrial burners may provide a purge period for air to flow through the appliance prior to starting fuel flow. The control also prohibits main burner fuel from being admitted until the control has proved the existence of the ignition flame.

The discharge of unburned oil from pot-and-sleeve-type vaporizing burners should be prevented by a constant level valve or by barometric feed. Each of these provisions maintains a predetermined level of oil in the burner below any point of overflow.

Air/Fuel Interlocks

If the safe operation of a burner depends upon a forced or induced draft fan or air compressor to supply combustion air, an interlock is needed to shut off fuel in case the air supply is interrupted.

Atomizer/Fuel Interlocks

In an oil-burning system, where air, steam, or other means for atomization can be interrupted without stopping oil delivery to the burner, an interlock must be provided to immediately shut off the oil if atomization fails.

Pressure Regulation and Interlocks

Gas burners and pressure-type oil burners require uniform fuel pressure so fuel burns safely and efficiently. Pressure regulators are used to maintain uniform fuel pressures. In situations where fuel pressure fluctuations may be expected, pressure interlocks are provided, which shut off the fuel to the main burners when the pressure is too low or too high for safe operation.

Oil Temperature Interlocks

Oil burners requiring heated oil for satisfactory operation are equipped with a low oil temperature switch to prevent the burner from starting or to shut it down if the oil temperature falls below the required minimum.

Manual Restart

If a burner is not equipped to provide safe automatic restarting after shutdown, its control system must be arranged to require manual restarting after any control operates to extinguish the burner flame.

Remote Shutoff

It is good practice to provide a remote control for manually stopping the flow of fuel to the burner. An identified switch in the burner supply circuit, placed near the entrance to the room where the burner is located, can be used with electrically powered equipment. A valve in the fuel supply line, operable from a location reached without passing near the burner, may also be used.

Safety Shutoff Valves

Safety shutoff valves prevent the abnormal discharge of fuel. They should be constructed so they cannot be readily restrained or blocked in the open position. Such valves should automatically close when de-energized, regardless of the position of any damper-operating lever or reset handle. Electrically operated valves should not need to be energized to close. Pressure-operated valves should close upon failure of the operating pressure.

Safety Control Circuits

Safety control ac circuits are of the two-wire type with one side grounded. They must not exceed a nominal 120 V and must be protected with suitable fuses or circuit breakers. All switches should be in the hot ungrounded line. The accidental grounding of such a circuit will not cause a required safety control to be bypassed.

Appliances and equipment fired with fuel burners require some additional controls to avoid excessive pressure, temperature, or other abnormal conditions. These controls are covered later in this chapter.

Appliances listed by testing agencies are equipped with all the required safety controls. The application of controls to fuel-burning systems assembled in the field or intended for specific applications should be entrusted only to people specializing in this work. The standards pertaining to fuel-burning equipment referenced in the bibliography include specific information on controls essential to safe operation of fuel-burning equipment.

Safety controls are generally preset by the manufacturer or the installer. Any alteration of these settings or bypassing any safety control could lead to serious consequences. When a faulty control is replaced, a similar model and setting should be used. Figures 3-5L and 3-5M illustrate typical arrangements of safety valves and interlocks for industrial gas and oil burners, respectively.

HEATING APPLIANCES AND THEIR APPLICATIONS

Central heating appliances, room heating and cooking appliances, and miscellaneous heat-producing devices, which are fired by the various fuels and methods discussed, are described here.

Central Heating Appliances

Central heating appliances are divided into four broad categories: (1) boilers, (2) central furnaces, (3) floor furnaces, and (4) wall furnaces.

Boilers: These are either of the steam or hot water type and are constructed of cast iron or steel. (See Figure 3-5N.) The *ASME Boiler and Pressure Vessel Code* defines low-pressure boilers (for low-pressure steam heating, hot water heating, and hot water supply) as those steam boilers that operate at pressures not exceeding 15 psi (100 kPa), and hot water boilers as those that operate at pressures not exceeding 160 psi (1,100 kPa) and at temperatures not exceeding 250°F (120°C). The *ASME Boiler and Pressure Vessel Code* also covers the construction of power boilers and locomotive boilers.

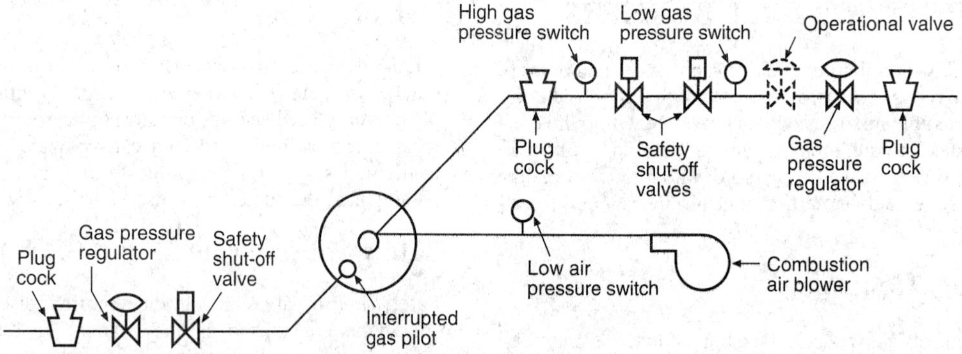

FIG. 3-5L.   *Typical safety valves and interlock arrangement for industrial gas burner.*

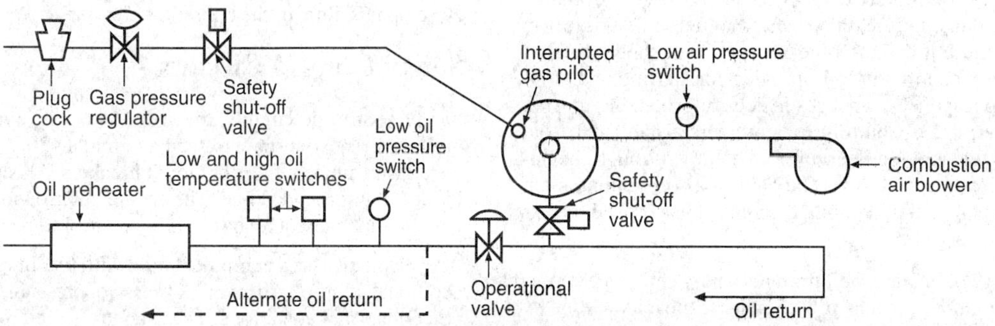

FIG. 3-5M.   *Typical safety valves and interlock arrangement for industrial oil burner.*

Boilers are equipped with safety devices to prevent overpressure conditions. Automatically fired boilers also have controls that will shut off the burner or electric heating elements under low water conditions or when a predetermined pressure or temperature has been reached.

The fire problems of boilers involve their mountings and clearances to combustibles. Explosions resulting from unburned fuel accumulations in the combustion chamber or due to overpressure conditions could rupture fuel lines, thus contributing to the potential for subsequent fire or explosion damage.

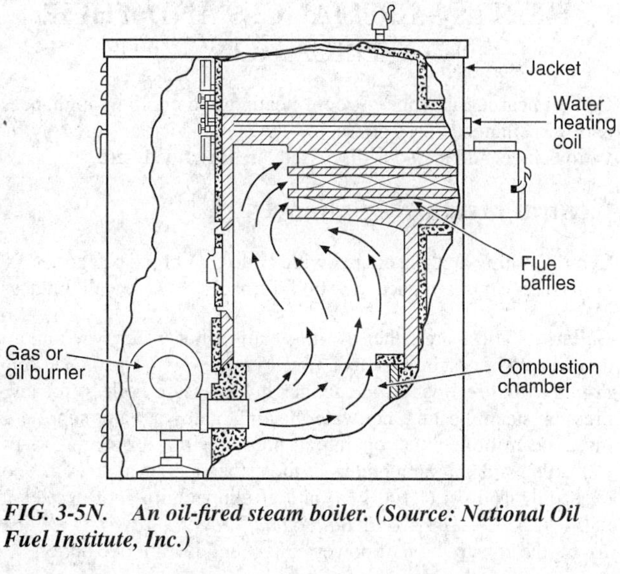

Central warm air furnaces: The residential types currently installed in the United States are usually equipped with circulating fans, filters, and other features that make them, in effect, air-conditioning systems.

Central warm air furnaces are of the following types:

1. Gravity—depends primarily upon the circulation of air by gravity.
2. Gravity with integral fan—fan is an integral part of the furnace construction and is used to overcome internal resistance to airflow.
3. Gravity with booster fan—fan does not restrict flow of air by gravity when fan is not operating.
4. Forced air—equipped with fan that provides the primary means of circulating air. (See Figure 3-5O.)

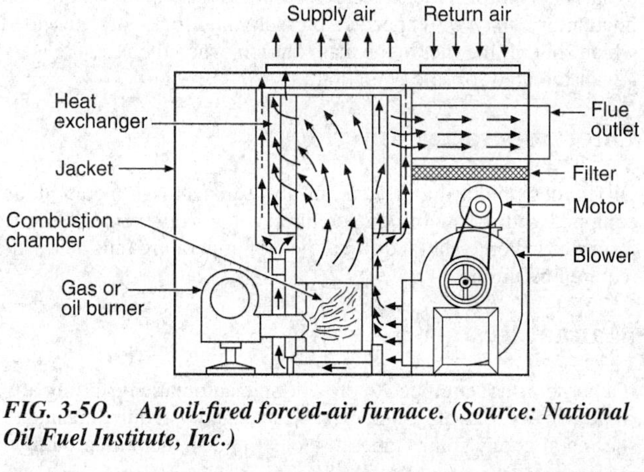

FIG. 3-5N.   *An oil-fired steam boiler. (Source: National Oil Fuel Institute, Inc.)*

FIG. 3-5O.   *An oil-fired forced-air furnace. (Source: National Oil Fuel Institute, Inc.)*

Forced-air central warm-air furnaces may be classified according to the direction of airflow through them, i.e., as horizontal, up-flow, or downflow.

Gravity furnaces are floor mounted and can heat only spaces above them. Forced-air furnaces may be floor mounted or suspended and may be found on most any floor of a building, including the attic and roof.

Under some conditions of operation, plenums of warm-air heating furnaces may become hot enough to ignite adjacent unprotected woodwork. Clearances or insulation are necessary in these instances. [Complete requirements for installation of systems in residences can be found in NFPA 90B, *Standard for the Installation of Warm Air Heating and Air Conditioning Systems* (hereinafter referred to as NFPA 90B).]

Automatic controls, also called high-limit controls, can shut off the fuel or electric supply whenever the air in the furnace warm air plenum or at the beginning of the main supply duct—at a point not affected by radiated heat—reaches 250°F (120°C) or less, depending upon the type of furnace and its installation. (See Figure 3-5P.) Automatic controls cannot be set above prescribed temperature limits. Some furnaces used only with special duct systems are factory equipped to permit temperatures above 250°F (120°C) at the inlet of the supply ducts.

Central warm air furnaces also cause fires due to inadequate clearances to combustible construction, lack of proper limit controls, heat exchanger burnouts, and other causes associated with lack of proper installation, servicing, and maintenance.

Pipeless furnaces: These are essentially gravity warm air furnaces that are not connected to ducts. The furnace is mounted directly under the space to be heated. All the air heated in its outer jacket is discharged through a register above the heat exchanger. They are used for heating small homes and other small one-story properties.

Pipeless furnaces may produce dangerously high temperatures, particularly if the air circulation is restricted. For example, if the register in the floor of a small hallway heats adjoining rooms by air circulating through opened doors, the temperature of the entire hall may be raised to a dangerous degree if the doors are closed tightly at a time when the furnace is in full operation. Fires also have occurred when clothing left to dry above a central floor register was ignited. It is good practice to equip automatically fired pipeless furnaces with limit controls.

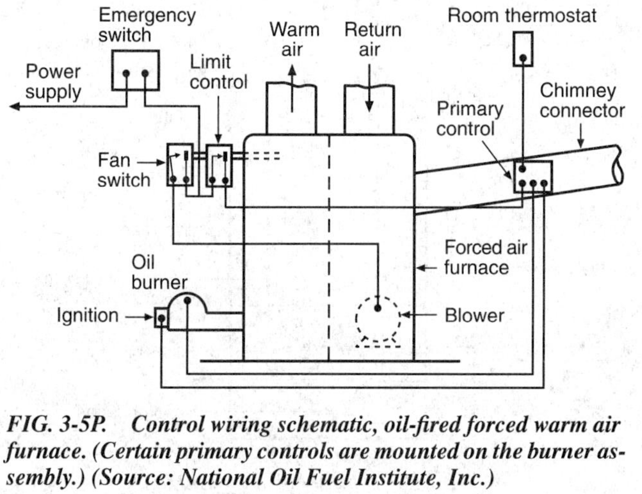

Floor furnaces: Most oil- and gas-fired floor furnaces are designed so they may be installed in combustible floors (see Figure 3-5Q); however, they should not be installed in combustible floors if they have not been listed by a testing laboratory for such use. Clearances should be sufficient not only for protection of combustible floors and walls, but also to keep the furnace casing and the piping connected to it out of contact with earth or damp materials.

Temperature limit controls are provided on floor furnaces to shut off the fuel supply when the temperature of the discharged air reaches a predetermined level.

In auditoriums, public halls, or assembly rooms, floor furnaces should not be installed in the floor of any aisle, passageway, or exitway. Floor furnace registers may become hot enough to cause burns under some conditions; therefore, registers in dwellings should not be located in passageways to bathrooms, where persons would be likely to step with bare feet. Users should be warned against covering registers or placing clothing on them to dry. Older types of floor furnaces do not have a temperature limit control, and may become dangerously hot if the air passages are blocked with dust and lint. Inspection and vacuum cleaning or hand removal of this blockage is essential for continued safe operation.

Wall furnaces: These are self-contained indirect-fired gas or oil heaters installed in or on a wall. They supply heated air directly to the space to be heated, either by gravity or a fan through grilles on openings or boots in the casing supplied by the manufacturer. The furnaces may be of the direct-vent type or are vent- or chimney-connected, depending upon the fuel. Limit controls limit outlet air temperature. The fire problems with recessed wall furnaces are similar to those encountered with most warm air furnaces. A wall furnace installation is shown in Figure 3-5R.

Duct furnaces: These furnaces are installed directly in ducts of some warm air and air-conditioning systems and depend upon air circulation from a blower not furnished as a part of the furnace. They may be oil or gas fired or electrical, and are equipped with a limit control to shut off the fuel supply or electricity at excessive temperatures.

When duct furnaces are used in conjunction with refrigeration coils in a combined heating and cooling system, the furnace should

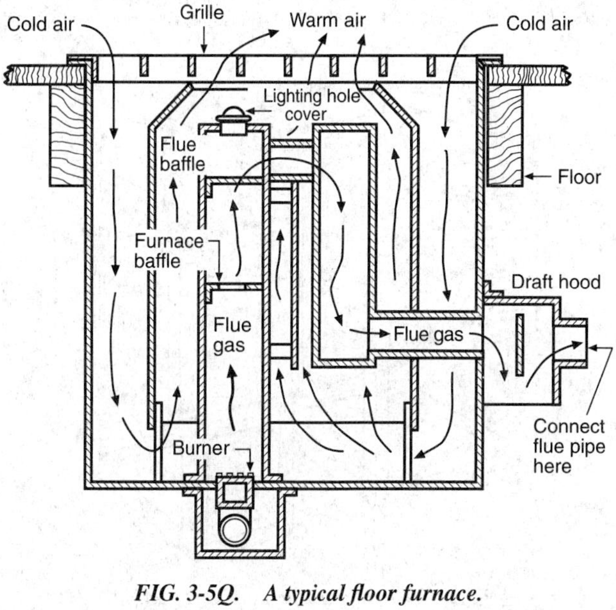

FIG. 3-5P.   *Control wiring schematic, oil-fired forced warm air furnace. (Certain primary controls are mounted on the burner assembly.) (Source: National Oil Fuel Institute, Inc.)*

FIG. 3-5Q.   *A typical floor furnace.*

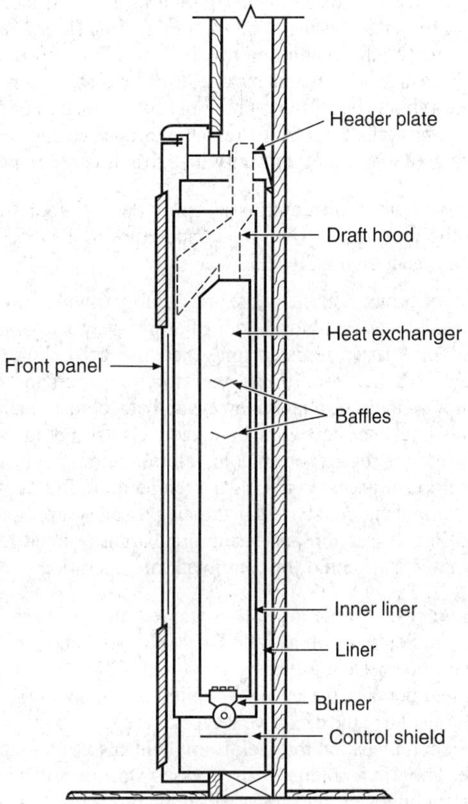

FIG. 3-5R. *Typical gas-fired wall furnace.*

be located upstream from the refrigeration coil or parallel with it to prevent condensation from corroding the furnace. There are, however, furnaces made of corrosion-resistant material, which may be installed downstream from the refrigeration coil. When the furnace is located upstream from the refrigeration coil, the coil should be designed so that excessive pressures and temperatures will not develop in the coil. A duct furnace installation is shown in Figure 3-5S.

Warm air heating panels: These are used in low-temperature, forced-air systems to circulate warm air through plenums or chambers that have one or more surfaces exposed to the space to be heat-

ed. NFPA 90B gives recommendations for the use, construction, and installation of these panels.

Radiant heating: This type of heating utilizes panels of hot water piping or electric heating elements that usually operate at moderate temperatures. The panels are embedded in plaster walls or ceilings or in cement floors. Hot water pipes or electrical equipment in radiant systems should conform to standard installation practices. NFPA 70 contains specific recommendations on the installation of electrical space heating equipment.

Heat pump: This term is applied to a type of forced-air heating system in which refrigeration equipment is used in such a way that heat is taken from a heat source and given up to the conditioned space when heat service is wanted and is removed from the space and discharged to a heat sink when cooling and dehumidification are desired. These systems frequently have supplemental heating units. In such cases, the units are equipped with an interlock to prevent the unit from operating, unless the indoor air circulating fan on the system is running. The units also have temperature limit controls. Hazards are those presented by power, refrigeration equipment, and heat units, if these are part of the system.

Unit Heaters

These are classed as self-contained, automatically controlled, chimney- or vent-connected air heating appliances with an integral means for air circulation. They may be floor mounted or suspended and are equipped with temperature limit controls. The term *unit heater* as used in NFPA standards is intended to cover appliances for heating nonresidential properties. Thus, it excludes room heaters, floor furnaces, and similar devices. More specifically, a unit heater is an appliance consisting of a heating element and fan housed in a common enclosure and placed within or adjacent to the space to be heated. Unit heaters should not be confused with heat exchangers, which are equipped with fans to circulate heated air. In the latter type, hot water or steam is piped from a heating unit to the heat exchanger in the area to be heated.

Unit heaters that are designed for connection to a duct system may be considered as central heating furnaces and should be provided with the same safeguards. A typical unit heater is shown in Figure 3-5T.

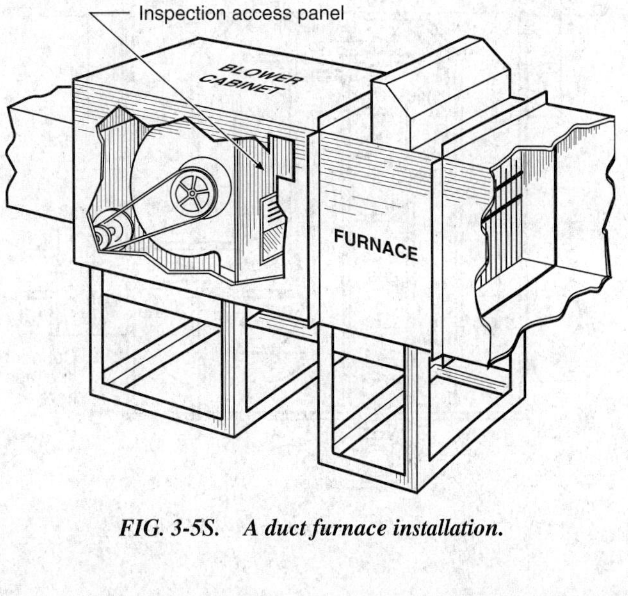

FIG. 3-5S. *A duct furnace installation.*

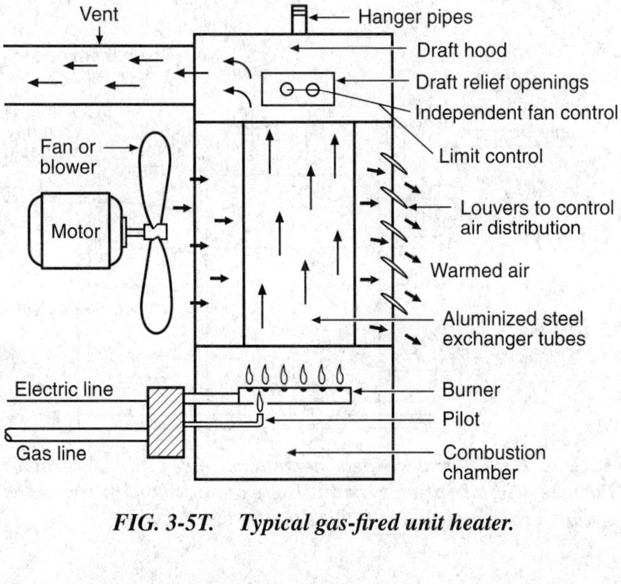

FIG. 3-5T. *Typical gas-fired unit heater.*

Room Heaters and Cooking Appliances

A room heater differs from a central heating furnace in that it is a self-contained air heating appliance designed for the direct heating of the space around it. External pipes or ducts are not used to distribute the heat. Room heaters fall into two general types: (1) circulating and (2) radiant. A fuel-burning circulating heater has an outer jacket surrounding the combustion chamber. It is arranged with openings at the top and bottom so that air circulates between the inner and outer shell. It does not have openings in the sides of the outer jacket in order to permit direct radiation. From a safety point of view, room heaters other than those specifically designed as circulating type are classified as being of the radiant type. They include such appliances as wood and coal stoves, electric and gas logs, open-front heaters, wall heaters, gas coal baskets, and fireplace inserts. Care must be taken, especially with heaters that can be moved easily, to place the radiant type so that radiating elements and surfaces are not directed toward walls, drapes, or furniture in close proximity to the heater.

Fuel-burning heaters and residential cooking appliances should be connected to a chimney or vent, where appropriate. The exceptions are gas room heaters and gas cooking appliances listed for unvented use.

Fuel-burning room heaters are not to be installed in sleeping quarters for use of transients, as in hotels and motels, nor in institutions such as homes for the aged, sanitariums, convalescent homes, orphanages, etc. It is good practice to install direct-vent-type heating appliances in sleeping quarters where heaters are permitted and in rooms generally kept closed.

Solid-Fuel Room Heaters

Fires involving coal and wood stoves or room heaters have dropped dramatically over the last few years. (See Figure 3-5U.) Much of the improvement can be attributed to increased vigilance in the correct installation and operation of the appliances. Fires in solid-fuel room heaters are due to two underlying conditions: (1) inadequate clearances to combustible materials and other installation deficiencies and (2) improper or inadequate maintenance. Many wood stoves need frequent charging, often involving much regulating of draft controls to prevent overheating. Sometimes, too, weather conditions can affect the chimney draft, causing improper combustion.

A solid-fuel heater essentially is a fire chamber surrounded by a decorative enclosure. The chamber may be equipped with manual draft controls or a thermostatic control and provided with a chimney connector and other parts that may be required for installation. Instructions that come with heaters that have been listed by testing laboratories are quite precise on the type of heat shielding and the clearances from combustibles that are required.

The vast majority of solid-fuel room heaters are designed to burn only wood, which is easily burned on the bottom of the fire chamber, resting on a bed of firebrick, ashes, or sand. For coal, a grate or other provision to supply air below the fuel bed is needed.

Unless a special chimney is called for by the listing or installation instructions (as in the case of room heaters for use in mobile homes), the typical solid-fuel heater can be connected to either a standard masonry chimney or any listed factory-built chimney. Laboratory tests for listing heaters establish the dimensions required for adequate clearances. Because clearances from large heaters may exceed 36 in. (0.9 m), the tests may be performed with supplemental heat protection or shielding for walls. For proper clearances and floor protection, see the section of this chapter on installation.

Hazards with Solid-Fuel Room Heaters

1. *Overfiring.* If an appliance is overheated deliberately or through neglect, there is a severe and immediate hazard if the unit or its connector glows even a very dull red. Control is to simply close doors and air dampers until things cool down.
2. *Careless handling of ashes containing live coals.* Ashes should never be collected in combustible containers, nor should the containers ever be placed on a combustible floor.
3. *Inadequate connector clearance.* This includes violations of the required 18-in. (457-mm) clearance or lack of adequate protection where a single wall connector must penetrate a combustible wall to join a masonry chimney. Reduced connector clearances are permissible in some instances, such as the entry of single wall connectors into listed supports for factory-built chimneys. The reduced clearances have been validated by tests and are described in the installation instructions for listed heaters. (See Figure 3-5V.)
4. *Creosote accumulation and fires.* Creosote is a collection of tars and other hydrocarbons produced by the pyrolytic distillation of wood. Under ideal conditions, these products are consumed in the flames of a wood fire. When wood is burned in a stove under air-limited conditions, it may not be entirely burned and may deposit on surfaces of the stove, connector, and chimney. Creosote is combustible and, if ignited, can fuel a dangerous fire in portions of the venting system not designed to withstand combustion. The fire can damage the system components or spread to the adjacent structure.

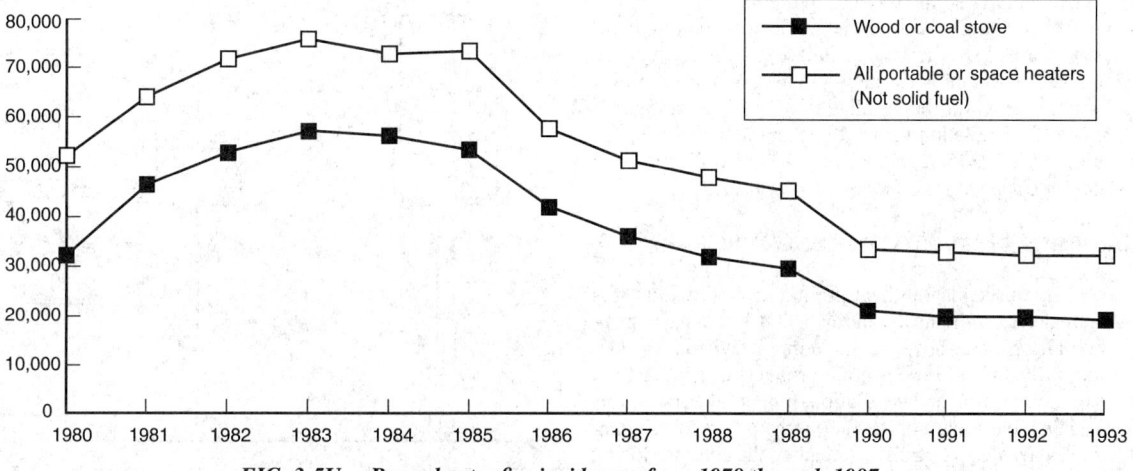

FIG. 3-5U.   *Room heater fire incidences from 1978 through 1987.*

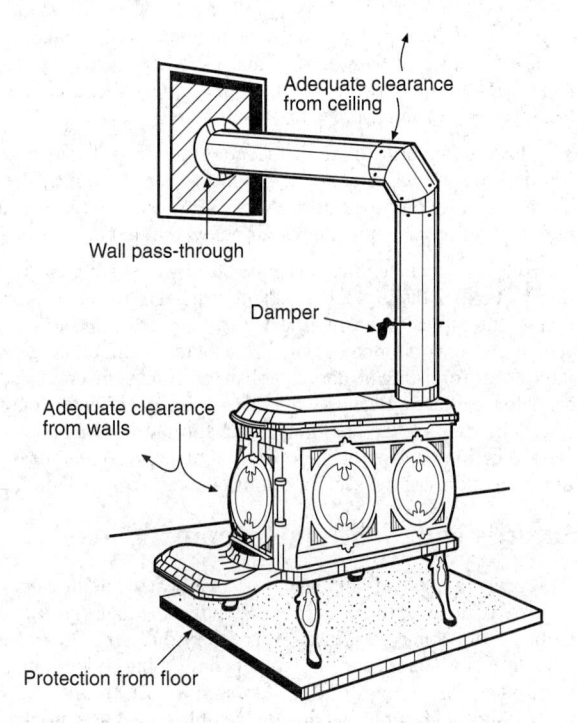

FIG. 3-5V. Adequate clearances should be maintained in order to ensure safe operation of coal and wood stoves.

Airtight wood stoves, when operated with draft controls restricted or with large fuel loads, can cause a significant buildup of creosote in a short period of time. Therefore, frequent inspection, and, when required, cleaning of the chimney system is in order when these appliances are used.

The species and moisture content of the wood are not major determinants of the rate of creosote buildup, but habits of stove operation are. Proper sizing of the stove, with respect to the heating need, will help minimize the production of creosote by reducing the need for air-limited, low-temperature operation. The user should avoid long, extended burn cycles and large loads of fuel. The control of creosote and chimney fires is discussed in more detail under "Hazards of Chimneys," later in this chapter.

Restaurant Cooking Appliances

These include a variety of devices such as ranges, deep fat fryers, steamers, broilers, hot plates and griddles, portable ovens, and others. These devices, like other heat-producing devices, require careful installation to avoid overheating of adjacent combustibles. In some of them, grease accumulates, adding to the fire hazard. NFPA 54 contains requirements for the installation of gas-fired restaurant cooking appliances. Ventilation of fixed restaurant cooking equipment is covered in NFPA 96, *Standard for Ventilation Control and Fire Protection of Commercial Cooking Operations.*

Miscellaneous Heat-Producing Devices

There are so many miscellaneous heat-producing devices that it is unrealistic to treat each individually in this text. Certain basic principles apply to all of them, however, including provision for adequate clearances to combustibles, venting products of combustion, adequate air for combustion and ventilation, proper burner controls to safeguard against the hazard of the fuel, proper storage and handling of the fuel, and safety controls to safeguard the heat-producing device.

Testing laboratories test and list many miscellaneous types of heat-producing appliances, such as direct-fired heaters, incubators, lanterns, forges, torches, etc.

Kerosene stoves: These may be defined as self-contained, self-supporting, kerosene-burning ranges, room heaters, or water heaters that are not connected to chimneys but are equipped with integral fuel supply tanks not exceeding 2-gal (7.6-L) capacity. Terms often applied to kerosene room heaters are cabinet heaters and space heaters. Since they are not connected to chimneys, they can be moved rather easily although they are not considered portable. This feature of kerosene stoves has frequently resulted in creation of severe fire and life hazards due to improper placement of the stoves in relation to nearby combustibles and paths of exitways. Kerosene stoves listed by testing laboratories incorporate fire protection features that may be missing from those not listed. Among the more important features of listed stoves are the types of construction materials, primary control valves, and drip pans used in them.

Portable kerosene heaters: These are similar in hazard to kerosene stoves because they are not connected to a chimney. The hazard in their use, however, is increased by their even greater portability and subsequent misuse and misplacement. They are also subject to misfueling; substitution of gasoline or other inappropriate liquid; and refueling when hot, which can produce flammable vapors. The safest heaters incorporate fire safety features not included in others, such as special types of latching devices and integral sheet-metal trays under the burners to catch oil drips. They usually employ wick-type burners integral with the oil reservoir. Portable kerosene heaters should include a tip-over switch that automatically snuffs the burner if the unit is jostled or tipped.

Conversion range oil burners: These consist essentially of a single- or double-sleeve-type burner assembly, regulating valves, and an oil supply assembly with a suitable supporting stand and seamless connecting tubing. A thermal valve located in the burner compartment of the stove adjacent to the burner is installed in the oil supply line. (See Figure 3-5W.)

Range oil burners, which are found most frequently in the northeastern section of the United States, are designed to burn kerosene, range oil, or similar fuel. They are primarily installed only in stoves or ranges originally designed to use solid fuel. Range oil burners should not be mistaken for conversion oil burners of the vaporizing-pot type designed for conversion of central heating appliances.

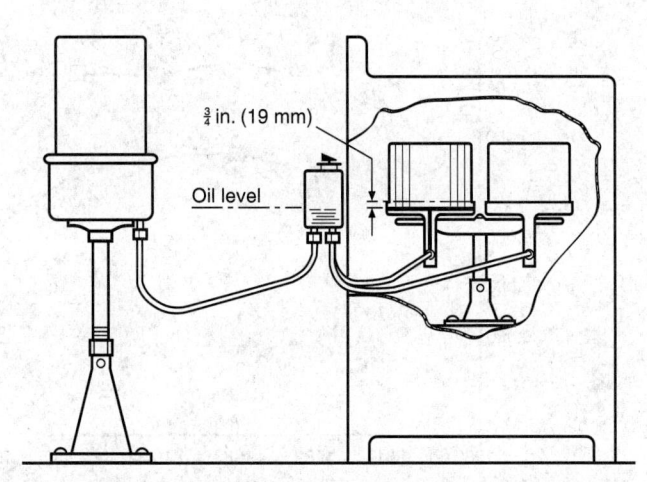

FIG. 3-5W. Typical installation of conversion range oil burner.

Salamanders: These are typical miscellaneous portable heating devices. They are used at construction sites and in unheated buildings. Too frequently, a salamander consists of only a metal drum with some holes punched in it for draft, and wood scrap and other construction waste materials for fuel. These devices present an acute spark hazard and the hazard of carbon monoxide poisoning because they are not chimney connected. Crude salamanders of this type, and gas- and oil-fired salamanders that have not been tested and listed by a laboratory, should not be used.

UL and American Gas Association (AGA) laboratories test and list portable gas- and oil-fired heaters, such as salamanders. One way to use salamanders is to carry heated air indoors through flexible ducts from a heating unit located outdoors. NFPA 58, *Standard for the Storage and Handling of Liquefied Petroleum Gases* (hereinafter referred to as NFPA 58), has extensive material on the use of LP-Gas-fired portable heaters.

DISTRIBUTION OF HEAT BY DUCTS AND PIPES

Warm Air Distribution Systems

Central warm air heating systems consist basically of a heat exchanger with an outer casing or jacket that is connected to a system of ducts, air passages, or plenums that carry heated air (the supply side) to the spaces to be heated and return air (the return side) from the heated spaces to the heat exchanger.

NFPA 90B contains specific requirements for systems that are installed in one- and two-family dwellings and other occupancies not exceeding 25,000 cu ft (700 m^3) in volume. Systems that are installed in spaces exceeding 25,000 cu ft (700 m^3) are covered in NFPA 90A, *Standard for the Installation of Air Conditioning and Ventilating Systems* (hereinafter referred to as NFPA 90A). The text of this portion of the Handbook is confined to a discussion of ducts for warm air heating systems installed in dwellings and other relatively small spaces.

Warm air supply ducts: Warm air ducts must be installed with clearances as given in Table 3-5B. They are substantially constructed of metal or of Class 0 or Class 1 duct materials* and are properly supported and protected against injury.

Supply air plenums: It is common practice in some localities to use the crawl space as the supply plenum for warm air heating systems when a one-story single-family house does not have a basement. These systems utilize a downflow furnace that discharges warm air through directional ducts into the crawl space. An opening cut in the floor of each room of the house and covered with a grille serves as the supply register. The hazards of these systems include: use of combustible construction as a plenum, use of the crawl space for storage, presence of a large enclosed combustible area with no limitations on fire spread and with direct openings to every room of the house, and the possibility of reverse flow in the downflow furnace (the cold air return then becomes the supply duct). NFPA 90B gives specific recommendations and limitations applicable to the use of under-floor space as supply plenums, and it should be consulted for detailed guidance.

---

*Duct materials are tested and listed by UL in accordance with UL 181, *Standard for Safety Factory-Made Air Ducts and Air Connectors*, and are classified as follows: Class 0—Air duct materials having a fire hazard classification of zero (flame spread and smoke developed). Class 1—Air ducts that have a flame spread rating of not more than 25 without evidence of continued progressive combustion and a smoke developed rating of not more than 50.

TABLE 3-5B. *Installation Clearances for Horizontal Warm Air Ducts*

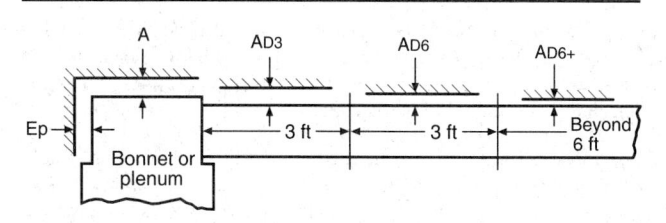

A—Clearance above top of casing, bonnet, plenum, or appliance determined by Table 3-5E.
A$_{D3}$—Clearance from horizontal warm air duct within 3 ft (0.9 m) of plenum.
A$_{D6}$—Clearance from horizontal warm air duct between 3 and 6 ft (0.9 and 1.8 m) of plenum.
A$_{D6+}$—Clearance from horizontal warm air duct beyond 6 ft (1.8 m) of plenum.
E$_P$—Clearance from any side of bonnet or plenum.

If A Equals	A$_{D3}$ Equals	A$_{D6}$ Equals	A$_{D6+}$ Equals	E$_P$ Equals	Method of Firing
1	1	0	0*	1	Automatic oil, comb. gas-oil, or gas
2	2	0	0*	2	Automatic oil, comb. gas-oil, or gas
6	6	6	0*	6	Automatic oil, comb. gas-oil, or gas
6	6	6	1*	6	Automatic stoker fired†
18	18	6	1*	18	Any fuel or control.

*Clearance A$_{D6+}$ to be maintained to a point where there is a change in direction equivalent to 90 degrees or more.
†Furnace must be equipped with limit control that cannot be set higher than 250°F (120°C) and must also have a barometric draft control operated by draft intensity and permanently set to limit draft to a maximum intensity of 0.13 in. water gage (32 kPa), otherwise clearances should be as indicated for any fuel.

Horizontal supply ducts and vertical ducts, risers, boots, and register boxes within certain limits of the warm air furnace can reach temperatures that can become hazardous; therefore, safe clearances to combustibles must be maintained. The clearances required for the many possible configurations of warm air ductwork are specifically provided in NFPA 90B, which should be consulted.

Registers: To prevent excessive heat from developing in the duct system, one register or grille must be installed without a shutter and without a damper in the duct to it. The exceptions are automatic oil- or gas-fired systems that have approved temperature limit controls or systems with dampers and shutters designed so that they cannot shut off more than 80 percent of the duct area.

Where registers are installed in the floor over the furnace, as in pipeless furnaces or floor furnaces, the register box should consist of a double wall with an air space of not less than 4 in. (102 mm) between walls. Furnaces with a cold air passage around the warm air passage comply with this requirement.

Return air ducts and plenums: The best way to conduct return air to the furnaces and duct heaters is through continuous ducts. Return ducts do not need to be made of the same materials as supply ducts except for those portions directly over the heating surface or within 2 ft (0.6 m) of the outer casing or jacket of the furnace or duct heater. They should not, however, be made of materials more combustible than 1-in. (25.4-mm) nominal wood boards. Under-floor spaces may be used for return ducts from rooms directly above, but such spaces should be used only if they are not more than 2 ft (0.6 m) in depth from the bottom of floor joists, are clean of all combustible material, and are tightly and substantially enclosed. A vertical stack for return air should not be connected to registers on more than one floor.

The interior of combustible ducts must be lined with metal at points where there is danger from smoldering or burning incandescent particles dropped through a register, such as directly under floor registers and at the bottom of vertical ducts.

Air filters: Air filters should be the kind that will not burn freely or emit large volumes of smoke or other objectionable products of combustion. Filters qualifying as Class 1 and Class 2, as defined in NFPA 90B, are accepted as meeting these requirements. Only liquid adhesive coatings with a flash point of 325°F (163°C) or higher (Pensky-Martens closed-cup tester) are acceptable. Filters are not installed in ducts of heating systems unless the system design calls for them. Otherwise the filters may restrict the flow of air and cause dangerous overheating.

Solid fuel add-on furnaces: Solid fuel add-on furnaces are designed to connect to the existing air distribution ductwork of a house and supply heated air as a supplement to the operation of the regular furnace. Most, though not all, are independently controlled, with a separate thermostat. A wood or coal fire is established in the unit, and the thermostat regulates the flow of air to the combustion chamber, and thereby the amount of heat produced. Properly installed and operated, add-on furnaces can supply all or most of the central heating needs of a home.

However, add-on furnaces can also present some unique hazards. They tend to have large combustion chambers that can be filled with a large volume of wood. When the heating need is not great, the thermostat tends to keep the fire low and smoldering, with consequent production of a large amount of creosote. Add-on furnaces and their venting systems need frequent inspection and cleaning.

NFPA 211, *Standard for Chimneys, Fireplaces, Vents, and Solid Fuel-Burning Appliances* (hereinafter referred to as NFPA 211), specifically prohibits the use of a chimney flue to vent both a solid fuel appliance and an appliance burning other fuels. Therefore, a properly installed solid fuel add-on furnace must have its own chimney flue, separate from that serving the conventional furnace or water heater.

The method by which the add-on is connected to existing ductwork is critical. Properly configured, the add-on will have its own distribution blower and not rely on the blower of the existing furnace. This allows the supply (hot air) plenum of the add-on to be connected to the supply plenum of the furnace; and the return (cold air) side of the add-on to draw air directly through the return ducting. This is called a *parallel* connection and allows the add-on to function identically with the conventional furnace.

A parallel connection can also be accomplished without connection of the return side of the add-on to the return ducting system. In this configuration, the add-on's blower is simply open and draws air from the surrounding area. This can cause significant depressurization of the basement, which could result in flow reversal of the add-on flue or other appliances in the area, with consequent spillage of combustion products. Such open installations are strongly discouraged.

A connection method that has been frequently used for unlisted and homemade add-on furnaces is a *series* connection. This configuration is definitely hazardous. In this form, the add-on does not have a blower, but attempts to use the existing furnace blower for air distribution. Both the return side and supply side of the add-on are connected to the existing return plenum or ductwork, with a baffle to direct the flow of return air through the add-on. As a result, hot air produced through the add-on is circulated through portions of the furnace and duct system, which were not designed for hot air. Furthermore, in a power failure the add-on will continue to produce

heat, but the distribution blower will not run. Very hot air will back up and be concentrated in the return ducting system, which may be constructed of combustible materials. Numerous fires have resulted from such installations.

Heating panels: NFPA 90B recommends that heating panels be used only with automatically fired gas- or oil-burning or electric forced warm air systems that will limit furnace outlet air temperature to 200°F (93°C) or systems equipped with steam or hot-water heat exchangers utilizing steam that cannot exceed 15 psig (103 kPa) or hot water that cannot exceed 250°F (120°C).

Panels used with automatically fired forced warm air systems must be of noncombustible material or at least of material with a flame spread rating of not more than 25, as determined in accordance with NFPA 255, *Standard Method of Test of Surface Burning Characteristics of Building Materials* (hereinafter referred to as NFPA 255).

Where the warm air supply is from a steam or hot-water heat exchanger, the panels may be made of material not more flammable than 1-in. (25.4-mm) nominal wood boards. No single vertical heating panel should serve more than one story of any building.

Steam and Hot-Water Pipes

The temperature of saturated steam varies with the pressure. For example, at atmospheric pressure the temperature of steam is 212°F (100°C), while at 25 psig (172 kPa) its temperature is 266.7°F (130.4°C). Other values of temperature of saturated steam at various pressures are given in Table 3-5C.

Water at high temperature and pressure is also used for heating. The system pressure must always exceed the pressure at the saturation temperature to prevent the water from flashing to steam. This means that the pressure in high-temperature hot-water systems must exceed the values shown in Table 3-5C at the specific temperature at

TABLE 3-5C. *Temperature of Saturated Steam*

Gage Pressure		Temperature		Gage Pressure		Temperature	
psi	kPa	°F	°C	psi	kPa	°F	°C
0*		212.0	100.0	40	275	286.7	141.5
1	6.895	215.4	101.8	45	310	292.3	144.6
2	13.79	218.5	103.3	50	345	297.7	147.6
3	20.69	221.5	105.3	75	517	320.0	160.0
4	27.58	224.4	106.9	100	690	337.8	169.9
5	34.48	227.1	108.4	125	862	352.9	178.3
6	41.37	229.6	109.8	150	1034	365.9	185.5
7	48.26	232.3	111.3	200	1379	387.9	197.7
8	55.16	234.7	112.6	300	2068	421.7	216.5
9	62.06	237.0	113.9	400	2758	448.2	231.2
10	68.9	239.4	115.2	500	3448	470.1	243.4
11	75.84	241.5	116.4	600	4137	488.8	253.8
12	82.74	243.7	117.6	700	4826	505.6	263.1
13	89.64	245.8	118.8	800	5516	520.3	271.3
14	96.53	247.8	119.9	900	6205	534.0	278.9
15	103.4	249.8	121.0	1000	6895	546.4	285.8
20	137.9	258.8	126.0	1500	10,342	597.6	314.2
25	172.4	266.7	130.4	2000	13,790	636.8	336.0
30	206.8	274.1	134.5	3211	22,139	706.1†	374.5
35	241	206.8	138.1				

*Zero gage pressure corresponds to an absolute pressure of 14.696 psi (101.329 kPa).

†Critical point, from Keenan's Tables.

Note: Superheated steam has a higher temperature than saturated steam at the same pressure, depending upon the degree of superheat.

which the system operates. Most systems operate at temperatures between 250 and 430°F (120 and 220°C).

Recommended clearances for steam and hot-water pipes and radiators are given in Table 3-5D.

Radiators and pipes should not be used as racks for drying purposes. Where radiators are placed in window recesses or concealed spaces, such spaces should be lined with noncombustible material, have ample air circulation, and be kept clean.

INSTALLATION OF HEATING APPLIANCES

Any source of heat is a potential fire hazard unless it is arranged to prevent the possibility of dangerous temperatures developing on adjacent combustible materials. It is possible for wood and certain other combustible materials to ignite at temperatures far below their usual ignition temperatures if they are continually exposed to relatively moderate heat over long periods of time. To safeguard against this possibility, heat-producing appliances must be installed with adequate clearance between the appliance and combustibles. Then, under conditions of maximum heat (long and continued exposure), the temperature of exposed combustibles will not exceed dangerous limits. Minimum clearances for various types of heat-producing devices, which should prevent the temperature from exceeding the maximum in exposed combustibles, have been determined by testing laboratories and through field experience. Table 3-5E gives the required standard clearances around and above gas- and oil-fired and electrical residential, commercial, and industrial heat-producing appliances. Table 3-5F gives clearances for solid-fuel burning appliances. The term "listed appliances," which appears in the subheads of Tables 3-5E and 3-5F, refers to appliances that have been tested by testing laboratories. The proper clearances for these appliances, as determined by tests, are published in a listing. The recommended clearances are also indicated either on the appliance or in the manufacturer's installation manual included with the appliance.

Examples of laboratories that test and list appliances are the Factory Mutual Research Corporation, Underwriters Laboratories Inc., Underwriters Laboratories of Canada, and American Gas Association Laboratories.

TABLE 3-5D. Installation Clearances for Steam and Hot-Water Pipes and Radiators

	Description	Clearances (in.)	(mm)
A.	Hot-water pipes and radiators supplied by automatically fired gas, gas-oil, or oil-burning boilers equipped with limit control that cannot be set to permit a water temperature above 150°F (66°C).	None	
B.	Hot-water and steam pipes and radiators supplied with hot water at not more than 250°F (120°C) except as permitted in A above, and with steam at not over 15 psig (103 kPa).	1*	25*
C.	Steam pipes carrying steam at pressures above 15 psig (103 kPa), but not over 500 psig (3450 kPa)	6	15

*At points where pipes emerge from a floor, wall, or ceiling, the clearance of the opening through the finish floor boards or wall or ceiling boards may not be less than 1/2 in. (13 mm). Each such opening must be covered with a plate of noncombustible material.

Modifications of Clearances

Listings of heat-producing appliances and methods for setting clearances between appliances and combustible surfaces are based upon criteria in Underwriters Laboratories Inc. test standards:[1]

- Maximum temperature rise of 117°F (65°C) above room temperature on exposed surfaces; and

- Maximum temperature rise of 90°F (50°C) above room temperature on unexposed surfaces, such as beneath the appliance, floor protector, or wall-mounted protective device.

These requirements are based upon the fact that, while the ignition temperature of wood products is generally quoted to be on the order of 392°F (200°C),[2] wood that is exposed to constant heating over a period of time may undergo chemical change, resulting in a much-lowered ignition temperature and increased potential for self-ignition.

Mitchell[3] presents data on wood fiberboard exposed to temperatures as low as 229°F (109°C) that resulted in ignition after prolonged exposure. MacLean[4,5] reports charring of wood samples at temperatures as low as 200°F (93°C). He concludes that wood should not be exposed to temperatures appreciably higher than 151°F (66°C) for long periods. McGuire[6] suggests that the maximum safe temperatures on the surface of a combustible material adjacent to a constant heat source should be no more than 212°F (100°C).

Clearly, the ignition of wood at moderately elevated temperatures is a complex phenomenon; the length of exposure is indeed an important parameter.[7,8] While exact limits recommended in the literature vary due to exposure time and details of the tests conducted, the numerous documented fires involving the ignition of wood members near low-pressure steam pipes[9] suggest an upper temperature limit for combustible materials exposed to long-term, low-level heating should not be appreciably higher than 212°F (100°C).

The clearances given in Tables 3-5E and 3-5F will provide reasonable protection for exposed surfaces, including those of considerable area, against assumed conditions of continuous operation of heat-producing appliances at maximum temperatures. However, where the appliance is very large, increased clearances to combustible materials may be required. This is a good reason why all appliances should be tested and installed in accordance with the terms of their listing. On the other hand, there are many installations that have not had fires where the clearances are less than that specified. Careful operation of the devices in many such cases has limited temperatures to below the maximum. Yet every year there are thousands of heating-plant fires when cold weather forces long operation at maximum temperatures. Therefore, where the appliance is used more frequently and for longer periods of time in very cold weather, more frequent checks should be made of the temperature of exposed combustibles and the chimney system.

Installing heating appliances, which under normal operation will not expose adjacent combustible materials to dangerous temperatures, is not sufficient. Clearances and protection also must be designed for reasonable protection when the heating device is operated at its maximum temperature. However, under any conditions of operation the temperature of the exposed combustible materials should not exceed 180°F (82°C). Some provision for ventilation, air circulation, or other method of cooling is necessary to dissipate heat.

In large rooms, a suitable clearance between a heating appliance and combustible material is all that is necessary to prevent ignition. In small rooms or poorly ventilated spaces, particularly where the size of the heating device is large in proportion to the room, dangerous temperatures may be built up no matter how great

TABLE 3-5E. Standard Installation Clearances for Nonsolid Fuel Heat-Producing Appliances*

These clearances apply unless otherwise shown on listed appliances. Appliances should not be installed in alcoves or closets, unless so listed. For installation on combustible floors, see †.

Residential-Type Appliances For Installation in Rooms that Are Large (See ‡)	Above Top of Casing or Appliance		From Top and Sides of Warm-Air Bonnet or Plenum		Appliance					
					From Front (See §)		From Back		From Sides	
	(in.)	(mm)	(in.)	(mm)	(in.)	(mm)	(in.)	(mm)	(in.)	(mm)
Boilers and Water Heaters Steam boilers—15 psi (103 kPa) Water boilers—250°F (120°C) Water heaters—200°F (93°C) All water walled or jacketed										
Automatic oil or comb. gas-oil	6	152	—		24	610	6	152	6	152
Automatic gas	6	152	—		18	457	6	152	6	152
Electric	6	152	—		18	457	6	152	6	152
Furnaces—Central Gravity, upflow, downflow, horizontal and duct. warm-air—250°F (120°C) max.										
Automatic oil or comb.gas-oil	6¶	152	6¶	152	24	610	6	152	6	152
Automatic gas	6¶	152	6¶	152	18	457	6	152	6	152
Electric	6¶	152	6¶	152	18	457	6	152	6	152
Furnaces—Floor For mounting in combustible Floors										
Automatic oil or comb.gas-oil	36	914	—		12	305	12	305	12	305
Automatic gas	36	914	—		12	305	12	305	12	305
Electric	36	914	—		12	305	12	305	12	305
Heat Exchanger Steam—15 psi (103 kPa) max. Hot water—250°F (120°C) max. —	1	25	1	25	1	25	1	25	1	25
Room Heaters Circulating type Vented or unvented										
Oil	36	914	—		24	610	12	305	12	305
Gas	36	914	—		24	610	12	305	12	305
Radiant or other type Vented or unvented Oil	36	914	—		36	914	36	914	36	914
Gas	36	914	—		36	914	18	457	18	457
Gas with double metal or ceramic back	36	914	—		36	914	12	305	18	457
Radiators Steam or hot water										
Gas	36	914	—		6	152	6		6	
Ranges—cooking stoves Vented or unvented							Firing Side		Opp. Side	
							24 in.		18 in.	
							6 in.		6 in.	
Oil	30	762			—		9	229	10	254
Gas	30	762			—		6	152	6	152
Electric	30	762			—		6	152	6	152
(See**)										
Clothes dryers Listed types										
Gas	6	152	—		24	610	6	152	6	152
Electric	6	152	—		24	610	0		0	
Incinerators Domestic types See §§										
—	36	914	—		48	1219	36	914	36	914

Appliance

		Above Top of Casing or Appliance (See ††) (in.)	(mm)	From Top and Sides of Warm-Air Bonnet or Plenum (in.)	(mm)	From Front (in.)	(mm)	From Back (See ††) (in.)	(mm)	From Sides (See ††) (in.)	(mm)
Low-Heat Appliances Any and All Physical Sizes, Except as Noted											
Boilers and Water Heaters											
100 cu ft (2.8 m³) or less Any psi steam	All fuels	18	457	—	—	48	1219	18	457	18	457
50 psi (345 kPa) or less Any size	All fuels	18	457	—	—	48	1219	18	457	18	457
Unit Heaters											
Floor mounted or suspended—any size	Steam or hot water	1	25	—	—	—	—	1	25.4	1	25.4
Suspended—100 cu ft (2.8 m³) or less	Oil or comb., gas-oil	6	152	—	—	24	610	18	457	18	457
Suspended—100 cu ft (2.8 m³) or less	Gas	6	152	—	—	18	457	18	457	18	457
Suspended—Over 100 cu ft (2.8 m³)	All fuels	18	457	—	—	48	1219	18	457	18	457
Floor mounted, any size	All fuels	18	457	—	—	48	1219	18	457	18	457
Ranges—Restaurant Type											
Floor mounted	All fuels	48	1219	—	—	48	1219	18	457	18	457
Other Low-Heat Industrial Appliances											
Floor mounted or suspended	All fuels	18	457	18	457	48	1219	18	457	18	457
Commercial-Industrial Type Medium Heat Appliances											
Boilers and Water Heaters											
Over 50 Psi (345 kPa) Over 100 cu ft (2.8 m³)	All fuels	48‡‡	1220	—	—	96	2440	36‡‡	910	36‡‡	910
Other Medium-Heat Industrial Appliances											
All sizes	All fuels	48‡‡	1220	36	910	99	2440	36‡‡	910	36‡‡	910
Incinerators											
All sizes	All fuels	48‡‡	1220	—	—	96	2440	36‡‡	910	36‡‡	910
High-Heat Industrial Appliances											
All sizes	All fuels	180‡‡	4570	—	—	360	9140	120‡‡	910	120‡‡	910

*Standard clearances may be reduced by affording protection to combustible material in accordance with Table 3-5G.

†An appliance may be mounted on a combustible floor if the appliance is listed for installation on a combustible floor, or if the floor is protected in an approved manner. For details of protection, reference may be made to NFPA 211.

‡Rooms that are large in comparison to the size of the appliance are those having a volume equal to at least 12 times the total volume of a furnace and at least 16 times the total volume of a boiler. If the actual ceiling height of a room is greater than 8 ft (2.4 m), the volume of a room must be figured on the basis of a ceiling height of 8 ft (2.4 m).

§The minimum dimension should be that necessary for servicing the appliance, including access for normal maintenance, care, tube removal, etc.

¶For a listed oil, combination gas-oil, gas, or electric furnace, this dimension may be 2 in. (51 mm) if the furnace limit control cannot be set higher than 250°F (120°C), or this dimension may be 1 in. (25.4 mm) if the limit control cannot be set higher than 200°F (93°C).

***To combustible material or metal cabinets. If the underside of such combustible material or metal cabinet is protected with asbestos millboard at least ¼ in. (6.4 mm) thick covered with sheet metal of not less than 24 in. (610 mm).

††If the appliance is encased in brick, the 18-in. (457-mm) clearance above and at sides and rear may be reduced to not less than 12 in. (305 mm).

‡‡If the appliance is encased in brick, the clearance above may be not less than 36 in. (914 mm), and at sides and rear may be not less than 18 in. (457 mm).

§§Clearance above the charging door should be not less than 48 in. (1219 mm).

For SI units: 1 in. = 25.4 mm.

TABLE 3-5F. *Standard Clearances for Solid-Fuel-Burning Appliances*
(For Reduced Clearances, see Table 3-5G.)
These clearances apply to listed appliances installed in rooms that are large in comparison with the size of the appliances.

Appliance Type	Above Top of Casing or Appliance, Above Top and Sides of Furnace Plenum or Bonnet (in.)	(mm)	From Front (in.)	(mm)	From Back‡ (in.)	(mm)	From Sides‡ (in.)	(mm)
Residential Appliances Steam boilers — 15 psi (103 kPa) Water boilers — 250°F (120°C) max. Water boilers — 200°F (93°C) max. All water-walled or jacketed	6	152	48	1219	6†	152†	6†	152†
Furnaces Gravity and forced air§	18	457	48	1219	18	457	18	457
Room heaters, fireplace stoves, combinations	36	914	36	914	36	914	36	914

					Firing Side		Opposite Side	
Ranges Lined fire chamber Unlined fire chamber	30* 30*	762* 762*	36 36	914 914	24 36	610 914	18 18	457 457

*To combustible material or metal cabinets. If the underside of such combustible material or metal cabinet is protected with sheet metal of not less then 24 gage [0.024 in. (0.61 mm)], spaced out 1 in. (25.4 mm), the distance shall be permitted to be reduced to not less than 24 in. (610 mm).
†Adequate clearance for cleaning and maintenance must be provided.
‡Provisions for fuel storage must be located at least 36 in. (914 mm) from any side of the appliance.
§For clearances from air ducts, see NFPA 90B, *Standard for the Installation of Warm Air Heating and Air Conditioning Systems*.
Source: NFPA 211.

the clearance unless some provision is made for cooling by air circulation. Only appliances designed and tested for installation in confined spaces, such as alcoves or closets, are to be installed in such spaces. The clearances specified for them are observed whether the enclosing walls are combustible or noncombustible.

Table 3-5G and Figures 3-5X through 3-5CC show how clearances can be reduced by installing protection between the heat-producing device and the combustible material. This information is especially helpful in situations of inadequate clearance where it would be impractical to move the heat-producing appliance.

Clearance reduction systems: If the air space between heating appliances and combustible material is small, a barrier of metal, metal and insulating material, or masonry heat-resistive material should be installed. Metal and masonry shields, or clearance reduction systems, tend also to distribute the heat, preventing in some measure its building up at one location. Except for specific systems described in Table 3-5G, an air space must be left between the shields and the heat-radiating surface on one side and between the barrier and the combustible material on the other. The air currents set up by this procedure will prevent the combustible material from attaining a dangerous temperature, even with the smaller clearance.

Limitations of Insulation

To make certain that insulating material will provide adequate protection, one needs to understand its limitations. The insulation may be used on the heating appliance or in combination with sheet metal and air spaces (Table 3-5G) if its value under such circumstances has been verified by tests. Many heat-producing appliances have built-in insulation, which makes it safer to reduce the required clearances to combustible material.

Insulation alone, however, is not sufficient. Regardless of the thickness of insulation, long continued heat may eventually penetrate it. However, if some method is used to conduct the heat away before it reaches the combustible material, the insulation will provide adequate protection.

Continued high temperatures over long periods of time have caused many fires under apparently safe conditions. Figure 3-5DD shows fires caused under conditions that the layperson would consider safe. Solid masses of brick or concrete or plaster finish may not provide any fire protection under some circumstances, especially where masonry coverings are attached directly to the surface of the combustible material. Heat transfer to the combustible material is usually more efficient as the density of the covering material increases. Therefore, without an air space between the masonry and the combustibles, the heat transfer can actually increase, resulting in the ignition of the combustibles. However, since the masonry products do have a relatively large mass, it will take some time of operation of the appliance to transfer sufficient energy to ignite combustibles. This, however, is entirely dependent upon the weight and dimensions of the masonry and the rate of heat transfer from the appliance. The same rationale would also apply to products other than masonry attached directly to the surface of combustibles.

Mountings

The limitations of masonry, concrete, metal, and other materials as insulating mediums apply particularly to the underside of stoves, heaters, boilers, furnaces, and other similar heat-producing appliances. Tables 3-5H and 3-5I give a list of mountings for various classes of heat-producing appliances.

The testing laboratories indicate in their listings whether the appliances tested may be installed on combustible or noncombustible floors. In some cases, there is a statement on the appliance itself to the effect that it may be installed on a combustible floor. If no such indication appears, however, it is advisable to check the listing for the particular appliance.

TABLE 3-5G. *Reduction of Appliance Clearance with Specified Forms of Protection* (see footnotes.)
(Source: NFPA 211, *Standard for Chimneys, Fireplaces, Vents, and Solid Fuel-Burning Appliances*)

Clearance Reduction System Applied to and Covering All Combustible Surfaces within the Distance Specified as Required Clearance with No Protection (See Table 3-5E.)	Maximum Allowable Reduction in Clearance (%)		Where the required clearance with no protection is 36 in. (914 mm), the clearances below are the minimum allowable clearances. For other required clearances with no protection, calculate minimum allowable clearance from maximum allowable reduction.§§¶¶			
	As Wall Protector (%)	As Ceiling Protector (%)	As Wall Protector (in.)	(mm)	As Ceiling Protector (in.)	(mm)
(a) 3 1/2-in. (90-mm) thick masonry wall without ventilated air space	33	—	24	610	—	—
(b) 1/2-in. (13-mm) thick noncombustible insulation board over 1-in. (25.4 mm) glass fiber or mineral wool batts without ventilated air space	50	33	18	457	24	610
(c) 0.024-in. (0.61-mm), 24-gage sheet metal over 1-in. (25.4-mm) glass fiber or mineral wool batts reinforced with wire, or equivalent, on rear face with ventilated air space	66	50	12	305	18	457
(d) 3 1/2-in. (90-mm) thick masonry wall with ventilated air space	66	—	12	305	—	—
(e) 0.024-in. (0.61-mm), 24-gage sheet metal with ventilated air space	66	50	12	305	18	457
(f) 1/2-in. (13-mm) thick noncombustible insulation board with ventilated air space	66	50	12	305	18	457
(g) 0.024-in. (0.61-mm), 24-gage sheet metal with ventilated air space over 0.024-in. (0.61-mm), 24-gage sheet metal with ventilated air space	66	50	12	305	18	457
(h) 1-in. (25.4-mm) glass fiber or mineral wool batts sandwiched between two sheets 0.024-in. (0.61-mm), 24-gage sheet metal with ventilated air space	66	50	12	305	18	457

*Spacers and ties must be of noncombustible material. No spacers or ties are to be used directly behind appliance or conductor.

†With all clearance reduction systems using a ventilated air space, adequate air circulation must be provided as described in Figure 3-5Z. There must be at least 1 in. (25.4 mm) between the clearance reduction system and combustible walls and ceilings for clearance reduction systems using a ventilated air space.

‡Mineral wool batts (blanket or board) must have a minmum density of 8 lb/ft³ (128.7 kg/m³) and have a minimum melting point of 1500°F (816°C).

§Insulation material used as part of clearance reduction system must have a thermal conductivity of 1.0 (Btu-in.)/(sq ft–hr-°F) or less. Insulation board must be formed of noncombustible material.

¶If a single-wall connector passes through a masonry wall used as a wall shield, there must be at least 1/2 in. (13 mm) of open, ventilated air space between the connector and the masonry.

**There must be at least 1 in. (25.4 mm) between the appliance and the protector. In no case must the clearance between the appliance and the wall surface be reduced below that allowed in the table.

††Clearances in front of the loading door or ash removal door, or both, of the appliance must not be reduced from those in Section 10-1.

‡‡All clearances and thicknesses are minimums; larger clearances and thicknesses shall be permitted.

§§To calculate the minimum allowable clearance, the following formula can be used: $C_{pr} = C_{un} \times (1 - R/100)$, where C_{pr} is the minimum allowable clearance, C_{un} is the required clearance with no protection, and R is the maximum allowable reduction in clearance.

¶¶Refer to Figures 3-5AA and 3-5BB for other reduced clearances using materials found in (a) through (h) of this table.

Air for Combustion and Ventilation

In many locations, combustion-type heat-producing appliances have ample sources of air for efficient combustion in addition to the ventilation required to prevent undue temperature rises. In basements of dwellings, for example, sufficient air comes in through the cracks around doors and windows.

In relatively tight rooms, such as furnace and boiler rooms, a means to supply air for combustion and ventilation must be provided. Because there are so many variables involved, there is no universally accepted formula for calculating the size of openings necessary to provide adequate air for combustion and ventilation. For example, the tightness and size of the furnace or boiler room, and the operation of exhaust fans and other equipment, would affect the static air pressure in the building. Both NFPA 31 and NFPA 54, as well as other standards pertaining to heat-producing appliances, include specific recommendations on how to supply the air for combustion and ventilation.

For certain laboratory-tested appliances, notably gas- and oil-burning appliances for installation in closets, the minimum size necessary to provide enough air for the operation of the unit and ventilation of the room or space is specified for an opening to the heater room or space.

In recent years there has been a clear trend toward tighter, more energy-efficient buildings, particularly houses. Such buildings have less ability to supply the necessary combustion air through natural infiltration. At the same time there has been an increase in the use of high-capacity exhaust and ventilation devices, such as attic fans and down-draft-powered cooking ranges. The operation of such devices tends to forcibly remove more air than can easily be replaced by infiltration. Such buildings are prone to depressurization, where the pressure inside the house, or portions of it, is significantly lower than the atmospheric pressure outside.

Depressurization can cause constant or intermittent spillage of flue gases from combustion appliances. If the drop in building pressure is greater than the draft created in the chimney or vent, flue gases and air will be drawn down through the venting system and into the building—a phenomenon called backdrafting. This has a serious negative effect on indoor air quality. The moisture produced by combustion will raise indoor humidity levels and promote the growth of molds and fungus. If combustion is not complete, carbon monoxide and other products may be introduced to the indoor environment and can become concentrated to dangerous levels.

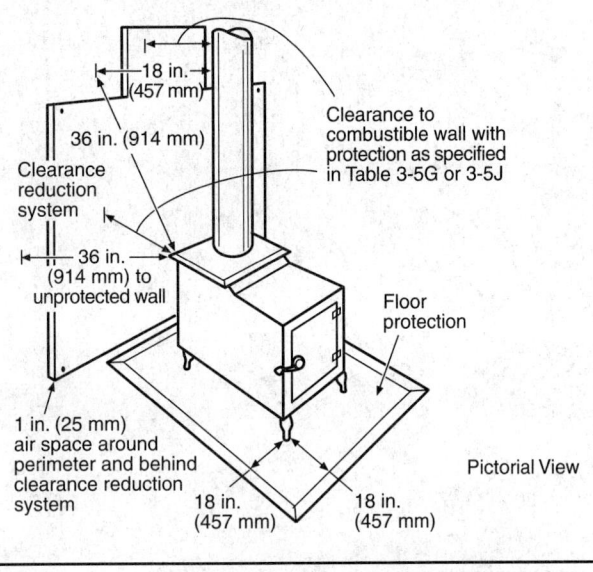

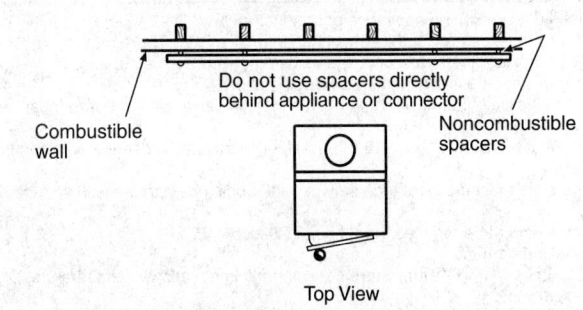

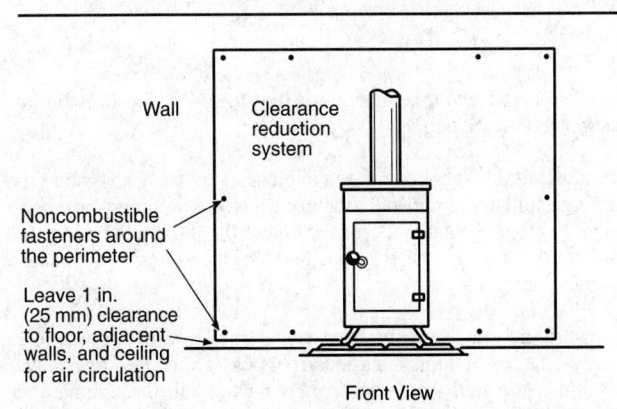

FIG. 3-5X. *A clearance reduction system for a heat-producing appliance. The system is a barrier of metal, metal and insulating material, or masonry heat-resistive material with an air space between the barrier and a combustible surface.*

Direct-vent, "sealed combustion" appliances are essentially immune to such problems, since they take combustion air directly from the outdoors and do not communicate with the indoor atmosphere. Appliances with powered exhausts are also less likely to suffer from backdrafting. For natural draft, conventionally vented appliances, methods of predicting and diagnosing the risk of backdrafting in any particular house are being developed and popularized. Codes are increasingly requiring positive provisions for makeup air in tight, mechanically ventilated houses, so that indoor pressure does not become exceedingly low.

Installation of a natural draft appliance should always include an empirical test of the sufficiency of draft and vent flow. Such tests should be conducted with other exhaust devices operating, including fans and other appliances, to simulate worst-case conditions under which the appliance must properly operate and be fully vented.

Clearances for Servicing

Clearances must also be provided for servicing and maintenance of equipment. Lack of proper servicing and good maintenance can result in fires. If lack of space around an appliance makes accessibil-

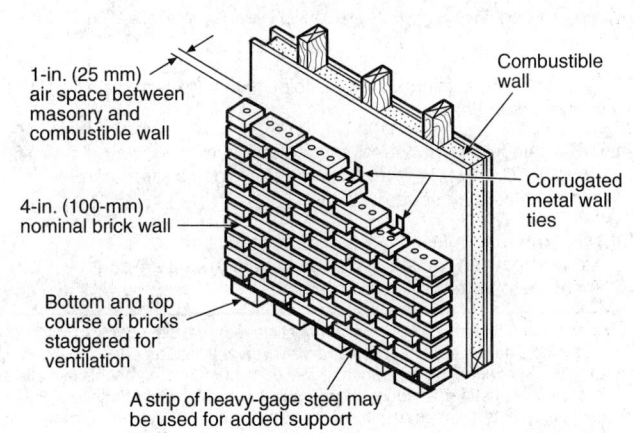

Note: Do not place masonry wall ties directly behind appliance or connector.

FIG. 3-5Y. *Details of a masonry clearance reduction system.*

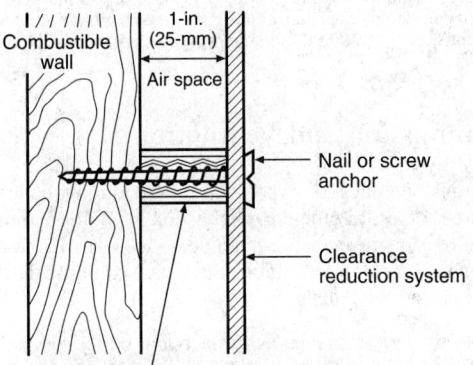

1-in. (25-mm) noncombustible spacer, such as stacked washers, small-diameter pipe, tubing, or electrical conduit.

Masonry walls may be attached to combustible walls using wall ties.

Do not use spacers directly behind appliance or connector.

FIG. 3-5Z. *Details of anchoring a clearance reduction system to a combustible wall.*

ity difficult, the appliance, or at least some parts of it, will be neglected. This same reasoning applies to appliances located in out-of-the-way places and places difficult to reach. Suspended furnaces and furnaces in attics and underfloor crawl spaces are typical examples of such installations.

CHIMNEY AND VENT CONNECTORS

Chimney and vent connectors are specifically the pipe or breeching used to connect fuel-burning appliances with the required chimney or vent, unless the chimney or vent is attached directly to the appliance. NFPA 211 includes requirements for chimney and vent connectors and should be consulted for information on materials to be used and the sizing and installation of connectors.

Chimney and vent connectors must be installed with adequate clearances to combustibles. Table 3-5J provides guidance for appropriate clearances. Table 3-5K shows how clearances can be reduced

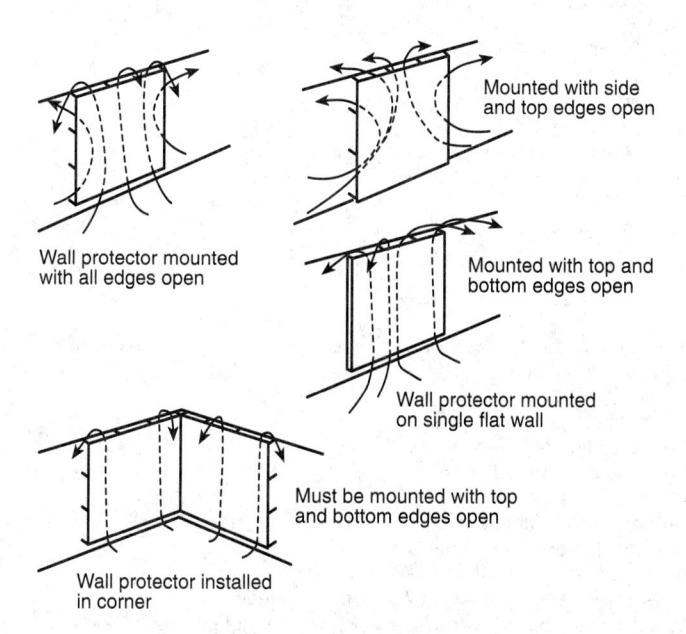

FIG. 3-5AA. *Airflow patterns for various configurations of clearance reduction systems having an air space between the barrier and the combustible surface it protects.*

by installing protection between the connector and combustible material.

Connectors for residential-type appliances may pass through walls or partitions constructed of combustible material if the connector is either listed for wall pass-through or is routed through a device listed for wall pass-through. Connectors for residential-type appliances with inside diameters less than or equal to 10 in. (254 mm) may pass through walls or partitions constructed of combustible material to a masonry chimney if the connector system selected or fabricated is installed as illustrated in Table 3-5L. Figures 3-5EE and 3-5FF illustrate two example installations in detail.

Appliances to Be Chimney or Vent Connected

All oil-fired appliances are chimney connected, except direct-fired heaters, listed kerosene stoves, and portable kerosene heaters. All solid-fueled appliances are chimney connected.

NFPA 54 is quite specific on the types of gas appliances that are required to be vented through vents or chimneys and those that are not so required. Generally, it is the larger types of heating appliances, e.g., boilers and furnaces, that require venting, while smaller appliances, such as listed cooking stoves and hot plates, do not. However, if several smaller appliances that normally do not require venting are located within the same space or room, some may require venting if the aggregate input for all the appliances exceeds 20 Btu per hr per cu ft (0.2 kW/m^3) of the space in which they are installed. NFPA 54 should be consulted for the specifics on the types of gas appliances requiring venting, the conditions under which vents may be omitted or required, and the type of chimney, gas vent, or venting system that can be used and its installation.

Listed unvented room heaters may be used, but care should be taken to follow the manufacturer's instructions. Such heaters are equipped with an oxygen depletion safety shutoff system designed to shut off the gas supply to the heater if the oxygen in the surrounding atmosphere is reduced below about 18 percent. Unvented room heaters are not to be used in institutions, convalescent homes, orphanages, etc.

A change to the 1996 edition of the *National Fuel Gas Code* allows listed unvented wall-mounted heaters to be used in bedrooms and bathrooms that meet the following limitations:

1. The installation of unvented space heaters in bedrooms and bathrooms must be approved by the Authority Having Jurisdiction.
2. Heaters in bathrooms must have an input no greater than 6,000 Btu/hr (1,760 W/hr).
3. Heaters in bedrooms must have an input no greater than 10,000 Btu/hr (2,930 W/hr).

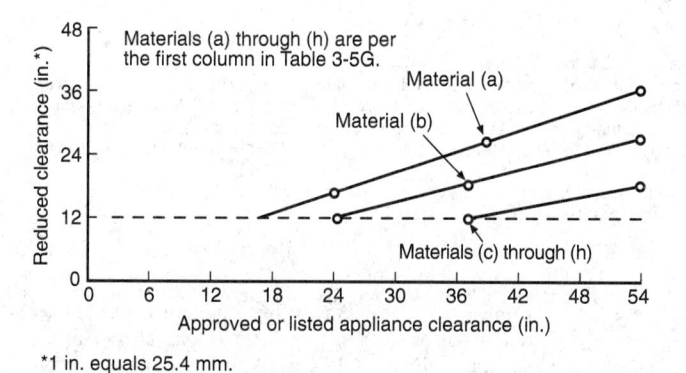

*1 in. equals 25.4 mm.

FIG. 3-5BB. *Wall protection using materials specified in Table 3-5G.*

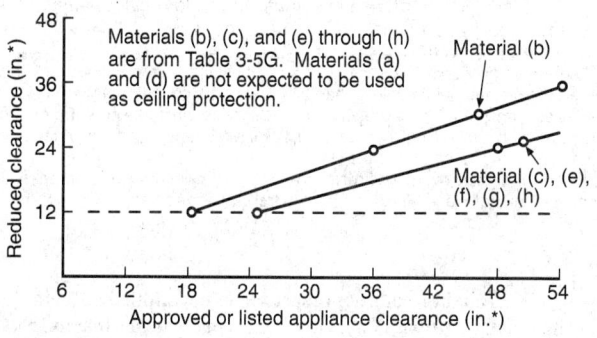

*1 in. equals 25.4 mm.

FIG. 3-5CC. *Ceiling protection using materials specified in Table 3-5G.*

TABLE 3-5H. *Floor Mountings for Residential Nonsolid Fuel Heat-Producing Appliances*
(See Table 3-5I for residential solid fuel appliances.)

Type of Mounting	Required for the Following Types of Heaters and Furnaces
No Floor Protection: Combustible floors.*	Residential-type central furnaces so arranged that the fan chamber occupies the entire area beneath the firing chamber and forms a well-ventilated air space of not less than 18 in. (457 mm) in height between the firing chamber and the floor, with at least one metal baffle between the firing chamber and the floor Low-heat appliances (see Table 3-5E for examples) in which flame and hot gases do not come in contact with the base, on legs that provide not less than 18 in. (457 mm) open space under the base of the appliance, with at least one sheet metal baffle between any burners and the floor Other appliances for which there is evidence that they are designed for safe operation when installed on combustible floors
Metal: Sheet metal not less than 24 gage (0.024 in./0.61 mm) or other approved noncombustible material, laid over a combustible wood floor.*	Heating and cooking appliances set on legs or simulated legs that provide not less than 4 in. (100 mm) open space under the base Ordinary residential stoves with legs Residential ranges with legs Residential room heaters with legs Water heaters with legs Laundry stoves with legs Room heaters with legs†
Hollow Masonry: Hollow masonry not less than 4 in. (100 mm) in thickness laid with ends unsealed and joints matched in such a way as to provide free circulation of air through the masonry.	Downflow furnaces
Hollow Masonry and Metal: Hollow masonry not less than 4 in. (100 mm) in thickness covered with sheet metal not less than 24 gage (0.024 in./0.61 mm), laid over a combustible floor. The masonry will be laid with ends unsealed and joints matched in such a way as to provide a free circulation of air from side to side through the masonry.	Heating furnaces and boilers in which flame and hot gases do not come in contact with the base Floor-mounted heating and cooking appliances Residential stoves without legs Residential ranges without legs Room heaters without legs Water heaters without legs Laundry stoves without legs Residential-type incinerators Restaurant ranges on 4-in. (100-mm) legs Other low-heat appliances on 4-in. (100-mm) legs Medium-heat appliances on legs which provide not less than 24 in. (610 mm) open space under the base
Two Courses Masonry and Plate: Two courses of 4-in. (100-mm) hollow clay tile covered with steel plate not less than ³/₁₆ in. (4.76 mm) in thickness, laid over a combustible floor. The courses of tile will be laid at right angles with ends unsealed and joints matched in such a way as to provide a free circulation of air through the masonry courses.	Heating furnaces and boilers in which flame and hot gases come in contact with the base Restaurant ranges Other low-heat appliances

*Where an appliance is mounted on a combustible floor, and solid fuel is used or the appliance is a domestic-type incinerator, a sheet of ¼-in. (25-mm) asbestos covered by sheet metal not less than 24 gage (0.024 in./0.61 mm) will be required, extending at least 18 in. (457 mm) from the appliance on the front or side where ashes are removed. (The sheet of asbestos may be omitted where the protection required under the appliance is sheet metal only.) For residential-type incinerators, the protection must also extend at least 12 in. (305 mm) beyond all other sides. If the appliance is installed with clearance less than 6 in. (152 mm), the protection for the floor should be carried to the wall.
†Floor protection for radiating-type gas-burning room heaters that make use of metal, asbestos, or ceramic material to direct radiation to the front of the device should extend at least 36 in. (914 mm) in front when the heater is not of a type approved for installation on a combustible floor.

4. The bedroom or bathroom must be an unconfined space as defined in the *National Fuel Gas Code*, which means that the room volume must be at least 50 cu ft for each 1,000 Btu/hr of heater input.

This change was first made as a tentative interim amendment to the 1992 edition of the *National Fuel Gas Code* that was adopted in 1994. It resulted from problems in the southern United States where gas-fired unvented space heaters have a long history of use due to the moderate climate. Old heaters, many installed before 1940, though in need of repair, could not be repaired because parts were not available, nor could they be replaced due to the pre-1994 *National Fuel Gas Code* prohibition of unvented space heaters in bedrooms and bathrooms.

TABLE 3-5H. *(Continued)*

Fire-Resistive Floors Extending 6 in. (152 mm):

Floors of fire-resistive construction with noncombustible flooring and surface finish and with no combustible material against the underside thereof, or on fire-resistive slabs or arches having no combustible material against the underside thereof. Such construction will extend not less than 6 in. (152 mm) beyond the appliance on all sides, and where solid fuel is used, it will extend not less than 18 in. (457 mm) at the front or side where ashes are removed.

Floor-mounted heating and cooking appliances
Residential-type room heaters
Residential-type water heaters

Fire-Resistive Floors Extending 12 in. (305 mm):

Floors of fire-resistive construction with noncombustible flooring and surface finish and with no combustible material against the underside thereof, or on fire-resistive slabs or arches having no combustible material against the underside thereof. Such construction will extend not less than 12 in. (305 mm) beyond the appliance on all sides, and where solid fuel is used, it will extend not less than 18 in. (457 mm) at the front or side where ashes are removed.

Heating furnaces or boilers
Restaurant-type cooking appliances
Residential-type incinerators.
Other low-heat appliances.

Fire-Resistive Floors Extending 3 ft (0.9 m):

Floors of fire-resistive construction with noncombustible flooring and surface finish and with no combustible material against the underside thereof, or on fire-resistive slabs or arches having no combustible material against the underside thereof. Such construction will extend not less than 3 ft (0.9 m) beyond the appliance on all sides, and where solid fuel is used, it will extend not less than 8 ft (2.4 m) at the front or side where ashes are removed.

Medium-heat appliances and furnaces (See Table 3-5G for examples.)

Fire-Resistive Floors Extending 10 ft (3.1 m):

Floors of fire-resistive construction with noncombustible flooring and surface finish and with no combustible material against the underside thereof. Such construction will extend not less than 10 ft (3.1 m) beyond the appliance on all sides, and where solid fuel is used, it will extend not less than 30 ft (9.1 m) at the front or side where hot products are removed.

High-heat appliances and furnaces. (See Table 3-5G for examples.)

*Where an appliance is mounted on a combustible floor, and solid fuel is used or the appliance is a domestic-type incinerator, a sheet of ¹/₄-in. (25-mm) asbestos covered by sheet metal not less than 24 gage (0.024 in./0.61 mm) will be required, extending at least 18 in. (457 mm) from the appliance on the front or side where ashes are removed. (The sheet of asbestos may be omitted where the protection required under the appliance is sheet metal only.) For residential-type incinerators, the protection must also extend at least 12 in. (305 mm) beyond all other sides. If the appliance is installed with clearance less than 6 in. (152 mm), the protection for the floor should be carried to the wall.

†Floor protection for radiating-type gas-burning room heaters that make use of metal, asbestos, or ceramic material to direct radiation to the front of the device should extend at least 36 in. (914 mm) in front when the heater is not of a type approved for installation on a combustible floor.

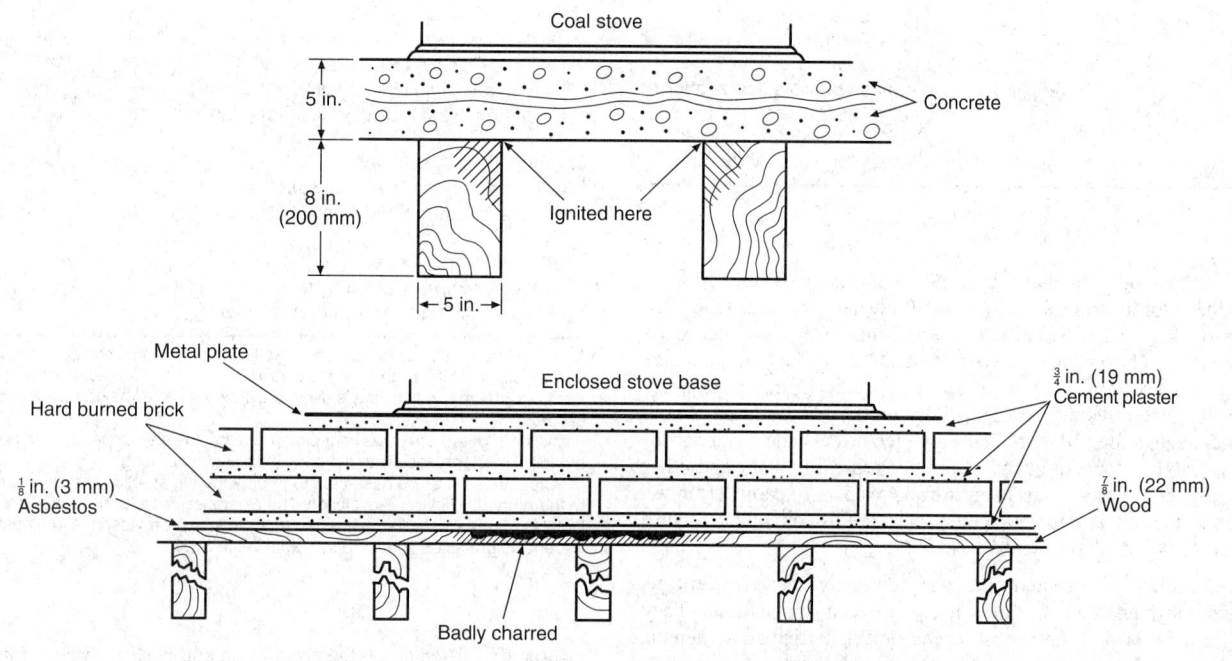

FIG. 3-5DD. *Typical fires due to improperly installed heating appliances.*

TABLE 3-5I. *Floor Mountings for Residential Solid Fuel-Burning Appliances*

Kind of Appliance	Allowed Mounting
(1) All forced air & gravity furnaces, steam and water boilers.	Floors of fire-resistive construction with noncombustible water heaters, fireplace flooring, and surface finish, or fire-resistive arches or slabs. These constructions must have no combustible material against the underside thereof. Such construction must extend not less than 18 in. (457 mm) beyond the appliance on all sides.
(2) Residential type ranges, stoves, room heaters, and combination fireplace stove/room heaters, having less than 2 in. (51 mm) of ventilated open space beneath the fire chamber or base of the unit.	These appliances must not be placed on combustible floors. On combustible floors when such floors are protected by 4 in. (100 mm) of hollow masonry, laid to provide air circulation through the masonry layer. Such masonry must be covered with 24 gage (0.024 in./0.61 mm) sheet metal.
(3) Residential-type ranges, water heaters, fireplace stoves, room heaters, and combination stove/room heaters having legs or pedestals providing 2 to 6 in. (51 to 152 mm) of ventilated open space beneath the fire chamber or base of the heater.	The required floor protection must extend not less than 18 in. (457 mm) on all sides of the appliance. On noncombustible floors, such floors must extend not less than 18 in. (457 mm) on all sides of the appliance.
(4) Residential-type ranges, water heaters, fireplace stoves, room heaters, and combination fireplace stove/room heaters having legs or pedestals providing over 6 in. (152 mm) of ventilated open space beneath the fire chamber or base of the covered heater.	On combustible floors when such floors are protected by closely spaced masonry units of brick, concrete or stone, which provide a thickness of not less than 2 in. (51 mm). Such masonry must be covered by 24-gage (0.024 in./0.61-mm) sheet metal. The required floor protection must extend not less than 18 in. (457 mm) on all sides of the appliance. On noncombustible floors, such floors must extend not less than 18 in. (457 mm) on all sides of the appliance.

TABLE 3-5J. *Chimney Connector and Vent Connector Clearances from Combustible Materials*

(Source: NFPA 211, *Standard for Chimneys, Fireplaces, Vents, and Solid Fuel-Burning Appliances*)

Description of Appliance	Minimum Clearance*	
	(in.)	(mm)
Residential-Type Appliances		
Single-Wall Metal Pipe Connectors		
Gas Appliances without Draft Hoods	18	457
Electric, Gas, and Oil Incinerators	18	457
Oil and Solid-Fuel Appliances	18	457
Unlisted gas Appliances with Draft Hoods	9	229
Boilers and Furnaces Equipped with Listed Gas Burners and with Draft Hoods	9	229
Oil Appliances Listed as Suitable for Use with Type L Vents	9	229
Listed Gas Appliances with Draft Hoods and Other Category I Gas Appliances Listed for Use with Type B Vents‡	6	152
Type L Vent Piping Connectors		
Gas appliances without Draft Hoods	9	229
Electric, Gas, and Oil Incinerators	9	229
Oil and Solid-Fuel Appliances	9	229
Unlisted Gas Appliances with Draft Hoods	6	152
Boilers and Furnaces Equipped with Listed Gas Burners and with Draft Hoods	6	152
Oil Appliances Listed as Suitable for Use With Type L Vents†		
Listed Gas Appliances with Draft Hoods and Other Category I Gas Appliances Listed for Use with Type B Vents‡		
Type B Gas Vent Piping Connectors		
Listed gas appliances with draft hoods and other category I gas appliances listed for use with type B vents‡		
Low-Heat Appliances		
Single-Wall Metal Pipe Connectors		
Gas, Oil, and Solid-Fuel Boilers, Furnaces, and Water Heaters	18	457
Ranges, Restaurant-Type	18	457
Oil Unit Heaters	18	457
Unlisted Gas Unit Heaters	18	457
Listed Gas Unit Heaters with Draft Hoods	6	152
Other Low-Heat Nonresidential Appliances	18	457
Medium-Heat Appliances		
Single-Wall Metal Pipe Connectors		
All Gas, Oil, and Solid-Fuel Appliances	36	914
High-Heat Appliances		
Masonry or Metal Connectors		
All Gas, Oil, and Solid-Fuel Appliances§		

*These clearances apply, except if the listing of an appliance specifies a different clearance, in which case the listed clearance takes precedence.

†If listed Type L vent piping is used, the clearance shall be permitted to be in accordance with the vent listing.

‡If listed Type B or Type L vent piping is used, the clearance shall be permitted to be in accordance with the appliance and vent listing.

§Clearances shall be based on good engineering practice and acceptable to the Authority Having Jursidiction. The clearances from connectors to combustible materials shall be permitted to be reduced, provided the combustible material is protected in accordance with Table 3-5K.

The change was made in the *National Fuel Gas Code* following the introduction of wall-mounted unvented heaters. One concern of the *National Fuel Gas Code* Committee regarding space heaters in bedrooms and bathrooms had been the lack of clearance to combustibles in these small rooms, specifically the probability that the heaters would be used to dry towels and clothing, which could create a fire problem. Wall-mounted heaters address this concern. Another concern of the committee regarding these and other space heaters was a means to ensure adequate air of combustion. As a result, the standard to which the heaters are listed requires the use of an oxygen depletion sensor.

Ventilating hoods and exhaust systems may be used to vent gas utilization equipment installed in commercial applications. They may also be used to vent industrial equipment, particularly when the process itself requires fume disposal. If industrial gas utilization equipment is located in a large and well-ventilated space, it may be operated by discharging the products of combustion directly into the atmosphere.

TABLE 3-5K.   *Reduction of Connector Clearance with Specified Forms of Protection**, †, ‡, §, ¶
(Source: NFPA 211, *Standards for Chimneys, Fireplaces, Vents, and Solid Fuel-Burning Appliances*)

Clearance Reduction Applied to and Covering All Combustible Surfaces within the Distance Specified as Required Clearance with No Protection (See Table 3-5J.)	Maximum Allowable Reduction in Clearance (%)		Where the required clearance with no protection is 18 in. (457 mm), the clearances below are the minimum allowable clearances. For other required clearances, calculate minimum allowable clearance from maximum allowable reduction.††			
	As Wall Protector (%)	As Ceiling Protector (%)	As Wall Protector		As Ceiling Protector	
			(in.)	(mm)	(in.)	(mm)
3 1/2-in. (90-mm) thick masonry wall without ventilated air space*	33	—	12	305	—	—
1/2-in. (13-mm) thick noncombustible insulation board over 1-in. (25.4-mm) glass fiber or mineral wool batts without ventilated air space†	50	33	9	229	12	305
0.024-in. (0.61-mm), 24-gage sheet metal over 1-in. (25.4-mm) glass fiber or mineral wool batts reinforced with wire, or equivalent, on rear face with ventilated air space‡	66	50	6	152	9	229
3 1/2-in. (90-mm) thick masonry wall with ventilated air space§	66	—	6	152	—	—
0.024-in. (0.61-mm), 24-gage sheet metal with ventilated air space¶	66	50	6	152	9	229
1/2-in. (13-mm) thick noncombustible insulation board with ventilated air space#	66	50	6	152	9	229
0.024-in. (0.61-mm), 24-gage sheet metal with ventilated air space over 0.024-in. (0.61-mm), 24-gage sheet metal with ventilated air space**	66	50	6	152	9	229
1-in. (25.4-mm) glass fiber or mineral wool batts sandwiched between two sheets 0.024-in. (0.61-mm), 24-gage sheet metal with ventilated air space††	66	50	6	152	9	229

*Spacers and ties shall be of noncombustible material. No spacers or ties shall be used directly behind appliance or connector.

†With all clearance reduction systems using a ventilated air space, adequate air circulation must be provided. There shall be at least 1 in. (25.4 mm) between the clearance reduction system and combustible walls and ceilings for clearance reduction systems using a ventilated air space.

‡Mineral wool batts (blanket or board) shall have a minimum density of 8 lb/cu ft (128.7 kg/m³) and have a minimum melting point of 1500°F (816°C).

§Insulation material used as part of clearance reduction system shall have a thermal conductivity of 1.0 (Btu-in.)/(ft²-hr-°F) or less. Insulation board shall be formed of noncombustible material.

¶If a single-wall connector passes through a masonry wall used as a wall shield, there shall be at least 1/2 in. (13 mm) of open, ventilated air space between the connector and the masonry.

#There shall be at least 1 in. (25.4 mm) between the connector and the protector. In no case shall the clearance between the connector and the wall surface be reduced below that allowed in the table.

**All clearances and thicknesses are minimum; larger clearances and thicknesses shall be permitted.

††To calculate the minimum allowable clearance, the following formula can be used:
$C_{pr} = C_{un} \times (1 - R/100)$, where C_{pr} is the minimum allowable clearance, C_{un} is the required clearance with no protection, and R is the maximum allowable reduction in clearance.

VENTS

Vents are laboratory-tested, factory-built units used to vent fuel-burning, heat-producing appliances. Specific types of vents have specific uses. (See Table 3-5M.)

Types of Vents

Type B gas vents: These vents are used to vent listed gas appliances that have draft hoods or are specifically listed to be used with Type B gas vents.

Type BW gas vents: These vents are used with listed gas wall furnaces having capacities not greater than that of Type BW gas vents. (See Figure 3-5GG.)

Type L vents: These vents are used to vent oil-burning appliances listed for use with Type L vents and gas appliances listed for use with Type B vents.

Special gas vents: These are listed vents specifically for venting Category II, III, and IV gas appliances. Special gas vents must be specified by the appliance manufacturer.

Pellet vents: These vents are used for venting listed pellet fuel-burning appliances. They are similar in construction to Type L vents and may be additionally listed as such.

Single-wall metal pipe: Single-wall metal pipe, constructed of galvanized sheet steel not lighter than No. 20 galvanized sheet gage [0.020 in. (0.51 mm)] or other noncombustible corrosion-resistant material, may be used to vent residential-type and low-heat gas appliances equipped with draft hoods. They may also be used to vent incinerators used outdoors, such as in open sheds, breezeways, or carports. (See Table 3-5M.) Single-wall metal pipe is not used with solid-fuel appliances, which require either masonry or factory-built chimneys.

Plastic pipe: Unlisted PVC or other plastic pipe may be used to vent Category IV gas appliances, subject to the specifications and limitations in the appliance manufacturer's instructions.

TABLE 3-5L. *Connector Systems and Clearances from Combustible Walls for Residential Heating Appliances*

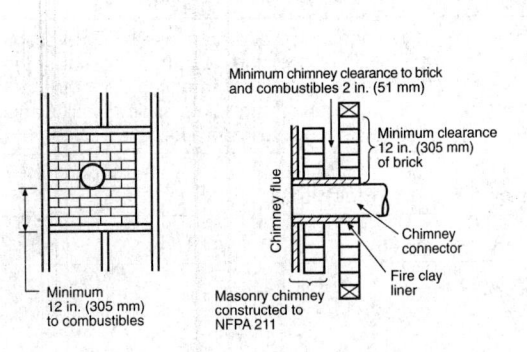

System	Clearance (in./mm)
A. Minimum 3.5-in. (90-mm) thick brick masonry wall framed into combustible wall with a minimum of 12 in. (305 mm) brick separation from clay liner to combustibles. Fire clay liner (ASTM C315, *Standard Specification for Clay Fire Linings,* or equivalent), minimum 5/8-in. (16-mm) wall thickness, must run from outer surface of brick wall to, but not beyond, the inner surface of chimney flue liner and must be firmly cemented in place.	12/305
B. Solid insulated listed factory-built chimney length of the same inside diameter as the chimney connector and having 1 in. (25 mm) or more of insulation with a minimum 9-in. (229-mm) air space between the outer wall of the chimney length and combustibles. The inner end of the chimney length must be flush with the inside of the masonry chimney flue and must be sealed to the flue and to the brick masonry penetration with nonwater-soluble refractory cement. Supports must be securely fastened to wall surfaces on all sides. Fasteners between supports and the chimney length must not penetrate the chimney liner.	9/229
C. Sheet steel chimney connector, minimum 24 gage [0.024 in. (0.61 mm)] in thickness, with a ventilated thimble, minimum 24 gage [0.024 in. (0.61 mm)] in thickness, having two 1-in. (25 mm) air channels, separated from combustibles by a minimum of 6 in. (152 mm) of glass fiber insulation. Opening must be covered and thimble supported with a sheet steel support, minimum 24 gage [0.024 in. (0. 61 mm) in thickness. Support must be securely fastened to wall surfaces on all sides and must be sized to fit and hold chimney section. Fasteners used to secure chimney sections must not penetrate chimney flue liner.	6/152
D. Solid insulated listed factory-built chimney length with an inside diameter 2 in. (51 mm) larger than the chimney connector and having 1 in. (25 mm) or more of insulation, serving as a pass-through for a single-wall sheet steel chimney connector of minimum 24 gage [0.024 in. (0.61 mm)] thickness, with a minimum 2-in. (51 mm) air space between the outer wall of chimney section and combustibles. Minimum length of chimney section must be 12 in. (305 mm). Chimney section concentric with, and spaced 1 in. (25 mm) away from, connector by means of sheet steel support plates on both ends of chimney section. Opening must be covered and chimney section supported on both sides with sheet steel supports of minimum 24 gage [0.024 in. (0.61 mm)] thickness. Supports must be securely fastened to wall surfaces on all sides and must be sized to fit and hold chimney section. Fasteners used to secure chimney sections must not penetrate chimney flue liner.	2/51

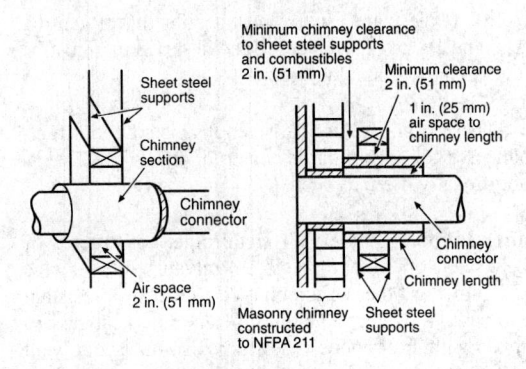

Additional requirements:

1. Insulation material used as part of wall pass-through system must be of noncombustible material and must have a thermal conductivity of 1.0 Btu • in./sq ft • °F (3.88 kg • cal/hr • °C) or less.

2. All clearances and thicknesses are minimums; larger clearances and thicknesses are acceptable.

3. Any material used to close up an opening for the connector must be of noncombustible material.

4. A connector to a masonry chimney, except for System B, must extend to in piece through the wall pass-through system and the chimney wall to the inner face of the flue liner, but not beyond.

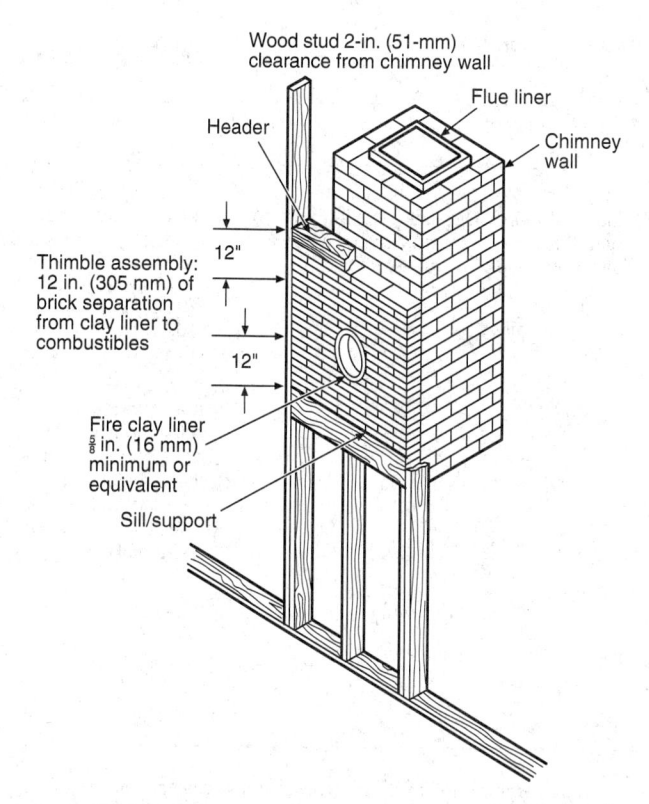

FIG. 3-5EE. *Detail of System A in Table 3-5L.*

New Considerations for Venting Gas Appliances

Prior to the 1980s most gas furnaces were of similar design. They typically included an atmospheric burner, supplemented by excess air introduced in and above the combustion zone. Flue gases passed though a clamshell heat exchanger into a draft hood incorporated into or directly attached to the outlet of the furnace. The draft hood allowed for the relief of backdrafts, which would otherwise interfere with combustion, and also allowed a significant amount of dilution air to mix with the flue gases. The flue gas/air mixture was then drawn up the chimney or vent by draft created by the buoyancy of warm gases. Gas furnaces of this design typically operated in the range of 65 to 75 percent annual fuel utilization efficiency (AFUE).

During the 1970s and 1980s gas appliance manufacturers responded to both market and regulatory pressures by developing more efficient appliances. The National Appliance Energy Efficiency Act of 1987 (NAEEA) now requires that all gas furnaces achieve a minimum AFUE of 78 percent. Most standard gas furnaces now fall in the range of 78 to 83 percent AFUE. Such furnaces are referred to as "mid-efficiency" appliances, to distinguish them from conventional "low-efficiency" designs.

In order to achieve these higher efficiencies, both on-cycle and off-cycle heat losses up the flue needed to be reduced. Conventional furnace design was altered in several ways:

1. The standing pilot was eliminated, replaced by some form of intermittent ignition device (IID).
2. In most designs the draft hood was eliminated, thus reducing on-cycle dilution air and the off-cycle loss of heated room air.
3. The draft hood was replaced by a small fan at the flue outlet of the appliance. This fan provides for a controlled flow of flue gases and excess air through the heat exchanger, and inhibits the flow of heat from the furnace into the flue during the off-cycle.

Mid-efficiency furnaces of this design are known as "fan-assisted" furnaces.

4. In some designs the heat exchanger was made somewhat more efficient.

Another class of furnaces, i.e., "high-efficiency" condensing furnaces, has evolved. These achieve efficiencies well above 90 percent. Such appliances contain a second heat exchanger that is designed to cool the flue gases below their dewpoint, thereby capturing the latent heat of the water vapor created by combustion. Since condensation takes place intentionally, within the appliance, these furnaces must be constructed of corrosion-resistant materials and designed for suitable disposal of the condensate.

Conventional appliance design (below 75 percent AFUE) allowed for simple and mostly trouble-free venting. Relatively high flue gas temperatures provided adequate draft in most chimney and vent configurations, while excess and dilution air helped control condensation by lowering the dewpoint temperature (the temperature below

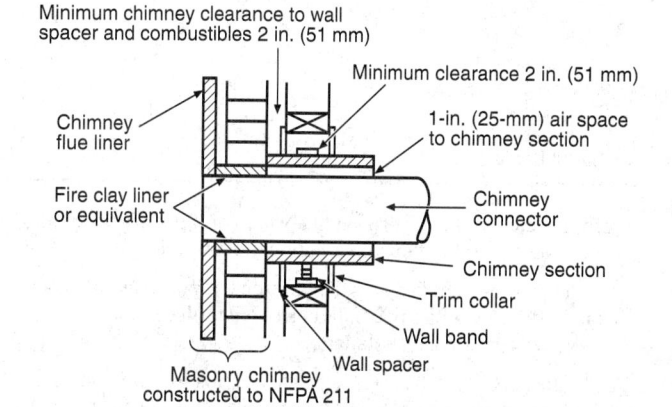

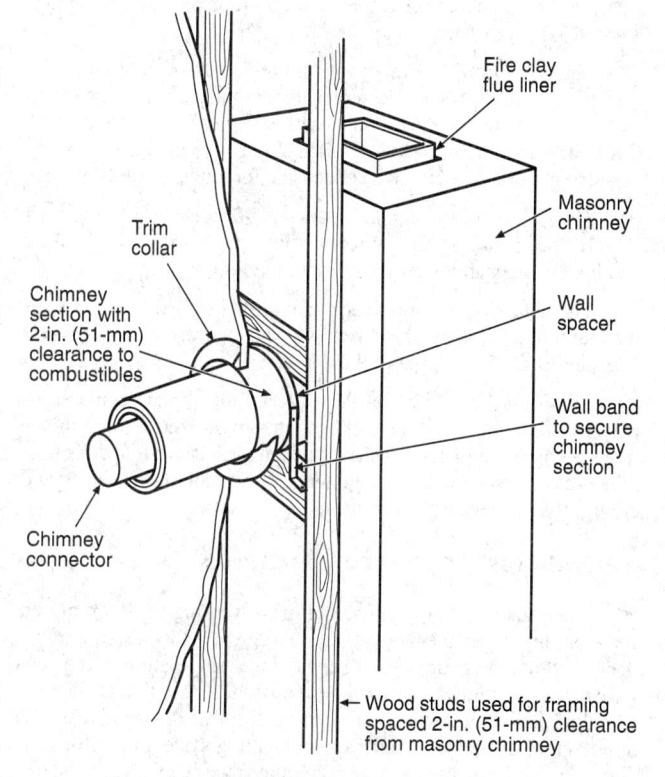

FIG. 3-5FF. *Detail of System D in Table 3-5L.*

TABLE 3-5M. *Vent Selection Chart*

Type of Vent					
Type B—Gas	Type BW—Gas	Type L—Oil	Special Gas Vent Systems	Pellet Vent	Metal Pipe
Column I All listed gas appliances with draft hoods and other Category I gas appliances listed for use with Type B vents, such as: 1. Central furnaces 2. Duct furnaces 3. Floor furnaces 4. Heating boilers 5. Ranges 6. Built-in ovens 7. Vented wall furnaces listed for use with Type B vents 8. Room heaters 9. Water heaters 10. Horizontal furnaces 11. Unit heaters	Column II 1. Vented wall furnaces listed for use with Type BW vents only.	Column III 1. Low-temperature flue gas appliances listed for use with Type L vents. 2. Gas appliances shown in Column I.	Column IV 1. Listed Category II, III, and IV gas appliances only.	Column V 1. Listed pellet-burning appliances listed for use with pellet vents.	Column VI 1. Incinerators used outdoors, such as in open sheds, breezeways, or carports. 2. Gas appliances shown in Column I. 3. Listed residential and low-heat gas appliances without draft hoods, and unlisted residential and low-heat gas appliances with or without draft hoods.

which moisture will condense on flue walls) of the flue gas mixture, and providing a drying flow of air during the off-cycle.

The new mid-efficiency and high-efficiency designs create significantly different venting conditions and introduce new considerations in vent design and installation.

Reduced loss of heat through the venting system and lower vent gas flow rate result in lower venting system temperatures, with a consequent reduction in available draft and in the surface temperature of venting system components.

Elimination of the draft hood and reduction of dilution air increases the humidity and dewpoint temperature of the vent gases, making them more prone to condensation in the venting system. Because they dry out more slowly, vent system components will be exposed to potentially corrosive condensate for longer periods of time.

Elimination of the standing pilot and off-cycle flow of room air inhibits the drying of vent system components during the off-cycle and lowers the standby temperature of the flue.

The reduction in dilution air and use of fan assistance increases the maximum capacity of the venting system; a vent of a given size can handle more gas input to the appliance.

The presence of a fan at the outlet of the appliance increases the likelihood that the vent will become pressurized. Where this is intended by the appliance design, the vent must be designed for positive pressure. Where this is not intended, the vent must be sized and configured to prevent pressurization.

Vented Gas Appliance Categories

From the standpoint of the general type of venting system required for a gas appliance, the two key variables are: (1) the pressure of the gases within the venting system and (2) the temperature of the vent gases. The vent pressure can either be positive (relative to the surrounding atmosphere) or non positive (neutral or negative). The temperature of the gases entering the venting system can either be above or below a critical temperature that is likely to result in an excessive volume or duration of condensation in the vent.

The combination of these two variables creates a matrix of four possible appliance types, or categories, with respect to their venting needs. These vented gas appliance categories are defined in NFPA 54, as follows:

Category I: An appliance that operates with a non-positive vent static pressure and with a vent gas temperature that avoids excessive condensate production in the vent.

Category II: An appliance that operates with a non-positive vent static pressure and with a vent gas temperature that may cause excessive condensate production in the vent.

Category III: An appliance that operates a positive vent static pressure and with a vent gas temperature that avoids excessive condensate production in the vent.

Category IV: An appliance that operates with a positive vent static pressure and with a vent gas temperature that may cause excessive condensate production in the vent.

The category of a gas appliance can be determined only in the laboratory, using criteria from the appropriate ANSI appliance test standard. The vent pressure that is produced by the appliance is determined using a special test apparatus. The determination of whether the vent gas temperature will cause or avoid "excessive condensate production in the vent" is unique for each type of gas appliance and is based on an efficiency profile of the appliance type. It is not possible to reliably determine the category of an appliance through testing in the field.

Categorization criteria for every gas appliance type have been or are being developed. The category of an appliance is part of its certification and will be shown on the rating plate. There is no further need to evaluate the category in the field.

Both Category I and III appliances generally fall within the mid-efficiency range. Since both typically have a small fan at the outlet, the major difference between them is whether or not the fan is used to force vent gases through the vent, or simply to pull gases

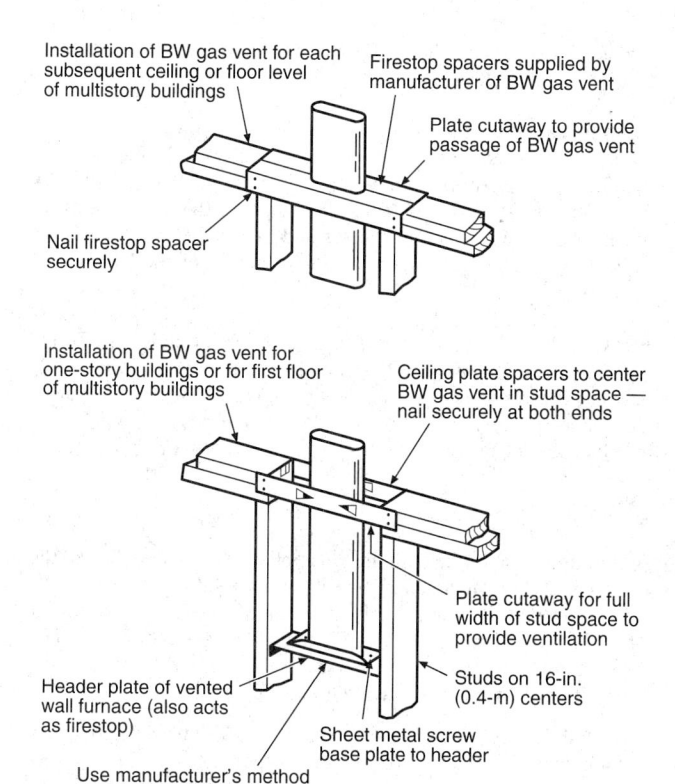

FIG. 3-5GG. Installation of Type BW gas vents for vented wall furnaces.

through the heat exchanger, delivering them at neutral pressure to the beginning of the vent. Category IV represents high-efficiency, condensing appliances. Category II appliances are problematic, since they need to operate under natural draft, but without sufficient vent gas temperature to produce much draft. There are few, if any, Category II appliances on the market.

The category of a gas appliance determines the general characteristics of the venting system that must be used. Category III and IV furnaces, which are designed to operate with positive vent gas pressure, must use a vent that can withstand the pressure and be sealed against leakage. Category I and II appliances are expected to be vented by natural draft and must use a vertical venting system sized and configured to develop sufficient draft.

Further, Category I and III appliances have relatively high vent gas temperatures. The vent may not need to withstand severe exposure to condensate, but must be able to withstand relatively high temperatures. Category II and IV appliances need not withstand such high temperatures, but must be corrosion resistant and designed to contain and dispose of condensate.

Venting Category I Appliances

Category I appliances and low-efficiency, uncategorized appliances are vented through "conventional" venting systems: mainly Type B vents and masonry chimneys. However, venting systems for mid-efficiency, fan-assisted appliances call for careful attention to sizing and factors of vent geometry. For instance, long connector runs between the appliance and vertical vent can cause excessive heat loss and unacceptable condensation. Masonry chimneys, due to their mass, heat up slowly and have limited use with mid-efficiency ap-

pliances. Frequently, masonry chimneys must be relined with a lightweight aluminum or stainless-steel liner, in order to provide adequate capacity and minimize condensation. The use of chimneys and vents on the exterior of the building is severely restricted or prohibited in most climate zones by NFPA 54.

Procedures for sizing venting systems for Category I appliances have undergone significant changes in the past decade. NFPA 54 has provided vent sizing tables since the 1950s. These tables were developed for draft-hood-equipped appliances operating in the low-efficiency range. Modern mid-efficiency appliances, without draft hoods, bring different conditions to the vent, which must be anticipated in the sizing regimen.

In-depth research into the venting needs of mid-efficiency appliances, much of it sponsored by the Gas Research Institute (GRI), has led to the updating and expansion of the original NFPA 54 sizing tables. The original tables addressed only the *maximum* capacity of vents and chimneys, which corresponded to a minimum size that could be used for an appliance of a given input rate and a given vent geometry to prevent spillage. With the advent of fan-assisted appliances, the maximum capacity (or minimum size) also becomes important to prevent pressurization of the vent by the fan.

The new tables introduce the concept of a *minimum* vent capacity, which is the smallest appliance input rating that can be connected to a given vent. A vent connected to an appliance with an input lower than the vent's minimum capacity will warm up slowly, and condensation will continue until the vent wall is warmer than the dewpoint temperature of the vent gases. This longer "wet time," or exposure to condensate, may lead to corrosion and premature failure of the vent. The minimum capacities are designed to limit the wet time to acceptable levels. The concept of minimum vent capacity results in a maximum vent size that can be used for the given input rate and vent geometry.

For any proposed system, a chimney or vent size must be found that lies between the minimum and maximum. The new tables in NFPA 54 also add a number of conditions and adjustments to the former table values that may further restrict the applicability of a vent. Depending on the details of the system, the choice of acceptable sizes may be quite narrow, or even nonexistent. Combinations of appliance input and vent configuration that were, in the past, used routinely must now be carefully scrutinized and may not be acceptable for mid-efficiency, fan-assisted appliances.

Venting Category III Appliances

Like Category I appliances, Category III appliances are mid-efficiency appliances. However, they are designed to exhaust the products of combustion under positive pressure rather than natural draft. Since the "work" will be done by a fan, such systems can typically be vented horizontally through a sidewall. Thus, they avoid the need for a chimney or other vertical vent running up through, or beside, the building.

Vents for Category III appliances are not designed with reference to standard installation requirements and sizing tables found in codes. Instead, the materials, size, and installation guidelines for each appliance are found in the manufacturer's instructions. Such instructions must be adhered to in order to produce a proper installation.

Category III appliances have relatively high vent gas temperatures, which are under positive pressure within the vent. Therefore, the vent must be of a temperature-resistant material that can be constructed with sealed joints to prevent leakage. A new type of listed vent system, special gas vents, has been developed to meet these requirements. (See Table 3-5M.) Such systems must comply with UL 1738, *Standard for Safety Venting Systems for Gas-Burning Appliances.*

They are typically formed of a high-temperature plastic resin or stainless steel. Sections of pipe are joined together, using a manufacturer-specified sealant cement.

Since the vent will be subjected to elevated temperature, some degree of condensation, and mechanical stress induced by thermal expansion, it is extremely important that the vent be installed with care. Hangers and other securement of the pipe must be done according to instructions. The specific sealant required by the manufacturer must be used, and joints fitted together according to the specified procedure. Some systems require condensate disposal systems and a specific pitch or rise back to, or away from, the appliance. Both appliance and vent manufacturer's instructions must be consulted and followed.

Venting Category IV Appliances

To obtain efficiencies approaching 90 percent or better, the flue gases from high-efficiency, Category IV appliances have been reduced to temperatures as low as 100°F (38°C). Because these temperatures are too low to allow atmospheric venting, mechanical venting is required to exhaust the flue gases under positive pressure. Furthermore, the temperature of the flue gases is low enough to cause condensation of the water vapor in flue gases. Thus, special treatment is needed to provide a venting system that will retain and dispose of the condensate and, in cases of mechanical venting, to prevent leakage of flue gases from the venting system.

Unlike Category I appliances, high-efficiency Category IV appliances do not use standardized, code-specified venting systems. Instead, the type of venting system to be used, and any limitations on size and configuration, must be specified explicitly in the appliance manufacturer's instructions. Such limitations include maximum length and number of turns in the vent. Typically, manufacturers' instructions specify the use of PVC (or other plastic pipe) or a high-alloy, corrosion-resistant type of stainless steel, such as AL29-4C. These can frequently be run horizontally through a side wall, with an appropriate termination. Many designs use a parallel pipe to bring in combustion air, creating a closed combustion, direct vent system.

Manufacturer's instructions will also specify the method to be used to dispose of condensate, as well as necessary clearances from the vent. The instructions must be followed exactly to ensure a safe and properly performing system.

Installation of Vents

Types B, BW, and L, and special gas vents: Installations must be made in compliance with the terms of their laboratory listings and the manufacturers' instructions, making certain that the required clearances are maintained.

Single-wall metal pipe: The use of single-wall metal pipe is restricted to runs directly from the space in which the gas appliance is located through the roof or exterior wall to the outside. The pipe is not to originate in any attic or concealed space and is not to pass through any attic or inside space, nor through any floor or ceiling. The minimum clearances from combustible material for single-wall metal pipe are as given for connectors in Table 3-5J. (See Figure 3-5HH.) These clearances may be reduced if appropriate protection is provided for combustible materials. (See Table 3-5K.) Where a single-wall metal pipe passes through an exterior wall constructed of combustible material, or the pipe passes through a roof constructed of combustible material, it is protected at the point of passage as specified for connectors in Table 3-5L. If the appliance the pipe serves is a gas appliance for use with a Type B gas vent, protection is by a noncombustible, nonventilating thimble not less than 4 in. (100 mm) larger in diameter than the pipe and extending not less than 18 in. (457 mm) above and

6 in. (152 mm) below the roof, with annular space open at the bottom and closed at the top.

Venting capacities: The venting capacities of various sizes of Type B gas vents, chimneys, and single-wall metal pipe used for venting gas appliances are given in tables in NFPA 54. NFPA 54 also contains data that may be used to calculate the size of vents required under various circumstances and to establish configurations of venting arrangements. Approved engineering methods may also be used.

Firestopping: Vents that pass through floors of buildings requiring the protection of vertical openings are enclosed within walls having a fire-resistance rating of not less than 1 hr if the vents are in a building less than four stories in height, and not less than 2 hr if the vents are located in a building four stories or more in height.

Draft Hoods

A draft hood is a device built into an appliance, or made a part of the vent connector from an appliance. It is designed to: (1) ensure the ready escape of the flue gases in the event of no draft, backdraft, or stoppage beyond the draft hood; (2) prevent a backdraft from entering the appliance; and (3) neutralize the effect of stack action of the chimney or gas vent upon the appliance operation.

Draft hoods are an important safety feature of gas appliances and are generally required on all vented appliances, except incinerators, dual-oven-type combination ranges, direct-vent appliances,

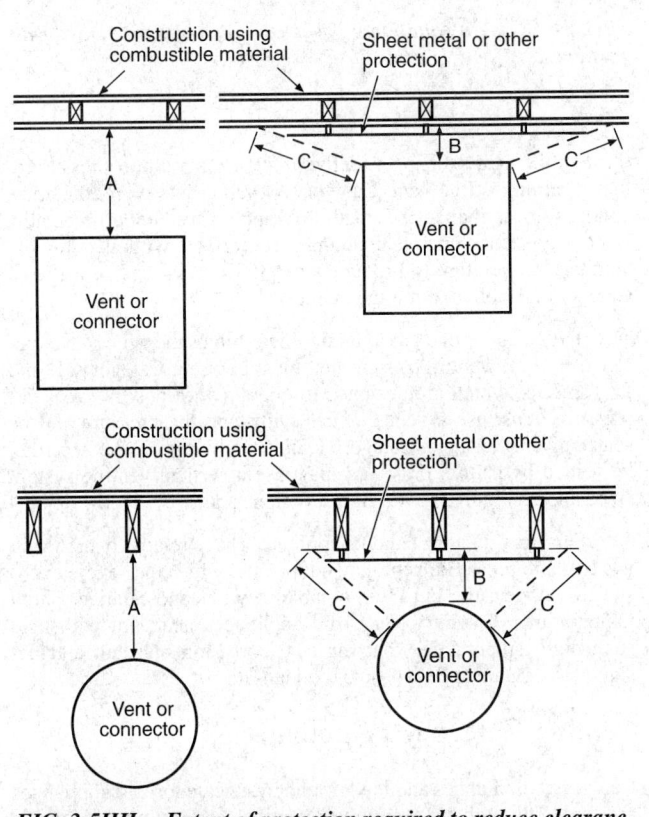

FIG. 3-5HH.   *Extent of protection required to reduce clearances from combustibles for vents and connectors. "A" equals the required clearance with no protections as specified in Table 3-5J; "B" equals the reduced clearance permitted by the types of protection specified in Table 3-5K. The protection applied to construction using combustible material should extend far enough in each direction to make "C" equal to "A."*

and units designed for power burners or for forced venting by mechanical means. Many newer mid-efficiency appliances replace the draft hood with a fan at the appliance outlet. Such fans do not necessarily pressurize the vent, since they may be designed only to pull the gases through the heat exchanger and deliver them at neutral pressure to the beginning of a natural draft venting system.

CHIMNEYS AND FIREPLACES

Chimneys are classified into three major types: (1) factory built, (2) masonry, and (3) unlisted metal (smokestacks). In addition to the major chimney classes, there are subclasses that involve type of building and surroundings. For example, small, well-insulated, factory-built chimneys may be installed at close clearances in residential frame construction, while larger chimneys for the same type of appliances are suitable mainly for noncombustible surroundings or in well-ventilated open-room use at much greater clearance to combustibles. Details of installation for chimneys, fireplaces, and vents are provided in NFPA 211.

Factory-Built Chimneys

A factory-built chimney is an assembly of manufactured components that form a completed chimney. Factory-built chimneys are tested for compliance with safety standards. (See bibliography.)

Factory-built chimneys for use with residential-type appliances are tested fully enclosed to establish a minimum clearance to combustibles not greater than 2 in. (51 mm). The test is conducted using a flue gas generator that simulates an appliance steadily producing 1,000°F (540°C) gases. A typical factory-built chimney for residential appliances is shown in Figure 3-5II.

For building heating equipment [gases to 1,000°F (540°C)], commercial ovens and furnaces [gases to 1,400°F (760°C)], and medium-heat appliances [gases to 1,800°F (980°C)], factory-built chimneys are available in sizes from 10 in. (250 mm) to several feet (1 ft equals 0.3 m) in diameter. These chimneys are tested for the appropriate temperature and are used mainly in noncombustible surroundings, although tested thimbles are available for penetration of combustible roofs. None of these are suitable for close clearance to combustible enclosures; however, this is seldom a problem in the usual large building of masonry or metal construction. A typical unenclosed type of large factory-built chimney for serving building boilers is shown in Figure 3-5JJ.

Current information on types, usage, and installation limitations of all UL-listed factory-built chimneys is given in the *UL Gas and Oil Equipment Directory*, published annually with a supplement.

Masonry Chimneys

Field construction requirements for masonry chimneys are detailed in NFPA 211, as well as in building codes. Field-erected masonry chimneys, which are subject to the know-how and ability of available labor, are not tested, except perhaps for a smoke test. A minimum residential masonry chimney consists of a refractory fire clay tile liner and an air space of roughly 1/2 in. (13 mm) between the liner and brick, with all liner joints grouted both to prevent leakage and to center and support the tile liners. (See Figure 3-5KK.) One course of common brick around the liner suffices for usual residential chimneys. For higher temperature classifications, firebrick is used as the liner, with additional courses of brick for larger size or greater strength and security. All chimneys should, of course, be supported by footings suitable for the chimney size and weight and be provided with cleanouts between the points of entry and the lowest chimney level.

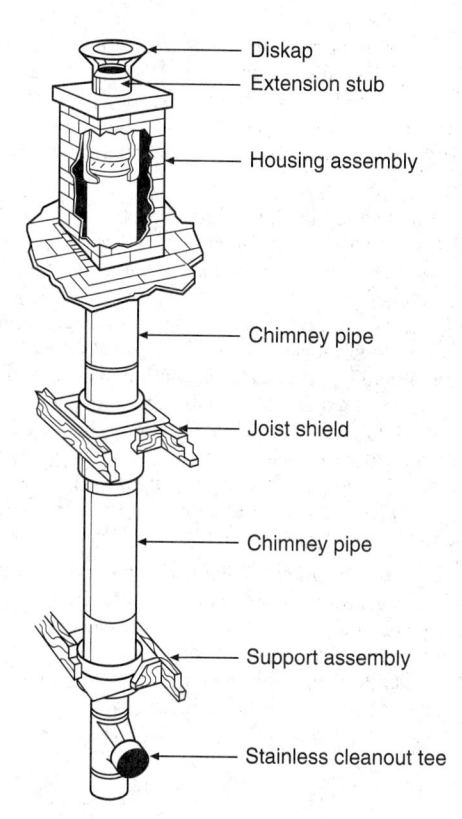

FIG. 3-5II. *A typical factory-built chimney. (Source: Metalbestos Div., Wallace-Murray Corp.)*

For residential and low-heat chimneys located entirely or partially inside the building, a minimum of 2 in. (51 mm) of air space is required between the chimney and surrounding combustibles. For residential and low-heat chimneys located entirely outside the building, a minimum of 1-in. (25.4 mm) clearance is required. For higher heat chimneys, larger clearances are appropriate. All clearances must consist of an air space. Solid material, including building insulation, whether noncombustible or not, must not be allowed to impinge into the required air space clearance.

As a general rule, the best and safest chimney operation and appliance control is obtained with only one appliance attached to a chimney flue. Two or three gas appliances, however, may be connected into a masonry chimney flue at one level of a building, provided that the connector size, its vertical rise, and chimney size and height are in accordance with tabulated capacity data in NFPA 54. The construction of masonry chimneys for open-front fireplaces is the same as for other residential appliances and is specified in NFPA 211.

Flue liners for masonry chimneys: All masonry chimneys must have flue liners. The liner serves several functions that are essential for chimney safety.

1. The liner provides a smooth corrosion- and erosion-resistant surface that is conducive to the flow of flue gases, and that protects the brick and mortar flue wall from direct exposure to combustion products.

2. The liner forms a sealed conduit that contains the products of combustion, including gases, moisture, heat, and solids such as soot and creosote, and conducts flue gases to the outside atmosphere.

3. The liner is an important element in the thermal performance of the chimney. The liner, with its surrounding air space, forms a thermal break between the warm flue gases and the chimney wall. This helps maintain the temperature of the gases for good draft and reduces temperatures on the exterior of the chimney wall.

Traditionally, the flue lining for residential and low-heat type chimneys has been a clay flue lining, which has been fired to produce a vitreous, relatively non-porous wall. Manufacturing specifications for clay flue linings are set in ASTM C315, *Standard Specification for Clay Flue Linings*. The general requirements for clay flue linings and their installation are specified in NFPA 211. Clay linings should be installed with small joints struck smooth on the inside, bedded in a non-water soluble refractory cement, such as a calcium aluminate mixture. All tile sections should be carefully aligned and should never be offset more than 30 degrees from the vertical. Liners should start from a point at least 8 in. (203 mm) below the lowest inlet, and extend continuously to a point at least 2 in. (51 mm) above the chimney crown (splay or wash).

In recent years a variety of alternative flue lining systems has been developed. These can be used either as the original liner in the chimney, in place of clay lining, or retrofitted into an existing chimney. Relining of chimneys is most often done when the original liner has become damaged or deteriorated, or when a smaller flue is needed to serve a new connected appliance.

Although NFPA 211 does provide for the installation of unlisted linings, most alternative lining systems have been tested and listed to UL 1777, *Standard for Safety Chimney Liners*. As with all listed systems, they should be used only for the intended purpose and installed in strict compliance with their manufacturer's instructions.

Most alternative linings fall into one of two types: (1) stainless steel and (2) cast-in-place cementitious mixtures. Stainless steel liners are usually made of a 300 series alloy and can be composed of rigid pipe sections or a continuous flexible tube that can be passed through chimney offsets. Stainless steel linings must be insulated, either with a ceramic blanket wrapped around the pipe before insertion, or with a backfill of a moist vermiculite/cement mixture, depending on the liner listing. Loose, unbound backfills are not permitted. Stainless steel lining systems come with a rain cap tested to exclude most precipitation, as well as other parts that must be used to form a complete, listed system.

Cast-in-place liners are usually composed of a proprietary mixture of perlite or vermiculite, and Portland or other cement as a binder. One method involves insertion of a properly sized, sausage-shaped inflated bladder within the chimney, held off the walls by spacers. A very wet slurry of liner mix is then pumped around the tube and allowed to harden. After several days the bladder is deflated and removed, leaving a smooth, formed flue. In another method, a damp mixture is poured in the chimney while a bell-shaped slipform is pulled up, forcing the mixture against the chimney walls. With either method, a second coating of a glazing material may be applied to the flue to reduce moisture penetration.

Special versions of either stainless steel or cast-in-place liner systems can be designed and listed to obviate the need for air space clearances to combustibles normally required for masonry chimneys. These "zero-clearance" liners typically contain more insulating material or wall thickness to reduce the amount of heat reaching the outside of the chimney wall. They are useful for lining existing chimneys that were constructed without proper clearances. In order to be eligible for this application, the liner will be explicitly labeled for zero-clearance and will have special installation requirements that must be followed.

Other specialized subsets of chimney liner systems are those designed only for use with a specific type or class of appliance. The primary example of this are liners for use with Category I gas appli-

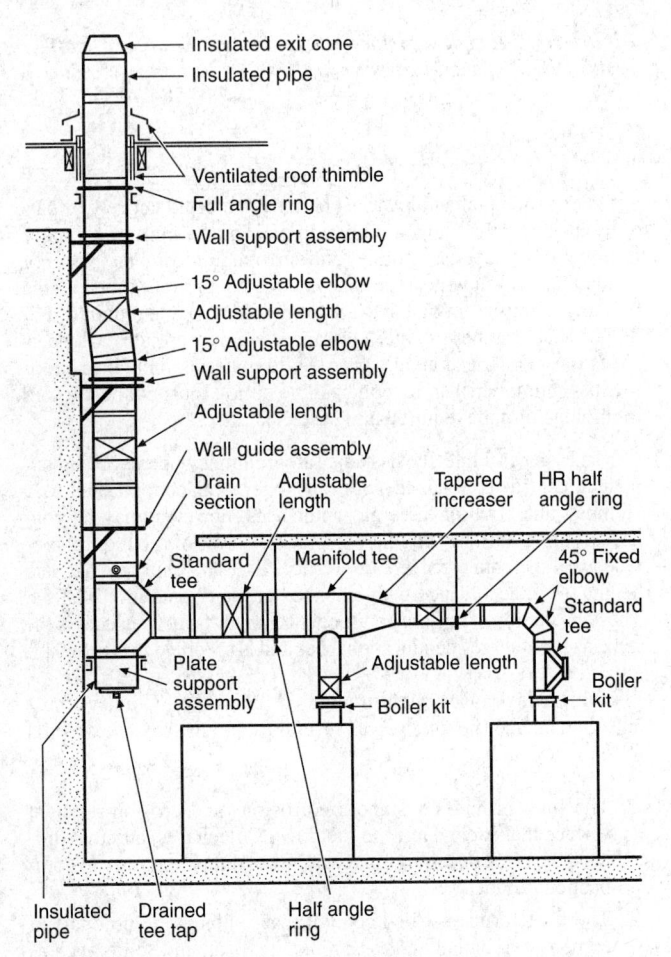

FIG. 3-5JJ. *A typical unenclosed factory-built chimney for serving building boilers.*

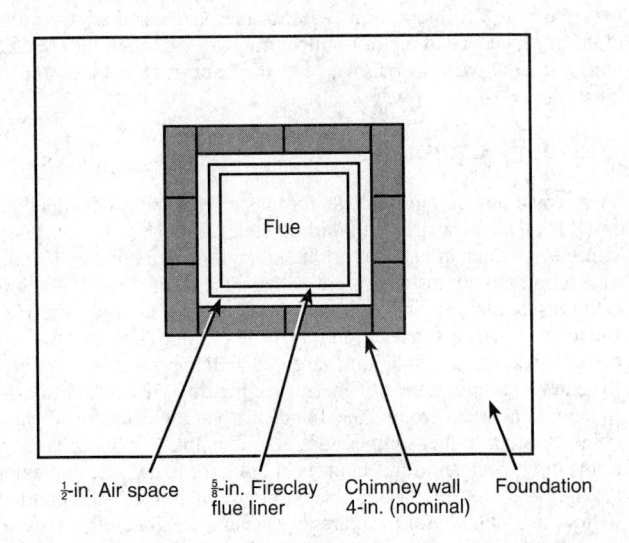

FIG. 3-5KK. *A minimum residential masonry chimney.*

ances. Many such systems are made of aluminum, which would not be suitable for the temperatures produced by general residential appliances. They have been tested under the temperature conditions of UL 441, *Standard for Safety Gas Vents*, for Type B gas vents, and can be used for the same applications. They are, however, still chimney liners listed under UL 1777 and must be installed under the conditions appropriate for chimney liners, according to their own manufacturer's instructions. NFPA 54 and NFPA 211 both require that a notice warning against connection of inappropriate appliances be posted near the chimney inlet.

Unlisted Metal Chimneys

Metal chimneys are suitable for all the classes of appliances, but are not subjected to safety testing of any kind. They are distinct from factory-built chimneys, which may be made of metal but are tested and listed assemblies. The major hazard with metal chimneys is inadequate clearance to combustibles where they penetrate ceilings and roofs. (See Figure 3-5LL.)

Metal chimneys may be of single-wall metal for low-temperature use, such as with gas appliances, or metal lined with firebrick or refractory mortar for medium- and high-heat service. They may be located inside or outside of buildings, but not inside or outside of one- and two-family dwellings or buildings of wood-frame construction. The conditions under which metal chimneys may be used are quite limited and are spelled out in detail in NFPA 211. The standard is also quite specific on acceptable materials of construction.

Regardless of permitted clearances, metal chimneys should not be enclosed with combustible construction. It is particularly important that the metal used be resistant to corrosion if the gases are 350°F (177°C) or below, because exposed outdoor portions of the chimney walls may be below the dew point. The resulting continuous condensation can lead to rapid corrosion and very short chimney life.

Chimney Functions

The purpose of a chimney is to create draft or negative pressure to provide combustion air for the fuel or fuel bed and to remove products of combustion from the appliance and building. For open fireplaces or for gas appliances having draft hoods, chimney draft has very little influence on the combustion process; the chimney serves only as a conduit for carrying away the products of combustion. For forced-draft or packaged boilers, the chimney may be under slight positive pressure and, again, serves only as a duct to carry away

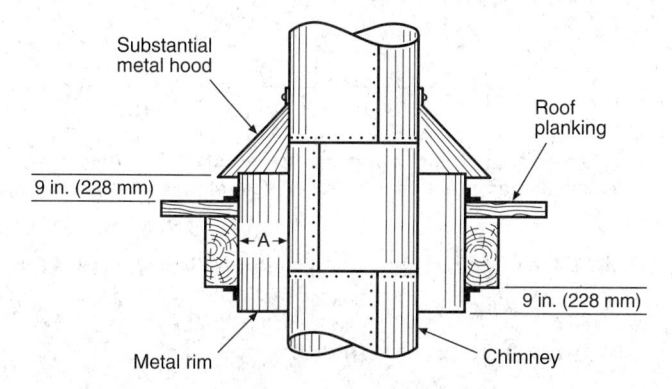

FIG. 3-5LL. *Arrangement of a metal chimney through a wood roof. For chimneys serving residential-type or low-heat appliances, dimension "A" should be at least 6 in. (152 mm). For chimneys serving medium-heat appliances, dimension "A" should be at least 18 in. (457 mm).*

combustion products. The advantage of the slight positive pressure, commonly from 0.5 to 1.5 in. of water column (120 to 370 Pa), is that it permits use of much smaller chimneys and eliminates some concerns about the draft-producing ability or height of the chimney.

The technology of matching the size, height, and configuration of a chimney to the temperature, type, and quantity of fuel for its correct capacity is covered in the ASHRAE *Handbook and Product Directory*.

Height of Chimneys

A variety of factors determines the height of any chimney, including the type of appliance, need for draft, roof construction, and building height. The minimum height of a chimney (or gas vent) for gas appliances having draft hoods is based on the height specified in the applicable test standard for that appliance. For example, gas furnaces, water heaters, boilers, and room heaters must pass draft hood spillage tests with 5 ft (1.5 m) of chimney height measured from the draft hood to chimney outlet, while wall furnaces require a vent outlet at least 12 ft (3.7 m) above the floor.

In contrast to these low heights for draft hood appliances, some older types of solid fuel boilers require up to 50 or 70 ft (15 to 20 m) of chimney to produce the draft needed to overcome fuel bed and internal flow resistance and to develop rated heat output.

Residential oil furnaces and boilers with pressure atomizing burners are designed to function with negative outlet draft settings in the range of 0.04 to 0.06 in. of water column (10 to 15 Pa). Manufacturers of factory-built residential chimneys have prepared draft chimney height tables that allow for such factors as the number of connector elbows, fuel input rating, and outlet size, so that an adequate height can be selected.

Building or dwelling height and configuration also govern chimney height. The generally cited rule for minimum height of the chimney outlet is illustrated in Figure 3-5MM. The chimney height selected must be adequate for proper appliance operation and for adequate chimney flow, as well as for proper height above the roof. Owing to the aerodynamic complexity of buildings, roofs, and chimneys, the dimensions suggested in Figure 3-5LL sometimes are not adequate for efficient operation, which is attested to by the proliferation of special vent caps, chimney extensions, and other devices used in attempting to cure fireplace or draft problems.

Airflow over the roof must be visualized to analyze a structure for the correct chimney top location and to minimize backdrafts due to wind. Only the pressure or windward side of pitched roofs needs to be considered, because the flow moving up a building wall or toward the ridge separates from the downward side at the ridge or building edge and creates a zone of negative pressure, which circulates and eddies in this protected or cavity zone, as shown in Figure 3-5NN. The building thus may aid or impede chimney flow by using the wind to create a zone of negative or positive pressure at the chimney outlet.

Spark Arresters on Chimneys

Spark arresters are required on refuse burners or incinerators, but are desirable for all solid fuels. If sparks are expected from a fireplace or solid-fuel appliance, a chimney of any kind should have a spark screen and/or the roof surface material should be noncombustible. The usual recommendation for an arrester is for a wire mesh cloth or expanded metal having approximately $1/2$-in. (13-mm) square openings. Smaller openings clog rapidly, while larger ones could allow passage of some sparks. Stainless-steel material is recommended, with galvanized steel hardware cloth a poor second choice because the coating soon burns off and frequent replacement is necessary. Factory-built chimney manufacturers offer a variety of chimney cap options with integral and accessory spark-arrester screens.

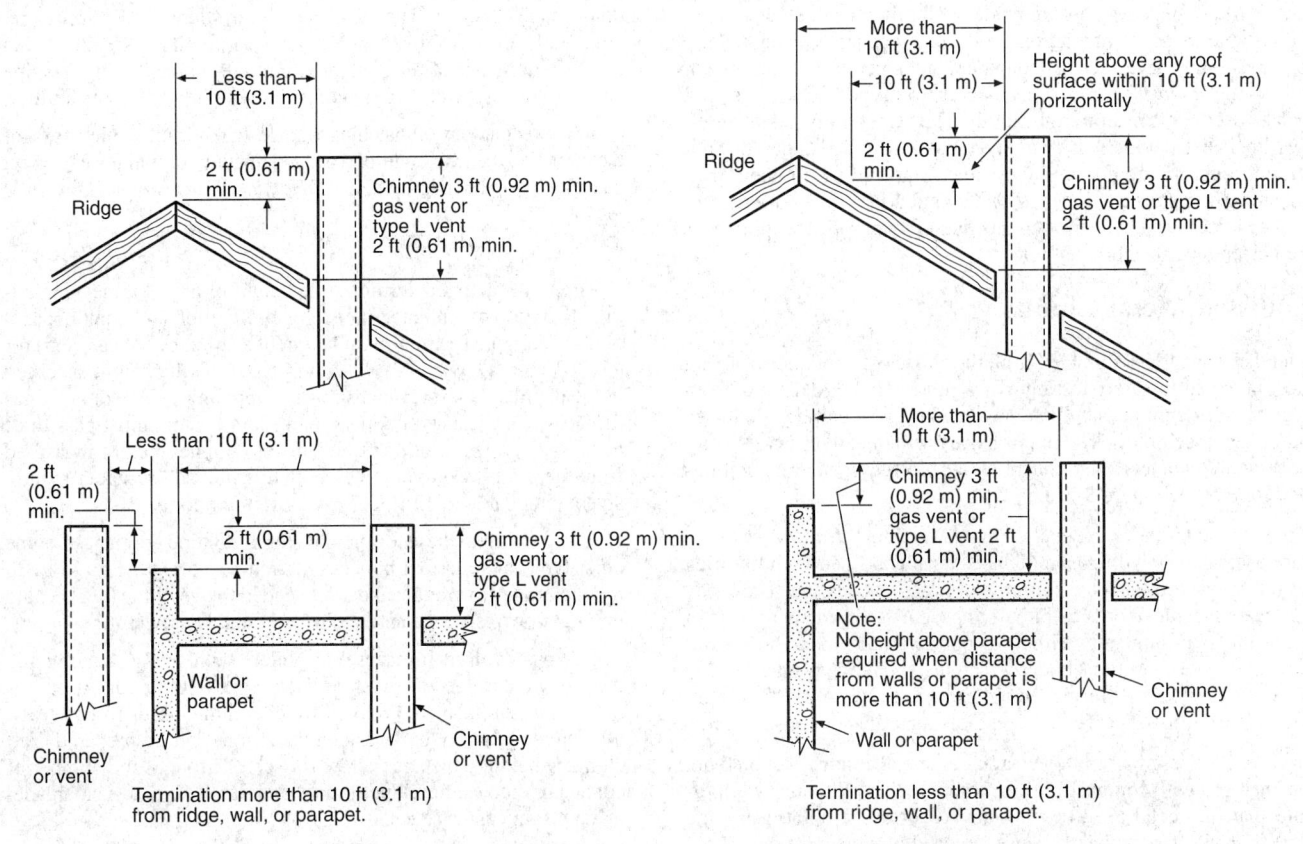

FIG. 3-5MM. *Termination of vents and chimneys above buildings.*

Selection of Chimneys

As an aid to selecting chimneys, heat-producing appliances have been graded by temperatures developed in the heating media or material being heated, and also by size. The grades of low-, medium-, and high-heat appliances can then be used to choose the chimney. (See Table 3-5N.) This selection method provides help if the appliance outlet gas temperature is consistent with its grade; however, a study of several hundred types of process appliances, ovens, and furnaces has indicated the precepts of the grading system do not always hold true. With the availability of modern temperature instrumentation and controls and the need to conserve energy, the outlet flue gas temperature and other conditions can and should be known more precisely than heretofore for the vast majority of fuel-burning equipment. This information, which should be readily obtainable from equipment and appliance manufacturers, can offer a logical basis for chimney selection and for determining safe conditions of installation and use.

As an example of a grading inconsistency, a steam boiler operating at more than 50 psig (350 kPa) is classified in some codes as a medium-heat appliance. In selecting a chimney, it could be assumed that the outlet temperature was as high as 1,500 to 1,800°F (815 to 980°C). Modern multiple-pass boilers, however, operate at controlled outlet gas temperatures only 100 to 200°F (38 to 93°C) above steam temperature. Thus, the correct chimney for even a 150 psig (1,000 kPa) steam boiler would be selected for an outlet gas temperature of 365.9°F +200°F, or 565.9°F (185°C +93°C, or 278°C). This is well below the 600°F (316°C), which is one minimum temperature at which the medium heat grade begins. Thus, in this instance, neither the grade of appliance based on temperature of the process [2,400°F (1,300°C) gas in the flame] or material

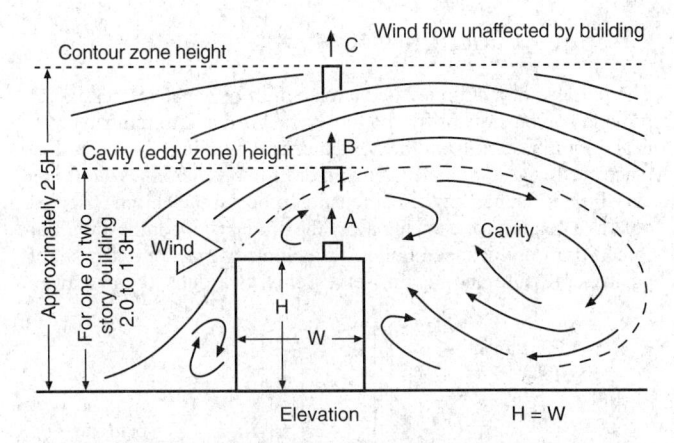

FIG. 3-5NN. *Eddy and contour zones due to airflow for one- or two-story buildings and their effect on chimney gas discharge.*

[365.9°F (185°C) steam] provides a conclusive basis for selecting the boiler's chimney.

Chimney Selection Checklist

Selection of the correct chimney may be aided by the following two checklists. (In those areas with adopted building codes, the code requirements take precedence over the generalities in the checklists; however, if special engineering is needed for a particular application, this approach should generally also meet the intent of any modern performance code.)

TABLE 3-5N. *Chimney Selection Chart*

Chimney Type (Construction Requirements)				
1. Factory Built—Residential Type and Building Heating Appliance	1. Factory Built—Residential Type and Building Heating Appliance	1. Factory Built—1,400°F (760°C)	1. Factory Built—Medium Heat Appliance	
2. Masonry, Residential Type	2. Masonry, Low Heat Type	2. Masonry, Low-Heat Type	2. Masonry, Medium-Heat Type	1. Masonry, High-Heat Type
	3. Metal, Low-Heat Type*	3. Metal, Low-Heat Type*	3. Metal, Medium-Heat Type	2. Metal, High-Heat Type

Maximum Continuous Appliance Outlet Flue Gas Temperature				
1000°F (538°C)	1000°F (538°C)	1400°F (760°C)	1800°F (982°C)	Over 1800°F (982°C)

Types of Appliances to Be Used with Each Type Chimney†

Column I	Column II	Column III	Column IV	Column V
A. Residential-type appliances, such as: 1. Ranges 2. Warm-air furnaces 3. Water heaters 4. Hot-water heating boilers 5. Low-pressure steam heating boilers 6. Incinerators 7. Floor furnaces 8. Wall furnaces 9. Room heaters 10. Fireplace stoves 11. Fireplace stove-room heater B. Fireplaces: 1. Factory built 2. Masonry 3. Freestanding (see fireplace stove)	A. All appliances shown in Column I B. Nonresidential-type building heating appliances for heating a total volume of space exceeding 25,000 cu ft (708 m³) C. Steam boilers operating at not over 1000°F (538°C) flue gas temperature; pressing machine boilers	All appliances shown in Columns I and II, and appliances such as: 1. Class A ovens or furnaces operating at temperatures below 1400°F (760°C) as defined in NFPA 86‡ 2. Annealing baths for hard glass (fats, paraffin, salts, or metals) 3. Bake ovens (in bakeries) 4. Candy furnaces 5. Core ovens 6. Feed drying ovens 7. Forge furnaces (solid fuel) 8. Gypsum kilns 9. Hardening furnaces (below dark red) 10. Lead melting furnaces 11. Nickel plate (drying) furnaces 12. Paraffin furnaces 13. Restaurant-type cooking appliances using solid or liquid fuel 14. Sulphur furnaces 15. Tripoli kilns (clay, coke, and gypsum) 16. Wood-drying furnaces 17. Zinc amalgamating furnaces	All appliances shown in Columns I, II, and III and appliances such as: 1. Alabaster gypsum 2. Annealing furnaces (glass or metal) 3. Charcoal furnaces 4. Cold stirring furnaces 5. Feed dryers (direct-fire heated) 6. Fertilizer dryers (direct-fire heated) 7. Galvanizing furnaces 8. Gas producers 9. Hardening furnaces (cherry to pale red) 10. Incinerators, commercial and industrial 11. Lehrs and glory 12. Lime kilns 13. Linseed oil boiling 14. Porcelain biscuit kilns 15. Pulp dryers (direct-fire heated) 16. Steam boilers operating at over 1000°F (538°C) flue gas temperature 17. Water-glass kilns 18. Wood-distilling furnaces 19. Wood-gas retorts	All appliances shown in Columns I, II, III, and IV and appliances such as: 1. Bessemer retorts 2. Billet and bloom furnaces 3. Blast furnaces 4. Bone calcining furnaces 5. Brass furnaces 6. Carbon point furnaces 7. Cement brick and tile kilns 8. Ceramic kilns 9. Coal and water gas retorts 10. Cupolas 11. Earthenware kilns 12. Glass blow furnaces 13. Glass furnaces (smelting) 14. Glass kilns 15. Open-hearth furnaces 16. Ore-roasting furnaces 17. Porcelain baking and glazing kilns 18. Pot-arches 19. Puddling furnaces 20. Regenerative furnaces 21. Reverberatory furnaces 22. Vitreous enameling ovens (ferrous metals)

*Single-wall metal chimneys or unlisted metal chimneys must not be used inside one- and two-family dwellings.

†For appliance types not listed in Columns I through V, the appropriate chimney must be selected on the basis of the appliance outlet flue gas temperature when the appliance is fired at its normal maximum input, and type of surroundings.

‡NFPA 86, *Standard for Ovens and Furnaces*.

Code Factors:

Applicable building code
Type or grade of appliance
Appliance manufacturers' instructions
Chimney manufacturers' instructions
Listed, certified appliance
Unlisted appliance
Earthquake zone—for masonry types
Air quality and external discharge factors
Type, occupancy, and use of building
Availability of fire protection service

Engineering Factors:

Temperature of gases from appliance
Characteristics of fuel and combustion gases
Corrosives, particulates, deposits, dewpoint
Quantity of combustion products
Operating needs and cycle of appliance
Method of draft control
Location of chimney in or outside of structure
Materials adjacent to chimney
Space available for chimney
Length and insulating ability of chimney

Hazards of Chimneys

For chimneys serving oil, wood, or coal appliances, the possibility of fire involves one or more of the following elements: (1) operator error or ignorance, (2) control failure, (3) improper installation, (4) use of a defective or unlisted appliance, (5) ignition of combustible soot or creosote deposits in the chimney, (6) serious cracks or internal collapse of the masonry, (7) defective construction, and (8) failure to secure joints of factory-built products.

Even masonry chimneys for gas appliances operating at very low temperatures have their problems. Condensation of flue product moisture may eventually disintegrate the mortar, leaving the chimney weak and easily toppled—a hazard to passersby.

In masonry chimneys, the following conditions may cause a fire in adjacent combustible materials:

1. Filling the space between the chimney and wood framing with insulation, or placing framing into or against the chimney wall. Note that a fill of even noncombustible insulation can be hazardous. Though the insulation itself may not ignite, it interferes with the dissipation of heat away from the chimney and framing. Temperatures may build to the point where framing separated from the chimney by insulation is exposed to potential ignition. All clearances from chimneys must be maintained as an air space, without a fill of any kind.
2. Using a restrictive chimney cap in combination with cracks or failed mortar in a chimney, which may cause excessive leakage of hot gas at roof level, thus endangering rafters and roofing.
3. Prolonged overfiring, which may cause excessive heating on adjacent walls. At temperatures of 500°F (260°C) on the brick exterior surface, 2-in. (51-mm) clearance to nearby combustibles is of little protective value, particularly if there is little ventilation airflow.
4. Creosote or soot fires, insofar as masonry chimneys are susceptible to collecting creosote deposits. When thick or heavy enough, these deposits can burn long enough to cause outer chimney surface temperatures that can ignite the structure around them. Further, upward gas flow velocity may be sufficient to carry sparks up from the fire or burning creosote and to drop them on the roof. After a severe chimney fire, both the tile liner and bricks may have serious cracks due to thermal expansion.

Factory-built chimneys provide many opportunities for installation errors and shortcuts, which can lead to fire hazards. Among these are:

1. Installation of the chimney ceiling support above the ceiling framing. This necessitates passing the single-wall connector through ceiling framing at what could be dangerously close clearance; thus, the hazard might more properly be blamed on the connector.
2. Failure to secure joints or align gas-carrying parts. Many designs of factory-built chimneys allow very little tolerance for misalignment. Failure to make sure that the inner pipes of a multi-wall chimney are aligned and connected can allow flames to enter external passages reserved for cooling airflow.
3. Use of mismatched chimney parts. A given design or brand of chimney is intended to be assembled only with identical parts. There is no interchangeability and it is extremely bad practice to mix multi-wall and insulation-filled chimney sections.
4. Failure to maintain proper air space clearances. As with masonry chimneys, close proximity of framing, or filling of the air space with insulation, can cause a fire hazard. Factory-built chimneys are listed for specific clearances, which are shown in the instructions and on labels. When properly used, the required manufacturer's support and spacing parts will enforce proper clearance.

The foregoing installation errors may cause fires at relatively low flue gas temperature; however, even a correctly assembled and installed chimney lacks the capability of repeatedly acting as a combustion chamber.

Chimney or Creosote Fires

The term *creosote* as applied to chimney systems refers to the tarry brown or hard black internal deposits produced by burning wood. Combustion of wood produces varying amounts of complex hydrocarbons that can condense and accumulate inside connector and chimney surfaces. These combustible deposits ignite when temperature is sufficiently high and will burn at temperatures in the range of 1,200 to 2,000°F (650 to 1,100°C). The temperature and duration of the fire both depend upon the accumulated thickness of creosote, chemical composition of the deposit, availability of oxygen, type of heating appliance, type of chimney, etc.

In masonry chimneys, intense fires can cause cracking of tiles and brick masonry, as well as excessive heat transfer through chimney walls and hazardous overheated external surface temperatures. Clay flue liners are quite resistant to high temperatures, when applied gradually. However, chimney fires typically cause a very sudden rise in internal chimney temperature. This results in thermal shock as the hot inner surface of the liner expands against the cooler, inflexible outer portions. Longitudinal cracks in one or more liner sections are a frequent result of chimney fires. Such cracks can appear closed when the chimney is cool, but can open up to measurable dimensions when the chimney is heated by subsequent use.

The likelihood of fire spread to the adjacent structure is related to the duration of the chimney fire. Because of their mass, masonry chimneys produce significant thermal inertia. A short, though intense, chimney fire is less likely to overheat the chimney structure than a longer-lasting fire. It is therefore important that, after initial steps are taken to control a chimney fire, the fire department be called to make sure that it is completely extinguished.

In factory-built chimneys, as with masonry, the effect of chimney fires varies depending upon attained temperature. Factory-built chimneys are tested with repeated application of flue gas temperatures up to 2,100°F (1,149°C) for periods of up to 30 min. They may, therefore, be expected to survive a chimney fire in usable condition. This, however, should never be simply assumed; chimney fires can exceed the test temperatures, and differential heating of flue surfaces can create unmanageable stresses in the metal. It is not uncommon for the liner of a factory-built chimney to exhibit buckling after a fire. If the buckling is minor, such that flue gases or other combustion products cannot pass beyond the liner, the chimney can continue to be used. If buckling is more significant, the affected section, at minimum, should be replaced. In general, if the outer chimney is not discolored and remains shiny, it is unlikely that any building structural damage has occurred, since shiny surfaces are unlikely to radiate damaging amounts of heat if the chimney is installed at correct clearance.

If it is suspected that a chimney fire has occurred, both masonry and factory-built chimneys should be thoroughly inspected. The flue surface, in particular, should be closely examined for cracks, holes, missing sections, or distortion. Unless the chimney is very short, such inspections should be done using closed-circuit video inspection equipment, which is the only practical way to view the inaccessible flue surface. The outer chimney surface should be examined as well, for cracks, smoke leaks, or major discoloration. Combustibles surrounding the chimney should be checked for evidence of charring and any insulation removed from required air space clearances. Damaged chimney caps and other appurtenances should be replaced. The chimney should be placed back in service only after any necessary repairs have been made.

Control of chimney fires: In wood-burning stoves and heaters having air dampers and doors, reducing the air supply by closing them is the best option; then the combustion rate in the chimney depends upon its air supply. When other appliances are connected to the chimney or a barometric damper is used for draft control, these can supply air and must also be closed to control a chimney fire. Flare-type chimney fire extinguishers (similar to automotive safety flares) are available that, when activated and placed in the appliance, release a cloud of fire extinguishing agent into the chimney. For the agent to be effective, it must be retained in the chimney. Therefore, the stove should be closed to prevent the agent from being drawn quickly up and out of the chimney. Open-front fireplaces or fireplace stoves without doors may not accumulate creosote rapidly, but under fire conditions air control is difficult. The best policy with fireplaces is routine chimney inspection and mechanical cleaning (brushes and scraping).

Because of the possibility of fire spread to the building structure, occupants should, at minimum, be gathered and prepared to leave the building. The fire department should be immediately called, both as a precaution against fire spread and to ensure that the chimney fire is completely extinguished.

Maintenance and Inspection of Chimneys

In order to remain safe and effective, chimneys need periodic inspection and maintenance. NFPA 211 requires all chimneys (regardless of the fuel utilized in connected appliances) to be inspected at least once a year. Such inspections should include, at minimum, examination of the inner flue surface for the presence of combustible deposits and freedom from cracks, gaps, softening, or other damage or deterioration. The outer wall should be examined for missing bricks or mortar and evidence of leakage. Although clearances to the building structure are usually determined at the time of construction, inspection should verify that no insulation or loose combustible material has fallen into the required air space.

For chimneys serving solid-fuel equipment, cleaning of creosote and other combustible deposits from the flue is the most frequent maintenance activity. A common rule of thumb suggests that a buildup of $1/4$ in. (6.4 mm) in thickness indicates a need for cleaning. This may be most applicable to soft, powdery, or crusty deposits. Deposits of smooth, gummy, or hard tar-glaze, on the other hand, may contain significantly more potential energy, and can fuel a more intense and longer-lasting fire than less dense material. Therefore, when any thickness of tarry deposits is detected, immediate cleaning is in order.

Cleaning of loose deposits is usually done with a stiff wire brush, fitted to the flue dimensions and scrubbed up and down the chimney with rods. More resistant deposits may require the use of a flat wire brush or specialized manual scrapers. There are various mechanical devices available that use rotating chains or brush heads to flail and chip hard creosote deposits on the flue walls. In extreme cases, there are chemical treatments that can soften or loosen deposits. Mechanical and chemical cleaning methods should be used only by trained professional chimney service specialists.

NFPA 211 also includes requirements for corrective action when inspection shows that the chimney is no longer suitable for the intended application. Any damage or deterioration, from whatever cause, that impairs the ability of the chimney to conduct the prod-

ucts of combustion directly to the outdoors, while protecting the structure and occupants from heat, leakage, or spillage, is cause for repair or replacement of the chimney. Specifically, with respect to flue lining, NFPA 211 requires repair, replacement, or relining if the lining has "softened, cracked, or otherwise deteriorated such that it no longer has the continued ability to contain the products of combustion, i.e., heat moisture and flue gases."

FIREPLACES AND FIREPLACE STOVES

Fireplaces of masonry or factory-built construction are primarily open fire-chamber appliances with no controls on the air supply to the fire. These require a properly sized chimney to carry out the smoke and combustion gases. Fireplace stoves are free-standing, factory-built appliances operating with an open front, but, if these have doors, they may also operate as stoves or heaters with controlled combustion air. The addition of glass doors to a masonry fireplace, or the installation of a closed-front fireplace insert, will transform its combustion characteristics to that of a heater.

Special techniques are required for connection of a heating appliance to a fireplace. The appliance must be directly connected to masonry fireplace flues with a suitable connector extending through the damper and smoke chamber. Appliances must not be inserted in, or connected to, factory-built fireplaces unless the appliance is listed for that fireplace, and then only in strict accordance with manufacturer's instructions.

Masonry Fireplaces

Field-constructed masonry fireplaces must conform to local building codes or provisions of applicable standards (e.g., NFPA 211) to ensure safe operation. They may be served by a masonry residential chimney or may use a factory-built chimney with a suitable factory-furnished transition between smoke chamber and chimney. Prefabricated designs of masonry fireplaces and masonry fireplace chimney combinations are available in some localities. When a lining of low-duty firebrick is provided, masonry fireplaces require at least 8 in. (203 mm) of brick thickness or equivalent for the back and sides, with 2-in. (51-mm) clearance to combustibles at the sides and 4-in. (102-mm) minimum clearance from the back. When the lining described above is not used, the thickness of the masonry at the back and sides should not be less than 12 in. (305 mm). The entire fireplace must be supported on a noncombustible footing, with structural floor framing well away from direct heat conduction from the fire zone or external hearth. (See Figure 3-5OO.)

Hearth extensions of approved noncombustible material must be provided for all fireplaces. The hearths must extend at least 16 in. (406 mm) in front and 8 in. (203 mm) beyond the sides of fireplaces with an opening of less than 6 sq ft (0.56 m²). If the opening is 6 sq ft (0.56 m²) or larger, the hearth must extend at least 20 in. (508 mm) to the front and 12 in. (305 mm) to the sides.

Masonry fireplaces may be built around steel fireplace units or heat circulating forms incorporating an air chamber. A total thickness at back and sides of not less than 8 in. (203 mm) must be provided, of which not less than 4 in. (102 mm) is solid masonry. The same clearances to surrounding combustibles apply to fireplaces with circulators.

Factory-Built Fireplaces

Factory-built fireplaces are mainly of metal construction using multiple air spaces, refractory hearths and liners, and insulation to obtain the required level of safety. Factory-built fireplaces are tested and listed to ANSI/UL 127, *Standard for Safety Factory-Built*

Fireplaces. Each fireplace must be installed with the specific chimney and other parts called for by the manufacturer's instructions and the terms of its listing.

Combustion Air Inlets

Some fireplaces incorporate a duct to bring combustion air directly to the fire chamber from the outdoors. This theoretically reduces the consumption of room air and minimizes the possibility of excessive depressurization of the house. Because such ducts communicate with the combustion zone, care must be taken to ensure that they do not create a hazard. Building fires have been reported where combustion products, such as embers, have entered the duct in proximity to combustible construction.

NFPA 211 does not allow combustion air ducts to terminate within the fire chamber of a fireplace unless the device is listed for such an application. Otherwise, the duct must terminate outside the fireplace; if within 6 in. (152 mm) of the opening, they must be equipped with a guard to prevent the entry of embers, ashes, and flame, etc. The duct itself must be made of 26-gage galvanized steel or equivalent, and must maintain a 1-in. (25.4-mm) clearance to combustibles.

Factory-built fireplaces may use combustion air ducts that are incorporated into the original design of the listed fireplace, and, if optional, must be a listed accessory to the particular fireplace. They will be accompanied by specific installation instructions and must be installed accordingly.

No combustion air duct should draw air from an attic, basement, garage, or other interior space, or an area where combustible vapors might be present. Any duct that connects directly to the fire chamber should not extend vertically above the fireplace, to avoid the possibility of the duct acting like a chimney and drawing in combustion products.

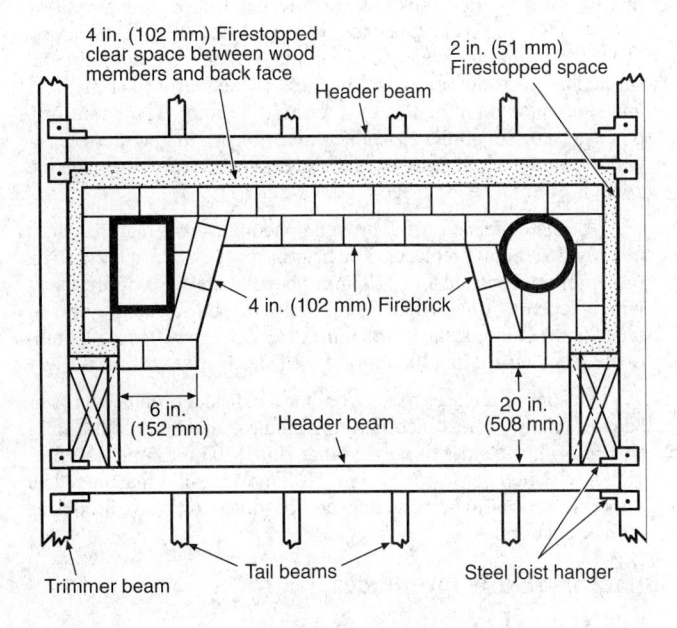

FIG. 3-5OO. *Floor framing around a fireplace, when the fireplace opening is 6 sq ft (0.56 m²) or larger.*

Chimney Sizes for Fireplaces

For masonry fireplaces, the correct size and height of chimney will ensure that all smoke, fire, and gas goes up the chimney. Although imprecise, there are several rules of thumb for sizing fireplace flues. For very short chimneys [less than about 15 ft (4.6 m)], the actual cross-sectional area of the flue should be at least one-eighth of the area of the frontal opening of the fireplace. For chimneys of normal height, rectangular flues should be at least one-tenth of the fireplace opening; round flues should be at least one-twelfth of the opening. The nominal size of tile liner must be carefully checked to ensure that the internal cross-section area is adequate. In addition, the area of opening through the damper throat must be approximately twice the required flue area or greater, or it may reduce flow and possibly cause smoking.

For factory-built fireplaces, the chimney and, therefore, the flue size, is an integral part of the appliance design. Only the chimney specified or supplied by the fireplace manufacturer should be used. There are also factory-built fireplaces made of masonry material that are assembled on site from pre-cast components. Some of these come with an integral chimney, while others specify a particular size chimney of conventional construction. Some of these fireplaces utilize advanced principles of airflow through the fireplace and can use unconventionally small flue sizes; ratios of fireplace opening area to flue area of 20:1 are possible with some designs.

Charts are available for factory-built and masonry chimneys in the ASHRAE *Handbook and Product Directory* and in manufacturers' literature that indicate the correct chimney size and height in relation to frontal opening, based on an average room air velocity of not less than 0.8 ft per sec (0.24 m/s) into the opening. This minimum velocity rule is valid for a wide variety of open fire chamber combustion systems, including masonry fireplaces as well as combination room heaters and fireplace stoves.

Factory-Built Fireplace Stoves

These free-standing units are usually of metal or combination metal and refractory construction and may require a specific type of chimney and special connectors, or may be connected to any acceptable chimney by conventional means. Fireplace stoves are tested and listed (see ANSI/UL 737, *Standard for Safety Fireplace Stoves*), but if the stove has doors it becomes a combination fireplace and room heater, and to be listed for operation with doors closed it must also comply with ANSI/UL 1482, *Solid Fuel Type Room Heaters.*

Investigation of many combination fireplace stove heater designs has demonstrated the validity of the geometrical relationship between heater size and required clearance. Simply stated, larger units may require greater than standard clearance. To be sure of correct installation, the manufacturer's installation instructions for listed stoves and heaters should be scrupulously observed. These instructions have been reviewed for consistency with test results and are as essential to the safety of an installation as are the design features of the stove.

While untested fireplace stoves and related heaters can be installed and operated safely, they are seldom accompanied by carefully written instructions. Many imitations of listed designs can be found that are sold through retail outlets having little interest in the final installation. The lack of clear instructions and warnings poses many hazards to the consumer.

Hazards of Fireplaces and Their Chimneys

Open fireplaces, transferring heat primarily by radiation from the fire, can create several kinds of hazards:

1. Ignition by radiation of furnishings or walls located too close to the fire.

2. Ignition of furnishings or carpeting by sparks or from logs rolling out of the fire. For large, high fireplaces having a high fuel capacity, the 20-in. (508-mm) hearth width requirement may be grossly inadequate.

3. Ignition of combustible structural materials by excessive heat conducted through hearth or fireplace walls, which can make prolonged overfiring of a masonry fireplace dangerous.

4. Escape of gases from the opening into a construction gap between the fireplace shell and the front facing wall. This is a common installation fault with factory-built fireplaces, since a smoking unit—one with insufficient chimney flow—allows very hot gas to escape into this area.

5. Sparks and embers falling onto a combustible floor between the fireplace and its hearth extension. This can be prevented by placing the masonry or factory-furnished hearth extension flush against the front of the fireplace, or by using a sheet of metal at the floor surface under the joint between the fireplace and its hearth.

6. Burning highly combustible solid materials, such as a dried Christmas tree and wrappings. The high gas temperatures resulting may ignite creosote deposits, or cause damage to the masonry. The high rate of burning may cause the flames, heat, and smoke to billow out of the fireplace opening, possibly igniting nearby combustibles. Regardless of how flammable these materials are, burning them a little at a time obviously will cause no problem.

7. Use of any flammable liquid to start or rekindle a fire, or in any room with or near an open fire.

8. Failure to clean the chimney whenever creosote deposits build up to dangerous thickness. A layer of creosote $1/16$ in. (1.6 mm) thick in an 8-in. (200-mm) size chimney 15 ft (4.6 m) high contains as much as 122,000 Btu (128 MJ) of fuel. If this burns off in 15 min, the resulting temperature may reach 1,700°F (930°C) at a heat release rate as high as 488,000 Btu per hr (140 kW).

9. Use of an incorrect or mismatched chimney. For safe operation, many types of factory-built fireplaces require a chimney that interconnects with its air passages. Failure to connect the fireplace to this specific chimney could be extremely dangerous and clearly violates installation requirements and the listing.

10. Failure to install firestopping around chimneys, which might be termed an indirect installation hazard. This will not start a fire, but, if one starts elsewhere, the lack of firestopping may cause severe losses. Because of the numerous possibilities for operator or construction errors in the vicinity of the fireplace itself, every precaution should be taken to prevent rapid involvement of the entire structure in case of fire. Many modern single and multiple residences of frame construction utilize a chase or vertical enclosure, inside or external to the dwelling, containing one or more fireplace chimneys. The absence of floors or other internal framing makes it easy to install all the chimneys, but also provides an ideal passage for rapid vertical spread of flame to upper levels and the roof. The shape and size of firestopping for this sort of installation cannot readily be anticipated by the fireplace manufacturer, but carefully fitted sheet metal or lath and plaster may limit the spread and allow time for effective fire fighting.

BIBLIOGRAPHY

References Cited

1. UL 1482, *Standard for Safety Room Heaters, Solid-Fuel Type*, Underwriters Laboratories Inc., Northbrook, IL, 1994.
2. Schaffer, E. L., "Smoldering Initiation in Cellulosics Under Prolonged Heating," *Fire Technology*, Vol. 16, No. 1, Feb. 1980, pp. 22–28.
3. Mitchell, N. D., "New Light on Self-Ignition," *NFPA Quarterly*, Vol. 45, No. 2, Oct. 1951, pp. 165–172.
4. MacLean, J. D., *"Effect of Heat on Properties and Serviceability of Wood: Experiments on Thin Wood Specimens,"* Report No. R1471, 1945, Forest Products Laboratory, Madison, WI.
5. MacLean, J. D., *Rate of Disintegration of Wood Under Different Heating Conditions*, American Wood-Preservers Association, 1951.
6. McGuire, J. H., "Limiting Safe Temperature of Combustible Materials," *Fire Technology*, Vol. 5, No. 3, August 1969.
7. "Ignition and Charring Temperatures of Wood," Report No. 1464, 1958, Forest Products Laboratory, Madison, WI.
8. Shelton, J. W., *Wood Heat Safety*, Garden Way Publishing, Charlotte, VT, 1979.
9. Matson, A. F., Dufori, R. E., and Breen, J. F., "Performance of Type B Gas Vents for Gas-Fired Appliances, Part II, Survey of Available Information on Ignition of Wood Exposed to Moderately Elevated Temperatures," *UL Bulletin of Research*, No. 51, May 1959, Underwriters Laboratories Inc., Northbrook, IL.

References

ANSI/UL 103, *Standard for Safety Chimneys, Residential Type and Building Heating Appliance Factory-Built*, Underwriters Laboratories Inc., Northbrook, IL, 1994.
ANSI/UL 127, *Standard for Safety Factory-Built Fireplaces*, Underwriters Laboratories Inc., Northbrook, IL., 1992.
ANSI/UL 441, *Standard for Gas Vents*, Underwriters Laboratories Inc., Northbrook, IL, 1994.
ANSI/UL 641, *Standard for Safety Low-Temperature Venting Systems, Type L*, Underwriters Laboratories Inc., Northbrook, IL, 1994.
ANSI/UL 737, *Standard for Safety Fireplace Stoves*, Underwriters Laboratories Inc., Northbrook, IL, 1991.
ANSI/UL 959, *Standard for Safety Medium-Heat Appliances Factory-Built Chimneys*, Underwriters Laboratories Inc., Northbrook, IL, 1994.
ANSI/UL 1738, *Standard for Safety Venting Systems for Gas-Burning Appliances, Categories II, III, and IV*, Underwriters Laboratories Inc., Northbrook, IL, 1990.
ANSI/UL 1777, *Standard for Chimney Liners*, Underwriters Laboratories Inc., Northbrook, IL, 1995.
ASME Boiler and Pressure Vessel Code, American Society of Mechanical Engineers, New York, 1995.
ASTM Standards on Gaseous Fuels, (Parts 23, 24, 25, and 26), American Society for Testing and Materials, W. Conshohocken, PA, 1988.
ASTM C315, *Standard Specification for Clay Flue Linings*, American Society for Testing and Materials, W. Conshohocken, PA, 1991.
ASTM D396, *Standard Specification for Fuel Oils, American Society for Testing and Materials*, W. Conshohocken, PA, 1992.
1996 ASHRAE Handbook—HVAC Systems and Equipment, American Society of Heating, Refrigerating and Air Conditioning Engineers, Atlanta, GA, 1996.
Combustion Handbook, North American Mfg. Co., Cleveland, OH, 1978.
UL 181, *Standard for Safety Factory-Made Air Ducts and Air Connectors*, Underwriters Laboratories Inc., Northbrook, IL, 1994.
UL 441, *Standard for Safety Gas Vents*, Underwriters Laboratories Inc., Northbrook, IL, 1994.
UL 1738, *Standard for Safety Venting Systems for Gas-Burning Appliances*, Underwriters Laboratories Inc., Northbrook, IL, 1993.
UL 1777, *Standard for Safety Chimney Liners*, Underwriters Laboratories Inc., Northbrook, IL, 1992.
UL Gas and Oil Equipment Directory, Underwriters Laboratories Inc., Northbrook, IL, 1994.

NFPA Codes, Standards, and Recommended Practices

Reference to the following NFPA codes, standards, and recommended practices will provide further information on the heating systems and appliances discussed in this chapter. (See the latest version of *The NFPA Catalog* for availability of current editions of the following documents.)

NFPA 30, *Flammable and Combustible Liquids Code*
NFPA 31, *Standard for the Installation of Oil-Burning Equipment*
NFPA 54, *National Fuel Gas Code*
NFPA 58, *Standard for the Storage and Handling of Liquefied Petroleum Gases*
NFPA 70, *National Electrical Code®*
NFPA 86, *Standard for Ovens and Furnaces*
NFPA 86C, *Standard for Industrial Furnaces Using a Special Processing Atmosphere*
NFPA 90A, *Standard for the Installation of Air Conditioning and Ventilating Systems*
NFPA 90B, *Standard for the Installation of Warm Air Heating and Air Conditioning Systems*
NFPA 96, *Standard for Ventilation Control and Fire Protection of Commercial Cooking Operations*
NFPA 97, *Standard Glossary of Terms Relating to Chimneys, Vents, and Heat-Producing Appliances*
NFPA 211, *Standard for Chimneys, Fireplaces, Vents, and Solid Fuel-Burning Appliances*
NFPA 255, *Standard Method of Test of Surface Burning Characteristics of Building Materials*
NFPA 8501, *Standard for Single Burner Boiler Operation*
NFPA 8502, *Standard for the Prevention of Furnace Explosions/Implosions in Multiple Burner Boilers*
NFPA 8503, *Standard for Pulverized Fuel Systems*
NFPA 8504, *Standard on Atmospheric Fluidized-Bed Boiler Operation*
NFPA 8505, *Recommended Practice for Stoker Operation*
NFPA 8506, *Standard on Heat Recovery Steam Generator Systems*

Additional Reading

AGA Directory of Certified Appliances and Accessories, American Gas Association, Cleveland, OH, 1993.
Alternative Heater Fires: A Critical Review of Safety Issues, Federal Emergency and Management Agency, Emmitsburg, MD, 1988.
ANSI Z83.3, *Gas Utilization Equipment in Large Boilers*, American Gas Association Laboratories, Cleveland, OH, 1989.
Code for the Installation of Heat-Producing Appliances, American Insurance Association, NY.
Corry, B., "Understanding and Investigating Chimney Fires," *Fire and Arson Investigator*, Vol. 42, No. 2, Mar. 1992, pp. 21–23.
Domanski, P. A., *et al.*, "Simulation Model and Study of Hydrocarbon Refrigerants for Residential Heat Pump Systems," *Proceedings* of a Conference on New Applications of Natural Working Fluids in Refrigeration and Air Conditioning, May 10-13, 1994, Hanover, Germany, 1994.
Fang, J. B., "Quantification of Heat Losses Through Structural Supports for Shallow Trench Heat Distribution Systems," National Institute of Standards and Technology, Gaithersburg, MD, NISTIR 89-4134, Oct. 1989, 107 pages.
Fang, J. B., "Thermal Analysis of Directly Buried Conduit Heat Distribution Systems," National Institute of Standards and Technology, Gaithersburg, MD, NISTIR 4365, Aug. 1990, 96 pages.
Fanney, A. H., "Field Monitoring of a Variable-Speed Integrated Heat Pump/Water Heating Appliance," National Institute of Standards and Technology, Gaithersburg, MD, NIST BSS 171, R&D Project RP 89-60, June 1993, 59 pages.
Fanney, A. H., "Field Monitoring of a Variable-Speed Integrated Heat Pump/Water-Heating Appliance," *ASHRAE Transactions*, Vol. 101, No. 2, 1995, pp. 1–15.
Galloway, F. M. and Hirschler, M. M., "Experiments for Hydrogen Chloride Transport and Decay in a Simulated Heating, Ventilating and Air Conditioning System and Comparison of the Results With Predictions From a Theoretical Model," *Journal of Fire Sciences*, Vol. 9, No. 4, July/Aug. 1991, pp. 259–275.
Galloway, F. M., and Hirschler, M. M., "Hydrogen Chloride Transport and Decay in a Simulated Heating, Ventilating and Air conditioning System," Sixteenth International Conference on Fire Safety, Volume 16, Product Safety Corp., Sunnyvale, CA, 1991, 40–53 pp.
Hall, J. R., "Update on the Auxiliary Heating Fire Problem," *Fire Journal*, Vol. 80, No. 2, 1986, p. 52.
Harris, R.J., *Gas Explosions in Buildings and Heating Plant*, British Gas Corporation, E & F N Spon, Ltd., London, England, 1983.
Johnston, D., *et al.*, "Chimney Fires: Causes, Effects & Evaluation," Chimney Safety Institute of America, Gaithersburg, MD, 1992.
Lemoff, T. C., ed., *National Fuel Gas Code Handbook*, 2nd ed., National Fire Protection Association., Quincy, MA, 1992.
Lentini, J. J., "Gasoline and Kerosene Don't Mix—At Least Not in Kerosene Heaters," *Fire Journal*, Vol. 83, No. 4, 1989, p. 13.
Liu, S. T., Kelly, G. E. and Terlizzi, C. P., "Experimental Study on the Performance of a Combination Appliance for Domestic Hot Water and Space Heating," Department of Energy, Washington, DC, NISTIR 4356, Aug. 1990, 32 pages.
Liu, S. T., Kelly, G. E. and Terlizzi, C. P., "Performance Testing of a Family of Type I Combination Appliances," Department of Energy, Washington, DC NISTIR 5626, Apr. 1995, 30 pages.
Miller, A., "U.S. Home Product Report, 1985-1989 (Appliances and Equipment)," National Fire Protection Association, Quincy, MA, Home Product Report, Mar. 1992, 89 pages.
Ohlemiller, T., and Sharb, W., "Products of Wood Smolder and Their Relation to Wood Burning Stoves," NBSIR 88-3767, May 1988, National Bureau of Standards, Gaithersburg, MD.
Parsons, R. A., *ASHRAE Handbook—Fundamentals*, American Society of Heating, Refrigerating, and Air-Conditioning Engineers, Atlanta, GA, 1989.
Paul, D. D., *et al.*, "Venting Guidelines for Category I Gas Appliances with Fan-Assisted Combustion Systems," GRI-89/0016, 1992, Gas Research Institute, Chicago, IL.
Peacock, R. D., "Chimney Fires: Intensity and Duration," *Fire Technology*, Vol. 22, No. 3, 1986, pp. 234-252.
Peacock, R. D., "Wood Heating Safety Research: An Update," *Fire Technology*, Vol. 23, No. 4, 1987, pp. 292-312.
"Pumps for Oil-Burning Appliances. Standard for Safety," Underwriters Laboratories, Inc., Northbrook, IL, UL 343, 7th Edition, Apr. 29, 1993, 28 pages.
Rayment, R. and Whittle, G. E., "Domestic Warm-Air Heating Systems Using Low-Grade Heat Sources," Fire Research Station, Borehamwood, England, BRE IP 01/89, Jan. 1989, 4 pages.
Shepherd, T. A., "Spillage of Flue Gases From Open-Flued Combustion Appliances," Building Research Establishment, Garston, England, BRE IP 21/92, Dec. 1992, 4 pages.
"Standard for Safety Tests for Flammability of Plastic Materials for Parts in Devices and Appliances," Underwriters Laboratories, Inc., Northbrook, IL, UL 94, 4th Edition, June 18, 1991, 34 pages.
Toy, D. A., "Take Those Hazards out of Your Gas Supply System," *Semiconductor International*, Vol. 12, No. 8, July 1989, pp. 66–70.
Treado, S. J., and Bean, J. W., "Interaction of Lighting, Heating and Cooling Systems in Buildings," National Institute of Standards and Technology, Gaithersburg, MD, NISTIR 4701, Mar. 1992, 115 pages.
UL 372, *Standard for Safety Primary Safety Controls for Gas- and Oil-Fired Appliances*, 5th ed., Underwriters Laboratories Inc., Northbrook, IL, 1994.
Ulley, L., "Chem Baggies for Chimneys," Fire *Fighting in Canada*, Vol. 39, No. 5, June 1995, p. 23.
Using Coal and Wood Stoves Safely, A Hazard Study, National Fire Protection Association, Quincy, MA, 1974.
Vickers, A. K., "Kitchen Ranges in Fabric Fires," NBS Tech Note—817, April 1974, National Bureau of Standards, Gaithersburg, MD.
Voigt, G. Q., "Fire Hazard of Domestic Heating Installations," NBS Research Paper RP596, Sept. 1933, National Bureau of Standards, Gaithersburg, MD.

"Voimalaitosten ja lampokeskusten sammutusjarjestelmien valinta helpott-uu. [Choice of Extinguishing Systems for Power and Heating Plants To Be Make Easier.]," *Palontorjuntatekniikka*, Vol. 1, 1994, p. 34.

Walton, W. D., "Fire Testing of Roof-Mounted Solar Collectors by ASTM E108," NBSIR 81-2344, August 1981, National Bureau of Standards, Gaithersburg, MD.

Walton, W. D., "Solar Collector Fire Incident Investigation," NBSIR 81-2326, August 1981, National Bureau of Standards, Gaithersburg, MD.

Welsh, P.A., "Testing the Performance of Terminals for Ventilation Systems, Chimneys and Flues," Building Research Establishment, Garston, England, BRE IP 05/95, Mar. 1995, 4 pages.

Wieder, M., "Fire Department Response to Chimney Fires," *Firehouse*, Vol. 21, No. 2, Feb. 1996, pp. 112, 114–116.

Wood Heating Seminar Proceedings, Documents 1-1977, 2-1977, 3-1978, 4-1979, 5-1979, Wood Energy Institute, Camden, ME.

BOILER-FURNACES

Revised by Shelton Ehrlich

More than any other single invention, the steam engines of New-comen and Watt led to the modern age. The primitive boilers that produced the steam for these early engines consisted of a furnace, a space for burning the fuel, and a separate boiler, with a large copper pot heated from beneath. Improvements in iron fabrication techniques made possible the integration of the furnace and boiler.

In fire-tube boilers the furnace is a large central horizontal tube surrounded by the water-filled boiler. The hot exhaust gases pass out of the furnace and then back through tubes that traverse the boiler. Failure to keep water in fire-tube boilers led to catastrophic boiler explosions, resulting in many deaths in the mid-19th century (and such explosions still occur occasionally). Inventors were kept busy trying to solve this problem, as well as automating the feeding of coal via the use of mechanical stokers. Another important "invention" of 1859 was the Steam Boiler Assurance Company in the U.K., the beginning of the boiler inspection and insurance industry.

Further, material improvements led to the development of the water-tube boiler. Here the water is in the tubes and the hot gases pass across the tubes. While individual water tubes could still burst, boiler safety was much improved. With the development of pulverized coal firing, water-tube boilers would grow to enormous size. The increase in size led to the possibility of catastrophic furnace explosions. Steady improvements in design, materials, instrumentation, operator training, and supervision [addressed by American Society of Mechanical Engineers (ASME), insurance, National Fire Protection Association (NFPA), Authorities Having Jurisdiction, etc.] have led to a great reduction in fire-tube boiler explosions and water-tube boiler furnace explosions.

Though this chapter discusses water-tube boiler furnaces, the safety principles outlined are also applicable to fire-tube boilers. Fuel-burning systems and related control equipment for single- and multiple-burner boiler furnaces are discussed here. Furnaces for fuels burned in suspension and in fluidized beds are included; other solid-fuel-burning units, such as stokers, are not.

Further, it is not the intent of this chapter to include startup, shutdown, or operating procedures or a detailed description of interlock and trip systems. This information is contained in the applicable NFPA standards listed at the end of this chapter.

Shelton Ehrlich, retired from Electric Power Research Institute, is now principal consultant of Ehrlich Associates, Palo Alto, CA, and a member of NFPA's technical committee on boiler combustion systems hazards.

THE COMBUSTION PROCESS

Combustion is the rapid combination of oxygen with combustible elements of a fuel. There are three important combustible elements: (1) carbon, (2) hydrogen, and (3) sulfur. When burned to completion with oxygen, these elements unite according to the following:

$$C + O_2 = CO_2 + 14,100 \text{ Btu/lb (or 32 797 kJ/kg) of carbon}$$

$$2H_2 + O_2 = 2H_2O + 61,100 \text{ Btu/lb (or 142 119 kJ/kg) of hydrogen}$$

$$S + O_2 = SO_2 + 3,980 \text{ Btu/lb (or 9 258 kJ/kg) of sulfur}$$

Waste gases rich in carbon monoxide (CO) are often burned in boilers and release 4,347 Btu/lb (10 111 kJ/kg) of CO.

To compete effectively, designers must carefully balance the need to provide for complete combustion against the capital cost of the boiler and the auxiliary power needs. The key factors, often referred to as the "three Ts" of combustion, are: (1) *time*, set by the size of the furnace; (2) *temperature*, set by the arrangement of heat-removal surfaces; and (3) *turbulence*, set by the skill of the designer in manipulating flows of air and fuel and the interaction between all the burners in the furnace. A fourth critical factor is *continuous,* i.e., the need to supply fuel, air, and ignition source on an uninterrupted basis.

Not only must the fuel and air be supplied continuously but the ratio of fuel-to-air must be kept constant within a relatively narrow range. The burner designer and the limits built into the controls provide that the ignition source (feedback from the fuel just burned) be available for the fuel just entering. When any of these factors go awry (e.g., the fuel is interrupted for an instant; the air flow suddenly increases, say because a blockage is removed; or the burner does not hold the flame close), then ignition can be lost and the controlled flammable mixture becomes an uncontrolled explosive mixture. A large boiler-furnace is always just a few seconds away from catastrophe. It is only through the skill of the many people responsible for ensuring that events are always correct that large furnace explosions are so remarkably rare.

Figure 3-6A shows how fuel and air are introduced and burned in a furnace through a modern low-NO$_x$ burner. Pulverized coal, carried by a small stream of air termed *primary air,* is injected through the center of the burner and meets with the swirling bulk of the air, termed *secondary air.* The several functions of this burner can be seen in Figure 3-6A: a zone just at the burner outlet in which fuel is ignited and partially burned, but with a strong deficiency of air (A); and a second zone in which thermal NO$_x$ is reduced to N$_2$ (C). The mixing and burning of the remaining air and residual fuel (the *char*

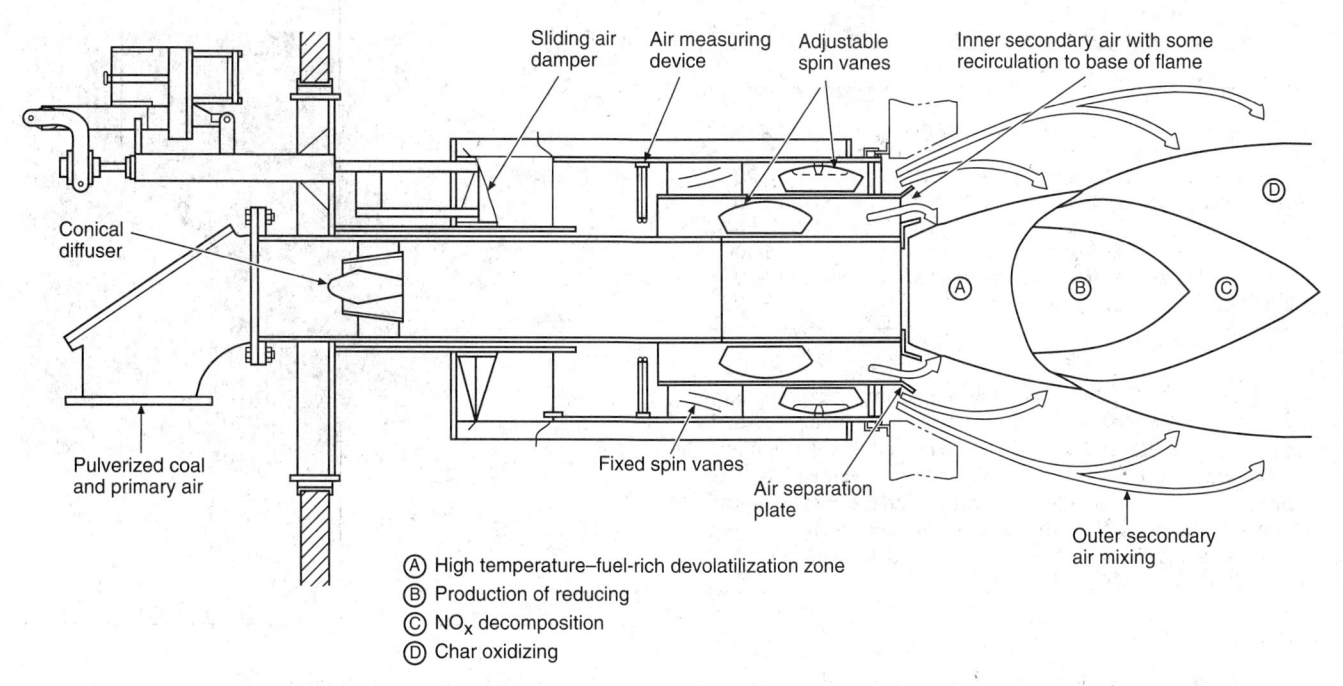

A. High temperature–fuel-rich devolatilization zone
B. Production of reducing
C. NO_x decomposition
D. Char oxidizing

FIG. 3-6A. Schematic of burner flame conditions. (Source: Babcock & Wilcox)

that remains after the volatile fuel components have been driven off near the burner) takes place in the furnace. Radiation from the brightly luminous flame and the recirculation of hot gas due to turbulence provide a stable ignition source. The combustion process proceeds toward completion through a wide range of fuel-air ratios, fuel-rich just at the burner exit and fuel-lean at the furnace exit.

FUELS

After two periods of rapidly rising oil prices, coal became the least expensive boiler fuel almost everywhere in the world. Some large boilers, built when fuel prices were more competitive, still burn residual fuel oil because: (1) they may be incapable of burning coal or (2) use of very low sulfur fuel oil is virtually mandated. Other boilers, particularly in gas-producing regions, were designed to burn natural gas when gas prices were maintained low via federal controls. In winter, when gas is urgently required by pipelines, many of these boilers will burn distillate oil. Most boilers in commercial installations and in small industries also use natural gas and clean oil, although all fuels are, to some extent, used.

In recent years, with the commercialization of fluidized bed combustion and a federal law making the use of waste fuels especially profitable for independent power production, a wide array of waste fuels are now used. Included are wood wastes, petroleum coke, coal-mining residues, and many others. Still, the dominant fuel worldwide remains coal, and its dominant form is as a finely pulverized powder with a mean particle size under 50μ. There are also waste fuels produced in oil refineries, chemical plants, paper mills, etc., which are burned in specially designed furnaces that may demand special precautions in addition to those described herein for the more common fuels.

Natural gas is gaining market share; combined cycle gas turbine power plants are now the most common new power plant worldwide. See Section 3, Chapter 4, "Emergency and Standby Power Supply"; and Section 3, Chapter 10, "Stationary Combustion Engines," for some discussion of gas turbines. In a combined cycle the hot turbine exhaust generates steam in a boiler, which is termed a *heat recovery steam generator* (HRSG). To maintain steam tem-

perature and pressure at all loads, many of these HRSGs are also fired with natural gas or light oil, and they have many features in common with other boilers. However, the profusion of HRSG designs and supplemental firing schemes has led to a number of serious accidents. For more information on HRSGs, see NFPA 8506, *Standard on Heat Recovery Steam Generation Systems.*

OIL- AND GAS-BURNING SYSTEMS

The burner is the principal component of oil- or gas-firing systems. Its purpose is to introduce fuel and air into the furnace in the proper proportion, and swirl, droplet size, etc., to sustain efficient exothermic chemical reactions. The combustion efficiency of a burner (and furnace) should be as high as possible to minimize energy lost as either unburned fuel or excess air.

Normal use of a boiler requires operation at different outputs to meet the varying load. The specified operating range for a burner is the ratio of full load heat production of the burner to the minimum load at which the burner is capable of stable ignition and reliable operation. For example, with a multiple burner-boiler rated at 100,000 lb/hr (45 359 kg/hr) steam capacity, a load range of 4:1 on the burners means that the unit can provide stable ignition and complete combustion at any load from 100,000 lb/hr (45 359 kg/hr) down to 25,000 lb/hr (11 340 kg/hr) without requiring one burner to be taken out of service.

Because of the chemistry of combustion, more than the theoretical air quantity is necessary to achieve essentially complete combustion. The amount of excess air needed for complete combustion with minimum stack loss must be balanced against compliance with NO_x emission requirements. The excess air normally required with coal, oil, and gas, expressed as a percent of theoretical air, is given in Table 3-6A. Many boilers operate with substantially higher excess air levels; some because of the particular fuel, the furnace design, or poor attention to optimum performance.

A frequently used burner for gas and oil is the circular burner. (See Figure 3-6B.) Normally, the capability of an individual circular burner is limited to about 200 million Btu/hr (58 620 kW). The

TABLE 3-6A. *Usual Amount of Excess Air Required for Complete Combustion*

Fuel	Type of Furnace or Burner	Excess Air (wt. %)
Pulverized coal	Completely water-cooled	15-20
Crushed coal	Cyclone furnace	10-15
	Fluidized bed	15-20
Fuel oil	Oil burner, register type	5-10
	Multifuel burners and flat flame	10-20
Natural gas	Register-type burner	5-10

tangentially displaced doors built into the air register provide the turbulence necessary to mix the fuel and air and to produce a stable flame. While the fuel is introduced to the burner in a fuel-rich mixture in the center, the direction and velocity of the air and the fuel-dispersion pattern thoroughly mix the fuel with the combustion air. A burner specially designed for low NO_x emissions is shown in Figure 3-6C. An air-flow monitor is incorporated in this burner to allow for balancing air flow from burner to burner.

Oil Burners

A main function of an oil burner is to atomize the oil, i.e., turn the thick stream of oil into a very fine mist of oil droplets. This provides the large oil surface area needed for rapid ignition and, as each droplet is surrounded by air, complete combustion. There are a number of types of oil atomizers, but the discussion here is limited to the two most popular types: (1) steam or air atomizers and (2) mechanical atomizers.

For proper atomization, oil heavier than No. 2 must be heated to reduce viscosity to between 100 and 150 SUS.* Steam or electric heaters are used to raise the oil temperature to the required level—approximately 135°F (57.2°C) for No. 4 oil, 185°F (85°C) for No. 5, and 200°F (93.3°C) to 220°F (104.4°C) for No. 6. With certain oils, better combustion is obtained at somewhat higher temperatures than are required for atomization. However, oil must not be heated to the point where vapor binding might occur in the fuel pump; this could cause flow interruptions and loss of ignition.

It is also important that oil be free of acid, grit, and other foreign matter that might clog or damage burners or their control valves. In order to prevent a loss of fuel, flow strainers installed to remove particles must be cleaned as often as an increasing pressure drop indicates.

Steam or air atomizers: Generally, the steam or air-oil atomizer operates by producing a steam-fuel (or air-fuel) emulsion that atomizes the oil through the rapid expansion of the steam when released into the furnace. (See Figure 3-6D.) The atomizing steam must be dry because moisture causes pulsations that can lead to loss of ignition. Where steam is not available, moisture-free compressed air would be substituted.

The steam atomizer performs efficiently over a wider load range than other types of atomizers. Normally, it properly atomizes oil down to 20 percent of rated capacity (turndown = 5:1); in some instances, steam atomizers have been successfully operated at 5 percent of capacity. This high level of turndown may be inconsistent with high efficiency because the total air flow must not be permitted to fall below 25 percent of full capacity rate (the purge rate). Also,

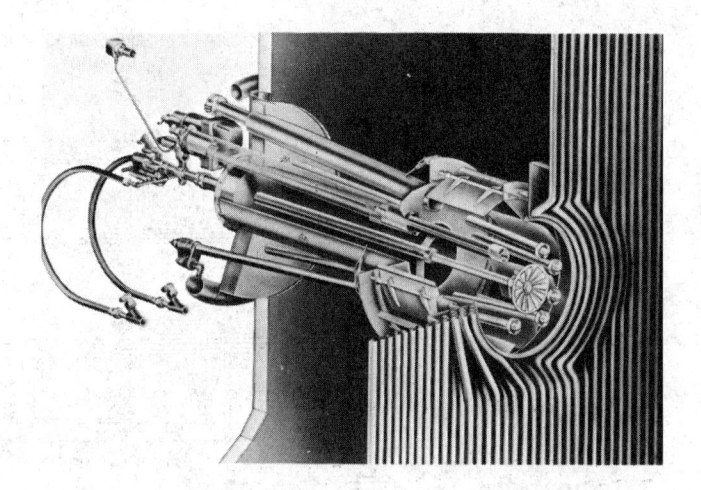

FIG. 3-6B. Circular register for gas and oil firing. (Source: Babcock & Wilcox)

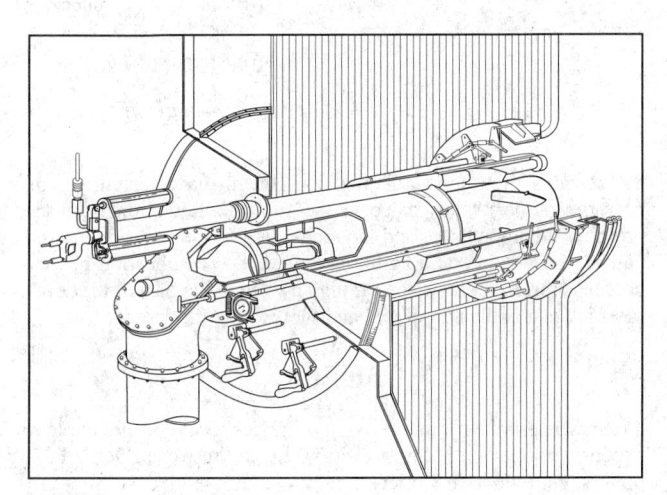

FIG 3-6C. Burner for low emissions of NO_x. (Source: Babcock & Wilcox)

these extremes in range cannot be fully utilized because the temperature of the furnace falls below that needed to complete combustion despite the excellent quality of atomization.

Mechanical atomizers: In the mechanical atomizer, the pressure of the fuel itself is used as the driving force for atomization. A wide variety of mechanical atomizer designs have been developed over the years. Excessive maintenance is required to keep those having rotating mechanical parts close to the furnace operational.

Return-flow atomizers are used for many marine installations and some stationary boilers where the use of steam or air is inappropriate, impractical, or uneconomic. (See Figure 3-6E.) Oil pressure required for these atomizers ranges from 600 to 1,000 psi (4.14 to 6.90 MPa), depending on capacity, load range, and fuel.

Mechanical atomizers are also available in sizes up to 200 million Btu/hr (58 620 kW) input. The acceptable operating range may be as much as 10:1 or as little as 3:1, depending on the maximum oil pressure used, the furnace configuration, air temperature, and burner throat velocity. Return-flow atomizers are ideally suited for standard grades of fuel oil where it is desired to meet load variations without changing sprayer plates, or cutting burners in and out of service.

---

*Abbreviation for Saybolt Universal Seconds. This is the efflux time in seconds of 60 ml (approximately 2 oz) of sample flowing through a calibrated universal orifice in a Saybolt viscometer under specified conditions.

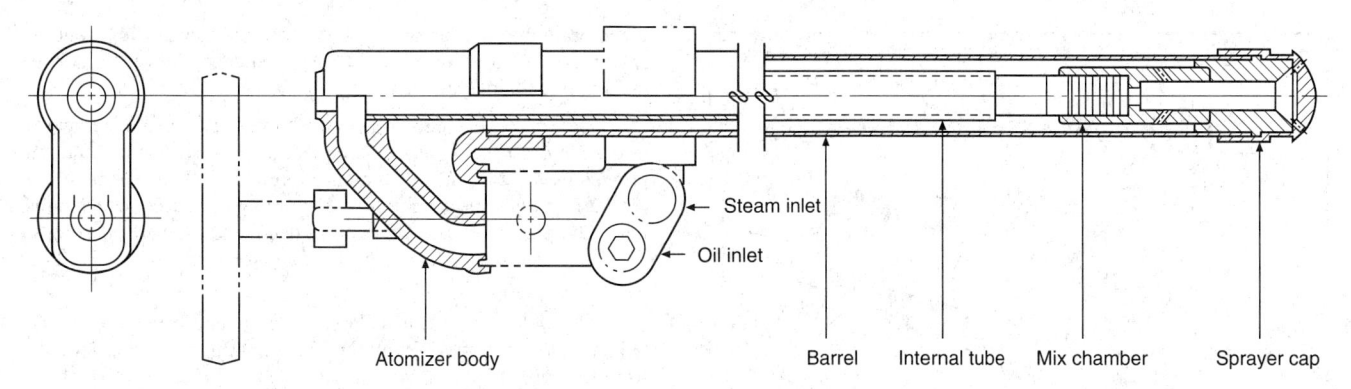

FIG. 3-6D.   Steam-fuel atomizer assembly. (Source: Babcock & Wilcox)

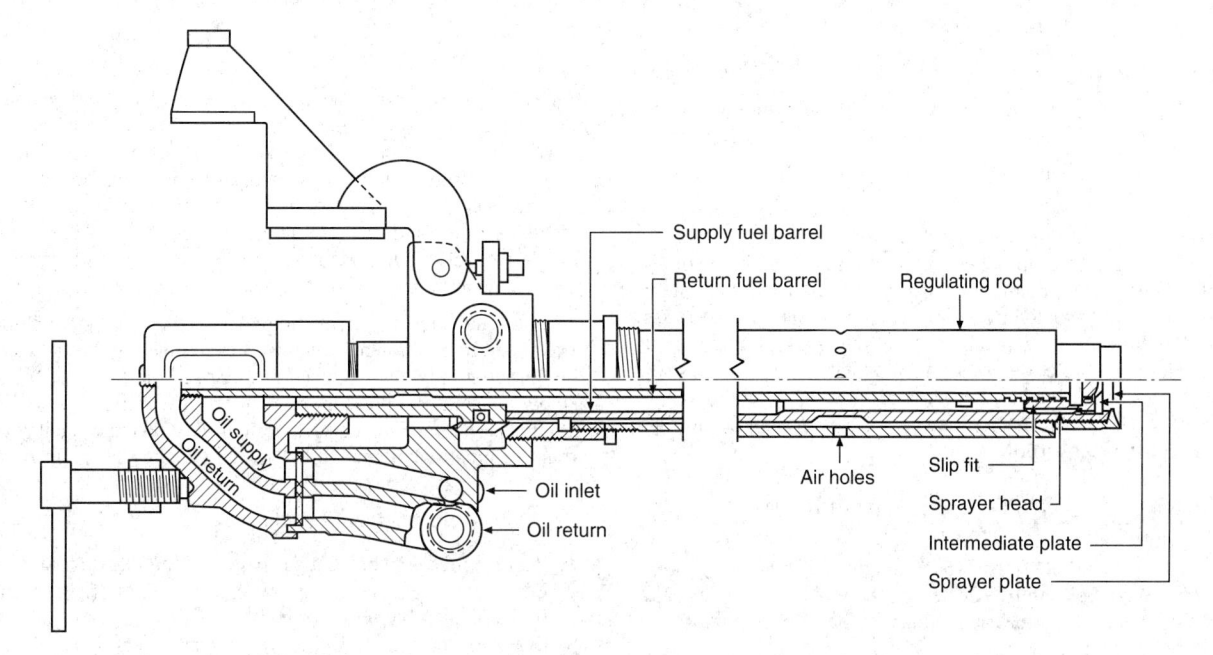

FIG. 3-6E.   Mechanical return-flow oil atomizer assembly. (Source: Babcock & Wilcox)

Natural Gas Burners

Natural gas is the ideal boiler fuel, requiring no preparation for combustion. Basically, gas burners mix fuel and air in either of two ways: (1) premix or (2) external mix. With the advent of stringent NO_x emission regulations, changes were required in the design and operation of large gas burners.

In a premix burner, gas and a portion of the air are mixed before the mix is introduced to the burner nozzle. A common method involves mixing gas and air in the suction side of a mechanical blower. Another method uses the venturi effect, wherein a gas jet creates a negative pressure at the air input orifice and draws air into the mixer. This design is familiar on kitchen stoves and home heating furnaces. The third premix method also uses the venturi effect but, in this case, an air jet draws fuel gas into the mixer, thus requiring a gas regulator to reduce fuel gas pressure to atmospheric before gas and air are mixed.

In external-mix gas burners, the fuel and air are mixed external to the nozzle. When external-mix gas burners have individual elements or spuds (see Figure 3-6F), part or all of the gas discharges in front of the impeller. This provides a local fuel-rich zone and thus

serves as an ignition stabilizer at high gas flow rates. A burner contains several spuds, each a gas pipe with a number of precisely drilled holes near the tip, to discharge gas for ignition at the burner throat. Depending on the manufacturer, each spud may be fitted with a flame holder or may share a common flame stabilizer to provide a local fuel-rich zone to stabilize ignition at low inputs. Gas ports are relatively large in order to minimize plugs. The spuds shown in Figure 3-6F may be withdrawn for service while the burner is in operation. Gas spuds are located so that oil will not impinge on them in multifuel burners.

With the proper selection of control equipment, a multifuel-fired furnace with a multiple spud-type burner is capable of changing from one fuel to another without a drop in steam pressure or output and is also capable of simultaneously firing natural gas and oil in the same burner. The control system's primary function is to achieve complete and efficient combustion by balancing fuel in the proper relationship with burner air flow. This type of element is designed for use with natural gas or other gaseous fuels containing at least either 70 percent methane, 70 percent propane, or 25 percent hydrogen by volume. The element is also designed for a maximum input of 200 million Btu/hr (58 620 kW) per burner.

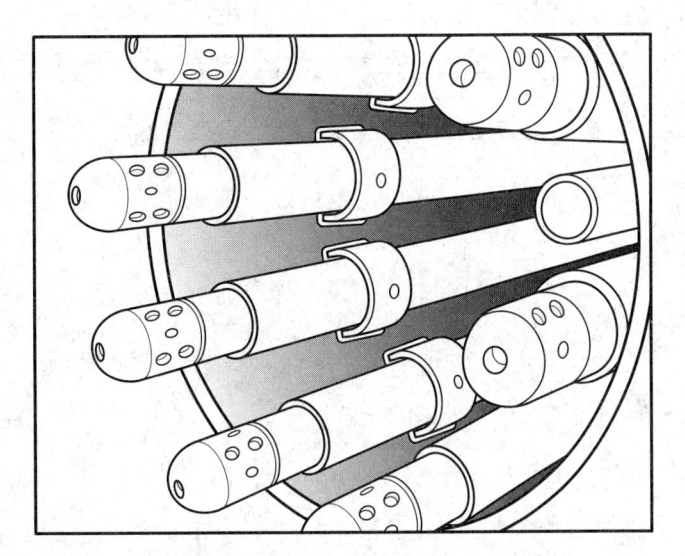

FIG. 3-6F. External-mix gas burner with individual spuds.

To provide safe operation, ignition of a gas burner should remain close to the burner impeller(s) throughout the full range of allowable gas pressures, at both normal airflows, and abnormally high airflows. Ideally, it should be possible, at the minimum load, to pass full-load airflow through the burner without loss of ignition. It should also be possible to pass as much as 25 percent in excess of theoretical air at full load without loss of ignition. With this range of airflow, it is not likely that ignition can be lost, even momentarily, during some upset in airflow.

PIPING AND CONTROL DEVICES

Fuels must be essentially free of contamination and be under control to ensure combustion safety. The design and reliability of fuel-handling systems are, therefore, important factors in minimizing the risk of fire or explosion.

Oil-Fired Systems

The fuel-oil supply system should be designed to ensure a continuous, steady flow of fuel that will meet the requirements of the boiler over the entire load range. This includes coordinating the main fuel control valve, burner safety shutoff valve, and associated piping volume to prevent fuel pressure transients that might exceed stable flame burner limits as a result of placing burners in or taking them out of service. (See Figure 3-6G.)

Fill and recirculation lines should be connected to storage tanks so that they always discharge below the liquid level. This will avoid excessive evaporation and the generation of static electrical charges in free-falling fuel.

Strainers, filters, traps, and sumps are some of the devices that may be used to remove contaminants from oil systems. Examples of contaminants are salt, sludge, water, and a variety of abrasive particles such as sand and corrosion products. It is good practice to route piping and locate valves to minimize their exposure to: (1) temperature extremes that may alter the oil's viscosity or pressure, and (2) physical damage in the event of an explosion.

Burner shutoff valves should be located as close to the burner as possible to minimize the volume of oil that may remain in the burner line downstream of the valve. This will minimize the oil that

could drain into the furnace following an emergency trip or burner shutdown. Positive means must be provided to prevent oil from leaking into an idle burner.

Fuel oil must be delivered to the burners at a specific temperature and pressure for proper atomization. Adequate recirculation provisions must be made for controlling the viscosity of the oil at the burners for initial lightoff and subsequent operation. Systems must also prevent excessively hot oil from entering fuel-oil pumps; otherwise, the pumps may vapor-bind and interrupt the supply of oil to the burners.

Positive means are needed to prevent fuel oil from entering the burner header system through the recirculating valves, particularly from the fuel supply system of another boiler. Check valves have been undependable for this function in heavy oil service. Provisions should also be made for clearing (scavenging) the passages of an atomizer into the furnace when a burner is taken out of service.

Gas-Fired Systems

Natural gas fuel-supply systems must be able to provide a continuous, steady flow over the entire load range. This includes coordination of main fuel control valve, burner safety valves, and the associated piping volume to avoid large gas pressure transients. Transients could cause burner limits to be exceeded and result in an unstable flame condition as burners are placed in or taken out of service. A well-designed system will also be capable of stable gas supply, despite anticipated furnace pressure pulsations.

The fuel supply system components located outside the boiler room need to be arranged to prevent excessive fuel gas pressure in the burner supply system, even in the event of failure of the main supply constant gas pressure regulator(s). (See Figure 3-6H.) Usually this is accomplished by providing full relieving capacity vented to a safe location. Where full relieving capacity is not installed, a high-supply gas pressure trip should be installed.

The gas-fired system should have:

1. Positive means to prevent gas from leaking into an idle furnace.
2. Piping vents located upstream of the last shutoff valve in any line to a main burner or igniter.
3. Provision to permit piping system leak tests and subsequent repair.
4. Permanent provisions to include means for making easy and accurate tightness tests of the main gas supply safety shutoff valves as well as the individual burner gas safety shutoff valves.
5. Vents located so that there is no possibility of vented gas entering the boiler room or adjacent buildings.
6. Vents placed high enough that escaping gas will not be a fire hazard.
7. Header vent lines that run independently.
8. Igniter vent subsystem that runs independently of the burner vent subsystem so there is no chance of backflow into the burner.
9. No cross-connection between venting systems of different boilers.

PULVERIZED COAL SYSTEMS

One of the variables in the design of a pulverized coal supply system is the location of the primary air fan relative to the air heater and pulverizer. The combination shown in Figure 3-6I is known as a "cold" system, and includes a large primary air fan supplying drying air via the air heater to a group of pulverizers. The other basic system provides hot air to individual primary air fans for each pulverizer located downstream of the air heaters.

Pulverized-coal-fired systems capable of burning a wide range of coal fuels have several necessary functions. Raw stored coal must

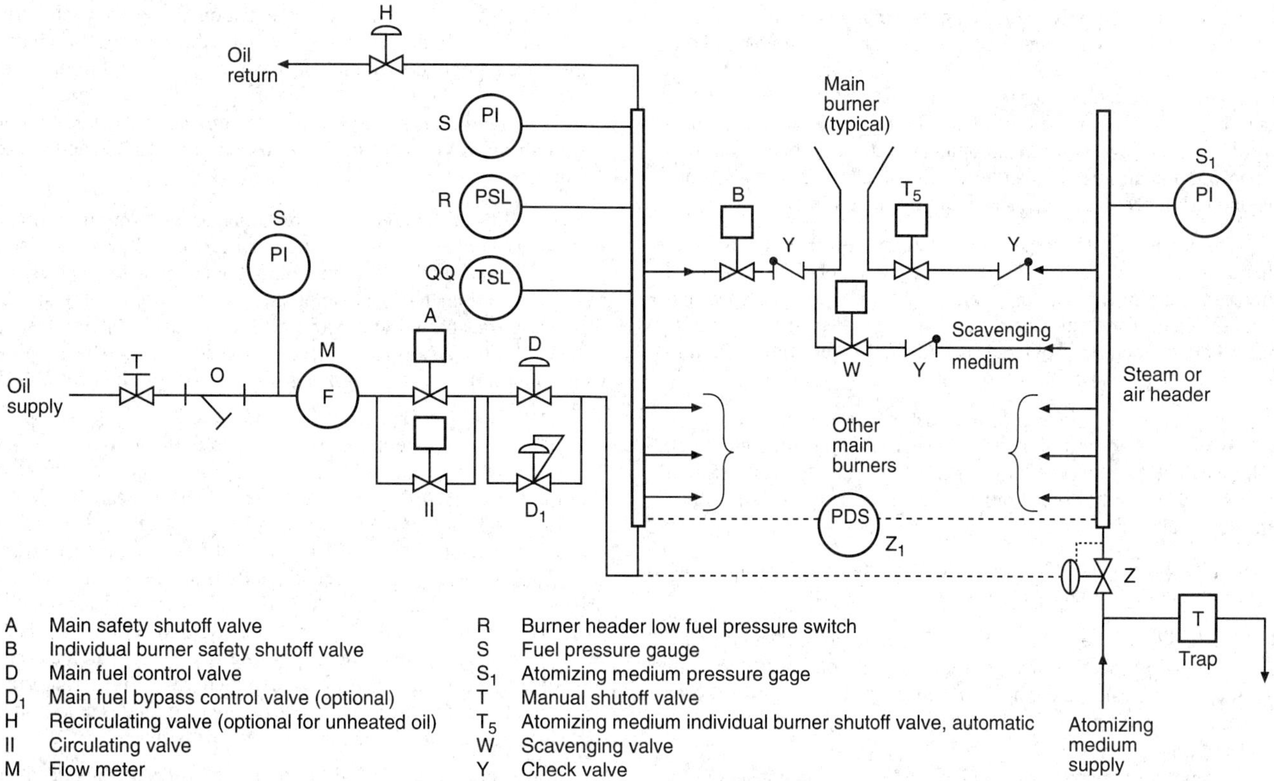

A	Main safety shutoff valve	R	Burner header low fuel pressure switch
B	Individual burner safety shutoff valve	S	Fuel pressure gauge
D	Main fuel control valve	S_1	Atomizing medium pressure gage
D_1	Main fuel bypass control valve (optional)	T	Manual shutoff valve
H	Recirculating valve (optional for unheated oil)	T_5	Atomizing medium individual burner shutoff valve, automatic
II	Circulating valve	W	Scavenging valve
M	Flow meter	Y	Check valve
O	Cleaner or strainer	Z	Differential pressure control valve
QQ	Low temperature or high viscosity alarm switch	Z_1	Differential pressure alarm and trip switch

FIG. 3-6G. *Schematic of typical main oil burner—steam atomizing.*

A	Main safety shutoff valve	O	Cleaner or strainer
B	Individual burner safety shutoff valve	P	Restricting orifice
C_1	Burner header atmospheric vent valve	Q	Burner header high fuel pressure switch
C_2	Individual burner atmospheric vent valve	Q_2	High fuel supply pressure switch
D	Main fuel control valve	R	Burner header low fuel pressure switch
D_1	Main fuel bypass control valve (optional)	R_1	Burner header low fuel pressure switch (alternate location)
I	Charging valve (optional) (must be self-closing)	R_2	Low fuel supply pressure switch
J	Constant fuel pressure regulator	S	Fuel pressure gauge
K	Pressure relief valve	T	Manual shutoff valve
M	Flow meter		

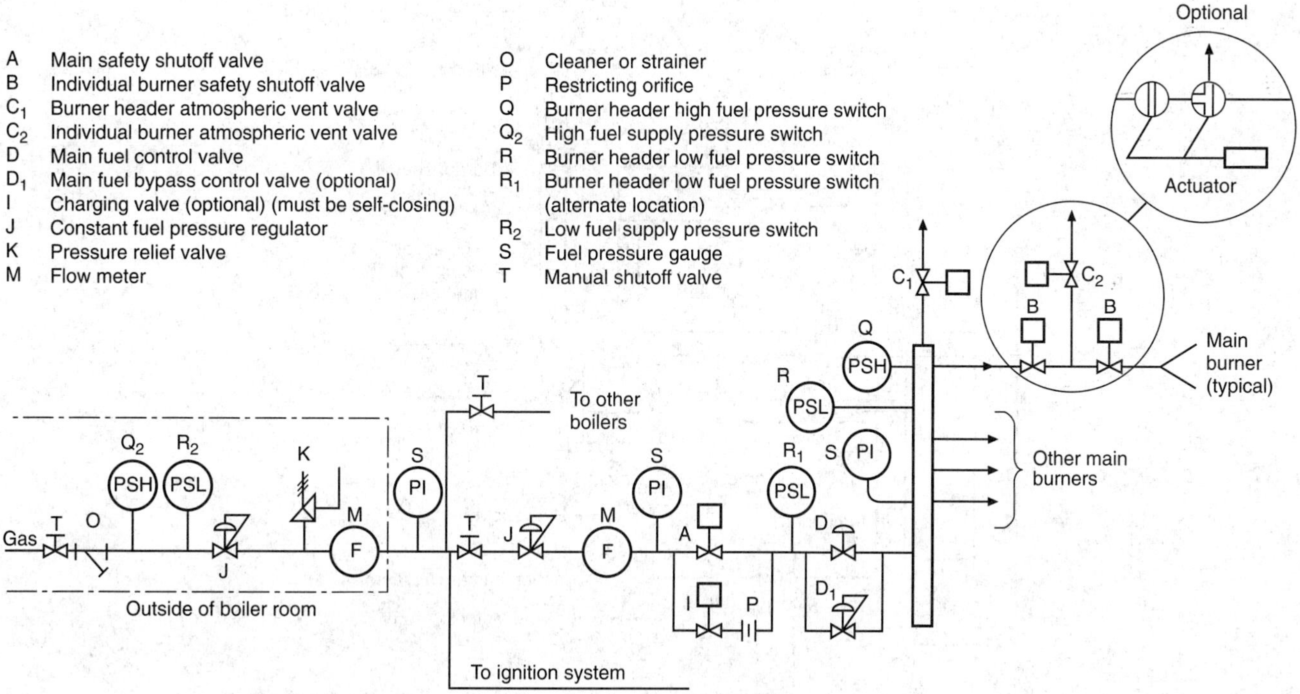

FIG. 3-6H. *Typical fuel supply system for natural gas-fired multiple-burner boiler-furnace.*

be transferred to the pulverizer in measured and controlled quantities. Hot air mixed with tempering cool air is blown into the pulverizer to dry the coal. (See Figure 3-6I.) The mix temperature is controlled by the temperature of the air-coal stream at the outlet of the pulverizer based on the type of coal being pulverized. A temperature that is too low, as when particularly wet coal is being fed, will impair pulverization. A temperature that is too high may cause coking and risk a fire in the pulverizer and burner piping.

The advent of NO_x emission regulations has increased the likelihood of unburned carbon in the fly ash. It is therefore even more important than in the past that coal be pulverized to the specified fineness. Testing and pulverizer maintenance will ensure proper sizing. Where needed, improved classifiers can be retrofit on existing pulverizers.

The primary air is designed to transport the fine coal from each pulverizer to its several associated burners in controlled amounts, well distributed among the burners. It is important to establish that the transport velocity in each pulverized coal pipe is equal. Periodic retesting is also required.

At the burner, the coal-air mixture is combined with a measured amount of secondary air to serve several purposes:

1. The fuel must be volatilized and completely consumed in a continuous process starting with an initial zone of stable ignition at the burner and continuing through the combustion process with minimum unburned combustibles in the stack gas and ash hoppers.
2. The total air to each burner and to the burner zone must contain the lowest level of excess air possible to minimize stack losses and NO_x while achieving a high level of combustion efficiency.

3. The air should surround the combustion process at each burner to minimize reducing atmospheres at tube surfaces. Some boilers are equipped with an "air wash" to protect tubing near the burners.
4. Stable ignition is necessary at each burner to ensure continuous combustion. A means of flame detection is also needed at each burner.

Products of combustion are transported from the furnace through platen and pendant superheater and reheater surfaces, through the convection pass including the economizer surface, and through the air heater to accomplish the necessary heat transfer before being transported to the air pollution control system(s) by the induced draft (ID) fan. The boiler is designed, based on the specified coal, to minimize slag and other solid deposits on the heat transfer surfaces.

The largest modern boilers are capable of supplying steam to 1300-MW reheat turbines. This large capacity coupled with poorer fuel quality created the need for large pulverizers capable of processing 50 tons (45.36 metric tons) or more of coal per hr. The pulverizer shown in Figure 3-6J is a roll-and-race type that utilizes three large-diameter rolls spaced equally around the mill. The rolls are mounted on axles; and, in turn, the roll assemblies are attached by a pivoted connection to a stationary overhead frame that keeps them in their roll path while permitting limited radial freedom of movement. Grinding pressure is supplied by springs that apply force to the axles of the rolls. The grinding ring rotates at low speed and is shaped to form a race in which the rolls run.

Raw coal is fed into the mill either inside or outside of the grinding race. It immediately mixes with the partially dried and ground coal that is circulated within the grinding zone by the airflow

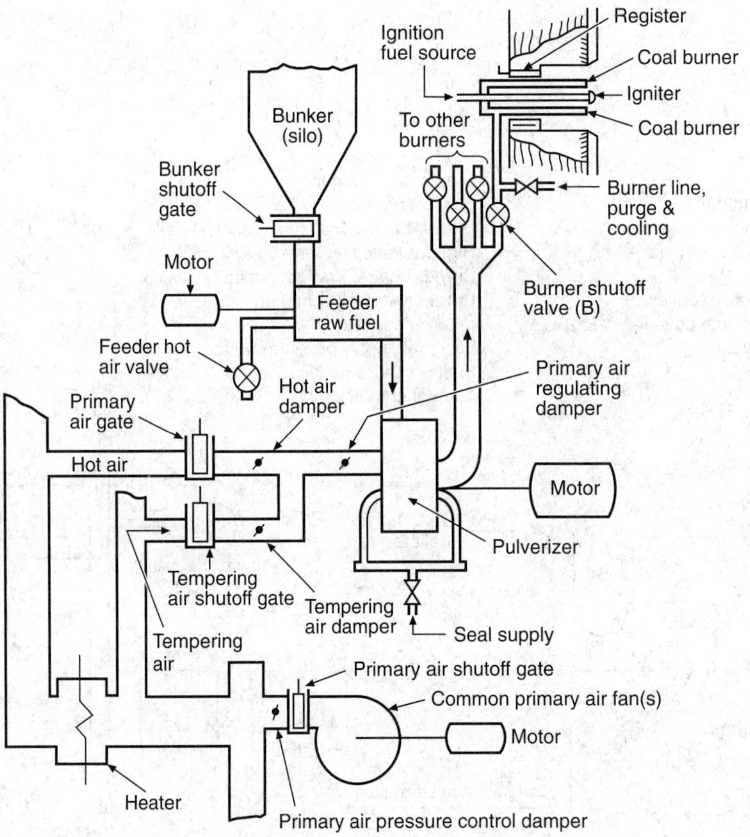

FIG. 3-6I.   Direct-firing system for pulverized coal. (Source: Babcock & Wilcox)

through the pulverizer. Coal fines are carried by the air to the classifier. The finest fraction of the coal, along with the air, leaves the pulverizer through the outlet pipes. Oversize coal is returned to the grinding zone through a seal at the bottom of the classifier.

FLUIDIZED BED COMBUSTION

The world's first fluidized bed combustion (FBC) boiler, a pilot unit burning 600 lb (272.2 kg) of coal per hr, was placed in service in 1967. In 1995 a 350-MW FBC was placed in commercial service in Japan. Many units of moderate size, 10 to 100 MW, have been built around the world to exploit the availability of low-cost, waste fuels. There are many diverse FBC designs for smaller units.

The two central characteristics of the fluidized bed boiler, when compared to stokers and pulverized coal, are: (1) the bed and (2) the bed temperature. Bed refers to the pile of granular solid particles in which the fuel is burned. Air flowing up from beneath and through the bed causes the individual granules to separate and move about. Fuel added to the bed is ignited by the hot bed particles. In turn, the burning fuel transfers the heat of combustion to the inert bed particles.

The bed, cooled by a variety of techniques, moderates the combustion and keeps the *bed temperature* in the range of 1,500°F (815°C) to 1,650°F (899°C). (See Figure 3-6K.) This, in turn, leads to reduced slagging and fouling, lower emissions of NO_x, and an environment where added limestone (primarily $CaCO_3$) can react with sulfur. The moderate combustion temperature and turbulence also provide an ideal environment for the combustion of a wide variety of low-grade fuels, e.g., anthracite culm to wood wastes, as well as high-grade coal.

To control the bed temperature at about 1,500°F (815°C), roughly 50 percent of the heat of combustion must be removed as rapidly as it is released. The bed solids play this role by giving up heat to boiler, superheater, and reheater surfaces, both inside and outside the furnace boundaries. In bubbling fluidized beds, water tubes are often placed in the combustion bed itself. Particles also recycle within the furnace, giving up heat to the water walls. In a circulating fluidized bed (CFB), heat is also transferred from the bed particles to the water walls and, in some designs, to a bubbling bed, external to the furnace. (See Figure 3-6K.) In another variation, the FBC furnace is enclosed within a pressure vessel and the uncooled flue gas, cleaned of dust, is used to drive a gas turbine. About 10

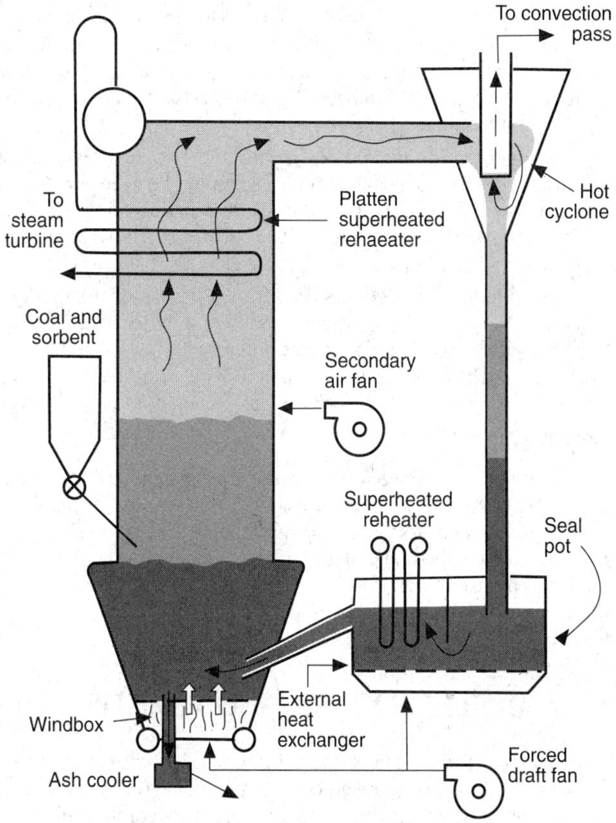

FIG. 3-6K. Circulating fluidized bed boiler.

FBCs were operating or under construction around the world in 1995—most in the 80-MW size range.

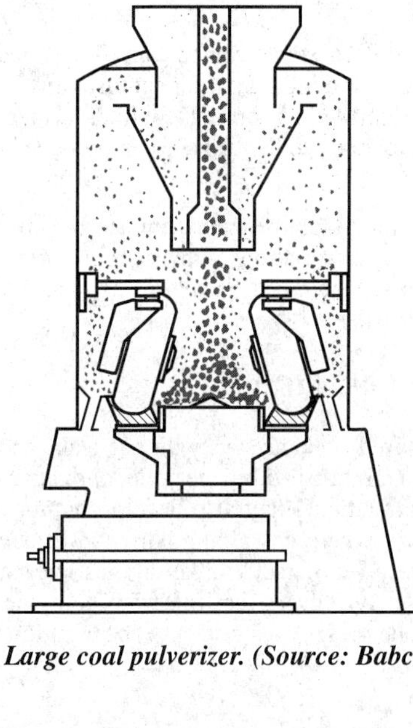

CFB systems currently have the dominant market share. While combustion and other system hazards are similar for bubbling and circulating FBCs, the remainder of this discussion will describe CFBs.

To control NO_x, air is admitted both below and above the dense bed. Although the split of fluidizing (primary) and secondary air varies between designs, the values are roughly 50/50. Thus, the dense bed operates with an air deficiency, i.e., it is fuel rich, and more than half the combustion occurs above the dense bed. This mode of operation is analogous to that of the low-NO_x burner shown in Figure 3-6A. The temperature of the gas-solids stream leaving the furnace is roughly at the same temperature as the dense bed. Further, 99+ percent of the solids are removed by the large cyclone for recycle to the furnace.

Special FBC Hazards

Most potential FBC hazards are similar to those of suspension-fired boilers. However, there are some hazards unique to FBCs. For this new technology, it is important that operators be trained to deal with relatively rare but serious possibilities. The special hazards are:

1. *Hot solids spills.* A large FBC contains many tons of very hot aerated solids. Bottom ash-removal pipes have failed in several plants, leading to a large spills. Plant designers should avoid placing critical components and controls in the vicinity of a potential spill point.

2. *Lime burns.* Removal of most of the fuel's sulfur requires limestone feed rates that exceed the stoichiometric ratio of calcium-to-sulfur [40 lb/32 lb (18 kg/15 kg) = 1:1]. Ratios of 2:1 or more are not unusual, which means that the flyash and bottom ash are

FIG. 3-6J. Large coal pulverizer. (Source: Babcock & Wilcox)

rich in CaO, i.e., quicklime. High temperature can be generated when quicklime is mixed with water deliberately, as in an ash conditioner; or inadvertently, as when in-furnace maintenance is required. Removal of the bed before furnace entrance is now the common practice.

3. *Steam generation after trip.* The mass of hot particles contain sufficient sensible heat to continue steam production at near full capacity for several minutes after the fuel source is tripped if the fluidizing air flow is continued. Thus, a loss of feedwater should lead to a fan trip as well as a fuel trip.

4. *Operation at very high fuel-air ratios.* Because of the mass of hot solids, an FBC's inherent combustion stability permits operation with as much as four times the normal fuel-to-air ratio. In several such cases, the accumulation of fuel in the bed led to a rapid increase in bed temperature when some operational change occurred.

While FBCs do have some special hazards that need to be considered by designers and operators, the safety record over FBCs' short commercial history has been quite good. To start an FBC, it is necessary to heat the bed material to above the ignition point of the main fuel by use of oil or gas burners. Just as in any boiler, FBCs are most vulnerable to explosion during startup. (See hazards discussed for oil and gas firing.)

BOILER-FURNACE HAZARDS

Explosion and fire are the most serious potential hazards in boiler-furnaces and their associated pipes, ducts, fans, and fuel bunkers. Explosions occur when an ignition source contacts a combustible mixture of fuel and air. When a combustible mixture has accumulated in a large, confined space of the equipment, a very damaging explosion can occur. Such accumulations may result from an equipment malfunction or an operator error associated with an inadequate or improper purge or incorrect operation of burner equipment. Because even a minute ignition source may initiate a substantial explosion and ignition sources may abound in a furnace, e.g., static electricity, hot slag, or glowing fly ash in a hopper, safe operation must emphasize avoiding creation of a large combustible mixture.

A temporary loss of flame caused by an interruption in fuel, air, or ignition energy may allow a combustible fuel-air mixture to accumulate before ignition is re-established. Any momentary fuel-rich operation may also lead to a combustible accumulation. For example, leaking fuel may vaporize and collect in an idle furnace only to ignite explosively when the first burner is lighted. In multiburner units, the undetected loss of flame at a burner may allow an explosive mixture to accumulate somewhere in the volume of the furnace, only to be ignited by other burners. The classic example of operator error is the failure to purge the furnace of combustible mixtures between repeatedly unsuccessful attempts to lightoff the burner. Another is to ignore the implications of small explosions, i.e., *puffs*, that do no damage but may be symptoms of a serious defect.

Hazards of Oil Firing

Fuel oil is a complex blend of hydrocarbons having a wide variety of molecular weights, and boiling, freezing, and flash points. At a sufficiently high temperature fuel oil will partially volatilize thereby creating new and unpredictable liquids, gases, and solids. Boilers have been fired with unrefined crude oil containing very volatile light ends, such as propane, butane, and pentane. However, with the rise of oil prices this is no longer common. Refined boiler fuels have had the light ends removed to produce gasoline and other valuable products, and are thus less volatile. A new fuel form termed *Orimulsion*™ is a blend of bitumen (e.g., tar) and water. Any water-

containing fuels can pose a hazard when separated water, rather than fuel, is pumped through the burner.

No. 1 and No. 2 fuel oil, and fuels commonly known as kerosenes, range oil, furnace oil, and diesel oil, are called *distillates*, while No. 5 and No. 6 fuel oil are referred to as *residuals*. Most large boilers designed for oil are designed for Nos. 5 or 6 oil. Distillates are used in boilers designed for natural gas as the primary fuel. Very volatile oils (sometimes used in very cold regions), crude oil, and many byproduct fuels require that electrical motors, switches, and other components be designed to explosionproof standards.

For heavy oils (Nos. 4, 5, and 6), preheating is necessary to reduce the viscosity of the oil flowing to the burners and ensure proper atomization. This preheating is an important part of an integrated safety system designed to avoid flame interruptions. Also, if sludge is allowed to accumulate in storage tanks, it may plug strainers or burner tips. Water normally accumulates below the stored oil, which is less dense; it is therefore important to design tank suction lines so that they draw from above any potential water level.

Oils from different sources, even with similar specifications, may sometimes cause problems when stored in the same tank due to chemical reactions. When oil shipments having widely different viscosities or gravities are stored in the same tank, a significant change in fuel input rate (without a corresponding change in airflow) may occur and impair complete combustion. In any event, fuel oils should be tested for their conformance to specifications soon after receipt.

Where the same burner is used for heavy and light oil (depending on price or availability, for example), special care must be taken that the correct valve settings and interlocks are selected. When very light oils with low electrical conductivity are used, static electricity can build up and provide an uncontrolled ignition source. Light or heavy, oil leaks pose a potential fire hazard. Assembly and disassembly of pipes and equipment should be done with care to avoid leaks. Leaks that do occur should be promptly stopped and cleaned up.

Combustion efficiency of mechanical atomizing burners may also be affected by a wear-induced change in orifice size. Improper cleaning may also damage burner tips. Individual burners operating fuel rich or with very low excess air may cause combustibles to accumulate in the furnace. Periodic flow testing may be necessary. Unsafe operating conditions can also be created by the failure of personnel to reinstall a tip, gaskets, or sprayer plate during burner reassembly.

Major oil flow transients can be caused by rapid operation of an oil supply valve, individual burner shutoff valve, or the regulating valve in the return oil line from the burner header. Uncontrolled changes in the fuel input to the furnace can produce hazardous conditions.

Oil flow to individual oil guns can also be adversely affected by such conditions as variations in burner elevation, distance from the regulating valve, and pipe size, all of which can be hazardous on low-pressure burners.

Hazards of Gas Firing

Hazards in gas-fired systems start with gas leaks and the development of fuel-rich mixtures within the boiler or ductwork. Potentially hazardous accumulations may also develop within other building areas, particularly where gas piping is routed through spaces that are not adequately ventilated. Unlike oil, gas leaks are not visible and plant noise may prevent the leak from being heard. Odorants may or may not be present at sufficient concentration to be detected via the nose.

Within the furnace it is possible for air-fuel ratios to be altered severely without producing any visible evidence at the burners, furnace, or stack, thus allowing a combustible mixture to accumulate over time. Combustion control systems, when responding to falling steam flow or pressure due to an increase in demand for fuel, are potentially dangerous unless protected or interlocked to prevent the creation of a fuel-rich mixture and a loss of ignition.

While gas liquids are normally removed in or near the gas fields, natural gas may be *wet*. A wet gas implies the presence of liquid fuel components, which, if carried into the burners, can result in momentary increases in fuel input and/or a flameout, followed by possible reignition and explosion. For this reason, systems using wet gas require special attention.

Gases supplied from one or more sources can introduce unacceptable hazards if they have significant differences in volumetric heating value due to the ratios of propane, butane, etc., to methane. With such variable supplies, it is necessary to provide instruments that are responsive to variations in heat value (e.g., specific gravity or heating value meters), and appropriate alarm and combustion compensation devices. Similar problems will be encountered if proper recharging of the gasline does not follow line maintenance.

Discharges from relief valves will produce a particular hazard unless vent lines are carefully designed with all potential weather conditions considered to avoid reentry of gas into the boiler room. Purging preceding maintenance can also lead to a hazard.

Burner pulsation or combustion-driven oscillations remain a mystery associated with gas firing and, to a lesser degree, with oil firing. When one or more burners on a large unit begin to pulsate, there is a feedback mechanism that leads to variations in furnace pressure, fuel input, and heat release. The action may become violent as each pulse adds energy to the next oscillation. At times, the whole boiler may shake. Adjusting only one burner may start or stop pulsation. At times, only minor burner adjustments can eliminate pulsation. In other instances, it may be necessary to make physical alterations to the burners. Alterations may include modifying the gas ports, impinging gas streams on one another, or installing a device that changes the fuel-air mixing characteristics.

Hazards of Pulverized Coal Firing

When coal is crushed, fresh reactive surfaces become available for oxidation. The coal in the storage bunkers ahead of any pulverizer is subject to self-heating and, in cases where the heat is not dissipated, spontaneous combustion. The three conditions that can lead from moderate self-heating to uncontrolled combustion are: (1) excessive fines, (2) a source of air, and (3) storage for an extended period. Once a mass of coal is burning, quenching with water will lead to a reaction between water and carbon, producing combustible CO and H_2. Dealing with a bunker fire usually requires that the burning coal be run into the boiler via the feeders, pulverizers, and burners. This "solution" can lead to pulverizer fires or explosions. Self-heating can be detected by monitoring the CO content of the air in the empty space above the coal. Thermocouples may also be useful, but they can easily miss a hot spot.

Very wide variations in the size of raw coal, moisture, or clay content may cause erratic feeding of the coal to the pulverizer, and, often, complete stoppages. As-received coal may contain foreign matter, such as scrap iron, blasting wire, railroad ballast, wood, rags, excelsior, etc., which may interrupt the coal feed, damage or jam equipment, or cause a spark within a pulverizer.

A large boiler may burn 100 lb (45.4 kg) of pulverized coal per sec, but as little as 0.05 oz per cu ft (0.05 kg/m³) of air forms an explosive mixture; therefore, as little as 3 lb per 1,000 cu ft (1.0 kg in 20 m³) can be hazardous. Thus, an explosive mixture will develop very quickly if a momentary flameout occurs. A special hazard is methane gas released from freshly crushed or pulverized coal, which may accumulate in enclosed spaces, e.g., bunkers, and within pulverizers and burner piping.

Pulverized coal, entrained in the primary air stream, is conveyed through pipes from the pulverizer to the burners. To prevent flashback or settling of coal particles in the burner pipes and preignition, it is necessary to maintain the design air velocity at all loads. Pulverizer wear or broken classifiers may introduce larger particles into the burner pipes that will settle at even higher velocities. Provisions should be made for cooling and emptying pulverizers and/or burner lines when the burners they supply are shut down. This will avoid spontaneous combustion and explosion in the pulverizer or burner lines.

Pulverizer fires and explosions are serious hazards. A sudden, large increase in temperature of the mixture leaving the pulverizer or in the temperature of the casing indicates that a pulverizer fire has started. Fires are caused by: (1) feeding burning fuel from the bunkers or (2) by spontaneous combustion of fuel in the pulverizer or piping. While pulverizers are now designed to withstand a substantial overpressure, a fire is still serious and should be dealt with promptly.

The general steps for extinguishing the fire are:

1. Remove the pulverizer from service without emptying to avoid creating a sudden air-rich (explosive) mixture condition.
2. Prevent air infiltration by closing all burner line valves and inlet air dampers, and sealing the air valve.
3. Admit steam or inert gas into the pulverizer through the connection provided to:
   (a) Smother the fire.
   (b) Maintain an inert atmosphere until the coal has been removed and the housing has cooled.
4. Dump the coal stored in the pulverizer through the pyrite gates.
5. Rotate the pulverizer without operating the primary air fan to complete the swirling and cleaning procedure.

Various pulverizer manufacturers have different requirements.

HAZARDS OF LOW NO$_X$ FIRING

Regulatory pressures to reduce emissions of existing units could lead to furnace safety problems.

1. NO$_x$ reduction methods may reduce the margins formerly available to prevent accumulation of unburned fuel during upsets.
2. Flue gas recirculation, a common method of reducing NO$_x$ production, requires that operators be aware of the oxygen content of the combustion "air" and/or have knowledge of the distribution of air and of recycled flue gas.
3. Unburned combustibles often increase with low-NO$_x$ firing. This means that horizontal ducts and hoppers may contain smoldering, carbon-rich ash—a hidden ignition source. A CO analyzer will allow avoidance of fuel-rich operation, but good practices are still evolving.

HAZARDS OF FLUIDIZED BED FIRING

Attempting to start main-fuel flow before the bed is at the proven temperature for that particular fuel can lead to a furnace explosion. Interlocks are to be provided to prevent such maloperation and must be reset, after testing, when a very different coal is to be fired. Hot restarts without a purge, e.g., following an emergency trip, are permitted only if the temperature of the fluidized bed is high enough to ensure proper ignition of fuel.

While flame detectors are useful for proving a pulverized coal flame, they are not useful in FBC. Instead, the bed temperature is proved safe by a number of in-bed thermocouples. The thermocouple readings are only reliable indicators of an average "safe" bed temperature when the bed is actively fluidized and thus thoroughly mixed.

OPEN REGISTER LIGHTOFF AND CONTINUOUS PURGE PROCEDURE

Furnace explosions in large water-tube boilers in the 1960s led to the codification of the open-register lightoff and continuous purge procedure for all fuels. This procedure maintains airflows at or above the prescribed minimum of 25 percent of full-load *mass* rate airflow during all operations of the boiler. Until the 1995 edition of NFPA 8502, *Standard for the Prevention of Furnace Explosions/Implosions in Multiple Burner Boilers*, the *volumetric* airflow rate was specified in recognition that mass flow rate must be increased at low (during startup and shutdown) air temperature to maintain air and gas velocities. In any event, the continuous purge procedure has been shown to prevent the inadvertent accumulation of a hazardous air-fuel combustible mixture in the furnace or setting.

The continuous purge procedure is based on the concept that three basic conditions will significantly improve the margin of safety, particularly during startup. These conditions are:

1. Minimum number of required equipment manipulations, thus reducing the likelihood of operator errors or equipment malfunctions;
2. Means for establishing the desired fuel-rich condition at individual burners during lightoff; and
3. A very air-rich furnace atmosphere during lightoff and warmup, created by maintaining total furnace airflow at the same rate as that required for furnace purge.

In its simplest form, the basic procedure that satisfies these three operating objectives is as follows:

- Place all or most of the burner air registers in a predetermined open position.

- Purge the furnace and boiler settings with the burner air registers in that same predetermined position. The total airflow for purge must not be less than 25 percent or more than 40 percent of full-load mass airflow.

- Light the first burner or group of burners without any change in airflow setting or in burner air register position.

At initial startup a boiler should be tested to determine whether any modifications are required to obtain satisfactory burner ignition, or to satisfy other design limitations during lightoff and warmup. For example, some boilers will be purged with the registers in the normal operating position. In this event, it may be necessary to momentarily close the registers of the burner being lighted to establish ignition. However, unnecessary modifications in basic procedure are to be avoided, thereby satisfying to the greatest degree possible the three basic objectives given above, particularly that of minimizing equipment manipulations.

Suggestions have been made to modify the continuous purge requirement to allow improved NO_x control and increased efficiency at low loads. No changes have been approved by a NFPA committee, since tests and data do not exist that prove reducing airflow below 25 percent (or purge) can be done safely.

FIRE AND EXPLOSION PROTECTION

Prevention is clearly the key to fire and explosion protection. Good facility design, selection of reliable equipment, system monitoring, and malfunction alarm instrumentation are key elements in preventing fires and explosions in boiler-furnace systems. With the cost of a major boiler explosion ranging to $100 million, replacement power costs included, operator training and maintenance critical to prevention are obviously worth the cost.

The entire system—boiler, furnace, fuel supply, air supply, vents, piping, and ducts—is designed to specific parameters and for specific operating limits. At no time should these limits be exceeded without a detailed engineering analysis and the modifications indicated. This is particularly relevant when new emission standards are imposed on an operational plant.

SPECIAL CONSIDERATIONS FOR SMALL BOILERS

Large boilers will be constructed as separate structures, but small boilers [under 12,500,000 Btu/hr (3,663 kW)] are often installed in existing buildings. Good facility design requires the boiler be installed in a separate room or structure, preferably of noncombustible construction. Boilers should be set on concrete floors or platforms that extend beyond the equipment for a distance of at least 4 ft (1.2 m) in each direction. If they must be set on combustible floors, there should be sufficient air circulation beneath the furnace to keep the temperature of the combustible floor below 160°F (71°C). (See Section 3, Chapter 5, "Heating Systems and Appliances," for an explanation of the temperature limits for protection of combustible surfaces.) This protection can be provided, for example, by laying 4-in. (102-mm) thick hollow masonry block covered with sheet metal, on the floor.

If possible, metal chimneys or smokestacks should not pass through combustible floors, ceilings, or walls. If such chimneys must pass through combustible roofs, sufficient clearance and/or insulation should be provided to keep the temperature of the combustible materials below 160°F (71°C). Generally, minimum clearance is considered to be 18 in. (457 mm).

INTERLOCKS, ALARMS, AND OPERATOR COMPETENCE

A system of interlocks should be provided to prevent improper sequencing by operating personnel and to shut down operations if certain critical malfunctions occur. Both audible and visual alarms serve to warn operators to take certain corrective steps; others may indicate what automatic functions have occurred to reduce a hazard. New boilers should not be fired until adequate safeguards have been installed and tested. Bypassing safety interlocks with electrical jumpers, for example, should *never* be permitted by the contractor or the owner.

Safe operation cannot be ensured solely by equipment design and adherence to the manufacturer's operating instructions. Competent operators who understand the processes involved are also needed. Technical competence should be maintained by a program of periodic retraining so that increasing familiarity with the system does not tempt operating personnel to take shortcuts in operating procedures or to bypass safety devices. Where operator error has been responsible for a serious accident, inadequate training has often been found to be the root cause.

A program of preventive maintenance is needed for the equipment and control devices. Poor preventive maintenance will lead to the need for costly corrective maintenance. A comprehensive equipment history that records, for each date, the conditions found, work done, and the changes made, will provide the records needed to plan the next maintenance effort. Forms for these records are often available from the boiler's insurer. Cleanliness and good housekeeping will also contribute to the prevention of fire and explosion. Where pulverized coal is the fuel, care should be taken in cleaning dust settled on ledges (and other spills), so as not to create a combustible dust cloud.

For smaller units, automatic sprinklers are commonly required in combustible boiler rooms; however, in noncombustible boiler rooms, where the potential for sustained fire is slight, portable fire extinguishers or small hose lines might be adequate. Additional guidance on fire protection for fuel-supply systems can be found in Section 3, Chapter 21, "Storage of Flammable and Combustible Liquids"; and Section 3, Chapter 22, "Storage of Gases."

BIBLIOGRAPHY

References

Cote, A. E., ed., *Industrial Fire Hazards Handbook*, 3rd ed., National Fire Protection Association, Quincy, MA, 1990.

Factory Mutual Research Corporation, "High-Pressure Steam Boilers—Care and Operation," Factory Mutual Research Corp., Norwood, MA.

Factory Mutual Research Corporation, "Elements of Combustion, Controls, and Safeguards in Industrial Heating Equipment," Loss Prevention Data 6-0, Factory Mutual Research Corp., Norwood, MA.

Factory Mutual Research Corporation, "Oil- and Gas-Fired Single-Burner Boilers," Loss Prevention Data 6-4, Factory Mutual Research Corp., Norwood, MA.

Factory Mutual Research Corporation, "Oil- or Gas-Fired Multiple-Burner Boilers," Loss Prevention Data 6-5, Factory Mutual Research Corp., Norwood, MA.

National Board of Boiler and Pressure Vessel Inspectors, *Boiler Owner and Operator's Guide,* The National Board of Boiler and Pressure Vessel Inspectors.

NFPA Codes, Standards, and Recommended Practices

Reference to the following NFPA codes, standards, and recommended practices will provide further information on the safeguards for boiler-furnaces discussed in this chapter. (See the latest version of *The NFPA Catalog* for availability of current editions of the following documents.)

NFPA 30, *Flammable and Combustible Liquids Code*
NFPA 54, *National Fuel Gas Code*
NFPA 68, *Guide for Venting of Deflagrations*
NFPA 8501, *Standard for Single Burner Boiler Operation*
NFPA 8502, *Standard for the Prevention of Furnace Explosions/Implosions in Multiple Burner Boilers*
NFPA 8503, *Standard for Pulverized Fuel Systems*
NFPA 8504, *Standard on Atmospheric Fluidized-Bed Boiler Operation*
NFPA 8506, *Standard on Heat Recovery Steam Generator Systems*

Additional Reading

ANSI/ASME, *Boiler and Pressure Vessel Code*, American Society of Mechanical Engineers, NY, 1995.

Davis, G. D., "Investigation and Repair of an Auxiliary Boiler Explosion," *Plant/Operation Progress*, Vol. 6, No. 1, Jan. 1987, pp. 42-45.

Ehrlich, S., "Safe Design and Operation of Fluidized Bed Systems," 13th International Conference on Fluidized Bed Combustion, May 1995, ASME, New York.

Graham and Trotman, Ltd., eds., *Boiler Operator's Handbook*, 2nd ed., State Mutual Book and Periodical Service, Ltd., New York, 1989.

Katzel, J., "Focus on Boilers," *Plant Engineering*, Vol. 44, No. 10, May 24, 1990, p. 67.

Laursen, T. A., *et al.*, "Results of the Low NO_x Cell Burner Demonstration at Dayton Power & Light Company's J. M. Stuart Station Unit No. 4," EPRI/EPA Joint Symposium on Stationary Combustion NO_x Control, May 1993, Miami Beach, FL.

Nikitenko, G., *et al.*, "Test Results of a Low NO_x Combustion System for Boston Edison's New Boston Unit 1," International Joint Power Generation Conference, Oct. 1993, Kansas City, MO.

Nikitenko, G., *et al.*, "Results of Long-Term Operation of Babcock & Wilcox DRB-XCL Burners and High-Spin Classifiers at Alabama Power Company Gaston Steam Plant," International Joint Power Generation Conference, Oct. 1993, Kansas City, MO.

HEAT TRANSFER SYSTEMS (NONWATER MEDIA)

Revised by Peter J. Gore Willse

The term "heat transfer fluid" refers to the broad spectrum of either liquid or vapor media used to transfer heat energy at a controlled rate from one place to another. While practically all industrial processes involve the exchange of heat in one form or another, the use of heat transfer fluids is especially associated with the chemical, paint, textile, plastics, food, petroleum, and paper industries. The fluids are used to heat or cool process and reaction vessels, to heat distillation stills, heat rollers for drying paper, to calender films and textiles, and for many other forms of processing.

For additional information relevant to heat transfer systems, see Section 3, Chapter 8, "Industrial and Commercial Heat Utilization Equipment"; Section 3, Chapter 9, "Oil Quenching and Molten Salt Baths"; Section 3, Chapter 19, "Chemical Processing Equipment"; Section 3, Chapter 21, "Storage of Flammable and Combustible Liquids"; and Section 4, Chapter 6, "Flammable and Combustible Liquids."

TYPES OF TRANSFER FLUIDS

If the temperature required by the transfer process is above the freezing point of water and below about 350°F (177°C), the choice is usually between these two media. On the other hand, if the temperature needed is above or below these two points, it is desirable, if not necessary, to consider other fluids. For temperatures below the freezing point of water, the most common heat transfer fluids are air; refrigerants, such as halogenated hydrocarbons; ammonia; brines; or solutions of ethylene glycol and water.

When temperatures increase above 350°F (177°C), the vapor pressure of water increases rapidly, requiring expensive high-pressure processing equipment. Therefore, a more practical and inexpensive means of heating or temperature control at higher temperatures is to use a fluid or vaporizing fluid with a high boiling temperature. Care should be taken to use high flash point, thermally stable, and noncorrosive fluids designed for this purpose. The high boiling point liquids that are commonly used can be generally categorized as either heat transfer oils, molten salts, or specially formulated heat transfer media.

There are two types of heat transfer systems: (1) liquid and (2) vapor phase. With liquid-phase systems, the temperature of the fluid changes as heat is transferred, resulting in nonuniform temperatures

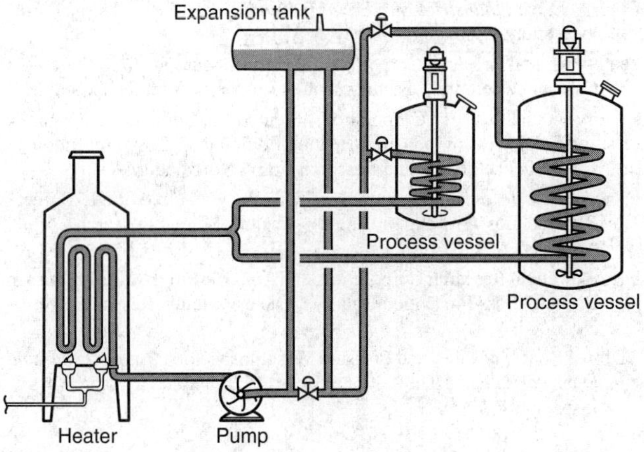

FIG. 3-7A. *Liquid-phase heat transfer system with forced circulation-type heater.*

even if large circulation rates are employed. (See Figure 3-7A.) With vapor-phase systems, heat is transferred at the saturation temperature of the vapor, which affords controlled temperatures. Combinations of these systems can be used as well. (See Figures 3-7B and 3-7C.)

Heat Transfer Oils

These are specially refined petroleum oils for use at temperatures up to approximately 600°F (315°C). At higher temperatures, the oils undergo thermal cracking to produce light hydrocarbons and polymers. Petroleum oils are also subject to oxidation at temperatures above 392°F (200°C) and in contact with air. These oils will provide a long service life and good heat transfer when the manufacturer's recommended guidelines are followed.

Molten Salts

These salts are generally used as heat absorbers or cooling media, and thus serve to control exothermic reactions that occur in the manufacture of certain chemicals. The most frequently used salt is a mixture of sodium and potassium salts having a relatively low melting point. Since these salts are solid at room temperature, it is obvious that care must be exercised to keep the temperature of the

Peter J. Gore Willse, P.E., is a research consultant at Industrial Risk Insurers (IRI), Hartford, CT. He is a member of the NFPA Technical Committee on Ovens and Furnaces.

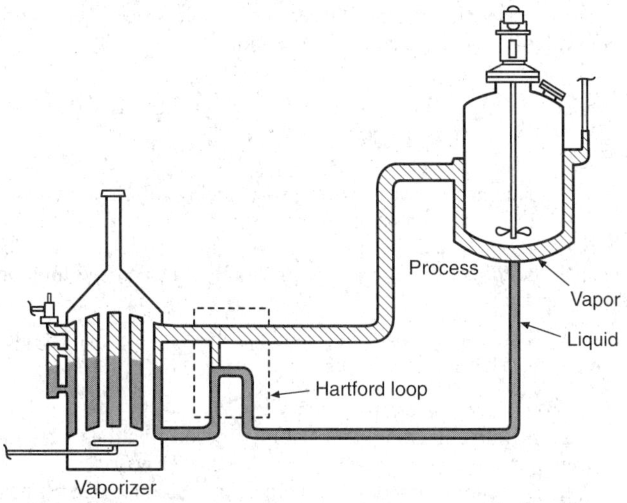

FIG. 3-7B.   *Vapor-phase gravity condensate return heat transfer system. This system has a vertical fire-tube-type vaporizer with natural circulation.*

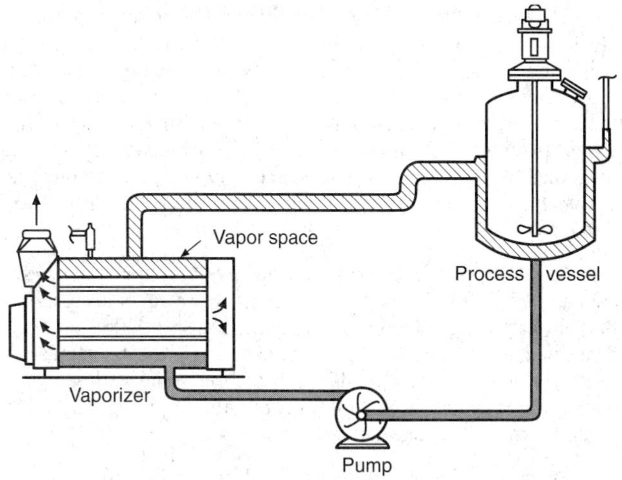

FIG. 3-7C.   *Vapor-phase heat transfer system with pumped condensate return. This system has a horizontal fire-tube vaporizer with natural circulation.*

salt above its melting point to prevent solidification that can plug the system. Also, care must be taken during startup to control the heat input so the mixture will melt slowly. Many molten salts are strong oxidizers, and this should also be considered where these salts are used.

Synthetic Fluids

A variety of specially formulated synthetic fluids are used as a heat transfer fluid. Synthetic fluids are subject to thermal cracking and oxidation; however, they are safe, noncorrosive, essentially non-toxic, and thermally stable when used under the manufacturer's recommended guidelines. The synthetics create a fire hazard if released into the firebox of a fuel-fired heater. Table 3-7A gives the physical properties of some common heat transfer fluids and oils.

When these fluids are used in the liquid state, the unit is referred to as a heater, and when the fluid is vaporized the unit is called a vaporizer. Two types of equipment are manufactured: (1) fuel-fired heaters, which are generally designed for heat duty use in excess of 1 million Btu per hr (29 280 kW), and (2) electric heaters, which are usually designed for heat duty use of less than 1 million Btu per hr (29 280 kW). Capacities of vaporizers and heaters range from a few thousand to over 175 million Btu per hr (524 694 kW).

HAZARDS OF ORGANIC HEAT TRANSFER FLUIDS

The basic hazards of organic heat transfer liquids are those associated with the heating and transfer of a combustible liquid near or above its fire point in a closed system.

TABLE 3-7A.   *Physical Properties Typical of Heat Transfer Fluids*

Compound	Freezing Point		Boiling Point		Flash Point		Fire Point	
	°F	°C	°F	°C	°F (C.O.C.)	°C	°F (C.O.C.)	°C
1,2,4-trichlorobenzene	63	17	417	213	210	99	*	*
Tetrachlorobenzene (isomer mixture)	170	77	480	249	None	—	*	*
Diphenyl ether—diphenyl eutectic	54	12	495	257	255	124	275	135
Biphenylyl phenyl ether (isomer mixture)	99	37	680	360	370	188	410	210
O-biphenylyl phenyl ether	122	50	670	354	370	188	410	210
Di- and triaryl ethers	<0	−17	572	300	305	152	315	157
Dimethyl-diphenyl ether (isomer mixture)	−40**	−40	554	290	—	—	—	—
Tetramethyl diphenyl ether (isomer mixture)	—	—	590	310	—	—	—	—
Di-sec-butyl diphenyl ether (isomer mixture)	—	—	705	374	380	193	400	204
Dicyclohexyldiphenyl ether (isomer mixture)	—	—	785	418	—	—	—	—
Dodecyldiphenyl ether (isomer mixture)	45**	7	>800	>427	410	210	440	227
Ethyldiphenyl (isomer mixture)	<−60**	<−51**	536	280	—	—	—	—
Partially hydrogenated terphenyl	−15**	−26**	690	366	335	168	375	191
Aliphatic oil	15	−9	720–950	382–510	425	218	475	246
Alkylaromatic oil	20	−6	~650	342	350	177	390	199

*None to boiling point
**Pour point

The type of vaporizer shown in Figure 3-7D is designed for relatively low heat release. The lower drum provides a space for the collection of any sludge or foreign substances within the system, preventing such materials from fouling the heating surface. Precautions must be taken to prevent the liquid level from falling below the tops of the tubes and impairing circulation. In direct-fired units, impingement of heater burner flames on empty tubes can produce excessive local temperatures, resulting in deposits and ultimate failure of the tubes.

Some organic heat transfer media have very low surface tension and viscosity. To prevent leakage, special care is necessary in the fabrication and erection of equipment. Construction in accordance with the *ASME Boiler and Pressure Vessel Code* is mandatory in nearly all U.S. and Canadian jurisdictions. The wide range in temperature requires provision for expansion of piping. High temperatures may make ordinary relief valve springs unreliable.

Because of the cost of organic fluids, it is common practice to set the blowoff points of safety valves so high that there is very little likelihood of the valve opening under normal operating conditions. To allow for this, the *ASME Boiler and Pressure Vessel Code* requires that reservoirs be designed for a working pressure of at least 40 psi (276 kPa) above the normal operating pressure, but the reservoirs should not be intentionally fired at rates above rated capacity and temperature.

Most corrosive problems result from contamination of the system by foreign chemicals. Contamination by leakage from the process side can lead to attack in the heat transfer loop. The severity and nature of attack in these instances depends on the degree and specific type of contamination. For example, chloride contamination can cause failure in the 300 series of stainless steels because of embrittlement or stress corrosion cracking.

SAFEGUARDS FOR FLAMMABLE HEAT TRANSFER SYSTEMS

Where a flammable heat transfer system having a flash point below 300°F (149°C) is heated above its atmospheric boiling point, the vaporizer or heater and the user facility should be located in a detached building or in a cutoff room of damage-limiting construction.

If located in a processing building, the vaporizer and heater room should have the following features:

1. Automatic sprinkler protection
2. Cutoff from other important facilities by construction with minimum one-hr, fire-resistance rated, fire barrier walls
3. Piping securely supported and otherwise protected against mechanical injury, with adequate clearance from combustible materials
4. Floors curbed and drained to protect nearby facilities

Equipment should always be operated within the temperature and pressure limits specified by the manufacturer of the heat transfer medium and by the manufacturer of the equipment. Good maintenance is a key factor to any accident-free operating system and should include periodic testing of the heat transfer medium, heater fire surfaces, safety valves, and control devices.

System vaporizers and heaters should be under automatic control and regulated by temperature and pressure switches in the outlet.

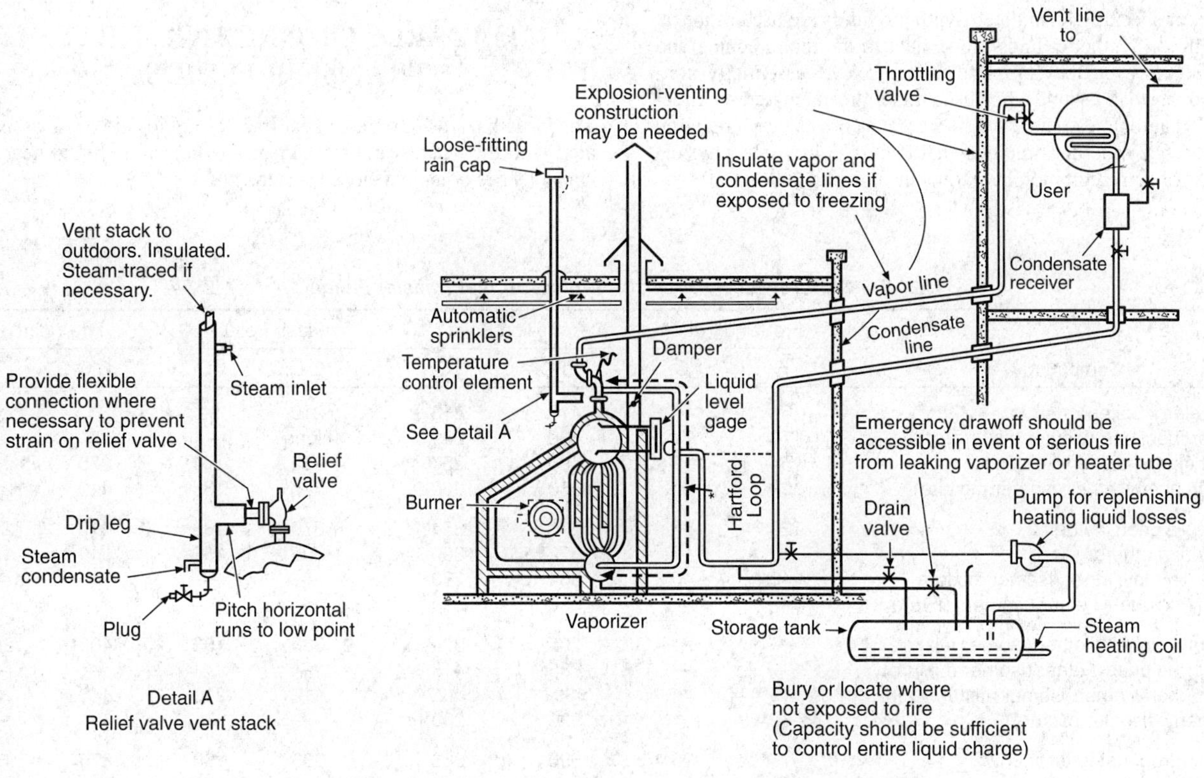

* Hartford loop prevents forcing liquid out of the vaporizer up through the condensate line after the liquid level in the upper drum drops below point C.

FIG. 3-7D. Gravity return vaporizer arrangement. (Source: Factory Mutual Research Corp.)

On vapor systems, separate safety limit switches should be provided to shut down the heater automatically and sound an alarm either when the vaporizer level becomes dangerously low, or when the vapor pressure or temperature is excessive. On circulating liquid systems, limit switches, in addition to the low liquid level cutoff, should be provided to shut off the heater automatically if liquid temperature is excessive or if circulation rate is low.

A manually controlled steam or inert gas smothering system for the firebox is advisable where a supply of steam or inert gas is available. If fire occurs in the vaporizer, follow these steps:

1. Shut off the heating units.
2. Open steam or inert gas smothering systems or use portable Class B extinguishers at openings to the firebox.
3. Unless a leak is near the bottom of the vaporizer, do not drain the liquid until after the fire has been extinguished and the unit is cooled down—otherwise, vaporizer drums and tubing may become overheated.
4. If necessary, drain liquid in the vaporizer back to the storage tank. If a storage tank is not available, drain to a safe location.

To minimize the hazard of fires from organic heat transfer media leaking and soaking into pipe insulation, heating spontaneously, and igniting, eliminate the source of leakage when it occurs by repacking valve stems, replacing leaky gaskets, etc. Cover insulation in areas of potential leakage with oil-resistant cement or use a nonabsorbent insulation, such as cellular glass or reflective aluminum foil.

BIBLIOGRAPHY

Reference

ASME Boiler and Pressure Vessel Code, American Society of Mechanical Engineers, NY, 1995.

NFPA Codes, Standards, and Recommended Practices

Reference to the following NFPA codes, standards, and recommended practices will provide further information on heat transfer systems discussed in this chapter. (See the latest version of *The NFPA Catalog* for availability of current editions of the following documents.)

NFPA 10, *Standard for Portable Fire Extinguishers*
NFPA 12, *Standard on Carbon Dioxide Extinguishing Systems*
NFPA 13, *Standard for the Installation of Sprinkler Systems*
NFPA 86, *Standard for Ovens and Furnaces*

Additional Reading

Bownan, J. W., and Perkins, R. P., "Preventing Fires with High-Temperature Vaporizers," *Plant/Operations Progress*, Vol. 9, No. 1, 1990, pp. 39-43.

Chapman, K. S., Ramadhyani, S., and Viskanta, R., "Modeling and Analysis of Heat Transfer in a Direct-Fired Continuous Reheating Furnace," American Society of Mechanical Engineers (ASME), Heat Transfer in Combustion Systems, Winter Annual Meeting, HTD-Vol. 122, December 10–15, 1989, San Francisco, CA, 1989, pp. 35–43.

Domanski, P. A, *et. al.*, "Simulation Model and Study of Hydrocarbon Refrigerants for Residential Heat Pump Systems," *Proceedings* of a Conference on New Applications of Natural Working Fluids in Refrigeration and Air Conditioning, May 10–13, 1994, Hanover, Germany, 1994.

Factory Mutual Research Corporation, "Heat Transfer by Organic Fluids," *Loss Prevention Data Sheet 7-99/12-19*, Factory Mutual Research Corp., Norwood, MA.

Fang, J. B., "Quantification of Heat Losses Through Structural Supports for Shallow Trench Heat Distribution Systems," National Institute of Standards and Technology, Gaithersburg, MD, NISTIR 89-4134, Oct. 1989, 107 pages.

Fang, J. B., "Thermal Analysis of Directly Buried Conduit Heat Distribution Systems.," National Institute of Standards and Technology, Gaithersburg, MD, NISTIR 4365, Aug. 1990, 96 pages.

Fanney, A. H. and Klein, S. A., "Thermal Performance Comparisons for Solar Hot Water Systems Subjected to Various Collector and Heat Exchanger Flow Rates," *Solar Energy*, Vol. 40, No. 1, 1988, pp. 1–11.

Fanney, A. H., "Field Monitoring of a Variable-Speed Integrated Heat Pump/Water Heating Appliance," National Institute of Standards and Technology, Gaithersburg, MD, NIST BSS 171, R&D Project RP 89-60, June 1993, 59 pages.

Fanney, A. H., "Field Monitoring of a Variable-Speed Integrated Heat Pump/Water-Heating Appliance," *ASHRAE Transactions*, Vol. 101, No. 2, 1995, pp. 1–15.

Harris, R.J., *Gas Explosions in Buildings and Heating Plant*, British Gas Corporation, E & F N Spon, Ltd., London, England, 1983.

Industrial Risk Insurers, "Organic Heat Transfer Fluids and Equipment," IRI Information IM.7.1.5, Industrial Risk Insurers, Hartford, CT, 1994.

Lee, B. T., and Walton, W. D., "Fire Experiments and Flash Point Critieria for Solar Heat Transfer Liquids," Final Report, National Bureau of Standards, Gaithersburg, MD, Department of Energy, Washington, DC, NBSIR 79-1931, Nov. 1979, 40 pages.

Peavy, B. A., and Dressler, W. E., "Transpiration Heat Transfer in Thermal Energy Storage Devices," National Bureau of Standards, Washington, DC, NBSIR 77-1237, May 1977.

Thomas, W. C., "Effects of Test Fluid Composition and Flow Rates on the Thermal Efficiency of Solar Collectors," Final Report, Virginia Polytechnic Institute and State University, Blacksburg, National Bureau of Standards, Washington, DC, Department of Energy, Washington, DC, NBS-GCR-80-254; VPI/SU-80-03; Aug. 1980.

"Voimalaitosten ja lampokeskusten sammutusjarjestelmien valinta helpottuu. [Choice of Extinguishing Systems for Power and Heating Plants To Be Made Easier.]," *Palontorjuntatekniikka*, Vol. 1, 1994, p. 34.

INDUSTRIAL AND COMMERCIAL HEAT UTILIZATION EQUIPMENT

Revised by Raymond F. Ostrowski

Heat utilization equipment has potential hazards involving both heat generation and the process materials. Fuel-fired systems, electrically heated systems, and heat transfer systems all have their individual hazards. The hazards of exposure of adjacent materials and overheating may be common to any type of heating system. The hazards and control methods are discussed in subsequent paragraphs of this chapter.

The process hazards within the equipment may involve combustible materials, flammable liquids, or flammable gases; such hazards would occur in the curing of solvent-based coating materials, roasting or drying of combustible agricultural products, or heat treating in a flammable special atmosphere. The principal method of fire or explosion prevention is to prevent overheating or an accumulation of a gas or vapor-air mixture in the explosive range. Subsequent paragraphs in this chapter describe the various classes of ovens, furnaces, and dryers, and present general process hazards. The prevention and protection methods that usually are applied to those processes and devices also are described.

Guidelines, rules, and methods applicable to the safe operation of such equipment are covered by NFPA standards and those of other organizations. These should be referred to for details that could not be included in this text. Additional information is available in this handbook in the following chapters: Section 4, Chapter 6, "Flammable and Combustible Liquids"; and Section 4, Chapter 7, "Gases." The grading of heat-producing appliances is discussed in these chapters: Section 3, Chapter 5, "Heating Systems and Appliances"; Section 3, Chapter 6, "Boiler Furnaces"; and Section 3, Chapter 7, "Heat Transfer Systems (Nonwater Media)."

Heat utilization equipment is so varied in size, complexity, location, and use that it has been difficult to develop rules that would apply to every type of oven, furnace, or dryer. Users and designers must use engineering and supervisory skills to bring together the proper combination of controls, protective devices, and operator training necessary for proper equipment operation.

Heat utilization equipment failures have resulted because someone either ignored safety procedures and designs or was unaware they existed, due to inadequate training of operators and maintenance technicians, faulty equipment design, complacency on the part of users, or improper selection of combustion safeguards.

This chapter is not a guide to solving all of these problems. However, if the guidelines and principles it presents are used, a number of problems either will be mitigated or solved.

INDUSTRIAL HEAT UTILIZATION EQUIPMENT

Industrial heat utilization equipment includes a variety of forges, furnaces, kettles, kilns, ovens, and retorts that are heated by gas, oil, solid fuels, or electricity. They may be fired directly, with the products of combustion entering the process space, or indirectly, with radiant tubes or other heat exchanger methods. Heat transfer media, such as organic fluids, also are used where steam or hot water does not provide the temperature or thermal efficiencies desired for the equipment. Figures 3-8A and 3-8B illustrate typical industrial heat utilization equipment.

Industrial ovens and furnaces, special atmosphere furnaces, vacuum furnaces, after-burners and catalytic combustion systems, dehydrators, dryers, and lumber kilns are discussed in this chapter.

Fire and Explosion Problems

When heat utilization equipment is installed, the following factors need to be considered: (1) the proximity and combustibility of the

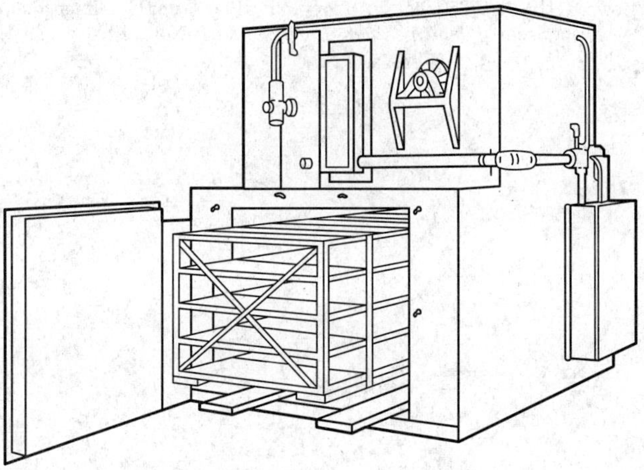

FIG. 3-8A. *A typical industrial oven.* (*Source: Electronics Corp. of America*)

Raymond F. Ostrowski works at Protection Controls, Inc., Skokie, IL. He is a member of NFPA's Technical Committee on Ovens and Furnaces.

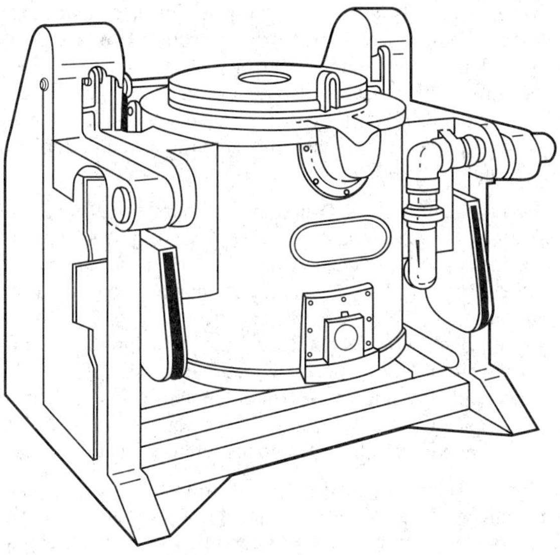

FIG. 3-8B.   A fuel fire melting furnace. (Source: Electronics Corp. of America)

contents of the building where the equipment is located; (2) construction of the building; (3) setting; (4) ventilation; (5) location within the building; (6) heat, gas, and smoke disposal; (7) maximum temperature required; and (8) handling of heated materials in connection with equipment. Fire in combustibles can be prevented by insulation or by separating them from the source of heat. Overheating can be prevented by temperature controls.

Explosion hazards exist where there are flammable vapor-air mixtures from gas or oil fuel or from the volatiles released from the material being dried. Explosions or fires may be prevented by ventilation and controls that keep the flammable vapor content below 25 percent of the lower flammable limit (LFL) of the vapor-air mixture.

Some special process ovens and furnaces contain hydrogen or other flammable gases for such purposes as annealing copper and heat treating other metal shapes. To operate these devices safely, it is necessary to prevent air from entering the processing enclosure under normal operating conditions. Both normal and unscheduled starts and stops of these devices require special safety procedures that depend heavily upon a skilled operator and adequate control equipment.

OVENS AND FURNACES

This discussion covers the location, design, construction, operation, protection, and maintenance of the industrial heating enclosures known as ovens or furnaces. It does not cover small-cabinet or stove-type ovens for domestic use. The source of heat for industrial heating may be gas burners, oil burners, electric heaters, infrared lamps, induction heaters, or steam radiation systems. In practically all cases, there are fire or explosion hazards from either the fuel used, volatiles produced by material in the oven, or by a combination of both.

There is no clear distinction between an oven and a furnace. The dictionary definition of an oven is "a compartment or receptacle for heating, baking, or drying by means of heat"; a furnace is "an enclosed chamber or structure in which heat is produced for heating a building, reducing ores and metals, baking pottery, etc." It has been an industry rule of thumb to classify heating devices that do not "indicate color" [i.e., operate at temperatures of less than approximately 1,000°F (540°C)] as ovens. This rule does not always

apply, e.g., coke ovens operate at temperatures in excess of 2,000°F (1,093°C), and some furnaces operate at temperatures below 1,000°F (540°C).

NFPA Classification of Ovens and Furnaces

The classification system for heat processing equipment as set forth in NFPA 86, *Standard for Ovens and Furnaces* (hereinafter referred to as NFPA 86), is as follows:

Class A ovens or furnaces are heat utilization equipment operating at approximately atmospheric pressure wherein a potential explosion and/or fire hazard may be occasioned by the presence of flammable volatiles or combustible material processed or heated in the oven. Such flammable volatiles and/or combustible material may, for instance, originate from paints, powder, or finishing processes including dipped, coated, or sprayed-on impregnated materials; or wood, paper and plastic pallets; or spacers; or packaging materials. Polymerization or similar molecular rearrangements and resin curing are processes that may produce flammable residues and/or volatiles. Potentially flammable materials, such as quench oil, waterborne finishes, cooling oil, etc., in sufficient quantities to present a hazard, are ventilated according to Class A standards. Ovens may also utilize a low-oxygen atmosphere to evaporate solvent.

Class B ovens or furnaces are heat utilization equipment operating at approximately atmospheric pressure wherein there are no flammable volatiles or combustible material being heated.

Class C furnaces are those in which there is a potential hazard due to a flammable or other special atmosphere being used for treatment of material in process. This type of furnace may use any type of heating system and includes the special atmosphere supply system(s). Also included in the Class C standards are integral quench and molten salt bath furnaces. See NFPA 86C, *Standard for Industrial Furnaces Using a Special Processing Atmosphere* (hereinafter referred to as NFPA 86C).

Class D furnaces are vacuum furnaces that operate at temperatures above ambient to over 5,000°F (2,760°C) and at pressures below atmospheric using any type of heating system. These furnaces may include the use of special processing atmospheres. See NFPA 86D, *Standard for Industrial Furnaces Using Vacuum as an Atmosphere* (hereinafter referred to as NFPA 86D).

Classification by Type of Handling System

Ovens and furnaces may be further classified according to the way a material is handled. The two principal types are: (1) the batch oven or furnace, sometimes referred to as "in and out," "intermittent," or "periodic" type; and (2) the continuous oven or furnace.

Batch type: In the batch oven or furnace the temperature is practically constant throughout the interior. The material to be heated is placed in a predetermined position and remains there until the process is complete. Material is then removed, generally through the opening by which it entered. Figures 3-8C and 3-8D are examples of a batch-type furnace and oven, respectively.

Continuous type: In the continuous furnace or oven, the material moves through the furnace while being heated. The straight-line furnace is probably the most common of this type. The material passes through a continuous oven or furnace on a conveyor or on rollers. The oven or furnace may operate at a constant temperature throughout, or it may be divided into zones maintained at different temperatures. Figure 3-8E is an example of a continuous-type furnace.

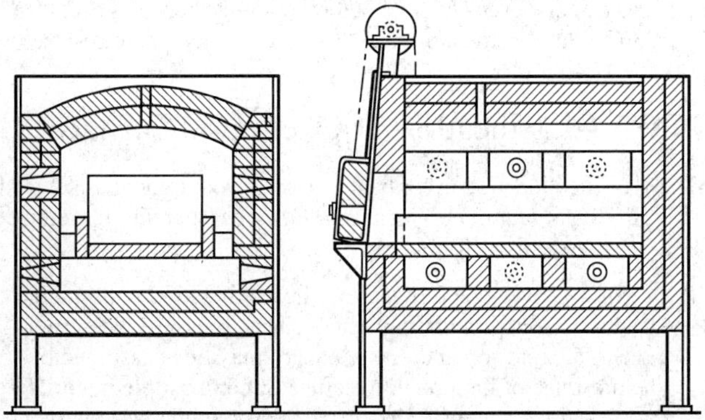

FIG. 3-8C. *A large batch-type, under- and over-fired, semi-muffle, heat treating furnace.*

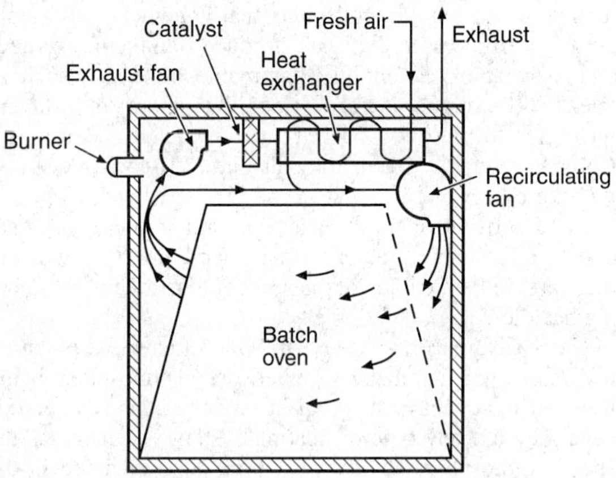

FIG. 3-8D. *Batch oven with a catalytic heater for air pollution control.*

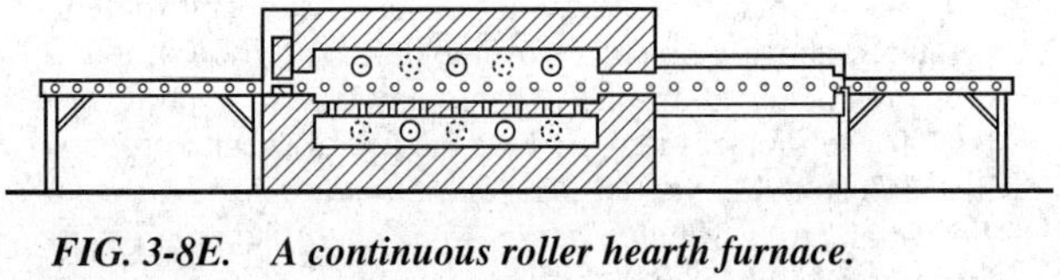

FIG. 3-8E. *A continuous roller hearth furnace.*

A common variation of the continuous-type furnace or oven is one known as a rotating hearth or rotating table furnace. In it, the material is placed on the hearth and is removed after the hearth has completed one revolution. Another design feeds the material being processed through the revolving hearth furnace or tube by means of a stationary internal screw thread.

Location and Construction of Ovens and Furnaces

Ovens and furnaces should be located where they will present the least possible hazard to life and property. To prevent or minimize damage from a fire or explosion, they may need to be surrounded by walls or partitions and located either at or above grade, because basements below grade are difficult to ventilate and offer severe obstacles to proper explosion release.

The oven or furnace and the building that houses it need to be of noncombustible construction, and explosion relief venting should be provided where required. Combustibles in the vicinity of the oven

should be adequately separated or properly insulated and each oven or furnace should have its own venting facilities. When gas or oil fuel is used, the heater and the oven or furnace should have separate venting unless the products of combustion discharge directly into the oven. Except in special cases, separate mechanical means are needed to provide air for the combustion and ventilation; natural draft is usually inadequate.

Furnaces that exhaust directly outdoors sometimes may be necessary, depending upon the heating process, type of combustion, and the hazard to personnel. If a furnace exhausts directly into a room, the room must have a balanced mechanical ventilation system to bring in fresh air and carry the exhaust outdoors. The supply inlets and exhaust outlets should be arranged to provide a uniform flow of air throughout the area without any dead-air pockets. The system should also remove any toxic contaminants at their maximum anticipated rate of release and keep concentrations below the established maximum allowable concentration (MAC) values.

Some furnaces are made of metal with a brick or masonry covering; some are made of metal with metal supports but no covering; and others have an inner lining of ceramic fibers or fire clay products. Occasionally furnaces have air spaces or noncombustible fillers between the outer and inner walls.

Ovens and furnaces should be well separated from valuable stock, important power equipment, machinery, and automatic sprinkler risers, so there will be minimum interruption to production and protection if there are accidents to the oven or furnace. They should be readily accessible with adequate space above them for automatic sprinklers, the proper use of hose streams, the proper functioning of explosion vents, and routine inspection and maintenance. Roofs and floors of ovens should be insulated. The space above and below the ovens should be ventilated to keep temperatures at combustible ceilings and floors below 160°F (71°C).

Oven and Furnace Heating Systems

For the purpose of this chapter, the term "furnace heating system" includes the heating source, associated piping and wiring used to heat the furnace, auxiliary quenches, and the work therein.

Note 1: For the protection of personnel and property, careful consideration should be given to the supervision and monitoring of conditions that could cause, or could lead to, a real or potential hazard on any installation.

Note 2: The presence of safety equipment on an installation cannot, in itself, ensure absolute safety of operation.

Note 3: There is no substitute for a diligent, capable, well-trained operator.

Note 4: Highly repetitive operational cycling of any safety device can reduce its life span.

The three most common methods of transferring heat to materials in ovens or furnaces are: (1) direct contact with the products of combustion, (2) convection and direct radiation from the hot gases, and (3) reradiation from the hot walls of the furnace. In muffle furnaces, the products of combustion are separated from the material being heated by a metal or refractory muffle, and heat transfer occurs by radiation. (See Figure 3-8C.) In liquid bath furnaces, i.e., salt baths or molten metal used for tempering, hardening, galvanizing, tinning, etc., a metal pot containing a liquid is heated and the heat is transferred through the liquid to the material placed in the pot.

Oven heaters: The two general types of oven heaters are: (1) direct fired and (2) indirect fired. In direct-fired heaters, the products of combustion enter the work chamber and contact the work in process; this is not the case in indirect-fired oven heaters. Instead, heating comes from radiation from tubes or from air passing over tubes and into the oven. Dangerous fuel-air mixtures cannot readily fill the work space of an indirect-fired oven. Nevertheless, explosions may still occur from vapors given off by a flammable liquid drying process.

There are several arrangements of these two types of oven heaters—direct-fired internal, direct-fired external, indirect-fired internal, and indirect-fired external heaters. Figure 3-8F shows three variations of direct-fired external heaters, two of indirect-fired internal heaters, and three of indirect-fired external heaters, with the advantages and disadvantages for each type. The exhaust ventilation arrangements in these designs provide for air movement through the oven with no recirculation through the exhaust fan. The direct-fired external-type

oven may have a single burner or a relatively small number of burners that simplify the completing of automatic safety controls.

Furnace heaters: Furnace heaters are usually arranged like ovens—direct-fired internal, indirect-fired external, etc., though sometimes the terms are different. If products of combustion are under a hearth and then carried up and into the heating chamber, the furnace is said to be underfired. When the same thing occurs in a chamber at one side of the furnace and passes over a bridge wall into the heating chamber, the furnace is referred to as side-fired. A furnace in which the products of combustion are produced in a space above the heating chamber and pass through a perforated arch into the heating chamber is called an over-fired furnace. If combustion occurs at some distance above the heating chamber and hearth, and the products of combustion are deflected onto the hearth by an arched roof,

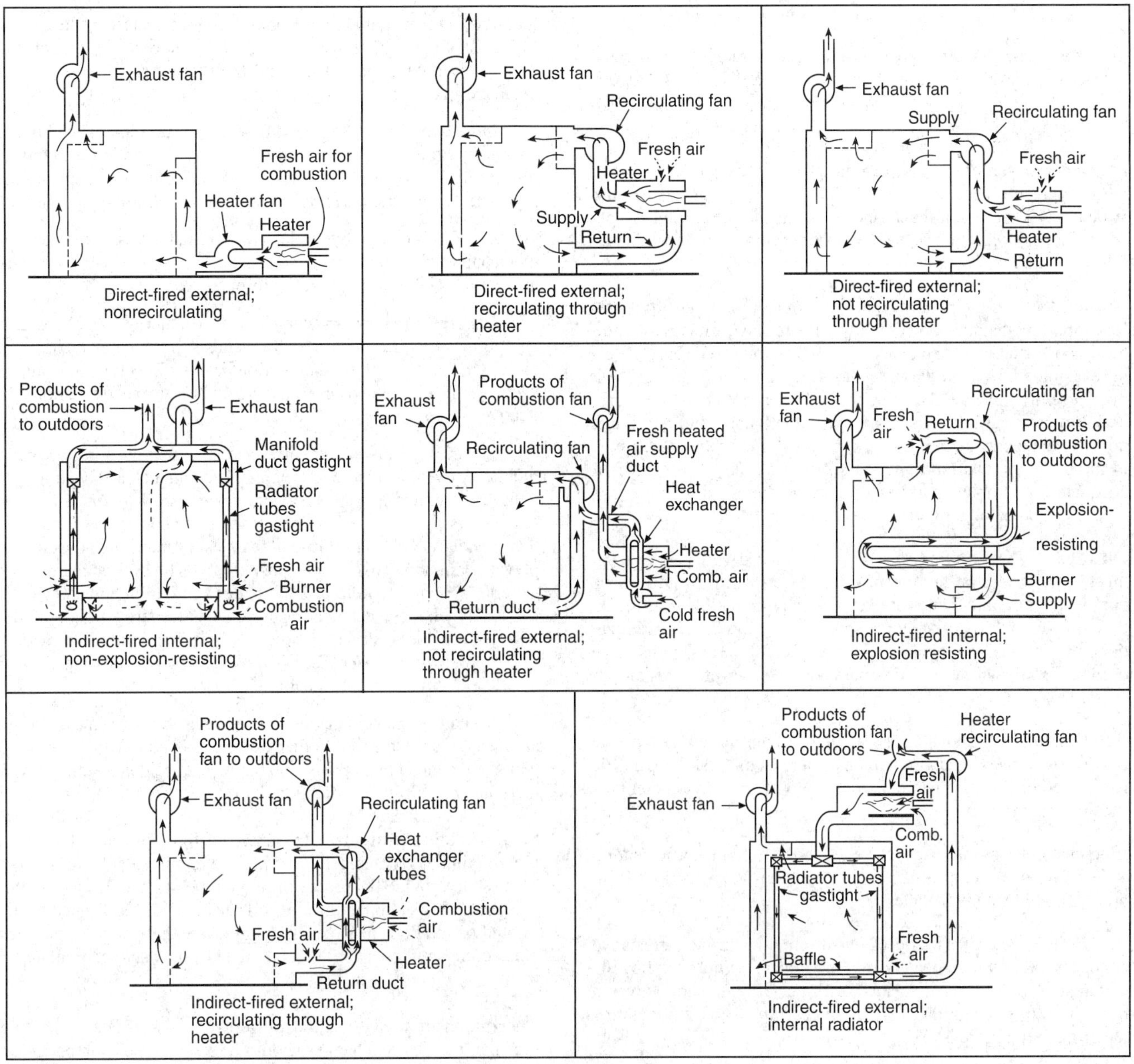

FIG. 3-8F.   Types of oven heating systems.

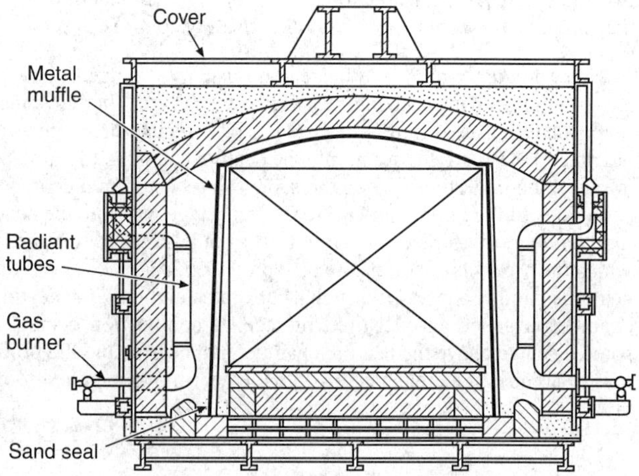

FIG. 3-8G. *An annealing muffle furnace.*

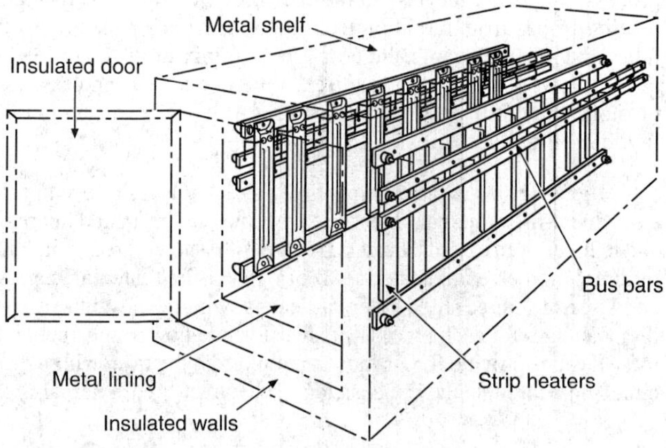

FIG. 3-8H. *Strip heaters mounted in a small oven.*

the furnace is called a reverberatory furnace. A radiant-tube-heated furnace is an arrangement for indirect firing. (See Figure 3-8G.)

Sources of Heat

Heat for an oven or furnace may be provided by gas burners, oil burners, electric heaters, steam radiation, or heat transfer medium systems. It is important that flames, heating surfaces, or other possible sources of ignition be located where drippings or dust cannot fall or accumulate on them.

Gas fired: Gas fuel is any type of gas in common industrial use. It is important that the burner, its adjustment, and the means of combustion be suitable for the type of gas to be burned. The burner may have a single nozzle, with burners located singly or in groups, or it may have multiple nozzles in perforated-pipe, ribbon, or slot burners. Burners must light easily, have a stable flame at all ports, and not have the tendency to flash back or blow off over the entire range of turndown under all draft conditions in the oven. A supply of air adequate for complete combustion may be premixed with the gas, nozzle-mixed at the burner, or otherwise provided.

Oil fired: Oven heating systems may be fired with fuel oil. Oil must be vaporized before it can be burned. Most oven and furnace burners atomize and then vaporize oil while mixing it with combustion air. Heat for vaporization is generated by the flame. Atomization may be caused by pressure, mechanical spraying into fine droplets, sonic vibration, low-pressure air, high-pressure air, or steam.

Combination gas- and fuel-oil fired systems use either separate gas and oil burners, or configurations where an oil atomizing nozzle is centered on a gas burner and arranged to use a common combustion air source.

Electric heating systems: There are five types of electric heating systems for ovens and furnaces: (1) resistance, (2) infrared, (3) induction, (4) arc, and (5) dielectric.

Resistance heat is produced by current flow through a resistive conductor. Resistance heaters may be "open," with bare heating conductors, or "insulated sheath," with heater conductors covered by a protective sheath that may be filled with electrical insulating material. (See Figure 3-8H.)

Infrared heat is transmitted as electromagnetic waves from incandescent lamps with filaments that operate at temperatures lower than the filament temperature of ordinary incandescent lamps, so that most of the radiation occurs in the infrared part of the spectrum. These waves pass through air and transparent substances but not opaque objects, and release their heat energy to these objects.

Induction heat is developed by electric currents induced in the charge by an externally applied alternating magnetic field. Induction heaters have an electric coil surrounding the oven space, and heating is by electric currents induced in the work being processed.

Arc heat is caused by an electric current that passes between either a pair of electrodes or between electrodes and the work, causing an arc that releases energy in the form of heat.

Dielectric heat occurs when dielectric materials are exposed to an alternate electric field. The frequencies are generally 3MHz or more—higher than those in induction heating. This type of heater is useful for heating materials that are commonly considered nonconductive.

Electric systems can be arranged so that processing does not require an oven enclosure. "Ovenless," i.e., unenclosed heating systems, can employ lamps, resistance-type electric elements, or infrared heaters to vaporize flammable, toxic, or corrosive liquids and their residues. Enclosures around ovenless systems are advisable to prevent flammable, toxic, or corrosive vapors from escaping into the general area, and to help provide better ventilation and safeguards for personnel. However, heating systems with energy input of under 100 kW may be excluded if adequate area ventilation is provided. (NFPA 70, *National Electrical Code®*, gives guidance for electrical installations in hazardous locations.)

All parts of heaters that operate at elevated temperatures within an oven or furnace and all other energized parts must be protected to prevent contact by persons, and to prevent accidental contact with materials being processed and with drippage from the materials.

Steam heating systems: In steam heating systems, the steam pressure in heat exchanger coils must be regulated at the minimum required to provide the proper drying temperatures. This avoids unnecessarily high temperatures at coil surfaces. The coils must not be located on the floor of the oven or anywhere that paint drippings or other combustibles, such as recirculated lint, may accumulate on them.

Thermal heat transfer fluid systems: As with steam heating systems, the heat exchanger should be kept free of combustibles, such as lint. The heat exchanger may be located external of the work chamber and a recirculating heated air system used.

Fuel Hazards

Gas or vapors from unburned or incompletely burned fuel may, when mixed with air, be within the explosive range. These hazards develop during lighting off, firing, and shutting down the oven or furnace; therefore, it is necessary to treat each operation as a separate condition requiring certain specific operating procedures to avoid mishaps.

Lighting off: Before torches, sparks, or other ignition sources are introduced, and until all burners are properly lighted, the operator must take every precaution, using all practical automatic safety controls, to avoid producing dangerous unburned fuel accumulations. The following precautions should be followed:

1. Purge possible accumulations of unburned fuel.
2. Ignite burners promptly with substantial igniters.
3. If another ignition attempt is necessary, purge before introducing the ignition source and fuel again.

Firing: In the firing phase, safety requires continuous ignition and complete burning of the fuel before it passes beyond its normal combustion zone. To maintain this, the mixer and burner assembly must proportion fuel and air properly throughout the combustion zone, and the mixture velocity in the combustion zone must be neither too high, causing extinguishment by blowoff, nor too low, causing the flame to flashback or "go out." Thus good burner mixer design will be one of the main factors in safety during firing.

Air for combustion is obtained from the primary and secondary air supplied at the burner. Partial or total failure of the combustion air supply can cause an unstable flame, which in turn may lead to flame failure and the introduction of unburned fuel into the combustion chamber. When too little air is supplied, the result is an over-rich mixture and incomplete combustion. The flammable incomplete products of combustion that leave the burner at concentrations too high for prompt ignition may later, within the oven or ductwork, become diluted by air into the flammable range and ignite, causing an explosion. Over-rich combustion also may produce rapid smothering and extinguishment of the burner flame. The flammable products of incomplete combustion followed by raw fuel likewise may become explosive when later diluted by air in another part of the system. Therefore, precautions must be taken to shut off the fuel and to require manual reset in the event that the air supply for combustion fails.

Liquid fuel, like fuel oil, must be atomized so that it will ignite easily and burn quickly. This can be accomplished by ejecting the liquid fuel at high pressure, or by directing steam or air jet into the oil stream. Improper oil temperature or high viscosity can prevent proper flow, and can create partial obstructions in burner tips and loss of oil or atomizing medium pressures, causing improper atomization. Failure to atomize properly will usually result in an unstable flame, which in turn can lead to flame failure. Other conditions that may cause flame failure are stoppage of fuel supply by an improperly closed fuel valve or other pipe obstruction, and the presence of water in a fuel oil line.

Shutting down: Following a shutdown, a dangerous accumulation of unburned fuel may occur in an oven and heating system if any manual fuel valves are left open or are leaking, and/or safety shutoff valves are not tight-closing. If the leaking fuel subsequently is ignited by hot refractory material or by another ignition source when starting up, an explosion may occur.

Supervisory Controls for Ovens and Furnaces

It is essential that all ovens and furnaces processing flammable materials, involving flammable vapors, or heated with combustible fuels be provided with adequate supervisory devices that ensure sufficient pre-ventilation, adequate ventilation during operation, and proper operating conditions that will not permit fires or explosions to develop. All safety devices must be listed for the service intended. While it is true that a competent operator is essential, it is also true that assistance is needed because the operator cannot continually supervise everything. The operator is, however, responsible for the proper maintenance and testing of the oven's operating and control equipment.

The type of supervisory control equipment used with oven or furnace assemblies depends upon the requirements of the particular operation. A list of the principal types of supervisory controls is given in Table 3-8A. Figure 3-8I shows a typical arrangement of the devices for a gas-fired unit. Most of these units are tested by testing laboratories and listed according to the appropriateness of the devices for varying situations.

Equipment for Ovens and Furnaces

Full supervisory control of a gas- or oil-fired oven, especially one that is direct fired, may require the following: (1) determining that all fuel valves are closed and not leaking, (2) establishing and maintaining required ventilation, (3) turning on and igniting the gas pilot only at the conclusion of a required pre-ventilation period, and (4) opening the fuel valves (including the safety shutoff valve) that supply the burner only after the combustion safeguard shows that the pilot flame has been ignited, and that fuel and air (or other atomizing agent) pressures are correct. If the burner flame is not promptly established, the safety shutoff valve will close and the entire cycle must be repeated, starting from the beginning. Once started, operation continues only as long as the supervisory controls

TABLE 3-8A. *Supervisory Control Equipment for Ovens and Furnaces*

Ventilation controls
  Airflow switches
  Pressure switches
  Fan shaft rotation detectors
  Dampers
  Position limit switches
  Electrical interlocks
  Preventilation time-delay relays

Fuel supervisory controls
  Safety shutoff valves
  Supervising cock (FM cock)
  Flame detection units (combustion safeguards)
  Flow meters
  Firechecks
  Reliable ignition sources
  Pressure switches
  Program relays

Temperature controllers
Excess temperature limit controller
Continuous vapor concentration indicators and controls
Conveyor interlocks (with steam and electric-resistance heating equipment)
Electrical overload protection (with resistance and induction heating equipment)
Low oil-temperature limit controls (on oil-burner equipment using heavy residual fuel oil, such as No. 5 and No. 6, which require preheating)

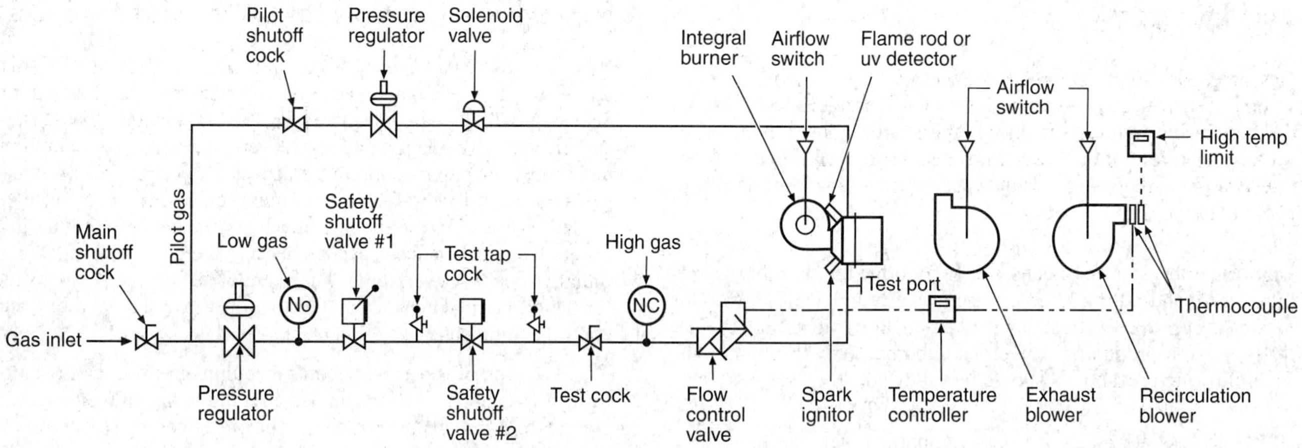

FIG. 3-8I. *Example of a gas piping diagram.*

indicate normal conditions; failure in any respect shuts down the entire system, and the full cycle must begin again.

NFPA 86, NFPA 86C, and NFPA 86D specify in considerable detail the safeguards needed for different types of ovens and furnaces.

When installations contain a large number of burners, the usual supervisory controls for each pilot and each burner may not be considered practical; in such cases, special devices and arrangements may have to be employed. A special continuous line pilot for multiple burners is illustrated in Figure 3-8J.

Controls for a large number of cup burners with flame propagation between individual burners may be provided as shown in Figure 3-8K.

Multiburner combustion safeguards are now available, making it possible to install flame supervision for each burner, where previously the supervisory cock system was the only safeguard available.

When the capacity of a gas-fired burner system exceeds 400,000 Btu/hr (117 kW), two safety shutoff valves are required. A manually operated plug cock and test tap must be provided downstream of the safety valves. This arrangement ensures positive fuel shutoff when testing the valves for leakage.

Safety ventilation is not needed in an oven that never contains flammable or noxious vapors, and it is unnecessary to guard against flammable fuel-air mixtures in an electric or steam-heated oven.

Operator Training

Alert and competent operators are essential to safe operations. New operators should be thoroughly trained and tested in the use of the equipment. Regular operators should be retrained at intervals to maintain proficiency and effectiveness. Operators must have access to operating instructions at all times.

Operating instructions should be provided by the equipment manufacturer. These instructions should include schematic piping and wiring diagrams, as well as: (1) light-up procedures, (2) shutdown procedures, (3) emergency procedures, and (4) maintenance procedures.

Operator training should include information on: (1) combustion of air-fuel mixtures, (2) explosion hazards, (3) sources of ignition and ignition temperature, (4) atmosphere analysis, (5) handling of flammable atmosphere gases, (6) handling of toxic atmosphere gases, (7) functions of control and safety devices, and (8) purpose and basic principles of the gas atmosphere generators.

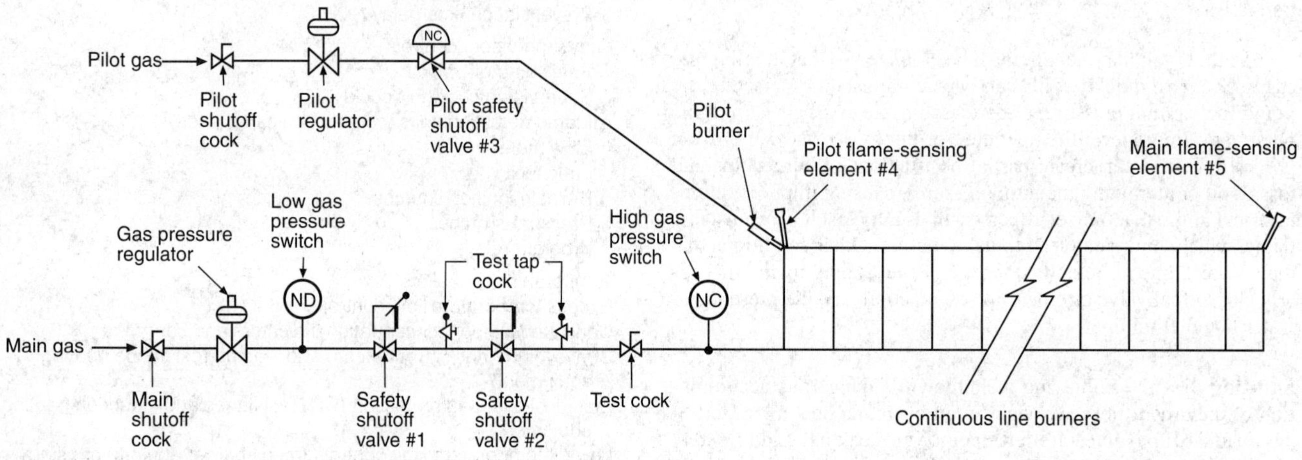

FIG. 3-8J. *Example of an approved combustion safeguard supervising a pilot for a continuous line burner during lighting-off and the main flame alone during firing.*

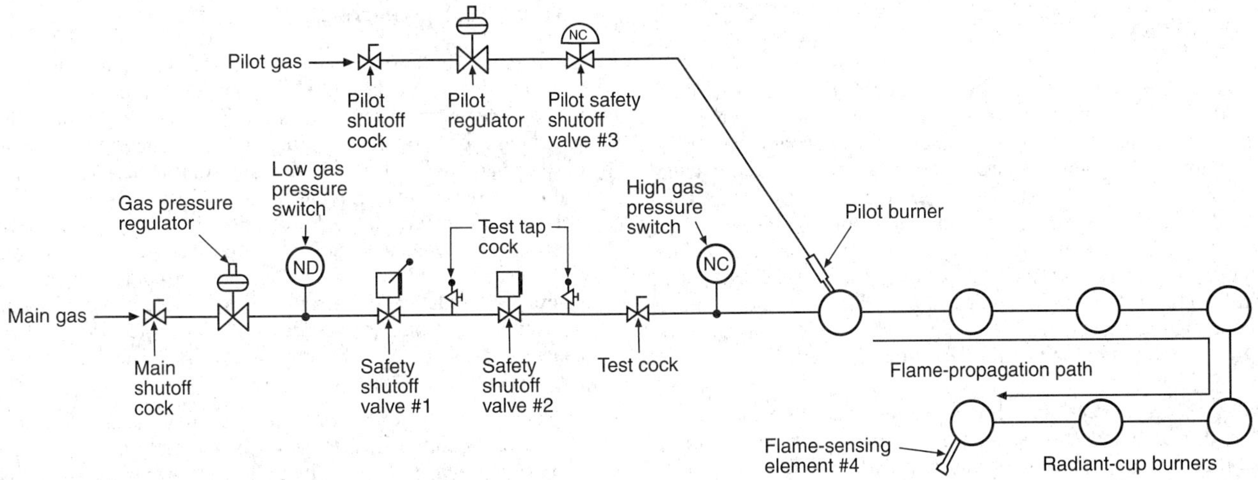

FIG. 3-8K. *Example of an approved combustion safeguard supervising a group of radiant-cup burners having reliable flame-propagation characteristics from one to the other by means of flame-propagation devices.*

Testing and Maintenance for Ovens and Furnaces

The operating and supervisory control equipment of each oven should be checked and tested regularly, preferably once a week. At less frequent intervals, probably annually, a more comprehensive test and check must be performed by an expert. All deficiencies must be corrected promptly, and a regular cleaning program followed to cover all portions of the oven and its attachments. Access openings for cleaning the oven enclosure and the connecting ducts must be provided.

A program for inspecting and maintaining oven safety controls is given in the appendices of NFPA standards for ovens and furnaces.

CLASS A OVENS AND FURNACES

Adequate ventilation must be provided in the operation of Class A ovens and furnaces where there is an explosion potential because of the presence of flammable vapors or fuel-air mixtures. Such fires or explosions may, in general, be prevented by good ventilation and supervisory controls that keep the flammable vapor content well below the LFL.

Ventilation

Ventilation is required while an oven is in operation and flammable vapors are given off. Control devices ensure that the ventilating and pre-ventilating systems are operating. Failure of the ventilating fan causes shutdown of the heating system and the conveyor that carries material into a continuous oven. The following discussion refers only to that ventilation required for safe operation, and not that required for combustion and recirculation. It also does not apply to ovens operating in conjunction with solvent recovery systems, which may use a low-oxygen atmosphere.

Proper ventilation includes a sufficient supply of fresh air, proper exhaust to outdoors, and properly distributed air circulation sufficient to ensure that the flammable vapor concentration throughout the oven is safely below the LFL at all times. The quantity of fresh air required for safe ventilation is determined by the amount of vapor released during the process.

In general, mechanical ventilation to outdoors is required on all ovens in which flammable or toxic vapors are liberated, as well as for ovens heated by direct-fired gas or oil heaters.

Ovens in which flammable or toxic vapors are never released do not require ventilation for safety, if heated by steam or electric energy, or by gas- or oil-fired indirect heating equipment. On new ovens of every size, ventilation provided by a separate exhauster is advisable whenever appreciable amounts of flammable vapors are given off by the work.

Continuous conveyor ovens: The general rule for ventilating continuous conveyor-type ovens is to provide not less than 10,000 cfm (283 m³/min) of fresh air at 70°F (21°C) for each 1 gal (3.79 L) of common solvent introduced into the oven. The basis for this rule is that 1 gal (3.79 L) of common solvent produces a quantity of flammable vapor that will diffuse in air to form roughly 2,500 cu ft (71 m³) of the leanest explosive mixture. Because a considerable portion of the ventilating air may pass through the oven without completely traversing the zone in which vapors are given off, the ventilation air may be distributed unevenly. To provide a margin of safety, four times this amount of air, or 10,000 cu ft (283 m³) [referred to 70°F (21°C)], is required for each 1 gal (3.79 L) of solvent evaporated. In certain solvents, however, the volume of air rendered barely explosive exceeds 2,500 cu ft (71 m³), and the safety factor decreases proportionately.

When a continuous-type oven is designed to operate with a particular solvent and the ventilating air can be accurately controlled, the required ventilation can be determined by calculation. [See NFPA 86 for information on how to calculate the volume of vapor produced by 1 gal (3.79 L) of solvent and the volume of air required to provide sufficient dilution to prevent an ignitable mixture.] As with the general rule for oven ventilation, the calculated rate of air change includes a factor of safety four times the volume of air required to prevent an ignitable mixture.

Batch process (box) ovens: The nature of the work being processed is the basis for determining the ventilation rate in batch process ovens. Because of the wide variations in the materials, rate of evaporation, and coating thickness, it is preferable to make tests and calculations to figure the proper ventilation rate. However, years of testing and experience have shown that approximately 380 cfm (10.7 m³/min) [referred to 70°F (21°C)] of ventilation for each 1 gal (3.79 L) of flammable volatiles released from a batch of sheet metal or metal parts being baked after dip coating is a reasonably safe rate of air change.

For other types of work, the figure of 380 cfm (10.7 m³/min) [referred to 70°F (21°C)] is also used, unless the required ventilation rates can be calculated from reliable previous experience, or the maximum evaporation rate is determined by tests run under actual operating conditions. In the latter case, a safety margin requires a rate of air change equaling four times the volume of air needed to produce an ignitable mixture. In any event, caution is needed when applying this estimating method to work of low mass, e.g., paper, textiles, etc., which will heat quickly, or work coated with materials containing highly volatile solvents. Either condition may give too high a peak evaporation rate for the estimating method.

Temperature correction: Temperature corrections must be made when using the above rules, since the volume of a gas varies in direct proportion to its absolute temperature. (0°F is equivalent to approximately 460° abs, and 0°C is equivalent to approximately 273K).

For example, to supply 10,000 cu ft of fresh air referred to 70°F (530° abs) to an oven operating at 300°F (760° abs), it is necessary to exhaust:

$$\frac{760}{530} \times 10,000 = 14,320 \text{ cu ft of } 300°F \text{ air.}$$

To supply 283 m³ of fresh air referred to 21°C (294 K) to an oven operating at 149°C (422 K), it is necessary to exhaust:

$$\frac{422}{294} \times 283 = 406 \text{ m}^3 \text{ of } 149°C \text{ air.}$$

In some cases, process requirements call for more ventilation than needed to maintain safe conditions in an oven. When this is true, an approximate method of figuring ventilation may be adequate for checking safety requirements. Except in these cases, all factors, including solvent characteristics, type of oven, material being processed, oven temperatures, and effect of temperature on the LFL, must be carefully considered so that an adequate safety factor is ensured.

Low-oxygen ovens: Low-oxygen ovens are used for evaporation of solvent in an atmosphere where the oxygen concentration is below the flammability level of the solvent. For some solvents this operating value would be approximately 8 percent oxygen, which includes a safety factor of approximately 3 percent. There is a need for oxygen analysis to verify the low oxygen concentrations. Inert gas is used to displace the oxygen (air) during startup, and combustibles during shutdown. These systems are used for solvent recovery from the coating cure process in the oven as shown in Figure 3-8L. The treated product passes into and out of the oven enclosure through the oven openings. The oven atmosphere consists of an inert carrier gas that is continuously recirculated through the oven enclosure (line 1). Solvents evolve from the treated product and build up to an equilibrium vapor level that is much higher than the allowable levels in an air atmosphere. A bleed stream (line 2), which is typically 1 percent of the recirculation flow, is processed by a solvent recovery system. A liquid (line 3), which provides cooling through vaporization, acts to condense the solvents that are pumped to storage (line 4). After solvents are stripped from the inert gas stream, it is returned directly to the oven enclosure (line 5) to resume its role as a solvent vapor carrier.

Fire and Explosion Protection

Automatic sprinklers and water spray systems should be considered for heat processing equipment that contains or processes sufficient combustible materials to sustain a fire. The amount of protection required depends upon the construction and arrangement of the oven and the materials handled in it. If combustible material is processed, or if trucks or racks are combustible (or subject to loading with excess finishing material), fixed protection must extend as far as necessary into the enclosure and exhaust ducts. It also should be present where an appreciable amount of flammable drippings from finishing materials accumulates within the oven. If desired, supplementary carbon dioxide, foam, dry chemical, or halon protection may be permanently installed, but such protection is not a substitute for automatic sprinklers.

The use of steam for fire protection in ovens and dryers generally is not recommended. However, when there is no alternative, steam smothering systems may be allowed when oven temperatures exceed 225°F (107°C) and large supplies of steam are available at all times. Complete standards have not been developed for the use of steam as an extinguishing agent. Steam is not as dependable as water, carbon dioxide, dry chemical, halon, or foam.

Portable extinguishers are needed near the oven, oven heater, and related equipment, including dip tanks or other finishing processes operated in conjunction with the oven. Small-hose stations with combination nozzles also should be provided so all parts of the oven structure can be reached.

Ovens that may contain flammable gas or vapor mixtures must be equipped with unobstructed relief vents to release internal explosion pressures. These vents, panels, or doors may be secured with explosion-relieving hardware or gravity-retained panels that provide adequate insulation and possess the necessary structural strength. The weight of the panel should be minimal to permit movement at the lowest practical pressures. Explosion-relief panels must be proportioned according to the ratio of area to the explosion-containing volume of the oven, with due allowance made for openings or access doors equipped with approved explosion-relieving hardware. The preferred ratio is 1 sq ft (0.09 m²) of relief panel area to every 15 cu ft (0.42 m³) of oven volume.

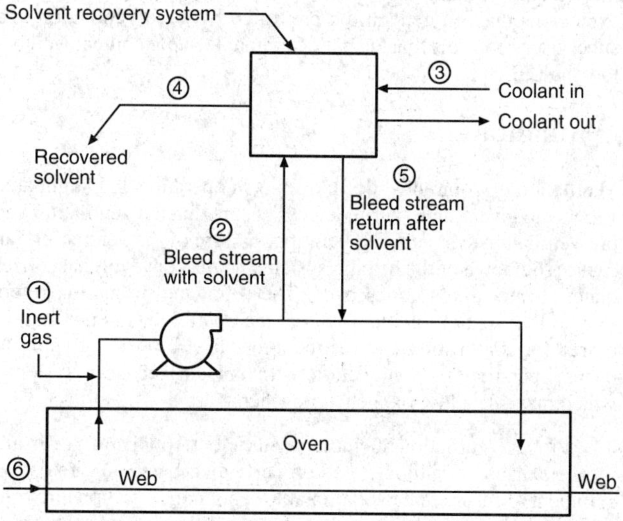

FIG. 3-8L. An example of a low-oxygen oven with a solvent recovery system.

CLASS B INDUSTRIAL FURNACES

In many ways, Class B furnaces are similar to Class A ovens and furnaces. But, because no flammable volatiles or residues are present, there is nothing combustible in the construction or contents of the Class B furnace.

In many Class B furnaces, little or no effective explosion-relief venting can be provided because of the weight and strength of the walls. Pre-ventilation is important before a source of ignition is introduced into the furnace. In some cases, completely automatic purging may be practical; in others, purging may be partly manual. In either case, supervisory controls are needed to interlock the ventilation, fuel supply, combustion air, safety shutoff valve, and flame failure devices.

In multiburner installations, combustion safeguards with flame supervision should be applied. When furnaces have zones operating at different temperatures, the zones may have to be treated as separate units.

Changes in the usual safety requirements may be permitted in some ovens or furnaces, but if there is any possibility of explosion from unburned fuel, adequate safeguards should be provided. An audible alarm may be installed to indicate unsafe conditions.

CLASS C INDUSTRIAL FURNACES USING A SPECIAL PROCESSING ATMOSPHERE

Special atmosphere furnaces are used to improve the quality of metals and metal alloys by heating them in an atmosphere in which air has been replaced by other gases, some of which are combustible. In most cases, the atmosphere gas is used to prevent oxidation of the metal during heating, but it also may prevent the removal or addition of carbon. Some processes that use atmospheric gases are bright annealing of copper and steel, scale-free hardening and annealing of castings, brazing, and sintering. Examples of protective gases used in these processes are hydrogen, charcoal gas, and dissociated ammonia. Also used are various hydrocarbon gases produced by equipment that processes the gas used for firing, generally in the presence of a catalyst. During the past few years, synthetic atmospheres that include methanol and other stored gas components have been used.

Some heat-treating furnaces contain an inert atmosphere (carbon dioxide, helium, argon, nitrogen) so they present no special fire hazard; however, these gases could present a health hazard. Other gases produce a flammable atmosphere (hydrogen, dissociated ammonia, incompletely burned hydrocarbon gas, carbon monoxide, and methane), which presents an explosion hazard; therefore their use requires special safeguards. Atmosphere furnaces are not limited to flammable gases but also include acid gases, such as chlorine and anhydrous hydrochloric acid. When the latter are used in a special atmosphere, extreme care must be taken to keep air from entering the furnace. Regardless of the type of special atmosphere, Class C furnaces have fuel hazards similar to those of the Class A and B ovens and furnaces discussed earlier (although most Class C units are indirectly fired or electrically heated). The hazards in Class C furnaces exist chiefly at three times: (1) before the process starts and the flammable atmosphere is replacing air in the furnace; (2) when the process is finished and air is being readmitted; and (3) when, for some reason, the special atmosphere supply is interrupted and air is permitted to enter. If at each of these times sufficient inert gas can be introduced into the furnace to prevent a combustible mixture, there is no danger of an explosion. Automatic introduction of inert gas upon failure of the special atmosphere supply is desirable; at the least, an audible alarm should notify the oven operator of atmosphere supply failure or other upset conditions.

Inert Gas Purge Procedure

Begin by verifying the adequacy of the inert gas supply. After all doors (if any) are closed, make sure that the flammable atmosphere gas, flame curtain, and other valves are closed. The furnace may then be heated to operating temperature. If so, introduce the inert gas at a rate capable of maintaining a positive pressure in the furnace. Sample the furnace atmosphere until two consecutive readings indicate the oxygen content is below 1 percent. With at least one furnace zone above 1,400°F (760°C), ignite the pilots at the outer doors and effluent vents. The flammable special atmosphere then may be introduced. When the flammable atmosphere is flowing, the inert gas should be turned off immediately. When flames appear at the vestibule effluent ports, the atmosphere introduction has been completed. The curtain burners then may be ignited.

To remove the flammable atmosphere with the furnace at operating temperature, the doors (if any) should remain closed, the inert gas purge actuated, and a positive furnace pressure maintained. Shut off the special atmosphere, flame curtain, and other valves; and sample the atmosphere until two consecutive analyses indicate the atmosphere is below 50 percent of its lower explosive limit (LEL). The furnace is then purged, and the doors may be opened, and the inert gas turned off.

The normal procedure for startup or shutdown of most Class C furnaces when the temperature is above 1,400°F (760°C) is to burn out the air at the start of the process and the flammable atmosphere at the finish. In furnaces with an operating temperature above 1,400°F (760°C), burning may be performed at the start by bringing the temperature up to 1,400°F (760°C) before the special atmosphere is introduced. Automatic means are required to prevent the introduction of flammable fluids into a furnace before the furnace temperature has risen to 1,400°F (760°C). At this temperature the flammable gas will burn in the oven until the oxygen is used up and the hazard removed, allowing the operation to be started. At the end of the process, the heating should be continued so the temperature stays above 1,400°F (760°C); then the flammable gas supply is shut off and air gradually admitted. When burning stops, the flammable gas has been consumed. When an alarm indicates failure of the flammable gas supply or of the heating system, the oven operator must immediately initiate an inert gas purge, or start the admission of air to burn the flammable gas in the furnace. This must be done before the furnace cools to below 1,400°F (760°C). Where operators are not present and flammable gas flow is interrupted because of insufficient temperature inside the furnace, a flow control unit should automatically admit a flow of inert gas that will restore positive pressure without delay.

Special Atmosphere Generators

Special atmosphere generators are a source for the atmospheric gases used in some Class C furnaces. One type of generator (i.e., exothermic) produces the atmospheric gas by completely or partially burning fuel gas at a controlled ratio, usually at 60 to 100 percent aeration, while another type (i.e., endothermic) produces the atmosphere at a controlled ratio of less than 50 percent. Atmospheres from an exothermic generator may be either inert or flammable, depending upon the generator design and operating range, while those from endothermic generators are always flammable.

Another type of generator is an ammonia dissociator, which, by temperature reaction with a catalyst in an externally heated vessel, produces dissociated ammonia (25 percent nitrogen and 75 percent hydrogen) from ammonia.

The special atmosphere generator must be provided with adequate supervisory controls. These would normally include interlocking of raw gas, air (if needed), burners for heating or processing, feed and discharge pressures, etc. Safety shutoff valves are usually provided in feed and discharge piping, and devices also are provided to indicate the pressure or rate of flow of processed gas to the furnace, and its analysis. This can vary. The operator is thus assisted in his/her efforts to supply the furnace with the desired special atmosphere.

The best location for the generator and its auxiliary equipment, such as surge tank, compressor, aftercooler, storage tank, etc., is in a separate, detached building of light, noncombustible construction.

CLASS D VACUUM FURNACES

Vacuum furnaces are used for heat treating metals; they are not, however, limited to this industry. In a vacuum furnace a vacuum pump is used to displace oxygen, and, in most cases, to reduce the water vapor content or dew point, as well.

Vacuum furnaces are usually batch furnaces. Batch furnaces are further classified into hot-wall and cold-wall furnaces; the latter are in greater use at the present time. Examples of hot-wall and cold-wall vacuum furnaces are shown in Figures 3-8M and 3-8N, respectively. In the hot-wall furnace, the entire vacuum vessel is heated, though usually not above 1,800°F (982°C) because of the reduction in the strength of materials at elevated temperatures. However, installation of a second vacuum vessel outside the vacuum retort (within which a roughing vacuum is maintained during the heating cycle) permits construction of larger hot-wall furnaces with higher operating temperatures.

Cold-wall furnaces consist of a water-cooled vacuum vessel. Usually the heating elements are inside the vacuum vessel. The walls can be maintained at near ambient temperature during high temperature operations; thus large units operating at high temperatures [4,000 to 5,000°F (2,204 to 2,760°C)] may be constructed. The two most common methods of heating cold-wall furnaces are by resistance and induction, with the heating elements located within the vacuum vessel. The heating elements are usually water-cooled, though occasionally they may be air-cooled. Insulation may be affected by radiation shields constructed of low-emissivity, oxidation-resistant metals with proper vapor pressure characteristics. Refractory insulation can be used in some instances; but this is difficult due to out-gassing of entrapped air in the refractory.

Mechanical-type pumps may achieve vacuum pressures of 10^{-3} to 10^{-2} torr. (Torr is a unit of pressure equal to 1/760 of an atmosphere.) For greater vacuum, booster oil-diffusion pumps are used to achieve pressures in the range of 10^{-5} to 10^{-7} torr; they must be "backed" by a supplementary pump.

Fractionating oil-diffusion pumps also are used and should always be backed by a rotary pump, or rotary and mechanical booster combination. They produce pressures from below 10^{-3} torr down to about 5×10^{-7} torr. The accidental admission of air into a heated diffusion pump has resulted in an explosion of the pump fluid. The explosion forces damaged both the pump and the furnace hot zone.

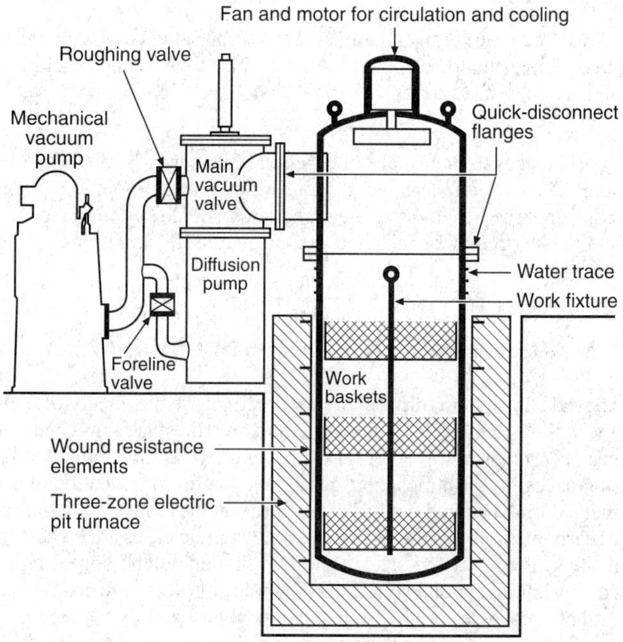

FIG. 3-8M. Example of a hot-wall, single-pump retort vacuum furnace.

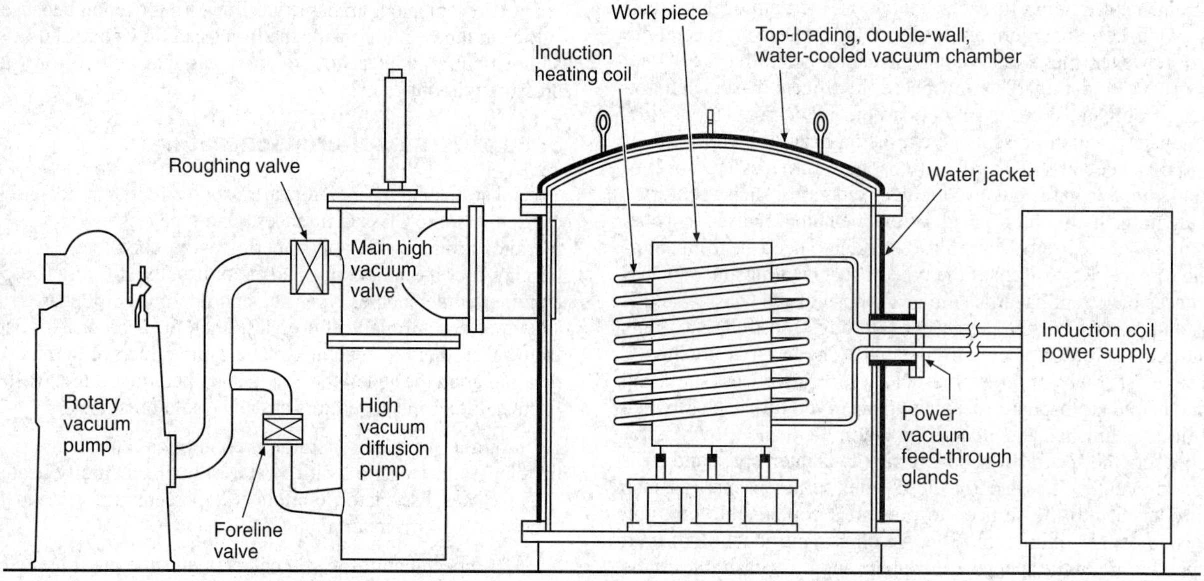

FIG. 3-8N. Example of a cold-wall, induction-heated vacuum furnace.

Vacuum furnaces have the same hazards as most Class C furnaces. Besides those hazards associated with the heat sources, the additional hazards of vacuum furnaces are:

1. Water leaks in either heating elements or vessel jackets can cause explosions. If water enters the furnace at operating temperatures, it will cause more than just a steam explosion.
2. Collapse of the furnace wall if a relief valve on the water jacket fails.
3. Collapse of the vacuum retort in a hot-wall furnace, if the vessel uses materials that have inadequate strength at high temperatures. If this occurs in a gas-fired unit, the flame can be pulled into the vacuum pump and ignite the oil in the pumps.
4. Vacuum pumps that pull fluids (water or oil) from hydraulic seal pots.
5. The condensed metallic vapors on electrical insulators, which can cause short circuiting, because this kind of furnace operates at pressures that can vaporize metals.
6. Short circuiting, if improperly supported heat shields sag at high temperatures and contact heating elements.
7. Hot spots on furnace walls that can weaken the furnace wall, if heat shields sag.
8. Temperature control problems not present in other types of furnaces and ovens. Optical pyrometers must have a line of sight to the work and their accuracy may be seriously impaired by gases, smoke, or discoloration of the sight glass.
9. Heat transfer—unless the thermocouple is actually attached to the part being measured, the heat transfer is based wholly on radiation. In air, a thermocouple receives heat by conduction and convection; therefore, with no air (or gas) in the furnace, the thermocouple response is slower. A gap of as little as 0.001 in. (0.025 mm) between the thermocouple and part or surface being measured can change significantly the response time of the thermocouple. A thermocouple on the heating element could mean that the parts would not reach the desired temperatures because the heating elements would of necessity have a temperature higher than the part (center of furnace), at least until equilibrium has been reached.
10. Induction heating—keeping piping conduits, building columns, beams, etc., out of the induction field, if the furnace is heated by induction. Any one of these items near an improperly shielded furnace can be heated by the induction coil inside the furnace. For instance, a steel bar placed so that it touches the furnace and a metal floor will be visibly hot in a matter of minutes. Of course, this also can happen on a Class B induction furnace.

AFTERBURNER AND CATALYTIC COMBUSTION SYSTEMS

Fume incinerators are combustion oxidation chambers designed to destroy process exhaust vapors or fumes by heat. Both direct flame and catalytic oxidation are used to reduce fumes, odors, vapors, and gases to acceptable exhaust products, such as carbon dioxide and water vapor. Exhaust that contains something besides plain hydrocarbons and oxygenated species may require additional special treatment, scrubbing, and filtration to remove particulate matter as well as halogens, hydroxides, sulfur oxides, and nitrogen oxides.

The advantages of fume incinerators can include reduced cleaning costs, reduced equipment downtime, reduced fire and explosion hazards, compliance with federal, state, and local pollution regulations, and savings in plant heating costs. Process fuel consumption can be reduced by the heat recovery of the burned exhaust fumes.

Installations have been damaged by fires in ducts between the process units and the incinerator, explosions of accumulated vapors in the ducts before or during start-up, improper operation of gas-fired incinerators, and overheating of the catalytic element or combustion chambers. The causes of these fires have been inadequate duct cleaning; inadequate duct design; incomplete prepurging of the ducts, process unit, and incinerator; failure by the operator to follow proper operating procedures; and malfunction of burner and temperature controls.

Afterburner (Direct-Flame) Incineration

Direct-flame fume incinerators can be used for a wide range of organic solvent vapors, organic dusts, and combustible gases. In order to burn, the fumes must be heated to their autoignition temperatures with sufficient oxygen present to complete the chemical reaction. Quenching the burner flame may occur if the burner capacity is insufficient or the flame pattern and mixing are inadequate. The fumes must reach the autoignition temperature and remain there long enough (dwell time) for the chemical reaction to occur (the dwell time is usually 0.4 to 0.8 sec). Ample oxygen, more than 16 percent, is necessary for complete combustion. The step sequence for successful incineration is shown in Figure 3-8O.

Operating temperatures in the combustion chamber are usually designed for a range of 1,200 to 1,500°F (650 to 816°C). Tests on some units have reported approximately 92 percent conversion efficiency to CO_2 at 1,300°F (704°C) and 96 percent conversion efficiency at 1,450°F (788°C). These conversion percentages of fumes to CO_2 are frequently required by air pollution codes. In any case, complete combustion requires that the fumes have sufficient air, proper mixing with the air, adequate dwell time, and adequate combustion chamber temperature.

Direct-flame combustion chambers are usually heavy refractory lined with external burners, such as tunnel burners, or light refractory with sectional line burners or line burners with mixing plates. (See Figure 3-8P.)

When metal construction is used, the design must take into account high thermal stresses and possible overheating of the metals. The combustion chambers, kiln, and boiler burner flames sometimes are used as fume incinerators.

Contaminated process waste streams might be inert gases with a low combustible hydrocarbon content. This mixture may be mixed

FIG. 3-8O. *Steps required for successful incineration of dilute fumes.*

with sufficient air to ensure combustion, then oxidized in a direct-flame or catalytic incinerator.

Special precautions must be taken where concentrated fumes are exhausted from the process. For safety, they are usually diluted with air to below 50 percent LEL for transfer to the incinerator.

Concentrated combustible fumes above the LEL are normally burned in flare stacks or as fuel in various types of heating equipment. The latter requires special burner design and combustion control supervision.

Catalytic Combustion Systems

Air pollution from furnace exhaust often is removed or reduced by catalytic combustion systems. A catalytic heater employs catalysts to accelerate the oxidation or combustion of air-fuel or air-fume mixtures for eventual release of heat to an oven or other process. Catalytic heaters may be used to burn a fuel gas, with substantial portions of the energy released as radiation to the processing zone. Alternately, catalytic heaters may be installed in the oven exhaust stream to release heat from evaporated oven byproducts, with available energy returned by a heat exchanger for recirculation through the oven processing zone.

Three types of catalytic combustion elements are available. The first is an all-metal mat used either as a fuel-fired radiant heater or alternately to oxidize combustible materials in air-fume mixtures. The second type is of ceramic or porcelain construction arranged in various configurations for gas fuel or fume oxidation with catalyst media, including a variety of "rare earth" elements, e.g., platinum, or metallic salts. Both types of these elements are classified as "fixed-bed" catalysts since they are normally held rigidly in place by clamps, cement, or other means. A third type of element consists of a bed, pellets, or granules supported or retained between screens in a fixed position, but with the individual members free to migrate within the bed.

Heating systems that employ catalysts are widely used to conserve oven fuel and control air pollution emissions. (See Figures 3-8Q and 3-8R.) Catalytic heaters cannot, however, oxidize or

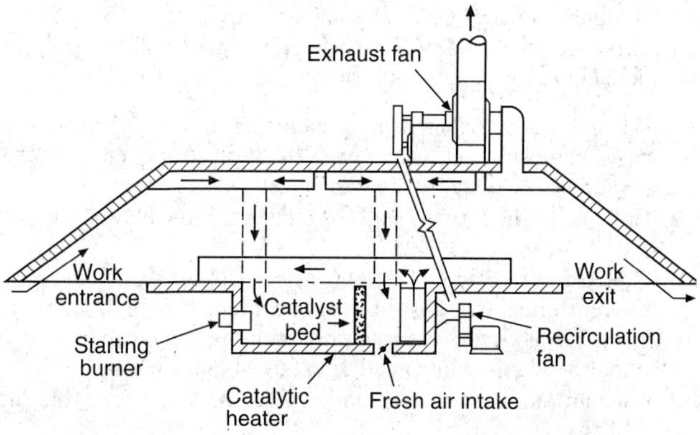

FIG. 3-8Q. *A direct-type catalytic oven heater for partial air pollution control.*

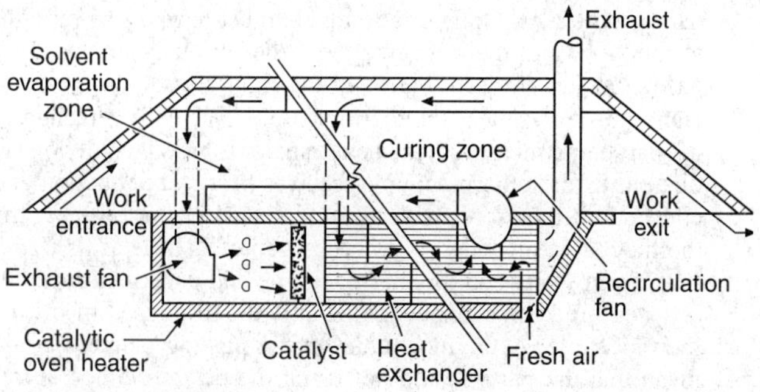

FIG. 3-8R. *An indirect-type catalytic oven heater for full air pollution control.*

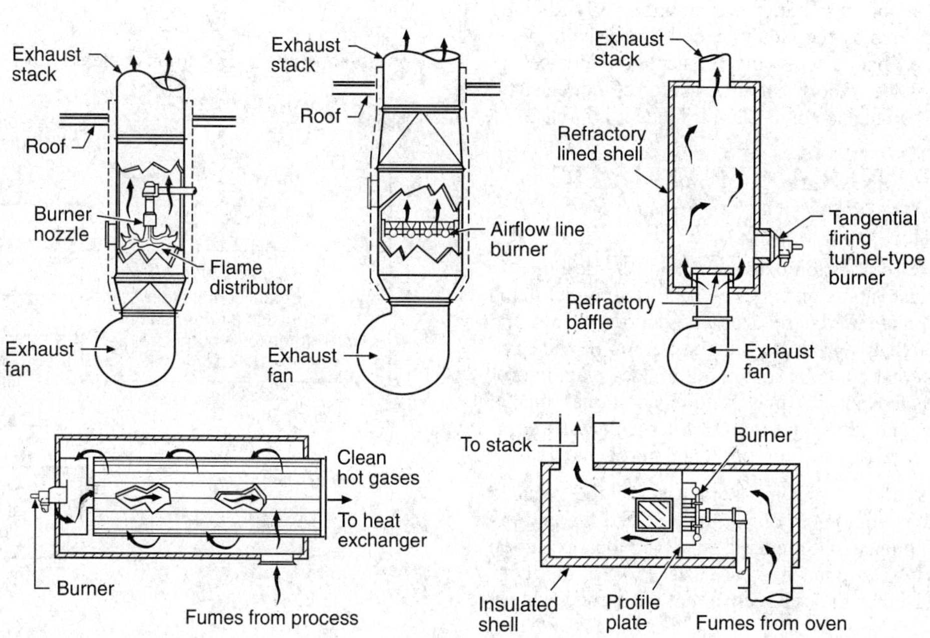

FIG. 3-8P. *Typical direct-flame fume incinerators. (Source: Maxon Premix Burner Co., Inc.)*

consume silicones, chlorine compounds, and metallic vapors as from tin, mercury, and zinc; these elements and various inorganic dusts may retard or paralyze the catalysts.

Installation

All components of the afterburners and catalytic combustion system, related process equipment, and interconnecting ducts must be provided with controls and safeguards to supervise conditions during start-up, operation, and shutdown. In some installations, many different concentrations of fumes may develop. The fume collection and delivery to the incinerator is an important part of the system. A careful investigation must be made for the incinerator and the associated equipment, including the appropriateness of the design and operating procedures.

HEAT RECOVERY

Heat exchangers and direct recirculation methods of heat recovery are often used to make process and fume incineration more economical. Some plants have calculated that if heat recovery methods are applied to heat generating processes, including fume incineration, the recovered heat can supply a substantial portion of the entire plant's heat or process demands. (See Figure 3-8S.)

Recovered heat may be used for: (1) the process as either a sole or supplementary heat source, (2) the process for some zones in a multizone unit, (3) other nearby processes, (4) preheating fumes to incinerators, (5) heating plant makeup air, and (6) a waste heat boiler serving multiple plant services. However, dirty stream deposits in the heat exchanger may make it inoperable. Combustible deposits and flammable liquids heated to high temperatures within the heat exchanger could cause a fire or explosion.

LUMBER KILNS

Drying lumber to a predetermined moisture content is accomplished in a variety of structures called kilns. Kilns make it possible to turn freshly cut, green wood into dry, accurately dimensioned lumber in a much shorter time than seasoning wood in open air. Because large quantities of combustible material are exposed to temperatures that can approximate ignition temperatures, kilns present a high degree of fire hazard.

Although called dry kilns, wood dryers usually employ moisture to maintain a uniform content within the wood during the drying, thus eliminating warping, checking, and splitting. The amount of moisture will vary with the species, and significant variations may be found within trees of the same species. The length of time required to bring the moisture content down to an optimum of about 2 percent depends upon the species, its original moisture content,

the dimensions of the pieces being seasoned, the type of kiln, and the volume of material.

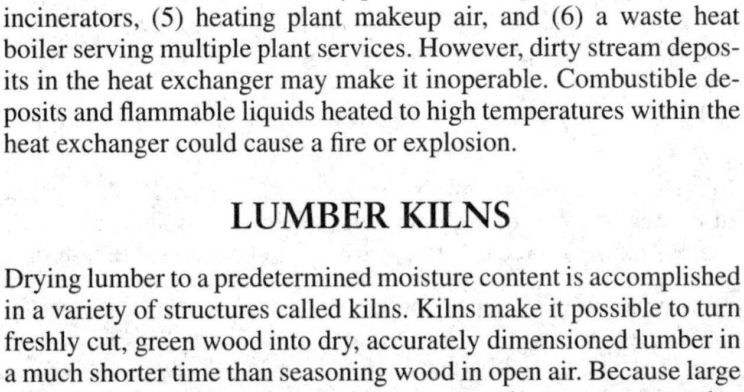

Types of Kilns

A dry kiln is basically an oven with controlled heat and humidity. It can be either a batch dryer or a progressive dryer. The wood being dried in a batch or compartment kiln remains stationary throughout the process. Temperature and humidity in all parts of the kiln are maintained as uniformly as possible and are adjusted as the wood dries.

Progressive kilns permit green lumber to be introduced in one end while dried lumber is being removed from the other end. Such kilns are designed so that somewhat higher temperatures are maintained at the dry, or discharge, end rather than at the loading end.

In natural circulation kilns, heated air rises up through the stacked lumber by convection. (See Figure 3-8T.) Losing its heat, air travels down to the heating device and is reheated. During the first part of the drying cycle, the air flows up along the sides, over the stacked lumber, and down through air passages in the stack. When the moisture content has been reduced to somewhere between 20 and 10 percent, heated air is directed up through the center of the stack to equalize the drying. Vents in the walls or roof of the kiln exhaust the hot, moisture-laden air.

Forced circulation kilns move air through the stacked lumber by either internal or external blowers. (See Figure 3-8U.) Figure 3-8V shows an internal fan kiln with fans located beneath the floor. Internal fans are reversible, changing the airflow for optimum drying. Internal fan kilns normally require cloth or metal baffles to eliminate turbulence and keep the air flowing in the desired direction.

Much the same airflow systems are used in both batch and progressive kilns. There are some minor differences, however. In progressive kilns, the air flows through the length of the chamber and is discharged at the green end. Because the air has picked up moisture and is cooler by the time it reaches the loading end of the kiln, the drying rate there is much slower.

Kiln Construction

Unlike most other buildings, kilns are subjected to extreme variations in internal temperature and humidity. Such variations cause unusual expansion and contraction that reduce structural integrity of

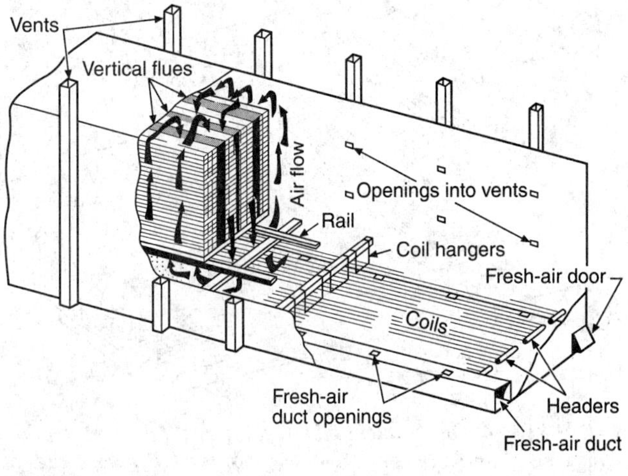

FIG. 3-8S. Typical incineration system incorporating waste heat recovery with fume and process air preheating.

FIG. 3-8T. A natural circulation, steam-heated compartment kiln. Arrows indicate air movement during the early stages of the drying cycle.

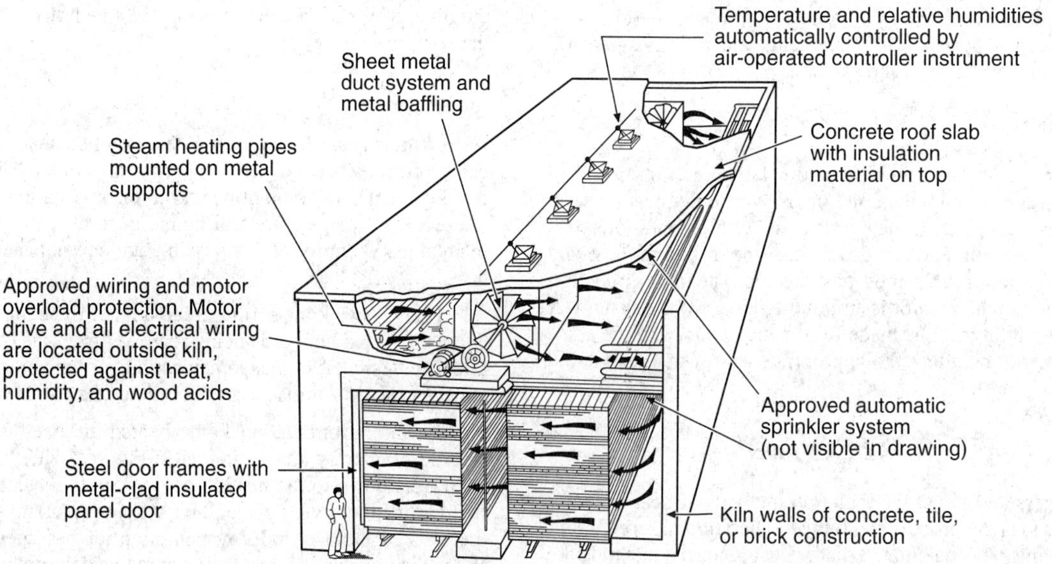

FIG. 3-8U. *A forced circulation double-track compartment kiln. Note that automatic sprinklers are installed above and below the platform between the kiln and the overhead fan room.*

the kiln and increase heat loss. Untended structural defects also can lead to premature failure of the structure. To reduce the fire hazard, kilns should be of fire-resistant or heavy-timber construction.

Heat Sources

Dry kilns require a constant source of heat, provided either directly or indirectly, to vaporize the water content of the wood. Direct heating is accomplished by circulating hot gases (produced by burning gas, oil, sawdust, or other fuels) through the stacked lumber. It also can be done by heating large metal surfaces with an open gas or oil flame as in Figure 3-8W. Air circulated by internal fans passes over the metal, which acts as a heat exchanger, then flows over the lumber, carrying off moisture vapor.

Steam is a common source of heat for indirectly heated kilns, but hot gases and electrical resistance heaters also are used. Steam is circulated through pipes and hot gases through ducts.

The Fire Hazards

Lumber kilns present high-fire hazards. This is especially true when direct heat systems or high-pressure steam systems are used, and in those instances where the structure itself is of combustible materials.

Direct-fired kilns present the greatest hazard because open ignition sources are close to the wood. (See Figure 3-8V.) The kilns should be considered analogous to Class A ovens in that they should be equipped with all the combustion controls normally required for drying ovens where the heating fuel is introduced into the heating

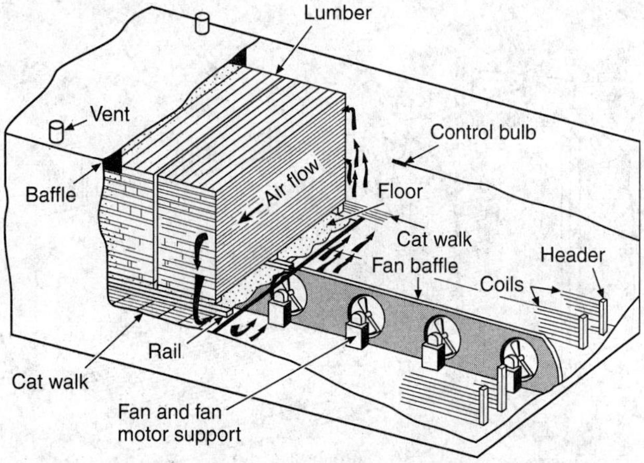

FIG. 3-8V. *A compartment kiln with internal fans and steam coils located under the grating at the floor level. Broken pieces of stickers and sawdust can fall through the grates and collect around the coils and fan motors to become a hazard, particularly if high-pressure steam coils are used.*

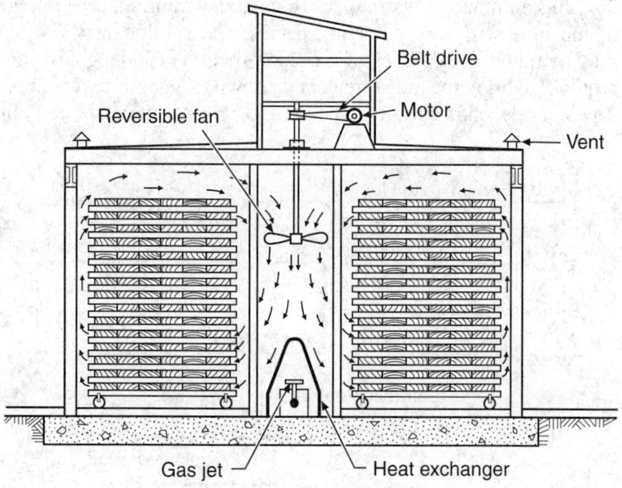

FIG. 3-8W. *A double-track, internal fan compartment kiln that is heated directly by a gas burner.*

enclosure itself. Indirectly heated kilns utilizing steam coils as heat exchangers and having controlled humidity are usually considered to be of low hazard.

The Safeguards

The important requirements for fire safety are automatic sprinkler protection, sound construction, good housekeeping, automatic humidity control, good air circulation, and proper ventilation. Kilns should be situated at safe distances from storage yards, sheds, and mill buildings. Ideally, they should be of fire-resistant construction and equipped with a complete automatic sprinkler system connected to an adequate water supply. The sprinkler protection should extend to fan houses and control rooms. Hydrants or hose connections should be located on the exterior for manual fire fighting.

DEHYDRATORS AND DRYERS

Dehydrators and dryers for agricultural products, commonly referred to as dryers, use heat to reduce the moisture content of products. The hazards of dryers are: (1) the possibility of igniting combustible materials near them, (2) the use of fuel or electricity as a heat source, and (3) the ignition of stock being dried.

Types of Dehydrators and Dryers

The three types of agricultural product dryers are: (1) continuous, (2) batch, and (3) bulk. They differ in the arrangement and operation of the drying chamber.

Continuous dryers include:

1. Drum dryers for milk, puree, and sludges.
2. Spray dryers for milk, eggs, and soup.
3. Flash dryers for chopped forage crops.
4. Gravity dryers (may also be batch type) for small grains, beans, and seeds. (See Figure 3-8X.)
5. Tunnel dryers (may also be batch type, and may be further classified according to airflow and whether or not intermediate heating is used) for fruits, vegetables, grains, seeds, nuts, fibers, and forage crops. (See Figure 3-8Y.)
6. Rotary dryers for milk, puree, and sludges. (See Figure 3-8Z.)

Batch dryers may be either fixed or portable and include pan dryers for sugar, puree, sludges, and other products. (See Figure 3-8AA.)

Bulk dryers dry the product in a bin, crib, or compartment in which it is to be stored. They are used to dry seeds, grains, nuts, tobacco, hay, and forage. (See Figure 3-8BB.)

Methods of Heating

Dryers for agricultural products may be direct fired (where products of combustion contact the material being dried) or indirect fired. The heaters may be oil fired, gas fired, solid-fuel fired, electrical, or heated by a heat transfer medium, such as steam. In general, the requirements for burner installation and fuel storage are the same as those for other heat-producing devices.

If gas-fired infrared heaters or lamps are used, their focal length should be ample so the surface of the drying product does not reach unsafe temperatures. If electrical infrared lamps are used in dryers, the lamps should be located where they cannot collect combustible dust.

Solid-fuel furnaces (other than those burning coke and anthracite coal) should not be used where the products of combustion can enter the drying chamber. Indirect solid-fuel dryers need temperature-controlled heat relief openings to the outside.

Dryer Controls

Some suggested controls for dryers, except those on the heating equipment, include:

1. A method for automatically shutting down the dryer in the event of fire or excessive temperature.

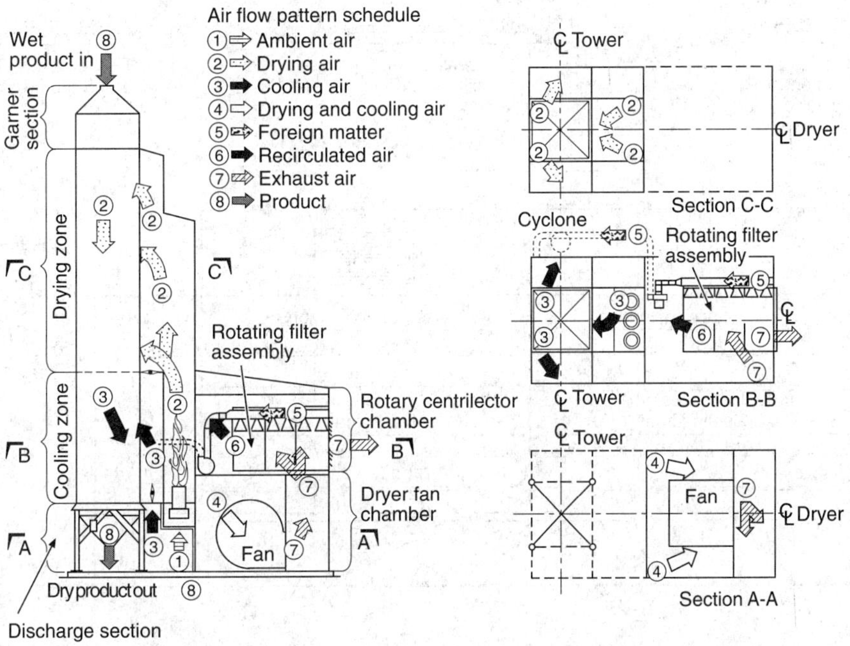

FIG. 3-8X.   A tower-type gravity dryer. (Source: Aeroglide Corp.)

2. A thermostat in the exhaust air when the product is fed automatically from the dryer to a storage building. In the event of excessive temperature, the thermostat:
   (a) Shuts off heat to the dryer and stops airflow (except when the product being dried is in suspension)
   (b) Stops the flow of the product
   (c) Sounds an audible alarm.
3. A thermostat in combustible dryers that, when the temperature of the combustible reaches 165°F (74°C), shuts off heat to the dryer but permits unheated air to pass through, and activates an audible alarm.
4. A device to shut off heat to the dryer if air movement through the dryer stops.
5. A high-limit thermostat located between the heat-producing device and the dryer.

Burner Controls

In general, the burner controls for dryers are the same as those for other automatically fired devices. A manual, quick-closing shutoff valve should be installed in the supply line of gas- and oil-fired burners, and controls should be arranged so that following automatic shutdown, manual restart will be necessary. Other control safeguards include flame failure protection and pre-ventilation of the combustion chamber. All safety devices must be listed for the service intended.

Construction and Installation of Dryers

Because dryers operate at elevated temperatures, they should be made of fire-resistant or noncombustible materials. If combustible materials must be used, dryers must not be subjected to sustained temperatures in excess of approximately 165°F (74°C). Expansion joints should be provided to prevent damage from expansion and contraction.

Secondary air openings for direct-fired dryers are screened with 1/2-in. (12.7-mm) mesh screen so materials cannot enter the combustion chamber. Primary air openings require screens with mesh 1/4 in. (6.4 mm) or smaller. An ample supply of easily opened access panels is necessary for inspection, cleaning, and fire fighting.

When stock is moved through the dryers so that it generates static electricity, all conductive parts of the dryers should be electrically bonded and grounded.

Like any heat-producing device, a dryer must have adequate clearance from nearby combustibles to prevent overheating. If there is a combustible dust hazard in the same building as the dryer, the heating device and blowers are installed in a dust-free room or area separated from the rest of the building. Ducts to convey heated air to the dryer and exhaust air from the dryer to the outside should be noncombustible.

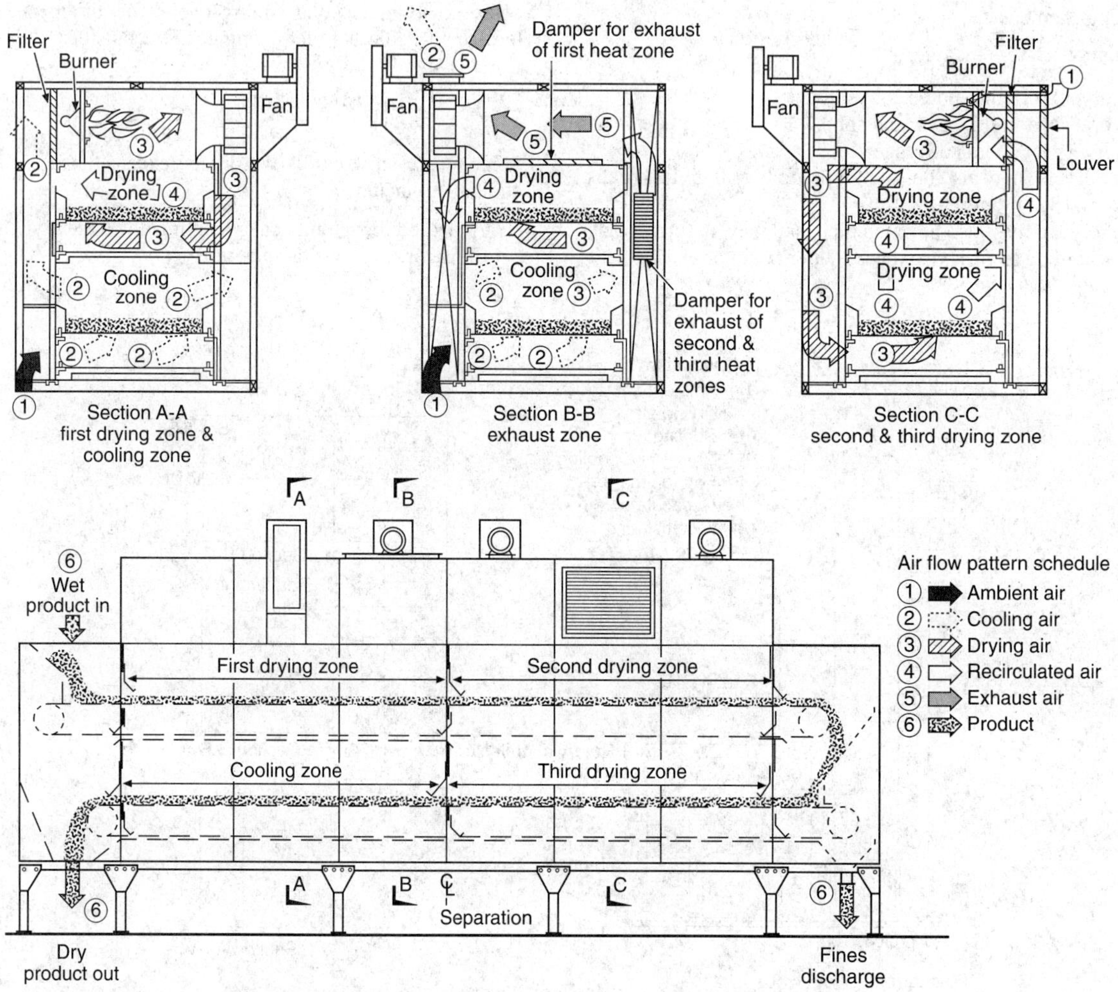

FIG. 3-8Y. A tunnel dryer. (Source: Aeroglide Corp.)

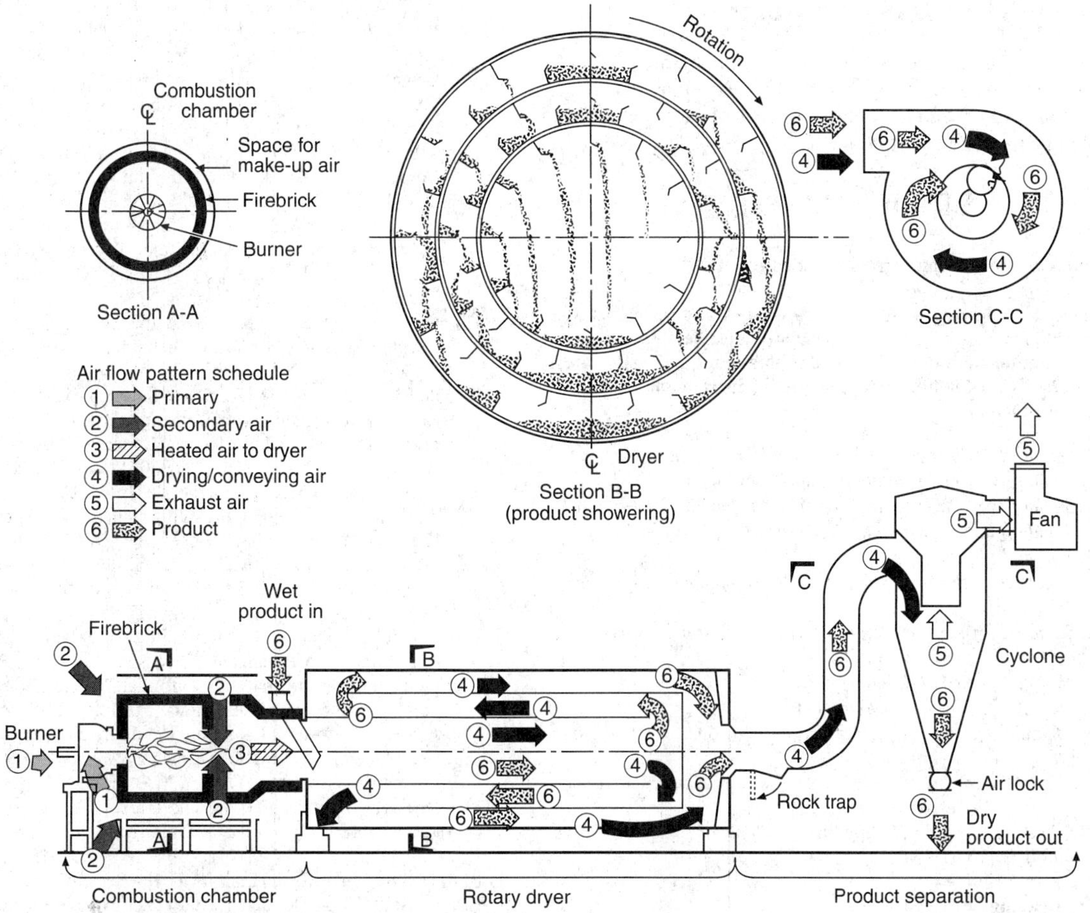

FIG. 3-8Z.   A rotary dryer. (Source: Aeroglide Corp.)

Extinguishing Equipment

The best way to protect a dryer enclosure is to install a water spray or automatic sprinkler system within the enclosure where possible. An exception to this is the direct-fired rotary dryer, which may be damaged by the internal application of water. A carbon dioxide system is satisfactory protection for this type dryer.

Fire protection may be needed in the product and exhaust dust collecting system, bins, chutes, ducts, cyclone collectors, and dust collectors. In addition to automatic sprinkler or deluge systems, high-speed infrared detectors with water spray may be needed to extinguish ignition sources, sparks, or embers in the ducts and to prevent propagation to the bins or dust collectors.

To extinguish small fires in and around most dryers, standpipe hoses are most useful, though water-type portable fire extinguishers also may be used.

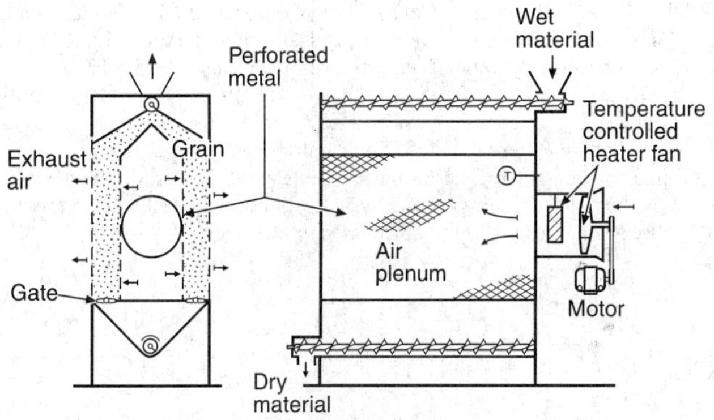

FIG. 3-8AA.   A batch-type grain dryer.

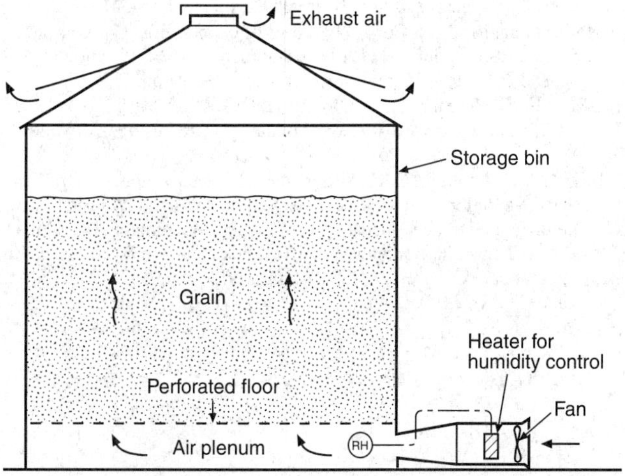

FIG. 3-8BB.   A bulk-type grain dryer.

Cooling of Dehydrated Products

A product being dried requires adequate cooling before it is packaged or stored. The amount of cooling required to prevent subsequent ignition will depend upon the properties of each material and how it is to be packaged or stored.

BIBLIOGRAPHY

NFPA Codes, Standards, and Recommended Practices

Reference to the following NFPA codes, standards, and recommended practices will provide further information on industrial and commercial heat utilization equipment discussed in this chapter. (See the latest version of *The NFPA Catalog* for availability of current editions of the following documents.)

NFPA 12, *Standard on Carbon Dioxide Extinguishing Systems*

NFPA 13, *Standard for the Installation of Sprinkler Systems*

NFPA 15, *Standard for Water Spray Fixed Systems for Fire Protection*

NFPA 31, *Standard for the Installation of Oil-Burning Equipment*

NFPA 54, *National Fuel Gas Code*

NFPA 58, *Standard for the Storage and Handling of Liquefied Petroleum Gases*

NFPA 61, *Standard for the Prevention of Fires and Dust Explosions in Agricultural and Food Products Facilities*

NFPA 70, *National Electrical Code®*

NFPA 86, *Standard for Ovens and Furnaces*

NFPA 86C, *Standard for Industrial Furnaces Using a Special Processing Atmosphere*

NFPA 86D, *Standard for Industrial Furnaces Using Vacuum as an Atmosphere*

NFPA 325, *Guide to Fire Hazard Properties of Flammable Liquids, Gases, and Volatile Solids*

Additional Reading

Chapman, K. S., Ramadhyani, S. and Viskanta, R., "Modeling and Analysis of Heat Transfer in a Direct-Fired Continuous Reheating Furnace," American Society of Mechanical Engineers (ASME), Heat Transfer in Combustion Systems, Winter Annual Meeting, HTD-Vol. 122, December 10–15, 1989, San Francisco, CA, 1989, pp. 35–43.

Cooke, G. M. E., "Use of Plate Thermometers for Standardising Fire Resistance Furnaces," Fire Research Station, Borehamwood, England, BRE OP 58, March 1994, 146 pages.

Crowhurst, D., "Explosion Hazards in Dryers," *Fire Surveyor*, Vol. 8, No. 3, June 1989, pp. 17–23.

Eckoff, R.K., *Dust Explosions in the Process Industries*, Butterworth—Heinemann, Ltd., Oxford, England, 1991.

Factory Mutual Research Corp., "Process Furnaces," *Loss Prevention Data Sheet 6-10*, Factory Mutual Research Corp., Norwood, MA.

Factory Mutual Research Corp., "Elements of Combustion, Controls, and Safeguards in Industrial Heating Equipment," *Loss Prevention Data Sheet 6-0*, Factory Mutual Research Corp., Norwood, MA.

Factory Mutual Research Corp., "Industrial Ovens and Dryers," *Loss Prevention Data Sheet 6-9*, Factory Mutual Research Corp., Norwood, MA.

Filius, K. D., Whitworth, C. G., "Emissions Characterization and Off-Gas System Development for Processing Simulated Mixed Waste in Plasma Centrifugal Furnace," Department of Energy, Washington, DC, University of California, Berkeley, Toxic Toxic Combustion Byproducts, 4th International Congress, ABSTRACTS ONLY, June 5–7, 1995, Salt Lake City, UT, 1995, 24 pages.

Gibson, T. O., "Loss Prevention Features Perform in Fuel Heater Loss Incident," *Plant/Operation Progress*, Vol. 7, No. 4, Oct. 1988, p. 258.

Gullett, B. K., *et al.*, "Formation of Polychlorinated Dioxins and Furans in Cement Kiln Operations," University of California, Berkeley, Toxic Toxic Combustion Byproducts, 4th International Congress, ABSTRACTS ONLY, June 5–7, 1995, Salt Lake City, UT, 1995, 79 pages.

Harris, R.J., *Gas Explosions in Buildings and Heating Plant*, British Gas Corporation, E & F N Spon, Ltd., London, England, 1983.

Hoerning, J. M. and Ragland, K. W., "Aromatic Emissions From Incineration of Selected Wastes Using a Laboratory Scale Rotary Kiln," University of California, Berkeley, Toxic Toxic Combustion Byproducts, 4th International Congress, ABSTRACTS ONLY, June 5-7, 1995, Salt Lake City, UT, 1995, 107 pp.

Holmes, J. G., Hedman, B. A., and Salama, S. Y., "Overview of Industrial Drying Needs and Competing Technologies," *Plant/Operation Progress*, Vol. 7, No. 3, July 1988, pp. 199-203.

Inculet, I. I., *et. al.*, "Ignition Studies of Selected Explosive Mixtures of Gases and Dusts Emitted From Cement Kilns," *IEEE Transactions on Industry Applications*, Vol. 29, No. 1, January/February 1993, pp. 82–87, IUSD 90–88, IEEE Log Number 9204185, IEEE Industry Applications Society Annual Meeting, October 2–7, 1990, Seattle, WA, 1990.

Koo, J., *et. al.*, "Formation of Toxic Byproducts From Pilot-Scale Rotary Kiln Incinerator for Mixed Polyethylene Waste," University of California, Berkeley, Toxic Toxic Combustion Byproducts, 4th International Congress, ABSTRACTS ONLY, June 5–7, 1995, Salt Lake City, UT, 1995, 38 pages.

Lemieux, P. M., Linak, W. P. and Wendt, O. L., "Waste and Sorbent Parameters Affecting Mechanisms of Transient Emissions From Rotary Kiln Incineration," University of California, Berkeley, Toxic Toxic Combustion Byproducts, 4th International Congress, ABSTRACTS ONLY, June 5–7, 1995, Salt Lake City, UT, 1995, 39 pages.

Linville, J. L., ed., *Industrial Fire Hazards Handbook*, 3rd ed., National Fire Protection Association, Quincy, MA, 1990.

McCoy, C. S., Dillenback, M. D., and Triax, D. J., "Major Fire in a Steam-Methane Reformer Furnace," *Plant/Operation Progress*, Vol. 5, No. 3, July 1986, pp. 165-168.

Miron, Y., Smith, A. C. and Lazzara, C. P., "Sealed Flask Test for Evaluating the Self-Heating Tendencies of Coals," Bureau of Mines, Pittsburg, PA, RI 9330, 1990, 22 pages.

"Pumps for Oil-Burning Appliances. Standard for Safety," Underwriters Laboratories, Inc., Northbrook, IL, UL 343, 7th Edition, April 29, 1993, 28 pages.

Rasmussen, E. F., *Dry Kiln Operator's Manual*, Harwood Research Council, Memphis, TN, 1988.

Reid, R. N., "Rotary-Kiln Furnaces for Disposal of Hazardous Waste," *Proceedings* for the Pacific Rim Conference of Building Officials, April 9–13, 1989, Honolulu, HI, Intl. Conf. of Building Officials, Whittier, CA, 1989, pp. 189–195.

Schenck, H. W., Wendt, O. L. and Kerstein, A. R., "Mixing Characterization of Transient Puffs in a Rotary Kiln Incinerator," University of California, Berkeley, Toxic Toxic Combustion Byproducts, 4th International Congress, ABSTRACTS ONLY, June 5–7, 1995, Salt Lake City, UT, 1995, 86 pp.

Shepherd, T. A., "Spillage of Flue Gases From Open-Flued Combustion Appliances," Building Research Establishment, Garston, England, BRE IP IP 21/92, December 1992, 4 pages.

Sidhu, S. and Dellinger, B., "Contribution of Cement Kiln Raw Meal Organics to PIC Emissions," University of California, Berkeley, Toxic Toxic Combustion Byproducts, 4th International Congress, ABSTRACTS ONLY, June 5–7, 1995, Salt Lake City, UT, 1995, 113 pp.

Sparrow, R. E., "Firebox Explosion in a Primary Reformer Furnace," *Plant/Operation Progress*, Vol. 5, No. 2, April 1986, pp. 122-128.

"Voimalaitosten ja lampokeskusten sammutusjarjestelmien valinta helpottuu. [Choice of Extinguishing Systems for Power and Heating Plants To Be Made Easier.]," *Palontorjuntatekniikka*, Vol

OIL QUENCHING AND MOLTEN SALT BATHS

Revised by Raymond Ostrowski

Oil-quenching operations are heat treatment processes that both harden and temper metals. Molten salt baths are used for heat treatment of metals, ceramics, or polymers to increase their hardness.

The first part of the chapter addresses the various elements of oil quenching, safeguards, and fire protection. The second part of the chapter describes salt bath types and applications, hazards, and safeguards.

OIL QUENCHING

One process in the heat treatment of metals is a controlled cooling or quenching of heated materials by immersion in a liquid-quenching medium. This process hardens and tempers the metal by imparting metallurgical changes to its surface. Due to the combustible nature of quench oils, the process presents serious fire hazard potentials.

Additional hazard and protection information relevant to oil quenching can be found in Section 3, Chapter 13, "Fluid Power Systems." Fire suppression information can be found in Section 6, "Suppression."

Contributing to the hazards of oil quenching are:

1. Requirements for a special atmosphere (a flammable gas blanketing the surface of the oil).
2. Temperature requirements of the quench medium.
3. Physical properties of the quench medium.
4. Volume limitations of the quench medium.
5. Size and configuration of process materials.
6. Locations of furnaces and quench tanks.
7. Mutual exposure between quenching and other processing or storage facilities.

Whether the quenching is an automatic and continuous, semi-automatic, or batch operation, it will involve elevators, conveyors, hoists, and cranes, either individually or in combination, to immerse the work in, move it through, and remove it from the oil bath. (See Figure 3-9A.) Although all three steps are important to the process, the one that is critical to safety is the entrance of the work into the quench.

Raymond Ostrowski is a sales manager with Protection Controls, Inc., Skokie, IL and a member of NFPA's Technical Committee on Ovens and Furnaces.

QUENCHING OILS

In most cases, mineral oils are used for quenching, but specific metallurgical requirements may dictate the use of mixtures with animal or vegetable oils. In addition, wetting agents may be blended with certain oils or oil mixtures. In any case, it is important to use a quenching oil with a low viscosity.

Essential quench oil properties include the ability to remain stable over periods of extended usage and to retain fluidity. For unheated quenching, operating temperatures between 100 and 200°F (38 and 93°C) are considered normal. Standard quench oils for use in this temperature range usually have flash points somewhat above 300°F (149°C). For heated quenching, operating temperatures between 200 and 400°F (93 and 204°C) are common. Quench oils used at these temperatures generally have a flash point above 500°F (260°C).

POLYMER QUENCHING

A rather recent development is the use of polymer quenchant in integral quench furnaces. A furnace designed for oil is not designed for a quick change to polymer quenching. Caution should be observed and the original furnace manufacturer should be contacted prior to changing from oil to polymer quenchant.

QUENCH TANKS

A quench tank should allow proper quenching under normal conditions, and provide for minor variations in equipment control functions and operator error. The design of the tank freeboard, overflow drains, and liquid level control are all critical. A fire that is confined to the liquid surface within a tank is much more readily controlled and extinguished than a fire involving quench oil that has overflowed the tank.

The distance from the quench surface to the top of the tank, with the tank loaded to capacity, is known as the freeboard. Freeboard design should take into account the splashing to be expected when the maximum workload is immersed with maximum speed. The distance between the liquid level with the workload submerged and any openings in the tank wall should be not less than 6 in. (152 mm) below the door or any opening into the furnace.

Adequately sized, fully trapped overflow drains are important safety features for all oil quench tanks. They should direct the overflow to a safe location outside of buildings or into special tanks. As

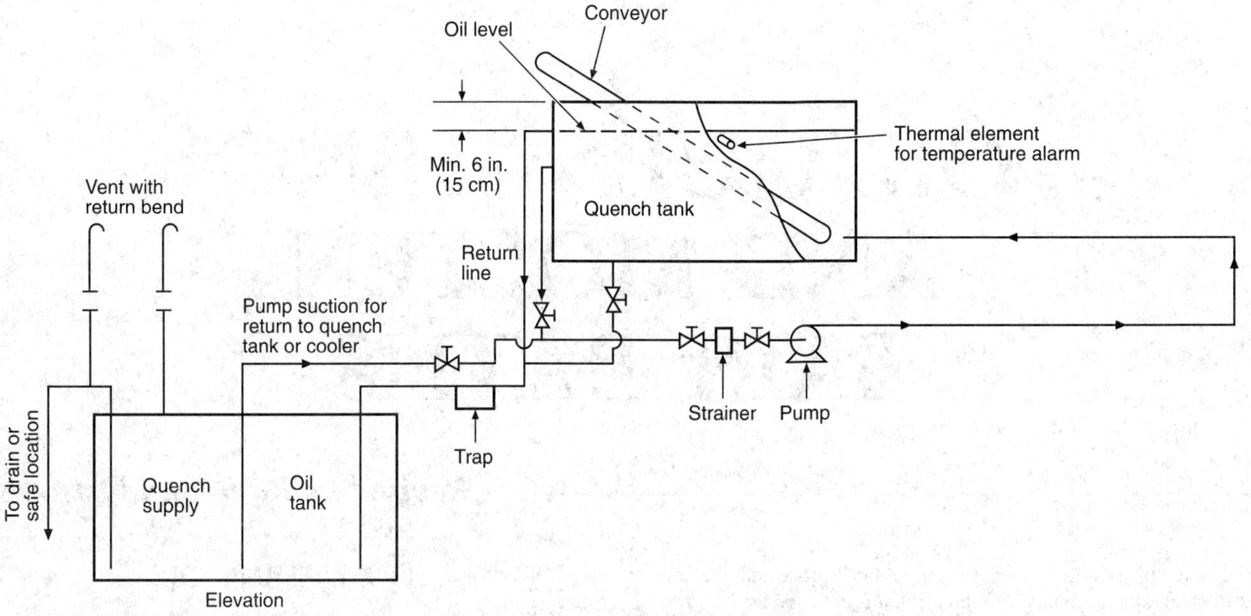

FIG. 3-9A. A schematic of a typical continuous-type oil quench tank. (Source: Factory Mutual Research Corp.)

a practical matter, small quench tanks frequently may be installed without overflow drains. However, quench tanks with a liquid capacity of 150 gal (0.57 m³) or a liquid surface area of 10 sq ft (0.93 m²) or larger should be provided with overflow drains.

While overflow drains should be specifically designed for each tank, certain minimum sizes have been established and accepted. (See Table 3-9A.) For large quench tanks, multiple overflow pipes are preferable to a single large pipe, provided the aggregate cross-sectional area is equivalent to that of a single pipe. Piping connections on drains and overflow lines must be designed to permit ready access for inspection and cleaning.

Emergency Drains

Under serious fire conditions, it may be necessary to empty a quench tank in order to reduce the amount of quench fluid available. This can be performed readily through adequately sized, fully trapped, and valved bottom drains directed to safe locations. Where gravity drainage is not possible, special pumps may be provided for oil removal. Drains and pumps should be sized so that the quench oil can be removed within 5 min. Gravity drains should be sized according to the tank capacities and drain diameters given in Table 3-9B.

Emergency drains must be used only under the guidance and control of well-trained and experienced personnel, as improper usage can result in greater hazards. Whenever a flammable gas processing atmosphere is maintained above the quench oil, removal of the oil can create a negative pressure that can result in explosion, an increase in fire severity, or both.

Tank Location

Heat-treating operations involving combustible quench oils should be housed in fire-resistive buildings and should be well separated from exit areas, combustible materials, valuable stock, power equipment, and important process equipment.

The safest location for quench tanks is at grade level. Boilover from tanks above grade can be expected to spread fire to floors below, thereby making fire control more difficult and significantly in-

TABLE 3-9A.   Quench Tank Overflow Drains, Minimum Pipe Sizes

Liquid Surface Area		Minimum Pipe Diameter (I.D.)	
sq ft	m²	in.	mm
10–75	0.93–6.7	3	76
75–150	6.7–14	4	101
150–225	14–21	5	127
225–325	21–30.2	6	152
325+	30.2+	8	203

TABLE 3-9B.   Gravity Drain Pipe Diameters, I.D.

Tank Capacity		Pipe Diameter (I.D.)	
gal	m³	in.	mm
500–750	1.9–2.8	3	76
750–1000	2.8–3.8	4	101
1000–2500	3.8–9.5	5	127
2500–4000	9.5–15	6	152
Over 4000	Over 15	8	203

creasing the potential fire loss. Fires in below-grade locations will make manual fire fighting difficult and result in a significant increase in fire loss.

MATERIAL TRANSFER

Rapid and complete immersion of the work in process is essential to safe and proper heat-treating metallurgical processes. The method of immersion must result in minimal splashing and no overflow of the quench oil outside of the tank. It is also essential that the possibilities for partial immersion be eliminated or minimized. Partial immersion of the work is the most common cause of quench oil fires.

Chutes

Many furnaces are designed so that the work in process drops off the end of a conveyor, through a chute, and into the quench oil. The chute design and construction must allow the work to fall freely under all normal furnace feed conditions. This will involve proper chute sizing to accommodate the work, proper pitch to ensure continued stock motion, and smooth surfaces to prevent the work from being stuck in the chute. (See Figure 3-9B.)

Elevators

When stock in process is handled on trays or in baskets, elevators are commonly used to immerse the work in the quench oil. (See Figures 3-9C and 3-9D.) Partial immersions are caused more by trays or baskets and elevators than by any other method. The following are three critical concerns:

1. The elevating mechanism must be adequately supported by structural members to prevent its falling unevenly.
2. Adequate guides must be provided to ensure uniform movement within the quench tank and to prevent an elevator from being wedged, as this could result in partial immersion.

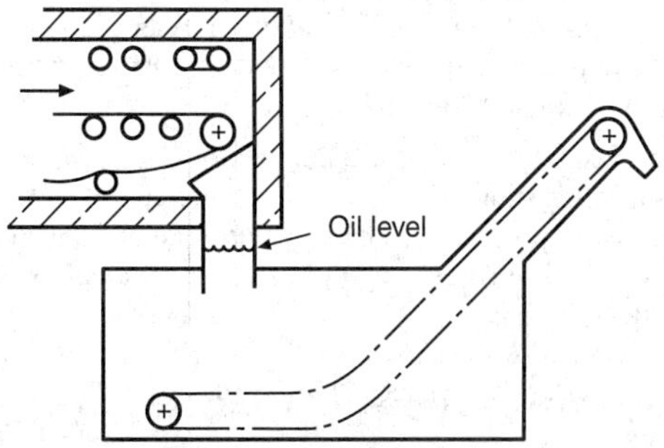

FIG. 3-9B.   Bottom chute-type quench.

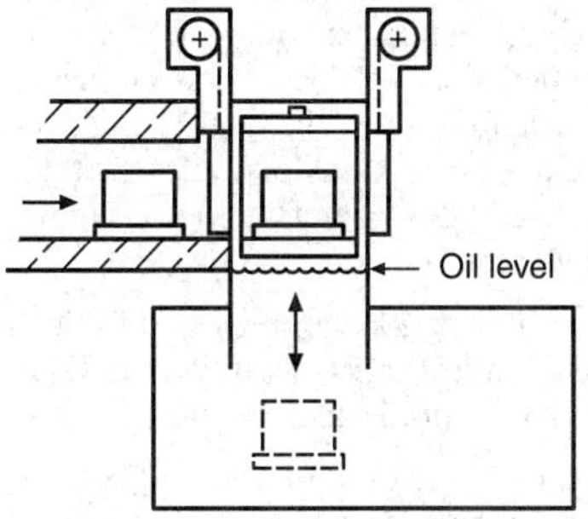

FIG. 3-9C.   Dunk-type elevator quench.

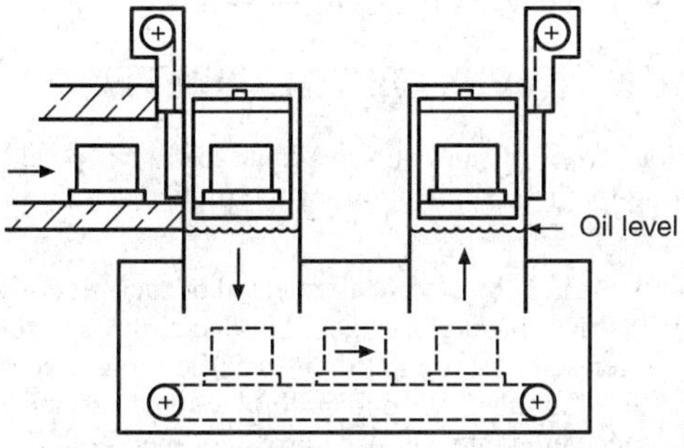

FIG. 3-9D.   Transfer-type elevator quench.

3. Suitable guides and stops must be provided to prevent shifting of the workload, as this can cause elevator jamming and partial immersion.

Hoists and Cranes

A hoist or crane is usually required for moving large, specialized workloads into and out of the quench tank. Proper positioning of this equipment can be accomplished by stops and/or limit switches. Mechanical guides may be required to ensure that the work is properly positioned as it enters the quench tank.

Unless the oil drains off the work at the end of the quench period, an excessive amount of oil will be lost. Usually the work will still be warm at this time and some oil may vaporize. Any ignition source can produce a fire at this point in the process. Oil vapors will condense on cool surfaces above the drain area and thus contribute significantly to the potential fire loss.

OIL TEMPERATURE CONTROL

The control of quench oil temperature within specified design limits is essential to the heat-treating process as well as for safety. Failure of a cooling system can cause the oil to overheat. What is more hazardous, however, is a cooling system failure that allows water to enter the quenching oil. An excessive amount of water can produce a boilover when a hot workload is immersed in the quench. The exact critical water volume varies somewhat with the specific oil or oil mixtures being used, but if the water content reaches 0.50 percent by volume, the oil is no longer considered safe to use.

Quench oil can be cooled by water circulating through coils in the quench tank, by external heat exchangers, or by water jackets for the quench tank. Water jackets and internal water coils *must not* be used with combustible quenching oils. In these designs, any mechanical failure of the coil or the quench tank shell will result in water entering the quench medium. Such failures are not readily detectable; the first indication may be a boilover or a steam explosion when a hot workload is immersed in the quench.

If an external heat exchanger is used, quench oil must be circulated through the exchanger at a higher pressure than the exchange medium (water). If a leak should develop, oil will enter the water and be wasted. If the water pressure is higher than the oil pressure, such a leak would result in water entering the oil, creating a potential for boilover or steam explosion.

The continuous flow of cooling water is essential for temperature control. This can be properly supervised by observing the discharge from an open drain on the water side of the heat exchanger. Waterflow indicators are needed where a completely closed system must be used.

An oil quench tank is built as an integral part of many special atmosphere furnaces. In these cases, the vestibule above the quench tank is water-cooled, usually by water jackets. Many failures of the interior jacket walls have released cooling water into the quench tank below. As a result, safety considerations have dictated the use of special materials for vestibule jackets or the use of external coils. With special atmosphere furnaces, the use of a combustible gas for the atmosphere also will result in moisture development when the exit door is opened and the atmosphere is burned out.

If the oil level is too low and a large workload is immersed in the quench, the oil can overheat. Therefore, low-oil-level detectors should be used to sound an alarm and shut down quenching operations before overheating occurs.

Agitation is critical to maintaining safe quench oil temperature and uniformity of temperature throughout the bath. If the agitation mechanism fails, localized overheating will occur, which may cause a fire at the surface. Agitation failure and subsequent overheating can cause excessive vaporization, which can raise the pressure inside an enclosed special atmosphere furnace. Gas and oil vapor may then be forced out of the chamber around doors and through vents. Frequently, these escaping gases ignite, damaging the facilities. Agitation systems should be supervised automatically and their failure should result in the safe shutdown of operations.

Where the process requires that the quench be heated, i.e., fuel fired, electrically heated, or steam heated, immersion units are used as the heat source. All three methods of heating will create excessive temperatures at the interface of the heating unit and the quenching oil. A quench heating system should be prevented from operating if the oil level is too low, the agitation equipment is not functioning, or the oil temperature is too high. Where combustible oils are used, a separate excess high-limit temperature controller should be interlocked in the system to prevent quenching and to shut off the oil heating system when the oil temperature exceeds a specified maximum limit.

CENTRAL OIL SYSTEM

Many heat-treating plants utilize a quenching oil that is common to several operations. Properly designed central oil systems can contribute to the maintenance of reasonably dependable and problem-free oil-quenching operations. A proper design includes filtering to remove particulate contamination, water removal for safety of operations and prevention of boilover, and cooling to deliver the quenching medium at an acceptable temperature.

Water separation can be accomplished by settling and the use of centrifuges. However, these systems are not dependable enough to eliminate the need for the oil in quench tanks to be tested periodically.

Since central systems involve a constant removal and replacement of oil from the quench tanks, it is essential for safety that oil flow be supervised. To avoid foaming and boilover, replenishment oil should not be added while the quench temperature is 212°F (100°C) or higher. Regardless of the oil being used, the temperature must never be permitted to rise to a value that is less than 50°F (28°C) below its flash point.

THE SAFEGUARDS

All automatic shutdowns of quenching operations should result in the workload being completely immersed in, or removed from, the quench. Partial immersion always must be considered hazardous.

Whenever movement of the workload into, or out of, the quench has been stopped by a malfunction, qualified operating personnel must be permitted to override the safety interlocks so that manual attempts can be made to complete immersion or remove the workload. If exit doors must be opened, a fire condition should be anticipated. Extinguishing facilities adequate to protect the operator and prevent property damage should be available.

All safety controls and their interlocking functions should be tested on a regular schedule. The time interval between tests should be adequate for each installation but should not exceed six months. Refer to manufacturers' guidelines.

Hydraulic control systems add to the fire hazard in the vicinity of high-temperature equipment. Fire-resistant hydraulic fluids should be used. When combustible hydraulic oils must be used, proper maintenance of the hydraulic equipment is vital to safety. See NFPA 86C, *Standard for Industrial Furnaces Using a Special Processing Atmosphere*, for detailed safeguards.

FIRE PROTECTION

Fire protection for combustible oil quenching operations requires a detailed evaluation of the inherent fire potentials. One of the most effective forms of area protection is the automatic water spray system. Experience has shown that these ceiling-mounted sprinkler systems will limit building and equipment damage from quench oil fires whether they are confined to the quench tank surface or spread over a large area by a quench oil boilover.

Specific protection for open quench tank surfaces and oil drainage areas is also important. Most oil fires can be extinguished by fixed carbon dioxide or dry chemical systems. In some operations, various foam systems can also be effective. However, the suitability of foam will be determined by the quenching oil used and the temperatures involved. These systems are designed to operate automatically, well in advance of any potential sprinkler system discharge.

Those situations that may require operating personnel to manually release jammed workloads and close furnace doors will dictate the provision of fire control and/or extinguishing facilities for the protection of the operator. In these instances, carbon dioxide or fixed water spray systems are the most effective. The most important fire protection features in every heat-treatment shop are the manual fire extinguishing and control equipment and plant personnel properly trained in its use. In addition to an adequate supply of portable, hand-carried fire extinguishers, the larger, wheeled extinguishers should also be available.

Appropriately spaced hose connections with water fog or water spray nozzles should be considered an essential part of all heat-treating area fire protection. These can provide prolonged periods of fire control and life safety beyond the limited supplies of portable extinguishers.

MOLTEN SALT BATHS

Molten salt baths have become increasingly popular for the heat treatment of metals, ceramics (e.g., glass), or polymers (e.g., Teflon®) because they provide rapid and precise heat transfer and the required equipment is relatively inexpensive. A molten salt bath is defined as any heated container that holds a melt or fusion of one or more chemical salts in a fluid state. The work to be treated is immersed in the bath. There are many kinds of salt baths that use various salts and salt mixtures and that operate at various temperatures, depending on the results required.

Additional information can be found in Section 3, Chapter 8, "Industrial and Commercial Heat Utilization Equipment," and Section 3, Chapter 23, "Storage and Handling of Chemicals."

As work is immersed fully into a salt bath, the salt serves as a protective environment. Heat is transferred more rapidly than by any other type of furnace heating. Moreover, because of the constant convection currents in the heated liquid medium, uniform temperature distribution is more easily attained.

TYPES OF SALT BATHS

The various types of salt baths available today can be classified by the heating method, application, or the particular salt mixture utilized.

Salt baths may be electrically heated or fuel-fired. Electrically heated baths may have electrodes introduced through the top surface of molten salt or through the sides of the furnace below the salt surface. (See Figures 3-9E and 3-9F.) In some instances, ribbon-type resistance elements are used around a metal pot or container. (See Figure 3-9G.)

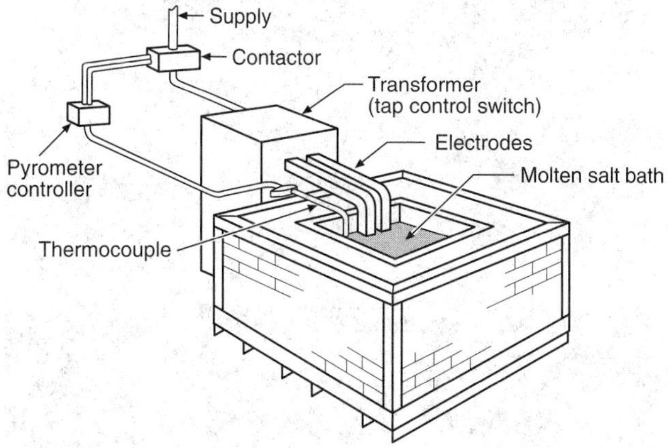

FIG. 3-9E.   Typical electrode immersion-type electrical heating arrangement.

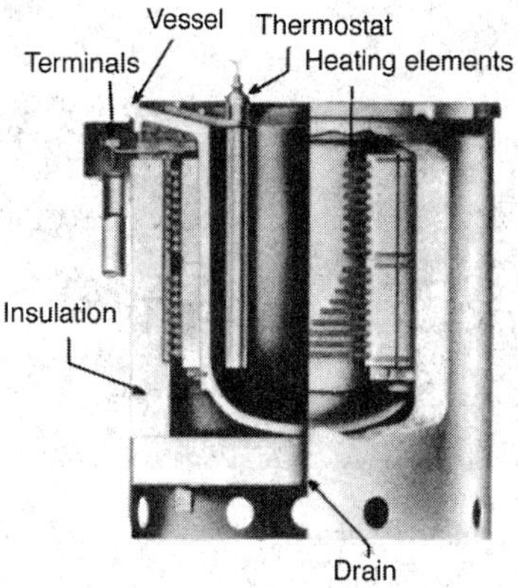

FIG. 3-9G.   Resistance heated salt bath with pot. (Source: Sunbeam Equipment Co.)

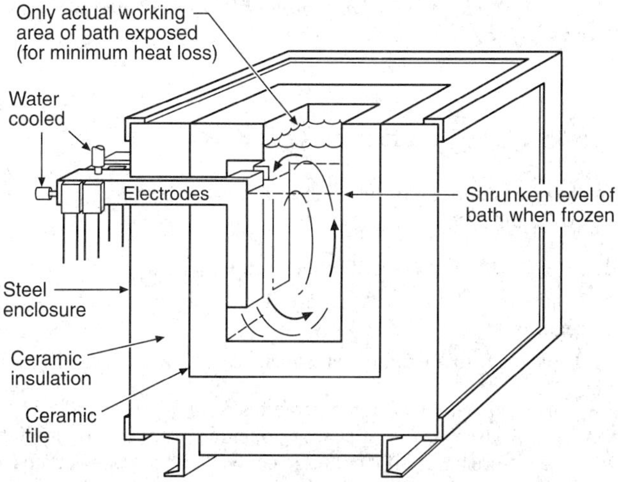

FIG. 3-9F.   Submerged electrode-type salt bath. (Source: Ajax Electric Co.)

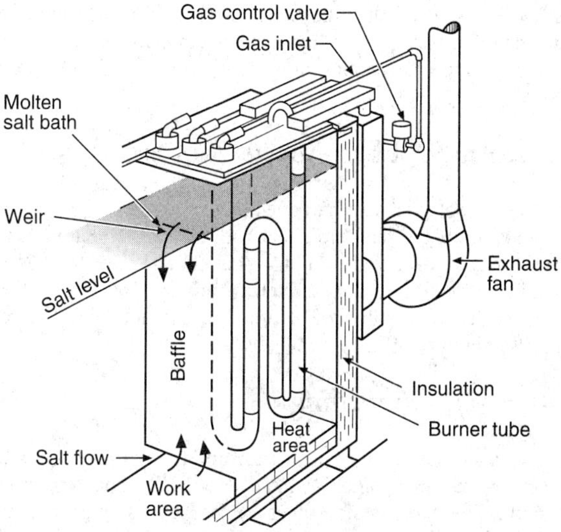

FIG. 3-9H.   Gas-fired radiant tube salt bath heating section.

There are two types of fuel-fired salt baths. In one, the burners are fired directly outside a retort containing the salt. In some instances the burners are fired in metal tubes, known as radiant tubes, and the tubes are immersed in the molten salt. (See Figure 3-9H.)

APPLICATIONS

Major uses of molten salt baths are descaling, liquid carburizing (case hardening), cyaniding and nitriding, neutral hardening, tool steel hardening, annealing, brazing tempering, and isothermal quenching. Each of these operations requires a different salt or salt mixture at different temperatures.

Descaling

For the removal of oxides that are difficult to remove by normal means, e.g., pickling, grit blasting, polishing, etc., there are three different descaling processes. These processes are: (1) the reducing sodium hydride process, (2) the oxidizing sodium hydroxide process, and (3) the electrolytic process. In the reducing sodium hydride process, a fused melt of sodium hydroxide and sodium hydride at temperatures between 650 and 780°F (343 and 415°C) is used. For oxidizing, a melt of sodium carbonate, sodium chloride,

and sodium hydroxide is used at temperatures from 700 to 950°F (371 to 510°C). In the electrolytic process, a fused melt of sodium carbonate, sodium chloride, and sodium hydroxide is used at about 900°F (482°C). Electrolytic oxidation reduction is accomplished by passing direct current through the work. A reversing switch changes the polarity of the work from positive to negative which effectively cleans the surfaces.

Liquid Carburizing

Case hardening is accomplished by diffusing carbon and small amounts of nitrogen into steel surfaces. The salts used are molten mixtures of sodium cyanide, sodium chloride, and barium chloride at temperatures between 1,450 and 1,750°F (788 and 955°C), depending upon the depth of case required. (See Figure 3-9I.)

Cyaniding and Nitriding

These two operations are also performed with either sodium or potassium cyanides as the chief ingredient. Usually the operating temperatures are lower, usually 950 to 1,050°F (510 to 566°C), for nitriding.

FIG. 3-9I.   *Liquid carburizing and isothermal heat treating line. (Source: Ajax Electric Co.)*

Neutral Hardening

The hardening of ferrous alloys without harmful surface effects is performed in mixtures of sodium, potassium, and barium chlorides. Operating temperatures may range from 1,400 to 2,350°F (760 to 1,288°C).

Hardening of High-Speed Tool Steels

High-speed tool steels are hardened in a series of molten salt baths at various temperatures. A preheat bath at approximately 1,400°F (760°C) may be followed by a high-heat bath at 2,350°F (1,288°C), which in turn is followed by a quench bath at 1,100°F (593°C). All of these baths are mixtures of barium, sodium, and potassium chlorides. (See Figure 3-9J.)

Isothermal Quenching

To achieve various levels of physical properties in the heat treatment of steels, isothermal quenching rapidly cools metal parts in a molten salt bath. The three major types of isothermal quench are: (1) austempering, (2) martempering, and (3) cyclic annealing. Each produces different hardness properties in metal alloys. Bath temperatures vary from 400 to 1,300°F (205 to 705°C). The salts used are nitrate/nitrite salts for the lower temperatures and neutral chloride types for the higher temperatures.

Annealing

Process annealing of carbon steels is usually carried out in carbonate and chloride salts at a temperature of 1,250 to 1,300°F (677 to 704°C). For stress relief of carbon steels, the melt is usually a nitrate salt at approximately 1,000°F (538°C). Chloride salts are used to anneal stainless steel products as well as nickel-chrome alloys at temperatures of 1,550 to 2,150°F (843 to 1,177°C).

Brazing

Salt bath brazing of ferrous and nonferrous alloys with silver and aluminum alloys, brass, and copper is another popular application for molten salt baths. Depending on the alloy and type of brazing, temperatures and salt mixtures vary considerably. For aluminum dip brazing, the salt is a mixture of chlorides with smaller percentages of sodium and/or aluminum fluorides, which act as fluxes. In copper brazing, barium chloride mixtures are used. Chlorides of sodium and potassium are used for silver alloy and brass brazing applications. Carburizing salts are used for a combination of brass brazing and carburizing.

FIG. 3-9J.   *Salt bath for hardening of high-speed tool steel. (Source: Ajax Electric Co.)*

THE HAZARDS

The hazards common to salt bath furnaces can be divided into the following three types:

1. Fire caused by contact of molten salt with combustibles.
2. Explosion of the salt mixture due to physical or chemical reaction.
3. The danger to operating personnel.

Since salts are used at temperatures from 300 to 2,400°F (149 to 1,316°C), the ejection of salt by popping, spattering, or spilling will create a fire hazard if it comes in contact with anything combustible, including wood floors. Since molten salts have relatively little surface tension and low viscosities, any minor physical disturbance or chemical reaction can result in salt being ejected. When liquids (water, oil, etc.) or materials reactive to the particular salt utilized penetrate the surface of the salt bath, a violent ejection can result. Table 3-9C lists common salts and their melting points. Table 3-9D lists common salt mixtures and their melting points. Proprietary mixtures of these salts are formulated for specific applications and temperatures. Intermixing of certain salts is hazardous, e.g., nitrates and cyanide. Consult suppliers for recommendations.

Nitrates are particularly hazardous due to their ability to start and support combustion. When nitrates are overheated to temperatures in excess of 1,100°F (593°C), rapid dissociation with the release of nitrous oxide fumes occurs. These fumes are injurious to operating personnel and are corrosive to the adjacent equipment. Explosions may occur if nitrate salt leaks from an externally heated container (pot) into a superheated combustion chamber. The overheating of nitrates can be caused by temperature controls that malfunction, an accumulation of sludge in the bath, or by operator errors.

If the operating temperature of a nitrate or nitrite bath exceeds the maximum limit, the following emergency action should be taken:

1. Shut off the heat supply.
2. Remove all work from the bath.
3. Move employees to a safe location.

TABLE 3-9C. *Melting Points of Common Chemical Salts*

Salt	Melting point °F	Melting point °C
Barium chloride*	1764	963
Barium fluoride	2336	1280
Boric oxide (anhydride)	1071	578
Calcium chloride*	1422	773
Calcium fluoride	2480	1360
Calcium oxide	4662	2572
Lithium chloride*	1135	613
Lithium nitrate	491	255
Magnesium fluoride	2545	1396
Magnesium oxide	5072	2800
Potassium carbonate*	1636	891
Potassium chloride*	1429	776
Potassium cyanide*	1174	634
Potassium fluoride	1616	880
Potassium hydroxide*	716	380
Potassium nitrate*	631	333
Potassium nitrite*	567	297
Sodium carbonate*	1564	851
Sodium chloride*	1479	804
Sodium cyanide*	1047	564
Sodium fluoride*	1796	980
Sodium hydroxide*	605	318
Sodium metaborate	1771	966
Sodium nitrate*	586	308
Sodium nitrite*	520	271
Sodium tetraborate	1366	741
Strontium chloride	1603	873

*Most common salts.

Heated nitrate salts will react strongly with carbonaceous materials, such as oil, soot, tar, graphite, and cyanides—all of which may be utilized in some other metal-treating process. Accidental mixing of cyanides with molten nitrates can cause explosions of considerable magnitude.

The introduction of aluminum parts into a nitrate bath approaching or exceeding the melting points of these metals can cause fire and explosion. Magnesium alloys should never be immersed in a nitrate salt.

The storage of salts can cause serious problems. Chemicals, such as nitrates and cyanides, may mix and interact. Many salts are hydroscopic and, when stored in damp areas, will absorb moisture. When these salts are subsequently heated, moisture will be released below the surface and an explosion will ensue.

Intensive external reheating and remelting of a solidified salt bath may result in sufficiently rapid expansion to bulge or rupture the container. Such reheating can generate enough pressure to fracture the surface crust of the salt and scatter hot salts over a wide area.

When exposed to air having even a slight moisture content, the fumes from many salt baths become corrosive because chlorides, fluorides, and other salts will form acids by hydrolysis. These fumes are highly corrosive to adjacent building structures, wiring, and equipment and are, of course, detrimental to the bath operator.

SAFETY CONTROL EQUIPMENT

A molten salt bath furnace must be equipped with control instrumentation and interlock systems that provide protection against the various and known types of potential equipment malfunctions.

TABLE 3-9D. *Melting Points of Common Salt Mixtures**

Mixture and Proportion	Melting point °F	Melting point °C
Lithium nitrate 23.3, sodium nitrate 16.3, potassium nitrate 60.4	250	121
Potassium hydroxide 80, potassium nitrate 15, potassium carbonate 5	280	138
Potassium nitrate 53, sodium nitrate 7, sodium nitrite 40†	285	141
Potassium nitrate 56, nitrite 44	295	146
Potassium nitrate 51.3, sodium nitrate 48.7	426	219
Sodium nitrate 50, sodium nitrite 50†	430	221
Sodium hydroxide 90, sodium nitrate 8, sodium carbonate 2	560	293
Lithium chloride 45, potassium chloride 55	666	352
Barium chloride 31, calcium chloride 48, sodium chloride 21†	806	430
Calcium chloride 66.5, potassium chloride 5.2, sodium chloride 28.3†	939	504
Calcium chloride 67, sodium chloride 33	941	505
Potassium chloride 35, sodium chloride 35, lithium chloride 25, sodium fluoride 5†	960	516
Potassium chloride 40, sodium chloride 35, lithium chloride 20, sodium fluoride 5†	990	532
Barium chloride 48.1, potassium chloride 30.7, sodium chloride 21.2†	1026	552
Sodium chloride 27, strontium chloride 73	1049	565
Potassium chloride 50, sodium carbonate 50†	1086	586
Barium chloride 35.7, calcium chloride 50.7, strontium chloride 13.6	1110	599
Barium chloride 50.3, calcium chloride 49.7	1112	600
Potassium chloride 61, potassium fluoride 39	1121	605
Sodium carbonate 56.3, sodium chloride 43.7†	1177	636
Calcium chloride 81, potassium chloride 19	1184	640
Barium chloride 70.3, sodium chloride 29.7†	1209	654
Potassium chloride 56, sodium chloride 44†	1220	660
Sodium chloride 72.6, sodium fluoride 27.4	1247	675
Barium fluoride 70, calcium fluoride 15, magnesium fluoride 15	1454	790
Barium chloride 83, barium fluoride 17	1551	844
Calcium fluoride 48, magnesium fluoride 52	1738	948

*Lowest constant melting points given; proportions are percentages by weight.
†Most common salt mixtures.

Gas- and oil-fired salt bath furnaces must be provided with an approved flame safeguard device for each burner and must be interlocked to: (1) shut off the fuel supply to the affected burner(s) and (2) activate an alarm.

An excess temperature control instrument must be provided that: (1) has its own temperature-sensing element, (2) can be interlocked to shut off the heating system, and (3) activates an alarm when an excess temperature condition is detected.

When nitrate salts are used (regardless of the type of heating system), a "heat rate" controller should be installed to prevent a too-rapid heatup, thus preventing localized overheating and ignition of the salt.

THE SAFEGUARDS

While there appear to be many hazards connected with the operation of molten salt bath furnaces, fire, explosion, and injury can be prevented by the constant observance of ordinary precautions.

Clean dry sand may be used for diking purposes to confine and prevent the spread of the escaped melt. Carbon dioxide or dry chemical extinguishers may be used to extinguish burning carbonaceous material in the immediate vicinity of the salt bath.

The preferred location for a salt bath is a noncombustible area. The bath may be housed in a cement-lined pit or curbed area large enough to contain the contents in case of leakage or spill. It should be protected against leakage of liquids from other sources and be provided with baffles to prevent splashover from one tank to another. Salt bath furnaces provided with steel enclosures greatly reduce risks of leakage.

All salts should be shipped and stored in tightly covered containers designed to prevent the absorption of liquids or moisture. Nitrate salts should be stored in a fire-resistive, damp-free room and separated a reasonable distance from heat, liquids, and reactive chemicals. Nitrate and/or nitrite salts should never be stored in the vicinity of cyanide salts. Only the actual amount required to recharge the bath should be removed from the storage area, and this amount should be melted immediately. Salts should be transported from the storage area to the furnace in suitable containers to prevent loss during transport.

The safety precautions that apply to the heating systems of salt baths are identical to those for the heating systems of other industrial furnaces. Some additional safety measures are necessary. Flame should be tangential to the wall of a salt pot or container when a gas- or oil-fired heating system is used. The products of combustion should be vented. Radiant tubes and electrodes should be of materials resistant to the corrosive action of the salts used.

The buildup of sludge in any type of salt bath should be avoided, and the sludge periodically removed. Sludge is caused by the deterioration of salt, the introduction of foreign material into the bath, the dropping of small parts into the bath, or the freezing of salt due to a cold furnace bottom. Several methods of removing sludge and foreign objects are available. Some equipment has been designed to alleviate the buildup problem. Properly applied mechanical agitation greatly aids in the elimination and/or removal of the contaminants. Agitation prevents temperature stratification and freezing on the furnace bottom. Agitation also suspends and directs the contaminating particles into a settling pan or through a filter basket. (See Figures 3-9K and 3-9L.)

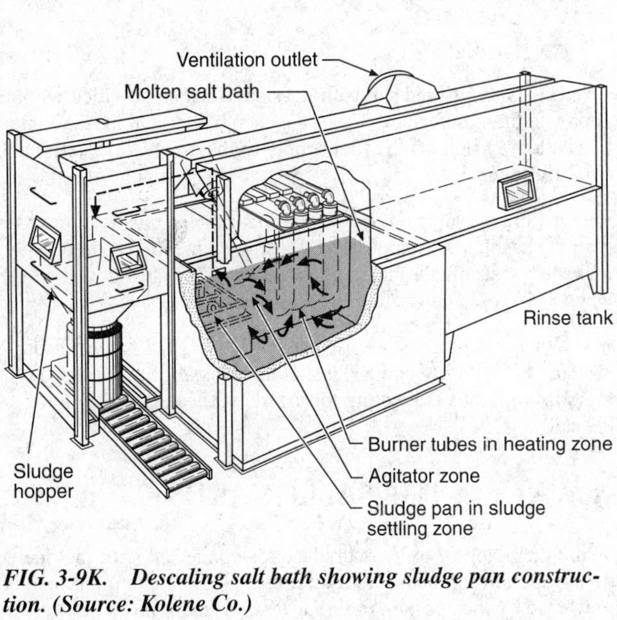

FIG. 3-9K. *Descaling salt bath showing sludge pan construction. (Source: Kolene Co.)*

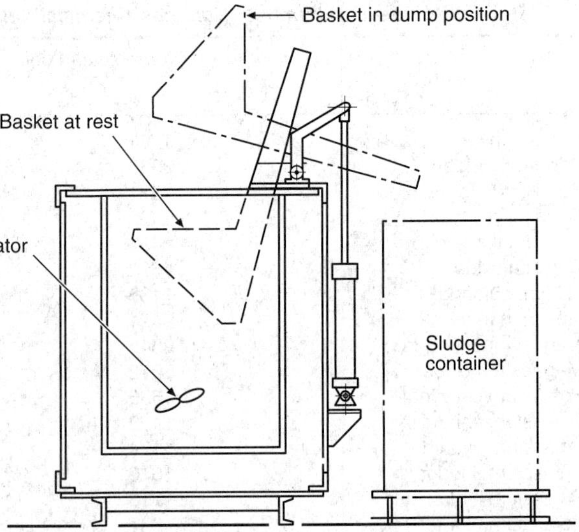

Features
- Continuously filters out suspended particles
- Automatically lifts for cleanout
- Suitable for retrofitting
- Replaces manual desludging operations
- Does not interrupt production

FIG. 3-9L. *Automatic basket dumper. (Source: Ajax Electric Co.)*

Two thermocouples properly located in the bath will provide safe control and protection from overheating. An overtemperature control should be arranged to automatically shut off the heat source and actuate visual and audible alarms in the event of a malfunction of the normal operating controls. All sensing couples must be protected from the corrosive effects of the particular salt used.

For the removal and control of any toxic, corrosive, or heated fumes, all salt bath furnaces should be equipped with hoods made of material that will not be affected by the corrosive nature of such fumes. In many cases, chemical scrubbers or suitable filters should be installed to remove the particulates. (See Figure 3-9M.) Removal or neutralization of fumes may require special equipment.

All fixtures used for dipping parts in baths, such as hooks, ladles, or baskets, should be of a solid, closed design without corners that could retain water. Parts, such as closed piping or other hollow metallic articles in which air or moisture may be trapped, should not be inserted into a molten bath without making provisions for venting the heated, expanded air or moisture. Immersion of hollow parts in salt baths should be very gradual. Any fixture used in one salt bath should be thoroughly cleaned and dried before it is immersed in another bath.

When electrodes are water cooled, an instrument should be provided to detect failure of the water cooling system.

If a salt bath is used as an internal quench in a furnace with a combustible atmosphere, measures should be provided to prevent carbon precipitation onto the salt surface. In addition, adequate salt circulation should be provided to prevent localized hot spots whenever the salt is exposed to furnace temperatures.

A cold salt bath that has frozen over should be remelted or liquefied by applying initial heat as near to the top as possible. The application of heat to the external bottom surfaces will cause melting near the bottom and leave the top solidified. Expanding volume causes excessive pressure and failure of the container, with the subsequent ejection of hot salts. Breaking the surface crust will result in an explosive ejection of the salt. This hazard can be avoided by

FIG. 3-9M. Typical hooded enclosure, overhead hoist, and work platform greatly improve safety together with improved productivity and ease of maintenance.

inserting a tapered solid bar in the molten salt adjacent to the immersed electrodes. After the salt has frozen, the bar is removed. The cavity created by the removal of the bar will serve as a vent duct for the escape of the gases formed. Gas burners that heat the top layers and gradually heat downward may be used if they are used carefully. For the submerged type of electrodes, a resistance heating coil set in between electrodes is recommended. Pumps are available to remove salt while it is in the molten state. Considerable care should be exercised in removing any molten salt; it should be pumped directly into steel drums where it can freeze over immediately. Salt and drums may then be properly discarded. For all types of heating, a "soft start" with reduced power input is recommended until melting is established from top to bottom.

All operators and workers in the vicinity of molten salt baths should be provided with corrosion-resistant and heat-resistant shoes, gloves, aprons, hard hats, and face shields. Solutions for cleansing eyes and supplies for treating minor burns should be available at all times. Breathing apparatus should also be provided for emergency use against oxides of nitrogen or corrosive fumes, such as chlorides and fluorides.

Constant vigilance with safety in mind and good housekeeping by intelligent and well-trained personnel can keep a salt bath operation as safe as any other metallurgical process conducted in an industrial furnace.

BIBLIOGRAPHY

NFPA Codes, Standards, and Recommended Practices

Reference to the following NFPA codes, standards, and recommended practices will provide further information on the safeguards for oil quenching and molten salt baths discussed in this chapter. (See the latest version of *The NFPA Catalog* for availability of current editions of the following documents.)

NFPA 10, *Standard for Portable Fire Extinguishers*
NFPA 11, *Standard for Low-Expansion Foam*
NFPA 12, *Standard on Carbon Dioxide Extinguishing Systems*
NFPA 13, *Standard for the Installation of Sprinkler Systems*
NFPA 14, *Standard for the Installation of Standpipe and Hose Systems*
NFPA 15, *Standard for Water Spray Fixed Systems for Fire Protection*
NFPA 17, *Standard for Dry Chemical Extinguishing Systems*
NFPA 34, *Standard for Dipping and Coating Processes Using Flammable or Combustible Liquids*
NFPA 70, *National Electrical Code®*
NFPA 86, *Standard for Ovens and Furnaces*
NFPA 86C, *Standard for Industrial Furnaces Using a Special Processing Atmosphere*
NFPA 101®, *Life Safety Code®*
NFPA 220, *Standard on Types of Building Construction*

Additional Reading for Oil Quenching

Benedetti, R. P., ed., *Flammable and Combustible Liquids Code Handbook*, 5th ed., National Fire Protection Association, Quincy, MA, 1993.
Cote, A. E., ed., "Oil Quenching," *Industrial Fire Hazards Handbook*, 3rd ed., National Fire Protection Association, Quincy, MA, 1990.
Factory Mutual Research Corp., "Oil Quenching of Metals," *Handbook of Industrial Loss Prevention*, 2nd ed., McGraw-Hill, New York, 1967.

Additional Reading for Molten Salt Baths

Cote, A. E., ed., *Industrial Fire Hazards Handbook*, 3rd ed., National Fire Protection Association, Quincy, MA, 1990.
"Flammable Liquid Drainage and Containment," *Record*, Vol. 69, No. 4, July/Aug.1992, pp. 24–31.
"Flammable Liquids—Risk, Regulations and Protection Measures to Safeguard Flammable Liquids," *Record*, Vol. 69, No. 1, Jan./Feb. 1992, pp. 15–21.
Olsson, S., and Ryderman, A., "Extinguishment of Oil Spray Fires With Water: Experimental Procedures and Test Data," Swedish National Testing and Research Institute, Boras, Sweden, SP REPORT 1990:32, CIB W14/92/07 (S), 1990, 67 pages.
"Proper Flammable and Combustible Liquid Handling," *Record*, Vol. 68, No. 5, Sept./Oct. 1991, pp. 16–17.
Whiteley, B., "Foaming Sprinklers and Flammable Liquid Fires," *Fire Prevention*, No. 262, Sept. 1993, pp. 32–35.

STATIONARY COMBUSTION ENGINES

Revised by James B. Biggins

Stationary combustion engines are prime movers (e.g., as internal combustion engines, external combustion engines, gas turbine engines, rotary engines, and free piston engines) that are installed for use in a location as permanent equipment. Portable engines that remain connected for use in the same location for a period of one week or more are also considered to be stationary combustion engines. The various types of engines noted above can be grouped into two primary types: (1) reciprocating engines and (2) gas turbine engines.

Stationary engines are used for a wide number of applications in industry. The most common use is in tandem with a generator for the production of electricity in commercial and emergency situations. Larger engines are frequently used to drive large pumps and compressors, such as those used in the petrochemical and gas-pipeline transmission industries. Engine-driven pumps and compressors handle a variety of fluids, including flammable liquids and gases.

This chapter discusses the potential fire hazards associated with such engines, and addresses the location, installation, and protection of these engines.

More in-depth discussions of the hazards that may be associated with the use and storage of fuels for these engines can be found in: Section 3, Chapter 21, "Storage of Flammable and Combustible Liquids"; Section 3, Chapter 22, "Storage of Gases"; Section 4, Chapter 6, "Flammable and Combustible Liquids"; and Section 4, Chapter 7, "Gases". Additional information on engines used to drive fire pumps is provided in Section 6, Chapter 4, "Fire Pumps."

PRIMARY ENGINE TYPES

Reciprocating Engines

Reciprocating engines can range in size from small portable gasoline engines to very large diesel engines used for power generation. These engines are typically of two main types: (1) otto cycle engines, which use a spark plug to ignite a fuel-air mixture, and (2) diesel cycle engines, in which high-pressure compression raises the air temperature to the ignition temperature of the injected fuel oil. The characteristics of, and hazards associated with, all reciprocating engines are similar.

James B. Biggins, P.E., C.S.P., is consultant and assistant vice president with Marsh & McLennan Protection Consultants in Chicago, IL. He is a member of NFPA's Technical Committee on Internal Combustion Engines.

Gas Turbine Engines

Gas turbine engines are typically larger engines, ranging in size from 100 to 100,000 hp (75 to 75,000 kW). Stationary gas turbine engines are used mainly for power generation and gas-pipeline transmission. Gas turbine engines are either open or closed cycle.

In an open cycle, the working fluid passes through the engine only once. In a closed cycle, the working fluid is continually recycled through the engine. The majority of gas turbine engines in use are of the open cycle design.

POTENTIAL FIRE HAZARDS

The fire hazards associated with stationary combustion engines are: (1) combustible liquids and flammable gases (used as fuel), hydraulic fluids, lubricants, and in some cases, the material being pumped or compressed; (2) the temperature of exposed engine parts, particularly exhaust manifolds and pipes, which can be hot enough to ignite flammable or combustible materials; (3) the potential arcing of electrical components, such as starters, generators, distributors, and magnetos, which can ignite flammable vapors; (4) combustible materials stored within the engine room or used in the engine installation, such as air filters; and (5) engine disintegration.

ENGINE INSTALLATIONS

NFPA 37, *Standard for the Installation and Use of Stationary Combustion Engines and Gas Turbines*, and local codes provide minimum requirements for the safe installation of engines, and include specifications for the engine installation, engine room compartmentation, fuel supply piping, room ventilation, and engine exhaust systems. (See Figure 3-10A.)

Engine Locations

Stationary combustion engines may be located either outdoors, inside structures, or on the roofs of structures. The selection of an engine location should consider the distance from combustible walls and ceilings, exterior wall openings, and the availability of sufficient air for combustion and ventilation. Engine room compartmentation should comply with applicable standards and local codes, which may specify minimum fire resistance ratings.

Combustible materials should not be stored in rooms containing stationary engines, or within the compartments of a gas turbine.

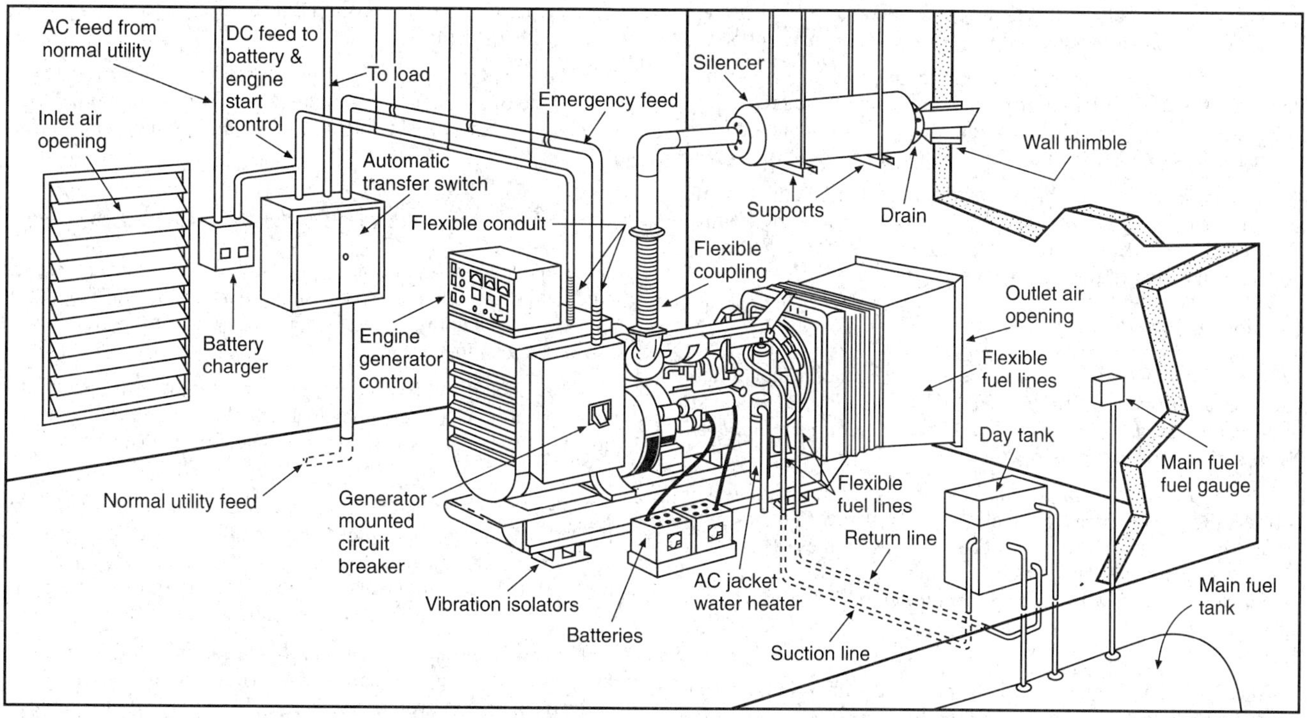

FIG. 3-10A.   A typical generator installation used to supply emergency power. (Source: Caterpillar)

Engines should be isolated from locations where combustible dust or airborne flyings can be produced. Engine exhaust manifolds and piping can reach high temperatures, and combustible dusts and flyings deposited on these surfaces may ignite. For ignition to occur, conditions must be favorable for prolonged contact between the combustible material and the hot surface. A good maintenance and housekeeping program will lessen the chance of such an occurrence.

Exhaust Systems

Fires can be caused by failure to allow sufficient clearance between exhaust piping and combustible walls, ceilings, or roofs. Engine exhaust systems (e.g., manifolds, mufflers, exhaust piping, etc.) can reach temperatures in excess of 1,000°F (538°C), and special precautions are needed when these items penetrate combustible construction.

Engine exhaust systems should be designed based on the expected flue gas temperatures. Exhaust pipes should be constructed of sufficient strength to withstand the service. Provisions are needed in the exhaust system to prevent damage from the ignition of unburned fuel. Low points, with suitable means to allow the drainage, should also be incorporated into the system design.

It may be necessary to install a flexible connector from the engine to the exhaust pipe to minimize the potential for a break in the exhaust system due to engine vibration or heat expansion.

Exhaust systems should maintain a minimum clearance of 9 in. (229 mm) from combustible materials. Special precautions are required for cases where it is necessary to pass the exhaust system through walls or roofs of combustible construction. The exhaust system should terminate outside the structure at a point where the hot gases or sparks will be discharged harmlessly and not come in contact with combustible material.

Electrical Installations

Arcing electrical engine components should be safeguarded by flashover shielding, flame arresters, purging, or ventilation, if flammable liquids or gases are being pumped or compressed.

The electrical installation for the engine should be in conformance with the provisions of NFPA 70, *National Electrical Code®*. Engine rooms or other locations are not to be classified as hazardous locations, as defined in the *National Electrical Code*, solely by reason of the engine fuel.

FUEL SUPPLIES

Stationary combustion engines have been designed to operate with a number of different fuels. The two most common types are: (1) flammable and combustible liquids, such as gasoline or diesel fuel, and (2) flammable gases, such as propane or natural gas.

Many engine applications require on-site fuel supplies for reliability. The fuel supply may be stored either outdoors or within a building. Local codes and standards may restrict fuel tank size, construction, and location, and should be consulted.

Liquid Fuels

Diesel fuel and gasoline are the two most widely used liquid fuels for stationary engines. Gasoline, due to its low flash point [–45°F (–43°C)], presents a greater fire hazard than other liquid fuels. Gasoline storage in buildings is limited to the capacity of the integral tank on the engine, but not more than 25 gal (95 L). When larger quantities of liquid fuel are required in buildings, a diesel-fueled engine generally is used.

Liquid fuel systems should be designed to minimize the potential for an accidental release of fuel into a structure. Alarms, float-controlled valves, gages, and sight glasses should be installed to aid personnel in operating the fuel system. Diking, or some other method of spill containment, should be considered for fuel storage tanks installed either inside structures, or aboveground outdoors.

Liquid fuels are supplied to engines by pumps. An exception is that gravity feed is permitted from an integral tank. The fuel supply to tanks above the lowest level of a building should be pumped in a manner acceptable to the Authority Having Jurisdiction.

An overflow line, high-level alarm, and high-level automatic shutoff should be provided for all tanks that are filled by pumps. Care should be taken to avoid locating fuel lines, vents, and overflow lines near the engine combustion air intake.

Gaseous Fuels

Gas-fueled engines should be installed in accordance with applicable codes and standards, such as NFPA 54, *National Fuel Gas Code*, or NFPA 58, *Standard for the Storage and Handling of Liquefied Petroleum Gases*.

Some gas fuels, such as propane, are heavier than air and will collect in low-lying areas. Special precautions need to be taken to ensure proper ventilation is provided when using these fuels. Gas pressure regulators located inside a structure should be provided with a vent outdoors, discharging a minimum of 5 ft (1.5 m) from the structure. An alternative is the provision of a listed vent limiting device.

Gas piping to engines should be provided with a shutoff valve located remote from the engine, preferably outside the structure for those engines located indoors. Each gas-fueled engine should be provided with a means to automatically shut off the flow of gas to the engine, in case the engine stops from any cause.

LUBRICATION AND HYDRAULIC SYSTEMS

Engine lubrication and hydraulic systems generally utilize a combustible oil. In some larger gas turbine units, the combustible oil may be replaced with a synthetic, fire-resistant fluid. This type of lubricating fluid has a higher flash point than lubricating oil. The use of a fluid with a higher flash point decreases the potential for ignition of a spill from a heat source, such as hot engine parts.

An oil mist is often present in the crankcases of large reciprocating engines which, upon ignition, may result in a combustion explosion within the crankcase. As a result, many such engines are equipped with crankcase explosion vents. In other cases, the crankcase is maintained under a nonflammable atmosphere.

Glass gages and sight glasses should be protected against physical damage, if their breakage could permit the escape of oil.

INSTRUMENTATION AND CONTROL

Total mechanical failure or disintegration of an engine occurs very infrequently. Gas turbines are at greater risk to this phenomenon than reciprocating engines, because of the turbine's very high rotational speeds.

This hazard is controlled by the provision of automatic engine speed governors. An overspeed shutdown device should be provided on large engines [greater than 100 hp (75 kW)] of both types. Furthermore, the design of gas turbines for stationary installations usually incorporates construction features to contain the compressor, turbine blades, and rotor parts.

Devices should be provided on all engines greater than 10 hp (7.5 kW) to shut down the engine for high-jacket water temperature, high-cylinder temperature, or low lubricating oil temperature. Additional shutdowns should be provided for high-oil temperature on larger engines. Gas turbines should also be arranged to shut down on high-exhaust temperatures.

The starting sequence for gas turbines should incorporate a purge cycle of adequate duration to ensure that a nonflammable atmosphere is present in the turbine and exhaust system prior to ignition.

INSTALLATION SAFEGUARDS

Larger installations should have provision for a remote emergency shutdown of the engine. Most installations also should have a remote means of shutting down fuel and lubricating oil pumps not directly driven by the engine.

Care should be taken in the selection of inlet air filters, particularly for larger gas turbine installations. Whenever possible, filters should be of materials that will not burn freely when exposed to fires.

Diesel engines installed in hazardous locations, as defined by NFPA 70, *National Electrical Code*, should be installed such that the engine is shut down by isolating both the fuel and combustion air supplies. This will prevent the diesel engine from "running on," utilizing the hazardous atmosphere as the fuel source.

Backfires through gas air mixers and carburetor air intakes occur as a result of engine malfunction. These can ignite grease and oil deposits on and around the engine. Flame arresting equipment should be provided at these points if flammable liquids or gases are being pumped or compressed.

OPERATING INSTRUCTIONS

Instructions for normal starting, stopping, operation, and routine maintenance procedures should be posted near the equipment.

Emergency procedures for the safe shutdown of the engine should be developed. These instructions typically include location and operation of remote shutdown stations, shutdown of lubricating oil and hydraulic oil pumps, and closing of the fuel shutoff valve.

FIRE PROTECTION

Manual fire protection equipment, such as portable fire extinguishers, should be provided in accordance with applicable standards and codes. Larger installations, particularly those utilizing an appreciable amount of lubricating or hydraulic oils, and those utilizing liquid fuel, should be protected by automatic fire suppression systems. Sprinkler systems are typically the protection provided for reciprocating engines installed within structures.

Gas turbine installations may be protected by a number of different system applications, dependent on the preference of the engine's manufacturer, operator, or both. Fire protection system options include pre-action waterspray systems, carbon dioxide systems, clean agent systems, and water mist extinguishing systems. Dry chemical extinguishing systems are sometimes utilized to protect the exhaust compartments of these engines. The interlocking of fire protection system operation with the automatic closing of the fuel valves can be of assistance in limiting the size of a fire.

BIBLIOGRAPHY

NFPA Codes, Standards, and Recommended Practices

Reference to the following NFPA codes, standards, and recommended practices will provide further information on the safeguards for stationary combustion engines discussed in this chapter. (See the latest version

of *The NFPA Catalog* for availability of current editions of the following documents.)

NFPA 30, *Flammable and Combustible Liquids Code*
NFPA 37, *Standard for the Installation and Use of Stationary Combustion Engines and Gas Turbines*
NFPA 54, *National Fuel Gas Code*
NFPA 58, *Standard for the Storage and Handling of Liquefied Petroleum Gases*
NFPA 68, *Guide for Venting of Deflagrations*
NFPA 70, *National Electrical Code®*
NFPA 99, *Standard for Health Care Facilities*
NFPA *101®*, *Life Safety Code®*
NFPA 211, *Standard for Chimneys, Fireplaces, Vents, and Solid Fuel-Burning Appliances*
NFPA 220, *Standard on Types of Building Construction*

NFPA 850, *Recommended Practice for Fire Protection for Electric Generating Plants and High Voltage Direct Current Converter Stations*

Additional Reading

"Engine Power Test Code-Spark Ignition and Compression Ignition—Net Power Rating," ANSI/SAE J1349, Society of Automotive Engineers, Inc., Warrendale, PA, 1990.
"Power Piping," ANSI/ASME B31.1, American Society of Mechanical Engineers, New York, 1992.
"Procurement Standard for Gas Turbine Ratings and Performance," ANSI B133.6, American National Standards Institute, New York, 1978 (reconfirmed 1994).
"Welded Steel Tanks for Oil Storage," API 650, American Petroleum Institute, Washington, DC, 1993.

METALWORKING PROCESSES

Paul G. Dobbs

This chapter describes the various methods used to shape, dimension, and finish metals. These include tool, electrical discharge, and electromechanical machining. The fire and explosion hazards of the various processes, metals, and fluids are indicated and the proper safeguards discussed.

Additional information on the characteristics of metals is included in Section 4, Chapter 16, "Metals." Extinguishing agents for metals are discussed in Section 6, Chapter 26, "Combustible Metal Agents and Application Techniques." For more information, the Bureau of Mines publishes a bulletin on the explosibility of metal dusts.[1]

THE METALWORKING PROCESS

The metalworking process includes shaping, dimensioning, or surface finishing of a wide variety of metals. The operations include turning, planing, shaping, slotting, milling, sawing, boring, drilling, grinding, abrasive cutting, filing, threading, reaming, broaching, deburring, lapping, chamfering, spinning, etc. Some of these same operations and equipment are also used to work on other-than-metal materials, such as plastics or wood. These other materials likely would increase the hazard of the operation; however, the extent of that increase, or the additional protection that may be necessary, is not covered in this chapter.

Upon cursory observation of a machining operation, it would appear there are few, if any, fire hazards about which to be concerned. This is deceptive for there are aspects of machining that require careful and constant attention. The fire record is full of experiences that point to the need for considerable pre-planning and preventive action. The principal hazards involve the coolants/lubricants used for the lubrication of the cutting tools and the possible combustion of the cuttings (chips) and fine particles (fines) that are produced as the workpiece is shaped and cut. The three principal sources of ignition are: (1) the heat generated by the work being done, (2) the friction created by the cutting tool, and (3) spontaneous oxidation of materials used. It is important to remember that nearly all metals will burn in air under certain conditions, depending on size, shape, and quantity.

The machining operation can be broken down into five components for the purpose of analyzing the hazards present: (1) machine tool; (2) cutting tool; (3) raw materials; (4) oils, including cutting oils, lubricating oils, and hydraulic oils; and (5) machine cuttings and their collection systems.

The Machine Tool

The modern industrial machine tool may be a single-motor machine, such as a drill press, or it may be a very large multi-motored automatic machine with a highly complex electronics control system. Stock can be brought as needed to an individual machine where a worker then mounts it to the machine and initiates the operation manually. More likely there is some degree of automation in these operations, and, in many cases, several machines can be located in close proximity to facilitate the easy movement of stock from one machining operation to the next. Almost all machines are electric-motor operated and likely contain some lubricating oils and grease. Electrical supply to this equipment is generally 480 volts or less; however, some larger equipment will operate at higher voltages.

In almost all cases, the lubricating oil for the machine is self-contained, i.e., in the unit. Although the amount may become significant [50 to 100 gal (189 to 378 L)], it is generally not a significant hazard. The containment structures for these oils are usually substantial, and leaks are rare.

The electric motors and the lubricating oils systems on the machines do not alone constitute a hazard sufficient to warrant any automatic protection. However, portable Class BC extinguishers should be available in the area.

Traditionally, the machine tool is a stand-alone piece of equipment, usually operated manually by a trained operator, who sets up the stock, then controls the cutting operation. In the modern industrial setting, this arrangement has gradually given way to numerically and digitally controlled equipment, leading to more sophisticated electronic equipment in the vicinity of these operations. While this equipment does not itself present significant fire hazards, it does present an additional loss potential from fire exposure. It can also be a factor in the damage and downtime of a piece of equipment involved in a fire. In addition, modern electronic equipment is more susceptible to damage from exposure fires. This potential for increased loss due to exposure fires should be included in the evaluation of automatic protection in occupancies that include such equipment.

Cutting Tool

The cutting tool can either be fixed in place while the stock is moved past it, such as on a lathe, or the tool itself can be moving, such as

---

Paul G. Dobbs is district engineering manager for American Risk Consultants Corp., based in Plymouth, MI.

in a mill, surface grinder, or boring machine. The tools may be pointed or bladed and may be of hardened high-carbon steel, high-speed steel, hard ceramics, or steel tipped with hard metal carbides of tungsten, cobalt, tantalum, or other hard metallic elements. Grinding abrasives are usually alundum or silicon carbide grains bonded with resin or ceramics and pressed into the shape of wheels. These cutting tools shape the raw materials into the finished part. They can also represent a point of ignition, as they can develop significant heat from the friction of the work they are performing. Dull tools not only increase the requirement for power and influence the quality and tolerance of the work being done, but also are often the ignition source of fires in machines. Dull tools tend to produce finer particles that are more readily combustible or explosive than the larger chips, which are generally produced by sharp tools. Cutting tools normally do not require any special protection. Of course if they are transported on wood pallets or in cardboard cartons or other combustible containers, those materials may necessitate the provision of sprinkler protection.

Raw Materials

The raw materials arriving at a machining plant may be in the form of clean castings, forgings, rough-cut plate, rolled or drawn bar or merchant mill stock, fabrications, tubes, or rolled or extruded shapes. In general, these raw materials have significant mass and so can be considered as noncombustible unless exposed to massive fire. This applies even to the light metals that are known to be more readily combustible when they have a large surface-to-mass ratio. In most cases, though, the raw materials do not present any significant combustible loading. However, the packaging materials used in the transport of these materials must be considered, and, if sufficient combustibles are present, automatic sprinkler protection should be provided as necessary.

Other raw materials include cutting, lubricating, and hydraulic oils; cutting tools; etc. Bulk shipments and storage of combustible oils should be handled in accordance with accepted standards. Further discussion of these hazards can be found in Section 3, Chapter 21, "Storage of Flammable and Combustible Liquids."

Hazard potentials can change significantly once the raw materials are subjected to the metalworking processes. The type of operation the stock is subject to determines the size of pieces of that material that will be removed in the form of chips or fines. Chips and fines significantly increase the surface-to-mass ratio of the material, and, in some cases, they may become subject to spontaneous ignition or to ignition from outside sources. Table 3-11A shows the combustibility of fine powders of various metals, and the amount of

energy needed to initiate combustion. Almost all metals have some degree of combustibility, and, in the case of such materials as aluminum, magnesium, and titanium, they can represent a very significant hazard. As mentioned, some chips in cuttings may be subject to spontaneous ignition. Zirconium and uranium powders tend to ignite spontaneously. However, spontaneous ignition in such apparently noncombustible materials as iron and steel borings and turnings has been recorded in scrapyards and in the holds of ships carrying scrap. In these cases, fires have occurred where the heat of oxidation cannot be dissipated sufficiently to the surrounding atmosphere to prevent the development of ignition temperatures in these materials. In most normal situations, though, this shouldn't be a problem.

Metals, e.g., magnesium and titanium, resist the normal extinguishing methods, such as automatic sprinklers, fire hoses, or regular portable extinguishers. When these materials are being used, special dry-powder extinguishers rated for Class D fires should be available nearby for use, if needed.

Oils

Oil is used during metalworking operations. On the machines, lubricating oils and/or greases may be used to facilitate the operation of the equipment. Coolant/lubricants are often used to cool and lubricate the tool and stock at the point of operation, and to facilitate removal of chips and cutting fines; these oils vary from alkaline aqueous solutions to light petroleum distillates and compounded oils. They are usually liberally applied to the point of operation for the mentioned purposes and then collected, separated from the chips, filtered, and reused. The cleaning and filtering equipment can either be self-contained to the machine itself, or part of the larger central system that interconnects many machines. Essentially all of the oil-based and compounded cutting fluids are combustible to some degree.

Compounded oils that are acidic or contain organic additives that may oxidize and then become acidic with use should not be used for the machining of magnesium and certain other metals, because the acid can react with the metal to release hydrogen. In these cases, only uncompounded mineral oils should be used. Water-oil emulsions are also used as cutting fluids, especially where only light machining operations are taking place. In the emulsion state these oils are nearly noncombustible. Before being mixed with water or if they come out of emulsion, as might happen if the water evaporates from the emulsion, the hazard becomes that of the oil itself. Nevertheless, water-oil emulsions or other noncombustible oils should be used wherever possible to reduce hazards. Where regular cutting

TABLE 3-11A. *Ignition and Explosibility of Metal Powders*[1]

	Ignition temp. cloud layer		Min. explosive concentration		Min. ignition energy dust cloud	Maximum pressure		Maximum rate of pressure rise		Explosibility† Index
	°F	°C	oz/ft³	g/m³	mJ	psig	kPa	psi/sec	Pa/s	
Aluminum	650	343	0.045	45	50	73	503	20000	508	>10
Magnesium	620	327	0.040	40	40	90	620	9000	228	>10
Zirconium	20	7	0.045	45	15*	55	379	6500	146	>10
Titanium	330	165	0.045	45	25	70	482	5500	140	>10
Uranium	20	7	0.060	60	45*	53	365	3400	86	>10
Iron	440	227	0.200	200	72–305	45	310	600	15	0.1
Zinc	680	360	0.500	500	960	48	331	1800	45	<0.1
Bronze	370	188	1.000	1000	—	44	303	1300	33	—
Copper	700	371	—	—	—	—	—	—	—	<0.1

*In this test, 1 g of powder was used. Larger quantities ignited spontaneously.
†Index of Explosibility: none 0, weak 0.1, moderate 0.1–1.0, strong 1.0–10, severe 10.

oils are used, the quantity and extent of use should be evaluated and automatic sprinkler protection considered.

In some cases where additional cooling is required, mineral oil, kerosene, or other materials having lower flash points are sometimes used. If these fluids are utilized within close proximity of their flash point [50°F (10°C)], special extinguishing systems, such as local applications of CO_2 or dry chemical systems, should be considered.

Cutting oils have a tendency to form fine mists around the cutting operations that can drift from the equipment and can condense on other machines, floors, and building structural members. Housekeeping in the surrounding areas is important in these instances, as severe fires can develop and spread over wide areas utilizing these oils. This is most significant where oil deposits have accumulated on vertical surfaces, such as the sides of equipment and building columns. Where such accumulations occur, periodic cleaning of these areas is important. Adequate enclosures and exhaust ventilation to remove this oil spray and mist should be provided.

Where central oil collection and cleaning systems are provided, the dirty oil is often collected in floor trenches that can interconnect numerous machines. If these systems are used, they should be tightly covered, or, preferably, fixed piping systems should be utilized. Where floor trenches are utilized, the possibility of spreading fire from one piece of equipment to the next is greatly enhanced, and, therefore, some form of automatic protection, such as sprinklers or carbon dioxide systems, should be considered. The central cleaning system will generally have quantities of oil present sufficient to warrant at least automatic sprinkler protection in the area. Provisions for overflow of the oil tanks in these areas caused by water from sprinklers or fire hose should be provided, channeling to a safe area.

As one of the purposes of the cutting fluid is to remove heat from the work area, the temperature of the return oil to the work station should be checked regularly to ensure that it does not approach its flash point. Where oil from a central system is pumped back to individual work stations, provisions should be made for manual or, preferably, automatic shutoff of the oil supply. Automatic shutoff can be actuated by sprinkler waterflow detection, or other automatic fire detection systems.

Automatic ceiling sprinkler protection should be provided where combustible cutting oils are used on a general basis. This protection should be installed on an ordinary-hazard basis, with adequate water supplies provided.

Many machine tools employ hydraulic devices for various control actuators. Oil pressures used can vary from 400 to several thousand psi (400 psi equals approximately 42 kPa). The reservoirs and pumps for the hydraulic oil can either be self-contained, i.e., within the machine itself, or part of a larger central system.

Nearly all hydraulic oils are combustible to some extent, even though there are some less-hazardous hydraulic fluids that significantly reduce the fire hazard. These less hazardous fluids will burn when released at high pressure and then brought into contact with a source of ignition. However, they do not sustain combustion when that ignition source is eliminated.

The resulting fire from a high-pressure hydraulic oil leak is usually a very intense torch-like flame that characteristically has a very high rate of heat release. Causes of hydraulic oil release can be deteriorated flexible hoses; failure of pipe joints, especially if threaded connections are utilized; or failure of valve packing or gaskets.

One of the easiest ways to extinguish hydraulic oil leak fires is simply shutting off the oil supply. However, experience has shown that if there is a failure in a pressurized oil system, especially one that results in an intense fire, personnel in the area are likely to run for safety without shutting down the pump. For this reason, remotely located emergency shutoff switches should be provided, or, as a more positive step, automatic shutoff should be provided. Automatic shutdown can be provided by low-level switches in the main oil reservoir; an excess flow switch; a fixed-temperature release, such as a fusible link or thermostat provided in the vicinity of the equipment; or through interconnection to a sprinkler waterflow switch. In most cases where significant hydraulic oil systems are utilized, automatic sprinkler protection should be provided. If the building is not sprinklered completely, the system should extend for at least 20 ft (6 m) beyond all areas of hydraulic usage. Suitable Class BC portable extinguishers are also necessary throughout these areas.

Welded connections should be used wherever possible to eliminate the possibility of leaks from threaded joints. Tubing should be used in preference to pipe. Piping or tubing should be securely fastened in place, and flexible hoses should be kept clear of all objects and not allowed to bend in excessively tight turns. Routine maintenance checks of all hydraulic system components should be made, and necessary repairs or corrective action taken on a timely basis. Leaks should not be allowed to continue, and good housekeeping should be maintained around all equipment.

Most petroleum-based hydraulic oils have a flash point between 350 and 600°F (177 and 315°C). Less hazardous hydraulic fluids also have flash points in a similar range; however, they do not sustain combustion when the ignition source is removed. These fluids are generally of the oil-water emulsion type, water and glycol type, or purely synthetic fluids. Each type has characteristics that cover nearly the entire range of operating temperatures and gasket and seal materials used in conjunction with petroleum-based oils. (See Section 3, Chapter 13, "Fluid Power Systems," for further information on the hazards and safeguards associated with hydraulic pressure systems.) It is important to ensure that the fluid being used is compatible with the seals, gaskets, and packing in the system. If a less hazardous fluid is being retrofilled to an existing system, this compatibility factor becomes especially important.

Machine Cuttings

All machining operations result in the creation of chips, which will vary widely in character, size, and configuration, depending on the operation being performed, the feed and speed of cutting, and the intrinsic characteristics of the metal being cut. Frequently, cuttings and chips appear to be quite large and heavy, and, for this reason, it is often an erroneous assumption that they do not constitute a fire hazard or an explosion hazard. However, a significant amount of fines may be obscured by the more obvious heavy chips.

Combustibility depends upon the combustible character of the metal and upon the ratio of the exposed surface of the cutting or chip to its mass. The thinner the cutting and the smaller the particles, the greater the likelihood of fire. Table 3-11A indicates that relatively small amounts of energy are needed to initiate combustion in some fine metal powders.

The accumulation of chips, cuttings, or fine particles should never be permitted at the machine. They should be removed and placed in a noncombustible container and suitably stored. In small, single-machine operations, the removal of chips is often done manually. In the larger integrated and automated operations, central collection systems are often used.

Where heavy chips and cuttings are generated, they are often mechanically separated from the cutting fluids and stored in some suitable container until disposed. Heavy chips represent the hazard attributable to the particular metal involved. In addition, there is the likelihood that cutting oil residue remains on these chips, and that hazard must also be considered in determining protection for the storage and handling of these materials.

Where high-speed operations, harder metals, or grinding and more precise finishing operations are used, much smaller chips and finer particles of the metal are often generated. The removal and collection of these particles is often accomplished with a pneumatic transport system. These systems usually consist of hoods over the work area, connected by ductwork to a central material/air separator. These separators can be in the form of cyclone collectors, bag collectors, wet collectors, and electrostatic precipitators.

Because of the flow of a significant amount of fine particles in the system, it should be designed with explosion hazards in mind, especially when dealing with the more combustible metals. The system should be designed so that at no time will the fine dust concentration approach the lower explosion limit (LEL) of the dust. The ductwork should be designed and constructed so that it can contain the pressure buildup from an explosion without failing. Bag-type collectors are usually not utilized for collection of fine metal dust, because they inherently provide the optimum condition for rapid and complete combustion. Any combustible filter medium would tend to catch the particles of combustible or explosive metals and hold them in an envelope of oxygen; therefore, all that would be necessary to initiate a fire or explosion would be a very small amount of energy, such as a static discharge.

Wet collectors for fine metal dusts should be used only after careful consideration. Many light-metal dusts react with water to produce highly explosive hydrogen gas. Wetted sludge of these metals is also highly explosive, unless it is kept submerged in the water. Figure 3-11A is a schematic diagram of a water-precipitation-type collector. Further discussions of this aspect of metalworking operations can be found in the following chapters: Section 4, Chapter 15, "Dusts," and Section 4, Chapter 19, "Air-Moving Equipment."

ALTERNATE MEANS OF MACHINING

There are other processes used for the shaping and finishing of metals that require no cutting edge as such.

Electric Discharge Machining

Most important and widely used of the alternate processes is the electric discharge machine (EDM). Electric discharge machining is a process for metal removal by electric arc discharge. In this machine, an electrode is placed in close proximity of the workpiece and is bathed in a continuous flow of dielectric fluid. A direct cur-

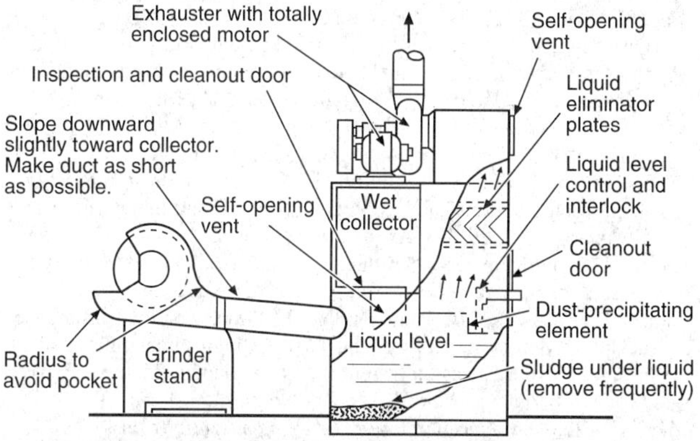

FIG. 3-11A. *A schematic diagram of a water-precipitation-type collector for use in collecting dry combustible metal dust without creating explosive dust clouds or dangerous deposits in ducts or collection chambers. The diagram is intended only to show some of the features that should be incorporated in the collector design. The collector may be used for all combustible metal dusts.*

rent of up to 400 amperes at 400 volts is pulsed from the control unit, and a current flows from the electrode through the dielectric fluid to the workpiece. Each pulse partially ionizes the dielectric fluid and causes a submerged spark that melts into and dislodges a small metal particle. Thus, the required shape is created by this eroding process. A continuous flow of the dielectric fluid is maintained to remove the metal particles. This fluid is filtered before being recirculated to the work area. This filtration system can either be proprietary to an individual machine or a part of a larger central system. The principal hazard associated with this operation is the dielectric fluid, which almost always is combustible. Flash points can be as low as 200°F (93°C). Under normal work conditions, there is a sufficient level of oil above the electrode to quench the spark and prevent the ignition of the oil. However, if the oil level is allowed to become too low, the electrode arc can become a ready source of ignition.

Several large losses have occurred in recent years that have been associated with EDM operations. These losses highlight the importance of maintaining proper fluid levels in the machine and the need for proper precautions when central oil filtration systems are utilized. Because fires within the individual machine work areas are generally confined to a particular piece of equipment, it is important to keep some separation between these machines. At least 5 ft (1.5 m) is considered the minimum clearance to prevent fire spread from one machine to the next. Due to the quantities and combustible nature of the dielectric fluid, automatic sprinkler protection on an ordinary-hazard basis should be provided throughout areas of this occupancy. Suitable Class BC portable extinguishers are also necessary and should be provided within close proximity of these machines.

In order to prevent operation of the machine when there is insufficient fluid to cover the electrode, a low-level interlock should be provided within the work tank. If gravity flow of fluid to the work tank is provided, then automatic-closing fusible-link-operated shutoff valves should be provided to close off the flow of oil in the event of fire.

Significantly greater hazard is created in EDM operations where there is a central oil filtration system. Where these systems are used, at least a 1-hr fire-rated cutoff should be provided to separate these operations from the remaining occupancy. A manual emergency shutoff switch controlling the central oil system should also be provided and located in an area that would be easily accessible during a fire at either the central oil system or the EDM work station. Oil flow between the filtration system and EDM should only be within rigid or flexible metal piping. No plastic or rubber hoses should be utilized, unless specifically listed for such application.

Electromechanical Machining

Electromechanical machining (ECM) is a process in limited specialized use. It operates by anodic dissolution in an electric cell in which the workpiece is the anode and the tool is the cathode. A direct current of 50 to 2,000 amperes at 4 to 30 volts is imposed. At 1,000 amperes, the rate of metal removal is approximately 1 cu in. per min (16 387 mm³/min). Current densities of 100 to 2,000 amperes per sq in. (one square inch equals 64.5 mm²) are employed.

FIRE HAZARDS

The principal hazards encountered in a machining operation are:

1. Chip fires at the machine, where ignition is caused by the heat of metalworking, friction of the chip against the tool, or both;
2. Spontaneous combustion of cuttings;
3. Combustion of coolant/lubricants;
4. Fine particles that are either combustible or explosible;

5. Reaction of certain metals with water or other agents;
6. Combustion of pressurized hydraulic fluids used for the actuation of machine tools and/or their accessories;
7. Combustion of oil vapors deposited on building structures;
8. Combustion of oil-saturated floors; and
9. Combustion of dielectric fluid in EDM equipment.

The principal sources of ignition are:

1. Smoking materials;
2. Heat of cutting;
3. Spontaneous oxidation;
4. Hot particles from grinding, dressing of grinding wheels, and welding and cutting;
5. Hot surfaces, such as furnaces, torches, etc.;
6. Electrical sparking or arcing; and
7. Impact ignition of certain pyrophoric surface compounds, which sometimes form during the earlier stages of fabrication. (An example of this is magnesium nitride, which sometimes appears on the surface of castings and can explode under the impulse of a very minor impact.)

SAFEGUARDS

Building Construction and Protection

Buildings that house metalworking processes should be of fire-resistive construction with a noncombustible roof deck. If the building is large, it should be subdivided by fire walls to limit the spread of fire.

The need for sprinkler protection should be carefully considered. It will depend largely upon the combustible characteristics of the building structure, especially the floors and roof deck, and the extent of the hazard posed by the metalworking machine or raw material as previously discussed. Small fires at machines can be best controlled with portable extinguishers of the correct classification. If proper housekeeping is practiced, the fire hazard from cuttings and turnings can be minimized.

Buildings where machining operations are performed should be regularly inspected for the accumulation of oily deposits and/or fine combustible metal particles. Cleaning should be undertaken before hazardous accumulations develop. All electrical control equipment, particularly those for numerical control or computer-directed control of machines, should be housed in vaportight enclosures. The equipment itself should be inspected at regular intervals and cleaned when necessary. If possible, control equipment should be located in a separate room, cutoff by 1-hr-rated walls.

Fire Protection

The more combustible metals should never be machined unless a suitable fire extinguisher is immediately available to the machine operator. It is important to remember that extinguishers for Class A, B, and C fires are generally ineffective for burning metal or metal dust. Only extinguishers rated for Class D fires should be used on burning metal. When they are used, extreme care must be exercised to ensure that the emergent force of the extinguishing agent or the action of its application to the fire is not sufficient to create an explosive metal dust cloud.

For their own safety, it is most important that all workers thoroughly understand the hazards involved and be properly instructed so they will fully understand the possible consequences of unconsidered or improper action under stress. In many instances, it is highly desirable that local fire fighters, who would normally respond to an alarm, be similarly instructed.

BIBLIOGRAPHY

Reference Cited

1. U.S. Bureau of Mines, "Explosibility of Metal Dusts," Bulletin RI-6516, 1964, U.S. Bureau of Mines, Washington, DC.

NFPA Codes, Standards, and Recommended Practices

Reference to the following NFPA codes, standards, and recommended practices will provide further information on metalworking processes discussed in this chapter. (See the latest version of *The NFPA Catalog* for availability of current editions of the following documents.)

NFPA 10, *Standard for Portable Fire Extinguishers*
NFPA 13, *Standard for the Installation of Sprinkler Systems*
NFPA 30, *Flammable and Combustible Liquids Code*
NFPA 34, *Standard for Dipping and Coating Processes Using Flammable or Combustible Liquids*
NFPA 65, *Standard for the Processing and Finishing of Aluminum*
NFPA 70, *National Electrical Code®*
NFPA 70B, *Recommended Practice for Electrical Equipment Maintenance*
NFPA 72, *National Fire Alarm Code*
NFPA 75, *Standard for the Protection of Electronic Computer/Data Processing Equipment*
NFPA 79, *Electrical Standard for Industrial Machinery*
NFPA 91, *Standard for Exhaust Systems for Air Conveying of Materials*
NFPA 325, *Guide to Fire Hazard Properties of Flammable Liquids, Gases, and Volatile Solids*
NFPA 480, *Standard for the Storage, Handling, and Processing of Magnesium Solids and Powders*
NFPA 481, *Standard for the Production, Processing, Handling, and Storage of Titanium*
NFPA 482, *Standard for the Production, Processing, Handling, and Storage of Zirconium*
NFPA 505, *Fire Safety Standard for Powered Industrial Trucks Including Type Designations, Areas of Use, Maintenance, and Operation*
NFPA 651, *Standard for the Manufacture of Aluminum Powder*
NFPA 801, *Standard for Facilities Handling Radioactive Materials*

Additional Reading

Eckoff, R.K., *Dust Explosions in the Process Industries*, Butterworth—Heinemann, Ltd., Oxford, England, 1991.
Hertzberg, M., Zlochower, I. A. and Cashdollar, K. L., "Explosibility of Metal Dusts," Short Communication, *Combustion Science and Technology*, Vol. 75, No. 1–3, 1991, pp. 161-165.
Kedzierski, M. A. and Worthington, J. L., III, "Design and Machining of Copper Specimens With Micro Holes for Accurate Heat Transfer Measurements," *Experimental Heat Transfer*, Vol. 6, 1993, pp. 329–344.
Long, M. H. and Croce, P. A., "Process Industry: Protection Based on Risk Quantification," Worcester Polytechnic Institute, Firesafety Design in the 21st Century, Conference, Part 1, Conference Report, Part 2, Conference Papers, May 8–10, 1991, Worcester, MA, 1991, 236–245.
Marks, L. S., *Mechanical Engineers Handbook*, 9th ed., McGraw-Hill, New York, 1988.
Rhein, R.A. and Carlton, C.M., "Extinction of Lithium Fires: Thermodynamic Computations and Experimental Data From Literature," *Fire Technology*, Vol. 29, No. 2, 1993, pp. 100–130.
Sharma, T. P. and Kumar, S., "Products of Combustion of the Metal Powders," *Fire Science and Technology*, Vol. 12, No. 2, 1992, pp. 29–38.
Sharma, T. P., Varshney, B. S. and Kumar, S., "Studies on the Burning Behavior of Metal Powder Fires and Their Extinguishment. Part 2. Magnesium Powder Heaps on Insulated and Conducting Material Beds," *Fire Safety Journal*, Vol. 21, No. 2, 1993, pp. 153–176.
"Tactics for Large Magnesium Fires," *Fire Engineering*, Vol. 148, No. 4, April 1995, pp. 38–47.
Varshney, B. S., Kumar, S. and Sharma, T. P., "Studies on the Burning Behavior of Metal Powder Fires and Their Extinguishment. Part 1. Mg, Al, Al-Mg Alloy Powder Fires on Sand Bed," *Fire Safety Journal*, Vol. 16, No. 2, 1990, pp. 93–117.
Walker, J. R., *Machining Fundamentals*, Goodheart-Willcox Co., Inc., South Holland, IL, 1989.

AUTOMATED PROCESSING EQUIPMENT

John F. Bloodgood

Like many other facets of industry, automation has benefited greatly from the technological developments of the last few decades. Thirty years ago, automation was accomplished by use of large relay control panels. In 1969, programmable controllers (PCs) began to replace relays in the automotive industry. By 1971, other industries were using PCs. In 1977, microprocessor-based PCs were controlling automated processes. Today, automation has become very complex and sophisticated. A great deal of care is needed to design safe automated systems.

Programmable controllers have introduced new fire protection problems into the processes that they automate. This is because the PC has been given control over such things as flammable liquids flow, the position of combustible materials, the initiation of sparks and flames, the motion of friction-causing parts, and other potential sources of fuel and ignition. As a result, any kind of failure in the PC can cause a fire or explosion. This chapter addresses how to analyze and deal with the increased potential for fire or explosion with automated equipment. The chapter does not cover the hazards and protection of individual pieces of equipment that are included in other chapters of this handbook. Many of the issues raised also are germane to independently operated machines.

Modern automated machinery usually consists of many pieces of equipment acting in a coordinated manner without operating personnel control. These systems are commonly referred to as manufacturing cells or flexible manufacturing systems. Common examples include numerically controlled machine tools (NC), computerized numerically controlled machine tools (CNC), industrial robots, work-handling machines or systems, material/workpiece transfer equipment, and associated control and information-handling equipment. (See Figure 3-12A.) Automated systems also control many other types of processes.

Each application of automation must be reviewed for any hazardous situations it may introduce. This is done by reviewing both the PC and the process being controlled.

GENERAL CONSIDERATIONS

Automated processing systems consist of two or more machines grouped together to perform a series of operations on materials or parts. Typically, these machines run automatically without an operator, or sometimes one person may run more than one machine. A

John F. Bloodgood, P.E., is president of JFB Enterprises, Consulting Engineers, Fond du Lac, WI and a member of NFPA's Technical Committee on Industrial Machinery.

FIG. 3-12A. *Industrial robots are increasing productivity, lowering manufacturing costs, and improving product quality. The robots pictured here are spot-welding automobile frames. (Source: Robotic Industries Association)*

supervisory controller (usually a computer) directs the interrelations between these machines, transmits information to the individual machine controllers, and monitors the process. The individual machine units (machine and controller) may be supplied by different vendors, thus requiring compatibility between the individual units and the supervisory controller. The issue of compatibility places additional responsibility on the manufacturer for proper installation, maintenance, and training of the end user. In addition, the machine controllers and the supervisory controllers utilize sophisticated computer equipment, operating in a real-time mode, that transmits large amounts of information at low levels and high speeds—conditions that require greater attention to the design, installation, operation, and maintenance of these systems.

AUTOMATED PROCESSING EQUIPMENT

Analysis of the Controller

Microprocessor-based PCs introduce failures never before dealt with in relay-controlled automation. The most common of these failures are described in Table 3-12A.

TABLE 3-12A. *Programmable Computer Failures*

CPU STALL	Processor ceases to execute the program.
INPUT/OUTPUT SCAN FAILURE	Processor fails to scan input/output signals for change in status.
INPUT FAILURE	Input module locks in "on" position and does not respond to further input changes.
ADDRESSING FAILURE	Processor fails to correctly consult input/output intelligence.
OUTPUT FAILURE	Output module locks in "on" position and does not respond to the controller.
MEMORY FAILURE	Incorrect bit in memory causes an improper instruction to be given.
WATCHDOG TIMER FAILURE	Timer controlling execution of instructions fails.
PROGRAM CARD FAILURE	Any problem with the program card.
MOMENTARY POWER OUTAGE	Temporary loss of power to controller.

Problems of personnel safety, power networks and distribution, grounding, electrical noise interference, software application and maintenance, and fail-safe requirements are more evident in these integrated manufacturing systems. Another way PCs introduce fire or explosion potential is by their introduction into areas where relay control panels didn't fit, but which may contain flammable vapors. When this is done, the controller should be intrinsically safe as described in UL 913, *Standard for Safety for Intrinsically Safe Apparatus and Associated Apparatus for Use in Class I, II, and III, Division l Hazardous (Classified) Locations.*

Safety

The operation of integrated manufacturing systems is more automated than individual machines, in terms of starting motions and the transferring of material or parts, without operator intervention. Integrated systems can also cover more complex operations on larger areas of a factory floor. It is therefore necessary to provide additional safety measures beyond those for conventional single-machine work stations. Typical measures include barriers and gates, presence-sensing devices, and visual or audible warning devices to announce pending machine and/or material movement.

Movement of a machine or workpiece and/or a material-handling device that occurs at unannounced times could catch personnel in a "pinch point" and result in a serious injury. Thus, the primary safety consideration is to keep personnel out of the work area while the system is operating. Enclosing the system in a cage or fence is one solution to the problem, but it is not always practical, since many systems require personnel in the work area during operation. Often, it is not possible to determine if anyone is in the work area when the equipment is operating. One alternate solution in particularly hazardous work areas is to provide barriers and/or guards, interlocked with the individual controllers or the supervisory controller. All hazardous motions should be automatically stopped if the interlock is interrupted.

Another alternative is to provide presence-sensing devices, such as floormats or light-beam curtains, that will either stop all dangerous motions immediately or stop the automatic cycle once the motions in progress have been completed. Visual or audible annunciators that warn of pending axis motions or the cycling of auxiliary equipment also can be used if personnel must be in the work area during automatic operation.

Failsafe travel limits, such as mechanical stops or hard-wired travel switches, always should be used in systems where software controllers for individual machines or system monitoring are utilized. The failsafe travel limits ensure that, if the controller should fail or a runaway condition should exist in either an axis motion or an auxiliary cycle, the motion would be physically restrained or the power would be removed from that motion.

ANALYSIS OF THE PROCESS

For some processes, failure of a PC will merely result in inconvenience. It is not unusual, however, for more serious problems to be introduced. For example, PC-controlled robot welders may destroy parts if the arc is not turned off at the right time. They could also ignite combustible materials that mistakenly end up in the area.

PCs controlling hazardous processes, such as spray painting, solvent drying, and fuel firing, also present many potential fire and explosion hazards. A PC failure could create hazardous conditions by not stopping paint guns, solvent introduction, or fuel flow.

Another complication of automated processing is the potential for interruption of production. In automated processes that cannot be run manually, PC failures will stop production. The length of time needed to repair or replace controllers must be considered. Off-the-shelf controllers are easier to replace than custom-made ones. Since off-the-shelf controllers would require reprogramming, software backup should be maintained in a safe location.

Power Networks and Distribution

Careful consideration must be given to the design of the automatic factory power network when large systems are installed. Busbars, which supply electrical power to machines or equipment that are highly inductive or utilize power converters, should be isolated from those that supply power to sensitive electronic equipment. Electrical noise on common power distribution lines can cause malfunctions or permanent damage to electronic equipment. Separate power buses, isolation transformers, and filtering devices are some of the techniques used to minimize the problems associated with power distribution. In some cases, it is necessary to monitor bus loading and automatically schedule the sequencing of particularly heavy electrical loads on the network.

Grounding: The grounding of electrical equipment has always been a consideration for the safety of personnel to guard them from electrical shock. In systems using sensitive electronic equipment, grounding is a primary concern to protect the controller and its input/output connections from the effects of electrical noise. The grounding of control equipment in industrial environments should be integrated by the equipment user into a coordinated ground system. When control equipment is grounded, it is interconnected into the system ground by building columns or ground rods.

In electrical/electronic controllers, there are usually subsystems that are tied to the ground system. It is essential that the noncurrent-carrying metallic enclosures be grounded, including all internal frames and equipment support structures. A specific ground point is provided within the enclosure for the equipment ground to make connection to the ground system. A separate tie point is provided to connect the control common (the zero potential reference point for the electrical and electronic equipment) to the ground system. Having two separate ground reference points separates the two ground systems within the controller, eliminating ground loop problems. Providing two separate ground reference points within the enclosure allows the equipment ground to be connected to the building, while the control ground is connected to a separate ground rod independent of the building structure.

DESIGN SOLUTIONS

To ensure a high degree of reliability in programmable controller control functions, and particularly those that are safety related, these functions should be either periodically checked for proper operation, continuously monitored, or be installed with either redundant or diversified circuits using independent circuitry. Input circuits, input/output addressing, and input/output scanning should be verified with a dynamic safety check, on the order of several times per sec. Output devices, particularly those that control hazardous conditions, should be monitored and, in critical areas, designed using independent redundant circuits. The central processing unit (CPU) as well as output devices are monitored using a watchdog timer, which itself is provided with a dynamic safety check. Sum and parity checks are used on memories.

Above all else, the entire process control system must be designed to fail safely. The response to detection of PC abnormalities can be an alarm, the shutdown of chosen process functions, or both. However, if a PC failure can result in unsafe process conditions, the process should be safely shut down upon this failure. In other words, the system should be failsafe.

Process control devices that may fail in a shorted condition (e.g., rectifiers) can be arranged to be disabled manually. Critical controller functions can be independently hard-wired to ensure the ability to operate properly in the event of failure.

Electrical Noise Interferences

Because entire books have been written on this complicated subject, only the most important considerations are covered here. The effects of coupling electrical noise into control equipment become more pronounced as the operating speed and equipment complexity increase, while the immunity to disturbances decreases. This is certainly the case with controllers applied to industrial automation equipment. The microelectronic circuits operate above 1 MHz, with harmonics in the range of 2,100 MHz to 1 GHz. Signal levels are typically below 15 V. Noise coupling from disturbing sources, such as power inverters, motors, contactors, and relays, and associated wiring runs can be detrimental to the operating system, unless proper techniques are used to reduce the disturbances or minimize coupling into the sensitive circuits. Commonly used methods of reducing the disturbing effects of electrical noise are: (1) employing transformers or filters to isolate power wiring from signal and control wiring, and (2) shielding sensitive control and signal wiring by using shielded conductors or coaxial cable. Proper grounding of the shield (usually at only one end) is a must. For extremely sensitive circuits, the use of double shielding (one shield tied to the equipment ground and the other tied to the control common) may be required. Some circuits or subsystems may have to be entirely enclosed in a grounded shield.

Sensitive signal conductors should be separated to the greatest degree practical from noise sources. One example of this problem involves power leads that are run in the same cable or raceway as digital transmission or analog signal lines. Wherever possible, these lines should be in separate cables or raceways. When this is not possible, it is necessary to use shielding or additional conductors that are tied to the ground. Remember that all unused conductors should be grounded at one end to minimize coupling.

The effect of electrostatic discharge is another potentially serious problem that is often overlooked. Microelectronic devices (memory and logic chips) must be handled by personnel who are grounded by either a wrist strap or conductive plates located on the soles of their shoes. However, nonconductive materials (clothing, paper, etc.) in close proximity to the chips also carry an electrostatic charge that can be neutralized by ionized air. The precautions taken for handling microelectronic parts are typically discontinued once the chips are inserted into a printed circuit assembly. Although a completed printed circuit assembly will dissipate some of the charge, the chips are still subject to damage from electrostatic discharge. The grounding or discharging of personnel who are required to work with integrated circuits, including those who assemble, inspect, test, ship, or maintain the equipment, is important to ensure reliable operation of the electronic equipment. An electric charge of thousands of volts can be developed by simply getting up from a chair or paging through a set of schematic drawings. With the present technology, as little as 50 V of electric charge can permanently damage, degrade the performance of, or shorten the life of these components. In industrial systems employing thousands or tens of thousands of these devices, the potential for a malfunction or failure due to electrostatic discharge is very high and should be considered as serious a problem as shock hazard.

Software Application and Maintenance

The use of programmable electronic controllers and software programming has introduced a new series of problems that were not present with hardware systems. Failure modes and malfunctions are harder to detect and correct since the controllers must first verify the proper operation of the hardware and then test with application software. Since the software tends to be unique for each controller (one of the major advantages of programmable electronic controllers), elaborate testing and verification methods must be devised to ensure that the software functions properly under all dynamic conditions. This problem is compounded when the controllers must interact in a manufacturing system. Even if the software is properly certified before shipment to the end user, it is often modified in the field to correct for unexpected conditions of actual use. Many times, the corrected software is put into operation without a complete recertification, or, even worse, it is corrected at the job site without being retested. In a large, complex system, the results of this "on-the-fly" modification method can be disastrous.

The typical failure mode being "fail unsafe" is another problem created by the introduction of electronic circuitry. Silicone-controlled rectifiers, triacs, and electronic switching devices usually fail in a shorted condition, resulting in such hazards as axis or auxiliary cycle runaway. Short of removing power to the portion of the system that has failed, the condition cannot be corrected. Typical methods of overcoming this problem are redundant circuits, hard-wired contacts that can break the connection to the drives or motors that control motions or auxiliary cycles, or control circuits that monitor and automatically disable equipment that can create hazardous conditions.

The transmission of large amounts of digital information between system elements presents the potential of system malfunctions if the data are not properly received. Wrong axis commands, machine conditions, or even entirely incorrect part programs may be sent or received. Since there is no operator intervention before machine or controller action is initiated, injury can result to personnel or damage can occur to the machine or the work in progress. Redundant transmission or error checking and correction are means of reducing the effects of this problem. Proper shielding and protection of these transmission lines is mandatory if these problems are to be kept to a minimum.

STANDARDS RELATED TO AUTOMATED PROCESSING EQUIPMENT

A number of standards have been or are in the process of being developed that relate to factory automation and address subjects previously covered in this chapter. ANSI B11.1, *Construction, Care, and Use of Machine Tools*, correlated the hazards to persons created by various types of machine tools and the measures to be used to ensure minimum exposure to these hazards. A standard on safety issues related to manufacturing systems/cells (ANSI B11.20, *Safety Requirements for Construction, Care, and Use of Machine Tools, Manufacturing Systems/Cells*) has been developed. This standard deals with the hazards and their associated risks created from the interaction of a number of machines working in a coordinated manner to produce discrete parts or assemblies. There is also a safety standard for robots, developed by a committee sponsored by the Robotics Industries Association (RIA). In turn, these standards incorporate the electrical requirements found in NFPA 79, *Electrical Standard for Industrial Machinery.*

Internationally, safety standards have been developed to encompass industrial manufacturing systems and robots, as well as electrical equipment of industrial machines. International Organization for Standardization (ISO) Technical Committee 199 is responsible for basic safety standards related to machinery. International Electrotechnical Commission (IEC) Technical Committee 44, electrical standard IEC 204–1, *Electrical Equipment of Industrial Machines—Part 1: General Requirements*, has been adopted as a European standard as part of a series of standards used to support the European Directive (law) 89/392/EEC, on machinery safety. The Machinery Safety Directive has been in full effect since January 1, 1995, and affects all machinery manufacturers who are marketing their products in the European Union.

Within Europe itself, a number of standards related to general requirements for machines and machining systems have been or are in the process of being developed. These include requirements for machines, such as machine tools, rubber and plastic machines, and material handling systems, as well as various types of devices, such as light screens, two-hand control, and interlocking devices. It is anticipated that much, if not all, of the research in Europe related to safety of machinery will be elevated into the international standards community (ISO and IEC) within the next few years, making it vital for the United States to participate in these activities, if U.S. industry does not wish to be severely affected in its ability to sell machinery and related products.

Standards related to programmable controllers are being developed by an IEC technical committee. These standards are divided into two series. The first covers such topics as operating conditions, input/output structures, and programming languages. The second is directed toward safety issues, including both hardware and software. There are no corresponding U.S. standards on programmable controllers, since all effort in this area has been focused on IEC activities.

There is also a new European directive on electromagnetic compatibility (89/336/EEC), which went into effect on January 1, 1996. This directive requires that all electrical equipment (with few exceptions) must not generate sufficient electrical fields to disturb other electrical equipment, and, at the same time, must not be susceptible to effects of electromagnetic interference (both electric fields and common mode) and electrostatic discharge in such a manner that a malfunction or failure of the equipment will result. Both European and IEC standards are being developed to provide guidance on the levels of emission and immunity, means for reducing the effects of electromagnetic interference, and testing methods.

The Institute of Electrical and Electronics Engineers (IEEE) has produced a standard on electrical interferences: IEEE 518, *Guide for the Installation of Electrical Equipment to Minimize Electrical Noise Inputs to Controllers from External Sources*, which covers various types of noise that can affect the operation of electrical equipment (e.g., common mode, electromagnetic, and electrostatic) and aims to reduce the effects of these interferences.

Considerable effort both in the United States and internationally has been expended on the encoding and transmission of digital information over industrial networks. The most prominent of these activities is the manufacturing automation protocol (MAP) specification, which is based on the international open system interconnection (OSI) model. Coupled with this work are the series of international standards on the manufacturing message specification (MMS). Adjunct to this work is the soon to be released standard on the digital transmission of encoded serial information between machine controllers and drives.

BIBLIOGRAPHY

References

ANSI B11.20, *Safety Requirements for Construction, Care, and Use of Machine Tools, Manufacturing Systems/Cells*, American National Standards Institute, New York, 1991.

IEC 204-1, *Electrical Equipment of Industrial Machines, Part I*: General Requirements, International Electrotechnical Commission, Geneva, 1984.

IEEE 518, *Guide for the Installation of Electrical Equipment to Minimize Electrical Noise Inputs to Controllers from External Sources*, Institute of Electrical and Electronics Engineers, Piscataway, NJ, 1990.

NFPA Codes, Standards, and Recommended Practices

Reference to the following NFPA codes, standards, and recommended practices will provide further information on automated processing equipment discussed in this chapter. (See the latest version of *The NFPA Catalog* for availability of current editions of the following documents.)

NFPA 79, *Electrical Standard for Industrial Machinery*

NFPA 497A, *Recommended Practice for Classification of Class I Hazardous (Classified) Locations for Electrical Installations in Chemical Process Areas*

Additional Reading

Bell, R. and Smith, S., "Overview of Draft IEC International Standard: Functional Safety of Programmable Electronic System," Third International Conference for Management and Engineering of Fire Safety and Loss Prevention: Onshore and Offshore, Elsevier Applied Science, New York, 1991, pp. 269–289.

Haley, C., "Fundamentals of Digital Electronic Systems and Combinational Logic Circuits," *Fire Engineers Journal*, Vol. 53, No. 170, Sept. 1993, pp. 23–24.

May, W. B. and Park, C., "Building Emulation Computer Program for Testing of Energy Management and Control System Algorithms," Department of Energy, Washington, DC, NBSIR 85-3291, Dec. 1985.

Merrick, D., "Fire Protection for Robotics—A Systems Approach," *Industrial Fire World*, Vol. 2, No. 4, August 1987, pp. 18-21.

Stone, W. C., "Development of a Fast-Response Variable-Amplitude Programmable Reaction Control System," National Institute of Standards and Technology, Gaithersburg, MD, NISTIR 5118, Jan. 1993, 238 pages.

FLUID POWER SYSTEMS

Revised by Paul K. Schacht

Fluid power systems are used for transmitting power or motion to various parts of equipment and machines. The function of the fluid may be power multiplication as in a hydraulic press, actuation of automatic equipment as in die casting, or remote actuation or pressure control of machines or instruments. (See Figure 3-13A.)

Hydraulic fluids are generally petroleum-based and present fire hazards. For additional information on allied topics, consult the following chapters: Section 4, Chapter 6, "Flammable and Combustible Liquids"; and Section 3, Chapter 21, "Storage of Flammable and Combustible Liquids."

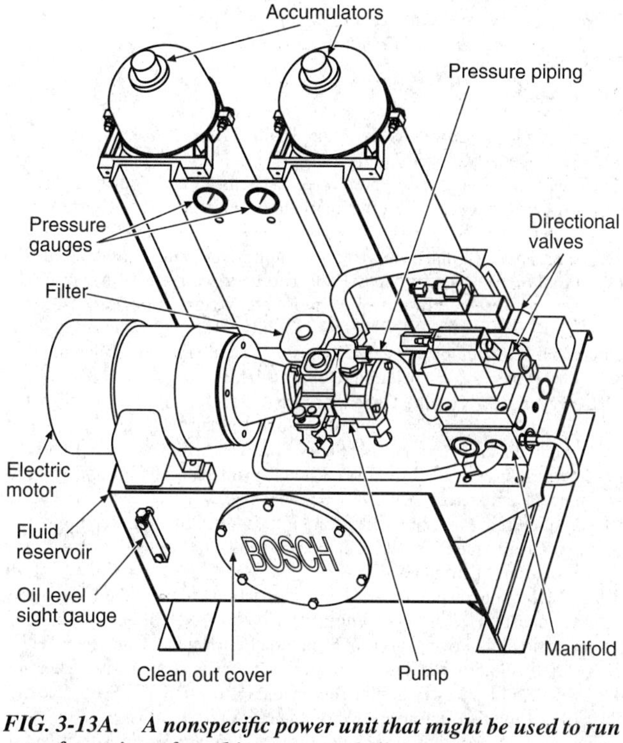

FIG. 3-13A. A nonspecific power unit that might be used to run any of a variety of machines, e.g., a drill or press.

Paul K. Schacht is manager of technical services, Robert Bosch Corporation, Racine, WI, and chairman of the fluids coordinating committee for the National Fluid Power Association.

Early fluid power systems used water as the hydraulic medium. Because of its corrosive effect on the metal parts of the machine, and lack of lubricity, it was replaced by petroleum oil. Except for its fire hazard, oil is an ideal hydraulic fluid. It is noncorrosive, is compatible with a wide variety of materials of construction, is readily available, is relatively inexpensive, and can be recycled. Flash points range from 300 to 600°F (148 to 316°C), and autoignition temperatures from 500 to 750°F (260 to 399°C).

Although petroleum oils are by far the most commonly used hydraulic fluids, synthetic lubricants are being used more frequently in industrial and mobile applications. There are several classes of synthetic lubricants. The most common are synthesized hydrocarbon fluids (SHF), such as polyalphaolefins (PAO) and dialkyl benzenes. Flash points of these fluids range from about 225 to 500°F (106 to 260°C).

Seals and hoses made from nitrile or fluorocarbon polymers and elastomers are commonly used with petroleum oils and the synthesized hydrocarbons. Butyl rubber cannot be used, as it will swell and soften when in contact with these fluids. As with petroleum oil, a wide variety of materials are compatible with synthetic fluids.

FLUIDS UNDER PRESSURE

Pressurized oils or synthesized hydrocarbons in fluid power systems present a considerable fire hazard, particularly where ignition sources are constantly present as in die casting, plastic molding, automatic welding, heat treating of metals, and the engines of mobile equipment.

High-pressure pipe with welded and screwed joints, steel tubing, and metal-reinforced rubber hose is used to conduct oil at pressures ranging up to 10,000 psi (68 950 kPa). Failure of piping, particularly at threaded sections, failure of valves and gaskets or fittings, pulling out of tubing from fittings, and rupture of flexible hose have been the principal causes of oil being released from a fluid power system. Lack of adequate supports or anchorage to prevent vibration or movement of piping has often been a factor in these failures. Repeated flexing and abrasion of rubber hose against other hoses or parts of machines have created weak spots that eventually failed. Tubing under pressure has failed due to being accidentally cut by torches or stepped on during maintenance procedures.

Guidelines to be followed when selecting hydraulic tubing can be found in the National Fluid Power Association and Society of Automotive Engineers (SAE) standards listed in the bibliography.

FIRE CHARACTERISTICS

When oils or synthesized hydrocarbons under pressure are released by failure of equipment, the usual result is an atomized spray of mist or oil droplets which, depending on pressure, may encompass large areas. The oil spray is readily ignited, and the resulting fire is usually torchlike with a very high rate of heat release.

If systems contain oil, the following recommendations should be met in addition to those recommended for less hazardous fluids:

1. Automatic sprinklers provided in the area.
2. Automatic or remote manual switches to shut off the hydraulic pumps in the event of fire, if over 100 gal (378 L) of fluid are involved.
3. Fire extinguishers provided suitable for Class B fires.
4. Threaded pipe connections avoided wherever possible.
5. Machines isolated from positive ignition sources, such as metal casting.
6. Armored hose or pressure hose enclosed in a second tube to contain escaping fluid.
7. Regular inspections of the entire hydraulic system.

LESS HAZARDOUS HYDRAULIC FLUIDS

Less hazardous fluids, which are typically referred to as fire-resistant fluids, have been developed to replace petroleum-based fluids in applications where there is a potential source for fluid ignition. Although these fluids represent a lower fire hazard than petroleum oils or synthesized hydrocarbons, all will ignite under some conditions. Since some types of fluid offer much greater margins of safety than others, the degree of safety required should be considered at the time the hydraulic medium is selected. Despite being less hazardous, the above itemized recommendations for petroleum-based fluids are reasonable precautions to be employed when using fire-resistant fluids.

When converting to less hazardous fluids, and to perform required routine maintenance, consult with component and fluid suppliers, so that suitable equipment and procedures can be selected.

Precautions and recommendations to be used in changing from one type of hydraulic fluid to any other are published by the National Fluid Power Association.

Water-Glycol Fluids

Water-glycol fluids normally consist of 35 to 50 percent water, and ethylene or propylene-glycol, thickeners, and additives. Recommended temperature limits are 0 to 150°F (–18 to 66°C), with normal operating temperatures of 120 to 130°F (49 to 54°C). At higher temperatures, the rate of water evaporation is such that addition of makeup water is frequently required. As the water evaporates, the viscosity increases, and the fire resistance decreases.

Water-glycol fluids are compatible with most types of seals, gaskets, and hoses made from nitrile, fluorocarbon, butyl, or ethylene-propylene, but are incompatible with certain types of leather, cork, and paper. These fluids generally have a solvent action on most petroleum-compatible paints and coatings.

Synthetic Fluids

Synthetic fluids are nonwater-containing man-made materials, such as phosphate esters, blends of phosphate esters with petroleum oil, and polyol esters. Recommended temperature limits are 20 to 150°F (–7 to 66°C), with normal operating temperatures of 120 to 130°F (49 to 54°C).

Phosphate esters and phosphate ester blends are not compatible with seals, gaskets, or hoses made from nitrile, neoprene, or natural rubber, and these elements must be replaced with fluorocarbon or other compatible materials.

Polyol esters are compatible with seals, gaskets, or hoses made from nitrile or fluorocarbon.

All synthetics typically attack paints, coatings, and electrical insulation.

Water-in-Oil Emulsions

Water-in-oil emulsions consist of 35 to 40 percent water, petroleum oil, emulsifiers, and additives. The water is dispersed in fine droplets in the oil phase. Recommended temperature limits for the emulsion are 50 to 150°F (10 to 66°C), with normal operating temperatures of 120 to 130°F (49 to 54°C). At higher temperatures, frequent addition of water is required. Loss of water tends to reduce viscosity and increase flammability. Water-in-oil emulsions are compatible with most seals and gaskets, and with hoses made from nitrile, fluorocarbon, or neoprene. Seals or gaskets made from cork, paper, leather, or butyl are unacceptable. These fluids generally have a solvent action on most petroleum-compatible paints and coatings.

BIBLIOGRAPHY

NFPA Codes, Standards, and Recommended Practices

Reference to the following NFPA codes, standards, and recommended practices will provide further information on the safeguards for fluid power systems discussed in this chapter. (See the latest version of *The NFPA Catalog* for availability of current editions of the following documents.)

NFPA 10, *Standard for Portable Fire Extinguishers*
NFPA 13, *Standard for the Installation of Sprinkler Systems*
NFPA 30, *Flammable and Combustible Liquids Code*

Additional Reading

"Approving Less Hazardous Hydraulic Fluids," *Record*, Vol. 69, No. 5, Sept./Oct.1992, pp. 17–19.
Corbin, E., and Sale, J. W., "Fire Protection at the Grand Coulee Hydro Project," *Proceedings* of the International Conference on Hydropower: Waterpower '87, ASCE, New York, 1988, pp. 2091-2099.
Davis, K., *et. al.*, "Fluids, Lubricants, Coatings, and Elastomer Materials," Final Report, September 1, 1989–February 28, 1991, University of Dayton Research Inst., OH, Wright Lab., Wright-Patterson AFB, OH, UDR-TR-91-103, WL-TR-91-4097, July 1991, 387 pages.
Friedman, R., *Principles of Fire Protection Chemistry*, 2nd ed., National Fire Protection Association, Quincy, MA, 1989.
Hathaway, L. R., "Turbine-Generator Fire Prevention Programs Cut Costs," *Fire Journal*, Vol. 84, No. 3, May/June 1990, p. 47.
Holmstedt, G., and Persson, H., "Spray Fire Tests with Hydraulic Fluids," *Proceedings* of the First International Symposium on Fire Safety Science, Hemisphere, NY, 1986, pp. 869-879.
"Hydraulic Fluid Power-Fire Resistant Fluids—Information Report on Company Trade Names," T2.13.2R2-1989, National Fluid Power Association, Milwaukee, WI.
Khan, M. M. and Tewarson, A., "Characterization of Hydraulic Fluid Spray Combustion," *Fire Technology*, Vol. 27, No. 4, Nov. 1991, pp. 321–333.
"Pneumatic Fluid Power—Use of Synthetic Lubricants—Guidelines," ANSI/B93.50-79, National Fluid Power Association, Milwaukee, WI.
"Pressure Ratings for Hydraulic Tubing and Fittings," SAE J1065, Society of Automotive Engineers, Warrendale, PA.
Saunders, J., "Testing: Fluids Under Fire," *Lubricants World*, Vol. 5, No. 6, June 1995, pp. 15–23.
"Standard Practice for the Use of Fire Resistant Fluids in Industrial Hydraulic Fluid Power Systems," ANSI/B93.5M-1988, National Fluid Power Association, Milwaukee, WI.
Stewart, H. L., *Hydraulic and Pneumatic Power for Production*, 4th ed., Industrial Press Inc., New York, 1977.

"Tube Fittings for Flammable and Combustible Fluids, Refrigeration Service, and Marine Use," Standard for Safety, Underwriters Laboratories, Inc., Northbrook, IL, UL 109, 5th Edition, Apr. 28, 1993, 13 pages.

VanDam, J., Laruelle, L. and Daenens, P., "Qualitative and Quantitative Determination of Thermal Degradation Products of Type C Hydraulic Fluids," *Journal of Fire Sciences*, Vol. 12, No. 4, July/Aug. 1994, pp. 376–387.

Vásquez, I., *et. al.*, "Suppression of Elevated Temperature Hydraulic Fluid and JP-8 Spray Flames," *Proceedings* of the Fourth International Symposium on Fire Safety Science, International Association for Fire Safety Science, 1994, pp. 1255–1266.

Yule, A. J. and Moodie, K., "Method for Testing the Flammability of Sprays of Hydraulic Fluid," *Fire Safety Journal*, Vol. 18, No. 3, 1992, pp. 273–302.

WELDING AND CUTTING

Revised by August F. Manz

In its largest sense, welding includes any materials joining process. As used in this chapter, welding means only fusion welding, i.e., melting together. The American Welding Society (AWS) defines welding as "a joining process that produces coalescence* of materials by heating them to the welding temperature, with or without the application of pressure or by the application of pressure alone, and with or without the use of filler metal." Similarly, melting is the significant feature in reference to thermal cutting. The AWS defines thermal cutting as "a group of cutting processes that severs or removes metal by localized melting, burning, or vaporizing of the workpiece." Both welding and thermal cutting require high-intensity energy sources—usually an electric arc or the heat of combustion of a fuel gas. In this text it is not possible to cover all welding and cutting processes; therefore only basic information about the more common processes is discussed. AWS terms will be used, where possible, in the remainder of this chapter.

For additional information, consult Section 3, Chapter 11, "Metalworking Processes"; Section 3, Chapter 22, "Storage of Gases"; and Section 4, Chapter 7, "Gases."

Torch fires accounted for about 5 percent of fires in 1988–1992 in nonresidential structures and 8 percent in industrial and manufacturing properties.[1,2] Typically, three-fourths of torch fires are due to cutting or welding torches, which amounts to thousands of fires a year. However, education, training, and on-the-job practice can significantly reduce the potential for welding and cutting fires. There are four factors that must be kept in mind:

1. Two sides of the fire triangle are always present, i.e., a source(s) of ignition and air to support combustion. The other controllable element is the combustible material.
2. Familiarity with cutting or welding as an activity. A shop that performs welding or cutting operations on a routine basis in a fixed location can develop confidence and reinforce safe habits more than a shop that performs welding or cutting only occasionally and in a variety of different temporary locations.
3. The kind of process and equipment used and their potential fire effect. Cutting, as well as certain arc welding operations, produces literally thousands of ignition sources in the form of sparks and hot slag. Other types of torches, such as arcs and oxy-fuel gas flames, have only themselves as ignition sources.

4. The supervision and training of the welder. Is the welder trained in the proper use of the equipment and mindful of the exposure? Who has knowledge of and has assessed the risks involved in bringing a torch into the work area and has authorized its use there?

PROCESSES USING ELECTRICITY

Arc Welding (AW)

This term applies to a number of processes that use an electric arc as the heat source for melting and joining metals. The arc is a useful tool because its heat can be concentrated and controlled quite effectively. Frequently, but not always, a filler metal must be used to obtain a good joint. The arc is struck between the metals to be welded and an electrode, which is maneuvered along the joint or which may remain stationary while the work is moved beneath it. The electrode may be consumable or nonconsumable. If the latter, a separate rod or wire may be used as the filler metal. Consumable electrodes supply their own filler metal by melting. The molten metal and weld zone may be protected from atmospheric contamination by use of a supplementary shielding gas, or a shielding atmosphere produced by decomposition of a flux which is used with certain types of electrodes and wires.

Shielded metal arc welding (SMAW): This simple well-known process is widely used with ferrous-based metals. (See Figure 3-14A.) It produces coalescence of metals by heating them with an arc between a covered metal electrode and the weld pool. The process is used with shielding gases from the decomposition of the electrode covering, without the application of pressure, and with filler metal from the electrode. The process requires an alternating or direct current power supply, power cables, and an electrode holder. Shielded metal arc welding can be performed readily in remote, unusual, or confined locations. Consequently, this process is widely used in such industrial applications as construction, shipbuilding, and pipeline erection. Fairly simple, portable units are frequently used in maintenance and field construction work.

Gas metal arc welding (GMAW): This process uses a continuous filler metal electrode, which functions as one terminal of the arc, and a gas to shield the arc and the weld metal. The shielding gas depends on the base metal and process variations. It may be an inert

---

August F. Manz is president of A. F. Manz Associates, welding technology and safety consultants, Union, NJ and a member of NFPA's Technical Committee on Cutting and Welding Practices.

*This is the preferred American Welding Society term used to describe the flowing together into one body of the materials being welded.

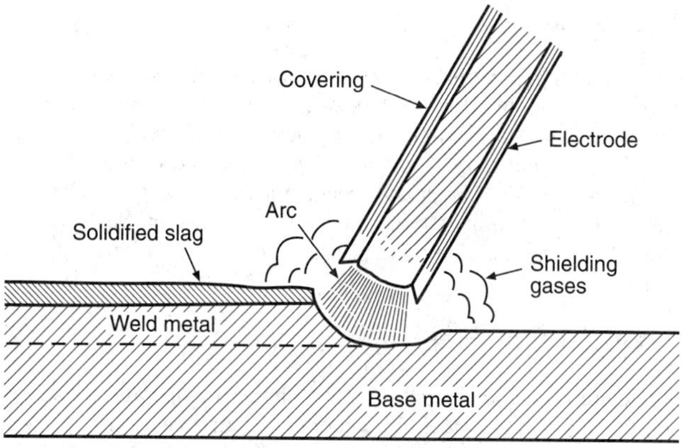

FIG. 3-14A.   Schematic diagram of a typical shielded metal arc welding operation. (Source: Lincoln Electric Co.)

gas, such as argon or helium; an active gas, such as carbon dioxide; or mixtures involving these gases with additions of hydrogen, nitrogen, or oxygen. Overall, this process can be used to join just about any metal in any configuration of joint.

Flux cored arc welding (FCAW):   Although this process is similar to gas metal arc welding in equipment used and types of applications, it uses tubular rather than solid electrodes. Minerals and alloys in the core assist in weld protection, and many such electrodes are intended for use with shielding gases. Other kinds of cored electrodes are self-shielding and produce their own protective envelope without an auxiliary shielding gas.

Gas tungsten arc welding (GTAW):   This process is similar to gas metal arc welding and flux cored arc welding, but it uses a nonconsumable tungsten electrode for one pole of an inert gas-shielded arc. Filler metal may be used. The most common inert shielding gases are argon and helium. Equipment required comprises a power supply, welding torch, source of inert gas, suitable pressure regulators and flow meters, and connecting hoses.

Plasma arc welding (PAW):   This is characterized by the plasma state, wherein the temperature of a gas is raised until the gas becomes at least partly ionized, enabling it to conduct an electric current. It uses a constricted arc between a nonconsumable electrode and the weld pool (transferred arc) or between the electrode and the constricting nozzle (nontransferred arc). A plasma arc torch incorporates an electrode and an orifice through which a flow of gas forms the arc plasma. Secondary gas, which flows through an outer nozzle cup arrangement that encircles the arc plasma orifice, shields the arc and the weld. Typical shielding gas may be argon, helium, or mixtures of argon with hydrogen or helium. Filler metal may or may not be used. Equipment required is not unlike that for gas tungsten arc welding, with necessary differences in power supply and control, torch design, and gas supply and control.

Submerged arc welding (SAW):   In this process the arc and molten metal are shielded by molten flux and a layer of unmelted flux granules. A continuously fed electrode is submerged in the flux. The arc is submerged below the flux. The process is used without pressure. Radiation and fumes are minimal, and fire hazard potential is likewise relatively minor compared to other processes.

Resistance Welding (RW)

Welding heat for these processes is created by resistance to flow of current through the parts being joined. Resistance welding is gener-

ally used to join two metal pieces that may have different thicknesses and composition. Electrodes conduct current through the pieces, which are clamped or rigidly held together to provide good contact and pressure at the joint.

Flash Welding (FW)

Although defined by the AWS as resistance welding, flash welding is considered a process unto itself. Heat is created by a flashing action at the interface. Pressure is applied after heating, resulting in the expulsion of metal, formation of a flash, and usually a significant shower of sparks. Applications usually involve butt welding rods or bars end to end or edge to edge. The process is almost always automatic. Flash welding machines vary significantly in size, some being large enough to weld sheets in a steel mill.

Electroslag Welding (ESW)

Electroslag welding is a process used primarily for vertical position welding. This process uses a slag, which is conductive while molten, to protect the weld and melt base-metal edges and filler metal. An arc is needed to start the process by melting the slag and preheating the work because the unfused, solid slag is nonconductive. Once the process is started, the arc is no longer necessary, and resistance to current flow through the molten slag produces the heat to sustain the process.

Arc Cutting (AC)

This term applies to a group of thermal cutting processes that, like the kindred welding processes, melt the metals to be cut by heating with an arc struck between an electrode and the workpiece.

Air carbon arc cutting (CAC-A):   This process uses electrodes comprising a mixture of graphite and carbon, and which may be coated with a copper layer to improve current-carrying capacity. Metal melted by the arc is blown away by a jet of compressed air supplied by conventional compressors, usually at about 80 psi (550 kPa). Current is provided by standard electric welding power supply units rated for CAC-A duty, or special power units.

Plasma arc cutting (PAC):   This cutting process achieves cutting action with an extremely hot jet of ionized gases at high velocity. The high-velocity hot jet is produced by forcing an arc and gas through a small orifice. Arc energy confined to a small area melts the metal, and the jet of highly heated, expanding gases forces the molten metal from the cutting area. High-arc voltages, special power supplies, and water-cooled torches are required. Air, nitrogen, argon-hydrogen, argon-nitrogen, and argon-helium gases are commonly used.

OXYFUEL GAS PROCESSES

Oxyfuel Gas Welding (OFW)

This process uses the heat of combustion of a fuel gas and oxygen flame at high temperatures to melt the workpiece base metal and filler metal, if these are used. The flame (except for oxygen-hydrogen) is readily adjusted to provide a "neutral," oxidizing (excess oxygen), or carburizing[†] (excess fuel gas) environment. The type of flame used depends on the materials being welded.

Acetylene remains the pre-eminent welding fuel because it has the highest flame temperature of all the fuel gases. Other fuel gases or mixtures, such as methyl acetylene-propadiene (stabilized), have found some acceptance. Other hydrocarbons (propane, propylene,

---

†Sometimes carburizing flames are called reducing flames.

butane, natural gas-methane) are not suitable for welding ferrous metals. Hydrogen has a low flame temperature and heat content, and its flame, which is colorless, is difficult to adjust for oxygen-fuel gas ratio. Oxy-hydrogen welding does have application for welding lead, which is frequently but erroneously called lead burning. Table 3-14A shows neutral flame temperatures of some fuel gases.

TABLE 3-14A. *Flame Temperatures of Fuel Gases with Oxygen*

Gas	Neutral Flame Temperature	
	°F	°C
Acetylene	5589	3087
Methyl acetylene-propadiene (stabilized)	5301	2927
Propylene	5250	2900
Hydrogen	4820	2660
Natural gas-methane	4600	2538
Propane	4579	2526

With the exception of acetylene, neutral flame temperatures are significantly lower than the maximums, so it is not possible to melt and control the weld, except perhaps for thin sheet metal. At maximum temperatures, the flames are strongly oxidizing and not usable because of deleterious metal oxide formation in the weld metal. Limited use of methyl acetylene-propadiene (stabilized) requires special procedures.

Brazing and Braze Welding (BW)

Brazing (B): This is broadly defined by the AWS as a group of welding processes in which the base metal is heated but not fused, and the joining is accomplished with a filler metal having a liquidus temperature above 840°F (450°C) and below the solidus temperature of the base metal. Moreover, brazing is characterized by the filler metal being distributed through a closely fitted joint by capillary action.

Braze welding: Braze welding is of more significance as it relates to fire prevention because of its frequent use in repair and maintenance activities. It differs from brazing in that capillary action in the joint is not a factor; the filler metal is laid down in a groove or fillet at the point of application. The edges of the joint are heated to a dull red color, flux is used, and filler metal is supplied by a bronze rod. However, the base metal is not melted, only the filler metal melts. The process is widely used on ferrous metals, as well as on copper and nickel alloys. Sometimes it provides a solution to the problem of joining dissimilar metals.

Hard soldering: This common but incorrect term is used to describe brazing with a silver-base filler metal. The operation is often performed with an oxyfuel gas torch, as in assembling copper pipe joints with silver brazing alloy. Many times, however, the oxygen is not pure gas but is supplied as air, as in the air-acetylene torch. Thus, only one small gas supply cylinder may be needed, enhancing portability.‡

---

‡The air-fuel gas flame, particularly air-acetylene, also finds significant use in soldering, commonly called "soft soldering," which by definition involves a filler metal that melts below 840°F (450°C). Plumbing jointwork, refrigeration piping, and heating and ventilating duct assemblies are applications where the air-fuel gas flame is used. Equipment portability is an advantage—but it may also signify access to places where fire prevention is not easily accomplished and is easily overlooked because a single-joint operation can be done in a matter of minutes.

Surfacing

The oxyfuel gas flame (usually oxyacetylene) is used not in a true welding sense but rather to deposit a layer of filler metal on a base metal to obtain certain surface properties or dimensions. Bronze may be used, but more frequently the process is one of hard facing, where the deposited layer is a special alloy that will greatly prolong the life of parts subject to extreme wear or abrasion. Surfacing can also be accomplished by use of arc welding and thermal spraying processes.

Oxyfuel Gas Heating Operations

There are numerous industrial and commercial operations that regularly employ an oxyfuel or air-fuel flame but do not involve a joining or severing operation. The flame and the nature and location of these operations nonetheless may represent an appreciable fire potential. Some such operations are:

Forming: Heating (ironwork, piping, etc.) to facilitate bending, shaping, or straightening.

Annealing, flame hardening, flame softening: Use of the flame in a controlled manner to obtain specific properties.

Flame priming: Heating to remove scale and rust in preparing metal surfaces for painting.

Flame descaling: Heating (generally a steel mill application) to remove scale from bars, billets, slabs, etc., to facilitate machining or inspection.

Other applications: Paint burning, glass finishing, leather edging, babbitting, or antiquing of wood.

Oxyfuel Gas Cutting (OFC)

This term describes a group of oxygen cutting processes named by the specific fuel gas used [e.g., oxy-acetylene cutting (OFC-A), oxy-natural gas cutting (OFC-N)] for severing metals by the reaction of high-purity oxygen with the metal at elevated temperatures.

Burning iron in oxygen produces iron oxide, normally a solid, but the oxide melts at a temperature below the melting point of iron or steel and runs off as slag. A variation in the process is oxyfuel gas gouging, wherein a relatively low-velocity oxygen jet permits gouging or grooving of a metal surface in a reasonably smooth, well-defined manner.

Oxyfuel Gas Welding and Cutting Equipment

Basic elements are fuel gas and oxygen supplies, pressure regulators, conduits (hoses or piping) to convey the gases, and a torch to mix and burn the gases in controlled fashion and provide the jet used in cutting operations.

In its simplest form, the equipment comprises a cylinder each of fuel gas and oxygen, a pressure regulator on each cylinder, hoses, and a torch. (See Figure 3-14B.) A steel mill, however, might have a major installation for cutting and welding operations. (See Figure 3-14C.) Fuel gas might be supplied from large multicylinder manifolds (possibly truck mounted and replaceable), from storage tanks for liquefied fuel, or from a public utility natural gas main. Oxygen may be supplied from the mill's own on-site oxygen-generating plant, from storage tanks for liquid oxygen or high-pressure gas or both, or from multicylinder manifolds. A major installation may also have an extensive fixed-piping distribution network and consuming devices running the gamut from simple hand torches to sophisticated steel-conditioning machines.

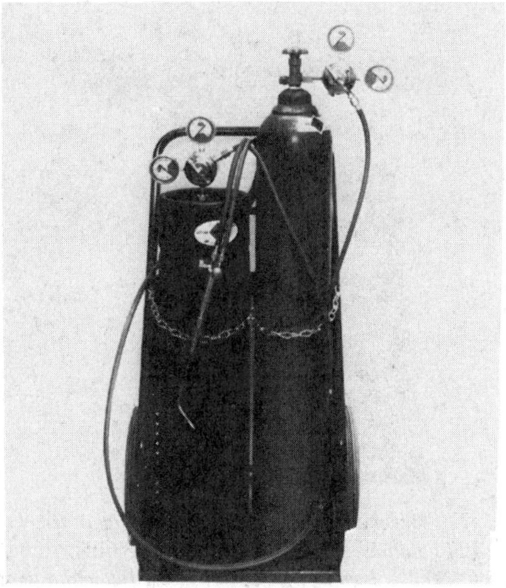

FIG. 3-14B. A portable welding outfit with oxygen and acetylene outfits chained to an easy-rolling cylinder truck (cutting attachments not shown). (Source: The ESAB Group, Inc.)

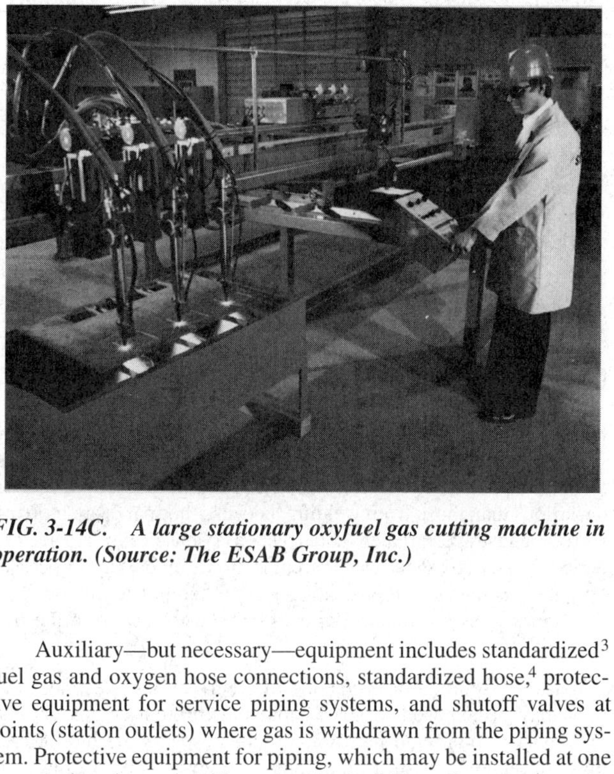

FIG. 3-14C. A large stationary oxyfuel gas cutting machine in operation. (Source: The ESAB Group, Inc.)

Auxiliary—but necessary—equipment includes standardized[3] fuel gas and oxygen hose connections, standardized hose,[4] protective equipment for service piping systems, and shutoff valves at points (station outlets) where gas is withdrawn from the piping system. Protective equipment for piping, which may be installed at one or more locations, should prevent the backflow of oxygen into the fuel gas supply system, the passage of flame flashback into the fuel gas supply system, and the development of pressures in excess of system components ratings. In some systems, backflow prevention devices also may be required at station outlets. (See NFPA 51, Standard for the Design and Installation of Oxygen-Fuel Gas Systems for Welding, Cutting, and Allied Processes.)

Welding torches have inlet connections and valves for each gas at the rear of the handle. (See Figure 3-14D.) The gases are controlled by inlet valves and thoroughly mixed before issuing from the torch tip. Cutting torches are similar but provide passageways and

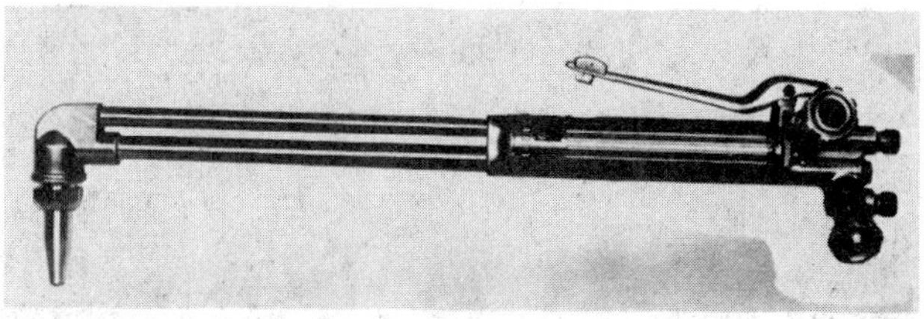

FIG. 3-14D. A typical cutting torch. All control valves are located at the rear of the body. (Source: The ESAB Group, Inc.)

separate valving for supplying and controlling the cutting oxygen jet at the center of the tip.

Mechanized cutting is common. Equipment varies from relatively simple, portable machines, used for straightline work and perhaps circles and some irregular shapes, to highly sophisticated multi-torch machines that can trace (by photocell, laser beams, or other electronic means) intricate shapes and accurately and simultaneously produce a number of parts of the same shape.

THERMAL SPRAYING (THSP)

Closely allied to welding are several processes, known collectively as thermal spraying,[§] in which finely divided metallic or nonmetallic materials are deposited in molten or near-molten condition to form coatings. Special "guns" are used. (See Figure 3-14E.) In flame spraying (FLSP), an oxyfuel gas flame is used to melt the coating material, and an auxiliary compressed gas may be used to assist in atomizing and propelling coating material to the workpiece.

Arc spraying (ASP) employs an arc between two consumable electrodes of coating material, plus auxiliary compressed gas to atomize and propel the coating particles. Plasma spraying (PSP) uses a nontransferred plasma arc as the heat source and for propelling the coating material. In detonation flame spraying, the coating material is melted and propelled by the controlled explosion of fuel gas and oxygen.

SAFEGUARDS

Equipment Preparation and Condition

Although it is beyond the scope of this chapter to describe all the fire prevention considerations that are tied into proper equipment design, installation, and maintenance, some significant items have been selected for specific attention.

Oxyfuel Gas Equipment

Equipment meeting recognized standards (e.g., torches, cylinder manifolds, pressure regulators, pipeline protective devices, etc.) should be used.

Since oxygen is a far more powerful oxidizer than air, oxygen equipment must be kept clean (i.e., free of oil, grease, and other combustible contaminants). Materials that burn in air will burn violently in pure oxygen at normal pressure and explosively in pressurized oxygen. Also, many materials that do not burn in air will do so in pure oxygen, particularly under pressure. Consequently, it is widely recognized good practice to reserve equipment for oxygen service only.

§American Welding Society preferred terminology for "metal spraying" or "metalizing."

FIG. 3-14E. *Thermal spray gun. (Source: American Welding Society)*

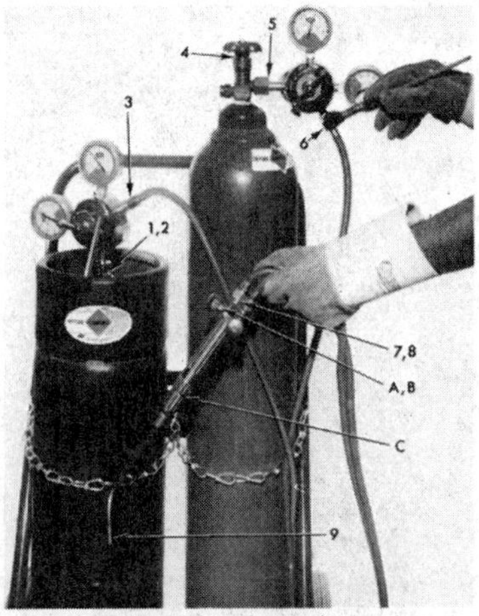

FIG. 3-14F. *Before operating an oxyfuel gas outfit, the equipment should be checked for leaks. After pressurizing both hose lines (with torch valves tightly closed), test for leakage at the following points, using an approved leak-test solution: (1,2) acetylene cylinder connection and acetylene cylinder valve spindle, (3) acetylene regulator-to-hose connection, (4) oxygen valve spindle, (5) oxygen cylinder connection, (6) oxygen regulator-to-hose connection, (7,8) hose connections at the torch, and (9) torch tip (for leakage past the torch valves). Later, after lighting the torch, check for leakage at the throttle valve stems (A,B) and at the welding head-to-torch handle connection (C). (Source: The ESAB Group, Inc.)*

Proper storage of gas cylinders is important. (See NFPA 51, *Standard for the Design and Installation of Oxygen-Fuel Gas Systems for Welding, Cutting, and Allied Processes.*)

Cylinders should be moved and handled in accordance with recognized practices, e.g., those of the American National Standards Institute (ANSI) and the Compressed Gas Association (CGA). (See bibliography.) There are many such recognized practices, and their importance may vary with working locations and conditions. However, cylinders should always be supported or located in such a way that they cannot be knocked over. Ruptured cylinders can explode.

Oxyfuel gas cutting and welding equipment should be tested periodically for leaks. (See Figure 3-14F.) Testing frequency depends upon the specific kind of equipment involved and how often it is used. Hose connections are known possible trouble spots. Also, experience has shown that fires occur at fuel gas cylinder-to-regulator connections simply because someone failed to tighten the joint properly. The ignition of leaking gas at this point may in turn cause the release of cylinder safety devices, especially with acetylene cylinders fitted with fusible plugs, thereby releasing more gas and increasing the size of the fire.

Only standard welding hose should be used.[4] It should be frequently inspected for burns, cuts, worn places, abrasions, and similar defects. Taped repairs are unacceptable. Replace the damaged hose, or, if feasible, cut out the affected area and insert a proper splice.

When repairs to equipment, e.g., torches and regulators, are required, they should be carried out by trained, skilled mechanics.

Arc Welding Equipment

When using arc welding equipment, the following should be addressed to avoid accidents, electric shock, or equipment damage:

1. Use equipment meeting recognized criteria, such as that provided by the American National Standards Institute, National Electrical Manufacturers Association (NEMA), Underwriters Laboratories Inc. (UL), etc. Installation, including incoming power lines and grounding of the machine frame or case, should comply with NFPA 70, *National Electrical Code®*, with particular attention to Article 630, "Electric Welders"; and with ANSI/AWS Z49.1, *Standard for Safety in Welding and Cutting.*
2. Proper storage and handling procedures for cylinders of shielding gases should be observed.

3. At each work location, cylinders should be supported or located in such a way that they cannot be knocked over accidentally. Precautions must be taken that the cylinders are not grounded.
4. Cable sizes should be adequate for current and anticipated duty cycles. Sustained overloading of inadequate cables may burn away insulation. Cables should be inspected frequently for wear and damage, and properly repaired or replaced when necessary.

Precautions for the Work Area

Hazardous sparks, e.g., globules of molten, burning metal or hot slag, are produced by both welding and cutting operations. Sparks from cutting, particularly oxyfuel gas cutting, are generally more hazardous than those from welding, because the sparks are more numerous and travel greater distances. In a sense, they are jet propelled by the oxygen or airstreams used in cutting processes. Oxyfuel gas flames and electric arcs are inherent and obvious ignition sources, as are hot workpieces or sections cut from the base workpiece.

Either isolation or protection of combustibles is essential, for they may be exposed to sparks that fall through cracks or other openings in floors and partitions. If those sparks are of sufficient mass to retain heat for a time, they may ignite combustibles. The recommended requirements for combustible control in the cutting or welding work area are:

1. Move all combustibles a safe distance away—at least 35 ft (10.6 m) horizontally—and be sure that there are no openings in walls or floors within 35 ft (10.6 m).
2. Move the work to a safe location.
3. If neither of the foregoing steps is possible, protect the exposed combustibles with suitable fire-resistant guards and provide a trained fire watcher with extinguishing equipment readily available.

These steps are only a partial solution to the problem of preventing cutting and welding fires. There are other important factors to consider. Are there any inconspicuous combustibles in an area proposed for cutting and welding operations? What conditions must be met before cutting and welding operations can take place? Who has the responsibility for authorizing the work to proceed? Are cutters, welders, and their supervisors properly trained in the use of their equipment and in emergency procedures should a fire occur? If an outside firm is engaged to do cutting and welding work, chances are that its employees will be unfamiliar with the premises and its contents. Have they been briefed on the conditions in the areas where they will work?

Based on the fundamental but necessary understanding that welders, their supervisors, and facility management share the responsibility for fire safety, the following abridged version of NFPA 51B, *Standard for Fire Prevention in Use of Cutting and Welding Processes*, should be helpful. It is adapted here for convenience, and should not be used in place of NFPA 51B.

1. Management must establish areas designed and authorized for cutting and welding and/or designate a knowledgeable person to authorize welding or cutting in areas not specifically designed for such processes. This management designee must require trained fire watchers where the potential exists for a significant fire to develop. Fire watchers must also be required where appreciable quantities of shielded combustibles are less than 35 ft (10.6 m) away; where wall or floor openings within 35 ft (10.6 m) expose combustibles in adjacent areas; or where combustibles adjacent to opposite sides of partitions, ceiling, or roofs are likely to be ignited by heat from the work. Management must select contractors with a view to their awareness of risks and the quality of their personnel. Management must also advise contractors of the presence of flammable materials or other hazardous conditions on the property work site.
2. The supervisor of welding and cutting operations, e.g., the plant manager, plant maintenance foreman, contractor, or contractor's foreman, in areas not designed for such processes must be assigned the following responsibilities:
   (a) Determine what combustible materials are present at the work site.
   (b) If necessary, have the work or the combustibles moved, or have the combustibles shielded.
   (c) Secure authorization from management, preferably in the form of a written permit.
   (d) See that the welder is aware of the authorization and conditions.
   (e) See that fire watchers are available when required.
   (f) Make a final check for fires one-half hour after completion of welding or cutting operations in cases where a fire watcher was not required.
3. Welders must have their supervisor's approval before starting to cut or weld, must handle their equipment safely, and must continue to work only so long as approval conditions do not change.
4. There are a number of precautions to be observed during welding and cutting operations:
   (a) Welding and cutting must not be permitted in flammable (explosive) atmospheres; near large quantities of exposed, readily ignitable materials; in areas not authorized by management; or on metal partitions, walls, or roofs with combustible covering or with combustible sandwich-type panel construction.
   (b) Floors must be free of combustibles, such as wood shavings. If the floor is of combustible material, it must be kept wet or otherwise protected.
   (c) If combustibles are closer than 35 ft (10.6 m) to the welding or cutting process and the work cannot be moved or the com-

bustibles relocated at least 35 ft (10.6 m) away, they must be protected with flame resistant covers or metal guards or curtains. This also applies to walls, partitions, ceilings, or roofs of combustible construction. (See Figure 3-14G.)
   (d) Openings in walls, floors, or ducts must be covered if within 35 ft (10.6 m) of the work.
   (e) Cutting or welding on pipes or other metal in contact with combustible walls, partitions, ceilings, or roofs must not be performed if close enough to cause ignition by heat conduction.
   (f) Charged and operable fire extinguishers must be readily available. Trained fire watchers must be posted. In the absence of fire watchers, an important minimum step is to check the work area and adjacent areas carefully for at least one-half hour after completion of welding or cutting to detect possible smoldering fires.

SPECIAL SITUATIONS AND ADDITIONAL PRECAUTIONS

Containers

Welding or cutting on containers can cause—and has caused—explosions or fires with potential for deaths or serious injuries. Any container can contain or may have contained a combustible. Therefore, all containers must be considered hazardous unless they have been tested and found safe, cleaned, or rendered inert. Therefore it is essential that flammable liquids, solids, or vapors be removed from containers by some type of adequate cleaning procedure before welding or cutting.

Depending upon the application, it may be necessary or desirable to supplement the cleaning with inerting, water flooding, or periodic testing (for flammables) of the atmosphere within the container. With some materials, water washing or flushing may be sufficient; others may respond to steam cleaning; some may require alkaline cleaners (trisodium phosphate or caustic soda); some may call for more specialized cleaning methods. Heavy, viscous liquids and solids can be especially difficult to remove completely. Residues are especially dangerous because they may be volatilized or decomposed into flammable products by the heat of the torch or arc.

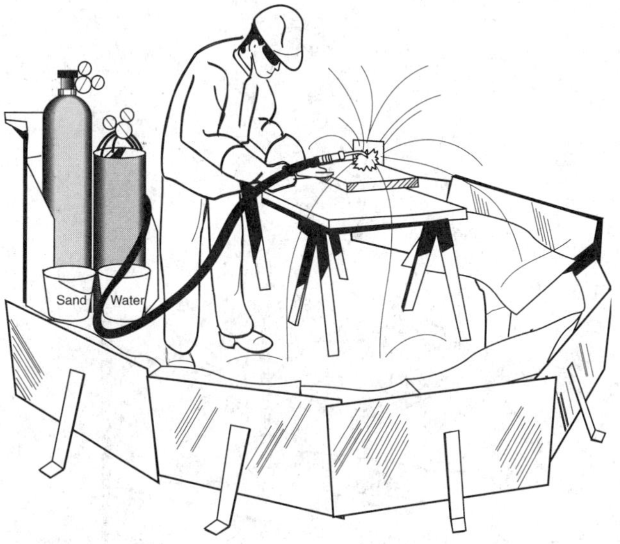

FIG. 3-14G. *Safe work procedures. (Source: ESAB Group, Inc.)*

For details concerning these hazards, see UL 551, *Standard for Safety Transformer-Type Arc-Welding Machines;* and AWS F4.1, "Recommended Safe Practices for the Preparation for Welding and Cutting of Containers and Piping."

Jacketed Containers and Hollow Parts

There are some similarities between the hazards of cutting or welding jacketed containers or other hollow workpieces, and hazards of cutting or welding containers. Air confined inside an unvented hollow part will expand when heated, and pressure will increase. Since hot metal rapidly loses its strength, the container or hollow piece may burst with explosive force at the focus of the cutting or welding work. It is well to be suspicious of closed metal parts that seem unusually light. Drilling a vent hole before heating the part may be necessary. (See Figure 3-14H.)

Hot Tapping

Occasionally, there are situations where emergency repair or the complete impracticality of emptying and cleaning demand welding or cutting on a container while it holds flammable gas or liquid (for example, a natural gas transmission pipeline or utility distribution system). Schemes to accomplish such hot tapping with relative safety have been developed. (See ANSI B31.8, *Gas Transmission and Distribution Piping Systems;* and API PUBL 2201, "Procedures for Welding or Hot Tapping on Equipment Containing Flammables.") Needless to say, any such work must be performed only by specially trained and qualified staff, using recognized and authorized methods.

Public Exhibitions and Demonstrations

Special and enhanced fire safeguarding is necessary when welding or cutting is performed at trade shows and exhibitions. Characteristically, places of public assembly, such as auditoriums and hotel exhibit halls, and concentrations of people are involved. When welding or cutting work is planned in such exposures, the fire department must have prior notification, operations must be under the control of a qualified person, gas cylinders must be charged to only one-half their maximum permissible content, storage sites for gas cylinders must have special restrictions, and fire extinguishing equipment must be appropriate to the situation.

Personnel Protection and Ventilation

One aspect of welding and cutting fire protection that is sometimes overlooked or inadequately considered is the safety of the operator, helpers, or nearby workers. Flame-resistant gloves, wool clothing, aprons of leather or other durable flame-resistant material, cape sleeves or shoulder covers with skull caps under helmets or with goggles for overhead work, leggings for heavy work, and high-top safety shoes are generally recommended. Trousers should not be turned up or cuffed on the outside, front pockets on clothing should be eliminated, and sleeves and collars kept buttoned to prevent sparks from entering and lodging in such places. Outer clothing should be free of oil and grease. Cotton instead of wool clothing may be worn if the cotton is chemically treated to reduce its combustibility. (See Figure 3-14I.)

Adequate ventilation must be provided wherever welding and cutting are performed to protect the operator from inhaling noxious gases and fumes. Potentially hazardous materials may exist in certain fluxes, coatings, and filler metals. In some cases, general natural draft ventilation may be adequate. Other operations may require forced-draft ventilation, local exhaust hoods or booths, or personal respirators or air-supplied masks.

Oxygen, since it accelerates combustion, must never be used to cool the welder, ventilate a confined space, or dust off clothing. Tests and experience have shown that oxygen-saturated clothing or clothing in an oxygen-enriched atmosphere will literally burn in a flash, with extremely serious and sometimes fatal results.

Manufacturers' Recommendations

Procedures given in manufacturers' instructions for setting up, connecting, lighting or starting, adjusting, and maintaining equipment have specific, safety-oriented purposes. These procedures should be followed. In addition, precautions and safe practices publications with valuable information about fire and personnel safety are available from equipment suppliers.

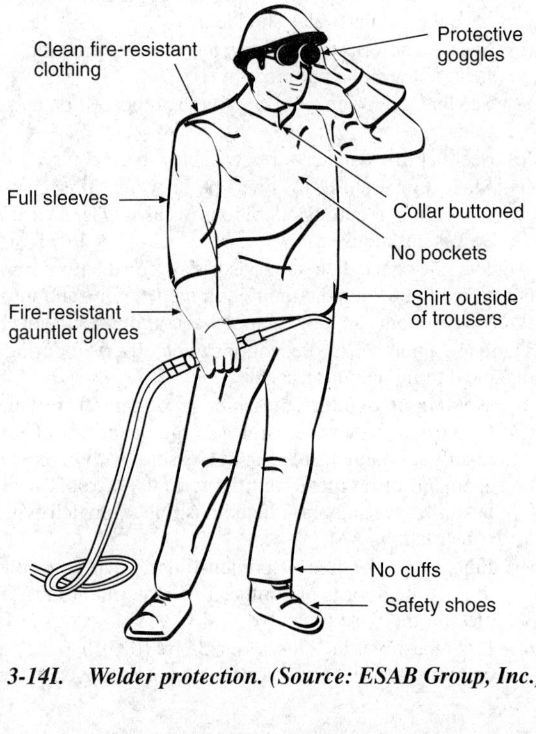

FIG. 3-14H.   Vent hole drilling. (Source: ESAB Group, Inc.)

FIG. 3-14I.   Welder protection. (Source: ESAB Group, Inc.)

BIBLIOGRAPHY

References Cited

1. Miller, A. L., "Special Data Information Package-Torches," NFPA Fire Analysis and Research Division, Quincy, MA, Dec. 1994.
2. Hall, John R., Jr., "The U.S. Fire Problem Overview Report Through 1993," NFPA Fire Analysis and Research Division, Quincy, MA, Jan. 1995.
3. CGA E-I, "Standard Connections for Regulator Outlets, Torches and Fitted Hose for Welding and Cutting Equipment," 3rd ed., Compressed Gas Association, Inc., Arlington, VA, 1994.
4. *Specifications for Rubber Welding Hose*, 4th ed., Rubber Manufacturers Association, Washington, DC, and Compressed Gas Association, Arlington, VA, 1976.

References

ANSI/AWS Z49.1, *Standard for Safety in Welding and Cutting*, American National Standards Institute, New York, 1988.

ANSI B31.8, *Gas Transmission and Distribution Piping Systems*, Paragraph 841.27, American National Standards Institute, New York, 1992.

API PUBL 2201, "Procedures for Welding or Hot Tapping on Equipment Containing Flammables," American Petroleum Institute, Washington, DC, 1985.

AWS F4.1, "Recommended Safe Practices for the Preparation for Welding and Cutting of Containers and Piping," American Welding Society, Inc., Miami, FL, 1994.

CGA P-1, "Safe Handling of Compressed Gases in Containers," 8th ed., Compressed Gas Association, Inc., Arlington, VA, 1991.

NEMA EW-1, "Electric Arc Welding Power Sources," National Electrical Manufacturers Association, Washington, DC, Rev. Nov. 1991.

UL 551, *Standard for Safety Transformer-Type Arc-Welding Machines*, Underwriters Laboratories Inc., Chicago, IL, 1994.

NFPA Codes, Standards, and Recommended Practices

Reference to the following NFPA codes, standards, and recommended practices will provide further information on the safeguards for welding and cutting operations discussed in this chapter. (See the latest version of *The NFPA Catalog* for availability of current editions of the following documents.)

NFPA 10, *Standard for Portable Fire Extinguishers*

NFPA 50, *Standard for Bulk Oxygen Systems at Consumer Sites*

NFPA 51, *Standard for the Designs and Installation of Oxygen-Fuel Gas Systems for Welding, Cutting, and Allied Processes*

NFPA 51B, *Standard for Fire Prevention in Use of Cutting and Welding Processes*

NFPA 70, *National Electrical Code®*

NFPA 241, *Standard for Safeguarding Construction, Alteration, and Demolition Operations*

NFPA 306, *Standard for the Control of Gas Hazards on Vessels*

NFPA 327, *Standard Procedures for Cleaning or Safeguarding Small Tanks and Containers Without Entry*

Additional Reading

"Acetylene Cylinders—Hazards and Procedures," *Fire Prevention*, No. 202, Sept. 1987, pp. 31-32.

ANSI Z117.1, *Safety Requirements for Confined Spaces*, American National Standards Institute, NY, 1989.

API PUBL 2009, "Safe Welding and Cutting Practices in Refineries, Gas Plants, and Petrochemical Plants," American Petroleum Institute, Washington, DC, 1988.

API PUBL 2013, "Cleaning Mobile Tanks in Flammable or Combustible Liquid Service," 6th ed., American Petroleum Institute, Washington, DC, 1991.

API PUBL 2015, "Safe Entry and Cleaning of Petroleum Storage Tanks," 5th ed., American Petroleum Institute, Washington, DC, 1994.

Ashmore, F., and McNally, M., "Dealing with Acetylene," *Fire Prevention*, No. 204, Nov. 1987, pp. 37-38.

Bartknecht, W., "Ignition Capabilities of Hot Surfaces and Mechanically Generated Sparks in Flammable Gas and Dust/Air Mixtures," *Plant/Operation Progress*, Vol. 7, No. 2, 1988, pp. 114-121.

CGA Bulletin SB-4, "Handling Acetylene Cylinders in Fire Situations," Compressed Gas Association, Arlington, VA, 1990.

CGA C-4, "Method of Marking Portable Compressed Gas Containers to Identify the Material Contained," 4th ed., Compressed Gas Association, Arlington, VA, 1990.

CGA G-1, "Acetylene," 8th ed., Compressed Gas Association, Arlington, VA. 1990.

CGA Pamphlet E-2, "Hose Line Check Valve Standards for Welding and Cutting," 3rd ed., Compressed Gas Association, Arlington, VA, 1991.

CGA Pamphlet E-3, "Pipeline Regulator Inlet Connection Standards," 3rd ed., Compressed Gas Association, Arlington, VA, 1991.

CGA Pamphlet E-4, "Standard for Gas Pressure Regulators," 3rd ed., Compressed Gas Association, Arlington, VA, 1994.

CGA Pamphlet E-5, "Torch Standard for Welding and Cutting," 2nd ed., Compressed Gas Association, Arlington, VA, 1991.

CGA Pamphlet G-4, "Oxygen," 8th ed., Compressed Gas Association, Arlington, VA. 1987.

CGA Pamphlet G-4.4, "Industrial Practices for Gaseous Oxygen Transmission and Distribution Piping Systems," 3rd ed., Compressed Gas Association, Arlington, VA, 1993.

CGA V-1, "Compressed Gas Cylinder Valve Outlet and Inlet Connections," 6th ed., Compressed Gas Association, Arlington, VA, 1987, with 1991 addendum.

Compressed Gas Association, *Handbook of Compressed Gases*, 3rd ed., Van Nostrand Reinhold, NY, 1990.

Connor, L. P., Chapter 16, "Safe Practices," *Welding Handbook*, 8th ed., American Welding Society, Miami, FL, 1987.

Hoelemann, H., and Worpenberg, R., "Investigations on Fires Caused by Welding, Cutting, and Related Procedures—Damage Evaluation," *Schweissen und Schneiden*, Vol. 38, No. 4, 1986, pp. E60-E63.

Kennebeck, M. E., "Eight Points to Ponder for Safe Welding and Cutting," *Welding Journal*, Vol. 67, No. 9, 1988, pp. 67-69.

Nightingale, P. J., "Investigation into an Explosion that Occurred During a Welding Operation," *Plant/Operation Progress*, Vol. 8, No. 1, Jan. 1989, pp. 29-32.

NSC, *Accident Prevention Manual for Business and Industry*, 10th ed., National Safety Council, Itasca, IL, 1992.

OSHA, 29 CFR 1910. 252, Subpart Q, *Welding and Cutting*, Government Printing Office, Washington, DC.

Safe Practices, American Welding Society, Inc., Miami, FL, 1987.

Safety and Health Fact Sheets, American Welding Society, Inc., Miami, FL, 1990.

UL 123, *Standard for Safety Oxyfuel Gas Torches*, 8th ed., Underwriters Laboratories Inc. Northbrook, IL, 1992.

UL 407, *Standard for Safety Manifolds for Compressed Gases*, 5th ed., Underwriters Laboratories Inc., Northbrook, IL, 1993.

WOODWORKING FACILITIES AND PROCESSES

John M. Cholin

After thousands of years of civilization and technological progress, wood, humanity's first structural material, remains the favorite material for the construction of homes and furnishings. The demand for wood products has not diminished with the introduction of plastics as some had predicted; it has increased. With the ever increasing demand for wood products, a continued growth in the forest products industry has been noted. Many forest products firms are housed in timber-frame structures that are hundreds of years old while others are ultramodern facilities with automated processes, controlled by computers in new fire-resistant structures. Some woodworking facilities are one-person shops while others employ as many as 2,000 people. Some woodworking shops produce custom-designed high-quality furnishings while others produce dimension lumber, paneling, engineered wood products, rifle stocks, knobs, millwork, picture frames, cutting boards, or musical instruments. It is the wide size, age, business volume, and product range coupled with the inherently combustible nature of wood that makes the fire protection for any woodworking facility a challenge.

This chapter will address the fire protection problems that are unique to wood product manufacturing facilities and processes. It will not study wood as a material. That is covered in Section 4, Chapter 3, "Wood and Wood-Based Products." Nor will fire protection concerns over wood construction and timber-frame construction under fire exposure be emphasized. These topics are covered elsewhere. This chapter is devoted to the woodworking facility and the production processes commonly employed in those facilities.

"Primary forest products," including structural timbers, dimension lumber, shingles, and veneers, are produced from a wide variety of tree species. Only cutting and drying are required to derive these products from the raw log. Primary forest products are usually produced in facilities that are dedicated to the specific product, situated close to the forest source. The primary product is then shipped as a commodity to various producers of either secondary or tertiary forest products. There are exceptions to this general scheme of things. Some "secondary forest products," such as plywood, particleboard, flake board, oriented strandboard, and fiberboard, are produced in large centralized facilities that consume logs in their raw form. Other secondary forest products, including millwork, knobs, spindles, paneling, glue-laminated beams, composite wood joists and dowels, are made from the primary dimension stock in

separate facilities or portions of larger entities that produce the tertiary product. Each of these forms of wood has specific structural characteristics that makes it the material of choice for specific applications. These products are referred to as "secondary," because processes subsequent to cutting and drying are necessary to result in the product, yet the product is still not in a form that is normally consumed on a retail level.

The majority of the primary and secondary forest products are consumed as raw material for "tertiary forest products," i.e., forest products that are consumed on the retail level. The primary examples of these are housing and furnishings. Lumber, a primary forest product, is cut and shaped into wood components that are then assembled with other components to ultimately become a finished piece of furniture.

WOODWORKING PROCESSES

Virtually all woodworking processes involve reducing the size of a piece of wood through mechanical means. Whenever mechanical work is done on matter, heat is liberated. This is a principle of physics called the work/heat equivalence. Consequently, one must anticipate the liberation of heat whenever wood is processed. If this heat is allowed to accumulate in a sufficiently small mass of wood, the nominal ignition temperature of 536°F (280°C), where exothermic pyrolysis begins, will be attained and the wood will continue charring. (See Section 4, Chapter 3, "Wood and Wood-Based Products.") Such temperatures can easily be achieved from frictional heating with powered cutting and shaping tools. If conditions conducive to sustained combustion are maintained, wood brought to this temperature will ultimately ignite. Consequently, every mechanized woodworking process represents a very real ignition hazard.

Furthermore, most woodworking processes produce waste wood as a by-product of the size reduction to the desired size and shape. Wood waste exists over a wide range of particle size, from pieces the layman would consider a board to fine powdery dust. Wood waste management is a widespread problem throughout the forest products industries. Fine wood particulate is very easily ignited. When suspended in air it burns very aggressively, producing a deflagration. When wood dust exists in a mass, such as an accumulation on a floor, it can smolder for extended periods of time (days), producing a charred mass that can flare up as a deflagration as soon as sufficient air at sufficient velocity is provided to suspend the particulate in the air stream. The combustibility of wood waste particulate is a recurring problem and must be expected in every forest products facility.

John M. Cholin, P.E., is an independent fire protection consultant and engineer with J. M. Cholin Consultants, Inc., Oakland, NJ and a member of NFPA's Technical Committee on Wood, Paper, and Cellulosic Dusts.

The combustibility of wood waste dusts is the basis for concern with the generation of sparks by woodworking machinery. The probability of spark generation is related to both the wood species and the woodworking process. In general, species that exhibit wide variations in density are prone to produce more sparks. The feed rate of a saw or milling machine is generally adjusted for the average density of the species. When an abnormally hard portion of a board hits the cutter, excessive heat is generated. The other important species-dependent factor is the average hardness of the wood. The harder the wood the more likely it is to generate sparks when maximum recommended feed rates are exceeded.

Some generalities can be made regarding the relative tendency of various species to present a spark generation problem, but only in the context of the process. The hardwoods maple, hickory, cherry, and white oak are commonly viewed as representing a high probability of spark generation in the context of milling and cutting. Other hardwoods, including beech, red oak, birch, yellow poplar, and ash, have represented less of a problem. The softwoods tend to be less prone to spark generation in sawing and milling. The theory is that the resin in the softwood chips can evaporate with a cooling effect, whereas hardwoods lack these resins and, consequently, ignite. However, in sanding operations the resinous softwoods tend to load the abrasive grit much more aggressively than the nonresinous hardwoods. This results in a tendency of resinous softwoods to generate sparks at rates several orders of magnitude higher than that of hardwoods in the sanding operation.

Dimension Cutting

Dimension cutting usually occurs at a specialized facility called a sawmill. The raw logs are first debarked and then cut to one of several standard working lengths. The bark is collected beneath the debarker and usually transported to a collection silo by a belt conveyor. The log is then squared on at least two sides. The slabs taken from the log to square it up are waste and are either diverted to a pile for residential fire wood or introduced into a wood waste system via a hog. The squared-up log is then run through the saw and converted into rough green dimension stock.

When raw logs are reduced to green dimension stock, the moisture content in the wood is sufficiently high to make ignition of the stock being cut unlikely. However, any sawdust that has accumulated around the saw will rapidly air dry to a moisture content of approximately 10 percent, making it readily ignitable. Rough sawing does introduce an ignition hazard from "inclusions" in the log. These inclusions run the gamut from arrowheads, bullets, nails, and sugar-taps to stones and knots. While larger mills may have either radio frequency or ultrasonic metal detectors monitoring the feed to the saw, such an inclusion is still occasionally hit by the saw. This generally produces large quantities of heat, resulting in the ejection of sparks and embers into the surrounding area.

The major source of ignition in the rough sawing operation is the ignition of sawdust as the result of an inclusion or broken sawtooth. The removal of wood waste, including sawdust, is of paramount importance. The management of this fire risk necessitates a wood waste removal system to prevent accumulations of wood dust in the sawmill.

Drying

Green wood is between 30 and 50 percent water, by weight. As it dries, it shrinks. Consequently, green wood is generally deemed unusable as a structural material. Therefore, before a piece of wood can be used it must be dried. Most wood is kiln dried to a moisture content of 7 percent.

In the lumberyard, boards are individually inspected and built into hacks for drying and handling. Lumber hacks are about 14 ft (4.3 m) long, 6 to 8 ft (1.8 to 2.4 m) wide, and vary in height from 4 to 6 ft (1.2 to 1.8 m). Each layer of boards in the hack is separated by spacers known as kiln sticks. Hacks are stacked by forklift trucks back to back and as high as 16 ft (4.9 m). The honeycomb construction and the relatively easy access to oxygen make it almost impossible to extinguish a fire once it is underway in a lumber hack. Aisle widths should allow fire-fighting equipment to maneuver effectively.

The kiln removes the moisture from the lumber in a manner that minimizes the differences in rate of shrinkage throughout the board. The rough green dimension stock is stacked with "stickers" between adjacent boards to maximize the flow of air through the stack. It is then introduced into the kiln where the temperature is raised, often as high as 300°F (149°C) while the humidity is maintained at 100 percent. Once the entire stack of lumber has attained the kiln temperature, the humidity is slowly allowed to decrease while the temperature is maintained. This drives the moisture from the wood without causing cracks (called "checks"), wrapping, twisting, or other defects. The stack of lumber emerges from the kiln after a period of usually weeks at a nominal 7 percent moisture content, the ideal moisture level for use in construction, furniture, and other tertiary forest products.

The drying operation does not involve reducing the size of the lumber, but does involve heating a combustible material to a temperature close to 392°F (200°C), where endothermic pyrolysis can begin occurring. (See Section 4, Chapter 3, "Wood and Wood-Based Products.") Furthermore, the productivity of the kiln is measured by the number of board feet of lumber that exit the kiln per unit of time and heating fuel consumed. This creates a financial incentive to run the kiln as hot as possible.

Drying kilns are often heated with wood waste. Wood particulate is conveyed from a storage silo into the boiler via a pneumatic conveyance system. Wherever wood is handled as a particulate solid, the fire and deflagration potential exists. This is covered in detail later, in the subsection "Wood Waste Management."

Veneer Cutting

Veneer mills are unique, in that the size reduction is performed with knives rather than saws. Veneer logs are often air dried for an extended time period prior to processing. They are then cut to length prior to slicing. Slicing is performed with knives rather than saws. Veneer is cut either by a rotary or rift machine. The rotary machine turns the log and slices a thin layer from the circumference as it rotates. The rift cut veneer is produced by slicing along the length of the log which has been sawn down its middle. As the veneer comes off the log, it is transported via conveyor to a dryer where it is rapidly brought to the nominal 7 percent moisture content. Most commercial hardwood veneer is a nominal $1/24$ to $1/32$ in. (1.1 to 0.8 mm) thick, whereas softwood veneers and plies vary from $1/24$ in. (1.1 mm) for face ply to $1/8$ in. (3.2 mm) for core plies. Consequently, drying occurs quite rapidly. The dried veneer is either stacked flat and baled for shipping or transported via belt conveyor to a paneling facility where it is laminated into multi-ply material or where it is laminated onto a composite wood core.

The object of veneering is to obtain more square feet (m²) of good-looking or usable wood from a log. In general, there is less waste generated in veneering than in the dimension sawmill. However, there is still waste in the form of sawdust, veneer flakes, veneer strands, bark, flattening slabs, and cut-offs. This waste represents a similar fire hazard to that in the sawmill. Management of the fire risk is dependent upon a wood waste removal system to minimize the accumulation of wood waste in the site. This is covered in detail later in the subsection "Wood Waste Management." Furthermore, veneer cutting operations require very large mechanical forces. These forces are generated by large electrical motors or hydraulic systems that introduce additional ignition mechanisms and attendant fire hazards.

Milling

The rough sawn dimension stock serves as a raw material for the production of lumber. Most dimension stock is milled to standard dimensions, such as "2 by 4" or "1 by 6." Milling involves passing the stock through edgers, jointers, and planers, all of which have hardened steel cutters rotating at high speed that cut away enough wood to ensure a smooth, straight, and flat surface on all four faces. Finger-joiners produce interlocking fingers on the ends of boards allowing them to be joined, end-to-end, with glue to make longer pieces. Other secondary milling operations further reduce the lumber to pieces with specific cross sections for door stiles, door frames, window frames, window sills, moldings, tongue-and-groove flooring and paneling, and other special uses. These specific cross-sections are produced by mills with cutters that have been precisely ground to produce the desired shape.

Milling produces large quantities of wood chips and dust. The high speeds of the cutters and the high feed rates of modern production facilities make the generation of sparks and embers the rule rather than the exception. This reality complicates the problems in the wood waste management for the site. The wood waste must be removed from the milling machine to prevent an accumulation of chips. The wood waste stream is likely to contain embers.

Storage

Finished secondary wood products are generally stacked solid and banded together into a single unit that can be easily handled with a forklift. This method of storage minimizes the access to air in the event of an ignition and, hence, mitigates the fire hazard to some degree. Nevertheless, storage areas for forest products contain large concentrations of large quantities of very combustible material. This is a high-challenge fire hazard.

Among the main fire hazards around lumber storage areas, which usually have only hydrant protection, are lightning, discarded cigarettes, and boiler room sparks. Portable heaters and bonfires or barrel fires lit for warmth also pose hazards during the winter.

The fire problems become worse when units of dimension lumber are broken down and the lumber is stacked in racks. Random horizontal stacking allows greater surface exposure to atmosphere. Under no circumstance should lumber be stored in a vertical orientation. This orientation creates chimneys between adjacent planks, rendering the lumber virtually unextinguishable.

Cutting

As lumber is used as a raw material in the tertiary forest products facility, one of the most common processes is cutting to component dimension. Cutting across the grain of the board is called cross-cutting, while cutting parallel to the grain is called ripping or resawing. The byproducts of cutting are sawdust, block scraps from cross-cutting, and kerf scraps from resawing.

The cutting creates a severe fire ignition exposure as dry stock is processed by a machine that has the ability to invest large quantities of energy into the fine wood waste created by the machine. The sawdust is easily ignited and is produced at elevated temperatures. This is especially true of facilities that process woods with large variations in density in a single board or billet. When the wood density increases, the feed rate must be reduced to keep operating temperatures below the autoignition temperature for the sawdust. In mechanized facilities, the feedback from the saw to the feeding unit may not be sufficiently responsive to ensure adequate control over cutting temperature in the workpiece. The result is sparks and embers in the sawdust stream from the unit when knots, crotches, burls, and butts are encountered in the cut. This is especially common in resawing.

The high-saw speeds and high-feed rates of modern production facilities make the generation of sparks and embers in the cutting operations the rule rather than the exception. This reality complicates the problems in the wood waste management for the site. The wood waste must be removed from the saw to prevent an accumulation of chips. The wood waste stream is likely to contain embers.

Turning

Turning has historically represented somewhat less of a fire risk than other shaping methods. However, when billets of wood that exhibit wide variations in density or glue-ups are used there is an ignition potential. Part of the difficulty with turning is that the roughing, final shaping, and sanding are often done at the same location without removing the workpiece from the lathe. This creates a dust stream having a wide range of particle size and introduces it into the pneumatic wood waste removal system. This is very undesirable, as it increases the likelihood of a dust-collector fire.

Laminating

Laminating covers a very wide range of processes. There is an incredible range of surfaces, substrates, and adhesives in use today. Some lamination surfaces are themselves engineered materials consisting of multiple layers of different materials all integrated into a single layer with advanced adhesives. Recent amendments to the Clean Air Act have placed pressure on laminators to reduce volatile organic compound (VOC) emissions. Consequently, many of the toluene- and xylene-based glues have been replaced with casein, acrylic, and urethane systems. Most of these are heat activated and are applied at a very high solids content.

Application of the adhesive is usually performed at a spray station. Adhesive application does not involve the complexities of spray finishing. Over-spray is easily controlled and electrostatic spraying is rarely employed due to the poor electrical properties of the wood-based substrates. Once the adhesive is applied, the laminate facing is applied and the adhesive is activated by applying both heat and pressure, either by passing the laminate through heated rollers or placing the laminate in a heated press.

The heat source for the heated rollers or press is either steam or heat transfer oil (e.g., Therminol™). This introduces a fire source, as combustible heat transfer fluids under pressure have been known to provide an ignition source when lines have ruptured while at elevated temperature and pressure.

Sanding

Each component that is to become an exposed part of a finished assembly must be smooth and free of sharp edges. The final form and surface quality of virtually all wood products is achieved by sanding. A wide variety of sanding machines are used in the tertiary forest products industries, from handheld random orbital sanders for finish sanding to widebelt abrasive planers capable of taking an eighth of an inch off (1 in. = 25.4 mm) in a single pass. (See Figure 3-15A.)

Sanding operations probably represent the greatest hazard to the modern forest products plant. The hazard is caused by the large quantities of fine dust produced during sanding. The larger sanding machines also are capable of producing substantial quantities of

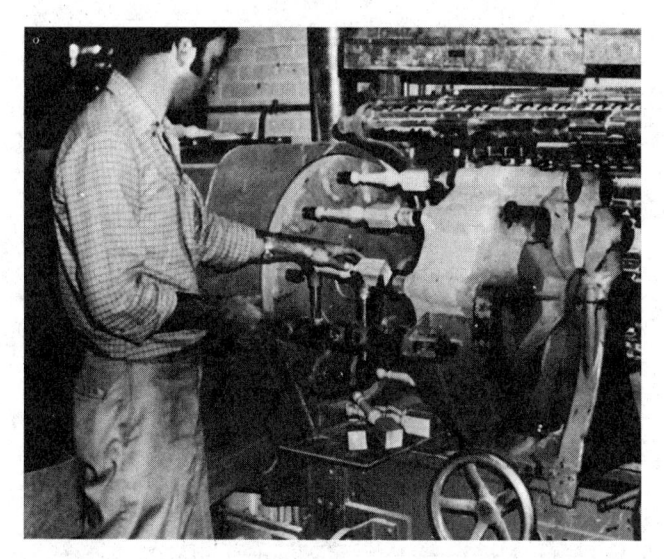

FIG. 3-15A.   A sanding machine for applying a smooth finish to furniture legs prior to finishing. The sanding operation is particularly hazardous because of the fineness of the dust created.

heat. The heat from the sanding operation vaporizes resins and extractives in the wood which then condense on the interiors of the dust collection system ducts, coating the interior surface over time. The grit on the sanding belt can become "loaded" with compressed wood dust, forming a composite material, which can achieve red-hot temperatures as it is rubbed across the stock. Sander dust, vaporized resins, and red-hot bits of wood/resin composite are all picked up by the dust collection system. The pneumatic waste wood conveyance ducts running from the sanders to the dust collector are frequent sources of fire.

Glue-up/Assembly

The glue-up and assembly of tertiary forest products is the least hazardous portion of the production facility. While glues have traditionally been dissolved in mineral spirits, toluene, or other volatile organic compounds, the new polyurethane and acrylic adhesives are not.

Finishing

Years ago finishing operations represented a more serious concern than they usually do today. The concern over VOC emissions has made finishes using nitrocellulose, toluene, xylene, and methylethylketone (MEK) vehicles far less common than they once were. New acrylic, urethane, and water-based finishes are replacing the older finishes in most of the large-volume facilities.

Many of the larger, more modern woodworking plants employ as many as 20 spray booths in a conveyorized finishing operation, using 20,000 to 40,000 sq ft (1,858 to 3,716 m²) of floor space. Ovens for curing the finished pieces operate at approximately 250°F (121°C). Due to the competitive pressure to produce a flawlessly finished product, multicoat finishes are customary, necessitating curing ovens with high temperatures and airflows and overnight storage between coats to ensure finish quality.

The lower-volume smaller facilities are often found to still be using older finishing procedures based upon volatile vehicles. Wherever volatile flammable liquids are used there is the possibility

of a fire. The principal flammable solvents used in wood finishing are listed below with their approximate flash points:

Acetone	-4°F (-20°C)
Methylethylketone	16°F (-9°C)
Naphtha (VM&P)	50°F (10°C)
Xylene	81°F (27°C)
Toluene	40°F (4.4°C)

All of these materials are prone to electrostatic ignition under certain circumstances. Fires are occasionally caused when flammable vapors come in contact with lighting fixtures, heaters, sparking motors, electrical switches, etc.

Finished Products Storage

The finished products storage of forest products naturally varies with the type of product. However, the storage areas should be protected in a manner consistent with the combustible loading that can be achieved in a storage area used for the storage of furnishings or other cellulosic combustibles.

Wood Waste Management

Most forest products production processes generate wood waste. By the time parts have progressed through rough milling and finish machining, a large part of the original material has been discarded. Large chips, chunks, board ends, resaw kerfs, and lumber scraps account for a large part of the discarded material. The larger material is usually conveyed by belt conveyor or wheelbarrow to a hog, while the finer material is generally introduced into a pneumatic conveyance system called the "dust collection system."

A wood hog is a high-power mill that reduces large wood waste to chips, splinters, or flakes, depending upon the intended subsequent use. This particulate can be conveyed pneumatically. Larger facilities operate hogs that can consume entire wood pallets and tree trunks, reducing them to a shower of chips. The outlet of the hog is usually connected via a high-power fan to a pneumatic conveyance duct that conveys the chips to a silo for intermediate storage. The chips are usually used for two purposes: (1) raw feed stock for the manufacture of wood composite materials, such as particleboard and fiber board (see Section 4, Chapter 3, "Wood and Wood-Based Products"), or (2) as a fuel for generating process steam and heating.

All of the previously described wood-shaping and cutting processes generate wood chips, shavings, and dusts. These materials are usually removed from the machine producing them via a pneumatic conveyance system, often referred to as the "dust collection system." This system consists of a system of ducts extending to the wood-cutting part of each process machine. The chips and dust are sucked up by the dust collection system and are conveyed suspended in air to a dust collector where the wood chips and dusts are separated from the conveyance air. In many installations, about 80 percent of the air is returned to the plant. Return of conveyance air into the plant atmosphere can cut down on the air makeup requirement and reduce plant heating fuel needs. However, it introduces serious fire protection concerns that must be addressed.

There are two general designs of dust collection systems common in the forest products industries. The first is the "single-stage" system. (See Figure 3-15B.) These systems consist of a single dust collector either in the form of a cyclone separator or a combination

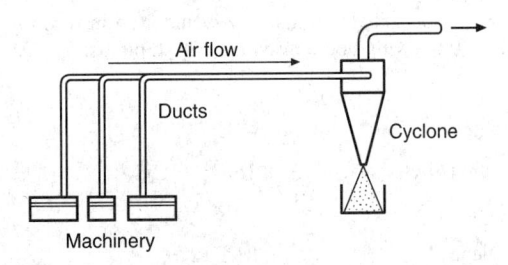

FIG. 3-15B. A single-stage collection system consisting of a cyclone separator.

cyclone/baghouse unit. The second is a two-stage system. (See Figure 3-15C.) These usually consist of a cyclone separator followed by a bag-type filter house, usually referred to as a "baghouse." Cyclone separators are intended for removing the larger particulate from the air stream. However, they cannot remove fine dusts efficiently enough to be practical where conveyance air is being returned to the workplace. If conveyance air is to be returned to the workplace, a design decision that is usually elected in order to save on the cost of heating the facility, a filter must be used to remove the fine, dusty particulates. The filter is usually implemented with an array of cloth bags in a steel enclosure, called a baghouse. The baghouse removes wood dust having particle sizes larger than the weave of the cloth.

Most facilities have two-stage systems. Historically, they started out as a single-stage system using only a cyclone to separate the larger, usable wood waste from the air stream, exhausting the conveyance air to the outside. Subsequently, a baghouse was added allowing the return of the heated conveyance air back to the facility. Newer facilities are now often designed with a single-stage cyclone/baghouse combined unit, where the bottom of the unit acts as the cyclone separator and the upper portion contains the bag-type filters. (See Figure 3-15D.)

When wood dust suspended in air in a concentration above the minimum combustible concentration (MCC) is ignited, it burns very violently, producing a deflagration. The likelihood of a deflagration in wood particulates suspended in air is dependent upon the concentration of fuel per unit volume. However, the proportion of the fuel that has an effective particle size below approximately 420 microns in equivalent diameter is also an important factor. When the effective particle size of the wood dust is greater than 420 microns in equivalent diameter, the material will still burn but will generally not result in a deflagration that is capable of producing a pressure within a containment enclosure (such as a building, cyclone, baghouse, or storage silo) sufficient to cause structural damage.

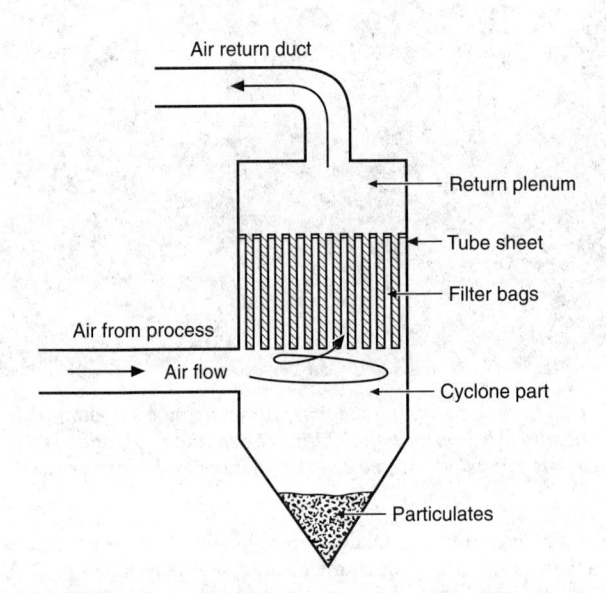

FIG. 3-15C. A two-stage dust collection system consisting of a cyclone and baghouse.

Since both cyclones and baghouses separate wood waste from the conveyance air, there is a continuous concentration gradient within these units as a normal part of their operation. The concentration gradient effectively guarantees that some portion of the inte-

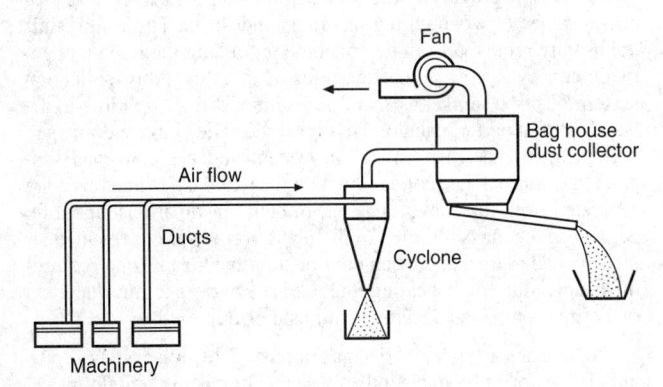

FIG. 3-15D. (Top) Single-stage dust collector with a combination cyclone and baghouse. (Bottom) Example of a single-stage dust collector installation.

rior volume of the dust collector has an environment conducive to supporting a deflagration. It is true that most cyclones are not designed to separate the fine "deflagrable" dust from the conveyance air under normal operating conditions, but they will produce a deflagrable environment during startup and shutdown of the conveyance system. The baghouse is designed to collect fine particulate, and conditions conducive to a deflagration exist during startup, normal operations, and shutdown. Furthermore, since no filtering system is 100 percent efficient, the finest wood dusts often pass through the filter media of the baghouse and are conveyed back into the production space. This results in an accumulation of ultra-fine wood dust on many of the undisturbed horizontal surfaces in most production shops.

Wherever a cloud of wood dust of the critical average particle size is ignited, it produces a deflagration, a ball of flame. The deflagration causes flame and combustion gases to expand spherically from the point of ignition through the cloud. The expanding deflagration forces the air within the enclosure to move in front of the ball of flame and hot gases, suspending additional dust in the air, making much larger dust clouds of varying density throughout the facility. Once started, a succession of ignitions may occur, eventually attaining sufficient force to cause the enclosure to fail. The rupture of the enclosure is an explosion.

Unfortunately, the finer the dust particles, the easier they are to ignite and the more violently they will burn. To varying degrees, all dust collection systems remove the larger, less combustible wood particles, allowing the finer particles to pass through the filter medium and back into the production facility. As the efficiency of the dust collection system is improved in an effort to keep the facility cleaner, the combustibility of the dust in the collector increases. The combustibility of the residual dust that passes through the filter, returning to the occupied space, also increases. In short, the better the dust collection system the greater the fire hazard.

Note: *The fire hazard does not go away if there is no dust collection system.* If the wood chips and dusts are not collected, the employees are exposed to the health risks associated with inhaling the dust. Furthermore, the dust is free to migrate throughout the workplace. In this circumstance, whenever someone drops a plank or does something else that causes the dust to billow up, there is the possibility that the dust cloud will be ignited. The choice is between a fire hazard distributed throughout the facility or a fire hazard concentrated in a specific spot. *The fire hazard is always there.*

FIRE PREVENTION

Design

The fire safety of a facility can be improved substantially by addressing the fire prevention concerns during the layout of the production area. Significant fire safety improvements can be made in most existing facilities by reorganizing the layout. Processes that are likely to cause ignition should be moved away from processes or areas where there is significant fuel loading or fuels of elevated combustibility. Appliances that use electric heating, such as glue pots and bending irons, should be restricted to specific dedicated areas that are dust free and physically separated from cutting, shaping, and sanding machines as well as inventory storage and finishing areas. Whenever possible, fire-resistant separations equipped with listed fire doors at each penetration should be installed between the wood shaping/cutting, finishing, assembly, and storage areas.

Since most of the commonly used wood finishes have a solvent of flammable liquid, there is a fire hazard wherever finishes are stored or applied. The shop, regardless of size, must be organized to keep flammable vapors out of confined spaces and away from igni-

tion sources. All flammable materials must be stored and used in compliance with NFPA 30, *Flammable and Combustible Liquids Code.*

The use of portable electric space heaters to supplement the main heating plant distributes an ignition source throughout the facility. The heating plant should be designed to eliminate any need for portable space heaters.

Properly grounded electrical outlets should be installed where the electricity is needed. Extension cords, 3- to 2-wire adapters ("cheater plugs"), cube taps, and outlet extenders should be replaced with permanent electrical service in compliance with NFPA 70, *National Electrical Code®.*

The dust collection system should be designed pursuant to the appropriate codes and standards. NFPA 1, *Fire Prevention Code,* stipulates that combustible material must not pass through a fan. (See Section 35-2.1 of NFPA 1.) This requirement was adopted due to the long history of ignitions occurring as a result of combustible particulate solids moving through a fan in a pneumatic conveyance system. The impact of the fan blade on the combustible particulate or tramp iron produces heat that can raise particles to their autoignition temperature. This is particularly true when particles of a broad range of particle sizes are transported in a common duct. In-line fans also can ignite the conveyed material if their bearings overheat or if a fan blade strikes the housing. Consequently, pneumatic conveying systems should be designed as negative pressure systems with the fan located on the clean air side of the dust collector. If this is not possible, or if the pneumatic conveying system already exists, the ignition hazard represented by an in-line fan should be managed with a spark detection and extinguishment system downstream from the fan, as discussed below.

NFPA 650, *Standard for Pneumatic Conveying Systems for Handling Combustible Materials,* and NFPA 664, *Standard for the Prevention of Fires and Explosions in Wood Processing and Woodworking Facilities,* establish a number of requirements on the design of the facility, especially as it relates to the design of the pneumatic conveying system. One of the requirements is that dust collectors be of noncombustible construction and situated outside the structure. (See Section 4-3 of NFPA 650, and Chapter 8 of NFPA 664.) Furthermore, if cleaned conveyance air is to be returned to the facility interior, a listed spark detection system activating an abort gate is required. (See Section 8-4 of NFPA 664.) (See Figure 3-15E.)

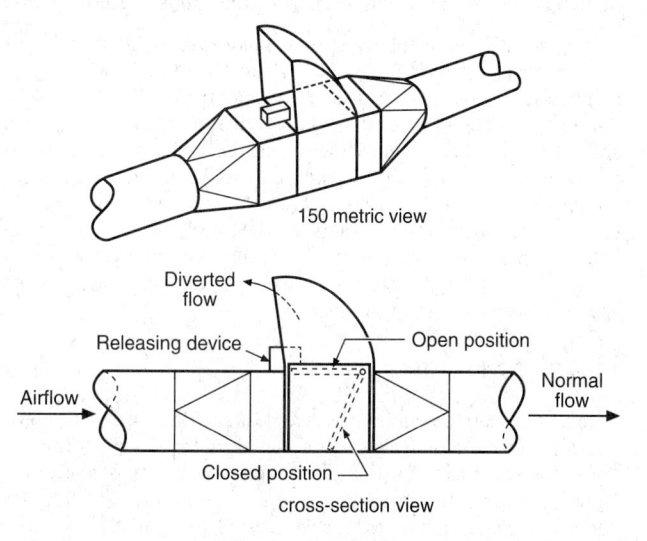

150 metric view

Diverted flow

Releasing device — Open position

Airflow — Normal flow

Closed position

cross-section view

FIG. 3-15E.   *An abort gate diverts the air stream to the outside, preventing the passage of flames and combustion gases into the facility via the clean air return duct.*

There are several other design methods that can reduce the likelihood of a dust collection system fire. They do not eliminate the possibility of a fire but do reduce the probability. They are derived from the fundamentals of spark physics.

A small wood particle (sander dust) is easily ignited, but it burns out very quickly since it represents very little fuel. Large wood particles (router chips) are harder to ignite but will burn for a longer period of time, sometimes long enough to survive until they reach the dust collector (baghouse). Serious fires can occur when fine sander dust and larger chips from routers, planers, and saws are conveyed together in the same duct. The likelihood of a dust collector fire can be reduced by conveying fine dust and large chips in separate ducts until they reach the collector. The same strategy holds true for separating the lines that come from the sanding of finished surfaces or resinous woods and the sanding of unfinished surfaces or nonresinous woods. The resins in the finishing materials tend to load-up the grit on the belts. When the resin/wood composite reaches ignition temperature it can ignite the nonresinous dust in the line. Finally, longer ducts should increase the probability that an ember will burn out before it reaches the dust collector where the continuous concentration gradient exists that is so conducive to an ignition. These steps will only reduce the probability of a dust collector fire; they will not eliminate the possibility.

Equipment Maintenance

There are ignition hazards associated with the electrically powered woodworking machines, both stationary and portable. Switches wear out and start "arcing." Power cords become worn and frayed. Bearings fail and overheat. All of these are examples of the electrically generated ignition sources. Lighting can also be the cause of excessive heat or catastrophic electrical ignition. Frayed power cords, arcing switches, clogged ventilation ports, worn bearings, worn belts, misaligned pulleys, and misaligned guards and fairings around moving parts have all been identified as sources of ignition.

Cutting tools should also be maintained as sharp as practical. This improves both the productivity and the fire safety of a facility. Sharper tools cut more cleanly, producing less dust and chips with smoother surfaces, both factors that reduce the fire risk. Furthermore, a sharp tool produces less heat during the cutting process, making it less likely that the cutting process will raise particles to the autoignition point. Keep cutters sharp and have a specific, quantitative method of determining when sharpening is needed.

Finally, abrasive cutters, such as sanding belts and discs, must be kept clean, in good repair, and free of "loading" consisting of compressed wood dust. A specific quantitative method for determining when abrasive belts should be replaced must be developed and enforced. When a belt becomes loaded, it generates tremendous frictional heating that readily ignites the dust produced by the sanding operation. Excessive abrasive belt temperatures cause the cloth backing of the belt to fray, resulting in the catastrophic failure of the belt. When a belt fails, shards of sanding belt material are often ejected into the dust collection system where they often strike sparks as they pass down the duct.

Heating Systems

Furnaces and other sources of space heat should utilize a combustion air source that is isolated from the air in the woodworking facility. Most furnaces use ambient interior air as a source of combustion air. The furnace burns a fuel to produce hot flue gas which passes through a heat exchanger on its way up the chimney. Interior cool air is passed across the heat exchanger, absorbing heat in the process, and is then forced into the interior space. (See Figure 3-15F.)

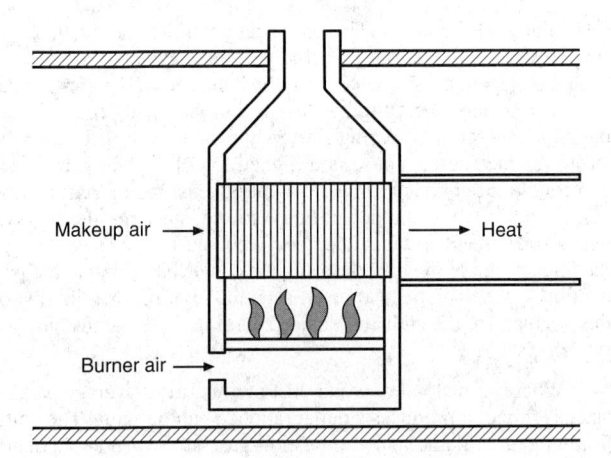

FIG. 3-15F.   A simplified hot-air furnace.

If wood dust is accidentally mixed with the combustion air, it will ignite, potentially sending flames into the production area. This is especially true of gas-fired space heating units commonly found in industrial spaces. This concern can be addressed by installing a fresh combustion air inlet duct, running from the outside to the furnace air inlet, as shown in Figure 3-15G. This will prevent wood dust from invading the combustion air stream of the furnace.

Regular inspection of the heating system is advisable. It is all too common to make revisions to support systems, especially the heating plant, over the years as additional staff necessitate them. In some cases, additional heating appliances have been added under crisis conditions. Portable space heaters are placed where an employee gets cold, often without adequate consideration of the proximity of combustibles. This can create an environment conducive to accidental fire ignition.

Personnel/Procedures

A second major source of fire ignition is accidental ignition by employees. Safe operating procedures must be established and reviewed on a regular basis: control feed rates on cutting and sanding machines; make sure that the staff is properly trained in the correct method of installing and achieving proper alignment of abrasive belts.

There are procedural ways to manage the ignition risk represented by the electrically driven machinery. There have been occasions where a switch has failed into the "on" position, starting a

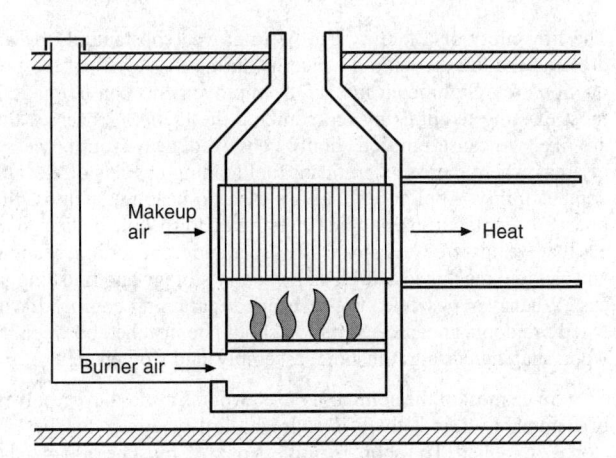

FIG. 3-15G.   A modified furnace/heater.

machine when no one was present. The facility should be equipped with a dedicated set of breakers for the machinery, enabling those responsible for the facility to turn off the source of electrical energy before leaving for the night. An alternative where small machinery operating at 110 V ac is used, is an "unplug when you're done" policy that is monitored by management to ensure that all equipment is disconnected when not in use.

A specific location should be reserved for lunch breaks. This limits the probability that small appliances, such as coffee makers, are placed in uncontrolled locations. Furthermore, small electrical appliances, such as coffeepots, microwave ovens, and hotplates, should be powered through switched outlets that become de-energized when the lights are turned off.

Smoking should not be allowed in a woodworking facility. Smoking materials account for a distressingly high percentage of the accidental fire ignitions. Improper disposal of smoking materials will represent a pressing threat to any woodworking facility.

Welding: Welding for routine maintenance and repair should be performed per NFPA 51B, *Standard for Fire Prevention in Use of Cutting and Welding Processes.* It should be controlled and carefully monitored. When welding is performed, small pellets of molten metal fly off from the material being welded. If these land on combustible material, a fire can result. *It can take several hours for the fire to develop.* Welding should not be performed unless and until the area is thoroughly cleaned of all wood and wood waste and the appropriate fire extinguishing equipment is in place. A fire watch, maintained for at least two (2) hours after the welding has ended should be required whenever welding is performed, regardless of how small the needed repair is. (See Section 3-3 of NFPA 51B.)

Housekeeping: The wood dust that is an inescapable part of a woodworking facility is a combustible material that represents a serious and extensively distributed fire hazard. Every effort must be made to minimize the accumulation of wood dust on upward facing horizontal surfaces and to routinely remove all dust that evades the preventive measures.

An efficient, properly constructed dust collection system is the first step in keeping dust levels under control. However, no system is 100 percent efficient. Routine vacuum cleaning of the entire facility is of paramount importance. Under no circumstance should high-pressure air be used to "blow-down" a work area. This merely stirs up the dust and ensures that the finest, most combustible fraction of the dust is deposited in the highest, most difficult to clean spaces.

Keep the heat exchanger in the furnace/heaters *clean.* The surface of the heat exchanger will get hot enough to ignite surface dust or dust in the air. This is especially true in the autumn when the furnace/heaters are turned on for that first cold morning of the fall.

FIRE PROTECTION

General

In general, woodworking facilities are filled with wood and/or fabric-covered finished and semifinished goods, as well as other raw materials. Automatic sprinkler coverage is generally considered the most appropriate general-area protection for woodworking facilities. However, in some older buildings, it may be difficult to achieve full protection. NFPA 13, *Standard for the Installation of Sprinkler Systems,* requires design of sprinkler systems to be based on Ordinary Hazard-Group 3 design rules for the machine room. Other areas, such as finishing rooms and upholstery rooms, may need designs based upon a higher density, involving larger pipe sizes and closer spacing of sprinkler heads. Whenever possible, these processes should be located in separate buildings or separated from each other by fire-rated construction with openings protected by automatic fire doors.

The degree of protection needed can vary from facility to facility, but a basic recommendation is that a yard system of mains and hydrants capable of supplying at least 1,000 gal/min (3.79 m³/min), enough to sustain four 2¹/₂-in. (64-mm) hose streams simultaneously, should be installed per NFPA 24, *Standard for the Installation of Private Fire Service Mains and Their Appurtenances,* as appropriate. If conditions warrant, expanded supplies and larger fire stream appliances may be needed for effective fire control. Good unobstructed fire lanes are needed in the yards so that fire equipment can approach the lumber piles. NFPA 46, *Recommended Safe Practice for Storage of Forest Products,* provides guidance on protection for lumber in storage.

Portable Fire Extinguishers

Portable fire extinguishers should be installed per the requirements of NFPA 10, *Standard for Portable Fire Extinguishers,* and relevant local codes. There is a wide diversity in the types of fires that are apt to occur in a woodworking plant. Class A, B, and C fuels are abundant.

There should be water or aqueous film-forming foam (AFFF) extinguishers for the wood storage and general production areas. Dry chemical agent extinguishers should be used for the finishing areas and wherever combustible or flammable liquids are stored or used. Dry chemical extinguishers are also preferable for use on electrical equipment.

Gaseous extinguishers, including both the halon and carbon dioxide (CO_2) units, should not be used in a woodworking shop. When these units are discharged, the gas rushes out of the extinguisher, causing a great deal of turbulence in the surrounding air. This can cause any wood dust in the vicinity of the fire to billow up in a cloud. This is an extremely dangerous situation. If the suspended wood dust contacts the flame, a deflagration will probably result.

ABC-type dry chemical is often called a "universal dry chemical." However, keep in mind that a given extinguisher's effectiveness varies greatly with the fuel. While a given extinguisher could easily extinguish a 10-ft (3.1 m) diameter combustible liquid fire, it would not be able to extinguish a 10-ft (3.1 m) diameter stack of wood pallets. Nevertheless, ABC-type dry chemical is effective on wood, paper, and combustible liquid fires, yet is electrically nonconductive. It is an excellent choice as a general-purpose fire extinguisher for most of the areas in the woodworking environment.

In the finishing areas there is a concentration of combustible liquids, and it is appropriate to use a fire extinguisher that is optimized for that fuel. The BC sodium bicarbonate dry chemical extinguisher is generally recommended for this type of service. The ABC dry chemical unit will extinguish flammable liquid fires, but may not be as effective as the BC sodium bicarbonate dry chemical units. The sodium bicarbonate dry chemical units are not very effective on wood and paper fires. Consequently, they should only be used on combustible and flammable liquids or gases.

AFFF is excellent on wood and paper, and combustible or flammable liquids. However, foams are not as effective on three-dimensional fires, i.e., fire where the fuel is stacked up, vertically. AFFF is also electrically conductive. These factors may limit its applicability. However, AFFF is generally considered a good choice for a stockroom, veneer storage, lumber storage, and corrugated cardboard storage areas.

The diversity of combustibles introduces some risks associated with fire extinguishers. Their placement and signage should be designed to minimize the likelihood that the wrong type of unit will be used.

Finishing Process

The principal hazard in the spray application of finishes involves flammable and combustible liquids and their vapors or mist, and combustible residues in spray booths. Properly constructed booths with adequate mechanical ventilation are the ideal way to discharge vapor to a safe location. Minimizing all sources of ignition in the spraying area and constantly supervising the overall finishing process are essential to a safely conducted operation.

NFPA 33, *Standard for Spray Application Using Flammable or Combustible Materials*, provides the minimum compliance design criteria and operating procedures for spray application of flammable coatings. This standard should be applied to all finishing areas. Mechanical ventilation of the spraying area should remove solvent vapors and control overspray. The ventilation system should maintain an average velocity of not less than 100 ft/min (30.5 m/min) over the entire open face of the booth. This is generally considered adequate to entrain and confine overspray to the booth interior. In some larger facilities, the spray booths exhaust over 25,000 cu ft/min (708 m³/min). This necessitates careful management of the makeup air demand.

The curing ovens associated with the finishing process are generally classified per NFPA 86, *Standard for Ovens and Furnaces,* as continuous, Class A ovens. These are enclosures that operate at nominal atmospheric pressure and in which there is an explosion or fire hazard due to the presence of flammable volatiles and combustible residues. The oven exhaust system is usually vented to a solvent recovery system prior to discharge to the outside.

Spark Detection for the Wood Waste/Dust Collection System

Sprinkler systems are intended to provide fire suppression over the area of the facility production floor. However, their usefulness in protecting the facility from a dust deflagration originating in the wood waste/dust collection system is limited; they can protect the remains of a dust collector after an explosion, but they are of little use in protecting the dust collection system from fire and explosion.

Spark detection and extinguishing systems can make very significant reductions in the hazard associated with pneumatic wood waste conveyance systems, subject to one important caveat. Spark detection and extinguishment systems cannot be used where deflagrable conditions exist in the conveyance ducts. When deflagrable conditions exist in the conveyance ducts, explosion prevention systems outlined in NFPA 69, *Standard on Explosion Prevention Systems*, must be employed. Spark detection and extinguishing systems should be thought of as fire suppression systems that prevent the introduction of an ignition source into the dust collector.

In this context, the "fire" is a spark the size of a BB produced by the woodworking machinery and picked up by the dust collection system. Once within the conveyance duct it travels at speeds of thousands of feet per minute (m/min). When a spark arrives in the collector it encounters a continuous concentration gradient of wood dust and, hence, a region where the conditions are conducive to ignition of the dust suspended in the air within the dust collector. If the spark has sufficient energy and is capable of delivering the threshold power (energy per unit time) to the dust/air mixture, the dust within the dust collector will ignite and a deflagration will result. Sprinkler systems cannot respond to this type of a fire scenario in a meaningful manner.

A deflagration does not result every time a spark is introduced into a dust collector. A second scenario often occurs. The spark ignites the wood dust adhering to the surface of the filter bags. With the air movement providing plenty of atmospheric oxygen, the fire develops on the filter bag rapidly, involving the entire collector interior. In this case, the heat release is very rapid and could actuate a sprinkler head. Unfortunately, the operation of the sprinkler head might complicate matters. There is concern that the operation of the sprinkler head might suspend dust and cause a deflagration. In either case, the dust collector sustains serious damage and the return air duct has conveyed smoke, flame, and combustion gases into the facility, spreading the fire throughout the plant.

The fact that many dust collectors have been operating for decades without having sustained a fire cannot be used as a justification for not protecting both the employees and the facility from a dust collector fire or explosion. The machinery in the modern forest products facility generates an endless succession of sparks. In some facilities, as many as dozens of sparks per hour have been detected. Under normal circumstances, most of these sparks burn out. They are quenched as they collide with the walls of the duct, with large wood chips, or with resinous material in the duct. They don't survive sufficiently long to reach a location where they can ignite a dust cloud. Perhaps only one in a million sparks generated will survive the journey from its source to the dust collector. There are some management options that can reduce the probability of a spark reaching the collector.

Three documents of the *National Fire Codes®* address the dust collection systems found in the forest products industries.

1. NFPA 1, *Fire Prevention Code*, states: Dust shall not pass through a fan or blower.
2. NFPA 650, *Standard for Pneumatic Conveying Systems for Handling Combustible Materials*, states: Air-material separators shall be located outside of buildings.
3. NFPA 664, *Standard for the Prevention of Fires and Explosions in Wood Processing and Woodworking Facilities*, states: Filtered air shall not be recycled back into the building unless the following conditions are met.
   (a) The system shall be equipped with an approved spark detection and suppression system,
   (b) The recycled air duct shall be fitted with an abort damper that would be actuated by the spark detector bypassing the air to atmosphere, away from the plant, and
   (c) The abort damper shall be provided with a manual reset so that, after it has aborted, it can only be returned to the closed position at the damper. Automatic or remote reset shall not be allowed.

These requirements are made with the intent of achieving reasonable levels of life safety and property protection. These requirements do not address the need to secure the productive capacity of the site.

NFPA 650 and NFPA 664 require that the dust collector be of noncombustible construction and located outside the building. NFPA 664 requires that a spark detector be placed on the return air ducts that bring air from the collector back into the occupied space. The specifics relating to how spark detection systems are designed and installed are addressed in NFPA 72, *National Fire Alarm Code.* The design, use, and installation of the spark detectors are addressed in Chapter 5 of NFPA 72. However, the activation of an abort gate is a remedial action necessary to secure the life safety and property of the site. A listed local fire protective signaling system control panel is required for the activation of the abort gate. Consequently, the minimum spark detection system consists of: (1) spark detectors in sufficient number to comply with the requirements of Chapter 5 of NFPA 72 and (2) a control panel with secondary power supply that

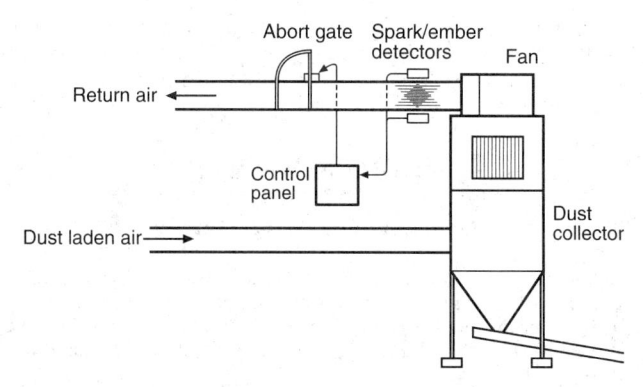

FIG. 3-15H.   *The minimum code-compliant implementation of spark detection.*

is listed for releasing device service pursuant to Chapter 2 of NFPA 72. Thus, the minimal system is represented in Figure 3-15H.

NFPA 650 and 664 require the spark detection system to actuate an abort damper on the return air duct to divert flame and smoke out of the building. This system does nothing to protect the dust collector, itself. Once the abort gate has operated, the occupants of the building and the building itself are considered adequately protected and the dust collector is allowed to burn. Since it is of noncombustible construction it represents, in theory, a limited fire threat.

This represents the minimum code-compliant protection for the woodworking facility dust collector. This does not represent the majority of the spark detection systems. Most systems are installed to address the losses in production due to a fire in the dust collector. Spark detection is usually used in conjunction with an extinguishing system to quench sparks before they enter the dust collector.

The typical spark detection and extinguishment system (see Figure 3-15I) comprises three basic components, which are listed as a system:

1. A set of spark/ember detectors mounted on the duct.
2. A water spray extinguishing unit mounted downstream from the detectors, and
3. A control panel to provide power to, and receive signals from, the detectors and the extinguisher.

The distance between the detector and the extinguishing unit must be carefully computed, based upon the duct diameter, air speed, and the response time parameters of the spark detectors, extinguishing unit, and control panel. Likewise, there is usually a minimum distance between the dust collector and the extinguishing unit. This distance is determined by the air speed, duct diameter, and water pressure. Failure to compute these factors properly renders the system of limited value as it then becomes possible for the spark to pass the extinguishing unit before the water discharge has established an effective extinguishing concentration in the duct.

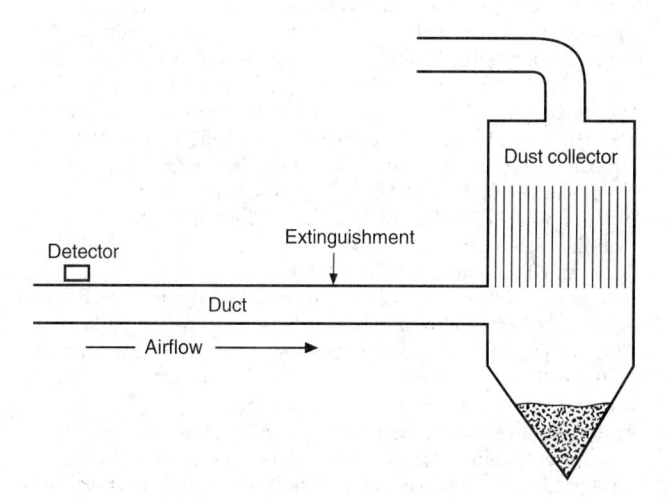

FIG. 3-15I.   *The basic spark detection and extinguishment system design.*

Spark detection and extinguishment is a listed system that is normally mounted on the inlet pipes that bring dust to the collector. Each line entering the dust collector must be monitored. The principal concern is the protection of the collector from sparks entrained in the airflow. Spark detection and extinguishment systems have an excellent record of protecting the collector. Each time a spark is detected, the extinguishing unit is energized and a spray of water is established in the duct. The control panel is equipped with a timer that controls the duration of the water discharge to a period between 3 and 15 sec. After this brief spray, the extinguishing unit is de-energized and the system waits for the next spark. The air movement is not shut down. The whole dust collection system continues to operate. The water discharge is normally a gallon or two (1 gal = 3.785 L), which evaporates in seconds. (See Figures 3-15J and 3-15K.)

This approach is appropriate for the majority of applications. The spark detection and extinguishment on the inlet lines detect and quench the sparks in the wood waste stream, preventing burning material from entering the collector. The spark detection on the clean air return and the abort damper prevent smoke, flame, and burning material from entering the facility from the collector. Some Authorities Having Jurisdiction do not require the detection on the clean air side if detection and extinguishment are located in the inlet lines. However, spark detection on the inlet side of the collector should not be used to actuate the abort damper on the outlet side. Most properly working spark detection systems will detect several

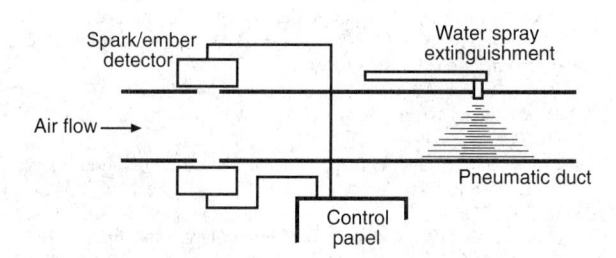

FIG. 3-15J.   *Spark detection and extinguishment protecting an inlet line of a dust collector.*

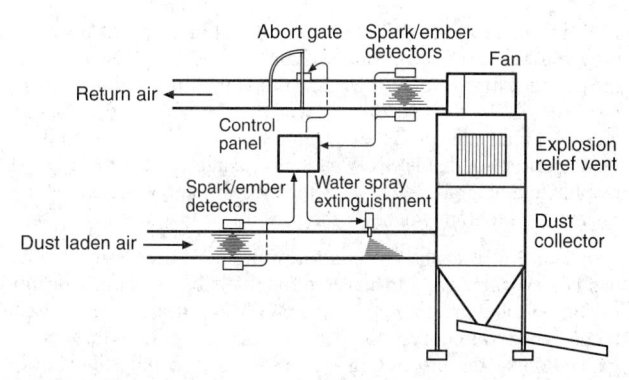

FIG. 3-15K.   *Typical use of the spark detection/extinguishment system protecting the dust collector, and complying with the requirements for diversion of the return air.*

sparks per day under normal conditions. Once the abort damper has operated, the air movement must be shut down to enable the manual reset of the damper. This is a serious inconvenience if it happens too frequently. Furthermore, the interruption of the air movement through the dust collector allows all of the dust that has adhered to the filter bags to fall down into the bottom of the collector. If there is any burning material within the collector, the air shutdown could cause a deflagration.

In general, a spark detection system does not "completely protect" a dust collection system, the dust collector, or the facility it serves. While such a system, with extinguishment covering the inlet lines and detection activating an abort gate on the air return lines, addresses much of the fire risk, a properly protected dust collection system will include either deflagration pressure-relief vents sized using the calculation procedures in NFPA 68, *Guide for Venting of Deflagrations*, or explosion prevention and protection systems that meet NFPA 69, *Standard on Explosion Prevention Systems*, to address the possibility of a deflagration in the collector producing a flame front that travels too fast to be diverted by the abort gate. Furthermore, a sprinkler system in the dust collector installed per NFPA 13, *Standard for the Installation of Sprinkler Systems*, should be considered, as it will probably limit the structural damage to the dust collector in the event of a fire and, hence, facilitate the return to operation.

There are some fire protective design concepts that should be carefully scrutinized before using them in protecting dust collectors for combustible particulate solids. The use of heat detectors to actuate a gaseous extinguishing system may increase the hazard rather than reduce it. In a 50,000 cu ft/min (1,415 m³/min) dust collection system at standard temperature and pressure (STP), it takes a fire of 15.3 Btu/sec (16.14 kW) to produce a 1°F (0.555°C) increase in the temperature of the air flowing through the collector. Fixed temperature heat detectors generally require temperature increases on the order of 50°F (28°C) before operating. Differential temperature detection systems are rarely stable when the threshold differential is less than 10°F. Using the 10°F (5.5°C) temperature differential, such a system could be expected to initiate response when the fire had achieved a heat output of approximately 153 Btu/sec (161.4 kW). This is equivalent to a pool of burning gasoline of approximately 1 sq ft (0.0929 m²). If the air movement through the collector is disrupted (by the discharge of an extinguishing agent and the shutdown of the air movement necessitated to attain and maintain extinguishing concentration) while a fire of this magnitude exists within the collector, a deflagration is likely. A similar set of concerns should be addressed in the event that water deluge actuated by heat detection is contemplated.

SUMMARY

Both the nature of wood as a raw material and the methods employed to convert that raw material into a finished product contribute to a very high level of fire risk in woodworking facilities. This is compounded by the structures in which woodworking facilities are housed. Finally, the production methods produce large quantities of wood particulate that is easily ignited and, in many cases, truly vulnerable to deflagration. These factors necessitate an aggressive loss prevention program whenever these facilities are addressed.

The fire risk inherent in a forest products processing facility can be very effectively managed through deliberate facility design, operations, and fire protective systems. Woodworking facilities can be designed to minimize the opportunities for ignition through organization of the production process to keep combustibles away from ignition sources, compartmentation of the facility, proper design of the wood waste management system, and utilization of automatic sprinklers. Operational policies can contribute to the

reduction of fire risk, especially through a strictly enforced housekeeping program. Finally, protection of the wood waste/dust collection system can reduce the probability of an explosion to very acceptable levels.

BIBLIOGRAPHY

NFPA Codes, Standards, and Recommended Practices

Reference to the following NFPA codes, standards, and recommended practices will provide further information on the safeguards for woodworking facilities and processes discussed in this chapter. (See the latest version of *The NFPA Catalog* for availability of current editions of the following documents.)

NFPA 1, *Fire Prevention Code*
NFPA 10, *Standard for Portable Fire Extinguishers*
NFPA 13, *Standard for the Installation of Sprinkler Systems*
NFPA 24, *Standard for the Installation of Private Fire Service Mains and Their Appurtenances*
NFPA 30, *Flammable and Combustible Liquids Code*
NFPA 33, *Standard for Spray Application Using Flammable or Combustible Materials*
NFPA 46, *Recommended Safe Practice for Storage of Forest Products*
NFPA 51B, *Standard for Fire Prevention in Use of Cutting and Welding Processes*
NFPA 68, *Guide for Venting of Deflagrations*
NFPA 69, *Standard on Explosion Prevention Systems*
NFPA 70, *National Electrical Code®*
NFPA 72, *National Fire Alarm Code*
NFPA 86, *Standard for Ovens and Furnaces*
NFPA 101®, *Life Safety Code®*
NFPA 231, *Standard for General Storage*
NFPA 600, *Standard on Industrial Fire Brigades*
NFPA 650, *Standard for Pneumatic Conveying Systems for Handling Combustible Materials*
NFPA 664, *Standard for the Prevention of Fires and Explosions in Wood Processing and Woodworking Facilities*

Additional Reading

Aycock, B. B., "Furniture Manufacturing," *Industrial Fire Hazards Handbook*, 3rd ed., Cote, A. E., ed., National Fire Protection Association, Quincy, MA, 1990.
Cholin, J. M., "Everything You Didn't Know You Wanted to Know About Spark/Ember Detection but Are About to Learn," *The Moore-Wilson Signaling Report*, Focus Publishing Enterprises, Bloomfield, CT. (Serialized over 5 issues, March 1993.)
Cholin, J. M., "Fire Protection for the Midsized Woodworking Shop," J. M. Cholin Consultants, Inc., Oakland, NJ, 1993.
Janssens, M., "Rate of Heat Release of Wood Products," *Fire Safety Journal*, Vol. 17, No. 3, 1991, pp. 217–238.
Ohtani, H., Ohno, T., and Uehara, Y., "Heat Release Measurement During Smoldering Combustion of Wood Sawdust," *Journal of Applied Fire Science*, Vol. 4, No. 3, 1994/1995, pp. 161–169.
Parker, W. J., "Wood Materials. Part A. Prediction of the Heat Release Rate From Basic Measurements," Chapter 11, *Heat Release in Fires*, Elsevier Applied Science, NY, 1992, pp. 333–356.
Tran, H. C., "Experimental Heat Release Rate of Wood Products," First U. S. Symposium for Heat Release and Fire Hazard, Abstracts, Interscience Communications Limited, 1991, pp. 1–2.
Tran, H. C., "Wood Materials. Part B. Experimental Data on Wood Materials," Chapter 11, *Heat Release in Fires*, Elsevier Applied Science, New York, 1992, pp. 357–372.
White, R. H., and Nordheim, E. V., "Charring Rate of Wood for ASTM E 119 Exposure," *Fire Technology*, Vol. 28, No. 1, Feb. 1992, pp. 5–30.
Yudong, L. and Drysdale, D., "Measurement of the Ignition Temperature of Wood," *Fire Safety Science*, Vol. 1, No. 1, Sept. 1992, pp. 25–30; First Asian Conference for Fire Science and Technology (ACFST), October 9–13, 1992, Hefei, China, International Academic Publishers, China, 1992, pp. 380–385.

SPRAY FINISHING AND POWDER COATING

Don R. Scarbrough

Spray application of coatings is a process basic to the manufacture of a broad variety of fabricated products, and a high percentage of factories operate at least one paint spray booth. Regardless of the purpose for which the coating is applied, flammable or combustible materials are commonly used in the process. Fires in paint operation areas develop very quickly, have high heat release rates, and produce large volumes of toxic smoke. Preventive and protective measures for control of these hazards must be undertaken with special attention to the peculiar hazards of each process.

TYPES OF COATINGS

Fluid Coatings

Flammable and combustible fluid coatings have been in industrial use for many years, and their use involves familiar fire risks, discussed here at some length.

Waterborne coatings, which are enjoying expanded use, present somewhat of an enigma, i.e., many formulations that exhibit a flash point do not have a fire point and will therefore not sustain a flame. While these materials do not present fire hazards typical of flammable liquids, their dried residues are combustible and must be managed as such. ASTM D4206, *Standard Test Method for Sustained Burning of Liquid Mixtures by the Seta Flash Apparatus (Open Cup)*, sets forth a procedure useful to determine whether or not a given material, as a liquid, will support flame. If it does, it should be treated as a flammable liquid.

The most familiar atomizing device for fluid coatings is the air spray gun, which uses jets of high-pressure air to break up fluid into a fine mist. Another device commonly seen in high-volume production processing is the airless atomizer, which generates a spray of fluid by hydraulic means without using compressed air. Fluid pressures used with this type of atomizer range from 300 psi (2,068 kPa) to approximately 7,000 psi (48,265 kPa).

Air and airless atomizers are also used in electrostatic spray operations. For this process, the appropriate atomizer is built into a spray gun that has a high-voltage electrical input. Voltages applied to the charging electrode range from approximately 35,000 to somewhat over 100,000 V. There is less overspray with the electrostatic

method than with the air or airless spray methods because the charged atomized particles are attracted to the grounded workpiece.

A third type of atomizer depends upon electrostatic forces for its operation. In its most common configuration, a sharp-edged disc with a diameter of 6 to 12 in. (152 to 305 mm) is mounted with its axis oriented vertically and is then charged electrically to about 100 kV. The disc is spun about its axis while the coating fluid is poured slowly onto the surface. Centrifugal force spreads the fluid into a thin film and carries it to the sharp edge of the disc, where the film is disrupted by electrostatic forces and sprayed in a 360-degree pattern. Workpieces are carried to the process zone by a conveyor and collect coating as they pass through. A form of this device that has gained popularity has the disc developed into the form of a cup or bell 2 or 3 in. (50 or 76 mm) in diameter that rotates at a high speed—sometimes approaching 60,000 rpm. Bell atomizers are frequently seen mounted in banks of 6 to 12 units at a single spray station.

Powder Coatings

A coating process that has gained broad acceptance is the application of organic coatings in the form of dry powder. In this process, the powder is first suspended in air and then charged electrostatically from a dc power supply operating between 60 and 120 kV. The powder is then directed toward the workpiece and held in place by electrostatic forces. The powder is formed into a continuous coating as it melts during passage through a process oven. The powder coating process differs substantially from the fluid coating processes because no organic solvents are used and no readily ignitable residues are produced.

Electrostatic application of powder is most commonly accomplished with spray guns. These spray guns are simple devices to which a mixture of powder and air is fed through a single tube. A separate cable connected to the gun provides the high voltage necessary for operation of "corona-type" guns, while those that operate by frictional charging (referred to as "tribo-charging" guns) require no external source of high voltage.

Workpieces having dimensions less than approximately 4 in. (102 mm) or that are configured as long ribbons or cables can be powder coated with a device known as an electrostatic fluidized bed. In this process, powder is held in a container having an open top and a porous bottom through which an upward flow of air causes the powder mass to levitate or "fluidize." Charging electrodes near the surface or beneath the fluidizing plate impart an electrical charge to the powder. Electrically grounded workpieces pass over the surface

Don R. Scarbrough is product safety staff consultant for the Nordson Corp., Westlake, OH, and has been an active member of the NFPA Technical Committee on Finishing Processes since 1972.

of the bed and collect a coating of powder. The coating is then cured in a baking oven.

Although they are not widely used, there are techniques for applying dry powder coatings without the use of electrostatics. These processes typically involve preheating the workpiece to a temperature substantially above the melting point of the powder and then applying the powder, either by dipping the workpiece into a fluidized bed or by spraying air-suspended powder directly onto the hot surface of the workpiece. The powder melts immediately upon contact and flows to form a film, which is subsequently cured in a bake oven.

SPRAY PROCESS EQUIPMENT AND COMPONENTS

Fluid Supply

Air spray guns draw their coating fluid from either small cups mounted on the guns, or through hoses connected to larger pressurized containers called paint tanks or pressure pots that have capacities varying from 10 to 60 gal (38 to 227 L). In very high production arrangements, the coating may be fed by a pump from a bulk tank. Fluid pressures between 3 and 75 psig (21 and 517 kPa) are customary.

To ensure satisfactory results, industrial coating mixtures are usually modified shortly before spraying by the addition of solvents to adjust viscosity. A safety tank or container is used to reduce the possibility of a flammable liquid spill while handling or transporting this blended material.

Spray Guns and Devices

Among the various forms of industrial air spray guns, the two most commonly used are the pistol grip hand spray gun and the machine-mounted automatic air spray gun. Air is supplied to the hand spray gun through one hose, while fluid is either hose fed or drawn from a gun-mounted container. (See Figure 3-16A.)

Electrical grounding of airless spray guns to drain off static electricity generated during spraying is provided through a wire or conductive layer built into the fluid hose. A manual safety lock is usually provided on hand units to secure the trigger and prevent accidental actuation.

Handheld and automatic electrostatic spray guns have electrically nonconducting extensions at the front end to insulate the energized components from the grounded parts. In addition to being connected to coating fluid and air supplies, these guns are also connected to high-voltage power supplies. Voltage at the atomizer is in the 30 to 75 kV range for hand-held units and in the 30 to 120 kV range for automatic units. In hand spray guns of this type, the cable carrying the high-voltage power to the gun also supplies an electrical ground for the pistol grip and trigger. To prevent electrical sparking from accidental contact of the charged elements of the gun with a grounded object, the cable is usually terminated at the gun in an electrical resistance on the order of 75 to 250 MΩ (megohm). In a more recent development, the high-voltage power supply has been integrated into the gun, thereby eliminating the high-voltage cable. Some automatic electrostatic spray guns are not equipped with a high-impedance termination, but rely on the process control approach of maintaining separation between gun and workpiece to prevent electrical sparking.

Figure 3-16B shows an electrostatic disc surrounded by chairs hanging from a conveyor loop. Immediately above the disc is its motor drive. Partially obscured, at the top of the illustration, is a device that moves the disc up and down, enabling it to coat workpieces over the entire height of the carriers. Separate electrical and fluid in-

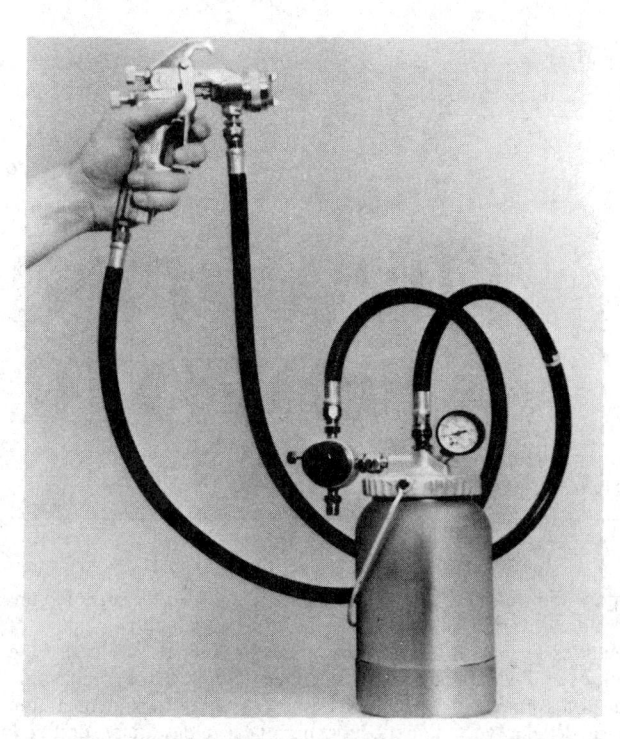

FIG. 3-16A. *Hand air spray gun with fluid supplied from 2-qt (1.9-L) pressure pot. (Source: DeVilbiss)*

puts are remotely controlled. Discs commonly are equipped with variable voltage power supplies that can be adjusted over a range of approximately 50 to 120 kV. Since considerable electrical energy is stored on the disc, adequate distance must be maintained between it and the workpiece to prevent sparking.

Spray Booths

A spray booth is a power-ventilated structure used to enclose a spraying operation. While in most cases the booth structure physi-

FIG. 3-16B. *Electrostatic disc. (Source: Ransburg)*

cally surrounds the spray operation, some configurations leave the spray operation unenclosed and surround the operation with a controlled stream of air drawn into the booth.

The most popular type of spray booth is the "open face" or "open front" arrangement. It is a boxlike structure that has one open side. The ventilation system associated with the booth may provide an airflow horizontal to the floor (cross draft) or vertical to the floor (down draft) as demanded by process requirements.

For some operations, such as production finishing of automotive bodies, a tunnel spray booth is used. (See Figure 3-16C.) This structure is virtually always arranged with vertical down draft ventilation and a horizontal floor-mounted conveyor running the length of the tunnel. Makeup air is introduced through special ductwork and air diffusers in the ceiling.

Special-Purpose Enclosures

Continuous coaters: These enclosures are individually engineered for a specific coating process. Within the coater enclosure is an array of spray guns that apply coatings to workpieces that pass through the coater at conveyor speeds between 100 and 600 ft/min (30 and 183 m/min). The entry and exit vestibules are equipped with exhaust shrouds to capture any vapors or minor amounts of overspray that drift out from the main enclosure. Most of the overspray is collected in a sump and drawn off through a pumping system, which recycles the coating into the coating process.

Decorating machines: These machines are used in conjunction with masking devices to paint stripes or other patterns on workpieces, such as automobile side moldings. The mask is mounted in the opening beneath the row of cylinders at the front of the machine, and the workpiece is placed face down on the mask. When the operating cycle is started, air-driven pistons in the row of cylinders clamp the work to the machine. Spray guns inside the cabinet automatically tilt in one direction and travel the length of the workpiece. At the end of the workpiece, the spray guns tilt in the opposite direction and travel back to the starting point. The air-driven clamps automatically release the work at the end of the operation.

An exhaust system draws fresh air in through the grill at the front of the machine, and internal filters remove overspray from the air vapor stream before it is exhausted. Since the process is totally

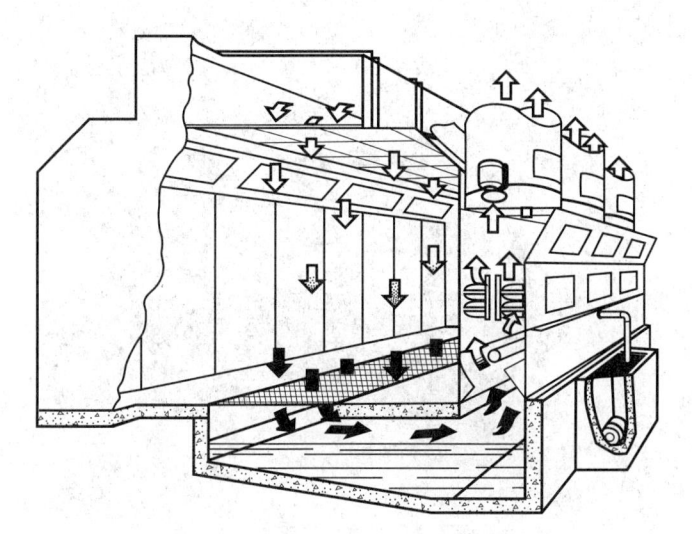

FIG. 3-16C. Tunnel booth. (Source: Binks)

enclosed, the risk of escaping vapors or overspray and subsequent ignition is significantly diminished.

Spray Rooms

The size, shape, and weight of some workpieces may make the use of spray booths impractical. When this is the case, an entire room dedicated to the spray process is a legitimate alternative. Such rooms, called spray rooms, are classified as hazardous areas. Power ventilation removes the combustible vapors that are released during the process, but velocities are not high enough to capture particulate matter. Therefore, overspray is permitted to settle to the floor where it accumulates as a combustible residue until it is removed by a mechanical cleaning process. Vapor removal systems are most effective when the exhaust intake is located along one wall within 1 ft (0.3 m) of the floor and the makeup air is introduced along the opposite wall.

Open-Floor Spraying

A spray operation conducted without a spray booth and in an area not separated from general factory operations by a partition is called open-floor spraying. Depending upon the quantity of volatiles to be released in the operation and existing ventilation within the building, forced ventilation may or may not be provided. Whether forced or not, adequate ventilation must be provided to prevent accumulation of flammable concentrations of vapor. Particulate overspray materials are allowed to settle to the floor and their residues are removed by mechanical means. Ignition sources incidental to other factory operations are a great concern whenever open-floor spray techniques are used, and it is common to establish a buffer zone surrounding the spray area for about a 20-ft (6-m) radius to separate processes.

Overspray Collectors

Vapor and overspray removal systems associated with spray booths or enclosures typically include a fan to create an airflow, and a collection system that separates and collects particulate matter from the airstream then exhausts vapors to the exterior of the building. Overspray collectors can generally be placed into one of four categories, as follows.

Baffle maze: The baffle maze consists of a series of flat panels arranged in a staggered pattern through which the airstream is directed. A substantial portion of its particulate burden is removed by direct impaction and collection upon the surface of the dry baffle panels where it remains until removed by mechanical cleaning. Such systems are of limited efficiency and permit appreciable amounts of fine particulates to pass through.

Dry filter: Collectors using paper or fiberglass filter elements are called dry filters and are popular for low-to-intermediate volume spray operations. The typical filter used in this type of collector is a replaceable element approximately 20 in. (508 mm) square and perhaps 2 in. (51 mm) thick. Secondary cloth filters may be found mounted in the plenum behind the primary filters to provide more efficient capture of fine particles. Particulate residues are permitted to accumulate on the filter until a significant obstruction of the airflow through the filter is noted, then the fouled filters are discarded and replaced.

Waterfall and cascade scrubbers: Where high-volume spray coating operations are conducted for several hours a day, waterfall or cascade scrubbers are commonly used. The exhaust airstream is either scrubbed directly by sprays of water coming from nozzles or it follows a path that takes it through several stages of waterfall. Figure 3-16D is an example of one variation of this type of scrubber.

Particulate matter is accumulated in a water tank from which it is removed either manually or automatically by a sludge removal device.

Chemical compounds must be added to the water to prevent paint residues from adhering to the walls or piping thereby creating open channels through the water curtain. Air passing through these open channels will not be adequately cleansed and will carry particulate matter into the exhaust system; the particulate matter may be deposited on the lining of the exhaust stack, on exhaust fan blades, or on the roof of the building.

Venturi scrubbers: Perhaps the most efficient overspray collector is the venturi scrubber. This device directs the exhaust airflow through a narrow throat (venturi) through which a high-velocity spray of water is also directed. Virtually all particulate matter is extracted from the airstream and trapped in the water. The water is processed through a tank where residues are removed by settling and skimming. The same chemical compounds mentioned in the discussion of the waterfall-type scrubber must be added to the water used in this system to prevent plugging and blocking of nozzles. Figure 3-16E is a schematic of a venturi scrubber.

POWDER COATING PROCESS EQUIPMENT AND COMPONENTS

Spray Process

The powder coating spray process utilizes a device called a feeder that mixes powder with air and supplies the mixture to the spray guns. The powder may be contained in a fluidized bed or in a hopper having an inverted cone-shaped bottom. Feeders may supply a single gun or several guns with a separate ejector for each gun.

The most common coating powder is epoxy powder, but others include acrylic, polyester, vinyl, nylon, butyrate, polyolefin, and

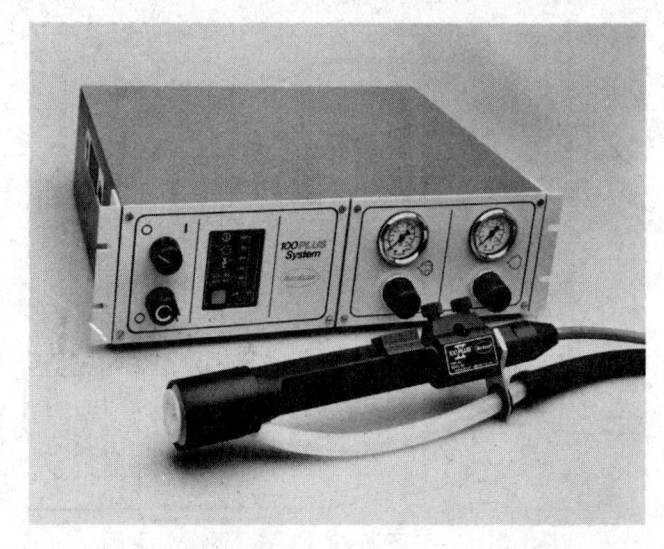

A Contaminated air inlet
B Scrubber water flow to nozzle
C Venturi throat
D Water/air separator
E Cleaned air exhaust
F Water and sludge drain

FIG. 3-16E.   Venturi scrubber. (Source: Nordson)

alkyd. These powders are shipped from the manufacturer to the user in containers of 25 to 1,000 lb (11.34 to 453.6 kg) capacity. They may be classified as ordinary combustibles and, as such, may be stored without requirements for extra-hazard protection.

Spray guns used for the electrostatic application of dry powders do not differ greatly in appearance from guns used for the electrostatic application of fluids. (See Figure 3-16F.) Since no atomization is required, the functions of the spray gun are simply to control the shape of the spray pattern and to impress a high-voltage charge upon the powder cloud. Powder and electrostatic power input connections may be made at the rear of the barrel or from below to the base of the grip and the forward portion of the barrel. Connections to automatic guns typically are made at the rear. Guns having integrated power supplies have no high-voltage cable, but are commonly powered through a lightweight cable carrying about 12 V.

Electrostatic power supplies rectify stepped-up common ac line voltage inputs, either 115 or 230 V, to produce dc output ranging from 30 to 100 kV. Several automatic guns may be connected to a single power supply. For handgun applications, however, a separate power supply is provided for each gun, and each supply has a

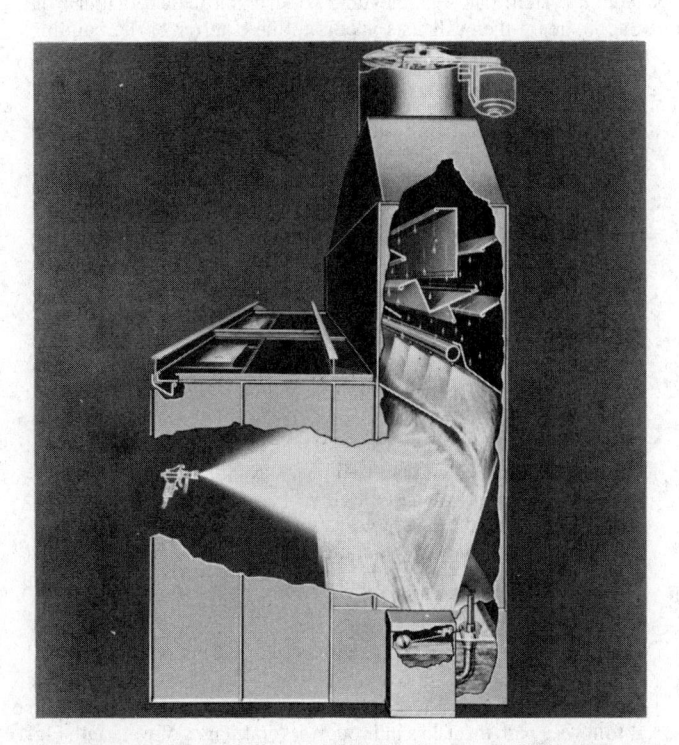

FIG. 3-16D.   Water wash booth. (Source: DeVilbiss)

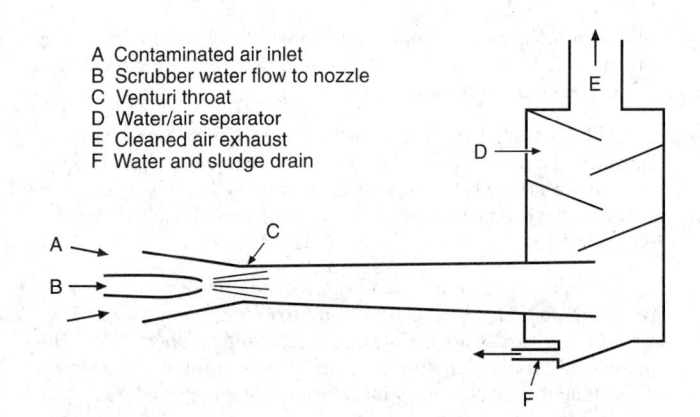

FIG. 3-16F.   Automatic powder gun. (Source: Nordson)

control circuit that prevents the pack from being energized unless the trigger is actuated.

The interconnection between power supply and spray gun is usually made through a high-voltage coaxial cable. At the termination of the cable within the spray gun, the charging circuit is connected to the gun electrode through a 75 to 250 megohm resistor. In operation, a faint blue glow, called a corona, is developed around the end of the electrode which projects through the front of the gun. Individual powder particles are charged as they pass through the corona.

In a class of equipment commonly referred to as "tribo-charging guns," no external power supply or cable is used. These devices generate static charge directly upon the surface of the powder granules through frictional contact between the high-velocity airborne granules and internal surfaces of the apparatus.

A common facility in which electrostatic powder spray operations are conducted is a spray booth with a hopper bottom through which exhaust air is drawn into a ductwork that leads to a powder collector. A cloth or fabric filter within the collector separates the powder from the airstream. Powder separated within the collector falls to the bottom of the hopper from which it is extracted and is subsequently reintroduced to the feeder for the spray guns. A basic outline of this arrangement is shown in Figure 3-16G.

In some installations, a cyclone is placed between the booth and the filter collector. This device uses centrifugal force to separate larger particles from the airstream prior to filtration.

A more recent development is the integration of the spray booth and powder recovery system into a single structure. (See Figure 3-16H.) In this equipment, cartridge-type filters and the exhaust fan are built into the structure of the spray booth. The ductwork, which conventionally would separate the booth from the powder collector, is eliminated. Collected powder is pneumatically pumped from the hoppers immediately beneath the recovery filters directly back to the spray gun feeders. Within the last 16 years this arrangement has gained dominance in new installations. A further development that has gained broad acceptance is the use of flame-retardant plastics for the spray enclosure. Many of the powder booths currently manufactured have plastic walls and roof panels.

Electrostatic powder coating can also be conducted within an enclosure commonly referred to as a pipe coater. The device consists of a steel enclosure, a ring of automatic electrostatic powder spray guns, and a conveyor. The workpiece enters at one end, passes through a cloud of coating powder, and leaves the coater at the opposite end. The workpiece is preheated to a temperature above the melting point of the powder, so the powder will fuse upon contact. The coater is provided with an exhaust system and powder recovery filter system similar to that used with the booth arrangement described in connection with Figure 3-16G.

Fluidized Bed

Fluidized beds may be constructed in a wide variety of sizes. They have vertical walls (typically steel) and a porous floor (textile or plastic). Air flows upward through the floor and through the powder. When the airstream is adjusted properly, the powder will behave as a liquid, flowing readily around, and contacting all surfaces of any object that is dipped into it. It will immediately fuse to a workpiece that has been preheated above the melting point of the powder. After the workpiece has been removed from the bed, it is placed in an oven to complete the curing of the coating.

An electrostatic fluidized bed has a series of high-voltage electrodes near the surface of the fluidized powder mass or, alternatively, below the fluidizing plate. Grounded workpieces, which may or may not be preheated, pass over the electrodes and are coated electrostatically, then cured in a bake oven.

Both regular and electrostatic fluidized bed arrangements usually are provided with peripheral air exhaust systems that collect any powder grains escaping the bed and prevent their distribution into the general factory area.

FLUID SPRAY PROCESS HAZARDS AND CONTROL

Materials and supplies used in organic spray finishing processes are usually flammable or combustible, are often toxic, and, in some cases, may be highly reactive or unstable. For these reasons, spray finishing operations are considered hazardous, and suitable preventive and protective measures should be taken to minimize the hazards.

Fire Prevention

Hazard identification: Materials' hazards should be identified and marked upon containers as soon as they arrive in the facility. A

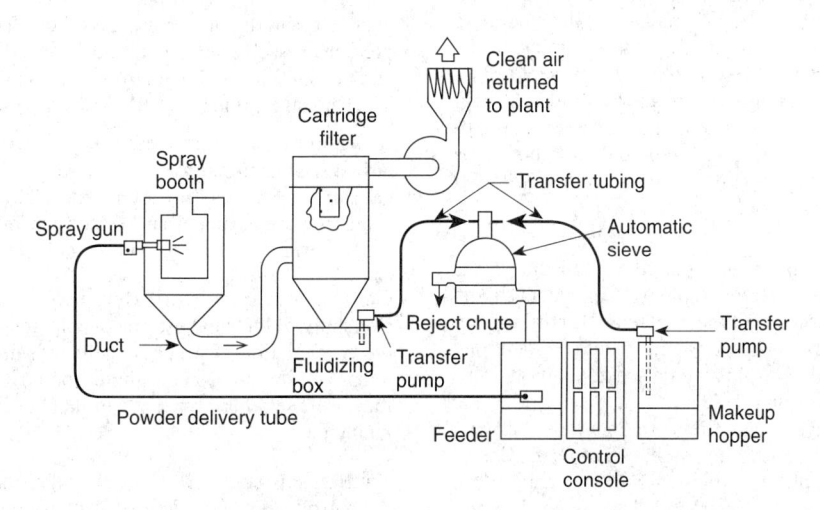

FIG. 3-16G. Diagram of automatic recycle of a single-color powder-coating system showing the booth, collector, fan, and filter. (Source: Nordson)

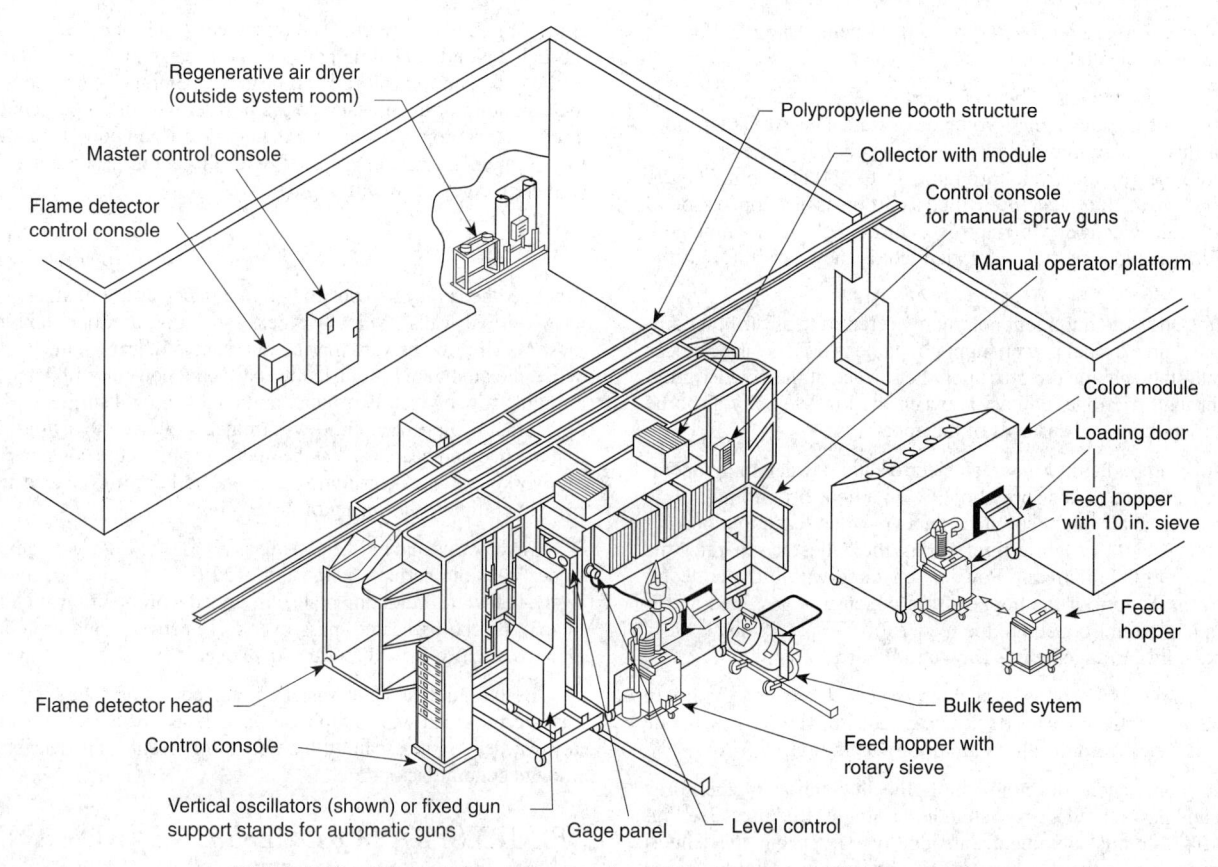

Regenerative air dryer
(outside system room)

Master control console

Flame detector
control console

Polypropylene booth structure

Collector with module

Control console
for manual spray guns

Manual operator platform

Color module

Loading door

Feed hopper
with 10 in. sieve

Feed
hopper

Bulk feed sytem

Feed hopper with
rotary sieve

Flame detector head

Control console

Vertical oscillators (shown) or fixed gun
support stands for automatic guns

Gage panel

Level control

FIG. 3-16H. *Integrated powder spray booth/recovery system. (Source: Nordson)*

detailed method of identification is contained in NFPA 704, *Standard for the Identification of the Fire Hazards of Materials*. The NFPA 704 system provides for the identification of five degrees of health, flammability, and reactivity hazards.

Storage and handling: Bulk supplies of flammable liquids should be stored outdoors, away from buildings or in special indoor storage facilities. Smaller quantities are then brought indoors to a mixing room where they are prepared for use. The mixing room should be located adjacent to an outside wall provided with explosion-relief vents, and should be isolated from the rest of the building by fire-resistant construction. The room should have sufficient mechanical ventilation to prevent the development of flammable vapor concentrations in the explosive range. To prevent rupture of the container due to fire exposure, all flammable liquid containers of greater than 5-gal (19-L) capacity that are kept indoors should be equipped with a special plug that incorporates a pressure-relief valve, a vacuum-relief valve, and a flame arrester.

Prepared coating materials are transported from the mix room to the spray area in containers or through piping. Containers should be equipped with tightly clamped lids that will retain vapors and liquids in the event the container is upset. Piping should be identified as process pipe containing flammable materials, and a shutoff valve should be installed at each point where a hose is connected to the system. Emergency stop controls and fire alarm interlocks are very important to keep runaway pumps from continuing to feed a fire after hoses have been burned off. All piping, pumps, and supply and receiving vessels must be electrically grounded and bonded to prevent accumulation of dangerous static electrical charges during transfer of fluids.

Quantities of flammable or combustible liquids kept in or near process areas should be limited to the amount needed for one working shift at a time. The materials should be put in closed containers equipped with appropriate pressure-relief devices and flame arresters at all openings. These materials are best kept outside the spray booth and separated from it by several feet (1 ft = 0.305 m) to prevent them from becoming involved in a spray booth fire.

Flammable or combustible liquids should never be transferred from one container to another by the application of air pressure to the original shipping container. Pressurizing such containers may cause them to rupture, creating a serious flammable liquid spill. Any pressure vessels used should be designed specifically for such use and be manufactured in conformance with the American Society of Mechanical Engineers' (ASME) *Code for Unfired Pressure Vessels.*

Fluid hoses connected to some electrostatic equipment must have a special structure to resist both fluid and high-voltage electrical stresses that result from use of electrically conductive paint. If ordinary hose or tube is substituted for this use, electrical failure of the hose wall will lead to pinhole failures and ignition of fires.

Control of vapors and overspray: Limitation of vapors and overspray to the smallest practicable area is accomplished through the combination of process enclosures and power ventilation systems. Ventilation systems should be checked frequently to ensure they are operating properly and that specified flow rates are being maintained.

Ignition sources: All sources of ignition should be barred from, or controlled in, areas defined as Division 1 or Division 2 in Article 516 of NFPA 70, *National Electrical Code®.* The careless use of smoking materials and improperly supervised welding or flame cutting operations pose significant threats to a fluid-coating applications system. A hotwork permit system and prohibition of smoking should be in practice.

To minimize chances of ignition, open flames, spark-producing equipment, and any exposed surfaces exceeding ignition temperature of the material being sprayed should be prohibited in either Division 1 or Division 2 areas. No equipment or process capable of producing sparks or hot particles should be located above or adjacent to those areas classified as hazardous unless partitions or other means of separation are provided.

All electric wiring and equipment, unless specially designed and manufactured for use in hazardous areas, may be regarded as potential ignition sources. Only those devices and wiring types listed for this use may be installed or used in areas classified as hazardous. To accommodate failure, electric lighting fixtures located above a classified area should be totally enclosed or provided with a guard that will prevent hot particles from dropping into the hazardous area. Exhaust fans should be nonsparking and conform to Air Movement and Control Association (AMCA) Class C requirements.

Static electricity remains a common source of flammable vapor ignition in electrostatic spray finishing operations. Energized electrodes send generous amounts of electrical charge into the air, and it then collects upon surrounding surfaces. Accumulation of this charge may raise a surface to a high voltage. Electrostatic spray systems should be equipped with electrical interlocks to de-energize the electrostatic power supply any time the spray guns are not actually in use. Most modern electrostatic spray guns that are listed for use with flammables include a charging circuit that limits the energy of a discharge arc to a value below that which is sufficient to cause ignition.

Electrostatic ignition can be prevented by electrically bonding together all the elements of a fluid transfer or spray system along with the workpiece and all other electrically conductive objects located within 10 ft (3 m) of the charged elements, and then grounding the bond. Additional information on bonding and grounding can be found in NFPA 77, *Recommended Practice on Static Electricity*.

Human beings are also conductors of electricity; thus, the spray operator cannot be overlooked in the grounding process. Gripping the grounded spray gun with the bare hand or wearing shoes with electrically conductive soles, provided the floor is grounded, are two workable methods for grounding the operator. Even in the absence of electrostatic equipment, dangerous static electrical charges can be generated by the simple act of walking across a spray area floor that is covered with sticky paint residue. The charging mechanism is identical to that seen in the more familiar act of walking across a wool or nylon carpet on a dry day: a spark and shock results when a doorknob or switchplate is touched. That spark is quite capable of igniting flammable vapors.

Static electrical charge is also generated by the operation of nonelectrostatic air and airless spray equipment, resulting in discharge sparks from either the spray apparatus or the object being sprayed. To prevent ignition of fires from such discharges, electrical bonding and grounding of apparatus and the sprayed object is necessary.

Housekeeping: Residues of spray materials are a solid form of readily ignitable fuel. A routine maintenance program should provide for the periodic removal of overspray residue from walls, floor, and ceiling of the spray booth, room, or area, as well as from conveyors and the interior of ventilation ducts. Residue buildup on conveyors may become particularly hazardous by acting as electrical insulation, which interferes with workpiece grounding. Contaminated spray booth filters should be either removed from the building

as soon as they have been replaced, or kept immersed in water until disposal, as they present a serious spontaneous heating hazard. Only a few hours of "fermenting" is required for a waste container holding paint-fouled filters or rags to become so hot that it bursts into flame. Alternately, the spontaneous heating hazard may be eliminated by curing the paint in fouled filters or rags by processing them through a paint bake oven before disposal.

Operator training: Operating personnel should be exposed to thorough initial training when they begin the job and should be given periodic additional training to maintain their appreciation of the hazards involved, to teach them hazard control procedures, and to educate them in new processes. They should be aware of the inherent hazards of the spray materials being used, particularly if the materials are toxic, chemically unstable, or reactive. They should also be familiar with the system of hazard identification that is employed. Operators should be drilled in proper operating, bonding and grounding, emergency, and maintenance procedures.

Fire Protection

Areas in which fluid spray processes are conducted and areas where coatings and solvents are mixed or stored should be separated from other plant operations by appropriate distance or partitions. Automatic sprinklers inside spray booths, ventilation ducts, and at the ceiling contribute considerably to the control of fire. In open areas, draft curtains extending down from the ceiling around spraying operations, along with smoke and heat vents, will slow the "mushrooming" of hot combustion gases along the ceiling and thus limit the number of sprinklers that will activate. This will concentrate sprinkler discharge into the area where it is most needed and will minimize smoke and heat interference with fire-fighting efforts.

Research within the last year has shown that fire in the paint spray pattern radiates heat at an extremely high rate.[1] Temperature of surfaces 3 ft (0.9 m) from the spray gun have been shown to rise more than 150°F (66°C) per second. Residues exposed to such radiant heating ignite within a few seconds. Fast-acting optical flame detectors have been shown capable of interrupting spray within a small fraction of a second, thereby preventing ignition of surrounding residue. Such detectors are now required by NFPA 33, *Standard for Spray Application Using Flammable or Combustible Materials* (hereafter referred to as NFPA 33), for automatic electrostatic installations.

Any fluid transfer or supply system for flammables, whether driven by pump or by pressure vessel, should be arranged with emergency shutdown provisions and an interlock with the fire alarm and flame detector to interrupt flow in the event of fire or accidental spill, thereby limiting available fuel. "Panic button" actuators should be located at the operator control station and at the exit door.

Portable fire extinguishers and standpipe hoses are useful against fires involving small spray operations. The rate of fire development across residue accumulations can be very high, however. It must be recognized that in only a few seconds a fire can grow to a size that will be beyond portable extinguisher control, particularly in spray booths of the baffle- or dry-filter type or in operations where paint residues are permitted to accumulate on the floor. For enclosed or semi-enclosed coating processes, such as in continuous coaters or automatic decorating machines, automatic fire extinguishing systems may be used to flood the interior of the machines.

A spray booth and its exhaust ductwork should maintain structural integrity (i.e., not collapse) during a fire of nominal proportions so that the major portion of flame, heat, and fire gases produced will vent from the building through the exhaust ductwork. This diminishes the chances of fire spread in the building that will interfere with egress and with access to the seat of the fire by members of the fire service. Exhaust fans and air makeup units should remain in operation during fire conditions to maximize smoke removal. Spray booths and their exhaust ductwork are required by various codes and standards to be constructed of steel, masonry, or other material of equivalent fire resistance. Survivability of booth structure is considerably enhanced by the cooling effect of water from sprinklers inside the stack and booth. The probability that heat from a fire within a booth and stack will ignite nearby combustible materials can be reduced to an acceptable level by reasonable clearances between any combustible structure or stored materials and the surfaces of the booth or stack. (For recommended clearances, see NFPA 33.) Spray rooms are usually required to be separated from other occupancies within the same building by walls, floors, and ceiling structures with a minimum 2-hr fire-resistance rating.

POWDER COATING PROCESS HAZARDS AND CONTROL

Powder coating operations using combustible organic powders are classified as hazardous processes because the powders can burn vigorously when suspended as airborne dusts and can explode when confined. There are no flammable vapors in the powder coating process, and the energy required to ignite a cloud of air-suspended coating powder is from 100 to 1,000 times higher than that required to ignite flammable vapors. Powder deposited onto workpieces or booth walls, or lying on the floor is not readily ignitable.

Fire Prevention

Storage and handling: Coating powders that are kept in shipping containers are not customarily classified as hazardous materials and are commonly stored under conditions appropriate for ordinary combustible materials. Reasonable care is required in the handling and movement of the powder containers because a fire hazard could be created should the containers be broken and the powder distributed as an air-suspended cloud of dust.

Waste powder is commonly packaged into fiber drums or cardboard cartons lined with plastic bags and shipped to a landfill for disposal in accordance with Environmental Protection Agency (EPA) regulations. If the waste container has been processed through a baking cycle, the waste powder will have fused into a solid block of plastic that is no longer vulnerable to scattering if the container is broken.

Spilled powder pickup and routine cleanup is usually accomplished with an industrial vacuum approved for use with combustible dust. Explosions have occurred when unapproved vacuums (with open-frame universal motors and sparking commutators) were used.

Coating operations: Virtually all spray powder coating processes are conducted in spray booths that are more tightly enclosed and permit considerably less overspray to escape than those booths used for fluid coating processes. All connections in the hoses and piping associated with pneumatic transfer equipment should be routinely checked to confirm that they are secure and that ground connections are in place. Breakage of any such connection during operation could create a substantial hazard by blowing powder into the workspace air outside the process enclosure.

To prevent the escape of powder, the spray booth must have adequate ventilation that provides effective capture velocity at all its openings. Minimum average velocities are considered to be in the vicinity of 60 fpm (18 m/min). Interlocks should be installed to permit operation of the powder application equipment only while the ventilation equipment is in operation and, similarly, to permit oper-

ation of the electrostatic power supply only during the time the powder is actually being applied.

Ignition sources: The control methods for ignition sources for powder coating are similar to those for spray finishing.

Static accumulations on electrically isolated conductive objects are a major cause of fires in electrostatic powder coating installations. Metal workpieces suspended from conveyor hangers that do not provide an effective electrical circuit between the workpiece and ground pass through the coating process zone while discharging hot electrical sparks. This sparking can ignite the powder cloud, and flaming will be sustained by the continuing flow of powder and air. In an electrostatic fluidized bed, the entire volume of the bed can become involved in fire.

Some application equipment will discharge incendiary sparks when approached too closely by a grounded object. Proper use of this type of equipment requires safeguards to prevent any grounded object from approaching the electrically charged high-voltage elements closer than twice the sparking distance. Discharge sparks that do occur during use of this type of equipment may be traceable to improper racking of parts on a conveyor or swinging of the conveyor racks. To reduce the probability of a small fallen part being drawn into a duct and producing mechanical sparking, grillwork at the intake of any exhaust duct is highly recommended, and magnetic devices can be added to catch tramp metal.

Spray booths connected to collectors by ductwork should be provided with a means of preventing entry of tramp iron to the duct. Screens and magnetic traps are effective.

A hotwork permit system and prohibition of smoking should be in practice.

Fire Protection

Protective measures for powder coating installations are: (1) keep the powder collector from becoming involved in a fire that has originated in the spray operation, (2) equip the collector with pressure-relief vents and ductwork in order to prevent its rupture in an explosion, and (3) protect the building from heat and pressure effects that would be generated by an explosion.

To prevent a fire in a powder spray booth from spreading through connecting ductwork, the ventilation system is usually arranged to keep powder concentrations in the ductwork below one-half the minimum explosible concentration (MEC). In those cases where fire has persisted in the spray booth for periods ranging beyond approximately 15 sec, however, the powder collector is threatened by glowing embers that have formed in the spray fire. These glowing embers are independent of powder concentration and are capable of igniting the powder collector.

Powder collectors can be protected by a fast-acting flame detector within the spray booth, interlocks to shut down equipment, and a fast-acting damper to interrupt airflow in the duct between the spray booth and the powder collector. In operation, this apparatus will recognize any flame that has ignited in the spray booth and respond quickly (within 1/2 sec) to the flame by shutting down all process equipment energy supplies (including electricity and compressed air) and by closing the ductwork damper. This same equipment can be arranged to extend the response sequence by discharging an inerting gas into the dust collector.

Use of multispectral "smart" flame detectors from one manufacturer has been monitored for seven years (millions of hours of operation) in industrial service. Although fires ignited many times, not a single case of damage resulted.[1]

For those application systems that require operation with combustible concentrations of powder in the ductwork between the application area and the powder collector, a potential exists for almost instantaneous involvement of the collector whenever a fire occurs in the application area. Such systems should either be equipped with appropriately engineered suppression systems or designed in such a manner that the powder application and collection equipment will be fire and explosion resistant.

Airborne clouds of powder in a confined space will burn at far higher rates than in unconfined space. Since the interior of a conventional dust collector is a confined space, steps must be taken to accommodate the sudden pressurization and volumes of smoke and fire gases that will evolve in the event of ignition. The usual protective technique used is to install automatic pressure-relief vents in the collector housing. These vents will open should an ignition occur and allow the combustion products to escape, thereby limiting pressure development. When the collector is located within a building, ductwork is usually required to direct the vented combustion products to the exterior of the building or to a safe location. Vent ductwork must be kept short. Lengths in excess of 10 ft (3 m) are of questionable value.

When a collector explosion occurs, fire, clouds of unburned powder, and molten globules of partially burned powder are projected from pressure-relief openings and from spray booth openings. If vented into a building, a major secondary explosion may occur, resulting in catastrophic damage.

If powder escapes from the collector into the building, it will collect on horizontal surfaces and create a potential for a secondary explosion. The primary explosion may throw loose powder into the air in suspension where it could be ignited by flames or embers that persist from the initial combustion. If a substantial quantity of powder is thrown into suspension by the primary explosion, the secondary explosion could be considerably more serious than the primary. Protection is achieved primarily through: (1) maintenance of process equipment, (2) procedures, and (3) scrupulous housekeeping to collect any powder that does escape.

The dust explosion hazard is virtually eliminated (through elimination of the confined space necessary to form an explosion) in an integrated spray booth/collector of the type shown in Figure 3-16H and in other apparatus incorporating the same design principles. This class of equipment, since it operates without explosion risk, needs no explosion vent or ductwork. However, flame detectors are still needed—when automatic spray guns are used—to quickly interrupt powder feed in the event of fire and thereby limit heat damage and formation of toxic smoke.

After ignition of a spray gun powder cloud or a fluidized bed, the flame can usually be extinguished almost instantly by simply turning off the spray gun feeder or the air supply to the fluidized bed. The normal velocity of the feed systems exceeds typical flame-front propagation velocities and flame therefore does not back-stream in hoses from guns to feeders, even though powder concentrations in hoses may exceed MEC.

Automatic sprinkler systems are commonly required in factory areas where powder coating operations take place, as well as within the booth and collector. Because sprinklers have rather long operating delays relative to the duration of a powder fire (in a system equipped with automatic flame detectors), it is common for powder fires to be extinguished by shutting down the equipment before sprinklers can activate.

In those cases where small fires have persisted in spray booth residues, portable water fire extinguishers or sprinklers arranged to protect the filter cartridges or bags have been found to be adequate.[1]

BIBLIOGRAPHY

Reference

1. Unpublished research by the author.

NFPA Codes, Standards, and Recommended Practices

Reference to the following NFPA codes, standards, and recommended practices will provide further information on the safeguards for spray finishing and powder coating discussed in this chapter. (See the latest version of *The NFPA Catalog* for availability of current editions of the following documents.)

NFPA 10, *Standard for Portable Fire Extinguishers*
NFPA 13, *Standard for the Installation of Sprinkler Systems*
NFPA 30, *Flammable and Combustible Liquids Code*
NFPA 33, *Standard for Spray Application Using Flammable or Combustible Materials*
NFPA 68, *Guide for Venting of Deflagrations*
NFPA 69, *Standard on Explosion Prevention Systems*
NFPA 70, *National Electrical Code®*
NFPA 77, *Recommended Practice on Static Electricity*
NFPA 704, *Standard System for the Identification of the Fire Hazards of Materials*

Additional Reading

ASME, *Code for Unfired Pressure Vessels*, American Society of Mechanical Engineers, New York.

ASTM D92, *Standard Test Method for Flash and Fire Points by Cleveland Open Cup*, 1990 ed., American Society for Testing and Materials, W. Conshohocken, PA.

Benedetti, R. P., ed., *Flammable and Combustible Liquids Code Handbook*, 5th ed., National Fire Protection Association, Quincy, MA, 1993.

Benrashid, R. and Nelson, G. L., "Synergistic Fire Performance of Zinc and Zinc Compounds as Coatings or Fillers in Engineering Plastics," Customer Demands for Improved Total Performance of Flame Retarded Materials, October 26–29, 1993, Tucson, AZ, Fire Retardant Chemicals Assoc., Lancaster, PA, 1993, pp. 47–70.

"Coatings for Plastic Glazing," *Automotive Engineering*, May 1993, pp. 24–28.

Davis, K., "Fluids, Lubricants, Coatings, and Elastomer Materials," Final Report, September 1, 1989-February 28, 1991, Wright Lab., Wright-Patterson AFB, OH, UDR-TR-91-103, WL-TR-91-4097, July 1991, 387 pages.

Dehart, M. G., Musser, D. A., and Krajkowski, E., "Elimination of Rolled Powder Fires. Phase 2," Final Report, March 1991-March 1992, Hercules Inc., Radford, VA, ES-1AA-6-855364, ARAED-CR-92003, Sept. 1992, 248 pages.

Deil, D. G., "Magnesium Coating Effects in Boron Combustion," *Proceedings* of the Fall Technical Meeting, 1995, for Chemical and Physical Processes in Combustion, 1995, pp. 329–332.

Evans, B., "Paint Finishes Can Aggravate the 'Fireball Effect'," *Architects Journal*, Vol. 197, No. 26, June 30, 1993, p. 9.

Factory Mutual Research Corporation, "Spray Applications of Flammable and Combustible Materials," Loss Prevention Data 7-27, Factory Mutual Research Corporation, Norwood, MA.

Grand, A. F., "Fire Evaluations of Coatings for Glass-Reinforced Polymeric Composites," Second International Conference on Fire and Materials, Interscience Communications Limited, 1993, pp. 143–159.

Ingason, H., "Response Characteristics of Glass Bulb Sprinkler Heads Mounted in a Paint Spray Booth," *Fire Technology*, Vol. 29, No. 4, Nov. 1993, pp. 317–331.

Ingason, H., and Persson, H., "Brandrisker vid sprut-malning-hogtrycks-sprutning betydligt brandfarligare an kon-ventionell sprutmalning. [Fire Hazard of Spray Painting, Flash Point, Heat of Combustion, Combustion Efficiency.]," (Abstract in English), SP Report 1989:09, 1989, 36 pages.

"Instantaneous Fire Detection/Suppression System for Jaguar Paint Spray Electrostatic Plant," *Product Finishing*, Vol. 40, No. 7, 1987, pp. 14-15.

Kaetzel, L. J., and McKnight, M. E., "Enhancing Coatings Diagnostics, Selection, and Use Through Computer Based Knowledge Systems," *Proceedings* for SSPC 95 Protective Coatings Blazing New Trails,

Balancing Economics and Compliance for Maintaining Protective Coatings Steel, Structures Painting Council (SSPC), SSPC 95-09, 1995, pp. 287–295.

Kimbrough, R. "Paint Spray Booth Fire System Applications," *Fire Systems*, Vol. 5, No. 2, 1991, pp. 33–41.

Langford, T. H., *et al.* "Elimination of Rolled Powder Fires," Final Report, Nov.1987-Mar. 1990, Army Ammunition Plant , Radford, VA, RAD-140.10, ARAED-CR-90013, ES-1A-6-8553, September 1990, 336 pages.

Martin, J. W., *et al.*, "Methodologies for Predicting the Service Lives of Coating Systems," E. I. du Pont de Nemours and Co. (Inc.), Philadelphia, PA, NIST BSS 172, Oct. 1994, 73 pages.

Scarbrough, D. R., "Spray Finishing and Powder Coating," *Industrial Fire Hazards Handbook*, Cote, A.E., ed., 3rd ed., National Fire Protection Association, Quincy, MA, 1990.

Sharma, T. P., Varshney, B. S., and Kumar, S., "Studies on the Burning Behavior of Metal Powder Fires and Their Extinguishment. Part 2. Magnesium Powder Heaps on Insulated and Conducting Material Beds," *Fire Safety Journal*, Vol. 21, No. 2, 1993, pp. 153–176.

Varshney, B. S., Kumar, S., and Sharma, T. P., "Studies on the Burning Behavior of Metal Powder Fires and Their Extinguishment. Part 1. Mg, Al, Al-Mg Alloy Powder Fires on Sand Bed," *Fire Safety Journal*, Vol. 16, No. 2, 1990, pp. 93–117.

Woodruff, F. A., "Developments in Coating and Electrostatic Flocking," *Journal of Coated Fabrics*, Vol. 22, Apr. 1993, pp. 290–297.

DIPPING AND COATING PROCESSES

Revised by J. William Sheppard

This chapter discusses the processes, equipment, hazards, and fire protection of dipping and coating operations that use flammable and combustible liquids. Such processes often involve liquids having tremendous heat energy and heat release capabilities, which can cause rapid property destruction when involved in a fire if not properly protected. The heat of combustion of flammable liquids is approximately two and one-half times that of an equivalent weight of wood.[1] In addition, heavy concentrations of smoke and toxic products of combustion often develop and make fire fighting extremely difficult.

THE PROCESSES

Applications

Dipping and coating processes include, but are not limited to, finishing, impregnating, priming, cleaning, quenching, and other similar operations during which materials are immersed in, passed through, or coated by flammable or combustible liquids.

Processes vary widely and range from the cleaning of small parts in small quantities to an automated part-finishing process that uses a tank containing several thousand gallons (1,000 gal equals 3,780 L) of a flammable or combustible liquid. As is the case with any potentially hazardous operation, a complete evaluation of the process and an understanding of the properties of the liquid(s) in use are essential.

Equipment

The basic process components, e.g., dip tanks, flow coaters, curtain coaters, and roll coaters, present similar hazards. Related process equipment often used in conjunction with dipping and coating operations includes conveyors, pumps, and piping systems, flammable or combustible liquid storage tanks, ovens, liquid heaters or heat exchangers, agitators, detearing equipment, and ventilation and exhaust systems. Exhaust systems may include energy recovery and pollution control devices. Support equipment must be selected and installed with full consideration of the inherent process hazards involved.

Dip tanks: Dip tanks are liquid containers of various sizes and shapes designed for the process involved. Tank sizes vary from a few gallons with small exposed surface area to several thousand gallons with large exposed surface area (1 gal = 3.785 L). Figure 3-17A is an example of a typical dip tank process system. Dip tanks, their drainboards, and their covers should be constructed of noncombustible material, such as heavy metal, reinforced concrete, or masonry. They should be designed for the process and liquid involved, with consideration given to static head of the liquid, corrosion, mechanical damage, and ease of maintenance and repair. Spill or exposure fires could weaken tank supports, possibly causing tank collapse. Therefore, supports for large tanks should be made of reinforced concrete or protected steel.

Flow coaters: In flow coaters, the coating material is applied from nozzles or slots in an unatomized state onto the material being coated. (See Figure 3-17B.) The excess is collected in a trough or sump below the workpiece, returned to the reservoir, and recirculated to the nozzles by a pump. The nozzles may be fixed or oscillating. The more nozzles used, the larger the reservoir and pump capacities required, and the greater the solvent loss will be. The workpiece usually enters and leaves through conveyor openings, and a drip tunnel is used in place of the drainboards that are used with dip tanks. The tunnel is enclosed on all sides, except for the conveyor opening, and is sloped toward the trough or sump. Airborne solvent concentration in the tunnel is sometimes used to keep the coating from drying before it gets to the oven.

Curtain coaters: Curtain coaters are used to apply coating material to flat or slightly curved workpieces. (See Figure 3-17C.) Coating material is pumped from a reservoir to a coating head, which has a small reservoir with a dam or weir forming one side. Coating material pumped to the head overflows the dam and forms a continuous vertical stream, which drops down onto the workpiece. Excess coating material drops into a trough and returns to the reservoir to be recirculated.

Roll coaters: In a roll coater, coating material is applied by bringing the workpiece into contact with one or more liquid-coated rollers. (See Figure 3-17D.) The coating material may come from an open pan or a nip formed by two rollers. Coating material is supplied from a reservoir by a pump and piping system.

PROCESS HAZARDS

Dipping and coating articles or materials by passing them through flammable and combustible liquids involves the danger of fire and explosion of vapor-air mixtures. Generally, the severity of the hazard

J. William Sheppard is an administrator of corporate security and fire prevention, General Motors, Detroit, MI, and a member of NFPA Technical Committees on Private Water Supplies, Ovens and Furnaces, and Water Based Fire Prevention Systems.

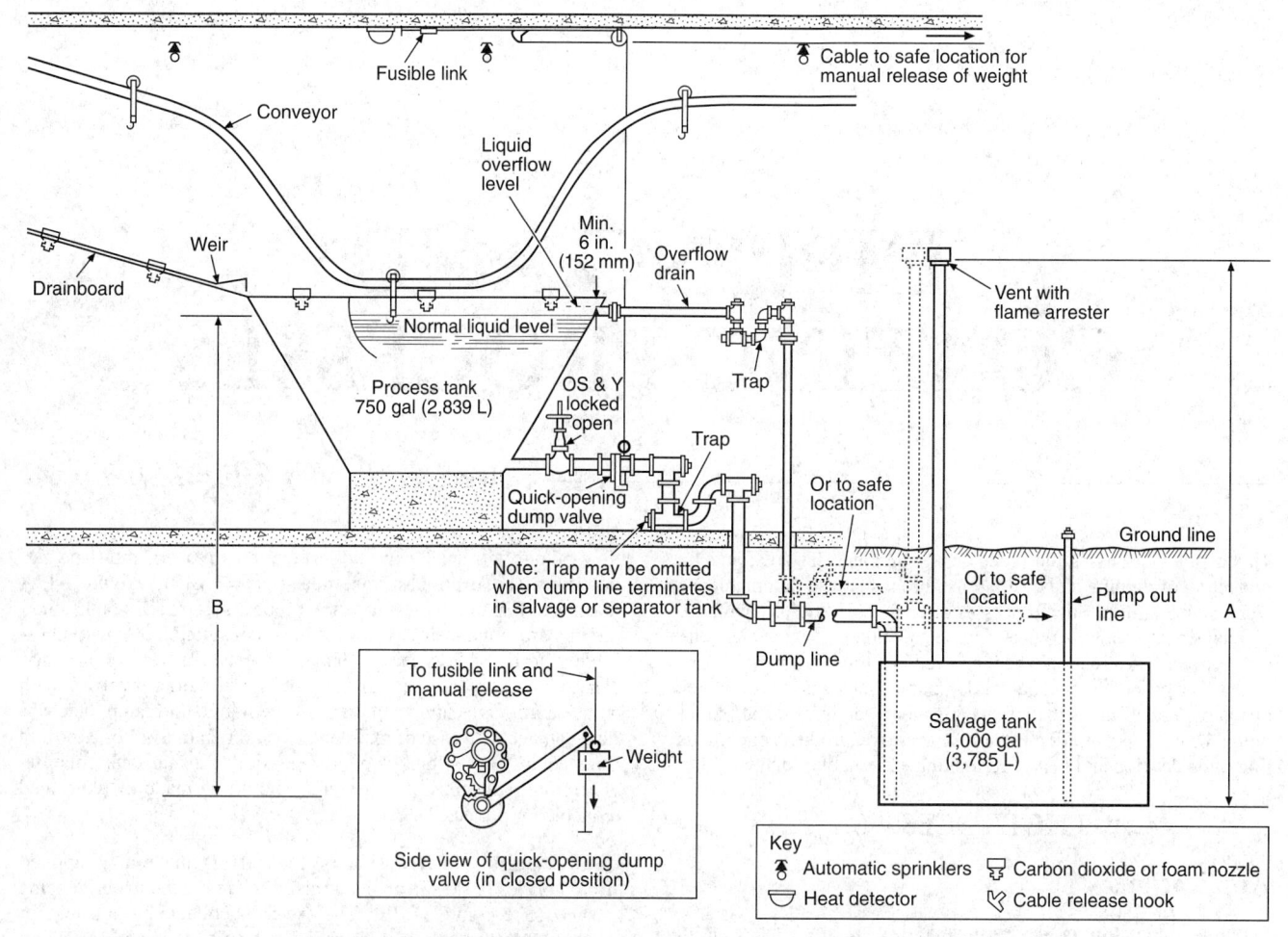

Note: A is greater than B but discharge is at least 12 ft (3.6 m) above ground.

FIG. 3-17A. Dip tank, drainboard, and conveyor system.

depends on the character and flammability of the liquids used, the quantities present, and the rate of vapor generation. (Characteristics of flammable and combustible liquids are discussed in detail in Section 3, Chapter 21, "Storage of Flammable and Combustible Liquids.")

Extent of the hazard is defined by: (1) ease of ignition and (2) rate of flammable vapor generation from the liquid surface and from surfaces of freshly coated articles, floors, drain or drip boards, and other related equipment. Two further hazards are: (1) the intensity and burning characteristics of the flammable vapor evolved and (2) the probability of fire spread from radiated heat or from the flow of burning liquids because of a container rupture, boilover, or overflow.

Flammable liquid vapors are usually heavier than air and, therefore, flow to low points. They may travel great distances before being exposed to ignition sources that can cause flashback to the process area.

Flammable liquids are often lighter than, and immiscible in, water. During fire-fighting operations, water applied to such liquids may cause overflow and float the burning liquid away from the process to other areas.

HAZARD REDUCTION

Hazards inherent in dipping and coating operations can be reduced by properly locating the processes, installing ventilation and ex-

haust systems, eliminating ignition sources, and providing proper maintenance and periodic inspections. Special equipment design, employee training, and proper fire protection will also help to reduce process hazards.

Process Location

The principle of segregating hazards to confine fires and limit damage is fundamental in the selection of locations for dipping and coating operations. The personal safety of workers and other building occupants, as well as the potential exposure of large values, must also be considered.

A preferable location for dipping and/or coating operations is in a detached or cutoff one-story sprinklered building of fire-resistive construction. When operations are located on upper stories, floors should be made waterproof and equipped with drains to a safe location. Dipping and coating operations should not be located below grade, directly over basements, or near pits or trenches, because liquid drainage and vapor removal will be difficult to achieve. Curbs and trapped drains should be provided to control liquid flow where necessary.

When, due to its nature, the process cannot be cut off, it should be located in an area that is clear of combustibles, means of egress, and other important processes. Combustible floors, ceilings, and surrounding walls should be protected. Noncombustible

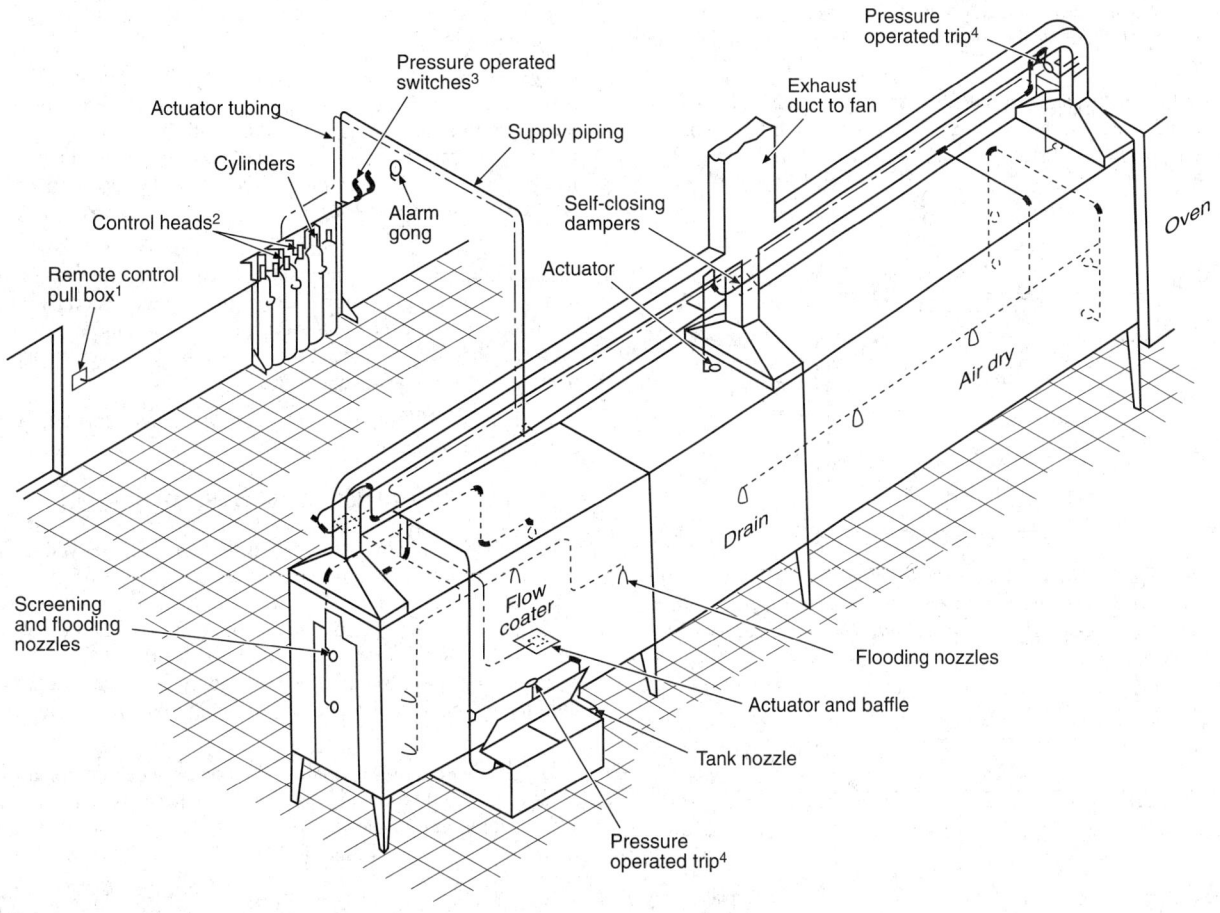

Notes:

1. Normal manual pull-box control and cable.
2. Pneumatic control heads with emergency manual controls.
3. Pressure-operated switches to shut down exhaust fan, pumps, conveyor, etc., and sound alarm.
4. Pressure-operated trips to release self-closing dampers in exhaust duct and self-closing cover on paint tank.

FIG. 3-17B. *Flow coater. (Source: NFPA 34, Standard for Dipping and Coating Processes Using Flammable or Combustible Liquids)*

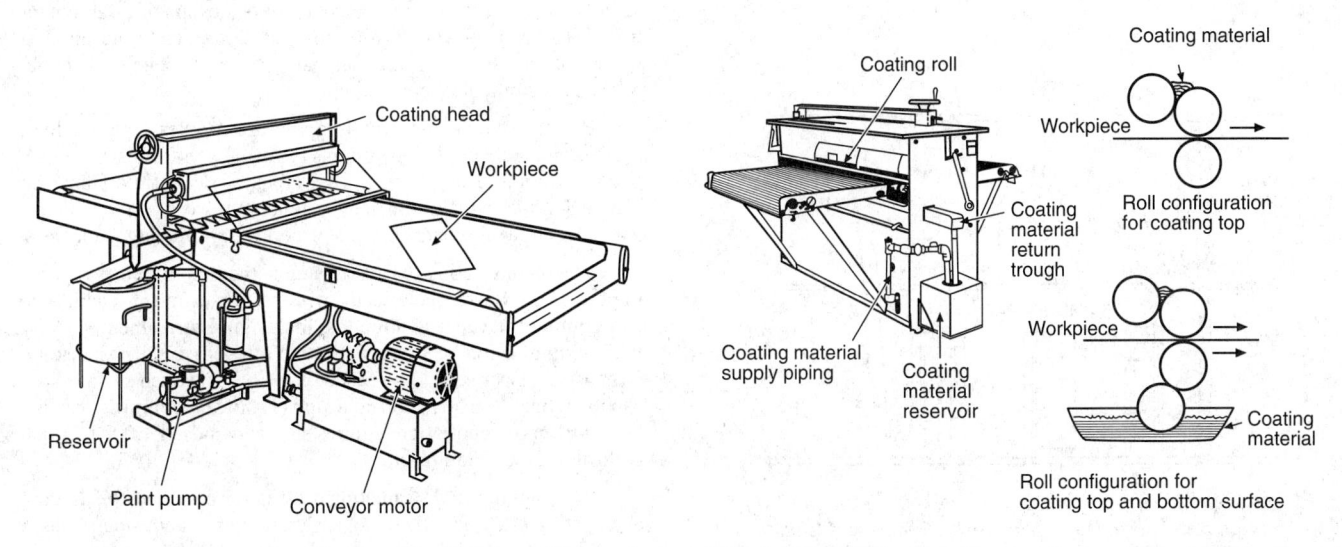

FIG. 3-17C. *Curtain coater. (Source: Factory Mutual Research Corp.)*

FIG. 3-17D. *Roll coater. (Source: Factory Mutual Research Corp.)*

curtainboards should be installed around the perimeter of the process or protected construction area and extended downward as far as is practical.

As flammable and combustible liquid fires generally develop rapidly and release large amounts of heat, automatic roof vents are extremely desirable. Roof vents allow heat and smoke to escape, improving the chances of fire control by automatic sprinklers or the fire department.

When processes are located in confined areas, and highly flammable or unstable liquids are used, explosion relief venting may be necessary to lessen potential losses. Venting can be in the form of lightweight walls and roofs or explosion-relieving wall panels, roof latches, and windows. Explosion relief is discussed in detail in NFPA 68, *Guide for Venting of Deflagrations.*

Careful evaluation of the process is necessary in developing location requirements for dipping and coating operations. Unfortunately, small dipping and coating operations generally are not considered in the same light as the larger operations, and, therefore, little consideration is given to their locations. However, even a small operation can produce a major fire loss.

Figures 3-17E (a), (b), and (c) show common location features of dipping and coating operations.

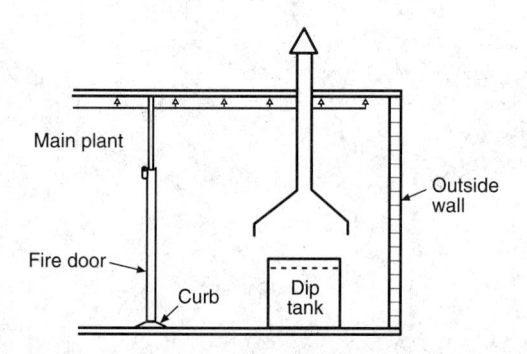

FIG. 3-17E(a). Most satisfactory arrangement for location of processes. Tank located near an outside wall, cut off from main plant areas by fire-resistive construction. (Source: Factory Mutual Research Corp.)

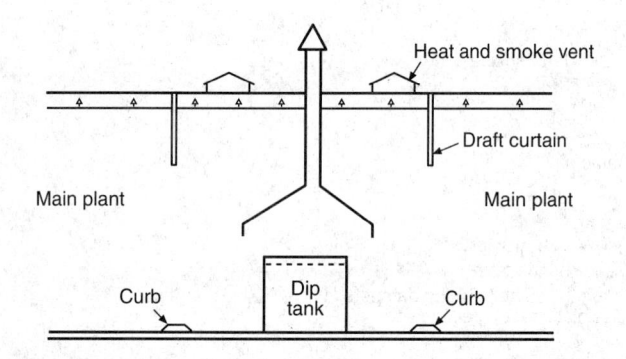

FIG. 3-17E(b). Satisfactory arrangement. Tank located in main plant area cut off by draft curtains and curbing. Heat and smoke vents should limit opened sprinklers to those near the fire and reduce smoke damage. (Source: Factory Mutual Research Corp.)

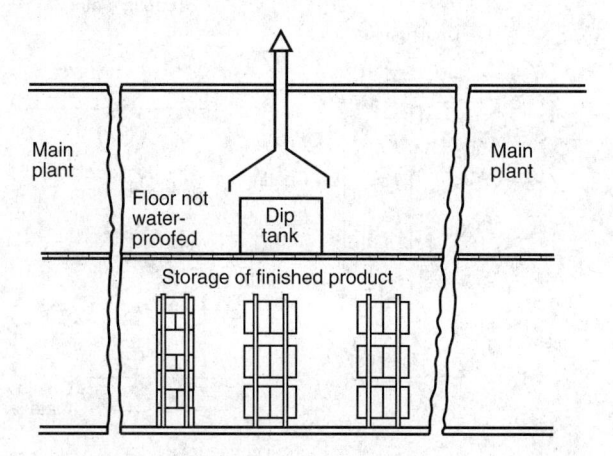

FIG. 3-17E(c). Unsatisfactory arrangement. Tank not cut off from main plant area. Water from hose streams will leak down through floor and damage finished product. (Source: Factory Mutual Research Corp.)

Ventilation and Exhaust Systems

When processes involve liquids that produce vapors, ventilation is necessary to limit the vapor source to the smallest practical space possible. The vapor source created by dipping and coating processes is defined as "the liquid exposed in the process and on the drain board. Also, any dipped or coated object from which it is possible to measure vapor concentrations exceeding 25 percent of the lower flammable limit at a distance of 1 ft (0.305 m) in any direction from the object."[2]

The extent of the vapor area of a process depends upon the properties of the liquid, such as vapor pressure, flash point, boiling point, and evaporation rate, along with the characteristics of the process or wetted surface area exposed. Figure 3-17F clearly shows the difference in exposed wetted surface area between two dip tank operations using the same quantity of liquid. Note that, with all else being the same, the process shown in the upper part of the figure would generate a greater volume of vapor and have a higher rate of heat release than that in the lower part; thus it would offer a greater fire hazard.

Vapor areas for each process should be kept as small as is practical but should not extend more than 5 ft (1.5 m) from the vapor source. The extent of a vapor area can be determined with a combustible gas analyzer. A theoretical ventilation rate can be determined when the liquid usage rate is known.[3] Since vapor concentrations vary widely from one process to another, a single safety exhaust ventilation system cannot be designed for all operations. The prime objective is to limit the vapor area to the smallest space possible and to prevent dead air spaces or pockets where vapors can accumulate.

Because flammable liquid vapors are heavier than air, low-point peripheral ventilation systems are usually more desirable than overhead hood arrangements. Two types of design are shown in Figure 3-17G. The least efficient system is the overhead, or canopy, hood that allows discharge of some vapors to surrounding work areas.

When considering exhaust systems, the ratio of air to vapor is important. However, the effectiveness of exhaust air in picking up and diluting the vapor is often lost in design. Entrainment velocity with mixing action becomes the key to efficient use of exhaust air. Exhaust rates per sq ft (1 sq ft = 0.0929 m^2) of wetted surface area usually range from 50 to 200 cu ft/min (1.4 to 5.7 m^3/min). Effective entrainment of vapor often requires slot velocities of 1,000 to 2,000 cu ft/min (28 to 56 m^3/min).[4]

In designing and evaluating ventilation systems for dipping and coating processes, there are several important considerations:

1. Automatic processes should be interlocked to shut down the operation in the event of an exhaust system failure.

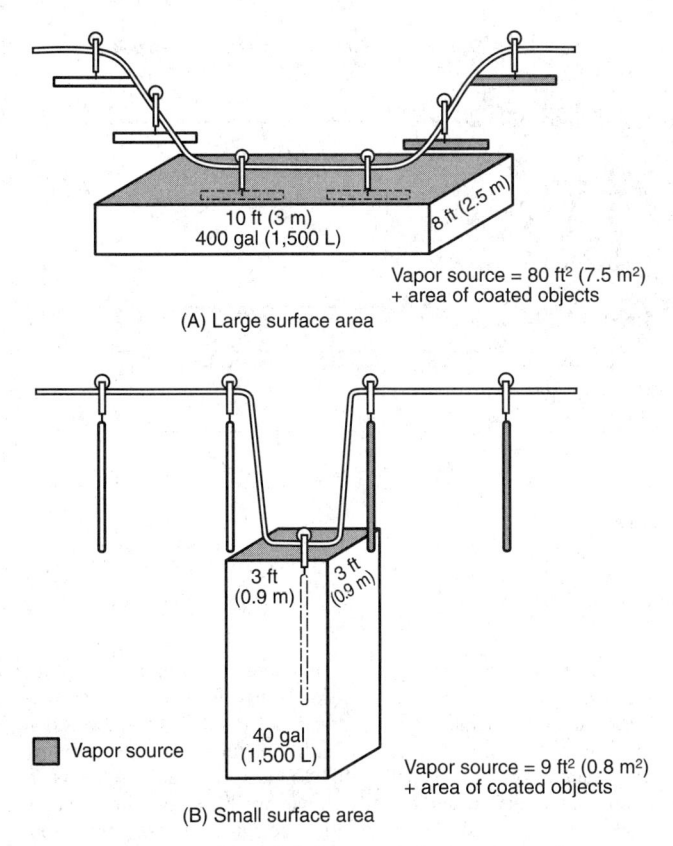

FIG. 3-17F.   *The difference in vapor source area between two dip tanks with the same coating volume. For SI units: 1 ft = 0.305 m; 1 sq ft = 0.0929 m²; and 1 gal = 3.785 L. (Source: Factory Mutual Research Corp.)*

2. The supply of makeup air should be sufficient to allow efficient operation of the exhaust fans and to minimize dead air spaces.
3. Ideally, each exhaust system should discharge directly outdoors. When exhaust gases must be treated to satisfy environmental protection requirements or when energy conservation measures are used, individual, direct discharge of each system may not be practical. Manifold systems increase the hazard; therefore, all such systems should be specifically approved and strict preventive maintenance measures taken. Reactive materials and coatings should not be used when ducts are connected to a manifold.
4. Exhaust air should not be used for makeup air in occupied spaces. It can be used in unoccupied areas, but only if it has been decontaminated to safe levels.
5. Exhaust ducts should not discharge near air intakes, nor should they discharge less than 6 ft (1.8 m) from a combustible wall or roof or 25 ft (7.6 m) from combustible construction or an unprotected opening in a noncombustible exterior wall.
6. Exhaust ducts should be equipped with ample access doors to facilitate cleaning.
7. Ducts should be constructed of fire-resistive material, such as steel or masonry, and should be substantially supported. They should be installed with adequate clearance from combustibles. Dampers should not restrict exhaust ventilation to below minimum safe levels.

Ignition Sources

The key to minimizing the hazard of fire or explosion from vapor-air mixtures in dipping and coating processes is maintaining concentrations well below the lower flammable limit (LFL) through

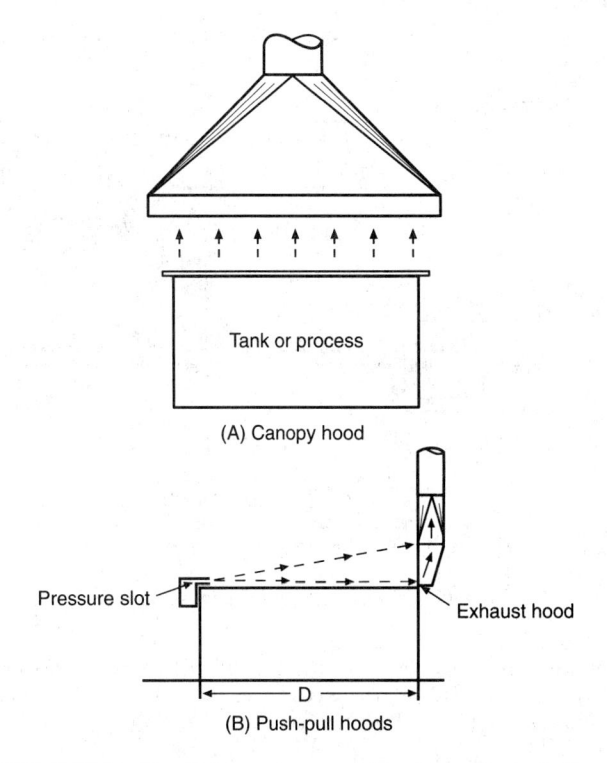

FIG. 3-17G.   *Ventilation systems used for open process tanks. (Source: American Conference of Governmental Industrial Hygienists)*

properly designed safety exhaust ventilation systems. Without vapor (fuel) and air (oxygen) properly mixed, ignition cannot take place even though sources of ignition are available. As it is almost impossible to keep processes free of flammable vapor-air mixtures in the explosive range, sources of ignition must be eliminated or equipment must be designed for use in hazardous areas containing flammable vapors and/or combustible residue.

Electrical equipment: Where flammable liquids are used or where combustible liquids are used at or above their flash point temperatures, electrical equipment should conform to the requirements for hazardous locations as outlined in NFPA 70, *National Electrical Code®*. NFPA 70 classifies areas in which special types of electrical equipment must be used. In the case of dipping and coating processes, the classified areas are measured from the vapor source, which may be the liquid surface (open tanks) or wetted surfaces (freshly coated workpieces, drainboards, floors). The space containing hazardous vapor concentrations normally extends laterally and upward a short distance from the vapor source and, since the vapors are usually heavier than air, downward to the floor.

Classified areas are shown in Figure 3-17H. Class I, Division 1, electrical equipment is required within a radial distance of 5 ft (1.5 m) from the vapor source and within a horizontal distance of 25 ft (7.6 m) from the vapor source when the equipment is in pits. If a pit is not vapor-stopped at a point 25 ft (7.6 m) from the vapor source, the entire pit is considered a classified area. Class I, Division 2, electrical equipment is required 3 ft (0.91 m) beyond Division 1 in all areas down to 3 ft (0.91 m) above the floor. In the space between the floor and 3 ft (0.91 m) above the floor, the Division 2 area extends 20 ft (6.1 m) beyond the Division 1 area. An exception to the electrical requirements above is tanks containing 5 gal (19 L) or less and having 5 sq ft (0.46 m²) or less of exposed surface area. Vapor concentrations during operating and shutdown periods cannot

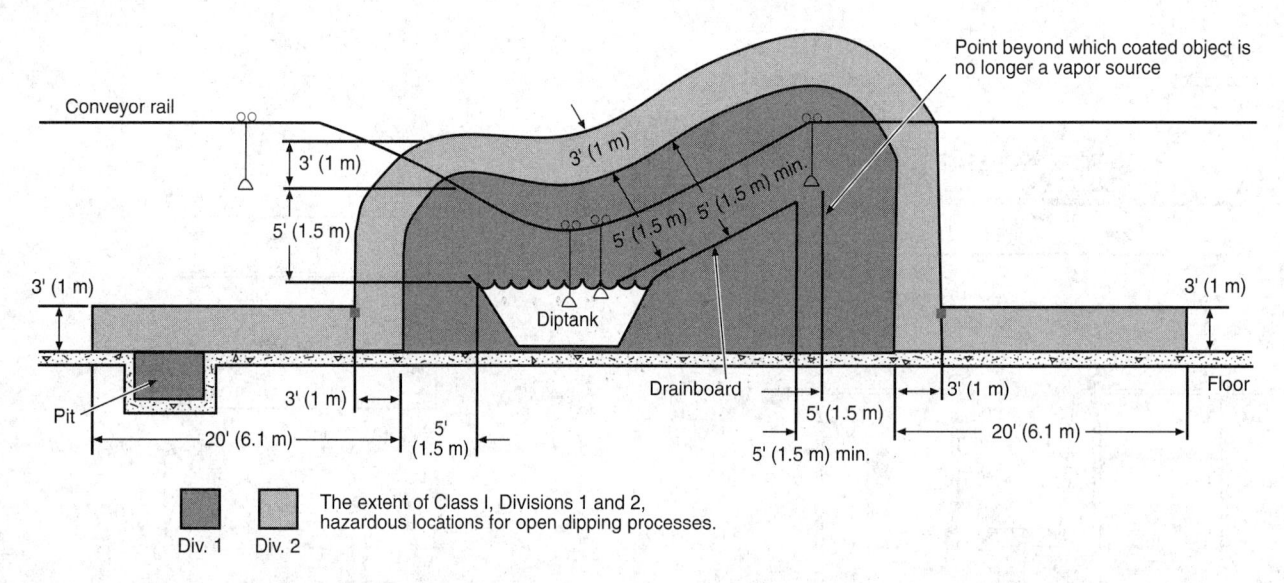

FIG. 3-17H.   The Class I, Divisions 1 and 2, hazardous locations for a dipping operation. For SI units: 1 ft = 0.305 m.

exceed 25 percent of the LFL. These installations generally present less of a hazard and ordinary electrical equipment may be accepted more than 8 ft (2.4 m) from the vapor source.

It is usually undesirable to install electrical equipment in the vicinity of dipping and coating operations. In such locations, the equipment would be subject to deposits of combustible residue, which would complicate maintenance and cleaning procedures. However, if electrical equipment must be installed in such locations, it should be a type that can be exposed safely to flammable vapors and combustible residues. Wiring in rigid conduit or in threaded boxes or fittings without taps, splices, or terminal connections is permitted in both residue and vapor areas.

Fixed lighting fixtures located outside of, but above, classified areas should be protected against physical damage, which may allow them to become sources of ignition of flammable vapors.

In areas where the electric service is supplied from overhead wires and lightning is a fairly common occurrence, protection against lightning-induced surge voltages should be provided. In the absence of lightning protection, high voltage surges could enter vapor areas and create high-energy ignition sources. Suitable protection includes lightning arresters, interconnection of all grounds, and surge protection capacitors.

Under abnormal conditions, overheating and arcing can occur in electrical circuits and equipment. Therefore, it is important that electrical circuits and equipment servicing dipping and coating processes have the proper overcurrent protection, are grounded, and receive regular maintenance.

Dipping and coating processes are frequently changed when new coating materials are introduced, production demand is increased, and new equipment is installed. Quite often, unsafe electrical installations and poor maintenance habits creep into the production process. It is common to find extensions connected to additional fans and lighting that are not suitable for use in classified hazardous areas.

Other ignition sources: The presence of open flames, spark-producing processes or devices, and heated surfaces having temperatures high enough to ignite flammable vapors should be prohibited in flammable vapor areas.

Ovens located directly above or adjacent to dipping and coating operations have been the ignition source for fires. This equipment should be located as far as practical from dipping and coating operations, with noncombustible partitions provided when reasonable distances cannot be maintained. Exhaust ventilation systems of ovens and dryers located directly above, or adjacent to, dipping and coating processes should be interlocked with process ventilation systems, so that heating and process equipment cannot function unless all ventilation equipment is in operation. Indirect heating systems, which do not present ignition sources, are usually preferred in drying operations. (For more information on heating systems, see Section 3, Chapter 8, "Industrial and Commercial Heat Utilization Equipment.")

Dipping and coating processes often involve the use of conveyors, rollers, collectors, festoon dryers, and other devices that move the workpiece or liquid through the process. Flowing liquids and materials moving through a process can generate static electricity. A static charge can accumulate on process equipment and develop a difference in electrical potential sufficient to cause a spark containing enough energy to ignite flammable vapor-air concentrations. Static elimination by bonding and grounding process equipment or by humidification or ionization of the local atmosphere is essential in reducing the hazard of fire or explosion.

Of equal importance is the banning in hazardous classified areas of equipment and processes that are not essential to dipping and coating operations but that are potential ignition sources. Included in this category are: cutting and welding operations, portable heaters, spark-producing (ferrous) tools, and smoking. Nonferrous tools are available as are nonferrous rotating parts for exhaust fans. If hazardous maintenance operations, such as cutting or welding, must be performed in the vicinity of dipping or coating processes, they should be performed in accordance with recognized good practices by competent personnel under strict supervision. "No smoking" signs in process areas must be properly posted.

Special Design Considerations

Some hazards of dipping and coating operations can be reduced by special design features in the process equipment.

Many flammable and combustible coatings are less dense than, and immiscible with, water. In the event of fire, water from hose streams or sprinkler systems could cause an overflow of a large amount of liquid from the tank unless certain design precautions have been taken.

For example, the top of the tank should be at least 6 in. (152 mm) above the floor, and the liquid surface should be at least 6 in. (152 mm) below the top of the tank. Tanks having a capacity of more than 150 gal (568 L) or a liquid surface area of 10 sq ft (0.93 m^2) or more should be equipped with trapped overflow pipes that discharge to a safe location. Overflow pipes should be capable of handling the maximum delivery of the tank's fill pipes or of the automatic sprinkler discharge. In any case, overflow pipes should never be less than 3 in. (76.2 mm) in diameter.

Larger tanks—those having a capacity of 500 gal (1.9 m^3) or more—should be equipped with a bottom drain of sufficient size to empty the tank in 5 min. The flammable liquid should be drained to a vented salvage tank or other safe location. If gravity flow is not practical, pumps may be used to empty the tank. The drain may be operated manually or automatically. However, if a manual drain release is used, it should be accessible and at a safe location. If, due to increased hazard or by nature of the operation, bottom drains cannot be provided, the process should be cut off or fully diked, and both an automatic special extinguishing system and sprinklers should be installed.

Conveyor systems used in dipping and coating processes should be arranged to stop automatically in the event of a fire or failure of the exhaust ventilation system.

In processes where there is a possibility of flammable liquids being heated above their boiling points or to within 100°F (37.8°C) of their autoignition temperatures, suitable excess temperature limit controls are needed to prevent rapid vapor buildup and possible autoignition.

Controls should limit the surface temperatures of heated workpieces to at least 100°F (37.8°C) below the autoignition temperature of the coating material used. This limitation does not apply to quenching tanks, which are discussed in Section 3, Chapter 9, "Oil Quenching and Molten Salt Baths."

Limit controls, such as liquid level devices, meters, and timers, should be used to prevent overfilling of tanks where automatic filling is designed into the process. If pumps are used, they should be interlocked to shut down in the event of fire.

Maintenance, Training, and Inspection

Location, ventilation, equipment design, and the elimination of ignition sources are key to the reduction of the hazards in dipping and coating processes using flammable and combustible liquids. Further, maintenance, training, and regular inspection of equipment are essential to safe operations. Unfortunately, due to production schedules, changes in processes, and turnover of personnel, these essentials may be neglected, resulting in unsafe operating conditions.

Areas in the vicinity of dipping and coating processes should be kept free of accumulations of combustible residues and unnecessary combustible materials. Spontaneous ignition, due to oxidation or exothermic reaction between various coating components, often occurs when excess residue accumulates in work areas, ducts, duct discharge points, or other adjacent areas. When excess residues accumulate in such locations, operations should be discontinued until conditions are corrected.

Combustible coverings (e.g., thin paper or plastic) and strippable coatings are often used to facilitate cleanup of drippings and residues. The increased amount of combustibles introduced by the use of these materials is offset by the improved housekeeping and ease of cleaning that they provide. Suitable containers should be provided for the disposal of waste and rags used for cleaning.

The size and complexity of dipping and coating operations may vary widely but, in all operations, the personnel involved should be instructed in the potential safety and health hazards inherent in the particular process employed. The nature of the process and operational, maintenance, protection, and emergency procedures should be included. It is desirable to record such training and provide refresher instruction from time to time to ensure continuity and proper action in the event of an emergency.

Depending upon the size and nature of the process, periodic (usually at least monthly) inspection should be made of dipping and coating processes. Inspections should include noting the condition of equipment covers, overflow pipe inlets and outlets, discharge, bottom drains and valves, electrical wiring and equipment, grounding and bonding connections, ventilation equipment, and extinguishing equipment. Inspections and protection equipment tests should be documented and conducted using an appropriate inspection form. Defective equipment and unsafe conditions should be corrected promptly.

FIRE PROTECTION

In spite of the considerable care taken in the operation of dipping and coating processes that use flammable and combustible liquids, fires and explosions do occur, and fire protection and suppression equipment are necessary.

Fire loss experience has shown that damage in small processes can be as severe as in large operations. Less consideration may be given to the location of small processes, and they may be in areas that expose high values or combustible construction. Some form of protection is usually required for all dipping and coating processes.

By far the simplest device for extinguishing a fire in a process tank is a self-closing cover, which cuts off the fire's air supply. Normally the cover is held open by a cable and fusible link. When exposed to the heat of a fire, the solder in the fusible link melts, releasing the cover and allowing it to drop tightly in place over the tank. (See Figure 3-17I.) Automatic sprinklers are often provided at

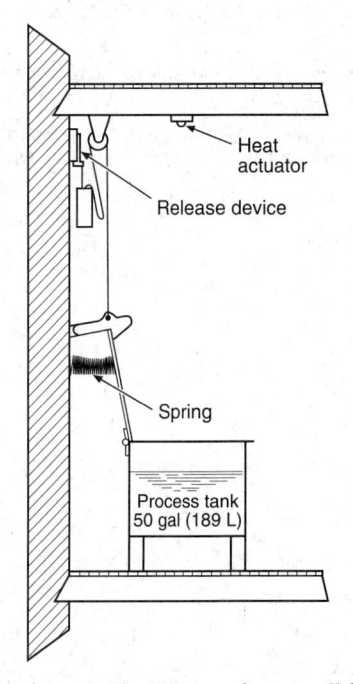

FIG. 3-17I. Automatic closing cover for a small fixed dip tank. For SI units: 1 gal = 3.785 L.

ceiling level and can provide adequate protection for small processes and where combustible coatings are used with limited liquid surface area exposure. Sprinklers are intended to help control the fire, prevent damage to the building, and allow effective manual fire fighting. Automatic sprinklers are often required as backup to special extinguishing systems.

Where processes involve considerable volumes, large liquid surface areas, or large wetted areas of low flash point materials, special extinguishing systems should be provided. In the case of dip tanks, special systems should be provided for tanks having capacities in excess of 150 gal (569 L) or surface areas greater than 10 sq ft (0.93 m^2). Suitable protection systems use foam, carbon dioxide, or dry chemical as the extinguishing medium. For liquids having flash points above 140°F (60°C), water spray systems may be used. All systems should operate automatically and conform to fire protection standards. The complexity and value of some processes and the absence of sprinklers as backup often make it desirable to provide redundancy in protection systems.

Dipping and coating process areas should be equipped with portable fire extinguishers suitable for use on flammable and combustible liquids fires.

Due to the flash fire conditions that may occur in flammable liquids fires, emergency personnel or fire brigades should be trained to act immediately to effect both automatic and manual protection methods to fight the fire to achieve rapid control and limit possible injury and damage.

BIBLIOGRAPHY

References Cited

1. Factory Mutual Research Corporation, "Flammable Liquids—General Safeguards," Loss Prevention Data 7-32, July 1993, Factory Mutual Research Corp., Norwood, MA, p. 41.

2. NFPA 34, *Standard for Dipping and Coating Processes Using Flammable or Combustible Liquids*, National Fire Protection Association, Quincy, MA, 1995.

3. NFPA 86, *Standard for Ovens and Furnaces*, National Fire Protection Association, Quincy, MA, 1995.

4. *Industrial Ventilation—A Manual of Recommended Practice*, 22nd ed., Committee on Industrial Ventilation, Cincinnati, OH, pp. 10-93 to 10-118.

NFPA Codes, Standards, and Recommended Practices

Reference to the following NFPA codes, standards, and recommended practices will provide further information on the safeguards for dipping and coating processes discussed in this chapter. (See the latest version of *The NFPA Catalog* for availability of current editions of the following documents.)

NFPA 10, *Standard for Portable Fire Extinguishers*
NFPA 11, *Standard for Low-Expansion Foam*
NFPA 12, *Standard on Carbon Dioxide Extinguishing Systems*
NFPA 13, *Standard for the Installation of Sprinkler Systems*
NFPA 15, *Standard for Water Spray Fixed Systems for Fire Protection*
NFPA 25, *Standard for the Inspection, Testing, and Maintenance of Water-Based Fire Protection Systems*
NFPA 34, *Standard for Dipping and Coating Processes Using Flammable or Combustible Liquids*
NFPA 35, *Standard for the Manufacture of Organic Coatings*
NFPA 68, *Guide for Venting of Deflagrations*
NFPA 69, *Standard on Explosion Prevention Systems*
NFPA 70, *National Electrical Code®*
NFPA 77, *Recommended Practice on Static Electricity*
NFPA 86, *Standard for Ovens and Furnaces*
NFPA 91, *Standard for Exhaust Systems for Air Conveying of Materials*
NFPA 327, *Standard Procedures for Cleaning or Safeguarding Small Tanks and Containers Without Entry*
NFPA 497M, *Manual for Classification of Gases, Vapors, and Dusts for Electrical Equipment in Hazardous (Classified) Locations*

Additional Reading

Burke, R., "Flammable Liquids," *Firehouse*, Vol. 20, No. 5, May 1995, pp. 36–37, 110–111.
Evans, B., "Paint Finishes Can Aggravate the 'Fireball Effect'," *Architects Journal*, Vol. 197, No. 26, June 30, 1993, p. 9.
"Flammable Liquid Drainage and Containment," *Record*, Vol. 69, No. 4, July/Aug. 1992, pp. 24–31.
"Flammable Liquids—Risk, Regulations and Protection Measures to Safeguard Flammable Liquids," *Records*, Vol. 69, No. 1 Jan./Feb. 1992, pp. 15–21.
Lentini, J. J. and Waters, L. V., "Behavior of Flammable and Combustible Liquids," *Fire and Arson Investigator*, Vol. 42, No. 1, Sept. 1991, pp. 39–45.
"Proper Flammable and Combustible Liquid Handling," *Record*, Vol. 68, No. 5, Sept./Oct. 1991, pp. 16–17.
Tabar, D. C., "Coatings Industry Responsibility in Safety and Loss Prevention," *Coatings*, Vol. 41, No. 2, Washington DC, 1989.
Whiteley, B., "Foaming Sprinkler and Flammable Liquid Fires," *Fire Prevention*, No. 262, Sept. 1993, pp. 32–35.
Woodruff, F. A., "Development in Coating and Electrostatic Flocking," *Journal of Coated Fabrics*, Vol. 22, April 1993, pp. 290–297.

EXTRUSION AND MOLDING PROCESSES

Revised by Ken Adams

The multi-billion-dollar plastics industry is one of the largest industries in the United States. Plastics are a basic material of use—on a par with metals, glass, wood, and paper. They have become increasingly important to advanced concepts in architecture, aerospace, communications, and transportation, as well as medicine and the arts. The plastics industry originated in the United States in 1868 when John Wesley Hyatt developed celluloid—the first American plastic.

At least 6,000 companies in the United States make plastics; i.e., they produce basic material, process or fabricate plastics into products or parts, or finish their goods. Companies that use plastics or these materials and services number well into the tens of thousands.

Plastic is any one of a large and varied number of materials consisting wholly or in part of combinations of carbon with oxygen, hydrogen, nitrogen, and other organic and inorganic elements. While it is solid in the finished state, at some stage in its manufacture it is molten or softened, and thus capable of being formed into various shapes, most through the application, either singly or together, of heat and pressure. Whatever their properties or form, however, plastics fall into one of two groups: (1) thermoplastics or (2) thermosets. (See Table 3-18A.)

Thermoplastic resins can be repeatedly softened and hardened by heating and cooling, without a chemical change taking place.

Thermoset resins, when heat treated, undergo a chemical reaction and cannot be softened by further heating without causing a chemical change that would forever alter the material.

There are approximately 30 major classes of plastics or polymer groupings, with numerous variations of individual plastic products. This diversity has made plastics applicable to an extremely broad range of end-uses and products. Yet, paradoxically, this diversity has also made it difficult to grasp the concept of a single family of materials encompassing such a far-reaching span of physical and chemical properties and fire hazard characteristics.

Relevant material can also be found in Section 3, Chapter 33, "Housekeeping Practices," and Section 4, Chapter 6, "Flammable and Combustible Liquids."

Ken Adams, assistant technical director for standards and technical support, The Society of the Plastics Industry, Inc., revised the chapter for this edition.

PLASTICS PROCESSING

The plastics industry as a whole has three broad areas of processing: (1) manufacturing, (2) conversion, and (3) fabricating. Although each is basically distinct, they may be conducted in the same or at different locations.

Manufacturing, or synthesizing, of the basic plastic or feedstock sometimes includes compounding with colorants or other additives. The hazards of the synthesis plant are basically those of a chemical plant. (See Section 3, Chapter 23, "Storage and Handling of Chemicals.")

Conversion of the plastics materials into useful articles by molding, extrusion, or casting usually involves heating the plastic so it will flow into a shape that is retained when the plastic is cooled. Although chemical reactions are not often a significant part of these operations, the plastics industry refers to them as "processing" or "converting."

Fabricating encompasses the largely mechanical operations of bending, machining, cementing, decorating, and polishing plastics.

These three areas, i.e., manufacturing, conversion, and fabricating, accurately categorize the fire hazards that may be encountered in the plastics industry. It is important to recognize, though, that the synthesis, or manufacturing, of a given plastic may be much more hazardous than its molding or extrusion. It is also important to recognize that some plants nominally doing only converting and/or fabricating may also be conducting chemical operations with flammable or reactive materials. One example is the overlapping of molding and synthesis operations in a plant that impregnates liquid polyester resins into reinforcing glass fibers for molding boat hulls or containers.

This chapter discusses the processing, or converting, phase of the plastics, which includes recognizing, of course, that in some instances hazards associated with all three phases of the industry are carried on in one location.

The Basic Hazards

Most plastics are combustible organic compounds that can burn under certain conditions. But aside from the inherent combustibility of plastic formulations influenced by the basic polymers used, the nature of plastic additives, the form the final product takes, and the conversion of feedstock plastics into finished articles involve the hazards associated with combustible dusts, flammable solvents, electrical faults, hydraulic fluids, and the storage and handling of

TABLE 3-18A. *Examples of Plastics Materials*

Thermoplastic		Thermoset	
ABS (acrylonitrile-butadiene-styrene)	Polyethylene	Epoxy	Silicones
A combination of three different monomers—acrylonitrile, butadiene, and styrene gas—suitable for tough products; has good electrical insulating qualities; used to make automobile front grills, household appliance components, business machine and telephone components.	Tough; excellent chemical resistance and electrical insulating quality; near zero moisture absorption; used for packaging, molded housewares, and toys.	Excellent for chemical resistance, electrical properties, and dimensional stability; will bond metals, glass, hard rubber, etc.; used for aircraft components, tanks, and electrical and electronic assemblies.	Long-term heat resistant, water repellent, electrical insulator, and mineral acid resistant; used in the electrical and electronics industry; insulation in meters and generators and induction heating apparatus.

large quantities of combustible raw materials and finished products. (See the chapters in this handbook that specifically discuss these hazards.)

Terminology

Before discussing the converting of plastic feedstocks into finished products, and the associated hazards, certain process terms and definitions are provided.

Additive: Any material added in a minor amount to basic resins or compounds to alter properties. Additives may be dyes, pigments, powdered fillers for stiffening, plasticizers for flexibility, fibers to reinforce, antioxidants, lubricants to aid flow into or release from a mold, or fire retardants.

Binders: The resins in a plastic mixture that hold together all of the other ingredients.

Blowing agent: Any substance that alone or in combination with other substances is capable of producing a cellular structure in a plastic.

Blown tubing: A thin film made by extruding a tube and simultaneously inflating it with air while hot; distention may be 20 times the diameter of the extruded tube.

Casting: Flowing liquid material into a mold or against a substrate with little or no pressure, followed by solidification and removal of the formed object.

Extrusion: The process of passing molten plastic under pressure through a die to form a continuous profile; the equipment is called an extruder.

Fabrication: The making of articles by machining, cementing, heat-sealing or thermoforming of preformed sheets, rods, or tubes. The term is used in contrast to "processing."

Fillers: Materials added to plastic to modify physical properties and/or reduce its cost. For example, they may be used to increase heat resistance and alter the dielectric strength. A wide variety of products are used, e.g., wood flour, cotton, sisal, glass, and clay.

Film: A general term for sheets of plastic not more than 0.01 in. (0.25 mm) thick, regardless of the process used to make them.

Finishing: The removal of burrs, flash, gates, and defects from plastic articles, and also the development of a desired surface texture.

Flash: Unwanted projections from molded articles resulting from flow of plastic into space between matching parts of a mold. (The term has no fire connotation.)

Foam (cellular plastic): A plastic with numerous cells dispersed throughout its mass. Rigid foams have rapidly reached large-volume production as thermal insulation boards for construction, cups for hot and cold drinks, trays for prepackaged meats, and shock-resistant packaging; flexible foams are used for furniture padding and insulation of outer garments.

Foaming agent: See "blowing agent."

Forming: The process in which the shape of plastic pieces, such as sheets, rods, or tubes, are changed into a desired configuration.

Laminate: A composition of several layers of material(s) firmly held together by heat and pressure, bonding, or impregnation.

Lubricant: Material added to improve feeding of powder or granules into molding or extrusion machines, to improve flow of molten plastic through machines and into molds, or to prevent adhesion of plastic to molds; the last of these uses is called "mold-release." Typical lubricants are zinc stearate, carnauba wax, and silicone oil.

Molding: The process of forcing plastic into a cavity to achieve a desired shape.

Monomers: A simple compound usually containing carbon and of low molecular weight, which can react to form a polymer by combination with itself or with other similar molecules or compounds.

Plasticizers: Materials added to plastic to make the finished product more flexible or to facilitate compounding. Some plasticizers increase the combustibility of the plastic while others serve as flame retardants.

Plastics: Materials that contain as an essential ingredient an organic substance of large molecular weight, are solid in their finished state, and at some stage in their manufacture or processing into finished articles can be shaped by flow.

Polymerization: The process by which molecules of a monomer are made to respectively combine with other monomers. The result is a much longer chain-like molecule.

Processing: The converting of polymers into useful articles by molding or extrusion from granules, depositing film from solvent, or laminating resin and reinforcement. Most often the molding operation uses heat, but it is largely or entirely a physical rather than a chemical process.

Reinforced plastic: A plastic with a fibrous filler that significantly increases flexural, impact, or tensile strength. Additives are usually glass, cotton, or nylon fibers. Reinforced plastics may be thermoplastic granules for injection molding, or may be materials using large areas of reinforcement as in layup molding or pulp molding.

Resin: A solid, semisolid, or pseudosolid organic material that often has high molecular weight, and that exhibits a tendency to flow when subjected to stress and usually has a softening or melting range. However, common usage of the term in the plastics industry does not always conform to the definition. The term is often used for uncured fluid thermosetting materials, some chemically modified material resins, and is often used synonymously with the terms "plastic" and "polymer."

RAW MATERIALS

Many plastic feedstocks are derivatives of petroleum or gas recovered during the refining process; e.g., ethylene monomer (a widely used feedstock) is derived in gaseous form from petroleum refinery gas, liquefied petroleum gases, or liquid hydrocarbons. (See Figure 3-18A.)

The monomer is subjected to a chemical reaction, known as polymerization, that causes the small molecules to link together into increasingly longer molecules. Chemically, the polymerization reaction has turned the monomer into a polymer (the reason names of so many plastics materials begin with the prefix "poly"). The polymer, or plastic resin, must next be prepared for use by the processor, who will turn it into a finished product. In some instances, it is possible to use the plastic resin or feedstock as it comes out of the polymerization reaction. More often, however, it is transformed into a form that can be more easily handled by the processor and more easily handled in processing equipment. The most common solid forms for the plastic resin are pellets, granules, flakes, and powder. Some feedstocks are also available as semisolids, e.g., pastes, or as liquids (for casting).

THE PRODUCTION PROCESS

The ways in which plastics can be processed into useful end-products are as varied as the plastics themselves.

Though the processes differ, there are common elements. In the majority of cases, thermoplastic compounds in the forms of pellets, granules, flakes, and powder must be melted by heat so they can flow. Pressure is usually involved in forcing the molten plastic into a mold cavity or through a die, and cooling must be provided to allow the molten plastic to harden. With thermosets, heat and pressure also are most often used. In their case, however, heat serves to trigger a chemical reaction and causes the thermosetting resin to cure (set) in the mold under pressure. When thermoplastic or thermoset resins are in liquid form, heat and/or pressure need not necessarily be used; however, in many casting techniques intended for high-speed production, they do play a role.

The following descriptions of processes cover the basics of the major manufacturing systems. It should be recognized, however, that there are variations in virtually every process.

Blow molding: This process is generally used only with thermoplastics. It is applicable to the production of hollow plastics products, such as bottles and gas tanks. Blow molding involves the melting of the thermoplastics resin and then forming it into a tube-like shape (known as a parison). The ends of the tube are then sealed, and air is injected into the tube. The tube, in a softened state, is inflated inside the mold and forced against the walls of the mold. (See Figure 3-18B.)

Calendering: This process can be used to convert thermoplastics into film and sheeting, and to apply a plastic coating to textiles or other supporting materials. In calendering film and sheeting, the plastic compound is passed between a series of three or four large heated revolving rollers which squeeze the material between them into a sheet or film.

Casting: This process may be used both with thermoplastics and thermosets to make products, shapes, rods, and tubes, by pouring a liquid monomer-polymer solution into an open mold or a closed mold where it finishes polymerizing into a solid. Pressure need not be used with the casting process, unlike the molding process. The starting material in the casting process is usually in liquid form rather than solid form.

Coating: Coating may be accomplished by using either thermoplastic or thermosetting materials. The coating may be applied to metal, wood, paper, glass, fabric, leather, or ceramics. (See Figure 3-18C.)

Compounding: Mixing additives into previously formed resin, using kneading mixers, screw extruders of varied design, masticating rolls, or calenders, is called compounding. Rolls and kneaders are heated by high-pressure steam or heat transfer fluids.

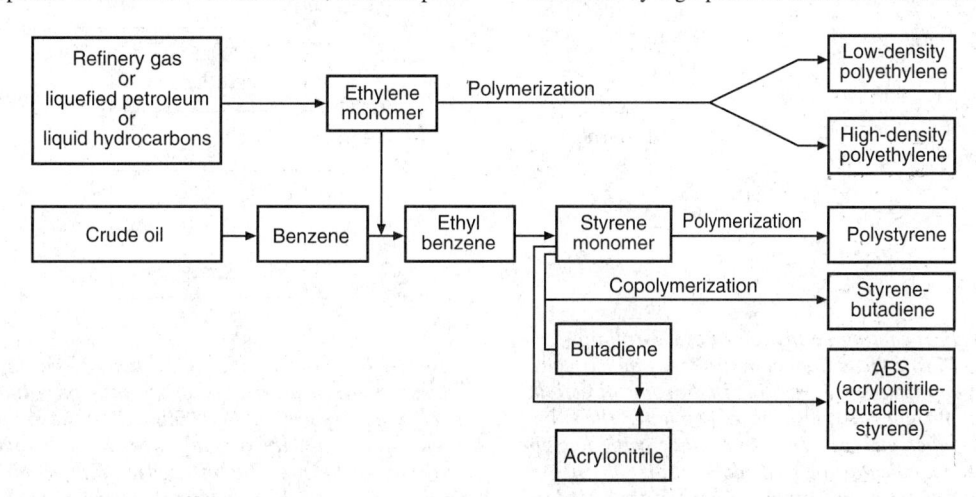

FIG. 3-18A.   Flow chart shows ethylene's role in the manufacture of polyethylene, polystyrene, and styrene copolymers. (Source: The Society of the Plastics Industry, Inc.)

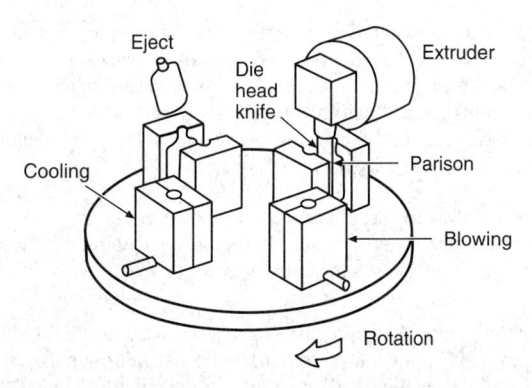

FIG. 3-18B. *Diagram of continuous extrusion blow-molding setup, using a rotating horizontal table. Plastic parison (in cylindrical shape) is extruded from die into mold which closes on the parison (knife cuts the parison off from the extrudate). Mold then rotates to second station where air is injected into the parison (still hot and therefore formable) to blow it out to the shape of the inside mold cavity. At third station, the blown part is allowed to cool and set. Mold finally rotates to last station where finished part (in this case, a bottle) is ejected from mold. (Source: The Society of the Plastics Industry, Inc.)*

Screw extruders have largely displaced rolls and kneaders for compounding, because they provide better control, continuous output, and less exposure of hot plastic to air.

Compounding for flexible polyvinyl chloride (PVC) packaging uses high-intensity mixers and ribbon blenders. The mixer spins the resin at high speed. This generates heat, which allows the resin to accept plasticizers, stabilizers, etc. After absorption at 190 to 240°F (88 to 116°C), the batch drops to a blender to cool for storage or extrusion.

Compression molding: This is the most common method of forming thermosetting materials. Compression molding is simply the squeezing of material into a desired shape by applying heat and pressure to the material in a mold. (See Figure 3-18D.)

Extrusion: This method is employed to form thermoplastic materials into continuous sheeting, film, tubes, rods, profile shapes, or filaments and to coat wire, cable, and cord. (See Figure 3-18E.)

Foam plastics molding: This is the manufacturing process where foams can be used in casting, calendering, coating, rotational molding, flow molding, and even injection molding and extrusion. (See Figure 3-18F.) The manufacturing of foamed products is discussed in more detail later in this chapter in the subsection "Flammable Solvents."

High-pressure laminating: A process that uses thermosetting plastics to hold together reinforcing materials, such as cloth, paper, wood, or glass fibers. Heat and pressures of at least 1,000 psi (6,894 kPa) are used to produce the laminated product.

Injection molding: The method of forming objects from granular or powdered plastics, most often of the thermoplastic type, in which the material is fed from a hopper to a heated chamber in which it is softened, after which a ram or screw forces the material into a mold. Pressure is maintained until the mass is hardened sufficiently for removal from the mold. (See Figure 3-18G.)

Reaction injection molding (RIM): This is a technique used primarily for molding polyurethane elastomers of foams into end-products with solid integral skins and cellular cores. Basically, two or more pressurized reactive streams are impinged together under high pressure in a mixing chamber. The resulting mixture is then injected, under low pressure, into the mold. There, the reaction begins and continues until the liquid mixture has set up into a solid or cellular finished product.

Reinforced plastics processing: In this process, resins (acting as binder material) are combined with reinforcing materials (usually in fibrous form) to produce composite products having exceptional strength-to-weight ratios and outstanding physical properties. The resins may be either thermosets or thermoplastics. (See Figure 3-18H.)

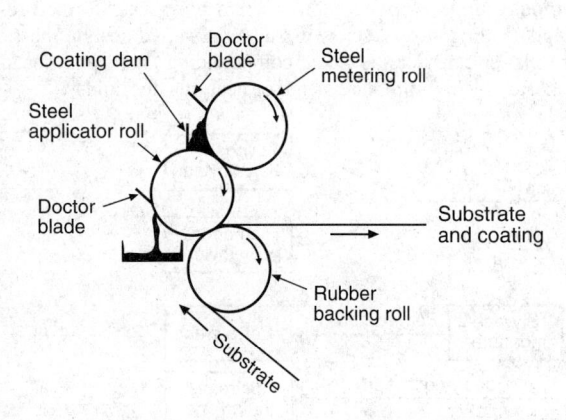

FIG. 3-18C. *A typical coating setup, known as a 3-roll nip-fed reverse roll coater. Plastic feeds from dam through nip between steel metering and applicator rolls, rotating in the same direction. At bottom of the applicator roll, plastic is laid on top of the substrate (e.g., fabric, paper, etc.) as it comes in contact with the substrate at nip between applicator roll and backing roll (which carries the substrate up from the bottom of setup). Doctor blade is used to scrape off excess plastic from applicator roll. (Source: The Society of the Plastics Industry, Inc.)*

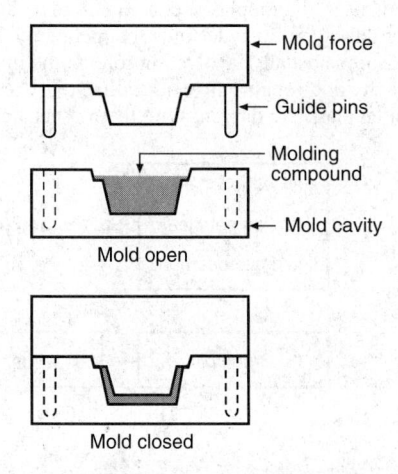

FIG. 3-18D. *Basics of a simple two-piece compression mold. Plastic molding material is loaded into lower half (cavity) of the heated mold (shown at top). Top half of the mold (mold force) is then lowered, and the two halves are brought together under pressure (shown below). The softened molding material is thus formed into the shape of the cavity and allowed to harden with further heating. Mold is then opened and part is removed. (Source: The Society of the Plastics Industry, Inc.)*

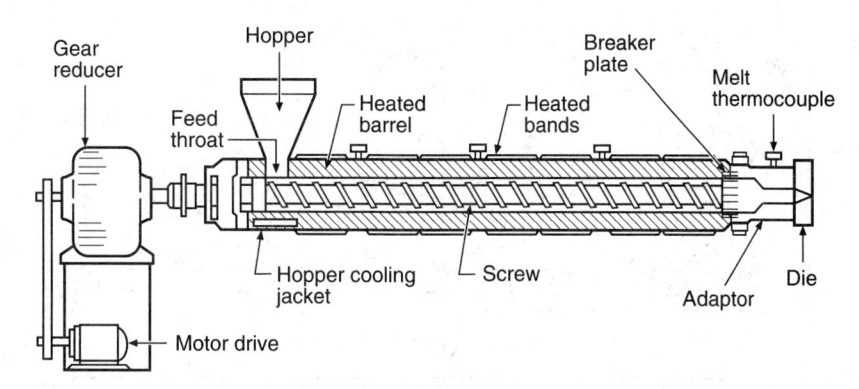

FIG. 3-18E.   In a basic single-screw extruder, plastics pellets (or powders) are fed into the hopper, through the feed throat, and into a screw that rotates in a heated barrel. The rotation of the screw (which is powered by the drive motor) conveys the plastic forward for melting and delivery through the breaker plate (reduces the rotary motion of the melt), through the adapter, and into the die, which dictates the shape and size of the final extrudate. (Source: The Society of the Plastics Industry, Inc.)

Rotational molding: A process used to make hollow, one-piece parts. Finely powdered plastic or molding granules are placed in a cavity that is rotated about two axes to distribute the plastic. While rotating, the mold is heated to melt the plastic and then cooled to solidify the part.

Thermoforming: Thermoplastic sheeting is heated to its softening temperature and forced against the contours of a mold by mechanical means, e.g., tools, plugs, solid molds, etc., or by pneumatic means, e.g., differentials in air pressure created by pulling a vacuum between sheet and mold (vacuum forming) or by using the pressures of compressed air to force the sheet against the mold (pressure forming). (See Figure 3-18I.)

Transfer molding: Most generally used for thermosetting plastics, this is a process similar to compression molding. The molding material is placed in a pot at the top of a closed mold and a plunger

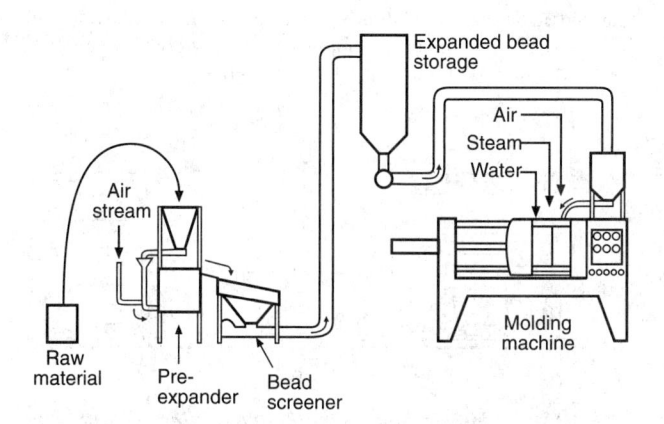

FIG. 3-18F.   Among the many variations in molding foamed plastics is this setup for steam-chest molding expandable styrene beads into products such as foam cups, novelties, building products, etc. In this operation, the expandable beads, containing a blowing agent, are pre-expanded with steam, then screened to remove large clumps. The expanded beads are next blown into a storage hopper and allowed to dry and stabilize. From here, they feed into the final mold where steam is again used to complete expansion of the beads so that they fill the mold and fuse together. Water is used for cooling, prior to opening the mold and removing the finished foamed styrene part. (Source: The Society of the Plastics Industry, Inc.)

is used to force the material into the gates, runners, and cavities of the mold. Heat is used to cure the material, and the part is then removed.

THE FIRE HAZARDS

Plants converting feedstocks into finished products are subject to a variety of hazards that can result in explosions and fire. The broad area of hazard involves the presence of combustible dusts, flammable and combustible liquids, high-heat elements, hydraulic and heat transfer fluids, static electricity, and failure to observe good storage and housekeeping practices.

Dusts

Although when in solid massive form many plastics can be difficult to ignite and will not continue to burn on removal of exterior fuel, nearly all will burn rapidly in the form of dust, and if dispersed in air can be explosively ignited by a spark, flame, or metal surface above 700°F (371°C). Dust explosions should be considered possible when operations use pulverized plastic, convey larger granules through pneumatic conveying systems, or produce dust by machining or sanding in finishing work. Additives, such as wood flour or finely ground dye stuffs, also require safeguarding when being added to plastics.

Plastic pellets for injection or extrusion molding are commonly known as molding powder. In reality they are not pulverized material but cubes or cylindrical pellets. They are usually screened to remove finer particles to permit more uniform feeding to machines, and they are free from hazard of dust explosion. However, dust can be generated by abrasion of these particles when conveyed in a long pneumatic system.

Trimmings from injection molding are cut to small size for reuse alone or for blending with fresh (virgin) molding pellets. This cutting is called "regrinding," although it is a shearing and impacting action deliberately intended to avoid making dust and fine particles. Regrinding usually generates some fine powder, and dust hazards should be considered when much regrinding is done.

Compounding of resins with such additives as dyes, pigments, fillers, mold-release or flow-improving lubricants, plasticizers for flexibility, ultraviolet or heat stabilizers, or modifying resins is also a source of dust hazards. For rapid mixing, most of these are charged to mixing equipment as fine powders, and dust explosions are possible with any of these ingredients. Some compounding, especially for

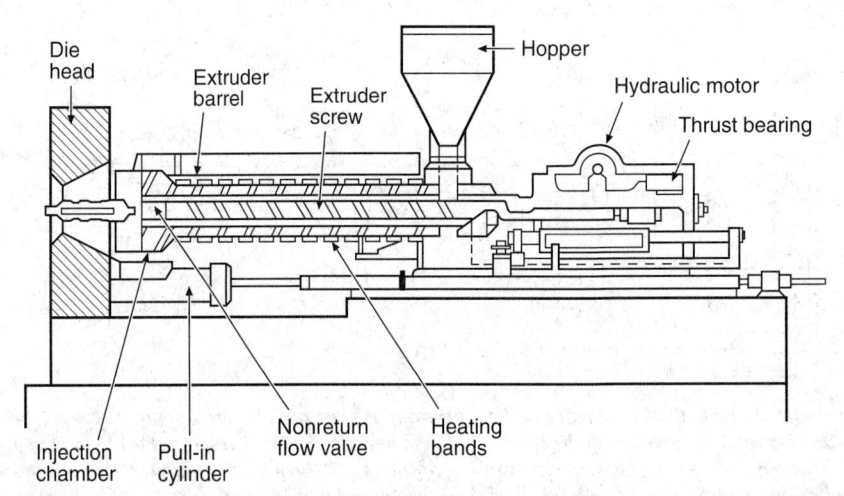

FIG. 3-18G. *Diagram of injection molding machine (reciprocating screw). Plastics pellets feed through the hopper into the screw (much like the screw in an extruder) where they are compacted, melted, and pumped by the rotation of the screw past the nonreturn flow valve to the front of the screw where they are allowed to accumulate. At the proper time, the rotation of the screw is stopped and the amount of molten plastic in front of the screw is injected into the mold, using the screw as a plunger activated by the hydraulic injection cylinders. In the mold, the molten plastic flows throughout the cavity, completely filling it. The plastic is then allowed to cool and harden, the mold is opened, and the finished part removed. The back end of the machine shown in this figure contains the motors and drives needed to power the machine. (Source: The Society of the Plastics Industry, Inc.)*

reclaiming of once-processed materials, is done at molding machines in plastics plants of all types.

Incorporation of wood flour, cotton flock, or other combustible fillers can increase the explosibility of dust. Incorporation of low percentages of fire retardants has but little effect on explosibility of dusts.

It is important that the escape and dispersion of dust into the atmosphere of a converting plant be kept to a minimum. Equally important is that provisions are made to reduce the possibility of ignition, relieve explosion pressure, and confine and control fire.

Guidance in controlling dust explosion hazards in plastics plants is found in NFPA 654, *Standard for the Prevention of Fire and Dust Explosions in the Chemical, Dye, Pharmaceutical, and Plastics Industries.* Other applicable standards are NFPA 68, *Guide for Venting of Deflagrations;* NFPA 69, *Standard on Explosion Prevention Systems;* and NFPA 650, *Standard for Pneumatic Conveying Systems for Handling Combustible Materials.* NFPA 650 is useful as a guide for evaluating pneumatic conveying systems.

Flammable Solvents

Flammable organic solvents are found in nearly every plastics plant. They may be used in very small quantities to apply adhesives, lacquers, or paints to molded or fabricated items; in large amounts to

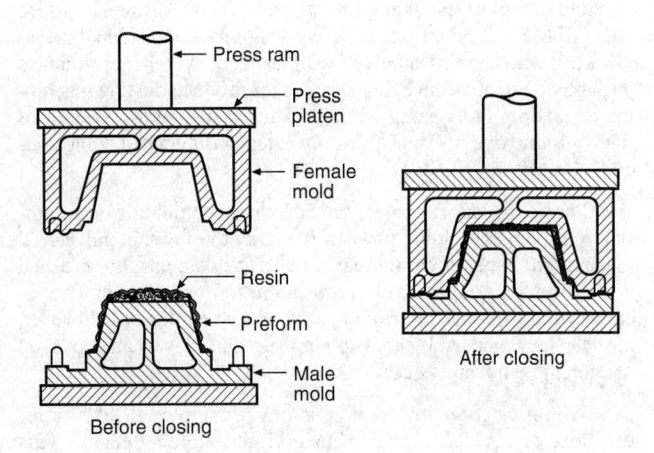

FIG. 3-18H. *Matched-die molding is one technique for producing reinforced plastic parts. Basically, it is a compression molding process in which resin and glass fibers are shaped into the finished product under heat and pressure between the two halves of a mold (male and female halves). The glass-fiber reinforcements are laid over the male mold in the form of a "preform," a combination of glass and resin preformed before molding to the basic shape of the part to be molded. Additional liquid resin mix is added before mold is closed. (Source: The Society of the Plastics Industry, Inc.)*

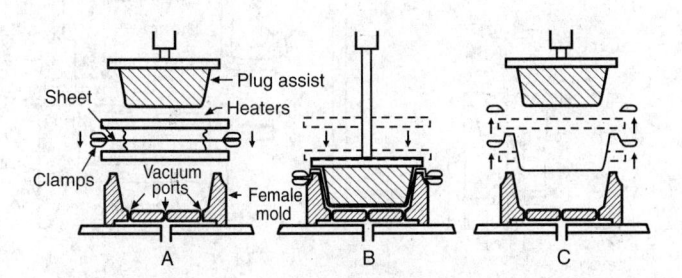

FIG. 3-18I. *This variation on the thermoforming of plastic sheet is known as plug-assist vacuum forming. In operation, the plastic sheet is clamped in place and heaters move in to heat the sheet top and bottom to soften it (A). Heaters are then withdrawn, and the frame holding the sheet is lowered down to contact the mold. At this point, the plug-assist is lowered into the softened sheet, stretching it down to the bottom of the mold cavity (B). After the plug-assist has reached its closed position, a vacuum is drawn through the ports to pull the stretched sheet completely into the cavity and finish the forming. Next, the plug-assist is withdrawn, the formed sheet is cooled, and the clamps are opened to remove the formed part from the frame (C). (Source: The Society of the Plastics Industry, Inc.)*

coat plastic on cloth, paper, leather, or metal; or on metal belts from which a dried film will be stripped. There may be increased hazard when solvents are applied to plastics, particularly when printing or coating on fast-moving films, because plastics usually have high electrical resistivity; they generate and retain static charges more readily than paper or cotton fabric.

Small plants are increasing their use of solvents for the preparation of both rigid and flexible foam plastic. The plastic is moistened with solvent and heated above the boiling point of the solvent in a closed mold or extruder. Upon release of pressure, the boiling solvent expands the resin and produces a bubbled structure. Another type of foaming process heats a resin containing a chemical additive that evolves gas, usually nitrogen, carbon dioxide, or steam. Either the solvent or the chemical agent is called a blowing agent. The most common rigid foam is made from beads of polystyrene moistened with about 10 percent pentane or a similar hydrocarbon. The pentane does not soften and expand the resin at room temperature, but does so at the low-pressure steam temperature used in molding to shape, or in extruding as sheet. It is imperative that sources of electric spark or flame be avoided. Resin is usually shipped from the manufacturer with the hydrocarbon blowing agent already mixed with it. Containers will have free vapor of pentane, and should be opened outdoors or with good exhaust ventilation. Expanded articles should be aged under forced hot air to remove nearly all flammable blowing agent before shipping.

Flexible urethane foams for upholstery and garments, or rigid foams for pour-in-place construction or refrigeration insulation, are blown by steam and carbon dioxide generation, which are by-products of the formation of the urethane resin from basic ingredients. Urethane foams of lower density may be made with a low-boiling hydrocarbon in the resin mixture to increase the expansion.

Foams of polyvinyl chloride are used in garments and upholstery. These are usually called "expanded vinyl," because the extent of foaming is limited to provide an essentially continuous outer surface. These are usually made with the chemical blowing agent azodicarbonamide, also called azobisformamide. Above 300°F (149°C) it provides nitrogen gas at a controlled rate, is essentially nonhazardous in the proportions used, and does not give off flammable vapors. Drums of the reagent should be cooled with water when exposed to fire.

Improper handling of flammable liquids has caused serious fires in plastics plants. Failure to recognize the importance of static spark prevention, explosionproof electrical equipment, and vapor removal systems has been the most frequent cause of flammable liquid fires. NFPA Standards 30, 33, 35, 70, 77, and 91 (see bibliography for titles) contain further information on safeguards to observe in the handling and use of flammable liquids as they apply, or may apply, to the plastics industry.

Heating Elements

Molding and extrusion operations, for both shaping and compounding, have hazards associated with local overheating of electrical components. Operating temperatures normally range from 300° to 650°F (149° to 343°C), depending on the plastic being processed. The upper temperature range is beyond that practical for heat transfer fluids, so electric resistance heating is almost universally employed. Heater bands are required to fuse the resin in the feed section at the upstream end of the extruder barrel. It is not uncommon for controllers to stick, permitting resistance heaters to exceed the temperature set for the thermocouple controller. In most cases, the character of extrudate will markedly change—and thus warn the operator—well before heater bands get hot enough to be a source of ignition.

Some areas within equipment may not be regularly purged by flow of the plastic feedstock. Material remaining in such areas can be subject to exclusively high temperature or be kept too long at a normally acceptable temperature. Decomposition may then take place, resulting in the release of gases that may be combustible. It is good practice to start heating such equipment first at the downstream end to ensure fluidity of material and hence relief of pressure.

Cleanliness in molding and extruding areas is vital to reduce the hazard of ignition from overheated bands where flammable vapors may be generated.

Electrical wiring in heat processing equipment on plastics machines should be installed in accordance with applicable provisions of NFPA 70, *National Electrical Code®*.

Static Electricity

Many operations in plastics plants generate static electricity. Because plastics are such good electrical insulators, static electricity on them can rapidly build up to spark discharge—a hazardous condition if dust or flammable vapors are present. Operations that can generate static are stripping of films from production or printing equipment, or rapid passage of films across rolls or guides. Belts for power transmission also are a significant source of static discharge. Because of their low water-absorption and high resistivity, plastics cannot have their static charge dissipated by high ambient humidity as practiced in cotton, wool, and paper mills. Attention should be given to grounding of equipment and ensuring that tinsel conductors firmly contact moving films or filaments. Care should also be given to separating vapor and dust hazards from machines where static electricity ignition sources could develop. NFPA 77, *Recommended Practice on Static Electricity*, is a good source for the safeguards to use in protecting against static electricity hazards.

Hydraulic Pressure Systems

Hydraulic systems are used to clamp molds and to provide pressure to rams or screws that force molten plastic into molds by compression, transfer, or injection molding. The molten plastic may be at pressures up to 20,000 psi (138,000 kPa), but the hydraulic systems are normally less than 2,000 psi (13,800 kPa). Petroleum fluids have been used in plastics operations where heating elements are generally below 600°F (315°C). (The same fluids have a poor record in die-casting metals parts because molten metal is at a much higher temperature.)

Fire-resistant water-glycol or water-oil emulsions or "synthetic" fluids are available for hydraulic systems. Substituting fire-resistant fluids for petroleum fluids should follow the recommendations of the manufacturers of hydraulic system components.

Storage Arrangements

The fire hazards of plastics in storage, whether as feedstocks for the conversion process or as finished articles, are determined by their chemical composition, physical form, and storage arrangement. The physical form may be foam, solid sheet, pellets, flakes, random-packed small objects, bags, or cartons. The storage of plastics generally should not exceed a maximum height of approximately 20 ft (6.1 m). The hazard of a particular plastic in any form of storage arrangement is the same whether it is encapsulated or nonencapsulated. If fire occurs, large quantities of smoke are usually generated, making manual fire fighting difficult and venting desirable in building construction consideration.

Plastics, such as fluorocarbons, unplasticized polyvinyl chloride, and phenolics, can be protected the same way as any Class III commodity (wood, paper, and natural fiber cloth), regardless of their physical form or storage arrangement. Pellets and small objects can be protected the same as Class IV commodities. (Commodity classifications are based on NFPA 231, *Standard for General Storage*.)

Thermoplastics that present a fire hazard include polystyrene and acrylonitrile-butadiene-styrene (ABS), as well as polyurethane, polyethylene, plasticized polyvinyl chloride, and thermosets such as polyesters, although all of these to a lesser extent. These plastic materials will melt and break down (depolymerize) into their monomers, acting and burning like flammable liquids. High sprinkler discharge densities over relatively large areas are necessary to protect these types of plastics. In the form of foamed material, these plastics present the most severe fire hazard.

One-story buildings without basements are preferable for storage of plastics materials because of greater efficiency for fire fighting, ventilation, and salvage operations. Plastics should be stored, handled, and piled according to their fire characteristics.

Housekeeping Practices

Housekeeping is basic to good fire safety. Good housekeeping practices reduce the danger of fire simply because they are time-proven methods of controlling the presence of unwanted fuels, obstructions, and sources of ignition. Approved containers should be provided and properly maintained for the disposal of refuse and rubbish. Spills of flammable liquids or combustible materials should be promptly cleaned up and properly disposed. Removal of combustible dust and lint accumulations from walls, ceilings, and exposed structural members is necessary. Smoking control, in addition to sensible regulations, also requires receptacles for spent smoking materials. Areas containing flammable liquids, vapors, or combustible plastics materials, as well as areas used for processing, manufacturing, or storage, must be clearly identified to prohibit smoking or the use of open-flame devices.

THE SAFEGUARDS

Good fire protection starts with the fire safe design of the plant or warehouse, or inspection and modification of the existing facilities. Sprinkler-protected, noncombustible construction is appropriate for buildings occupied for storage, processing, and manufacturing of combustibles such as those involved in the plastics industry. Automatic sprinklers, standpipe and hose systems, and water-type portable extinguishers should be supplemented by fire extinguishers and special automatic systems suitable for flammable liquid fires and electrical fires, where these hazards exist.

Consideration should be given to the provision of roof vents, particularly in large one-story warehouses of manufacturing plants.

Building Construction

Long, narrow buildings provide greater ease in protection and fire fighting than large, square buildings. One-story buildings without basements are preferable to multistory buildings which may be subject to the spread of fire from lower to upper floors.

Large properties are best subdivided into fire areas to limit the spread of fire. Storage and manufacturing areas particularly need to be separated from each other by walls with sufficient fire rating to protect each area and occupancy from the other in case of fire. Preferably, fire walls should be without openings, but if openings are necessary, protection can be provided by self-closing or automatic fire doors suitable for openings in fire walls. Generally, a single fire area should not exceed 50,000 sq ft (4,645 m²).

Fire Control Systems

An alarm system that alerts building occupants, notifies fire suppression departments, and activates automatic suppression equipment is a necessary protection feature. Rapid extinguishment may be achieved by providing plastics processing equipment, conveyors, and manufacturing machinery subject to ignition or explosion with automatic fire-, smoke-, or explosion-detecting devices to initiate an alarm and to activate automatic suppressing systems (water spray, foam, dry chemical, carbon dioxide, or halogenated extinguishing agents).

Sprinklers are the most important single system for automatic control of fires in plastics plants. Among the advantages of automatic sprinklers is the fact that they operate directly over a fire and that smoke, toxic gases, and reduced visibility do not affect their operation. Automatic sprinklers, standpipes, and fire hose connections depend upon an adequate water supply delivered with the necessary pressure to control fires.

BIBLIOGRAPHY

References

Cote, A. E., ed., *Industrial Fire Hazards Handbook*, 3rd ed., National Fire Protection Association, Quincy, MA, 1990.
1988 Facts and Figures of the Plastics Industry, The Society of the Plastics Industry, Inc., Washington, DC, 1988.
The Story of the Plastics Industry, 13th ed., The Society of the Plastics Industry, Inc., Washington, DC, 1971.

NFPA Codes, Standards, and Recommended Practices

Reference to the following NFPA codes, standards, and recommended practices will provide further information on the safeguards for extrusion and molding processes discussed in this chapter. (See the latest version of *The NFPA Catalog* for availability of current editions of the following documents.)

NFPA 10, *Standard for Portable Fire Extinguishers*
NFPA 11, *Standard for Low-Expansion Foam*
NFPA 11A, *Standard for Medium- and High-Expansion Foam Systems*
NFPA 12, *Standard on Carbon Dioxide Extinguishing Systems*
NFPA 12A, *Standard on Halon 1301 Fire Extinguishing Systems*
NFPA 13, *Standard for the Installation of Sprinkler Systems*
NFPA 14, *Standard for the Installation of Standpipe and Hose Systems*
NFPA 15, *Standard for Water Spray Fixed Systems for Fire Protection*
NFPA 16, *Standard for the Installation of Deluge Foam-Water Sprinkler and Foam-Water Spray Systems*
NFPA 17, *Standard for Dry Chemical Extinguishing Systems*
NFPA 30, *Flammable and Combustible Liquids Code*
NFPA 33, *Standard for Spray Application Using Flammable or Combustible Materials*
NFPA 35, *Standard for the Manufacture of Organic Coatings*
NFPA 40E, *Code for the Storage of Pyroxylin Plastic*
NFPA 68, *Guide for Venting of Deflagrations*
NFPA 69, *Standard on Explosion Prevention Systems*
NFPA 70, *National Electrical Code®*
NFPA 77, *Recommended Practice on Static Electricity*
NFPA 91, *Standard for Exhaust Systems for Air Conveying of Materials*
NFPA 101®, *Life Safety Code®*
NFPA 231, *Standard for General Storage*
NFPA 650, *Standard for the Pneumatic Conveying Systems for Handling Combustible Materials*
NFPA 654, *Standard for the Prevention of Fire and Dust Explosions in the Chemical, Dye, Pharmaceutical, and Plastics Industries*

Additional Reading

Babrauskas, V., "Plastics. Part B. The Effects of FR Agents on Polymer Performance," Chapter 12, *Heat Release in Fires*, Elsevier Applied Science, New York, 1992, pp. 423–446.

Bresle, A., "The Hazards of PVC—A Second View," *Fire Prevention*, No. 217, March 1989, pp. 19-23.

Brown, J. E., "Plastics. Part C. Composite Materials," Chapter 12, *Heat Release in Fires*, Elsevier Applied Science, New York, 1992, pp. 447–459.

Crawford, R. J., *Plastics Engineering*, 2nd ed., Pergamon Press, New York, 1987.

Hirschler, M. M., "Heat Release from Plastic Materials," First U. S. Symposium for Heat Release and Fire Hazard, Abstracts, Interscience Communications Limited, 1991, pp. 3–4.

Johnson, D. G., "Combustion Properties of Plastics," *Journal of Applied Fire Science*, Vol. 4, No. 3, 1994/1995, pp. 185–201.

Takahashi, S., "Extinquishment of Plastics Fires with Plain Water and Wet Water," *Fire Safety Journal*, Vol. 22, No. 2, 1994, pp. 169–179.

CHEMICAL PROCESSING EQUIPMENT

Revised by Richard F. Schwab

This chapter discusses chemical reactions, the means of reaction control, and the equipment used to produce chemicals and synthetics, such as fibers, plastics, and some drugs. Operations to chemically change the properties of materials, as in the chemical pulping of wood and in the tanning of leather, are also covered. However, in discussing fire and explosion prevention and loss control, the hazards of each system must be considered as a whole when appropriate separation distances and other means of minimizing damage are examined.

Additional relevant information can be found in the following chapters: Section 1, Chapter 6, "Explosions"; Section 3, Chapter 7, "Heat Transfer Systems (Nonwater Media)"; Section 3, Chapter 14, "Welding and Cutting"; Section 3, Chapter 31, "Waste Handling and Control"; Section 3, Chapter 32, "Hazardous Waste Control"; and Section 4, Chapter 14, "Explosion Prevention and Protection."

PLANT SITING

Loss Prevention in the Process Industries[1] notes that safety is a prime consideration in plant siting. The most important feature is distance between the site and areas of high population density. Distance tends to reduce casualties; i.e., if a toxic release occurs, the effect of distance reduces the concentration of gas and buys time to permit evacuation. The disaster at Bhopal, India, in December 1984 was undoubtedly exacerbated by the very high population density immediately adjacent to the plant. If the hazard is fire or explosion, distance will reduce the intensity of heat or blast overpressure as well as damage done by missiles. The density of population adjacent to the liquefied petroleum gas storage tanks involved in a fire and explosion in Mexico City, Mexico, in September 1984 led to very high loss of life. These two incidents resulted in 3,000 to 5,000 fatalities and rank among the worst industrial accidents in history.[2]

A plant with serious toxic gas release potential or major fire or explosion hazards must be designed to high standards of safety and loss prevention. Distance and plant siting cannot be relied upon to prevent disasters but must be considered as key features in limiting the potential magnitude of an incident.[3]

Richard F. Schwab, P. E., S.F.P.E., AIChE, is manager of process safety and loss prevention, AlliedSignal Inc., Morristown, NJ. Mr. Schwab is Chairman of NFPA's Technical Committee on Electrical Equipment in Chemical Atmospheres.

Separation Distances

Single or allied processes are usually located on individual blocks of land surrounded by access roads. The theory behind the block approach is that fires or unignited spills can then be attacked from all directions. This approach provides maximum latitude to take advantage of both wind direction and the shelter afforded by peripheral structures in the block. If the right of way for these roads is 50 ft (15.25 m) wide, a minimum distance of 50 ft (15.25 m) between structures on adjacent blocks is ensured. This distance has proved adequate to prevent the spread of fires through highly protected properties.

Chemical manufacturing plants pose a special danger of explosion, especially in a large-loss incident. During 1971 to 1988, for example, there were 15 chemical plant fires each involving $5 million in damage reported to NFPA. These 15 incidents collectively caused 20 deaths and $418.9 million in direct property damage. Nine of the 15 incidents (62 percent of the dollar loss, 90 percent of deaths) began with explosions, another two incidents (26 percent of dollar loss, 10 percent of deaths) had explosions after fire began, a twelfth had explosions that may have preceded or followed the fire (5 percent of dollar loss), and two had reports that did not indicate whether explosions occurred (4 percent of dollar loss). Only one of the 15 (3 percent of dollar loss) definitely did not involve an explosion.

Where an explosion is possible, the 50-ft (15.25-m) distance does not provide sufficient protection. One solution has been to place any unit involving an explosion hazard at least 50 ft (15.25 m) away from the block perimeter. This gives a minimum distance (in the adjacent blocks) of 100 ft (30.5 m) between explosion hazards. In the case of equipment explosions resulting from an increase in internal pressure, the shock wave would probably be adequately attenuated by that distance. This is true whether the explosion is of the thermal type, represented by an ordinary runaway reaction and failure of the container, or of the shock type, where the pressure release is caused by a reaction at supersonic speed. In the latter case, the rapid pressure increase could produce shrapnel, so a containment barricade would be needed. If the material that might cause the explosion is a true explosive, whether formed intentionally or accidentally, and whether inside or outside of the equipment, the spacing should be in accordance with the American Table of Distances for Storage of Explosives. (See also NFPA 495, *Explosive Materials Code*.)

Conservative views on spacing are taken by the American Oil Company. Its *Engineering for Safe Operations* states: "Generally, a distance of 250 ft (76.2 m) between units or between tankage has been used as a desirable spacing."[4] Also, Figures 3-19A (1), (2), and (3) give the recommendations of the Industrial Risk Insurers (IRI), with respect to petrochemical plants. Their booklet also suggests distances for separation of units in refineries, gasoline plants, terminals, oil pump stations, and offshore property.[5]

The greater distance suggested by the American Oil Company and IRI are justified by the large amounts of flammables usually present in equipment and piping in oil refineries and petrochemical plants. They are probably adequate for controlling even large flammable liquid fires.

Another method for plant spacing was presented using *The Mond Fire, Explosion, and Toxicity Index*.[6] This resulted from an official inquiry into the Flixborough disaster in 1974. It requires a ranking of hazards of various units at an early stage of design. The rankings are then used to lay out a plant to minimize loss potential. The method has undergone further study, and a revision was suggested in 1989.[7]

For another type of explosion hazard, no spacing standards exist. This is the type of explosion that results when a flammable gas or super-heated flammable liquid is released into the atmosphere and mixes with air to form a cloud with a volume in the explosive range. Such a cloud can release 2 to 10 percent of its heat of combustion in the form of explosive energy. In estimating potential losses from vapor clouds, IRI uses the 2 percent figure. This decision was reached after studying the results of 79 cases of vapor cloud formation.[8] In 7 cases, there was no ignition; in 17 there was ignition but no pressure was generated.

If a hydrocarbon releases 20,000 Btu per lb (46.5 mJ/kg) when burned and 2 percent of that, or 400 Btu per lb (930 kJ/kg) were released explosively by ignition of its mixture with air, then 5 lb (2.3 kg) of hydrocarbon would release 2,000 Btu per lb (4.65 mJ/kg) explosively—the same as 1 lb (0.45 kg) of TNT. Where such releases are possible, spacing is not a reliable defense against damage, since no one can predict where the cloud will drift before it is ignited.

Inter-unit spacing requirements (ft). `/` = no spacing requirements; 1 ft = 0.305 m

Columns (left to right):
1 = Service buildings
2 = Motor control centers and electrical substations
3 = Utilities areas
4 = Cooling towers
5 = Control rooms
6 = Compressor buildings
7 = Large pump houses
8 = Process units moderate hazard
9 = Process units intermediate hazard
10 = Process units high hazard
11 = Atmospheric storage tanks
12 = Pressure storage tanks
13 = Refrigerated storage tanks dome roof
14 = Flares
15 = Unloading and loading racks
16 = Fire water pumps

Row item	1	2	3	4	5	6	7	8	9	10	11	12	13	14	15	16
Motor control centers and electrical substations	/															
Utilities areas	/	/														
Cooling towers	50	50	/													
Control rooms	50	50	100	50												
Compressor buildings	/	/	100	100	/											
Large pump houses	100	100	100	100	100	30										
Process units moderate hazard	100	100	100	100	100	30	30									
Process units intermediate hazard	100	100	100	100	100	30	30	50								
Process units high hazard	200	100	100	100	200	50	50	100	100							
Atmospheric storage tanks	400	200	200	200	300	100	100	200	200	200						
Pressure storage tanks	250	250	250	250	250	250	250	250	300	350						
Refrigerated storage tanks dome roof	350	350	350	350	350	350	350	350	350	350						
Flares	350	350	350	350	350	350	250	250	350	350						
Unloading and loading racks	300	300	300	300	300	300	300	300	300	300	300	400	400	/		
Fire water pumps	200	200	200	200	200	200	200	200	200	300	250	250	350	500	50	
Fire stations	50	50	50	50	50	200	200	200	300	300	350	350	350	300	200	/
Fire stations	50	50	50	50	50	200	200	200	300	300	350	350	350	300	200	/

1 ft = 0.305 m
/ = no spacing requirements

FIG. 3-19A(1). Inter-unit spacing requirements for oil and chemical plants.

FIG. 3-19A(2). Intra-unit spacing requirements for oil and chemical plants.

Column headers (diagonal): Compressors, Intermediate hazard pumps, High hazard pumps, High hazard reactors, Intermediate hazard reactors, Moderate hazard reactors, Columns, accumulators, drums, Rundown tanks, Fired heaters, Air cooled heat exchanger, Heat exchangers, Pipe racks, Emergency controls, Unit block valves, Analyzer rooms.

Compr	Int pumps	High pumps	High react	Int react	Mod react	Columns	Rundown	Fired	Air cooled	Heat exch	Pipe racks	Emerg	Unit valves	Analyzer
30														
30	5													
50	5	5												
50	10	15	25											
50	10	15	25	15										
50	10	15	25	15	15									
50	10	15	50	25	25	15								
100	100	100	100	100	100	100	100							
50	50	50	50	50	50	50	100	25						
30	15	15	25	15	15	15	100	50	/					
30	10	15	25	15	10	10	100	50	15	5				
30	10	15	25	15	10	10	100	50	/	10	/			
50	50	50	100	50	50	50	100	50	50	50	50	/		
50	50	50	100	50	50	50	100	50	50	50	50	/	/	
50	50	50	50	50	50	50	100	50	50	50	50	/	/	/

1 ft = 0.305 m
/ = no spacing requirements

Procedures used to estimate damage from explosion pressure resulting from ignition of a flammable vapor cloud are:[9,10]

1. Estimate the volume of the vapor produced by the release.
2. Assume that a percentage of the heat of combustion of this volume will be released as explosion energy.
3. Equate this energy to an equivalent amount of TNT.
4. Estimate the damage, based on the explosion of that amount of TNT.

This is discussed briefly by Davenport.[8,11] More extensive comments, with a table of pressure ranges at which different units of equipment may be expected to fail, have been made by Nelson.[12]

Van Wingerden et al.[9] extensively discuss the problems in evaluating vapor cloud explosions.

Experimental work has been performed to develop an improved method to predict the blast effect from vapor clouds.[13] The method recognizes the importance of partial confinement, with regard to the strength of vapor cloud explosions. The experiments were made to quantify the influence of obstacles and partial confinement.[9]

Other Means of Loss Control

As chemical plants process larger quantities of materials, it becomes impractical to provide ever-increasing separation of units. Where toxic, flammable, reactive, or otherwise hazardous materials may be spilled, the logical approach is to:

1. Minimize the possibility of uncontrolled spills.
2. Minimize the size of a possible spill. (If spills do occur, the design features should keep them small.)
3. Minimize the spread of a possible spill. (If spills are large in volume, design features should keep them confined.)
4. Prepare alarm and evacuation plans if toxic releases can occur.
5. Control sources of ignition.
6. Provide protection for exposed property if ignition does occur.

A number of means are available to control spills. One is the obvious technique of providing diking or curbing to restrict the spread of a liquid spill.

Another is the use of foam to cover the liquid spill so as to restrict the vaporization of the liquid, carefully considering the physical and chemical properties of the liquid to make sure that the foam is compatible.

FIG. 3-19A(3).   *Storage tank spacing requirements for oil and chemical plants.*

Column headers (diagonal, left to right):
1. Floating and cone roof tanks <3000 barrels
2. Floating and cone roof tanks >3000 <10,000 barrels
3. Floating roof tanks >10,000 <300,000 barrels
4. Jumbo floating roof tanks >300,000 barrels
5. Cone roof tanks class II, III product >10,000 <300,000 barrels
6. Cone roof tanks inerted class I product >10,000 <150,000 barrels
7. Pressure storage vessels spheres and spheroids
8. Pressure storage vessels drums and bullets
9. Refrigerated dome roof storage tanks

1	2	3	4	5	6	7	8	9
0.5 D*								
0.5 D	0.5 D							
1 x D	1 x D	1 x D						
1 x D	1 x D	1 x D	1 x D					
0.5 D	0.5 D	1 x D	1 x D	0.5 D				
1 x D	1 x D	1 x D	1 x D	1 x D	1 x D			
1.5 D 100' min	1.5 D 100' min	1.5 D 100' min	2 x D	1.5 D 100' min	1.5 D 100' min	1 x D 50' min		
1.5 D 100' min	1.5 D 100' min	1.5 D 100' min	2 x D	1.5 D 100' min	1.5 D 100' min	1 x D 100' min	1 x D	
2 x D 200' min	2 x D 200' min	2 x D 200' min	2 x D	2 x D 200' min	2 x D 200' min	1 x D 100' min	1 x D 100' min	1 x D 100' min

D = Largest tank diameter
1 barrel = 42 gallons = 159 L
°C = (°F-32) x 0.555
1 ft = 0.305 m

* For Class II, III products, 5 ft spacing is acceptable
** Or Class II or III operating at temperatures > 200 F

Another is the use of water spray, using either a fixed sprinkler or water spray system with monitor nozzles (in some cases remotely controlled) to wash out the vapor from the atmosphere. This method is appropriate for vapors that are water soluble and also for liquids that can be expected to be released as a mist. It also requires a large water supply at high pressure to be immediately available to respond to the emergency. In many cases, fire water systems have been tapped, since they are usually available at the required pressure and amount.

Steam fogging systems have also been used to help dissipate vapor releases in plants where a large steam supply is always immediately available for this purpose.

In all cases, drainage systems are required and must be designed to accommodate the expected water or foam discharge along with the expected spill.

In 1988 the AIChE published *Guidelines for Vapor Release Mitigation.*[14] This reference provides valuable information that can mitigate situations that can cause major fire, health, and environmental problems.

EXPOSURE PROTECTION

Automatic fire control and extinguishing systems are the first line of defense against fire and explosion emergencies. However, if an explosion hazard exists, system design would require barricades for the protection of important components, such as deluge valves and dry pipe valves.[15] If they might be exposed to explosion pressure, however improbably, process control houses, including areas where people congregate (e.g., change houses), should be of explosion-resistant design.[16-20] The procedure offered by Bradford and Culbertson[16] is based on resistance to static pressure; that in the Chemical Manufacturers Association's (CMA) Safety Guide SG-22[17] is based on elastic response (dynamic response) to a pressure impulse. Lawrence and Johnson[19] give a concise exposition of the design theory.

Reliably available water at pressure and volume sufficient to supply the maximum foreseeable fire-fighting and exposure protection demand for a minimum of four hours is desirable. Of course, built-in or added fire resistance can give excellent exposure protection.

Protection for a chemical plant generally requires a specialized fire department where each fire fighter is equipped with full protective clothing, including self-contained breathing apparatus (SCBA), as well as acid-resistant hose and special extinguishing equipment where available. NFPA 49, *Hazardous Chemicals Data,* is an excellent source for information on personal protective measures needed to fight fires involving different chemicals, as well as the tactics that should be used in emergencies. Appropriate symbols from NFPA 704, *Standard System for the Identification of the Hazards of Materials for Emergency Response,* posted at locations on the property where chemicals are present, can give helpful guidance on inherent hazards of materials present and on planning effective fire-fighting operations.

IGNITION SOURCES

Figure 3-19A(2) recommends that open flames be kept at least 50 ft (30.5 m) from any hazardous area. This contemplates fixed equipment; however, it could also apply to maintenance work. A better approach is to prohibit open flames or other hotwork in all but predetermined specified areas, except where the hotwork is to be done under a permit that limits scope and duration. It must be remembered that piping for high-pressure steam, heat-transfer media, and hot process streams may need insulation to keep it from becoming a source of ignition. When piping operating at high temperatures is insulated, it may be necessary to leave flanges uninsulated. If the bolts joining flanges get too hot, their thermal expansion will relax the compression on the joints and may lead to leakage.

CONTROL OF SPILLS

Large flammable liquid spills should be confined where they may occur. Although diking is often used, it may not prevent liquid around the spill source from burning. Equipment may be damaged further or the spill aggravated unless efficient, prompt, and preferably automatic fire control measures are taken. A better approach is to provide drainage to an impounding basin located where a fire, if not extinguished, can burn out harmlessly or a nonburning toxic material can be appropriately neutralized. A spill can be ignited some time during the drainage process, so care must be taken to avoid damage during drainage. If the sewerage system is underground and is not provided with liquid seals at all entrance points, explosions in the system could result. Fire in open drainage ditches could damage anything without adequate protection that is located alongside the ditches or that crosses them, such as utility or piping systems. Problems associated with such drainage can be avoided by using the trench system described in NFPA 15, *Standard for Water Spray Fixed Systems for Fire Protection.*

Total confinement of a large flammable gas or vapor spill is not possible, due to the gaseous nature of the spill. Hose lines or monitors with spray nozzles can be used, because of the air entrained by the spray, to dilute the gas cloud and to move it in a desired direction until it dissipates. Fixed water spray nozzles may be used to set up water curtains, which prevent a cloud from drifting to a furnace or other ignition source. Units that have been effective in experiments with small, low-level vapor releases were discussed in a number of articles in *Loss Prevention.*[18,21] The use of steam nozzles to dilute and direct vapors upward at an installation has been reported.[22] There are problems with this approach. Large steam capacity must always be available; and there is need for meticulous grounding of all metallic material that may be charged by steam passing over it or condensing on it, in order to avoid static sparks that might ignite the material being dispersed.

Keeping Spills Small

Hazardous spills can be kept small by keeping the amount of hazardous material used in the process small. Where this is impractical, valves to isolate all large quantities of material should be provided. These valves should be installed at each outlet of any large container through which material might escape (except at relief devices). They can also be used to subdivide long runs of large-diameter pipe. The valves should be firesafe, i.e., they should not leak appreciably when exposed to fire after being closed, and they should be arranged to operate both remotely and at the valve.

If operated by electric motor, both the power wiring to the motor and the signal wiring to the starting switch should be arranged to remain functional during any fire or other emergency requiring valve operation, for at least as long as it might take to discover the emergency, transmit the alarm, and operate the valve. A minimum of 15 minutes for this fire protection sequence is suggested. If valves are held open by instrument air pressure and closed by spring or gravity when air pressure is lost, plastic tubing can be used in the air line, at least near the valve. The plastic will melt quickly in a fire, and the valve will close before it is damaged. Since it is necessary to detect spills before ignition, flammable vapor detectors of the diffusion-head type may be used for that purpose.[23]

The distinction between a small spill and a large one must be fixed individually for each plant. A small spill is one that can be easily handled by the exposure protection and confinement means. All others are large spills. In the case of jets of liquefied flammable gas[24,25] or hot flammable liquid,[26] where spacing is not a reliable defense against explosion damage, various quantities have been proposed as limits that must not be exceeded. One organization suggests 10,000 lb (4,536 kg) of gas or vapor as a maximum permissible leakage rate that would permit formation of no more than 1,000 lb per min (454 kg/min) of flammable gas or vapor.

Preventing Spills

Because many chemicals cause unusual corrosion or abrasion problems, or their processing requires unusually high or low temperatures, chemical plants should be constructed of appropriate materials and should follow good maintenance practices. Assuming that there are no problems in these areas, the reaction systems themselves may cause trouble.

In general, reactions are used to produce desired commodities that differ chemically from the raw materials from which they are made. The differences are produced by controlled chemical changes. Uncontrolled changes can produce unwanted materials and violent system failures that result in spills. The common denominator in these failures is overheating, i.e., overheating of a container so that it softens and fails, or overheating of the contents so that the container bursts from too much pressure.

Accidents of these types result from a failure to move heat adequately from a heat source to a heat-absorbing system or to a coolant from a heat-releasing system. Heat is usually produced when materials dissolve, crystallize, condense, or are adsorbed. Heat is also produced mechanically by crushing, grinding, milling, compressing, and pumping, and by exothermic chemical reactions. Heat transfer is also required when heat is needed for evaporation, melting, decreasing viscosity, initiating exothermic chemical reactions, and driving endothermic reactions. In any case, loss prevention requires an understanding of heat transfer.

HAZARDS OF HEAT TRANSFER

When a hot fluid, gas, or liquid transfers heat to a cooler fluid through a barrier such as the metal wall of a vessel, the resistance to heat flow, as shown by a temperature drop, is mainly due to the fluid films on the metal surfaces rather than the thermal conductivity of the metal. (See Figure 3-19B.) Flow in the heating and heated fluids is usually turbulent, with invisible swirls and eddies; but in the films, it is laminar with thin layers of molecules sliding slowly over each other.

The film layer nearest the metal moves slowest and its temperature is nearest to that of the metal. This is important because high temperature causes decompositions of most organic and some inorganic fluids. Although a large temperature differential between the fluids moves heat faster than a small one, the hottest part of the heated film must be kept cool enough so that it does not decompose and form a solid film. When such a solid film forms, the temperature of the metal on the heated side will increase, so that the metal is damaged (softened or chemically altered) and a "burnout" occurs.

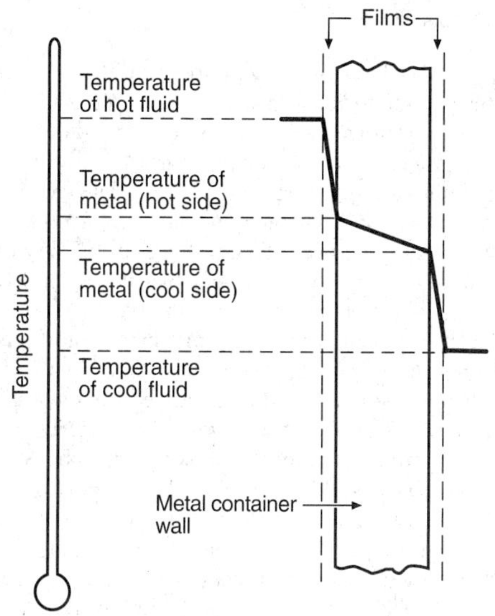

FIG. 3-19B.   *The effect of fluid films on temperatures at the sur-faces of the metal wall of a vessel.*

Precautions consist of methods to detect the fluid breakdown, such as checking the fluid for tar, solids, or significant change in viscosity or volatility. Where the heater is of a single tube design, an internal buildup will cause an easily observable increased pressure drop as the fluid flows through the tube. For this reason, single-pass tube heaters are to be preferred over those of multiple fluid path designs.

When the breakdown of the fluid is intentional, as in the case of cracking a petroleum fraction, the internal deposits are removed on a scheduled basis before they reach dangerous thickness. This is usually done by burning them out with air or a mixture of air and steam.[27] Requirements for safety involve:

1. Purging the unit of flammables, usually with steam, before introducing the oxidizing atmosphere.
2. Purging after burning out the deposits.
3. Safe disposal of the materials purged.
4. A system of valving, usually involving double valves with a purged or vented space between, to ensure against mixing air and combustibles.

If the heated fluid is a liquid and starts to boil, heat transfer improves as the film is thinned by the action of the bubbles on the metal, growing and breaking away. This formation of bubbles on the heated surface is called "nucleate boiling." If the temperature of the heat source is then raised in an attempt to increase the rate of boiling, the temperature may become too high. Vapor bubbles then coalesce into a vapor film, instead of breaking away individually. This is called "film boiling." Vapor films offer more resistance to heat flow, so, again, "burnout" may occur.

Endothermic Reactions

Endothermic reactions are most commonly supplied with heat by combustion processes. This energy may be obtained by partial oxidation of the reacting stream, e.g., cracking natural gas to acetylene by burning part of it with oxygen and then quenching the partially oxidized stream suddenly; or concentrating a liquid by submerged combustion (burning a premixed gaseous fuel beneath the surface of a liquid); or roasting ore or cementitious material in a kiln. The energy may also be supplied through the walls of a container via a liq-

uid heat transfer media or by hot gases from a combustion process or another hot gas stream.

The same general hazards exist whether the material being heated is a reacting system or one being decomposed or heated. All cases require consideration of the metallurgical and construction characteristics of the equipment. These problems are accentuated in the case of direct-fired equipment. The heater must resist corrosion, pitting, and scaling from the materials being heated as well as from the stream supplying the heat.

Construction must be such that heated and cooled portions can expand and contract freely. This is particularly important if the alloys used enter or pass through a brittle phase at any stage of heating or cooling.

Two more important, but sometimes overlooked, points are:

1. The metal of the heater and the heat transfer system and the fluids with which they are in contact must be compatible. The common molten salt heat transfer medium, a mixture of sodium nitrate and sodium nitrite used in indirect-fired heaters, will support the combustion of steel if the temperature is high enough and ignition occurs[28]; steel and copper in thin sections, as on the fins of heat transfer tubes, can ignite easily in hot chlorine.
2. The heat transfer medium should not create a hazard if it leaks into the material being heated or *vice versa*. Where leakage one way but not in another can be temporarily tolerated (e.g., sulfuric acid into water as opposed to water into sulfuric acid), pressures of the coolant and the cooled material must be suitably different. Alternatively, annular tubes with the annulus separating the two fluids may be used with means to detect leakage into the annulus from either direction. To promote heat transfer, the annulus may be filled with a conducting liquid inert to both fluids or, as in "still" tubes, intermittently bridged with heat-conducting metal wires or perforated discs.

The preceding discussion applies equally well when the heat transfer is from an exothermic reaction to a fluid used for cooling.

It is impossible to be all inclusive about the hazards of heat transfer. Two examples, however, may indicate the scope of the problem. In the first example, electric inductive heat was being used on a cast-iron vessel with thermostatic controls. Cast-iron growth caused intercrystalline cracking. The cracks interrupted the flow of the induced electric currents throughout the vessel, the thermostats called for more heat, and the vessel eventually failed from localized overheating. In another instance, sodium silicate was condensed on a stainless-steel pipe where it acted as a flux to remove the alloying elements so that the pipe lost its corrosion resistance and failed.

Exothermic Reactions

Proper application of heat transfer principles is essential to the control of exothermic reactions. Such use requires knowledge of the reaction system. First, it is necessary to know the thermal stability and shock sensitivity of the system and its components, raw materials, intermediates, products, and by-products. Second, the reaction kinetics must be known, i.e., the total amount of the heat released and how fast it is released.

For process design, it is necessary to have all of this data under conditions that would exist if the reactants were supplied in grossly incorrect proportions, in the wrong order, without mixing (both with and without a subsequent late start of the mixing system), at the wrong temperatures, if power failed, if the coolant line broke, or if any other upset condition imaginable should occur.

STABILITY AND SHOCK SENSITIVITY

Most chemicals are reasonably stable and insensitive. Where this is not true, data usually is available from the supplier or in the literature. NFPA 49 discusses specific chemicals, and NFPA 491M, *Manual of Hazardous Chemical Reactions*, covers hazardous systems. *The Handbook of Reactive Chemical Hazards*[29] is another excellent reference in this field.

The aforementioned publications provide many references. When data are not available, thermodynamic calculations and laboratory-scale stability testing should be used.[30-35] Many such test procedures have been developed by Committee E27 on Hazard Potential Chemicals and are published by the American Society for Testing and Materials (ASTM). Where explosive materials are involved, the hazard may be reduced to an acceptable level by diluting the explosive material in an inert chemical or by providing barricades around the explosive materials.

If the instability hazard is less than explosive, there are various ways to reduce it. Chemical reactions progress faster as the temperature rises, so the methods used involve keeping the material from being heated to an unsafe level. For example, the boiling point of a liquid is lowered by vacuum. Vacuum or freeze drying can help avoid overheating. A high boiling material that becomes unstable at or near its boiling point may have steam bubbled through it at a temperature below the (normal) boiling point.

Vapor from the material goes along with the steam, and separation can be made after condensing the steam-vapor mixture. Viscous materials can be concentrated by running them down the heated interior walls of a wiped film evaporator. Mechanically operated blades keep wiping the interior surface and the film cannot thicken, so the film layer nearest the hot wall cannot overheat. (See Figure 3-19B.) Liquid nitrogen or dry ice (solid CO_2) may be added to material going to a grinder. This not only takes care of the heat produced mechanically but also tends to inert the atmosphere and reduce chances of a dust explosion.

In some cases, pressure may affect stability. Liquids, such as propargyl bromide and nitromethane, will simply boil away when heated at atmospheric pressure. If heated under pressure, however, the boiling point might be raised to the point where the liquid decomposes violently, with resulting explosion of the container. Pure gases, such as acetylene, ethylene, and nitrous oxide, will decompose violently under pressure. The testing described earlier should be thorough enough to detect such a possibility. Also, when considering systems, one should keep in mind that pressure may cause a phase change. For example, chlorine gas liquefies fairly easily, and a system easy to control when using the gas may be hard to control if liquid is accidentally formed.

REACTORS

Reaction systems are of two types: (1) continuous and (2) batch. A continuous reacting system may be thought of as a pipe, although it may contain pumps, compressors, elongated vessels, etc., with the raw materials and possibly an inert carrier flowing in at one end, or at appropriate intervals along the pipe and products and by-products flowing out the other end. In a batch process, the chemicals are added to a single unit all at once or at appropriate feed rates; then the reaction takes place, the unit is emptied, and the process is repeated. Batch process systems have more flexibility, in that, generally speaking, the equipment can be used to make several different products. Continuous process equipment is generally limited to the production of one product.

Continuous Reactors

It is common to have one reactant, pure or diluted, circulating continuously throughout the reactor with another being added, reacting, and its products removed by condensation, washing, filtering, or similar appropriate means. In normal operation, the major safety control is automatically cutting off the appropriate feed stream if the heat transfer system fails. Arrangements are often made for alternate means of supplying heat transfer fluids. Reliability of the shut-off system may be enhanced by using two valves in series with a vent between (known as double block and bleed system). Depending on the complexity of the system, this same shutoff may be actuated by changes in the process conditions. An example of a process condition change might be too low a temperature, which would indicate that the reaction was not proceeding properly. This signals that the concentrations of unreacted materials may be building up to the point where they could react with violence and overpower the means of heat removal, or react in a part of the system designed to recover product rather than control the reaction. Also, when starting up or shutting down, the continuous reactor may go through situations that are more hazardous than its normal operation.

Batch Reactors

The essential parts of a batch reactor are:

1. A vessel to contain the reaction at its maximum pressure.
2. A heat transfer system to control the temperature of the reacting mixture.
3. A stirrer or agitator to keep the temperature and composition of the reacting mixture uniform.
4. A relief system to protect the vessel against overpressure.

The vessel: This may range from an open wood or rubber-lined steel tank in which bauxite ore reacts with sulfuric acid to produce a solution of alum to a glass-lined or stainless-steel pressure vessel. Regardless of the size or configuration, the vessel used as a batch reactor will probably have to be entered for cleaning and inspection, while the continuous reactor can be cleaned automatically with a cleaning fluid. Cleaning or entering vessels will require special precautions if flammable cleaning agents are used or if vapors or residues make entry hazardous.

Heat transfer methods: Common methods of heat transfer use jackets or coils on the exterior of the vessel or coils in the interior. Others circulate the reacting mixture from the vessel through an external heat exchanger and back to the vessel. A third method allows the reacting mixture to boil. The heat of vaporization is then removed by condensing the vapor externally, which allows the condensed liquid to run back and thus continue the cycle. This last method does not provide means to heat the material in the vessel, which may be needed to start the reaction, even though cooling is needed as the reaction progresses. The methods outlined may be used in combination.

Mixing: This is usually performed with an agitator mounted on a shaft driven by an external motor. If an external heat exchanger is used, the mixture leaving it may be recirculated back into the vessel to accomplish the needed mixing. Where heat-sensitive materials are involved and it is necessary to avoid frictional heating by moving parts, mixing can be accomplished by "sparging" (bubbling) an inert gas or possibly air through the reactants.

Pressure relief of batch reactors: Batch reactors in which a flammable atmosphere might sometimes exist are usually designed to contain the pressure that might result from ignition, or they are purged or inerted to eliminate the hazard. If vessels are constructed according to the *Pressure Vessel Code* of the American Society of

Mechanical Engineers (ASME), they are required to have a relief device adequate to keep the vessel from being overpressured because of heating from an external source, including an exposure fire, normal heating, or the introduction of fluids at pressures higher than those for which the vessel is designed. The *Code* gives no guidance for the relief of pressure caused by heat of reaction or as it is more commonly known, a runaway reaction.

Most chemical reactions double in speed with each rise in temperature of 18°F (10°C). This means that a heat removal system would have to remove twice as much heat at 140°F (60°C) as at 122°F (50°C), four times as much at 158°F (70°C), etc. For this reason, it is unsafe to run a batch reaction more than 32°F (14°C) above that of the cooling medium.[36] If the cooling system is overpowered, the reaction keeps increasing in speed, the retained heat builds up pressure, and the pressure must be relieved or the vessel will rupture. Austin has given a good explanation of this problem.[37] The Dow Chemical Company's *Guidelines for Process Scale-Up* asks the following questions about runaway reactions:[38]

Can you prevent runaway reactions?	Yes	No
By adequate heat transfer	___	___
By quenching the reaction	___	___
By stopping feed streams	___	___
By dilution of the reactor contents	___	___

Score: Two or more methods—You're in good shape. Only one method—Better find a second line of defense. None of the above—You'd better stop right now and reconsider your design for safety.

However, since even multiple safeguards may fail, emergency venting is supplied as a last ditch protection for the vessel. For example, loss of mixing could make all the suggested preventive measures inoperative or dangerously inefficient. Materials that have leaked into the hollow shaft of an agitator have created pressure and spewed materials into the main reaction system, which speeds up the main reaction so that it overpowers the cooling system. Other problems were discussed earlier in this chapter in the subsection "Endothermic Reactions."

The problem of adequate sizing of emergency venting was recognized by the AIChE, and a committee was organized to study the problem. This became known as the Design Institute for Emergency Relief Systems (DIERS). Twenty-nine companies sponsored the work and spent $1.6 million on research. The project was successfully completed in 1985. The key contributions developed to provide the chemical process industry with the tools necessary to evaluate emergency vent requirements are:

1. Emphasizing the importance of proper determination of the source term (i.e., energy release rate and gas generation rate for vapor and gassy systems, respectively) under upset conditions, and development of a new bench-scale apparatus in which such determination can be obtained and extrapolated directly to full-scale application.
2. Emphasizing the importance of two-phase flow in the emergency relief discharge process, and development of vent sizing techniques that are valid for multi-phase behavior.

The DIERS methodology can be complex and time consuming and may be beyond the capability of many small plants. As a result, some inexpensive and easily employed techniques have been developed. These commercially available tools should make it possible for all facilities to design adequate emergency venting for processing plants. A number of articles have appeared in AIChE publications, *Chemical Engineering Progress,* and *Plant/Operations Progress* (and its successor publication *Process Safety Progress*)

concerning the DIERS program and the ways to use it.[39-48] The theoretical advances allowing a rational design procedure to be used for handling potentially runaway reaction conditions are best summarized in reference 49. Symposia on this subject, primarily sponsored by DIERS, AIChE, are held periodically.[50]

Safety Instrumentation

Safety instrumentation will vary from process to process and plant to plant. There are, however, a few general rules:[51]

1. Measure, as directly as possible, the variables of interest. For example, if an agitator is driven by an electric motor, a relay can be used to show that the switch to the motor is closed. An ammeter would be better because that would show that current is flowing; a wattmeter would be best because it would show that work is being done and that the impeller has not fallen off the agitator shaft.
2. Reliability and maintenance must be of high quality; e.g., if an external fire could upset the reaction, the warning instrument should not be disabled by the fire. Since safety instruments will not have to operate often, frequent checks must be made to ensure they are in operating condition.
3. Provide redundancy or diversity in the system. If temperature is critical, provision of two thermocouples may lead to two different readings. It is best to provide three and trust the majority of the readings. However, since a temperature increase is usually accompanied by a pressure increase, one pressure sensor and one thermocouple may be used, with credence given to the more pessimistic report.

CHEMICAL PLANT OPERATIONS AND EQUIPMENT

The generally accepted unit physical operations require equipment to carry out the following functions:

1. Heat transfer
2. Fluid flow
3. Crushing and grinding
4. Mixing
5. Mechanical separation
6. Distillation
7. Evaporation
8. Crystallization
9. Filtration
10. Absorption
11. Adsorption
12. Drying

Descriptions of individual pieces of equipment can be found in most engineering handbooks.

Heaters and Coolers (Heat Transfer)

These devices are either: (1) direct (flue gas, kilns, spray cooling, vaporization) or (2) indirect (coils and jackets). Heat transfer media usually work by boiling to absorb heat or condensing to give it up but also may work just by becoming hotter or cooler without phase change. Hazards of heat transfer have been discussed previously in this chapter.

Fluid Flow

Fluid flow is accomplished by: (1) fans and compressors, (2) pumps, or (3) vacuum jets. Fans, compressors, and pumps heat the materials they move because of mechanical work. Compressors require good aftercoolers. The liquid in pumps can boil, and the pump bearings fail from overheating if a pump operates against a closed discharge valve. If a centrifugal pump suction is inadequate, cavitation may cause the pump rotor to fail. Failure of vacuum devices may permit unwanted air to enter the equipment and cause the temperature in the equipment to rise.

Crushing and Grinding

Equipment for these operations include: (1) mills (impact, ball, hammer, roller, disk), and (2) crushers (cone, gyratory, jaw). All of these devices produce heat and may produce fine material capable of a dust explosion. Ball mills may become overpressured as liquid in their contents are heated. If gaskets fail and the liquids are flammable, a dangerous situation will result.

Mixing

Devices used in the mixing process include: (1) tumblers; (2) venturi mixers; (3) kneaders, rolls, and mullers; and (4) propellers, blades, and turbines. These devices also produce heat to varying degrees. When it is necessary to produce a hazardous mixture, e.g., a blend of starch and an oxidizer for use as a flour aging material, the mixer should be vented to prevent explosions. Any dust collector should be of the water-wash type.

Mechanical Separation

Mechanical separation includes: (1) cyclones, (2) bag filters (shake or blow back), (3) ore tables, (4) screens, (5) electrostatic precipitators, (6) flotation separators, and (7) centrifuges. The first five are usually dry type and the latter two wet, although there are liquid cyclones and wet screening processes. Devices used in dry service for combustible dusts should have deflagration (explosion) venting or an explosion suppression system. If collectors have internal combustible bags, an internal sprinkler system is desirable. It may not save the bags, but it will save the housing and the mechanical devices. When materials wet with flammable liquids are centrifuged, the operation should be conducted under inert gas or with an explosion suppression system.

Distillation

The types of stills are: (1) batch, (2) continuous, (3) pressure, (4) vacuum, and (5) steam. All distillations involve heat transfer. The major hazard of this operation is that a flammable vapor may be released to the atmosphere.[52] The direct use of steam is discussed in the "Stability and Shock Sensitivity" section of this chapter.

Evaporation

The three types of evaporators are: (1) multiple effect, (2) vacuum, and (3) wiped film. All three use heat transfer. The multiple-effect system condenses vapor from the first unit to heat the second, etc. The initial heat transfer medium is used in the first, where the material is most concentrated and boils at the highest temperature. Vacuums and wiped films have been discussed previously in this chapter.

Crystallization

The two types of crystallizers are: (1) vacuum and (2) pan. Both involve cooling to remove heat of crystallization.

Filtration and Agglomeration

Filtration and agglomeration involve: (1) plate and frame filters, (2) Nutsche vacuum filters, (3) drum type, (4) rotary cell type, (5) ag-

glomerating tables, and (6) pelletizers. No unusual hazards are involved other than possible exposure of flammable liquids to the air.

Adsorption

The two types of adsorbers are: (1) activated carbon and (2) zeolites (molecular sieves). Adsorption produces heat just as absorption does. However, when the material being adsorbed is a flammable vapor or gas, the heat is retained on the surface of the adsorbent. Heat may build up enough to cause fire in the activated carbon or initiate the polymerization of the adsorbed reactive materials, such as ethylene.

Drying

The seven general types of dryers are (1) spray, (2) fluidized bed, (3) vacuum, (4) tray, (5) belt, (6) drum, and (7) azeotropic. All these involve heat transfer. In the case of azeotropic dryers, where heat is removed by boiling, condensation and return of the cool liquid, the unwanted component, usually water, can be discarded if the condensate forms a two-phase system. All other types of dryers may present problems of mixing flammable vapors with air. Spray and fluid-bed dryers may expose thermally unstable dry materials to hot heat-transfer surfaces.

CONCLUSION

A summary of necessary knowledge and procedures to ensure safe chemical processing has been suggested, as follows:[53]

1. Know the total reaction energy in the system.
2. Know the rate of energy release.
3. Evaluate thermal and shock sensitivity data.
4. Design the process to control the rate of energy release, and in doing so:
   (a) Be alert for trace compounds or catalytic impurities that may accelerate reaction rates.
   (b) Prevent buildup or concentration of high-energy materials in the system. Calculate a material balance.
   (c) Design into the process the ability to safely accommodate inadvertent releases.

Before beginning operations, a hazard analysis of the proposed or modified process is generally required.[54,55]

BIBLIOGRAPHY

References Cited

1. Lees, F. P., *Loss Prevention in the Process Industries,* Vol. 1, Butterworth & Co., Ltd., London, 1980, pp. 210–230.
2. Marshall, V. C., *Major Chemical Hazards*, Halsted Press, Div. of John Wiley & Sons, New York, 1987.
3. Melancon, C. L., "Improving Emergency Control and Response Systems," *Loss Prevention*, Vol. 13, AIChE, New York, 1980, pp. 43–49.
4. *Engineering for Safe Operations*, No. 8, American Oil Company (now Standard Oil of Indiana), Chicago, IL, 1964.
5. *General Recommendations for Spacing*, IM.2.5.2, Industrial Risk Insurers, Hartford, CT, March 2, 1992.
6. Lewis, D. J., "The Mond Fire, Explosion, and Toxicity Index Applied to Plant Layout and Spacing," *Loss Prevention,* Vol. 13, AIChE, New York, 1980, pp. 20–26.
7. Lewis, D. J., "Plant Layout and Storage Area Spacing," *Proceedings* of the 6th International Symposium on Loss Prevention and Safety Promotion in the Process Industries, Vol. 1, European Federation of Chemical Engineers, Oslo, Norway, 1989, pp. 7-1, 7–16.
8. Davenport, J. A., "A Survey of Vapor Cloud Incidents," *Loss Prevention,* Vol. 11, AIChE, New York, 1977, pp. 39–49.

9. van Wingerden, K., *et al.*, *Guidelines for Evaluating the Characteristics of Vapor Cloud Explosions, Flash Fires, and Bleves,* AIChE-CCPS, New York, 1994.

10. Decker, D. A., "An Analytical Method for Estimating Overpressure from Theoretical Atmospheric Explosions," presented at the 1974 Annual Meeting of NFPA-SFPE, May 23, 1974.

11. Davenport, J. A., and Lenoir, E. M., "A Survey of Vapor Cloud Incidents—2nd Update," 26th Loss Prevention Symposium (New Orleans), AIChE, New York, March 30–April 2, 1992 (paper 74d).

12. Nelson, R. W., "Know Your Insurers' Expectations," *Hydrocarbon Processing,* 1977, pp. 103–108.

13. van Wingerden, C. J. M., "Experimental Investigation into the Strength of Blast Waves Generated by Vapor Cloud Explosions in Congested Areas," *Proceedings* of the 6th International Symposium on Loss Prevention and Safety Promotion in the Process Industries, Vol. 1, European Federation of Chemical Engineers, Oslo, Norway, 1989, pp. 26-1 to 26-16.

14. Prugh, R. W., *Guidelines for Vapor Release Mitigation,* American Institute of Chemical Engineers, New York, 1988.

15. Rinder, R. M., and Wachtell, S., "Establishment of Design Criteria for the Safe Processing of Hazardous Materials," *Loss Prevention,* Vol. 7, AIChE, New York, 1967, pp. 28–30.

16. Bradford, W. J., and Culbertson, T. L., "Design of Control Houses to Withstand Explosive Forces," *Loss Prevention,* Vol. 1, AIChE, New York, 1967, pp. 28–30.

17. "Siting and Construction of New Control Houses for Chemical Manufacturing Plants," Safety Guide SG-22, Chemical Manufacturers Association, Washington, DC.

18. Eggleston, L. A., Herrera, W. R., and Pish, M. D., "Water Spray to Reduce Vapor Cloud Spray," *Loss Prevention,* Vol. 10, AIChE, New York, 1976, pp. 31–42.

19. Lawrence, W. E., and Johnson, E. E., "Design for Limiting Explosion Damage," *Chemical Engineering,* Vol. 81, No. 1, 1974, pp. 96–104.

20. Jones, D. W., Guidelines for Evaluating Process Plant Buildings for Explosion or Fire, AIChE-CCPS, New York.

21. Watts, J. W., Jr., "Effects of Water Spray on Unconfined Flammable Gas," *Loss Prevention,* Vol. 10, AIChE, New York, 1976.

22. *The Safe Dispersal of Large Clouds of Flammable Heavy Vapors,* ed., Imperial Chemical Industries, Ltd., Heavy Organic Chemicals Division, Billingham, England, 1971.

23. Johanson, K. A., "Gas Detectors by the Acre," *Instrumentation Technology,* Vol. 21, No. 8, 1974, pp. 33–37.

24. Burgess, D. S., and Zabetakis, M. G., "Detonation of a Flammable Cloud Following a Propane Pipeline Break," RI 7752, USDI Bureau of Mines, Washington, DC.

25. Goforth, C. P., "Functions of a Loss Control Program," *Loss Prevention,* Vol. 4, AIChE, New York, 1970, pp. 1–5.

26. Kletz, T., "Lessons to Be Learned from Flixborough" *Loss Prevention,* Vol. 9, AIChE, New York, 1975.

27. Armistead, G., Jr., *Safety in Petroleum Refining and Related Industries,* 1st ed., John G. Simmonds & Co., Inc., New York, 1950, pp. 118–119.

28. "Potential Hazards in Molten Salt Baths for Heat Treatment of Metals," NBFU Research Report RR-2, 1954, American Insurance Association, New York.

29. Bretherick, L., *The Handbook of Reactive Chemical Hazards,* 5th ed., edited by P. G. Curben, Butterworth-Heinemann Boston, 1995.

30. Coffee, R. D., "Hazard Evaluation Testing," *Loss Prevention,* Vol. 3, AIChE, New York, 1969, pp. 18–21.

31. Coffee, R. D., "Hazard Evaluation: The Basis for Chemical Plant Design," *Loss Prevention,* Vol. 7, AIChE, New York, 1973, pp. 58–60.

32. Davis, E. J., and Ake, J. A., "Equilibrium Thermochemistry Computer Programs as Predictors of Energy Hazard Potential," *Loss Prevention,* Vol. 7, AIChE, New York, 1973, pp. 67–73.

33. Stull, D. R., "Linking Thermodynamics and Kinetics to Predict Real Chemical Hazards," *Loss Prevention,* Vol. 7, AIChE, New York, 1973, pp. 67–73.

34. Trewick, D. N., Claydon, C. R., and Seaton, W. H., "Appraising Energy Hazard Potentials," *Loss Prevention,* Vol. 7, AIChE, New York, 1973, pp. 21–27.

35. Way, D., "Fire Protection Engineering," *Loss Prevention,* Vol. 3, AIChE, New York, 1969, pp. 23–25.

36. Boynton, E. D., Nichols, W. B., and Spurline, H. M., "Control of Exothermic Reactions," *Industrial & Engineering Chemistry,* Vol. 51, No. 4, 1959, pp. 489–494.

37. Austin, G. T., "Hazards of Commercial Chemical Reactions," *Safety and Accident Prevention in Chemical Operations,* Fawcett, H. H., and Wood, W. S., eds., John Wiley & Sons, New York, 1965.

38. Kline, P. E., *et al.*, "Guidelines for Process Scale-Up," *Chemical Engineering Progress,* Vol. 70, No. 10, 1974, pp. 67–70.

39. Fauske, H. K., "A Quick Approach to Reactor Vent Sizing," *Plant/Operations Progress,* Vol. 3, No. 3, 1984, pp. 145–146.

40. Fauske, H. K., and Leung, J. C., "New Experimental Technique for Characterizing Runaway Chemical Reactions," *Chemical Engineering Progress,* Vol. 81, No. 8, 1985, pp. 39–45.

41. Fauske, H. K., "Emergency Relief Systems (ERS) Design," *Chemical Engineering Progress,* Vol. 81, No. 8, 1985, pp. 53–58.

42. Fauske, H. K., "Emergency Relief System Design for Reactive and Non-Reactive Systems: Extension of the DIERS Methodology," *Plant/Operations Progress,* Vol. 7, No. 3, 1988, pp. 153–158.

43. Fauske, H. K., Grolmes, M. A., and Clare, G. H., "Process Safety Evaluation Applying DIERS Methodology to Existing Plant Operations," *Plant/Operations Progress,* Vol. 8, No. 1, 1989, pp. 19–24.

44. Fauske, H. K., *et al.*, "An Easy and Inexpensive Approach to DIERS Methodology," presented at AIChE Loss Prevention Symposium, Houston, TX, April 1989.

45. Fisher, H. G., "DIERS Research Program on Emergency Relief Systems," *Chemical Engineering Progress,* Vol. 81, No. 8, 1985, pp. 36–38.

46. Grolmes, M. A., Leung, J. C., and Fauske, H. K., "Large-Scale Experiments of Emergency Relief Systems," *Chemical Engineering Progress,* Vol. 81, No. 8, 1985, pp. 57–62.

47. Grolmes, M. A., and Leung, J. C., "Code Method for Evaluating Integrated Relief Phenomena," *Chemical Engineering Progress,* Vol. 81, No. 8, 1985, pp. 47–52.

48. Leung, J. C., and Stepaniuk, N. J., "Emergency Relief Considerations Under Severe Segregation Scenarios," presented at AIChE Loss Prevention Symposium, New Orleans, LA, March 1988.

49. Fisher, H. G., *et al.*, "Emergency Relief System Design Using DIERS Technology" (Design Manual), DIERS-AIChE, New York, 1992.

50. Melhem, G. A., and Fisher, H. G., eds., *Proceedings* International Symposium on Runaway Reactions and Pressure Relief Design: Aug. 2–4, 1995, AIChE, New York, 1995.

51. Doyle, W. H. "Instrument Connected Losses in the CPI," *Instrumentation Technology,* 1972, pp. 38–42.

52. Doyle, W. H. "Minimizing Serious Fires and Explosions in the Distillation Process," Technology Report 74-2, 1974, Society of Fire Protection Engineers, Boston, MA.

53. *The ABCs of Reactive Chemical Processing,* Dow Chemical Company, Midland, MI.

54. Hendershot, D., *et al.*, *Guidelines for Hazard Evaluation Procedures,* 2nd ed., AIChE-CCPS, New York, 1992.

55. Ormsby, R. W., *et al.*, *Guidelines for Chemical Process Quantitative Risk Analysis,* AIChE-CCPS, New York, 1989.

References

Annual Book of Standards, American Society for Testing and Materials, W. Conshohocken, PA. (Updated annually.)

Heemskerk, A. H., *et al., Guidelines for Chemical Reactivity Evaluation and Application to Process Design,* AIChE-CCPS, New York, 1995.

Pressure Vessel Code, American Society of Mechanical Engineers, New York. (Continuously updated.)

NFPA Codes, Standards, and Recommended Practices

Reference to the following NFPA codes, standards, and recommended practices will provide further information on the safeguards for chemical processing equipment discussed in this chapter. (See the latest version of *The NFPA Catalog* for availability of current editions of the following documents.)

NFPA 15, *Standard for Water Spray Fixed Systems for Fire Protection*
NFPA 49, *Hazardous Chemicals Data*
NFPA 68, *Guide for Venting of Deflagrations*
NFPA 306, *Standard for the Control of Gas Hazards on Vessels*
NFPA 491M, *Manual of Hazardous Chemical Reactions*
NFPA 495, *Explosive Materials Code*

NFPA 704, *Standard System for the Identification of the Hazards of Materials for Emergency Response*

Additional Reading

Atallah, S., and Allan, D. S., "Safe Separation Distances from Liquid Fuel Fires," *Fire Technology*, Vol. 7, No. 1, Feb. 1971, pp. 47–55.

Bartknecht, W., *Explosions*, Springer-Verlag, Berlin, Heidelberg, New York, 1981.

Bowen, J. E., *Emergency Management of Hazardous Materials Incidents*, National Fire Protection Association, Quincy, MA 1994.

"Chemical Laws Improve Safety and Loss Control.," *Record*, Vol. 69, No. 4, July/August 1992, pp. 15–20.

Davenport, J. A., "Explosion Losses in Industry," *Fire Journal*, Vol. 75, No. 1, Jan. 1981, pp. 52–56, 71–73.

Donahue, M. L., "CHEMTREC: A Vital Link in Chemical Emergencies," *NFPA Journal*, Vo. 87, No. 3, May/June 1993, pp. 86–88, 90.

Doran, P., and Greig, R., "Mond Index Assesses Hazard Poetntial of Chemical Plant," *Process Engineering*, Vol. 70, No. 2, 1989, pp. 67–68.

"Engineering Special Chemical Risks," *Record*, Vol. 70, No. 3, May/June 1993, pp. 3–8.

Eversole, J. M., "Planning, Preparation and Response to Explosions and Fires at Chemical Plants," International Fire Conference and Exhibition in Tokyo for Firesafety Frontier '94, Creating a Safe Tomorrow, 1994, pp. 63–67.

"Explosion and Fire Hazards in the Storage and Handling of Organic Peroxides in Plastic Fabricating Plants," SPI-FPC 19, June 1964, The Society of the Plastics Industry, New York.

Fawcett, H. H., and Wood, W. S., eds., *Safety and Accident Prevention in Chemical Operations*, 2nd ed., Wiley—Interscience, New York, 1982.

Fire Inspectors Guide to NFPA 1031, National Fire Protection Association, Quincy, MA, 1983.

Grace, C., "Fluid Choice Takes the Steam out of Unsafe Process Heaters," *Process Engineering*, 5, 1977, pp. 85, 87, 88.

Halpaap, W., "Special Appliance for the Chemical Industry," *Fire International*, 5, 56, 1977, pp. 44–50.

Holmes, N., "Respiratory Protection in the Chemical Industry," *Fire International*, No. 144, August/September 1994, pp. 27–28.

Howard, H. A., "Chemical Plant Fire," *Fire Engineering*, Vol. 142, No. 2, 1989, pp. 28–34, 37–43.

Isman, W. E., "Chemical Properties Influence Decisions," *NFPA Journal*, Vol. 85, No. 6, November/December 1991, pp. 94–95.

Joschek, H. I., "Risk Assessment in the Chemical Industries," *Plant/Operations Progress*, Vol. 2, No. 1, Jan. 1983, pp. 1–5.

Kearney, J., "Bulk Chemical Incidents: Learning the Lessons From Past Emergencies," *Fire*, Vol. 86, No. 1057, July 1993, pp. 41–42.

Kirk, R. E., and Othmor, D. F., *Encyclopedia of Chemical Technology*, 3rd ed., Interscience Encyclopedia, Inc. New York, 1983.

Kletz, T. A., "Fires and Explosions of Hydrocarbon Oxidation Plants," *Plant/Operation Progress*, Vol. 7, No. 4, Oct. 1988, pp. 226–230.

Kuchta, J. M., *Investigation of Fire and Explosion Accidents in the Chemical, Mining, and Fuel Related Industries: A Manual*, U.S. Department of Interior, Bureau of Mines, Washington, DC, 1986.

Lewis, B., and Von Elbe, G., *Combustion, Flames, and Explosions of Gases*, 3rd ed., Academic Press, New York, 1987.

Lewis, R. J., *Dangerous Properties of Industrial Material*, 8th ed., Von Nostrand-Reinhold Company, New York, 1996.

MacDarmid, J. A., and North, G. J. T., "Lessons Learned from a Hydrogen Explosion in a Process Unit," *Plant/Operation Progress*, Vol. 8, No. 2, April 1989, pp. 96–99.

Malone, G. and Hession, M., "Failure of Water Supplies Hampered Firefighting at Cork Chemical Plant," *Fire*, Vol. 86, No. 1062, December 1993, pp. 23–24.

McElroy, F. E., ed., *Accident Prevention Manual for Industrial Operations*, 9th ed., National Safety Council, Chicago, IL, 1988.

Mikkola, E., "TOXFIRE: Toimintaohjeet kemikaalivarastojen tuli-palojen hallintaan—projekti kaynnistynyt. [TOXFIRE: New Project on Operational Guidelines For Managing Fires in Chemicals Stores.]," *Palontorjuntatekniikka*, March 1993, pp. 28–29.

Morehart, J., "QR-AS Sprinkler Test in Chemical Laboratories," VHS Video Tape, National Institutes of Health, Bethesda, MD, Federal Fire Forum, Current Technical Fire Issues, November 1, 1993, Gaithersburg, MD, 1993.

Norstrom, G. P., II, "Fire/Explosion Losses in the CPI," *Chemical Engineering Progress*, Vol. 78, No. 8, August 1982, pp. 80–87.

"Occupational Exposure to Hazardous Chemicals in Laboratories: Chemical Hygiene Plan," *Health and Safety Instruction*, Number 20, January 1991, 15 pages.

Pilborough, L., *Inspection of Chemical Plants*, Gulf Publishing Co., Houston, TX, 1982.

Piatt, J. A., "Chemical Process Safety Management Within the Department of Energy," *Proceedings* for the Thirteenth International System Safety Conference on Hazard Control Methodologies for the Future, 1995, pp. 1–6.

Rho, S. K., Measures Taken for Fire Protection in Place of Work (Focused on Chemical Factories)," International Fire Conference and Exhibition in Tokyo for Firesafety Frontier '94, Creating a Safe Tomorrow, 1994, pp. 313–318

"Safe and Efficient Plant Operation and Maintenance," *Chemical Engineering*, McGraw-Hill, New York, 1980.

Staggs, K. J., et al., "Development of Flammable Liquid Storage Wooden Cabinets for Chemical Laboratories," Department of Energy, Washington, DC, UCRL-ID-115605, November 1993, 39 pages.

Stull, J. O., et al., "Developing and Selecting Test Methods for Measuring Protective Clothing Performance in Chemical Flashover Situations," Performance of Protective Clothing: Challenges for Developing Protective Clothing for the 1990s, Fourth (4th) Volume, ASTM STP1133, ASTM, W. Conshohocken, PA, 1991, pp. 908–923.

Tokle, Gary, *Hazardous Materials Response Handbook*, 2nd ed., National Fire Protection Association, Quincy, MA, 1992.

Uehara, Y., "Fire Safety Assessments in Petrochemical Plants," *Proceedings* of the Third International Symposium on Fire Safety Science, Elsevier Applied Science, London, 1991, pp. 83–96.

Wackerlig, H., "Risk Assessment of Chemical Warehouse," *Fire Protection*, Vol. 17, No. 4, December 1990, pp. 4–6.

Yates, J., "Phillips Petroleum Chemcial Plant Explosion and Fire, Pasadena, Texas, October 23, 1989," USFA Fire Investigation Technical Report Series, Federal Emergency Management Agency, Washington, DC, Report 035, 1990.

Zabetakis, M. D., "Flammability Characteristics of Combustible Gases and Vapor," Bulletin 627, 1965, U. S. Bureau of Mines, Pittsburgh, PA.

AEROSOL CHARGING OPERATIONS

Byron L. Briese

A wide variety of products are packaged in aerosol form for consumer and industrial uses. Applications of this packaging technique include lubricants, cleaning agents, personal care products, food products, paints, and pesticides.

Prior to the issuance of the first edition of NFPA 30B, *Code for the Manufacture and Storage of Aerosol Products* (hereinafter referred to as NFPA 30B), in 1990, fire protection of aerosol products was governed by NFPA 30, *Flammable and Combustible Liquids Code* (hereinafter referred to as NFPA 30). Under this code, aerosol products were considered to be equivalent to Class IA flammable liquids. This classification often led to fire protection strategies, particularly in aerosol storage facilities, that were considered by many users to be excessive.

Fire testing efforts, chiefly conducted in the early 1980s by a consortium of aerosol manufacturers, warehouse operators, and interested insurance firms, indicated that the fire characteristics of aerosol products were more complicated than provided for under NFPA 30. Some products presented severe fire hazards, others presented a lower fire challenge than a Class IA flammable liquid. The improved understanding led to the development of an aerosol product fire hazard classification system, improved fire protection strategies for aerosol storage facilities, and later to the development of NFPA 30B.

This chapter will review the construction of aerosol products, the fire hazard classification system associated with aerosol products, the fire behavior of aerosol products, and fire protection strategies for manufacturing or storage occupancies.

CONTAINERS

The most common aerosol containers are high-strength, metal units with capacities ranging from a few ounces up to a quart. (For SI units: 1 oz. = 28.4 g; 1 qt = 0.946 L.) The top and base of the container are generally domed. Unit working pressure ranges between 240 to 400 psi (1,655 to 2,758 kPa).

A finished aerosol product consists of the basic elements shown in Figure 3-20A. A base product is dispensed into the aerosol container, an actuator assembly is fitted to the container body, and the container is pressurized with propellant. If a nozzle or tip assembly is not part of the previously installed actuator assembly, one is installed. Often, a protective outer cap (not shown in Figure 3-20A) is also provided to prevent damage to the nozzle or premature discharge of product.

PROPELLANTS

A variety of gas propellants are available to aerosol manufacturers. Chiefly, these propellants are hydrocarbons that are chosen, by the packager, for the characteristics that they provide for discharge of the base product. Characteristics of concern to aerosol packagers can include odor, purity, stability, and toxicity. Often, to satisfy particular needs, blends of hydrocarbon gases are selected by the aerosol formulator. All hydrocarbon aerosol propellant gases are flammable gases. Certain special formulations might use noncombustible gases such as carbon dioxide.

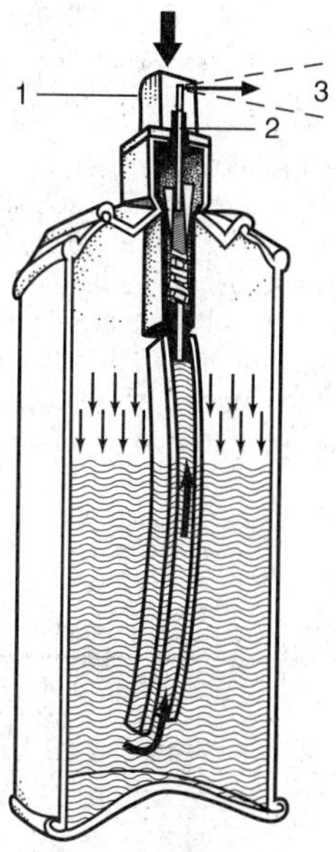

FIG. 3-20A.   *An aerosol can (cutaway view). When the plunger (1) is pressed, a hole in the valve (2) allows a pressurized mixture of product and propellant (3) to flow through the plunger's exit orifice.*

Byron L. Briese, P.E., is a fire protection consultant based in Knoxville, TN, and vice president of American Risk Consultants Corp. He is a member of NFPA's Technical Committee on Flammable and Combustible Liquids.

Propellants are typically blends of propane and butane. Pentane may also be used to produce a specific propellant. Specific blends of these gases are provided to aerosol fillers by propellant suppliers, and are marketed as "aerosol"-grade gases. As with most consumer and industrial products, these gas blends are sold under specific manufacturer's product names.

Hydrocarbon aerosol propellants are gaseous at ambient temperatures and pressures. They will condense to liquid form under moderate pressure or at low temperature. They maintain relatively constant pressure, at a given temperature, until all of the liquid propellant has been converted to gas.

Table 3-20A contains a summary of the physical and fire properties of the common propellant gases. The vapor pressure of these materials and the large quantities of gas they produce when vaporized makes them suitable propellants. A particular gas, or gas blend, may be referenced by its vapor pressure at 70°F (21°C). For example, A-31 refers to aerosol-grade isobutane, which has a vapor pressure of 31 psig (214 kPa). A-108 refers to propane, with a vapor pressure of 108 psig (745 kPa) at 70°F (21°C). A particular, desired vapor pressure can be achieved by blending gases.

BASE PRODUCTS

The material dispensed from the aerosol container is referred to as the "base product," or as the "concentrate." These liquid products run the full range of consumer and industrial products. The fire hazards of these materials can be minimal, e.g., a water-based liquid, to highly flammable, such as with many lubricants or coatings. An understanding of the physical properties and fire hazards of these materials is important. Each plays an important role in developing the proper fire plans for facilities that manufacture or store aerosol products.

CLASSIFICATION OF AEROSOLS

In order to determine appropriate fire protection measures for use with aerosol products, aerosol products are classified by NFPA 30B into one of three hazard categories. The 1994 edition of the standard recognized a revised approach to aerosol hazard classification developed by Factory Mutual Research Corporation (FMRC).

Prior to 1994, the fire hazards of aerosol products were characterized using a number of methods. These methods were: full-scale fire tests, small-scale fire tests; or the use of a decision tree. The decision tree, included in Appendix A of the 1990 edition of NFPA 30B, characterized an aerosol product's flammability, based upon the percentage of flammable material in the base product, the miscibility of the flammable material with water, and the percentage of flammable propellant compared to total container contents. (See Figure 3-20B.)

The 1994 edition recognizes two classification methods: (1) full-scale testing using a 12-pallet test array and (2) classification based upon the chemical heats of combustion of the aerosol products. The chemical heats of combustion calculation method accounts for the chemical heats of combustion for the propellant and the base product.

Based upon the aerosol product's aggregate chemical heat of combustion, the following hazard categories have been developed:

Level 1: Those aerosol products with total chemical heat of combustion equal to or less than 8,600 Btu/lb (20 kJ/g).

Level 2: Those aerosol products with total chemical heat of combustion greater than 8,600 Btu/lb (20 kJ/g) and less than or equal to 13,000 Btu/lb (30 kJ/g).

Level 3: Those aerosol products with total chemical heat of combustion greater than 13,000 Btu/lb (30 kJ/g).

Proper fire hazard classification of aerosol products is particularly important in storage areas used for aerosol products. The expected fire behavior, and associated protection requirements, will vary greatly with the class of aerosol product.

FIRE BEHAVIOR OF AEROSOLS

Fire tests of large amounts of filled aerosol containers confirms what has often been reported about actual fire incidents involving aerosol products. Level 2 aerosols produced fires so intense that ruptured containers rocketed through a simulated storage facility, occasionally trailing flaming liquid. Sometimes flaming packaging material or plastic caps were propelled beyond the test area.

Level 3 aerosols burned more intensely and rocketed with trailing burning liquid more frequently than Level 2 products. In addition, they produced dense black smoke that completely obscured visibility in the simulated storage area within 4 to 5 minutes.

In rack storage tests, the rupture of both Level 2 and Level 3 aerosols spanned the aisle in the test array. Ignition of the "target" rack within the test array occurred only in Level 3 tests. In both rack and pallet storage, rupturing cans dislodged surrounding containers and exposed others inside the load, increasing the fire severity.

EXAMPLE FIRE INCIDENTS

Aerosol products have been directly involved in large-scale, costly, and occasionally fatal fire incidents. In one fire incident, an aerosol can, improperly disposed of in an incinerator, blew off the incinerator door, igniting adjacent rubbish. Aerosols improperly stored near a heater produced a fire scenario that rendered manual firefighting efforts futile. At an aerosol filling facility, an explosion involving hydrocarbon propellant killed 4 employees and injured 18 others.

TABLE 3-20A. *Fire Hazard Properties of Hydrocarbon Propellants*

	Flammable Limits*		Flash Point		Autoignition Temperature	
	Lower	Upper	°F	°C	°F	°C
Propane	2.1	9.5	−156	−104	940	504
Isobutane	1.8	8.4	−117	−83	890	477
n-Butane	1.8	8.4	−101	−74	860	460
iso-Pentane	1.4	7.6	−60	−51	788	420
n-Pentane	1.5	7.8	−40	−40	500	260

* Percent in air by volume

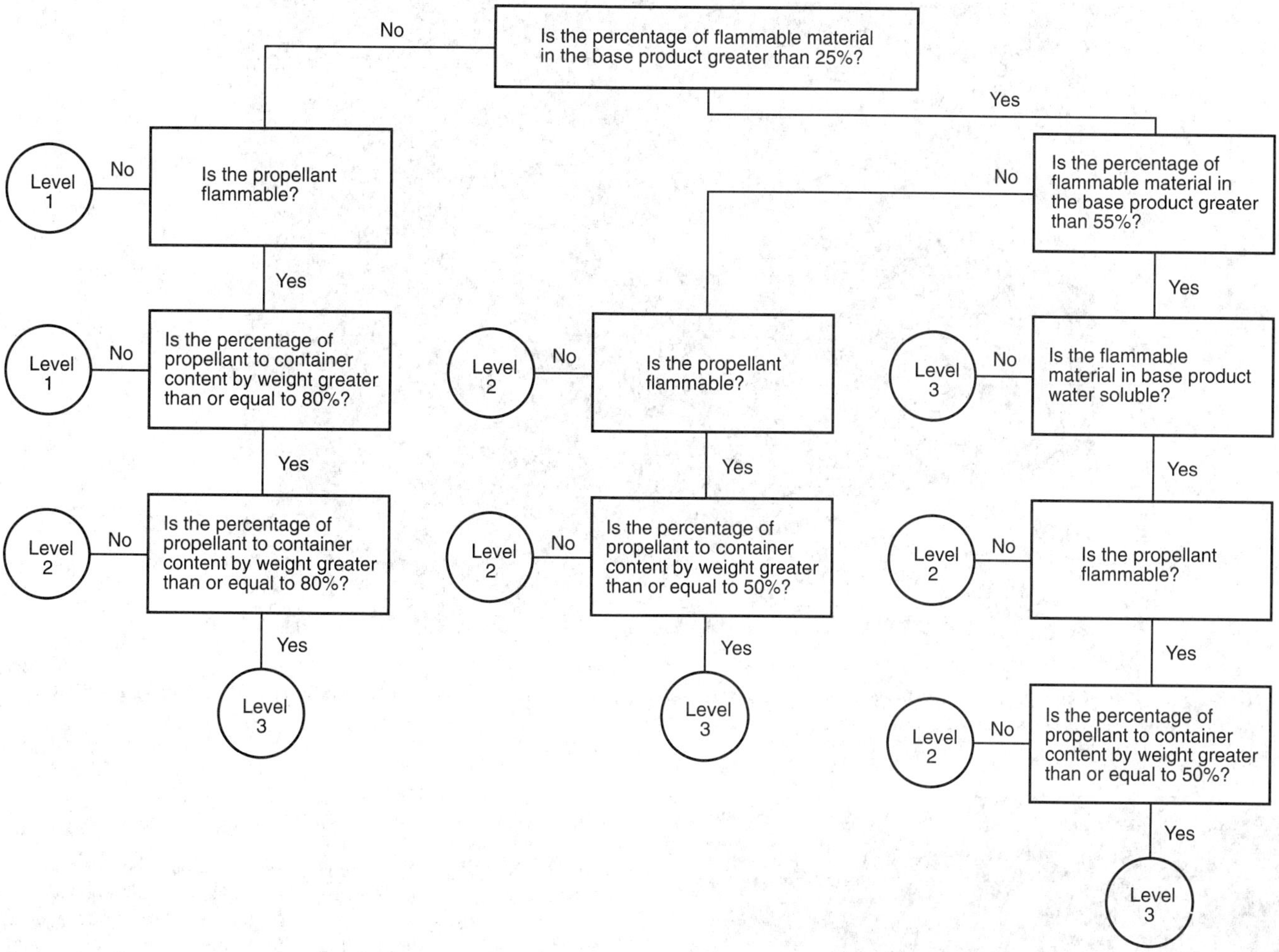

FIG. 3-20B. *Decision tree for determining level classification of aerosol products.*

One of the most notable fires involving aerosol products destroyed a 27-acre (10.9 ha) distribution center operated by the Kmart® Corporation in Falls Township, PA, in 1982. A fire in palletized storage of Level 3 aerosol products overwhelmed the automatic sprinkler protection. Rocketing aerosol cans quickly spread the fire beyond the aerosol storage area. The large volume of fire and failure of the automatic sprinkler protection led to the loss of structural integrity within the building. The loss, expressed in 1982 dollars, was estimated to be in excess of $100 million. (See Figures 3-20C and 3-20D.)

Aerosol products also were present in a fire that resulted in the loss of a large storage facility in Elizabeth, NJ, in 1985, and the loss of a coatings product distribution facility in Dayton, OH, in 1987. These, and similar incidents, indicate that aerosol storage arrangements should be maintained in configurations that are within the capability of the level of fire protection provided. Fire incidents involving aerosol filling facilities can expose workers and properties to fast-growing fires or explosions. Proper controls, interlocks, and safeguards should be provided to prevent a catastrophic fire scenario.

AEROSOL FILLING PLANTS

Fire and explosion hazards within aerosol filling operations are controlled using a number of layers of protection. These layers are:

(1) incident prevention, (2) fire/explosion suppression, and (3) incident magnitude mitigation.

Aerosol filling facilities, properly designed to mitigate fire and explosion hazards, are subdivided into several areas. Propellant gases are stored in remotely located, bulk storage facilities. Further subdivisions can occur in multiple-line filling operations and filling components, such as propellant pumps. The general manufacturing area and the propellant filling and charging operation require major separation.

Filling of aerosol cans with base product and charging with propellant must take place in an area separated from the main manufacturing area. Many modern designs use a propellant charging house that is physically separate from the main aerosol production building.

Empty aerosol cans are prepared in the manufacturing area. The containers are then conveyed to the charging area. Base product is added to the container, the actuator assembly is sealed into place, and the container is charged with propellant. After the charging and filling operation is completed, the container is conveyed back to the general manufacturing area for inspection and placement into storage.

The initial preparatory steps within the general manufacturing area typically present no unusual fire hazards. Combustibles in the area might include packaging materials, such as cardboard and

FIG. 3-20C. *Aerial view from the northwest shows total destruction of the warehouse areas. This photo was taken three days after the fire. (Jay Crawford, Bucks County Courier Times)*

FIG. 3-20D. *These aerosol cans were located in the area of fire origin of the Kmart fire.*

plastic wrap. The area will also customarily include large quantities of metal containers.

Within the charging facility, several severe fire hazards may exist. The aerosol base product will be present. This material may be a combustible or flammable liquid. Most importantly, propellant gas will be present in the area. Small quantities of propellant are lost to the atmosphere during charging operations. The propellant gas losses and any vapor evolved from the base product must be controlled by a properly maintained, dedicated ventilation system.

Several active prevention and protection systems should be provided. For prevention purposes, a combustible gas detection sys-

tem should be installed and interlocked to the charging area's ventilation system, equipment controls, and alarm system.

The charging area ventilation system should be designed by experienced professionals and should account for anticipated propellant losses during filling operations. Further, it must have the capability to provide high-volume air removal under emergency operations. When the combustible gas detection system senses conditions above 20 percent of the propellant's lower explosive limit (LEL), the ventilation rate in the charging area should be increased to at least 150 percent of the design airflow or two air changes per minute, whichever is greater. An alarm should also sound to alert operators to the condition.

Automatic equipment controls should also be provided. When the combustible gas detection system senses 40 percent of the propellant's LEL, equipment, such as the main propellant supply, should be shut down automatically. Another audible alarm should alert operators to this condition.

At 100 percent of the propellant's LEL, all charging equipment in the area should be shut down.

In the event that prevention measures fail, an explosion suppression system should also be provided. Such a system will respond quickly to suppress an explosion in its earliest stages. Wet-pipe automatic sprinkler protection should also be provided in the area. All protective systems should be interlocked to aerosol production equipment, so that equipment is shut down in the event of protective system activation.

In the event that both prevention and suppression measures fail, the design of the charging facility should provide for relief of deflagrations. Structural elements in the charging area, located adjacent to major buildings or assets, should be designed to withstand the anticipated overpressure of a deflagration. Deflagration vents, directed away from buildings and personnel, should be provided at the charging area exterior walls.

As filled aerosol cans are returned from the filling area to the main production area, quality testing is performed to identify leaking cans. Cans are conveyed through warm test baths, and defective units removed to disposal containers. Reject-can containers should be provided with active ventilation to remove propellant gas and/or base-product vapors.

PRODUCT STORAGE

Storage arrangements for aerosol products depend, to a great degree, on the amount of material stored by the manufacturer or distributor. NFPA 30B provides guidance for the storage of limited amounts of aerosol products in assembly, business, educational, industrial, or institutional occupancies. Storage in a single fire area within these occupancies for Level 2 and Level 3 products should be limited to 1,000 lb (454 kg) and 500 lb (227 kg), respectively. If larger amounts of aerosols are required, storage should be in separate inside storage areas or flammable liquid cabinets.

Large amounts of aerosol products should be stored in dedicated storage facilities, such as an aerosol warehouse. An aerosol warehouse is separated from other occupancies by either distance or a fire-rated assembly, or both. Fire suppression features in an aerosol warehouse must be appropriate for the storage configuration of the facility.

Wet-pipe, automatic sprinkler protection is recommended for aerosol storage. The design basis for sprinkler protection principally depends upon: (1) the classification of the aerosol product, and (2) the method of storage, i.e., rack storage or palletized storage. Live fire testing has produced substantial knowledge of appropriate sprinkler design. Standard spray sprinklers, large orifice sprinklers, extra-large orifice sprinklers, and early suppression, fast response (ESFR) sprinklers may be selected for the protection of stored aerosol products.

For the design and evaluation of sprinkler protection, Level 1 aerosol products are considered to be a Class III commodity, as defined by NFPA 231, *Standard for General Storage,* and NFPA 231C, *Standard for Rack Storage of Materials.* Protective criteria for these commodities can be drawn from the standard appropriate for the storage configuration.

Sprinkler protection for palletized or solid-pile storage of aerosol products should be provided in accordance with Table 3-20B or Table 3-20C.

Sprinkler protection for rack storage of aerosol products should be provided in accordance with Tables 3-20D, 3-20E, 3-20F, or 3-20G.

TABLE 3-20B. *Arrangement and Protection of Palletized and Solid-Pile Level 2 Aerosol Storage**

Max. Ceiling Ht (ft)	30	30	25	25
Maximum Pile Ht (ft)	5	15	18	20
Sprinkler	Standard or large orifice	ESFR (K = 13.5 to 14.5)‡	Large drop 0.64 in.	ESFR (K = 13.5 to 14.5)‡
Temp. Rating†	High	Ordinary	Ordinary	Ordinary
Sprinkler Spacing (ft²)	100 max.	80-100	80-100	80-100
Sprinkler Demand	0.30 gpm/ft² over 2500 ft²	12 sprinklers at 50 psi	15 sprinklers at 50 psi	12 sprinklers at 50 psi
Hose Stream Demand (gpm)				
Duration (hr)	2	1	2	1

*All fire tests on which this table is based were conducted with spray, large drop, or ESFR sprinklers. This does not include spray or large drop sprinklers equipped with quick-response links. The Response Time Index (RTI) of spray and large drop sprinklers shall not be less than 181 (ft/sec)$^{1/2}$ [100 (meter/sec)$^{1/2}$].
†When sprinklers having higher temperature ratings are used, such as near unit heaters, refer to NFPA 13, *Standard for the Installation of Sprinkler Systems.*
For SI units: 1 ft = 0.305 m; 1 ft² = 0.0929 m; 1 gpm/ft² = 40.743 L/min/m²; 1 psi = 6.895 kPa; 1 gpm = 3.785 L/min.
‡Sprinkler discharge coefficient

TABLE 3-20C. *Arrangement of Protection for Palletized and Solid-Pile Level 3 Aerosols**

Maximum Ceiling Ht (ft)	30	30	25	20	20
Maximum Pile Ht (ft)	5	15	15	10	5
Sprinkler	Standard, large, or extra-large orifice‡	ESFR (K = 13.5 to 14.5)§	ESFR (K = 13.5 to 14.5)§	Large drop 0.64 in.	Standard orifice
Temp. Rating†	High	Ordinary	Ordinary	Ordinary	High
Sprinkler Spacing (ft²)	100 max.	80-100	80-100	80-100	100 max.
Sprinkler Demand	0.60 gpm/ft² over 2500 ft²	12 sprinklers at 75 psi	12 sprinklers at 50 psi	15 sprinklers at 75 psi	0.30 gpm/2500 ft²
Hose Stream Demand (gpm)					
Duration (hr)	2	1	1	2	2

*All fire tests on which this table is based were conducted with spray, large drop, or ESFR sprinklers. This does not include spray or large drop sprinklers equipped with quick-response links. The Response Time Index (RTI) of spray and large drop sprinklers shall not be less than 181 (ft sec)$^{1/2}$ [100 (meter/sec)$^{1/2}$].
†When sprinklers having higher temperature ratings are used, such as near unit heaters, refer to NFPA 13, *Standard for the Installation of Sprinkler Systems.*
‡Extra-large orifice sprinklers shall have a minimum operating pressure of 10 psi (69 kPa).
For SI units: 1 ft = 0.305 m; 1 ft² = 0.0929 m²; 1 gpm/ft² = 40.743 L/min/m²; 1 psi = 6.895 kPa; 1 gpm = 3.785 L/min.
§Sprinkler discharge coefficient

TABLE 3-20D. *ESFR (K = 13.5 to 14.5) Arrangement and Protection of Level 2 Rack Storage**

Max. Ceiling Ht (ft)	30	25
Max. Storage Ht (ft)	15	20
Temp. Rating†	Ordinary	Ordinary
Sprinkler Spacing (ft²)	80-100	80-100
Sprinkler Demand	12 sprinklers at 50 psi	12 sprinklers at 50 psi
Hose Stream Demand (gpm)	250	250
Duration (hr)	1	1

*Single and double-row racks only.
†When sprinklers having higher temperature ratings are used, such as near unit heaters, refer to NFPA 13, *Standard for the Installation of Sprinkler Systems.*
For SI units: 1 ft = 0.305 m; 1 ft² = 0.0929 m²; 1 gpm/ft² = 40.743 L/min/m²; 1 psi = 6.895 kPa; 1 gpm = 3.785 L/min.

TABLE 3-20E. *ESFR (K = 13.5 to 14.5) Arrangement and Protection of Level 3 Rack Storage**

Max. Ceiling Ht (ft)	30	25
Max. Storage Ht (ft)	15	15
Temp. Rating†	Ordinary	Ordinary
Sprinkler Spacing (ft²)	80-100	80-100
Sprinkler Demand	12 sprinklers at 75 psi	12 sprinklers at 50 psi
Hose Stream Demand (gpm)	250	250
Duration (hr)	1	1

*Single and double-row racks only.
†When sprinklers having higher temperature ratings are used, such as near unit heaters, refer to NFPA 13, *Standard for the Installation of Sprinkler Systems.*
For SI Units: 1 ft = 0.305 m; 1 ft² = 0.0929 m²; 1 gpm/ft² = 40.743 L/min/m²; 1 psi = 6.895 kPa; 1 gpm = 3.785 L/min.

TABLE 3-20F. *Protection of Rack Storage of Level 2 Aerosols with Spray Sprinklers*

Level	Ceiling Sprinkler Arrangement	In-Rack Sprinkler Arrangement*	Clearances: Storage to Sprinklers†	Ceiling Demand	In-Rack Sprinkler Demand	Duration: Sprinklers and Hose Stream
2	High temperature, 100 ft² max. spacing; standard or large orifice	Ordinary temperature sprinklers 8 ft apart max. One line at each tier except top. Locate in longitudinal flue spaces double-row racks.	15 ft max. If clearance exceeds 15 ft (4.6 m), a barrier with face sprinklers below is required.	0.30 gpm/ft² over 2500 ft²	30 gpm per sprinkler minimum. Base on operation of hydraulically most remote: (1) 8 sprinklers if one level. (2) 6 sprinklers each of 2 levels if only 2 levels. (3) 6 sprinklers on top 3 levels if 3 or more levels.	2 hr

*For multiple-row storage, refer to Figure 4-2(a) of NFPA 30B where distance between transverse flues does not exceed 6 ft (1.8 m); refer to Figure 4-2(b) of NFPA 30B where distance between transverse flues exceeds 6 ft (1.8 m).
†Provide at least 6 in. (150 mm) between sprinkler deflectors and top of storage in tier. If clearance exceeds 15 ft (4.6 m), a barrier with face sprinklers below is required.
For SI units: 1 ft = 0.305 m; 1 ft² = 0.0929 m²; 1 gpm/ft² = 40.743 L/min/m²; 1 psi = 6.895 kPa; 1 gpm = 3.785 L/min.

TABLE 3-20G. *Protection of Rack Storage of Level 3 Aerosols with Spray Sprinklers*

Level	Ceiling Sprinkler Arrangement	In-Rack Sprinkler Arrangement	Clearance: Storage to Sprinklers†	Ceiling Demand	In-Rack Sprinkler Demand	Duration Sprinklers and Hose Stream
3	High temperature, large or extra-large* orifice, 100 ft² max. spacing	Ordinary temperature sprinklers 8 ft apart max. Install in longitudinal flue and on face of each tier except top tier.	5 ft (1.5 m) or less	0.30 gpm/ft² over 2500 ft²	30 gpm per sprinkler minimum. Base on operation of hydraulically most remote: (1) 8 sprinklers if 1 level. (2) 6 sprinklers each of 2 levels if only 2 levels. (3) 6 sprinklers on top 3 levels if 3 or more levels.	2 hr
			More than 5 ft (1.5 m) to 15 ft (4.6 m)	0.60 gpm/ft² over 1500 ft² to 2500 ft². Interpolate for clearances between 5 ft and 15 ft.	30 gpm per sprinkler minimum. Base on operation of hydraulically most remote: (1) 8 sprinklers if 1 level. (2) 6 sprinklers each of 2 levels if only 2 levels. (3) 6 sprinklers on top 3 levels if 3 or more levels.	2 hr

*Extra-large orifice sprinklers shall have a minimum operating pressure of 10 psi (69 kPa).
†At least 6 in. (150 mm) shall be provided between sprinkler deflectors and top of storage.
For SI units: 1 ft = 0.305 m; 1 ft² = 0.0929 m²; 1 gpm/ft² = 40.743 L/min/m²; 1 psi = 6.895 kPa; 1 gpm = 3.785 L/min.

TABLE 3-20G *(Continued)*

Level	Ceiling Sprinkler Arrangement	In-Rack Sprinkler Arrangement	Clearance: Storage to Sprinklers†	Ceiling Demand	In-Rack Sprinkler Demand	Duration Sprinklers and Hose Stream
			More than 15 ft (4.6 m) or more than 5 ft (1.5 m) where barriers are used	0.30 gpm/ft² over 2500 ft² plus a barrier above top tier of storage with face sprinklers below	30 gpm per sprinkler minimum. Base on operation of hydraulically most remote: (1) 8 sprinklers if 1 level. (2) 6 sprinklers each of 2 levels if only 2 levels. (3) 6 sprinklers on top 3 levels if 3 or more levels.	2 hr
3 maximum ceiling height: 30 ft (9 m)	High temperature, large or extra-large* orifice, 100 ft² max. spacing	Ordinary temperature sprinklers 8 ft (2.4 m) apart max. One line at each except top. Locate in longitudinal flue. For multiple-row racks.	Up to 15 ft (4.6 m)	0.60 gpm/ft² over 2500 ft²	30 gpm per sprinkler minimum. Base on operation of hydraulically most remote: (1) 8 sprinklers if 1 level. (2) 6 sprinklers each of 2 levels if only 2 levels. (3) 6 sprinklers on top 3 levels if 3 or more levels.	2 hr
			More than 15 ft (4.5 m)	60 gpm/ft² over 2500 ft² plus a barrier above top tier with face sprinklers below	30 gpm per sprinkler minimum. Base on operation of hydraulically most remote: (1) 8 sprinklers if 1 level. (2) 6 sprinklers each of 2 levels if only 2 levels. (3) 6 sprinklers on top 3 levels if 3 or more levels.	2 hr

*Extra-large orifice sprinklers shall have a minimum operating pressure of 10 psi (69 kPa).
†At least 6 in. (150 mm) shall be provided between sprinkler deflectors and top of storage.
For SI units: 1 ft = 0.305 m; 1 ft² = 0.0929 m²; 1 gpm/ft² = 40.743 L/min/m²; 1 psi = 6.895 kPa; 1 gpm = 3.785 L/min.

BIBLIOGRAPHY

References

Best, R., "$100 Million Fire in Kmart® Distribution Center," *Fire Journal*, Vol. 77, No. 2, March 1983, pp. 36–42.
Factory Mutual Research Corporation, "Storage of Aerosol Products—ESFR Protection," Technical Advisory Bulletin, Filed with LPDS 7–29S, Factory Mutual Research Corp., Norwood, MA.
Factory Mutual Research Corporation, "Storage of Aerosol Products," *Loss Prevention Data Sheet 7–29S*, 1983, Factory Mutual Research Corp., Norwood, MA.
Factory Mutual Research Corporation, "Aerosol Flammability Test," *Technical Advisory Bulletin*, filed with *Loss Prevention Data Sheet 7–29S*, Factory Mutual Research Corp., Norwood, MA.
Hydrocarbon Propellants: Considerations for Effective Handling in the Aerosol Plant and Laboratory, Chemical Specialties Manufacturers Association, Washington, DC, 1979.
Sanders, P.A., PhD., *Handbook of Aerosol Technology*, 2nd ed., New York, 1979.
UL 913, *Intrinsically Safe Apparatus and Associated Apparatus for Use in Class I, II, and III, Division I, Hazardous (Classified) Locations*, Underwriters Laboratories Inc., Northbrook, IL, 1988.

NFPA Codes, Standards, and Recommended Practices

Reference to the following NFPA codes, standards, and recommended practices will provide further information on aerosol charging operations discussed in this chapter. (See the latest version of *The NFPA Catalog* for availability of current editions of the following documents.)

NFPA 11A, *Standard for Medium- and High-Expansion Foam Systems*

NFPA 12, *Standard on Carbon Dioxide Extinguishing Systems*
NFPA 12A, *Standard on Halon 1301 Fire Extinguishing Systems*
NFPA 13, *Standard for the Installation of Sprinkler Systems*
NFPA 15, *Standard for Water Spray Fixed Systems for Fire Protection*
NFPA 30, *Flammable and Combustible Liquids Code*
NFPA 30B, *Code for the Manufacture and Storage of Aerosol Products*
NFPA 45, *Standard on Fire Protection for Laboratories Using Chemicals*
NFPA 58, *Standard for the Storage and Handling of Liquefied Petroleum Gases*
NFPA 68, *Guide for Venting of Deflagrations*
NFPA 69, *Standard for Explosion Prevention Systems*
NFPA 70, *National Electrical Code®*
NFPA 77, *Recommended Practice on Static Electricity*
NFPA 91, *Standard for Exhaust Systems for Air Conveying of Materials*
NFPA 231, *Standard for General Storage*
NFPA 231C, *Standard for Rack Storage of Materials*
NFPA 704, *Standard System for the Identification of the Fire Hazards of Materials*

Additional Reading

"Aerosol Storage: The Problems and Solutions," P8207, Factory Mutual Research Corporation, Norwood, MA.
"Assessing the Flammability of Aerosols in Warehouses," *Fire Prevention*, No. 201, July/August 1987, pp. 24–28.
Benedetti, R. P., ed., *Flammable and Combustible Liquids Code Handbook*, 5th ed., National Fire Protection Association, Quincy, MA, 1993.
DeHaan, J. D., and Howard, W. A., "Combustion Explosions Involving Household Aerosol Products," *Proceedings*, Twentieth International Conference on Fire Safety, Volume 20, Product Safety Corp., Sunnyvale, CA, 1995, pp. 67–71.

"ESFR Sprinklers: Your Way to Fight Aerosol Fires," *Sentinel*, Second Quarter 1991, p. 6.

Gewain, R. G., "Fire Protection of Aerosol Products," *Industrial Fire Safety*, Vol. 2, No. 1, Jan./Feb. 1993, pp. 27–31.

"Protecting Aerosol Production Supplies," *Sentinel*, Second Quarter 1991, p. 7.

Rizzo, J. J., "The Development of the duPont Aerosol Flammability Test," *Fire Journal*, Vol. 81, No. 2, 1987, p. 32.

Roberts, D. J., "Aerosols and Fire Prevention," *Fire Prevention*, No. 217, March 1989, pp. 24–27.

"Storing Aerosol Products Safely," *Sentinel*, Second Quarter 1991, pp. 3–9.

STORAGE OF FLAMMABLE AND COMBUSTIBLE LIQUIDS

Revised by Anthony M. Ordile

The first requirement for the storage of flammable and combustible liquids is properly designed containers that are liquidtight and from which release of vapors, if required, is carefully controlled. The containers range from large, upright, outdoor storage tanks of millions of gallons (1,000,000 gals = 3785 m³) capacity down through drums to small cans containing ounces of liquid.

This chapter discusses the various types of containers used in the storage and transportation of flammable and combustible liquids and the precautions that should be observed in handling the liquids as they are loaded, unloaded, or dispensed.

Additional information relevant to the storage of flammable and combustible liquids can be found in Section 1, Chapter 6, "Explosions"; Appendix A, "Tables and Charts"; Section 4, Chapter 1, "Fire Hazards of Materials: An Overview"; Section 3, Chapter 2, "Control of Electrostatic Ignition Sources"; Section 5, Chapter 8, "Gas and Vapor Testing"; and Section 6, "Suppression."

TANK STORAGE

Tanks may be installed aboveground, underground, or, under certain conditions, inside buildings. Openings and connections to tanks, for venting, gaging, filling, and withdrawing, can present hazards if they are not properly safeguarded.

Given substantially constructed, properly installed, and well-maintained tanks, storage of flammable and combustible liquids is less dangerous than transferring these liquids. The severity of the storage hazard might seem to depend upon the quantity stored. As a practical matter, however, the size of the tank or the number of tanks is less important than such factors as the characteristics of the liquid stored, the design of the tank and its foundation and supports, the size and location of vents, and the piping and its connections.

Flammable liquids expand when heated. Gasolines expand about 0.07 percent in volume for each 10°F (5.5°C) increase in temperature within ordinary atmospheric temperature ranges. The effect of temperature increase on the volume of acetone, ethyl ether, and certain other flammable liquids with higher coefficients of expansion is greater than in the case of gasolines. To avoid danger of overflow, tanks should not be filled completely, particularly where

cool liquid is placed in a tank in a warm atmosphere—for example, when filling an automobile tank from an underground gasoline tank on a warm day.

Several methods are used to prevent storage evaporation loss and loss of vapors as the tank is filled. Underground tanks reduce evaporation losses, since there is less fluctuation of temperature. Aboveground tanks often are painted with aluminum or white paint to reflect heat, thus decreasing the temperature rise of liquid contents and slowing evaporation of tank contents. Floating roof tanks minimize vapor loss and tend to reduce the fire hazard. Storage of gasoline in pressurized tanks reduces the loss of vapor. In some cases, vapors are conserved by the use of lifter roof or vapor dome tanks, or the vents from several cone roofed tanks may be connected through manifolds to a vapor dome or pressure-type tank.

The vapor space in tanks storing flammable liquids with vapor pressures above 4 psia (28 kPa), such as gasoline, is normally too rich to burn. The ratio of vapor to air is above the upper flammable (explosive) limit (UFL). However, if the temperature of the liquid gasoline is in the range of –10 to –50°F (–23 to –46°C), the vapor space may be within the flammable range. When a tank is being off-loaded, or when there is a sudden cooling rainstorm on a hot day, a portion of the tank vapor spaceway may be within flammable limits, resulting from introduction of air. This condition may remain for several hours or even days, due to stratification of the vapors.

The vapor space in tanks storing low vapor pressure liquids (below approximately 2 psia or 14 kPa), such as kerosene, is normally too lean to burn. The ratio of vapor to air is below the lower flammable (explosive) limit (LFL). However, if the entire body of liquid is heated to its flash point, as may happen during refining processes or by exposure to fire, the vapor space may enter the flammable range. It should be noted that it is the temperature of the liquid and not the temperature of the vapor space that determines the presence of a flammable vapor-air mixture. The oil vapors driven off by the heated air in the vapor space are condensed back to liquid by the cooler body of oil. Therefore, the vapors are only flammable for a very short distance above the liquid surface despite the fact that the air within the tank may be considerably above the flash point temperature.

Tank storage of ethyl and methyl alcohol, JP-4 or Jet B turbine fuel, and other liquids of similar vapor pressure (approximately 2 to 4 psia at 100°F or 14 to 28 kPa at 38°C) presents an unusual hazard, since the vapors are normally in the flammable range. Storage in floating roof or similar tanks or the addition of inert gas in the vapor space is desirable to reduce the possibility of an explosion in

Anthony M. Ordile, P.E., is a consulting engineer employed by Loss Control Associates, Inc., a fire protection and process safety consulting firm in Langhorne, PA. He is a member of the NFPA Technical Committee on Flammable and Combustible Liquids.

the vapor-air mixture in the tank. Floating roof tanks, cone roof tanks with internal floating roofs, lifter roof tanks, and vapor dome tanks are also used for vapor conservation purposes for Class I liquids.

Aboveground Storage Tanks

Storage tanks come in a variety of design and construction. They are divided into three general categories of pressure design: (1) atmospheric tanks, for pressures of 0 to 1.0 psig (0 to 6.9 kPa); (2) low-pressure storage tanks, for pressures from 1.0 to 15 psig (6.9 to 103 kPa); and (3) pressure vessels, for pressures above 15 psig (103 kPa). Some of the more common types of aboveground storage tanks are shown in Figures 3-21A and 3-21B. Pressure tanks and pressure vessels normally are used for vapor conservation purposes, particularly for liquids with high vapor pressures.

In recent years, aboveground secondary containment-type tanks have become increasingly more popular for the atmospheric storage of flammable and combustible liquids. The popularity of these tanks has been aided by federal regulations now imposed on underground tanks. Various types of aboveground secondary containment tanks are manufactured and approved for use. These tanks incorporate a primary steel tank that is either completely encased by another steel or fiberglass tank or a geomembrane shell so that any leakage from the primary tank is contained by the outer tank or shell. For protection against exposure fires and vandalism, many aboveground secondary containment tanks are encased with 6 in. (152 mm) of reinforced concrete. Requirements covering size limitations, pipe connections, and overfill protection are contained in NFPA 30, *Flammable and Combustible Liquids Code* (hereinafter referred to as NFPA 30).

Construction: All aboveground storage tanks should be built of steel or concrete, unless the character of the liquid necessitates the use of other materials. Both steel and concrete tanks resist heat from exposure fires. Tanks built of materials less resistant to heat (e.g., low melting point materials) might result in tank failure and fire spread.

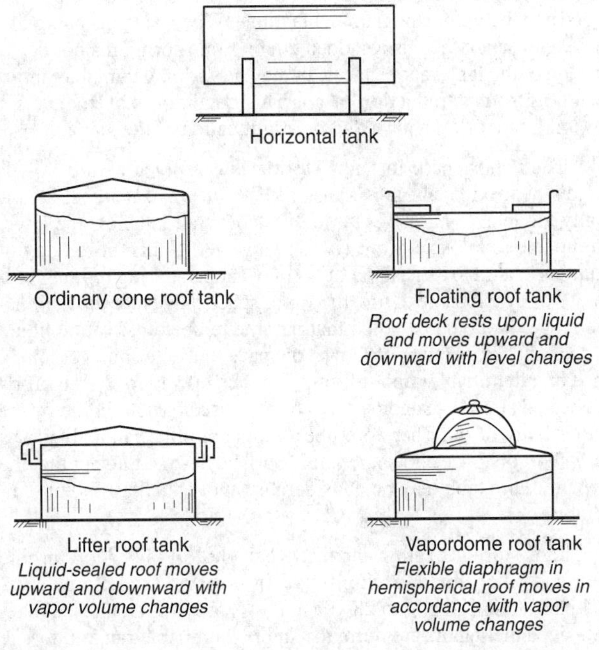

FIG. 3-21A. *Common types of atmospheric storage tanks.*

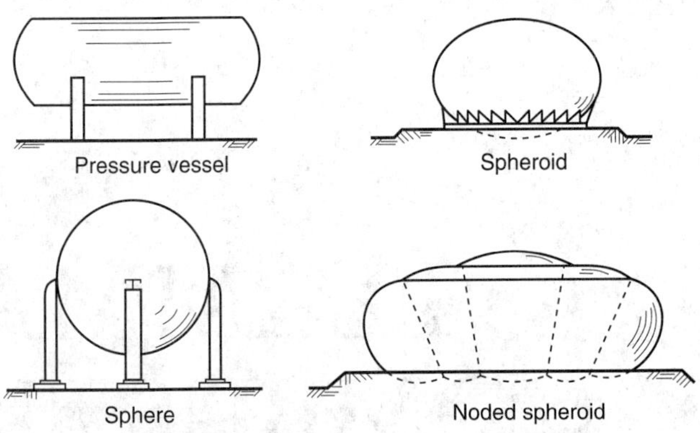

FIG. 3-21B. *Common types of low-pressure tanks or pressure vessels.*

The thickness of the metal used in tank construction is based not only on the strength required to hold the weight of the liquid, but also on an added allowance for corrosion. When intended for storing corrosive liquids, the specifications for the thickness of the tank shell are increased to provide additional metal and allow for the expected service life of the tank. In some cases, special tank linings are used to reduce corrosion. Periodic inspection should be made to ascertain metal thickness of the tank, establish safe operating limits, and avoid overstressing the tank. The inspection of tanks for corrosion may be through the use of visual, magnetic particle, ultrasonic, and radiographic methods; or by experience gained from the storage of similar materials. Ultrasonic devices operate on a principle of the length of time it takes for sound waves to reflect from a surface. Any difference in metal thickness is quickly disclosed by these instruments, which are particularly effective when large areas with many potential corrosion spots are involved.

For atmospheric, vertical, cylindrical, and aboveground welded tanks, API Standard 650, *Welded Steel Tanks for Oil Storage*,[1] may be used in calculating the minimum thickness of shell plates. The required nominal thickness of shell plates must be the greater of the design shell thickness, including any corrosion allowance, or the hydrostatic test shell thickness, but in no case less than the values given in Table 3-21A.

TABLE 3-21A. *Shell Plate Thickness*

Nominal Tank Diameter (ft)	Nominal Thickness (in.)
Smaller than 50	$3/16$
50 to, but not including, 120	$1/4$
120 to 200, inclusive	$5/16$
Over 200	$3/8$

For SI units: 1 ft = 0.304 m; 1 in. = 25.4 mm.

Those tanks provided with the label of Underwriters Laboratories Inc. (UL), or built in accordance with American Petroleum Institute (API) standards, meet the required construction specifications.

Concrete tanks require special engineering. Unlined concrete tanks should be used only for the storage of liquids with a specific gravity of 40 degrees API or heavier. Concrete tanks with special linings are permitted for other services, provided the design is in accordance with appropriate engineering practices. Tanks built of material other than steel should be designed to specifications embodying equivalent steel safety factors.

Installation: Aboveground tanks should be installed in accordance with NFPA 30, which specifies distances from tanks to property lines that can be built upon, and to public ways or important buildings. These distances will vary depending upon whether the pressures in the tanks, including those attained during fire exposure, can be under or over 2.5 psig (17 kPa), as well as whether the contents are stable or unstable liquids, or liquids having boil-over characteristics. Other factors affecting distances are the size and design of the tank, protection for exposures, fire extinguishing or control systems provided, and other protection features. The spacing between tanks is also specified in NFPA 30.

When a tank is located in an area that may be subjected to flooding, applicable provisions in NFPA 30 should be followed.

Venting and flame arresters: An appropriate vent must be provided for normal operation of any tank to permit filling and emptying and for the maximum expansion or contraction of the tank contents with changes in temperature. API Standard 2000, *Venting Atmospheric and Low-Pressure Storage Tanks,*[2] provides information on sizing vents for product movement and breathing. Clogged or inadequately sized vents may result in the rupture of tanks from internal pressure, or collapse due to internal vacuum. During the filling operation, the vents discharge flammable vapors. If the mixture is sufficiently rich or if the vent location is such that the released vapor is a hazard, the vapors should be piped to a safe outside location at least 12 ft (3.7 m) above grade. The vapor release should not occur under eaves, or close to doors, windows, or possible sources of ignition.

Venting devices are normally closed when the tank is not under pressure or vacuum. Devices installed on vent pipes can be approved pressure-vacuum conservation (breather) vent valves or flame arresters to prevent flashback into tanks. (See Figures 3-21C and 3-21D.) These devices are required when a flammable mixture is present, and for the storage of all Class I liquids. Arresters constructed of banks of parallel metal plates or tubes having a large surface of metal to dissipate heat are more effective than screens for larger openings and are less subject to clogging and corrosion. Heat is absorbed by the metal plates or tubes, which lower the temperature of the vapor below its self-ignition point. However, if the flame arrester is exposed to long burning periods, the metal plates can become sufficiently hot on the bottom side to ignite any flammable vapor-air mixtures in the tank. Approved flame arresters are cellular arresters that include perforated-plate arresters, parallel-plate arresters, crimped-metal-ribbon arresters, and sintered metal arresters. The two critical parameters that govern the effectiveness of cellular arresters are: (1) diameter or width and (2) length of the flame passages.

Vent design should be suitable for venting tank vapor space in accordance with API 2000[2] and NFPA 30 and should not restrict vapor relief under design conditions.

Where the liquids stored have flash points in the range of normal summer temperatures, the vapor space above the liquid in the tank will normally contain vapors in the flammable range. Flame arresters have their most important application on such tanks. However, condensation and crystallization of certain liquids and freezing of moisture in winter may make conservation vents and flame arresters impractical. Steam tracing (steam heating) is provided in some cases to prevent freezing or crystallization of contents.

Although a wire screen of 40 mesh* ordinarily will prevent the passage of flame through small openings, it cannot be relied upon as an effective flame arrester because of possible physical damage to the wires or clogging of the mesh by dirt or other residues.

Pressure-relief devices: Appropriate venting of aboveground tanks and vessels may also be achieved by the use of pressure-relief

---

*A woven wire fabric in which there are 40 wires per in. (25.4 mm) in each direction and 1,600 interstices per sq in. (645 mm²). The size of the openings depends upon the diameter of the wires and upon the mesh of the screens, but the size of wire is approximately uniform for any given mesh, and specifying a screen in terms of its mesh determines the size of the openings with sufficient accuracy for practical purposes. The individual opening in a 40-mesh screen has an area of about 0.00022 sq in. (0.14 mm²).

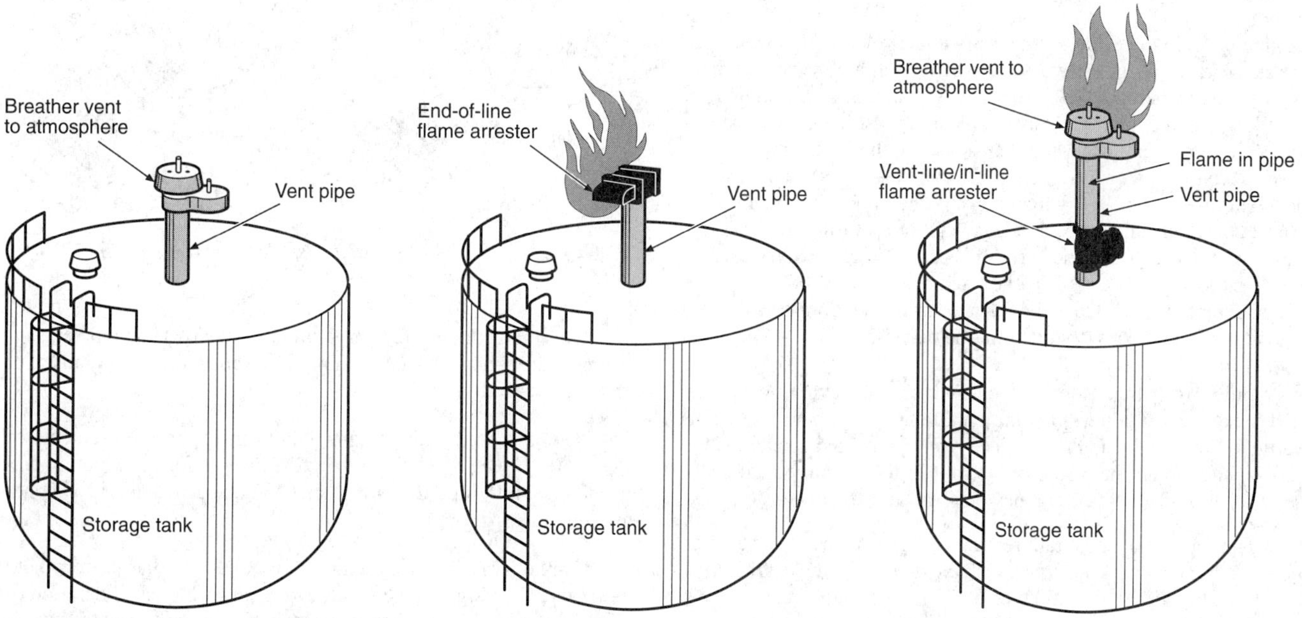

FIG. 3-21C. *Typical arrangement of pressure-vacuum conservation (breather) vent valves and flame arresters. (Source: The Protecto-seal Company.)*

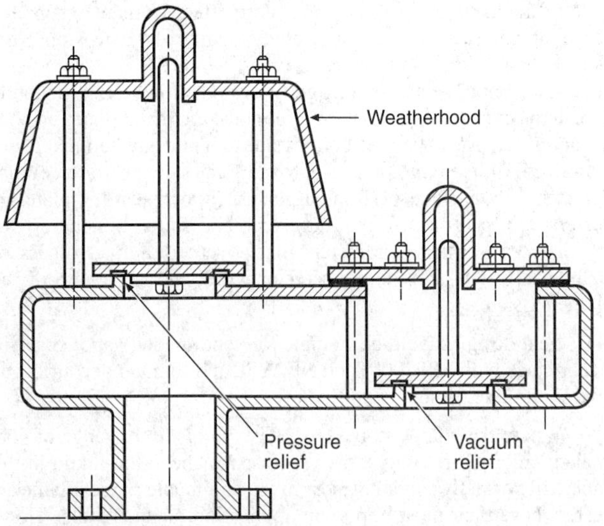

FIG. 3-21D. *Typical pressure-vacuum conservation (breather) vent. (Source: The Protectoseal Company.)*

devices. These devices are actuated by inlet static pressure and are designed to open during an emergency or abnormal conditions to prevent a rise of internal pressure in excess of a specified value within the tank or vessel.

Pressure-relief valves are frequently of the spring-loaded type, which are designed to automatically reclose after actuation and prevent the further release of gas, vapor, or liquid. (See Figure 3-21E.)

Other types of pressure-relief devices are: (1) pilot-operated pressure-relief valves that include a pressure-relief valve combined with and controlled by an auxiliary pressure-relief valve and (2) rupture disk devices that are non-reclosing differential pressure-relief devices actuated by inlet pressure and designed to function by bursting the rupture disk. (See Figure 3-21F.)

Emergency venting: In addition to the normal operating vents, emergency relief of internal pressure is required for most aboveground tanks in the event a fire occurs under or around the tank. In the absence of a proper provision for such relief, high pressures may be generated when the tank is exposed to external fire, resulting in a boiling liquid expanding vapor explosion (BLEVE). Such explosions are infrequent, but when they do occur, the results may be disastrous to life and property. BLEVEs can be avoided by providing adequate pressure relief, which permits the vapors to escape and burn at the vents and prevents the tank from rupturing. Emergency relief venting may take the form of loose manhole covers that lift under pressure, weak roof-to-shell seams, floating roofs, rupture disks, remotely actuated vapor depressuring, or commonly used pressure-relief devices designed for the purpose.

Unless they are adequately vented, horizontal cylindrical tanks under excessive internal pressure commonly fail at the ends. For vertical cone-roofed tanks designed with weakened seams at the roof-to-shell joint, the lifting of the roof or top of the roof-to-shell seam affords adequate emergency pressure relief.

The danger of tank failure from internal pressure when exposed to fire depends to a considerable extent upon the characteristics of the liquid, the size and type of the tank, and the intensity and duration of the fire. The smaller the tank or the smaller the volume of the liquid in the tank, the shorter the time under fire exposure before a BLEVE occurs.

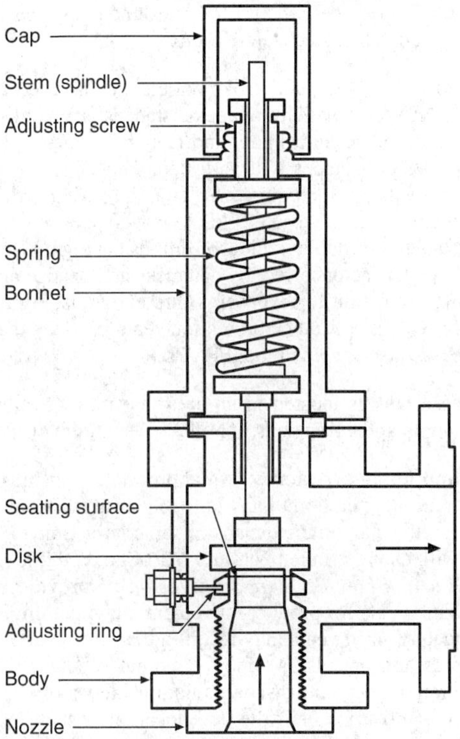

FIG. 3-21E. *Typical pressure-relief valve device of the spring-loaded type. (Source: API 520)[3]*

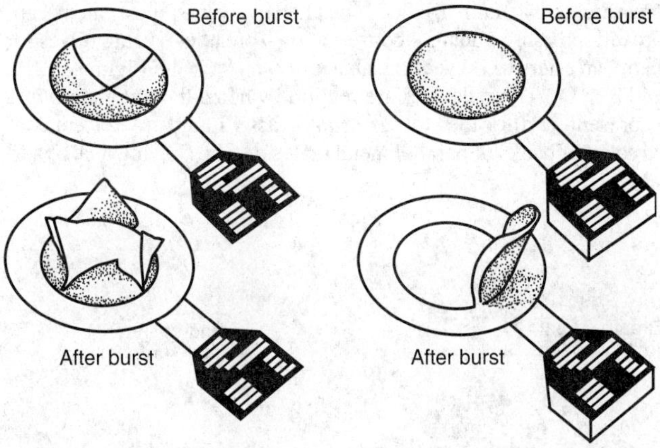

FIG. 3-21F. *Types of rupture disk devices. (Source: API 520)[3]*

On small cone-roofed tanks, under 20 ft (6 m) in diameter, supplemental venting may be needed to prevent preferential failure of the shell-to-bottom seam.

Dome-roofed tank designs should meet the requirements of API 650, *Welded Steel Tanks for Oil Storage,*[1] for construction and venting. Dome-roofed designs should not be mistaken for cone-roofed designs.

Table 3-21B is based upon the required discharge from both normal and emergency vents. It is derived from a consideration of the probable maximum rate of heat transfer per unit area; the size of tank and the percentage of total area likely to be exposed; the time required to bring tank contents to boil; the time required to heat unwet portions of the tank shell or roof to a temperature where the metal will lose strength; and the effect of drainage, insulation, and the application of water in reducing fire exposure and heat transfer. Due to the wide variance in flow characteristics of manufactured vents, each vent design under 8 in. (203 mm) should be flow tested rather than relying upon vent size. The flow capacity of larger vents may be flow tested or calculated in accordance with the formula in NFPA 30.

The wetted area referred to in Table 3-21B is the internal portion of the tank in contact with the liquid (assuming a full tank) that is expected to be exposed to the flame of an external ground fire. The requirements for emergency venting are based on the heating of the liquid in the tank. Therefore, the wetted area (exposed to an external ground fire, such as a spill) varies with different tank designs: 55 percent of the total exposed area of a sphere or spheroid, 75 percent of the total exposed area of a horizontal tank, and the first 30 ft (9 m) abovegrade of the exposed shell area of a vertical tank.

For tanks and storage vessels designed for pressures over 1 psig (6.9 kPa), the total rate of venting is given in Table 3-21B. When the exposed wetted area or the surface is greater than 2,800 sq ft (260 m²), the total rate of venting is as given in Table 3-21C, or it can be calculated by the following formula:

$$CFH = 1.17A^{0.82}$$

where

CHF = venting requirement, in cu ft or m³ of free air per hour, and

A = exposed wetted surface, in sq ft or m².

TABLE 3-21B. *Wetted Area Vs. Amount of Free Air per Hour Required for Relief Venting for Aboveground Tanks*
(For purposes of calculating the capacity of venting devices, "free air" is defined as air at 14.7 psia and 60°F or 101 kPa and 15.5°C)

ft²	m²	ft³/hr	m³/hr
20	1.86	21,100	597
30	2.78	31,600	895
40	3.71	42,100	1,192
50	4.64	52,700	1,492
60	5.57	63,200	1,790
70	6.50	73,700	2,037
80	7.43	84,200	2,384
90	8.36	94,800	2,675
100	9.29	105,000	2,973
120	11.15	126,000	3,568
140	13.00	147,000	4,163
160	14.86	168,000	4,757
180	16.72	190,000	5,380
200	18.58	211,000	5,975
250	23.23	239,000	6,768
300	27.87	265,000	7,504
350	32.51	288,000	8,155
400	37.16	312,000	8,834
500	46.45	354,000	10,024
600	55.74	392,000	11,100
700	65.03	428,000	12,120
800	74.32	462,000	13,082
900	83.61	493,000	13,960
1,000	92.90	524,000	14,838
1,200	111.48	557,000	15,772
1,400	130.06	587,000	16,622
1,600	148.64	614,000	17,387
1,800	167.22	639,000	18,095
2,000	185.80	662,000	18,746
2,400	222.96	704,000	19,935
2,800	260.12	742,000	21,011

Note: Interpolate for intermediate values.

For SI units:

$$m^3/hr = 220\,A^{0.82}$$

The foregoing formula is based on $Q = 21,000A^{0.82}$.

The total emergency relief venting capacity for any specific liquid may be determined by the following formula:

$$\text{Cu ft of free air per hour} = V\frac{1,337}{L\sqrt{M}}$$

For SI units:

$$m^3/hr = \frac{3,107}{L\sqrt{M}}$$

where

V = cu ft or m³ of free air per hour from Table 3-21B,

L = latent heat of vaporization of specific liquid in Btu per lb or kJkg, and

M = molecular weight of specific liquids.

The required flow rate may be multiplied by the appropriate factor listed in the following schedule when protection is provided as indicated (only one factor may be used for any one tank):

- 0.5 for approved drainage for tanks over 200 sq ft (18.6 m²) of wetted area
- 0.3 for approved water spray and drainage
- 0.3 for approved insulation
- 0.15 for approved water spray with approved insulation and drainage

The above factors may be reduced by 50 percent for water-miscible liquids with heats of combustion and rates of burning equal to or less than those of ethyl alcohol (ethanol), but in no case can the factors be reduced to less than 0.15.

When pressure tanks or vessels are exposed to fire, a violent BLEVE may result if the steel in the vapor space is softened by heat. Localized overheating of the tank shell has been caused by vent fires impinging on tank surfaces. The outlets of all vents and vent drains on aboveground tanks designed for 2.5 psi (17 kPa) or greater should be arranged to prevent localized overheating of any part of the tank from a vent fire.

TABLE 3-21C. *Rate of Venting for Tanks with an Exposed Wetted Area Larger than 2800 Sq Ft (260 m²)*

ft²	m²	ft³/hr	m³/hr
2,800	260	742,000	21,011
3,000	279	786,000	22,257
3,500	325	892,000	25,258
4,000	372	995,000	28,175
4,500	418	1,100,000	31,149
5,000	464	1,250,000	35,396
6,000	557	1,390,000	39,360
7,000	650	1,570,000	44,457
8,000	743	1,760,000	49,838
9,000	836	1,930,000	54,652
10,000	929	2,110,000	59,749
15,000	1,393	2,940,000	83,252
20,000	1,858	3,720,000	105,339
25,000	2,322	4,470,000	126,577
30,000	2,787	5,190,000	146,965
35,000	3,251	5,900,000	167,070
40,000	3,716	6,570,000	186,043

Foundations and supports: Tanks should be set on firm foundations. They should be adequately supported to minimize the possibility of uneven settling and to minimize the corrosion in any part of the tank resting on the foundation. Vertical tanks are normally set on a slightly elevated concrete pad to provide a sound base, and normally above the adjacent ground level to protect the bottom of the tank from any water in the area. Exposed pilings or steel supports under any flammable liquid tank should be protected by fire-resistive materials with a fire resistance rating of not less than 2 hours.

In certain situations, the use of water spray protection on tank supports may be approved for use, in lieu of providing fire-resistive materials.

Drainage and dikes: Where important facilities, waterways, or adjoining properties would be endangered by release of liquids stored in tanks, it is necessary to provide a means to control any spillage. The most desirable method is to locate the tanks on sloping ground. Directional dikes or drainage ditches could then divert spillage away from all tanks and into an impounding basin where the liquid could burn safely without exposing other tanks, property, or waterways. Where impounding basins cannot be used, dikes built around the tanks can prevent the spread of liquid. Dikes may be constructed of earth, concrete, solid masonry, or steel built to withstand the lateral pressure of a full liquid head. When several large tanks are in a single diked enclosure, it may be desirable to place drainage channels or intermediate dikes between tanks. Intermediate dikes, which should be at least 18 in. (457 mm) in height, are effective in preventing small spills from exposing other tanks in the diked enclosure. Such spills may result from a leaking valve or connection or from overfilling the tank.

Diked enclosures should be designed to contain the greatest amount of liquid that can be released from the largest tank within the enclosure, assuming a full tank. When calculating the volumetric capacity of the diked enclosure, the volume of tanks within the diked area to the height of the dikes must be considered as unavailable space for the liquid to flow.

Where needed, diked areas are provided with trapped drains to remove rain water or water used for fire fighting. The best practice is to keep drain valves normally closed, and open them at intervals as needed, since permanently open drains would discharge liquids in case of leakage from the tank. The drain valves should also be accessible under fire conditions, i.e., they should be located outside the dike. Oil separators are effective in skimming off oil flowing on the surface of water, but separators normally will not control oil discharge when the entire flow through the drain is oil.

Dikes higher than 6 ft (1.8 m) are not desirable without additional safety measures. Where circumstances require high, close dike walls, means should be provided for access to tanks, valves, and other equipment, and for safe egress from the diked enclosure. Where the average height of a dike containing Class I liquids is over 12 ft (3.6 m) high or where the distance between a tank and the inside edge of the dike wall is less than the height of the dike wall, provisions should be made for normal operation of valves and for access to tank roofs without entering below the top of the dike. These provisions reflect a concern for the accumulation of Class I liquid vapors reaching harmful levels when confined in the narrow space between the high dike wall and the tank.

Aboveground secondary containment-type tanks meeting the requirements of NFPA 30 are not required to meet the provisions for drainage or diking.

Fire hazards of aboveground tanks: A commonly encountered type of aboveground tank installation includes tanks on unprotected steel supports. Tank truck loading can be accomplished by gravity when done immediately adjacent to these tanks. The common fire at such installations originates at the tank vehicle, and, in short order, the unprotected steel supports fail and drop the tanks to the ground. The tanks may break the piping or rupture on hitting the ground, releasing their entire contents. To prevent tank collapse, steel supports should be protected by 2-hr fire- resistive coverings. The loading rack should be at least 25 ft (7.6 m) from the tanks if Class I flammable liquids are handled, and 15 ft (4.6 m) with Class II and Class III liquids.

Improper and inadequate emergency venting have been factors in tank failures during exposure to fire. Since inadequate and improperly designed vents have been the cause of severe BLEVEs and have resulted in deaths and injuries to fire fighters and spectators, it is of vital importance that the standards on venting be followed. In fighting tank fires, it is essential to cool the shell above the liquid level to keep the steel from overheating. Failure to do so could cause the tank to bulge and rupture, regardless of the adequacy of the venting.

Also, a factor in fires involving storage tanks has been the failure of piping and valves. Such failures have resulted in the addition of their contents to the ground fire. Pipe systems may be located either aboveground or underground. Piping from aboveground tanks is normally placed aboveground to avoid corrosion problems and to aid in detecting pipe leaks. If piping is in open trenches, suitable fire barriers should be placed at certain intervals to prevent the flow of liquid from one section of the facility to another. Underground piping is not subject to fire exposure. All piping must be protected against physical damage and excessive stresses that arise from expansion, contraction, vibration, and settlement. Materials that are subject to failure due to thermal shock, such as cast iron, and low melting point materials such as aluminum, copper, and brass, should not be used. Steel or nodular iron pipe and valves should be used for external tank connections through which liquid would normally flow, unless the chemical characteristics of the liquid stored are incompatible with steel. Welded pipe connections or flanged joints are preferred for aboveground piping, particularly for large-size pipe. Screwed pipe connections larger than 3 in. (76 mm) are subject to disengagement in a long-exposure fire unless the connections are back-welded. Pipe joints dependent upon the friction characteristics of combustible materials for mechanical continuity of piping are subject to failure under fire exposure conditions.

Many storage tank fires originate with an internal explosion or by a spill fire exposing the tanks. In a prolonged tank fire, the shell of a large vertical tank is likely to fold into the tank above the burning liquid level, without splitting the tank shell.

Internal explosions of storage tanks can occur when the vapor space above the liquid reaches the flammable range. Liquids with flash points near the stored temperature are most susceptible to ignition. Liquids with low flash points will have a flammable vapor-air mixture in the vapor space if the temperature is markedly reduced, and high-flash-point liquids will form a flammable vapor-air mixture when heated. During fire exposures, internal explosions have occurred in tanks containing liquids with flash points above the stored liquid temperature. Such explosions occur because the liquid is heated by the exposure fire and the vapor passes into the flammable range when a part of the tank shell in the vapor space is hot enough to ignite the vapors.

Floating roof tanks provide greater fire safety for aboveground installations and, as a result, are permitted to be located closer to a property line than are other types of tanks. Explosions may occur, however, when floating roof tanks are virtually empty of liquid and the roof has been allowed to rest on the low-level supports, thus creating a flammable vapor space. When fires occur in floating roof tanks, they are normally restricted to the seal space between the roof and the shell and often can be extinguished by portable extinguishers or handheld foam lines. However, there have been a few cases of full-surface fires that were the result of the roof sinking. Roof sinking can occur as a result of the roof hanging up due to overfill or tank obstructions, excessive snow loading or rain water, and excessive water or foam from fire-fighting activities. A good tank roof and seal maintenance program, along with proper tank operating procedures, is the best way to prevent problems resulting in fires. (See Figure 3-21G.)

Fire protection equipment should be installed on tanks where needed. (See NFPA 11, *Standard for Low-Expansion Foam.*)

Underground Storage Tanks

Underground tanks are generally considered the safest form of storage. However, recent concern about groundwater contamination has increased concern about use of underground storage tanks. Environmental safety regulations should be reviewed when planning installation of underground storage tanks.

Underground tanks must be designed to withstand safely the service to which they are subjected, including the pressure of the earth, concrete, or possible aboveground vehicle traffic. Typical sizes for underground tanks are shown in Table 3-21D.

TABLE 3-21D. *Typical Sizes of Underground Flammable Liquid Tanks*

Capacity		Diameter			Length		
gal	L	ft	in.	m	ft	in.	m
300	1,136	3	0	0.91	6	0	1.83
560	2,120	4	0	1.22	6	0	1.83
1,000	3,785	4	0	1.22	11	0	3.35
1,000	3,785	5	4	1.62	6	0	1.83
1,000	3,785	4	0	1.22	11	0	3.35
1,000	3,875	5	4	1.62	6	0	1.83
1,500	5,678	5	4	1.62	9	0	2.74
2,000	7,571	5	4	1.62	12	0	3.66
2,500	9,463	5	4	1.62	15	0	4.57
3,000	11,356	5	4	1.62	18	0	5.49
3,000	11,356	6	0	1.83	14	0	4.27
4,000	15,142	5	4	1.62	24	0	7.32
4,000	15,142	6	0	1.83	19	0	5.79
5,000	18,927	6	0	1.83	24	0	7.32
6,000	22,712	6	0	1.83	29	0	8.84
6,000	22,712	8	0	2.44	16	0	4.88
7,500	28,390	8	0	2.44	20	0	6.10
10,000	37,854	8	0	2.44	27	0	8.23
10,000	37,854	9	0	2.74	21	0	6.40
10,000	37,854	10	0	3.05	17	0	5.18
10,000	37,854	10	6	3.20	15	7	4.75

Construction: Rigid specifications established by UL ensure reasonable safety for those tanks bearing its label. Underground tanks also may be of unlined concrete for the storage of liquids having a specific gravity of 40 degrees API or heavier. Lined concrete tanks may be used for liquids having a lighter specific gravity, provided the lining is suitable for the liquid being stored and has satisfactory adherence to the concrete.

Installation: Underground tanks may be buried outside or under buildings. Tanks that are buried underneath buildings should have fill and vent connections outside the building walls. Tanks should be set on firm foundations and surrounded with soft earth or sand that is well tamped into place.

Excavation for underground storage tanks should be made with due care to avoid undermining of foundations of existing structures.

FIG. 3-21G. Double-seal system for floating roof tanks. (Source: NFPA 11, Standard for Low-Expansion Foam.)

Underground tanks must be protected against damaging loads imposed by the cover over the tanks and such factors as building foundations and vehicular traffic. Normally, no special protection is needed if the tanks are well supported underneath and buried sufficiently deep. However, if tanks are located in areas where higher-than-normal loads may be imposed, concrete, paving, or additional earth coverage may be necessary. Piping subject to possibly damaging loads or vibrations is frequently protected by sleeves, casings, or flexible connectors to ensure the integrity of the line.

Workmanship is the largest cause of failure and leaks in underground tanks. Generally, poorly installed piping systems connected to underground tanks will show evidence of leakage in the first year. The normal life expectancy of a properly installed underground steel tank with either a fiberglass outer wall or cathodic protection is 25 to 30 years.

Environmental regulations or a hazard analysis may require the use of a double-walled tank design and/or the use of groundwater monitoring for tank and piping leaks. The soil in which the tank is buried is very important, since some soils, such as red clay, may be highly corrosive because of their chemical composition or moisture content. This is also true if construction debris, cinders, shale, or other foreign matter is mixed, even in very small quantities, with otherwise "clean" backfill. The use of homogeneous, clean backfill and protective coatings prolongs the life of steel tanks and piping. Cathodic protection of buried tanks and piping is often necessary. NFPA 30 requires that tanks and their piping be protected either by using corrosion-resistant materials of construction, such as special alloys and fiberglass, or by using a properly engineered, installed, and maintained cathodic protection system in accordance with recognized standards of design.

Electrolytic corrosion in an electrically conductive tank and piping system may occur at points where metals having different electromotive qualities, such as steel and brass, are connected. Connections of two dissimilar metals should be avoided to prevent galvanic corrosion.

Stray electrical currents may initiate corrosive action, but their presence may be difficult to determine until after the action has progressed to the point where damage has been done to the tank or piping. Cathodic protection or insulation is sometimes used to protect underground tanks from stray currents.

Fiberglass-reinforced plastic (FRP) tanks for underground installation eliminate the corrosion problem encountered with steel tanks. However, it is vitally important that the manufacturer's instructions for proper installation of these tanks be followed closely. Care should be taken in the use of FRP tanks to ensure that liquids stored in the tanks are not destructive to the plastic tank construction.

Tanks should be anchored or weighted to prevent floating in locations where the groundwater level is high or may rise in case of flood. Details of installation and protection are covered in NFPA 30.

The proximity of underground tanks to a building foundation is not a direct measure of the potential danger to the building if a leak should develop in the tanks. Leaking contents from underground tanks have been known to travel several miles underground (1 mile equals 1.6 km) or to penetrate 24 in. (610 mm) of waterproofed concrete before appearing in a building. Tanks suspected of leaking should be tested hydrostatically with the same liquid stored in the tank. (See Figure 3-21H.) Air tests or testing with liquids other than that stored in the tank have proved to be dangerous and inconclusive for detecting suspected leaks in underground tanks. Additional details about the causes of corrosion, locating leaking tanks, and removal of liquids in the ground are covered in NFPA 329, *Recommended Practice for Handling Underground Releases of Flammable and Combustible Liquids.*

Tanks Inside Buildings and Storage Tank Buildings

Tank installations storing flammable and combustible liquids are permitted inside buildings when meeting the requirements of NFPA 30 for storage tank buildings. A tank installation that has a canopy or roof that does not limit the dissipation of heat or dispersion of vapors and does not restrict fire-fighting access and control is considered by NFPA 30 as an outside aboveground tank installation and not a storage tank building.

Design: Tanks designed for installation within buildings have the same shell thickness and design features as required for outside aboveground storage tanks. In certain specialized processing operations, large tanks are required and are designed as low-pressure storage tanks or pressure vessels.

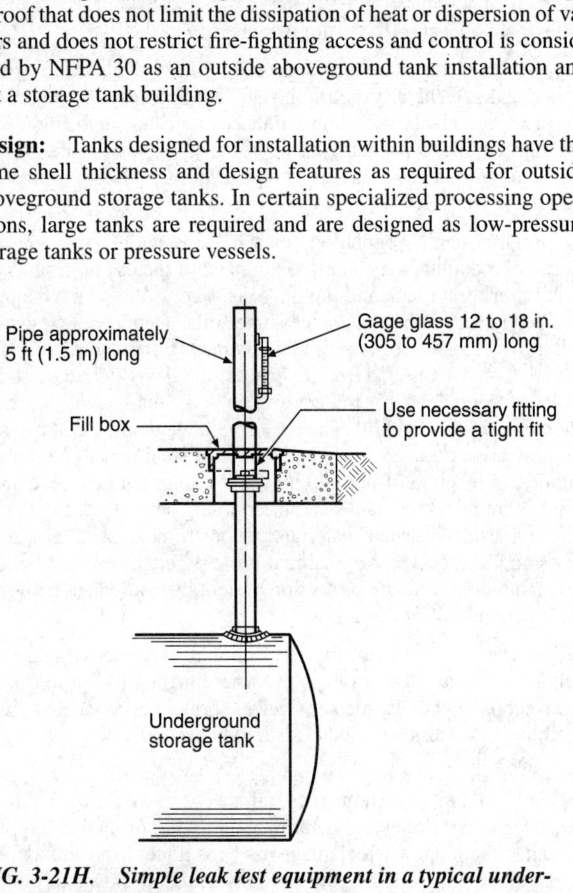

FIG. 3-21H. Simple leak test equipment in a typical underground tank installation.

For the storage of flammable and combustible liquids in accordance with NFPA 30, the capacity of any individual storage tank within a building is limited to 100,000 gal (378,500 L), unless the Authority Having Jurisdiction permits the use of larger tanks. Storage tank buildings should be constructed so as to maintain their structural integrity for 2 hr under fire exposure conditions and to provide adequate access and egress for unobstructed movement of all personnel and fire protection equipment.

For the storage of liquid petroleum products in accordance with NFPA 30A, *Automotive and Marine Service Station Code,* tanks enclosed in substantially liquid- and vapor-tight enclosures without backfill are permitted inside buildings where it is impractical because of property or building limitations to install the tanks aboveground or underground in accordance with NFPA 30. Such enclosures are constructed of 6 in. (152 mm) of reinforced concrete, with access through the top of the enclosure. Means are also provided for using portable ventilating equipment to discharge flammable vapors outside the building.

For the storage of fuel oil in accordance with NFPA 31, *Standard for the Installation of Oil-Burning Equipment,* unenclosed supply tanks in buildings connected to oil-burning appliances are limited to one 660-gal (2,500-L) tank or two tanks of an aggregate capacity of 660 gal (2,500 L). Fuel oil tanks larger than 660 gal (2,500 L) may be installed in buildings, provided the tanks are within a 3-hr fire-resistive enclosure.

Installation: Tanks and any associated equipment within a storage tank building should be located so that a fire in the area should not constitute an exposure hazard to adjoining buildings or tanks for a period of time consistent with the response and suppression capabilities of the fire-fighting operations available to the location.

For specific requirements covering the installation of storage tank buildings, with respect to exposed property lines, public ways, and important buildings, see NFPA 30.

Storage tank buildings storing Class I liquids or Class II or Class IIIA liquids at temperatures above their flash points are required to be ventilated at a rate sufficient to maintain the concentration of vapors within the building at or below 25 percent of the lower flammable limit. The tank vents should be arranged to ensure that flammable vapors are not released inside the building. Also, electrical equipment and wiring specified and installed in accordance with NFPA 70, *National Electrical Code®,* is required to prevent a source of ignition for any flammable vapors that may be present under normal operations or during a spill.

Drainage systems should be provided in storage tank buildings to minimize fire exposure to other tanks and to prevent the discharge of flammable or combustible liquids to public waterways, public sewers, and adjacent properties.

Storage tank buildings should also have fire prevention control systems and methods for: (1) life safety, (2) minimizing property loss, and (3) reducing fire exposure to adjoining operations and property. Such systems and methods include emergency alarm systems; automatic sprinkler, foam-water, water spray, or deluge systems; and thermally or remotely actuated valves at liquid transfer connections, along with programs to control ignition sources and static electricity.

Gaging Tanks

Tank openings for gaging or measuring the quantity of liquid may permit the escape of vapors during the gaging operation. Such openings are particularly undesirable in tanks located in buildings or buried under basements and are prohibited by NFPA standards unless protected by a spring-loaded check valve or other approved device.

Substitutes for manual gaging include heavy-duty flat-gage glasses; magnetic, hydraulic, or hydrostatic remote reading devices; and sealed float gages. These devices, however, must be maintained in reliable operating condition. Ordinary gage glasses should not be used, since their breakage may permit the escape of liquid.

Cleaning Tanks

Cleaning tanks for the purpose of making repairs requires great care. Precautions must be taken to avoid igniting flammable vapors and to protect personnel against toxic vapors.[4,5]

Work on empty tanks must be performed only under the supervision of persons well versed in fire and explosion hazards and in the procedures required to properly safeguard the operation. Unless the work can be done with the tank and all connected piping and fittings completely filled with water, all vapors should be removed by cleaning with steam or chemical solutions, or by displacement with water, air, or inert gas. In some instances, the tank vapor space may be filled with inert gas by specially qualified personnel to provide a safe atmosphere. The selection of the method of rendering a tank safe will depend upon several factors, such as the character of the liquid, the size of the tank, the flammability and reactivity of residues, and the type of work to be performed. With many reactive materials, it will be necessary to obtain information from the manufacturer regarding recommended practices for safe cleaning.

Removal of flammable vapors by displacement: This is sometimes referred to as "purging" and may be accomplished by one of several methods.

1. *Displacement with water:* Where the flammable liquid previously contained is known to be readily displaced by, or soluble in, water, it can be removed completely by using water to alternately fill and drain the tank. The operation should be repeated several times until tests with a combustible gas indicator show that the vapors are no longer present. Acetone and ethyl alcohol are examples of water-soluble liquids.
2. *Displacement with air:* Frequently, flammable vapors may be removed by purging with air through the use of venturi-type air movers or low-pressure blowers, and a safe atmosphere is sustained by continuous ventilation. Air movers should be restricted to those operated by steam, by air, or by electric motors approved for use in the atmosphere involved. When steam operates the air mover, the air mover should be bonded or in electrical contact with the tank. When small tank openings cannot accommodate an air mover, the contents can sometimes be purged with compressed air connected to a metallic pipe bonded to the tank. Care should be exercised to prevent overpressuring a small tank if compressed air is used. Irregularly shaped tanks may not be thoroughly purged by this method because the tank may contain pockets that cannot be reached effectively by the air stream. In air purging, the concentration of flammable vapors in air in the tank may go through the flammable range before safe atmospheres are obtained. Therefore, all precautions must be taken to minimize the hazards of ignition by static electricity. Air movers should be clamped or bolted, and thereby inherently bonded, to the vessel being ventilated. (See Figure 3-21I.)
3. *Displacement with inert gas:* When nitrogen in cylinders, or carbon dioxide in low-pressure containers or in solid form (dry ice), is available in sufficient quantity, it may be used to purge the flammable vapors from tanks without the hazards of having the vapor-air mixture in the tank vapor space pass through the flammable range. This procedure should be followed by air ventilation of the tank. High-pressure carbon dioxide, such as that from a fire extinguisher, should not be used because static electricity is generated. Several explosions or fires have occurred when CO_2 extinguishers were used to inert tanks or vessels.

Inerting the vapor space: Inerting is a means of safeguarding a tank so that work can be performed on it. Inerting a tank can be accomplished by reducing the oxygen content to the point where combustion cannot take place in the vapor space. However, individuals in direct charge of the work must be thoroughly familiar with the limitations and characteristics of the inert gas being used. Attempting such work without proper knowledge or equipment can be hazardous because a false sense of security is engendered. The oxygen content should be maintained at substantially zero during the entire period work is in progress. Inerting gases include carbon dioxide and nitrogen. Both may be obtained in pressurized cylinders, and carbon dioxide may also be obtained in solid form.

Removal of residues: The removal of liquid and solid residues is necessary, as they may release flammable vapors during repair operations involving "hotwork." The removal of residues can be accomplished by steam or chemical cleaning or by other recognized methods. In steam cleaning, the rate of steam supply should be sufficient to exceed the rate of condensation, and the steam nozzle should be electrically bonded to the container shell. During the steam cleaning process, the entire container is, of course, heated close to the boiling point of water [212°F (100°C)]. Chemical cleaning may be necessary to remove some residues, but this may present personnel health hazards that should be guarded against.

Although a tank's atmosphere is safe for entry, the removal of residues may cause the release of additional hazardous and/or flammable vapors. During such work conditions, continuous ventilation can maintain the vapor concentration at a safe limit. Tanks can be continuously ventilated by air movers at the roof manholes. Continuous monitoring of the tank's atmosphere will provide notification of potentially hazardous vapor levels.

Special care must be taken to eliminate any source of ignition in the vicinity of the tank or in the path of vapors being displaced. Bonding, either with special bond wires or by metal-to-metal contact with clamped or bolted connections providing inherent bonding, must be used where static-producing devices are used. All electrical equipment, such as inspection lights and motors, used in connection with the cleaning operations must be designed for such use.

Tests for the presence of flammable vapors constitute the most important phase of the cleaning or safeguarding procedure and must be made before commencing any alterations or repairs; immediately

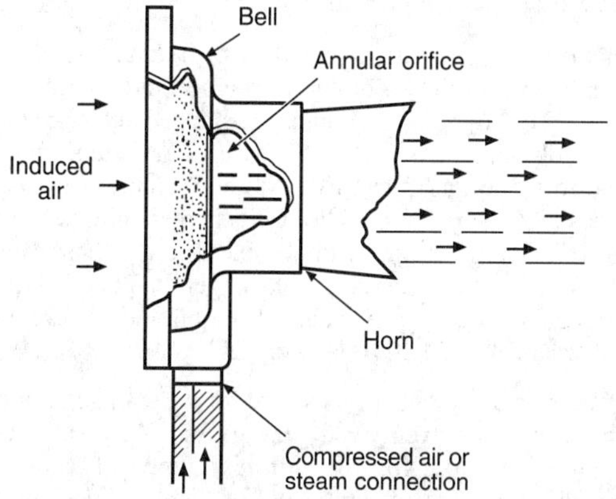

FIG. 3-21I. Schematic drawing of the operation of an air remover. Compressed air or steam is admitted to the bell of the air remover and enters the horn through the annular orifice around the base. As the steam or compressed air passes through the horn, its expansion induces a rapid flow of air, equal to approximately ten times the volume of the steam or compressed air.

after starting any welding, cutting, or heating operations; and frequently during the course of such work. The tests made with a combustible-gas indicator in good working order will normally produce reliable readings. Considerable care should be exercised to ensure that the indicator is calibrated for the vapors involved, correctly scaled, and properly used and read, and that the readings are interpreted and applied properly. Where an inert gas is used, the oxygen content of the tank is measured to determine whether a hazardous condition exists. This may be determined directly by the use of an oxygen indicator, or indirectly by the use of an indicator showing the concentration of the inert gas being used. All testing must be conducted by persons experienced in the operation and in the limitations of the indicators used.

When hotwork is to be done on small tanks or containers that cannot be entered, the combustible gas indicator should show no appreciable indication of the presence of flammable vapors. Extra precautions are necessary if the container last contained a high-flashpoint liquid, because no vapor reading will show on the indicator. If welding or cutting is done, the heat may vaporize some of the liquid, create a flammable vapor-air mixture in the drum, and explode it. Such containers require special precautions.

Hotwork may be performed safely, under qualified supervision, in tank cars, tank trucks, and tanks that can be entered when the flammable vapors are under 20 percent of the lower flammable limit. Additional precautions, such as protective clothing and self-contained breathing apparatus, are necessary to protect the health of persons entering tanks that have contained leaded gasoline or other highly toxic residues.

Where toxic materials have been stored in a tank, additional tests may be required to determine if the atmosphere is safe from the standpoint of health. For example, leaded gasoline tanks need to be lead-free before they can be entered without breathing equipment and specialized protective clothing.

More complete instructions for cleaning tanks or containers are covered in NFPA 327, *Standard Procedures for Cleaning or Safeguarding Small Tanks and Containers Without Entry;* NFPA 306, *Standard for the Control of Gas Hazards on Vessels;* API RP 2013, *Cleaning Mobile Tanks in Flammable or Combustible Liquid Service;*[4] and API RP 2015, *Cleaning Petroleum Storage Tanks.*[5]

OTHER STORAGE OF FLAMMABLE LIQUIDS

Flammable and combustible liquids are packaged, shipped, and stored in bottles, drums, and other containers ranging in size up to 60 gal (227 L). Additionally, liquids are shipped and stored in intermediate bulk containers up to 793 gal (3,000 L) and in portable tanks up to 5,500 gal (20,818 L). Storage requirements for these containers are covered in the NFPA 30 chapter entitled "Container and Portable Tank Storage," with the exception of those intermediate bulk containers and portable tanks larger than 660 gal (2,498 L) that are required to meet the applicable requirements covered in the NFPA 30 chapter entitled "Tank Storage."

Examples of container types used for the storage of liquids include glass, metal, polyethylene (plastic), and fiberboard. The maximum allowable size for the different types of containers is governed by the class of flammable or combustible liquid to be stored in it. (See NFPA 30.)

The principal hazard of closed container storage is the possibility of overpressure failure of the container when exposed to fire. This release of liquid adds to the intensity of a fire and may cause the rupture of other containers, resulting in a rapidly spreading fire. Fire tests have shown that ordinary automatic sprinkler systems

may be inadequate to control a fire involving drums of flammable liquids, or to prevent overpressuring of drums if flammable liquid containers are piled too high.

Container Storage in Buildings

Container storage of flammable and combustible liquids may exist in mercantile and industrial occupancies and in general-purpose and flammable liquid warehouses. Any liquid-storage building or room or any portion of a building or room where containers are stored should be designed to protect the containers from exposure fires. This may require the installation of fire walls or partitions.

The life hazard to the occupants of buildings, exposure to other buildings, building construction, and the degree of fire protection provided are factors to be considered when evaluating the amount of container storage in buildings. Details and limitations on closed container storage are given in NFPA 30.

Flammable liquid storage cabinets: Specially designed storage cabinets are available for storing not more than 120 gal (454 L) of Class I, Class II, and Class IIIA liquids, of which not more than 60 gal (227 L) should consist of Class I and Class II liquids. These cabinets are typically constructed of No. 18 gage sheet steel consisting of a double wall with $1^1/_2$-in. (38-mm) air space, although wood cabinets constructed in accordance with the requirements of NFPA 30 are permitted. (See Figure 3-21J.) No more than three such cabinets should be located in a single fire area, except in industrial occupancies where additional cabinets are permitted if separated by 100 ft (30.5 m), or where six cabinets are permitted if the building is sprinklered. (See NFPA 30.)

Hazardous Material Storage Lockers

Hazardous material storage lockers have become very popular in recent years, as they meet the environmental concerns dictated by the special handling of hazardous chemicals, some of which are flammable and combustible liquids. These lockers are movable, modular prefabricated storage buildings that provide a safe and cost-effective means of storing hazardous materials. The lockers are equipped with spill containment features and can be provided with electricity, mechanical ventilation, and fire suppression systems. Generally, the lockers are located outside, but NFPA 30 permits the lockers to be utilized as inside liquid storage rooms, as long as the lockers are constructed with the fire-resistive ratings and applicable requirements listed in NFPA 30. NFPA 30 limits the gross floor area of the lockers to 1,500 sq ft (139 m²). (See Figure 3-21K.)

Drum Storage Outdoors

Outdoor drum storage should be located in such a manner as to reduce the spread of fire to other materials in storage, adjacent buildings, or neighboring properties. (See Table 3-21E.) The storage area should be graded in a manner to divert possible spills away from buildings or other exposures, or curbing should be used to contain spills. Areas used for drum storage should be kept free of ignition sources, such as open flames, and combustibles, such as weeds and debris. The area should also be protected against tampering or trespassers, and smoking should be prohibited in the area.

HANDLING OF FLAMMABLE AND COMBUSTIBLE LIQUIDS

Loading and Unloading

Spills may occur at loading and unloading stations for tank cars and tank vehicles, which if ignited could endanger neighboring tanks

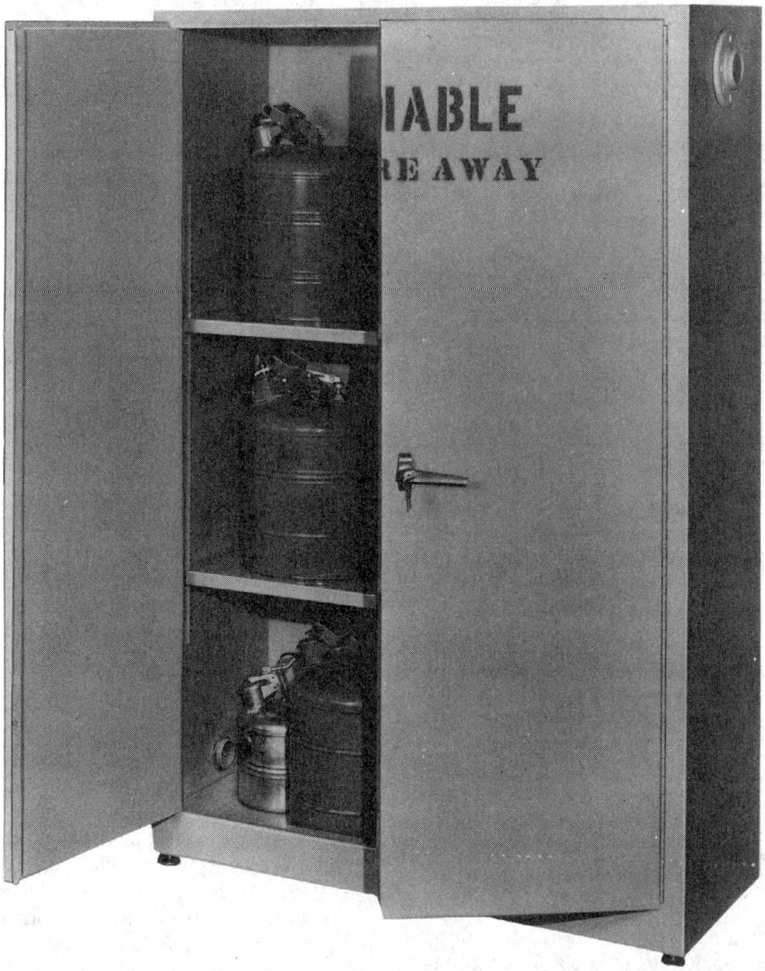

FIG. 3-21J. *Typical metal flammable liquid storage cabinet. Metal cabinets are constructed of at least No. 18 gage steel and double walled with 1½-in. (38-mm) air space on all sides. The door should have a three-point latch, with a sill raised to at least 2 in. (51 mm) above the bottom of the cabinet. The cabinet should be marked in conspicuous lettering: "FLAMMABLE — KEEP FIRE AWAY." (Source: Flammable and Combustible Liquids Code Handbook)*

FIG. 3-21K.   *Typical hazardous material storage lockers. [Source: JUSTRITE (left); The HAZ-STOR Company (right)]*

and buildings. Tank vehicle and tank car stations for Class I liquids should be located a minimum of 25 ft (8 m) from storage tanks, other plant buildings, and the nearest line of property that can be developed. For Class II and Class III liquids, the separation distance is 15 ft (4.6 m). Level ground is desirable, and drains, diversionary curbs, or natural ground slope can be utilized to prevent spills from spreading to other parts of the plant or to other property.

Many liquids, including gasoline, jet fuels, toluene, and light fuel oil, can build up dangerous static electrical charges on the surface of the liquid. If a flammable vapor-air mixture is present at the surface of the liquid when high static electrical discharges occur, an explosion or fire can result. Excessive turbulence, pumping two dissimilar materials, the free fall of liquid through the vapor space, and the use of filters capable of removing micron-sized particles

TABLE 3-21E. *Outdoor Liquid Storage in Containers and Portable Tanks*

	1				2				3		4		5	
Class	Container Storage Max. per Pile*§				Portable Tank Storage Max per Pile (gal)*				Distance Between Piles or Racks		Distance to Property Line That Can Be Built Upon†‡		Distance to Street, Alley, or Public Way‡	
	gal	m³	Height		gal	m³	Height							
	gal	m³	ft	m	gal	m³	ft	m	ft	m	ft	m	ft	m
IA	1,100	4.16	10	3.05	2,200	8.32	7	2.1	5	1.5	50	15.2	10	3.0
IB	2,200	8.32	12	3.66	4,400	16.65	14	4.3	5	1.5	50	15.2	10	3.0
IC	4,400	16.65	12	3.66	8,800	33.30	14	4.3	5	1.5	50	15.2	10	3.0
II	8,800	33.30	12	3.66	17,600	66.62	14	4.3	5	1.5	25	7.6	5	1.5
III	22,000	88.38	18	5.49	44,000	166.56	14	4.3	5	1.5	10	3.0	5	1.5

*See 4-7.1.1 of NFPA 30 regarding mixed class storage.
†See 4-7.1.3 of NFPA 30 regarding protection for exposures.
‡See 4-7.1.4 of NFPA 30 for smaller pile sizes.
§For storage in racks, the quantity limits per pile do not apply, but the rack arrangement must be limited to a maximum of 50 ft (15.2 m) in length and two rows of 9 ft (2.7 m) in depth.

are the most common causes of explosions or fires originating from static electricity in aboveground tanks, tank vehicles, tank barges, and tank ships. Where flammable vapors may exist, and to reduce static-caused ignitions, it is recommended that the delivery rate be slowed until the fill pipe is covered. A minimum 30-sec relaxation time should be provided downstream of a filter.

Bonding provisions for protection against static sparks must be provided at all loading stations when Class I liquids are loaded. The bonding must be provided between the fill pipe or piping and the tank vehicle. Bonding connections should be made before the dome covers are opened and should remain in place until filling is complete and all dome covers have been closed and secured. Closed metal piping systems (eliminating exposure to air) for loading and unloading may eliminate the need for special bonding provisions, since the piping system is inherently bonded to the vehicle. Bonding provisions are not required when Class II and Class III liquids are being loaded and when vehicles are loaded with products not having a static-accumulating tendency, such as asphalts, crude oils, and water-soluble liquids.

Stray electrical current protection is necessary for ship and tank car loading and unloading in those areas where stray currents may be present. To protect against stray currents at tank car stations, the fill pipe can be permanently bonded to at least one rail and to the rack structure. In addition, all pipes entering the rack area should be provided with insulating sections to electrically isolate the rack piping from the pipelines.

For further details on control of static, refer to NFPA 77, *Recommended Practice on Static Electricity*, and API Standard 2003, *Protection Against Ignitions Arising Out of Static, Lightning, and Stray Currents*.

Tank ships and barges are loaded and unloaded at oil piers that may or may not be used exclusively for this purpose. The location and arrangement of the oil piers is governed by direction and velocity of the waterway, the range of tides, and the direction and frequency of prevailing high-velocity winds. Hazards related to the loading and unloading of vessels are comparable to those for tank vehicles and tank cars, except the quantities of liquid involved are generally larger.

Piping and Valves

Piping systems protected against physical damage are preferable to portable containers for conveying quantities of liquids throughout buildings. All piping systems should be designed so that, in case of pipe breaks, liquid will not continue to flow by gravity or by siphoning. Valves should be provided at accessible points to control or stop the flow. Emergency remote controls are frequently provided for valves or pumps, particularly at dispensing locations. Other remotely controlled valves installed for normal operating procedures can frequently be used during fire emergency procedures to control or stop the flow of liquids. Valves are available that close automatically if subjected to fire conditions. Dispensing outlets for systems under gas pressure or gravity head should be equipped with self-closing valves.

Pipe materials should be used that: (1) are resistant to the corrosive properties of the liquid handled; (2) have adequate design strength to withstand the maximum service pressure (including shock and surge pressures that may be expected) and temperature; and (3) when possible, are resistant to physical damage and thermal shock. Where low melting point materials, such as aluminum and brass; materials that soften on fire exposure, such as plastics; or nonductile material, such as cast iron, are necessary, special consideration should be given to their behavior on exposure to fire. After installation, piping systems should be tested at 150 percent of the maximum anticipated pressure of the system, or pneumatically tested to 110 percent of the maximum anticipated pressure of the system, but not less than 5 psig (35 kPa) at the highest point of the system. The test should be maintained for a sufficient time to complete visual inspection of all joints and connections, but not for less than 10 min.

Since some valves are used infrequently, it is necessary to make periodic maintenance inspections to ensure that the valves will operate during emergency conditions. All aboveground piping and connections should be inspected periodically to detect and prevent leakage. Internal corrosion in pipes is a particular problem when liquids are corrosive, or where piping is used for the continuous flow of liquid under high-pressure operations common at refineries or chemical plants. Maintenance inspections of the thickness of pipe walls can be performed in several ways similar to the inspection of tanks. Underground piping is also subject to external corrosion and should be protected against such corrosion by suitable coatings and cathodic protection. Tests for possible leaks in underground piping can be performed hydrostatically, using the liquid normally handled in the piping system.

Dispensing and Handling Methods

Large quantities of flammable or combustible liquids are best transferred through piping by pumps. Gravity flow is not desirable, except as required in process equipment. If positive displacement pumps are used, they should be provided with a pressure relief that discharges back to the tank or to the pump section. Only inert gas should be used to transfer Class I liquids. The use of air pressure to transfer Class II and Class III liquids is acceptable, unless the liquids are heated above their flash point, in which case inert gas should be used. Whenever air or inert gas pressure systems are used for transfer, the vessels, containers, tanks, and piping systems should be capable of withstanding the operating pressures. In addition, safety and operating controls, such as pressure relief devices, should be provided to prevent overpressuring any part of the system.

One of the safest methods for handling flammable or combustible liquids is to pump the liquid from underground storage tanks, through an adequately designed piping system protected from physical damage, to the dispensing equipment located outdoors or in specially designed inside dispensing rooms. Such rooms should have at least one exterior wall for explosion relief and accessibility for fire fighting, interior walls with appropriate fire-resistance ratings, and adequate ventilation and drainage, and be free of sources of ignition.

Where solvents are pumped from storage tanks to the point of use in an industrial building, emergency switches should be located in the dispensing area, at the normal exit door or at other safe locations outside the fire area, and at the pumps to shut down all pumps in case of fire.

Where dispensing is by gravity flow, such as in the filling of containers in an industrial operation, a shutoff valve should be installed as close as practical to the vessel being unloaded. A control valve should be located near the end of the discharge pipe. Additionally, in some filling operations, a heat-actuated valve is desirable to shut off the flow of liquid.

The preferred method of dispensing flammable and combustible liquids from a drum is by use of an approved hand-operated pump drawing through the top. In certain cases, an approved drum faucet may be used for dispensing from a drum. However, hand-operated pumps are safer than faucets, because the hazard of leakage is reduced.

For handling small quantities of flammable and combustible liquids, safety cans are preferred. Safety cans are substantially constructed to avoid the danger of leakage and are designed to minimize the likelihood of spillage or vapor release and of container rupture under fire conditions. (See Figure 3-21L.) Liquids can also be dispensed from the original shipping containers. Open pails or open buckets should never be used for storage.

Flammable liquids should always be handled and dispensed in a well-ventilated area free of sources of ignition, and bonding should be provided between the dispensing equipment and the container being filled. (See Figure 3-21M.)

TRANSPORTATION OF FLAMMABLE AND COMBUSTIBLE LIQUIDS

Tank Vehicles

NFPA 385, *Standard for Tank Vehicles for Flammable and Combustible Liquids* (hereinafter referred to as NFPA 385), provides for

FIG. 3-21L.   *Typical safety cans having pouring outlets with tightfitting caps or valves normally closed by springs, except when held open by hand, so that contents will not be spilled if a can is tipped over. The caps also provide an emergency vent when the cans are exposed to fire.*

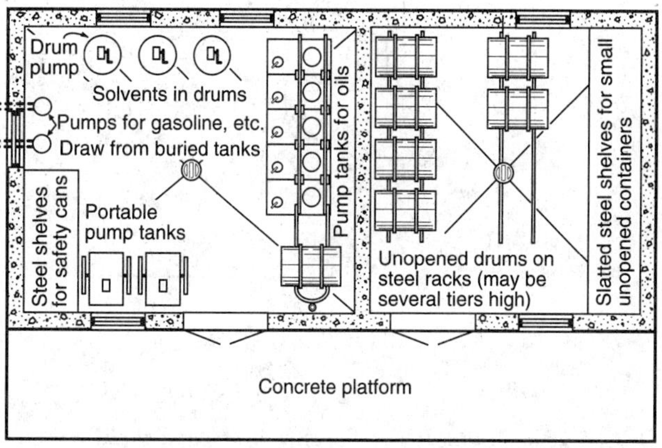

FIG. 3-21M.   *Layout of a storage and dispensing house.*

substantially constructed vehicles and tanks that present relatively little danger of fire when involved in minor traffic accidents. Under these recommendations, tanks are constructed to withstand all but the most violent impact without rupturing and releasing liquid. Vents are provided so that, if a tank is subjected to fire, vapor will burn at the vent, avoiding the danger of rupture from excessive internal pressure.

For all but viscous liquids, a shutoff valve located inside the shell of the cargo tank is required. It is kept closed, except during loading or unloading operations. A shear section is provided in the piping connected to the internal valve. Therefore, in an accident that could damage the piping, the piping breaks at the shear section, leaving the internal valve undamaged and closed.

The operating mechanism for the valve is also required to have a secondary control, remotely located, that can be used to shut off the valve in case of fire or severe spillage during unloading. In addition, a fusible element [to operate at not more than 250°F (121°C)] is required in the control mechanism for the valve, to permit the valve to close automatically in case of fire. (See Figures 3-21N and 3-21O.) Other important fire protection features are detailed in NFPA 385.

In the past, many municipalities have attempted to minimize fire problems by specifying the maximum sizes of cargo tanks on trucks. The wisdom of this sort of provision is dubious. Assume, for example, that it is necessary to transport 4,000 gal (15,000 L) of gasoline over city streets to make local deliveries. A legal restriction as to the maximum size of the tank might require the use of four 1,000-gal (3,800-L) tank trucks in place of one 4,000-gal (15,000-L) tank. This quadruples the tank-truck traffic, with four times the accident potential. The danger from a gasoline fire is not directly proportionate to the quantity of gasoline involved. For this reason, NFPA standards have not recommended any limitation on the maximum size of tank trucks.

Tank-truck traffic through congested districts of cities should be avoided as much as possible, since any fire is likely to have more serious consequences in congested districts than in sparsely settled areas. Bypass routes, such as those commonly specified for all kinds of through truck traffic, are the obvious solution to this particular problem.

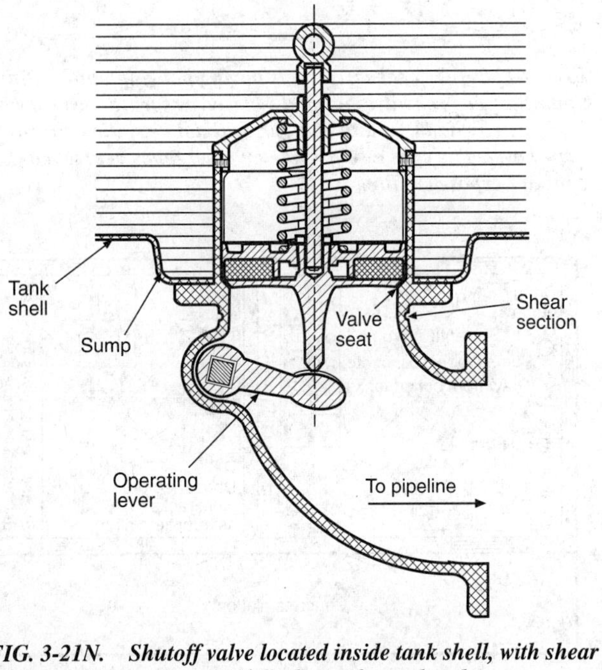

Parking of loaded tank vehicles is another feature that can be appropriately regulated by municipalities. Such parking can be prohibited on city streets. Similarly, tank vehicles should not be parked in public garages. Permissible parking locations for tank vehicles may be specified, if necessary. Any property zoned for aboveground oil storage, e.g., the ordinary bulk oil plant, should be a location where tank vehicles can be parked without any undue increase in hazard to the public.

To prevent the hazards associated with "switch loading," no tank vehicle compartment that has been utilized for transporting Class I liquids should be loaded with Class II or Class III liquids until the compartment and all piping, meters, and hoses have been completely drained. When Class II or Class III liquids are loaded into a tank vehicle that previously contained a Class I liquid that was not adequately drained out, there exists the potential for the generation of a static spark to occur in a mixture that is within the flammable range, resulting in an explosion.

Transportation of flammable liquids in tank vehicles in interstate commerce is governed by regulations of the U.S. Department of Transportation (DOT), as contained in the Code of Federal Regulations, Title 49, "Transportation."[6] There are comprehensive specifications and labeling requirements for shipping containers for various types of products. (Absence of a label does not necessarily mean that the material is nonhazardous, however.)

Rail, Ship, and Pipeline

The transportation of flammable liquids by railroad tank cars is under the jurisdiction of the DOT.[6] The design of tank cars is rigidly controlled and is governed by the nature of the contents being carried. Emergency venting requirements are included in DOT regulations.

The transportation of flammable and combustible liquids in bulk on board vessels in the United States is under the jurisdiction of the Commandant of the U.S. Coast Guard. These transportation requirements include the design of the vessel, whether it be a tank ship or tank barge, as well as requirements for extinguishing systems or portable extinguishers. Further requirements include limitations on container storage of drums or portable tanks, restrictions for passenger-carrying vessels, and many other requirements for the safe transportation of flammable and combustible liquids by water.

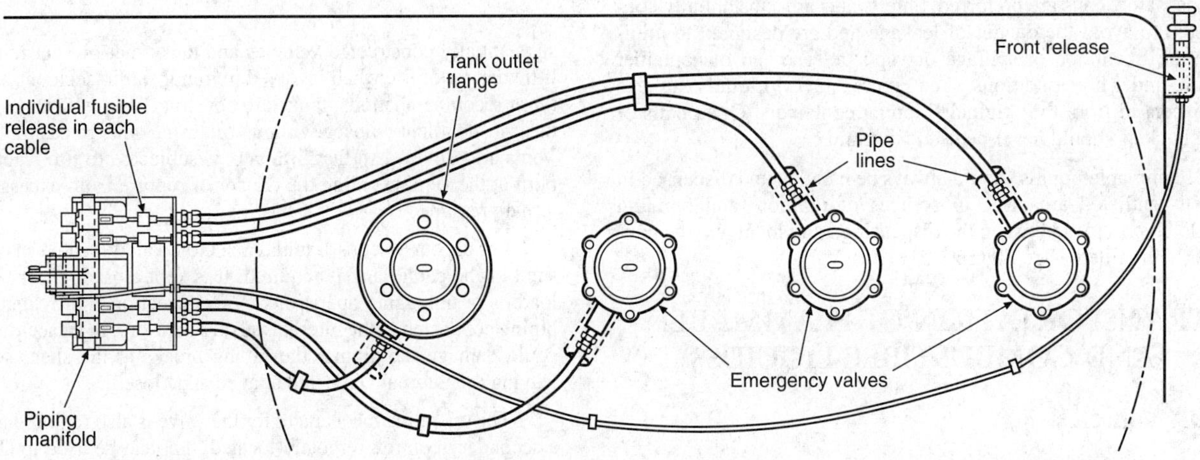

FIG. 3-21N. *Shutoff valve located inside tank shell, with shear section to leave valve seat if discharge faucet breaks.*

FIG. 3-21O. *Bottom view of tank vehicle showing typical installation of emergency valves and controls.*

Details of these requirements can be secured at the nearest U.S. Coast Guard Merchant Marine Inspection Office.

Standards for pipelines transporting liquids are published by the American Society of Mechanical Engineers (ASME). These standards include piping requirements and installation recommendations.

BIBLIOGRAPHY

References Cited

1. API Standard 650, *Welded Steel Tanks for Oil Storage*, American Petroleum Institute, Washington, DC, May 1993.
2. API Standard 2000, *Venting Atmospheric and Low-Pressure Storage Tanks*, American Petroleum Institute, Washington, DC, Sept. 1992.
3. API 520, *Sizing, Selection, and Installation of Pressure-Relieving Devices in Refineries*, Part I, *Sizing and Selection*, American Petroleum Institute, Washington, DC, July 1990.
4. API RP 2013, *Cleaning Mobile Tanks in Flammable or Combustible Liquid Service*, 5th ed., American Petroleum Institute, Washington, DC, Jan. 1991.
5. API RP 2015, *Cleaning Petroleum Storage Tanks*, American Petroleum Institute, Washington, DC, Dec. 1993.
6. *Code of Federal Regulations*, Title 49, "Transportation," U.S. Government Printing Office, Washington, DC.

References

API Standard 2003, *Protection Against Ignitions Arising Out of Static, Lightning, and Stray Currents*, American Petroleum Institute, Washington, DC, Dec. 1991.

Atkinson, G. T., "Fire Spread in a Pallet Load of Bottles of Flammable Liquid," Short Communication, *Fire Safety Journal*, Vol. 22, No. 4, 1994.

Back, G. G., "Experimental Evaluation of Water Mist Fire Suppression System Technologies Applied to Flammable Liquid Storeroom Applications," *Proceedings* for the International Conference on Fire Research and Engineering, Society of Fire Protection Engineers (SFPE), Boston, MA, 1995, pp. 127–132.

Bielen, R. P., "Fire Research With Flammable and Combustible Liquids in Bulk Merchandising Retail Facilities," *Fire Technology*, Vol. 29, No. 1, February 1993, pp. 80-81.

El-Iskandarani, B. M. K., "Emergency Relief Venting Analysis for Flammable Liquid Storage Tanks," Worcester Polytechnic Institute, Thesis, May 1994, 160 pages.

"Flammable Liquid Drainage and Containment," *Record*, Vol. 69, No. 4, July/Aug. 1992, pp. 24–31.

Flammable Liquids—Risk, Regulations and Protection Measures to Safeguard Flammable Liquids," *Record*, Vol. 69, No. 1, Jan./Feb. 1992, pp. 15–21.

Goodman, D., "Fire Retarded Cardboard Cartons—An Approach to Allow General Warehousing of Flammable Liquids in Plastics Containers," Fire Safety Developments and Testing: Toxicity—Heat Release—Product Development—Combustion Corrosivity, October 21–24, 1990, Ponte Vedra Beach, FL, Fire Retardant Chemicals Assoc., Lancaster, PA, 1990, pp. 157–165.

Hall, J. R., Jr., "Fire Risk Analysis Model for Assessing Options for Flammable and Combustible Liquid Products in Storage and Retail Occupancies," *Fire Technology*, Vol. 31, No. 4, 4th Quarter, Nov. 1995, pp. 290–308.

Kokkala, M., "Extinquishment of Liquid Fires With Sprinklers and Water Sprays: Analysis of the Test Results," VTT-Technical Research Center of Finland, Espoo, VTT Research Reports 696, Project PAL9002, Aug. 1990, 54 pages.

Lentini, J. J., Waters, L. V., "Behavior of Flammable and Combustible Liquids," *Fire and Arson Investigator*, Vol. 42, No. 1, Sept. 1991, pp. 39–45.

Mulhaupt, R., "Fire Protection for Flammable and Combustible Liquids Warehouse," First International Conference for Fire Suppression Research, May 5–8, 1992, Stockholm, Sweden, 1992, pp. 393–399.

"National Wholesale/Retail Occupancy Fire Research Project. Task 1. Protection of Flammable Liquids," Technical Report, Underwriters Laboratories, Inc., Northbrook, IL, National Fire Protection Research Foundation, Quincy, MA, Nov. 1992, 167 pages.

Nugent, D. P., "Directory of Fire Tests Involving Storage of Flammable and Combustible Liquids in Small Containers," Schirmer Engineering Corp., Deerfield, IL, Directory, Feb. 1994, 75 pages.

Nugent, D. P., "Fire Tests Involving Storage of Flammable and Combustible Liquids in Small Containers," *Journal of Fire Protection Engineering*, Vol. 6, No. 1, 1994, pp. 1–10.

Omans, L. P. "Fighting Flammable Liquids Fires: A Primer. Part 1. The Family of Foams," *Fire Engineering*, Vol. 146, No. 1, Jan. 1993, pp. 50–54, 57–61.

Omans, L. P., "Fighting Flammable Liquids Fires: A Primer. Part 2," *Fire Engineering*, Vol. 146, No. 2, Feb. 1993, pp. 50–52, 56–58.

Omans, L. P., Fighting Flammable Liquid Fires: A Primer. Part 3," *Fire Engineering*, Vol. 146, No. 3, Mar. 1993, pp. 93–94, 96–98, 100, 102, 104.

"Outline of Investigation for Fire Test of Packaging Systems of Class 1 and 2 Liquids in Plastic Containers," Underwriters Laboratories, Inc., Northbrook, IL, Subject 2019, Feb. 1993, 15 pages.

"Proper Flammable and Combustible Liquid Handling," *Record*, Vol. 68, No. 5, Sept./Oct. 1991, pp. 16–17.

Sharma, T. P., *et al.*, "A New Particulate Extinquishant for Flammable Liquid Fires," *Proceedings* of the Second International Symposium on Fire Safety Science, Hemisphere Publishing Corporation, New York, 1989, pp. 667–680.

Sharma, T. P., *et al.*, "Experiments on Extinction of Liquid Hydrocarbon Fires by a Foam Technique," *Proceedings* of the Fourth International Symposium on Fire Safety Science, International Association for Fire Safety Science, 1994, pp. 865–876.

Staggs, K. J., *et al.*, "Development of Flammable Liquid Storage Wooden Cabinets for Chemical Laboratories," Department of Energy, Washington, DC, UCRL-ID-115605, Nov. 1993, 39 pages.

Vaivads, R., *et al.*, "Flammability of Alcohol—Gasoline Blends in Fuel Tanks," *Proceedings* of the Fourth International Symposium on Fire Safety Science, International Association for Fire Safety Science, 1994, pp. 575–588.

Venart, J. E. S., *et al.*, "To BLEVE Or Not To BLEVE: Anatomy of a Boiling Liquid Expanding," New Brunswick Univ., Fredericton, Canada, American Institute of Chemical Engineers, AIChE Meeting on Loss Prevention, Mar./Apr. 1992, New Orleans, LA, 1992, pp. 1–21.

NFPA Codes, Standards, and Recommended Practices

Reference to the following NFPA codes, standards, and recommended practices will provide further information on storage of flammable and combustible liquids discussed in this chapter. (See the latest version of *The NFPA Catalog* for availability of current editions of the following documents.)

NFPA 11, *Standard for Low-Expansion Foam*
NFPA 30, *Flammable and Combustible Liquids Code*
NFPA 30A, *Automotive and Marine Service Station Code*
NFPA 31, *Standard for the Installation of Oil-Burning Equipment*
NFPA 70, *National Electrical Code®*
NFPA 77, *Recommended Practice on Static Electricity*
NFPA 306, *Standard for the Control of Gas Hazards on Vessels*
NFPA 325, *Guide to Fire Hazard Properties of Flammable Liquids, Gases, and Volatile Solids*
NFPA 327, *Standard Procedures for Cleaning or Safeguarding Small Tanks and Containers Without Entry*
NFPA 329, *Recommended Practice for Handling Underground Releases of Flammable and Combustible Liquids*
NFPA 385, *Standard for Tank Vehicles for Flammable and Combustible Liquids*

Additional Reading

American Conference of Governmental Industrial Hygienists, *Threshold Limit Values*, Cincinnati, OH, 1989.

American Industrial Hygiene Association, *Hygienic Guide Series*, Detroit, MI.

Benedetti, R., *Flammable and Combustible Liquids Code Handbook*, 5th ed., National Fire Protection Association, Quincy, MA, 1993.

Dunbar, J., "Packaging and Storing Flammable Liquids," *Products Finishing*, Vol. 52, No. 9, June 1988, pp. 118-121.

Flick, E. W., ed., *Industrial Solvents Handbook*, 4th ed., Noyes Data Corp., Park Ridge, NJ, 1990.

Fox, R. B., "Tank Fires: A Case Study History," *Industrial Fire World*, Vol. 1, No. 9, 1986.

Fullam, B. W., "Fire Protection of LPG Vessels," *Proceedings* of the Gastech 86 LNG/LPG Conference, Gastech, 1986, pp. 152-156.

Hadjisophocleous, G. V., Sousa, A. C. M., and Venart, J. E. S., "Fire Engulfment of Horizontal Cylinder Tanks," *Proceedings* of the 1989 National Heat Transfer Conference: Heat Transfer Phenomena in Radiation, Combustion, and Fires, American Society of Mechanical Engineers: Heat Transfer Division, New York, 1989, pp. 339-347.

Hawthorne, E., *Petroleum Liquids—Fire and Emergency Control*, Prentice-Hall, Englewood Cliffs, NJ, 1987.

Hughes, J. R., *Storage and Handling of Petroleum Liquids*, 3rd ed., Chas. Wiley, New York, 1987.

Hunt, D. L. M., and Ramskill, P. K., "A Description of ENGULF—A Computer Code to Model the Thermal Response of a Tank Engulfed in Fire," Report No. SRD/HSE/R 354, March 1987, United Kingdom Atomic Energy Authority, Warrington, UK.

Isner, M. S., "$49 Million Loss in Sherwin-Williams Warehouse Fire," *Fire Journal*, Vol. 82, No. 2, March/April 1988, p. 65.

James, D., "Highly Flammable Liquids and LPG," *Fire Prevention Handbook*, Butterworths, Boston, 1986, pp. 84-95.

Johnson, M. R., "Fire Protection for Railroad Tank Cars," *Proceedings* of the ASME Pressure Vessels and Piping Conference, ASME, 1986, p. #8.

Martinsen, W. E., Johnson, D. W., and Millsap, S. B., "Determining Spacing by Radiant Heat Limits," *Plant/Operation Progress*, Vol. 8, No. 2, Jan. 1989, pp. 25-28.

Masters, R. W., "Large-Area Hydrocarbon Firefighting," *Industrial Fire World*, Vol. 2, No. 4, August 1987, pp. 7-12.

National Safety Council, *Accident Prevention Manual for Industrial Operations*, 9th ed., Chicago, IL, 1988.

Phillips, H., and Pritchard, D. K., "Performance Requirements of Flame Arresters in Practicle Applications," *Proceedings* of the Conference on Hazards in the Process Industries, Institution of Chemical Engineers, 1986, pp. 47-61.

Potapov, A. Y., "Calculation of the Allowable Probability of Failure of an Acetylene Flame Arrestor," *Chemical and Petroleum Engineering* (English Translation), Vol. 23, Nos. 11-12, 1988, pp. 569-571.

Presdee, B., "Computerized Oil Rig Fires Could Save Lives," *Fire International*, Vol. 12, No. 113, Oct./Nov. 1988, p. 49, 52-53.

Przybyla, L., and Gandhi, P., "The Results Are in on Flammable Liquids in Plastic Containers," *Fire Journal*, Vol. 84, No. 3, May/June 1990, p. 38.

Safety Digest of Lessons Learned, Vols. 3, 6, 7, 8, 9, American Petroleum Institute, Washington, DC, 1986.

Sax, N. I., *Dangerous Properties of Industrial Materials*, 7th ed., Van Nostrand Reinhold, New York, 1989.

Schoen, W., and Droste, B., "Investigations of Water Spraying Systems for LPG Storage Tanks by Full-Scale Fire Tests," *Journal of Hazardous Materials*, Vol. 20, Dec. 1988, pp. 73-82.

Schwab, R. F., "Consequences of Solvent Ignition During a Drum Filling Operation," *Plant/Operation Progress*, Vol. 7, No. 4, Oct. 1988, p. 242.

"Shipping," Title 46, Parts 146 to 149; Title 49, "Transportation," Parts 171 to 178, *Code of Federal Regulations*, U.S. Government Printing Office, Washington, DC.

Siddle, J. F., "Fire Hazards of Flammable Liquids in Small Containers," *Proceedings* of the Conference on Hazards in the Process Industries, Institute of Chemical Engineers, 1986, pp. 111-120.

Steinbrecher, L., "Analysis of a Tank Fire 'Classic,'" *Fire Engineers Journal*, Vol. 47, No. 144, 1987, pp. 15-17.

Tabor, D. C., "Foam-Water Sprinkler Protection for Flammable Liquids," *Plant/Operation Progress*, Vol. 8, No. 1, 1989, pp. 218-224.

"Which of the Following Presents the Greatest Challenge to Sprinklers? Flammable Liquids in: (a) aerosol cans, (b) metal containers, (c) glass containers, (d) plastic containers," *Record*, Jan./Feb. 1987, pp. 4-9.

Williams, D., "Extinguishing Flammable Liquid Fires," *Fire International*, Vol. 11, No. 107, Oct./Nov. 1987, pp. 126-127.

STORAGE OF GASES

Revised by Theodore C. Lemoff

This chapter describes essential fire safety characteristics of containers in which gases are stored, as well as some fundamental precepts for storage arrangements.

A gas must be stored in a container that is gastight for: (1) the range of temperature and pressure conditions present at the storage location and (2) the conditions represented by the transportation environment prior to arrival at a storage site. Thus, it is preferable that the storage container contain the source of energy needed to remove the gas from storage and convey it to the point of use. This energy is represented by the pressure of the gas in the container.

In the case of containers in which the gas is entirely in the gaseous phase (compressed gases), the pressure is applied at the charging plant or, in the case of pipelines, by compressors spaced along the pipeline. In the case of containers of gas in which the gas is partly in the gaseous phase and partly in the liquid phase (i.e., liquefied gases, including cryogenic liquefied gases), the pressure is obtained as a result of heat stored in the liquid, which is directly related to how much the temperature of the liquid is above its normal boiling point.

Therefore, gas containers, whether shaped as tanks or as pipelines, are closed pressure vessels containing considerable energy per unit of volume, and requiring careful design, fabrication, and maintenance. Furthermore, because the containers are closed and excessive pressures can develop when they are exposed to rather nominal heat sources and fire (or by overfilling, in the case of liquefied gas containers), overpressure protection is usually needed.

GAS CONTAINERS

Most countries have regulations for the design, fabrication, and maintenance of gas containers. Although basically similar, the regulations differ in detail. This chapter concentrates on North American practices.

In North America, there are two types of gas containers: (1) cylinders and (2) tanks. Originally, the distinction between cylinders and tanks was based upon size. Cylinders, being the smaller, were considered portable; the tanks, being the larger, were essentially used in stationary service. There was also a distinction that reflected the pressures in the containers. Cylinders were thought of as being used for high pressures and tanks for low or moderate pressures. Over the years, these distinctions have lessened so that, today,

the only real distinction lies in the regulations or codes under which the container is built.

Practically all gases must be transported from the manufacturer to the user, making the safety of the container in transportation a matter of primary concern. As a result, criteria for many gas containers reflect transportation safety conditions. Because it could be hazardous as well as uneconomical to require gases to be transferred from a shipping container to a separate container for other use, every effort has been made to utilize the same container whenever feasible. This is generally the procedure used for the smaller containers.

Gas Cylinders

Gas cylinders are fabricated in accordance with regulations and specifications of the U.S. Department of Transportation (DOT) in the United States and Transport Canada (TC) in Canada. The requirements are the same in both countries. Prior to April 1, 1967, DOT regulations were promulgated by the Interstate Commerce Commission (ICC), and many cylinders are still in service with ICC markings. Cylinders are generally limited to a maximum water capacity (i.e., the capacity when completely filled with water) of 1,000 lb (454 kg), or approximately 120 gal (450 L) of water.

The regulations cover the service pressure the cylinder must be designed for, the gas or group of gases that it can contain, safety devices, and requirements for in-service (transportation) testing and requalification. The specifications cover such criteria as metal composition and physical testing, wall thickness, joining methods, nature of openings in the container, heat treatments, proof testing, and marking.

Gas Tanks

Customarily, gas tanks are fabricated in accordance with Section VIII, "Unfired Pressure Vessels," of the *Boiler and Pressure Vessel Code*, published by the American Society of Mechanical Engineers (ASME), or tank fabrication standards of the American Petroleum Institute (API). ASME tanks are usually smaller tanks under moderate pressure, and API tanks are usually very large tanks under low pressure.

Cylinders or tanks that are part of transportation units, such as cargo vehicles or railcars, are subject to additional criteria in the regulations, primarily to reflect the fact that the containers are on wheels. Tanks for cargo vehicles are basically ASME code tanks,

---

Theodore C. Lemoff, P.E., is the senior gases engineer on the staff of NFPA.

while tanks for railcars are covered by specific DOT and TC specifications.

In the late 1960s and early 1970s, a series of serious railcar derailments occurred in the United States, reflecting the deterioration of tracks and track beds due to inadequate maintenance. These incidents were characterized by both impact and fire-caused boiling liquid expanding vapor explosions (BLEVEs) of tank cars containing liquefied flammable gases. As a result, DOT regulations were revised in 1977 to require that these tank cars be protected thermally and equipped with a type of coupler that prevents vertical disengagement. Unless the insulation is contained in a steel outer jacket, the car also must be equipped with heavy steel-plate head shields for impact protection. Existing cars were required to be retrofitted by the end of 1980. In the few derailments involving tank cars outfitted in this fashion, these safeguards have proved effective—a BLEVE either has not occurred at all or has been delayed for several hours.

A number of such containers are exempted by the regulations. These include certain very small containers and containers with nonflammable cryogenic gases, including oxygen, where the pressure in the container as shipped is below 40 psia (an absolute pressure of 276 kPa).

There is usually a lag in time between the need to ship a gas that is new to commerce and the promulgation of specifications for a transportation container. In such cases, DOT or TC exemptions, which spell out safety criteria agreed upon by the authorities, are issued.

In themselves, the DOT and TC regulations apply only to cylinders and tanks in transportation in interstate or interprovincial commerce. However, many consensus codes and standards extend these regulations to transportation in intrastate commerce and also extend applicable criteria to use and storage at consumer sites.

The aforementioned requirements for "in-service" reinspection and requalification of DOT specification containers are also extended to use and storage at consumer sites by the consensus standards and codes. However, the ASME *Boiler and Pressure Vessel Code* and API standards do not contain such provisions. If not covered by consensus standards and codes, in-service inspection and requalification become a matter of owner judgment, and of such state or local regulations as may apply, or conditions set forth in an insurance contract.

Regardless of the degree of structural integrity incorporated into gas containers, abuse of the container must be avoided. This is especially true of portable cylinders, which are subject to mishandling. General handling of safety precautions for gas cylinders is contained in the Compressed Gas Association's "Safe Handling of Compressed Gases in Containers."

Pipelines

Gases used in large volumes are often transported by pipelines. Natural gas is customarily transported by pipeline, as is much LP-Gas and some industrial gases, such as anhydrous ammonia, oxygen, hydrogen, and ethylene.

Since 1968, most of the transmission and distribution of flammable gases by pipeline has been regulated in the United States by the DOT Office of Pipeline Safety and is covered by federal regulations. These regulations cover such items as pipe materials; design for pressure and other stresses (which, among other criteria, require stronger piping as population density increases); piping components, including valves (emergency flow control valve spacing in the pipeline system is also affected by population density); joining methods; installation of meters, service regulators, and service lines; corrosion control; test requirements; certain operating requirements; and maintenance, including leakage surveys.

Normally, DOT regulations do not apply to piping on consumer premises. Such piping for the more common gases is covered by consensus codes and standards, notably NFPA 54 (ANSI Z223.1), *National Fuel Gas Code*, and ANSI/ASME B31, *Code for Pressure Piping*.

STORAGE SAFETY CONSIDERATIONS*

Fire protection safeguards for gas storage reflect: (1) the hazards of the container/gas combination and (2) the hazards of the gas when it escapes from the container. Discussions of the basic hazards of gases, gas emergency control, and specific gases are especially pertinent.

During at least some period, any fire incident can manifest both of these hazards at the same time. However, the container/gas combination hazard always exists. The fire hazard of escaping gas, on the other hand, may be negligible if the gas is nonflammable.

Container/Gas Hazard Safeguards

The major container/gas hazard is the BLEVE. The BLEVE hazard is restricted to containers of liquefied gases, and the major cause of such BLEVEs in storage is fire exposure. BLEVEs resulting from corrosion of a container are far less frequent, and impact-caused BLEVEs even more so for containers in storage.

Container insulation: The BLEVE hazard is greatly affected by container features that restrict the opportunity for the container metal to be overheated. Especially notable in this respect is the presence of insulation between an exposing fire and the portion of the container subject to internal pressure. All cryogenic liquefied gas containers are insulated as a matter of functional necessity. The BLEVE hazard can essentially be eliminated by also considering insulation system behavior under fire exposure. North American standards for such systems on storage containers reflect this situation. For example, insulation for flammable cryogenic gas containers is required to be noncombustible and (if in a form such that loss of an enclosing jacket could cause a serious loss in insulating capacity, e.g., powder or granules), the container jacket is required to be steel or concrete rather than aluminum.

Containers of noncryogenic gases are not required to be insulated for operational reasons and are seldom insulated for fire safety reasons. However, insulation of larger liquefied flammable gas tanks is becoming more frequent in specific installations where other BLEVE prevention measures, e.g., water cooling, are deficient.

A basic BLEVE safeguard is to reduce the chances of fire exposure to the container. This philosophy is also applicable to containers of nonliquefied gases (compressed gases) because, although by definition they are not subject to a BLEVE, they can still fail explosively from fire exposure.

Limiting combustibles in the area: Essential to this safeguard is limiting the quantity of combustibles in the vicinity of gas containers. This is valid whether the storage is indoors or outdoors. Except where small quantities of gas are involved, e.g., one or two cylinders of compressed gas, it is highly desirable that storage rooms or docks be of noncombustible or limited combustible construction. (See NFPA 220, *Standard on Types of Building Construction*, for these definitions.) If the building that houses the storage room presents a substantial fire load, the storage room walls should have a suitable fire-resistance rating.

---

*The philosophy of this analytical approach is developed in Section 4, Chapter 7, "Gases"; it would be worthwhile to study that chapter before proceeding further in this chapter.

Containers of flammable gases should not be stored with containers of nonflammable gases. This is especially true with respect to gases that, if released, could intensify combustion of the flammable gases. Notable in this respect are oxidizing gases, such as oxygen or nitrous oxide. However, it is best also to avoid storage of inert gases near flammable gases because the container hazard reflects only pressure. Storage of nonflammable and oxidizing gases in the same area is safe.

Separation: As a general rule applicable to cylinders and small liquefied gas tanks, separation of flammable gases from nonflammable gases by 20 ft (6 m) is appropriate. Provision of a noncombustible barrier as high as the containers [usually 5 ft (1.5 m)] and having a fire-resistance rating of at least 1/2 hr is an acceptable substitute for this distance where cylinders or small tanks [less than 1,000 lb (454 kg) water capacity] are involved.

The use of barriers around larger containers of flammable gases is questionable unless they permit application of cooling water, prevent pocketing of escaping gas, and allow unrestricted egress of personnel from the area in an emergency.

For restrictions over the proximity of combustibles to oxidizing gases, it must be recognized that the degree of combustibility has little practical significance in the presence of these gases. For example, asphalt paving is not normally considered to be combustible. However, in contact with oxygen, especially liquid oxygen, it is not only combustible, but can be explosive.

Application of water: During a fire exposure, the application of water is a basic safeguard to prevent a BLEVE or a compressed gas container failure. In indoor locations, automatic sprinkler protection can greatly limit pressure rise from heat and high metal temperatures from fire exposure. However, for maximum effectiveness, the system should be capable of furnishing a discharge density of at least 0.25 gpm per sq ft [10 (L/min)/m^2] over an area of at least 3,000 sq ft (279 m^2), and the sprinklers should be located not more than 20 ft (6 m) above the floor where the containers are stored. Further characterization of sprinkler protection may be needed if the storage is in racks or in high piles.

A water spray fixed system of similar capacity is also effective and can be installed outdoors. For the protection of larger containers of liquefied flammable gases (chemically not self-reactive), a density in excess of 0.25 gpm of water per sq ft [10 (L/min)/m^2] of container surface is needed to prevent failure, although a 0.25 gpm per sq ft density [10 (L/min)/m^2] may prevent a crack from proceeding to the point of container rupture.

Water application may not always be appropriate for tanks containing cryogenic liquids. Keep in mind that these containers are usually not pressure vessels, and the cold liquid is at its normal atmospheric boiling point. Thus, the container cannot BLEVE. Application of water could result in water entering the pressure-relief device and freezing, rendering it inoperative and resulting in pressure vessel failure. Also, water is at a temperature well above the temperature of the liquid and is a heat source that may promote vaporization of the liquid.

Overpressure limiting devices: The satisfactory operation of container overpressure limiting devices is vital to controlling the BLEVE or compressed gas container failure hazard. Even though these devices cannot by themselves always prevent container failure, they do extend the prefailure time in all cases and often can prevent failure under many fire exposure conditions. It is essential that the device not be blocked closed by corrosion, paint deposits, etc., and not be damaged mechanically. Portable containers should be checked for this every time they are taken into the facility and whenever they are connected to consuming equipment or are filled.

Spring-loaded pressure-relief valves on the larger stationary storage containers should be tested at five- to ten-year intervals (more often if in corrosive or reactive gas service).

Relief devices on liquefied gas containers should always directly contact and monitor the vapor space, because their relieving capacity is determined under vapor discharge conditions and is severely restricted if liquid is discharged. With the exception of some LP-Gas containers used in mobile engine fuel systems and 1-ton (907-kg) chlorine containers, portable liquefied gas containers should be stored in an upright position to attain this effect.

Care in handling: Also reflecting the container/gas hazard, it is important that the containers not be subjected to physical abuse. Although quite sturdy as a result of their design as pressure vessels, any dent or gouge can reduce safety factors and, at the least, shorten failure times from fire exposure or lead to impact failures upon subsequent movement. If the valve is broken off on some smaller portable compressed gas containers, the nozzle reaction from escaping gas can be sufficient to propel the cylinder violently. Where the container is designed for a valve-protecting cap or collar, these always should be in place during storage or movement.

DOT requalification: DOT containers must be requalified at intervals specified in the federal regulations. These intervals vary with the kind of gas and type of container. In most cases, this interval is five years, but can be as long as twelve years, depending on the retest method used. A requalified DOT container must be marked with the date of requalification. If such a marking is not present on a compressed gas DOT specification container and the manufacture date marking is more than five years old, it should be brought to the attention of whoever last charged the container.

The container owner is responsible for requalification. In many instances, the user/storer of portable gas containers is not the owner. In others, however, the user/storer is the owner, such as for LP-Gas containers used in industrial truck service and filled on site. Those responsible for such containers should be aware of this and of the fact that the requalification procedures are rigorous and require considerable technical expertise.

Safeguards for Escaping Gas

The major escaping gas hazard is combustion of flammable gas and is, in turn, manifest as either a fire or a combustion explosion. Fire, of course, can also lead to the previously discussed explosive container failures. Again, the reader is referred to Section 4, Chapter 7, "Gases," for a discussion of the elements of this hazard.

Gases in storage do not present many sources of escaping gas beyond those already discussed under container/gas hazard safeguards. The container itself, its valves, and overpressure protection devices are the only sources of escaping gas, provided that the gas is not being transferred into and out of storage.

Inspection for leakage: All storage containers should be inspected periodically for leakage from container appurtenances or from the container itself. Portable containers should not be placed into storage if they are leaking. To prevent this, incoming shipments should always be inspected upon arrival.

The senses of sight, sound, and smell are invaluable leak detectors. Although most gases are invisible, some do have color, e.g., chlorine. Liquefied gases escaping as liquids can lead to the formation of a visible cloud of condensed water vapor. Because they are stored under pressure, a leak can be accompanied by a hissing sound. Although many gases are odorless, some do possess strong odors, e.g., chlorine and anhydrous ammonia. Natural gas and LP-Gas have an odorant added to them. Many gas detection instruments are manufactured. Leak detection solutions are available that, when applied to small leaks, show bubbles.

One source of escaping gas in storage locations is the operation of overpressure protection devices. Because the pressure in a gas container is directly related to the gas temperature, and the gas temperature in uninsulated containers reflects the temperature of the surroundings, room temperature and solar radiation will affect the pressure. In general, this pressure will result in the operation of overpressure devices on fully charged containers if the temperature reaches 130 to 140°F (54 to 60°C). Therefore, storage areas should not be allowed to reach such temperatures, and reflective paint should be maintained on outdoor containers subject to solar radiation.

Leakage from containers themselves is less common and develops slowly, so detection is facilitated prior to the development of a hazardous situation. Leakage from the inner container of an insulated container for a cryogenic gas is indicated by the formation of water condensate or frost on the outside container due to contact with air on the cold surface. While this often only indicates a void in the insulation rather than leakage, this appearance should be investigated.

Overpressure protection: Overfilled liquefied gas containers can lead to overpressure device operation at much lower temperatures. (See discussion of Charles' Law in Section 4, Chapter 7, "Gases.") Portable liquefied gas containers should be checked for proper quantities before being placed into storage. Many such containers may be checked by weighing them. The weight of the empty container (tare weight) must be marked on the container in accordance with DOT regulations. The weight of gas must be determined by the use of data that include the specific gravity of the gas and its temperature when placed into the container. In other cases, the quantity is checked by the use of a gage that measures the liquid level. The most accurate gage requires that a small amount of gas be released. This should be done outside under carefully controlled conditions. These gages are known as fixed-tube gages or try-cocks. They will release gas only if the level is safe and will release liquid if there is too much in the container.

Ventilation of spaces: Indoor storage areas should be ventilated, regardless of the chemical hazard of the gas. In areas used solely for storage (no filling of containers), the amount of ventilation need not be great—one-half to one air change per hour or so. However, even inert gases, because they are often odorless, tasteless, and colorless, can present asphyxiation hazards in unventilated areas.

Controlling ignition sources: Ignition sources should be controlled in flammable gas storage areas. The vapor density of the gas, in part, determines the extent of the area in which ignition sources should be eliminated or controlled. Many gases are heavier than air at all times. Others will be temporarily heavier than air when they are released in liquid form and vaporize. In general, flammable gas storage areas with no container filling are classified as Division 2 locations for purposes of installing electrical equipment because of the ventilation provided and the nominal leakage potential.

Most flammable gases are Group C or D materials, in the context of NFPA 70, *National Electrical Code®*. Acetylene (Group A) and hydrogen (Group B) are common exceptions. Because of the limited variety of Group A and B electrical equipment available, electrical equipment may have to be avoided; be purged and ventilated in accordance with NFPA 496, *Standard for Purged and Pressurized Enclosures for Electrical Equipment;* or be intrinsically safe.

Fire protection and control of gas fires are discussed in Section 4, Chapter 7, "Gases." Sprinkler and water spray protection are also discussed in that chapter. An essential aspect is that the extinguishment of gas fires by the application of extinguishing agents should be restricted to small leaks where the consequences of reignition can be tolerated.

BIBLIOGRAPHY

References

ANSI/ASME B31, *Code for Pressure Piping*, American National Standards Institute, New York.
ASME *Boiler and Pressure Vessel Code*, American Society of Mechanical Engineers, New York.
CGA, "Safe Handling of Compressed Gases in Containers," *Pamphlet P-1*, Compressed Gas Association, Arlington, VA.

NFPA Codes, Standards, and Recommended Practices

Reference to the following NFPA codes, standards, and recommended practices will provide further information on storage of gases discussed in this chapter. (See the latest version of *The NFPA Catalog* for availability of current editions of the following documents.)

NFPA 50, *Standard for Bulk Oxygen Systems at Consumer Sites*
NFPA 50B, *Standard for Liquefied Hydrogen Systems at Consumer Sites*
NFPA 51, *Standard for the Design and Installation of Oxygen-Fuel Gas Systems for Welding, Cutting, and Allied Processes*
NFPA 51B, *Standard for Fire Prevention in Use of Cutting and Welding Processes*
NFPA 54, *National Fuel Gas Code*
NFPA 58, *Standard for the Storage and Handling of Liquefied Petroleum Gases*
NFPA 59, *Standard for the Storage and Handling of Liquefied Petroleum Gases at Utility Gas Plants*
NFPA 59A, *Standard for the Production, Storage, and Handling of Liquefied Natural Gas (LNG)*
NFPA 70, *National Electrical Code®*
NFPA 99, *Standard for Health Care Facilities*
NFPA 220, *Standard on Types of Building Construction*
NFPA 496, *Standard for Purged and Pressurized Enclosures for Electrical Equipment*

Additional Reading

Atallah, S., and Borows, K. A., "Survey of Fire Protection Systems at LNG Facilities," Topical Report, July 1990–Nov. 1990, Gas Research Institute, Chicago, IL, GRI-91/0028, Apr. 5, 1991.
Bettis, R. J., Makhviladze, G. M., and Nolan, P. F., "Behavior of Heavy Gas and Particulate Clouds," *Proceedings* of the First International Symposium on Fire Safety Science, Hemisphere, 1986, pp. 919-930.
Colton, W., and Mayer, P. E., "Fire Detection and Suppression in Natural Gas Pipeline Compressor Stations," *Papers* presented at the ASME Gas Turbine Conference and Exhibition, American Society of Mechanical Engineers, 87-GT-103, 1987, p. 7.
Compressed Gas Association, *Handbook of Compressed Gases*, 3rd ed., Van Nostrand Reinhold, New York, 1990.
Cowley, L. T., and Pritchard, M. J., "Large-Scale Natural Gas and LPG Jet Fires and Thermal Impact on Structures," Third International Conference for Management and Engineering of Fire Safety and Loss Prevention: Onshore and Offshore, Elsevier Applied Science, New York, 1991, pp. 209–218.
Davenport, J. A., "Insurer's View of Gas Plant and Fuel Handling Facilities," *Proceedings* of the American Institute of Chemical Engineers 1986 Annual Meeting, AIChE, 1986, p. 19.
DeNeuvers, N., "Propane Over-Filling Fires," *Fire Journal*, Vol. 81, No. 5, 1987, p. 80.
Fullam, B. W., "Fire Protection of LPG Vessels," *Proceedings* of the Gastech 86 LNG/LPG Conference, Gastech, 1986, pp. 152-156.
Kirby, R. E., "Major Propane Gas Explosion and Fire, Perryville, Maryland, July 6, 1991," USFA Fire Investigation Technical Report Series, Federal Emergency Management Agency, Washington, DC, Report 053, 1992.
Lewis, B., and Von Elbe, G., *Combustion, Flames and Explosions of Gases*, 3rd ed., Academic Press, Troy, MI, 1987.

"LPG: Storage of Liquefied Petroleum Gas," P7717, Factory Mutual Research Corp., Norwood, MA.

Nettleton, M. A., *Gaseous Detonations: Their Nature, Effects, and Control,* Chapman and Hall, New York, 1987.

Sato, K., "Extinquishment of Liquefied Gas Fires by Carbon Dioxide and Liquid Nitrogen," *Bulletin of Japanese Association of Fire Science and Engineering,* Vol. 40, No. 1, 1990, pp. 29–35.

Schoen, W., and Droste, B., "Investigations of Water Spraying Systems for LPG Storage Tanks by Full-Scale Fire Tests," *Journal of Hazardous Materials,* Vol. 20, Dec. 1988, pp. 73-82.

"The Design and Construction of Liquefied Petroleum Installations at Marine and Pipeline Terminals, Natural Gas Processing Plants, Refineries, Petrochemical Plants and Tank Farms," *API Standard 2510,* American Petroleum Institute, Washington, DC.

STORAGE AND HANDLING OF CHEMICALS

William J. Bradford

Safe storage of chemicals requires a knowledge of all of the hazardous properties of the chemicals. Many materials exhibit more than one potential for fire and explosion, and detailed knowledge of hazard potential is best obtained from the manufacturer's detailed literature on the chemical. All manufacturers are required by law to prepare and distribute material safety data sheets (MSDS) for all products they make. The MSDS gives information on toxicity, flammability, reactivity, and stability, as well as safe handling and storage practices.

The quantity, size, and nature of the container, as well as its storage arrangement, affect the safety of storage. In addition to knowledge of the hazard potential, it is also necessary for safe handling to have details on the process in which the chemical is to be used. The temperature, pressure, and concentration of all chemicals involved in the process must be known. Other factors to be considered include potentially hazardous byproducts due to normal operating conditions, and byproducts that could be formed by contamination or operation outside the normal parameters. Only with complete information can safe storage and handling of chemicals be achieved.

PRINCIPLES OF GOOD STORAGE

Segregation

The first principle of good storage practice for chemicals is segregation, including separation from other materials in storage, from processing and handling operations, and from incompatible materials. Segregation can take the form of isolation in a separate detached structure or isolation in the same building (by means of fire walls). Segregation can also be achieved by separating chemicals within the same building by an intervening empty area or by intervening storage of inert or nonhazardous materials. The extent of segregation depends upon the quantity of materials being stored, the physical state of the chemicals, and their degree of incompatibility. In addition, the known behavior of materials under fire conditions will affect the extent and type of segregation needed. Rupture of vessels and possible mixing of incompatible materials under fire conditions must be considered.

William J. Bradford, P.E., is a loss prevention consultant. He is a registered professional engineer in Connecticut and Rhode Island, is a member of AIChE and SFPE, and serves on several NFPA technical committees.

Protection Against Physical Damage

Containers for chemicals are designed to be compatible with the materials they contain. However, when containers are subjected to physical abuse, containers may be damaged by the release of chemicals, which can considerably increase fire and explosion hazards. For this reason, protection of the containers against physical damage in shipment, transfer, and storage is very important. In addition, the container may represent a potential hazard under fire conditions. For example, storage of oxidizers in fiber packs or even plastic containers may be safe under normal conditions, but if a fire occurs, the oxidizers and the combustible packages can magnify fire intensity.

Hazard Identification

All storage areas should conspicuously display signs to identify the material stored in the area. The hazard identification system described in NFPA 704, *Standard System for the Identification of the Hazards of Materials for Emergency Response*, should be used. When materials having different hazard identifications are stored in the same area, the area should be marked to indicate the most severe health, flammability, and reactivity hazard present. Hazard identifications for many materials can be found in NFPA 49, *Hazardous Chemicals Data*, and NFPA 325, *Guide to Fire Hazard Properties of Flammable Liquids, Gases, and Volatile Solids*.

Fire Control

The selection of extinguishing agents is determined by the reactivity of the chemical, the physical state of the chemical (i.e., solid, liquid, or gas), the toxicity of the chemical, and its expected products of combustion. Water is normally used as the extinguishing agent, except on chemicals that may be dangerously water-reactive. However, the toxicity of the material in water must be considered as an additional factor. The contamination of potable water supplies by toxic runoff from fire fighting could create a major health hazard.

Automatic protection for storage should be employed wherever possible. The protection must be specifically designed based upon the extent of fire hazard, the arrangement of material, and the maximum quantity that can be stored.

Manual fire fighting to supplement the automatic protection may be severely limited by the toxicity of the material, obscuration due to smoke, the nature of the products of combustion, or the possibility that fires may lead to explosions in storage areas for reactive chemicals. Fire fighters should wear self-contained breathing apparatus (SCBA) to avoid the hazard of toxic materials. Where explosion is a possibility, manual fire fighting should be conducted from a remote location. In some cases, the hazard will be sufficiently high so that fire fighters should not approach the area to attempt any type of manual fire fighting.

TOXICITY OF CHEMICALS

The toxicity of chemicals is particularly important from the point of view of fire protection, regardless of the toxic material's fire hazard. A fire or explosion may subject fire fighters to a severe life hazard if the toxic material is accidentally released while they are present. In situations where fire officers are aware of a real or potential toxicity hazard, the decision may be made to forgo effective manual fire fighting.

Before using a chemical, one should obtain and evaluate information about its toxicity. When the toxicity problem appears to be severe, an effort should be made to find a suitable less toxic substitute. If there is no practical way to eliminate the toxic material, protection should be provided for those subject to daily exposure and for fire department personnel. Those who may be exposed during a fire or other emergency should be informed of the potential hazard and advised of the proper protective clothing and breathing apparatus to wear. Automatic fire protection should always be provided when a fire hazard is present in conjunction with a toxicity hazard.

Information on the toxicity of chemicals can be obtained from various sources, including the manufacturers of the chemicals; NFPA 49, *Hazardous Chemicals Data;* Patty's *Industrial Hygiene and Toxicology;* and other references.

There are two ways to protect against the toxic effects of chemicals during handling. First, use the most practical of the available methods for controlling and confining the chemical so that the toxic material cannot be contacted, swallowed, or inhaled in dangerous quantities during normal operations. Second, in areas where toxic chemicals are handled, educate all persons about the hazards, precautionary procedure, danger signals, and emergency procedures to be followed.

Of the methods used to control and confine toxic chemicals, handling them in a closed system is absolutely necessary. However, a leak in the system is always possible and could subject persons to a severe health hazard. If the gas is not an irritant, personnel must be warned of the exposure by automatically operated toxic gas indicators or other alarm devices. In some installations, it may be possible to maintain a slight negative pressure on the closed system to prevent the escape of toxic materials in case of a minor leak. Generally, such systems are ducted to a scrubber where the toxic gas can be neutralized. Wherever possible, processes should be installed outdoors where natural air movement will dilute and dissipate toxic gases or vapors.

OXIDIZING CHEMICALS

In considering storage facilities for oxidizing chemicals, remember that oxidizing chemicals usually are not themselves combustible, but may provide oxygen to accelerate the burning of other combustible materials. Combustible materials and flammable liquids, therefore, should not be stored in the same storage areas with oxidizing chemicals. Storage buildings should be of noncombustible or fire-resistive construction. Combustible packaging and wood pallets may represent a severe hazard and should be eliminated. Because certain oxidizing materials undergo dangerous reactions with specific noncombustible materials, the possibility of dangerous reactions should be considered when deciding on acceptable storage facilities. Chlorates, for example, should not be stored with acids or

combustible materials. Inorganic peroxides also react with acids, yielding hydrogen peroxide. A suitable storage facility for inorganic peroxides would be a dry, fire-resistive storage room in which there are no combustible contents or acids. When classifying chemicals for storage purposes, one should keep in mind that the label on the container required for interstate shipment classifies the chemical by hazard for transportation purposes only; a knowledge of the fire, health, and reactivity hazards of the chemical should be the guide when arranging storage.

Because of the probability that spilled material will become mixed with combustible refuse, it is important to clean up all spills immediately and thoroughly. Combustible linings of barrels should be removed from the storage area and destroyed as soon as the barrels are empty.

NFPA 430, *Code for the Storage of Liquid and Solid Oxidizers,* addresses storage of oxidizing chemicals.

Nitrates

Awareness of the fire hazard properties of the inorganic nitrates is important because these materials are widely used in fertilizers, salt baths, and other industrial applications. Under fire conditions, inorganic nitrates may melt and release oxygen, causing the fire to intensify. Molten nitrates react with organic materials with considerable violence, usually releasing toxic oxides of nitrogen. When solid streams of water are used for fire fighting, they may produce steam explosions upon contact with molten material. Common nitrates are discussed in this section. Other nitrates have somewhat similar properties.

Sodium nitrate: Noncombustible sodium nitrate promotes combustion of other materials. It evolves oxygen when heated to about 700°F (370°C), thereby increasing the intensity of any fire in its vicinity. It is soluble in water and is hygroscopic. This water solubility (common to most nitrates) is indirectly responsible for many serious fires. Paper, burlap, or cloth bags of nitrates that become moist during shipment or storage retain an impregnation of nitrate after drying and are thus highly combustible. For this reason, nitrates should be transferred from bags or wood barrels to noncombustible bins for storage. The bags or barrels should be thoroughly washed. For the same reason, storage of bulk sodium nitrate on wood floors or against wood walls or posts is hazardous. An intimate mixture of sodium nitrate and organic material can be exploded by a flame.

Potassium nitrate: The properties and hazards of potassium nitrate are similar to those of sodium nitrate; however, potassium nitrate is less moisture absorbent.

Ammonium nitrate: Like other inorganic nitrates, ammonium nitrate is an oxidizing agent and will increase the intensity of a fire. The oxidizing gas it gives off is nitrous oxide rather than oxygen. Chemical- and fertilizer-grade ammonium nitrate must not be confused with certain ammonium nitrate combustible material mixtures used as explosives. All grades of ammonium nitrate can be detonated if they are in the proper crystalline form, if the initiating source is sufficiently large, or if heated under sufficient confinement (the purest material needing the greatest confinement).

As is evident from the shipboard explosions at Texas City, Texas, and Brest, France (both in 1947), and the Red Sea in 1953; the explosion following a freight train wreck at Traskwood, Arkansas, in December 1960; and the explosion in a bulk warehouse near Pryor, Oklahoma, in January 1973, certain conditions other than initiation by explosives can cause ammonium nitrate to detonate. An extensive series of tests to determine the explosion hazards of fertilizer-grade ammonium nitrate under fire exposure was conducted by the U.S.

Bureau of Mines,[1] and from that study it would appear that the following conclusions can be reached:

1. Although detonation initiation in ammonium nitrate as a result of fire exposure cannot be ruled out completely, a direct burning-to-detonation transition in commercial fertilizer-grade ammonium nitrate (referred to, hereafter, as "AN") appears to be possible only in a pile of extremely large dimensions, with the ignition at the bottom or center of the pile. (The ammonium nitrate involved in the Texas City, Brest, and Red Sea explosions was organic coated, with substantially different burning characteristics from those of the AN manufactured today.)

2. Projectiles derived from nearby explosions can initiate reactions leading to detonations, particularly in hot AN. Ordinary sporting arms bullets are incapable of initiating detonations of AN under normal storage conditions.

3. When AN is intimately mixed with fuel oil, ground polyethylene, or ground paper, transition from burning to detonation is possible, although quite unlikely, in pile sizes that are typical of those found in storage and transportation. Such mixtures are properly classified as blasting agents and should be stored according to the requirements in NFPA 495, *Explosive Materials Code*.

4. Gas detonations are incapable of initiating detonation of AN.

5. Hot AN can be detonated by projectile impact. (Initiation of detonation of AN in the Traskwood freight train wreck and fire may have been by projectiles derived from a gasoline-nitric acid detonation, because tank cars of gasoline and of nitric acid were also in the wreck. Nitric acid may have become mixed with the burning gasoline.)

Tests conducted by Underwriters Laboratories Inc. indicate that mixtures of ammonium nitrate and ammonium sulfate containing not more than 40 percent by weight of ammonium nitrate, and mixtures of ammonium nitrate and calcium carbonate (calcium ammonium nitrate) containing not more than 61 percent ammonium nitrate, are not explosive under conditions met in practice. When exposed to fire, calcium ammonium nitrate forms calcium nitrate and ammonium carbonate which absorbs heat while decomposing to ammonia, carbon dioxide, and water. However, the contamination problem has not been thoroughly investigated.

Ammonium nitrate in water solution is not hazardous unless spilled into combustible material and permitted to dry. Research, however, has shown that solutions containing up to 8 percent water can be detonated.[2]

Cellulose nitrate: For hazards and properties of cellulose nitrate, see Section 4, Chapter 10, "Plastics and Rubber."

Nitric Acid

(See section entitled "Corrosive Chemicals," in this chapter.)

Nitrites

Nitrites should not be confused with nitrates. Nitrites contain one less oxygen atom than nitrates but are more active oxidizing agents since they melt and release oxygen at lower temperatures. Because nitrites in mixtures with combustible substances are hazardous, such mixtures should not be subjected to heat or flame. Certain nitrites, notably ammonium nitrite, are by themselves explosive. Nitrites should be treated like nitrates with respect to storage, handling, and fire fighting.

Inorganic Peroxides

Sodium, potassium, and strontium peroxide: Although these chemicals are themselves noncombustible, they react vigorously with water and release oxygen as well as large amounts of heat. Large quantities of sodium and potassium peroxides may react explosively with water, and heat from reaction with just a little water may cause the contents of an entire container to decompose. If organic or other oxidizable material is present when this reaction takes place, fire is likely to occur.

Barium peroxide: Heat releases oxygen from barium peroxide. Intimate mixtures of barium peroxide and combustible or readily oxidizable materials are explosive and easily ignited by friction or by contact with a small amount of water.

Hydrogen peroxide: In contrast to the four abovementioned peroxides, which are white powders, hydrogen peroxide is a syrupy liquid. In pure form, it is relatively stable. When heated to and kept at a temperature of 212°F (100°C), 99.2 percent hydrogen peroxide decomposes at the rate of 4 percent per year, and 50 to 90 percent hydrogen peroxide at a rate of something less than 2 percent a day. An increase in temperature increases the decomposition rate of hydrogen peroxide about $1^1/_2$ times for each 10°F (–12°C). Near the boiling point, the rate of decomposition is very rapid, and, if adequate venting is not provided, the pressure in the container may cause it to rupture. In general, the dilute material is less stable than the concentrated.

Solutions at concentrations of between 86 and 90.7 percent hydrogen peroxide have been demonstrated to be detonable.[3] At a concentration above about 92 percent, the liquid can be exploded by shock. Concentrated hydrogen peroxide vapors can be exploded by a spark. At atmospheric pressure, the boiling material must be 74 percent hydrogen peroxide or higher to produce explosive vapors.

Decomposition of hydrogen peroxide produces water, oxygen, and heat. At concentrations above 35 percent, the heat is sufficient to turn all the water into steam, assuming that the decomposition began at room temperature [72°F (22°C)]. This means that a steam explosion is possible with the sudden decomposition of concentrated material. Decomposition may be caused by contamination with iron, copper, chromium, and many other metals (except aluminum) or their salts. Decomposition can also be caused by combustible dust, or by contact with a rough surface, such as ground glass.

Hydrogen peroxide is a strong oxidizing agent and may cause ignition of combustible material with which it remains in contact. This possibility is remote, except in the case of concentrations greater than 35 percent.

Chlorates

An adequate understanding of the fire hazard properties of the chlorates can be gained from the best-known member of the group, potassium chlorate.

Potassium chlorate: This white crystalline substance is water soluble, noncombustible, and a strong oxidizing agent. When heated, it gives up oxygen even more readily than do nitrates. Mixture with combustible materials, e.g., floor sweepings, should be prevented because under such conditions potassium chlorate may ignite or explode spontaneously. Drums containing chlorates may explode when heated. Sodium chlorate has properties similar to potassium chlorate.

Chlorites

Sodium chlorite: This is a powerful oxidizing agent that forms explosive mixtures with combustible materials. In contact with strong acids, it releases explosive chlorine dioxide gas. At 347°F (175°C), sodium chlorite decomposes with the evolution of heat.

Dichromates

Among the dichromates, all of which are noncombustible, ammonium dichromate is most readily decomposed. It begins to decompose at 356°F (180°C); above 437°F (225°C) the decomposition becomes self-sustaining and is accompanied by swelling and the release of heat and nitrogen gas. Closed containers rupture at the decomposition temperature.

The other dichromates, such as potassium dichromate, react with readily oxidizable materials which, in some cases, may cause them to ignite. They release oxygen when heated.

Hypochlorites

Calcium hypochlorite at a concentration above 50 percent by weight may cause combustible or organic materials to ignite on contact. When heated, it gives off oxygen. With acids or moisture, it freely evolves chlorine, chlorine monoxide, and some oxygen at ordinary temperatures. It is sold as bleaching powder or, when concentrated, as a swimming pool disinfectant.

Perchlorates

Perchlorates contain one more oxygen atom than chlorates. They have roughly similar properties, but are more stable than chlorates. They are explosive under some conditions, such as when in contact with concentrated sulfuric acid.

Ammonium perchlorate: This chemical has great explosive sensitivity when contaminated with such impurities as sulfur, powdered metals, carbonaceous materials, and reducing agents. The pure material in finely divided form can detonate if involved in a fire. A mixture with a chlorate may form spontaneously explosive ammonium chlorate.

Potassium perchlorate, sodium perchlorate, and magnesium perchlorate: Each of these chemicals forms explosive mixtures with combustible, organic, or other easily oxidizable materials. Magnesium perchlorate is sometimes used in laboratories in place of calcium chloride as a dessicant. Such use requires vigilance to avoid dangerous contamination.

Permanganates

Mixtures of inorganic permanganates and combustible material are subject to ignition by friction or may ignite spontaneously if an inorganic acid is present. Explosions may occur whether the permanganate is in solution or is dry.

Potassium permanganate: This chemical reacts violently with finely divided oxidizable substances. On contact with sulfuric acid or hydrogen peroxide, it is explosive.

Persulfates

Persulfates, e.g., potassium persulfates, are strong oxidizing agents that may cause explosions during a fire. Oxygen released by the heat of a fire may cause an explosive rupture of the container, or the explosion may follow an accidental mixture of the persulfate with combustible material.

Fire Control Methods

With one or two exceptions, water appears to be the only suitable extinguishing agent for fires involving inorganic oxidizing agents. Water in large quantities should be used to control fires involving nitrates, nitrites, and chlorates. Carbon dioxide and other smothering agents are of little or no extinguishing value because the oxidizing material furnishes its own oxygen for combustion. Although inorganic peroxides decompose when moist and liberate oxygen, water should still be used on fires in combustible materials that are located in the vicinity of the peroxide.

It may be possible to extinguish a fire involving a peroxide spill with a dry chemical extinguisher or by smothering with dry sand or soda ash. If these methods fail, the area should be flooded with water from hose streams.

Self-contained breathing apparatuses should be worn by fire fighters. During a fire involving nitrates, one of many dangers is inhalation of oxides of nitrogen. As much ventilation as possible should be provided to permit rapid dissipation of the products of combustion and heat. When water in solid streams strikes molten nitrate, steam explosions may cause a violent eruption of the molten material.

COMBUSTIBLE CHEMICALS

Essentially, all organic chemicals are combustible. Storage practices for such chemicals closely follow those for the more common solid combustible materials discussed in other chapters. Sulfur and sulfides of sodium, potassium, and phosphorus may be considered principally as combustible chemicals. They do have some properties that warrant special mention. Combustion of these materials produces sulfur dioxide, an irritating gas. Storage should be away from oxidizing materials and, in the case of sulfides, should be separated from strong acids.

Water should be used for fighting sulfur and sulfide fires. For fighting fires of sulfur dust, water spray is recommended, principally to avoid stirring up dust clouds and causing a dust explosion. Fires in closed spaces, e.g., tanks of molten sulfur, can best be fought by closing the container and allowing the heavy sulfur dioxide produced by combustion to smother the fire.

Carbon Black

Carbon black may be formed by the decomposition of acetylene, by incomplete combustion of natural gas, or a mixture of natural gas and a liquid hydrocarbon, or by cracking hydrocarbon vapor in the absence of air. It is most hazardous immediately following manufacture when bags of the finished product may contain red-hot carbon particles. While carbon black adsorbs oxygen, slow smoldering may develop. To prevent this hazard, carbon black is stored in an observation warehouse before shipment or final storage. It is well established that carbon black will not heat spontaneously after thorough cooling and airing, although it may generate heat in the presence of oxidizable oils.

Tests show that a dust explosion hazard does not exist. Red-hot metal, electric sparks, and burning magnesium ribbon will not cause carbon black dust clouds to ignite explosively, but ignition has been obtained by using 1 oz. (28 g) of gunpowder.

Carbon black comprises 98 percent of the world's production of powdered carbon.

Lamp Black

Lamp black is formed by burning low-grade heavy oils or similar carbonaceous materials with insufficient air. It adsorbs gases to a marked degree and often ignites spontaneously when freshly bagged. It has great affinity for liquids and heats in contact with drying oils. The possibility of lamp black dust explosions is increased by the presence of unconsumed oil that adheres to the carbon. U.S. Bureau of Mines' tests indicate that a dust explosion hazard may exist if the oil content exceeds 13 percent. Lamp black should be thoroughly cooled before it is bagged and then stored in a cool, dry area away from oxidizing materials.

Lead Sulfocyanate

Lead sulfocyanate burns slowly. It decomposes when heated, yielding among its decomposition products highly toxic and flammable carbon disulfide and highly toxic but nonflammable sulfur dioxide.

Nitroaniline

This combustible solid melts at 295°F (146°C), and its flash point is 390°F (199°C). In the presence of moisture, it nitrates organic materials, which may result in their spontaneous ignition.

Nitrochlorobenzene

Nitrochlorobenzene is a solid material at ordinary temperatures and gives off flammable vapors when heated.

Sulfides

Antimony pentasulfide: This is readily ignited and hazardous in contact with oxidizing materials. On contact with strong acids, antimony pentasulfide yields hydrogen sulfide.

Phosphorus pentasulfide: This ignites readily, and in the presence of moisture may heat spontaneously to its ignition temperature [287°F (142°C)]. The products of combustion include highly toxic sulfur dioxide and phosphorus pentoxide. Reaction of phosphorus pentasulfide with water yields hydrogen sulfide.

Phosphorus sesquisulfide: With an ignition temperature of only 212°F (100°C), this chemical is easily ignited and is considered highly flammable. Toxic sulfur dioxide is a product of combustion.

Potassium sulfide and sodium sulfide: These are moderately flammable solids that form toxic sulfur dioxide when burning, and hydrogen sulfide on contact with acids.

Sulfur

Sulfur at ordinary temperatures is a yellow solid or powder consisting of rhombic crystals that melt in the vicinity of 234°F (112°C), depending on their purity. Sulfur boils at about 832°F (444°C). Except in small quantities, it is shipped and stored as a liquid at a temperature below 300°F (149°C). It is combustible, and its vapor forms explosive mixtures with air. [Its flash point is 405°F (207°C).] Finely divided sulfur dust likewise possesses an explosion hazard that requires control during storage and handling. Ignition temperatures of dust clouds vary upward from 374°F (190°C). Sulfur contains varying amounts of hydrocarbons, depending on the source. These hydrocarbons gradually react with the molten material to form combustible and highly toxic hydrogen sulfide. Storage tanks and pits for molten sulfur must be ventilated to prevent accumulation of this gas. Except in the presence of lamp black, carbon black, charcoal, and a few less common substances, spontaneous ignition of sulfur is practically nonexistent. It melts and flows when burning and evolves large quantities of toxic, irritating, and suffocating sulfur dioxide. This gas attacks the eyes and throat, and complicates fire fighting. Sulfur also forms highly explosive and easily detonated

mixtures with chlorates and perchlorates, and forms gunpowder when mixed with potassium nitrate and charcoal.

Naphthalene

Naphthalene is combustible both in solid and in liquid form. Naphthalene vapors and dusts form explosive mixtures with air.

UNSTABLE CHEMICALS

Special precautions must be taken for storage of chemicals that are subject to spontaneous decomposition or other dangerous reactions. The precautions should be planned to minimize the possibility of a dangerous reaction and to prevent injuries and extensive property damage if one should occur. Steps to be taken to protect against the hazard will depend upon the conditions that affect the stability of the chemical being stored. Points peculiar to unstable chemicals to be considered are: (1) the catalytic effect of containers, (2) materials in the same storage area that could initiate a dangerous reaction, (3) the presence of inhibitors, and (4) the effect of direct sunlight or temperature changes. There is a need for pressure-relief vents for containers and explosion venting for the storage area in addition to the usual considerations, such as automatic sprinkler or water spray protection, and elimination of all combustible material from the storage area. Whenever possible, unstable chemicals should be stored in a detached outside location.

Fire fighters should be thoroughly briefed on the proper procedures to be followed if called upon to fight a fire involving or exposing a storage area containing unstable chemicals.

Acetaldehyde

Acetaldehyde is a highly reactive compound. Under suitable conditions, the oxygen or any hydrogen can be replaced. Acetaldehyde undergoes numerous condensation, addition, and polymerization reactions, which may take place with explosive violence.

Ethyl acrylate, methyl acrylate, methyl methacrylate, and vinylidene chloride: These are flammable liquids that may polymerize at elevated temperatures, as in fire conditions. If the polymerization takes place in a closed container, the container may rupture violently. These liquids usually contain an inhibitor to prevent polymerization.

Ethylene Oxide

Ethylene oxide may polymerize violently when catalyzed by anhydrous chlorides of iron, tin, or aluminum; oxides of iron (i.e., iron rust) and aluminum; and alkali metal hydroxides. Violent polymerization of ethylene oxide may be initiated by heat or shock. Ethylene oxide reacts with alcohols, organic and inorganic acids, ammonia, and many other compounds. The heat liberated by these exothermic reactions may cause polymerization of the unreacted ethylene oxide. Ethylene oxide vapors may detonate if some initiating heat source is present. Ethylene oxide vapors in a tank exposed by fire may be rapidly heated to their ignition temperature unless the tank is kept wet with water spray. The flammable range of ethylene oxide in air is 3 to 80 percent. Although the upper limit is frequently reported to be 100 percent, explosions of mixtures containing greater than 80 percent ethylene oxide are the result of chemical decomposition.

Hydrogen Cyanide

Hydrogen cyanide is flammable and poisonous. In either liquid or vapor states, it has a tendency to polymerize. The reaction is catalyzed by alkaline materials, and since one of the products of the polymerization reaction is alkaline (ammonia), an explosive reaction will eventually take place. By adding sulfuric, phosphoric, or some other acid to neutralize the ammonia, the rate of polymerization in the liquid can be held down to a safe speed. Potassium cyanide and sodium cyanide on contact with acids release poisonous and flammable hydrogen cyanide vapor.

Nitromethane

Nitromethane is a combustible liquid. At 599°F (315°C) and 915 psig (6308 kPa), it decomposes explosively. Noteworthy detonations of nitromethane in railroad tank cars occurred at Niagara Falls, New York, and at Mt. Pulaski, Illinois, in 1958. Although undiluted nitromethane may detonate under certain conditions of heat, pressure, shock, and contamination, the causes of these two tank car explosions were not definitely determined. A possible explanation is contamination of the nitromethane by some material previously carried in the tanks. For a discussion of the hazards of nitromethane and other nitroparaffins, see AIA Research Report No. 12.[4]

Organic Peroxides

Organic peroxides are an important group of chemicals widely used in the plastics industry as polymerization reaction initiators, in the milling industry as flour bleaches, and in the chemical and drug industries as catalysts. All organic peroxides are combustible and, as in the case of inorganic peroxides, increase the intensity of a fire. Many organic peroxides can be decomposed by heat, shock, or friction. The rate of decomposition depends on the particular peroxide formulation and the temperature. Some, such as methyl ethyl ketone peroxide, are detonatable. Organic peroxides may be liquid (e.g., t-butyl perbenzoate) or solid (e.g., benzoyl peroxide) and are often dissolved in flammable or combustible solvents. The peroxides are often found wet with water or diluted with stable liquids. With solutions, sensitive crystals can be formed by freezing.

This widely used group of unstable chemicals deserves special attention. Many organic peroxides are diluted to the point where hazardous reactions are impossible; others require special precautions. The manufacturer of the peroxide should be consulted for specific recommendations.

NFPA 43B, *Code for the Storage of Organic Peroxide Formulations*, gives quantity and arrangement specifications for various formulations.

Benzoyl peroxide: In undiluted form, benzoyl peroxide ignites very readily and burns with great rapidity, similar to burning an equal amount of black powder. Decomposition by heat is rapid, and, if the benzoyl peroxide is confined when heated, explosive decomposition will occur. Decomposition can also be initiated by heavy shock or frictional heat. Most benzoyl peroxide is shipped diluted or water-wet to reduce the fire hazard.

Ether peroxides: During storage, practically all ethers form ether peroxides. When the ether peroxide mixture is heated or concentrated, the peroxide may detonate. With some ethers, the quantities of peroxide formed are too small to be of significance. Peroxide formation is a hazardous property of diethyl ether, ethyl tertiary butyl ether, ethyl tertiary amyl ether, and the isopropyl ethers. Isopropyl ether is said to be considerably more susceptible to peroxide formation than other ethers. When pure, dry ether is stored under laboratory conditions in a clear and colorless bottle, it will develop detectable amounts of peroxides in one month. Light seems to be a more important factor than heat, although peroxides have been known to form in amber bottles. Ether sealed in copper-plated cans or in otherwise inhibited cans is not likely to form peroxides. Although there is apparently no means yet available to completely

eliminate peroxide formation, using any one of numerous patented inhibitors, or copper or iron along with storage in metal or opaque amber glass containers, provides sufficient stability for all practical purposes. Ether should not be dry-distilled unless peroxides have been proved absent.

Styrene

Styrene polymerizes slowly at ordinary temperatures, and the rate increases as the temperature increases. Since the polymerization reaction is exothermic, the reaction will eventually become violent as it is accelerated by its own heat. Inhibitors are added to styrene to prevent dangerous polymerization.

Vinyl Chloride

Vinyl chloride is a toxic and flammable gas that may polymerize at elevated temperatures, as in fire conditions, and cause violent rupture of the container. Vinyl chloride usually contains an inhibitor to prevent polymerization.

Fire Control Methods

Water is the recommended extinguishing agent for fires involving unstable chemicals, including organic peroxides. Automatic sprinklers are the best protection because fires involving such materials can lead to explosions. Manual fire fighting must be conducted from a distance where the fire fighters will be protected from an explosion.

WATER- AND AIR-REACTIVE CHEMICALS

Water-reactive and air-reactive chemicals present significant fire hazards. Significant quantities of heat are liberated during the reactions. If the chemical is combustible, it is capable of self-ignition; if noncombustible, the heat of reaction may be sufficient to ignite nearby combustible materials.

Water-Reactive Chemicals

Anhydrides, carbides, hydrides, alkali metals (lithium, sodium, potassium), and similar chemicals should be stored in dry areas in waterproof and airtight containers that are kept off the floor on skids. The storage area should not contain combustible material and/or incompatible chemicals.

Air-Reactive Chemicals

Aluminum hydride, aluminum alkyls, yellow phosphorous, and similar chemicals must be stored to avoid contact with air. Yellow phosphorous, for example, is stored under water. Other materials, such as aluminum alkyls, which react with both water and air, must be stored under inert liquids or inert gas. Storage areas should contain no incompatible materials.

Alkalies (Caustics)

Caustic soda (sodium hydroxide or lye) and caustic potash (potassium hydroxide) are the most common alkalies. Although caustics are noncombustible, they generate heat when mixed with water. In contact with water, dry solid caustics will react. The heat generated (heat of solution) may be sufficient to ignite combustible material. Caustic solutions may generate hydrogen on contact with zinc, galvanized metals, or aluminum.

Aluminum Trialkyls

Most of these organic metal compounds are pyrophoric, i.e., they ignite spontaneously on exposure to air, and react violently with water and certain other chemicals. Triethylaluminum, the most common member of this group, ignites spontaneously in air and on contact with water. When it is mixed with strong oxidizing agents or with halogenated hydrocarbons, violent reactions or detonations may occur.

Anhydrides

Acid anhydrides are compounds of acids from which water has been removed. They react with water, usually violently, to regenerate acids. Organic acid anhydrides are combustible and usually are more hazardous than their corresponding acids, since their flash points are lower. Acetic anhydride has a flash point of 129°F (54°C); propionic anhydride, 165°F (74°C), open cup; butyric anhydride, 190°F (88°C), open cup; and maleic anhydride, 218°F (103°C). Inorganic acid anhydrides, e.g., chromium anhydride and phosphorus anhydride, are not combustible.

Carbides

The carbides of some metals, such as sodium and potassium, may react explosively on contact with water. Many, such as calcium carbide, lithium carbide, potassium carbide, and barium carbide, decompose in water to form acetylene. In addition to the hazard of the formation of flammable gas, another fire hazard of certain carbides is the generation of heat in contact with water. When one-third its weight of water is added to a water-reactive carbide, the temperature may be raised sufficiently to ignite the gas generated. Sodium carbide becomes heated to incandescence when placed in chlorine, carbon dioxide, or sulfur dioxide. The carbides of silicon and tungsten are very stable.

Charcoal

Under certain conditions, charcoal reacts with air at a sufficient rate to cause the charcoal to heat spontaneously and ignite. Charcoal made from hard wood by the retort method appears to be particularly susceptible. Spontaneous heating occurs more readily in fresh charcoal than in old material; the more finely divided it is, the greater the hazard. The principal causes of spontaneous heating of charcoal appear to be: (1) lack of sufficient cooling and airing before shipment; (2) charcoal becoming wet; (3) friction in grinding of finer sizes, particularly of material insufficiently aired before grinding; and (4) carbonizing of wood at too low a temperature, leaving the charcoal in a chemically unstable condition.

Coal

Under some conditions, virtually all grades of coal (except high-grade anthracite) are subject to spontaneous heating and ignition. Although the basic causes for the spontaneous heating of coal are not well defined, it is believed that adsorption of oxygen or oxidation of finely divided particles is the main cause. While instances of spontaneous heating of coal are numerous, the number of fires from this cause is insignificant when the large number of coal storage piles is considered.

The principal conditions believed to affect the susceptibility of coal to spontaneous heating are: (1) the fineness of the particles, (2) the oxygen adsorptive abilities of the particles, (3) the trapped and confined moisture content of the coal, (4) air trapped in voids in coal piles, (5) the presence of sulfur in the form of pyrites or marcasites, (6) free gases in the pile, (7) foreign substances in the pile, (8) the method and depth of piling, (9) the temperature of the containing walls and floor or of surrounding or surrounded surfaces, and (10) the type and amount of ventilation.

Hydrides

Most hydrides are compounds of hydrogen and metals. Metal hydrides react with water to form hydrogen gas.

Sodium hydride: A gray-white crystalline, free-flowing powder, sodium hydride will ignite with explosive violence on contact with water. When exposed to air, absorption of moisture may cause ignition.

Lithium hydride: When this combustible solid reacts vigorously with water, hydrogen gas and heat are evolved. Lithium hydride dust is likely to explode in humid air. Static electricity may cause the dust to explode in dry air.

Lithium aluminum hydride: Like lithium hydride, this chemical is a combustible solid that reacts rapidly with water to form hydrogen gas and heat. The heat will probably cause the hydrogen to ignite. When lithium aluminum hydride is in a solution with ether, a fire involving the solution is essentially an ether fire. Small amounts of water cause the burning to intensify. Combustible materials on which the solution has spilled may ignite spontaneously or be ignited by light friction.

Oxides

Oxides of some metals and nonmetals react with water to form alkalies and acids, respectively. This reaction takes place violently with the infrequently used sodium oxide. Calcium oxide, more commonly known as quicklime or unslaked lime, also reacts vigorously with water (slaking) with the evolution of enough heat to ignite paper, wood, or other combustible material under some conditions.

Phosphorus

Two forms of phosphorus, i.e., white and red, are in common use.

White (or yellow) phosphorus: This type is the more dangerous because of its ready oxidation and spontaneous ignition in air. It is common practice to ship and store white phosphorus under water, usually with the mixture in a hermetically sealed metal container. Periodic checks should be made to ensure that containers do not leak. White phosphorus is very toxic and should not be permitted to come in contact with the skin. On ignition, dense white clouds of toxic fumes are evolved which attack the lungs.

Red phosphorus: This type is less hazardous than white, does not oxidize and burn spontaneously at ordinary temperatures, and can be shipped and stored without the protection of water, although it should be kept in closed containers away from oxidizing agents. It is formed by heating white phosphorus. The solid form is not toxic but, once vaporized, it takes on all the fire hazards of white phosphorus to which it reverts on condensation. Care should be taken in opening containers of red phosphorus because spontaneous ignition has been known to take place with exposure to air.

Sodium

(See Section 4, Chapter 16, "Metals.")

Sodium Hydrosulfite

Sodium hydrosulfite burns slowly and produces sulfur dioxide as a combustion product. On contact with moisture and air, sodium hydrosulfite heats spontaneously and may ignite nearby combustible materials.

Fire Control Methods

The value of automatic sprinklers depends upon the type of chemical reaction that will occur when the chemical comes in contact with water. Fires involving hydrides, for example, can be smothered with a special graphite-base powder or with inert material, such as dry, finely divided calcium magnesium carbonate (dolomite). Hydride fires in an enclosed space should not be put out because the continued evolution of hydrogen after extinguishment will create an explosion hazard. For the fires involving air-reactive white (or yellow) phosphorus, water will solidify the phosphorus melted by the heat of the fire, after which it can be covered with dry sand or dirt. All of the chemical must be disposed of before it can dry out and reignite.

Several of these chemicals are both air- and water-reactive and the water reaction can be particularly violent, e.g., aluminum alkyls. Fires of such materials can be contained by dry chemical while the material burns out under control. Metallic sodium, which is both water- and air-reactive, can be extinguished by the use of dry chemical. The residual sodium should then be submerged in oil.

CORROSIVE CHEMICALS

The term "corrosive" refers to those chemicals that have a destructive effect on living tissues. Although some corrosive chemicals are strong oxidizing agents, they are separately classified to emphasize their injurious effect upon contact or inhalation. It should not be inferred, however, that a chemical is not injurious because it is otherwise classified. For example, caustics, classified as water- and air-reactive chemicals, are also corrosive. *Care should be taken to prevent inhalation, ingestion, and contact with all chemicals unless they are known to be harmless.*

Inorganic Acids

Concentrated aqueous solutions of the inorganic acids are not in themselves combustible. Their chief hazard lies in the danger of leakage and possible mixture with other chemicals or combustible material stored in the vicinity, since fire or explosions could occur in some cases.

Hydrochloric acid: In concentrated solution, hydrochloric acid is hazardous because its reaction with certain metals (including tin, iron, zinc, aluminum, and magnesium) forms hydrogen gas. Strong oxidizing agents mixed with hydrochloric acid cause the release of chlorine gas. A mixture of nitric and hydrochloric acids generates chlorine and nitrous oxide.

Hydrofluoric acid: Either anhydrous or aqueous hydrofluoric acid is noncombustible and does not cause ignition of combustible materials with which it comes in contact. However, it is highly toxic, irritating to the eyes, and inflicts severe skin burns. When it comes in contact with metals, hydrogen is generated.

Nitric acid: Under certain conditions nitric acid nitrates cellulose material. Thus, wood that comes in contact with the acid or its vapor may ignite much more easily. Spontaneous heating follows if strong solutions of the acid mix with organic material. In general, concentrated nitric acid nitrates organic materials; dilute acid oxidizes them, giving off oxides of nitrogen during the process. These oxide fumes (colorless to brown) are usually present in fires in buildings where nitric acid is used. A concentration of this gas (actually a mixture of several gases) so small that it is not objectionable at the time of inhalation can result in serious illness or death to the victim, although no effects may be felt for some time. If white fuming nitric acid (more than 97.5 percent nitric acid) is spilled into burning gasoline, it will detonate.[1]

Perchloric acid: When misused or used in concentrations greater than 72 percent, perchloric acid can be extremely dangerous. At the normal commercial strength (72 percent), it is a strong oxidizing and dehydrating agent when heated, but a strong nonoxidizing acid at room temperature. Because of this and other advantageous properties, it is widely used in analytical laboratories. The rate of burning of organic substances is greatly increased by contact with perchloric acid. Explosions have occurred at wood and plastic laboratory hoods after long exposure of the hoods to perchloric acid vapors. Strong dehydrating agents, such as concentrated sulfuric acid or phosphorus pentoxide, convert perchloric acid solution to anhydrous perchloric acid, which decomposes even at room temperature and explodes with terrific violence. It also explodes on contact with many organic substances. For these reasons, dehydrating agents should never be mixed with perchloric acid.

Sulfuric acid: This chemical has the added hazardous property of absorbing water from organic material with which it may come in contact. Charring takes place and sufficient heat may be evolved to cause ignition. Particular care must be taken to avoid painful skin burns inflicted by bodily contact. Dilute sulfuric acid will dissolve metals with the evolution of hydrogen.

The Halogens

The members of the halogen (salt-producing) group are all chemically active and have similar chemical properties. The individual elements, i.e., fluorine, chlorine, bromine, and iodine, differ from each other in decreased chemical activity in the order named. Bromine and iodine have the least fire hazard. They are noncombustible, but will support combustion of certain substances. Turpentine, phosphorus, and finely divided metals ignite spontaneously in the presence of the halogens. The fumes are poisonous, as well as corrosive and irritating to the eyes and throat.

Bromine: This is a dark reddish-brown corrosive liquid which may cause fire when in contact with combustible materials.

Chlorine: This is a heavy, greenish-yellow, highly toxic gas given off in some manufacturing processes and by bleaching powder (chloride of lime), especially in the presence of strong acids. It is not flammable itself, but may cause fires or explosions, especially if it comes in contact with acetylene, ammonia, turpentine, hydrocarbons, or finely powdered metals. At 484°F (251°C), mild steel ignites and burns in chlorine. Adequate ventilation should be provided in any process where this gas is generated.

Fluorine: A greenish-yellow gas, fluorine is one of the most reactive elements known. It combines, in most cases spontaneously, with practically all known elements and compounds under suitable conditions. Fluorine reacts violently with hydrogen and many organic materials. It explodes in contact with metallic powders, attacks glass and most metals, and reacts explosively in contact with water vapor. It may be safely handled in nickel or Monel™ cylinders. Fluorine reacts with these two metals, but forms a protective nickel fluoride layer that prevents further action. However, moisture or other impurities within the cylinder may cause such violent reaction that the metal will melt and ignite in the fluorine. The tank will then burst and scatter molten metal.

Iodine: This chemical is usually in the form of purplish-black volatile crystals that are corrosive. Reports indicate that iodine is explosive when diffused with ammonia (it forms explosive nitrogen triiodide) and when mixed with turpentine or lead triethyl.

Fire Control Methods

Water in spray form is recommended for fighting fires in acid and alkali storage areas. Water from straight streams mixing with concentrated acid or a caustic agent will heat and spatter the corrosive chemicals. A fire involving perchloric acid can cause an explosion if the acid becomes mixed with organic material. Ample precautions should be taken to protect fire fighters from possible explosions. Fire fighters should avoid contact with spilled acids and inhalation of their toxic fumes by the use of self-contained breathing apparatuses.

Chlorine and fluorine do not represent fire hazards, but the release of dangerous toxic gases should be expected if cylinders are present in a fire. It is mandatory that a self-contained breathing apparatus be used when fighting a fire where chlorine or fluorine cylinders are located.

RADIOACTIVE MATERIAL

Radioactive elements and compounds have fire and explosion hazards identical to those of the same material when not radioactive. An additional hazard is introduced by the various types of radiation emitted, all of which are capable of causing damage to living tissue. Under fire conditions, vapors and dusts (smoke) may be formed that could contaminate not only the building of origin but neighboring buildings and outdoor areas. The fire protection engineer's main concern is to prevent the release or loss of control of these materials by fire during fire extinguishment.

Fire Protection for Radioactive Materials

The life hazard introduced by an escape of radioactive dusts and vapors during a fire makes it vitally important to take all practical steps to prevent a fire from involving these materials. Radioactivity is not detectable by any of the human senses. Special instruments and measuring techniques are required to identify and evaluate it. The hazard is affected by the form of the material, i.e., whether solid, liquid, or gas, and by the container in which it is kept or handled.

Radioactivity can cause loss of life, injuries, and damage to and extended loss of materials used, as well as damage to equipment and buildings. Manual fire fighting may be limited by the danger to fire fighters from exposure to radioactivity. Salvage work and resumption of normal operations at a property may be delayed where a fire or explosion causes loss of control over radioactive substances. The need to decontaminate buildings, equipment, and materials presents a serious and complicated problem.

Smoke and products of combustion from fires in places where there are radioactive materials must be controlled. The runoff of water used in fighting fires must also be controlled. Fire fighters require protective clothing and respiratory protection equipment. Fire control must be thoroughly preplanned. With radiation hazards, automatic sprinklers are preferable to measures requiring manual fire fighting. This lessens the amount of radioactive smoke or products of combustion and water runoff to be dealt with manually.

Handling Radioactive Materials

The possibility of accidental release of radioactive material because of a fire or explosion, with the resultant health hazard to fire fighters and others, is a strong argument for careful attention to methods of fire prevention and control in laboratories and other occupancies handling radioactive materials.

More information can be obtained from NFPA 801, *Standard for Facilities Handling Radioactive Materials.* This publication calls attention to basic information concerning radiation protection methods and provides some guidance on fire protection practices to those who design and operate such facilities. Most radioactive materials introduce little or no fire or explosion hazard, so the fire hazard of facilities handling radiation usually can be determined by a knowledge of the combustibility of the building and its furnishings, and the fire and explosion hazards of the nonradioactive chemicals.

The fire and explosion hazard can be substantially reduced by use of a fire-resistive building, noncombustible interior finish and furnishings wherever possible, and enforcement of strict controls to minimize the hazards of flammable liquids and other chemicals that may be necessary for laboratory work or for other occupancies. In any facility handling radioactive materials, one feature that can cause trouble unless the building has been properly designed is the duct system usually required for the safe disposal of contaminated vapors, gases, and dusts. Unless this system is properly arranged, it can spread radioactive contaminants to noninvolved parts of a facility during a fire.

Due to the need to immediately control any fire that might eventually release radioactive materials and the potential health hazard to those who could be exposed to radiation during manual fire control, there can be no question as to the desirability of automatic sprinkler protection for radiation laboratories and other areas where radioactive materials are handled. Special precautions to be followed by fire-fighting personnel are described in *Handling Radiation Emergencies.*[5] The procedures stress steps that should be taken to protect fire fighters at all times, including the use of radiation monitoring devices, self-contained breathing apparatuses, and regular fire department protective clothing.

MATERIAL SUBJECT TO SELF-HEATING

Charcoal

Spontaneous heating hazards with charcoal can be controlled by thorough cooling and ventilation before bagging and storage. It is important to keep the charcoal dry, to prevent contamination with foreign combustibles, and to avoid contact with heat sources. Small quantities of charcoal are normally stored in heavy paper bags. The spontaneous heating hazard of individual small bags, as may be found in a dwelling, is not serious.

When fire occurs in charcoal, water should be used, but directed only on the fire, without wetting nonburning material. The damaged and wet materials should be removed from the storage building at once, because wet charcoal is even more susceptible to self-heating than when dry.

Agricultural Products

Spontaneous heating in agricultural products can be prevented by control of moisture. Proper curing (as for hay) and adequate aeration will prevent heat buildup. Where the moisture content cannot be controlled, or where any suspicion of spontaneous heating exists, thermocouples may be used in stacks or bales. Pointed, hollow metal rods or pipes with holes drilled in the lower ends often are used to permit insertion of temperature-recording instruments into the subsurface areas of the stored material. Regular checks can then be made for the development of any hazardous temperature conditions. Where evidence of dangerous spontaneous heating is noted, the material should be removed quickly from storage.

When spontaneous heating occurs in agricultural products, it is important to have ample fire-fighting facilities available. Water hoses are needed to combat a fire that may occur spontaneously on exposure of the hot spots to air. Hay, in and around hot spots, should be wetted thoroughly before complete uncovering or attempted removal.

MIXTURES OF CHEMICALS

This chapter has presented several examples of dangerous reactions that can occur when certain chemicals are mixed. Examples have been given of chemicals that can increase the ease of ignition or the intensity of burning of the combustible materials with which they are mixed. In order to recognize the innumerable combinations of so-called incompatible chemicals, it is necessary to have a knowledge of the potentially dangerous reactions of individual chemicals. NFPA 491M, *Manual of Hazardous Chemical Reactions*, contains more than 3,400 dangerous reactions that have been reported in chemical literature and elsewhere.

TRANSPORTATION OF CHEMICALS

The safe transportation of chemicals depends upon a knowledge of the hazardous properties of the chemicals, the normal and abnormal conditions to which the chemicals may be exposed during shipment, and the conditions of packing and shipping that will minimize the possibility of accidental release or reaction of chemicals.

In the United States, shipments of hazardous chemicals in interstate or foreign commerce by land, water, or air must comply with U.S. Department of Transportation (DOT) regulations. Among items covered by these regulations are: (1) construction of containers; (2) methods of packing; (3) weight of chemical per package; (4) marking and labeling; (5) loading, placarding, and movement of railroad cars; and (6) regulations for motor vehicle equipment and motor vehicle operation on the highway.

State regulations for intrastate transportation usually agree with DOT interstate commerce regulations. In Canada, shipment of explosives and other dangerous chemicals are regulated by Transport Canada (TC). Chemicals shipped in accordance with Canadian regulations may be shipped to a destination in the United States or through the United States en route to a point in Canada.

Principles governing the safe transportation of chemicals include:

1. Constructing the container of material that will not react with or be decomposed by the chemical.
2. Excluding chemicals that can react dangerously with each other from the same outside container.
3. Packaging toxic and radioactive chemicals so they will not present a health hazard during normal transportation conditions and will not be released in an accident or under other abnormal conditions.
4. Providing sufficient outage for maximum expansion of liquids under conditions to be expected during transportation.
5. Limiting the amount of chemical that can be released through container breakage or leakage by limiting maximum size of individual containers.
6. Cushioning containers to minimize the possibility of breakage.

WASTE CHEMICAL DISPOSAL

The safe disposal of hazardous chemical waste is controlled by a 1976 federal law—the Resource Conservation and Recovery Act (RCRA PL 94-590, Subtitle C, Hazardous Waste Management). This act provides for cradle-to-grave regulation of potentially dangerous byproducts spawned by industrial technology. Specifically, the law covers any solid, liquid, or gaseous waste that exceeds federal criteria established for the following characteristics: (1) toxicity, (2) persistence and degradability in nature, (3) potential for accumulation in tissue (bioaccumulation), (4) reactivity, (5) flammability, (6) corrosivity, and (7) radioactivity.

Standards have been published that regulate generators of hazardous waste in several different areas, i.e., recordkeeping practices, labeling and containerization, disclosure of waste composition to haulers and disposers, use of a manifest to monitor the life cycle of a waste stream, and reporting requirements to applicable local, state, and federal agencies. Additionally, standards have been developed to control the storage, transportation, and final disposition to ensure that these compounds are handled in properly designed facilities. For example, secured chemical landfills must have a sufficiently impervious base to prevent contaminated leachate from entering adjacent groundwater, and incinerators must achieve a 99.9 percent destruction efficiency.

Recognizing the hazards involved, it is advisable to verify the capabilities of any firm utilized for disposal of hazardous wastes. At the very minimum, operating permits should be reviewed to confirm that a particular waste can be legally processed by any company under consideration for this service.

BIBLIOGRAPHY

References Cited

1. Van Dolah, R. W., *et al.*, "Explosion Hazards of Ammonium Nitrate Under Fire Exposure," RI 6773, 1966, U.S. Bureau of Mines, Washington, DC.
2. Fukuyama, I., "Sensitive Ammonium Nitrate," *Journal of Industrial Explosives Society* (Kogyo Kayaku Kyohaishi), Vol. 18, No. 1, 1957, pp. 64-66.
3. U.S. Bureau of Mines, "Research and Technologic Work on Explosives, Explosions, and Flames: Fiscal Year 1967," IC 8387, August 1968, U. S. Bureau of Mines, Washington, DC.
4. AIA, "Nitroparaffins and Their Hazards," NBFU Research Report No. 12, 1959, American Insurance Association (formerly National Board of Fire Underwriters), New York.
5. Purington, R., and Patterson, W., *Handling Radiation Emergencies*, National Fire Protection Association, Quincy, MA, 1977.

NFPA Codes, Standards, and Recommended Practices

Reference to the following NFPA codes, standards, and recommended practices will provide further information on storage and handling of chemicals discussed in this chapter. (See the latest version of *The NFPA Catalog* for availability of current editions of the following documents.)

NFPA 43B, *Code for the Storage of Organic Peroxide Formulations*
NFPA 49, *Hazardous Chemicals Data*
NFPA 325, *Guide to Fire Hazard Properties of Flammable Liquids, Gases, and Volatile Solids*
NFPA 430, *Code for the Storage of Liquid and Solid Oxidizers*
NFPA 490, *Code for the Storage of Ammonium Nitrate*
NFPA 491M, *Manual of Hazardous Chemical Reactions*
NFPA 495, *Explosive Materials Code*
NFPA 655, *Standard for Prevention of Sulfur Fires and Explosions*
NFPA 704, *Standard System for the Identification of the Fire Hazards of Materials for Emergency Response*
NFPA 801, *Standard for Facilities Handling Radioactive Materials for Emergency Response*

Additional Reading

Barry, I., "Vermiculite Cements and Passive Fire Protection in the Hydrocarbon and Chemical Industries," *Fire Surveyor*, Vol. 20, No. 5, Oct. 1991, pp. 4–7.
Brethrick, A. L., *Handbook of Reactive Chemical Hazards*, 4th ed, Butterworths, Boston, 1990.
Carroll, T. R., and Schwope, A. D., "Non-Destructive Testing and Field Evaluation of Chemical Protective Clothing," Final Report, Federal Emergency Management Agency, Washington, DC, FA-106, Dec. 1990, 64 p.
"Chemical Laws Improve Safety and Loss Control," *Record*, Vol. 69, No. 4, July/Aug. 1992, pp. 15–20.

"Engineering Special Chemical Risks," *Record*, Vol. 70, No. 3, May/June 1993, pp. 3–8.

Eversole, J. M., "Planning, Preparation and Response to Explosions and Fires at Chemical Plants," International Fire Conference and Exhibition in Tokyo for Firesafety Frontier '94, Creating a Safe Tomorrow, 1994, pp. 63–67.

Fischer, S., *et al.*, "Vadautslapp av brandfarliga och giftiga gaser och vatskor. [Toxic and Inflammable/Explosive Chemicals: A Swedish Manual for Risk Assessment.]," National Defence Research Establishment, Stockholm, Sweden, FOA-D-95-00099-4.9-SE, Mar. 1995, 323 pages.

Goydan, R., *et al.*, "Computer System to Predict Chemical Permeation Through Fluoropolymer-Based Protective Clothing," Performance of Protective Clothing: Challenges for Developing Protective Clothing for the 1990s, Fourth (4th) Volume, ASTM STP1133, ASTM, W. Conshohocken, PA, 1991, pp. 956–971.

Gulliani, D., Wilusz, E., and Galezewski, A., " Flame Retardant Elastomers for Chemical Protective Gloves," *Journal of Fire Sciences*, Vol. 12, May/June 1994, pp. 246–256.

Harwood, P., "Warwickshire's Changing Philosophy on Chemical Protection Suits," Fire and Rescue Service, Warwickshire, England, *Fire Research News*, No. 18, Winter 1994, pp. 19–20.

Holmes, N., "Respiratory Protection in the Chemical Industry," *Fire International*, No. 144, Aug./Sept. 1994, pp. 27–28.

Howard, H. A., "Chemical Plant Fire," *Fire Engineering*, Vol. 142, No. 2, 1989, pp. 28-34, 37-43.

Isman, W. E., "Chemical Properties Influence Decisions," *NFPA Journal*, Vol. 85, No. 6, Nov./Dec. 1991, pp. 94–95.

Kearney, J., "Bulk Chemical Incidents: Learning the Lessons From Past Emergencies," *Fire*, Vol. 86, No. 1057, July 1993, pp. 41–42.

Linville, J. L., ed., *Industrial Fire Hazards Handbook*, 3rd ed., National Fire Protection Association, Quincy, MA, 1990.

Marzi, W. B., "Europaische Normung von Chemikalienschutzkleidung und vfdb-Richtlinie 0801. [European Standardization of Protective Clothing Against Chemicals and VFDB Buide 0801.]," *VFDB*, Jan. 1993, pp. 6–7.Noll, G.C., CSP, "Handling Haz Mats: Subsurface Spills and Releases into a Sewer System,"*Fire Engineering*, Vol. 148, No. 4, Apr. 1995, pp. 54–60.

"Nitromethane: Storage and Handling," NP Series TDS No. 2, 3rd ed., IMC Chemical Group, Inc., Terre Haute, IN.

Piatt, J. A., "Chemical Process Safety Management Within the Department of Energy," *Proceedings* for the Thirteenth International System Safety Conference on Hazard Control Methodologies for the Future, 1995, pp. 1–6.

Rho, S. K., "Measures Taken for Fire Protection in Place of Work (Focused on Chemical Factories)," International Fire Conference and Exhibition in Tokyo for Firesafety Frontier '94, Creating a Safe Tomorrow, 1994, pp. 313–318

Say, D. J., "Chemical Protective Clothing: Still a Long Way To Go," *Fire Engineering*, Vol. 144, No. 8, Aug. 1991, pp. 86–88,90–92.

Sax, N. I., *Dangerous Properties of Industrial Materials*, 7th ed., Van Nostrand Reinhold, New York, 1989.

Sax, N. I., and Lewis, R. J., *Hazardous Chemical Desk Reference*, VanNostrand Reinhold, New York, 1989.

"Occupational Exposure to Hazardous Chemicals in Laboratories: Chemical Hygiene Plan," *Health and Safety Instruction*, Number 20, Jan. 1991, 15 pages.

Schwope, A. D., and Renard, E. P., "Estimation of the Cost of Using Chemical Protective Clothing," Performance of Protective Clothing: Challenges for Developing Protective Clothing for the 1990s, Fourth (4th) Volume, ASTM STP1133, ASTM, W. Conshohocken, PA, 1991, pp. 972–981.

Staggs, K. J., *et al.*, "Development of Flammable Liquid Storage Wooden Cabinets for Chemical Laboratories," Department of Energy, Washington, DC, UCRL-ID-115605, Nov. 1993, 39 pages.

Storment, S., and Veghte, J. H., "Qualitatively Evaluating the Comfort, Fit, Function and Integrity of Chemical Protective Suit Ensembles. Evaluation of ASTM Standard F-1154," Task 2, Final Report, Federal Emergency Management Agency, Washington, DC, FA-107, Sept. 1991, 29 p.

Stull, J. O., "New Comprehensive Performance Standards for Chemical Protective Gloves, Boots, and Other Types of Protective Clothing," Performance of Protective Clothing: Challenges for Developing Protective Clothing for the 1990s, Fourth (4th) Volume, ASTM STP1133, ASTM, W. Conshohocken, PA, 1991, pp. 322–338.

Stull, J. O., *et al.*, "Developing and Selecting Test Methods for Measuring Protective Clothing Performance in Chemical Flashover Situations," Performance of Protective Clothing: Challenges for Developing Protective Clothing for the 1990s, Fourth (4th) Volume, ASTM STP1133, ASTM, W. Conshohocken, PA, 1991, pp. 908–923.

Stull, J. O., White, D. F., and Heath, C. A., "Selection and Development of Chemical-Resistant, Flame-Resistant Protective Gloves for U.S. Navy Shipboard Use," Performance of Protective Clothing: Challenges for Developing Protective Clothing for the 1990s, Fourth (4th) Volume, ASTM STP1133, ASTM, W. Conshohocken, PA, 1991, pp. 867–884.

Uehara, Y., "Fire Safety Assessments in Petrochemical Plants," *Proceedings* of the Third International Symposium on Fire Safety Science, Elsevier Applied Science, London, 1991, pp. 83–96.

"Use and Care of the Coat and Trousers, Chemical Protective, Aircrew, Flame Resistant," Army Natick Research Development and Engineering Center, MA, 1990, 22 pages.

Veghte, J. H., "Field Evaluation of Chemical Protective Suits," Final Report, Task 1, Federal Emergency Management Agency, Washington, DC, FA-108, Sept. 1991, 38 pages.

Wackerlig, H., "Risk Assessment of Chemical Warehouses," *Fire Protection*, Vol. 17, No. 4, Dec. 1990, pp. 4–6.

STORAGE AND HANDLING OF SOLID FUELS

Revised by John G. Puder

This chapter discusses the practices and fire hazards associated with the storage and handling of the two principal solid fuels—coal and wood—particularly as they apply to electric generation and industrial environments. The hazards associated with the actual use of these fuels in firing boilers and furnaces are covered in Section 3, Chapter 6, "Boiler-Furnaces." This chapter confines itself to the hazards associated with the storage and handling of the fuels to the point of delivery into the firing system.

The following chapters may be of interest to readers of this chapter: Section 9, Chapter 19, "Outdoor Storage Practices," which contains information about the outdoor storage of wood products; Section 9, Chapter 20, "Materials-Handling Equipment," and Section 4, Chapter 19, "Air-Moving Equipment," which both offer information about conveying systems; Section 4, Chapter 13, "Deflagration (Explosion) Venting," which discusses explosion venting; Section 9, Chapter 32, "Motor Vehicles," which contains information about the use of motor vehicles in storage facilities; and Section 9, Chapter 29, "Electric Generating Plants."

COAL AS A FUEL

Coal presents hazards between the time it is mined and its eventual consumption in boilers and furnaces. These hazards are associated with its transportation, storage, and bunkering. They can be classified as: (1) spontaneous heating, (2) dust explosions, and (3) gas generation and explosions.

Oxidation

ASTM D388, *Standard Classification of Coals by Rank*, ranks coal by its tendency to absorb oxygen. The lower the rank, the greater the tendency for coal to absorb oxygen (i.e., oxidize).

The simple downward classification of coals is: (1) anthracite, (2) bituminous, (3) subbituminous, and (4) lignite. Some coals oxidize far more readily than others. Anthracite, the highest ranked coal, has very low spontaneous heating tendencies.

Generally, the following statements are true of coal:

1. The higher the inherent (equilibrium) moisture, the higher the oxidizing tendency.

2. The lower the moisture and ash-free Btu, the higher the oxidizing tendency. The higher the oxygen in the coal, the higher the heating tendency.

3. Sulphur, once considered a major factor, is now thought to be a minor factor in the spontaneous heating of coal. There are many very-low-sulphur western subbituminous and lignite coals that have very high oxidizing characteristics, and there are high-sulphur coals that exhibit relatively low oxidizing characteristics.

4. Spontaneous oxidation is a solid-to-gas reaction, which happens initially when air (a gas) scrubs past a coal surface (a solid). Oxygen from the air is absorbed by the coal, raising the temperature of the coal. As the reaction proceeds, the moisture in the coal is liberated as a vapor, and then some of the volatile matter that normally has a distinct odor is released. The amount of surface area of the coal that is exposed is a direct factor in its heating tendency. The finer the size of the coal, the more surface is exposed per unit of weight and the greater the oxidizing potential, all other factors being disregarded.

5. Many times, segregation of the coal particle sizes is the major cause of heating. The coarse sizes allow the air to enter the pile at one location and react with high surface area fires at another location. Coals with a large top size, e.g., ≥4 in. (100 mm), will segregate more in handling than those ≤2 in. (50 mm) in size.

6. It is generally believed that the rate of reaction doubles for every 15 to 20°F (8 to 11°C) increase in temperature.

7. Freshly mined coal has the greatest oxidizing characteristic, but a hot spot in a pile may not appear before one or two months. As the initial oxidation takes place, the temperature gradually increases and the rate of oxidation accelerates.

8. There is a critical amount of airflow through a portion of a coal pile that maximizes the oxidizing or heating tendencies of coal. If there is no airflow through a pile, there is no oxygen from the air to stimulate oxidation. If there is a plentiful supply of air, any heat generated from oxidation will be carried off, and the pile temperature will reach equilibrium with the air temperature; this is considered a ventilated pile.

9. When there is just sufficient airflow for the coal to absorb most of the oxygen from the air and an insufficient airflow to dissipate the heat generated, the reaction rate increases, and the temperatures may eventually exceed desirable limits.

The general ranges of densities of coal are given in Table 3-24A to illustrate the significance of the density of the coal at various stages in its utilization.

John G. Puder is president of F. E. Moran, Inc., Special Hazards Systems, Northbrook, IL.

TABLE 3-24A. *General Ranges of Densities of Coal*

Physical Description of Coal	Range of Density, lb/cu ft (kg/m³)	Approximate Percentage of Coal Pile Occupied by Air
Coal in the seam before mining	80–85 (12.8–13.6)	0%
Coal size 2" × 0 normal shipping size as loaded in RR car, truck, etc., no compaction	50–55 (8–8.8)	36%
Coal size 2" × 0 compacted with track-mounted dozers	60–65 (9.6–10.4)	24%
Coal size 2" × 0 compacted with rubber-tired carryall or sheepsfoot roller	70–75 (11.2–12)	12%

STORAGE PRACTICES FOR COAL

The methods used to build outside storage piles and to fill silos, bins, and bunkers are significant in controlling the oxidation hazard of coal. Coal is usually stored either in an uncompacted [50 to 55 lb/cu ft (801 to 881 kg/m³)] or compacted [70 to 75 lb/cu ft (1,121 to 1,201 kg/m³)] form. When coal is not compacted, air occupies approximately 35 percent of the volume of the pile, so it is relatively easy for air to move through the pile. Coal is at this density when it is handled on conveyors or placed in trucks, railroad cars, bunkers, or silos, or in any form of "live" pile. Coal is uncompacted whenever it is moved and falls on itself.

Compacted Coal Storage

Generally, when coal is compacted, it is placed in relatively thin layers from 6 to 12 in. (152 to 305 mm) thick. Each layer is then rolled over with a device that has a relatively high loading in lb/sq ft (kg/m²) of surface area (1 lb/sq ft equals 4.9 kg/m²). Rubber-tired carryalls are normally used for this purpose, since they compact the coal while they convey the coal to the desired location.

A coal pile should be compacted on all sides to eliminate any low-density areas. When coal is compacted, it has a density of 70 to 75 lb/cu ft (1,121 to 1,201 kg/m³), and air occupies only 12 percent of the pile. This makes it far less likely that there would be appreciable air movement through the pile. Virtually all types of coal have been stored successfully in this manner, including piles exceeding 1 million tons (1 ton equals 1,016 kg) and at depths approaching 100 ft (30.5 m).

Large users of coal, e.g., utilities and large industries, will generally have compacted permanent storage piles and uncompacted "live" storage piles, while the smaller users, e.g., small industry, commercial, and residential, normally have uncompacted storage. All coal users have some form of uncompacted storage, and this form of storage may require special fire protection considerations.

Uncompacted Coal Storage

Ideally, all uncompacted coal is stored in properly designed bunkers, silos, bins, and even outside piles so all of the coal moves through the storage system with no dead or unmoving portion of the coal pile. Although the coal is still subject to the slow oxidation, it will move through the system before any appreciable heating can occur. Many new bunkers, silos, and bins are designed for mass flow of coal, and heating problems have been minimized. Coal piles may be worked in such a manner that the piles are rotated and the first coal in is the first coal out. This achieves the same results as the mass flow conditions in bins where none of the coal remains for prolonged periods of time.

When bunkers, silos, or bins have not been, or cannot be, designed to achieve mass flow, heating of the coal may occur and create problems. As much as practical, sealing off the bottom of the storage system can minimize the airflow through the coal; however, space above the coal in the storage system should not be sealed. This will prevent the accumulation of gases that may become liberated from freshly mined or crushed coal. In these cases, the area above the coal should be ventilated to carry off any accumulation of gases. Some storage systems, particularly ones where pulverized coal is stored, are designed to prevent the formation of a hazardous mixture of gases and coal dust with air.

It may be necessary to routinely draw down the storage system and clean the static deposits of coal from the bunker to minimize problems. Because cleaning bunkers, silos, or bins can be hazardous, special safety precautions should be taken.

If heating occurs in a bunker, silo, or bin, it is generally in an area where static deposits of coal have remained for a period of time. Once spontaneous heating develops to the fire stage, it becomes very difficult to extinguish the fire, short of emptying the bin, bunker, or silo. Therefore, provisions for emptying the bunker should be provided. Fire fighting in coal silos is a difficult, long-duration activity, and may take several days to completely extinguish the fire.

Carbon dioxide vapor has proved to be an effective fire-suppression agent when forced through the coal by pressure injection in the upper and lower parts of the silo. The rate of application should be sufficient to compensate for absorption by the coal. Inerting can normally be accomplished within an 8-hr shift, if a supply of carbon dioxide is available.

Since carbon dioxide is stored as a liquid, it must be vaporized before injection into the silo. This is accomplished using an external, in-line vaporizer sized such that the anticipated application rate can be increased if the maximum anticipated leakage rate is exceeded.

Normal inerting procedure involves starting the carbon dioxide flow at a preset volume from the flow-control valve. Initially, the carbon dioxide vapor should be injected into the top of the silo, into the space above the coal, to mitigate the potential of developing an explosive atmosphere. Once this inerting is established, the flow can be reduced and the primary flow directed into the bottom (while continuing trickle injection into the top) of the silo. Silo geometry creates a chimney effect, which will prevent the carbon dioxide from settling throughout the silo. Therefore, it is necessary to add injection points near the bottom. In addition, high-expansion foam can be added to the top.

Experience indicates that carbon dioxide inerting is considered successful when the carbon dioxide concentration reaches 65 percent. However, the logistics of injecting carbon dioxide might result in almost 100 percent saturation throughout most of the silo in order to achieve at least 65 percent concentration in all portions of the silo. Since leakage in the silo is inevitable, it will be necessary to anticipate using a large quantity of carbon dioxide. Experience at one utility suggests that it is advisable to order a quantity of carbon dioxide equivalent to about 3 times the gross volume of the silo when using a low-pressure system.

When carbon dioxide is used, there is a risk of oxygen depletion in the area above, around, or below a silo, bin, or bunker. Areas where gas could collect and deplete oxygen should be identified with appropriate barriers and warning signs.

Nitrogen has been used successfully to inert silo fires. It is applied in a manner very similar to carbon dioxide. A notable difference is that nitrogen has about the same density as air (whereas carbon dioxide is significantly more dense). Therefore, it must be applied at numerous injection points around the silo to ensure that it displaces available oxygen. This results in the need for more injection equipment and a larger quantity of agent.

Another method is the use of Class A foams and penetrants. These agents have found some success, but it is difficult to predict the resources required for successful fire control. The agents generally require mixing with water prior to application, usually in the range of 1 percent by volume. The typical application of Class A foam is 1 percent to fight wetland fires. However, coal plants have reported success with using Class A foams at 0.1 percent. This causes the agent to act as a surfacant. Higher percentages have caused excessive bubble accumulation, which impedes penetration into the coal.

Water and steam are not recommended, because the potential for explosion exists when water reaches the hot spot and the expansion of the steam is trapped by the water. The introduction of steam can actually increase spontaneous heating because of the introduction of additional heat and moisture.

Agent application can be enhanced by using portable infrared heat detectors to search for hot spots, either on the sides or top of the silo, to facilitate injection of the agent as close as possible to the fire area. The infrared imagery can be used to evaluate performance and monitor progress of the attack. The agent solution must penetrate to the seat of combustion to be effective. This can be affected by the degree of compaction, voids, rate of application, evaporation rate, etc. Runoff must be drained through feeder pipes, and will require collection, cleanup, and disposal.

Detection and location of hot spots may be achieved by using flammable gas monitors, portable infrared heat detection, or thermography. A long thermocouple [e.g., 10 ft (3.1 m)] connected to a portable instantaneous readout monitor can be employed. Pushing the thermocouple into the coal storage can detect developing hot areas or strata at different depths. Periodic monitoring of temperature change in these areas will help predict spontaneous combustion development and aid in response preplanning.

Regardless of the type of suppression approach selected, prefire planning is an important element of successful fire control and extinguishment. If all necessary fire-fighting resources are not stockpiled onsite, suppliers should be contacted in advance to ensure that equipment and supplies are available on relatively short notice.

Personnel requirements for this fire-fighting activity should be identified in advance. Personnel should be trained and qualified for fire fighting in the hot, smoky environment that might accompany a silo fire. This includes the use of self-contained breathing apparatus (SCBA) and personal protective equipment. Personnel engaged in this activity should be minimally trained and equipped to the structural fire brigade level, as defined in NFPA 600, *Standard on Industrial Fire Brigades*.

Selecting an outside storage site: Many plants have uncompacted storage piles on the ground, and some precautions must be taken in selecting and preparing a storage site. First, it should be determined that the proposed site is not over open sewers or other devices that may bring air into the base of the pile. Neither should the site contain any steam lines or systems that could raise the temperature of a portion of the pile beyond normal ambient temperatures.

Once selected, the site should be cleaned of all vegetation and foreign materials, particularly materials that may have heating tendencies. Then it should be graded smoothly and contoured so that water will not drain into the base of the pile. The periphery of a pile should be given special attention in the grading process to make sure that all of the water draining from the pile will be carried away.

Recognizing that uncompacted piles may tend to heat, a good practice is to size and/or shape a pile for easy access to areas that have heat so that they can then be handled appropriately. It is also good practice to limit the height of an uncompacted pile to approximately 15 ft (4.6 m) if the coal is known to have high heating tendencies. Greater pile heights can be tolerated if the coal has low heating characteristics and if the pile can be rotated with ease.

Weatherproofing of Coal Piles

In very special cases, e.g., when coal piles are to be retained for long periods or coking properties are of special importance, the exclusion of air can be further improved by covering the completed pile with road tar in the same manner as the same dressing is applied to roads. Experience has shown that road tar is an efficient and economical method and, in addition, can be used on heated heaps which, in the complete absence of air, would then stop oxidizing and cool off. The sealing of a coal pile has some added advantages, such as protection against wind loss and reduction of channeling due to heavy rain. A secondary effect, therefore, is reduction of atmospheric and river pollution.

Special Storage Precautions

It is beneficial to handle coal going to storage in a manner that prevents segregation of coarse coal from the fines. Segregation occurs when coal is placed in a conical pile; the coarse coal runs to the edge and outside of the pile, and the fines collect in the center. The air can then move readily through the coarse coal to the fines containing the large surface area, greatly encouraging heating of the pile.

Placing coal in storage near a vertical wall presents special problems, because coal tends to be less compacted when it is placed near these vertical surfaces unless special precautions are taken. Preferably, coal should not be stored around supporting beams or similar members. Less dense coal provides a natural flue for air ventilation. When vertical walls or surfaces of any kind cannot be avoided, the coal should be stored along the wall or parallel to the wall, as opposed to pushing the coal up to the vertical surface.

Selecting Coal for Storage

Commercial and domestic users select the size of coal best suited for their burning equipment, and, in most cases, it is usually a double-screened coal with a $1\frac{1}{4}$-in. (32-mm) top size, a $\frac{1}{4}$-in. (6-mm) bottom size, and a relatively small percentage of fines. This size coal, if handled reasonably well, will not segregate, and a pile can be considered fully ventilated. There will be a minimum tendency of heating because of fewer fines and because any heat generated will be readily removed from the pile.

Industries and utilities normally use a coal with a larger percentage of fines, and it is advantageous to use a coal that can be handled and stored with ease. Normally, a good shipping size is 2-in. (50-mm) top size or less, but if a larger size is used, extra care should be exercised to minimize segregation and oxidizing in storage.

Detection of Heating

Heating in a coal storage pile is usually detectable by smelling or seeing the vapor. Vapor will travel through the path of least resistance in a storage pile and may emerge from the pile at a location remote from the hot spot. This vapor should not be confused with a moisture vapor that may be noticeable on cool mornings.

When heating occurs in coal, the vapor generally emitted has a distinct odor. It is part of the volatile matter of the coal and may contain some sulphur compounds that are particularly noticeable. As stated, in many cases this vapor is visible, and it should alert the operator to start planning a procedure to eliminate a potential hazard. Depending on the materials of construction, and whether the heating has occurred in bunkers, bins, or silos, the operator has several choices. One alternative is to select a convenient time to empty the storage vessel to the point where the hot material can be removed, or cooled and removed. In some cases the vessel can be sealed and inerted with an inert gas; this will minimize further heating. Care may still need to be taken in ultimate utilization of the coal.

Sometimes small-diameter pipes of approximately 1 in. (25 mm) in diameter are placed into a pile to record temperatures at several elevations; however, it is easy to be misled, because temperatures within a pile can vary, even within a few feet of each other. Normally, if the pile is less than the temperature the human body can tolerate, e.g., 125°F (51.6°C), it is reasonable to leave the pile as is, but to continue monitoring. If the temperature is over 125°F (51.6°C), plans should be made to remove the hot area. A hot area usually can be dug out and spread out to cool on the surface of the pile if the heated area is not too large and hot. At other times, the coal can be dug out and sprayed with water and mixed with cooler coal and then consumed as soon as possible. When hot coal is fed to pulverized fuel-fired equipment, the operator should use extreme caution and be familiar with the potential hazards. (See NFPA 850, *Recommended Practice for Fire Protection for Electric Generating Plants and High Voltage Direct Current Converter Stations*, for information on fire prevention and fire protection for coal.)

DUST EXPLOSIONS WITH COAL

The conditions that create dust explosions usually occur when dry coal is being transferred from place to place or when an air current disturbs dust that has accumulated on surfaces. An ignition source must also be present at the location of the coal dust/air mixture to create the explosion. Although these types of explosions are rare, when they do occur, injury and loss of life and damage to equipment can be substantial.

There are several precautions that can be taken to help eliminate this problem. Commercial and domestic coals, for example, can be double-screened to minimize the number of fines, and they may also be treated with oil. Providing sufficient surface moisture to the coal to get the fine dust to stick to the larger sizes can be very effective in many installations.

Wetting agents sometimes are used with water to improve dust control. Many industrial and utility plants use vacuum hoods or other sophisticated systems to minimize the dust problem. Eliminating ignition sources is desirable, but is not always possible; therefore, it is more effective to eliminate coal dust/air mixtures that may explode. Some of the ignition sources that can be controlled are smoking, cutting, and welding. Some of the causes that are much more difficult to eliminate are static electricity and occasional electrical sparks.

GAS GENERATION AND EXPLOSIONS WITH COAL

Methane and other gases can be liberated from freshly mined and freshly crushed coal. An accumulation of a dangerous proportion of gas from this source is rare, but when explosions do occur, severe and fatal losses may result.

Dangerous accumulations of gas can occur over the tops of bins, silos, and bunkers, and in underground coal reclaiming facilities. Adequate ventilation is the best solution to prevent dangerous accumulation. The ventilation system should sweep airflow across the tops of bins, silos, and bunkers, and through the conveyor gallery of underground reclaiming systems. Care should be taken in the design and operation of the storage system to minimize any airflow through the stored coal, thus preventing spontaneous heating. In addition, any airflow through the pile should be in a direction that will minimize any dangerous accumulations of gas.

If there is any suspicion of gas accumulations, a methane monitor can be permanently installed, and portable detectors can be used by personnel working in these areas.

WOOD AS A FUEL

Wood is becoming widely used because of the cost and occasional shortage of crude oil and because of improved technology for whole tree chipping on site. States, e.g., Maine, New Hampshire, Vermont, and other northern tier states, that have reasonably abundant supplies of wood are continuing to research how wood can be converted into usable form to produce energy for heat and power generation. In Maine alone, more than 20 wood-to-energy projects have been proposed that would have a total capacity of 479 mW.

The increasing use of wood as an energy source competes with the use of wood as raw material for pulp, paper, and building products. Consequently, much of the research into wood use today is concentrated upon the less commercially useful wood species and upon wood waste from the forest products industry. The emphasis is upon total use of the tree, either in the production of the product or of the waste as fuel.

Hazards of Wood Fuels

Given a source of ignition, wood in all its forms will burn. As the tree is cut and processed into its eventual use, the hazard of fire increases from two directions: (1) reduction of moisture content from air drying, and (2) reduction of the wood particle size from the standing tree to finished lumber, to chips for fuel and chemical pulp, to waste wood particles the size of finely divided dust and flour. As the drying and reduction process proceeds from the relatively large particles down to the dust/flour stage, the hazard changes from that of a Class A combustible material subject to typical burning characteristics to one of such rapid combustion that it is termed an explosion. In considering the storage of wood as a fuel, the hazards encountered will range from those for solid wood to those for wood dust/flour.

Solid wood: The hazard for solid wood is Class A combustible material not readily ignited but capable of generating a large amount of heat if allowed to burn unimpeded. The rapidity with which fire will spread depends upon pile configuration; moisture of the wood and surrounding atmosphere; physical characteristics of the wood, i.e., size of individual pieces; whether wood is peeled or with bark; and species of wood. Pile size and configuration will determine the amount of radiant heat released.

Chips: This form of wood is being used on an ever-increasing scale as fuel for heat and power generation because it can be mechanized on the scale needed. The hazard is again one of a Class A combustible, and because of its small size [$5/8$ to $1\ 1/4$ in. (16 to 32 mm)], fires in piles tend to be surface type with no more than a few inches (mm) of fire penetration into the pile itself. Internal fires do occur from spontaneous heating as a result of excessive internal heat buildup. These fires are usually from chip fines that cause too much compaction or humus material within the pile, which is subject to more rapid oxidation than clean wood chips.

The chip fines can present a flash fire or explosion hazard if handled inside and dispersed in a manner that allows the wood dust to collect on structural members or to be suspended in air at concentrations above the lower explosive limit (LEL).

Sawdust and shavings: Sawdust and shavings, a byproduct of the lumber mill and wood processing plants, are generally used as a fuel on site. If strategically located, the usable material, along with other waste wood, may be shipped to other locations for use in manufacturing pulp, particleboard, or pellet fuel. The hazards are Class A combustible material and possible explosion in storage and handling from wood dust buildup.

Other wood waste: Wood scrap and edgings are another waste byproduct, particularly of lumber mills and furniture manufacturing plants. This material is generally put through a hog to reduce the size for conveying or blowing to storage bins and boiler rooms for fuel. If the larger pieces of scrap are recoverable for use as chips in pulp manufacturing, they are conveyed to a chipper and pneumatically blown to a loading station for transfer into rail cars or tractor trailers for shipment. The hazards are similar to those for chips and sawdust. The larger wood particles are readily ignited, while the fines generated by the sawing, planing, and chipping processes may contain particles small enough to be an explosion hazard.

Pellets: The process uses pulverized wood waste from the forest products industry, which is formed under high temperature and pressure into pellets approximately $1/4$ by $3/4$ in. (6 by 19 mm). The finished pellets are stored in bulk form inside to keep them dry; the raw stock of wood waste is stored outside. The finished pellet storage presents little dust hazard but will burn as a Class A combustible material.

Bark: Bark is generated at pulp mills, sawmills, and chip plants as a waste material and used to a limited extent as a soil conditioner and as fuel for limited-size boilers (when mixed with other wood scrap from the process), or it is simply disposed of by burning in teepee-type bark burners. Until recent years, bark was basically a disposal problem. Some pulp mills, however, generated too much bark over the years to be disposed of by the above means, and it was simply piled at a detached site convenient to the plant. Over many years of operation, these piles have grown into miniature mountains. The sheer height and bulk of the piles has created a fire hazard from internal heating leading to spontaneous ignition, and from external sparks from heavy equipment used to build the pile, particularly during dry periods.

A method has been outlined for eliminating fires in wood bark piles by controlling the flow of air (oxygen) horizontally into the pile.[1] Oxygen control is achieved by sloping the pile at a 30- to 40-degree angle from the horizontal. In Figure 3-24A, dimension "A" is the depth at which pressure and moisture allow the chemical reaction to generate enough heat and volatile material to provide two sides of the fire triangle, i.e., heat and fuel. Dimension "B" is the distance air (oxygen) must travel to complete the fire triangle and is approximately three times greater, with a 30-degree slope than with a 60-degree slope. The upward flow of heated gases within the pile

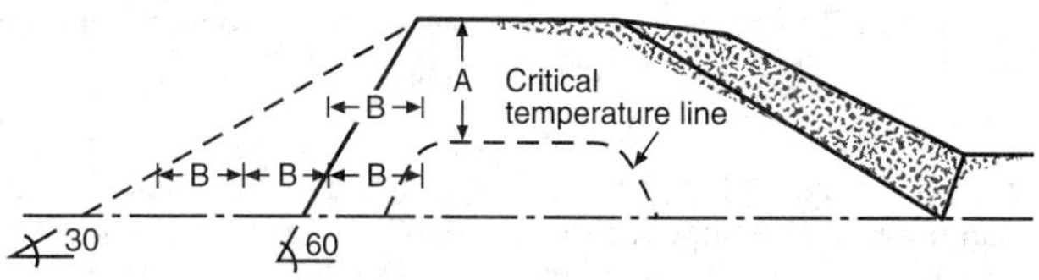

FIG. 3-24A.   A method for eliminating fires in wood bark piles by controlling the flow of oxygen horizontally into the pile.

prevents oxygen from reaching the point of combustion except from the nearly horizontal direction, which means the resistance to oxygen flow increases as dimension "B" increases. Where increasing the slope is not practical, sealing the slope to air penetration is an alternative.

What was formerly a waste disposal problem has become a valuable commodity as the price of fuel oil has gone up and the technology for burning bark on a large scale has improved. For those with such piles, they can be reclaimed as fuel for relatively large bark boilers. The reclaimed bark is mixed with the bark generated daily in the pulp manufacturing process.

STORAGE PRACTICES FOR WOOD FUELS

Logs that are to be brought to the use site for industrial and commercial conversion to chips for fuel should be stored in ranked piles (i.e., piled in parallel form), because such piles are lower in height with less volume and less radiant heat exposure than stacked piles (i.e., large circular "matchstick" piles). The use of long logs, which is likely, will dictate the ranking method of storage. The piles should be 100 ft (30.5 m), and preferably more, from the nearest important building. The length and width of piles and pile height should be kept as small as practical to make fire fighting easier and to limit the amount of wood subject to a fire. The storage area should be clean and on solid ground, with good access for fire-fighting purposes.

Residential use of cordwood and long logs is generally in the five to fifteen cord range and presents a relatively light hazard. Common sense dictates maintaining as much clear space as possible between the rough storage and dwelling. After cutting and splitting for use, the wood often is moved into the basement or garage for seasoning where there can be a fire potential from heating appliances, electrical wiring, and smoking. Care should be taken to store the wood away from the furnace it is to be burned in, electrical switchboxes, and wiring. A better method is to store the wood in a pile detached from the residence and bring in a week's supply as needed.

The use of wood chips for a residential fuel is still in the developmental stage. Furnaces and the means of automatically stoking them with the chips as well as chip storage are being developed and marketed. Wood chips for fuel, often mixed with other biomass, has forged ahead in industrial and independent energy projects as illustrated by the numerous wood-to-energy projects proposed in the state of Maine.

Where chips for fuel are used on a large scale, it likely will be in a relatively isolated area close to the wood source and unfortunately away from municipal fire protection facilities and water supplies. Only the larger operations under these conditions can economically justify an independent water supply distribution system, and it will be doubly important to limit outdoor chip pile size and height to dimensions even less than those recommended in NFPA 46, *Recommended Safety Practice for Storage of Forest Products* (hereinafter referred to as NFPA 46). The plant should be located, if possible, near a pond or stream so water is available for hose streams. Any outside storage should be well isolated from processing buildings and surrounding forests.

Storage of chips, sawdust, shavings, and sander wood waste, in silos, bins, and buildings, may present fire and explosion hazards. Therefore, storage should be arranged to minimize these hazards by proper location, preferably outside the buildings, and by physical construction of the bin or silo and associated ducts with explosion-relief venting and explosion-isolating chokes.

Large-scale bark-burning boilers require a continuous supply of fuel to avoid fluctuations in steam output or the need to fire oil burners to take up the slack. Consequently, it is common practice to provide an enclosed surge storage area capable of providing a one- or two-day supply of fuel. One design used for such a storage building is the "A" frame style with storage capacities up to 2,000 short tons (1,814 metric tons) at 50 to 60 percent moisture. These storage buildings should be designed to minimize the possibility of wood dust buildup on structural building members by minimizing surface areas on which the wood dust can collect. If possible, the walls should be sloped at angles approaching 60 degrees. Structural members that cannot be sloped should be covered with sheet metal designed to eliminate flat surfaces upon which the dust can settle.

HANDLING WOOD FUELS

Logs, either tree length or cordwood, are loaded and piled using mechanical equipment mounted either on the log truck or on a separate vehicle. Logs present minimal fire hazard exposure to the wood-handling operation.

Wood chips and wood waste material are generally transferred either pneumatically or by belt conveyor from one place to another. The fans, ducts, cyclones, and collectors associated with air movement of wood fuel constitute a fire and explosion hazard to the process of which they are a part; therefore, they should be located so as to minimize this exposure, preferably outside the buildings whenever possible.

Magnetic separators should be provided to reduce the possibility of a spark igniting the dust. The fans should be downstream of the collectors, if possible, to prevent passage of the wood fuel through the fan.

Belt conveyors present a fire hazard from the belt material as well as from the wood material being conveyed. Sources of ignition are frictional heat from belt slippage, cutting and welding, smoking, and overheated bearings. Belt conveyor fires are sometimes complicated by the need to run the belts overhead, particularly in large bark-burning boilers. Overhead conveyors present an access problem for hose stream use.

Screw conveyors and rotary airlocks should be utilized to provide chokes for fire and explosion isolation where the wood material being handled is fine enough to present an explosion hazard. It is necessary to remove a revolution (helix) of the screw in order to provide the choke. Backdraft dampers can also be used in high-velocity systems. (See Figure 3-24B.)

FIRE PREVENTION FOR WOOD FUELS

Where there is outside bulk storage of wood for fuel in any form, the first line of defense should be to limit the pile size to the absolute minimum required for economical operation. The homeowner and small commercial stick wood consumer will likely limit inventory to one or possibly two years' supply, whereas the larger commercial and industrial users who consume wood fuel to generate heat and power will likely restrict inventory to one day or even hours since, in many cases, they are consuming the wood waste as it is generated in the process.

Logs should be ranked rather than stacked and otherwise arranged in accordance with NFPA 46. If large chip piles are necessary, they should be arranged as outlined in NFPA 46. Two or more smaller piles are better than one large pile.

There is indication that foreign material in piles of chips or bark accelerates the buildup of heat, possibly from chemical reaction or simply the increased porosity that allows freer movement of air through the pile. Keeping the piles free of contamination is effective fire prevention.

Pile site selection, whether for chips or bark, should take into consideration the topography of the ground, which should be clean and free of combustibles. Clearance from surrounding forest, brush, or grass should be adequate to prevent an exposure fire from reaching the storage pile. Piles should be detached from buildings and located sufficiently apart from each other to allow fire-fighting efforts to control an exposing fire.

Bark piles stored outside, whether rough or hogged bark, create heat which can lead to spontaneous ignition under certain conditions. The piles should be limited in height and contoured to restrict the flow of air (oxygen) through the pile. Heat that is generated within the pile rises to the surface and is dissipated to the surrounding air, which tends to keep the internal temperature below the spontaneous ignition point. However, in northern climates a blanket of ice and snow can impede heat release to the extent that combustion does occur. Thus, restricting the pile height to limit heat from pressure and restricting airflow through the side of the pile become doubly important under these conditions. (See Figure 3-24A.)

Storage bins, silos, and vaults for waste wood fine enough to be an explosion hazard should be outside plant buildings, if possi-

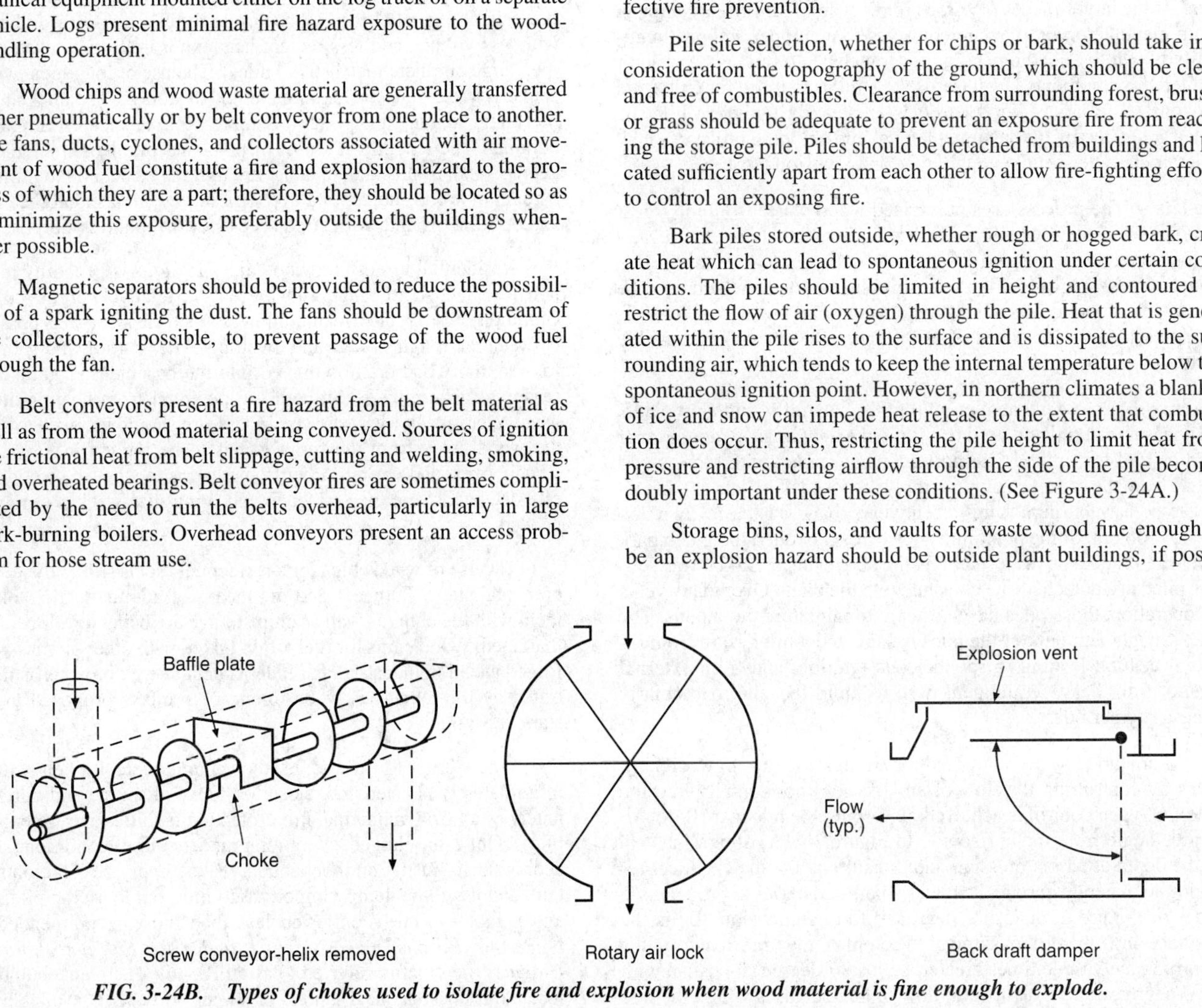

FIG. 3-24B. *Types of chokes used to isolate fire and explosion when wood material is fine enough to explode.*

ble, or, if inside, should be arranged to vent explosions to the outside. For further details, see NFPA 664, *Standard for the Prevention of Fires and Explosions in Wood Processing and Woodworking Facilities.*

The need to control moisture content of wood waste fuel for more efficient combustion may require use of lightly constructed metal buildings on a concrete slab to protect the fuel from the weather. Such buildings should be designed to minimize dust buildup on structural members. Lighting, if required, should conform to applicable provisions of NFPA 70, *National Electrical Code®.*

If the wood waste being stored contains a considerable amount of fines in the wood/flour range (such as sander fines), it may be necessary to incorporate explosion relief in the building design. A partially open structure may be a feasible design.

FIRE PROTECTION FOR WOOD FUELS

The bins, silos, or vaults in which chips or other wood waste are collected should be constructed of metal and located outside the boiler building, where possible. Collector and storage units handling fines that present an explosion hazard should be constructed with explosion-relief venting arranged to operate well below the design strength of the equipment. If bulk storage must be in or adjacent to the boiler building, it should be cut off with a fire-rated masonry wall.

Ranked log piles, arranged as previously described, should be protected with a yard hydrant system suitable for the storage arrangement and size, as defined in NFPA 46. Where the plant location and size preclude having an adequate water supply and a plant fire brigade, smaller, more segregated piles will be necessary, with greater clear space between the wood storage and plant.

Chip and bark piles also require large amounts of water to control and extinguish fires. While it is likely that storage piles of chips and bark to be used as fuel would not be the same size as piles of these materials at pulp mills, conscious effort should be made to keep the piles limited in size in order to reduce the volume of water and the fire-fighting effort needed to combat a fire. (See NFPA 46 for details of yard fire protection.)

Automatic sprinkler or water spray protection should be provided in the collecting, conveying, and storage areas. Where fuel storage buildings are utilized in the system, they should be provided with automatic sprinkler protection in accordance with NFPA 13, *Standard for the Installation of Sprinkler Systems;* NFPA 15, *Standard for Water Spray Fixed Systems for Fire Protection;* and NFPA 850.

Where bark or chips are transported from the outside storage pile or silo to an inside bin for feeding to the boiler by enclosed conveyors, the structure should be noncombustible and arranged to meet applicable safety codes. Regardless of the system type used, the conveyor should be interlocked to shut down upon actuation of the sprinkler or detection system. As a result of the increased interest in resource recovery and the expansion of independent, non-utility power generation, the use of nontraditional (alternative) solid fuels has become common.

Municipal solid waste (MSW), refuse-derived fuel (RDF), biomass, culm, gob, and rubber tires are examples of such fuels that have been utilized in the production of electric power.

While biomass exhibits most of the same challenges as wood; and culm and gob are similar to coal; MSW, RDF, and rubber tires present unique fire and explosion hazards due to the potential for unsuitable waste (e.g., flammable liquids, explosives, acetylene, etc.). Refer to NFPA 850 for prevention and protection details.

BIBLIOGRAPHY

Reference Cited

1. Halsey, K., "Controlling Bark Pile Fires," *Pulp and Paper Magazine,* Dec. 1980.

References

ASTM D388, *Standard Classification of Coals by Rank,* American Society for Testing and Materials, W. Conshohocken, PA (Rev. A, 1992).

Carini, R. C., *et al., The Relative Effectiveness of Different Agents in Dealing with Coal Pulverizer Fire and Explosion Prevention,* Coal Technology Conference, Houston, TX, Nov. 1984, p. 9.

"Coal Fires and Explosions: Prevention, Detection, and Control," preliminary draft, RP-1883-1, Oct. 1984, Electric Power Research Institute, Palo Alto, CA, pp. 2-11, A-14, Fig. 2-4a, Fig. 2-5a.

Fisher, J. E., and Wakeman, J. F., *Detecting Fires and Preventing Explosions in Coal Pulverizers,* 2nd International Coal Utilization Exhibition and Conference, Houston, TX, Nov. 1979, p. 1.

Rigsby, L. S., "Self-Heating Can Be Understood and Controlled," *Coal Quality,* Spring 1983, pp. 16–20.

Schwab, R., "Dusts," *Fire Protection Handbook,* 17th ed., National Fire Protection Association, Quincy, MA, 1991, pp. 3-133–3-142.

Smoot, L. D., *et al., Pulverized Coal Power Plant Fires and Explosions, Summary Report,* Part 1. Prepared for Research and Development Department, Utah Power and Light, Salt Lake City, UT, 1979, p.18.

NFPA Codes, Standards, and Recommended Practices

Reference to the following NFPA codes, standards, and recommended practices will provide further information on the storage and handling of solid fuels discussed in this chapter. (See the latest version of *The NFPA Catalog* for availability of current editions of the following documents.)

NFPA 13, *Standard for the Installation of Sprinkler Systems*

NFPA 15, *Standard for Water Spray Fixed Systems for Fire Protection*

NFPA 46, *Recommended Safe Practice for Storage of Forest Products*

NFPA 61, *Standard for the Prevention of Fires and Dust Explosions in Agricultural and Food Products Facilities*

NFPA 68, *Guide for Venting of Deflagrations*

NFPA 69, *Standard on Explosion Prevention Systems*

NFPA 70, *National Electrical Code®*

NFPA 91, *Standard for Exhaust Systems for Air Conveying of Materials*

NFPA 231D, *Standard for Storage of Rubber Tires*

NFPA 600, *Standard on Industrial Fire Brigades*

NFPA 664, *Standard for the Prevention of Fires and Explosions in Wood Processing and Woodworking Facilities*

NFPA 850, *Recommended Practice for Fire Protection for Electric Generating Plants and High Voltage Direct Current Converter Stations*

Additional Reading

Bland, K. E., "Behavior of Wood Exposed to Fire: A Review and Expert Judgment Procedure for Predicting Assembly Failure," Worcester Polytechnic Inst., MA, Thesis, May 1994, 159 pages.

Braun, E., "Self-Heating Properties of Coal," NBSIR 87-3554, August 1987, Center for Fire Research, Gaithersburg, MD.

Castleberry, J. C., and Smith, P. A., "Pulp and Paper Processing," *Industrial Fire Hazards Handbook,* 3rd ed., National Fire Protection Association, Quincy, MA, 1990.

Chen, J. C., "Distributed Activation Energy Model of Heterogeneous Coal Ignition," *Proceedings* of the Combustion Institute/Eastern States Section Fall Technical Meeting, 1995 on Chemical and Physical Processes in Combustion, 1995, pp. 345–348.

Chen, X. D., and Stott, J. B., "Calorimetric Study of the Heat of Drying of a Sub-Bituminous Coal," *Journal of Fire Sciences,* Vol. 10, No. 4, July/Aug. 1992, pp. 352–361.

Collins, P. K., *et al.,* "Flammability Characteristics of Treated Coals," Proceedings of the Fossil Fuels Combustion Symposium 1989, Twelfth Annual Energy-Sources Technology Conference and Exhibition, American Society of Mechanical Engineers, Petroleum Division, Vol. 25, New York, pp. 19-24.

Eisfeld, D., "Activation Process in the Spontaneous Ignition of Coal," *Glueckauf-Forschungshefte*, Vol. 49, No. 6, Dec. 1988, pp. 298-308.

Essenhigh, R. H., *et al.*, "Ignition of Coal Particles—A Review," *Combustion and Flame*, Vol. 77, No. 1, July 1989, pp. 3-30.

Factory Mutual Research Corporation, *Handbook of Industrial Loss Prevention*, Chapters 66 and 70, Norwood, MA.

Fernandez-Pello, A. C., "On Solid Fuel Ignition and Flame Spread," NIS-TIR 5499, Sept. 1994, National Institute of Standards and Technology, Annual Conference on Fire Research: Book of Abstracts, October 17–20, 1994, Gaithersburg, MD, 1994, 149–150 pp.

Fredlund, B., "Modeling of Heat and Mass Transfer in Wood Structures During Fire," *Fire Safety Journal*, Vol. 20, No. 1, 1993, pp. 39–69.

Gouws, M. J., *et al.*, "Adiabatic Apparatus to Establish the Spontaneous Combustion Propensity of Coal," *Mining Science and Technology*, Vol. 13, No. 3, Dec. 1991, pp. 417–422.

Herbert, M., "Beating Fire and Explosion Risks in Coal Mines," *Fire Prevention*, No. 231, July/Aug. 1990, pp. 25–28, 30–31.

Janssens, M., "Piloted Ignition of Wood: A Review," *Fire and Materials*, Vol. 15, No. 4, Oct./Dec. 1991, pp. 151–167.

Janssens, M., "Rate of Heat Release of Wood Products," *Fire Safety Journal*, Vol. 17, No. 3, 1991, pp. 217–238; QMC Fire and Materials Center in association with Fire Research Station, International Conference for Fire: Control the Heat...Reduce the Hazard, October 24–25, 1988, London, England, 1988, pp. 18/1–10.

Janssens, M., "Thermal Model for Piloted Ignition of Wood Including Variable Thermophysical Properties," *Proceedings* of the Third International Symposium on Fire Safety Science, Elsevier Applied Science, New York, 1991, pp. 167–176.

Janssens, M. L., "Fundamental Thermophysical Characteristics of Wood and Their Role in Enclosure Fire Growth," National Forest Product Assoc., Washington, DC, Thesis, Sept.1991, 492 pages.

Janssens, M. L. and White, R. H., "Short Communication: Temperature Profiles in Wood Members Exposed to Fire," *Fire and Materials*, Vol. 18, No. 4, July/Aug. 1994, pp. 263–265.

Jones, J. C., "Anomalies in the Self-Heating Temperature Histories of a Higher Rank Coal," *Journal of Fire Sciences*, Vol. 13, No. 5, Sept./Oct. 1995, pp. 378–385.

Jones, J. C., Hughes, K. C., and Wearing, H., "On the Action of Wetting Agents in the Extinction of Wood Fires," *Journal of Fire Sciences*, Vol. 10, No. 1, Jan./Feb. 1992, pp. 20–27.

Jones, J. C., "Steady Behaviour of Long Duration in the Spontaneous Heating of a Bituminous Coal," *Journal of Fire Sciences*, Vol. 14, No. 2, Mar./Apr. 1996, pp. 159–166.

King, J., "Improved Standards for Solid Fuel Heaters," Building Research Association of New Zealand, Judgeford, *BUILD*, Feb. 1991, pp. 17–18.

Kubler, H., "Indicators and Significance of Air Supply in the Combustion of Wood for Heat," *Wood and Fiber Science*, Vol. 23, No. 2, 1991, pp. 153–164.

McIntosh, A. C. and Tolputt, T. A., "Critical Heat Losses to Avoid Self-Heating in Coal Piles," *Combustion Science and Technology*, Vol. 69, No. 4–6, 1990, pp. 133–145.

Mikkola, E., "Charring of Wood Based Materials," *Proceedings* of the Third International Symposium on Fire Safety Science, Elsevier Applied Science, New York, 1991, pp. 547–556.

Mikkola, E., "Charring of Wood," VTT-Technical Research Center of Finland, Espoo, VTT Research Reports 689, Project PAL7003, June 1990, 35 pages.

Mutanen, K., Nissinen, K., and Linna, V., "Improvement of Safety in Peat Handling," *Proceedings of the Symposium on Low-Grade Fuels*, Part 2, Technical Research Center of Finland, 1989, pp. 31-42.

NFPA Industrial Fire Protection Section, Pulp & Paper-Wood Products Industry Committee, "Wood Products," *Industrial Fire Hazards Handbook*, 3rd ed., National Fire Protection Association, Quincy, MA, 1990.

Ohlemiller, T. J., "Smoldering Combustion Propagation on Solid Wood," *Proceedings* of the Third International Symposium on Fire Safety Science, Elsevier Applied Science, New York, 1991, pp. 565–574.

Ohtani, H., Miyazawa, S., and Nakaya, I., "Experimental Study on Bottom Surface Combustion of Woods, Japanese Association of Fire Science and Engineering, Fire Research Annual Conference, May 17–18, 1990, pp. 103–106.

Ohtani, H., Ohno, T., and Uehara, Y., "Heat Release Measurement During Smoldering Combustion of Wood Sawdust," *Journal of Applied Fire Science*, Vol. 4, No. 3, 1994/1995, pp. 161–169.

Park, S. H., and Tien, C. L., "Radiation Induced Ignition of Porous Solid Fuels," *Combustion and Flame*, Vol. 95, No. 1–6, 1994, pp. 173–192.

Parker, W. J., " Wood Materials. Part A. Prediction of the Heat Release Rate From Basic Measurements," Chapter 11, *Heat Release in Fires*, Elsevier Applied Science, NY, 1992, pp. 333–356.

Pasek, E. A., and McIntyre, C. R., "Heat Effects on Fire Retardant-Treated Wood.," *Journal of Fire Sciences*, Vol. 8, No. 6, Nov./Dec. 1990, pp. 405–420; Fire Retardant Chemicals Association. Fire Safety Developments and Testing: Toxicity—Heat Release—Product Development—Combustion Corrosivity, October 21–24, 1990, Ponte Vedra Beach, FL, 1990, pp. 181–192.

"Prevention, Detection, and Control of Coal Pulverizer Fires and Explosions," Report EPRI CS No. 5069, 1987, Electric Power Research Institute, Palo Alto, CA.

Schultz, H. E., III and Richards, R. C., "Suppression Methods for Deep Seated Coal Fires," Final Report, Department of Transportation, Washington, DC, CG-M-2-90, GC-MFSRD-49, Mar. 1990, 122 pages.

Smith, A. C., Miron, Y., and Lazzsara, C. P., "Large-Scale Studies of Spontaneous Combustion of Coal," Bureau of Mines, Pittsburgh, PA, RI 9346, 1991, 32 pages.

Smith, A. C., Miron, Y., and Lazzara, C. P., "Inhibition of Spontaneous Combustion of Coal," Report of Investigations No. 9196, 1988, U.S. Bureau of Mines, Washington, DC.

Smith, A.C., *et al.*, "Method to Evaluate the Performance of Coal Fire Extinguishants," Bureau of Mines, Pittsburgh, PA, RI 9392, 1991, 15 pages.

Takahashi, S., " Minimum Oxygen Concentration to Extinguish Stack Solid Fuel Fires by Means of Air, Diluted by Carbon," *Bulletin of Japanese Association of Fire Science and Engineering*, Vol. 41, No. 1, 1992, pp. 21–28.

Thompson, C., "Fire Safety in Bulk Sulphur Handling Locations," *Fire Prevention*, No. 208, 1988, pp. 28-31.

Tran, H. C., "Experimental Heat Release Rate of Wood Products," Abstracts of the First U. S. Symposium on Heat Release and Hazard, Interscience Communications Limited, Dec. 1991, San Diego, CA, 1991, pp. 1–2.

Tran, H. C., "Wood Materials. Part B. Experimental Data on Wood Materials," Chapter 11, *Heat Release in Fires*, Elsevier Applied Science, NY, 1992, pp. 357–372.

Tuominen, J., and Ihatsu, R., "Handling of Solid Fuels in Multi-Fuel Power Plants," *Bulk Solids Handling*, Vol. 6, No. 5, 1986, pp. 855-858.

Tzeng, L., and Atreya, A., "Theoretical Investigation of Piloted Ignition of Wood." National Institute of Standards and Technology, Gaithersburg, MD, NIST-GCR-91-595, Aug. 1991, 230 pages.

Woodside, A., and Cunningham, M. J., "Open Fireplaces and Insert Solid Fuel Stoves: An Experimental and Analytical Study," Building Research Association of New Zealand, Judgeford, BRANZ Study Report SR26, 1990, 48 pages.

Yudong, L., and Drysdale, D., "Measurement of the Ignition Temperature of Wood," *Fire Safety Science*, Vol. 1, No. 1, Sept. 1992, pp. 25–30; University of Science and Technology of China, First Asian Conference on Fire Science and Technology, (ACFST), October 9–13, 1992, Hefei, China, International Academic Publishers, China, 1992, pp. 380–385.

Zoufal, R., "Determination of the Thermal Conductivity Coefficient and the Specific Thermal Capacity of Wood at High Temperatures," Third International Symposium on Fire Protection of Buildings, May 10–12, 1990, Eger, Hungary, 1990, pp. 45–51.

STORAGE AND HANDLING OF RECORDS

Revised by Thomas Goonan

The information explosion brought about by such modern technology as telecommunications and computers has created a massive increase in the total volume of records that are generated. This increase has intensified the problems involved in protecting records from fire. Traditional fire-resistive containers, such as insulated file cabinets, safes, and vaults, have not lost any of their capabilities to protect paper records. In fact, recent advances have introduced devices that can safeguard magnetic or photographic records. In many cases, however, the volume of records material has become so great that it is impractical to invest in either the devices or the space that would be required to safeguard all the records to the level of protection that is justified. New methods of records storage have been developed that use the maximum cubic capacity of the area assigned to records storage. Some of these storage arrangements not only place the records at risk, but also contain sufficient fire potential to be a danger to the structure and all other operations housed in it.

This chapter discusses ways of identifying and classifying valuable records to determine the amount of protection, justified by their value, against fire and its associated perils. The relative susceptibility of various records to flame, heat, smoke, and water exposure is discussed. Because water damage is a serious byproduct of fire containment efforts, information on salvaging watersoaked documents is provided. The different ways in which the risk of extensive loss of valuable records can be mitigated are considered. Protection against perils not associated with fire is not discussed.

Detailed discussions of the best ways to provide maximum reasonable protection for records are contained in NFPA 232, *Standard for the Protection of Records* (hereinafter referred to as NFPA 232), and NFPA 232A, *Guide for Fire Protection for Archives and Records Centers* (hereinafter referred to as NFPA 232A). Further information can be found in "Protecting Federal Records Centers and Archives from Fire,"[1] and the reference book *Protecting the Library and Its Resources.*[2]

Adequate provisions must be maintained for the safety of persons in records storage facilities. This can be a particularly difficult task when the facility is designed to provide maximum security of its contents against illegal entry. However, emergency exits are often necessary to ensure life safety for persons in the records area in case of fire, even at some sacrifice in security. Alternate egress from

catwalks in records centers often requires a common sense design approach. General good practice recommendations for means of egress should be followed.

STORAGE OPTIONS

Bulk Storage

Bulk storage of records creates a fire hazard in itself. Bulk storage describes any sizeable collection of records not contained in vaults, safes, or insulated cabinets. This includes collections of records ranging from small file rooms to the largest archives or records centers. Storage methods include, but are not limited to, file cabinets; various types of shelving, including open shelves and mobile shelves; palletized cardboard boxes; transport cases; miscellaneous cardboard boxes; and devices for unusually shaped records, such as blueprints, magnetic tapes, photographic film, and other media. Locations may range from an area within a general office complex to specially built records facilities. It is not uncommon to house record collections in basements or attics of public buildings, in office buildings, in converted factory or warehouse buildings of various constructions and levels of quality, in public warehouses, underground, or in facilities protected against war-time disasters.

Although the desktop personal computer industry continues to promise the "paperless office," the outcome so far has resulted in an acceleration of paper records *plus* an acceleration of electronic records.

Open-Shelf Storage

The trend toward making the maximum use of available space in buildings has often resulted in using open-shelf storage methods, normally with the records held in either file folders or various kinds of cardboard boxes. Typically, the racks of records face each other across 25- to 30-in. (0.6- to 0.75-m) aisles. The aisle exposure presents a wall of paper made up of the sides of boxes or loose ends of paper sticking out of file folders. They can be easily ignited accidentally by any ignition source, such as a match, a cigarette, a portable heater, faulty fluorescent ballast, or by an exposed incandescent light bulb. Paper ignites at a relatively low temperature. Ignition of a few pieces of paper, on a filing cart, for instance, can readily transmit ignition to the boxes.

Attempts have been made to develop economical methods for increasing the flame resistance of the typical records center cardboard boxes. The most frequently attempted method is coating the

Thomas Goonan, P.E., FSFPE, is a principal of Tom Goonan Associates, Springfield, VA. He is chairman of the NFPA Records Protection Committee and a member of the NFPA Electronic Computer / Data Processing Equipment and General Storage Committees.

box with an intumescent type of fire-retardant paint. Tests of records boxes protected by such paint, properly applied, show that the coating will substantially delay actual ignition of the cardboard box material. However, since intumescent paint does not effectively react to heat under about 400°F (204°C), the temperature of any modest exposure fire (such as might occur on a file cart) will weaken the paper in the box to the point where the box will break open under the weight of the paper it contains and expose the ordinary combustible contents of the box to ignition. In a small-scale test, in the 1960s, conducted as a joint effort by the NFPA Committee on Records Protection and the U.S. General Services Administration (GSA), a fire-retardant paint coating on boxes only briefly delayed ignition and the spread of fire up and across the face of the records in storage.

Full-scale fire tests were conducted in December 1974 in untreated boxes of paper records stored on open steel shelves 13 ft (4 m) high.[1] Properly designed sprinklers were adequate to control the fire, but the fire developed very rapidly anyway. From a small ignition source, flames reached the top of the rack in 4½ min. Eight sec later, the facing array of boxes across the 30-in. (0.75-m) aisle flashed into flame from top to bottom. The first sprinkler had a 286°F (141°C) rating and activated in a little more than 5 min. By then, the ceiling temperature directly over the fire had reached 1,900°F (1,038°C). The direct cause of the intense fire was rupture of the walls of the cardboard cartons and the release of the contents. Paper records exfoliated into the aisle and many burned while falling. A test was conducted with an identical arrangement, except that unboxed files were oriented perpendicular to the aisle, which exposed the edges of paper to easy ignition. In this test, fire spread at a leisurely pace and did not develop into an intense fire. The basic difference in the rate of fire growth between these two fire tests was attributed to whether or not paper records exfoliated into the aisles. The progress of the fire can also be dramatically changed by changing sprinkler temperature ratings, response time, and sprinkler drop size characteristics.[3] Some sprinklers especially designed for fast response to fire exposure would be expected to reduce fire spread, heat rise, and extent of fire damage under similar conditions.

In other tests of high-piled storage conducted by Underwriters Laboratories Inc. (UL)[4] and tests of 6-ft (1.5-m) high archival shelving arrangements conducted by the GSA, the fire, at the end of an early and relatively short development stage, preheated a sufficient amount of the exposed boxes so that fire development characteristics changed suddenly, temperatures rose quickly, and the flame enveloped large areas. Neither of the tests involved exposed loose paper typical of systems with open-shelf filing.

In open-shelf storage, the close proximity of the opposing sides of the aisle can result in increased radiant heat feedback and flaming ignition of evolved gases across the narrow aisle. The severity and rapidity of fire development is accelerated by increased stack height and slowed by increased aisle width.

Properly designed automatic sprinklers will control the fire, but factors that speed fire development and penetration increase the amount of information loss and the amount of irrecoverable damage.

Mobile Shelving

A class of storage devices known variously as mobile shelving, track files, compaction files, moveable files, etc., consists of open-shelf units mounted on tracks. In use, all of the shelves are pushed together except that one aisle is left open for access to two units. To access another unit, shelf units are moved on the tracks until access to the desired unit is gained. Units may be arranged to move electrically or manually. The shelves that are pushed together form continuous horizontal tunnels, except where shelves are built with continuous transverse dividers. Mobile shelving containing combustibles should be protected by overhead sprinklers.

A fire in the open aisle of mobile shelving is similar to a fire in open-shelf storage and is readily controllable by overhead sprinklers. When a fire elsewhere in the mobile shelving array is sheltered from sprinkler application, it is a deep-seated burrowing fire. Full-scale fire tests conducted in 1978 indicate that a fire in a tunnel formed by shelving will develop and spread very slowly. When shielded from direct sprinkler discharge, fire is likely to continue spreading through the entire array unless the shelf unit has internal metal dividers.[5] Although protection by sprinklers alone may permit fire involvement of the entire array, the fire is very likely to be confined to that single array. Overhead sprinklers provide adequate protection for a building containing mobile shelving, and they prevent fire from jumping to other arrays.[6]

In addition to sprinkler protection, a smoke detection system is suggested in mobile shelving areas not continuously occupied to provide a fire warning well in advance of sprinkler operation. Early-warning smoke detectors may help limit the extent of fire damage in mobile shelving if skilled forces are quickly available during non-operating hours.

The U.S. National Archives and Records Administration recently opened a very large facility in College Park, MD, employing state-of-the-art mobile shelving. The facility is about 1.9 million sq ft (176,510 m²) in size, protected throughout with wet-pipe automatic sprinklers, augmented in records storage areas with smoke detection. Approximately 520 miles (837 km) of mobile shelves are provided for records storage. Typical shelving units are 8 ft (2.4 m) high and about 20 ft (6.1 m) in length. Shelves are electrically propelled, opening at the selected aisle, but during closed hours and during a fire alarm, shelves are automatically parked a few inches apart. Each shelf is directly accessible to sprinkler water from overhead sprinklers that will operate by heat from a fire in the array. Smoke detectors ensure prompt response to an incipient fire and open the array to sprinkler control and extinguishment.

Automated Files

Program libraries (e.g., open-reel tapes, tape cassettes, etc.) are normally stored in a separate room adjoining a computer room. A code-conforming library has its own fire separation walls, protected openings, and separate fire extinguishing module(s). Program tapes are manually accessed. Automated file systems are coming into use in which a relatively unlimited number of cassette-type programs or record tapes are stored in a series of modules within the computer room. At the command of the computer, a desired tape is mechanically accessed from storage, transported and installed in the proper slot in the computer, removed, and restored. NFPA 75, *Standard for the Protection of Electronic Computer/Data Processing Equipment*, requires that these tape libraries, being a severe unprotected hazard to the computer and peripherals, be sprinklered internally.

Plastic Media

The combustibility or flammability of magnetic media and its containers is a prime concern when such media are stored in bulk quantities. Generally, acetate and polyester-based tapes do not present a hazard any more severe than paper. Polystyrene cases and reels, however, present a severe fire hazard condition because they have a high fuel value, a high heat release rate, and shed water. UL tests conducted in the 1960s have shown that storage systems designed to safeguard materials of cardboard or paper composition would not adequately protect materials involving polystyrene. If the containers are made of another plastic, the condition may vary, although most thermosplastics exhibit similar properties. In any case, it is necessary to limit the height and extent of the storage of reels encased in plastic or to design special protection systems for them.

Fortunately, it appears that open-reel data storage is a declining technology, but huge quantities remain in storage and constitute a major self-exposure and exposure to other records. Computer records are being stored on an increasing variety of magnetic and optical media in widely varying formats. The main trend is the dramatic increase in information density each change involves. This factor has serious implications for the archivist, for fire prevention/ control, and for the owner of the information. Loss of a records box or its volume equivalent filled with records on paper would approximate the contents of a file drawer. Loss of an equal volume of CD-ROMs might represent the loss of an entire vault filled with paper records. The increase in information density also implies that minor dimensional changes, e.g., due to slight overheating, could make the information forever unreadable. Even minor smudges can make this media unreadable until cleaned. It is evident that reasonable protection can be achieved only by storing duplicate records in widely separated locations.

DAMAGEABILITY AND SALVAGE

Ordinary paper records survive reasonably well when exposed to most of the effects of fire, except direct exposure to flaming. With rare exceptions, burned paper records represent a total loss, while a high recovery rate is practical where the records have been exposed only to water, high humidity, smoke, and moderately high temperatures of about 350°F (177°C). While exfoliated pages burn nearly instantaneously, packed paper files burn slowly at the edges. Prompt extinguishment of packed files leaves the information largely intact, and near total recovery is possible. Nonpaper records media tend to be more susceptible to damage than paper.

Photographic Records

Photographic records, whether they are on traditional acetate, glass base, or any special base, consist of an image held in place by an emulsion. The image can be distorted or destroyed under any condition that loosens the emulsion. Tests have demonstrated that these emulsions will not withstand high temperature and humidity conditions, and that they are particularly susceptible to steam.[7,8] The traditional insulated safe or insulated filing device depends upon the water of crystallization in the insulation material to limit the internal temperature, and may be vented into the interior of the device. Therefore, it is expected that a steam atmosphere at 212°F (100°C) or higher will exist within the records protection equipment under severe fire exposure. The tests demonstrated that such exposure would damage or destroy the information on photographic media. A high degree of safety, however, may be achieved by placing the photographic media inside a steel can and sealing it with a moisture-resistant tape before storage within a record container.

The fire problem with photographic records that are stored in bulk facilities is similar to that encountered with paper records, except that in salvage efforts it is important to give priority to the photographic records. While the cold water used in extinguishing the fire will not immediately attack the photographic emulsion as would steam, it will tend to cause some softening. If salvage efforts are not undertaken immediately, there is a tendency for the emulsion to stick to adjacent material and damage the image.

Magnetic Media

The more common magnetic records consist of magnetic impulses retained in an iron oxide or similar deposit held to an acetate, polyester, or similar plastic-based material. The tapes are usually wound on polystyrene or similar plastic reels and contained in polystyrene cases. This is particularly true of many tapes used with electronic computer systems. The security of the record is primarily related, as with photographic records, to the stability of the emulsion. Even minor distortion is severe in the case of magnetic records, since machines cannot make subjective judgments regarding distortion. If the emulsion is softened by the heat from fire, for example, there is a tendency for the layers of tape to stick to each other on the reel and to be destroyed in efforts to unreel. A tape can be considered safeguarded only up to temperatures in the range of 125°F (52°C) and in relative humidity not exceeding 85 percent.

Salvage

The problem of recovering wet records is the same whether they are damaged by a fire, or some other source, such as flood, hurricane, heavy rainstorm, roof leakage, spillage from operations located above, or a breakdown of any of the numerous water or steam systems in the building. It is generally recognized that virtually any wet paper records can be recovered, provided prompt and proper action is taken.

Good sources for further information on salvaging records are contained in NFPA 910, *Recommended Practice for the Protection of Libraries and Library Collections*, "Salvaging and Restoring Records Damaged by Fire and Water,"[9] *Managing the Library Fire Risk*,[10] "After the Water Comes,"[11] and *Procedures for Salvage of Water-Damaged Library Materials*.[12]

RECORDS STORAGE

Factory-Built Devices

Limited quantities of valuable records may be stored in factory-built records protection equipment, such as insulated records containers, fire-resistant safes, insulated filing devices, and insulated file drawers. These devices are available in varying degrees of resistance to fire, heat, and impact, and the degree of protection to be specified for a particular application will depend upon the severity of the exposure and the items to be stored. The test procedure applicable to this equipment and the ratings employed are described in UL 72, "Tests for Fire Resistance of Record Protection Equipment."[13]

Records protection equipment is classified in terms of two elements: (1) an interior temperature limit and (2) a time in hours. Three temperature limits are employed: 350°F (177°C), regarded as a suitable limit for paper records; 150°F (66°C), regarded as a limiting temperature for photographic records; and 125°F (52°C), regarded as suitable for most magnetic media. The time limits employed are $1/2$, 1, 2, 3, or 4 hr. The complete rating, consisting of the two elements, indicates that the specified interior temperature limit is not exceeded when the record container, safe, or filing device is exposed to a standard test fire as described in UL 72,[13] for the length of time specified.

Ratings are assigned to the various categories, as follows:

Insulated Records Containers

    Class 125—4 hr
    Class 125—3 hr
    Class 125—2 hr
    Class 125—1 hr
    Class 150—4 hr
    Class 150—3 hr
    Class 150—2 hr
    Class 150—1 hr

Class 350—4 hr

Class 350—3 hr

Class 350—2 hr

Class 350—1 hr

Fire-Resistant Safes

Class 125—4 hr

Class 125—3 hr

Class 125—2 hr

Class 125—1 hr

Class 150—4 hr

Class 150—3 hr

Class 150—2 hr

Class 150—1 hr

Class 350—4 hr

Class 350—3 hr

Class 350—2 hr

Class 350—1 hr

Insulated Filing Devices

Class 125—1 hr

Class 125—$1/_2$ hr

Class 150—1 hr

Class 150—$1/_2$ hr

Class 350—1 hr

Class 350—$1/_2$ hr

Insulated File Drawers

Class 125—1 hr

Class 150—1 hr

Class 350—1 hr

Insulated records containers and fire-resistant safes must protect contents from heat to an extent described in the requirements, before and after a fall from 30 ft (9 m). Insulated filing devices, which are not drop-tested, are not required to have the strength to endure such an impact.

Insulated records containers, fire-resistant safes, insulated filing devices, and insulated file drawers must sustain sudden exposure to high temperatures to an extent described in the requirements without the unit exploding as a result of such exposure.

Ordinary, uninsulated steel files and cabinets provide only a limited measure of protection from an exposure fire, as heat sufficient to char contents is quickly transmitted to the interior. However, they can be very useful where the major fire exposure is from the records themselves. They are commonly used for records having no extraordinary value. They are also used for the organized storage of valuable records in protected facilities, including vaults, file rooms, and document buildings. A records vault or file room in which all records are kept in metal file cabinets is much safer than one in which the records are kept in cardboard boxes or on open shelves. If an ignition occurred in an open file drawer, fire development would likely be very slow. An ignition in open-shelf files, however, could be beyond manual control within a few minutes.

Files and cabinets made of wood, fiberboard, or other combustible materials can release their contents in a fire, adding to the fire hazard. They should not be used where they will expose valuable records storage.

Vaults and File Rooms

Standards for the construction of fire-resistive vaults and file rooms are contained in NFPA 232. The term "vault" refers to a completely fire-resistive enclosure up to 5,000 cu ft (142 m³) in volume, used exclusively for records storage, with no work to be carried on inside. It is equipped, maintained, and supervised to minimize the possibility of a fire starting within, and to prevent a severe fire of long duration on the outside from entering, provided the vault door is closed. Vaults are designed to fire resistance classifications of 2, 4, and 6 hr, indicating that, under standard test conditions, heat will not rise above a specified temperature inside and the construction will withstand both the exposure fire and the application of fire hose streams during that period. Vaults are constructed in the field and do not carry a testing laboratory label. Vault doors, however, are laboratory tested and carry ratings conforming to the vault construction in which they are to be used. (See NFPA 232.) Unlike some other fire doors, vault doors limit the temperature on the interior face to 350°F (177°C) during fire exposure, so that paper could be stored in contact with the door without danger of ignition. Provisions in NFPA 232 require sprinklers for oversize vaults up to 25,000 cu ft (708 m³) in volume.

Vaults usually contain a substantial fuel load, and by their nature contain only vital and important records. The internal contents of some vaults are more of a fire hazard than any external exposure. In recognition of this situation, vault standards permit sprinkler systems, lighting, and low-energy circuits, with suitable safeguards. Installation of smoke detectors and/or sprinklers to detect and extinguish a vault fire with a minor increase in ignition sources is permitted. A fire within an unprotected vault can be disastrous unless it is immediately discovered and extinguished with portable fire-fighting devices. Fire extinguishers of the water type or fire hose, or both, should be outside the vault in an accessible location near the door.

Additional protection may be gained by storing particularly sensitive documents in protected filing devices *inside* vaults. Be careful that fire exposure is not increased thereby, such as installing the safe in a vault occupied by open-shelf filing.

A standard fire-resistive file room is defined as an enclosure not exceeding 50,000 cu ft (1416 m³) in volume and 12 ft (3.66 m) in height, designed to a fire resistance classification of 1, 2, 4, or 6 hr. The volume and height limitations restrict the quantity of records exposed to destruction by fire in a single enclosure and reduce the possibility of fire originating within the enclosure.

File room doors are labeled as to their fire resistance following a laboratory test under the procedures of UL 155, "Tests for Fire Resistance of Vault and File Room Doors,"[14] but the ratings are only for $1/_2$ or 1 hr. These ratings anticipate that paper and other combustibles will not be stored nearer than 3 ft (0.9 m) from the unexposed face of the door, nor 6 in. (152 mm) to the side from the door jambs.

Lighting, heating, ventilation, etc., are permitted inside file rooms, as are filing cabinets and furniture, provided they are noncombustible. Although it is not to be used as a working space for other than filing purposes, the file room is more susceptible to a fire start than a vault. Due to this vulnerability, installation of automatic fire detection and fire control systems should be considered in relation to the values involved. Automatic sprinklers are required in file rooms having open-shelf filing.

Records Centers and Archives

Bulk storage of records in buildings set aside for the specific purpose or in major portions of other buildings [in rooms exceeding the 50,000 cu ft (1416 m^3) fire-resistive file room limitation] should comply with the recommendations set forth in NFPA 232A. For facilities of this size, where the most severe hazard may not be from a fire exterior to the facility but from a fire starting and spreading entirely within the records storage area, the level of fire-resistive construction and protection must be individually determined. Within such a facility, secondary operations may have to be segregated from the actual storage by fire-resistive construction. Full-scale fire tests of 14-ft (4.3-m) high open-shelf records storage in a sprinklered facility indicated the likelihood of ceiling temperatures in excess of 1,200°F (649°C) for periods of up to 8 min. when using standard 286°F (141°C) sprinklers.[1] Lightweight bar joist roofs often present in records centers are vulnerable to early collapse unless additional precautions are taken, such as might be provided by developments in large-drop sprinklers, quick-response sprinklers, or fireproof coatings for the steel. A thorough fire risk analysis by qualified personnel should be conducted due to the high fire potential from the concentration of combustibles and the probability of loss of large volumes of valuable documents.

FIRE RISK ANALYSIS

Maximum possible protection is not desirable for the bulk of what comprises records storage. Most records can be reconstructed, and duplicates are often available in other locations, and their loss may not create any substantial hardship. The value of certain, irreplaceable documents warrants especially sophisticated protective measures to preserve a single sheet of paper. The great majority of records in a particular collection, however, may dictate consideration of less-than-optimum methods of records protection and anticipation of potential small fire losses, yet still provide a high degree of assurance against a significant loss.

A records protection program should be based on an inventory that determines type, volume, rate of acquisition, duplication of data, rate of disposal, and class of importance of the records. Records can be generally classified in one of two ways to assess their value:

1. *Vital records.* These records are irreplaceable; records where reproduction does not have the same value as an original; records that give direct evidence of legal status, or ownership; records needed to sustain a business or avoid delay in restoration of production, sales, or service; and records of accounts receivable.
2. *Important records.* These are necessary records that can be reproduced from original sources only at considerable expense or loss of time.

Other records may be useful, but should be kept well separated from vital and important records because they may constitute a fire exposure in and of themselves.

Fire Risk Evaluation Factors

In considering the protection of valuable records, four basic items must be evaluated. They are:

1. The probability that the building containing the records, as well as neighboring operations, will have a fire outside the records storage activity that will spread to the records. Determining the probability of such fire hazards will involve consideration of building construction, contents, possible fire causes, and general features of fire protection that might stop a fire that originated

elsewhere before it reaches the records. (See Section 7, Chapter 4, "Structural Integrity During Fire.")
2. The probability of a fire starting within the records storage activity. This includes assessing the ignition sources, vulnerability to arson, and the susceptibility of the records, or their containers, to ignition.
3. The quantity of fuel the records represent, particularly as it relates to available or proposed capability for fire extinguishment, and the structural stability of the storage arrangement and the building enclosure.
4. The susceptibility of the records to damage from fire, fire effects (heat, smoke, vapors, etc.), and fire extinguishing efforts (principally water damage and impact damage from hose streams and other extinguishing devices, and physical disruption from manual fire fighting).

When records must be housed in a building that may burn around them, properly rated vaults and containers can give reasonable protection against the external fire exposure. The hazard of the records themselves, such as those stored within a vault, file room, segregated floor or section of a fire-resistive building, or records center, can be substantial. For example, a fire involving military records of 19 million U.S. servicemen occurred in an unsprinklered, six-story, fire-resistive building. The top floor, where the fire started, was completely destroyed. Fire-fighting efforts prevented the fire from spreading downward. Had the fire started on a lower floor, it is likely that the entire building would have been lost.[1]

The protection methods described in this chapter yield protection guidelines commensurate with the hazard and the sophistication of the protection systems. The degree of fire risk and the potential for loss in collections not suitable for cabinet or vault storage may require evaluation by a person knowledgeable in this type of analysis.

FIRE RISK REDUCTION

Since most records media are combustible, 100 percent effective protection is not feasible and efforts should be directed to reducing the risk of fire and its associated effects. One of the most effective means for limiting the disastrous effects of a fire in a records storage facility is to prepare duplicates and store them away from the originals where they will not be subject to the same incident. Often the duplicate copy is on microfilm, which is inexpensive, easily transported, and easily stored at a remote facility. Once a duplicate has been prepared, the value of the original is reduced considerably, unless the original document is required for legal purposes or is of intrinsic or historic value. If the duplication is complete, up-to-date, and easily accessible, then reduced protective measures may be justified for both the original and the duplicate collection.

If the prior generation of computer records is stored remote from the current generation, the effect of loss of the current records in use at the computer center would be reduced. Although they are not exact duplicates, the earlier records still greatly simplify the task of reconstructing current records.

Protective Containers

Heavily insulated, massive vaults, safes, and filing cabinets are the traditional way to safeguard valuable records against the effects of fire. The practice of encapsulating the records so that they are fully protected against the greatest potential fire exposure is still viable. Containers are available that will keep the internal temperature and humidity as low as necessary to retain the data on magnetic tape and other temperature-sensitive media.

Vaults are usually used where the volume of valuable records is large. They are often the only practical method of protection in non-fire-resistive buildings where the probable fire severity exceeds the rated endurance of 4-hr safes. Safes are used for smaller volumes of records where it is necessary to have records near their point of use, where the cost of vault construction would be prohibitive, or where the building does not lend itself to vault construction.

Protective containers for records are rated by tests under standard fire conditions. They are rated according to the time elapsed before the interior of the container reaches 350°F (177°C). This time factor provides a measure of safety, since the ignition temperature of most paper is somewhat higher.

Records containers rated for interior temperatures that do not exceed 150°F (66°C) and relative humidity that does not exceed 85 percent at temperatures above 120°F (49°C) are available. Containers rated with these more stringent requirements are intended to provide protection for films, magnetic tapes, plastic media, and similar materials. The development of new and different records materials, however, requires a careful evaluation of their characteristics to ensure proper protection. An important limitation on this type of container is that it is not subjected to the traditional drop test while containing magnetic or similar media records. The device should not be considered capable of safeguarding magnetic and similar records in any building that could collapse in a fire and cause debris to fall on the container, nor should the container itself be subject to falling several floors.

Table 3-25A indicates the approximate fire resistance that can be safely expected from various types of records containers exposed to a fully developed fire. The protection provided varies considerably, depending upon design and materials. Thickness of insulation alone does not give a reliable indication of fire resistance, which can be determined accurately only by tests.

TABLE 3-25A. *Fire Resistance of Records Containers*

Insulated records vault doors	2, 4, and 6 hrs
Insulated file room doors	1/2 and 1 hr
Steel-plate vault doors with inner doors	About 15 min
Steel-plate door without inner doors	Less than 10 min
Modern safes	1, 2, and 4 hr
"Old-line," "Iron," or "Cast-iron" safes, 2 to 6 in. wall thickness	Uncertain
Insulated records containers (files and cabinets, etc.)	1/2, 1, and 2 hr
Containers with air space or with cellular or solid insulation less than 1 in. thick	10 to 20 min
Uninsulated steel files, cabinets, wood files, wood or steel desks	About 5 min

For SI units: 1 in. = 25.4 mm.

Patterns of records container performance in real fires were analyzed in depth in a special five-year study done by NFPA over four decades ago.[15] While the statistics from that era can no longer be considered representative, the leading factors of contents destruction are probably the same ones one would find today. They are listed below in order of frequency from the old study.

• Container exposed to fire beyond its capacity to withstand (mostly unlabeled containers, some 1-hr containers, a few 2-hr containers, no 4-hr containers)

• Obsolete design or construction of unlabeled container

• Container submerged in water

• Container door left open

The first two of these were cited much more often than the second two, which underlines the importance of selecting the right containers for the fire hazard in the first place. As one might expect, the longer the time rating, the more likely the container was to preserve the contents.

The Concept of Early Warning

Records administrators usually consider the sprinkler system as optional when they plan records protection. A smoke detection system coupled with manual response, once highly favored by archivists and librarians as a full substitute for sprinkler protection, is now only considered to be a useful supplement to sprinklers under favorable conditions. In a slowly developing fire, a well-designed smoke detection system can give an advance warning before heat-responsive devices, such as automatic sprinklers, operate. If manual fire extinguishment is mobilized by a manned emergency center, it may be able to attack the fire before it gains headway. Whether manual extinguishment causes more or less damage than sprinklers depends upon the skill of the responding force and how the fire progresses while the responding force is mobilizing. In any case, although automatic detection and manual extinguishment can be valuable adjuncts to automatic sprinklers, they would not constitute an acceptable alternative to automatic sprinkler protection under most conditions found in records storage.

Smoke detectors in records storage are particularly useful when the area is unoccupied. Spot-type detectors on high ceilings are unlikely to be activated until smoke from a smoldering fire completely permeates the air, or flaming fire gains sufficient headway to drive a column of smoke and heat to the ceiling. Therefore rack detection on projected beam detectors should be considered.

Smoke detection systems may not be appropriate in open-shelf records storage, where fire development is likely to be very rapid. They may provide supplementary protection in high-value collections, particularly if measures have been taken to slow down development of a fire (such as storage in totally closed metal containers), or burning material is prevented from falling off shelves. Smoke detection equipment can be useful in unoccupied areas of mobile shelving or other arrangements in which combustible contents are sheltered from sprinklers, or storage arrangements slow the spread of fire. Heat detection, either spot type or line type, may be located near or in the racks. This type of detection may be more suited where a rapidly spreading fast burning fire is expected.

Smoke detection systems are also used to activate extinguishing systems, such as carbon dioxide, halon, or high-expansion foam. They are used to activate preaction sprinkler systems, which are sometimes installed to prevent accidental water damage.

Duct detectors are used to shut down fans, activate dampers, or initiate smoke control measures. Detectors are also used to cause held-open doors to close and to stop conveyors, in a variety of single-purpose actions.

Heat detectors can perform most of the same duties as smoke detectors, but usually with a much longer time lag. An extended time lag is sometimes desirable, such as in the opening of a sprinkler, or the release of halon gas.

Hand-Held Extinguishing Appliances

Even if the facility is equipped with an automatic extinguishing system, manual fire control devices should be provided for staff use. Fire extinguishers are most often encountered. The preferred types are those containing clear water as the extinguishing agent. Although carbon dioxide extinguishers do not leave a residue or dampen documents, they are not effective on the deep-seated fires likely to occur in records storage. Although the use of halon extinguishing agents is being phased out worldwide, they are still used extensively in record storage systems. Halon extinguishers have the same limited extinguishing ability as carbon dioxide, in that they are ineffective on deep-seated fires. The user of halons should also be aware of the toxicity of the agent and avoid high concentrations of it. Although national standards rightly limit human exposure to the various halons, there is no mechanism, except the knowledge and alertness of the extinguisher operator, to manage the extent of exposure to halon. Most extinguisher users, however, are unaware of this hazard and are at risk of overdose. Multipurpose dry chemical extinguishers may be provided, but they leave considerable residue and have only limited effectiveness on deep-seated fires.

Water hose lines may supplement, or be used instead of, fire extinguishers. To minimize water damage, rubber-lined hose with an adjustable water spray, straight stream, and shutoff nozzle is preferable in records storage areas. Its use greatly expands the firefighting capabilities of the staff. Hose lines connected to carbon dioxide or halon storage tanks or cylinders have been used. However, these systems should not be undertaken without extensive training and appropriate protective equipment because the danger to the fire brigade is much greater than with the use of hand extinguishers containing the same agent.

AUTOMATIC EXTINGUISHING SYSTEMS

A variety of automatic fire extinguishing systems are suitable for records storage protection; the choice depends upon the economics of a particular situation and the degree of sophistication desired. All of the extinguishing systems described are loss initiated; the system does not detect and begin extinguishment until a fire is established and a fire loss has occurred. Systems that are more sensitive and reactive are available for highly specialized requirements. A large increase in cost and rate of false operation should be expected. Increased maintenance is usually required in more sophisticated systems to attain the reliability inherent in standard sprinkler systems.

Automatic Sprinklers

Automatic sprinkler systems are the most common type of automatic protection. They have proved effective in records storage fire incidents in limiting both fire and water damage. Water is discharged only in the immediate vicinity of the fire, and salvage techniques have been perfected to the point where recovery of wetted items is no longer unusual. Accidental opening of a sprinkler is rare. The probability of water damage from an accidental discharge can be further reduced using a preaction sprinkler system where water does not enter the piping until a fire detector operates. Preaction systems introduce additional failure modes, which may be unacceptable. A more sophisticated variation is a system where the water supply valve cycles on and off, depending upon the actual presence of fire as sensed by a detector in the fire area. Also available are sprinklers that cycle on and off individually, but the piping remains filled with water. However, the increased cost of these systems must be weighed against the value of the records being stored.

Water Mist Systems

"Mist" nozzles, long used in cotton processing factories to raise humidity, reduce static electricity discharge, and reduce incidence of flash fire in accumulations of fine fibers, are being used in a new type of sprinkler system. Parallel developments are being pursued for two types of application. In one model, a small array of open mist nozzles are supplied through a valve controlled by heat or smoke detector. Many arrays are installed in a large area. In another model, individual nozzles are activated by an internal valve controlled by an integral heat detector.

Fire tests are being conducted on a room-size scale to determine application limitations for development and refinement of installation standards. Applicability to large libraries and records centers is yet to be determined.

Foam Systems

Tests have shown high-expansion foam to be an effective fire control system for records, though few high-expansion foam systems are known to have been installed in records centers.[16] Rapid fire control comparable to, or possibly even faster than, that achieved by automatic sprinkler protection may be expected. Water damage to any single item will be low, but all items within the room or fire area involved will be affected. (A side effect unrelated to fire was observed in a large-scale test, which may be a disaster in itself: The records box contents labels exposed to foam loosened and slid to the floor.)

Carbon Dioxide Systems

A few automatic total flooding carbon dioxide systems having a high rate of discharge have been installed in records storage or library areas. A properly designed system should promptly control fire with limited damage. To be successful, a high concentration of gas must be maintained for an extended period. The concentration must be maintained until a deep-seated fire cools below its ignition temperature. The operation of such systems, however, involves a hazard to life and must be delayed until the area to be flooded is cleared of people.

Halon Systems

The manufacture of Halon 1301 (bromotrifluoromethane) and various related Freon™ gases (used in fire protection, refrigeration, and air conditioning) are being phased out rapidly by international treaty because of concern for ozone depletion in the upper atmosphere. Several substitute gases have been proposed for fire extinguishment, none of which has so far received universal acceptance.[17] Remaining Halon 1301 stocks are expected to remain in service until discharged and dispersed in the atmosphere, without replacement.

Existing automatic total flooding halon systems utilize flame-inhibiting liquefied gas under pressure. Halon 1301 is being marketed to use in specially engineered systems for the protection of valuable records. Primary advantages are that the gaseous agent leaves no residue and is only mildly toxic in the concentrations in which it is used. Occupants, however, should not be unnecessarily exposed to Halon 1301. Deep-seated smoldering fires will not be extinguished by halon, as it inhibits flaming only.

System Reliability

All fire control systems are subject to failure when various conditions are exceeded or not met. In Australia, where sprinklers are closely supervised, the success rate is more than 98 percent. It should not be assumed that sprinklers, let alone other fire control systems, have the same success rate in the United States.

Gaseous systems have a number of possible failure modes that are unique to them, although this does not necessarily imply an unusually high failure rate in practice. An independent detection system must work and must be correctly linked to the gas release. Usually, electric power is required for the detection system, linkage, and release to operate. The gas containers must be filled and connected to distribution piping. If the room has been enlarged since the system was installed, the gas volume applied may be inadequate. If any of the doors fail to close properly, a critical amount of gas may escape. If the ventilating system continues to operate, critical duct dampers fail to close, a window remains open or gets broken, or walls are breached for construction, a critical amount of gas may escape. If fire fighters open doors to assist in extinguishment, a critical amount of gas may escape. If deep-seated fires are not extinguished prior to gas dissipation, rekindling of the fire may be beyond the control of manual forces if no connected reserve gas supply is available for reapplication.

High-expansion foam systems have similar detection and activation vulnerabilities. They are especially vulnerable to loss of power because substantial power is required to operate the large fans used to generate the foam. Foam is also subject to blockage by doors, walls of stock, and fabric curtains. It is also somewhat vulnerable to loss of containment, although not nearly to the extent of gaseous systems.

Fire Prevention and Emergency Planning

Although detection and extinguishing systems are important, particularly for the bulk storage of valuable records, the first line of defense remains a good fire prevention and risk reduction program. Good housekeeping, orderliness, maintenance of equipment, and prohibition of smoking in records storage and handling areas are fundamental principles of good records management.

An emergency action plan that is kept current and practiced is essential to limit damage in case a fire occurs. Staff personnel cannot be expected to approach a fire and use extinguishers or hose lines effectively without adequate training. Fire-resistive containers and vaults are of much greater value if there has been training in procedures for fire emergencies. The following suggestions apply:

1. Records should be returned to their places of safety accurately, quickly, and without confusion or oversight. If there is no standard records protection, the best plan is to have the most important records carried out of the building. Drills are valuable to train employees to meet emergencies.
2. Records belonging in vaults or safes should never be left out overnight.
3. Important materials that belong under protection should not be allowed to accumulate on desks.
4. Because records normally safeguarded are often unprotected while temporarily on loan, copies instead of the originals should be loaned, and the originals retained in safe storage.

BIBLIOGRAPHY

References Cited

1. General Services Administration, "Protecting Federal Records Centers and Archives from Fire," General Services Administration, Washington, DC, 1977.
2. Gage-Babcock & Associates, *Protecting the Library and Its Resources*, American Library Association, Chicago, IL, 1963.
3. Chicarello, P. J., and Troup, J. M., *Fire Tests of Records Storage in a Fixed Storage Module Under the Protection of Large-Drop Sprinklers*, Factory Mutual Research Corp. (sponsored by U.S. General Services Administration), West Glocester, RI, 1980.
4. Jensen, R. H., *Report on Fire Tests in High-Piled Combustible Stock*, Underwriters Laboratories Inc. (sponsored by FIA and NBFU), Northbrook, IL, 1963.
5. Chicarello, P. J., *et al.*, *Fire Tests in Mobile Storage Systems for Archival Storage*, Factory Mutual Research Corp. (sponsored by U.S. General Services Administration), West Glocester, RI, 1978.
6. Longheed, G. D., Mawhinney, J. R., O'Neill, J., "Full-Scale Fire Tests and the Development of Design Criteria for Sprinkler Protection of Mobile Shelving Units," at National Research Council of Canada, Ottawa, Ontario; Gage-Babcock & Associates, Vienna, VA, *Fire Technology*, Vol. 30, No. 1, first quarter 1994, pp. 98-132.
7. McCrea, J. L., *et al.*, "Report to the Committee on Protection of Records of the National Fire Protection Association on Fire Tests on Microfilms Stored in Insulated Records Containers," Eastman Kodak Co., Rochester, NY, 1956.
8. McCrea, J. L., and Adelstein, P. Z., "Fire Tests on Microfilm in Insulated Records Containers," Second Report to the Committee on Protection of Records of the National Fire Protection Association, Eastman Kodak Co., Rochester, NY, 1958.
9. Fire Council, "Salvaging and Restoring Records Damaged by Fire and Water," *Federal Fire Council Recommended Practice No. 2*, Federal Fire Council, Washington, DC, 1963.
10. Morris, J., *Managing the Library Fire Risk*, University of California, Office of Insurance and Risk Management, Berkeley, CA, 1975.
11. Spawn, W., "After the Water Comes," *Bulletin of the Pennsylvania Library Association*, Vol. 28, No. 6, 1973, pp. 242-251.
12. Waters, P., *Procedures for Salvage of Water-Damaged Library Materials*, Library of Congress, Washington, DC, 1975.
13. Underwriters Laboratories Inc., "Tests for Fire Resistance of Record Protection Equipment," UL 72, Underwriters Laboratories Inc., Northbrook, IL, 1991.
14. Underwriters Laboratories Inc., "Tests for Fire Resistance of Vault and File Room Doors," UL 155, Underwriters Laboratories Inc., Northbrook, IL, 1990.
15. National Fire Protection Association, "Records Container Performance," *NFPA Quarterly*, Vol. 47, No. 2, 1953, pp. 159-170.
16. Atomic Energy Commission, "High-Expansion Foam Fire Control for Records Storage Centers," Report IDO-12050, 1966, Atomic Energy Commission, Idaho Falls, ID.
17. Su, J. Z., Kim, A. K., and Mawhinney, J. R., "Review of Total Flooding Gaseous Agents as Halon 1301 Substitutes," *Journal of Fire Protection Engineering*, Vol. 8, No. 2, 1996.

NFPA Codes, Standards, and Recommended Practices

Reference to the following NFPA codes, standards, and recommended practices will provide further information on the safeguards for records storage discussed in this chapter. (See the latest version of *The NFPA Catalog* for availability of current editions of the following documents.)

NFPA 10, *Standard for Portable Fire Extinguishers*
NFPA 11, *Standard for Low-Expansion Foam*
NFPA 12, *Standard on Carbon Dioxide Extinguishing Systems*
NFPA 12A, *Standard on Halon 1301 Fire Extinguishing Systems*
NFPA 13, *Standard for the Installation of Sprinkler Systems*
NFPA 14, *Standard for the Installation of Standpipe and Hose Systems*
NFPA 72, *National Fire Alarm Code*
NFPA 75, *Standard for the Protection of Electronic Computer/Data Processing Equipment*
NFPA 232, *Standard for the Protection of Records*
NFPA 232A, *Guide for Fire Protection for Archives and Records Centers*
NFPA 910, *Recommended Practice for the Protection of Libraries and Library Collections*

Additional Reading

Artim, N., "Fire Protection Strategy for Library Bookstacks," *Proceedings of the Annual Conference for inFIRE (international network for Fire Information and Reference Exchange.)*, Apr. 28–30, 1993, Norwood, MA, 1993, pp. 67–78.
Gage Babcock and Associates, "Report of Fire Tests, Mobile Compact Shelving Systems, Archives II—Phase Z," Underwriters Laboratories Inc., (Sponsored by National Archives and Records Administration and Hellmuth Obarta Kassabaum-Ellerbe Becket, a joint venture), Northbrook, IL, May 1996.
Hagglund, B., Fransson, C., and Bengtson, S., "Use of Zone and Field Model to Recommend Fire Protection Measures in New Stores of the Royal Library in Stockholm," National Defense Research Establishment, Sundbyberg, Sweden, FOA Report C20981-2.4, June 1994.

Jones, N., "Arson-for-Profit Investigations, Success or Failure? Recovering Water Damaged Business Records," *Fire and Arson Investigator*, Vol. 40, No. 3, Mar. 1990, pp. 50–52.

Lee, S., "National Library of Scotland: Protecting a Nation's Heritage," *Fire Prevention*, No. 249, May 1992, pp. 22–26.

Milke, J. A., and Gerschefski, C. E., "Overview of Water Mist Research for Library Applications," *Proceedings* for the International Conference on Fire Research and Engineering, Sept. 10–15, 1995, Orlando, FL, SFPE, Boston, MA, 1995, pp. 133–138.

Morris, J., "Protecting Libraries and Museums from Fire," *Fire Science and Technology*, Vol. 11, No. 1–2, 1991, pp. 35–43.

O'Neill, J. G., and Hahl, C. E., "Securing Our Nation's History with Improved Fire Protection," *Consulting-Specifying Engineer*, Vol. 17, No. 5, May 1995, pp. 36–38, 40, 42, 44.

Scoones, K., "Serious Fires in Libraries and Museum, 1986–1991," *Fire Prevention*, No. 254, Nov. 1992, pp. 20–21.

STORAGE AND HANDLING OF GRAIN MILL PRODUCTS

James E. Maness

The grain industry includes the movement, storage, and processing of grains, such as wheat, corn, oats, and sorghums, and oilseeds, such as soybeans and sunflower seeds. The industry can be separated into two broad segments: (1) grain elevator operations, and (2) millers or processors. Grain elevators are facilities that receive, store, and distribute whole grains and oilseeds for direct use, process manufacturing, or export. Grain elevators receive directly from the farmer and can be classified as either "country" or "terminal" elevators, with terminal elevators further categorized as inland or export facilities. Their end product is virtually identical to their raw material, with only minimal cleaning, drying, and blending and, in some cases, fumigation to preserve and meet product quality standards. Grain millers and processors, on the other hand, take whole grains as their raw material and process these into products, such as animal feeds and feed ingredients, cereals, flour, starch, meal, vegetable oils, corn sugars, brewery products, etc. The discussion that follows is primarily aimed at the grain elevator segment of the industry, and those portions of grain milling and processing that handle and store grains or dry milled products.

The growth of the grain industry in the United States has paralleled the phenomenal growth in production, consumption, and export of grains and oilseed products. This growth has been the result of many factors—the most important of which is the ideal temperate climate of the United States for producing these agricultural commodities. The favorable climate combined with the mechanization of farming, the development of high-yielding hybrids, the use of fertilizers, the free market system, and uniquely efficient storage and transportation systems have affected the growth. Table 3-26A indicates the dramatic rate of growth in harvested grain and oilseed crops.[1]

In the last two decades there has been much consolidation of the industry, resulting in fewer facilities operating at greater efficiency handling and processing larger crops at a faster rate. Table 3-26B gives the structure and productivity of various sectors of the U.S. industry, in terms of the number of facilities and their capacity and/or production.[2] All sectors shown in Table 3-26B handle grain and grain products and generally have a grain elevator portion in their operations. Figure 3-26A shows a major export elevator in the New Orleans, LA, area.

TABLE 3-26A. *Production of U.S. Grains and Oilseeds in Bushels*

Year	Corn	Wheat	Soybeans	Others	Total (1,000 bu)
1950	3,131,009	1,026,755	287,010	2,105,195	6,549,969
1955	3,184,836	938,159	371,276	2,548,501	7,042,772
1960	4,352,668	1,363,443	558,778	2,317,196	8,592,085
1965	4,084,342	1,315,613	845,608	2,227,321	8,472,884
1970	4,151,938	1,351,588	1,127,100	2,272,590	8,903,216
1975	5,828,961	2,122,459	1,547,383	2,092,432	11,591,235
1980	6,644,841	2,374,306	1,792,062	1,767,121	12,578,330
1985	8,875,453	2,424,115	2,099,056	2,228,974	15,627,598
1990	7,933,068	2,738,574	1,921,787	1,834,555	14,427,984
1995	7,374,000	2,186,000	2,152,000	981,000	12,693,000

Source: *Statistical Reporting Service*, U.S. Dept. of Agriculture.[1]
For SI units: 1 bushel = 0.0352 m³.

The awareness of dust explosion hazards has caused a transformation in the design and construction of grain elevator facilities during the last 15 years, from the totally enclosed concrete structured headhouse (which contains much of the handling and cleaning equipment) to construction of open towers. This awareness has greatly increased because of: (1) the severity and frequency of explosions that occurred in the late 1970s and early 1980s, (2) grain industry efforts to research the causes and prevention of grain dust fires and explosions, and (3) the promulgation of the Occupational Safety and Health Administration (OSHA) rule 29 CFR 1910.272 regarding grain-handling operations. Figure 3-26B shows the results of a dust explosion at a Houston, TX, area export elevator in 1976. Figures 3-26C and 3-26D show the results of other dust explosions.

Information and new concepts learned about grain dust fires are reflected in the 1995 edition of NFPA 61, *Standard for the Prevention of Fires and Dust Explosions in Agricultural and Food Products Facilities* (hereinafter referred to as NFPA 61), which combines four previous NFPA standards, i.e., NFPA 61A, NFPA 61B, NFPA 61C, and NFPA 61D.

RAW MATERIALS

Grain and oilseeds consist primarily of starch or carbohydrates, protein, fiber, and various vegetable oils, all capable of burning under

James E. Maness is assistant vice president and safety director of the Bungee Corporation, St. Louis, MO. Mr. Maness is a member of NFPA's Technical Committee on Agricultural Dusts.

TABLE 3-26B.   *U.S. Grain-Handling Facilities and Capacities*[2]

Type of Facility (Commercial)	Estimated Number of Facilities	Annual U.S. Characteristics (Estimated for each industry segment)
Grain Elevators		
Country elevators	10,835	198.8 million tons storage capacity
Terminals	700	42 million tons storage capacity
Export elevators	60	9.3 million tons storage capacity; 120,000 tons/hr handled; Exports 115 million tons of commodities
Soybean Crushing	63	Crushes 44 million tons of soybeans; daily crush cap 132,000 tons
Cottonseed Crushing	35	Crushes 4.93 million tons
Sunflower Crushing	7	Crushes 4,755 tons/day capacity
Feed Mills	6,723	Produces 164.7 million tons of feed
Wheat Flour Mills	234	Mills 31 million tons of wheat; daily production capacity 75,000 tons of flour
Dry Corn Mills		
Degerminating	15	Grind capacity 3.78 million tons of corn
Full seed	55	Grind capacity 280,000 tons of corn
Wet Corn Mills	24	Processes 33.6 million tons of corn; produces 15 million tons of corn sweeteners
Total No. Facilities	18,751	

For SI units: 1 ton U.S. = 0.907 ton metric.

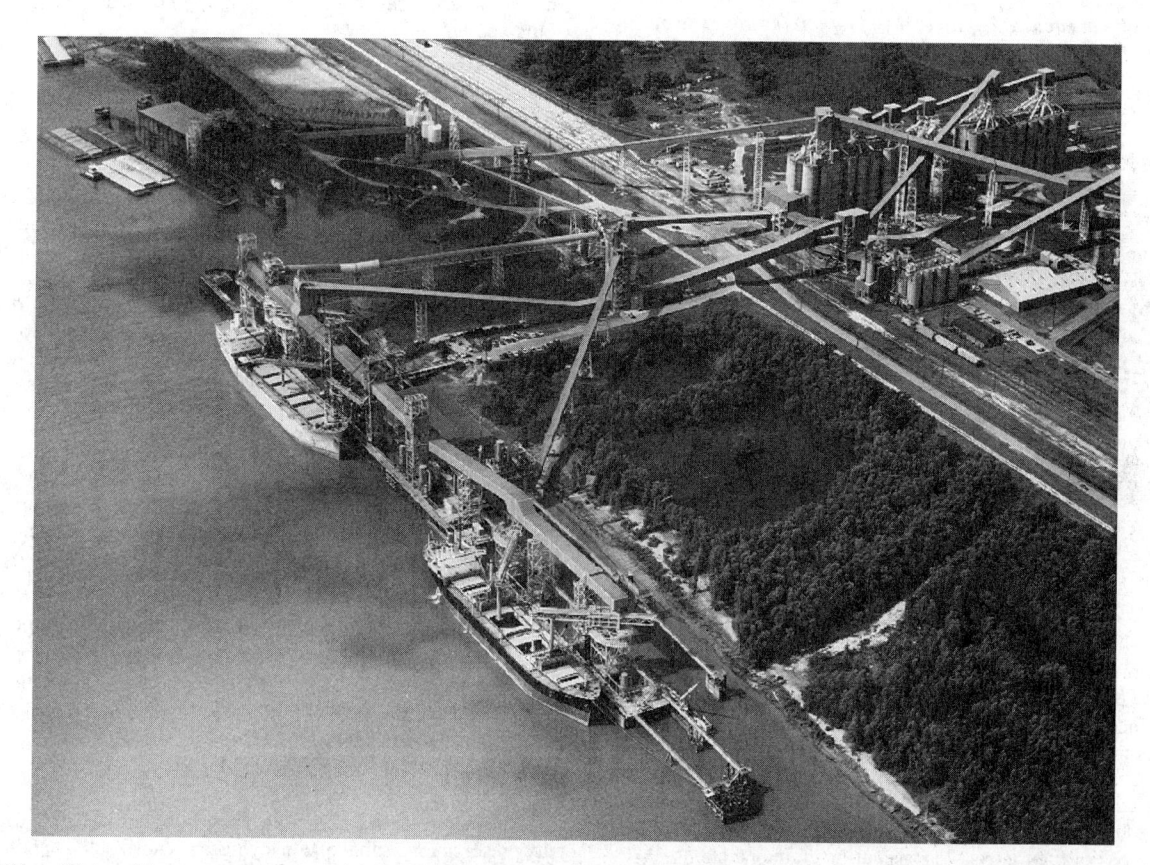

FIG. 3-26A.   *Pictured is an export elevator in the New Orleans, LA, area. The facility was built in the early 1980s to replace a facility that had a major explosion. Many new design concepts were used, including elimination of bucket elevators and use of inclined belt conveyors, and elimination of the headhouse and use of turnheat distributors rather than bin galleries.*

very specific conditions. In their raw, whole-kernel states, these commodities are stable and not readily subject to combustion if protected from moisture, insects, and fungi. So-called spontaneous combustion is a product of microbiological spoilage created by fungi combined with certain levels of moisture and temperature.[3] The incidence of this source of ignition of fires and explosions in grain elevators is rare, although it has been known to occur. Good grain industry practice is to store grains at moisture and temperature levels that keep the grain from deteriorating and, thus, minimize the possibility of spontaneous combustion.

FIG. 3-26B. *Pictured is the resulting damage to a major export elevator in the Houston, TX, area. The facility had two head-houses—one at each end of an array of concrete silos. The building in the foreground was a bagging shed.*

FIG. 3-26C. *Pictured is a very destructive explosion and fire that occurred at this Des Moines, IA, facility in the early 1980s.*

The probability of fires and explosions at grain-handling facilities is directly related to the increase in the amount of starch and other particles that are released when grain kernels are broken either during handling or processing. For example, artificial drying of corn can introduce stress cracks into the corn kernel, which leads to increased breakage during subsequent handling. Unless controlled, this breakage can result in increased levels of grain dust in the grain and can increase the risk of fires and explosions during grain operations.

As size reduction of the grain kernel proceeds, the susceptibility to fires and explosions increases dramatically.[4] The frequent handling of grain from farm to consumer progressively creates more broken kernels and, hence, more dust. This reduction in particle size also occurs as an integral part of the milling and processing operations.

Since grain is essentially handled as a bulk commodity, the economies of scale continue to dictate larger and more automated handling units. The grain is normally transported by truck, rail, barge, and ship, in bulk. Only a small fraction of the grain is bagged,

FIG. 3-26D. *This explosion at a Burlington, IA, facility in 1987 caused extensive damage, leveling the headhouse.*

consisting primarily of seed stocks and bagged commodities for export to a few emerging nations that are not equipped to discharge bulk vessels.

STORAGE

There are numerous types and configurations of bulk grain storage facilities. Structures that consist of upright concrete silos, wood bins, steel silos, and steel tanks can generally be unloaded using gravity flow. (See Figure 3-26E.) In addition to conventional warehouse structures, there are structures, referred to as flat storage, which do not totally empty by gravity and require manual labor or special power equipment to remove residual grains. Further, outdoor temporary piles are used during harvest periods to store the grain. Many facilities serve a multipurpose function, including the operation of other related agricultural businesses, such as seed and fertilizer distribution.

The size of the storage structure is determined by the nature of an individual facility's operation. Individual bins may range from 1,000 bushels (35.2 m³) to 2,000,000 bushels (70,400 m³) or even larger. The largest single facility has a total storage capacity exceeding 45,000,000 (1,524,000 m³) bushels.

Grain temperature systems are often used to monitor grain temperature for grain quality and spontaneous heating purposes at facilities that store grains for long time periods. In addition, storage systems are often equipped with aeration systems to help maintain grain temperature, moisture content, and quality.

HANDLING

The primary function of a grain handler, other than storage and quality preservation, is that of conveying grain horizontally and vertically into and out of the storage facility. Conveyors common to most bulk handling industries are used in this process. The following discussion shows how these conveyors are used in the grain industry and highlights modified designs tailored to grain industry needs.

Belt Conveyors

The most common conveyor employed to move grain horizontally from point to point is the trough-belt conveyor. By virtue of the wide choices of speeds and belt widths available, any desired volume,

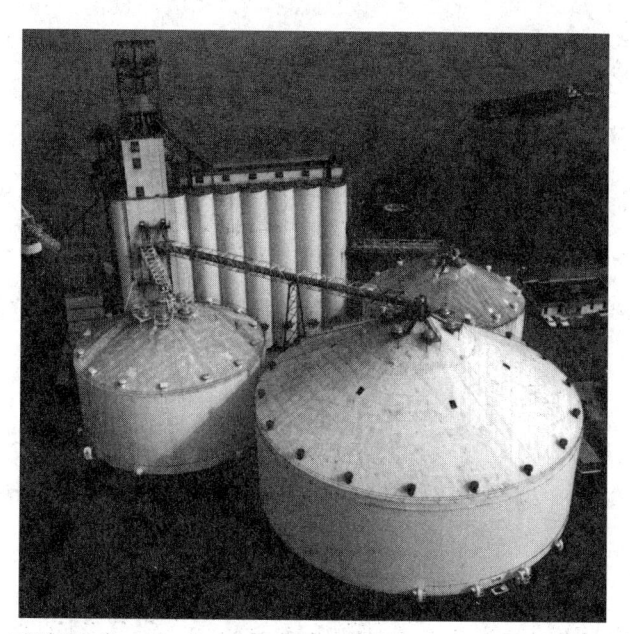

FIG. 3-26E. *Pictured here is a combination of traditional elevator design with new concepts to help minimize explosions and limit damage. The concrete elevator is a traditional headhouse with concrete silo complex. However, bucket elevators have been extended outside and vented. Steel tanks with conveyors are used to store grain. The gallery above the bins houses enclosed drag conveyors.*

within reason, can be accommodated in this manner. Belt conveyors are used to meet large production capacity needs, often exceeding 20,000 bushels per hr (bph) upwards to 80,000 bph at large export elevators. (For SI units: 1 bushel = 0.0352 m³.) Belt conveyors are generally highly reliable with low maintenance requirements. The fire hazards associated with these conveyors are the combustible rubber belting material, the tendency of dust to be liberated at grain transfer points (belt to belt or to other equipment or storage), and mechanical failure due to poor maintenance. The fire hazard can be reduced with the use of flame-resistant belting[5] and good maintenance procedures. The dust hazard can be mitigated with the use of enclosures, aspiration, belt speed control, or a combination of these. Newer belt designs often totally enclose the belt with critical bearings kept outside of the enclosure. These new designs control the dust by containment and aspiration.

Chain Conveyors

Alternatives to the belt conveyor include the *en masse,* drag, or chain conveyors, which are totally encased in a housing that prevents the escape of dust. These are usually of more limited capacity and convey at reduced linear speeds with much deeper grain depths. Normally, only the loading point or discharge point needs to be aspirated. Their higher energy use (than belt conveyors), increased maintenance, and installation costs are major factors when considering drag conveyors for use.

Screw Conveyors

The helical screw conveyor is a standard for low-volume conveying for short distances. The grain industry generally uses screw conveyors for handling materials, such as dust or small product streams. A major drawback to screw conveyors is that they can cause damage to whole grain, adding to grain breakage. Bearing failure is also a prime concern which requires periodic inspection and lubrication.

Pneumatic Conveyors

The milling portion of the industry uses pneumatic conveying systems extensively. These are especially effective for confinement and movement of finely ground commodities in a complex processing operation requiring multipoint pickup or distribution. There can be some concern about static electricity buildup and discharge if the systems are not properly grounded and maintained.

Bucket Elevators

The common principle used in grain elevator design is that of elevating the commodity to the highest optimum point, then permitting the material to flow by gravitational force down through the various garners, hopper weighing scales, cleaners, and finally through spouts into storage compartments. (See Figure 3-26F.) Use of gravity flow not only makes efficient use of the energy required to move the mass, but eliminates re-elevating and rehandling, which can cause increased breakage and damage to the kernel or seed. Grain quality is often monitored by using automatic grain samplers to collect representative samples during gravitational grain flow. Each subsequent rehandling contributes to reducing the quality of the grain, and generation of additional quantities of fine particles and dust. The bucket elevator (or "leg" as it is referred to in the industry) is the primary equipment used to gain these elevations. Inclined belts have been used in several large terminals, but are often impractical for most smaller facilities because of the large amount of land they require.

The bucket elevator is the workhorse of the industry and requires special attention as an explosion hazard. The dust concentration within a bucket elevator is likely to be above the lower explosive limit (LEL) during normal operations. This combined with the pumping action of the buckets moving in a confined enclosure and the high amount of mechanical energy inherent in its operation have led to its identification as the principal ignition source in explosions.[6,7] For these reasons, it is becoming industry practice to locate new bucket elevators outside of totally enclosed headhouse structures. (See Figure 3-26G.) Where "legs" cannot be installed outside, they can be explosion vented to the outside or be equipped with explosion suppression in addition to other safety devices. (See Figure 3-26H.)

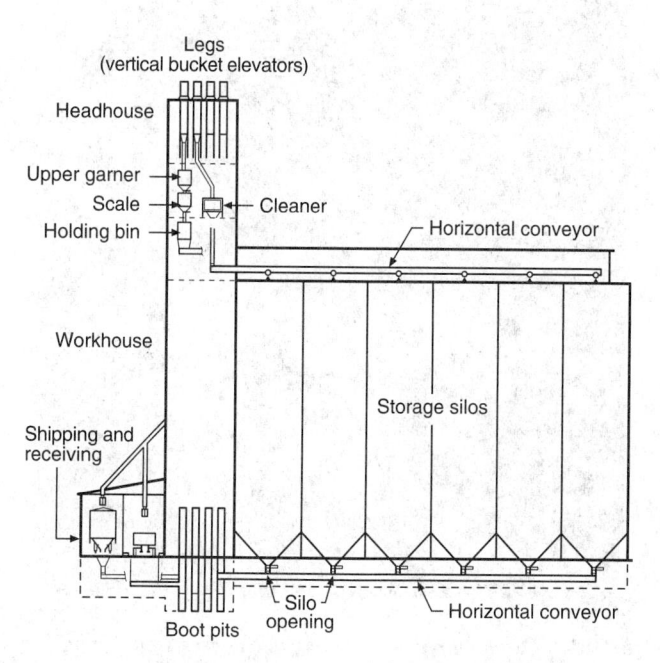

FIG. 3-26F. *Traditional concrete elevator design.*

Spouting and Lining

Grain and oilseeds are very abrasive commodities that can rapidly erode steel conveying spouts used to channel the flow through elevators. For very long grain drops, a grain-retarding device (dead box) can be used to slow down the grain and reduce breakage and dust generation.

The abrasive nature of grain creates an almost constant demand for maintenance to eliminate leaks and dust emissions. While patching and repair are adequate temporary measures, they seldom fully restore a spout. An alternative has been the wide use of abrasion-resistant liners that can be totally replaced without disruption of the outer spout. Materials most commonly used are abrasion-resistant alloy steel plates and high-density synthetic plastics (most frequently used), such as polyethylene or urethane and certain ceramics. These can usually be formed or molded to the contour of the spouts and bolted into place without the need for welding or other heating devices. Although the high-density plastics are not easily ignited, they can burn and must be removed or protected whenever welding or cutting is conducted on the spout.

Receiving and Shipping

The first and last adjuncts to a grain storage elevator are the machinery and structures needed to unload or load the grain from and into a carrier. Wherever a truck, railcar, barge, or ocean vessel shipping/receiving system is used, these operations are at ground level, partly or completely in the open, and usually connected to the main storage structure with an underground tunnel beneath the discharge receiving hopper or with an overhead bridge.

The free-fall of grain through open spaces into receiving hoppers presents a unique dust control problem influenced by surface winds and the lack of sufficient enclosures to contain the dust emission. The problem is less of an explosion or fire hazard than a nuisance, as the dust can fly about, since the operation is essentially an open-air one. Environmental laws and regulations have required the

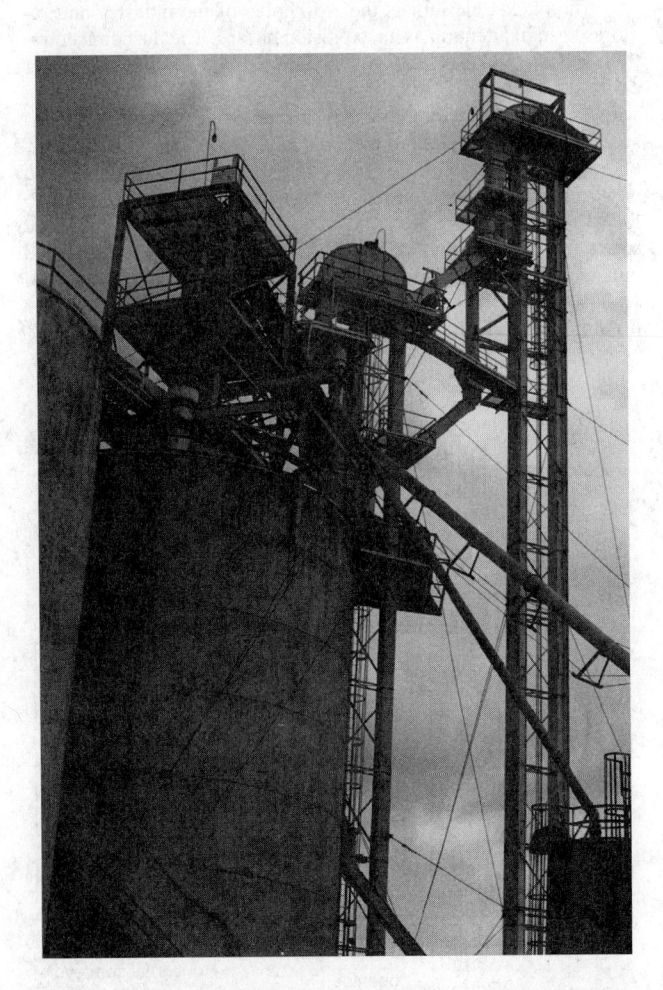

FIG. 3-26G. *Pictured is the use of outside free-standing bucket elevators with steel tank storage.*

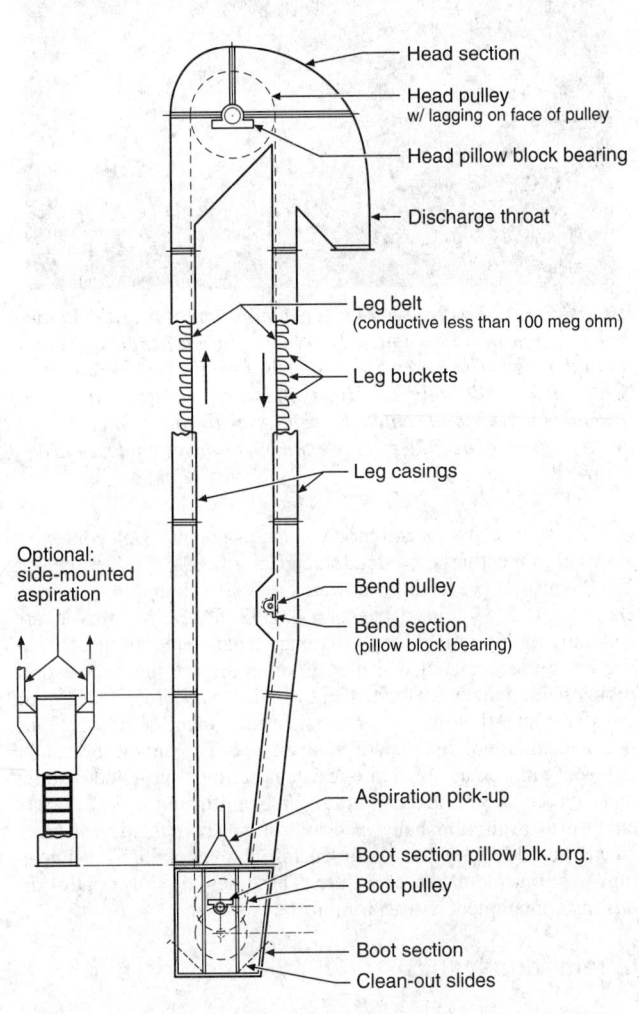

Notes:
1. Capacity can be a few hundred bushels per hour to 60,000 bushels per hour.
2. Belts can have multiple rows of buckets (metal or plastic, which is preferred) across them.
3. Take up can be gravity or screw style.
4. Capacity is belt speed x bucket capacity.
5. Head pulley is normally lagged.
6. Legs can be in excess of 200 ft tall.
7. Legs interconnect the entire headhouse when placed inside.
8. Slow motion devices are used to detect overload for belt slippage.
9. Bearing and alignment rub blocks with thermocouple establishes temperature monitoring.
10. Casings and head sections can be vented when leg is outside. Head sections of inside legs can often be vented to the outside.

FIG. 3-26H. *Bucket elevator. (For SI units: 1 bushel = 0.0352 m³; 1 ft = 0.305 m.)*

same attention to these ground-level emissions as to elevated sources of emissions, which can present greater concerns.

Grain Drying

The principal processing operation at most grain elevators is that of drying the grain to moisture levels low enough to preserve quality during storage or to meet grain standards. There are many types of dryers used in the industry, with batch dryers or continuous flow dryers being the most common. Continuous flow dryers are either column (which allow continuous flow of grain through a screened column while hot air is passed perpendicular to the grain flow) or rack dryers (where grain flows down an air plenum with hot air flowing up through the cascading grain flow). (See Figure 3-26I.)

The typical modern grain dryer is direct-fired, i.e., the heat of the burned fuel is directed into a stream of air that is passed directly through the moist grain. The fuels used are principally natural gas, fuel oil, or vaporized liquid propane. Although dryers are designed to minimize their fire hazards, inadequate maintenance and im-

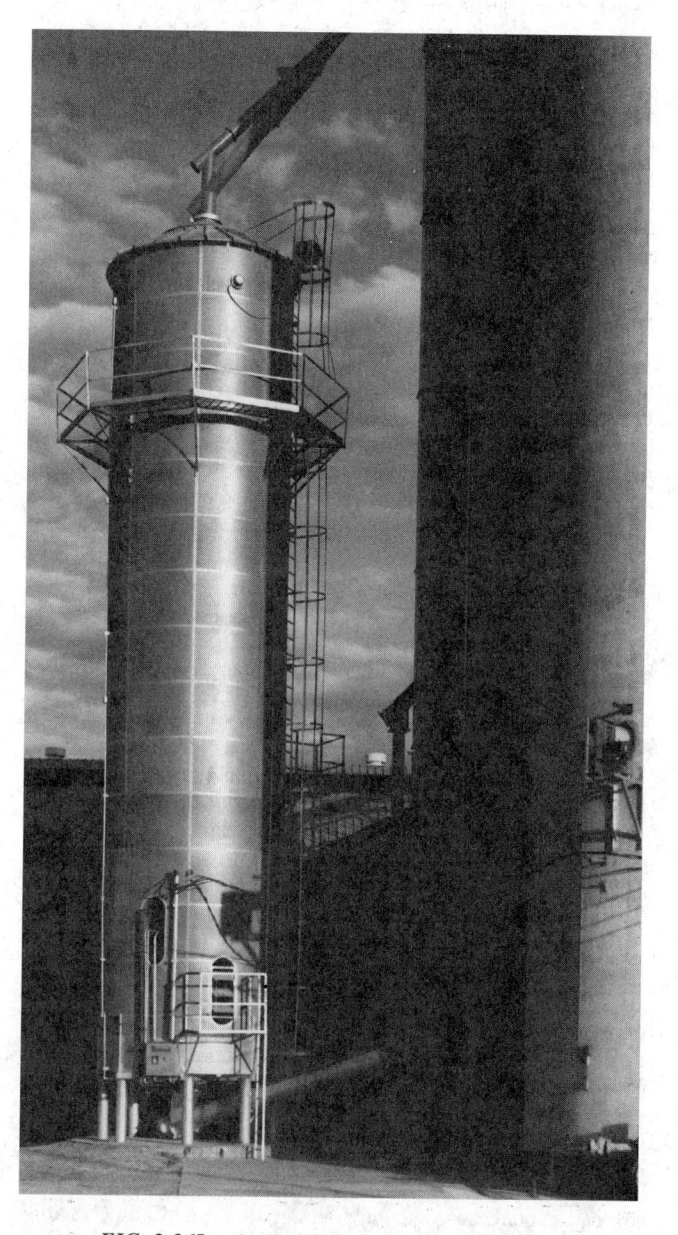

FIG. 3-26I. A commercial column grain dryer.

proper operations can result in ignition of the grain being dried. Examples of causes of these fires are the failure to properly and frequently remove excessive accumulations of dust and combustible materials from the inside of the dryer or the occurrence of blockages to the flow (such as a broken bucket from a "leg"), allowing the grain to overheat.

The resulting "hang-ups" and over-dried materials and particles readily ignite when burner temperature gets too hot or small embers are entrained in the plenum air. The multiple severe grain dryer fires and explosions of 15 to 20 years ago have not recurred, reflecting the installation of a number of safety devices and improved operating procedures.

Ideally, a grain dryer should be housed in a structure completely removed from the storage facility. Inside-located bulk grain dryers are prohibited by OSHA in 29 CFR 1910.272(p), unless located in a separate fire area or protected by explosion suppression. Dryers must also be equipped with certain automatic safety controls that shut off fuel for power failure, provide flame or exhaust air movement interruption, and stop the feed to the dryer if excess temperature occurs in the exhaust section of the dryer. The grain is fed to the dryer from an elevated source, and returned from the dryer to storage in a second separate conveyance for distribution into silos. Locating a direct-fired dryer away from the storage unit itself, as well as good operating procedures, minimizes the risk of more serious fires or explosions within the silo structure. In cases where a dryer serves solely to heat grain without cooling, the hot grain is often cooled in the storage silos by fans forcing ambient air up through the grain mass until air-temperature equilibrium is reached.

The grain temperature typically should not exceed 150°F (65.5°C) in a drying operation [corresponding to a hot-air drying temperature between 180 and 200°F (82 and 93°C)], thus avoiding heat damage to the kernels.[3] As the wet grain is heated, the evaporating water at the surface of the kernels has a cooling effect that prevents the kernels from overheating. The heated air leaving the dryer is in a near-saturated state, and some coarse dust and fine particles can escape with it.

Grain Cleaning

Another important activity in some grain-handling facilities is scalping and cleaning of grain through the use of aspiration and/or screening systems to remove extraneous material from the grain, such as bits of stalks, stems, seed pods, husks, corn cobs, weed seeds, or fine broken grain particles. These materials not only affect the quality of the grain, but are also more prone to ignition than the grain itself by virtue of their extremely dry state and high fiber content. In some cases, green weed seeds and plant particles can cause additional heating concerns. The grain is cleaned by passing it over vibrating, rotating, or gyrating screening devices or stationary gravity screens for simple size separation. A positive air aspiration system can be used prior to the screening system. Aspiration is often used to remove dust generated by the grain movement within screen-type cleaning systems. Fires associated with cleaners can occur due to lack of maintenance and components coming loose and rubbing, causing frictional heat.

Grinding and Cracking

Some grain facilities serve specialty industries that require grain to be cracked or ground. This entails the use of hammer mills or grinders to accomplish the size reduction, particularly in feed, processing, or milling operations.

Hammer mills have been sources of fires and dust explosions, which can be transmitted through aspiration connections. The hammer mill is frequently used to grind corn and other feed grains in preparing feeds. Care must be taken to remove foreign objects, especially stones and metallic objects, from entering the grinding mechanisms with the use of screening and magnetic devices.

Dust Control

Supplementing the conveying, elevating, drying, screening, sampling, weighing, and storage activities is the dust control process. Each point of grain transfer and handling can produce suspended dust. A complete dust control system at an elevator consists of a combination of methods, such as proper and periodic housekeeping, good maintenance to prevent grain leaks in equipment and spouting, the enclosure of grain-handling equipment to prevent the escape of dust, the reduction of grain-handling speeds, and the use of active dust control provided by mechanical dust collection systems to reduce fugitive emissions. In some cases, oil additives have been sprayed on grain to help reduce grain dust emissions during handling.

Mechanical dust collection continues to be the primary means used to control fugitive dust emissions. The dust collector most commonly employed is the "baghouse" or fabric filter, wherein the dust-laden air is passed through filter media and exits virtually dust-free (up to 99.99 capture efficiency). The dust is recovered in the filter housing, stored in a remote bin or tank, and can sometimes be sold as a by-product of grain for use by the animal feed industry. Often the dust must be given away to avoid disposal and freight costs. Most often the dust is returned to the grain stream in such a manner so as not to create additional airborne dust. However, at export facilities, dust returned to the grain is prohibited once it has been placed in a storage bin or tank. Mechanical dust systems are complex processing operations with self-cleaning devices for the bags, automatic discharge mechanisms, and continuous conveying to disposal points or storage tanks.

The cyclone collector was commonly used prior to clean-air laws, but is used less frequently for controlling dust emissions. Nevertheless they continue to be relied upon for preventing product losses and serve as a separator for dust aspirating systems. Cyclones are not as efficient in capturing smaller particle emissions. Collection efficiencies vary greatly, ranging up to 80 percent for low-efficiency cyclones and up to 95 percent for high-efficiency cyclones. For fine particles, efficiency is only 50 to 60 percent. The cyclone is still often used in conjunction with some dust systems by being located ahead of bag filters to remove and recover very large particles for re-entry into the grain or product streams. The remaining fine dust particles are then collected in bag filters for disposal or for re-admission into the operation where circumstances and practices permit.

Dust collection systems can be a fire hazard, because they concentrate the dust in specific pieces of equipment and can serve as a path for flame or explosion transmission to interconnecting equipment and areas. For this reason, dust filters should be located outside and should be equipped with explosion vents to minimize possible damage, should an explosion occur. Alternatively, dust systems can be located in separate explosion-vented rooms or be equipped with explosion suppression systems.

Dust systems must be periodically inspected for proper operation. Pressure gages are often installed to monitor the performance of the unit. Excessive pressure drop across the bags can indicate that filter bags need to be changed or cleaned out. Low pressure can indicate a broken bag or improper operation of the unit.

THE FIRE HAZARD

Although dust explosions receive the most attention and awareness regarding hazards at grain-handling facilities, there are a number of other fire hazards at these facilities. Fires are much more numerous than dust explosions. However, no reliable information is available regarding the number and types of fires in grain facilities. Fires often precede a dust explosion; ignition sources include improper welding and cutting, failed bearings, overheated or failed electrical equipment, and mechanical failures.

Fires may occur in nearly any area of a grain facility, particularly where potential ignition sources are present, such as motors, drives, bearings, and electrical equipment. Areas of grain facilities where fires might occur include the dryer, motor control rooms, bucket elevators, conveyors, storage bins, grinding or milling areas, and offices. Potential fuel sources for fires in grain-handling facilities vary depending upon construction, equipment, and types of operations.

Generally speaking, grain facility structures are constructed of noncombustible materials, such as concrete and steel. Older facilities may have some wood construction.

The following materials may serve as fuel sources at grain handling facilities: dust, grain, grain screenings, plastics, lubricants and grease, fabric cloth materials, rubber belts, polyvinyl chloride (PVC), conveyor and bucket elevator belting, roofing materials, propane and natural gas, and electrical wiring coverings. Many of these materials are not peculiar to the grain industry and, in fact, are common in most industrial facilities. However, grain dryer fires, conveyor belting fires, bucket elevator fires, and grain fires provide unique fire challenges. Fire challenges can be dealt with as described in the National Grain and Feed Association's publication, entitled *Emergency Preplanning and Firefighting Manual: A Guide for Grain Elevator Operators and Fire Department Officials*. Some typical considerations and factors to consider for several types of grain facility fires are discussed below.

The grain dryer's most significant fire hazard is that of igniting the grain, or grain dust and extraneous material that has accumulated on ledges and surfaces inside the dryer. The large amount of heat needed—in some dryers, more than 20,000,000 Btu/hr (21.12 million kJ/hr)—combined with the large size—up to 80 ft (24.4 m) high—requires good operating procedures and proper maintenance of safety devices, such as temperature sensors, UV flame detectors, limit switches, and fuel cutoff devices.

In addition to the preventive measures used to reduce the possibility of a grain dryer fire, each elevator must have an emergency plan for dealing with a fire. The size and unique design of a grain dryer makes fighting the fire with hose streams difficult. Probably the most effective consideration in fighting a dryer fire is moving the burning grain to a safe area away from the elevator where water can be selectively applied. In some cases, small smoldering fires in dryers can be removed manually by shoveling the smoldering material into a bucket and removing it to a safe location. For modern grain dryers, a means can be provided for unloading (emergency dumping) of the dryer contents to a safe location such that a fire exposure is not created for adjacent buildings, structures, or equipment. One common method is to place a door or chute on the dryer discharge conveyor for emergency dump purposes.

The hazard of grain itself igniting must also be addressed from a preventive, as well as a fire fighting, standpoint. Preventive measures are basically to keep open flame (e.g., welding, torch cutting, cigarettes) and hot surfaces or objects away from the grain. Once grain ignites, it becomes quite difficult to extinguish, and, if the extinguishing method creates a dust cloud, an explosion is likely. A permit system should be utilized to control hotwork, such as welding and cutting. The application of large amounts of water in a storage bin will not only deteriorate the grain, but, in the case of soybeans, could cause enough swelling of the soybeans to rupture the bin itself. A bucket elevator fire is of great concern, since internal dust concentrations can result in a dust explosion. Elevator management should consider meeting with local fire departments to review emergency procedures should a grain fire occur.

Conveyor belting fires are difficult to deal with because of the intensity of the fire and the amount of smoke and vapors given off. Belting fires once ignited are difficult to put out and require large amounts of water to control. If a belting fire cannot be extinguished, consideration can be given to cutting the belt and dropping it to isolate the fire and prevent it from spreading.

Grain bin fires are particularly difficult to deal with depending upon the fires' location and the quantity of grain involved. Bin fires should be approached as a potential explosion hazard. Air sources, such as aeration fan openings and portals, should be sealed off to limit the source of oxygen. If water is applied into the bin, a fog spray is recommended to wet down the dust on top of the bin, including any accumulations under the bin top roof, being careful not to create a dust cloud. Small surface fires can be extinguished using a fog nozzle and removed from the top manually or with the use of a pneumatic evacuator. Fires on the top surface of grain in concrete silos may be controlled by using carbon dioxide to blanket the top of the grain. However, the use of carbon dioxide to extinguish deep-bedded fires is often not successful. If burning or suspect grain is allowed to flow from the bin, it should be directed to the outside by spouting or conveyed to the outside using portable augers or evacuators. Suspect grain should never be handled in bucket elevators or other enclosed equipment. Deep-bedded fires in grain bins or inside large grain storage structures are the most difficult to deal with because of the amount of grain to be handled and the difficulty of locating and accessing the fire. If a pipe or hose can be probed to the fire area, it is possible to apply water to that portion of the grain mass. It is nearly impossible to apply water to the grain surface to put out deep-bedded fires, and the application of excessive amounts of water can create structural hazards and can destroy grain that can otherwise be reclaimed. In large flat structures, it may be possible to

remove the grain with mobile equipment (front-end loaders) or to selectively unload portions of the grain mass to access the fire.

In all cases of dealing with a grain fire in storage structures, a careful plan should first be established to deal with such fires, following these principles: (1) shut off the grain flow; (2) seal off all openings that allow oxygen to reach the fire, including shutting down aeration or roof exhaust fans; (3) use fogging nozzles to wet down dust and grain; (4) do not allow high-pressure water streams to be applied that could create dust clouds; (5) carefully consider how suspect or burning grain will be removed from storage; and (6) do not needlessly expose anyone to the fire or a dust explosion hazard.

Another potential fire hazard in grain elevator operations is the use of propane and natural gas. Both are used for grain drying or comfort heating. They present a serious fire hazard in grain operations because of the presence of below-grade areas in grain elevators and because the heavy rail or truck traffic can cause deterioration in underground pipes. If a break occurs, the gas or propane vapors can seep into the below-grade tunnels until discovered or ignited. Elevator management should insist that such gas and propane lines be installed according to NFPA codes, and maintained regularly.

THE EXPLOSION HAZARD

Dust explosions certainly must be considered the number one hazard in the grain industry. The first recorded grain dust explosion incident is reported to have occurred in 1785.[4] However, recognition of dust explosions was not common until the late 1800s. Figure 3-26J shows the number of dust explosions, injuries, and deaths that have occurred during the last 20 years. Figure 3-26K indicates a trend that the incidents of dust explosions and their severity has continued to decrease during this time period. Table 3-26C shows the locations of primary dust explosions, indicating that bucket elevators, storage bins, hammer mills, dust collectors, and enclosed equipment are the

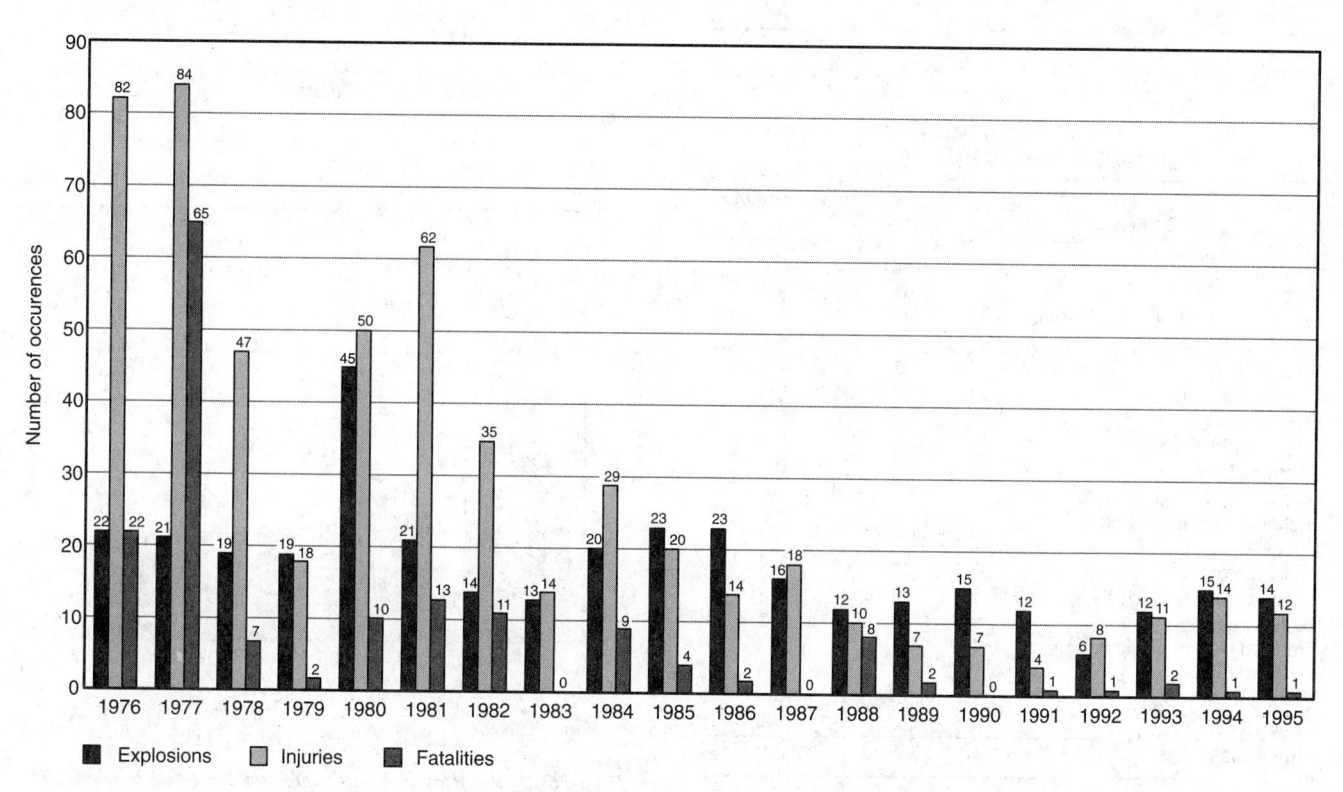

FIG. 3-26J. *Agricultural dust explosions, 1976–1995. (Source: See reference 12.)*

areas where explosions often initiate. Figures 3-26B through 3-26D show the severity and type of damage that can occur from grain dust explosions. Secondary explosions occur when the primary explosion pressure wave disturbs dust layers, creating a dust cloud that is ignited by the flame front, causing a more severe and damaging explosion. The flame front travels much slower than the pressure wave, which may allow reaction time to avoid fire and explosion hazards. Table 3-26D indicates the probable ignition sources that have been identified in the past explosions. Welding and cutting operations, fires, overheated bearings, and electrical failures are among the most frequent explosion ignition sources. An analysis of available information for dust explosions during the period from 1958 through 1995 is presented in Table 3-26E.

TABLE 3-26C. *Probable Location of Primary Dust Explosion (1958-78)* and (1984-85, 1988-89, 1991, 1992-95)†*

Location	No. of Facilities	Percent of Facilities
Unknown	146	41.2
Bucket elevator	105	29.6
Storage bins or tanks	19	5.4
Hammer mills, roller mills, or other grinding equipment	18	5.1
Dust collector	14	3.9
Other areas inside elevator	12	3.3
Other areas inside equipment	11	3.0
Headhouse	9	2.5
Adjacent or attached feed mill	8	2.2
Grain dryer	3	0.8
Inside electrical equipment	2	0.6
Outside and adjacent to facility	2	0.6
Pellet collector	2	0.6
Tunnel	2	0.6
Other	2	0.6
Sample size	355	100.0

*Source: reference 11.
†Source: reference 12.

TABLE 3-26D. *Probable Ignition Sources (1958-78)* and (1984-85, 1988-89, 1991, 1992-95)†*

Source	No. of Facilities	Percent of Facilities
Unknown	154	43.4
Welding	51	14.4
Fire other than welding or cutting	22	6.1
Overheated bearings	17	4.7
Miscellaneous	15	4.3
Tramp metal	13	3.6
Friction sparks	13	3.6
Other spark	12	3.3
Electrical failure	12	3.3
Unidentified foreign objects	10	3.0
Friction from choked "leg"	8	2.3
Lightning	7	2.0
Faulty motors	5	1.5
Extension cords in "legs"	4	1.2
Static electricity	4	1.2
Fire from friction of slipping belt in "leg"	3	0.8
Rubbing-pulley	3	0.8
Smoking material	2	0.5
Sample size	355	100.0

*Source: reference 11.
†Source: reference 12.

As a result of increases in grain dust explosion incidents and fatalities in the late 1970s and early 1980s, the industry, through the National Grain and Feed Association, established a Fire and Explosion Research Council, to increase the industry's understanding of known causes of explosions and to conduct research into the basic factors contributing to grain dust explosions. This research has revealed a number of options for helping to prevent explosions and created a better understanding of their causes. Major accomplishments of research included better guidelines for explosion venting of bucket elevators and the use of explosion suppression systems to control bucket elevator explosions. (See Figure 3-26L.) Explosion venting of grain bins and grain galleries was also studied, resulting in a better understanding of explosions of these structures. (See Figure 3-26M.) Because of their large length to cross-sectional width or diameter ratio, these structures are difficult to explosion vent, except for mild or limited explosions.

Bearings were identified as a significant factor in dust explosions, requiring good maintenance practices. Metal sparking and electrostatic sparking were also studied. The research showed that electrostatic sparks were not generally a problem, except under special circumstances where there is high charge generation and poor grounding. The major problem with metal sparking was determined to be high-speed grinding equipment and frictional sparks from mechanical failures or objects becoming lodged in equipment. In addi-

TABLE 3-26E. *Grain Dust Explosion Factors*

Type of Facility Involved in Explosions		Commodity Handled at Time of Explosion	
Facility	Percent	Commodity	Percent
Grain elevators	65.5	Corn	49.5
Feed mills	15.3	Sorghum	7.3
Corn processors	6.2	Wheat	7.3
Flour mills	3.1	Soybeans	5.2
Rice mills	2.8	Rice	4.7
Others	7.1	Corn starch	3.6
		Barley	2.6
		Others	19.8
Total	100.0		100.0

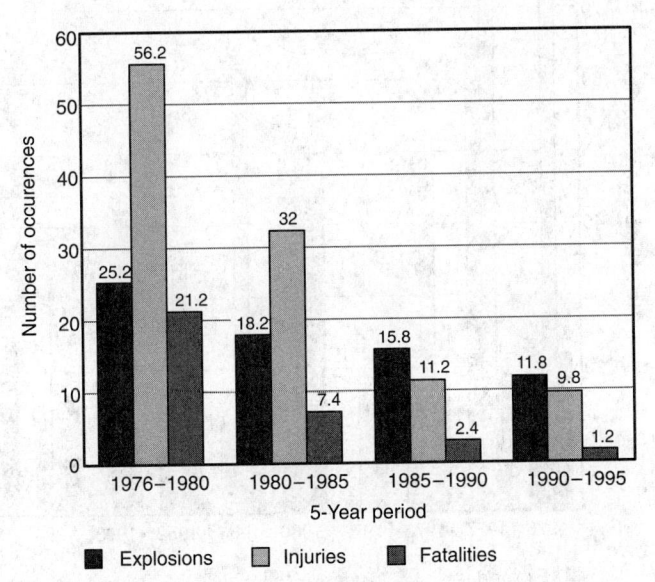

FIG. 3-26K. *Agricultural dust explosion decreases presented in 5-year segments. (Source: See reference 12.)*

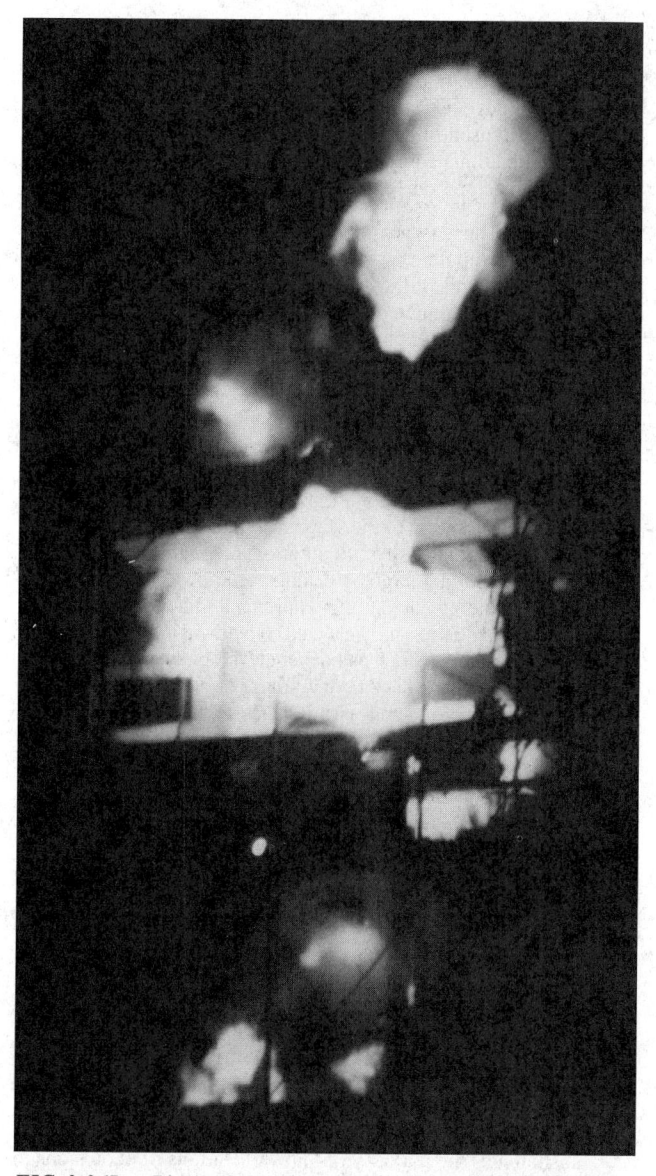

FIG. 3-26L. *Pictured is a dust explosion test of a bucket elevator conducted by Fenwalls, Inc., for the National Grain and Feed Association in 1979. Vents were located along the "leg" casing and in the head section. The testing determined venting criteria to be used on bucket elevators. Also tested was the use of explosion suppression systems.*

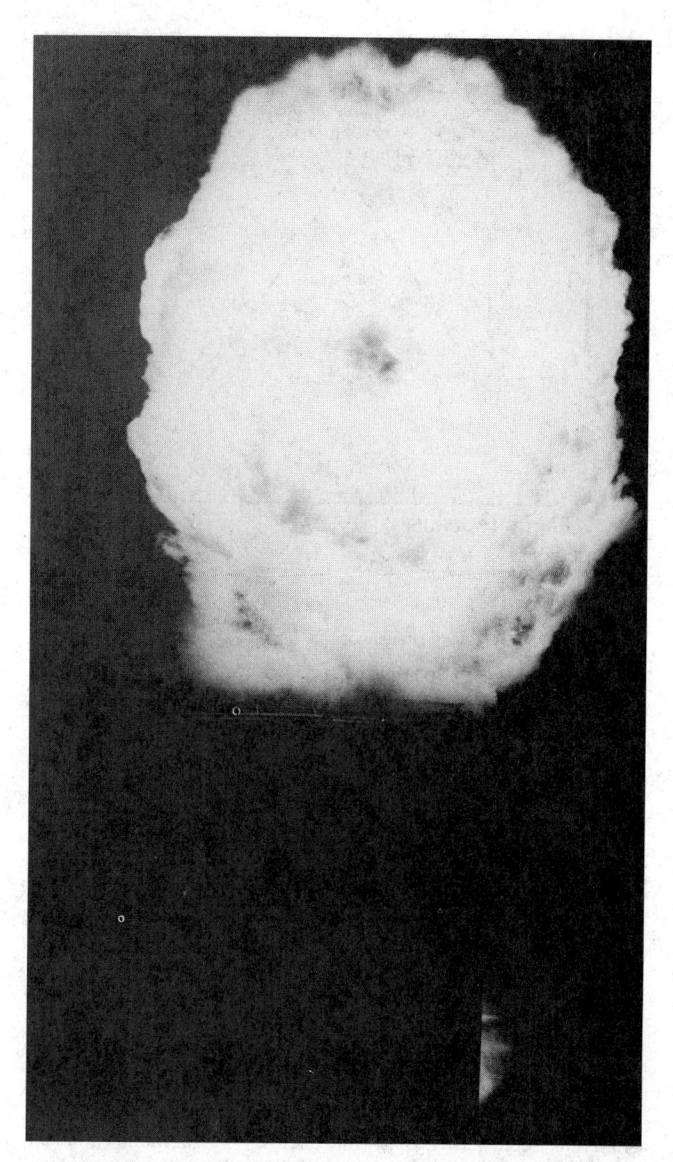

FIG. 3-26M. *A dust explosion test conducted in Norway by Dr. Rolf Eckhoff. The research tested explosion venting of grain bins. The large fireball, at the top, shows the initial venting. The flame beginning to come out of the right side shows the bin splitting open on the side. This illustrates the difficulty of venting elongated grain bins.*

tion to the above, the research focused on identifying design methods that could be used to help minimize grain dust explosions.

Elements of a Dust Explosion

The elements of a grain dust fire or explosion are almost axiomatic; namely, that to be initiated and sustained there must be fuel, oxygen, and an ignition source. To have an explosion, a fourth element, i.e., confinement (which confines the rapidly expanding heated gases of combustion within a constraining enclosure until the pressures exceed the ultimate strength of the enclosure), is necessary. (See Figure 3-26N.)

The precise circumstance under which an explosion of a grain dust will occur is a complex combination of dust particle size; concentration in the air [oz/ft³ (g/m³)]; the energy of the ignition source;

and less easily determined factors, such as the moisture content of the dust (or percent relative humidity of the air) and the actual composition of the dust. Dust from each agricultural commodity has its own explosion characteristics. While researchers have not agreed precisely on the limits of the various characteristics of a particular dust, their conclusions are generally within an acceptably narrow range. Studies of explosibility of agricultural dust show that lower explosion limit for dust concentrations needed for an explosion ranges from 25 to 500 g/m³. Grain dust generally varies from 0.025 to 0.055 oz/ft³ (25 to 55 g/m³). (See Table 3-26F.) Suspended dust concentrations of this type represent a severely dense cloud through which a 100-watt light bulb cannot be seen at a distance of approximately 9.8 ft (3 m) or one's hand cannot be seen at arm's length. Such dust concentrations make breathing extremely difficult. Some of the highest dust concentrations occur inside bucket elevators,

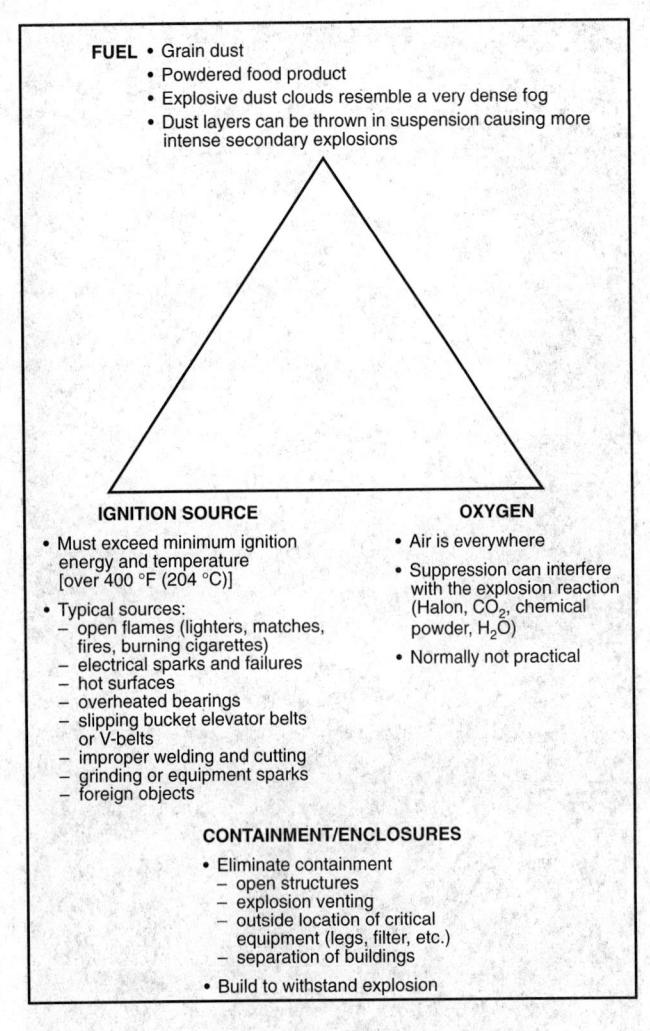

FUEL • Grain dust
• Powdered food product
• Explosive dust clouds resemble a very dense fog
• Dust layers can be thrown in suspension causing more intense secondary explosions

IGNITION SOURCE
• Must exceed minimum ignition energy and temperature [over 400 °F (204 °C)]
• Typical sources:
  – open flames (lighters, matches, fires, burning cigarettes)
  – electrical sparks and failures
  – hot surfaces
  – overheated bearings
  – slipping bucket elevator belts or V-belts
  – improper welding and cutting
  – grinding or equipment sparks
  – foreign objects

OXYGEN
• Air is everywhere
• Suppression can interfere with the explosion reaction (Halon, CO_2, chemical powder, H_2O)
• Normally not practical

CONTAINMENT/ENCLOSURES
• Eliminate containment
  – open structures
  – explosion venting
  – outside location of critical equipment (legs, filter, etc.)
  – separation of buildings
• Build to withstand explosion

FIG. 3-26N. Dust explosion elements.

grain bins, scale garners, and other enclosures. Maximum explosive pressures can exceed 100 psi (690 kPa), with the maximum rate of rise of pressure approaching up to 8,500 psi per sec (58 608 kPa/ sec), which is an indication of the intensity and speed of a grain dust explosion. When the rate of rise of pressure is higher, explosions are more severe.

The fuel: Dust particles emanating from various emission points within a grain elevator are of varying composition and of a wide range of sizes. It is generally agreed by researchers that particle sizes below 100 microns constitute the greatest hazards.[8] A considerable portion of the dust within the elevator environment is smaller than 100 microns. Larger particles tend to settle out rapidly. (See Table 3-26G.)

While it seems improbable that such a dense cloud would exist within the ambient space of an elevator structure where personnel are present, such concentrations have been measured within the confines of bucket elevators and may also occur in conveyor housings, bins being loaded, silos, dust collecting systems, and connecting spout work.

The mechanism of an explosion depends upon the immediate heat release of a burning particle to ignite and support the burning of adjacent particles.[4] As this rapid spread of flame proceeds from particle to particle, pressure waves and thermal expansion of the air

can create an intense shock sufficiently strong to rupture the typical reinforced concrete structure. In studies performed by the U.S. Bureau of Mines,[9] the maximum pressures for corn dust are greater than 100 psig (gage pressure of 690 kPa). Concrete structures found in elevators can usually withstand no more than 25 psig (172 kPa). Most grain-handling equipment enclosures fail at pressures less than 6 psig (gage pressure of 41.4 kPa).

Suspended dust is not the only fuel with which to be concerned. Dust accumulation on floors, walls, rafters, and equipment may become suspended if disturbed by vibration, fires, or small explosions. If this accumulated dust is suspended in sufficient concentration, the resulting explosive dust concentration can become ignited and progress into an explosion. This re-suspended dust can encompass large volumes and propagate explosions through an entire elevator. This illustrates the importance of keeping overhead surfaces and ledges clean of excessive dust accumulations.

Ignition sources: The second most important element of a grain dust explosion is ignition of the suspended dust cloud by a source of energy of sufficient intensity and duration. One ignition source that has been identified in a large percentage of known instances is improper use of welding and cutting equipment. Other identified ignition sources include fires and heat caused by the frictional energy of mechanical equipment, such as bucket elevators, bearings, and belt drives. Heat or arcing caused by the failure of electrical equipment, such as lighting, motors, and wiring, has also been identified as an ignition source. Miscellaneous ignition sources include open flames from matches or smoking, space heaters, lightning, and internal combustion engines on vehicles such as industrial trucks.

Torches and electric arcs used for hotwork represent one of the most intense energy sources used at grain facilities. These sources can quickly ignite dust and construction materials. Welding and cutting is also a concern in the grain industry, because many elevators do not have full-time maintenance personnel. This results in unsupervised contractors performing welding operations within the grain elevators. These contractors may be unfamiliar with the fire and explosion potential of grain dust and need to be monitored by elevator management. A formalized hotwork welding and cutting work permit should be established to help control hotwork operations.

Inasmuch as the majority of known locations of explosions in grain elevators stem from the bucket elevator, it would follow that this single piece of operational equipment presents the most serious ignition hazard to the grain handler.[6,7] This conveyance produces ignition energy in a number of ways. Overloading or stalling of the belt generates intense frictional heat on the revolving drive pulley. This has been known to burn the belting to the point of failure, allowing the severed and flaming pieces to drop within its housing. Failure of belt splices could lead to the same results.

Misalignment of belts can cause the "leg" casing to become hot enough to ignite combustible materials, such as dust or lubricants. The misaligned belt itself probably does not get hot enough to ignite while it is moving, since it has the chance to cool as it moves, but may ignite after the equipment is turned off and the belt is stationary next to the hot casing.

Another potential ignition source associated with bucket elevators is that of overheating bearings. Some older conveyor designs have tail or head pulley bearings located inside of the casings. If these bearings overheat, they can provide sufficient heat for ignition of static or suspended dust. Even bearings located outside of casings can ignite layered dust which, in turn, can be drawn into the "leg" casing by the dust collection aspiration or the "air pumping" action of the "leg" itself.[6]

TABLE 3-26F. *Dust Explosibility Properties**

Material	Lower Explosive Limit (LEL)	Minimum Ignition Energy (Joules)†	Minimum Ignition Temperature		Maximum Explosive Pressure (psi)‡	Maximum Rate of Pressurization (psi/sec)‡
			Cloud	Layer		
Coffee	85	0.16	410	240	44	500
Corn	45	0.04	400	250	95	6,000
Corncobs	30	0.04	400	190	110	5,000
Grain (mixed)	55	0.03	430	230	115	5,500
Soy	35	0.05	520	190	99	6,500
Sugar	35	0.03	350	220	91	5,000
Wheat	55	0.06	480	220	103	3,600
Wheat starch	25	0.02	380	210	105	8,500
Wheat flour	50	0.05	380	360	95	3,700
Coal, Pittsburgh	55	0.06	610	ND	83	2,300

*Source: Reference 8.
†1 cal = 4.187 J.
‡1 psi = 6.895 Kpa.
ND = Not Determined

TABLE 3-26G. *Size Distribution**

Dust size (µ)	Dump Pit (mostly beeswings)		Belt Loading (mostly starch dust)		Main Elevator (60:40 mixture beeswings, starch)		Beans Dust		Mesh	In.	mm
	Retained on %	% cum	Retained on %	% cum	Retained on %	% cum	Retained on %	% cum			
+150	94.8	94.8	—	—	56.0	56.0	16.0	16.0	+100	—	—
150–100	3.7	98.5	—	—	11.3	67.3	12.1	28.1	100	0.0059	0.150
100–74	1.1	99.6	—	—	7.0	74.0	13.4	41.5	150	0.0041	0.104
74–38	0.4	100.0	—	—	6.0	80.0	9.2	50.7	200	0.0029	0.074
38–21	—	—	31	31	6.0	86.0	16.3	67.0	450	0.0017	0.043
21–16	—	—	28	59	5.0	91.0	16.0	83.0	630	—	—
16–8	—	—	22	81	4.0	95.0	11.0	94.0	937	—	—
8–6	—	—	10	91	3.0	98.0	4.0	98.0	1,875	—	—
6–4	—	—	3	94	2.0	100.0	2.0	100.0	2,300	—	—
4–2	—	—	2	96	—	—	—	—	4,500	—	—
2–1	—	—	3	99	—	—	—	—	6,250	—	—
–1	—	—	1	100	—	—	—	—	12,500	—	—
Total	100.0	—	100.0	—	100.0	—	100.0	—	—	—	—

*Source: reference 10.

Extraneous foreign material, such as scrap metal, tools, wood, stones, or pieces of concrete, is also a concern within a bucket elevator. Such material may or may not produce sparks with sufficient energy to ignite dust, but certainly can result in plugged spouts, damage to the belting, or deformation of the elevating buckets. When this occurs, the chances of belts stalling or continuous frictional rubbing is increased. The use of grating to reduce foreign objects from entering the grain receiving pit and subsequent handling equipment is common at most facilities. Most grates are set 2.5 in. (63.5 mm) apart, with varying lengths of the opening. The grate size can be adjusted for the commodity; e.g., a smaller opening can be used for wheat than can be used for receiving soybeans in southern locations or for receiving corn cobs. Many processors, millers, and grain export facilities rely upon magnets on the receiving grain stream to capture tramp metal.

Nonconductive belting moving over the pulleys in bucket elevators can create significant static electricity on the buckets. Research indicates that sufficient energy to initiate a dust explosion is not released from this static discharge, but prudent practice is to reduce this static accumulation by using electrostatically conductive belting and good grounding practices.[13] OSHA calls for belting to have a surface resistivity less than 100 megohms. Rubber belting has typical resistivity of 1 megohm, and PVC belts are typically 15 megohms in resistivity.

The energy inherent in electrical systems is extremely high, but can be minimized as an ignition source by strict adherence to the provisions of NFPA 70, *National Electrical Code*® (hereinafter referred to as NFPA 70), for the appropriate hazardous area classification. For areas that have an exposure to higher levels of airborne and layer dust, electrical equipment is normally rated as explosion proof for Class II, Division 1, Group G areas. Other areas can be rated for electrical equipment that is suitable for Class II, Division 2 (dusttight), Group G locations. Many in the industry choose to utilize electrical equipment that has a dual rating for Class II, Division 1 and 2, Group G areas to avoid the problem of the wrong equipment getting in an area. The importance of proper electrical equipment applies to both fixed and portable equipment. This consideration is important for portable electrical devices, lighting, low-voltage control circuitry, extension drop-lights, and communications equipment.

Oxygen: The oxygen concentration required for a dust explosion is that found in the atmosphere. Using inert gases in equipment to reduce the oxygen concentration enough to prevent explosions has been suggested, but has limited application due to the large size and volume of grain-handling equipment. Processing equipment, such as hammer mills, may be more likely candidates for inerting.

Confinement: As with all dust explosions (except detonations), the pressures generated after ignition of a grain dust explosion increase until the fuel or oxygen is consumed or until the explosion is vented. If there is no confinement (i.e., unlimited venting), explosion pressures are minimal, and the incident should more properly be called flash fire. On the other hand, as confinement increases, the explosion pressures can build up to experimentally observed levels over 100 psig (690 kPa). Grain elevator structures and equipment cannot withstand pressures anywhere near these levels, so considerable damage will occur, unless these pressures can be vented. Due to structural considerations, most existing grain elevators would be nearly impossible to retrofit with adequate venting, but newly designed elevators are adaptable to increased venting and the resulting reduction in explosion confinement. Explosion venting must be done to the outside of the facility to avoid secondary explosions or endangerment to personnel. Although the levels of recommended venting are not always possible, some venting is better than no venting when feasible to help eliminate pressure development during an explosion.

An alternative to explosion venting is explosion suppression, during which a chemical is injected into equipment during the early stages of explosion development to stop the process. The chemical can interfere with the combustion reaction (halon), or inert the environment (CO_2), or cool the process with a water spray. Suppression systems can be specifically engineered for a piece of fixed equipment.

The construction of structures and equipment that can totally contain an explosion is generally not practical, with the exception of hammer mills, which can often withstand the explosive pressures. However, the transmission of pressures and flames to connecting equipment can present explosion problems.

The only other means of dealing with confinement is to provide separation between buildings and structures to help minimize the propagation of an explosion. NFPA 61 recommends that a separation of 100 ft (30.5 m) be provided between personnel-intensive areas and concrete elevator headhouses and silos. Some exceptions are allowed, based on the location of bucket elevators (outside), type of construction (damage limiting), or property size. The concept of separation can be taken into account for major modifications or new facilities, but is often impractical for existing facilities.

THE SAFEGUARDS

Building Design

NFPA 61 has been prepared as a standard to assist designers of grain-handling facilities. The standard devotes considerable attention to the physical structure. NFPA 61 is not intended to apply to existing facilities, which may have been constructed prior to its publication; however, major modification or expansion made to an existing facility should enhance safety to life and property. The following discussion summarizes the current industry trends concerning the design of new facilities.

The traditional design approach to grain elevator facilities has been to construct upright storage silos of reinforced concrete or steel. The facility has a high contiguous structure known as a headhouse, with inside-located bucket elevators to elevate the grain high enough to pass through necessary scales, samplers, garners, cleaners, and distributors by gravity into the storage bins. Figure 3-26F illustrates the traditional grain elevator design. Structures are constructed of noncombustible materials. There has been a constant evolution of facility design to meet the larger grain productions and to be able to handle higher flow rates with more automation. Higher

throughput facilities have evolved to load unit trains in excess of 100 railcars [3,300 bushels (119 m³) per railcar] or several barges [up to 60,000 bushels (2160 m³) each] a day. Equipment has also changed to meet these new demands. Facility owners have also been concerned with safety and the elimination of fire or dust explosion hazards and make changes, as resources permit.

Modern design has moved toward the elimination of the enclosed elevator headhouse.[14,15] Bucket elevators are placed outside the facility or inside structural steel or reinforced concrete frameworks to eliminate the element of confinement. (See Figure 3-26O.) The bucket elevators then elevate the grain to the top of storage to distributors that connect to turnheads, which spout directly into bins or to enclosed conveyors, which move the grain to bins, eliminating the enclosed gallery above the bins. (See Figures 3-26P, 3-26Q, and

FIG. 3-26O. Open tower structure and explosion relief venting are evident in the new Corpus Christi, TX, public elevator.

FIG. 3-26P. The headhouse concept is eliminated in the new design used in this Superior, WI, elevator.

3-26R.) Some large terminals, particularly export facilities, have utilized inclined belt conveyor systems in lieu of bucket elevators. (See Figure 3-26A.) These incline designs convey the grain to separate weighing and cleaning towers. Enclosures, such as dust collectors, bucket elevator housings, and ancillary structures, are often designed to relieve explosion pressure waves as much as possible, should such an incident happen. This is not always physically possible, because of the very nature of the configuration of the enclosure.

According to NFPA 68, *Guide for Venting of Deflagrations,* numerous methods have been proposed for calculating the vent area for an enclosure. Some venting models have used the surface area of the enclosure as a basis for determining vent area. Analysis of available data shows that such methods overcome certain deficiencies of previous methods of calculating vent area. The recommended venting equation is:

$$A_v = \frac{C(A_S)}{(P_{red})^{1/2}}$$

where

A_v = vent area (ft² or m²),

C = venting equation constant (commonly 0.10 for grain dust),

A_s = internal surface area of enclosure (ft² or m²), and

P_{red} = maximum internal overpressure that can be withstood by the weakest structural element (psi or kPa).

The use of doors, windows, multiple explosion-relief panels, and light-gage structural coverings on steel structures is a practical approach to an acceptable design. Stairwells and elevator shafts, where they are enclosed, should be protected with approved fire doors.

Fire walls of approved design can be provided to separate the grain-handling function from adjoining grain-processing operations, such as flour milling, preparation for oil extraction, feed milling, or grinding. Where tunnels, basements, or other underground structures can be avoided, the alternate use of ground-level or aboveground structures can give opportunity for maximum openings to the atmosphere.

Where sufficient land area is available, the placement of structures for weighing, cleaning, and other operational functions can be remote from silo storage and interconnected with inclined belt systems.

All surfaces, bin bottoms, and spout inclinations should be designed to be self-cleaning when the grain or grain products flow. Structural ledges, beams, or other horizontal surfaces should be designed, where practicable, with an inclination of a minimum of 60 degrees from the horizontal to prevent gradual buildup of static dust. Where possible, design should accommodate the washing down of accumulated dust on vertical walls and overhead structures with water.

Where possible, the placement of bucket elevators, in open air, apart from the structure should be considered. Additionally, office buildings, employee break rooms, maintenance shops, inspection

FIG. 3-26Q.   The Naples, IL, Terminal Elevator Company, designed and built by Borton, Inc., incorporated explosion venting technology on bucket elevators with the headhouse structure left open on one side.

FIG. 3-26R. A country elevator in O'Neill, NE, features an outside located bucket elevator and elimination of gallery.

labs, and control rooms should be located remote from concrete structures when constructing new elevators.

Mechanical Design

Many devices are available to detect, warn, and control faulty operation of mechanical equipment or their components. Many of these systems were neither available nor sufficiently developed for use in Class II, Group G, dust atmospheres until recent years. Among them are hot-bearing sensors, speed indicators, alignment devices, level-sensing gages, slowdown detectors, overflow alarms, and pressure gages, all of which can serve to indicate or react to malfunctioning units. Particularly important among these are the speed indicator and alignment devices for monitoring bucket elevators. If properly designed, installed, and monitored, these devices can prevent ignition from stalled and rubbing belting.

A combination of mesh screens, grating, scalpers, or magnets will prevent large extraneous material and metal from entering into the grain handling machinery. Depending on individual elevators,

permanent or electromagnets, large-mesh screening, grizzlies, and specific gravity separators may be appropriate. Their use is particularly important just prior to the point where the grain enters hammermills or other types of grinders.

Dust Control

The need for adequate dust control has been established. Dust control in elevators, however, should not be limited to installation of dust collection systems, but should be a systematic approach to dust control. This systematic approach should include designs that limit the amount of dust liberated from the grain, mechanical dust collection at grain transfer points where dust occurs, containment of dust to selected points where mechanical dust collection can be effective, reduction of dust concentrations within equipment, maintenance of dust collection equipment, and manual housekeeping to complement mechanical dust control.

The design of dust collection systems must provide sufficient capture velocity at the point of emission, particularly at conveyor loading and discharge points. Design must also provide sufficient air velocity [typically between 3,500 and 4,000 ft per min (1068 to 1220 m/min)] within the ducts to prevent settlement of the particles and subsequent plugging. Blast gates and fresh-air inlet dampers are necessary to balance the airflow to all points served by the system so that starving of some remote emission points is avoided. Operators must understand the dust collection systems, so that they can monitor and maintain their effectiveness on a daily basis.

If a baghouse or fabric filter is used, the porosity of the filter media must be maintained to avoid diminishing the airflow. High humidity and fine dust particles often form a cake on the bags and reduce the air passage. In such instances, replacement or laundering of the bags is necessary. (See NFPA 91, *Standard for Exhaust Systems for Air Conveying of Materials,* and NFPA 650, *Standard for Pneumatic Conveying Systems for Handling Combustible Materials*, for further information on the application and design of air-moving systems.)

Containment of dust to selected points or the proper return of the dust downstream of the capture point not only makes the mechanical dust collection system more effective, but also limits the amount of manual housekeeping effort needed to reduce accumulated dust.

Typical methods to reduce dust liberation from grain include reducing the distance and velocity grain is allowed to free fall onto conveyors or from spouting. (Choke feeding, where possible, is the best approach to limit grain velocity.) Reduction of conveyor belt speeds and adequate belt tension can both be effective dust control methods where horizontal belt conveyors are used. Well-designed transition points that do not have abrupt changes in direction can also limit the amount of dust liberated from the grain. Additives, such as soybean or mineral oil, have been proposed as an additional method to control dust. Research is being conducted to determine whether these additives deteriorate grain quality for the end-user and whether grain so treated can meet the grading standards of the U.S. Department of Agriculture (USDA).

Regardless of the facility design or amount of mechanical dust collection, manual housekeeping efforts are important in grain elevator facilities. Although safe levels of accumulated dust are impossible to determine, housekeeping programs at each facility should be designed to keep the dust accumulation at the lowest feasible level. This housekeeping program should focus on not only grain dust but also any type of combustibles to reduce the possibility of their providing an ignition source for a dust explosion. Good maintenance to spouts and other equipment leaking grain dust is also an essential element to explosion prevention. In addition, proper safety procedures must be followed to avoid ignition sources while blowing down overhead ledges. Some in industry use vacuum systems to clean overhead and floor areas in lieu of blow down, or they use periodic wash down of some areas, such as reclaim tunnels. Further, new techniques are sometimes used to pressurize key areas of a facility to prevent dust from escaping the equipment.

Electrical Design

Electrical wiring and equipment in grain dust environments should be installed in accordance with Articles 500 and 502 and, where applicable, Article 504 of NFPA 70. These articles need to be thoroughly understood by those installing or maintaining electrical equipment in a grain elevator facility. Individual paragraphs in these articles cover equipment requirements for maximum surface temperature, transformers and capacitors, surge protection, wiring methods, switches/breakers, etc.; motors; ventilating piping; utilization equipment; lighting equipment; flexible cords; receptacles and plugs; signal and communication systems; live parts; and grounding. No attempt is made here to paraphrase these articles. The reader is encouraged to review the most current edition of NFPA 70.

Away from the technical requirements of NFPA 70, the development of economical and miniaturized electronic components for use in Class II, Group G atmospheres has made the operation and control of grain elevators more sophisticated and less labor intensive. Remote control of the entire facility from a centralized control center is now a reality in new installations. This technological advance has dictated the need for redundant detection of malfunctions, since people are no longer in an operating area to observe the minute-by-minute activity.

The more common safeguards include electrical interlocking to simultaneously shut down all activities in a sequenced fashion when one unit of a sequentially connected operation fails. Annunciator panels can display the exact cause of shutdown to the operator. Ammeters and load indicators can depict the exact weight being carried by conveying equipment. Spring-return pneumatic devices can activate shutdown conditions even in a total power failure. Fully electronic scales now safely transmit millivolt differentials, detected by a load cell or strain gage, to a remote amplifier for direct conversion into information on weight.

Maintenance

A preventive maintenance program is essential not only to ensure trouble-free operation and to minimize breakdowns, but, in grain elevator facilities, it is also a critical safeguard for eliminating ignition sources. This preventive maintenance program should include lubrication of bearings, checking of belt splices, replacement of bent or missing bucket elevator cups, and the early remedy of leaking, worn-out spouting.

Dust filters should be checked frequently for plugging, excessive pressure drop, and worn or poorly sealed rotary airlocks. Ductwork in the aspiration system should be free of plug-ups and bent ducts, and blast gates should be opened or closed consistent with system balancing.

Electrical junction boxes should be kept closed and lighting fixture protective globes kept in place. Whenever maintenance work on the electrical systems is performed, qualified personnel should ensure that the power is disconnected or that hazardous dust concentrations are not present.

Welding and any other hotwork should be rigidly controlled by facility management. The use of a written "hotwork permit" is generally the best system to ensure that hotwork is not performed in the presence of explosive dust concentrations. This is particularly im-

portant with contractors whose employees may not be aware of the dust explosion hazard.

Improved maintenance diagnostic equipment can also be effective in monitoring equipment that could become an ignition source for an explosion. This diagnostic equipment includes bearing temperature monitors, bearing vibration monitors, and infrared analysis of electrical and mechanical equipment.

Fire Control Systems

Grain elevators have few applications where traditional automatic sprinkler systems would provide much protection. The nature of the explosion hazard is such that, even if sprinklers are installed, the lines would likely be ruptured by the shock waves. Additionally, explosions are normally initiated so rapidly that fusible links would have insufficient time to react. Combine this ineffectiveness with the lack of combustibles present and the catastrophic damage that could result from water leaks sweeping into grain storage bins, and it becomes clear that general-purpose automatic sprinkler systems would do little to mitigate damage from an explosion hazard. There are, however, several specific applications where sprinklers, fixed water spray nozzles, and explosion suppression systems can be effective. Strategic location of fire extinguishers throughout the facility and near major potential ignition sources, such as major drive units, is important to allow personnel to deal with small incipient fires before they create more serious concerns.

Wood elevators, which have a significant fire hazard potential, as well as the explosion hazard, are the major exception where automatic sprinkler installation can be justified. Additional specific applications where sprinklers or fixed water spray nozzles can have some benefit are: (1) within the drive pulley enclosure of bucket elevators; (2) at remote, elevated belt gallery structures where manual fire fighting would be impossible; (3) in some tunnel areas where access is difficult; and within some grain dryers in which, because of design, (4) it would be impossible to use hose streams effectively.

Explosion suppression offers the grain industry a possible tool to reduce the number of explosions. Explosion suppression technology (i.e., rapid detection of incipient explosion and immediate introduction of suppressing agent) has been used effectively in other industries for some time. Recent research efforts directed at applying the technology to the grain industry have resulted in the development of self-contained explosion suppression devices that are suitable for installation on bucket elevators.[16] Application of hose streams within an elevator where the possibility of a dust explosion exists must be a well-considered endeavor. Any application of high-pressure water or use of a fire extinguisher that disperses accumulated dust must be avoided to prevent an airborne dust concentration that could be ignited.

SUMMARY

Much of the above discussion is a reiteration of concepts and principles that have been known by the grain-handling industry for a number of years or recently learned through its research efforts. The devices, systems, and machinery now available are much more extensive than in earlier years. There are over 10,000 grain elevators in the United States, ranging from crossroad country stations to newly constructed export facilities. Many of these newer facilities have incorporated the devices mentioned above.

In spite of this, grain dust explosions still occur although at a lower incidence rate than in past years. In many instances, the exact location of the primary explosion and ignition source are not precisely known. However, this much is known: the initial explosion has the characteristic of transmitting pressure waves, vibration, and air movement throughout the facility, which could result in other deposited dust being thrown into suspension. This additional dust might have been lodged on the ledges, beams, walls, floors, and inaccessible places and is exposed to the progressing flame front. The result is a series of additional explosions known as secondary explosions.[17]

In summarizing the potential hazards associated with the handling and storage of grain, one must remember that dust is a powerful fuel. Constant awareness and proper training of all who work in the facilities is essential to continue the recent trend of reduced numbers of explosions. Prudent consideration of the inherent dangers requires respect, care, and adherence to building codes and regulations in the design and construction of each facility. Once the facility is built and operating, each day requires positive management appreciation of the inherent dangers, and action to ensure that all those working in the facility understand the basics of proper operations, maintenance, safety, housekeeping, and training.

Investigations are continuing into the causes of grain dust explosions, in hope of more clearly defining the practices and methods needed to reduce the hazard. It would appear that the most fruitful field for concentration is the reduction in the potential of bucket elevators as ignition sources. The outside location of bucket elevators, explosion venting, and explosion suppression are getting increased usage. New developments in computer technology are making the use of programmable logic controllers at grain facilities more commonplace; these can monitor safety devices, bearing temperatures, and equipment loads. This trend, coupled with further use of new design techniques, will do much to help the industry cope with dust explosions.

If it is assumed that some dust will always be present within the grain elevator facility, the elimination of the explosion hazard can only be accomplished by controlling the dust and removing its ignition sources.

BIBLIOGRAPHY

References Cited

1. *Statistical Reporting Service*, U.S. Department of Agriculture, Washington, DC, 1996.
2. *Industry Trade Associations & 1996 North American Grain & Milling Annual Publication*, Sosland Publishing Company, Kansas City, MO.
3. Christensen, C.M., ed., *Storage of Cereal Grains and Their Products*, American Association of Cereal Chemists, St. Paul, MN, 1982.
4. Palmer, K.N., *Dust Explosions and Fires*, Chapman and Hall, London, 1973.
5. "Mineral Resources," 30 CFR 18.65, U.S. Government Printing Office, Washington, DC.
6. "Prevention of Grain Elevator and Mill Explosions," Publication NMAB 367-2, 1982, National Materials Advisory Board, National Academy of Science, National Academy Press, Washington, DC.
7. Anderson, R., *Proceedings* of International Symposium on Grain Dust Explosions, Session III, Grain Elevator and Processing Society, Minneapolis, MN, 1977.
8. Lilienfeld, P., "Special Report on Dust Explosibility," GCA-TR-78-17-6, EPA Contract No. 68-01-4143, Technical Service Area 1, Task Order No. 24, March 1978, GCA Technology Division, Environmental Protection Agency, Washington, DC.
9. Jacobsen, M., *et al.*, "Explosibility of Agricultural Dust," RI 5753, 1961, U.S. Bureau of Mines, Washington, DC.
10. Matkovic, I.. M., "Dust Composition, Concentration, and Its Effects," *Proceedings* of International Symposium on Grain Dust Explosions, Grain Elevator and Processing Society, Minneapolis, MN, 1977.
11. "Prevention of Grain Elevator Explosions—An Achievable Goal," U.S. Department of Agriculture, Washington, DC, 1979.
12. Schoef, R., unpublished data released by Kansas State University, in cooperation with USDA's Federal Grain Inspection Service, compiled along with reference 11 data.
13. Dahn, J., *Electrostatic Grounding Characteristics of Grain Facilities*, National Grain and Feed Association, Fire and Explosion Research Council, Washington, DC, 1982.
14. Gordon, R. C., and Maness, J. E., *A Practical Guide to Elevator Design*, National Grain and Feed Association, Washington, DC, 1979.
15. Gordon, R. C., Young, C. R., and Goedde, M. G., *Retrofitting and Constructing Grain Elevators for Increased Productivity and Safety*, National Grain and Feed Association, Washington, DC, 1984.
16. Gilles, J., *Explosion Venting and Suppression of Bucket Elevators*, National Grain and Feed Association, Fire and Explosion Research Council, Washington, DC, 1980.
17. Tamanini, F., *Dust Explosion Propagation in Simulated Grain Conveyor Galleries*, National Grain and Feed Association, Fire and Explosion Research Council, Washington, DC, 1983.

NFPA Codes, Standards, and Recommended Practices

Reference to the following NFPA codes, standards, and recommended practices will provide further information on the safeguards for grain and grain milling discussed in this chapter. (See the latest version of *The NFPA Catalog* for availability of current editions of the following documents.)

NFPA 10, *Standard for Portable Fire Extinguishers*
NFPA 13, *Standard for the Installation of Sprinkler Systems*
NFPA 14, *Standard for the Installation of Standpipe and Hose Systems*
NFPA 15, *Standard for Water Spray Fixed Systems for Fire Protection*
NFPA 25, *Standard for the Inspection, Testing, and Maintenance of Water-Based Fire Protection Systems*
NFPA 30, *Flammable and Combustible Liquids Code*
NFPA 31, *Standard for the Installation of Oil-Burning Equipment*
NFPA 36, *Standard for Solvent Extraction Plants*
NFPA 51B, *Standard for Fire Prevention in Use of Cutting and Welding Processes*
NFPA 54, *National Fuel Gas Code*
NFPA 58, *Standard for the Storage and Handling of Liquefied Petroleum Gases*
NFPA 61, *Standard for the Prevention of Fires and Dust Explosions in Agricultural and Food Products Facilities*
NFPA 68, *Guide for Venting of Deflagrations*
NFPA 69, *Standard on Explosion Prevention Systems*
NFPA 70, *National Electrical Code*®
NFPA 72, *National Fire Alarm Code*
NFPA 77, *Recommended Practice on Static Electricity*
NFPA 80, *Standard for Fire Doors and Fire Windows*
NFPA 86, *Standard for Ovens and Furnaces*
NFPA 91, *Standard for Exhaust Systems for Air Conveying of Materials*
NFPA 101®, *Life Safety Code*®
NFPA 496, *Standard for Purged and Pressurized Enclosures for Electrical Equipment*
NFPA 505, *Fire Safety Standard for Powered Industrial Trucks Including Type Designations, Areas of Use, Maintenance, and Operation*
NFPA 650, *Standard for Pneumatic Conveying Systems for Handling Combustible Materials*
NFPA 780, *Lightning Protection Code*

Additional Reading

Bartknecht, W., "Effectiveness of Explosion Venting as a Protective Measure for Silos," *Plant/Operation Progress*, Vol. 4, No. 1, 1985, pp. 4-13.
Bartknecht, W., *Dust Explosions Course Prevention Protection*, Springer Verlag, Berlin, Heidelberg, NY, 1989.
Bonney, M. J., "Suppressing Dust Explosions," *Fire*, Vol. 88, No. 1086, Dec. 1995, p. 40.
Britton, L. G., "Systems for Electrostatic Evaluation in Industrial Silos," *Plant/Operation Progress*, Vol. 7, Jan. 1988, pp. 45-50.
Britton, L. G., and Kuby, D. C., "Analysis of a Dust Deflagration," *Plant/Operation Progress*, Vol 8, No. 3, July 1989, pp. 177-180.
"Dust Collectors," *Loss Prevention Data Sheet 7-73*, Factory Mutual Research Corp., Norwood, MA.
Gillis, J. P., "Rapid Fire Detection in Dust Process Equipment Utilizing High Air Flows," *Proceedings* of the American Institute of Chemical Engineers National Meeting, AIChE, 1988, p. 25.

Goforth, K., and Ostrowski, C., *Emergency Pre-planning and Firefighting Manual—A Guide for Grain Elevator Operators and Fire Department Officials*, National Grain and Feed Association, Washington, DC.

Goyer, D., "Grain Dust Explosion Averted Dockyard Blaze," *Fire Fighting in Canada*, Vol. 39, No. 1, Feb. 1995, pp. 4–6.

"Grain Storage and Grain Milling," *Loss Prevention Data Sheet 7-15,* Factory Mutual Research Corp., Norwood, MA.

Gray, T. A., "Firefighting Hazards at Grain Facilities," *Fire Engineering*, Vol. 147, No. 11, Nov. 1994, pp. 56–57.

Halpin, T., Jr. and Shattuck, D., " Fires in Agricultural Silos," *Fire Engineering*, Vol. 147, No. 3, Mar. 1994, pp. 12–13, 16.

Hertzberg, M., *A Critique of the Dust Explosibility Index*, U.S. Department of the Interior, Bureau of Mines, Washington, DC, 1987.

Hoenig, S. A., "Reducing the Hazards of Dust Explosions," *Plant/Operation Progress*, Vol. 8, No. 3, July 1989, pp. 119-128.

Isner, M. S., "$16 Million Fire Destroys Yuma Food Plant," *NFPA Journal*, Vol. 87, No. 4, July/Aug. 1993, pp. 33–39.

Lunn, G. A., "Note on the Lower Explosibility Limit of Organic Dusts," *Journal of Hazardous Materials,* Vol. 17, No. 2, Feb. 1988, pp. 207-213.

Maddison, N., "Dangers in the Dust," *Fire Prevention*, No. 264, Nov. 1993, pp. 22–25.

Moore, P. E., "Industrial Dust Explosions," Cashdollar, K. L., and Hertzberg, M., eds., ASTM Special Publication 958, American Society for Testing and Materials, W. Conshohocken, PA, 1987.

NFGA, *Fire and Explosion Research Council Research Reports* (series), National Grain and Feed Association, Washington, DC.

"*Prevention of Grain Elevator and Mill Explosions,*" Publication NMAB-2, 1982, National Materials Advisory Board, National Academy of Science, National Academy Press, Washington DC.

"Recipe for a Dust Explosion," *Record*, Vol. 72, No. 4, Fourth Quarter 1995, pp. 3–12.

"Risks in Handling Combustible Dusts," *Fire Prevention,* No. 213, Oct. 1988, pp. 32-33.

Steer, P., "Feed Mill Destroyed," *Fire Fighting in Canada*, Vol. 37, No. 6, Aug. 1993, pp. 42–43.

Winward, J. and Campanaro, P., " Fire in Mill Construction: Breeding Ground for Conflagration," *Fire Engineering*, Vol. 145, No. 1, Jan. 1992, pp. 50–52, 54, 56, 58, 60.

Zeeuwen, J. P., "Dust Explosions—How to Assess the Risk," *Fire Prevention*, No. 228, Apr. 1990, pp. 26–28.

GRINDING PROCESSES

Revised by Delwyn D. Bluhm

This chapter discusses the fire hazards of grinding processes. Grinding (or pulverizing and/or milling) is an industrial process in which materials are reduced to very small particles. Sometimes the grinding process of combustible and some normally noncombustible materials produces a highly explosive dust that is hazardous when dispersed in critical concentrations in air. A serious dust fire or explosion can begin either in the grinding equipment itself or in the environment surrounding the equipment. Examples of materials having combustible dust hazards are wheat flour, wood flour, sulfur, starch, coal, some plastics, aluminum, and magnesium.

According to experts[1] in coal mining, electric power generation, grain handling, plastics, chemicals, wood products, and metal powders industries, dust explosions and fires have been a continuing problem. Various preventive and protective measures must be taken to ensure safety, utilizing both existing technologies and new data for practical designs of safe new industrial equipment.

It is important to distinguish the processes used to reduce the size of materials by grinding from those processes that employ abrasive wheels, disks, or drums to prepare surfaces or shape articles of wood, metal, or plastic. The latter processes could produce explosive dusts under certain conditions. However, those operations normally are conducted in environments where there is little likelihood that dust concentrations will reach the lower explosive limit (LEL). This chapter, therefore, is concerned only with the hazards of milling or grinding operations, i.e., the former grinding processes, involving large quantities of potentially explosive materials.

Section 1, Chapter 6, "Explosions," and Section 4, Chapter 14, "Explosion Prevention and Protection," present guidelines for grinding hazards control. Section 3, Chapter 15, "Woodworking Facilities and Processes," and Section 9, Chapter 23, "Food Processing Facilities," offer additional insight into the hazards of organic dusts. See also Section 3, Chapter 26, "Storage and Handling of Grain Mill Products."

GENERAL

Processes

Grinding operations are performed either wet or dry. Water is an excellent medium for wet grinding, though other liquids are some-

times used. Kerosene, for example, is used for the wet milling of magnesium. Because dust fires or explosions are the major hazards of grinding processes, dry grinding is emphasized in this chapter.

Materials

Agricultural products, such as wheat, corn, and other grains, present explosion hazards both during storage and processing into flour or starch. Wood flour, finely divided sawdust, sulfur, coal, and some plastics present the same hazards as magnesium and aluminum. An explosion or fire can originate in the process equipment itself or in the ambient environment into which dust may escape and accumulate.

Actual occurrences, reinforced by laboratory tests, have demonstrated that finely divided particles of almost any readily oxidized material suspended in proper concentration in air will ignite and explode or burn. Table 3-27A classifies 59 different materials into three groups according to the maximum rate of pressure rise during an explosion. The maximum rate of pressure rise is indicative of the violence of a dust explosion. The table includes eight different plastic dusts, but does not cover the entire range of explosive plastic dusts. Table 3-27B gives the explosion characteristics of a variety of agricultural and industrial dusts. Materials other than those mentioned also may be explosive under proper conditions of particle size and oxygen availability.

The maximum rate of pressure rise is calculated from the steepest part of a pressure *vs.* time curve generated during a test explosion in a closed vessel. As with the maximum explosion pressure, appropriate adjustments need to be made if the dispersing air effectively increases the ambient pressure. The average rate of pressure rise is also calculated from the pressure *vs.* time curve. An approximate correlation often exists between the average and maximum rates of pressure rise, but the latter is preferred when assessing practical needs. The average rate of pressure rise can be affected by slow development of the explosion and is usually one-half to one-third the value for the maximum rate of pressure rise.

The U.S. Department of Interior Bureau of Mines has developed three additional measures of relative explosion hazard: the ignition sensitivity (IS), the explosion severity (ES), and the explosibility index (EI). Each of these is a dimensionless value derived by comparing the measured explosibility parameters for a given dust, as described above, with those of Pittsburgh coal dust.

Delwyn D. Bluhm, Ph.D., P.E., is senior engineer and manager of research and development engineering for Ames Laboratory and Institute for Physical Research and Technology, operated by Iowa State University, Ames, IA.

The relative explosion hazards may be derived using the following formulas:

$$IS = \frac{\min T_1 \times \min E_1 \times \min C_e}{\min T_1 \times \min E_1 \times \min C_e}$$

$$ES = \frac{\max T_2 \times \max P'}{\max T_e \times \max p}$$

$$EI = IS \times ES$$

where

IS = Ignition sensitivity

ES = Explosion severity

EI = Explosibility index

T_i = Ignition temperature

E_i = Ignition energy

C_e = Explosive concentration

P_e = Explosive Pressure

p = Rate of pressure rise

(Prime parameters represent Pittsburgh coal dust; unprimed parameters represent dust samples.)

Known values for these quantities are included in Table 3-27B. In the absence of a sound theoretical basis for predicting explosion hazards of dusts, the explosibility index provides a useful evaluation of relative explosibility. An empirical correlation between the U.S. Bureau of Mines indices and a descriptive categorization, i.e., the explosion hazard rating, is shown in Table 3-27C.

Materials are ground either in batches or continuously. Continuous processes may be either open or closed circuit. (See Figures 3-27A and 3-27B.) In air-swept mills, air is blown in at one end and the ground material is removed at the other end in air suspension. Batch mills are generally used only where small quantities are processed, due to the high labor costs for charging and discharging the mill.

Grinding equipment is classified according to the manner in which force is applied to the material. The material may be ground: (1) between two solid surfaces, (2) by being forced against one solid surface (jet mill method), (3) by the action of the surrounding medium, or (4) by the nonmechanical introduction of energy, such as thermal shock, explosive shattering, or electrohydraulic processes. This chapter is concerned mainly with the first two methods. Also associated with grinding equipment are classifiers which separate out the fine product and return the coarse mill for regrinding.

Fire and Explosion Safety

All grinding equipment (e.g., mills) that produce combustible dusts should be isolated. If possible, they should be in separate, detached, one-story, noncombustible buildings entirely above grade. Adequate explosion venting is essential. The surfaces of interior walls and equipment should be smooth to minimize dust accumulations and facilitate cleaning. If a milling operation must be located in a multi-purpose building, it should be in a room with at least one exterior wall that can be arranged to relieve explosion pressures, and the interior walls around the mill area should be of explosion-resistant construction.

TABLE 3-27A. *Classification of Explosive Dusts Based on Maximum Rates of Pressure Rise*

Low maximum rates of pressure rise†

Metal Dusts	Miscellaneous Dusts
Antimony	Anthracite
Cadmium	Carbon black
Chromium	Coffee
Copper	Coke, low volatile
Iron (impure)	Graphite
Lead	Leather
Tungsten	Tea

Moderate maximum rates of pressure rise ‡

Metal Dusts or Powders

Iron (carbonyl, electrolytic or H2, reduced)	Polyethylene
	Polystyrene
Manganese	Urea—melamine
Tin	Vinyl butryal
Zinc	

Grains, Spices, etc., Dusts

Alfalfa	Miscellaneous Dusts
Cocoa	Bituminous Coal
Grain Dust and Flour	Cork
Mixed Grains	Calcium lignosulfonic acid
Rice	Coumarone indene
SoyBean	Dextrin
Spices	Lignin
Starch	Lignite
Yeast	Peat
	Powdered drugs

Plastic Dusts

Cellulose acetate	Pyrethrum
Methyl methacrylate	Shellac
Phenolformaldehyde	Silicon
Phthalic anhydride and its resins	Sulfur
	Tung
	Wood flour

High maximum rates of pressure rise§

Metal Dusts

Aluminum	*Sorbic acid
*Stamped aluminum	*Titanium
Magnesium	*Zirconium
Magnesium—aluminum alloys	Some metal hydrides

*These are exceptionally fast.
†<7300 psi/sec (500,000 kPa/sec) measured via the Hartmann apparatus.
‡±7300 to 22,000 psi/sec (500,000 to 151,000 kPa/sec) measured via the Hartmann apparatus.
§>22,000 psi/sec (151,000 kPa/sec) measured via the Hartmann apparatus.

Although the mill is the source of combustible dust and may also provide an ignition source, explosions generally develop somewhere in the system downstream of the mill. Also, mills are generally constructed substantially enough to withstand explosion pressures. Explosion venting is needed where this is not the case.

As much as practical, nonsparking construction materials should be used to minimize sparks. Magnetic separators should be installed in front of mills to remove foreign ferrous metal, and screens should be used to remove rock and other nonferrous foreign material. Mills should be grounded to minimize the possibility of ignition by static sparks. Open flames and smoking should not be permitted, and welding and cutting equipment should be used only when the mill is shut down and the area has been made entirely dust free.

Good housekeeping is essential. Although well-designed grinding mills minimize dust leakage and reduce necessary cleaning, it is

TABLE 3-27B. Explosion Characteristics of Various Dusts

(Compiled from the following reports of the U.S Department of Interior, Bureau of Mines: RI 6516, "Explosibility of Agricultural Dusts"; RI 5753, "The Explosibility of Agricultural Dusts"; RI 5971, "Explosibility of Metal Powders"; RI 5971, "Explosibility of Dusts Used in the Plastics Industry"; RI 6597, "Explosibility of Carbonaceous Dusts"; RI 7132, "Dust Explosibility of Chemicals, Drugs, Dyes, and Pesticides"; and RI 1200, "Explosibility of Miscellaneous Dusts".)

Type of Dust	E_i Explosibility Index	I_s Ignition Sensitivity	E_s Explosion Severity	P_e Max. Explosion Pressure Rise (psig)	(kPa)	Max. Rate of Pressure Rise (psi/sec)	(kPa/sec)	T_c Ignition Temp. Cloud (°C)	Cloud (°F)	Layer (°C)	Layer (°F)	E_i Min. Cloud Ignition Energy (J)	C_e Min. Explosion Conc. (oz/cu ft)	(g/m³)	Limiting Oxygen Percentage* (Spark Ignition)
Agricultural Dusts															
Alfalfa meal	0.1	1.2	6.6	455	1,100	7,585	530	986	—	—	—	0.105	—	105	—
Cocoa bean shell	13.7	3.6	3.8	77	531	3,300	22,754	470	878	370	698	0.030	0.040	40	—
Coffee, raw bean	<0.1	0.1	0.1	33	228	150	1,034	650	1,202	280	536	0.320	0.150	150	C17
Cornstarch, commercial product	9.5	2.8	3.4	106	731	7,500	51,713	400	752	—	—	0.040	0.045	45	—
Cork dust	>10	3.6	3.3	96	662	7,500	51,713	460	860	230	446	0.035	0.035	35	—
Grain dust, winter wheat, corn, oats	9.2	2.8	3.3	131	903	7,000	48,265	430	806	240	464	0.030	0.055	55	—
Peat, sphagnum, sun dried	2.0	2.0	1.0	104	717	2,200	15,169	460	860	210	410	0.050	0.045	45	—
Pyrethrum, ground flower leaves	0.4	0.6	0.6	95	655	1,500	10,344	460	860	—	—	0.080	0.100	100	—
Rice	0.3	0.5	0.5	47	324	700	4,827	510	950	450	842	0.100	0.085	85	C15
Soy flour	0.7	0.6	1.1	94	648	800	5,116	550	1,022	340	644	0.100	0.060	60	—
Wheat flour	4.1	1.5	2.7	97	669	2,800	19,306	440	824	440	824	0.060	0.050	50	—
Yeast, torula	2.2	1.6	1.4	123	848	3,500	24,133	520	968	260	500	0.050	0.050	50	—
Carbonaceous Dusts															
Charcoal, hardwood mixture	1.3	1.4	0.9	83	572	1,300	8,964	530	986	180	356	0.020	0.140	140	—
Coke, petroleum	0.1†	0.1†	—	—	—	200	1,379	670	1,238	§	—	‡	1.000	1005	—
Graphite	0.1§	0.1§	—	—	—	—	—	§	—	580	1,076	—	—	—	—
Lignite, California	>10	5.0	3.8	94	648	8,000	55,160	450	842	200	392	0.030	0.030	30	—
Metals															
Cadmium, atomized (98% Cd)	—	—	—	7	48	100	690	570	1,058	250	482	4,000	—	—	C10
Iron, carbonyl (99% Fe)	1.6	3.0	0.5	43	296	2,400	16,548	320	608	310	590	0.020	0.105	105	C10
Lead, atomized (99% Pb)	—	—	—	—	—	—	—	710	1,310	270	518	—	—	—	—
Magnesium, milled, Grade B	>10	3.0	7.4	116	800	15,000	103,425	560	1,040	430	806	0.040	0.030	30	C10
Manganese	0.1	0.1	0.7	53	365	4,900	33,786	460	860	240	464	0.305	0.125	125	—
Tantalum	0.1	0.1	0.7	55	379	4,400	30,338	630	1,166	300	572	0.120	<0.200	<200	—
Thermoplastic Resins and Molding Compounds															
Cellulose acetate	>10	8.0	1.6	85	586	3,600	24,822	420	788	—	—	0.015	0.040	40	C14
Methyl methacrylate polymer	6.3	7.0	0.9	84	579	2,000	13,790	480	896	—	—	0.020	0.030	30	C11
Polyethylene, low-pressure process	>10.0	22.4	2.3	80	552	7,500	51,713	450	842	—	—	0.010	0.020	20	—
Polystyrene molding compound	>10.0	6.0	2.0	77	531	5,000	34,475	560	1,040	—	—	0.040	0.015	15	C14
Thermosetting Resins and Molding Compounds															
Phenolformaldehyde	>10.0	9.3	1.4	77	531	3,500	24,133	580	1,076	—	—	0.015	0.025	25	C17
Urea formaldehyde molding compound, Grade II, fine	1.0	0.6	1.7	89	614	3,600	24,822	460	860	—	—	0.080	0.085	85	C17
Special Resins and Molding Compounds															
Lignin, hydrolized-wood-type, fines	>10.0	5.6	2.7	102	703	5,000	34,475	450	842	—	—	0.020	0.040	40	C17
Rubber, synthetic, hard, contains 33% sulfur	>10.0	7.0	1.5	93	641	3,100	21,375	320	608	—	—	0.030	0.030	30	C15
Shellac	>10.0	25.2	1.4	73	503	3,600	24,822	400	752	—	—	0.010	0.020	20	C14

*Numbers in this column indicate oxygen percentage while the prefix indicates the diluent gas. For example, the entry "C17" means dilution to an oxygen content of 17 percent with carbon as the diluent gas.

†0.1 designates materials presenting primarily a fire hazard, as ignition of the dust cloud is not obtained by the spark or flame source but only by the intense heated surface source.

‡Guncotton ignition source.

§No ignition.

TABLE 3-27C.   *Correlation Between Descriptive Categories for Dust Explosions and U.S. Bureau of Mines Indices**

Type of Explosion	Ignition Sensitivity	Explosion Severity	Explosibility Index
Weak	<0.2	<0.5	<0.1
Moderate	0.2–1.0	0.5–1.0	0.1–1.0
Strong	1.0–5.0	1.0–2.0	1.0–10
Severe	>5.0	>2.0	>10

*From: Jacobson et al., U.S. Bureau of Mines, RI5753, "The Explosibility of Agricultural Dusts."

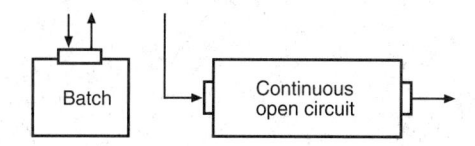

FIG. 3-27A.   *Batch and continuous grinding systems.*

not always possible to maintain absolutely tight systems. Vacuum cleaning is the best way to remove any dust which escapes into the room, as it prevents the dust from forming explosive clouds.

For grinding operations where ignition sources are difficult to control, the equipment may be protected by introducing a continuous flow of inert gas, such as carbon dioxide, nitrogen, or flue gas. This will keep the normal oxygen content within the equipment sufficiently low to prevent an explosion. See section "Protection Against Fires or Explosions" later in this chapter.

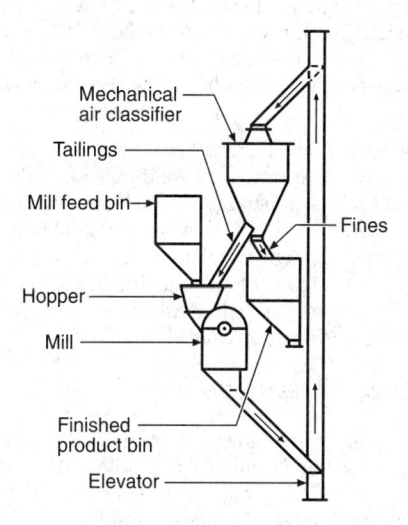

FIG. 3-27B.   *A hammer mill in a closed circuit with air classifier.*

CHARACTERISTICS OF DUST EXPLOSIONS

Dust explosions in some respects are similar to vapor and gas explosions, but they do differ in some important ways. Like a gas, dust must be mixed with air or another supporter of combustion, and a source of ignition is generally required to cause an explosion. Rarely have dust explosions resulted from spontaneous oxidation and heating. Reaction rates and rates of pressure rise are usually higher in vapor and gas explosions than in dust explosions. However, complete combustion of dust in a given volume of air will frequently de-

velop energy and pressure greater than that developed by the combustion of a gas. Dust explosions, then, are sometimes more disastrous than gas explosions. This, in part, is due to their slower rate of development and longer duration. The slower rate of development results from the fact that the combustion of dust is a surface reaction, and the diffusion of oxygen toward the reacting surface is necessarily slower and less complete than it is in a flammable gas.

Requisite Conditions for Dust Explosions

In order for a dust explosion to occur, four conditions must be satisfied simultaneously:

1. A combustible solid in a finely divided state must be dispersed in an oxidizing medium—usually oxygen in air.
2. The concentration of the dust in air must be within the explosible range.
3. An external source of ignition of sufficient energy and duration to initiate the explosive chain reaction for that particular dust must be present.
4. The combustion must occur in a confined volume.

The rapid chemical reaction, or flash fire, characteristic of explosions will occur if only the first three conditions are satisfied. However, the rapid buildup of excessive pressures, inherent in the working definition of dust explosions, will result only when the reaction occurs in an enclosed space.

Characterization of Explosion Hazards of Dusts

Parameters employed to describe the relative explosion hazards of various dusts include:

1. Lower and upper limit of dust concentrations within which an explosion is possible.
2. Minimum ignition energy, i.e., minimum electric spark energy required for ignition of the dust cloud.
3. Minimum ignition temperature as measured by a furnace apparatus.
4. Maximum oxygen concentration permissible to prevent ignition.
5. Maximum explosion pressure attained during the course of the explosion.
6. Maximum rate of pressure rise (sometimes also the average rate of pressure rise).

Values for these parameters are not fixed but depend on various factors—namely, particle size and shape, ambient temperature and pressure, moisture content of the dust, degree of turbulence in the suspension, and size of the ignition source. (See Table 3-27B.) Experimental work has made possible some qualitative observations of the effects of these factors, but quantitative relations are not available.

Definitions of dust in terms of particle size vary, but normally dust is defined as particles with diameters of 0.1 to 1,000 microns (1 micron equals 10^{-6} m).

The minimum explosible concentration and the minimum ignition energy generally tend to decrease with a decrease in average particle size (i.e., the explosion hazard increases with a decrease in particle size). For average particle sizes less than 50 microns, the effect is much less pronounced. According to Palmer,[2] uniform dispersion of the dust in laboratory test equipment may become more difficult as particles become very small. Due to the greater cohesiveness of very fine particles, some may exist as agglomerates rather than as individual particles, leading to an apparent reduction in explosibility. A different method of dispersion or a more vigorous ignition source may break up the agglomerations, resulting in an increased explosion hazard. Tests performed in large-scale coal mines are said to confirm this.

Little experimental evidence is available concerning the effect of particle shape on explosibility. Particles may resemble fibers, needles, or flakes, as well as spheres. Studies employing atomized spherical particles and flaked aluminum particles indicate that a significant difference in explosibility can occur with changes in particle shape. Size and shape, which are generally measured before ignition, may be altered during preignition stages, and subsequently affect propagation of the explosion. Changes may occur as a result of melting, vaporization, expansion to form hollow spheres, and fragmentation.

The effect of ambient temperature on explosibility would be particularly applicable to industrial situations, such as dryers, but no information is available on measurements in plant-scale units. Theoretically, it would be expected that the minimum ignition energy would decrease as the ambient temperature increased, other factors remaining constant. If the final temperature reached by the combustion products is limited by the molecular dissociation of product gases, then the maximum explosion pressure would be expected to decrease as the ambient temperature increases, and the rate of pressure rise would increase as the ambient temperature increased. Laboratory tests employing coal dust revealed that the minimum ignition energy decreased with increasing ambient temperature as expected, but only over a limited temperature range; thereafter, it increased. Palmer[2] also notes that the minimum ignition energy would depend on previous exposures to high temperatures.

If the pressure in the reaction vessel is atmospheric when the explosion is initiated, rises in pressure during the course of the explosion are not considered to be changes in ambient pressure. Laboratory tests employing methane/air mixtures resulted in proportional increases in maximum explosion pressures with increased initial pressure, but smaller increases were obtained for coal dust/air mixtures. The maximum explosible concentration decreased with increasing ambient pressure, while the minimum explosible concentration remained relatively unchanged. Explanations for these effects have not yet been explored.

An increase in moisture content of dust tends to increase the minimum explosible concentration, the minimum ignition energy, and the minimum ignition temperature when measured in small-scale apparatus. These effects are apparently due to the absorption of heat in vaporizing water and to the decrease in dispersibility of the dust. Nineteen percent water content (on a weight basis) was found to prevent ignition of starch dust.

Tests with coal dust/air suspensions have indicated that explosions tend to increase in severity (i.e., maximum explosion pressure and/or maximum rate of pressure rise increases) with increases in the size of the ignition source. The nature of the ignition source also affects the explosibility of dusts. Organic dusts tend to be ignited more readily by heated coils, but metal dusts react more readily to spark ignition.

Experiments indicate that some dusts produce stronger explosions than others. Metallic dusts, such as stamped aluminum powder, milled and stamped magnesium, and atomized aluminum, produce the most violent dust explosions. Phenolformaldehyde resin, cornstarch, soybean protein, wood flour, and coal dust, respectively, followed the metallic dusts in explosive intensity.

The character and severity of any dust explosion will be affected by several factors. One of these is particle size. For any given material, the finer the particle size, the more violent the explosion. Less energy will be required to ignite the dust, and it will remain in suspension for a longer time, increasing the total force exerted. Figure 3-27C shows the relationship of particle size to the explosibility index.

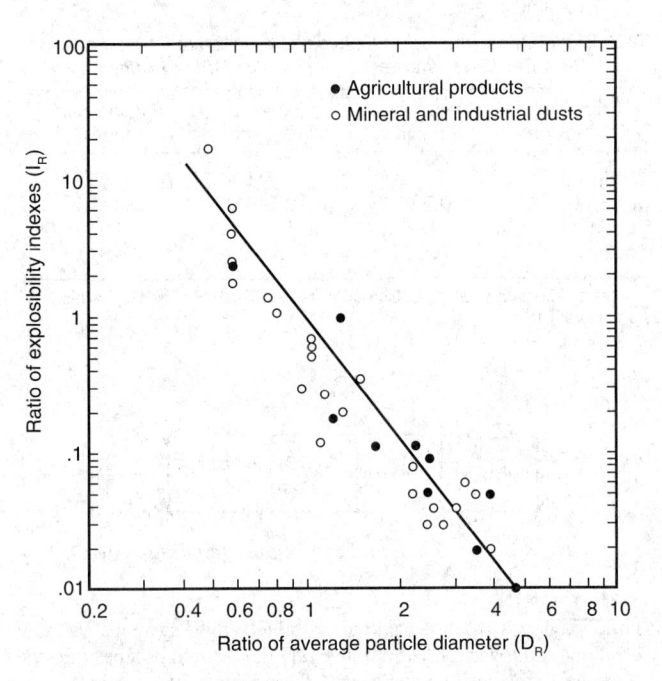

FIG. 3-27C. Effect of particle diameter on relative explosibility index. Relative average particle diameter is the ratio of the mean particle size of the dust to the mean size of the through No. 220 sieve sample. Relative explosibility index is the ratio of the index computed for the dust to that computed for a through No. 200 sieve sample. (Source: U.S. Bureau of Mines RI 5753)

Turbulence is another factor that contributes to the severity of a dust explosion. Turbulence speeds up the diffusion of oxygen to the reacting surfaces and promotes stronger explosions. The smaller the particle size and the greater the turbulence, the more the dust resembles a gas or vapor in its explosive characteristics.

Relatively little investigation has been undertaken with regard to the effects of turbulence on explosibility. Tests with coal dust/air suspensions revealed an increase in maximum explosion pressure and maximum rate of pressure rise with increased turbulence. Both parameters passed through maxima and then decreased, although more slowly in the case of the maximum explosion pressure. Sources of turbulence can include the presence of obstacles and rapid volume expansion due to flame.

Particle Size Classifiers

Classifiers separate out the fine product and return the coarse material (circulating load) to the mill for regrinding, along with new material being fed to the mill. If the fines are continuously removed, a mill performs much more efficiently. Closed circuit grinding with size classifiers provides even more uniform size distribution and is also more economical.

Wet classifiers are generally used for large-scale wet milling operations, as in cement and ore processing plants. The simplest type of wet classifier is a settling basin arranged so that the fines do not have time to settle out and are drawn off, while the coarse material is raked to a central discharge. Wet classifiers of this type do not present an explosion hazard.

Dry classifiers may be installed external to the mill in a closed circuit (see Figure 3-27B), or they may be internal as an integral part of the mill (see Figure 3-27F).

Air classifiers are used for most dry milling operations. There are a number of different types, but all these classifiers are based

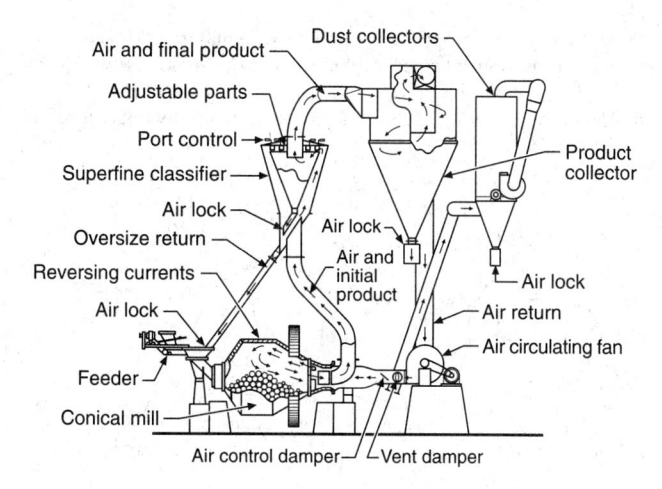

FIG. 3-27D. *A Hardinge conical (tumbler type) mill with reversed current air classifier.*

upon the principles of air drag and particle inertia. One type directs an airstream across a stream of the particles to be classified. Another has adjustable flow baffles, and still another changes the direction of air flow. (See Figure 3-27D.) The double-cone classifier uses centrifugal action, induced by flow through vanes, which cause coarse particles to move outward and down the wall of the inner cone and thus return to the grinding zone, while the upward-moving air stream entrains the fines. (See Figure 3-27E.)

Rotating blades are the main elements of several types of classifiers. The centrifugal motion established by the rotating blades tends to throw the coarser particles outward and returns them to the grinding zone, while the fines are carried off in the airstream. (See Figure 3-27F.)

GRINDING HAZARDS

The hazards of grinding operations lie in the fact that the process produces very fine particles of readily oxidized materials that may be mixed with process or environmental air in flammable or explosive concentrations. To separate the properly sized material from the coarser particles being fed into the mill, a system of classifiers is used. For the most part, classifiers are based on the force of gravity and the principles of air drag and particle inertia. A continuous flow of air passes through the mill, regulated so that particles of the desired fineness are carried through to a collecting bin or compartment, while coarser particles fall out and return to the mill for further grinding. This principle is illustrated in Figure 3-27B.

There are several different types of air classifiers, some of which may be external to the mill as in Figure 3-27B. Others may be internal as in the ring-roller mill illustrated in Figure 3-27F. Here the whizzers are rotating blades whose centrifugal action throws the coarser particles outward, permitting them to drop back onto the grinding surfaces. The finished product is carried through an outlet by the airstream.

Theoretically, the process airstream should be fully and tightly enclosed so that it does not carry the finished product or unwanted dust into the atmosphere surrounding the mill. In practice, this is rather difficult to accomplish, and dust builds up in the structure

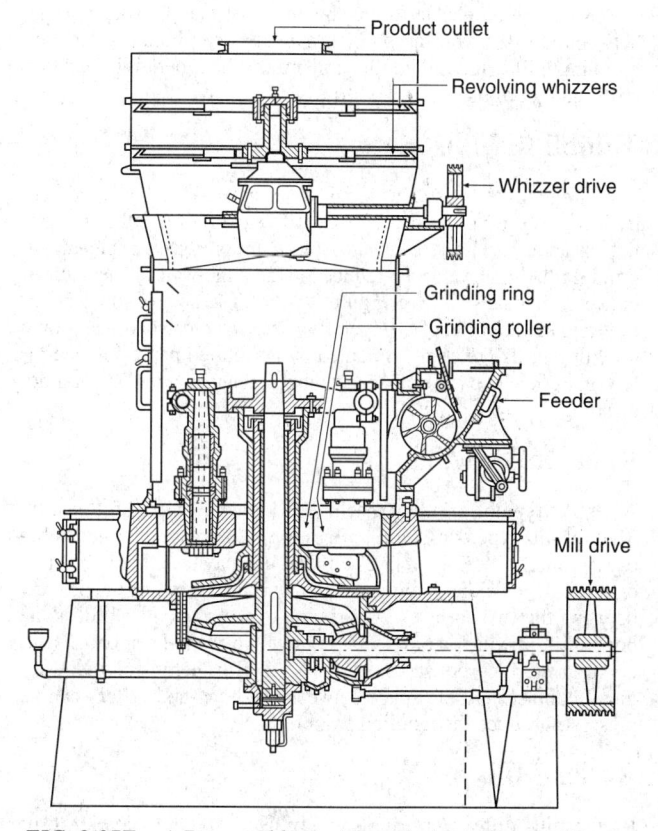

FIG. 3-27E. *A bowl mill (variation of a ring roller mill). (Source: C. E. Raymond, Combustion Engineering, Inc.)*

FIG. 3-27F. *A Raymond high-side mill with internal whizzer classifier.*

housing the mill. There are, then, two distinct hazards. One is that an explosion may occur within the milling system, and the other is that an explosion may occur in the structure. All that is required is a critical concentration of the material in air and an ignition source.

The lower limit of flammability will vary from one material to another. For some, such as phthalic anhydride, shellac, aluminum stearate, and phenothiazine, it may be as low as 0.015 oz/cu ft (15 g/m^3). For zinc, it may be as much as 0.5 oz/cu ft (500 g/m^3). Upper flammability limits for dusts, i.e., the concentration above which an explosion will not occur, are poorly defined, not usually reproducible, and not yet determined for many dusts. Therefore, any concentration above the lower limit should be considered potentially explosive. The minimum electric spark energy required for ignition of a dust cloud can vary from as little as 10 mJ (millijoule) to as much as 4000 J (joules).

Ignition sources, too, can vary widely. Since a bit of metal accidentally entering the mill may cause a spark as it strikes against one of the grinding surfaces, mills should be equipped with magnetic or mechanical separators to remove any tramp metal from the feed system. Sparking may also be caused by tools made of ferrous metals. Other possible ignition sources are static electricity, hot surfaces, friction, open flames, welding arcs, and personnel smoking in the area.

GRINDING EQUIPMENT

Equipment for reducing particle size of a given material can be classified into four groups, according to the method of grinding used: (1) a material can be reduced between two solid surfaces; (2) it can be reduced by impact on one solid surface; (3) it can be reduced by action of the surrounding medium; or (4) it can be reduced by nonmechanical means, such as thermal shock, explosive shattering, or electrohydraulic processes.

Most grinding is performed by machines that pass the material between two solid surfaces. Such machines can be classified into tumbling mills, ring roller mills, roller mills, hammer mills, attrition mills, and jet mills.

Tumbling Mills

Tumbling mills consist of a cylindrical or conical shell charged with balls of steel, flint, or porcelain, or with steel rods. As the shell revolves about its horizontal axis, the balls or rods tumble about, grinding the material to be reduced against the wall of the shell or between themselves. (See Figure 3-27D.) The size of the balls or rods and the duration of the operation determine particle size. Some tumbling mills are compartmented by perforated partitions that allow material to pass from one compartment to another for finer grinding.

Ring Roller Mills

A second type of grinding machine is known as a ring roller mill. Mills of this type consist of a grinding ring or plate that moves between rollers. (See Figure 3-27F.) The ring may be either horizontal or vertical, and either it or the roller may rotate, grinding the product between the two surfaces. A variation of the ring roller mill is the bowl mill, in which a bowl and ring revolve around stationary rollers to grind the product. (See Figure 3-27E.) There is no metal-to-metal contact, and the space between the bowl and rollers can be preset to produce the required particle size.

Roller Mills

Roller mills differ from ring roller mills in that the material to be ground passes between two or more rollers revolving in opposite directions at different speeds. (See Figure 3-27G.) A scraper blade at the discharge end removes the finely ground material. Most dry materials are ground between rollers having corrugations that determine the final particle size. Corrugations may be either sharp or dull, and they may be used in various combinations to achieve the desired results.

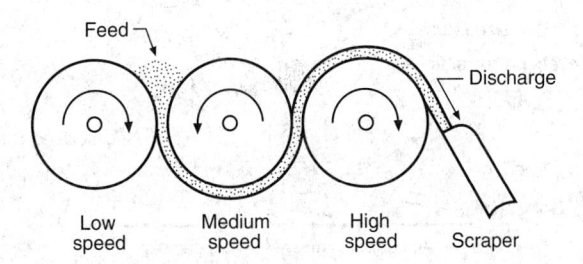

FIG. 3-27G. *A roller mill for paint grinding.*

Hammer Mills

Hammer mills have hammers, or beaters, attached to a rotating shaft. (See Figure 3-27H.) The hammers may be of almost any shape and hinged or fixed to the shaft. The fineness of the finished product is determined by the speed of the rotating shaft, the clearance between the hammers and grinding plates, the number and size of the hammers, the feed rate, and the size of the discharge openings. Two variations of the hammer mill are: (1) the disintegrator and (2) the pin mill.

The disintegrator has a vertical rotating shaft with hammers that run at close tolerances to a cylindrical screen. (See Figure 3-27I.) Material is fed parallel to the rotating shaft, and the centrifugal action of the hammers discharges the ground material into a primary chute.

The pin mill is a high-speed mill with two disks in which pins are set in alternating circular rows. Either one or both of the disks may rotate. If both rotate, they do so in opposite directions. The material to be ground is broken up between the pins. (See Figure 3-27J.)

Attrition Mills

Attrition mills use metallic or abrasive grinding plates that rotate at high speed on either a horizontal or vertical plane. One disk may be stationary, or the two may rotate in opposite directions. The material enters at the axis and is discharged at the periphery of the grinding

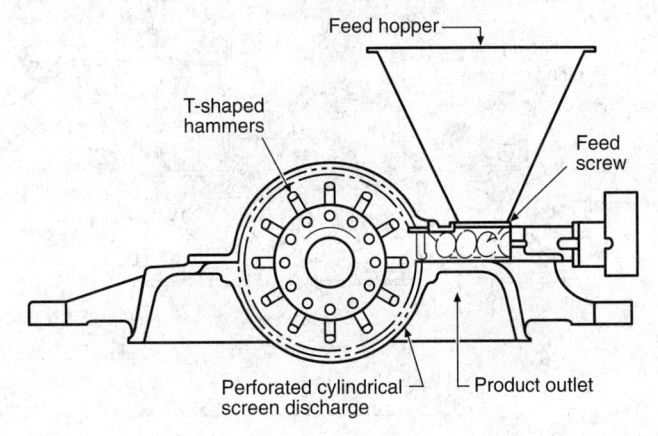

FIG. 3-27H. *A Milro-Pulverizer hammer mill. (Source: Pulverizing Machinery Co.)*

plates. One type of attrition mill, known as the Buhrstone mill, uses two circular stones instead of metal or abrasive disks between which the material is ground.

Jet Mills

Another type of mill, the jet mill, differs from the others in that the material is not ground against a hard surface. Instead, a gaseous medium is introduced. The gas may convey the feed material at high velocity in opposing streams, or it may move the material around the periphery of the grinding and classifying chamber. The high turbulence causes the particles of feed material to collide and grind upon themselves.

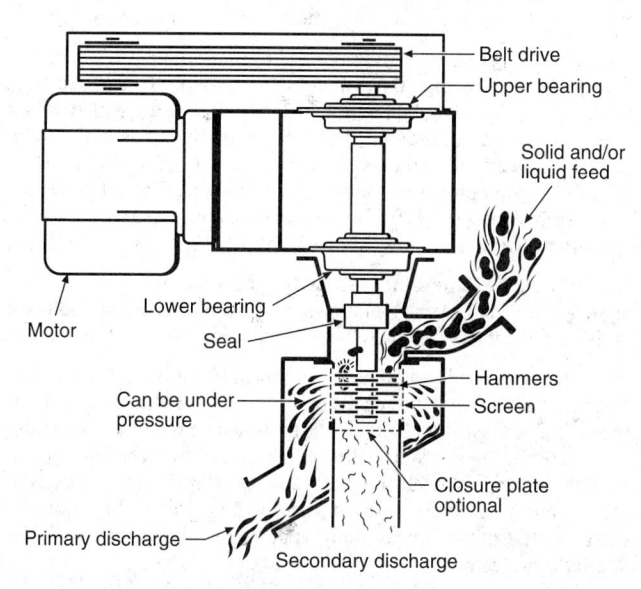

FIG. 3-27I.   A Reitz disintegrator (variation of a hammer mill). (Source: Reitz Manufacturing Co.)

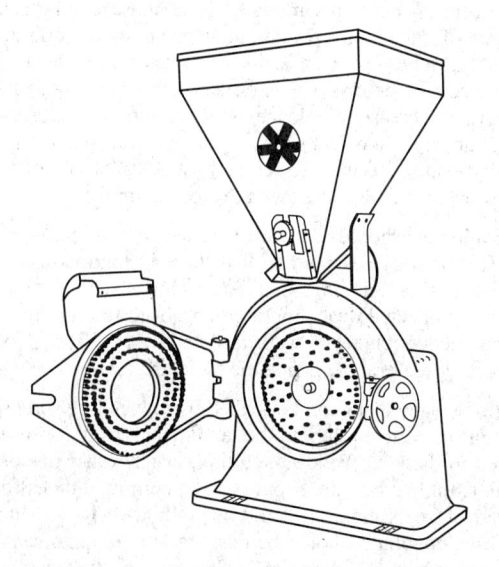

FIG. 3-27J.   An Alpine-Kolloplex pin mill (variation of a hammer mill). (Source: Alpine-American Corp.)

APPLICATION OF GRINDING EQUIPMENT

Agricultural Products

The roller mill is the traditional machine for grinding wheat and rye into high-grade flour. Generally, rollers with dull corrugation are used. For very tough wheat, however, a sharp roller is used against a sharp roller, and for other grades, combinations of dull and sharp rollers are used. Rollers with sharp corrugations are used for grinding corn and feed. High-speed hammer mills or pin mills are used to produce flour with controlled protein content. Disk attrition mills are also used for grinding wheat.

After the oil has been extracted, soybeans or soybean cake is ground in attrition mills or flour rollers, depending upon whether the product is to be a feed meal or flour. In some cases, a hammer mill may be used as a preliminary disintegrator for pressed cakes, including linseed and cottonseed cake.

Where only medium fineness is required, a hammer mill is used to produce starch, potato flour, tapioca, and similar flours. For finer flour products, a high-speed impact mill, such as a pin mill, is used.

Carbon Products

Bituminous coal and pitch are used as fuel for industrial furnaces, boilers, and rotary kilns. Pulverized coal is either blown directly into the furnace as it is pulverized, or pulverized in a central grinding system and stored in a bin until it is used. Ball-, tube-, ring-roller-, bowl-, and ball-and-ring-type mills are used for direct firing of large installations. Ring-roller mills are also used to pulverize coal for bin systems.

Anthracite coal is harder to reduce than bituminous coal. Ball or hammer mills are used to pulverize anthracite coal for foundry facing mixtures. Calcined anthracite, used in the manufacture of electrodes, is generally pulverized in ball-and-tube mills, or ring-roller mills.

The grinding characteristics of coke vary from petroleum coke, which is relatively easy to grind, to certain foundry and retort coke, which is difficult to grind. Where uniform size of particles with a minimum of fines is required, rod or ball mills are used in a closed circuit with screens.

Natural graphite is classified in three grades: (1) flake, (2) crystalline, and (3) amorphous. Flake is the most difficult to grind to a fine powder, and crystalline is the most abrasive. Ball, tube, ring-roller, and jet mills with or without air classification are used for grinding graphite. For handling large capacities, ball and tube mills are used, especially for the flake and crystalline grades. Graphite for pencils is ground in a jet pulverizer. Ball mills in a closed circuit with air classifiers have been used for grinding artificial graphite. Charcoal and Gilsonite® are ground in hammer mills with air classifiers.

Chemicals

Hammer mills are generally used to pulverize dry colors and dyestuffs, with pebble mills used for small lots. Hammer or jet mills with air classifiers for size limitation are used for dyes that are coarsely crystalline. A ring-roller mill is used for fine grinding of sulfur, with inert gas injected into the mill.

Pulverizing of metallic soaps, such as stearates, requires certain provisions to keep the material cool and in rapid motion. Since these materials tend to cake, batch grinding is not practicable.

Stearates are pulverized in multi-cage mills, screen mills, and hammer mills with air classification.

Organic Polymers

The grinding characteristics of various resins, gums, waxes, hard rubbers, and molding powders are such that when a finely ground product is required, it may be necessary to use a water-jacketed mill or a pulverizer with an air classifier in which cooled air is introduced into the system. Hammer mills are generally used for this purpose. Some resins with low softening temperatures can be ground by mixing dry ice with the material before grinding or by introducing refrigerated air into the mill.

Most gums and resins used in the paint, varnish, or plastics industries do not require very fine grinding, so hammer mills or roll crushers will produce a satisfactory product. Some resins used in the phenolic resin industries that require very fine pulverization are ground in a pebble mill and cooled with water or brine in a closed circuit with an air classifier. A ring-roller mill with an internal air classifier is used to pulverize phenolformaldehyde resins.

Hard rubber is ground on heavy steam-heated rollers. The materials pass through a series of rollers in a closed circuit with screens and air classifiers. Usually the rollers are of different sizes, and the machines operate at relatively low speeds so as not to generate too much heat.

Molding powders are produced with hammer or attrition mills in closed circuits equipped with either screens or air classifiers.

Cryogenic Grinding

Although cooling is required for some mills because of the material being ground, cryogenic grinding can be applied in any mill to produce a smaller-sized particle than could be obtained otherwise. Existing mills can be converted by adding a liquid nitrogen storage tank, a piping system, and a properly designed hopper. In this way, the material can be cooled before it is ground.

The material can be precooled in the hopper by immersing it in a nitrogen bath, it can be cooled in the grinding chamber by spraying it with liquid nitrogen, or both methods can be used simultaneously. Gaseous nitrogen has been used for years to provide an inert atmosphere in mills, particularly in jet mills. Cryogenic grinding can be designed to provide a protective inert atmosphere within the mill. In spice grinding, freezing the spice before grinding gives it a superior appearance and it retains the aroma and flavor usually lost when it is not precooled.

Cryogenic grinding has been applied to the production of powdered coatings used for insulation and protective coatings, and for powders used in the manufacture of bearings, with improved wear-life and increased lubrication properties. Other areas where cryogenic grinding may have application are in recycling scrap materials, grinding existing materials for newer applications and processes (rotational molding, spray powder for textile stiffener), and grinding protein concentrations.

PROTECTION AGAINST FIRES OR EXPLOSIONS

Combustion and explosion, or more properly deflagration, relative to dusts, are practically identical processes. The difference lies in the speed with which the oxidation reaction takes place. NFPA 68, *Guide for Venting of Deflagrations*, defines deflagration as "burning which takes place at a flame speed below the velocity of sound in the unburned medium." It defines explosion as "the bursting of a building or container as a result of development of internal pressure beyond the confinement capacity of the building or container." Inasmuch as this is usually the result of a deflagration, the term explosion, as generally used herein, refers to the entire process.

Because an explosion takes place almost instantaneously, it cannot be brought under control by containment as can many fires. The effects, however, can be minimized by certain protective measures; these are prevention, venting, inerting, and suppression.

Prevention

The simplest means of preventing an explosion is to keep a critical concentration of dust from developing and to eliminate any potential sources of ignition.

In grinding operations, there are two types of locations where dust accumulation may become critical. One is within the milling or grinding equipment itself. The other is in the surrounding environment, i.e., the structure in which the equipment is housed.

The design and construction of the structure may vary considerably, depending upon the product being ground. It is impractical to construct a building that will withstand the pressure generated by a dust explosion. An ordinary 12-in. (0.3-m) thick brick wall can be destroyed by an internal pressure of less than 1 psi (6.9 kPa). Most dust explosions produce much higher pressures. Typical pressures range from 13 psi (90 kPa) for zinc to 89 psi (614 kPa) for stamped aluminum, with most being above 30 psi (207 kPa).

If a dust explosion hazard exists, the structure should be designed with panels, vents, windows, or other closures that will open at the lowest practical pressures and minimize structural damage.

Materials used should be noncombustible or fire resistive. Any interior walls intended to serve as fire walls should be capable of providing at least three hours of fire resistance under standard fire test methods. Interior stairs, lifts, or elevators should be enclosed in shafts of noncombustible materials and have fire-resistive ratings of at least one hour. Such enclosures should be protected by automatically closing fire doors, and any openings in fire walls should be similarly protected.

The number of horizontal surfaces that might collect dust and are difficult to reach or are inaccessible for cleaning should be minimized. Wherever practical, such surfaces should be built up to an angle of at least 60 degrees from the horizontal so that dust will tend to slide off rather than accumulate. Good housekeeping is a necessary part of explosion prevention. It means both equipment and structure must be kept clean so dust cannot accumulate. Dust should be removed by vacuum systems. Brushing or sweeping will disperse the dust and increase turbulence, which in turn increases the possibility of an explosion occurring and increases its potential severity. In grain elevators, recently developed grain dust probes may be used to alarm when dust levels exceed explosive limits.[1]

Equipment should be made of metal and be as dusttight as possible. It should be designed so that there is continuous suction at openings during grinding, dumping, transfer, and similar operations. The collected dust should be conveyed through tightly constructed ducts or chutes to well-designed dust collectors located in a safe place, preferably outside the structure.

Since many dusts can be ignited by low-energy sparks, potential ignition sources should be eliminated from or adequately shielded in the area. Welding, cutting, or any other operation that uses an open flame or arc should not be permitted unless the work area is dust free. Smoking should be prohibited. Torque-limiting or fluid-drive couplings should be designed to dissipate heat readily. Moving equipment, elevators, belts, and conveyors should be grounded or of nonconductive material to eliminate the possibility of static sparks. All electrical wiring should conform to the requirements of NFPA 70, *National Electrical Code®*, for hazardous locations containing combustible dusts.

Extinguishing

Adequate fire-extinguishing equipment, both fixed and portable, should be provided. Hose nozzles should be of the spray type, since solid streams can cause turbulence and dispersion of dust into the air and increase the explosion hazard. Automatic sprinklers should be provided to protect the building(s).

Venting

Vents are openings in the equipment or structure that allow heated explosion gases to escape more readily. Venting does not prevent explosions, but it does serve to limit the maximum pressure resulting from a deflagration. The most effective vents are free and unrestricted openings; however, these are not always practical.

Vents should be closed in such a manner that they will open under the lowest practical pressure. Typical venting arrangements include rupture diaphragms, hinged or blowout windows and panels, and weakly constructed walls and roofs.

The venting area required, as calculated from empirical formulas, depends upon the expected pressure and its rate of rise, the type of closure, and the volume and strength of the enclosure. Determination of vent area for an enclosure is mostly empirical. Vents should be located where minimum damage will result from the shattering or blowing out of the vent closure.

Inerting

One method of preventing a dust explosion is inerting, i.e., the replacement of oxygen in the grinding process with an inert gas, such as nitrogen or carbon dioxide. Such equipment as grinders, pulverizers, mixers, dryers, conveyors, dust collectors, and filling machines can be protected by this method.

The amount of oxygen that must be replaced to provide a safe concentration depends upon the type of dust, particle size, concentration, turbulence, diluent gas, and intensity of the ignition source. To prevent ignition of carbonaceous dusts by a spark, for example, oxygen should be reduced to 8 percent by nitrogen or to 11 percent by carbon dioxide. However, if a stronger ignition source is present, the oxygen should be reduced to 3 and 4 percent, respectively.

Many factors will affect the use of inerting to prevent explosion. Among them are protection of personnel, the hazard to be protected, the required reduction in oxygen concentration, the availability and cost of the inert gas supply, and the necessary control equipment.

Suppression

Explosion suppression is a technique of stopping an explosion before it develops destructive pressures. The factors involved here are much the same as those used in the extinguishment of fire, namely cooling, limiting the supply of oxygen, and inhibiting flame spread.

Despite the rapidity with which combustion proceeds in an explosion, there is a short period of time before which the destructive force is evolved. During this time, the initial pressure rise can be detected by suitable sensing devices, which automatically trigger the release of the suppressing agent, normally an inert gas or liquid, that inhibits the combustion process.

Explosion suppression systems can be used in confined spaces, such as reactors, mixers, pulverizers, mills, dryers, storage bins, ovens, bucket elevator transport systems, and pneumatic transport systems. The effective application of such systems requires careful

consideration of many factors. Among them are characteristics of the dust, rate of pressure rise, ignition sources, and characteristics of the suppressant. Though the principles of explosion suppression apply to all installations, each suppression system must be individually designed to cover the wide range of variables that will be encountered.

BIBLIOGRAPHY

References Cited

1. Cashdollar, K. L., and Hertzberg, M., eds., "Industrial Dust Explosions," ASTM Special Technical Publication 958, 1986, American Society for Testing and Materials, W. Conshohocken, PA.
2. Palmer, K. N., *Dust Explosions and Fires*, Chapman and Hill, London, 1973.

References

Johnson, *et al.*, "Explosibility of Agricultural Dusts," U.S. Bureau of Mines RI 5753, 1961, U.S. Bureau of Mines, Washington, DC.

NFPA Codes, Standards, and Recommended Practices

Reference to the following NFPA codes, standards, and recommended practices will provide further information on grinding processes discussed in this chapter. (See the latest edition of *The NFPA Catalog* for availability of current editions of the following documents.)

NFPA 61, *Standard for the Prevention of Fires and Dust Explosions in Agricultural and Food Products Facilities*
NFPA 68, *Guide for Venting of Deflagrations*
NFPA 69, *Standard on Explosion Prevention Systems*
NFPA 70, *National Electrical Code®*
NFPA 651, *Standard for the Manufacture of Aluminum Powder*
NFPA 654, *Standard for the Prevention of Fire and Dust Explosions in the Chemical, Dye, Pharmaceutical, and Plastics Industries*
NFPA 655, *Standard for Prevention of Sulfur Fires and Explosions*
NFPA 664, *Standard for the Prevention of Fires and Explosions in Wood Processing and Woodworking Facilities*

Additional Reading

Bartknecht, W., "Ignition Capabilities of Hot Surfaces and Mechanically Generated Sparks in Flammable Gas and Dust/Air Mixtures," *Plant/Operation Progress*, Vol. 7, No. 2, 1988, pp. 114–121.
Bluhm, D. D., "Grinding and Milling Operations," *Industrial Fire Hazards Handbook*, 3rd ed, National Fire Protection Association, Quincy, MA, 1990.
Bonney, M. J., "Suppressing Dust Explosions," *Fire*, Vol. 88, No. 1086, Dec. 1995, p. 40.
Britton, L. G., and Kuby, D. C., "Analysis of a Dust Deflagration," *Plant/Operation Progress*, Vol. 8, No. 3, July 1989, pp. 177–180.
Bruderer, R. E., "Ignition Properties of Mechanical Sparks and Hot Surfaces in Dust/Air Mixtures," *Plant/Operation Progress*, Vol. 8, No. 3, July 1989, pp. 152–164.
Davenport, J. A., "Explosion Losses in Industry," *Fire Journal*, Vol. 75, No. 1, pp. 52–56, 71–73.
Friedman, R., *Principles of Fire Protection Chemistry*, 2nd ed., National Fire Protection Association, Quincy, MA, 1989.
Goyer, D., "Grain Dust Explosion Averted Dockyard Blaze," *Fire Fighting in Canada*, Vol. 39, No. 1, Feb. 1995, pp. 4–6.
Gray, T. A., "Firefighting Hazards at Grain Facilities," *Fire Engineering*, Vol. 147, No. 11, Nov. 1994, pp. 56–57.
Halpin, T., Jr. and Shattuck, D., "Fires in Agricultural Silos," *Fire Engineering*, Vol. 147, No. 3, Mar. 1994, pp. 12–13, 16.
Hertzberg, M., *et. al.*, "Explosives Dust Cloud Combustion," Twenty-fourth Symposium (International) on Combustion, Combustion Institute, Pittsburgh, PA, 1992, pp. 1837–1843.
Hertzberg, M., Zlochower, I. A. and Cashdollar, K. L., "Explosibility of Metal Dusts," Short Communication, *Combustion Science and Technology*, Vol. 75, No. 1–3, 1991, pp. 161–165.
"Historic Blazes: Blaze in Steam-Powered Mill Curbs Industrial Revolution," *Fire Prevention*, No. 260, June 1993, pp. 31–32.

Hwang, C. C., and Litton, C. D., "Ignition of Combustible Dust Layers on a Hot Surface," *Proceedings* of the Third International Symposium of Fire Safety Science, Elsevier Applied Science, New York, 1991, pp. 187–196.

Kauppinen, E. I., "Burning Question at the Pulp Mill," *Industrial Horizons*, Vol. 2, 1995, pp. 15–16.

Maddison, N., "Dangers in the Dust," *Fire Prevention*, No. 264, Nov. 1993, pp. 22–25.

Mangs, J., "Prosessiteollisuuden polyrajahdyksien paineenalentamisaukkojen mitoitus II. [Dimensioning Pressure Outlets to Avert Dust Explosions in the Processing Industry.]," *Palontorjuntatekniika*, Mar. 1992, pp. 30–31.

Mintz, K. J., "Ignition Temperatures of Dust Layers: Flaming and Non-Flaming," *Fire and Materials*, Vol. 15, No. 2, Apr./June 1991, pp. 93–96.

Mintz, K. J., "Upper Explosive Limit of Dusts: Experimental Evidence for Its Existence Under Certain Circumstances," *Combustion and Flame*, Vol. 94, No. 1/2, 1993, pp. 125–130.

Mittal, M., and Guha, B. K., "Study of Ignition Temperature of a Polyethylene Dust Cloud," *Fire and Materials*, Vol. 20, No. 2, Mar./Apr. 1996, pp. 97–105.

Moore, P. E., "Industrial Dust Explosions," K. L. Cashdollar and M. Hertzberg eds, *ASTM Special Publication 958*, American Society for Testing and Materials, W. Conshohocken, PA, 1987.

Palmer, K. N., "Dust Explosions: Initiation, Characteristics, and Protection," *Chemical Engineering Progress*, Vol. 86, No. 3, Mar. 1990, pp. 33–36.

"Prevention, Detection, and Control of Coal Pulverizer Fires and Explosions," Electric Power Research Institute, Report EPRI CS No. 5069, 1987, Palo Alto, CA, p. 280.

"Recipe for a Dust Explosion," *Record*, Vol. 72, No. 4, Fourth Quarter 1995, pp. 3–12.

Steer, P., "Feed Mill Destroyed," *Fire Fighting in Canada*, Vol. 37, No. 6, Aug. 1993, pp. 42–43.

Winward, J. and Campanaro, P., "Fire in Mill Construction: Breeding Ground for Conflagration," *Fire Engineering*, Vol. 145, No. 1, Jan. 1992, pp. 50–52, 54, 56, 58, 60.

Zalosh, R. G., "Review of Coal Pulverizer Fire and Explosion Incidents," *Symposium on Industrial Dust Explosions*, STP 958, 1987, American Society for Testing and Materials, W. Conshohocken, PA.

Zeeuwen, J. P., "Dust Explosions—How to Assess the Risk," *Fire Prevention*, No. 228, Apr. 1990, pp. 26–28.

REFRIGERATION SYSTEMS

Henry L. Febo, Jr.

Industrial refrigeration systems usually are designed to provide a means for cooling specific locations to temperatures below ambient. Although there are many types of refrigeration systems, they are all heat exchangers differing only in mechanical and thermal design, employing different fluids for refrigerants.

This chapter outlines recent regulatory changes affecting refrigeration systems. There is also a review of the new refrigerant identification and hazard classification schemes. It will discuss basic operating principles and identify some of the more common types of refrigeration systems and associated hazards. It excludes systems designed exclusively for air cooling building occupants.

EFFECTS OF REGULATORY CHANGES

In 1978 chlorofluorocarbons (CFCs) were banned from use in nonessential aerosols by the United States in an effort to reduce depletion of the ozone layer. Increased use of CFCs in manufacturing operations resulted in a continuation rather than a reduction of the problem.

The United Nations began to address the problem and called a conference in Vienna in 1985 that led to the adoption of the "Montreal Protocol on Substances that Deplete the Ozone Layer." The Protocol became effective worldwide in 1989. This led the United States to enact the Clean Air Act of 1990 that required the total phaseout of CFC production by the year 2000. Recently, this date was revised to January 1, 1996. At this time, recycled and reclaimed stocks are the only source of CFC refrigerants.

The Clean Air Act also mandates reduction of CFC emissions to the "lowest achievable levels." This includes a ban on intentional venting and limitations on purge units expelling CFCs into the atmosphere.

Alternative refrigerants are being rapidly developed and commercialized. At present there are two on the market: HFC 134a and HCFC 123. These replace R-12 or R-500, and R-11, respectively. These are not "drop in" replacements. There would be incompatibilities with lubricants and residual refrigerant in existing equipment. These issues need to be properly addressed on any changeover.

Further, the regulations require that all personnel handling refrigerants be certified. This includes service, operating, and installation personnel.

In addition to the above, there has also been some discussion regarding the possible phasing out of all halogenated refrigerants

containing chlorine, such as hydrochlorofluorocarbons (HCFCs), by the year 2030.

Available alternatives to the use of halogenated refrigerant systems are:

1. Ammonia systems,
2. Absorption systems,
3. Steam-jet systems,
4. Systems employing CFC substitutes, and
5. Hydrocarbon refrigeration systems.

The use of ammonia as a refrigerant is affected by a number of regulations of the U.S. Occupational Safety and Health Administration (OSHA) and the Environmental Protection Agency (EPA). Application of these various regulations is based on specific quantity limits, typically 500 to 1,000 lb (227 to 454 kg). One regulation in particular that might have to be considered is OSHA's 29 CFR 1910.119, "Process Safety Management Standard." This regulation applies to any system with over 10,000 lb (4,540 kg) of ammonia, quite common in industrial systems.

Consult a regulatory specialist prior to proceeding with a new or modified installation.

APPLICATIONS

The following brief outline includes many of the principal applications of refrigeration systems.

1. Industrial, refining, and chemical:
   (a) Controlling vapor pressure of volatiles during distillation, separation, or processing.
   (b) Shifting solubility relationships to permit segregation and removal of undesired constituents, such as asphalt or wax, in lubricating oils.
2. Manufacture, freezing, preservation, and distribution of food products.
3. Air conditioning.
4. Manufacture, preservation, and distribution of medicine and drugs.
5. Environmental testing chambers.
6. Cold treatment of metals.
7. Industrial testing.
8. Miscellaneous:
   (a) Cold storage of flowers and furs.
   (b) Ice making and skating rinks.

Henry L. Febo, Jr., P.E., is an engineering specialist at Factory Mutual Research Corp. in Norwood, MA.

REFRIGERANT CLASSIFICATIONS

Numerical Designation

One does not usually associate low temperatures with fire hazards, but refrigeration systems are of concern for two reasons: (1) some of the fluids used as refrigerants are flammable, and (2) some are explosive in critical combination with air. Others are toxic and, if they escape during a fire, can cause injury or death or interfere with fire-fighting operations.

Refrigerants are identified by a numerical designation as specified in ANSI/ASHRAE 34, *Number Designation and Safety Classification of Refrigerants* (hereinafter referred to as ANSI/ASHRAE 34). The number can be up to 4 digits with each representing the number of fluorine, hydrogen, and carbon atoms as well as the number of unsaturated carbon-carbon bonds, respectively. The number of chlorine atoms can be inferred, and the number of bromine atoms is designated by an uppercase "B" in the numerical designation.

Commercialized refrigerant blends are designated in either the 400 or 500 series. The 400 series is for zeotropic* blends, with the number identifying the compounds in the blend but not the amount of each. The 500 series is for azeotropic† blends, with the number indicating both the compounds and, by definition, the amounts of each.

The 600 series has been assigned to miscellaneous organic compounds, and the 700 series to inorganic compounds.

Safety Designation

ANSI/ASHRAE 34 also assigns safety groups to the various refrigerant compounds. The classification consists of two alphanumeric characters (e.g., A2 or B1). The capital letter indicates toxicity, and the Arabic numeral indicates flammability. The safety group classification can be represented as in Figure 3-28A. Complete details of the classification criteria and methods can be found in ANSI/ASHRAE 34.

The safety class changed with the 1992 edition of ANSI/ASHRAE 34. The pre-1989 system evolved over the years and attempted to classify these two independent parameters, i.e., toxicity and flammability, with a simple (or modified single) designator, i.e., Group 1, 2, 3a, 3b, 4a, and 4b. The new system is less arbitrary in assignment of the letter for toxicity and number for the flammability index.

A list of the more commonly used refrigerants, some selected properties, and their safety class can be found in Table 3-28A.

BASIC OPERATING PRINCIPLES

All refrigeration systems take heat from one space or medium and transfer it to another. To accomplish the transfer, fluids are used that will readily absorb, transport, and release heat under controlled conditions. These fluids used as refrigerants perform in both the liquid and gaseous states. When a refrigerant in a liquid state changes to a gas, it absorbs heat from its surroundings. Conversely, when a refrigerant in a gaseous state condenses back to a liquid, it releases

*Zeotropic: These blends comprise multiple components of different volatilities that, when used in refrigeration cycles, change volumetric composition and saturation temperatures as they evaporate (boil) or condense at constant pressure.

†Azeotropic: These blends comprise multiple components of different volatilities that, when used in refrigeration cycles, *do not* change volumetric composition and saturation temperatures as they evaporate (boil) or condense at constant pressure.

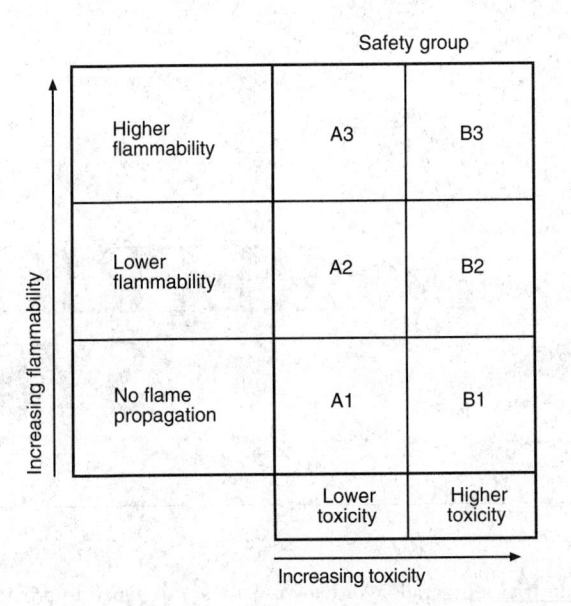

FIG. 3-28A.   Refrigerant safety group classification. (Source: ANSI/ASHRAE 34)

heat to its surroundings. The most effective refrigerant is a material that can vaporize and condense at the desired temperatures and pressures.

The unit of refrigeration in the United States is known as the ton. It is historically developed from the heat required to melt one ton [2,000 lbs (907 kg)] of ice at 32°F (0°C) during a period of 24 hours. Since the heat of fusion of water is approximately 144 Btu/lb (3.4×10^5 J/kg), a U.S. ton of refrigeration equals 12,000 Btu/hr (3.5 kW).

A refrigeration system consists of a circulating refrigerant, condenser and receiver, a control device, an evaporator, and a compression device. (See Figure 3-28B.) At the start of a cycle the refrigerant passes from the receiver at high pressure to the control device, which is an expansion valve where the high-pressure liquid is allowed to expand (thus reducing its pressure and temperature), and then to the evaporator with many coils or tubes that will remove

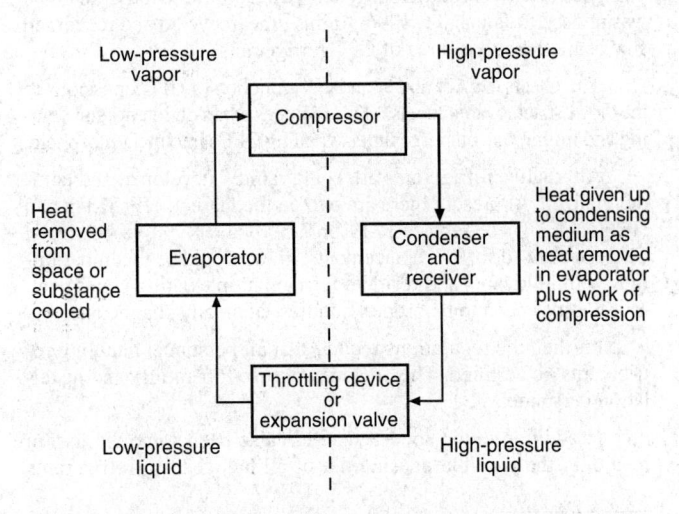

FIG. 3-28B.   Schematic of basic mechanical refrigeration system. (Source: Factory Mutual Research Corp.)

TABLE 3-28A.   *Refrigerant Classification and Selected Properties*

Refrigerant	Name	Chemical Formula	Boiling Point °F	Boiling Point °C	Specific Gravity (Air=1)	Autoignition Temperature °F	Autoignition Temperature °C	Flammable Limits % by Volume in Air Lower	Flammable Limits % by Volume in Air Upper	Safety Class*
R-11	Trichlorofluoromethane	CCl_3F	76	24	4.79	—	—	—	—	A1
R-12	Dichlorodifluoromethane	CCl_2F_2	−22	−30	4.17	—	—	—	—	A1
R-13	Chlorotrifluoromethane	$CClF_3$	−114	−81	3.60	—	—	—	—	A1
R-13B1	Bromotrifluoromethane	$CBrF_3$	−72	−58	—	—	—	—	—	A1
R-14	Tetrafluoromethane	CF_4	−198	−127	—	—	—	—	—	A1
R-22	Chlorodifluoromethane	$CHClF_2$	−42	−41	2.98	1170	632	Very weakly flammable		A1
R-23	Trifluoromethane	CHF_3	−116	−82	—	—	—	—	—	NR†
R-40	Methyl chloride	CH_3Cl	−11	−24	1.8	1170	632	7	19.1	B2
R-113	Trichlorotrifluoroethane	CCl_2FCClF_2	118	48	6.46	1256	680	Very weakly flammable		A1
R-114	Dichlorotetrafluoroethane	$CClF_2CClF_2$	39	4	5.89	—	—	—	—	A1
R-115	Chloropentafluoroethane	$CClF_2CF_3$	−38	−39	—	—	—	—	—	A1
R-123	2,2-dichloro-1,1,1-trifluoroethane	$CHCl_2CF_3$	81	27	—	—	—	—	—	A1
R-134a	1,1,1,2-tetrafluoroethane	CH_2FCF_3	−15	−26	—	—	—	‡	‡	A1
R-142b	1-chloro-1,1-difluoroethane	CH_3CClF_2	14	−10	1.19	—	—	9.0	14.8	A2
R-152a	1,1-difluoroethane	CH_3CHF_2	−13	−25	—	—	—	3.7	18.0	A2
R-170	Ethane	C_2H_6	−128	−89	1.04	959	515	3	12.5	A3
R-290	Propane	C_3H_8	−44	−42	1.56	871	466	2.2	9.6	A3
R-C318	Octafluorocyclobutane	C_4F_8	21	6	—	—	—	—	—	A1
R-400	R-12 and R-114 (composition varies)	$CCl_2F_2/$ $C_2Cl_2F_4$	—	—	—	—	—	—	—	A1/A1
R-500	R-12, 73.8%, & R-152a, 26.2%	$CCl_2F_2/$ CH_3CHF_2	−27	−33	—	—	—	—	—	A1
R-502	R-22, 48.8%, & R-115, 51.2%	$CHClF_2/$ $CClF_2CF_3$	−49	−45	—	—	—	—	—	A1
R-503	R-23, 40.1%, & R-13, 59.9%	$CHF_3/CClF_3$	−126	−88	—	—	—	—	—	NR†
R-600	Butane	C_4H_{10}	31	−1	2.06	761	405	1.9	8.5	A3
R-600a	Isobutane	$CH(CH_3)_3$	11	−12	2.0	860	460	1.8	8.4	A3
R-611	Methyl formate	$HCOOCH_3$	90	32	2.07	853	456	4.5	23	B2
R-717	Ammonia	NH_3	−28	−33	0.59	1204	651	16	25	B2
R-744	Carbon dioxide	CO_2	−109	−79	1.52	—	—	—	—	A1
R-764	Sulfur dioxide	SO_2	14	10	2.21	—	—	—	—	B1
R-1150	Ethylene	C_2H_4	−155	−104	0.98	842	450	2.7	36	A3
R-1270	Propylene	C_3H_6	−53	−47	1.5	851	455	2.0	11	A3

*Safety class in accordance with ANSI/ASHRAE 34.
†NR = not rated.
‡Exhibits combustibility at 5.5 psig (140 kPa) and 350°F (177°C).

heat from the area or secondary fluid. The compression device, where an outside energy source is used for operation, takes the low-pressure gas, increases its pressure, and then discharges the hot, high-pressure gas into a condenser. Here the gas gives up the heat obtained in the evaporator and compressor, and condenses into a liquid, returning to the receiver where the cycle began.

TYPES OF SYSTEMS, BASIC HAZARDS

Refrigeration systems are classified in two ways: (1) by the type of compression system (mechanical, absorption, steam-jet ejectors) and (2) by the method of cooling the space or substance (direct or indirect).

Mechanical System

Description: A mechanical refrigeration system, where the compression step is achieved by a mechanical compressor, is probably the most common refrigeration system type in use. The compressor types include reciprocating [horizontal, vertical (Figure 3-28C), or "V" design], centrifugal (Figure 3-28D), or rotary screw type. Refrigerants can include any of the fluids shown in Table 3-28A.

Figure 3-28E illustrates a typical ammonia refrigeration system showing the relationship of the major components, such as the compressor, expansion valve, and evaporator.

Hazards: The hazards are primarily related to the refrigerant fluid. Those with a flammability rating of 1 present no fire hazard but, if released into a confined space, present oxygen depletion hazards to personnel, as would those rated 2 or 3.

Those with a 2 flammability rating pose limited fire and explosion hazard, according to the ANSI/ASHRAE 34 classification method. The potential fire and explosion hazards of these fluids should be evaluated before any installation decision is made.

Those with a 3 flammability rating are highly flammable. These are used primarily in the petroleum refining and petrochemical industries. Use of these refrigerants requires special design and construction to control the fire and explosion hazards, and to minimize the consequences of an unexpected release.

Refrigerants carrying a B toxicity rating present additional personnel hazards that must be recognized and addressed in design and emergency planning.

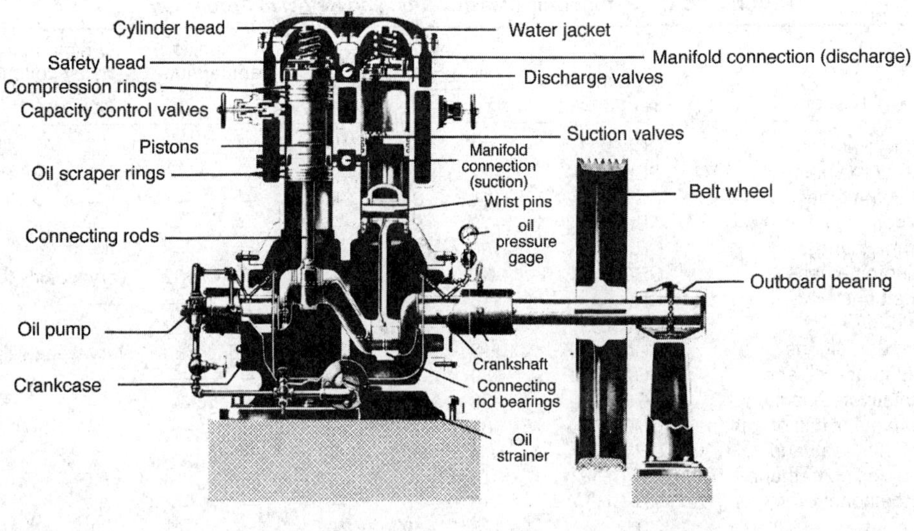

FIG. 3-28C.   *Reciprocating compressor, vertical single acting. (Source: Factory Mutual Research Corp.)*

The equipment used for the compression step will present various hazards not directly addressed in this chapter. These include potential for overspeed, overpressure, leakage of refrigerants or lubricant, and personnel safety issues. These should be recognized in the design and installation phase and addressed in a preventive maintenance program.

Absorption System

Description: The absorption cycle is the oldest known for producing a refrigeration effect. Mechanically, it is the simplest of the systems in common use today.

The simple absorption system consists of the basic components shown in Figure 3-28F, with the absorber, heat exchanger, and generator system serving as the compression step.

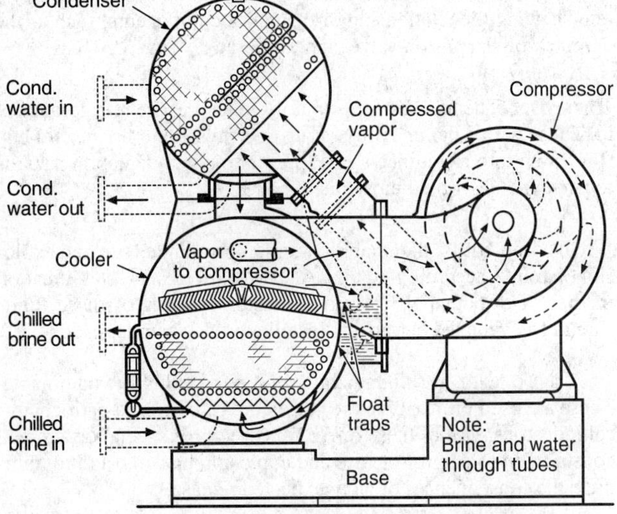

FIG. 3-28D.   *Cross-section of centrifugal refrigeration system showing refrigeration cycle. (Source: Factory Mutual Research Corp.)*

The absorption cycle uses an absorbent as a secondary fluid to absorb the primary fluid, which is a gaseous refrigerant that has been vaporized in the evaporator. The evaporation process absorbs heat, providing the needed refrigeration.

The absorption and mechanical compression cycles have in common the evaporation and condensation of a refrigerant liquid, occurring at two pressure levels within the unit. The two cycles differ in that the absorption cycle uses a heat-operated generator to produce the pressure differential, whereas the mechanical compression cycle uses a compressor. Both cycles require energy for operation, i.e., heat and a small amount of mechanical energy (a pump) in the absorption cycle and mechanical energy in the compression cycle.

The two most common working fluids presently in use are: (1) water-lithium bromide (H_2O-LiBr), and (2) ammonia-water (NH_3-H_2O), where the components are given as refrigerant-absorbent. Note, in one system, water is the refrigerant, while in the other it is the absorbent.

The water-lithium bromide is used primarily for air-conditioning applications, while the ammonia-water is for large tonnage industrial applications requiring low temperatures for process cooling.

Water-lithium bromide is a brine that has a subatmospheric vapor pressure at typical cycle conditions. Machines that use water-lithium bromide operate under a vacuum. Conversely, ammonia-water solutions have vapor pressures up to 295 psi (2,034 kPa) at cycle conditions. As a result, equipment used for ammonia-water systems requires much stronger vessels.

The heat energy for the generator can come from various sources. These include steam or hot fluids (called indirect fired); clean, hot waste gases (heat recovery, also indirect fired); or flame (direct fired).

Hazards: Lithium bromide is a relatively stable compound (nontoxic, nonflammable, and nonexplosive). If contaminated it can generally be reclaimed.

Ammonia systems have inherent hazards, as described in the section entitled "Hazard Control and Emergency Response." Similar safeguards in that section would apply.

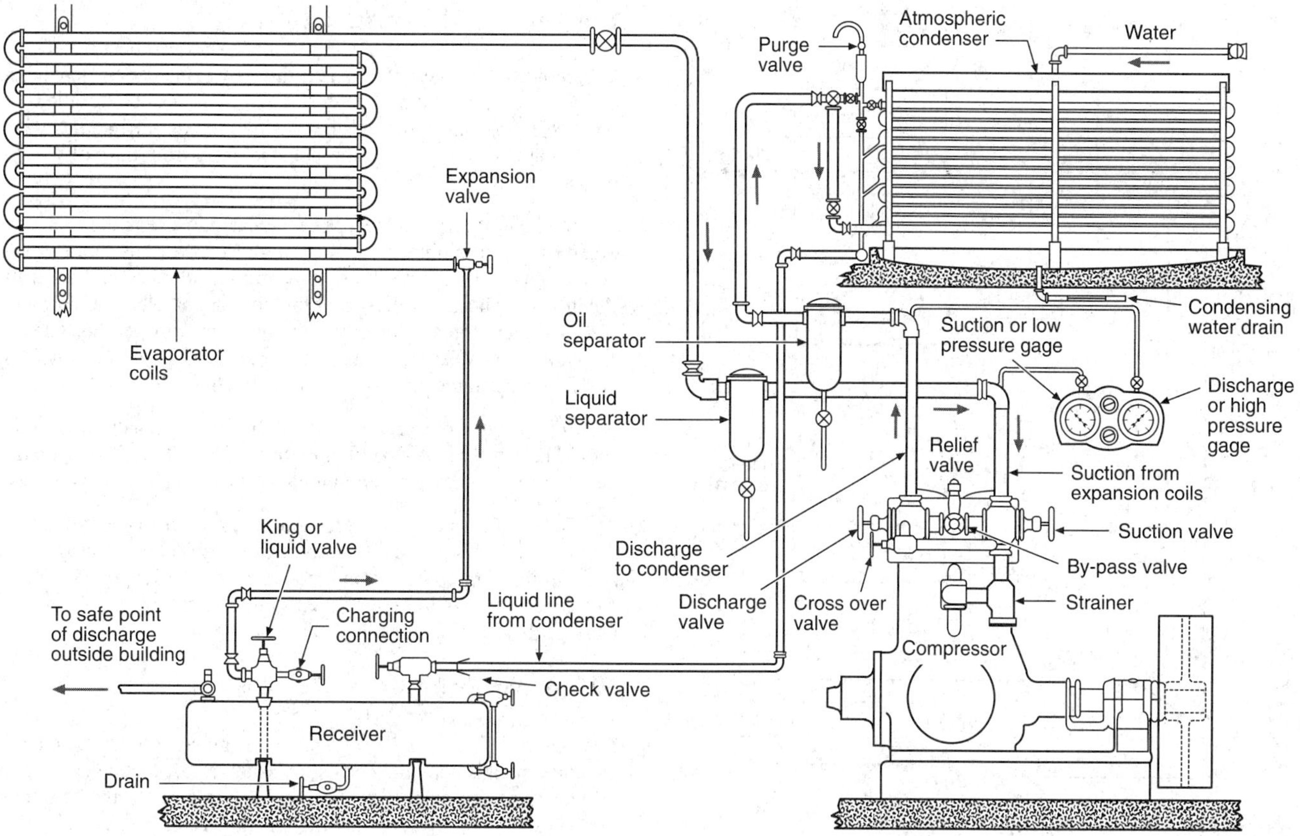

FIG. 3-28E.   Typical ammonia refrigeration plant. (Source: Factory Mutual Research Corp.)

Steam-Jet Ejector System

Description: The steam-jet (also known as the steam vacuum, steam-ejector, and chill-vactor) refrigeration cycle has been used for over 70 years, yet it is one of the least common refrigeration cycles in use today. Its lack of popularity stems, at least in part, from some inherent performance limitations. Several hundred of these units are still in use in such industries as pulp and paper, petrochemical, food processing, and pharmaceuticals where they are used for air conditioning and process cooling.

The steam-jet refrigeration cycle is similar to the more conventional refrigeration cycles having basic system components, such as an evaporator, compression device, condenser, and refrigerant. Instead of a mechanical compressor, the steam-jet system employs a steam ejector or booster to compress the refrigerant to the condenser pressure level.

In the steam-jet refrigeration cycle, water is used as the refrigerant and the cooling effect is produced by the continuous vaporization of a part of the water in the evaporator at a low absolute pressure level.

Figure 3-28G shows a simple open, chilled-water refrigeration system. In this open system, refrigerant water is circulated directly to the cooling load. It is possible to have a closed chilled-water system in which a secondary fluid, usually water or brine, is used to serve the cooling load.

Hazards: Due to the simple nature of this system, no unique hazards are presented. Because water is the refrigerant, care is needed to prevent freezing in the evaporator. This includes operational controls and low-refrigerant-temperature alarms.

Direct System

In addition to being classified according to the type of compression stage employed, refrigeration systems are further differentiated by how the heat is removed from the cooling load. In direct systems, the evaporator is in direct contact with the product or space to be cooled, or it is in air-circulating passages connected to the area or product. A household refrigerator is an example of a direct system.

Indirect System

An indirect system uses an intermediate heat transfer fluid, which is first cooled by the refrigerant and then circulated to the material or space. The intermediate fluid is called "brine," because early refrigerating systems used a mixture of sodium or calcium salt and water. Brine has been defined as any liquid cooled by a refrigerant and used for the transmission of heat energy without a change in state. Organic chemical solutions, such as ethylene glycol, as well as brine, are employed in indirect systems.

HAZARD CONTROL AND EMERGENCY RESPONSE

Detailed criteria for installation and arrangement of refrigeration systems are addressed in various national standards and codes. (See bibliography.) This section summarizes some basic considerations and safeguards. Some of the following criteria may not be practical where small quantities of refrigerants are involved.

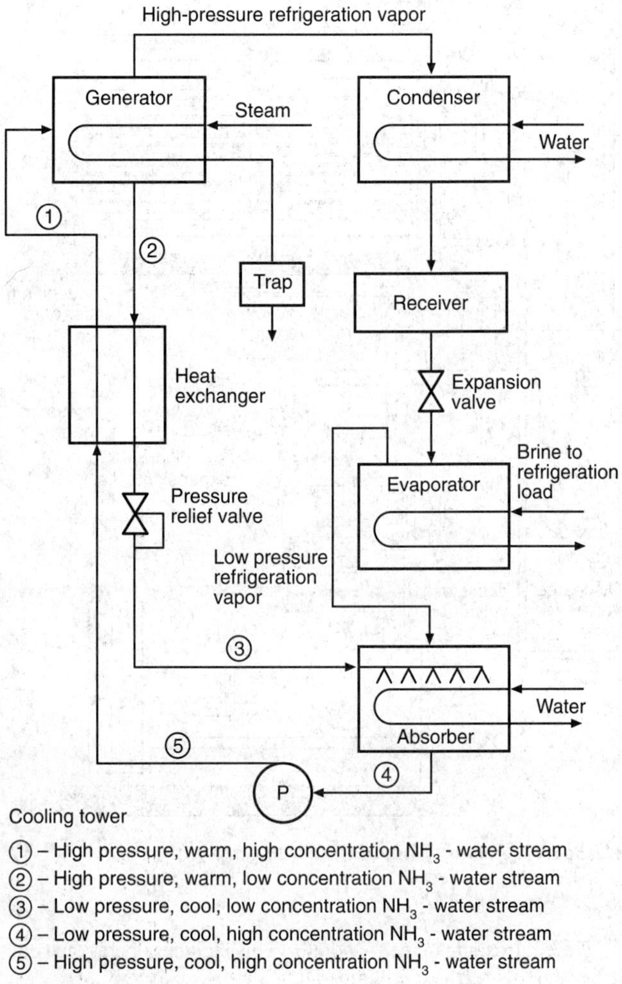

Cooling tower

① – High pressure, warm, high concentration NH_3 - water stream
② – High pressure, warm, low concentration NH_3 - water stream
③ – Low pressure, cool, low concentration NH_3 - water stream
④ – Low pressure, cool, high concentration NH_3 - water stream
⑤ – High pressure, cool, high concentration NH_3 - water stream

FIG. 3-28F. Basic absorption refrigeration system. (Source: Factory Mutual Research Corp.)

Location and Construction

The choice of refrigerant type is limited by some codes, depending on the occupancy of the area where the system is to be installed. Indirect systems can sometimes be used with toxic refrigerants when the machine room can be safely located away from normally occupied areas.

Refrigeration machinery is best located in a separate room, cut off from any adjoining occupancy. In most cases, this room should be limited to containing the refrigeration machinery and controls. Electric switchgear, boilers, fuel fire equipment, etc., should be located in other rooms. Even the noncombustible refrigerants (Classes A1 and B1) can be decomposed by arcs, flames, and hot surfaces, possibly releasing corrosive chlorides and fluorides.

The room construction should be noncombustible. Cutoffs should be 1-hr fire rated for flammability Class 1 and 2 refrigerants, and 2-hr fire rated where flammability Class 3 refrigerants are used.

Where Class 2 and 3 refrigerants are used, the potential for room explosion must be considered. Outdoor locations or rooms of damage-limiting construction (see NFPA 68, *Guide for Venting of Deflagrations*) would be preferred to minimize damage from a release of these refrigerants.

Electrical Equipment

For flammability Class 1 refrigerants, the major concern for electrical equipment is potential corrosion issues. Best practice is to locate all electrical equipment not directly associated with the refrigeration system to areas other than the machine room.

For flammability Class 2 and 3 refrigerants, the machine room should conform to Class I, Division 2, of NFPA 70, *National Electrical Code®*.

For ammonia, some codes and Authorities Having Jurisdiction allow use of ordinary electrical equipment, if:

1. The machine room is continuously ventilated, and failure of the ventilation system sounds an alarm, or

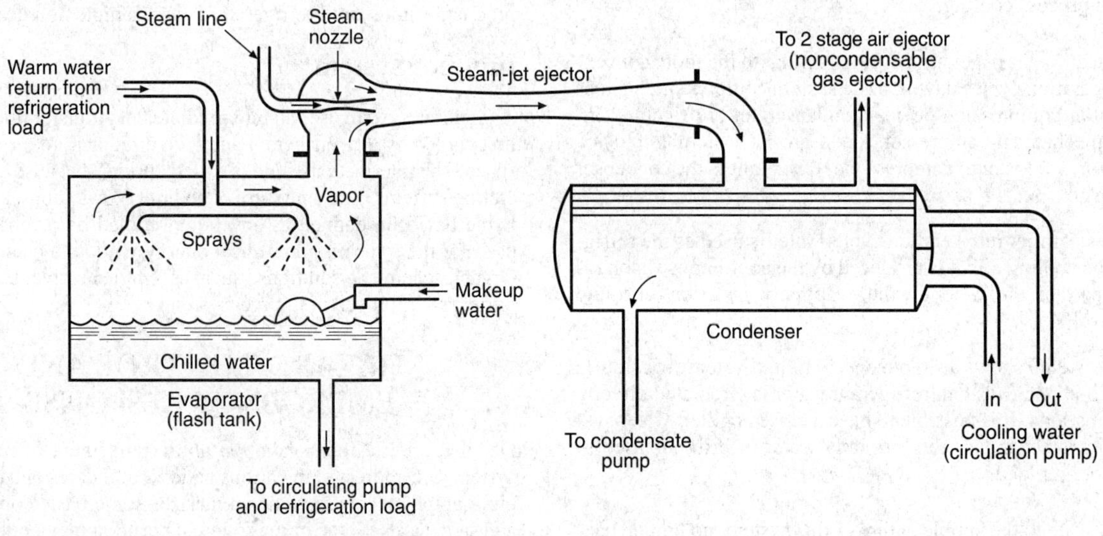

FIG. 3-28G. Open, chilled-water steam-jet refrigeration system with surface condenser. (Source: Factory Mutual Research Corp.)

2. The machine room is equipped with a vapor detector that will automatically start the ventilation system and sound an alarm at a detection level at or below 1,000 ppm.

Ventilation

For Class A1 refrigerants, natural or mechanical ventilation is necessary for equipment cooling and personnel comfort only. For all other refrigerants, mechanical ventilation is needed. Where the hazard is only toxicity, appropriate guidelines should be followed. Where refrigerant flammability is an issue, continuous mechanical ventilation of 1 cu ft/min/sq ft (0.3 m³/min/m²) of floor area will minimize the possibility of developing an explosive concentration in the event of normal or minor leakage.

Fire Protection

Control of fire emergencies generally consists of reducing the heat of the fire with water and, if possible, shutting off the gas supply. Water should be applied as a spray from hoses or fixed nozzles.

Where the construction is combustible, or flammable refrigerants are used, automatic sprinklers provide the best and most reliable form of fire protection.

All machine rooms should be provided with portable extinguishers suitable for at least Class B and C fires.

Miscellaneous

To minimize refrigerant release from equipment and piping, a high level of maintenance should be provided, especially where refrigerants are flammable or toxic. In addition, the initial installation should ensure that piping and components are compatible with the refrigerant, and installed according to appropriate codes. ANSI/ASHRAE 15, *Safety Code for Mechanical Refrigeration*, details the requirements for piping, pressure relief, etc. Special care should be taken to protect refrigerant piping from mechanical damage.

Control of the nonfire emergency is accomplished by directing, diluting, and dispersing the gas to prevent it from coming in contact with people or goods. Simultaneously, steps should be taken to stop the flow of gas. Air, water, and steam are practical media for accomplishing dilution and dispersion. Water, in the form of a spray from hoses or fixed nozzles, is most commonly used for this purpose.

Ammonia presents hazards of fire, explosion, toxicity, and contamination of goods. It may be most severe in the refrigerated area and machine room, but can spread to adjoining areas. Emergency planning should address control of the release, protection of personnel, and decontamination or preservation of the stored commodity.

BIBLIOGRAPHY

References

ANSI/ASHRAE 15, *Safety Code for Mechanical Refrigeration*, American Society of Heating, Refrigeration, and Air-Conditioning Engineers, Inc., Atlanta, GA, 1994.

ANSI/ASHRAE 34, *Number Designation and Safety Classification of Refrigerants*, American Society of Heating, Refrigerating, and Air-Conditioning Engineers, Inc., Atlanta, GA, 1992.

ANSI/ASME B31.5, *Refrigeration Piping*, American Society of Mechanical Engineers, New York, NY, 1992.

ANSI/IIAR 2, *Equipment Design and Installation of Ammonia Refrigeration Systems*, International Institute of Ammonia Refrigeration, Chicago, IL, 1992.

Loss Prevention Data 12-53R, *Absorption Refrigeration Systems*, Factory Mutual Research Corp., Norwood, MA, 1991.

Loss Prevention Data 12-54R, *Steam-Jet Refrigeration Systems*, Factory Mutual Research Corp., Norwood, MA, 1992.

Loss Prevention Data 12-61R, *Mechanical Refrigeration*, Factory Mutual Research Corp., Norwood, MA, 1993.

NFPA Codes, Standards, and Recommended Practices

Reference to the following NFPA codes, standards, and recommended practices will provide further information on the safeguards for refrigeration equipment discussed in this chapter. (See the latest version of *The NFPA Catalog* for availability of current editions of the following documents.)

NFPA 10, *Standard for Portable Fire Extinguishers*

NFPA 12, *Standard on Carbon Dioxide Extinguishing Systems*

NFPA 17, *Standard for Dry Chemical Extinguishing Systems*

NFPA 30, *Flammable and Combustible Liquids Code*

NFPA 68, *Guide for Venting of Deflagrations*

NFPA 70, *National Electrical Code®*

NFPA 325, *Guide to Fire Hazard Properties of Flammable Liquids, Gases, and Volatile Solids*

Additional Reading

ASHRAE Handbook, *1993—Fundamentals*, American Society of Heating, Refrigerating, and Air-Conditioning Engineers, Inc., Atlanta, GA.

Benedetti, R. P., ed., *Flammable and Combustible Liquids Code Handbook*, 5th ed., National Fire Protection Association, Quincy, MA, 1993.

James, R. W., "Safe Use of Refrigerants for Process Applications," *Journal of Loss Prevention in the Process Industries*, Vol. 7, No. 6, Nov. 1994, pp. 492–500.

Olson, C., "New Standards Cool Concerns over Alternative Refrigerants," *Building Design and Construction*, Vol. 33, No. 11, Nov. 1992, pp. 62–64.

LASERS

Revised by Carol Caldwell

Lasers have become a common part of life. To increase fire engineers' understanding of lasers, this chapter discusses how lasers work, their fire hazards, and some hazard protection measures. Lasers have applications in many different fields. The most frequent exposure to lasers is when buying merchandise. They are often used to scan product codes for pricing. In business they are used in laser printers and CD-ROMs. In the entertainment industry, lasers have opened up many opportunities. They are used in music, CD movies, for movie effects, and in public displays at theme parks.

The medical field has taken advantage of lasers' properties, and they are used frequently in surgery. Additionally, veterinarians have used them on their patients. In the construction industry, lasers are used for measuring distances and alignment. The alignment use is for pipes, tunnels, ceilings, and floors. Manufacturers use lasers to cut, weld, and drill, in items as diverse as metal, paper, and fabric. They also have application in research facilities and the military.

A primary benefit of lasers can also be a significant fire hazard. Lasers that are powerful enough to be used to cut, weld, and drill are an ignition hazard. Without proper attention and controls, they can start and have started fires. The laser itself also has potential fire concerns from the use of flammable liquids and gases, as well as high-voltage electrical components. Ideally, laser fire protection is achieved by physical controls. However, in certain fields, particularly the medical industry, administrative controls are the primary fire protection measure.

Further information on lasers can be found in Section 4, Chapter 8, "Medical Gases." The bibliography references other applicable sources of information.

LASER PROPERTIES

The term *laser* is actually an acronym for "light amplification by stimulated emission of radiation." The radiation referred to in the acronym *laser* should not be confused with ionizing radiation associated with radioactive materials; it refers to the broad spectrum of electromagnetic radiation that includes visible light, ultraviolet light, and infrared light.

Visible light is one form of electromagnetic radiation. Electromagnetic radiation consists of electrical and magnetic energy trav-

eling together in wave form. Other forms of electromagnetic radiation are not visible to the human eye, such as radio waves, ultraviolet light, and microwaves.

All forms of electromagnetic radiation taken together produce the electromagnetic spectrum. Different portions of this spectrum describe the frequency and wavelength of the various electromagnetic wave forms. (See Figure 3-29A.)

Wavelength refers to the distance between peaks on a wave or the distance for the wave to complete one cycle. Wavelength is measured in meters. Frequency refers to the number of cycles the wave completes in one second. A "hertz" is one cycle per second and is the normal term of expression for frequency.

Figure 3-29A shows that low-frequency, long wavelength radiation is found at the low end of the spectrum, while high-frequency, short wavelength radiation is found at the high end of the spectrum. Examples of low-frequency, long wavelength radiation are radio waves, television waves, and microwaves. Examples of high-frequency, short wavelength radiation are ultraviolet light and x-rays. Radiation visible to the human eye in the form of light makes up a small portion of the entire spectrum. Figure 3-29B illustrates the visible light range within the electromagnetic spectrum.

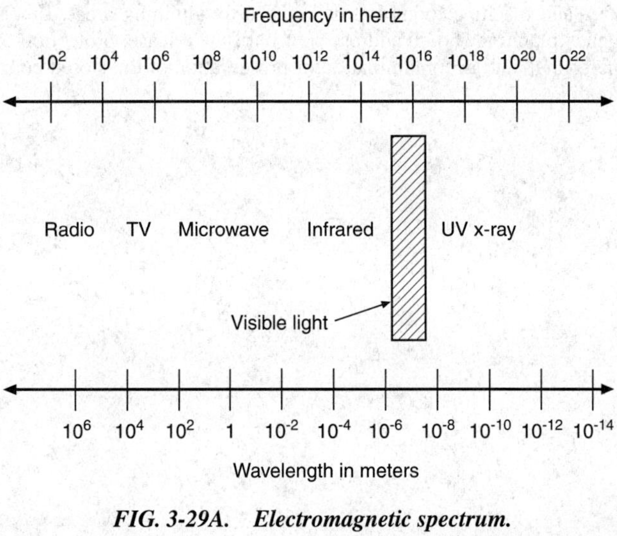

FIG. 3-29A. *Electromagnetic spectrum.*

Carol Caldwell is president of Caldwell Consulting, Ltd. in Christchurch, New Zealand. She is the chairman of the NFPA Technical Committee on Laser Fire Protection.

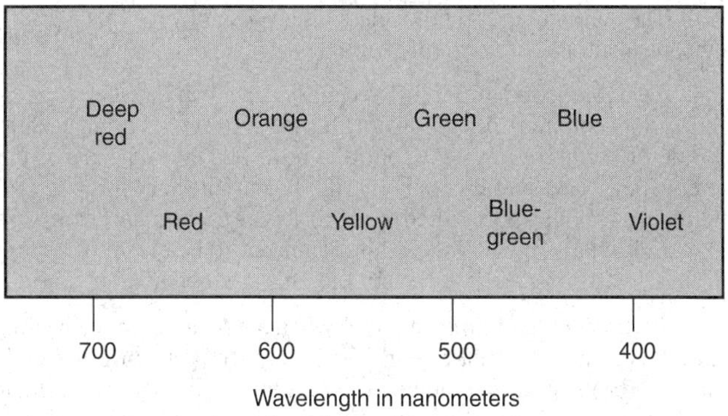

FIG. 3-29B. Visible light spectrum.

The study of laser light is generally concentrated in the optical spectrum, or that portion of the electromagnetic spectrum from infrared to ultraviolet light, which includes visible light.

Infrared radiation is radiation with wavelengths ranging between 10^{-4} and 10^{-6} m. The name *infrared* comes from the fact that the wavelengths are longer than the visible red end of the spectrum. Visible light is electromagnetic radiation that is visible to the human eye. Visible light has a wavelength between 400 and 700 nanometers (a nanometer is one billionth of a meter). The eye is most sensitive to green and yellow colors in the range of 555 nanometers and least sensitive to blue and red near their respective ends of the electromagnetic spectrum. Ultraviolet light is that portion of the electromagnetic spectrum with wavelengths between 10 and 400 nanometers, just beyond the visible portion of the electromagnetic spectrum at the violet end.

Electromagnetic radiation is produced whenever an electron gives up energy into an electric field. When the electron gives up energy, it transitions to a lower energy state and gives up a photon, or a packet of energy.

In order to get the electrons to give up energy, the electrons must be raised to a higher energy level. When raised to a higher energy level, the electron is said to be excited, or in an excited state. When electrons randomly give up photons by transitioning to a lower energy state, the light or electromagnetic radiation is diffuse. However, if the atoms can be excited and then influenced to produce the emission of a photon of a certain wavelength and in a certain direction, then the resulting light can be utilized in unique ways. Obviously, to make the light emitted useful, a great many photons of the proper type must be produced.

Laser light has several unique properties. One of those properties is that laser light consists of only a single color. In addition, the wavelengths of this single color are within a very narrow range. This property is called monochromaticity. A second property is that of directionality. Normal light is radiated in all directions in a random fashion. Laser light, however, is radiated in a narrow beam in a single direction. The third unique property of laser light is that of coherence. Laser light is radiated in an orderly pattern known as being "in phase." When all the separate waves are in phase, the resultant combination wave is much stronger. Thus, laser light is a very intense light of a single color, being radiated in a single direction, with little divergence. It is these properties that give laser light its diverse applications and its hazards.

A laser, then, must be constructed of components that can produce the necessary photons in sufficient quantities to produce a useful light beam. Lasers consist of generally three components: (1) a lasing medium, (2) an energy source (or pumping system), and (3) a feedback mechanism. Other accessories, such as lenses, mirrors, and shutters may be added to the system to produce special effects, but only the basic three are needed for lasing action.

Lasing Medium

The lasing medium is a collection of atoms that can be excited to a state to produce a stimulated emission of photons. The active medium may be a solid, liquid, or gas, or semiconductor material.

Energy Source

The pumping system is a source of energy that can be used to raise the electrons in the lasing medium to their excited state. The pumping system may take one of several forms: (1) optical pumping systems use strong sources of light, such as xenon flashtubes or flashlamps, or a continuous arc lamp; (2) electron collision pumping is created when an electric current is passed through the laser material; or (3) chemical pumping is created by the making or breaking of chemical bonds. All three forms of pumping are commonly in use.

Feedback Mechanism

The feedback mechanism is an apparatus that returns a portion of the laser light created in the lasing medium back to the lasing medium. This feedback creates a situation where the photons created in the lasing medium are passed back through the lasing medium, creating more photons. If sufficient feedback is created, the creation of laser light can be made continuously and at high levels. The normal feedback mechanism is a pair of mirrors aligned to reflect the laser light back and forth. One of the mirrors is only partially reflective, permitting some light to leave the laser. The portion of the light that is not emitted continues to be reflected, creating more photons.

LASER OPERATION

Laser operation can best be described in three steps:

1. Energy is supplied to the laser material by the pumping system. This energy is stored by the laser material in the form of electrons in their excited state. The pumping system must continue until a population inversion (i.e., more electrons in the excited state than in their normal state) takes place for laser light to be produced.
2. Once the population inversion is achieved, the decay of a few electrons to a lower energy level will start a chain reaction. Photons released will strike electrons, creating more photons of the same wavelength, phase, and direction.
3. When the photons reach one of the mirrors, they will be reflected back into the laser material, supporting continued chain reaction and the production of even more photons. Some photons will strike the partially reflective mirrors; when that occurs, some of them will be reflected back into the laser material and some will be emitted as laser light.

Modes of Operation

The modes of laser operation are differentiated by the rate at which energy is delivered. Some lasers produce a steady flow of light. These lasers are known as continuous wave lasers, or CW lasers. Pulsed lasers are lasers operated to create repetitive laser pulses. A special type of pulsed laser is a Q-switched laser. In the Q-switched laser, a shutter-like device is placed between the mirrors of a laser to interrupt the lasing action. The shutter can act as a mirror in itself, allowing the continued reflection of photons and buildup of excited electrons, but not allowing any photons to escape through the partially reflective mirror. When the shutter opens, a very intense pulse is released.

ANSI CLASSIFICATION OF LASERS

The American National Standards Institute (ANSI) has adopted a classification system for lasers, based on the ability of the primary laser beam or reflected primary laser beam to cause biological damage to the eye or skin during intended use:

Class 1: A low-powered laser device that cannot, under normal circumstances, emit laser radiation that creates an optical hazard. These are sometimes called exempt lasers. (See note after Class 4 lasers.)

Class 2: Low-powered laser devices operating in the visible spectrum that cannot injure a person accidentally, but which may injure the eye when viewed directly for an extended period of time.

Class 3: Medium-powered laser devices that are capable of causing eye damage with short-duration exposures to the direct or specularly reflected beam. Occasionally, some Class 3 lasers are considered a fire hazard.

Class 4: Lasers that can injure if viewed directly or if reflections are viewed. These lasers can also cause severe skin damage and are to be considered a fire hazard.

NOTE: Some Class 4 lasers are called "embedded lasers" and given a lower classification. They still have the same inherent fire risk due to the power of the beam. However, because of engineering controls, they do not have the same risk of laser radiation hazards as the normal Class 4 laser.

HAZARDS AND CONTROL MEASURES

In addition to the biological hazards classified by the ANSI system, lasers produce several other hazards that must be considered.

Electrical Hazards

The most common hazard associated with lasers of all types is the electrical hazard. Almost all lasers contain high-energy power supplies. These power supplies can produce both electrical shock hazards and, as in all electrical pieces of equipment, fire hazards. Normal electrical safety devices, such as fuses, circuit breakers, insulation, grounding devices, and interlocks, can effectively control these hazards.

Many lasers use capacitors. Capacitors store electrical charges, with the intent to release high electrical charges in a very short period of time. Accidental discharge of capacitors is, therefore, extremely dangerous. Shorting or internal electrical faults can cause capacitor rupture, fires, or explosions. Because of all these possible hazards, capacitors require certain safety precautions. Capacitors should be isolated within screens, shields, barriers, or uninhabited rooms designed to protect against shock, burns, and fire, and to contain fragments from an explosive capacitor failure. Covers or doors to capacitors should be interlocked to prevent charging (when open) by dumping and grounding the capacitors.

Laser Beam Hazards

The laser beam from Class 4, and some Class 3 lasers, is powerful enough to ignite combustible materials. The ignition hazard applies in all applications of laser use, e.g., health care, industrial, manufacturing, commercial, research, and military. Some lasers are "embedded lasers" and can be called Class 1 lasers by the ANSI classification system. They are, in fact, a higher class laser (generally a Class 4 laser), and the beam is a fire hazard.

There are many different precautions that can be taken to minimize the potential for fire from the laser beam ignition hazard. Some of them are physical controls; however, the majority of the precautions are administrative. The beam intensity profile and alignment should be verified prior to use. Appropriate beam stop materials should be in place. Attention needs to be given to the materials adjacent to the laser beam; they must not be combustible.

Particular attention is required when using lasers in the health-care field. The laser beam (consider it an open flame or other ignition source) is used in conjunction with many combustible materials. There have been fires using lasers in the health care industry. This is recorded in the data base on laser fire incidents, established by the NFPA Committee on the Recommended Practice on Laser Fire Protection. With the exclusion of some metals, no material is "fire safe" around lasers. Potential fuels in health care facilities include: patients' hair; gastrointestinal gases (e.g., methane, hydrogen, hydrogen sulfide); prepping agents; and fabric products, such as drapes, towels, gowns, and dressings. Additional associated combustible materials to consider are plastic and rubber products, tracheal tubes, gloves, anesthesia masks, petroleum-based ointments, and the laser circuitry itself.

Also complicating the situation in the health care industry is the potential for an oxygen-enriched atmosphere. Many materials' potential for ignition is enhanced when present in an oxygen-enriched atmosphere. (See Section 4, Chapter 9 "Oxygen-Enriched Atmospheres," for more information on oxygen-enriched atmospheres.) Flammable gastrointestinal gases also require special attention. The gases need to be eliminated or managed in the presence of lasers. It is assumed that the anesthetic gas itself is nonflammable; otherwise, it is a significant concern during laser surgery.

Flammable Liquid Hazards

Lasers can use flammable liquids as part of the process. In particular, dye lasers use a lasing medium of a complex fluorescent organic dye dissolved in an organic solvent. Very often the organic solvent is a flammable liquid. The dye is used as part of the process of exciting selected atoms. Many of these lasers use only 1 L of solvent. Others use from 5 gal to hundreds of gal of solvent. In the event of a spill, there are ignition sources present, such as the laser beam, and electrical components.

Many of the protective measures for handling flammable liquids are already well documented. Some general precautions will be discussed herein. It is important to incorporate into the design or installation a means of containing a spill. Pumps, motors, and other electrical components should be appropriately rated for the application or designed as intrinsically safe. Use of metal tubing instead of plastic tubing for the flammable liquids is suggested.

Compression-type fittings are recommended for flammable liquid lines instead of slip-on friction fittings without clamps. The integrity of tubing and connections should be checked periodically. Good housekeeping is necessary in these areas. Warning signs appropriate for the flammable liquids should be posted. Storage should follow accepted practice for flammable liquids. During cleanup and disposal, it needs to be recognized that the waste still has the properties of a flammable liquid.

For larger volume laser systems, the fire protection measures are similar to any industrial process using large quantities of flammable liquids. Methods to monitor the volume level, flow, and pressure of the systems should be installed. Consideration of the adequacy of the ventilation in the area is required. Remote shutdown capability should be incorporated into the system design.

Flammable and Reactive Gas Hazards

Occasionally, lasers use flammable or reactive gases. The gases should be checked to determine if they are used in their pure form or mixed with an inert gas. The smallest quantity and percentage of gas appropriate for the system should be used. Whenever possible, the gases should be stored outside the building or in a ventilated gas cabinet designed for that purpose. The ventilation in the area where the gases are used should be assessed to determine the consequences of a gas leak, i.e., is it possible for the area to be within the flammable range for the particular gas? The integrity of the piping systems should be checked at regular intervals.

Consideration should be given to the installation of combustible gas detectors designed to activate at 25 percent of the lower flammable limit. Exhaust vents from gas piping should be located away from air intakes. The exhaust gases might need to be mixed with an inert gas, depending upon the hazard. Additionally, care is required to ensure that the exhaust gases do not stagnate in the exhaust piping system. The regulator for the gas should not be used as a flow control valve; a supplemental valve should be installed. All piping should be labeled to identify specific contents. The general vicinity of the gases should have signs indicating the hazard. Emergency shutdown of the gases should be built into the system at the gas supplies and at another remote location.

Operations

It is recommended that any sites using lasers have individuals familiar with lasers and who are responsible for adherence to fire safety practices. Only personnel trained on the lasers should be permitted to use them. Other documents establish training requirements for laser use. Maintenance should only be performed by knowledgeable personnel. A log of the maintenance activity should be kept.

Emergency Response

Prior to using a laser that is considered a fire hazard, appropriate personnel should be familiar with emergency procedures. Issues to consider are the location of exits, laser shutdown procedures, notification procedures, and the location and use of fire extinguishers. Personnel in the vicinity of lasers generally have duties different from fire fighting, so it is important that training of what to do in a fire situation takes place. Efforts should be made to ensure that the responding fire fighters are aware of the hazards involving lasers.

SUMMARY

Lasers are a part of everyday life. They are quite beneficial for humanity. However, adequate precautions must be taken to minimize the fire hazards. In particular, Class 4 and some Class 3 lasers have significant fire safety concerns. The laser beam itself is an ignition hazard, and other products used in the process to generate the laser are fire hazards. It is important that those involved with using lasers are aware of the fire hazards and take appropriate control measures. With proper attention, lasers can be used in a fire safe manner.

BIBLIOGRAPHY

References

ANSI B57.1 (CGA Pamphlet V-1-1987) *Standard for Compressed Gas Cylinder Valve Outlet and Inlet Connections*, American National Standards Institute, New York, NY.

ANSI Z136.1-1993, *Standard for the Safe Use of Lasers*, American National Standards Institute, New York, NY.

ANSI Z136.3-1988, *Standard for the Safe Use of Lasers in Health Care Facilities*, American National Standards Institute, New York, NY.

"Control of Hazards to Health from Laser Radiation," Technical Bulletin TB MED 279, 1975, Department of the Army, Washington, DC.

21 CFR (FDA), Part 1040, Chap. 1, "Performance Standards for Light Emitting Products," Apr. 1994, Food and Drug Administration, Washington, DC.

OSHA "Guidelines for Laser Safety and Hazard Assessment," Instruction Publication 8-1.7, Aug. 1991, Occupational Safety and Health Administration, Washington, DC.

NFPA Codes, Standards, and Recommended Practices

Reference to the following NFPA codes, standards, and recommended practices will provide further information on the safeguards for lasers discussed in this chapter. (See the latest version of *The NFPA Catalog* for availability of current editions of the following documents.)

NFPA 13, *Standard for the Installation of Sprinkler Systems*

NFPA 30, *Flammable and Combustible Liquids Code*

NFPA 70, *National Electrical Code®*

NFPA 75, *Standard for the Protection of Electronic Computer/Data Processing Equipment*

NFPA 99, *Standard for Health Care Facilities*

NFPA 115, *Recommended Practice on Laser Fire Protection*

Additional Reading

Bennet, H. E., ed., "Laser Induced Damage in Optical Materials," *Proceedings* of the 16th Symposium on Optical Materials for High Powered Lasers, 1984, Boulder, CO, National Institute for Standards and Technology, Gaithersburg, MD, 1986.

"Decreasing Airways Fire Risk During Laser Surgery Is Aim of New Subcommittee," *ASTM Standardization News*, Vol. 18, No. 12, December 1990, pp. 20–22.

deRichemond, A. L., "Laser Resistant Endotracheal Tubes: Protection Against Oxygen-Enriched Airway Fires During Surgery?," ASTM STP 1111, American Society for Testing and Materials, Flammability and Sensitivity of Materials in Oxygen-Enriched Atmospheres, Vol. 5, Symposium Sponsored by ASTM Committee G-4 on Compatibility and Sensitivity of Materials in Oxygen Enriched Atmospheres, May 14–16, 1991, Cocoa Beach, FL, ASTM, W. Conshohocken, PA, 1991, pp. 157–167.

Lavid, M., *et al.* "IR Laser Ignition of Natural Gas," Combustion Institute/Eastern States Section, *Proceedings* of the 1991 Fall Technical Meeting of Chemical and Physical Processes in Combustion, October 14–16, 1991, Ithaca, NY, 1991, pp. 1–4.

SEMICONDUCTOR MANUFACTURING

Revised by William B. Marshall

Semiconductor devices, such as computer chips, are commonplace in the modern world. Applications range from the digital wristwatch to the most advanced supercomputers under development. In between, uses include televisions, VCRs, communication devices, electronic controls on appliances, and the control module in most automobiles.

Semiconductors are made from a variety of basic materials. The most common material is silicon. Other types include gallium arsenide, germanium, gallium phosphide, gallium arsenide phosphide, indium phosphide, and mercury cadmium telluride. A single manufacturing plant may use any combination of these materials as the base crystal. All are prepared in a similar manner.

Semiconductor devices have extremely small circuit patterns. A single particle of microscopic dust can ruin a chip. Because of this, the devices are fabricated in a cleanroom. A room can be considered to be a cleanroom if it can maintain a particle level of fewer than 100,000 particles larger than 0.5 micrometer in size per cu ft (0.028 m³) of air. Such a room would be designated a Class 100,000 cleanroom. The most advanced semiconductors today require cleanrooms of Class 1 design. This means the room has fewer than 1 particle larger than 0.5 micrometer (0.0000196 in.) in size per cu ft (0.028 m³) of air. In contrast to this, normal urban air has particle counts in the range of 5,000,000 particles per cu ft (0.028 m³), while rural air averages 1,000,000 particles per cu ft (0.028 m³). The human body can release up to 2,000,000 particles per minute into the atmosphere.

Most materials either conduct electricity (conductors) or do not conduct electricity (insulators). Semiconductor materials do not fall into either of these classifications, since they are neither good conductors nor good insulators. Their electrical conductivity can be altered by minute changes in their chemical makeup or by creating changes in electrical fields inside the device. Many different ways exist to create such changes, and much of the research in the field has been directed toward improving present methods and developing new ones. This capability to change the electrical properties of a material, or of selected regions of the material, is what makes semiconducting materials so valuable in modern electronics.

In the manufacturing process, selected areas of the devices are made to be either conductors or insulators, thus creating electrical circuits, transistors, capacitors, etc. Highly sophisticated manufac-

turing techniques allow circuit patterns as narrow as 5 millionths of an inch (¹/₈ micrometer) in width. The circuit patterns are applied in sequential layers, one upon the other. Forty or more layers of circuit patterns may be combined on a single chip, depending on chip design. A modern 4-Mb (4 million bits of information) memory chip typically contains over 8 million individual transistors and capacitors.

The loss potential in a semiconductor manufacturing plant can be enormous. Typical plants cost more than $100 million to construct and $1 billion to equip. Add to this the value of completed and in-process wafers, which are very susceptible to damage from fire and smoke, and the potential disruption in production schedules, product distribution, and business interruption, the potential for loss can be staggering.

The in-process or finished devices generally present no unique hazard. In contrast to this, the manufacturing process presents several unique hazards to production workers and fire fighters. Acids, alkalis, flammable liquids, flammable gases, pyrophoric gases, and toxic gases are common in semiconductor manufacturing facilities.

THE PRODUCTION PROCESS

Numerous steps are involved in the manufacturing process of an integrated circuit, or chip. In general, these can be broken down into pre-production, wafer fabrication, and post-production. The pre-production steps include circuit design, mask production, crystal growth, slicing, and preparation. The wafer fabrication process includes a number of different operations that are used to create the intricate pattern of the circuit in successive layers. Figure 3-30A shows the general process. Post-production steps include testing, sawing, bonding, encapsulating, retesting, and packaging the devices. This chapter will emphasize pre-production and wafer fabrication, as it occurs in the cleanroom environment.

Pre-Production

The pre-production processes center on the preparation of materials needed in the fabrication of the chip. These materials include the base crystal or wafer and the masks necessary to develop the various circuits. A chip is constructed of millions of individual transistors, capacitors, and the paths connecting them. These components are constructed on the chip in layers. Each layer must align perfectly with the layers below. Failure to attain this perfect alignment results in chips that do not function.

William B. Marshall is senior engineer, AT&T Company, based in Orlando, FL. He is a member of the NFPA Technical Committee on Cleanrooms.

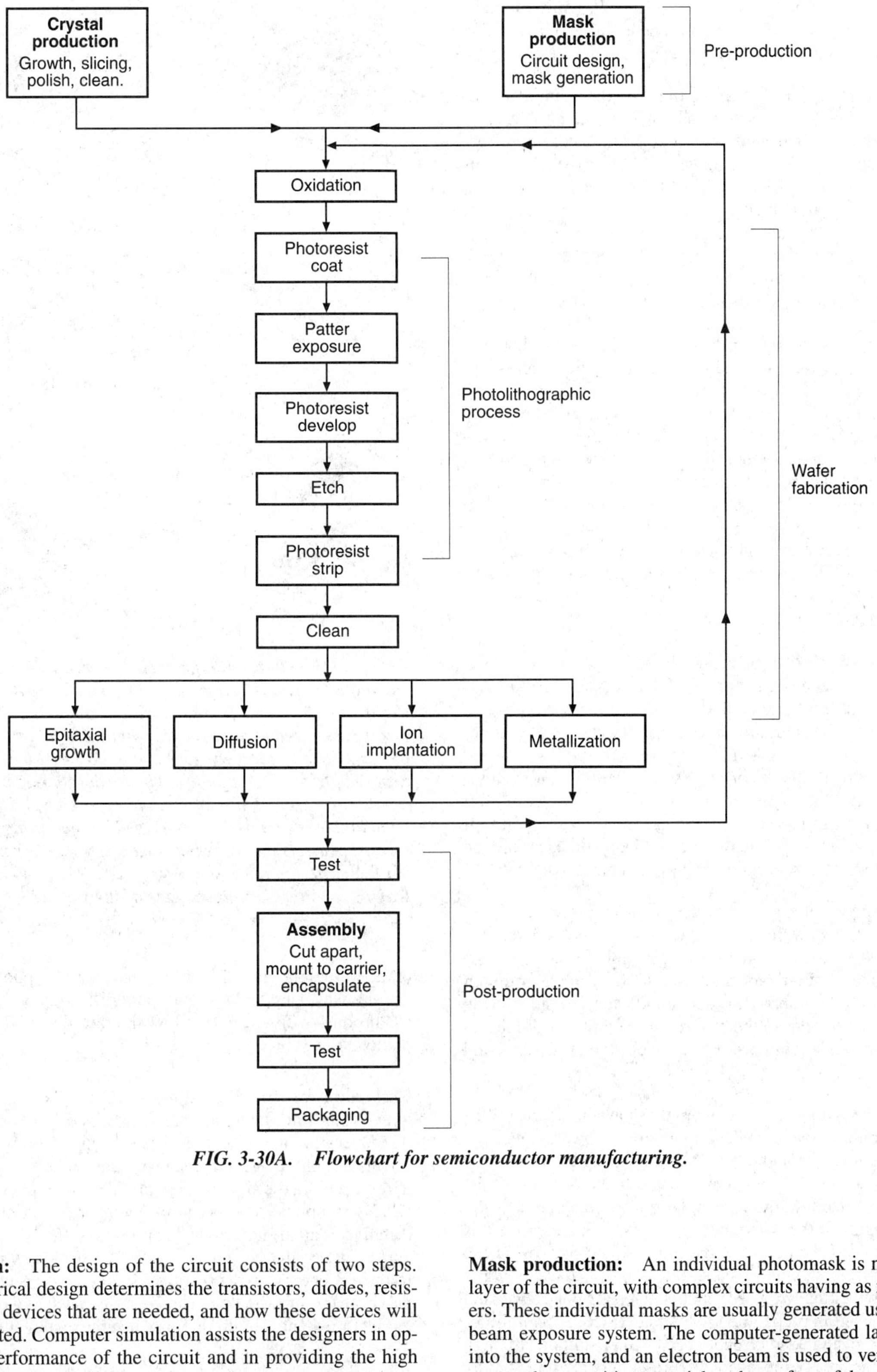

```
┌──────────────────────┐              ┌──────────────────────┐
│   Crystal            │              │    Mask              │
│   production         │              │    production        │    Pre-production
│ Growth, slicing,     │              │ Circuit design,      │
│   polish, clean.     │              │ mask generation      │
└──────────────────────┘              └──────────────────────┘
```

```
                        ┌──────────────────┐
                        │    Oxidation     │
                        └──────────────────┘

                        ┌──────────────────┐
                        │  Photoresist     │
                        │     coat         │
                        └──────────────────┘

                        ┌──────────────────┐
                        │    Patter        │
                        │   exposure       │              Photolithographic
                        └──────────────────┘                  process

                        ┌──────────────────┐
                        │  Photoresist     │
                        │    develop       │
                        └──────────────────┘

                        ┌──────────────────┐
                        │      Etch        │
                        └──────────────────┘

                        ┌──────────────────┐
                        │  Photoresist     │
                        │     strip        │
                        └──────────────────┘

                        ┌──────────────────┐
                        │     Clean        │
                        └──────────────────┘
```

Epitaxial growth	Diffusion	Ion implantation	Metallization

Wafer fabrication

```
                        ┌──────────────────┐
                        │      Test        │
                        └──────────────────┘

                        ┌──────────────────┐
                        │    Assembly      │
                        │   Cut apart,     │
                        │ mount to carrier,│              Post-production
                        │   encapsulate    │
                        └──────────────────┘

                        ┌──────────────────┐
                        │      Test        │
                        └──────────────────┘

                        ┌──────────────────┐
                        │    Packaging     │
                        └──────────────────┘
```

FIG. 3-30A.   Flowchart for semiconductor manufacturing.

Circuit design: The design of the circuit consists of two steps. First, the electrical design determines the transistors, diodes, resistors, and other devices that are needed, and how these devices will be interconnected. Computer simulation assists the designers in optimizing the performance of the circuit and in providing the high level of reliability required.

The second step is the layout of the circuit. The designer uses computerized tools to translate the electrical design to physical patterns on the chip. Computer-aided layout permits precise, error-free positioning of the hundreds of thousands of geometric elements required to form a single chip design.

Mask production: An individual photomask is needed for each layer of the circuit, with complex circuits having as many as 40 layers. These individual masks are usually generated using an electron beam exposure system. The computer-generated layout is entered into the system, and an electron beam is used to very precisely expose a photosensitive material on the surface of the chromium-coated glass mask. This selectively etches off portions of the chromium in the locations exposed to the electron beam. The result is an array of geometric openings in the chromium layer.

The design geometry on these imprinted plates, or masks, is photolithographically printed onto (silicon) wafers during the wafer fabrication process. Although the chips are small, many identical patterns can be placed on a single mask. The ability to make masks and use them is a complicated factor in chip manufacturing.

Crystal growth, slicing, and preparation: In the manufacture of silicon-based devices, polycrystalline silicon is placed into the crystal grower, an electrically heated, water-cooled furnace. Impurities, also known as dopants, are added to the grower in precise amounts to control the electrical resistivity of the final crystal. To form a single crystal, the charge of polycrystalline silicon and dopants is heated in the crystal grower until the material is completely molten at 2,552°F (1,400°C). A rotating starter crystal, or seed, is dipped into the molten silicon, and withdrawn very slowly. As the seed is withdrawn, a single crystal ingot grows around it.

The final ingot is ground into a cylinder of the precise diameter required for further processing, and sliced into very thin wafers, approximately 0.03 in. (0.76 mm) thick. The resulting wafers are lapped, or ground, to obtain perfectly flat, parallel surfaces. Next, the wafers are polished to remove any residual surface damage, using abrasives and chemical etchants, such as nitric, hydrofluoric, and sulfuric acid, and sodium hydroxide. The wafers are then cleaned by rinsing with deionized (DI) water and washing with an organic solvent, such as acetone, isopropanol, methanol, or 1,1,1-trichloroethane. Also, some combinations of acid may be utilized. At this stage, the wafers are ready for the fabrication process.

Wafer Fabrication

The steps used in the fabrication of specific chips vary widely. Generally, the steps are repeated a number of times during the fabrication process, in differing orders, to define each successive layer of the circuit. Most of the steps are pattern-creating steps that first determine a pattern on the chip and then create portions of the circuit in the pattern defined. The major processes usually include photolithography, epitaxial growth, deposition, diffusion, oxidation, ion implantation, metallization, etching, and cleaning. As noted in Figure 3-30A, the process steps can be repeated many times before the chip is completed. Each process is discussed in more detail below.

Photolithography: This process step is, in many respects, the most critical process in wafer fabrication, as it is used to define the circuit components and paths on the wafer. Photolithography is similar to the simple photography processes used to make common snapshots. Photolithography in semiconductor manufacturing projects the image or defines the pattern where portions of the circuit will be created in operations, such as deposition, diffusion, and ion implantation.

A typical photolithographic sequence as shown in Figure 3-30B, using negative photoresist, includes application of the photoresist (a light-sensitive emulsion) on the wafer, alignment of the mask over the wafer, exposure of the uncovered portions of photoresist through the mask using ultraviolet light, and development of the photoresist to remove the unexposed photoresist. Removal of this unexposed photoresist creates a three-dimensional pattern on the chip exposed for the etching process that follows. After etching is complete, the remaining (exposed) photoresist is stripped away and the wafer is cleaned in preparation for the next step. A photolithographic sequence using positive photoresist is very similar, although in that process the areas exposed to ultraviolet light are removed by the developing process.

Equipment used in photolithography includes wafer spin coaters, wafer aligners, ovens, steppers, and wafer developers. Either positive or negative photoresist may be used. (Note: Generally, positive resists are not flammable and negative resists are flammable solutions.) Various chemicals are used in the development and processing

FIG. 3-30B. Photolithography. A pattern is etched onto a silicon dioxide layer that has been (1) grown on to the surface of the wafer. (2) The wafer is coated with photoresist. (3) The coated wafer is exposed to ultraviolet light through the patterned mask. (4) Exposure renders the exposed photoresist insoluble. The unexposed photoresist is removed in the developing process, leaving a patterned area of silicon dioxide exposed. (5) The exposed silicon dioxide is removed by etching. The remaining hardened photoresist protects the unexposed silicon dioxide from the etchant. (6) Remaining photoresist is stripped away in a chemical bath leaving a pattern on the surface of the wafer.

of the wafers for photolithography. Because the photoresist reacts to certain wavelengths of light, photolithography is performed in cleanrooms with special yellow lighting that will not expose the photoresist.

Epitaxial growth: Epitaxial growth usually occurs at the beginning of the wafer fabrication process. Epitaxy is the deposition of a thin layer of material on the surface of the wafer with exact crystal orientation of the wafer. It can be used to make a very abrupt change in resistivity in the silicon substrate. The chemical vapor deposition (CVD) of epitaxy can be atmospheric, reduced pressure, or by a technique called molecular beam epitaxy (MBE). In conventional epitaxy, the wafer is heated to 1,472°F (800°C) by impinging radio frequency energy on the surface. Hydrogen gas is used to clean and passivate the surface, then reactive gases, such as dichlorosilane and tetrachlorosilane, are used to produce the epitaxial layer.

Deposition: Deposition is used to introduce dopants, or impurities, into the exposed portions of the crystal structure of the wafer to change the electrical characteristics of that area. Typical dopant sources are: arsine, arsenic trichloride, arsenic trioxide, boron tribromide, boron trichloride, boron trioxide, diborane, boron nitride, phosphorus oxychloride, phosphorus tribromide, phosphorus pentoxide, or phosphine. These dopants are passed over the wafer surface at temperatures ranging from 1,472 to 1,832°F (800 to 1,000°C) causing the atoms to reside near the surface of the wafer.

Diffusion: Diffusion is the process of redistributing the dopant atoms deeper into the wafer bulk. Tube furnaces, using electric heating elements to bring the wafers to 1,652 to 2,192°F (900 to 1,200°C), are used to provide the energy necessary to redistribute the dopants to a predetermined depth.

Oxidation: Oxidation is used to grow a layer of oxide on the surface of the wafer. Oxidation is frequently performed in a tube furnace, as is diffusion, at temperatures between 1,472°F (800°C) and 1,832°F (1,000°C) (lower temperatures than diffusion). In this process, steam and either wet oxygen or dry oxygen are introduced into the furnace to cause growth of the layer. Alternate techniques for depositing an oxide layer utilize the decomposition of silane, or other silicon-containing gases, in either a high- or low-pressure environment. The thickness of the oxide layer is controlled by the time and temperature in the furnace and the rate of gas flow.

Ion implantation: Ion implantation is used to place atoms of a particular dopant into the crystal structure of the wafer. The wafer is loaded into the ion implanter, and a controlled high-energy beam of ions of the selected dopant gas is directed at the wafer. The beam is directed at the entire wafer, implanting ions into all exposed regions. Common ion implant gases are arsine, boron trifluoride, diborane, hydrogen, and phosphine. (Note: Some processes may utilize solid sources of dopants in substitution of gaseous dopants.)

Metallization: Metallization is used to interconnect the various elements to form a completed circuit. In metallization, the exposed regions of the wafer are coated with the metal to create circuit paths. Typical metallization processes include flash evaporation and sputtering. The process is conducted in a vacuum where the metal is evaporated by an electron beam or radio frequency (RF) induction. The evaporated metal condenses on the wafer in the desired film thickness. Among the materials used are the metals aluminum, gold, silver, nickel, chromium, and titanium, and metal silicides.

Etching: Most wafers go through the etching process several times at different stages of fabrication. Etching is used to remove selected portions of the chip pattern in preparation for further process steps. Two etching processes are used: (1) wet etching and (2) dry etching.

Wet etching is frequently performed in an open bath, commonly called a wet bench, in the cleanroom, although automated enclosed robotic units are available. Ammonium fluoride and hydrofluoric acid are used as oxide etches; mixtures of acetone, nitric, and phosphoric acids are used as metal etches. Wet benches also typically contain baths for neutralizing or rinsing the etchant from the wafer and for cleaning the wafer.

Dry etching is performed in an evacuated chamber where an etchant gas is passed over the wafer. Boron trichloride, chlorine, nitrogen trifluoride, oxygen, trifluoromethane, and tetrafluoromethane are common etchant gases.

Cleaning: Wafer cleaning is necessary at many stages in the fabrication process to ensure proper preparation for critical operations that follow. Most cleaning is performed in open baths in the wet benches, using sulfuric acid, hydrogen peroxide, nitric acid, hydrofluoric acid, hydrochloric acid, and ammonium hydroxide.

Post-Production

Test and repair: Once the wafer fabrication is completed, the wafer is cleaned and ready for testing. Each of the chips on the wafer (from a few to over 600 chips on a single wafer, depending on the chip and wafer size) is tested for proper operation by a computerized test apparatus. Chips failing the test may be repaired using a laser repair system, depending on the chip design. Defective chips are marked for identification, and the wafer is sent on to the assembly area.

Assembly and test: Following testing, the wafer is sawed with a diamond saw into individual chips. The good chips are sorted out and bonded into packages, or chip carriers. Circuitry on the chip is connected to the leads on the carrier by a computerized machine using gold wires thinner than a human hair. The package is then encapsulated in an epoxy material for protection and tested once again. Chips passing the final quality control checks are packaged for shipment to the customer.

HAZARD CONTROL IN SEMICONDUCTOR MANUFACTURING

Several classes of hazards are encountered in semiconductor manufacturing plants. Most are process-related and include flammable liquids and gases, pyrophoric and toxic gases, toxic liquids and solids, radio frequency fields, and ionizing radiation.

Most of these potential hazards are commonly found in other industries. Plating operations, crystal growing, flammable liquids, and flammable gases all present hazards that are widely recognized and for which effective loss prevention controls are established.

There are, however, several hazards that are not widely encountered. Pyrophoric gases, toxic materials (gases, liquids, and solids), radio frequency fields, ionizing radiation, and special handling techniques required for certain chemicals due to purity constraints are examples of uncommon potential hazards that may be encountered.

General Safeguards

Semiconductor manufacturing cleanrooms should be located in fire-resistant construction and must be equipped with a full automatic fast-response sprinkler system. The cleanroom should be cut off from other areas of the plant by fire-rated construction having a minimum rating of one hour. Because of the high values present, additional compartmentation is highly desirable. Ventilation systems, including recirculating and exhaust systems, should be independent of ventilation systems in adjoining compartments.

Recirculating air systems should be designed with the flexibility to shut down in the event of a fire or chemical spill. Exhaust systems serving processes using toxic materials should continue running during such incidents. This will allow continued removal and treatment of potentially dangerous fumes and vapors from the cleanroom, as well as assist in smoke removal. Most of these exhaust systems contain treatment processes to remove or neutralize the chemical(s) in the exhaust stream. Caution should be used in placing automatic sprinklers inside ducts used for fume exhaust, because tests have demonstrated that sprinklers can significantly reduce the exhaust flow through the duct when operating. This may be undesirable in exhausts handling toxic fumes or vapors.

Fire extinguishers and hose stations should be provided. The minimum extinguisher capability should be distributed on an extra-hazard schedule in fabrication areas. (See NFPA 10, *Standard for Portable Fire Extinguishers*.) Carbon dioxide (CO_2) fire extinguishers are the most commonly used in cleanrooms. Dry chemical extinguishers are not recommended for use in cleanrooms. The powders used in these extinguishers will contaminate much of the cleanroom when discharged, causing any wafers in the room to be contaminated beyond salvage. Most of the equipment will require meticulous and expensive cleaning to remove the powder, and the high-efficiency particulate air (HEPA) filters will probably have to be replaced at great expense. In specific isolated situations, combustible metal extinguishing powders, such as MET-L-X, may be appropriate at combustible metal processes.

Selected process equipment or wet benches may be considered for protection with CO_2 extinguishing systems. However, process constraints should be evaluated when considering this option.

FIRE HAZARDS

Flammable Liquids

A number of fairly common flammable liquids are used in the fabrication process. Most are alcohols, or alcohol-based mixtures. When these liquids are to be used in direct wafer processing, i.e., in contact with the wafers, absolute purity of the liquid must be maintained. As in the cleanroom atmosphere itself, a single microscopic particle can contaminate the chip, rendering it useless.

Storage and handling of flammable liquids should generally follow the principles outlined in NFPA 30, *Flammable and Combustible Liquids Code*. Special precautions are necessary to handle flammable materials safely. These include carts designed specifically to transport the containers safely from the flammable liquid storage room to the cleanroom. These carts hold the containers firmly upright and can even contain the contents in the event of leakage. Regardless of container construction, flammable liquids not being used should be stored in listed flammable liquid storage cabinets. Personnel training and restricted travel routes for the carts between storeroom and cleanroom should be provided.

In some cases, the flammable liquid is piped into the process equipment. The bulk supply will vary in size and location according to process requirements. Appropriate safeguards must be taken to ensure stability of the pipe system. Bulk dispensing in closed, dual-contained systems is becoming more prevalent throughout the industry.

Flammable and Pyrophoric Gases

Some of the gases used in semiconductor manufacturing are either flammable or pyrophoric (i.e., ignite spontaneously upon contact with air). The gases may be pure, or they may be found as mixtures, with a carrier gas such as hydrogen or nitrogen. Pyrophoric and toxic gases should be placed outside the plant. Gas may be piped to the cleanroom in welded, double-walled, stainless steel tubing. The gas travels in the inner tube(s), while the annular space between the outer tube and the inner tube(s) is used to confine any leak from the inner tube(s). This annular space is usually inerted or evacuated, and monitored by pressure sensors and/or gas detectors to automatically shut off the flow of gas in the event of a leak.

Process equipment using high concentrations of pyrophoric gases, such as silane, should be equipped with exhaust gas conditioning in the exhaust stream. These exhaust systems should be equipped with continuous monitoring devices for abnormal conditions.

The burning characteristics of pyrophoric gases, such as silane, dichlorosilane, diborane, and phosphine, are not fully understood. This is mostly because the engineering controls provided for these processes have been so successful in eliminating uncontrolled releases. Experience has shown that when a leak occurs in an unconfined area (e.g., outdoors), the pyrophoric gases usually ignite upon release, and subsequently burn in a manner similar to other flammable gases. When released into a confined area, such as a gas cabinet, silane has been known to detonate.

Cleanroom Furnishings

Work stations, wet benches, and other equipment used in fabrication areas are specially designed to reduce dust accumulation and release of particulates (from the finish treatment) into the room, and for compatibility with the chemicals used in or around the equipment. Fiberglass-reinforced plastic ducts for exhaust of corrosive fumes are usually present.

HEALTH HAZARDS

Health hazards are also present in semiconductor manufacturing, but are beyond the scope of this handbook. They are mentioned briefly only to acknowledge their existence. For additional information concerning any particular hazard, consult an industrial hygienist or safety engineer. Semiconductor plants should have persons on staff trained in these disciplines.

Several of the gases used, such as arsine, phosphine, and diborane, are highly toxic. The toxicity hazards are controlled and monitored on a continual basis. In addition to the piping and control systems, the exhaust systems from the process tools should include treatment processes for these materials to avoid release into the atmosphere.

Many of the liquid and solid chemicals used in semiconductor manufacturing are also toxic. Examples include arsenic, various acids, and solvents. Safety considerations including ventilation and employee training are required.

In addition to chemical hazards present in the fabrication area, there are sources of ionizing radiation, laser radiation, RF, and UV exposure. The machines should have appropriate safety devices and controls incorporated into them, and follow the Semiconductor Equipment and Materials International standards (SEMI) for cleanroom equipment.

BIBLIOGRAPHY

NFPA Codes, Standards, and Recommended Practices

Reference to the following NFPA codes, standards, and recommended practices will provide further information on the processes and protective systems for semiconductor manufacturing discussed in this chapter. (See the latest version of *The NFPA Catalog* for availability of current editions of the following documents.)

NFPA 10, *Standard for Portable Fire Extinguishers*
NFPA 12A, *Standard on Halon 1301 Fire Extinguishing Systems*
NFPA 13, *Standard for the Installation of Sprinkler Systems*
NFPA 24, *Standard for the Installation of Private Fire Service Mains and Their Appurtenances*
NFPA 25, *Standard for the Inspection, Testing, and Maintenance of Water-Based Fire Protection Systems*
NFPA 30, *Flammable and Combustible Liquids Code*
NFPA 70, *National Electrical Code®*
NFPA 75, *Standard for the Protection of Electronic Computer/Data Processing Equipment*
NFPA 86, *Standard for Ovens and Furnaces*
NFPA 90A, *Standard for the Installation of Air Conditioning and Ventilation Systems*
NFPA 91, *Standard for Exhaust Systems for Air Conveying of Materials*
NFPA 231, *Standard for General Storage*
NFPA 231C, *Standard for Rack Storage of Materials*
NFPA 255, *Standard Method of Test of Surface Burning Characteristics of Building Materials*
NFPA 318, *Standard for the Protection of Cleanrooms*

Additional Reading

Anderson, M., "Fire Protection Problems—Electronics and Semiconductor Industries," *Papers* presented at the Summer National Meeting of the American Institute of Chemical Engineers, AIChE, 1987, p.13.

Bolmen, R. A., "Designing for Fire Safety," *Semiconductor International*, Vol. 9, No. 11, Nov. 1986, pp. 112-114.

Brown, A. R., "Computer Chip Clean Rooms: A New Fire Protection Challenge," *Fire Prevention*, No. 287, Mar. 1996, pp. 12–14.

Cider, L., "Cleaning and Reliability of Smoke-Contaminated Electronics," *Fire Technology*, Vol. 29, No. 3, 1993, pp. 226–245.

Fisher, F. L., *et al.*, "Fire Protection of Flammable Work Stations in the Clean Room Environment of a Microelectronic Fabrication Facility," *Fire Technology*, Vol. 22, No. 2, 1986, pp. 148–166.

Fogarty, K., "Responding to Semiconductor Facilities," *Fire Engineering*, Vol. 147, No. 11, Nov. 1994, pp. 46–48.

Nakajima, T. and Miyawawa, I., "Systematization of the Distinction of Mixed Gases Using Semiconductor Gas Sensors," *Bulletin of Japanese Association of Fire Science and Engineering*, Vol. 39, No. 2, 1990, pp. 19–31.

Ohtani, H., "Combustion Behavior of Gases for Semiconductor Production," Yokohama National Univ., Japan, Science University of Tokyo, Japan, '93 Asian Fire Seminar, October 7-9, 1993, Tokyo, Japan, 1993, pp. 59–65.

Okayama, Y., *et al.*, "Carbon Monoxide Sensor Element Made of SnO_2-Sb_2O_3-Pt Semiconductor," 1990, 2 pages.

Peterson, D., and Schwalbe, K., "Preventing Fire Disasters in Computer Rooms," *Consulting/Specifying Engineer,* Vol. 8, No. 3, Sept. 1990, pp. 96–98, 100, 102.

Price, B. J., "Computer Room Fire Protection," *Library Hi Tech*, Vol. 29, No. 1, Apr./June 1990, pp. 43–56.

Simmons, F., "Computer Room Protection—The Debate Continues," *Fire Prevention*, No. 232, Sept. 1990, pp. 19–21.

Singer, P. H., "Wet Bench Fire Suppression," *Semiconductor International*, Vol. 10, No. 10, 1987, pp. 154–157.

Toy, D. A., "Take Those Hazards out of Your Gas Supply System," *Semiconductor International*, Vol. 12, No. 8, July 1989, pp. 66–70.

Werman, L., *et. al.*, "Building a Clean Room One Lab at a Time," *Consulting-Specifying Engineer*, Vol. 18, No. 4, Oct. 1995, pp. 26–28, 30, 32.

WASTE HANDLING AND CONTROL

Lawrence G. Doucet

SOLID WASTE MANAGEMENT SYSTEMS

Although typical solid wastes are rarely a source of ignition, they are a potential source of fuel for fires or ignition from other sources. Therefore, a proper and efficient waste management system incorporating prompt treatment and disposal is essential for good fire safety. This part of the chapter covers the fire safety aspects of solid waste management, treatment, and disposal systems and equipment.

Waste management generally comprises the collection, internal transport, interim storage, treatment, and final disposal of wastes. A great variety of alternate processes and equipment are available for providing these functions as part of a total waste management system. Depending upon the specific types, forms, quantities, and characteristics of the waste, management systems can range from a few simple processes to a complex combination of many processes.

Waste handling generally refers to those functions associated with the movement of wastes after creation, excluding storage, processing, treatment, and final disposal. It includes collection of waste at points of generation, transport of the materials, and unloading from the transport system.

Waste-handling systems also comprise equipment and devices for loading or charging waste into treatment systems and processes. For example, mechanical loading devices may be part of a system for charging solid waste into an incinerator. Waste-handling systems and equipment include conveyors, chutes, carts, transport vehicles, elevators, and lifts. These systems can range in sophistication from fully manual to fully automatic operation.

Waste storage includes the interim containment of accumulated materials prior to subsequent handling, processing, treatment, or disposal. Waste storage, depending on the waste type and form, can be loose, compacted, shredded, or another processed form. Waste can also be stored in bulk or individual containers, such as drums or cartons.

Waste processing includes those functions that prepare or alter the waste by changing its shape, size, uniformity, or consistency. Such processes are usually provided to facilitate handling, storage, treatment, or disposal processes. Waste-processing systems and equipment include compactors, shredders, crushers, pulpers, pulverizers, baggers, encapsulators, extruders, and de-watering devices.

Waste treatment and disposal systems include incinerators, steam sterilizers or autoclaves, and various technologies combining shredding with biomedical-waste disinfection or sterilization processes, such as chemical injection, microwaving, infrared radiation, and gamma radiation.

WASTE STORAGE ROOMS

Rooms in a building or structure that are used for the storage or handling of waste in any form and in amounts exceeding one cu yd (0.765 m³) uncompacted measure are subject to the requirements of NFPA 82, *Standard on Incinerators and Waste and Linen Handling Systems and Equipment* (hereinafter referred to as NFPA 82). The walls, ceilings, and floors of such rooms are required to have fire resistance ratings of at least 2 hr, and room openings must be protected by self-closing fire doors suitable for Class B openings. In addition, such rooms require automatic sprinklers, in accordance with NFPA 13, *Standard for the Installation of Sprinkler Systems* (hereinafter referred to as NFPA 13).

WASTE CHUTES AND HANDLING SYSTEMS

Waste chutes are generally fixed systems for transporting waste from points of generation or interim storage to centralized areas for processing, treatment, or disposal. There are basically five types of chute systems:

1. *General access, gravity-type system.* This type of chute consists of an enclosed vertical passageway through which waste is transferred by gravity alone. All occupants of the building have unlimited access to the chute.

2. *Limited access, gravity-type system.* This type of chute is similar to a general access type except that access is restricted. Entry is gained through locked chute doors or locked service opening room doors. These chutes are used primarily in hospitals and health care institutions.

3. *Pneumatic system.* This type of system uses air flow to transport waste from chute service openings to a central collection area. Such systems have chutes that may run horizontally, vertically, or inclined, depending upon blower and design characteristics. High-velocity air, which typically transports waste materials at speeds of about 60 mph (79 km/hr), is aspirated through the top of the chute and exhausted at its termination. These systems are usually found in hospitals, but they can be

Lawrence G. Doucet, P.E., DEE, is president of the environmental consulting and engineering firm of Doucet & Mainka, P.C., Peekskill, NY.

used in any size or type of building. Figure 3-31A shows a diagrammatic view of a pneumatic chute system.

4. *Gravity-pneumatic systems.*   This type of system uses a conventional gravity chute to feed a collecting chamber which, in turn, feeds a pneumatic chute system.

5. *Multiloading pneumatic systems.*   This type of system is similar to a pneumatic type except that entry is gained through automatic or self-closing, positive-latching inlet doors that are interlocked so that only one can be opened at a time and not when the material damper is open.

Waste Chute Design and Construction

Waste chute design and construction are critical to fire safety in waste-handling systems. Chutes could readily serve as channels capable of carrying smoke and flames throughout a building or facility. They are often considered as waste storage areas since they could become clogged with waste during use and present fire and smoke hazards. Criteria relative to chute construction, chute enclosures, fire dampers, sprinklers, service openings, and the like are all directed toward minimizing and controlling potential fire hazards. In this regard, strictly adhere to the requirements of NFPA 82.

Gravity-type chutes may be constructed of unlined steel, refractory-lined steel, or masonry. Pneumatic waste-handling systems are constructed only of unlined steel. Unlined chutes are fabricated of stainless, galvanized, or aluminized steel for corrosion resistance. Each type of chute construction has separate design and construction criteria, as detailed in NFPA 82, for providing structural integrity combined with adequate fire safety measures.

Chute size: Gravity-type chutes are required to be not less than 22.5 by 22.5 in. (570 by 570 mm) or not less than 24 in. (610 mm) inside diameter. Pneumatic waste-handling system chutes are required to be not less than 16 in. (406 mm) inside diameter, but they may be smaller if all waste materials are first shredded. However, these sizing criteria are strictly minimum requirements. Waste types and volumes, waste package sizes, user habits, and specific applications must be carefully considered before accepting minimum sizing for chutes and service openings. For example, a kitchen compactor may produce a typical package measuring $16 \times 16 \times 9$ in. ($406 \times 406 \times 230$ mm), which would have a recommended chute diameter of about 36 in. (915 mm).

Chute service openings: Properly sized chute service openings are necessary to prevent chute clogging. Oversized openings increase the probability that overly large or excessive quantities of waste will be loaded into the chute and become clogged. Standards

Suction generator

Discharge collectors

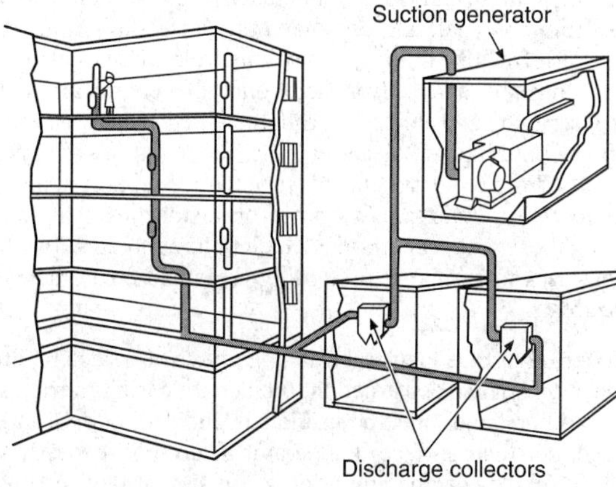

FIG. 3-31A.   A diagrammatic representation of a buildingwide pneumatic chute system.

for maximum service opening sizes, developed to prevent such conditions, are as follows:

1. General access gravity chute service openings are limited to one-third of the cross-sectional area of a square chute or 44 percent of the cross-sectional area of a round chute.

2. Limited access gravity-type chutes are limited to two-thirds of the cross-sectional area of the chute.

3. Pneumatic chute service openings may be equivalent to or greater than the cross-sectional area of the chute. However, pneumatic system service entrances, or charging stations, comprise a compartment with electrically interlocked inner and outer doors. The capacity of the charging compartment, with only one door being allowed open at a time, limits the volumes or sizes of waste that can be loaded into the chute.

Chute doors: Chute service openings are required to be equipped with fire-rated door enclosures to prevent or minimize the spread of chute fires. Otherwise, natural airflows and pressures could cause flames and smoke to be exfiltrated into other building areas. According to NFPA 82, chute doors must also be self-closing and positive-latching types. Also, pneumatic system service openings are required to have two electrically interlocked doors. For structural integrity, door frames must be securely fastened to both the chute and chute enclosure walls.

Chute enclosures and penetrations: In order to ensure fire integrity, chute system enclosures, including floor and wall penetrations, require fire-rated protection. Metal chutes, which in themselves offer no fire resistance, must be totally enclosed within walls of masonry or other noncombustible, fire-rated material. However, masonry chutes meeting the fire resistance requirements of NFPA 82 are not required to be enclosed. If a masonry chute penetrates a floor without closing the opening entirely, the space between the floor and chute must be filled with material equivalent to the floor in fire resistance.

Wherever a waste chute penetrates a fire-rated floor-and-ceiling assembly, fire wall, or other fire-rated assembly, a fire damper is required. However, a fire damper may not be needed if the chute is enclosed within a fire-rated shaft. Schematic diagrams of typical chute system fire rating requirements are shown in NFPA 82.

Chute discharge and terminal rooms: Waste chutes are required to discharge or terminate in an enclosed room or compartment separated from other parts of the building. Since a chute-terminal enclosure area is subject to potential ignition sources, the walls, ceiling, and floor of the room or compartment must have a fire-resistance rating of 1 or 2 hr, depending on the number of stories in the building and whether a gravity or pneumatic system is used. Terminal enclosures are required to have automatic fire doors, automatic sprinklers, and a ventilation system designed to exhaust fire and smoke to the outdoors. These requirements are in recognition of the seriousness of fires from waste, which are generally difficult to control and typically generate large quantities of smoke.

In gravity-pneumatic-type chute systems, the collecting chambers, which can be considered interim discharge or terminal points, are vulnerable to potential fires. They should be adequately sprinklered and provided with ready access for fire-fighting measures. Also, the valve room of the system should be sprinklered to aid in cooling the space.

Some older gravity chute systems discharged directly into an incinerator, and, in some systems, the waste chute also served as the incinerator chimney. Because of the tremendous potentials for fire and smoke hazards, most of these older installations have been discontinued or abandoned. Today, waste chutes are not permitted to discharge directly into an incinerator. Furthermore, because of the inherent difficulties in maintaining fire integrity, systems for automatically transferring waste materials from chute terminal enclosures directly into an incinerator are not recommended.

Chute sprinklers: According to NFPA 82, chute systems must be either sprinklered or they must be fully capable of containing a fire and venting products of combustion. Metal chutes require automatic sprinklers at the top and at alternate floor levels. Specific requirements for chute sprinklers are in NFPA 13. A good practice is to locate sprinklers where they can be inspected and maintained and yet be out of reach of vandals and beyond the range of falling waste. Masonry chutes or chutes constructed in accordance with requirements for factory-built, medium-heat appliance-chimney sections are not required to be sprinklered.

Chute service opening enclosures: For many years it was general practice to put chute service openings in public corridors for convenience sake. However, aside from the problems of unsightliness and poor sanitation, fire hazards were seriously increased and the fire integrity of a building reduced. Fatalities occurred when chute fires spilled into corridors and cut off means of escape. NFPA 82 requires every service opening to be located in a room or compartment separated from other parts of the building by an enclosure or room having fire-rated wall-and-ceiling assemblies and with the opening into the enclosure protected by a self-locking fire door.

INCINERATORS

Incineration is an engineered process using high-temperature combustion for the treatment and disposal of waste. Properly designed and operated systems can reduce the weight and volume of most residential, commercial, and institutional solid waste by as much as 95 percent or more, thereby greatly reducing off-site haulage and disposal costs. Incinerators are often used for the destruction and sterilization of biomedical and pathological wastes at health care facilities, universities, research facilities, and the like. Some facilities use on-site incineration for the destruction of hazardous and toxic wastes. In addition, incineration with waste heat recovery is often an attractive and cost-effective alternative energy source for many facilities. In terms of fire safety, incineration provides prompt and complete waste destruction, which substantially reduces interim, on-site waste storage requirements and associated potential fire hazards.

Many early incinerator designs comprised little more than enclosed, single-chamber fire boxes in which uncontrolled burning took place. Such systems provided poor waste destruction and were notorious sources of smoke and odor problems. However, modern technology offers systems that are readily capable of meeting demanding waste destruction and performance requirements in compliance with stringent environmental regulations. In addition, properly designed, constructed, and operated systems, equipment, and support facilities can be completely safe in terms of fire safety.

Incineration Fire Safety

The very nature of incineration presents potential fire hazards: storage and handling of combustible, sometimes highly volatile waste; the presence of high-temperature, flaming combustion; exhaust and ducting of high-temperature combustion gases; the handling of hot ashes; and the presence of fuel-handling and combustion systems.

Incineration fire safety must take into account not only the obvious measures such as the presence of sprinklers and room fire-rating but also less apparent, indirect measures as related to facility

planning and design, such as system layout, equipment orientation, waste flow patterns, system operations, and maintenance practices.

Incinerator Categories

There are several different incinerator technologies and designs. There is also a wide variation in the types and characteristics of waste that can be incinerated. This section is not intended to cover or include all of the design and construction details for each type of solid waste incinerator technology and application. However, all design, construction, control, and other features as needed to reduce or minimize fire hazards are required for all incinerator facilities.

The basic types of solid waste incinerators that are typically used in residential, institutional, and commercial facilities are as follows.

Multiple-chamber incinerator: Consists of a primary and one or more secondary chambers, and operates with high-excess combustion air—typically 200 to 300 percent. Wastes are loaded into the primary chamber for combustion, and most combustibles entrained in the flue gases are burned in the secondary chambers. These are designed and constructed in accordance with standards developed in the mid-1950s. However, very few multiple-chamber incinerators have been built over the last 10 to 15 years. (See Figure 3-31B.)

Controlled-air incinerator: A two-stage combustion process: (1) wastes in the primary chamber, or first stage, are burned under starved air, or oxygen-deficient conditions; and (2) smoke and volatiles from the first stage are combusted in the secondary chamber under excess air conditions—typically 100 to 200 percent. It is estimated that more than 90 percent of the incinerators built over the last 20 years have been of the controlled-air type. (See Figure 3-31C.)

Rotary kiln incinerator: Comprises a cylindrical combustion chamber, or kiln, oriented at a slight incline to horizontal, which rotates slowly on its axis. Wastes are loaded at one end, and kiln rotation provides turbulence for thorough waste destruction as well as for the discharge of ash residues from the opposite end. Combustibles entrained in the flue gases from the kiln are burned in a secondary chamber. Over the last 5 to 10 years, these systems have been used in an increasing number of institutional and commercial applications and facilities.

Design and Construction

Incineration systems and equipment, including breachings and stacks, are subject to extremely severe operating conditions. These include very high and widely fluctuating temperatures, thermal shocks from wet material, residues that cause slag, clinkers and spalling of refractory linings, explosions from items such as aerosol cans, corrosive attacks from acids and other chemicals, and mechanical scraping and abrasions from metallics in the waste stream, as well as operating and cleanout tools. Incineration systems must be suitably designed, constructed, structurally supported and reinforced to resist cracking, warping, or distortions that may be caused by these severe conditions. Such failures could allow flames and hot gases to escape into building areas and create dangerous fire hazards.

Exterior surface incinerator temperatures must also be limited by proper design and construction. Incinerator casings normally accessible to personnel must be insulated, shielded, or appropriately shrouded so that surface temperatures do not exceed 160°F (71.1°C). Handles of operating doors and similar devices must be designed for a maximum 40°F (10°C) rise if metallic, and 60°F (15°C) rise if nonmetallic.

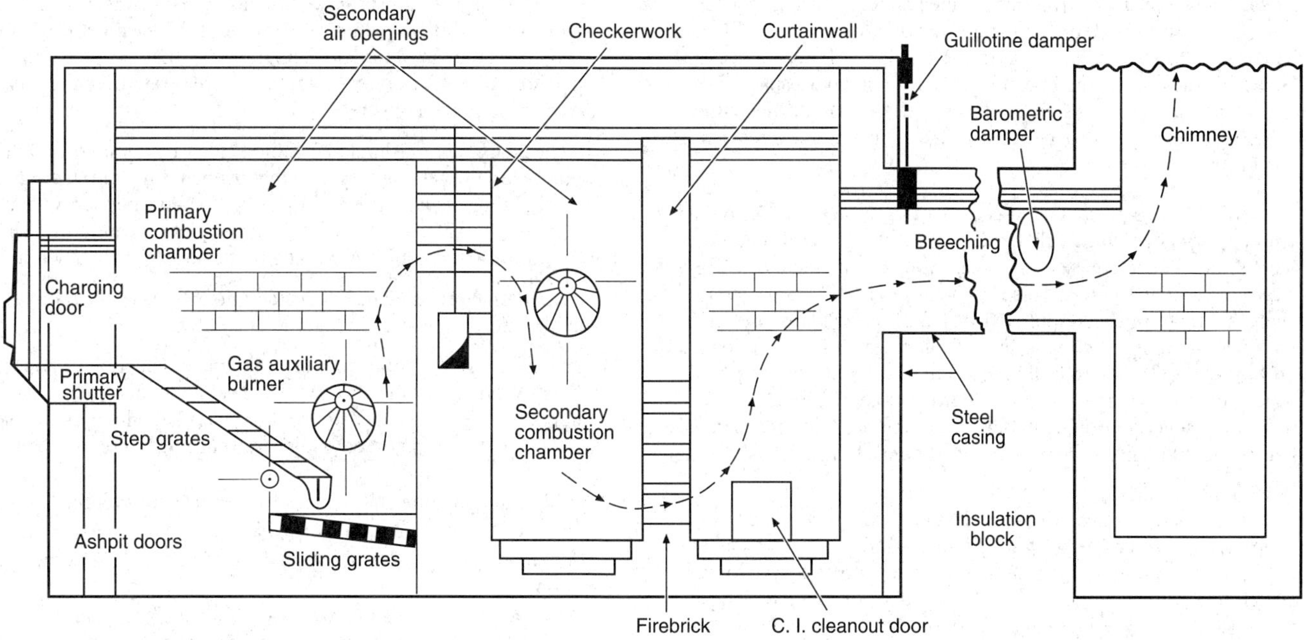

FIG. 3-31B.   *A typical in-line type multiple-chamber incinerator.*

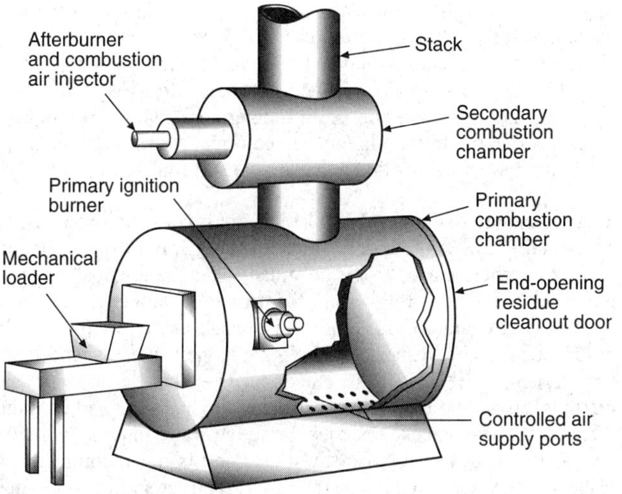

FIG. 3-31C.   *A schematic of a typical controlled-air incinerator.*

Explosion-relief protection is also a requirement. Incinerators must be equipped with a suitable relief door or panel with an effective area of 1 sq ft per 100 cu ft (0.0929 m^2/2.83 m^3) of primary combustion chamber volume.

Location and Arrangement

Incineration systems should be well planned and implemented to ensure that: (1) waste and residue containers do not block charging and cleanout operations or access to work areas or passageways; (2) waste can be charged in a smooth, efficient manner; (3) all parts of the incinerator, including the ash pit, combustion chambers, and breechings are easily and safely accessible for cleaning, repair, and servicing; and (4) clearances above the charging door and between the incinerator top and sides to combustible materials are in compliance with codes and standards.

Charging Systems

Loading of waste into an incinerator has a particularly high fire-hazard potential, i.e., it provides a direct interface between high-temperature, flaming combustion conditions within the incinerator and surrounding building areas. Inadequate design and improper operation could permit radiant heat, flames, and combustion products to escape from the incinerator into building areas.

Two methods of incinerator charging are: (1) manual loading directly through a charging door and (2) mechanical loading with an automatic-airlock charging system.

Direct manual loading: This method is often used on incinerators that charge less than about 500 lb per hr (227 kg/hr), as well as on systems that are batch loaded or that require loading at very infrequent intervals. Fire-prevention measures include minimizing charging-door opening times and eliminating spillage and blockage. The following should be provided:

1. Charging-door openings large enough to accept the largest or bulkiest waste items without blockage or obstruction. Clear opening areas should not be less than 24 by 24 in. (610 by 610 mm). However, excessively oversized openings must be avoided.
2. Charging-door openings and frames that have smooth edges and are free of projections that may catch or hang up waste materials and cause spillage.
3. Charging doors that operate easily and have positive latching devices and handles that remain relatively cool. Guillotine charging doors should be counterweighted or motor operated.
4. Interlocks to shut off primary chamber burners when the charging door is opened.
5. Loading trays, where appropriate, in front of charging doors to assist the operators in handling and loading heavy and unwieldy waste materials.

Mechanical loading: Automatic, mechanical charging systems are almost always used on incinerators greater than about 500 lb per hr (227 kg/hr) and where controlled charging rates are needed. However, they are sometimes employed on smaller-capacity systems where a continuous airlock interface is required. Mechanical loaders are also used to protect against incinerator overcharging conditions.

NFPA 82 requires that waste-charging systems be provided with appropriate controls that will prevent the direct discharge of flames, combustion gases, and heat from the incinerator during waste-loading operations. This is to include a hopper/ram-type mechanical loader or other approved system. The standard also stipulates that the only exceptions are those incinerators that are loaded on a batch basis, whereby the charging door is not opened while waste combustion is taking place. Furthermore, it should be noted that many state environmental agencies now legislate mechanical loaders on all incinerators as a means of controlling stack and fugitive emissions.

The most common type of mechanical charging system is the "hopper/ram loader," as shown in Figure 3-31D. Typically, waste is put into the charging hopper and, upon actuation of a system-start switch, the hopper cover closes, the fire door opens, the charging ram injects the waste into the incinerator, the charging ram returns, the fire door closes, and the hopper cover opens for the next loading cycle. Hopper volume and a cycle timer help limit the amount of waste that can be loaded into the incinerator per hour, which helps prevent overcharging. An airlock interface is provided between the incinerator and ambient conditions, because the fire door and hopper cover are not both opened simultaneously.

Two potential causes of fire and smoke problems that must be guarded against in hopper/ram loader systems are:

1. *Partial closing of fire door.* Waste can lodge under the door and prevent it from closing tightly. This waste could ignite and spread fire back into the hopper and beyond. The partially closed door could also significantly affect incinerator draft conditions, which in turn could result in heavy smoke escaping under the partially closed door. The fire door should be closed by power, not gravity, in order to seal the opening as tightly as possible should any waste become lodged beneath.
2. *Ram ignition hazards.* When the charging ram injects waste into the incinerator, its face is directly exposed to high temperatures. Eventually, it could heat up enough to be a potential fire

hazard. The hot ram face could cause plastic waste bags and similar materials to melt and adhere to it. If these items do not drop from the ram during its loading cycle, they become ignited and are carried back into the hopper where they will ignite waste remaining in the hopper.

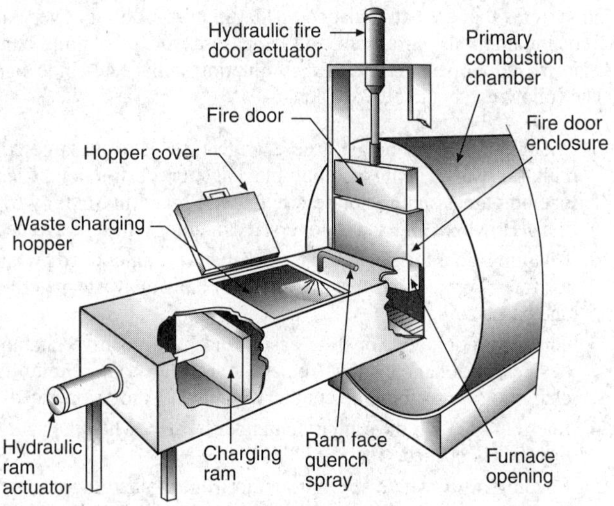

The charging ram should be cooled either by an internal water circulation or a water-spray system that quenches the ram face after every charging stroke. In addition, the loading-hopper volume should be such that a single ram stroke completely empties it. This will minimize the risk of waste being in the hopper to catch on fire.

Additional recommended considerations to protect against accidental ignition of wastes in loading hoppers are:

1. Provision of a flame scanner for immediate detection of hopper fires. Actuation of the scanner could sound an alarm, activate a hopper water-spray system, and/or automatically initiate the loading sequence in order to inject the burning materials into the incinerator.
2. A high-temperature automatic sprinkler directly over the hopper and independent of the room sprinkler system.
3. A manually actuated water spray located above the loading hopper.
4. An emergency switch that would override the normal automatic loading-cycle-control sequence and cause immediate injection of hopper contents into the incinerator.

Residue Handling and Removal

Residue, or ash, handling and removal systems range from fully manual to fully automatic. Because of high costs, space limitations, operational complexities, and technological limitations, most existing older incinerators are cleaned out manually. However, most newer incinerators, and almost all incinerators charging more than about 1000 lb per hr (454 kg/hr), are equipped with an automatic or semiautomatic ash-handling and removal system.

Manual residue removal: This cleanout method involves raking or shoveling ashes into barrels, bins, or tubs. Small-capacity incinerators can be cleaned from the outside, but large-capacity units may require the operator to enter the primary chamber for cleanout operations. Manual residue removal is objectionable and potentially hazardous in many respects, including: (1) intensive, difficult labor requirements; (2) dangers to personnel from heat, ash, dust, and noxious gases; and (3) fire hazards from the handling and containment of unquenched ash residues. Spraying water into the incinerators to facilitate ash removal by quenching is not recommended, because the water will damage the furnace refractory due to thermal shock. Water quenching is limited to ash already removed from the furnace.

Ash cleanout without containment or quenching could be a potential fire hazard if the ashes are hot or contain embers and smoldering materials. Because of this, NFPA 82 requires that a system and appropriate measures be provided to adequately quench and/or fully contain ash residues removed from the incinerator during cleanout operations. This could include such features as water sprays, a wet quench pit, or a special containment enclosure to enable ash cleanout with minimal exposure to ambient conditions.

Automatic residue removal: These systems utilize grates, transfer rams, or pulse hearths to transfer and discharge ashes from the primary chamber. In semi-automatic-type systems, ashes are collected in residue carts. In fully automatic systems, ashes are discharged into a residue conveyor from which they are continuously transferred to a large container for removal. Residue conveyor systems are equipped with a water trough, or sump, which fully satu-

FIG. 3-31D. A typical mechanical loader.

(Labels in figure: Hydraulic fire door actuator; Primary combustion chamber; Fire door; Fire door enclosure; Hopper cover; Waste charging hopper; Hydraulic ram actuator; Charging ram; Ram face quench spray; Furnace opening)

rates the ashes. With cart-type systems, the ashes are quenched by water sprays surrounding the discharge chute from the incinerator.

For fire safety protection, it is recommended that ash carts used in both manual and semi-automatic systems be handled with care and, prior to off-site disposal, stored in a fire-rated area in a location that is isolated from stored waste materials. Even when quenched by water sprays or hoses, the containers may still be very hot, and residues in the ash carts may tend to ignite spontaneously if disturbed and exposed to room air. It is also recommended that considerations be given for dumping ash carts at a landfill rather than within the facility building areas.

Incinerator Chimneys and Breeching Systems

The design, construction, termination, and clearances of incinerator chimneys, and the flue-gas-ducting, or breeching, systems must be in strict accordance with NFPA 82 and NFPA 211, *Standard for Chimneys, Fireplaces, Vents, and Solid Fuel-Burning Appliances*, as well as applicable building codes.

Chimneys and breeching systems handling low-temperature, wet, and saturated incinerator flue gases downstream from a scrubber must be designed and constructed to withstand acidic corrosion conditions. Chimneys and breeching systems handling hot and high-temperature flue gases should be refractory lined. It should be noted that there is no single chimney-refractory lining, or construction material, or family of materials, suitable for ensuring chimney integrity and fire safety for all incinerator operating conditions and waste types. Such materials must be carefully evaluated and selected.

NFPA 82 stipulates that chimneys for incinerators serve no other purpose and they must not be connected to boiler chimneys unless they are of suitable construction and properly designed and controlled for serving both systems.

Auxiliary Fuel Systems

Incinerators require auxiliary, or supplemental, fuel for ignition, warm-up, and maintaining proper combustion temperatures. Most incinerators fire either natural gas or distillate fuel oil. It is imperative that auxiliary fuel systems, including storage and handling systems, burners, and controls, be installed and tested in accordance with all applicable requirements of local codes and NFPA standards. Burners should be of a type specifically designed and constructed for use in incineration systems. Burners must also be equipped with automatic ignition systems, and flame safeguard and management systems.

Combustion and Ventilation Air

Depending on the types and heating values of wastes burned, incinerators typically require on the order of 10 to 20 lb (4.5 to 9 kg) of combustion air per pound (0.45 kg) of waste incinerated. Ventilation air is also required to remove heat that is normally transmitted, or radiated, through incinerator casings and associated equipment, in order to maintain comfortable operating conditions and acceptable incinerator-room temperature levels. Insufficient supply of combustion and ventilation air may result in operational problems and unsafe conditions. Incinerators located in confined spaces or areas enclosed by tight-fitting partitions must be provided an air-supply louver or duct in accordance with NFPA 82 requirements. Louvers and ducts providing direct outside air must be sized for 1 sq in. (645.2 mm^2) of free area per 4000 Btu per hr (1172 W/hr) of total incinerator heat input.

Electrical Service

Power and control wiring, including electrical components and control devices, must be in accordance with applicable local codes and NFPA 70, *National Electrical Code®*. It is recommended that wiring and conduit not be attached directly to incinerator casings. Wiring within about 1 ft (0.30 m) of incinerator casings should have high-temperature insulation. It is also recommended that incinerator system control panels be located less than 3 ft (0.92 m) from the casing. Wiring, relays, and transformers for auxiliary burners should be either remotely mounted or protected from high-temperature damage by the burner-blower fans.

Incinerator Rooms

Incinerators, including associated waste- and residue-handling systems, must be located in a separate, fire-rated room or compartment with an automatic fire door suitable for a Class B opening. Such incinerator rooms may also be contained in the same enclosure with the building heating system and equipment.

Incinerator Operations

To properly manage the severe and complex operating conditions typical of incineration systems, well-trained operating personnel are required. Training programs and comprehensive operating and maintenance manuals and instructions are highly recommended and mandated in the environmental regulations of many states. Such instructions should not only ensure proper incinerator operations, but also the continued implementation of safe operating conditions.

WASTE COMPACTORS

Waste compactors are devices that use electro-mechanical-hydraulic means to reduce the volume of waste and package it in the reduced condition. The two types of compactors regulated by NFPA 82 are: (1) domestic and (2) commercial-industrial.

Domestic Compactors

Commonly called kitchen compactors, domestic compactors are designed for use in dwellings and apartment units for compaction of single-family residential waste. This appliance reduces the fire hazard of stored waste by retaining it in a metal container under compaction. Units are of the undercounter and moveable types. They must be capable of manual opening such that waste can be easily removed in the event of equipment electrical failure.

Commercial-Industrial Compactors

These compactors are used in multiple-family dwellings and many other classes of occupancies as the prime waste-handling system. They may be located either indoors or outdoors and can be chute or hand fed.

There are four types of compactor systems.

1. *Bulkhead compactor.* Waste is compacted in a chamber against a bulkhead. When a compacted block is ready for removal, a bag is installed and filled with the compacted block.
2. *Extruder.* Waste is forced through a cylinder that has a restricted area. Driving forces compact the material and extrude it into a "slug," which is broken off and bagged or placed in a container. The operation of this type of compactor is shown in Figure 3-31E.
3. *Carousel bag packer.* Waste is compacted into a container in which a bag has been inserted. Various configurations are used. One method utilizes a compaction chamber from which the

FIG. 3-31E. *The sequence of operation of an extruder-type refuse compactor. Waste falling from the chute (A) trips a sensor and starts the compacting cycle. The ram moves backward (B) to let refuse fall into the compactor chamber and then moves forward (C) with more than 40,000 lb (18,144 kg) of force to push the waste through a system of restrictors that compacts the waste and pushes it out of the machine.*

compacted plug is pushed into the bag. Another uses a cylinder that acts as a compaction chamber that is inserted into the bag. Upon compaction, the cylinder is removed while a compacting head holds the material in place.

4. *Container packers.* Waste is compacted directly into a bin, cart, or container. When full, the container can be either manually or mechanically removed from the compactor and compaction area.

Compaction containers must be provided with either a 2$\frac{1}{2}$-in. (63.5-mm) hose connection suitable for standard fire-fighting equipment (and located near the top of the container) or an access door that can be opened without disconnecting the container from the compactor.

Chute Termination Bin

Most compactor waste chutes do not feed directly into the compactor but into a small storage chamber, chute connector, or an impact area that is usually large enough to temporarily hold small quantities of waste. Potential fire hazards can be minimized by installing sprinklers and providing doors large enough to allow access in the event of fire. If a compactor is charged manually from a large bin with an open top, the bin need not be sprinklered. It is sufficient to rely on the sprinklers protecting the compactor room for protection.

The bottom closure of a storage bin (or area) and its ability to be opened under fire conditions deserves attention. If, for instance, there is equipment failure, waste will not only build up in the chute's termination chamber but also in the chute itself. In the event of a fire and sprinkler discharge, the weight of soggy waste could become excessive and jam simple slide devices. Sufficient strength should be built into these closure devices to allow opening them without breakage under such conditions. The most satisfactory situation is one that allows discharge into other receptacles, bins, or equipment without excessive chute collection.

Chute-fed compactors should be provided with an automatic, special, fine-water-spray sprinkler in the compaction chamber. However, hand-fed compactors do not require such protection.

Compactor Rooms

Compacted waste has a wide range of densities and can burn to produce large amounts of smoke. Therefore, storage of compacted waste in buildings should be minimized.

Compactor systems are required to be enclosed within fire-rated rooms with fire doors suitable for Class B openings. Automatic sprinklers are also required in compactor rooms.

SHREDDERS

Shredders, as well as granulators, grinders, pulpers, chippers, and the like, are sometimes used to process waste for volume reduction in order to minimize storage-space requirements and to facilitate systems where waste must be mechanically transferred before storage. They are also used in combination with other processes for disinfecting or sterilizing biomedical waste. It is good practice to sprinkler shredder feed bins. If they are serviced by chutes, arrangements must be made to bypass the shredder, since shredders have frequent periods of repair. All bypass areas and storage areas should also be sprinklered and enclosed within fire-rated rooms with fire doors suitable for Class B openings. The discharge from shredding devices also requires sprinkler protection.

The possibility of explosion is a particular hazard in connection with shredder operations. This can result from ignition of the dust-laden air mixtures that normally surround shredders during operation. A recommended method of protection is to provide an explosion-suppressive system within the shredder room and a vent, preferably above the shredder, to safely relieve the impact of a rapid pressure buildup or explosion.

INDUSTRIAL WASTE SYSTEMS EQUIPMENT

The term *industrial waste* generally describes waste emanating from manufacturing facilities, processing plants, factories, and the like. However, it essentially has no meaning without further definition, because it encompasses an almost infinite variety of materials. This part of the chapter addresses industrial waste systems and equipment.

According to 1992 estimates, a total of about 6.5 billion tons of industrial waste are generated annually in the United States, of which about 200 million tons are designated as hazardous, according to the U.S. Environmental Protection Agency (EPA).[2] These hazardous wastes must be managed and disposed in compliance with specific EPA regulations.

Approximate distribution of hazardous wastes generated by some major industries is shown in Figure 3-31F. Table 3-31A lists annual waste generation rates for prominent manufacturing industries.

Under the Resource Conservation and Recovery Act of 1976 (RCRA), the EPA has instituted a complex set of regulations for controlling hazardous waste from generation through final disposal. This is known as "cradle-to-grave" control. To deal with dumpsite cleanup costs and liabilities not covered by RCRA, the EPA enacted legislation known as the "Superfund" bill. The Toxic Substances Control Act of 1976 (TSCA) was established to regulate chemicals not controlled under other regulations. This act, for example, specifies disposal requirements for polychlorinated biphenyl compounds (PCBs).

For most industries, particularly those generating hazardous waste, the evaluation and selection of waste management and disposal systems is difficult and complex. First, there are numerous options and alternate technologies available, each with differing benefits, risk factors, and costs. Second, the forms and properties of waste materials vary widely and greatly impact the feasibilities and cost-effectiveness of the alternate technologies. Third, regulatory requirements and restrictions may limit the suitability of various technologies. These complexities and multiplicities are also site specific, making it necessary to evaluate options and alternatives for each facility on an individual, case-by-case, basis.

All other[1]

SIC 22 Textile mill products
SIC 24 Lumber and wood products
SIC 25 Furniture and fixtures
SIC 27 Printing and publishing
SIC 30 Rubber and miscellaneous plastic products
SIC 31 Leather and leather tanning
SIC 32 Stone, clay, and glass products
SIC 35 Machinery except electrical
SIC 38 Instruments and related products
SIC 39 Misc. manufacturing industries

Non-manufacturing categories[2]

SIC 5085 Drum reconditioners
SIC 07   Agricultural services
SIC 5161 Chemical warehouses
SIC 40   Railroad transportation
SIC 55   Automotive dealers and
         gasoline service stations
SIC 72   Personal services
SIC 73   Business services
SIC 76   Misc. repair services
SIC 80   Educational services
SIC 82   Health services

FIG. 3-31F. Pie chart showing the distribution of hazardous waste generation by Standard Industrial Classification (SIC).[2]

WASTE MATERIALS

Industrial waste is as varied as the raw materials used in production operations. However, the nature of the industry generally determines the nature of the waste. A furniture plant, for example, discards wood scraps, sawdust and shavings, fabric scraps, glues, stains, and varnishes. Within any specific industry there may be many different types and forms of waste materials, ranging from plant trash to spent chemicals and objectionable fumes. Other items may include production residues and by-products, packaging scraps, reject products, returned goods, and even animal carcasses from biomedical and veterinary-type research activities. Such waste can be in the form of solids, sludges, liquids, or gases. The waste can be hazardous or nonhazardous, and it can be combustible or noncombustible. Obviously, the combinations are virtually endless.

Hazardous Waste

Hazardous waste is generally defined as any waste or combination of wastes that poses a substantial danger, now or in the future, to human, plant, or animal life, and which, therefore, cannot be handled or disposed of without special precautions. Hazardous waste categories include toxic chemical, flammable, corrosive, reactive, explosive, radioactive, and biomedical. If improperly handled or disposed, such waste can contaminate surface and ground waters as well as the ambient atmosphere. It can poison, burn, maim, blind, and kill people and other living organisms. Some are nondegradable and persist in nature indefinitely. Some may accumulate in living

TABLE 3-31A.   Industrial Solid-Waste Production Rates[1]

SIC* Code	Industry	Waste Production Rate (tons/employee/year)
201	Meat Processing	6.2
2033	Cannery	55.6
2037	Frozen foods	18.3
Other 203	Preserved foods	12.9
Other 20	Food processing	5.8
22	Textile-mill products	0.26
23	Apparel	0.31
2421	Sawmills and planing mills	162.0
Other 24	Wood products	10.3
25	Furniture	0.52
26	Paper and allied products	2.00
27	Printing and publishing	0.49
281	Basic chemicals	10.00
Other 28	Chemical and allied products	0.63
29	Petroleum	14.8
30	Leather and plastic	2.6
31	Leather	0.17
32	Stone, clay	2.4
33	Primary metals	24
34	Fabricated metals	1.7
35	Nonelectrical machinery	2.6
36	Electrical machinery	1.7
37	Transportation equipment	1.3
38	Professional and scientific instruments	0.12
39	Miscellaneous manufacturing	0.14

*Standard Industrial Classification

things, and some may work their way into the food chain. Also, some may catch fire or explode at normal temperatures and pressures when exposed to air or water, or by being jarred or dropped.

The EPA, under RCRA, defines and regulates wastes that are hazardous because of either ignitibility, corrosivity, reactivity, or toxicity. The RCRA regulations not only identify specific hazardous criteria, but also list specific waste types, constituents, and sources generating wastes with these hazardous characteristics. Radioactive waste is regulated by the Nuclear Regulatory Commission (NRC). Biomedical, or infectious, waste is currently regulated by state and local environmental agencies and health departments.

Biomedical waste management and disposal are currently not regulated on a federal level; however, the EPA is in the process of promulgating regulations for new and existing medical waste incinerators, and the Department of Transportation (DOT) has recently promulgated regulations for the off-site transport of biomedical waste.

Hazardous Characteristics

Ignitibility: Liquid wastes are considered ignitibly, or flammably, hazardous if their flash point is less than 140°F (60°C). These mainly consist of contaminated organic solvents, but they may include oils, plasticizers, complex organic sludges, and off-specification chemicals.

Nonliquid wastes are considered ignitibly hazardous if they are capable, under standard temperatures and pressures, of causing fires through friction, absorption of moisture, or through spontaneous chemical changes. These include pyrophoric materials, such as phosphorous and aluminum alkyls, which ignite when exposed to air.

Organic vapors and fumes may be flammably hazardous if released in combustible concentrations. These may travel considerable distances, reach an ignition source, and ignite. NFPA 86, *Standard for Ovens and Furnaces*, identifies specific requirements for the safe venting of such gases.

Corrosivity: Corrosive wastes can severely damage skin and other living tissues, as well as construction materials, such as metals and finished surfaces. Wastes are defined as corrosively hazardous if their pH values are 2 or less, 12 or more, or if they corrode steel at a rate greater than $1/4$ in. (6.35 mm) per year. These wastes generally comprise strong acids and alkalis.

Reactivity: Hazardous wastes characterized as reactive are normally unstable and readily undergo violent changes without detonation. Some reactive wastes are capable of detonation or explosive decomposition at standard temperatures and pressures. Also, some reactive wastes react violently with water, and some form potentially explosive mixtures or toxic vapors if mixed with water. Reactive materials, such as sodium, potassium, and aluminum alkyls, react violently with water and burn fiercely. Strong oxidizers in contact with organic materials may cause rapid combustion or explosion.

Toxicity: Toxic waste poisons or produces injury, upon contact with or through accumulation in or on the body of a living organism. This may occur by inhalation or through contaminated food or water. It is both a short-term and long-term problem. Most toxic waste contains either metals, such as arsenic, barium, cadmium, chromium, and lead, or synthetic organics, such as various pesticides. RCRA specifies an extraction procedure test to determine whether such materials are present in significant concentrations to be a toxic hazard.

Explosives: These mainly consist of obsolete ordnance and manufacturing wastes from commercial explosives industries. They may also consist of waste from the manufacturing of various propellants and pyrotechnics. However, some waste, which may be generated or which are formed as carbon and organic dusts, may be potentially explosive in various air concentrations when exposed to ignition sources, such as open flames or sparks.

Radioactivity: Most radioactive wastes consist of conventional, nonradioactive materials contaminated with radionuclides in concentrations ranging from parts per billion to as much as 50 percent. The biological hazards from radioactive materials are due to the effects of penetrating and ionizing radiation rather than chemical toxicity. The long-term hazards associated with specific radioactive waste are not necessarily proportional to the nominal level of radioactivity, but rather to the specific toxicity and decay rate of each radionuclide which may be present. Radioactive industrial waste is primarily generated from biomedical-type research activities. These generally comprise items such as animal carcasses and liquid scintillation counting (LSC) vials with tracer level, or low-level, concentrations of radioactivity.

Infectiousness: Waste characterized as infectious contains pathogens capable of producing disease in humans. Pathogens are disease-producing microorganisms, and includes bacteria, fungi, viruses, viroids, rickettsiae, and protozoa. Industrial wastes that are potentially infectious are almost exclusively generated from biomedical research activities.

Waste Characterization

Waste stream characterization provides the starting point for designing and selecting industrial waste management systems. This involves identification of average, or representative, waste compositions and properties, deviations from the averages, and the forecasting of likely changes.

A characterization program usually begins with a detailed inventory and classification of all waste stream sources, components, and constituents. These are then organized into various groupings, or categories, to facilitate further evaluations. Typical groupings include form, such as solids, sludges, and liquids; combustible *vs.* noncombustible; and hazardous *vs.* nonhazardous. The waste classification system shown in Table 3-31B has proved particularly useful to the author.

The next step in a characterization program usually involves determination of the physical and chemical properties of the waste materials, or groups of waste materials. Physical properties of concern typically include physical state, size, shape, weight, volume, temperature, viscosity, and pressure. Chemical properties typically include elemental constituents, toxicity, corrosivity, explosivity, flash point, heat of combustion, and products of combustion. Table 3-31C is a comprehensive checklist of parameters that may be significant in a waste characterization program.

Depending on the specific waste, sufficient characterization data may be readily available in published literature, from faculty records, or engineering handbooks. When such data are not available or are inadequate, laboratory testing and analytical work may be required. Waste testing methodologies are typically done in accordance with American Society for Testing and Materials (ASTM) procedures. Hazardous waste sampling and analysis procedures are usually in accordance with "Test Methods for Evaluating Solid Waste—Physical/Chemical Methods," SW-846, as issued by the EPA.

WASTE MANAGEMENT SYSTEMS

As with solid waste management systems (see the first part of this chapter), a great variety of alternate processes and equipment is available as part of a total industrial waste management system. Depending upon the specific types, forms, quantities, and hazardous characteristics of the waste materials, management systems can range from a few simple processes to a complex combination of many processes.

Waste-handling systems and equipment include conveyors, chutes, carts, and transport vehicles, elevators and lifts, as well as pumps and piping for liquids, and ducts and blowers for gases and fumes. These systems can range in sophistication from fully manual to fully automatic operation.

Systems for storing and handling liquid chemical wastes often require special designs and safety provisions. Problems of concern typically include viscosity, freeze protection, abrasion, corrosion, slagging, and adverse chemical reactions, such as spontaneous ignition, rapid heat release, foaming, precipitation, solidification, and vaporization of low-boiling-point compounds. Safety provisions include spill and runoff containment, static electricity prevention, flame arresters, and systems for handling the vapors or fumes vented from the storage tanks. In addition, special precautions are sometimes necessary to ensure that incompatible wastes are segregated to prevent undesirable reactions.

For safely handling fumes or vapors from industrial ovens and furnaces, dilution air is required such that the combustible concentrations are below 50 percent, and usually less than 25 percent, of lower explosive limits (LEL). Potentially explosive dusts require controls to prevent combustible air mixtures from forming and being ignited.

See Table 3-31D for applicability of various waste management and disposal methods for different waste classes.

TABLE 3-31B.   *Waste Classification System[2]*

Class	Remarks	Example
I. Solid wastes		

A.   Putrescibles
      Household garbage
      Vegetable and fruit processing
         wastes
      Animal manure
      Dead animals
      Meat, poultry, and seafood
         processing wastes
      Others, not elsewhere classified*		

B. Bulky combustibles	Bulky is defined to mean a material of a size to present problems by jamming in a compaction truck hopper, an incinerator feed chute, or other such problems in handling or disposal. Its dimensions may not be defined, except by the size and nature of handling or disposal equipment.	
Wood		Timbers, pallets, cross-ties
Paper and paper products		Large cardboard packing boxes, boxcar linings
Cloth		Filter cloths, mattresses
Plastics		Styrofoam™ logs, polystyrene sheets, garden hose
Rubber		Belting, tires
Leather		Conveyor belting
Yard and street wastes		Tree limbs
C. Bulky noncombustibles		
Metal		Drums, bedsprings
Mineral		Carboys, bathroom fixtures
D. Small combustibles	Small is defined to mean a piece of waste material of a small enough size to present no danger of jamming equipment, or otherwise causing problems because of its size. It refers to the size of a piece of the material, not the size of the delivered load.	
Wood		
Paper and paper products		
Cloth		Gloves
Leather		Shoes
Plastics		Milk cartons
Rubber		Galoshes, butyl rubber crumbs
Yard and street wastes		Street sweepings, leaves
E. Small noncombustibles		
Metal		Cans
Mineral		Bottles
Ashes		Furnace ashes
F. Nonempty cans, bottles, and drums	This functional class will require precise definition of the contents by one of the other classes. This establishes the quantity of material delivered in this manner instead of in bulk form.	Bottles of wastes from laboratories
G. Gas cylinders		Oxygen, acetylene
H. Powders and dusts Organic		Pesticides, grain dusts, chemical powders, coal dusts

Metallic inorganic
      Nonmetallic inorganic
      Explosive		

*Other (not elsewhere classified)—represents the "cleanest" of material and is not intended as a catch-all grouping.

TABLE 3-31B. *(Continued)*

Class	Remarks	Example
I. Pathological wastes Cloth, paper, and plastic Animal and human wastes Instruments and utensils		
J. Sludges	Materials that are solid in appearance, but are wet, either from water or liquid organics. The solids portion is classified.	

Chlorinated
    Brominated
    Fluorinated
    Acid
    Alkaline
    Water-reactive (unhydrolyzed)		

Air-reactive	May represent a material highly reactive with the moisture in the air, or with the oxygen in the air	
Putrescible	May represent the wet form of any of the materials listed under the functional class of the same name, but also includes such materials as wastewater treatment plant sludges	

Miscellaneous organic
    Metallic inorganic
    Nonmetallic inorganic
	
	Refers to metals in the uncombined form
	Inorganic compounds
		
		Particles of metal in oil
		Filter cakes, $CaCO_3$ precipitate

K. Demolition and construction	This functional class is subdivided into the analytical classes shown for bulky and small, combustibles and noncombustibles. It is included as a separate class to quantify materials from this source and to recognize that a delivery may contain a wide mixture of large and small combustibles and noncombustibles.	
L. Abandoned vehicles		
M. Radiological wastes	This functional class will require further definition by one of the other classes. It must be considered to define this type of contamination to refuse, and to consider the special measures involved.	
II. Liquid wastes		
A. Wastewaters	Waste liquids composed almost entirely of water, but containing contaminants in low enough concentration (usually much less than 1.0 percent) to be handled through a sewer system to a wastewater treatment plant. This definition is proposed for exclusion purposes, since these wastes are not normally considered as refuse.	
B. Contaminated waters	Waters containing contaminants in a concentration too high, or of a nature, that handling through a wastewater treatment plant is unacceptable	

Chlorinated
    Brominated
    Fluoridated
    Acid
    Alkaline
    Putrescibles
    Insoluble oils
    Soluble oils
    Toxic organics
    Toxic inorganics
    Soluble metals
    Others (not elsewhere classified)*
		Blood, grease

*Other (not elsewhere classified)—represents the "cleanest" of material and is not intended as a catch-all grouping.

TABLE 3-31B.   *(Continued)*

Class	Remarks	Example
C. Liquid organics 　　Chlorinated 　　Brominated 　　Fluoridated 　　Sulfurated 　　Acid 　　Alkaline 　　Water reactive (unhydrolyzed) 　　Shock reactive 　　Toxic and hazardous 　　Soluble metals 　　Others (not elsewhere 　　　classified)*	Liquid at all ambient temperatures	Many solvents Pesticides
D. Tars 　　Chlorinated 　　Brominated 　　Fluorinated 　　Sulfurated 　　Acid 　　Alkaline 　　Water-reactive 　　Chemically reactive 　　Self-reactive (monomers) 　　Toxic and hazardous 　　Soluble metals 　　Others (not elsewhere 　　　classified)*	Stiff materials that are semi-solid at low ambient temperatures.	Chlorinated-substituted hydrocarbons Brominene-substituted hydrocarbons Fluorine-substituted hydrocarbons Sulfur-substituted hydrocarbons Low pH, corrosive solvents Unhydrolyzed materials that react violently 　　with water Sodium, calcium
E. Slurries 　　Organic in water 　　Inorganic in water 　　Organic in liquid organic 　　Inorganic in a liquid organic	Liquid materials that contain solids, but which readily flow or pump. The liquid and solid materials both are classified.	Lime slurry Metallic sodium in oil
III. Gaseous wastes	This classification is restricted to those gaseous materials that have been or might become the responsibility of a disposal group or department to treat, burn, or otherwise alter before discharge to the atmosphere.	
A. Odorous		Mercaptans, H_2S
B. Combustible particulate 　　Solids 　　Mists		
C. Organic vapors		Volatile solvents
D. Acid gases		SO_2, HCl

*Other (not elsewhere classified)—represents the "cleanest" of material and is not intended as a catch-all grouping.

TREATMENT AND DISPOSAL SYSTEMS

Because of the high costs and liabilities typically associated with waste treatment and disposal, initial efforts are usually directed to reducing waste quantities. Techniques include reuse and recycling of materials that would otherwise be waste, changes in manufacturing and production operations, and substitutions of raw materials to others generating less waste. Also, substitutions and reductions in the usage of hazardous chemicals may result in reduced hazardous waste disposal requirements.

TABLE 3-31C.   General, Physical, and Chemical Parameters of Possible Significance in the Characterization of Solid Wastes[3]

General Parameters

Compositional weight fractions
A. Domestic, commercial, and
    institutional
    Paper (broken into subcategories)
    Food wastes
    Textiles
    Glass and other ceramics
    Plastics
    Rubber
    Leather
    Metals
    Wood (limbs, sawdust)
    Bricks, stones, dirt, ashes
B. Other municipal
    Dead animals
    Street sweepings
    Catch-basin cleanings
C. Agricultural
    Field
    Processing
D. Industrial
E. Mining/metallurgical
F. Special
    Radioactive
    Munitions, etc.
    Pathogenic
    Moisture

Process weight fractions
Combustible
Compostable
Processable by
    landfill
Salvageable
Having instrinsic
    value

Chemical Parameters

General
pH
Alkalinity
Hardness ($CaCO_3$)
MBAS (methylene-blue
    active substances)
BOD (biochemical oxygen
    demand)
COD (chemical oxygen
    demand)
Rate of availability of
    nitrogen
Rate of availability of
    phosphorus
Crude fiber
Organic (%)
Combustion parameters
    Heat content
    Oxygen requirement
    Flame temperature
    Combustion products
        (including ash)
    Flash point
    Ash-fusion characterization
    Toxicity
    Corrosivity
    Explosivity
    Other safety factors
    Biological stability
    Attractiveness to vermin
Inorganic and elemental
Moisture content
Carbon
Hydrogen
(P_2O_5 and phosphate)
Sulfur content
Alkali metals
Alkaline-earth metals
Precious metals
Heavy metals
    Especially mercury
    Lead
    Cadmium
    Copper
    Nickel
Toxic materials
    Chromium
    Especially arsenic
    Selenium
    Beryllium
    Asbestos
Eutrophic materials
    Nitrogen
    Potassium
    Phosphorus

Organic
Soluble (%)
Protein nitrogen
Phosphorus
Lipids
Starches
Sugars
Hemicelluloses
Lignins
Phenols
Benzene oil
ABS (alkyl benzene sulfonate)
CCE (carbon chloroform
    extract)
PCB (polychlorinated biphenyls)
PNH (polynuclear hydrocarbons)
Vitamins (e.g., B-12)
Insecticides (e.g., Heptochlor, DDT,
    Dieldrin, etc.)

Physical Parameters

Total wastes
Size
Shape
Volume
Weight
Density
Density stratification
Surface area
Compaction
Compactability
Temperature
Color
Odor
Age
Radioactivity
Physical state
    Total solids
    Liquids
    Gas
Solid wastes
Soluble (%)
Suspendable (%)
Combustible (%)
Volatile (%)
Ash (%)
    Soluble (%)
    Suspendable (%)
Hardness

Particle characteristics
Shape distribution
    Shape
    Surface
    Porosity
    Sorption
    Density
    Aggregation
Liquid wastes
Turbidity
Color
Taste
Odor
Temperature
Viscosity data
    Specific gravity
    Stratification
Total solids (%)
    Soluble (%)
    Suspended (%)
    Settleable (%)
Dissolved oxygen
Vapor pressure
Effect of shear rate
Effect of temperature
Gel formation
Gaseous wastes
Temperature
Pressure
Volume
Density
Particulate (%)
Liquid (%)

One EPA study reported 43 potentially feasible waste treatment processes. These processes include physical treatment, chemical treatment, biological treatment, and thermal treatment. They provide such functions as volume reduction, separations of waste stream com-ponents, and destruction and detoxification of hazardous constituents. Often several treatment processes are linked in series to provide a required degree of treatment. Residues from these processes usually require disposal, such as land burial.

TABLE 3-31D. Waste Management and Disposal Methods for Different Waste Classes

Waste Classification Utilization Chart	Toxicity	Explosiveness/ Flammability	Pathogenicity	Radioactivity	Std. Storage	Spec. Storage	Std. Coll. Vehicle	Spec. Coll. Vehicle	Std. Pipeline	Spec. Pipeline	Compaction	Grinding	Pulping (w/H_2O)	Composting	Incineration	Incineration (Liquid)	Incineration (Spec.)	Chemical Treat. or Alteration	Sanitary Landfill	Ocean Disposal§
Solid wastes																				
Putrescible			X		X		X				X	X	X	X	X				X	
Bulky combustible						X		X			X	X			X			X	X	
Bulky noncombustible						X		X			X	X							X	
Small combustible					X		X				X	X	X	X	X				X	
Small noncombustible					X		X				X	X							X	
Nonempty cans, bottles, drums*	X	X	X	X	X	X	X											X	X	
Gas cylinders		X				X		X										X	X	
Powders and dusts		X				X		X										X	X	
Pathological wastes	X		X	X		X		X									X	X	X	
Sludges	X					X		X									X	X	X	X
Demolition and construction						X		X				X						X	X	
Abandoned vehicles						X		X			X	X						X	X	
Radiological waste				X		X		X										X	X	X
Liquid wastes†																				
Wastewaters									X							X		X		
Contaminated waters‡	X			X		X		X	X	X						X		X		
Liquid organics‡	X		X			X		X	X	X							X		X	X
Tars	X	X				X		X	X	X							X	X	X	
Slurries	X					X		X	X	X							X		X	X
Gaseous wastes†																				
Odorous	X	X				X		X										X	X	
Combustible particulate		X				X		X										X	X	
Organic vapors	X	X				X		X										X	X	
Acid gases	X	X				X		X										X	X	

*For contents treatment, see proper classification; this class for container only.
†Final disposal of these wastes requires an initial transformation to the solid class (except as noted).
‡Pipeline selection dependent upon specific contaminant.
§Ocean disposal as locally permitted or approved—not recommended procedure.

Selection of treatment processes depends on the type, form, and quantities of waste, required performance to satisfy environmental requirements, and overall system economics. Table 3-31E summarizes the currently available hazardous waste treatment and disposal processes discussed below.

Physical Treatment

Physical treatment processes generally provide for separation of waste stream components or phases. These are particularly useful for concentrating specific hazardous constituents within dilute waste mixtures, thus reducing the overall quantities requiring disposal as a hazardous waste. These also provide for the recovery of specific materials in resource recovery operations.

The following physical treatment processes are potentially feasible for wastes:

Adsorption
Centrifugation
Dialysis
Electrodialysis
Electrolysis
Electrophoresis
Filtration
Flocculation
Flotation
Freeze crystallization
Freeze drying
Freezing
High-gradient magnetic separation
Reverse osmosis
Stripping
Ultrafiltration
Zone refining

Chemical Treatment

Chemical treatment processes are particularly useful for the detoxification of hazardous wastes. They also provide for the separation of specific waste stream components. However, they are generally limited to liquid forms of waste materials.

TABLE 3-31E. *Hazardous Waste Treatment and Disposal Processes*

Processes	Functions Performed*	Types of Waste†	Forms of Waste‡	Resource Recovery Capacity
Physical treatment				
Carbon sorption	VR, Se	1, 3, 4, 5	L, G	Yes
Dialysis	VR, Se	1, 2, 3, 4	L	Yes
Electrodialysis	VR, Se	1, 2, 3, 4, 6	L	Yes
Evaporation	VR, Se	1, 2, 5	L	Yes
Filtration	VR, Se	1, 2, 3, 4, 5	L, G	Yes
Flocculation/setting	VR, Se	1, 2, 3, 4, 5	L	Yes
Reverse osmosis	VR, Se	1, 2, 4, 6	L	Yes
Ammonia stripping	VR, Se	1, 2, 3, 4	L	Yes
Chemical treatment				
Calcination	VR	1, 2, 5	L	
Ion exchange	VR, Se, De	1, 2, 3, 4, 5	L	Yes
Neutralization	De	1, 2, 3, 4	L	Yes
Oxidation	De	1, 2, 3, 4	L	
Precipitation	VR, Se	1, 2, 3, 4, 5	L	Yes
Reduction	De	1, 2	L	
Thermal treatment				
Pyrolysis	VR, De	3, 4, 6	S, L, G	Yes
Incineration	De, Di	3, 5, 6, 7, 8	S, L, G	Yes
Biological treatment				
Activated sludges	De	#	L	No
Aerated lagoons	De	3	L	No
Waste stabilization ponds	De	3	L	No
Trickling filters	De	3	L	No
Disposal/storage				
Deep-well injection	Di	1, 2, 3, 4, 6, 7	L	No
Detonation	Di	6, 8	S, L, G	No
Engineered storage	St	1, 2, 3, 4, 5, 6, 7, 8	S, L, G	No
Land burial	Di	1, 2, 3, 4, 5, 6, 7, 8	S, L	No
Ocean dumping	Di	1, 2, 3, 4, 7, 8	S, L, G	No

*Functions: VR, volume reduction; Se, separation; De, detoxification; Di, disposal; St, storage.
†Waste types: 1, inorganic chemical without heavy metals; 2, inorganic chemical with heavy metals; 3, organic chemical without heavy metals; 4, organic chemical with heavy metals; 5, radiological; 6, biological; 7, flammable; 8, explosive.
‡Waste forms: S, solid; L, liquid; G, gas.

The following chemical treatment processes are potentially feasible for wastes:

Chemical oxidation
Chemical reduction
Hydrolysis
Liquid-liquid solvent extraction
Neutralization
Ozonation
Photolysis

Thermal Treatment

Thermal treatment processes include pyrolysis and incineration. Incineration is high-temperature oxidation, or combustion, while pyrolysis is thermal decomposition in the absence of oxygen. Thermal treatment processes can be designed to dispose of virtually all types and forms of waste materials, but only their organic constituents can be actually destroyed by thermal treatment.

Depending on waste type and composition, thermal treatment processes can provide the greatest weight and volume reduction. In addition, they can destroy or detoxify hazardous organics, sterilize infectious materials, and provide for waste heat recovery. Some industries utilize thermal treatment processes in conjunction with their production operations for the conversion, reclamation, and reuse, or recovery, of residues and byproducts. Examples include thermal distillation for spent solvent reclamation and hydrochloric acid recovery from the incineration of chlorinated hydrocarbons in conjunction with a quench/neutralization system.

The following are specific thermal treatment process types.

Pyrolysis: As indicated, pyrolysis is a thermal treatment process similar to incineration except that decomposition occurs in the absence of oxygen. Heat usually is added externally, and temperatures typically are maintained much lower than incineration temperatures.

The pyrolysis process basically distills the volatile fraction from the waste, leaving inorganics, or ash, and fixed carbons, or char, behind. The volatiles can be combusted as fuel in a boiler or furnace. If the inorganics have non-slagging characteristics and comprise only a small percentage of the solid residues, the char can also be burned as fuel. Pyrolysis is a potentially useful process for treating wastes containing salts, metals, and other contaminants that would otherwise inhibit conventional incineration systems. However, several types of incinerators, including rotary kilns, multiple-hearth incinerators, and special designs, can be operated in a pyrolytic mode.

Incineration: Incineration is an engineered process using high-temperature combustion. Variables most affecting incineration system design, construction, and operations include waste combustibility and heating values, operating temperatures, retention times for both waste and products of combustion, turbulence within the combustion zone, and emission requirements. There are many basic types of industrial incinerators. Each is generally suitable for specific waste types, forms, and quantities; and each has limitations, advantages, and disadvantages.

The standard, basic types of industrial incinerators include the following:

Open burning. Basically limited to the detonation of explosive waste in remote, open areas on a flat, gravel base. Open burning for other waste is prohibited in most of the country.

Open-pit or air curtain destructor. Single chamber, above or below ground, with an open top for burning explosive or reactive materials, such as nitrocellulose. A high-velocity air stream across the open top provides a "curtain" of air to promote combustion turbulence within the pit.

Multiple-chamber incinerator. Similar to that described in the beginning of this chapter for general solid-waste type incinerators. Usually limited to solid wastes, but hearth designs can accommodate sludges. Liquid wastes can be burned in suspension.

Controlled-air incinerator. Similar to that described in the beginning of this chapter for general solid-waste type incinerators. Usually limited to solid wastes and limited quantities of sludges. Liquid wastes can be burned in suspension.

Rotary kiln incinerator. Similar to that described in the beginning of this chapter for general solid-waste type incinerators. Rotary kilns are highly versatile systems suitable for most organic wastes, including solids and sludges. Liquid wastes can be burned in suspension or loaded in drums or containers.

Multiple-hearth incinerator. Wastes move slowly downward through vertically stacked refractory hearths within a vertically oriented, cylindrical chamber. Rotating rabble arms with plow blades move wastes across and downward through each hearth level. Usually limited to organic sludges, but can accommodate granulated solid wastes. Liquids and gases can be injected between various hearths.

Rotary hearth incinerator. Comprises a flat, horizontally oriented, doughnut-shaped hearth that rotates around its central axis within a combustion chamber. Sludges are fed onto the hearth at one location, they are combusted as the hearth rotates, and resultant ash residues are scraped off at a discharge point just prior to a full rotation. System is basically limited to sludges, and it operates predominantly in a pyrolytic mode.

Fluidized-bed incinerator. Wastes are injected into hot agitated bed of granular particles that are suspended within cylindrical chamber by a high-pressure blower. Rapid combustion and heat transfer between wastes and bed. Usually limited to organic sludges, but can accommodate granulated solid wastes. Liquids and gases can be injected into the bed.

Infrared incinerator. Wastes are loaded onto a slowly moving conveyor belt that passes below a series of infrared heaters within a combustion chamber. Ash residues are discharged at the end of the belt. System is limited to sludges, granulated solids, and contaminated soils.

Liquid injection incinerator. Refers generically to those systems designed to burn liquid wastes in suspension within a combustion chamber. Depending on heating value, wastes can be injected through nozzles or combusted through a burner system.

Table 3-31F summarizes the basic parameters of the most common industrial waste incinerators.

Gaseous waste incinerators: Refers to those systems designed for the combustion of fumes, vapors, odors, and the like. Three basic types are:

1. Flares: Open burning of combustible gases that are either near or above their lower explosive limits (LEL) of concentration or above their upper explosive limits (UEL) of concentration.
2. Thermal burners: Combustion of dilute, or low-combustible, gases in the presence of a burner flame within a chamber.
3. Catalytic burners: Combustion of dilute gases that are preheated, exposed to a catalyst material, and oxidized at relatively lower temperatures. Catalyst materials are subject to damage or deactivation from gas stream contaminants.

Other types of industrial incinerators that are either highly specialized, unique, or still under development include:

1. Molten-salt incineration,
2. Wet air oxidation or zimmerman process,
3. Plasma destruction,
4. High-temperature fluid-wall reactor (HTFW) or thagard process,
5. Supercritical fluid technology or MODAR process, and
6. Critical fluid system (CFS).

Table 3-31G shows the incinerator technologies that are best suited or recommended for different waste types and forms.

Biological Treatment

Biological treatment processes utilize microorganisms for the decomposition of organic compounds in the waste. These are applicable to aqueous waste streams with solid or solvent organics. Composting is a biological treatment process used for solid wastes.

The following biological treatment processes are potentially feasible for wastes:

1. Activated sludge,
2. Aerated lagoons,
3. Anaerobic digestion,
4. Composting,
5. Enzyme treatment,
6. Trickling filters, and
7. Stabilization ponds.

Other Technologies

Other thermal treatment technologies and variations to conventional incineration used for waste combustion include:

1. Burning in conventional boilers,
2. Cement kilns,
3. Other industrial furnaces or ovens,
4. At-sea incineration, and
5. Mobile incineration.

Steam sterilization, or autoclaving, is a thermal treatment process for infectious waste. Autoclave equipment is designed to kill pathogens in waste by exposing them to relatively high temperatures—between 240 and 280°F (116 to 138°C)—for periods ranging from about 15 minutes to several hours. Retention times for effective treatment are dependent on steam temperatures and pressures, types and forms of the waste, quantities of waste, and specific type and design of the autoclave system.

ULTIMATE DISPOSAL

Ultimate disposal is basically the final treatment or disposition of treated and untreated waste materials and residues. For example, with solid waste incineration, a substantial percentage of residues

TABLE 3-31F. *Summary of Waste Incinerators[4]*

Type	Process Principle	Application	Combustion Temp.	Residence Time
Rotary kiln	Slowly rotating cylinder mounted at slight incline to horizontal. Tumbling action improves efficiency of combustion	Most organic wastes; well suited for solids and sludges; liquids and gases	1500–3000°F (815–1650°C)	Several seconds to several hours
Multiple hearth	Solid feed slowly moves through vertically stacked hearths; gases and liquids fed through side ports and nozzles	Most organic wastes, largely in sewage sludge; well suited for solids and sludges; also handles liquids and gases	1400–1800°F (760–980°C)	Up to several hours
Liquid injection	Vertical or horizontal vessels; wastes atomized through nozzles to increase rate of vaporization	Limited to pumpable liquids and slurries (750 SSU or less for proper atomization)	1200–3000°F (650–1650°C)	0.1 to 1 sec
Fluidized bed	Wastes are injected into a hot agitated bed of inert granular particles; heat is transferred between the bed material and the waste during combustion	Most organic wastes; ideal for liquids, also handles solids and gases	1400–1600°F (760–870°C)	Seconds for gases and liquids; longer for solids
Pyrolysis	Thermal decomposition in the absence of oxygen; transforms organic materials into solids, liquids, and gaseous organic materials of simpler structure	Primary as fuel; useful with solids and sludges; potential for resource recovery of breakdown products	900–1500°F (480–815°C)	Normally 12 to 15 min

TABLE 3-31G. *Matrix for Matching Waste Type with Incineration Process[5]*

Waste Type	Rotary Kiln*	Multiple Hearth*	Fluidized Bed*	Liquid Incinerator	Multiple-Chamber Incinerator
Solids					
Granular homogeneous	X	X	X		X
Irregular bulky (Pallets, etc.)	X				X
Low melting point (Tars, etc.)	X		X	If material can be melted and pumped	
Organic compounds with fusible ash constituents	X	X			
Gases					
Organic vapor laden			X		
Liquids					
High organic strength aqueous wastes, often toxic	If equipped with auxiliary liquid injection nozzles		X		
Organic liquids	If equipped with auxiliary liquid injection nozzles		X	X	
Solids/liquids					
Waste contains halogenated aromatic compounds [2,200°F (1204°C) minimum]	X	X	If liquid		
Aqueous organic sludges	Provided waste does not become sticky upon drying	X	X		

*Suitable for pyrolysis operation.

typically remain as ash and/or sludges from the air pollution control system and need disposal. If the waste being incinerated is regulated as hazardous, the ash residues must be analyzed to determine whether they require disposal as a hazardous waste in a permitted facility. Likewise, the ash residues from incinerating radioactive waste may also require disposal in a permitted site if radionuclide concentrations exceed NRC levels. In some states, even ash residues from infectious and nonhazardous wastes must be analyzed to determine whether their heavy metal concentrations will not exceed safe toxicity concentration levels when placed in landfills. Alternate disposal processes include land burial, deep-well injection, and ocean dumping.

Land Burial

Two types of land burial are: (1) sanitary landfills and (2) secure, or industrial, landfills. These are engineered systems as opposed to indiscriminate dumping. Sanitary landfills are used for the disposal of nonhazardous solid wastes. Wastes are covered daily with earth to minimize health vector problems, blowing of debris, and open burning. However, unless specially designed, sanitary landfills have the potential for surface and groundwater pollution from leaching, as well as air pollution from gas venting.

Secure landfills for hazardous wastes are regulated under RCRA. These require special liner materials to prevent leaching, special caps to prevent surface exposures, venting systems for handling gases, and monitoring and test wells for verifying the integrity of the liner structure.

Deep-Well Injection

Deep-well injection involves the pumping of liquid wastes into underground cavities or reservoirs that have been identified as geologically secure. Since a prime concern is to protect all usable underground water, the reservoirs must be located far below potable water aquifers and isolated by thick, nearly impermeable strata, such as shale, limestone, or dolomite. Average well depths are about 5,000 ft (1,525 m), with some as deep as 10,000 ft (3,050 m). Most are located in the U.S. Southwest where the geology can accommodate this disposal technology.

Ocean Dumping

Sewage sludges, industrial wastes, and explosives have been dumped at sea for years in designated and approved locations. The U.S. Army Corps of Engineers and the U.S. Coast Guard are basically responsible for the sea transportation and disposal areas for industrial wastes. Of course, the primary concern with this type of disposal is the threat to marine life and coastal areas from improperly sealed containers of hazardous wastes. Because of recent washups of sewage sludge and medical waste on beaches in the U.S. Northeast, national legislative measures have been taken to outlaw ocean dumping of all waste.

Table 3-31H compares key parameters of primary hazardous waste treatment and disposal technologies.

TABLE 3-31H. *Comparison of Some Hazard Reduction Technologies*[6]

	Disposal		Treatment		
	Landfills and Impoundments	Injection Wells	Incineration and Other Thermal Destruction	Emerging High-Temperature Decomposition*	Chemical Stabilization
Effectiveness: how well it contains or destroys hazardous characteristics	Low for volatiles, questionable for liquids; based on lab and field tests	High, based on theory, but limited field data available	High, based on field tests, except little data on specific constituents	Very high, commercial-scale tests	High for many metals, based on lab tests
Reliability issues	Siting, construction, and operation Uncertainties: long-term integrity of cells and cover, liner life less than life of toxic waste	Site history and geology; well depth, construction, and operation	Monitoring uncertainties with respect to high degree of DRE; surrogate measures, PIC incinerability †	Limited experience mobile units; on-site treatment avoids hauling risks; operational simplicity	Some inorganics still soluble; uncertain leachate test; surrogate for weathering
Environmental elements most affected	Surface and groundwater	Surface and groundwater	Air	Air	Groundwater
Least compatible wastes‡	Liner reactive; highly toxic, mobile, persistent, and bioaccumulative	Reactive; corrosive; highly toxic, mobile, and persistent	Highly toxic and refractory organics, highly heavy metals concentration	Some inorganics	Organics
Costs: low, mod, high	L to M	L	M to H (Coincin. = L)	M to H	M
Resource recovery potential	None	None	Energy and some acids	Energy and some metals	Possible building material

*Molten salt, high-temperature fluid wall, and plasma arc treatments.
†DRE = destruction and removal efficiency. PIC = product of incomplete combustion.
‡Wastes for which this method may be less effective for reducing exposure, relative to other technologies. Wastes listed do not necessarily denote common usage.

CODES, REGULATIONS, AND STANDARDS

The proper and safe management and disposal of industrial waste require in-depth knowledge of the material properties, alternate treatment technologies, and applicable regulations. There are also various codes and standards dealing with waste handling and disposal. The more significant of these are listed below.

Federal Codes and Regulations for Waste Management and Disposal

1. Resource Conservation and Recovery Act of 1976 (RCRA), Subtitle C, Hazardous Waste Regulations, 40 CFR, parts 260-265 and 122-124.
2. Nuclear Regulatory Commission (NRC), Part 20, Standards for Protection Against Radiation, 10 CFR.
3. Toxic Substances Control Act of 1976 (TSCA).
4. Clean Air Act of 1963 (CAA) and Clean Air Act Amendments of 1970 and 1990, including:
   (a) National Ambient Air Quality Standards (NAAQS),
   (b) National Emission Standards for Hazardous Air Pollution (NESHAP),
   (c) Prevention of Significant Deterioration (PSD), and
   (d) Nonattainment Regulations (NA).
5. Federal Pesticide Control Act of 1972; also known as Federal Insecticide, Fungicide, and Rodenticide Act (FIFRA).
6. Oil Pollution Act of 1961.
7. Federal Water Pollution Control Act of 1948 (FWPCA).
8. Occupational Safety and Health Act of 1970 (OSHA), including National Institute of Occupational Safety and Health (NIOSH).
9. Hazardous Substances and Hazardous Waste Response, Liability and Compensation Act; also known as "Superfund."

General Standards for Waste Materials and Their Management

NFPA 49, *Hazardous Chemicals Data.*

NFPA 69, *Standard on Explosion Prevention Systems.*

NFPA 325, *Guide to Fire Hazard Properties of Flammable Liquids, Gases, and Volatile Solids.*

NFPA 430, *Code for the Storage of Liquid and Solid Oxidizers.*

NFPA 480, *Standard for the Storage, Handling, and Processing of Magnesium Solids and Powders.*

NFPA 481, *Standard for the Production, Processing, Handling, and Storage of Titanium.*

NFPA 482, *Standard for the Production, Processing, Handling, and Storage of Zirconium.*

NFPA 491M, *Manual of Hazardous Chemical Reactions.*

NFPA 654, *Standard for the Prevention of Fire and Dust Explosions in the Chemical, Dye, Pharmaceutical, and Plastics Industries.*

"Biosafety in Microbiological and Biomedical Laboratories," U.S. Department of Health and Human Services—Public Health Service: Center for Disease Control and National Institutes of Health, Atlanta, GA, May 1983.

"EPA Guide for Infectious Waste Management," PB 86-199130, May 1986, Office of Solid Waste, U.S. Environmental Protection Agency, Washington, DC.

"Finding the R_x for Managing Medical Waste," OTA-O-459, Sept. 1990, U.S. Congress, Office of Technology Assessment, Washington, DC.

"Sampling and Analysis Methods for Hazardous Waste Incineration," 1st ed., Contract 68-02-3111, Feb. 1982, Industrial Environmental Research Laboratory, U.S. Environmental Protection Agency, Washington, DC.

"Test Methods for Evaluating Solid Waste—Physical/Chemical Methods," SW-846, Revision B, July 1981, Industrial Environmental Research Laboratory, U.S. Environmental Protection Agency, Washington, DC.

General Standards for Waste Treatment and Disposal Systems

NFPA 30, *Flammable and Combustible Liquids Code.*

NFPA 82, *Standard on Incinerators and Waste and Linen Handling Systems and Equipment.*

NFPA 86, *Standard for Ovens and Furnaces.*

NFPA 801, *Standard for Facilities Handling Radioactive Materials.*

"Draft Manual for Infectious Waste Management," SW-957, Sept. 1982, Office of Solid Waste, U.S. Environmental Protection Agency, Washington, DC.

General Standards for Burners and Equipment

NFPA 31, *Standard for the Installation of Oil-Burning Equipment.*

NFPA 54, *National Fuel Gas Code.*

NFPA 850, *Recommended Practice for Fire Protection for Electric Generating Plants and High Voltage Direct Current Converter Stations.*

UL 296, *Standard for Safety Oil Burners*

UL 372, *Standard for Safety Primary Safety Controls for Gas- and Oil-Fired Appliances.*

UL 795, *Standard for Safety Commercial–Industrial Gas Heating Equipment.*

The above codes and standards are not all-inclusive. Other agencies, such as Industrial Risk Insurers (IRI), American National Standards Institute (ANSI), and American Society for Testing and Materials (ASTM), may have standards applicable to specific waste management and disposal operations. In addition, state and local regulations and requirements may be more stringent than federal regulations or standards.

BIBLIOGRAPHY

References Cited

1. Wilson, D. G., *Handbook of Solid Waste Management*, Van Nostrand Reinhold Co., New York, 1977.
2. "Hazardous Waste Generation and Commercial Hazardous Waste Management," SW-894, Dec. 1980, U.S. Environmental Protection Agency, Washington, DC, p. 111-3.
3. Ulmer, N. S., "Physical and Chemical Parameters and Methods for Solid-Waste Classification," Open-File Progress Report RS-03-68-17, 1970, U.S. Environmental Protection Agency, Washington, DC.
4. Shen, T. T., Chen, M., and Lauber, J., "Incineration of Toxic Chemical Wastes," *Pollution Engineering*, Oct. 1978, pp. 45-50.
5. Hitchcock, D., "Solid Waste Disposal: Incineration," *Chemical Engineering*, May 21, 1979, pp. 185-194.

6. "Technologies and Management Strategies for Hazardous Waste Control," OTA-M-197, March 1983, Congressional Office of Technology Assessment, Washington, DC.

Reference

"Serious Reduction of Hazardous Waste," OTA-ITE-318, Sept. 1986, Congressional Office of Technological Assessment, Washington, DC.

NFPA Codes, Standards, and Recommended Practices

Reference to the following NFPA codes, standards, and recommended practices will provide further information on waste handling and control discussed in this chapter. (See the latest version of *The NFPA Catalog* for availability of current editions of the following documents.)

NFPA 13, *Standard for the Installation of Sprinkler Systems*
NFPA 30, *Flammable and Combustible Liquids Code*
NFPA 31, *Standard for the Installation of Oil-Burning Equipment*
NFPA 49, *Hazardous Chemicals Data*
NFPA 54, *National Fuel Gas Code*
NFPA 58, *Standard for the Storage and Handling of Liquefied Petroleum Gases*
NFPA 68, *Guide for Venting of Deflagrations*
NFPA 69, *Standard on Explosion Prevention Systems*
NFPA 82, *Standard on Incinerators and Waste and Linen Handling Systems and Equipment*
NFPA 86, *Standard for Ovens and Furnaces*
NFPA 97, *Standard Glossary of Terms Relating to Chimneys, Vents, and Heat-Producing Appliances*
NFPA 101®, *Life Safety Code®*
NFPA 211, *Standard for Chimneys, Fireplaces, Vents, and Solid Fuel-Burning Appliances*
NFPA 325, *Guide to Fire Hazard Properties of Flammable Liquids, Gases, and Volatile Solids*
NFPA 480, *Standard for the Storage, Handling, and Processing of Magnesium, Solids and Powders*
NFPA 481, *Standard for the Production, Processing, Handling, and Storage of Titanium*
NFPA 482, *Standard for the Production, Processing, Handling, and Storage of Zirconium*
NFPA 491M, *Manual of Hazardous Chemical Reactions*
NFPA 801, *Standard for Facilities Handling Radioactive Materials*

Other Standards

UL 296, *Standard for Safety Oil Burners*
UL 372, *Standard for Primary Safety Controls for Gas- and Oil-Fired Appliances*
UL 795, *Standard for Commercial-Industrial Gas Heating Equipment*

The above codes and standards are not all-inclusive. Other agencies, such as Industrial Risk Insurers (IRI), American National Standards Institute (ANSI), and American Society for Testing and Materials (ASTM), may have standards applicable to specific waste management and disposal operations. In addition, state and local regulations and requirements may be more stringent than federal regulations or standards.

Additional Reading

Barton, R. G., *et al.*, "Mechanistic Analysis of the Vaporization of Metals During Waste Incineration," Utah Univ., Salt Lake City, University of California, Berkeley, Toxic Toxic Combustion Byproducts, 4th International Congress, ABSTRACTS ONLY, June 5-7, 1995, Salt Lake City, UT, 1995, 51 pages.
Beyler, C. L., and Hunt, S. P., "Fire Development in and Hazard Analysis of Concrete Hazardous/Radioactive Waste Storage Facilities," *Proceedings* of the Society of Fire Protection Engineers Engineering Seminars on Performance-Based Fire Safety Engineering, Nov. 15–17, 1993, Phoenix, AZ, SFPE, Boston, MA, 1993, pp. 45–55.
Bloom, J. M., Mason, B. J., and Spier, R., "Fire Investigation, Hazardous Waste, and Liability," Fire and Arson Investigator, Vol. 41, No. 2, Dec. 1990, pp. 33–36.
Doucet, L. G., *State-of-the Art Hospital and Institutional Waste Incineration: Selection, Procurement, and Operations*; Technical Document

No. 055940, Jan. 1991, American Society for Hospital Engineering, Chicago, IL.
"EMC Offers a Solution to the Problem of Hazardous Household Waste," Fire Chief, Vol. 35, No. 7, July 1991, pp. 52–54.
"Fire Protection for Belt Conveyors," FMRC Loss Prevention Data Sheet 7-11, Factory Mutual Research Corp., Norwood, MA.
Filius, K. D. and Whitworth, C. G., "Emissions Characterization and Off-Gas System Development for Processing Simulated Mixed Waste in Plasma Centrifugal Furnace," MSE, Inc., Butte, MT, Department of Energy, Washington, DC, University of California, Berkeley, Toxic Toxic Combustion Byproducts, 4th International Congress, ABSTRACTS ONLY, June 5–7, 1995, Salt Lake City, UT, 1995, 24 pages.
"Flammable and Combustible Waste Disposal in the Plastics Industry," Bulletin No. 13, Society of the Plastics Industry, New York.
Hayes, W. K., "Recycle And/Or Disposal of Flame Retardant Plastics," Ethyl Corp., Baton Rouge, LA, Fire Retardant Chemicals Association, Customer Demands for Improved Total Performance of Flame Retarded Materials, Oct. 26–29, 1993, Tucson, AZ, Fire Retardant Chemicals Assoc., Lancaster PA, 1993, pp. 283–287.
Hoerning, J. M. and Ragland, K. W., "Aromatic Emissions From Incineration of Selected Wastes Using a Laboratory Scale Rotary Kiln," Wisconsin Univ., Madison, University of California, Berkeley, Toxic Toxic Combustion Byproducts, 4th International Congress, ABSTRACTS ONLY, June 5–7, 1995, Salt Lake City, UT, 1995, 107 pages.
Hunt, S. P., "Concrete Hazardous/Radioactive Waste Storage Facilities: Fire Development and Hazard Analysis, Selected Readings in Performance-Based Fire Safety Engineering. 1995, Boston, MA, Society of Fire Protection Engineers, Boston, MA, 1995, pp. 117–128.
Huotari, J. and Vesterinen, R., "PCDD/F Emissions From Cocombustion of RDF With Peat, Wood Waste and Coal in FBC Boilers," VTT Energy, Jyvaskyla, Finland, University of California, Berkeley, Toxic Toxic Combustion Byproducts, 4th International Congress, ABSTRACTS ONLY, June 5–7, 1995, Salt Lake City, UT, 1995, 78 pages.
Johnson, G. D., "Fiscal Year 1992 Program Plan for Evaluation and Remediation of the Generation and Release of Flammable Gases in Hanford Site Waste Tanks," Westinghouse Hanford Co., Richland, WA, Department of Energy, Washington, DC, WHC-EP-0537, Jan. 1992, 85 pages.
Kephart, W., Eger, K. and Clemens, M. K., "Toxic Combustion Byproducts: Generation, Containment, Separation, Cleansing for Hazardous, Mixed, and Transuranic Waste Processing," Foster Wheeler Environmental Corp., Argonne National Laboratory, IL, University of California, Berkeley, Toxic Toxic Combustion Byproducts, 4th International Congress, ABSTRACTS ONLY, June 5–7, 1995, Salt Lake City, UT, 1995, 25 pages.
Koo, J., *et al.*, "Formation of Toxic Byproducts From Pilot-Scale Rotary Kiln Incinerator for Mixed Polyethylene Waste," Korea Advanced Institute of Science and Technology, Kolon Engineering Ltd., Co., University of California, Berkeley, Toxic Toxic Combustion Byproducts, 4th International Congress, ABSTRACTS ONLY, June 5–7, 1995, Salt Lake City, UT, 1995, 38 pages.
Larsen, F. S., *et al.*, "Hydrocarbon and Formaldehyde Emissions From the Combustion of Pulverized Wood Waste," Combustion Science and Technology, Vol. 85, No. 1–6, 1992, pp. 259–269.
Law, C. K., "Considerations of Droplet Processes in Liquid Hazardous Waste Incineration," Combustion Science and Technology, Vol. 74, No. 1–6, 1990, pp. 1–15.
Lemieux, P. M., Linak, W. P., and Ryan, J. V., "Use of Surrogate Performance Indicators to Predict Emissions of Trace Organics From Hazardous Waste Incinerators," Air and Energy Engineering Research Lab., Research Triangle Park, NC, Acurex Environmental Corp., Research Triangle Park, NC, University of California, Berkeley, Toxic Toxic Combustion Byproducts, 4th International Congress, ABSTRACTS ONLY, June 5–7, 1995, Salt Lake City, UT, 1995, 127 pages.
Lemieux, P. M., Linak, W. P. and Wendt, O. L., "Waste and Sorbent Parameters Affecting Mechanisms of Transient Emissions From Rotary Kiln Incineration," Environmental Protection Agency, Research Triangle Park, NC, Arizona Univ., Tucson, University of California, Berkeley, Toxic Toxic Combustion Byproducts, 4th International Congress, ABSTRACTS ONLY, June 5–7, 1995, Salt Lake City, UT, 1995, 39 pages.
Leonard, J. T., *et al.*, "Fire Hazard Assessment of Shipboard Plastic Waste Disposal Systems," Naval Research Laboratory, Washington, DC, Hughes Associates, Inc., Columbia, MD, NRL/MR/6184-94-7452, Feb. 28, 1994, 55 pages.

Levendis, Y. A., *et al.*, "Comparative Study of Combustion and Organic Emissions of Waste Tire Crumb and Pulverized Coal," Northeastern Univ., Boston, MD, University of California, Berkeley, Toxic Toxic Combustion Byproducts, 4th International Congress, ABSTRACTS ONLY, June 5–7, 1995, Salt Lake City, UT, 1995, 66 pages.

McKnight, M. E., "Review of Current Research and Activities Involving Characterization, Abatement and Disposal of Lead-Containing Paint Films," National Institute of Standards and Technology, Gaithersburg, MD, NISTIR 90-4285, May 1990, 13 pages.

Nollet, A. R., "How to Prevent Shredding Plant Explosions," *World Wastes*, Vol. 32, No. 7, 1989, pp. 56-60.

Rhoeds, B. T., *et al.*, "Evaluation of Solid Waste Drum Fire Performance," *Proceedings* of the International Conference on Fire Research and Engineering, Sept.10–15, 1995, Orlando, FL, SFPE, Boston, MA, 1995, pp. 403–406.

Serbin, S. I., Sacchi, G. F. and Sarofim, A. F., "Application of Plasma-Chemical Technology in Hazardous Waste Incinerators," Massachusetts Institute of Technology, Cambridge, University of California, Berkeley, Toxic Toxic Combustion Byproducts, 4th International Congress, ABSTRACTS ONLY, June 5–7, 1995, Salt Lake City, UT, 1995, 29 pages.

Silva, M., "Assessment of the Flammability and Explosion Potential of Transuranic Waste," Environmental Evaluation Group, Albuquerque, NM, DOE/AL/58308-48, June 1991, 96 pages.

Snyder, K. A. and Clifton, J. R., "4SIGHT Manual: A Computer Program for Modelling Degradation of Underground Low Level Waste Concrete Vaults," National Institute of Standards and Technology, Gaithersburg, MD, NISTIR 5612, June 1995, 73 pages.

Tsang, W., "Chemistry of Hazardous Waste Incineration," Combustion Institute/Eastern States Section, 1990 Fall Technical Meeting on Chemical and Physical Processes in Combustion, Dec. 3–5, 1990, Orlando, FL, 1990, pp. B/1–11.

Turner, D. A. and Miron, Y., "Testing of Organic Waste Surrogate Materials in Support of the Hanford Organic Tank Program," Final Report, Department of Energy, Washington, DC, WHC-MR-0455, UC-600, Jan. 1994, 175 pages.

Wei, M. S., Savin, T. J., and Fisher, G., "Air Quality Modeling Analysis for Hazardous Waste Sites," Parsons Engineering Science, Inc., Oak Ridge, TN, Parsons Engineering Science, Inc., Fairfax, VA, University of California, Berkeley, Toxic Toxic Combustion Byproducts, 4th International Congress, ABSTRACTS ONLY, June 5–7, 1995, Salt Lake City, UT, 1995, 84 pages.

HAZARDOUS WASTE CONTROL

Revised by Sami Atallah

Waste is a natural by-product of many human activities. The varieties, quantities, and hazards of today's waste appear to be directly proportional to the degree of technological development of a nation. The Environmental Protection Agency (EPA) estimated that, in 1993, municipal solid waste alone totaled 207 million tons (188 metric tons).[1] This is but a small fraction of the total solid waste produced by commercial, mining, construction, agricultural, manufacturing, and other industrial activities. During the course of a year, over 3.5 billion tons (3.3 billion metric tons) of solid waste is produced—roughly 100 lb (45 kg) per day for every man, woman, and child.[2]

In addition to solid waste, many gaseous products, such as those produced in combustion processes (e.g., CO_2, NO_x, SO_2, soot, and hydrocarbons), and, to a lesser extent, chlorofluorocarbons (CFCs), are released into the atmosphere. These latter chemicals had been extensively used as aerosol propellants, refrigerants, and fire-extinguishing agents (Halon 1301 and Halon 1211). However, under the Montreal Protocol on Substances that Deplete the Ozone Layer, signed by the United States and many other countries on September 16, 1987, the production and use of CFCs is expected to be phased out by the year 2000. The pollution of air with such gaseous-waste products is not only detrimental to human health, it reduces visibility and contributes to the corrosion and deterioration of clothing, art treasures, historical structures, metals, rubber, and paint. Acid rain, the product of further atmospheric oxidation and dissolution of NO_x and SO_x in rain water, adversely affects the growth of forests and agricultural products, and the survival of aquatic life in streams, ponds, and lakes. The accumulation of CO_2 in the atmosphere is expected to have long-range effects on the warming of the planet with dire climatological, agricultural, and ocean-level implications. CFCs destroy the ozone layer in the stratosphere; ozone protects us against harmful ultraviolet radiation from the sun. The depletion of this layer increases the risk of skin cancer.

Domestic, commercial, and industrial liquid wastes invariably find their way to bodies of water and add to the problem of the dwindling potable water supply. Industrial liquid waste may contain metallic salts that have immediate or long-term effects on humans. High concentrations of these salts may also build up in fishes, crustaceans, and plants with similar effects on people who consume them. Some organic wastes from commercial, household, and industrial sources reduce the concentration of dissolved oxygen in water supplies. This may lead to the extinction of aquatic life, and the growth of disease-carrying bacteria and parasites. Other wastes may contain known or suspected carcinogens, such as chlorinated hydrocarbon solvents, benzene, and polychlorinated biphenyls (PCBs).

Although all waste is a nuisance to society, not all waste is necessarily hazardous. Hazardous waste may be defined as that which poses a threat to human health, the environment, and public property if handled or disposed of improperly. The hazard is created by virtue of the toxicity, flammability, explosivity, reactivity, radioactivity, corrosivity, or etiologic (disease-causing) potential of the waste.

Under Section 313 of the Emergency Planning and Community Right-to-Know Act of 1986, which was expanded under the Pollution Prevention Act of 1990, EPA collects annual data on the emission rates of over 300 listed toxic chemicals and 20 chemical categories from over 23,000 qualifying companies. These companies employ 10 or more full-time employees and process more than 25,000 lb (11 340 kg) or use more than 10,000 lb (4 536 kg) of the listed chemicals.

These data show[3] that, in 1993, the reporting facilities released more than 2.8 billion lb (1.27 billion kg) of the listed chemicals into the environment. Air emissions constituted 59.5 percent of all toxic releases. Surface-water releases, which include releases to rivers, lakes, oceans, and other water bodies, accounted for 9.7 percent of all releases. Releases to land, which include landfills, surface impoundments, and other types of land disposal, accounted for 10.3 percent of all releases. Underground injection accounted for the remainder (20.5 percent). In addition to these releases, more than 4.7 billion lb (2.14 billion kg) of toxic chemical waste were transferred to off-site locations for recycling (69.1 percent); energy recovery (10.3 percent); treatment at publicly owned treatment works (6.7 percent); other treatment, such as by biological degradation, incineration, neutralization, or physical separation (7 percent); and disposal (6.9 percent).

Numerous federal, state, and local laws and regulations have been enacted to protect the public and the environment against exposure to hazardous waste. The federal laws attempt to specify and control the quantities, concentrations, and types of chemicals that may be released and the manner in which they may be packaged, stored, transported, and disposed of safely.

Table 3-32A is a summary of pertinent federal laws and related terms used in this chapter. In addition to the federal acts listed in Table 3-32A, parts of other laws and amendments to existing

---

Sami Atallah is president of Risk & Industrial Safety Consultants, Inc., Des Plaines, IL.

TABLE 3-32A. *Summary of Pertinent Federal Laws and Terminology*

CAA:	Clean Air Act (1955, 1970, 1977, 1990)—a federal law that mandates and enforces toxic emissions standards for stationary sources and motor vehicles. The CAA amendments of 1990 added restrictions on air toxics, ozone-depleting chemicals, stationary and mobile emission sources, and emissions implicated in acid rain and global warming. They also required industries that handle threshold quantities of listed hazardous chemicals to prepare a risk management plan.
CERCLA:	Comprehensive Environmental Responsibility, Compensation and Liability Act (1980, 1986), sometimes referred to as Superfund—a federal law that authorizes the identification and remediation of abandoned hazardous waste sites.
CWA:	Clean Water Act (1972, 1987)—a federal law that regulates the discharge of pollutants into surface waters and to publicly owned and municipal water treatment facilities.
EPA:	Environmental Protection Agency, established under the Reorganizational Plan 3 of 1970 as an independent federal agency to ensure the protection of the environment by abating and controlling pollution. It coordinates all governmental action; conducts research; and monitors, sets, and enforces environmental standards.
FEMA:	Federal Emergency Management Agency—a federal agency that is accountable for all federal emergency preparedness, mitigation, and response activities for all natural, man-made, or nuclear emergencies; also responsible for administering certain training funds under SARA.
FIFRA:	Federal Insecticide, Fungicide, and Rodenticide Act (1972, 1988)—a federal law that mandates toxicity testing and registration of pesticides, and establishes tolerance levels for residues of these agents in food.
OSHA:	Occupational Safety and Health Act (1970)—a federal law that called for the establishment of the Occupational Safety and Health Administration, which oversees and regulates workplace health and safety. OSHA standards require facilities that process specified quantities of listed toxic and flammable chemicals to comply with 29 CRF 1910.119, "Process Safety Management." It also requires employers to inform and train employees in the hazards of the chemicals they handle, under 29 CFR 1910.1200, "Hazard Communication."
RCRA:	Resource Conservation and Recovery Act (1976, 1984)—enacted as an amendment to the Solid Waste Disposal Act, RCRA protects human health and the environment and encourages the conservation of valuable material and energy sources. It provides assistance to state and local governments for prohibiting open dumping; regulating the management of hazardous wastes; and encouraging recycling, reuse, and treatment of hazardous wastes. RCRA provides for "cradle-to-grave" tracking of hazardous waste, from generator to transporter to treatment, storage, or disposal.
SARA:	Superfund Amendments and Reauthorization Act (1986)—a federal law that reauthorizes and expands the jurisdiction of CERCLA and designates requirements mandated in Title III of SARA for public disclosure of chemical information and the development of emergency response plans.
TSCA:	Toxic Substances Control Act (1976)—a federal law that authorizes EPA to gather information on new and existing chemical risks to human health and the environment.
UST:	Underground Storage Tanks—regulated by RCRA, Subtitle I, these are tanks with 10 or more percent of their volume underground, used to store CERCLA-regulated hazardous chemicals or petroleum products.

laws regulate specific waste-disposal and handling operations. For example, the Marine Protection, Research, and Sanctuaries Act regulates ocean dumping and incineration on the high seas; the Hazardous Materials Transportation Act regulates the packaging and containerization of hazardous materials, including hazardous wastes; the Hazardous and Solid Waste Amendments to RCRA (Subtitle I) provide for the development and implementation of a comprehensive regulatory program for underground storage tanks; and the Safe Drinking Water Act establishes maximum contaminant levels for drinking water and regulates underground injection of hazardous wastes.

The safe disposal of hazardous waste introduces many problems not only to the industry generating the waste, but also to the waste-transportation and disposal agency, and to fire fighters and other response forces who may be called upon to clean up an accidental release. These problems arise because the contents of waste are usually not well defined—sometimes not even known. Waste is inherently a mixture of many components with differing hazardous properties. One component, for example, may be a flammable solvent while another is a highly toxic pesticide. Disposing of the solvent by burning may release the pesticide (or its toxic combustion products) to the atmosphere. Additional problems arise in handling and storing waste products. Self-heating and ignition, corrosion of pipe and tank-wall materials, and reaction with other noncompatible wastes have been known to cause major catastrophic events. Fire fighters may also be accidentally exposed to a hazardous-waste material released as a result of an unrelated fire or an explosion to which they have responded.

This chapter will: (1) discuss the sources and characteristics of waste materials with particular emphasis on hazardous waste, (2) summarize waste handling, disposal, and recovery procedures, and (3) examine the various approaches to fire prevention and protection in cases involving hazardous waste.

SOURCES OF WASTE

There are many sources of waste materials in present-day society. To a great extent, the source determines what the composition of the waste may be, its bulk, density, quantity, and the type and degree of hazard that it may present. These factors, in turn, determine how the waste should be collected, processed, or disposed of in a practical and safe manner.

Domestic, Commercial, and Industrial Sources

Table 3-32B shows typical breakdowns of the contents of collected solid waste from domestic, commercial, and industrial sources. Obviously, these breakdowns will vary among sources within each category. For example, suburban homes are expected to produce more yard waste than city residences. Restaurants generate more food waste and less paper waste than office buildings. Hospitals dispose of cotton and paper products that may be contaminated with hazardous chemicals and etiologic agents.

The nature of an industry determines the content of the waste generated. A furniture plant, for example, discards wood scrap, sawdust, and shavings as well as oily rags, stain, varnish, lacquer, glue,

TABLE 3-32B. *Typical Contents of Solid Waste (Percentage by Weight)*

	Domestic	Commercial	Industrial
Paper products	51.5	69.0	17.4
Glass	15.0	7.0	3.4
Cans and metals	7.0	10.0	8.0
Plastics	2.0	10.0	
Cloth, leather, rags	4.0		0.4
Food	10.0	4.0	8.4
Wood	2.0		17.9
Yard waste	8.5		
Minerals			36.7
Chemical waste			3.3
Rubber			1.6
Other			3.1

etc., in addition to paper, metal, and other commercial items. The waste products of chemical processing plants vary considerably depending on the raw materials used and the end products processed. These plants may discard substantial amounts of off-grade or reject products, by-products, tars, spent catalyst, precipitates, etc. On the other hand, nuclear power plants produce radioactive waste.

At any processing plant, some materials will be accidentally spilled in storage and handling. Cleanup operations from filling equipment, fabrication operations, and general plant maintenance will generate some waste. Rejected, aged, or damaged packages; off-specification materials; and overruns will also be discarded.

Transportation Sources

The movement of bulk or packaged goods may also generate waste. Broken packages, leaking drums, and overheating or freezing of cargo are contributing sources. Washing tank cars or tank trucks produces substantial amounts of waste material that may be emulsified or dissolved in water. A spill of a hazardous liquid or solid-chemical cargo as a result of a railway, truck, ship, or pipeline accident contaminates soil and/or nearby bodies of water and constitutes a public hazard and another waste disposal problem. This is particularly true if the accident occurs in a remote area where experienced personnel are not readily available to assess the potential hazard to the public and the environment and to provide guidance in controlling the hazard.

Storage Sources

Warehouse inventory control necessitates that unmarketable or undesirable materials of all kinds, forms, and hazards be discarded occasionally. Tanks must be cleaned as storage needs change. The washing of transfer lines and tanks results in a waste product.

Materials released as a result of plant accidents, such as when packages are damaged by forklifts, require cleanup and disposal. Water, fire, or explosion damage can also result in substantial amounts of undesired product.

Other Sources

Several other human activities produce waste. Harvesting of agricultural products, mining operations, and demolition of old structures are examples of operations that generate large quantities of waste products with various characteristics.

CHARACTERIZATION OF WASTE

The safe handling, collection, and disposal of hazardous waste can be accomplished only if the physical, chemical, and hazardous properties of its components are known and that information is properly applied. Quite often though, the composition is not known because waste is usually a mixture of many components with differing properties. This section will discuss the physical, chemical, and hazardous properties that may influence the selection of response procedures to accidental releases and the appropriate disposal techniques for wastes.

Physical State

The ease of handling and disposal of waste is highly dependent on its physical state.

Solid waste: This category includes most domestic, commercial, and industrial trash, metal, wood, plastic scrap, chemicals, food, etc. (See Table 3-32B.) It also includes agricultural waste products, demolition waste, and spent ore from mining operations. The bulk density of these wastes differs considerably depending on the source, mix of materials, and the degree of compactness of the waste. For example, uncompacted household waste has a bulk density of 200 to 250 lb per cu yd (119 to 148 kg/m³), while waste in a landfill may have a bulk density of 750 to 1,000 lb per cu yd (445 to 594 kg/m³). Solid waste also includes combustible metal waste (e.g., magnesium shavings) and semisolids, such as waxes, soaps, and elastomers.

Liquid waste: This category includes aqueous and nonaqueous solutions and suspensions generally produced as a result of cleaning operations. Nonaqueous solvents may be flammable (e.g., toluene and xylene) or noncombustible (e.g., tricholorethylene). Viscous liquids, such as pastes and tars, and suspensions, such as acid sludges, require special handling procedures and equipment.

Gaseous waste: Most gaseous waste is generated from industrial operations and is disposed of by venting into the atmosphere. Hazardous gaseous-waste products are regulated by the Environmental Protection Agency (EPA) and must be properly disposed of, such as by flaring, removal, or recovery of the hazardous components by chemical means. Of particular concern to fire fighting and other response personnel are old unidentifiable gas cylinders with walls that may have become weakened due to age and corrosion. Discarded small butane and LP-Gas cylinders in household trash may also pose an explosion hazard to trash collectors and compactors and to operators of shredding equipment.

Properties of Hazardous Waste

Few discarded materials are so compatible with the environment or so inert as to have no short-term or long-term impact. Hazards that appear minor may have unexpected impacts long after disposal. When two or more hazards pertain to a material, the lesser may not receive the necessary consideration. Mixing of two discarded substances may result in a chemical reaction with severe and unexpected consequences.

Since waste is generally a mixture of many components, its physical and chemical properties cannot be defined with any degree of accuracy. Whenever possible, the approximate composition of a hazardous waste should be ascertained from the originating source or from the manifest accompanying the waste being transported. Generally, when one component predominates, the physical and chemical properties of the waste mixture are nearly those of the major component. This is not true for the hazardous properties of waste mixtures consisting of a relatively harmless major component and small amounts of highly toxic, radioactive, or etiologically active components. The hazard, in this case, is determined by the smaller components.

Hazardous Solid Waste

The Environmental Protection Agency defines hazardous solid waste as that which may: "(i) cause, or significantly contribute to, an increase in mortality or an increase in serious irreversible, or incapacitating reversible, illness; or (ii) pose a substantial present or potential hazard to human health or the environment when it is improperly treated, stored, transported, disposed of, or otherwise managed." The determining characteristic of a hazardous solid waste must be "measurable by an available standardized test method . . . or reasonably detectable by generators of solid waste through their knowledge of their waste."[4]

Basically, EPA standards require that a solid waste be listed and treated as hazardous if it meets the ignitibility, corrosivity, reactivity, and/or toxicity characteristics prescribed by the standard. Highlights of the standard are given below.[5]

Ignitibility: A solid waste exhibits this characteristic if a representative sample of the waste has any of the following properties:

1. It is a liquid, other than an aqueous solution containing less than 24 percent alcohol by volume, and has a closed-cup flash point less than 140°F (60°C) (Class I and Class II liquids, as defined in NFPA 30, *Flammable and Combustible Liquids Code*), as determined by an approved test method.
2. It is not a liquid and is capable under standard temperature and pressure of causing fire through friction, absorption of moisture, or spontaneous chemical changes and, when ignited, burns so vigorously and persistently that it creates a hazard.
3. It is a flammable compressed gas as defined by another federal standard.[6] A flammable compressed gas is defined as one that forms flammable mixtures with air at concentrations less than 13 percent (by volume) or has a flammability range with air which is wider than 12 percent regardless of the lower flammable limit. The standard specifies additional criteria for defining flammable vapors and aerosols.
4. It is an oxidizer, such as a chlorate, permanganate, inorganic peroxide, or a nitrate, that yields oxygen readily to stimulate the combustion of organic matter.[7]

Corrosivity: A solid waste exhibits this characteristic if a representative sample of the waste has either of the following properties:

1. It is aqueous and has a pH of less than or equal to 2 (i.e., highly acidic) or greater than or equal to 12.5 (i.e., highly alkaline, or basic).
2. It is a liquid and corrodes steel at a rate greater than 0.25 in. (6.35 mm) per year at a test temperature of 130°F (55°C).

Reactivity: A solid waste exhibits this characteristic if it behaves in any of the following ways:

1. It is normally unstable and readily undergoes violent change without detonating.
2. It reacts violently with water.
3. It forms potentially explosive mixtures with water.
4. When mixed with water, it generates toxic gases, vapors, or fumes in a quantity sufficient to present a danger to human health or the environment.
5. It is a cyanide-bearing or sulfide-bearing waste that, when exposed to pH conditions between 2 and 12.5, can generate toxic gases, vapors, or fumes in a quantity sufficient to present a danger to human health and the environment.

6. It is capable of detonation or explosive reaction if it is subjected to a strong initiating source or if heated under confinement.
7. It is readily capable of detonation or explosive decomposition or reaction at standard temperature and pressure.
8. It is a known forbidden or a Class A or B explosive.

Toxicity: The EPA defines toxic wastes as those that have been found to be fatal to humans in low doses or, in the absence of data on human toxicity, it has been shown in studies to have an oral LD_{50} toxicity (rat) of less than 50 mg/kg, an inhalation LC_{50} toxicity (rat) of less than 2 mg/L, or a dermal LD_{50} toxicity (rabbit) of less than 200 mg/kg or is otherwise capable of causing or significantly contributing to an increase in serious irreversible, or incapacitating reversible, illness. A toxic waste is also one that contains or degrades into a listed toxic material in large enough concentrations to pose a potential hazard to the public and/or the environment. The standard for toxicity provides descriptions of laboratory tests for extracting and analyzing the potentially toxic components of wastes.

Hazardous Medical Waste

During the summer of 1988, a number of incidents were reported where medical waste, such as blood vials and bags, syringes, pill bottles, and bandages, washed ashore at various public Atlantic coast beaches. The publicity associated with these incidents, and others involving the disposal of medical waste in dumpsters, hastened the promulgation of the Medical Waste Tracking Act of 1988. The law required EPA to develop a two-year demonstration "cradle-to-grave" tracking system for four participating states. This law has since expired. Individual states have developed their own environmental standards in which they defined what constitutes hazardous medical waste and identified acceptable disposal methods. Further, the Occupational Safety and Health Administration (OSHA) also promulgated a standard[8] for the protection of employees against exposure to bloodborne pathogens. The Department of Transportation (DOT) has also developed standards for the proper packaging and shipment of hazardous medical waste.[9]

Medical waste may be defined as that which is generated during the diagnosis, treatment, or immunization of human beings or animals; in research pertaining thereto; or in the production or testing of regulated biologicals. It is generated at such facilities as hospitals, health care centers, nursing homes, funeral homes, blood banks, veterinary clinics, and medical-research laboratories.

Medical waste is a likely source of pathogens (infectious agents). Some of the diseases that may be transmitted through improper disposal and handling of infectious medical waste are acquired immune deficiency syndrome (AIDS), typhoid, cholera, dysentery, anthrax, salmonellosis, tuberculosis, and burcellosis. Development of a strict medical-waste management plan would reduce the risk of infection to staff, patients, and the general public.

Medical waste may be categorized as follows:

1. *Cultures and stocks.* Cultures and stocks of infectious agents and associated biologicals, including cultures from medical and pathological laboratories; cultures and stocks of infectious agents from research and industrial laboratories; wastes from the production of biologicals; discarded live and attenuated vaccines; and culture dishes and devices used to transfer, inoculate, and mix cultures.
2. *Pathological wastes.* Human pathological wastes, including tissues, organs, and body parts and body fluids that are removed during surgery or autopsy, or other medical procedures, and specimens of body fluids, and their containers.
3. *Human blood and blood products.* Liquid-waste human blood; products of blood; items saturated and/or dripping with human blood; items that were saturated and/or dripping with hu-

man blood that are now caked with dried human blood, including serum, plasma, and other blood components, and their containers, which were used or intended for use in either patient care, testing, and laboratory analysis, or the development of pharmaceuticals. Intravenous bags are also included in this category.

4. *Sharps.* These are objects that have been used in animal or human patient care or treatment, or in medical, research, or industrial laboratories. They include hypodermic needles, syringes (with or without the attached needle), pasteur pipettes, scalpel blades, blood vials, needles with attached tubing, and culture dishes (regardless of presence of infectious agents). Also included are other types of broken or unbroken glassware that were in contact with infectious agents, such as used slides and cover slips.

5. *Animal waste.* Contaminated animal carcasses, body parts, and bedding of animals that were known to have been exposed to infectious agents during research (including research in veterinary hospitals), production of biologicals, or testing of pharmaceuticals.

6. *Isolation waste.* Biological waste and discarded materials contaminated with blood, excretion, exudates, or secretions from humans who are isolated to protect others from certain highly communicable diseases, or isolated animals known to be infected with highly communicable diseases.

7. *Unused sharps.* Unused, discarded sharps, e.g., hypodermic needles, suture needles, syringes, and scalpel blades.

COLLECTION, HANDLING, AND DISPOSAL OF WASTE

The disposal of waste may be achieved through a variety of techniques and processes, depending on the physical, chemical, and hazardous properties of the waste. These techniques range from improvements in operations that reduce the total quantity of waste, to the introduction of process changes that reduce the concentration of hazardous components to acceptable levels. Recycling and the recovery of energy from waste streams are other approaches that have become economically acceptable as the cost of energy and raw materials continues to increase.

Treatment of Gaseous Waste

Industrial gaseous waste may be released in the atmosphere, provided that the concentrations of the contaminants meet allowable levels set by EPA and the Clean Air Act. Table 3-32C summarizes the primary ambient air quality standards set by EPA for typical contaminants. Also shown are contaminant background levels in nature.

Gaseous waste streams can be purified by removing undesirable components or by reducing their concentrations to acceptable levels. This may be achieved by chemical conversion, absorption, adsorption, or particulate removal.[12]

Chemical conversion: An undesirable component may be allowed to react chemically with other compounds to form a harmless component or one that may be removed easily. Sulfur dioxide, for example, may be removed by reaction with lime or magnesia. Catalytic converters are used to convert nitrogen oxides and hydrocarbons into nitrogen, CO_2, and H_2O.

Absorption: Several gaseous pollutants may be removed by absorption. Ammonia and hydrogen chloride, for example, may be removed by water scrubbing. Sulfur dioxide may be absorbed in an alkaline solution.

Adsorption: Certain solids with a large pore-surface area selectively adsorb small concentrations of specific gases. Activated carbon, silica or alumina gel, fuller's earth and other clays are typical adsorbents. The operator is left with the problem of disposing of the contaminated adsorbent or the concentrated stream of undesirable gas released upon adsorbent regeneration.

Particulate removal: Many industrial gaseous streams contain undesirable dusts, mist, or smoke. These suspensions may be removed by filtration, sedimentation, centrifugal separation, electrostatic precipitation, and wet scrubbing.

Liquid Waste Treatment

The volume and strength of industrial aqueous waste may be reduced by various process modifications. Although the hazard is not completely eliminated, a smaller volume of contaminated water or a stream with a lower concentration of a hazardous component is easier to handle and dispose of. Neutralization of acidic or alkaline waste may be achieved by chemical reaction with the appropriate compounds. Inorganic salts may be removed by precipitation, ion exchange, and adsorption by carbon beds. Holding ponds may be used to retain industrial waste so that it may be disposed of with domestic waste at a constant rate rather than overloading the sewage-disposal system at any one time of the day.

Suspended solids may be separated by such processes as sedimentation (allowing heavy particles to sink to the bottom), flotation (skimming floating materials off the surface), drying followed by incineration, centrifuging, and filtration. Colloidal solids are more difficult to remove because their particle size range falls between that of suspended and dissolved solids. Colloidal solids will not settle down unless chemicals are added that help coagulate the solid

TABLE 3-32C. *Ambient Air Quality Standards*[10,11]

Contaminant	Background Level	EPA Primary Level	Allowable Exposure
SO_2	<0.002 ppm	0.03 ppm	AAM
		0.14 ppm	Max. 24-hr conc., N
CO	0.03 ppm	9 ppm	8-hr AC, N
		35 ppm	1-hr AC, N
O_3	0.01–0.05 ppm	0.12 ppm	1-hr AC, N
NO_2	4 ppb	52 ppb	AAM
Particulates	10–60 $\mu g/m^3$	150 $\mu g/m^3$	24-hr AC
		50 $\mu g/m^3$	AAM

AAM = annual arithmetic mean,
AC = average concentration
hr = hour
N = not to exceed once per calendar year

into larger masses that would sink. Many toxic pesticides, insecticides, and herbicides may also be removed by adsorption in certain clays.

Many food-processing operations as well as the manufacture of textiles and paper produce a waste that is high in organic matter. Biological degradation is generally achieved by lagooning, treatment with biologically activated sludge, the use of trickling filters, wet combustion, deep-well injection, and other means.

Wastewater from tanneries, slaughterhouses, and canneries may contain etiologic agents. These waters must be sterilized by chlorination, ozonization, or ultraviolet radiation before they are discharged into public water supplies.

Organic liquid waste components may be recovered by fractional distillation. The remaining components may be disposed of by incineration or burial in environmentally acceptable locations.

Disposal and Recovery of Solid Waste

Table 3-32B shows that a large fraction of domestic, commercial, and industrial solid waste is combustible. Thus, the burning of scrap and refuse under controlled conditions is one of the most satisfactory methods of disposal. Municipal incinerators operated primarily for domestic and commercial trash will usually accept industrial refuse as long as it does not introduce additional hazards or operating problems.

In recent years, the recovery of glass and metal from these wastes and the use of the combustible component as a fuel supplement has become economically attractive. The basic steps involved in these recovery operations are waste collection, transportation to the disposal site, storage, shredding, separation and recovery of glass and metal, burning of the combustible component, and ash removal. (See Figure 3-32A.)

Air-pollution-control equipment is nearly always needed on industrial incinerators. Environmental control devices depend upon the composition of the stack effluent. Such devices include the following:

1. Afterburners to ignite combustible particles,
2. Cyclones to remove larger particles,
3. Low-energy scrubbers to remove larger particulates,
4. High-energy venturi scrubbers to remove fine particulates and/or water-soluble acid gases,
5. Electrostatic precipitators to remove fine particulates and mists, and
6. Fabric filters (bag houses) to remove fine particulates.

Principal hazards of operating an incinerator are the following:

1. Overheating due to feeding a large amount of volatiles,
2. Overheating due to an increase in the heat of combustion,
3. Structural failure due to overheating and/or corrosion,
4. Corrosion of scrubber due to acid, and
5. Failure of scrubber system.

Compacted trash on transportation vehicles has been known to result in fires. Ignition sources may be smoldering ashes from fireplaces or unextinguished cigarettes. Shredding operations have also been the scene of occasional explosions and fires; in some cases, as a result of discarded explosive materials. Small butane and LP-Gas cylinders and gasoline containers with some fluid remaining in them may be the source of the flammable vapor. The ignition sources may be frictional sparks between the grinding or shredding teeth, the walls of the shredder, and other solid components of the refuse.

Another approach for recovering the heating content of combustible waste is by pyrolysis. When polymers and organic materials are heated, they yield oils and other recoverable compounds. Although pyrolysis has not yet been widely adopted, it has promise as a conservation measure. A retort is charged with selected scrap and then heated to cause decomposition and distillation of the volatiles and oils. The process is somewhat similar to a coke oven, and similar operating hazards should be expected. (See Figure 3-32B.)

Most nonhazardous domestic, commercial, and industrial waste today is disposed of in sanitary landfills. In a landfill operation, discarded material is mixed with earth and then compacted in a wide trench or other depression. Layer upon layer of mixture is added and then 2 ft (0.6 m) of earth is spread and compacted on top. This process helps control odors, rodents, and scavenger birds.

Successful landfill operation depends heavily upon biological activity within the fill material. There is some tendency for methane gas to be generated during biological activity. There have been some fires and explosions in buildings constructed over old landfills due to the seepage of biologically generated methane into the building and its eventual ignition. The recovery of natural gas from waste landfills is under way in various parts of the United States.

Hazardous wastes cannot be disposed of in a sanitary landfill. They are either chemically treated or incinerated in properly designed furnaces that ensure the complete degradation of the hazardous components of the waste. Otherwise, they may be disposed of in secure landfills. These are properly designed and engineered containment facilities that ensure that no waste can ever seep into the underground water system to pollute natural bodies of water or the public water-supply system. Transporting the hazardous waste from the source to the disposal site usually requires federal, state, and local permits. The waste must be packaged or containerized in approved containers. A manifest describing the contents, their hazards, and emergency-response procedures must accompany the

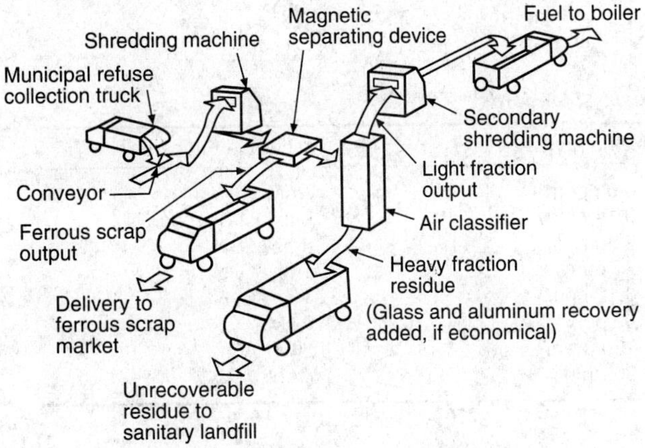

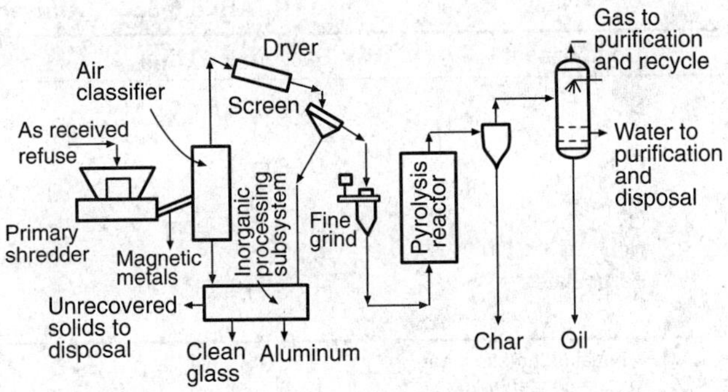

FIG. 3-32A. *Typical refuse recovery process.*

FIG. 3-32B. *Flow diagram of a pyrolysis recovery system.*

shipment. The operator of the secure landfill generally requires a complete chemical analysis of the hazardous waste so that it may be handled and sited properly.

Treatment of Hazardous Medical Waste

State standards governing the packaging, segregation, and identification of hazardous medical waste differ. Nevertheless, these standards generally require reporting, recordkeeping, and tracking of medical waste. The standards apply to waste generators, transporters, on-site incineration facilities, and off-site treatment, destruction, and disposal facilities.

Noninfectious medical waste generated at any medical facility typically constitutes up to 70 percent of all medical waste. To reduce disposal costs, the noninfectious waste is generally segregated from infectious medical waste at the source. This reduces the potential for cross-contamination and allows disposal at sanitary waste-disposal sites without further treatment. Segregation at the source also reduces the volume of medical waste that needs to be treated before disposal.

A standard operating procedure (SOP) should be developed by all facilities that store, handle, transport, or dispose of infectious waste.[13] Unsightly pathological waste should be placed in opaque plastic bags. All containers and bags should be red and labeled with the word "infectious" and/or with the international biohazard symbol. (See Figure 3-32C.) Wastes that are to be transported should be double-bagged and placed in a semi-rigid container to prevent leakage. Wastes should be loaded and unloaded manually, and never compacted, to prevent compromising the integrity of the packaging. These wastes should be arranged chronologically to meet state regulations for storage periods. All storage facilities should have limited access with the entrances marked by the international biohazard symbol and the word "infectious."

Infectious medical waste must be treated and rendered harmless prior to disposal at a sanitary waste disposal site. Several treatment methods are available, including the following six methods:

1. Steam sterilization,
2. Incineration,
3. Thermal inactivation (dry heat sterilization),
4. Gas/vapor sterilization,
5. Sterilization by irradiation, and
6. Chemical disinfection.

Steam sterilization is the most commonly used method for destroying pathogens in medical waste. It is preferred for contaminated solid surfaces such as sharps, and low-density waste, which may be easily penetrated by steam. Incineration is the preferred method for treating AIDS-contaminated and combustible infectious waste. Proper incineration of waste reduces the volume that needs to be disposed of by up to 95 percent, while rendering the waste totally harmless. Like steam sterilization, thermal inactivation is used

for the sterilization of contaminated surgical instruments and equipment. The dry heat has the advantage of not corroding instruments, thereby increasing their life.

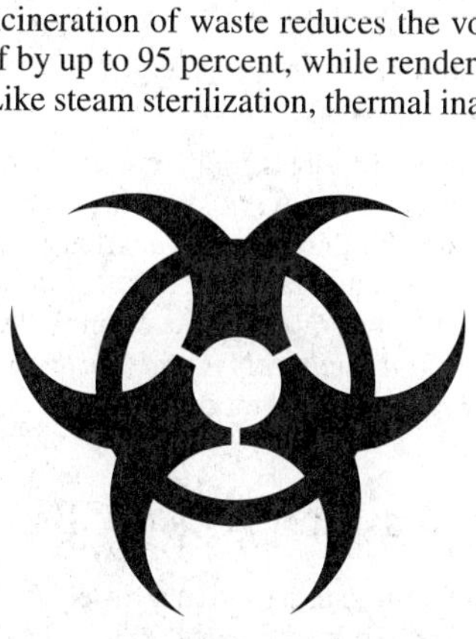

Although the remaining three methods may be used to treat medical waste, they tend to be more costly and require special precautions to protect the operator. Gas/vapor sterilization, for example, requires the use of ethylene oxide or formaldehyde, both of which are highly flammable and hazardous to human health. Irradiation with ultraviolet (UV) and gamma radiation is a newly developed technology that is used for pre-sterilization of medical products and may be used to treat waste water. Chemical disinfection is carried out by applying hydrogen peroxide, acids, or alcohols to exposed surfaces, utensils, and medical supplies.

In order to ensure the effectiveness of any treatment, it is important that an SOP be established for each waste stream to be treated. Each SOP should address the type of waste, variations in the waste stream, and the limitations of the treatment apparatus. The SOP must also specify procedures for biologically monitoring and testing of the treated waste to ensure that it is no longer hazardous.

Protection of Hazardous Waste Operators

OSHA standard "29 CFR 1910.120: Hazardous Waste Operations and Emergency Response" protects workers involved in hazardous-waste operations, and in responding to accidental releases of hazardous materials and wastes. This law[14] requires the following:

1. The development and implementation of a written safety and health program for employees (and subcontractors' employees) who are involved in hazardous-waste operations. The program must be designed to identify, evaluate, and control safety and health hazards, and provide for emergency response for hazardous-waste operations.
2. Hazardous-waste-site characterization, analysis, and control in order to identify specific site hazards, determine appropriate safety and health control procedures, and to implement these procedures to control employee exposure to the identified hazards.
3. Extensive training and retraining of employees before they are permitted to engage in hazardous-waste operations, and informational programs to ensure that employees and subcontractors are aware of the degree and nature of safety and health hazards specific to the work site.
4. Medical surveillance of employees exposed to (or potentially exposed to) hazardous substances, health hazards, or who wear respirators.
5. The implementation of engineering controls, work practices, and personal protective equipment, and the continuous monitoring of the work area to ensure that employees are not exposed to levels that exceed permissible exposure limits for hazardous substances at the site.
6. The development of a written emergency-response plan to handle anticipated emergencies prior to the commencement of hazardous-waste operations.
7. Additional requirements for the development and implementation of decontamination, and materials-handling procedures, and ensuring the presence of adequate illumination and sanitation facilities at the work site, and that proper excavation procedures are followed.

HAZARD PREVENTION AND CONTROL

Prevention Measures

The probability of exposure to hazardous waste may be reduced through preventive actions achieved through pre-planning, proper

FIG. 3-32C.   The international biohazard symbol.

waste-handling, and storage-handling procedures. Preventive measures include the following:

1. Segregating chemically incompatible wastes, for example, by ensuring that combustible waste does not contain self-heating or oxidizing materials.
2. Disposing of hazardous waste promptly so that only small quantities are allowed to accumulate at any one time.
3. Locating combustible waste piles away from buildings, roadways, and ignition sources.
4. Maintaining flammable and corrosive liquid wastes in approved metal containers.
5. Inspecting hazardous-waste containers continuously to ensure their integrity.
6. Properly identifying and marking waste containers. The hazards associated with each waste should be shown clearly on the exterior of each container, storage tank, transport vehicle, or building.
7. Training personnel in proper waste-handling procedures, use of personal protective equipment, and emergency response measures to an accidental release of hazardous waste. Training should include realistic drills simulating all credible release scenarios.

Emergency Response Planning

A fire department should coordinate and preplan its approach to the handling of hazardous-waste emergencies with all waste-generating industries or disposal facilities within its jurisdiction. The fire department should be aware of the types, quantities, and hazards of all raw materials, products, and wastes stored at each plant site as well as those that are transported to and from the plant. The fire department should then ensure the availability to its personnel of any necessary specialized handling equipment and materials, detection devices, and personal protective equipment.

In preparing fire-response plans, fire fighters should recognize that water used to control or extinguish fires at chemical plants or warehouses where hazardous materials are stored will most likely become contaminated with these materials, and thus constitute a hazardous waste. If the contaminated water is allowed to reach natural bodies of water, such as through the rain-water drainage system, the hazard will spread outside the plant, and endanger the public and the environment. A good example of such an event is the fire that occurred in November, 1986, at a warehouse for agricultural chemicals located along the Rhine River in Basel, Switzerland. The contaminated water resulted in a massive fish kill, and a water-pollution problem that spread all the way to Germany and Holland.

It is very important, therefore, that the fire department, together with plant management, predetermine the locations of storm-water drains and sewers, and the general layout and slope of the land on which the plant is built. Precautions must then be built into the emergency-response plan to ensure that fire-fighting water will not pollute any natural water bodies or drinking-water supplies.

Response Procedures

The potential severity of an accident involving the release of a hazardous waste may be reduced through response measures developed and tested under realistic drill conditions. In the absence of such plans, or when the waste components are not known, it is advisable to assume that an industrial waste is hazardous and to treat it with due care until more information becomes available from reliable sources. In general, response procedures include the following:

1. Wearing full protective clothing and breathing apparatus;

2. Using the appropriate detection devices to check for the flammability, toxicity, or radioactivity of the released waste or its products of combustion;
3. Stemming the flow of the waste at the source;
4. Diking the area around a liquid spill to limit its spread to sewage lines and drains and to reduce the vapor release rate;
5. Staying upwind when fumes are being generated;
6. Controlling ignition sources; and
7. Evacuating people who are downwind.

The decision to extinguish burning waste should be made very carefully after a full assessment of the consequences. Questions to be considered include the following:

1. Is the extinguishing agent to be used compatible with the waste?
2. Are sealed drums or containers of flammable liquids or compressed gases in danger of exposure to the fire?
3. Are there any radioactive, explosive, or oxidizing materials in the burning waste or near the fire?
4. Will extinguishment exacerbate the problem by allowing flammable or toxic vapors to be released once the fire is extinguished?
5. Will the use of water result in a severe pollution problem?

SPECIFIC WASTES

Combustible Refuse

Refuse of a highly combustible nature, such as dry waste paper, excelsior, etc., should be collected in metal containers and not allowed to accumulate. Quantity storage of these materials should be separated from buildings, roadways, and ignition sources by a distance of 50 ft (15 m) or more. Transport to an incinerator or landfill on a frequent schedule will minimize the fire hazard.

Drying Oils

Rags or paper that have absorbed drying-type oils, such as linseed oil or turpentine, are subject to spontaneous heating. They should be kept in well-covered metal cans and thoroughly dried before collection or transport.

Flammable Liquids and Waste Solvents

Waste solvents have variable flash points, hence varying levels of hazard, depending upon composition. Some may contain solids, tars, waxes, and other materials that impede flow. Chlorinated solvents and water may also be present.

The two most common methods of disposal for such wastes are by: (1) incineration or (2) the use of a disposal contractor. In either case, the solvents are collected in 5-gal (19-L) cans or 55-gal (208-L) drums for disposal. If a contractor is used, the approximate composition of the waste must be provided.

Installations having substantial amounts of waste solvent may install their own solvent-recovery system. Alternatively, they may install an incinerator. In that case, an auxiliary gas or oil burner is needed to maintain adequate firebox temperature. Usually a scrubber section is used to remove acid gases and other pollutants. See NFPA 82, *Standard on Incinerators and Waste and Linen Handling Systems and Equipment*, for further information on industrial incinerator installations.

Polychlorinated Biphenyls (PCBs)

PCBs were manufactured and used in a variety of electrical and mechanical applications from the early 1930s until regulated by the Toxic Substances Control Act in 1976. They had been extensively used as a dielectric fluid in capacitors and transformers and as a lubricant in compressors and other machinery. They continue to be used today, but only in enclosed systems.

PCBs have been known to cause temporary skin irritations and nervous-system symptoms in humans and are suspected carcinogens. Improper combustion may produce highly toxic and carcinogenic products.

Extreme care must be taken in handling PCBs and fires involving PCBs. Waste PCBs from old electrical equipment must be disposed of by incineration at an approved site or at sea or buried in approved containers in a secure landfill.

Liquid and Solid Oxidizing Materials

Spilled oxidizing materials and leaking or broken containers should be immediately removed to a safe area to await disposal in conformance with applicable regulations and manufacturers' instructions. (See NFPA 430, *Code for the Storage of Liquid and Solid Oxidizers*.) Most, if not all, oxidizers can be rendered harmless by dilution with water. However, some solutions cause pollution of streams and rivers, and thus require pretreatment.

Combustible and Reactive Metals

Small scraps of sodium from laboratory use can be dissolved in ethyl alcohol (not isopropyl alcohol) and the resulting alkoxide neutralized with acid. Large quantities can be offered to a vendor for purification and recycling. Oil-contaminated dispersions or other moderate quantities can be burned in dry pans if provision is made to control the oxide fumes, which are highly alkaline. Disposal of sodium by throwing it directly into water is spectacular but extremely hazardous and will contaminate the water with sodium hydroxide. Equipment contaminated with small amounts of sodium may be cleaned by a remotely controlled introduction of water or steam. Potassium, sodium-potassium alloy (NaK), and lithium are handled in the same manner as sodium. Lithium reacts somewhat less violently with water.

It is usually feasible to reclaim magnesium fines and scrap in the form of defective castings, clippings, and coarse chips. If this is to be done, either locally or through a smelting firm, the scrap should be thoroughly dried before remelting. Scrap in the form of fine chips and dust-collector sludge can be disposed of in a specially constructed incinerator or by burning in a safely located outdoor area paved with fire-brick or hard-burned paving brick. By spreading the scrap or sludge in a layer about 4 in. (102 mm) thick, on which ordinary combustible material is placed and ignited, the combustion of magnesium can be safely conducted. Another method of deactivating magnesium sludge is to treat it with a 0.5 percent solution of ferrous chloride. Since hydrogen is generated by this reaction, this method of disposal is conducted in an open container in an outdoor location where the hydrogen can be safely dissipated. (See NFPA 480, *Standard for the Storage, Handling, and Processing of Magnesium Solids and Powders*.)

Spontaneous combustion of titanium has occurred in fines, chips, and swarf that were coated with water-soluble oil. Disposal containers should be tightly closed and segregated so that spontaneous burning would not involve other exposures. Incineration in the open, as for magnesium scrap, may be feasible. (See NFPA 481, *Standard for the Production, Processing, Handling, and Storage of Titanium*.)

While small amounts of zirconium fines can be mixed with sand or other inert material and buried, incineration of this combustible metal is generally preferred for larger quantities. Because of the spontaneous ignition potential, personnel carrying scrap should wear flame-protective clothing and equipment and carry the scrap in

buckets on a yoke between them. A layer may be ignited with excelsior. (See NFPA 482, *Standard for the Production, Processing, Handling, and Storage of Zirconium*.)

Lithium hydride and lithium aluminum hydride can ignite upon contact with water or moist air. Alkaline oxides resulting from incineration may be collected and neutralized.

Radioactive Waste

Radioactive waste is generated at nuclear power plants in the form of spent fuel and contaminated instruments, equipment, piping, and other objects such as clothing, gloves, and cleaning rags. Other sources include nuclear-physics research laboratories and tailings from uranium mines.

Recent advances in the use of radioactive materials in medical diagnostics and therapy has led to an increase in the sources of radioactive medical waste. Radioactive isotopes of iodine, cesium, iridium, phosphorus, and gallium are widely used in hospitals and laboratories for diagnostics and patient treatments. Contaminated vials, containers, and capsules are growing in quantity. All radioactively contaminated materials must be properly disposed of.

The packaging of all radioactive waste, including radioactive medical waste, is regulated by the Nuclear Regulatory Commission (NRC) while the Department of Transportation (DOT) regulates its transportation. The EPA has developed standards for protecting the public against exposure to nuclear-reactor wastes in general, and for the proper storage and disposal of such waste. The specific requirements vary greatly, and depend upon the type of radioactive waste that is generated.

For most installations, a contractor licensed to dispose of radioactive materials can be employed. (See NFPA 801, *Standard for Facilities Handling Radioactive Materials*.) Radioactive materials that are flammable or combustible must be handled with particular care. Combustible radioactive wastes should not be allowed to accumulate. Storage of such wastes near an air intake is particularly undesirable. If the products of combustion of waste materials containing long-lived radioactive materials are dispersed through air-conditioning or compressed-air systems, extensive decontamination may become necessary.

Liquid radioactive waste may be concentrated and retained until the radioactivity has decayed to a safe level. Combustible radioactive waste, such as absorbent paper used to wipe contaminated surfaces, should be placed in approved containers for disposal.

The treatment of radioactive medical wastes is less extensive than that for other medical wastes. For radioactive medical waste with a half-life of less than 65 days, the waste may be stored in a secure area for a minimum period of 10 half-lives and discarded as sanitary wastes.

Most "spent" or used nuclear fuel is currently being stored in specially designed deep pools of water at nuclear reactor sites.[15] Some radioactive military waste is being stored at federal reservations aboveground in heavy, thick-walled metal or concrete structures. All of these storage facilities are temporary. In 1982 the Office of Civilian Radioactive Waste Management was charged by Congress to develop a mined geologic repository for the disposal of these wastes. As of 1995, no permanent site had been chosen.

Oil Spills on Water

Booming of spills has been proved effective in containing spills of liquids on relatively calm and current-free waters. Various makeshift designs of booms have been tried with relative ineffectiveness, but commercially available booms that recognize the hydrodynamics and aerodynamics involved in the confinement of spills on water have proved quite effective. Because of ecological considerations, this has become an important means of containing an oil spill, and more effective equipment is now available.

Following the confinement of spills on water, various ways of removing the confined liquid have been used, including skimming devices or absorbents. Absorbents, such as straw, plastics, sawdust, and peat moss, have been spread on the surface of the spill and then collected and burned on shore. Skimming devices operate on several principles, including pumps and separators. Power boats with skimmers on the bow can scoop up oil and water, sending it through an oil separator and rollers to which oil adheres. The oil is then removed by scraping or compression.

Sinking agents to which oil may adhere and sink to the bottom have been used in the past. Several proprietary sinking agents have been developed, but such products as carbonized sand, brick dust, crushed clinkers, and cement also have been used. However, sinking agents are objectionable because they contribute to water pollution and are no longer recommended, except possibly for use in the ocean far from land.

Regardless of whether a spill occurs on land or on water, preplanning is imperative and should include prompt notification of all local, state, or federal agencies dealing with the problem of water pollution.

Underground Storage Tanks

The EPA estimates that there are more than two million underground-storage-tank (UST) systems in the United States located at more than 700,000 operating facilities nationwide.[16] In addition, there are numerous abandoned tanks. Although most of these tanks are (or were) used for the storage of motor fuels, some are (or were) used for the storage of other hazardous chemicals. Tests on a random sample of more than 10,000 USTs have shown that about 25 percent of these systems are leaking, either because they lack corro-

sion protection, or because of improper installation, piping failure, and overfilling.[16]

Underground liquid leaks invariably find their way to groundwater, which is the primary source of drinking water for more than 50 percent of the U.S. population. Leaking petroleum products contaminate the soil and the air. Their vapors may accumulate in confined spaces, such as septic tanks, sewers, and building basements. The vapors are hazardous to human health. If ignited, they may result in a fire or an explosion.

To protect the public and the environment against leaks from USTs, the federal government extended and strengthened the provisions of the Solid Waste Disposal Act as amended by RCRA. Subtitle I of RCRA regulates USTs containing "regulated substances" and releases of these substances to the environment.

Subtitle I defined a UST as a tank system, including its piping, that has at least 10 percent of its volume underground. Regulated substances include those defined as hazardous under CERCLA, and petroleum products. This standard requires that USTs installed after 1988 be properly built, certified, equipped with overfill protection and detection systems, and, if made of steel, protected against corrosion. By 1998, all USTs installed before 1988 must have corrosion protection for steel tanks and piping, and devices to prevent overfilling. Existing tanks must also be tested for leaks in accordance with a schedule determined by their age.[17] (See Figure 3-32D.)

Leaks from USTs must be detected by one (or a combination) of the following monitoring methods:[18]

1. Automatic tank gaging,
2. Monitoring for vapor in the soil,
3. Interstitial monitoring between the storage tank and an external protective tank, and
4. Monitoring for the presence of contaminants in groundwater.

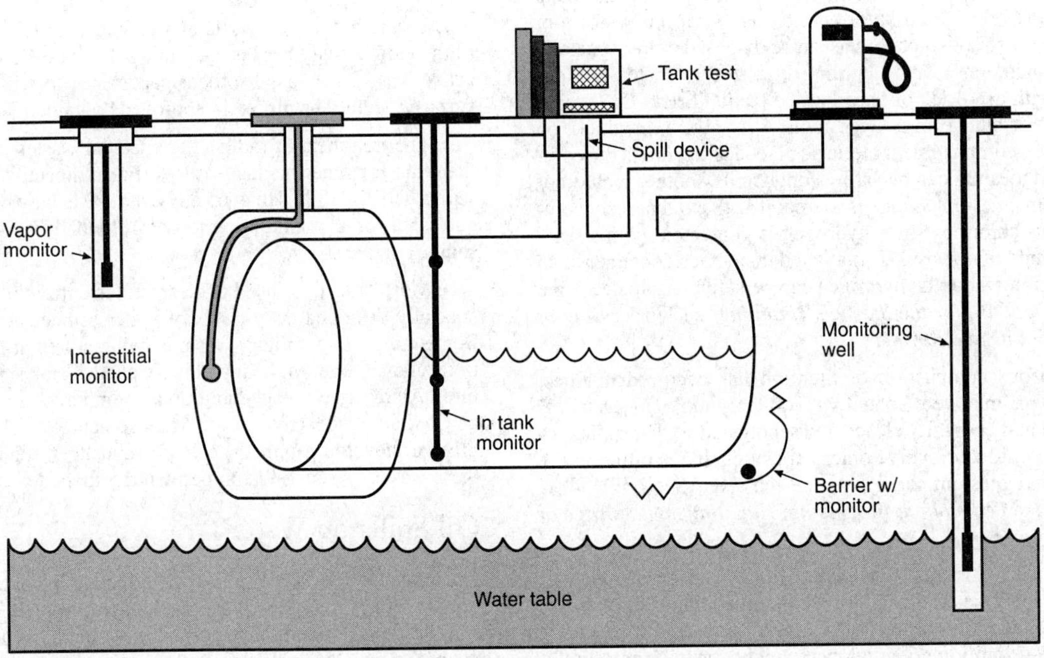

FIG. 3-32D. UST leak detection alternatives.

BIBLIOGRAPHY

References Cited

1. EPA, "Characterization of Municipal Solid Waste in the United States: 1994 Update," Executive Summary, EPA 530-S-94-042, Nov. 1994, Office of Solid Waste and Emergency Response, U.S. Environmental Protection Agency, Washington, DC.
2. Reindl, J., "Interrelationships Within the Solid Wastes System," *Solid Wastes Management*, April 1977, pp. 22-23, 54-56.
3. EPA, "1993 Toxics Release Inventory, Public Release Data," EPA 745-R-95-010, March 1995, U.S. Environmental Protection Agency, Office of Pollution Prevention and Toxics, Washington, DC.
4. EPA, "Criteria for Identifying the Characteristics of Hazardous Waste and Listing Hazardous Waste," *Federal Register*, Vol. 45, No. 98, Subpart B, May 19, 1980, pp. 33121–33122.
5. CFR, "Title 40, Section 261, Subpart C Characteristics of Hazardous Waste," *Code of Federal Regulations*, Washington, DC, 1995.
6. CFR, "Title 49, Section 173.300, Subpart G, Transportation," *Code of Federal Regulations*, Washington, DC, 1995.
7. CFR, Section 173.151, Subpart E, Transportation, *Code of Federal Regulations*, Washington, DC, 1995.
8. CFR, "Title 29, Section 1910:1030: Bloodborne Pathogens," *Code of Federal Regulations*, Washington, DC, 1995.
9. CFR, "Title 49, Section 173.386: Etiologic Agents, and Section 173.387: Regulated Medical Waste," *Code of Federal Regulations*, Washington, DC, 1995.
10. World Bank, *Environmental Considerations for the Industrial Development Sector*, Office of Environmental and Health Affairs, The World Bank, Washington, DC, 1978.
11. CFR, "Title 40, Subchapter C: Air Programs, Part 50: National Primary and Secondary Ambient Air Quality Standards," *Code of Federal Regulations*, Washington, DC, 1995.
12. Reindl, J., "Examining Disposal and Recycling Techniques for Solid Wastes," *Solid Wastes Management*, May 1977, pp. 60, 68, 70, 92, 94.
13. Meaney, J. G., and Cheremisinoff, P.N., "Medical Waste Strategy," *Pollution Engineering*, Vol. 21, No. 11, Oct. 1989, pp. 92-106.
14. CFR, "Title 29, Section 1910.120: Hazardous Waste Operations and Emergency Response," *Code of Federal Regulations*, Washington, DC, 1995.
15. Department of Energy, "Locations of Spent Nuclear Fuel and High-Level Radioactive Waste Ultimately Destined for Geologic Disposal," DOE/RW-0447, Sept. 1994, DOE's Office of Civilian Radioactive Waste Management, Washington, DC.
16. CFR, "Title 40, Sections 280 and 281: Underground Storage Tanks," *Code of Federal Regulations*, Washington, DC, 1995.
17. EPA, "Musts for USTs," EPA/530/UST-88/008, 1988, U.S. Environmental Protection Agency, Washington, DC.
18. Niaki, S., and Broscious, J. A., *Underground Tank Leak Detection Methods*, Noyes Publications, Park Ridge, NJ, 1987.

NFPA Codes, Standards, and Recommended Practices

Reference to the following NFPA codes, standards, and recommended practices will provide further information on hazardous waste control discussed in this chapter. (See the latest version of *The NFPA Catalog* for availability of current editions of the following documents.)

NFPA 30, *Flammable and Combustible Liquids Code*
NFPA 40, *Standard for the Storage and Handling of Cellulose Nitrate Motion Picture Film*
NFPA 40E, *Code for the Storage of Pyroxylin Plastic*
NFPA 43D, *Code for the Storage of Pesticides*
NFPA 49, *Hazardous Chemicals Data*
NFPA 55, *Standard for the Storage, Use, and Handling of Compressed and Liquefied Gases in Portable Cylinders*
NFPA 61, *Standard for the Prevention of Fires and Dust Explosions in Agricultural and Food Product Facilities*
NFPA 82, *Standard on Incinerators and Waste and Linen Handling Systems and Equipment*
NFPA 91, *Standard for Exhaust Systems for Conveying of Materials*
NFPA 325, *Guide to Fire Hazard Properties of Flammable Liquids, Gases, and Volatile Solids*

NFPA 329, *Recommended Practice for Handling Underground Releases of Flammable and Combustible Liquids*
NFPA 430, *Code for the Storage of Liquid and Solid Oxidizers*
NFPA 471, *Recommended Practice for Responding to Hazardous Materials Incidents*
NFPA 472, *Standard for Professional Competence of Responders to Hazardous Materials Incidents*
NFPA 480, *Standard for the Storage, Handling, and Processing of Magnesium Solids and Powders*
NFPA 481, *Standard for the Production, Processing, Handling, and Storage of Titanium*
NFPA 482, *Standard for the Production, Processing, Handling, and Storage of Zirconium*
NFPA 490, *Code for the Storage of Ammonium Nitrate*
NFPA 491M, *Manual of Hazardous Chemical Reactions*
NFPA 495, *Explosive Materials Code*
NFPA 498, *Standard for Safe Havens and Interchange Lots for Vehicles Transporting Explosives*
NFPA 651, *Standard for the Manufacture of Aluminum Powder*
NFPA 654, *Standard for the Prevention of Fire and Dust Explosions in the Chemical, Dye, Pharmaceutical, and Plastics Industries*
NFPA 655, *Standard for Prevention of Sulfur Fires and Explosions*
NFPA 664, *Standard for the Prevention of Fires and Explosions in Wood Processing and Woodworking Facilities*
NFPA 704, *Standard System for the Identification of the Fire Hazards of Materials*
NFPA 801, *Standard for Facilities Handling Radioactive Materials*

Additional Reading

Barkenbus, B. D., *et al.*, "Environmental Protection for Hazardous Materials Incidents. Volume 1," Hazardous Materials Incident Management System, Final Report, October 1986-June 1990, Air Force Engineering and Services Center, Tyndall AFB, FL, ESL-TR-89-15 VOL 1, Nov. 1990, 349 pages.
Barkenbus, B. D., *et al.*, "Environmental Protection for Hazardous Materials Incidents. Volume 2," Appendices, Final Report, Oct. 1986-June 1990, Air Force Engineering and Services Center, Tyndall AFB, FL, ESL-TR-89-15 VOL 2, Nov. 1990, 254 pages.
Barton, R. G., *et al.*, "Mechanistic Analysis of the Vaporization of Metals During Waste Incineration," Utah Univ., Salt Lake City, University of California, Berkeley, Toxic Toxic Combustion Byproducts, 4th International Congress, ABSTRACTS ONLY, June 5–7, 1995, Salt Lake City, UT, 1995, 51 pages.
Beecher, N., and Rappaport, A., "Hazardous Waste Management Overseas," *Chemical Engineering Progress*, Vol. 86, No. 5, May 1990, pp. 30-39.
Beyler, C. L. and Hunt, S. P., "Fire Development in and Hazard Analysis of Concrete Hazardous/Radioactive Waste Storage Facilities," *Proceedings* of the Society of Fire Protection Engineers Engineering Seminars on Performance-Based Fire Safety Engineering, November 15–17, 1993, Phoenix, AZ, SFPE, Boston, MA, 1993, pp. 45–55.
Bloom, J. M., Mason, B. J. and Spier, R., "Fire Investigation, Hazardous Waste, and Liability," Fire and Arson Investigator, Vol. 41, No. 2, Dec. 1990, pp. 33–36.
Brunner, C. R., *Handbook of Hazardous Waste Incineration*, TAB Professional and Reference Books, Blue Ridge Summit, PA, 1989.
Carroll, T. R. and Schwope, A. D., "Non-Destructive Testing and Field Evaluation of Chemical Protective Clothing," Final Report, Federal Emergency Management Agency, Washington, DC, FA-106, Dec. 1990, 64 p.
Cheremisinoff, P. N., Casana, J. G., and Ouellette, R. P., *Underground Storage Tanks Guidebook*, Pudvan Publishing Co., Northbrook, IL, 1987.
"Engineering Handbook for Hazardous Waste Incineration," SW-889, Sept. 1981, U.S. Environmental Protection Agency, Washington, DC.
"EMC Offers a Solution to the Problem of Hazardous Household Waste," Fire Chief, Vol. 35, No. 7, July 1991, pp. 52–54.
Evans, D. D., *et al.*, "Burning, Smoke Production, and Smoke Dispersion from Oil Spill Combustion," NISTIR 89-4091, May 1989, National Institute for Standards and Technology, Gaithersburg, MD.
Evans, D. D., *et al.*, "Combustion of Oil on Water," NBSIR 86-3420, Nov. 1987, National Bureau of Standards, Gaithersburg, MD.
Evans, D. D., *et al.*, "Environmental Effects of Oil Spill Combustion," NISTIR 88-3822, Sept. 1988, National Institute for Standards and Technology, Gaithersburg, MD.

Filius, K. D. and Whitworth, C. G., "Emissions Characterization and Off-Gas System Development for Processing Simulated Mixed Waste in Plasma Centrifugal Furnace," MSE, Inc., Butte, MT, Department of Energy, Washington, DC, University of California, Berkeley, Toxic Toxic Combustion Byproducts, 4th International Congress, ABSTRACTS ONLY, June 5–7, 1995, Salt Lake City, UT, 1995, 24 pages.

Fire Protection Guide to Hazardous Materials, National Fire Protection Association, Quincy, MA 1995.

Gaglierd, A.M. and Hilinski, R.H., "Response to a Medical Emergency Involving Radioactive Materials," *Fire Engineering*, Vol. 148, No. 7, July 1995, pp. 32–39.

Gronow, J. R., Schofield, A. N., and Jain, R. K., eds., Land Disposal of Hazardous Waste, Wiley, New York, 1988.

Gulliani, D., Wilusz, E. and Galezewski, A., "Flame Retardant Elastomers for Chemical Protective Gloves," *Journal of Fire Sciences*, Vol. 12, May/June 1994, pp. 246–256.

Harwood, P., "Warwickshire's Changing Philosophy on Chemical Protection Suits," Fire and Rescue Service, Warwickshire, England, *Fire Research News*, No. 18, Winter 1994, pp. 19–20.

Hayes, W. K., "Recycle And/Or Disposal of Flame Retardant Plastics," Ethyl Corp., Baton Rouge, LA, Fire Retardant Chemicals Association, Customer Demands for Improved Total Performance of Flame Retarded Materials, October 26–29, 1993, Tucson, AZ, Fire Retardant Chemicals Assoc., Lancaster PA, 1993, pp. 283–287.

"Hazardous and Industrial Waste," *Proceedings* of the Mid-Atlantic Industrial Waste Conference, Hazardous Materials Control Research Institute, Silver Springs, MD, 1988.

Henry, M. F., *Hazardous Materials Response Handbook*, National Fire Protection Association, Quincy, MA 1990.

Henry, M. F., "Update on the Underground Leakage Problem," *Fire Journal*, Vol. 80, No. 1, 1986, p. 26.

Hoerning, J. M. and Ragland, K. W., "Aromatic Emissions From Incineration of Selected Wastes Using a Laboratory Scale Rotary Kiln," Wisconsin Univ., Madison, University of California, Berkeley, Toxic Toxic Combustion Byproducts, 4th International Congress, ABSTRACTS ONLY, June 5–7, 1995, Salt Lake City, UT, 1995, 107 pages.

Holmes, N., "Respiratory Protection in the Chemical Industry," *Fire International*, No. 144, Aug./Sept. 1994, pp. 27–28.

Hunt, S. P., "Concrete Hazardous/Radioactive Waste Storage Facilities: Fire Development and Hazard Analysis, Selected Readings in Performance-Based Fire Safety Engineering. 1995, Boston, MA, Society of Fire Protection Engineers, Boston, MA, 1995, pp. 117–128.

Huotari, J. and Vesterinen, R., "PCDD/F Emissions From Cocombustion of RDF With Peat, Wood Waste and Coal in FBC Boilers," VTT Energy, Jyvaskyla, Finland, University of California, Berkeley, Toxic Toxic Combustion Byproducts, 4th International Congress, ABSTRACTS ONLY, June 5–7, 1995, Salt Lake City, UT, 1995, 78 pages.

Johnson, G. D., "Fiscal Year 1992 Program Plan for Evaluation and Remediation of the Generation and Release of Flammable Gases in Hanford Site Waste Tanks," Westinghouse Hanford Co., Richland, WA, Department of Energy, Washington, DC, WHC-EP-0537, Jan. 1992, 85 pages.

Johnson, N. P., and Cosmos, M. G., "Thermal Treatment Technologies for Hazardous Waste Remediation," *Pollution Engineering*, Vol. 21, No. 11, Oct. 1989, pp. 66-85.

Kephart, W., Eger, K. and Clemens, M. K., "Toxic Combustion Byproducts: Generation, Containment, Separation, Cleansing for Hazardous, Mixed, and Transuranic Waste Processing," Foster Wheeler Environmental Corp., Argonne National Laboratory, IL, University of California, Berkeley, Toxic Toxic Combustion Byproducts, 4th International Congress, ABSTRACTS ONLY, June 5–7, 1995, Salt Lake City, UT, 1995, 25 pages.

Klunk, D. and Blaine, B., "Santa Fe Springs' Model Haz Mat Plan. How One City Protects Its Environment," *American Fire Journal*, Vol. 45, No. 2, Feb. 1993, pp. 28–31.

Koo, J., "Formation of Toxic Byproducts From Pilot-Scale Rotary Kiln Incinerator for Mixed Polyethylene Waste," Korea Advanced Institute of Science and Technology, Kolon Engineering Ltd., Co., University of California, Berkeley, Toxic Toxic Combustion Byproducts, 4th International Congress, ABSTRACTS ONLY, June 5–7, 1995, Salt Lake City, UT, 1995, 38 pages.

Larsen, F. S., *et al.*, "Hydrocarbon and Formaldehyde Emissions From the Combustion of Pulverized Wood Waste," Combustion Science and Technology, Vol. 85, No. 1–6, 1992, pp. 259–269.

Law, C. K., "Considerations of Droplet Processes in Liquid Hazardous Waste Incineration," Combustion Science and Technology, Vol. 74, No. 1–6, 1990, pp. 1–15.

Lemieux, P. M., Linak, W. P., and Ryan, J. V., "Use of Surrogate Performance Indicators to Predict Emissions of Trace Organics From Hazardous Waste Incinerators," Air and Energy Engineering Research Lab., Research Triangle Park, NC, Acurex Environmental Corp., Research Triangle Park, NC, University of California, Berkeley, Toxic Toxic Combustion Byproducts, 4th International Congress, ABSTRACTS ONLY, June 5–7, 1995, Salt Lake City, UT, 1995, 127 pages.

Lemieux, P. M., Linak, W. P., and Wendt, O. L., " Waste and Sorbent Parameters Affecting Mechanisms of Transient Emissions From Rotary Kiln Incineration," Environmental Protection Agency, Research Triangle Park, NC, Arizona Univ.,Tucson, University of California, Berkeley, Toxic Toxic Combustion Byproducts, 4th International Congress, ABSTRACTS ONLY, June 5–7, 1995, Salt Lake City, UT, 1995, 39 pages.

Leonard, J. T., *et. al.*, "Fire Hazard Assessment of Shipboard Plastic Waste Disposal Systems," Naval Research Laboratory, Washington, DC, Hughes Associates, Inc., Columbia, MD, NRL/MR/6184-94-7452, February 28, 1994, 55 pages.

Levendis, Y. A., *et. al.*, "Comparative Study of Combustion and Organic Emissions of Waste Tire Crumb and Pulverized Coal," Northeastern Univ., Boston, MD, University of California, Berkeley, Toxic Toxic Combustion Byproducts, 4th International Congress, ABSTRACTS ONLY, June 5–7, 1995, Salt Lake City, UT, 1995, 66 pages.

Linville, J., ed., *Industrial Fire Hazards Handbook*, 2nd ed., National Fire Protection Association, Quincy, MA, 1990.

Marzi, W. B., "Europaische Normung von Chemikalienschutzkleidung und vfdb-Richtlinie 0801. [European Standardization of Protective Clothing Against Chemicals and VFDB Buide 0801.]," *VFDB*, Jan. 1993, pp. 6–7.

McKnight, M. E., "Review of Current Research and Activities Involving Characterization, Abatement and Disposal of Lead-Containing Paint Films," National Institute of Standards and Technology, Gaithersburg, MD, NISTIR 90-4285, May 1990, 13 pages.

Melvold, R. W., Gibson, S. C., and Scarberry, R., *Sorbents for Liquid Hazardous Substance Cleanup and Control*, Noyes Publications, Park Ridge, NJ, 1988.

"Occupational Exposure to Hazardous Chemicals in Laboratories: Chemical Hygiene Plan," *Health and Safety Instruction*, Number 20, Jan. 1991, 15 pages.

Perry, B., ed., *Handbook of Hazardous Waste Regulation*, 3rd ed., Business and Legal Reports, Madison, CT, 1989.

Pocket Guide to Chemical Hazards, DHEW (NIOSH) Publication No. 85-114, 1987, U.S. Department of Health, Education and Welfare, Washington, DC.

Rhoeds, B. T., *et. al.*, "Evaluation of Solid Waste Drum Fire Performance," *Proceedings* of the International Conference on Fire Research and Engineering, September 10–15, 1995, Orlando, FL, SFPE, Boston, MA, 1995, pp. 403–406.

Say, D. J., "Chemical Protective Clothing: Still a Long Way To Go," *Fire Engineering*, Vol. 144, No. 8, Aug. 1991, pp. 86–88, 90–92.

Schwope, A. D. and Renard, E. P., "Estimation of the Cost of Using Chemical Protective Clothing," Performance of Protective Clothing: Challenges for Developing Protective Clothing for the 1990s, Fourth (4th) Volume, ASTM STP1133, ASTM, Philadelphia, PA, 1991, pp. 972–981.

Serbin, S. I., Sacchi, G. F. and Sarofim, A. F., "Application of Plasma-Chemical Technology in Hazardous Waste Incinerators," Massachusetts Institute of Technology, Cambridge, University of California, Berkeley, Toxic Toxic Combustion Byproducts, 4th International Congress, ABSTRACTS ONLY, June 5–7, 1995, Salt Lake City, UT, 1995, 29 pages.

Silva, M., "Assessment of the Flammability and Explosion Potential of Transuranic Waste," Environmental Evaluation Group, Albuquerque, NM, DOE/AL/58308-48, June 1991, 96 pages.

Snyder, K. A. and Clifton, J. R., "4SIGHT Manual: A Computer Program for Modelling Degradation of Underground Low Level Waste Concrete Vaults," National Institute of Standards and Technology, Gaithersburg, MD, NISTIR 5612, June 1995, 73 pages.

Stull, J. O., "New Comprehensive Performance Standards for Chemical Protective Gloves, Boots, and Other Types of Protective Clothing," Performance of Protective Clothing: Challenges for Developing Protective

Clothing for the 1990s, Fourth (4th) Volume, ASTM STP1133, ASTM, Philadelphia, PA, 1991, pp. 322–338.

"Threats to the Environment," *Fire Prevention*, No. 231, July/Aug. 1990, pp. 15–16.

Travis, J.R., *et al.*, "An HMS/RRAC Analysis of a High-Level Radioactive Waste Tank Farm," *Fire Science and Technology*, Vol. 13, Suppl., 1993, pp. 135–167.

Tsang, W., "Chemistry of Hazardous Waste Incineration," Combustion Institute/Eastern States Section, 1990 Fall Technical Meeting on Chemical and Physical Processes in Combustion, Dec. 3–5, 1990, Orlando, FL, 1990, pp. B/1–11.

Turner, D. A. and Miron, Y., "Testing of Organic Waste Surrogate Materials in Support of the Hanford Organic Tank Program," Final Report, Department of Energy, Washington, DC, WHC-MR-0455, UC-600, Jan. 1994, 175 pages.

Veghte, J. H., "Field Evaluation of Chemical Protective Suits," Final Report, Task 1, Federal Emergency Management Agency, Washington, DC, FA-108, Sept. 1991, 38 pages.

Watts, J. M., Jr., "Environmental Protection: A Fire Safety Objective?," Editorial, *Fire Technology*, Vol. 28, No. 3, Aug. 1992, pp. 193–194.

Wei, M. S., Savin, T. J. and Fisher, G., "Air Quality Modeling Analysis for Hazardous Waste Sites," Parsons Engineering Science, Inc., Oak Ridge, TN, Parsons Engineering Science, Inc., Fairfax, VA, University of California, Berkeley, Toxic Toxic Combustion Byproducts, 4th International Congress, ABSTRACTS ONLY, June 5–7, 1995, Salt Lake City, UT, 1995, 84 pages.

York, K. G., and Grey, G. L., "Hazardous Materials/Waste Handling for the Emergency Responder," *Fire Engineering*, New York, 1989.

HOUSEKEEPING PRACTICES

John T. Higgins

The Occupational Safety and Health Administration (OSHA) requires that every employer maintain a safe and healthful place of employment. This means the care and maintenance of the property. This is the definition of good housekeeping. It is basic to fire safety and should be a major concern in every type of occupancy, from the simplest dwelling to the most complex industrial facility.

This chapter covers the fundamentals of good housekeeping practices. It discusses in some detail the specifics of practices to be observed in providing fire safe and effective housekeeping in connection with the care and maintenance of buildings.

In today's society, each facility should have a written, effective safety and health program in place. To be effective, this program should include the involvement of all facility employees. A strong part of this program should be a fire safety plan, and an important part of the fire safety plan is housekeeping practices that include inspections, equipment layout, storage and handling practices, and an effective preventive maintenance program to limit or eliminate potential sources of ignition. This chapter presents suggestions on where employees can interact with management on housekeeping.

Good storage practices for various materials and the different storage configurations used for them can be found in the following chapters: Section 3, Chapter 21, "Storage of Flammable and Combustible Liquids"; Section 3, Chapter 22, "Storage of Gases"; Section 3, Chapter 23, "Storage and Handling of Chemicals"; Section 3, Chapter 24, "Storage and Handling of Solid Fuels"; Section 3, Chapter 25, "Storage and Handling of Records"; and Section 3, Chapter 32, "Hazardous Waste Control."

GOOD HOUSEKEEPING THEORY

A good housekeeping program concerns itself with the less complex aspects of tidiness and order, waste control, and the regulation of such personal practices as smoking, which, without reasonable controls, could lead to hazardous conditions.

In looking at the theory of accidents, one can find information that can help in the prevention of fire. For each serious fire that is reported, there might be 10 minor fires; 100 "close calls" or near misses that could have led to fires; and 1,000 unsafe conditions, recognized or unrecognized, that could have led to near misses. These

illustrative numbers may be high or low, but the bottom line is that many conditions do exist in the workplace that can cause a fire. And, in fact, a facility can get away with having these conditions over a long period of time without a fire occurring. However, a time may come when conditions are just right and these same unsafe conditions are the cause of the fire or are the factors that allow the fire to spread uncontrolled. If one can recognize these conditions and eliminate them, the potential for a disastrous fire in our workplace will be substantially reduced. This can be accomplished through good inspection procedures and good housekeeping practices.

Poor housekeeping contributes to loss potential by increasing fire and explosion hazards in several ways:

1. It provides more places for a fire to start.
2. It creates a greater continuity of combustibles, which makes it easier for fire to spread.
3. It provides a greater combustible loading for the initial fire to feed upon.
4. It creates the potential for flash fire or dust explosions when layers of lint or dust are allowed to accumulate.
5. It allows spills or drips of flammable or combustible liquids to accumulate, which could catch fire (including spontaneous ignition in some situations).
6. When not properly addressed, friction, static, or electrical connections can be sources of ignition.
7. Poorly controlled smoking policies can lead to a source of ignition.
8. It increases the potential for spontaneous ignition.

In addition to the increased hazard, poor housekeeping can have a negative effect on production. Quality is hard to maintain when the workspace is crowded and messy. Efficiency suffers because people normally tend to work faster and more accurately if their surroundings are clean. Maintenance of good workspace and adequate aisles to allow not only the free flow of materials into and out of the work area but also good exit capabilities is as important as keeping things clean and well laid out. Thus, good housekeeping will not only prevent fires and possibly save lives, it can also improve production and employee morale as well.

The Essentials of Good Housekeeping

The degree of effort and attention needed for proper housekeeping is influenced, of course, by the type of building and the overall size of the facility involved. But most significant are the specific occupancies of the facility. Some processes produce more waste, leakage,

---

John T. Higgins, P.E., C.S.P., is a safety and loss prevention consultant in the corporate safety, industrial hygiene, and loss prevention department of Dow Corning Corporation, Midland, MI.

and vapors than others, thus contributing to the extent of housekeeping problems. In addition, the acceptable level of cleanliness varies from occupancy to occupancy. What is satisfactory in a foundry would probably not be tolerable in an office building. And the cleanliness of the average office would hardly be satisfactory for an electronic "clean room."

In order to have good housekeeping, the program must start at the top. In today's business atmosphere, all sorts of titles are used, from the traditional manager and supervisor to the new title of team leader or crew leader. Whatever the title, the top facility manager/team leader must set the tone. This is not to say that an individual supervisor/team leader/crew leader at the department level cannot go it alone, but his/her success will be much improved if the top management of the facility is behind the program. To begin, team leaders (at all levels) must define what is meant by good housekeeping. To do this, each level of leadership should review the plant and then determine:

1. Is the present level of housekeeping acceptable?
2. Are the appropriate tools provided for general plant cleanup? For emergency cleanup?
3. Are adequate storage areas provided for both incoming raw materials and outgoing products?
4. Are leadership's offices reflective of the good housekeeping required of the facility areas?
5. Are adequate waste receptacles provided?

Proper housekeeping does not just happen. It requires the leadership and wholehearted support and direction of leadership of the facility and the cooperation of all employees. Leadership cannot merely decree that good housekeeping is a desired goal. Good housekeeping requires effort from everyone. First management must set the tone, i.e., the level of acceptable housekeeping. Next, management must enlist the aid of everyone to help. Employee ideas should be solicited on how to maintain good housekeeping. Leadership must either act on the employee ideas or explain why it cannot.

It is important that all employees accept the responsibility for housekeeping in their respective area. The facility will probably either hire a contract janitorial service or create a janitorial department, or have a combination of both. However, janitors will only do the general cleaning. Ensuring that materials, tools, wastes, etc., are placed in their proper locations is the job of each employee who handles them.

Safety inspections are important. When doing so, leadership must demonstrate with their actions and words the level of housekeeping they will accept.

Where proper housekeeping does not exist, it is usually because inadequate attention is paid to, or inadequate action taken in, one or more of the following areas.

Communications: Leadership must publicize its commitment to good housekeeping. This publicity must be reinforced periodically, and recognition given whenever notable improvement or outstanding performance takes place. Team leaders must provide feedback to leadership on how the program is working and on what they need to meet the program goals. This can be done through safety inspection reports prepared by team or crew leaders and employees, and periodic inspections by facility leadership. In addition, team or crew leaders must be able to meet with facility leadership periodically to review performance, to revise goals, and to offer and substantiate recommendations. These team/crew leaders should also encourage feedback from employees, in the form of suggestions and constructive criticism. One important place to discuss these issues is at monthly safety meetings. Assign employees to teams to find innovative methods to identify solutions to problems that have been brought forth either through inspections or by employee suggestions. More efficient housekeeping methods can often be more readily identified by the worker who uses the area than by facility leadership. These new methods are also more readily accepted by workers.

Equipment: Housekeeping efforts should not founder from lack of necessary tools or equipment. This includes the tools normally used by the janitorial department. The simple step of placing a sufficient number of easily accessible wastebaskets or trash receptacles at points of need can reduce the amount of waste deposited on the floor or in the product.

In some production and handling areas, dust, lint, and other waste may be produced constantly. In these cases, vacuum pick-up stations tied into an exhaust and collection system may be needed at specific points of waste generation. Keep in mind that this dust or lint may be combustible, and care must be taken to ensure that there is no possibility to ignite this material. For area cleaning, powered floor sweepers or rail-mounted traveling cleaners may be necessary. Areas in which large quantities of scrap or discarded packaging materials continuously accumulate may need not only large trash containers, but also motorized equipment that can be emptied or replaced frequently. In areas where grease or oil may accumulate, a high-pressure steam floor-washing system may be required.

A well-planned preventive maintenance program on all equipment will find and eliminate leaks of liquid, electrical hazards, static buildup, and friction caused by lack of lubrication.

Layout and storage: Overcrowding is a major impediment to proper housekeeping. Blocked or restricted aisles limit access and in so doing, hamper efficient cleaning and trash pickup. Lack of sufficient workspace and storage capacity results in inefficient operations, an inability to maintain order, and worker frustration. The creative use of racks, shelving, and bins, or marking aisles and storage areas by painting lines on the floor can provide solutions.

With its negative influence on good housekeeping, disorganized and haphazard storage is usually a detriment to effective fire protection, as well. Fire extinguishers, small hose stations, and sprinkler system control valves can become blocked and inaccessible, while other fire equipment, such as fire doors, may be made inoperable. Last, in a fire emergency, it will be more difficult for the fire department to attack and extinguish a fire even in a sprinklered occupancy.

Environment: Equally as important as the natural environment is the artificial environment created in the workplace. The control of process fumes, vapors, dusts, and flyings required today for the well-being of the worker has also had a beneficial impact on overall cleanliness in many work areas. Consideration should be given to totally enclosed dust- and/or vaportight process equipment in addition to the ventilation equipment mentioned above.

Personnel: Employees must be informed on the necessity of keeping their respective areas clean. This is a critical part of the program. To do this, leadership must constantly communicate to the employees what is acceptable and unacceptable about the way the facility is being maintained, from a housekeeping standpoint. In addition, employees should be encouraged to discuss ways to eliminate hazards and improve working conditions.

BUILDING CARE AND MAINTENANCE

The three basic requirements for good housekeeping are: (1) proper layout and equipment; (2) correct materials handling and storage; and (3) cleanliness and order. Any facility that implements these basics has laid the foundation for good housekeeping. Using them, the facility can develop special housekeeping practices to deal with its own specific problems.

In pursuing cleanliness and order, effective care and maintenance of buildings require special housekeeping practices to reduce the fire danger to buildings.

Floors

The general care, treatment, cleaning, and refinishing of floors may present a fire hazard if flammable solvents or finishes are used or if combustible residues are produced in quantity. For example, fires have resulted from the use of a flammable liquid, such as gasoline, to clean a greasy spot on the floor. In general, cleaning solvents with flash points below room temperature are too dangerous to use to clean floors. When selecting a cleaning agent, care should be taken about its toxicity to employees and to the environment if it is eliminated through the sewer system. Safe materials are available for most of the following purposes.

Sweeping compounds: Many excellent compounds exist on the market for sweeping floors and absorbing oily materials. Noncombustible materials should be used for this purpose. Sawdust and similar combustible materials should be avoided or, if used, should be disposed of in covered metal containers.

Floor oils: Compounds containing oils and low-flash-point solvents are a hazard, particularly when freshly applied. In addition, component oils may be subject to spontaneous heating. To reduce the fire hazard, suitable attention must be given to the safe storage of oily mops, sponges, and wiping rags in metal or other noncombustible containers. Any combustible oil used to excess increases the combustibility of the floor. Oil-soaked floors, the product of years of use, also show increased combustibility.

Floor waxes: Low-flash-point solvents are hazardous, especially when used with electric polishers. In such instances, ignition might result from friction and sparking. Water-emulsion waxes are preferable.

Furniture polishes: Furniture polishes containing oils subject to spontaneous heating become hazardous when rags that are saturated with these polishes are not disposed of properly. Such oil-soaked rags should be placed in metal or other noncombustible containers.

Nonhazardous cleaning agents: Flammable cleaning solvents need not be used since a number of nonhazardous cleaning agents are available. These relatively safe materials are stable and have high flash points and low toxicity. There are several commercial stable solvents available that have flash points ranging from 140 to 190°F (60 to 88°C) and have a comparatively low degree of toxicity. Safe materials are available for most of the preceding purposes.

Dust and Lint

A necessary procedure in many occupancies is the removal of combustible dust and lint accumulations from walls, ceilings, and exposed structural members. Unless this procedure is performed safely, as by vacuum cleaners or air-moving (blower and exhaust) systems, this procedure may present a fire or explosion hazard. In some cases, vacuum cleaning equipment must be equipped with dust-ignitionproof motors to ensure safe operation in dust-ladened atmospheres.

Care should be taken not to dislodge into the atmosphere any appreciable quantities of combustible dust or lint which might ignite or form an explosive mixture with air. A lot of work can be eliminated by applying suction at locations where dust may escape from processing machinery, and conveying the aspirated dust to safely located collectors. Blowing down dust with compressed air may create dangerous dust clouds, and such cleaning should be done only when other methods cannot be used and after all possible sources of ignition have been eliminated. In most localities, it is possible to obtain the services of reliable professional industrial cleaning specialists to remove dust accumulations safely.

Exhaust Ducts and Related Equipment

The exhaust ducts from the hoods over cooking ranges, such as those found in facility cafeterias, present troublesome problems because grease condenses inside the ducts and on exhaust equipment. Grease accumulations can be ignited by sparks from the range or, more often, by small fires in overheated cooking oil or fat. Without these grease accumulations in the hood and duct, stovetop fires can often be extinguished or allowed to burn out without causing appreciable damage. Fires occur frequently in frying because cooking oils and fats are heated to their flash points and may reach their self-ignition temperatures when accidentally overheated or spilled on the hot stovetop.

Grease removal devices: All exhaust systems for kitchen cooking equipment must be equipped with grease removal devices. These include such items as grease extractors, grease filters, or special fans designed to remove grease vapors effectively and provide a fire barrier. Grease filters, including frames, and other grease removal devices should be made of noncombustible materials.

Ducts: There is no practical method for preventing all kitchen duct fires, but the danger can be minimized through a combination of precautions as outlined in NFPA 96, *Standard for Ventilation Control and Fire Protection of Commercial Cooking Operations*. It is good practice to clean hoods, grease removal devices, fans, ducts, and associated equipment frequently. The exhaust system should be inspected daily or weekly, depending on its use, to determine if grease or other residues are accumulating in it.

Clean ducts are essential to fire safety, but they often remain dirty because cleaning them is a difficult and unpleasant job. One source of help is a commercial firm that specializes in this sort of work. In any case, never try burning the grease out; it is a dangerous practice, even though duct systems installed according to NFPA standards are designed to withstand burnout.

In cleaning the exhaust system, avoid using flammable solvents or other flammable cleaning aids. Do not start the cleaning process until all electrical switches, detection devices, and extinguishing system supply cylinders have been turned off or locked in a "shut" position. This will prevent both the exhaust fan and the fire extinguishing system (if the exhaust duct is equipped with one) from actuating accidentally. Once the cleaning process is completed, the switches and other controls should be returned to normal operating positions.

Satisfactory cleaning results have been obtained with a powder compound consisting of one part calcium hydroxide and two parts calcium carbonate. This compound saponifies the grease or oily sludge (i.e., converts it to soap), thus making it easier to remove and clean. The process requires proper ventilation. Another cleaning method is to loosen the grease with steam and then scrape the residue out of the duct. This has proved to be quite effective.

Spraying duct interiors with hydrated lime after cleaning is a fire prevention method used commercially. This procedure tends to saponify the grease and may facilitate subsequent cleaning, but it does not provide permanent fire retardancy.

Other duct systems: All duct systems can accumulate dirt and whatever other material is dispersed in the facility. Frequent cleaning of these systems is necessary for good housekeeping. Filters that are changed frequently can help. Particular attention should be given to building ventilation systems, including fire cutoff devices. In today's tight building design, it has been shown that microorganisms can build up and cause problems in these systems, or cause illness. Dirty outlets on ceilings and walls are evidence of the effects of neglect.

The use of a professional cleaning company or specially trained employees should be considered to ensure proper handling of the dust and dirt from ducts. This is especially true if the dust is combustible or explosive, since special equipment is needed to clean this type of system.

OCCUPANCY AND PROCESS HOUSEKEEPING

Housekeeping programs must give special consideration to disposal of rubbish, control of smoking habits, and other housekeeping hazards. It is a good idea to have an inspection of facilities by security officers after employees have left a facility for the day or weekend. This inspection should occur about 1 hr after the facility has been vacated, and should be repeated on a regular basis while the facility is vacated.

Disposal of Rubbish

The key in this area is not to give fire a place to start. The proper handling and disposal of rubbish is an integral part of the housekeeping process, and its success depends primarily upon having and observing a satisfactory routine. The proper and regular disposal of combustible waste products is of the utmost importance.

In both industrial and commercial properties, the removal of combustible waste products at the end of each workday or at the end of each work shift is a common practice. In some properties, more frequent waste disposal is necessary. In others, the collection, storage, and disposal routine vary with the nature of the property use. In all cases, however, an adequate program for dealing with this problem is a fire safety essential. Keeping a place tidy also depends on providing enough wastebaskets, bins, cans and other proper containers so that building users will find tidiness convenient.

Receptacles: Noncombustible containers should be used for the disposal of waste and rubbish. This is true even of such small receptacles as ashtrays and wastebaskets and applies, of course, to the larger units found in commercial and industrial properties. Industrial waste barrels should be made of metal and equipped with fitted covers. Care should be taken to avoid mixing waste materials where such mixing introduces hazards of its own. (See Figure 3-33A.)

Plastic wastebaskets of varying sizes are readily available and are popular because they are quiet, attractive, and scratch and dent resistant; however, not all plastic baskets have the same burning characteristics. Some melt and burn readily, adding fuel to a fire and creating a comparatively serious fire exposure problem by collapsing and spilling their burning contents. This is also true of many plastic liners used for wastebaskets and receptacles. Other baskets may contribute relatively little fuel to the fire while maintaining their shape fairly well.

Segregation of waste: It is not good housekeeping practice to dump all manner of dry waste down refuse chutes or to place it in a common bin or storage receptacle. For example, combustible metal dusts and metal powders dumped into chutes may explode if ignit-

FIG. 3-33A. *Waste containers designed to snuff out accidental fires in their contents and to limit external surface temperatures to no more than 175°F (80°C) above room temperature. (Source: Justrite Mfg. Co.)*

ed. Mercury batteries and pressurized containers, such as the ubiquitous aerosol can, may also explode when incinerated or mixed with rubbish that is subsequently burned. Precautions should be taken to keep combustible items separate from each other and from noncombustible items.

Requirements of the U.S. Environmental Protection Agency (EPA) and the state waste control departments must also be considered. Most chemicals must be handled in a specific prescribed manner. These rules are contained in the Resource Conservation and Recovery Act (RCRA) and the Hazardous Waste Amendment to that act. Contact the EPA for further information.

Control of Ignition Sources

Control of smoking: While society in general has made many changes in the smoking habits of employees, there are still a significant number of employees who do smoke. The best policy is to prohibit smoking altogether. This eliminates the possibility of smoking materials being improperly discarded. It is critical that such a policy be strictly enforced and have the support of all employees. Where this is not a possible consideration, smoking regulations must be specific as to location and, preferably, time. Areas in which smoking is permissible, as well as those in which it is limited or prohibited entirely, must be clearly marked by appropriate signs that leave no question as to what is allowed where. (See Figure 3-33B.)

FIG. 3-33B. *Examples of signs permitting or forbidding smoking in designated areas. (Source: Factory Mutual Research Corp.)*

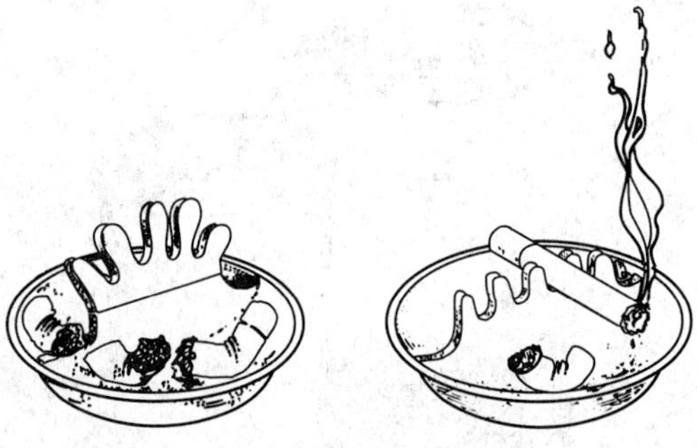

FIG. 3-33C.   Two types of well-designed ashtrays.

In addition to sensible regulations, smoking control also requires adequate receptacles for spent smoking materials. Properly designed ashtrays are essential to safe smoking. They should be made of noncombustible materials with grooves or snuffers that hold cigarettes securely. Their sides should be steep enough to force smokers to place cigarettes entirely within the ashtray. (See Figure 3-33C.) In industrial buildings, large containers of sand are often used to conveniently and safely extinguish and dispose of spent smoking materials.

Improperly designed ashtrays may constitute a hazard, particularly if they allow a lit cigarette or cigar to fall or roll away. A lighted butt also may easily come in contact with combustible materials and start a fire under certain circumstances.

The contents of ashtrays must be disposed of carefully because a live butt may well be mixed in with apparently innocuous ashes. If lighted smoking materials were to be dumped into an ordinary wastebasket, they could set paper or some other piece of combustible rubbish on fire. To prevent this from happening, reserve special covered metal containers for discarded smoking materials only.

Control of static: Static can be produced by the flow of dissimilar materials past each other. For example, liquid or dust conveyed through a pipe or duct produces an electrical potential. Under the right conditions, if adequate oxygen is present, a static discharge can occur that will ignite the flammable vapor or combustible dust. It is critical that adequate grounding and bonding be provided to decrease the risk of this happening. Part of the site-preventive maintenance program should be annual inspection and testing of all grounds, including building grounds and bonding.

Control of friction: A preventive maintenance program must be in place to identify and eliminate potential sources of friction. Lubrication is one important part of this program.

Control of electrical hazards: Routine inspections should identify overloaded electrical circuits, excess electrical extension cords, frayed cords, missing grounding plugs, etc.

Industrial Housekeeping Hazards

Some industrial occupancies have special housekeeping problems inherent to the nature of their operations. For these particular problems, specific planning and arrangements are necessary.

Clean waste and rags: Clean cotton waste or wiping rags are generally considered to be mildly hazardous, chiefly because they are readily flammable when not baled, and there is always the likelihood that dirty waste may become mixed with them. The presence of dirty waste or small amounts of certain oils may lead to spontaneous heating. Reclaimed waste is considered somewhat more hazardous than new waste. It is common practice to handle clean waste in the same manner as dirty waste, although the fire hazard is relatively small.

Large supplies of clean waste are best kept in bins made entirely of metal, or of wood lined with metal and provided with covers that are normally kept closed. Several bins may be provided where the supplies are large or where different kinds of waste are kept. The covers on such bins should be counterweighted so that they may be readily raised and lowered. The counterweight ropes can have fusible links to ensure that the covers are closed automatically in the event of fire.

Local supplies of clean waste are usually kept in small, properly marked waste cans. Providing local supply points for clean waste can help eliminate the practice of keeping waste in clothes lockers, drawers, benches, and similar locations. If clean waste is put in such places, workers may mistakenly believe that other, more oily waste is also allowed there when, in fact, the combination of the two may result in fire. These same policies should apply to reused (clean) cotton gloves and may also apply to cotton coveralls and uniforms.

Coatings and lubricants: Paints, greases, and similar combustibles are widely used at industrial occupancies, and a good housekeeping program will make sure that their combustible residues are collected and disposed of safely. The discharge of vapors from spray booths should be so arranged that the vapors are conducted directly to the outside and the residues accumulate safely. Use of wash-water spray booths for painting should also be considered.

Drip pans: Drip pans are essential at many locations, notably under motors, machines using cutting oils, and bearings. They are also used with borings and turnings that may contain oil. Drip pans should also be used wherever flammable and combustible liquids are dispensed. Drip pans should be made of noncombustible material and contain an oil-absorbing compound. At many industrial occupancies, commercial oil-absorbing compounds consisting largely of diatomaceous earth are used instead of sawdust or sand. The regular removal of oil-soaked material is recommended.

Flammable and corrosive liquids waste disposal: The disposal of combustible liquid waste often presents a troublesome problem. Any waste material that is a corrosive liquid (pH <2 or >12.5), or is a liquid with a flash point of 140°F (60°C) or less, is considered a hazardous waste under the Federal Resource Conservation and Recovery Act (RCRA). Drums of this waste must be labeled in accordance with this act. This includes labeling the liquid as hazardous waste and including the proper EPA codes. These waste products must be disposed of in a facility that is licensed to handle this waste. Care should be taken to ensure that the facility taking the waste will properly dispose of it.

Flammable liquid spills: Flammable liquid spills may be anticipated wherever such products are handled or used, and some means of coping with these spills must be kept on hand. These include a supply of suitable absorptive material and special tools to help limit the spill. Workers should understand and promptly take the steps needed to cut off sources of ignition, ventilate the area, and safely dissipate any flammable vapors.

Flammable liquid storage: Flammable liquids should be stored in a segregated area. (See Section 3, Chapter 21, "Storage of Flammable and Combustible Liquids"). Good housekeeping practices will ensure that only limited quantities of flammable and combustible liquids are kept on the production or work area. These should be protected in suitable containers. No storage of flammable liquids should be allowed in the general storage area, except as allowed in

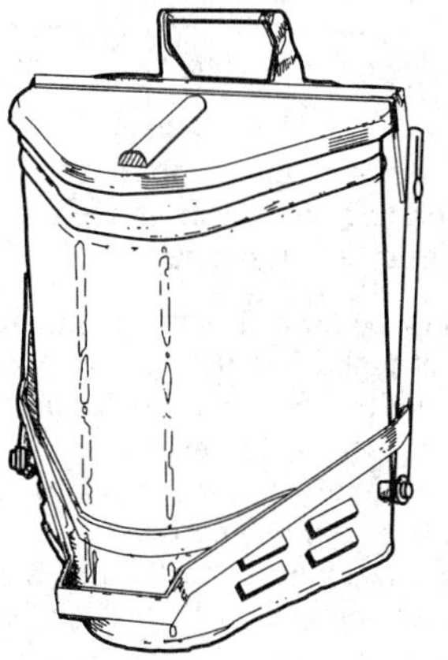

FIG. 3-33D.   A portable waste can that is equipped with a self-closing cover. This type of container can be used to store oily waste materials, particularly if they are subject to spontaneous heating. (Source: The Protectoseal Co.)

NFPA 30, *Flammable and Combustible Liquids Code.* Combustible liquid storage should be limited and protected as allowed in NFPA 30.

Oil puddling: Accumulations of oil can present a housekeeping problem at industrial locations where a considerable amount of oil is used. Poor maintenance of industrial hydraulic elevator installations may result in oil leaks that eventually form puddles on the elevator machine room floors and in the bottoms of hoistway pits. Although most oils used in hydraulic elevator systems have high flash points, any combustible oil can be a source of fire, particularly when it is found in puddles that contain accumulations of debris. Puddled oil and materials used to absorb oil spills should be disposed of in metal barrels.

Oily waste: Oily wiping rags, sawdust, lint, clothing, and other items can be highly dangerous, particularly if they contain oils subject to spontaneous heating. To dispose of all such materials in ordinary quantities, a standard waste can that has been listed and labeled by testing organizations is best. For large amounts, heavy metal barrels with covers are ideal. Good practice calls for cans containing oily waste to be emptied daily and for wiping rags to be kept in covered metal containers until they can be laundered. (See Figure 3-33D.)

Packing materials: Almost all packing materials used today are combustible and, consequently, hazardous. Plastic pellets and rigid forms, excelsior, shredded paper, sawdust, burlap and other such materials should be treated as clean waste. However, large quantities may have to be kept in special vaults or storerooms. Automatic sprinklers are the best protection for areas where considerable quantities of packing materials are stored or handled.

Used or waste packing materials and the crating materials from receiving and shipping rooms must be removed and disposed of as promptly as possible in order to minimize the danger of fire. Ideally, the packing and unpacking processes should be conducted in an orderly manner so that excessive quantities of packing materials do not become strewn about the premises. A specially marked (or identified) area should be provided to accumulate this material. This area should be cleaned frequently and the debris removed to an outside storage receptacle.

Pallet storage: NFPA 231, *Standard for General Storage,* specifies that empty wood pallet storage be severely limited in storage ar-

eas. Storage of empty wood pallets in excess of these limits gives a fire an excellent place to grow beyond the limits of the available fire protection.

Lockers and Cupboards

Many industrial facilities provide their employees with lockers in which to put their personal belongings. These lockers may present a fire hazard if they are untidy or are used as general storerooms for such waste material as oily rags and cloths or paint-smeared clothing. These items may ignite spontaneously or they may be accidentally ignited by matches and imperfectly extinguished pipes and cigars that employees may inadvertently leave in their lockers.

Wood lockers can ignite and spread fire. Metal ones are preferred, but they should be inspected regularly. Metal lockers may contain a fire if they are of solid construction, including fronts and bottoms, partitions, and backs. Backs and partitions made of expanded metal or wire screen may allow fire to spread unchecked and should not be used.

Lockers that are arranged in two tiers, i.e., one upon the other, are generally unsatisfactory. They do not hold clothes without mussing them, and it often becomes the habit of those using such lockers to keep their clothes outside the locker or to throw them haphazardly into the locker, thus increasing the danger of spontaneous heating when such clothing is spotted with oil or paint.

Some lockers are provided with mechanical exhaust ventilation. Where this is true, NFPA 91, *Standard for Exhaust Systems for Air Conveying of Materials,* should be followed to avoid spreading a fire that originates in a locker.

Industrial plants that furnish and wash their employees' protective clothing may use a system of wire baskets, one per employee, suspended from the ceiling by a small chain running over a pulley, instead of lockers. This method has proved successful in maintaining cleanliness and thereby reducing the fire hazard.

Where automatic sprinklers are installed, lockers must have expanded metal or screen tops to enable water from the sprinklers to reach the contents. Paper can be pasted on the tops to keep dust out. Sloping tops are also advisable to help prevent both fires and accidents. Material cannot be placed on top of a locker designed this way.

Wood supply cupboards constitute a fire hazard in places such as machine and paint shops where woodwork becomes oil- or paint-soaked and where clothes or oily waste may be left in them. Wood cupboards should be inspected regularly to ensure that they are always clean. The ideal cupboards for tools and similar items are made entirely of steel.

OUTDOOR HOUSEKEEPING PRACTICES

Good housekeeping practices are as essential out-of-doors as they are indoors. Failure to comply with good housekeeping practices out-of-doors may threaten the security of exposed structures and goods stored outside. The accumulation of rubbish and waste and the growth of tall grass and weeds adjacent to buildings or stored goods are probably the most common hazards. A regular program for policing the grounds is essential.

Grass and Weed Control

Tall grass, dry weeds, and bushes around buildings, along highways, on railroad properties, and along the streets of large industrial and commercial complexes present a definite fire hazard. To reduce this hazard, those responsible for maintaining these properties have tried to control or destroy such vegetation.

One way to get rid of unwanted vegetation is to apply a chemical solution which poisons the weeds. Among the chemicals used are chlorate compounds, particularly sodium chlorate. Unfortunately, those who use chlorate compounds do not always realize that they are oxidizing agents. When these compounds come in contact with combustible materials, they virtually prime those materials for a fire or explosion. During hot periods in the summer, large numbers of fires have resulted from the use of sodium chlorate solutions on dry grass and weeds. Fires have also been reported in buildings and other structures in which such solutions have been spilled. Chlorate compounds spilled on clothing may present a danger to personnel, too.

There are weed killers that are not toxic and do not pose a fire hazard. Calcium chloride and agricultural borax, applied dry or in solution, are effective nonhazardous weed killers, as are various proprietary solutions. A number of commercial chemical weed killers, such as ammonium sulfamate, have little or no fire hazard and only a slight toxic hazard. Sodium arsenite and other compounds containing arsenic are efficient herbicides, but they are poisonous and not generally recommended.

The amounts of various chemicals needed to effectively kill weeds and the duration of their effect vary, depending on the weed-killing agent used, the character of the vegetation, the climate, and the soil. Manufacturers' directions indicate the proper amounts to be used under various conditions.

Another way to remove vegetation is to burn it. This may be done only where environmental regulations permit outdoor burning and must be carefully controlled. Adequate fire-extinguishing equipment must be readily available at all times. Using this method, grass and weeds are usually cut down, collected in piles, and ignited. When the grass is too damp to propagate fire easily, flame-throwing torches may be used on the piles. However, these torches introduce a hazard if not carefully operated.

All burning introduces a hazard. Grass fires may spread out of control and ignite nearby buildings. To avoid this hazard, controlled burning should only be done at certain times of the year and then under the direct supervision of the fire department.

Fire authorities issue fire permits, where permissible, to help them control burning. These permits provide the authorities with an opportunity to educate the public in safe burning. They also help the fire department limit burning to nonhazardous periods of the year and better manage burning done during the hazardous periods.

Outdoor Storage

Goods stored outdoors should be properly separated from buildings of combustible construction and from other combustible storage that might constitute an exposure hazard. These separations should be maintained by the housekeeping staff, who must see to it that they are never blocked, even temporarily, by such things as contractors' shacks, discarded crates, pallets, or other combustibles. Obstructed aisles could hamper fire-fighting operations if the need for them ever arises. Passageways between storage piles should also be unobstructed and clear of combustibles.

Proper housekeeping also requires that smoking in outdoor storage areas be controlled. Suitable signs should be posted and large noncombustible receptacles should be provided for the disposal of smoking materials before entering a "no smoking" area.

Outdoor Rubbish Disposal

Combustible waste materials stored outdoors to await subsequent disposal as rubbish should be placed not less than 20 ft (6 m), and preferably 50 ft (15 m), from buildings, and at least 50 ft (15 m)

from public roadways and sources of ignition, such as incinerators. They should also be enclosed with a secure noncombustible fence of adequate height.

The most satisfactory solution to the rubbish disposal problem is regular collection and removal from the premises. Burning rubbish is generally unsafe and is not permitted in built-up urban areas.

If rubbish must be burned outdoors, it should be done in the early morning or at night, because the night moisture reduces the chance that sparks will ignite combustibles in the surrounding area. This is the reasoning behind certain fire department and forest service regulations that limit outdoor burning to certain days or times of day. Of course, there are some times when outdoor burning is strictly prohibited. Most parts of the United States and Canada experience days when things are so dry that any burning is dangerous.

Even when rubbish is not burned outdoors, it can present a fire hazard. All rubbish dumps, even landfills, are susceptible to fire. And sparks and flying brands from dump fires can carry the fire long distances. This is also true of sparks and brands produced by bonfires and incinerators that lack adequate spark arresters.

INSPECTION

Housekeeping inspections are an important part of an overall housekeeping program. This type of program should be combined with a complete safety inspection program. This type of inspection has four main objectives:

1. Maintain a safe work environment.
2. Control unsafe actions of employees.
3. Maintain operations profitability (and product quality).
4. Maintain operations to meet or exceed acceptable safety and government standards.

Although the word "housekeeping" is never mentioned, it does not take much imagination to see good housekeeping is involved in each objective. Good housekeeping is an essential part of any successful safety program and that includes the fire safety program.

There are several types of inspections. Some involve preventive maintenance of equipment, such as lubrications of equipment or static grounding continuity. Others involve fire equipment inspections. But most important is a regular inspection by the team/crew leader and facility management.

These inspections must be well defined. The program should include what areas are to be inspected, how often the inspections should be made, what is acceptable performance, and who will conduct the inspections.

One of the most time-consuming parts of an inspection is writing a report. Here, computers and bar coding can help. Bar coding is commonly used today in identification of products. However, bar codes can also be used in the identification of hazards.

The facility inspector can carry a bar-code reader and recorder. Bar codes identify the department, date, and time of the inspection, as well as the inspector. Common hazards for the facility are entered on a bar-code list carried by the inspector. For each item or hazard on the list, a recommendation is also keyed into the computer program. The inspector simply scans the bar code for the desired item. At the completion of the inspection, the information is downloaded into a computer and a report is printed.

This format also allows positive comments to be made where employees have made positive contributions to the housekeeping program. This positive reinforcement will encourage employees to continue their good work.

Fire Prevention Checklist

Electrical equipment

No makeshift wiring
Extension cords serviceable
Motors and tools free of dirt and grease
Lights clear of combustibles
Circuits properly fused or otherwise protected
Equipment approved for use in hazardous areas (if required)
Ground connections clean and tight and have electrical continuity

Friction

Machinery properly lubricated
Machinery properly adjusted and/or aligned

Special fire-hazard materials

Storage of special flammables isolated
Nonmetal stock free of tramp metal

Welding and cutting

Area surveyed for fire safety
Combustibles removed or covered
Permit issued

Open flames

Kept away from spray rooms and booths
Portable torches clear of flammable surfaces
No gas leaks

Portable heaters

Set up with ample horizontal and overhead clearances
Safely mounted on noncombustible surface
Secured against tipping or upset
Not used as rubbish burners
Combustibles removed or covered
Use of steel drums prohibited

Hot surfaces

Hot pipes clear of combustible materials
Ample clearance around boilers and furnaces
Soldering irons kept off combustible surfaces
Ashes in metal containers

Smoking and matches

"No smoking" and "smoking" areas clearly marked
No discarded smoking materials in prohibited areas
Butt containers available and serviceable

Spontaneous ignition

Flammable waste material in closed, metal containers
Piled material, cool, dry, and well ventilated
Flammable waste material containers emptied frequently
Trash receptacles emptied daily

Static electricity

Flammable liquid dispensing vessels grounded or bonded
Proper humidity maintained
Moving machinery grounded

Housekeeping

No accumulations of rubbish
Safe storage of flammables
Passageways clear of obstacles
Automatic sprinklers unobstructed
Premises free of unnecessary combustible materials
No leaks or dripping of flammables and floor free of spills
Fire doors unblocked and operating freely with fusible links intact

Extinguishing equipment

Proper type
In working order
In proper location
Service date current
Access unobstructed
Personnel trained in use of equipment
Clearly marked

FIG. 3-33E.   A sample checklist can serve as a guide for the supervisor in drawing up an inspection list. It should be reviewed regularly to keep it up to date.

This computer format also allows for followup to housekeeping recommendations. It is very important that, when recommendations are made, positive change results in the area. This change can either be the action recommended or another action that provides equal protection.

When a bar code system does not fit the inspection program, one may find that a checklist will be a valuable aid. An example of a checklist is shown in Figure 3-33E.

In order to prevent inspections from taking considerable time, the inspector should:

1. Have a definite schedule.
2. Inspect only one part of the department at a time.
3. Rotate inspection responsibilities among department members.
4. Make a point of looking for housekeeping problems as part of a daily routine.

Trading inspections between departments can help increase inspection effectiveness; monotony can distort evaluation of real problem areas.

SUMMARY

A well-planned housekeeping program will yield the following benefits:

- Reduced operating costs
- Increased production
- Improved production control
- Conservation of material and parts
- Reduced production time
- Better use of space
- Improved traffic flow (i.e., permits faster travel)
- Reduced accident rate
- Reduced fire hazards
- Compliance with OSHA Act

BIBLIOGRAPHY

NFPA Codes, Standards, and Recommended Practices

Reference to the following NFPA codes, standards, and recommended practices will provide further information on housekeeping practices discussed in this chapter. (See the latest version of *The NFPA Catalog* for availability of current editions of the following documents.)

NFPA 30, *Flammable and Combustible Liquids Code*
NFPA 82, *Standard for Incinerators, and Waste and Linen Handling Systems and Equipment*
NFPA 91, *Standard for Exhaust Systems for Air Conveying of Materials*
NFPA 96, *Standard for Ventilation Control and Fire Protection of Commercial Cooking Operations*
NFPA 231, *Standard for General Storage*

Additional Reading

Anderson, R., "Safety Inspection and Bar Codes; A Perfect Match," *Chemical Health and Safety*, Vol. 2, No. 4, July/August, 1995, pp. 7-9.
Clinton, W. J. & Gore, A. J. "The New OSHA: Reinventing Worker Safety & Health," *National Performance Review*, May, 1995.
Cote, A. E., ed., "Industrial Housekeeping Practices," *Industrial Fire Protection Handbook*, 3rd ed., National Fire Protection Association, Quincy, MA, 1990, pp. 1049-1063.
Garside, R., "Maintaining Maintenance in Hazardous Areas," *Control and Instrumentation*, Vol. 18, No. 11, 1986, pp. 35, 37.

"Good Housekeeping," *National Safety and Health News*, Vol. 134, No. 3, Sept. 1986, pp. 28-32.

Higgins, L. R., and Morrow, L. C., eds., "Sanitation and Housekeeping," *Maintenance Engineering Handbook*, 4th ed., McGraw-Hill, New York, 1988.

Higgins, L. R., *Maintenance Engineering Handbook*, 44th ed., McGraw Hill, New York.

Holden, P., "The Inspection Process," *Chemical Processing*, July 1995, p. 82.

Lee, R., *Building Maintenance Management*, 3rd ed., Collins, London, 1987.

"Prevention of Housekeeping-Related Fires," Factory Mutual Research Corp., Norwood, MA, P9217, Mar. 1994; Factory Mutual Research Corp., *Business Under Fire! A Manager's Guide to Fire Prevention Programs*, 1994, pp. 1–4.

Rosaler, R. C., Rice, J. O., and Hicks, T. G., eds., *Industrial Maintenance Reference Guide*, McGraw-Hill, New York 1987.

MATERIALS, PRODUCTS, AND ENVIRONMENTS

If prevention fails, the next opportunity to reduce fire losses comes in the selection and arrangement of everything that can burn in the space where the fire started. A small fire in a trash can, for example, will pose no serious threat if the trash can is the only object in the middle of an enormous concrete floor in some vast storage room. However, that same burning trash can could start a devastating fire if it were near an upholstered chair in a densely furnished room in an occupied residential property, against an untreated plywood wall in the lobby of an old hotel, or adjacent to a solvent dispensing area in a furniture manufacturing facility.

The differences among these scenarios lie in the burning properties of the materials near the point of initial ignition, the way those materials are used in products or assemblies, and other features of the environment at the point of ignition. The first item ignited in an unwanted fire, if it is small enough, might be considered merely the "fuse" to the first large fuel package, which is the real key to creating a serious fire hazard. Section 4 addresses the burning properties of materials and products in particular environments, paying special attention to those which may promote such rapid and severe early fire growth that the effectiveness of later-acting fire-protection strategies would be seriously impaired.

Chapters 1 and 2 of Section 4 are about the basic science of this part of fire protection. Chapter 1 addresses burning properties more generally, with special attention given to thermal properties, such as fast or slow development and high or low rates of heat release. Chapter 2 focuses on the range of combustion products—toxic gases and other dangerous effects—that can become components in the smoke and gas produced by fire. Chapters 1 and 2 help to place the other chapters of Section 4 in context.

Chapters 3 through 5 address wood and wood-based products, cellulosic materials, and fibers and textiles—all commonly found in

fire scenarios as large fuel packages of ordinary combustibles—and methods of making them fire-retardant or flame-resistant. Chapter 4 addresses modification of the burning properties of some classes of products through fire retardancy or flame resistance. Chapter 17 addresses large fuel packages in residential fires, namely, upholstered furniture and mattresses.

Chapters 6 through 8, 12, 15, and 16 address certain types of materials that are of special concern because of their potential for extremely rapid or explosive fire development. Chapter 11 addresses pesticides and their unusual toxic properties. Chapter 10 addresses plastics and rubber, which constitute large full packages of ordinary combustibles.

Chapters 9 and 18 address certain specialized environments—those created by oxygen-enriched atmospheres and energy conservation programs—that may affect the burning properties of many different materials. Energy conservation may lead to less oxygen flow to a fire, while an oxygen-enriched atmosphere may lead to more.

Finally, as Chapter 1 of this section says, "It is becoming possible to predict how a material, burning in a given scenario, will influence the hazard to life from that fire. This emerging field of fire protection, known as fire hazard analysis, requires information on the burning properties of a material and its conditions of use, as well as smoke obscuration and smoke toxicity data." Measurement and modeling of systematic, state-of-the-art consideration of materials, products, and environments and their impact on early fire development is described in Section 11, especially in Chapters 4 through 8, 10, 12, and 13.

—Martha H. Curtis
December 1996

FIRE HAZARDS OF MATERIALS: AN OVERVIEW

Frederic B. Clarke

This section of the Handbook is devoted to the specific fire problems posed by various kinds of materials. Solving such problems involves preventing ignition or, if ignition occurs, minimizing the size of the fire and mitigating its effects. In addition to heat, fire's effects often include those from smoke, such as toxicity or corrosivity, as well as its ability to hamper vision.

This chapter addresses the various kinds of fire hazards, ways of characterizing those hazards, and some of the methods used to reduce them. The most commonly encountered classes of materials, and their special fire problems, are dealt with in detail in the succeeding chapters of this section.

BURNING OF MATERIALS

Ordinary burning means oxidation, i.e., the chemical reaction of a material with oxygen to form relatively simple compounds called oxides. This is generally accompanied by the release of energy as heat and light. The first descriptor of the hazard of a material might be the amount of heat it produces when it burns, i.e., the heat of combustion. The greater the heat of combustion, the greater its potential as a fuel, and the hotter it burns.

The amount of heat produced on burning is determined almost entirely by the chemical composition of the material. The speed with which the heat is produced, on the other hand, is determined by the physical form of the material. For example, excelsior burns more quickly than a solid block of wood, although equal weights of the two will produce essentially the same amount of heat.

Organic Materials

Any discussion of the fire hazards of materials must center upon those materials that are organically derived, i.e., those that are based on carbon. This is simply because organic materials as a class are ubiquitous. The simplest organic compounds, such as methane and propane, are common gaseous fuels. Organic liquids are fuels, solvents, and chemical intermediates, but the richest variety of materials are organic solids. They include everything from specific compounds, such as aspirin, to common but complex everyday materials, such as wood, paper, textiles, and plastics. These materials have carbon as their principal constituent; almost all contain some

hydrogen; and many contain oxygen, nitrogen, and other elements in varying amounts.

The molecules that compose most common organic materials consist of many thousands of atoms linked together to form the chains and networks necessary for useful mechanical properties. These very large molecules are often of natural origin, such as the cellulose in wood or cotton and the protein-based structure of wool. They can also be man-made, such as plastics. Polyethylene, for example, is manufactured as the name suggests: by chemically linking together thousands of molecules of ethylene, a gas, into each large "macro" molecule of polyethylene, a nonvolatile solid with many everyday uses. The process is called polymerization and is the origin of almost all plastics or synthetic polymers.

Energy Production

Most organic materials burn readily. The common products of combustion of organic materials are water, which is the oxide of hydrogen, and carbon dioxide. A tabulation of the heats of combustion of organic materials is contained in Appendix A. Those figures are based on complete burning. In practice, there may be insufficient oxygen available to convert all of the carbon to carbon dioxide; some carbon monoxide and soot may be formed, so the heats are upper limits of the actual amount of heat to be expected for the combustion of materials under normal conditions.

Examination of the heats of combustion tables in Appendix A will show that, while the heat of combustion is quite different for different organic materials, the heat produced per unit weight of oxygen consumed is the same within about 10 percent. This fact, sometimes called Thornton's Rule,[1] allows one to use oxygen consumption as a reasonable measure of the heat produced by a burning organic material.

Organic materials fall into two broad classes: (1) hydrocarbon based and (2) cellulose based. The former are derivatives of the unoxidized hydrocarbon building blocks: $—CH_2—$, or $—CH—$. The latter are based on a partly oxidized carbon unit: $—CH(OH)—$. In a sense, cellulose-based material is already partly oxidized in its natural state, so that when the two classes are combusted to carbon dioxide and water, cellulose consumes less oxygen and produces less heat.

For equivalent amounts of oxide formed, the hydrocarbon-based materials consume 50 percent more oxygen and thus produce about 50 percent more heat. On a weight basis, the difference is greater; hydrocarbons produce over twice the heat, pound for pound

---

Frederic B. Clarke, Ph.D., is president of Benjamin/Clarke Associates, Inc., of Kensington, MD, a firm specializing in fire hazard analysis.

(Kg for Kg) as do cellulosics. Hence, from an energy viewpoint, oil or natural gas is a better fuel than wood, which is composed largely of cellulose.

Rate of Burning

The rate at which the fire products are produced is always an important component of fire hazard, whether the concern is for heat, smoke, or toxic gases. Later sections of this chapter address the measurement of this factor, but it is instructive to examine some of the general physical factors that contribute to the burning rate.

The burning material may be gaseous, liquid, or solid, but the oxygen (normally free oxygen in the air) is usually in a gaseous state. For the necessary chemical reactions to occur, the fuel and oxygen must be brought into contact at a molecular level, and this, in turn, means that burning is generally a vapor-phase phenomenon.

The rate of burning is a function of how fast the chemical reaction of oxidation occurs. In premixed flames, i.e., those in which mixing has occurred before the combustion is initiated, the burning rate is controlled only by the inherent rate at which the substances combine. This is generally quite fast; flames propagate under premixed conditions at several feet per second (one ft per sec equals 0.3 m/s). It is for this reason that the contact of air and combustible vapors is so dangerous; the process, once started, is virtually impossible to interrupt, except in closed spaces specifically equipped for that purpose.

A more common mode of burning is the diffusion flame, shown schematically in Figure 4-1A. In this case, vaporized fuel mixes with oxygen in the combustion zone. The rate of burning is essentially controlled by the rate at which the two components arrive in the heated combustion zone. Once they arrive, combustion is, by comparison, instantaneous.

Since gases mix with one another readily, the burning of a gaseous fuel, such as hydrogen or methane in air, is a rapid process. However, the burning of a liquid or solid requires first that the fuel be converted to the gaseous state (volatilization). This process consumes heat energy to produce more volatile fragments. As a practical matter, therefore, the rate of volatilization of a material strongly affects its rate of burning. Unlike smaller molecules, macromolecules are too large to be transported "as is" into the vapor phase. Ac-

tual chemical bonds must first be broken, and this consumes energy as well.

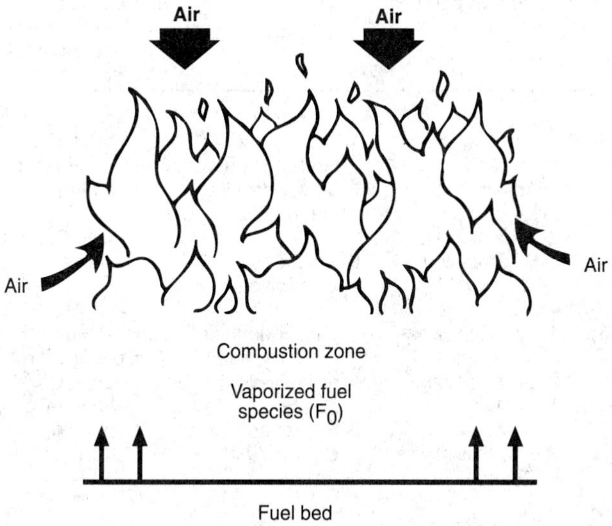

Liquids and solids are more concentrated fuels than are gases. Propane, for example, burns completely according to the following expression:

$$C_3H_8 + 5O_2 \rightarrow 3CCO_2 + 4H_2O$$

At any given pressure, one volume of propane gas consumes five volumes of oxygen, or about 25 volumes of air. By contrast, liquefied propane is some 300 times denser than propane gas under normal conditions. This means that one volume of the liquid will produce 300 times as much heat, but it will also require 300 times as much air. Thus, the burning of highly volatile liquids and solids may also be affected by the rate of delivery of oxygen to the combustion zone, especially under poorly ventilated conditions.

Except for hypergolic combinations of materials (i.e., those that ambient temperatures are sufficient to ignite), the initiation of the combustion reaction requires some added heat. This heat must be sufficient to vaporize enough of the fuel to initiate reaction and to accelerate the chemical combustion reaction to a rate where it can sustain itself. The heat required for ignition is greatly dependent upon the physical state of the materials and the heat transfer properties of its environment. Thus, combustible materials present in a dust, i.e., with a large surface-to-volume ratio, may burn very readily or explode, while the same material in a solid block may be very difficult to ignite at all.

EXTREMES OF BURNING BEHAVIOR

The extremes of burning behavior differ principally from normal burning in the time scale over which burning occurs. At the fast end of the spectrum are explosions, while at the slow end are the phenomena of auto-oxidation and smoldering. Explosions are a source of substantial hazard and are discussed in detail in several chapters in this handbook. The slow processes, however, are less well known. Auto-oxidation is the combination of a material with oxygen at a rate too slow to produce the observable heat and light normally associated with fire. Metals are particularly prone to auto-oxidation, demonstrated by the rusting of iron, the anodization of aluminum, and the tarnishing of silver. Many organic materials are also prone to auto-oxidation. The deterioration of rubber and plastics over time is often the result of slow oxidative processes. Antioxidants can be added to such materials in order to increase their durability and working life. Such additives operate either by being themselves oxidized in preference to the organic material or by chemically inhibiting the oxidation reaction.

Self-heating, and eventual spontaneous ignition, can occur if the heat produced by the auto-oxidation is not removed. This is especially true in porous solids, such as coal, where air can diffuse into the interior of the material, and yet the heat produced is effectively contained by the insulating properties of the material.

Smoldering is a distinct burning process, quite different from flaming. Only a small fraction of combustible materials, those which can produce a porous char during the course of combustion, will smolder. This char allows air to reach the smoldering region, which is on or below the surface of the fuel. Compared to flaming, smoldering is a slow process. Temperatures in the smolder zone are typically 900 to 1600°F (483 to 871°C); temperatures in a flame typically reach 2700°F (1482°C). When materials smolder, the degree of combustion is usually less complete than when the same materials burn under flame. Most smolderers are organic and all are solids, which can yield both volatile fuel species and a rigid, porous char structure.

FIG. 4-1A.   A schematic view of a burning surface.

Air   Air

Air   Air

Combustion zone

Vaporized fuel
species (F_0)

Fuel bed

MEASURING AND CONTROLLING HAZARDS

The vastly different burning characteristics of gases, liquids, and solids pose different types of hazards. The methods for measuring and controlling them are also different. Table 4-1A shows the different kinds of measurement techniques available for different components of hazard. It is most convenient to think of hazard control in terms of controlling first the likelihood of ignition, then the control or containment of the fire spread, and finally management of the fire impact if ignition and spread cannot be prevented.

Fire Hazards of Gases and Dusts

Flammable gases and combustible dusts burn rapidly once ignited, so avoidance of ignition is very important. The principal tools for doing this are: (1) the determination of flammability limits, i.e., the concentration range of a particular gas or dust (usually in air) within which combustion will occur, and (2) operating in a manner such that the concentration is maintained beyond those limits. It is also possible to reduce the likelihood of ignition of a gas or dust by adding chemical inhibitors to raise the lower flammability limit. In

practice, however, this is rarely done, either because of expense or because it alters otherwise desirable properties.

Ignition control is generally effected by stringent storage and handling safeguards; the bulk of the discussion in the chapters on flammable gases are concerned with these procedures. It is also possible in closed environments to reduce the amount of oxygen available and thereby to raise the effective flammability limit. Conversely, atmospheres that are enriched with oxygen beyond normal atmospheric concentrations offer special explosion hazards for vapors and dusts. The intensity of a fire resulting from gases, vapors, or dust is a function of the density of gas or the concentration and particle size of the dust. Very light gases, e.g., hydrogen, and very finely divided dust are rapidly dispersed and can lead to the onset of explosive conditions in a correspondingly shorter period of time than more dense materials. Fire and explosion control efforts concentrate on containing the products and, under specialized conditions, providing for automatic suppression of an incipient explosion. This is done by rapidly introducing a suppressant, such as water vapor or a chemical inhibitor, into the exploding cloud. To be effective, automatic suppression must occur in a time scale of a few hundredths of a second.

In recent years there has been concern over the possibility of large flammable gas fires in populated areas. With vast quantities of

TABLE 4-1A. *Means of Measuring and Controlling Fire Hazards of Materials*

Hazard Component Material	Ignition		Spread and Growth		Fire Impact	
	Measurement	Ignition Control	Measurement	Control of Spread and Growth	Measurement	Control of Fire Impact
Gases	Flammability Limits	1. Storage and Handling Safeguards 2. Inerting of Atmospheres	Flammability Limits, Density, Diffusion Coefficient	Tank Venting Procedures	Placarding, NFPA Hazardous Materials Identifiers	1. Emergency Response and Evacuation 2. Placarding, NFPA Hazardous Materials Identifiers
Liquids	Flash Point	1. Handling Safeguards 2. Hazard Classification Systems	Volatility	1. Ventilation and Flame Arrest 2. Storage and Tank Separation	Placarding, NFPA Hazardous Materials Identifiers	1. Emergency Response and Evacuation 2. Placarding, NFPA Hazardous Materials Identifiers
Solids 1. Textiles, Cushioning Materials	1. Ease of Ignition Tests and Small-Scale Flame Spread Tests	1. Flame-Resistive Materials and Treatments 2. Protective Layers	1. Supported Flame Spread 2. Rate-of-Heat Release	1. Material Selection and Assembly 2. Detection and Suppression	1. Toxicity of Combustion Products (under development) 2. Smoke Generation	1. Breathing Apparatus for Fire Fighters 2. Smoke Control Systems
2. Structural Materials, Building Components and Finishes	Same	Same	Flame Spread, Rate-of-Heat Release	1. Fire Resistivity 2. Low Combustibility 3. Fire-Resistive Coatings 4. Detection and Suppression	1. Fire Endurance (Stability, Integrity, Insulation) 2. Smoke and Toxic Combustion Products	Same, Plus Construction Design: Compartmentation; Separation

gaseous fuels now being transported either in bulk or by pipeline, the likelihood of such occurrences commands attention. The principal means by which the impact of such accidents is minimized is the use of hazardous material identification systems, e.g., placarding and NFPA 704, *Standard System for the Identification of the Hazards of Materials for Emergency Response*. In the absence of a major breakthrough in the prevention or suppression of such fires, the most appropriate means of minimizing fire impact is the advanced planning, preparedness, and response of trained emergency personnel.

Fire Hazards of Liquids

Since burning actually occurs in the vapor phase, the most hazardous combustible liquids are those with a high vapor pressure, or volatility. An empirical measure that combines volatility with the heat-producing capabilities of the vapor is the flash point determination. The flash point is simply the temperature at which a liquid gives off vapors that can be ignited under specified laboratory conditions. Flash point determinations give rise to hazard classification systems, the most severe hazard being afforded by those liquids with the lowest flash points. As in the case of gases, however, the principal means of controlling the ignition of combustible liquids is in handling safeguards. Elaborate procedures exist to minimize the escape of flammable vapors in the handling of volatile liquids and to avoid sources of ignition.

If a fire involving such materials is initiated, means are also available to prevent the supply of additional combustible fuel to the fire. These include designs for the venting of storage tanks in order to minimize explosions, and flame arrestors to guard against the travel of the flame into the storage area itself. Storage and separation criteria for bulk storage of flammable liquids are also used to minimize the likelihood that additional supplies of flammable liquids will come in contact with the fire. The intensity of the fire once begun is controlled by: (1) the fuel's volatility, and (2) the amount of heat released when the fuel burns. Thus, heavy oils or tars may be difficult to ignite, but can burn readily once underway. A portion of the heat produced in the flame radiates back to the fuel surface and vaporizes more fuel. For most common organic liquids, the heat required to vaporize a given amount of material is a small percentage of the heat of combustion.

Some fire-fighting techniques interfere with the passage of fuel from the liquid to the vapor phase. These include cooling the liquid to slow vaporization and the use of foam to cover the liquid surface.

Fire Hazards of Solids

The part of Table 4-1A relating to solids concentrates on practical everyday materials; it ignores the large group of chemicals that are solid but which are primarily found in industrial environments. Unless these materials are present as dusts, foams, or in other forms that present high surface areas, their fire hazards are similar to those of liquids.

In Table 4-1A, solid materials are divided into two major classes: (1) flexible materials, such as textiles and cushioning; and (2) structural materials, which can include everything from steel and concrete to wood and synthetic structural plastic foams. For both of these classes of materials there is a variety of tests to determine their susceptibility to ignition. However, since ignition requires the volatilization of some of the solid fuel, ignition behavior is strongly dependent upon the amount of heat applied to the surface. Therefore, different ignition tests often give different results, depending on the size of the ignition source. The same principle applies to tests that attempt to measure flame spread over small samples. In many such tests, a sample receives no radiant heat load except that available from its own burning. The larger the amount of sample burning, the greater the heat transferred to the unburned area

ahead of the flame, and a large sample will appear to have a higher flame spread rate than a small one.

The fire hazards posed by inorganic structural materials are most likely to be passive. For example, steel can lose its strength, concrete can crack and spall, and glass can break and melt on exposure to high temperature. Such materials are therefore rated on their ability to withstand such high-temperature effects.

Flame-Resistant Treatments

Improvements in ignition control of materials have come about as a result of flame-resistant treatments developed for both natural and man-made materials. It is also possible to design polymeric systems in which a substantial fraction of the material is unavailable as fuel. One example is the incorporation of an inert or energy-absorbing filler, such as calcium carbonate or alumina trihydrate, into a plastic. Another example is the inclusion of flame-inhibiting agents, often containing chlorine, bromine, or fluorine, in the monomers from which plastics are made. A third approach is the development of polymers based on materials other than carbon. These include silicone polymers and phosphonitrilics, whose composition is principally phosphorus and nitrogen.

Flame spread tests are probably the best-known fire performance tests. The most widely used of these are the Steiner Tunnel Test, contained in NFPA 255, *Standard Method of Test of Surface Burning Characteristics of Building Materials* (ASTM E84); and the radiant panel test, included in ASTM E162, *Standard Test Method for Surface Flammability of Materials Using a Radiant Heat Energy Source*. These tests attempt to simulate spread of fire across a plane surface and, because flame spread under actual conditions often occurs over preheated surfaces, the tests may include radiant heating of the sample from an imposed external heat flux.

The role of heat release rate in determining fire hazards has recently been recognized. Two standard test methods for heat release rate have been developed: (1) NFPA 263, *Standard Method of Test for Heat and Visible Smoke Release Rates for Materials and Products* (ASTM E906); and (2) ASTM E1354, *Standard Test Method for Heat and Visible Smoke Release Rates for Materials and Products Using an Oxygen Consumption Calorimeter*. These tests measure the rate at which heat is produced, generally as a function of the imposed radiant energy. Because of the importance of an imposed flux in real fire hazard, these measurements promise to be of wide utility in predicting fire hazards. Ideally, the effects of fire-resistant treatments should be detectable as diminished heat release rate measurements.

It is important to recognize that claims for flame resistance are tied closely to the test method that is used to evaluate fire performance. Materials that resist ignition or behave acceptably in small-scale tests may be wholly inappropriate for use when more severe or substantially different fire conditions are encountered. Figure 4-1B, for example, shows that the burning rate depends heavily on imposed heat load (external flux). The relative performance of materials under a radiant heat load of 1 or 2 W/cm^2 may be different from their performance at 5 W/cm^2, an external flux representative of what might be imposed on an exposed surface in a well-developed room fire. This makes it important that materials be tested under conditions that duplicate those expected in real fire scenarios whenever possible.

The steepness of this dependence can be estimated by knowing the ratio of the heat of combustion to the heat required for vaporization: the higher the ratio, the more burning rate will increase with increasing external flux.

Allusions are occasionally made to the hazards of man-made polymers, such as polyethylene, as being similar to those

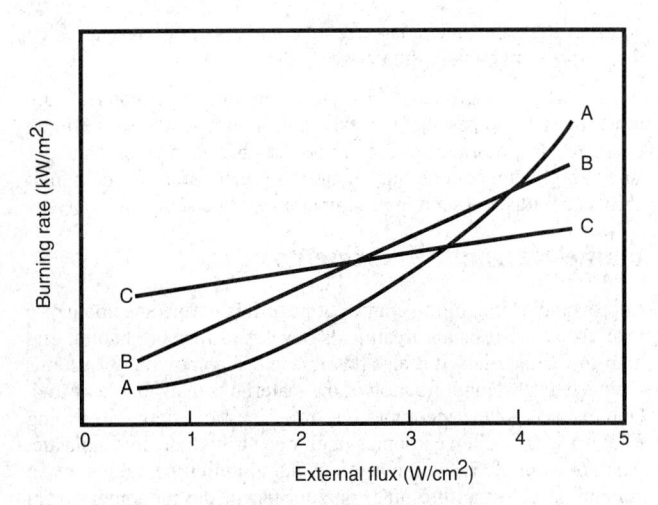

FIG. 4-1B. ***Burning behavior of various materials as a function of external flux (W/cm²).***

of hydrocarbon fuels, such as gasoline, because they both have similar heats of combustion. In fact, however, such polymers require much more heat for vaporization than do the small volatile molecules to which they are compared, so the polymer's burning rate does not increase nearly so rapidly and, in most situations, it burns much more slowly.

SMOKE

Because smoke* is usually the primary threat to life safety, it receives a great deal of attention. Traditionally, smoke has been characterized simply by its ability to obscure light. However, there has been increasing interest in the identity and effects of combustion products themselves, as synthetic materials have become common components of building construction and furnishings. Because of the variability of their chemical composition, smoke from synthetic materials is more likely to contain constituents not often encountered when natural materials were the rule. The real question, however, is not whether new materials produce toxic combustion products—they do—but whether these products represent a kind different from, or a more serious hazard than, those of traditional materials. Although there is no general answer to this question—it depends on the material in question and the use to which it is put—the toxic hazard from most large life-threatening fires is dominated by carbon monoxide, which is produced by virtually all organic materials.[2] The amount of smoke produced and the chemistry of the combustion products (i.e., the smoke's toxicity) are greatly influenced by the manner in which the smoke is generated. The heating rate or combustion temperature of the sample, the availability of oxygen, whether the sample smolders or flames—all of these are important factors.[3]

Although these measurement techniques are far from perfect, their development means that it is becoming possible to predict how a material, burning in a given scenario, will influence the hazard to life from that fire. This emerging field of fire protection, known as fire hazard analysis, requires information on the burning properties of a material, and its conditions of use, as well as smoke obscuration

and smoke toxicity data. Fire hazard analysis is described in more detail in Section 11, Chapter 7, "Fire Hazard Analysis," and Section 11, Chapter 8, "Fire Risk Analysis."

BIBLIOGRAPHY

References Cited

1. Thornton, W. M., *Philosophical Magazine*, Vol. 33, 1917, pp. 196–203.
2. Babrauskas, V., Levin, B., Gann, R., Paabo, M., Harris, R., Peacock, R., and Yusa, S., "Toxic Potency Measurement for Fire Hazard Analysis," NIST Special Publication 827, Dec. 1991, National Institute of Standards and Technology, Gaithersburg, MD.
3. Clarke, F., "Toxicity of Combustion Products: Current Knowledge," *Fire Journal*, Vol. 77, No. 5, 1983, p. 84ff.

NFPA Codes, Standards, and Recommended Practices

Reference to the following NFPA codes, standards, and recommended practices will provide further information on the fire hazards of materials discussed in this chapter. (See the latest version of *The NFPA Catalog* for availability of current editions of the following documents.)

NFPA 49, *Hazardous Chemicals Data*
NFPA 255, *Standard Method of Test of Surface Burning Characteristics of Building Materials*
NFPA 258, *Standard Research Test Method for Determining Smoke Generation of Solid Materials*
NFPA 259, *Standard Test Method for Potential Heat of Building Materials*
NFPA 263, *Standard Method of Test for Heat and Visible Smoke Release Rates for Materials and Products*
NFPA 701, *Standard Methods of Fire Tests for Flame-Resistant Textiles and Films*
NFPA 704, *Standard System for the Identification of the Hazards of Materials for Emergency Response*

Additional Reading

Aronstein, J., "Fire Hazard Due to High Resistance Connections," Consultant, Poughkeepsie, New York, 1990, 9 pages.
ASTM E162, *Standard Test Method for Surface Flammability of Materials Using a Radiant Heat Energy Source*, American Society for Testing and Materials, W. Conshohocken, PA, 1994.
ASTM E1354, *Standard Test Method for Heat and Visible Smoke Release Rates for Materials and Products Using an Oxygen Consumption Calorimeter*, American Society for Testing and Materials, W. Conshohocken, PA, 1994.
Babrauskas, V., "Burning Rates," *The SFPE Handbook of Fire Protection Engineering*, 2nd ed., DiNenno, P. J., *et al.*, eds., National Fire Protection Association, Quincy, MA, 1995.
Babrauskas, V., "Cone Calorimeter—A Versatile Bench-Scale Tool for the Evaluation of Fire Properties," *Proceedings* from the Conference on New Technology to Reduce Fire Losses and Costs, Elsevier, New York, 1986, pp. 78–87.
Babrauskas, V., "Toxic Fire Hazards: Control by Limiting Toxic Potency or Control by Limiting Burning Rate?," Flame Retardants ' 94, Sixth International Conference, January 26–27, 1994, London, England, Interscience Communications Ltd., London, England, 1994, pp. 239–250.
Babrauskas, V., "Toxicity, Fire Hazard and Upholstered Furniture," Third European Conference on Furniture Flammability (EUCOFF' 92), Nov. 1992, Brussels, Interscience Communications Ltd., London, England, 1992, pp. 125–133.
Babrauskas, V. and Peacock, R. D., "Heat Release Rate: The Single Most Important Variable in Fire Hazard," *Fire Safety Journal*, Vol. 18, No. 3, 255–272, 1992.
Beever, Paula, "Self-Heating and Spontaneous Combustion," 2nd ed., DiNenno, P. J., *et al.*, eds., National Fire Protection Association, Quincy, MA, 1995.
Bryan, J., "Damageability of Buildings, Contents and Personnel from Exposure to Fire," *Fire Safety Journal*, Vol. 11, 1986, p. 15.
Bukowski, R. W., "On the Central Role of Fire Calorimetry in Modern Fire Hazard Assessment," National Institute of Standards and Technology, Gaithersburg, MD, DOT/FAA/CT-95/46, AAR-423; National Institute

---

*Smoke is defined here as the total airborne effluent from heating or burning a material. Thus, expressions such as "smoke and toxic gases" are, by this definition, redundant.

of Standards and Technology, Fire Calorimetry, *Proceedings*, July 27–28, 1995, Gaithersburg, MD, 1995, 81 pages.

Cleary, T. G., Ohlemiller, T. J. and Villa, K. M., "Influence of Ignition Source on the Flaming Fire Hazard of Upholstered Furniture," *Fire Safety Journal*, Vol. 23, 1994, pp. 79–102.

Coaker, A. W. and Hirschler, M. M., "New Low Fire Hazard Vinyl Wire and Cable Compounds," *Fire Safety Journal*, Vol. 16, No. 3, 1990, pp. 171–196.

Emmons, H. W., "Analyzing Far Field Effects," *Fire Safety Journal*, Vol. 12, No. 3, 1987, p. 183.

"Fire Hazard Assessment," *Fire Prevention Design Guide*, Fire Protection Association, London, England, FPDG 7, Aug. 1990, 6 pages.

Fire and Smoke: Understanding the Hazards, National Academy Press, Washington, DC, 1986.

Fowell, A. J., "Developments Needed to Expand the Role of Fire Modeling in Material Fire Hazard Assessment," National Institute of Standards and Technology, Gaithersburg, MD, DOT/FAA/CT-93/3; Federal Aviation Administration (FAA), International Conference for the Promotion of Advanced Fire Resistant Aircraft Interior Materials, February 9–11, 1993, Atlantic City, NJ, 1993, pp. 255–262.

Goransson, U., "Model for Predicting the Hazard of Fire Growth," Statens Provingsanstalt, Boras, Sweden, EUREFIC (European Reaction to Fire Classification), International Seminar on Performance Based Reaction to Fire Classification, September 11–12, 1991, Copenhagen, Denmark, Interscience Communications Ltd., London, England, 1991, pp. 15–22.

Goss, J., "Forensic Scientist's Experiments Have Revealed Fire Hazard of Brake Fluids," *Fire*, Vol. 85, No. 1044, June 1992, p. 8.

Hirschler, M. M., "Fire Retardance, Smoke Toxicity and Fire Hazard," Flame Retardants '94, Sixth International Conference, January 26–27, 1994, London, England, Interscience Communications Ltd., London, England, 1994, pp. 225–237.

Hirschler, M. M., "How to Measure Smoke Obscuration in a Manner Relevant to Fire Hazard Assessment: Use of Heat Release Calorimetry Test Equipment," *Journal of Fire Sciences*, Vol. 9, No. 3, May/June 1991, pp. 183–222.

Hirschler, M. M., "Smoke and Heat Release and Ignitability as Measures of Fire Hazard From Burning of Carpet Tiles," *Fire Safety Journal*, Vol. 18, No. 4, 1992, pp. 305–324.

Hirschler, M. M., "Measurement of Smoke in Rate of Heat Release Equipment in a Manner Related to Fire Hazard," *Fire Safety Journal*, Vol. 17, No. 3, 1991, pp. 239–258.

Lyons, J., *The Chemistry and Uses of Fire Retardants*, R. E. Krieger Pub. Co., Malabar, FL, 1987. [Reprint of 1970 edition]

Mulholland, G. W., "Smoke Production and Properties," *The SFPE Handbook of Fire Protection Engineering*, 2nd ed., DiNenno, P. J., *et al.*, eds., National Fire Protection Association, Quincy, MA, 1995.

Nelson, H. E. and Forssell, E. W., "Use of Small-Scale Test Data in Hazard Analysis," *Proceedings* of the Fourth International Symposium on Fire Safety Science, Intl. Assoc. for Fire Safety Science, Boston, MA, 1994, pp. 971–982.

"New Hazard Uncovered—Rolled Nonwoven Fabrics," *Record*, Vol. 68, No. 6, Nov./Dec. 1991, pp. 3–8.

Newman, J.S., "Technological Breakthroughs in Material Flammability Evaluations," *Fire Technology*, Vol. 30, No. 4, 1994, pp. 445–457.

Peacock, R. D. and Bukowski, R. W., "Prototype Methodology for Fire Hazard Analysis," *Fire Technology*, Vol. 26, No. 1, Feb. 1990, pp. 15–40; American Society of Safety Engineers (ASSE) Twenty-ninth Professional Development Conference and Exposition, Building Our Professional Heritage—Protecting Our Nation's Resources, June 24–27, 1990, Washington, DC, American Society of Safety Engineers, Des Plaines, IL, 1990 pp. 20–37.

Prugh, R. W., "Quantitative Evaluation of "BLEVE" Hazards," *Journal of Fire Protection Engineers*, Vol. 3, No. 1, 1991, pp. 9–24.

Purser, D. A., "Toxicity, Toxic Hazard and Furniture in Fires," Paper 8, Second European Conference on Furniture Flammability, June 27–28, 1990, Brussels, Belgium, 1990, pp. 8/1–17.

Quintiere, J. G. and Cleary, T. G., "Fire Growth Hazard Analysis From Property Tests," CFR Video Seminar, National Institute of Standards and Technology, Gaithersburg, MD, Video, July 31, 1990.

Quintiere, J. G., Haynes, G. and Rhodes, B.T., "Applications of a Model to Predict Flame Spread Over Interior Finish Materials in A Compartment," *Journal of Fire Protection Engineering*, Vol. 7, No. 1, 1995, pp. 1–14.

Quintiere, J. G., "Application of Flame Spread Theory to Predict Material Performance," *Journal of Research of the National Bureau of Standards*, Vol. 93, No. 1, Jan.-Feb. 1988, pp. 61–70.

Spearpoint, M. J. and Shipp, M. P., "Measurement of Heat Release From Burning Automobiles for Channel Tunnel Hazard Assessment," Fire Research Station, Borehamwood, England, Federal Aviation Administration, Fire Calorimetry, *Proceedings*, July 27–28, 1995, Gaithersburg, MD, 1995, pp. 135–147.

Tewarson, A., "Fire Properties of Materials for Model-Based Assessments for Hazards and Protection," Chapter 30, ACS Symposium Series 599, 208th National Meeting of the American Chemical Society for Fire and Polymers II: Materials and Tests for Hazard Prevention, August 21–26, 1994, Washington, DC, American Chemical Society, Washington, DC, 1995, pp. 450–497.

Tewarson, A., "Generation of Heat and Chemical Compounds in Fires," *The SFPE Handbook of Fire Protection Engineering*, 2nd ed., DiNenno, P. J., *et al.*, eds., National Fire Protection Association, Quincy, MA, 1995.

Zhang, J., Shields, T. J. and Silcock, G. W. H., "Fire Hazard Assessment of Polypropylene Wall Linings Subjected to Small Ignition Sources," *Journal of Fire Sciences*, Vol. 14, No. 1, Jan./Feb. 1996, pp. 67–84.

COMBUSTION PRODUCTS AND THEIR EFFECTS ON LIFE SAFETY

Gordon E. Hartzell

The thermal decomposition or combustion of every combustible material or product produces a smoke atmosphere that is toxic. In sufficiently high concentrations, these smoke atmospheres present hazardous conditions to exposed humans. Predominant among the hazards are impaired vision due to eye irritation, narcosis from inhalation of asphyxiants, and irritation of the upper and/or lower respiratory tracts.

These effects, often occurring simultaneously in a fire, contribute to physical incapacitation, loss of motor coordination, faulty judgment, disorientation, restricted vision, and panic. The resulting delay or prevention of escape may lead to subsequent injury or death from further inhalation of toxic gases and/or the suffering of thermal burns. Survivors from a fire also may experience postexposure pulmonary (lung) complications that can lead to delayed death.

FIRE AND COMBUSTIBLE MATERIALS

Almost all polymeric materials can undergo pyrolysis and/or combustion. The process of pyrolysis is defined as a "process of simultaneous phase and chemical species change caused by heat," with combustion being defined as "a chemical process of oxidation that occurs at a rate fast enough to produce temperature rise and usually light either as a glow or flame."[1] The processes of pyrolysis and combustion have both physical and chemical properties.[2]

Polymeric materials, upon application of sufficient thermal energy, often first undergo phase change, such as melting in the case of thermoplastics, followed by chemical decomposition. The chemical decomposition processes may involve a number of mechanisms, such as chain scission, chain stripping, and crosslinking. These are endothermic processes resulting in the production of volatile low-molecular-weight products, which may or may not then undergo actual combustion.

If smoke from pyrolysis contains gases that are themselves combustible, combustion may occur if an oxidizing agent (air) and an ignition source are present, provided that they are heated to their piloted ignition temperatures. In order for the process of combustion to be self-sustaining, it is necessary for the burning gases to feedback sufficient heat energy to the material to continue the production of gaseous fuel vapors or volatiles. The process is a continuous feedback loop; heat transferred to the material causes the generation of flammable volatiles; these volatiles react with the oxygen in the air

above the material to generate heat; and a part of this heat is transferred back to the material to continue the process. The chemical oxidation processes for the combustion of organic materials are generally quite exothermic, with more than ample energy being produced to continue the pyrolysis and bond-breaking processes.

It is difficult to generalize the flammability properties of materials, since fire performance is influenced by a number of factors, including the chemical composition and structure of the material, the use of additives in formulated systems, and even the conditions of the fire. Highly crosslinked thermoset polymers normally burn less readily than thermoplastics. Cellular plastics (foams) generally burn quite readily due to their large surface area and good thermal insulating properties, which prevent dissipation of heat. Polymer systems containing halogen [e.g., polyvinyl chloride (PVC)] burn with difficulty; however, the addition of plasticizers increases their propensity to burn. Fire-retardant additives, which include organic-halogen and phosphorus-containing compounds, as well as a number of metal oxides and hydrates, are employed to increase resistance to ignition and/or lower burning rates of polymer systems.[3] However, inherently fire-resistant polymers and polymer systems containing fire-retardant additives will still burn under sufficiently severe thermal conditions.

SMOKE

Smoke is defined as "the airborne solid and liquid particulates and gases evolved when a material undergoes pyrolysis or combustion."[1] It is important in the science of combustion toxicology to first understand the formation of the toxic species present in the smoke produced from the burning of materials.

Carbon Monoxide

Carbon monoxide (CO) is produced from both smoldering and flaming combustion. The production of CO from smoldering fires is a special and very complex combustion phenomenon, the chemistry and physics of which remain insufficiently understood. Compared with flaming combustion, smoldering is a very slow process. Lethal concentrations of CO are generally attained within one to three hours after initiation of smoldering; however, transition to flaming combustion usually occurs much sooner.[4]

The production of CO from the flaming combustion of materials is largely dependent upon the supply of oxygen. Oxygen available to a fire can be limited either by lowering the oxygen

---

Gordon E. Hartzell, Ph.D., is a consultant on toxic hazards in fires, with Hartzell Consulting, Inc., San Antonio, TX.

concentration in the incoming air supply or by a reduction in the volume flow of air to a fire.

The formation of CO is related to the fuel-to-air equivalence ratio,[5] which is defined as:

$$\varphi = \frac{(kg\ fuel\ /\ kg\ air)}{(kg\ fuel/kg\ air)stoich} \qquad (1)$$

where "stoich" denotes conditions at which the ratio between fuel and oxygen is that required for complete combustion, with no excess oxygen. (For U.S. conversion: 1 lb = 0.4536 kg.) Thus, when $\varphi = 1$, exact stoichiometric conditions exist between fuel and air. For values of $\varphi < 1$, the fire is well ventilated; for values of $\varphi > 1$, the fire is fuel-rich and ventilation controlled. These latter conditions are ones that favor the formation of CO.

Studies have shown that, under well-ventilated conditions ($\varphi \ll 1$), yields of CO are very low.[5] These yields remain negligible until φ reaches about 0.5. Thereafter, the yields of CO increase rapidly as values of φ increase, reaching a plateau for most materials in the range of from 0.1 to 0.2 kg CO/kg fuel. (For U.S. conversion: 1 lb = 0.4536 kg.) These yields of CO correspond to an equivalence ratio of somewhat greater than 1.0.

For rooms of moderate size, flashover occurs at about the time when φ reaches 0.5.[5] This is also the point at which CO yields begin to rise dramatically. Thus, a rapid increase in CO yield occurs almost simultaneously with flashover. At flashover, the rate of CO production also increases significantly, since it is dependent on the mass burning rate which increases rapidly at flashover. Thus, the rapid rise in CO production at flashover is due both to the fire becoming ventilation controlled and to the very rapid rise in the mass burning rate. Once φ reaches a value greater than 1.0, the CO yield becomes rather constant, with the production rate being dependent only on the mass burning rate. For a sufficiently large fire whose flame radiative heat flux and mass generation rate of material vapor have approached their asymptotic values, the mass burning rate remains rather constant and, thus, the CO production rate is also constant.[6] These phenomena have been observed with large-scale fire tests involving fully furnished rooms.[7]

Carbon Dioxide

The ratio of carbon dioxide/carbon monoxide (CO_2/CO), often used as a descriptive characteristic of a fire, also depends more upon the ventilation conditions of the fire than on the nature of the materials being burned.[5] Studies on the dependence of CO_2/CO ratios on the equivalence ratio show that, for well-ventilated fires, i.e., $\varphi \ll 1$, essentially all the fuel carbon is oxidized to CO_2. Once the equivalence ratio has exceeded that associated with flashover ($\varphi > 1$), indicating a fuel-rich or ventilation-controlled fire, two plateaus are seen. For fuels that do not contain oxygen in the molecule, the CO_2/CO ratio reaches a plateau at about 20, while for those containing oxygen in their molecular structure, the ratio reaches a plateau at about 12.5. Thus, there appears to be a small fuel chemistry effect when considering postflashover fires.

Hydrogen Cyanide

The generation of hydrogen cyanide (HCN) is both material dependent and temperature dependent. Only nitrogen-containing materials yield HCN, with relatively high temperatures being required. In contrast to CO, there have been insufficient studies on HCN to enable its formation in fires to be quantified. If sufficient oxygen is present, low concentrations of oxides of nitrogen (NO_x) may also be formed from nitrogen-containing materials.[8] Although one study reported NO_x production from nitrogen-containing fuels to be far less than that for HCN, there are conflicting data in the literature.

Halogen Acids

Polymer systems containing halogen (fluorine, chlorine, or bromine) result in the formation of the halogen acids (HF, HCl, and HBr), the production of which is largely material dependent as long as thermal decomposition temperatures are reached. For example, HCl is evolved quantitatively from polyvinyl chloride (PVC) at temperatures of about 437 to 527°F (225 to 275°C).[2] The halogen acids are formed in the pyrolysis component of the combustion process and are not oxidized further. Thus, the halogen acids are produced even if flaming combustion does not occur. Since the production efficiencies for the formation of HF, HCl, and HBr are close to being quantitative, maximum yields would be expected in fires. It should be noted, however, that halogen acid concentrations decay rather quickly in the presence of sorptive surfaces[9] and water droplets present in most fire effluents at temperatures below 212°F (100°C). Thus, maximum possible concentrations are rarely actually encountered.

Organic Irritants

Pyrolysis and/or incomplete combustion of organic materials can lead to a wide variety of organic irritant species. The most important of these are formaldehyde, unsaturated aldehydes (especially acrolein), and isocyanates (from polyurethanes). Acrolein, in particular, has been demonstrated to be present in many fire atmospheres.[10] It is formed from the smoldering of all cellulosic materials and from the pyrolysis of polyethylene.[11]

Combustion Physics

In addition to chemical processes, the physics of combustion must also be considered in order to fully appreciate the rapid development of toxic smoke atmospheres in a fire. Fire size is generally described in terms of its rate of energy or heat release in kilowatts (kW). A small fire confined to an object of origin is one of about 30 to 100 kW, while a large fire progressing beyond its compartment of origin (flashover) is perhaps in excess of 1000 kW (one megawatt). The mass loss rate of a fire (i.e., the rate at which smoke is produced) is related to its rate of heat release through the fuel's effective heat of combustion. Using a typical effective heat of combustion for many polymeric materials of about 25 MJ•kg•s^{-1}, a fire size of about 1000 kW translates to a mass loss rate in the order of 0.04 kg•s^{-1}. (For U.S. conversion: 1 lb = 0.4536 kg.) This mass loss rate represents approximately 254 oz (7200 g) of smoke being evolved from a 1000 kW fire in about 3 min. If the smoke is assumed to be of only "average" toxicity, it could be fatal to humans exposed in a typical 15- × 18- × 8-ft room (approximately 60 m^3) within just a few minutes. Incapacitation might well occur within only 2 to 3 min. The toxicological effects of smoke produced from fires usually involve relatively high concentrations of toxicants inhaled over quite short periods of time. Thus, many criteria commonly used in conventional toxicology to describe hazardous conditions, such as threshold limit value (TLV) and industrial long-term exposure value, are normally of little use in combustion toxicology.

TOXICITY OF SMOKE

Fire gas toxicants are usually considered as belonging to one of three basic classes: (1) asphyxiants or narcosis-producing toxicants; (2) irritants, which may be sensory/upper respiratory or pulmonary; and (3) toxicants exhibiting other or unusual effects. Although always a possibility, the third class has few documented examples.

Asphyxiants

In combustion toxicology, the term "narcosis" refers to the effects of asphyxiant toxicants that are capable of resulting in central nervous system depression, with loss of consciousness and ultimately death. Effects of these toxicants depend upon the accumulated doses, i.e., both concentration and the time or duration of the exposure. The severity of the effects increases with increasing doses.

Carbon monoxide: The toxic effects of carbon monoxide are those of anemic hypoxia.[12] Hypoxia refers to the condition in which there is an inadequate supply of oxygen (O_2) to body tissue, with anemic hypoxia being characterized by a lowered oxygen-carrying capacity of the blood even when the arterial partial pressure of O_2 and the rate of blood flow are normal. This is due to competition between O_2 and CO for the heme-binding sites of hemoglobin, with the affinity of hemoglobin for CO being over 200 times greater than for O_2. Even partial conversion of hemoglobin to carboxyhemoglobin (COHb) significantly reduces the oxygen-transport capability of the blood such that serious toxic signs and symptoms are produced. Additionally, partial conversion of hemoglobin to COHb causes oxygen that is bonded as oxyhemoglobin (HbO_2) to be more tightly held and less available to body tissues. The extent to which blood hemoglobin is converted to COHb can readily be measured in a clinical laboratory and is expressed as percent COHb saturation.

The signs and symptoms produced by exposure to CO are directly related to the percent of blood hemoglobin that is converted to COHb.[13] Deaths of humans exposed to CO produced in fires have been associated with COHb saturations ranging from 1 percent to 99 percent.[14] Although most deaths have been reported to be in the 50 percent to 70 percent COHb range,[15,16] both lower and higher values are not unusual. Low COHb values may be due to preexisting physiological conditions, such as ingestion of alcohol or medication, pulmonary insufficiency, cardiovascular disease, etc., which may have already compromised an individual. The high oxygen demand resulting from physical exertion may also cause a person to succumb to low COHb saturation. High COHb values at death may be attributed to the low oxygen demand of a person either at rest or at quite low physical activity. There is no specific COHb saturation below which one would be expected to live and above which death would occur. As with essentially all biological responses, COHb saturations associated with death are best represented as a statistical distribution.

Binding of O_2 and CO with hemoglobin occurs in equilibria, with both association and dissociation taking place.[12] Thus, CO and hemoglobin are in equilibrium with COHb, with the concentration of COHb being dependent on the partial pressure (or concentration) of CO in the atmosphere. Each concentration of CO is associated with an equilibrium COHb saturation, or, stated another way, for any given value of COHb saturation there is a minimum concentration of CO that will attain it. At any given O_2 concentration, higher equilibrium COHb saturations can be obtained only by increasing the concentration of CO.

The competition between O_2 and CO for bonding sites on hemoglobin is defined by the Haldane equation[12]:

$$\frac{(CO\,Hb)}{(Hb\,O_2)} = M \frac{(P_{CO})}{(P_{O_2})} \qquad (2)$$

where

$(COHb)/(HbO_2)$ = ratio of carboxyhemoglobin to oxyhemoglobin,

$(P_{CO})/(P_{O_2})$ = ratio of the respective partial pressures (or concentrations), and

M = constant that is, to some extent, species dependent.

The value of M for humans is about 245. The Haldane equation is commonly used to calculate the CO concentration that would be in equilibrium with any given COHb saturation.

The time required for a human subject to attain a given COHb saturation can be approximated from the Stewart-Peterson equation.[17]

$$\% CO\,Hb = (3.317 \times 10^{-5})(ppm\,CO)^{1.036}(RMV)(t) \qquad (3)$$

where

ppm CO = CO concentration (ppm),

RMV = respiratory minute volume (L/min), and

t = exposure time (min).

At rest, the respiratory minute volume (RMV) for humans averages approximately 8.5 L/min,[15] but may increase during strenuous activity to as much as 100 L/min. Derived from the Stewart-Peterson equation, curves showing the relationship between CO concentration and time of exposure for various COHb saturation values for humans can be constructed.[16]

Studies using rats have shown that elevated temperatures have an exacerbating effect on CO intoxication,[18] possibly through increased rate of respiration. This is likely to occur with humans, as well.

Hydrogen cyanide: Hydrogen cyanide (HCN), approximately 25 times more toxic than carbon monoxide, is a very rapidly acting toxicant.[19] The action of HCN is due to the cyanide ion, which is formed by hydrolysis in the blood. Unlike CO, which remains primarily in the blood, cyanide ions are distributed throughout the body water and make contact with the cells of tissues and organs. Cyanide ions readily react with the enzyme cytochrome oxidase to form a cytochrome oxidase-CN complex, and also with methemoglobin to form cyanomethemoglobin. If the concentration of cyanide ions is not sufficiently great as to cause death, the ions are slowly released from the complexes with cytochrome oxidase and methemoglobin and converted to thiocyanate ions by the enzyme rhodanese. This detoxification process, along with the fact that cyanide is distributed throughout the body, makes correlation of blood cyanide content with actual exposure to HCN difficult.

Cytochrome oxidase occupies a central role in the utilization of oxygen in practically all cells. Its inhibition rapidly leads to loss of cellular functions (cytotoxic hypoxia), then to cell death. In contrast to CO, cyanide ion does not decrease the availability of oxygen but, rather, prevents the utilization of oxygen by cells. The heart and brain are particularly susceptible to this inhibition of cellular respiration, with bradycardia, cardiac arrhythmia, and EEG brain wave activity indicative of central nervous system depression having been reported in studies using monkeys.[20] Although cardiac irregularities are often noted in HCN intoxication, the heart invariably outlasts respiration, and death is usually due to respiratory arrest of central nervous system origin.

A complication resulting from the inhalation of HCN is one of respiratory stimulation or hyperventilation,[15] presumably due to the body's regulating chemoreceptor cells responding to a need for oxygen. This causes a more rapid uptake of not only HCN but also other toxicants that may be present. The hyperventilatory effect of

HCN may also account for its rather steep dose-response relationship.

Data relating symptoms in humans to various concentrations of HCN are very limited. One widely referenced descriptive account of hydrogen cyanide intoxication of humans reports that 50 ppm may be tolerated for 30 to 60 min without difficulty, 100 ppm for that same period is likely to be fatal, 130 ppm may be fatal after 30 min, and 181 ppm may be fatal after 10 min.[21]

The role of hydrogen cyanide as a causative agent in human fire fatalities is considerably less clear than that of carbon monoxide. Documented cases are rare in which hydrogen cyanide alone can be shown to be the primary toxicant. Blood can be analyzed for cyanide, but the procedure is more complex than that for carbon monoxide, and the reliability of the results is dependent on proper blood storage conditions prior to analysis. Analyses should be interpreted with caution, with consideration given to the analytical method used, the blood storage history, and the reputation of the responsible laboratory. It is generally held that blood cyanide concentrations greater than 1.0 µg/mL are indicative of possibly significant toxicological effects due to inhalation of hydrogen cyanide.[14,22] Blood cyanide levels greater than 3.0 µg/mL are generally considered to be lethal. Significant concentrations of blood cyanide are normally found associated with carboxyhemoglobin saturation due to inhalation of CO.[22,23] The contribution of each to death often cannot be determined with confidence.

There is no evidence for synergism between hydrogen cyanide and carbon monoxide; however, these two toxicants are generally agreed to be additive in their effects.[24,25]

Carbon dioxide: Carbon dioxide (CO_2) is quite low in its own toxicological potency and is not, by itself, normally considered to be significant as a toxicant in combustion toxicology. However, it does stimulate both the rate and depth of breathing, thereby increasing the respiratory minute volume (RMV). The RMV, increasing about 50 percent with only 2 percent CO_2, may be as much as 8 to 10 times normal in the presence of 10 percent CO_2. Such respiratory stimulation causes accelerated inhalation of toxicants leading, for example, to an increased rate of formation of blood COHb from inhalation of CO. However, the same equilibrium COHb saturation of blood is reached as in the absence of CO_2.[26]

Increased incidence of lethality, particularly postexposure, of rats has also been reported with combinations of CO and CO_2.[27] This effect may be associated with the combined insult of respiratory acidosis (caused by CO_2) and metabolic acidosis (caused by CO), a condition from which the rodent has difficulty recovering postexposure. Whether or not this latter effect of CO_2 also occurs with primates has not been determined.

Oxygen depletion: Since oxygen is consumed in the combustion process, oxygen depletion or vitiation must also be considered a toxic component of smoke. When oxygen drops from its usual level of 21 percent in air to approximately 17 percent, a person's motor coordination is impaired. When oxygen drops into the range of 14 to 10 percent, a person is still conscious but may exercise faulty judgment and will be quickly fatigued. In the range of 10 to 6 percent oxygen, a person loses consciousness and must be revived with fresh air or oxygen within a few minutes to prevent death.[28] During periods of exertion, increased oxygen demands may result in oxygen deficiency symptoms at higher oxygen levels.

Irritants

Irritant effects, produced in essentially all fire atmospheres, are normally considered by combustion toxicologists as being of two types: (1) sensory irritation, including irritation of the eyes and the upper respiratory tract; and (2) pulmonary irritation affecting the lungs. Most fire irritants produce signs and symptoms characteristic of both sensory and pulmonary irritation.

Eye irritation, an immediate effect that depends primarily upon the concentration of an irritant, may significantly impair escape of a victim from a fire situation. Nerve endings in the cornea are stimulated, causing pain, reflex blinking, and tearing. Severe irritation may also lead to subsequent eye damage. Victims may shut their eyes, partially alleviating these effects; however, this action may also impair their escape from a fire.

Airborne irritants enter the upper respiratory tract, causing burning sensations in the nose, mouth, and throat, along with the secretion of mucus. These sensory effects are also primarily related to the concentration of the irritant and do not normally increase in severity as the exposure time is increased.

Following signs of initial sensory irritation, significant amounts of inhaled irritants may also be quickly taken into the lungs, with the symptoms of pulmonary or lung irritation being exhibited. Lung irritation is often characterized by coughing, bronchoconstriction, and increased pulmonary flow resistance. Tissue inflammation and damage, pulmonary edema, and subsequent death may follow exposure to high concentrations, usually within 6 to 48 hrs. Inhalation of pulmonary irritants also appears to increase susceptibility to postexposure bacterial infection. Unlike sensory irritation, the effects of pulmonary irritation are dependent both on the concentration of the irritant and on the duration of the exposure.

Both inorganic irritants (e.g., halogen acids, nitrogen oxides) and organic irritants (e.g., aldehydes) can be formed in fires.

Halogen acids: The halogen acids have received particular attention, with the most important being hydrogen chloride (HCl) resulting from the decomposition of polyvinyl chloride (PVC). Hydrogen chloride is both a potent sensory irritant and also a strong pulmonary irritant. Concentrations as low as 75 to 100 ppm are extremely irritating to the eyes and the upper respiratory tract, suggesting possible impairment of physical activity, such as escape.[29] However, hydrogen chloride was found not to be physically incapacitating to baboons subjected to concentrations up to 17,000 ppm for 5 min.[30] Postexposure deaths were reported from doses that did not appear to incapacitate. Comparable studies have not been conducted using actual PVC smoke.

Another study, using baboons exposed for 15 min to PVC smoke containing 5000 ppm HCl, did not show any smoke-related impairment of pulmonary function when tested at 3, 90, 180, and 360 days postexposure.[31] Lesions to mucosal surfaces of the mouth were observed, however, along with evidence of bronchoconstriction and lowered PaO_2 levels during the exposure.

Considerable controversy has existed as to what concentration of HCl is hazardous to humans. It is questionable whether lethality data from numerous studies with rodents can be directly extrapolated to humans because of significant anatomical differences between the respiratory tracts of the rodent and the primate which, in turn, result in very different responses to HCl. Rodents exhibit a decrease in respiratory rate and minute volume upon exposure to HCl, whereas these values have been shown to increase for primates.[32] Interestingly, exposure doses (concentrations × time) of HCl causing postexposure lethality in rats are actually in the same range as those that have resulted in postexposure deaths of baboons.[33] However, the data for baboons are very limited and the comparison made is rather subjective. It is prudent to recognize that HCl is likely to be dangerous to humans at concentrations well below those indicated by its lethal toxic potency. Hydrogen chloride would not appear to be physically incapacitating, nor would it seem to produce significant chronic respiratory complications after short exposures to concentrations of 5,000 ppm and below. Although acute symptoms, ranging from severe discomfort to noticeable upper respiratory tissue damage, may be experienced, significant pulmonary function changes would not be expected, except at higher concentrations. Because of the effects of bronchoconstriction and lowered PaO_2 levels resulting from inhalation of HCl, persons having preexisting compromised pulmonary function may be more susceptible than healthy individuals.

Fire-retardant additives based on chlorine or bromine are also a source of halogen acids in fires, with fluoropolymers being the major source of HF. Although studies using the other halogen acids, HF and HBr, have been very limited, it would appear that these acid gases exhibit irritant effects similar to those of HCl at comparable concentrations.[15]

Nitrogen oxides: Nitrogen dioxide (NO_2) and nitric oxide (NO) are the major components of a mixture of nitrogen oxides usually referred to as NO_x. Nitric oxide is only about one-fifth as potent as NO_2[8] Studies on rats exposed to NO_2 under the conditions of smoke toxicity test methods indicate NO_2 to have a lethal toxic potency comparable to HCN.[34] In contrast to HCN, the toxicity of NO_x is primarily due to its properties as a pulmonary irritant, with lethality of rats being postexposure, usually within one day.

Organic irritants: Although numerous organic irritants may be produced in fires, only acrolein has received significant attention in combustion toxicology. Concentrations of acrolein as low as a few parts per million are extremely irritating to the eyes and upper respiratory tract. Interestingly, studies with baboons showed that concentrations up to 2,780 ppm for 5 min were not physically incapacitating during exposure.[30] Pulmonary complications caused by that and even lower concentrations resulted in death within hours after the exposure, however.

In the absence of other information, a mass loss of organic material produced from the burning of 1 g for each m^3 of volume has been proposed as potentially incapacitating due to sensory irritation, with a mass loss of 10 g for each m^3 presenting highly dangerous conditions with regard to pulmonary irritation. (For U.S. conversions: 1 oz = 28.36 g; 1 cu ft = 0.0283 m^3.)[15]

QUANTIFICATION OF TOXIC POTENCY

The standard in combustion toxicology for quantifying the toxic potency of individual fire gases or of smoke is generally regarded to be the LC_{50} for 30 min exposure of rats.[35,36] The LC_{50} is defined as "the concentration of toxic gas in ppm or smoke in g/m^3 statistically calculated from concentration-response data to produce lethality in 50 percent of test animals within a specified exposure and postexposure time." (For U.S. conversions: 1 oz = 28.36 g; 1 cu ft = 0.0283 m^3.) It is significant that consensus among experts has been to recognize that rats are reasonably acceptable animal models for human exposure to smoke when the principal effects are due to inhalation of asphyxiant toxicants.[16,37]

Over the past 15 years, LC_{50} values have been experimentally determined for individual fire gases, as well as for many materials and products. This extensive data bank has become extremely useful in the development of a strategy for the calculation of smoke LC_{50} values from combustion analytical data without the need for exposure of animals. This strategy is commonly referred to as fractional effective dose (FED) methodology.[24,38,39] The FED is "the ratio of the Ct product (concentration × time) for a gaseous toxicant produced in a given test to that Ct product of the toxicant that has been statistically determined from independent experimental data to produce lethality in 50 percent of test animals within a specified exposure and postexposure time."[36]

Although individual fire gas toxicants may exert quite different physiological effects through different mechanisms, when present in a mixture each may result in a certain degree of compromise experienced by an exposed subject. It should not be unexpected that varying degrees of a partially compromised condition may be roughly additive in contributing to incapacitation or death. This has been demonstrated in a number of studies using rodents. As a result, it is now fairly well agreed that the asphyxiants carbon monoxide and hydrogen cyanide are additive when these toxicants are represented as fractional effective doses.[24,25] Furthermore, analysis of data from experiments exposing rats to mixtures of hydrogen chloride and carbon monoxide, and also mixtures of hydrogen chloride and hydrogen cyanide, suggest that FEDs leading to lethality may also be approximately additive.[40,41]

The additivity of FEDs has become a useful tool in combustion toxicology for the calculation of toxic potencies from combustion product analysis data, as well as for the assessment of potential toxic hazards to humans.[42] Mathematically, the principle is expressed as

$$FED = \sum_{i=1}^{n} \int_{0}^{t} \frac{C_i}{(Ct)_i} \Delta t \qquad (4)$$

where

C_i = concentration of the toxic component "i," and

$(Ct)_i$ = specific exposure dose (concentration × time)

required to produce the toxocological effect.

When the accumulated FED = 1, it is expected that the mixture of gaseous toxicants would be lethal to 50 percent of exposed animals, i.e., the LC_{50}. Use of this principle has also been termed the "N-Gas Model" by the National Institute of Standards and Technology (NIST).[25,42] Examination of data from a series of pure gaseous toxicant experiments in which various percentages of animals died indicated that the mean FED value corresponding to the LC_{50} was 1.07, using the "N-Gas" calculation, with 95 percent confidence limits of ± 0.20.[43]

TEST METHODS FOR TOXICITY OF SMOKE

Tests for the toxicity of smoke produced by a burning material rely on determination of the lethal toxic potency, which is based on concentration/response relationships obtained by exposing groups of rodents over a fixed time to different concentrations of smoke. This is accomplished by conducting a series of experiments in which the quantity of combusted material or the flow rate of diluting air is varied in order to produce different concentrations. The number of animals that die increases as the exposure concentration is increased. Recent test methods, now adopted by American Society for Testing and Materials (ASTM),[36] National Fire Protection Association (NFPA),[36] and International Standards Organization (ISO),[35] minimize or even eliminate the use of animal exposures by using FED methodology to calculate LC_{50} values from analytical data.

Smoke toxicity test data are generally reported as 30-min LC_{50} values in units of grams of material per unit exposure chamber volume or total airflow, i.e., g/m^3. (For U.S. conversions: 1 oz = 28.36 g; 1 cu ft = 0.0283 m^3.) The weight of material may be expressed either as that subjected to the test conditions or as that consumed in the test. (These values will be identical if the specimen is completely combusted with no residue.) However, one test method, i.e., UPITT, commonly used in the United States, expresses the LC_{50} only as grams of material subjected to the test protocol. Thus, the LC_{50} reported with the UPITT test is not a true expression of concentration. Care should be taken in comparing LC_{50} values obtained from different test methods to ensure that the basis for reporting data is the same.

The major test methods for assessing materials with regard to the toxic potency of combustion products can be summarized as follows.

NBS

The NBS test is a static or closed system using a cup-type furnace as the combustion device.[44] Data for both nonflaming and flaming combustion are obtained by conducting tests just below and just above the autoignition temperature of the specimen, respectively. Concentration-response (lethality) relationships for 30-min exposures (14-day postexposure observation) are determined using rats held in tubular restrainers for head-only exposure. Six rats are used per test. Reported LC_{50} values are in units of g/m^3 and are based on specimen weight charged to the cup. (For U.S. conversions: 1 oz = 28.36 g; 1 cu ft = 0.0283 m^3.)

The NBS test, commonly used in the United States during the 1980s, has largely been abandoned in favor of a newer NIST test method,[5] now adopted, in part, as ASTM E1678.[36] Considerable data on common generic materials were obtained over the years during which the NBS test was used. These data are still often referenced and used in the assessment of toxic hazards.

UPITT

The UPITT test (University of Pittsburgh) exposes four mice per test to fire effluents produced in a dynamic (flow-through) system from the 68°F (20°C) per min ramped heating of a specimen in a box or muffle furnace.[45] Exposures are for 30 min (plus 10 min postexposure), starting when 1 percent of the sample weight has been lost. Bioassays include concentration-response and time-to-death for lethality. In the UPITT test, the concentration term in the LC_{50} refers only to the weight of test specimen charged to the furnace, rather than to actual concentration. Thus, LC_{50} values obtained from this test are not directly comparable to those from other methods. This test is currently required in the United States by the State of New York for certain construction and finish materials. No acceptance levels or criteria for classification of materials have ever been set.

LC_{50} values have been reported and filed with New York State for over 15,000 products.[46] It is of interest that 63.3 percent of the LC_{50} values fall between 5 and 12.5 g, with another 32.7 percent being in the range of 12.5 to 28.1 g. (For U.S. conversion: 1 oz = 28.36 g.) Thus, only about 4 percent of products reported are outside about a six-fold range of LC_{50} values.

DIN 53436

The DIN 53436 Test (German Standards Institution) is characterized by the use of a moving annular tube furnace operating at a constant temperature in the range of 392 to 1832°F (200 to 1000°C).[47] The furnace is programmed to travel the length of a quartz tube containing the sample. Decomposition, taking place in an air stream counter current to the direction of furnace travel, is intended to result in the continuous flow of fire effluents of constant composition. Radiant heat is the major source of energy transfer. This test device, used in six European countries, offers a rather wide range of well-controlled combustion conditions. Ratios of CO_2/CO can be made to vary widely, depending on the airflow rate used; thus, both freely ventilated and ventilation-limited fires may be simulated. Specimens may undergo either flaming or nonflaming combustion, depending on the imposed heat flux level and/or the presence of an ignition device.

Combustion product analytical data obtained from the DIN 53436 test have been used for the calculation of LC_{50} values using the FED principle, as well as the N-Gas method described in connection with the ASTM E1678 test.[48] Comparisons between calculated and experimental LC_{50} values for over 30 materials tested revealed fairly good correlation and predictability ($r > 0.8$), giving additional credibility to the concept that lethal toxic potency can largely be attributed to a small number of major toxic gases.

ASTM E1678/NFPA 269

With the experimental details originally developed jointly at NIST[5] and at Southwest Research Institute,[49] this test method subjects a specimen to ignition while exposed for 15 min to a radiant heat flux of 50 kW/m^2.[36] The smoke produced is collected for 30 min within a 53-gal (200-L) chamber communicating through a connecting chimney with the combustion assembly. Concentrations of the major gaseous toxicants, CO, CO_2, and O_2, and, if present, HCN, HCl, and HBr, are monitored over the 30-min period.

The lethal toxic potency, LC_{50}, of the test specimen is predicted from the combustion atmosphere analytical data by first calculating the fractional effective dose (FED) for the test. Using Equation 5, the LC_{50} is then calculated as that specimen mass loss that would yield a FED = 1 within a chamber volume of 35 cu ft (1 m^3).

The 30-min LC_{50} for a test specimen is calculated from equation 5.

$$LC_{50} = \frac{\text{specimen mass loss}}{\text{FED} \times \text{chamber volume}} \quad (5)$$

where the specimen mass loss is in grams, and the chamber volume is 0.2 m^3. The resulting LC_{50} has the units of g/m^3. (For U.S. conversions: 1 oz = 28.36 g; 1 cu ft = 0.0283 m^3.)

The predicted LC_{50} is then confirmed in limited tests, using rats, to ensure that the monitored toxicants account for the observed toxic effects.

The experimental procedures of ASTM E1678, *Standard Test Method for Measuring Smoke Toxicity for Use in Fire Hazard Analysis*, and NFPA 269, *Standard Test for Developing Toxic Potency Data for Use in Fire Hazard Modeling*, are identical. The two standards differ only in that NFPA 269 provides a mathematical adjustment of experimental LC_{50} values, in order to make them specifically appropriate for the toxic hazard assessment of oxygen-vitiated, post-flashover fires.

ISO 13344

This method[35] subjects a test specimen to the combustion conditions of a specific laboratory combustion device, chosen to have demonstrated relevance to one or more of the specific classes or stages of fires.[50] This provision allows for use of the DIN 53436 tube furnace in Europe and use of the radiant furnace/53-gal (200-L) exposure system of ASTM E1678 in the United States. Concentrations of the major gaseous toxicants in the smoke atmosphere are monitored over a 30-min period, with Ct products for each being determined from integration of the areas under the respective concentration-time plots. The Ct product data, along with either the mass charge or the mass loss of the test specimen during the test, are then used in FED calculations to predict the 30-min LC_{50} of the test specimen. The resulting calculated LC_{50} has the units of g/m^3. (For U.S. conversions: 1 oz = 28.36 g; 1 cu ft = 0.0283 m^3.)

If considered necessary, the calculated LC_{50} may then be experimentally confirmed as precisely as toxicologically relevant. Confirmation ensures that the monitored toxicants account for the observed toxic effects. ISO 13344 makes optional any bioassay confirmation of the calculated LC_{50}. A purpose of this provision is to minimize or even avoid the use of live animals for testing purposes, unless absolutely necessary and properly justified under various countries' regulations.

SIGNIFICANCE OF SMOKE TOXICITY TEST DATA

Combustion toxicology in the 1970s was clouded with considerable controversy over issues such as the probability of "super toxicants," whether or not to assess incapacitation, the validity of data obtained from tests using rodents in assessing toxicity to humans, the precision and accuracy of smoke toxicity data, the relevance of laboratory-scale tests to real fires, and, ultimately, how such data should be used.

The issue of potential "super toxicants" in fire atmospheres has, with more experience in the testing of materials, been generally dismissed since there have been few documented examples. One involved the formation of a neurotoxin from the thermal decomposition of a noncommercial rigid polyurethane foam,[51] while another concerned the unusual toxic potency exhibited by polytetrafluoroethylene in some laboratory tests. The latter case would appear to be largely an artifact of the test method.[52] It is now widely accepted that smoke toxicity can, to a large extent, be explained both qualitatively and quantitatively on the basis of a small number of toxic gases.[53] Experience has shown that the LC_{50} values for most polymeric materials fall within a range of approximately one order of magnitude from about 5 g/m^3 to 60 g/m^3.[15] (For U.S. conversions: 1 oz = 28.36 g; 1 cu ft = 0.0283 m^3.) Values for most varieties of wood and other cellulosic materials also fall within that range.

Although considerable attention was once directed toward incapacitation studies, current smoke toxicity tests make no such assessment. Incapacitation is simply inferred from lethality data, with combustion toxicologists generally regarding incapacitating exposure doses to be about one-third to one-half of those required for lethality. Although tests have been developed for assessment of incapacitation due to sensory irritation,[54] they have not received significant use. This is due, at least in part, to a lack of validation between the tests and actual effects on humans.

The propensity for smoke to have the same effects on humans in fires can only be inferred to the extent that the rat or mouse can be correlated with humans as a biological system. As a result of comparisons of rodent data with limited human and non-human primate data, the use of rats as an acceptable model for human exposure, at least to the asphyxiant gases, has received general endorsement among smoke toxicologists.[37] Although the 30-min exposure time for animals in laboratory smoke toxicity tests is considered by some to be rather long compared to most human exposures to smoke in fires, reasonable extrapolation of toxic potency test data for shorter exposures can be made through the use of Haber's Rule (concentration × time = constant).

It was once felt that every polymeric material and formulation would need to be tested, since the toxicity of its smoke could well be unique to the material, i.e., highly material-dependent. However, the formation of the major toxicants in smoke has now been recognized to be often more dependent on the conditions of the fire, i.e., highly fire-dependent. With the production of common toxicants having both material- and fire-dependent aspects, some doubt is cast on the significance of data obtained using laboratory tests that may not simulate real-fire conditions. It is important to recognize that the results obtained with smoke toxicity tests relate only to the specific conditions of the test itself and may not be relevant to *any* real-scale fire.

A special problem exists with CO as the major toxicant in most real-scale fires. It is one of changing ventilation or air supply throughout the course of a fire. Combustion consumes oxygen from the air, causing increasingly vitiated conditions that favor increased production of carbon monoxide. These conditions are generally not attained in small-scale laboratory tests. A correction for CO production under post-flashover conditions has been proposed to enable laboratory LC_{50} values to be used in real-fire toxic hazard assessment.[5]

Toxic hazard, most commonly defined as the potential for harm from toxic products of combustion,[1] can, at times, be the primary concern in a fire. At other times, it could be irrelevant. Smoke toxicity test data are not, in themselves, an indication of toxic hazard, or even relative toxic hazard. However, they may be used as an element of fire hazard assessment which takes into account all of the factors that are pertinent to an assessment of the fire hazard of a particular end use.[36]

Toxic Hazards in Fires*

After about 20 years of studies in combustion toxicology, it has been observed that researchers have perhaps been overly obsessed with smoke toxicity test methods for materials and products and may have lost sight of the real objective—life safety of humans. The science of fire safety engineering now addresses this objective by providing for fire hazard analysis using a variety of fire parameters, including smoke toxicity, in mathematical models.

The key element in the analysis of threat of life due to toxic fire hazards involves the estimation of the occupant's condition (e.g., incapacitation) at each increment of exposed time, i.e., whether or not safe escape is possible. In doing so, the effects of asphyxiant toxicants and sensory/upper respiratory irritants must be considered separately. This is because effects due to asphyxiants are related to their accumulated dose (concentration × time) whereas effects due to sensory/upper respiratory irritation are related only to their concentration.

The asphyxiant component of toxic hazard analysis can take either, or both, of two forms, depending upon the nature of available input data.[55] The first of these is termed a toxic gas model which requires as input the concentration-time profile of each asphyxiant gas present, along with an effective concentration-time value expected to cause incapacitation by each. This model is most useful in consideration of fires where concentration-time data can either be measured or be estimated from fire growth models. Alternatively, a mass-loss model may be used. The mass-loss model requires as input the rate of mass loss of materials or products in a fire, along with effective concentration-time values expected to cause incapacitation due to smoke produced from those materials or products. These latter values are derived from laboratory test methods (e.g., ASTM E1678, NFPA 269, ISO 13344, *Determination of the Lethal Toxic Potency of Fire Effluents*). Estimations of mass-loss rate can often be made from fire size and energy considerations.

The basic principle for modeling the asphyxiant component of toxic hazard involves the same FED concept used to calculate LC_{50} values from analytical data.[39] However, rather than determining the 30-min-accumulated FED for an exposure by integrating from 0 to 30 min, the time at which the accumulated FED value becomes equal to 1 is determined. This time represents the mode, or the most frequently expected time for incapacitation to occur.

---

*For more information on the techniques of fire and risk analysis, see Section 11, Chapter 7, "Fire Hazard Analysis," and Section 11, Chapter 8, "Fire Risk Analysis."

The form of Equation 6 for hazard fatality due to asphyxiants is

$$FED = \int_0^t \frac{[CO]}{35000} dt + \int_0^t \frac{\exp([HCN]/43)}{220} dt \qquad (6)$$

where CO and HCN concentrations are in ppm. Equation 6 is derived from actual incapacitation experiments using nonhuman primates.[15,30] It is most applicable when oxygen concentrations are not lower than 13 percent, and CO_2 concentrations do not exceed 2 percent.

The sensory/upper respiratory irritant component of toxic hazard is addressed in a somewhat analogous manner, with the exception that concentrations, rather than accumulated doses, are utilized. With irritants, the time at which the total fractional effective concentration (FEC) becomes equal to 1 represents the mode, or the most frequently expected time for incapacitation to occur.

The sensory/upper respiratory irritant component of toxic hazard is difficult to assess since very few controlled studies have been made on the effects of human exposure to irritants; most information is only anecdotal. Also the severity of such effects has been highly subjective, covering a range from mild eye and upper respiratory irritation to severe pain. Despite the scarcity of hard data, it has been proposed that sensory/upper respiratory irritation will at least impair or even prevent escape by incapacitating those exposed. However, it has also been claimed that the effects of sensory/upper resipiratory irritants peak at moderate concentrations and that, although the effects may be unpleasant, they would not significantly impair the ability to escape. Most of the evidence obtained under controlled experimental conditions actually suggests that sensory/upper respiratory irritants, although irritating and even painful, may not be incapacitating at all. Their impact on escape capability may be largely psychological, with either impairment or even enhancement being possible, depending upon other factors.

PROBABILISTIC NATURE OF TOXIC HAZARD ANALYSIS

The methodologies just described involve either the accumulated dose or the instantaneous concentration associated with the occurrence of a biological response, such as incapacitation, of people exposed to fire effluents. However, the accumulated dose or concentration for a response actually represents the *maximum likelihood* in a statistical distribution of people's responses surrounding that dose or concentration, i.e., the mode, or most frequently expected dose or concentration for the response to occur for exposure of a number of people. Doses or concentrations for the responses of individuals would be statistically distributed around the mode in a frequency distribution curve.

The maximum of the frequency distribution for a response occurs, by definition, at the time at which either the total accumulated FED or the total instantaneous FEC value for asphyxiant or irritant toxicants is equal to 1. However, the setting of criteria for life safety and the predicted ability of occupants to escape from a fire situation must also take into consideration the expected frequency distribution of people's responses relative to the accumulated FED values and instantaneous FEC values. Target FED and FEC values are, therefore, chosen which are more appropriate to a desired acceptable incidence of human incapacitation.

The traditional approach for establishing safe levels or tolerances for chemical agents is to reduce their effective dose or concentration values by factors that take into consideration variation in susceptibility within the human population. With respect to this variability, evidence derived from limited studies on the pharmacokinetics of chemicals has indicated that a 10-fold difference in these parameters corresponds to 7 to 9 standard deviations in populations of normal adults. Thus, significant conservatism is implied with application of a safety factor of 10 in order to accommodate even more susceptible human subpopulations exposed acutely to fire effluents. A target factor of 0.1 FED or 0.1 FEC is, therefore, now being suggested for the safe escape of nearly all exposed individuals. Other target values may be chosen, however, depending upon the objectives of the user.

It should be noted that this quantitative assessment of fires' threat to life, due to toxic hazards, is currently under study within the International Standards Organization (ISO), Technical Committee 92 (Life Safety), Subcommittee 3 (Toxic Hazards in Fires). The interested reader should keep abreast of future developments emanating from this international standards group.

A noteworthy consequence of the probabilistic nature of the prediction of biological responses resulting from exposure of people to smoke atmospheres is that statistical distributions for two such responses, such as incapacitation and death, may overlap. This gives rise to a probability that some may die without others even being incapacitated. It is known, of course, that this commonly occurs in real fires.

HEAT

The burning of most materials is an exothermic chemical oxidation process. Energy from the process is evolved as heat, which possesses both convective (hot gases) and radiative components. The radiative component represents energy released in the visible and infrared portions of the spectrum and is seen as flame or luminosity of a fire.

Heat poses a significant physical danger to humans. If the total heat energy reacting with the body surpasses the capability of the physiological defense processes to compensate, a series of events will occur, ranging from minor injury to death. The effects of exposure to heated air are greatly augmented by the presence of moisture in the fire atmosphere. With higher moisture content, transfer of heat energy is more efficient and the body is less able to rid itself of the heat burden. Moisture can be present in a fire environment as the result of natural humidity, from the combustion itself, and from the application of water for extinguishment.

If excessive heat is conducted rapidly to the lungs, a serious decline in blood pressure may result, along with capillary blood vessel collapse, leading to circulatory failure. Severe heat also may cause fluid buildup in the lungs. In fire tests conducted by the National Research Council of Canada (NRCC), 300°F (149°C) was taken as the maximum survivable breathing air temperature.[57] A temperature this high can be endured for only a short period and not at all in the presence of moisture. It has been suggested that fire fighters should not enter any hostile atmosphere without full protective clothing and masks.

In school fire tests conducted in Los Angeles in 1959, a temperature of 150°F (66°C) at the 5-ft (1.5-m) level was selected as the temperature beyond which teachers and children could not be expected to enter a corridor from a relatively cool room.[58] This selection assumed exposure to dry air, and only for the brief period of time necessary to reach exits.

Skin tissue burns commonly are classified as first-, second-, or third-degree burns. First-degree burns involve only the outer layer of the skin and are characterized by abnormal redness, pain, and sometimes a small accumulation of fluid underneath the skin. Second-degree burns penetrate more deeply into the skin. The burned area is moist and pink, the skin blisters, and there is usually a considerable amount of subcutaneous fluid accumulation. Third-degree burns usually are dry, charred, or pearly white. If a large percentage of the body skin tissue has suffered third-degree burns, postexposure consequences may be extremely critical.

Studies have shown that a skin surface temperature as low as 160°F (71°C) will result in second-degree burns if contact is maintained for 60 sec.[59] If the surface temperature is increased, the time to produce second-degree burns decreases. For example, burns result in 30 sec at 180°F (82°C), and in 15 sec at 212°F (100°C). A tolerance limit for radiant heat of 2.5 kW/m² has been suggested.[60]

Before human skin can absorb sufficient heat to raise its surface temperature, the body's heat-dissipating capabilities must be defeated. The body dissipates heat by evaporative cooling (perspiration) and by circulation of blood. The cooling effect of evaporation of skin moisture may compensate for the effect of heat on skin up to 140°F (60°C) or more in dry air. This limit is lower in moist air. The time for the skin temperature to increase depends upon the exposure temperature, which, in most fires, increases very rapidly. Under these conditions, the skin temperature may increase faster than the defense mechanisms can function. Thermal tolerance data for unprotected skin of humans at rest would suggest a limit of about 248°F (120°C), above which considerable pain would quickly be incurred.[15] Exposure to convective heat below this temperature may still result in hyperthermia, without the production of actual burns.

Hyperthermia occurs if the body absorbs heat faster than it can be dissipated by evaporation of surface moisture and outward radiation. The entire body temperature is thereby elevated sufficiently above normal to cause damage to the central nervous system and, possibly, death.

VISIBLE SMOKE

In addition to fire gases, smoke also consists of finely divided particulate matter and suspended liquid droplets known as aerosols. These are formed from the burning of most materials under the conditions of incomplete combustion normally present in fires. Since the average size of the particulates and aerosols is about the same as the wavelength of visible light, the light is scattered and vision through smoke is obscured. Petroleum-derived materials, especially those from aromatic hydrocarbons, are particularly prone to the formation of dark, sooty smoke. There is, however, no relationship between the color of visible smoke and the toxicity of fire gases that may be present.

Since smoke obscures the passage of light, visibility to exits may be blocked, thus impairing escape of victims from a fire. The development of quantities of smoke sufficient to impair egress can be very rapid and usually is the first hazard to occur in a fire. As evidenced in nearly every one of the Los Angeles school fire tests, smoke in the corridors arising from fires in the basement reached untenable levels before temperatures attained hazardous condition.[58] In the tests, smoke, as it pertained to visibility, was the principal hazard. Although smoke frequently provides an early warning of fire, it also contributes to creation of panic because of its blinding and irritating effects.

Obscuration of vision due to smoke is related to its concentration and is usually expressed either as optical density per meter (OD/m) or as the extinction coefficient K (K = OD/m × 2.3). The response to obscuration of vision due to smoke, and the subsequent inability to escape, is highly variable. Although visibility (m) can be calculated as the reciprocal of the OD/m, actual requirements for escape depend to a large extent on the size of the enclosure and the occupants' familiarity with escape routes. Suggested tenability limits for optical density have ranged from 0.5/m[60] (2 m visibility), for occupants of small rooms who are familiar with escape routes, down to about 0.065/m[61] (15 m visibility), for large enclosures in which occupants are unfamiliar with their surroundings.

Smoke particulates and aerosols also can be harmful when inhaled, and long exposure may cause damage to the respiratory system. Smoke particulates often are of a sufficiently small size that enable them to be inhaled deeply into the lungs, where absorbed toxicants may produce damage to the respiratory system. These effects have not been sufficiently studied as to enable a full understanding of their consequences.

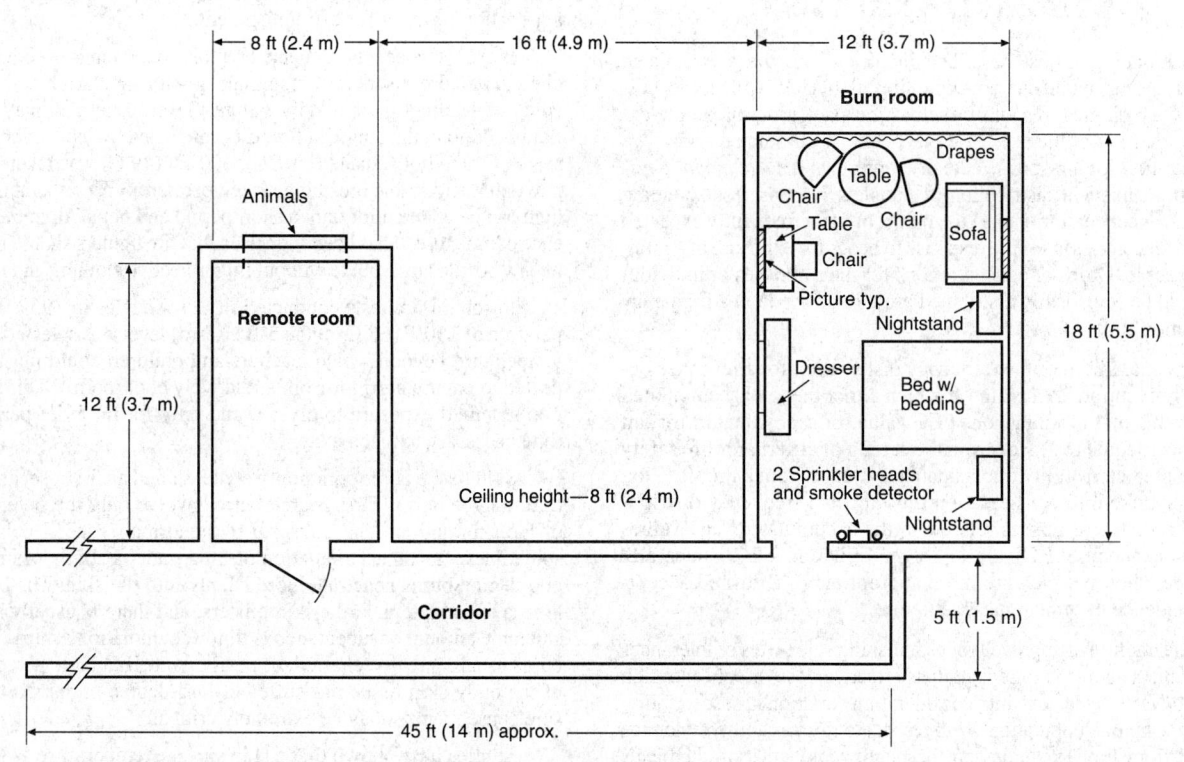

FIG. 4-2A.   Burn room, corridor, and remote room full-scale test facility at Southwest Research Institute.

DEVELOPMENT OF TOXIC HAZARDS

Full-scale fire tests are useful for placing the development of toxic hazard in perspective. Four such tests were reported by Southwest Research Institute (SWRI) in which fires were set in a fully furnished simulated hotel room with attached corridor and remote room.[7] (See Figure 4-2A.) During the tests, the door from the fire room to the corridor was fully open. The door between the corridor and the remote room was open approximately 1 in. (25.4 mm) until about 3 min after flashover, when it was closed to prevent the development of excessive concentrations of toxicants.

The sequence of events that occurred in the burn room is shown as a composite illustration in Figure 4-2B. The tests were initiated with a smoldering fire in the chair closest to the sofa. During the smoldering phase of approximately 19 min, no conditions were attained that would be considered hazardous to life. Following flaming ignition of the chair, the fire progressed rapidly to flashover within about 8 min, accompanied by life-threatening conditions of temperature, carbon monoxide, hydrogen cyanide, and oxygen depletion in the room of origin. After flashover in the burn room, tenability in the remote room also quickly deteriorated, as shown in Figure 4-2C, first with visual obscuration by smoke, followed by rapidly increasing concentrations of toxic gases. Test animals (rats) in the remote room became incapacitated within about 2 min, with death occurring from carbon monoxide asphyxiation approximately 11 min after flashover. From present knowledge of carbon monoxide intoxication, it is

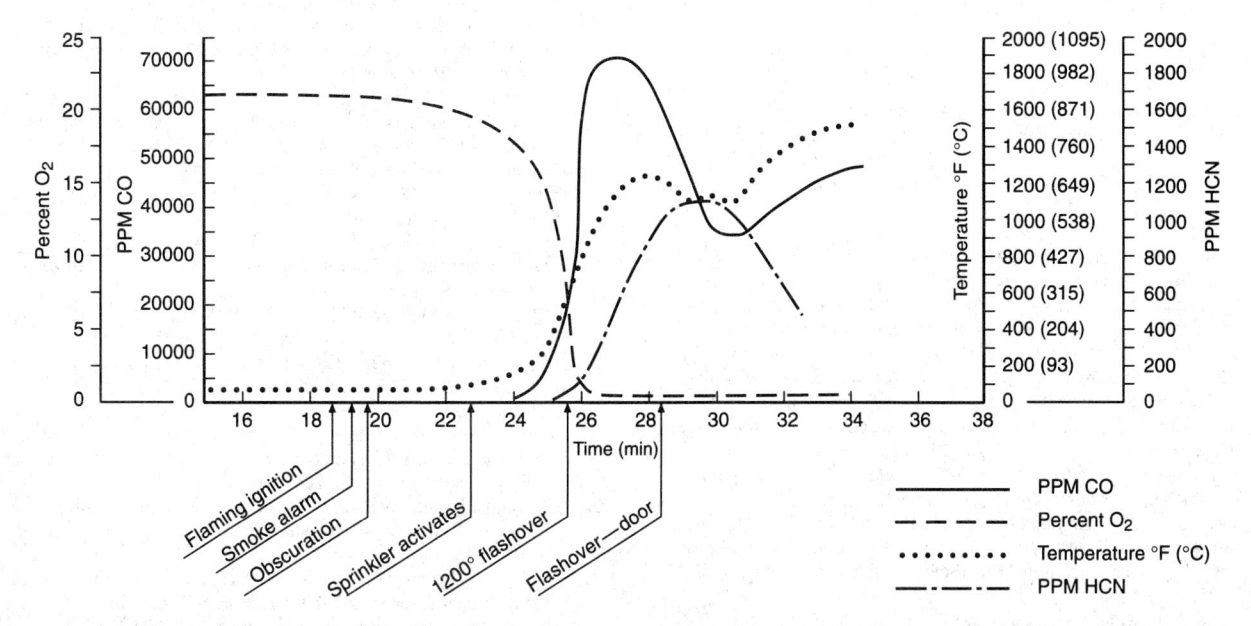

FIG. 4-2B. *Composite illustration of events and development of hazardous conditions at 5¹/₂-ft (1.7-m) level in the burn room during fully furnished room fires conducted at SWRI.*

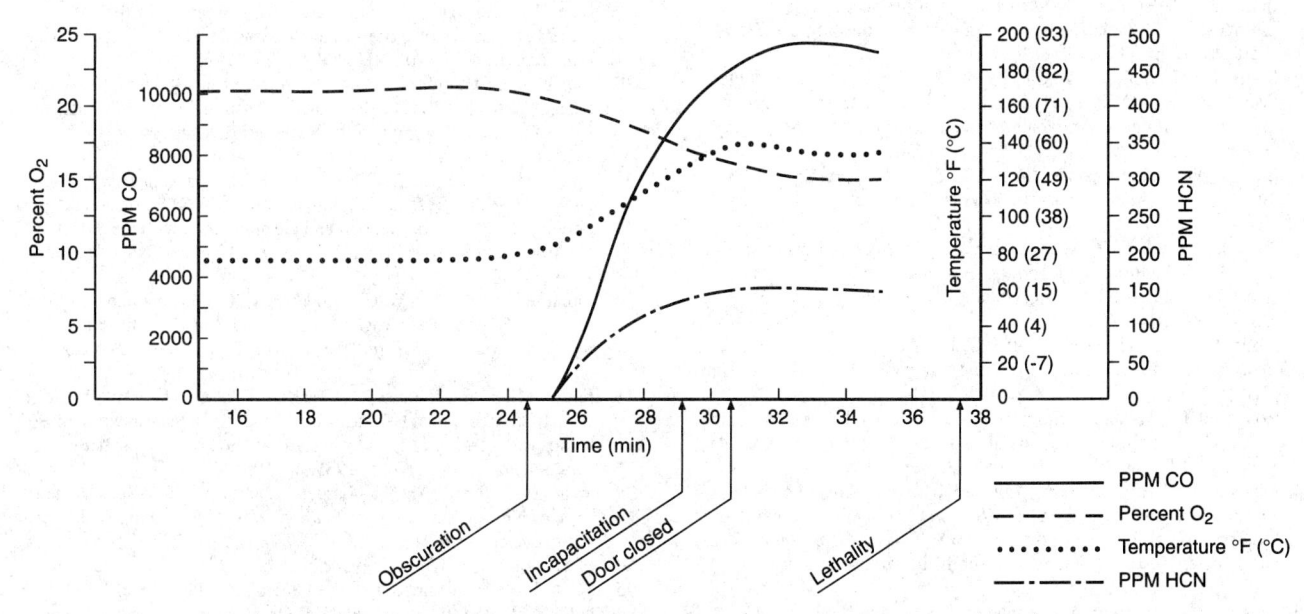

FIG. 4-2C. *As in Figure 4-2B, but for remote room.*

likely that humans in the remote room would have been incapacitated and killed in approximately the same time intervals.

It was considered significant in these tests that the toxic hazard did not develop until flashover was approached in the burn room. After that point, smoke obscuration and life-threatening conditions increased very rapidly in the burn room, corridor, and the remote room. The smoke detectors employed in these tests provided adequate warning (5 min) prior to the development of acutely toxic concentrations of combustion products. The sprinkler activation times demonstrated that a properly installed and working sprinkler system, by controlling the fire at an early stage, would have prevented any significant toxic threat from developing.

BIBLIOGRAPHY

References Cited

1. ASTM E176, *Standard Terminology of Fire Standards*, American Society for Testing and Materials, W. Conshohocken, PA, 1993.
2. Beyler, C. L., and Hirschler, M. M., "Thermal Decomposition of Polymers," *The SFPE Handbook of Fire Protection Engineering*, 2nd ed., Dinenno, P. J., ed., National Fire Protection Association, Quincy, MA, 1995, pp. 1-99–1-119.
3. Babrauskas, V., Harris, R. H., Jr., Gann, R. G., Levin, B. C., Lee, B. T., Peacock, R. D., Paabo, M., Twilley, W., Yoklavich, M. F,. and Clark, H. M., "Fire Hazard Comparison of Fire-Retarded and Non-Fire-Retarded Products," NBS Special Publication 749, 1988, National Bureau of Standards, U.S. Department of Commerce, Gaithersburg, MD.
4. Ohlemiller, T. J., "Smoldering Combustion," *The SFPE Handbook of Fire Protection Engineering*, 2nd ed., DiNenno, P. J., Ed., National Fire Protection Association, Quincy, MA, 1995 pp. 2-171–2-179.
5. Babrauskas, V., Levin, B. C., Gann, R. G., Paabo, M., Harris, R. H. Jr., Peacock, R. D., and Yusa, S., "Toxic Potency Measurement for Fire Hazard Analysis," NIST Special Publication 827, 1991, National Institute of Standards and Technology, Gaithersburg, MD.
6. Tewarson, A., "Generation of Heat and Chemical Compounds in Fires," *The SFPE Handbook of Fire Protection Engineering*, 2nd ed., DiNenno, P. J., *et al.*, eds., National Fire Protection Association, Quincy, MA, 1995, pp. 3-53–3-124.
7. Grand, A. F., Kaplan, H. L., Beitel, J. J., Switzer, W. G., and Hartzell, G. E., "An Evaluation of Toxic Hazards from Full-Scale Furnished Room Fire Studies," Special Technical Publication 882, 1986, American Society for Testing and Materials, W. Conshohocken, PA, pp. 330-353.
8. Tsuchiya, Y., "Relative Significance of NO*x* and HCN in Fire Gas Toxicity," *Fire Safety Science—Proceedings* of the Second International Symposium, Wakamatsu, T., ed., Hemisphere Publishing Co., New York, 1989, pp. 381–390.
9. Beitel, J. J., Bertello, C. A., Carroll, W. F., Jr., Grand, A. F., Hirschler, M. M., and Smith, G. F., "Hydrogen Chloride Transport and Decay in a Large Apparatus: II, Variables Affecting Hydrogen Chloride Decay," *J. Fire Sciences*, Vol. 5, No. 2, 1987, pp. 105–145.
10. Burgess, W. A., Treitman, R. D., and Gold, A., *Air Contaminants in Structural Firefighting*, Harvard School of Public Health, Boston, MA, 1979.
11. Potts, J. W., Lederer, T. S., and Quast, J. F., "A Study of Inhalation Toxicity of Smoke Produced Upon Pyrolysis and Combustion of Polyethylene Foams, Part I: Laboratory Studies," *J. Comb. Tox.*, Vol. 5, 1978, pp. 408–433.
12. Smith, R. P., "Toxic Responses of the Blood," *Casarett and Doull's Toxicology*, 3rd ed., Macmillan Publishing Company, New York, 1986, pp. 228–231.
13. Roughton, F. J. W., and Darling, R. C., "The Effect of Carbon Monoxide on the Oxyhemoglobin Dissociation Curve," *Am. J. Physiol.*, Vol. 141, 1944, pp. 17–31.
14. Esposito, F. M., and Alarie, Y., "Inhalation Toxicity of Carbon Monoxide and Hydrogen Cyanide Gases Released During the Thermal Decomposition of Polymers," *Advances in Combustion Toxicology*, Vol. III, Hartzell, G. E., ed., Technomic Publishing Co., Inc., Lancaster, PA, 1992, pp. 78–125.
15. Purser, D. A., "Toxicity Assessment of Combustion Products," *The SFPE Handbook of Fire Protection Engineering*, 2nd ed., DiNenno, P.

J., *et al.*, eds., National Fire Protection Association, Quincy, MA, 1995, pp. 2–85—2–146.
16. Kaplan, H. L., and Hartzell, G. E., "Modeling of Toxicological Effects of Fire Gases: I. Incapacitating Effects of Narcotic Fire Gases," *J. Fire Sciences*, Vol. 2, No. 4, 1984, pp. 286–305.
17. Stewart, R. D., Peterson, J. E., Fisher, T. N., Hosko, M. J., Baretta, E. D., Dodd, H. C., and Herrmann, A. A., *Arch. Environ. Health*, Vol. 26, 1973, p. 1.
18. Sanders, D. C., and Endecott, B. R., "The Effect of Elevated Temperature on Carbon Monoxide-Induced Incapacitation," *Advances in Combustion Toxicology*, Vol. III, Hartzell, G. E., ed., Technomic Publishing Co., Inc., Lancaster, PA, 1992, pp. 343–357.
19. Swinyard, E. A., "Noxious Gases and Vapors: Carbon Monoxide, Hydrocyanic Acid, Benzene, Gasoline, Kerosene, Carbon Tetrachloride, and Miscellaneous Organic Solvents," *The Pharmacological Basis of Therapeutics*, 3rd ed., Goodman, L., and Gilman, A., eds., Macmillan Publishing Co., New York, 1965, pp. 915–928.
20. Purser, D. A., Grimshaw, P., and Berrille, K. R., "Intoxication by Cyanide in Fires: A Study in Monkeys Using Polyacrylonitrile," *Arch. Env. Health*, Vol. 39, 1984, p. 394.
21. Kimmerle, G., "Aspects and Methodology for the Evaluation of Toxicological Parameters During Fire Exposure," *J. Fire and Flammability/Combustion Toxicology Supplement*, Vol. 1, 1974, p. 42.
22. Halpin, B. M., and Berl, W. G., "Human Fatalities from Unwanted Fires," Report NBS-GCR-79–168, 1978, U.S. National Bureau of Standards, Washington, D.C.
23. Harland, W. A., and Woolley, W. D., "Fire Fatality Study," Building Research Establishment Information Paper, IP 18/79, 1979, University of Glasgow, Glasgow, Scotland.
24. Hartzell, G. E., Switzer, W. G., and Priest, D. N., "Modeling of Toxicological Effects of Fire Gases: V. Mathematical Modeling of Intoxication of Rats by Combined Carbon Monoxide and Hydrogen Cyanide Atmospheres," *J. Fire Sciences*, Vol. 3, No. 5, 1985, pp. 330–342.
25. Levin, B. C., Paabo, M., Gurman, J. L., and Harris, S. E., "Effects of Exposure to Single or Multiple Combinations of the Predominant Toxic Gases and Low-Oxygen Atmospheres Produced in Fires," *Fund. and Appl. Tox.*, Vol. 9, 1987, pp. 236–250.
26. Hartzell, G. E., and Switzer, W. G., "On the Toxicities of Atmospheres Containing Both CO and CO_2," *J. Fire Sciences*, Vol. 3, No. 5, 1985, pp. 307–309.
27. Levin, B. C., Paabo, M., Gurman, J. L., Harris, S. E., and Braun, E., "Toxicological Interactions Between Carbon Monoxide and Carbon Dioxide," *Proceedings* of 16th Conference of Toxicology, Air Force Medical Research Laboratory, Dayton, OH, 1986; *Toxicology*, Vol. 47, 1987, pp. 135–164.
28. Reinke, R. W., and Reinhardt, C. F., "Fire, Toxicity, and Plastics," *Modern Plastics*, Feb. 1973, pp. 94–98.
29. Kane, L. E., Barrow, C. S., and Alarie, Y., "A Short-Term Test to Predict Acceptable Levels of Exposure to Airborne Sensory Irritants," *Amer. Ind. Hyg. Assoc. J.*, Vol. 40, 1979, pp. 207–229.
30. Kaplan, H. L., Grand, A. F., Switzer, W. G., Mitchell, D. S., Rogers, W. R., and Hartzell, G. E., "Effects of Combustion Gases on Escape Performance of the Baboon and the Rat," *J. Fire Sciences*, Vol. 3, No. 4, 1985, pp. 228–244.
31. Kaplan, H. L., Switzer, W. G., Hinderer, R. K., and Anzueto, A., "Acute and Long-Term Effects of Polyvinylchloride (PVC) Smoke on the Respiratory System of the Baboon and a Comparison with the Effects of Hydrogen Chloride (HCl)," *J. Fire Sciences*, Vol. 11, No. 6, 1993, pp. 485–511.
32. Kaplan, H. L., Switzer, W. G., Hinderer, R. K,. and Anzueto, A., "Studies of the Effects of Hydrogen Chloride and Polyvinylchloride (PVC) Smoke in Rodents," *J. Fire Sciences*, Vol. 11, No. 6, 1993, pp. 512–552.
33. Hartzell, G. E., Packham, S. C., Grand, A. F., and Switzer, W. G., "Modeling of Toxicological Effects of Fire Gases: III. Quantification of Postexposure Lethality of Rats from Exposure to HCl Atmospheres," *J. Fire Sciences*, Vol. 3, No. 3, 1985, pp. 195–207.
34. Levin, B. C., Paabo, M., Highbarger, L., and Eller, N., "Synergistic Effects of Nitrogen Dioxide and Carbon Dioxide Following Acute Inhalation Exposures in Rats," NISTIR 89–4105, 1989, National Institute of Standards and Technology, Gaithersburg, MD.
35. ISO 13344, "Determination of the Lethal Toxic Potency of Fire Effluents," 1995, International Standards Organization, Geneva, Switzerland.

36. ASTM E1678, *Standard Test Method for Measuring Smoke Toxicity for Use in Fire Hazard Analysis*, American Society for Testing and Materials, W. Conshohocken, PA, 1995.
37. ISO TR 9122, Part 5, "Prediction of the Toxic Effects of Fire Effluents," 1993, International Standards Organization, Geneva, Switzerland.
38. Hartzell, G. E., Priest, D. N., and Switzer, W. G., "Modeling of Toxicological Effects of Fire Gases: II. Mathematical Modeling of Intoxication of Rats by Carbon Monoxide and Hydrogen Cyanide," *J. Fire Sciences,* Vol. 3, No. 2, 1985, pp. 115–128.
39. Hartzell, G. E., and Emmons, H. W., "The Fractional Effective Dose Model for Assessment of Toxic Hazards in Fires," *J. Fire Sciences,* Vol. 6, No. 5, 1988, pp. 356–362.
40. Hartzell, G. E., Grand, A. F., and Switzer, W. G., "Modeling of Toxicological Effects of Fire Gases: VI. Further Studies on the Toxicity of Smoke Containing Hydrogen Chloride," *J. Fire Sciences,* Vol. 5, No. 6, 1987, pp. 368–391.
41. Hartzell, G. E., Grand, A. F., and Switzer, W. G., "Toxicity of Smoke Containing Hydrogen Chloride," *Fire and Polymers,* Nelson, G. L., ed., American Chemical Society, Washington, DC, 1990, pp. 12–20.
42. Levin, B. C., Paabo, M., Bailey, C., Harris, S. E., and Gurman, J. L., "Toxicological Effects of the Interactions of Fire Gases and Their Use in a Toxic Hazard Assessment Computer Model," *The Toxicologist,* Vol. 5, 1985, p. 127.
43. Levin, B. C., Paabo, M., Gurman, J. L., Clark, H. M., and Yoklavich, M. F., "Further Studies of the Toxicological Effects of Different Time Exposures to the Individual and Combined Fire Gases: Carbon Monoxide, Hydrogen Cyanide, Carbon Dioxide, and Reduced Oxygen," *Polyurethanes '88, Proceedings* of the 31st Society of the Plastics Industry, Polyurethane Division Meeting, Technomic Publishing Co., Inc., Lancaster, PA, 1988, pp. 249–252.
44. Levin, B. C., Fowell, A. J., Birky, M. M., Paabo, M., Stolte, A., and Malek, D., "Further Development of a Test Method for Assessment of the Acute Inhalation Toxicity of Combustion Products," NBSIR 82–2532, 1982, National Institute of Standards and Technology, Gaithersburg, MD.
45. "New York State Uniform Fire Prevention and Building Code," Article 15, Part 1120, *Combustion Toxicity Testing,* Office of Fire Prevention and Control, Department of State, New York State, Albany, NY, 1986.
46. Chien, W. P., Delain, G. E., and Richards, R. S., "New York State Fire Gas Toxicity Requirements: Past, Present, and Future," *J. Fire Sciences,* Vol. 9, No. 3, 1991, pp. 173–182.
47. German Standards Institution, DIN 53436, "Erzeugung Thermischer Zersetzungsproducte von Werkstoffen unter Luftsufuhr und ihre Votikologische Prufung. Teil 1, Zersetzungsgerat und bestimmung de Versuchstemperatur. Teil 2, Verfahren zur Thermischen Zersetzung. Teil 3, Entwurf-Verfahren zur Inhalations-Toxikologischen Untersuchung Thermischer Zersetzungsproducte," 1981.
48. Pauluhn, J., "A Retrospective Analysis of Predicted and Observed Smoke Lethal Toxic Potency Values," *J. Fire Sciences,* Vol. 11, No. 2, 1993, pp. 109–130.
49. Grand, A. F., "Development of a Product Performance Smoke Toxicity Test," Final Report to the National Institute of Building Sciences, Project 01–1744, 1990, Southwest Research Institute, San Antonio, TX.
50. ISO TR 9122, Part 4, "The Fire Model," 1993, International Standards Organization, Geneva, Switzerland.
51. Petajan, J. H., Voorhees, K. J., Packham, S. C., Baldwin, R. C., Einhorn, I. N., Grunnet, M. L., Dinger, B. G., and Birky, M. M., "Extreme Toxicity for Combustion Products of a Fire-Retarded Polyurethane Foam," *Science,* Vol. 187, 1975, pp. 742–744.
52. Baker, B. B., Jr., and Kaiser, M. A., "Understanding What Happens in a Fire," *Analytical Chemistry,* Vol. 63, 1991, pp. 79–83.
53. Purser, D. A., and Woolley, W. D., "Biological Studies of Combustion Atmospheres," *J. Fire Sciences,* Vol. 1, No. 2, 1983, pp. 118–144.
54. Barrow, C. S., Alarie, Y., and Stock, M. F., "Sensory Irritation Evoked by the Thermal Decomposition Products of Plasticized Poly-Vinyl Chloride," *Fire and Materials,* Vol. 1, 1976, pp. 147–153.
55. Hartzell, G. E., "Smoke Toxicity and Toxic Hazards in Fires," *ASTM Standardization News,* Vol. 22, No. 11, 1994, pp. 38–43.
56. Schaper, M., "Development of a Database for Sensory Irritants and Its Use in Establishing Occupational Exposure Limits," Am. Ind. Hyg. Assoc. J., 54, 1993, pp. 488–544.
57. Shorter, G. W., *et al.,* "The St. Lawrence Burns," *NFPA Quarterly,* Vol. 53, No. 4, Apr. 1960, pp. 300–316.
58. *Operation School Burning,* National Fire Protection Association, Quincy, MA, 1959, p. 25.
59. Woodson, W. E., *Human Factors Design Handbook,* McGraw-Hill, Inc., New York, 1981, p. 812.
60. Babrauskas, V., Ph.D., "Full-Scale Burning Behavior of Upholstered Chairs," *Technical Note 1103,* National Institute of Standards and Technology, Gaithersburg, MD, 1979.
61. Jin, T., *Journal of Fire and Flammability,* Vol. 12, 1981, p. 130. (No longer published.)

Additional Readings

Purser, D. A., "Toxicity Assessment of Combustion Products," *The SFPE Handbook of Fire Protection Engineering,* 2nd ed., DiNenno, P. J., et al., eds., National Fire Protection Association, Quincy, MA, 1995, pp. 2-85–2-146.
Hall, J. R., and Harwood, B., "Smoke or Burns—Which Is Deadlier?" *NFPA Journal,* Jan./Feb. 1995, pp. 38–43.

WOOD AND WOOD-BASED PRODUCTS

Revised by John M. Cholin

Wood, arguably the first structural material to be utilized by pre-historic man, is still the single most versatile structural material in use today. It is light in weight, strong, durable, flexible, and resilient, yet it possesses an intrinsic beauty. These characteristics have made it a persistent first choice construction material. Furthermore, once dwelling spaces are constructed, those spaces are filled with furnishings that are often constructed of wood and wood products. The fact that this material is truly ubiquitous in modern society necessitates a thorough understanding of both its nature as a fuel and its behavior as a structural material under conditions of fire exposure.

One reason wood and wood products have been so universally adopted as a structural material is the fact that it can be processed into so many different forms that take specific advantage of the unique chemical and physical characteristics of wood as a raw material. The wood from a wide variety of trees can be transformed into structural timbers, dimension lumber, shingles, and veneers. These are called "primary forest products," as only cutting and drying are required to derive these products from the log.

Raw wood can be used to make a variety of wood-based building products, including multiple-ply glue-laminations, fiberboard, insulation board, ceiling tile, plywood, flakeboard, strandboard, and particleboard. Each of these forms of wood has specific structural characteristics that makes it the material of choice for specific applications. These products are often referred to as "secondary forest products," since secondary processes are necessary over and above cutting and drying to result in the product.

Primary and secondary forest products are usually not consumed directly. They are the material from which directly consumed products are constructed. The primary examples of these are housing and furnishings. Lumber, a primary forest product, is cut and shaped and assembled with other components to ultimately become a finished piece of furniture. Occasionally these are referred to as "tertiary forest products."

The unique chemical composition of wood makes it possible to process it into materials that have no resemblance to the log from which they came. Wood is the major raw material source for paper and paper-based products, such as cardboard, box board, and corrugated board used in containers. Paper, used alone or combined with metal foils and plastics, is employed in the manufacture of many

hundreds of products, including writing and wrapping papers, magazines, books, bags, wall coverings, and labels. Furthermore, when exposed to the appropriate chemical treatment, wood yields a material called cellulose, which is the fundamental raw material for a wide variety of products, including transparent films (e.g., Saran Wrap®), motion picture film, textile fibers, explosives and propellants, coatings, plastics, and foodstuffs.

Because wood and its cellulose derivatives are found in such a wide range of forms, it should come as no surprise that they are frequently involved in fires. The discussion of wood in this chapter is limited to solid wood and wood-based materials, including paper and paper derivatives. This chapter will address the chemistry of cellulose, the process of ignition of wood and cellulosic materials, and the behavior as a structural component under conditions of fire exposure. Those interested in a more detailed treatment on the subject should consult the list of references at the end of this chapter.

There are several other chapters in this handbook that specifically deal with various wood-derived cellulosic materials. Cellulose-based lacquers and paints are covered in Section 3, Chapter 16, "Spray Finishing and Powder Coating." Textiles, including acetates and various rayons, are discussed in Section 4, Chapter 5, "Fibers and Textiles." Plastics and films made from cellulose are covered in Section 4, Chapter 10, "Plastics and Rubber." Other reconstituted cellulose materials used in explosives and propellants are discussed in Section 4, Chapter 12, "Explosives and Blasting Agents." The fire hazards of finely divided wood particles, such as sander dust, are a special topic and are discussed in Section 4, Chapter 15, "Dusts." In addition, there are several chapters that deal with occupancy classes where wood and wood-derived materials are predominant influences.

Manufacturing and process information on various wood products can be found in Section 3, Chapter 15, "Woodworking Facilities and Processes." Fire hazards of wood fuels are covered in Section 3, Chapter 24, "Storage and Handling of Solid Fuels." Products of combustion, their importance on the fire scene, their physiological effects on persons trapped in the fire environment, and the degree to which they affect rescue and fire-fighting efforts are discussed in Section 4, Chapter 2, "Combustion Products and Their Effects on Life Safety." Fire-retardant treatments are discussed in Section 4, Chapter 4, "Fire-Retardant and Flame-Resistant Treatments of Cellulosic Materials." Biomass and standing timber hazards are discussed in Section 10, Chapter 7, "Wildland Fire Management."

John M. Cholin, P.E., is an independent fire protection consultant and engineer with J.M. Cholin Consultants, Inc., Oakland, NJ.

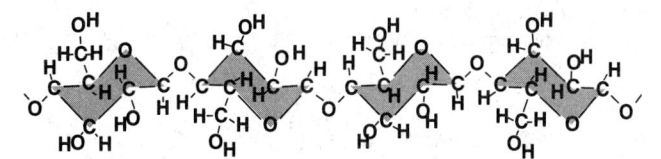

FIG. 4-3A.   *Cellulose is a polymer of glucose units produced by plants.*[1]

FIG. 4-3B.   *A schematic representation of a cellulose fibril.*

NATURE OF WOOD AND WOOD-BASED PRODUCTS

To understand the behavior of forest products, both as fuel and as structural material under exposure to fire, one must start with an understanding of the material called "wood." This understanding starts with the chemistry of the principal component of wood, i.e., cellulose.

Chemical Composition of Wood

Cellulose is a naturally occurring polymer constructed from the simple sugar, glucose. Its chemical structure is shown in Figure 4-3A.[1]

The individual molecules of glucose are chemically linked together during the growth process of the living cell to produce a long-chain polymer, cellulose. In most cases, the average cellulose molecule is between 300 and 3,000 glucose units long.[2] The long-chain polymer cellulose molecules are organized in bundles of parallel chains called fibrils. The fibrils are linked together via hydrogen bonding through a water molecule. (See Figure 4-3B.)

Adjacent fibrils are cemented together by the living cell with special linking polymeric molecules consisting of hemicelluloses, pectins, also called lignin. These polymeric chemicals bind adjacent fibrils together into a mat or sheet that forms the cell wall of the plant cell. The cells of the living plant are organized into long strands, like a string of pearls, and this gives rise to the cellulose fiber.

A collection of complex polymeric aromatic alcohols, called lignin, binds adjacent cellulose fibers together, forming a matrix around and between adjacent cellulose fibers. This matrix can be imagined as being similar to a microscopic plate of spaghetti, where the noodles are the cellulose fibers and the sauce is the lignin polymers, binding the fibers together. It is a structure that is not unlike fiber-reinforced plastic (fiberglass), and results in the wood fiber.

The cellulose molecule also has numerous hydrogen bonding sites along its length.[2] Wherever an –OH group exists along the length of the cellulose molecule, there exists a site where hydrogen bonding via an adsorbed water molecule can serve to bind adjacent cellulose fibers together. The combined effect of the hydrogen bonding via adsorbed water molecules and the chemical linkage formed by the lignin polymers binds adjacent wood fibers together into the material known as wood.

Obviously, wood is a complex structure with tremendous strength for its weight. It consists of bundles of hollow fibers, ce-

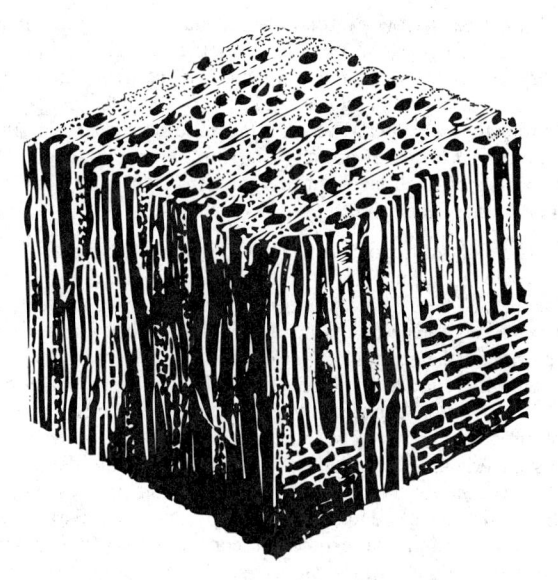

FIG. 4-3C.   *A photomicrograph of red oak.*

mented together along their length, forming a rigid, yet flexible, structure. (See Figure 4-3C.)

Although wood is a complex of many substances, cellulose [chemical formula $(C_6H_{10}O_5)_x$] represents its major component. Hemicellulose is a five-carbon sugar consisting of carbon, oxygen, and hydrogen. Lignins are aromatic alcohols, also consisting of carbon, oxygen, and hydrogen. Consequently, the majority of the mass of any sample of dry wood consists primarily of carbon, hydrogen, and oxygen, with a small percentage of nitrogen from cellular proteins. There are small quantities of other elements that remain as an ash when wood is burned. It is interesting to note that untreated wood contains virtually no sulfur, distinguishing it from most other solid fuels. (See Table 4-3A.)

TABLE 4-3A.   *Chemical Composition of Woods, Exclusive of Water*

Species	Constituents, Percent by Weight				
	Carbon	Hydrogen	Oxygen	Nitrogen	Ash
Oak	50.16	6.02	43.26	0.09	0.37
Ash	49.18	6.27	43.19	0.07	0.57
Elm	48.99	6.20	44.25	0.06	0.50
Beech	49.06	6.11	44.17	0.09	0.57
Birch	48.88	6.06	44.67	0.10	0.29
Pine	50.31	6.20	43.08	0.04	0.37
Poplar	49.37	6.21	41.60	0.96	1.86
Calif. redwood	53.50	5.90	40.30	0.10	0.20
Western hemlock	50.40	5.80	41.40	0.10	2.20
Douglas fir	52.30	6.30	40.50	0.10	0.80

A sample of wood, minus all water, contains, on average, 50 percent cellulose by weight; approximately 24 percent hemicellulose, which is also a form of sugar; 23 percent lignin; 1 percent extractives (in the form of gums, resins, oils, and other materials); and the remainder ash. Different tree species yield slightly different percentages, but these values can be thought of as the average.

Some woods, especially the soft woods (pines, spruces, and true firs), are quite resinous, with commensurately higher levels of the extractives and lower levels of hemicellulose and lignin. In these woods, the resins can become quite volatile, especially at elevated temperatures. In contrast, most hardwoods contain such small quantities of extractives that they are of little, if any, consequence to the fire protection specialist. The ash is the mineral content of a sample of wood and is primarily potash (potassium carbonate). The remainder consists of trace levels of manganese, magnesium, iron, zinc, and other transition elements.

The major difference between various woods depends largely on the density (specific gravity) of the wood produced by a particular species of tree. This variation in density is related to cell wall thickness and translates into the dimensional stability and strength properties for that particular species of wood. The other species-dependent difference is in the quantity and chemical composition of the extractives that contribute to the color, scent, rot resistance, and other properties of wood. In some species, these extractives are resinous materials, which produce both flammable and combustible vapors. These vapors can produce hazardous atmospheres under the right conditions of temperature and confinement. Furthermore, some resins undergo an exothermic condensation reaction, which has been suspected to be the cause of spontaneous ignition in certain wood products, e.g., piled chips of southern pine used in making paper.

Moisture Content

Wood exhibits a natural affinity for water and always contains significant quantities of moisture. Water plays an important part in the physical properties of wood through its contribution to the bonding between adjacent cellulose molecules. When the tree is alive, much of the space within the cells is occupied by water. However, when the log is cut, this moisture begins leaving the wood. The water in the cell interiors is slowly replaced by air. The moisture content varies with species and the degree that the sample has been dried. Regardless of prior processing, wood will always come to equilibrium with its environment. The energy liberated during the combustion of wood is dependent upon this moisture content.

The liquid water in the living tree (green wood) is called "free water." The free water is that water absorbed within the cellular structure of the wood in excess of the fiber saturation point (FSP). The fiber saturation point is the maximum moisture content the cellulose can retain as adsorbed water. The free water occupies cell cavities and readily flows through the tubular veins along the length of the sample. Free water readily leaves the log through air drying. As the log dries, air replaces the free water.

The log begins with the same FSP moisture content as the species of tree from which it came. This moisture remains in the form of vapor in the cell cavity, and as adsorbed water in the cell wall. At the FSP, a sample may still contain between 25 and 30 percent water, by weight. This moisture is called "bound water," since it is hygroscopically bound to the cellulose molecules through a weak chemical bond called a "hydrogen bond." Bound water will slowly leave the log or board at a rate determined by the temperature and relative humidity. This process is usually accelerated by kiln drying.

As the moisture content approaches approximately 7 percent, by weight, very little additional moisture is lost with increases in temperature or duration of heating. The moisture content below the range of 5 to 7 percent is called the "moisture of constitution." This water forms bonds between adjacent cellulose molecules and contributes to the structural strength of the wood fiber. This water cannot be removed without the physical destruction of the wood. Consequently, even "dry" wood always contains some moisture.

Construction lumber is considered to be dry at 19 percent moisture content (MC), but wood used for furniture is dried to 7 per-

cent moisture content or less. Moisture content in wood is calculated as a percentage of the oven dry weight by the formula:

$$MC = \left(\frac{\text{Original Weight - Dry Weight}}{\text{Dry Weight}} \right) 100\%$$

Wood reaches an equilibrium moisture content (EMC) with its surroundings at a given temperature and relative humidity. In a heated building, the EMC of structural wood members is usually slightly less than 10 percent moisture content in the winter months and about 12 percent moisture content in the summer. The lumber in a newly constructed building is generally wetter than that in a building that has existed for some time. The moisture in wood is not uniformly distributed, and wood is often drier or wetter at the surface than in its interior. Surface layers, which control ignition and flame spread, respond much faster to changes in atmospheric conditions than the interior; thus, bulk moisture content measurement must be considered with this fact in mind. Figure 4-3D illustrates conditions under which equilibrium will be established at three different temperatures.

Air-dried wood usually attains a moisture content of approximately 10 percent water by weight. If wood is dried to a moisture content below this nominal 10 percent in a kiln, for example, it will immediately begin absorbing moisture from the air when it is removed from the kiln. The overwhelming majority of the wood found in structures, either as part of the structure or contents, will be dried, usually kiln dried, and can be assumed to be at a nominal 8 to 10 percent (by weight) moisture content. A small variation about the nominal 10 percent moisture content has very little influence on the ignitability and heat release of wood. It does, however, affect the dimensions and strength of a given piece of wood substantially. This creates the incentive to dry wood to a minimum moisture content and maintain that level, especially with woods used for interior furnishings.

The level of moisture in a wood sample can have an impact on the ignition and combustion of that sample.[7] "Green" wood, that which has not been dried, often contains as much as 50 percent water by weight. This free water boils off as the sample burns, absorbing a significant portion of the heat generated by the combustion of the cellulosic content. Dry wood is devoid of the free water and much of the bound water. Furthermore, the bound water effectively exists as vapor, not liquid, and does not have the ability to absorb heat through vaporization. This results in a combustible material with a significantly greater heat output during a fire.

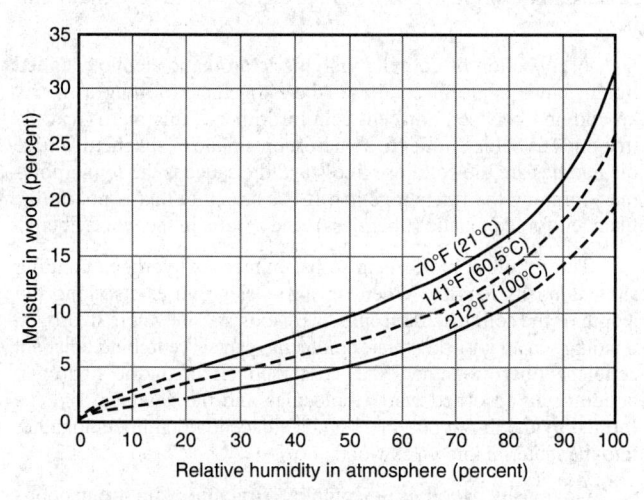

FIG. 4-3D. *Equilibrium moisture content of wood.*

Moisture content has an important bearing on the electrical properties of wood. Dry wood is a good insulator, but wet wood is a conductor. The electrical resistance of wood increases in approximate proportion to the moisture content of wood, from 25 percent to about 6 percent moisture content where it becomes too high to measure with a resistance moisture meter.

The rate of moisture loss or gain depends a great deal upon the size and shape of the wood product. Some finely divided cellulosic materials absorb or lose moisture very rapidly, and their combustibility can vary from low to high within just a few hours. Examples are shredded newspaper, tissues, and loose forest litter (duff). Dense and compact combustibles, such as books or large timbers, change moisture content much more slowly. Rate of moisture loss depends on the relative humidity, the temperature, and the air velocity. Moisture loss is quite slow in air drying where there is no control over air velocity, temperature, or relative humidity; lumber may require 30 days or more to dry in such an environment. But in a kiln, where humidity, temperature, and air velocity can be adjusted, lumber can be dried in less than 24 hours. As a general rule of thumb, the amount of time required to dry lumber varies with the square of the thickness. Thus, a 1-in.-(25.4-mm-) thick board can be air dried in a few weeks, but a railroad crosstie or large timber requires 24 to 30 months. If wood is dried below the EMC it will reach in service, it will regain moisture slowly until it reaches that EMC. (See Figure 4-3D.)

Wood Derivatives

The chemical nature of wood and its cellulose is the basis for the diverse range of materials that are directly derived from wood. Most papers are made by a chemical process that removes the lignin and hemicellulose from the wood, yet leaves the cellulose fibers intact. This de-lignined cellulose is called "α-cellulose." These fibers are suspended in water and spread out in a thin layer. The layer is passed through sets of rollers that squeeze out the free water, and then through a dryer that drives off much of the remaining interstitial water, resulting in a sheet of paper. The cellulose fibers are held together, forming the sheet of paper, by the hydrogen bonds and the water bound to the cellulose. Newsprint is made by a process in which the fibers are separated by a mechanical or grinding process, i.e., no chemicals are used. Some papers are made without any performance-modifying additives, e.g., absorbent papers. However, most paper products contain performance-modifying additives, such as binders (wet-strength resins), fillers (clay), pigments (titanium dioxide), colorants (dyes), sizes (rosin, starch), and coatings (paraffins or plastics).

Wood is used as the primary feed stock for the production of purified cellulose. Purified wood cellulose fiber is used in disposable sanitary applications, such as baby diapers and feminine protection products. Purified cellulose is a common ingredient of many food products where it is used to thicken liquids and to improve the texture of foods. It is commonly found in diet aids where it serves as a bulking agent and a source of dietary fiber. Many cellulose-based products are not discernible as being made from wood or paper. Chemically modified cellulose gives rise to such materials as nitrocellulose lacquers, films, and plastics; acetate films; textiles and plastics; rayon textiles; cellophane; and smokeless powder used in small-arms cartridges.

Wood cellulose can be broken down to glucose, using appropriate chemical techniques, to make molasses suitable for cattle feed supplements, or fermented to make ethyl alcohol (grain alcohol) for use as a gasoline additive. The hemicellulose portion of wood (composed of five-carbon sugars, such as xylose) can be fermented to make methyl-ethyl-ketone (MEK) or butyl alcohol (butanol), used as solvents.

Destructive distillation of wood releases the extractives and the products of the pyrolytic decomposition of the wood. This process produces charcoal, methanol (wood alcohol), acetic acid, formic acid, and wood creosote, as well as a low-heat-output fuel that can be used to power internal combustion engines. Other processes are used to produce turpentine and rosin, known as naval stores, from standing pine trees, old stumps, or from the byproducts of the manufacture of kraft paper by the sulfate process.

Wood-based products, paper, and paper-based products generally contain much less moisture than the wood from which they are derived, making them easier to ignite and more combustible than solid wood.

Wood Byproducts

Bark is often encountered in the context of wood-waste utilization and specialty construction materials. Bark is an important byproduct of forest products processing. It is often used for heating fuel in many forest products facilities. Bark differs both structurally and chemically from wood. It consists of a cellulosic material called suberin, which is utilized in the form of cork. Commercial cork, obtained from the bark of the cork oak, is resistant to high temperatures and ignition, and burns with difficulty. Cork contains a natural adhesive that allows small particles to be bonded under heat and pressure to make composition cork. Bark and cork generally have lower moisture contents than wood. Bark is of commercial significance in that it contains tannin which is extracted and utilized in the curing of leather.

There are many other fibrous cellulosic materials that behave in a manner similar to that of wood, and consequently can be included in this chapter. These materials include bagasse (spent sugar cane stalks), jute, hemp, and sisal. These materials have burning characteristics similar to wood-based materials. These and other similar products are often used as a source of fiber for construction materials, papers, and insulating materials. As with more traditional wood-derived materials, these fibers may be impregnated, pressure treated, or coated with a variety of chemicals, potentially altering the ignition and burning characteristics of such materials.

PHYSICAL NATURE OF WOOD

The cellular structure of wood provides a material that is light and strong. It appears to be an ideal structural material. However, it is not without its shortcomings, and it is these shortcomings that have precipitated the development of the engineered wood-based materials.

Mechanical Properties

While wood is strong, its strength is very dependent upon the orientation of the stress relative to the grain of the piece under consideration. The modulus of compressibility is good along the grain, but poor across the grain. The structural strength characteristics of wood are dependent upon the moisture content. If the moisture content of a sample is allowed to stay above the fiber saturation point, removal of moisture has little effect on wood, other than to reduce its weight. However, once the moisture content is brought below the fiber saturation point, the strength properties improve considerably.

Solid wood is preferred where the nature of wood is consistent with the use. These are applications where the strength vs. weight properties of wood, the ease of fastening, and environmental endurance make it the most economical approach. Wood products made from solid wood include lumber, timbers, mine timbers, railroad ties, poles, and pilings. Other applications are driven by aesthetic concerns, such as upscale furnishings. Other solid-wood products are barrels, boxes, crates, and other containers. Each of these uses takes advantage of specific and often unique characteristics of wood or a species of wood, and the economies gained from those characteristics. Hardwoods are used primarily for selected interior finish applications, the manufacture of furniture, and as feed stock for cellulose fiber for engineered wood products and specialty papers. Softwoods are more commonly used for structural building components, such as timbers, beams, joists, and studs. They are also the wood of choice for mine timbers, railroad ties, and pilings, and other uses where large dimensions are required.

Secondary wood products are designed to address one or more of the shortcomings of primary stock. Lumber dimensions vary with moisture content, and the dimensional variation is dependent upon the orientation of the grain. This dimensional instability can be addressed by using wood products made from thin sheets of wood (veneer) bonded together with adhesives. These materials are termed plywood and laminated veneer lumber (LVL). Present-day large structural wood beams are often made from laminations of smaller dimension lumber.

The utilization percentage of a log is increased when it is converted to smaller particles that are bonded together with an adhesive resin. This yields an engineered wood composite material. Consequently, both the cost and dimensional properties of primary wood can be improved upon by conversion to a wood composite material. These materials are produced from wood that has been mechanically reduced in size to form flakes, strands, splinters, fibers, or particles. They are then mixed with thermoset plastic resins and processed to form a wood composite material with specific mechanical properties. This category includes a wide range of materials, such as flakeboard, oriented strand board (OSB), medium-density fiberboard (MDF), particleboard, fiberboard, and hardboards.

Thermal (Heat) Conductivity

Thermal conductivity (K factor—the inverse of insulating value, R factor) of a particular material greatly influences how a material will burn. Thermal conductivity is a measure of the rate at which absorbed heat flows through the mass of material. For instance, wood is a poor heat conductor and thus has a high insulating value. Steel exhibits a thermal conductivity approximately 350 times that of an "average" specimen of wood. Aluminum exhibits a thermal conductivity 1,000 times greater than an average wood specimen.

The thermal conductivity of wood is dependent on the direction of heat flow, with respect to the axis of grain orientation in wood, the moisture content of the wood, and the density (specific gravity) of the wood. Thermal conductivity is two to three times greater along the grain in the longitudinal direction than across the grain. Thermal conductivity of wood in any direction is about one-third greater for moisture content above 30 percent than for drier wood.

The poor thermal conductivity of wood is a significant factor in the notably good performance heavy timber construction provides during a fire. When the exterior surface of a timber is exposed to the heat of a fire, the insulating nature of the wood slows the temperature rise of the core of the timber. Furthermore, the surface of the timber forms a char layer, which improves the R-factor of the timber and prevents atmospheric oxygen from penetrating into the timber, where it can support combustion.

Effect of Fire Retardants

Fire-retardant chemicals can be, and frequently are, added or applied to wood and wood-based products. When properly formulated and used, fire-retardant chemicals reduce the combustibility of these products. Depending on the formulations used, flame and surface combustion can be reduced once the outside source of heat is removed. Some fire-retardant treatments (FRT) may adversely affect the strength properties of plywood sheathing and structural lumber, causing them to become brittle and fail prematurely or under loads that one would normally expect them to bear without difficulty. There are a number of theories regarding the mechanism of this loss of strength, including chemical attack on the cellulose fibers in the wood by the fire-retardant chemicals and the high temperatures used to redry FRT plywood and lumber after treatment. Regardless, the impact of the use of fire retardants on the strength of wood members must be addressed when using these materials. The current building codes rightfully recognize the contribution these materials make in improving the fire safety of the structures where they are used and provide for derating to accommodate the possible loss of strength.

The Effect of Preservatives

To resist rot and insect damage many wood products are treated chemically with biocidal materials. These include railroad crossties, utility poles, pilings, landscaping ties, and exterior dimension stock for decks and other wood structures. These wood products are usually pressure treated with various wood preservatives containing creosote (a mixture of single and conjugated ring aromatic compounds made from coal tar) or pentachlorophenol in solutions of oil similar to diesel or fuel oil, or with water-soluble salt solutions containing arsenic, chromium, copper, or phenol. All of these materials are toxic to humans, in varying degrees, if ingested or applied directly to the skin. Waterborne salts are usually used to treat dimension decking lumber and plywood, whereas the oil-based preservatives are more commonly used for treating utility poles, railroad ties, and pilings.

These treatments introduce two concerns for the fire protection specialist. First, they can affect the ignitability and burning rates of the treated wood products. Wood products freshly treated with creosote or pentachlorophenol in oil solutions are often more easily ignited because of the presence of the oil vehicle and can be expected to burn with a greater heat output. Second, these treatments introduce toxic chemicals into the wood that will be released if involved in a fire. These materials add to the toxicity of the smoke produced from such a fire.

Building Fashions

The prevailing trends in housing design have a significant impact on the types of wood products used in a particular structure. This affects the fire safety considerations for that structure, since the selection of construction materials affects the behavior of the structure in the event of a fire. During the 1970s, due primarily to a desire for a more "natural" appearance in homes and enactment of new legislation designed to encourage conservation and save energy, more external wood materials (wood shingles or siding) were used. In many parts of the country, a return to wood as a primary fuel (burned in fireplaces and stoves) and to log homes with shingle roofs created greater fire risks. The 1980s saw increased use of vinyl-clad fiberboards used for building sidings, and substitution of oriented strandboards (OSB) or flakeboards for plywood in construction. Use of wood shingles declined with a return to fiberglass and asphalt-derived materials. There has also been a marked increase in the manufacture and use of treated lumber and plywood in construction. More insulation products made from various organic and inorganic materials are now being used, while asbestos has been virtually eliminated in sidings and flooring materials. All of these factors have important bearing on the ignitability, fire growth rate, heat output, and structural performance of the building during a fire.

VARIABLES INFLUENCING COMBUSTIBILITY

Whether one is primarily concerned with mere ignition or with the energies obtained from sustained combustion, there are several factors that consistently enter into the analysis. These include the physical form of the fuel wood, the thermal conductivity of the fuel wood (as discussed earlier), and the moisture content of the wood.

Physical Form

The influence of physical form on the ignition and burning characteristics of wood and wood-based products is illustrated by the fact that wood kindling will flame and burn from relatively small heat sources while heavier wood logs will resist ignition to a considerable degree. Paper is easier to ignite than small pieces of wood. The reason for this is that, as the size of the particle diminishes, the ratio of surface area to volume (mass) increases. Thus there is both greater exposure of fuel to air and less mass to conduct heat away from the particle; consequently, heat does not readily dissipate within the material.

It follows that small, thin forms of wood (such as excelsior), or other combustible solids (such as paper), will burn more readily than larger objects of the same material. This fact is most important when dealing with wood-derived combustible particulate solids. Wood dusts, chips, and shavings represent a serious fire risk, especially when pneumatically conveyed. Very finely divided wood particles, such as sander dust or wood flour (often made from nut shells), are prone to deflagration similar to other carbonaceous dusts. These materials create a serious threat of explosion in the pneumatic conveying and dust collection systems commonly used to handle them. Sawdust, router chips, shavings, and other wood waste are often thought of as less hazardous due to the much larger particle sizes. However, the pneumatic conveying of these particulates causes attrition of the particles, which results in fine dusts capable of undergoing a deflagration. Fires in pneumatic conveying and dust-control systems are a constant hazard in most wood manufacturing operations due to ignitability and pervasive presence of finely divided wood particulates.

Although paper is considered to be easy to ignite, large compact masses of paper, such as books or cut and packaged stacks of paper, do not burn readily due to charring and limited availability of combustion air in the interior of the mass. Additives, such as binders, colorants, sizes, etc., in the paper also serve to inhibit combustion. However, large rolls of paper or paper-based products stacked on end may unwind (exfoliate) during a fire, exposing long, single sheets of paper which are easily ignited and burn readily.

Moisture Content

Ignition and burning tests in the laboratory show that the behavior of combustible solids of the same size, shape, and chemical composition varies markedly, depending upon the moisture content. In research conducted by the British Department of Scientific and Industrial Research (now the Ministry of Technology), the influence of the moisture content of oak and western red cedar on the time necessary for these woods to ignite was studied; the results are shown in Figure 4-3E.[3] Radiation intensity in this figure is expressed in terms of the number of calories falling on one square centimeter of the exposed wood in one second, using a test apparatus developed by the Ministry.

Once a fire is well under way, the significance of the moisture factor decreases, since the heat radiation and therefore the rate of pyrolysis increases. Under such conditions, wood with a moisture content of 50 percent or more will burn.

There is a time lag between changes in the fuel moisture content of cellulosic materials and changes in atmospheric humidity. Most solid cellulosic combustibles are not immediately absorptive of airborne water vapor; conversely, the moisture content of most solids will be retained for a longer period of time than air retains its moisture in suspension. Thus, the humidity in the ambient air is not always a reliable index of the moisture content of ordinary combustibles.

IGNITION OF WOOD AND WOOD PRODUCTS

The ignition of a piece of wood is not a simple process; it is actually quite complex and subject to a number of variables. Furthermore, the process of ignition does not always follow the models used. Nevertheless, to a first order approximation, the ignition mechanism for wood and wood-based products begins with the application of heat with sufficient intensity and power to raise the fuel to its pyrolytic temperature. Pyrolysis is the thermal decomposition of the cellulose and related compounds. Usually, the pyrolysis produces combustible gases and vapors, which are released into the surrounding air. When these combustible gases and vapors are produced at a sufficient rate, they produce the visible flame usually associated with combustion.

The thermal decomposition of wood is a complex process. As the temperature is raised, different sets of physical and chemical reactions become the predominant factors. However, in general, this process can be thought of as occurring in four stages, corresponding to four temperature regions, as the sample is heated.[4]

The initial stage occurs between room temperature and 392°F (200°C). As the sample passes through this stage, the bound water is driven off as water vapor. Carbon dioxide evolution has also been observed at these temperatures. These products are noncombustible. However, some combustible materials, e.g., the low molecular weight extractives, if present, formic acid, and acetic acid, also vaporize and leave the sample as gases at this temperature. These materials are combustible. However, the heat necessary to drive them off from the sample is greater than the heat that can be derived from their combustion. Consequently, this phase is an energy-consuming (endothermic) phase and is somewhat reversible, in that water will be reabsorbed if the sample is allowed to cool.

The second phase or region occurs between 392°F (200°C) and 536°F (280°C). By the time the sample has reached this temperature, the majority of the water has been driven off. The principal pyrolytic product is carbon monoxide. Carbon monoxide is combustible, but

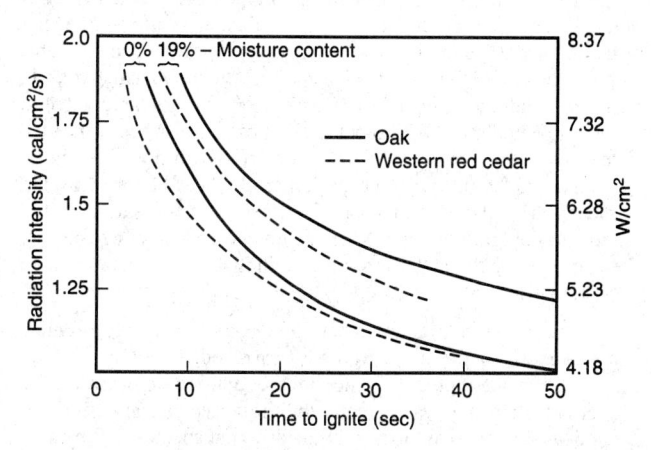

FIG. 4-3E. *Effect of moisture content on the ignition of wood.*

the quantities produced are relatively small and, as with the earlier phase, this phase is still endothermic.

The third phase or region occurs between 536°F (280°C) and 932°F (500°C). In this temperature region, the pyrolytic decomposition of the wood becomes energy producing (exothermic) and can proceed without the investment of additional energy from an external source of heat. This is the critical phase of the ignition process. If the heat liberated due to the pyrolytic decomposition of the sample is retained by the sample, its temperature continues to increase and ignition proceeds. However, if the heat liberated by the pyrolysis is dissipated more rapidly than it is produced, the sample's temperature decreases and the process becomes endothermic once again. During this phase, the pyrolysis products include a variety of gases and vapors that are primarily single-carbon and two-carbon hydrocarbon compounds. These compounds are extremely combustible and can produce a flame. The process of pyrolysis does not convert cellulose entirely into volatile products. Much of the carbon in the cellulose remains as unburned charcoal.

The fourth region is where the temperature exceeds 932°F (500°C). At these temperatures, the wood is essentially gasified, rapidly forming volatile and flammable gases of one to four hydrocarbon units. This phase more completely utilizes the carbon in the cellulose, resulting in a smaller charcoal residue.

The ease of ignition is related to the ignition temperature and the radiant power necessary to raise the temperature of a sample to that temperature. If wood were a better conductor of heat, it would take far more energy to reach the ignition temperature. The insulating qualities of wood are critical, allowing attainment of ignition temperatures with relatively small quantities of ignition energy.

Ignition temperatures may vary, depending on the analytical method and test equipment used. With proper test equipment, some exothermic reaction can be observed in the initial stage. In the third phase, ignitable gases are produced and the pilot ignition temperature is reached. When the decomposition has progressed to the point where the evolved gases no longer insulate the charcoal layer from oxygen, spontaneous ignition may result. However, this ignition mechanism requires a significant investment of heat energy into the fuel before ignition occurs.

Common experience leads one to believe that ignitability is dependent upon the species. However, this is only partially true. Ignition temperature does not appear to be well correlated to species, as seen in Tables 4-3B and 4-3C. However, temperature is only half of the equation; power (joules of energy per unit time) is the other. Resinous softwoods often require greater ignition power than equivalent masses of nonresinous hardwoods. As a specimen of softwood is being heated, the resins are free to melt and vaporize before the exothermic pyrolysis temperature is attained. This actually consumes energy and cools the sample. At this stage of ignition, the softwood actually requires more energy per unit time from the ignition source due to this evaporative cooling effect. Once the volatile resins have left the sample, its temperature rises more rapidly due to its now lowered mass. The accumulation of resin vapors at the surface of the sample can accelerate ignition if the ignition heat source is a flame. Once the resin vapors ignite, the heat liberated by the combustion of these vapors rapidly raises the temperature of the sample to the ignition point. When samples are slowly "cooked" in the absence of a flame, these variations in ignition temperatures are less dramatic.

Specific ignition temperatures of wood are difficult to determine because of the many variables involved. Leading research groups have attempted to pinpoint the ignition temperature of wood, but the results vary greatly. In one series of tests using untreated, air-dried wood blocks of nine different species, 1¹/₄ by 1¹/₄ by 4 in. (32 × 32 × 102 mm), the Forest Products Laboratory in

TABLE 4-3B. *Ignition Temperatures of Woods*[5]

Wood 2¹/₂ in. (63.5 mm) long and weighing 3 g (0.10 oz)	Self-Ignition Temperature	
	°F	°C
Western red cedar	378	192
White pine	406	207
Long leaf pine	428	220
White oak	410	210
Paper birch	399	204

TABLE 4-3C. *Ignition Temperatures of Wood Shavings*[*]

Wood Shavings	Self-Ignition Temperature	
	°F	°C
Short leaf pine	442	228
Long leaf pine	446	230
Douglas fir	500	260
Spruce	502	261
White pine	507	264

*Source: National Bureau of Standards [now the National Institute for Standards and Technology (NIST)].

Madison, WI, found that the specimens could be heated to temperatures varying from 315 to 385°F (157 to 196°C) for 40 min in a stream of heated air (held at a constant temperature) without igniting. In another series of tests by the same laboratory, using the same test procedures but higher temperatures, 34 untreated softwood and hardwood species ignited in 4 to 6 min at temperatures varying from 608 to 660°F (320 to 349°C) for the softwood, and from 595 to 740°F (313 to 393°C) for the hardwood. It was noted that a fairly close relationship existed between the specific gravity of the wood under test and the ignition temperature. In general, low-density species ignited at lower temperatures than high-density species.[6] The National Bureau of Standards [now the National Institute for Standards and Technology (NIST)] tested three conifers and two hardwoods and reported ignition temperatures from 378 to 428°F (192 to 220°C) in small samples varying from shavings to match size.

The ignition and charring temperatures of wood were recorded at the Forest Products Laboratory.[7] The findings, represented in Table 4-3D, show the length of time before wood specimens maintained at the specified constant temperatures evolved combustible gases in sufficient quantities to be ignited by a pilot flame located about ¹/₂ in. (12.7 mm) above the specimen. The specimens, 1¹/₄ by 1¹/₄ by 4 in. (32 × 32 × 102 mm) in size and oven-dry, were heated in a vertical quartz chamber 3 in. (76 mm) in diameter by 10 in. (254 mm) long that was maintained at constant temperature by an electric furnace.

C. R. Brown reported on ignition temperatures of some materials, as shown in Table 4-3B.[5] In his investigations, specimens about 2¹/₂ in. (63.5 mm) long and weighing 3 g (0.10 oz.) were suspended in a vertical glass container 2¹/₂ in. (63.5 mm) in diameter and 4¹³/₁₆ in. (122 mm) high. The container was heated by an electric furnace, the temperature being increased at a predetermined rate. Air preheated to the temperature of the furnace was passed through the container at measured rates of flow. The results were found to be affected by the size of the sample, the rate of heating, and the rate of air flow over the specimens.

TABLE 4-3D. *Time Required to Ignite Wood Specimens*[7]

Wood 1¼ by 1¼ by 4 in. (32 × 32 × 102 mm)	No Ignition in 40 min		Exposure Before Ignition, By Pilot Flame, Minutes						
	°F	°C	356°F (180°C)	392°F (200°C)	437°F (225°C)	482°F (250°C)	572°F (300°C)	662°F (350°C)	752°F (400°C)
Long leaf pine	315	157	14.3	11.8	8.7	6.0	2.3	1.4	0.5
Red oak	315	157	20.0	13.3	8.1	4.7	1.6	1.2	0.5
Tamarack	334	167	29.9	14.5	9.0	6.0	2.3	0.8	0.5
Western larch	315	157	30.8	25.0	17.0	9.5	3.5	1.5	0.5
Noble fir	369	187	—	—	15.8	9.3	2.3	1.2	0.3
Eastern hemlock	356	180	—	13.3	7.2	4.0	2.2	1.2	0.3
Redwood	315	157	28.5	18.5	10.4	6.0	1.9	0.8	0.3
Sitka spruce	315	157	40.0	19.6	8.3	5.3	2.1	1.0	0.3
Basswood	334	167	—	14.5	9.6	6.0	1.6	1.2	0.3

In 1947, the National Bureau of Standards (now NIST) reported ignition temperatures of wood as shown in Table 4-3C. In these tests, the specimens of wood in the form of shavings were heated in a glass test tube in the presence of a measured flow of heated air at optimum velocity. Ignition temperature was determined to be the lowest temperature at which the exothermic oxidation reaction culminated in an ignition, as indicated by flaming or glowing combustion.

Wood in contact with steam pipes or a similar constant temperature source over a very long period of time may undergo a chemical change, resulting in the formation of charcoal, which is capable of heating spontaneously. It has been suggested that 212°F (100°C) is the highest temperature to which wood can be continually exposed without risk of ignition.[8]

Spontaneous Heating and Ignition

While coal and other organic materials have a reputation for "self-heating" and "self-igniting," there is little evidence to support the claims of self-ignition of wood products. In general, wood, wood-based products, and cellulosic materials that are clean and dry will not self-heat and spontaneously ignite. Normally a spark, open flame, contact with hot surfaces, or exposure to some other thermal radiator is necessary in order to attain the ignition temperatures of wood-based fuels. In those episodes where it has appeared that wood products did ignite spontaneously, there were polymerizing resins involved that provided the exothermic reaction necessary to heat the wood to its ignition temperature.

It has been shown that cellulosic materials contaminated with certain drying oils (e.g., linseed oil, tung oil, and turpentine) can self-heat when stored in unventilated spaces. The actual source of the heat is the exothermic chemical reactions involving the oils. In such cases if self-heating is allowed to continue, exothermic pyrolysis can be achieved and ignition can result.

Similarly, external heating can raise certain wood-based materials to a temperature where the exothermic pyrolytic reactions begin. Once this temperature is achieved, the material can then self-heat as long as the conditions are conducive to the retention of the heat generated by the pyrolysis. For example, if wood fiberboards are not subjected to a direct, artificial heat source either during manufacture or when installed in structures, they will not heat spontaneously. On the other hand, tests have shown that heating such fiberboards at relatively moderate temperatures can initiate an exothermic reaction leading to their ignition. In one such test, a 22-inch (559-mm) high stack of wood fiberboards ⅛-in. (3-mm) thick, subjected to a heat source of 228°F (109°C), was ignited in 96 hours.

Wood and wood-based products are not susceptible to the same type of self-heating from microbiological action that is prevalent in certain agricultural products.

COMBUSTION OF WOOD AND WOOD-BASED MATERIALS

The factors that affect the ignitability of wood are also important factors governing the combustion of wood. The moisture content is an important factor in determining the heat output of a given sample and, hence, the heat available to perpetuate the fire. Likewise, the chemical composition is an important determinant of the combustion of a sample. Some chemical additives will reduce the heat output and yield a less violent flame, whereas other additives may actually increase the heat content of the fuel. Finally, the resin content of the fuel is important.

There is a characteristic odor of wood smoke. The smoke is bluish or dark when the wood is green or wet and low temperature or insufficient air is present; but, with high temperature and excess air, the smoke is often almost invisible. This can be used to infer the range of size for the particulates that make up the smoke. Since particulates diffract light that is of a wavelength shorter than the particle diameter, the bluish smoke implies smoke particulates with an average diameter of approximately 0.4 microns, the wavelength of blue light. As the efficiency of combustion is improved, the particle diameters become even smaller resulting in "invisible" smoke. Burning wood produces smoke containing glowing particles and large amounts of a gray ash. Wood burned under conditions of incomplete combustion in which little or no air is present produces charcoal, carbon monoxide, and a number of condensible yet combustible vapors which are called, in the aggregate, "creosote."

Air Availability

An adequate amount of air (oxygen) is essential for the complete combustion of cellulosic materials, regardless of physical form. Combustion engineers have suggested a formula for the minimum volume of air required for complete combustion of solid fuels with air at 62°F (17°C):

$$V_m = 147\left[C + 3\left(H - \frac{O}{8}\right)\right] \text{cu ft}$$

where

V_m = minimum volume of air in cu ft,

C = part by weight of carbon in 1 lb of fuel,

H = parts by weight of hydrogen in 1 lb of fuel, and

O = parts by weight of oxygen in 1 lb of fuel.

(For SI units: 1 cu ft = 0.0283 m^3; 1 lb = 0.454 kg.)

This formula provides a basis for determining whether a given fire of given fuel load is underventilated and yielding less than its optimum heat release or fully ventilated and yielding the anticipated heat release for the fuel mass involved. The underventilated fire does produce "destructive distillates," such as carbon monoxide, formic acid, aldehydes, ketones, and carboxylic acids, which can deflagrate when exposed to a source of fresh air. This is a serious concern for fire fighters and should not be overlooked.

The residue after a fire provides valuable insight into the combustion processes of the fire. The products of combustion and the residues remaining after combustion vary according to the degree of ventilation of the fire. When excess air is supplied to a wood fire, the solid residue remaining after combustion is in the form of a gray ash. When black charcoal is the combustion residue, the indication is that the fire was underventilated.

Rate of Combustion

Since the combustion of forest products is a surface phenomenon, the rate of combustion can be influenced by the physical state and orientation of the fuel and any factor affecting the availability of combustion air. Increasing the volume of air available to the fire, regardless of the means used, accelerates the rate of combustion until the air velocity reaches a point where it begins diluting the pyrolysis gas concentration in air to levels below that necessary for optimum combustion efficiency. Then airflow becomes excessive and the air begins acting as a heat transfer fluid, robbing the fire of the heat needed to continue the pyrolysis necessary to maintain the supply of gaseous distillates that fuel the flame. If the air supply velocity is further increased, the volume of air may exceed the ability of the fuel mass to supply sufficient gaseous distillates, and the flame ceases. The combustion will continue but in a solid phase, evidenced by the glowing surface of the fuel, rather than the gaseous phase evidenced by the flame.

In most freely ventilated fires involving combustible solids, once ignition is achieved, the fire follows the t^2 growth curve until the entire surface of the fuel is involved. Once there is total surface involvement, the rate of combustion remains essentially constant until a limitation in fuel or combustion air is encountered. The rate of combustion is significantly influenced by the physical form of the combustible, the air supply present, the moisture content, and allied factors; but there is the fundamental need for progressive vaporization of the solid by heat exposure for combustion to proceed.

Speed of Flame Propagation

Speed of flame propagation over the surface of a combustible critically affects the severity of a fire in many cases. The 1959 Los Angeles School Fire Tests[10] showed that, under controlled test conditions, flames spread over combustible surfaces with such speed that observers had to flee the sudden release of heat and flame. In 1949, the British Department of Scientific and Industrial Research conducted full-scale dwelling fire tests on the influence of combustible wallboard *vs.* the use of plaster finish.[11] The record of the British tests graphically illustrates the observed time-temperature curves in the living rooms of the two houses tested. Originating in an easy chair in the living room, the fires in their initial stages were identical in both dwellings. The loading of combustible materials was roughly proportional to the loadings observed in occupied dwellings in England. It should be noted that, in the United States and Canada, earlier ignition and faster flame spread on the combustible surfaces could be expected, since winter heating in these coun-

tries results in lower relative humidities and, thus, lower moisture content of the combustibles.

Currently, the comparative assessment of flame spread characteristics of building materials is performed per ASTM E84 (NFPA 255) using the Steiner Test Tunnel.[12] Red oak was originally used as a reference standard to establish the performance rating equal to 100. Today the test method is referenced to specific performance parameters on a noncombustible surface. Nevertheless, red oak still achieves a rating of approximately 100, and most other wood species fall into the range between 90 and 160. Flame spread indexes for selected wood species are listed in Table 4-3E.

TABLE 4-3E. *Flame Spread Indexes for Selected Wood Species, 1-in. (25.4-mm) Nominal Solid Dimension Lumber*[13]

Species	Flame Spread Index	Source
Bald cypress	145-150	UL
Eastern red cedar	110	HUD/FHA
Alaska yellow cedar	78	CWC
Western red cedar	70	HPMA
Douglas fir	70-100	UL
Western hemlock	60-75	UL
Western white pine	75, 72	UL, HPMA
Eastern white pine	120-215, 85	UL, CWC
Lodgepole pine	65-110	CWC
Ponderosa pine	105-230	UL
Red pine	142	CWC
Southern yellow pine	130-195	UL
Redwood	70	UL
Eastern white spruce	65	CWC, UL
Sitka spruce	100	UL
Yellow birch	105-110	UL
Cottonwood	115	UL
White (sugar) maple	104	CWC
Red oak	100	UL
White oak	100	UL
Sweetgum	140-155	UL
Walnut	130-140	UL
Yellow poplar	170-185	UL

CWC: Canadian Wood Council.
HPMA: Hardwood Plywood Manufacturers Association.
HUD/FHA: U.S. Dept. of Housing and Urban Development.
UL: Underwriters Laboratories Inc.

Flame spread ratings developed for virtually all wood-based panel products are based upon ASTM E84,[12] UL 723,[14] or NFPA 255 flame spread rating tests, using the Steiner Test Tunnel.

Fire Resistance

The fire resistance behavior of wood panels has been reviewed using the ASTM E119 (NFPA 251) test apparatus and procedure.[15] When wood panels are exposed to this test, surface ignition occurs in approximately 2 min.[13] Charring proceeds at a rate dependent upon the species and moisture content. However, the average char rate is approximately $1/30$ in. (0.85 mm) per min for the next 8 min. The insulating effect of the char layer slows the subsequent char rate to approximately $1/40$ in. (0.63 mm) per min. ASTM E119[15] test data has yielded a set of empirical relationships for predicting the rate of char in the direction transverse to the grain of the wood.[13]

For Douglas fir, $R = 1/[(57.4 + 1.16M)\rho + 8.4]$

For southern yellow pine, $R = 1/[(11.7 + 0.24M)\rho + 25.7]$

For white oak, $R = 1/[(40.1 + 0.81M)\rho + 15.0]$

where

R = rate of char in in. per min,

M = moisture content in percent, and

ρ = dry specific gravity.

(For SI units: 1 in. = 25.4 mm)

Table 4-3F lists the burn-through rates for vertically fire-exposed wood panels of 1.0-in. (25.4-mm) thickness, under the ASTM E119[15] (NFPA 251) fire exposure test method. The specific gravity and rate are for samples at moisture content of 6.0 percent.

TABLE 4-3F. *ASTM E119 Fire Resistance Data for Selected Woods*[13]

Species	Specific Gravity (in./hr)	Rate
Bald cypress	0.44	1.7
Basswood	0.42	2.4
Yellow birch	0.63	2.0
Chestnut	0.45	1.7
Douglas fir	0.45	1.6
Eastern hemlock	0.40	1.6
White (sugar) maple	0.64	2.1
Northern red oak	0.61	1.8
White oak	0.67	1.5
Eastern white pine	0.39	1.6
Ponderosa pine	0.42	2.1
Southern yellow pine	0.55	2.2
Sugar pine	0.32	2.0
Redwood	0.38	1.6
Sitka spruce	0.43	1.7
Sweetgum-sapwood	0.55	2.4
Sweetgum-heartwood	0.52	1.6
Yellow poplar	0.44	2.1

(For SI units: 1 in. = 25.4 mm.)

The temperature at the base of the char layer is usually in the neighborhood of 550°F (287°C), whereas 1/4 inch beneath the char the temperature is approximately 360°F (182°C).[13] This steep temperature gradient is typical of most wood species, and demonstrates the insulating effect of the char layer. This allows wood structural members to retain their load-bearing capacity during a fire until the charring has reduced the residual cross-section of the member below the minimum necessary for the design load.

Heat Release and Heat Release Rate

The heat release rate of wood and wood-based fuels becomes an important consideration when analyzing the behavior of a space under fire conditions. The heat release per unit of mass and the quantity of material present predict the total heat release rate attainable and, hence, the probability of flashover.[16]

The heat of combustion of most species of wood falls into the range of 8,000 to 12,000 Btu per lb (18 600 to 27 900 kJ/kg).[13] Under fire conditions, the heat release rate of the wood component of the fuel load depends on the incident radiant flux, moisture content, thickness, orientation, boundary conditions at the rear face, and the oxygen concentration of the ambient air. It is also a function of time, characterized by an early peak followed by a slow decay until flaming ceases. A second peak will also occur if the rear face of the specimen is insulated.

Heat release rates are determined with the ASTM E906 test method, which defines the conditions precisely.[17] For an oven-dry vertical pine board of nominal 1-in. (25.4-mm) thickness exposed to a thermal radiation level of 60 kW/m^2 in normal air with the rear

face exposed to a water-cooled plate, a peak heat release rate of 250 kW/m^2 in 19 min was reached when the flame went out.[18] A series of wood products with moisture contents of about 8 percent exhibited peak heat release rates between 90 and 170 kW/m^2 under the same conditions. The heat release rate of wood products increases linearly with the density.[19] The total heat release rate of the materials of construction and furnishings is an important factor in determining whether a room will reach flashover conditions.[20]

Table 4-3G shows some figures for the heats released by various wood products and also gives comparative figures for some other materials. It should be noted that the heat of combustion of most dry wood products is less per pound (kg) than that of the other fuels listed. The trend of replacing wood furnishings with furnishings made of plastics actually increases the fire load. This is an important factor when estimating the quantity of combustible materials exposed to a fire scenario. Such estimates must be adjusted for the nature of the combustibles when determining the needed fire resistance of structural elements or the optimum quantity of water required per unit of floor area for sprinkler discharge.

TABLE 4-3G. *Heat of Combustion of Various Wood and Wood-Based Products and Comparative Substances*

Substance	Heating Value	
	Btu per lb	kJ/kg
Wood sawdust (oak)	8,493*	19,755
Wood sawdust (pine)	9,676*	22,506
Wood shavings	8,248*	19,185
Wood bark (fir)	9,496*	51,376
Corrugated fiber carton	5,970*	13,866
Newspaper	7,883*	18,336
Wrapping paper	7,106*	16,529
Petroleum coke	15,800	36,751
Asphalt	17,158	39,910
Oil (cottonseed)	17,100	39,775
Oil (paraffin)	17,640	41,031

*Dry

The amount of fuel contributed by one combustible *vs.* another, or different forms of the same combustible, can also be used to evaluate the desirability of utilizing the material in a given situation. This is particularly true for measuring the fuel contributed by certain forms of building materials.

Engineered Wood-Composite Products

Engineered wood-composite products include laminated veneer lumber, glu-lam beams, composite/wood joists and beams, and other wood-composite materials, such as oriented strandboard (OSB), flakeboard, medium density fiberboard (MDF), and particleboard. The use of engineered wood-composite materials facilitates the production of lightweight structural components, such as composite/wood joists and beams, with substantially reduced mass and cross-sectional area. For example, composite/wood joists and beams provide superior structural strength with reduced mass, allowing greater design flexibility and economy. Nevertheless, their performance under fire conditions has become a source of some concern. There may be occasions where these engineered wood-composite products behave very differently than their primary wood precursers. (See Figure 4-3F.)

The introduction of glues and resins into the structural materials changes the chemistry of those materials. The relatively simple

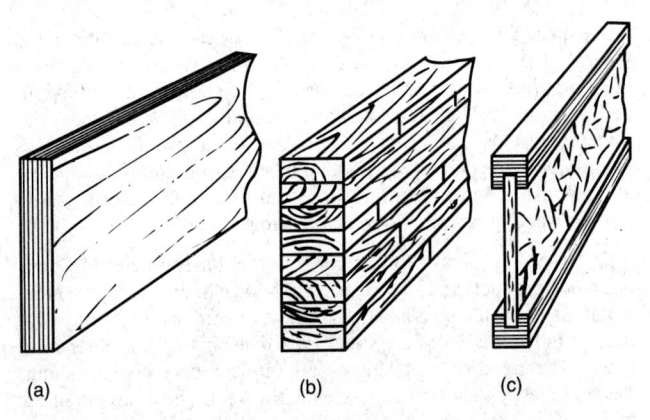

FIG. 4-3F.   (a) Laminated veneer lumber, (b) glu-lam beam, and (c) composite wood joist.

combustion chemistry of cellulose is complicated by the presence of petroleum-derived plastic resins, some of which may contain aromatic rings, halogens, epoxies, and other unsaturated hydrocarbon constituents. This affects heat output, flame spread rate, smoke composition, and smoke density during a fire involving these materials.

The more serious concern is the structural performance of engineered wood products, especially composite/wood joists and beams, under exposure to a fire. The reduced cross-sectional area of a composite/wood joist or beam provides a larger surface area for a given mass (weight) of material. Since both pyrolysis and combustion are surface phenomena, the increased surface area provides for a more rapid consumption of the fuel. As a result, the engineered wood-composite components burn faster than their solid, sawn lumber equivalents.[21] (See Figure 4-3G.)

The elevated rate of combustion on a mass consumed per unit time basis would predict earlier structural failure of a building supported by composite/wood components than one constructed of equivalent solid sawn lumber. However, there is the additional factor of the engineered distribution of mass in the composite/wood component. The mass of the composite/wood beam is concentrated in those areas where it is structurally significant, resulting in a cross section similar to an I-beam. This places the web and top flange in compression, and the bottom flange in tension. (See Figure 4-3H.) As the wood-composite material begins to char, the pyrolysis of the wood fibers leads to a more rapid deterioration of the structural properties than one encounters with dimension sawn lumber of equivalent design capacity. This effect combined with the greater rate of combustion due to the increased surface area and the greater heat re-

lease rate due to the petroleum-derived resins has been used to explain the earlier-than-expected failure of structures during fires.

When engineered wood-composite components were first introduced, there were also concerns that the heat of a fire might cause the adhesives to soften, allowing the component to fail. This notion might have come from anecdotal experience with casein adhesives. This does not seem to be the case with current wood-composite components using modern phenolic thermoset resins. The available research has not identified structural failures that are attributable to adhesive/resin failure.[21]

The National Fire Protection Research Foundation has been instrumental in the investigation of these questions. As the use of composite/wood beams and joists continues, the fire service will probably find it necessary to rethink the length of time a structure can be involved before it is deemed structurally unsafe for interior fire fighting.

The implications of the use of engineered lightweight construction components on the fire safety and fire endurance of structures continues to be an area of research. As wood has become more expensive and as structural engineers continue to make structures more efficient and lighter in weight, lightweight wood trusses may become more common. This demands that their behavior in a fire be better understood.

A solid sawn lumber joist or beam endures both compression and tension stresses under design loads, as shown in Figure 4-3I. Since the mechanical properties of wood vary with the orientation of the stress to the grain of the wood fibers, greater stresses can be supported by an engineered wood assembly of less weight if the orientation of the grain of the wood fibers is "adjusted" to employ the wood in an orientation relative to the applied stress, which takes maximum advantage of its mechanical properties. This approach yields a structure shown in Figure 4-3I.

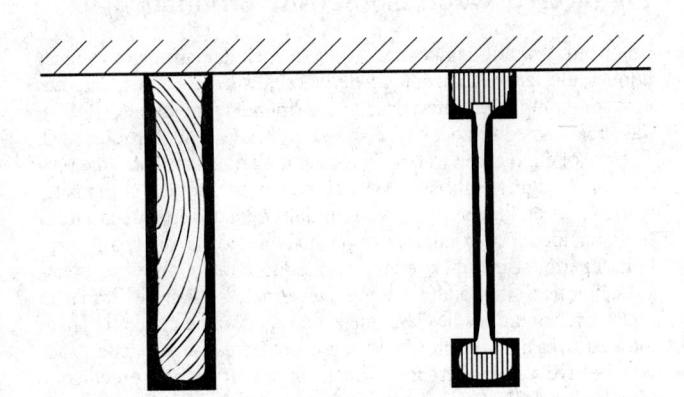

FIG. 4-3H.   The orientation of load forces on a beam or joist.

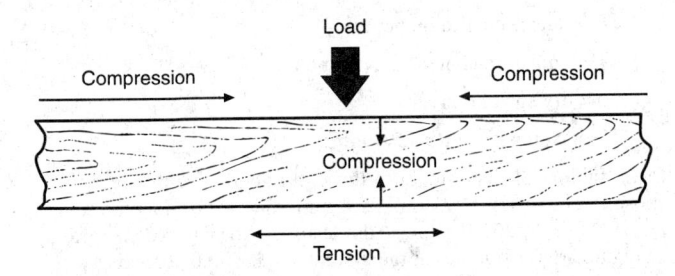

FIG. 4-3G.   The charring of a composite/wood joist compared with a sawn joist after the same fire exposure.

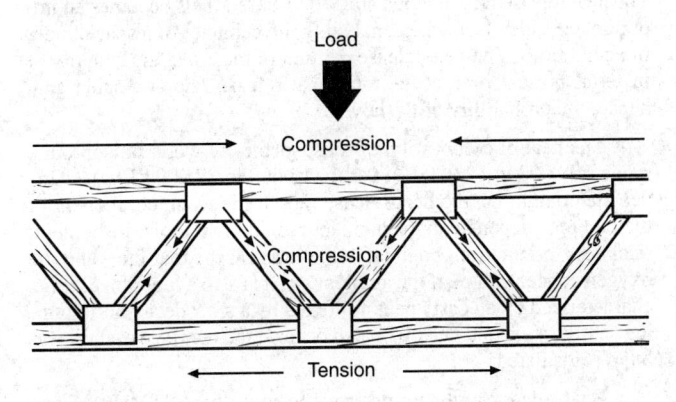

FIG. 4-3I.   The stresses in the lightweight wood truss assembly with both compression and tension stresses oriented along the wood grain.

These lightweight assemblies have two features that may prove to be a concern as the research is completed. First, the surface area-to-mass ratio of the lightweight truss assembly is much greater than that of the solid sawn plank. Since burning rate is proportional to the surface area, this could mean that the lightweight truss assembly will fail sooner under fire conditions. Second, lightweight truss assemblies are assembled from individual members that are attached to each other with steel gusset plates. There is currently some concern that the transmission of heat through the steel gusset plate will accelerate charring beneath the plate, leading to early failure of the attachment. Since the loss of an attachment threatens the load-bearing capacity of the entire truss, this could provide a mechanism for premature failure of a roof supported by lightweight truss assemblies.[22]

The National Fire Protection Research Foundation has been the coordinator of research into these questions. There is considerable research yet to be conducted to develop the answers addressing the above concerns.[22]

BIBLIOGRAPHY

References Cited

1. Roberts, J., and Caserio, M., *Modern Organic Chemistry*, W.A. Benjamin, Inc., New York, 1967, pg. 443.
2. Lehninger, A. L., *Biochemistry*, Worth Publishers, Inc., New York, 1970, pp. 231–232.
3. FOC, "Fire Research Report for the Year 1952," Department of Scientific and Industrial Research and Fire Offices' Committee, Her Majesty's Stationery Office, London.
4. Beall, F. C., and Eichner, H. W., "Thermal Degradation of Wood Components: A Review of the Literature," Report No. 130, 1970, Forest Products Laboratory, Madison, WI.
5. Brown, C. R., "The Ignition Temperatures of Solid Materials," *NFPA Quarterly*, Vol. 28, No. 2, 1934, pp. 135–145.
6. Fleischer, H. O., "The Performance of Wood in Fire," Report No. 2202, 1960, Forest Products Laboratory, Madison, WI.
7. McNaughton, G. C., "Ignition and Charring Temperatures of Wood," Mimeo No. R1464, 1944, Forest Products Laboratory, Madison, WI.
8. McGuire, J. H., "Limited Safe Surface Temperatures for Combustible Materials," *Fire Technology*, Vol. 5, No. 3, 1969, pp. 237–241.
9. Mitchell, N. D., "New Light on Self-Ignition," *NFPA Quarterly*, Vol. 45, No. 2, 1951, pp. 165–172.
10. LAFD, *Operation School Burning No. 1*, Los Angeles Fire Department, National Fire Protection Association, Quincy, MA, 1959.
11. FOC, "Fire Research Report for the Year 1950," Department of Scientific and Industrial Research and Fire Offices' Committee, Her Majesty's Stationery Office, London.
12. ASTM E84, *Standard Test Method for Surface Burning Characteristics of Building Materials*, American Society for Testing and Materials, W. Conshohocken, PA, 1994.
13. *Wood Handbook: Wood as an Engineering Material*, USDA Forest Service Handbook, No. 72, Government Printing Office, Washington, DC, 1987.
14. UL 723, *Standard for Safety Test for Surface Burning Characteristics of Building Materials*, Underwriters Laboratories Inc., Northbrook, IL, 1993.
15. ASTM E119, *Standard Test Methods for Fire Tests of Building Construction and Materials*, American Society for Testing and Materials, W. Conshohocken, PA, 1988.
16. Babrauskas, V., "Estimating Room Flashover," *Fire Technology*, Vol. 16, No. 2, May 1980.
17. ASTM E906, *Standard Test Method for Heat and Visible Smoke Release Rates for Materials and Products*, American Society for Testing and Materials, W. Conshohocken, PA, 1983.
18. Parker, W. J., and Long, M. E., "Development of a Heat Release Rate Calorimeter at NBS," *Ignition, Heat Release, and Noncombustibility of Materials*, ASTM STP 502, American Society for Testing and Materials, W. Conshohocken, PA, 1972, pp. 135–151.
19. Chamberlain, D. L., "Heat Release Rate Properties of Wood-Based Materials," NBSIR 83-2597, 1983, National Bureau of Standards, Gaithersburg, MD.
20. McCaffrey, B. J., Quintiere, J. G., and Harkelroad, M. F., "Estimating Room Temperatures and the Likelihood of Flashover Using Fire Data Correlations," *Fire Technology*, Vol. 17, No. 2, 1980.
21. Grundahl, K., P.E., *National Engineered Lightweight Construction Fire Research Project*, National Fire Protection Research Foundation, Quincy, MA, 1992.
22. Malanga, R., P.E., "Fire Endurance of Lightweight Wood Trusses in Building Construction," *Fire Technology*, Vol. 31, No. 1, National Fire Protection Association, Quincy, MA, 1995.

References

Bierberdorf, F. W., and Yuill, C. H., "An Investigation of the Hazards of Combustion Products in Building Fires," U.S. Public Health Service Contract No. PH86-62-208, Final Report, 1963.
BOCA, *The BOCA Basic Building Code*, 5th ed., Sec. 922, Building Officials and Code Administrators International, Chicago, 1970.
Brenden, J. J., "Determining the Utility of a New Optical Test Procedure for Measuring Smoke from Various Wood Products," Research Paper 137, 1970, Forest Products Laboratory, Madison, WI.
Ferguson, G. E., "Fire Gases," *NFPA Quarterly*, Vol. 27, 1933.
ICBO, *Uniform Building Code*, Vol. 1, Sec. 4202b, International Conference of Building Officials, Whittier, CA, 1970.
Imaizumi, K., "Stability in Fire of Protected and Unprotected Glued Laminated Beams," *Transactions*, No. 263, Chalmers University of Technology, Gotenburg, Sweden, 1963, pp. 46–71.
NFPA, "Fire Gas Research Report," *NFPA Quarterly*, Vol. 45, No. 3, 1952.
NSF, "Answers to Burning Questions—Reducing the Hazards of Fire," *Mosaic*, National Science Foundation, Washington, DC, 1973.

NFPA Codes, Standards, and Recommended Practices

Reference to the following NFPA codes, standards, and recommended practices will provide further information on wood and wood-based products discussed in this chapter. (See the latest version of *The NFPA Catalog* for availability of current editions of the following documents.)

NFPA 251, *Standard Methods of Tests of Fire Endurance of Building Construction and Materials*
NFPA 255, *Standard Method of Test of Surface Burning Characteristics of Building Materials*
NFPA 258, *Standard Research Test Method for Determining Smoke Generation of Solid Materials*

Additional Reading

Atreya, A., and Emmons, H. W., "Wood Ignition and Pyrolysis," 9th Joint Panel Meeting of the UJNR Panel on Fire Research and Safety, NBSIR 88-3753, Apr. 1988, National Bureau of Standards, Gaithersburg, MD.
Bland, K. E., "Behavior of Wood Exposed to Fire: A Review and Expert Judgment Procedure for Predicting Assembly Failure," Worcester Polytechnic Inst., MA, Thesis, May 1994, 159 pages.
Carling, O., "Brandteknisk dimensionering av massiva trakonstruktioner. (Fire Engineering Design of Timber Structures.)," TrateknikCentrum, Stockholm, Sweden, Report I 9004018, Apr. 1990, 140 page.
Eboatu, A. N., Garba, B. and Akpabio, I. O., "Flame-Retardant Treatment of Some Tropical Timbers," *Fire and Materials*, Vol. 17, No. 1, 1993, pp. 39–42.
Egan, M. R., and Litton, C. D., "Wood Crib Fires in a Ventilated Tunnel," No. 9045, 1986, U.S. Department of the Interior, Bureau of Mines, Washington, DC.
Emmons, H. W., and Atreya, A., "The Science of Wood Combustion," *Indian Acad. Sci. Eng. Sci.*, Vol. 5, pp. 259–269.
Fire Retardent Chemicals Association, *Fire Safety Problems Leading to Current Needs and Future Opportunities*, Technomic, Lancaster, PA, 1989.
"Flame Retardant Treatments for Wood and Its Derivatives. Fire Prevention Design Guide," Fire Protection Association, London, England, FPDG 13, Aug. 1990, 4 pages.
Fredlund, B., "Modeling of Heat and Mass Transfer in Wood Structures During Fire," *Fire Safety Journal*, Vol. 20, No. 1, 1993, pp. 39–69.

Fredlund, B., "Calculation of the Fire Resistance of Wood Based Boards and Wall Constructions," Lund Institute of Science and Technology, Sweden, LUTVDG/TVBB-3053, 1990, 93 pages.

Gardner, W. D., and Thomson, C. R., "Flame Spread Properties of Forest Products: Comparison and Validation of Prescribed Australian and North American Flame Spread Test Methods," *Fire and Materials*, Vol. 12, No. 2, June 1988, pp. 71–85.

Handbook of Wood and Wood-based Materials for Engineers, Architects, and Builders, Forest Products Laboratory, Hemisphere, New York, 1989.

Isner, M. S., "Untreated Wood Shingles Spread Apartment Fire," *Fire Journal*, Vol. 83, No. 2, Mar. 1989, p. 77.

Janssens, M., "Piloted Ignition of Wood: A Review," *Fire and Materials*, Vol. 15, No. 4, Oct./Dec. 1991, pp. 151–167.

Janssens, M., "Rate of Heat Release of Wood Products," *Fire Safety Journal*, Vol. 17, No. 3, 1991, pp. 217–238; QMC Fire and Materials Center in association with Fire Research Station, International Conference for Fire: Control the Heat...Reduce the Hazard, October 24–25, 1988, London, England, 18/1–10 pp., 1988.

Janssens, M., "Thermal Model for Piloted Ignition of Wood Including Variable Thermophysical Properties," *Proceedings* of the Third International Symposium on Fire Safety Science, Elsevier Applied Science, New York, 1991, pp. 167–176.

Janssens, M. L., "Fundamental Thermophysical Characteristics of Wood and Their Role in Enclosure Fire Growth," National Forest Product Assoc., Washington, DC, Thesis, Sept. 1991, 492 pages.

Janssens, M. L. and White, R. H., "Short Communication: Temperature Profiles in Wood Members Exposed to Fire," *Fire and Materials*, Vol. 18, No. 4, July/Aug. 1994, pp. 263–265.

Janssens, M., "Rate of Heat Release of Wood Products," *Fire Safety Journal*, Vol. 17, No. 3, 1991, pp. 217–238.

Jones, J. C., Hughes, K. C. and Wearing, H., "On the Action of Wetting Agents in the Extinction of Wood Fires," *Journal of Fire Sciences*, Vol. 10, No. 1, Jan./Feb. 1992, pp. 20–27.

Kashiwagi, T., Werner, K., and Ohlemiller, T. J., "Products of Wood Gasification," NBSIR 85-3127, Apr. 1985, National Bureau of Standards, Gaithersburg, MD.

Kotoyori, T., "Critical Ignition Temperatures of Wood Sawdusts," *Proceedings* of the First International Symposium on Fire Safety Science, Hemisphere, New York, 1986, pp. 463–471.

Kubler, H., "Indicators and Significance of Air Supply in the Combustion of Wood for Heat," *Wood and Fiber Science*, Vol. 23, No. 2, 1991, pp. 153–164.

LeVan, S. L. and Winandy, J. E., "Effects of Fire Retardant Treatments on Wood Strength: A Review," *Wood and Fiber Science*, Vol. 22, No. 1, 1990, pp. 113–131.

McNeil, R. R., "Fire Impact on Wood: The Next Five Years," *Wood and Fiber*, Vol. 9, No. 1, pp. 2–12.

Mikkola, E., "Charring of Wood Based Materials," *Proceedings* of the Third International Symposium on Fire Safety Science, Elsevier Applied Science, New York, 1991, pp. 547–556.

Mikkola, E., "Charring of Wood," VTT-Technical Research Center of Finland, Espoo, VTT Research Reports 689, Project PAL7003, June 1990, 35 pages.

Ohlemiller, T. J., "Smoldering Combustion Propagation on Solid Wood," *Proceedings* of the Third International Symposium on Fire Safety Science, Elsevier Applied Science, New York, 1991, pp. 565–574.

Ohlemiller, T., and Sharb, W., "Products of Wood Smolder and Their Relation to Wood Burning Stoves," NBSIR 88-3767, May 1988, National Bureau of Standards, Gaithersburg, MD.

Ohlemiller, T. J., Kashawagi, T., and Werner, K., "Wood Gasification at Fire Level Heat Fluxes," *Combustion and Flame*, Vol. 69, 1987, pp. 155–170.

Ohtani, H., Miyazawa, S. and Nakaya, I., "Experimental Study on Bottom Surface Combustion of Woods, Japanese Association of Fire Science and Engineering, Fire Research Annual Conference, May 17–18, 1990, pp. 103–106.

Ohtani, H., Ohno, T. and Uehara, Y., "Heat Release Measurement During Smoldering Combustion of Wood Sawdust," *Journal of Applied Fire Science*, Vol. 4, No. 3, 1994/1995, pp. 161–169.

Ostman, B. A. L. and Tsauntaridis, L. D., "Heat Release and Classification of Fire Retardant Wood Products," *Fire and Materials*, Vol. 19, No. 6, 1995, pp. 253–258.

Parker, W. J., "Prediction of the Heat Release Rate of Wood," *Proceedings* of the First International Symposium on Fire Safety Science, Hemisphere, New York, 1986, pp. 207–216.

Parker, W. J., "Prediction of the Heat Release Rate of Douglas Fir," *Proceedings* of the Second International Symposium on Fire Safety Science, Hemisphere Publishing Corporation, New York, 1989, pp. 337–346.

Parker, W. J., " Wood Materials. Part A. Prediction of the Heat Release Rate From Basic Measurements," Chapter 11, *Heat Release in Fires*, Elsevier Applied Science, NY, 1992, pp. 333–356.

Pasek, E. A. and McIntyre, C. R., "Heat Effects on Fire Retardant-Treated Wood.," *Journal of Fire Sciences*, Vol. 8, No. 6, November/December 1990, pp. 405–420; Fire Retardant Chemicals Association. Fire Safety Developments and Testing: Toxicity—Heat Release—Product Development—Combustion Corrosivity, October 21–24, 1990, Ponte Vedra Beach, FL, 1990, pp. 181–192.

Sarvaranta, L. and Mikkola, E., "Mortars Containing Wood-Based Fibers Under Thermal Exposure Using Cone Calorimeter Heating," *Fire and Materials*, Vol. 19, No. 1, Jan./Feb. 1995, pp. 35–41.

Tillott, R. J., "Preserving Timber From Fire," *Fire Prevention*, No. 240, June 1991, pp. 18–22.

Tran, H. C., "Experimental Heat Release Rate of Wood Products," First U. S. Symposium for Heat Release and Fire Hazard, Abstracts, Interscience Communications Limited, 1991, pp. 1–2.

Tran, H. C., "Wood Materials. Part B. Experimental Data on Wood Materials," Chapter 11, *Heat Release in Fires*, Elsevier Applied Science, NY, 1992, pp. 357–372.

Tran, H. C. and Janssens, M. L., "Wall and Corner Fire Tests on Selected Wood Products," *Journal of Fire Sciences*, Vol. 9, No. 2, Mar./Apr. 1991, pp. 106–124.

Tran, H. C., "Modelling of Wall Fires of Selected Wood Products," Forest Products Lab., Madison, WI, Nineteenth IUFRO World Congress, Aug. 1990, Montreal, Canada, 1990, pp. 1–12.

Tzeng, L. and Atreya, A., "Theoretical Investigation of Piloted Ignition of Wood." National Institute of Standards and Technology, Gaithersburg, MD, NIST-GCR-91-595, Aug. 1991, 230 pages.

White, R. H., "Analytical Methods for Determining Fire Resistance of Timber Members," *SFPE Handbook of Fire Protection Engineering*, 2nd ed., National Fire Protection Association, Quincy, MA, 1995.

White, R. H., "Analytical Methods for Determining Fire Resistance of Timber Members," *The SFPE Handbook of Fire Protection Engineering*, Second Edition, Section 4, Chapter 12, National Fire Protection Association, Quincy, MA, June 1995, pp. 217–229.

White, R. H., "Fire Resistance of Wood-Frame Wall Sections," Forest Service, Madison, WI, Sept. 1990, 26 pages.

White, R. H. and Nordheim, E. V., "Charring Rate of Wood for ASTM E 119 Exposure," *Fire Technology*, Vol. 28, No. 1, Feb. 1992, pp. 5–30.

Winandy, J. E., Ross, R. J. and LeVan, S. L., "Fire-Retardant-Treated Wood: Research at the Forest Products Laboratory," *TRADA*, Vol. 4, 1991, pp. 4.69–4.74, International Timber Engineering, *1991 Proceedings*, September 2–5, 1991, London, England, 1991.

"Wood Handbook: Wood as an Engineering Material," *USDA Forest Service Handbook*, No. 72, Government Printing Office, Washington, DC, 1987.

Yates, J. M., "Firefighters Push for Mandatory Flammability Testing," *Wood and Wood Products*, Vol. 92, No. 11, Oct. 1987, p. 38.

Yudong, L. and Drysdale, D. "Measurement of the Ignition Temperature of Wood," *Fire Safety Science*, Vol. 1, No. 1, Sept. 1992, pp. 25–30; First Asian Conference for Fire Science and Technology (ACFST), October 9–13, 1992, Hefei, China, International Academic Publishers, China, 1992, pp. 380–385.

Zoufal, R., "Determination of the Thermal Conductivity Coefficient and the Specific Thermal Capacity of Wood at High Temperatures," Third International Symposium on Fire Protection of Buildings, May 10–12, 1990, Eger, Hungary, 1990, pp. 45–51

FIRE-RETARDANT AND FLAME-RESISTANT TREATMENTS OF CELLULOSIC MATERIALS

Revised by James R. Shaw

This chapter discusses the basic principles of fire-retardant and flame-resistant treatments, their uses and limitations, and methods of treatment for wood, wood-based composite panels, paper, and miscellaneous decorative materials.

Related information can be found in the following chapters: Section 4, Chapter 3, "Wood and Wood-Based Products"; Section 4, Chapter 5, "Fibers and Textiles"; Section 4, Chapter 10, "Plastics and Rubber"; Section 7, Chapter 3, "Interior Finish"; and Section 10, Chapter 7, "Wildland Fire Management."

Cellulosic materials are an integral part of the living environment. Homes, schools, factories, offices, shops, etc., utilize large quantities of cellulosic products in construction, furnishings, and decorations. Fire-retardant and flame-resistant treatments for these cellulosic materials can increase fire protection.

Popular usage has created a variety of terms associated with this subject, resulting in frequent misconceptions, misunderstandings, misuses, and an undesirable degree of ambiguity. Terms such as "fire resistant," "fire retardant," "flame resistant," "flame retardant," and "flameproof" are often used indiscriminately and incorrectly.

The term "fire resistant" should not be used in reference to the fire-retardant treatment of combustibles. Fire resistant is used properly only to signify the ability of a structure, material, or assemblage to resist the effects of a large-scale severe fire exposure. The American Society for Testing and Materials (ASTM) defines fire resistance as "the property of a material or assembly to withstand fire or give protection from it."[1] Fire-resistance rating or fire endurance is the time, in minutes or hours, that materials or assemblies have withstood a fire exposure, as established in accordance with the test procedures of NFPA 251 (ASTM E119), *Standard Methods of Tests of Fire Endurance of Building Construction and Materials.* "Flameproof" and its derivatives are also misleading and subject to abuse and misunderstanding; their use should be discouraged.

"Fire retardant" signifies a lesser degree of protection than fire resistant. It should be used in reference to chemicals, treatments, or coatings used to reduce the combustibility of building materials, and other such treated materials. ASTM suggests "fire retardant" be used as a modifier only with defined compound terms, such as fire-retardant barrier, fire-retardant chemical, fire-retardant coatings, and fire-retardant treatments.[1] Flame spread rating refers to indices or classifications obtained according to NFPA 255 (ASTM E84), *Standard Method of Test of Surface Burning Characteristics of Building Materials* [hereinafter referred to as NFPA 255 (ASTM E84)], or other flame spread tests. "Flame retardant" and "flame resistant" may be used more or less interchangeably and denote decorative materials that, due to chemical treatment or inherent properties, do not ignite readily or propagate flaming under small-to-moderate fire exposure. "Flame retardant" is the preferred term to denote chemicals, processes, or coatings used for the treatment of decorative materials, including fabrics, foliage, Christmas trees, and similar materials used as decorations or furnishings.

Fire retardants and flame-retardant treatments frequently have been misused. They have a widespread appeal due perhaps, in part, to a general lack of understanding about the limitations of the treatments, and a failure to differentiate between those that are effective and those that are not. One popular misconception is that all fire-retardant treatments give a fire-resistance rating. This is not true. While some fire-resistant coatings may improve the fire endurance of combustible materials and assemblies, other treatments are only effective in retarding the rate of burning and the rate at which fuel is contributed by the treated material. This action, however, can reduce the fire intensity of some materials that otherwise might be very fast burning. This burning rate property is measured and reported as the "rate of heat release."

In other words, there are practical limits to fire-retardant treatments. They are not all permanent, they are not capable of making cellulosic materials noncombustible, they affect other properties of the treated material, and they add expense. They do, however, provide protection on a competing basis with non-fire-retardant treatments and coatings. For example, fire-retardant paints have been improved to the point where many provide a colorful, protective coating with comparable wear resistance and strength of normal oil- and water-based paints.

BASIC PRINCIPLES

Treatments to reduce the combustibility of wood, fabrics, and other natural materials have a long history reaching back to at least the time of Moses. The early use and development of chemical retardants was limited to materials that were accidentally found to be effective. Research in basic combustion processes and action of retardants was greatly expanded during World War II due to increased military needs for flame-resistant fabrics and fire-retardant wood. This expansion continued for several years, leading to new-generation fire-retardant formulations for wood. Because cotton

James R. Shaw, Ph.D., is a senior scientist in the corporate research and development division of Weyerhaeuser Company, Tacoma, WA.

textiles and wood products have by far the longest history of use and are of great economic importance, much of the research has been directed toward these cellulosic materials. In the past few years the research activity in the United States has stabilized.

Burning of Cellulose

Since cellulose is the major component of wood and other cellulosic materials and the major source of combustible volatiles that fuel the flaming combustion, it is appropriate to focus on the cellulose reactions in the thermal degradation process.

The chemical reactions and the mechanisms involved in the combustion of untreated cellulose are, by themselves, highly complex and not yet fully understood.[2] The complexity is magnified by the addition of fire-retardant chemicals.

Wood or cellulose normally does not burn directly. The initial reaction to high heat exposure is pyrolysis, in which some of the products given off are gases, vapors, or mists that can form flammable mixtures with air. On ignition, the combustion of these products is evidenced by visible flames. The solid residues include charcoal, which can undergo a different form of combustion by combining directly with oxygen, as evidenced by glowing. This is an important distinction, since many chemicals that are effective flame inhibitors have little or no effect in retarding flameless combustion, or afterglow. For most applications, flame retardance is more important, but there are some instances where glow resistance is equally necessary.

Theories of Fire-Retardant Mechanisms

Four general theories have evolved to explain the mechanisms by which chemicals retard flaming or glowing of cellulosic materials. It is agreed that no single theory is adequate to explain these mechanisms and that, in most cases, more than one mechanism is involved. The four theories that offer the most satisfactory explanations for the function of fire-retardant chemicals are as follows.

Thermal: There are three possible ways that fire-retardant materials work to reduce thermal buildup of the treated combustible: (1) they increase thermal conductivity to dissipate the heat of combustion, (2) they increase thermal absorption to reduce the amount of heat available for pyrolysis, and/or (3) they provide thermal insulation to reduce heat access to the substrate. Thermal insulation is the most important mechanism and is also discussed below, under coating theories.

Coating: Some retardants melt or fuse at relatively low temperatures, and it is believed that they may form an insulating coating over the fibers of the treated material, acting to exclude oxygen and inhibit the escape of combustible gases. More effective retardants exhibit a bubbling or foaming action, creating an insulating barrier. Intumescent paints and coatings protect substrates by this mechanism.

Gas: Some chemicals function by releasing nonflammable gases, such as water, ammonia, and carbon dioxide, under heat exposure that dilute the combustible gases sufficiently to render the mixture nonflammable in air. Other retardants catalytically inhibit the free radical chain reactions that occur in the gas phase (flaming combustion). Halogen-containing retardants exhibit this type of mechanism and, when combined with antimony, a synergistic (more than additive) effect is observed.

Chemical: Simplified representations of the chemical mechanism are shown in Figures 4-4A and 4-4B. These diagrams illustrate the different effects of fire on untreated and treated cellulose and are diagramatically explained in "Flameproofing Textile Fabrics."[3] Untreated cellulose breaks down into liquids and a smaller amount of

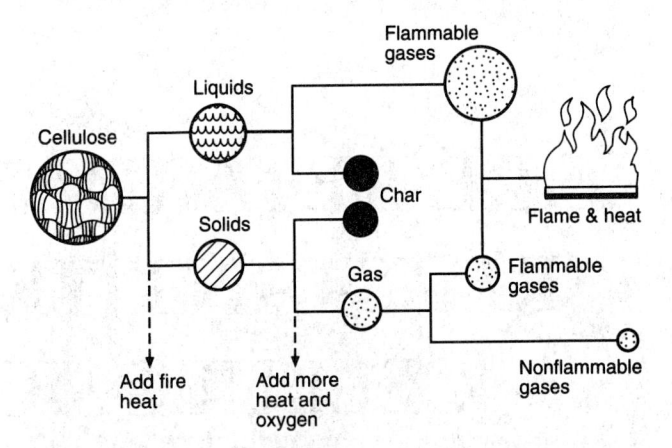

FIG. 4-4A. *Schematic diagram of the burning sequence of untreated cellulose. The size of the circle is representative of the volume of product. Note predominance of liquid and flammable gas products.*

solids. The liquids, in turn, break down into flammable gases and char, and the solids decompose into char and gases, some of which are flammable, some not. The flammable gases volatilized from the liquids combine with those from the solids to cause flaming and the emission of additional heat, which continues the reaction.

With treated cellulose, fewer liquids and more solids are formed, and more of the solids become char. The reduced amount of liquids creates some char and some flammable gases. Because the amount of flammable gases from both the liquid and solid phases is now significantly less with treated cellulose, ignition is delayed and there is less heat released. Therefore, flame resistance and a reduced burning rate is achieved. Another postulate, an expansion of the chemical theory known as the catalytic dehydration theory, has been advanced.[4]

Studies at the U.S. Forest Products Laboratory[5] have shown that pyrolysis reactions (without the presence of oxygen) for the lignin component of wood start at approximately 428°F (220°C), while that for the cellulose fraction starts near 518°F (270°C). However, once pyrolysis for the cellulose fraction is started, its decomposition is primarily an endothermic reaction and is nearly completed when

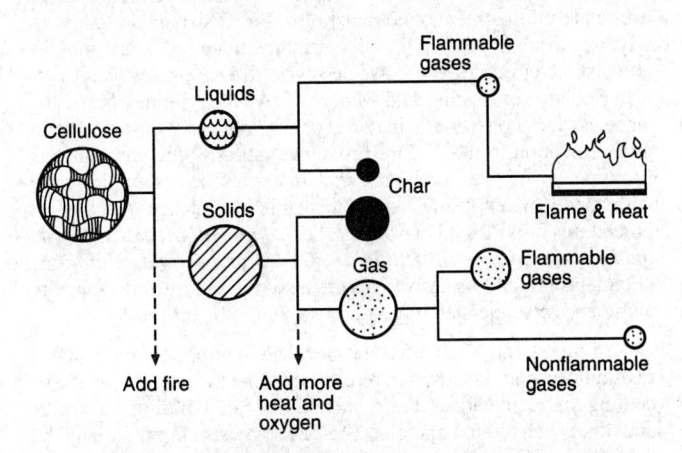

FIG. 4-4B. *Schematic diagram of the burning sequence of treated cellulose. The size of the circle is representative of the volume of product. Note the predominance of solids and the severe reduction in liquids and flammable gases emanating from the liquids, as compared with the same burning sequence for untreated cellulose as illustrated in Figure 4-4A.*

752°F (400°C) is reached. The lignin decomposes more slowly, almost in an entirely exothermic reaction, and loses less than half its weight in pyrolysis up to 1112°F (600°C).

Fire-retardant chemicals greatly influence the pyrolysis reaction of the cellulose fraction, frequently reducing the temperature at which decomposition starts from 518 to 446°F (270 to 230°C). As a result, the decomposition of the cellulose is promoted directly to form char and water, instead of forming intermediate flammable gases and tar products; and the char yield from the cellulose at 680°F (360°C) is increased from 15 percent up to 30 to 35 percent. The chemical treatment also increases the char yield from the lignin component from 49 percent up to 57 percent, but here the significance is less.

It was further shown that the combustion reactions for wood can be divided primarily into two exothermic stages: (1) one near 626°F (330°C) associated with flaming as the cellulose fraction of the wood undergoes decomposition into flammable gases and tars and becomes ignited, and (2) a second reaction near 806°F (430°C) associated with glowing as the lignin fraction of the wood, which has been changed to charcoal, undergoes oxidation. Fire-retardant chemicals greatly reduce the intensity of the first exotherm, and some chemicals also reduce the intensity of the second exotherm, thus reducing the heat released by the treated wood over a wide temperature range. Other chemicals reduce the first exotherm and intensify the second, thus requiring a second chemical to result in maximum fire-retardant performance. Heat release during the pyrolysis reactions for wood was found to be less than 5 percent of that released during the combustion reactions.

All of this data tended to prove that one of the principal reactions of fire-retardant chemicals impregnated into wood is to accelerate chemical reactions at certain temperature ranges. This reduces the release of certain intermediate products, such as levoglucosan tars and flammable gases, which may contribute to the flaming of wood, and also results in the formation of greater percentages of charcoal and water. Some chemicals are then further effective in reducing the oxidation rate for the charcoal residue. An excellent discussion of these basic principles is contained in the Forest Products Laboratory report, "Theories of the Combustion of Wood and Its Control."[6]

FIRE-RETARDANT TREATING METHODS

There are four basic methods of treatment to produce fire retardance in materials: (1) chemical changes (substitutions, admixtures), (2) impregnation (saturation, absorption), (3) pressure impregnation, and (4) coating. Chemical changes and pressure impregnation are limited to manufacturing processes and procedures, and are not adaptable for field use. Impregnation and coating methods may be used in the field. Generally, methods employed in manufacturing processes are to be preferred, because they provide greater uniformity, permanence, and dependability. Applications in the field, or subsequent to manufacture of the basic material, are more subject to variation in skill and integrity of the processor or applicator. Treatments of the coating- and pressure-impregnation type are covered in NFPA 703, *Standard for Fire Retardant Impregnated Wood and Fire Retardant Coatings for Building Materials.*

Chemical Changes

Fire-retardant chemical changes are primarily effective with plastics and with synthetic fibers, because different elements and compounds can be chosen that will alter the burning characteristics of the final product. Since the fire-retardant elements are an integral part of the chemical composition, these materials are said to be "in-

herently" fire retardant. Naturally occurring polymers, such as cellulose, cannot be made fire retardant by this method. However, another type of chemical fire-retardant treatment, known as chemical grafting, can be applied to natural polymers. In chemical grafting, the material is treated by a multistage process in which the material is chemically activated so that molecules of fire-retardant chemical are bonded to molecules of the material being treated. A follow-up curing step ensures permanence. This approach has not developed into a commercially viable method of producing fire-retardant-treated cellulosic products. Research efforts on chemical modification of wood continue, but the primary focus is on wood preservation and enhanced dimensional stability of wood.[7]

Impregnation

Impregnation refers to a technique for treating absorbent materials. The flame-retardant chemicals are either dissolved or dispersed in a solvent, usually water, and the material to be treated is thoroughly wetted or saturated with the solution. The solution may be applied by spraying or by immersion of the material. The latter process is used for treatment of large volumes of fabric yardage. The excess solution is extracted by passing the saturated material through squeeze rollers before drying.

In the simple water-soluble salt impregnation processes, the result is merely to deposit minute crystals of the salts within or on the surface of the fibers. The process is purely physical, i.e., no chemical reaction is involved. With durable treatments, insoluble deposits are formed by chemical reactions or other means.

Certain cellulose-based products, such as paper, acoustical tile, and wood-based compositions, incorporate a wet pulp stage during the manufacturing process. It is possible to add fire-retardant chemicals at the wet pulp stage, resulting in even distribution of the chemicals through the entire mass of the finished product. This practice is prevalent in manufacturing treated papers of many types, and is currently in use by several producers of building materials. Wood-fiber acoustical tile and building panels processed by this method have a NFPA 255 (ASTM E84), flame spread rating of less than 25.

Pressure Impregnation

Fire-retardant pressure impregnation is used for treating relatively dense nonabsorbent materials, such as wood. This process replaces the air in the wood cells with fire-retardant chemical solutions. (The chemicals are deposited as the solutions dry.) The treating solution is usually forced into the wood by the standard vacuum pressure methods used in the wood preserving industry.[8] Compared to simple impregnation techniques, pressure impregnation provides for deeper penetration and much greater chemical retention.

Coating

There are several types of fire-retardant coatings that are useful in treating many materials. Coatings may be applied at any stage from manufacture to use. They may either actively inhibit flame spread to some degree or present a noncombustible surface over which flame cannot spread. They are used predominantly on nonabsorbent building materials that cannot be treated by any other method; on Christmas trees and similar decorative materials; and, to a limited extent, on paper and fabrics that for various reasons cannot be treated effectively by impregnation.

The effectiveness of a coating depends on the chemical and physical properties of the material to which the coating is applied, the effectiveness of the coating on this material, the ability of the applier, and the thoroughness of the treatment. Since types of coatings and their effects vary depending on the material treated, they are discussed in the subsections of this chapter that cover materials.

PRACTICAL LIMITATIONS OF TREATMENTS

Ideally, a fire-retardant treatment would permanently eliminate all flame spread and smoke generation characteristics from a material; the color, feel, strength, weight, and other key characteristics of the basic material would not suffer from the treatment; and the method would be inexpensive, easy, and safe. There are no present-day treatments meeting all of these qualifications. As the different materials are discussed, the limitations of fire-retardant treatments will be identified.

WOOD

Wood can be made fire retardant by either pressure impregnation or coatings. While both of these treatments will reduce the flame spread of wood, only some fire-retardant coatings are significantly effective in increasing the resistance of wood to degradation under sustained fire exposure or in delaying reduction of its load-bearing capacity.

While the fire-retardant treatment that would make wood noncombustible does not exist, the spread of flame from the immediate area of an incipient fire can be retarded and, in some cases, prevented by present-day treatments, with a corresponding substantial reduction in smoke contribution. Also, the fire-resistance rating of wood structural members in buildings may be improved through the use of fire-resistant coatings.[9,10,11]

Fire-Retardant Pressure Impregnation

Pressure impregnation treatments deposit chemicals within the fibrous passages of wood. Many chemicals exhibit fire-retardant properties, but because of cost limitations or various other objectionable characteristics, comparatively few are considered practical. Chemicals commonly used in the past include monobasic and dibasic ammonium phosphates, ammonium sulfate, borax, boric acid, zinc chloride, and sodium dichromate, usually in various combinations. Newer formulations tend to exclude hygroscopic salts, such as ammonium sulfate and zinc chloride, and, in some cases, include organophosphorus compounds and amino resins. Figure 4-4C shows the effect of some inorganic salts on the flame spread of Douglas fir. The most significant improvements in the new formulations are the high tolerance for high relative humidity (RH) conditions (up to 95 percent RH) and exterior application due to their nonleachability. The impregnation operation can be controlled to secure a predetermined absorption of solution. The important considerations are the depth of penetration and the amount of chemical deposited per unit volume of the wood.

Since the fire-retarding effect of the treatment is directly related to the amount of the chemical deposited within the wood, it is interesting to compare the effect of an ordinary brush treatment with a pressure-impregnation treatment. A 10-ft (3-m) high fence 50 ft (15.25 m) long of nominal 1-in. (25-mm) pine boards would absorb about 4 gal (15 L) of treating solution if applied by brush. The effect on flame spread would be minor [flame spread index (FSI) = 130 to 195, Class III]. When the same boards are pressure impregnated to refusal (i.e., the point at which no more solution is absorbed by the wood), they absorb about 225 gal (852 L) of chemical and the flame spread is cut to one-fourth or less of the original (FSI < 25, Class I).

A load or "charge" of wood is locked in a large cylinder that can be up to 8 ft (2.5 m) in diameter and over 140 ft (43 m) long. The air is then withdrawn, and the resulting vacuum draws much of

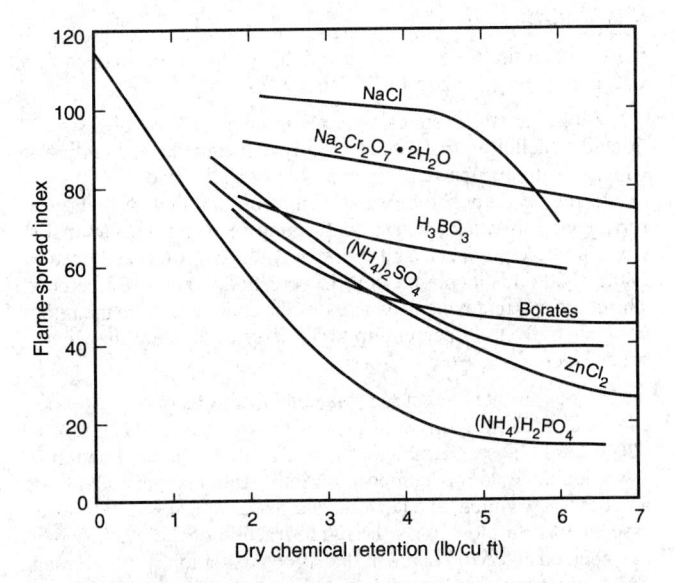

FIG. 4-4C. *Effect of inorganic additives on flame-spread index of Douglas fir, measured in 8-ft (2.4-m) tunnel.*[12]

the entrapped air from the wood cells. After a delay to allow the vacuum to stabilize, the fire-retardant chemical solution is bled into the cylinder. After the cylinder has been filled with the treating solution, additional chemicals are forced into it under pressure. As the pressure is raised [up to 200 psi (1,379 kPa)], the chemicals are forced into the empty wood voids. The process can be run to refusal and the unabsorbed salt solution drained from the cylinder. A slight vacuum may then be drawn to remove surface chemicals that would cause dripping and slow drying. After the charge is removed and dried, it is ready for shipment and use. Additional details are available from the American Wood Preservers Association.

This treating process has become an integral part of the definition of fire-retardant treated wood (FRTW) which is, per NFPA 255 (ASTM E84),

> any wood product impregnated with chemicals by a pressure process or other means during manufacture, and which, when tested in accordance with ASTM E84 for a period of 30 minutes, shall have a flame spread of not over 25 and show no evidence of progressive combustion. In addition, the flame front shall not progress more than 10 1/2 ft (3.2 m) beyond the centerline of the burner at any time during the test.

Advantages: Pressure impregnation permits a large amount of fire-retardant chemical to be deposited in the wood under close control and with uniform and predictable results. Samples for tests are readily available. Of primary importance, the treatment can be considered permanently effective under normal and proper conditions of use. Since both water-soluble and nonleachable fire retardants are available, it is important that nonleachable treatments be chosen for exterior use. Treatments with low hygroscopicity should be used for interior applications where high humidity conditions exist.

Disadvantages: In recent years, the fire-retardant chemical producers changed to "new generation," low-hygroscopicity formulations that dramatically improved the resistance to corrosion of metal fasteners. However, there have been reports of structural failures in fire-retardant treated plywood roofs where high temperature and humidity conditions exist. Research efforts at the Forest Products Laboratory have shown that acidic fire-retardant chemicals, such as monoammonium phosphate, can cause premature degradation and simultaneous strength loss at elevated temperatures [>150°F (66°C)] and humidities. These conditions may be found in attics and affect fire-retardant treated plywood sheathing and other treated structural components. It is important to note that not all fire-retardant formulations currently being used for wood show the same degree of strength losses under the chosen laboratory conditions. Therefore, it is necessary for the user of fire-retardant-treated wood and fire-retardant-treated plywood to review the available literature on this subject, including the information available from the American Plywood Association (APA): The Engineered Wood Association and the fire-retardant chemical producers. Several technical reports on this problem have been published by the Forest Products Laboratory.[13]

Another disadvantage of pressure impregnation is that the wood must be treated before use and after dimensioning.

Fire performance characteristics: The pressure impregnation process can substantially improve the fire performance characteristics of wood, as measured by the tunnel test, the room fire test, and heat-release rate tests. The effect can be illustrated by data shown in Tables 4-4A through 4-4D.

TABLE 4-4A. *Surface Burning Characteristics of Pressure-Treated and Untreated Wood**

Species	FSI[†] (untreated)	FSI[†] Pressure-Impregnated	Smoke-Developed Pressure Impregnation
Douglas fir (lumber)	70-100	≤25	≤25
Hemlock (lumber)	60-70	≤25	≤25
Southern yellow pine (lumber)	130-195	≤25	≤25
Western red cedar (lumber)	70	≤25	≤25
Redwood (lumber)	65-70	≤25	≤25
Red oak (lumber)	91	≤25	≤25
Western spruce (lumber)	100	≤25	≤25
Douglas fir (plywood)	115-130	≤25	≤25
Southern yellow pine (plywood)	95	≤25	≤25
Particleboard	140-200	≤25	≤25

*As tested by NFPA 255 (ASTM E84), selected species only. (See bibliography.)
† Flame spread index.

TABLE 4-4B. *Heat Release Rates of Pressure-Treated and Untreated Wood**

Species (Product)	Irradiance (kW/m²)	Average Peak HRR[†] (kW)	Average HRR (kW)
Southern yellow pine (plywood)*	50	237	156
Southern yellow pine (fire-retardant treated plywood)*	50	151	55

*Unpublished data generated at Weyerhaeuser Company.
†HHR–Heat release rate

TABLE 4-4C. *Effect of Pressure Impregnation on Douglas Fir Plywood in the Room Fire Test**

	Untreated	Pressure Impregnated
Flashover time	3.3 min	None
Maximum heat release rate	7.3 mW	0.6 mW
Maximum smoke generation (O.D.[†])	1.11	0.76

*The ignition exposure was 40 kW (net) for 5 min., 160 kW (net) for 5 min., and 0 kW (net) for 5 min.
†Optical density.

TABLE 4-4D. *Variations in Effect of Different Fire-Retardant Coatings on Unprimed Douglas Fir*

	Untreated	Painted with Coating X[†]	Painted with Coating Y[†]	Painted with Coating Z[†]
Total coverage	—	100	100	200
[sq ft per gal (L/m²)]	—	2.45 L/m²	2.45 L/m²	1.23 L/m²
Flame spread*	100	60	10	25
Fuel contributed*	100	35	15	10
Smoke developed*	100	40	5	10
Rate per coat	—	300	200	200
[sq ft per gal (L/m²)]	—	0.82 L/m²	1.23 L/m²	1.23 L/m²
Number of coats	—	3	2	1

*Where a range of values was obtained, the figure given here is the highest obtained.
†Each coating is listed by Underwriters Laboratories Inc., as meeting the minimum requirements for a fire-retardant coating.

Testing: The effectiveness of fire-retardant pressure impregnation treatments is tested by several methods. The tunnel test [NFPA 255 (ASTM E84)] has been used to determine the flamespread and smoke characteristics of many pressure impregnation treatments. A standard room fire test of wall and ceiling materials was developed at the National Institute of Standards and Technology Center for Fire Research and is used to evaluate treated panel products. This room test method has been standardized by the International Standards Organization (ISO) as ISO 9705, *Fire Tests—Full-Scale Room Test for Surface Products*. The test utilizes the oxygen depletion method for calculating the rate of heat release. Data collected by this method are used in the validation of fire mathematical models. In recent years, rate of heat release tests, such as the cone calorimeter (ASTM E1354, *Standard Test Method for Heat and Visible Smoke Release Rates for Materials and Products Using an Oxygen Consumption Calorimeter*) and the intermediate scale calorimeter (ICAL) [ASTM E1623, *Standard Test Method for Determination of Fire and Thermal Parameters of Materials, Products, and Systems Using an Intermediate Scale Calorimeter (ICAL)*] have been used successfully in distinguishing burning rate characteristics of fire-retardant-treated and untreated wood products.

The City of New York originally established three tests: (1) crib (ASTM E160, *Test Method for Combustible Properties of Treated Wood by the Crib Test*), (2) shavings, and (3) timber, for evaluating fire-retardant-treated wood. The fire-tube test (ASTM E69, *Test Method for Combustible Properties of Treated Wood by the Fire-Tube Apparatus*), developed at the Forest Products Laboratory, has been used extensively both in the United States and in several other countries to measure the effectiveness of fire-retardant treatments. Other tests, such as fire penetration and ignition tests, have been used by different organizations and individuals.

Model codes allow the use of fire-retardant-treated wood in noncombustible construction, with some limitations. As shown in Table 4-4A, lumber and plywood for construction can be treated by pressure impregnation to achieve a flame spread rating as low as 15 and to perform well in rated assemblies.

Plywood: Fully treated plywood is satisfactory only when surface appearance is unimportant. For decorative purposes where the face veneer must have the beauty of natural wood, untreated face veneers of various species are laminated to treated cores. The flame spread classification of these treated plywood products varies in approximately inverse proportion with the density of the face veneer, and ranges generally from 20 to 75, satisfying all but the most restrictive requirements. Fully treated plywood generally has a flame spread index less than 25 (Class I).

Relative toxicity: Combustion toxicity testing of treated and untreated wood suggests that smoke from treated wood is "no more toxic than untreated Douglas fir."[14] This is true even though minor amounts of toxic vapors, such as HCN and nitrogen oxides, are expected to be liberated from some treatments upon degradation.

Smoke toxicity values are now required by the State of New York on all wood products to be sold in-state that are used for interior finish and interior floor finish. These values are generated using the University of Pittsburgh Test Method. Treated and untreated wood products tested to date range in LC_{50} values of 7 to 72 g.[15] Many treated and untreated wood products have also been tested by the radiant furnace method. These results also indicate that smoke from treated wood is slightly less toxic than from untreated.[16]

Fire-Retardant Coatings

In existing structures of untreated wood, or in cases where fire-retardant pressure impregnations are impracticable, fire-retardant coatings can be used. Coatings can be applied to any exposed surface of structural members, interior finish, or contents. They may be applied by means of brush, roller, sprayer, or trowel. As in the case of pressure impregnations, the degree of the flame spread reduction secured by a coating is dependent upon the original combustibility of the surface to be protected, the effectiveness of the coating material, the amount of coating used, the thoroughness of the application, and the size and severity of the exposing fire.

With normal paints, the usual aim is to get maximum coverage with minimum material. The approach is different in the application of fire-retardant coatings. Here, the objective is to provide the needed amount of coating per unit area required to ensure a definite degree of protection. Film thickness for intumescent coatings is a very important factor in their ultimate performance.

Among the many proprietary fire-retardant coatings, several have been listed by testing laboratories. Beware of any so-called fire-retardant or "fireproof" paint that has not been certified as to its flame spread characteristics by an independent testing laboratory. There are four types of fire-retardant coatings for wood, as follows.

1. *Intumescent paints.* These coatings, upon the action of heat, expand from a thin, paint-like coating to a thick, puffy, burnt-marshmallow-type coating. The puffy coating results in one or all of the following effects: insulation of fuel from heat, exclusion of oxygen from fuel, production of diluent gases, and reduction of flammable gases. It will retain its effectiveness until broken up either by high heat or by sustained heat.
2. *Mastics.* These materials are applied by trowel or by heavy-duty spray equipment and form a thick coating on the surface of the combustible material. They vary from hard, ceramic-like materials to soft, tar-like coatings. All withstand significant amounts of heat and inhibit flame spread by the impervious noncombustible membrane they form.
3. *Gas-forming paints.* These coatings, when heated, release quantities of noncombustible gas that dilute the oxygen at the protected surface so there is not sufficient oxygen to support combustion.
4. *Cementitious and mineral-fiber coatings.* Limited research has been conducted on the potential use of cementitious and mineral-fiber coatings on wood to improve its fire endurance characteristics. These coatings are commonly used on structural steel to enhance fire endurance by protecting the steel from the intense heat of a fire. Some researchers feel that structural wood systems may be protected in the same manner.

The customary finishing materials for wood, such as ordinary stains, oil paints, enamels, varnishes, shellac, lacquer, and waxes, are of no appreciable value in protecting wood against fire, and, in some cases, increase combustibility. Paint, however, may be of value in preventing absorption of oil and in promoting cleanliness. It should also be noted that dry, decayed wood is more susceptible to ignition by small sparks from cigarettes than is sound wood and that paint coatings tend to prevent surface decay.

Advantages: Fire-retardant coatings have the following advantages: (1) they can be used on combustible materials already in place, (2) they are relatively inexpensive, (3) they may be easily applied, and (4) they do not cause strength loss at high temperatures and humidities.

Disadvantages: Fire-retardant coatings have the following disadvantages: (1) ease of application can lead to thin and ineffective treatments, (2) unexposed surfaces cannot be treated, (3) limited life and durability, (4) susceptibility to damage, and (5) required maintenance.

Fire performance characteristics: The fire performance of wood can be improved through the use of fire-retardant coatings. The amount of improvement is determined by the effectiveness of the treatment. Underwriters Laboratories Inc. has established a minimum standard of effectiveness for listed coatings to be a flame spread rating of 50 or less in the tunnel test, or a 50 percent reduction in flame spread as compared to the untreated surface, whichever is lower.

Listed coatings, when properly applied, are effective in improving fire performance characteristics, some to a point well below the limits noted above. The variations in effect of listed fire-retardant coatings on Douglas fir are shown in Table 4-4D.

Permanence: Some fire-retardant paints and coatings have good permanence when used in interior locations of relatively normal humidity and temperatures. Water-resistant, oil-based, intumescing fire-retardant coatings have been developed.[18] But the use of fire-retardant coatings in exterior locations, in areas of relatively high humidity, and in areas of high temperature has not been thoroughly tested. One difficulty in testing these exterior paints is the problem of weathering samples. Samples for small-scale tests can be placed in a weather accelerator or weatherometer, and the results measured. These small-scale tests, however, do not produce significant flame spread results. There is also doubt as to the comparability of weatherometer exposures and actual long-term weathering. The samples required for the acceptable large-scale tests are too large for accelerated weathering and thus must be exposed to actual weathering, which requires long periods of time. Until this testing problem can be solved, the use of fire-retardant coatings in exterior locations is questionable.

Strength and deterioration: Strength and deterioration of both the materials coated and their fastenings is not a problem with most fire-retardant coatings.

Moisture: Only those coatings that use hygroscopic fire-retardant chemicals are vulnerable to absorbing moisture, thus reducing their effectiveness.

Toxicity: The reactive toxicity of the total products of combustion of the coated material as compared to the uncoated material should be considered when examining fire-retardant coatings.

Preservative Treatments

Where subject to decay or insect attack, or to attacks by marine borers, timber and lumber commonly are pressure treated with coal tar creosote; pentachlorophenol; chromated copper arsenate; ammonical copper arsenate; or many new, environmentally friendly, chemicals. Such treatments are used in piers, bridges, trestles, decks, and some locations in buildings.

Oil-type preservatives, such as creosote, cause an increase in fire hazard as compared with untreated wood. Fire retardants utilizing the proper ratio of phosphorous-halogen mixture are effective in reducing the combustibility of the treated wood to that of untreated wood; however, there is an initial flash of "light ends" not observed in untreated wood.[19]

Fire may spread more rapidly and generate higher surface temperatures in structures containing newly creosoted wood than in structures of untreated wood, particularly where general fire temperatures are sufficiently high to vaporize the creosote oils. The adverse fire experience in fires in creosoted pier substructures, however, is due primarily to the excessive areas of wood construction without firebreaks, which are inaccessible for manual fire fighting and lack automatic sprinkler protection. As the construction

industry shifts to new waterborne treatments this concern will lessen in new structures.

Waterborne salt preservatives can provide some fire-retardant effect, but normal preservative concentrations are not recognized as effective fire-retardant treatments. Conversely, some fire-retardant formulations combine fire retardance and resistance to decay or to attack by insects. Borates are a good example of this dual effect.

Christmas Trees

Christmas trees become extremely flammable when cut long in advance of use and then brought indoors where heat and low humidity accelerate drying. To reduce the hazard, the tree should be kept indoors only as long as absolutely necessary, the trunk sawed off at an angle at least 1 in. (25.4 mm) above the original cut end, and the tree placed in water while it is in the house. Water should be added at intervals to the pail, tub, or tree stand in which the tree is placed so that the water level is always above the cut. To achieve a satisfactory degree of flame resistance in any combustible material, including Christmas trees, it is essential to get a certain minimum quantity of effective flame-retardant chemical either into or on the surface of the material to be treated. Since Christmas trees by their nature are not absorbent, the only way to effectively treat them is to apply a surface coating. Treating Christmas trees with simple solutions of water-dissolved chemicals, such as borox-and-boric acid, diammonium phosphate, or ammonium sulfate, is not effective. In order to provide a significant flame-retardant value, the testing solution or coating must be thick or syrupy enough to form a fairly heavy coating.

Much misunderstanding in this area results from tests of fresh trees treated with simple water-thin solutions of the type described above. Such tests lead to the erroneous conclusion that the treatment is effective, when the fact is that the tree was naturally flame resistant due to its water content.

Pressurized aerosol containers of flame-retardant Christmas tree coatings may be convenient, but their value is questionable due to the very limited quantity of the contents. Many cans would be necessary to adequately treat all parts of larger trees, but the high cost leads the user to apply only a token (and useless) coating.

WOOD-BASED COMPOSITE PANELS

The group of materials referred to here as "wood-based composite panels" includes products such as insulation boards, hardboards, particleboards, oriented strand boards, medium-density fiberboard, and laminated paper boards. They consist of a mixture of fibers, particles, or flakes and binder formed under pressure into board-like panels or tiles. These board or panel materials in final form retain some of the properties of the original wood but, because of the manufacturing methods, gain new and different properties from those of the wood.

Wood-based composite panels can be fire-retardant treated by chemical admixtures, impregnations, and coatings. These treatments vary in their effectiveness and are somewhat limited in their use by their effect on the physical properties of the untreated panel.

Admixtures

When mineral fibers, such as glass or rock wood, are added to a mixture of organic fibers, they will reduce the combustibility of the mixture, depending upon the amount of mineral fiber used. If mineral fibers completely replace the organic fibers, then the product becomes noncombustible, provided the binder used to join the fibers is noncombustible or has a sufficiently high ignition temperature. Mixtures of alumina trihydrate, boric acid, and wood fibers have been successfully combined into panel products with very low flame spread and smoke-developed values. Fire-retardant particleboard and oriented strand board are prepared by adding the fire-retardant chemicals to the furnish prior to board making. These products give very consistent flame spread and smoke-developed values if chemical retentions are consistent.

Impregnations

Attempts have been made to reduce the combustibility of organic composite boards by pressure impregnation of the finished product and by treatment of the raw fiber product itself. Treatments have included soluble salts, insoluble complexes, and chemical reactions with the organic materials themselves. The effect of these treatments on combustible composite boards varied with respect to the type of treatment and type and quantity of chemicals used. Treatment by pressure impregnation has not yet proved to be commercially feasible, due in part to losses in physical properties and inadequate fire retardancy. At least one manufacturer has developed a successful wood pulp treatment process for both composite boards and acoustical tile, and similarly treated wood fiber acoustical tile is produced by other manufacturers.

Coatings

The treatments previously discussed are made during the manufacture of the material. Coatings can be applied during manufacture or to material already in use. Factory-applied coatings are usually fire-retardant paints; however, many are of questionable effectiveness when exposed to severe fire conditions. Their main effect is to cover the fuzzy surface of the untreated material.

The application of fire-retardant paints of the intumescent variety to existing composite board will provide good protection for the exposed side of the board, provided the coatings are maintained. Some composite boards exhibit considerable dimensional instability, expanding when damp and contracting when dry. This movement may expose cracks and gaps in the material that will open untreated areas to the action of heat and rapid flame spread.

Fire Performance Tests of Wood-based Composites

The effectiveness of the treatments of composite boards can be shown with a fire test. Most of the methods listed under the section on wood can be used to evaluate their fire performance characteristics.

PAPER

Paper can be treated by impregnation to make it flame and glow retardant after the ignition source has been removed. In addition to treating various forms of paper, e.g., building paper, wrapping, packing, and decorative paper, these same treatments can be used to preserve foliage, leaves, grass skirts, broom straw, excelsior, and similar materials if they are sufficiently absorbent. Adequate wetting may be achieved by dipping or spraying with some materials, while others may require soaking for several hours or even days.

The development of flame-retardant products has been given impetus by the requirements for nonflammable display and decorative material in hotels, theaters, schools, hospitals, nursing homes, and other high-occupancy buildings. Decorative crepe paper, corrugated display cardboard, textiles, and window-shade materials may be made flame resistant during manufacture, which is the most reliable method. Field treating paper is difficult due to variations in paper finishes, sizing, and color fastness. It is preferable to purchase factory-treated materials.[20]

MISUSE OF FIRE RETARDANTS

Fire-retardant treatments are used for reduction of flame spread on structural wood members, and combustible interior finish, such as wood paneling, acoustical tile, and wood-based composite panels. Flame-retardant treatments are used for reducing the flammability of contents, furnishings, and decorative materials, such as curtains, draperies, upholstery, crepe and corrugated papers, Christmas trees, and miscellaneous dried or artificial plants and foliage. In all these applications, there are opportunities for misuse or misrepresentation. Some of the more prevalent kinds of misuse are described, as follows.

1. Promoting pressure-impregnated wood for uses where noncombustible materials are absolutely required. Such substitutions are appropriate in certain limited applications, but even the most effective treatments cannot entirely eliminate some contribution of fuel and smoke. The concept of a "limited combustibility classification" is being considered on the basis of heat release rate characteristics. Fire-retardant-treated wood is an example of a material that could be so classified.
2. Using pressure-impregnated or coated materials that are not intended for use in locations that are exposed to the weather or unusually high humidity conditions.
3. Attempting to treat the newer synthetic fiber fabrics with common water-soluble-salt flame-retardant solutions. Such solutions are effective only on cellulose fibers, such as cotton and rayon, and the common animal fibers, such as wool and silk. They are not effective for the treatment of nylon, acetate, Orlon®, Dacron®, Acrilan®, or any of the similar synthetics presently in wide use.
4. Marketing flame-retardant solutions in small aerosol spray containers for home use. This practice is subject to many abuses, including the lack of product labels to indicate the limitations of the contents and instructions for proper application. Some advertising contains wildly exaggerated claims for the product's utility, giving the impression that the contents of one can will "flameproof" virtually everything in the home. The product itself may be excellent and convenient to use, but there is a great risk that the average user will not achieve an effective degree of flame resistance. With few exceptions, flame-retardant treatments should be left to skilled persons.
5. Advocating the use of ordinary water-soluble-salt solutions for the treatment of nonabsorbent materials. Effective treatment of nonabsorbent materials can be achieved only by means of a substantial surface coating, requiring that the treating solution be of a syrup- or paint-like consistency.

The technology of flame-retardant treatment is highly useful and can do much to increase fire safety. But because it is also a field vulnerable to entry by unskilled entrepreneurs and charlatans, the buyer should be alert to its limitations. Products that have been tested, rated, and labeled by reputable, independent testing laboratories are the consumers' best safeguard against the pitfalls listed above.

BIBLIOGRAPHY

References Cited

1. ASTM E176, *Standard Terminology of Fire Standards*, American Society for Testing and Materials, W. Conshohocken, PA, 1993.
2. Shafizadeh, F., "The Chemistry of Pyrolysis and Combustion," ACS Monography 207, 1984, American Chemical Society, Washington, DC.
3. Little, R. W., Church, J. M., and Coppick, S., "Flameproofing Textile Fabrics," ACS Monography 104, 1947, American Chemical Society, Washington, DC.

4. Schuyten, H. A., Weaver, J. W., and Reid, J. D. , "Some Theoretical Aspects of the Flameproofing of Cellulose," *Advances in Chemistry Series No. 9*, American Chemical Society, Washington, DC, 1954.

5. U.S. Forest Products Laboratory, 1962–64: Eickner, H. W., "Basic Research on the Pyrolysis and Combustion of Wood," *Forest Products Journal*, Apr. 1962; Browne, F. L., and Tang, W. K., "Thermogravimetric and Differential Thermal Analysis of Wood and of Wood Treated with Inorganic Salts During Pyrolysis," *Fire Research Abstracts and Reviews*, Vol. 4, Nos. 1 and 2, 1962, pp. 76–91; Tang, W. K., and Neill, W. K., "Effect of Flame Retardants on Pyrolysis and Combustion of Alpha-Cellulose," *Journal of Polymer Sciences*, Part C6, 1964, pp. 65–81; Brenden, J., "The Influence of Inorganic Salts on the Products of Pyrolysis of Ponderosa Pine," Master's thesis, University of Wisconsin, 1963; Tang, W. K., "Effect of Inorganic Salts on the Pyrolysis, Ignition, and Combustion of Wood, Cellulose, and Lignin," Ph.D. thesis, University of Wisconsin, 1964; Browne, F. L., and Tang, W. K., "Effect of Various Chemicals on the Thermogravimetric Analysis of Ponderosa Pine," U.S. Forest Service Research Paper RPL-6, May 1963, Forest Products Laboratory, Madison, WI.

6. Browne, F. L., "Theories of the Combustion of Wood and Its Control," Report No. 2136, 1958, Forest Products Laboratory, Madison, WI.

7. Rowell, R. M., "Chemical Modification of Wood," *Wood and Cellulosic Chemistry*, Marcel Dekker, Inc., New York, 1991, pp. 703–756.

8. American Wood Preservers, "All Timber Products—Preservative Treatment by Pressure Processes," C1-77, 1977, American Wood-Preservers Association, Vienna, VA.

9. White, R. H., "Use of Coatings to Improve Fire Resistance of Wood," *Fire-Resistive Coatings: The Need for Standards*, ASTM STP 826, Lieff, M., and Stumpf, F. M. eds., American Society for Testing and Materials, W. Conshohocken, PA, 1983, pp. 24–39.

10. Lieff, M., "Fire-Resistant Coverings," *Fire-Resistive Coatings: The Need for Standards*, ASTM STP 826, Lieff, M., and Stumpf, F. M., eds., American Society for Testing and Materials, W. Conshohocken, PA, 1983, pp. 3–13.

11. Stumpf, F. M., "Spray-Applied Fibrous Material Fire-Resistive Coatings," *Fire-Resistive Coatings: The Need for Standards*, ASTM STP 826, Lieff, M., and Stumpf, F. M., eds., American Society for Testing and Materials, W. Conshohocken, PA, 1983, pp. 14–23.

12. Eickner, H. W., and Schaffer, E. L., *Fire Technology*, 1967, 3(2), pp. 90–104.

13. LeVan, S., and Collet, M., "Choosing and Applying Fire-Retardant-Treated Plywood and Lumber for Roof Designs," General Technical Report FPL-GTR-62, 1989, U.S. Department of Agriculture, Forest Service, Forest Products Laboratory, Madison, WI; LeVan, S., and Winawdy, J.E., "Effects of Fire-Retardant Treatments on Wood Strength: A Review," *Wood and Fiber Science*, 22 (1), 1990, pp. 113–131.

14. Leiu, P. J., Magill, J. H., and Alarie, Y. C., "An Evaluation of Some Polyphosphazenes and Commercial Fire Retardants for Wood," *Journal of Combustion Toxicology*, Vol. 9, May 1982, p. 65.

15. *New York State Uniform Fire Prevention and Building Code*, Article 15, Lenz and Riecker, Albany, NY.

16. Alexeeff, G. V., and Packham, S.C., "Use of Radiant Furnace Fire Model to Evaluate Acute Toxicity of Smoke," *Journal of Fire Sciences*, Vol. 2, 1984, pp. 306–320.

17. Memorandum to the National Institute of Building Sciences Com/Tox Steering Committee from Wayne Ellis, June 15, 1989.

18. Verburg, G. B., *et al.*, "Water-Resistant, Oil-Based, Intumescing Fire-Retardant Coatings, Part I, Developmental Formulations," *Journal of the American Oil Chemists' Society*, Vol. 41, Oct. 1964.

19. Gooch, R. M., Kenaga, D. L., and Tobey, H. M., "The Development of Fire Retardants for Wood Treated with Oil-Type Preservatives," *Forest Products Journal*, Vol. 9, No. 10, Oct. 1959.

20. Segal, L., "Flameproofing: Facts and Fallacies," *Fire Journal*, Vol. 60, No. 6, Nov. 1966, pp. 41–45.

NFPA Codes, Standards, and Recommended Practices

Reference to the following NFPA codes, standards, and recommended practices will provide further information on fire-retardant and flame-resistant treatments of cellulosic materials discussed in this chapter.

(See the latest version of *The NFPA Catalog* for availability of current editions of the following documents.)

NFPA *101®*, *Life Safety Code®*
NFPA 220, *Standard on Types of Building Construction*
NFPA 251, *Standard Methods of Tests of Fire Endurance of Building Construction and Materials*
NFPA 255, *Standard Method of Test of Surface Burning Characteristics of Building Materials*
NFPA 703, *Standard for Fire-Retardant Impregnated Wood and Fire-Retardant Coatings for Building Materials*

Additional Reading

Amundson, F. J., "Fire Retardant Cedar Shakes and Shingles," Fire Retardant Chemicals Association, 1992 Spring Conference on Technical and Marketing Issues Impacting the Fire Safety of Building and Construction and Home Furnishings Applications, March 29-April 1, 1992, Orlando, FL, Technomic Publishing Co., Lancaster, PA, 1992, pp. 241–248.

ASTM E69, *Standard Test Method for Combustible Properties of Treated Wood by the Fire-Tube Apparatus,* American Society for Testing and Materials, W. Conshohocken, PA, 1995.

ASTM E84, *Standard Test Method for Surface Burning Characteristic of Building Materials*, American Society for Testing and Materials, W. Conshohocken, PA, 1995.

ASTM E1354, *Standard Test Method for Heat and Visible Smoke Release Rates for Materials and Products Using an Oxygen Consumption Calorimeter,* American Society for Testing and Materials, W. Conshohocken, PA 1994.

ASTM E1623, *Standard Test Method for Determination of Fire and Thermal Parameters of Materials, Products, and Systems Using an Intermediate Scale Calorimeter (ICAL)*, American Society for Testing and Materials, W. Conshohocken, PA, 1994.

ASTM D5516, *Standard Test Method for Evaluating the Flexural Properties of Fire-Retardant Treated Softwood Plywood Exposed to Elevated Temperatures*, American Society for Testing and Materials, W. Conshohocken, PA, 1994.

Badr, O. and Karim, G. A., "Experimental Study of Self-Ignition and Smoldering of Moist Cellulosic Materials," *PD-Emerging Energy Technology*, Vol. 36, 1991, pp. 15–21.

Bland, K. E., "Behavior of Wood Exposed to Fire: A Review and Expert Judgment Procedure for Predicting Assembly Failure," Worcester Polytechnic Inst., MA, Thesis, May 1994, 159 pages.

Borg, L. B., "Christmas Tree Fire Scars Utah's Historic Governor's Mansion," *Sprinkler Age*, Vol. 14, No. 7, July 1995, pp. 10, 12.

Camino, G., Costa, L., and Martinasso, G., "Intumescent Fire-Retardant Systems," *Polymer Degradation and Stability*, Vol. 23, No. 4, 1989, pp. 359–376.

Carling, O., "Brandteknisk dimensionering av massiva trakonstruktioner. (Fire Engineering Design of Timber Structures.)," TrateknikCentrum, Stockholm, Sweden, Report I 9004018, Apr. 1990, 140 pages.

Chen, Y., "An Experimental Study of the Pyrolysis of Pure and Fire-Retarded Cellulose," NIST-GCR-89-566, June 1989, National Institute for Standards and Technology, Gaithersburg, MD.

Chen, Y., *et. al.*, "Combustion Properties of Pure and Fire-Retarded Cellulose," *Combustion and Flame*, Vol. 84, No. 1–2, 1991, pp. 121–140.

Collier, P. C. R., "Design of Loadbearing Light Timber Frame Walls for Fire Resistance. Part 1," Study Report, Building Research Association of New Zealand, Judgeford, BRANZ Study Report SR36, Dec. 1991, 52 pages.

Collier, P., "Design of Light Timber Framed Walls and Floors for Fire Resistance," Technical Recommendation, Building Research Association of New Zealand, Judgeford, BRANZ Technical Recomm. 9, Aug. 1991, 19 pages.

Daikoku, M., Venkatesh, S. and Saito, K., "The Use of Cellulose Sample for Material's Flammability and Pyrolysis Tests," *Journal of Fire Sciences*, Vol. 12, No. 5, Sept./Oct. 1994, pp. 424–441.

Damant, G. H. and Nurbakhsh, S., "Christmas Trees: What Happens When They Ignite?," *Fire and Materials*, Vol. 18, No. 1, Jan./Feb. 1994, pp. 9–16.

David, J., "Seven-Fatality Christmas Tree Fire, Canton, Michigan, December 22, 1990," USFA Fire Investigation Technical Report Series,

TriData Corp., Arlington, VA, Federal Emergency Management Agency, Washington, DC, Report 046, 1991, 24 pages.

Day, T., "Surface Coatings: Will Time Affect Their Performance?" *Fire Prevention*, No. 222, Sept. 1989, pp. 40–41.

Eboatu, A. N., Garba, B. and Akpabio, I. O., "Flame-Retardant Treatment of Some Tropical Timbers," *Fire and Materials*, Vol. 17, No. 1, 1993, pp. 39–42.

Evans, P. D., *et al.*, "Fire Resistance of Preservative-Treated Slash Pine Fence Posts," *Forest Products Journal*, Vol. 44, No. 9, Sept. 1994, pp. 37–39.

Favro, P. C., "Fire-Retardant Treatments for Wood Roofs: A Permanent Solution," American Technologies International, Fire Safety '91 Conference, Volume 4, November 4–7, 1991, St. Petersburg, FL, Am. Technologies Intl., Pinellas Park, FL, 1991, pp. 136–145.

Fire Retardant Chemicals Association, *Fire Safety Problems Leading to Current Needs and Future Opportunities*, Technomic, Lancaster, PA, 1989.

"Flame Retardant Chemicals—The Current Picture," *Fire Prevention*, No. 199, May 1987, pp. 33–35.

"Flame Retardant Treatments for Wood and Its Derivatives. Fire Prevention Design Guide," Fire Protection Association, London, England, FPDG 13, Aug. 1990, 4 pages.

Fredlund, B., "Calculation of the Fire Resistance of Wood Based Boards and Wall Constructions," Lund Institute of Science and Technology, Sweden, LUTVDG/TVBB-3053, 1990, 93 pages.

Gann, R. G., "Flame Retardants," *Kirk-Othmer Encyclopedia of Chemical Technology*, 4th Edition, Volume 10, John Wiley and Sons, Inc., NY, 1994, pp. 930–936.

Garba, B., Eboatu, A. N. and, Atta-Elmannan, M. A., "Note: Effect of Flame Retardant Treatment on Energy of Pyrolysis/Combustion of Wood Cellulose," *Fire and Materials*, Vol. 18, No. 6, 1994, pp. 381–383.

Gillon, R. D., "Intumescent Coatings: Their Uses and Future Development," Fire Surveyor, Vol. 18, No. 2, Apr. 1989, pp. 13–20.

Giudice, C. A., and delAmo, B., "Flame Retardant Paints. Part 1," *European Coatings Journal*, No. 11, 1991, pp. 740, 745, 748, 750, 752, 754–755.

Giudice, C. A., and delAmo, B., "Flame Retardant Paints. Part 2," *European Coatings Journal*, No. 1–2, 1992, pp. 8–10, 12–14.

Goodman, D., "Fire Retarded Cardboard Cartons—An Approach to Allow General Warehousing of Flammable Liquids in Plastics Containers," Occidental Chemical Co., Pottstown, PA, Fire Retardant Chemicals Association, Fire Safety Developments and Testing: Toxicity—Heat Release—Product Development--Combustion Corrosivity, October 21–24, 1990, Ponte Vedra Beach, FL, Fire Retardant Chemicals Assoc., Lancaster, PA, 1990, pp. 157–165.

Hall, J. N., "Paper Mill Fires," *Fire Engineering*, Vol. 143, No. 9, Sept. 1990, pp. 48–50, 52–53, 56.

Hall, R. W., "Investigation of Melt Stick Characteristics of Flame Retardant Treated Polyester-Cotton Blend Fabrics," Fourth Annual Conference on Protective Clothing: A New Review of Applications, Materials Used, Needs, Concerns, and Trends, April 24–26, 1990, Charlotte, NC, 1990, pp. XX/1–11.

Harper, R. J., and Ruppenicker, G. F., "Flame Retardancy of Cotton: Current Perspective," Recent Advances in Flame Retardancy of Polymeric Materials: Materials, Applications, Industry Developments, Markets, Volume 2, May 14–16, 1991, Stamford, CT, Business Communications Co., Inc., Norwalk, CT, 1991, pp. 141–149.

Hirata, T., *et al.*, "Combustion Toxicity of Woods Treated With Retardants and CCA Preservatives," Forestry and Forest Products Research Institute, Ibaraki, Japan, NISTIR 4449; U. S./Japan Government Cooperative Program on Natural Resources (UJNR), Eleventh Joint Panel Meeting on Fire Research and Safety, October 19–24, 1989, Berkeley, CA, 1990, pp. 72–79.

Hirato, T., and Kawamoto, S., "Toxic Gases from Combustion of Wood Treated with Retardants," 9th Joint Panel Meeting of the UJNR Panel on Fire Research and Safety, NBSIR 88-3753, Apr. 1988, Center for Fire Research, pp. 417–427.

Hirano, T., and Nakaya, I., "Strategy of Fire Suppression by Using Fire Retardants," 9th Joint Panel Meeting of the UJNR Panel on Fire Research and Safety, NBSIR 88-3753, Apr. 1988, Center for Fire Research, pp. 397–404.

Horrocks, A. R., Akalin, M. and Price, D., "Smoke, CO2 and CO Evolution From Cotton and Flame Retarded Cotton. Part 2. Behavior of Single Layer Fabrics in Air at Elevated Temperatures," *Journal of Fire Sciences*, Vol. 8, No. 2, Mar./Apr. 1990, pp. 135–151.

Horrocks, A. R., Allen, J. and Price, D., "Testing and Performance of Flame Retardant Textiles," Fire Retardant Chemicals Association, International Conference on Fire Safety--Fire Retardant Technology and Marketing. March 25–28, 1990, New Orleans, LA, Fire Retardant Chemicals Assoc., Lancaster, PA, 1990, pp. 181–198.

Horvat, S. G., "Use of a Brominated Organic (Not a Brominated Diphenyloxide) in a Flame Retardant Chlorinated Paraffin Dispersion for Cellulosic and Cellulosic Blend Fabrics for Textile Applications," 1992 Spring Conference on Technical and Marketing Issues Impacting the Fire Safety of Building and Construction and Home Furnishings Applications, March 29-April 1, 1992, Orlando, FL, Technomic Publishing Co., Lancaster, PA, 1992, pp. 213–223. "Intumescent Fire Seals Product Review," *Fire Surveyor*, Vol. 17, No. 2, Apr. 1988, pp. 17–19, 22–29.

Hume, S. F., "Christmas Tree Burns: Raising Public Safety Awareness," *Fire Engineering*, Vol. 145, No. 12, Dec. 1992, pp. 46–47.

ISO 9705, Fire Tests—Full-scale Room Test for Surface Products, International Standards Organization, 1, rue de Varembé, Case postale 56, CH-1211 Genève 20, Switzerland, 1993.

James, R., "Intumescent Fire Seals," *Fire Surveyor*, Vol. 17, No. 2, Apr. 1988, pp. 13–15.

Koffel, W. E., "Fire-Retardant Wood Shakes: Are They Effective?," *NFPA Journal*, Vol. 85, No. 1, Jan./Feb. 1991, pp. 24–25.

Kumar, V., "Flame Retardancy in Carpets," Special Conference on Recent Advances in Flame Retardancy of Polymeric Materials—Materials, Applications, Industry Developments, Markets, May 15–17, 1990, Stamford, CT, 1990, pp. 1–8.

LeVan, S. L., and Winandy, J. E., "Effects of Fire Retardant Treatments on Wood Strength: A Review," *Wood and Fiber Science*, Vol. 22, No. 1, 1990, pp. 113–131.

Levan, S. L., "Effect of Cyclic Exposure on the Mechanical Properties of Fire Retardant Treated Plywood," *Mogjae Gonghak*, Vol. 18, No. 1, 1990, pp. 31–38.

LeVan, S. L., Ross, R. J. and Winanady, J. E., "Effects of Fire Retardant Chemicals on Bending Properties of Wood at Elevated Temperatures," Forest Products Lab., Madison, WI, FPL-RP 498, 1990, 24 pages.

Lucas, C. L., "Improved Intumescent Fire Protection for Hazardous Environments," *Proceedings* of the Conference on Offshore Passive Fire Protection, Plastics and Rubber Institute, 1987, p. 8.

Lyon, D. E., *et al.*, "Evaluation of Strength Properties of Fire-Retardant Treated Wood Using the 1983 National Forest Products Association Protocol," *Forest Products Journal*, Vol. 38, No. 6, 1988, pp. 13–18.

Lyons, J. W., *The Chemistry and Uses of Fire Retardants*, reprint, R. E. Krieger, Malabar, FL, 1987.

Mischutin, V., "Flame-Retardant Home Furnishings," *American Dyestuff Reporter*, Vol. 76, No. 3, 1987, pp. 38–42, 49.

Minichshofer, H., "New Developments in Flame-Retardant Protective Clothing," *Technical Textiles*, No. 2, Mar. 1990, p. 7.

Mischutin, V., "Flame Retarding Cellulosic and Blended Textiles," Amspec Chemical Corp., Glouster City, NJ, Fire Retardant Chemicals Association, Fire Safety Developments and Testing: Toxicity—Heat Release—Product Development—Combustion Corrosivity, October 21–24, 1990, Ponte Vedra Beach, FL, Fire Retardant Chemicals Assoc., Lancaster, PA, 1990, pp. 193–203.

Needles, H. L., *et al.*, "Assessment of Flame-Retardant Coated Fabric Flammability by Small-Scale Test Methods," Journal of Coated Fabrics, Vol. 17, Oct. 1987, pp. 129–134.

Olesen, F. B., "Fire Engineering Design of Timber Structures," Aalborg Univ., Sweden, ISSN 0902-7513-R9329, Aug. 1993, 7 pages.

Ostman, B. A. L., and Tsauntaridis, L. D., "Heat Release and Classification of Fire Retardant Wood Products," *Fire and Materials*, Vol. 19, No. 6, 1995, pp. 253–258.

Pasek, E. A., and McIntyre, C. R., "Heat Effects on Fire Retardant-Treated Wood.," *Journal of Fire Sciences*, Vol. 8, No. 6, Nov./Dec. 1990, pp. 405–420; Fire Retardant Chemicals Association. Fire Safety Developments and Testing: Toxicity—Heat Release—Product Development—Combustion Corrosivity, October 21–24, 1990, Ponte Vedra Beach, FL, 1990, pp. 181–192.

Pellejero, A. L., and Tapia, P. H., "Effect of Fire-Retardants on Textile Fiber blends and Their Application in Fire Behavior. [Influencia del producto ignifugo y su aplicacion en el comportamiento al fuego en mezclas de fibras textiles.]," *Mapfre Seguridad*, Vol. 14, No. 53, 1994, pp. 91–92.

Price, D., Akalin, M. and Horrocks, A. R., "Mechanistic Information From DTA and Evolved Gas Analysis Studies of Flame Retarded Cotton and

Polyester/Cotton Blend Fabrics," Fire Retardant Chemicals Association, International Conference on Fire Safety—Fire Retardant Technology and Marketing. March 25-28, 1990, New Orleans, LA, Fire Retardant Chemicals Assoc., Lancaster, PA, 1990, pp. 199–211.

Rowland, L., "Case of the Exploding Christmas Tree," *Fire and Arson Investigator*, Vol. 45, No. 2, Dec. 1994, pp. 37–38.

Saville, N., "Overview of Fire Blocking Fabrics," *Proceedings* of the 32nd International SAMPE Symposium and Exhibition: Advanced Materials Technology ' 87, SAMPE, Covina, CA, 1987, pp. 1347–1359.

Seyed-Yagoobi, J., Bell, D. O., and Asensio, M. C., "Heat and Mass Transfer in a Paper Sheet During Drying," *Journal of Heat Transfer*, Vol. 114, May 1992, pp. 538–541; National Heat Transfer Conference, July 28–31, 1991, Minneapolis, MN, 1992.

"Sprayed Passive Fire Protection," *Insulation Journal*, Vol. 30, No. 8, 1986, p. 14.

Suzuki, M., Dobashi, R., and Hirano, T., "Burning Process of Cellulosic Fibers Composing Filter Paper During Flame Spread," First Asian Conference on Fire Science and Technology (ACFST), October 9-13, 1992, Hefei, China, International Academic Publishers, China, 1992, pp. 499–504.

Sweeney, M., "Maintaining Textile Flame Retardancy," *Fire Prevention*, No. 269, May 1994, pp. 15–17.

Sweet, M. S., "Fire Performance of Wood: Test Methods and Fire Retardant Treatments," Recent Advances in Flame Retardancy of Polymeric Materials: Materials, Applications, Industry Developments, and Markets, Volume 4, May 18–20, 1993, Stamford, CT, Business Communications Co., Inc., Norwalk, CT, 1993, pp. 36–43.

Tempesti, A., "Inherently Flame Retardant Fibers for Furniture Applications," SNIA Fiber, Italy, Paper 12 (Abstract Only); Interscience Communications Limited. Second European Conference on Furniture Flammability, June 27–28, 1990, Brussels, Belgium, 1990, 12 pages.

Tillott, R. J., "Preserving Timber From Fire," *Fire Prevention*, No. 240, June 1991, pp. 18–22.

Topf, R., "New Developments in Intumescent Materials," Proceedings of the Conference on New Technologies to Reduce Fire Losses and Costs, Elsevier, New York, NY, 1986, pp. 155-161.

Underwriters Laboratories Inc., *Building Materials Directory*, Underwriters Laboratories Inc., Northbrook, IL. (Published annually.)

White, R. H., "Fire Resistance of Wood-Frame Wall Sections," Forest Service, Madison, WI, Sept. 1990, 26 pages.

Winandy, J. E., "Fire-Retardant-Treated Wood: Effects of Elevated Temperature and Guidelines for Design," *Wood Design Focus*, Summer 1990, pp. 8–10.

Winandy, J. E., Ross, R. J., and LeVan, S. L., "Fire-Retardant-Treated Wood: Research at the Forest Products Laboratory," *TRADA*, Vol. 4, 1991, pp. 4.69–4.74, International Timber Engineering, *1991 Proceedings*, September 2–5, 1991, London, England, 1991.

Winandy, J., *et al.*, "Thermal Degradation of Fire-Retardant Treated Plywood. Development and Evaluation of a Test Protocol," Forest Products Lab., Madison, WI, FPL-RP-501, June 1991, 25 pages.

Youngs, R. W., "Fire Retardant Paperboard Packaging," FRC Technologies, Inc., Fire Retardant Chemicals Association, Fire Safety Developments and Testing: Toxicity—Heat Release—Product Development—Combustion Corrosivity, October 21-24, 1990, Ponte Vedra Beach, FL, Fire Retardant Chemicals Assoc., Lancaster, PA, 1990, pp. 167–180.

Zoufal, R., "Determination of the Thermal Conductivity Coefficient and the Specific Thermal Capacity of Wood at High Temperatures," Third International Symposium on Fire Protection of Buildings, May 10-12, 1990, Eger, Hungary, 1990, pp. 45–51.

FIBERS AND TEXTILES

Revised by Salvatore A. Chines

Fibers and textiles are an integral part of daily life. Clothing, furniture, carpets, and curtains are examples of textile materials produced with fibers. Almost all fibers are combustible. This characteristic results in problems when the fibers are manufactured, while they are stored, while they are made into fabrics, while they are collected as waste, and while they are worn as apparel or used as furnishings. Fiber combustibility combined with the prevalence of textile products explains the frequency of textile-related fires and the many deaths and injuries that result.

This chapter discusses commonly used fibers and textiles, their flammability, their end use, and their fire-retardant treatments. Also detailed in this chapter are special environments where textiles are used and the testing methods employed in evaluating and creating standards for the use of fibers and textiles.

Studies of fatal fires continue to show that textiles provide the initial fuel for the largest share of fatal structure fires. From 1989 through 1993, for example, 41.0 percent of fatalities in structural fires came in fires that began with the ignition of a textile product. These fires also accounted for 15.5 percent of 1989 through 1993 structure fire incidents and 13.2 percent of direct property damage from 1989 through 1993 structure fires. Many fires start with the ignition of textile products then rapidly involve other combustibles, such as foam plastics used as filling materials in textile-covered upholstered furniture.

Table 4-5A shows the leading fabric products, ranked by their share of textile fires, and the number of fires and civilian deaths and dollars of direct property damage occurring in fires where the first fuel was textile in that product. Table 4-5B provides a similar display by type of textile.

More information can be found elsewhere in this handbook, especially in the following chapters: Section 4, Chapter 8, "Medical Gases"; Section 4, Chapter 9, "Oxygen-Enriched Atmospheres"; Section 4, Chapter 17, "Upholstered Furniture and Mattresses"; Section 7, Chapter 3, "Interior Finish"; and Section 11, Chapter 3, "Use of Fire Incident Data and Statistics."

FIBERS

Fibers are the basic components of all textiles. Fibers may be either natural or man-made. Natural fiber sources include cellulosic, pro-

tein, and mineral. Cellulosic fibers come from a variety of plants, the most common being cotton. Protein fibers come from a variety of animal hairs. Mineral fibers include mineral wool and fiberglass. Man-made fibers are produced by chemical processes. All fibers come in a variety of lengths. Although natural fiber length can be slightly manipulated genetically, man-made fiber can be produced in any length desired.

Natural Fibers

The most common cellulosic fiber used in textile manufacture is cotton. Cotton consists of more than 90 percent cellulose $(C_6H_{10}O_5)_x$. Other plant fibers, such as jute, flax (linen), hemp, and sisal, are primarily cellulosic in composition. Cellulosic fibers are combustible; the ignition temperature of cotton fiber is 752°F (400°C), for example. When ignited, these fibers produce heat, smoke, carbon dioxide, carbon monoxide, water, and numerous other compounds. Plant fibers char but do not melt. Many cellulosic fibers are hollow. Each fiber has its own air supply. This allows fire to burrow deeply into bales.

Protein fibers, such as silk, wool, and other hair fibers derived from animals, are chemically different from cellulosics because they consist of complex protein molecules containing high percentages of nitrogen as well as carbon, hydrogen, oxygen, and small amounts of sulfur. Wool, for example, supports combustion with difficulty, and other factors being equal, is more difficult to ignite, burns more slowly, and is easier to extinguish than cotton. This is seen in wool's very small share of the fire problem, as evidenced in Table 4-5B.

Man-Made Fibers

Man-made fibers are produced by processing liquids through small holes (spinnerets) and allowing the emerging filaments to be immersed in a coagulating medium. Spinneret orifices generally range from 0.002 to 0.005 in. (0.05 to 0.13 mm) in diameter. Most spinnerets are made of iridium and platinum. The resulting fibers can be used as "filament yarns," or can be cut into shorter lengths and spun into "staple yarns." The staple yarns can contain more than one type of fiber, e.g., blends of polyester and cotton. Many different textile properties can be attained by blending various fibers.

Man-made fibers can be divided into two groups: (1) "regenerated" or "reconstituted" fibers and (2) fully synthetic fibers. The major regenerated fiber in use today is viscose rayon. Rayon is made from dissolved cellulose and is similar to cotton in its charring and

---

Salvatore A. Chines is a research consultant with Industrial Risk Insurers, Hartford, CT.

TABLE 4-5A.   *Reported U.S. Structure Fires Involving Textiles as the Type of Material First Ignited, by Form of Material, Annual Average 1989–1993**

Fabric Product	Fires		Civilian Deaths		Direct Property Damage (Millions)	
Mattress, bedding, or pillow	33,200	(33.3%)	625	(36.8%)	$ 327.2	(33.6%)
Clothing on or off a person	17,800	(17.8%)	222	(13.0%)	$ 153.3	(15.7%)
Upholstered furniture	14,300	(14.4%)	631	(37.1%)	$ 210.0	(21.5%)
Linen, other than bedding	5,300	(5.3%)	38	(2.2%)	$ 29.9	(3.1%)
Carpet or rug	5,000	(5.0%)	52	(3.1%)	$ 58.0	(5.9%)
Curtain, drape, or tapestry	4,400	(4.5%)	28	(1.7%)	$ 37.9	(3.9%)
Fiber or lint	4,400	(4.5%)	0	(0.0%)	$ 12.6	(1.3%)
Other known	15,200	(15.3%)	103	(6.1%)	$ 145.7	(15.0%)
Total	99,600	(100.0%)	1,699	(100.0%)	$ 974.6	(100.0%)

*Includes proportional share of fires with unknown type of material first ignited, and allocation of textile fires with unknown form of material first ignited.
Source: NFPA analysis of NFIRS and NFPA Survey.

TABLE 4-5B.   *Reported U.S. Structure Fires Involving Textiles as the Type of Material First Ignited by Type of Material, Annual Average 1989–1993**

Textile Type	Fires		Civilian Deaths		Direct Property Damage (Millions)	
Cotton, rayon, or cotton-polyester blend	57,200	(57.5%)	912	(53.7%)	$ 492.6	(50.5%)
Manufactured fabric	35,900	(36.1%)	708	(41.7%)	$ 424.9	(43.6%)
Wool or wool mixture	1,600	(1.6%)	19	(1.1%)	$ 16.5	(1.7%)
Fur, silk, or other fabric	400	(0.4%)	67	(0.4%)	$ 6.1	(0.6%)
Other or unclassified	4,400	(4.4%)	53	(3.1%)	$ 34.4	(3.5%)
Total	99,600	(100.0%)	1,699	(100.0%)	$ 974.6	(100.0%)

*Annual average of U.S. fire experience reported to fire departments, 1989-1993.
Source: NFIRS and NFPA Survey.

burning properties. Acetate is made by the reaction of cellulose and acetic acid; it melts before burning. Its burning characteristics are more like thermoplastic fibers (nylon, olefin, and polyester) than cellulose.

The major fully synthetic fibers are acrylics (which char, intumesce, and burn when exposed to heat) and the thermoplastic fibers, such as nylon, olefin (including polypropylene and the less popular polyethylene fibers), and polyester. Acrylics and nylon contain nitrogen atoms and, like wool, produce hydrogen cyanide when burned. The thermoplastics do not char but shrink away from heat sources and melt, form holes near the heat source, and/or ablate. If such fabrics burn at all, the upward flame spread tends to be slow. Molten polymer runs downward, moving the flames sideways and downward, or forms drops that may or may not flame. Compared to fabrics containing cellulose, thermoplastics are generally more difficult to ignite, and the flames tend to cover smaller areas, travel more slowly, or self-extinguish.

TEXTILE MANUFACTURING

Many fibers are spun into yarns and then woven or knitted into a variety of fabrics, e.g., plain or patterned fabrics, or pile, such as flannels, terry cloth towels, artificial furs, and carpets. Fibers also can be processed directly into useful structures, such as nonwoven fabrics or battings. Fabrics are usually scoured, bleached, dyed, and/or finished with chemicals that give them a stiffer or softer feel, permanent press characteristics, luster, or flame or weather resistance.

Figure 4-5A shows the steps in manufacturing woven cotton and cotton blend materials.

Opening, Picking, Cleaning, and Blending

Cotton received from the gin arrives in bales tightly matted and highly compressed. Debris in the form of dirt, leaf matter, grass,

stones, baling wire, machine dust from the ginning process, and tool parts from the handling process can be present. When bales of cotton are brought from the warehouses into the opening rooms, bale ties and coverings are removed and bales are allowed to sit for several hours until they reach room temperature and adjust to the humidity of the area. To ensure uniformity in the end product, cotton must be blended from several bales to make the size, length, and color of the cotton as uniform as possible. This blending process, starting with opening, continues through most processing operations. Opening machinery loosens and separates the cotton into tufts, so it can be effectively cleaned and further processed.

Removing impurities, short fibers, and "dust" from the cotton fiber is essential to ensure product quality and to allow the use of high-speed spinning machines. The cleaning operations are performed pneumatically and mechanically, while the fibers are further separated by spiked and saw-toothed rollers. The dust and light impurities are sucked away, while heavy waste is removed by inertial traps. The efficiency and high speeds of these cleaning machines critically depend upon the airflow within the equipment and within the dust-removal system.

In the chute-feed system, bales are automatically fed into openers connected to various mixers and blenders. The material is then air-driven through ductwork to the carding process, where card lines are fed from individual mixers/reserve hoppers. Usually, hand-fed waste-opener hoppers reintroduce usable waste to the process.

These chute-fed card systems could easily transport fires throughout the entire process, particularly with the high speed of the equipment. Causes of these fires are foreign material in the stock being processed, friction, and defective electrical equipment. Fire in cotton tends to spread fast and flash over an entire group of open bales before it can be extinguished. Tramp metal in the cotton is a notable cause of fires in opener and picker rooms. The development of strong and reliable permanent magnets in the 1940s brought

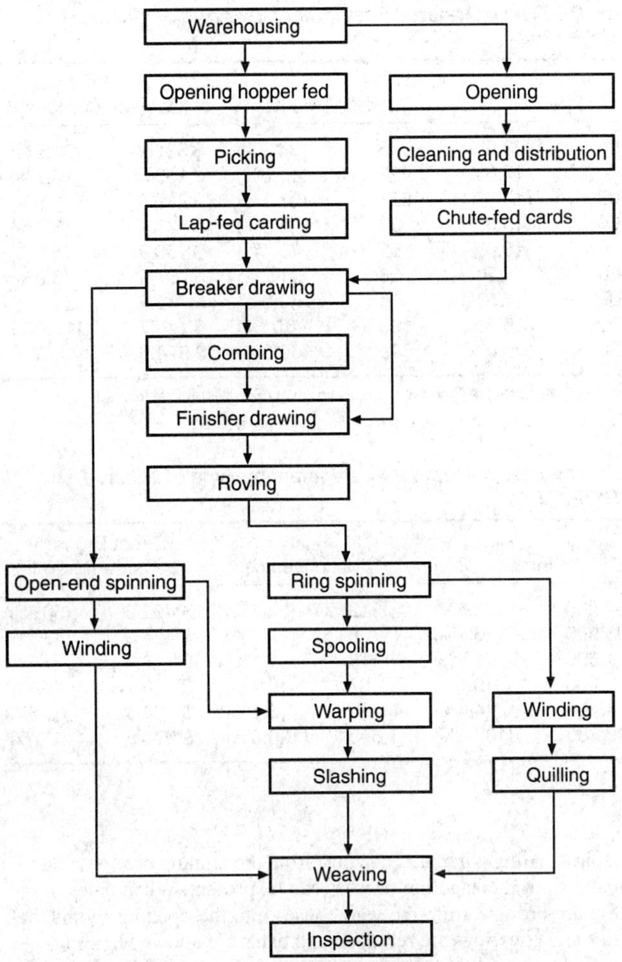

FIG. 4-5A. Flowchart of cotton and cotton blend woven material manufacturing.

about the use of magnetic separators, which significantly contributed to reducing the number of fires caused by foreign materials. Magnets, however, are not the complete answer to the problem. Non-ferrous objects, such as stones, can still produce sparks if they strike any rapidly moving steel part of the opening or cleaning machinery. Gravity traps, which use gravity and inertia to separate heavy foreign materials from the cotton, can be strategically installed to remove these foreign materials.

Carding

Carding machines separate and lay the cotton fibers parallel to each other. During the carding process, two surfaces of wire or metallic card cloth, moving in the same direction but at different speeds on the card machine, comb the cotton into a fine web of generally parallel fibers. A rapidly vibrating vertical comb and a doffer roll remove the fibers from the main drum in a thin film known as the "web." The web is drawn through a tapered cylindrical device called a "trumpet," which condenses the web into a soft, fleecy, rope-like strand of cotton known as a "sliver." The sliver is coiled into storage cans and moved to subsequent equipment.

Fire hazards in a card room are not as great as in opener rooms. However, the frequency of card room losses and the magnitude of the average loss have increased over the years, due to higher operating speeds, insufficient maintenance, or cleaning.[1]

Combing

Combing, an optional process used before drawing, produces a fine, high-quality product. Revolving or oscillating combs equipped with needles perform the combing operation. The purpose is to separate the shorter fibers from the longer ones for removal. The resulting combed fibers are more uniform.

Drawing and Roving

The purpose of "drawing" or "drafting" is to further straighten the cotton fibers, reduce the size of the sliver, and accomplish some degree of blending and additional paralleling of fibers. Usually six or more strands of "card sliver" are fed into the drawing frame and passed through many sets of rolls. The roving process is similar to drawing, except the end result is a condensed sliver about $1/8$ in. (3.2 mm) thick.

Spinning

Two main types of spinning frames are: (1) "ring" and (2) "open end." Open-end spinning eliminates the need for the slubbing process, because the cotton fed into the machine is in the form of a sliver rather than a roving. It is about ten times faster than ring spinning. The yarn discharging from the spinning head can be wound directly onto spools to eliminate the need of a winding operation. Ring-spinning is performed on a spinning frame by reducing the roving strand to the desired size, imparting twist. The finished yarn or thread is wound on a bobbin. Ring-spun yarn has somewhat greater strength than open-end spun yarn.

Many mills are installing suspended ceilings above the spinning frames to better maintain temperature and humidity control. Unfortunately, cotton fiber and lint tend to accumulate above the ceiling tiles and create a potential fire hazard.

Warping

Warping is the first step in preparing for the weaving process. The end threads are tied into continuous lengths that are then threaded through guides and eyelets and wound in parallel on the "section beam." Warper beams are sometimes stored in a high-piled rack configuration consisting of cantilever beams.

Slashing

Before the warp yarns can be used in a loom, they must be treated and coated to prevent the yarns from breaking from tension forces and rubbing or chaffing during the weaving process. In this process, yarn is run through a hot solution of starch, gum, and a softening agent, then dried and wound on the "loom beam." The coating applied is called "sizing."

Loss exposure from fire in these areas is moderate due to combustible loading. However, improperly handled starch can result in a dust explosion. The use of starch has diminished greatly. Polyvinyl alcohol (PVA) is being used in its place.

Weaving

Weaving converts yarn into cloth by interlacing two sets of yarn on a loom. One set is the "warp," the long longitudinal threads; the other is the "filling" or "weft," that runs the width of the loom.

Losses in plain weave rooms, which are second only to cotton opening rooms, are usually small and contained in one machine. Loss experience reveals the two outstanding causes are: (1) defective electrical equipment and (2) excessive friction.

Nonwoven Fabrics

Nonwoven fabrics are textile structures produced by bonding and/or random interlocking of fibers. This can be accomplished mechanically, chemically, thermally, or by using solvents or adhesives. There are two common methods of making nonwoven fabrics: (1) resin bonding and (2) thermoplastic bonding. Resin bonding produces fabric in web form directly from aqueous dispersion equipment. The newly formed web is dried, heat cured, and may be calendered. Some iron-on fabric adhesives are made in this manner.

Thermoplastic bonded nonwoven fabrics contain either cut staple or filament fibers with low melting points. Blends with natural fibers can be used. The fabric is produced by needle punching or other mechanical means. Fibers are laid into a mat formation and numerous overhead needles mechanically entangle the fibers. Heated calendars or high-speed embosser calendars cause the fibers to bond by controlled melting and softening. Some materials use adhesives or latex to further stiffen the fabric.

Nonwoven fabrics are widely used in many ways, e.g., as HAZMAT protective clothing, medical clothing, disposable clothing, stiffeners, and backings in clothing and rugs.

Nonwoven fabrics generally do not include paper or wool felt materials, even though the processes are somewhat similar. Loose, high-loft fibers are used as filling materials and personal hygiene products.

TEXTILE COMBUSTIBILITY

The ignitability, flame spread and heat release rates, total heat produced, heat shrinkage, and ease of fire extinction in textiles depends on the type of fibers used, their percent content, the configuration of the fabric, weight, and construction. The finish generally does not alter these properties significantly, unless the fabric is treated with a flame-retardant. Flame-retardant treatments can be either durable or nondurable. Normal laundering and dry cleaning may reduce flame-retardancy.

Common textile fibers are listed in Table 4-5C, along with some of their burn hazard properties, major trade names, and end

TABLE 4-5C.   *Fire Hazard Properties of Common Textile Fibers*

Fiber Designation	Decomp. Temp. °F (°C)	Melting Temp. °F (°C)	Ignition Temp. °F (°C)	Burning Temp. °F (°C)	Burning Behavior	Typical Trade Names	End Uses
A. *Natural Fibers*							
Cellulosic Cotton, hemp, jute, linen, sisal, etc.	580-610 (305-320)	— —	490-750 (255-400)	1560 (850)	Char, burn, sometimes afterglow	—	Apparel, furnishings, towels, cordage
Protein Wool, mohair, cashmere, camel hair, etc.	450 (230)	— —	1060-1110 (570-600)	1720 (940)	Char, intumesce, burn less readily than cellulosics	—	Apparel, blankets, carpets, furniture covers
B. *Man-made Fibers*							
Acetate	570 (300)	500 (260)	820-1000 (440-545)	1720 (940)	Melts and burns	Ariloft, Celanese, Chromspun, Estron	Apparel, lingerie, furnishings
Acrylic	540-590 (280-310)	— —	860-1050 (460-565)	1560 (850)	Chars, intumesces, burns	Acrilan, Creslan, Orlon, Zefran	Apparel, furnishings, carpets, blankets, pile fabrics
Nylon	600-790 (315-420)	420-490 (215-255)	840-1060 (450-570)	1600 (870)	Melts, ablates, burns	ACE, Anso, Antron, Blue "C," Caprolan, Celanese, Cordura, Courtaulds, Cumnloft, Ultron	Apparel, lingerie, furnishings, carpets, cordage, industrial uses
Olefin (polypropylene)	750 (400)	330 (165)	930-1060 (500-570)	1540 (840)	Melts, ablates, burns	Herculon, Marquesa, Marvess, Patlon	Knitted sportswear, carpets, cordage, furniture covers, industrial uses
Polyester	680-750 (360-400)	480-570 (250-300)	840-1040 (450-560)	1290-1330 (700-720)	Melts, ablates, burns	A.C.E., Avlin, Dacron, Fortrel, Holofil, Kodel, Trevira	Apparel, lingerie, furnishings, carpets, blankets, fiberfill, cordage, industrial uses
Rayon (viscose)	550 (290)	—	790 (420)	1560 (850)	Chars, burns	Avril, Coloray, Fibro, Zantrel	Apparel, lingerie, furnishings
Spandex	581-671 (305-355)	446-482 (230-250)	780 (415)	NA*	Melts, burns	Lycra	In apparel and lingerie where stretch is desired

*NA Not available

uses. The data have been collected from a variety of sources.[2,3,4] The descriptions of burning behavior in this table and the discussions throughout this chapter are based on small-scale tests and may be misleading when large fires or strong radiation are encountered. In such cases, even fabrics that are considered flame-retardant may burn readily.

The orientation of fabrics also influences their combustibility. Fabrics generally burn faster in the vertical position than in the horizontal position. Some fabrics melt when burned, form a pool, and require other combustibles to act as a wick before their full combustibility can be exhibited.

All common natural fibers, except glass, contain organic compounds and the elements carbon and hydrogen, and often also oxygen and nitrogen. Coating added to textile fabrics can greatly alter their combustibility. Chlorine is found in vinyl-coated fabrics.

Thermoplastics are hard to ignite. There are, however, many situations in which thermoplastics can burn readily. This occurs when they are stiffened by finishes or heavy dye application, causing them to lose their ability to evade flames and ablate. It also occurs in blends of thermoplastic and char-forming fibers (even when the latter are flame-retardant), e.g., in the popular polyester/cotton blends, which burn more like 100 percent cotton than 100 percent polyester. A thermoplastic fiber fabric in contact with a char-forming fabric, e.g., a polyester/cotton skirt in contact with a nylon slip,[5] or a polyester curtain in contact with a foam lining, can burn readily. If a burning outer garment is not in contact with a nylon or polyester undergarment, the latter may not ignite, but it may become limp, shrink, make close contact with, and possibly stick to, the skin, causing injury even without burning. Finally, if an ablated thermoplastic fabric still flames when it falls, it can ignite garments or furnishings near it.

However, with such noted exceptions, thermoplastics can be considered relatively safe with respect to small-flame ignition, as compared to cellulosics and if used as single-layer apparel. Thus, accidents in which children's nightwear was the first item to ignite have been essentially eliminated since the introduction of the children's sleepwear standard,[6] and the subsequent change from cotton to primarily thermoplastic nightwear,[7,8] at no additional cost of the garment.[9] On the other hand, thermoplastics, even if they are hard to ignite, should not be used in garments that are intended to protect against heat, such as fire fighter or industrial uniforms, because of their tendency to shrink, melt, and stick when exposed to elevated temperatures.

In addition to the burn behavior, Table 4-5C lists temperatures that have been cited for onset of decomposition, melting, ignition, and the burning temperature of the flames. The ranges are often quite wide for any one fiber because of the differences in methods for measuring such temperatures, as well as the variety of polymer compositions used in each fiber group for various end uses. In the case of nylon, there are two major types, i.e., Nylon 6 and Nylon 66, which vary in the structure of the base molecule. For the ignition temperature of cotton, one finds two values in the literature, i.e., 490 and 750°F (255 and 400°C); it is possible that the lower value was obtained with aged cotton. (It is well known that the ignition temperature of wood decreases with aging, and that may hold for cotton as well.) The ignition temperature of blends tends to be similar to that of the component with the lower ignition temperature. Thus, measured by one method, the ignition temperature for 14 different fiber types ranged from 750°F (400°C) for cotton, to 1,110°F (600°C) for wool; 50/50 blends of cotton and another fiber ranged from 770 to 860°F (410 to 460°C).

Noncombustible Textiles

Noncombustible fabrics include those made entirely from inorganic materials. Table 4-5D lists noncombustible fibers. Glass-based fabrics listed by testing laboratories for use as draperies are woven from uncoated glass yarns that do not burn or propagate flame. If a sufficient quantity of combustible coating or decorative material is applied to glass fabrics or to other noncombustible fabrics, however, the fabrics will support continued flaming. The inherent brittleness of glass and the other fibers listed in Table 4-5D eliminates them from use in clothing fabrics. They are used as reinforcement for plastics and, in the case of relatively low-priced glass fibers, for curtains and draperies.

TABLE 4-5D. *Noncombustible Fibers*

Beta Fiber	Stainless Steel
E-glass	Super alloy
Quartz	Refractory-whiskers
Carbonaceous residue	Alumina
Carbon	Zirconia
Graphite	Boron

High-Temperature and Flame-Retardant Textiles

Several of the fibers that resist high temperatures and are used in textiles are listed in Table 4-5E[10,11] and in Table 4-5F, along with their ignition temperatures, limiting oxygen index (LOI), and some other physical properties.

The limiting oxygen index is a way to measure the tendency of a fabric, once ignited, to continue burning after the ignition source is removed. LOI is defined as the minimum volume concentration of oxygen in a mixture of oxygen and nitrogen that will just support sustained combustion of the material when ignited in a vertical position at its uppermost edge.[12] Cotton fabrics, for example, have a LOI of 19 percent, which means that, if the oxygen is reduced below 19 percent, the cotton will not continue to burn after the ignition source is removed. The higher the LOI of a fabric, the greater the probability that it will cease to burn once the ignition source is removed. Fabrics with high ignition temperatures and high LOIs can be expected to be safe for clothing and furnishings, since they will not ignite readily and will not continue to burn after an accidental ignition source is removed. However, it should be noted that the use of the LOI test for textiles has abated in recent years; full-scale tests of the various items are considered more reliable.

As previously discussed, many synthetic fabrics shrink when exposed to temperatures approaching their melting or decomposition temperatures. When shrinkage brings the fabric into contact with the skin, the insulating layer of air is eliminated and the amount of heat transferred to the skin is increased significantly. Thermal shrinkage is a particularly important property to consider when selecting fabric for use in protective clothing. Another important property of fabrics used in protective clothing is the ability to maintain integrity during a fire. Fabrics that weaken by charring may split apart and fall away, leaving the skin directly exposed to heat.

FLAME-RETARDANT TREATMENTS

The flame-retardant treatment of theater scenery, curtains, and draperies in places of public assembly is commonly required by local fire authorities. Treated fabrics are used in hotels, hospitals, and other occupancies in the interest of preservation of lives and property.

Many businesses specialize in the flame-retardant treatment of theater scenery, draperies, and other fabrics, using standard chemicals. However, some of these businesses may not treat fabrics prop-

TABLE 4-5E. *High-Temperature, Flame- and Chemical-Resistant Fibers*

	Ryton	Ultrate	Nomex §,#	Kevlar §,#	PBI#	T-84 Polyimide	Kynol II,# Phenolic
Softening Point °F (°C)	545 (285)	>600 (>315)	Decomposes†	Decomposes†	Decomposes†	Decomposes†	NA
Upper Use Temp. (50% tenacity) °F (°C)	400 (205)	500 (260)	450 (230)	500 (260)	>600 (>315)	>550 (>290)	NA
Strength (g/denier)	2–4	4–6	3–5	18–20	3–5	3–4	1.5
Elongation (%)	20–40	20–40	20–40	3–4	25–40	30–40	35
Chemical Resistance							
Hot acid	VG	G	P	P	VG	VG	P
Hot base	VG	VG	P	F	G	P	P
Hydrolysis	VG	VG	F	F–P	VG	VG	NA
Solvents	VG	VG	G	VG	VG	VG	VG
Dry heat resistance	G	VG	G	G	VG	VG	VG
Flammability (LOI)‡	34	26	30	29	38–49	37	29–31
Smoke density*	NA	NA	5	5	3	7	low

*Smoke density is a relative scale where, for example, wool = 46 and cotton = 26.
†Decomposes rather than melts.
‡LOI: Limiting oxygen index; for example, non-fire-retardant wool = 28 and cotton = 19.
Generic Name:
§Aramid
II Novoloid
#Presently used in structural fire fighter's protective gear.

VG = very good; G = good; F = fair; P = poor; NA = not available.
Table, with minor additions, reproduced by courtesy of Albany International Research Co., Mansfield, MA.

TABLE 4-5F. *Flame-Retardant (FR) Fibers*

Fiber Type	LOI	Ignition Temp °F (°C)	Burning Behavior °F (°C)	Trade Names
PVC	37	930 (500)	Chars, shrinks at low temperatures	Rhovyl, Leavyl, Teviron
FR rayon	31	—	Chars	PFR
Matrix	29-32	930-1000 (500–540)	Chars at 430 (220), shrinks	Kohjin, Cordelan
Modacrylic	27-30	600 (315)	Chars, shrinks at 280-360 (140-180)	SCF
FR polyester	28	—	Melts at 480 (250)	Trevira 271, Heim
FR wool	32-34	1060-1100 (570–595)	Chars, intumesces	—
FR cotton	28-32	450 (230)	Chars	—

erly or may use relatively ineffective or nonpermanent chemicals. It is therefore advisable to deal only with businesses of known reliability, or, if dealing with an unknown concern, to have treated fabrics tested for adequacy of treatment.

The resistance of such treatments to aging, laundering, and dry cleaning must be ensured. Also, it must be emphasized that such treatments are only expected to make fabrics resistant to ignition by small flames, and that they may burn readily in larger fires.

The effects of chemical treatments in reducing the flammability of combustible fabrics are varied and complex, and all phases are not fully understood. There are five different ways in which chemicals or mixtures retard spread of flame and afterglow: (1) they generate noncombustible gases that tend to exclude oxygen from the burning surface; (2) radicals or molecules from degradation of the flame-retardant chemical react endothermically and interfere with chain reactions in the flame; (3) the flame-retardant chemical decomposes endothermically; (4) a nonvolatile char or liquid is formed by the chemical, which reduces the amount of oxygen and heat that can reach the fabric; and (5) finely divided particles are formed that change the combustion reactions. Usually a flame-retardant chemical or mixture affects flammability in more than one of

these ways. A review of flame-retardant treatments for textiles can be found in *Fire Safety Journal*.[13] Flame-retardant treatments can increase the LOI of cotton from 19 to 28 or 32, and of wool from 28 to 32 or 34.

Hundreds of different chemicals have been suggested as flame-retardants for textiles. It must be emphasized that no one system works for all fibers, as can be seen from Table 4-5G. In addition, some chemical systems change textile properties, e.g., resistance to sunlight, absorption, color, and flexibility, or lose some of their effectiveness when exposed to high temperatures or repeated launderings. Much progress has been made in overcoming most of these undesirable effects. Loss of strength, for example, has become a minor problem in correctly treated flame-retardant cotton fabrics. However, concern that some flame-retardants may be skin irritants, carcinogens, or too expensive can deter wider use.

It must be emphasized that treatments that protect against ignition of textiles by small flames do not necessarily impart resistance to smoldering ignition by cigarette. For example, flame-retardant treatment of cotton generally lowers the charring temperature of cotton but often decreases the cigarette ignition resistance.

TABLE 4-5G. *Flame-Retardant Chemicals for Textiles*

Fiber	Typical Flame-Retardant Chemical
Cotton	Tetrakis (hydroxymethyi) phosphonium salt insolubilized with ammonia gas
Cotton-rayon (disposable, nondurable finish)	Diammonium phosphate; ammonium sulfamate; boron compounds
Rayon (modified fiber)	Hexapropoxy phosphazene
Polyester (modified fiber)	Oligomeric phosphonate decabromodiphenyl ether (DBDPE) and antimony oxide
Polyester, acetate, nylon	Oligomeric phosphonate decabromodiphenyl ether (DBDPE) and antimony oxide
Nylon (nondurable finish)	Thiourea
Wool	Ti, Zr compounds, dibromoterephthalic acid
Modacrylics (modified fibers)	Vinylchloride, vinylidene chloride, vinylbromide as comonomer

RELATIONSHIP OF FLAMMABILITY AND END USE

So far this discussion of textile flammability has been confined primarily to the textiles themselves. The end use of the textile, e.g., ordinary or heat-protective clothing, curtains, carpets, etc., is very important when evaluating the flammability of a textile.

Clothing

Congress passed the *Flammable Fabrics Act of 1953*,[14] after a rash of serious fire accidents in which rayon-pile fabric made into children's cowboy chaps or sweaters was involved. All wearing apparel sold since 1953 has to pass a test described in the Consumer Product Safety Commission's (CPSC) "Standard for the Flammability of Clothing Textiles."[15] The test, however, excluded the rayon-pile fabrics and other extremely flammable fabrics, such as very high-pile cotton flannels, stiffened nylon nets, and very light fabrics. However, all normal apparel fabrics must pass this test.

Several studies indicated that additional protection of young and old populations would be desirable. For example, in a study of 463 fire fatalities that occurred from 1972 through 1977 in the State of Maryland,[16] deaths from clothing ignitions comprised 5.5 percent of the fatalities. Of those victims, 21.6 percent were 9 years old or younger, and 20.5 percent were 60 years old or older. Persons aged 50 years and older were more likely to become fire casualties. The number of fire deaths in this age group was approximately twice the predicted number, based on 1970 census data for Maryland. Factors that contributed to the increased proportion of fire fatalities at both ends of the age spectrum were inability to rapidly and successfully cope with the fire situation and lowered tolerance to toxic combustion products. This applied especially to the older population, where increased cardiopulmonary weakness contributed to lower tolerance to toxic exposures. In addition, it was shown that many victims of garment fires wore sleepwear at the time of the incident.

Consequently, a children's sleepwear standard was promulgated for sizes 0 through 6x in 1971, followed by a standard for sizes 7 through 14, three years later.[6] At the time of promulgation, this standard strained the state of the art, but thermoplastic and flame-retardant garments were soon produced, which passed this rather severe test. The results have been unexpectedly beneficial.

Since the introduction of the children's sleepwear standard, the number of burn deaths among children under age 15, due to ignitions of any type of clothing, have dropped to five or fewer a year. However, a serious problem still exists for older adults. From 1983 through 1985, for example, clothing-ignition burn deaths accounted for roughly one fire death of every 20 fire deaths overall, but for one of every eight fire deaths for people age 65 and over. Since people age 65 and over also have twice the fire death rate of the whole population, this means that they have five times the rate of death due to burns from clothing ignition fires.[17]

The style or type of garment is an important factor in evaluating its fire hazards. For example, a full-length gown containing several yards of flammable fabric could be a death trap, whereas a tightly fitting blouse or shirt made of the same material might be far less dangerous. Not only would the gown present much more fuel for burning, but its size and loose shape would make it far more vulnerable to contact with typical ignition sources. The wearer is unaware that an ignition source is in contact with the garment. Flared skirts, large bows, and full-sleeved kimonos are other examples of easily ignitable garments. The availability of oxygen on both sides of loosely fitting garments accelerates flame spread.

Heat-Protective Clothing

Heat-protective clothing should not burn, melt, or disintegrate on exposure to heat or flame. It should be an effective thermal barrier (consistent with the ability of the wearer to tolerate the weight, heat load, bulkiness, and stiffness of highly insulative garments, such as fire fighters' coats and pants), should not shrink excessively when heated, and should be durable. Detailed reviews of research and development on heat-protective garments and methods to measure their properties can be found in other publications.[18,19] Abbott and Schulman have made an analysis of exposure conditions encountered by fire fighters, and protective materials for routine and emergency conditions.[20]

For structural fire fighting, turnout coats are generally worn in the United States. These coats consist of a water- and abrasion-resistant outer shell, a vapor barrier to prevent penetration of water and chemicals, and an inner liner to provide further insulation against heat and cold. The common materials for the outer shell are Nomex®, blends of PBI and Kevlar®, and flame-retardant cotton. Among the vapor barriers used are neoprene-coated polyester/cotton fabrics or Goretex® fabrics, which are water vapor permeable but do not permit liquid water permeation. Nomex® needled or batting materials, flame-retardant cotton or wool fabrics, etc., are popular inner liners. Similar materials are used in bunker pants, gloves, and hoods used to protect the ears and neck. European fire departments frequently use wool outer shells in their turnout coats; wool tends to char at relatively low temperatures.

Curtains, Draperies, and Wall Coverings

Curtains, draperies, and wall coverings can ignite from such ignition sources as electrical malfunction or wastepaper-basket fires. Flames rapidly spread to walls and ceilings. It has been shown that large differences exist between the potential of various domestic curtain materials to ignite wall-covering materials and the production of heat and smoke this creates.[21] Curtains in public occupancies are regulated by local authorities, and conforming materials do not seem to present major problems during the initial stages of a fire. The test most prescribed by regulatory authorities for curtains and draperies is NFPA 701, *Standard Methods of Fire Tests for Flame-Resistant Textiles and Films* (hereinafter referred to as NFPA 701). NFPA 701 recently addressed that, even if single layers of curtain fabrics pass the requirements of the standard, the combination of two or three layers, often found in motels, etc., may burn vigorously. This was illustrated in a full-scale test.[22] Attempts are presently being made to develop a test method for multiple layers.

Carpets and Rugs

Until 1960 there was very little fire experience to indicate that carpets were a significant factor in fires. Consequently, floor coverings were customarily exempted from coverage in building codes regulating life safety from fire. Since 1960, although there were a small number of fires reported in which carpets spread fire, those that have been reported indicate that carpets can be the prime material involved in the development or spread of a fire. In 1970 a fire in a California dwelling was first seen by an occupant when it was a few feet in diameter on carpeting in front of the fireplace.[23,24] As the occupant applied water with a garden hose to the burning carpet, the fire spread rapidly on the carpet and enveloped the entire room within ten minutes after discovery, including furniture and wood paneling. Because of such experiences, carpet flammability is currently being regulated at the federal, state, and municipal levels. The persistence of fatal carpet and rug fires noted in Table 4-5A primarily occurs in the much less regulated home environment.

Among the many factors in carpet construction that may affect ease of ignition and ability to spread fire are: fiber type, area unit weight density (or, loops per square inch), pile height, backing material, padding, type of adhesive, single *vs.* double back, and type of dye. (Carpet flammability tests for residential and special occupancies are discussed later in this chapter.)

TEXTILE FLAMMABILITY TESTS

The severity of the hazard of a fabric in a particular type of product, such as a dress, curtain, or carpet, depends greatly upon the ease of ignition of the product, the speed with which flame spreads, the amount of heat generated and the rate at which it is released, susceptibility to melting, and the quantity and composition of the smoke and gaseous products of combustion. No individual test method gives a complete evaluation of the hazard, and there are no nationally recognized test methods to measure some of the hazards. Many of the more important test methods for textile products have been summarized.[25,26,27] With the exception of the test methods for mattresses and upholstered furniture, those in common use in the United States are described below.

Clothing

The federal *Flammable Fabrics Act of 1953* was signed into law in 1953 and amended in 1967.[14] The act prohibits the sale of all wearing apparel and fabrics subject to the act that are classified "highly flammable" when tested according to 16 CFR Part 1610, which was formerly designated Commercial Standard CS 191-53.[15] The specified test is designed to measure ease of ignition and rate of flame spread. It does this by impinging a 5-in. (127-mm) long gas flame for 1 sec against the lower end of the upper face of a 2- by 6-in. (51- by 152- mm) specimen that is suspended at a 45-degree angle from the horizontal in a test chamber. If the fabric is ignited by the 1-sec exposure, the time required for the flame to spread 5 in. (127 mm) on the fabric is used to classify the fabric. This test has been successful in excluding inordinately flammable garments from interstate com-

merce. It does not discriminate among other apparel fabrics, even though large differences may exist between them. For example, pile fabrics, such as cotton/polyester terry fabrics and cotton flannels cut into gowns, robes, and pajamas, gave considerably higher burn injury estimates (percentage of body surface burned weighted by depth of burn) than estimates obtained with almost 60 other common apparel fabrics when burned on an instrumented mannequin.[28,29] The same fabrics ranked high on the flammability scale when tested in 12 laboratory fabric flammability tests, but easily passed the Part 1610 requirements. One-hundred percent nylon and polyester fabrics also passed the test, as well as some blends containing wool or modacrylic fibers, which produced very low injury estimates on the mannequin and low flammability results in the other laboratory tests.

The test method in the 1980 edition of NFPA 702,* *Classification of the Flammability* of *Wearing Apparel*, is the same as that in Part 1610,[14] with two important exceptions. The committee that drafted the NFPA-wearing apparel standard noted that many fabrics, particularly those without raised fiber surfaces, were not ignited when the top surface was touched by a small test flame for 1 sec. Those fabrics could not be evaluated for rate of flame spread. Furthermore, the committee concluded that the lower edge of a fabric should be easier to ignite than the surface. Consequently, the test flame in NFPA 702 is brought in contact with the lower edge of all samples and, in the case of those without raised fiber surface, is held there until ignition occurs. Many fabrics that do not ignite in Part 1610[14] ignite in the NFPA 702 test.[29]

Stringent standards were introduced for children's sleepwear in the early 1970s; Parts 1615 and 1616[6] are much more restrictive than Part 1610.[15] Fabric samples 10 by 3 in. (254 by 89 mm) are suspended vertically in a cabinet and subjected to a test flame along the bottom edge. A fabric passes the test if the average char length of five specimens does not exceed 7 in. (178 mm), and no specimen has a char length of 10 in. (254 mm). The barrel of the test flame burner is $^3/_4$ in. (19 mm) below the lower edge of the sample and the test flame is adjusted to extend $1^1/_2$ in. (38 mm). The sample is exposed for 3 sec. Char length is the distance from the fabric's lower edge to the end of the void or tear in the charred, burned, or damaged area. The tear is made by lifting the sample by one of the corners near the lower edge after weights had been attached to the opposite corner near the lower edge. The small-scale test method in NFPA 701 and *Federal Test Method Standard No. 191*, "Flame Resistance of Cloth; Vertical Method 5903.2,"[30] are similar to the children's sleepwear test.[6]

The United States thus has two apparel flammability standards: (1) the children's sleepwear standard, which has greatly reduced the incidence of accidents but restricted the choice of fabrics, and (2) the general apparel standard, which has eliminated inordinately flammable, limited-use fabrics, but does not significantly affect fabric choice. However, it permits many highly flammable fabrics to be used without regard to garment configuration. Attempts to further categorize fabric flammability hazards have been made; some have based such categorization on flame-spread rate measurements.[29,31] Flame-spread rate measured on remainders of garments involved in fire accidents, however, did not correlate with the size of the victim's burn injury.[32] Researchers have developed methods for measuring the rate of heat release from burning fabrics.[33,34,35] Braun *et al.*[35] suggested that fabrics with a low rate of heat release could be used in all garments; those with short ignition times and higher rate of heat release could be used only in tightly fitting garments; and finally suggested the elimination of fabrics with short ignition time ($^1/_2$ sec or less) and high-heat release rate. The International Standards Organization (ISO) is also proposing a standard based on garment fit.

---

*NFPA 702 was withdrawn as an NFPA document and is no longer available.

Heat-Protective Clothing

NFPA 1971, *Standard on Protective Clothing for Structural Fire Fighting*, employs three methods to evaluate the fire safety of turnout coats:

1. An ignition test: "Flame Resistance of Cloth: Vertical Test Method 5903.2,"[30] which is quite similar to the children's sleepwear test, except that it uses a different gas, with a flame temperature of 1550°F (843°C), and a 12-sec ignition time. Maximum char length is 4 in. (102 mm), afterflame time 2 sec.

2. A heat stability test: The fabrics used in the ensemble must not char, separate, or melt when exposed in a forced-air oven at 482°F (250°C) for 5 min.

3. A heat protection test: The thermal protection performance (TPP) test, which is described below.[36]

The third method measures actual heat transfer through the ensemble and relates it to burn injury potential, rather than relying on the ensemble thickness. A horizontal specimen cut from the ensemble is placed over a heat source delivering 2 cal/cm^2 per sec [7.4 Btu/sq ft per sec (84 kW/m^2/s)]. The heat source consists of two gas flames and nine quartz heaters adjusted to obtain a 50:50 ratio of convective and radiative heat. The time until an incipient second-degree burn to skin would occur in contact with the inner layer is measured; this is based on work correlating the rate of heat and the total heat delivered to skin with burn injury severity.[37] This time measure, multiplied by two because of the exposure to 2 cal/cm^2 per sec, is called the thermal protection performance. A minimum TPP rating of 35, well within the state of the art, has been proposed as the acceptance criterion.

The Health and Safety Division of the International Association of Fire Fighters has written guidelines for procurement of fire fighter protective clothing, citing, among other items of interest, Occupational Safety and Health Administration (OSHA) standards and military procurement specifications as of 1983.[38,39] NFPA also has standards for station/work uniforms, helmets, gloves, and protective footwear for structural fire fighters.

The protection in real fires provided by turnout coats varying widely in construction and ranging in TPP from 33 to 53 was investigated. It was concluded that the TPP is predictive of protection afforded if the coats are exposed to rapidly increasing heat. A coat with a TPP of 35 will provide protection for only a few seconds in a flashover situation.[40,41]

A TPP test like the one described above, but using a 7.4 Btu/sq ft per sec (84 kW/m^2/s) gas flame rather than a combination of flame and radiant heat, is listed by the American Society for Testing and Materials (ASTM).[42] Fabrics exposed to other heat sources on one side and in contact with artificial skin or other thermosensors can be ranked according to protective value.[42,43,44] Tests for the tendency of molten metals to adhere to protective clothing, for thermal protective performance of materials when in contact with hot surfaces, and for resistance to penetration by chemicals have been developed or are under development.

Blankets

In an ASTM blanket flammability test,[45] a small flame is brought down on the center of a specimen for 1 sec. The specimen is covered with a paper monitor with a 2-in. (5-cm) circular cutout. If the paper monitor does not char or become discolored, the material is Class 1, and is considered acceptable.

Curtains, Draperies, and Wall Coverings

Curtains and draperies for public occupancies are generally regulated by local or state fire authorities, and the most widely used test is described in NFPA 701. The small-scale part of this method excludes certain thermoplastic fabrics that develop unacceptable char length, because in the specimen frames they cannot shrink freely away from the flames and ablate as they can in real-life situations. The American Society for Testing and Materials' ASTM D3659, *Standard Test Method for Flammability of Apparel Fabrics by Semi-Restraint Method*,[46] has been suggested to replace the small-scale NFPA 701 test. In this test, the 15- by 6-in. (381- by 152- mm) specimens are attached to a bar on top of the test apparatus and are somewhat restrained from movement by chains that loosely connect the bottom corners of the specimens to fixed points. With the specimen frames removed, more thermoplastic materials pass the test. According to the present version of NFPA 701, materials that fail the small-scale test can still be qualified if they pass the large-scale test.

Curtain fabrics that can pass NFPA 701 as single layers can burn vigorously in combination with other curtain fabric layers.[22] Such multi-layer curtain types are often found in motels.

Many jurisdictions require that wall coverings including fabrics be tested by ASTM E84[47] or NFPA 255, *Standard Method of Test of Surface Burning Characteristics of Building Materials* (hereinafter referred to as NFPA 255). A study was undertaken to compare the burning characteristics of a number of textile wall coverings in full-scale tests, yielding results of two bench-scale tests—the cone calorimeter and the lateral ignition flamespread test (LIFT).[48] In this test, specimens approximately 32 by 6 in. (80 by 15 cm) are mounted on an inert insulating material, with the longer dimension horizontal. A radiant panel is mounted at an angle to the specimen so that the heat flux over the specimen is varied. Flame spread and heat flux at which ignition occurs are measured. Use of such tests may eventually obviate the use of the more complex tunnel tests in which the materials—unrealistically for wall coverings—are mounted horizontally in a tunnel.

Carpets and Rugs

Carpets and rugs are currently regulated in the United States by federal regulations.[49] Separate parts of the regulations apply to particular sizes. In testing these carpets and rugs, a 9- by 9-in. (229- by 229- mm) sample is placed on the floor of a test chamber, and a flat steel plate with an 8-in. (203-mm) diameter hole in the center is placed on the sample. A methanamine tablet is placed in the center of the hole and ignited. The sample passes the test if the charred portion does not extend beyond 1 in. (25.4 mm) of the edge of the hole at any point. Seven of eight specimens of other-than-small carpets and rugs must pass if carpets made of the tested material are to be sold. In the case of small carpets and rugs, those that do not pass the test may be sold if they carry a permanent label stating that they failed to pass and should not be used near an ignition source. This requirement eliminated only a small percentage (usually low quality) of the carpets on the market at the time the standard was introduced.

Hospitals and other institutions receiving federal financial aid are required to comply with flammability limits for carpets and other floor coverings used on patient-occupied areas and exit ways. The flame spread rating of floor coverings must not exceed 75, as determined by ASTM E84.[47] This test is similar to NFPA 255 and is commonly referred to as the Steiner Tunnel Test.

The Appendix of NFPA *101®*, *Life Safety Code®*, suggests that where limitations are to be placed on the flammability of floor coverings, the test method in NFPA 255 is to be used. Objections have been raised regarding its use to evaluate floor coverings, because positioning the test specimen on top of the tunnel bears no resemblance to its position when in place in a building. The method is also criticized because it may result in delamination, melting, and dripping of test specimens, which removes combustible material from the reach of the burner. The Canadian carpet test places the specimen on the floor of a tunnel.[50]

In an effort to more correctly assess the flame spread hazard of flooring materials, the concept of critical radiant flux was developed. In their work on model corridor fire tests, the National Bureau of Standards (now NIST) found that the radiant energy impacting on a floor covering had a significant influence on the propagation of flame across the floor covering. This work led to the development of the flooring radiant panel test (FRPT).

The flooring radiant panel test as embodied in NFPA 253, *Standard Method of Test for Critical Radiant Flux of Floor Covering Systems Using a Radiant Heat Energy Source*, subjects a horizontal floor-covering specimen to a radiant energy flux from a gas-fired panel at a 30-degree angle. The radiant energy flux decreases along the length of the specimen according to a standard energy flux *vs.* distance profile. A pilot flame ignites the specimen at the end with the higher radiant energy flux. The flame front is allowed to travel until it is no longer reinforced by the radiant energy flux and extinguishes. The heat flux at this point of the specimen is called the "critical radiant flux." For corridors and exitways in such occupancies as hospitals and nursing homes, a minimum of 0.45 W/cm^2 is recommended (except one- and two-family houses—where the recommendation is 0.22 W/cm^2).[51] The critical heat flux of carpets then is believed to have aggravated the initial fire conditions; e.g., the Baptist Tower fire was found to be 0.1 W/cm^2.[52] However, the above requirements are well within the state of the art of carpet manufacture.

Textiles in Aircraft

The flammability of textiles in the crew and passenger compartments of aircraft is regulated by the Federal Aviation Administration (FAA). The vertical test method 5903.2 is used;[30] average char length for three specimens must not exceed 8 in. (203 mm), afterburn time must be less than 15 sec, and the time that drippings may continue to flame must be less than 5 sec. In addition, the FAA has specified a new test for passenger seats.[53] For the purpose of this test, a unique oil burner that has the following specifications is used: (1) kerosene consumption, 2 gal per hr (7.57 L/h); (2) nozzle diameter, 6 in. (152 mm); (3) distance of nozzle from side of seat to be tested, 4 in. (102 mm); (4) heat delivery to seat specimen, 10 Btu per sq ft per sec (115 kW/m^2 per sec); and (5) temperature at specimen, 1850°F (1010°C). At least three seats must be tested. For at least half of the number tested, the flame spread must not reach the side of the cushion opposite the burner, the average weight loss must not exceed 10 percent, and there must be no flaming melt drip below the cushion. Among the seat assemblies that pass this test are: (1) those that contain the widely used wool/nylon fabric; (2) interlayers consisting of aluminized or ordinary Nomex® or PBI fabrics, which prevent pyrolysis gases from venting toward the flames; and (3) polyurethane foam. An interliner made from Vonar® neoprene-type foam sheet also makes it possible to pass this test, but the lighter fabrics are more cost effective in this application. Other interliner materials are under development, and the use of interliners has spread to institutional furniture as well.

Textiles in Motor Vehicles

Materials used in the occupant compartments of passenger cars, multipurpose passenger vehicles, trucks, and buses in the United States must not burn or transmit a flame front across the surface at a rate of more than 4 in. (102 mm) per min when tested according to the motor vehicle safety standard of the Department of Transportation (DOT).[54] In that standard, a specimen is suspended in a horizontal position in a test chamber, and one end is ignited by a Bunsen burner flame. The specimen is held in a U-shaped clamp and the assembly is suspended so that the top of the barrel of the burner is $3/4$ in. (17 mm) below the exposed edge of the specimen. The burner flame is adjusted to $1 1/2$ in. (38 mm) high, and the flame temperature to that of a natural gas flame. The specimen is exposed to the test flame for 15 sec. Timing begins when the flame from the burning specimen reaches a point $1 1/2$ in. (38 mm) from the open end of the specimen, and timing ends when the flame has progressed to 10 in. (254 mm) or self-extinguishes. As previously mentioned, burn rate must not exceed 4 in. (102 mm) per min. This test can be considered to provide only minimum protection, and work is in progress to develop a more realistic test, especially for school buses.

Tents, Tarpaulins, Air-Supported Structures

NFPA 102, *Standard for Grandstands, Folding and Telescopic Seating, Tents, and Membrane Structures* (hereinafter referred to as NFPA 102), requires that the textiles used in these structures meet the appropriate flame-resistance requirements when tested in accordance with NFPA 701. Covered by NFPA 102 are: (1) tents and air-supported structures used for assembly; (2) tents and air-supported structures in which animals are stabled; (3) tents and air-supported structures located in a portion of the premises used by the public; (4) tents and air-supported structures in places of assembly in or about which fuel-burning devices are located; and (5) all tarpaulins used in connection with (1) through (4).

The small-scale test in NFPA 701 was mentioned earlier. The large-scale test in NFPA 701 uses specimens 7 ft (2.135 m) long and 5 or 25 in. (127 or 635 mm) wide, depending upon whether the specimen is to be tested as a flat sheet or is to be hung in folds. The flat specimen is suspended vertically in an open-ended metal stack and its lower edge is exposed to a 11-in. (280-mm) high flame from a Bunsen burner, the top of whose barrel is 4 in. (102 mm) below the bottom of the specimen. A folded specimen is tested with its folds $1/2$ in. (12.7 mm) apart in the same metal stack and with the same burner arrangement. For both single-sheet and folded specimen, the test flame is applied for 2 min, then withdrawn, and the duration of flaming combustion recorded. Length of char, i.e., the distance from the top of the test flame to the top of the charred area resulting from spread of flame or afterglow, is also recorded. To pass the test, afterflame time cannot exceed 2 sec, and the char length cannot exceed 10 in. (254 mm). For specimens in folds, maximum afterflame time cannot exceed 2 sec, and maximum char length cannot exceed 35 in. (890 mm), but afterglow may spread in the folds.

In addition, the Canvas Products Association International (CPAI) has developed CPAI, *A Specification for Flame-Resistant Materials Used in Camping Tentage*.[55] The test for wall and top material is essentially the same as the 5903.2 test[30]; and for flooring, the methanamine tablet carpet test is prescribed. Maximum char length for walls and tops varies from 4.5 in. (114 mm) for light fabrics to 9 in. (228 mm) for heavy fabrics. For floors, the Part 1630 criteria apply.[49] Provisions for leaching, weathering, and simulated sunlight exposure are part of this voluntary standard. Conforming tents carry the CPAI label.

TEXTILES IN SPECIAL ENVIRONMENTS

Certain ambient atmospheric conditions influence the behavior of treated textiles in fire situations. They are: (1) oxygen-enriched atmospheres, (2) compressed-air atmospheres, (3) atmospheres where flammable inhalation anesthetizing agents may be present, and (4) exposure to weather.

Action of Treated Textiles in Oxygen-Enriched Air

Some industrial processes require work in atmospheres enriched above the normal oxygen content. Such atmospheres present a high fire hazard, particularly to the clothing worn by the workers. The effect of flame-retardant treatments on the combustibility of the fabrics has been studied.[56]

An indication of the resistance to burning of flame-retardant textiles in oxygen-enriched atmospheres can be inferred from the limiting oxygen indices in Table 4-5F.

The textile fibers department of the Du Pont Company has conducted research on their Nomex® high-temperature-resistant nylon fiber in both high-pressure air environments and oxygen-enriched atmospheres at greater-than-normal pressures. Nomex®-based fabric was found to retain its flame-retardant qualities under air pressure of four atmospheres (59 psig or 405 kPa). At this pressure and in higher oxygen concentrations (40 percent and greater), Nomex® did support combustion, but at a substantially slower rate than did untreated, readily combustible fabrics. NASA has further improved the state of the art in this area, and uses materials such as PBI, flame-retardant cotton, etc., in the oxygen-rich atmospheres.

Fabrics used for hospital operating room garments, or in other atmospheres with flammable gas contamination, should be electrically conductive to prevent accumulation of static electricity. Such accumulations, if discharged to ground, could provide a spark of sufficient energy to ignite the gas.

Weather-Exposed Textiles

Flame-retardant treatments for tents, awnings, tarpaulins, and other fabrics exposed to the weather must not leach out over a period of time. If water-soluble chemicals are used, the treatments must be renewed at frequent intervals. Both the NFPA 701 and CPAI[55] tests for tentage specify methods for determination of the effects of weathering.

A wide variety of chemicals can be used for treating fabrics exposed to the weather. Many of the formulas used include chlorinated paraffin, chlorinated synthetic resins, or chlorinated rubber, in combination with various water-insoluble metallic salts, plasticizers, stabilizers, synthetic resins, pigments, binders, mildew inhibitors, etc. Such mixtures are not water soluble and are used with hydrocarbon solvents, or are suspended in aqueous emulsions. The effective application of such treatments calls for techniques and equipment ordinarily available only for factory processing or to businesses specializing in flame-retardant applications. Application of paint to these treated fabrics may lessen their flame resistance.

BIBLIOGRAPHY

References Cited

1. "Cotton Processing," *IRInformation®*, Section IM.17.21.1, Industrial Risk Insurers, Hartford, CT.
2. MMFPA, *Man-Made Fibers Fact Book*, Man-Made Fibers Producers Association, Inc., Washington, DC, 1988.
3. Gottlieb, I.M., and Beck, L. R., Jr., "Thermal Stability, Flammability, and Melting Characteristics of Textile Substrates," *Annals NY Academy of Science*, Vol. 82, 1959, pp. 724–743.
4. Einsele, U., "Uber das Brennverhalten und den Brennmechanismus von Synthesefasern," *Melliand Texilberichte*, Vol. 53, 1972, pp. 1395–1402.
5. McCullough, E.A., and Noel, C. J., "Flammability Characteristics of Layered Fabric Assemblies," *Proceedings* Information Council on Fabric Flammability, 1978, pp. 175–184.
6. CPSC, *Code of Federal Regulations*, Part 1615, "Standard for the Flammability of Children's Sleepwear: Size 0 Through 6x (FF3–71"; Part 1616, "Sizes 7 Through 14 (FF5-74)." U.S. Consumer Product Safety Commission, Washington, DC, 1974.
7. Bolieu, S., "Children's Sleepwear Flammability Standards: Have They Worked?" *Proceedings* Information Council on Fabric Flammability, 1977, pp. 1–5.
8. Crikelair, G. F., "A Win for the Team," *Proceedings* Information Council on Fabric Flammability, 1975, pp. 218–233.
9. Beckwith, O.P., "Current Status of Children's Sleepwear Manufacturing and Marketing," *Proceedings* Information Council on Fabric Flammability, 1979, pp. 114–122.
10. AIRC, *Chemical, High-Temperature, and Fire-Resistant Fibers*, Albany International Research Co. (formerly FRL), Newsletter, Nov. 1, 1988, Mansfield, MA.
11. Carborundum, *Kynol® Flame-Resistant Fibers*, The Carborundum Co., Sanborn, NY 1975.
12. ASTM D2863, *Method for Measuring the Minimum Oxygen Concentration to Support Candle-Like Combustion of Plastics (Oxygen Index)*, 1991, American Society for Testing and Materials, W. Conshohocken, PA.
13. Gordon, B. F., "Flame Retardants and Textile Materials," *Fire Safety Journal*, Vol. 4, 1981, pp. 109–123.
14. CPSC, *Flammable Fabrics Act of 1953* (as amended and revised 1967), U.S. Consumer Product Safety Commission, Washington, DC, 1967.
15. CPSC, *Code of Federal Regulations*, Part 1610, "Standard for the Flammability of Clothing Textiles (formerly CS 191–53)," and Part 1611, "Standard for the Flammabilty of Vinyl Plastic Film (formerly CS 192-53)," U.S. Consumer Product Safety Commission, Washington, DC, 1988.
16. Berl, W. G., and Halpin, B. M., "Human Fatalities from Unwanted Fires," *Fire Journal*, Vol. 73, No. 5, Sept. 1979, pp. 105–123.
17. Harwood, B., and Hall, J.R., Jr., "What Kills in Fires: Smoke Inhalation or Burns," *Fire Journal*, May/June 1989, pp. 28–34.
18. Brewster, E. P., and Barker, R. L., "A Summary of Research on Heat-Resistant Fabrics for Protective Clothing," *American Industrial Hygiene Association Journal*, Vol. 44, 1983, pp. 123–130.
19. ASTM, "International ASTM Symposium on the Performance of Protective Clothing," *ASTM Special Technical Publication*, American Society for Testing and Materials, W. Conshohocken, PA, 1984.
20. Abbott, N.J., and Schulman, S., "Protection from Fire: Nonflammable Fabrics and Coatings," *Journal of Coated Fabrics*, Vol. 6, July 1976, pp. 48–64.
21. Moore, L. D., "Full-Scale Burning Behavior of Curtains and Draperies," NBSIR 78–1448, 1978, U.S. National Bureau of Standards, Washington, DC.
22. Belles, D. W., and Beitel, J. J., "Do Multi-Layer Draperies Pass the Single-Layer Fire Test?" *Fire Journal*, Sept./Oct. 1988.
23. "Burning Incense Fell on Carpet," *Fire Journal*, Vol. 64, No. 4, July 1970, p. 50.
24. "Polyester Carpet Fire, Suisun, CA," *Fire Journal*, Vol. 64, No. 5, pp. 88, 89.
25. Hilado, C. J., *Flammability Test Methods Handbook*, Technomic Publishing Company, Inc., Westport, CT, 1973.
26. Krasny, J. F., "Flammability Evaluation Methods for Textiles," *Flame-Retardant Polymeric Materials*, Vol. 3, Lewin, M., Atlas, S. M., and Pearce, E. M., eds., Plenum, NY, 1982.
27. Troitzsch, J., *International Plastics Flammability Handbook: Principles—Regulations—Testing and Approval*, Macmillan, New York, 1983.
28. Bercaw, J. R., "Cooperative Program on General Apparel Flammability: Status Report," *Proceedings* Information Council on Fabric Flammability, 1977, pp. 175–190.
29. Krasny, J. F., "Apparel Flammability: Accident Simulation and Bench-Scale Tests," *Textile Research Journal*, Vol. 56, 1986, pp. 287–303.
30. GSA, "Flame Resistance of Cloth: Vertical Test Method 5903.2," Federal Test Method Standard No. 191, 1971, General Services Administration, Washington, DC.
31. Nielsen, E.B., and Richards, H. R., "A Proposed Method for Measuring Rate-of-Burn," *Journal of the American Association of Textile Chemists and Colorists*, Vol. 1, 1969, pp. 270–277.
32. Vickers, A., Krasny, J., and Tovey, H., "Some Apparel Fire Hazard Parameters," *Proceedings* Information Council on Fabric Flammability, 1973, pp. 205–226.

33. Miller, B., "A New Concept for Monitoring the Burning Behavior of Fabric," *Proceedings* Information Council on Fabric Flammability, 1976, pp. 198–208.

34. Miller, B., and Meiser, C. H., Jr., "Heat Emission from Burning Fabrics, Potential Harm Ranking," *Textile Research Journal*, Vol. 48, 1978, pp. 238–243.

35. Braun, E., *et al.*, "Back-up Report for the Proposed Standard for the Flammability of General Wearing Apparel," NBSIR 76–1072, 1976, National Bureau of Standards, Washington, DC.

36. Behnke, W. P., "Predicting Flash Fire Protection of Clothing from Laboratory Tests Using Second-Degree Burn to Rate Performance," *Fire and Materials*, Vol. 8, No. 2, 1984, pp. 57–63.

37. Stoll, A. M., and Chianta, M. A., "Method and Rating Systems for Evaluation of Thermal Protection," *Aerospace Medicine Journal*, Vol. 40, 1969, pp. 1232–1238.

38. Smith, M. J., *et al., Procuring Fire Fighter Protective Clothing*, Health and Safety Division, Department of Research, International Association of Fire Fighters, AFL-CIO-CLC, Washington, DC, 1983.

39. Veghte, J. H., *Firefighter's Protective Clothing*, Janesville, Dayton, OH, 1988.

40. Krasny, J. F., Rockett, J. A., and Dingyi, H., "Protecting Fire Fighters Exposed in Room Fires: Comparison of Results of Bench-Scale Tests for Thermal Protection and Conditions During Room Flashover," *Fire Technology*, Vol. 24, 1988, pp. 5–19.

41. Krasny, J. F., and Peacock, R. D., "Protecting Fire Fighters Exposed in Room Fires: Part 2. Performance of Turnout Coat Materials under Actual Fire Conditions" (unpublished, 1989).

42. ASTM D4108, "Thermal Protective Performance of Materials for Clothing by Open-Flame Method," 1987, American Society for Testing and Materials, W. Conshohocken, PA.

43. Schoppe, M. M., Welsford, J. M., and Abbott, N. J., "Resistance of Navy Shipboard Work Clothing Materials to Extreme Heat," Tech. Rep. No. 148, 1982, Navy Clothing and Textile Research Facility, Natick, MA.

44. Braun, E., *et al.*, "Measurement of the Protective Value of Apparel Fabrics in a Fire Environment," *Journal of Consumer Product Flammability*, Vol. 7, 1980, pp. 15–25.

45. ASTM D4151, "Standard Test Method for Flammability of Blankets," 1992, American Society for Testing and Materials, W. Conshohocken, PA.

46. ASTM D3659, "Standard Test Method for Flammability of Apparel Fabrics by Semi-Restraint Method," 1993, American Society for Testing and Materials, W. Conshohocken, PA.

47. ASTM E84, "Standard Test Method for Surface Burning Characteristics of Building Materials," 1994, American Society for Testing and Materials, W. Conshohocken, PA.

48. Hackleroad, M.F., "Fire Properties Data Base for Textile Wallcoverings," NISTIR 89–4065, May 1989, National Institute of Science and Technology, Washington, DC.

49. CPSC, *Code of Federal Regulations*, Part 1630: "Standard for the Surface Flammability of Carpets and Rugs (FFI-70)," Part 1631: "Standard for the Surface Flammability of Small Carpets and Rugs (FF2-70)," U.S. Consumer Product Safety Commission, Washington, DC, 1970.

50. CGSB, "Standard Method of Test for Surface Burning Characteristics of Flooring and Floor Covering Materials," B.54.9-1972, Canadian Government Specification Board, Ottawa, Ontario.

51. Benjamin, I.A., and Davis, S., "Flammability Testing of Carpet," NBSIR 78-1436, 1978, National Bureau of Standards, Washington, DC.

52. Willey, A. E., "Fire, Baptist Towers Housing for the Elderly," *Fire Journal*, Vol. 67, No. 3, 1973, pp. 15–21, 103.

53. FAA, "Flammability Requirements for Aircraft Seat Cushions," 14 CFR, Parts 25, 29, and 1231, Oct. 11, 1983, Federal Aviation Administration, Washington, DC.

54. DOT, *Code of Federal Regulations*, Part 571, "Motor Vehicle Safety Standard No. 302, Flammability of Interior Materials—Passenger Cars, Multi-purpose Passenger Vehicles, Trucks, and Buses," U.S. Department of Transportation, Washington, DC.

55. CPAI, *A Specification for Flame-Resistant Materials Used in Camping Tentage*, Canvas Products Association International, St. Paul, MN, 1984.

56. Turner, H. L., and Segal, L., "Fire Behavior and Protection in Hyperbaric Chambers," *Fire Technology*, Vol. 1, No. 4, 1965, pp. 269–277.

NFPA Codes, Standards, and Recommended Practices

Reference to the following NFPA codes, standards, and recommended practices will provide further information on fibers and textiles discussed in this chapter. (See the latest version of *The NFPA Catalog* for availability of current editions of the following documents.)

NFPA *101®, Life Safety Code®*
NFPA 102, *Standard for Grandstands, Folding and Telescopic Seating, Tents, and Membrane Structures*
NFPA 253, *Standard Method of Test for Critical Radiant Flux of Floor Covering Systems Using a Radiant Heat Energy Source*
NFPA 255, *Standard Method of Test of Surface Burning Characteristics of Building Materials*
NFPA 701, *Standard Methods of Fire Tests for Flame-Resistant Textiles and Films*
NFPA 1971, *Standard on Protective Ensemble for Structural Fire Fighting*
NFPA 1975, *Standard on Station/Work Uniforms for Fire Fighters*

Additional Reading

"Baled Fibers," *Loss Prevention Data Sheet 8-7*, Factory Mutual Research Corp., Norwood, MA.

Baitinger, W. F., "Wash Durability of FR Cotton," Technical Westex Inc., Chicago, IL, ASTM STP 1133; American Society for Testing and Materials (ASTM), Performance of Protective Clothing: Challenges for Developing Protective Clothing for the 1990s, Fourth (4th) Volume, ASTM STP 1133, June 18–20, 1991, Montreal, Canada, ASTM, Philadelphia, PA, 1991, pp. 766–774.

Beninate, J. V., *et al.,* "Wool: Its Effect on Flame Retardant Properties of Blend Fabrics," *Journal of Fire Sciences,* Vol. 1, No. 2, Mar./Apr. 1983, pp. 145–154.

Benisek, L., *et al*, "Latest Developments in Wool Aircraft Textile Furnishings," *Fire Safety Journal*, Vol. 17, No. 3, 1991, pp. 187–203.

Bhatnagar, V. M., ed., "Advances in Fire Retardant Textiles," and "Flammability of Apparel," Vols. 5 and 7, respectively, *Progress in Fire Retardancy Series,* Technomic, Westport, CT, 1975.

Buchanan, M. S., "Bountifil DB: Flammability Characteristics of a 100% Polyester Fiber Cushion," Product Safety Corporation Fifteenth International Conference on Fire Safety, Volume 15, January 8–12, 1990, Millbrae, CA, Product Safety Corp., Sunnyvale, CA, 1990, pp. 102–107.

Carroll-Porczynski, C. Z., *Flammability of Composite Fabrics*, 1st American ed., Chemical Publishing Co., New York, 1984.

Christian, S. D., "Current and Developing European and International Standards for Furniture and Textiles," Home Office, London, England, Paper 4, Interscience Communications Limited, European Conference on Furniture Flammability, 2nd June 27–28, 1990, Brussels, Belgium, 1990, pp. 4/1–6.

Fire Safety Progress in Regulations, Technology, and New Products, Technomic, Lancaster, PA, 1987.

"The Flammability of Floor Coverings," *Journal of Fire and Flammability,* Vol. 1, 1970, pp. 64–70.

Fukatsu, K., "Kinetic and Thermogravimetric Analysis of Thermal Degradation of Polychlal Fiber/Cotton Blend," *Journal of Fire Sciences*, Vol. 8, No. 3, May/June 1990, pp. 194–206.

Hall, R. W., "Investigation of Melt Stick Characteristics of Flame Retardant Treated Polyester-Cotton Blend Fabrics," Fourth Annual Conference on Protective Clothing: A New Review of Applications, Materials Used, Needs, Concerns, and Trends, April 24–26, 1990, Charlotte, NC, 1990, pp. XX/1–11.

Harper, R. J., and Ruppenicker, G. F., "Flame Retardancy of Cotton: Current Perspective," Business Communications Co., Inc. (BCC), Recent Advances in Flame Retardancy of Polymeric Materials: Materials, Applications, Industry Developments, Markets, Volume 2, May 14–16, 1991, Stamford, CT, Business Communications Co., Inc., Norwalk, CT, 1991, pp. 141–149.

Hasselbrack, S. A., "Flame-Retarding Wool Textile Materials and the Evaluation of Thermally Stable Polymers for Commercial Airplanes," *Proceedings* of Improved Fire- and Smoke-Resistant Materials for Commercial Aircraft Interiors, November 8–10, 1994, Washington, DC, National Academy Press, Washington, DC, 1994, pp. 165–174.

Horrocks, A. R., Akalin, M., and Price, D., "Smoke, CO_2 and CO Evolution From Cotton and Flame Retarded Cotton. Part 2. Behavior of Single

Layer Fabrics in Air at Elevated Temperatures, *Journal of Fire Sciences*, Vol. 8, No. 2, Mar./Apr. 1990, pp. 135–151.

Horrocks, A. R., Allen, J., and Price, D., "Testing and Performance of Flame Retardant Textiles," Fire Retardant Chemicals Association International Conference on Fire Safety—Fire Retardant Technology and Marketing. March 25–28, 1990, New Orleans, LA, Fire Retardant Chemicals Assoc., Lancaster, PA, 1990, pp. 181–198.

Horrocks, A. R., Anand, S. C., and Sanderson, D., "Fibre-Intumescent Interactive Systems for Barrier Textiles," British Plastics Federation and Institute of Materials, Sixth International Conference, Flame Retardants ' 94, January 26–27, 1994, London, England, Interscience Communications Ltd., London, England, 1994, pp. 117–128.

Horrocks, A. R., *et. al.*, "Influence on Laundering on Durable Flame Retarded Cotton Fabrics. Part 1. Effect of Oxidant Concentrations and Detergent Type," *Journal of Fire Sciences*, Vol. 10, No. 4, July/Aug. 1992, pp. 335–351.

Horvat, S. G., "Use of a Brominated Organic (Not Brominated Diphenyloxide) in a Flame Retardant Chlorinated Paraffin Dispersion for Cellulosic and Cellulosic Blend Fabrics for Textile Applications," 1992 Spring Conference on Technical and Marketing Issues Impacting the Fire Safety of Building and Construction and Home Furnishings Applications, March 29-April 1, 1992, Orlando, FL, Technomic Publishing Co., Lancaster, PA, 1992, pp. 213–223.

Hshieh, F. Y., and Beeson, H. D., "Flammability Testing of Pure and Flame Retardant-Treated Cotton Fabrics," *Fire and Materials*, Vol. 19, No. 5, Sept./Oct. 1995, pp. 233–239.

"Ignition Sources for Determination of Burning Behavior of Textiles and Textile Products," International Organization for Standardization, Geneva, Switzerland, ISO/TC 38/SC 19 N395, May 8, 1990, 26 pages.

Jackson, J., "Textile Flammability Testing," Society of Plastics Engineers and Fire Retardant Chemicals Association, Fire Retardant Blends, Alloys and Thermoplastic Elastomers: Material Source—Processing—Applications and Testing, March 17–20, 1991, Hilton Head Island, SC, Fire Retardant Chemicals Assoc., Lancaster, PA, 1991, pp. 191–197.

Johnson, R. F., "Camping Tent Flammability—What the Record Shows," *Fire Journal*, Vol. 71, No. 4, July 1977, pp. 50–55.

Kirkland, K. M., "High Performance Polyethylene Fibers for Protective Clothing Applications," Fourth Annual Conference on Protective Clothing: A New Review of Applications, Materials Used, Needs, Concerns, and Trends, April 24–26, 1990, Charlotte, NC, 1990, pp. VI/1–25.

Krasny, J. F., "Textile Item Fire Safety: Accomplishments and Failures," First International Conference on Fire and Materials, September 24–25, 1992, Arlington, VA, 1992, pp. 109–116.

Krasny, J. F., Singleton, R. W., and Pettengill, J., "Performance Evaluation of Fabrics Used in Fire Fighter' s Turnout Coats," *Fire Technology*, Vol. 18, No. 4, 1982, pp. 309–318.

Lilani, H. N., "Comparative Evaluation of Heat/Flame Resistant Hybrid Textiles for Protective Clothing Applications," Product Safety Corporation, Seventeenth International Conference on Fire Safety, Volume 17, January 13–17, 1992, Millbrae, CA, Product Safety Corp., Sunnyvale, CA, 1992, pp. 225–241.

Martin, J. R., and Miller, B., "The Thermal and Flammability Behavior of Polyester-Wool Blends," *Textile Research Journal*, Vol. 48, 1978, pp. 97-103.

Michener, J. W., "Textile Industry and Flammability Regulations: An Update," Business Communications Co., Inc. (BCC), Recent Advances in Flame Retardancy of Polymeric Materials: Materials, Applications, Industry Developments, Markets, Volume 2, May 14–16, 1991, Stamford, CT, Business Communications Co., Inc., Norwalk, CT, 1991, pp. 213–219.

Mischutin, V., "Flame Retarding Cellulosic and Blended Textiles," Amspec Chemical Corp., Glouster City, NJ, Fire Retardant Chemicals Association, Fire Safety Developments and Testing: Toxicity—Heat Release—Product Development—Combustion Corrosivity, October 21-24, 1990, Ponte Vedra Beach, FL, Fire Retardant Chemicals Assoc., Lancaster, PA, 1990, pp. 193–203.

Morin, C.J., "Protective Clothing Applications for Flame Resistant Cotton," Fourth Annual Conference on Protective Clothing: A New Review of Applications, Materials Used, Needs, Concerns, and Trends, April 24–26, 1990, Charlotte, NC, 1990, pp. III/1–4.

Pellejero, A. L., and Tapia, P. H., "Effect of Fire-Retardants on Textile Fiber blends and Their Applications in Fire Behavior. [Influencia del producto ignifugo y su aplicacion en el comportamiento al fuego en mezclas de fibras textiles.]," *Mapfre Seguridad*, Vol. 14, No. 53, 1994, pp. 91–92.

Pike, R., "High Performance Fibers For Protective Applications," Fourth Annual Conference on Protective Clothing: A New Review of Applications, Materials Used, Needs, Concerns, and Trends, April 24–26, 1990, Charlotte, NC, 1990, pp. V/1–10.

Price, D., and Tunc, M., "Textile Flammability," *Chem. Br.*, Vol. 23, No. 3, 1987, pp. 235–236.

Quintiere, J. G., Hopkins, M., and Hopkins, D., Jr., "Room-Corner Fire Prediction for Textile Wall Materials," *Proceedings* of the International Conference on Fire Research and Engineering, September 10–15, 1995, Orlando, FL, SFPE, Boston, MA, 1995, pp. 272–277.

Saito, N. and Yanai, E., "Classification of Flammability of Non-Melting Flame Retardant Textiles," Report of Fire Research Institute of Japan, No. 70, 9–16, September 1990.

Saville, N., "Overview of Fire Blocking Fabrics," *Proceedings* of the 32nd International SAMPE Symposium and Exhibition: Advanced Materials Technology ' 87, SAMPE, Covina, CA, 1987, pp. 1347-1359.

Saville, N., and Squires, M., "Latest Developments in Fire Resistant Textiles," *Textile Month*, May 1990, pp. 47–50, 52.

Saxena, N. K., Sharma, S. K., and Gupta, D. R., "Development and Evaluation of Fire Retardant Cotton Fabrics," *Fire Science and Technology*, Vol. 12, No. 2, 1992, pp. 1–10.

Suzuki, M., Dobashi, R., and Hirano, T., "Burning Process of Cellulosic Fibers Composing Filter Paper During Flame Spread," First Asian Conference on Fire Science and Technology (ACFST), October 9–13, 1992, Hefei, China, International Academic Publishers, China, 1992, pp. 499–504.

Sweeney, M., "Maintaining Textile Flame Retardancy," *Fire Prevention*, No. 269, May 1994, pp. 15–17.

Taday, H., "Widening Market for Flame-Retardant Textiles," *Fire Prevention*, No. 243, October 1991, p. 9.

Tempesti, A., "Inherently Flame Retardant Fibers for Furniture Applications," SNIA Fiber, Italy, Paper 12 (Abstract Only), Interscience Communications Limited, European Conference on Furniture Flammability, 2nd June 27–28, 1990, Brussels, Belgium, 1990, 12 pages.

"Textiles: Design of Apparel for Reduced Fire Hazard," Technical Report, International Organization for Standardization," Geneva, Switzerland, ISO TR 9240, October 1, 1992, 10 pages.

Tomann, J., "Puuvillakankaalla verhoillun pehmustetun huonekalun sytyminen kytevaan paloon. [Padded Furniture With Cotton Upholstering Ignites Into a Smoldering Fire.]," *Palontorjuntatekniika*, March 1993, pp. 22–23.

Villa, K. M., "Textile Test Methods for Protective Clothing Standards," Fourth Annual Conference on Protective Clothing: A New Review of Applications, Materials Used, Needs, Concerns, and Trends, April 24–26, 1990, Charlotte, NC, 1990, pp. X/1–5.

Woolliscroft, D., "Flame and Fire Protection," *Textile Horizons*, Vol. 7, No. 11, 1987, pp. 42–43.

Yanai, E., and Saito, N., "Classification of Flammability of Thermal Shrinkable and/or Melting Flame Retardant Textiles," Fire Research Institute, Tokyo, Japan, Report of Fire Research Institute of Japan, No. 75, 1993, pp. 1–8.

Yates, J. M., "Firefighters Push for Mandatory Flammability Testing," *Wood and Wood Products*, Vol. 92, No. 11, Oct. 1987, p. 38.

Yuill, C. H., "Floor Coverings: What is the Hazard?" *Fire Journal*, Vol. 61, No. 1, Jan. 1967, pp. 11–19.

Zhiyi, Z., "Preparatory Manufacture of the Blended Dyeable Flame-Retarded Polypropylene Fiber," Fire Retardant Chemicals Association International Conference on Fire Safety—Fire Retardant Technology and Marketing. March 25–28, 1990, New Orleans, LA, Fire Retardant Chemicals Assoc., Lancaster, PA, 1990, pp. 33–41.

FLAMMABLE AND COMBUSTIBLE LIQUIDS

Revised by Orville M. Slye, Jr.

The discussion of flammable and combustible liquids presented in this chapter focuses on the fire hazards of these materials. After these fire hazards are discussed, the NFPA classification system for flammable and combustible liquids is presented. The physical and fire characteristics of these materials are discussed next, including a special section on the burning characteristics of liquids. Methods to prevent fires while using flammable and combustible liquids complete the chapter.

For more information about flammable and combustible liquids not contained in this chapter, see the following chapters: Section 1, Chapter 5, "Chemistry and Physics of Fire," which includes information about the physical and fire characteristics of liquids and other materials; Section 3, Chapter 2, "Control of Electrostatic Ignition Sources," which discusses the control of static electricity as an ignition source; Section 3, Chapter 21, "Storage of Flammable and Combustible Liquids"; Section 3, Chapter 23, "Storage and Handling of Chemicals," which discusses the storage and handling of chemicals and flammable and combustible liquids; Section 4, Chapter 14, "Explosion Prevention and Protection," which discusses explosion venting; and Section 4, Chapter 19, "Air-Moving Equipment," which discusses blower and exhaust systems for flammable and combustible liquids.

Flammable and combustible liquids fires can be controlled and extinguished using foam, dry chemical agents, carbon dioxide, halogenated agents, or water. Section 6, Chapter 12, "Water Spray Protection," reviews use of water spray for cooling and extinguishment; Section 6, Chapter 18, "Halogenated Agents and Systems," discusses halon agents; Section 6, Chapter 20, "Carbon Dioxide and Application Systems," reviews use of carbon dioxide agents; Section 6, Chapter 21, "Dry Chemical Agents and Application Systems," discusses use of dry chemical agents; and Section 6, Chapter 22, "Foam Extinguishing Agents and Systems," contains information on use of foam agents for extinguishment of flammable and combustible liquids fires.

HAZARDS OF FLAMMABLE AND COMBUSTIBLE LIQUIDS

Flammable and combustible liquids do not burn or explode by themselves. Instead, flammable vapors resulting from the evaporation from those liquids elevated to temperatures above their flash points and exposed to an ignition source, such as a spark, will burn. Since, by definition, most flammable liquids are normally stored and handled above their flash points, they continually give off vapors when the vapor-air mixture is within the flammable range.

Ignitable mixtures occur when concentrations of vapors in air are within a definite percentage range, commonly referred to as the flammable (explosive) range. The lower limit of the range is known as the lower flammable limit (LFL). The upper limit of the range is known as the upper flammable limit (UFL). For example, the proportions (flammable range) for carbon disulfide are from about 1 to 44 percent carbon disulfide vapors in air by volume; for ethyl alcohol, from about 4 to 19 percent by volume; and for gasoline, from about 1.4 to 7.6 percent by volume. Therefore, storing flammable and combustible liquids in the proper closed containers and minimizing the exposure of the liquid to air while in use are fundamental safety measures.

Explosions of flammable vapor-air mixtures near the lower or upper limits of the flammable range are less intense than those occurring in well-mixed "stochiometric" concentrations of the same mixture. Flammable vapor-air explosions occur most frequently in confined spaces, such as containers, tanks, rooms, or buildings. The violence of flammable vapor explosions depends upon the nature of the vapors and enclosure containing the mixture, as well as on the quantity of vapor-air mixture.

Distinct from an explosion of a flammable vapor-air mixture inside a tank or container is the overpressuring of a tank which results in a rupture. As with vapor explosions, pressure ruptures vary in intensity. Any closed container may rupture violently when exposed to a severe fire.

Fire and explosion prevention measures are based on one or more of the following techniques or principles: (1) exclusion of sources of ignition, (2) exclusion of air (oxygen), (3) storage of liquids in closed containers or systems, (4) ventilation to prevent the accumulation of vapor within the flammable range, and (5) use of an atmosphere of inert gas instead of air. Extinguishing methods for flammable and combustible liquid fires involve shutting off the fuel supply, excluding air by various means, cooling the liquid to stop evaporation, or a combination thereof.

Gasoline is the most widely used flammable liquid. It is commonly known that gasoline generates flammable vapors at ambient temperatures. There are many other volatile flammable products. A comprehensive list of the more commonly used flammable or combustible liquids, including the characteristics of each, appears in

Orville M. Slye, Jr., P.E., is president of Loss Control Associates, Inc., Langhorne, PA.

NFPA 325, *Guide to Fire Hazard Properties of Flammable Liquids, Gases, and Volatile Solids.*

Flash point, though commonly accepted as the most important criterion of the relative hazard of flammable and combustible liquids, is not the only factor used to evaluate the hazard. The ignition temperature, flammable range, rate of evaporation, reactivity when contaminated or exposed to heat, density, and rate of diffusion of the vapor also affect the hazard. The flash point and other factors that determine the relative susceptibility of a flammable or combustible liquid to ignition have comparatively little influence on the liquid's burning characteristics after fire has burned for a short time.

Although many flammable and combustible liquids can be classed as normal or stable liquids, others introduce the problem of instability or reactivity.

The storage, handling, and use of unstable (reactive) flammable or combustible liquids require special attention. It may be necessary to increase the distances to property lines from storage tanks and between adjacent tanks or to provide extra fire protection. For example, it would be poor practice to locate heat-reactive and water-reactive flammable or combustible liquids tanks adjacent to each other. In the event of a ground fire, water applied to the heat-reactive tank for protection might penetrate the tank of water-reactive liquid and cause a violent reaction.

CLASSIFICATION OF FLAMMABLE AND COMBUSTIBLE LIQUIDS

For fire protection purposes, an arbitrary division between liquids and gases has been established, based upon the definition of a flammable liquid found in NFPA 30, *Flammable and Combustible Liquids Code* (hereinafter referred to as NFPA 30). Liquids are defined as those fluids with a vapor pressure not exceeding 40 psia (2,068 mm Hg), which is approximately 25 psi gage pressure at 100°F (37.8°C). Another arbitrary division also has been established between liquids and solids for the purposes of this classification system. A liquid is additionally defined as one which has a fluidity greater than that of 300 penetration asphalt.

The following classification system for flammable and combustible liquids (from NFPA 30) is based on the division of flammable liquids into three categories. It is anticipated that, in most areas, the indoor temperature could reach 100°F (37.8°C) at some time during the year. Therefore, all liquids with flash points below 100°F (37.8°C) are called Class I liquids. In some areas, the ambient temperature could exceed 100°F (37.8°C), requiring only a moderate degree of heating to bring the liquid to its flash point. Based on this concept, an arbitrary division of 100 to 140°F (37.8 to 60°C) was established for liquids in this flash point range, to be known as Class II liquids. Since liquids with flash points higher than 140°F (60°C) would require considerable heating from a source other than ambient temperatures, they have been identified as Class III liquids.

Flammable Liquids

1. Flammable liquids have flash points below 100°F (37.8°C) and vapor pressures not exceeding 40 psia at 100°F (2,068 mm Hg at 37.8°C).
2. Class I liquids include those with flash points below 100°F (37.8°C) and may be subdivided as follows:
   (a) Class IA liquids includes those with flash points below 73°F (22.8°C) and with boiling points below 100°F (37.8°C).
   (b) Class IB liquids includes those with flash points below 73°F (22.8°C) and with boiling points at or above 100°F (37.8°C).

(c) Class IC liquids includes those with flash points at or above 73°F (22.8°C) and below 100°F (37.8°C).

Combustible Liquids

Liquids with flash points at or above 100°F (37.8°C) are referred to as combustible liquids and may be subdivided as follows:

1. Class II liquids have flash points at or above 100°F (37.8°C) and below 140°F (60°C).
2. Class IIIA liquids have flash points at or above 140°F (60°C) and below 200°F (93°C).
3. Class IIIB liquids have flash points at or above 200°F (93°C).

Other Classification Systems

There are other classification systems for flammable and combustible liquids. In some, the break points between classes of liquids are different, or the open-cup flash-point tester is used to determine the flash point. In some classification systems, the solubility of the liquid with water is considered.

Solids with Flash Points

Many combustible chemicals that are solids at 100°F (37.8°C) or above are classified as solids. When heated, the solid becomes liquid and gives off flammable vapors; flash points can then be determined. In their liquid state, these solids should be treated as liquids with similar flash points. Some manufactured solids, such as paste waxes or polishes, may contain varying amounts of flammable liquids. The flash point and amount of liquid in such manufactured solid materials will indicate the degree of hazard.

PHYSICAL PROPERTIES OF LIQUIDS

In liquids, molecules move freely among themselves, but they do not tend to separate from one another in the way that molecules in gases do. Liquids, unlike gases, are only slightly compressible and are incapable of indefinite expansion. They differ from solids in the ease with which their molecules move against one another, causing them to adapt to the shape of the containing vessel. Some liquids are very viscous, however, and no sharp definition can be drawn between these liquids and solids. Neither is there a sharp line of demarcation between liquids and gases. Materials can exist in either state, depending upon temperature and pressure conditions. Liquids tend to become gases as their temperatures increase or their pressures decrease. On the other hand, gases tend to become liquids as their temperatures decrease or their pressures increase. However, a material can only exist as a gas, regardless of pressure, if its temperature is above the so-called critical temperature. (The critical temperature is an intrinsic property of a material and is the temperature above which the material can exist only in the gaseous state.) In all cases, the actual state of a liquid is determined by the combination of temperature and pressure.

PHYSICAL CHARACTERISTICS OF FLAMMABLE AND COMBUSTIBLE LIQUIDS

Specific Gravity

Since the specific gravity of water equals one, a liquid with a specific gravity of less than one will float on water (unless it is water soluble). A specific gravity greater than one indicates that water will float on the liquid. Because fire-fighting foams are water-based extinguishing agents, this is an important consideration in fighting a flammable or combustible liquids fire.

Vapor Density Ratio

Vapor density is the weight per unit volume of a pure gas or vapor. For fire protection purposes, vapor density is reported in terms of the ratio of the relative weight of a volume of vapor to the weight of an equal volume of air, under the same conditions of temperature and pressure. In this respect, the vapor density ratio is similar to the specific gravity of a liquid, except that air is used as the standard in place of water. Air is taken as unity and the vapor density is reported as a ratio. A vapor density of 3 (i.e., 3:1) indicates that the vapor is three times as dense, or heavy, as air.

$$\text{Vapor density ratio} = \frac{\text{molecular weight (MW) of the material}}{\text{composite molecular weight of air}}$$

$$= \frac{\text{MW}}{29}$$

Vapor density ratios are reported at equilibrium temperature and atmospheric pressure conditions. Unequal or changing conditions will appreciably change the density of any vapor.

Generally, pure vapors are seldom found, except when the liquid is normally stored above its boiling point. For those liquids that are stored or handled below their boiling points, the vapor located above the liquid will be a mixture of vapor and air.

For liquids with boiling points above the prevailing ambient temperature at atmospheric pressure, the vapor density may be a misleading figure. For example, the boiling point of ethyl acetate is 171°F (77°C). At 70°F (21°C) its vapor pressure is about 2.35 psi (16.2 kPa) or 0.16 atm (atmospheres). Thus, an equilibrium mixture of ethyl acetate vapor and air at 70°F (21°C), as might be found in a closed vessel, would have a composition of 16 percent ethyl acetate vapor and 84 percent air. The "theoretical" vapor density of ethyl acetate is 3.0. The "actual" density of the "vapor-air mixture" that is produced by ethyl acetate at 70°F (21°C) and at atmospheric pressure is:

$$0.16 \times 3.0 + 0.84 \times 1 = 1.32$$

Vapor densities are ordinarily used as an indication of the settling or rising tendency of a vapor.

Vapor Pressure

Where a liquid is present in a closed container with an atmosphere of vapor-air mixture above the liquid surface, the percentage of vapor in the mixture may be determined from the vapor pressure. The percentage of vapor is directly proportional to the relationship between the vapor pressure of the liquid and the total pressure of the mixture. For example, acetone at a temperature of 100°F (37.8°C) has a vapor pressure of 7.6 psia (52 kPa). Assuming total pressure at 14.7 psia (101 kPa), the proportion of acetone vapor present will be 7.6 divided by 14.7, or 52 percent. Vapor pressures of petroleum liquids usually are determined by the Reid method, as recommended by the American Society for Testing and Materials (ASTM D323).[1] This method gives the vapor pressure in psia (kPa) at 100°F (37.8°C), which differs slightly from the true vapor pressure. Vapor pressure figures for many substances can be found in chemical handbooks. If the closed-cup flash point of a liquid and the vapor pressure at the flash point temperature are known, the lower flammable limit for the vapor (at the flash point temperature), in percent

by volume at normal atmospheric pressure, can be calculated as follows:

$$\text{LFL} = \frac{V}{0.147} \left(\text{in SI}: \ \text{LFL} = \frac{V}{1.01} \right)$$

where LFL is percent vapor by volume at the lower flammable limit, and V is vapor pressure in psia (kPa) at flash point temperature. At other pressures,

$$\text{LFL} = \frac{100V}{P}$$

where P is the ambient pressure in psia (kPa).

In a mixture of volatile flammable liquids, the effects of the two vapor pressures, one upon the other, generally depend upon whether the two liquids are completely miscible, partly miscible, or completely immiscible. If the two liquids are completely miscible, they lower each other's vapor pressures; if they are nearly completely immiscible, the vapor pressure of the mixture is the sum of the partial pressure of each liquid layer (Dalton's Law of Partial Pressure); if both liquids are partially miscible, the relationships are more complex.

Evaporation Rate

Evaporation rate is the rate at which a liquid is converted to the vapor state at any given temperature and pressure. The differing rates of mixture evaporation is of primary concern in fire protection. In general, as the boiling point decreases, the vapor pressure and the evaporation rate increases.

Viscosity

The viscosity of a liquid is a measure of its resistance to flow resulting from the combined effects of adhesion and cohesion, i.e., a measure of internal friction in a fluid. Although there are several recognized devices for determining viscosity, the measurement principles are generally the same. These involve measuring the time required for a predetermined quantity of liquid at a specified temperature to flow into a receiving flask or through an orifice of a prescribed size.

Latent Heat of Vaporization

Latent heat of vaporization is the heat that is absorbed when one gram of liquid is transformed into vapor at the boiling point under one atmosphere pressure, and is expressed in Btu per lb or in calories per gram (cal/g).

Solubility in Water and Surface Tension

Solubility in water and surface tension are other characteristics of a liquid that concern the fire protection field. Fires in liquids that are water soluble can be extinguished by dilution of the liquid with water, or by use of "alcohol-type" foams for extinguishment. The use of wetting agents affects the surface tension of a liquid and, in some cases, aids in fire extinguishment.

FIRE CHARACTERISTICS OF LIQUIDS

Flash Point

A vapor mixed with air in proportions below the lower limit of flammability may burn at the source of ignition, i.e., in the zone immediately surrounding the source of ignition, without propagating (spreading) flame away from the ignition source. The flash point of a liquid corresponds roughly to the lowest temperature at which the vapor pressure of the liquid is just sufficient to produce a flammable mixture at the lower limit of flammability. (See Figure 4-6A.)

There are several types of test apparatus used for determining flash point. (See Figure 4-6B.) The tag closed tester is described in ASTM D56, *Standard Test Method for Flash Point by Tag Closed Tester*. The Pensky-Martens closed tester, as described in ASTM D93, *Standard Test Methods for Flash Point by Pensky-Martens Closed Tester*, is specified for testing liquids with flash points above 200°F (93°C), or for certain viscous or film-forming materials.

ASTM D3278, *Standard Test Methods for Flash Point of Liquids by Setaflash Closed-Cup Apparatus*, is used for testing paints, enamels, lacquers, varnishes, and related products and their components with flash points between 32 and 230°F (0 and 18°C).

Open-cup flash points are sometimes used in grading flammable liquids in transportation and are determined by ASTM D1310, *Standard Test Method for Flash Point and Fire Point of Liquids by Tag Open-Cup Apparatus*, or ASTM D92, *Standard Test Method for Flash and Fire Points by Cleveland Open Cup*. Open-cup flash points represent conditions with the liquid in the open and are generally higher than the closed-cup flash point figures for the same substances. However, closed-cup flash point figures are generally higher than the actual lower temperature limit of flammability (flash point) when tested in a vertical tube utilizing the principle of upward flame propagation.

It is essential to realize that flash point varies with pressure and with the oxygen content of the atmosphere, as well as with the purity of the product being tested, the method of test, and the skill of the operator conducting the test. It is quite possible to have a flammable mixture far below the stated flash point value if the pressure is considerably less than one atmosphere.

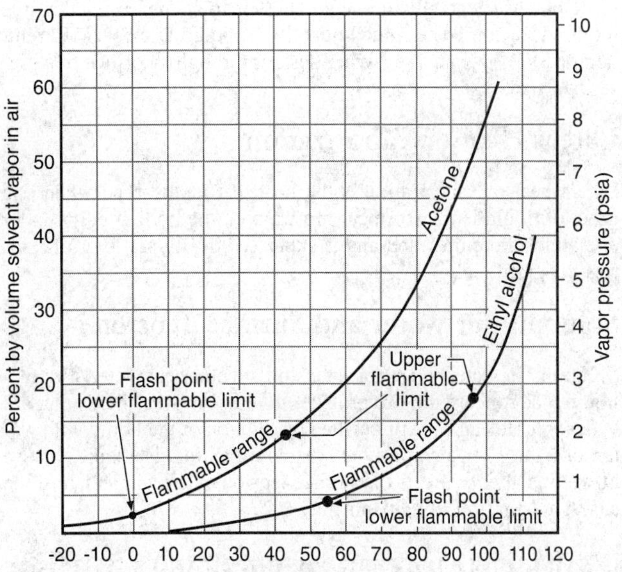

Ignition Temperature (Autoignition Temperature, Autogenous Ignition Temperature)

The autoignition temperature reported for a flammable liquid is generally the temperature to which a closed or nearly closed container must be heated in order that the liquid in question, when introduced into the container, will ignite spontaneously and burn. Time lags of a minute or more are frequently involved. The standard testing method for autoignition temperatures of petroleum products is described in ASTM E659, *Standard Test Method for Autoignition Temperature of Liquid Chemicals*.

Ignition Temperatures and Molecular Weights

Within a given hydrocarbon series, such as the straight chain series running from normal methane down through normal decane, igni-

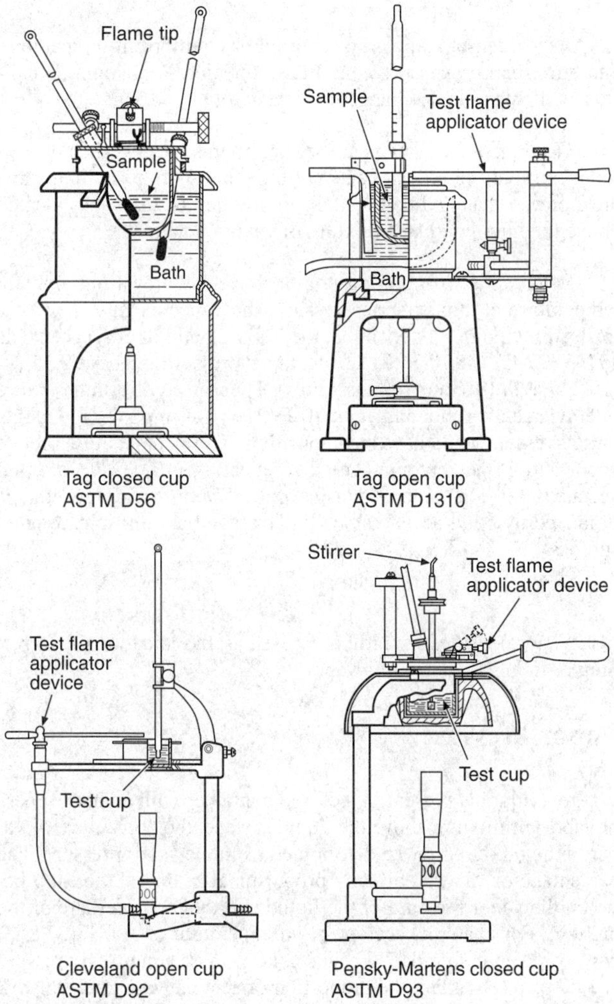

FIG. 4-6B. *Four of the commonly used testers for determining flash points of flammable or combustible liquids. The material to be tested is slowly heated, and at periodic intervals a test flame is applied to the vapor space. Flash point is the temperature at which a flash of fire is seen when the test flame is applied. The full details of conducting tests for each type of testing apparatus are given in the applicable standards of the American Society for Testing and Materials, or by the manufacturer.*

FIG. 4-6A. *Relationship among flash point, flammable limits, temperature, and vapor pressure for acetone and ethyl alcohol. Liquid, vapor, and air in equilibrium in a closed container at normal atmospheric pressure. [One psia = psi + 14.7; psi × 6.89 = kPa; 5/9 (°F −32) = °C.]*

tion temperature decreases as molecular weight or carbon chain length increases, all other factors being equal. (See Figure 4-6C.) Thus, pentane [$CH_3(CH_2)_3CH_3$; molecular weight 72.1] has a higher ignition temperature than hexane [$(CH_3CH_2)_4CH_3$; molecular weight 86.2].

Boiling Point

The temperature at which the equilibrium vapor pressure of a liquid equals the total pressure on the surface is known as the boiling point. The boiling point is entirely dependent on the total pressure. (Boiling point increases with an increase in pressure.) Theoretically, any liquid may be made to boil at any desired temperature by sufficiently altering the total pressure on its surface. Similarly, unless decomposition takes place, any liquid may be made to boil at any desired pressure by sufficiently changing its temperature.

The temperature at which a liquid boils when under a total pressure of one atmosphere (14.7 psia or 101 kPa) is termed the normal boiling point.

The above applies to pure materials or constant boiling mixtures. Most flammable liquids and gases on the market today are mixtures and do not follow the physical laws governing pure materials. The boiling points of mixtures are reported as distillation curves. The 10 percent point of a distillation performed in accordance with ASTM D86, *Standard Test Method for Distillation of Petroleum Products*, is used as the boiling point for most NFPA standards.

Flammable (Explosive) Limits

The term "lower flammable limit" (LFL) describes the minimum concentration of vapor to air below which propagation of a flame will not occur in the presence of an ignition source. The "upper flammable limit" (UFL) is the maximum vapor to air concentration above which propagation of flame will not occur. If a vapor-to-air mixture is below the lower flammable limit, it is described as being "too lean" to burn, and if it is above the upper flammable limit, it is "too rich" to burn.

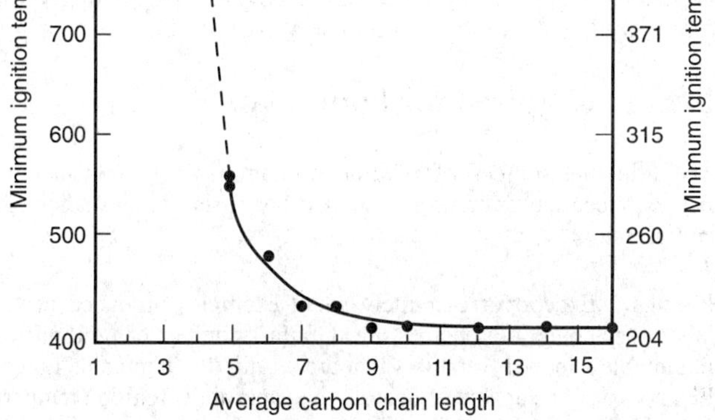

FIG. 4-6C. Minimum ignition temperatures of hydrocarbons of various carbon chain lengths.

When the vapor-to-air ratio is somewhere between the lower flammable limit and the upper flammable limit, fires and explosions can occur. The mixture is then said to be within its flammable or explosive range. When the mixture happens to be in the intermediate range between the LFL and UFL [synonymous with lower explosive limit (LEL) and upper explosive limit (UEL), respectively], the ignition is more intense and violent than if the mixture were closer to either the upper or lower limits.

Calculating Vapor Volume and Flammable Mixtures

It is frequently helpful to calculate the volume of air required to provide dilution sufficient to prevent the formation of an ignitible mixture as, for example, in the design of a ventilating system for a drying oven. This can be readily done when the quantity of solvent supplied is known or can be reasonably estimated.

For example, consider a process in which acetone vapor is released. The flammable range for acetone vapor in air is from 2.6 to 12.8 percent by volume. Any mixture containing more air than 100 minus 2.6, or 97.4 percent of air—equivalent to 97.4 divided by 2.6, or approximately 37 volumes of air to one volume of vapor—will be too lean to be ignitible. Hence, if the volume of vapor can be estimated, it is a simple matter to determine the required air volume.

The volume of vapor produced from 1 gal of solvent can be calculated from the specific gravity of the liquid and the vapor density as follows:

Cubic feet of vapor from 1 gal of liquid =

$$\frac{8.33 \times \text{specific gravity of liquid } (H_2O = 1)}{0.075 \times \text{vapor density of vapor (air = 1)}}$$

where 8.33 is the weight of 1 gal of water in lb, and 0.075 is the weight of 1 cu ft of air in lb.

Stated more simply:

$$\text{Vapor equivalent of 1 gal is}: \ 111 \times \frac{\text{Sp. G.}}{\text{V.D.}}$$

If the vapor density is not known, it can readily be calculated from the molecular weight.

Again, taking acetone as an example [specific gravity (Sp. G.) = 0.792; vapor density (V.D.) = 2], yields:

$$\text{Vapor equivalent of 1 gal is}: \ 111 \times \frac{0.792}{2} = 44 \text{ cu ft}$$

The volume of air required to dilute vapor from 1 gal of acetone to below the lower flammable limit is:

$$44 \times 37 = 1,628 \text{ cu ft}$$

If the rate of evaporation of acetone into a given space (oven, etc.) should be 1 gpm, it would require 1,628 cfm of uncontaminated ventilating air to keep the vapor concentration below the lower flammable limit.

For SI units, the volume of vapor (m^3) produced by 1 L of solvent can also be calculated from the specific gravity of the liquid and the vapor density, as follows:

The cubic meter vapor equivalent of 1 liter (L) of liquid is:

$$L = 0.83 \times \frac{\text{Sp. G.}}{\text{V.D.}}$$

For acetone, vapor equivalent of 1 liter (L) is:

$$L = \frac{0.83 \times 0.792}{2} = 0.33 \text{ m}^3$$

In SI units, the volume of air required to dilute 1 L of acetone to below the lower flammability limit is:

$$0.33 \times 37 = 12.2 \text{ m}^3$$

For different evaporation rates, the air quantity would be proportional. In practice and as a safety factor, a substantial excess of air ventilation would be applied because of the inevitable nonuniformity of the atmosphere within the enclosure.

Where it is desirable to keep a space "too rich" to be ignitible, a similar procedure (using the upper flammable limit) may be used to determine the maximum air quantity that can be tolerated. This is not, however, a recommended procedure, since it is necessary to enter and pass through the flammable range during the operation.

Variation in Hazard with Temperature and Pressure

The temperature and pressure to which a particular liquid is subjected must often be considered in the practical application of the flammable limits of flammable liquids. It can be shown that these factors have a pronounced effect on the fire and explosion hazard and are as important in safe handling as are the flammable limits themselves. A liquid that has a flash point above room temperature and, therefore, is not expected to produce flammable vapors, will do so if its temperature is raised to its flash point or higher.

The extent of liquid vaporization varies as pressure in the vapor space over the liquid is increased or lowered. More vapor is released as pressure is lowered, and less vapor will result from increased pressure above the liquid. In the same manner, at a higher temperature the liquid will have a higher vapor pressure and will tend to vaporize in greater amounts. When at a given temperature and pressure the liquid has vaporized to the point where no further vaporization will occur unless conditions are changed, a state of equilibrium is said to have been reached. It is evident that equilibrium cannot exist in other than a closed system. In the open air, a vaporizing liquid would continue to vaporize until the supply was exhausted. The pressure-temperature effect, therefore, is applicable only in tanks, pipes, and processing equipment, in which the liquid and vapor-air mixture approach equilibrium.

Figures 4-6D and 4-6E show the effect of temperature and pressure on flammable limits, as determined by Mitchell and Vernon.[2] In using these charts, it should be noted that they represent equilibrium conditions and that, prior to the establishment of equilibrium, values may be different.[2] For example, when a liquid is introduced into a pressure container with air at a temperature such that the resultant vapor-air mixture will be above the upper flammable limit (according to the chart), some time will elapse before sufficient evaporation has taken place to reach this condition; meanwhile, the mixture may be flammable or explosive.

It should also be noted that the pressures developed, or the intensity of an explosion, will vary with the initial pressure of the vapor-air mixture. With high initial pressures, the pressures are greater, as in the familiar example of the gasoline engine. With low initial pressures, the pressures are relatively lower.

The lower limits of flammability of solvent vapors are affected by the temperature. This is a safety factor in installations, such as industrial ovens, where pressure is not a factor but where temperature needs to be taken into account in hazard evaluation. In such in-

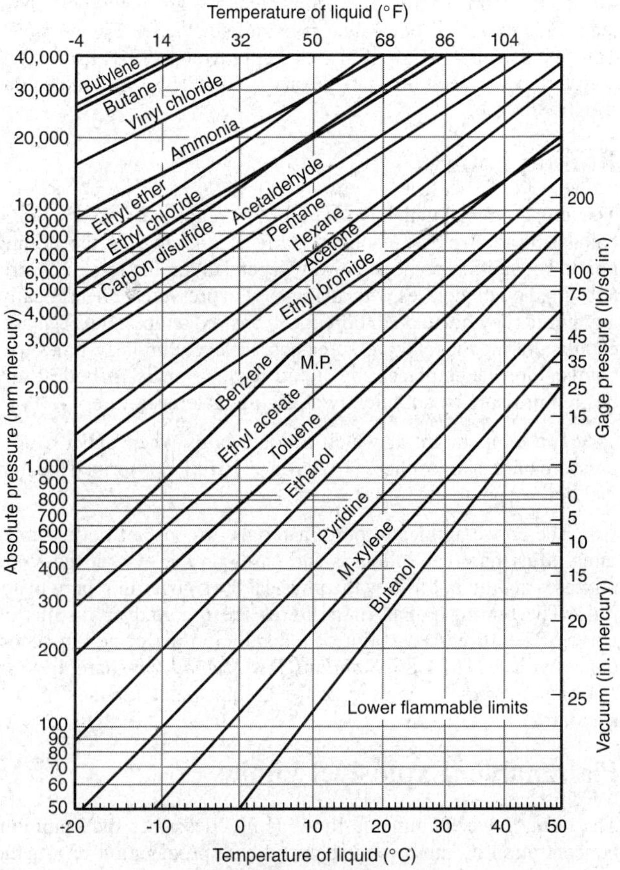

FIG. 4-6D. *Variation of lower flammable limits with temperature and pressure. This chart is applicable only to flammable liquids or gases in equilibrium in a closed container. Mixtures of vapor and air will be too "lean" to burn at temperatures below and at pressures above the values shown by the line on the chart for any substance. Conditions represented by points to the left and above the respective lines are accordingly nonflammable. Points where the diagonal lines cross the zero gage pressure line (760 mm of mercury, absolute pressure) indicate flash-point temperatures at normal atmospheric pressure.*

dustrial ovens, mechanical or forced ventilation is usually employed to dilute the concentration of the vapors below their lower flammable limits in air and to remove them from the oven. Information on the lower flammable limits of the vapors at elevated temperatures is needed in order to determine the amount of ventilation required for safe operation of an oven.[3] (See Table 4-6A.)

Energy Required for Ignition of Vapors

The principal sources of ignition of flammable liquids include flames, electrical or frictional sparks, hot surfaces, and adiabatic compression.

Flames: Except for extremely small examples produced under laboratory conditions, flames are unfailing sources of ignition for flammable vapor-air mixtures that are within the flammable range. Flames must be capable of heating the vapor to its ignition temperature in the presence of air in order to be a source of ignition. For some liquids and solids, it will be necessary for the flame to be of sufficient duration and heat to volatilize the fuel and to ignite the re-

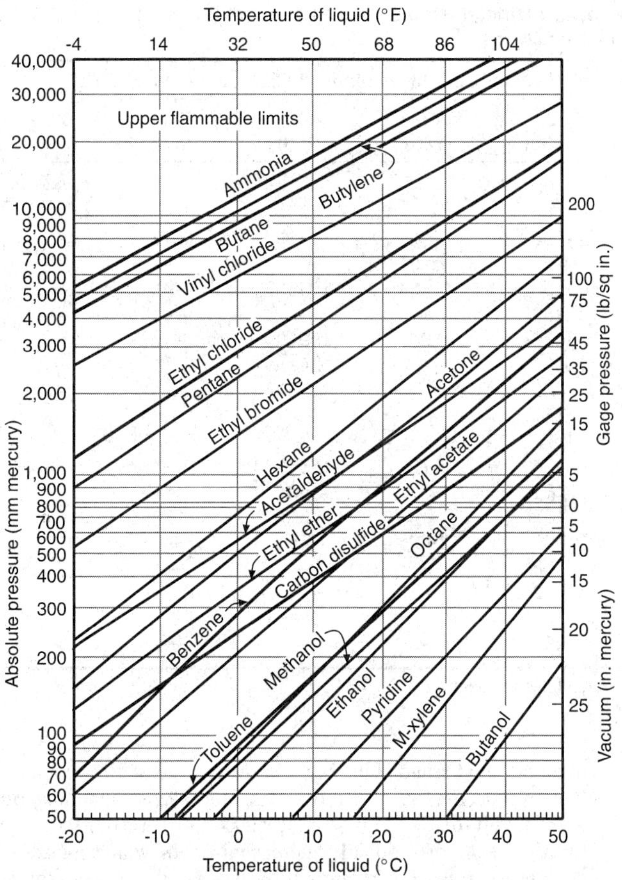

FIG. 4-6E.   Variation of upper flammable limits with temperature and pressure. This chart is applicable only to flammable liquids or gases in equilibrium in a closed container. Mixtures of vapor and air will be too "rich" to be flammable at temperatures above and pressures below the values shown by the line on the chart for any substance. Conditions represented by points to the right of and below the respective lines are accordingly nonflammable.

leased vapors. Once ignited, the radiated heat from the burning vapors perpetuates the burning process.

Electrical, static, and frictional sparks: These must have sufficient energy to ignite flammable vapor-air mixtures. Electrical sparks from most commercial electrical supply installations are above flame temperatures and will usually ignite flammable mixtures. However, friction sparks may fail to ignite a flammable mixture because the spark may be of such short duration as to fail to heat the vapor to its ignition point. Also, not only must an electrical spark have sufficient intensity and length, but some variable conditions must also exist within certain limits before ignition will occur. The nature of the ignition points and surfaces, as well as the composition, temperature, and pressure of vapor-air mixtures, are the principal variables in calculating ignition. Most of these same factors and variables also apply to static electrical sparks and frictional sparks as a source of ignition because, in either case, the spark must be of sufficient duration, intensity, or both, to create sufficient heat to cause ignition.

Hot surfaces: These can be a source of ignition if they are large enough and hot enough. The smaller the heated surface, the hotter it

must be to ignite a mixture. The larger the heated surface in relation to the mixture, the more rapidly ignition will take place and the lower the temperature necessary for ignition. A flammable liquid, however, must remain in contact with a hot surface for a sufficient length of time to form a vapor-air mixture within its flammable range. For example, a single drop of a low-viscosity, highly volatile flammable liquid that falls on the surface of an electric hotplate at 2000°F (1093°C) may ignite. A hot exhaust pipe in the open seldom ignites a flammable mixture, even though the surface temperature may be considerably above the environmental ignition temperature test method.

Adiabatic compression: This has been the cause of several destructive explosions. When controlled, adiabatic compression is the basis of diesel engine operation. A rapidly compressed flammable mixture will be ignited when the heat generated by the compressing action is sufficient to raise the flammable vapor to its ignition temperature.

Behavior of Mixed Liquids

The behavior of mixed liquids varies considerably, depending upon physical characteristics of the liquid and environmental conditions. However, the vapor pressure or evaporation rate of mixed liquids is particularly important in the prevention of fires. These factors are significant when a mixture of liquids is listed as nonflammable or has a high flash point but becomes flammable under conditions of use. As a specific example, sufficient carbon tetrachloride can be added to gasoline so that the mixture has no flash point. However, while standing in an open container, the carbon tetrachloride will evaporate more rapidly than the gasoline. Over a period of time, therefore, the residual liquid will first show a high flash point, then a progressively lower one, until the flash point of the final 10 percent of the original sample will approximate that of the higher boiling fractions of gasoline.

In order to evaluate the fire hazard of such liquid mixtures, fractional evaporation tests can be conducted at room temperature in open vessels. After evaporation of appropriate fractions, such as 10, 20, 40, 60, and 90 percent of the original sample, flash-point tests can be conducted on the residue. The results of such tests indicate the class into which the liquid should be placed if the conditions of use are such as to make it likely that appreciable evaporation will take place.

With other liquid mixtures, the flash point may increase as evaporation occurs.

Noncompatible Materials

Noncompatible materials of construction or equipment are those materials with which an unstable flammable liquid or its normal reaction products will combine, or whose products of corrosion are contaminants. For example, if mercury is a harmful contaminant, then mercury thermometers should not be used. If a chlorinated compound slowly releases hydrochloric acid, aluminum should not be used. The aluminum chloride formed by this acid may act as a catalyst. There are many materials that will chemically react with each other; when there is a question, reference should be made to NFPA 491M, *Manual of Hazardous Chemical Reactions.*

BURNING CHARACTERISTICS OF LIQUIDS

Since it is the vapors from flammable liquids which burn, the ease of ignition as well as the rate of burning can be related to such properties as the vapor pressure, flash point, boiling point, and evaporation rate. Liquids with vapors in the flammable range above the

TABLE 4-6A. *The Effect of Elevated Temperature on the Lower Flammable Limit of Combustible Solvents as Encountered in Industrial Ovens*

Solvent	Flash Point Closed Cup	Lower Flammable Limit Percent Vapor by Volume at Initial Temperature, °F (°C)						
		Room Temp.	212 (100)	392 (200)	437 (225)	482 (250)	572 (300)	662 (350)
Acetone	3	2.67	2.40	2.00*	—	—	—	—
Amyl Acetate, Iso	77	—	1.00	0.82	—	0.76*	—	—
Benzene	−4	1.32	1.10	0.93	—	—	0.80*	—
Butyl Alcohol, Normal	100	—	1.56	1.27	1.22*	—	—	—
Cresol, Meta-Para	202	—	1.06†	0.93	—	0.88*	—	—
Cyclohexane	−4	1.12	1.01	0.83*	—	—	—	—
Cyclohexanone	111	—	1.11	0.96	0.94	0.91*	—	—
Ethyl Alcohol	54	3.48	3.01	2.64	—	2.47	2.29*	—
Ethyl Lactate	131	—	1.55	1.29	—	1.22*	—	—
Gasoline	−45	1.07	0.94	0.77*	—	—	—	—
Hexane, Normal	−15	1.08	0.90	0.72*	—	—	—	—
High-Solvency Petroleum Naphtha	36	1.00	0.89	0.74	0.72	0.69*	—	—
Methyl Alcohol	52	6.70	5.80	4.81	—	4.62	4.44*	—
Methyl Ethyl Ketone	21	1.83	1.70	1.33*	—	—	—	—
Methyl Lactate	121	—	2.21	1.86	1.80	1.75*	—	—
Mineral Spirits, No. 10	104	—	0.77	0.63*	—	—	—	—
Toluene	48	1.17	0.99	0.82	—	—	0.72*	—
Turpentine	95	—	0.69	0.54*	—	—	—	—
V. M. & P. Naptha	28	0.92	0.76	0.67*	—	—	—	—

* Rapid and extensive thermal decomposition and oxidation reactions in vapor-air mixture at this temperature.
† Lower limit determined at 302°F (150°C).

liquid surface at the stored temperature will have a rapid rate of flame propagation. Flammable and combustible liquids with flash points above temperatures at which they are stored have a slower rate of flame propagation, because it is necessary for the heat of the fire to sufficiently heat the liquid surface and form a flammable vapor-air mixture before the flame will spread through the vapor. There are many variables affecting the rate of flame propagation and burning, including environmental factors, wind velocity, temperature, heat of combustion, latent heat of vaporization, and barometric pressure.

Liquid hydrocarbons normally burn with an orange flame and emit dense clouds of black smoke. Alcohols normally burn with a clean blue flame and produce very little smoke. Certain terpenes and ethers burn with considerable ebullition (boiling) of the liquid surface, making it difficult to extinguish fires involving these substances.

Boilover, Slopover, and Frothover

Three specific conditions—boilover, slopover, and frothover—deserve special mention in regard to fires in open-topped tanks containing various types of oils.

Boilover: Boilover is a phenomenon that may occur spontaneously during a fire in an open-top tank containing certain types of crude oils. This may occur when the roof of a tank is blown off by an explosion, or when a floating roof is sunk to expose the crude. After a long period of quiescent burning, there is a sudden overflow or ejection of some of the residual oil in the tank. It is caused by boiling water that forms a quickly expanding steam-oil froth. The frothing results from the existence of the following three conditions, the absence of any one of which will prevent the occurrence:

1. The tank must contain free water or water-oil emulsion at the tank bottom. This situation will normally prevail in tanks storing crude oil.
2. The oil must contain components having a wide range of boiling points, such that when the lighter components have been distilled off and burned at the surface, the residue, with a temperature of 300°F (149°C) or over, is more dense than the oil immediately below. This residue sinks below the surface and forms a layer of gradually increasing depth which advances downward at a rate substantially faster than the rate of regression of the burning surface. This sets up the so-called "heat wave," which is the result of localized settling of a part of the hot surface oil until it reaches the colder oil below. There is no heat conduction from the burning surface downward.
3. Sufficient content of heavier oils are present in the oil to produce a residue that can form a tough persistent froth of oil and steam.

NFPA records indicate the most serious loss-of-life tank fire in history occurred on December 19, 1982, in Tacoa, Venezuela. The storage tank, located a few miles northwest of Caracas in the small seaside village of Tacoa, supplied fuel to an electric generating plant. Two workers died upon ignition of the tank. After six hours of intense burning, the contents of the tank erupted into an extremely violent boilover. The final toll—160 people dead, hundreds more injured, others still unaccounted for, and damage estimated at $50 million. Fifty of those who died were fire fighters, which also makes the Tacoa incident the gravest of its kind in terms of loss of life of fire personnel. (See Figure 4-6F.)

Slopover: Slopover can result when a water stream is applied to the hot surface of a burning oil, provided the oil is viscous and its temperature exceeds the boiling point of water. Since only the surface oil is involved, a slopover is a relatively mild occurrence.

Frothover: Frothover refers to the overflowing of a container not on fire when water boils under the surface of a viscous hot oil. A typical example would be when hot asphalt is loaded into a tank car containing some water. The first asphalt is cooled by contact with the cold metal, and, at first, nothing may happen. But when the water becomes heated and starts to boil, the asphalt may overflow the tank car.

A similar situation can arise when a tank containing a water bottom or wet emulsion is used for storing oil slops or residium at

FIG. 4-6F.   Emergency personnel on the scene of the fuel-oil storage tank fire boilover in Tacoa, Venezuela, December 19, 1982. At this moment, the fire had been declared under control. Note the fire fighters on the top of the dike, and the other personnel along the roadway fence. The boilover occurred minutes after this photo was taken, and all of the people shown above were reportedly killed.

temperatures below 200°F (93°C), and the tank receives a substantial addition of hot residium at a temperature of 300°F (149°C) or higher. After enough time has elapsed for the effect of the hot oil to reach the water in a tank, a prolonged boiling action can take place, which can remove a tank roof and spread froth over a wide area.

Burning Rates of Liquids

The burning rate of flammable liquids will vary somewhat similar to the rate of flame propagation. Gasoline, a compound of light and heavy fractions, will burn more rapidly at first while the lighter fractions are burning, while the heavier fractions will burn at a rate approaching that of kerosene. The burning rate for gasoline is 6 to 12 in. (150 to 300 mm) of depth per hr, and for kerosene the rate is 5 to 8 in. (130 to 200 mm) of depth per hr. For example, a pool of gasoline, $1/_2$ in. (12.7 mm) deep, could be expected to burn itself out in $2^1/_2$ to 5 min.

In a series of tests conducted at the U.S. Bureau of Mines,[4] the burning rates of several liquids and gases were determined and found to approach "a maximum and constant value with increasing pool diameter. This constant burning rate is proportional to the ratio of the net heat of combustion to the sensible heat of vaporization."[4]

Based upon these observed burning rates in petroleum tank fires, an estimate can be made of the extent of area that will be involved in fire from a petroleum spill. When a spill is burning, the fire area will first be small and then spread to a point of equilibrium where it will burn as fast as it is released. At a burning rate of 1 ft per hr (0.3 m/hr), 1 gal per min (3.8 L/min) of spillage will reach an equilibrium burning area of about 8 sq ft (0.74 m²). Thus, a 10 gpm (38 L/min) spill rate will be in equilibrium with 80 sq ft (7.4 m²) of burning area; 100 gpm at 800 sq ft (380 L/min at 74 m²); etc. For liquids having higher burning rates, the area would be smaller; and for those liquids having lower burning rates, the area would be larger. Also, the terrain will have an influence on the shape of the burning area.

FIRE PREVENTION METHODS

Whenever flammable and combustible liquids are stored or handled, the liquid is usually exposed to the air at some stage in the operation, except where the storage is confined to sealed containers that are not filled or opened on the premises or where handling is in closed systems and vapor losses are recovered. Even when the storage or handling is in a closed system, there is always the possibility of breaks or leaks, which permit the liquid to escape. Therefore, ventilation is of primary importance to prevent the accumulation of flammable vapors. It is also good practice to eliminate sources of ignition in places where low-flash-point flammable liquids are stored, handled, or used, even though no vapor may ordinarily be present.

Whenever possible in manufacturing processes involving flammable or combustible liquids, equipment, such as compressors, stills, towers, and pumps, should be located in the open. This will lessen the fire potential created by the escape and accumulation of flammable vapors. Gasoline and almost all other flammable liquids produce heavier-than-air vapors, which tend to settle on the floor or in pits or depressions. Such vapors may flow along the floor or ground for long distances, be ignited at some remote point, and flash back. The removal of such vapors at the floor level (including pits) is usually the proper method of ventilation. Convection currents of heated air or normal vapor diffusion may carry even heavy vapors upward, and, in such instances, ceiling ventilation may also be desirable. Ventilation to eliminate flammable vapors may be either natural or artificial. Although natural ventilation, which depends upon temperature and wind conditions, has the advantage of not being dependent on manual starting or on power supply, it is not so easily controlled as is mechanical ventilation. Mechanical ventilation should be used wherever there are extensive indoor operations involving flammable and combustible liquids.

Explosion Venting

In rooms or buildings where possible explosions of flammable vapors may occur, it is recommended that relief through explosion venting be provided for at least Class IA liquids and unstable liquids.

Substitution of Nonflammable Liquids

The hazard created by flammable liquids may be avoided or reduced by the substitution of relatively safe materials. Such materials should be stable, have a low toxicity, and be either nonflammable or have a high flash point. For example, trichloroethylene, though higher in price, is nonflammable at ordinary temperatures, and may for some uses be substituted for a more hazardous flammable solvent. Tetrachloroethylene (perchloroethylene) is another nonflammable liquid. However, even though these products are less toxic than carbon tetrachloride, they should be used only in well-ventilated areas, with full vapor recovery and controls to prevent release to the atmosphere.

There are several commercial-type stable solvents available that have flash points from 140 to 190°F (60 to 88°C) and that have a comparatively low degree of toxicity.

Specially refined petroleum products, first developed as "Stoddard Solvent" but now sold by different companies under a variety of trade names, have solvent properties approximating gasoline, but with fire hazard properties similar to those of kerosene. A danger in their use lies in the possibility that persons believing that they are using a safe solvent without fire hazard may neglect ordinary precautions that would be observed with a liquid such as kerosene. When heated to above their flash point [about 100°F (37.8°C)], these solvents produce vapors as flammable as those of gasoline at its flash-point temperature.

Several other types of commercial solvents which are mixtures of liquids with differing rates of evaporation are available. Some are mixtures of gasoline or one of the various naphthas and a chlorinated solvent having a different and often higher evaporation rate than the flammable solvent which, over a period of time, would leave the original low-flash-point solvent. These mixtures create a toxicity hazard, as well as a fire hazard, and their use in open containers should be discouraged.

Solvents and solvent vapors are toxic in varying degrees, and ventilation is almost always necessary to keep vapor concentration within safe limits.

BIBLIOGRAPHY

References Cited

1. ASTM D323, *Standard Test Method for Vapor Pressure of Petroleum Products (Reid Method)*, American Society for Testing and Materials, W. Conshohocken, PA, 1994.
2. Mitchell, F. C., and Vernon, H. C., "Effect of Pressure on Explosion Hazards," *NFPA Quarterly*, Vol. 31, No. 4, Apr. 1938, pp. 306–313.
3. Matson, A.F., and DuFour, R.E., "Flammable Limits at Oven Temperatures," *NFPA Quarterly*, Vol. 43, No. 4, Apr. 1950, pp. 260–264.
4. Burgess, D. S., Strasser, A., and Grumer, J., "Diffusion Burning of Liquid Fuels in Open Trays," *Fire Research Abstracts and Reviews*, Vol. 3, 1961, pp. 177–192.

References

ASTM D56, *Standard Test Method for Flash Point by Tag Closed Tester*, American Society for Testing and Materials, W. Conshohocken, PA, 1993.
ASTM D86, *Standard Test Method for Distillation of Petroleum Products*, American Society for Testing and Materials, W. Conshohocken, PA, 1993.
ASTM D92, *Standard Test Method for Flash and Fire Points by Cleveland Open-Cup*, American Society for Testing and Materials, W. Conshohocken, PA, 1990.
ASTM D93, *Standard Test Methods for Flash Point by Pensky-Martens Closed Tester*, American Society for Testing and Materials, W. Conshohocken, PA, 1990.
ASTM D1310, *Standard Test Method for Flash Point and Fire Point of Liquids by Tag Open-Cup Apparatus*, American Society for Testing and Materials, W. Conshohocken, PA, 1990.
ASTM D3278, *Standard Test Methods for Flash Point of Liquids by Setaflash Closed-Cup Apparatus*, American Society for Testing and Materials, W. Conshohocken, PA, 1989.
ASTM E659, *Standard Test Method for Autoignition Temperature of Liquid Chemicals*, American Society for Testing and Materials, W. Conshohocken, PA, 1978 (reconfirmed 1994).
"Bulletin of Research No. 43," Underwriters Laboratories Inc., Northbrook, IL.

NFPA Codes, Standards, and Recommended Practices

Reference to the following NFPA codes, standards, and recommended practices will provide further information on flammable and combustible liquids discussed in this chapter. (See the latest version of *The NFPA Catalog* for availability of current editions of the following documents.)

NFPA 30, *Flammable and Combustible Liquids Code*
NFPA 30A, *Automotive and Marine Service Station Code*
NFPA 30B, *Code for the Manufacture and Storage of Aerosol Products*
NFPA 31, *Standard for the Installation of Oil-Burning Equipment*
NFPA 68, *Guide for Venting of Deflagrations*
NFPA 69, *Standard on Explosion Prevention Systems*
NFPA 325, *Guide to Fire Hazard Properties of Flammable Liquids, Gases, and Volatile Solids*
NFPA 327, *Standard Procedures for Cleaning or Safeguarding Small Tanks and Containers Without Entry*
NFPA 385, *Standard for Tank Vehicles for Flammable and Combustible Liquids*
NFPA 491M, *Manual of Hazardous Chemical Reactions*

Additional Reading

Atkinson, G. T., "Fire Spread in a Pallet Load of Bottles of Flammable Liquid," Short Communication, *Fire Safety Journal*, Vol. 22, No. 4, 1994, pp. 409–415.
Accident Prevention Manual for Industrial Operations, 9th ed., National Fire Safety Council, Chicago, 1988.
Babrauskas, V., "Pool Fires: Burning Rates and Heat Fluxes," *SFPE Handbook of Fire Protection Engineering*, National Fire Protection Association, Quincy, MA, 1988.
Back, G. G., "Experimental Evaluation of Water Mist Fire Suppression System Technologies Applied to Flammable Liquid Storeroom Applications," *Proceedings* for the International Conference on Fire Research and Engineering, Society of Fire Protection Engineers (SFPE), Boston, MA, 1995, pp. 127–132.
Bielen, R. P., "Fire Research With Flammable and Combustible Liquids in Bulk Merchandising Retail Facilities," *Fire Technology*, Vol. 29, No. 1, Feb. 1993, pp. 80–81.
Benedetti, R. P., ed., *Flammable and Combustible Liquids Code Handbook*, 3rd ed., National Fire Protection Association, Quincy, MA, 1987.
Burke, R., "Flammable Liquids," *Firehouse*, Vol. 20, No. 5, May 1995, pp. 36–37, 110–111.
"Controlling the Power of Flammable Liquids," Pamphlet No. P7045, Factory Mutual Research Corp., Norwood, MA.
Davis, K., "Fluids, Lubricants, Coatings, and Elastomer Materials," Final Report, September 1, 1989-February 28, 1991, Wright Lab., Wright-Patterson AFB, OH, UDR-TR-91-103, WL-TR-91-4097, July 1991, 387 pages.
Dunbar, J., "Packaging and Storing Flammable Liquids," *Products Finishing*, Vol. 52, No. 9, June 1988, pp. 118–121.
El-Iskandarani, B. M. K., "Emergency Relief Venting Analysis for Flammable Liquid Storage Tanks," Worcester Polytechnic Inst., Thesis, May 1994, 160 pages.
"Flammable Liquid Drainage and Containment," *Record*, Vol. 69, No. 4, July/Aug. 1992, pp. 24–31.
"Flammable Liquids—Risk, Regulations and Protection Measures to Safeguard Flammable Liquids," *Record*, Vol. 69, No. 1, Jan./Feb. 1992, pp. 15–21.
Goodman, D., "Fire Retarded Cardboard Cartons—An Approach to Allow General Warehousing of Flammable Liquids in Plastics Containers," Fire Safety Developments and Testing: Toxicity—Heat Release—Product Development—Combustion Corrosivity, October 21–24, 1990, Ponte Vedra Beach, FL, Fire Retardant Chemicals Assoc., Lancaster, PA, 1990, pp. 157–165.
Hall, J. R., Jr., "Fire Risk Analysis Model for Assessing Options for flammable and Combustible Liquid Products in Storage and Retail Occupancies," *Fire Technology*, Vol. 31, No. 4, 4th Quarter, Nov. 1995, pp. 290–308.
Handbook of Organic Industrial Solvents, 6th ed., Alliance of American Insurers, Schaumburg, IL, 1987.
Hird, D., "Fire-Fighting Foams for Flammable Liquid Fires," *Fire Prevention*, No. 202, Sept. 1987, pp. 20–22, 24, 26.
Holmstedt, G., and Persson, H., "Spray Fire Tests with Hydraulic Fluids," *Proceedings* of the First International Symposium on Fire Safety Science, Hemisphere, New York, 1986, pp. 869–879.
"IRI Promotes Safer Metal Container for Flammable Liquids Storage," *Sentinel*, Vol. 50, No. 3, Third Quarter 1993, p. 9.
Kanury, M. A., "Ignition of Liquid Fuels," *SFPE Handbook of Fire Protection Engineering*, National Fire Protection Association, Quincy, MA, 1988.
Kokkala, M., "Extinguishment of Liquid Fires With Sprinklers and Water Sprays: Analysis of the Test Results," VTT-Technical Research Center of Finland, Espoo, VTT Research Reports 696, Project PAL9002, Aug. 1990, 54 pages.
Kokkala, M., Andsten, T., and Bjorkman, J., "Extinguishment of Liquid Fires with Sprinklers and Water Sprays: Description of Tests," *Valtion Teknillinen Tutkimuskeskus*, No. 593, 1989, Technical Research Institute of Finland, Espo, Finland.
Kukkola, T., "Palavan nesteen varastoalueet. [Storage Areas for Flammable Liquids.]," *Palontorjuntatekniikka*, Apr. 1995, pp. 26–27.
Lentini, J. J. and Waters, L. V., "Behavior of Flammable and Combustible Liquids," *Fire and Arson Investigator*, Vol. 42, No. 1, Sept. 1991, pp. 39–45.

Linville, J., ed., *Industrial Fire Hazards Handbook*, 3rd ed., National Fire Protection Association, Quincy, MA, 1990.

Martinsen, W. E., Johnson, D. W., and Terrell, W. F., "BLEVES: Their Causes, Effects, and Prevention," *Hydrocarbon Processing*, Vol. 65, No. 11, 1986, pp. 141-148.

Mudan, K. S., and Croce, P. A., "Fire Hazard Calculations for Large Open Hydrocarbon Fires," *SFPE Handbook of Fire Protection Engineering*, National Fire Protection Association, Quincy, MA, 1988.

Mulhaupt, R., "Fire Protection for Flammable and Combustible Liquids Warehouses," First International Conference for Fire Suppression Research, May 5–8, 1992, Stockholm, Sweden, 1992, pp. 393–399.

"National Wholesale/Retail Occupancy Fire Research Project. Task 1. Protection of Flammable Liquids," Technical Report, Underwriters Laboratories, Inc., Northbrook, IL, National Fire Protection Research Foundation, Quincy, MA, Nov. 1992, 167 pages.

Northrup, S. D., "Evaluation of Less Flammable Transformer Liquids," *Proceedings of the 5th BEAMA International Electrical Insulation Conference*, Cavanagh Associates, Leatherhead, England, 1986, pp. 101–106.

Nugent, D. P., "Directory of Fire Tests Involving Storage of Flammable and Combustible Liquids in Small Containers," Schirmer Engineering Corp., Deerfield, IL, Directory, Feb. 1994, 75 pages.

Nugent, D. P., "Fire Tests Involving Storage of Flammable and Combustible Liquids in Small Containers," *Journal of Fire Protection Engineering*, Vol. 6, No. 1, 1994, pp. 1–10.

Omans, L. P., "Fighting Flammable Liquid Fires: A Primer. Part 1. The Family of Foams," *Fire Engineering*, Vol. 146, No. 1, Jan. 1993, pp. 50–54, 57–61.

Omans, L. P., "Fighting Flammable Liquid Fires: A Primer. Part 2," *Fire Engineering*, Vol. 146, No. 2, Feb. 1993, pp. 50–52, 56–58.

Omans, L. P., "Fighting Flammable Liquid Fires: A Primer. Part 3," *Fire Engineering*, Vol. 146, No. 3, Mar. 1993, pp. 93–94, 96–98, 100, 102, 104.

"Outline of Investigation for Fire Test of Packaging Systems of Class 1 and 2 Liquids in Plastic Containers," Underwriters Laboratories, Inc., Northbrook, IL, Subject 2019, Feb. 1993, 15 pages.

"Proper Flammable and Combustible Liquid Handling," *Record*, Vol. 68, No. 5, Sept./Oct. 1991, pp. 16–17.

"Proper Flammable and Combustible Liquid Handling," *Record*, Vol. 68, No. 5, Sept./Oct. 1991, pp. 16–17.

Przybyla, L. and Gandhi, P., "Results Are In On Flammable Liquids in Plastic Containers," *Fire Journal*, Vol. 84, No. 3, May/June 1990, pp. 38–38, 41–43.

Sax, N. I., *Dangerous Properties of Industrial Materials*, 7th ed., Van Nostrand Reinhold Co., NY, 1989.

Sharma, T.P., *et al.*, "A New Particulate Extinguishant for Flammable Liquid Fires," *Proceedings* of the Second International Symposium on Fire Safety Science, Hemisphere Publishing Corporation, New York, 1989, pp. 667–680.

Sharma, T.P., *et al.*, "Experiments on Extinction of Liquid Hydrocarbon Fires by a Foam Technique," *Proceedings* of the Fourth International Symposium on Fire Safety Science, International Association for Fire Safety Science, 1994, pp. 865–876.

"Sherwin-Williams Takes Lead in Flammable Liquids Fire Research," *The Sentinel*, Fourth Quarter 1990, pp. 8–9.

Staggs, K. J., *et al.*, "Development of Flammable Liquid Storage Wooden Cabinets for Chemical Laboratories," Department of Energy, Washington, DC, UCRL-ID-115605, Nov. 1993, 39 pages.

Sundin, D. W., and Northrup, S. D., "Relationship Between Liquid Flammability Tests and Transformer Fire Situations," Conference Record—Industrial and Commercial Power Systems Technical Conference 1987, *IEEE*, 1987, pp. 117–121.

Threshold Limit Values, American Conference on Governmental Industrial Hygienists, Cincinnati, OH, 1989.

Vaivads, R., *et al.*, "Flammability of Alcohol—Gasoline Blends in Fuel Tanks," *Proceedings* of the Fourth International Symposium on Fire Safety Science, International Association for Fire Safety Science, 1994, pp. 575–588.

Venart, J. E. S., *et al.*, "To BLEVE Or Not To BLEVE: Anatomy of a Boiling Liquid Expanding," New Brunswick Univ., Fredericton, Canada, American Institute of Chemical Engineers, AIChE Meeting on Loss Prevention, Mar./Apr. 1992, New Orleans, LA, 1992, pp. 1–21.

Weast, R. C., ed., *Handbook of Chemistry and Physics*, 69th ed., CRC Press, Boca Raton, FL, 1988.

Whiteley, B., "Foaming Sprinklers and Flammable Liquid Fires," *Fire Prevention*, No. 262, Sept. 1993, pp. 32–35.

Williams, D., "Extinguishing Flammable Liquid Fires," *Fire International*, Vol. 11, No. 107, Oct.-Nov. 1987, pp. 126–127.

GASES

Revised by Theodore C. Lemoff

The term "gas" describes the physical state of a substance that has no shape or volume of its own but will take the shape and fill the entire volume of whatever container or other enclosure it occupies. This is in contrast to a liquid, which has no shape of its own but does have volume, and to a solid, which has both its own shape and volume. Gases are composed of extremely minute particles in constant motion. This motion affects the properties and behavior of gases; e.g., the higher the temperature, the more rapid the motion.

This chapter presents a definition of the substance known as gas and then outlines the various ways gases are classified. The hazards that gases present in storage differ from those presented when gases escape—both kinds of hazards are discussed along with the control measures to follow in an emergency. The chapter concludes with a summary of the properties and behavior of some common gases. For each gas, information is provided about its classification, chemical properties, physical properties, usage, hazards inside containers, hazards when released from containment, and emergency control.

Additional information about gases can be found in the following chapters: Section 1, Chapter 6, "Explosions"; Section 3, Chapter 22, "Storage of Gases"; Section 4, Chapter 9, "Oxygen-Enriched Atmospheres"; Section 4, Chapter 14, "Explosion Prevention and Protection"; and Section 6, Chapter 22, "Foam Extinguishing Agents and Systems."

GASES DEFINED

Because all substances can exist as gases, depending upon the temperature and pressure applied to them, the term "gas" as used in this chapter is applied only to substances that exist in the gaseous state at so-called "normal" temperature and pressure (NTP) conditions—approximately 70°F (21°C) and 14.7 psia (an absolute pressure of 101 kPa). However, even at normal or near normal temperatures and pressures, many substances can exist as either liquids or gases. The term "gas" is not precisely defined in NFPA standards. However, NFPA standards partially define a flammable liquid as a liquid having a vapor pressure not exceeding 40 psia at 100°F (an absolute pressure of 275 kPa at 38°C). For comparative purposes, any substance, or mixture of substances, that in its liquid state exerts a vapor pressure greater than 40 psia at 100°F (an absolute pressure of 275 kPa at 38°C) can be considered a gas.

Theodore C. Lemoff, P.E., is the senior gases engineer on the staff of NFPA.

CLASSIFICATION OF GASES

Effective handling of the great number and variety of gases in commerce and in the environment requires that gases be classified. These classifications are based on certain common denominators that reflect the chemical and physical properties of gases and the primary uses of gases.

Classification by Chemical Properties

The chemical properties of gases are of primary fire protection concern due to their ability to react chemically with other materials (or within themselves) to produce potentially hazardous quantities of heat or reaction products, or to produce physiological effects hazardous to humans.

Flammable gases: In NFPA usage, any gas that will burn in the normal concentrations of oxygen in the air is considered a flammable gas. Flammable gases will burn in air the same way flammable liquid vapors burn in air, i.e., each gas will burn only within a certain range of gas-air mixture compositions (the flammable or combustible range) and will ignite only at or above a certain temperature (the ignition temperature).

In a few instances, consideration of the width of the flammable range or the magnitude of the lower limit of flammability or both has resulted in classification of an ostensibly flammable gas as nonflammable because the chances of fire are low under certain conditions. Anhydrous ammonia is notable in this respect. While it is a flammable gas in the context of the previous paragraph, it is classified by the U.S. Department of Transportation (DOT) as a nonflammable gas for purposes of transportation in interstate commerce.

Although flammable liquid vapors and flammable gases exhibit similar combustion characteristics, the term "flash point," which describes a common and useful combustion property of flammable liquids, has no practical significance for flammable gases. The flash point is a measure of the temperature at which a flammable liquid produces sufficient vapors for combustion, and this temperature is always below the normal boiling point. A flammable gas exists normally at a temperature exceeding its normal boiling point, even when the gas is in the liquid state (as it often is in shipment and storage). Thus, the gas exists at a temperature that not only exceeds its flash point, but which is usually well above it.

Nonflammable gases: Nonflammable gases are those that will not burn in any concentration of air or oxygen. A number of these gases, however, support combustion while others suppress combustion. Those gases that support combustion are often referred to as oxidizers or oxidizing gases. They are generally either oxygen or mixtures of oxygen and other gases, such as oxygen-helium or oxygen-nitrogen mixtures, or certain gaseous oxides, such as nitrous oxide. These mixtures contain considerably more oxygen than is present in the oxygen-nitrogen mixture that comprises air.

Gases that will not support combustion are generally known as inert gases. Among the most common are nitrogen, argon, helium, and other rare gases in the atmosphere, as well as carbon dioxide and sulfur dioxide. (There are some metals, however, that are combustible and can react vigorously in carbon dioxide or nitrogen atmospheres, e.g., magnesium.)

Reactive gases: Most gases can be made to react chemically with some other substance under some conditions. Therefore, the term "reactive gas" is used to distinguish gases that will either react with other materials or within themselves (producing potentially hazardous quantities of heat or reaction products) by a chemical reaction other than burning (combustion) and under reasonably anticipated initiating conditions of environmental heat, shock, etc.

Fluorine is an example of a highly reactive gas, because it reacts with practically all organic and inorganic substances at normal temperatures and pressures, often fast enough to result in flaming. Another example is the reaction between chlorine (a nonflammable gas) and hydrogen (a flammable gas), which can produce flames.

Several gases can rearrange themselves chemically when subjected to reasonably anticipated conditions of environmental heat and shock, including fire exposure to their containers, with the production of potentially hazardous quantities of heat or reaction products. Examples of these gases are acetylene, methyl acetylene, propadiene, and vinyl chloride. For transport or storage, these gases are usually mixed with other substances or placed in special containers to stabilize them against reasonably anticipated reaction initiators.

Toxic gases: Certain gases can present a serious life hazard if they are released into the atmosphere. These gases, which are poisonous or irritating when inhaled or contacted, include chlorine, hydrogen sulfide, sulfur dioxide, ammonia, and carbon monoxide, among others. The presence of these gases may complicate fire fighting efforts by exposing fire fighters to toxic hazard.

Classification by Physical Properties

These properties are of primary fire protection concern because they affect the physical behavior of gases while they are inside containers and after any accidental release from containers. Until they are used, gases must be completely confined in containers during transportation, transfer, and storage. In any situation, the quantity (weight) of gas is important, since gases are inherently lighter than liquids or solids. It is a matter of economic necessity and the ease of usage that gases be packaged in containers that contain as much gas as is practical. These requirements have resulted in transportation and storage of gases in the liquid as well as the gaseous state. Distinguishing between gases in a liquid or gaseous state is important for the application of sound fire prevention and protection practices.

Compressed gases: For purposes of this chapter, a compressed gas is one that exists solely in the gaseous state under pressure at all normal atmospheric temperatures inside its container. The pressure depends on the pressure to which the container was originally charged, and on how much gas remains in the container, although the gas temperature will have some effect. There are no universally defined lower or upper limits to the container pressure. In the United States, the lower limit is customarily considered to be 25 psi (273

kPa) at normal temperatures [70 to 100°F (21 to 38°C)]. The upper limit is restricted only by the economics of container construction and is usually in the range of 1,800 to 3,600 psi (12,512 to 24,923 kPa).

A container of compressed gas is still rather limited in the weight of gas it can hold. For example, the largest common portable cylinder of compressed oxygen contains only about 20 lb (9 kg) of oxygen, or about 245 cu ft (6.9 m³) of oxygen measured at NTP of 70°F (21°C) and 14.7 psia (101 an absolute pressure of kPa).

Liquefied gases: A liquefied gas is one that, at 70°F (21°C) inside its closed container, exists partly in the liquid state and partly in the gaseous state, and under pressure as long as any liquid remains in the container. The pressure depends on the temperature of the liquid, although the quantity of liquid can affect this under some conditions.

A liquefied gas is much more concentrated than a compressed gas. For example, the aforementioned compressed oxygen cylinder could hold about 116 lb (53 kg) of liquefied oxygen, or about 1,400 cu ft (39.6 m³) of oxygen measured at NTP, which is about six times greater. This comparison is valid only for illustrative purposes, because compressed oxygen and liquefied oxygen will not be found in the same type of container.

Cryogenic gases: A cryogenic gas is a liquefied gas that exists in its container at temperatures below –130°F (–90°C). A principal reason for this distinction with respect to a liquefied gas is that a cryogenic gas cannot be retained indefinitely in a container. Heat from the atmosphere, which can be slowed but not prevented from entering the container, tends to continually raise the container pressure. If the gas is confined, the resulting pressure could greatly exceed any feasible container strength.

Some pertinent physical properties of liquefied gases, including cryogenic gases, are given in Table 4-7A. It is emphasized that these descriptions of gas classifications by physical properties pertain only to usage in context of this chapter. Somewhat different descriptions of the terms compressed gas, liquefied gas, and cryogenic gas can be found in codes and regulations, particularly the DOT hazardous materials regulations applicable to interstate transportation. For example, in the United States, many liquefied gases are classified as compressed gases by DOT.

Classification by Usage

Standard- and code-making organizations and general industry often classify gases by their principal use. This classification scheme is not as precise as the preceding scheme, and there is much overlap in uses with the gases.

Fuel gases: These gases are flammable gases customarily used for burning with air to produce heat, which in turn is used as a source of heat (comfort and process), power, or light. By far the principal and most widely used fuel gases are natural gas and the liquefied petroleum gases (butane and propane).

Industrial gases: These include the entire range of gases classified by chemical properties customarily used in industrial processes, for welding and cutting, heat treating, chemical processing, refrigeration, water treatment, etc.

Medical gases: By far the most specialized usage classification, medical gases are used for medical purposes, such as anesthesia and respiratory therapy. Oxygen and nitrous oxide are common medical gases.

TABLE 4-7A. *Physical Properties of Cryogenic Gases*
(Compiled from data in the *Handbook of Compressed Gases,* Compressed Gas Association)

Name	Normal Boiling Point Temp °F	Temp °C	Liquid Density lb/cu ft	Liquid Density kg/m³	Gas Density lb/cu ft	Gas Density kg/m³	Latent Heat Btu/lb	Latent Heat kJ/kg	Normal Conditions 70°F, 14.7 psi (21°C, 101 kPa) Gas Density lb/cu ft	Gas Density kg/m³	cu ft Gas from 1 cu ft Liquid	m³ Gas from 1 m³ Liquid
Air	−317.8	−194.3	54.6	87.5	—	—	88.2	205.1	0.075	0.120	728	726
Argon	−302.6	−158.9	86.98	139.3	0.356	0.570	70.2	163.3	0.103	0.165	842	842
Carbon monoxide	−312.7	−191.5	51.12	81.9	—	—	92.8	215.8	0.073	0.117	706	706
Ethylene	−154.8	−103.7	35.4	56.7	0.130	0.208	208.0	483.8	0.072	0.115	487	487
Fluorine	−306.6	−188.1	94.1	150.7	0.363	0.581	74.1	172.3	0.098	0.160	961	961
Helium	−452.1	−268.9	7.8	12.5	0.106	0.170	10.3	23.9	0.01	0.016	754	754
Hydrogen	−423.0	−252.8	4.43	7.1	0.084	0.134	192.7	448.2	0.005	0.008	850	850
Methane	−258.7	−161.5	26.5	42.4	0.111	0.178	219.2	509.8	0.042	0.067	636	636
Nitrogen	−320.4	−195.8	50.46	80.8	0.288	0.461	85.7	199.3	0.072	0.115	696	696
Oxygen	−297.4	−183.0	71.27	114.2	0.296	0.474	91.7	213.3	0.083	0.133	861	861

BASIC HAZARDS OF GASES

A systematic evaluation of gas hazards distinguishes between the hazards presented by a gas when it is confined in a container and the hazards presented when it escapes from a container—even though both of these hazards may be simultaneously present in a single incident.

Hazards within Containers

Gases expand when heated, producing an increase in pressure on a container, which can result in gas release and/or cause container failure. In addition, containers can lose their strength and fail during a fire. Compressed and liquefied gases are affected somewhat differently when heated.

A compressed gas (solely in the gaseous state) simply attempts to expand and follow the classic gas behavior laws. No actual gas follows these laws exactly, but the laws of Boyle and Charles are sufficiently accurate to predict the behavior of compressed gases under commonly encountered conditions. It is essential, however, that a consistent set of U.S. customary or SI units be used in calculations using the "absolute" values for temperature and pressure. In the formulas, T = absolute temperature (deg F + 459; or deg C + 273), P = absolute pressure (gage pressure in lb per sq in. + 14.7 psi; or kPa), and V = volume in cu ft or m³ or any other suitable unit of volume.

Boyle's Law

Boyle's Law states that the volume occupied by a given mass of gas varies inversely with the absolute pressure if the temperature is not allowed to change, or

$$PV = \text{constant}$$

Charles' Law

Charles' Law states that the volume of a given mass of gas is directly proportional to the absolute temperature if the pressure is kept constant. Thus:

$$\frac{V}{T} = \text{constant}$$

Therefore, for most gases, within practical working limits, the relation among temperature, pressure, and volume may be closely approximated by the following formula:

$$\frac{T_1}{T_2} = \frac{P_1 V_1}{P_2 V_2}$$

T_1, P_1, and V_1 refer to initial conditions; T_2, P_2, and V_2 refer to changed conditions to be determined.

EXAMPLE 1: Assume a 10 cu ft cylinder of compressed gas at 70°F has a gage pressure of 1000 psi. Assume that the temperature is increased to 150°F. What will be the pressure?

$T_1 = 70 + 459 = 529$

$P_1 = 1,000 + 14.7 = 1,014.7$

$V_1 = 10$

$T_2 = 150 + 459 = 609$

$P_2 = $ to be determined

$V_2 = 10$

Substituting in the formula:

$$\frac{529}{609} = \frac{1,014.7 \times 10}{P_2 \times 10}$$

$$P_2 = 1,014.7 \times \frac{609}{529} = 1.170$$

Therefore, gage pressure will be 1,170 − 14.7, or 1,155.3 psi.

EXAMPLE 2: Assume a 2.0 m³ cylinder of compressed gas at 20°C has a gage pressure of 1,200 kPa. Assume that the gas is further compressed into a 1.0 m³ cylinder and maintained at 50°C. What will be the gage pressure?

$T_1 = 20 + 273 = 293$

$P_1 = 1,200 + 101 = 1,301$

$V_1 = 2.0$

$T_2 = 50 + 273 = 323$

$P_2 = $ to be determined

$V_2 = 1.0$

Substituting in the formula:

$$\frac{293}{323}=\frac{1,301\times2.0}{P_2\times1.0}$$

$$P_2 = 2,868$$

A liquefied gas, including a cryogenic gas (which is partly in the liquid state), exhibits a more complicated behavior because the net end result of heating is the combination of three effects. First, the gas phase is subject to the same effect as for a compressed gas. Second, the liquid attempts to expand, compressing the vapor. Third, the vapor pressure of the liquid increases as the temperature of the liquid increases. The combined result of these effects is an increase in pressure when the container is heated.

A most serious pressure rise can occur if the liquid expansion results in the container becoming full of liquid (the initial gas-phase condensing). If this happens, a small amount of additional heating results in a large increase in pressure. For this reason, it is *vital* never to place more liquefied gas in the liquid phase into a container than can be accommodated—leaving a vapor space—if the liquid temperature is raised to a level commensurate with the expected ambient temperatures. The proper quantity varies considerably with the liquefied gas and the factors that affect expected temperature rises, such as the temperature of the liquid when placed into the container, the size of the container, and whether the container is insulated or installed above or below ground. The quantities permitted are commonly expressed as "filling densities" or "loading densities" and are specified for specific gases (in some cases, groupings of similar gases) by codes, standards, and regulations. Filling densities expressed in terms of weight are absolute values, i.e., the designated weight of gas may always be placed into the container. Filling densities expressed in terms of volume, however, must always be qualified by specifying the liquid temperature.

Table 4-7B shows an example of how filling densities are expressed for liquefied gases—in this case, liquefied petroleum gases stored at normal temperatures in uninsulated containers. Note that larger quantities of higher specific gravity materials are permitted (reflecting the fact that these liquids tend to expand less) and that larger containers can be filled more than smaller ones can (reflecting the fact that it takes longer for them to absorb heat from atmospheric temperatures or solar radiation). Also, underground containers can be filled even more (reflecting the fact that their ambient temperatures are relatively constant and well below summer atmospheric temperatures).

Containers of compressed or liquefied gases can represent high levels of potential energy release due to the concentration of matter by compression or liquefaction. Container failure releases this energy—often extremely rapidly and violently—with a simultaneous release of gas to the surroundings and propulsion of the container or container pieces. Compressed gas container failures are distinguished more by the flying missile hazard than by the results of gas release, because these containers contain lesser quantities of gas. Liquefied gas container failures can release larger quantities of gas.

Overpressure Relief Devices

Spring-loaded pressure-relief valves or bursting discs, or sometimes both, are provided on most compressed and liquefied gas containers to limit container pressure to a level the container can safely withstand, although fusible plugs are sometimes used on smaller containers. The start-to-discharge pressure settings of these devices are related to the strength of the container. In most cases, the relieving capacity (in terms of gas flow rate through them) is based upon consideration of heat input rates resulting from fire exposure, because this is generally the largest anticipated source of heat. In some instances, such as underground or insulated containers, other sources of overpressure may be the dominant effect of fire exposure.

As noted, the relieving capacity of these devices is based upon discharge of gas. In the case of liquefied gas containers exposed to fire, it is possible to have conditions such that liquid will be discharged instead of gas, i.e., when the container is tipped over. Under such conditions, the relieving capacity will be reduced—in some cases as much as 60 to 70 percent. With the possible exception of the fire exposure condition, this reduction is of little practical significance. Even in the case of fire exposure, this situation is not as critical as it may appear, because prevention of container failure under fire exposure conditions requires other safeguards in addition to the overpressure relief device. Inspection, maintenance, and replacement of these devices are essentially unregulated and they are subject to potentially harmful influences.

TABLE 4-7B. *Maximum Permitted Filling Density (LP Gas Stored at Normal Temperatures in Uninsulated Containers)*

Specific Gravity at 60°F (15.6°C)	Aboveground Containers				All Underground Containers	
	0 to 1,200 gal*		Over 1,200 gal*			
	% of WWC†	Vol. % at 60°F‡	% of WWC†	Vol. % at 60°F‡	% of WWC†	Vol. % at 60°F‡
0.496—0.503	41	82.0	44	88.0	45	90.0
0.504—0.510	42	82.6	45	88.5	46	90.5
0.511—0.519	43	83.5	46	89.4	47	91.3
0.520—0.527	44	84.3	47	90.0	48	91.8
0.528—0.536	45	84.7	48	90.3	49	92.2
0.537—0.544	46	85.2	49	90.7	50	92.5
0.545—0.552	47	85.6	50	91.2	51	93.0
0.553—0.560	48	86.4	51	91.8	52	93.6
0.561—0.568	49	87.0	52	92.3	53	94.1
0.569—0.576	50	87.5	53	92.7	54	94.4
0.577—0.584	51	87.9	54	93.1	55	94.8
0.585—0.592	52	88.3	55	93.5	56	95.1
0.593—0.600	53	88.9	56	94.0	57	95.6

* Total water capacity, U.S. gallons (1,200 U.S. gallons equals 4.54 m³).
† WWC—water weight capacity.
‡ 60°F = 15.6°C.

Containers of certain poisonous and highly toxic gases do not have overpressure relief devices, because the overall hazard of a prematurely operating or leaking relief device outweighs the hazard of container failure from overpressure. It is also the practice in some countries, especially in Asia and the Mediterranean area, not to provide overpressure protection on liquefied gas containers—a practice that reflects the widespread use and installation of such containers inside buildings, including homes, and commercial and institutional buildings. In such cases, the hazard of gas release indoors due to operation of the relief device is felt to outweigh the hazard of container failure due to overpressure.

The Liquefied Gas BLEVE

Instances where containers of liquefied gases fail and break into two or more pieces are common enough to warrant treatment in some detail. This failure is termed "boiling liquid expanding vapor explosion" (BLEVE), a type of pressure-release explosion.

All liquefied gases are stored in containers at temperatures above their boiling points at NTP and remain under pressure only so long as the container remains closed to the atmosphere. This pressure ranges from less than 1 psi (6.9 kPa) for some cryogenic gas containers to several hundred psi for noncryogenic liquefied gas containers at normal storage temperatures. If the pressure is reduced to atmospheric, such as through container failure, the substantial heat that is in effect "stored" in the liquid causes very rapid vaporization of a portion of the liquid, to a degree directly proportional to the temperature difference between that of the liquid at the instant of container failure and the normal boiling point of the liquid. For many liquefied flammable gases, this temperature difference at normal atmospheric temperatures can result in vaporization of about one third of the liquid in the container.

Because overpressure relief devices are set to discharge at pressures corresponding to liquid temperatures above normal atmospheric temperatures (to prevent premature operation), the liquid temperature is higher than this if container failure occurs when a relief device is functioning. Therefore, more liquid is vaporized under these conditions—often over one half of the liquid in the container. This is the usual situation when a container fails from fire exposure. The remaining liquid unvaporized is cooled by the "self-extraction" of heat when the pressure is reduced to atmospheric and is cooled to near its normal boiling point.

Liquid vaporization is accompanied by a large liquid-to-vapor expansion. It is this expansion process that provides the energy for propagation of cracks in the container structure, propulsion of pieces of the container, rapid mixing of the vapor and air resulting in the characteristic fireball upon ignition by the fire which caused the BLEVE, and atomization of the remaining cold liquid. Many of the atomized droplets burn as they fly through the air. However, it is not uncommon for the cold liquid to be propelled from the fire zone too quickly for ignition to occur and fall to Earth still in liquid form. In one case, dissolved spots in asphalt paving were noted up to $1/2$ mile (0.8 km) from the site of an LP-Gas BLEVE. In other BLEVEs, fire fighters have been cooled by cold liquid passing in their vicinity.

Reduction of internal pressure to atmospheric level in a container results from structural failure of the container. Failure is most often due to weakening of the container metal from flame contact; however, this will happen if the container is punctured or fails for any other reason.

As shown in Figure 4-7A, the strength of carbon steel steadily decreases with temperature increases above about 400°F (204°C). Figure 4-7A is based upon a typical low-carbon steel. The curves will vary quantitatively with other steels, but the loss of strength

with increasing temperature is valid for all common metals, and the critical temperatures are well below those attainable in a fire.

Figure 4-7A also shows why entirely satisfactory performance of a spring-loaded relief valve to design parameters cannot prevent a BLEVE. By its nature, such a valve cannot reduce the pressure to atmospheric but only to a point somewhat below its start-to-discharge pressure. Therefore, the liquid will always be at a temperature above its normal boiling point, pressure will remain inside the container, and the container structure will be stressed in tension. This stressed area is shown as a shaded area in Figure 4-7A for a common type of LP-Gas container having a pressure-relief valve set for 250 psi (172 kPa). Again, while this area will vary for different steels or pressure vessel relief valve design characteristics, it is evident that if the metal is heated above this range (which is quite possible in the event of direct flame contact), the metal will not withstand the stress and the container will fail.

It is extremely difficult to significantly heat the container metal where it is in contact with liquid because the liquid conducts the heat away from the metal and acts as a heat absorber. For example, when the relief valve cited in this example is discharging, the propane liquid stays in the 120 to 140°F (49 to 60°C) range. As a result, the metal temperature is well within safe limits. This situation does not exist for the metal in the vapor space of the container, as vapor is relatively nonheat conductive and has little heat-absorbing capacity.

In most BLEVEs that have been studied where the failure was due to metal overheating, it originated in the metal of the vapor space and was characterized by both the metal stretching and thinning out and the appearance of a longitudinal tear which progressively got larger until a critical length was reached. At this time, the failure became brittle in nature and propagated at sonic velocity through the metal in both longitudinal and circumferential directions. As a result, the container came apart in two or more pieces.

Magnitude of a BLEVE: Although most liquefied gas BLEVEs that involve container failure result from fire exposure, a few BLEVEs have occurred due to container failures from other causes,

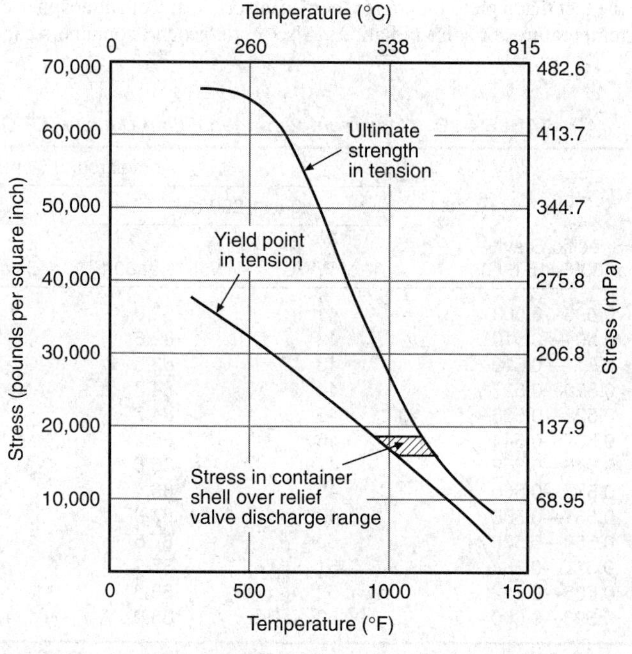

FIG. 4-7A. *Behavior of liquefied gas container metal (carbon steel) when exposed to fire.*

such as corrosion or impact from an outside force. Impact failures are particularly noticeable in transportation accidents involving rail-cars and cargo vehicles. In these cases, the BLEVE generally occurs simultaneously with impact. In one instance, however, a 30,000-gal (113.5 m³) tank car of LP-Gas was only severely weakened by impact during derailment and did not BLEVE until more than 40 hr later. The tank car had been lifted and moved without incident in the interim. At the time of failure, however, the internal pressure was increasing as the ambient temperature was rising.

The size of a BLEVE depends upon the weight of the container pieces and upon how much liquid vaporizes when the container fails. This is analogous in many respects to the performance of rockets, as far as propulsion of container parts is concerned. Most liquefied gas BLEVEs occur when containers are from slightly less than $^1/_2$ to about $^3/_4$ full of liquid. The liquid vaporization-expansion-energy to container-piece weight ratio is such that pieces are propelled for distances up to approximately $^1/_2$ mile (0.8 km). Deaths from such missiles have occurred up to 800 ft (244 m) from larger containers. Fireballs several hundred feet in diameter are not uncommon, and deaths from burns have occurred to persons as much as 250 ft (76 m) from the larger containers.

A major LP-Gas disaster with multiple BLEVEs occurred on November 19, 1984, near Mexico City, Mexico. According to a Swedish investigator who cooperated with the NFPA, and contrary to earlier published reports, this incident occurred at about 5:00 a.m. on a Monday while the terminal was being filled by pipeline from a refinery in another part of the country. A failure in a component of the 12-in. (25.4-mm) fill line resulted in a line break and the development of a large LP-Gas cloud which ignited upon reaching a ground-level flare about 20 min later. The resulting explosion and fire is believed to have been the major factor in the 500 deaths reported. The flame issuing from the broken pipeline impinged on one of the four 10,000-barrel (1,590-m³) spheres which BLEVEd about 10 min later. In all, approximately 15 BLEVEs were noted on seismic charts among the four 10,000-barrel (1,590 m³) spheres and 48 horizontal tanks.

The two 15,000-barrel (2,384 m³) spheres remained on their foundations, but were severely damaged. (See Figure 4-7B.) The four 10,000-barrel (1,590 m³) spheres were found in pieces both on and off the terminal site. All 48 horizontal tanks were damaged in varying degrees, with major sections as far as 3,600 ft (1.1 km) from

the site. Figure 4-7C illustrates the spectacular dimensions a BLEVE can reach.

Time intervals for fire-caused BLEVEs: The time between initiation of flame contact and a BLEVE varies because it depends upon such widely varying factors as the size and nature of the fire as well as the container itself. Uninsulated containers located aboveground can BLEVE in the absence of water cooling in a matter of a very few minutes in the case of small containers to a few hours for very large containers. Data on insulated containers is meager because only cryogenic containers and some reactive-gas containers are usually insulated. However, there is no doubt that insulation designed for fire-exposure conditions can delay BLEVE times significantly. In one case involving an insulated LP-Gas railroad tank car, the BLEVE did not occur until $20^1/_2$ hr of fire exposure—undoubtedly an extreme example. In comparison, using fire tests on LP-Gas railroad tank cars, a BLEVE occurred in 93 min in the insulated case, as opposed to 25 min for the uninsulated tank.

Protection against a BLEVE: Protection for an uninsulated liquefied gas container that may be exposed to fire is provided by the application of water so that a film of water coats portions of the container not in internal contact with liquid. The methods used can range from hose streams to installation of water spray fixed systems.

The preceding data on BLEVE magnitude, time intervals, and protection are applicable only to unreactive liquefied flammable gases. The additional chemical energy in reactive gases introduces chemical factors not present in the purely physical and combustion phenomena just described, even though the outward appearance and general effects are similar. The hazard can be increased by evolution of internal heat of chemical reaction (which adds to the heat from the exposing fire) and the inability of externally applied cooling water to cool the liquid inside the container.

Combustion within Containers

A less frequent but significant hazard of gas stored in containers is the danger of container failure from overpressure, resulting from combustion of the gas while inside the container. Flammable gas-air

FIG. 4-7B. Two of the 15,000-barrel (2,384 m³) spheres involved in the Mexico City LP-Gas disaster. Note how far out of vertical they are. This is due to partial failure of the container supports.

FIG. 4-7C. The fireball formed in a BLEVE involving LP-Gas railroad tank cars at Crescent City, IL, on June 21, 1970. The elevated water tank at center is a point of reference in visualizing the dimensions of the fireball.

or gas-oxygen mixtures are seldom intentionally provided in containers, but can be established accidentally. Most of these explosions have occurred in industrial and medical gas applications where oxygen or compressed air is often used in conjunction with flammable gases. Where such a possibility is inherent in a process, e.g., an oxygen-fuel gas metal cutting system, provisions are made to prevent the occurrence of such mixtures in containers. More generally, most explosions can be prevented by education and training in proper container filling procedures. Consumers of industrial and medical gases seldom have the expertise needed to fill containers safely.

Combustion when Released from Containers

When gases are released from their containers, the hazards vary according to the chemical and physical properties of the gas and the nature of the environment into which they are released. All gases, with the exception of oxygen and air, present a hazard to life if they displace the breathing air. Nitrogen, helium, argon, and other odorless and colorless gases are particularly hazardous, because they are not detectable by the human senses. The minimum oxygen concentration in air for survival is about 6 to 10 percent (compared with the normal 21 percent) by volume, but even at higher concentrations judgment and coordination are affected.

Toxic or poisonous gases: These gases present obvious life hazards. They are especially dangerous when released during a fire because they can impede fire-fighting efforts by either preventing or delaying access to the site by fire fighters.

Oxygen and other oxidizing gases: Although these gases are nonflammable, they can make combustibles ignite at lower temperatures, accelerate combustion, and start fires by causing flames in fuel-burning appliances to extend beyond their combustion chambers.

Liquefied gases, including cryogenic gases: Because of their low temperatures, these gases present a hazard to persons and property when they escape as liquids. Contact with cold liquid can cause frostbite which can be severe if the exposure is prolonged. The properties of many structural materials, particularly carbon steel and plastics, are affected by low temperatures and embrittlement, which could lead to structural failure.

Flammable gases: Because of their prevalence, the behavior of flammable gases when released from their containers is of major interest. Released flammable gases present two basic hazards: (1) combustion explosions and (2) fire. Failure to distinguish between the circumstances surrounding these two hazards can result in misapplication of protective measures.

Combustion Explosions

It is useful to consider a flammable gas combustion explosion as occurring in the following sequence:

1. A flammable gas or the liquid phase of a liquefied flammable gas is released from its container, piping, or equipment (including the normal operation of an overpressure relief device). If liquid escapes, it rapidly vaporizes and produces the potentially large quantities of vapor associated with this liquid-to-vapor transition.
2. The gas mixes with the air.
3. With certain proportions of gas and air (in the flammable or combustible range), the mixture is ignitable and will burn. (See Table 4-7C.)

TABLE 4-7C. *Combustion Properties of Common Flammable Gases*

Gas	Btu per cu ft (Gross)	mJ/m³ (Gross)	Limits of Flammability Percent by Volume in Air		Specific Gravity (Air = 1.0)	Air Needed to Burn 1 cu ft of Gas cu ft	Air Needed to Burn 1 m³ of Gas m³	Ignition Temp.	
			Lower	Upper				°F	°C
Natural gas									
High inert type (*Note 1*)	958–1051	35.7–39.2	4.5	14.0	.660–.708	9.2	9.2	—	—
High methane type (*Note 2*)	1008–1071	37.6–39.9	4.7	15.0	.590–.614	10.2	10.2	900–1170	482–632
High Btu type (*Note 3*)	1071–1124	39.9–41.9	4.7	14.5	.620–.719	9.4	9.4	—	—
Blast furnace gas	81–111	3.0–4.1	33.2	71.3	1.04–1.00	0.8	0.8	—	—
Coke oven gas	575	21.4	4.4	34.0	.38	4.7	4.7	—	—
Propane (commercial)	2516	93.7	2.15	9.6	1.52	24.0	24.0	920–1120	493–604
Butane (commercial)	3300	122.9	1.9	8.5	2.0	31.0	31.0	900–1000	482–538
Sewage gas	670	24.9	6.0	17.0	0.79	6.5	6.5	—	—
Acetylene	1499	55.8	2.5	81.0	0.91	11.9	11.9	581	305
Hydrogen	325	12.1	4.0	75.0	0.07	2.4	2.4	932	500
Anhydrous ammonia	386	14.4	16.0	25.0	0.60	8.3	8.3	1204	651
Carbon monoxide	314	11.7	12.5	74.0	0.97	2.4	2.4	1128	609
Ethylene	1600	59.6	2.7	36.0	0.98	14.3	14.3	914	490
Methylacetylene, propadiene, stabilized (*Note 4*)	2450	91.3	3.4	10.8	1.48	—	—	850	454

Note 1: Typical composition CH₄ 71.9–83.2%; N₂ 6.3–16.2%
Note 2: Typical composition CH₄ 87.6–95.7%; N₂ 0.1–2.39
Note 3: Typical composition CH₄ 85.0–90.1%; N₂ 1.2–7.5
Note 4: MAPP® Gas

4. When ignited, the flammable mixture burns rapidly and produces heat rapidly.

5. The heat is absorbed by everything in the vicinity of the flame and by the very hot gaseous combustion products.

6. Nearly all materials expand when they absorb heat. The one material in the vicinity of the flame or hot gaseous combustion products that expands most when it is heated is air. Referring to the "gas laws" cited earlier in this chapter, it will be noted that air expands to double its original volume for every 459°F (237°C) it is heated.

7. If the heated air is not free to expand because, for example, it is confined in a room, the result is a rise in pressure in the room.

8. If the room structure is not strong enough to withstand the pressure, some part of the room will suddenly and abruptly move and depart from its original position, and a bang, woosh, boom, or other noise will be heard. This activity, in part, describes an explosion. Because the source of pressure is combustion, this kind of explosion is called a combustion explosion. It is also called less accurately a room explosion or vapor-air explosion.

A combustion explosion requires that a quantity of flammable gas-air mixture accumulates in an enclosure. In addition, the quantity/strength-of-enclosure relationship must be such that the strength of some part of the enclosure must be exceeded by the pressure-building potential of the mixture. If the enclosure is strong enough to withstand the pressure, a combustion explosion could not occur, because it is the manner in which the enclosure performs that basically determines whether such an explosion occurs. However, few enclosures are strong enough to withstand such pressure.

If an enclosure was full of a flammable gas-air mixture at atmospheric pressure, the enclosure would have to withstand about 60 to 110 psi (400 to 750 kPa) in order to remain intact to prevent a combustion explosion. If a reactive flammable gas is involved or oxygen enrichment occurs, even higher pressures are possible. Conventional structures are capable of withstanding pressures only in the order of $1/2$ to 1 psi (3.5 to 7 kPa). This wide pressure disparity shows clearly that conventional structural enclosures are vulnerable even if they are not full of a flammable gas-air mixture. Experience supports this conclusion, and it has been estimated that most combustion explosions of conventional structures occur with less than 25 percent of the enclosure occupied by the flammable mixture. This fact should not be overlooked—it is erroneous to assume that a flammable gas-air mixture needs to completely fill a room or building before an explosion can occur.

The mechanics of gas accumulation in a structure are affected by the rate of gas release, whether the gas is in the liquid or gas phase, the density of the gas, and the ventilation in the structure. Classic diffusion laws are of little significance under actual conditions, because the combination of extremely slow release rates and airtight structures seldom happens. For this reason, and the fact that most flammable gas-air mixtures are about 90 percent air, the density of the gas itself (i.e., whether it is lighter or heavier than air) is less often a significant factor in gas combustion explosions involving structures than one might expect.

Due to the large and rapid flammable gas-air mixture potential of liquefied gases, codes and standards impose severe limitations on the handling of such gases indoors. Considering the fireball aspect of the BLEVE, it is evident that an indoor fireball behaves similarly to an ignited gas-air mixture accumulation; the results of a BLEVE of a container located indoors can be very similar to a combustion explosion.

Combustion Explosion Safeguards

Basic combustion explosion prevention safeguards are designed to limit flammable gas-air mixture accumulation in a structure. Fundamental safeguards are the use of rugged containers and equipment that minimize the chance of leakage, the minimal quantity of gas released by emergency flow control devices, and the use of limiting orifices to minimize the quantity released. Burners are often equipped with flame failure devices that shut off the flow of gas if the flame is extinguished for any reason.

Many gases are colorless and odorless, and it is common to odorize the more widely used fuel gases to increase chances of leak detection. This is particularly true for the common fuel gases, e.g., natural gas and LP-Gas. Codes, standards, and regulations customarily require that these gases be odorized so that they are detectable by humans at gas concentrations in air not exceeding one fifth of the lower limit of flammability. The odorants are usually volatile organic liquids containing chemically combined sulfur and have a characteristic "gassy" odor.

While it is an effective safeguard, odorization has limitations. The functioning detector, i.e., a human's sense of smell, is not always present, such as when the premises are vacant or the occupants are asleep. In addition, all odors deaden the sense of smell if inhaled long enough, and senses of smell vary. Odorants can be scrubbed from the gas by some soils, a factor where leakage stems from underground piping. Mixture accumulations can also be limited by ventilation systems in structures. These systems, however, are usually in industrial operations, and even then only rather nominal release rates can be handled by practical ventilation systems because of the necessity to condition the makeup ventilation air for comfort purposes.

Ignition source control is also fundamental to combustion explosion prevention. However, this safeguard is also limited mainly to industrial operations where the many flammable gases burned in heat-producing equipment provide an inherent ignition source.

In addition to combustion explosion prevention safeguards, the severity of the explosion can be reduced by special structural design whereby some elements of the structure are designed to dislocate at lower pressures, and other elements are designed to stay in place at the lower pressures that result. This is known as "explosion venting" and is covered in some detail in NFPA 68, *Guide for Venting of Deflagrations*. Such building design is not practical for most ordinary buildings, however, so this practice is essentially restricted to certain industrial structures.

Flammable gas fires: The flammable gas fire can be considered as an aborted combustion explosion wherein an explosive quantity of flammable gas-air mixture does not accumulate because the mixture is either ignited too quickly or a confining structure is not present. As would be expected when flammable gas escapes outdoors, fires are more likely than explosions. However, if a massive release occurs, it is possible for the air or surrounding buildings to comprise enough confinement to lead to a type of combustion explosion known as an open-air explosion or space explosion. Liquefied noncryogenic gases are subject to this phenomenon, as are hydrogen, ethylene, and some reactive gases that have an extremely rapid rate of flame propagation. An example of the intentional provision of prompt ignition to achieve a fire instead of a combustion explosion is the use of pilots in gas-burning equipment, such as range ovens, water heaters, boilers, and furnaces.

Gas fire prevention safeguards include many of the combustion explosion prevention safeguards. However, unlike the combustion explosion, the destructive effects of the gas fire can be minimized by control measures applied after the fact.

GAS EMERGENCY CONTROL

Controllable gas emergencies resulting from the escape of gas from containers present two basic forms of hazard: (1) toxic, inert, or oxidizing gases can present hazards to persons or property and unignited flammable gases can be ignited, possibly explosively ("no fire" emergencies); and (2) gas fires can present thermal hazards to persons or property ("fire" emergencies) and, if such fires expose gas containers, introduce the possibility of container failure and the BLEVE. A fire in any combustible material also presents the hazard of container failure from fire exposure.

Control of "No Fire" Emergencies

Control generally consists of directing, diluting, and dispersing the gas to prevent contact with persons, preventing it from infiltrating structures if release is outdoors, and avoiding its contact with ignition sources while, if possible, simultaneously stopping the flow of escaping gas. Gas direction, dilution, and dispersion require the use of a carrier fluid; air, steam, and water have proved to be practical. The use of air, for all practical purposes, is limited to indoor situations and is an extension of the ventilation safeguards against a combustion explosion. Steam has been distributed through systems of fixed nozzles around outdoor ethylene processing equipment and where the large quantities of needed steam are available.

Water in the form of a spray, applied from hoses or monitor nozzles or by fixed water spray systems, is the most common carrier fluid. The use of hose streams is largely a fire department operation because of the personnel requirements. Fixed water spray systems for this purpose are designed differently from the more conventional spray systems used to control fire. To date, such systems have been provided at a few outdoor ethylene and liquefied natural gas (LNG) facilities. When water is used on cryogenic liquid containers, care must be taken to ensure that water does not contact the pressure-relief valve, as it can easily freeze and defeat the operation of this important safety device.

The physical properties of the escaping gas affect control techniques. With compressed gases, the density of the gas is an important factor. When such gases are colorless and odorless, control tactics may be complicated because instruments may be needed to define the extent of the hazardous area.

Liquefied gases possess an inherent visible indicator of their location, because the refrigerating effect of their vaporization condenses water vapor from the air and produces a visible fog. The fog roughly defines the gas area, but invisible ignitible gas-air mixtures often extend for several feet (1 ft = 0.3 m) beyond the extremities of the visible fog.

Because noncryogenic liquefied gases contain considerable heat for vaporization, they will often vaporize so rapidly from contact with the air or ground that they will not exist in the liquid phase once they escape—at least not to the extent that a pool will form. The lower vapor pressure noncryogenic liquefied gases, such as butane and chlorine, and those with high latent heats of vaporization, such as anhydrous ammonia, are exceptions. Even the higher vapor pressure gases, such as propane, will pool when ambient temperatures are well below freezing.

The cryogenic liquefied gases, on the other hand, must obtain nearly all the heat for vaporization from ground or air contact and, therefore, will characteristically form a pool if the leak continues for a long enough time. In such cases, application of a carrier fluid will increase the vaporization rate if applied to the liquid—an often undesirable effect.

The gas produced near the source of vaporization of a liquefied gas is always heavier than air at normal temperatures because of the low temperature of the gas at that point. This, together with the associated water fog, tends to cause even normally lighter-than-air gases to hug the ground for some distance.

The use of foam, particularly high-expansion foam, to control the flow of gas produced by a vaporizing cryogenic gas which is lighter than air at ambient temperatures, has been investigated to some extent. A sufficiently thick blanket of foam can warm the gas to the point where it will rise above the blanket instead of flowing horizontally near the ground. However, this is applicable only to established pools of vaporizing liquid and not to the initial high vaporization rate escape phase.

Some gases will react chemically with the carrier fluid—particularly if the carrier is steam or water. Chlorine is a notable example—the reaction produces hydrochloric acid. However, the principal problem in this respect is enlargement of the leak as the metal container reacts with the acid.

A special type of "no fire" control for indoor application to flammable gases is the explosion suppression system. Actually, such systems do permit ignition to occur but arrest the combustion of the flammable gas-air mixture before the pressure needed for a combustion explosion is obtained.

Control of "Fire" Emergencies

Control of fire emergencies is generally the control of heat from the fire by the application of water while, if possible, stopping the flow of escaping gas. Many gas fires can be extinguished by conventional extinguishing agents, including carbon dioxide, dry chemical, and the halogenated agents. However, the potential conversion of a gas fire into a combustion explosion if gas continues to escape after extinguishment must be recognized. The generally accepted practice is to limit actual extinguishment by agent application to small leaks that will not present a hazard if they reignite. In addition, the use of halogenated agents is being phased out because of their detrimental effect on Earth's ozone layer.

Methods of water application are the same as described for the "no fire" emergencies, i.e., hose streams, monitor nozzle streams, and fixed water spray systems, as well as sprinkler systems. The selection of the particular application method, or combination of methods, requires a sound fire protection analysis of the existing conditions. This is particularly true where water is used to prevent a BLEVE because of the limited time available if uninsulated containers are involved. This, together with the specialized tactics needed for safety of emergency personnel, has been shown to severely tax the capabilities of hose stream application in many instances. Manual actuation of fixed systems (water spray or sprinkler) that have nozzles and dry piping in the fire area is questionable, because the immediate intensity of a gas fire can quickly damage the piping before the water is turned on.

Conventional automatic sprinkler protection is limited to indoor or roofed-over areas. However, sprinklers have been effective in greatly reducing the number of overpressure relief devices on cylinders that operate during a fire. In turn, the number of cylinders that may fail from contact with the burning relief device effluents is reduced; sprinkler spacing and water density must be tailored to this hazard.

Foam can control a fire in a pool or tank of cryogenic gas, but cannot extinguish it. The degree of control depends upon the extent to which the foam can cover the liquid and the length of time the foam application can be maintained.

SPECIFIC GASES
Acetylene

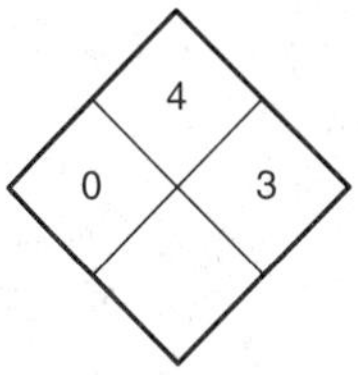

UN 1001

Classifications: Reactive, flammable, compressed, industrial.

Note: The health hazard rating of acetylene has been revised from 1 to 0. This revision is based on the 1990 edition of NFPA 704, *Standard System for the Identification of the Fire Hazards of Materials* (hereinafter referred to as NFPA 704), where qualitative health criteria was added to the standard. The revised health hazard rating is based on toxicological data.

Chemical properties: Acetylene is comprised of carbon and hydrogen joined by a triple chemical bond which is responsible for its reactivity. In the liquid or solid states, or in the gaseous state at moderate or high pressures, acetylene can decompose rapidly with the formation of carbon and hydrogen and evolution of heat. Decomposition can be initiated by heat. Decomposition of liquid or solid acetylene also can be initiated by mechanical impact. In a confined space, the heated decomposition gases can cause overpressure and container, piping, or equipment failure.

As an indication that a hazardous, heat-producing reaction can occur in the absence of air, an upper flammable limit of 100 percent is often found in the literature, rather than the 81 percent given in Table 4-7C. This technical misapplication of the flammable range concept has caused confusion.

Acetylene can react with certain metals to produce metallic acetylides, extremely shock-sensitive explosive compounds which, if detonated even in small quantities, can initiate acetylene decomposition. Copper and some copper alloys are notable in this respect, and their use must be avoided in most acetylene piping and equipment. However, these metals can be used in some components of acetylene systems under certain conditions that reflect the reaction kinetics involved. Torch tips, for example, fall into this category. Brass containing less than 65 percent copper can be used for acetylene piping systems.

Acetylene is not toxic and many years ago was mixed with oxygen for use as an anesthetic. Pure acetylene is odorless, but acetylene in general use has a characteristic odor due to minor impurities inherent in its generation from calcium carbide or derivation from other hydrocarbons.

Physical properties—autoignition temperature [581°F (305°C)]: Although acetylene is considered a compressed gas, the usual acetylene container used in transportation and storage does not exclusively contain acetylene in the gaseous phase and is unique in this respect. To ensure stability under reasonably anticipated thermal and mechanical impact conditions, acetylene cylinders are filled with a porous-mass packing material containing very small pores or cellular spaces so that any volume of gas therein is correspondingly small. This limits the decomposition energy available and restricts communication between spaces. In addition, the mass is saturated with acetone or dimethylformamide (DMF)—a flammable liquid in which acetylene is very soluble. In this manner, acetylene gas can be compressed in solution in a manner similar to that of carbon di-

oxide in water to produce carbonated water. When the pressure is reduced, e.g., by opening a cylinder valve, the gas escapes from solution and thus leaves the container in the gaseous state. Currently, a charging pressure maximum of 250 psig at 70°F (1724 kPa at 21°C) is recognized by DOT and the Transport Canada (TC) regulations. There will be a variation of about 2.5 psig rise or fall per °F of temperature change (9 kPa per 1°C). For each atmosphere (14.7 psi or 101.3 kPa) of pressure, acetone or DMF will dissolve about 25 times its own volume of acetylene so that at 250 psig (1724 kPa) acetone will dissolve about 425 volumes of acetylene. An acetylene cylinder is illustrated in Figure 4-7D.

Usage: Acetylene is used primarily in chemical processing and as a fuel gas in oxygen-fuel gas cutting and welding operations. Occasionally it is manufactured at the consuming location by reacting calcium carbide (a solid) and water in acetylene generators and piped directly to gas holders or to the point of use.

Hazards inside containers: Acetylene is shipped in DOT and TC cylinders as described previously and stored in these or in low-pressure gas holders. It is either handled in hoses or piping systems.

Acetylene cylinders are indirectly protected against overpressure by heat-actuated devices, usually fusible metal plugs of eutectic alloys which have low melting points similar to those alloys employed in automatic sprinklers. These may be threaded plugs, or the fusible metal can be incorporated into the construction of the cylinder valve. Unlike spring-loaded pressure relief valves, operation of fusible plugs results in complete reduction to atmospheric pressure, because both the compressed gas and acetone or DMF are released at the same time. Due to this mode of overpressure protection, acetylene cylinders are not subject to a BLEVE, which by definition cannot occur if the liquid contents are not above their boiling points at NTP. Under fire exposure conditions and during charging operations in acetylene cylinder charging plants, however, internal decomposition has occurred which has resulted in BLEVE effects, such as cylinder rupture and propulsion, accompanied by small fireballs.

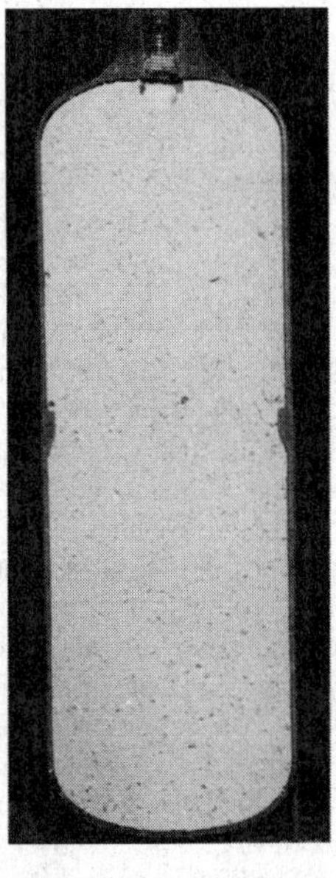

FIG. 4-7D. Cutaway of a typical acetylene cylinder showing porous filler material. Cylinder valve and packing for well at top of cylinder are not shown.

Because susceptibility to decomposition is directly related to pressure, i.e., the higher the pressure, the easier it is to initiate decomposition and the more violent the effects, an acceptable degree of stability in most piping systems containing gaseous acetylene is achieved by limiting or regulating the pressure. In general, this pressure must not exceed 15 psi (103 kPa). In applications requiring higher pressures, particularly in acetylene cylinder charging plants, special piping design features can be employed to control this hazard.

The use of acetylene in either the liquid or solid phases is extremely hazardous, and is prohibited by standards and codes covering conventional applications.

Hazards when released from containment: When acetylene is released from containment, it presents combustion explosion and fire hazards. Because of its reactivity, it is easier to ignite than most flammable gases and burns more rapidly. The latter effect increases the severity of combustion explosions and increases the difficulty of providing explosion venting. Acetylene is only slightly lighter than air and does not readily dissipate. (See Table 4-7C.)

Because of acetylene's reactivity, the design of electrical equipment for use in acetylene atmospheres is unique to this material, and Group A electrical equipment (see NFPA 70, *National Electrical Code®*, Article 500) is devoted exclusively to acetylene. In practice, electrical equipment is avoided in the more likely acetylene release areas or the release potentials are reduced to a level where Group C or D equipment can be used. As a result, available Group A equipment is extremely limited in variety.

Emergency control: Because of the limited quantities of acetylene in conventional shipping modes and storage practices, emergencies seldom occur. Fire control measures are similar to those involved with any flammable, nontoxic gas—i.e., application of water to containers for cooling and dissipation of leakage and stopping the flow of escaping gas if possible.

Hazards of acetylene cylinders exposed to fire: Special precautions are needed when acetylene cylinders are exposed to fire. If an acetylene cylinder is exposed to fire, the fusible-metal pressure-relief devices will function when the metal reaches about 212°F (100°C). The nature of acetylene presents hazards that may not be present with other gases. Acetylene will decompose into hydrogen and carbon at temperatures in excess of 570°F (300°C). This property can present a danger when cylinders are involved in fire and heated locally above 570°F (300°C). When this occurs, cylinders can become overpressured and can, in some cases, rupture violently. The decomposition of acetylene can be so rapid that the pressure-relief provided by a plug(s) cannot always prevent rupture. Rupture can occur in as soon as a few minutes, or up to several hours after exposure.

Acetylene cylinders in the United States and Canada have fusible-metal pressure-relief devices incorporated into the cylinder or cylinder valve. Small cylinders incorporate the fusible metal into the valve. Medium and large cylinders incorporate the fusible metal in a plug(s) in the top head or bottom head. Most acetylene cylinders in Europe and other parts of the world do not have pressure-relief devices and are more susceptible to overpressure and rupture. When part of an acetylene cylinder is exposed to fire and is heated over 570°F (300°C), the acetylene will begin to decompose into carbon and hydrogen. Decomposition is rapid and liberates large quantities of energy. Cylinders exposed to fire should be sprayed with water from a protected location, and persons should not approach an acetylene cylinder that has been exposed to fire. Cylinders exposed to fire can explode even 24 hr after the exposure.

A common hazard of acetylene cylinders happens when a small leak occurs and ignites, resulting in a small flame. This small flame can cause heating of the fusible metal, resulting in the release of a torch-like flame which can be as much as 12 ft (3.6 m) in length. This sudden flame can be a danger to people in the area and to other cylinders.

Anhydrous Ammonia

UN 1005

Classifications: Flammable, liquefied (including cryogenic), industrial.

Chemical properties: Anhydrous ammonia is comprised of nitrogen and hydrogen. While often referred to simply as "ammonia," this term is more appropriately used to describe a solution of anhydrous (which means "without water") ammonia in water. The nitrogen component is inert in the combustion reaction and accounts for the somewhat limited flammability of anhydrous ammonia as manifest by its high lower-flammability limit and low heat of combustion. (See Table 4-7C.)

Moist ammonia will vigorously attack copper and zinc and many of their brass or bronze alloys. This has caused some problems in agricultural areas where, because of their similar vapor pressures, anhydrous ammonia and propane are often handled in the same storage tanks, transport containers, and associated equipment at different times of the year. LP-Gas equipment customarily utilizes brass, bronze, copper, and zinc extensively for valves, gages, piping, regulators, etc. This equipment, if converted to anhydrous ammonia service, must be changed to steel. Subsequently, anhydrous ammonia must be carefully purged from a container when it is reconverted to LP-Gas service.

The characteristic pungent odor and irritant properties of anhydrous ammonia serve as warnings for this relatively toxic gas. However, the effectiveness of the warning depends upon the rate of release. Large clouds of ammonia gas have been rapidly produced from large liquid leaks, so that people have been trapped and killed before they could evacuate the area.

Physical properties: At its normal boiling point of –28°F (–33°C), anhydrous ammonia has a liquid density of 42.6 lb per cu ft (682.4 kg/m³), a gas density of 0.055 lb per cu ft (0.88 kg/m³), and a latent heat of vaporization of 589.3 Btu/lb (1371 kJ/kg). At NTP, the gas has a density of about 0.045 lb per cu ft (0.72 kg/m³), and vaporization of 1 cu ft (0.3 m³) of liquid will produce about 885 cu ft (25 m³) of gas.

When vaporized it becomes a colorless gas, with a penetrating suffocating characteristic odor. Liquid released under pressure floats and boils on water.

Usage: Anhydrous ammonia is used primarily as an agricultural fertilizer, as a refrigerant, and as a source of hydrogen for metal heat treating and semi-conductor manufacturing special atmospheres.

Hazards inside containers: Anhydrous ammonia is shipped in DOT and TC cylinders, and DOT-specification cargo trucks, railroad tank cars, and barges. It is stored in cylinders, American Society of Mechanical Engineers (ASME) code tanks, and in cryogenic form in insulated American Petroleum Institute (API) tanks.

Anhydrous ammonia containers with a capacity of less than 165 lb (79 kg) are not required to be equipped with overpressure protection devices. This reflects the toxicity of the gas and is a safety tradeoff between the hazards of container rupture from overpressure and gas release due to operation of an overpressure protective device, especially indoors where the smaller containers are often found.

Few BLEVEs of uninsulated anhydrous ammonia containers have been recorded, reflecting the limited flammability of the gas which minimizes the probability of it supplying the exposing fire. Fires that have been reported have been in other combustibles.

Hazards when released from containment: Anhydrous ammonia presents combustion explosion and fire hazards (as well as a toxicity hazard) when released from containment. However, its high lower limit of flammability and low heat of combustion reduce these hazards substantially.

If anhydrous ammonia is released outdoors, it is difficult for it to reach the lower flammability limit concentration except for small zones in the immediate vicinity of the leak. Even where large quantities of liquid are released, ignitable concentrations tend to be in discontinuous pockets and this, together with the low heat of combustion, reduces the possibilities of sustained burning. Experience indicates that similar circumstances apply when the release occurs indoors in conventional and reasonably well-ventilated buildings. In unusually tight rooms, such as refrigerated process or storage areas, however, the release of liquid or large quantities of gas can result in the accumulation of hazardous quantities of a flammable mixture and result in a combustion explosion. In such cases, even though the low heat of combustion produces lower pressures than most flammable gases, the pressure is enough to do major structural damage. As a result of these factors, anhydrous ammonia fires are infrequent and, where ignition does occur, a combustion explosion is the likely result. In a NFPA fire record analysis of 36 incidents from 1929 through 1969 where released gas or liquid was ignited, 28 incidents, or about 78 percent, resulted in a combustion explosion. All occurred indoors.

Emergency control: At normal temperatures, anhydrous ammonia gas weighs about 0.6 times as much as air. Because of its pronounced solubility in water, the spread of escaping anhydrous ammonia gas can be readily controlled by water spray. If hose streams are used, the toxic and inerting properties of anhydrous ammonia require fire fighters to use protective breathing apparatus. Where liquid contact is likely, full protective clothing should be worn. If release of liquid in cryogenic form occurs, pooling is possible and application of water to pools should be avoided, unless the vapors can be controlled, in order to prevent increasing vaporization rate.

Carbon Dioxide

UN 2187
Cryogenic Liquid

UN 1013
Compressed Gas

Classifications: Nonflammable (inert), liquefied, industrial.

Chemical properties: Carbon dioxide is composed of carbon and oxygen. The ability of carbon to chemically link with as much oxygen as possible prohibits its further oxidation (combustion), making it nonflammable. Although nontoxic, carbon dioxide can cause asphyxiation due to the displacement of air.

Physical properties: Carbon dioxide exhibits some unique physical properties. At a temperature of –69.9°F (–56.6°C) and a pressure of 60.4 psig (416.4 kPa), it can exist in a container simultaneously as a liquid, solid, and gas (the triple point). At temperatures and pressures above these and below 87.8°F (30.8°C), it exists as both liquid and gas. Above 87.8°F (30.8°C), it exists only in the gaseous phase. At normal temperatures, the gas is about 1 $1/2$ times heavier than air.

Usage: Carbon dioxide is used primarily to carbonate beverages, to provide an inert atmosphere, and to extinguish fires.

Hazards inside containers: Carbon dioxide is shipped in DOT and TC cylinders and insulated DOT-specification cargo trucks and railroad tank cars. It is stored in cylinders at a pressure of about 850 psi at 70°F (5861 kPa at 21°C) or in insulated ASME-code tanks at pressures of about 200 to 312 psi (1380 to 2,150 kPa) and temperatures of –20 to +40°F (–28.8 to +4.4°C).

Cylinders are provided with overpressure protection in the form of frangible (bursting) discs. Insulated tanks are equipped with pressure-relief valves. Carbon dioxide cylinders have BLEVEd as a result of corrosion, with enough energy to nearly demolish small cargo vehicles. The NFPA has no reports of BLEVEs of other types of containers.

Hazards when released from containment: In addition to the hazard of asphyxiation, contact with cold vaporizing carbon dioxide can cause frostbite. For these reasons, actuation of automatic room-flooding extinguishing system applications should incorporate a time delay if personnel occupy the protected area.

Emergency control: Carbon dioxide emergencies are "no fire" emergencies involving potential asphyxiation hazards. Structure ventilation indoors and water spray control outdoors are applicable. Self-contained breathing apparatus should be used.

Chlorine

UN 1017
Chlorine

Classifications: Reactive, nonflammable, liquefied, toxic, industrial.

Note: The health hazard rating of chlorine has been revised from 3 to 4. This revision is based on the 1990 edition of NFPA 704 where qualitative health criteria was added to the standard. The revised health hazard rating is based on toxicological data.

Chemical properties: Chlorine is a basic chemical element. It is a strong oxidizer. Although it is nonflammable, it can react with many organic materials corrosively and, in some instances, explosively, particularly with acetylene, turpentine, ether, gaseous ammonia, hydrocarbons, most fuel gases, and finely divided metals.

The reactivity of chlorine necessitates special attention to the container, piping, and equipment materials of construction. Below 230°F (110°C), steel, copper, iron, and lead are widely used. In contact with water, chlorine forms hypochlorous and hypochloric acids which are corrosive to most metals.

Chlorine has been used in warfare as a poison gas. Liquid chlorine can burn skin. Its sharp odor serves as a warning.

Physical properties: The normal boiling point of chlorine is about –30°F (–34.4°C). At 32°F (0°C), the liquid density is 91.7 lb per cu ft (1468 kg/m^3) and the gas density is 0.2 lb per cu ft (3.2 kg/m^3). The latent heat of vaporization at the normal boiling point is 123.7 Btu/lb (287.7 kJ/kg). At 32°F (0°C), the vaporization of 1 cu ft of liquid will produce about 458 cu ft of gas (1 m^3 of liquid will produce about 458 m^3 of gas). Chlorine has a greenish-yellow color.

Usage: Chlorine is used primarily in chemical processing, bleaching, purification of drinking water and swimming pools, and sanitation of industrial and sewage wastes.

Hazards inside containers: Chlorine is shipped in DOT and TC cylinders, in DOT-specification portable tanks (so-called "one-ton containers"), and in insulated railroad tank cars.

Chlorine cylinders and 1-ton (1,016-kg) containers are provided with overpressure protection by fusible plugs. Tank cars are protected by pressure-relief valves. Because the operation of fusible plugs results in complete depressurization, the probability of a BLEVE is low, and the NFPA has no record of a BLEVE of a tank car containing chlorine, even though such an occurrence is theoretically possible.

Steel is a suitable material for a chlorine container at normal temperatures; however, it is rapidly attacked at temperatures above approximately 230°F (110°C), and leaks can develop upon fire exposure. For this reason, as well as for BLEVE protection, water should be applied to fire-exposed containers. Insofar as possible, application of water directly upon a leak should be avoided, because it may enlarge the leak from acid corrosion.

Hazards when released from containment: Chlorine presents primarily toxicity and corrosion hazards when released from containment. Since chlorine gas is about $2^1/_2$ times heavier than air, it will hug the ground.

Emergency control: The gas can be controlled by water spray. If hose streams are used, fire fighters should wear full protective clothing. Water applied to liquid chlorine will accelerate vaporization.

The chlorine industry has developed an extensive leak-control program that includes strategic siting of trained emergency personnel and special equipment kits for stopping the more common leaks in cylinders, ton containers, and tank cars. Information on this program is available from The Chlorine Institute, 2001 L St. NW, Ste 506, Washington, DC 20036.

Ethylene

Classifications: Flammable, compressed, cryogenic, industrial.

Chemical properties: Ethylene is comprised of carbon and hydrogen and contains a double chemical bond that imparts a degree of reactivity. Except at very high pressures normally encountered only in chemical processing, it is a stable material. It has a wide flammable range (see Table 4-7C) and high burning velocity, reflecting its reactivity. Although it is nontoxic, ethylene is an anesthetic and asphyxiant.

Physical properties: See Table 4-7A.

Usage: Ethylene is principally used in chemical processing, e.g., manufacture of polyethylene plastic. It is also used to ripen fruit.

Hazards inside containers: Ethylene is shipped as a compressed gas in DOT and TC cylinders and as a cryogenic gas in insulated cargo trucks and railroad tank cars in accordance with DOT specifications and regulations. It is stored in cylinders or in insulted ASME-code or API tanks.

Ethylene cylinders (except for medical cylinders which may have fusible plugs or combination safety devices) are protected against overpressure by frangible (bursting) discs. Insulated truck tanks and tank cars are protected by pressure-relief valves. Cylinders are subject to failure from fire exposure, but not to BLEVEs because they do not contain liquid. Insulated containers are subject to BLEVEs, but few incidents have been recorded. One such incident occurred in Spain; however, the cargo tank and insulation system did not comply with North American standards. A few instances of cryogenic truck container overpressure rupture have occurred due to freezing of relief valves that closed due to improper installation of the valves.

Hazards when released from containment: Ethylene presents combustion explosion and fire hazards when released from containment. A wide flammable range and high burning rate accentuate these hazards. In a number of instances involving rather large outdoor releases, open-air or open-space explosions have occurred.

Emergency control: Escaping ethylene presents both "no fire" and "fire" emergency situations. At normal atmospheric temperatures, ethylene gas is very slightly lighter than air. Ethylene gas vaporizing from the cryogenic liquid near its normal boiling point is heavier than air at 70°F (21°C) and can spread along the ground. Escaping liquid will also pool on the ground. Although the visible fog created is a rough indication of the extent of the hazardous area, the hazard can extend beyond the visible area.

Escaping gas can be controlled by water spray. Contact between water and pooled ethylene should be avoided to prevent increased vaporization, unless the vapors can be controlled. Water should be applied to fire-exposed containers and the flow of escaping gas should be stopped if possible.

Hydrogen

UN 1038
Cryogenic Liquid

UN 1962
Compressed Gas

UN 1066
Cryogenic Liquid

UN 1049
Compressed Gas

Classifications: Flammable, compressed, cryogenic, industrial.

Chemical properties: Hydrogen is a basic chemical element. Hydrogen has an extremely wide flammable range and the highest burning velocity of any gas. Its ignition temperature is reasonably high, but its ignition energy is very low. Because hydrogen contains no carbon, it burns with a nonluminous flame which is often invisible in daylight. Hydrogen is nontoxic.

Physical properties: See Table 4-7A.

Usage: Hydrogen is principally used in chemical processing, for hydrogenation of edible oils, in welding and cutting, as a metal heat-treating special atmosphere, as a coolant in large electrical generators, and in the manufacture of semiconductors.

Hazards inside containers: Hydrogen is shipped as a compressed gas in uninsulated DOT and TC cylinders, as a cryogenic gas in insulated DOT and TC cylinders, and in insulated cargo trucks and railroad tank cars according to DOT specifications and regulations. It is stored in cylinders or in ASME-code insulated tanks.

Compressed hydrogen cylinders are protected against overpressure by frangible (bursting) discs or such discs in combination with fusible plug devices. Insulated cylinders, trucks and railcar tanks, and storage tanks are protected by pressure-relief valves and frangible disks. Compressed gas cylinders are subject to failure, but not to BLEVEs because they do not contain liquid. Insulated cryogenic containers are theoretically subject to BLEVEs, but NFPA has no record of such an occurrence.

Hydrogen is also shipped as a refrigerated liquid at about its normal atmospheric boiling point of –423°F (–253°C) in insulated containers. These are pressure vessels with reclosing pressure-relief valves, which will open to relieve the pressure that develops when heat leakage into the container results in boiling of some of the stored hydrogen. When this occurs, a visible cloud will develop as the air is cooled. Do not spray water in the vicinity of the discharge point, as it can freeze and block this important safety procedure.

Hazards when released from containment: Hydrogen presents both combustion explosion and fire hazards when released from containment. Although its wide flammable range and high burning rate accentuate these hazards, its low ignition energy, low heat of combustion on a volume basis, and its nonluminous (low thermal radiation level) flame exert counteracting influences in many instances.

Because of its low ignition energy, when gaseous hydrogen is released at high pressure, nominally small heat-producing sources, e.g., friction and static generation, can result in prompt ignitions. On this basis, hydrogen may be frequently thought of as self-igniting under these circumstances. The record of releases in high-pressure applications reveals that fires rather than combustion explosions occur. When hydrogen is released at low pressures, however, self-ignition is unlikely. Rather, hydrogen combustion explosions occur that are characterized by very rapid pressure rises which are extremely difficult to vent effectively. Open-air or open-space explosions have occurred from large releases of gaseous hydrogen.

Because of its very low boiling point, contact between liquid hydrogen and air can result in condensation of air and its oxygen and nitrogen components. A mixture of hydrogen and liquid oxygen is potentially explosive, even though the quantities involved are likely to be small. Reported accidents from this source have been generally restricted to the interiors of liquefaction equipment and to small containers of liquid hydrogen that are handled in the open atmosphere.

At ordinary temperatures, hydrogen is very light, weighing only about $1/15$ as much as air. The accordingly high diffusion rate makes it difficult for hydrogen to accumulate in conventional structures, unless the escape rate is high. This tends to reduce its combustion explosion hazard.

Emergency control: Escaping gaseous hydrogen seldom presents a "no fire" emergency situation, because it either ignites promptly or rises in the atmosphere rapidly. Hydrogen gas vaporizing from the cryogenic liquid near its normal boiling point is slightly heavier than air at 70°F (21°C); this fact, together with the visible fog of condensed water vapor created, causes it to spread along the ground for sizable distances (depending upon leak size and meteorological conditions). Because of the low gas density of vapors produced from vaporizing cryogenic hydrogen liquid, impounding or diked areas have not been required.

Ignitible mixtures can extend well beyond the visible cloud. Such escapes can be controlled by water spray. Contact between water and pooled hydrogen should be avoided to prevent increased vaporization, unless the vapor can be controlled.

Water should be applied to fire-exposed containers and the flow of gas stopped, if possible. Because hydrogen burns with a flame that is often invisible in daylight and because its flames produce low levels of thermal radiation, people have actually walked into its flames. When approaching hydrogen fires, a useful technique is to hold a broom out in front, while progressing slowly and at the same time throwing dirt ahead. The combustibles in the dirt will incandesce and locate the edge of the fire.

Liquefied Natural Gas (LNG)

UN 1972

Classifications: Flammable, cryogenic, fuel.

Chemical properties: LNG is a mixture of materials all comprising carbon and hydrogen. The principal component is methane, with lesser amounts of ethane, propane, and butane. The composition will vary, depending principally upon whether the source of the natural gas (which has been liquefied) is a transmission pipeline or gas wells. In the former instance, more of the propane and butane is removed prior to introduction into the pipeline. (See Table 4-7D.)

LNG is nontoxic but is a simple asphyxiant.

Physical properties: See Table 4-7D.

Usage: LNG is used as a source of natural gas to augment pipeline supplies during periods of extreme demand (peak shaving), to supply gas distribution systems in areas remote from central distribution systems, and as a basic supply of natural gas. To a small extent, it is used as a vehicle propulsion fuel.

Hazards inside containers: LNG is shipped as a cryogenic gas in insulated cargo trucks built to DOT specifications and regulations and in marine vessels under DOT authorization. It is stored in insulated ASME-code or API tanks. LNG containers are protected against overpressure by pressure-relief valves. Such containers are

TABLE 4-7D. *Approximate Properties of LNG*

Composition	
Methane	83–99%
Ethane	1–13%
Propane	0.1–3%
Butane	0.2–1.0%
Physical Properties	
Normal boiling point (NBP)	−250° to −260°F (−160 to −164°C)
Density liquid at NBP	$3^1/_2$ to 4 lb per gal (0.42 to 0.48 g/cm^3)
Density vapor at NBP (compared with air at 70°F or 21°C)	1.47
Liquid to vapor expansion	600 to 1
Heat of vaporization	220–248 Btu/lb (512–577 kJ/kg) 770–990 Btu/gal (215–276 mJ/m^3)
Combustion Properties	
Flammable range	5–14% (methane at normal temperatures) 6–13% (methane near −260°F or −127°C)
Heat of combustion	22,000 Btu/lb (51.2 mJ/kg)
Burn rate, steady-state pool	0.2–0.6 in. (5 to 15 mm) per minute

theoretically subject to BLEVEs, but NFPA has no record of such an occurrence.

Hazards when released from containment: LNG presents both combustion explosion and fire hazards when released from containment. At the present time, LNG is seldom used indoors and when so used the structure is designed for a combustion explosion hazard in accordance with national standards and regulations. Available test and experience data indicates that escaping LNG is not subject to open-air or open-space explosions.

Emergency control: Escaping LNG presents both "no fire" and "fire" emergency situations. LNG gas vaporizing from the cryogenic liquid near its normal boiling point is about $1^1/_2$ times heavier than air is at 70°F (21°C) and will spread along the ground accompanied by the visible fog created by condensed water vapor. This distance will depend upon the leak size and meteorological conditions and upon the geometry of the required liquid-impounding area provided. Ignitible areas are roughly defined by the visible cloud, but can extend beyond the visible area. Such escape can be controlled by water spray. Contact between water and pooled LNG should be avoided to prevent increased vaporization, unless the vapor can be controlled. Water should be applied to fire-exposed containers and the flow of gas stopped, if possible.

Liquefied Petroleum Gases (LP-Gas)

UN 1075

Classifications: Flammable, liquefied (including cryogenic), fuel.

Chemical properties: LP-Gas is a mixture of hydrocarbon materials. In commerce, LP-Gas is predominantly either propane or normal butane or mixtures of these with smaller amounts of ethane, ethylene, propylene, iso-butane, and butylene (including isomers). Principal variations in composition occur depending upon whether the source is gas wells or petroleum refineries. LP-Gas is nontoxic but is a simple asphyxiant.

Physical properties: See Table 4-7E.

Usage: LP-Gas is used principally as a domestic, commercial, agricultural, and industrial fuel gas, in chemical processing, and as an

TABLE 4-7E. *Approximate Properties of LP-Gases*

	Commercial Propane	Commercial Butane
Vapor Pressure in psig at:		
70°F	127	17
100°F	196	37
105°F	210	41
130°F	287	69
Vapor Pressure in kPa at:		
20°C	895	103
40°C	1,482	285
45°C	1,672	345
55°C	1,980	462
Specific Gravity Liquid at 60°F (15.5°C)	0.509	0.582
Initial Boiling Point at 14.7 psia	−44°F	15°F
Initial Boiling Point at 101 kPa	−42°C	−9°C
Weight per gal of Liquid at 60°F, lb	4.20	4.81
Weight per m^3 of Liquid at 15.5°C	504 kg	582 kg
Specific Heat of Liquid, Btu/lb 60°F	0.630	0.549
Specific Heat of Liquid, kJ/kg at 15.5°C	1.46	1.28
Cu ft of Vapor per gal at 60°F	36.38	31.26
m^3 of Vapor per L at 15.5°C	0.271	0.235
Cu ft of Vapor per lb at 60°F	8.66	6.51
m^3 of Vapor per kg at 15.5°C	0.534	0.410
Specific Gravity of Vapor (Air = 1) at 60°F (15.5°C)	1.50	2.01
Ignition Temperature in Air	920–1120°F (493–549°C)	900–1000°F (482–538°C)
Maximum Flame Temperature in Air	3595°F (1980°C)	3615°F (1990°C)
Limits of Flammability in Air, Percent of Vapor in Air-Gas Mixture:		
(a) Lower	2.15	1.55
(b) Upper	9.60	8.60
Latent Heat of Vaporization at Boiling Point:		
(a) Btu per lb	184	167
(b) Btu per gal	773	808
(c) kJ/kg	428	388
(d) kJ/L	216	226
Total Heating Values after Vaporization:		
(a) Btu per cu ft	2,488	3,280
(b) Btu per lb	21,548	21,221
(c) Btu per gal	91,547	102,032
(d) kJ/m^3	92,430	121,280
(e) kJ/kg	49,920	49,140
(f) kJ/m^3	25,140	28,100

engine fuel. In domestic and recreational applications, it is sometimes known as bottled gas.

Hazards inside containers: LP-Gas is shipped as a liquefied gas in uninsulated DOT and TC cylinders and in ASME tanks and in DOT-specification cargo trucks, railroad tank cars, and marine vessels. It is also shipped at low temperatures near its normal boiling point at atmospheric pressure in insulated marine vessels. It is stored in cylinders, ASME-code tanks, and insulated API tanks.

LP-Gas containers are generally protected against overpressure by pressure-relief valves, although some cylinders are protected by fusible plug devices and, occasionally, by a combination of these. Most containers are subject to BLEVEs.

Hazards when released from containment: LP-Gas presents both combustion explosion and fire hazards when released from containment. Based on reported incidents, the combustion-explosion experience dominates. This hazard is accentuated when LP-Gas is used indoors—reflecting the fact that 1 gal (3.78 L) of liquid propane or butane will produce about 245 to 275 gal (927 to 1,041 L) of gas. For this reason, standards and codes severely restrict the uses of liquid-phase LP-Gas indoors. Large releases of LP-Gas outdoors have led to a few reported open-air detonations.

Emergency control: Escaping LP-Gas presents both "no fire" and "fire" emergency situations. LP-Gas vapor is normally $1^{1}/_{2}$ to 2 times heavier than air, and the vapor produced as LP-Gas vaporizes from the liquid at its normal boiling point is even heavier. Therefore, it will tend to spread along the ground accompanied by the created visible fog of condensed water vapor. Ignitible mixtures extend beyond the visible area. Such escape can be controlled by water spray. When propane is stored and handled at atmospheric temperatures, it is unlikely to pool, except under very low ambient temperature conditions. Butane stored and handled at atmospheric temperatures and low-temperature LP-Gas are likely to pool. Contact between water and pooled LP-Gas should be avoided to prevent increased vaporization, unless the vapor can be controlled. Water should be applied to fire-exposed containers and the flow of gas stopped, if possible.

Methylacetylene-Propadiene, Stabilized (MPS)

UN 1060

Classifications: Flammable, liquefied, industrial.

Chemical properties: Methylacetylene-propadiene, stabilized, is a mixture of materials all comprised of carbon and hydrogen. In commerce, and in compliance with standards covering its use as a metal-cutting gas, MPS has composition per Table 4-7F.

Methylacetylene (also known as "propyne") and propadiene (also known as "allene") contain double and triple chemical bonds and are unstable, reactive materials presenting hazards similar to acetylene. The achievement of an adequate degree of stability for this product depends upon the existence of a sufficient quantity of the propane, butane, and iso-butane components of the mixture. MPS is marketed under various trade names, including "MAPP

GAS" (Monsanto Chemical Co.) and "APACHE GAS" (Air Products and Chemicals, Inc.).

MPS has a degree of chemical reactivity with metals, similar to acetylene—especially as it pertains to the use of copper and copper alloys. (See "Acetylene" in this chapter.)

MPS is nontoxic but has some anesthetic and narcotic effects. It has a strong odor detectable at concentrations as low as 100 ppm in air. (See Tables 4-7C and 4-7G.)

Physical properties: See Table 4-7G.

Usage: MPS is principally used as a metal-cutting fuel gas with oxygen.

Hazards inside containers: MPS has physical properties very similar to propane and is shipped as a liquefied gas in uninsulated DOT and TC cylinders and in DOT-specification cargo trucks and tank cars. It is stored in cylinders and ASME-code tanks. Its hazards inside containers are analogous to propane. The pressure of MPS in piping systems is not restricted.

Hazards when released from containment: MPS presents combustion explosion and fire hazards when released from containment. It is easier to ignite (due to a lower ignition energy) than propane

TABLE 4-7F. *MAPPS Composition*

1. Methylacetylene-propadiene (in combination with a maximum ratio of 3.0 moles of methylacetylene per mole of propadiene in the initial liquid-phase in a storage container)	–68 mole percent *maximum*
2. Propane, butane, isobutane (in combination)	–24 mole percent *minimum*, of which at least one-third (8 mole percent of total mixture) shall be butane and/or isobutane
3. Propylene	–10 mole percent *maximum*
4. Butadiene	–2 mole percent *maximum*

TABLE 4-7G. *Selected Properties of MPS**

Molecular Weight (average-stabilized mixture)	42.3
Heat of Vaporization	227 Btu/lb (528 kJ/kg)
Gross Heat of Combustion	21,700 Btu/lb (50.5 mJ/kg)
Normal Boiling Range, 1 Atm.	–36 to –4°F (–37.8 to –20°C)
Flame Temperature in Oxygen	5301°F (2927°C)
Specific Heat, Liquid Cp—70°F, (21°C) 1 Atm.	0.362 Btu/lb°F [1.5 J/(kg/°K)]
Specific Heat, Liquid Cv—70°F, (21°C) 1 Atm.	0.277 Btu/lb°F [0.9 J/(kg/°K)]
Critical Temperature	245°C (118°C)
Critical Pressure	752 psi (5.2 MPa)
Burning Velocity	15.4 ft/sec (4.7 m/sec)
Explosive Limits in Oxygen	2.5 to 60%
Vapor Pressure at 70°F (21°C)	94 psi (648 kPa)
Specific Gravity of the Liquid (60/60°F)	0.576
Specific Volume of the Gas at 60°F (21°C), 1 Atm.	8.85 cu ft/lb (0.55 m3/kg)
Heat of Formation	1075 Btu/lb (2.5 mJ)
Minimum Ignition Energy	0.2–0.25 mJ

* MAPP GAS (Dow Chemical Company).

and burns more rapidly; thus, it is similar to ethylene in these respects.

MPS is classified as a Group C material for purposes of suitability of electrical equipment installed in classified hazardous areas.

Emergency control: See "Liquefied Petroleum Gases" in this chapter.

Oxygen

UN 1073   UN 1072

Cryogenic Liquid   Compressed Gas

Classifications: Nonflammable (oxidizing), compressed, cryogenic, industrial, medical.

Chemical properties: Oxygen is a basic chemical element. It reacts with practically all materials and the general reaction is known as oxidation. Combustion is a particular type of oxidation reaction. In most combustion reactions, the oxygen is accompanied by nitrogen, in a mixture known as air. Nitrogen contributes nothing to the combustion reaction and actually inhibits it. Therefore, concentrations of oxygen in excess of its concentration in air increase most combustion hazards to a degree directly related to the concentration. This affects all basic combustion parameters, except for the heat of combustion. For example, ignition temperatures and energies are lowered, the flammable range is widened, and the burning rate is increased as the oxygen concentration increases—the ultimate effects occurring at an oxygen concentration of 100 percent. In consideration of these properties, the design of systems containing 100 percent oxygen requires particular attention to these factors from a compatibility standpoint.

Physical properties: See Table 4-7A.

Usage: Oxygen is primarily used in the production of steel, in metal welding and cutting, in medical applications, in life-support systems, in aeronautic and aerospace applications, and in chemical processing.

Hazards inside containers: Oxygen is shipped as a compressed or cryogenic gas in DOT and TC cylinders and in tank cars and trucks in accordance with DOT specifications and regulations. It is stored in DOT-specification cylinders (insulated or uninsulated) or in ASME-code insulated tanks.

Compressed-oxygen cylinders are protected against overpressure by frangible (bursting) discs or a combination of these with fusible plugs. Insulated cryogenic containers are protected with pressure-relief valves. Metals for containers and piping must be carefully selected, depending on service conditions. The various steels are acceptable for many applications, but some service conditions may call for other materials (usually copper or its alloys) because of their greater resistance to ignition and lower rate of combustion.

Similarly, materials that can be ignited in air have lower ignition energies in oxygen. Many such materials may be ignited by friction at a valve seat or stem packing or by adiabatic compression produced when oxygen at high pressure is rapidly introduced into a system initially at low pressure.

There are few reported cases of oxygen container failures from fire exposure—probably reflecting standards that require separation of containers from flammable gases and other combustibles. Most oxygen system components have failed due to accumulations of grease, oil, etc., on the surfaces of these components in contact with oxygen, reflecting poor housekeeping practices. Since materials such as grease are easily ignited in air, they are especially ignitible in 100 percent oxygen. Their ignition often results in combustion of system components that are noncombustible in air, including metallic components. Such incidents usually involve only small portions of such components, but may occur in a spectacular way, causing property damage and personal injury. They are referred to conventionally as "flashes."

Hazards when released from containment: Release of compressed or liquefied oxygen is usually observed by acceleration of the fire in whatever is burning. Release of liquid oxygen in the absence of fire presents an increased possibility that an oxygen-fuel mixture may occur prior to ignition. If ignition delay occurs under these circumstances, an explosion may result. Almost any combination of liquid oxygen and a combustible material is potentially explosive due to the rapid combustion circumstances produced. Some commercial explosives have been derived in this manner.

Oxygen is slightly heavier than air at the same temperature.

Emergency control: Escaping oxygen presents basically a "no fire" situation, but can create a "fire" situation in the vicinity of fired equipment or arcing electrical equipment by causing these to get out of control. For example, metallic components of internal combustion engines have burned in the oxygen-rich atmospheres produced by escaping oxygen. Oxygen gas vaporizing from the cryogenic liquid near its normal boiling point is approximately four times heavier than air is at 70°F (21°C) and will spread along the ground accompanied by the visible fog of condensed water vapor that has been created. Such escape can be controlled by water spray. Contact between water and pooled liquid oxygen should be avoided to prevent increased vaporization, unless the vapor can be controlled. Water should be applied to fire-exposed containers and the flow of gas stopped, if possible.

Liquid oxygen spilled from containers will remain in liquid form for some time while slowly boiling. If it is in contact with any organic material, such as asphalt paving, the combination is shock sensitive and can detonate. The hazard exists only when liquid oxygen is present, and extreme caution is advised.

Utility Gases

Classifications: Flammable, fuel.

Chemical properties: The term "utility gas" is applicable to any flammable gas distributed by gas utilities as a fuel gas. Natural gas dominates this field today. Some LP-Gas is distributed by gas utilities. Most utilities use LP-Gas to augment natural gas supplies during short periods of peak demand which occur in unusually cold weather.

The creation of natural gas is the result of decomposition of organic material by heat, pressure, and bacteriological action in the absence of air, usually below ground. As it is evolved underground, natural gas consists of both flammable and nonflammable gases. The flammable gases are comprised of carbon and hydrogen and are principally methane and ethane with some propane, butane, and pentane. The nonflammable gases are principally nitrogen and carbon dioxide. In commerce, most of the nonflammable gases are removed prior to distribution so that the natural gas consists of about 70 to 90 percent methane, with the remainder mostly ethane. (See Table 4-7C.)

Natural gas is nontoxic but is an asphyxiant. This is in marked contrast to the manufactured gas formerly widely distributed as utility gas and which contained quantities of poisonous carbon monoxide.

Utility natural gas has no odor of its own and is generally odorized as distributed.

Physical properties: With the exception of liquefied natural gas (LNG), utility natural gas is distributed in piping as a compressed gas at pressures ranging from approximately 0.25 to 1,000 psi (1.72 to 6,895 kPa) and in cylinders at pressures up to 3,600 psi (24.8 MPa). The gas has a density of about $2/_3$ that of air.

Usage: Utility natural gas supplies about one third of the total fuel energy of the United States for domestic, commercial, and industrial heating and power. It is also being used as a motor fuel and is known as "compressed natural gas (CNG)." Natural gas (usually supplied from a gas utility distribution piping system) is compressed to 2,400 to 3,600 psi (16.5 to 24.8 MPa), stored in a fueling station, and dispensed into vehicle containers at the same pressure.

Hazards inside containers: Utility natural gas is distributed nearly exclusively by a network of more than 1 million miles (1 million miles equals 1.6 million km) of underground pipeline in the United States and Canada. It is transported from gas wells in producing areas in large-diameter transmission pipelines at pressures up to approximately 1,000 psi (6,895 kPa). These pipelines are tapped by utility gas companies, and pressures are reduced for distribution—generally to around 0.25 to 60 psi (1.72 to 414 kPa). Regulators and safety relief devices are used to control pressures.

Steel and cast-iron pipe are used extensively. In recent years, thermoplastic and thermosetting plastic piping have been used extensively in distribution piping at pressures up to approximately 100 psi (689.5 kPa).

BIBLIOGRAPHY

Reference Cited

1. CGA SB-4, *Handling Acetylene Cylinders in Fire Situations*, Compressed Gas Association, Inc., Arlington, VA, 1994.

NFPA Codes, Standards, and Recommended Practices

Reference to the following NFPA codes, standards, and recommended practices will provide further information on gases discussed in this chapter. (See the latest version of *The NFPA Catalog* for availability of current editions of the following documents.)

NFPA 12, *Standard on Carbon Dioxide Extinguishing Systems*
NFPA 49, *Hazardous Chemicals Data*
NFPA 50, *Standard for Bulk Oxygen Systems at Consumer Sites*
NFPA 50A, *Standard for Gaseous Hydrogen Systems at Consumer Sites*
NFPA 50B, *Standard for Liquefied Hydrogen Systems at Consumer Sites*
NFPA 51, *Standard for the Design and Installation of Oxygen-Fuel Gas Systems for Welding, Cutting, and Allied Processes*
NFPA 51A, *Standard for Acetylene Cylinder Charging Plants*
NFPA 51B, *Standard for Fire Prevention in Use of Cutting and Welding Processes*
NFPA 52, *Standard for Compressed Natural Gas (CNG) Vehicular Fuel Systems*
NFPA 54, *National Fuel Gas Code*

NFPA 58, *Standard for the Storage and Handling of Liquefied Petroleum Gases*
NFPA 59, *Standard for the Storage and Handling of Liquefied Petroleum Gases at Utility Gas Plants*
NFPA 59A, *Standard for the Production, Storage, and Handling of Liquefied Natural Gas (LNG)*
NFPA 68, *Guide for Venting of Deflagrations*
NFPA 70, *National Electrical Code®*
NFPA 99, *Standard for Health Care Facilities*
NFPA 302, *Fire Protection Standard for Pleasure and Commercial Motor Craft*
NFPA 328, *Recommended Practice for the Control of Flammable and Combustible Liquids and Gases in Manholes, Sewers, and Similar Underground Structures*
NFPA 704, *Standard for the Identification of the Fire Hazards of Materials*

Additional Reading

"Acetylene," Pamphlet G-1, Compressed Gas Association, Inc., Arlington, VA, 1990.
ANSI B31.8-1986, *Gas Transmission and Distribution Piping Systems*, American Society of Mechanical Engineers, NY, 1992.
Atallah, S., and Borows, K. A., "Survey of Fire Protection Systems at LNG Facilities, Tropical Report, July 1990-November 1990, Risk and Industrial Safety Consultants, Inc., Des Plaines, IL, Gas Research Inst., Chicago, IL, GRI-91/0028, April 5, 1991, 34 pages.
Barry, T. F., "Application of Fire and Explosion Risk Assessment to an LPG Bulk Storage Facility," ASTM STP 1150, American Society for Testing and Materials. Fire Hazard and Fire Risk Assessment, Sponsored by ASTM Committee E-5 on Fire Standards, December 3, 1990, San Antonio, TX, ASTM, Philadelphia, PA, 1990, pp. 183–206.
"BLEVE Causes Massive Damage in Gas Depot, Sydney, Australia, April 1, 1990," *Fire Protection*, Vol. 18, No. 3, Sept. 1991, pp. 13–15.
Burger, L. W., "Computer Modelling: Predicting Accidental Gas Releases to the Atmosphere," *Fire Protection*, Vol. 18, No. 3, Sept. 1991, pp. 18–19.
CGA G2.1/ANSI K61.1-1989, *Safety Requirements for the Storage and Handling of Anhydrous Ammonia*, Compresses Gas Association, Arlington, VA.
Cloud Explosions by Water Sprays," *Gas Engineering and Management*, Vol. 31, No. 7-8, July/Aug. 1991, pp. 166–172.
Code of Federal Regulations, Title 49, Parts 171–190.
Compressed Gas Association, *Handbook of Compressed Gases*, 3rd ed., Van Nostrand Reinhold, New York, 1990.
Cowley, L.T., and Pritchard, M. J., "Large-Scale Natural Gas and LPG Jet Fires and Thermal Impact on Structures," Third International Conference for Management and Engineering of Fire Safety and Loss Prevention: Onshore and Offshore, Elsevier Applied Science, New York, 1991, pp. 209–218.
Cowley, L. T. and Johnson, A. D., "Oil and Gas Fires: Characteristics and Impact," Work Package No. FL1, Shell Research Ltd., Chester, UK, Feb. 1991, 244 pages.
DeNevers, N., "Propane Over-Filling Fires," *Fire Journal*, Vol. 81, No. 5, 1987, p. 80.
Droste, B., "Fire Protection of LPG Tanks With Thin Sublimation and Intumescent Coatings," *Fire Technology*, Vol. 28, No. 3, Aug 1992, pp. 257–269.
Dunn, V., "BLEVE: The Propane Cylinder," *Fire Engineering*, Vol. 141, No. 8, 1988, pp. 64–70.
Fire, F. L., "Ethylene," *Fire Engineering*, Vol. 140, No. 8, 1987, p. 53.
Fire, F. L., "Liquified Petroleum Gas," *Fire Engineering*, Vol. 142, No. 1, 1989, pp. 51–55.
Friedman, D. M., Williams, T. A. and Farkhondehpay, D., "Safety Assessment for Indoor Refueling of Compressed Natural Gas Mass Transit Buses," *Industrial Fire Safety*, Vol. 2, No. 1, Jan./Feb. 1993, pp. 56–60.
Gratz, R., and Wagenknecht, M., "Hei be Oberflachen als Zundquelle fur reins Acetylen. [Hot Surfaces as Ignition Point for Pure Acetylene.]," *VFDB*, No. 3, February 1994, pp. 103–111.
Gubin, E. I., and Dik, I. G., "Flame Propagation in a Dusty Gas," *Combustion, Explosion and Shock Waves*, (English Translation), Vol. 23, No. 6, May 1988, pp. 685-690.
Holemann, H., "Wo gibt es brandschutztehnische Defizite im Bereich der offentlichen Gasversorgung? [Where are Fire Protective Deficits in Publics Gas Supplies?]," *VFDB*, No. 1, Feb. 1991, pp. 6–10.

Hustad, J. E., and Sonju, O.K., "Experimental Studies of Lower Flammability Limits of Gases and Mixtures of Gases at Elevated Temperatures," *Combustion and Flame*, Vol. 71, No. 3, Mar. 1988, pp. 283–294.

"Hydrogen," Pamphlet G-5, Compressed Gas Association, Inc., Arlington, VA, 1991.

Imai, T., "Catalytic Combustion of Combustible Gases Mainly Containing Methane," Mitsubishi Heavy Ind. Ltd., Japan, Japan Kokai Tokkyo Koho JP, November 10, 1995, pp. 1–4.

Isner, M. S., "Natural Gas Explosion in Motel, Hagerstown, Maryland, February 18, 1990," Fire Investigation Report, National Fire Protection Association, Quincy, MA, 1990, 28 pages.

Johannsson, O., "The Disaster of San Juanico," *Fire Journal*, Jan. 1986, p. 32.

Jones, B., "Acetylene Cylinders: A Swedish Approach," *Fire Engineers Journal*, Vol. 56, No. 180, January 1996, pp. 15–16.

Kirby, R. E., "Major Propane Gas Explosion and Fire, Perryville, Maryland, July 6, 1991," USFA Fire Investigation Technical Report Series, Federal Emergency Management Agency, Washington, DC, Report 053, 1992, 65 pages.

Lamkie, A. J., and Davis, D., "Night Into Day: The Edison, New Jersey, Gas Pipeline Explosion," *Fire Engineering*, Vol. 148, No. 5, May 1995, pp. 34–42.

Leiker, D., and Mesch, T., "Natural Gas Explosions Level Westminster Homes. On the Job: Colorado," *Firehouse*, Vol. 20, No. 7, July 1995, pp. 50–52, 54.

Lewis, B., and Von Elbe, G., *Combustion, Flames and Explosions of Gases*, 3rd ed., Academic Press, Troy, MD, 1987.

Liquified Petroleum Gas, John Wiley & Sons, Chichester, 1987.

Martinsen, W. E., *et al.*, "BLEVEs: Their Causes, Effects, and Prevention," *Hydrocarbon Processing*, Nov. 1986.

"Mapp Industrial Gas," Loss Prevention Data Sheet 7–94, Factory Mutual Research Corp., Norwood, MA.

McRae, M. H., "Anhydrous Ammonia Explosion in an Ice Cream Plant," *Plant/Operations Progress*, Vol. 6, No. 1, Jan. 1987, pp. 17–19.

Nat, A. G. S., "Computer Modelling: Watching Gas Releases Like a HAWK," *Fire Protection*, Vol. 18, No. 3, Sept. 1991, pp. 20–21.

National Transportation Safety Board, several reports of investigations of utility natural gas explosions and fires, National Transportation Safety Board, Washington, DC.

Nettleton, M. A., *Gaseous Detonations: Their Nature, Effects, and Control*, Chapman and Hall, New York, 1987.

Ohtani, H., Gon, L. S. and Uehara, Y., "Combustion Characteristics of Several Flammable Gases With Chlorine Trifluoride," *Proceedings* of the Fourth International Symposium on Fire Safety Science, Intl. Assoc. for Fire Safety Science, Boston, MA, 1994, pp. 493–502.Powell, F., "Can Non-sparking Tools and Materials Prevent Gas Explosions?," *Gaz-Eaux-Eaux Usees*, Vol. 66, No. 6, 1986, pp. 419–428.

Patrick, K. A., "Balboa Gas Line Rupture, ' Up Close and Personal '," *Fire Engineering*, Vol. 147, No. 8, Aug. 1994, pp. 62–64, 66–68.

Prugh, R. W., "Mitigation of Vapor Cloud Hazards," *Plant/Operations Progress*, Vol. 5, No. 3, July 1986, pp. 169–174.

Prugh, R. W., "Quantitative Evaluation of "BLEVE" Hazards," *Journal of Fire Protection Engineers*, Vol. 3, No. 1, 1991, pp. 9–24.

Ramachandran, G., "Risk of Explosion From Trains Carrying LPG and Other Dangerous Goods," *Fire Surveyor*, Vol. 19, No. 3, June 1990, pp. 8–13.

"Safety Relief Valves for Anhydrous Ammonia and LP-Gas," Standard for Safety, Underwriters Laboratories, Inc., Northbrook, IL, UL 132, 5th Edition, April 28, 1993, 15 pages.

Sato, K., "Extinguishment of Liquefied Gas Fires by Carbon Dioxide and Liquid Nitrogen," *Bulletin of Japanese Association of Fire Science and Engineering*, Vol. 40, No. 1, 1990, pp. 29–35.

Shebeko, Y. N., *et al.*, "Fire and Explosion Risk Assessment for LPG Storages," *Fire Science and Technology*, Vol. 15, No. 1/2, 1995, pp. 37–45.

Sidebotham, G., *et al.*, "Nitrous Oxide Anesthesia Creates Flammable Intentinal Gases," *Proceedings* of the 1995 Fall Technical Meeting of the Combustion Institute/Eastern States Section on Chemical and Physical Processes in Combustion, October 16–18, 1995, Worcester, MA, 1995, pp. 463–466.

Sumathipala, U. K., *et al.*, "Fire Engulfment of Pressure-Liquefied Gas Tanks: Experiments and Modeling," New Brunswick Univ., Fredericton, Canada, National Research Council of Canada, Ottawa, Ontario, Atomic Energy of Canada, Ltd., Manitoba, Canada, University of Science and Technology, Beijing, People' s Republic of China, ASTM STP 1150; American Society for Testing and Materials. Fire Hazard and Fire Risk Assessment, Sponsored by ASTM Committee E-5 on Fire Standards, ASTM STP 1150, December 3, 1990, San Antonio, TX, ASTM, Philadelphia, PA, 1990, pp. 100–115.

U.S. Dept. of Energy Research Reports, Contract DE-AC06-76 RLO 1830, Pacific Northwest Laboratory, Battelle Memorial Institute (NTIS)

Van den Berg, A. C., *et al.*, "Current Research at T.N.O. on Vapor Cloud Explosion Modeling," *Proceedings* of the International Conference on Vapor Cloud Modeling, Boston, 1987.

Van Wingerden, C. J. M., *et al.*, "Vapor Cloud Explosion Blast Prediction," *Plant/Operation Progress*, Vol. 8, No. 4, 1989, pp. 234–238.

VanTiggelen, P. J. and Thill, L., "Halon 1301 and Gaseous Detonations," *Proceedings* of the 1993 Halon Alternatives Technical Working Conference, May 11–13, 1993, Albuquerque, NM, 1993, pp. 481–491.

Weiss, E. S. and Sapko, M. J., "Oxygen Balanced Emulsion-Anfo Blends for Use in Flammable Atmospheres," Sixth Annual Symposium on Explosive and Blasting Research, 1990, pp. 193–204.

Wighus, R., "Active Fire Protection: Extinguishment of Enclosed Gas Fires With Water Sprays," Norwegian Fire Research Lab., Trondheim, Norway, STF25 A91028, June 1989, 97 pages.

Wighus, R., "Extinguishment of Enclosed Gas Fires With Water Spray," *Proceedings* of the Third International Symposium for Fire Safety Science, Elsevier Applied Science, New York, 1991, pp. 997–1006.

Woodward, F. P., Hansen, L. E. and Melton, D. A., "Cryogenic Liquid Oxygen Container Explosion: An Investigation," *Fire Engineering*, Vol. 147, No. 1, Jan. 1994, pp. 59–63.

PNL-4401, LNG Annotated Bibliography

PNL-4398, LNG Fire and Vapor Control System Technologies

PNL-3991, Assessment of Research and Development

MEDICAL GASES

Revised by Gerald L. Wolf, MD

The use of oxidizing gases in hospitals represents a fire hazard different from that of any other environment. Since many patients cannot be readily moved because of dependency on life-support equipment, immobility becomes lethal in a fire emergency.

The recognition of these hazards, particularly extreme in the anesthetized patient, instituted a gradual shift to nonflammable volatile anesthetic agents after World War II, which was partially offset by the proliferation of other potential fuels, such as disposables made of paper and plastics. The increased use of oxidizers, e.g., oxygen and nitrous oxide (N_2O), and the increased reliance on potential ignition sources, e.g., electrical appliances, continued to put the hospitalized patient at increased risk. Thus, while practice has changed, the fire triangle (i.e., ignition source, oxygen, and fuel) persists, albeit in altered form, and the hospital represents a fire safety problem unique in society. This chapter discusses those parts of the health care fire safety problem related to the storage and use of medical gases.

The safe practices and hazards discussed in this chapter apply to all health care facilities that use medical gases, e.g., hospitals, ambulatory health care centers, nursing homes, dental offices, etc., and also have relevance to the home health care environment.

Pertinent information for those concerned with the problem and hazard of medical gases and equipment can be found in the following chapters: Section 3, Chapter 2, "Control of Electrostatic Ignition Sources"; Section 3, Chapter 21, "Storage of Flammable and Combustible Liquids"; Section 3, Chapter 22, "Storage of Gases"; Section 4, Chapter 1, "Fire Hazards of Materials: An Overview"; Section 4, Chapter 6, "Flammable and Combustible Liquids"; Section 4, Chapter 7, "Gases"; Section 4, Chapter 9, "Oxygen-Enriched Atmospheres"; and Section 9, Chapter 26, "Fire Protection of Laboratories Using Chemicals."

ANESTHETIC AGENTS

The introduction of diethyl ether as an anesthetic in 1846 led to dramatic changes in surgical practice. Its benefits were so obvious that general anesthesia rapidly became an accepted technique. Diethyl ether had many drawbacks, however, including flammability, and the search for better agents led to the development and introduction

of many. Initially, the agents developed were also volatile liquids, i.e., agents stored in the liquid phase and converted to the vapor phase in vaporizers. With the advent of practical methods for compressing, storing, and delivering gases in regulated amounts, the use of so-called true gases (e.g., nitrous oxide and cyclopropane) became popular.

Enrichment of anesthetic atmospheres with oxygen occurred nearly simultaneously with the introduction of nitrous oxide and allowed anesthetists to administer appropriate concentrations of anesthetic to patients while, at the same time, maintaining adequate levels of oxygenation. The practice of using oxygen as a component of inhalation anesthesia persists to this day.

Flammability of Anesthetics

Anesthesiologists in the Western world have nearly completely abandoned the use of flammable anesthetic agents. During earlier times, however, most anesthetic agents were flammable, and resulted in a number of fires and explosions, accompanied by injuries and fatalities. Diethyl ether, premixed with air, burns briskly, while diethyl ether premixed with oxygen explodes violently. Cyclopropane-oxygen mixtures create even more violent explosions, since the concentration used for anesthesia is frequently close to stoichiometric.

Former flammable agents included the true gases, e.g., cyclopropane and ethylene; and the volatile liquids, e.g., diethyl ether, divinyl ether, and ethyl chloride. Nonflammable anesthetic agents include the true gas, nitrous oxide; and the volatile liquids, halothane, enflurane, isoflurane, desflurane, and sevoflurane.

The method to reduce the flammability of a hydrocarbon involves substituting halogens (usually fluorine or chlorine) for hydrogen atoms. Methoxyflurane and halothane were the first nonflammable halogenated hydrocarbon anesthetics developed, followed by enflurane, isoflurane, and then desflurane and sevoflurane.

When methoxyflurane, halothane, or trichloroethylene are heated, they burn. However, under usual anesthetic conditions of use, and in the absence of heating to ignition temperature, these agents are considered nonflammable. This is also true for the newer agents, sevoflurane and desflurane. Sevoflurane is reported to have a lower flammability limit of 11 percent in oxygen and 10 percent in nitrous oxide. In oxygen, desflurane is reported to have a lower flammability limit of 11 percent and an upper flammability limit of 21 percent. In nitrous oxide, the reported lower flammability limit is 18 percent, and the upper flammability limit is 27 percent.

Gerald L. Wolf, MD, is an anesthesiologist at SUNY/Health Science Center at Brooklyn in Brooklyn, NY. He is chairman of the Gas Delivery Equipment Committee and member of the Laser Fire Protection and Oxygen-Enriched Atmospheres Committees.

Nitrous oxide and oxygen, both oxidants, vigorously support combustion. The oxygen in the nitrous oxide molecule is released under flame conditions, contributing to the oxidation-reduction reaction, and the released heat of dissociation contributes to the damage created by the fire.

Safe Practices in Using Anesthetics

A consideration of safe practices in anesthesia begins with an understanding of the basic chemistry and physics of oxygen and the anesthetic gases, as well as of the vapors produced by the volatile liquid agents.

A gas consists of individual molecules moving in a linear direction at high velocity. The molecular movement increases with a rise in temperature and decreases with cooling. This movement produces a parallel increase or decrease in the pressure exerted by that gas, dependent upon its temperature. When a gas is compressed, its individual molecules are forced closer together, thus increasing molecular movement and temperature. Gases possess definite mass, but neither definite shape nor definite volume (i.e., they assume the shape and volume of their containers, if any).

The relationship between pressure, volume, and temperature is defined by various gas laws. From these gas laws the general gas law can be established; i.e., the pressure of any given quantity of gas is proportional to its absolute temperature and inversely proportional to its volume. Algebraically, pressure times volume divided by temperature equals a constant. These gas laws assume the gas to be ideal, while in actuality no gas is ideal.

The anesthetic agent nitrous oxide is stored in cylinders under pressure sufficient to liquefy the gas; therefore, cylinders of compressed nitrous oxide contain both gas and liquid. The anesthetic is dispensed as a gas through needle valves and flowmeters. In contrast, liquid anesthetic agents, such as halothane and isoflurane, are stored as liquids that then volatilize in vaporizers during use.

A liquid possesses volume and mass but no definite shape. At any gas-liquid interface, molecules are continually escaping from the liquid state into the gaseous state. The pressure created by the molecules that escape into the gaseous state is called the vapor pressure. As the temperature of a liquid is increased, its vapor pressure increases and reaches atmospheric pressure at its boiling point.

The use of vaporizers during anesthesia delivery decreases the temperature of the liquid anesthetic. The carrier gases (nitrous oxide and oxygen) used to deliver inhalation anesthetic agents to patients pass through the vaporizer in a steady flow. These carrier gases carry the vaporized molecules of the volatile anesthetic to the anesthesia circuit. In the absence of applied heat, the temperature of the volatile liquid remaining in the vaporizer, and the temperature of the reservoir itself, falls as the volatile liquid vaporizes (i.e., latent heat of vaporization). As the temperature of the remaining liquid falls, so does its volatility. Some anesthetic vaporizers are designed to maintain the temperature of the liquid at or near room temperature so that constant rates of vaporization are attained. Other vaporizers control the flow of diluent gases by means of a bimetallic flow controller that alters flow as temperature changes, so that constant anesthetic concentration is maintained.

Desflurane, an anesthetic with a vapor pressure of 26 in. Hg at 68°F (664 mm Hg at 20°C), is back pressurized to 61 in. Hg (1550 mm Hg) in the vaporizer in order to prevent boiling. The vaporizer is also heated to 73 to 77°F (23 to 25°C) by means of a heating coil that supplies the required calories for heat of vaporization of the high-vapor-pressure anesthetic. The electrically powered heating unit must not achieve a temperature in any of its components approaching the ignition temperature of any potential fuel in the oxi-

dant atmosphere of oxygen and nitrous oxide. That temperature has been empirically set at 572°F (300°C).

Anesthesia Workstation

Anesthesia workstation refers to the combination of an anesthesia machine and incorporated monitors. Anesthetic agents are administered using an anesthesia machine. (See Figures 4-8A and 4-8B.) The machine consists of a number of components. A source of oxygen and anesthetic nitrous oxide gas is required, from either gas cylinders, connection to a central piping system, or, as is usually the case, from both of these sources. The medical gas cylinders contain gases that are either nonliquefied or liquefied. A pressure regulator to reduce high-pressure gases to a working pressure, usually about 50 psi (345 kPa), is also required. Gases are piped to needle valves, which control the flow of the gas delivered to the anesthesia circuit. To measure the flow of these gases, each gas passes through a flowmeter.

Rotameters (Thorpe tubes) are the most common flowmeter used in anesthesia machines. A Thorpe tube is a tapered glass tube containing a bobbin or rotor which floats in the gas stream. The space between the outside of the rotor and the inside of the glass tube is an annular orifice which increases as the rotor rises in the tube. When the gas passes across the annular orifice, a fall in pressure occurs. The difference between the upstream and downstream pressures is equal to the weight of the rotor or bobbin at equilibrium. Flowmeters are calibrated for specific gases by passing the specific gas initially through a master flowmeter, then through the flowmeter

FIG. 4-8A. A typical operating room with anesthesia equipment on both sides of the O/R table.

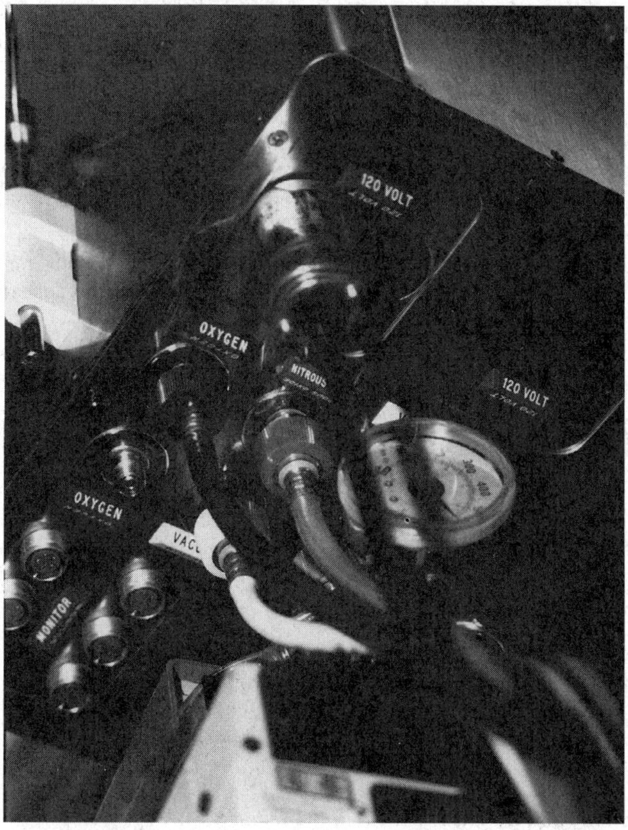

FIG. 4-8B.   A typical overhead utility supply station that provides electricity, gases, and vacuum for anesthesia equipment.

to be calibrated. Flows are indicated by noting the position of the rotor in the flowmeter being calibrated.

From the flowmeters, gases enter a common mixing manifold. The mixture of gases then enters the anesthesia circuit. The anesthesia circuit is the portion of the anesthesia machine that administers the anesthetic directly to the patient by inhalation and into which the patient breathes during exhalation. During exhalation, carbon dioxide, a product of metabolism, enters the anesthesia circuit and is absorbed so that rebreathing of carbon dioxide does not occur. Absorption is by chemical reaction, employing an absorber that contains 1.3 to 0.3 gal (0.5 to 1 L) of granular soda lime (a mixture of sodium and calcium hydroxides, water, a color indicator, and a binder).

Valves in the anesthesia circuit permit the patient to breathe in unidirectional fashion, i.e., from the rebreathing bag to the lungs for inhalation, and from the lungs through the absorber to the rebreathing bag for exhalation. The vaporizer is located between the gas delivery of the anesthesia machine and the anesthesia circuit.

In an attempt to minimize ignition of combustibles within the piping and circuitry of the anesthesia machine, it has been suggested that the maximum temperature of potential ignition sources be limited to below the ignition temperature of those combustibles, at the oxidizing atmosphere present at both normal- and single-fault condition.

Anesthesia machines may incorporate various add-on monitoring devices to monitor function of the machine, and patient physiologic parameters. The monitoring devices, electronically based, present potential electric hazards, such as shock, burn, and ignition.

OXYGEN

Oxygen as a Medical Agent

In the early days of inhalation anesthesia, diethyl ether was administered by the open-drop technique. Dropped onto a gauze-covered mask, the ether volatilized in the air of the operating room atmosphere. Oxygen enrichment was later introduced. In the presence of an oxygen-enriched atmosphere, the lower limit of flammability and the ignition temperature of flammable agents are reduced. If such a mixture approaches stoichiometric concentration, i.e., the proportion of oxygen and fuel at which complete combustion of fuel occurs with neither fuel nor oxygen in excess, the most violent of combustion events occurs.

From a practical standpoint, oxygen is routinely used for many therapeutic purposes. In anesthesia, it is used to create a wider margin of safety for the patient undergoing anesthesia and surgery. In respiratory therapy, oxygen-enriched atmospheres are commonly used. Hyperbaric chambers, for various forms of hyperbaric medicine, also utilize oxygen-enriched atmospheres either in the form of compressed oxygen or air. Thus, recognizing that oxygen-enriched atmospheres are present in various therapeutic applications, fire or explosion prevention in health care facilities must take into account the control of potential fuels and ignition sources.

The concept of the problem of oxygen enrichment in relation to the medical gas fire problem must be extended to encompass the other commonly used oxidant, nitrous oxide. Since nitrous oxide vigorously supports combustion and also releases heat of decomposition, the fire hazard in N_2O is compounded. The widespread use of oxygen and nitrous oxide in operating rooms, in conjunction with the "wet environment" of the operating room, along with a wide array of potential electrical ignition sources, makes the operating room a particularly hazardous fire location.

Oxygen Supplies

Most oxygen for medical use is prepared by compression and liquefaction of air, followed by fractional distillation. Once purified, it is transported to and stored at the site of consumption as a compressed gas or as a liquid. Because of the economies in bulk oxygen use, almost all medium- to large-size hospitals rely on bulk oxygen storage on site. NFPA 50, *Standard for Bulk Oxygen Systems at Consumer Sites*, covers safety standards for the design and operation of such bulk oxygen storage units.

Oxygen concentrators separate air into its major components, nitrogen and oxygen, and then deliver oxygen of purity approaching 90 percent. Such oxygen concentrators are particularly useful when used on-site in facilities such as mobile military hospitals, hospitals in inaccessible geographic areas, and home health care environments.

Distribution Systems for Medical Gases

Distribution systems for medical gases consist of central supply systems with appropriate controls, piping systems, and terminal units at end-use points.

Gases are distributed within the hospital by piping systems dedicated for the individual gases, such as oxygen, compressed air, and nitrous oxide. Oxygen and nitrous oxide pipeline systems installed in health care facilities require the use of copper or brass tubing or pipe with screwed, brazed, or otherwise high-melting-point joints. The system must be pressure tested with oil-free dry air or nitrogen before use. [See NFPA 99, *Standard for Health Care Facilities* (hereinafter referred to as NFPA 99), for details.]

Since it is extremely difficult to extinguish a fire supported by a continuous source of oxygen or nitrous oxide, all piping systems containing an oxidizing gas for use in health care facilities require a remote shutoff valve to shut off the flow of gas to a station outlet or series of station outlets in the event of a fire.

When a system is installed that pipes more than one medical gas, care must be taken to check all station outlets for delivery of the correct gas *before* the system is put in use. Cross connections have accidentally been made, resulting in injury or death of patients from anoxia.

Specifications for the design, installation, testing, and operation of nonflammable medical piped gas systems for health care facilities are covered in Chapter 4 of NFPA 99. Certain Compressed Gas Association (CGA) publications cover requirements related to the installation of systems. (See bibliography for titles.)

Station outlets for medical gas supply systems must be equipped with noninterchangeable fittings, keyed to the specific gas. Station outlets also require a label with the name of the gas supplied to the outlet. Medical gas cylinder outlets are also keyed to the gas supplied. The Diameter Index Safety System is utilized for large-size gas cylinders and mating connectors. For small-size cylinders, the Pin Index System is employed. This system uses two holes drilled along a radius beneath the outlet orifice of the valve body, keyed to the specific gas. The cylinder yokes are equipped with pins that mate with the holes in the valve body of the cylinder. The holes and pins are located in such a manner that it is impossible to insert a cylinder in the yoke for a different gas.

The American standard for color coding of high-pressure gas cylinders includes green for oxygen, blue for nitrous oxide, brown for helium, and gray for carbon dioxide. Color combinations are available for gas mixtures of certain percentages. However, color coding of the contents of a cylinder is considered *secondary* to labeling (per NFPA 99). One rationale for this requirement is that persons who are color blind may not distinguish cylinder contents if color code is the only distinguishing requirement. Color code standardization does not exist internationally.

Cylinder weights, sizes, pressures, and contents are shown in Table 4-8A from Appendix C-12.5 of NFPA 99. Data for this appendix was supplied by the Compressed Gas Association.

In some health care facilities, oxygen is supplied to the piping system by a manifold to which cylinders are connected rather than by a liquid oxygen bulk source as described above. Another method occasionally used is individual oxygen cylinders at the site of use.

Storage of medical gas cylinders must be such that oxidizing gas cylinders, i.e., oxygen, compressed air, and nitrous oxide, are *not* stored with flammable gas cylinders.

Cylinder handling requires safety precautions. Large cylinders are heavy; mechanical damage, as well as personnel injury, may result if cylinders are dropped. Cylinders should be transported only on approved cylinder carts. When large cylinders are moved, the cylinder cap must be tightly affixed. Cylinders in use or in storage must be chained to a wall; affixed to a cylinder stand; or, in the case of small cylinders, securely attached to the equipment in use. Valves

TABLE 4-8A *Typical Medical Gas Cylinders—Volume and Weight of Available Contents.*‡*

All Volumes at 70°F (21.1°C)

Cylinder Style & Dimensions	Nominal Volume cu in./liter	Contents	Air	Carbon Dioxide	Cyclo-Propane	Helium	Nitrogen	Nitrous Oxide	Oxygen	Mixtures of Oxygen Helium	CO$_2$
B 3$^1/_2$" od × 13" 8.89 × 33 cm	87/1.43	psig		838	75				1900		
		Liters		370	375				200		
		lb-oz		1–8	1–7$^1/_4$				—		
		kilograms		.68	.66				—		
D 4$^1/_2$" od × 17" 10.8 × 43 cm	176/2.88	psig	1900	838	75	1600	1900	745	1900	**	**
		Liters	375	940	870	300	370	940	400	300	400
		lb-oz	—	3–13	3–5$^1/_2$	—	—	3–13	—	**	**
		kilograms	—	1.73	1.51	—	—	1.73	—	**	**
E 4$^1/_4$" od × 26" 10.8 × 66 cm	293/4.80	psig	1900	838		1600	1900	745	1900	**	**
		Liters	625	1590		500	610	1590	660	500	660
		lb-oz	—	6–7		—	—	6–7	—	**	**
		kilograms	—	2.92		—	—	2.92	—	**	**
M 7" od × 43" 17.8 × 109 cm	1337/21.9	psig	1900	838		1600	2200	745	2200	**	**
		Liters	2850	7570		2260	3200	7570	3450	2260	3000
		lb-oz	—	30–10		—	—	30–10	122 cu ft	**	**
		kilograms	—	13.9		—	—	13.9		**	**
G 8$^1/_2$" od × 51" 21.6 × 130 cm	2370/38.8	psig	1900	838		1600		745		**	**
		Liters	5050	12,300		4000		13,800		4000	5330
		lb-oz	—	50–0		—		56–0		**	**
		kilograms	—	22.7		—		25.4		**	**
H or K 9$^1/_4$" od × 51" 23.5 × 130 cm	2660/43.6	psig	2200			2200	2200	745	2200		
		Liters	6550			6000	6400	15,800	6900		
		lb-oz	—			—	—	64	244 cu ft		
		kilograms	—			—	—	29.1	—		

*These are computed contents based on nominal cylinder volumes and rounded to no greater variance than ±1%.
**The pressure and weight of mixed gases will vary according to the composition of the mixture.
‡This table reprinted with permission from the Compressed Gas Association, Inc.

on cylinders must not be damaged. If a valve breaks, the cylinder may be propelled by the release of the high-pressure gas. Serious injury or even death of personnel may result.

Oxygen is utilized in emergency service vehicles. Generally, small cylinders are employed. The same precautions must be exercised in emergency service vehicles as in a permanent facility.

Transfilling gas from cylinder to cylinder is a hazardous procedure and should be performed only by qualified personnel. It is not permitted in patient care areas of health care facilities. (See NFPA 99.) The Compressed Gas Association publishes safety guides for transfilling cylinders. (See bibliography.)

In addition to bulk storage of liquid oxygen, liquid oxygen within a hospital can also be distributed by portable liquid oxygen units.

RESPIRATORY THERAPY

Respiratory therapy began with the use of oxygen-enriched atmospheres in oxygen tents. Subsequently, various methods for the use of oxygen in medical care were developed, including oxygen hoods, incubators for newborns, oro-nasal or nasal masks, nasal catheters, and devices that mechanically assist or control breathing while delivering medical gases via tracheal tubes. This equipment, which intermittently generates pressure, delivers oxygen, air-oxygen mixtures, or oxygen mixed with therapeutic agents. These devices may be powered by a pressurized oxygen supply or by an electrically operated motor and pump system. Controls for such ventilators are either pneumatic, fluidic, or electronic. One type of respiration unit is shown in Figure 4-8C.

The above-mentioned combination of oxygen-enriched atmosphere and an electrically operated power supply, in addition to the proximity of potential fuels, has contributed to respiratory therapy equipment fires.

In oxygen-enriched atmospheres, combustible substances possess a lower ignition temperature and burn more rapidly than in air. Flame spread can be extremely rapid, especially if bedding or clothing is saturated with oxygen. The safe practices for the use of respiratory therapy equipment are detailed in Chapter 8 of NFPA 99.

Helium/oxygen mixtures are occasionally used in treating certain forms of respiratory distress and in anesthesia for laser airway surgery. This gas mixture should be treated as an oxygen-enriched atmosphere, and precautions noted above should be followed. It should also be noted that helium delays ignition of combustibles due to its higher thermal conductivity than that of nitrogen. However, if combustion occurs, this higher thermal conductivity creates a more rapid flame spread rate in helium than in nitrogen. In addition, since helium possesses a lower specific heat than nitrogen, the flame temperature will be higher in helium dilution than nitrogen. Thus, while helium delays ignition, if ignition does occur, the fire in helium atmosphere is more destructive than in nitrogen atmosphere.

COMPRESSED AIR

Compressed air has a wide variety of uses in medicine. It is used in hyperbaric chambers. If pure, it can be mixed with oxygen for ventilation of patient's lungs, thus avoiding the complications of pure oxygen breathing over prolonged periods of time. Compressed air is also used in various laboratory apparatus. Chapter 4 of NFPA 99 includes provisions for the installation of a central piping system for compressed air for medical purposes.

Atmospheric air compressed to greater than ambient pressure [i.e., greater than 14.7 psi (101 kPa)] constitutes an oxygen-enriched atmosphere. Although not as hazardous as pure oxygen, the

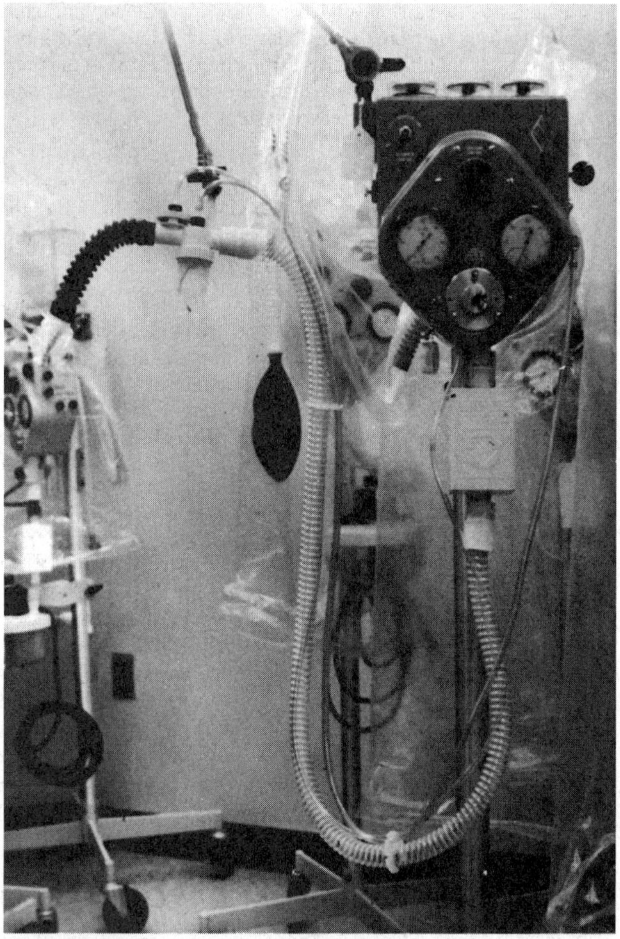

FIG. 4-8C. A respirator typically used for intensive-care patients. The unit is connected to the hospital compressed air and oxygen distribution system via quick-connect couplers on hoses at rear. Desired air-oxygen mixture is delivered to patient via bacteria filter and through the attached corrugated tubing.

ease of ignition of combustible substances is nevertheless increased, and flame spread rate is much greater than in air at ambient pressures. The fire hazard created by compressed air is thus significant. Fire extinguishment in compressed air atmosphere is difficult unless a water-deluge system is employed. Fire blankets and dry chemical agents are ineffective, since the burning continues in gas pockets that require the burning material be cooled by water.

If a fire occurs in electrically powered equipment in a compressed air atmosphere, it is necessary to first deenergize the equipment. The fire can then be extinguished with water.

Compressed Air Supplies

Compressed air is obtained from gas cylinders containing compressed air, from mixers that mix nitrogen and oxygen from cylinders, and from cylinders previously filled from separate oxygen and nitrogen sources at 1:4 proportion. If a compressor is used as a source of air for a piped medical air system in a health care facility, the system must not add impurities to the air, such as hydrocarbons, carbonaceous particulates, or carbon monoxide. To meet these specifications, at minimum the air supply must not come from the building in which the compressor is located.

In addition, the air must be free of any contaminant lubricants or products resulting from the breakdown of lubricants. Some lubricants release contaminants such as carbon monoxide under the pressure and temperature generated in an air compressor, which could be distributed in the compressed air system.

In a health care facility using a compressor to furnish air to the operating room and as an adjunct to respiratory therapy, electric power to the compressor must be wired to the critical branch of the essential electrical system of the facility. The specifications for this electric system are set forth in Chapter 3 of NFPA 99.

If a health care facility is located in an area in which the atmosphere is highly contaminated, it may be desirable to utilize compressed air from cylinders rather than compressing it on site. Some suppliers of cylinders may not have a pure source of atmospheric air to compress, and oxygen-nitrogen mixtures may then be utilized. This so-called "synthetic air" is purer than the ambient air, because it is made up of pure oxygen and nitrogen, and so contains no atmospheric pollutants.

Because of the problems of back firing, the medical compressed air system should not be used for any nonpatient purpose.

VACUUM SYSTEMS

In addition to central piped gas systems for compressed air, oxygen, and nitrous oxide, central vacuum systems are a standard feature in most health care facilities. Station inlets for the vacuum system are commonly installed in emergency rooms, operating rooms, recovery rooms, intensive care units, delivery rooms, nurseries, and patient rooms. While portable individual vacuum pump units have been used in the past, most health care facilities presently use central systems.

A central vacuum system requires both collection bottles and trap bottles at the station inlets. The collection bottles collect the fluids. The trap bottles ensure that none of the material so collected enters the suction line. Current practice dictates retaining the suction material at the patient care site, preventing contamination and cross-infection within the premises.

A central vacuum system requires adequate piping capacity and pump units of sufficient size to meet peak needs. At a minimum, two pumps are required, sized so that if one fails the other(s) can meet the need. The pumps are connected so that they will run alternately to provide even wear. Central vacuum systems are also required to be wired to the critical branch of the essential electrical system, since the system provides life support function in many situations.

A central vacuum system may also be used in laboratories and other hospital facilities. Such systems should not be connected to the medical-surgical vacuum system piping. However, a common receiver tank is permitted.

In the operation of a central vacuum system, it is important to avoid drawing flammable gases or liquids into the system. At the discharge end, the system should avoid contaminants returning into the facility. Requirements for vacuum systems in health care facilities are included in Chapter 4 of NFPA 99.

Waste Anesthetic Gas Disposal (Scavenging)

Accepted anesthesia practice dictates that waste anesthetic gas disposal (WAGD) be incorporated in anesthetizing locations, since pollution of the operating room may create an occupational hazard for personnel. Anesthesia techniques may utilize high gas flows that result in a significant portion of the excess gases being vented to the operating room atmosphere, thus creating the alleged hazard.

Waste anesthetic gas disposal systems include the collection system for the excess anesthesia gases, the system that transports the gases, and the discharge system. If a central vacuum system is to be used for WAGD, it must have a capacity greater than that for conventional operating room requirements (i.e., suctioning only surgical patients).

For more details on WAGD, ANSI Z79.11, *Anesthetic Equipment—Scavenging Systems for Excess Anesthetic Gases*, should be reviewed.

Housekeeping Vacuum Systems

In some health care facilities, a vacuum system for housekeeping purposes may exist. It is used in conjunction with vacuum equipment for cleaning corridors, drapes, and other related housekeeping chores. This system *must* be separate from the central vacuum system used for medical purposes.

HEALTH CARE FACILITY LABORATORIES

Health care facility laboratories may utilize a wide variety of gases and gaseous mixtures for chemical analysis, the generation of flames, and other purposes. Chapter 10 of NFPA 99 sets forth the design features and practices required for the safe operation of such a laboratory.

GAS STERILIZING

For many years, sterilization by either dry heat or steam was standard practice for the eradication of bacteria and other pathogens from reusable medical equipment. Ethylene oxide sterilization was then introduced, which employs ethylene oxide, an extremely volatile liquid that vaporizes readily at room temperatures. It is an effective bactericidal and virucidal agent.

To sterilize objects, the sterilizer chamber is first evacuated of air, then filled with ethylene oxide gas. The gas penetrates the material contained in the chamber and destroys bacteria, viruses, and other pathogenic organisms.

Ethylene oxide is highly flammable and chemically reactive, and proper operation of the equipment is required in order to avoid the danger of accidental ignition or explosion. The liquid is delivered to the facility in cylinders that must be stored properly, as is the case with any container of flammable liquid or gas. It also can be delivered mixed with other gases to reduce flammability. Ethylene oxide is toxic to humans, and proper venting of the ethylene oxide sterilizer to the outside of the building is mandatory.

Other chemical sterilizing and disinfecting agents have been used in the past. These have included chlorine, ozone, sulfur dioxide, and formaldehyde gas. Because of the problems in handling these agents, their corrosive nature, and the greater efficacy and safety associated with heat and ethylene oxide sterilization, these agents are no longer used for medical sterilization purposes.

FIRE PREVENTION AND FIRE RESPONSE

Fire and Explosion Prevention

Potential ignition sources in hospital oxygen-enriched atmospheres include cigarette smoking, electrosurgical units, lasers, fiberoptic light sources, heat of adiabatic compression of gases, electric sparks, friction sparks, heated objects (including open flame), static electricity, short circuits, defibrillators, and others. In view of the hazards involved, fire prevention methods generally strive for redundancy to establish built-in safeguards of a multifaceted nature so that an explosion or fire will not occur even if one element of the safety recommendations fails or is omitted.

To prevent dangerous adiabatic compression heating of oxidizing gases, it is recommended that cylinder valves be opened *slowly* to allow only gradual introduction of the high-pressure gas downstream from the cylinder valve. This allows a slow buildup of pressure and, hence, temperature buildup is minimized, thus allowing the mass of the metal of the regulator or other gas-containing component to dissipate the heat rapidly enough to prevent the buildup of dangerously high temperatures.

The elimination of electric sparks is directed toward the proper design, application, and maintenance of electrical equipment. Specifications for this electrical equipment, as well as for line cords and power supplies, are set forth in NFPA 99. Recommendations for the proper use of this equipment by facility personnel are also spelled out in NFPA 99.

Cylinders, regulators, pipeline systems, and other systems containing compressed air, oxygen, and nitrous oxide must be cleaned specifically for such use and maintained free of hydrocarbon contaminants (solvents, lubricants, etc.) that may act as fuels. Combustion of such fuels contributes to the fire hazard and creates increased pipeline pressure and the release of potentially toxic pyrolysis and combustion products.

Other hospital fire fuels include mattresses, bed and operating-room table sheets and drapes, patient hair, anesthesia and respiratory care equipment, cleaning and prepping solutions, rubber gloves, oxygen-delivery tubing and masks, gauze pads and sponges, and the intestinal gases hydrogen and methane. Under the conditions of oxygen-enrichment and high pressure, certain metals can burn. Aluminum is particularly susceptible to ignition under such conditions.

Fire-Fighting Problems

Oxygen-enriched atmospheres are encountered during practically every inhalation anesthetic (nitrous oxide, an anesthetic, is also considered an oxygen-enriched atmosphere) and respiratory therapy procedure. When bedding and clothing become saturated with oxygen and a fire occurs, flame spread rate will be extremely rapid and large quantities of water are required for extinguishment. Dry chemical fire extinguishers and fire blankets are ineffective. If the fire involves electrical equipment, such equipment must first be deenergized. The fire may then be extinguished with water. Halon and carbon dioxide fire extinguishers are also appropriate. Safe practices for the prevention of fires during the use of respiratory therapy are covered in Chapter 8 of NFPA 99. Included is a recommended response in the event of a fire involving respiratory therapy equipment.

Flammable solvents are present in most health care facilities: in pharmacy storage, hospital laboratories, housekeeping, and other areas. Special fire problems are created if any of these substances are allowed to enter drains, the central suction system, or ventilation ducts.

Mechanical hazards are created by the potential energy contained in compressed gas cylinders.

For some surgical procedures, an operating room meets the definition of a "wet location." While high-voltage electrical equipment (e.g., radiological) and electronic equipment (e.g., medical monitoring and computer) do not alone usually create a fire hazard, combustibles stored in their vicinity may be involved in a fire. The free use of water is inadvisable because of the electric shock hazard to the fire fighter or damage to the equipment. The presence of radioisotopes may also hamper fire suppression methods. Fire service

personnel must be made aware of all special problems created by equipment or supplies in any health care area.

An aid to the identification of hazards associated with medical gases and agents is the NFPA hazards identification symbol system. (See NFPA 704, *Standard System for the Identification of the Hazards of Materials for Emergency Response.*) This identification scheme, based on the "704 diamond," can visually present information on flammability, health, and self-reactivity hazards, as well as special information associated with the hazards of the materials being identified. Chapter 10 of NFPA 99 requires application of the NFPA hazards identification symbols to laboratories. It is also prudent policy to apply the system throughout a facility in such areas as general stores, housekeeping, pharmacy, operating rooms, gas and flammable liquid storage rooms, and any other location where volumes of combustible or flammable materials are stored or used, or where certain special fire-fighting problems exist. Fire-fighting personnel are trained to recognize and decipher the meaning of the "704 diamond."

Adoption of the 704 symbol system affords enhanced fire protection and some measure of prevention and education of employees, as well. Fire personnel are instantly apprised of the hazards they face. In addition, employees are educated about the hazardous nature of the materials they employ daily. It is also possible that, in deploying the system, personnel discover hazardous materials no longer in use that should be removed from the hospital premises, thus eliminating potential problems.

HYPERBARIC CHAMBERS

Another use of an oxygen-enriched atmosphere is in the operation of hyperbaric chambers. Some chambers may be small and accommodate only a single patient; others may be designed to allow the performance of an operative or other procedure, and are thus large enough to accommodate one or more patients and several attendants. Because the pressure of the air in the chamber is elevated, an increased partial pressure of oxygen results (unless means are taken to limit the oxygen content). On occasion, the percent of oxygen is deliberately increased in the chamber. Safety standards for hyperbaric chambers and facilities are detailed in Chapter 19 of NFPA 99.

AMBULATORY CARE FACILITIES

Many requirements for the safe use of oxygen-enriched atmospheres in ambulatory care facilities, dental offices, and other facilities are identical to those for hospitals. Individual vacuum units are commonly utilized, rather than a central system, because of the relatively small amount of suction needed in dental offices. In larger dental offices, or ambulatory care facilities in which surgery is performed under general anesthesia, it is necessary to incorporate a central vacuum system for the operating rooms and the recovery rooms, as well as for central piping of oxygen and nitrous oxide.

Requirements for inhalation anesthesia in such facilities are contained in Chapters 13, 14, and 15 of NFPA 99.

BIBLIOGRAPHY

NFPA Codes, Standards, and Recommended Practices

Reference to the following NFPA codes, standards, and recommended practices will provide further information on medical gases discussed in this chapter. (See the latest version of *The NFPA Catalog* for availability of current editions of the following documents.)

NFPA 10, *Standard for Portable Fire Extinguishers*
NFPA 30, *Flammable and Combustible Liquids Code*

NFPA 49, *Hazardous Chemicals Data*
NFPA 50, *Standard for Bulk Oxygen Systems at Consumer Sites*
NFPA 53, *Guide on Fire Hazards in Oxygen-Enriched Atmospheres*
NFPA 77, *Recommended Practice on Static Electricity*
NFPA 99, *Standard for Health Care Facilities*
NFPA 101®, *Life Safety Code®*
NFPA 115, *Recommended Practice on Laser Fire Protection*
NFPA 325, *Guide to Fire Hazard Properties of Flammable Liquids, Gases, and Volatile Solids*
NFPA 491M, *Manual of Hazardous Chemical Reactions*
NFPA 704, *Standard System for the Identification of the Hazards of Materials for Emergency Response.*
NFPA 801, *Standard for Facilities Handling Radioactive Materials*

Additional Reading

The following documents are available from the American National Standards Institute, 11 W. 42nd St., New York, NY 10036.

ANSI Z79.11, *Anesthetic Equipment—Scavenging Systems for Excess Anesthetic Gases*
ANSI Z136.1, *Safe Use of Lasers*
ANSI Z136.3, *Safe Use of Lasers in Health Care Facilities*

The following documents are available from the American Society for Testing and Materials, 100 Barr Harbor Dr., W. Conshohocken, PA 19428.

ASTM F1161, *Standard Specification for Minimum Performance and Safety Requirements for Components and Systems of Anesthesia Gas Machines*

The following documents are available from the Canadian Standards Association, 178 Rexdale Blvd., Rexdale, Ontario, Canada.

CSA Z305.1-M 84, *Nonflammable Medical Gas Piping Systems*
CSA Z305.2-M80, *Low-Pressure Connecting Assemblies for Medical Gas Systems*
CSA Z305.4-M77, *Qualification Requirements for Agencies Testing Nonflammable Medical Gas Piping Systems*

CSA Z305.5-M86, *Medical Gas Terminal Units*

The following documents are available from the Compressed Gas Association, Inc., 1725 Jefferson Davis Hwy., Arlington, VA 22202.

ANSI/CGA C-4, *Method of Marking Portable Compressed Gas Containers to Identify Material Contained Therein*
CGA C-9, *Standard Color Marking of Compressed Gas Containers Intended for Medical Use*
CGA G-4, *Oxygen*
CGA G-4.3, *Commodity Specification for Oxygen*
CGA G-6, *Carbon Dioxide*
CGA G-6.2, *Commodity Specification for Carbon Dioxide*
CGA G-7, *Compressed Air for Human Respiration*
CGA G-7.1, *Commodity Specification for Air*
CGA G-9.1, *Commodity Specification for Helium*
CGA G-10.1, *Commodity Specification for Nitrogen*
CGA P-2, *Characteristics and Safe Handling of Medical Gases*
CGA P-2.5, *Transfilling of High-Pressure Gaseous Oxygen to Be Used for Respiration*
CGA P-2.6, *Transfilling of Low-Pressure Liquid Oxygen to Be Used for Respiration*
CGA P-2.7, *Guide for the Safe Storage, Handling, and Use of Portable Liquid Oxygen Systems in Health Care Facilities*
CGA S-1.1, *Pressure Relief Device Standards—Cylinders for Compressed Gases*
CGA V-1, *Compressed Gas Cylinder Valve Outlet and Inlet Connections*
CGA V-5, *Diameter—Index Safety System (Non-Interchangeable Low-Pressure Connections for Medical Gas Applications)*
CGA V-6, *Cryogenic Liquid Transfer Connections*
Compressed Gas Association, *Handbook of Compressed Gases*, 3rd ed., Van Nostrand, Reinhold, New York

The following documents are available from the International Organization for Standardization, 1 Rue de Varembé , Geneva, Switzerland.

ISO 8359:1988(E), *Oxygen Concentrators for Medical Use—Safety Requirements*
ISO 9179:1990(E), *Terminal Units for Use in Medical Pipeline Systems*

OXYGEN-ENRICHED ATMOSPHERES

Revised by Coleman J. Bryan and Joel M. Stoltzfus

This chapter discusses the fire hazards that may be associated with oxygen-enriched atmospheres. Methods that can reduce and protect against fire hazards are discussed. Oxygen is a clear, odorless, tasteless, colorless element. It most commonly is found in the gaseous state and comprises approximately 21 percent of the atmosphere. As the most common oxidizing agent in the atmosphere, oxygen can react with many materials in highly exothermic reactions. An oxygen-enriched atmosphere (OEA) is defined as an atmosphere in which the concentration exceeds 21 percent by volume, or the partial pressure of oxygen exceeds 160 torr (21.3 kPa), or both. (See Table 4-9A.)

Oxygen supports combustion and is necessary for plant and animal life. The concentration of oxygen available in the atmosphere is generally sufficient for human needs during normal conditions of health and for most combustion applications. Further, it has many uses under different circumstances. Oxygen-enriched atmospheres are commonly used in medical applications, industrial processes, underwater tunneling and caisson work, deep-sea exploration, space flight operations, and commercial and military aviation. Oxygen-enriched atmospheres may develop inadvertently at any time when oxygen or compressed air is transported, stored, or utilized.

The preparation of oxygen conventionally involves the compression of air, followed by cooling and reexpansion, that is repeated until the temperature of the air falls below the critical pressure of oxygen, causing it to liquefy. Fractional distillation of the liquid air then separates its various gaseous components. The oxygen is then collected and may be stored as a compressed gas or a liquid. Oxygen is also separated from air by absorption and membrane systems and by electrolysis.

Further information concerning oxygen-enriched atmospheres can be found in the following chapters: Section 3, Chapter 13, "Fluid Power Systems"; Section 4, Chapter 1, "Fire Hazards of Materials: An Overview"; Section 4, Chapter 6, "Flammable and Combustible Liquids"; Section 4, Chapter 7, "Gases"; Section 4, Chapter 8, "Medical Gases"; Section 6, Chapter 25, "Special Systems and Extinguishing Techniques"; Section 9, Chapter 8, "Health Care Occupancies"; and Section 9, Chapter 26, "Fire Protection of Laboratories Using Chemicals."

Coleman J. Bryan is chief of the Failure Analysis and Physical Testing Branch at the NASA John F. Kennedy Space Center, FL. Joel M. Stoltzfus is an aerospace engineer at the NASA White Sands Test Facility, Las Cruces, NM.

FIRE HAZARDS

The phenomenon known as "fire" is a chemical reaction between a fuel and oxygen. The initiation and rate of the reaction is subject to the laws of chemical kinetics. The degree of fire hazard varies with the concentration of oxygen present, the diluent gas, and the total pressure. In most commonly encountered oxygen-enriched atmospheres, an increased fire hazard is produced by the increased partial pressure of oxygen (e.g., in an atmosphere of compressed air). However, an oxygen-enriched atmosphere may not necessarily produce an increased fire hazard. Certain oxygen-enriched atmospheres can exhibit combustion-supporting properties similar to ambient air, while others are incapable of supporting the combustion of normally flammable materials. For example, a 4 percent oxygen mixture in nitrogen or helium with a total pressure of 12 atm (1216 kPa) will not support the combustion of paper, even though the partial pressure of oxygen is 365 torr (48.7 kPa). However, different concentrations and pressures may accelerate the combustion of materials; facilitate ignition; and, in general, increase the fire hazard. Most commonly encountered oxygen-enriched atmospheres fall into this last category. (See Table 4-9A.)

Fires and explosions have occurred in many diverse circumstances involving intentional and unintentional oxygen-enriched atmospheres. Among them are:

1. Transport, transfer, and production of oxygen. Typical are truck delivery of liquid oxygen, leakage or venting from storage or dispensing equipment, and recharging of containers.
2. Medical applications involving inhalation therapy, life-support systems, resuscitation procedures, hyperbaric chambers, and anesthesia.
3. Industrial applications in petrochemical processing, steel manufacturing, mining, secondary treatment of municipal and industrial wastewaters, manufacturing microchips, paper and pulp bleaching procedures, fireflooding in tertiary oil recovery, and caisson work and underwater tunneling.
4. Aerospace activities that involve the use of liquid oxygen in liquid-fueled rockets and aviator breathing systems. Gaseous oxygen is used in the NASA space shuttle orbiter at concentrations as high as 25.9 percent oxygen under normal conditions, and the space station is designed to use a 10.2 psia (70.3 kPa) 30 percent oxygen atmosphere.
5. Test systems, ground support equipment, laboratory piping systems, and aceteylene/oxygen torches using oxygen-enriched and high-pressure oxygen environments.

TABLE 4-9A. *Partial Pressure of Oxygen in a Rarified or Compressed-Air Atmosphere*

Total Absolute Pressure			Altitude Above or Depth Below Sea Level	Partial Pressure of Oxygen if Atmosphere Is Air	Concentration of Oxygen if Partial Pressure of Oxygen Is 160 mm Hg or torr
Atmospheres	mm Hg or torr	psia (kPa)	Meters (Feet) Air or Sea Water	mm HG or torr*	% by Volume
1/5	152	2.9 (20)	11,735 (38,500)	32	100.0†
1/3	253	4.9 (33.8)	8,832 (27,500)	53	62.7†
1/2	380	7.3 (50.3)	5,486 (18,000)	80	42.8†
2/3	506	9.8 (67.6)	3,353 (11,000)	106	31.3†
1	760	14.7 (101.4)	Sea Level	160	20.9
2	1,520	29.4 (202.7)	−10 (−33)	320†	10.5
3	2,280	44.1 (304.1)	−20 (−66)	480†	6.9
4	3,040	58.8 (405.4)	−30 (−99)	640†	5.2
5	3,800	73.5 (506.8)	−40 (−132)	800†	4.2

*This column shows the increased available oxygen in compressed-air atmospheres.
†Oxygen-enriched atmosphere.
(Source: NFPA 53, *Guide on Fire Hazards in Oxygen-Enriched Atmospheres*)

6. Inadvertent oxygen-enriched atmospheres that may result from improper design, malfunction, or improper use of oxygen storage or dispensing equipment. Oxygen is more soluble in aircraft fuel than nitrogen and, if such fuel is exposed to air for a significant period of time, enough oxygen may dissolve in the fuel and release during flight because of the decreased atmospheric pressure to create a hazardous oxygen-enriched atmosphere.

7. Inadvertent substitution of oxygen for air or nitrogen.

Further, an oxygen-enriched atmosphere may develop in situations where nitrous oxide is employed. Two common nitrous oxide applications are: (1) medical, e.g., as a mild anesthetic agent; and (2) industrial, as a propellant in aerosol products. The use of other oxidants, e.g., chlorine, chlorates, nitric oxide, and ozone, may result in enhanced combustion. Appropriate safety literature, such as materials safety data sheets (MSDS), should be consulted.

IGNITION AND COMBUSTION OF MATERIALS

Flames involve strongly exothermic reactions between oxidants and fuels, producing combustion products at high temperatures. In general, for a fuel molecule and an oxygen molecule to interact chemically, sufficient energy has to be imparted to these molecules to enable a collision between the two to result in a chemical transformation. The minimum energy for the molecules to react is referred to as the "activation energy." If the energy released by this reaction is imparted to adjacent molecules in the form of heat, the temperature of the gas will increase. An increase in gas temperature increases the number of molecules with energy equal to the activation energy and increases the reaction rate. As the temperature further increases, enough fuel and oxygen molecules react with enough additional energy released to enable the reaction to become self-sustaining until one or both of the reactants have been consumed and the combustion ceases.

The minimum ignition energy for combustion will vary with the type of ignition source, the specific chemical nature and physical character of the fuel, and the composition and pressure of the atmosphere. Though most combustion is accompanied by a gas- or vapor-phase combustion reaction, certain materials, e.g., metals, can burn in liquid or solid phase; i.e., a condensed-phase reaction.[1-5] If the reaction is to continue in the vapor phase, as in the case of solids or liquids, sufficient thermal energy first needs to be supplied to convert a part of the fuel to vapor. In all cases, for the combustion to proceed, the ignition source has to impart energy to the fuel at a faster rate than the fuel loses energy. The ignition sources of principal concern for oxygen-enriched atmosphere application can be categorized into six types:

1. Electrical sources, such as electrostatic and break (arc) sparks;
2. Hot surfaces, such as friction sparks and heated wires;
3. Heated gases, independent of surfaces, generated by adiabatic compression or jets of hot gas (includes pilot flames);
4. Exothermic chemical reactions;
5. Mechanical sources, such as frictional heating and particle impact; and
6. Laser sources.

In general, the greater the oxygen concentration, the lower the energy required for ignition and the faster the flame spread rate. Figure 4-9A depicts, in an oversimplified manner, the effects of variations in oxygen concentration and environmental pressure. For a fixed volume percent oxygen, the minimum ignition varies inversely with the square of the pressure. There exists a minimum pressure below which ignition will not occur. As the temperature of a given system increases, less and less energy is required to ignite the mixture until it reaches a sufficiently high temperature to ignite

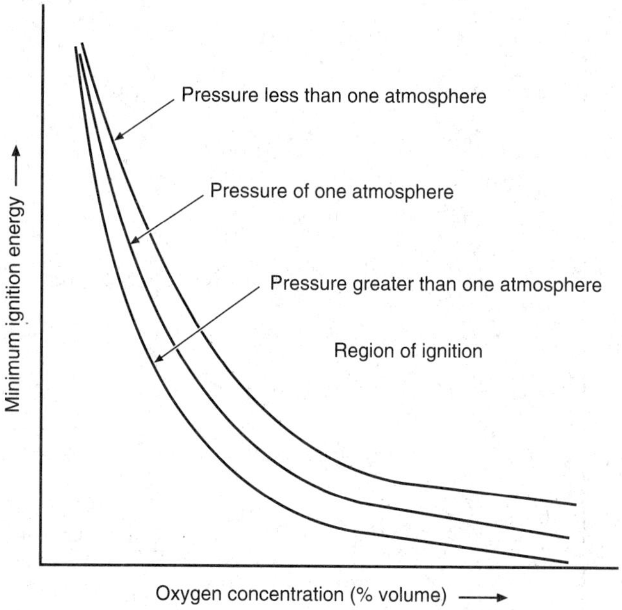

FIG. 4-9A. *Effects of variations in oxygen concentration and environmental pressure on the minimum ignition energy. (Source: NFPA 53, Guide on Fire Hazards in Oxygen-Enriched Atmospheres)*

spontaneously. This minimum temperature is known as the auto-ignition or spontaneous ignition temperature.

The likelihood of ignition and the rate of flame propagation of a combustible are primarily influenced by the oxygen content in the environment. Thus, a minimum amount of oxygen must be present for a flame to propagate, regardless of the ratio of the fuels to inert gases present. Effects on ignition energy requirements by typical ignition sources will vary with the particular inert gas present and are a function of the heat capacity and thermal conductivity of these gases in the atmosphere.

Combustion is a complex sequence of chemical reactions between a fuel and an oxidant, accompanied by the evolution of heat and, usually, by the emission of light. The rate of combustion process depends on the chemical nature of the fuel and oxidant, their relative concentrations, environmental pressure and temperature, and other physical parameters, e.g., geometry and ventilation.

Combustible Gases, Vapors, and Liquids

Once ignition occurs, the propagation of combustion requires a continued supply of fuel and oxidant. The limits of flammability represent the extreme concentration limits of a combustible in an oxidant through which a flame, once initiated, will continue to propagate at the specified pressure and temperature. In the case of combustible gases, vapors, and liquids, two types of mixtures, i.e., homogeneous or heterogeneous, can exist within the atmosphere. A homogeneous mixture is one in which the components are intimately and uniformly mixed so that any small volume is representative of the whole. If the mixture is not homogeneous, it is necessarily heterogeneous.

Figure 4-9B shows that, for liquid fuels in equilibrium with their vapors in air (or in oxygen), a minimum temperature exists for each fuel above which sufficient vapor is released to form a flammable vapor-air (or vapor-oxygen) mixture. The experimentally determined value of this minimum temperature is commonly referred to as the "flash point."

The total environmental pressure also has an effect on the limits of flammability. (See Figure 4-9C.) For a given atmospheric composition, an increase in pressure generally broadens the flammability range. The low-pressure limit is dependent on the particular fuel and oxidant, as well as the temperature, size, geometry,

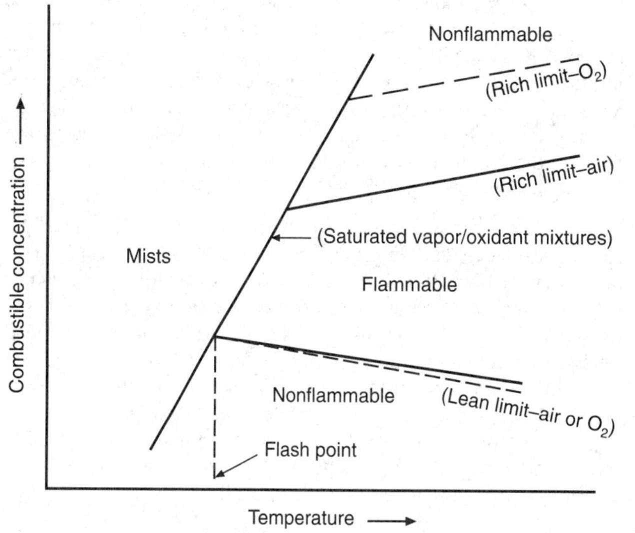

FIG. 4-9B. *Minimum temperatures for flammable vapor mixtures. (Source: NFPA 53, Guide on Fire Hazards in Oxygen-Enriched Atmospheres)*

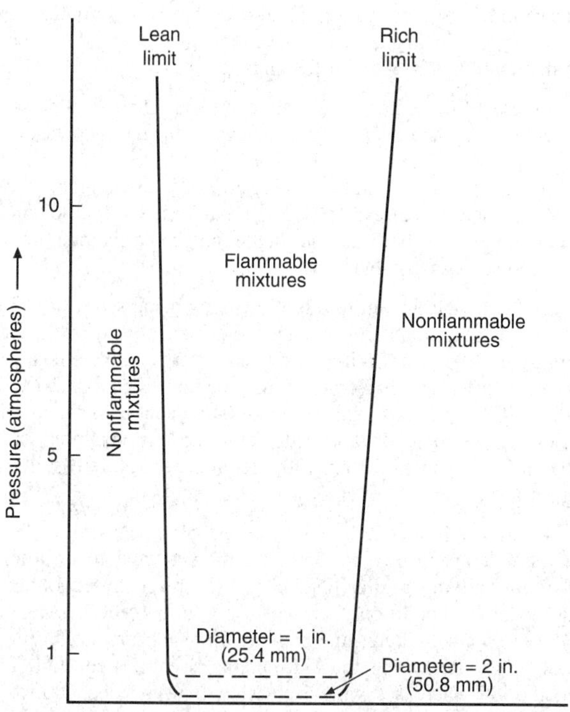

FIG. 4-9C. *Effects of pressure on limits of flammability. (Source: NFPA 53, Guide on Fire Hazards in Oxygen-Enriched Atmospheres)*

and attitude of the confining vessel. The quenching or low-pressure limits are represented in Figure 4-9C by broken lines to indicate their dependency on surroundings.

The lower limit of flammability is of great interest, since it defines the minimum combustible concentration required for the propagation of flame throughout the particular mixture. The minimum temperature at which a lower limit concentration can exist depends on the volatility of the combustible and corresponds roughly to the flash point of the combustible. Flammable liquids have flash points in air of less than 100°F (38°C). (See Table 4-9C.) Also, fuel-air mixtures formed at or above their flash point temperature will propagate flame if ignited. In oxygen, flash points of combustibles are slightly lower than those in air. Where fuel mists or foams are formed, the mixtures may propagate flame at temperatures far below the flash points of the fuels.

Typically, the lower flammability limits for most hydrocarbon fuels, anesthetics, and solvents are between 1 to 5 percent by volume in air or oxygen at 1 atm pressure. In comparison, the values for most lubricants are less than 1 percent by volume because of the higher molecular weights of such fluids. Further, lubricants need much higher temperatures, e.g., 200 to 700°F (93 to 371°C) to form lower-limit flammable mixtures than do paraffins or other hydrocarbons.[6] Although most lower flammable limits in oxygen do not differ greatly from those in air, the upper flammable limits usually require much higher oxygen concentration, which tends to be about 50 percent for many materials. (See Table 4-9C.)

Combustible Solids—Nonmetals

The burning of solid combustibles requires the consideration of only heterogeneous fuel-oxidant systems. As in the case of flammable liquids and gases, the flame reaction occurs in the gas phase. Once a particular solid combustible has been ignited, propagation of flame requires that a portion of the heat of combustion be fed back to the solid fuel to cause its vaporization or pyrolysis, or both, thereby making additional gaseous fuel available to mix with the oxidant. The flame process is of the diffusion type.

Figure 4-9D depicts the effects of oxygen content and pressure on the flame spread rate. Note that increasing the partial pressure of oxygen at a constant environmental pressure may change the classification of a material from the nonflammable to the flammable category. For materials already in the flammable category based on 21 percent oxygen, further increases in the oxygen partial pressure result in higher flame spread rates.

Organic materials in the form of finely dispersed dust clouds are extremely susceptible to combustion. Typically, spark ignition energies are less than 0.1 J. Single fibers of organic materials, e.g., lint, cotton tufts, and fluffy fabrics, may be ignited by a 0.02 J static spark in 100 percent oxygen but not in 64 percent oxygen.[7] Textile fabrics, e.g., cotton and wool used in clothing, can be ignited with a spark energy as low as 2.3 J in 100 percent oxygen, whereas, in normal air, a spark energy of at least 193 J is required.[8]

Combustible materials, when heated, may self-ignite at relatively low temperatures that approach the spontaneous or autogenous ignition temperatures [(SIT) or (AIT), respectively] obtained under ideal test conditions. As shown in Figure 4-9E, the AIT rapidly decreases with increasing pressure up to approximately 7 MPa (1,000 psig), after which, the AIT remains essentially constant. Further, the AIT increases with decreasing oxygen concentration at constant pressure down to the lower flammability limit, as shown in Table 4-9D.

The majority of the materials commonly used in the home and workplace are flammable in air. Chemical additives to plastics and textile fabrics, such as halogens, borax, phosphates, and various

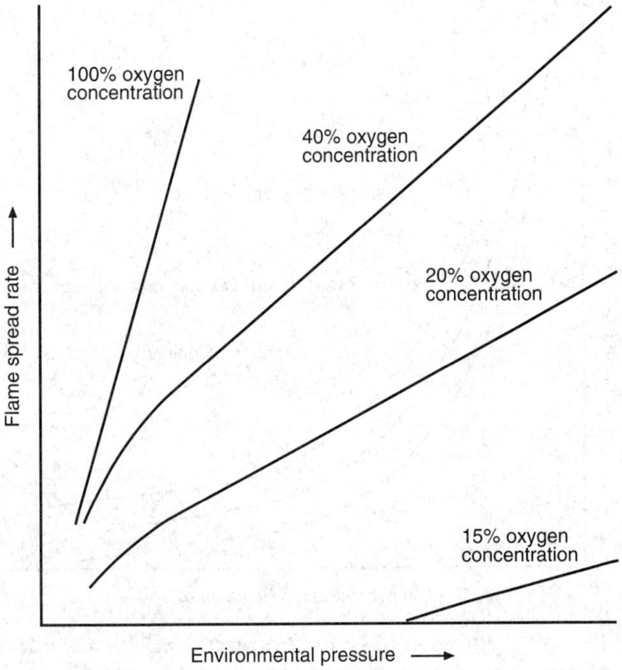

FIG. 4-9D.   Effects of oxygen content and pressure on flame spread rate. (Source: NFPA 53, Guide on Fire Hazards in Oxygen-Enriched Atmospheres)

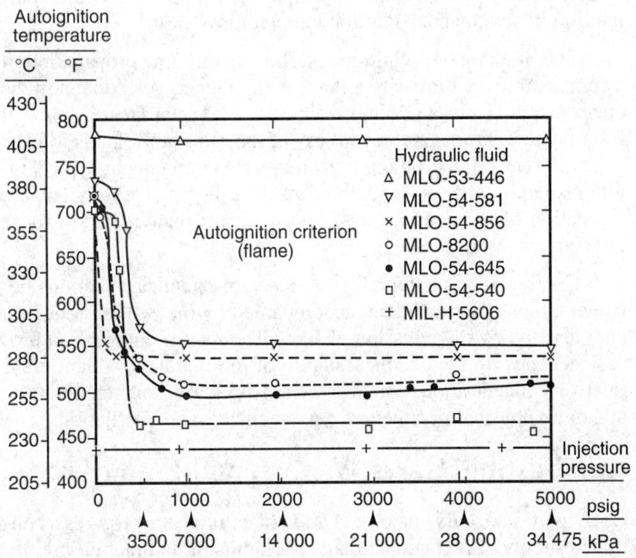

FIG. 4-9E.   Effect of pressure on autogenous ignition temperatures. (Source: NFPA 53, Guide on Fire Hazards in Oxygen-Enriched Atmospheres)

metal oxides, are effective in reducing both ignitability and flammability and have been found to be effective for many materials in atmospheres containing up to 30 to 35 percent oxygen. However, it should be mentioned that materials treated with some fire retardants will burn differently than nontreated materials and may generate corrosive or toxic vapors during the combustion process. The flame resistances of several materials, treated and untreated, are shown in Table 4-9E.

TABLE 4-9C. *Ignition and Flammability Properties of Combustible Liquids and Gases in Air and Oxygen at Atmospheric Pressure*

Combustible	Flash Point[1] Air °C (°F)		Min. Ign. Temperature[2] Air °C (°F)		Oxygen °C (°F)		Min. Ign. Energy[3] Air mJ	Oxygen mJ	Flammability Limits[4] Vol. % Air LFL	Air UFL	Oxygen LFL	Oxygen UFL
Hydrocarbon Fuels												
Methane	Gas		630	(1166)	—		0.30	0.003	5.0	15	5.1	61
Ethane	Gas		515	(959)	506	(943)	0.25	0.002	3.0	12.4	3.0	66
n-Butane	−60	(−76)	288	(550)	278	(532)	0.25	0.009	1.8	8.4	1.8	49
n-Hexane	−3.9	(25)	225	(437)	218	(424)	0.288	0.006	1.2	7.4	1.2	52*
n-Octane	13.3	(56)	220	(428)	208	(406)	—	—	0.8	6.5	≤0.8	—
Ethylene	Gas		490	(914)	485	(905)	0.07	0.001	2.7	36	2.9	80
Propylene	Gas		458	(856)	423	(793)	0.28	—	2.4	11	2.1	53
Acetylene	Gas		305	(581)	296	(565)	0.017	0.0002	2.5	100	≤2.5	100
Gasoline (100/130)	−45.5	(−50)	440	(824)	316	(600)	—	—	1.3	7.1	≤1.3	—
Kerosene	37.8	(100)	227	(440)	216	(420)	—	—	0.7	5	0.7	—
Anesthetic Agents												
Cyclopropane	Gas		500	(932)	454	(849)	0.18	0.001	2.4	10.4	2.5	60
Ethyl Ether	−28.9	(−20)	193	(380)	182	(360)	0.20	0.0013	1.9	36	2.0	82
Vinyl Ether	−30	(<−22)	360	(680)	166	(331)	—	—	1.7	27	1.8	85
Ethylene	Gas		490	(914)	485	(905)	0.07	0.001	2.7	36	2.9	80
Ethyl Chloride	−50	(−58)	516	(961)	468	(874)	—	—	4.0	14.8	4.0	67
Chloroform	———————— Nonflammable ————————											
Enflurane	>200	(93)	NA	NA	NA	NA	NA	NA	NA	NA	9.8	NA
Isoflurane	>200	(93)	NA	NA	NA	NA	NA	NA	NA	NA	8.8	NF
Desflurane	NF		NA	NA	NA	NA	NA	NA	NA	NA	17.2	20.8
Nitrous Oxide	———————— Nonflammable ————————											
Solvents												
Methyl Alcohol	12.2	(54)	385	(725)	—		0.14	—	6.7	36	≤6.7	93
Ethyl Alcohol	12.8	(55)	365	(689)	—		—	—	3.3	19	≤3.3	—
n-Propyl Alcohol	15	(59)	440	(824)	328	(622)	—	—	2.2	14	≤2.2	—
Glycol	111	(232)	400	(752)	—		—	—	3.5*	—	≤3.5	—
Glycerol	160	(320)	370	(698)	320	(608)	—	—	—	—	—	—
Ethyl Acetate	−4.4	(24)	427	(800)	—		0.48	—	2.2	11	≤2.2	—
n-Amyl Acetate	24.4	(76)	360	(680)	234	(453)	—	—	1.0	7.1	≤1.0	—
Acetone	−17.8	(0)	465	(869)	—		1.15	0.0024	2.6	13	≤2.6	60*
Benzene	−11.1	(12)	560	(1040)	—		0.22	—	1.3	7.9	≤1.3	30
Naphtha (Stoddard)	37.8	(~100)	232	(~450)	216	(~420)	—	—	1.0	6	≤1.0	—
Toluene	4.4	(40)	480	(896			2.5	—	1.2	7.1	≤1.2	—
Butyl Chloride	−6.7	(20)	240	(464)	235	(455)	0.332	0.007*	1.8	10	1.7	52*
Methylene Chloride	—		615	(1139)	606	(1123)	—	0.137	15.9*	19.1*	11.7*	68
Ethylene Chloride	13.3	(56)	476	(889)	470	(878)	2.37	0.011*	6.2	16	4.0	67.5
Trichloroethane	—		458	(856)	418	(784)	—	0.092	6.3*	13*	5.5*	57*
Trichloroethylene	32.2	(90)	420	(788)	396	(745)	—	18*	10.5*	41*	7.5	91*
Carbon Tetrachloride	———————— Nonflammable ————————											
Miscellaneous Combustible												
Acetaldehyde	−27.2	(−17)	175	(347)	159	(318)	0.38	—	4.0	60	4.0	93
Acetic Acid	40	(104)	465	(869)	—		—	—	5.4*	—	≤5.4	—
Ammonia	Gas		651	(1204)	—		>1000	—	15.0	28	15.0	79
Aniline	75.6	(168)	615	(1139)	—		—	—	1.2*	8.3	≤1.2	—
Carbon Monoxide	Gas		609	(1128)	588	(1090)	—	—	12.5	74	≤12.5	94
Carbon Disulfide	−30	(−22)	90	(194)	—		0.015	—	1.3	50	≤1.3	—
Ethylene Oxide	<17.8	(<0)	429	(804)	—		0.062	—	3.6	100	≤3.6	100
Propylene Oxide	−37.2	(−35)	—		400	(—)	0.14	—	2.8	37	≤2.8	—
Hydrogen	Gas		520	(968)	400	(752)	0.017	0.0012	4.0	75	4.0	95
Hydrogen Sulfide	Gas		260	(500)	220	(428)	0.077	—	4.0	44	≤4.0	—
Bromochloromethane	—		450	(842)	368	(694)	—	—	NF[5]	NF	10.0	85
Bromotrifluoromethane	Gas		>593	(>1100)	657	(1215)	—	—	NF	NF	NF	NF
Dibromodifluoromethane	Gas		499	(930)	453	(847)	—	—	NF	NF	29.0	80

[1] Data from Footnotes 1 and 2; open cup method.
[2] Data from Footnotes 3, 4, 5, and 6.
[3] Data from Footnotes 7, 8, 9, 10, and 3.
[4] Data from Footnotes 11, 12, 3, 4, and 6.
[5] NF — No flammable mixtures found in Footnote 4.
* Data at 200°F (93°C).
(Source: NFPA 53, *Guide on Fire Hazards in Oxygen-Enriched Atmospheres*)

TABLE 4-9D. *Minimum Hot Plate Ignition Temperatures of Six Combustible Materials in Oxygen-Nitrogen Mixture at Various Total Pressures*

Material	Oxidant	Ignition Temperature, °C Total Pressure, Atmospheres			
		1	2	3	6
Cotton	Air	465	440 (425)†	385	365
sheeting	42% O$_2$, 58% N$_2$	390	370	355	340
	100% O$_2$	360	345	340	325
Cotton	Air	575	520 (510)	485 (350)	370 (325)
sheeting	42% O$_2$, 58% N$_2$	390 (350)	335	315	295
treated*	100% O$_2$	310	—	300	285
Conductive	Air	480	395	370	375
rubber	42% O$_2$, 58% N$_2$	430	365	350	350
sheeting	100% O$_2$	360	—	345	345
Paper	Air	470	455	425	405
drapes	42% O$_2$, 58% N$_2$	430	—	400	370
	100% O$_2$	410	—	365	340
Nomex® fabric	Air	>600	>600	>600	560
	42% O$_2$, 58% N$_2$	550	540	510	495
	100% O$_2$	520	505	490	470
Polyvinyl	Air	>600	—	495	490
chloride	42% O$_2$, 58% N$_2$	575	—	370	350
sheet	100% O$_2$	390	—	350	325

*Cotton sheeting treated with Du Pont X-12 fire retardant; amount of retardant equal to 12 percent of cotton specimen weight.
†Values in parentheses indicate temperature at which material glowed.
(Source: NFPA 53, *Guide on Fire Hazards in Oxygen-Enriched Atmospheres*)

The oxygen index test is commonly used to evaluate the ignition and propagation properties of materials in oxygen-enriched atmospheres. Many commonly used materials will ignite and burn in as little as 16 percent oxygen; however, fluorocarbon plastics did not propagate flames in 100 percent oxygen. The oxygen index values for a number of materials are shown in Table 4-9F.

Electrical wiring insulation is a critical application of polymeric materials. Commonly used polymers are Teflon, Nomex, Kapton, Kynar, silicone, polyolefin, and poly(vinyl chloride). Of these, Teflon is the most ignition and flame resistant in oxygen-enriched atmospheres. Kapton was found to be resistant to ignition in oxygen-enriched atmospheres, but has been found to be susceptible to arc tracking. Potting compounds, circuit boards, and other components of the electrical system may also contribute to the fuel supply.[9,10]

The rate at which flame spreads under a given set of conditions in an oxygen-enriched atmosphere is the single most important property to fire hazard development in such an atmosphere. The rate of flame spread over the surface of a given combustible material provides an indication of the speed at which a fire involving the combustible will develop. (See Table 4-9G.) Flame spread rates are much greater in an upward direction, due to the buoyant convection, than in the horizontal and downward directions. The rate of flame spread increases with: (1) an increase in the oxygen concentration at constant pressure, or (2) an increase in total pressure at a constant percentage of oxygen. Table 4-9H shows typical data for the burning of filter paper over a wide range of pressure and atmospheric conditions.

Several test methods for determining flame spread rate include the following:

1. NFPA 255, *Standard Method of Test of Surface Burning Characteristics of Building Materials.*
2. ASTM E162, *Standard Test Method for Surface Flammability of Materials Using a Radiant Heat Energy Source.*
3. ASTM D568, *Standard Test Method for Flammability of Flexible Plastic.*
4. ASTM D1230, *Standard Test Method for Flammability of Apparel Textiles.*
5. ASTM D2863, *Standard Test Method for Measuring the Minimum Oxygen Concentration to Support Candle-like Combustion of Plastics (Oxygen Index).*
6. ASTM G125, *Standard Test Method for Measuring Liquid and Solid Material Fire Limits in Gaseous Oxidants.*

For each method, selected materials are tested under conditions of intended use prior to use in oxygen-enriched atmospheres.

Combustible Solids—Metals

The burning of metals can occur in either the vapor phase or in a condensed-phase reaction and, therefore, can require the consideration of both homogeneous and heterogeneous fuel-oxidant systems. Once a particular metallic combustible has been ignited, propagation of the combustion, whether burning in the vapor phase or condensed phase, requires that a portion of the heat of combustion (assuming the ignition source has been removed) be fed back to the solid fuel to cause it to heat past the ignition point.

All metals, with the possible exception of the noble metals—gold and platinum—can be expected to ignite in oxygen at some elevated temperature and pressure. Metals most subject to ignition hazards are those configured with high surface-to-volume ratios, such as dusts, thin sheets, wires, and wire meshes. When the structural elements of systems containing pressurized oxygen ignite and burn, the results are often catastrophic, due to the explosion-like release of high-pressure gases and ejection of burning debris. Ignition mechanisms include mechanical impact, particle impact, friction, electrical arc and spark, resonance, rupture, exposure of fresh metal surfaces, and promoted ignition. The most ignitable common metals are titanium, magnesium, and lithium; the least ignitable are nickel,

TABLE 4-9E. Flame Resistance of Materials Held Vertically at One Atmosphere Pressure in O₂/N₂ Mixtures

NRL Sample Number	Material	Combustion in O₂/N₂ Mixtures		
		21% O₂	31% O₂	41% O₂
FM-1	Rosin-impregnated paper	Burned		
FM-3	Cotton terry cloth	Burned		
FM-28	Cotton cloth, white duck	Burned		
FM-4	Cotton terry cloth, Roxel®-treated	No	No	Burned
FM-5	Fleece-backed cotton cloth, Roxel®-treated	Surface only	Burned	Burned
FM-14	Cotton O.D. Sateen, Roxel®-treated	No	Burned	—
FM-15	Cotton green whipcord, Roxel®-treated	No	Burned	—
FM-16	Cotton white duct, Roxel®-treated	No	Burned	—
FM-17	Cotton, King Kord, Roxel®-treated	No	Burned	—
FM-29	Cotton white duck, treated with 30% boric acid-70% borax	No	Burned	Burned
FM-30	Cotton terry cloth, treated with 30% boric acid-70% borax	No	Burned	Burned
FM-6	Fire resistant cotton ticking	No	Burned	
FM-7	Fire resistant foam rubber	No	No	Burned
FM-9	Nomex® temperature resistant Nylon	No	Burned	—
FM-10	Teflon® fabric	No	No	No
FM-11	Teflon® fabric	No	No	No
FM-12	Teflon® fabric	No	No	No
FM-13	Teflon® fabric	No	No	No
FM-19	Verel fabric	No	Burned	Burned
FM-22	Vinyl-backed fabric	No	Burned	Burned
FM-23	Omnicoated Dupont high-temperature fabric	No	Burned*	Burned
FM-24	Omnicoated glass fabric	No	No	Burned*
FM-20	Glass fabric, fine weave	No	No	No
FM-21	Glass fabric, knit weave	No	No	No
FM-25	Glass fabric, coarse weave	No	No	No
FM-26	Glass fabric, coarse weave	No	No	No
FM-27	Aluminized asbestos fabric	No	No	Burned
FM-32	Rubber from aviator oxygen mask	Burned	Burned	Burned
FM-33	Fluorolube® grade 362	No	No	No**
FM-34	Belco no-flame grease	No	No	No**

*Burned only over igniter.
**White smoke only.
(Source: NFPA 53, *Guide on Fire Hazards in Oxygen-Enriched Atmospheres*.)

TABLE 4-9F. Oxygen Index for Selected Materials[46]

Material	Description	OI[a]
Polyacetal		16
Loctite pipe sealant		
Nuclear grade PST®	anaerobic sealant (cured), cup test[b]	17
Type PS/T	anaerobic sealant (cured), cup test[b]	20
Poly(methylmethacrylate)		
Plexiglas®		18.5±0.5
ECO/Rubber	epichlorohydrin rubber	18.5
Silicone® rubber		
RTV 102		23
Silastic® 732		25
SMS 2454		25
RTV 60		28.5
RTV 560		29
RTV 560 mixture	user-added 50% glass	36
Silicone® grease	cup test[b]	26±1
Rectorseal® #15 thread sealant		<30.0
Durabla gasket		28.0±.5
Fluorosilicone grease #822	cup test[b]	30
Blue Gard® gaskets		
Blue Gard® 3000	nonasbestos gasket	30.5±0.5
Blue Gard® 3200	nonasbestos gasket	31
Blue Gard® 3400	nonasbestos gasket	52
Blue Gard® 3200	nonasbestos gasket	60
Blue Gard® 3000	nonasbestos gasket	62
Blue Gard® 3300	nonasbestos gasket	68
Nylon		
Nylon 66		30.5
Nylon 66 (glass-filled)		23.5
CYL-SEAL thread sealant		38
Polyvinylidene fluoride		
Kynar®		39
Fluorocarbon rubber		
Viton®	brown O-ring	40.5±0.5
Viton®, green	green O-ring	42
Viton A®		57, 57.5
Viton E-60C®		60.5
Viton B®, #V494-70		DNP
Balston® filters		
Type Epoxy	cut from cylinder	42.5
Type Q-fluorocarbon	cut from cylinder	47±1
Type H-inorganic	cut from cylinder	DNI
Molykote® Z powder	MOS₂ cup test[b]	45
Bel-Ray Greases	cup test[b]	57
FC1260	cup test[b]	DNP
FC1245	halocarbon oil/graphite	
Key Abso-Lute®	cup test[b]	67
CTFE Lubricants		
Fluorolube GR362 grease	cup test[b]	67±4
Halocarbon 25-20 oil	cup test[b]	75
Halocarbon 11-14S oil	cup test[b]	DNP
Fluorocarbon FEP	tubing	77
Fluorocarbon PFA		100
Fluorocarbon TFE		DNP
PFPE grease		
Fomblin RT15®	cup test[b]	DNP
Krytox 283AC®	cup test[b]	DNP

[a] DNP (Did not propagate); DNI (Did not ignite).
[b] Cup test performed basically as described by Nelson, G.L., and Webb, J.L. "Oxygen Index of Liquids, Techniques and Application." *Journal of Fire and Flammability*, Vol. 4, July 1973, pp. 210-226.

copper, and cobalt. Increase in oxygen pressure and concentration promotes the ignition of metals at lower temperatures.[11]

As with nonmetals, the extent of combustion and flame propagation for metals depends on a number of factors including the absolute pressure, ambient temperature, fuel and oxidizer composition, geometric shape and temperature of the fuel sample, and direction of combustion front. Depending on these factors, the combustion front

TABLE 4-9F. *(Continued)*

Material	Description	OI[a]
Krytox GPL 225®	cup test[b]	DNP
Krytox GPL 205®	cup test[b]	DNI
Tribolube 13C®	cup test[b]	DNP
PFPE fluid		
Fomblin Y25®	cup test[b]	DNI
Krytox GPL 105®	cup test[b]	DNP
CTFE plastic		
Kel-F 81®	15% glass filled	DNP
Kel-F 81®	nonplasticized	DNP
Perfluoroelastomer		
Kalrez® 1045	O-ring	DNP
Kalrez® 1050	O-ring	DNP
Kalrez® 4079	O-ring	DNP
Silica gel	cup test[b]	DNI
Blue Drierite	cup test[b]	DNI
Kaowool Insulation	alumina-silica	DNI
Cerawool paper		DNI
Fiberglass/cement board		DNI
Kwik Flux #54®	cup test[b]	DNI
Asbestos cement board		
Transite®		DNI
Sindanyo CS51®		DNI
Turnalite TI 150®		DNI
Asbestos paper	32 lb/100 ft²	DNI

[a] DNP (did not propagate); DNI (did not ignite).
[b] Cup test performed basically as described by Nelson, G. L., and Webb, J. L. "Oxygen Index of Liquids, Techniques and Application." *Journal of Fire and Flammability*, Vol. 4, July 1973, pp. 210–226.
(Source: NFPA 53, *Guide on Fire Hazards in Oxygen-Enriched Atmospheres.*)

in metals can propagate at greatly varying rates. For example, a 0.13-in. (3.3-mm) diameter 316 stainless steel rod burning upward in 7 MPa (1000 psig) oxygen will propagate at about 0.43 in. per sec (11 mm/s), whereas a 0.13-in. (3.3-mm) diameter 6061 aluminum rod will burn at 2.5 in. per sec (64 mm/s).[12] In general, the combustion front propagation rate increases with increasing ambient pressure, oxidizer concentration, ambient temperature, and decreasing sample dimensions.[13,14]

Flammability limits per se do not exist for most structural metal alloys, since they burn mostly in the liquid rather than the vapor phase. However, two measures of the relative flammability of metals exist that are of practical value. They are: (1) the minimum oxygen pressure required to support combustion of a standard sample (threshold pressure) and (2) the minimum oxygen concentration required to support combustion of a standard sample at a given pressure (oxygen index). Data on the threshold pressures and oxygen indices of several metals and alloys are provided in Tables 4-9I (A) and (B) and Figure 4-9F, parts (a), (b), and (c), respectively.

DESIGN OF SYSTEMS FOR OXYGEN-ENRICHED ATMOSPHERES

In the design of systems associated with oxygen-enriched atmospheres, e.g., gas and compressed air supplies, welding applications, spacecraft propulsion and ground support equipment, self-contained underwater breathing apparatus (SCUBA), medical applications (including home assisted-breathing apparatus), etc., the following fire hazards should be considered:

TABLE 4-9G. *Effect of Oxygen on Flame Spread Rates over Various Materials (Edges Not Inhibited)*

Material	Flame Spread Rate (mm/s)	
	In Air	In 258 mm Hg Oxygen
Aluminized mylar tape	—	49.53
Aluminized vinyl tape	NI	78.74±10.16
Asbestos insulating tape	NI	2.03
Butyl rubber	0.152	0.40±0.04
Canvas duck	NP	6.35
Cellulose acetate	0.305	7.1
Chapstick	NI	46.23
Cotton shirt fabric	NP	38.1±1.27
Electrical insulating resin	NI	6.86
Electrical terminal board	NI	1.524±.254
Fiberglass insulating tape	NI	106.68±15.24
Foam cushion material	4.83	314.96
Foamed insulation	0.051	55.88±5.08
Food packet, aluminized paper	NI	7.112±1.27
Food packet, brown aluminum	NI	17.78±7.62
Food packet, plastic	8.38	13.97
Glass wool	NI	NI
Kel-F®	NI	NI
Masking tape	4.32	46.228
Natural rubber	0.254	15.49
Neoprene rubber	NI	(8.13±1.0)
Nylon 101	NI	(4.83±1.27)
Paint, Capon, ivory	NI	9.652±1.016
Paint, Pratt & Lambert, grey	NI	15.24±5.08
Plexiglas®	0.127	(8.89±0.25)
Polyethylene	0.356	(6.35±1.27)
Polypropylene	0.254	(8.89±0.25)
Polystyrene	0.813	(20.32±5.08)
Polyvinyl chloride	NI	(2.54±0.25)
Pump oil	NI	122.606
Refrigeration oil	NI	20.828±1.778
Rubber tubing	0.76	6.096
Silicone® grease	NI	23.368
Solder, rosin core	NI	4.572
Sponge, washing	1.78	205.74±2.54
Teflon® pipe sealing tape	NI	NI
Teflon® tubing	NI	NI
Tygon tubing	4.57	12.7±1.27
Viton A	NI	.076±.051
Wire, Mil W76B, orange	NI	14.478±1.27
Wire, Mil W76B, blue	NI	—
Wire, Mil W76B, yellow	NI	—
Wire, Mil W16878, green	NI	NI
Wire, Mil W16878, black	NI	NI
Wire, Mil W16878, yellow	NI	NI
Wire, Mil 16878, white	NI	NI
Wire, misc., white, ³/₃₂	NI	8.382
Wire, misc., black, ³/₁₆	NI	—
Wire, misc., brown, ⁷/₃₂	NI	12.954±1.27
Wire, misc., yellow ⁷/₆₄	NI	22.606
Wire, misc., yellow, ⁵/₃₂	NI	10.414

NP — No sustained propagation of flame.
NI — No ignition of material.
(Source: NFPA 53, *Guide on Fire Hazards in Oxygen-Enriched Atmosphere*)

1. Characteristics of materials of construction (e.g., flame spread rate, ignition susceptibility),

2. Risk of fire initiation (expectancy),

TABLE 4-9H.   *Typical Measured Burning Rates for Strips of Filter Paper at 45° Angle*

Total pressure		atm. abs. ft. of seawater	Burn Rate, cm/s					
Gas Composition (dry basis)								
%O₂	%N₂ (a)	%He	0.21 atm. —	0.53 atm. —	1.00 atm. 0 ft.	4.03 atm. 100 ft.	7.06 atm. 200 ft.	10.09 atm. 300 ft.
99.6	0.4	0.0	2.32	3.13	4.19	d	d	d
50.3	49.7	0.0	1.13	1.44	2.36	3.72	5.10	6.34
			1.17			3.77	4.06	
20.95,b	79.05	0.0	c	0.80	1.17	1.82	2.80	3.13
					1.17	1.78	2.28	3.25
					1.10			
49.5	0.0	50.5	1.24	1.87	2.96	4.06	4.90	d
				1.90	2.89		4.82	
					2.89			
20.3	0.0	79.7	c	c	c	2.23	2.61	2.49
47.0	24.6	28.4	d	d	2.74	3.66	4.41	5.53
					2.68		4.64	6.78
20.9	39.6	39.5	d	d	1.38	2.28	2.71	3.72
					1.38	2.28	2.83	3.13
					1.35	1.97	2.74	3.56
					1.27	2.28		3.33
						1.81		3.00
						1.72		

a. Includes any argon that was present.
b. Compressed air.
c. Sample would not burn, even with brightly glowing igniter grid.
d. No run was made under these conditions.
(Source: NFPA 53, *Guide on Fire Hazards in Oxygen-Enriched Atmospheres*)

3. Presence of potential energy (e.g., compressed gases, etc.), and
4. Personnel escape routes from occupied areas.

In addition to evaluating the fire-safe characteristics of all materials involved in oxygen-enriched environments under the end-use performance conditions, oxygen service durability is evaluated by: (1) accelerated time tests for deterioration and (2) high-energy tests for degradation. The acceptability of candidate materials for oxygen-enriched atmosphere systems applications is, in part, based on data developed by tests, such as:

1. ASTM D2512, *Test Method for Compatibility of Materials with Liquid Oxygen (Impact Sensitivity Threshold and Pass-Fail Techniques).*
2. ASTM G72, *Standard Test Method for Autogenous Ignition Temperature of Liquids and Solids in a High-Pressure Oxygen-Enriched Environment.*
3. ASTM G74, *Standard Test Method for Ignition Sensitivity of Materials to Gaseous Fluid Impact.*
4. ASTM G86, *Standard Test Method for Determining Ignition Sensitivity of Materials to Mechanical Impact in Pressurized Oxygen Environments.*
5. ASTM G124, *Standard Test Method for Determining the Combustion Behavior of Metallic Materials in Oxygen-Enriched Atmospheres.*
6. ASTM G125, *Standard Test Method for Measuring Liquid and Solid Material Fire Limits in Gaseous Oxidants.*

Other tests are available for the determination of flash and fire points; upward/downward combustion propagation rates; calorific fuel values (heat of combustion); electrical wire insulation, coating, and accessory flammability; electrical overload and hot-wire ignition; odor; toxicity; and off-gassing.

Electrical equipment used in oxygen-enriched atmospheres is limited to that approved at the maximum anticipated oxygen pressure and concentration. Since most commonly used metals burn freely in oxygen-enriched atmospheres, electrical contacts that do not burn in normal atmospheres could burn away and initiate insulation fires in an oxygen-enriched atmosphere. Fire-resistive encapsulation, electrical mineral insulation, and noncombustible organometallic or ceramic (cermet) have been found to be effective for power cables.

The following "fire-stopping" design techniques can minimize ignition potential and fire spread for systems in oxygen-enriched atmospheres:

1. Avoid mass concentration of combustible materials near potential heat or ignition sources.
2. Use spatial separation and configuration to minimize or eliminate flame propagation paths.
3. Create "thermal damping" by judicious placement of fire-resistant heat-sink masses.
4. Install flashover barriers.
5. Create sealed packaging, e.g., inerted compartmentations, fire-resistant encapsulation, etc.
6. Install automatic fire monitoring, e.g., infrared thermography, fire detectors, etc.
7. Use hose line flashback arresters [consult Compressed Gas Association, Inc., (CGA) TB-3, *Hose Line Flashback Arresters.*]

The design of systems that will be used in oxygen-enriched atmospheres should only be undertaken by experienced personnel. All materials that come into contact with oxygen-enriched atmospheres should be cleaned to remove all foreign material. To aid in the education of designers, ASTM Committee G4 has developed a standards technology training course, "Controlling Fire Hazards in Oxygen Handling Systems," which is accompanied by a course

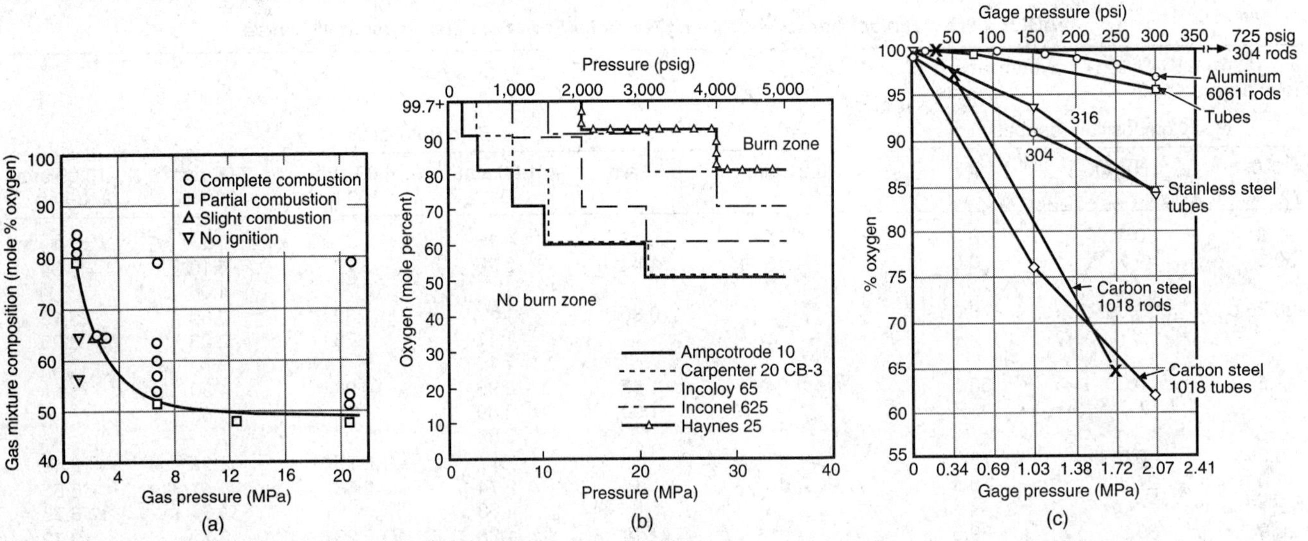

FIG. 4-9F. *Oxygen indices. (Source: NFPA 53, Guide on Fire Hazards in Oxygen-Enriched Atmospheres)*

textbook, *Fire Hazards in Oxygen Systems.* This course was developed by a committee comprising members from the oxygen manufacturers, government, equipment suppliers, and users. The course is based on the experience of committee individuals and their respective organizations. Course content was derived from many of the following documents:

1. CGA G-4.1, *Cleaning Equipment for Oxygen Service.*
2. CGA G-4.4, *Industrial Practices for Gaseous Oxygen Transmission and Distribution Piping Systems.*
3. CGA Video AV-8, "Characteristics and Safe Handling of Cryogenic Liquid Gaseous Oxygen."
4. CGA P-14, *Accident Prevention in Oxygen-Rich and Oxygen-Deficient Atmospheres.*
5. CGA E-2, *Hose Line Check Valve Standards for Welding and Cutting.*
6. European Industrial Gas Association 33/86/E, *Cleaning Equipment for Oxygen Service.*
7. European Industrial Gas Association 5/75/E, *Code of Practice for Supply Equipment and Pipeline Distributing Non-Flammable Gases and Vacuum Services for Medical Purposes.*
8. European Industrial Gas Association 6/77, *Oxygen-Fuel Gas Cutting Machine Safety.*
9. European Industrial Gas Association 8/75/E, *Prevention of Accidents Arising from Enrichment or Deficiency of Oxygen in the Atmosphere.*
10. European Industrial Gas Association 12/80/E, *Pipelines Distributing Gases and Vacuum Services to Medical Laboratories.*
11. European Industrial Gas Association 10/81/e, *Reciprocating Compressors for Oxygen Service.*
12. European Industrial Gas Association 27/82/E, *Turbo Compressors for Oxygen Service, Code of Practice.*
13. NFPA 53, *Guide on Fire Hazards in Oxygen-Enriched Atmospheres.*
14. NFPA 50, *Standard for Bulk Oxygen Systems at Consumer Sites.*
15. NFPA 51, *Standard for the Design and Installation of Oxygen-Fuel Gas Systems for Welding, Cutting, and Allied Processes.*
16. NFPA 99, *Standard for Health Care Facilities.*

17. ASTM G63, *Standard Guide for Evaluating Nonmetallic Materials for Oxygen Service.*
18. ASTM G88, *Standard Guide for Designing Systems for Oxygen Service.*
19. ASTM G94, *Standard Guide for Evaluating Metals for Oxygen Service.*
20. ASTM G128, *Guide for the Control of Hazards and Risks in Oxygen-Enriched Atmospheres.*

FIRE EXTINGUISHMENT IN OXYGEN-ENRICHED ATMOSPHERES

Fire extinguishing systems for use in oxygen-enriched atmospheres face many requirements in addition to those required for conventional systems because of ignition susceptibility, increased flame spread rate and burning intensity, and the flammability of normally flame-resistant materials. In general, these requirements cannot be satisfied by the simple extension of traditional extinguishment techniques. Special instruction and training of emergency personnel is required, as well as careful selection of extinguishing agents and systems.

Most common materials exhibit increased burning rate in oxygen-enriched atmospheres. To protect occupants and real property, fire extinguishants must: (1) act rapidly to be effective, (2) be inherently nontoxic, and (3) not produce toxic or corrosive decomposition products.

Water has proved to be an effective extinguishant in oxygen-enriched atmospheres when applied in sufficient quantities. Water at a spray density of $1\frac{1}{4}$ gpm per sq ft (50 L/min/m^2) applied for 2 min will extinguish cloth burning in 100 percent oxygen at atmospheric pressure. The application method of the water is all-important. Water extinguishing systems must be carefully designed such that: (1) all protected space is covered by the minimum spray density, and (2) water is distributed to a depth sufficient to extinguish stratified fires in nonhomogeneous materials, e.g., layers of cloth in clothing.

The use of conventional halogenated extinguishants has been restricted by the Montreal Protocol, due to the ozone-depleting po-

TABLE 4-9I(A). *Threshold Pressures^a in Oxygen of 3.2-mm (0.13-m) Diameter Rods Ignited at the Bottom*

Material	Threshold Pressure^a	
	(MPa)	(psia)
Monel® K-500	>68.9	>10,000
Inconel MA754	>68.9	>10,000
Monel® 400	>68.9	>10,000
Brass 360 CDA	>68.9	>10,000
Nickel 200	>55.2	>8,000
Copper 102	>55.2	>8,000
Hastelloy C276	20.7	3,000
Inconel® 600	17.2	2,500
Inconel® 625	17.2	2,500
Hastelloy C22	13.8	2,000
Inconel 718	6.9	1,000
440C SS	6.9	1,000
316 SS	6.9	1,000
304 SS	6.9	1,000
17-4 PH SS	6.9	1,000
Weldalite 049	2.1	300
Aluminum 8090	2.1	300
Aluminum 2219	0.2	35
HSLA steel^b	0.17	25
Aluminum 99.9%	0.17	25
Ti-6Al-4V	0.007	1

^aThreshold pressure is the minimum test pressure required to support complete combustion of the test sample.
^bDenotes high-strength low-alloy steel.
(Source: NFPA 53, *Guide on Fire Hazards in Oxygen-Enriched Atmospheres*)

TABLE 4-9I(B)[54] *Threshold Pressures^a in Oxygen of 60 × 60 Wire Meshes Rolled into 6.4-mm (0.25-in.) Diameter Cylinders Ignited at the Bottom*

Material	Threshold Pressure^a	
	(MPa)	(psia)
Nickel 200	>69	>10,000
Copper 100	0.3	47
Monel® 400	≤0.085	≤12.4
316 stainless steel	≤0.085	≤12.4
304 stainless steel	≤0.085	≤12.4
Carbon steel	≤0.085	≤12.4

^aThreshold pressure is the minimum test pressure required to support complete combustion of the test sample.
(Source: NFPA 53, *Guide on Fire Hazards in Oxygen-Enriched Atmospheres*)

tential of the agents themselves as well as their decomposition products. Halon® 1301 (bromotrifluoromethane) has proved to be effective in extinguishing fires in oxygen-enriched atmospheres. However, its use has been restricted, i.e., existing systems can remain in place, but replacement due to use or leakage is no longer allowed.

Use of other extinguishants has been examined. Diluents, such as carbon dioxide and nitrogen, have been employed. Carbon dioxide systems have become widely used in unoccupied computer rooms. Note, however, their use must be restricted in occupied areas. No reliance should be placed on blankets made of Nomex®, wool, or fiberglass.

New technology in the use of high-expansion, water-based foams has resulted in improved extinguishing properties. However, the applicability of these agents to each particular candidate system must be carefully evaluated with regard to the available space, required time, and application methods. More data has become available on the use of low-expansion foam and dry chemical extinguishants in normal atmospheres, but their applicability in oxygen-enriched atmospheres is still unknown.

Rapid flame spread in oxygen-enriched atmospheres requires that fire extinguishing systems be capable of fast automatic actuation by fire detectors, as well as by manual actuation. Fixed systems utilize an extinguishing agent acceptable for use on fires in oxygen-enriched atmospheres, and automatic actuation occurs in less than 1 sec of the perception of detectable flame. In addition to the automatic fixed system in occupied areas, a manually operated water hose not less than $1/_2$-in. (13-mm) inside diameter, and with an effective nozzle pressure of not less than 50 psi (345 kPa) above the ambient pressure, is usually provided.

BIBLIOGRAPHY

References Cited

1. Sircar, S., Gabel, H., Stoltzfus, J. M., and Benz, F. J., "The Analysis of Metals Combustion Using Real-Time Gravimetric Technique," *Symposium on Flammability and Sensitivity of Materials in Oxygen-Enriched Atmospheres*, Vol. 5, ASTM STP 1111, J. M. Stoltzfus and K. McIlroy, eds., American Society for Testing and Materials, W. Conshohocken, PA, 1991, pp. 313–325.
2. Steinberg, T. A., Mulholland, G. P., Wilson, D. B., and Benz, F. J., "The Combustion of Iron in High-Pressure Oxygen," *Combustion and Flame 89*, Elsevier Science Publishing Co., Inc., 1992, pp. 221-228.
3. Wilson, D. B., and Steinberg, T. A., "The Combustion Phase of Burning Metals," *Combustion and Flame*, Vol. 91, No. 2, Nov. 1992, pp. 200–208.
4. Mellor, A. M., "Heterogeneous Ignition of Metals: Model and Experiment," Final Report, NASA Grant NSG-641, Princeton University, Princeton, NJ, 1967.
5. Monroe, R. W., Bates, C. E., and Pears, C. D., "Metal Combustion in High-Pressure Flowing Oxygen," *Symposium on Flammability and Sensitivity of Materials in Oxygen-Enriched Atmospheres*, Vol. 5, ASTM STP 812, B. L. Werley, ed., American Society for Testing and Materials, W. Conshohocken, PA, 1991.
6. Kuchta, J. M., and Cato, R. J., "Review of Ignition and Flammability Properties of Lubricants," Tech. Report AFAPL-TR-67-126, pub. no. AD665121, Jan. 1968, Air Force Aero Propulsion Laboratory, Pasadena, CA.
7. Guest, P. G., "Oily Fibers May Increase Oxygen Then Fire Hazard," *The Modern Hospital*, May 1965.
8. Voigtsberger, P., "The Ignition of Oxygen-Impregnated Tissues by Electrostatic and Mechanically Produced Sparks," SMRE Translation No. 5005, June 1964, British Ministry of Power, London.
9. NRL Memorandum Report 5508, "Flashover Failures from Wet Wire Arcing and Tracking," National Technical Information Service, Atlanta, GA.
10. "Kapton® Wire Arc Track Testing," NASA Report CR 185642, National Technical Information Service, Atlanta, GA.
11. White, E. L., and Ward, J. J., "Ignition of Metals in Oxygen," DMIC Report 224, Feb. 1966, Defense Metals Information Center, Battelle Memorial Institute, Columbus, OH.
12. Benz, F. J., Shaw, R. C., and Homa, J. M., "Burn Propagation Rates of Metals and Alloys in Gaseous Oxygen," *Flammability and Sensitivity of Materials in Oxygen-Enriched Atmospheres*, Vol. 2, ASTM STP 910, M. A. Benning, ed., American Society for Testing and Materials, W. Conshohocken, PA, 1986, pp. 135–152.
13. Zawierucha, R., McIlroy, K., and Mazzarella, R. B., "Promoted Ignition Combustion Behavior of Selected Hastelloys in Oxygen Gas Mixtures," *Flammability and Sensitivity of Materials in Oxygen-Enriched Atmospheres*, Vol. 5, ASTM STP 1111, J. M. Stoltzfus and K. McIlroy, eds., American Society for Testing and Materials, W. Conshohocken, PA, 1991, pp. 270–287.

14. Sato, J., "Fire Spread Rates along Cylindrical Metal Rods in High-Pressure Oxygen," *Flammability and Sensitivity of Materials in Oxygen-Enriched Atmospheres*, Vol. 4, ASTM STP 1040, J. M. Stoltzfus, F. J. Benz, and J. S. Stradling, eds., American Society for Testing and Materials, W. Conshohocken, PA, 1989, pp. 162–177.

References

"ASRDI Oxygen Technology Survey, Vol. IX: Oxygen Systems Engineering Review," NASA SP-3090, 1978, National Technical Information Service, Springfield, VA.

"Design Guide for High-Pressure Oxygen Systems," N8332990, 1983, National Technical Information Service, Springfield, VA.

"Flammability and Sensitivity of Materials in Oxygen-Enriched Atmospheres," ASTM STP 812, B. L. Werley, ed., American Society for Testing and Materials, W. Conshohocken, PA, 1983.

"Flammability and Sensitivity of Materials in Oxygen-Enriched Atmospheres," Vol. 2, ASTM STP 910, M. A. Benning, ed., American Society for Testing and Materials, W. Conshohocken, PA, 1986.

"Flammability and Sensitivity of Materials in Oxygen-Enriched Atmospheres," Vol. 3, ASTM STP 986, D. W. Schroll, ed., American Society for Testing and Materials, W. Conshohocken, PA, 1988.

"Flammability and Sensitivity of Materials in Oxygen-Enriched Atmospheres," Vol. 4, ASTM STP 1040, J. M. Stoltzfus, F. J. Benz, and J. S. Stradling, eds., American Society for Testing and Materials, W. Conshohocken, PA, 1989.

"Flammability and Sensitivity of Materials in Oxygen-Enriched Atmospheres," Vol. 5, ASTM STP 1111, J. M. Stoltzfus and K. McIlroy, eds., American Society for Testing and Materials, W. Conshohocken, PA, 1991.

"Flammability and Sensitivity of Materials in Oxygen-Enriched Atmospheres," Vol. 6, ASTM STP 1197, D. D. Janoff and J. M. Stoltzfus, eds., American Society for Testing and Materials, W. Conshohocken, PA, 1993.

"Flammability and Sensitivity of Materials in Oxygen-Enriched Atmospheres," Vol. 7, ASTM STP 1267, D. D. Janoff, W. T. Royals, and M. V. Gunaji, eds., American Society for Testing and Materials, W. Conshohocken, PA, 1995.

"Materials Safety Data Sheet for Gaseous Oxygen," L-4638A, Praxair, Inc., Linde Division Communications Dept., Danbury, CT.

"Materials Safety Data Sheet for Liquid Oxygen," L-4637B, Praxair, Inc., Linde Division Communications Dept., Danbury, CT.

"Precautions and Safety Practices for Gas Welding, Cutting, and Heating," L-2035JJ, Praxair, Inc., Linde Division Communications Dept., Danbury, CT.

"Safety Precautions for Oxygen, Nitrogen, Argon, Helium, Carbon Dioxide, Acetylene, FG-2, Ethylene Oxide, Sterilent Mixtures, and Speciality Gases," L-3499H, Praxair, Inc., Linde Division Communications Dept., Danbury, CT.

UL 252, *Standard for Safety Compressed Gas Regulators*, 6th ed., Underwriters Laboratories, Inc., Northbrook, IL, 1991.

UL 407, *Standard for Safety Manifolds for Compressed Gases*, 5th ed., Underwriters Laboratories Inc., Northbrook, IL, 1993.

NFPA Codes, Standards, and Recommended Practices

Reference to the following NFPA codes, standards, and recommended practices will provide further information on oxygen-enriched atmospheres discussed in this chapter. (See the latest version of *The NFPA Catalog* for availability of current editions of the following documents.)

NFPA 50, *Standard for Bulk Oxygen Systems at Consumer Sites*

NFPA 51, *Standard for the Design and Installation of Oxygen-Fuel Gas Systems for Welding, Cutting, and Allied Processes*

NFPA 53, *Guide on Fire Hazards in Oxygen-Enriched Atmospheres*

NFPA 99, *Standard for Health Care Facilities*

NFPA 255, *Standard Method of Test of Surface Burning Characteristics of Building Materials*

NFPA 325, *Guide to Fire Hazard Properties of Flammable Liquids, Gases, and Volatile Solids*

NFPA 491M, *Manual of Hazardous Chemical Reactions*

Additional Reading

Abbud-Madrid, A., *et al,* "Ignition of Bulk Metals by a Continuous Radiation Source in a Pure Oxygen Atmosphere," University of Colorado, Boulder, CO, Center for Combustion Research, Mechanical Engineering Dept. ASTM STP 1197, American Society for Testing and Materials Flammability and Sensitivity of Materials in Oxygen-Enriched Atmospheres, Volume 6, Symposium Sponsored by ASTM Committee G-4 on Compatibility and Sensitivity of Materials in Oxygen Enriched Atmospheres, May 11–13, 1993, Noordwijk, The Netherlands, ASTM, Philadelphia, PA, 1993, pp. 211–222.

Barthelemy, H., Delode, G., Vagnard, G., "Ignition of Materials in Oxygen Atmospheres: Comparison of Different Testing Methods for Ranking Materials," ASTM STP 1111, American Society for Testing and Materials, Flammability and Sensitivity of Materials in Oxygen-Enriched Atmospheres, Vol. 5, Symposium Sponsored by ASTM Committee G-4 on Compatibility and Sensitivity of Materials in Oxygen Enriched Atmospheres, May 14–16, 1991, Cocoa Beach, FL, ASTM, Philadelphia, PA, 1991, pp. 506–515.

Christianson, R., and Stoltzfus, J. M., "Selecting Materials for Use in Oxygen-Enriched Atmospheres," Lockheed, Las Cruces, NM, 93-WA/PID-2; American Society of Mechanical Engineers (ASME) Winter Annual Meeting, November 28–December 3, 1993, New Orleans, LA, 1993, pp. 1–9.

de Richemond, A. L., and Bruley, M. E., "Insidious Iatrogenic Oxygen-Enriched Atmospheres as a Cause of Surgical Fires," ECRI Plymouth Meeting, PA, ASTM STP 1197, American Society for Testing and Materials, Flammability and Sensitivity of Materials in Oxygen-Enriched Atmospheres, Vol. 6, Symposium Sponsored by ASTM Committee G-4 on Compatibility and Sensitivity of Materials in Oxygen Enriched Atmospheres, May 11–13, 1993, Noordwijk, The Netherlands, ASTM, Philadelphia, PA, 1993, pp. 66–73.

Feiereisen, T. J., *et al.,* "Gravity and Pressure Effects on the Steady-State Temperature of Heated Metal Specimens in Pure Oxygen Atmosphere, University of Colorado, Boulder, Cleveland, OH, ASTM STP 1197, American Society for Testing and Materials, Flammability and Sensitivity of Materials in Oxygen-Enriched Atmospheres, Vol. 6, Symposium Sponsored by ASTM Committee G-4 on Compatibility and Sensitivity of Materials in Oxygen Enriched Atmospheres, May 11–13, 1993, Noordwijk, The Netherlands, ASTM, Philadelphia, PA, 1993, pp. 196–210.

Glassman, I., "Combustion Fundamentals of Low Volatility Materials in Oxygen-Enriched Atmospheres," ASTM STP 1111, American Society for Testing and Materials, Flammability and Sensitivity of Materials in Oxygen-Enriched Atmospheres, Vol. 5, Symposium Sponsored by ASTM Committee G-4 on Compatibility and Sensitivity of Materials in Oxygen-Enriched Atmosphere, May 14–16, 1991, Cocoa Beach, FL, ASTM, Philadelphia, PA, 1991, pp. 7–25.

Gunaji, M. V., Steinberg, T. A., and Beeson, H. D., "Test Methods for Selecting Materials for Use in Oxygen-Enriched Atmospheres," *Proceedings* of the ASME Winter Conference, American Society of Mechanical Engineers, New York, NY, 1993, pp. 1–9.

Hshieh, F. Y., *et al.,* "Flammability Testing of Materials in Oxygen-Depleted and Oxygen-Enriched Environments Using a Controlled-Atmosphere Cone Calorimeter," Lockheed, Las Cruces, NM, National Aeronautics and Space Administration, White Sands Testing Facility, NM, CIB, W14/93/2 (J); Tsukuba Building Test Laboratory, Center for Better Living, *First (1st) Proceedings* of Japan Symposium on Heat Release and Fire Hazard, Session 3, Scope for Next-Generation Fire Safety Testing Technology. May 10–11, 1993, Tsukuba, Japan, 1993, pp. III/15–24.

Richardson, T. F., Santoro, R. J., and Kirby, M. J., "Effect of Inert Diluent Addition on Diffusion Flame Height in an Oxygen Atmosphere," 1990 Fall Technical Meeting of the Combustion Institute/Eastern States Section on Chemical and Physical Processes in Combustion, December 3–5, 1990, Orlando, FL, 1990, pp. 30/1–4.

Sato, J., Ohtani, H. and Hirano, T., "Ignition Process of a Heated Iron Block in High-Pressure Oxygen Atmosphere," *Combustion and Flame*, Vol. 100, No. 3, 1995, pp. 376–383.

Steinberg, T. A., and Benz, F. J., "Iron Combustion in Microgravity," *Symposium on Flammability and Sensitivity of Materials in Oxygen-Enriched Atmospheres*, Vol. 5, ASTM STP 1111, J. M. Stoltzfus and K.

McIlroy, eds., American Society for Testing and Materials, W. Conshohocken, PA, 1991, pp. 298–312.

Steinberg, T. A., Wilson, D. B., and Benz, F. J., "Burning of Metals and Alloys in Microgravity," *Combustion and Flame*, Elsevier Publishing Co., Vol. 88, 1992, pp. 309–320.

Steinberg, T.A., Wilson, D.B., and Benz, F.J., "The Combustion Phase of Burning Metals," *Combustion and Flame*, Elsevier Publishing Co., Vol. 91, 1992, pp. 200–208.

Weiss, E. S., and Sapko, M. J., "Oxygen Balanced Emulsion-Anfo Blends for Use in Flammable Atmospheres," Sixth Annual Symposium on Explosive and Blasting Research, 1990, pp. 193–204

PLASTICS AND RUBBER

Revised by Bert M. Cohn

Since World War II, the types, variations, and uses of plastics have proliferated at a tremendous rate, producing a family of materials that is extraordinarily complex and varied. Each of the variations can be considered combustible. Some burn very rapidly, and under certain conditions present extreme hazards to life and property. When such hazardous characteristics are identified, modifications can be made in manufacture and application that can reduce the hazard to acceptable levels. For example, cellulose nitrate motion picture film presents the hazard of flash burning. This hazard was controlled by phasing out the cellulose nitrate film and replacing it with slow-burning "safety" film. More recently, the hazards of cellular plastics, such as polyurethane and polystyrene, which were used for exposed interior finish in buildings, have been controlled by using flame-retardant additives during manufacture combined with thermal barriers installed in the field.

This chapter focuses on the fire behavior of plastics that are used as building materials and on the characteristics of polymers in general. The tests that are used to identify the relative hazards from fire and smoke of plastics used in building applications are also discussed. Plastics in fibers and textiles (including clothing, upholstery fabrics, and mattresses) are discussed in Section 4, Chapter 5, "Fibers and Textiles." The fire hazards associated with the manufacturing, processing, and warehousing of plastics and elastomers are discussed in Section 3, Chapter 18, "Extrusion and Molding Processes."

While being a part of the family of polymers, natural and synthetic rubbers, technically, are not plastics. Yet, because of the many similarities and the formulations that intermix the traditional characteristics of plastics and elastomers, rubbers are often included in discussions of plastics. In this chapter, the general term "plastic" is intended to include rubber, but it should be understood that rubber has a whole complex technology of its own—one that predates the modern plastics industry.

PLASTICS TERMINOLOGY

The following terms are used in this chapter or are commonly encountered in the application, use, and storage of plastics products.

Additive: Any material mixed with a resin to modify its processing or end-use properties. The resulting mix usually is called a "plastic" to distinguish it from the "resin" or principal ingredient. Additives may be dyes, pigments, powdered fillers for stiffening, plasticizers for flexibility, fibers to reinforce, antioxidants, lubricants to aid flow into or release from a mold, or flame retardants.

Binders: The resins in a plastic mixture that hold together all of the other ingredients. They are usually a product of polymerization or a naturally occurring substance of high molecular weight.

Blowing agent: A material that releases gas upon heating so that the plastic in which it is mixed will expand into foam. The gas may result from boiling of a liquid or from decomposition of the blowing agent (foaming agent).

Copolymers: High molecular weight compounds produced when two or more monomers are involved in a polymerization process.

Cross-linking: The establishing of chemical bonds between adjacent molecules to make a network that will reduce solubility or resist flow on heating. The amount of cross-linking may be enough to render the plastic highly resistant to flow at temperatures up to thermal decomposition, in which case the plastic is said to be a thermoset.

Fillers: Materials that modify the strength and working properties of a plastic. They may be used to increase heat resistance or alter the dielectric strength. A wide variety of products are used, including wood flour, cotton, sisal, glass, and clay.

Film: A general term for plastic not more than 0.01 in. (0.25 mm) thick, regardless of the process used to make it. "Foil" is used today only to describe metal.

Foam: Plastic with many small gas bubbles; cellular plastic.

Foaming agent: The same as a blowing agent.

Inhibitor: A material added to prevent or retard an undesired chemical reaction.

Laminate: A composition of two or more layers of plastic firmly adhered either by partly melting one or more of the layers, by an adhesive, or by impregnation. High-pressure laminates are rigid stock made in presses at pressures above 400 psi (2760 kPa).

Monomers: The small starting molecules, usually gaseous or liquid, used to produce the polymer resins.

Bert M. Cohn, P.E., is president of Bert Cohn Associates, Inc., Elmhurst, IL.

Plastic (ASTM definition): A material that contains as an essential ingredient one or more organic polymeric substances of large molecular weight, is solid in its finished state, and at some stage in its manufacture or processing into finished articles can be shaped by flow.

Plasticizers: Organic materials added to plastics to make the finished product more flexible or to facilitate compounding. Some plasticizers increase the combustibility of the plastic while others serve as flame retardants.

Polymerization: The process by which molecules of a monomer are made to add to themselves or to other monomers in a repetitive way, forming a much longer chainlike molecule.

Polymers: Large molecules made by combining smaller molecules by chemical reaction.

Reinforced plastic: A plastic with a filler that significantly increases flexural, impact, or tensile strength. Additives are usually glass, asbestos, cotton, or nylon fibers.

Resin (ASTM definition): A solid or pseudosolid organic material often of high molecular weight, which exhibits a tendency to flow when subjected to stress, usually has a softening or melting range, and usually fractures conchoidally. Note: In a broad sense, the term is used to designate any polymer that is a basic material for plastics.

Sheet: A general term for plastics 0.01 in. (0.25 mm) and thicker, regardless of the method of manufacture.

Stabilizer: A material added to prevent or retard an undesired chemical reaction. When used to prevent polymerization, the preferred term is "inhibitor." Stabilizers usually are present in less than 1 percent. They may be antioxidants, screening agents for ultraviolet radiation, or chemicals to neutralize decomposition products during hot processing.

Thermoforming: The process of stretching or bending a plastic sheet that has been softened by heating.

Thermoplastic (ASTM definition): A plastic that can be repeatedly softened by heating and hardened by cooling through a temperature range characteristic of the plastic and that in the softened state can be shaped by flow into articles by molding or extrusion.

Thermoset (ASTM definition): A plastic that, after having been cured by heat or other means, is substantially infusible and insoluble.

THE FAMILY OF POLYMERS

In the classification of engineering materials, polymers are one division under the general category of nonmetals, which also includes ceramics, glasses, and materials such as oils, greases, lubricants, papers, etc. Polymers are further subdivided into three classes—elastomers, thermosets, and thermoplastics—as shown in Figure 4-10A. Basically, all polymers are characterized by their very large molecules, formed in chemical "polymerization" reactions, which link two or more molecules into larger molecules. These larger molecules contain repeating structural units of the original molecules. If the linkages result in a rigid "cross-linked" molecular structure, a thermosetting plastic results. If the molecular structure is more flexible, including side chains not rigidly linked to the other molecules, the resulting material is a thermoplastic.

The products known as "plastics" include the common types of thermosetting and thermoplastic polymers included in Figure

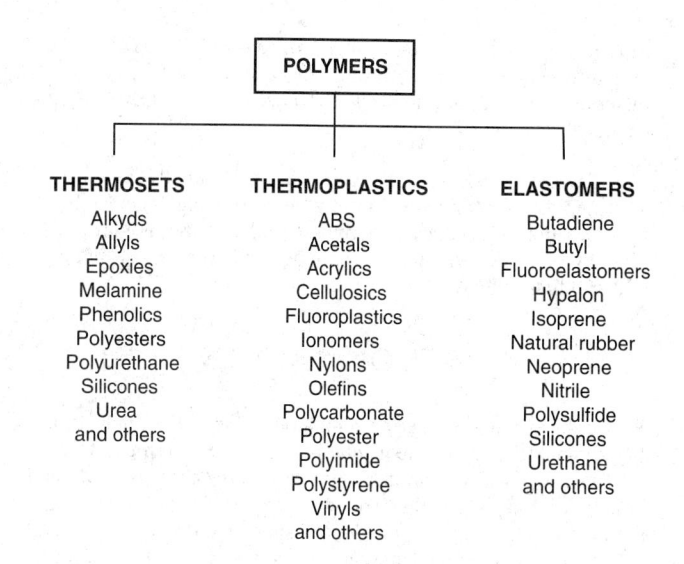

FIG. 4-10A.   The family of polymers.

4-10A, plus many more. In addition to the polymer constituent, most finished plastics contain plasticizers, colorants, fillers, stabilizers, reinforcing agents, lubricants, or other special additives. The variation of individual plastics-product formulations runs into the thousands. The ultimate form of each product, whether it appears in solid sections, films and sheets, foams, molded forms, synthetic fibers, pellets, or powders, has a significant effect on its fire properties. Additionally, the fire properties are influenced by the environment, and the resultant hazard by the quantity of the product and the conditions of its use. For these reasons, it is important to conduct fire tests of products in their end-use configuration and under conditions that simulate the end-use environment.

Elastomers

Elastomers are commonly referred to as "synthetic rubbers" and are characterized by their elastic or rubberlike properties. As a whole, this class of polymers is intended to reproduce or surpass the best properties of natural rubber for particular applications. The fire characteristics of elastomers are usually similar to those of natural rubber. A discussion of both natural and synthetic rubbers is included in this chapter.

Thermosetting Plastics

Thermosetting plastics, or thermosets, are hardened into a permanent shape in the manufacturing process. These materials cannot usually be softened by heating. For those that can be softened, the parts cannot be remelted or restored to their original state before curing. The components of thermosets can be purchased in liquid form as either monomer/polymer mixtures or as partially polymerized molding compounds. Thermosets are usually formed into their final shape with heat (sometimes under pressure), although some are cured at room temperature.

Thermoplastics

Thermoplastics do not cure or set into a permanent shape under the application of heat since heat will cause them to soften. They can be softened and rehardened repeatedly, although eventual thermal aging will cause a degradation of the material and limit the number of reheat cycles. When heated to a flowable state, thermoplastics can be transferred to a mold and cooled into the shape of the mold. Thermoplastics can be extruded through dies and solidified around other materials, such as insulation over electrical wiring. Generally, no chemical reaction takes place.

There is some overlap between thermosets and thermoplastics. Thermoplastics, after extrusion, can be cross-linked chemically or by irradiation to produce a thermoset product. The molding techniques typical of thermoplastics have been modified to permit injection molding of thermosets, such as phenolics.

USES OF PLASTICS

Plastics have wide ranging and continuously expanding applications. In building construction applications, plastics can be found in thermal and acoustical insulation, roofing materials, decorative trims and imitation wood parts, ceiling and wall panels, resilient flooring, light fixture diffusers, electrical conduits, insulation on wiring, water and gas piping, bathtubs and sinks, and heating ducts. Plastics are found in the housings and parts of equipment and appliances. Entire pieces of furniture, seat cushions, mattresses, carpeting, wall coverings, and draperies are often plastics, as are countertops, desktops, and tabletops. Textiles in clothing today often are partially or totally plastics products. Other uses of plastics include major parts, trim, and interiors of automobiles, buses, airplanes, and trains; packaging materials, including plastic wrap; housewares; sports equipment; toys; photographic films; magnetic tapes; phonograph records; and luggage.

Lack of stability under high temperatures and inherent combustibility so far have ruled out the use of plastics for applications where fire-resistance ratings are a requirement. One exception is the use of special silicone plastic foams for sealing openings in fire walls; the material retains its sealing and insulating properties while it slowly burns away during fire exposure.

ADDITIVES AND BLENDS

The physical properties of the basic types of polymers can be altered substantially during manufacturing by a variety of techniques. An extraordinary number of combinations of special properties are obtainable through the blending, or alloying, of one polymer with another. Additives, such as carbon black, glass fibers, flame retardants, and fillers ranging from cotton to clay, can alter not only the strength and other properties needed for the finished product but also the response to heat and flame.

Fiberglass and other threads added to plastics (usually, polyethylene) are used in the manufacture of fiber-reinforced plastic (FRP) products, such as boat hulls, automobile body parts, fishing rods, skis, motor homes, swimming pools, bathtubs, shower enclosures, building panels, and translucent roofing and siding. While flame retardants can be added to lower the ASTM E84, *Standard Test Method for Surface Burning Characteristics of Building Materials*, flamespread index to a range of 20 to 75, general-purpose FRP panels and other products are likely to have much higher surface flamespread ratings, in the range of 150 to 400.[1]

MAJOR GROUPS OF PLASTICS

The major groups of plastics are briefly described in the following paragraphs. Hundreds of variations are sold under thousands of individual trade names. More detailed information can be obtained from consulting the standard reference works of the industry. (See references at the end of this chapter.)

Positive identification of the plastic used in a product is often very difficult. Guidelines are available to assist in the identification of the more common types. Braun[2] sets forth simple laboratory tests using heat (pyrolysis), flame, and melting, compiled in a small, easy-to-read reference volume. Saechtling's plastics identification material is included. (See Figures 4-10B through 4-10E.)

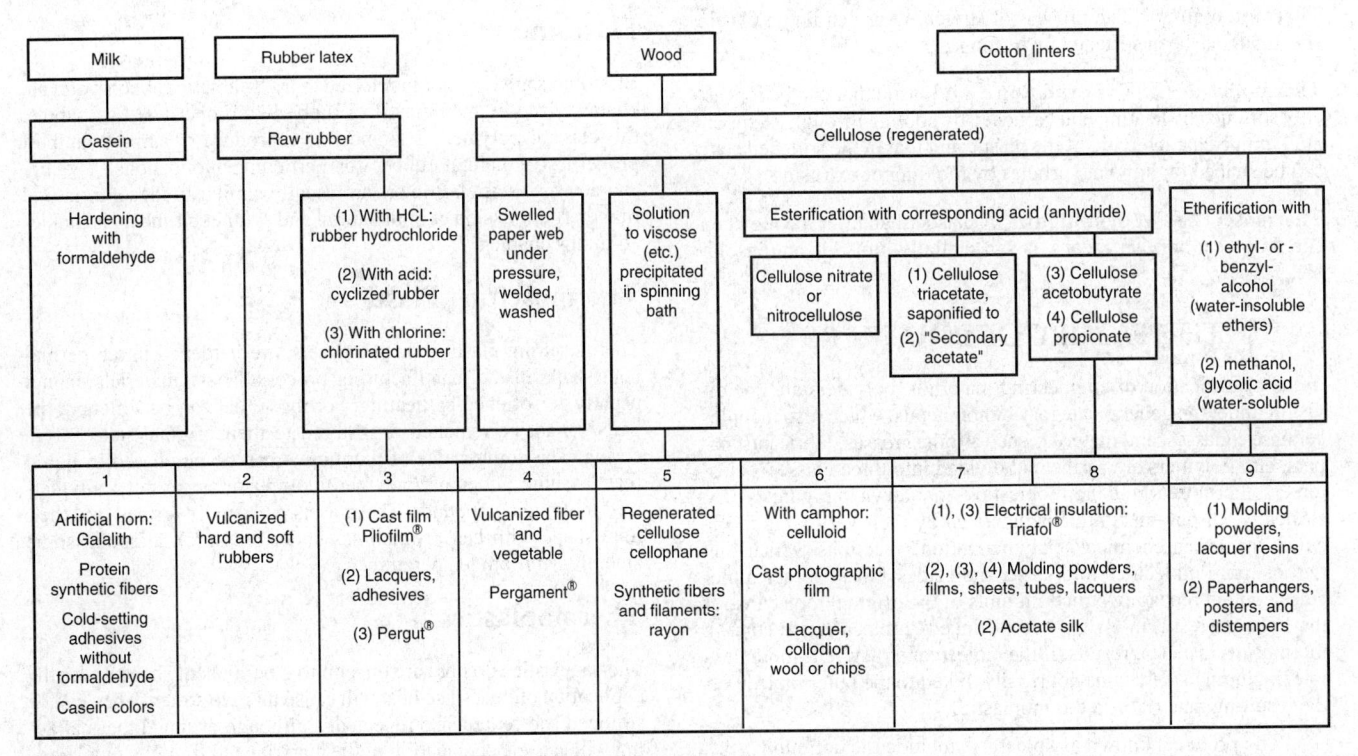

FIG. 4-10B. *Plastics from natural products.*[2,3]

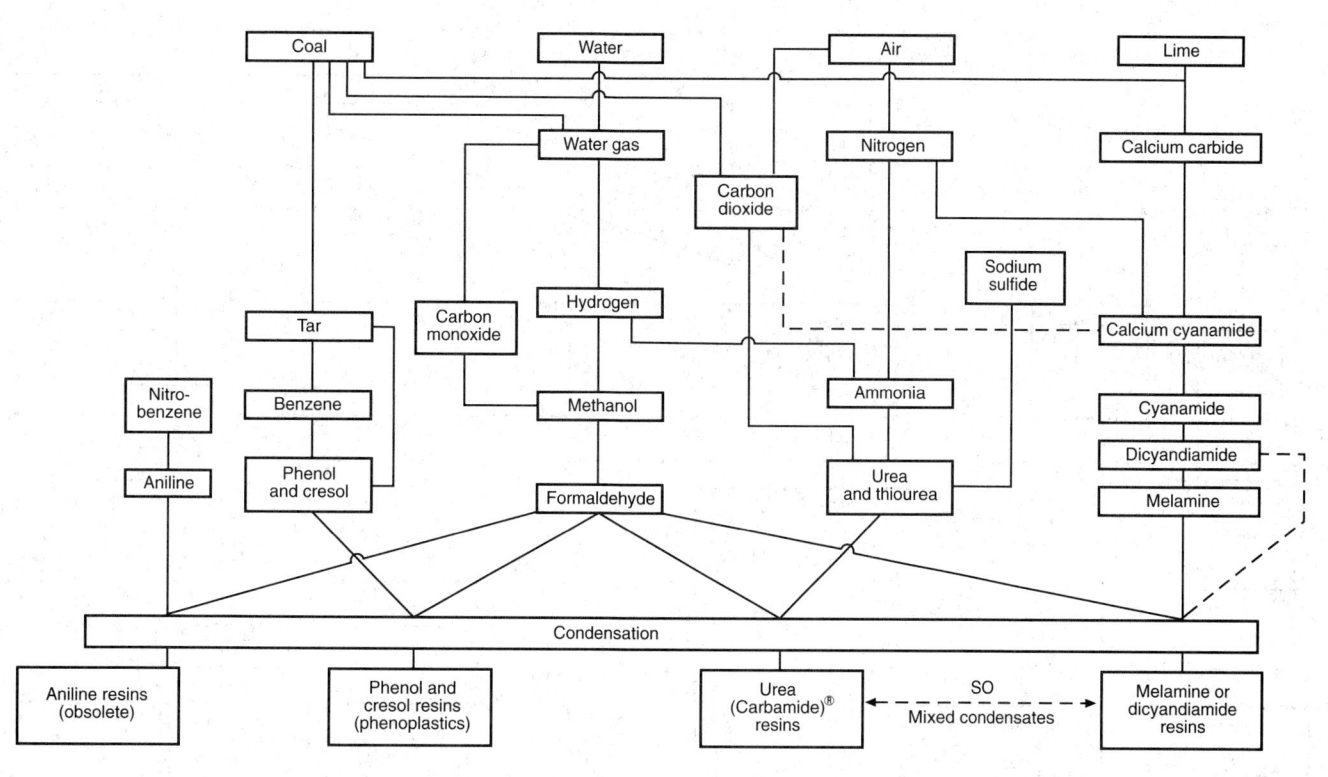

FIG. 4-10C.   *Classical condensation resins—thermosets.*²,³

Figure 4-10F presents a number of simple tests that take advantage of the unique characteristics that many common plastics possess.

According to Saechtling,³ plastics can be divided into four groups that reflect their derivation, method of synthesis, and historical development: (1) plastics from natural products, starting with vulcanized fiber in 1859, celluloid in about 1870, and artificial horn in 1897 as the oldest plastics (see Figure 4-10B); (2) "classical" condensation resins, including thermosets stemming from the discovery of Bakelite in about 1910 (see Figure 4-10C); (3) Staudinger polymerized plastics, deriving from foundations in 1922 for the polymerization of unsaturated monomers into long-chain macromolecules (see Figure 4-10D); and (4) plastics from multifunctional aliphatic and aromatic intermediate products, derived from work since the 1950s in combining various intermediates into a series of general-purpose and specially tailored engineering plastics (see Figure 4-10E.)

Commonly accepted abbreviations have been developed for many plastics materials, and some of these have been standardized by American Society for Testing and Materials (ASTM D1600, *Standard Terminology for Abbreviated Terms Relating to Plastics*).⁵ Manufacturers are being urged to offer the abbreviation along with trademark names to aid in the rapid identification of the material. This is particularly important for plastics products used in building construction. For the purposes of this text, the abbreviation is noted either in parentheses in the heading or after the name of the plastic in the descriptions that follow.

ABS

Developed to offset some of the shortcomings of polystyrene, ABS plastics are copolymers of acrylonitrile, butadiene, and styrene monomers in varying proportions. They are balanced for mechanical toughness, good service temperature for other-than-high temperature applications, and ease of fabrication. ABS materials have relatively good electrical insulating properties, good resistance to a wide range of chemicals, and good dimensional stability. Typical applications include automobile dashboards, pump impellers, refrigerator parts, appliance housings, luggage, protective helmets, battery cases, telephones, pipe and fittings, shoe heels, knobs, and tool handles. Typical tradenames include Abson, Cycolac™, GravoFLEX™, Kralastic, Ravikral, Styrophane, and Terluran.

Acetals (AR)

The acetal resins are made primarily by the polymerization of formaldehyde. Copolymers of acetal are also produced. These materials form one of the strongest and stiffest thermoplastics. The products show good resistance to organic solvents; however, they do react with strong acids or bases. These plastics are widely used for equipment gears, automotive equipment, plumbing parts, and appliances. Two basic types are: (1) a homopolymer (e.g., DuPont's Deltrin) and (2) a copolymer (e.g., Celanese's Celcon®).

Acrylics

The acrylics are basically polymethyl methacrylate (PMMA). Marketed under the common trade names Lucite™ and Plexiglas™, PMMA is known for its excellent clarity, high transparency, good resistance to sharp blows, strength, and rigidity. It is odorless, tasteless, and nontoxic, but the surface scratches easily, and contact with grit and harsh scouring agents must be avoided. Because it is a thermoplastic, it loses its shape in boiling water, although some cast acrylics have higher heat resistance. In sheets of varying thickness and color, PMMA is used for glazing, building panels, advertising displays, and protective shields. Solutions and emulsions of PMMA are produced for use as surface coatings, adhesives, and finishes. In rods and tubes, it is used for electric lighting and medical illuminating equipment, which utilize its capacity to transmit light along its length and around curves. In its molded form, it is used for lenses, dentures, toilet articles, and telephones.

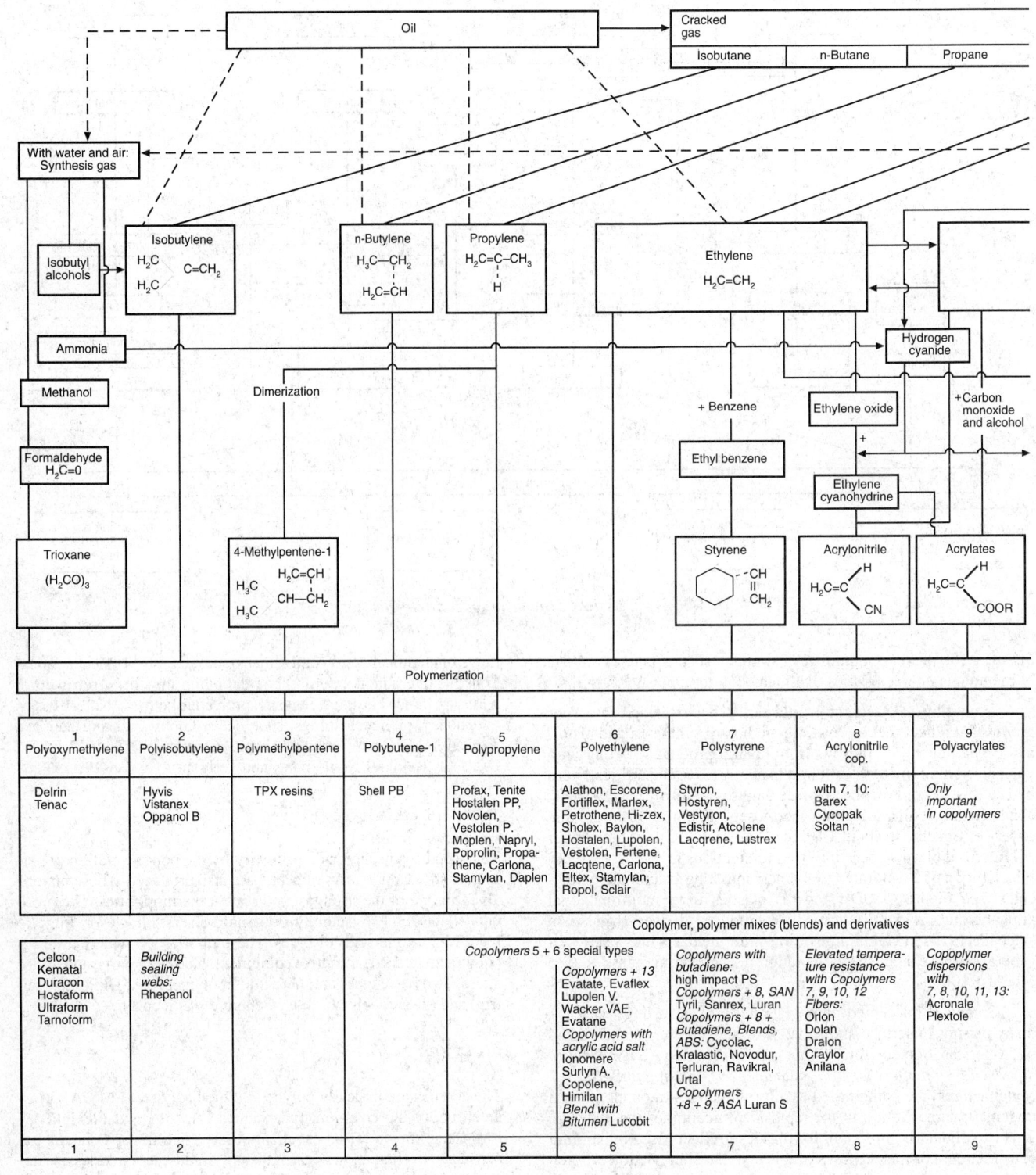

FIG. 4-10D. Double-bond polymerization thermoplastics.[2,3]

1 Polyoxymethylene	2 Polyisobutylene	3 Polymethylpentene	4 Polybutene-1	5 Polypropylene	6 Polyethylene	7 Polystyrene	8 Acrylonitrile cop.	9 Polyacrylates
Delrin Tenac	Hyvis Vistanex Oppanol B	TPX resins	Shell PB	Profax, Tenite Hostalen PP, Novolen, Vestolen P. Moplen, Napryl, Poprolin, Propa-thene, Carlona, Stamylan, Daplen	Alathon, Escorene, Fortiflex, Marlex, Petrothene, Hi-zex, Sholex, Baylon, Hostalen, Lupolen, Vestolen, Fertene, Lacqtene, Carlona, Eltex, Stamylan, Ropol, Sclair	Styron, Hostyren, Vestyron, Edistir, Atcolene Lacqrene, Lustrex	with 7, 10: Barex Cycopak Soltan	*Only important in copolymers*
Celcon Kematal Duracon Hostaform Ultraform Tarnoform	*Building sealing webs:* Rhepanol			*Copolymers 5 + 6 special types* *Copolymers + 13* Evatate, Evaflex Lupolen V. Wacker VAE, Evatane *Copolymers with acrylic acid salt* Ionomere Surlyn A. Copolene, Himilan *Blend with Bitumen* Lucobit	*Copolymers with butadiene:* high impact PS *Copolymers + 8, SAN* Tyril, Sanrex, Luran *Copolymers + 8 + Butadiene, Blends, ABS:* Cycolac, Kralastic, Novodur, Terluran, Ravikral, Urtal *Copolymers +8 + 9, ASA* Luran S	*Elevated tempera-ture resistance with Copolymers 7, 9, 10, 12* *Fibers:* Orlon Dolan Dralon Craylor Anilana	*Copolymer dispersions with* 7, 8, 10, 11, 13: Acronale Plextole
1	2	3	4	5	6	7	8	9

Alkyds

Good insulating properties, excellent moldability, short cure time, and low cost are primary reasons for the use of thermosetting alkyds in electrical materials, such as housings and cases for switchgear, encapsulation of capacitors and resistors, and parts for fuses, light switches, and insulators. Maximum use temperature of the molded parts is 250 to 300°F (120 to 150°C). Liquid resins are used for enamels and lacquers for cars, stoves, and refrigerators.

Allyls (Allylics)

Allyl resins and monomers are thermosetting plastics that are com-mercially available as the diallyl esters of phthalic (DAP) and iso-

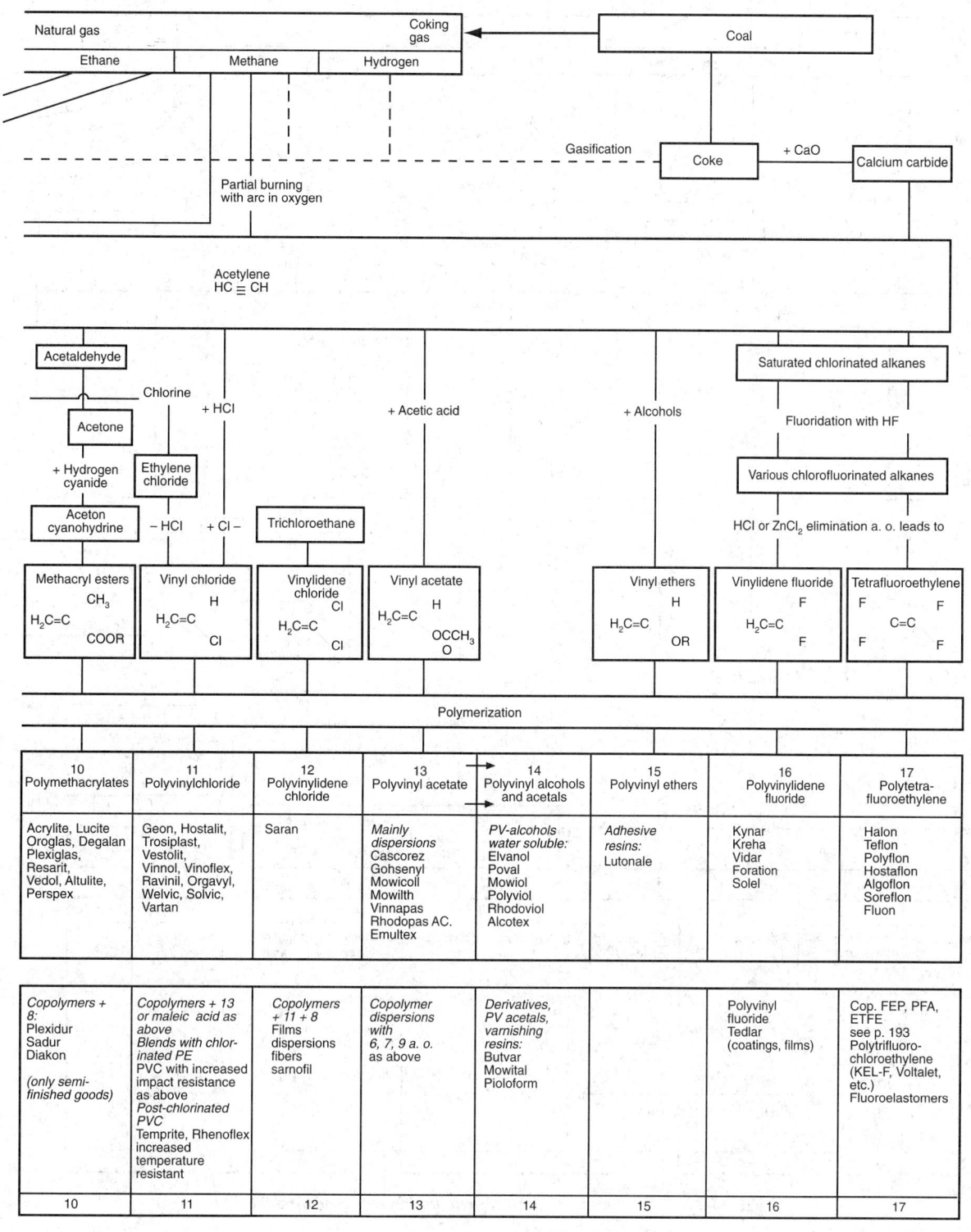

FIG. 4-10D.   (Continued)

phthalic (DAIP) acids. Other monomers in this grouping are diallyl maleate (DAM) and diallyl chlorendate (DAC). The latter is used for flame-resistant formulations because of the high chlorine content of the resins. They can be polymerized by raising their temperature or by the use of peroxide initiators. The molding and electrical properties of these materials are excellent. They are primarily used for electrical components, decorative laminates, sealants, and coatings.

Cellulosics

Cellulosic plastics are produced by the chemical modification of cellulose, a natural polymer that is a major constituent of plant life. The primary sources of cellulose for this purpose are cotton linters and wood pulp. The term "cellulosic plastics" applies only to plastics whose resinous content is an ester or ether of cellulose. Included

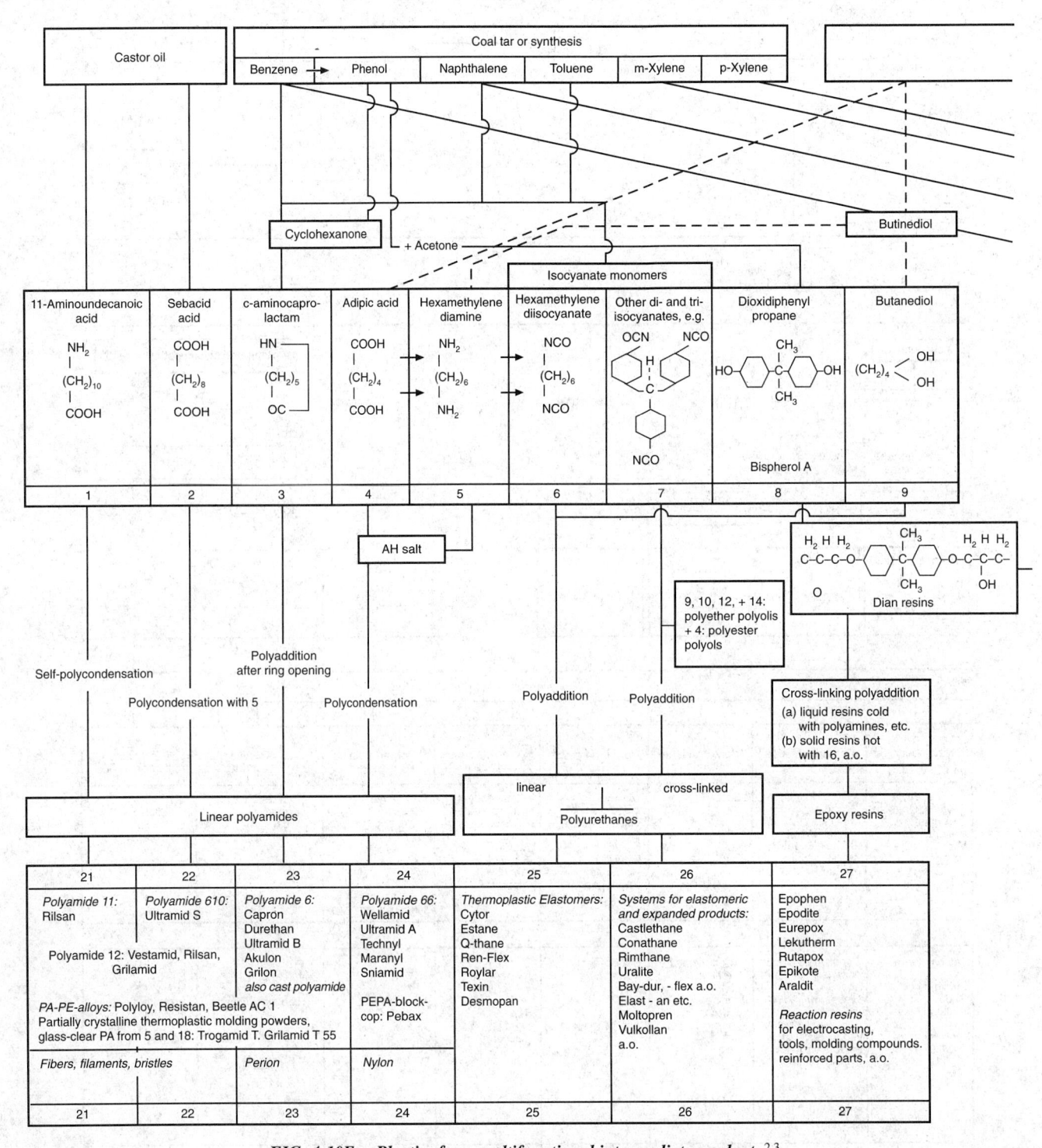

FIG. 4-10E. *Plastics from multifunctional intermediate products.*[2,3]

in this group are cellulose acetate, cellulose acetate butyrate, methyl cellulose, cellulose triacetate, cellulose propionate, ethyl cellulose, cellulose nitrate, cellulose films, and fibers.

Cellulose acetate (CA): A thermoplastic made by treating cellulose with acetic acid and acetic anhydride, with sulfuric acid as a catalyst. It can be formed into molding powder, foamed by rapid evaporation of a solvent, dissolved for lacquers, cast into film, and spun from solution into fibers. It has replaced hazardous cellulose nitrate for photographic use under the name "safety" film. (See also

cellulose triacetate.) Cellulose acetate molding compositions are tough and of good color and gloss; hardness is controlled by varying the ratio of plasticizer and ranges from rigid to flexible. Primary uses are in electronic components, insulation tapes, food packaging, toys, eyeglass frames, and sound recording and computer tapes. Rigid cellulose acetate foam is used as the core in sandwich panels and marine floats. Variations in composition include the degree of esterification of cellulose and amount and choice of plasticizer, which is usually a phthalate ester. Molding stocks for general purposes ignite easily, give off burning drips, burn with a dark yellow

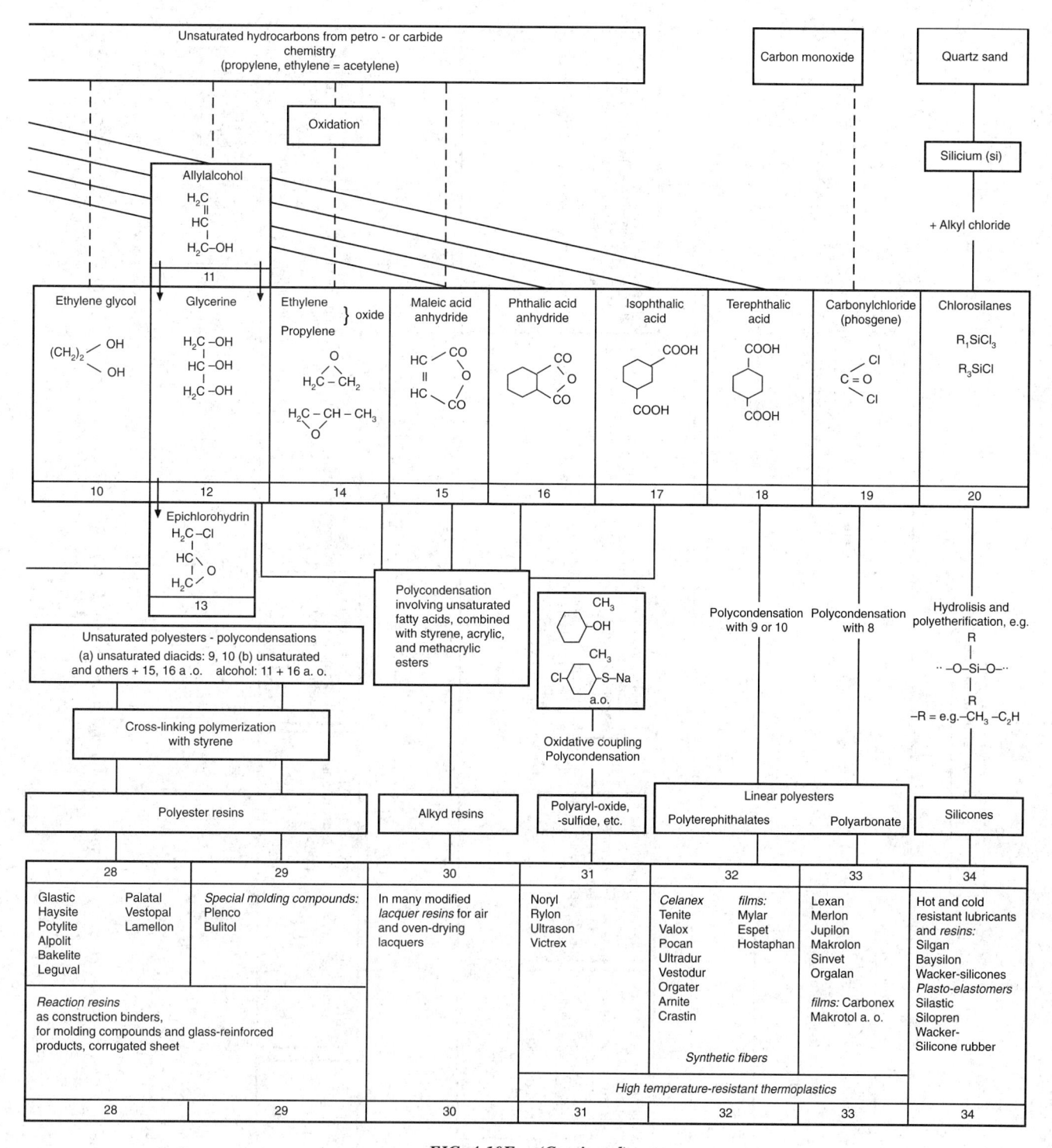

Unsaturated hydrocarbons from petro - or carbide chemistry (propylene, ethylene = acetylene)

Carbon monoxide

Quartz sand

Oxidation

Silicium (si)

Allylalcohol

H_2C
$\|$
HC
$|$
$H_2C–OH$

11

+ Alkyl chloride

Ethylene glycol	Glycerine	Ethylene / Propylene } oxide	Maleic acid anhydride	Phthalic acid anhydride	Isophthalic acid	Terephthalic acid	Carbonylchloride (phosgene)	Chlorosilanes
$(CH_2)_2$ $\langle$ OH / OH	$H_2C–OH$ / $HC–OH$ / $H_2C–OH$		HC $-CO$ $\|$ HC $-CO$ $\rangle O$	$\langle$ CO / CO $\rangle O$	COOH / COOH	COOH / COOH	Cl / C=O / Cl	R_1SiCl_3 / R_3SiCl
10	12	14	15	16	17	18	19	20

Epichlorohydrin
$H_2C–Cl$
HC
H_2C $\rangle O$

13

Unsaturated polyesters - polycondensations
(a) unsaturated diacids: 9, 10 (b) unsaturated and others + 15, 16 a .o.   alcohol: 11 + 16 a. o.

Polycondensation involving unsaturated fatty acids, combined with styrene, acrylic, and methacrylic esters

CH_3 / OH
CH_3
Cl — S – Na
a.o.

Polycondensation with 9 or 10

Polycondensation with 8

Hydrolisis and polyetherification, e.g.
R
·· –O–Si–O–··
R
$–R = e.g.–CH_3 –C_2H$

Cross-linking polymerization with styrene

Oxidative coupling Polycondensation

Linear polyesters

Polyester resins	Alkyd resins	Polyaryl-oxide, -sulfide, etc.	Polyterephithalates	Polyarbonate	Silicones
28	29	30	31	32	33	34

28	29	30	31	32	33	34
Glastic Palatal / Haysite Vestopal / Potylite Lamellon / Alpolit / Bakelite / Leguval	*Special molding compounds:* / Plenco / Bulitol	In many modified *lacquer resins* for air and oven-drying lacquers	Noryl / Rylon / Ultrason / Victrex	*Celanex* *films:* / Tenite Mylar / Valox Espet / Pocan Hostaphan / Ultradur / Vestodur / Orgater / Arnite / Crastin	Lexan / Merlon / Jupilon / Makrolon / Sinvet / Orgalan / / *films:* Carbonex / Makrotol a. o.	Hot and cold resistant lubricants and *resins:* / Silgan / Baysilon / Wacker-silicones / *Plasto-elastomers* / Silastic / Silopren / Wacker-Silicone rubber
Reaction resins as construction binders, for molding compounds and glass-reinforced products, corrugated sheet				*Synthetic fibers*		

High temperature-resistant thermoplastics

| 28 | 29 | 30 | 31 | 32 | 33 | 34 |

FIG. 4-10E.   (Continued)

flame, produce moderate smoke, and have an odor of vinegar, unless masked by the odor of the plasticizer. Flame-resistant stocks are widely used for molding.

Cellulose acetate butyrate (CAB):   A plastic made by treating cellulose with a mixture of acetic and butyric acids and anhydrides in the presence of a catalyst. The ratio of acetic and butyric components can be varied to produce the desired flexibility. It absorbs less water than cellulose acetate. Compositions with ultraviolet stabiliz-

ers are used for small outdoor and indoor signs. Major uses are for automobile steering wheels and taillights, telephone handsets, toothbrush handles, business machine keys, appliance housings, piping, and tubing.

Cellulose propionate (CP):   A thermoplastic molding composition made from cellulose and propionic acid. Molding compositions are usually supplied as granules or pellets, and finished articles can be fabricated by injection molding or extrusion. It has similar uses

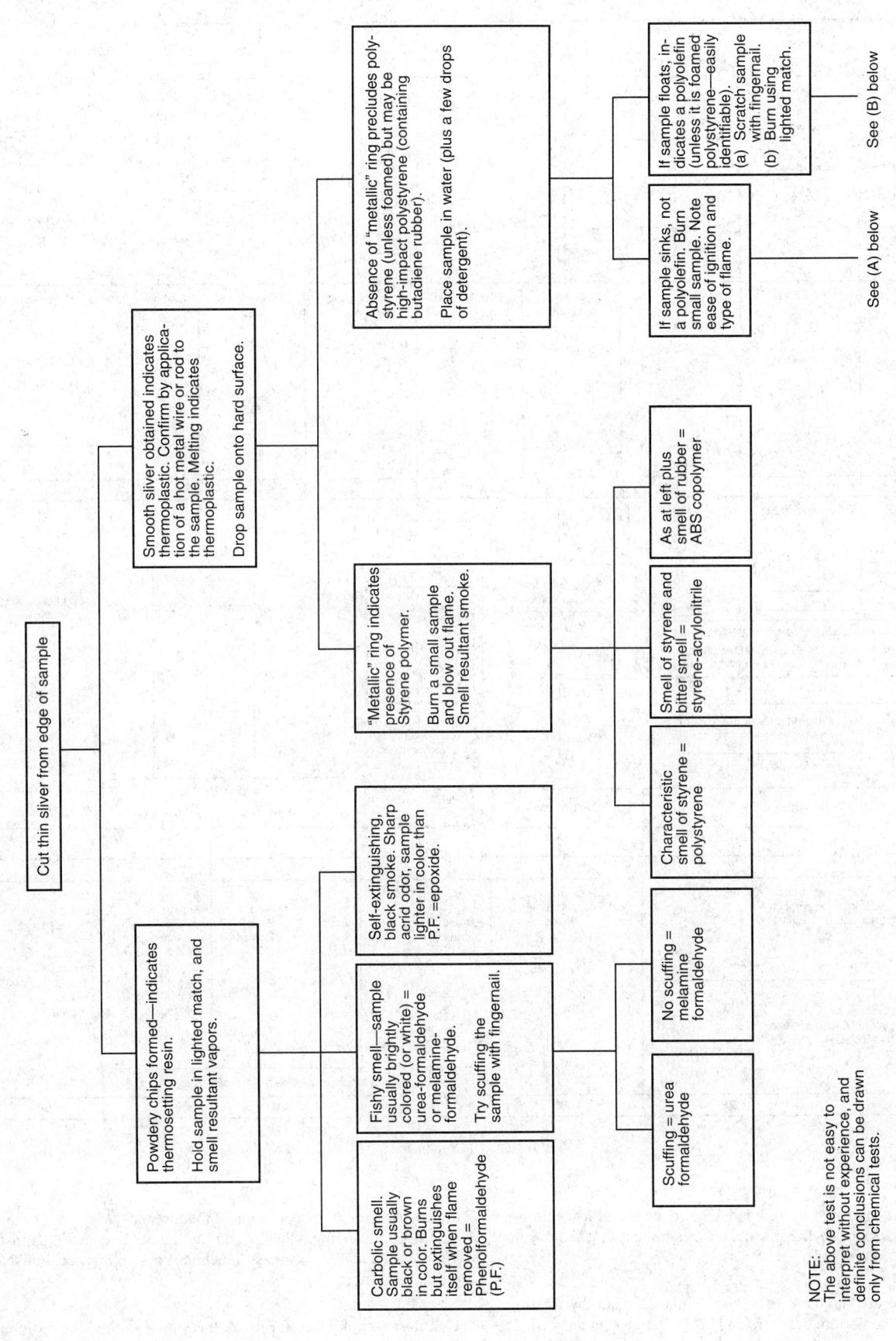

Cut thin sliver from edge of sample

Powdery chips formed—indicates thermosetting resin.

Hold sample in lighted match, and smell resultant vapors.

Smooth silver obtained indicates thermoplastic. Confirm by application of a hot metal wire or rod to the sample. Melting indicates thermoplastic.

Drop sample onto hard surface.

Carbolic smell. Sample usually black or brown in color. Burns but extinguishes itself when flame removed = Phenolformaldehyde (P.F.)

Fishy smell—sample usually brightly colored (or white) = urea-formaldehyde or melamine-formaldehyde.

Try scuffing the sample with fingernail.

Self-extinguishing, black smoke. Sharp acrid odor, sample lighter in color than P.F. =epoxide.

"Metallic" ring indicates presence of Styrene polymer.

Burn a small sample and blow out flame. Smell resultant smoke.

Absence of "metallic" ring precludes polystyrene (unless foamed) but may be high-impact polystyrene (containing butadiene rubber).

Place sample in water (plus a few drops of detergent).

Scuffing = urea formaldehyde

No scuffing = melamine formaldehyde

Characteristic smell of styrene = polystyrene

Smell of styrene and bitter smell = styrene-acrylonitrile

As at left plus smell of rubber = ABS copolymer

If sample sinks, not a polyolefin. Burn small sample. Note ease of ignition and type of flame.

If sample floats, indicates a polyolefin (unless it is foamed polystyrene—easily identifiable).
(a) Scratch sample with fingernail.
(b) Burn using lighted match.

See (A) below

See (B) below

NOTE:
The above test is not easy to interpret without experience, and definite conclusions can be drawn only from chemical tests.

FIG. 4-10F. Simple tests for identifying the more common plastics. (Adapted from Briston, J. H., and Grosselin, C. C., Introduction to Plastics, Newnes-Butterworth, London, 1968.[4])

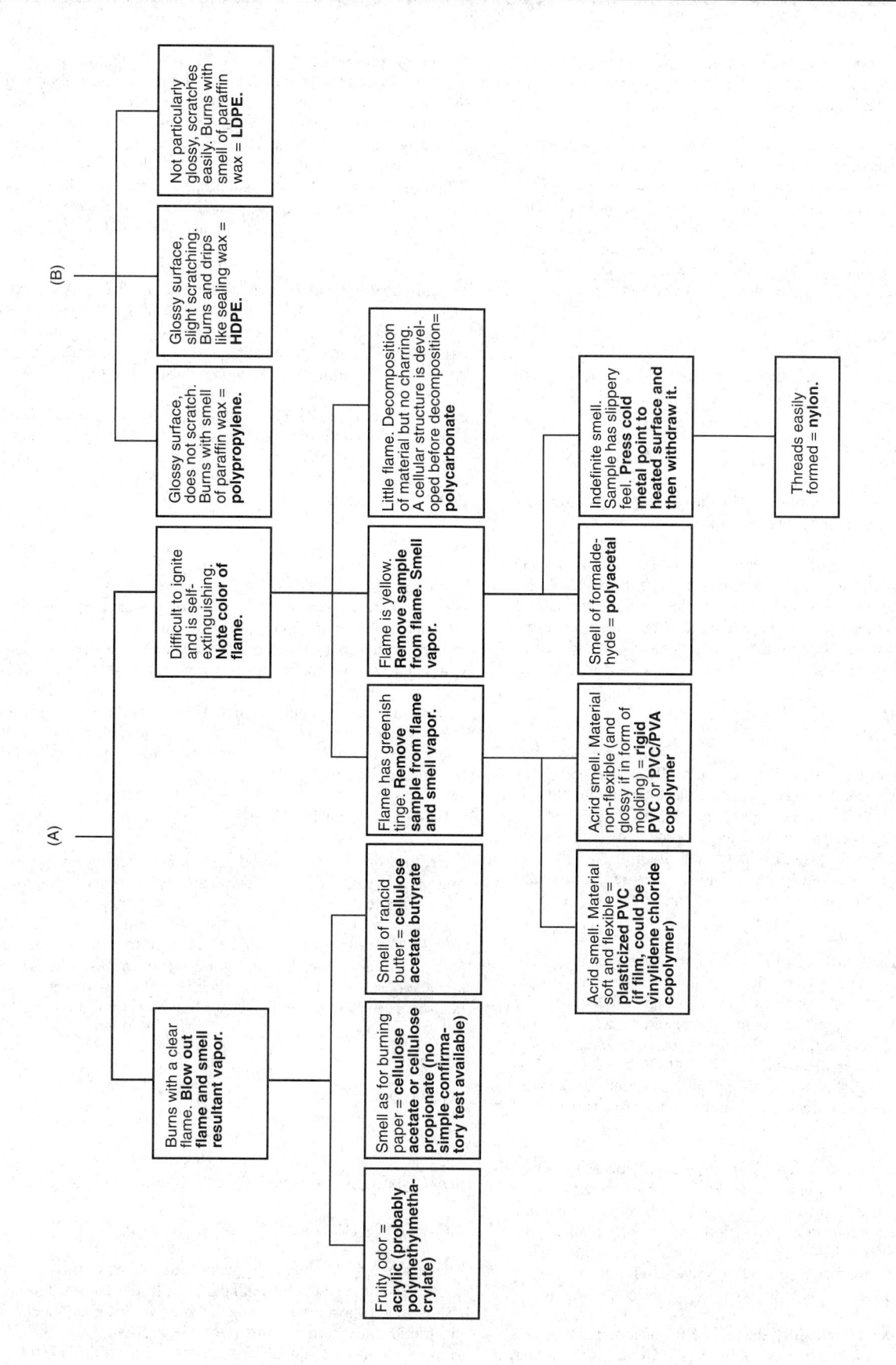

(A)

(B)

Burns with a clear flame. **Blow out flame and smell resultant vapor.**

Difficult to ignite and is self-extinguishing. **Note color of flame.**

Glossy surface, does not scratch. Burns with smell of paraffin wax = **polypropylene.**

Glossy surface, slight scratching. Burns and drips like sealing wax = **HDPE.**

Not particularly glossy, scratches easily. Burns with smell of paraffin wax = **LDPE.**

Fruity odor = **acrylic (probably polymethylmetha-crylate)**

Smell as for burning paper = **cellulose acetate or cellulose propionate (no simple confirma-tory test available)**

Smell of rancid butter = **cellulose acetate butyrate**

Flame has greenish tinge. **Remove sample from flame and smell vapor.**

Flame is yellow. **Remove sample from flame. Smell vapor.**

Little flame. Decomposition of material but no charring. A cellular structure is devel-oped before decomposition= **polycarbonate**

Acrid smell. Material soft and flexible = **plasticized PVC (if film, could be vinylidene chloride copolymer)**

Acrid smell. Material non-flexible (and glossy if in form of molding) = **rigid PVC or PVC/PVA copolymer**

Smell of formalde-hyde = **polyacetal**

Indefinite smell. Sample has slippery feel. **Press cold metal point to heated surface and then withdraw it.**

Threads easily formed = **nylon.**

FIG. 4-10F: (Continued)

to CA and CAB, but due to better surface hardness than CAB and less moisture sensitivity than CA, it is also used for mechanical pencils, telephones, and shoe heels.

Cellulose triacetate (CAT):   A thermoplastic made by reacting cellulose with acetic anhydride. It differs from cellulose acetate in chemical structure in that it consists of three acetate groups (instead of two) attached to each glucose unit of the cellulose molecule. Because of its high-softening temperature, it is not suitable as molding stock. It has lower water absorption than CA and is now the prevalent base for photographic safety film, drawing and printing film, magnetic tape, and transformer and capacitor dielectrics.

Cellulose Nitrate (CN)

Because of the extreme fire hazards associated with cellulose nitrate in its processing, use, and storage, this plastic is discussed in detail later in this chapter.

Epoxies (EP)

Epoxy resins are primarily produced by the reaction of epichlorohydrin and bisphenol-A in the presence of a catalyst. By varying the ratio of the ingredients, the resultant product may be a low-viscosity fluid or a high-melting solid. Two primary classes are available: (1) the liquid resins that are combined with curing agents for adhesives, potting, and tooling compounds, and (2) the solid resins that are modified with other resins to make coating materials. The epoxy resins also are used as "body solders" for automobile repair, as dies for forming sheet metal, as piping and tubing, and as foamed blocks for construction insulation.

Fluoroplastics

Fluoroplastics or fluorocarbons are commercially available in several major groups under the headings polyethylene-tetrafluoroethylene (ETFE), polyethylenechlorotrifluoroethylene (ECTFE), polytetrafluoroethylene (TFE), polyfluoroethylene-propylene (FEP), polychlorotrifluoroethylene (CTFE), polyvinyl fluoride (PVF), and polyvinylidene fluoride (PVF_2). All of these resins are characterized by resistance to solvents and chemical attack and good-to-excellent weatherability. Six are described below.

Polytetrafluoroethylene (PTFE):   A representative of a group of completely fluorinated resins. It is handled as general-purpose molding powders, fine granules, and aqueous dispersions. The methods of processing include compression molding, extruding, dip coating, and film casting. PTFE has a very extended range of serviceable temperatures, from –450 to 500°F (–265 to 260°C). It is used as chemical gasketing, high-frequency and high-temperature electrical insulation, lubricants, and on molded bearings and coated cookware.

Polyfluoroethylene-propylene (FEP):   A copolymer of tetrafluoroethylene and hexafluoropropylene. FEP can be handled through the melt-flow processes involved in extrusion and injection molding. It is widely used in industrial applications, particularly as insulation for process industries involving heat and corrosive atmospheres. It is also used as a lining for vessels, pipes, valves, and fittings handling corrosive fluids.

Polyethylene-tetrafluoroethylene (ETFE):   A copolymer of ethylene and tetrafluoroethylene. It is processed by the conventional method of extrusion and injection molding. Its applications include use as gears, pump components, automotive parts, labware, valve linings, electrical connectors, and wire coatings. ETFE can be reinforced with glass fiber.

Polychlorotrifluoroethylene (PCTFE):   A thermoplastic resin produced in various formulations of fluoroplastics. It differs from the usual fluoroplastic in that its molecular structure contains chlorine. It flows like FEP for injection molding or extrusion, but is less thermally stable. It is chemically resistant to many corrosive liquids. Its serviceable temperatures range from –400 to 390°F (–240 to 200°C). Its electrical and thermal properties make it valuable for insulation, cable assemblies, printed circuits, and electronic components. It also is used in the chemical process industries in valves, fittings, and gaskets.

Polyethylene-chlorotrifluoroethylene (E-CTFE):   A 1:1 copolymer of ethylene and chlorotrifluoroethylene. E-CTFE is processed by extrusion, molding, rotocast, and powder coating techniques. Applications include chemically resistant linings, coatings, containers, labware, moldings, wire coatings, fibers, and films.

Polyvinyl fluoride (PVF):   A plastic produced through the polymerization of vinyl fluoride. Films are cast from selected hot solvents, and the major use is as a weather-resistant surface on building siding made of aluminum, wood, composite boards, or polyester glass fiber laminates.

Ionomers

The term "ionomer" is used to describe the class of thermoplastic polymers in which ionized carboxyl groups create ionic crosslinks in the intermolecular structure. Ionomer resin formulations are available in a wide variety of compositions. They are outstanding plastics, with regard to toughness, transparency, and solvent resistance, and they resist impact at temperatures as low as –160°F (–107°C). Articles made from these plastics include housewares, toys, containers, safety shields, tool handles, electrical insulation, and packaging. These plastics are also used in the lamination of various building construction materials.

Melamines (MF)

Melamine, one of the members of the amino family, is a thermosetting plastic noted for its extreme hardness, permanent colors, and good electrical insulating characteristics. Alpha-cellulose filled, general-purpose grades are used for dishes and kitchenware, where their lack of taste and odor and resistance to food stains make them particularly desirable. Mineral-filled grades are used in electrical applications. The resin may be bonded with fabrics, paper, wood veneers, etc., to produce laminates for building applications.

Nylon

See polyamides.

Olefins

See polybutylene, polyethylene, polyallomers, and polypropylene.

Phenolics (PF)

The discovery of phenol formaldehyde plastics (Bakelite™) in about 1910 marked the beginning of the plastics industry. Many varieties are now available, marked by low cost and good mechanical, electrical, and thermal properties. General-purpose, wood-filled grade phenolics are suitable for use up to about 300°F (150°C), but glass- or mineral-filled compounds have long-term retention of properties up to approximately 500°F (260°C). Some special, heat-resistant types are usable for up to 1 hr at temperatures as high as 1000°F (540°C).

Paper or fabric impregnated with phenolic resin in a water or organic solvent can be wound into tube form or pressed into flat sheets to make tough thermoset "laminates" with low water absorption and high service temperature capabilities. Laminates are used for electrical insulation and bearings for marine propellers and heavy rolling mills. Similar compositions are used as laminating adhesives for exterior-grade plywood and as impervious surfacing materials. Paper impregnated with phenolic resin and printed with wood grain is frequently the top surface for wall panels or furniture, with backing of other phenolic laminates or plywood. Phenolics are standard binders for brake shoes, sandpaper, grinding wheels, sand-casting foundry molds, and wood chip particleboard.

Two types of phenolic foam—reaction foam and syntactic foam—are used in construction or for thermal insulation. Reaction foam is made by curing the phenolic resin with agitation so rapid that water is produced as a byproduct, and is vaporized and trapped to make the voids in the foam. Syntactic foam is made from hollow spheres of phenolic resin with a phenolic or other resin binder. Each type is used for cores in sandwich panels, and as buoyancy or a stiffening means in marine and aircraft structures. Cast phenolic resin is largely used for decorative applications.

Polyallomers

Polyallomers are made by polymerizing two monomers with catalysts so that blocks of each polymer are formed in the polymer chain. The most common polyallomer contains ethylene and propylene. Applications for polyallomers include films, bottles, appliance parts, automotive parts, toys, and closures.

Polyamides

Two aliphatic polyamides—nylon-6,6 and nylon-6—find their greatest uses in home furnishings, apparel, and tire cord, and account for most of the production in this family of polymers. The manufacture of nylon-6,6 started in 1938, and nylon-6 in 1939. The first major application was in the hosiery market. Nylon has become one of the leading polymers due to its high-impact strength, versatility, and ease of processing. Other forms of nylons are also commercially available and find application in ropes, fishlines, cable jacketing, gears, slide fasteners, boat propellers, and appliance parts.

Aromatic polyamides are more heat resistant. Their commercial production started in 1961 with DuPont's Nomex™ and, more recently, Kevlar™ fiber. Nomex™ is found in items such as fire fighter's protective clothing, while Kevlar™ is ideally suited for tires and ballistic vests because of its strength. Kevlar™, together with graphite, is finding a use in superior composite materials, such as fiber-reinforced plastics (FRP). Polyamide imides and polyimides are other members of this grouping, typically used for heat-resistant electrical wire and cable insulation.

Polybutylene (PB)

Polybutylene is flexible, has excellent creep resistance at room and elevated temperatures, and has higher long-term temperature resistance than other polyolefins. Other advantages include good toughness, high tear strength, and good moisture barrier and electrical insulation characteristics. Applications include film and sheeting, hot-melt coatings and adhesives, flexible pipe and tubing, and shrink film.

Polycarbonates (PC)

Polycarbonates are thermoplastic polyesters of carbonic acid. The resin is available in several different forms, including pellets, powder, film, sheet, rod, plate, and tubing. High-impact strength, good electrical properties, clarity, and resistance to creep have caused this material to find many applications, such as in appliances, electrical and electronics equipment, food-handling systems, and sporting goods. Its transparency makes it useful for lenses, safety shields, and glazing.

Polyester, High-Temperature Aromatic

Important characteristics of this material are its dimensional stability at high temperatures [over 600°F (315°C)], self-lubricating properties, good stiffness, electrical insulating properties, and solvent resistance. It decomposes rapidly at approximately 1000°F (540°C). An unusual aspect of this thermoplastic is its very high thermal conductivity, which makes the material useful in applications, such as bearings and electrical insulation, where failure from localized hot spots is a problem with most plastics. Stick-free coatings for frying pans and encapsulation of electronic parts are among its other uses. Trade names include Ekonol™ and Ekcel™.

Polyesters, Thermoplastic

Unsaturated polyesters show good weathering resistance, light stability, and heat resistance. Their primary application is in the production of reinforced plastics. They are used to fabricate structures ranging in size from electrical components to large boat hulls.

Saturated polyesters do not have the double bond that would permit further polymerization by treatment with initiators, such as organic peroxides. Saturated polyesters of high molecular weight, such as ethylene glycol terephthalate, are used principally in fiber and film production, e.g., magnetic recording tapes, tough or boil-in-bag food packaging, and satellite balloons. They can be back-plated with metals for decorative films and nameplates. The low molecular weight esters are used as plasticizers for vinyl and acrylic resins.

Thermoplastic polyester molding compositions are available. Besides poly (ethylene-terephthalate)—PET, there are poly (1,4-butylene-terephthalate)—PBT, poly (tetramethylene-terephthalate)—PTMT, and poly (cyclohexylenedimethylene-terephthalate-isophthalate)—PCDT. PBT and PTMT are the same chemically.

Thermoplastic polyesters have engineering applications, including automotive parts, appliance parts, business machine parts, electrical and electronic parts, gears, bearings, pulleys, and pump components. The polyester PET, discovered in Britain during World War II, can be spun into a fiber that is commonly used in textiles (e.g., Dacron™, etc.).

Polyethylene (PE)

Three major types of polyethylene are marketed: the low-, medium-, and high-density products. In addition, copolymers of ethylene-ethyl acrylate, ethylene-vinyl acetate, and ethylene-butylene are produced. The three density ranges are 56.8 to 57.7, 57.8 to 58.7, and 58.8 to 60.2 lb per cu ft (90.9 to 92.4 kg/m³, 92.5 to 94.0 kg/m³, and 94.1 to 96.4 kg/m³).

Low-density stocks are waxy, relatively limp, and are the toughest of the polyethylenes, with an upper service temperature of approximately 160°F (70°C). As density increases, so does surface hardness, stiffness, resistance to permeation by oils and water, and softening temperature. For polyethylenes of the highest density, the upper service temperature is 240°F (115°C) under no-load conditions.

The outstanding properties of polyethylenes are: (1) good moldability, (2) good mechanical properties, (3) excellent electrical resistance, (4) low moisture vapor transmission, (5) resistance to solvents and chemicals, and (6) lightness in weight. Polyethylene is widely used for film, sheeting, plastic bottles, piping, tubing, dish pans, garbage containers, laundry baskets, electronic components, insulation, and textiles. Boat and motor vehicle fuel tanks are a recent application for high-density polyethylene. Polyethylene can be crosslinked chemically or by irradiation to form a thermoset with high heat resistance.

Polyphenylene Oxide (PPO)

Polyphenylene oxide is a thermoplastic resin formed by the catalyzed oxidation of 2,6-xylenol. It is a specialized, highly engineered plastic with a wide temperature range, good mechanical and electrical properties, and resistance to corrosive materials. It has many electrical and electronic applications, and is used for packaging frozen foods and as a replacement for glass and stainless steel in hospital utensils and medical instruments.

Polyphenylene Sulfide (PPS)

Excellent chemical resistance and high stiffness at elevated temperatures make this material useful as a coating for various metals, including aluminum, steel, cast iron, brass, etc., and in the petroleum and chemical processing industries. Other uses include cookware, pump impellers, pumps, conveyor rollers, and machine parts.

Polypropylene (PP)

Polypropylene possesses the lowest density of the commercially available thermoplastics. In addition, it exhibits good rigidity, high yield strength, good surface hardness, exceptional flexural properties, resistance to chemicals, and excellent dielectric properties. Like all olefins, it has excellent resistance to water solutions that are highly damaging to many metals.

Primary areas of use include the usual categories of molding, as well as in film and piping. It is used for luggage, housewares, automobile and aviation components, and appliance parts. Because of its superior insulating properties, it is also used to make electronic components. Carpeting made of polypropylene fibers also is available. Polypropylene filled with asbestos offers high-impact strength, stiffness [up to 250°F (120°C)], and low friction.

Polystyrene (PS)

A glassy transparent thermoplastic, primary characteristics of polystyrene are hardness, rigidity, clarity, and heat and dimensional stability. By varying the polymerization reaction, polystyrenes have been developed with higher heat distortion temperatures. Styrene polymers are marketed in granular and powder form for extrusion or molding. Beads of polystyrene that contain the flammable gas pentane as a blowing agent are sold for the production of polystyrene foam.

General-purpose and impact grades of polystyrene are used for refrigerator liners, appliance housings, automotive applications, films for wrapping, sporting goods, toys, and novelties. Heat-resistant polystyrene is used for television sets, radios, and illuminated signs. Glass-filled polystyrene is used for business machines, electronic equipment, and military hardware. Polystyrene foam is used to make toys, insulation, displays, shipping containers, and molded furniture. Construction panels, consisting of cores of polystyrene foam between sheets of plywood, plastic, or aluminum, and foamed-in-place polystyrene insulation are used in the building industry.

Polysulfone (PSU)

Polysulfone is a thermoplastic resin that contains a diphenylene group. This group imparts certain desirable characteristics to the polymer, including thermal stability [over 300°F (150°C)], oxidation resistance, and rigidity. Glass fibers may be added for improved environmental stress crack resistance. Commercial applications include electronic units, electrical circuit breaker parts, automobile engine parts, aircraft interiors, kitchen range hardware, electroplating equipment, and appliance hardware.

Polyurethane (PUR)

The polyurethanes consist of a group of polymers that are produced in the following two general forms: (1) foams, e.g., flexible, semiflexible, and rigid; and (2) elastomers, e.g., casting compounds and elastoplastic resins, adhesives, coatings, and spandex fiber. The basic reaction used to produce the urethane polymers involves isocyanates and reactive hydrogen-bearing materials, such as polyethers, castor oils, amines, carboxylic acid, and water. By varying the number of branchings, it is possible to make polyurethanes that are thermoplastic or thermosetting.

The polyurethane foams are widely used in the production of upholstered furniture, bedding, sponges, toys, wearing apparel, and medical dressings. Rigid urethane foams are used for insulation in building construction. Substantial quantities of the polyurethanes are used to produce coatings and adhesives. Its rubberlike qualities aid in the production of specialty fillers and elastomers. Gears, sprockets, and rolls are produced from urethanes. Very thin films are used to produce containers impermeable to gas and moisture.

Silicones (SI)

The silicone resins comprise a large group of polymers consisting of chains of alternating silicone and oxygen atoms with organic groups, e.g., methyl groups attached to the silicone atoms. They are stable at high and low temperatures, have good dielectric properties, resist weathering, do not react with most chemicals, and repel water.

Silicone emulsions are used as mold release agents, defoamers, and waterproofing materials. A compound of liquid silicone and finely divided filler produces a grease suitable for high-temperature lubrication applications. Silicone paints, varnishes, and enamels have many uses, including cloth and wire coating. With inorganic fillers, such as glass or asbestos, silicones are molded into such products as connector plugs, coil bobbins, insulators on induction furnaces, and heat barriers in jet engine afterburners. Silicone glass laminate applications include rigid hot air ducts for aircraft, terminal boards, and transformer spools and spacers. Rigid silicone foams can be preformed or foamed in place. High-temperature insulation is the principal use of rigid silicone foams.

Styrene Acrylonitrile (SAN)

Produced through the copolymerization of styrene and acrylonitrile, styrene acrylonitrile is a transparent thermoplastic with improved heat resistance, stiffness, and scratch and impact resistance. It finds application in decorative panels, food packages, "dishwasher-safe" housewares, battery cases, lenses, glasses, appliance parts, and automobile components.

Ureas (UF)

Urea formaldehyde, as a thermosetting plastic, is especially suitable for any application that must resist heat. The material has excellent resistance to oils and greases, but is not recommended for high humidity environments. Because urea formaldehyde is hard, strong, odorless, and tasteless, it has been frequently used in tableware, crockery, and decorative articles for the home. Other uses are for electrical parts, such as switches and outlet covers, bottle tops, toys, and as a laminate in building panels. Plaskon™ and Beetle™ are two well-known trade names.

Vinyls

The vinyl resins are thermoplastics that exist in a wide variety of polymers and copolymers. Commercial production of the vinyl chloride monomer starts with acetylene or ethylene. Most vinyl plants utilize an oxychlorination process with ethylene to produce vinyl chloride. Copolymers of vinyl chloride and vinyl acetate account for a large part of vinyl production.

The major applications for flexible and rigid vinyls include: automobile seat covers, floor mats, and moldings; shower curtains; window shades; wrappings; bottles; adhesives; flooring; siding; piping; wiring; swimming pools; records; and medical appliances.

Polyvinyl chloride (PVC): This type of plastic is made as a rigid product for a number of building components and as a flexible plasticized stock for upholstery and wearing apparel. It exists in hundreds of individual formulations. It has good abrasion resistance.

The unplasticized PVC softens as it burns, producing white smoke and acrid fumes, which can be corrosive. Most of the chlorine content of PVC is released as hydrogen chloride. The fire properties of plasticized PVC are determined to a large extent by the nature of the plasticizer. Building products should be tested individually.

Vinyl chloride—vinyl acetate copolymers (polyvinyl acetate) (PVAC): These contain from 5 to 20 percent vinyl acetate in the polymer. They are used for more flexible stocks than PVC and are more readily softened by plasticizers. Their properties are essentially those of PVC.

Polyvinyl dichloride (PVDC): Because this plastic is made by chlorinating PVC, its properties are essentially those of rigid PVC. It is used most where more toughness, stiffness, and heat resistance are needed than is offered by PVC.

Vinyl acetate polymer (VAC): A thermoplastic that is used in coatings, as a wood primer, and as an adhesive in a milky aqueous dispersion. It is not used as a molding or sheet plastic, as its upper service temperature is below 120°F (50°C).

Vinyl alcohol resins (PVA): Because they are made by hydrolysis of polyvinyl acetate, PVA products swell or dissolve completely in water. They are highly impervious to oils and many lacquer solvents. They are used in hoses for spray painting and as water-soluble film for dose-packaging detergents, etc.

Polyvinyl butyral (PVB): Polyvinyl butyral is produced by condensing butyraldehyde with polyvinyl alcohol. It has remarkable adhesive qualities and is primarily used as plasticized film for laminating automotive safety glass.

Polyvinyl formal (PVFM): This plastic is produced by condensing formaldehyde with polyvinyl alcohol to make a hard resin. Polyvinyl formal is then mixed with small amounts of phenolic resin to improve the bond and reduce creep. It is not to be confused with polyvinyl fluoride. (See Fluoroplastics.) Its principal use is as standard insulation enamel for magnet wire in small motors and electronic equipment.

Vinylidene chloride polymer: This is a thermoplastic that as a film has low permeability to water vapor. A principal use is as a coating on cellophane. Filaments are used for webbing on garden furniture and for window screening.

Vinylidene chloride-vinyl chloride copolymer: This plastic is less rigid than PVC and flows more readily in extrusion and molding. Its major use is as a coating for metals and molded electrical parts.

NATURAL AND SYNTHETIC RUBBER

Like plastics, rubber appears in many shapes and forms, from very soft and pliable products to hard rubber that looks and feels like a molded plastic. Many polymers can be manufactured in elastomeric grades and are classified in that form as synthetic rubbers. Crepe soles fall in that category.

Natural Rubber (NR)

Natural rubber is a constituent of latex, the white milky sap collected from the rubber tree. The processed rubber may be used in sheet form for items like shoe soles, extruded into products like inner tubes, or compression molded with heat to make items like tires and hot-water bottles. Liquid latex has a variety of applications. It is often coated on fabrics for waterproofing, mixed with cement for superior bonding qualities, included in water-based paints to produce a cohesive film after the paint dries, and expanded into the familiar foam latex, or foam rubber. The density of the foam is controlled by varying the ratio of foam to air and the size of the cavities.

Vulcanization is a curing process that changes the isoprene molecule to a polyisoprene and produces a thermoset polymer that cannot be reshaped or remolded when reheated. By prolonging the vulcanization process, the degree of hardness can be varied, eventually resulting in "hard rubber" (known as ebonite or vulcanite in England). Hard-rubber tubes and rods have been popular products, valued for their resiliency, electrical insulating quality, and chemical inertness. Other common products have included automobile battery cases, combs, telephone receivers, castors, wheels, and bushings. In many of these applications, synthetics have replaced natural rubber.

Synthetic Rubber (SR)

These materials are intended to imitate the elasticity of natural rubber, and they can be vulcanized. They have similar long-chain molecular structures, built up from copolymerized synthetic monomers, such as butadiene with styrene, or butadiene with acrylonitrile. The former method produces styrene-butadiene rubber (SBR), such as the early Buna™ and Buna-S™ rubbers; styrene-butadiene rubber is used extensively in automobile tires. Nitrile rubber (NBS) is the product of the butadiene-acrylonitrile reaction, while butyl rubber is produced from an isobutylene isoprene copolymer. Neoprene (GR) is a very strong and durable synthetic made from chloroprene.

Synthetic foam rubber usually means a soft flexible product of the polyurethane type and is either polyester (AU) or polyether (EU). However, new types are continually being developed, including ethylene-propylene rubber (EPR), polybutadiene (BR), and polyisoprene, the synthetic equivalent of natural rubber.

Silicone rubber (SI) incorporates the inorganic, inert element silicone, which gives it a number of unique properties. It is the only satisfactory soft material for human implants, and it is commonly used for breast implants and many other implants, including artificial ears, skull plates, and jawbones. In building applications, self-curing, water-repellent silicone rubber is used as a sealant, and it finds many applications where a nonstick surface is desired. Some formulations have been used successfully as firestops and to seal openings in fire walls, where its flexibility gives performance superior to grout or plaster.

Additional information sources on synthetic rubbers are: *Plastics: Design and Materials*,[6] *Introduction to Industrial Polymers*,[7] and *Rubber Technology Handbook*.[8]

CELLULOSE NITRATE (CN)

Cellulose nitrate (also referred to as nitrocellulose and pyroxylin) is made by the action of nitric and sulfuric acids on cellulose materials, such as cotton. Pyroxylin plastic comprises the lower nitrated products and contains from 11 to 12 percent nitrogen.

Cellulose nitrate material that has been subjected to heat, such as that salvaged from a fire, may be so altered in composition that it is subject to spontaneous ignition.

When cellulose nitrate products are heated to temperatures above 300°F (150°C), decomposition starts, which generates further heat and soon raises the material to its ignition temperature. Some researchers have reported decomposition after long-continued exposure to temperatures not much above that of boiling water. There are a number of cases on record of ignition from contact with steam pipes and electric light bulbs. Decomposition of cellulose nitrate generates heat and does not depend upon an external air supply.

Some of the gases produced by decomposition are highly toxic. The effect of carbon monoxide is well known. The toxic effect of the oxides of nitrogen is often delayed; persons exposed to these gases may show no immediate ill effects, but fatalities may follow some hours or days after exposure.

Cellulose nitrate, when in solution in acetone or any of the various solvents, has no greater hazard than that of the solvent. If the solvent is lost, the hazard reverts to that of cellulose nitrate.

Most articles formerly made of cellulose nitrate are now made from less hazardous plastics, but it is still used to some extent in industrial applications, for items such as shoe heels, some housewares, and as a pigment in lacquers.

Processing Cellulose Nitrate

Fire safety in plants processing cellulose nitrate plastic consists essentially of four parts: (1) segregation of hazards, (2) elimination of ignition sources, (3) good housekeeping, and (4) strong fire control facilities. In some plants, finished cellulose nitrate plastic parts may be assembled with other articles that do not involve any special fire hazard, e.g., toilet seats. In such cases, the hazard and necessary protection depend upon the quantity of cellulose nitrate used. If the quantities used are large, the same safeguards should be followed as for plants in which the product is made entirely from cellulose nitrate.

Cellulose Nitrate Storage

Storage of cellulose nitrate plastics requires special facilities. Depending on the quantity to be stored, these facilities may be cabinets, vaults, storage rooms, or isolated storage buildings. Storage facilities are described in detail in two NFPA standards: NFPA 40, *Standard for the Storage and Handling of Cellulose Nitrate Motion Picture Film* (hereinafter referred to as NFPA 40), and NFPA 40E, *Code for the Storage of Pyroxylin Plastic.*

Cellulose Nitrate Photographic Film

All photographic film made in the United States and Canada since 1952 is the safety-base type, usually cellulose acetate or triacetate or polyester. An engraving process used in some printing plants engraves on cellulose nitrate plates.

There is comparatively little storage of cellulose nitrate X-ray film, although some hospitals may still store X-rays made on cellulose nitrate film many years ago. However, appreciable quantities of old nitrate motion picture and aero film still exist. Small amounts of portrait, industrial, and amateur nitrate film probably exist, but the usual storage of this material (in paper envelopes) poses no serious hazard. Nitrate-based film manufactured in other countries may occasionally be found.

The existence of old nitrate motion picture and aero film does present a hazard, since persons handling this material may not be familiar with its serious hazards and the elaborate precautions necessary to handle it safely. Although it is unlikely that nitrate film will be projected in a theater handling present-day motion pictures, such material may be shown in small "art" or "film society" theaters. Unfortunately, projectionists may not be aware of the fact that a nitrate print is to be shown. For this reason, theater projection standards for nitrate film should still be kept in effect. Storage of nitrate film should be in accordance with NFPA 40.

The amounts of scrap-nitrate motion picture film have greatly decreased, and most of the agencies that formerly handled this material are no longer in existence. This has given rise to the possibility that scrap could accumulate in dangerous quantities. Prior to environmental concern with air pollutants from outdoor burning, the best procedure for disposing of scrap nitrate film was to burn it at a remote location in the open, with all personnel at a safe distance upwind (because toxic oxides of nitrogen are among the products of combustion of this film). If accumulations of scrap-nitrate film are a problem, the best course to follow would be to check with environmental authorities to learn if such open burning is permitted locally.

Transportation of Cellulose Nitrate

The U.S. Department of Transportation (DOT) regulates the interstate transportation of cellulose nitrate plastic by rail, land, or water. These regulations permit dry shipment of cellulose nitrate (pyroxylin) scrap when specially packaged, but if there is evidence, or the possibility of decomposition, the cellulose nitrate must be packaged under water. Shipment of cellulose nitrate sheets, rolls, rods, and tubes is not subject to special DOT regulation, except when shipped by rail express or water.

Articles manufactured entirely of or containing cellulose nitrate plastic do not require special packaging when shipped by land, rail, or water. Special shipping regulations apply to cellulose nitrate wet with alcohol, solvent, or water.

Shipments of cellulose nitrate motion picture and X-ray films are permitted when certain packaging requirements are met, including inside containers made of light metal, cardboard, or fiberboard for individual reels of film. NFPA 40 places restrictions on the transportation of nitrate film in public vehicles so as to protect the public against the consequences of a possible fire involving the film.

Air transportation of cellulose nitrate plastic scrap is not permitted by regulations of the International Air Transport Association. Cellulose nitrate plastic rods, sheets, rolls, tubes, manufactured articles, and motion picture and X-ray films are permitted only when specially packaged.

Extinguishing Cellulose Nitrate Fires

Large volumes of water are required to extinguish cellulose nitrate fires. Automatic sprinklers constitute the best general protection for areas where cellulose nitrate materials are stored or handled. Water supplies should be particularly strong. Closer sprinkler spacing than for ordinary occupancies is required in order to supply sufficient water to absorb the heat. The water from sprinklers, in addition to its extinguishing and cooling effects, tends to absorb the poisonous oxides of nitrogen fumes but has no effect on carbon monoxide.

If hose streams are required, the objective should be to get large quantities of water on the fire as soon as possible. Fire fighters should be located upwind or protected by self-contained breathing apparatus (SCBA) to prevent inhalation of the products of combustion.

FIRE BEHAVIOR OF PLASTICS

While there is no special fire hazard for most accepted usages of plastics, some exhibit burning characteristics that are considerably different from those encountered with the more traditional cellulosic building materials.

Special Fire Behavior Problems

Test methods, which in the past have been adequate to indicate the relative hazard of materials under actual use conditions, have failed to predict the fire behavior of some plastics. In addition, different fire conditions cause significantly different burning characteristics. Of principal concern has been fire behavior that poses unusual hazards to life and property, including the following.

Ignitability and rate of burning: Although plastics tend to have a higher ignition temperature than wood and other cellulosic products, some are easily ignited with a small flame and burn vigorously. Very high surface flame spread rates have been reported—up to approximately 2 ft per sec (0.6 m/s), or 10 times the rate of flame spread across most wood surfaces.

Smoke produced: The burning of some plastics is characterized by the rapid generation of large amounts of very dense, sooty, black smoke. Chemicals added to inhibit flammability may increase smoke production.

Toxic gases: Any fire will generate lethal products of combustion, principally carbon monoxide. Depending on the plastic and the particular fire conditions, highly toxic gases, such as hydrogen cyanide and hydrogen chloride, may also evolve.

Flaming drips: Thermoplastic articles tend to melt and flow when heated. In a fire situation, this characteristic may cause the material to melt away from the flame front and inhibit further burning, or it may produce flaming and tar-like dripping, which is difficult to extinguish and may start secondary fires.

Deviation from test results: Small-scale "Bunsen burner" tests are used for product development and laboratory control purposes. In the past they were used to indicate that certain plastics were "self-extinguishing" or "nonburning" and, presumably, safe for use. Unfortunately, in real-life situations the same materials have shown flash burning characteristics. Somewhat larger-scale tests, such as the NFPA 255, *Standard Method of Test of Surface Burning Characteristics of Building Materials*, "tunnel test" also fail to adequately predict flash burning characteristics under some end-use conditions.

Corrosion: Severe corrosion damage to sensitive electronic equipment and to metal surfaces has been reported following fires involving commonly used plastics, such as polyvinyl chloride.

The fire behavior problems summarized above can occur under every conceivable condition of burning, from complete to partial combustion or smoldering and destructive pyrolysis. When the plastics and their constituent modifying agents, including flame-retardant additives, burn, they can produce a wide variety of noxious and toxic byproducts in varying concentrations. In this respect, plastics are similar to most ordinary combustibles, such as wood, leather, wool, silk, etc., in that they are capable of thermal degradation into volatile and gaseous products of combustion able to cause harmful physiological effects when breathed. In general, carbon monoxide is generated more rapidly than other toxic gases and tends to be the principal factor in fire fatalities. Nevertheless, unusually high burning rates, unusually heavy smoke production, and a higher heat content per unit weight are responsible for a greater concern about the fire behavior of certain plastics.

Smoke Generation

Smoke generation from a given polymer may vary widely depending upon the nature of the polymer, the additives present, whether fire exposure was flaming or smoldering, and the type of ventilation that was present. Several comments from Gaskill's[9] comprehensive study on smoke density are especially pertinent:

1. Woods and most polymeric material pyrolytically degrade, yielding smokes that are dense to very dense. Ventilation tends to clear the smoke, but in most cases it does not reduce the intensity to the point of satisfactory visibility. With fire retardants incorporated into the polymer (at least in the case of solid materials), heat and flame generally produce very dense smokes, and this occurs rapidly.
2. Woods and those polymeric materials that burn cleanly will yield smokes, under conditions of heat and flame, of somewhat less density.
3. Urethane foams under flaming or nonflaming exposures generally yield dense smokes and, with few exceptions, obscuration occurs in a fraction of a minute. Flaming exposure usually causes the smokes to be generated in less than 15 sec, and the intensities in this case are usually higher than in the nonflaming case.

Cellular Plastics

Cellular plastics (or foamed or expanded plastics) have been in widespread use since the 1960s. Without adequate flame retardants and without barriers to resist ignition, however, these plastics in flexible form, e.g., mattresses, seat cushions, carpet padding, etc., and in rigid form, e.g., building insulation, have created an unacceptable fire hazard. Some reports tell of fast-spreading, high-intensity fires and voluminous smoke production. The low resistance of cellular plastics to excessive heat and their ease of ignition contrast sharply with their most valuable property in building construction—their excellent resistance to heat transfer. In a 1969 study of the flammability of cellular plastics,[10] Underwriters Laboratories Inc. (UL) reported on an analysis of 34 fires where rigid cellular plastics had been used as building materials. The report indicated that:

1. The fire performance of cellular plastics can vary considerably depending upon whether they are exposed or protected and/or flame resistant, along with the degree of flame resistance. But the reduction in flammability through the use of inhibitors or protective coverings is not conclusively demonstrated by the study.
2. In 8 of 21 fires, a high rate of flame spread was reported and appeared to be especially typical of sprayed-on polyurethane and rigid polyurethane boards. Flame spread ratings from NFPA 255, *Standard Method of Test of Surface Burning Characteristics of Building Materials* (hereinafter referred to as NFPA 255), tunnel tests of some of the materials ranged from 5 to 1375 (red oak, 100).

3. The fire behavior of cellular plastics appears to be affected by the extent of foam application, i.e., whether the foam is applied to the walls and ceiling or only to the walls.
4. The fire behavior of the flexible foam plastics appears to be similar to that of the rigid plastic foams. (Peculiar to the flexible foams, though, is the pool of liquid product that can be generated in a fire in the bulk storage of the foam "buns." The liquefied plastic burns like a high flash-point flammable liquid.)

Whereas untreated rigid polyurethane burns rapidly and is readily ignited by common ignition sources found in the environment, fire-retardant grades have substantial resistance to localized, temporary ignition sources under most fire environments. Sustained application of heat and flame causes such foam to burn, but the burning rate tends to be retarded. Often, burning will cease entirely when the external flame source is removed. The production of dense smoke is a characteristic of both treated and untreated urethane.

Untreated rigid polystyrene foam, likewise, ignites easily and burns vigorously with the production of dense, black smoke and a very black, viscous melt product which can burn with the intensity of a flammable liquid. Fire-retardant treatments inhibit ignition; the plastic tends to shrink away from fixed heat sources without igniting. The heat of combustion of styrene is about 50 percent higher than that of urethane foam [18,000 Btu/lb *vs.* 12,000 Btu/lb (42 mJ/ kg *vs.* 28 mJ/kg)].

At times, cellular plastics that exhibit good resistance to burning under most exposure conditions have shown a markedly different behavior when surfaces of the exposed plastic are arranged to cause a feedback of radiant heat between the surfaces.[11] The combination of low thermal conductivity, high rate of heat release, and rapid degradation under heat exposure appear to be responsible for a rapid upward propagation of the flame front occurring in a room corner configuration. When the ceiling is also insulated with the exposed foamed plastic, the corner configuration encourages rapid extension of the fire across the underside of the ceiling. This phenomenon has occurred even with ignition sources as small as a wastebasket fire. The laboratories of Factory Mutual Research Corporation devised the "corner test" to simulate this exposure condition.

Rubber

Burning rubber is notable for its smell and dense smoke. The material, as generally used, is not easily ignited and does not burn with unusual rapidity.

The bulk storage of rubber items, particularly rubber tires, is subject to hot, difficult-to-control, and very smoky fires. Surface application of water often is only marginally effective because the fire tends to burrow into the material. Special automatic sprinkler protection is required to help ensure control if the storage is more than 5 ft (1.52 m) high. (See NFPA 231D, *Standard for Storage of Rubber Tires.*)

Rubber or plastic jacketing on grouped electrical cables has provided the fuel for fires in the electrical and communications industries. Considerable work has been done to raise the operating temperature limits and ignition temperatures of insulation for electric power and control cables. An applicable standard is IEEE 383, *Standard for Type Test of Class IE Electric Cables, Field Splices, and Connections for Nuclear Power Generating Stations.* Butyl/ neoprene and neoprene rubber insulation tend to burn readily and fail such small-scale tests, while Hypalon™, for instance, is a jacketing material with more resistance to heat and ignition.

No significant differences appear to exist in the ignition or combustion characteristics of natural rubber and the various synthetics. Silicone rubbers can be formulated to reduce combustibility to a minimum under ordinary conditions, but not all silicone rubbers

have that quality: the material failed the requirement of no flashover in a quarter-scale room fire test conducted at the National Bureau of Standards.[12]

Test Results

The variation of the fire behavior of plastics when compared to other building materials was further demonstrated in tests conducted at Underwriters Laboratories Inc.[13] Thirty-one materials were subjected to the Steiner tunnel test (NFPA 255; ASTM E84, *Standard Test Method for Surface Burning Characteristics of Building Materials*; and UL 723, *Standard for Safety Test for Surface Burning Characteristics of Building Materials*) to establish flame spread and smoke-developed classifications. These materials were also subjected to a 20-lb (9.1-kg) wood crib exposure fire in a full-size corner configuration. Testing established that the 20-lb (9.1-kg) wood crib presented fire severity conditions about equal to the 5,000 Btu per min (88 kW) burner in the tunnel test, generating a heat flux of about 0.97 Btu/sec • ft² (1.1 W/cm²). Four materials in the test series achieved an identical flame-spread classification of 18 (cement asbestos, 0; red oak, 100), but the smoke that developed and the extent of burning were markedly higher for the flame-retardant grade of plastic foam. (See Table 4-10A.)

TABLE 4-10A. *Test Results (Cellular Plastics) for a 20-lb (9.1-kg) Wood Crib Fire Test*

Code Letter	Material	FSC*	SDC†	Affected Area‡
E	3¹/₂-in. (89-mm) thick glass fiber blanket insulation, 6% resin by weight, unfaced, 0.65 pcf (10.4 kg/m³) density	18	3	24%
I	¹/₂-in. (12.7-mm) thick wood fiberboard acoustic ceiling panels, integrally flame retardant treated, decorative finished, 18.8 pcf (301.2 kg/m³) density	18	4	37%
T	¹/₂-in. (12.7-mm) thick wood particleboard, integrally flame retardant treated	18	37	25%
Z	2-in. (51-mm) polyurethane spray cellular plastic on ¹/₄-in. (6-mm) asbestos cement board; 2.2 pcf (35.2 kg/ m³) density, flame retardant	18	935	85%

* Flame spread classification, tunnel test result.
† Smoke developed classification, tunnel test result.
‡ Percent of 192 sq ft (17.8 m²) total wall and ceiling area affected in full-scale corner test (7% affected in control fires).

These four building materials were also subjected to exposures in the heat-release rate calorimeter test developed by the National Bureau of Standards.[14] As reported by UL[13] in the document on the 1973-74 test series, the first three specimens, code letters E, I, and T, yielded identical heat release rates of 0.26 Btu/sec • ft² (0.3 W/ cm²) at an exposure energy level of 2.6 Btu/sec • ft² (3 W/cm²). However, a heat release rate 25 times higher [6.6 Btu/sec • ft² (7.5 W/cm²)] was recorded for the flame-retardant plastic, code letter Z.

Another cellular plastic, 2-in. (51 mm) extruded polystyrene cellular plastic boardstock [1.8 lb/ft³ (0.03 g/cm³) density, flame retardant], obtained a flame spread classification of 3 in the tunnel test, but its smoke developed classification was 785 (red oak, 100), and 146 sq ft (13.6 m²), or 76 percent, of the wall-ceiling area of the corner test structure was adversely affected by the burning.

The fire problem associated with plastics was studied by the Products Research Committee (PRC), formed to administer a trust fund established by a number of companies engaged in the manufacture and sale of cellular plastics and their ingredients. The PRC issued its final report in 1980[15] after a five-year effort, which included sponsoring a number of research programs. Highlights of conclusions reached by the PRC included the following:

1. Provisions of current building codes, requiring a thermal barrier over cellular plastic insulation in wall cavities, should remain in force until further advances are realized in fire safety engineering.
2. Properly designed interlayers between furniture fabric and padding are an effective means of reducing the fire risk of upholstered furniture.
3. The use of results of many different small-scale fire tests to obtain an estimate of behavior of a material in fire is appropriate; however, the present state of the art does not allow regulatory decisions to be made without considerable expert judgment and, preferably, data from full-scale fire tests in which the most probable fire scenarios are simulated.

FIRE TESTS FOR PLASTICS

Combustibility characteristics and suitability for use should be determined for any material or assembly on the basis of tests that simulate the end-use condition as realistically as possible. The tests should be designed to show how the product will perform when it is subjected to the type of fire exposure that can be anticipated under the conditions of use. The most desirable tests, from the standpoint of fire protection, engineering, and codes, are those that are suitable for all products for a particular application and that treat all products equally. Separate tests for plastics, or any other group of materials, that do not allow comparison of the fire behavior under like test conditions are undesirable.

As explained previously, small-scale tests are now used mostly for product development and production control. Such tests are also suitable for determining the ignitability of such items as upholstery, where the ignition source is small and of low intensity, but are unsuitable for evaluating the fire characteristics of materials under room fire or other full-scale fire conditions.

Somewhat greater-intensity fire exposures are produced by standard fire tests, such as the Steiner tunnel test and the radiant panel test, both of which are used to determine the suitability of all kinds of materials as interior finish for walls and ceilings in buildings. As previously indicated, these tests have been proved inadequate to accurately predict the burning characteristics of some plastics under certain conditions. Higher-intensity exposures encountered under actual-use conditions may cause fire behavior substantially more hazardous than indicated by the results of these tests. The Factory Mutual Research Corporation corner test tends to reproduce these more severe exposure conditions. It is now common to require materials, such as plastics, to be qualified under more than one test in order to ensure their suitability in building construction. Factory Mutual has written about its corner test as follows:

The characteristic of rapid flame spread across the exposed surface of polyurethane, even with automatic sprinkler protection, prompted the development of the Factory Mutual corner test. The test determines the performance of various types of foamed plastic insulation, in combination with various surface treatments, under full-scale fire conditions. The test procedure is designed to simulate the exposure that would be expected in an essentially noncombustible occupancy (e.g., an isolated stack of pallets). When sprinklers are provided, the exposure is increased to simulate a fire in a combustible occupancy.[16]

The full-scale corner test is conducted in a 25-ft (7.6-m) high test building, and the initiating fire is a 750-lb (340-kg) wood crib. A scaled-down version, 1/12 full scale, produces similar results at lower cost.

Other tests that may be required by the Authority Having Jurisdiction are used to determine, separately, ignition temperature, surface flame spread index, and the rate of smoke production. Potential heat (caloric value) of the plastic material may have to be determined, but this parameter is of little significance in fire protection engineering; more important is the rate at which heat is liberated and contributes to the intensity of the fire. A multiplicity of tests to qualify plastics, or any other material, for end-use applications may not be required in the future. Test apparatus that measure heat release and smoke development rates at varying radiant heat flux exposures provide more realistic assessments of probable fire response under real-life situations than some of the older test methods. The first standardized test for this analysis is ASTM E906, *Standard Test Method for Heat and Visible Smoke Release Rates for Materials and Products*, for use with the Ohio State University (OSU) release-rate apparatus. (See Figure 4-10G.) Measurements include heat and smoke particles released by the test specimen per unit time and total during the test, toxic gases given off, ease of ignition, and flame spread rate. Procedures for similar analyses using an oxygen-consumption cone calorimeter developed at National Institute of Science and Technology (NIST) are standardized in ASTM E1354, *Standard Test Method for Heat and Visible Smoke Release Rates for Materials and Products Using an Oxygen Consumption Calorimeter*. These two tests have also been published as NFPA 263, *Standard Method of Test for Heat and Visible Smoke Release Rates for Materials and Products,* and NFPA 264, *Standard Method of Test for Heat and Visible Smoke Release Rates for Materials and Products Using an Oxygen Consumption Calorimeter*, respectively.

In addition to the fire test methods for building construction materials and assemblies in general (i.e., not product related), a number of test methods have been developed to analyze flammability and other fire response characteristics of plastics, specifically. Almost all are small-scale tests intended primarily for research, product development, and manufacturing quality control. They are not intended for, and should not be used for, building or room fire hazard assessments, unless their scope is so noted. A number of such test methods have been issued by the American Society for Testing and Materials (ASTM); descriptions can be found in ASTM D3814, *Standard Guide for Locating Combustion Test Methods for Plastics*. Underwriters Laboratories Inc. (UL) has several standards for measuring ease of ignition and other characteristics of plastics that are used in products that have been tested and listed or classified by UL. One common test is UL 94, *Standard for Safety Tests for Flammability of Plastic Materials for Parts in Devices and Appliances*. (See Figure 4-10H.) It includes procedures for hot-wire and arc ignition and for rates of burning while the sample is positioned horizontally and vertically. Results of this test, like the ASTM plastics flammability tests, cannot be compared to the fire behavior of the materials under full-scale fire exposures.

A number of organizations have cooperated in producing test methods for evaluating the hazard of plastics and textiles in specific applications. Working with the U.S. Consumer Product Safety Commission, the Upholstered Furniture Action Council, NIST, and others, NFPA has prepared test methods NFPA 260, *Standard Methods of Test and Classification System for Cigarette Ignition Resistance of Components of Upholstered Furniture*, and NFPA 261,

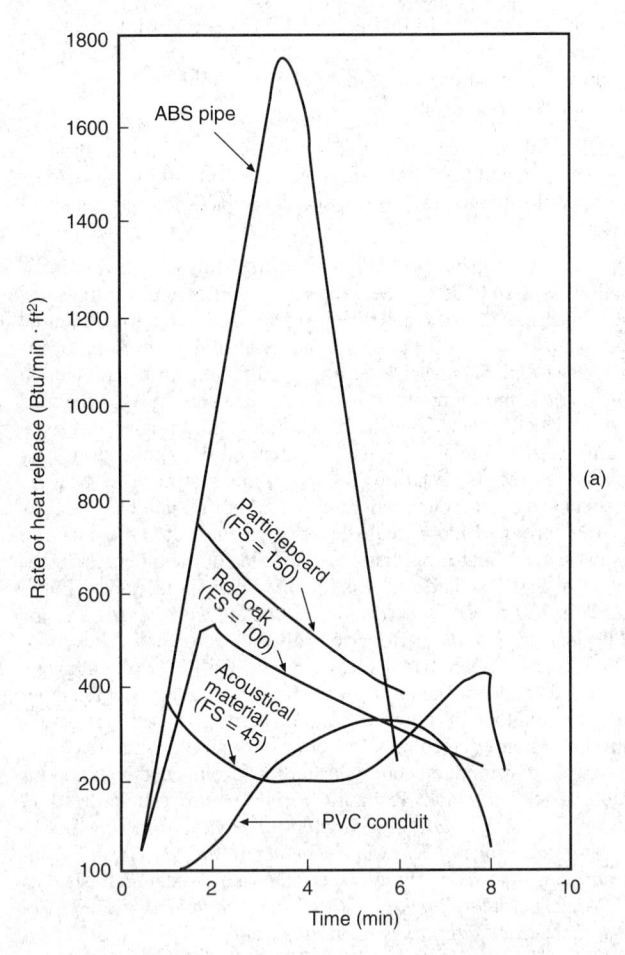

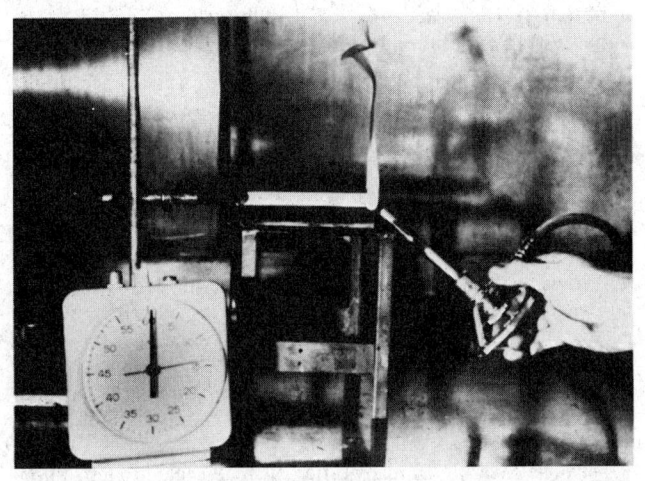

FIG. 4-10H. *Specimen being tested under the horizontal burning test of UL 94 for flammability of plastics. (Source: Underwriters Laboratories Inc.)*

ing Ignition Source, for the fire response of upholstered furniture exposed to flaming ignitions. Similarly, NFPA 267, *Standard Method of Test for Fire Characteristics of Mattresses and Bedding Assemblies Exposed to Flaming Ignition Source,* covers the fire response of mattresses and bedding exposed to flaming ignition. NFPA 262, *Standard Method of Test for Fire and Smoke Characteristics of Wires and Cables,* used for determining fire and smoke characteristics of plastic and other insulations on wires and cables, was derived from one of the UL tests. NFPA 264A, *Standard Method of Test for Heat Release Rates for Upholstered Furniture Components or Composites and Mattresses Using an Oxygen Consumption Calorimeter,* addresses fire characteristics of upholstered furniture, mattresses, and bedding when material samples are tested in the cone calorimeter.

Some U.S. government agencies have promulgated regulations to control the flammability of plastics and textiles in certain applications. Under MVSS-302, the DOT regulates interior materials for automobiles, using a test for burning rate in the horizontal mode. The U.S. Coast Guard, under Subchapter T, regulates materials used as resins for hulls in boats-for-hire carrying more than six passengers. The U.S. Coast Guard also uses a modified version of ASTM E136, *Standard Test Method for Behavior of Materials in a Vertical Tube Furnace at 750°C,* to establish the "noncombustibility" of materials used in the interior of passenger-carrying ships. Guidelines for the flammability of carpets, rugs, children's sleepwear, mattresses, and mattress pads are issued by the U.S. Consumer Product Safety Commission. The U.S. Nuclear Regulatory Commission has issued guidelines for the ignitability and flammability of plastics used in the insulation of electrical cables requiring qualification under test procedures in IEEE 383, *Standard for Type Test of Class IE Electric Cables, Field Splices, and Connections for Nuclear Power Generating Stations.*

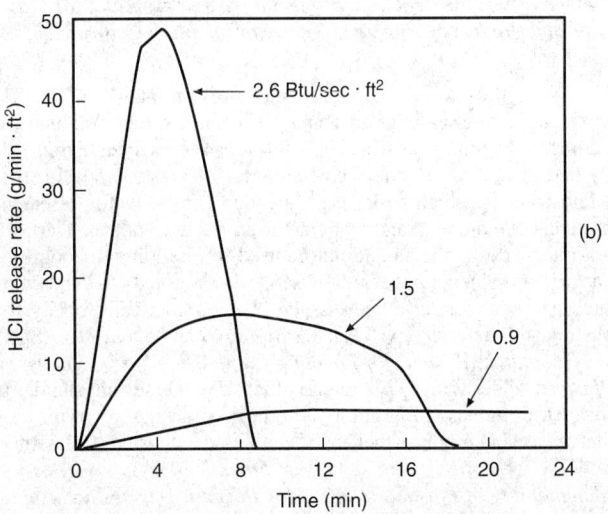

FIG. 4-10G. *Typical of the curves generated by the OSU release rate apparatus are: (a) heat release rates for several specimens at an exposure level of 2.6 Btu/sec • ft², and (b) hydrogen chloride release rates from a sample of conduit at different exposure levels. FS ASTM E84 flamespread rating. (One Btu/sec • ft² = 0.053 W/m².)*

Standard Method of Test for Determining Resistance of Mock-Up Upholstered Furniture Material Assemblies to Ignition by Smoldering Cigarettes, for determining the resistance to cigarette ignitions in upholstered furniture; and NFPA 266, *Standard Method of Test for Fire Characteristics of Upholstered Furniture Exposed to Flam-*

INSTALLATION SAFEGUARDS

In the current state of the art, large areas of cellular plastics left exposed in interior building construction or decoration must be considered to be extremely hazardous. Exposed foam on ceilings or walls, in air-handling plenums, in shafts, in other flue-like vertical spaces inside walls, or above hung ceilings should be avoided. A low flame-spread classification (e.g., 25 or less) does not ensure that a product has limited combustibility under such conditions, nor can it be assumed that automatic sprinklers will control the hazard. Coating such surfaces with a fire-retardant paint, intumescent or otherwise, is not likely to produce satisfactory results.

Model building codes require that foamed plastic used as interior wall insulation be covered with a thermal barrier, such as $1/2$-in. (12.7-mm) thick ordinary gypsum board, or other method that reduces the risk of ignition and the subsequent flash-fire propensity. The use of foamed plastics in the cavities of hollow masonry walls, such as perimeter insulation around the foundation of a building, insulation below concrete slabs on the ground, and for roof insulation under certain conditions, is generally accepted without thermal barrier protection. All such plastics in the thickness and density used must have a smoke-developed rating no greater than 450 and a flame-spread rating no greater than 75, under ASTM E84 (NFPA 255) tunnel test conditions.

In other than noncombustible or fire-resistant building types, foamed-plastic insulated steel or aluminum exterior building panels are permitted by the model building codes, provided the foam core has a flame-spread classification of 25 or less and the space is protected by automatic sprinklers.

When used as interior wall and ceiling finish, plastic materials other than foamed plastic generally are not subject to any special requirements. As for any other materials, plastics are subject to limitations on surface flame spread and often on the smoke generated, as measured by standard test procedures, such as the Steiner tunnel test or radiant panel test. These or other special limitations may be applied to plastics used as diffusers in lighting fixtures, where it is often acceptable to have a plastic that deforms and drops out of the fixture at an elevated temperature still well below its ignition point. Special limitations may also be set forth by building codes for the use of plastic glazing instead of glass in exterior walls of buildings, where the normal requirement for conformance to the limitations for interior finish may be waived.

Plastic laminates for countertops, kitchen cabinets, tabletops, etc., are not usually included in the definition of interior finish, as regulated by building codes. Even when it is not regulated by local code, the flammability level should be limited to that encountered when natural products are used in these applications. Plastic laminates are commonly available with low surface flame-spread classifications, in the range of 25 to 75.

Further guidelines for the installation of plastics in various applications can be found in publications of the Society of the Plastics Industry[17] and Factory Mutual Research Corporation.[18]

The packaging, storage, and shipping precautions that are customarily observed for ordinary combustible materials apply generally to most plastics. No special shipping requirements for plastics (other than cellulose nitrate) have been adopted by the DOT.

FIRE- AND FLAME-RETARDANT TREATMENTS

All organic and some inorganic materials burn under proper conditions. Fire and flame retardants can make a given material more difficult to ignite, and when ignited, cause it to burn more slowly. Retardants can be very effective in preventing ignition from low-level, short-duration flame and heat exposures. Even under moderate exposures, the burning characteristics can be greatly improved by fire retardants, depending upon the product and the additive used. For severe exposures, such as from a fully developed room fire, treated and untreated plastics typically show similar fire behavior.

Fire- and flame-retardant chemicals are added to plastics during the manufacturing process. Because of the widely varying properties of plastics, no single, universal additive is available, nor is there only one effective mechanism for imparting flame resistance. Fire- and flame-retardant additives may be blended into the polymer mix during processing, or they may be reactants that change the molecular structure of the polymer. The additives can be chemically reacted throughout the polymer substrate or on the surface of finished or semifinished products. Surface treatments added after the manufacturing process, such as fire-retardant paints or coatings, are unreliable without test data to prove the effectiveness of the combination.

Additives blended into the mix may be organic or inorganic materials. Organic additives typically are chlorinated and brominated hydrocarbons or halogenated and nonhalogenated organophosphorous compounds. Among the inorganic flame retardants are the salts of antimony, zinc, molybdenum, and aluminum. Those classed as reactants include brominated aromatics, brominated aliphatic polyols, and phosphorous-containing polyols.

Phosphorous additives appear to improve flame resistance by: (1) promoting formation of char to reduce concentration of carbon-containing gases; (2) forming a glassy, insulating layer; and (3) either entering into chemical reactions that remove heat from the combustion system, or chemically inhibiting the combustion process, or both. Compounds containing halogens, such as chlorine and bromine, tend to inhibit burning in the gas phase by interrupting chain reactions needed for continuous burning. Other flame retardants involve heat absorption reactions, e.g., by providing a heat sink through evaporation of water of hydration.[19]

A complicating aspect of treating plastics and other materials is that a fire-retardant chemical that works on one substrate system will not necessarily work on other substrates. Thus, fire-retardant chemicals for polystyrene will not necessarily work on polyethylene, nylon, wood, or cotton. Fire-retardant chemicals often impair some other material properties. For example, fire-retardant treatments for cotton fabrics generally reduce fabric strength and wear life. Fire retardants in plastics may reduce allowable processing temperatures and impair physical properties. Costs are generally increased.

Smoldering, rather than free-burning, combustion is likely to increase the amount of smoke produced, and there tends to be more smoke when fire retardants are added to polymers; but this is not true in all cases. Comparisons were made among five different types of products in fire-retardant and non-fire-retardant versions in a series of tests sponsored by the Fire Retardant Chemicals Association at the National Institute of Science and Technology.[20] Included were polystyrene TV cabinets, polyphenylene oxide business machine housings, polyurethane foam-padded chairs, electrical cable with polyethylene insulation and rubber jacketing, and polyester/glass electric circuit boards. Heat release, spread of flames, and the propensity for flashover were reduced when the fire-retardant products were tested. Smoke yields generally were similar, but fire-retardant TV cabinets produced twice the amount of smoke than the non-fire-retardant versions. None of the test specimens produced smoke of extreme toxicity. Potential escape times for occupants were reported to be 15 times greater when the fire-retardant products were used in the tests.

FIGHTING FIRES IN PLASTICS

Plastics other than cellulose nitrate are classified as ordinary combustibles. Consequently, extinguishing methods suitable for fires involving wood and other ordinary combustibles (Class A fires) should be used to extinguish a burning plastic. Some plastics, e.g., flexible polyurethane, may form a pool of burning liquid during a fire, similar to a high-flash-point combustible liquid (Class B) fire. Water from automatic sprinklers or spray nozzles on hoselines can be used for fire control, although the extent of a pool fire is likely to be overshadowed by the more generalized fire in the area.

Storage with a preponderance of plastics products or materials can develop into fires with extremely high temperatures that tremendously magnify the normal problem of radiant heat, structural damage, and exposure.[21] Rapid fire spread, high burning rates, and thick, dense smoke in heavy clouds can be anticipated. Factory Mutual Research Corporation recommends a two-fold attack: (1) properly designed sprinkler protection, and (2) appropriate manual fire fighting techniques. Large-orifice ($^{17}/_{32}$ in.) sprinklers have been found to be substantially better in controlling fires in plastics, and large-drop and early suppression fast response (ESFR) sprinklers were developed at Factory Mutual specifically to address the high-challenge plastics fire.

The physical form of plastics will greatly influence their fire behavior. Molding pellets, which are shipped in bags, drums, or large cartons, provide little surface for access to air. These same plastics in such shapes as flashlight cases will have much more access to air and may burn vigorously until heat causes the shapes to melt, thereby reducing the surface area exposed to air. Melting may be a hazard if it causes burning drips to carry flame to a lower floor or spreads fuel for later ignition.

While fire may develop more slowly in the more dense, unexpanded form than in foamed products, once established, the burning rate is comparable. For fires that have become well established in a large quantity of plastics products in a sprinklered warehouse, "buttoning up" the building to limit combustion air and letting the sprinklers do the work of fire control, for hours if necessary, before moving in for final mop-up by conventional means, is an approach recommended by Factory Mutual Research Corporation.[21] Shutting off sprinklers to allow the smoke to rise and vent the building is dangerous because the high combustibility of the storage can cause the blaze to rapidly flash out of control.

Toxicity

Possibly because of the long chemical names of some plastics, combustion and thermal decomposition products of plastics have been a cause for concern among fire fighters. For plastics now in commercial use, the hazard of carbon monoxide from partial combustion greatly outweighs the toxic effects of other fire gases, both as to nature and amount.

On burning, some plastics, e.g., polyvinyl chloride or ethylene sulfide rubbery caulking materials, generate hydrogen chloride or sulfur dioxide. Because these are strongly irritating, they ordinarily force evacuation long before their toxic effects become dangerous. These gases also are corrosive to metals and electrical equipment; therefore, such equipment should be ventilated, rinsed, or treated with dilute ammonia, and rinsed as soon as possible after exposure.

BIBLIOGRAPHY

References Cited

1. Gaylord, M. W., *Reinforced Plastics, Theory & Practice*, 2nd ed., Cahners, Boston, 1974.
2. Braun, D., *Identification of Plastics*, 2nd ed., Macmillan, New York, 1986.
3. Saechtling, Dr. H., *Plastics Handbook*, Hanser Press, Macmillan, New York, 1983.
4. Briston, J. H., and Grosselin, C. C., *Introduction to Plastics*, Newnes-Butterworth, London, 1968.
5. ASTM D1600, *Standard Terminology for Abbreviated Terms Relating to Plastics*, American Society for Testing and Materials, W. Conshohocken, PA (1994).
6. Katz, Cassell, and Collier, *Plastics: Design and Materials*, Macmillan Publishers, London, 1978.
7. Ulrich, H., *Introduction to Industrial Polymers*, Oxford University Press, New York, 1982.
8. Hoffman, W., *Rubber Technology Handbook*, Macmillan, New York, 1988.
9. Gaskill, J. R., "Smoke Development in Polymers During Pyrolysis or Combustion," *Journal of Fire and Flammability*, Vol. 1, July 1970, pp. 183–216.
10. UL, "Report on Flammability Study of Cellular Plastics," File NC 522, 1969, The Society of the Plastics Industry, Underwriters Laboratories Inc., Northbrook, IL.
11. Troup, W. W. J., *Cellular Plastics and the Building Fire Problem*, Factory Mutual Research Corp., Norwood, MA, 1969.
12. NBS, "Fire Hazard Evaluation of Shipboard Hull Insulation and Documentation of a Quarter-Scale Room Fire Test Protocol," NBSIR 83-2642, 1983, National Bureau of Standards, Washington, DC.
13. UL, "Flammability Studies of Cellular Plastics and Other Building Materials Used for Interior Finishes," Subject 723, 1975, Underwriters Laboratories Inc., Northbrook, IL.
14. Parker, W. J., and Long, M. E., "Development of a Heat Release Rate Calorimeter at NBS," ASTM Special Technical Publication 502, 1972, American Society for Testing and Materials, W. Conshohocken, PA.
15. PRC, "Fire Research on Plastics: The Final Report of the Products Research Committee," Products Research Committee, Lyons, J., Chairman, National Bureau of Standards, Washington, DC, 1980.
16. FMRC, "Rigid Foamed Polyurethane and Polyisocyanurate for Construction," Loss Prevention Data 1-57, 1978, Factory Mutual Research Corp., Norwood, MA.
17. Society of the Plastics Industry, *Expanded Polystyrene Division*, Washington, DC (various dates).
18. FMRC, Loss Prevention Data of the Factory Mutual Research Corporation, Norwood, MA:
    a. 1-57, "Rigid Foamed Polyurethane and Polyisocyanurate for Construction."
    b. 1-58, "Foamed Polystyrene for Construction."
    c. 1-59, "Reinforced Plastic Panels in Construction."
    d. 8-9, "Storage of Plastics and Elastomers."
19. *Modern Plastics Encyclopedia*, 1984-85, Vol. 56, No. 10A, McGraw-Hill, New York, 1985.
20. *Fire Safety Through Use of Flame-Retardant Polymers*, Fire Retardant Chemicals Association Conference Series, 1985, Technomic Publishing, Lancaster, PA.
21. "Stored Plastics—The High-Challenge Risk," Pamphlet P7742, 1987, Factory Mutual Research Corp., Norwood, MA.

References

American Society for Testing and Materials, W. Conshohocken, PA

ASTM D3814, *Standard Guide for Locating Combustion Test Methods for Plastics*
ASTM E84, *Standard Test Method for Surface Burning Characteristics of Building Materials*
ASTM E136, *Standard Test Method for Behavior of Materials in a Vertical Tube Furnace at 750°C*
ASTM E906, *Standard Test Method for Heat and Visible Smoke Release Rates for Materials and Products*
ASTM E1354, *Standard Test Method for Heat and Visible Smoke Release Rates for Materials and Products Using an Oxygen Consumption Calorimeter*

Institute of Electrical and Electronics Engineers, Inc.

IEEE 383, *Standard for Type Test of Class IE Electric Cables, Field Splices, and Connections for Nuclear Power Generating Stations* (Revised 1992).

The Society of the Plastics Industry, Inc., Washington, DC

Expanded Polystyrene Division:

"Fire Safety Guidelines for Use of Expanded Polystyrene in Building Construction," EPS 301, Cat. No. AH-101.
"Flammability of Expanded Polystyrene Insulation Used in Conjunction with Gypsum Wallboard."
"Flammability of Expanded Polystyrene Insulation when Used as Full-Scale Building Enclosure Ceiling Construction Material."

"Flammability of Expanded Polystyrene Insulation Used in Conjunction with Plywood Paneling."
"Flammability of Exposed Polystyrene Insulation (with and without Wooden Dividers)."
"Foamed Plastic, Full-Scale Corner Test, 7½-in. Thickness, EPS."
"Literature Review of the Combustion Toxicity of Expanded Polystyrene."

Polyurethane Division

U-100R2, "Fire Safety Guidelines for Use of Rigid Polyurethane Foam Insulation in Building Construction."
U-102R, "An Update Report on Findings of Fire Study of Rigid Cellular Plastic Materials for Wall and Roof Ceiling Insulation."
U-103, "Large-Scale Corner Wall Fire Tests of Spray-on Coatings over Rigid Polyurethane Foam Insulation."
U-105, "Evaluation of the Fire Performance of Carpet Underlayments."
U-106, "Fire Safety Guidelines on Flexible Polyurethane Foams Used in Upholstered Furniture and Bedding."
U-107, "Room-Scale Compartment Corner Tests of Spray-on Coatings over Rigid Polyurethane Foam Insulation."
U-110, "Flexible Polyurethane Foams and Combustibility."
U-111, "Using Flexible Polyurethane Foams Safely."
U-118, "Guide for Safe Handling and Use of Polyurethane and Polyisocyanurate Foam Systems."

Underwriters Laboratories Inc., Northbrook, IL

UL 94, *Standard for Safety Tests for Flammability of Plastic Materials for Parts in Devices and Appliances* (1991).
UL 723, *Standard for Safety Test for Surface Burning Characteristics of Building Materials* (1993).

NFPA Codes, Standards, and Recommended Practices

Reference to the following NFPA codes, standards, and recommended practices will provide further information on plastics and rubber discussed in this chapter. (See the latest version of *The NFPA Catalog* for availability of current editions of the following documents.)

NFPA 40, *Standard for the Storage and Handling of Cellulose Nitrate Motion Picture Film*
NFPA 40E, *Code for the Storage of Pyroxylin Plastic*
NFPA 231, *Standard for General Storage*
NFPA 231C, *Standard for Rack Storage of Materials*
NFPA 231D, *Standard for Storage of Rubber Tires*
NFPA 255, *Standard Method of Test of Surface Burning Characteristics of Building Materials*
NFPA 260, *Standard Methods of Tests and Classification System for Cigarette Ignition Resistance of Components of Upholstered Furniture*
NFPA 261, *Standard Method of Test for Determining Resistance of Mock-Up Upholstered Furniture Material Assemblies to Ignition by Smoldering Cigarettes*
NFPA 262, *Standard Method of Test for Fire and Smoke Characteristics of Wires and Cables*
NFPA 263, *Standard Method of Test for Heat and Visible Smoke Release Rates for Materials and Products*
NFPA 264, *Standard Method of Test for Heat and Visible Smoke Release Rates for Materials and Products Using an Oxygen Consumption Calorimeter*
NFPA 264A, *Standard Method of Test for Heat Release Rates for Upholstered Furniture Components or Composites and Mattresses Using an Oxygen Consumption Calorimeter*
NFPA 266, *Standard Method of Test for Fire Characteristics of Upholstered Furniture Exposed to Flaming Ignition Source*
NFPA 267, *Standard Method of Test for Fire Characteristics of Mattresses and Bedding Assemblies Exposed to Flaming Ignition Source*
NFPA 701, *Standard Methods of Fire Tests for Flame-Resistant Textiles and Films*

Additional Reading

ASTM, "Standard Method of Test for Density of Smoke from the Burning or Decomposition of Plastics," ANSI/ASTM D2843-88, American Society for Testing and Materials, W. Conshohocken, PA.

Alarie, Y. C. and Caldwell, D. J., "Toxicity of Plastic Combustion Products," Final Report," National Institute of Standards and Technology, Gaithersburg, MD, December 10, 1990.
Babrauskas, V., "Plastics. Part B. The Effects of FR Agents on Polymer Performance," Chapter 12, *Heat Release in Fires*, Elsevier Applied Science, New York, 1992, pp. 423–446.
Beyler, C. L., and Hirschler, M. M., "Thermal Decomposition of Polymers," *The SFPE Handbook of Fire Protection Engineering*, 2nd ed., DiNenno, P. J., *et al.,* eds., National Fire Protection Association, Quincy, MA, 1995.
Brown, J. E., "Plastics. Part C. Composite Materials," Chapter 12, *Heat Release in Fires*, Elsevier Applied Science, NY, 1992, pp. 447–459.
Cleary, T. G. and Quintiere, J. G., "Flammability Characterization of Foam Plastics," Society of Plastics Industries, Inc., Washington, DC, NISTIR 4664, Oct. 1991.
"Combustion Toxicology of Polymers," NMAB 318-3, 1978, National Materials Advisory Board, Washington, DC.
Cox, G., and Steven, G., eds., *Fundamental Aspects of Polymer Flammability*, American Institute of Physics, Chicago, 1987.
Dynamics of Current Developments in the Firesafety of Polymers, Technomic, Lancaster, PA, 1988.
Ellis, J. W., "Rigid Foam Plastic Insulation," *Fire Surveyor*, Vol. 19, No. 1, Feb. 1990, pp. 21–23.
Fire Safety Through Use of Flame Retardant Polymers, Fire Retardant Chemical Association's Conference Series, Technomic, Lancaster, PA, 1985.
Harmathy, T. Z., "Properties of Building Materials," *SFPE Handbook of Fire Protection Engineering*, 2nd ed., DiNenno, P. J., *et al.*, National Fire Protection Association, Quincy, MA, 1995.
Hilado, C. J., ed., *Flammability of Solid Plastics*, Technomic, Lancaster, PA, 1985.
Hilado, C., *Flammability Handbook for Plastics*, 3rd ed., Technomic Publishing Co., Westport, CT, 1982.
Hirschler, M. M., "Plastics. Part A. Heat Release From Plastic Materials," Chapter 12, *Heat Release in Fires*, Elsevier Applied Science, New York, 1992, pp. 375–422.
Hollins, L. T., "Plastic Gas Mains: Hazards and Tactics," *Fire Engineering*, Vol. 148, No. 12, Dec. 1995, pp. 49–52.
Johnson, D. G., "Combustion Properties of Plastics," *Journal of Applied Fire Science*, Vol. 4, No. 3, 1994/1995, pp. 185–201.
Kidder, R. C., "Effect of Fire Safety Regulatory Activities on the Plastics Scene in the USA," *Flame Retardants '85*, Plastics and Rubber Institute, 1985, p. 16.
Kidder, R. C., "Fire Safety Issues in the United States for Plastics in 1992," Fifth International Conference for Flame Retardants '92, Elsevier Applied Science, New York, 1992, pp. 21–32.
Larsen, J. B., *et al.*, "Toxicity of Combustion Products From Engineering Plastics," Sixteenth International Conference on Fire Safety, Volume 16, January 14–18, 1991, Millbrae, CA, Product Safety Corp., Sunnyvale, CA, 1991, pp. 189–202.
Masarik, I., "Ignitability and Burning of Plastic Materials: Testing and Research," Sixth International Fire Conference, Interflam '93, Fire Safety. Interscience Communications Ltd., London, England, 1993, pp. 567–577.
Nakagawa, Y., Komai, T. and Kohno, M., "Laboratory-Scale Gallery Fire Test on Rubber Conveyor Belts With Fabric Skeletons," *Fire and Materials*, Vol. 15, No. 1, 1991, pp. 17–26.
Nelson, G. L., "Recycling of Plastics: A New Challenge for Flame Retardants," *Proceedings* of the Twentieth International Conference on Fire Safety, Vol. 20, January 9–13, 1995, Millbrae, CA, Product Safety Corp., Sunnyvale, CA, 1995, pp. 189–199.
O'Neill, T. J., "Assessing the Corrosion Risk of Plastics in Fires," *Fire Safety Journal*, Vol. 15, No. 1, 1989, pp. 45–56.
Persson, H., "Evaluation of the RDD-Measuring Technique. RDD-Tests of the CEA and FMRC Standard Plastic Commodities," Swedish National Testing and Research Institute, Boras, Sweden, SP Report 1991:04, CIB W14/92/12 (S), 1991.
"Plastic Fire Hazard Classifications," Pamphlet P6221, Factory Mutual Research Corp., Norwood, MA.
Rubin, I. I., *Handbook of Plastic Materials and Technology*, Wiley, New York, 1988.
Saechtling, H., *International Plastics Handbook*, Macmillan, New York, 1984.

Scudamore, M. J., Briggs, P. J. and Praager, F. H., "Cone Calorimetry: A Review of Tests Carried Out on Plastics for the Association of Plastic Manufacturers in Europe," *Fire and Materials*, Vol. 15, No. 2, Apr./June 1991, pp. 65–84.

Smith, G. F., "Assessment of PVC Smoke," *Journal of Vinyl Technology*, Vol. 10, No. 2, June 1988, pp. 84-89.

Smith, S. R., "Plastics—The Hidden Danger in Our Surroundings," *Fire Safety Journal*, Vol. 45, No. 1, 1985, pp. 22–26.

Takahashi, S., "Extinguishment of Plastics Fires With Plain Water and Wet Water," *Fire Safety Journal*, Vol. 22, No. 2, 1994, pp. 169–179.

Thomson, H. E., and Drysdale, D. D., "Flammability of Plastics," *Fire and Materials*, Vol. 11, No. 4, Dec. 1987, pp. 163–172.

Tomes, W. J. and Golinveaux, J., "Fire Test Performance of Extra Large Orifice Sprinklers for Rack Storage of Group A Plastics in Warehouse-Type Retail Occupancies," *Proceedings* of the International Conference on Fire Research and Engineering, September 10–15, 1995, Orlando, FL, Society of Fire Protection Engineers (SFPE), Boston, MA, 1995, pp. 415–420.

UL 94, *Tests for Flammability of Plastic Materials for Parts in Devices and Appliances*, Underwriters Laboratories Inc., Northbrook, IL, 1982, 4th ed., 1991.

Williams, B. K., Larsen, J. B. and Nelson, G. L., "Combustion Product Toxicity of Engineering Plastics: Pulmonary Pathology," Seventeenth International Conference on Fire Safety, Vol. 17, January 13–17, 1992, Millbrae, CA, Product Safety Corp., Sunnyvale, CA, 1992, pp. 198–204.

Yung, D. and Mehaffey, J. R., "Fire Resistance Requirements for Rubber-Tire Warehouses," *Fire Technology*, Vol. 27, No. 2, May 1991, pp. 100–112.

PESTICIDES

William J. Keffer and Matthew Woody

Pesticides, as a class, are extremely diverse in chemical composition, in formulation and in relative hazard, reflecting the enormous diversity of pests, pest management problems, and application conditions. They are closely regulated under a wide variety of federal and state regulations and local codes and standards.

Pesticide products are widely distributed in retail and wholesale occupancies throughout the country. In many cases these occupancies are small businesses that also vend and store a multitude of other products, including food and clothing for human use.

Current industry trends, driven by increased pressures for waste reduction, costs of waste disposal and improvements in application equipment, have resulted in more concentrated pesticide formulations being routinely available throughout the industry.

Due to increasing awareness of the need for management of materials that may present unacceptable exposure hazards to workers, the general public, and the environment, a wide variety of "standard of care" regulations and guidelines have been promulgated and should be considered in management of pesticide occupancies. These new, more stringent and protective regulations present significant challenges to owners and managers of agricultural chemical facilities, due to the variety of serious potential hazards in these occupancies. Significant policy changes include: (1) creating formal comprehensive planning for product management, (2) improving requirements for training of employees and documenting competency, and (3) interfacing with the public to share information and management plans for hazardous chemical management.

The changing perspective on chemical management is evidence that prevention of release and exposure incidents has become a major concern. There have been numerous reports in recent years of facility release incidents, with and without fire, where the value of the product lost was insignificant compared to the cost of the incident mitigation, e.g., facility costs for replacement of responding fire service protective gear, hoses, cleanup, and incineration of large quantities of contaminated soil.

Local codes and standards incorporate changing risk perspectives rather slowly, because regulating must be mandatory and represent a minimum level of compliance to be met by all facilities within the jurisdiction. The current or potential owner or manager of a commercial occupancy dealing in pesticide storage must become familiarized with each updated requirement. This chapter provides pesticide facility owners and managers an overview of requirements and practices that reflect state-of-the-art facility location, construction, and administration.

REGULATIONS AND GUIDELINES

State and federal regulations affect the management of pesticides. The risk manager for a pesticide occupancy needs to be familiar with the requirements of the following applicable federal regulations, developed by U.S. Environmental Protection Agency (EPA), Department of Transportation (DOT), and Occupational Safety and Health Administration (OSHA), and their state counterparts.

Clean Air Act (CAA) (EPA)

This Act establishes emission performance standards for fertilizer and pesticide plants. Amendments require: (1) 75 percent reduction of toxic emissions, including several pesticides; and (2) regulations to prevent, detect, and respond to accidental releases of ammonia and several other pesticide active ingredients. A new risk management plan (RMP) includes detailed requirements and procedures for public and environmental exposure protection.

Clean Water Act (CWA) (EPA)

This Act prohibits the discharge of toxic pollutants into navigable waters without a permit. The definition of "navigable waters" is very broad. The CWA also establishes pre-treatment standards for waste waters sent to publicly owned treatment works (POTWs). New CWA regulations require the development and implementation of a comprehensive program to obtain permits and manage the discharge of storm water from privately owned facilities.

Commercial Motor Vehicle Safety Act/ Hazardous Materials Transportation Act (CMVSA/HMTA) (DOT)

In addition to setting minimum national standards for commercial drivers' licenses, recent amendments to these regulations establish minimum competencies for any person who handles, or prepares, hazardous materials for shipment.

William J. Keffer is senior engineering advisor, U.S. Environmental Protection Agency, Kansas City, KS. Matthew Woody is hazardous materials coordinator, Des Moines Fire Department, Des Moines, IA.

Comprehensive Environmental Response Compensation and Liability Act (CERCLA/Superfund) (EPA)

This act requires federal notification of spills or releases of hazardous substances that may reach the air, waters, groundwaters, or surface soil of the environment. CERCLA also regulates owner/operator liability for spill and release cleanup, including past disposals.

Emergency Planning and Community Right-to-Know Act (SARA Title III/EPCRA) (EPA)

This act mandates emergency planning and response capabilities, requires notification of local and federal authorities of inventoried hazardous chemicals, and dictates reporting of all environmental releases.

Endangered Species Act (ESA) (EPA)

As part of the rules developed under this act, a list of pesticide ingredients "which may affect" endangered species has been created. Management of these pesticides requires special consideration of potential environmental exposures.

Federal Insecticide, Fungicide, and Rodenticide Act (FIFRA) (EPA)

This act comprehensively regulates the distribution, sale, and use of pesticides. Under Section 19 of the 1988 amendments to this act, EPA's authority to regulate the storage, transportation, and disposal of pesticides; containers; rinsates; and contaminated materials has been greatly expanded. On August 13, 1992, EPA announced its final rule revising regulations regarding protection of workers from agricultural pesticides.

Occupational Safety and Health Act (OSHA)

The various provisions of this act require all facilities to take steps to provide for minimum worker protection. These requirements include hazard communication (HC) (29 CFR 1910.1200), fire and evacuation plans (FEP) (29 CFR 1910.37 and 1910.38a), hazardous waste operations and emergency response (HAZWOPER) (29 CFR 1910.120), and process safety management of highly hazardous chemicals (PSM) (29 CFR 1910.119). For a quick summary of the most pertinent new requirements, readers are referred to the nonmandatory appendix to the PSM regulation, and to the new, August 22, 1994, Appendix (e) to the HAZWOPER regulation.

Resource Conservation and Recovery Act (RCRA) (EPA)

The various provisions of this act control generation-to-disposal systems for managing hazardous wastes, provide certain exemptions for farmers discarding pesticide residues, and ban land disposal of hazardous waste.

Relatively brief summaries of the applicability of these regulations to specific operations can be found in the "Regulatory File" section of the current edition of the *Farm Chemicals Handbook*.[1] These concise summaries on the status of these regulations as they apply to the storage of pesticides can help facility owners and managers develop appropriate risk management strategies.

In recent years several other publications have been developed to provide concise guidance to facility risk managers and planners, especially those from small businesses with limited experience and time to study compliance requirements. The following are examples of the most useful of these documents.

1. Title III, "List of Lists."[2] This document presents a consolidated list of chemicals regulated by EPA under CERCLA/Superfund, SARA Title III/EPCRA, and the CWA, and provides a list of the threshold quantities in the various regulations.
2. *Comprehensive Planning Document*.[3] This document takes the key elements of 15 federal regulations and provides an organized matrix of where they apply to the development and management of a safe operation. The document organizes these regulations into a matrix that addresses:

Hazard identification	Community outreach
Safe storage and handling	Emergencies—evacuation
Normal process operations	Emergencies—response
Release prevention	Training

The organization of the topical list is the same as the sequence of tasks that a facility risk manager would follow to develop a single effective safety, health, and environmental management plan for a fixed facility.

3. "Risk Assessment Guidelines for Municipalities and Industry."[4] This document from the Major Industrial Accident Council of Canada (MIACC) provides a practical set of steps for the evaluation of relative hazard and the identification of vulnerable zones from potential accidents at a production and storage facility.
4. "SACS—Survey Form for Public Warehousing."[5] This American Crop Protection Association (ACPA) [formerly NACA] document takes eight key elements of facility location, structures, and management, and provides a series of questions with multiple answers to evaluate the relative risk of chemical accidents from a specific facility. This document provides the basis for much of this chapter.

Additional information relevant to the hazards of storage and use of pesticides can be found in this handbook. See the following sections and chapters:

Section 1

1-9 "Environmental Issues in Fire Protection"

Section 3

3-21 "Storage of Flammable and Combustible Liquids"

3-22 "Storage of Gases"

3-23 "Storage and Handling of Chemicals"

3-32 "Hazardous Waste Control"

3-33 "Housekeeping Practices"

Section 4

4-1 "Fire Hazards of Materials: An Overview"

4-6 "Flammable and Combustible Liquids"

4-7 "Gases"

Section 9

9-17 "Warehouse Operations"

9-18 "General Indoor Storage"

9-19 "Outdoor Storage Practices"

Section 10

Section 11

DEFINITION OF PESTICIDES

The definition of a pesticide as provided in FIFRA is comprehensive and includes a chemical or mixture of chemicals or substances used to repel or combat an animal or plant pest. This includes insects and other invertebrate organisms; all vertebrate pests, e.g., rodents, fish, pest birds, snakes, gophers; all plant pests growing where not wanted, e.g., weeds; and all microorganisms that may or may not produce disease in humans. Household germicides, plant growth regulators, and plant root destroyers are also included in the definition.

Pesticides are typed according to their primary or specific control purposes, or to reflect the manner in which they are used. Insecticides (insect control), fungicides (fungus and bacteria control), herbicides (unwanted plant control), nematocides (earth and water worm control), and rodenticides (rodent control) are examples of pesticides classified by control purposes. Fumigants are an example of pesticides that are classified by the manner in which they are used, in that a fumigant is a pesticide that acts in the gaseous, or "fume," state.

Pesticides are also classified by the hazard to users and/or the environment. Many concentrations and specific poisons are only available, legally, that is, to trained and licensed applicators. These are always identified by label information as "Restricted Use Pesticides."

Well over one billion pounds of pesticides are produced and sold annually in North America. In addition to their manufacturing facilities, pesticides can be found stored in commercial warehouses, agricultural chemical warehouses, farm supply stores, nurseries, farms, supermarkets, discount stores, hardware stores and other retail outlets, on the premises of commercial pest control applicators and lawn-care operators, and in the home. Large quantities are continually being transported as liquids, solids, and gases in highway and rail transportation.

HAZARDS AND HAZARD IDENTIFICATION

In addition to the obvious, but highly variable, health hazards of the pesticide active ingredient (i.e., a biological poison), pesticide formulation components (e.g., active ingredients, carriers, and adjuvants) can also be corrosive, reactive, flammable, or combustible. The products of combustion of some formulations may present a significant toxic hazard. Some formulations may act as oxidizing agents that can accelerate combustion. Runoff water from fires involving pesticide products may be highly contaminated and may present a human health and environmental hazard. It is not uncommon to find pesticides stored in close proximity to other agricultural products, swimming pool chemicals, and various fuels. This is especially common in large discount stores in urban and suburban areas.

Pesticide formulations are encountered in all three states of matter: liquid, solid, and gaseous. Flammable or combustible liquids are widely used to dissolve solid pesticides in order to facilitate application. Adjuvants are frequently added to reduce surface tension, increase uptake rates by target, and protect against wash off/abrasion, all of which may also increase the hazard during fire or other release incidents. The hazards of flammable or combustible liquids when released from containers can range from a simple spill fire to a combustible explosion of vapors, fogs, or mists during application. Solids are usually stored and handled as granules, dusts, and powders. If they are combustible, they can present a dust explosion hazard when dispersed into the air. Most pesticide powders, however, are specific formulations absorbed into an inert material, such as clay.

The intrinsic toxicity of the active ingredient in pesticides varies considerably. Table 4-11A lists the toxicity of the active ingredients of some pesticides routinely offered for retail sale to the general public and provides information on two common substances for comparative purposes.

TABLE 4-11A.   Example Pesticide Toxicities

	Oral LD_{50}* mg/kg	Dermal LD_{50}* mg/kg
Fungicides		
captan	9,000	—
chlorothalinol	>10,000	>10,000
Herbicides		
2,4-D	375	—
glyphosate	4,300	—
prometron (tech-Pramitol™)	2,980	>2,000
calcium acid methanearsonate	4,000	—
Insecticides		
acephate (tech-Orthene™)	945	>10,250
carbaryl (tech-Sevin™)	283	—
chloropyrifos (Dursban™)	270	2,000
diazanon (tech)	400	3,600
disulfuton (tech-Disyston™)	12	25
lindane	125	1,000
malathion	1,375	4,100
methoxychlor	6,000	—
rotenone	1,500	—
EPA definition of extremely hazardous substances	50	
Table salt	3,320	
Aspirin	1,200	

*LD_{50} is a measurement value used in toxicity tests to specify the dose of a substance that is most likely to cause death within 14 days in one half of the test animals.

Hazard information provided on pesticide labels is frequently based only on the active ingredient and not necessarily on the concentration in the specific formulation, and does not include fire, health, or reactivity hazards of the complete formulation consisting of active ingredients, carriers, and adjuvants. Information on the synergistic effects of fires or other releases involving a mixture of pesticide materials is not widely available. Concentrations of active ingredients vary widely in the various formulations. The system of hazard identification required by EPA is based on the active ingredient testing only and is therefore not fully compatible with NFPA 704, *Standard System for the Identification of the Hazards of Materials for Emergency Response* (hereinafter referred to as NFPA 704). Table 4-11B lists the current EPA-required warning systems for labeling pesticide containers based on active ingredients.

One additional pesticide categorization system may be helpful. There has been considerable effort on the part of the federal government to limit general public accessibility to many of the more toxic active ingredients and formulations of pesticide products. It may be assumed that those products and formulations that require licensed applicators may present a more serious health risk than the same

TABLE 4-11B. *EPA Toxicity Categories*

Category 1: POISON/DANGER
All pesticide products meeting the following criteria:
   Oral LD_{50} up to and including 50 mg/kg,
   Inhalation LD_{50} up to and including 0.2 mg/liter,
   Dermal LD_{50} up to and including 200 mg/kg,
   Eye effects—corneal opacity not reversible within 7 days, and skin effects corrosive.
Must bear on the front panel the signal word "Danger." In addition, if the product was assigned to Category 1 on the basis of its oral, inhalation, or dermal toxicity (as distinct from skin and eye local effects), the word "poison" must appear in red on a background of distinctly contrasting color, and the skull and crossbones must appear in immediate proximity to the word "Poison."

Category 2: WARNING
All pesticides meeting the following criteria:
   Oral LD_{50} from 50 through 500 mg/kg,
   Inhalation LD_{50} from 0.2 through 2.0 mg/liter,
   Dermal LD_{50} from 200 through 2,000 mg/kg,
   Eye effects—corneal opacity reversible within 7 days (irritation persisting for 7 days), and
   Skin effects—severe irritation at 72 hr.
Must bear on the front panel the signal word "Warning."

Categories 3 and 4: CAUTION
All pesticide products meeting the following criteria:
   Oral LD_{50} greater than 500 mg/kg,
   Inhalation LD_{50} greater than 2.0 mg/liter,
   Dermal LD_{50} greater than 2,000 mg/kg,
   Eye effects—no corneal opacity, and
   Skin effects—moderate irritation at 72 hr.
Must bear on the front panel the signal word "Caution."

chemicals formulated in more dilute concentrations for general retail sales. It should further be noted that health hazard threshold levels for restricted use of pesticide formulations are identical to at least the "3" level health hazard in NFPA 704.

Many of the more toxic and persistent pesticides have been canceled for any use by EPA but may still be found by fire and other emergency response personnel. These older pesticide stocks may present a very serious fire and/or health hazard for personnel involved in an emergency response to any occupancy containing these materials. Information on pesticide restrictions can be obtained from the nearest EPA office. Information on toxic effects for emergency personnel is available from poison control centers, the Agency for Toxic Substances and Disease Registry, health and safety groups, major manufacturers, and the National Pesticide Telecommunication Network (NPTN).

Effective risk management for general operations and for response is a joint responsibility of both the owner/operator and the emergency response organizations serving the facility. Adequate preplanning requires evaluation of material safety data sheets (MSDS) for each formulation. Manufacturers and formulators should be encouraged to adopt the proposed 16-element American National Standards Institute (ANSI) MSDS format, which is oriented toward response information. Poorly prepared and incomplete MSDS should be brought to the attention of preparers and, if necessary, OSHA. The emergency response organization must become familiar with MSDS and with interpretation of the physical and chemical characteristics that identify hazards from fire, corrosivity, reactivity, compressed gases, and various container types. Other key information that affects response tactics for releases includes:

1. Is the health hazard a level "3" or greater per NFPA 704? Is the material a "restricted-use" pesticide? Is the material listed as hazardous on EPA's "List of Lists"[2]? These chemicals may re-

quire substantial hazardous material (HAZMAT) skills, particularly if the significant route of exposure is dermal (i.e., skin contact). Conventional fire-fighting gear is totally ineffective against skin exposure to these HAZMATs.

2. Is the material listed as a "may affect" by the EPA? These chemicals include:

acephate	methyl bromide
aldicarb	naled
aluminum phosphide	parathion
azinphos-methyl	permethrin
bendiocarb	phorate
brodifacoum	Pival™
bromadiolone	potassium nitrate
bromethalin	sodium cyanide
carbofuran	sodium fluoroacetate
chlorophacinone	sodium nitrate
chlorpyriphos	terbufos
diphacinone	trifluralin
endosulfan	vitamin D-3
fenthion	warfarin
fenvalerate (and s-fenvalerate)	zinc phosphide
magnesium phosphide	

These chemicals may require special storage conditions and precautions against release of runoff to sensitive environments.

3. Does the material have high-leaching potential? If so, special precautions may be needed to prevent releases from reaching shallow groundwater. Some crop production chemicals of concern for high-leaching potential are:

Herbicides

Acclaim	Karmex™
Assure™	Linuron
Avenge™	MCPA ester
Balan™	Paarlan™
Betamix™	Pramitol™
Betanex™	Prefar™
Caparol™	Probe™ (discontinued 1993
Dacthal™	by Sandoz Agro, Inc.)
Devrinol™	Prowl™
Far-Go™	Roundup™
Fusilade™	Sonalan™
Goal™	Tandem™
Gramoxone™	Treflan™
Hoelon™ 3EC	Zorial™

Insecticides

Abate™	Dyfonate™
Ambush™, Pounce™	Ethion
(Permethrin)	Kelthane™
Amdro™	Lindane
Antor™	Lorsban™
Asana™XL	Mavrik™
Bolstar™	Morestan™
Carzol™	Omite™
Cymbush™, Ammo™	Thimet™
(Cypermethrin)	Vendex™
Diazinon	Zolone™
Dimilin™	

4. Other subject-specific regulatory lists should be checked to determine special concerns for environmental runoff. Two examples are:
   a. Maximum contaminate levels in water systems, and
   b. Proposed RCRA action levels.

5. The *Farm Chemicals Handbook*[1] provides an extensive list of human and wildlife toxicity ratings and routes of exposure for many currently used pesticides.

The information listed above should be: (1) collected in preplanning efforts and (2) evaluated by the emergency response organization before a release incident occurs. Collected information that is not accessible when needed or understood by the user represents a waste of time.

FACILITY PLANNING

The "SACS—Survey Form for Public Warehousing" is the source for most of the following specifications.[5] The following criteria represent ideal storage facility conditions for pesticide products with significant hazards.

Location and Site Factors

It is preferable that a pesticide facility be located in an area where exposure and risk is minimal should a fire or other disaster occur. Generally less-populated areas are preferred.

It is generally preferable that institutions housing sensitive populations, such as hospitals, retirement homes, day-care centers, etc., are not close to a pesticide facility. Actual distances should be determined by a realistic process hazards analysis developed on a site-specific basis. The potential impact of a facility emergency upon nearby populations can be approximated using materials such as: (1) computer-aided management of emergency operations (CAMEO), computer software developed by the National Oceanic and Atmospheric Administration (NOAA) and EPA; and (2) the document "Technical Guidance for Hazards Analysis," developed by EPA, Federal Emergency Management Agency (FEMA), and DOT. These materials allow site-specific factors to be incorporated into a credible worst-case estimate of the effects of a chemical release upon a "vulnerable zone" in proximity to a pesticide facility. By incorporating seasonal local weather data, such as prevailing wind directions, planners can devise zoning or siting guidelines for such facilities.

The preferred location of a facility relative to major highways is one of balance. It should be easily accessible to emergency responders and truckers, but yet not pose a major threat (e.g., fire, smoke, fumes) to traffic. Consideration should also be given to the property uses along selected routes to and from the facility to account for the potential damage from a transportation accident involving a vehicle containing hazardous pesticide products.

It is preferable for a facility to be located in an area where the probability of a natural disaster, such as a flood, earthquake, tornado, hurricane, etc., is relatively low.

Facility Construction

Single-story facilities without basements are preferred for the storage of hazardous chemicals, including pesticides. For fire protection reasons, noncombustible construction is preferred. NFPA 220, *Standard on Types of Building Construction*, provides details for approved noncombustible or limited-combustible materials. Insulation, when required, should not contribute to a fire, should not absorb pesticide materials, and should serve to contain a fire. Floors should be structurally sound and constructed of materials that are easy to maintain and clean. For hazardous pesticide storage, sealants should be considered that are compatible with the pesticide formulation components. Facilities, including storage areas, should be lit adequately to prevent accidents. Lighting levels should be adequate to read written instructions without an additional lighting source.

Ventilation in crop-production chemical storage is essential to eliminate: (1) odors and (2) smoke and fumes in the event of a fire or spill. Ventilation systems should be installed based on the vapor density of toxic vapors that may accumulate in the storage area. Ventilation designs should factor in handling of both materials and containers. As a general rule, a facility devoted solely to storage does not present the same ventilation demands as one with open-system mixing or container-filling operations, and, therefore, may safely operate utilizing passive ventilation.

Open-use, dispensing, or mixing operations are those that allow the liberation of potentially harmful vapors, gases, or particles to the atmosphere, and require exhaust provisions to safely carry them away from the workplace. Contingent upon the degree of hazard proposed by the materials exhausted, a means of collection or treatment may be necessary to avoid discharging harmful contaminants to the atmosphere. Normal operations at a storage facility do not pose such a hazard, since containers are stored closed awaiting shipment.

Climate must be factored into a ventilation system design. Outside ambient temperatures can create makeup air for a system that heats chemicals, resulting in increased generation of vapors. Severe winter cold temperature can solidify certain substances or freeze wet-type sprinkler systems. In addition, a continually operated ventilation system in severe climates can place substantial demands on building heating and air-conditioning systems and can be extremely costly to operate. Operational costs can be minimized through the utilization of an interior source of makeup air entering through approved dampers, when required. Continuously operated ventilation/exhaust systems may be most appropriate for open-system operations where the release of gases, vapors, or particles can be anticipated under normal circumstances.

For reasons of general risk reduction, and especially for personnel safety during a fire or other emergency, office and retail spaces should be of proper construction; have at least two exits, not through the storage area; and be provided with heating and ventilation separate from the storage space.

Fire Protection

Conventional precautions for fire protection of hazardous pesticide facilities must be balanced with concerns for off-site exposures due to toxic gases during a fire situation and the potential for environmental damage due to fire water runoff. (See subsection entitled "Environmental Impact and Containment.") A pesticide facility should ideally have 100 percent sprinkler coverage, adequate fire extinguishers for preliminary response by employees, and a fire detection system that includes smoke and heat detectors. National Fire Protection Association (NFPA) sprinkler, fire extinguisher, and fire hose standards are referenced at the end of this chapter; also review chapters in Section 6 of this handbook.

Due to the potential for contaminant spread via runoff from sprinkler systems, or the water reactivity of some materials, e.g., fumigants, consideration should be given to alternative means of fire extinguishment and control. Dry chemical, carbon dioxide, or other extinguishing systems may be effective means of fire protection in some chemical facilities. The compatibility of chemicals with alternative fire extinguishing systems should be given careful consideration.

If fire protection is to be provided by sprinkler systems, a detection system should alert facility personnel to the imminence of sprinkler discharge, allowing time for response that may control the incident before water flows from the system. Consideration should be given to both automatic and manual alarm systems that can provide rapid notification of incipient fire conditions and provide facility personnel with a means of giving prompt notification of fire and other types of emergency conditions. If the facility is not occupied on a 24-hr basis, supervision of these systems should be provided to ensure that necessary notifications are promptly made.

Warehouse personnel should be properly trained in the use of all types of fire extinguishers available in their location. At least annual training in fire extinguisher use should be practiced at the facility. It is recommended that these practice sessions be conducted with local fire personnel.

Careful location of valves controlling sprinkler system risers will provide emergency responders the means to control fire spread from an area of origin into adjoining areas of the facility, while allowing them to stop the discharge of water onto released chemicals.

Ideally the facility should have two independent water sources that are connected to an underground looped yard main with proper valving. A facility should ideally have an automatic alarm system with an Underwriters Laboratories/Factory Mutual Research Corporation (UL/FMRC)–approved installation, connected to a supervised UL/FMRC–approved central station or fire department, and tested at least bimonthly.

Interior fire walls of storage areas are necessary to restrict a fire from spreading from one area to another and should be 4-hr-rated, liquidtight, ramped at openings, and extend 3 ft above the roofline. The use of noncombustible walls or partitions can be an effective means of segregating incompatible materials within a building. Fire doors are necessary to prevent a fire from spreading from one fire area to another. Fire doors should: (1) be properly rated and (2) close automatically in the event of a fire. NFPA 80, *Standard for Fire Doors and Fire Windows*, provides criteria for interior fire wall construction and for fire doors.

All electric service should be in compliance with NFPA 70, *National Electrical Code®*.

Heat, if any, should be provided with the ignition sources outside the storage area.

Preferably, the fire department servicing the warehouse should: (1) have a low Insurance Services Office (ISO) rating, (2) operate a competent hazmat team, and (3) provide routine prevention inspection services using current codes.

Easy accessibility between the facility and responding fire department is essential in the event of a fire or chemical release. Onsite accessibility to all areas of the facility is important, i.e., the site plan should be such that all areas can be easily accessed by the fire department. Access considerations should include multiple gates to facilitate a safe, upwind entry to the facility for emergency response in the case of hazardous materials release.

Hazard warning signs, using NFPA 704, should be visibly posted on all facility structures and fixed tanks. NFPA 704 evaluations and installation locations for signs and placards should be coordinated with the servicing fire jurisdiction.

Site-planning efforts should establish separate storage areas. Preferably, flammable storage should be in a separate building. Consideration should be given to separating storage of flammable materials without a significant toxicity hazard from those materials that are flammable, produce toxic fumes, are skin-contact poisons, or represent significant environmental hazards. Nonflammable materials with NFPA 704 health hazard ratings of "3" or more should not be stored in these separate areas.

Routine inspections and tests should be conducted to ensure fire protection systems are operable. The inspections and tests should cover sprinkler systems, valves, fire pumps, extinguishers, hose systems, fire doors, runoff containment structures and vessels, and runoff treatment systems, if any. Particular attention should be paid to the location and management of RCRA-approved waste-storage areas and to areas and procedures for the management of returned and damaged goods. A "no-smoking" policy for the flammable storage area is essential.

Material-handling equipment must be properly rated and maintained for the desired occupancy, with the rating clearly marked on the side of each vehicle. The entrance to each storage area should be clearly posted with the class of vehicle permitted in that area. The refueling or recharging area should be properly protected. No other motor vehicles are allowed in the warehouse area.

Hotwork, e.g., welding, constitutes a fire hazard. Procedures for all hotwork should be documented and enforced to protect storage when hotwork is required.

Neighboring facilities may be engaged in activities that pose potential risks to the pesticide facility. Significant off-site hazards should be identified through EPCRA submittals to the local fire jurisdiction and discussions with fire prevention personnel.

Environmental Impact and Containment

In the event of a fire or atmospheric release of chemicals, it is important to assess and plan for the effect on sensitive areas downwind. Assessment for the pesticide facility should plan for identification of areas where the potential result is: (1) death, (2) reversible injury, and (3) inconvenience. Methods and procedures that provide real-time documentation of exposure levels during the release should be in place.

The location of the facility should be where spills and runoff from normal operations, liquid release incidents, and fire water runoff do not threaten the environment. The facility should ideally have the capability to contain spills and water runoff within the property in secondary containment that has an impervious liner to prevent groundwater contamination. If the nature of the material being contained permits, internal containment through the use of curbing and ramping of door openings provides an effective alternative to traditional methods of conducting spillage and runoff to an exterior bermed location. Advantages to internal containment include: (1) greater control over the materials and (2) prevention of rainwater intrusion into open exterior containment. The addition of rainwater could result in containment overflow or could multiply disposal costs by increasing liquid volume. Drawbacks to internal containment are: (1) potential workplace exposure to contaminants remaining in the building, (2) the presence of released chemicals that may pose the threat of flash fire, (3) vapor explosion or other unacceptable occurrence, and (4) increased difficulty for emergency responders whose work may be made more difficult by the retention of large quantities of spilled materials.

If spilled pesticides and contaminated water are conducted to an exterior containment system, drainage piping should be compatible with these materials and adequately sized to carry the anticipated flow. Exterior containment areas should be secured and located away from important valves, building exits, incompatible materials, or environmentally sensitive areas; if open to precipitation, they should be adequately sized to contain this added volume. Greater control over materials confined in exterior containment can be exercised through the use of containment tanks that are closed to the atmosphere and installed in a manner to ensure adequate drainage. Contingent upon the percentage of a containment vessel below grade, some jurisdictions view them as underground storage tanks. Local authorities should be consulted in regard to applicable codes and ordinances.

Adequate containment volume should be calculated, at a minimum, to include a reasonable/credible worst-case scenario release from: (1) a single tank or vessel and (2) potentially contaminated water from fire-fighting runoff and/or sprinkler operation. Water utilized during fire-control operations may dissolve chemicals, resulting in contaminated runoff from the fire scene; or light nonmiscible liquids may be borne on top of this runoff. The volume of fire-fighting water runoff to be contained should be determined by the Authority Having Jurisdiction, factoring in the water miscibility and specific gravity of the chemicals that may be released. Water discharged onto nonmiscible pesticides may not leach harmful contaminants from them; and if the specific gravity of these compounds is greater than that of water, containment of released chemicals alone may suffice. Overflow from secondary containment should not pose an immediate threat to people, property, or the environment.

Groundwater contamination from leaking underground storage tanks (UST) must be considered. Current UST monitoring requirements that require inventory reconciliation to within 1 percent of throughput may not be adequate to ensure that groundwater contamination is not taking place. In areas of shallow groundwater used for human or animal consumption, it may be desirable to consider a monitoring well network for early warning of groundwater releases.

Security

A pesticide facility should be provided with good exterior lighting. Fencing and/or gates can be effective deterrents to security breaches. Inbound and outbound shipments should be handled in a safe and secure area. Security alarms are an effective method to deter unauthorized entry. Patrolling the facility affords additional security and can provide early identification of chemical releases. Access to all hazardous chemical storage areas must be controlled. This is particularly significant for "restricted-use pesticides."

Pre-employment security checks serve to identify potential personnel problems. Security problems have often been related to the availability of pharmaceutical drugs. Routine testing for substance abuse serves to protect against product loss and careless handling of hazardous materials within the storage areas.

Routine physical inventories reduce pilferage and help identify problems early. Physical inventories must be thorough, with special attention given to visual inspection of each unit.

Security procedures that address preventive practices and reaction plans are important. Security procedures and access control for contractors working at the facility should be written and implemented.

Storage and Handling Practices

A facility should implement housekeeping practices on a routine basis. Special attention should be paid to accumulations of combustible materials and to the handling, storage, and prompt management of spills, leaks, and damaged and returned goods. Pests, e.g., rodents, bugs, and birds, should be controlled. NFPA 30, *Flammable and Combustible Liquids Code*, and NFPA 231, *Standard for General Storage*, provide information relative to aisle space in the storage area. Relevant NFPA standards should also be consulted as to pile size and height requirements. Storage area personnel and retail stock personnel should know the class and hazards of each product stored and displayed. Pesticide products should never be stacked to heights that exceed the structural strength of the packaging.

Flammable and restricted-use pesticides should be stored and displayed separate from other products. It is recommended that restricted-use pesticides not be displayed in retail areas accessible to the general public. A system that recognizes not only the chemical hazard classes of the materials involved but also the potential loss to the facility should a fire or release occur must be used. Stored substances must be segregated, according to accepted codes and standards. DOT tables for incompatible materials in highway cargo van shipments can be used as an initial guide.

Flammable aerosols constitute a major fire threat. If present, storage should be limited to less than 100 cu ft (2.83 m^3), caged, and segregated from other combustibles.

Facility management, assisted by safety and supervisory personnel, should develop and implement a routine system for safety, health, and environment walk-throughs of the facility. This creates a housekeeping system for maintenance, safe and compatible product storage, management of damaged and returned goods, and overall facility safety operation consistency. These tours should be supported by documentation.

Equipment in pesticide storage areas should be dedicated. If necessary to use this equipment in a food-stuff area, the equipment should first be decontaminated and the decontamination verified by appropriate analytical data. Wheel chocks are required by OSHA and provide a vital safety function. Forklifts should have both safety and loading/unloading lights, backup warning devices, and horn. The warehouse should have the capability to recoup, re-palletize, re-unitize, return to the manufacturer, or dispose by legal methods damaged and returned merchandise.

Empty pallet storage inside the facility should be kept to a minimum. The warehouse area should have floor markings indicating storage spaces, aisles, staging areas, routes, etc. Warehouse area temperatures should be in compliance with applicable government regulations and in accord with any guidelines noted on the MSDS or manufacturer's instructional materials. Indirect heating, such as steam, warm air, etc., is recommended. Heating systems should be safe and permanent. Airflow should not be directly on storage, and storage should be a safe distance from the heat source. Temperature regulation devices should be visible and accessible. Annual inspections of the heating and ventilation system by a competent contractor is recommended.

Administration

Pesticide facility administration, like that at any chemical facility, must be effective, concerned, and knowledgeable in all areas affecting their business and the business of their clients. The management and staff must be able to demonstrate a thorough knowledge of the regulations and standards, such as OSHA and NFPA, that impact the operation. Copies of relevant regulations, standards, and permits should be available at each facility. Initial and periodic refresher training should be provided as needed.

SARA Title III/EPCRA and local community right-to-know laws are becoming increasingly important in chemical operations. Management at chemical storage and retail facilities should: (1) have a thorough knowledge and understanding of the requirements, (2) file all necessary reports, (3) complete an effective emergency response preplan in concert with local fire authorities, and (4) be active in the local emergency planning committee.

Management must create, review, deliver, and keep records of effective training programs in all aspects of pesticide operations. The new Appendix (e) of the OSHA HAZWOPER regulation (29 CFR 1910.120) provides detailed guidance for the management of training. As requirements for "demonstrated competency" become more important, it will be increasingly necessary for facility management to coordinate training program development and delivery with local institutions, such as community colleges. Participation in outside trade councils may be another important source of effective training materials. Training program documentation must include comprehensive initial technical as well as safety/health training, refresher programs, on-the-job training, and drills and exercises with local responders.

Under SARA Title III/EPCRA, facilities may be required to prepare pre-emergency plans. Detailed guidance for such plans is available in NFPA 43D, *Code for the Storage of Pesticides*, and NFPA 471, *Recommended Practice for Responding to Hazardous Materials Incidents*, as well as in other documents. The pre-plan should address product spills, medical emergencies, and other types of disasters that may occur at the facility. The plan should be prepared with the cooperation and consultation of local response organizations and with assistance from state and federal organizations, if possible. The emergency response plan should be viewed as an integral component of an overall chemical management plan for the facility. Details of overall management systems for hazardous chemicals are contained in the OSHA *Process Safety Standard*. Once the plan has been completed, the various positions identified as participating actively in response, e.g., incident commander, technician, specialist, etc., must be trained to meet the requirements of the OSHA HAZWOPER regulations. NFPA 472, *Standard for Professional Competence of Responders to Hazardous Materials Incidents*, provides competency standards for such training.

The pre-plan should be exercised and reviewed annually to reflect changes in the influencing factors; this process should be documented. In the event of an emergency, a complete list of emergency contacts and phone numbers is necessary. These phone numbers should include both mandatory regulatory notification numbers as well as contacts for technical assistance and cleanup services. Examples of necessary phone numbers are: facility emergency coordinator, fire department, police, emergency medical transport and hospitals, CHEMTREC, chemical suppliers, among others. This phone list should be updated any time changes in contacts and personnel are made, and at least twice a year.

As part of the pre-emergency planning, the fire department and other involved agencies should identify under which scenarios they would continue to fight a fire *vs.* letting it burn. Factors to be evaluated when making this determination should include: (1) the presence of combustible loading in the form of flammable products, (2) combustible packaging materials, (3) the stage of the fire at the time emergency responders arrive, and (4) combustible construction. The intense heat generated by the burning of a relatively high volume of combustible materials is critical to the breakdown of pesticides during a fire. Other factors to be evaluated include: (1) the ability to contain fire-fighting water runoff and (2) the proximity of people, property, or sensitive environmental areas to the products of pesticide combustion.

Material safety data sheets (MSDS) should be available for reference in all emergencies. Additional and current sets of the MSDS should be in a location accessible during an emergency. The MSDS should be made available to the fire department, local hospitals, and other associated parties. The pesticide facility should also maintain an up-to-date site plan that shows the general classification and approximate quantities of products within the various storage areas. This site plan should also be available during an emergency. Approved "lock boxes" that can be mounted on a building exterior are available in sizes that accommodate MSDS binders and site plans. These boxes: (1) limit access to all but authorized personnel and (2) can protect against tampering when connected to security systems.

When a facility elects to provide in-house response capability for so-called "insignificant" releases, the following must be addressed: adequate training, protective equipment, exposure monitoring equipment, and containment hardware; supplies, such as sorbents and salvage containers, must be maintained.

BIBLIOGRAPHY

References Cited

1. *Farm Chemicals Handbook*. Published annually by Meister Publications, 37733 Euclid Ave., Willoughby, OH.
2. Title III, "List of Lists," EPA 500-B-94-002. Available free from the nearest EPA regional library.
3. *Comprehensive Planning Document*. Available free from Ecology and Environment, Overland Park, KS. Phone: (913) 432-9961.
4. "Risk Assessment Guidelines for Municipalities and Industry," MIACC, 265 Carling Ave., Suite 600, Ottawa, Ontario, Canada, K1S 2E1.
5. "SACS—Survey Form for Public Warehousing," American Crop Protection Association (ACPA), [Formerly National Agricultural Chemicals Association (NACA)], 1156 15th St. NW, Suite 400, Washington, DC.

NFPA Codes, Standards, and Recommended Practices

Reference to the following NFPA codes, standards, and recommended practices will provide further information on pesticides discussed in this chapter. (See the latest version of *The NFPA Catalog* for availability of current editions of the following documents.)

NFPA 10, *Standard for Portable Fire Extinguishers*

NFPA 13, *Standard for the Installation of Sprinkler Systems*

NFPA 14, *Standard for the Installation of Standpipe and Hose Systems*

NFPA 30, *Flammable and Combustible Liquids Code*

NFPA 43D, *Code for the Storage of Pesticides*

NFPA 70, *National Electrical Code*®

NFPA 80, *Standard for Fire Doors and Fire Windows*

NFPA 220, *Standard on Types of Building Construction*

NFPA 231, *Standard for General Storage*

NFPA 471, *Recommended Practice for Responding to Hazardous Materials Incidents*

NFPA 472, *Standard for Professional Competence of Responders to Hazardous Materials Incidents*

NFPA 505, *Fire Safety Standard for Powered Industrial Trucks Including Type Designations, Areas of Use, Conversion, Maintenance, and Operation*

NFPA 704, *Standard System for the Identification of the Hazards of Materials for Emergency Response*

Additional Reading/Software

CAMEO. Software developed by National Oceanic and Atmospheric Administration (NOAA), Hazardous Materials Response Branch N/CMS34, 7600 Sand Point Way NE, Seattle, WA, and U.S. Environmental Protection Agency (EPA), Washington, DC.

Designing Facilities for Pesticide and Fertilizer Containment, 1st ed., Midwest Plan Service, Agricultural and Biosystems Engineering Department, Iowa State University, Ames, IA, 1991.

"Standard Operating Safety Guides," EPA 9285.1-03. Available free from EPA by calling (908) 321-6740; ask for the safety group.

EXPLOSIVES AND BLASTING AGENTS

Revised Thomas P. Dowling

This chapter gives a short description of explosives and explosive materials, with emphasis on their fire and explosion hazards. Additional important data concerning the manufacture, transportation, storage, and use of explosive materials can be found in NFPA 495, *Explosive Materials Code* (hereinafter referred to as NFPA 495), and in other referenced material. NFPA 495 is a current and comprehensive guide that should be studied by all who have anything to do with explosives.

NFPA 495 provides a rule that must be obeyed always: *Do not fight fires involving explosive materials.* Look for the explosive signs. Do not fight the fire. Evacuate the area. In 1988, six Kansas City, MO, fire fighters were killed fighting a fire in an ammonium nitrate-based blasting agent storage facility.

The hazardous nature of explosives has long been recognized, and the rapid increase in their production and use makes it necessary to point out those properties that contribute most to the inherent dangers of these very important industrial and military products. In the industrial field alone, approximately 5 billion pounds of explosives and blasting agents were used in 1995, compared with 3.7 billion pounds in 1983, and 0.7 billion pounds in 1950 (1 lb = 0.45 kg). Of these totals, approximately 90 percent were used in mining, and the remainder principally in construction operations.[2]

Black powder was the first explosive known, although fireworks and pyrotechnics were known and used many centuries earlier.[1] The preparation and properties of black powder were described by Roger Bacon in the 13th century, and it was first used for military purposes in the 14th century. Black powder was first used for mining operations in the latter part of the 17th century, and it remained the dominant commercial explosive until the latter part of the 19th century. In 1846, the compound nitroglycerin was first prepared and its explosive properties studied. The extremely hazardous nature of this compound prevented wide use until 1867 when Alfred Nobel made the discovery that nitroglycerin absorbed in an inert material, such as diatomaceous earth, produced a mixture that was relatively safe to handle. This explosive mixture, with many modifications and improvements, remained the predominant commercial explosive until the introduction of modern blasting agents in the mid-1950s. Meanwhile, the use of black powder decreased to an insignificant quantity, largely because of its great fire hazard. In less than 20 years, blasting agents, based principally on the sensitization of am-

monium nitrate, have taken over approximately 90 percent of the market.[2] The change is based largely on: (1) economic considerations and (2) greatly improved safety of nonnitroglycerin explosive materials.

NATURE OF EXPLOSIVE MATERIALS

Terminology

An understanding of the nature of explosive materials is essential before one can understand the fire and explosion potential of these admittedly hazardous products. A few general terms are defined below to provide a background for future discussions.

Blasting agents (explosive 1.5): Materials that have been classified according to tests specified in Title 49 of the U.S. *Code of Federal Regulations*, Parts 170-189,[3] and found to be so insensitive that there is very little probability of accidental initiation to explosion or of transition from deflagration to detonation. Blasting agents (explosive 1.5), as mixed for use or shipment, cannot be detonated in the standard cap sensitivity test.

Detonator: Any device containing a detonating charge that is used for initiating detonation in an explosive. The term includes, but is not limited to, electric blasting caps of instantaneous and delay types, blasting caps for use with a safety fuse, instantaneous and delay blasting caps for use with shock tube or miniaturized detonating cord, and detonating cord delay connectors.

Explosion: An explosion is an effect produced by the sudden violent production or expansion of gases. It may be accompanied by heat, shock waves, and the disruption or enclosing of nearby structures or materials.

Explosive: An explosive is a substance or a mixture of substances that, when subjected to the proper stimuli, undergoes an exceedingly rapid self-propagating reaction, characterized by: (1) formation of more stable products (usually gases), (2) evolution of heat, and (3) development of a sudden pressure effect through the action of this heat on produced or adjacent gases. A somewhat simpler, but less accurate, definition states that an explosive is any chemical compound, mixture, or device whose primary or common use is to function by explosion. The term includes, but is not limited to, dynamite, black powder, initiating explosives, detonators, safety fuses, squibs, detonating cords, igniter cords, and igniters.

Thomas P. Dowling is manager of technical services for the Institute of Makers of Explosives. He is a member of NFPA's Technical Committee on Explosives.

Explosive material: Includes explosives, blasting agents, water gels, emulsions, slurries, and detonators. A list of explosive materials is published annually by the Bureau of Alcohol, Tobacco, and Firearms (ATF).

Propellant: An explosive material that normally functions by deflagration (burning) and is used for propelling purposes.

An explosion may result from: (1) chemical changes, such as those that accompany the detonation of an explosive or the combustion of a flammable gas-air mixture; (2) physical or mechanical changes, such as the bursting of a steam boiler, or a high-pressure reaction vessel; and (3) atomic changes, such as those that occur in a nuclear explosion.

Fires and Explosions

A distinction must be made between controlled and unwanted fires and explosions. Controlled combustion reactions are basic to the production of power and necessary for an industrial economy. Likewise, detonation reactions are essential for use by the mining and construction industries. On the other hand, the loss of life and natural resources from unwanted fires and explosions is a matter of national concern. Since fire is a major hazard associated with the utilization of explosives and blasting agents, this chapter discusses the basic nature of these materials and methods for reducing the loss from uncontrolled reactions in these materials.

TYPES OF EXPLOSIVE MATERIALS

Although this chapter mainly addresses commercial explosives, some of the similarities and differences between commercial and military explosives should be considered. Most military explosives have high shattering power and relatively high detonation velocities. Military explosives are often kept in storage for relatively long periods of time and therefore must have extremely good stability. It is essential that these explosives detonate reliably even after storage in a wide variety of conditions. While there are no truly military explosives equivalent to the blasting agents used in commercial operations, the military uses commercial-type explosives and blasting agents for operations, such as road building, airstrip preparations, and a wide variety of similar operations.

As stated previously, commercial explosives have a wide range of properties. Explosives to be used in underground operations must have relatively good fume characteristics. This requires formulations that are nearly oxygen-balanced and therefore produce a minimal amount of carbon monoxide and oxides of nitrogen. The most widely used military explosives, such as trinitrotoluene (TNT) and cyclonite (RDX), are quite oxygen-deficient and therefore produce large quantities of toxic gases. Commercial dynamites normally have a fairly long storage life, but blasting agents are more often used soon after manufacture, so storage is not as important a factor. However, prolonged storage of dynamite, slurries, or ammonium nitrate with fuel oil (ANFO) under adverse storage conditions, such as high heat and humidity, may severely reduce the shelf life of the products.

Commercial Explosive Types

Explosive materials can be divided into a number of specific types according to their characteristic properties.[4] The following types are generally recognized in industry.

Primary or initiating high explosives: These are quite hazardous, but they play a very important role in the utilization of explosive materials. Typical explosives of this type are mercury fulminate, lead styphnate, and lead azide. Because of their hazardous nature they are seldom, if ever, used alone, and their principal function is to initiate detonation in less-sensitive explosives. These primary explosives are readily detonated by heat or mechanical shock. They are used almost exclusively as the initiating agent in detonators and are classified by the Department of Transportation (DOT) as 1.1 (Class A). Unless shipped as a component of a manufactured article, initiating explosives normally must be wetted with water or alcohol for shipment.

Secondary high explosives: These are materials that are relatively insensitive to mechanical shock and heat and yet are readily detonated by the shock from a primary explosive. They can be used for both military purposes and in commercial blasting operations. Secondary high explosives include such products as TNT, RDX, dynamite, nitroglycerin, and pentaerythritol tetranitrate (PETN). Because of its sensitivity, nitroglycerin is normally not used alone. Secondary high explosives are generally cap-sensitive, which distinguishes them from blasting agents, and are classified by DOT as 1.1 materials.

Low explosives or propellants: These are used primarily for propulsion purposes. They normally function by deflagration (burning) rather than detonation, although some propellants are susceptible to detonation. Black powder, smokeless powder, and solid rocket fuels fall into this category. Fire constitutes the greatest hazard in the handling and use of propellant explosives. Propellants can be 1.1 or 1.3 materials (Class A or Class B explosives).

Blasting agents: Today, almost 90 percent of all blasting operations in the United States are carried out using non-nitroglycerin materials. The most commonly used explosive (1.5-blasting agent) is a fuel oxidizer system that consists primarily of ammonium nitrate and a fuel, such as No. 2 diesel fuel. The formulation is termed ANFO and is sold under a variety of trade names. The addition of approximately 6 percent fuel oil to ammonium nitrate prills (pellets) produces a very efficient blasting material. This combination has the advantage of low cost, satisfactory fume characteristics, and greatly increased safety over nitroglycerin dynamites. While the combination does not burn readily, it still may change in a well-established fire to produce a detonation. ANFO does have a disadvantage in that it is difficult or impossible to use it under very wet conditions. Many variations of the basic ANFO formulation have been developed and successfully used. Powdered aluminum or other fuels can be added to increase the energy and blasting efficiency of the basic ANFO formulation.

Water gels, slurries, and emulsions: Operations where a water-resistant explosive material is a necessity, and where higher densities than those that can be obtained with ANFO are desirable, have led to the development of water gels, slurries, and emulsions. These may be classed as either 1.5 (blasting agents) or 1.1 explosives, depending on their formulation and sensitivity. Ammonium nitrate is usually the basic oxidizer in this type of explosives, and it may be sensitized by a variety of materials, including paint-grade aluminum, monomethylamine nitrate, TNT, and includes various gelling agents. Approximately 15 percent of the market has been taken over by water gels, slurries, and emulsions.

Nuclear explosives: Nuclear explosives are used for military purposes only.

CLASSES OF EXPLOSIVES

In 1991 the Department of Transportation (DOT) issued new regulations for the transportation of hazardous materials based upon the *United Nations Recommendations on the Transportation of Dangerous Goods*. These new regulations revised DOT's hazardous material regulations, with respect to hazard communication, classification, and packaging of explosives.

Under the United Nations recommendations, hazardous materials are assigned to a class according to their specific characteristics. Explosives are assigned to Class 1; and the explosives in Class 1 are divided into six divisions, as follows:

Division 1.1 consists of explosives that have a mass explosion hazard. A mass explosion is one that affects almost the entire load instantaneously.

Division 1.2 consists of explosives that have a projection hazard but not a mass explosion hazard.

Division 1.3 consists of explosives that have a fire hazard and either a minor blast hazard or a minor projection hazard, or both, but not a mass explosion hazard.

Division 1.4 consists of explosives that present a minor explosion hazard. The explosive effects are largely confined to the package, and no projection of fragments of appreciable size or range is to be expected. An external fire must not cause virtually instantaneous explosion of almost the entire content of the package.

Division 1.5 consists of very insensitive explosives. This division comprises substances that have a mass explosion hazard but are so insensitive that there is very little probability of initiation or of transition from burning to detonation under normal conditions of transport.

Division 1.6 consists of extremely insensitive articles that do not have a mass explosion hazard. This division comprises articles that contain only extremely insensitive detonating substances and that demonstrate a negligible probability of accidental initiation or propagation.

Where the classification system in effect prior to January 1, 1991, is referenced in state or local laws, ordinances, or regulations not pertaining to the transportation of hazardous materials, Table 4-12A can be used to compare old and new hazard class names.

TABLE 4-12A. *Comparison of Old and New Hazard Class Names as of January 1, 1991*

Current Classification	Class Name Prior to January 1, 1991
Division 1.1	Class A explosives
Division 1.2	Class A or B explosives
Division 1.3	Class B explosives
Division 1.4	Class C explosives
Division 1.5	Blasting agents
Division 1.6	No applicable hazard class

Other Classification Systems

The Bureau of Alcohol, Tobacco, and Firearms (ATF), Department of the Treasury, has a classification system based on the type of explosive material involved. In its system, explosive materials are classified as high explosives, low explosives, and blasting agents. This classification is employed in the enforcement of Title 11, "Regulation of Explosives," of the *Organized Crime Control Act of 1970* (18 U.S.C., Chapter 40), which is intended to provide for the secure and safe storage of explosive materials.

Permissible Explosives

While the term "permissible explosive" is not exactly a classification, it does describe a type of explosive that has been tested and approved by the Department of Labor, Mine Safety, and Health Administration (MSHA) as meeting the minimum safety requirements for use in underground coal mines.[5] These materials are modified dynamites or water gels that exhibit reduced tendencies to ignite flammable gas-air or gas-air-coal dust mixtures.

Two-Component Explosives

Two-component, or binary, explosives consist of two or more unmixed, commercially manufactured, prepackaged chemicals, including oxidizing chemicals, flammable liquids, or solids that are not independently classified as explosives. Until mixed, they do not have to be stored, transported, and handled as explosives.

MANUFACTURE OF EXPLOSIVE MATERIALS

All manufacturers of explosive materials must obtain a license from ATF and must follow the ATF regulations as specified in Title 27 of the U.S. *Code of Federal Regulations*, Part 55. Additionally, state and local regulatory agencies may also have rules covering the manufacture of explosive materials and may license manufacturers.

Cartridged or packaged explosive materials are manufactured at fixed plant sites and are transported to the point of use or are shipped to storage/distribution facilities generally located in or near explosive-consuming areas. The production of explosive materials at fixed plant sites is under the close supervision of qualified personnel. The safety record at such plants is extremely good, largely because the hazards are well recognized and operating procedures are designed to keep potential hazards to a minimum. Fire is the chief concern during manufacture, and plant layout keeps this hazard within reasonable limits. Smoking is *never* allowed near the explosive operations, and the use of flame-producing equipment is allowed only in specific areas under strict safety regulations.

During the past 25 years there has been a marked increase in the use of bulk explosives at large mining, quarrying, and construction blasting operations. Bulk explosives are usually 1.5 materials (blasting agents) and can consist of ANFO, water gels, slurries, emulsions, or a blend of these materials. Bulk explosives can be plant-mixed and delivered to the blast site by bulk trucks, or they can be mixed at the blast site in a bulk truck. Bulk explosives are normally moved by auger or pumped from the bulk truck directly into the blastholes.

Fire is a safety concern in the operation of any type of bulk truck, and the driver/operator must be thoroughly trained in the operation of the mixing, pumping, and/or augering equipment, and the characteristics, safe handling, and emergency procedures for the explosive materials being handled. The driver/operator must have a DOT commercial driver's license (CDL), with a hazardous materials (H) endorsement.

TRANSPORTATION OF EXPLOSIVE MATERIALS

The DOT regulates the transportation of all explosive materials in accordance with regulations set forth in U.S. *Code of Federal Regulations*, Title 49, Chapter I, "Transportation," Parts 100–199.[6] Copies of these regulations can be obtained from the U.S. Government Printing Office, or from the Bureau of Explosives of the Association of American Railroads.

Most commercial explosives are now transported over public highways by truck. Military explosives are transported both by rail and by truck. Fire is the greatest concern in the transportation by motor vehicle over public highways. Although tire fires are less common since the introduction of tubeless tires, they are still a concern and represent a hazard that is difficult to control because the driver of a truck may be unaware of the fire until it has gained considerable headway. Detonators may be transported in limited quantities on the same vehicle with explosives and blasting agents, provided DOT regulations concerning quantity and type of container are observed. Truck transportation over public highways is of particular concern, because there is increased exposure of the public to fire and explosion hazards. Unfortunately, fires and other types of accident situations tend to draw spectators, which increases the danger of casualties. While rail transportation generally provides a lesser hazard to the public, there is an increased danger resulting from the mixing of different types of cargo in a given train load.

The DOT publishes the *North American Emergency Response Guidebook* (ERG), which covers emergency response procedures for transportation incidents involving explosives or other hazardous materials. The publication is widely distributed and is normally carried on all hazardous materials transport vehicles and by emergency responders, such as fire and police departments. Copies of the ERG can be purchased from the U.S. Government Printing Office or a number of commercial outlets.

STORAGE OF EXPLOSIVE MATERIALS

A primary objective for the storage of most nonexplosive industrial materials is to provide protection for such materials from their surroundings. Because of the hazardous nature of explosive materials, additional factors must be considered. Safety to the industrial workers involved and to the general public in the vicinity of such storage is of primary importance. Since most explosive materials are utilized by the mining and construction industries, these industries have primary concern in training their employees to provide maximum attainable safety. Employees of explosives manufacturing firms are well aware of the hazards of their occupation, and safety is an integral part of their work.

Another factor connected with storage is indirectly related to safety and involves the element of security. In recent years it has become increasingly important to protect the public from the illicit use of explosives, such as those obtained by individuals through illegitimate means. The worldwide concern with terrorism has made security an important consideration in the storage of explosives. Most explosives used for illegal purposes have been obtained from legitimate stores of the materials. To assist in preventing this misuse of explosives, federal law now provides that the Bureau of Alcohol, Tobacco, and Firearms, Department of the Treasury, regulate the manufacture, distribution, and storage of explosive materials. Copies of these regulations can be obtained from that agency.

Storage Magazines

To provide proper storage for the wide variety of explosive materials now being produced, five different types of magazines are recognized. The requirements for these five types of magazines are given in the regulations developed by the Bureau of Alcohol, Tobacco, and Firearms. The specifications for these magazines and types of materials to be put in them can be obtained from ATF or can be found in NFPA 495, *Explosive Materials Code.*

When trucks transporting explosives must stop in transit, or the load must be transferred in transit, terminals are now provided for motor vehicles carrying explosives. These terminals are well controlled and are operated under strict safety regulations. Requirements for these terminals are contained in NFPA 498, *Standard for Safe Havens and Interchange Lots for Vehicles Transporting Explosives.*

The probability of accidental initiation of an explosive while in an approved storage location is slight. To safeguard the general public, the *American Table of Distances for Storage of Explosive Materials* (ATD) has been prepared to indicate the isolation distances for stores of explosives from points of contact with the public. (See Table 4-12B.) This table is revised periodically by the Institute of Makers of Explosives. While 1.5 materials (blasting agents) are appreciably less sensitive than 1.1 materials (Class A explosives), once they are initiated to detonation, the damage produced by such an accident is comparable, and the ATD (Table 4-12B) is used to determine the safe distances of stores of 1.5 materials (blasting agents) from inhabited buildings, public highways, and passenger railways. However, because of the lower sensitivity of 1.5 materials (blasting agents), the separation distances between magazines of 1.1 materials (Class A explosives) and stores of 1.5 materials (blasting agents), or between stores of 1.5 materials (blasting agents), need not be as great as that required between stores of 1.1 materials (Class A explosives). (See Table 4-12C.)

The ATD applies to the manufacture and permanent storage of *commercial* explosive materials. The distances specified are those measured from the explosive materials storage facility to the inhabited building, highway, or passenger railway, irrespective of property lines.

The ATD covers all *commercial* explosive materials, including, but not limited to, high explosives, blasting agents, detonators, initiating systems, and explosive materials in process. The ATD is not designed to be altered or adjusted to accommodate varying explosive characteristics, such as blast effect, weight strength, density, bulk strength, detonation velocity, etc.

The ATD should not be used to determine safe distances for blasting work, the firing of explosive charges for testing or quality control work, or the open detonation of waste explosive materials. It may, however, be utilized as a guide for developing distances for the unconfined, open burning of waste explosive materials where the probability of transition from burning to high-order detonation is unlikely.

Table 4-12C shows the recommended separation distances and the thickness of artificial barricades for the separation of 1.1 materials (Class A explosives) from 1.5 materials (blasting agents), and the separation between stores of 1.5 materials (blasting agents). When barricades are *not* provided, the recommended distances must be increased by a factor of six, as shown in Table 4-12C (Note 2).

FIRE PROTECTION FOR EXPLOSIVE MATERIALS

Fire is a principal cause of accidents involving explosive materials. Although explosives may vary in their sensitivity to fire conditions, they are liable to produce a disastrous explosion when subjected to fire. The only effective method for fire protection with explosive materials is to eliminate the source of fire. This means meticulous housekeeping standards. A small fire in oily rags or wastepaper can trigger a major catastrophe. Smoking or the use of any fire-producing equipment should not be permitted where explosive materials are produced, handled, stored, or used. Because of the relative insensitivity of 1.5 materials (blasting agents), there is always a danger that personnel will become careless around these materials. A fire and subsequent explosion in a blasting agent mixing house near Norton, VA, illustrates the potential hazard from the careless use of heat-producing equipment.[8] Another fire and explosion in a blasting agent storage facility killed six Kansas City, MO, fire fighters who were fighting the fire.[9]

TABLE 4-12B. *The American Table of Distances for Storage of Explosives*

The American Table of Distances is reprinted from IME Safety Library Publication No. 2 with permission of the Institute of Makers of Explosives, and this was revised in June of 1991.

Quantity of Explosive Materials[1,2,3,4]		Distance in ft*							
		Inhabited Buildings[9]		Public Highways Class A to D[11]		Passenger Railways—Public Highways with Traffic Volume of More than 3000 Vehicles/Day[10,11]		Separation of Magazines[12]	
Pounds† Over	Pounds† Not Over	Barricaded [6,7,8]	Un-barricaded	Barricaded [6,7,8]	Un-barricaded	Barricaded [6,7,8]	Un-barricaded	Barricaded[6,7,8]	Un-barricaded
0	5	70	140	30	60	51	102	6	12
5	10	90	180	35	70	64	128	8	16
10	20	110	220	45	90	81	162	10	20
20	30	125	250	50	100	93	186	11	22
30	40	140	280	55	110	103	206	12	24
40	50	150	300	60	120	110	220	14	28
50	75	170	340	70	140	127	254	15	30
75	100	190	380	75	150	139	278	16	32
100	125	200	400	80	160	150	300	18	36
125	150	215	430	85	170	159	318	19	38
150	200	235	470	95	190	175	350	21	42
200	250	255	510	105	210	189	378	23	46
250	300	270	540	110	220	201	402	24	48
300	400	295	590	120	240	221	442	27	54
400	500	320	640	130	260	238	476	29	58
500	600	340	680	135	270	253	506	31	62
600	700	355	710	145	290	266	532	32	64
700	800	375	750	150	300	278	556	33	66
800	900	390	780	155	310	289	578	35	70
900	1000	400	800	160	320	300	600	36	72

*One ft equals 0.305 m.
†One lb equals 0.4536 kg.

Explanatory Notes Essential to the Application of the American Table of Distances for Storage of Explosives

NOTE 1: "Explosive materials" means explosives, blasting agents, and detonators.

NOTE 2: "Explosives" means any chemical compound, mixture, or device, the primary or common purpose of which is to function by explosion. A list of explosives determined to be within the coverage of Title 18, United States Code, Chapter 40, "Importation, Manufacture, Distribution and Storage of Explosive Materials," is issued at least annually by the Director of the Bureau of Alcohol, Tobacco, and Firearms of the Department of the Treasury. For quantity and distance purposes, detonating cord of 50 grains per foot should be calculated as equivalent to 8 lb (3.7 kg) of high explosives per 1000 ft (305 m). Heavier or lighter core loads should be rated proportionately.

NOTE 3: "Blasting agents" means any material or mixture consisting of fuel and oxidizer, intended for blasting, and not otherwise defined as an explosive, provided that the finished product, as mixed for use or shipment, cannot be detonated by means of a No. 8 test blasting cap where unconfined.

NOTE 4: "Detonator" means any device containing any initiating or primary explosive that is used for initiating detonation. A detonator may not be permitted to contain more than 10 g of total explosives by weight, excluding ignition or delay charges. The term includes, but is not limited to, electric blasting caps of instantaneous and delay types, blasting caps for use with safety fuses, detonating cord delay connectors, and nonelectric instantaneous and delay blasting caps that use detonating cord, shock tube, or any other replacement for electric leg wires. All types of detonators in strengths through No. 8 cap should be rated at 1¹/₂ lb (0.7 kg) of explosives per 1000 caps. For strengths higher than No. 8 cap, the manufacturer should be consulted.

NOTE 5: "Magazine" means any building, structure, or container, other than an explosives manufacturing building, approved for the storage of explosive materials.

NOTE 6: "Natural barricade" means natural features of the ground, such as hills, or timber of sufficient density that the surrounding exposures that need protection cannot be seen from the magazine when the trees are bare of leaves.

NOTE 7: "Artificial barricade" means an artificial mound or revetted wall of earth of a minimum thickness of 3 ft (0.9 m).

NOTE 8: "Barricaded" means the effective screening of a building containing explosive materials from the magazine or another building, a railway, or a highway by a natural or an artificial barrier. A straight line from the top of any sidewall of the building containing explosive materials to the eave line of any magazine or other building or to a point 12 ft (3.7 m) above the center of a railway or highway shall pass through such barrier.

NOTE 9: "Inhabited building" means a building regularly occupied in whole or part as a habitation for human beings, or any church, schoolhouse, railroad station, store, or other structure where people are accustomed to assemble, but does not include any building or structure occupied in connection with the manufacture, transportation, storage, or use of explosive materials.

NOTE 10: "Railway" means any steam, electric, or other railroad or railway that carries passengers for hire.

NOTE 11: "Highway" means any public street, public alley, or public road.

NOTE 12: Where two or more storage magazines are located on the same property, each magazine shall comply with the minimum distances specified from inhabited buildings, railways, and highways, and, in addition, they should be separated from each other by not less than the distances shown for "separation of magazines," except that the quantity of explosive materials contained in detonator magazines shall govern with regard to the spacing of said detonator magazines from magazines containing other explosive materials. If any two or more magazines are separated from each other by less than the specified "separation of magazines" distances, such magazines, as a group, shall be considered as one magazine, and the total quantity of explosive materials stored in such group shall be treated as if stored in a single magazine located on the site of any magazine of the group, and shall comply with the minimum specified distances from other magazines, inhabited buildings, railways, and highways.

NOTE 13: Storage in excess of 300,000 lb (136,200 kg) of explosive materials in one magazine generally is not necessary for commercial enterprises.

NOTE 14: This table applies only to the manufacture and permanent storage of commercial explosive materials. It is not applicable to the transportation of explosives or any handling or temporary storage necessary or incident thereto. It is not intended to apply to bombs, projectiles, or other heavily encased explosives.

NOTE 15: Where a manufacturing building on an explosive materials plant site is designed to contain explosive materials, the building shall be located at a distance from inhabited buildings, public highways, and passenger railways in accordance with the American Table of Distances based on the maximum quantity of explosive materials permitted to be in the building at one time.

TABLE 4-12B. *The American Table of Distances for Storage of Explosives (Continued)*

Quantity of Explosive Materials[1,2,3,4]		Distance in ft*							
		Inhabited Buildings[9]		Public Highways Class A to D[11]		Passenger Railways—Public Highways with Traffic Volume of More than 3000 Vehicles/Day[10,11]		Separation of Magazines[12]	
Pounds† Over	Pounds† Not Over	Barricaded[6,7,8]	Un-barricaded	Barricaded[6,7,8]	Un-barricaded	Barricaded[6,7,8]	Un-barricaded	Barricaded[6,7,8]	Un-barricaded
1000	1200	425	850	165	330	318	636	39	78
1200	1400	450	900	170	340	336	672	41	82
1400	1600	470	940	175	350	351	702	43	86
1600	1800	490	980	180	360	366	732	44	88
1800	2000	505	1010	185	370	378	756	45	90
2000	2500	545	1090	190	380	408	816	49	98
2500	3000	580	1160	195	390	432	864	52	104
3000	4000	635	1270	210	420	474	948	58	116
4000	5000	685	1370	225	450	513	1026	61	122
5000	6000	730	1460	235	470	546	1092	65	130
6000	7000	770	1540	245	490	573	1146	68	136
7000	8000	800	1600	250	500	600	1200	72	144
8000	9000	835	1670	255	510	624	1248	75	150
9000	10,000	865	1730	260	520	645	1290	78	156
10,000	12,000	875	1750	270	540	687	1374	82	164
12,000	14,000	885	1770	275	550	723	1446	87	174
14,000	16,000	900	1800	280	560	756	1512	90	180
16,000	18,000	940	1880	285	570	786	1572	94	188
18,000	20,000	975	1950	290	580	813	1626	98	196
20,000	25,000	1055	2000	315	630	876	1752	105	210

*One ft equals 0.305 m.
†One lb equals 0.4536 kg.

Explanatory Notes Essential to the Application of the American Table of Distances for Storage of Explosives

NOTE 1: "Explosive materials" means explosives, blasting agents, and detonators.

NOTE 2: "Explosives" means any chemical compound, mixture, or device, the primary or common purpose of which is to function by explosion. A list of explosives determined to be within the coverage of Title 18, United States Code, Chapter 40, "Importation, Manufacture, Distribution and Storage of Explosive Materials," is issued at least annually by the Director of the Bureau of Alcohol, Tobacco, and Firearms of the Department of the Treasury. For quantity and distance purposes, detonating cord of 50 grains per foot should be calculated as equivalent to 8 lb (3.7 kg) of high explosives per 1000 ft (305 m). Heavier or lighter core loads should be rated proportionately.

NOTE 3: "Blasting agents" means any material or mixture consisting of fuel and oxidizer, intended for blasting, and not otherwise defined as an explosive, provided that the finished product, as mixed for use or shipment, cannot be detonated by means of a No. 8 test blasting cap where unconfined.

NOTE 4: "Detonator" means any device containing any initiating or primary explosive that is used for initiating detonation. A detonator may not be permitted to contain more than 10 g of total explosives by weight, excluding ignition or delay charges. The term includes, but is not limited to, electric blasting caps of instantaneous and delay types, blasting caps for use with safety fuses, detonating cord delay connectors, and nonelectric instantaneous and delay blasting caps that use detonating cord, shock tube, or any other replacement for electric leg wires. All types of detonators in strengths through No. 8 cap should be rated at $1\frac{1}{2}$ lb (0.7 kg) of explosives per 1000 caps. For strengths higher than No. 8 cap, the manufacturer should be consulted.

NOTE 5: "Magazine" means any building, structure, or container, other than an explosives manufacturing building, approved for the storage of explosive materials.

NOTE 6: "Natural barricade" means natural features of the ground, such as hills, or timber of sufficient density that the surrounding exposures that need protection cannot be seen from the magazine when the trees are bare of leaves.

NOTE 7: "Artificial barricade" means an artificial mound or revetted wall of earth of a minimum thickness of 3 ft (0.9 m).

NOTE 8: "Barricaded" means the effective screening of a building containing explosive materials from the magazine or another building, a railway, or a highway by a natural or an artificial barrier. A straight line from the top of any sidewall of the building containing explosive materials to the eave line of any magazine or other building or to a point 12 ft (3.7 m) above the center of a railway or highway shall pass through such barrier.

NOTE 9: "Inhabited building" means a building regularly occupied in whole or part as a habitation for human beings, or any church, schoolhouse, railroad station, store, or other structure where people are accustomed to assemble, but does not include any building or structure occupied in connection with the manufacture, transportation, storage, or use of explosive materials.

NOTE 10: "Railway" means any steam, electric, or other railroad or railway that carries passengers for hire.

NOTE 11: "Highway" means any public street, public alley, or public road.

NOTE 12: Where two or more storage magazines are located on the same property, each magazine shall comply with the minimum distances specified from inhabited buildings, railways, and highways, and, in addition, they should be separated from each other by not less than the distances shown for "separation of magazines," except that the quantity of explosive materials contained in detonator magazines shall govern with regard to the spacing of said detonator magazines from magazines containing other explosive materials. If any two or more magazines are separated from each other by less than the specified "separation of magazines" distances, such magazines, as a group, shall be considered as one magazine, and the total quantity of explosive materials stored in such group shall be treated as if stored in a single magazine located on the site of any magazine of the group, and shall comply with the minimum specified distances from other magazines, inhabited buildings, railways, and highways.

NOTE 13: Storage in excess of 300,000 lb (136,200 kg) of explosive materials in one magazine generally is not necessary for commercial enterprises.

NOTE 14: This table applies only to the manufacture and permanent storage of commercial explosive materials. It is not applicable to the transportation of explosives or any handling or temporary storage necessary or incident thereto. It is not intended to apply to bombs, projectiles, or other heavily encased explosives.

NOTE 15: Where a manufacturing building on an explosive materials plant site is designed to contain explosive materials, the building shall be located at a distance from inhabited buildings, public highways, and passenger railways in accordance with the American Table of Distances based on the maximum quantity of explosive materials permitted to be in the building at one time.

TABLE 4-12B.   *The American Table of Distances for Storage of Explosives (Continued)*

Quantity of Explosive Materials[1,2,3,4]		Distance in ft*							
		Inhabited Buildings[9]		Public Highways Class A to D[11]		Passenger Railways—Public Highways with Traffic Volume of More than 3000 Vehicles/Day[10,11]		Separation of Magazines[12]	
Pounds† Over	Pounds† Not Over	Barricaded[6,7,8]	Un-barricaded	Barricaded[6,7,8]	Un-barricaded	Barricaded[6,7,8]	Un-barricaded	Barricaded[6,7,8]	Un-barricaded
25,000	30,000	1130	2000	340	680	933	1866	112	224
30,000	35,000	1205	2000	360	720	981	1962	119	238
35,000	40,000	1275	2000	380	760	1026	2000	124	248
40,000	45,000	1340	2000	400	800	1068	2000	129	258
45,000	50,000	1400	2000	420	840	1104	2000	135	270
50,000	55,000	1460	2000	440	880	1140	2000	140	280
55,000	60,000	1515	2000	455	910	1173	2000	145	290
60,000	65,000	1565	2000	470	940	1206	2000	150	300
65,000	70,000	1610	2000	485	970	1236	2000	155	310
70,000	75,000	1655	2000	500	1000	1263	2000	160	320
75,000	80,000	1695	2000	510	1020	1293	2000	165	330
80,000	85,000	1730	2000	520	1040	1317	2000	170	340
85,000	90,000	1760	2000	530	1060	1344	2000	175	350
90,000	95,000	1790	2000	540	1080	1368	2000	180	360
95,000	100,000	1815	2000	545	1090	1392	2000	185	370
100,000	110,000	1835	2000	550	1100	1437	2000	195	390
110,000	120,000	1855	2000	555	1110	1479	2000	205	410
120,000	130,000	1875	2000	560	1120	1521	2000	215	430
130,000	140,000	1890	2000	565	1130	1557	2000	225	450
140,000	150,000	1900	2000	570	1140	1593	2000	235	470

*One ft equals 0.305 m.
†One lb equals 0.4536 kg.

Explanatory Notes Essential to the Application of the American Table of Distances for Storage of Explosives

NOTE 1: "Explosive materials" means explosives, blasting agents, and detonators.

NOTE 2: "Explosives" means any chemical compound, mixture, or device, the primary or common purpose of which is to function by explosion. A list of explosives determined to be within the coverage of Title 18, United States Code, Chapter 40, "Importation, Manufacture, Distribution and Storage of Explosive Materials," is issued at least annually by the Director of the Bureau of Alcohol, Tobacco, and Firearms of the Department of the Treasury. For quantity and distance purposes, detonating cord of 50 grains per foot should be calculated as equivalent to 8 lb (3.7 kg) of high explosives per 1000 ft (305 m). Heavier or lighter core loads should be rated proportionately.

NOTE 3: "Blasting agents" means any material or mixture consisting of fuel and oxidizer, intended for blasting, and not otherwise defined as an explosive, provided that the finished product, as mixed for use or shipment, cannot be detonated by means of a No. 8 test blasting cap where unconfined.

NOTE 4: "Detonator" means any device containing any initiating or primary explosive that is used for initiating detonation. A detonator may not be permitted to contain more than 10 g of total explosives by weight, excluding ignition or delay charges. The term includes, but is not limited to, electric blasting caps of instantaneous and delay types, blasting caps for use with safety fuses, detonating cord delay connectors, and nonelectric instantaneous and delay blasting caps that use detonating cord, shock tube, or any other replacement for electric leg wires. All types of detonators in strengths through No. 8 cap should be rated at $1^{1}/_{2}$ lb (0.7 kg) of explosives per 1000 caps. For strengths higher than No. 8 cap, the manufacturer should be consulted.

NOTE 5: "Magazine" means any building, structure, or container, other than an explosives manufacturing building, approved for the storage of explosive materials.

NOTE 6: "Natural barricade" means natural features of the ground, such as hills, or timber of sufficient density that the surrounding exposures that need protection cannot be seen from the magazine when the trees are bare of leaves.

NOTE 7: "Artificial barricade" means an artificial mound or revetted wall of earth of a minimum thickness of 3 ft (0.9 m).

NOTE 8: "Barricaded" means the effective screening of a building containing explosive materials from the magazine or another building, a railway, or a highway by a natural or an artificial barrier. A straight line from the top of any sidewall of the building containing explosive materials to the eave line of any magazine or other building or to a point 12 ft (3.7 m) above the center of a railway or highway shall pass through such barrier.

NOTE 9: "Inhabited building" means a building regularly occupied in whole or part as a habitation for human beings, or any church, schoolhouse, railroad station, store, or other structure where people are accustomed to assemble, but does not include any building or structure occupied in connection with the manufacture, transportation, storage, or use of explosive materials.

NOTE 10: "Railway" means any steam, electric, or other railroad or railway that carries passengers for hire.

NOTE 11: "Highway" means any public street, public alley, or public road.

NOTE 12: Where two or more storage magazines are located on the same property, each magazine shall comply with the minimum distances specified from inhabited buildings, railways, and highways, and, in addition, they should be separated from each other by not less than the distances shown for "separation of magazines," except that the quantity of explosive materials contained in detonator magazines shall govern with regard to the spacing of said detonator magazines from magazines containing other explosive materials. If any two or more magazines are separated from each other by less than the specified "separation of magazines" distances, such magazines, as a group, shall be considered as one magazine, and the total quantity of explosive materials stored in such group shall be treated as if stored in a single magazine located on the site of any magazine of the group, and shall comply with the minimum specified distances from other magazines, inhabited buildings, railways, and highways.

NOTE 13: Storage in excess of 300,000 lb (136,200 kg) of explosive materials in one magazine generally is not necessary for commercial enterprises.

NOTE 14: This table applies only to the manufacture and permanent storage of commercial explosive materials. It is not applicable to the transportation of explosives or any handling or temporary storage necessary or incident thereto. It is not intended to apply to bombs, projectiles, or other heavily encased explosives.

NOTE 15: Where a manufacturing building on an explosive materials plant site is designed to contain explosive materials, the building shall be located at a distance from inhabited buildings, public highways, and passenger railways in accordance with the American Table of Distances based on the maximum quantity of explosive materials permitted to be in the building at one time.

TABLE 4-12B.   *The American Table of Distances for Storage of Explosives (Continued)*

Quantity of Explosive Materials[1,2,3,4]		Inhabited Buildings[9]		Public Highways Class A to D[11]		Passenger Railways—Public Highways with Traffic Volume of More than 3000 Vehicles/Day[10,11]		Separation of Magazines[12]	
Pounds† Over	Pounds† Not Over	Barricaded[6,7,8]	Un-barricaded	Barricaded[6,7,8]	Un-barricaded	Barricaded[6,7,8]	Un-barricaded	Barricaded[6,7,8]	Un-barricaded
150,000	160,000	1935	2000	580	1160	1629	2000	245	490
160,000	170,000	1965	2000	590	1180	1662	2000	255	510
170,000	180,000	1990	2000	600	1200	1695	2000	265	530
180,000	190,000	2010	2010	605	1210	1725	2000	275	550
190,000	200,000	2030	2030	610	1220	1755	2000	285	570
200,000	210,000	2055	2055	620	1240	1782	2000	295	590
210,000	230,000	2100	2100	635	1270	1836	2000	315	630
230,000	250,000	2155	2155	650	1300	1890	2000	335	670
250,000	275,000	2215	2215	670	1340	1950	2000	360	720
275,000	300,000	2275	2275	690	1380	2000	2000	385	770
150,000	160,000	1935	2000	580	1160	1629	2000	245	490
160,000	170,000	1965	2000	590	1180	1662	2000	255	510
170,000	180,000	1990	2000	600	1200	1695	2000	265	530
180,000	190,000	2010	2010	605	1210	1725	2000	275	550
190,000	200,000	2030	2030	610	1220	1755	2000	285	570
200,000	210,000	2055	2055	620	1240	1782	2000	295	590
210,000	230,000	2100	2100	635	1270	1836	2000	315	630
230,000	250,000	2155	2155	650	1300	1890	2000	335	670
250,000	275,000	2215	2215	670	1340	1950	2000	360	720
275,000	300,000	2275	2275	690	1380	2000	2000	385	770

*One ft equals 0.305 m.
†One lb equals 0.4536 kg.

Explanatory Notes Essential to the Application of the American Table of Distances for Storage of Explosives

NOTE 1: "Explosive materials" means explosives, blasting agents, and detonators.

NOTE 2: "Explosives" means any chemical compound, mixture, or device, the primary or common purpose of which is to function by explosion. A list of explosives determined to be within the coverage of Title 18, United States Code, Chapter 40, "Importation, Manufacture, Distribution and Storage of Explosive Materials," is issued at least annually by the Director of the Bureau of Alcohol, Tobacco, and Firearms of the Department of the Treasury. For quantity and distance purposes, detonating cord of 50 grains per foot should be calculated as equivalent to 8 lb (3.7 kg) of high explosives per 1000 ft (305 m). Heavier or lighter core loads should be rated proportionately.

NOTE 3: "Blasting agents" means any material or mixture consisting of fuel and oxidizer, intended for blasting, and not otherwise defined as an explosive, provided that the finished product, as mixed for use or shipment, cannot be detonated by means of a No. 8 test blasting cap where unconfined.

NOTE 4: "Detonator" means any device containing any initiating or primary explosive that is used for initiating detonation. A detonator may not be permitted to contain more than 10 g of total explosives by weight, excluding ignition or delay charges. The term includes, but is not limited to, electric blasting caps of instantaneous and delay types, blasting caps for use with safety fuses, detonating cord delay connectors, and nonelectric instantaneous and delay blasting caps that use detonating cord, shock tube, or any other replacement for electric leg wires. All types of detonators in strengths through No. 8 cap should be rated at 1¹/₂ lb (0.7 kg) of explosives per 1000 caps. For strengths higher than No. 8 cap, the manufacturer should be consulted.

NOTE 5: "Magazine" means any building, structure, or container, other than an explosives manufacturing building, approved for the storage of explosive materials.

NOTE 6: "Natural barricade" means natural features of the ground, such as hills, or timber of sufficient density that the surrounding exposures that need protection cannot be seen from the magazine when the trees are bare of leaves.

NOTE 7: "Artificial barricade" means an artificial mound or revetted wall of earth of a minimum thickness of 3 ft (0.9 m).

NOTE 8: "Barricaded" means the effective screening of a building containing explosive materials from the magazine or another building, a railway, or a highway by a natural or an artificial barrier. A straight line from the top of any sidewall of the building containing explosive materials to the eave line of any magazine or other building or to a point 12 ft (3.7 m) above the center of a railway or highway shall pass through such barrier.

NOTE 9: "Inhabited building" means a building regularly occupied in whole or part as a habitation for human beings, or any church, schoolhouse, railroad station, store, or other structure where people are accustomed to assemble, but does not include any building or structure occupied in connection with the manufacture, transportation, storage, or use of explosive materials.

NOTE 10: "Railway" means any steam, electric, or other railroad or railway that carries passengers for hire.

NOTE 11: "Highway" means any public street, public alley, or public road.

NOTE 12: Where two or more storage magazines are located on the same property, each magazine shall comply with the minimum distances specified from inhabited buildings, railways, and highways, and, in addition, they should be separated from each other by not less than the distances shown for "separation of magazines," except that the quantity of explosive materials contained in detonator magazines shall govern with regard to the spacing of said detonator magazines from magazines containing other explosive materials. If any two or more magazines are separated from each other by less than the specified "separation of magazines" distances, such magazines, as a group, shall be considered as one magazine, and the total quantity of explosive materials stored in such group shall be treated as if stored in a single magazine located on the site of any magazine of the group, and shall comply with the minimum specified distances from other magazines, inhabited buildings, railways, and highways.

NOTE 13: Storage in excess of 300,000 lb (136,200 kg) of explosive materials in one magazine generally is not necessary for commercial enterprises.

NOTE 14: This table applies only to the manufacture and permanent storage of commercial explosive materials. It is not applicable to the transportation of explosives or any handling or temporary storage necessary or incident thereto. It is not intended to apply to bombs, projectiles, or other heavily encased explosives.

NOTE 15: Where a manufacturing building on an explosive materials plant site is designed to contain explosive materials, the building shall be located at a distance from inhabited buildings, public highways, and passenger railways in accordance with the American Table of Distances based on the maximum quantity of explosive materials permitted to be in the building at one time.

TABLE 4-12C. *Table of Recommended Separation Distances of Ammonium Nitrate and Blasting Agents from Explosives or Blasting Agents*[1,6]

Donor Weight		Minimum Separation Distance of Acceptor when Barricaded[2](ft)		Minimum Thickness of Artificial Barricades[5] (in.)
Pounds Over	Pounds Not Over	Ammonium Nitrate[3]	Blasting Agent[4]	
	100	3	11	12
100	300	4	14	12
300	600	5	18	12
600	1000	6	22	12
1000	1600	7	25	12
1600	2000	8	29	12
2000	3000	9	32	15
3000	4000	10	36	15
4000	6000	11	40	15
6000	8000	12	43	20
8000	10,000	13	47	20
10,000	12,000	14	50	20
12,000	16,000	15	54	25
16,000	20,000	16	58	25
20,000	25,000	18	65	25
25,000	30,000	19	68	30
30,000	35,000	20	72	30
35,000	40,000	21	76	30
40,000	45,000	22	79	35
45,000	50,000	23	83	35
50,000	55,000	24	86	35
55,000	60,000	25	90	35
60,000	70,000	26	94	40
70,000	80,000	28	101	40
80,000	90,000	30	108	40
90,000	100,000	32	115	40
100,000	120,000	34	122	50
120,000	140,000	37	133	50
140,000	160,000	40	144	50
160,000	180,000	44	158	50
180,000	200,000	48	173	50
200,000	220,000	52	187	60
220,000	250,000	56	202	60
250,000	275,000	60	216	60
275,000	300,000	64	230	60

For SI Units: 1 lb = 0.454 kg; 1 ft = 0.305 m; 1 in. = 2.54 cm.

Notes to Table of Recommended Separation Distances of Ammonium Nitrate and Blasting Agents from Explosives or Blasting Agents

NOTE 1: Recommended separation distances are to prevent explosion of ammonium nitrate and ammonium nitrate-based blasting agents by propagation from nearby stores of high explosives or blasting agents referred to in the table as the "donor." Ammonium nitrate, by itself, is not considered to be a donor where applying this table. Ammonium nitrate, ammonium nitrate-fuel oil, or combinations thereof are acceptors. If stores of ammonium nitrate are located within the sympathetic detonation distance of explosives or blasting agents, 1/2 the mass of the ammonium nitrate shall be included in the mass of the donor.

NOTE 2: Where the ammonium nitrate or blasting agent, or both, is not barricaded, the distances shown in Table 4-12C must be multiplied by 6. These distances allow for the possibility of high-velocity metal fragments from mixers, hoppers, truck bodies, sheet-metal structures, metal containers, and the like that could enclose the donor. Where storage is in bullet-resistant magazines* recommended for explosives or where the storage is protected by a bullet-resistant wall, distances and barricade thicknesses in excess of those prescribed in the American Table of Distances (See Table 4-12B) are not required.

NOTE 3: The distances in the table apply to ammonium nitrate that passes the insensitivity test prescribed in the definition of ammonium nitrate fertilizer, promulgated by the Fertilizer Institute†; ammonium nitrate failing to pass said test must be stored at separation distances determined by competent persons and approved by the Authority Having Jurisdiction.

NOTE 4: These distances apply to blasting agents that pass the insensitivity test prescribed in regulations of the U.S. Department of Transportation and the U.S. Department of the Treasury, Bureau of Alcohol, Tobacco, and Firearms.

NOTE 5: Earth, sand dikes, or enclosures filled with the prescribed minimum thickness of earth or sand shall be permitted to be used as artificial barricades. Natural barricades, such as hills or timber of sufficient density that the surrounding exposures that need protection cannot be seen from the donor when the trees are bare of leaves, also shall be permitted to be used.

NOTE 6: For determining the distances to be maintained from inhabited buildings, passenger railways, and public highways, Table 4-12B, the American Table of Distances for Storage of Explosives, shall be used.

*For construction of bullet-resistant magazines, see NFPA 495.

†*Definition and Test Procedures for Ammonium Nitrate Fertilizer*, Fertilizer Institute, November 1964.

Some of the materials used in making explosives and blasting agents may also be hazardous. Materials likely to be found include aluminum powder, paint-grade aluminum, ammonium nitrate, ammonium perchlorate, sodium nitrate, fuel oil, gilsonite, bagasse, and various types of finely divided gums. NFPA codes and standards should be consulted for proper storage and protection of these materials.

For all explosive materials, whether in manufacture, storage, transportation, or use, housekeeping should be of the highest order.

FIGHTING FIRES INVOLVING EXPLOSIVES AND BLASTING AGENTS

No attempt should be made to fight a fire involving explosive materials. The area should be evacuated to a distance of at least 2000 ft (610 m).

The Norton, VA, fire mentioned previously is a good example of the necessity of retreating from an established fire involving blasting agents. The Luckenback Pier fire in Brooklyn, NY, in 1959, shows that even 1.4 (Class C) explosives represent a potential hazard. Several tons (one short ton equals 0.91 metric ton) of detonating cord were stored on a pier that became involved in a serious fire. After considerable delay, the entire mass detonated with many lives lost and property damage in the millions of dollars.

Truck fires involving explosive materials pose the greatest threat to the public, because the number of trucks in transit is so large. Immediate evacuation in such incidents is imperative.

DESTRUCTION OF EXPLOSIVE MATERIALS

At times it may be necessary to destroy commercial explosive materials. These may consist of explosives or blasting agents from containers that have been broken during transportation, or may be materials that have exceeded their recommended shelf life or are believed to be overage or are no longer needed.

Due to the many developments in explosives technology over the past few years, the appearance and characteristics of products have undergone marked changes. Before destroying any explosives product, one must be familiar with the properties of the product; consult the manufacturer of the product for the most current product information and the recommended method of disposal and/or destruction.

When improvised explosive devices or bombs, military ordnance, military explosives, or homemade or illegal explosive materials are involved, disposal should be under the supervision of trained bomb disposal experts from a governmental regulatory agency.

BIBLIOGRAPHY

References Cited

1. Dixon, W. T., and Fisher, A. W., eds., *Chemical Engineering in Industry*, Chapter 12, American Institute of Chemical Engineers, New York, 1958.
2. U.S. Dept. of the Interior, "Explosives," 1993, Mineral Industry Surveys, USDI U.S. Geological Survey, Washington, DC.
3. U.S. *Code of Federal Regulations*, Title 49, Chapter I, "Transportation," Parts 170-189, 1995, U.S. Government Printing Office, Washington, DC.
4. Cook, M. A., *The Science of High Explosives*, Reinhold Publishing Co., New York, 1958.
5. U.S. *Code of Federal Regulations*, Title 30, Chapter I, Subchapter C, Part 15, "Explosives and Related Articles," 1995, U.S. Government Printing Office, Washington, DC.
6. U.S. *Code of Federal Regulations*, Title 49, Chapter I, "Transportation," Parts 100-199, 1995, U.S. Government Printing Office, Washington, DC.
7. *American Table of Distances*, The Institute of Makers of Explosives, Washington, DC, 1991.
8. Van Dolah, R. W., and Malesky, J. S., "Fire and Explosion in a Blasting Agent Mix House Building, Norton, VA," RI 6015, 1962, USDI Bureau of Mines, Washington, DC.
9. Isner, M. S., "Blasting Agent Explosion—Six Fire Fighters Killed, Kansas City, MO," *NFPA Fire Investigation Report*, Nov. 29, 1988, National Fire Protection Association, Quincy, MA.

NFPA Codes, Standards, and Recommended Practices

Reference to the following NFPA codes, standards, and recommended practices will provide further information on explosives and blasting agents discussed in this chapter. (See the latest version of *The NFPA Catalog* for availability of current editions of the following documents.)

NFPA 490, *Code for the Storage of Ammonium Nitrate*
NFPA 495, *Explosive Materials Code*
NFPA 498, *Standard for Safe Havens and Interchange Lots for Vehicles Transporting Explosives*

Additional Reading

ANSI A10.7, *Construction and Demolition—Commercial Explosives and Blasting Agents—Safety Requirements for Transportation, Storage, Handling, and Use*, American National Standards Institute, New York, 1989.
"ATF Explosives Laws and Regulations," ATF Publication 5400.7, June 1990, Bureau of Alcohol, Tobacco, and Firearms, Washington, DC.
Construction Guide for Storage Magazines, IME Safety Library Publication No. 1, Institute of Makers of Explosives, Washington, DC, Aug. 1993.
Creighton, J. R., "Ignition Temperature of Solid Explosives Exposed to a Fire," *PSAM-II Proceedings*, An International Conference Devoted to the Advancement of System-Based Methods for the Design and Operation of Technological Systems and Processes, Vol. 2, Sessions 37–72, March 20–25, 1994, San Diego, CA, 1994, pp. 037/5–10.
Cudzilo, S., *et al.*, "Shock Initiation Studies of Ammonium Nitrate Explosives," *Combustion and Flame*, Vol. 102, No. 1/2, July 1995, pp. 64–72.
Department of Defense Explosive Hazard Classification Procedures, DLAR 8220.1, Defense Logistics Agency, Washington, DC, 1989.
DOT, *1996 North American Emergency Response Guidebook*.
Explosion and Explosives: Journal of the Industrial Explosives Society, Industrial Explosives Society, Japan, published periodically.
"Explosives Incidents Report, 1993," Bureau of Alcohol, Tobacco and Firearms, Washington, DC, ATF P 3320.4, July 1994.
"Explosives," Department of the Interior, Washington, DC, Annual Review, 1994.
"Fireworks: Spectacular but Dangerous," *Fire Command*, Vol. 56, No. 7, July 1989, pp. 11–14.
Fitzwater, T., "Introduction to Explosives," *American Fire Journal*, Vol. 45, No. 4, Apr. 1993, pp. 12–17.
Fitzwater, T., "Introduction to Explosives. Part 2. High Explosives," *American Fire Journal*, Vol. 46, No. 3, Mar. 1994, pp. 12–16.
Fitzwater, T., "Introduction to Explosives. Part 3. Slurries, Water Gels, Emulsions, Binary Explosives and Detonating Cord," *American Fire Journal*, Vol. 47, No. 11, Nov. 1995, pp. 12–14.
Glossary of Commercial Explosives Industry Terms, IME Safety Library Publication No. 12, Institute of Makers of Explosives, Washington, DC, Feb. 1991.
Hayes, R. W., "Computer Model for the Products of Combustion and Atmospheric Dispersion Resulting From Open Burning of Explosives and Propellants," CPIA Publication 543; Joint Army-Navy-NASA-Air Force (JANNAF) Safety and Environmental Protection Subcommittee Meeting, June 4, 1989–June 21, 1990, Livermore, CA, 199, pp. 269–277.

Jones, J. C., "Model for Critical Behavior in a Liquid Explosive Experiencing Dual Ambient Temperatures," *Journal of Fire Sciences*, Vol. 13, No. 5, Sept./Oct. 1995, pp. 399–406.

Jones, J. C. and Moore, A., "Model for Critical Behavior in a Liquid Explosive Experiencing Dual Ambient Temperatures. Part 2. Temperature Trajectories and Times to Explosion," *Journal of Fire Sciences*, Vol. 13, No. 5, Sept./Oct. 1995, pp. 407–413.

Jones, J. C., "A Model for Critical Behaviour in a Liquid Explosive Experiencing Dual Ambient Temperatures. Part 3. Examination of the Initial Conditions," *Journal of Fire Sciences*, Vol. 14, No. 2, Mar./Apr. 1996, pp. 124–127.

Pickett, M., "Explosives: What Firefighters Must Know. Part 1. Why Firefighters Should Know About Explosives and Hot To Identify Common Commercial Types," *Firehouse*, Vol. 20, No. 9, Sept. 1995, pp. 108, 110, 111.

Poulton, T. J. and Kosanke, K. L., "Fireworks and Their Hazards," *Fire Engineering*, Vol. 148, No. 6, June 1995, pp. 49–66.

Recommendations for the Safe Transportation of Detonators in a Vehicle with Certain Other Explosive Materials, IME Safety Library Publication No. 22, Institute of Makers of Explosives, Washington, DC, May 1993.

Safety in the Transportation, Storage, Handling, and Use of Explosive Materials, IME Safety Library Publication No. 17, Institute of Makers of Explosives, Washington, DC, July 1996.

United Nations Recommendations on the Transportation of Dangerous Goods, 9th rev. ed.

U.S. *Code of Federal Regulations*, Title 27, Part 55, "Explosive Materials Regulations," U.S. Government Printing Office, Washington, DC.

Warnings and Instructions (*Always and Never*), IME Safety Library Publication No. 4, Institute of Makers of Explosives, Washington, DC, Mar. 1992.

DEFLAGRATION (EXPLOSION) VENTING

Richard F. Schwab

The first document published by NFPA concerning deflagration venting was NFPA 68-T, *Explosion Venting Standard*, tentatively adopted by NFPA in 1945. The *Guide for Explosion Venting* was published in 1954. It brought together the best available information on the design and utilization of vents to limit the pressures developed by "explosions," and that information was sparse.

Unfortunately the data available at that time (1954) were very primitive, without any theoretical basis. It was customary at the time to express vent ratios in terms of vent area per 100 cu ft (2.8 m³) of enclosure volume. Very little regard was given to the strength of the enclosure and its effect on vent sizing, though it was recognized that a strong enclosure or a heavy building did not need as much vent area as a weak enclosure or building. It was recognized in a qualitative fashion that the rate of pressure rise had an effect on the size of the vent required for a given enclosure; i.e., a combustible dust or flammable gas showing test data with high rates of pressure rise required a larger vent ratio than those with lower rates of pressure rise. Also, increasing the mass of the vent closure was undesirable since it slowed down the opening of the closure, resulting in the buildup of destructive pressures within the enclosure.

This unfortunate state of affairs prevailed into the late 1960s and early 1970s when some work done at the U.S. Bureau of Mines and more extensive work in Great Britain, Germany, and Switzerland increased the pool of knowledge to the extent that the 1974 edition of NFPA 68, *Guide for Venting of Deflagrations* (hereinafter referred to as NFPA 68), was issued.

The 1974 edition of NFPA 68 incorporated information allowing the design of deflagration (explosion) vents for low-strength enclosures or buildings, utilizing the Runes equation which can be expressed as:

$$A_v = \frac{CL_1L_2}{P^{1/2}} \qquad (1)$$

where

A_v = necessary vent area (sq ft or m²);

C = constant, characteristic of fuel gas or dust;

L_1 = smallest dimension of the rectangular building enclosure to be vented (ft or m);

Richard F. Schwab, P.E., S.f.P.E., AIChE, is manager of process safety and loss prevention, AlliedSignal, Inc., Morristown, NJ.

L_2 = second smallest dimension of the rectangular building enclosure to be vented (ft or m); and

P = maximum internal building pressure that can be withstood by the weakest building member which is desired not to vent or break (lbf/sq in. or kN/m²).

Another equation, referred to as the Rasbash equation, was introduced for the design of vents for buildings:

$P_m = 1.5\,P_v + 0.5\,K$ (U.S. units)

$P_m = 1.5\,P_v + 3.5\,K$ (SI units)

where

P_m = maximum pressure reached during venting of the combustion (lbf/sq ft or kN/m²),

P_v = pressure within the building space at which the vent opens (lbf/sq ft or kN/m²),

K = venting ratio A_c/A_v (dimensionless),

A_c = smallest cross-sectional area of the building space, and

A_v = total area of the combustion vents.

Both of these equations were replaced in later editions of NFPA 68 by the Swift-Epstein equation, discussed later.

It was in the 1974 edition of NFPA 68 that the "cubic law" was introduced. This was developed by investigators in Europe and expressed in the general form by:

$$K = \left(\frac{dp}{dt}\right) V^{1/3} \qquad (2)$$

where

dp/dt = maximum rate of pressure rise during the combustion in the unvented vessel, and

V = volume of the unvented vessel.

This equation was the theoretical basis for the design of deflagration vents and was first introduced in the form of tables.

The 1978 version of NFPA 68 presented further developments, including the introduction of nomographs for gases and combustible dusts for use in sizing of vents in vessels with a strength of 1.5 psi (10.3 kPa or 0.1 bar gage) or more.

The 1988 version revised and updated the nomographs for both gases and dusts and introduced equations that were derived from the nomographs, which enabled the engineer to begin to computerize the data and facilitate design.

The 1994 edition of NFPA 68 is a complete rewrite and more effectively communicates the principles of venting deflagrations to users of the document, plus introduces newly developed information.

FUNDAMENTALS

It must be realized that deflagration (explosion) venting is not a preventive measure. It presupposes that the event will occur and that the equipment and/or building, properly designed to anticipate the event, will withstand the effects of the deflagration with a minimum amount of damage. That is all it is intended to accomplish.

For a deflagration to occur, it is necessary to have the fuel concentration, whether a flammable gas (vapor) or combustible dust aerosol or a mixture of both (hybrid mixtures), within the flammable range along with sufficient oxidant (usually air) reasonably well mixed with an ignition source sufficient to initiate the combustion. The heat plus the gases generated by the combustion will result in an increase in pressure. If the pressure increases to the point where the enclosure in which the deflagration occurs exceeds the ability of the enclosure to contain that pressure, the enclosure will rupture and an explosion will result.

If a vent closure is provided for the enclosure that is able to open and relieve the pressure and vent hot gases from the combustion process to a safe location, thus preventing the pressure buildup from exceeding the enclosure strength, one will have accomplished the mission of designing a proper deflagration vent.

Dusts smaller than 420 µm (passing through a 40-mesh screen) are classified as combustible dusts. When dispersed in air and ignited, they will propagate a flame. As with gaseous mixtures, there is a lower combustible limit that will support flame propagation. Dust clouds do not exhibit an upper limit that is of practical use in deflagration control.

The introduction of a flammable gas into a dust cloud usually reduces the lower combustible limit of the dust/air mixture and also markedly decreases the ignition energy of the dust/air mixture. The effect can occur even though the gas is well below its lower flammable limit when considering the gas/air mixture alone.

Introducing a flammable gas into a dust cloud generally increases the rate of combustion dramatically over what would normally be expected of the dust cloud alone. In some cases, it has been shown that introduction of a flammable gas into a dust cloud that normally would be considered a minimum hazard will result in a vigorous deflagration.

Mists of combustible liquids have upper and lower combustible limits very similar to that of flammable and combustible vapors, and act in a similar manner.

For the purposes of this chapter, the oxidant under consideration is air. The equations and nomographs are all based on air. Other oxidants can be expected to support combustion, but the rates of combustion will be different from air. It will be necessary to determine, by experiment, the effects of these differences on venting design.

Initial Temperature and Pressure

Increasing the initial pressure of the combustible mixture, whether it is a flammable gas/air or a combustible dust/air mixture, will produce a proportionate change in the maximum pressure developed by the mixture in a closed vessel.

An increase in the initial temperature of the flammable gas/air or dust/air mixture will increase the rate of pressure rise in a closed vessel over that seen at lower temperatures. The maximum pressure attained at the higher temperatures, however, will decrease over that seen at the lower temperatures.

Turbulence increases the rate of pressure rise and greatly enhances flame speed. In elongated enclosures, these increases in pressure can result in transition to detonation.

Generally speaking, moisture in the air (humidity) has no significant effect on a deflagration once ignition has occurred. However, moisture in dusts reduces the rate of pressure rise and the maximum pressure attained in the combustion. The moisture will also raise the ignition temperature and increase the ignition energy needed to ignite the dust aerosol.

Enclosure Strength

The term P_{red} is defined as the maximum internal pressure that can be resisted by the weakest structural element, and the term P_{stat} is defined as the static activation pressure. P_{red} can be selected up to two thirds the ultimate strength of the equipment, provided that deformation of the equipment can be tolerated. P_{red} should always exceed P_{stat} by at least 0.35 psig (0.02 bar gage) for low-strength enclosures and by at least 0.72 psig (0.5 bar gage) for high-strength enclosures.

Vent Variables

The maximum pressure developed within a vented enclosure decreases as the vent area increases. Location of multiple vents over the surface of an enclosure is recommended. There is no particular advantage to be gained from using circular vents instead of rectangular vents. The shape of the vent closure is of no significance.

The inertia of the vent should be as low as practical. Normally NFPA 68 recommends that the mass of the vent be 2.5 lb/sq ft (12.2 kg/m²). In cases where the rates of pressure rise are expected to be low, such as venting of St-1 dust explosions, the mass of the vent can be allowed to increase as much as 8 lb/sq ft (40 kg/m²).

The supporting structure of the enclosure being vented should be strong enough to withstand any forces of reaction developed as a result of the operation of the vent without vent ducts. A method of calculating those forces has been established from test results.

$$F_r = 1.2(A_v)(P_{red}) \qquad (3)$$

where

F_r = reaction force resulting from the combustion (lb),

A_v = vent area (in sq in.), and

P_{red} = maximum pressure developed during venting (psig).

An equation that approximates the duration of the thrust force of the dust deflagration has been developed that will aid in the design of the support structures for the enclosures with deflagration vents.[1]

$$t_f = \frac{10^{-2}(K_{st})(V^{1/3})}{(P_{red})(A_v)} \qquad (4)$$

where

t_f = duration of the pressure pulse (sec),

K_{St} = deflagration index for the dust [bar (m/sec)],

V = vessel volume (m³),

P_{red} = maximum pressure developed during venting (bar gage), and

A_v = area of vent (m²).

The equivalent static force that a structure supporting a vented enclosure will experience during deflagration venting is given by:

$$F_s = 0.62 (A_v)(P_{red}) \qquad (5)$$

where

F_s = equivalent static force experienced by supporting structure [lb (kg)]

A_v = vent area (sq in. and cm²), and

P_{red} = maximum pressure developed during venting [psig (kPa)].

When F_s is expressed in kilonewtons (KN), Equation 5 is written as

$$F_s = 119 (A_v)(P_{red})$$

where

A_v = vent area (m²), and

P_{red} = maximum pressure developed during venting (bar gage).

When a vent opens, it relieves the pressure along with the hot gases and unburnt fuel, and a fireball is formed. That fireball can extend outward from the vent and can be 8 times the cube root of the volume of the enclosure vented. The height of the fireball, H, could be the same dimension, with half the height below the center of the vent and half the height above. In some deflagrations, buoyancy effects can allow the fireball to rise to elevations well above these measurements.

$$D = 8 \left(V^{1/3} \right) = H \qquad (6)$$

The range of the flame is slightly shorter for the St-2 dusts as compared with the St-1 dusts.

The equation is valid for sizes of 0.3 m³ ≤ V < 10,000 m³

where

Static activation rupture disk gage pressure, P_{stat} < 0.1 bar

Maximum reduced explosion gage pressure, P_{max} ≤ 1 bar

Maximum explosion gage pressure, P_{max} ≤ 9 bar (closed vessel test), and

Dust characterization factor 200 bar m/s ≤ K_{St} < 300 bar m/s.

The VDI recommends another equation for heterogeneous dust/air mixture, as follows:

$$L_f = 15 V^{-0.25} \qquad (7)$$

This equation is valid for sizes between 10 m³ ≤ V < 10,000 m³, with the same conditions applying as for Equation 5.

Pressure Effects from a Vented Enclosure

When a vent closure opens, allowing burnt and unburnt hot fuel to vent to the atmosphere, there is a pressure buildup in the immediate vicinity. Two peaks generally occur. One is due to the venting process itself, i.e., the release of the hot gases from the enclosure into the outside atmosphere; and the second peak is the result of the combustion of the unburned fuel. Both are influenced by the K_{St} and

the location of the ignition within the enclosure. The maximum peak pressure can be calculated, as follows:

$$P_{max} = 200 P_{red} A^{0.1} V^{0.18} \qquad (8)$$

where

P_{max} = maximum value of peak pressure (bar gage),

A = vent area (m³), and

V = volume of the enclosure (m³).

The point where this pressure exists is at a distance from the vent closure and is calculated, as follows:

$$R_s = 0.25 L_f \qquad (9)$$

As one moves farther away from the vent at a distance r, the P_r decreases, as follows:

$$P_r = P_{max} \left(R_s / r \right)^{1.5} \qquad (10)$$

The equation is valid for:

Sizes between 0.3 m³ ≤ V and < 10,000 m³,

P_{stat} ≤ 0.1 bar static activation gage pressure of the rupture disk,

P_{red} ≤ 1 bar maximum reduced explosion gage pressure,

Maximum explosion gage pressure P_{max} = 9 bar gage, and

Dust-specific characteristic, K_{St} = 200 bar m/s.

LOW-STRENGTH ENCLOSURES

As far as NFPA 68 is concerned, low-strength enclosures are capable of withstanding pressures of no more than 1.5 psig (0.1 bar gage). This section of NFPA 68 mainly concerns itself with buildings, dust collectors, ovens, and similar equipment.

The basic equation that is used for sizing relieving vents for low-strength enclosures is the Swift-Epstein equation, expressed in its simplified form as:

$$A_v = \frac{C \times A_s}{P_{red}^{1/2}} \qquad (11)$$

where

A_v = vent area (sq ft or m²),

C = venting equation constant (see Table 4-13A),

A_s = internal surface area of the enclosure (sq ft or m²), and

P_{red} = maximum internal pressure that can be withstood by the weakest structural element not intended to fail (psig or bars, not to exceed 1.5 or 0.1 bar).

The internal surface area of the enclosure includes the floor, walls, and ceiling; in other words, the internal perimeter surfaces of the enclosure being protected. It does not include the surfaces of the pieces of equipment within the enclosure.

The values of C in Table 4-13A were determined by actual data. If suitable large-scale tests are conducted for a specific application, an alternate value of C can be determined.

No recommendations can presently be given for fast-burning gases, such as hydrogen, certain alkenes (such as ethylene), alkynes (acetylene), dienes, and epoxides. These gases are prone to translate into a detonation that cannot be properly vented. The recommended method allows for initial turbulence and turbulence-generating ob-

TABLE 4-13A. *Fuel Characteristic Constant for Venting Equation**

	C(psi)$^{1/2}$	C(bar)$^{1/2}$
Anhydrous ammonia	0.05	0.013
Methane	0.14	0.037
Gases with fundamental burning velocity less than 1.3 times that of propane	0.17	0.045
St-1 dusts	0.10	0.026
St-2 dusts	0.12	0.030
St-3 dusts	0.20	0.051

*See Table 4-3 of NFPA 68.

jects, and no data have been generated to address such fast-burning gases.

For elongated enclosures, the vent area should be applied as evenly as possible with respect to the longest dimension. If the available vent area is restricted to one end of an elongated enclosure, the ratio of length to diameter should not exceed 3. For cross sections other than circular or rectangular, the effective diameter can be taken as the hydraulic diameter, given by $4(A/p)$ where A is the cross-sectional area normal to the longitudinal axis of the space, and p is the perimeter of the cross section. Therefore, for enclosures with venting restricted to one end, the venting equation is:

$$L_3 \leq 12\,(A/p) \tag{12}$$

where

L_3 = longest dimension of the enclosure (ft or m),

A = cross-sectional area (sq ft or m^2), and

p = perimeter of cross-section (ft or m).

If the enclosure can contain a highly turbulent gas mixture and the vent area is restricted to one end, or if the enclosure has many internal obstructions and the vent area is restricted to one end, then the L/D of the enclosure should not exceed 2 or:

$$L_3 \leq 8(A/p)$$

The calculated vent area can be reduced by increasing the strength of the enclosure, P_{red}.

Equation 12 has been adapted and used in NFPA 68 since the 1988 edition.

Previous editions of NFPA 68, such as the 1974 and 1978 editions, used another equation somewhat similar in form, referred to as the Runes equation (see Equation 1), which was found to be unsatisfactory and as a result was replaced by the Swift-Epstein Equation.[2]

The Runes equation may appear in some of the older engineering literature on the subject of deflagration venting, but is no longer recommended. The Rasbash equation (discussed earlier) was also presented as an alternate to the Runes equation in the 1974 version, and is also no longer recommended.

HIGH-STRENGTH ENCLOSURES

High-strength enclosures are those capable of withstanding more than 1.5 psig (0.1 bar gage).

The design of deflagration vents is based on the cubic law, expressed in its general form as:

$$K = \frac{dp}{dt} V^{1/3} \tag{13}$$

In this case, K is not truly a constant, though many times it is referred to as one. It is more truly termed a characterization factor, which enables one to correlate the many variables needed for extrapolating test data to full-scale equipment.

There are two different characterization factors that are customarily used:

$$K_G = \frac{dp}{dt} V^{1/3} \tag{14}$$

or

$$K_{St} = \frac{dp}{dt} V^{1/3} \tag{15}$$

where

dp/dt = rate of pressure rise (bar gage per sec),

V = volume of the test enclosure (m^3),

K = characterization factor,

K_G = characterization factor (gases, vapors), and

K_{St} = characterization factor (dusts).

K_G and K_{St} are not interchangeable values nor can they be directly equated. Generally speaking, K_G values are obtained under initially quiescent test conditions, while K_{St} values are always obtained under turbulent test conditions. It should be noted that, if tests are made in enclosures where the gas or vapor is being maintained under turbulent conditions, then the K_G so obtained will be higher than that obtained under quiescent conditions.

The nomographs and equations in NFPA 68 are based on the correlation and extrapolation of data using numerous tests from small-test equipment to vessels as large as 250 m^3, based on the cubic law principle. (See Figure 4-13A.)

A typical nomograph for design of deflagration vents for dusts is shown in Figure 4-13B. The equation describing the venting nomograph for gases in equipment capable of withstanding 1.5 psig (0.1 bar gage) or more is:

$$A_v = aV^b e^{cP_{stat}} P_{red}^d \tag{16}$$

where

A_v = vent area (m^2),

V = enclosure volume (m^3),

e = 2.718 (base of natural logarithm),

P_{red} = maximum pressure developed during venting (bar gage), and

P_{stat} = vent closure release pressure (bar gage).

Exponent Values for use in Equation 16

	a	b	c	d
Methane	0.105	0.770	1.230	−0.823
Propane	0.148	0.703	0.942	−0.671
Hydrogen	0.279	0.680	0.755	−0.393
Coke gas	0.150	0.695	1.380	−0.707

Note: Since this equation is derived from nomographs, it is no more accurate than the nomographs themselves and should not be extrapolated beyond the nomograph limits. The equation is subject to the same limitations as the nomographs and, therefore, should not be used for indiscriminate extrapolation. Serious errors in the value of A_v will occur if this is done. (See Figure 4-13C.)

For, combustible dusts, the vent area can be calculated from the following equation

$$A_v = aV^{2/3}K_{St}{}^b P_{red}{}^c \qquad (17)$$

where

$$a = 0.000571e^{(2P_{stat})} \qquad (18)$$

$$b = 0.978e^{(-0.105P_{stat})} \qquad (19)$$

$$c = -0.687e^{(0.226P_{stat})} \qquad (20)$$

and

A_v = vent area (m²),

V = enclosure volume (m³),

e = 2.718 (base of natural logarithm),

P_{red} = maximum pressure developed during venting (bar gage),

P_{stat} = vent closure release pressure (bar gage), and

K_{St} = deflagration index for dust (bar-m/s).

Note: Since this equation is derived from nomographs, it is no more accurate than the nomographs themselves and should not be extrapolated beyond the nomograph limits. Serious errors in the value of A_v will occur if this is done. (See Figures 4-13D and 4-13E.)

A more recent equation for the calculation of deflagration (explosion) vents for flammable gases and vapors is provided by Bartknecht.[3]

$$A_v = \left[\begin{array}{l} (0.1265 \log K_G - 0.0567)P_{red,max}^{-0.5817} \\ +0.1754P_{red,max}^{-0.5722}(P_{stat} - 0.1) \end{array} \right] V^{2/3} \qquad (21)$$

where

K_G = specific gas characterization factor (bar•m/sec),

P_{stat} = vent closure release pressure (bar gage),

P_{red} = maximum pressure developed during venting (bar gage),

A_v = vent area (m³), and

V = enclosure volume (m³).

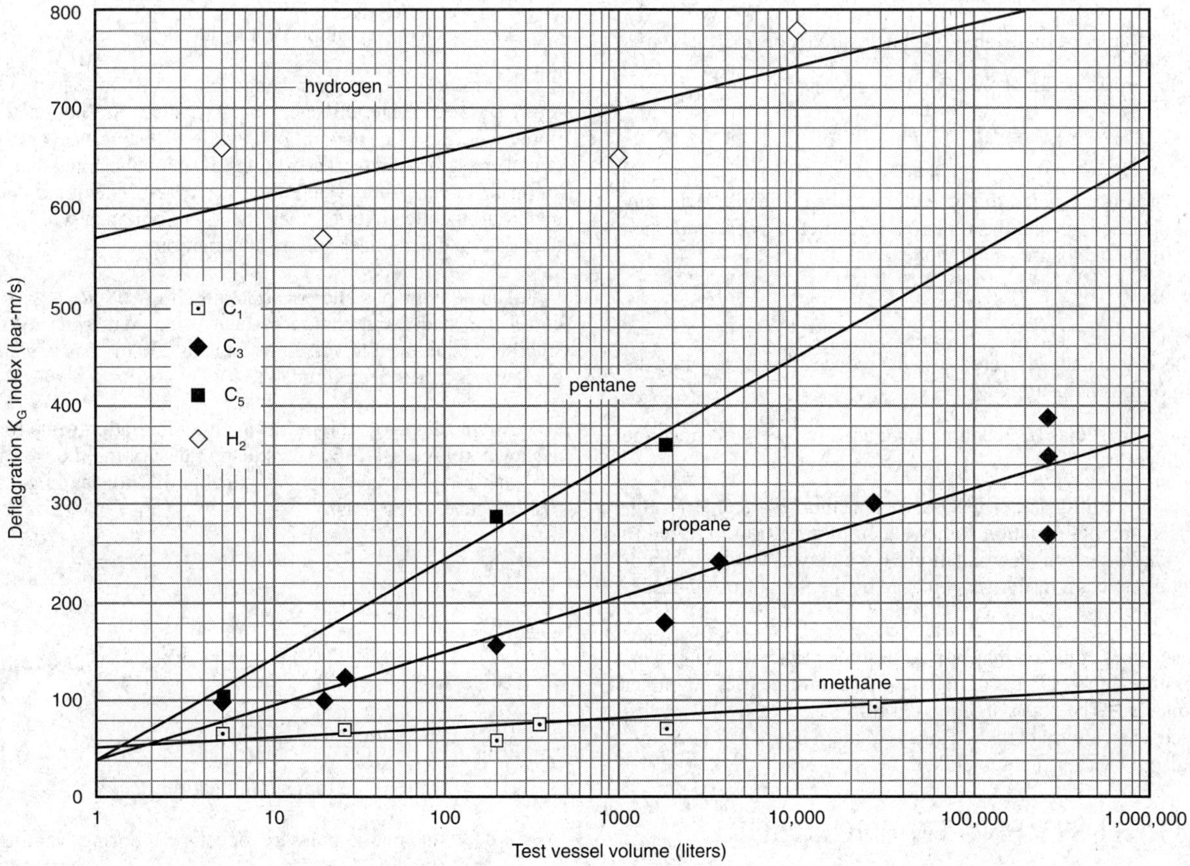

FIG. 4-13A.   Effect of test volume on K_G measured in spherical vessels. (Source: NFPA 68)

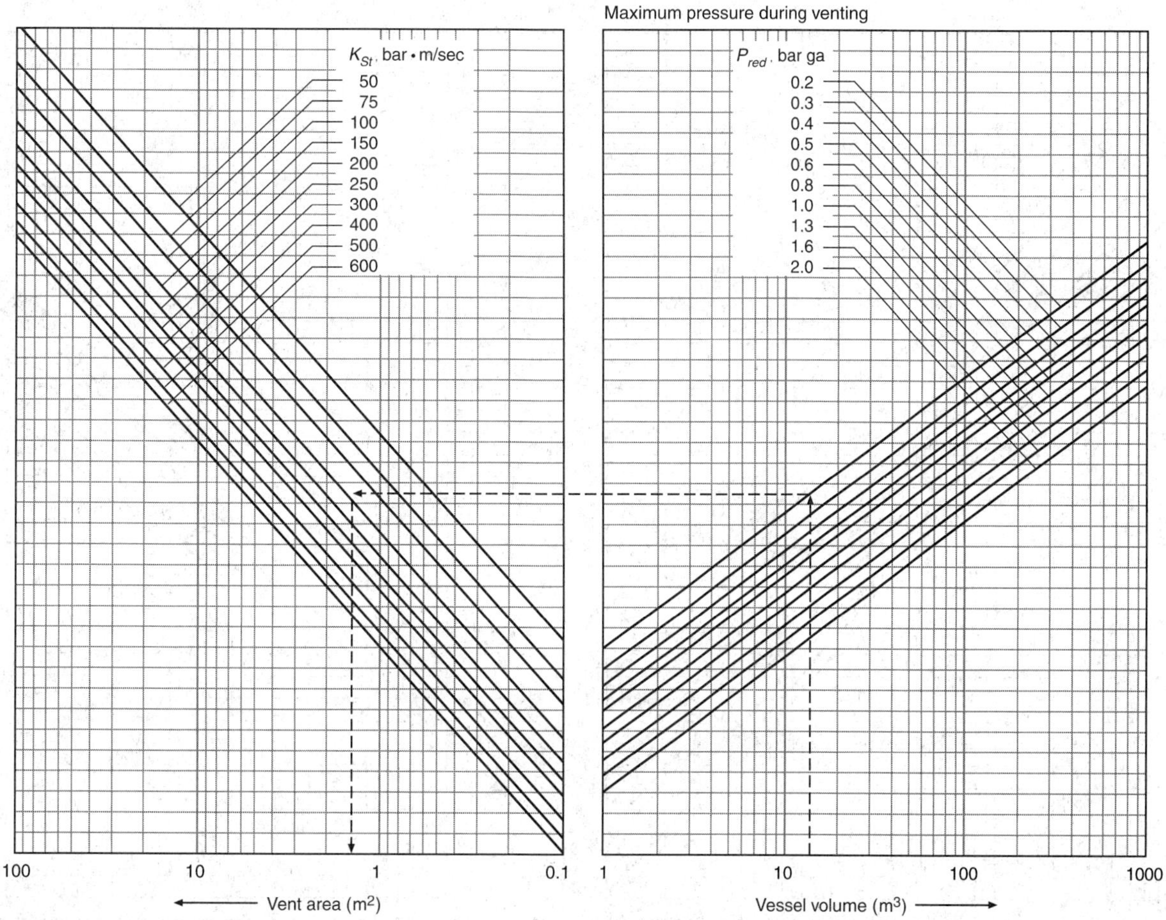

Maximum pressure during venting

Fig. 4-13B. *Venting nomograph for dusts (P_{stat} = 0.1 bar gage). (Source: NFPA 68)*

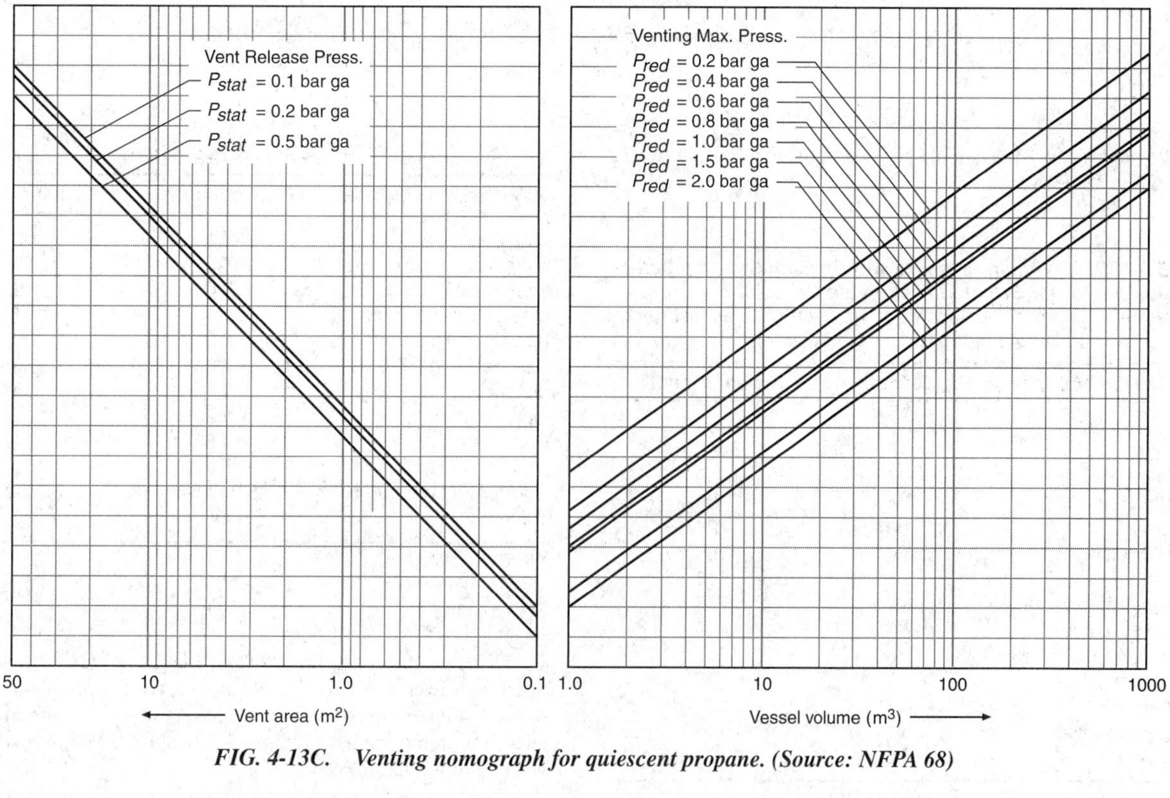

FIG. 4-13C. *Venting nomograph for quiescent propane. (Source: NFPA 68)*

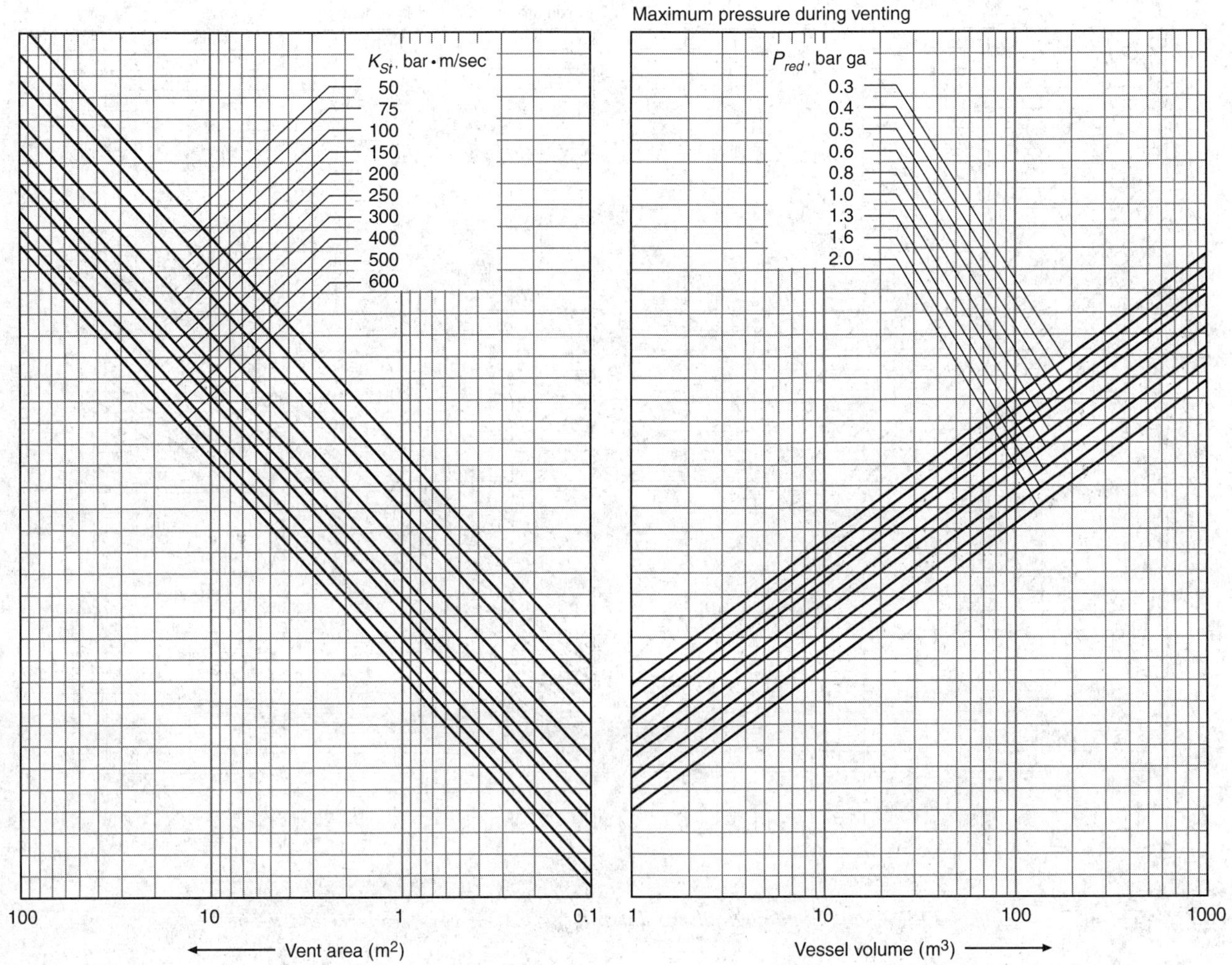

Maximum pressure during venting

FIG. 4-13D. *Venting nomograph for dusts (P_{stat} = 0.2 bar gage). (Source: NFPA 68)*

The values of K_G that should be used in the above Equation 21 are shown in Table 4-13B.

TABLE 4-13B. *Maximum Pressure and Specific Gas Characterization Factor for Selected Flammable Gases and Vapors.*
(5L sphere using an ignition source of 10 J at normal atmospheric pressure at normal room temperature.)

Flammable Material	P_{max} (bar gage)	K_G (bar m/sec)
Acetophenone[1]	7.6	109
Acetylene	10.6	1415
Ammonia[2]	5.4	10
β naphthol[3]	4.4	36
Butane	8.0	92
Carbon disulfide	6.4	105
Diethyl ether	8.1	115
Dimethyl formamide[1]	8.4	78
Dimethyl sulfoxide[1]	7.3	112
Ethane[1]	7.8	106
Ethyl benzene[1]	7.4	96
Hydrogen	6.8	550
Hydrogen disulfide	6.4	105
Isopropanol[1]	7.8	83

TABLE 4-13B. *(Continued)*

Flammable Material	P_{max} (bar gage)	K_G (bar m/sec)
Methane[1]	7.1	55
Methanol[1]	7.5	75
Methylene chloride[2]	5.0	5
Methyl nitrite	11.4	111
Mobiltherm light oil[1]	8.0	100
Neo Pentane	7.8	60
Octanol[1]	6.7	95
Octyl chloride[1]	8.0	116
Pentane	7.8	104
Propane	7.9	100
South African crude oil	6.8–7.6	35–62
Toluene[1]	7.8	94

1. Extrapolated value.
2. E = 100–200 J.
3. 200°C.

Dusts of the same hazard class that have maximum deflagration pressures not greater than 9 bar gage require less vent area than those that have a maximum deflagration pressure greater than 9 bar gage. The nomographs in Figures 4-13F and 4-13G are based on test work reported in NFPA 68.

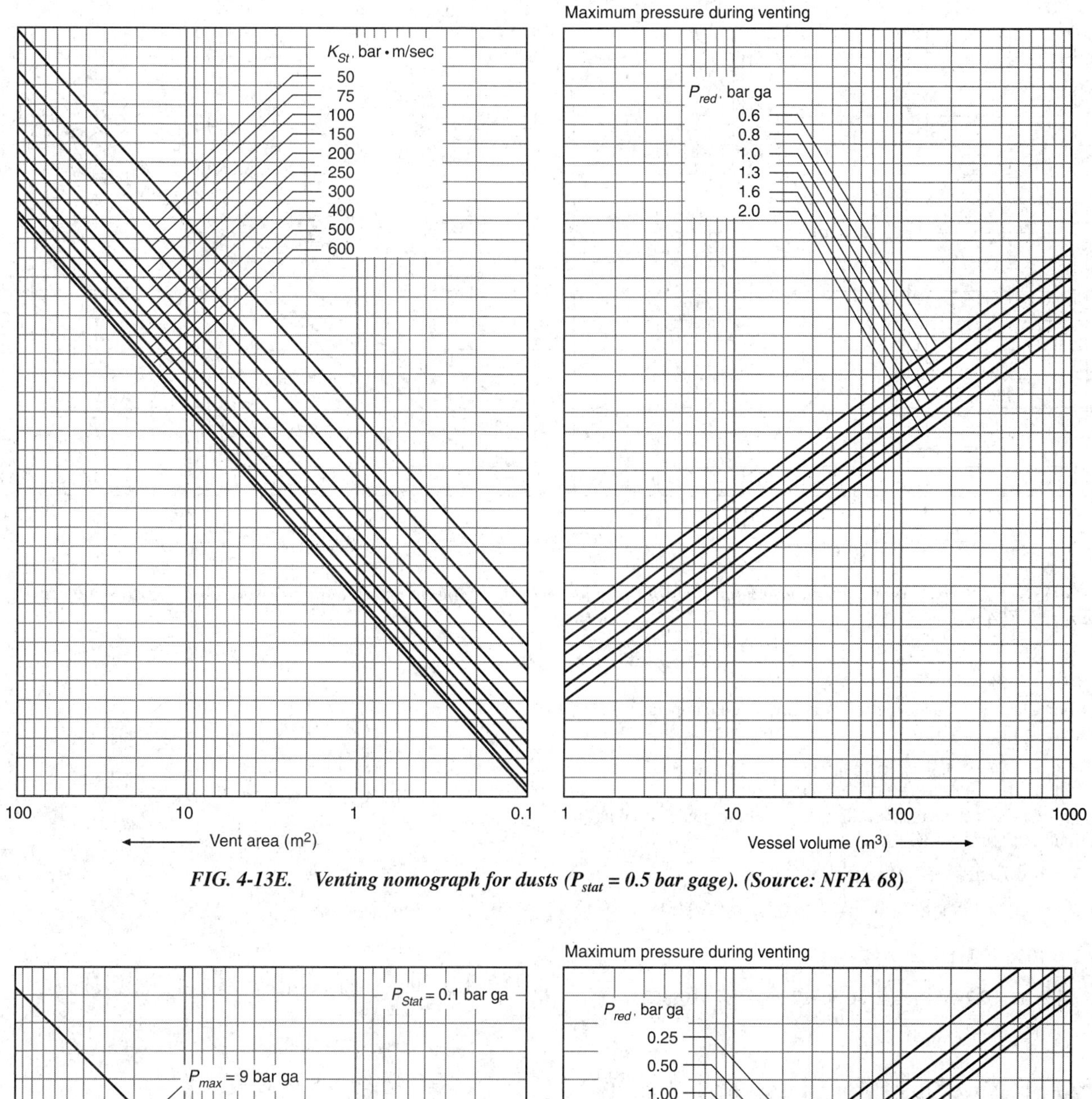

FIG. 4-13E. *Venting nomograph for dusts (P_{stat} = 0.5 bar gage). (Source: NFPA 68)*

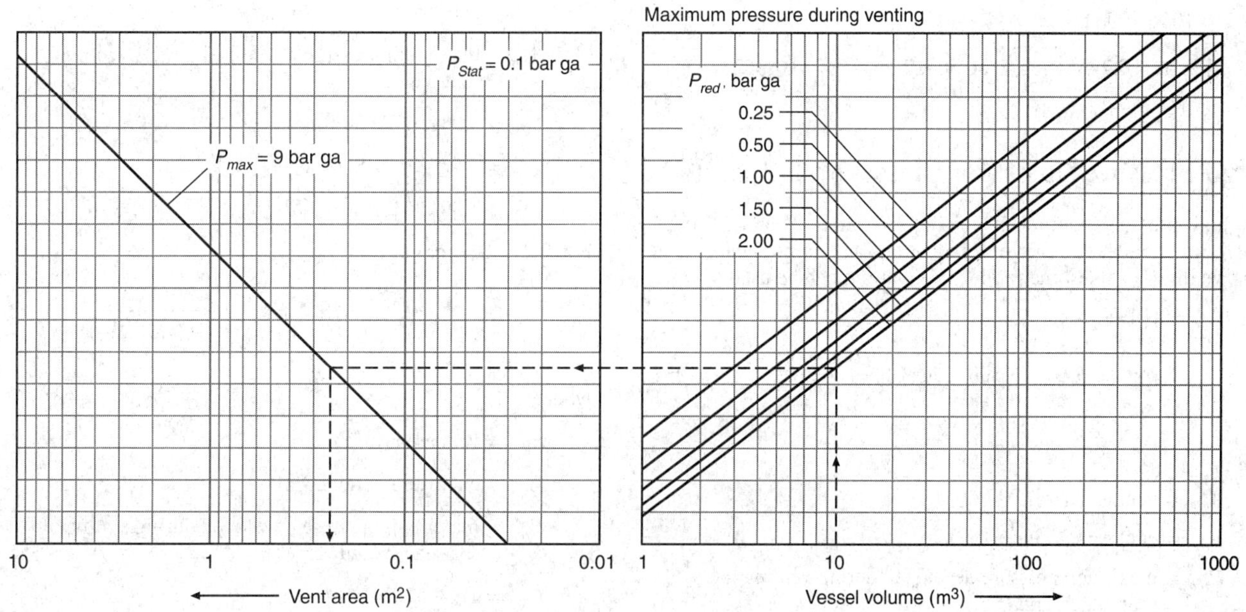

FIG. 4-13F. *Alternate venting nomograph for dusts of Class St-1 whose maximum deflagration pressure does not exceed 9 bar gage. (Source: NFPA 68)*

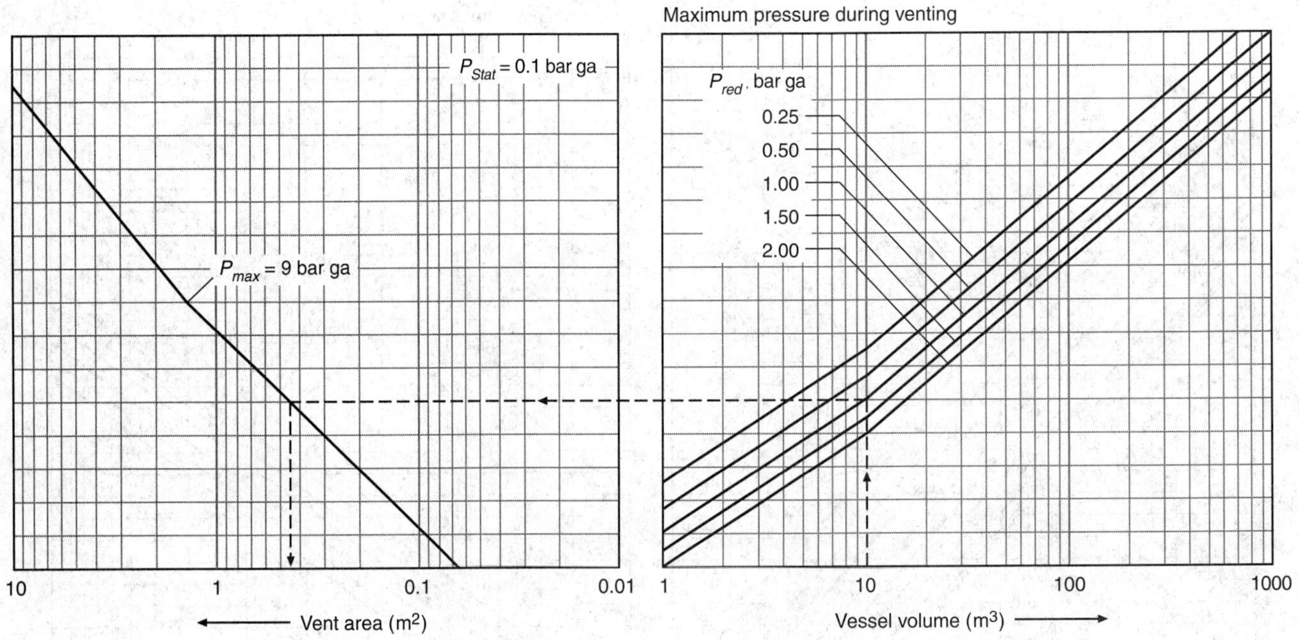

FIG. 4-13G. *Alternate venting nomograph for dusts of Class St-2 whose maximum deflagration pressure does not exceed 9 bar gage. (Source: NFPA 68)*

For Class St-1 dusts, the following equation can be used as an alternate to Figure 4-13F.

$$\log A_v = 0.77957 \log V - 0.42945 \log P_{red} - 1.24669 \quad (22)$$

For Class St-2 dusts, the following equations can be used as an alternate to Figure 4-13G.

For $V = 1$ to $10\ m^3$,

$$\log A_v = 0.64256 \log V - 0.46527 \log P_{red} - 0.99461 \quad (23)$$

For $V = 10$ to $1000\ m^3$,

$$\log A_v = 0.74461 \log V - 0.50017 \log \left(P_{red} + 0.18522\right) - 1.02406 \quad (24)$$

High-Pressure Dusts (for Classes St-1, 2, and 3)

For high-pressure equipment venting (for St-1, 2, and 3 dusts) where it is not desired to use the equations and nomographs using the specific K_{St} values, the following equations can be used.

For $P_{stat} = 0.1$ bar gage,

$$\log A_v + C = 0.67005 \log V + \frac{0.96027}{\left(P_{red}\right)^{0.2119}} \quad (25)$$

where

A_v = vent area (m^2),

V = enclosure volume (m^3),

P_{red} = maximum pressure developed during venting (bar gage), and

C = 1.88854 for St-1 dusts

= 1.69846 for St-2 dusts

= 1.50821 for St-3 dusts.

For $P_{stat} = 0.2$ bar gage,

$$\log A_v + C = 0.67191 \log V + \frac{1.03112}{\left(P_{red}\right)^{0.3}} \quad (26)$$

where

A_v = vent area (m^2),

V = enclosure volume (m^3),

P_{red} = maximum pressure developed during venting (bar gage), and

C = 1.93133 for St-1 dusts

= 1.71583 for St-2 dusts

= 1.50115 for St-3 dusts.

For $P_{stat} = 0.5$ bar gage,

$$A_v + C = 0.65925 \log V + \frac{1.20083}{\left(P_{red}\right)^{0.3916}} \quad (27)$$

where

A_v = vent area (m^2),

V = enclosure volume (m^3),

P_{red} = maximum pressure developed during venting (bar gage), and

C = 1.94357 for St-1 dusts

= 1.69627 for St-2 dusts

= 1.50473 for St-3 dusts.

VENT DISCHARGE DUCTS

Ideally, vents should discharge directly to the atmosphere to a safe location. However, this is not always practical and it becomes necessary to direct the hot gases and unburned fuel to a safe location as directly as possible via discharge ducts. Vent ducts will significantly increase the pressure developed in the enclosure during venting. The vent ducts should have a cross section at least as large as that of the vent itself. The increase in pressure developed during venting of gases is shown in Figure 4-13H; and, for dusts, in 4-13I. Other methods of predicting the effects of vent ducts have been presented.[4]

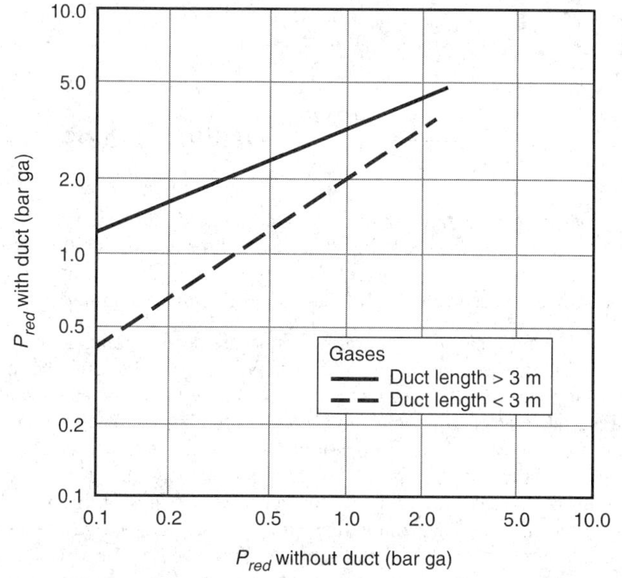

FIG. 4-13H.   *Maximum pressure developed during venting of gases, with and without vent ducts. (Source: NFPA 68)*

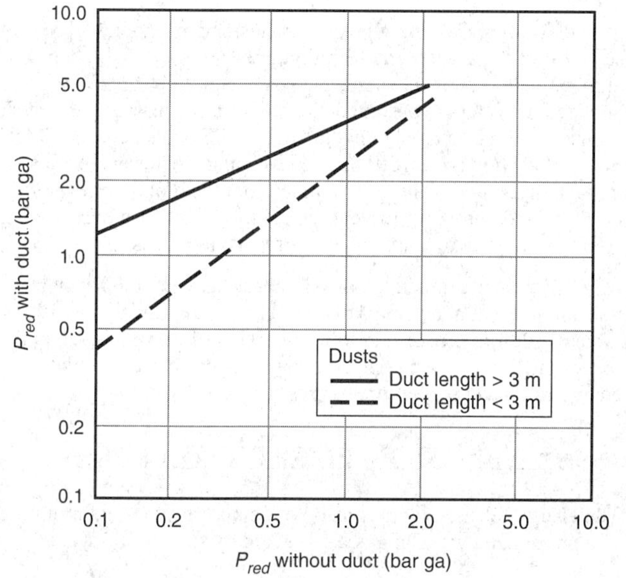

FIG. 4-13I.   *Maximum pressure developed during venting of dusts, with and without vent ducts. (Reference in NFPA 68)*

BIBLIOGRAPHY

References Cited

1. Faber, M., Symposium on Safety Against Explosions, Lucerne, Switzerland, June 5–7, 1984.
2. Swift, I., and Epstein, M., "Performance of Low-Pressure Explosion Vents," Paper 84d, 20th Annual Loss Prevention Symposium, AIChE Spring National Meeting, New Orleans, LA, Apr. 6–10, 1986.
3. Bartknecht, W., "Effectiveness of Explosion Venting as a Protective Measure for Silos," *Plant/Operations Progress,* Vol. 4, No. 1, Jan. 1985, pp. 4–13.
   ———, *Explosions-schutz; Grundlagen und Anwendung,* Springer-Verlag, Berlin, Germany, 1993.
   ———, *Explosions,* Springer-Verlag, Berlin, Germany, 1980.
   ———, *Dust Explosions,* Springer-Verlag, Berlin, Germany, 1989.
   ——— , "Pressure Venting of Dust Explosions in Large Vessels: Paper 82f," 20th Annual Loss Prevention Symposium, AIChE Spring National Meeting, New Orleans, LA, Apr. 6-10, 1986.
   ——— , "Pressure Venting of Dust Explosions in Large Vessels," *Plant/Operations Progress,* Vol. 5, No. 4, Oct. 1986, pp. 196–204.
4. Lunn, G. A., "Guide to Dust Explosion Prevention and Protection, Part 3: Venting of Weak Explosions and the Effects of Vent Ducts," Institution of Chemical Engineers, British Materials-Handling Board, London, England, 1988.
   Bruderer, R. E., "Advantages of the New Version of VDI 3673, Part 1, in Comparison with Chap. 7 of NFPA 68," *Process Safety Progress,* Vol. 14, No. 1, Jan. 1995, pp. 45–51.
   Howard, W. B., and Karabinis, A. H., "Tests of Explosion Venting of Buildings," *Plant/Operations Progress,* Vol. I, AIChE, NY, Jan. 1982, pp. 51–68.
   Simpson, L. L., "Equations for the VDI and Bartknecht Nomograms," *Plant/Operations Progress,* Vol. 5, No. 1, AIChE, NY, Jan. 1986, pp. 49–51.
   Siwek, R., "New Revised VDI Guideline 3673; Pressure Release of Dust Explosions: Paper 13a," Loss Prevention Symposium, Apr. 17-21, 1994, Atlanta, GA.
   Verein Deutscher Ingenieure VDI 3673, "Pressure Release of Dust Explosions," Nov. 1992.

NFPA Codes, Standards, and Recommended Practices

Reference to the following NFPA codes, standards, and recommended practices will provide further information on deflagration (explosion) venting discussed in this chapter. (See the latest version of *The NFPA Catalog* for availability of current editions of the following document.)

NFPA 68, *Guide for Venting of Deflagrations*

EXPLOSION PREVENTION AND PROTECTION

Joseph A. Senecal

A wide variety of materials are processed in industry. Many of these materials pose risks in handling due to their ability to sustain combustion in air. Of particular concern are those combustible materials that are processed as gases, mists, or finely divided solids, and have the possibility of being dispersed in air or other oxidizing gas. Flammable mixtures in air, if ignited, will deflagrate quickly, producing expanded volumes of hot gas in eight to ten times the volume of the original unburned gas. Ignition of a flammable fuel cloud within a process enclosure will result in a rise in pressure to a magnitude related to the burning characteristics of the fuel-air mixture and the ability of the enclosure to vent its contents. Such an event leading to damage to the process enclosure, injury of personnel, or other adverse effects is referred to as an explosion.

Hazards associated with deflagration explosions include catastrophic equipment failure, ejection of flame and unburned product (possibly hazardous in its own right) into the surroundings, possible secondary explosions leading to catastrophic facility damage, and personal injury. The purpose of this chapter is to provide basic guidance in the prevention of deflagrations and to outline the more commonly employed explosion protection methods.

The following elements must be simultaneously present in order for a deflagration explosion to occur:

1. A flammable mixture consisting of a fuel and oxygen, usually from air, or other oxidant;
2. A means of ignition; and
3. An enclosure.

A flammable mixture means that the oxygen and fuel components are intimately mixed and that they each are present at concentrations that fall inside a flammable composition boundary characteristic of each system of fuel, oxygen, and inert material (inert gas or solids).

Ignition of a flammable mixture will occur when a point source of sufficient energy content achieves a temperature above the ignition temperature of the mixture. All incandescent sparks (e.g., mechanical, electrical, electrostatic) have sufficiently high temperatures to cause ignition, but may lack sufficient energy to heat a minimal

propagating mass to its ignition temperature. A hot process surface may have a temperature below that required for prompt ignition, but may have large energy content. Dust deposits on such surfaces can be subject to accelerated self-heating and eventual ignition.[1,2]

Should ignition of a flammable mixture occur within an enclosure, whether or not the enclosure has points for ventilation, the internal pressure will rise as necessary to satisfy the non-steady-state material balance equation. Fully confined deflagrations commonly develop pressures in the range of eight to ten times the initial absolute pressure, or 100 to 140 psig (690 to 965 kPa). The time to achieve the maximum deflagration pressure depends on the vessel size and characteristics of the fuel, but is usually in the range of a few tens to a few hundreds of milliseconds. Some venting of expanding combustion gases will occur through normal process openings, but these are usually too small to prevent development of destructive pressures. Process enclosures are seldom designed for pressure containment. Pressures at which enclosure failure occurs can be quite low, particularly for enclosures of rectangular sheet metal construction. These can fail completely at internal pressures of a few pounds per square inch (psi). (For SI units: 1 psi = 6.895 kPa.)

In what follows, the subject of explosion prevention will be presented in the context of deflagration prevention through control or elimination of flammable atmospheres and ignition sources. Careful application of steps to prevent deflagrations is no guarantee that such events will not occur, but only that their likelihood is reduced. Additional protection is afforded through use of any of several explosion mitigation methods. These techniques come into play once ignition has occurred and serve to vent, contain, or redirect the expanding combustion gases, or suppress the combustion process itself.

The reader is urged to consult other chapters in this handbook relevant to the fundamentals of fires, explosions, and extinguishing systems. Before proceeding to a discussion of prevention and protection methods, some consideration is given to the magnitude of energy released upon failure of pressurized enclosures.

EXPLOSION ENERGY RELEASE

The energy released by expansion of compressed gas upon rupture of a pressurized vessel may be estimated by[3]

$$W_e = \frac{P_1 V_1}{k-1}\left[1-\left(P_2/P_1\right)^{\frac{(k-1)}{k}}\right] \qquad (1)$$

Joseph A. Senecal, Ph.D., P.E., M.B.A., is manager of the Combustion Research Center, Fenwal Safety Systems, Inc. He is a member of the NFPA Technical Committees on Explosion Protection Systems, Deflagration Venting and Explosion Protection Systems; the AICHE committee on Loss Prevention, and the ASTM committee on Hazards of Chemicals.

where

W_e = explosive energy released (J),

P_1 = vessel pressure at rupture (kPa),

P_2 = ambient pressure usually 101.3 kPa,

V = vessel volume (m^3), and

k = ratio of specific heats of vessel contents at time of failure (C_p/C_v).

(For U.S. conversions: 1 psi = 6.895 kPa; and 1 cu ft = 0.0283 m^3.)

Equation 1 assumes that the compressed gas expands isentropically from P_1 to P_2. It has been reported that this relation slightly overestimates the energy release for values of $P_1 - P_2$ less than about 690 kPa (100 psi) and underestimates the energy release at higher pressures. A more rigorous energy release estimate can be obtained by availability analysis.[4] However, Equation 1 can be used as a first approximation of energy release.

The calculated energy release may be expressed in terms of the explosive equivalent of a mass of TNT as:

$$m_{TNT} = 2.13 \times 10^{-7} W_e \text{ (kg)} \qquad (2)$$

EXAMPLE: What is the explosive energy release from a 10 m^3 vessel rupturing at pressure of 100 psig to atmospheric pressure due to a stoichiometric internal deflagration of propane and air?

The combustion reaction is (neglecting minor species, dissociation products, and radicals)

$$C_3H_8 + 5O_2 + 18.8N_2 = 3CO_2 + 4H_2O + 18.8N_2 \qquad (3)$$

The constant-pressure flame temperature of a stoichiometric (4.03 percent) propane-air flame is approximately 3497°F (1925°C).[5] The constant-volume flame temperature will be somewhat higher, as no PdV work was done by the heated gas. Heat losses would tend to decrease the temperature. In any case, this value is close enough to obtain reasonable heat capacity data from reference tables.[6] The heat capacities, C_p, of the principal combustion products at their elevated temperatures are:

$$N_2 : C_p = 36.3 \text{ J/mol} - \text{K}$$

$$CO_2 : C_p = 60.9 \text{ J/mol} - \text{K}$$

$$H_2O : C_p = 52.4 \text{ J/mol} - \text{K}$$

The molar average value of C_p of the combustion product mixture is 41.6 J/mol – K. Since $C_v = C_p - R$ (R is the gas law constant, 8.315 J/mol – K), the value of k is

$$k = C_p/C_v = 41.6/(41.6 - 8.315) = 1.25,$$

$$P_1 = 100 \text{ psig} = 114.7 \text{ psia} = 7.91 \times 10^5 \text{ Pa, and}$$

$$P_2 = 1 \text{ atm} = 1.013 \times 10^5 \text{ Pa.}$$

The explosive energy release is, therefore,

$$W_e = \frac{(791,000)(10)}{1.25 - 1} \left[1 - (101,300/791,000)^{\frac{(1.25-1)}{1.25}} \right]$$

$$= 1.07 \times 10^7 \text{ J}$$

$$m_{TNT} = (2.13 \times 10^{-7}) (1.07 \times 10^7)$$

$$m_{TNT} = 2.27 \text{ kg.}$$

(For U.S. conversions: 1 cu ft = 0.0283 m^3; 1 psi = 6.895 kPa.)

Thus, the explosive potential of an unprotected vessel can be readily appreciated in tangible terms. It is clear from Equation 1 that the explosive energy potential of an overpressurized vessel can be

minimized by preventing the buildup of excessive pressure within it or by relieving the excess pressure at as low a relief point as possible. Thus, the value of adequately designed and functioning systems to suppress pressure buildup or to relieve excess pressure is demonstrated.

DEFLAGRATION PREVENTION

Mixture Flammability Control Fundamentals

The notion of the "fire triangle" is that an ignition source must come together with both fuel and oxidant, the latter usually being atmospheric oxygen, in order to initiate and sustain a fire. This representation, while convenient, oversimplifies the more complex requirements needed to sustain a deflagration, particularly regarding fuel and oxygen requirements. A flammability diagram can be used to illustrate, in composition space, the boundary that divides mixtures of fuel, oxygen, and inert components into flammable and nonflammable regions. Figures 4-14A and 4-14B are two examples of two of the several types of such representations.

The three-component diagram is plotted as a triangular graph with each angle representing either fuel, oxidant, or inert gas. Where appropriate, the oxidant may be taken as air, or other separately managed stream containing the oxidizing species; and the inert gas as only added inert components not associated with atmospheric air. Only two components are plotted in some diagrams, the concentration of the third being determined by difference from 100 percent. The lower and upper flammable limits are two limiting points on such a diagram for compositions having no added inert component. (See references 5, 7, and 8 for more detailed discussions on flammability.)

An important part of deflagration prevention is in maintaining atmospheres in process volumes that are nonflammable. This may be achieved through use of any of several techniques, either singly or in combination. These include reduction of airborne concentrations of gaseous, dusty, or misted fuels. Alternatively, the concentration of

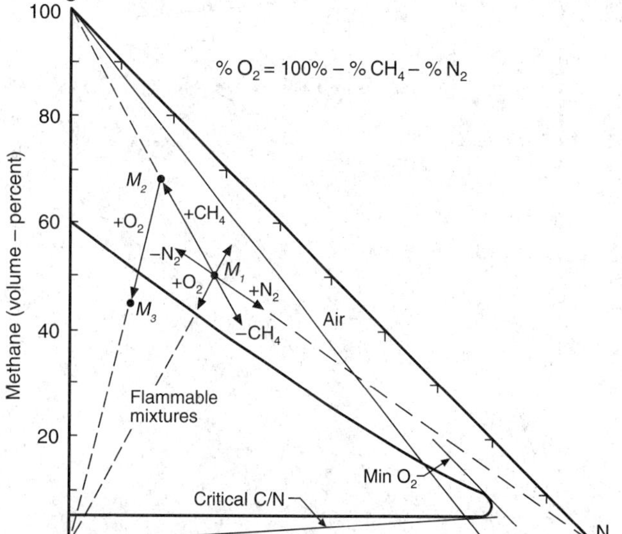

FIG. 4-14A. *Flammability diagram for methane-air oxygen system in triangular format. (Source: M.G. Zabetakis, Bulletin 627, U.S. Bureau of Mines)*

the oxidant may be reduced. Finally, a nonflammable component, such as inert gas, may be added to the atmosphere. Thus, the composition of the atmosphere within a process volume that is, or has the potential to be, in the flammable range is controlled to prevent this condition.

Reduction of Oxidant Concentration

The selection of a gas for inerting a process atmosphere will be influenced by several factors, such as inert gas effectiveness, availability, cost, process conditions, distribution requirements, and toxicity. The heat capacity of a gas is the most important inerting property. (See Table 4-14A for data on several inert gases.) This is because a nonflammable condition is achieved when the heat capacity of a fuel-oxygen-inert gas mixture is elevated above about 250 J/mol-K of oxygen in the mixture. Thus, the higher heat capacity of the inert gas, the lower the quantity required to achieve an inert mixture. Certain halogenated hydrocarbons, used in total flooding fire protection systems,[9] have also been shown to be useful in inerting flammable atmospheres.

TABLE 4-14A. *Heat Capacities of Inert Gases (Values at 25°C)*

Gas	C_p (J/mol-K)
Argon	20.9
Nitrogen	28.4
Steam	35.4 at 100°C
Carbon dioxide	37.5
HFC 23 (CHF_3)	51.6
HFC 227ea (CF_3CHFCF_3)	137
FC 3-1-10 (C_4F_{10})	191

For U.S. conversion: 77°F = 25°C.

FIG. 4-14B. *Flammability diagram of propane-air-nitrogen system in Cartesian format. (Source: F.T. Bodurtha, E.I. du Pont de Nemours & Co., Inc.)*

Nitrogen is commonly used as an inerting gas. It may be obtained in high-pressure cylinders or taken from stored cryogenically liquefied gas. Catalytic combustion of ammonia in air followed by water condensation results in a relatively pure grade of nitrogen for process inerting. Carbon dioxide is somewhat more effective than nitrogen due to its higher heat capacity. It can be stored as a liquefied compressed gas in cylinders or kept in low-pressure refrigerated tanks for use on demand. Flue gas from a combustion system is a mixture of carbon dioxide and nitrogen, usually saturated with water vapor, and having a low oxygen concentration. Flue gas is commonly used to inert storage tanks when off-loading oil from ships. Carbon dioxide should not be used as an inerting gas for certain refractory metals. Examples are magnesium, silicon, and zirconium. In such cases, it may be necessary to use a noble gas, such as argon, for inerting. Values of limiting oxygen concentration for a variety of gases, metals, and carbonaceous dusts are given in Tables 4-14B, 4-14C, and 4-14D.

Methods of Inert Gas Application

Batch methods: In *vacuum purging*, the pressure in the process unit is reduced from its initial value, P_1, to a lower pressure, P_2. Inert gas is then added to return the vessel to P_1. If the gas mixing within the process volume is assumed to be well mixed, then the fraction of original gas remaining after one such purge cycle will be

$$X = P_2/P_1 \qquad (4)$$

In *pressure purging*, a volume is pressurized from P_1, to a higher pressure, P_2, by addition of purge gas and then vented back to the initial pressure, P_1. The fraction of original gas remaining after each such cycle will be

$$X = P_1/(P_1 + P_2) \qquad (5)$$

The fraction of original gas remaining after n cycles by either batch purge method is

$$X(n) = X^n \qquad (6)$$

Continuous methods: Some enclosures, such as solvent storage tanks, are under continuous use or are not suited to batch purge methods due to structural considerations. Such tanks "breathe" upon addition or removal of liquid contents or when their internal temperature changes due to climatic variations. In *fixed rate* purging, inert gas is added continuously at the maximum anticipated demand rate. This method has the advantages of being simple to monitor and control but at the expense of purge gas consumption and, perhaps more important, results in carrying product vapors out of the enclosure.

Variable rate purging may involve elements of both pressure demand and load demand gas addition. A small rate of continuous inert gas addition may be used together with pressure monitoring to determine when automatic addition of gas is required to make up for vapor space cooling. Addition of inert gas may also be increased upon removal of liquid contents of a tank.

Purging enclosures of volatile liquids may be regulated under Title IV of the U.S. Clean Air Act Amendments of 1990. The capture and recovery, recycle, or incineration of process gas emissions has become much more common in the last 10 years. Vapor recovery systems need consideration with respect to possible explosion risks that may be present. The U.S. Coast Guard has promulgated rules on operation of marine vapor recovery systems pertaining to explosion prevention and control.[10]

TABLE 4-14B. *Maximum Permissible Oxygen Percentage to Prevent Ignition of Flammable Gases and Vapors, Using Nitrogen and Carbon Dioxide for Inerting*

	N_2-Air	CO_2-Air
	O_2 Percent Above Which Ignition Can Take Place	O_2 Percent Above Which Ignition Can Take Place
Acetone	13.5	15.5
Benzene (Benzol)	11	14
Butadiene	10	13
Butane	12	14.5
Butene-1	11.5	14
Carbon Disulfide	5	8
Carbon Monoxide	5.5	6
Cyclopropane	11.5	14
Dimethylbutane	12	14.5
Ethane	11	13.5
Ether (Diethyl)	10.5	13
Ethyl Alcohol	10.5	13
Ethylene	10	11.5
Gasoline	11.5	14
73–100 Octane	12	15
100–130 Octane	12	15
115–145 Octane	12	14.5
Hexane	12	14.5
Hydrogen	5	6
Hydrogen Sulfide	7.5	11.5
Isobutane	12	15
Isopentane	12	14.5
JP-1 Fuel	10.5	14
JP-3 Fuel	12	14
JP-4 Fuel	11.5	14
Kerosene	11	14
Methane	12	14.5
Methyl Alcohol	10	13.5
Natural Gas (Pittsburgh)	12	14
Neopentane	12.5	15
n-Heptane	11.5	14
Pentane	11.5	14.5
Propane	11.5	14
Propylene	11.5	14

Data in this table were obtained from publications of the U.S. Bureau of Mines.

Data were determined by laboratory experiments conducted at atmospheric temperature and pressure. Vapor-air inert-gas samples were placed in explosion tubes and exposed to a small electric spark or open flame.

In the absence of reliable data, the U.S. Bureau of Mines or other recognized authority should be consulted.

Concentration Reduction of Combustibles

Effective explosion prevention can be ensured if combustible materials are excluded from spaces to be protected. Three aspects of minimizing risks associated with combustibles are: (1) containment, (2) ventilation, and (3) purging.

Containment: Special considerations are necessary for each class of combustible material. For example, cylinders and tanks of flammable compressed gases must be properly stored or located to ensure valves and other appurtenances are not subject to accidental damage. Containers and tanks of flammable and combustible liquids need periodic inspection to verify tightness of closures and the absence of damage or corrosion. Filter bags in dust collectors need periodic in-

TABLE 4-14C. *Maximum Permissible Oxygen Concentration to Prevent Ignition of Combustible Metal/Dusts Using Carbon Dioxide and Nitrogen for Inerting*

	Oxygen Percent by Volume Above Which Ignition Can Occur	
	Carbon Dioxide-Air	Nitrogen-Air
Aluminum (atomized)	2	7
Antimony	16	—
Dowmetal	0	—
Ferrosilicon	16	17
Ferrotitanium	13	—
Iron, Carbonyl	10	—
Iron, Hydrogen reduced	11	—
Magnesium	0	2
Magnesium-Aluminum	0	5
Manganese	14	—
Silicon	12	11
Thorium	0	2
Thorium Hydride	6	5
Tin	15	—
Titanium	0	4
Titanium Hydride	13	10
Uranium	0	1
Uranium Hydride	0	2
Vanadium	13	—
Zinc	9	9
Zirconium	0	0
Zirconium Hydride	8	8

Data in this table were obtained principally from U.S. Bureau of Mines RI 6543.

In the furnace test, dust clouds of Zr, Th, U, and UH_3 also ignited in CO_2. During heating for several minutes, undispersed layers of samples of the following metal powders ignited (glowed) in CO_2: Stamped Al, Mg, ZnMg-Al, Dowmetal, Ti, TiH_2, Zr, ZrH_2, Th, ThH_2, U, and UH_3. Visible burning of dust layers was also observed in N_2 with powders of Mg, Sn, Mg-Al, Dowmetal, Ti, TiH_2, Zr, Th, ThH_2, U, and UH_3.

In the absence of reliable data for combustible dusts, the U.S. Bureau of Mines or other recognized authority should be consulted.

Data were obtained by laboratory experiments conducted at atmospheric temperature and pressure. An electric spark was the ignition source.

spection to locate and replace damaged elements that can allow passage of combustible dusts. It is recommended that conformance be maintained with materials-handling and storage requirements as outlined in NFPA codes (such as NFPA 30, *Flammable and Combustible Liquids Code*) and other codes and practices pertaining to specific flammable and combustible materials.

Ventilation: Where the presence of flammable gases or combustible dusts may be possible as a result of normal operations or upset conditions, their concentrations can be kept to a minimum or excluded altogether by general area or localized ventilation techniques. Continuous introduction of clean air to a room may be used to either continually sweep away contaminants or to maintain a room or process space under constant positive pressure to exclude entry of flammable materials. NFPA 69, *Standard on Explosion Prevention Systems* (hereinafter referred to as NFPA 69), Appendix D, contains guidance on the design of ventilation systems.

Purging: Purging is used to prevent development of ignitable atmospheres in enclosures or equipment. Specific methods are recommended for maintaining safe atmospheres inside of electrical equipment enclosures. Type X, Type Y, and Type Z purging pertain to reducing the classifications within enclosures from Division 1 to nonhazardous, Division 1 to Division 2, and Division 2 to nonhazardous,

TABLE 4-14D.   Maximum Permissible Oxygen Content to Prevent Ignition by Spark of Combustible Dusts Using Carbon Dioxide as the Atmospheric Diluent

Dust	Maximum Allowable Oxygen Concentration, Percent
Agricultural	
Clover seed	15
Coffee	17
Cornstarch	11
Dextrin	14
Lycopodium	13
Soy flour	15
Starch	12
Sucrose	14
Chemicals	
Ethylene diamine tetra acetic acid	13
Istoic anhydride	13
Methionine	15
Ortazol	19
Phenothiazine	17
Phosphorous pentasulfide	12
Salicylic acid	17
Sodium lingo sulfonate	17
Steric acid and metal stearates	13
Carbonaceous	
Charcoal	17
Coal, bituminous	17
Coal, subbituminous	15
Lignite	15
Metals	
Aluminium	2
Antimony	16
Chromium	14
Iron	10
Magnesium	0
Manganese	14
Silicon	12
Thorium	0
Titanium	0
Uranium	0
Vanadium	14
Zinc	10
Zirconium	0
Miscellaneous	
Cellulose	13
Lactalbumin	13
Paper	13
Pitch	11
Sewage sludge	14
Sulfur	12
Wood flour	16
Plastics Ingredients	
Azelaic acid	14
Bisphenol A	12
Casein, rennet	17
Hexamethylenetetramine	14
Isophthalic acid	14
Paraformaldehyde	12
Pentaerythritol	14
Phthalic anhydride	14
Polymer glyoxyl hydrate	12
Terephthalic acid	15

TABLE 4-14D.   (Continued)

Dust	Maximum Allowable Oxygen Concentration, Percent
Plastics—Special Resins and Molding Compounds	
Coumarone-indene resin	14
Lignin	17
Phenol, chlorinated	16
Pinewood residue	13
Rosin, DK	14
Rubber, hard	15
Shellac	14
Sodium resinate	14
Plastic—Thermoplastic Resins and Molding Compounds	
Acetal resin	11
Acrylonitrile polymer	13
Butadiene-styrene	13
Carboxymethyl cellulose	16
Cellulose acetate	11
Cellulose triacetate	12
Cellulose acetate butyrate	14
Ethyl cellulose	11
Methyl cellulose	13
Methyl methacrylate	11
Nylon polymer	13
Polycarbonate	15
Polyethylene	12
Polystyrene	14
Polyvinyl acetate	17
Polyvinyl butyral	14
Plastic—Thermosetting Resins and Molding Compounds	
Allyl alcohol	13
Dimethyl isophthalate	13
Dimethyl terephthalate	12
Epoxy	12
Melamine formaldehyde	17
Polyethylene terephthalate	13
Urea formaldehyde	16

Data in this table were obtained from U.S. Bureau of Mines RI 6543. The data were obtained by laboratory experiments conducted at room temperature and pressure, using a 24-watt continuous spark as the ignition source.

For moderately strong igniting sources, such as a low-current electrical arc or a heated motor bearing, the maximum permissible oxygen concentration is 2 percentage points less than the corresponding value for ignition by spark.

For strong igniting sources, such as an open fire, flame, or glowing furnace wall, the maximum permissible oxygen concentration is 6 percentage points less than the corresponding value for ignition by spark.

The maximum permissible concentration for ignition by spark, when nitrogen is used as the atmospheric diluent, can be calculated by the "rule-of-thumb" formula: $O_n = 1.30 O_c - 6.3$

where O_n = the maximum permissible oxygen concentration using nitrogen as the atmospheric diluent, and

O_c = the maximum permissible oxygen concentration using carbon dioxide as the atmospheric diluent.

Research data on the use of dry powders or water as inerting materials and on the effects of inerting on pressure development in a closed vessel are given in U.S. Bureau of Mines RI 6543, 6561, and 6811.

Ignition Source Control

The means by which flammable atmospheres can become ignited are diverse, indeed. While it is often true that the ignition source of a given event cannot be identified with certainty after the fact, due to the generalized destruction of the scene, a number of important mechanisms of deflagration initiation are recognized and should be carefully considered when conducting a hazard analysis.

respectively. The details of these purging methods can be found in NFPA 70, *National Electrical Code®* (hereinafter referred to as NFPA 70).

Mechanical sparks: A survey of 357 dust explosions between 1965 and 1980 revealed that, in 29 percent of the cases, the source of ignition was mechanically generated by sparks.[11] It has been shown[12] that sparks generated by grinding handheld steel can have sufficient energy to ignite organic dust clouds. This clearly points to the ignition energy potential when high-energy systems, e.g., coal pulverizers, mills of the rolling, hammer, and pin type, to name a few, cause grinding action on metal or other spark-inducing materials. Where tramp metal is a recognized hazard, the use of magnetic separators is recommended.

Fires: Unexpected open flaming due to ignition of accumulated dust can serve to ignite an ambient dust cloud. Mechanisms of causing such fires are several. Their risk of occurrence can be reduced by both understanding the ignition characteristics of the material being processed and by preventing the accumulation of materials susceptible to ignition by the given mechanism.

Hot surfaces: Accumulated dust on a motor housing, transformer, high-pressure steam pipe, or other like hot surface can lead to slow pyrolysis of the solids, which can then ignite. The ignition probability is related to the surface temperature, dust layer thickness, and the thermal stability of the dust. A method of characterizing surface ignition temperature for dusts is available.[13] Approved equipment will be marked to indicate the operating temperature. The temperature of equipment operating in Class I or II (see below) environments must not exceed the ignition temperature of the gas or dust to which it is exposed. NFPA 497M, *Manual for Classification of Gases, Vapors, and Dusts for Electrical Equipment in Hazardous (Classified) Locations*, provides ignition temperature data on dusts, vapors, and gases.

Autoignition: Dusts layers exposed to continuous uniform elevated temperature, or even ambient temperatures, can also undergo self-heating and ignite.[1,2] The temperatures at which autoignition of bulk dusts occurs are much lower than for surface ignition, above, and are dependent on the volume of the heated dust. A small-scale method for characterizing the self-heating tendency of dusts was developed by the U.S. Bureau of Mines.[14] Another method is recommended by the U.S. Department of Transportation for determination of packaging classification of self-heating materials for purposes of over-road shipment.[15]

Mechanical impact: Some materials may be subject to immediate ignition due to sudden mechanical impact. The ASTM E680 drop hammer test method can be used to characterize materials in this regard.[16]

Friction: Frictional heating of combustible materials is possible where moving components are found. Rotating drums and moving belts in bucket elevators is an example of a system that requires special consideration. Regular lubrication of bearings is essential. Use of belt-alignment devices is mandated in some cases.[17]

Embers: Wood processing operations, such as grinding, sanding, and cutting, can produce glowing wood particles that can be conveyed to a dust collector. Such an ember can lead to ignition of a wood-dust layer in which it becomes embedded. Spark detection and extinguishing systems may be appropriate in such operations.

Electrical equipment: Electrical motors, switches, lights, heaters, and similar high-current devices may be powerful ignition sources. Such equipment can be constructed for use in hazardous atmospheres, and is rated with respect to the type of atmosphere in which it is suited for use. A synopsis of the rating system is given below.

Class I	*Flammable gases.* Groups: A (acetylene); B (hydrogen); C (ethylene); and D (propane).
Class II	*Combustible dusts.* Groups: E (metal dusts); F (carbonaceous dusts, e.g., coal, charcoal, coke, etc.); and G (flour, grain, wood, plastics, chemicals).
Class III	*Combustible fibers.*
Division 1:	Flammable gases or combustible dusts may be present at ignitable concentrations, under normal operating conditions.
Division 2:	Where hazardous materials may be handled, processed, or used; ignitable atmospheres not normally present due to containment or ventilation of hazardous materials; areas adjacent to Division 1 locations.

The examples given in the groups above are representative only. Flammable gases are classified according to the maximum experimental safe gap (MESG), which prevents flame passage. MESG is determined by test.[18]

Thus, a motor rated as Class I, Division 1, Group D would be rated for use in an environment where a propane-like gas was likely to be present. See Article 500 of NFPA 70 for further information.

Unrated electrical equipment can be employed in hazardous areas, provided such equipment is contained in a pressurized or purged enclosure. The requirements for this approach are given in NFPA 496, *Standard for Purged and Pressurized Enclosures for Electrical Equipment*. Various purging methods are described that can render the classification within an enclosure from Division 1 to nonhazardous (Type X purging); from Division 1 to Division 2 (Type Y purging); and from Division 2 to nonhazardous (Type Z purging).

Electrostatic ignition: Conveyance of liquids or dusts leads to electron transfer at the boundary of the moving material and a duct or pipe. The flow of electric charge in this manner is called a *streaming current, I_s*. The discharge of the moving material from the conduit to a receiving point (tank or drum for liquids; bin or pile for solids), of capacitance C, results in charge separation and the development of an electric potential difference, V. The accumulated charge, Q, if not allowed to dissipate or *relax*, will have a stored energy of $E = QV/2 = CV^2/2$. Potentials exceeding 350 volts and stored energies exceeding 0.1 mJ are considered hazardous in flammable gas areas. Minimum ignition energies (MIE) of dust clouds vary widely but can be less than 10 mJ. Most dusts of organic composition have MIEs in the range of 10 to 100 mJ. Charge potentials of several thousand volts and stored energies sufficient to ignite dust clouds are readily achieved in some solids transfer operations.

Ignition of flammable atmospheres by electrostatic discharges can be avoided by first recognizing process operations that can lead to charge separation. One can then provide conduction paths to maintain process components at low or zero electric potential. Resistance to ground should not exceed 100,000 ohm from any component. It is important to consider all components in a system. A section of pipe may be connected to a system by bolted flanges, but electrically insulated by nonconductive gaskets. Such a component has the characteristics of a capacitor and can hold an electric charge. *Bonding* of electrically isolated process components is achieved by connecting a wire between them. (See NFPA 70 for guidance.) *Grounding* of the bonded components to an earth ground provides a conduction path that prevents charge accumulation. References 3,

19, 20, 21, and 22 should be consulted for a more complete discussion of electrostatic hazards.

Combined mechanisms: In any given ignition scenario it is likely that multiple ignition mechanisms were involved. For example, an unlubricated bearing (*friction, hot surface*) in the tension roll in the boot (bottom) section of a bucket elevator conveying grain could generate a smoldering mass of grain dust (*ember*). The burning matter could become dislodged and fall into the general mass of moving grain and be conveyed in a bucket into an upper section of the elevator during which time its extent of burning increases. At some moment the ember becomes a flaming mass, (*open flame*), and encounters a dust cloud with a concentration greater than its lower explosible limit, thus initiating a deflagration. The pressure developed by the primary deflagration could rupture the elevator leg enclosure. The escaping pressurized blast of gases could cause grain dust on nearby surfaces to become airborne and ignited, (*flame jet ignition*). An ensuing secondary deflagration would likely involve a much larger mass of fuel and air and lead to generalized pressure damage to the facility. Due to failure of a grease fitting, the plant could be substantially damaged. This example does not represent any particular event. Rather, it is a synthesis of many similar actual events that have occurred.

EXPLOSION PROTECTION

As used in the present context, explosion protection means those measures adopted to control and mitigate the effects of the ignition of the contents of a process volume that may be capable of propagating a deflagration. Thus, while appropriate explosion prevention measures may have been implemented, it is assumed at this point that these steps have only reduced the probability of initiating a deflagration and not eliminated such a possibility altogether. Whether the probability of the occurrence of a deflagration is now so small as to constitute a risk of acceptable magnitude is a matter for evaluation using the methods of risk and consequence analysis, which is beyond the scope of this chapter.

There are essentially only four strategies available as explosion protection options. Equipment may be designed for pressure containment. This approach may be required if no emission of vessel contents is tolerable (for concerns relating to toxicity, environmental, or public relations matters). In some cases, design for pressure containment (DPC) may be the most economical approach.

Deflagration venting is widely used to effectively direct expanding gases and overpressure out of a vessel to a safe location. This protection option may be appropriate when material toxicity and environmental contamination are not a concern and when a safe vent discharge location is available. When appropriate, venting is often the low-cost protection option. A discussion of the fundamentals and application of deflagration vents is given in Section 4, Chapter 13, "Deflagration (Explosion) Venting."

Explosion isolation is the strategy of preventing products of combustion produced in one location from passing to another location where personal injury or additional damage may occur. An example is the protection of operators at a manual dumping station from being injured by a backflash propagating through the material conveying system due to initiation of a deflagration in the receiving bin. Isolation is usually accomplished by use of fast-acting valves, deflagration interrupters, or chemical barriers.

Deflagration suppression may be the preferred approach when the consequences of employing the above options are unacceptable. Process equipment may be too large to economically construct for DPC; venting may not be possible due to the location or contents of the vessel; or mechanical isolation of the process system may have unacceptable secondary consequences in a complex plant.

Pressure Containment

The use of pressure relief vents on vessels in some processes may be impractical. Where the safety strategy is to allow a possible deflagration to occur, but be contained, the process vessel must be constructed to be able to contain the pressurized gases with no equipment damage or with limited acceptable damage. The design pressure to which the vessel must be fabricated is determined by the maximum anticipated pressure that may develop in the enclosure. This is the maximum deflagration pressure of the combustible material that is used. No credit is taken for pressure relief devices on the vessel.

Vessels designed and constructed for pressure containment must conform to the ASME *Boiler and Pressure Vessel Code*, Section VIII, Division I. The design pressure for deflagrations may be calculated, using Equation 1, to result in a vessel of sufficient strength to either not deform at the maximum deflagration pressure or to deform but not rupture. For vessels constructed of carbon steel or low alloy stainless steel, the design pressure is given by

$$P_{design} = 1.5 \left[R(P_i + 14.7) - 14.7 \right] / F_1 \qquad (7)$$

where

P_i = maximum initial pressure at ignition;

R = 9 for gases,

= 10 for dusts with $K_{St} < 300$ bar-m/s, and

= 13 for dusts with $K_{St} > 300$ bar-m/s;

F = 2 for "no deformation" design, and

= 4 for "no rupture" design.

The deflagration expansion ratio, R, is increased in inverse proportion to the absolute temperature for operating temperatures below 77°F (25°C), or

$$R(T_2) = 298 \, R / (T_2 + 273) \qquad (8)$$

where T_2 is the design temperature in Celsius, and R is either 9 or 10. See NFPA 69 for further details.

Design for pressure containment requires further consideration of the eventual release of the pressurized gases. If the vessel is truly leaktight, then pressure reduction will occur as the products of combustion cool. Alternatively, pressurized gases will flow through process openings or pressure relief devices. The essential point here is that hot combustion gases should be prevented from passing to other process units where adverse consequences may occur due to initiation of a fire or another deflagration. Techniques for process chemical isolation systems to prevent deflagration propagation are discussed later in this chapter.

Deflagration Venting

A detailed discussion of the design and use of deflagration vents is given in Section 4, Chapter 13, "Deflagration (Explosion) Venting," and in NFPA 68, *Guide for Venting of Deflagrations*. The following is a brief consideration of the utility and basis of selection of venting as an explosion protection strategy.

The simplest post-ignition method of explosion protection is deflagration venting. Pressure development within a process enclosure in which a deflagration occurs is prevented from attaining damaging levels by allowing the enclosure contents to be ejected, in a controlled way, to a safe location. It is the control and safety aspects of a vented deflagration that differentiate it from an explosion.

A deflagration venting system includes the mechanical aspects of the vent closure (disk, panel, door, etc.) and the opening in the wall of the process enclosure created when the closure bursts or otherwise opens. Vent systems come in a variety of designs, including single-use bursting types and resetable doors. Vents may be used on process vessels that are of relatively strong construction or on buildings that are relatively weak. Vent systems, where applicable, are relatively inexpensive solutions to the problem of deflagration pressure management. The following criteria should be satisfied to determine whether deflagration venting is a suitable method of explosion protection.

1. Sufficient vent area available. The total amount of vent area required to protect an enclosure may, in some cases, exceed the surface area available on the equipment.

2. Safe outdoor equipment location. Operation of a deflagration vent results in the discharge of burning material to the ambient environment. People, structures, and equipment need to be protected from high-velocity flames and pressure fronts. Indoor equipment may be able to be vented, provided a duct (the use of which affects the vent design) can be employed to direct the vent discharge to a safe location.

3. Environmental requirements are met. If the material discharged from a vented system is toxic, corrosive, otherwise hazardous, or noxious, then venting may not be allowed.

4. Good neighbor policy. Process plants in densely populated areas may want to avoid even safe discharge of benign materials due to public relations concerns.

Deflagration Suppression

Deflagration suppression is the only one of the four explosion protection methods that terminates the combustion process rather than redirecting or containing the energy liberated by combustion. The utility of this approach is found in that it provides protection where other methods fail to meet the requirements imposed by the hazard. Venting may be inappropriate where toxic or noxious materials are concerned or where equipment cannot be placed in a safe location. Isolation methods protect adjacent process units, not the unit where ignition has occurred. Containment is applicable only where the vessel has sufficient pressure resistance.

Deflagration suppression systems have been employed since the 1950s in industrial settings. The technology has undergone continual evolution, particularly in the areas of extinguishing agent type and effectiveness, extinguisher performance, detection systems, and control panel design and function.

Theory of deflagration suppression: Deflagration suppression systems are active systems that have a characteristic function time. The principle of operation is that, once detection of ignition has occurred, extinguishing agent is delivered to and distributed within the protected volume more rapidly than the rise in demand for the agent (to achieve extinguishment) by the growing flame ball. If the total rise in pressure that occurs in a vessel during suppression, P_{red} (reduced pressure), is less than the operating pressure limits of the equipment, then the suppression will be successful.

Explosibility: Combustion of a flammable mixture in a closed vessel results in a rapid rise in pressure. (See Figure 4-14C.) Key characteristics of closed-vessel combustion are the maximum pressure and the maximum rate of pressure rise, the $(dP/dt)_{max}$, developed during the event. The principal quantitative measure of the explosibility of a combustible material is computed as

$$K_{St} = (dP/dt)_{max} V^{1/3}, \text{ bar} - \text{m/s} \qquad (9)$$

where V is the volume of the test vessel in cubic meters, and $(dP/dt)_{max}$ is the maximum rate of pressure rise, in bar/s, attained over the range of fuel-air ratios tested. (For U.S. conversions: 1 cu ft = 0.0283 m^3; 14.5 psi = 1 bar.) A typical set of data points obtained in evaluating the explosibility of a dust is illustrated in Figure 4-14D. The term "index of explosibility" is sometimes applied to K_{St}. Explosibility is dependent on chemical composition and, for dusts, physical parameters, such as moisture content, particle size, and shape. The K_{St} value of a dust has been found to be nearly invariant with $V^{1/3}$ for vessels 5.3 gal (20 L) or larger in size. For this reason, it is important that K_{St} values be determined according to an approved standard that employs a vessel of at least 5.3 gal (20 L) volume.[23] It can be shown that the K_{St} value may be used to estimate the effective burning velocity in a flammable mixture as[1,2]

$$S = 0.21 \, K_{St} \, / \, [(P_{max}/P_0)^{1/k} (P_{max} - P_0)] \qquad (10)$$

where S is the apparent burning velocity in m/s. (For U.S. conversion: 1 ft = 0.305 m.) Thus, the explosibility of a mixture can be related to a burning rate that can then be compared to the agent delivery characteristics of a suppression system.

Deflagration suppression systems: Deflagration suppression is a special case of fire extinguishment and may be viewed as a competition between a rapidly rising rate of heat release from the combustion zone and the rate of agent delivery. The deflagration will be suppressed when the unburned fuel-air mixture has been rendered noncombustible due to the addition of agent, and the flame front has been cooled to the point of extinguishment.

At the onset of ignition there is some minimum uniform concentration of agent that renders the protected space noncombustible; that is the inerting concentration. As the deflagration evolves,

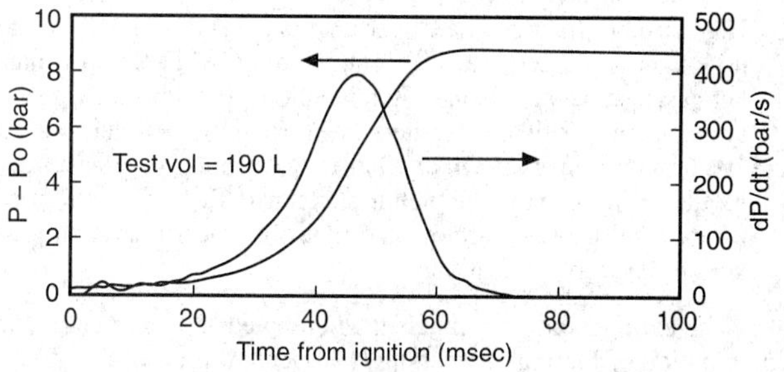

FIG. 4-14C. *Pressure and dP/dt versus time at 750 g/m^3 for closed-vessel deflagration of a proprietary corn starch product. (For U.S. conversions: 1 oz = 28.36 g; 1 cu ft = 0.0283 m^3; and 14.5 psi = 1 bar.) (Source: Fenwal Safety Systems)*

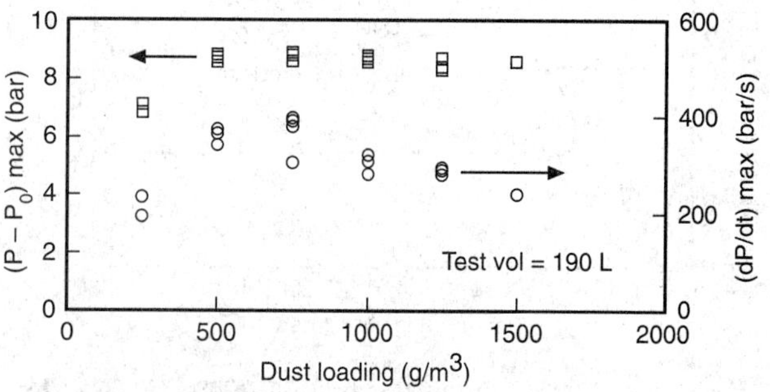

FIG. 4-14D. *Typical P_{max} and $(dP/dt)_{max}$ data obtained in determination of explosibility of a combustible dust (corn starch). (For U.S. conversions: 1 oz = 28.36 g; 1 cu ft = 0.0283 m^3; and 14.5 psi = 1 bar.) (Source: Fenwal Safety Systems)*

a proportionally larger quantity of agent is required, as the unburned material must be inerted and the burned volume must be quenched. The amount of agent that must be dispersed within the protected space to effect suppression increases with the progress of the deflagration. Should delivery of agent be delayed too long, the deflagration will pass the point of suppressibility. The result may be the attainment of normal deflagration pressures, or even higher pressures, depending on the choice of suppressant. Pressure-time traces of suppressed and unsuppressed deflagrations of a fast-burning dust are shown in Figure 4-14E.

Elements of a suppression system: A deflagration suppression system consists of components for detection, extinguishment, and control and supervision. Incipient deflagrations are detected using pressure detectors, rate-of-pressure-rise detectors, or optical flame detectors. Optical detectors, employing ultraviolet (UV) radiation sensors, are preferred in unenclosed environments with non-absorbing UV atmospheres. Examples of such environments are solvent storage and pump rooms and aerosol can filling rooms. Pressure and rate-of-pressure-rise detectors are employed in closed process equipment, and particularly where dusty atmospheres prevail.

Extinguishers are of the high-rate-discharge (HRD) type, charged with agent and propelling gas, usually nitrogen. The nitrogen overpressure is normally in the range of 300 to 900 psig (20 to 60 bar), depending on the manufacturer. Explosively opened valves, usually $1^3/_{16}$ to 5 in. (30 to 125 mm) in diameter, ensure the high rate of agent delivery necessary to effective suppression. Several types of extinguishing agents are employed, usually selected from among the following:

1. Water, with or without additives.
2. Dry chemical formulations, usually based on sodium bicarbonate or monoammonium phosphate. Other extinguishing powders have also been used.
3. Halocarbons (halogenated hydrocarbons). Halon 1301, while no longer manufactured, will continue to be available on a market basis, subject to changes in regulations, for suppression systems meeting critical-use criteria. Chlorobromomethane (Halon 1011) has been used widely, but may eventually come under regulation as an ozone-depleting substance. Research continues on the use of fluorocarbons and hydrofluorocarbons in explosion protection applications.

The extinguishing mechanism whereby each agent works is a combination of thermal quenching (100 percent in the case of water) and chemical inhibition, are beyond the scope of this chapter. The selection of agent is usually based on several considerations, such as effectiveness, toxicity, product compatibility, residual inerting, and volatility. Dry chemical agents are most commonly specified in deflagration suppression applications.

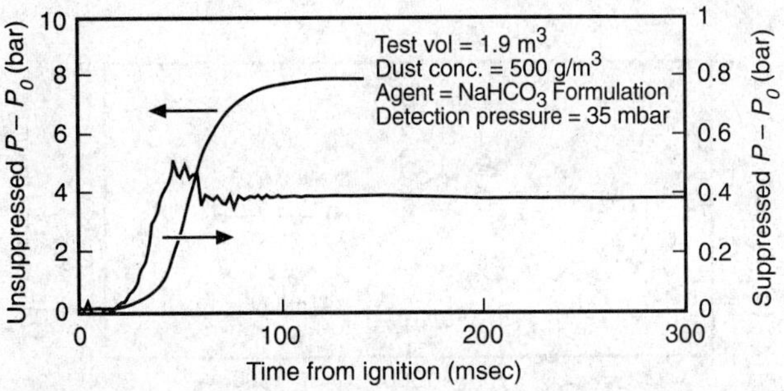

Distribution of agent within the protected volume is aided by use of nozzles. These may be fixed in position, or of the "pop-out" type which remains recessed until projected into the process volume by the momentum of the agent.

Control of a suppression system is achieved using an electronic power supply having battery backup power. This unit supervises the suppression system circuitry to ensure integrity of the system, and supplies the current to discharge explosive actuators employed on the extinguishers. Normally the process equipment being protected by the suppression system is automatically shut off upon detection of an incipient deflagration. An example of a process layout protected by a deflagration suppression system is shown in Figure 4-14F.

The protection afforded by a deflagration suppression system depends on several factors including the size and shape of the protected volume; the explosibility of the material being processed; the initial turbulence intensity; the number, size, and location of HRD extinguishers; and the means, sensitivity, and location of detection devices. The pressure-time profile of a suppressed deflagration of a combustible processed cellulosic dust is shown in Figure 4-14E.

Deflagration suppression systems are active systems that require periodic inspection and maintenance. The use of deflagration suppression systems is discussed in NFPA 69.

Prevention of Deflagration Propagation in Pipe Systems

Flammable gases and combustible dusts are commonly conveyed between process units in pipe or duct systems. For convenience, such systems will be referred to here simply as "pipes." Dust or powdered materials are frequently transported together with air in dilute or dense-phase conveying. Combustible vapors may be transported in an inerted condition, with nitrogen for example, or not inerted at all. A pipe carrying flammable mixtures can, should ignition occur at one end, act as a fuse. The ignited mixture can propagate flame through the pipe, leading to ignition of a process unit at the other end. Flame propagation, if possible, should be blocked near the point of ignition. Otherwise the speed at which a flame propagates in the pipe will steadily increase. If the mixture composition is in the detonable range the flame front can undergo deflagration to detonation transition, developing pressures of about 25 times the initial pressure and reflected pressures (that experienced by an object in the flow path) of up to 50 times the initial pressure.[24]

Prevention of flame propagation through pipe systems is an important aspect of total protection of a process plant. Several approaches are used in prevention of flame propagation in pipes. These include use of flame arresters, backflash interrupters, fast-closing valves, and chemical isolation systems.

Flame arresters: Flame arresters are devices that are constructed so as to cause the gas flow to pass through small passageways to prevent flame propagation due to thermal quenching. The maximum experimental safe gap (MESG), through which a flame cannot pass, can be determined using a device such as the Westerberg explosion test vessel.[25]

Flame arresters are normally placed at the end of a pipe line near a potential source of ignition. Special requirements are placed on flame arresters intended for use in a location that is a significant distance from the pipe end, e.g., mounted in-line. The U.S. Coast Guard has promulgated testing requirements for arresters that, when located in mid-pipe positions, may be challenged by a detonation. These requirements have been adopted in ANSI/UL 525.[26,29]

FIG. 4-14E. *Pressure-time traces of unsuppressed and suppressed deflagrations of a combustible processed cellulosic dust having a K_{St} value of 310 bar-m/s. The test was conducted in a 67 cu ft (1.9 m³) vessel. A proprietary sodium bicarbonate-based agent was used. (Source: Fenwal Safety Systems)*

Backflash interrupters: One method of preventing flame propagation in a pipe is to cause the pipe to open in such a manner as to discharge the burning material to the ambient environment while inhibiting flame propagation past the vent point. Before backflash interrupters are specified, they should be tested for the intended application.

Fast-closing valves: Positive mechanical isolation of a pipe system is another way to prevent passage of flames or pressure pulses from communicating between process units. Unlike flame arresters and backflash interrupters which are passive devices, valves are active components. They are used in conjunction with detection and control devices as part of an engineered system. One arrangement of an isolation valve system is shown in Figure 4-14G. A flame sensor is located at a sufficient distance from the valve such that, on detection of flame, the valve has adequate time to close. Fast-closing

valves have been used successfully in these applications to detect and intercept detonations in progress.[27]

Chemical isolation systems: Rather than completely blocking a pipe or duct by a valve, an alternative approach is to discharge a chemical extinguishing medium into the pipe, forming a non-flammable zone. Extinguishing compounds such as sodium bicarbonate or ammonium dihydrogen phosphate, chemicals commonly used in fire extinguishers, have been found to be very effective.[28] Halogenated hydrocarbons with low boiling points may be particularly useful, as they provide complete vapor blocks in pipes. Examples of chemicals of this type are HFC-23, HFC-125, HFC-227ea, and HFC-236fa. Halogenated agents leave no residue after discharge but may form halogen acids upon interacting with flame and hot products of combustion. These acids may be harmful to mechanical components as well as personnel. Both dry chemical and gaseous agents

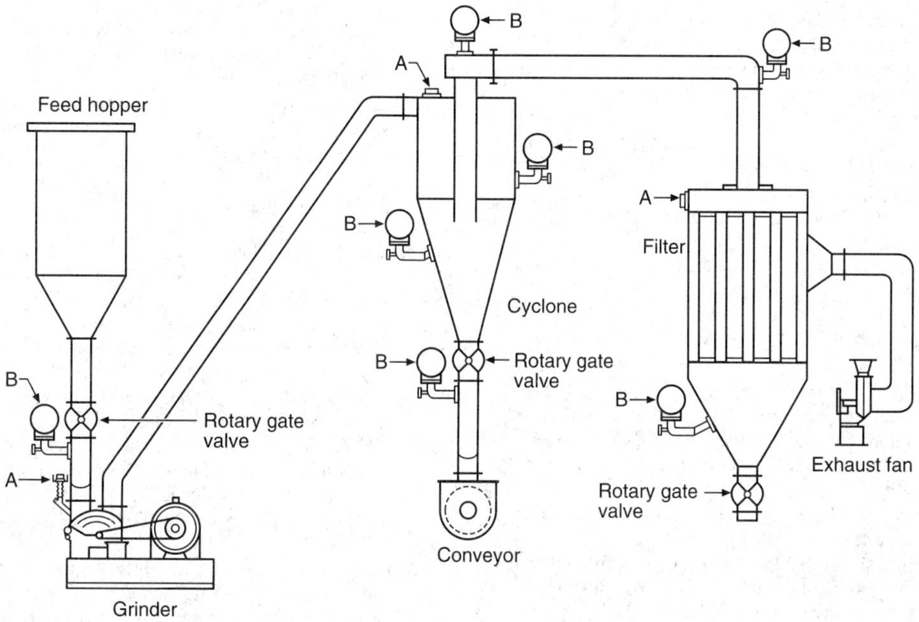

FIG. 4-14F. *Deflagration protection system for a typical grinding plant. Deflagration detectors (A) actuate high-rate-discharge extinguishers (B).*

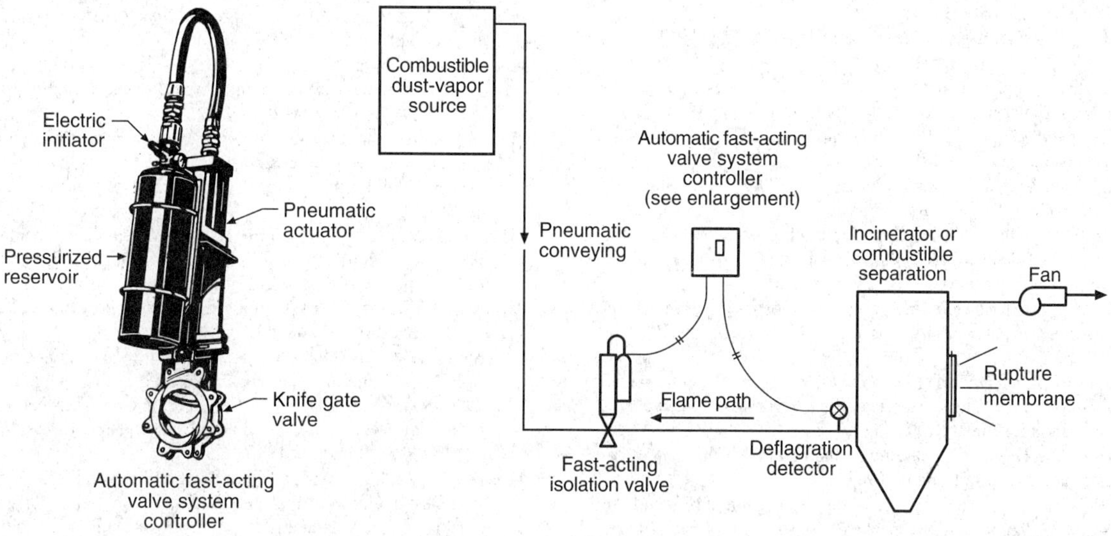

FIG. 4-14G. *Typical application of duct isolation system using a fast-acting valve.*

have been used successfully in industrial pipe and duct explosion protection systems applications. Applications where chemical barriers are used are often similar to the example shown for fast-acting valves in Figure 4-14G, where an agent bottle is used in place of a valve to create an inert chemical barrier in the duct. Chemical isolation systems must be specifically engineered for each application.

BIBLIOGRAPHY

References Cited

1. Beever, P. F., "Self-Heating and Spontaneous Combustion," *The SFPE Handbook of Fire Protection Engineering,* 2nd ed., DiNenno, P. J., *et al.* eds., National Fire Protection Association, Quincy, MA, 1995, pp. 2-180–2-189.
2. Zalosh, R. G., "Explosion Protection," *The SFPE Handbook of Fire Protection Engineering,* 2nd ed., DiNenno, P. J., *et al.* eds., National Fire Protection Association, Quincy, MA, 1995, pp. 3-312–3-329.
3. Crowl, D. A., and Louvar, J. F., *Chemical Process Safety: Fundamentals with Applications,* Prentice Hall, Englewood Cliffs, NJ, 1990.
4. Crowl, D. A., "Using Thermodynamic Availability to Determine the Energy of Explosion for Compressed Gases," *Plant/Operations Progress,* Vol. 11, No. 2, Apr. 1992, pp. 47–49.
5. Lewis, B., and von Elbe, G., *Combustion, Flames and Explosions of Gases,* 3rd ed., Academic Press, Orlando, FL, 1987.
6. *J. Phys. Chem.* Ref. Data, Vol. 14, Suppl. 1, 1985.
7. Bodurtha, F. T., *Industrial Explosion Prevention and Protection,* McGraw-Hill Book Company, New York, 1980.
8. Strelow, R. A., *Combustion Fundamentals,* McGraw-Hill Book Company, New York, 1984.
9. NFPA 2001, *Standard on Clean Agent Fire Extinguishing Systems,* National Fire Protection Association, Quincy, MA, 1994.
10. "Marine Vapor Control Systems," 33 CFR Part 154, *Federal Register,* June 21, 1990, pp. 25428–25451.
11. Bartknecht, W., *Dust Explosions,* Springer-Verlag, Berlin, 1989.
12. Bruderer, R. E., "Ignition Properties of Mechanical Sparks and Hot Surfaces in Dust/Air Mixtures," *Plant Operations Progress,* Vol. 8, No. 3, July 1989, pp. 152–164.
13. Miron, Y., and Lazzara, C., "Hot-Surface Ignition Temperatures of Dust Layers," *Fire and Materials,* 12, 1988, pp. 115–126.
14. Dorsett, H. G., *et al.*, "Laboratory Equipment and Test Procedures for Evaluating Explosibility of Dusts," RI 5624, 1960, U.S. Bureau of Mines, U.S. Department of the Interior, Washington, DC.
15. "Guidelines for the Classification and Packaging Group Assignment of Class 4 Materials," 49 CFR 173, App. E, Oct. 1, 1992, pp. 611–615.
16. ASTM E680 *Standard Test Method for Drop Weight Impact Sensitivity of Solid-Phase Hazardous Materials,* American Society for Testing and Materials, W. Conshohocken, PA, 1992.
17. "Grain-Handling Facilities, Inside Bucket Elevators," 29 CFR 1910.272(b)(2)(p)(6)(i), Apr. 1, 1991.
18. IEC, "Electrical Apparatus for Explosive Gas Atmospheres," *Publication 79-1A,* International Electrotechnical Commission, 1975.
19. Jones, T. B., and King, J. L., *Powder Handling and Electrostatics,* Lewis Publishers, Chelsea, MI, 1991.
20. Glor, M., *Electrostatic Hazards in Powder Handling,* John Wiley & Sons, New York, 1988.
21. Louvar, J. F., "Fundamentals of Static Electricity," *Proceedings* of the 28th Annual Loss Prevention Symposium, Paper 12a, American Institute of Chemical Engineers, Atlanta, GA, Apr. 1994.
22. Pratt, T. H., "Static Electricity in Pneumatic Transport Systems: Three Case Histories," *Process Safety Progress,* Vol. 13, No. 3, July 1994, pp. 109–113.
23. ASTM E1226, *Standard Test Method for Pressure and Rate of Pressure Rise for Combustible Dusts,* American Society for Testing and Materials, W. Conshohocken, PA, 1994.
24. Nettleton, M. A., *Gaseous Detonations: Their Nature, Effects, and Control,* Chapman and Hall, London, 1987.
25. Dufort, B. E., and Westerberg, W. C., "An Investigation of Fifteen Flammable Gases or Vapors with Respect to Explosion-Proof Electrical Equipment," Bulletin of Research No. 58, 1969, Underwriters Laboratories, Inc., Northbrook, IL.
26. ANSI/UL 525, *Standard for Safety Flame Arresters,* Underwriters Laboratories Inc., Northbrook, IL, 1994.
27. Senecal, J. A., and Meltzer, J. S., "Barrier Detonation Arresting Systems," American Petroleum Institute's Marine Technical-Environmental Conference, Chantilly, VA, Jan. 1992.
28. Chatrathi, K., and DeGood, R., "Explosion Isolation Systems Used in Conjunction with Explosion Vents," *Plant/Operations Progress,* Vol. 10, No. 3, July 1991, pp. 159–163.
29. "Vapor Control Systems," Appendix A to 33 CFR 154.820 *Fire, Explosion, and Detonation Protection,* Sept. 26, 1990.

References

Bonney, M. J., "Suppressing Dust Explosions," *Fire,* Vol. 88, No. 1086, Dec. 1995, p. 40.
Brenton, J. R., Thomas, G. O., and Al-Hassan, T., "Small-Scale Studies of Water Spray Dynamics During Explosion Mitigation Tests," *Proceedings* of the Institution of Chemical Engineers Symposium Series, Hazards XII: European Advances in Process Safety Conference, No. 134, 1994, Rugby, England, Institution of Chemical Engineers, Rugby, England, 1994, pp. 393–403.
Butz, J. R. and French, P., "Application of Fine Water Mists to Hydrogen Deflagrations," *Proceedings* of the 1993 Halon Alternatives Technical Working Conference, May 11–13, 1993, Albuquerque, NM, 1993, pp. 345–355.
"Confined Vented Explosions," Work Package No. BL2, British Gas Research and Technology, UK, Feb. 1991.
Cortese, R. A. and Sapko, M. J., "Flame-Powered Trigger Device for Activating Explosion Suppression Barrier," Bureau of Mines, Pittsburg, PA, RI 9365, 1991.
DiSanto, A. M., "Design Considerations for Fire Protection Systems Subjected to Explosions on Offshore Production Platforms," Worcester Polytechnic Inst., MA, Thesis, Apr. 1994.
Guidelines for Engineering Design for Process Safety, Center for Chemical Process Safety, American Institute of Chemical Engineers, New York, 1993. (See especially Ch. 11, "Sources of Ignition"; Ch. 14, "Pressure Relief Systems"; and Ch. 17, "Explosion Protection.")
"Interim Guidance Notes for the Design and Protection of Topside Structures Against Explosion and Fire," Steel Construction Inst., Berks, UK, SCI-P-112, 1991.
Isselhard, K., "Uberfuhrung der deutschen Explosionschutz-Richtlinien in eine europaische Norm-Erfahrungen und Ergebnis. [Adoption of German Explosion Protection Guidelines in a European Standard-Experience and Outcome.]," Berufsgenossenschaft der Chemischen Industrie, Heidelberg, Germany, *Stahl und eisen,* Vol. 115, No. 6, June 14, 1995, pp. 63–65.
Jones, A. and Thomas, G. O., "Mitigation of Small-Scale Hydrocarbon-Air Explosions by Water Sprays," *Transactions Institution of Chemical Engineers,* Vol. 70, No. A, 1992, pp. 197–199.
Jones, A. and Thomas, G. O., "Action of Water Sprays on Fires and Explosions: A Review of Experimental Work," *Process Safety Environment,* Vol. 71, No. 1, 1993, pp. 41–49.
Mangs, J., "Prosessiteollisuuden polyrajahdyksien paineenalentamisaukkojen mitoitus II. [Dimensioning Pressure Outlets to Avert Dust Explosions in the Processing Industry.]," *Palontorjuntatekniika,* Mar. 1992, pp. 30–31.
Mowrer, D. W., *et al.*, "Aircraft Fuel System Fire and Explosion Suppression Design Guide. Appendix C. Working Data Base. Part 1," Sept. 1986–Sept. 1989, SURVICE Engineering Co., Aberdeen, MD, Army Aviation Research and Technology Activity, Fort Eustis, VA, SURVICE-TR-89-008, Feb. 1990.
Mowrer, D. W., *et al.*, "Aircraft Fuel System Fire and Explosion Suppression Design Guide. Appendix C. Working Data Base. Part 2," Sept. 1986–Sept. 1989, SURVICE Engineering Co., Aberdeen, MD, Army Aviation Research and Technology Activity, Fort Eustis, VA, SURVICE-TR-89-008, Feb. 1990.
Mowrer, D. W., *et al.*, "Aircraft Fuel System Fire and Explosion Suppression Design Guide," Final Report, Sept. 1986–Sept. 1989, Survice Engineering Co., Aberdeen, MD, Army Aviation Research and

Technology Activity, Fort Eustic, VA, SURVICE-TR-89-008, US-AAVSCOM TR 89-D-16, JTCG/AS 89-T-005, Feb. 1990.

Senecal, J. A., "Explosion Protection in Occupied Spaces: The Status of Suppression and Inertion Using Halon and Its Descendants," 1993 International CFC and Halon Alternatives Conference, Stratospheric Ozone Protection for the 90's, October 20–22, 1993, Washington, DC, 1993, pp. 767–772.

Tamanini, F., "Explosion Suppression for Industrial Applications," Factory Mutual Research Corp., Norwood, MA, NISTIR 5766, Nov. 1995; National Institute of Standards and Technology, Solid Propellant Gas Generators: *Proceedings* of the 1995 Workshop, June 28–29, 1995, Gaithersburg, MD, 1995, pp. 208–217

Zabatakis, M. G., "Flammability Characteristics of Combustible Gases and Vapors," Bulletin 627, 1965, U.S. Bureau of Mines, Department of the Interior, Washington, DC.

NFPA Codes, Standards, and Recommended Practices

Reference to the following NFPA codes, standards, and recommended practices will provide further information on explosion prevention and protection discussed in this chapter. (See the latest version of *The NFPA Catalog* for availability of current editions of the following documents.)

NFPA 30, *Flammable and Combustible Liquids Code*
NFPA 68, *Guide for Venting of Deflagrations*
NFPA 69, *Standard on Explosion Prevention Systems*
NFPA 70, *National Electrical Code®*
NFPA 496, *Standard for Purged and Pressurized Enclosures for Electrical Equipment*
NFPA 497M, *Manual for Classification of Gases, Vapors, and Dusts for Electrical Equipment in Hazardous (Classified) Locations*
NFPA 2001, *Standard on Clean Agent Fire Extinguishing Systems*

DUSTS

Revised by Richard F. Schwab

Most finely divided combustible materials are hazardous. When suspended in air and ignited, they can cause severe explosions. This phenomenon has been known for over 200 years. The first recorded dust explosion occurred on December 14, 1785: a flour dust explosion in a warehouse in Turin, Italy. The entire industrial spectrum, including agricultural, chemical, metallurgical, mining, plastics, and woodworking industries, continues to be plagued by this problem.[1-4]

Although the basic principles for controlling dust explosions have been understood for many years, knowledge is becoming increasingly sophisticated as incidents continue to occur.[2, 4-7]

FACTORS INFLUENCING THE EXPLOSIBILITY OF DUSTS

The chance of a dust cloud igniting is governed by the size of its particles, dust concentration, impurities present, oxygen concentration, and the strength of the source of ignition.

Dust explosions usually occur as a series. Frequently, the initial deflagration is rather small in volume but intense enough to jar dust from beams, ledges, etc., or even rupture small pieces of equipment within buildings, such as dust collectors or bins. Subsequently, a much larger dust cloud develops, through which a secondary explosion can propagate. It is not unusual to have a series of explosions propagating from building to building.[7]

Hazards of Dusts

The hazard of any given dust is related to its ease of ignition and the severity of the ensuing explosion. The Bureau of Mines of the U.S. Department of the Interior has developed an arbitrary scale based on small-scale tests, which is quite useful for measuring the hazard. The ignition sensitivity (IS) is a function of the ignition temperature (of the dust cloud) and the minimum energy of ignition; the explosion severity (ES) is a function of the maximum explosion pressure and the maximum rate of pressure rise. To facilitate comparisons of explosibility data in U.S. Bureau of Mines tests, all test results[8] are related to a standard Pittsburgh coal dust taken at a concentration of 0.50 oz per cu ft (500 g/m^3), with the exception of some metal dusts.[9]

Richard F. Schwab, P.E., S.F.P.E., AIChE, is manager of process safety and loss prevention, AlliedSignal, Inc., Morristown, NJ, and serves as vice chairman of the NFPA Technical Committee on Explosion Protection Systems.

The ignition sensitivity and explosion severity of a dust are defined as:

$$IS = \frac{(T_i \times E_m \times C_m) \text{ Pittsburgh coal dust}}{(T_i \times E_m \times C_m) \text{ Sample dust}}$$

$$ES = \frac{(P_{max} \times R_{max}) \text{ Sample dust}}{(P_{max} \times R_{max}) \text{ Pittsburgh dust}}$$

where

T_i = ignition temperature of a dust cloud,

E_m = minimum ignition energy of a dust cloud,

C_m = minimum concentration for a combustible dust cloud,

P_{max} = maximum explosion pressure, and

R_{max} = maximum rate of pressure rise in the test apparatus.

The explosibility index (EI) is the product of the ignition sensitivity and explosion severity. This method allows one to rate the relative hazard of the dusts, as follows:

Type of Explosion	Ignition Sensitivity	Explosion Severity	Explosibility Index
Weak	<0.2	<0.5	<0.1
Moderate	0.2–1.0	0.5–1.0	0.1–1.0
Strong	1.0–5.0	1.0–2.0	1.0–10
Severe	>5.0	>2.0	>10

The concept of using this scale to measure the hazard of a combustible dust has in recent years been found to be outdated though it is still widely used.[10]

Table 4-15A gives the explosibility index, ignition sensitivity, explosion severity, maximum explosion pressure, maximum rate of pressure rise [not necessarily at the concentration of 0.50 oz per cu ft (500 g/m^3)], ignition temperature of both the dust cloud and a layer, the minimum ignition energy of the dust cloud, minimum explosion concentration, and the limiting oxygen concentration in a spark-ignition chamber.

TABLE 4-15A. *Explosion Characteristics of Various Dusts*

(Compiled from the following reports of the U.S. Department of Interior, Bureau of Mines: RI 5753, "The Explosibility of Agricultural Dusts"; RI 6516, "Explosibility of Metal Powders"; RI 5971, "Explosibility of Dusts Used in the Plastics Industry"; RI 6597, "Explosibility of Carbonaceous Dusts"; RI 7132, "Dust Explosibility of Chemicals, Drugs, Dyes, and Pesticides"; and RI 7208, "Explosibility of Miscellaneous Dusts.")

Type of Dust	Explosibility Index	Ignition Sensitivity	Explosion Severity	Maximum Explosion Pressure psig*	Max Rate of Pressure Rise psi/sec*	Ignition Temperature† Cloud °C	Ignition Temperature† Layer °C	Min Cloud Ignition Energy joules	Min Explosion Conc oz/cu ft‡	Limiting Oxygen Percentage§ (Spark Ignition)
Agricultural Dusts										
Cellulose, alpha	>10	2.7	4.0	117	8,000	410	300	0.040	0.045	—
Cornstarch commercial product	9.5	2.8	3.4	106	7,500	400	—	0.040	0.045	—
Lycopodium	16.4	4.2	3.9	75	3,100	480	310	0.040	0.025	C13
Wheat starch, edible	17.7	5.2	3.4	100	6,500	430	—	0.025	0.045	C12
Wood flour, white pine	9.9	3.1	3.2	113	5,500	470	260	0.040	0.035	—
Carbonaceous Dusts										
Charcoal, hardwood mixture	1.3	1.4	0.9	83	1,300	530	180	0.020	0.140	—
Coal, Pennsylvania, Pittsburgh (Experimental Mine Coal)	1.0	1.0	1.0	90	2,300	610	170	0.060	0.055	—
Chemicals										
Adipic acid	1.9	1.7	1.1	84	2,700	550	—	0.060	0.035	—
Bis-phenol A	>10	11.8	2.8	89	8,500	570	—	0.015	0.020	C12
Phthalic anhydride	6.9	11.9	1.4	72	4,200	650	—	0.015	0.015	C14
Stearic acid, aluminum salt (aluminum tristearate)	>10	21.3	1.9	87	6,300	420	440	0.015	0.015	—
Sulfur	>10	20.4	1.2	78	4,700	190	220	0.015	0.035	C12
Drugs										
Aspirin (acetylsalicylic acid), $C_9H_8O_4$	>10	2.4	4.3	88	>10,000	660	Melts	0.025	0.050	—
Vitamin C, ascorbic acid, $C_6H_8O_6$	2.2	1.0	2.2	88	4,800	460	280	0.060	0.070	C15, N12
Metals										
Aluminum flake, A 422 extra fine lining, polished	>10	7.3	10.2	127	20,000+	610	326	0.010	0.045	—
Magnesium, milled, Grade B	>10	3.0	7.4	116	15,000	560	430	0.040	0.030	—
Alloys and Compounds										
Aluminum-cobalt alloy (60—40)	0.4	0.1	3.5	92	11,000	950	570	0.100	0.180	—
Thermoplastic Resins and Molding Compounds										
Acrylamide polymer	2.5	4.1	0.6	85	2,500	410	240	0.030	0.040	—
Methyl methacrylate polymer	6.3	7.0	0.9	84	2,000	480	—	0.020	0.030	C11
Cellulose acetate	>10	8.0	1.6	85	3,600	420	—	0.015	0.040	C14
Polycarbonate	8.6	4.5	1.9	96	4,700	710	—	0.025	0.025	C15
Phenol formaldehyde molding compound, wood flour filler	>10	8.9	4.7	94	9,500	500	—	0.015	0.030	C14

*1 psi = 6.894 kPa.
† °F = 9/5 (°C + 32).
‡ 0.1 oz/cu ft = 100 g/m³.
§ Numbers in this column indicate percentage while the letter prefix indicates the diluent gas. For example, the entry "C13" means dilution to an oxygen content of 13 percent, with carbon dioxide as the diluent gas. The letter prefixes are: C = carbon dioxide, and N = nitrogen.
— No values listed.

Particle Size

The smaller the size of the dust particle, the easier it is to ignite the dust cloud, insofar as the exposed surface area of a unit weight of material increases as the particle size decreases. It is also true that particle size has an effect on the rate of pressure rise. For a given weight concentration of dust, a coarse dust will show a lower rate of pressure rise than a fine dust. The lower explosive limit (LEL) concentration, ignition temperature, and the energy necessary for ignition will decrease as the dust particle size decreases. Numerous studies by various investigators show this effect for a variety of dusts.[11]

Decrease in particle size also increases the capacitance of dust clouds, i.e., the size of electrical charges that can accumulate on particles in the cloud.[12,13] Because capacitance of solids is a function of surface area, the possibility of developing electrostatic charges of sufficient intensity to ignite a dust cloud increases as average particle size decreases. However, to produce such electrostatic charges requires, among other things, large quantities of dust in large volumes, with relatively high dielectric strengths and consequent long relaxation times. Because of the high ignition energies required for dust clouds to ignite in comparison with ignition energies of gases, attributing the cause of dust explosions to static electricity should be held suspect unless definite evidence exists to show that static electricity was a likely cause.[3,7]

Concentration

As with flammable gases and vapors, there is a specific range of dust concentration within which a dust explosion can occur. It is customary to express the concentration figures in terms of weight per unit volume, though, without knowledge of the particle size distribution of the sample, this expression is meaningless. The results presented in Table 4-15A were developed by the U.S. Bureau of Mines and were obtained with dusts small enough to pass through a 200 mesh screen (74 microns or smaller). In data published in Germany,[14,15] based on the data shown in Table 4-15A, the median particle size was shown for every test. These German tables were condensed from far more extensive tables,[14] which also included the particle size distribution for most of the reported tests.

Variations in minimum explosive concentration will occur with changes in particle diameter, i.e., the minimum explosive concentration is lowered as the diameter of particles decreases. Sample purity, oxygen concentration, strength of ignition source, turbulence of dust cloud, and uniformity of dispersion also affect the lower explosive limits (LEL) of dust clouds.

Upper explosive limits (UEL) for dust clouds have usually not been determined, mainly because of experimental difficulties. There is also a question of whether a clear-cut upper limit exists at all; from a practical point of view, this information is of questionable use.

Curves formed by plotting explosion pressures and rates of pressure rise against concentration show that explosion pressures and rates of pressure rise are at a minimum at the lower explosive limit, then rise to a maximum value at a given optimum concentration, then slowly decrease from this point. It is to be noted that the maximum pressure and the maximum rate of pressure rise do not always occur at precisely the same concentration. The destructive effect is determined primarily by the rate of pressure rise.

It appears that the most violent explosion occurs at a concentration slightly above that required for reaction with all of the oxygen in the atmosphere. At lower dust concentrations, less heat is generated and smaller peak pressures are developed. With dust concentrations greater than those causing the most violent explosions, absorption of heat by unburned dust is apparently the reason for less-than-maximum explosive pressures.

Moisture

Moisture in dust particles raises the ignition temperature of the dust because of the heat absorbed during heating and vaporization of the moisture. The moisture in the air surrounding a dust particle has no significant effect on the course of a deflagration once ignition has occurred. There is, however, a direct relationship between moisture content and minimum energy required for ignition, minimum explosive concentration, maximum pressure, and maximum rate of pressure rise. For example, the ignition temperature of cornstarch may increase as much as 122°F (50°C), with an increase of moisture content from 1.6 percent to 12.5 percent. As a practical matter, however, moisture cannot be considered an effective explosion preventive, since most ignition sources provide more than enough heat to vaporize the moisture and to ignite the dust. In order for moisture to prevent ignition of a dust by common sources, the dust would have to be so damp that a cloud could not be formed.

Inert Material

The presence of an inert solid powder reduces the combustibility of a dust because it absorbs heat, but the amount of inert powder necessary to prevent an explosion is usually higher than concentrations that would normally be found or could be tolerated as foreign material. The addition of inert material reduces the rate of pressure rise and increases the minimal dust concentration. Rock dusting of coal mines is a practical application of the use of inert dust to prevent explosion of combustible dust. For general rock dusting of coal mine entries, enough rock dust is usually added to provide an inert dust concentration of at least 65 percent of the total dust.[7]

Inert gas is effective in preventing dust explosions because it dilutes the oxygen to a concentration too low to support combustion. In selecting a suitable inert gas, care must be taken to choose one that is not reactive with the dust. Certain metallic dusts, for example, react with carbon dioxide or nitrogen.[9] Helium and argon are suitable diluents in such instances.

Oxygen Concentration, Turbulence, and Effect of Flammable Gas

Variations in oxygen concentrations affect the ease of ignition of dust clouds and the explosion pressures. With a decrease in the partial pressure of oxygen, the energy required for ignition increases, ignition temperature increases, and maximum explosion pressures decrease. The type of inert gas used as the diluent for reduction of the oxygen concentration also has an effect apparently related to molar heat capacity.

The combustion of dust takes place at the surfaces of the dust particles. The rate of reaction, therefore, depends on intimate mixing of the dust and oxygen. It is for this reason that turbulent mixing of dust and air results in more violent explosions than those obtained by ignition in more relatively quiescent mixtures.[1, 16, 17] The data presented in Table 3-12A were gathered by igniting the dust under violently turbulent conditions; thus, they cannot be compared to explosion data for flammable gases which were gathered under comparatively quiescent conditions.

Addition of a small amount of flammable gas to a dust cloud and igniting the resultant aerosol greatly increases the violence of the explosion, particularly at lower dust concentrations. The resultant rates of pressure rise are very much higher than usually anticipated. Without the dust, the remaining fraction of the total combustible in air, represented by the flammable vapor, would be in itself below the lower flammable limit (LFL). In certain drying operations, involving the evaporation of a flammable vapor from a combustible powder with entrainment of the combustible dust along with the flammable vapor, explosions occurred that were far more violent than anticipated when only the fraction represented by the flammable vapor alone was considered.[1, 3, 4, 18] Indeed, explosions occurred in such flammable vapor-combustible dust-air mixtures when the flammable vapor-air portion was below the LFL. When encountered, these situations need special safeguards, such as inert gas dilution, explosion suppression, very large explosion vents, and careful static electricity elimination designs.

DUST CLOUD IGNITION SOURCES

Dust clouds have been ignited by open flames, lights, smoking materials, electric arcs, hot filaments of light bulbs, friction sparks, high-pressure steam pipes and other hot surfaces, spontaneous heating, welding and cutting torches and sparks from these operations, and other common sources of heat for ignition. The dust cloud ignition temperatures given in Table 4-15A for the most part fall between 572 and 1,112°F (300 and 600°C), and a large majority of the reported minimum spark ignition energies were between 10 and 40 mJ. This can be contrasted to flammable vapor ignition energies with a range for the most part between 0.2 and 10 mJ. As a general rule, combustible dusts require 20 to 30 times the ignition energies of flammable vapors.[3]

Because ignition temperatures and ignition energies for dust explosions are much lower than the temperatures and energies of most common sources of ignition, it is not surprising that dust explosions have been caused by common sources of ignition. For this reason, the elimination of all possible sources of ignition is a basic principle of dust explosion prevention. Ignition sources in dust-handling operations are identified and recommendations for their elimination are described in various NFPA standards for the prevention of dust explosions. (See bibliography.)

DEFLAGRATION CHARACTERISTICS OF SELECTED COMBUSTIBLE DUSTS

Table 4-15B is based on information obtained from *Forschungsbericht Staubexplosionen: Brenn und Explosions Kenngrossen von Stauben*[14] and *Brenn und Explosions Kenngrossen von Stauben.*[15] For each dust, the tables show the median particle size of the material tested, as well as the following test results obtained in a 1 m^3 vessel: minimum explosible concentration, maximum pressure developed by the explosion (P_{max}), and the maximum rate of pressure rise $(dP/dt)_{max}$. Also shown are the K_{St} value, which is the equivalent to $(dP/dt)_{max}$ because of the size of the test vessel; dust hazard class; cloud and layer ignition temperatures; combustibility rating; and test number.

TABLE 4-15B. *Agricultural Products*

Material	Median Particle Size, μ	Minimum Explosive Concentration g/m²	$P_{max,}$ bar ga	$(dP/dt)_{max,}$ bar/sec	K_{St} bar-m sec	Dust Hazard Class
Cellulose	33	60	9.7	229	229	2
Cellulose,						1
pulp	42	30	9.9	62	62	
Cork	42	30	9.6	202	202	2
Corn	28	60	9.4	75	75	1
Egg white	17	125	8.3	38	38	1
Milk,						1
powdered	83	60	5.8	28	28	
Milk, non-						1
fat, dry	60	—	8.8	125	125	
Soy flour	20	200	9.2	110	110	1
Starch, corn	7	—	10.3	202	202	2
Starch, rice	18	60	9.2	101	101	1
Starch,						1
wheat	22	30	9.9	115	115	
Sugar	30	200	8.5	138	138	1
Sugar, milk	27	60	8.5	82	82	1
Sugar, beet	29	60	8.2	59	59	1
Tapioca	22	125	9.4	62	62	1
Whey	41	125	9.8	140	140	1
Wood flour	29	—	10.5	205	205	2

FACTORS INFLUENCING THE DESTRUCTIVENESS OF DUST EXPLOSIONS

While the destructiveness of a dust explosion depends primarily upon the rate of pressure rise, other contributing factors are: the maximum pressure developed, the duration of the excess pressure, the degree of confinement of the explosion (deflagration) volume, and the oxygen concentration.

Effect of Rate of Pressure Rise

The rate of pressure rise can be defined as the ratio of the increase in explosion pressure to the time interval of that increase. It is the most important single factor in evaluating the hazard of a dust and principally determines the destructiveness of a deflagration.

The rate of pressure rise is also an important consideration in the design of explosion (deflagration) vents, since it largely determines the size of the vent. In many cases, an extremely rapid rate of pressure rise indicates that an explosion vent design is either impractical or, at best, extremely difficult. In those cases, alternate protective schemes, such as use of inert gases, explosion suppression, or containment, should be considered. (See NFPA 69, *Standard on Explosion Prevention Systems.*)

Effect of Maximum Explosion Pressure

The maximum explosion pressures reported by recognized test procedures are, for the most part, in excess of 50 psig (345 kPa) and, sometimes, in excess of 100 psig (690 kPa). These figures are valid only for test conditions and are subject to changes by variation in particle size, concentration, and other variables. Considering that an ordinary brick wall 12 in. (0.3 m) thick can be destroyed by less than 1 psi (6.89 kPa) pressure, it is evident that it is not practical to construct a building strong enough to resist the maximum pressures resulting from a dust deflagration.

One reason why the degree of destruction is not greater in many dust explosions is that the dust is not uniformly dispersed throughout the explosion volume. A dust cloud is rarely ignited under optimum concentration conditions, with respect to the development of maximum explosion pressures.

Effect of Duration of Excess Pressure

Closely associated with maximum pressure and rate of pressure rise as indications of the destructiveness of dust explosions is the length of time the excess pressure is exerted on the surroundings. The area under the time-pressure curve determines the total impulse exerted. It is this total impulse rather than the force exerted at any moment that will determine the amount of destruction. The relationship between destructiveness and total impulse explains why dust explosions, which generally have slower average rates of pressure rise than gas explosions, may be more destructive than gas explosions.

Effect of Confinement

When a dust explosion occurs, gaseous products are formed and heat is released, which raises the temperature of the air in the enclosure. Since gases expand when heated, destructive pressures will be exerted on the surrounding enclosure, unless enough vent area is provided to release the gas pressure before dangerous pressures are reached.

Bartknecht,[1,19] Nagy and Verakis,[20] Field,[21] Palmer,[7] Cross and Farrer,[12] and Hartmann *et al.*[11] have examined the effect in some detail. The work of Bartknecht and his associates is generally considered to be the most comprehensive, leading to basic design methods for deflagration vents, and is the basis for the procedures shown in NFPA 68, *Guide for Venting of Deflagrations* (hereinafter referred to as NFPA 68). Generally his procedures relate to vessels or enclosures that are capable of withstanding 1.5 psig (10.4 kPa) or more. The work of Swift and Epstein[22] is the basis of the design procedures for low-strength structures [weaker than 1.5 psig (10.4 kPa)] in NFPA 68.

In some instances, it may be impractical to provide sufficient vent area to reduce pressures to a safe level. In these situations, the dust-producing operation can be conducted in the open, under an inert atmosphere, or protected by an explosion suppression system. The explosion suppression system consists of a flame or pressure detector and a flame-quenching agent that is expelled rapidly during the incipient stage of a deflagration.

Effect of Inerting on Explosion Pressures and Rates of Pressure Rise

Data published by the U.S. Bureau of Mines[23] and Bartknecht[19] show that reduction in the oxygen concentration in the atmosphere, and mixture of inert powder (Fuller's earth, or rock dust) or moisture with the combustible dust, reduces the maximum explosion pressure and rates of pressure rise. These data are shown in Figures 4-15A and 4-15B for explosions of 0.50 oz per cu ft (500 g/m^3) of cornstarch. A slight reduction in the atmospheric oxygen concentration or slight additions of inert powder or moisture have little effect on the explosion pressure. The effect of the inerting on the dust explosion becomes marked only when limiting values are approached. The maximum rate of pressure rise developed by the explosion (deflagration) appears to decrease almost linearly with use of inerting.

DUST EXPLOSION TEST APPARATUS AND PROCEDURES

There are a number of different types of dust explosion test chambers, and even more procedures for using them, each of which can profoundly affect the test results for any given dust sample. For example, the result of the dust explosion test in a closed chamber is affected by the geometry of the chamber, i.e., interior dimensions (cubical, spherical, cylindrical, etc.), and by the method of creating the dust cloud in the chamber (the method affects the turbulence of the cloud and uniformity of particle dispersion). This situation causes considerable difficulty in attempting to compare results obtained by various methods. In fact, unless several test results on the same dust sample are made in different chambers, and a consequent correlation is developed between the test methods, it can be said without reservation that any comparison of data obtained by different test methods is, at best, only qualitative. For example, attempts to correlate test results obtained in the Hartmann bomb with the spherical bomb developed by Bartknecht have been unsuccessful.

Lack of uniformity in test procedures also rules out any attempt to compare the pressures and rates of pressure rise resulting from tests on flammable vapors and dusts. By far, the largest amount of dust explosion test data has been obtained from tests conducted by the U.S. Bureau of Mines. The U.S. Bureau of Mines test equipment

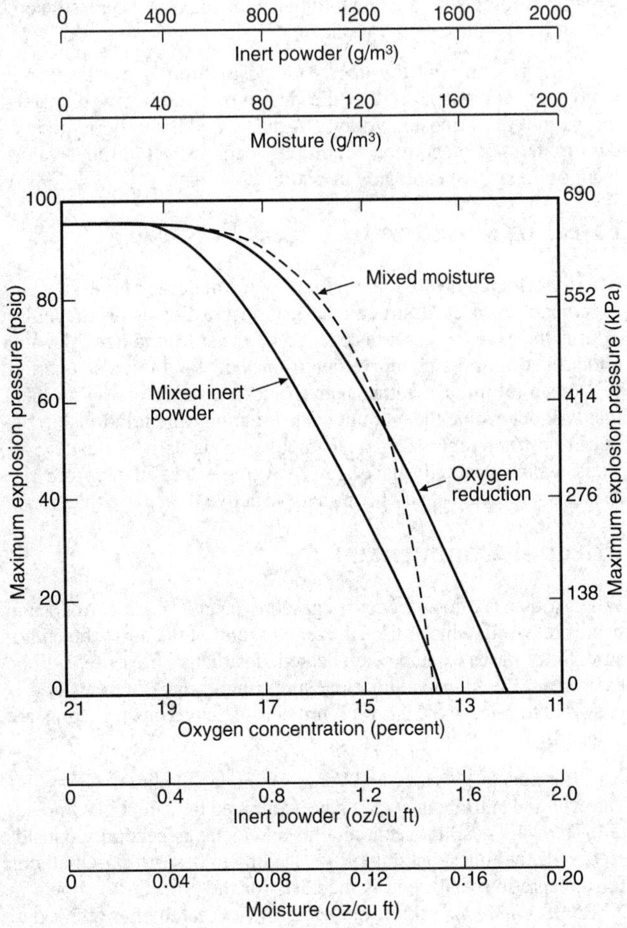

FIG. 4-15A. Effect of inerting on maximum pressure developed by explosion of 0.50 oz per cu ft (500 g/m^3) of cornstarch.

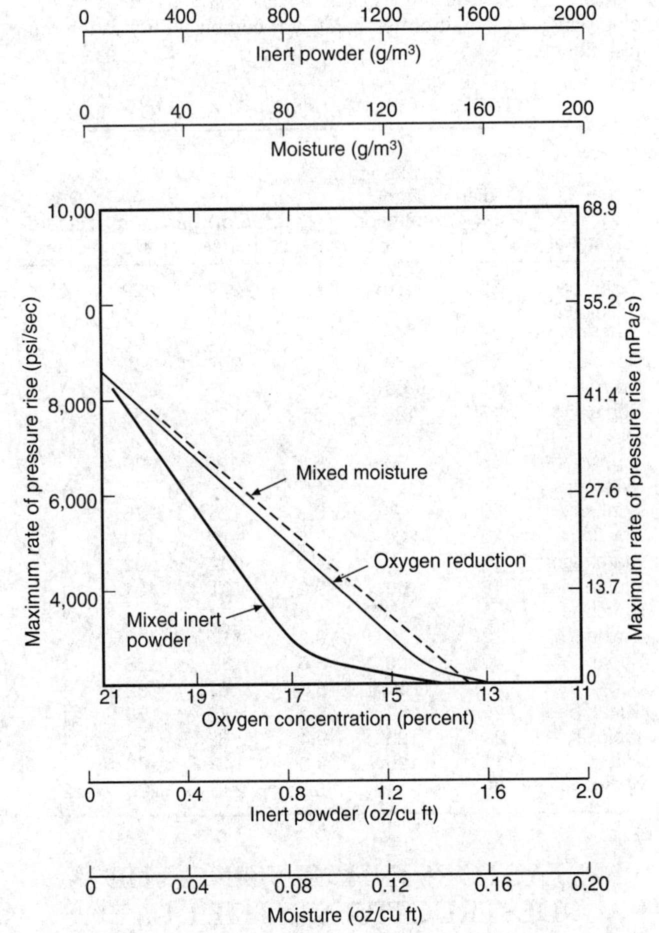

FIG. 4-15B. Effect of inerting on maximum rate of pressure rise developed by explosion of 0.50 oz per cu ft (500 g/m^3) of cornstarch.

and procedures are described in RI 5624.[24] The test equipment used by the U.S. Bureau of Mines is a cylindrical vessel of about 75 cu in. (1.2 L), standing about 13 in. (0.33 m) high, referred to as the Hartmann apparatus.[25] (See Figure 4-15C.) All of the data recorded in Table 4-15A have been collected in this type of test apparatus. This type of data is not satisfactory for designing explosion (deflagration) relief vents.

Recent data, particularly that intended for use in obtaining test data for use in designing explosion (deflagration) vents, were collected in spherical vessels. Bartknecht and his associates have developed a 20-L (about 5-gal) spherical bomb.[26, 27] (See Figure 4-15D.) Data collected in these containers tend to give more consistent results and follow the "cubic law" (described below). Data collected in a spherical bomb, properly calibrated, have been shown to generally follow the "cubic law" over a range from 20 L to 250 m³. This is the basis for the nomographs contained in NFPA 68.

Interpretation and Application of Dust Explosion (Deflagration) Test Data

Dust explosion test data are gathered in a variety of chambers, from a highly turbulent dust cloud to a practically stagnant cloud. Most modern test data are gathered in calibrated spherical test vessels that follow the "cubic law," which is represented for dusts as follows:

$$(dP/dt)_{max} \times (V)^{1/3} = K_{St}$$

where

P = pressure (bar),

t = time (sec),

V = volume (m³), and

K_{St} = normalized index (bar m/s).

This can also be expressed by the following relationship:

$$R_l/R_s = (V_s/V_l)^{1/3} = A_l/A_s$$

where

R_l = rate of pressure rise in the large chamber,

R_s = rate of pressure rise in the small chamber,

V_l = volume of the large chamber,

V_s = volume of the small chamber,

A_l = vent ratio of the large chamber, and

A_s = vent ratio of the small chamber.

(All variables must be in consistent units.)

This law indicates that the test data gathered in small chambers can be extrapolated to larger chambers with some degree of confidence, though care must be taken to keep the shape of small chambers and large chambers within reasonable limits. If the test is done in small spheres, the results cannot be expected to hold up in the design of explosion (deflagration) relief for a long narrow vessel.

Ignition Temperature

One piece of apparatus developed at the U.S. Bureau of Mines consists essentially of an electrically heated vertical cylindrical tube at the top of which is a small container holding a weighed amount of dust. The dust is projected down through the tube as a uniform cloud by a controlled blast of compressed air directed at the dust sample. The ignition temperature is considered to be the lowest temperature at which flame issues from the open bottom end of the tube. Differences in results are due to differences in the size and shape of test containers.[24] This test is used to determine the ignition temperature of the dust cloud.

The furnace is also used to determine the ignition temperature of a dust layer sample, 1/2 in. (12.5 mm) deep, 1 in. (25.4 mm) diameter, in a container suspended in the furnace heated by a stream of hot air flowing at 30 cu in. per min (1/2 L/min) past the sample.[24] Other recognized testing agencies use hotplates with the dust layers

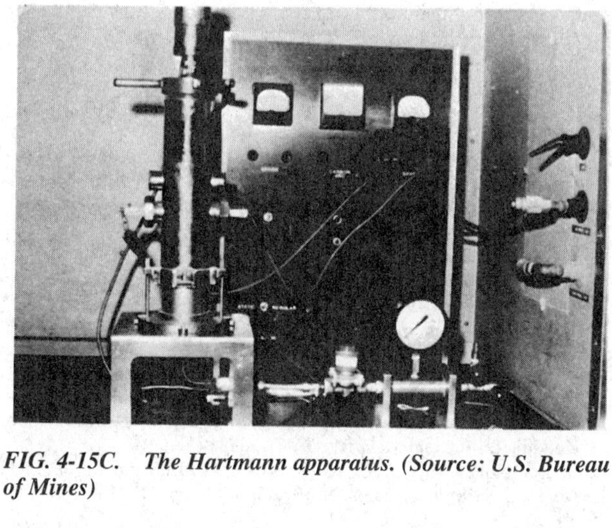

FIG. 4-15C. The Hartmann apparatus. (Source: U.S. Bureau of Mines)

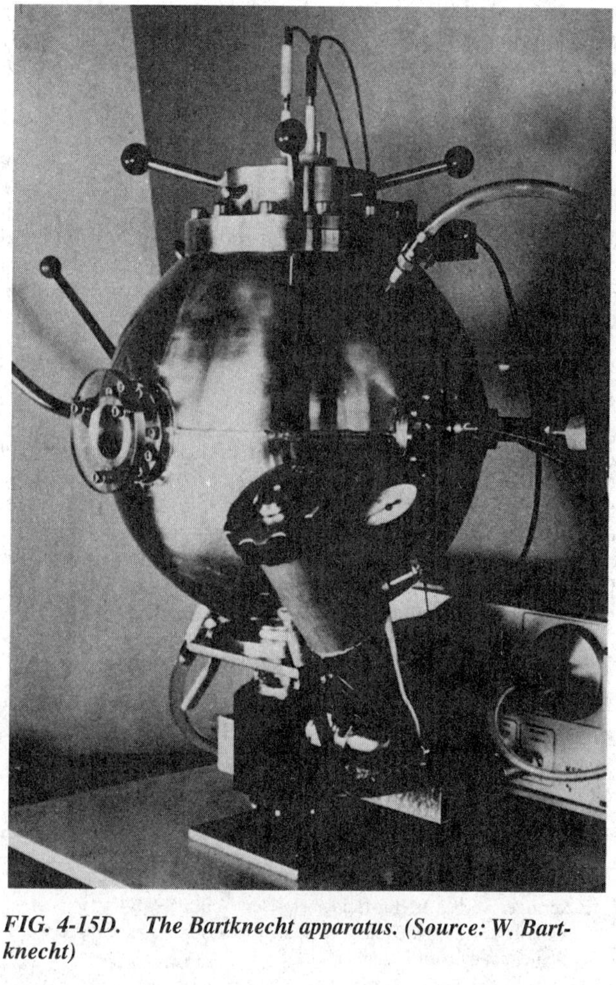

FIG. 4-15D. The Bartknecht apparatus. (Source: W. Bartknecht)

$3/_4$ in. (19 mm) deep.[14,15] Other researchers have reported test data using different test procedures[28,29] designed to investigate ignition sources encountered under industrial conditions.

Ignition in Inert Atmospheres

For the determination of the ignitability of dust in various inert gas-air mixtures, the apparatus consists of a vertical cylindrical tube containing tungsten electrodes to produce a spark for ignition. The tube is designed so that it can be flushed with the inert gas-air mixture prior to the test. The dust sample in a container near the top of the tube is projected downward as a cloud by a controlled blast of the inert gas-air mixture. Flame issuing from the open bottom of the tube signifies ignition.[28,29] Bruderer[28] has also investigated limiting oxygen concentrations for various ignition sources in his work.

Minimum Energies for Ignition

Apparatus developed at the U.S. Bureau of Mines, used to determine minimum energies needed to cause ignition of dust clouds, consists of a vertical cylindrical lucite tube. In this test, the dust is dispersed upward in the tube by controlled discharge of compressed air into the dust chamber at the bottom of the tube. A spark discharged by an electrical condenser is used as the ignition source, and by using condensers with different capacities, spark energy can be varied. Discharge of the condenser is synchronized with the formation of the dust cloud. Visible flame in the tube signifies ignition.

The data presented in Table 4-15A for cloud ignition energies has been gathered in this fashion. Recent work [24,30,31] has shown that, if the spark is generated in a different fashion with a high resistance in series with the capacitance discharge, it is possible to obtain ignition of the dust cloud considerably below that reported by the U.S. Bureau of Mines technique. Generally speaking, the data reported by the U.S. Bureau of Mines might be expected to be a factor of 3 to 5 times too high.[12]

Eckhoff[30] has managed to ignite some of the more sensitive dusts, such as aluminum ($5\mu m$) and sulfur, at energies of 1 mJ or less. The U.S. Bureau of Mines quotes 15 mJ for these materials. It is questionable whether a circuit that produces such a low-level spark of the nature required would occur naturally. Ignitability is not solely a function of spark energy but also a function of spark duration, with long-duration sparks igniting dusts more easily than short-duration sparks.[12]

Explosion Pressures

Apparatus described by the U.S. Bureau of Mines[24] for determining maximum explosion pressures and rates of pressure rise consists of a vertical cylindrical steel bomb into which a weighed sample of dust can be dispersed upward by a jet of compressed air. Pressures are measured by a strain gage transducer mounted in the top of the bomb. A pressure-time record is obtained on an oscillograph using a light beam galvanometer system.

BIBLIOGRAPHY

References Cited

1. Bartknecht, W., *Dust Explosions,* Springer-Verlag, New York, 1989.
2. Cashdollar, K. L., and Hertzberg, M., "Industrial Dust Explosions," ASTM Special Publication 958, 1987, American Society for Testing and Materials, W. Conshohocken, PA.
3. Bartknecht, W., *Explosion-Schutz: Grundlagen und Anwendung,* Springer-Verlag, Berlin-Heidelberg-New York, 1993.
4. Eckhoff, R. K., *Dust Explosions in the Process Industries,* Butterworth-Heinemann, Oxford, England, 1991.
5. van Laar, G. F. M., "Review of Incidents," European Information Centre for Explosion Protection (EuropEx), Antwerp, Belgium, 1989.
6. Best, R., "Two Grain Elevator Explosions Kill Five in Missouri," *Fire Journal,* Vol. 72, 1978, pp. 50–54.
7. Palmer, K. N., *Dust Explosions & Fires,* Halsted Press, Division of John Wiley & Sons, Inc., New York, 1973.
8. Jacobson, M., *et al.,* "Explosibility of Agricultural Dusts," RI 5753, 1961, U.S. Bureau of Mines, Pittsburgh, PA.
9. Jacobson, M., *et al.,* "Explosibility of Metal Powders," RI 6516, 1964, U.S. Bureau of Mines, Pittsburgh, PA.
10. Hertzberg, M., "A Critique of the Dust Explosibility Index: An Alternative for Estimating Explosion Probabilities," RI 9095, 1987, U.S. Bureau of Mines, Pittsburgh, PA.
11. Hartmann, I., *et al.,* "Recent Studies on the Explosibility of Cornstarch," RI 4725, 1964, U.S. Bureau of Mines, Pittsburgh, PA.
12. Cross, J., and Farrer, D., *Dust Explosions,* Plenum Press, New York, 1982.
13. Kunkel, W. G., "The State of Electrification of Dust Particles on Dispersion into a Cloud," *Journal of Applied Physics,* Vol. 21, 1950, pp. 820–832.
14. *Forschungsbericht Staubexplosionen: Brenn und Explosions Kenngrössen von Stauben,* Hauptverband der gewerblichen Berufsgenossenschaften e. V., Bonn, West Germany, 1980.
15. *Brenn- und Explosions-Kenngrössen von Stäuben,* BIA Berufsgenossenschaftaftiches Institute fur Arbeitssicherheit BIA Handbuch 7, Lfg. VI/87, Bonn, Germany, 1987.
16. FIA, *Dust Explosions Analysis and Control,* Factory Insurance Association (now Industrial Risk Insurers), Hartford, CT, 1966.
17. Amyotte, P. R., Chippett, S., and Pego, M. J., "Effects of Turbulence on Dust Explosions," *Prog. Energy Combust. Sci.,* Vol. 14, 1989, pp. 293–310.
18. Cotton, P. E., "New Test Apparatus for Dust Explosions," *NFPA Quarterly,* Vol. 45, No. 2, 1951, pp. 157–164.
19. Bartknecht, W., *Explosions: Course, Prevention, Protection,* Springer-Verlag, Berlin-Heidelberg-New York, 1981.
20. Nagy, J., and Verakis, H. C., *Development and Control of Dust Explosions,* Marcel Dekker, Inc., New York, 1983.
21. Field, P., *Dust Explosions,* Elsevier, Amsterdam, 1982.
22. Swift, I., and Epstein, M., "Performance of Low-Pressure Explosion Vents," *Plant/Operations Progress,* Apr. 1987, pp. 98–105.
23. Nagy, J., *et al.,* "Pressure Development in Laboratory Dust Explosions," RI 6561, 1964, U.S. Bureau of Mines, Pittsburgh, PA.
24. Dorsett, H. G., *et al.,* "Laboratory Equipment and Test Procedures for Evaluating Explosibility of Dusts," RI 5624, 1960, U.S. Bureau of Mines, Pittsburgh, PA.
25. ASTM E789, *Standard Test Method for Pressure and Rate of Pressure Rise for Dust Explosions in a 1.2 Liter Closed Cylindrical Vessel,* American Society for Testing and Materials, W. Conshohocken, PA.
26. Bartknecht, W., "Explosion Pressure Relief," AIChE Loss Prevention Symposium, Vol. 11, 1977, pp. 93–105.
27. Donat, C., "Pressure Relief as Used in Explosion Protection," AIChE Loss Prevention Symposium, Vol. 11, 1977, pp. 87–92.
28. Bruderer, R. E., "Ignition Properties of Mechanical Sparks and Hot Surfaces in Dust/Air Mixtures," *Plant/Operations Progress,* Vol. 8, No. 3, July 1989, pp. 152–164.
29. Bruderer, R. E., "Design Example: Explosion Selected for a Spray Drying Installation," *Plant/Operations Progress,* Vol. 8, No. 3, July 1989, pp. 141–146.
30. Eckhoff, R. K., "A Study of Selected Problems Related to the Assessment of Ignitability and Explosibility of Dust Clouds," *CHR,* Michelsens Institute for Vendenskapog Andsfrihet, Beretninger, XXXVIII, Bergen, Norway, 1976.
31. Hertzberg, M., *et al.,* "The Flammability of Coal Dust-Air Mixtures, Lean Limits, Flame Temperatures, Ignition Energies and Particle Size Effects," RI 8360, 1979, U.S. Bureau of Mines, Pittsburgh, PA.

References

ASTM E1226-88, *Standard Test Method for Pressure and Rate of Pressure Rise for Combustible Dusts (20-Liter Spherical Vessel),* American Society for Testing and Materials, W. Conshohocken, PA.

Britton, L. C., and Kirby, D. C., "Analysis of a Dust Deflagration," *Plant/Operations Progress,* Vol. 8, No. 3, July 1989, pp. 177–180.

Bruderer, R. E., "Design Example: Explosion Selected for a Spray Drying Installation," *Plant/Operations Progress*, Vol. 8, No. 3, July 1989, pp. 141–146.

Conti, R. S., Cashdollar, K. L., Hertzberg, M., and Liebman, I., "Thermal and Electrical Ignitability of Dust Clouds," RI 8798, 1985, U.S. Bureau of Mines, Pittsburgh, PA.

Field, P., and Abrahamsen, A. R., *Dust Explosion—The Respective Roles of the Hartmann Bomb and the 20-Liter Sphere in Prescribing the Size of Explosion Relief Vents: A Preliminary Study*, Building Research Establishment, Hertfordshire, UK, June 1981.

Field, P., *Dust Explosion Protection—A Comparative Study of Methods for Sizing Explosion Relief Vents*, Building Research Establishment, Hertfordshire, UK, July 1982.

Hoenig, S. A., "Reducing the Hazard of Dust Explosions: Agglomerate the Fines, Cool Off the Hot Spots, Collect the Fines at the Point of Origin, Neutralize Local Electrostatic Charges," *Plant/Operations Progress*, Vol. 8, No. 3, July 1989, pp. 119–128.

Lathrop, J. K., "54 Killed in Two Grain Elevator Explosions," *Fire Journal*, Vol. 72, No. 1, 1978, pp. 29–35.

Lunn, G. A., *Venting Gas and Dust Explosion—A Review*, Institution of Chemical Engineers, Rugby, UK, 1984.

Lunn, G. A., *Dust Explosion Prevention and Protection*, Part 1—Venting, 2nd ed., Institution of Chemical Engineers, Rugby, UK, 1992.

Lunn, G. A., *Guide to Dust Explosion Prevention and Protection, Part 3—Venting of Weak Explosions and the Effect of Vent Ducts*, Institution of Chemical Engineers, Rugby, UK, 1984.

Randant, S., "Dust Explosion Venting of Vessels and Silos," European Information Centre for Explosion Protection, Antwerp, Belgium, 1989.

Schofield, C., and Abbott, J. A., *Guide to Dust Explosion Prevention and Protection, Part 2—Ignition Prevention, Containment, Inerting, Suppression, and Isolation*, Institution of Chemical Engineers, Rugby, UK, 1988.

Senecal, J. A., "Deflagration Suppression of High K_{St} Dusts," *Plant/Operations Progress*, Vol. 8, No. 3, July 1989, pp. 147–151.

Siwek, R., "Dust Explosion Venting for Dusts Pneumatically Conveyed into Vessels," *Plant/Operations Progress*, Vol. 8, No. 3, July 1989, pp. 129–140.

Siwek, R., "New Knowledge about Rotary Air Locks in Preventing Dust Ignition Breakthrough," *Plant/Operations Progress*, Vol. 8, No. 3, July 1989, pp. 165–176.

Swift, I., "Designing Explosion Vents Easily and Accurately," *Chemical Engineering*, Apr. 1988, pp. 65–68.

Swift, I., "NFPA 68, Explosion Venting: What's New and How It Affects You," *Journal of Loss Prevention in Process Industries*, Vol. 2, Jan. 1989, pp. 5–15.

Tonkin, P. S., and Berlement, C. F. J., "Dust Explosions in a Large-Scale Cyclone Plant," Fire Research Note 942, July 1972, Dept. of the Environment and Fire Offices' Committee, Joint Fire Research Organization, London, UK.

NFPA Codes, Standards, and Recommended Practices

Reference to the following NFPA codes, standards, and recommended practices will provide further information on dusts discussed in this chapter. (See the latest version of *The NFPA Catalog* for availability of current editions of the following documents.)

NFPA 49, *Hazardous Chemicals Data*

NFPA 61, *Standard for the Prevention of Fires and Dust Explosions in Agricultural and Food Products Facilities*

NFPA 65, *Standard for the Processing and Finishing of Aluminum*

NFPA 68, *Guide for Venting of Deflagrations*

NFPA 69, *Standard on Explosion Prevention Systems*

NFPA 480, *Standard for the Storage, Handling, and Processing of Magnesium Solids and Powders*

NFPA 481, *Standard for the Production, Processing, Handling, and Storage of Titanium*

NFPA 482, *Standard for the Production, Processing, Handling, and Storage of Zirconium*

NFPA 651, *Standard for the Manufacture of Aluminum Powder*

NFPA 654, *Standard for the Prevention of Fire and Dust Explosions in the Chemical, Dye, Pharmaceutical, and Plastics Industries*

NFPA 655, *Standard for Prevention of Sulfur Fires and Explosions*

NFPA 664, *Standard for the Prevention of Fires and Explosions in Wood Processing and Woodworking Facilities*

Additional Reading

Amyotte, P. R., Mintz, K. J., Pegg, M. J., Sun, Yu-Hong, and Wilkie, K. I., "Effects of Methane Admixture, Particle Size, and Volatile Content on the Dolomite Inerting Requirements of Coal Dust," *Journal of Hazardous Materials*, 27, 1991, pp. 187–203.

Britain's Plastics Industry, Jordan and Sons, London, 1987.

Britton, L. G., and Kuby, D. C., "Analysis of a Dust Deflagration," *Plant/Operations Progress,* Vol. 8, No. 3, July 1989, pp. 177–180.

Bruderer, R. E., "Use Bulk Bags Cautiously," *Chemical Engineering Progress,* May 1993, pp. 28–31.

Bull, B., "Grain Elevator Explosions," *Firehouse,* Apr. 1978, pp. 40–41.

Cartwright, P., "Combating the Problem of Dust Explosions," *Process Engineering,* Vol. 70, No. 2, 1989, pp. 49–52.

Cashdollar, K. L., Weiss, E. S., Greninger, N. B., and Chatrathi, K., "Laboratory and Large-Scale Dust Explosion Research," *Plant/Operations Progress,* Vol. 11, No. 4, Oct. 1989.

"Combustible Dusts," Loss Prevention Data Sheet 7-76, Factory Mutual Research Corp., Norwood, MA.

Conti, R. S., Cashdollar, K. L., and Thomas, R. A., "Improved 6.8-L Furnace for Measuring the Autoignition Temperatures of Dust Clouds," Bureau of Mines, Pittsburg, PA, IC 9467, 1993.

Dahm, C. J., "How Safe Are Your Pneumatic Conveyers?" *Chemical Engineering Progress,* May 1993, pp. 17–22.

Dahm, C. J., Ashum, M., and Williams, K., "Contribution of Low-Level Flammable Vapor Concentration to Dust Explosion Output," *Plant/Operations Progress,* Vol. 5, No. 1, Jan. 1986, pp. 57–64.

Eckhoff, R. K., "Sizing of Dust Explosion Vents in the Process Industries: Advances Made During the 1980s," *Journal Loss Prevention of the Process Industries,* Vol. 3, July 1990, pp. 268–279.

Gubin, E. I., and Dik, I. G., "Flame Propagation in a Dusty Gas," *Combustion, Explosion and Shock Waves* (English transl.), Vol. 23, No. 6, May 1988, pp. 685–690.

Hertzberg, M., *A Critique of the Dust Explosibility Index,* U.S. Bureau of Mines, Pittsburgh, PA, 1987.

Hertzberg, M., *et. al.*, "Explosive Dust Cloud Combustion," Twenty-fourth International Symposium on Combustion, 24th, July 5–10, 1992, Sydney, Australia, Combustion Institute, Pittsburg, PA, 1992, pp. 1837–1843.

Hoenig, S. A., "Reducing the Hazard of Dust Explosions," *Plant/Operations Progress,* Vol. 8, No. 3, July 1989, pp. 119–128.

Hwang, C. C., and Litton, C. D., "Ignition of Combustible Dust Layers on a Hot Surface," Proceedings of the Third International Symposium on Fire Safety Science, Elsevier Applied Science, New York, 1991, pp. 187–196.

Lunn, G. A., "Note on the Lower Explosibility Limit of Organic Dusts," *Journal of Hazardous Materials,* Vol. 17, No. 2, Feb. 1988, pp. 207–213.

Mintz, K. J., "Problems in Experimental Measurements of Dust Explosions," *Journal of Hazardous Materials,* 42, 1995, pp. 177–186.

Mintz, K. J., "Upper Explosive Limit Dusts: Experimental Evidence for Its Existence Under Certain Circumstances," Combustion and Flame, Vol. 94, No. 1/2, 1993, pp. 125–130.

Opila, R. L., "Carefully Plan Dust Collection Systems," *Chemical Engineering Progress,* May 1993, pp. 22–27.

Palmer, K. N., "Dust Explosions: Initiation, Characteristics, and Protection," *Chemical Engineering Progress,* Vol. 86, No. 3, Mar. 1990, pp. 33–36.

"Risks in Handling Combustible Dusts," *Fire Prevention,* No. 213, Oct. 1988, pp. 32–33.

Skinner, S. J., "Case Histories of Dust Explosions," *Plant/Operations Progress,* Vol. 8, No. 4, 1989, pp. 211–217.

V.D., Guideline 2263, "Dust Fires and Dust Explosions Hazards–Assessment-Protective Measures," Beuth Verlog Gmb., Berlin, 1986.

Zeeurven, J. P., "Dust Explosions–How to Assess the Risk," *Fire Prevention,* No. 228, Apr. 1990, pp. 26–28.

METALS

Revised by Robert E. Tapscott

This chapter covers the fire hazard properties of combustible metals, and storage, handling, and transportation methods to prevent their ignition and to protect against other hazardous properties. Also discussed are the major fire problems that arise while combustible metals are processed in machine shops and foundries.

Nearly all metals will burn in air under certain conditions. Some oxidize rapidly in the presence of air or moisture, generating sufficient heat to reach their ignition temperatures. Others oxidize so slowly that heat generated during oxidation is dissipated before the metals become hot enough to ignite. Certain metals, notably calcium, hafnium, lithium, magnesium, plutonium, potassium, sodium, thorium, titanium, uranium, zinc, and zirconium, are referred to as combustible metals because of the ease of ignition of thin sections, fine particles, or molten metal. However, the same metals in massive solid form are comparatively difficult to ignite.

More information on the hazards of metal dusts is contained in Section 4, Chapter 15, "Dusts." Extinguishing agents suitable for use on combustible metal fires are discussed in Section 6, Chapter 26, "Combustible Metal Extinguishing Agents and Application Techniques."

Some metals, such as aluminum, iron, and steel, that are not normally thought of as combustible, can ignite and burn when finely divided; or, in the case of aluminum, when in contact with other burning materials. Clean, fine steel wool, for example, can be ignited. Particle size, shape, quantity, and alloy are important factors to be considered when evaluating metal combustibility. Dust clouds of most metals in air are explosive. Alloys, consisting of different metals combined in varying proportions, may differ widely in combustibility from their constituent elements. Metals tend to be most reactive when finely divided, and some powders may require shipment and storage under inert gas or liquid to reduce fire risks.

Hot or burning metals may react violently upon contact with other materials, such as any of the extinguishants used on fires involving ordinary Class A combustibles or flammable liquids. A few metals, such as thorium, uranium, and plutonium, emit ionizing radiations (e.g., gamma, beta, and alpha particles) that can complicate fire fighting and introduce a contamination problem.

Robert E. Tapscott, Ph.D., is Director, Center for Global Environmental Technologies, University of New Mexico, Albuquerque, NM. He is a member of the NFPA Technical Committees on Alternative Protection Options to Halons and Halogenated Fire Extinguishing Agents.

Temperatures produced by burning metals are generally much higher than temperatures generated by burning flammable liquids. Some hot metals can continue burning in carbon dioxide, nitrogen, or steam atmospheres in which ordinary combustibles or flammable liquids would be incapable of combustion.

Properties of metal fires cover a wide range. Burning titanium produces little smoke, while burning lithium smoke is dense and profuse. Some water-moistened metal powders, such as zirconium, burn with near-explosive violence, while the same powder wet with oil burns quiescently. Sodium melts and flows while burning; calcium does not. Some metals, e.g., uranium, acquire an increased tendency to burn after prolonged exposure to moist air, while prolonged exposure to dry air may make it more difficult to ignite the metal.

Inasmuch as the extinguishment of fires in combustible metals involves techniques not commonly encountered in conventional fire-fighting operations, it is good practice for those responsible for controlling combustible metal fires to gain experience in this area prior to the actual fire emergency. Fire fighters should practice extinguishing fires in those metals in an isolated outdoor location.

Where metals other than those described in this chapter are in use, it is most important that fire fighters gain some experience in extinguishing test fires involving the specific combustible metals.

MAGNESIUM

Properties

The ignition temperature of massive magnesium is very close to its melting point of 1202°F (650°C). (See Table 4-16A.) However, ignition of magnesium in certain forms may occur at temperatures well below 1202°F (650°C); magnesium ribbons and shavings can be ignited under certain conditions at about 950°F (510°C), and finely divided magnesium powder can ignite below 900°F (482°C).

Metal marketed under different trade names and commonly referred to as "magnesium" may be one of a large number of different alloys containing principally magnesium, but also significant percentages of aluminum, manganese, and zinc. Some of these alloys have ignition temperatures considerably lower than pure magnesium, and certain magnesium alloys will ignite at temperatures as low as 800°F (427°C). Flame temperatures can reach 2500°F (1371°C), although flame height above burning metal is usually less than 12 in. (305 mm).

Thin, small pieces of magnesium, such as ribbons, chips, and shavings, may be ignited by a match flame, whereas castings and

TABLE 4-16A. *Melting, Boiling, and Ignition Temperatures of Pure Metals in Solid Form*

Pure Metal	Melting Point °F	Melting Point °C	Boiling Point °F	Boiling Point °C	Solid Metal Ignition °F	Solid Metal Ignition °C
Aluminum	1,220	660	4,445	2,452	1,832*‡	1,000‡
Barium	1,337	725	2,084	1,140	347*	175*
Calcium	1,548	842	2,625	1,441	1,300	704
Hafnium	4,032	2,222	9,750	5,399	—	—
Iron	2,795	1,535	5,432	3,000	1,706*	930*
Lithium	367	186	2,437	1,336	356	180
Magnesium	1,202	650	2,030	1,110	1,153	623
Plutonium	1,184	640	6,000	3,316	1,112	600
Potassium	144	62	1,400	760	156*†	69†*
Sodium	208	98	1,616	880	239‡	115‡
Strontium	1,425	774	2,102	1,150	1,328*	720*
Thorium	3,353	1,845	8,132	4,500	932*	500*
Titanium	3,140	1,727	5,900	3,260	2,900*	1,593
Uranium	2,070	1,132	6,900	3,816	6,900*§	3,816*§
Zinc	786	419	1,665	907	1,652*	900*
Zirconium	3,326	1,830	6,470	3,577	2,552*	1,400*

*Ignition in oxygen.
†Spontaneous ignition in moist air.
‡Above indicated temperature.
§Below indicated temperature.

other large pieces are difficult to ignite even with a torch because of the high thermal conductivity of the metal. In order to ignite a large piece of magnesium, it is usually necessary to raise the entire piece to the ignition temperature. Magnesium melts as it burns and may form puddles of molten magnesium, which may present explosion hazards in the presence of water.

Scrap magnesium chips or other fines may burn as the result of ignition of waste rags or other contaminants. Chips wet with water, water-soluble oils, and oils containing more than 0.2 percent fatty acid may generate hydrogen gas. Chips wet with animal or vegetable oils may burn if the oils ignite spontaneously. Fines from grinding operations generate flammable hydrogen when submerged in water, but the metal particles themselves cannot be ignited if kept immersed. Grinding fines that are slightly wetted with water may generate sufficient heat to ignite spontaneously in air, burning violently as oxygen is extracted from the water with the release of hydrogen, which can also ignite.

Storage and Handling (Including Transportation)

The more massive a piece of magnesium, the more difficult it is to ignite. Once ignited, however, magnesium burns intensely and is difficult to extinguish. The storage recommendations in NFPA 480, *Standard for the Storage, Handling, and Processing of Magnesium, Solids and Powders* (hereinafter referred to as NFPA 480), take these properties into consideration. Recommended maximum quantities of various sizes and forms to be stored in specific locations are covered in this standard. The storage building preferably should be noncombustible, and the magnesium should be segregated from combustible material as a fire prevention measure.

With easily ignited lightweight castings, segregation from combustible materials is especially important. In the case of dry fines (fine magnesium scrap), storage in noncombustible covered containers in separate fire-resistive storage buildings or rooms with explosion venting facilities is preferable. For combustible buildings or buildings containing combustible contents, NFPA 480 recom-

mends automatic sprinkler protection to ensure prompt control of a fire before the magnesium becomes involved.

Because of the possibility of hydrogen generation and of spontaneous heating of fines wet with coolants (other than neutral mineral oil), it is preferable to store wet scrap fines outdoors. Covered noncombustible containers should be vented.

Magnesium in powder, pellet, or ribbon form is shipped as a flammable solid and must be appropriately labeled according to U.S. Department of Transportation (DOT) requirements. Massive pieces or castings do not require special shipping safeguards.

If dry magnesium scrap in the form of borings, shavings, and turnings is to be shipped in interstate commerce in less-than-carload lots, the shipper may use closed metal drums. In carload or truckload lots, four-ply paper bags may also be used.

Less-than-carload lots of scrap in the form of chips or sheets may be shipped in closed metal drums, wood barrels, or wood boxes. For bulk shipments, tight box cars, tightly closed steel-covered gondola cars, and closed or completely covered truck bodies may be used.

Process Hazards

In machining operations involving magnesium alloys, sufficient frictional heat to ignite the chips or shavings may be created if the tools are dull or deformed. If cutting fluids are used (machining of magnesium is normally performed dry), they should be mineral-oil types that have high flash points. Water or water-oil emulsions are hazardous, since wet magnesium shavings and dust liberate hydrogen gas and burn more violently than dry material when ignited. Machines and the work area should be frequently cleaned and the waste magnesium kept in covered, clean, dry steel, or other noncombustible drums, which should be removed from the building at regular intervals. Magnesium dust clouds are explosive if an ignition source is present. Grinding equipment should be equipped with a water-spray dust precipitator. (See Figure 4-16A.) A good grinder installation is designed to operate only if the exhaust blower and water spray are functioning properly. The equipment should be restricted to magnesium processing only.

Molten magnesium in the foundry presents a serious fire problem if not properly handled. Sulfur dioxide or melting fluxes are commonly used to prevent oxidation or ignition of magnesium

FIG. 4-16A. *A schematic diagram of a water-precipitation-type collector for use in collecting dry combustible metal dust without creating explosive dust clouds or dangerous deposits in ducts or collection chambers. The diagram is intended only to show some of the features that should be incorporated in the design of a collector. It may be used for all combustible metal dusts.*

during foundry operations. The action of sulfur dioxide is to exclude air from the surface of the molten magnesium; it is not an extinguishing agent.

Pots, crucibles, and ladles that may contact molten magnesium must be kept dry to prevent steam formation or a violent metal-water reaction. Containers should be checked regularly for any possibility of leakage or weak points. Steel-lined runoff pits or pits with tightly fitting steel pans should be provided, and the pans must be kept free of iron scale. Leaking metal contacting hot iron scale results in a violent thermite reaction. Use of stainless-steel pans or linings will eliminate this possibility.

Heat-treating ovens or furnaces, where magnesium alloy parts are subjected to high temperatures to modify their properties, present another special problem. Temperatures for heat treating needed to secure the desired physical properties are often close to the ignition temperatures of the alloys themselves, and careful control of temperatures in all parts of the oven is essential. Hot spots leading to local overheating are a common cause of these fires. Large castings do not ignite readily, but fine fins or projections on the castings, as well as chips or dust, more readily ignite. For this reason, castings should be thoroughly cleaned before heat-treating. Magnesium castings in contact with aluminum in a heat-treating oven will ignite at a lower temperature than when they are placed on a steel car or tray.

Magnesium should not be heat-treated in nitrate salt baths. Certain commonly used molten mixtures of nitrates and nitrites can react explosively with magnesium and magnesium alloys, particularly at temperatures over 1000°F (538°C).

Fighting Magnesium Fires

Magnesium and its alloys present special problems in fire protection. Magnesium combines so readily with oxygen that, under some conditions, water applied to extinguish magnesium fires may be decomposed into its constituent elements, i.e., oxygen and hydrogen. The oxygen combines with the magnesium, and the released hydrogen adds to the intensity of the fire. With the exception of the noble gases, such as helium and argon, none of the commonly available inert gases is suitable for extinguishing magnesium fires. The affinity of magnesium for oxygen is so great that it will burn in an atmosphere of carbon dioxide. Magnesium may also burn in an atmosphere of nitrogen to form magnesium nitride. For these reasons, any of the common extinguishing methods that depend on water, water solutions, or inert gases (other than the noble gases) are not effective on magnesium chip fires. Halogen-containing extinguishing agents (halons and many halon replacement agents) react violently with burning magnesium, the chlorine or other halogen combining with the magnesium. However, inerting with noble gases, e.g., helium or argon, will extinguish burning metal.

The method of extinguishing magnesium fires depends largely upon the form of the material. Burning chips, shavings, and small parts must be smothered and cooled with a suitable dry extinguishing agent, e.g., graphite and dry sodium chloride. Where magnesium dust is present, care must be taken to prevent a dust cloud from forming in the air during application of the agent because this may result in a dust explosion.

Fires in solid magnesium can be fought without difficulty if attacked in their early stages. Often, it may be possible to remove surrounding material, leaving the small quantity of magnesium to burn itself out harmlessly. Considering the importance of prompt attack on magnesium fires, automatic sprinklers are desirable because they provide automatic notification and control of fire. While the water from the sprinklers may have the immediate effect of intensifying magnesium combustion, it will serve to protect the structure and prevent ignition of surrounding combustible material. An excess of

water applied to fires in solid magnesium (avoiding puddles of molten metal) cools the metal below the ignition temperature after some initial intensification, and the fire goes out rapidly.

Magnesium fires in heat-treating ovens can best be controlled with powders and gases developed for use on such fires. By using melting fluxes to exclude air from the burning metal, fires in heat-treating furnaces have been successfully extinguished. Boron trifluoride gas is an effective extinguishing agent for small fires in heat-treating furnaces. Cylinders of boron trifluoride can be permanently connected to the oven or mounted on a suitable cart for use as portable equipment. Boron trifluoride is allowed to flow into the oven until the fire is extinguished, or, where large quantities of magnesium are well involved before discovery or where the furnace is not tight, the boron trifluoride will control the fire until extinguishment can be completed by the application of flux.

TITANIUM

Properties

Titanium, like magnesium, is classified as a combustible metal, but here again the size and shape of the metal determine to a great extent the ease of ignition. Castings and other massive pieces of titanium are not combustible under ordinary conditions; however, they can burn in the presence of other burning fuels or in high-oxygen atmospheres. Small chips, fine turnings, and dust ignite readily, and, once ignited, burn with the release of large quantities of heat. Tests have shown that very thin chips and fine turnings can be ignited by a match, and heavier chips and turnings, by a Bunsen burner. Coarse chips and turnings $^1/_{32}$ by $^3/_{28}$ in. (0.79 by 2.7 mm) or larger may be considered as difficult to ignite, but, unless it is known that smaller particles are not mixed with the coarser material in significant amounts, it is wise to assume that easy ignition is possible.

Dry titanium fines collected in cyclone separators can ignite spontaneously when allowed to drop freely through the air. Ignition temperatures of titanium dust clouds in air range from 630 to 1090°F (332 to 588°C) and of titanium dust layers from 720 to 950°F (382 to 510°C). Titanium dust can be ignited in atmospheres of carbon dioxide or nitrogen. Titanium surfaces that have been treated with nitric acid, particularly with red fuming nitric acid containing 10 to 20 percent nitrogen tetroxide, become pyrophoric and may be explosive.

Fine titanium chips coated with water-soluble oil have spontaneously ignited. The unusual conditions under which massive titanium shapes will ignite spontaneously include contact with liquid oxygen, in which case they can detonate on impact. It has been found that under static conditions, spontaneous ignition will take place in pure oxygen at pressures of at least 350 psi (2413 kPa). If the oxygen was diluted, the required pressure increased, but in no instance did spontaneous heating occur in oxygen concentrations less than 35 percent. Another requirement for spontaneous heating was a fresh surface that oxidizes rapidly and exothermically in an oxygen atmosphere.[1]

Storage and Handling (Includes Transportation)

Titanium castings and ingots are so difficult to ignite and burn that special storage recommendations for large pieces are not included in NFPA 481, *Standard for the Production, Processing, Handling, and Storage of Titanium*. Titanium sponge and scrap fines, on the other hand, do require special precautions, such as storage in covered metal containers and segregation of the containers from combustible materials. Because of the possibility of hydrogen generation in moist scrap and spontaneous heating of scrap wet with animal or vegetable oils, a yard storage area remote from buildings is recommended for scrap that is to be salvaged. Alternate recommended storage locations are detached scrap storage buildings and fire-resistant storage rooms. Buildings and rooms for storage of scrap fines should have explosion vents.

There are no special shipping requirements for titanium except when in powder form. When possible, titanium powder is shipped wet (not less than 20 percent water) in tightly closed metal shipping containers, which can be packed in outside wood boxes. The inside containers are limited to 10 lb (4.54 kg) net weight each, and the gross weight of the outside container cannot exceed 75 lb (34 kg). Cushioning of inside containers by noncombustible material, such as rock wool, is required. Single-trip metal barrels or drums with cushioned-inside metal drums are permitted by the DOT. Inside containers are flushed with inert gas before filling.

Process Hazards

Contact of molten metal with water is the principal hazard during titanium casting. To minimize this hazard, molds are usually thoroughly predried and vacuum or inert gas protection is provided to retain accidental spills.

The heat generated during machining, grinding, sawing, and drilling of titanium may be sufficient to ignite the small pieces formed by these operations or to ignite mineral-oil-based cutting lubricants. Consequently, water-based coolants should be used in ample quantity to remove heat, and cutting tools should be kept sharp. Fines should be removed regularly from work areas and stored in covered metal containers. To prevent titanium dust explosions, any operation that produces dust should be equipped with a dust collecting system discharging into a water-spray dust precipitator. (See Figure 4-16A.)

Descaling baths of mineral acids and molten alkali salts may cause violent reactions with titanium at abnormally high temperatures. Titanium sheets have ignited upon removal from descaling baths. This hazard can be controlled by careful regulation of bath temperatures.

There have been several very severe explosions in titanium melting furnaces. These furnaces utilize an electric arc to melt a consumable electrode inside a water-cooled crucible maintained under a high vacuum. Stray arcing between the consumable electrode and crucible, resulting in penetration of the crucible, permits water to enter and react explosively with the molten titanium. Indications are that such explosions approach extreme velocities. The design and operation of these furnaces require special attention in order to prevent explosions and to minimize damage when explosions do occur.

Fighting Titanium Fires

Tests conducted by Industrial Risk Insurers (IRI) on titanium machinings in piles and in open drums showed that water in coarse spray was a safe and effective means of extinguishing fires in relatively small quantities of chips.

Carbon dioxide, foam, and dry chemical extinguishers are not effective on titanium fires, but good results have been obtained with extinguishing agents developed for use on magnesium fires.

The safest procedure to follow with a fire involving small quantities of titanium powder is to ring the fire with a special powder suitable for metal fires and to allow the fire to burn itself out. Care should be taken to prevent formation of a titanium dust cloud.

ALKALI METALS: SODIUM, POTASSIUM, NaK, AND LITHIUM

Properties

Sodium: At room temperature sodium oxidizes rapidly in moist air, but spontaneous ignitions have not been reported except when the sodium is in a finely divided form. When heated in dry air, sodium ignites in the vicinity of its boiling point [1616°F (880°C)]. Sodium in normal room air and at a temperature only slightly above its melting point [208°F (98°C)] has been ignited by placing sodium oxide particles on its surface. This indicates the possibility of ignition at temperatures below the boiling point. Once ignited, hot sodium burns vigorously and forms dense white clouds of caustic sodium oxide fumes. During combustion, sodium generates about the same amount of heat as an equivalent weight of wood.

The principal fire hazard associated with sodium is its rapid reaction with water. It floats on water (density 0.97), reacting vigorously and melting. The hydrogen liberated by this reaction may be ignited by the heat of the reaction. Sodium (like other burning, reactive metals) reacts violently with halogenated hydrocarbons, with halogens, e.g., iodine, and with acids.

Potassium: The fire hazard properties of potassium are very similar to those of sodium, with the difference being that potassium is usually more reactive. For example, the reaction between potassium and the halogens is more violent, and, in the case of bromine, a detonation can occur. There is an explosive reaction with sulfuric acid. Unlike sodium, potassium forms some peroxides during combustion. These peroxides may react violently with organic contaminants.

NaK (sodium-potassium alloys): "NaK" is the term used when referring to any of several sodium-potassium alloys. The various NaK alloys differ from each other in melting point, but all are liquids or melt near room temperature. NaK alloys possess the same fire hazard properties as those of the component metals, except that the reactions are more vigorous. Under pressure, NaK leaks have ignited spontaneously.

Lithium: Lithium, like sodium and potassium, is an alkali metal. Lithium undergoes many of the same reactions as sodium. For example, both sodium and lithium react with water to form hydrogen, but whereas the sodium-water reaction can generate sufficient heat to ignite the hydrogen, the far-less violent lithium-water reaction does not. Lithium ignites and burns vigorously at a temperature of 356°F (180°C), which is near its melting point. Unlike sodium and potassium, it will burn in nitrogen. Noble gases, such as helium and argon, are used for inerting in equipment and storage containers involving lithium. The caustic (oxide and nitride) fumes accompanying lithium combustion are more profuse and dense than those of other alkali metals burning under similar conditions. Lithium is the lightest of all metals. During combustion, it tends to melt and flow.

Storage and Handling (Includes Transportation)

Because of their reactivity with water, alkali metals require special precautions to prevent contact with moisture. Drums and cases preferably are stored in a dry, fire-resistant room or building used exclusively for alkali metal storage. Since sprinkler protection would be undesirable, no combustible materials should be stored in the same area. However, if exposure fires could jeopardize equipment containing alkali metals, sprinkler protection should be considered. It is good practice to store both empty and filled alkali metal containers in the same area, and all containers should be on skids. There should be no water or steam pipes, but sufficient heat should be maintained to prevent moisture condensation due to atmospheric changes. Ventilation at a high spot in the room is desirable to vent any hydrogen that may be released by accidental contact of alkali metal with moisture.

Large quantities of alkali metal are often stored outdoors in aboveground tanks. In such installations weatherproof enclosures should cover tank manholes, and the free space within the tank should contain a nitrogen atmosphere. Argon or helium atmospheres should be substituted for nitrogen in the case of lithium.

Dry alkali metal is customarily shipped in watertight, steel barrels or drums. It is also shipped as a solid in tank car quantities. The DOT specifies the type of barrels, drums, and tank cars that are to be used in interstate transportation.

Transfer of an alkali metal to and from tank cars and storage tanks requires heating the metal to its melting point. This is accomplished by circulating hot oil through pipes within the tank. The alkali metal is transferred by evacuating a receiver with a vacuum pump, and as the metal is removed from the tank it is replaced by nitrogen. An alkali metal-level indicator should be provided and a low-pressure warning device should be part of the nitrogen system.

An alkali metal covered with an organic solvent may be shipped in 1-qt (946-mL) metal cans, each can placed in another can and cushioned with soda ash. The cans are shipped in wood boxes of 100-lb (45.4-kg) maximum gross weight.

Liquid alkali metal alloys may be shipped in pressure-tested metal cans, cushioned with noncombustible material and packed in wood boxes. Carload and truckload lots are permitted if special steel barrels and drums are used.

For small-scale transfer of solid alkali metal from a storeroom to the use area, a metal container with a tight cover is recommended. Alkali metal should be removed from storage in as small quantities as practicable. When stored on workbenches, it should be kept under kerosene or oil in a closed container. Alkali metal, with its great affinity for moisture, may react at the time it is sealed in a container with any atmospheric moisture. Due to the possible presence of hydrogen, containers should not be opened by hammering on the lid.

Process Hazards

Liquid alkali metals are valuable as high-temperature heat transfer media. For example, they are used in hollow exhaust valve stems in some internal combustion engines and in the transfer of heat from one type of nuclear reactor to a steam generator. In the latter process or other large-scale use of molten alkali metal, any equipment leak may result in a fire. Where molten alkali metal is used in process equipment, steel pans are located underneath to prevent contact and violent reaction of burning sodium with concrete floors.

Processing of alkali metal is essentially remelting it to form sticks or bricks or to add it as a liquid to closed transfer systems. During this handling, contact with moist air, water, halogens, halogenated hydrocarbons, and sulfuric acid must be avoided.

Fighting Fires of Sodium, Potassium, NaK, and Lithium

The common extinguishing agents, such as water, foam, and vaporizing liquids, should never be used because of the violent reactions of alkali metals with these agents. Special dry powders (essentially graphite) developed for metal fires, dry sand, dry sodium chloride, and dry soda ash can be effective. These finely divided materials blanket the fire while the metal cools to below its ignition temperature. Alkali metal burning in an apparatus can usually be extin-

guished by closing all openings. Blanketing with nitrogen is also effective. In the case of lithium, argon or helium atmospheres should be used. Copper powder is a very effective extinguishant for lithium fires.

ZIRCONIUM AND HAFNIUM

Properties

Zirconium: The combustibility of zirconium appears to increase as the average particle size decreases, but other variables, such as moisture content, also affect its ease of ignition. In massive form, zirconium can withstand extremely high temperatures without igniting, whereas clouds of dust in which the average particle size is 3 microns can spontaneously ignite at room temperature. Dust clouds of larger particle size can be readily ignited if an ignition source is present, and such explosions can occur in atmospheres of carbon dioxide or nitrogen, as well as in air. Zirconium dust will ignite in carbon dioxide at approximately 1150°F (621°C) and in nitrogen at approximately 1450°F (788°C). Layers of 3-micron-diameter dust are susceptible to spontaneous ignition. The depth of the layer and its moisture content are important variables for ignition. A cloud of zirconium dust in air presents a serious flash-fire hazard, in addition to a potential explosion hazard. Spontaneous heating and ignition are also possibilities with scrap chips, borings, and turnings if fine dust is present. Layers of 6-micron-diameter dust have ignited when heated to 374°F (190°C). Combustion of zirconium dust in air is stimulated by the presence of limited amounts (5 to 10 percent) of water. When very finely divided zirconium powder is completely immersed in water, it is difficult to ignite, but once ignited it burns more violently than in air.

Massive pieces of zirconium do not ignite spontaneously under ordinary conditions, but ignition will occur when an oxide-free surface is exposed to sufficiently high oxygen concentrations and pressure. The explanation for this reaction is the same as that cited for a similar titanium reaction. Zirconium fires (like fires involving titanium and hafnium) attain very high temperatures, but generate very little smoke.

Explosions can occur while zirconium is being dissolved in a mixture of sulfuric acid and potassium acid sulfate. Zirconium can explode during and following pickling in nitric acid, and also during treatment with carbon tetrachloride or other halogen-containing materials. Spontaneous explosions can occur during handling of moist, very finely divided, contaminated zirconium scrap.

Hafnium: Hafnium has combustion properties similar to those of zirconium. Hafnium burns with very little flame, but it releases large quantities of heat. Unless inactivated, hafnium in sponge form may ignite spontaneously.

Hafnium is generally considered to be somewhat more reactive than titanium or zirconium of similar form. Damp hafnium powder reacts with water to form hydrogen gas, but at ordinary temperatures this reaction is not sufficiently vigorous to cause the hydrogen to ignite. Under some conditions, however, ignition of the hydrogen may be expected to proceed explosively.

Storage and Handling (Including Transportation)

Special storage precautions are not required for zirconium castings because of the very high temperatures massive pieces of the metal can withstand without igniting. Zirconium powder, on the other hand, is highly combustible; consequently, it is customarily stored and shipped in 1-gal (3.78-L) containers with at least 25 percent water by volume. For specific details, refer to NFPA 482, *Standard for the Production, Processing, Handling, and Storage of Zirconium.*

Zirconium powder storerooms should be of fire-resistant construction equipped with explosion vents. Cans should be separated from each other to minimize the possibility of a fire at one can involving others and to permit checking of the cans periodically for corrosion. One plant handling zirconium has established the procedure of disposing cans containing powder that have been on the shelf for six months.

Dry zirconium powder or sponge is customarily shipped inside metal cans in wood or fiberboard boxes. The DOT does not require compliance with these shipping regulations if the particle size of the powder is greater than 20 mesh.

Wet zirconium powder and zirconium sludge may be shipped in glass or polyethylene containers within metal drums or wood boxes.

Hafnium transportation regulations and storage recommendations are the same as for zirconium.

Process Hazards

In general, processing recommendations for zirconium and hafnium are the same. Whenever possible, handling of zirconium powder should be under an inert liquid or in an inert atmosphere. An inert atmosphere of argon or helium in equipment and storage containers is a method used to prevent flash fires and explosions. If zirconium or hafnium powder is handled in air, extreme care must be used because the small static charges generated may cause ignition.

To prevent dangerous heating during machining operations, a large flow of mineral oil or water-base coolant is required. In some machining operations, the cutting surface is completely immersed. Turnings should be collected frequently and stored under water in cans. Where zirconium dust is a byproduct, dust-collecting equipment that discharges into a water-precipitation-type collector is a necessity. (See Figure 4-16A.)

Fighting Fires of Zirconium and Hafnium

Zirconium and hafnium fires can be fought in the same way. Fires exposing massive pieces of zirconium, for example, can be fought with water. Limited tests conducted by Industrial Risk Insurers have indicated that the discharge of water as a spray has no adverse effect on burning zirconium turnings. When a sprinkler opened directly above an open drum of burning zirconium scrap, there was a brief flareup, after which the fire continued to burn quietly in the drum. When a straight stream of water at a high rate of flow was discharged into the drum, water overflowed and the fire was extinguished.

Where small quantities of zirconium powder or fines are burning, the fire can be ringed with a special extinguishing powder to prevent its spread, after which the fire can be allowed to burn out. Special extinguishing powders have been effective in extinguishing zirconium fires. When zirconium dust is present, the extinguishing agent should be applied so that a zirconium dust cloud will not form. If the fire is in an enclosed space, it can be smothered by introducing argon or helium.

CALCIUM AND ZINC

Properties

Calcium: The flammability of calcium depends considerably on the amount of moisture in the air. If ignited in moist air, it burns without flowing at a somewhat lower rate than sodium. It decom-

poses in water to yield calcium hydroxide and hydrogen, which may burn. Finely divided calcium will ignite spontaneously in air. It should be noted that barium and strontium are similar to calcium in their fire properties.

Zinc: Zinc does not introduce a serious fire hazard in sheets, castings, or other massive forms because of the difficulty of ignition. Once ignited, however, large pieces burn vigorously. Moist zinc dust reacts slowly with the water to form hydrogen, and, if sufficient heat is released, ignition of the dust can occur. Zinc dust clouds in air ignite at 1110°F (599°C). Burning zinc generates appreciable smoke.

Storage, Processing, and Fighting Fires of Calcium and Zinc

The storage, handling, and processing recommended for magnesium is generally applicable to calcium and zinc.

METALS NOT NORMALLY COMBUSTIBLE

Aluminum

The usual forms of aluminum have a sufficiently high ignition temperature so that its burning is not a factor in most fires. However, aluminum can burn vigorously in the presence of other burning materials. In particular, there is evidence that aluminum in contact with magnesium burns readily, perhaps due to the formation of an alloy at the interface. Very fine chips and shavings are occasionally subject to somewhat the same type of combustion as described for magnesium. Powdered or flaked aluminum, however, can be explosive under certain conditions. (See Table 4-16A.)

When aluminum is substituted for iron, steel, or copper, the lower melting temperature is an important factor if resistance to fire temperatures is essential. Figure 4-16B shows the decrease in strength of an aluminum alloy (6061-T6) at elevated temperatures. Compare this with Figure 4-16C for steel.

The alloy 6061-T6 is typical of those used for architectural applications. Figure 4-16B also shows the strength of 6061-T6 alloy at various temperatures for five different heating times.[2]

Aluminum roofing laid directly on structural supports without a roof deck may be expected to melt through and vent a fire beneath it. For fire fighters working on such a roof, the hazard is greater than on a steel roof due to the strength loss under fire conditions.

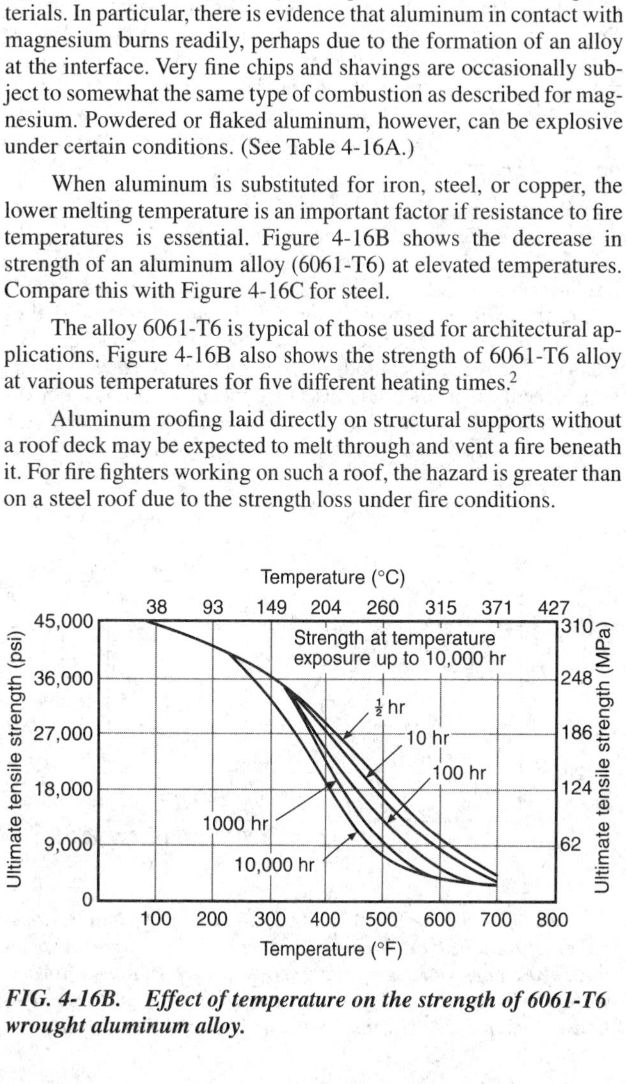

FIG. 4-16B.   *Effect of temperature on the strength of 6061-T6 wrought aluminum alloy.*

Aluminum tubing is used in place of copper in many applications for flammable liquids and gases. While in many cases aluminum is suitable for tubing, it should be used with respect for its lower melting temperature. A number of aircraft fires have been charged to the failure of thin-walled aluminum alloy tubing used in hydraulic lines containing flammable liquids. Under pressure, short circuits caused arcs to the tubing, melting the aluminum, and releasing the flammable liquid.

Iron and Steel

Iron and steel are not usually considered combustible; in a massive form (as in structural steel, cast-iron parts, etc.), they do not burn in ordinary fires. Steel in the form of fine steel wool or dust may be ignited in the presence of heat from, for example, a torch, sparking rather than flaming in most instances. Fires can occur in piles of steel turnings and other fines scrap which contain some oil and perhaps also other materials that facilitate combustion. Spontaneous ignition of water-wetted borings and turnings in closed areas, such as ship hulls, can also occur. The strength of structural steels (A7 and A36) when subjected to fire temperatures is shown in Figure 4-16C.

RADIOACTIVE METALS

Radioactivity does not influence nor is it influenced by fire properties of a metal. Radioactive metals include those few that occur naturally, e.g., uranium and thorium, and those produced artificially, e.g., plutonium and cobalt-60. Any metal can be made radioactive. Since radioactivity cannot be altered by fire, radiation will continue wherever the radioactive metal may be spread during a fire. Smoke from fires involving radioactive materials frequently causes more property damage than the fire. The "damage," however, is not physical but results from radioactive contamination that must be cleaned up to avoid exposure.

Naturally occurring radioactive metals consist of a mixture of atoms having slightly different mass. These are called isotopes. For example, the radioactivity of each uranium isotope is different, and it follows that the radiation hazards from any given piece of uranium will change with its isotopic composition, which can be varied by special processing. Change of isotopic content of a piece of uranium alters the nature and magnitude of radiation hazards, but does not alter the chemical or fire properties of the metal. The same is equally true of thorium and plutonium. Pre-fire planning is particularly important where radioactive materials are involved. Automatic sprin-

kler protection is generally provided for facilities containing radioactive materials.

In nuclear reactors, uranium, plutonium, and thorium may build up high levels of radioactivity. A few nuclear reactors utilize molten metallic sodium as a coolant, and the sodium may become radioactive. Fortunately, the requirements for fire protection are generally consistent with those independently necessitated for operational reasons.

Uranium

Normal uranium is a radioactive metal that is also combustible. Its radioactivity does not affect its combustibility, but can have a bearing on the amount of fire loss. Most metallic uranium is handled in massive forms that do not present a significant fire risk unless exposed to a severe and prolonged external fire. Once ignited, massive metal burns very slowly. A 1-in. (25.4-mm) diameter rod requires about one day to burn out after ignition. In the absence of strong drafts, uranium oxide smoke tends to deposit in the immediate area of the burning metal. Unless covered with oil, massive uranium burns with virtually no visible flame. Burning uranium reacts violently with carbon tetrachloride, 1,1,1 trichloroethane, the halons, and many halon replacements. For power reactors, uranium fuel elements are always encased in a metal jacket (usually zirconium or stainless steel).

Uranium in finely divided form is readily ignited, and uranium scrap from machining operations is subject to spontaneous ignition. This reaction can usually be avoided by storage under dry oil. Grinding dust has been known to ignite even under water, and fires have occurred spontaneously in drums of coarser scrap after prolonged exposure to moist air. Larger pieces generally have to be heated entirely to their ignition temperature before igniting. Moist dust, turnings, and chips react slowly with water to form hydrogen. Uranium surfaces treated with concentrated nitric acid are subject to explosion or spontaneous ignition in air.

Thorium

Thorium, like uranium, is a naturally occurring element. Both are known as "source materials," the basic materials from which nuclear reactor fuels are produced. The powdered form of thorium requires special handling techniques because of its low ignition temperature. It is handled dry in a helium or argon atmosphere. The dry metal powder should not be in air because the friction of the particles falling through the air or against the edge of a glass container may produce electrostatic ignition of the powder.

Powdered thorium is usually compacted into solid pellets weighing about 1 oz. (28 g) each. In this form it can be safely stored or converted into alloys with other metals. Improperly compacted thorium pellets have been known to slowly generate sufficient heat through absorption of oxygen and nitrogen from the air to raise a steel container to red heat.

Plutonium

Plutonium, like uranium, is radioactive as well as combustible. Plutonium is somewhat more susceptible to ignition than uranium and is normally handled by remote-control in an inert gas or "bone dry" air atmosphere. In finely divided form, such as dust or chips, plutonium is subject to spontaneous ignition in moist air.

Plutonium metal is never intentionally exposed to water, in part because of fire considerations. The massive metal ignites at about 1112°F (600°C) and burns in a manner quite similar to uranium, except that plutonium oxide smoke damage, i.e., airborne radioactive contamination, is more difficult to control and also more

FIG. 4-16C. *Strength of A7 and A36 structural steels at elevated temperature. The band indicates performance of a range of structural steels at elevated temperature. Actual performance will vary with the shape, size, and type of assembly exposed to fire conditions. (Data provided by the American Iron and Steel Institute.)*

hazardous. Because of certain nonfire hazard considerations, it is necessary to limit the quantity of plutonium kept at one location, thus limiting the maximum size of a fire in this metal. Plutonium, which ignites spontaneously, is normally allowed to burn under conditions limiting both fire and radiological contamination spread.

BIBLIOGRAPHY

References Cited

1. CEN, "Titanium Does a Fast Burn," *Chemical and Engineering News*, Vol. 36, No. 31, Aug. 1958, pp. 36–37.
2. DOD, "Metallic Materials and Elements for Aerospace Vehicles," *MIL-HDBK-5* (undated), U.S. Department of Defense, Washington, DC.

NFPA Codes, Standards, and Recommended Practices

Reference to the following NFPA codes, standards, and recommended practices will provide further information on metals discussed in this chapter. (See the latest version of *The NFPA Catalog* for availability of current editions of the following documents.)

NFPA 65, *Standard for the Processing and Finishing of Aluminum*
NFPA 480, *Standard for the Storage, Handling, and Processing of Magnesium Solids and Powders*
NFPA 481, *Standard for the Production, Processing, Handling, and Storage of Titanium*
NFPA 482, *Standard for the Production, Processing, Handling, and Storage of Zirconium*
NFPA 485, *Standard for the Storage, Handling, Processing, and use of Lithium Metal*
NFPA 651, *Standard for the Manufacture of Aluminum Powder*

Additional Reading

American Society of Metals International, *Metals Handbook*, 10th ed., Materials Park, OH, 1990.
Bryan, C. J., Stolzfus, J. M., and Gunaji, M. V., "Assessment of the Flammability Hazard of Several Corrosion Resistant Metal Alloys," John F. Kennedy Space Center, FL, National Aeronautics and Space Administration, Las Cruces, NM, Lockheed-ESC, Las Cruces, NM, ASTM STP 1197; American Society for Testing and Materials, Flammability and Sensitivity of Materials in Oxygen Enriched Atmospheres, May 11–13, 1993, Noordwijk, The Netherlands, ASTM, Philadelphia, PA, 1993, pp. 112–118.
Chan, S. H., and Tan, C. C., "Simplex Method and Duality Theory in Liquid Metal Combustion," 1990 Fall Technical Meeting of the Combustion Institute/Eastern States Section on Chemical and Physical Processes in Combustion, December 3–5, 1990, Orlando, FL, 1990, pp. 71/1–4.
Duclos, P., Binder, S., and Riester, R., "Community Evacuation Following the Spencer Metal Processing Plant Fire," *Journal of Hazardous Materials*, Vol. 22, No. 1, Sept. 1989, pp. 1–11.
"Fire Hazards in Aluminum Processing," *The Institution of Fire Engineers Quarterly*, Vol. 27, No. 65, Mar. 1967, pp. 94–97.
"Fire Protection for Combustible Metals," *National Safety News*, Vol. 119, No. 6, June 1979, pp. 75–82.
Hertzberg, M., Zlochower, I. A., and Cashdollar, K. L., "Explosibility of Metal Dusts," Short Communication, *Combustion Science and Technology*, Vol. 75, No. 1–3, 1991, pp. 161–165.

Jacobson, M., Cooper, A. R., and Nagy, J., *Explosibility of Metal Powders*, RI 6516, USDI, Bureau of Mines, Pittsburgh, PA, 1964.
Leihbacher, D., "Tactics for Large Magnesium Fires: Yonkers, New York Magnesium Incident," *Fire Engineering*, Vol. 148, No. 4, Apr. 1995, pp. 38–40.
Leonard, J. *et al.*, "Use of Copper Powder Extinguishers on Lithium Fires," Naval Research Laboratory, Washington, DC, Hughes Associates, Inc., Columbia, MD, Naval Sea Systems Command, Washington, DC, NRL/MR/6180-94-7490, Funding Numbers 61–M259-X3, July 8, 1994.
Markstein, G. H., "Combustion of Metals," *Journal of the American Institute of Aeronautics & Astronautics*, Vol. 1, 1963, pp. 550–562.
Ohlemiller, T. J. and Shields, J. R., "Effect of Suppressants on Metal Fires," National Institute of Standards and Technology, Gaithersburg, MD, NISTIR 5710, NIST SP 890, Volume 1, August 1995; Fire Suppression System Performance of Alternative Agents in Aircraft Engine and Dry Bay Laboratory Simulations, NIST SP 890, Vol. 1, Nov. 1995, 1995, pp. 97–119.
Peloubet, J. A., "Machining Magnesium—A Study of Ignition Factors," *Fire Technology*, Vol. 1, No. 1, Feb. 1965, pp. 5–14.
Peterson, W. S., ed., *Guidelines for Handling Molten Aluminum*, The Aluminum Association, Washington, DC, 1982.
Rhein, R. A., "Experimental Study of the Use of Liquid Argon and Argon-Filled Aqueous Foams for the Extinction of Lithium Fires," *Fire Technology*, Vol. 28, No. 4, Nov. 1992, pp. 290–316.
Rhein, R. A., Baldwin, J. C., and Beach, C. L., "Compositions for Extinguishing Titanium Fires," *Chemical Abstracts*, Vol. 103, 1985.
Rhein, R. A. and Carlton, C. M., "Extinction of Lithium Fires: Thermodynamic Computations and Experimental Data From Literature," *Fire Technology*, Vol. 29, No. 2, May 1993, pp. 100–130.
Riley, J. F., "Na-X, A New Fire Extinguishing Agent for Metal Fires," *Fire Technology*, Vol. 10, No. 4, Nov. 1974, pp. 269–274.
Shafirovich, E. Y. and Goldshleger, U. I., "Superheat Phenomenon in the Combustion of Magnesium Particles," *Combustion and Flame*, Vol. 88, No. 3–4, March 1992, pp. 425–432.
Sharma, T. P., Lal, B. B., and Singh, J., "Metal Fire Extinguishment," *Fire Technology*, Vol. 23, No. 3, 1987, pp. 205–229.
Sharma, T. P., Varshney, B. S. and Kumar, S., "Studies on the Burning Behavior of Metal Powder Fires and Their Extinguishment. Part 2. Magnesium Powder Heaps on Insulated and Conducting Material Beds," *Fire Safety Journal*, Vol. 21, No. 2, 1993, pp. 153–176.
Smithells, C. J., *Metals Reference Book*, 7th ed., Butterworths, Boston, 1991.
Steinberg, T. A., Wilson, D. B. and Benz, F., "Burning of Metals and Alloys in Microgravity," *Combustion and Flame*, Vol. 88, No. 3–4, March 1992, pp. 309–320.
Steinberg, T. A., Wilson, D. B. and Benz, F., "Combustion of Phase of Burning Metals," *Combustion and Flame*, Vol. 91, No. 2, Nov. 1992, pp. 200–208.
Twilt, L., "Strength and Deformation Properties of Steel at Elevated Temperatures: Some Practical Implications," *Fire Safety Journal*, Vol. 13, No. 1, 1988, pp. 1–8.
Varshney, B. S., Kumar, S. and Sharma, T. P., "Studies on the Burning Behavior of Metal Powder Fires and Their Extinguishment. Part 1. Mg, Al, Al-Mg Alloy Powder Fires on Sand Bed," *Fire Safety Journal*, Vol. 16, No. 2, 1990, pp. 93–117.
Weman, D. G., "Tactics for Large Magnesium Fires: St. Louis, MO, Magnesium Incident," *Fire Engineering*, Vol. 148, No. 4, April 1995, pp. 46–48.

UPHOLSTERED FURNITURE AND MATTRESSES

Vytenis Babrauskas

Upholstered items, e.g., upholstered furniture and mattresses, play a special role in fire hazard. This was first identified many years ago in a statistical study which determined that these items constitute the classes of burnable items accounting for the largest share of fire deaths in the United States.[1] In a number of subsequent reviews of fire loss statistics, the same basic observations continued to be made, although fire deaths from this cause—as from most causes—have decreased during the last two decades. It is estimated that there were approximately 47,200 home structure fires per year from 1989 to 1993 that were reported to U.S. fire departments as starting with the ignition of upholstered furniture or mattresses. These fires caused 1,369 civilian deaths, 5,199 civilian injuries, and $570.1 million in direct property damage per year. Approximately 35 percent of these fires were caused by cigarettes. Other major ignition sources were matches, lighters, fixed and portable space heaters, and cords and plugs. Lighted tobacco products led to approximately 23 percent of residential fire deaths in 1990–1994 and are the single largest cause of such deaths each year. However, they accounted for only 6 percent of the residential fires, and 12 percent of the injuries. Mortality and morbidity in cigarette-initiated fires are thus typically high. Victims of cigarette fires are not always the careless smokers themselves, and 42 percent of those killed by such residential fires are not in the same room as the fire when it begins.*

Upholstered furniture and mattresses may be ignited by cigarettes, by flaming sources, or by other sources, such as electric blankets. The latter is a factor only for mattress fires and will not be considered here.

The need for cigarette ignition resistance of upholstered items is unique—other furnishings, such as carpets and curtains, or apparel textile items have much smaller share of their fires due to cigarettes. Unfortunately, cigarette ignition resistance and resistance to flames are not necessarily compatible. Among common upholstery fabrics, medium and heavyweight thermoplastic fabrics (e.g., nylon, olefin, and polyester) generally resist cigarette ignition but do not resist flame ignition. Similarly, cellulosic materials (e.g., cotton, hemp, jute, linen, rayon fabrics, and batting) have poor cigarette ignition resistance, but over polyurethane foam they burn at a slower rate than the common combination of thermoplastic fabric and

*Laurence J. Stewart, "The U.S. Home Product Report, 1989–1993: Forms and Types of Materials First Ignited in Fires," Quincy, MA, NFPA Fire Analysis and Research, February 1996.

Vytenis Babrauskas, Ph.D., is president of Fire Science and Technology, Inc., Kelso, WA.

polyurethane foam. Only a smaller selection of fabric/padding combinations, e.g., wool fabric over advanced foams, combine good cigarette ignition resistance with relatively high resistance to flames.

UPHOLSTERED FURNITURE AND MATTRESSES: SIMILARITIES AND DIFFERENCES

Many aspects of flammability are similar for upholstered furniture and mattresses. However, some distinctions must be made: mattresses have predominantly flat surfaces that are not as likely to ignite by cigarettes as furniture crevices, i.e., the juncture of seat cushion and armrests and back of upholstered furniture. (In a similar vein, horizontal flat furniture surfaces are not as easily ignited by flame ignition sources as are vertical surfaces.)

On the other hand, cigarettes on mattresses may be inadvertently covered with sheets, blankets, or pillows. This increases the probability of ignition by cigarettes. Also, these intermediary materials may ignite from flames more readily than mattresses, and then expose the mattress to a much more severe fire. Consequently, it is more difficult to develop relevant ignition tests and standards for mattresses than for upholstered furniture.

Upholstered furniture can contain well over a dozen materials, e.g., cover fabrics and fabrics between the cover fabric and the padding (interliners); different padding materials in the seat, sides, and back; weltcords; another fabric and type of padding below the seat cushion (decking); and frames, springs, stiffening for the sides, etc. Mattresses are complex but contain fewer materials. In the ignition process, whether from a cigarette or small flame, the cover fabric and material immediately below it—one or two layers of padding—are important. As the fire progresses, other materials, including the frame and springs, become involved, which can affect the manner in which the burning item collapses.

Another difference stems from industry practices. It is possible to find older materials (cotton batting, latex foam, vegetable-origin fibers, etc.) in mattresses from certain manufacturers. By contrast, these materials have become quite rare in upholstered furniture, where polyurethane foam is by far the dominant padding material.

An important difference concerning mattresses is that radiant heat conditions can affect them more readily. This can mean that a mattress and a chair that show the same heat release rate when burned in a very large space will behave differently when placed in a small room. In the latter case, the mattress is likely to burn more severely than the upholstered chair. More technical discussions of differences in mattress and upholstered furniture behavior have been published.[2]

SMOLDERING AND FLAMING

Smoldering of upholstered items, which is the typical consequence of exposure to a burning cigarette, can produce enough smoke and toxic gases to incapacitate and kill, but rarely does so before the fire transitions to flaming or, in the absence of such a transition, before the fire self-extinguishes or is put out.[5] Transition from smoldering to flaming does not occur in all cases; when it does, times from placement of the cigarette to flaming of 20 minutes to several hours have been reported.[3,4] In a closed room, flaming can revert to smoldering when the oxygen supply is exhausted. With sufficient oxygen, a single upholstered item of sufficient burning rate can lead to room flashover.

Another feature of upholstered item fires is the difficulty of ensuring extinction, since wetting out or even immersion of the item in water may not ensure extinguishment. Items should be removed from the premises, torn apart, and the pieces thoroughly wetted. Many fire victims have wetted down a smoldering area in an upholstered chair or sofa, retired for the night, only to be faced several hours later with impenetrable smoke or a full-scale fire. This problem used to be very serious with latex foam items, but must be addressed with other types of constructions as well.

A monograph on upholstered furniture flammability contains an in-depth discussion of cigarette and small flame ignition resistance and post-ignition behavior.[6]

UPHOLSTERED ITEM FLAMMABILITY TESTS

Some cigarette and flame ignition standards and tests for upholstered items are listed in Table 4-17A. They can be classified as: (1) voluntary standards and tests, (2) federally enforced standards, and (3) state-imposed standards. A more extensive discussion of these standards and also of international standards was presented in 1989.[7]

Voluntary and Regulatory Standards

Voluntary standards are proposed by trade organizations, such as the Upholstered Furniture Action Council (UFAC) for residential furniture, and the Business and Institutional Furniture Manufacturers Association (BIFMA) for their products. These two standards have also been developed as NFPA 260, *Standard Methods of Tests and Classification System for Cigarette Ignition Resistance of Components of Upholstered Furniture*, and NFPA 261, *Standard Method of Test for Determining Resistance of Mock-Up Upholstered Furniture Material Assemblies to Ignition by Smoldering Cigarettes* (hereinafter referred to as NFPA 260 and NFPA 261, respectively). The BIFMA standard was originally developed by the

TABLE 4-17A. *Flammability Tests for Upholstered Furniture and Mattresses*

Test	Ignition Source		Status			Specimen		Occupancy	
	Cigarette	or Radiant Flame	Voluntary	Regulatory	Proposed	Small-scale	Actual Item or Large-Size Mockup	Residential	Special
Upholstered furniture									
ASTM E 1474; NFPA 264A†		x	x			x			x
NFPA 260 (UFAC)	x		x			x		x	
NFPA 261 (BIFMA)	x		x			x			x
Calif. T.B. 116†	x		x			x		x	
Calif. T.B. 117†	x	x		x		x		x	
UK BS 5852:1	x			x		x		x	
UK BS 5852:2		x		x		x		x	
ISO 8191	x			x			x	x	
Calif. T.B. 133; ASTM E 1537; NFPA 266		x		x			x		x
Nordtest NT FIRE 032†		x	x				x		x
ASTM stacking chair test		x			x		x		x
ASTM school bus seat test		x			x		x		x
ASTM D 3675‡		x		x		x			x
MVSS 302§		x		x		x			x
U.S. 13 CFR 25.1587 Appendix F Part II*				x			x		x
Mattresses									
U.S. 16 CFR 1632.4	x			x			x	x	
Calif. T.B. 129; ASTM E 1590; NFPA 267		x		x					
Calif. T.B. 121		x		x			x		x
UK BS 6807	x	x		x		x		x	
Nordtest NT FIRE 036	x	x	x			x		x	

†Also applicable to mattresses.
‡Only used for rail vehicle applications.
§Only used for road vehicle applications.
*Only used for aircraft seating.

National Bureau of Standards (NBS) for the Consumer Product Safety Commission (CPSC). CPSC decided to hold this standard, pending success of the UFAC voluntary approach. About 80 percent of the upholstered furniture on the market today carries a UFAC hangtag, and it has been reported that the UFAC standard has succeeded in improving the cigarette ignition resistance of furniture.[8]

The mattress cigarette ignition standard is enforced by CPSC on the federal level. The State of California provides for a voluntary standard, T.B. 116, which is similar to the UFAC standard for cigarette ignition resistance. California also requires on a mandatory basis compliance with its T.B. 117 test, which provides for small-scale tests on components. The latter contains provisions for both open-flame and smolder-resistance testing.[9] For certain occupancies, California also requires compliance with large-scale fire tests. The large-scale California tests T.B. 121, T.B. 129, and T.B. 133 are discussed below. Most of the other states do not have any regulatory requirements on a state level.

Some of the newer tests in Table 4-17A are more severe than the state standards, e.g., T.B. 116 and T.B. 117, in that they employ larger ignition sources or heating with radiative sources. These tests often include heat release rate, flame spread, and smoke measurements, and are frequently used for research, product development, and regulation of special occupancies by some state and local authorities.

In recent years, the fire-fighting community and others have expressed great concern about the rapid fire spread, high heat release rate, and copious smoke development from polyurethane foam, as compared to the materials used in older upholstered furniture and mattresses. In the United Kingdom, this concern has led to passage of a 1988 regulation that basically banned non-flame-retardant polyurethane foam[10]; its scope was enlarged in 1989.[11] The regulation specifies a number of tests for foams, fabrics, and mockups, as well as rules for labeling, etc. Foams would have to pass the requirements of British Standard BS 5852, Part 2, i.e., withstand ignition from a 0.6-oz (17-g) wood crib placed into the crevice of a mockup of polyester fabric and the foam to be tested. The largest effect of the British regulations was to replace standard or conventionally fire-retarded polyurethane foams with combustion-modified high-resilience (CMHR) foams. These are different from foams marketed as CMHR in the United States. In the United Kingdom, they are typically manufactured according to a process whereby the foam is melamine-treated. Interestingly, this has not resulted in a price increase. In addition, some changes in fabrics have also had to be made. One solution promoted in the United Kingdom has been back-coated fabrics that incorporate fire-retardant agents into a latex-type of rubbery backcoating. The Republic of Ireland has subsequently adopted the same regulations as in the United Kingdom. Other European regulations and standards exist, and these have been extensively reviewed.[12]

Component or Mockup Tests

Some tests are performed on mockups of furniture or on the actual furniture. Other tests prescribe testing of the individual components, i.e., the cover fabric, padding, weltcord, etc., separately. The component tests do not consider interaction between materials. Sometimes the tests are configured so as to employ "standard materials" for those components not being tested. If that standard material is not the worst-case selection, the test results may be difficult to interpret or apply; e.g., upholstered padding in the UFAC-NFPA tests (NFPA 260 and NFPA 261) is tested while covered with an approximately 14 oz per sq yd (475 g/m²) cotton fabric, but many fabrics with higher propensity to ignite from cigarettes are frequently used on furniture.

Cigarette tests specify placement of the cigarette in the most vulnerable configuration, which is the crevice formed by the horizontal and vertical upholstered surfaces except, of course, for mattress tests. The cigarette is covered with a piece of sheeting. This makes the test slightly more severe and more reproducible, presumably because of exclusion of air currents from the smoldering area.

For flame ignition component tests, ignition is often by a gas flame at the bottom of a vertical specimen, such as foam or fabrics.[10,13] Flaming ignition in furniture mockups is generally at the crevice location.

Two U.S. flammability tests for transportation seats present a wide range of severity. On the low end of the scale, the Motor Vehicle Safety Standard No. 302 requires a horizontal specimen and has minimal flame-spread requirements.[14] On the other end, the Federal Aviation Administration (FAA) provides for a large-scale test, discussed below.

Pass/fail criteria are most often expressed in terms of char length or obvious ignition (i.e., sustained, expanding smoldering) for cigarettes, and char length and afterflame and afterglow time for flame ignition. The newer test methods generally give quantitative results or ratings: i.e., heat release rate [Cone Calorimeter; ASTM E1354; NFPA 264, *Standard Method of Test for Heat and Visible Smoke Release Rates for Materials and Products Using an Oxygen Consumption Calorimeter* (hereinafter referred to as NFPA 264); NFPA 264A, *Standard Method of Test for Heat Release Rates for Upholstered Furniture Components or Composites and Mattresses Using an Oxygen Consumption Calorimeter* (hereinafter referred to as NFPA 264A)]; flame spread ratings (ASTM E162 and D3675); and smoke density [Cone Calorimeter; and NFPA 258, *Standard Research Test Method for Determining Smoke Generation of Solid Materials* (hereinafter referred to as NFPA 258)].

The cigarette used in all U.S. smoldering ignition tests is a filterless, 85-mm commercial cigarette (Pall Mall® brand). It is representative of many other cigarette brands in its propensity to ignite. In recent years, bills have been passed by the U.S. Congress and about a dozen state legislatures with the objective of reducing the tendency of cigarettes to ignite upholstered furniture. In 1984, Congress passed the "Cigarette Safety Act" (P.L. 98-567), which called for a study of the feasibility of making cigarettes with a low propensity to ignite, without increasing tar and nicotine. It was found that this propensity does not relate to the rate of burning or burning cone temperature of cigarettes, but to complex interactions among the tobacco column, the cigarette paper, and the surface on which the cigarette rests. This propensity can be greatly reduced by lowering tobacco packing density and cigarette diameter (both of which reduce fuel loading and tar and nicotine delivery), as well as by lowering cigarette paper porosity.[15] This was followed by a second law, the Fire Safe Cigarette Act of 1990 (P.L. 101-352), leading to a follow-up project at NIST to develop a standard test method for determining the propensity of cigarettes to ignite upholstered furniture.[16,17]

Large-Scale and Full-Item Tests

Testing for cigarette ignitability can be done very satisfactorily on small samples. However, for flaming behavior of fires the situation is different. A small Bunsen-burner or similar type of test cannot predict what will happen to real furniture. Thus, it had been recognized for a number of years that large-scale tests are necessary.

The first large-scale fire test for upholstered furniture to gain widespread acceptability was California Technical Bulletin 133.[18] The test comprises a room in which an actual furniture item (or a full-scale mockup) is fire tested. The original ignition source comprised 3.2 oz (90 g) of balled-up newspaper held in a wire cage placed into the crevice of the furniture. The criteria included room temperature rise, CO concentration, and smoke density in several locations, and the weight loss of the test item. Following a research program at the National Institute of Standards and Technology (NIST),[19] the test method was revised in 1990 in three fundamental respects: (1) for improved reproducibility, the ignition source was replaced with a gas burner; (2) the main pass/fail criterion was set as a heat release rate value, taken at the 80 kW level; and (3) a new option was added for running tests in the hood of a large-scale furniture calorimeter, rather than inside of a burn room.

The California T.B. 133 test was originally voluntary. In 1992 it was made mandatory in California for certain public occupancies (but not for residential use). It has also been adopted by other states. Its influence, however, greatly extends beyond these limited applications. Most of the larger manufacturers of contract furniture now offer a significant selection of T.B. 133-complying furniture. Much of this acceptance has come from concerns about product liability litigation. Thus, by conforming to the T.B. 133 test, manufacturers can establish that they are in conformance with the state of the art for fire safety of furniture. The test has been adopted by ASTM as ASTM E1537 and by NFPA as NFPA 266, *Standard Method of Test for Fire Characteristics of Upholstered Furniture Exposed to Flaming Ignition Source* (hereinafter referred to as NFPA 266). A UL method (UL 1056) was originally published and entailed quite different test conditions. Recently, however, the UL method was changed to make it rather similar to the methods of ASTM and NFPA. For actual testing, reference should be made to the ASTM or NFPA methods, since guidance to the laboratory is much more extensive in those documents than in the California Bulletin.

To meet T.B. 133 requirements, manufacturers have most often turned to using an interliner fabric, such as vinyl-coated glass fabric, or Nomex® or PBI nonwoven or woven fabrics, between the cover fabric and the padding. Foam improvements can be needed and care in selecting fabrics is also necessary. Especially, it is very difficult to pass the T.B. 133 test with polyolefin fabrics. It is important to realize that the test is a system test and not a materials test. Thus, using the same construction materials, it can be possible to pass the test with a smaller chair and fail it with a larger one. A number of other design variables, in addition to combustible mass, also enter into the picture.

For prison mattresses, California has long had a special large-scale test, T.B. 121. This is run in the same test room as for the T.B. 133 test, but with a different ignition source, which is a wastebasket placed underneath the test specimen.[20] More recently, the State of California developed a less severe mattress test suitable for other institutional occupancies. This test, T.B. 129,[21] uses a gas burner located alongside the edge of the specimen, with other test features being essentially identical to those of the T.B. 133, except for the pass/fail limit. For the California T.B. 129 test, this is set at 100 kW. The test has been adopted as ASTM E1590 and NFPA 267, *Standard Method of Test for Fire Characteristics of Mattresses and Bedding Assemblies Exposed to Flaming Ignition Source* (hereinafter referred to as NFPA 267).

A special category of concern is stacking chairs. These types of chairs are normally used in hotel ballrooms, churches, and other public assembly rooms. This chair type, when upholstered (some types are of solid molded plastic, without upholstery), usually has only a small amount of padding. Thus, it would seem that dangers associated with them would be small. After some significant fires, most notably the Cathedral Hill Hotel fire in San Francisco, CA, however, attention was focused on the fact that such chairs are often found in their stacked configuration, where they can be many units high. Under these conditions, the hazards are much increased. This concern has led the state of California to propose a large-scale fire test method for stacking chairs.[22] The method is also currently being considered by ASTM for adoption.

In the area of transportation seating, there is a large-scale regulatory test promulgated by the FAA for airplane seats.[23] It specifies a 115-kW, 6-in. (152-mm) diameter kerosene burner flame directed for 2 min. on the side of the seat. The flames must not spread across the seat, no flaming drips must fall, and a maximum weight loss is specified.

CHARACTERISTICS OF IGNITION SOURCES

The choice of the ignition source is crucial to fire testing. A test item that shows essentially no response under a given source may exhibit a runaway fire with just a slightly greater ignition source. Thus, the simulated source selected must be a realistic approximation of the actual fire conditions that may be encountered.

Table 4-17B shows the characteristics of some ignition sources that have been used in flammability experimentation with upholstered furniture items. The sources range over four orders of magnitude: from 5 to 50,000 W. The maximum heat fluxes for the sources, except the methenamine pill, range between 18 and 42 kW/m². This is a rather small range, and it can be generalized[24] that an "incidental" ignition source shows a typical flame flux of 35 kW/m². What varies when the ignition source strength is increased is not the flame flux, but rather the area over which the flux is applied. In the case of the polyethylene wastebasket, the area over which fluxes exceed 20 kW/m² is about 2.36 by 27.55 in. (60 by 700 mm); for a wood match this would be approximately 0.39 by 1.18 in. (10 by 30 mm). This relationship is generalized in Figure 4-17A. It is seen that there is a linear relationship between the power output of the ignition source and its flame coverage area. It should be noted that the above data are all for "incidental" fire sources. Heat fluxes from specialized sources, such as oxyacetylene torches, may be much higher, e.g., 140 kW/m².

The reason that the flame coverage area of the ignition source is important is because higher heat fluxes are required to raise small target areas to their ignition temperature than large areas. This is especially true for fire-retardant-treated materials. Unfortunately, no theoretical guidance is yet available on this point, which is one reason why ignition source selection must be done on an empirical basis. Too-small ignition sources can be misleading for an additional reason. Many materials shrink or melt away from a small ignition

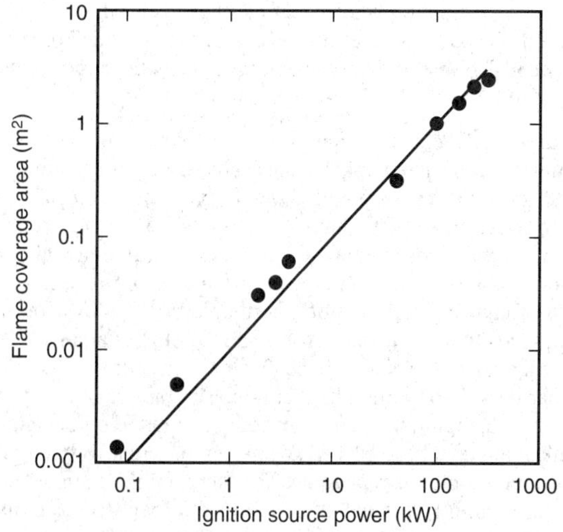

FIG. 4-17A.   *Flame area as a function of ignition source power. (For U.S. units: 1 ft = 0.305 m.)*

TABLE 4-17B. *Characteristics of Furniture Ignition Sources**

	Typical Heat Output (W)	Burn Time† (s)	Maximum Flame Height (mm)	Flame Width (mm)	Maximum Heat Flux kW/m^2
Cigarette 1.1 g (not puffed, laid on solid surface)					
Bone dry	5	1200	—	—	42
Conditioned to 50% R.H.	5	1200	—	—	35
Methenamine pill, 0.15 g	45	90		4	
Match, wood (laid on solid surface)	80	20–30	30	14	18–20
Wood cribs, BS 5852, Part 2					
No. 4 crib, 8.5 g	1000	190		15#	
No. 5 crib, 17 g	1900	200		17#	
No. 6 crib, 60 g	2600	190		20#	
No. 7 crib, 126 g	6400	350		25#	
Crumpled brown lunch bag, 6 g	1200	80			
Crumpled wax paper, 4.5 g (tight)	1800	25			
Crumpled wax paper, 4.5 g (loose)	5300	20			
Folded double-sheet newspaper, 22 g (bottom ignition)	4000	100			
Crumpled double-sheet newspaper, 22 g (top ignition)	7400	40			
Crumpled double-sheet newspaper, 22 g (bottom ignition)	17,000	20			
California T.B. 133, Crumpled newspaper, 95–105 g	15,000	110	710	610	80**
Polyethylene wastebasket, 285 g, filled with 12 milk cartons (390 g)	50,000	200‡	550	200	35§

*Sources: See reference 6.
†Time duration of significant flaming.
‡Total burn time in excess of 1800 sec.
§As measured on simulation burner.
#Measured from 25 mm away.
** Wide scatter.
For SI Units: 1 in. = 25.4 mm; 1 ft = 0.305 m; 1 oz = 28.4 g.

source; this can lead to passing of a Bunsen-burner-type test. When presented with a larger ignition flame area, however, the same item may be unable to shrink back and can ignite and burn vigorously. In general, extreme care should be used in interpreting any data from small, Bunsen-burner-type tests. Mostly such tests can only be considered for screening purposes and not for assessing actual product fire performance.

FLAME SPREAD

Flame spread has long been the basis of flammability evaluation of building materials; however, little quantitative work has been done on the flame spread over upholstered items. Cotton and rayon cover fabrics by themselves were shown to have flame spread rates inversely proportional to their weight;[25] however, when tested over polyurethane padding, the differences between flame spread rates were small. This is another case where component tests can give rise to misleading results, as mentioned earlier.

In another experimental series, the flame spread over seat cushions of chair mockups was measured by counting the number of 3.93-in. (100-mm) squares covered by flames as a function of time from ignition.[26] This flame-spread rate was found to correlate fairly well with bench-scale measurements on the same fabric/padding composites. There were major differences, however, in the behavior of charring and melting (thermoplastic) fabrics. Cotton and rayon, the cellulosic fabrics, charred ahead of the flame, and on the side of the chair opposite to the one which was burning. Eventually the char burst into flame. The thermoplastic fabrics initially shrank away from the ignition source, but being supported by the foam, the shrunken material bead burned and presented a relatively large ignition source for the foam padding. The fabric on the chair side opposite the burning side ablated, exposing the foam to the radiation. Thus, flames spread more rapidly over the thermoplastic fabric-covered mockups, even when the foam was neoprene and did not significantly flame during the test.

General flame-spread theory has not been much applied to upholstered items, primarily because of the variety of configurations and interaction between fabrics and padding. Dimensions in these items are sufficiently large so flame-spread and heat-release behavior are dominated by radiative, rather than convective, mechanisms. Recently, some studies of furniture flame spread in the "against the wind" mode, i.e., in the opposite direction to the draft induced by the buoyancy of the flame, have been conducted.[27] These may find an application in advancing the theory of furniture fires.

HEAT RELEASE RATE

In the last few years, it has been realized that heat release rate (HRR) is the single most important variable describing fire hazard.[28] The reason for this is not hard to understand. When confronted with a fire, normally the first question that arises is: How big is the fire? The HRR is simply the quantitative rephrasing of that question. It is important to note that by knowing the HRR of the full-scale object one does not separately need to inquire as to its flame spread characteristics. Items showing slow or incomplete flame spread reflect this in their HRR curves, which are low or slow growing.

In earlier days, it was not possible to discuss the HRR of burning items, simply because measuring tools did not exist for it. This situation was changed in 1982 when the present author and co-workers at NIST developed the furniture calorimeter.[29] The furniture calorimeter allowed simple, low-cost testing of whole furniture articles under realistic burning conditions. The HRR [measured in kilowatts (kW)] was measured, along with smoke and toxic gas evolution from the burning object. Subsequently, a European standard based on this was published.[30] This was followed by the U.S. standards discussed earlier, wherein the furniture calorimeter testing methodology is further developed (ASTM E1537 and E1590; NFPA 266 and 267). Furniture calorimeters are now available in many U.S. and foreign test laboratories and manufacturers' laboratories.

The furniture calorimeter is based on the *oxygen consumption* principle. This principle states that the combustion heat released is proportional to the amount of oxygen consumed to within about ±5 percent. The proportionality constant is 13.1×10^3 kJ per kg of O_2 consumed. Note that the *heat of combustion* of a fuel can vary widely and the oxygen consumption principle does *not* require that this be known or be constant. The details of this principle, along with extensive additional information on HRR applications, are available in a recent textbook.[31]

Measurements of HRR also use the identical oxygen consumption principle when performed in room fire tests, e.g., in the arrangement of T.B. 133.

The name "cone calorimeter" comes from the cone-shaped heater used to irradiate the specimen. The basic apparatus is standardized by ASTM as ASTM E1354, by NFPA as NFPA 264, and by the International Organization for Standardization as ISO 5660.

The above "cone calorimeter" standards all describe the basic equipment and test procedures. As such, they are pertinent to applications ranging from spacecraft to trash containers. Upholstered furniture, however, presents a special issue in that the samples are intrinsically composites, often comprising many layers. Thus, special specimen preparation instructions must be carefully followed in order to obtain reliable and reproducible results. These were first developed in the mid-1980s and were incorporated in the 1990 edition of NFPA 264A (the same material is in ASTM E1474). In subsequent years, the furniture on the marketplace came more to encompass fire-improved formulations and special constructions. Fire barriers (interliners) have come into widespread use in contract furniture as a consequence of T.B. 133 requirements. The earlier specimen preparation techniques proved to be insufficient for these new products. Also, with the widespread use of cone calorimeters in various laboratories, emphasis on quality assurance procedures needed to be increased, since testing was not always being done by specialists. Consequently, in 1994 a new, highly detailed specimen preparation protocol was developed.[33] This was extensively validated in round-robin testing and is forming the basis for updates to ASTM E1474 and NFPA 264A standards.

Predicting the HRR Performance of Furniture

The HRR of upholstered furniture can—in the worst circumstances—reach values of around 2000 to 3000 kW (2 to 3 MW) in a very short time, only 3 to 5 min. after ignition. In such a worst case, the hazard is extreme, since it only takes about 1 MW to flash over a room with a normal-sized door opening. Furthermore, there are few other consumer goods that can readily show such unfavorable fire behavior. Thus, it is very important that the HRR of furniture items be successfully measured and controlled.

In the T.B. 133 methodology, the furniture item is to be tested in full size, either in the furniture calorimeter or in a room fire test. Such an option should always be available to the manufacturer, but it does entail difficulties: costs and effort in full-scale testing can be much greater than for testing small specimens. Thus, as with the fire performance of many other categories of products, bench-scale equivalents have been sought. A *predictive* bench-scale test can provide nearly as good information as a full-scale test, yet be much easier and simpler to conduct. Figure 4-17B shows the general design flowchart. This figure also introduces the newest concept, that it may be possible to test *components*, not just *composites*.

An initial method for predicting the HRR performance of full-size furniture items on the basis of cone calorimeter results was presented in 1985.[6] Since then, additional research took place to enhance the state of the art. The most significant such advance was embodied in the European research program, CBUF (combustion behavior of upholstered furniture). This massive study,[34] completed

in 1995, yielded a new generation of predictive methods, based on very extensive testing and on advanced fire modeling work. The simplest model, based on direct correlation, can be summarized as follows.

Determination of propagating/non-propagating behavior: The most important factor in determining the flaming-combustion hazard of a furniture item is whether the item will or will not support the propagation of fire over its surface. *Propagating* chairs spread fire over most of the surface of the item. Such behavior means that the chair was largely consumed at the end of a fire, although substantial portions of hard combustibles (wood, particleboard) may remain. *Nonpropagating* chairs show a very different behavior. With such items, once the ignition source dies out or is removed, the fire does not sustain itself. Thus, most of the padding and fabric still remain after the fire is over. Combustible materials are usually gone directly under the fire source and in its immediate vicinity, but even a short distance farther the furniture item can appear essentially undamaged.

It is experimentally observed that non-propagating full-scale chair fires normally show peak HRR values of 20 to 100 kW. One can compare this to the 80 kW level that California has selected as the pass/fail limit for T.B. 133. The propagating/nonpropagating behavior is only slightly affected by small variations in the type of ignition source used.

For a nonpropagating fire, the total HRR (and the emissions of toxic gases) is essentially identical to what is measured for the chair

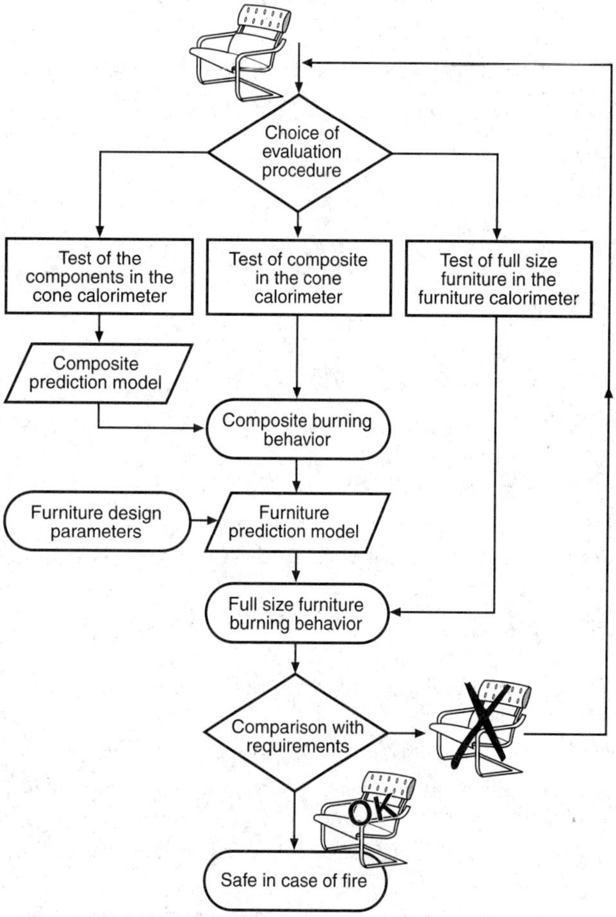

FIG. 4-17B. *Procedure for fire safety design of upholstered furniture.*

itself. For chairs that lead to propagating fires, however, the HRR (and toxic gas emissions) measured during a fire test only represents the *lower bound* to what may occur in a real fire. The reason is that for fires which are propagating fires it is readily possible to ignite additional items in the room, items that are not included in a test, such as the room-test alternative of T.B. 133. The heat and radiation levels coming from a 100 kW fire are not sufficient to cause injury to an occupant, unless the occupant also ignites his/her own clothing or unless the environment in question is a closed small room. Thus, it can be concluded that non-propagating fires are as *intrinsically safe* as a combustible item can ever be deemed to be.

Earlier studies[6] suggested that a propagation/nonpropagation criterion can be set as $\dot{q}''_{180} = 75$ kW/m^2, where $\dot{q}''_{180}$ is defined as the cone calorimeter HRR average for the period starting with ignition and ending 180 sec later. This limit was based on early experience with cone calorimeter testing, and both the irradiance and the test protocol were different from what was found necessary during the CBUF program. As a result of the CBUF study, a criterion value of $\dot{q}''_{180} = 65$ kW/m^2 was determined.

Prediction of the peak HRR: Analysis of the large amount of full-scale and bench-scale data collected within the CBUF program allowed equations to be developed to predict the peak HRR of the furniture (chair or sofa) item. Prediction of the peak HRR requires the use of Equations 1 through 5, below.

$$x_1 = (m_{soft})^{1.25}(style\ fac.\ A)(\dot{q}'' + \dot{q}''_{300})^{0.7}(15 + t_{ig})^{-0.7} \quad (1)$$

$$x_2 = 880 + 500(m_{soft})^{0.7}(style\ fac.\ A)(\Delta h_{c,eff}/q''_{tot})^{1.4} \quad (2)$$

If ($x_1 > 115$) or ($q'_{tot} > 70$ and $x_1 > 40$) or (style = {3,4} and $x_1 > 70$), then

$$\dot{Q} = x_2 \quad (3)$$

else,

If $x_1 < 56$, then

$$\dot{Q} = 14.4\ x_1 \quad (4)$$

else,

$$\dot{Q} = 600 + 3.77x_1 \quad (5)$$

For the above equations, the following cone calorimeter data are used:

$\dot{q}''_{pk}$ = peak HRR (kW/m^2) at 35 kW/m^2 exposure,

$\dot{q}''_{300}$ = 300-s average HRR (kW/m^2) at 35 kW/m^2 exposure,

t_{ig} = ignition time (s) at 35 kW/m^2 exposure,

$\Delta h_{c,eff}$ = effective heat of combustion (MJ/kg^1), and

q''_{tot} = total heat release (MJ/m^2) at 35 kW/m^2 exposure.

In addition, the total mass and the style factor of the full-scale item must be known:

m_{soft} = mass of upholstery (fabric, filling, interliner, etc.) materials (kg).

Style factors. Furniture style was already known to be an important variable affecting fire development from the earlier NIST studies, but the data available at that time were not adequate

to derive suitable numeric expressions covering the range of commercial furniture styles. The CBUF study, by contrast, was able to include a wide variety of commercially important furniture styles. Table 4-17C shows the furniture styles adopted. These are obviously not exhaustive of the market possibilities, but they do include the majority of common types. Style factors are used for two purposes: (1) for predictions of peak HRR values and (2) for predictions of time to peak. Style factor A is pertinent to calculations of peak HRR, while style factor B pertains to calculations of time to peak.

TABLE 4-17C. *Furniture Styles Defined in the CBUF Predictive Model*

Code	Style Factor A	Style Factor B	Type of Furniture
1	1.0	1.0	Armchair, fully upholstered, average amount of padding
2	1.0	0.8	Sofa, 2-seat
3	0.8	0.8	Sofa, 3-seat
4	0.9	0.9	Armchair, fully upholstered, highly padded
5	1.2	0.8	Armchair, small amount of padding
6	1.0	2.5	Wingback chair
7	—	—	Office swivel chair, plastic arms (unpadded), plastic, rear-back shell
8	—	—	High-back office swivel chair, plastic arms (unpadded), plastic back and bottom shell
9	—	—	Mattress, without innersprings
10	—	—	Mattress, with innersprings
11	—	—	Mattress and box spring (divan base) set
12	0.6	0.75	Sofa-bed (convertible)
13	1.0	0.8	Armchair, fully upholstered, metal frame
14	1.0	0.75	Armless chair, seat and back cushions only
15	1.0	1.0	Two-seater, armless, seat and back cushions only

The accuracy of the prediction of the above model (Equations 1 through 5) for peak HRR is shown in Figure 4-17C.

Prediction of total heat release: The total heat release is a measure of the effective fire load from the furniture item. The following equation was found to represent well the total heat:

$$Q = 0.9m_{soft} \cdot \Delta h_{c,eff} + 2.1(m_{comb,total} - m_{soft})^{1.5} \quad (6)$$

The effective heat of combustion comes from cone calorimeter tests of the composite (soft parts). For the full-scale article, $m_{comb,total}$ denotes the entire combustible mass, i.e., all except metal frame parts or other non-combustible pieces, and m_{soft} is the mass of the upholstery system. The results of this prediction are shown in Figure 4-17D.

Prediction of time to untenability: The time to untenability is defined as starting at the time when the full-scale HRR first exceeds 50 kW and ending when the smoke layer (100°C isotherm) is at 1.2 m above floor height. The measurement conditions correspond to the standard fire test room defined in ISO 9705. The untenability time, t_{UT}, is predicted by the following equation:

$$t_{UT} = 1.5 \times 10^5 (style\ fac.\ B)(m_{soft})^{-0.6}(\dot{q}''_{trough})^{-0.8}$$
$$\times (\dot{q}''_{ph\ \#2})^{-0.5}(t_{pk\ \#1} - 10)^{0.15} \quad (7)$$

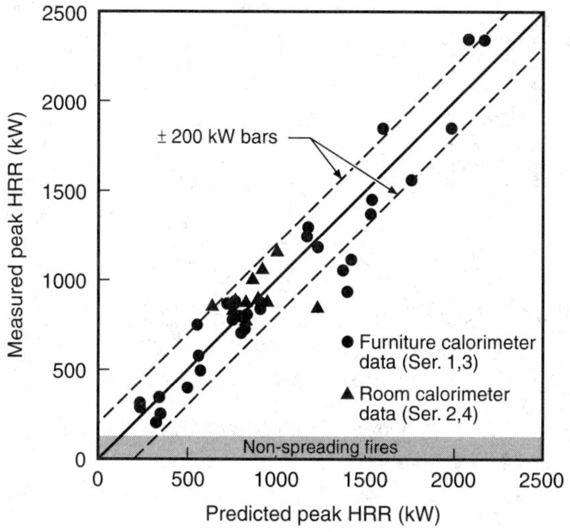

FIG. 4-17C. *Prediction of peak HRR of upholstered furniture, according to Equations 1 through 5.*

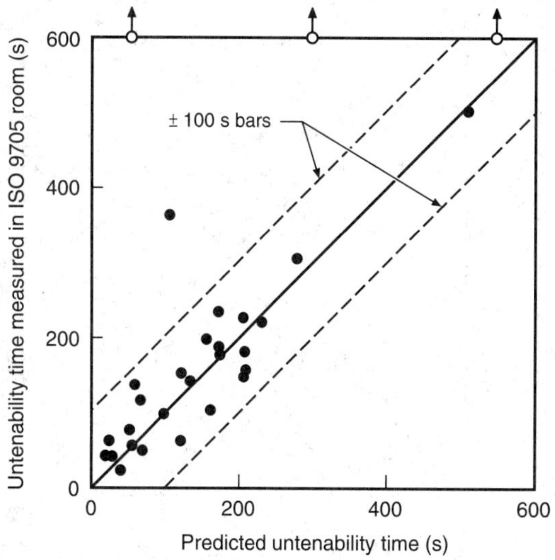

FIG. 4-17E. *Prediction of untenability time, according to Equation 7.*

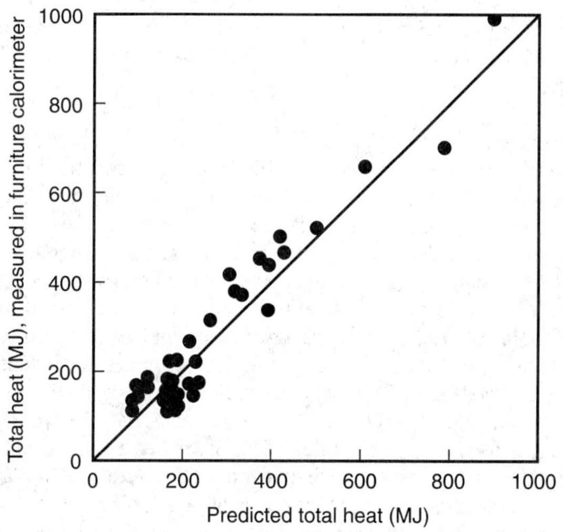

FIG. 4-17D. *Prediction of total heat release of upholstered furniture, according to Equation 6.*

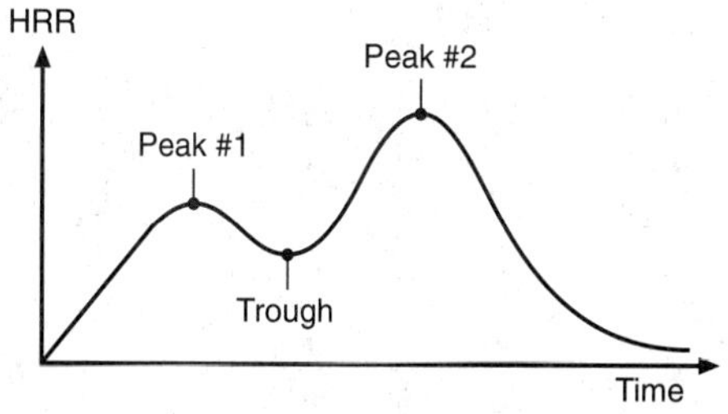

FIG. 4-17F. *Peak and trough values on the cone calorimeter HRR curve.*

The prediction results are indicated in Figure 4-17E. Here, $\dot{q}''_{trough}$ is the HRR in the cone calorimeter at the trough following the first peak; $\dot{q}''_{pk\#2}$ is the HRR at the second peak; and $t_{pk\#1}$ is the time to the first peak. These concepts are illustrated in Figure 4-17F.

Prediction of time to peak: The time to peak is considered an important modeling parameter of a fire. This time is defined to occur when the peak value of HRR is recorded. Generally, all other hazard variables are also at or near their peak values when the HRR peak occurs. Since the ignition period of a fire can be highly varied and refers only to individual circumstances, one can consider the starting point for analysis to be the time that sustained burning first occurs, i.e., when a fire level of 50 kW is first reached. Thus, it is of interest to be able to predict the time interval starting with the time that sustained burning first occurs, until the peak is reached. A predictive relationship was developed for estimating this time to peak as:

$$t_{pk} = 30 + 4900(style\ fac.\ B)(mass)^{0.3}\left(\dot{q}''_{pk\#2}\right)^{-0.5}$$
$$\times \left(\dot{q}''_{trough}\right)^{-0.5}\left(t_{pk\#1} + 200\right)^{0.2} \qquad (8)$$

where

t_{pk}	= time to peak, from start of sustained burning(s);
style fac. B	= style factor B, as given in Table 4-17C;
mass	= mass of upholstery (soft components only) of full-scale item (kg);
$\dot{q}''_{pk\#2}$	= second peak of cone calorimeter HRR curve (kW/m²);
$\dot{q}''_{trough}$	= trough of cone calorimeter HRR curve (kW/m²); and
$t_{pk\#1}$	= time to first peak of cone calorimeter HRR curve, from start of test(s).

The predictive results are shown in Figure 4-17G.

Prediction of smoke production: The following correlation was established for predicting the smoke:

if

$$p_2 < 1800 \qquad (9)$$

then,

$$SmP = p_2 \qquad (10)$$

else,

$$SmP = p_1 \qquad (11)$$

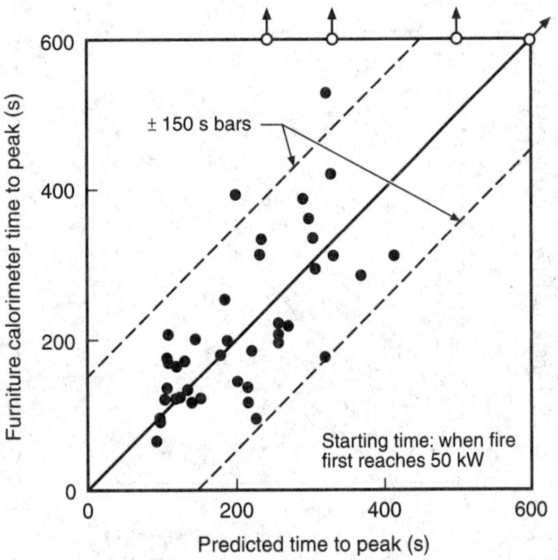

FIG. 4-17G. *Prediction of time to peak full-scale HRR, according to Equation 8.*

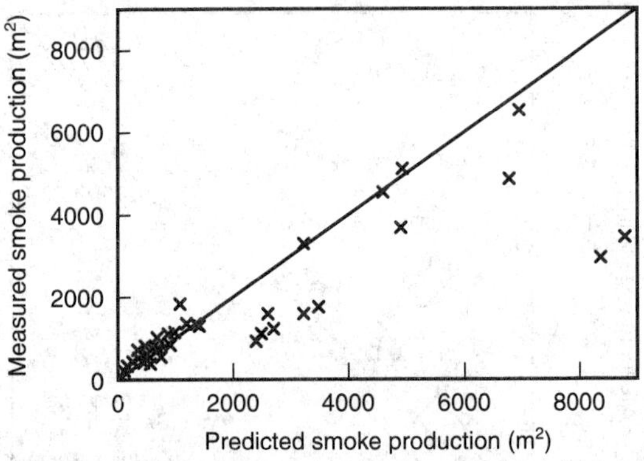

where

$$p_1 = 0.7 \cdot \frac{Q_{tot}\sigma_f}{\Delta h_{c,eff}} \qquad (12)$$

and

$$p_2 = s_o - 0.15s_o^{1.13} \qquad (13)$$

with

$$S_o = \frac{Q_{tot}\sigma_f\left(1.25 - 0.00188\dot{q}''_{35-60}\right)}{\Delta h_{c,eff}} \qquad (14)$$

and

$\dot{q}''_{35-60}$ = the 60 sec average HRR value from the cone calorimeter (kW/m²).

where

SmP = full-scale smoke production (m²),

Q_{tot} = full-scale total heat release (MJ),

σ_f = cone calorimeter average specific extinction area (m²/kg¹), and

$\Delta h_{c,eff}$ = cone calorimeter average effective heat of combustion (MJ/kg¹).

The results are plotted in Figure 4-17H. The agreement for fires showing moderate smoke development (less than about 1,500 m²) is remarkably good. Even for fires with seriously large smoke production, the actual measurements are either close to predicted values or lower; thus, the model is a conservative estimate under all circumstances.

The component model: Predictions of full-scale HRR from bench-scale data have normally required that the bench-scale test be done on a representative sample. For composite products, such as upholstered furniture, this means that testing is to be done on a bench-scale *composite*. Manufacturers have long sought to simplify matters by being able to test *components* and then to use some additional computational techniques in order to predict how a tested composite would have behaved. Until recently, such predictions

FIG. 4-17H. *Prediction of smoke production, according to Equations 9 through 14.*

were not possible to make. During the CBUF research, however, a technique was developed that allows a type of component testing to be utilized. In this procedure, foams are tested in the cone calorimeter with a special methane-flame burner, while fabrics are tested over a standard foam and with a different type of supplementary methane burner. A mathematical model is then used to predict the expected behavior of the composite. Further details are in the CBUF report.[34]

Toxic gases: Human incapacitation and fatalities in fires are much more commonly due to toxic gas effects than to direct burn injuries. This would seem to suggest that test methods should, first and foremost, measure toxic gases. This view, however, is over-simplified and does not reflect actual fire behavior realities. To understand this, it is best to consider the options for a manufacturer to reduce the hazard of upholstered furniture from toxic fire gases. The manufacturer could: (1) reduce the "per-gram" toxicity of the combustion products; or (2) reduce the *rate* at which combustion products are emitted. The effects of fire toxicity could be equally well mitigated by the first strategy or the second. In practice, it has been found that the first strategy is not feasible.[35] That is, materials are simply not to be found on the marketplace that show much reduced "per-gram" toxicities. In fact, most common materials available for constructing furniture exhibit very similar values. What the manufacturer can do, however, is to design the furniture in such a way that the burning *rate* is greatly reduced. This burning rate (which is a mass release rate) can be viewed as closely related to the HRR variable examined at length above. The range of peak HRR values for upholstered furniture has been seen to range from about 30 kW to 3,000 kW, which is an enormous variation. Thus, it can be concluded that toxic effects of upholstered furniture fires will be directly mitigated as the item's HRR is reduced. In addition, there are *no viable alternative strategies*. More detailed bases of these conclusions are also discussed in the CBUF study.[34]

PRODUCING FIRE-SAFE FURNITURE

The fire behavior of upholstered furniture can be optimized by careful material selection, design considerations, and other factors. The materials most suitable from a fire safety point of view are ones that are resistant to cigarette ignition, resistant to small-flame ignition, and produce only limited fire involvement when confronted with a large-source ignition. Some qualitative guidance on cigarette and small-flame ignition resistance is summarized in Table 4-17D. For fire performance once an item *is* ignited, some of the more important available strategies are summarized below. More comprehensive advice is available.[34]

TABLE 4-17D. *Upholstered Furniture Components Listed in Approximate Order of Descending Ignition Resistance*

Ignition Resistance	Cover Fabric	Padding	Interliner Fabric	Weltcords	Construction Parameters
A. Cigarette Ignition Resistance					
HIGH	Wool, PVC	Specialty foams†	Aluminized fabrics	Aluminized	Flat areas
▲	Heavy thermoplastics	Polyester batting	Neoprene sheets	PVC	Flat areas near
			Vinyl coated glass fab.	Thermoplastics	weltcord
	Cellulose/thermoplastics blends	SR PU	Novoloid felts	SR-treated	Tufts
	(depending on thermoplastic	SR cellulosic batting	Thermoplastic fabrics	cellulosics	Crevices
	percentages)	Untreated PU	Cellulosic fabrics	Cellulosics	
	Light thermoplastics	Mixed fiber batting			
▼	Light cellulosics	Cellulosic batting			
LOW	Heavy cellulosics				
B. Small-flame Ignition Resistance and Fire Growth					
HIGH	FR wool	Specialty foams‡	Aluminized, gas imperme-	Effect of	Flat areas
▲	Wool, PVC-coated cellulosics‡	FR cellulosic batting	able fabrics	weltcord has	Vertical areas
	Cellulosics‡	FR PU	Neoprene sheets‡	not been	Corner areas
		Cellulosic batting	Novoloid fabrics§	investigated,	
	Thermoplastics	Polyester batting	Aramid fabrics§‡	believed minor	
		Untreated PU	Vinyl-coated glass fabrics‡		
▼		Latex foam	FR cellulosic fabrics‡§		
LOW			Cellulosic fabrics†§		
			Thermoplastic fabrics		

SR—smolder resistant; FR—flame resistant; PU—polyurethane foam.
* Data on the behavior of acrylic are sparse but it seems to act more like cellulosics (smolder) than thermoplastics.
† Neoprene; combustion modified, high-resiliency PU.
‡ Heavier materials have higher flame ignition resistance and generally higher heat release and lower flame-spread rate.
§ Fabrics here include woven, knitted, and nonwoven structures.

Improving the Fabric

Fire behavior can be improved by selecting well-performing fabric categories. These include use of wool, modacrylic, FR-treated fabrics, and fabrics that do not show a strong melting tendency. For such fabrics, increased fabric weight usually corresponds to improved fire behavior. Thermoplastic fabrics are often troublesome; especially, the use of polyolefin (polypropylene) fabrics is to be avoided. Other thermoplastics usually perform better in flaming fires when they are of lighter weights than of heavier weights. Loosely woven or lofted fabrics are usually less desirable than flat, tight weaves.[36] In many cases, fabric backcoatings can be successfully used. Backcoatings may not be as effective as an interliner or as a substantially improved foam; nonetheless, they can significantly improve the fire performance of many fabrics without creating adverse usability or aesthetic effects. Backcoatings have been especially favored by U.K. manufacturers.

Fabrics comprising 100 percent cellulosic fibers (cotton, rayon, linen) usually show difficulties in passing cigarette-ignition tests, however. Cellulosic/thermoplastic blends, where the cellulosic component is less than 80 percent of the fabric weight, prove to be significantly more smolder resistant than pure cellulosics.[37] For a 100 percent cellulosic fabric to meet cigarette ignition requirements, it is very difficult if the fabric weight is greater than 440 g/m^2 (13 oz/yd^2), but is significantly less difficult[37] for fabrics lighter than 440 g/m^2. Thermoplastic fabrics of any weight rarely show any problems in meeting cigarette resistance requirements.

Improving the Foam

Essentially 100 percent of the upholstered furniture in the United States uses polyurethane foam, as do most (around 80 percent) of the mattresses. Because of the combustible nature of polyurethane foam, the simplest strategy to increase fire safety is to decrease the amount of foam used. As shown above, the peak HRR of a fire is highly dependent upon the mass of the upholstery. Another strategy involves substituting ordinary or (HR) grades of PU foam with an improved foam, possibly CMHR or one of the newer improved types. An innovative solution in this area has recently been commercialized, whereby polyols with substantially improved fire performance[38] are being seen as one way of significantly increasing furniture fire performance without requiring fundamental changes in furniture-making techniques. Composite foam strategies can be explored. An inexpensive base foam is in some way covered or coated with a foam that has FR properties. Early efforts in this vein have not yet shown significant improvement, but the general strategy is still considered promising. In rail-car seating, neoprene foam is often used. Neoprene foam can show excellent fire performance, but it is usually too costly and too dense for other applications.

Using Interliners

Fire-barrier types of interliners are among the most promising of the strategies that a manufacturer can adopt in order to increase the fire performance of upholstered furniture. Considered here are only modern interliners, of a type suitable for domestic or contract furniture use. There are lightweight (typically 70 to 170 g/m^2) interliners commercially available. Interliners can be woven, knitted, or nonwoven. Nonwoven interliners, however, often have the drawback that local density cannot be well controlled. Thus, the performance may not be as predictable as desired, due to existence of thin spots.

Several dozen different materials have been invented and proposed for interliner use. Commercially, the most common ones are: glass fiber, aramid (e.g., Kevlar®), and a proprietary glass fiber/FR cotton construction.

The interliner offers protection, since as the polyurethane foam underneath begins to melt, its upper surface moves steadily downward and away from the interliner which stays in place and acts as a thermal radiation shield. It also acts as a flame arrester, preventing the flame from reaching the foam. In a noninterliner system, the retreating surface of the foam is often "chased" by the burning fabric. By using an interliner, however, the burning fabric stays where it is while the foam melts away from it. This decreases the heat transfer to the foam and, thus, slows the melting and reduces the HRR. An interliner can also release flame-retardant chemicals into the gas phase and thus reduce flaming. Most of the common interliners, including the aramid and glass-fiber types, are purely nonreactive systems. At least one proprietary interliner, however, is heavily loaded with flame retardants. "Best quality" interliner-containing chairs show only limited burning of the fabric. The foam is fully protected. However, in some cases, the interliner provides protection to the foam only for a certain amount of time. Eventually, however, the foam burns up.

The most cost-effective way to utilize interliner technology is when the fabric supplier already laminates (attaches by fully gluing on) the interliner to the back of the upholstery fabric prior to delivering it to the furniture manufacturer. The average chair requires the use of about 4 m² of interliner.

The general term *interliner* sometimes is taken to encompass a polyester fiber batting *topper*, which is an intermediate layer placed below the fabric and on top of the foam. Such polyester toppers are often used to improve the cigarette-resistance of furniture. They can have undesirable effects on flaming-fire behavior, however, since the low-density polyester material is readily combustible.

Improvements Due to Design

A powerful strategy is the use of panel-mounting rather than webbing to support chair cushioning. Any such strategies that reduce the tendency to pool burning can cause a major reduction of fire hazard. Studies in England[39] demonstrated that about 50 percent of the peak HRR from an upholstered chair can be accounted for by pool burning.

The effects of button-tufting are not fully resolved,[34,40] but there is some evidence to suggest that they may, in certain cases, be problematic.

Fire performance will be improved if cushions are tightly upholstered and not sewn in a loose or floppy fashion.[40]

A fabric skirt along the lower edge of an upholstered chair is an easily combustible feature that can add to fire severity[40]; such skirts should be avoided where possible.

Wide gaps (>75 mm) between seat and back cushions and between arm cushions and seat cushions will improve the fire performance.[40]

Improvements Due to Frames

It is important to make sure that wood frame joints are strong and that metal joining plates are not used which could fail early in a fire. Thermoplastic material frames and frameless chair designs have generally unfavorable fire properties. For office chairs, make sure that the hard-plastic materials used do not melt and ignite readily and have adequate HRR properties.

CONCLUSIONS

For adequate fire safety of upholstered furniture, both cigarette ignition (smoldering) and flaming behavior must be considered. Assurance of hazard reduction cannot be had unless performance is demonstrated to be adequate on both scores. Separate assessment is required, since any given test method only examines flaming or smoldering performance, but not both.

Cigarette-ignition resistance of upholstered furniture is evaluated by test methods which, by now, have been long established as standard techniques and are published by both ASTM and NFPA. Cigarette-ignition resistance of mattresses has been regulated by the

federal government ever since the early 1970s. A newer issue is the possibility of reducing cigarette-ignition hazards by making safer cigarettes. This approach has been now comprehensively researched, but is not yet seen in the marketplace.

The marketplace for contract furniture has changed significantly since the introduction of the California T.B. 133 test, which examines the flaming-fire behavior of upholstered furniture. While this test's requirements are mandatory in only a limited number of situations, its effects have been widespread throughout the U.S. furniture industry. This has been due to manufacturers' awareness of the consequences of product liability litigation. A manufacturer whose furniture was involved in a fire is in a much stronger position if his/her furniture conformed to the state of the art in fire-safe design (as evidenced by passing the T.B. 133 test) than if he/she ignored such aspects of design.

The T.B. 133 test, in its current version, is based on evaluating the heat release rate (HRR) of the test item. The HRR is considered the single most important aspect of fire hazard. Separate testing for toxic gases is not considered necessary, since low HRR normally gives a good assurance that toxic effects are also well controlled.

The recently completed CBUF research program has resulted in a massive amount of recommendations to designers of furniture on how to design fire-safe furniture and how to test furniture in a comprehensive way. Additional modeling and testing information is contained in that report, which should be consulted by those wishing further guidance.

BIBLIOGRAPHY

References Cited

1. Clarke, F., II, and Ottoson, J., "Fire Death Scenario and Fire Safety Planning," *Fire Journal*, Vol. 70, May 1976, pp. 20–22, 117, 118.
2. Babrauskas, V., "Bench-Scale Predictions of Mattress and Upholstered Chair Fires," *Fire and Flammability of Furnishings* (ASTM STP 1233), American Society for Testing and Materials, W. Conshohocken, 1994, pp. 50–62.
3. Braun, E., *et al.*, "Cigarette Ignition of Upholstered Chairs," *Journal of Consumer Product Flammability*, Vol. 9, 1982, pp. 167–183.
4. Hafer, C. A., and Yuill, C. H., *Characterization of Bedding and Upholstery Fires*, Southwest Research Institute, San Antonio, TX, 1970.
5. Gann, R. G., Babrauskas, V., Peacock, R. D., and Hall, J. R., Jr., "Fire Conditions for Smoke Toxicity Measurement," *Fire and Materials*, 18, 1994, pp. 193–199.
6. Babrauskas, V., and Krasny, J. F., "Fire Behavior of Upholstered Furniture," Monograph MN-173, 1985, National Bureau of Standards, Gaithersburg, MD.
7. Babrauskas, V., "Flammability of Upholstered Furniture with Flaming Sources," *Cellular Polymers*, Vol. 8, 1989, pp. 198–224.
8. Fairall, P., "Analysis of CPSC 40 Chair Test Program," 1984, U.S. Consumer Product Safety Commission, Washington, DC.
9. Damant, G. H., "California's Flammability Standards for Residential Upholstered Furniture—Twenty Years Later," presented at the 19th Intl. Conf. on Fire Safety, Millbrae, CA, Product Safety Corp.,1994.
10. "The Furniture and Furnishings (Fire) (Safety) Regulations 1988," *Statutory Instruments 1988*, No. 1324, HMSO, London, 1988.
11. "The Furniture and Furnishings (Fire) (Safety) (Amendment) Regulations 1989," *Statutory Instruments 1989*, No. 2358, HMSO, London, 1989.
12. Vlot, I., "Fire Safety of Textile Products—Legislation and Standardization," Report 48, 1989, Consumer Safety Institute, Amsterdam.
13. "Flammability Information Package," California Bureau of Home Furnishings, North Highlands, CA, 1987.
14. Federal Register, "Flammability of Interior Materials—Passenger Cars, Multipurpose Passenger Vehicles, Trucks, and Buses," Motor Vehicle Safety Standard No. 302, Vol. 35, 1975.
15. Gann, R. G., Harris, R. H., Jr., Krasny, J. F., Levine, R. S., Mitler, H. E., and Ohlemiller, T. O., "The Effect of Cigarette Characteristics on the

Ignition of Soft Furnishings," NBS Technical Note 1241, 1988, National Bureau of Standards, Gaithersburg, MD.

16. "Final Report, Fire Safe Cigarette Act of 1990," Consumer Product Safety Commission, Washington, DC, 1993.

17. Ohlemiller, T. J., Villa, K. M., Braun, E., Eberhardt, K. R., Harris, R. H., Jr., Lawson, J. R., and Gann, R. G., "Test Methods for Quantifying the Propensity of Cigarettes to Ignite Soft Furnishings," NIST Special Publication 851, 1993, Natl. Inst. Stand. and Technol., Gaithersburg, MD.

18. "Flammability Test Procedure for Seating Furniture for Use in High-Risk and Public Occupancies," Technical Bulletin No. 133, 1984, Bureau of Home Furnishings, California State Department of Consumer Affairs, North Highlands, CA.

19. Parker, W. J., Tu, K.-M., Nurbakhsh, S., and Damant, G. H., "Furniture Flammability: An Investigation of the California Technical Bulletin 133 Test. Part III: Full-Scale Chair Burns," NISTIR 4375, 1990, U.S. Natl. Inst. Stand. and Technol., Gaithersburg, MD.

20. "Flammability Test Procedure for Mattresses for Use in High-Risk Occupancies," Technical Bulletin No. 121, 1980, Bureau of Home Furnishings, California State Department of Consumer Affairs, North Highlands, CA.

21. "Flammability Test Procedure for Mattresses for Use in Public Buildings," Technical Bulletin No. 129, 1992, Bureau of Home Furnishings and Thermal Insulation, California State Department of Consumer Affairs, North Highlands, CA.

22. Damant, G. H., "Development of a Test Method for the Flammability of Stacking Chairs," presented at the 19th Intl. Conf. on Fire Safety, Product Safety Corp., Millbrae, CA, 1994.

23. FAA, "Flammability Requirements for Aircraft Seat Cushions," Code of Federal Regulations 14 CFR, Parts 25, 29, 121, 1984, Federal Aviation Administration, Washington, DC.

24. Babrauskas, V., "Specimen Heat Fluxes for Bench-Scale Heat Release Rate Testing," *INTERFLAM '93*, Interscience Communications Ltd., London, 1993, pp. 57–74.

25. Lee, B. T., and Wiltshire, L. W., "Fire Spread Models of Upholstered Furniture," *Journal of Fire and Flammability*, Vol. 3, 1972, pp. 164–175.

26. Krasny, J. F., and Babrauskas, V., "Burning Behavior of Upholstered Furniture Mockups," *Journal of Fire Sciences*, Vol. 2, 1984, pp. 205–231.

27. Babrauskas, V., and Wetterlund, I., "The Role of Flame Flux in Opposed-Flow Flame Spread," *Procedures*, Fire and Materials, 3rd Intl. Conf., Interscience Communications Ltd., London, 1994, pp. 75–87.

28. Babrauskas, V., and Peacock, R. D., "Heat Release Rate: The Single Most Important Variable in Fire Hazard," *Fire Safety J.*, 18, 1992, pp. 255–272.

29. Babrauskas, V., Lawson, J. R., Walton, W. D., and Twilley, W. H., "Upholstered Furniture Heat Release Rates Measured with a Furniture Calorimeter," NBSIR 82-2604, 1982, U. S. National Bureau of Standards, Gaithersburg, MD.

30. "Upholstered Furniture: Burning Behaviour—Full-Scale Test," (NT FIRE) 032, 2nd ed., 1991, NORDTEST, Espoo, Finland.

31. Babrauskas, V., and Grayson, S. J., eds., *Heat Release in Fires*, Elsevier Applied Science Publishers, London, 1992.

32. Babrauskas, V., "Development of the Cone Calorimeter—A Bench-Scale Heat Release Rate Apparatus Based on Oxygen Consumption," *Fire and Materials*, 8, 1984, pp. 81–95.

33. Babrauskas, V., and Wetterlund, I., "Fire Testing of Furniture in the Cone Calorimeter—The CBUF Test Protocol," SP Report 1994:32, 1994, SP Swedish National Testing and Research Institute, Borås, Sweden.

34. Sundström, B., ed., "Fire Safety of Upholstered Furniture—The Final Report on the CBUF Research Programme," Report EUR 16477 EN, 1995, Directorate-General Science, Research and Development (Measurements and Testing), European Commission, Distributed by Interscience Communications Ltd., London.

35. Babrauskas, V., "Toxicity, Fire Hazard, and Upholstered Furniture," *Third European Conf. on Furniture Flammability*, Brussels, Interscience Communications Ltd., London, 1992, pp. 125–133.

36. Shelby-Williams, *TB-133 User-Friendly Selector*, Shelby-Williams Industries, Inc., Morristown, TN, 1993.

37. Damant, G. H., "Cigarette Ignition of Upholstered Furniture," presented at the 20th Intl. Conf. on Fire Safety, Product Safety Corp., Millbrae, CA, 1995.

38. Craig, T. A., Lear J. J., Motte, P., "New Generations Flame Retardant Polyols for Slabstock Applications," Conference *Proceedings*, Polyurethane World Congress, Oct. 10-13, Vancouver, BC, 1993, pp. 54-59.

39. Ames, S. A., FIRE: "Materials, Mechanisms and Measurements—Fire Chemistry Discussion Group Seminar," Bolton University, England, 14th April 1993.

40. Barile, P., "A Systematic Approach for Predicting Compliance with Technical Bulletin 133 for Untested Combinations of Chair Styles and Fabrics," presented at American Furniture Manufacturers Assn. Fire Safety Conf., Greensboro, NC, 1993.

References

ASTM D3675, *Standard Test Method for Surface Flammability of Flexible Cellular Materials Using a Radiant Heat Energy*, American Society for Testing and Materials, W. Conshohocken, PA, 1993.

ASTM E162, *Standard Test Method for Surface Flammability of Materials Using a Radiant Heat Energy Source*, American Society for Testing and Materials, W. Conshohocken, PA, 1994.

ASTM E1354, *Standard Test Method for Heat and Visible Smoke Release Rates for Materials and Products Using an Oxygen Consumption Calorimeter*, American Society for Testing and Materials, W. Conshohocken, PA, 1994.

ASTM E1474, *Standard Test Method for Determining the Heat Release Rate of Upholstered Furniture and Mattress Components or Composites Using a Bench Scale Oxygen Consumption Calorimeter*, American Society for Testing and Materials, W. Conshohocken, PA, 1992.

ASTM E1537, *Standard Test Method for Fire Testing of Real Scale Upholstered Furniture Items*, American Society for Testing and Materials, W. Conshohocken, PA, 1994.

ASTM E1590, *Standard Test Method for Fire Testing of Real Scale Mattresses*, American Society for Testing and Materials, W. Conshohocken, PA, 1994.

BIFMA, *Standard and Institutional Furniture Manufacturer's Association First Generation Voluntary Upholstered Furniture Flammability Standard for Business and Institutional Markets*, Business and Institutional Furniture Manufacturer's Association, Grand Rapids, MI, 1980.

BSI, "Fire Tests for Furniture, Part 1: Methods of Test for the Ignitability by Smokers' Materials of Upholstered Composites for Seating," BS 5852, Part 1: 1979, British Standards Institution, London.

BSI, "Fire Tests for Furniture, Part 2: Methods of Test for the Ignitability of Upholstered Composites for Seating by Flaming Sources," BS 5852, Part 2: 1982, British Standards Institution, London.

CGSB, "Method of Test for the Combustion Resistance of Mattress; Cigarette Test," 1977, Canadian Government Specification Board, Ottawa, Ontario, Canada.

CPSC, "Standard for the Flammability of Mattresses (and Mattress Pads)," Code of Federal Regulations, Part 1632, Final Rule, Oct. 10, 1984, U.S. Consumer Product Safety Commission, Washington, DC.

CPSC, "Proposed Standards for the Flammability (Cigarette Ignition) of Upholstered Furniture," Code of Federal Regulations, Part 1633, U.S. Consumer Product Safety Commission, Washington, DC.

GSA, "Flame Resistance of Cloth; Vertical Methods 5903.2," Federal Test Standard No. 191, 1971, General Services Administration, Washington, DC.

ISO, International Standard—Fire Tests—Reaction to Fire—Part 1: Rate of Heat Release from Building Products (Cone Calorimeter Method), ISO 5660-1:1993(E), 1993, International Organization for Standardization, Geneva, Switzerland.

ISO, International Standard—Fire Tests—Full-Scale Room Test for Surface Products, ISO 9705:1993(E), 1993, International Organization for Standardization, Geneva, Switerland.

Port Authority, "Specifications Governing the Flammability of Upholstery Materials and Plastic Furniture," 1981, The Port Authority of New York and New Jersey, New York, NY.

UFAC, *Important Consumer Safety Information from UFAC*, Upholstered Furniture Action Council, High Point, NC, 1984.

UL 1056, *Standard for Safety Fire Test of Upholstered Furniture*, Underwriters Laboratories Inc., Northbrook, IL, 1989.

NFPA Codes, Standards, and Recommended Practices

Reference to the following NFPA codes, standards, and recommended practices will provide further information on upholstered furniture and

mattresses discussed in this chapter. (See the latest version of *The NFPA Catalog* for availability of current editions of the following documents.)

NFPA 258, *Standard Research Test Method for Determining Smoke Generation of Solid Materials*

NFPA 260, *Standard Methods of Tests and Classification System for Cigarette Ignition Resistance of Components of Upholstered Furniture*

NFPA 261, *Standard Method of Test for Determining Resistance of Mock-Up Upholstered Furniture Material Assemblies to Ignition by Smoldering Cigarettes*

NFPA 264, *Standard Method of Test for Heat and Visible Smoke Release Rates for Materials and Products Using an Oxygen Consumption Calorimeter*

NFPA 264A, *Standard Method of Test for Heat Release Rates for Upholstered Furniture Components or Composites and Mattresses Using an Oxygen Consumption Calorimeter*

NFPA 266, *Standard Method of Test for Fire Characteristics of Upholstered Furniture Exposed to Flaming Ignition Source*

NFPA 267, *Standard Method of Test for Fire Characteristics of Mattresses and Bedding Assemblies Exposed to Flaming Ignition Source*

Additional Reading

Ames, S. A., "Measuring Fire Growth in Furniture," *Fire Surveyor*, Vol. 17, No. 2, Apr. 1988, pp. 4–10.

Ames, S. A., Babrauskas, V., and Parker, W. J., "Upholstered Furniture: Prediction by Correlations," Chapter 15, *Heat Release in Fires*, Elsevier Applied Science, New York, 1992, pp. 519–543.

Ames, S. A., and Rogers, S. P., "Large and Small Scale Fire Calorimetry Assessment of Upholstered Furniture," Fifth International Fire Conference on Fire Safety, Interflam '90, September 3–6, 1990, Canterbury, England, Interscience Communications Ltd., London, England, 1990, pp. 221–232.

Babrauskas, V., "Bench-Scale Predictions of Mattress and Upholstered Chair Fires: Similarities and Differences," National Institute of Standards and Technology, Gaithersburg, MD, National Institute of Justice, Washington, DC, NISTIR 5152, March 1993.

Babrauskas, V., "Flammability of Upholstered Furniture with Flaming Sources," European Conference on Furniture Flammability, London, 1988.

Babrauskas, V., "Toxicity, Fire Hazard and Upholstered Furniture," Third European Conference on Furniture Flammability (EUCOFF '92), Nov. 1992, Brussels, Interscience Communications Ltd., London, England, 1992, pp. 125–133.

Babrauskas, V., Damant, G., and Nubakhsh, S., "Heat Release Rate Testing of Mattresses: Full-Scale Measurements and Bench-Scale Predictions," Abstracts of the First U. S. Symposium on Heat Release and Fire Hazard, Dec. 1991, San Diego, CA, 1991, pp. 61–63.

Babrauskas, V., and Krasny, J. F., "Prediction of Upholstered Chair Heat Release Rates from Bench-Scale Measurements" (ASTM STP).

Babrauskas, V., and Walton, W. D., "A Simplified Characterization of Upholstered Furniture Heat Release Rates," *Fire Safety Journal*, Vol. 11, 1986, pp. 181–192.

Babrauskas, V., *et al.,* "Cone Calorimeter Used for Predictions of the Full-Scale Burning Behavior of Upholstered Furniture," *Proceedings* of the Fourth International Conference and Exhibition on Fire and Materials, November 15–16, 1995, Crystal City, VA, 1995, pp. 203–217.

Baroudi, D., and Kokkala, M., "Flame Spread and Heat Release Rate Model for a Burning Mattress," VTT-Technical Research Center of Finland, Espoo, ISBN 0–9516320–9–4; *Proceedings* of the Seventh International Interflam Conference, Interflam '96, March 26–28, 1996, Cambridge, England, Interscience Communications Ltd., London, England, 1996, pp. 37–46.

Bentley, R. L., and Buszard, D. L., "New Method for Measuring Post-Ignition Behavior of Upholstered Furniture Composites," Fifth International Conference, Flame Retardants '92, January 22–23, 1992, London, England, Elsevier Applied Science, New York, 1992, pp. 242–251.

Bernskiold, A., "Erfarenhetsaterforing fram brandprovningar av stoppade sittmobler. [Fire Tests of Upholstery Fabrics with Smoldering Cigarettes.] (Abstracts in English), Moabelinstitutet, Stockholm, Sweden, Report 63, 1991.

Brown, D., "Furniture Flammability—An Issue of Life and Death," European Bureau of Consumer Unions, European Commission DG 3, EGOLF

and CFPA Europe, Fire and Furnishing in Building and Transport, Nov. 6–8, 1990, Luxembourg, Belgium, 1990, pp. 1–11.

Chandler, R., and Schmitt, F., "Alternative Material for Non-Combustible Furniture," First International Conference on Fire Materials, September 24–25, 1992, Arlington, VA, 1992, pp. 89–107.

Christian, S. D., "Current and Developing European and International Standards for Furniture and Textiles," Home Office, London, England, Paper 4, Interscience Communications Limited, Second European Conference on Furniture Flammability, June 27–28, 1990, Brussels, Belgium, 1990, pp. 4/1–6.

Christian, S. D., "Regulatory Position in the United Kingdom in Respect of Upholstered Furniture Used in Both Contract and Domestic Applications," International Conference on Fire Safety—Fire Retardant Technology and Marketing, March 25–28, 1990, New Orleans, LA, Fire Retardant Chemicals Assoc., Lancaster, PA, 1990, pp. 87–105.

Cleary, T. G., Ohlemiller, T. J., and Villa, K. M., "Influence of Ignition Source on the Flaming Fire Hazard of Upholstered Furniture," *Fire Safety Journal*, Vol, 23, 1994, pp. 79–102.

Colver, C. P., and Colver, J. C., "Review of Progress Toward the Development of an Upholstered Furniture Flammability Standard," *Journal of Fire Sciences*, Vol. 5, No. 5, 1987, pp. 326–337.

D'Silva, A. P., and Sorensen, N., "Flammability Aspects of Decorative Trimmings. Part 1. Flammability of Trimmings Used on Upholstered Furniture," *Journal of Fire Sciences*, Vol. 14, No. 1, Jan./Feb. 1996, pp. 26–49.

Damant, G. H., "Cigarette Ignition of Upholstered Furniture," *Journal of Fire Sciences*, Vol. 13, No. 5, Sept./Oct. 1995, pp. 337–349.

Damant, G.H. "Cigarette Ignition of Upholstered Furniture," *Proceedings* of the Twentieth International Conference on Fire Safety, Jan. 9–13, Millbrae, CA, Product Safety Corp., Sunnyvale, CA, 1995, pp. 1–12.

Damant, G.H. "Cigarette Ignition of Upholstered Furniture," *Proceedings* of the Fourth International Conference and Exhibition on Fire and Materials, Nov. 15–16, 1995, Crystal City, VA, 1995, pp. 155–169.

Damant, G. H., and Nurbakhsh, S., "Heat Release Tests of Mattresses and Bedding Systems. Part 1. Full Scale Fire Tests," Bureau of Home Furnishings and Thermal Insulation, North Highlands, CA, October 1991.

Damant, G. H., and Nurbakhsh, S., "Heat Release Tests of Mattresses and Bedding Systems," Fifth International Conference, Flame Retardants'92, January 22–23, 1992, London, England, Elsevier Applied Science, New York, 1992, pp. 252–277.

Dietenberger, M. A., "Upholstered Furniture: Detailed Model," Chapter 14, *Heat Release in Fires*, Elsevier Applied Science, New York, 1992, pp. 479–518.

Dwyer, R. W., *et al.*, "Effects of Upholstery Fabric Properties on Fabric Ignitabilities by Smoldering Cigarettes," *Journal of Fire Science*, Vol. 12, May/June 1994, pp. 268–283.

Fitzgerald, W. E., "Quantification of Fires: 1. Energy Kinetics of Burning in a Dynamic Room-Size Calorimeter," *Journal of Fire and Flammability*, Vol. 9, Oct. 1978, pp. 510–527.

Fritz, T. W., and Hunsberger, P.L., "Testing of Mattress Composites in the Cone Calorimeter," *Proceedings* of the Fourth International Conference and Exhibition on Fire and Materials, November 15–16, 1995, Crystal City, VA, 1995, pp. 171–179.

"Furniture and Furnishings Regulations," *Fire Surveyor*, Vol. 18, No. 1, Feb. 1989, pp. 5–13.

Gallagher, J. A., "Interliner Effect on the Fire Performance of Upholstery Materials," *Journal of Fire Science*, Vol. 11, No. 1, Jan./Feb. 1993, pp. 87–105.

Gandhi, S. and Spivak, S. M., "Survey of Upholstered Furniture Fabrics and Implications for Furniture Flammability," *Journal of Fire Science*, Vol. 12, May/June 1994, pp. 284–312.

Goodall, C. A., *et al.*, "Effect of Ventilation on the Fire Behavior of Upholstered Furniture," Fire Research Station, Borehamwood, England, ISBN 0–9516320–9–4; *Proceedings* of the Seventh International Interflam Conference, Interflam '96, March 26–28, 1996, Cambridge, England, Interscience Communications Ltd., London, England, 1996, pp. 13–25.

Graham, R. A., "Upholstered Furniture Controls—Position of the Fire Authorities," Greater Manchester Fire Dept., UK, European Commission DG 3, EGOLF and CFPA Europe, Fire and Furnishing in Buildings and Transport, November 6–8, 1990, Luxemburg, Belgium, 1990, pp. 1–11.

Grand, A. F., "Real Scale Fire Evaluation of Upholstered Furniture," Omega Point Laboratories, San Antonio, TX, Fire Retardant Chemicals Association, International Conference on Fire Safety, 1993 Spring Conference, March 21–24, 1993, New Orleans, LA, Fire Retardant Chemicals Assoc., Lancaster, PA, 1993, pp. 223–235.

Grand, A. F., "Unusual Aspects of Upholstered Chair Flammability Performance," First International Conference on Fire Materials, September 24–25, 1992, Arlington, VA, 1992, pp. 63–72.

Grand, A. F., Priest, D. N., and Stansberry, H. W., II, "Burning Characteristics of Upholstered Chairs," Omega Point Laboratories, Inc., Elmendorf, TX, American Society for Testing and Materials (ASTM), Fire and Flammability of Furnishings and Contents of Buildings, ASTM STP 1233, Sponsored by ASTM Committee E5 on Fire Standards and Its Subcommittee E05.32 on Research, December 7, 1992, Miami, FL, ASTM, Philadelphia, PA, 1992, pp. 63–82.

Gravigny, L., "Draft European Directive on the Fire Behavior of Upholstered Furniture," Fifth International Conference, Flame Retardants '92, January 22–23, 1992, London, England, Elsevier Applied Science, New York, 1992, pp. 238–241.

Gravigny, L., "Le projet de directive sur le comportement au feu des meubles rembourres. [Policy Project Upon Fire Behavior of Padded Furniture.]," Commission of the European Community, Brussels, Belgium, European Commission DG3, EGOLF and CFPA Europe, Fire and Furnishing in Buildings and Transport, November 6–8, 1990, Luxembourg, Belgium, 1990, pp. 1–8.

Griggs, D., "Furniture Flammability—A Textile Specialists Viewpoint," Fire Prevention, No. 210, June 1988, pp. 28–30.

Hirschler, M. M., and Shakir, S., "Comparison of the Fire Performance of Various Upholstered Furniture Composite Combinations (Fabric/Foam) in Two Rate of Heat Release Calorimeters: Cone and Ohio State University Instruments," Journal of Fire Sciences, Vol. 9, No. 3, May/June 1991, pp. 223–248.

Hirschler, M. M., and Shakir, S., "Fire Performance of Fabric/Foam Combinations as Upholstered Furniture Composites in Rate of Heat Release Equipment (Cone and Ohio State University Calorimeters)," Sixteenth International Conference on Fire Safety, Volume 16, Jan. 14–18, 1991, Millbrae, CA, Product Safety Corp., Sunnyvale, CA, 1991, pp. 239–258.

Hirschler, M. M., and Smith, G. F., "Flammability of Sets of Fabric/Foam Combinations for Use in Upholstered Furniture," Fire Safety Journal, Vol. 16, No. 1, 1990, pp. 13–31.

Hulme, B., "Rate of Fire Growth Test for Furniture," Fire Research News, Vol. 13, Spring 1991, pp. 17–18.

Jervis, R. L., "Missing Link: Fire-Resistant Furnishings," Fire Journal, Vol. 81, No. 6, 1987, p. 56.

Jungensen, L., "Smoke and Toxic Gas Suppressants for Upholstered Polyurethane Foam," CIBA-GEIGY Corp., Hawthorn, NY, Business Communications Co., Inc. (BCC), Recent Advances in Flame Retardancy of Polymeric Materials: Materials, Applications, Industry Developments, Markets, Vol. 2, May 14–16, 1991, Stamford, CT, Business Communications Co., Inc., Norwalk, CT, 1991, pp. 335–336.

Krasny, F., and Huang, D., "Small Flame Ignitability and Flammability Behavior of Upholstered Furniture Materials," NBSIR 88-3771, June 1988, National Bureau of Standards, Gaithersburg, MD.

Krasny, J. F., et al., "Development of a Candidate Test Method for the Measurement of the Propensity of Cigarettes to Cause Smoldering Ignition of Upholstered Furniture and Mattresses," NBSIR 81-2363, 1986, National Bureau of Standards, Gaithersburg, MD.

Krasny, J. F., "Simple Method for Reducing Cigarette Caused Upholstery Fires," Textile Chemist and Colorist, Vol. 24, No. 11, Nov. 1992.

Krasny, J. F., and Gann, R. G., "Relative Propensity of Selected Commercial Cigarettes to Ignite Soft Furnishings Mockups," NBSIR 86-3421, 1986, National Bureau of Standards, Gaithersburg, MD.

Krasny, J. F., and Parker, W. J., "Impact of the Room Enclosure on the Peak Heat Release Rates of Upholstered Furniture," Proceedings of the Fourth International Conference and Exhibition on Fire and Materials, November 15–16, 1995, Crystal City, VA, 1995, pp. 181–190.

Lewis, L. S., et. al. "Effects of Upholstery Fabric Properties on Fabric Ignitabilities by Smoldering Cigarettes. Part 2," Journal of Fire Science, Vol. 13, No. 6, Nov./Dec. 1995, pp. 445–471.

Marchant, R., "Furniture Fire Hazards Coming Under Control," Fire Prevention, No. 226, Jan./Feb. 1990, pp. 28–30.

Messa, S., "EGOLF-Proposal for a Research Project: Fire Behavior of Upholstered Furniture," Ricerche sul fuoco (LSF), Italy, European Commission DG 3, EGOLF and CFPA Europe, Fire and Furnishing in Buildings and Transport, November 6–8, 1990, Luxembourg, Belgium, 1990, pp. 1–7.

Messa, S., "Undergoing Activities for European Harmonization of Building Products and Upholstered Furniture Reaction to Fire Tests," LSF Fire Laboratories, Italy, CIB W14/93/3 (J); First Proceedings of the Japan Symposium on Heat Release and Fire Hazard, Session 1, Development of Codes and Standards on Material Fire Safety Performance, May 10–11, 1993, Tsukuba, Japan, 1993, pp. I/41–79.

"Methods of Test for Assessment of Ignitability of Upholstered Seating by Smoldering and Flaming Ignition Sources," British Standards Institution, England, BS 5852:1990.

Migan, M., "Point de vue des autorites—position francaise. [Policy Project Concerning Fire Behavior of Padded Furniture.]," European Commission DG 3, EGOLF and CFPA Europe, Fire Furnishing in Building and Transport, Nov. 6–8, 1990, Luxembourg, Belgium, 1990, pp. 1–21.

Mischutin, V., "Flame-Retardant Home Furnishings," American Dyestuff Reporter, Vol. 76, No. 3, 1987, pp. 38–42, 49.

Mischutin, V., "New Clear FR Finish for Upholstery and Other Fabrics. Part 1," 1992 Spring Conference on Technical and Marketing Issues Impacting the Fire Safety of Building and Construction and Home Furnishings Applications, March 29-April 1, 1992, Orlando, FL, Technomic Publishing Co., Lancaster, PA, 1992, pp. 199–212.

Mitler, H. E. and Tu, K. M., "Effect of Ignition Location on Heat Release Rate of Burning Upholstered Furniture," National Institute of Standards and Technology, Gaithersburg, MD, NISTIR 5499, Sept.r 1994; National Institute of Standards and Technology, Annual Conference on Fire Research: Book of Abstracts, Oct. 17–20, 1994, Gaithersburg, MD, 1994, pp. 121–122.

Montagna, F. C., "Mattress Fires," Fire Engineering, Vol. 148, No. 4, Apr. 1995, pp. 18, 20.

Morotz-Cecei, K., and Beda, L., "Comparative Testing of the Flammability of Upholstery Textiles," Journal of Thermal Analysis, Vol. 32, No. 3, 1987, pp. 901–908.

Moulen, A. W., and Grubits, S. J., "Paper Ignition Sources Used to Examine the Fire Behavior of Seating and Bedding," Technical Record 474, (undated), Experimental Building Station, Department of Housing and Construction, Chatswood, N.S.W., Australia.

Myllymaki, J. and Baroudi, D., "Prediction of Heat Release Rate of Upholstered Furniture Using Integral Formulation," VTT-Technical Research Center of Finland, Espoo, ISBN 0-9516320-9-4; Proceedings of the Seventh International Interflam Conference, Interflam '96, March 26–28, 1996, Cambridge, England, Interscience Communications Ltd., London, England, 1996, pp. 27–35.

Natoli, G. S. and Calabrese, R. A., "Assessing Fire Performance of Furniture Cushioning Materials by BS-5852 vs California TB-133," Proceedings of the Polyurethanes World Congress Technical Session B on Furnishings, October 10–13, 1993, England, Technomic, Lancaster, PA, 1993, pp. 48–53.

NORDTEST, "Upholstered Furniture: Burning Behavior-Full Scale Test," NORDTEST, Espoo, Finland, NT FIRE 032, Edition 2, 1992.

Nurbakhsh, S., "Furniture Flammability Study. Part 1. Development of Full Scale Fire Testing Facility," Journal of Fire Sciences, Vol. 9, No. 5, Sept./Oct. 1991, pp. 390–405.

Nurbakhsh, S., Mikami, J. F. and Damant, G. H., "Full Scale Fire Tests of Upholstered Furniture With Rate of Heat Release Measurements," Journal of Fire Sciences, Vol. 9, No. 5, Sept./Oct. 1991, pp. 369–389.

Ohlemiller, T. J. and Villa, K. M., "Furniture Flammability: An Investigation of the California Bulletin 133 Test. Part 2. Characterization of the Ignition Source and a Comparable Gas Burner," National Institute of Standards and Technology, Gaithersburg, MD, NISTIR 4348, June 1990.

Parker, W. J., "Model to Predict the Burning Behavior of Upholstered Furniture Based on Test Data From Foam and Fabric Separately," Proceedings of the Fourth International Conference and Exhibition on Fire and Materials, November 15–16, 1995, Crystal City, VA, 1995, pp. 219–228.

Parker, W. J., et al., "Furniture Flammability: An Investigation of the California Technical Bulletin 133 Test. Part 3. Full Scale Chair Burns," National Institute of Standards and Technology, Gaithersburg, MD, Bureau of Home Furnishings and Thermal Insulation, North Highlands, CA, NISTIR 4375, July 1990, 43 pages.

Parrish, D. B. and Browning, S. A., "Flammability Testing of Upholstered Furniture Composites by the Oxygen Index Calorimeter,"

First International Conference on Fire and Materials, Sept. 24–25, 1992, Arlington, VA, 1992, pp. 125–142.

Paul, K., "Review of the Role and Uses of the Cone Calorimeter in Fire Tests of Upholstery and Other Material Composites," *Progress in Rubber and Plastics Technology*, Vol. 9, No. 4, 1993, pp. 286–320.

Paul, K. T., "Furniture and Fire Safety—The Need to Control," Paramount Exhibitions and Conferences, FIREX First International Conference on Building Products and Furniture—Assessing Their Fire Performance, Conference Papers, June 12–14, 1990, Birmingham, England, 1990, pp. 1–20.

Paul, K. T. and King, D. A., "Burning Behavior of Domestic Upholstered Chairs Containing Different Types of Polyurethane Foams," *Fire Safety Journal*, Vol. 16, No. 5, 1990, pp. 389–410; Plastics and Rubber Institute and British Plastics Federation Fourth International Conference, Flame Retardants 1990, Jan. 17–18, 1990, London, England, Elsevier Applied Science, New York, 1990, pp. 265–296.

Powell, D., "The Role of Combustion Modified Foams in Furniture: A Closer Look," *Fire Prevention*, No. 216, Jan./Feb. 1989, pp. 27–30.

Purser, D. A., "Toxicity, Toxic Hazard and Furniture in Fires," Huntingdon Research Center Ltd., UK, Paper 8, Interscience Communications Limited, Second European Conference on Furniture Flammability, June 27–28, 1990, Brussels, Belgium, 1990, pp. 8/1–17.

Quintiere, J. G., "Furniture Flammability: An Investigation of the California Technical Bulletin 133 Test. Part 1. Measuring the Hazards of Furniture Fires," National Institute of Standards and Technology, Gaithersburg, MD, NISTIR 4360, July 1990.

Rhyne, A. L. and Spears, A. W., "Sensitivity of a Cigarette Ignitability Index to Hypothetical Shifts in the Distribution of Upholstery Furniture Fabrics," *Journal of Fire Sciences*, Vol. 8, No. 1, Jan./Feb. 1990, pp. 3–22.

Rogowski, B. F. W., "Investigating the Contribution to Fire Growth of Combustible Materials Used in Building Components," *Proceedings* of the Conference on New Technology to Reduce Fire Losses and Costs, Elsevier, New York, 1986, pp. 88–105.

Rose, R. S., "Meeting California Bulletin 133 Criteria: A Guide to Selecting Polyurethane Foam, Fabric, and Backcoating for Furniture Cushions," Great Lakes Chemical Corp., West Lafayette, IN, Fire Retardant Chemicals Association, 1992 Fall Conference, Industry Speaks Out on Flame Retardancy: Coatings; Polymers and Compounding; Test Method Development; New Products, Oct. 27–30, 1992, Charleston, SC, Technomic Publishing Co., Lancaster, PA, 1992, pp. 23–27.

Ryynanen, T., "Asetukset patjojen ja pehmustettujen huonekalujen paloturvallisuudesta. [Decrees on the Fire Safety of Mattresses and Upholstered Furniture.]," *Palontorjuntateknikka*, Feb. 1992, pp. 26–27.

Smiecinski, T. M. and Grace, O. M., "Effect of Fabrics on Performance of Upholstered Furniture in the California Technical Bulletin 133 Test Procedure," Fifteenth International Conference on Fire Safety, Volume 15, Jan. 8–12, 1990, Millbrae, CA, Product Safety Corp., Sunnyvale, CA, 1990, pp. 23–26.

Smith, A. G. and Paul, K. T., "Fire Growth in Upholstered Furniture Towards a Practical Test System," Fifth International Fire Conference on Fire Safety, Interflam '90, Sept. 3–6, 1990, Canterbury, England, Interscience Communications Ltd., London, England, 1990, pp. 233–240.

Soderbom, J., "Combustion Behavior of Upholstered Furniture (CBUF) Work Package Report on Data Management," Swedish National Testing Institute, Boras, Sweden, SP REPORT 1995:50, ISBN 91–7848–577–0, ISSN 0284–5172, 1995.

Springer, G. S., "Behavior of Furniture Frames During Fire," NBS-GCR-86-512, Apr. 1986, National Bureau of Standards, Gaithersburg, MD.

Stiefel, S. W., et. al., "Fire Risk Assessment Method: Case Study 1, Upholstered Furniture in Residences," National Institute of Standards and Technology, Gaithersburg, MD, National Fire Protection Assoc., Quincy, MA, Benjamin/Clarke Associates, Inc., Kensington, MD, National

Fire Protection Research Foundation, Quincy, MA, NISTIR 90-4243, June 1990.

Sundstrom, B., "Combustion Behavior of Upholstered Furniture Tested in Europe," Swedish National Testing and Research Institute, Boras, Sweden, Chapter 37, ACS Symposium Series 599, American Chemical Society, 208th National Meeting on Fire and Polymers II: Materials and Tests for Hazard Prevention, Aug. 21–26, 1994, Washington, DC, American Chemical Society, Washington, DC, 1995, pp. 609–617.

Sundstrom, B., "Fire Safety Design of Upholstered Furniture: Application of the CBUF Project," Swedish National Testing and Research Institute, Boras, Sweden, ISBN 0–9516320–9–4; *Proceedings* of the Seventh International Interflam Conference, Interflam '96, March 26–28, 1996, Cambridge, England, Interscience Communications Ltd., London, England, 1996, pp. 3–12.

Thompson, S. L. and Apostolakis, G. E., "Response Surface Approximation for the Bench-Scale Peak Heat Release Rate From Upholstered Furniture Exposed to a Radiant Heat Source," *Fire Safety Journal*, Vol. 22, No. 1, 1994, pp. 1–24.

Tomann, J., "Puuvillakankaalla verhoillun pehmustetun huonekalun syttyminen kytevaan paloon. [Padded Furniture With Cotton Upholstering Ignites Into a Smoldering Fire.]," *Palontorjuntatekniikka*, Mar. 1993, pp. 22–23.

VanderVorm, J. K. J., "Position of Legislators on Fire Safety of Upholstered Furniture—A Netherlands View," Ministry of Welfare, Health and Cultural Affairs, The Netherlands, European Commission DG 3, EGOLF and CFPA Europe, Fire and Furnishing in Buildings and Transport, November 6–8, 1990, Luxembourg, Belgium, 1990, pp. 1–5.

VanWeperen, W. and Vlot, I., "Consumer Safety and Furniture Flammability," Consumer Safety Institute, Amsterdam, Paper 5 (Abstract Only), Interscience Communications Limited, Second European Conference on Furniture Flammability, 2nd June 27–28, 1990, Brussels, Belgium, 1990.

Villa, K. M. and Babrauskas, V., "Cone Calorimeter Rate of Heat Release Measurements for Upholstered Composites of Polyurethane Foam," National Institute of Standards and Technology, Gaithersburg, MD, NISTIR 4652, Aug. 1991.

Westerman, D., "Fire Risk Assessment of the Use of Upholstered Furnishings in Buildings and Means of Transport in Europe," CFPA Study, European Commission DG 3, EGOLF and CFPA Europe, Fire and Furnishing in Buildings and Transport, November 6–8, 1990, Luxembourg, Belgium, 1990, pp. 1–13.

Wilkinson, L., "Fire and Upholstery—New Regulations Are Not the End of the Road," *Fire Surveyor*, Vol. 19, No. 4, Aug. 1990, pp. 14–16.

Wu, E. C. and Ihrig, A. M., "Effects of Oxygen and Potassium Ions on the Ignition of Upholstery Materials by a Smoldering Cigarette," Fire Safety '91 Conference, Vol. 4, Nov. 4–7, 1991, St. Petersburg, FL, Am. Technologies Intl., Pinellas Park, FL, 1991, pp. 367–383.

Wujcik, S. E., et al., "Polyurethane Flexible Slab Stock Formulations for Meeting Flammability Tests in Residential and contract Furniture Applications," Seventeenth International Conference on Fire Safety, Volume 17, Jan. 13–17, 1992, Millbrae, CA, Product Safety Corp., Sunnyvale, CA, 1992, pp. 12–25.

Yates, J. M., "Firefighters Push for Mandatory Flammability Testing," *Wood and Wood Products*, Vol. 92, No. 11, Oct. 1987, p. 38.

Zicherman, J. B., and Allard, D. L., "Compartment Tests of Polyurethane Foam Seating Assemblies," *Fire Technology*, Vol. 24, No. 2, 1988, pp. 128–137.

EFFECTS OF ENERGY CONSERVATION IN BUILDINGS

T. T. Lie

Two decades of heightened sensitivity to uncertainty in both the cost and the availability of energy have led to continuing worldwide efforts to conserve energy. Energy conservation in buildings is an important objective in many countries, and various approaches have been taken to achieve it. Two well-known methods for conserving energy in buildings are: (1) to increase the amount of insulation used and (2) to seal air leaks. Another method is replacement of energy from nonrenewable sources, such as oil and gas, by energy derived from renewable ones, such as the sun and wind.

Although these various measures have contributed to a more efficient use of energy, they have also created problems. Some fire problems, for example, are an increased number of types of ignition sources and materials that may contribute to spread of fire and the production of smoke and toxic gases. The fraction of air recycled in heated and air-conditioned buildings is often increased, which tends to increase the hazard of combustion products because of confinement of these products. Other examples are increases of rate of fire growth and fire temperature (with better insulation), and the reduction of fire resistance of structural elements in specific cases (due to insulation).

This chapter discusses the role of building energy conservation measures, particularly insulation, on the fire behavior of structures. Closely allied with the subject matter of this chapter are two other chapters: Section 4, Chapter 2, "Combustion Products and Their Effects on Life Safety"; and Section 7, Chapter 3, "Interior Finish."

INSULATION

Most of the concern over fire problems associated with energy conservation measures in buildings is related to increased use of thermal insulation in the buildings. Insulation affects a fire in all phases of its development, from inception to decay. These phases will be discussed briefly because knowledge of the characteristics of the various phases of fire that will be considered in this chapter is important for understanding the influence of insulation on fire safety.

Phases of Fire

Normally, a room fire starts as a small fire caused by ignition of a material by a heat source, e.g., the flame of a match or the heat of a stove or hot electrical wire. Depending upon various factors, such as

the nature of the material that is exposed to the ignition source or the size of the ignition source, the fire may self-extinguish or grow to full development. Usually the fully developed stage of a fire is preceded by flashover, which is characterized by instantaneous ignition of materials in all parts of the room.

The course of a fire is usually divided into three periods: (1) a growth period, (2) a fully developed period, and (3) a decay period. (See Figure 4-18A.) In the growth period, i.e., the period before flashover, the temperatures in the room are relatively low, even when close to the burning materials, and evacuation from the fire area presents no problems. During flashover, however, the temperature rises very sharply to such a level that survival of the fire by persons still in the room at that stage becomes unlikely. Thus, the time interval between the start of a fire and the occurrence of flashover is a major factor in the time that is available for safe evacuation of the fire area.

There is negligible risk of structural failure in the growth period because of the low temperatures. Actual risk of failure of structural elements or of fire separations begins in the fully developed stage of the fire, after flashover. This risk also exists in the decay period, during which remnants of the combustible materials are consumed.

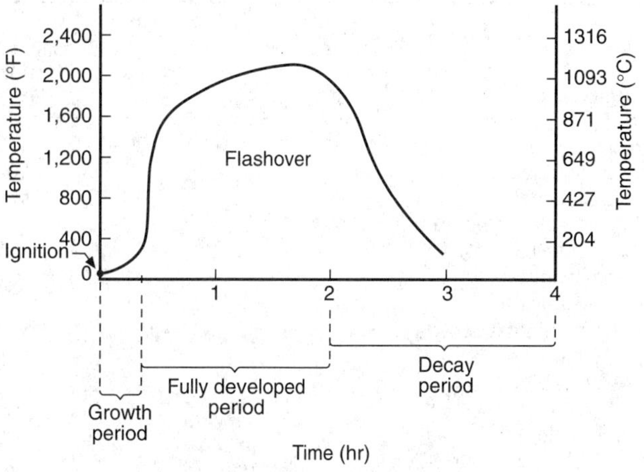

FIG. 4-18A. *Temperature course of a typical fire.*

T. T. Lie is research officer, Institute for Research in Construction, National Research Council of Canada.

Influence on Ignition of Materials

If thermal insulation is installed to surround heat-producing objects, such as electrical wires and light fixtures, those objects may become ignition sources. The insulation will impede the dissipation of heat, which will result in heat buildup in the objects. They may become hot enough to ignite combustible materials in contact with them.

Insulation also plays a role in the ignition of a material when it is applied to the back of a thin lining; it reduces heat losses at the back of the lining. The increase of surface temperature at the front side of an insulated interior finish material exposed to an ignition source may be significantly more rapid than that of an uninsulated material. Therefore, the insulated interior finish material may ignite more rapidly than the uninsulated one. The influence of insulation is small for thicker materials, i.e., thicker than $1/4$ in. (6.4 mm) for cellulosic linings, such as wood and board.[1]

In addition to the cases mentioned in which insulation indirectly contributes to the ignition of materials, the insulation itself may be a potential ignition hazard. The increased use of a wide variety of insulating materials, particularly foamed plastics, has created concern about this possibility. The plastics industry has been conducting extensive studies to investigate the fire performance of plastics to develop ways in which they can be used with reasonable safety.

Influence on Growth of Fire

During the growth of a fire, unburned material is successively ignited. The rate of growth of a fire depends upon many factors; one of the most important is the nature of the materials that contribute to fire growth. The faster the materials propagate fire and the more heat they develop, the faster the fire will grow. The heat developed by burning materials in the growth period is particularly important in enclosures of rather limited size, such as offices, living rooms, and classrooms. During the growth of the fire, the heat developed by the burning materials accumulates in the room. Because of this, other materials in the room may be heated so severely that after a short time they also ignite and, at this stage, flashover occurs.

Adding insulation may significantly affect the rate of growth of a fire. Because of the insulation, heat from fire will be conserved in the room in the same way as heat from heating systems is conserved. Consequently, the accumulation of heat in the room is accelerated, and the flashover stage may be reached much earlier than in a less-insulated room. According to estimates, the growth period reduces by a factor of about three by insulating a room of brick-plaster construction with a noncombustible lining of low thermal conductivity.[2]

In addition to accelerating fire growth by heat conservation, insulation may propagate flames itself, also affecting the growth of a fire.

Propagation of Flames

Once a material, e.g., interior finish in a room, is ignited, the flame may spread across the surface of the material. For flame to spread, heat transferred from the flame to the material must be sufficient to liberate flammable gases from the material near the flame front. Those gases will, in turn, be ignited. The gases are developed initially from the surface of the material; therefore, surface temperature is an important factor in determining the attainment of a condition favorable to flame spread. The faster the temperature rise of the surface, the faster the propagation of the flames.

The rate of temperature rise of the surface depends to a high degree upon the thermal properties of the material. In particular, the thermal conductivity (K) and the thermal capacity (ρc) of the material are important. The lower the thermal conductivity, the less of the heat received at the surface is transferred to the inside of the material. If, in addition, the thermal capacity of the material is small and thus relatively little heat is required to raise the temperature of the material, the temperature rise of the surface may be substantial in a short time. It can be shown for an idealized case that the time to reach a specific critical surface temperature, e.g., the temperature at which the surface ignites, is proportional to a quantity, $K\rho c$, which is often termed the thermal inertia of the material.[3]

In actual practice, however, more factors play a role in the determination of the surface temperature rise. For the purpose of illustrating the role of thermal inertia in flame propagation, the idealized case will be assumed in the following discussion.[4]

In Table 4-18A, typical values are given for the thermal inertia of some materials. The table also gives the time it takes to raise the surface temperature of the material to a level at which wood and plastic produce a significant amount of combustible gases, here assumed to be 400°F (204°C). It is further assumed that heat is supplied at a rate of 1050 Btu per sq ft/hr (3300 W/m²). At this rate,

TABLE 4-18A. *Thermal Inertia (Kρc) and Rate of Surface Temperature Rise of Various Materials*
[Rate of Heat Supply: 1050 Btu/(sq ft/hr) (3300 W/m²]

Material	Density range lb/cu ft (kg/m³)		Typical values of $K\rho c$ Btu²/(sq ft °F)² hr [W²/(m² °C)² s]		Time to reach a surface temperature of 400°F (204°C) (s)
Wood	30–40	(480–640)	1.7	(1.98×10^5)	600
Asbestos board	40–50	(640–800)	1	(1.16×10^5)	353
Gypsum board	40–50	(640–800)	0.6	(7.0×10^4)	212
Fiber insulation board	10–20	(160–320)	0.13	(1.51×10^4)	46
Cork board	5–10	(80–160)	0.06	(7.0×10^3)	21
Foam plastics:	1–3	(16–48)	0.01	(1.16×10^3)	4
Polyurethane					
Phenol-formaldehyde					
Polystyrene					
Cellular PVC					
Isocyanurate					
Fiber batts:	0.5–1.0	(8–16)	0.005	(5.81×10^2)	2
Glass fiber					
Mineral wool					

wood would attain this critical surface temperature in 600 sec, or 10 min. Under the same circumstances, fiber insulation board, which is known to propagate flames rapidly, would attain the critical surface temperature in 23 sec; for a foamed plastic it would take only a few seconds before its surface temperature reached the critical value.

The very rapid temperature rise of the surfaces of foamed plastics is probably the reason why they may propagate flames with extreme speed; this applies particularly to nonmelting or thermosetting foamed plastics. Thermoplastic foam may melt before its temperature reaches a critical value. In that case, the rate of temperature rise of the surface of the molten plastic slows, because the thermal inertia of the plastic surface increases and because heat is used in melting the plastic. Extremely fast propagation of fire is less likely to occur on thermoplastic foams that melt at relatively low temperatures [such as polystyrene and polyvinyl chloride, which melt at temperatures on the order of 250°F (121°C)] than on those that melt at higher temperatures. There are many thermoplastics, however, that do not melt before their surface reaches the critical temperature.

Protected Foamed Plastic

Several building codes recognize a material with a flame spread rating of 25 or less [as ascertained by the tunnel test in NFPA 255, *Standard Method of Test of Surface Burning Characteristics of Building Materials* (hereinafter referred to as NFPA 255)] as a low fire hazard material. Tests carried out on urethane foam, however, suggest that foams may accelerate the growth of fire much faster than expected on the basis of tunnel test results.

Concern about this possibility and the occurrence of a number of large-loss fires involving foamed plastic has led to more stringent requirements regarding the use of foams. One of these requirements is that plastic foams should be protected by a thermal barrier capable of preventing excessive heat transfer from heat source to plastic during a specified time.

Methods exist for evaluating the performance of a material as a thermal barrier. One of them is the method described in NFPA 251, *Standard Methods of Tests of Fire Endurance of Building Construction and Materials* (hereinafter referred to as NFPA 251), in which the area of the test specimen exposed to the test fire is at least 100 sq ft (9.3 m²). To pass the test, the protection must be capable of preventing excessive temperature rise at the unexposed side for 15 min or more when one side of the test specimen is exposed to standard heating. Because the temperatures during standard heating are of the same magnitude as those encountered in the post-flashover stage of a fire, the test method, which is intended to evaluate the performance of a protection in the pre-flashover stage of a fire, is a severe one. Temperatures in the pre-flashover stage of a house-burn test are shown in Figure 4-18B.[5] It can be seen that, up to approximately 15 min, the fire temperatures are substantially lower than the NFPA 251 standard curve.

Another method, which is less severe but sufficient to evaluate the performance of a protection in the pre-flashover stage of a fire and which will reasonably ensure that the foamed plastic will not contribute to the growth of a fire for at least the time measured in the test, is described in ULC-S124.[6] In this test, the thermal barrier is also exposed on one side to heating of a severity equal to that in the NFPA 251 fire test. The area of the specimen, however, is smaller, about 10 sq ft (0.93 m²), and the test also classifies thermal barriers that provide protection for less than 15 min.

The times during which materials provide protection against a post-flashover fire are given in Table 4-18B. Protection time in this table may be regarded as the approximate time it takes the temperature at the interface of the protective cover and the plastic foam to rise 250°F (121°C) above ambient. The total time before the plastic starts contributing significantly to the fire is thus equal to the time for

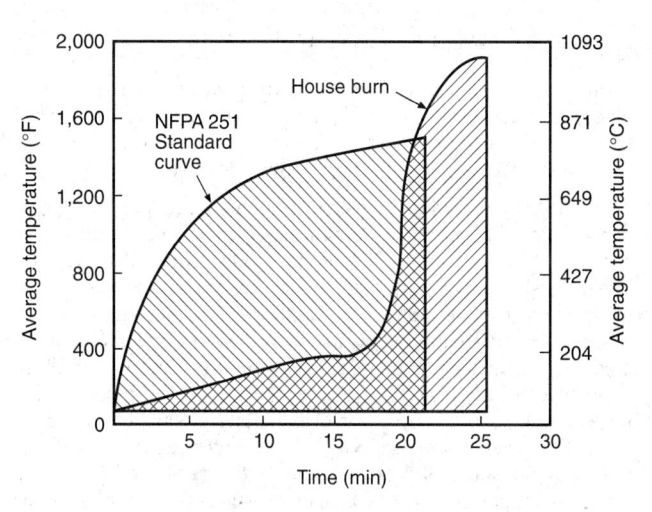

FIG. 4-18B.   *Temperature-time relationship according to NFPA 251, and for a house burn.*[5]

TABLE 4-18B.   *Time of Protection Against Post-Flashover Fire*

Material of Protection	Thickness of Protection in. (mm)		Protection Time (min)
Gypsum wallboard	3/8	(9.5)	11
Gypsum wallboard	1/2	(12.7)	15
Gypsum wallboard	5/8	(15.9)	20
Magnesium oxychloride	1/4–3/8	(6.4–9.5)	10–15
Magnesium oxychloride	1/2–5/8	(12.7–15.9)	20–25
Plywood	1/8	(4.9)	3
Plywood	1/4	(6.4)	5
Plywood	3/8	(9.5)	8
Plywood	1/2	(12.7)	11
Hardboard	1/8	(4.9)	3
Hardboard	1/4	(6.4)	5
Hardboard	3/8	(9.5)	8
Particleboard	1/4	(6.4)	4
Particleboard	3/8	(9.5)	6
Particleboard	1/2	(12.7)	8

the fire to grow from ignition to flashover, added to the protection time given in the table. If a combination of protecting materials is used, the total protection time may be assumed to be at least equal to the sum of the protection times of the individual protecting materials.

Because propagation of flame is a process that is determined mainly by the surface characteristics of the exposed materials, provision of a cover will reduce or remove the influence of the foamed plastic on fire growth. In this case, the propagation of flame will be determined by the surface characteristics of the cover material. If it has a low flame spread rating and its behavior is not significantly affected by the foamed plastic insulation, fast growth of a fire due to plastic insulation can be prevented in a room. It is possible, however, for fire to propagate in the cavity behind the cover, e.g., if the cover has a high thermal conductivity or openings are formed in it. It is also possible for fire to originate inside the cavity, e.g., because of overheating of electrical wires due to thermal insulation.

Foamed Plastic in Cavity Walls

Foamed plastic in cavity walls is usually sandwiched between two layers of material forming the wall. Sometimes there is an air space between the plastic and the wall, and sometimes the wall material is in direct contact with the plastic on both sides of it.

Depending upon the presence of an air space and the type of construction, i.e., whether the building is of noncombustible construction, a number of North American codes require that the insulation must satisfy specific flame spread ratings determined by using the test method described in NFPA 255. Because of doubt whether the contribution of insulation to the propagation of a fire in a cavity is related to its flame spread rating, in particular in those cases where there is no air space, tests were carried out to obtain more insight into the behavior of materials in cavities.[7] In these cases, the speed of propagation of the fire in the cavity is not determined by the speed of propagation over the surface of the plastic foam, but by the speed with which the plastic at the underside burns away.

The test results indicate that, for materials sandwiched between two layers of noncombustible materials without air space, the upward propagation of the fire is slow, even for materials with a high flame spread rating. Figure 4-18C shows a burned part of a urethane foam test specimen with a high flame spread rating (333 according to NFPA 255) after it has been exposed at the bottom to a standard fire (NFPA 251) for 6 hr. In this time, the fire had propagated to about 14 in. (356 mm) above the part of the specimen that was exposed to the fire.[7]

The fire may spread rapidly upward if there is an air space between the wall and foamed plastic. The speed of propagation depends not only upon the properties of the foamed plastic but also on several other factors, such as the properties of the wall material, width of the air space, and flow of hot gases through the cavity (due to chimney effect). Research is now in progress to examine the influence of a number of important factors on the speed of propagation of fire in cavity walls.

Irrespective of the circumstances, however, spread of fire from story to story in a cavity wall can be prevented by the use of well-designed fire stops. The installation of fire stops is prescribed in many codes. The role of fire stops has become very important due to the increase in the use of insulation in cavity walls. Therefore, much can be gained by measures that promote the installation of effective fire stops in multistory buildings.

Another consequence of the increasing use of combustible insulation in cavity walls is an increase in the chance of outbreak of fire in the cavity. Strictly speaking, however, this is not true. The chances of death in any given fire are roughly the same.

Insulation in Attics

The greatest concern for concealed-space fires is in situations where a large, undivided or unfirestopped space provides a route for fire growth that circumvents a building's compartmentation. Such spread has been observed in the common attics of rowhouses, for example. The hazard of a large, undivided space is increased if the combustible load is great.

Increase of building insulation and introduction of new insulating materials may have aggravated this situation. There is concern that widespread installation of thermal insulation may have adverse consequences on the performance of electrical wires, cables, and other electrical devices.[8] Since such devices generate heat, they may become overheated during operation if they are covered by building insulation. Tests have shown that electrical equipment located in insulated cavities operated at higher temperatures than in uninsulated cavities. However, many electrical devices are now listed for use with insulation covering them. Others are listed with specific clearances to the fixture indicated. Another cause of overheating is the liberation of exothermic heat by insulating materials as they cure.

A type of insulation known as loose-fill insulation is frequently used in attics. There are noncombustible insulating materials, such as mineral wool, but often combustible insulation made of cellulosic material is used. These cellulosic insulations are generally made of pulverized or finely chopped paper, frequently waste paper.[8] In this condition, the material is highly flammable and flame-retardant compounds have to be added to reduce its flammability. If the materials are properly treated with flame retardants, which can be determined by testing, and fire stops are installed in large attic areas, cellulosic insulation can be utilized without excessive risk.

SMOKE AND TOXIC GASES

Fire statistics show that thermal decomposition products (smoke and toxic gases) are responsible for most deaths.* Many new materials release harmful decomposition products very rapidly, and some of them produce much more smoke or toxic gas than traditional building materials. With the increased use of these new materials, the problem of smoke and toxic products has attracted even more attention.

Regulations to exclude the use of materials with excessive smoke production and methods to determine smoke density are specified in several codes.[9,10,11] NFPA, ASTM, and ISO recently adapted standard methods for analyzing the fire toxicity of products and materials. (See Section 4, Chapter 2, "Combustion Products and Their Effects on Life Safety.")

FIG. 4-18C. Test specimen after exposure to fire in a cavity wall (no air space, exposure time of 6 hr). (Source: National Research Council of Canada)

---

* Hall, John R., Jr., *Burns, Toxic Gases, and Other Hazards Associated with Fires*, Quincy, MA, NFPA Fire Analysis and Research Division, April 1996, p. 8.

Smoke

A major hazard of smoke is that it reduces visibility. It may impede the escape of occupants from a burning building and prolong exposure to the harmful effects of fire, such as the smoke itself, toxic products, and heat.

One of the most important factors that determines the reduction in visibility in a given space is the concentration of the smoke generated in that space. A measure of this concentration is the optical density of the smoke, which is a quantity that is directly proportional to the smoke concentration. In Table 4-18C, the second column lists relative optical densities of smoke generated by various materials under similar conditions. The third column lists times required to obscure a given room under the same conditions, assuming that wood smoke would obscure the room in 10 min. Although wood is considered here as a single material, there are differences in the smoke production among various species of wood. The figures, therefore, should be regarded as approximate and are intended to show the order of magnitude of the smoke production of various materials under one specific test condition.

TABLE 4-18C. *Smoke Production of Various Unprotected Materials*[12]

Material	Optical Density	Time of Obscuration (min)
Phenolformaldehyde foam	0.1	100
Wood	1	10
Boards	1–2	5–10
Cork	3	3.3
Polystyrene foam	5	2
Cellular PVC	10	1
Polyurethane foam	15	0.7

It can be seen that some materials used as insulation, e.g., cork and some foamed plastics, produce significantly more smoke than wood. If the insulation surface area is large in comparison with that of the other combustible materials in the room, and they are unprotected, smoke production will be dominated by the insulation. In this case, the time of obscuration, which is a measure of the time available for safe evacuation of the fire area, may drop to levels substantially lower than that if the smoke had been produced solely by wood. If, however, the insulation surface area is a small fraction of the surface area of the combustible materials in the room (not more than 10 percent of the surface area of the fire load) or the materials are protected by barriers sufficient to significantly delay the involvement of the insulation materials in fire, smoke contribution of the insulation will not be excessive in comparison with the total smoke produced by all burning materials.

If the insulation is protected, its smoke production may drop to very low values. Table 4-18D shows the influence of protection consisting of 1/2-in. (12.7-mm) Type X gypsum wallboard on the smoke production of a number of materials.

Another way to express optical density in practical terms is to relate it to visibility range. Results of various studies[13] showed that an optical density of approximately 4 corresponded to a visibility range of 80 ft (25 m), which was regarded as the critical range of visibility in multistory buildings. There is some variability in the critical range of visibility found in the studies, however. If one is familiar with the layout of an occupancy, the critical visibility range *for egress* is reduced by a factor of approximately 4.

TABLE 4-18D. *Influence of Protection on Smoke Production*[12]
[Protection of 1/2-in. (12.7-mm) Type X Gypsum Wallboard]

Material	Optical Density	Time of Obscuration (min)
Wood (unprotected)	1	10
Cork:		
Unprotected	3	3.3
Protected	0.06	180
Polystyrene foam:		
Unprotected	5	2
Protected	0.8	13
Polyurethane foam:		
Unprotected	16	0.6
Protected	4	2.5

Toxic Gases

Toxic gases can be fatal if they are present for a sufficient time and in sufficient quantities. In recent years, various new materials, especially synthetic polymers, have found increasing use in buildings, and their introduction has heightened the concern of fire authorities over toxic combustion products. Part of this increased concern has arisen because of the lack of information on toxic combustion products and the problem of assessing their potential hazard.

Comprehensive chemical analysis is necessary to identify unusual toxicants that may be produced during fires. For most practical applications, however, this is not necessary, since testing for a few of the most important known toxicants often gives sufficient information. Some compounds, such as carbon monoxide (CO), hydrogen chloride (HCl), hydrogen cyanide (HCN), and oxides of nitrogen (NO_x), are recognized as harmful products; others, such as water vapor and the hydrocarbons, contribute little or no toxic hazard. It is usually sufficient to decompose materials under specified conditions and determine the resulting concentrations of a few of the most important toxicants. From this information, a reasonable indication of the toxicity of the mixture of products can be obtained.

Table 4-18E shows the main harmful products of a number of materials and approximate values of the harmful concentrations for an exposure time of 30 min. According to the figures in this table, one of the most toxic products is HCN, which is harmful at concentrations of about 1/30th of the harmful concentration of CO. If the materials that produce HCN are present in small quantities, however, their contribution to toxicity may be less than that of the CO normally produced during a fire. This is also the case if the materials that generate very toxic products are properly protected. According to measurements made during tests in rooms with walls insulated with isocyanurate foam and protected with 1/4-in. (6.4-mm) Douglas fir plywood, no HCN was observed in the pre-flashover period, up to 21 min after the ignition.[5] The results indicate that, from the point of view of hampering safe evacuation of people by toxic combustion products, the materials in the room play a far greater role than the insulation in the walls if the insulation is adequately protected, for example, by boards or plywood of not less than 1/4-in. (6.4-mm) thickness.

HEATING SYSTEMS

The price of oil and uncertainties in its supply have stimulated the use of heating systems that do not operate on oil. Such heating systems, for example, utilize solar energy, wood, coal, and natural gas. Coal and gas have been used as the main fuels for heating buildings

TABLE 4-18E. *Main Harmful Products of Materials and Harmful Concentrations*[14,15,16]

Material		Harmful Product	Harmful Concentration in Parts per Million Parts of Air (30-min exposure)
210	210	Carbon monoxide (CO)	4,000 ppm
Polystyrene		CO—also styrene, but present in smaller quantities	
Polyvinyl chloride (PVC)		HCl—also corrosive CO	1,000–2,000 ppm
Plexiglas™ or Perspex™		CO—also methylmethacrylate, which is as toxic as CO but produced in smaller quantities	
Polyethylene		CO	
Ureaformaldehyde			
Isocyanurate			
Polyurethane		Hydrogen cyanide (HCN)	120–150 ppm
Acrylic fibres		CO	
Wool			
Nylon			

for a long time in several countries. The use of solar energy is relatively new, and wood has enjoyed renewed interest as a household heating fuel; in some areas, their use for heating buildings has increased tremendously.

Most of the fire hazard introduced by these systems and their proximity to building materials is related to the high temperatures that they may attain. For example, 400°F (204°C) for solar collectors is not abnormal, whereas wood-stove encasements may reach temperatures of over 1000°F (538°C). There is the risk that combustible materials may ignite spontaneously after prolonged exposure to these temperatures.

BUILDING STRUCTURE

Fire Severity

Actual risk of failure of structural elements or fire separations usually begins after flashover in the fully developed stage of a fire. The temperature of the fire in this stage will be affected by thermal insulation in the boundaries, such as walls and roofs, of the rooms; the greater the amount of insulation, the higher the fire temperature. Less of the heat generated by the burning materials in the room will be absorbed by the boundaries, and, as a consequence, the fire gas temperature will increase. The higher temperature, in turn, will accelerate the rate of decomposition of the materials in the room, which may result in an increase in the heat released in the room. According to existing information, however, this rate is a very weak function of the temperature of the fire gases.[15] Therefore the only significant effect on fire severity caused by adding insulation is expected to be a rise of fire temperatures, whereas the duration of the fire will be practically unchanged.

In addition to causing higher fire temperatures, insulation in roofs and walls also directly influences the temperatures reached in them by altering the heat transfer through the roof or walls. These temperature changes may affect the fire performance of structural elements and fire separations in various ways.

Roofs

Roof constructions in which fire performance may be affected by the addition of insulation are those supported by steel members, such as joists and beams, and those supported by steel wires, such as reinforcing and prestressing steel. Steel loses strength when it is heated to high temperatures; the higher its temperature, the lower its strength. In general, a critical steel temperature can be indicated at which the steel has lost so much strength that it can no longer support the load. Often a temperature of 1100°F (593°C) is assumed as the critical temperature of structural steel members and reinforcing steel, and a temperature of 800°F (427°C) as the critical temperature of prestressing steel.

Typical roof constructions that are supported by steel members are assemblies consisting of a steel deck supported by steel beams or joists and covered with concrete. Often the steel is protected from attaining excessive temperatures during a fire by a fire-resistant ceiling. At present, more roof insulation is placed on top of the concrete than in the past, to conserve energy. Because the insulation reduces heat losses to the air above the roof, all temperatures inside the structure, including the temperatures of the supporting beams, will be higher with increased insulation. If the concrete is sufficiently thick, however, the increase in temperature will be small. According to tests, the addition of insulation on top of the concrete has no significant effect on the fire resistance of the roof assembly if the thickness of the concrete is 2 in. (51 mm) or more.[17] If there is no concrete cover, and the insulation is placed directly on the steel deck, the fire resistance of the roof assembly may be reduced substantially. The influence on the fire resistance of the assembly depends upon the thermal properties and thickness of the insulation. A maximum thickness can be found beyond which there will be no further reduction in the fire resistance of the assembly. The order of magnitude of this thickness is 1.5 in. (38 mm), which was found during tests with a commonly used noncombustible insulation.

Typical roof constructions that are supported by steel wires are those consisting of reinforced or prestressed concrete slabs. If insulation is added on top of the concrete, the temperature of the reinforcing or prestressing wires may rise more rapidly than if the concrete is uninsulated. The influence of insulation on the temperatures of the wires depends upon various factors, such as the properties of the insulation, position of the wires in the concrete, and the thickness of the concrete. If the concrete is not very thin, say not less than 3 in. (76 mm) for normal-weight concrete, the influence is small. For lightweight concrete, the influence of adding insulation is less than for normal-weight concrete.

Walls

If walls are lined on the room side with an insulating material, the fire temperatures will increase. On the other hand, the temperatures in the wall will be lower, and, as a result, the fire resistance of the wall will improve. It is likely that lining a wall will have a favorable effect on its structural fire performance, particularly if the lining is a good insulator.

If the insulation is installed in a cavity, however, the effect on the structural fire resistance of the wall may be unfavorable. This is particularly true for thin walls in which unequal expansion of the wall layers on each side of the cavity may produce large deflections and cause the wall to buckle.

Insulation may reduce the structural fire performance of fire barriers, such as roofs, floors, and walls, but it increases their thermal fire resistance, i.e., their ability to prevent excessive heat penetration through the fire barriers.

BIBLIOGRAPHY

References Cited

1. Lie, T. T., *Fire and Buildings*, Applied Science Publishers Ltd., London, 1972, pp. 26–27.
2. Thomas, P. H., and Bullen, M. L., "The Effects of Insulation in Fires," *Fire*, 1980, p. 434.
3. Carslaw, H. S., and Jaeger, J. C., *Conduction of Heat in Solids*, 2nd ed., Oxford University Press, London, 1959.
4. Thomas, P. H., and Bullen, M. L., "On the Role of $K_\rho c$ of Room Lining Materials in the Growth of Room Fires," *Fire and Materials*, No. 3, 1979, pp. 68–73.
5. Holmes, C. A., *et al.*, "Fire Development and Wall Endurance in Sandwich and Wood-Frame Structures," Research Paper FPL 364, 1980, Forest Products Laboratory, U.S. Department of Agriculture.
6. ULC, "Standard Method of Test for the Evaluation of Protective Coverings for Foamed Plastic," ULC-S124, 1985, Underwriters Laboratories of Canada, Toronto, Ontario, Canada.
7. Lie, T. T., "Contribution of Insulation in Cavity Walls to Propagation of Fire," Fire Study No. 29, 1972, National Research Council of Canada, Ottawa, Ontario, Canada.
8. Degenkolb, J. H., "Energy Conservation and Its Effect on Buildings from the Fire Protection Viewpoint," *Fire Journal*, Vol. 72, No. 3, May 1978, pp. 86-90; and Vol. 72, No. 6, Nov. 1978, p. 8.
9. BOCA, *The BOCA Basic Building Code*, Building Officials Conference of America, Inc., Chicago, IL, 1984.
10. ICBO, *The Uniform Building Code*, International Conference of Building Officials, Pasadena, CA, 1994.
11. NRCC, *National Building Code of Canada*, National Research Council of Canada, Ottawa, Ontario, Canada, 1990.
12. Lie, T. T., "Smoke Contribution of Some Partition Components During a Fire Test," Technical Note 524, 1968, National Research Council of Canada, Ottawa, Ontario, Canada.
13. Tamura, G. T., *Smoke Movement and Control in High-rise Buildings*, National Fire Protection Association, Quincy, MA, 1994, pp. 27–28.
14. Sumi, K., and Tsuchiya, Y., "Evaluating the Toxic Hazard of Fires," *Canadian Building Digest*, National Research Council of Canada, Ottawa, Ontario, Canada, 1978.
15. Harmathy, T. Z., "Effect of the Nature of Fuel on Fully Developed Compartment Fires," *Fire and Materials*, No. 3, 1979, pp. 49-60.
16. Woolley, W. D., Ames, S. A., and Fardell, P. J., "Chemical Aspects of Combustion Toxicology of Fires," *Fire and Materials*, No. 3, 1979, pp. 110–120.
17. Stanzak, W. W., and Konicek, L., "Effect of Thermal Insulation and Heat Sink on the Structural Fire Endurance of Steel Roof Assemblies," *Canadian Journal of Civil Engineering*, No. 6, 1979, pp. 32–35.

Reference

Sumi, K., and Tsuchiya, Y., "Combustion Products of Polymeric Materials Containing Nitrogen in Their Chemical Structure," *Journal of Fire and Flammability*, 1973, pp. 15–22.

NFPA Codes, Standards, and Recommended Practices

Reference to the following NFPA codes, standards, and recommended practices will provide further information on effects of energy conservation in buildings discussed in this chapter. (See the latest version of *The NFPA Catalog* for availability of current editions of the following documents.)

NFPA 251, *Standard Methods of Tests of Fire Endurance of Building Construction and Materials*

NFPA 255, *Standard Method of Test of Surface Burning Characteristics of Building Materials*

Additional Reading

Babrauskas, V., "Wall Insulation Products: Full-Scale Tests versus Evaluation from Bench-Scale Toxic Potency Data," *Proceedings* of Interflam '96, Seventh International Interflam Conference, March 26–28, 1996, Cambridge, England, Interscience Communications Ltd., London, England, 1996, pp. 257–274.

Belles, D. W., "Loose-Fill Cellulose Insulation: An Agint [sic] Problem," *Journal of Applied Fire Science*, Vol. 3, No. 3, 1993/1994, pp. 295–303.

Briggs, P. J., and Murrell, J. M. "Do Insulated Sandwich Panels Pose a Fire Spread Threat?," Third International Conference and Exhibition on Fire and Materials, October 27–28, 1994, Arlington, VA, 1994, pp. 247–257.

Choi, K. K., "Effects of Foamed Plastic Insulation on Severity of Room Fires," *Fire Technology*, Vol. 22, No. 1, 1986, pp. 5–13.

Eisner, H., "How Safe Are Foam Roof Linings," *Fire Prevention*, No. 226, Jan.-Feb. 1990, pp. 32–33.

Ellis, J. W., "Rigid Foam Plastic Insulation," *Fire Surveyor*, Vol. 19, No. 1, Feb. 1990, pp. 21–24.

Kenneth M., "Roof Insulation Fire Safety. More Objectivity Required," *Fire Safety Engineering*, Vol. 2, No. 1, Feb. 1995, pp. 17–18.

Marshall, H. E., "Choosing Economic Evaluation Methods. Least-Cost Energy Decisions for Buildings. Part 3," Video Training Workbook, National Institute of Standards and Technology, Gaithersburg, MD, Department of Energy, Washington, DC, Federal Energy Management Program, NISTIR 5604, May 1995.

Mehaffey, J. R., "Fire Performance of Combustible Insulation in Buildings," *Journal of Thermal Insulation*, Vol. 10, Apr. 1987, pp. 256–269.

Mowrer, F. W., and Williamson, R. B., "Flame Spread Behavior of Char-Forming Wall/Ceiling Insulating Materials," *Proceedings* of the Third International Symposium on Fire Safety Science, Elsevier Applied Science, New York, 1991, pp. 689–698.

Murphy, J., and Mouton, D., "HCFC-14b in Rigid Insulating Polyurethane and Polyisocyanurate Foams: Decomposition Is Not a Concern," Alliance for Responsible CFC Policy; U. S. Environmental Protection Agency; Environment Canada; and United National Environmental Program, *Proceeding* of the 1992 International CFC and Halon Alternatives Conference, Stratosphere Ozone, Protection for the 1990s, Sept. 29–Oct. 1, 1992, Washington, DC, 1992, pp. 567–576.

Newman, J. S., and Tewarson, A., "Flame Spread Behavior of Char-Forming Wall/Ceiling Insulating Materials," *Proceedings* of the Third International Symposium on Fire Safety Science, Elsevier Applied Science, New York, 1991, pp. 679–688.

Ohlemiller, T. J., "Forced Smolder Propagation and the Transition to Flaming in Cellulosic Insulation," *Combustion and Flame*, Vol. 81, No. 3–4, Sept. 1990, pp. 354–365.

Petersen, S. R., and Fanney, A. H. "Technical and Economic Analysis of CFC-Blown Insulations and Substitutes for Residential and Commercial Construction," National Institute of Standards and Technology, Gaithersburg, MD, NBSIR 88–3829, July 1988; *Improved Thermal Insulation: Problems and Perspectives*, Ch. 4, 1991, Technical Publishing Co., Inc., Lancaster, PA, 1988, pp. 43–94.

Raeber, J. A., "Foamed Plastic Insulations: Three Controversies," *Construction Specifier*, Vol. 41, No. 11, 1988, pp. 31–32.

Ripley, D., "Insulation—A Costly Item," *Naval Architect*, 1988, p. 131.

Roisin, J. P., "Insulated Panels: Products and Hazards," *Fire Safety*, Vol. 3, No. 1, Feb. 1996, pp. 10–11.

Siddiqui, S. A., "Round Robin Fire Safety Tests on Cellulose Insulation," *Journal of Thermal Insulation*, Vol. 10, July 1986, pp. 11–21.

Sommerfeld, C. D., Walter, R., and Wittbecker, F. W., "Fire Performance of Facades and roofs Insulated With Rigid Polyurethane Foam: A Review of Full Scale Tests," ISBN 0-9516320-9-4; *Proceedings* of Interflam '96, Seventh International Interflam Conference, March 26–28, 1996, Cambridge, England, Interscience Communications Ltd., London, England, 1996, pp. 315–326.

Stirling, C. M., and Southern, J. R., "Stabilizing External Insulation in Fires," *Insulation Journal*, Vol. 33, No. 1, 1989, pp. 10–12.

Sultan, M. A., "Correlation of Small- and Full-Scale Fire Resistance on Insulated and Non-insulated Lightweight Fram Gypsum Board Wall Assemblies," National Research Council of Canada, Ottawa, Ontario, ISBN 0–6516320–9–4; *Proceedings* of Interflam '96, Seventh International Interflam Conference, March 26–28, 1996, Cambridge, England, Interscience Communications Ltd., London, England, 1996, pp. 933–937.

"Thermal Insulation and Fire Protection," *Insulation Journal*, Vol. 30, No. 8, 1986, pp. 13–14.

Wetterlund, I., "Thermal Stability of Insulating Materials at High Temperatures. A Proposal for a New Nordtest Method," Swedish National Testing Institute, Boras, Sweden, SP REPORT 1994:66, Nordtest Project 1114–93, 1994, 23 pages.

Woolhead, F., "Fire Test Report: Insulated Metal Roof Decks," *Fire Surveyor*, Vol. 19, No. 1, Feb. 1990, pp. 25–29.

AIR-MOVING EQUIPMENT

Revised by Jane I. Lataille

Many industrial and commercial activities and processes require the removal of smoke, fumes, and/or dusts. This necessitates equipment for moving air in and out of the processing area. Both the accumulation and moving of impurities, and the substitution of other atmospheres, present fire and explosion hazards. This chapter addresses these hazards and how to prevent them.

Additional information about hazards of air-moving equipment can be found in Section 3, Chapter 2, "Control of Electrostatic Ignition Sources"; Section 3, Chapter 8, "Industrial and Commercial Heat Utilization Equipment"; Section 3, Chapter 11, "Metalworking Processes"; Section 3, Chapter 15, "Woodworking Facilities and Processes"; Section 3, Chapter 19, "Chemical Processing Equipment"; Section 3, Chapter 27, "Grinding Processes"; Section 4, Chapter 1, "Fire Hazards of Materials: An Overview"; Section 7, Chapter 9, "Venting Systems for Commercial Cooking Equipment"; and Section 7, Chapter 14, "Airconditioning and Ventilating Systems."

Air-moving equipment includes all mechanical-draft duct systems, of both the pressure and exhaust types, for removal of dusts, vapors, and waste material, and for the conveying of materials. The hazards of such systems lie in the possibility of igniting flammable materials or vapors by sources such as sparks generated by fans or foreign material, e.g., tramp metal, or by overheated fan bearings. Also, such systems could inadvertently spread fires through the building.

Air-moving systems do, however, play an important role in fire protection. If the materials they remove were allowed to accumulate, the result could be an explosion from vapor-air mixtures, a flash fire due to accumulations of lint or other combustibles, and generally poor housekeeping conditions conducive to fires in general.

Air-moving systems usually contain axial flow fans or centrifugal blowers; several different designs of each kind are available. (See Figure 4-19A.) A straight-blade centrifugal fan is suitable for exhausting materials containing large particles, because it resists clogging and the blades can withstand considerable abrasion. Propeller fans will move large volumes of air, but only against small resistance. Since these fans could not operate efficiently against the friction a duct would introduce, they are not suitable for ducts of any appreciable length.

To reduce the hazard of fire, fans should be: (1) of noncombustible construction, (2) able to be shut down by remote control in case of fire, (3) accessible for maintenance, and (4) structurally sound enough to resist wear and overcome distortion and misalignment caused by structural weakness or overloading. Fans used for heat and smoke venting should be able to withstand temperatures of 932°F (500°C) for 4 hr. Construction materials for all fans must be appropriate for the intended use. A spark caused by a blade hitting the housing, due to distortion or misalignment, could result in ignition when the exhaust contains flammable solids or vapors. This possibility of friction sparks is minimized if the fan and fan housing are of nonsparking material. Aluminum and magnesium alloys are not compatible with ferrous alloys, which may rust.

Details on the installation of air-moving equipment are given in NFPA 91, *Standard for Exhaust Systems for Air Conveying of Materials*, and NFPA 96, *Standard for Ventilation Control and Fire Protection of Commercial Cooking Operations.*

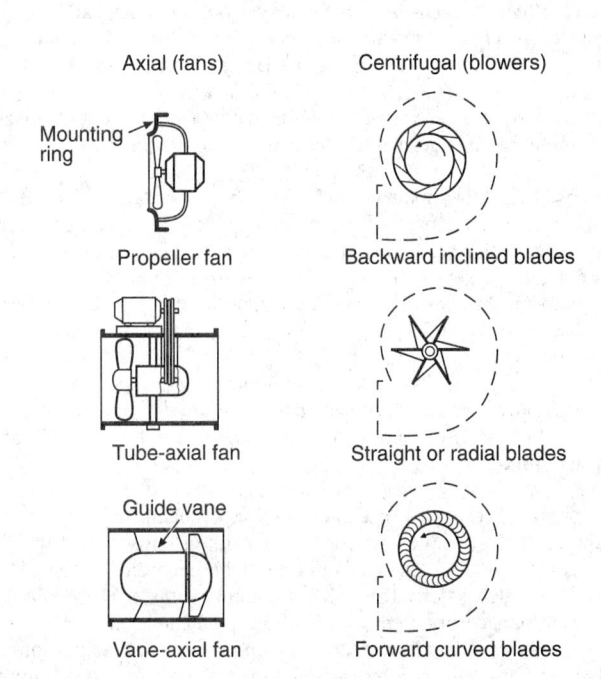

FIG. 4-19A.   *Fan and blower designs suitable for use in exhaust systems.*

Jane I. Lataille, P.E., ARM, is a research consultant with Industrial Risk Insurers, Hartford, CT.

HAZARDS OF SPECIFIC USES

Ventilation of Flammable Vapors

Because flammable vapors ignite easily, precautions must be taken to eliminate ignition sources in ducts that carry them. Using a single system to exhaust flammable vapors and particles from spark-producing processes is obviously dangerous. Similarly, drawing different vapors which individually may not be flammable, but which may be hazardous as a mixture in a single exhaust system, is extremely poor practice. For example, perchloric acid acts as an oxidizing agent at elevated temperatures. If perchloric acid vapors were drawn into a system exhausting organic materials, the combination could cause a fire or explosion.

It is important that vapors be withdrawn from the rooms or equipment in which they are generated and taken directly to the outside of the building or to suitably protected pollution-control equipment. Processes generating flammable vapors are best located along an outside wall of the building to facilitate efficient vapor removal.

When flammable vapors cannot readily be picked up at a specific source, general ventilation through a system of suction ducts with inlets to the room or area may be used. Because suction inlets have minimal directional effect beyond a few inches (1 in. equals 25.4 mm) from the face of the inlet, they should be located so that they produce a sweeping or purging effect that will eliminate pockets in which vapors may accumulate. A makeup air supply, properly located in relation to the point where vapors are generated, and openings for exhaust ducts will aid in the dilution and removal of vapors. The ventilation system should provide sufficient air movement to maintain the vapor concentration below the lower flammable limit (LFL) in the area where vapors are being liberated. If vapors are toxic, it may be necessary to maintain concentrations well below those required for flammability.

NFPA 86, *Standard for Ovens and Furnaces*, gives procedures for calculating the volume of air per minute necessary to dilute flammable vapors below their LFL.

Where heavier-than-air vapors or mixtures are handled, exhaust openings located near the floor line are generally the most effective. Conversely, for vapors or mixtures lighter than air, exhaust openings located near the top of the room, hood, or enclosure are the most effective. Caution in evaluating the vapor density is necessary where other than normal temperatures and pressures exist.

Fans: All fans should be suitable for moving vapors and air. If the vapor contains a material that could condense and build up on the fan blades, this possibility should be minimized by the selection of a fan with backward curved blades. Both the fan and the fan housing should be of nonferrous construction. The construction material chosen should not be adversely affected by the vapors handled.

Ducts: Ducts for systems handling flammable vapors need to be structurally capable of withstanding some fire exposure. Where vapors can condense, continuously welded joints in the ducts will prevent leakage.

Since flash fires or explosions can occur in ducts, ductwork should be installed away from combustibles, but where it will be readily accessible. Ducts should never be concealed in a wall or ceiling. Duct systems that exhaust flammable materials should terminate outside in an area where the vapors cannot be ignited and where they will not form deposits, such as overspray residue, on the building or other structures. An enclosure is sometimes used around the discharge end of a duct to prevent deposits on exposed property. These enclosures should be sprinklered.

Frequent cleaning of ducts, filters, and any enclosures will minimize both the intensity of a fire in a duct system and any spontaneous heating of deposits subject to this phenomenon.

Electrical equipment: All equipment that may be exposed to flammable vapors should comply with NFPA 70, *National Electrical Code®* (hereinafter referred to as NFPA 70) for such hazardous locations.

Ventilation of Corrosive Vapors and Fumes

The degree of expected corrosion is the governing factor in the construction of ducts exhausting corrosive vapors. In some cases, a heavier-gage metal is used, and in others, protective coatings may be sufficient. Occasionally, however, the corrosiveness is so extreme that a special lining is required. Stainless steel has been used very successfully in some cases.

Plastic duct systems are also used, but only if the corrosive vapors are nonflammable and the plastic has a flame-spread rating of 25 or less and a smoke-developed rating of 50 or less, when tested in accordance with NFPA 255, *Standard Method of Test of Surface Burning Characteristics of Building Materials*. Automatic fire protection is necessary in the ducts and also at the hood, canopy, or intake to plastic duct systems. A duct system of plastic materials for a typical industrial exhaust system is shown in Figure 4-19B, and a duct system for a typical hood exhaust system is shown in Figure 4-19C.

Ventilation of Kitchen Cooking Equipment

Exhaust systems for restaurant equipment require careful design, because grease condenses in the interior of the ducts. Grease accumulations may be ignited by sparks from the cooking appliance or, more often, by a small fire on the cooking appliance caused by overheated cooking oil or fat in a deep-fat fryer or on a grill. If the duct did not have a grease accumulation, cooking appliance fires could often be extinguished before causing appreciable damage. Fires occur so often in frying because cooking oils and fats are heated nearly to their flash points. Thus, fats and oils can reach their self-ignition temperatures by being accidentally overheated or by being ignited when spilled on a cooking appliance. Details of a typical kitchen range exhaust system are shown in Figure 4-19D.

Fans: Fans should be selected according to their ability to exhaust the required quantity of air against all calculated friction losses.

Ducts: The following should be considered when designing a duct system for commercial cooking equipment.

1. Design the system to minimize grease accumulations, with a minimum air velocity of 1,500 ft/min (458 m/min) through any duct.
2. Arrange ducts with ample clearance from combustible materials to minimize the danger of ignition, in case of fire in the duct.
3. Use ducts of substantial construction (not lighter than No. 16 Manufacturers Standard Gage steel or No. 18 Manufacturers Standard Gage stainless steel) with all seams and joints having a liquidtight, continuous weld.
4. Separate systems to ensure there is no connection with any other ventilating or exhaust system.
5. Lead ducts directly outside the building, without dips or traps, unless automatic grease removers are employed at the dips and traps.
6. Provide openings for inspection and cleaning. Do not install dampers in any duct system, unless required as part of a grease extractor or extinguishing system.

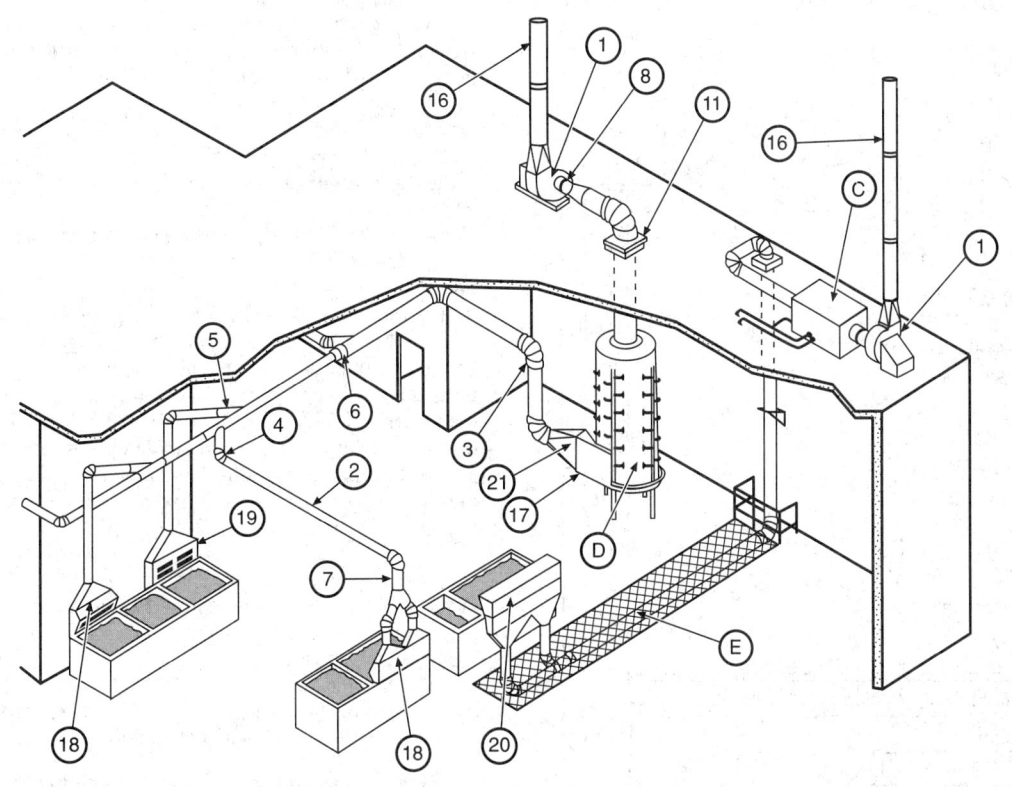

FIG. 4-19B. *Exhaust system for one-story building occupied by various-type fume hoods with vertical fume scrubber and service trench.*

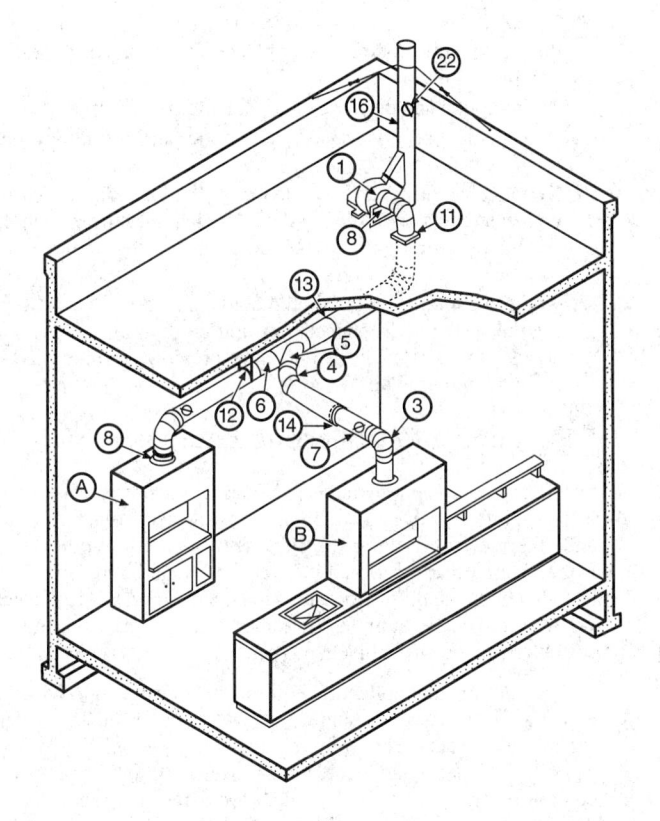

Nomenclature for Figures 4-19B and 4-19C:

I. Equipment
   A. Cabinet-type fume hood
   B. Bench-type fume hood
   C. Horizontal-type fume scrubber
   D. Vertical-type fume scrubber
   E. Service pit or trench

II. System components
  1. Air-moving equipment (centrifugal-type exhaust fan)
  2. Horizontal duct section
  3. 90-degree elbow
  4. Elbow (less than 90 degrees)
  5. Lateral entry
  6. Transition
  7. Manual balancing damper
  8. Flexible connection
  9. Fire damper
 10. Access door
 11. Counterflashing
 12. Duct hanger
 13. Circumferential girth joint (butt welded)
 14. Bell end duct seam
 15. Weather cap
 16. Fan discharge stack
 17. Flanged duct connection
 18. Open-face tank exhaust hood (updraft)
 19. Slotted-face tank exhaust hood (updraft)
 20. Open-face tank exhaust hood (downdraft)
 21. Round to rectangular (or square) transitional fitting
 22. Gravity-operated backdraft damper

FIG. 4-19C. *Rooftop exhaust system for one-story building occupied by a cabinet-type fume hood and bench-type fume hood.*

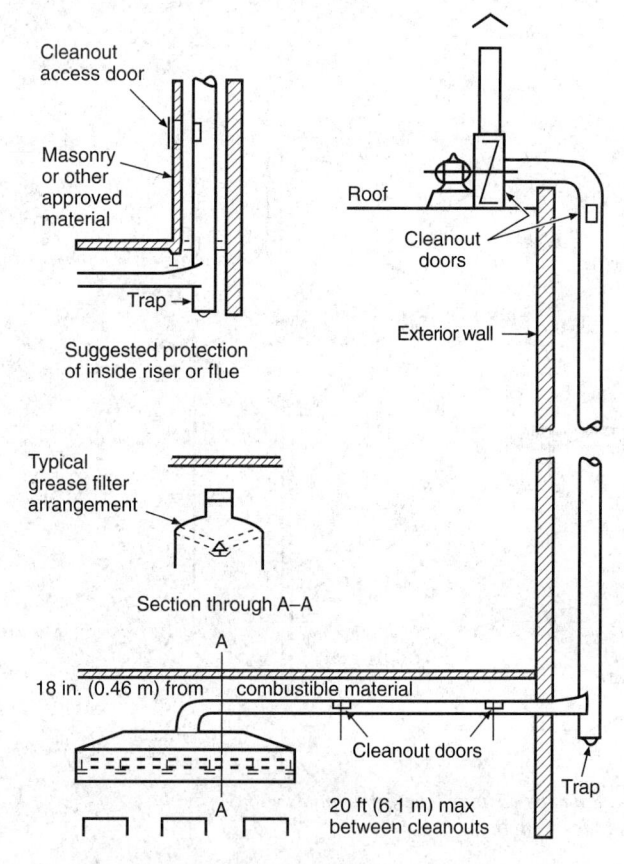

FIG. 4-19D. *Typical commercial cooking exhaust system for building with two stories or more. Duct riser should either be outside or enclosed in an approved fireproof material.*

Grease-removal devices: All kitchen exhaust systems require a means for removing grease. These may be grease extractors, grease filters, or other grease-removal devices, such as water-wash systems and special fans designed to remove grease vapors effectively and provide a fire barrier. Grease filters should be listed for use in commercial cooking operations. Mesh filters should not be used.

Electrical equipment: Manual control of the fan motor should be provided near the motor and also near the hood. In addition, it is desirable to have automatic shutdown of the motor by a heat-sensitive device in the hood close to the outlet.

All electrical equipment that may be exposed to vapors, grease, or heat should be installed in accordance with NFPA 70 requirements for those conditions. Fixtures should not be installed in ducts or hoods used for removal of cooking smoke or grease-laden vapors or located in the path of travel of such exhaust products, unless they are specifically designed and tested for such use. Electrical equipment may be placed outside the path of vapor travel by locating it on the outside of the hood with illumination through tight-fitting glass panels in the hood.

Dust Collecting and Stock and Refuse Conveying Systems

These systems consist of suction ducts and inlets, air-moving equipment, feeders, discharge ducts and outlets, collecting equipment, vaults, and other receptacles designed to collect powdered, ground, or finely divided material.

Collecting and conveying systems are often a very important part of the operations of a plant. When the material they handle is a combustible dust, there is danger of a fire or explosion. These factors make it a prime necessity that the design, construction, and operation of the systems conform to recognized standards. The general recommendations given in the first part of this chapter should be observed, as should the following more specific ones.

Fans: Systems conveying combustible dust should be arranged so that the fan is on the clean air side of the collector, so that the system operates under suction, and dust collecting equipment removes the dust before the air stream reaches the fan. This arrangement prevents combustible dust from passing through the fan where it may be ignited. If such an arrangement is not possible and the fan must be located between the dust-producing equipment and the collector, the blades and spider of the fan and the fan housing should be made of nonsparking material, with ample clearance between the blades and the housing. Fan bearings and motors should be outside of casings unless the fan and motor assembly are designed and tested for use in the dust atmosphere present.

Ducts: For conveying dusts or stock, ducts should be constructed of metal. Changes in direction in the duct system should be made with long bends or elbows. This minimizes accumulations of solid particles at the turns.

Following are some general design techniques recommended for ducts conveying dusts and other solid materials:

1. Every duct should be kept open and unobstructed throughout its length.
2. Not more than two branches should connect to any section of the main suction duct.
3. Branch ducts should be connected to the top or side of the main duct at an angle not exceeding 45 degrees, inclined in the direction of airflow.
4. Main suction and discharge ducts should be as short as possible.
5. Flexible duct sections should not restrict airflow and should be as short as possible.
6. Systems handling combustible dusts should, as far as possible, be located outside of the building. Branch ducts from each floor should pass through the wall and discharge into the main duct outside.
7. Additional branch ducts should not be added without redesigning the system. Disconnected or unused portions of the system should not be blanked off without providing orifice plates to maintain required airflow.

Separating and Collecting Equipment

Separating equipment includes cyclones, condensers, wet-type collectors, cloth screen and stocking arresters, centrifugal collectors, and other devices used for the purpose of separating solid material from the air stream in which it is carried. Then the material is collected in hoppers, bins, silos, and vaults. See Figures 4-19E through 4-19H for typical examples of some of this equipment and how sprinklers may be used in some types.

Separating and collecting equipment is designed to withstand anticipated explosion pressures, allowance being made for deflagration relief vents. This equipment should be constructed of steel or enclosed in steel and located outside the building. To prevent the equipment from collapsing (which could result in an explosible dust cloud), equipment should be well supported on steel, masonry, or concrete. Equipment and discharge ducts should be well separated from combustible construction and unprotected openings into buildings.

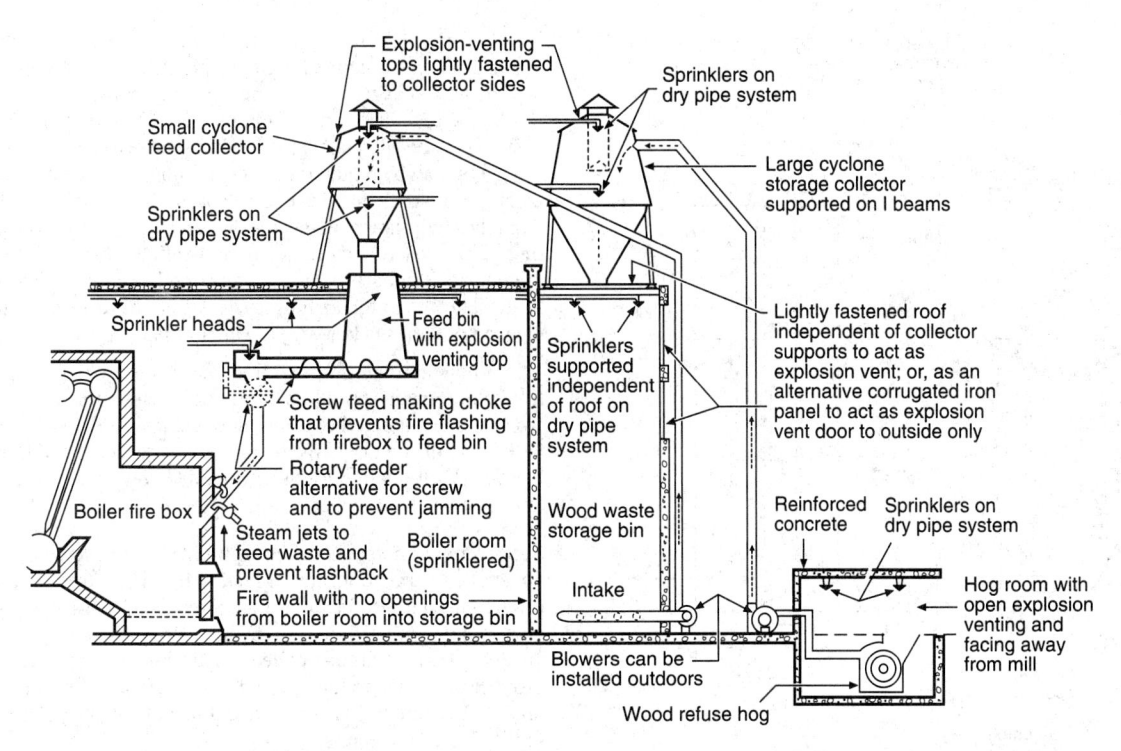

FIG. 4-19E. *Suggested arrangement for wood waste firing of boiler to minimize explosion and fire hazard.*

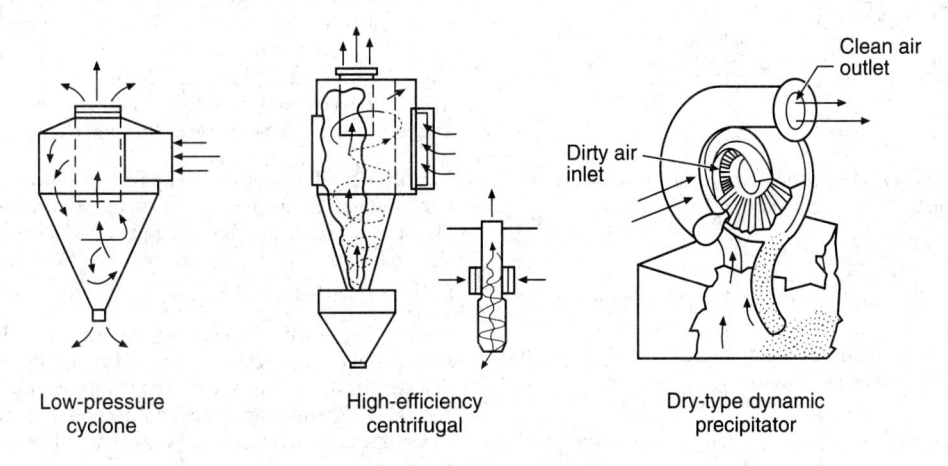

Low-pressure cyclone

High-efficiency centrifugal

Dry-type dynamic precipitator

FIG. 4-19F. *Examples of typical types of dust collecting equipment.*

Collectors that must be located indoors, and that cannot be constructed of sufficient strength to withstand anticipated explosion pressures, can be located in rooms near outside walls to facilitate deflagration venting.

Gravity feed through tightly fitted ducts is the best arrangement for delivering stock from separators, cyclones, or other collection equipment to storage receptacles. Delivery ducts from collection equipment should not convey refuse directly into the fireboxes of boilers, furnaces, refuse burners, or incinerators.

Where processes produce very fine dusts or where combustible metal dusts are handled, wet collectors provide efficient dust control. Dust is removed by passing the flow of air through a water curtain and is collected as a sludge submerged in water. Except where combustible metal dusts are collected, these wet collectors generally have no fire or explosion hazards.

Where refuse is to be used as fuel, the discharge system from the storage receptacle or intermediate feed bin to the furnace should be designed to prevent a flashback from the furnace. This may be accomplished by means of a choke feeder or choke conveyor so that a positive cutoff is provided. The installation of a steam spray in the duct to the furnace that blows steam in the direction of the fuel flow is recommended because it provides an added safety factor in preventing a flashback. See Figure 4-19E for a typical installation of this type.

The installation of screw conveyors or rotary feeds at appropriate points in stock or dust conveying systems has been used to good advantage in many plants by providing a "choke," which helps stop the spread of a flash fire. The installation of a deflagration vent also relieves the pressure of an explosion and prohibits its progress. This combination has often prevented small fires or explosions from becoming large and disastrous.

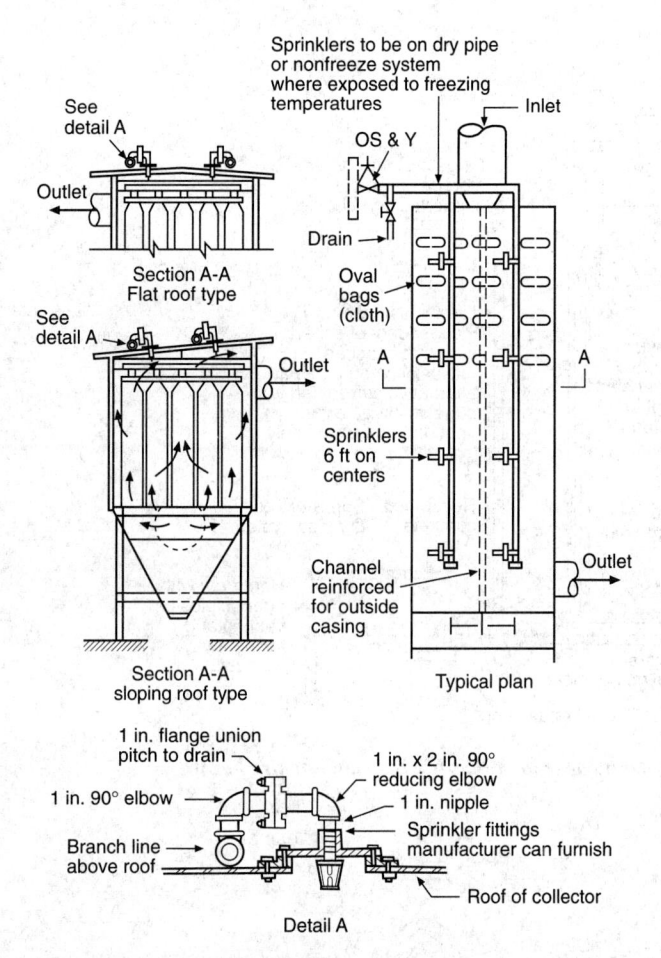

FIG. 4-19G. *Typical sprinkler protection for a bag-type dust collector. (For SI units: 1 in. = 25.4 mm; 1 ft = 0.305 m.)*

High-efficiency air filter units: Particulate air filter units are used to remove very fine particulate matter, i.e., to remove not less than 99.97 percent of 0.3-micron-diameter particles, from the air of industrial and laboratory exhaust systems.

The Nuclear Regulatory Commission (NRC) requires high-efficiency filters for filtering air exhausted from spaces where radioactive materials are handled. The filter units used are made of a filter medium of glass fiber, or equivalent inorganic material. Although the adhesive used is combustible, it contains a self-extinguishing additive. Fire of sufficient temperature and duration to melt glass fibers can damage the filter at points of contact, but will not stop the exhaust of filtered air through the undamaged area. The NRC regulations generally require large banks of these filters, limit the number in a fire area to 100, and require automatic sprinkler protection.

FIRE PROTECTION

Fire Extinguishing Systems

Because of the combustibility of either the ducts themselves, or of the material being transferred through them, the installation of automatic extinguishing systems is justified. The systems may be automatically or manually controlled and may utilize any of several appropriate extinguishing agents.

For example, if the material being transported is a flammable vapor or an easily ignited combustible dust, a deluge system actuated by heat-sensitive devices may be most appropriate. If the problem is the combustibility of the duct itself, an automatic sprinkler system may be adequate. In still other systems, a special extinguishing agent may be required or deemed more suitable. Sprinklers and discharge nozzles may have to be coated or covered to protect them from the environment inside the duct; they may have to be inspected on a regular basis and cleaned periodically to avoid serious buildup of deposits. Similarly, heat detectors or other types of detectors used to activate the extinguishing system may have to be periodically cleaned of deposits that might hamper their proper operation.

One important problem frequently overlooked is the obstruction of automatic sprinklers or other discharge nozzles by the ductwork. Care should be taken that ductwork does not seriously interfere with the discharge pattern. In many cases, additional sprinklers or nozzles will have to be extended down below the ductwork to provide satisfactory coverage.

Separating and collecting equipment: Cloth screen or bag-type dust collectors used for the collection of fine dusts present a special problem because of the use of a combustible fabric even where the dust is noncombustible. Automatic sprinkler protection may be needed where cloth dust collectors are important for continuity of production or where they provide exposure to other property. Figures 4-19G and 4-19H show typical layouts of sprinkler protection for cloth dust collectors.

Equipment of large volume, such as bins, or dust collectors in which pulverized stock is stored or may accumulate, should be protected by automatic sprinklers. Automatic carbon dioxide, dry chemical, or water spray extinguishing systems can be used effectively in dust collecting systems. Inert gas, such as nitrogen, may be effectively used to create safe atmospheres in conveying systems.

Ventilation of flammable vapors: Fixed fire-suppression systems suitable for the flammable material involved, as well as portable fire extinguishers, should be provided. The extinguishing agent should be water, foam, dry chemical, or CO_2 as appropriate for the hazard.

Ventilation of kitchen cooking equipment: For cooking installations, this equipment may consist of fixed-pipe carbon dioxide, dry chemical, wet chemical, foam-water sprinkler, water spray, or sprinkler systems, supplemented by portable alkaline (sodium bicarbonate or potassium bicarbonate) dry-chemical extinguishers.

Acidic base extinguishing agents, such as ammonium-phosphate-based multi-purpose types (ABC rated), impede saponification. Therefore, if the cooking equipment being protected involves exposed liquefied fat or oil in depth, such as deep-fat fryers, extinguishers employing these extinguishing agents are not recommended, as they may interfere with alkaline-based fixed systems.

Where fixed extinguishing systems are installed, they should: (1) be located so that they easily can be manually activated from the path of egress; (2) be arranged for simultaneous automatic operation of all systems in a single hazard area, upon operation of any one system; and (3) automatically shut off all sources of fuel and heat to all equipment protected, upon operation of the system.

Portable Extinguishing Equipment

Portable fire extinguishers of appropriate type and small hose with spray nozzles are helpful where fires may occur in ducts. This is true even where fixed extinguishing systems have been provided. Judicious placement of extinguishers and hoselines with respect to location of access panels will make them more effective.

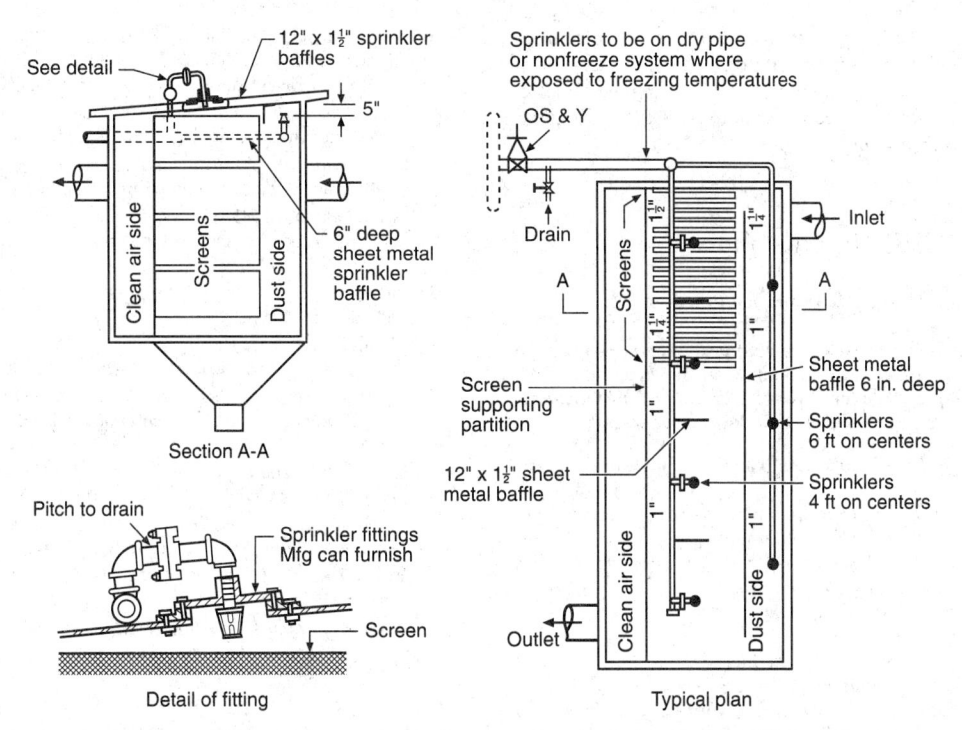

FIG. 4-19H. *Typical sprinkler protection for a screen-type dust collector. (For SI units: 1 in. = 25.4 mm; 1 ft = 0.305 m.)*

Magnetic Separators

As with flammable-vapor conveying systems, combustible-dust collecting systems should not be used with processes that may produce sparks. If there is a possibility of particles of ferrous materials ("tramp metal") entering the collecting systems, permanent magnetic separators should be installed at points where ferrous particles may enter. After collection, the dust passes into rooms or bins. These places of collection should be made of noncombustible construction and provided with deflagration vents terminating outside of the building.

Explosion Prevention and Deflagration Venting

When ignition sources are difficult to control, inerting may be used to create a safe atmosphere within the system, or an explosion suppression system or spark extinguishing system may be used. There are limits to these systems—inerting is generally practical for essentially closed systems, while explosion suppression is limited to those materials and enclosures that can be successfully protected. Efforts to suppress explosions in some materials, e.g., metal dusts, have not been successful. Spark extinguishing systems are suitable for fibers and flyings. Refer to NFPA 69, *Standard on Explosion Prevention Systems* for additional information.

Deflagration relief vents prevent or minimize damage to duct systems that carry explosible mixtures. The vents should lead by the most direct practical route to the outside of the building. NFPA 68, *Guide for Venting of Deflagrations*, gives information on designing deflagration vents. The use of suppression and inerting systems is limited by the configuration of the equipment and the physical and chemical properties of the material being conveyed.

Control of Static Electricity

When flammable vapors, dusts, or other materials pass through a duct, static charges on the duct are generated. If a charge is allowed to accumulate on an electrically insulated portion of a duct system, it could discharge to an adjoining duct section, at a joint for example, and ignite the material being conveyed. For this reason, exhaust systems carrying flammable vapors, dusts, gases, or other materials should be electrically bonded and grounded. Methods of eliminating dangerous static charges are treated in detail in NFPA 77, *Recommended Practice on Static Electricity*.

Dampers

Whenever ducts penetrate rated walls, floors, or partitions, dampers with equivalent fire-resistance ratings should be provided to preserve the integrity of the fire area. Dampers should also be used to prevent backdrafts in manifolded duct systems to keep undesirable vapors, dusts, or impurities from being sucked upstream of their collection point and being spread. Firestopping spool pieces are available for plastic ductwork. These pieces collapse when warm and block the duct, preventing fire spread.

Cleaning

Although clean ducts are essential in preventing fire spread, cleaning them is likely to be neglected because it is difficult and unpleasant, and can be dangerous if flammable solvents are used. Satisfactory cleaning results have been obtained with a powder compound consisting of one part calcium hydroxide and two parts calcium carbonate. This compound saponifies the grease or oily sludge, thus making it easier to remove and clean. Proper ventilation must be provided and safety precautions taken if cleaning is done inside the duct or fan housings.

Spraying the duct interiors with hydrated lime after cleaning is a fire prevention method in commercial use. Though this tends to saponify the grease and may facilitate subsequent cleaning, it does not provide any permanent fire protection or flame resistance, for grease eventually coats the compound.

BIBLIOGRAPHY

NFPA Codes, Standards, and Recommended Practices

Reference to the following NFPA codes, standards, and recommended practices will provide further information on air-moving equipment discussed in this chapter. (See the latest version of *The NFPA Catalog* for availability of current editions of the following documents.)

NFPA 33, *Standard for Spray Application Using Flammable or Combustible Materials*

NFPA 61, *Standard for the Prevention of Fires and Dust Explosions in Agricultural and Food Products Facilities*

NFPA 65, *Standard for the Processing and Finishing of Aluminum*

NFPA 68, *Guide for Venting of Deflagrations*

NFPA 69, *Standard on Explosion Prevention Systems*

NFPA 70, *National Electrical Code®*

NFPA 77, *Recommended Practice on Static Electricity*

NFPA 86, *Standard for Ovens and Furnaces*

NFPA 91, *Standard for Exhaust Systems for Air Conveying of Materials*

NFPA 96, *Standard for Ventilation Control and Fire Protection of Commercial Cooking Operations*

NFPA 255, *Standard Method of Test of Surface Burning Characteristics of Building Materials*

NFPA 318, *Standard for the Protection of Cleanrooms*

NFPA 654, *Standard for the Prevention of Fire and Dust Explosions in the Chemical, Dye, Pharmaceutical, and Plastics Industries*

Additional Reading

AMCA Supplement to January 1994 Issue of Heating, Piping, and Air Conditioning (64-pg. issue on fan applications and design), Air Movement and Control Association, Inc., Arlington Heights, IL 60004.

ANSI Committee on Ventilation, "Fundamentals Governing the Design and Operation of Local Exhaust Systems," ANSI Z9.2-1979 (reaffirmed 1991), American National Standards Institute, New York.

ASHRAE Handbook—Applications, American Society of Heating, Refrigerating, and Air-Conditioning Engineers, Inc., Atlanta, GA, 1991.

ASHRAE Handbook—Fundamentals, American Society of Heating, Refrigerating, and Air-Conditioning Engineers, Inc., Atlanta, GA, 1993.

ASHRAE Handbook—HVAC Systems and Equipment, American Society of Heating, Refrigerating, and Air-Conditioning Engineers, Inc., Atlanta, GA, 1992.

Bartknecht, W., *Explosions; Course, Prevention, Protection*, Springer-Verlag, New York, 1981.

Burchsted, C. A., "Basic Requirements for HEPA Filters," *Fire Technology*, Vol. 3, No. 4, Nov. 1967, pp. 271–280.

Committee on Industrial Ventilation, *Industrial Ventilation*, 20th ed., American Conference of Governmental Industrial Hygienists, Lansing, MI, 1988.

Frank, T. E., "Fire and Explosion Control in Bag Filter Dust Collection Systems," *Fire Journal*, Vol. 75, No. 2, Mar. 1981, pp. 75–80, 94.

Gill, K. E., and Sipila, O. E., "Kitchen HVAC Design," *Heating, Piping, and Air Conditioning*, Vol. 66, No. 10, Oct. 1994.

Gustafson, T., "Understanding Centrifugal Fans," *Plant Engineering*, February 6, 1995, pp. 50–52.

Heinsohn, R. J., *Industrial Ventilation*, Wiley & Sons, New York, 1990.

Levenback, G., "Grease Fires in Kitchens," *Fire Journal*, Vol. 66, No. 4, July 1972, pp. 69–72.

Reason, J., "Fans," Special Report, *Power*, Sept. 1982, pp. S.1–S.24.

Roberts, A. F., "Fire in Ducts Under Forced Ventilation Conditions," *Fire Technology*, Vol. 6, No. 1, Feb. 1970, pp. 13–21.

Schmitt, C. R., "Flammable Solvent Cleaning Operations in a Laminar Flow Cleanroom," *Fire Prevention and Suppression*, Hilado, C. J., ed., Technomic, Westport, CT, 1974, pp. 90–96.

Siconolfi, C. A., "Fire Retardancy in Chemical Process Ductwork," *Fire Technology*, Vol. 5, No. 3, Aug. 1969, pp. 217–224.

Smith, E. E., "Evaluation of the Fire Hazard of Duct Materials," *Fire Technology*, Vol. 9, No. 3, Aug. 1973, pp. 157–170.

Talbot, G., "Static Electricity," *Fire*, 70(871), 1979, pp. 397–398.

Wrenn, C., "Inerting for Safety," *Plant/Operations Progress*, Vol. 5, No. 4, 1986, pp. 225–227.

DETECTION AND ALARM

Time is the most important commodity in fire protection. The fire protection community's objective is to reduce reaction time, evacuation time, response time, and suppression time. Past fire events have shown that codes- and standards-complying fire alarm systems can provide the "window of safety" needed to meet fire protection goals. Coupled with automatic sprinkler systems, tools exist to significantly reduce life and property loss from fire.

Fire alarm systems have evolved through the years into sophisticated, computerized components capable of integrating entire building operations and control systems. However, one must remember that the fire alarm system is only part of the overall fire protection strategy of building and systems designers. The utilization of codes- and standards-complying fire alarm systems, along with proper fire barrier design and complete automatic suppression systems, will provide a strong basis for an effective fire safe design.

Fire alarm systems are used primarily for the protection of life, and secondarily for the protection of property. Many times these systems are used to protect the mission of the property owner or tenant. Mission protection simply means: Will a fire in my facility put me out of business? Unfortunately, over the years, partial detection coverage was assumed to be the maximum detection needed to provide "adequate" life safety protection. This is not true. As recent research and anecdotal fire evidence has shown, partial detection does not often, if ever, provide early warning of a fire condition; i.e., if the device installed to sense a fire condition is remote from the initial fire occurrence, it will not detect the fire in a timely fashion.

As the fire protection engineering and code-making professions move toward performance-based applications of fire alarm systems, it becomes increasingly important to design and install complete detection coverage for a facility, in order to ensure expected early-warning performance.

Section 5 provides the reader with an overview of fire alarm system concepts. The bibliographies provide the reader with additional references for more detailed fire alarm system design guidance.

In May of 1993, the membership of the National Fire Protection Association voted to adopt NFPA 72, *National Fire Alarm Code.* For the first time since the standards for fire alarm systems were introduced, all standards pertaining to fire alarm systems were combined in one document. The stated purpose of that effort included the goals of reducing duplicate requirements and making the standards more "user friendly."

The goal in the reorganization of Section 5 is the same. This section of the *Fire Protection Handbook* is designed for those individuals who desire a basic knowledge of fire alarm systems operational and design requirements. The following chapters are meant as background information and as a "road map" to direct the reader to the appropriate code, standard, or technical reference that must be fol-

lowed to provide codes- and standards-complying fire alarm system installations.

The chapters in this section are arranged in the following order:

Chapter 1, Fire Alarm Systems

Chapter 2, Automatic Fire Detectors

Chapter 3, Notification Appliances

Chapter 4, Fire Alarm System Interfaces

Chapter 5, Fire Alarm Systems, Inspection, Testing, and Maintenance

Chapter 6, Household Fire Warning Equipment

Chapter 7, Fire Protection Surveillance and Fire Guard Services

Chapter 8, Gas and Vapor Detection Testing

Chapters 1 through 4 have been arranged to provide the basic information of how a fire alarm system is "put together."

Chapter 1 addresses the background information regarding the types of fire alarm systems allowed in NFPA 72 and how a typical fire alarm system is classified based on its off-premises connection. The fundamentals of a codes- and standards-complying fire alarm system are established in this chapter.

Once the basic type of fire alarm system has been established, the type of automatic detection must be determined. Chapter 2 has been revised to include many of the latest automatic detectors that are now common in the marketplace.

Chapter 3 presents up-to-date information regarding notification appliances, and offers guidance with respect to the audibility and visibility requirements of codes- and standards-complying fire alarm systems.

Chapter 4 presents new information and requirements on an old subject: how to properly interface other fire protection systems to a fire alarm system. NFPA 72 has established new requirements in this area, and this chapter discusses some of the "how to comply" issues.

Chapter 5, for the first time in this section, introduces a basic primer on the reliability of fire alarm systems and the part that inspection, testing, and maintenance plays in ensuring that reliability. The reliability of fire alarm systems is made up of four basic components: (1) design, (2) equipment, (3) installation, and (4) maintenance. The basic long-term reliability of fire alarm systems depends on timely, effective maintenance. Fire alarm systems are often designed to meet NFPA 72, inspected during installation, and performance tested before the final acceptance is given. With the number of individuals who are supposed to review the fire alarm system, it would seem unlikely that an unreliable system could be installed. The fire professionals in the field indicate otherwise. The goal of this chapter is to provide the background to show how important each step of the process of commissioning a fire alarm system is to ensuring overall reliability of operation when the fire alarm system is called upon to do its job.

The revisions to Chapter 6 include the latest NFPA 72 requirements and a look at some of the new technologies available for household fire warning systems. Chapter 6 is a stand-alone chapter, with the exception that some of the information contained therein applies to commercial occupancies, such as apartments and hotels.

Chapter 7 provides guidance on the application of fire alarm systems in an industrial/commercial occupancy, and how these systems are integrated with other building management systems.

Although about specialized detection systems, Chapter 8 is included because gas and vapor detection is increasingly being connected to, and has become an integral part of, fire alarm systems.

Other areas in this handbook that emphasize concepts of integrated fire safety systems design include Chapters 2 and 3 of Section 1; and Sections 6 and 7, suppression and confinement, respectively. With the advent and popularity of fire models, it is recommended that the reader become familiar with those models that pertain to fire alarm systems. For a detailed treatment of calculation and performance-based methods of designing fire alarm systems, refer to *The SFPE Handbook of Fire Protection Engineering*.

—Wayne D. Moore
December 1996

FIRE ALARM SYSTEMS

Dean K. Wilson

A fire detection and alarm system is a key element among the fire protection features of any building. Properly specified, designed, manufactured, installed, maintained, tested, and used, a fire alarm system can help limit property fire losses in buildings, regardless of occupancy. And, since many of the fire deaths in the United States result from building fires, the use of fire detection and alarm systems in buildings also can help to significantly reduce the loss of life from fire.

Strictly adhering to the requirements of NFPA 72, *National Fire Alarm Code* (hereinafter referred to as NFPA 72), is fundamental to maintaining an appropriate level of quality control through the steps that lead to the successful provision of a fire detection and alarm system. This document, working in concert with NFPA 70, *National Electrical Code®* (hereinafter referred to as NFPA 70), particularly Article 760 on wiring for fire alarm systems, forms the basis for requirements that have been created through the consensus standards-making process. These requirements represent the combined experience and professional judgment of the members of the seven technical committees that write the various chapters of NFPA 72.

This chapter will discuss the philosophical basis for fire alarm system design, as well as the major operational characteristics of the various types of fire alarm systems currently in use.

SELECTING THE SYSTEM

In making the basic decision regarding what type of fire alarm system is needed for a particular building, certain questions must be answered. In answering these questions, one must bear in mind that the most successful approach to the overall protection of life and property is a holistic approach—an approach that relies on the fact that the protection provided by the sum of the individual components of the fire protection system will always be much greater than if one was to rely on each individual component alone. Thus, effective protection is a combination of interlocking, interrelated fire protection systems, designed to function in concert with each other to provide the desired level of protection.

It is not a matter of whether or not automatic sprinklers or a fire alarm system should be provided. But rather, by providing automatic sprinklers; a fire alarm system; proper building construction

features, such as fire cutoffs; along with a host of other valuable prevention and protection features, all built on a foundation of management commitment to fire prevention and control, only then can a building owner or occupant be assured that a facility is safe.

With this foundational truth in mind, fire alarm system selection must be based on answering the question: "What are the protection goals?" Is the building owner or occupant attempting to provide a level of life safety that will meet the needs of the people occupying the building and satisfy the requirements of the local building code and local Authority Having Jurisdiction? Or is the building owner or occupant intent on providing a needed level of property protection? Or is the building owner or occupant trying to ensure that the mission of the facility is not interrupted by fire?

The final answer may be "yes" to one or more of these questions. And, with each "yes" answer, the building owner or occupant will be guided to select the fire alarm system that will meet the needs of the facility.

MAKING THE SYSTEM WORK

Just as building fire protection is most effective when it consists of a series of interlocking, interrelated fire protection systems, ensuring the quality of a fire alarm system is the result of specific attention given to a series of interlocking, interrelated steps:

- Specify
- Design
- Manufacture
- Install
- Maintain
- Test
- Use

The user of the fire alarm system must define the reason why the system is being provided. This definition step will identify the protection goals of the user and also the external factors that may influence the fire alarm system installation. The user may determine that the protection goal is the preservation of the lives of the persons occupying the building. Such a *life safety goal* and the protection features it dictates will help determine the type of fire alarm system. Or the user may determine that the protection goal is the preservation of the physical building and its contents. Such a *property safety goal* and the protection features unique to this goal will also help determine the type of fire alarm system that can most effectively provide protection.

In analyzing the particular needs of a facility, the user may determine that the protection goal is the preservation of the ability to

Dean K. Wilson, P.E., is director of loss prevention training for Industrial Risk Insurers in Hartford, CT. He is chair of the Technical Correlating Committee on the *National Fire Alarm Code.®* He is publisher and co-editor of the fire alarm newsletter, *The Moore-Wilson Signaling Report.*

produce a product or provide a service. Such a *mission protection goal* will help determine the type of fire alarm system needed at the facility.

It may even be that the user determines that the protection goal actually embraces a combination of all three primary goals: (1) life safety, (2) property protection, and (3) mission protection. Once the protection goals are clearly defined, the features of the fire alarm system can be determined.

The feature selection will result in a clearly and concisely written set of specifications. These specifications will include the needed features and will include statements as to how the quality of the installation is to be ensured, including the initiating device, notification appliance, and signaling line circuit performance. This document will guide the system supplier in making recommendations regarding particular equipment, as well as guide the supplier by providing the quality control measures.

Based on the specifications, a fire protection engineer working either as a consultant to the user or as an employee of, or consultant to, the system supplier can develop a detailed system design. By exercising care to ensure that the design meets both the specifications and the requirements of NFPA 72, the designer will greatly enhance the overall quality of the fire alarm system.

Meanwhile, the manufacturer of fire alarm system components has been studying the needs of fire alarm users and has developed products with a wide variety of features. The manufacturer will provide products of high quality that can meet the requirements of NFPA 72 and pass the tests of the nationally recognized testing laboratories, such as Underwriters Laboratories Inc. (UL) and Factory Mutual Research Corporation (FMRC).

A qualified fire alarm system installer can now take the manufacturer's equipment that satisfies the user's specifications and the fire protection engineer's design, and install a fire alarm system. The process of installation must also meet the requirements of NFPA 72. Quality assurance during installation is critical to the long-term usefulness and stability of the fire alarm system. There are several ways of ensuring installation quality.

To help ensure installation quality, UL has developed two fire alarm certificate programs. One certificate is for central station fire alarm systems. The other is for local, auxiliary, remote station, proprietary, and emergency voice alarm/communication systems. UL-listed contractors may apply to UL for a certificate covering the installation of a fire alarm system at a specific protected premises. The certificate describes the system, and states that the contractor has met the requirements of NFPA 72 and the relevant UL standards. It further states that the contractor has a contract with the user for periodic testing and routine maintenance of the fire alarm system to ensure quality continuity. Periodically, UL representatives inspect a statistical sampling of certificated installations of each contractor to check on the accuracy of the information supplied on the certificate and to verify compliance with NFPA 72 and relevant UL standards.

FMRC has developed a placard that the contractor may post at the fire alarm installation, stating that the system has been installed to comply with the requirements of NFPA 72 and the relevant FMRC requirements. When FM representatives visit facilities insured within the FM System, they look for this placard as an indication that standard fire alarm service has been provided.

Some Authorities Having Jurisdiction (AHJs) require that installers demonstrate competency to install systems. Competency may be demonstrated by achieving certification through the National Institute for Certification in Engineering Technologies in fire protection, subfield fire alarms. Or, some AHJs may have developed an examination of their own that installers must pass.

FIG. 5-1A. *Technology increasingly impacts fire alarm systems, as evidenced by this solid-state, microprocessor-based fire alarm system control unit. An alpha-numeric visible display provides detailed information from the system. (Source: Fire Control Instruments)*

To ensure continuity of quality, a fire alarm system must be tested and maintained. NFPA 72 sets forth the requirements for maintenance and testing in Chapter 7. Once again, the competency of the persons performing the testing and maintenance is critical to achieving an appropriate level of quality. (See Chapter 5 of this section, "Fire Alarm Systems, Inspection, Testing, and Maintenance.")

Finally, just as the components of the fire alarm system must be listed by a nationally recognized testing laboratory for the specific use, the way the fire alarm system is used must conform to the requirements of NFPA 72. By carefully working through the steps, i.e., specify, design, manufacture, install, maintain, test, and use, the user of the fire alarm system will have a system that will serve the facility well for many years to come.

TYPES OF SIGNALS

Fire alarm systems may provide three types of signals: (1) alarm, (2) supervisory, and (3) trouble.

An *alarm signal* is a warning of fire danger that requires immediate action.

A *supervisory signal* indicates that action is needed in connection with the operation of other fire protection systems that are being monitored by the fire alarm system. Such systems may include extinguishing or suppression systems, such as automatic sprinkler systems, carbon dioxide systems, dry chemical systems, foam systems, and gaseous agent systems. They may also include the supervision of guards who make fire patrol tours throughout the protected premises. A supervisory signal may be either an "off-normal" signal, indicating a condition requiring attention; or a "restoration-to-normal" signal, indicating the condition that initiated the original off-normal signal has been resolved.

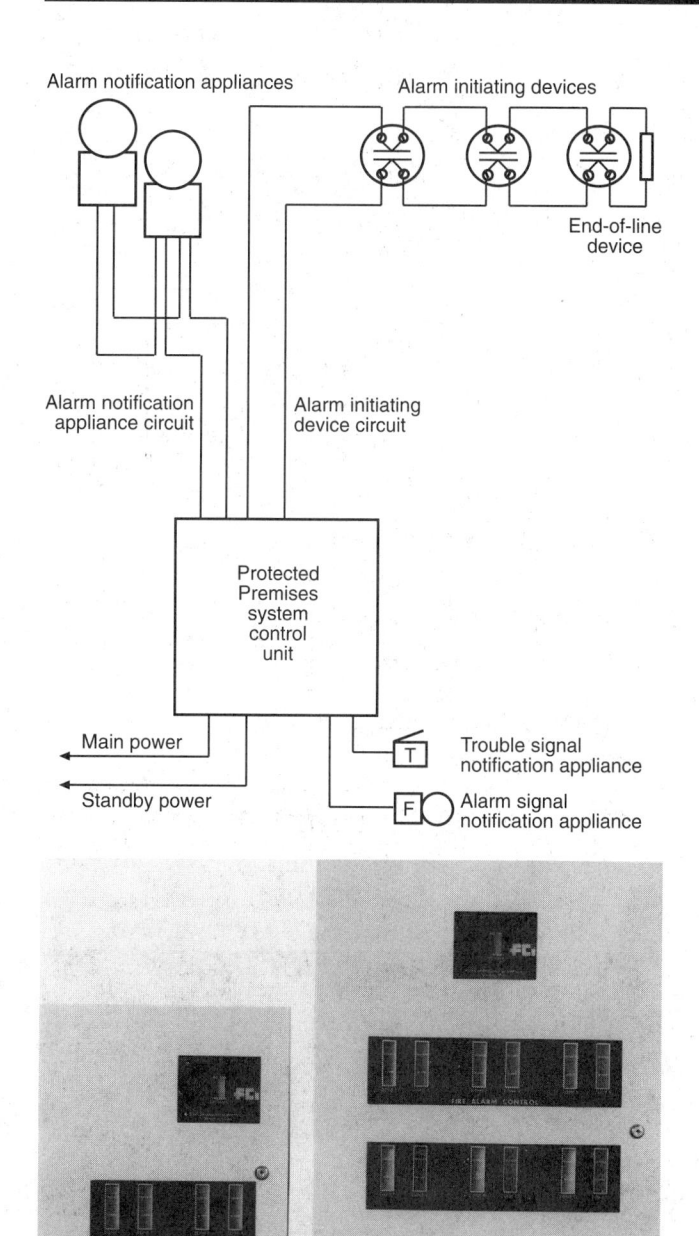

FIG. 5-1B. A typical arrangement of a local fire alarm system (top). A local system control unit (bottom). (Source: Fire Control Instruments)

A *trouble signal* indicates a fault in a monitored circuit or component of the fire alarm system, or the disarrangement of the primary or secondary power supply.

FIRE ALARM SYSTEM BASICS

Fire alarm systems are classified according to the functions they are expected to perform. Their installation, maintenance, and use are specified in NFPA 72.

The basic features of each system are:

1. A system control unit. (See Figure 5-1A.)
2. A primary, or main, power supply.

3. A secondary, or standby, power supply.
4. One or more initiating device circuits or signaling line circuits to which manual fire alarm boxes, sprinkler waterflow alarm initiating devices, automatic fire detectors, and other fire alarm initiating devices are connected.
5. One or more fire alarm notification appliance circuits to which audible and visible fire alarm notification appliances, such as bells, horns, stroboscopic lamps, and speakers, are connected.
6. Many systems also have an off-premises connection to a central station, proprietary supervising station, remote supervising station, or public fire service communication center by means of an auxiliary fire alarm system.

The primary and secondary power supplies at both the protected premises and at a subsidiary or supervising station must meet the requirements of NFPA 72. The primary power is usually supplied by a connection to utility-generated electric power. The connection must be from a branch circuit dedicated to the fire alarm system. The circuit and connections must be mechanically protected. The circuit disconnecting means must have a red marking, be accessible only to authorized personnel, and be identified as "Fire Alarm Circuit Control." Inside the fire alarm system control unit, a permanent legend must identify the location of the circuit disconnecting means.

Very remote areas may not have utility-generated power available. In these cases, electricity is usually supplied by generators on the site of the facility. Most often, more than one generator is provided, each operated for a number of hours at a time to help ensure the reliability of the electric power. Primary power for fire alarm systems at geographically remote facilities is, therefore, supplied by generators.

Secondary power supply for a fire alarm system is required to automatically supply the energy to the system within 30 sec whenever the primary power supply is not capable of providing the minimum voltage required for proper system operation.

The size of the secondary supply usually is measured in the amount of time that the secondary supply will operate the system, followed by a prescribed time period for the system to operate in an alarm condition. Local, central station, and proprietary systems must have 24 hr of standby power, followed by 5 min of alarm. Remote station and auxiliary systems must have 60 hr of standby power, followed by 5 min of alarm. Emergency voice/alarm communication systems must have 24 hr of standby power, followed by 2 hr of emergency operation. To allow calculation of the power required for 2-hr emergency operation, NFPA 72 specifies that the 2 hr of emergency operation is the equivalent of 15 min of operation, under full load, i.e., with all input devices and output appliances operating.

The required capacity of the secondary power can be supplied by storage batteries; storage batteries with 4 hr of capacity in conjunction with an engine-driven generator; or multiple engine-driven generators, each of which must be capable of supplying the entire standby power load. This generator is considered "dedicated" to the fire alarm system. This means that the reason the generator was installed was to supply secondary power to the fire alarm system. Of course, if the generator has a greater capacity than is needed by the fire alarm system, it may also be used to supply loads related to the fire alarm system, such as lighting or air-conditioning equipment serving the room where the fire alarm system is located.

In certain occupancies, an emergency power system is provided to ensure continuity of operations at the facility. When a fire alarm system receives its primary power from a buss connected to an emergency power system installed in accordance with NFPA 70, Article 700 or 701; or to an emergency power system installed in accordance with Article 702 that also meets the requirements of Article

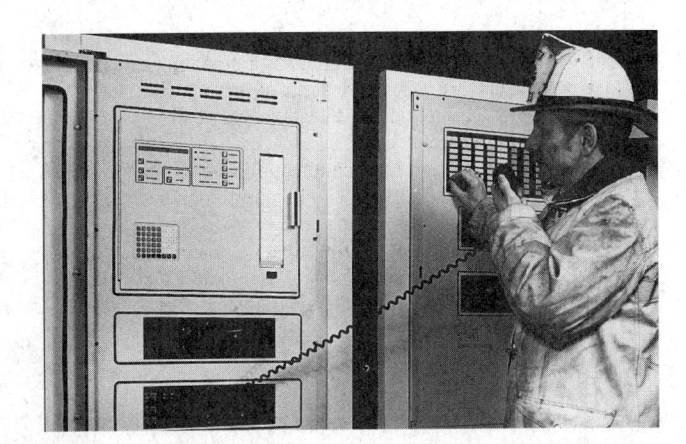

FIG. 5-1C. *Using an emergency voice/alarm communication system, a fire officer gives building occupants specific instructions during a fire. (Source: Simplex Time Recorder Company)*

700 or 701, then the fire alarm system is not required to have a secondary power supply. In actual applications, relatively few emergency power systems meet these stringent requirements.

All fire alarm system cable and the wiring installation must conform to Article 760 of NFPA 70, and the additional requirements of NFPA 72.

Fire alarm systems have three basic types of circuits: (1) initiating device circuits (IDC), (2) notification appliance circuits (NAC), and (3) signaling line circuits (SLC).

Initiating device circuits connect conventional (non-addressable) fire alarm and supervising initiating devices to the system control unit. Notification appliance circuits connect notification appliances (audible and visible) to the system control unit. The term *signaling line circuits* is used to define circuits over which two-way communications take place. This communication can take place between an addressable device and system control unit or the system control unit and the off-premises connection, such as a central station, a public fire service communications center, a remote station, or a proprietary central supervising station.

NFPA 72 provides the fire alarm system designer with various circuit configurations that provide different levels of operation and integrity. The circuit configuration can only be defined when both the fire alarm control unit operation and the wiring configuration are known.

A typical circuit used in many fire alarm systems consists of a two-wire circuit with an end-of-line resistor. Initiating devices, with normally open contacts, are connected in parallel. A small amount of electrical current flows through the wire to monitor the wiring integrity. A break in the wire causes a trouble condition at the system control unit. Everything electrically beyond the break in the wiring is out of service until repairs are made to the circuit. Operation of an initiating device shunts the resistor, thus increasing the current which causes the system control panel to respond in an alarm condition.

A circuit configuration that operates up to the single fault (open-circuit or ground-fault condition) is labeled a Class B circuit. A circuit configuration that requires the wiring to be returned to the system control unit, where the system control unit is configured to allow a device or appliance to operate on either side of a single open-circuit or ground-fault condition, is labeled Class A. Both of these designations can be applied to initiating device circuits, notification appliance circuits, and signaling line circuits.

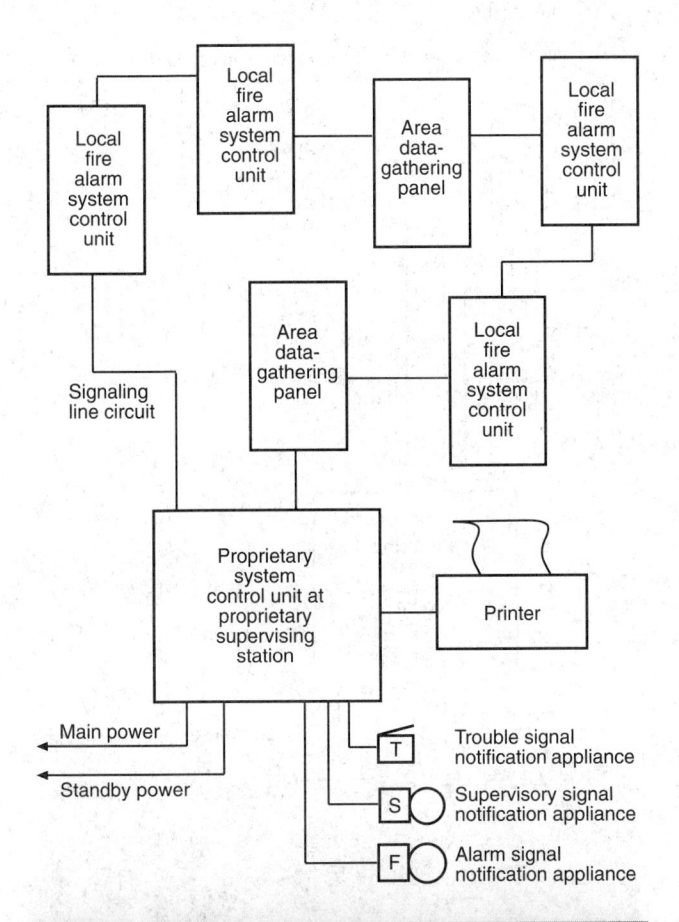

FIG. 5-1D. *A typical arrangement of a proprietary protective signaling system (top). At the central supervisory station, an attendant constantly monitors the proprietary protective signaling system (bottom). The printer on the desk provides a permanent record of signaling traffic. (Source: Simplex Time Recorder Company)*

Many fire alarm system control units have more than one initiating device circuit, so that the fire location can be indicated on an annunciator panel by floor, wing, subsection, or room. The annunciator can be built into the control unit or located in a lobby, maintenance area, telephone switchboard room, or some other location where it is accessible to building and fire service personnel.

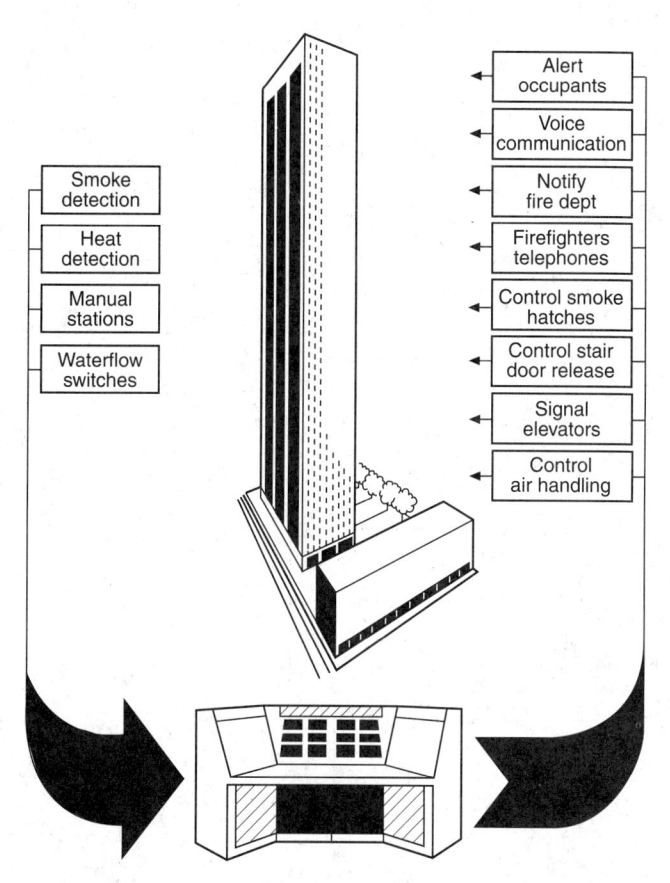

FIG. 5-1E.   A modern computer-controlled proprietary fire alarm system.

NFPA 72 significantly revised the manner in which requirements for fire alarm systems are stated. Chapter 1 covers the fundamentals of fire alarm systems. Chapter 2 is a completely stand-alone chapter that covers household fire warning systems. Chapter 3 covers the protected premises fire alarm system. Chapter 4 covers the supervising station fire alarm system. Chapter 5 covers initiating devices for fire alarm systems. Chapter 6 covers notification appliances for fire alarm systems and Chapter 7 covers the testing and maintenance of fire alarm systems.

The requirements for fire alarm systems that provide household fire warning have their own chapter (Chapter 2), and the requirements for fire alarm systems for all other occupancies are contained in Chapters 1, 3, 4, 5, 6, and 7.

A fire alarm system that meets the requirements of Chapters 1 and 3 forms the basic fire alarm system "building block." A protected premises fire alarm system, whether it is used to notify the occupants so they may safely evacuate, used to actuate a fire suppression or extinguishing system, or used to oversee conditions at a facility and report status changes to a supervising station, must meet the requirements of these two chapters. The initiating devices and notification appliances connected to the protected premises fire alarm system must meet the requirements contained in Chapters 5 and 6, respectively. A supervising station fire alarm system must meet the requirements contained in Chapter 4. And a protected premises fire alarm system, its initiating devices and notification appliances, along with a supervising station fire alarm system, must all meet the testing and maintenance requirements contained in Chapter 7.

Because of the familiarity and wide usage of industry terms that have evolved over the years, NFPA 72 has retained such termi-

nology. One example of this terminology is the system designations: protected premises system, emergency voice/alarm communication system, central station system, proprietary system, remote station system, public fire reporting system, and auxiliary system.

PROTECTED PREMISES SYSTEM

The main purpose of a protected premises fire alarm system is to activate local audible and visible alarm notification appliances to notify the occupants that they must evacuate the protected building. (See Figure 5-1B.) Such a system could be limited to the basic features described earlier. The system could also interface with other protective systems that would help make the building safer in case of fire. Such features might include, for example, the operation of a fire extinguishing or suppression system; the recall of elevators; the unlocking of doors; the closing of smoke barrier doors; the control of heating, ventilation, and air-conditioning (HVAC) equipment; or the operation of smoke control systems.

In a protected premises system, the alarm is not relayed automatically to a fire department. Instead, when the alarm sounds, someone must use some other means to notify the fire department. If the building is unoccupied at the time of the alarm, fire department response would depend upon a neighbor or passerby hearing or seeing the audible and/or visible fire alarm notification appliances and telephoning the fire department.

EMERGENCY VOICE/ALARM COMMUNICATION SYSTEM

This system is used in occupancies, such as high-rise buildings, where it is necessary to relocate occupants to areas of refuge, rather than evacuate them. (See Figure 5-1C.) The voice/alarm system consists of a series of high-reliability speakers located throughout the building. They are connected to, and controlled from, the fire alarm communication console located in an area designated as a building fire command station. From the building fire command station, individual speaker zones or the entire building can be selected to receive an alert tone followed by a voice message that gives specific instructions to the occupants. Some systems have fire warden telephone stations on each floor, or within each fire zone, to which a fire warden would report to assume local command and pass on specific evacuation instructions.

A trained building employee usually operates the fire command station until the fire department arrives, at which time the officer in charge takes over. The fire warden telephone stations may also be used during fire-fighting operations for communication with the fire fighters whose hand-portable two-way radios may not function properly in the high-rise building.

One important aspect of a voice/alarm communication system is that, since complete building evacuation is not always feasible, it can instruct occupants to relocate to areas of refuge where they can safely wait for the fire to be brought under control. In such cases, communication with these relocated occupants must be maintained to prevent panic and to facilitate further relocation if necessary.

CENTRAL STATION FIRE ALARM SYSTEM

A central station fire alarm system is designed to receive signals from a protected premises at a constantly attended location operated by a company whose purpose is providing central station service. This operating company must be listed by either UL or FMRC. Central station installations must have either the UL certificate or FMRC placard; whichever means of verifying the character of the

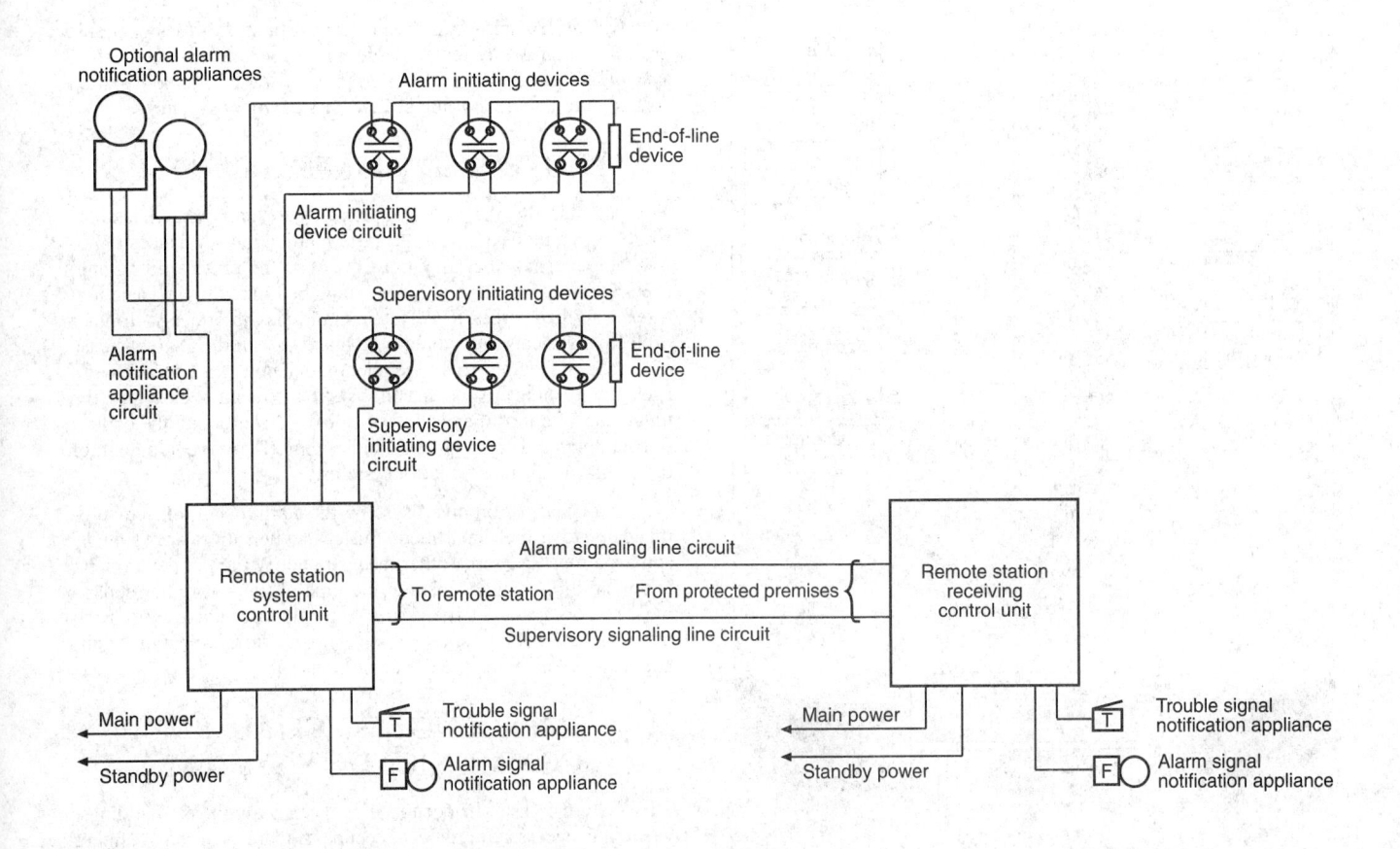

FIG. 5-1F. *A typical arrangement of a remote station fire alarm system. Note that, if a direct circuit connection is used, fire alarm and supervisory signals must be transmitted over separate signaling line circuits (top). Upon receipt of a remote station sprinkler waterflow alarm, the fire dispatcher prepares to retransmit the alarm to the fire companies that will respond (bottom).*

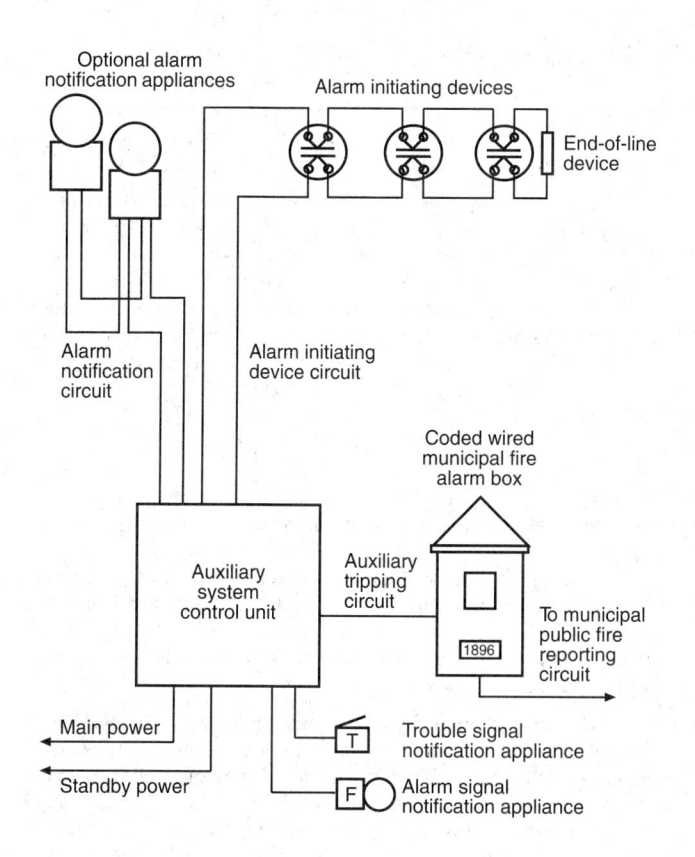

FIG. 5-1H.   *A local energy-type masterbox that connects the protected premises fire alarm system to the public fire reporting system. (Source: Gamewell Corp.)*

FIG. 5-1G.   *A typical arrangement of an auxiliary fire alarm system (top). An auxiliary system control unit (bottom). (Source: Simplex Time Recorder Company)*

installation is required by the Authority Having Jurisdiction. In the majority of cases, central station service is provided only for industrial and commercial facilities of relatively high value. In such cases, the central station fire alarm system is intended to enhance the property protection aspects of a facility in order to satisfy the recommendations submitted by a highly protected risk property insur-

ance company acting in its role of Authority Having Jurisdiction for matters relating to property insurance.

In some cases, a central station operating company may collect signals from a specific geographic area at a subsidiary station and then transmit the signals to a central station for handling. While the subsidiary station is not normally staffed, NFPA 72 requires that it be capable of being staffed in an emergency. Such an emergency might include the failure of the communication channels between the subsidiary station and the central station.

There are eight elements of central station service: four at the protected premises and four at the central station. The four at the protected premises are: installation, testing, maintenance, and runner service. The four at the central station are: management or oversight of the system, monitoring of signals from the protected premises, retransmission of fire alarm signals to the public fire service communication center, and recordkeeping.

Central station service may be provided entirely by a listed central station operating company, by a listed central station operating company which contracts one or more elements of central station service at the protected premises to a subcontractor, or by a listed local fire alarm installation company which will provide the four elements at the protected premises and subcontract the four elements at the central station to a listed central station operating company.

Signals may be transmitted from a protected premises using a variety of transmission technologies stated in NFPA 72, Chapter 4. Such technologies include active multiplex system encompassing leased line, optical fiber cable, private microwave radio system, and telephone-company-provided derived local channel system; digital alarm communicator system; digital alarm radio system; McCulloh system; two-way RF multiplex system; one-way private radio alarm system; or a directly connected noncoded system.

When they receive fire alarm, supervisory, or trouble signals, operators at the central station take actions prescribed by NFPA 72 In addition to notifying the public fire service communication center in the case of a fire alarm signal, the central station operators must notify designated persons either at the protected premises or at their homes. The operators must also dispatch a runner or service technician when equipment at the protected premises must be reset by someone from the central station. This runner must reach the premises within 1 hr.

Similarly, supervisory signals and trouble signals also require a runner response. Unresolved guard tour signals require a runner response within $1/_2$ hr of the expiration of a 15-min grace period. Other unresolved supervisory signals require a runner to reach the protected premises within 1 hr. In the case of a trouble signal, the service technician must arrive at the protected premises within 4 hr and begin making repairs to the fire alarm system.

PROPRIETARY SYSTEM

The proprietary fire alarm system is widely used in large commercial or industrial occupancies. (See Figure 5-1D.)

Signals transmitted over a proprietary system are received and automatically and permanently recorded at a constantly attended proprietary supervising station located either at the protected premises or at another location of the property owner. In very simplistic terms, a proprietary system is similar to a central station system, but with the supervising station owned and operated by the property owner and located at the protected premises or another location of the property owner.

Many existing proprietary systems have separate initiating device circuits for each building zone or subsection, similar to local, auxiliary, and remote station systems. However, due to the increasing use of electronics, newer proprietary systems for larger buildings often have signal multiplexing and built-in minicomputer systems. These systems receive all signals from the protected building over one or more communication paths, and determine the exact location of the fire by use of digitally coded information. Figure 5-1E shows the complexity and interrelationships among the many facets of a modern computer-controlled proprietary fire alarm system.

Proprietary control units are required to transmit alarm, supervisory, and trouble signals to the supervising station. Signals may be transmitted from a protected premises using a variety of transmission technologies stated in NFPA 72, Chapter 4. Such technologies include active multiplex system encompassing leased line, optical fiber cable, private microwave radio system, and telephone-company-provided derived local channel system; digital alarm communicator system; digital alarm radio system; McCulloh system; two-way RF multiplex system; one-way private radio alarm system; or a directly connected noncoded system.

Upon receipt of fire alarm signals, supervisory signals, or trouble signals, a runner must be dispatched to investigate. Fire alarm signals must also be retransmitted to the public fire service communication center.

REMOTE STATION SYSTEM

A remote station fire alarm system connects the outputs from a building fire alarm signal and transmits them to a remote location. (See Figure 5-1F.) NFPA 72 specifies that fire alarm signals must be received at the public fire service communication center, at a fire station, or at the location of the public agency that has the responsibility to receive fire alarms from the public.

If the public agency is unwilling to receive the remote station fire alarm signals, or if that agency is willing to allow another organization to receive those signals, then the signals may be received at a location acceptable to the Authority Having Jurisdiction that is attended by trained personnel 24 hr a day. Upon receipt of fire alarm signals, the personnel retransmit the signals to the public fire service communication center.

In some cases, the public agency receiving fire alarm signals may also be willing to receive supervisory signals and trouble signals. In most cases, supervisory signals and trouble signals are transmitted to a constantly attended location acceptable to the Authority Having Jurisdiction.

Signals may be transmitted from a protected premises using a variety of transmission technologies stated in NFPA 72, Chapter 4. Such technologies include active multiplex system encompassing leased line, optical fiber cable, private microwave radio system, and telephone-company-provided derived local channel system; digital alarm communicator system; digital alarm radio system; McCulloh system; two-way RF multiplex system; one-way private radio alarm system; or a directly connected noncoded system.

AUXILIARY SYSTEM

An auxiliary fire alarm system provides the interface between a protected premises fire alarm system and a public fire reporting system. (See Figure 5-1G.) NFPA 72 states requirements for public fire reporting systems in Chapter 4, Section 4-6. NFPA 72 also states requirements for auxiliary fire alarm systems in Chapter 4, Section 4-7. If a community does not have a public fire reporting system, then a protected premises within that community cannot have an auxiliary fire alarm system.

The transmission technology employed between the protected premises and the public fire service communication center is determined by the type of public fire reporting system, i.e., coded wired, coded radio, series telephone, or parallel telephone. The auxiliary fire alarm system interface may be a shunt-type or local energy-type connection to a coded wired, coded radio, or series telephone master fire alarm box. (See Figure 5-1H.) Or the auxiliary fire alarm system may be directly connected to a parallel telephone system circuit.

NFPA 72 states that only fire alarm signals may be transmitted by an auxiliary fire alarm system. In cases where the protected premises fire alarm system must also cover supervisory signals, these are usually transmitted by a remote station system. Trouble signals are either received only at the protected premises, or they are transmitted by a remote station system.

BIBLIOGRAPHY

NFPA Codes, Standards, and Recommended Practices

Reference to the following NFPA codes, standards, and recommended practices will provide further information on fire alarm systems discussed in this chapter. (See the latest version of *The NFPA Catalog* for availability of current editions of the following documents.)

NFPA 13, *Standard for the Installation of Sprinkler Systems*
NFPA 20, *Standard for the Installation of Centrifugal Fire Pumps*
NFPA 70, *National Electrical Code®*
NFPA 72, *National Fire Alarm Code*
NFPA 101®, *Life Safety Code®*

Additional Reading

Barnett, P., "Voice Alarm Systems: Loudspeakers—Fire, Flame or Fool Proof?," *Fire Surveyor*, Vol. 21, No. 6, Dec. 1992, pp. 7–9.
Barnett, P., "Merging of Expertise in Fire Alarm and Evacuation Systems Welcomed," *Fire*, Vol. 85, No. 1050, Dec. 1992, p. 18.
Bond, A., "Sound Guidance (Auditoria and Alarm Systems)," *Building Services*, Vol. 13, No. 2, Mar. 1992, pp. 28–30.
Bowden, G., "Integrity of Radio Activated Fire Alarm Systems: Just How Far Has the Technology Progressed?" *Fire Surveyor*, Vol. 22, No. 6, Dec. 1993, pp. 13–17.
Bowden, G., "Radio Fire Alarm Technology Moves into the Big League," *Fire Prevention*, No. 281, July/Aug. 1995, pp. 24–25.

Bryant, P., "British Standard Code of Practice For the Design, Installation and Servicing of Fire Detection and Alarm Systems in Buildings, BS 5839 PART 1:1988," Third International Symposium on Fire Protection of Buildings, May 10–12, 1990, Eger, Hungary, 1990, pp. 129–134.

Bukowski, R. W., and O'Laughlin, R. J., Fire Alarm Signaling Systems, 2nd ed., National Fire Protection Association, Quincy, MA, 1994.

Cholin, J. M., P.E., and Moore, W. D., P.E., "How Reliable Is Your Fire Alarm System?" NFPA Journal, Jan./Feb. 1995, p. 49.

Choppen, B., "Voice of Alarm in Public Areas," Fire Prevention, No. 266, Jan./Feb. 1994, pp. 28, 30.

Cordasco, G. M., "Modified Redundant Alarm Systems Increase Safety, Not Cost," NFPA Journal, Vol. 87, No. 6, Nov./Dec. 1993, pp. 68–69, 71–72.

"Digital Alarm Communicator System Units," Second Edition, Standard for Safety, Underwriters, Inc., Northbrook, IL, UL 1635, Feb. 21, 1991.

Donohue, P. A., "False Alarms from Automatic Fire Detection Systems," Building Research Establishment, Garston, UK, BRE IP 13/92, June 1992.

Faulkner, G., "Fire Alarm Systems Advancement—Meeting the Challenges . . . and More," Fire Engineers Journal, Vol. 51, No. 160, March 1991, pp. 9–10, 18.

Faup, J. J., "Alarm Systems Ensure Safe Elevator Recall," NFPA Journal, Vol. 87, No. 3, May/June 1993, pp. 28, 104.

Fowler, J., "Reducing False Alarms," Fire, Vol. 88, No. 1090, April 1996, pp. 27–28.

Fox, P., "Product Profile: Developments with the VESDA Aspirating Smoke Detectors," Fire Surveyor, Vol. 21, No. 3, June 1992, pp. 16–18.

Goedeke, A. D., et al., "Machine Vision Fire Detector System (MVFDS)," Final Report, May 1990–Nov. 1990, Donmar Limited, Newport Beach, CA, Air Force Engineering and Services Center, Tyndall AFB, FL, ESL-TR-91-02, Apr. 1991.

Hall, J. R., Jr., "U.S. Experience with Smoke Detectors and Other Fire Detectors: Who Has Them? How Well Do They Work? When Don't They Work?" National Fire Protection Assoc., Quincy, MA, June 1995.

Hicks, H. D., Jr., "Another Look at Design Guidelines for Voice Fire Alarm Systems," NFPA Journal, Jan./Feb. 1994, p. 50.

Hodge, S. G., "Fire Alarm Safety Improvement," Westinghouse Hanford Co., Richland, WA, WHC-SD-631-ATP-006, Oct. 13, 1994.

Ingrassia, E., "Fire-Alarm Systems: Seven Factors Influencing Selection," Consulting-Specifying Engineer, Vol. 12, No. 1, 60–62, July 1992, pp. 60–62.

Kirby, R., "From Manufacturer to Owner, Careful Software Management Is Crucial," NFPA Journal, Jan./Feb. 1995, p. 54.

Knox, R., "Redesigning the Alarm System at York Minster," Fire Prevention, No. 228, Apr. 1990, pp. 29–31.

Koffel, W. E., "Objectives Key to Alarm System Design," NFPA Journal, July/Aug. 1992, p. 12.

Korslund, S. M., "Technical Evaluation of Equipment Maintenance on Fire Alarm Detection, Suppression, and Signaling Systems on the Hanford Site," Westinghouse Hanford Co., Richland, WA, WHC-SC-GN-TEEM-10001, REV. 0, 1994.

Kozeki, D., et al., "Study of Fire Detection System Introducing AI Technology. Part 2. Role of Fuzzy Expert System to Reduce False Alarms," Matsushita Electric Works, Ltd., Osaka, Japan, Report of Research Institute of Japan, No. 71, Mar. 1991, pp. 21–31.

Letts, J., "Fire Detection and Alarm Systems—the Future," Fire Surveyor, Vol. 20, No. 6, Dec. 1991, pp. 48–50.

Meacham, B. J., "The Use of Artificial Intelligence Techniques for Signal Discrimination in Fire Detection Systems," Journal of Fire Protection Engineering, Vol. 6, No. 3, 1994, pp. 125–136.

Moore, W. D., and Wilson, D. K., The Moore-Wilson Signaling Report, published by Focus Publishing Enterprises, Bloomfield, CT.

Moore, W. D., ed., National Fire Alarm Code Handbook, National Fire Protection Association, Quincy, MA, 1993.

Ono, T., Ishii, H., and Muroi, N., "Early-Stage Fire Judgment Using Multi-Dimensional Measurement: A Fire Alarm System Using a Combination of Fire Judgment With Composite Fire Information and Fire Judgment With Time-Series Fire Information," Bulletin of Japanese Association of Fire Science and Engineering, Vol. 40, No. 1, 1990, pp. 7–17.

Papier, I. I., "NFPA 72, UL Standards Work Together," NFPA Journal, Sept./Oct. 1993, p. 25.

Pigott, B., "Computer Controlled Fire Alarms," Building Services, Vol. 13, No. 4, Apr. 1991, pp. 55–56.

Rosato, C., "Radio Revolution in Central Alarm Stations," Fire Prevention, No. 245, Dec. 1991, pp. 30–31.

Russo, W. P., "NFPA 72 Consolidates Signaling Standards," NFPA Journal, Sept./Oct. 1993, p. 71.

Salamone, R., "Retrofitting High-Rise Dorms with Alarm and Detection Systems," Fire Journal, Vol. 84, No. 1, Jan./Feb. 1990, pp. 37–39.

Sanderford, H. B., Jr., "Spread Spectrum Radio Alarms Give Hardwired Performance," Fire Journal, Vol. 84, No. 1, Jan./Feb. 1990, pp. 57–59.

Schifiliti, R. P., Meacham, B. J., and Custer, R. L. P., "Design of Detection Systems," SFPE Handbook of Fire Protection Engineering, 2nd ed., National Fire Protection Association, Quincy, MA, 1995.

Smith, R. L., "Performance Parameters of Fire Detection Systems," Fire Technology, Vol. 30, No. 3, 1994, pp. 326–337.

So, A. T. P., and Chan, W. L., "A Computer-Vision-Based and Fuzzy-Logic-Aided Security and Fire-Detection System," Fire Technology, Vol. 30, No. 3, 1994, pp. 341–356.

Stone, A., "Developments in Fire Alarm Control Equipment," Fire Safety Engineering, Vol. 1, No. 5, Oct. 1994, pp. 14–16.

Sultan, M. A. and Halliwell, R. E., "Optimum Location for Fire Alarms in Apartment Buildings," Fire Technology, Vol. 26, No. 4, Nov. 1990, pp. 342–356.

Terrett, D., "Voice Alarm Loudspeakers," Fire Surveyor, Vol. 22, No. 5, Oct. 1993, pp. 12–13.

Thompson, B., "Voice Alarm Systems: The Benefits of Integrated Voice Alarm Systems," Fire Surveyor, Vol. 21, No. 6, Dec. 1992, pp. 4–6.

Thuillard, M., "New Methods for Reducing the Number of False Alarms in Fire Detection Systems," Fire Technology, Vol. 30, No. 2, Second Quarter, 1994, pp. 250–268.

Todd, C., "Voice Alarms and Evacuation," Fire Prevention, No. 274, Nov. 1994, pp. 21–25.

Transue, R. E., "Integrating Fire Alarms into Building Design," Consulting-Specifying Engineer, Vol. 14, No. 7, Dec. 1993, pp. 36–38, 40.

Tucker, R. W., "Early Detection in Jails—While Minimizing Unwanted Alarms," SFPE Bulletin, Sep./Oct. 1990, pp. 6–8.

Training Manual on Fire Alarm Systems, National Electrical Manufacturers Association, Washington, DC, 1994.

AUTOMATIC FIRE DETECTORS

Revised by Wayne D. Moore

As soon as a fire starts, it produces a variety of environmental changes that help make its presence known. Human beings are excellent fire detectors—they possess the senses of smell, sight, hearing, taste, and touch. However, human senses can also be unreliable. Beginning in the mid-19th century, a number of mechanical, electrical, and electronic devices have been developed to mimic human senses in detecting the environmental changes created by fire.

The elements of a fire that are most often used as the basis for detection are heat, smoke (aerosol particulate), and light radiation. Complicating the detector's ability to detect a fire are two facts: (1) not all fires produce all of the elements, and (2) nonfire conditions can produce similar ambient conditions. The fire protection engineer must decide which of the elements produced by a fire might be expected from hostile fires and which similar ambient conditions might result from nonfire situations.

Most building codes require what amounts to partial coverage by detection systems, that is, some rooms or access are not covered by their own detectors. This becomes another limiting factor in the detector's ability to actually detect the fire. It must be understood that, if a fire is remote from a detector location, detection of that fire will be delayed.

In addition, placing a spot-type detector in a high-ceiling application will delay detection of all but the most rigorously burning fire.

Even if all of these elements—heat, smoke, and light—are present in a given fire, the magnitude of the different elements must exceed some theoretical basic level during fire development. It is also helpful to determine which element will appear first. This is especially true if life safety is involved. This chapter will discuss various automatic fire detectors and their principles of operation.

HEAT DETECTORS

Heat detectors are the oldest type of automatic fire detection device. They began with the development of automatic sprinklers in the 1860s and have continued to the present with a proliferation of var-

Wayne D. Moore, P.E., is vice president of engineering and marketing for MBS Fire Technology, Inc., and is the manager of the Atlanta, GA, office. He is currently chairman of the NFPA 72 Technical Committee for Chapter 3, "Protected Premises Fire Alarm Systems," of the *National Fire Alarm Code*. He is also editor of the first edition of the *National Fire Alarm Code Handbook* and co-editor of the national fire alarm systems newsletter, *The Moore-Wilson Signaling Report*.

ious types of devices. A sprinkler can be considered a combined heat-activated fire detector and extinguishing device when the sprinkler system is provided with waterflow indicators connected to the fire alarm control system. Waterflow indicators detect either the flow of water through the pipes or the subsequent pressure change upon actuation of the system.

Heat detectors that only initiate an alarm and have no extinguishing function are still in use. Although they have the lowest false alarm rate of all automatic fire detector devices, they also are the slowest in detecting fires. A heat detector is best suited for fire detection in a small confined space where rapidly building high-heat-output fires are expected, in areas where ambient conditions would not allow the use of other fire detection devices, or where speed of detection is not a prime consideration.

Heat detectors are generally located on or near the ceiling and respond to the convected thermal energy of a fire. They respond either when the detecting element reaches a predetermined fixed temperature or to a specified rate of temperature change. In general, heat detectors are designed to operate when heat causes a prescribed change in a physical or electrical property of a material or gas.

Operating Principles of Fixed-Temperature Heat Detectors

Fixed-temperature heat detectors are designed to alarm when the temperature of the operating element reaches a specified point. The air temperature at the time of alarm is usually considerably higher than the rated temperature because it takes time for the air to raise the temperature of the operating element to its set point. This condition is called thermal lag. Fixed-temperature heat detectors are available to cover a wide range of operating temperatures—from about 135°F (57°C) and higher. Higher temperature detectors are also necessary so that detection can be provided in areas normally subjected to high ambient (nonfire) temperatures, or in areas zoned so that only detectors in the immediate fire area operate.

Fusible-element-type: Eutectic metals—alloys of bismuth, lead, tin, and cadmium that melt rapidly at a predetermined temperature—can be used as operating elements for heat detection. The most common use is the fusible element in an automatic sprinkler. Fusing of the element allows the cover on the orifice to fall away, water to flow in the system, and the alarm to be initiated.

An eutectic metal may also be used to actuate an electrical heat detector. The eutectic metal is often used as a solder to secure a spring under tension. When the element fuses, the spring action closes contacts and initiates an alarm. (See Figure 5-2A.) Devices using eutectic metals cannot be restored; either the device or its operating element must be replaced following operation.

Continuous line-type: As alternatives to spot-type fixed-temperature detection, various methods of line-type detection have been developed. The detector shown in Figure 5-2B uses a pair of steel wires in a normally open circuit. The conductors are held apart by a heat-sensitive insulation. The wires, under tension, are enclosed in a braided sheath to form a single cable assembly. When the design temperature is reached, the insulation melts, the two wires contact, and an alarm is initiated. Following an alarm, the fused section of the cable must be replaced to restore the system.

A similar alarm device utilizing a semiconductor material and a stainless steel capillary tube has been used where mechanical stability is also a factor. (See Figure 5-2C.) The capillary tube contains a coaxial center conductor separated from the tube wall by a temperature-sensitive glass semiconductor material. Under normal conditions, a small current (i.e., below alarm threshold) flows in the circuit. As the temperature rises, the resistance of the semiconductor decreases, allows more current flow, and initiates the alarm.

Bimetallic-type: When two metals with different coefficients of thermal expansion are bonded together and then heated, differential expansion causes bending or flexing toward the metal having the lower expansion rate. This action closes a normally open circuit. The low-expansion metal commonly used is Invar™, an alloy of 36 percent nickel and 64 percent iron. Several alloys of manganese-copper-nickel, nickel-chromium-iron, or stainless steel may also be used for the high-expansion component of a bimetal assembly. Bimetals are used for the operating elements of a variety of fixed-temperature detectors. These detectors are generally of two types: (1) the bimetal strip and (2) the bimetal snap disc.

As it is heated, a bimetal strip deforms in the direction of the contact point. With a given bimetal, the width of the gap between the contacts determines the operating temperature: the wider the gap, the higher the operating point. The operating element of a snap disc device is a bimetal disc formed into a concave shape in its unstressed condition. (See Figure 5-2D.) Generally, a heat collector is attached to the detector frame to speed the transfer of heat from the room air to the bimetal. As the disc is heated, the stresses developed

cause it to suddenly reverse curvature and become convex. This provides a rapid positive action that closes the alarm contacts. The disc itself usually is not part of the electrical circuit.

All heat detectors using bimetal elements are automatically self-restoring after operation, when the ambient temperature drops sufficiently below the operating point.

Rate Compensation Detectors

A rate compensation detector is a device that responds when the temperature of the surrounding air reaches a predetermined level, regardless of the rate of temperature rise. (See Figure 5-2E.)

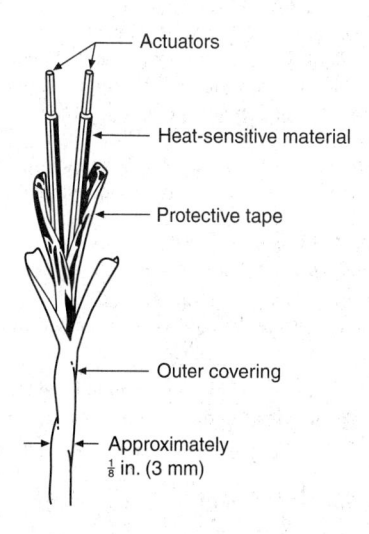

FIG. 5-2B.   Line-type heat detector. (Source: The Protectowire Company)

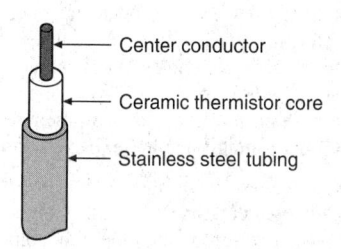

FIG. 5-2C.   View of the construction of the continuous thermal sensor showing outer tubing, ceramic thermistor core, and center wire. (Source: Alison Control, Inc.)

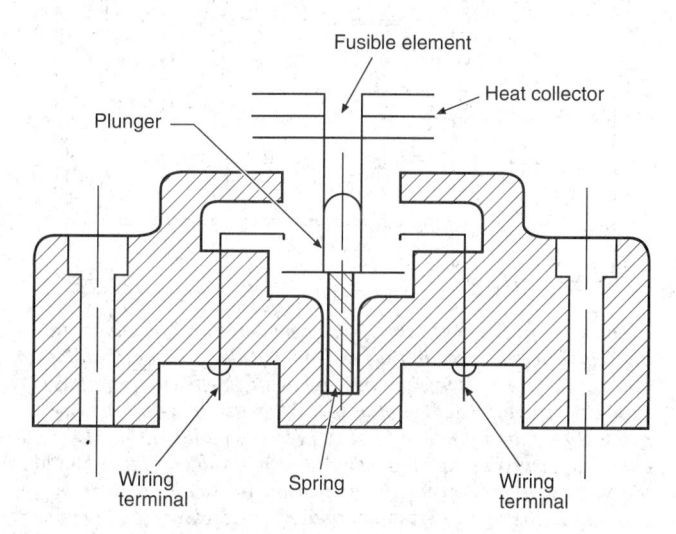

FIG. 5-2A.   Fixed-temperature heat detector, spot-type, fusible element.

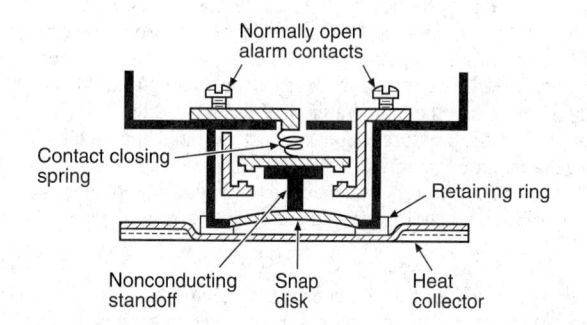

FIG. 5-2D.   Spot-type fixed-temperature snap disc detector. (Source: Edwards Systems Technology)

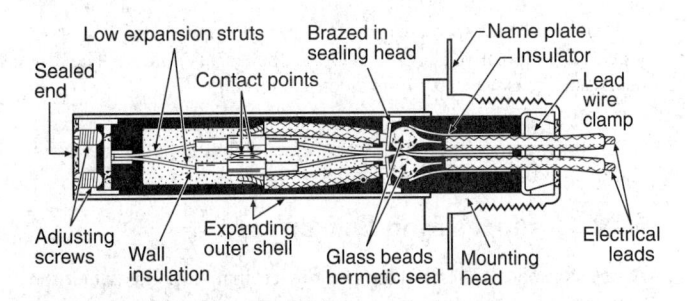

FIG. 5-2E. *Section of spot-type rate compensation detector. (Source: Fenwal, Inc.)*

A typical example is a spot-type detector with a tubular casing of metal that tends to expand lengthwise as it is heated, and an associated contact mechanism that will close at a certain point in the elongation. A second metallic element inside the tube exerts an opposing force on the contacts, tending to hold them open. The forces are balanced so that, with a slow rate of temperature rise, there is more time for heat to penetrate to the inner element. This inhibits contact closure until the total device has been heated to its rated temperature level. However, with a fast rate of temperature rise, there is less time for heat to penetrate to the inner element. The element therefore exerts less of an inhibiting effect, so contact closure is obtained when the total device has been heated to a lower level. This, in effect, compensates for thermal lag.

Thermal detectors using expanding metal elements are also automatically self-restoring after operation, when the ambient temperature drops to some point below the operating point.

Rate-of-Rise Detectors

One effect that a flaming fire has on the surrounding area is to rapidly increase air temperature in the space above the fire. Fixed-temperature heat detectors will not initiate an alarm until the air temperature near the ceiling exceeds the design operating point. The rate-of-rise detector, however, will function when the rate of temperature increase exceeds a predetermined value, typically around 12 to 15°F (7 to 8°C) per minute. Rate-of-rise detectors are designed to compensate for the normal changes in ambient temperature [less than 12°F (6.7°C) per minute] that are expected under nonfire conditions.

In a pneumatic fire detector, air heated in a tube or chamber expands, increasing the pressure in the tube or chamber. This exerts a mechanical force on a diaphragm that closes the alarm contacts. If the tube or chamber were hermetically sealed, slow increases in ambient temperature, a drop in the barometric pressure, or both, would cause the detector to initiate an alarm regardless of the rate of temperature change. To overcome this, pneumatic detectors have a small orifice to vent the higher pressure that builds up during slow increases in temperature or during a drop in barometric pressure. The vents are sized so that when the temperature changes rapidly, as in a fire, the rate of expansion exceeds the venting rate and the pressure rises. When the temperature rise exceeds 12 to 15°F (7 to 8°C) per minute, the pressure is converted to mechanical action by a flexible diaphragm. Pneumatic heat detectors are available for both line- and spot-type detectors. A schematic of a spot-type pneumatic heat detector is shown in Figure 5-2F, and a line-type is shown in Figure 5-2G.

Line-type: The line-type consists of metal tubing, in a loop configuration, attached to the ceiling or side wall near the ceiling of the area to be protected. (See Figure 5-2G.) Lines of tubing are normally spaced not more than 30 ft (9.1 m) apart, not more than 15 ft (4.5 m)

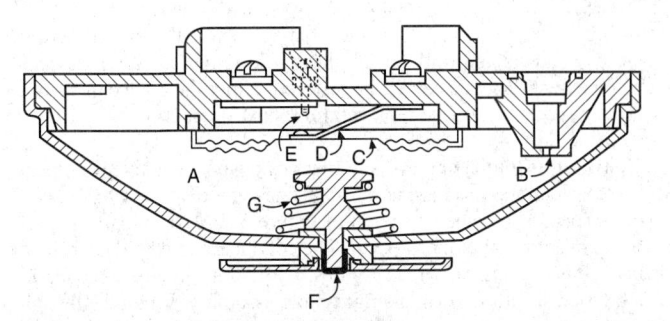

FIG. 5-2F. *A spot-type combination rate-of-rise, fixed-temperature device. The air in chamber A expands more rapidly than it can escape from vent B. This causes pressure to close electrical contact D between diaphragm C and contact screw E. Fixed-temperature operation occurs when fusible alloy F melts, releasing spring G which depresses the diaphragm closing contact points. (Source: Edwards Systems Technology)*

from a wall, and with no more than 1,000 ft (305 m) of tubing on each circuit. Also, a minimum of at least 5 percent of each tube circuit or 25 ft (7.6 m) of tube, whichever is greater, must be in each protected area. Without this minimum amount of tubing exposed to a fire condition, insufficient pressure would build up to achieve proper response.

In small areas where the line-type tube detectors might have insufficient tubing exposed to generate sufficient pressures to close the alarm contacts, air chambers or rosettes of tubing are often used. These units act like a spot-type detector by providing the volume of air required to meet the 5 percent or 25-ft (7.6-m) requirement. Since a line-type rate-of-rise detector is an integrating detector, it

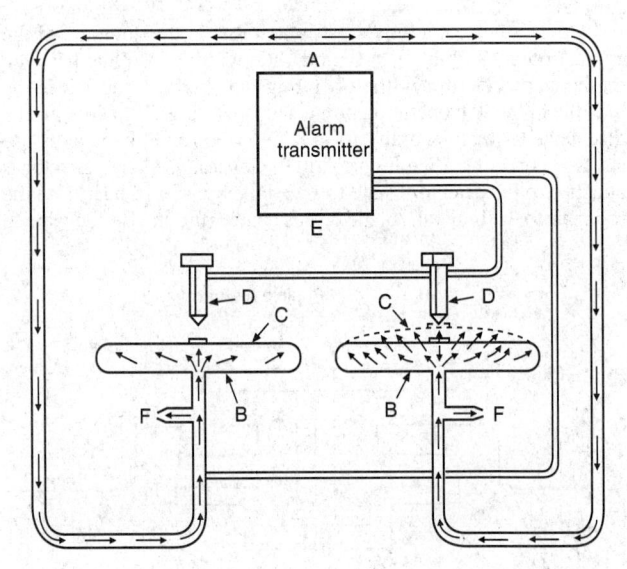

FIG. 5-2G. *Line-type rate-of-rise heat detector. The copper tubing A is fastened in a continuous loop to ceilings or walls and terminates at both ends in chambers B having flexible diaphragms C which control electrical contacts D. When air in the tubing expands under the influence of heat, pressure builds within the chambers, causing the diaphragms to move and close a circuit to alarm transmitter E. Vents F compensate for small pressure changes in the tubing brought about by small changes in temperature in the protected spaces. (Source: American District Telegraph Co. and its associated companies)*

will actuate either when a rapid heat rise occurs in one area of exposed tubing, or when a slightly less rapid heat rise takes place in several areas where tubing on the same loop is exposed.

Spot-type: The pneumatic principle is also used to close contacts within spot detectors. (See Figure 5-2F.) The difference between the line- and spot-type detectors is that the spot-type contains all of the air in a single container rather than in a tube that extends from the detector assembly to the protected area(s).

Combination Detectors

Combination detectors contain more than one element which responds to a fire. These detectors may be designed to respond from either element, or from the combined partial or complete response of both elements. An example of the former is a heat detector that operates on both the rate-of-rise and fixed-temperature principles. Its advantage is that the rate-of-rise element will respond quickly to a rapidly developing fire, while the fixed-temperature element will respond to a slowly developing fire when the detecting element reaches its set point temperature. The most common combination detector uses a vented air chamber and a flexible diaphragm for the rate-of-rise function, while the fixed-temperature element is usually leaf-spring restrained by an eutectic metal. (See Figure 5-2F.) When the fixed-temperature element reaches its design operating temperature, the eutectic metal fuses and releases the spring, which closes the contacts.

Electronic Spot-Type Thermal Detectors

A thermoelectric effect detector is a device that utilizes a sensing element consisting of one or more thermistors, which produce a change in electrical resistance in response to an increase in temperature. (See Figure 5-2H.) This resistance change is monitored by as-

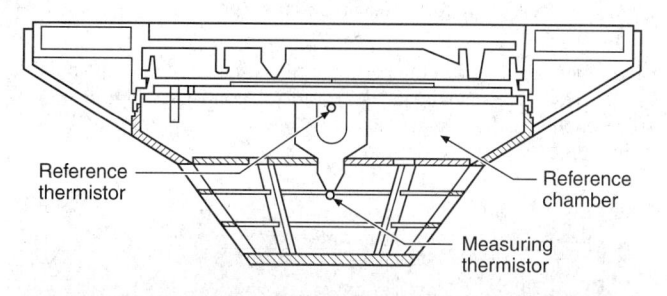

FIG. 5-2H.   Solid-state electronic thermal detector with dual thermistor sensing circuit. (Source: System Sensor/Division of Pittway)

sociated electronic circuitry, and the detector responds when the resistance changes at an abnormal rate (rate-of-rise type) or when the resistance reaches a specific value (fixed-temperature type).

Rate-of-rise detectors of this type use two thermistors. One is exposed to changes in atmospheric temperature. When the temperature rapidly changes as in a fire situation, the temperature of the exposed thermistor increases faster than the temperature of the unexposed reference thermistor, generating a net change in resistance causing the detector to go into alarm condition. Most rate-of-rise detectors are designed with a fixed-temperature backup feature so that, should the temperature rise be slower than 15°F (8°C) per min, the detector will operate when the exposed thermistor has reached a predetermined fixed temperature.

SMOKE DETECTORS

A smoke detector will detect most fires much more rapidly than a heat detector. This section will describe the various principles of smoke detector operation and their applications.

Smoke detectors are identified by their operating principle. Two of the operating principles are: (1) ionization and (2) photoelectric. As a class, smoke detectors using the ionization principle provide somewhat faster response to high-energy (open-flaming) fires, since these fires produce large numbers of the smaller smoke particles. As a class, smoke detectors operating on the photoelectric principle respond faster to the smoke generated by low-energy (smoldering) fires, as these fires generally produce more of the larger smoke particles. However, each type of smoke detector is subjected to, and must pass, the same test fires at testing laboratories in order to be listed.

Conventional smoke detectors provide a go, no-go form of detection. This means that other than alarm or no-alarm, no other information is transmitted to the fire alarm control unit. In order to provide a stable smoke detector, the system designer must ensure that the sensitivity level of the detector matches the worst environment in the facility to be protected. Some jurisdictions require that a minimum sensitivity level be used to help control the false or nuisance alarm issue.

Technological improvements in microprocessor use in fire alarm systems have led to the development of new smoke detector concepts. These new sensors use analog technology to measure the conditions in the area or space protected and transmit that information to the computer-based fire alarm control unit. This new sensor can report when it is too dirty to function properly or is getting too sensitive due to any number of conditions in the protected space.

Analog sensors provide an essentially false-alarm-free system from conditions normally found in a building. This sensor technology also allows the system designer to adjust the sensor's sensitivity to accommodate the ambient environment or use an extra-sensitive setting to protect a high-value or mission-sensitive area.

These sensors are available as photoelectric, ionization, or combination thermal, photoelectric, and ionization units. As fire alarm systems technology advances, analog sensors will be the sensor of choice for any system application, regardless of system size.

Ionization Smoke Detectors

Smoke detectors utilizing the ionization principle are usually of the spot type. An ionization smoke detector has a small amount of radioactive material that ionizes the air in the sensing chamber, rendering the air conductive and permitting a current flow through the air between two charged electrodes. This gives the sensing chamber an effective electrical conductance. When smoke particles enter the ionization area, they decrease the conductance of the air by attach-

AUTOMATIC FIRE DETECTORS

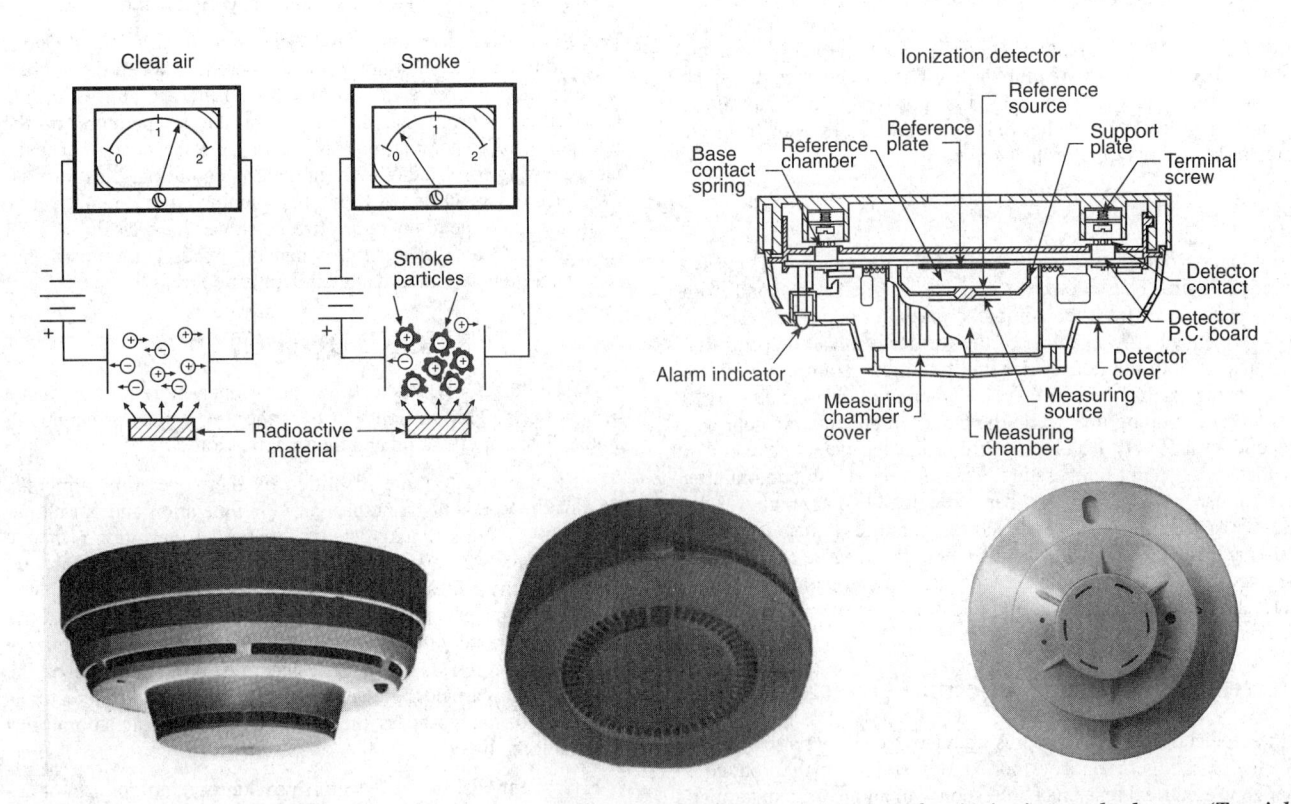

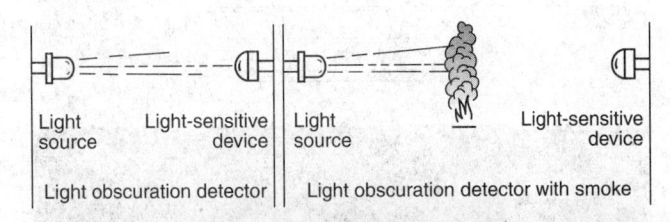

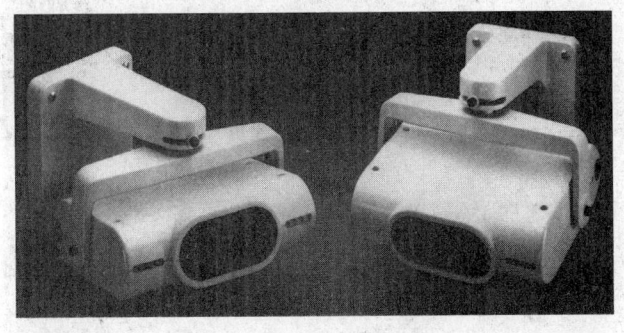

FIG. 5-2I. Principle of operation for ionization smoke detector (Top left); cross-section view of an ionization smoke detector (Top right) (Source: Pyrotronics); ionization smoke detector (Lower left) (Source: Pyrotronics); ionization smoke detector (Lower center) (Source: System Sensor/Division of Pittway); ionization smoke detector (Lower right) (Source: Fire Control Instruments, Inc.).

ing themselves to the ions, causing a reduction in ion mobility. When the conductance is below a predetermined level, the detector responds. (See Figure 5-2I.)

Photoelectric Smoke Detectors

The presence of suspended smoke particles generated during the combustion process affects the propagation of a light beam passing through the air. The effect can be utilized to detect the presence of a fire in two ways: (1) obscuration of light intensity over the beam path or (2) scattering of the light beam.

Light obscuration principle: Smoke detectors that operate on the principle of light obscuration consist of a light source, a light beam collimating system, and a photosensitive device. When dense smoke obscures part of the light beam, or less dense smoke obscures more of the beam, the light reaching the photosensitive device is reduced, and this initiates the alarm. (See Figure 5-2J.)

Most light obscuration smoke detectors are the beam type and are used to protect large open areas. They are installed with the light source at one end of the area to be protected and the photosensitive device at the other. Projected beam detectors are generally installed in accordance with the manufacturer's instructions.

Light-scattering principle: When smoke particles enter a light path, scattering results. Smoke detectors utilizing the photoelectric light-scattering principle are usually of the spot type. They contain a light source and a photosensitive device arranged so the light rays normally do not fall onto the device. When smoke particles enter the

light path, light strikes the particles and is scattered onto the photosensitive device, causing the detector to respond. (See Figure 5-2K.) The photosensitive device used in scattering detectors usually is a photodiode or a phototransistor.

FIG. 5-2J. Principle of operation for photoelectric obscuration smoke detector (Top). Typical photoelectric obscuration-type smoke detector (Bottom). (Source: System Sensor/Division of Pittway)

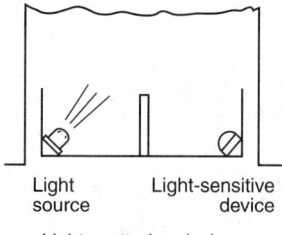

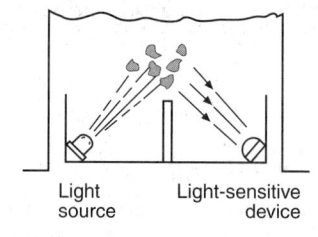

FIG. 5-2K.   Cross-section view of a photoelectric scattering smoke detector (Top) (Source: System Sensor/Division of Pittway); Photoelectric smoke detector (Center) (Source: System Sensor/Division of Pittway); Photoelectronic system smoke detector (Bottom) (Source: ESL, Sentrol, Inc.).

Air-Sampling-Type Smoke Detectors

Cloud chamber smoke detection principle: A smoke detector utilizing the cloud chamber principle is usually of the sampling type. An air pump draws a sample of air from the protected area(s) into a high-humidity chamber within the detector. After the air sample has been raised to a high humidity, the pressure is lowered slightly. If smoke particles are present, the moisture in the air condenses on them, forming a cloud in the chamber. The density of this cloud is then measured by a photoelectric principle. The detector responds when the density is greater than a predetermined level.

Continuous air-sampling smoke detection: In addition to the cloud chamber smoke detection devices, there are other smoke detection devices that actively and continuously sample the air from a protected space. The air-sampling system consists of sampling pipes spaced uniformly over the ceiling, together with two supplemental pipes arranged to sample the return air exiting from the monitored space. Each one of the ceiling pipe drops is capped and has a small sampling hole drilled in the cap to draw in a sample of air from that location. There are also sampling holes drilled in the sec-

tions of the supplemental pipes that extend across the return-air grilles.

This network of piping is connected to the detector/control unit where there is a fan, or aspirator, that creates a flow of air in the piping network, and this flow causes the pressure inside the pipe to be less than the local atmospheric pressure. The flow creates a slight vacuum; therefore, the piping network continually draws in air.

The sampled air is drawn through a filter to the detector assembly. Inside the detector is a very intense light source that irradiates the sampled air. If there are smoke particles in the sampled air, the device, which can sense smoke particles in extremely low concentrations, will activate the first of three levels of alarm conditions.

These systems are typically used in applications where dollar densities are very high, such as in electronic data-processing areas and museums, or where equipment survival is paramount to continuity of operations, such as in the communications industry.

GAS-SENSING FIRE DETECTORS

Many changes occur in the gas content of the environment during a fire. In large-scale fire tests, it has been observed that detectable levels of gases are reached after detectable smoke levels and before detectable heat levels. One of two operating principles, i.e., semiconductor and catalytic element, may be used in a gas-sensing fire detector.

Semiconductor Principle

Fire-gas detectors of the semiconductor type respond to either oxidizing or reducing gases by creating electrical changes in the semiconductor. The subsequent conductivity change of the semiconductor causes actuation of the detector.

Catalytic Element Principle

Fire-gas detectors of the catalytic element type contain a material which, in itself, remains unchanged, but which accelerates the oxidation of combustible gases. The resulting temperature rise of the element causes detector actuation.

FLAME DETECTORS

A flame detector responds either to radiant energy visible to the human eye (approximately 4000 to 7700 angstroms) or outside the range of human vision. Similar to the human eye, flame detectors have a "cone of vision," or viewing angle, that defines the effective detection capability of the detector. (See Figure 5-2L.) With this constraint, the sensitivity increases as the angle of incidence decreases. Such a detector is sensitive to glowing embers, coals, or flames which radiate energy of sufficient intensity and spectral quality to actuate the alarm. Each type of fuel, when burning, produces a flame with specific radiation characteristics. A flame detection system must be chosen for the type of fire that is probable. For example, an ultraviolet (UV) detector will respond to a hydrogen fire, but an infrared (IR) detector operating in the 4.4 micron sensitivity range will not. (See Figure 5-2M.) It is imperative, therefore, that a qualified fire protection engineer be involved in the design of these systems, along with assistance from the manufacturer's design staff.

Due to their fast detection capabilities, flame detectors are generally used only in high-hazard areas, such as fuel-loading platforms, industrial process areas, hyperbaric chambers, high-ceiling areas, and atmospheres in which explosions or very rapid fires may occur. Because flame detectors must be able to "see" the fire, they must not be blocked by objects placed in front of them. The infrared-type flame detector, however, has some capability for detecting radiation reflected from walls.

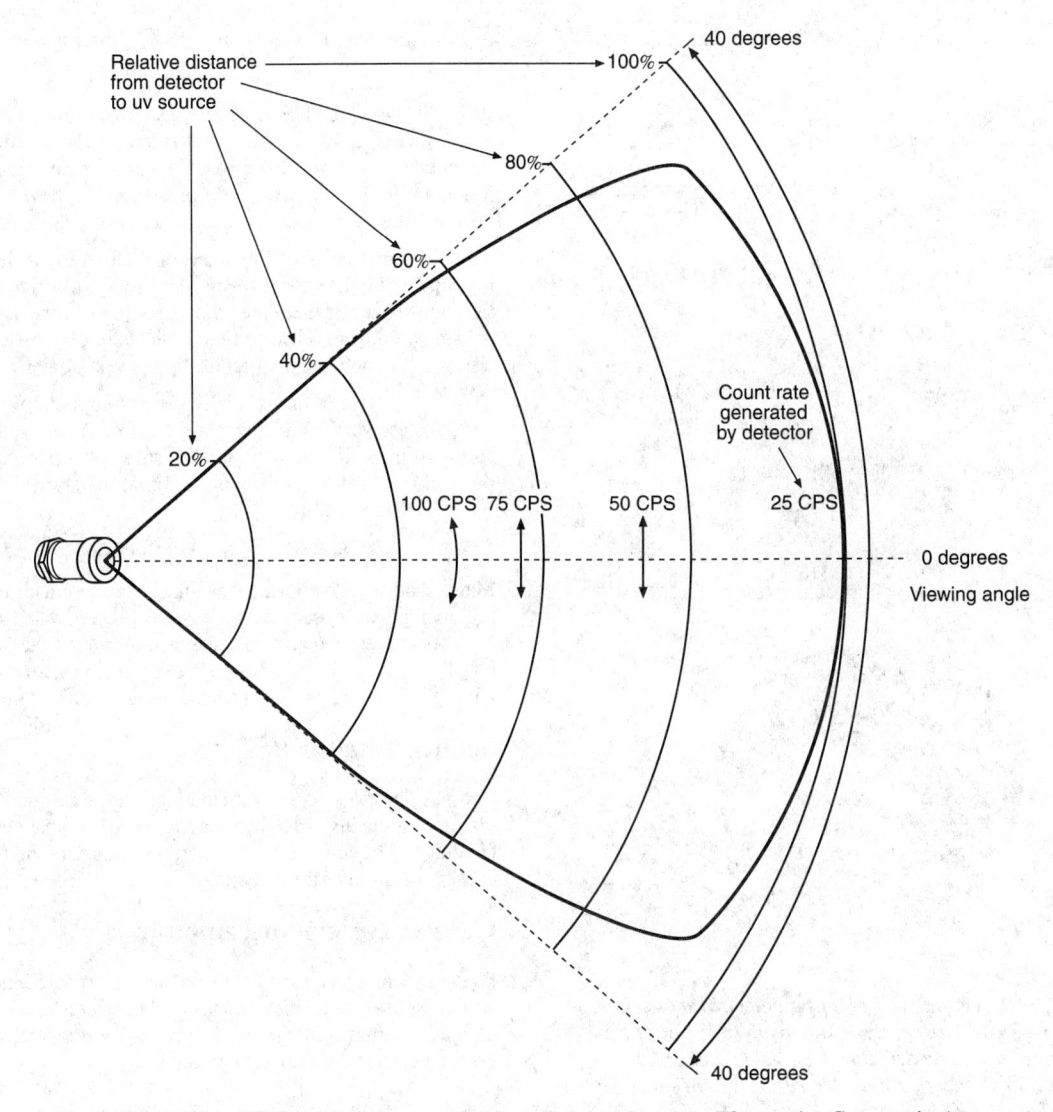

FIG. 5-2L. *UV fire detector cone of vision. (Source: Detector Electronics Corporation)*

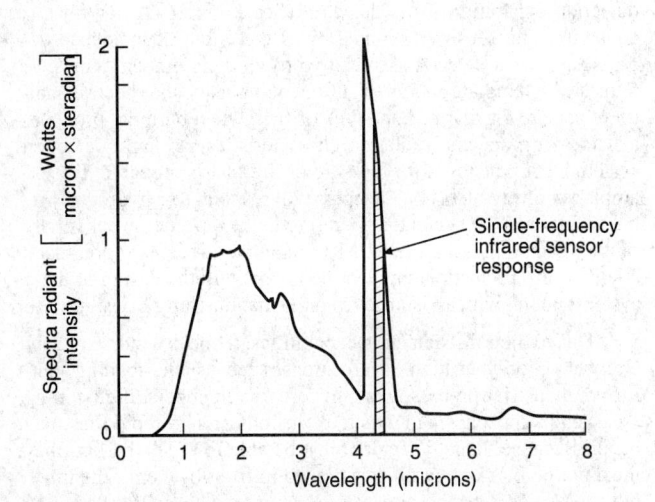

FIG. 5-2M. *Radiation spectra from a hydrocarbon fuel fire.*

Infrared Flame Detectors

An infrared (IR) detector basically is composed of a filter and lens system used to screen out unwanted wavelengths and focus the incoming energy on a photovoltaic or photoresistive cell sensitive to infrared energy. (See Figure 5-2N.) IR flame detectors can respond to the total IR component of the flame alone or in combination with flame flicker in the frequency range of 5 to 30 Hz.

A major problem in the use of infrared detectors receiving total IR radiation is the possible interference of solar radiation in the infrared region. When detectors are located in places shielded from the sun, such as vaults, filtering or shielding the unit from the sun's rays is unnecessary.

IR detectors are sensitive to most hydrocarbon fires (liquids, gases, and solids). Fires, such as burning metals, ammonia, hydrogen, and sulfur, do not emit significant amounts of IR radiation in the 4.4 micron sensitivity range of the detector. Thorough testing must be carried out prior to applying IR detectors to a non-hydrocarbon-fueled fire application.

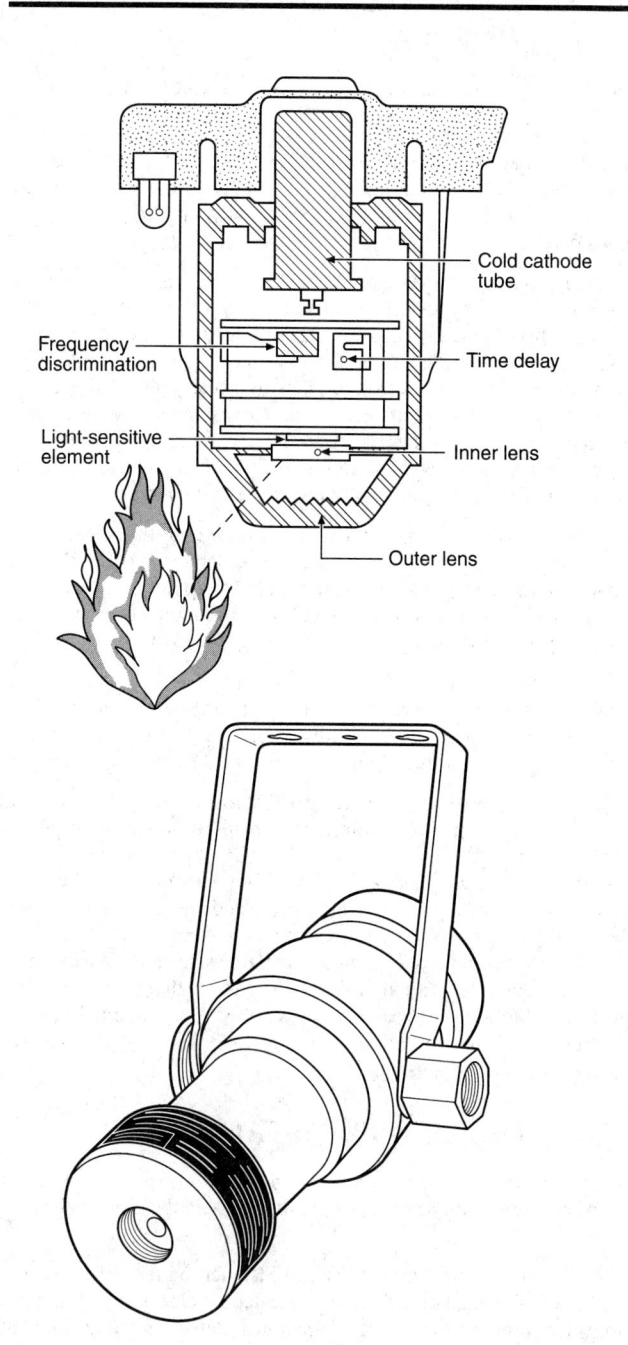

Cold cathode tube

Frequency discrimination

Time delay

Light-sensitive element

Inner lens

Outer lens

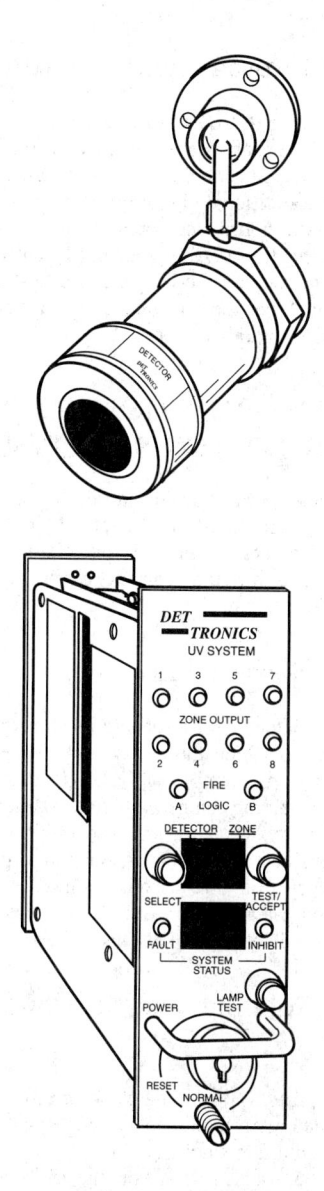

FIG. 5-2O. Typical UV detector assembly (Top). UV flame detector and controller (Bottom). (Source: Detector Electronics Corporation)

FIG. 5-2N. Cross-section of an infrared flame detector (Top) (Source: Pyrotronics); infrared flame detector (Bottom) (Source: Detector Electronics Corporation).

A buildup of ice or water film on the detector viewing window will greatly reduce its sensitivity. IR detectors are less affected by smoke, oil, and certain gases and vapors than UV detectors.

Ultraviolet Flame Detectors

Ultraviolet (UV) detectors generally use either a solid-state device, such as silicone carbide or aluminum nitride, or a gas-filled tube as the sensing element. (See Figure 5-2O.) UV detectors are essentially insensitive to both sunlight and artificial light.

A UV flame detector detects radiation emitted in the 1850 to 2450 angstrom range. Virtually all fires emit radiation in this band,

while the sun's radiation at this band is absorbed by the Earth's atmosphere. The result is that the UV flame detector is solar blind, meaning it will not cause an alarm in response to radiation from the sun. The implication of this feature is that it can easily be used both indoors and outdoors.

UV detectors are sensitive to most fires, including hydrocarbon (liquids, gases, and solids), metals (e.g., magnesium), sulfur, hydrogen, hydrazine, and ammonia.

Arc welding, electric arcs, lightning, X-rays used in nondestructive metal testing equipment, and radioactive materials can produce radiation levels that will activate a UV detection system.

The presence of UV-absorbing gases and vapors will attenuate the UV radiation from a fire, adversely affecting the ability of the detector to "see" a flame. Likewise, the presence of an oil mist in the air or an oil film on the detector window will have the same effect.

Ultraviolet/Infrared Flame Detectors

An ultraviolet/infrared (UV/IR) flame detector consists of a UV and a single-frequency sensor, paired together to form one unit. (See Figure 5-2P.) The two sensors individually operate the same as described above, but additional circuitry processes signals from both sensors. A fire alarm is produced only when both sensors detect a fire. The result is that the UV/IR system has better false alarm rejection capabilities than either the UV detection systems or IR detection systems alone. Since the UV/IR detector pairs two sensor types, it is subject to the limitations of both.

AMBIENT CONDITIONS AFFECTING DETECTOR RESPONSE

Ambient conditions need to be considered in the selection, placement, and response capability of detectors. The improper selection of a class of detector or the improper placement of a detector can create problems ranging from no alarm to excessive false alarms.

Ambient Background Level

When selecting a detector for a specific location, consider the ambient background to which the detector might be exposed under nonfire conditions. For example, an IR or UV detector used where there is gas or arc welding can generate false alarms due to the presence of radiant energy, which the unit was designed to detect. These responses are considered false alarms, since they do not result from a "hostile" fire. Detectors that respond to smoke particles are especially prone to false alarms from such sources as cooking fumes, cigarette smoke, and automobile exhaust fumes.

Heating, Ventilating, and Air Conditioning (HVAC)

In buildings or portions of buildings where forced ventilation is present, smoke detectors should not be located where air from supply diffusers could dilute smoke before it reaches the detector. Detectors should be located to favor the airflow toward return openings. This may require additional detectors, since placing detectors only near return air openings may leave the balance of the area with inadequate protection when the air-handling system is shut down.

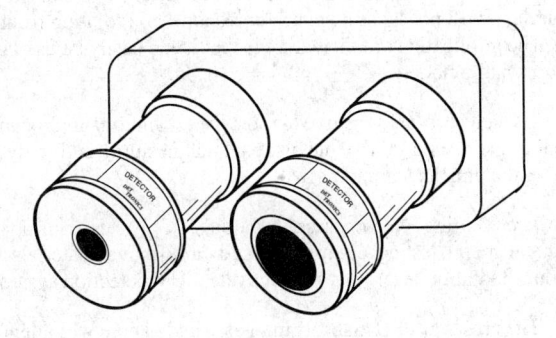

FIG. 5-2P. Ultraviolet/infrared flame detector. (Source: Detector Electronics Corporation)

SELECTION OF DETECTORS

When planning a fire detection system, the choice of detectors should be based on the kinds of potential fires expected. The type and quantity of fuel, possible ignition sources, ranges of ambient conditions, and the value of the protected property should all be considered.

In general, heat detectors have the lowest false alarm rate, but they also have the slowest response time. Since the heat generated by small fires tends to dissipate fairly rapidly, heat detectors are best used either to protect confined spaces or directly over hazards where flaming fires could be expected. They are usually installed in a grid pattern according to their recommended spacing schedule or with reduced spacing for faster response. The operating temperature of a heat detector should be a minimum of 25°F (14°C) above the maximum expected ambient temperature in the area protected.

Smoke detectors have a faster response time than heat detectors. They are better suited than heat detectors to protect large open spaces because smoke does not dissipate as rapidly as heat does in the same sized space. Smoke detectors are installed either according to prevailing air current conditions or in a grid layout.

Ionization smoke detectors are useful where flaming fires are a possibility. Photoelectric smoke detectors are best used in places with potential for smoldering fires or fires involving low-temperature pyrolysis of polyvinylchloride (PVC) wire insulation.

Flame detectors offer extremely fast response, but will warn of any source of radiation within their sensitivity range. False alarm rates can be high if this kind of detector is applied improperly. Because flame detectors are "line-of-sight" devices, care must be taken to ensure they can "see" the entire protected area and they will not be accidentally blocked by stacked material or equipment. The sensitivity of these units is determined by flame size and flame distance from the detector. Although fairly expensive, flame detectors are well suited to protecting areas where explosive or flammable vapors or dusts may be present because they are usually available in "explosionproof" housings.

DETECTOR INSTALLATION

After selection of the most suitable detectors for the job, the next step is to determine where to locate them within the space to be protected.

Spot-type detectors are usually installed on the ceiling or the wall, with the edge of the detector located no closer than 4 in. (100 mm) from the wall or ceiling. When heat detectors are installed at their listed spacing, detection times will approximately equal the operating times of standard 165°F (74°C) sprinklers. If a faster response is desired, detector spacing should be reduced. Where ceilings are high or ceiling construction is not smooth, spacing should be reduced accordingly. Specific information on the treatment of joisted, beamed, and sloped ceilings can be found in Appendix A of NFPA 72, *National Fire Alarm Code* (hereinafter referred to as NFPA 72).

Spacing of Heat Detectors on High Ceilings

Because hot air is diluted by colder air as it rises from a fire, it has always been assumed that heat detectors should be spaced closer together on a high ceiling to achieve the same response time that they would provide on an 8- to 10-ft (2.5- to 3-m) ceiling. To determine how much closer the detectors should be, a series of test fires were conducted using fires of various sizes and taking into consideration ceiling height, ambient temperature, and fire development. The data clearly show that heat detectors must be spaced closer together on a high ceiling to achieve the same response time they would have on a 10-ft (3-m) ceiling.

A typical chart from Appendix B of NFPA 72 shows a heat detector with a listed spacing of 30 ft (9.1 m) responding to a slowly developing fire. (See Figure 5-2Q.) The example shown indicates that, to detect a fire with a heat-release rate of 500 Btu/sec (527 kW) on a 10-ft (3-m) ceiling, a detector with a listing of 30 ft (9.1 m) should be placed on 17-ft (5.2-m) centers. If the ceiling is 20 ft (6.1 m) high, the same detector should be placed on 8-ft (2.4-m) centers to achieve the same response time it would have on a 10-ft (3-m) ceiling.

One might ask why a detector with a listing of 30 ft (9.1 m) is not spaced on 30-ft (9.1-m) centers when mounted on a 10-ft (3-m) ceiling. The answer is that the above example considered detecting a smaller fire. The test fires used by the testing laboratories more nearly approximated the 1000 Btu/sec (1055 kW) fire line on the chart. To detect a slowly developing fire with a heat release rate of 1000 Btu/sec (1055 kW), the chart shows that on a 10-ft (3-m) ceiling the recommended spacing would be 30 ft (9.1 m). On a 20-ft (6-m) ceiling, the recommended spacing would be 19 ft (5.8 m). NFPA 72, Chapter 5, requires the reduction of spacing for heat detectors mounted on ceilings higher than 10 ft (3 m). This reduction is shown in Table 5-2A.

TABLE 5-2A Heat Detector Spacing Reduction for High Ceilings

Ceiling Height (ft)		Percent of Listed Spacing
Above	Up To	
0	10	100
10	12	91
12	14	84
14	16	77
16	18	71
18	20	64
20	22	58
22	24	52
24	26	46
26	28	40
28	30	34

For SI units: 1 ft = 0.305 m.

Appendix B of NFPA 72 also contains charts for smoke detectors, which can be used as very expensive heat detectors to detect flaming fires only. The information regarding smoke detectors in Appendix B is intended only for flaming fires, not smoldering fires.

When installing any type of heat detector, consideration should be given to sources of heat within the protected space that might cause false alarms. For example, heat detectors should be located away from unit heaters and ovens where surges of hot air might be expected.

Proper installation is more critical for smoke detectors than it is for heat detectors because in a smoldering fire, smoke transport is strongly influenced by the convective airflow patterns within the protected area. For this reason, smoke detectors are not assigned a listed spacing by the testing laboratories. Although a grid pattern can be used as a starting point, care must be taken to appropriately locate smoke detectors with respect to the heating supply registers and return air registers. Smoke detectors should be located away from turbulence caused by supply air outlets. Their location should favor return air, because the return air will draw smoke toward the detector, and air velocity at the return tends to be lower.

Special Applications

Air duct smoke detectors are installed either in or on return air ducts or at the return air openings of HVAC systems in buildings. These detectors are used to prevent recirculation of smoke from a fire through the HVAC system within the building. Upon detection of a fire, the associated control system initiates an alarm and either shuts down the circulating blowers or switches them to a smoke exhaust mode. Detectors installed in air duct systems must never be considered substitutes for open-area protection. This is because of dilution of smoke-laden air by clean air from other parts of the building, and because the smoke may not be drawn into the HVAC system when the system is shut down.

Smoke-activated devices also are used to automatically close smoke doors in buildings to limit the spread of smoke in case of fire. Separate corridor ceiling-mounted smoke detectors can be connected to electrically operated hold-open devices on the doors, or smoke detectors can be built into the door closure units themselves.

Smoke stratification should also be considered when smoke detectors are installed. Smoke may stratify below a ceiling due to temperature gradients or airflow along the ceiling. Where stratification is possible, additional smoke detectors can be installed at alternate levels.

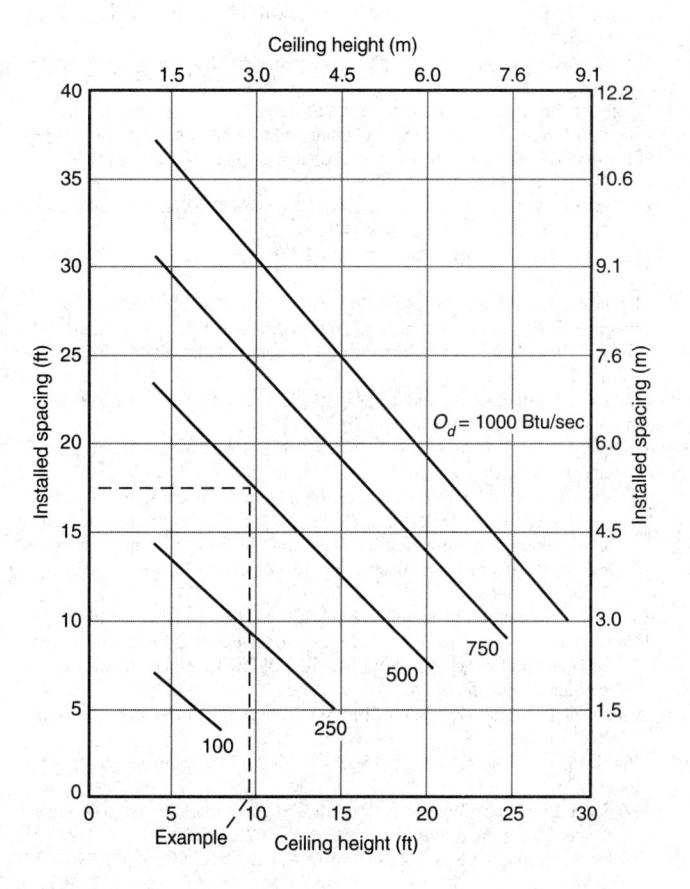

FIG. 5-2Q. Example of a design curve for a fixed-temperature heat detector with a listed spacing of 30 ft (9.1 m) for slow fires.

Installation of gas-sensing detectors is similar to that of smoke detectors. Fire gases tend to flow with smoke and are similarly affected by convected flows within the protected space. Gas detectors must also be located away from sources of oxidizable gases or vapors, such as aerosol sprays or hydrocarbon solvents, as these could cause false alarms.

The requirements of flame detectors are unlike those of heat or smoke detectors because spacings are not relevant for line-of-sight devices. Rather, flame detectors must be located so they can "see" light radiation emanating from any point within the protected space. Because the "cone of vision" for a given flame detector varies with the detector design, manufacturer's recommendations for area coverage must be followed. Flame detectors need to be shielded or located so they will not "see" radiant energy from nonfire sources that might cause false alarms.

DETECTOR MAINTENANCE AND TESTING

A thorough maintenance and testing program of all fire detectors is essential to provide continuous detector operation. A regular test of every smoke detector is of utmost importance because testing is the best way to determine that the detector has not failed. (See Section 5, Chapter 5, "Fire Alarm Systems, Inspection, Testing, and Maintenance.")

BIBLIOGRAPHY

NFPA Codes, Standards, and Recommended Practices

Reference to the following NFPA codes, standards, and recommended practices will provide further information on automatic fire detectors discussed in this chapter. (See the latest version of *The NFPA Catalog* for availability of current editions of the following document.)

NFPA 72, *National Fire Alarm Code.*

References

A Practical Guide to Fire Alarm Systems, Moore, W. D., P.E., ed., Central Station Alarm Association, Bethesda, MD, 1995.
Cholin, J. M., P.E., Moore, W. D., P.E., "How Reliable Is Your Fire Alarm System?" *NFPA Journal*, Vol. 89, No. 1, Jan./Feb. 1995, p. 49.
Fire Alarm Signaling Systems, Bukowski, R. W., P.E., and O'Laughlin, R. J., P.E., National Fire Protection Association and Society of Fire Protection Engineers, Quincy and Boston, MA, respectively, 1994.
Fire Detection Modeling: State of the Art, Schifiliti, R. P., and Pucci, W. E., Fire Detection Institute & RPSA, Inc., Reading, MA, 1995.
Guide for Application of System Smoke Detectors, National Electrical Manufacturers Association, Washington, DC, 1986.
International Fire Detection Literature Review & Technical Analysis, National Institute of Standards & Technology, Gaithersburg, MD, 1991. (Published and distributed by the National Fire Protection Research Foundation, Quincy, MA.)
International Fire Detection Research Project—Technical Report Year One.
Moore, W. D., P.E., *"Fire Detection Sensors and Alarm Systems,"* Chap. 38, *Handbook of Utilities and Services for Buildings*, C. Harris, ed., McGraw-Hill, New York, 1990.
National Institute of Standards & Technology, Gaithersburg, MD, 1993. (Published and distributed by the National Fire Protection Research Foundation, Quincy, MA.)
International Fire Detection Research Project—Technical Report Year Two, National Institute of Standards & Technology, Gaithersburg, MD, 1994. (Published and distributed by the National Fire Protection Research Foundation, Quincy, MA.)
The Moore-Wilson Signaling Report, D. K. Wilson, P.E., W. D., Moore, P.E., co-eds., Focus Publishing Enterprises, Bloomfield, CT.
National Fire Alarm Code Handbook, 1st ed., Moore, W. D., P.E., ed., National Fire Protection Association, Quincy, MA, 1993.

Additional Reading

Beever, P. F., "Estimating the Response of Thermal Detectors," *Journal of Fire Protection Engineering*, Vol. 2, No. 1, 1990, pp. 11–24.
Bertshinger, S., "Smoke Detectors and Alarms," *Fire Journal*, Vol. 81, No. 1, Jan.–Feb. 1988, pp. 43–53.
Breen, D. E., "Improved Life Safety Through More Reliable Smoke Detection Systems," *Proceedings* from the Conference on New Technology to Reduce Fire Losses and Costs, Elsevier, New York, 1986, pp. 227–234.
Brozovsky, E., Motevalli, V., and Custer, R. L. P., "A First Approximation Method for Smoke Detector Placement Based on Design Fire Size, Critical Velocity, and Detector Aerosol Entry Lag Time," *Fire Technology*, Vol. 31, No. 4, Nov. 1995, pp. 336–354.
Bukowski, R. W., "Fire Detection and Alarm Systems," NBSIR 86-3360, Apr. 1986, Center for Fire Research, Gaithersburg, MD.
Bukowski, R. W., "Studies Assess Performance of Residential Detectors," *NFPA Journal*, Vol. 87, No. 1, Jan./Feb. 1993, pp. 48–54.
Bukowski, R. W., "Techniques for Fire Detection," *Spacecraft Fire Safety*, NASA CP 2476, 1987, National Aeronautics and Space Administration.
Chohan, R. K., and Upadhyaya, B. R., "Safety and Fault Detection in Progress Control Systems and Sensors," *Fire Safety Journal*, Vol. 14, No. 3, Jan. 1989, pp. 167–177.
Cholin, J., "The Current State of the Art in Optical Fire Detection," *Plant/Operations Progress*, Vol. 8, No. 1, Jan. 1989, pp. 12–18.
Cholin, J., "Optical Fire Detection," *Chemical Engineering Progress*, Vol. 85, No. 7, July 1989, pp. 62–68.
Cholin, J. M., "Everything You Didn't Know You Wanted to Know about Spark/Ember Detectors, But Are about to Learn. Part 1," *Signaling Report*, Vol. 5, No. 2, Mar./Apr. 1993, pp. 14–15.
Cholin, J. M., "Everything You Wanted to Know about Flame Detectors, But Nobody Else Would Tell You. Part 3," *Signaling Report*, Vol. 3, No. 5, 14–15, Sep./Oct. 1991, pp. 14–15.
Chow, W. K., and Wong, W. C. W., "Experimental Studies on the Sensitivity of Fire Detectors," *Fire and Materials*, Vol. 18, No. 4, July/Aug. 1994, pp. 221–230.
Chow, W. K., and Yip, S. C., "Sensitivity Studies on Fire Detectors," *Journal of Applied Fire Science*, Vol. 3, No. 4, 1993–1994, pp. 359–371.
"Components of Automatic Fire Alarm Systems for Residential Premises. Part 1. Specification for Self-Contained Smoke Alarms and Point-Type Smoke Detectors, British Standards Institution, London, UK, BS 5446:Part 1:1990, Part 1, 1990.
Cooper, L., "Why We Need to Test Smoke Detectors," *Fire Journal*, Vol. 80, No. 6, Nov.–Dec. 1986, pp. 43–45.
Craig, J., "Smoke Alarms Are Not Enough," *Fire Prevention*, No. 233, Oct. 1990, pp. 22–23.
"Domestic Smoke Alarms," *Fire Prevention*, No. 283, Oct. 1995, p. 12.
Donohue, P. A., "False Alarms from Automatic Fire Detection Systems," Building Research Establishment, Garston, UK, BRE IP 13/92, June 1992.
Drake, A., "Detecting Incipient Fires—A New Approach," *Fire Surveyor*, Vol. 18, No. 5, Oct. 1989, pp. 5–11.
Dubivsky, P. M., "Underwriters Laboratories' Smoke Detector Standards and Tests," *Fire Journal*, Vol. 82, No. 1, 1988, p. 45.
Dyhring, F., "Trends in European Development of Efficient Equipment for Testing of Smoke Detectors," *Proceedings* of the Tenth International Conference on Automatic Fire Detection "AUBE '95", [Internationale Konferenz uber Automatischen Brandentdeckung.] April 4–6, 1995, Duisburg, Germany, 1995, pp. 250–260.
Endo, K., "Evaluation of Standard Fire Tests for Fire Detectors," 1990.
Evans, D. D., and Stroup, D. W., "Methods to Calculate the Response Time of Heat and Smoke Detectors Installed Below Large Unobstructed Ceilings," *Fire Technology*, Vol. 22, No. 1, 1986, pp. 54–66.
"Fire Detection and Suppression: Today's Technology," *Fire Safety Journal*, Vol. 14, Nos. 1 & 2, July 1988, pp. 1–125.
"Fire Detector Systems: The Range of Choice," *NFPA Journal*, No. 91, pp. 18–23.
"Focus on Fire Detection: Ionization Smoke Detectors—The High-Voltage Issues," *Fire Prevention*, No. 249, May 1992, pp. 35–36.
Fowler, J., "Reducing False Alarms," *Fire*, Vol. 88, No. 1090, Apr. 1996, pp. 27–28.
Fox, P., "Product Profile: Developments With the VESDA Aspirating Smoke Detectors," *Fire Surveyor*, Vol. 21, No. 3, June 1992, pp. 16–18.

Goedeke, A. D., "Characterization and Testing of Optical Fire Detectors and Immunity to False Alarm," Final Report, July 1989–Dec. 1989, Donmar Limited, Newport Beach, CA, Air Force Engineering and Services Center, Tyndall AFB, FL, ESL-TR-90-01, Apr. 1992.

Goedeke, A. D. and Gross, H. G., "Characteristics of Optical Fire Detector False Alarm Sources and Qualifications Test Procedures to Prove Immunity," Final Report, April 8, 1991–Oct. 8, 1992, Donmar Limited, Newport Beach, CA, Air Force Civil Engineering Support Agency, Tyndall AFB, FL, CEL-TR-92-62, Oct. 8, 1992.

Hall, J. R., "When Fire Detectors Don't Operate—A Growing Problem," *Fire Safety Journal*, Vol. 14, Nos. 1 & 2, 1988, pp. 25–32.

Hall, J. R., "U.S. Experience with Smoke Detectors: Who Has Them? How Well Do They Work? When Don't They Work?" National Fire Protection Association, Quincy, MA, 1988.

Hall, J. R., Jr., "U.S. Experience with Smoke Detectors and Other Fire Detectors: Who Has Them? How Well Do They Work? When Don't They Work?" National Fire Protection Association, Quincy, MA, June 1995.

Halliwell, R. E., and Sultan, M. A., "Attenuation of Smoke Detector Alarm Signals in Residential Buildings," *Proceedings* of the First International Symposium on Fire Safety Science, Hemisphere, Washington, DC, 1986, pp. 689–697.

Holker, J. R., and Lomax, G. R., "Sensory Early Warning Systems in Fire Detection," *Proceedings* of the Conference on New Technology to Reduce Fire Losses and Costs, Elsevier, 1986, pp. 246–255.

Hygge, S., "Installation and Reliability of a Free Smoke Detector," *Proceedings* of the First International Symposium on Fire Safety Science, Hemisphere, Washington, DC, 1986, pp. 739–748.

Jahr, F., "Focus on Fire Detection: Detector Success for Norway," *Fire Prevention*, No. 249, May 1992, pp. 37–38.

Jernigan, W., "Keeping the Smoke Detectors Operational: The Dallas Experience," *Fire Journal*, Vol. 81, No. 4, July 1987, p. 57.

Johnson, J. E., "Fire Detection: Past, Present, and Future," *Fire Journal*, Vol. 81, No. 5, Sept. 1987, p. 49.

Marriott, M., "Reliability and Effectiveness of Domestic Smoke Alarms," *Fire Engineers Journal*, Vol. 55, No. 177, June 1995, pp. 10–13.

Marriott, M., "Reliability of Domestic Smoke Alarms," *Fire Research News*, Vol. 13, Spring 1991, p. 8.

Meacham, B. J., "The Use of Artificial Intelligence Techniques for Signal Discrimination in Fire Detection Systems," *Journal of Fire Protection Engineering*, Vol. 6, No. 3, 1994, pp. 125–136.

Milke, J. A. and McAvoy, T. J., "Analysis of Signature Patterns for Discriminating Fire Detection with Multiple Sensors," *Fire Technology*, Vol. 31, No. 2, May 1995, pp. 120–136.

Moore, W. D., "Automatic Fire Detection: False Alarms," *NFPA Journal*, No. 88, pp. 146–148.

Neily, M. L., Smith, C. L., and Shapiro, J. I., "Residential Smoke Detector Performance in the United States," Consumer Product Safety Commission, Washington, DC, European Consumer Safety Association Conference, Nov. 22, 1993, Amsterdam, Netherlands, 1993, pp. 1–16.

Newman, J. S., "Prediction of Fire Detector Response," *Fire Safety Journal*, Vol. 12, No. 3, 1987, pp. 205–211.

Newman, J. S., "Principles of Fire Detection," *Fire Technology*, Vol. 24, No. 2, 1988, pp. 116–127.

Pati, V. B., *et al.*, "Simulation of Intelligent Fire Detection and Alarm System for a Warship," *Defense Science Journal*, Vol. 39, No. 1, Jan. 1989, pp. 79–94.

Pomroy, W. H., "Spontaneous Combustion Fire Detection for Deep Metal Mines," Report 9144, 1987, U.S. Bureau of Mines, Minneapolis.

Powell, B. D., Spring, D. J., and Adams, S. J., "Infrared Flame Detector with Fire-Ranging Ability," *Proceedings* of the Conference on Infrared Systems and Technology, International Society for Optical Engineering, Vol. 807, Bellingham, WA, 1987, pp. 69–72.

Salamone, R., "Retrofitting High-Rise Dorms with Alarm and Detection Systems," *Fire Journal*, Vol. 84, No. 1, Jan.–Feb. 1990, p. 37.

Scheidweiler, A., and Guttinger, H., "Reducing False Alarms From Smoke Detectors," *Fire International*, Vol. 14, No. 125, Oct./Nov. 1990, pp. 77–78.

Schierau, K., "Detection Equipment for Fire Prevention, A Standard Solution?" *Proceedings* of the Conference on New Technology to Reduce Fire Losses and Costs, Elsevier, 1986, pp. 256–263.

Shalna, A. J., "What a Combination: Beam Smoke Detectors and Hostile Environments," *Fire Journal*, Vol. 82, No. 4, Jul.–Aug. 1988, p. 53.

Shifiliti, R. P., Meacham, B. J., and Custer, R. L. P., "Design of Detection Systems," *SFPE Handbook of Fire Protection Engineering*, National Fire Protection Association, Quincy, MA, 1995.

Smoke Detectors: A Bibliographical Overview, Vance Bibliographies, Monticello, IL, 1985.

Stroup, D. W., Evans, D. D., and Martin, P. M., "Evaluating Thermal Fire Detection Systems," NBS SP 712 and NBS SP 713, Apr. 1986, Center for Fire Research, Gaithersburg, MD.

Stroup, D. W., and Evans, D. D., "Use of Computer Fire Models for Analyzing Thermal Detector Spacing," *Fire Safety Journal*, Vol. 14, Nos. 1 & 2, 1988, p. 33.

Tar, D., "Cerberus Infrared Flame Detectors," Third International Symposium on Fire Protection of Buildings, May 10–12, 1990, Eger, Hungary, 1990, pp. 173–174 .

Watanabe, A., Sasaki, H., and Unoki, J., "Overview of Fire Detection in Japan," *Proceedings* of the First International Symposium on Fire Safety Science, Hemisphere, Washington, DC, 1986, pp. 679–688.

Watkins, G., "Selection and Placement of Flame Detectors for Maximum Availability of the Detection System," Third International Conference on Management and Engineering of Fire Safety and Loss Prevention: Onshore and Offshore, Feb. 18–20, 1991, Aberdeen, Scotland, Elsevier Applied Science, New York, 1991, pp. 229–243.

"Will Your Smoke Detectors Operate?" *Fire Protection*, Vol. 17, No. 4, Dec. 1990, p. 7.

Wilton, R., "Recent Developments in the Fire Detection Field," *Fire Engineers Journal*, Vol. 47, No. 144, 1987, pp. 8–10.

Zallen, D. M., *et al.*, "Evaluation of Optical Fire Detectors," *Proceedings* of the 34th International Instrumentation Symposium, ISA, 1988, pp. 121–128.

NOTIFICATION APPLIANCES

Robert P. Schifiliti

Fire alarm systems and household fire warning equipment protect life by automatically indicating the need for the occupants to evacuate or relocate to a safe area. They may also notify emergency forces or other responsible persons who may then assist the occupants or assist in controlling and extinguishing the fire. Audible and visible notification appliances alert the occupants, and, in some cases, emergency forces, and convey information to them.

A fire alarm system that simply sounds an audible signal and flashes strobe lights in a space is conveying a single bit of information: Fire alarm. Systems that send voice announcements or that use text or graphic enunciators typically convey multiple bits of information. They may signal a fire alarm and give a specific location and information on how and where to evacuate or relocate. As the cost of emergency voice/alarm systems comes down closer to conventional systems, they are being used more and more by designers. They are almost always required by codes in high-rise buildings, but can also be effectively used in smaller buildings. When provided with detailed information about a fire emergency, people tend to evacuate more quickly and orderly. The audibility requirements for voice systems are the same as for conventional systems. In addition to audibility, the intelligibility of the signal content becomes important.

In addition to their use for fire alarm notification, audible and visible appliances may also be used to indicate a trouble condition in the fire alarm system. They may also be used as supervisory signals to indicate the condition or status of other fire protection systems, e.g., automatic sprinklers.

There are two operating modes for fire alarm notification appliances: (1) public operating mode and (2) private operating mode. In the public operating mode, audible and visible signaling is intended for the occupants or inhabitants of the area protected by the fire alarm system. In the private operating mode, audible and visible signaling is intended only for persons directly concerned with implementation and direction of emergency actions in the area protected by the fire alarm system. For example, in a bank or shopping mall, the fire alarm is intended to alert all occupants and convey the need for them to evacuate. This scenario describes a public mode system. In a prison or hospital, the system may be designed to notify only the trained staff. They will then initiate emergency procedures,

including assisting occupants who might not otherwise be able to evacuate or relocate. This scenario describes a private mode system. Similarly, a supervisory signal at a guard's station indicating an off-normal status of a fire pump is considered to be private mode signaling.

The operating mode of the system affects the selection and placement of notification appliances. In general, the requirements for private mode signaling are less stringent, because they are intended for a professional staff that has been trained to look and listen for that signal and to immediately take certain actions when it is seen or heard.

AUDIBLE SIGNALING

Audible notification appliances are the most common method for signaling a fire alarm condition in a building or area. There are two factors that affect the performance of audible notification systems. Here performance is defined as the ability to alert and convey information. The first is the rating of the audible appliance. Typically, audible appliances are rated by measuring the sound pressure level (SPL) at a fixed distance in a special room.

Without giving the exact mathematical definition, sound pressure level is a measurement of how loud something is. The measurement is expressed in decibels (dB). For almost all fire alarm applications, the sound pressure level is expressed in dBA, i.e., decibels, A-weighted. A sound pressure level expressed in dBA has been adjusted to account for the way the human ear perceives different frequencies. Typically, high-pitched sounds (high frequency) are heard better than low frequency, bass, sounds. The A-weighting adjustment corrects for this so that the loudness of sounds that are composed of different frequencies can be compared.

Audible appliance ratings are usually stated as a certain SPL at a distance of 10 ft (3.05 m). This is a starting point for the design or analysis of audible signaling systems. Because the rating is measured in a room with certain reverberation characteristics, the actual performance will vary depending on the field conditions. In a room or hallway with concrete block walls, a sheetrock ceiling, and a tile floor, the actual SPL at 10 ft (3.05 m) will probably be greater than the rating of the audible appliance. However, in an acoustically soft situation, such as a room with wood walls, carpet on the floors, and an acoustical tile ceiling, the SPL may be less than the appliance's rating.

The second factor affecting the performance of audible notification systems is the net sound pressure level produced throughout

Robert P. Schifiliti, P.E., is president of Robert P. Schifiliti & Associates, Inc., a fire protection engineering and consulting firm located in Reading, MA. He is chairperson of NFPA 72, *National Fire Alarm Code*, Chapter 6, "Notification Appliances for Fire Alarm Systems."

all of the building or area served by the fire alarm system. As one moves away from an audible appliance, the SPL decreases. As one gets closer, the SPL increases. In a wide-open space where the sound does not reflect off any surfaces (no reverberation), the SPL decreases by 6 dBA every time the distance from the source is doubled. This condition is called the free-field condition. An anechoic chamber, designed to absorb all sound waves, simulates a free-field condition. In almost all other situations there is some reverberation. Nevertheless, the 6-dBA rule can be used when designing a system or analyzing the expected performance of a system.

For example, if an appliance has a rating of 90 dBA at 10 ft (3.05 m) and the distance is doubled to 20 ft (6.10 m), one would expect to measure an SPL of approximately 84 dBA. (See Figure 5-3A.) At a distance of 40 ft (12.20 m), one would expect an SPL of 78 dBA. At 80 ft (24.40 m), the SPL would be approximately 72 dBA.

When there are multiple audible appliances in a space, their combined SPL is greater than any one, but the individual SPLs are not simply added together to get the result. For example, consider two appliances rated 90 dBA at 10 ft (3.05 m) that are spaced 40 ft (12.20 m) apart. Alone, each unit would produce about 84 dBA (90 – 6) in the center [20 ft (6.10 m) from each]. With both units operating, a measurement taken between the two at 20 ft (6.10 m) from each would be about 87 dBA (3 dBA greater than a single unit). Fire alarm designers usually ignore this effect so that their designs are conservative.

This type of analysis works well in open spaces. However, sound loss through walls and doors requires more complex calculation methods. Of course, in an existing building a temporary audible appliance can be used to study sound transmission and attenuation throughout the space and through the walls and doors. By measuring sound loss from one side of a door/wall assembly, designers can obtain representative values for a structure to be used in later design and analysis of the fire alerting system. *The SFPE Handbook of Fire Protection Engineering*[1] contains calculation methods that can be used in design and analysis of audible alerting systems.

Fire Alarm Audibility Design Goals

How loud should a fire alerting system be? Obviously, the audible signals must be louder than the ambient noise in a space. So, the first step is to measure the ambient noise. When designing new systems, tables of ambient noise levels may be used as an estimate. Appendix A of NFPA 72, *National Fire Alarm Code* (hereinafter referred to as NFPA 72), contains such a table. Two different ambient noise levels are required to complete a system design: (1) the 24-hr average SPL and (2) the maximum SPL that lasts 60 sec or more.

Fire alarm audibility goals are dependent on the mode of operation. In the public operating mode, the fire alarm system must produce a sound pressure level that is 5 dBA above any ambient noise

that lasts 60 sec or more, or 15 dBA above the 24-hr average, whichever is greater. In the private operating mode, the fire alarm system must produce a sound pressure level that is 5 dBA above any ambient noise that lasts 60 sec or more, or 10 dBA above the 24-hr average, whichever is greater. In either case, if the area will be used for sleeping, the minimum SPL required of the fire alarm system is 70 dBA.

Because building codes require residential- and other sleeping-type occupancies to have a minimum degree of soundproofing, they represent the most challenging scenarios for fire alarm system design. Given the high SPL that must be achieved to alert a sleeping person, it has been shown that audible appliances are almost always needed in each sleeping room.[1] In other occupancies, such as offices and schools, it may be sufficient to place audible appliances in the corridors only. In any case, careful design is needed to ensure that the audibility requirements are met.

Once the system is installed, a complete test, including SPL measurements, must be done to prove that the system meets the signaling requirements. Various types of sound level meters are available for use in testing system audibility. Some are very expensive and have many features for special system analysis, and others are very inexpensive, yet serve the purpose. It is important that the meter have an A-weighting scale for measurement of the ambient and fire alarm SPLs.

In addition to the minimum sound pressure levels that a fire alarm must produce, there is a maximum allowable SPL of 120 dBA at the minimum hearing distance. This is below the threshold of pain, which occurs at about 130 dBA, and is intended to reduce the chances of damaging someone's hearing. With appliances mounted about 7.5 ft (2.3 m) above the floor (to reduce the chances of being blocked by equipment or furnishings), the minimum hearing distance can be taken as 1.25 ft (0.38 m). This is equivalent to a 6-ft 3-in. (1.9-m) person standing directly below the unit. Using the 6-dBA rule, an appliance would have to have a rating of 102 dBA or less at 10 ft (3.05 m), so that the SPL would be 120 dBA or less at the minimum hearing distance.

It is often possible to achieve fire alarm audibility requirements by using a few very loud appliances or a larger number of units at a lower SPL. In general, system reliability is better with a larger number of lower SPL units. Also, with systems having a large number of appliances, the cost may be lower using a larger number of lower SPL units. If the system is being designed to transmit voice messages, the design should use a larger number of low-power units so that the voice message is intelligible. If only a few high-power speakers are used, reverberation of the sound may make it difficult to understand the message.

Effective July 1996, NFPA 72 requires that all audible evacuation signals comply with ANSI S3.41, *Audible Emergency Evacuation Signal.*[2] (Note: NFPA 72 does not itself require existing systems to comply with current requirements.) That signal is commonly referred to as a temporal code-three signal. The pattern consists of an "on" phase (a) lasting 0.5 sec +/– 10 percent, followed by an "off" phase (b) lasting 0.5 sec +/– 10 percent, for three successive "on" periods, which is then followed by an "off" phase (c) lasting 1.5 sec +/– 10 percent. The signal should be repeated for a period appropriate for the building, but not less than 180 sec. A single-stroke bell or chime sounded at "on" intervals lasting 1 sec +/– 10 percent, with a 2 sec +/– 10 percent "off" interval after each third "on" stroke, is acceptable. (See Figures 5-3B, 5-3C, and 5-3D.) The type of signal is not important; it may be a bell, horn, chime, slow whoop, or any other sound, provided it is not used for any purpose other than fire alarm. The intent is that the pattern of the sound will be standard from system to system. Systems intended to relocate

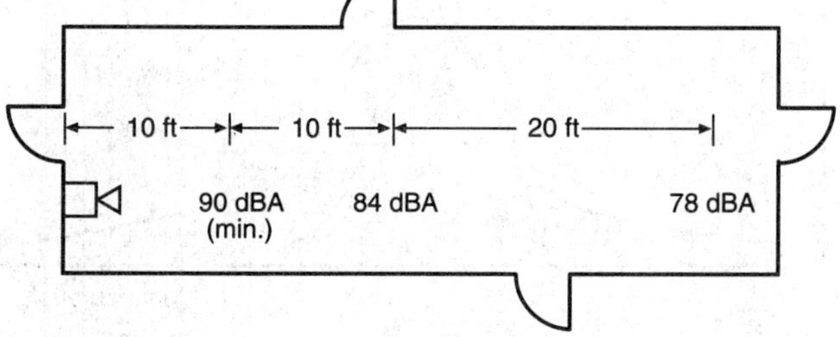

FIG. 5-3A.   Example of 6-dBA rule. (For SI units: 1 ft = 0.305 m.) (Source: Robert P. Schifiliti & Associates, Inc.)

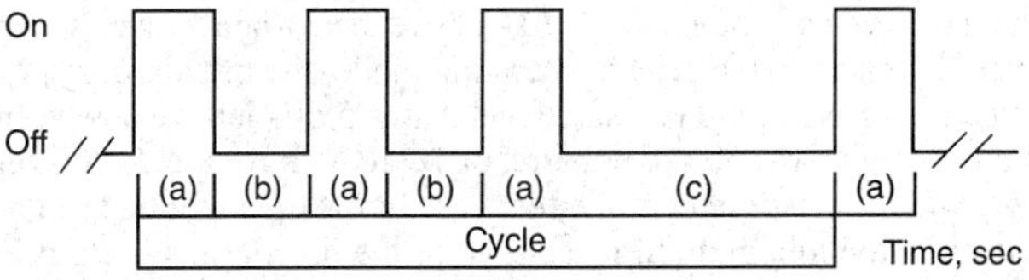

Key:
Phase (a) signal is "on" for 0.5 s ± 10%
Phase (b) signal is "off" for 0.5 s ± 10%
Phase (c) signal is "off" for 1.5 s ± 10% [(c) = (a) + 2(b)]
Total cycle lasts for 4 s ± 10%

FIG. 5-3B. Temporal pattern parameters.

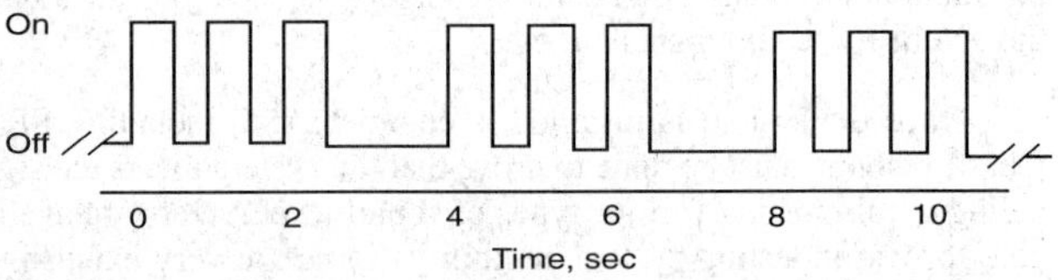

FIG. 5-3C. Temporal pattern imposed on signaling appliances that emit a continuous signal while energized.

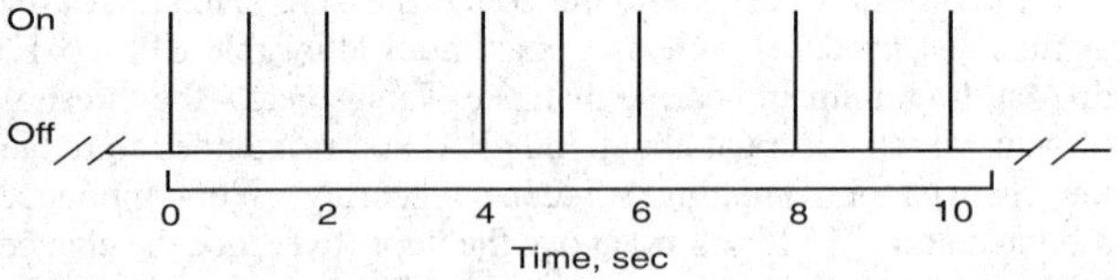

FIG. 5-3D. Temporal pattern imposed on a single-stroke bell or chime.

people or to simply alert them of a nonfire emergency must not use the standard evacuation signal.

VISIBLE SIGNALING

Visible fire alarm notification appliances are often intended only to augment audible appliances. However, when it is expected that hearing-impaired persons may be in the protected area, or when ambient noise levels are high, visible appliances may be the primary means of occupant notification that a fire emergency exists. Increasingly, visible appliances are being used as an integral part of occupant notification systems.

Building codes, NFPA *101*®, *Life Safety Code*®, and NFPA 72, as well as state and federal laws, all contain requirements for the use of visible notification appliances under certain circumstances. At present, only strobe lights are recognized as meeting the requirements of these codes and laws. The appliances must have a very short duration, high-intensity light output to meet stated performance requirements.

As with audible appliances, there are two factors affecting the performance of visible notification appliances: (1) source intensity and (2) illumination at some distance from the source.

Source intensity is a measure of the light output of the appliance. The unit of measure is the candela (cd). (This unit was formerly called "candlepower." There is a one-to-one relationship between candela and candlepower.) As one moves away from any light source, the illumination it provides decreases. Illumination is measured in units of lumens (lm) per sq ft, or lumens per sq m (also called "lux"). (Formerly, the unit used to describe illumination was the "footcandle." One footcandle equals one lumen per sq ft. One lumen per sq m equals 0.926 footcandles.) Figure 5-3E shows the relationship between source intensity and illumination.

The inverse square law is used to determine the illumination, E, at some distance, d, from a source of intensity, I, i.e., $E = I/d^2$.

Because strobe lights flash for a very brief period of time, the perceived brightness can vary depending on the actual peak source strength and duration of the flash. (See Figure 5-3F.) One appliance might reach a peak intensity of 1,000 candelas in 0.1 sec, while another might reach 750 candelas in 0.2 sec. Nevertheless, the human eye might perceive both as being equally bright. A complex mathematical relationship is used to relate the perceived brightness of a strobe light to that of a constantly burning light. The result is called the "effective intensity" [candela effective (cd eff)].

The design of visible notification systems involves many additional factors beyond those discussed herein. Many other factors, such as how much light is absorbed *vs.* how much is reflected by a surface, the angle that the light strikes a surface, and the color of the surface, affect the performance of the notification system. For instance, as light strikes a surface at some angle, only a portion of that light is reflected back to the human eye. The term used to describe the reflected light is "luminance." Thus, a certain intensity appliance produces a certain luminous intensity at some distance on a surface, but a person looking at that surface sees a certain luminance.

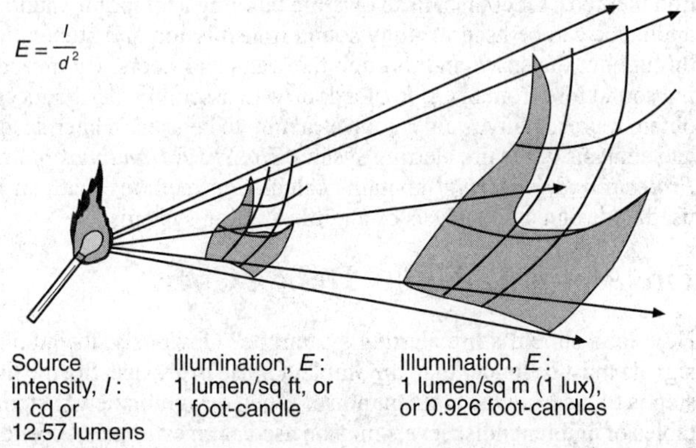

$$E = \frac{I}{d^2}$$

| Source intensity, I : 1 cd or 12.57 lumens | Illumination, E : 1 lumen/sq ft, or 1 foot-candle | Illumination, E : 1 lumen/sq m (1 lux), or 0.926 foot-candles |

FIG. 5-3E. Visible intensity vs. illumination. (Source: Robert P. Schifiliti & Associates, Inc.)

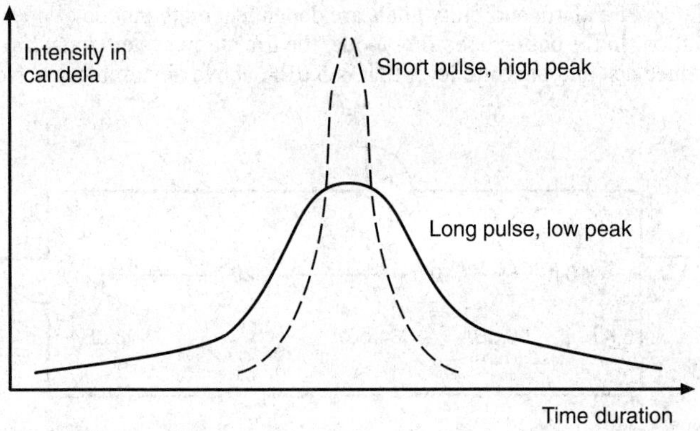

FIG. 5-3F. Visible peak vs. effective intensity. (Source: Robert P. Schifiliti & Associates, Inc.)

To simplify the process of designing and evaluating visible notification systems, NFPA 72 contains prescriptive requirements for the intensity and spacing of visible appliances that are sufficient in almost all applications to alert persons. In essence, NFPA 72 provides preset designs that can be used for a variety of actual field conditions.

The prescriptive requirements contained in the NFPA 72 are based, in part, on extensive tests performed by Underwriters Laboratories Inc. (UL).[3] The test program showed that virtually all subjects, including hearing-impaired persons, could be alerted by certain intensity strobe lights at certain maximum distances.

Fire Alarm Visibility Design Goals

NFPA 72 contains tables listing the minimum intensity strobe required in certain size rooms to alert the occupants. For instance, a single 15 cd eff strobe can be used when centered on one wall of a 20 ft × 20 ft (6.1 m × 6.1 m) room. There are also listings for use of multiple units in a single room. Using the 20 ft × 20 ft (6.1 m × 6.1 m) room as an example, the illumination on the opposite wall would be $15/20^2 = 0.0375$ lumens per sq ft. One could show that the distance from the appliance to the farthest corner is about 22.4 ft (6.8 m). The luminous intensity would then be $15/22.4^2 = 0.030$ lumens per sq ft. However, it should be kept in mind that strobe lights do not distribute their full-rated intensity in all directions. So, in the corner of a 20 ft × 20 ft (6.1 m × 6.1 m) room, the actual illumination may be 90 percent or less than that calculated, using the on-axis value for the intensity. By using the tables in NFPA 72, designers, without resorting to lengthy calculations, can be assured of providing a level of illumination that was found sufficient in the UL test series to alert hearing and hearing-impaired persons.

Occupancies where hearing-impaired persons may sleep must have higher strobe intensities than for non-sleeping areas to ensure that the hearing-impaired person can be awakened. Testing has shown that a 110 cd eff strobe, no more than 16 ft (4.9 m) from the pillow, was sufficient to awaken hearing-impaired persons. Other methods, such as bed shakers or fans, may be used to supplement visible signaling. In practice, a bedroom strobe light is often combined with a smoke detector. In that case, the strobe light must have an even higher intensity (177 cd eff) to overcome potential obscuration by a smoke layer at the ceiling.

Some spaces, such as corridors, can be treated differently from those rooms where a person may have his or her face turned away from a strobe light. In corridors, a person's view path is more concentrated, and the person is usually moving. Recognizing this, NFPA 72 allows the use of 15 cd eff strobes at longer distances than what would be required for rooms. Thus, the luminous intensity required is actually less, given that a person is likely to directly see the appliance or to see more surfaces illuminated by the strobe.

When designing the audible part of an occupant-notification system, it is best to use a greater number of lower power SPL units. The opposite is true for visible appliances. The use of a lower number of higher-intensity units will usually be more cost effective. Further, with fewer appliances, the distance between them is greater, and the chances of a person directly viewing more than two units at a time is reduced or eliminated. That, in turn, reduces the likelihood of triggering a seizure in a person who has photosensitive epilepsy.

When choosing and placing visible appliances, every effort should be made to place the unit in the concentrated viewing path of the target audience. For example, in a meeting room or classroom, the appliance should be on the wall that the majority of people face. Further, a mounting height of 6.5 ft (2.03 m) to 8 ft (2.44 m) ensures that the visible appliance is unlikely to be blocked by common furnishings.

In spaces such as warehouses, or large auditoriums, the use of high-intensity indirect signaling may be warranted. This method of indirect signaling uses much higher intensity strobes that must be located so that they cannot be directly viewed by the human eye (they could temporarily or permanently impair vision). The light from the strobes is used to illuminate large areas of walls and ceilings. The UL research used indirect illumination. However, the illumination was on a wall in front of the test subjects, and the maximum intensity was low enough to allow direct viewing of the appliance's element. Another possible method of visible signaling in such high-challenge spaces is to flash some or all of the building lights. While these methods may prove effective, at this time there is little test data to support specific performance requirements for these notification methods.

In the past, guidelines for meeting the visible signaling requirements of the Americans with Disabilities Act (ADA)[4] have cited strobe intensities and distances different from those specified in NFPA 72. However, through a process called "equivalent facilitation," the requirements of NFPA 72 have been shown to meet or exceed the original ADA "Accessibility Guidelines" (ADAAG). The ADAAG have been revised to follow NFPA 72 requirements.

SUMMARY

Once ANSI S3.41, *Audible Emergency Evacuation Signal*,[2] becomes required by NFPA 72 effective in July of 1996, some systems will require separate circuits for audible and visible appliances. Some combination audible/visible appliances may be available that can be installed on a common circuit. In any case, it will be important to size the circuit to handle the expected load. Greatest flexibility is obtained by putting fewer than the maximum number of units on a circuit when the system is first designed or installed. In this way, if the acceptance test shows deficiencies in some areas, additional units can be added without overloading the circuits or the system.

Most codes do not require audible and visible appliances to be on separately controlled circuits. However, if they are, the audible circuits can be silenced after a building has been completely evacuated, and the visible appliances can be left in operation. This is a way of telling people not to re-enter the building or space. Similarly, good design practice is to include an audible and a visible appliance on the outside of the building to assist arriving fire fighters in locating the building or building entrance. Placing appliances on each side of a building or at each entrance also helps inform persons that they should not enter the building or space.

When designing or installing occupant notification systems utilizing both audible and visible appliances, it is best to determine the locations of the visible units first. This is because their location is more dependent on viewing angles, intersections of corridors, and blockage by furnishings. Once the visible appliances have been placed, audible appliances can be located so that some or all of the units can be combination audible and visible appliances. The use of combination units may reduce installation costs slightly.

BIBLIOGRAPHY

References

1. Schifiliti, R. P., Meacham, B. J., and Custer, R. L. P., "Design of Detection Systems," *The SFPE Handbook of Fire Protection Engineering*, DiNenno, P. J., ed., 1995.
2. ANSI S3.41, *Audible Emergency Evacuation Signal*, American National Standards Institute, New York, 1990.
3. "Report of Research on Emergency Signaling Devices for Use by the Hearing Impaired," Mar. 1991, Underwriters Laboratories Inc., Northbrook, IL.

4. Americans with Disabilities Act, Public Law 101-336, Title III, Rule 36, Appendix A, "Accessibility Guidelines," 1990.

NFPA Codes, Standards, and Recommended Practices

References to the following NFPA codes, standards, and recommended practices will provide further information on notification appliances discussed in this chapter. (See the latest version of *The NFPA Catalog* for availability of current editions of the following documents.)

NFPA 72, *National Fire Alarm Code*
NFPA *101®*, *Life Safety Code®*

Additional Reading

ANSI S12.31, *Precision Methods for the Determination of Sound Power Levels of Broad-Band Noise Sources in Reverberation Rooms*, Acoustical Society of America, New York, NY, 1990.

ANSI S12.32, *Precision Methods for the Determination of Sound Power Levels of Discrete-Frequency and Narrow-Band Noise Sources in Reverberation Rooms*, Acoustical Society of America, New York, 1990.

Barnett, P., "Voice Alarm Systems: Loudspeakers—Fire, Flame or Fool Proof?" *Fire Surveyor*, Vol. 21, No. 6, Dec. 1992, pp. 7–9.

Barnett, P., "Merging of Expertise in Fire Alarm and Evacuation Systems Welcomed," *Fire*, Vol. 85, No. 1050, Dec. 1992, p. 18.

Bond, A., "Sound Guidance (Auditoria and Alarm Systems)," *Building Services*, Vol. 13, No. 2, Mar. 1992, pp. 28–30.

Bruck, D. and Horasan, M., "Non-arousal and Non-action of Normal Sleepers in Response to a Smoke Detector Alarm," *Fire Safety Journal*, Vol. 25, No. 2, 1995, pp. 125–140.

Bukowski, R. W., "Notification Appliances for the Hearing Impaired," VHS Video Tape, National Institute of Standards and Technology, Gaithersburg, MD, Video, Federal Fire Forum, Current Issues in Fire Protection, June 15, 1992, Gaithersburg, MD, 1992.

Cholin, J. M. and Moore, W. D., "How Reliable Is Your Fire Alarm System," *NFPA Journal*, Vol. 89, No. 1, Jan./Feb. 1995, pp. 49–53.

Choppen, B., "Voice of Alarm in Public Areas," *Fire Prevention*, No. 266, Jan./Feb. 1994, pp. 28, 30.

Craig, J., "Smoke Alarms Are Not Enough," *Fire Prevention*, No. 233, Oct. 1990, pp. 22–23.

DeVoss, F., "Bringing Alarms to Light—Signaling for the Hearing Impaired," Underwriters Labs., Inc., Northbrook, IL, *UL Lab Data*, Vol. 20, No. 1, 1990, pp. 4–7.

"Digital Alarm Communicator System Units," Second Edition, Standard for Safety, Underwriters Labs., Inc., Northbrook, IL, UL 1635, February 21, 1991.

Foster, J., "Fire Appliance Design: Options for the Future," *Fire Engineers Journal*, Vol. 52, No. 165, 14–17, June 1992, pp. 14–17.

Hicks, H. D., Jr., "Another Look at Design Guidelines for Voice Fire Alarm Systems," *NFPA Fire Journal*, Vol. 88, No. 1, 50–52, Jan./Feb. 1994, pp. 50–52.

Hodge, S. G., "Fire Alarm Safety Improvement," Westinghouse Hanford Co., Richland, WA, WHC-SD-631-ATP-006, Oct. 13, 1994.

Korslund, S. M., "Technical Evaluation of Equipment Maintenance on Fire Alarm Detection, Suppression, and Signaling Systems on the Hanford Site," Westinghouse Hanford Co., Richland, WA, WHC-SC-GN-TEEM-10001, REV. 0, 1994.

Moore, W. D., P.E., ed., *National Fire Alarm Code Handbook*, 1st ed., National Fire Protection Association, Quincy, MA, 1993.

Nober, E. H., Well, A. and Moss, S., "Alarms for the Hearing-Impaired," *Fire Prevention*, No. 233, Oct. 1990, pp. 28–31.

Nober, E. H., Well, A. D. and Moss, S., "Does Light Work As Well As Sound? Smoke Alarms for the Hearing-Impaired," *Fire Journal*, Vol. 84, No. 1, Jan./Feb. 1990, pp. 26–28, 30.

Rosato, C., "Radio Revolution in Central Alarm Stations," *Fire Prevention*, No. 245, Dec. 1991, pp. 30–31.

Scott, G., "Electronic Sirens or Air Horns?," *Fire Research News*, No. 15, Spring 1993, pp. 15–17.

Smith, P., "Smoke Alarms Fitted on Ground Floors May Not Be Loud Enough for Sleepers," *Fire*, Vol. 85, No. 1045, July 1992, p. 21.

Smith, R.L., "Performance Parameters of Fire Detection Systems," *Fire Technology*, Vol. 30, No. 3, 1994, pp. 326–337.

Sultan, M. A. and Halliwell, R. E., "Optimum Location for Fire Alarms in Apartment Buildings," *Fire Technology*, Vol. 26, No. 4, Nov. 1990, pp. 342–356.

Terrett, D., "Voice Alarm Loudspeakers," *Fire Surveyor*, Vol. 22, No. 5, Oct. 1993, pp. 12–13.

Thompson, B., "Voice Alarm Systems: The Benefits of Integrated Voice Alarm Systems," *Fire Surveyor*, Vol. 21, No. 6, Dec. 1992, pp. 4–6.

Todd, C., "Voice Alarms and Evacuation," *Fire Prevention*, No. 274, Nov. 1994, pp. 21–25.

UL 464, *Standard for Safety Audible Signal Appliances*, Underwriters Laboratories Inc., Northbrook, IL, 1990.

UL 1971, *Standard for Safety Signaling Devices for the Hearing Impaired*, Underwriters Laboratories Inc., Northbrook, IL, 1992.

FIRE ALARM SYSTEM INTERFACES

Revised by Wayne D. Moore

NFPA 72, the *National Fire Alarm Code* (hereinafter referred to as NFPA 72), requires that all fire protection systems in a protected property be connected to the protected premises fire alarm system. The integration of commercial fire and non-fire protection systems has existed for many years. In some instances, the combination of systems was an attempt to conserve valuable construction dollars through systems integration. In other cases, systems integration was seen as a necessity for ease of building management, e.g., combination security/access control and fire alarm systems or building automation and fire alarm systems. There are some commercially available combination systems that are tested, listed, and manufactured to serve as combination systems. However, there are many systems that provide some form of protection or management of a building, such as elevator control, door control, and heating, ventilating, and air-conditioning (HVAC) control, that often need to be integrated or interfaced with the fire alarm system. These systems are not manufactured, tested, or listed as a single integrated system. Their interface with fire protection systems will be presented in this chapter.

The most common fire protection system that is interfaced to a fire alarm system is the automatic sprinkler system. These important systems are monitored both for waterflow alarms and sprinkler system supervision.

The value of installing an automatic sprinkler system for life safety and property protection has been well documented over many years of service in a variety of specific applications. It is obviously necessary to ensure that the automatic sprinkler system is maintained in such a condition that it is always ready to discharge water on a hostile fire, holding the fire to its area of origin. One way of making certain that an automatic sprinkler system remains fully in service is to supervise its operational features. At the same time, it is necessary to initiate an alarm signal whenever the sprinkler system does operate so that personnel may respond promptly and take whatever action is needed. Thus, these two functions—waterflow alarm and automatic sprinkler system supervision—are vital elements of an overall fire protection program.

Wayne D. Moore, P.E., is vice president of engineering and marketing for MBS Fire Technology, Inc., and is the manager of the Atlanta, GA, office. He is currently chairman of the NFPA 72 Technical Committee for Chapter 3, "Protected Premises Fire Alarm Systems," of the *National Fire Alarm Code*. He is also the editor of the first edition of the *National Fire Alarm Code Handbook* and co-editor of the national fire alarm systems newsletter, *The Moore-Wilson Signaling Report*.

AUTOMATIC SPRINKLER SYSTEM WATERFLOW ALARMS

Waterflow alarm systems may be as simple as a hydraulically operated water motor gong serving a single sprinkler riser, or as complex as a multiplex proprietary fire alarm system covering several hundred sprinkler risers at a very large industrial facility. In either case, there are certain basic functions that must be provided by whatever waterflow alarm equipment is installed. These include:

1. The prompt detection of the discharge of a certain number of gallons per minute (L/min) at nominal pressure that would be equivalent to the amount of water discharged from a single operating sprinkler having the smallest orifice of those sprinklers installed in that particular system.
2. The reliable operation of a suitable audible alarm notification appliance. The exact nature of the particular appliance will depend on the purpose of the specific waterflow alarm system.
3. The provision of suitable facilities to test the operation of the waterflow alarm device, preferably by an actual flow of water equivalent to the discharge from a single automatic sprinkler having the smallest orifice of those sprinklers installed in the system.

Categorizing Waterflow Alarms

Waterflow alarm systems can be categorized according to their purpose and according to their method of operation. For example, the waterflow alarm system may be installed to provide an alarm signal that will alert someone outside the building. Or the waterflow alarm system may be installed to provide an alarm signal that will alert someone in the immediate area of the sprinkler riser. Such a signal would normally be produced by the operation of either a hydraulically operated or an electrically operated bell. The waterflow alarm system also may be installed to provide an alarm signal that will sound throughout the protected building. This will normally be achieved by connecting the waterflow alarm initiating device to a local fire alarm system. The waterflow alarm system may be installed to provide an alarm signal that will be transmitted off-premises to some constantly attended location by means of a signaling system for central station service, an auxiliary fire alarm system, or a remote station fire alarm system. The waterflow alarm system may be installed to provide an alarm signal that will be transmitted to a constantly attended location on the protected premises or to another location of the same owner. This will normally be accomplished by connecting the waterflow alarm initiating device to a proprietary fire alarm system.

Hydraulically Operated Waterflow Alarms

Hydraulically operated waterflow alarm notification appliances are called "water motor gongs." These devices operate very much like a miniature version of the water wheel that operated the mill shafting of a textile mill at the turn of the century. When an automatic sprinkler fuses in a fire, when a pipe breaks, or when a fitting develops a significant leak, water is diverted through an open alarm control valve to a nozzle assembly that directs a stream of water against the paddles of a water wheel. As the wheel turns, it operates a striking mechanism that repeatedly strikes the shell of a large gong. This produces a loud clanging sound that can be heard in the vicinity of the sprinkler riser. Most often, water motor gongs are mounted so that the gong shell is actually on the outside of the building. In countless instances of sprinkler operation, passersby have heard the water motor gong ringing and have telephoned the police or fire department. (See Figure 5-4A.)

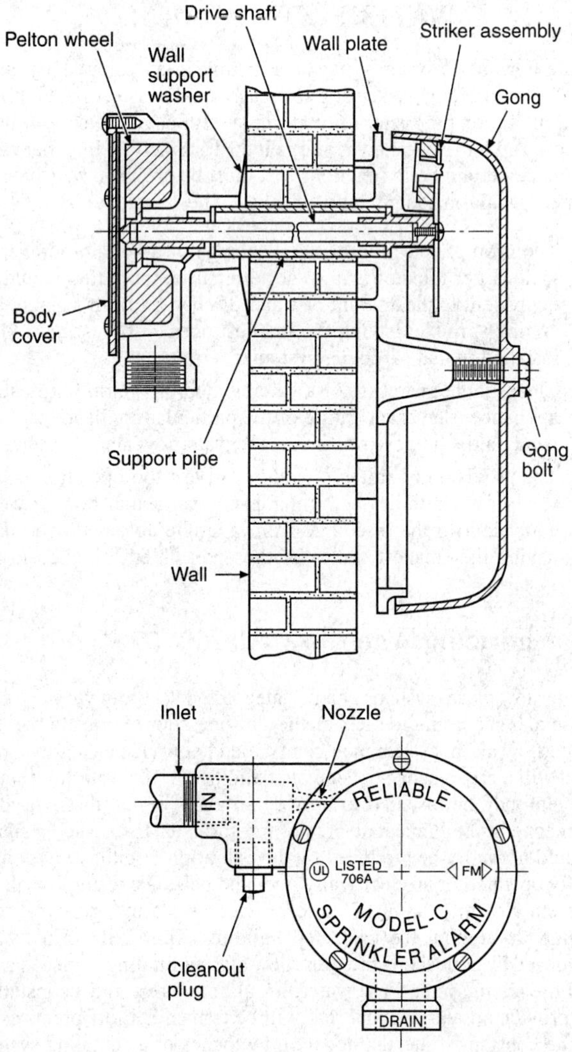

FIG. 5-4A. *A representative water motor gong. Shown is the Reliable Automatic Sprinkler Co., Inc., Model C. Other manufacturers have similar arrangements of proprietary equipment for water motor gong installations.*

Electrically Operated Waterflow Alarms

Waterflow alarms may be signaled using electrically operated fire alarm notification appliances, such as vibrating bells, horns, sirens, chimes, and several varieties of flashing lights. Such signals usually are initiated by electrical switches incorporated into some type of pressure- or flow-operated device.

Waterflow Alarms for Wet-Pipe Automatic Sprinkler Systems

For wet-pipe automatic sprinkler systems, the flow of water is generally detected either hydraulically by an alarm check valve or electrically by a vane-type waterflow alarm initiating device. Either of these would be installed at the point the water supply piping enters the building from the underground. Supplementary vane-type devices could also be installed at various points in the piping system, e.g., on each floor of a high-rise building.

Alarm Check Valves

Depending on the specific design, alarm check valves are generally referred to as being either "plain type" or "differential type." (See Section 6, Chapter 10, "Automatic Sprinkler Systems," for more information on alarm check valves.) The basic design of most alarm check valves incorporates a clapper that lifts from its seat when water is discharged from sprinkler system piping above the valve. The movement of the clapper is used in one of the following ways:

1. The valve seat ring can have a concentric groove with a pipe connection from the groove to an alarm outlet or port to which piping can be connected. Such valves are commonly called "differential" or "divided seat ring" alarm check valves. When the clapper of the alarm valve rises in response to discharge from the sprinkler piping above the valve, water also flows through the groove in the divided seat ring, out of the alarm port, and into the alarm piping. (See Figure 5-4B.)

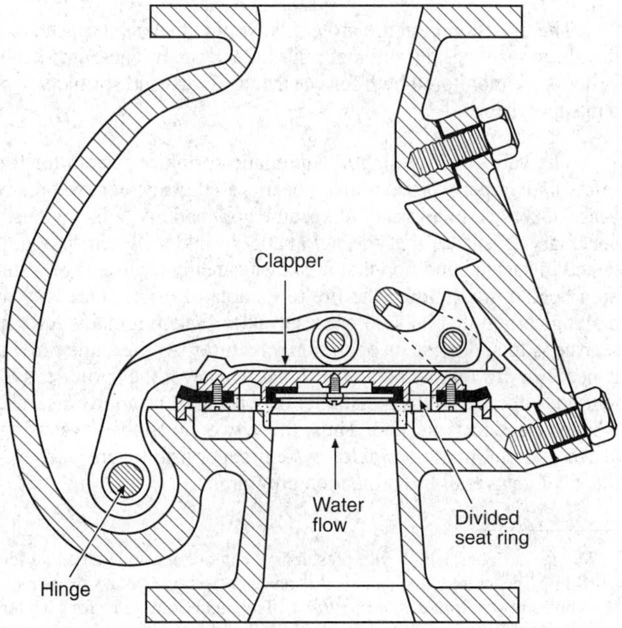

FIG. 5-4B. *A waterflow alarm check valve of the differential type. (Source: Firematic Sprinkler Devices, Inc.)*

2. The clapper of the alarm check valve may have an extension arm with a smaller clapper that covers the seat of a small pilot valve connected to an alarm outlet or port. Alarm piping may be connected to this port. When the clapper of the alarm valve rises in response to discharge from the sprinkler piping above the valve, the extension also rises, lifting the smaller clapper off the seat of the pilot valve. Water flows through the pilot valve, out of the alarm port, and into the alarm piping. (See Figure 5-4C.)

In either case, as the clapper lifts, it uncovers a passageway through which water can flow through an open alarm control valve to the water motor gong. To help reduce false alarms due to surges, the water may pass through a retard chamber after it passes through the open alarm control valve on its way to the water motor gong. This device is simply a large cast-iron or steel "bottle" that is equipped with a normally open automatic drain. As water flows into the retard chamber, the automatic drain shuts off. The time it takes for the water to fill up the chamber is the amount of retard time provided for the system. Once the chamber is filled, the water continues to flow to the water motor gong, operating that appliance. Once the water stops flowing, the automatic drain of the retard chamber returns to its normally open position, allowing the retard chamber to drain so that it is ready to be used the next time the clapper of the alarm check valve opens. (See Figure 5-4D.)

Pressure-Operated Switch Waterflow Alarm Initiating Devices

A pressure-operated switch may be mounted at the top of the retard chamber to provide a means of initiating the operation of an electric bell mounted in the vicinity of the riser. (See Figure 5-4E.) It is important to understand that such a switch mounted at the top of the retard chamber would not be permitted to initiate an alarm signal for a central station, local, auxiliary, remote station, or proprietary fire alarm system. Each of these systems requires that the waterflow alarm initiating device be located on the system in such a way that it cannot be tampered with or removed without initiating a signal. Because the water flowing to the retard chamber passes through a small alarm control valve that could be shut off—stopping the flow

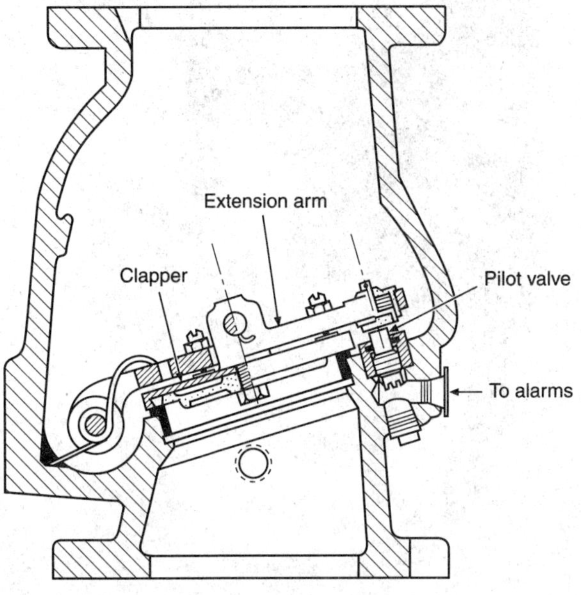

FIG. 5-4C. An alarm check valve with an auxiliary valve and pipe connection to alarms. Shown is the "Automatic" Sprinkler Co. Model 253 alarm check valve (no longer manufactured).

of water—this nontamper/nonremoval requirement of the signaling standards could not be met.

A pressure-operated switch could be mounted on a fitting connected to the alarm port of the alarm check valve, as long as there is no shutoff valve in the connection. In such a case, it would normally be necessary to provide an electronic or electro-pneumatic retard mechanism, either built into the pressure-operated switch or installed at the fire alarm system control unit, if one exists. A pressure-operated switch so located could be used to initiate a fire alarm signal on a central station, local, auxiliary, remote station, or proprietary fire alarm system.

Vane-Type Waterflow Alarm Initiating Devices

Far and away the most common electrically operated waterflow alarm initiating device is the vane-type waterflow switch. (See Figure 5-4F.) As a substitute for the hydraulically operated water motor gong, an electrically operated bell may be arranged to serve a wet-pipe automatic sprinkler system by connecting the bell to a vane-type waterflow alarm initiating device. These devices are designed to be inserted into the riser just above the point at which the water supply piping enters the building from underground. Even if the wet-pipe system has been equipped with an alarm check valve, a vane-type waterflow alarm initiating device may still be used. In this case, it normally would be installed just above the alarm check valve.

A small hole approximately 1 in. (25 mm) in diameter is drilled into the riser. The drill bit has a special piece of spring steel wire that captures the metal center of the hole, which is called the "coupon." Once the coupon is removed from the drill bit, it is usually attached to one of the mounting bolts of the waterflow alarm initiating device to show future inspectors that it was not left inside the pipe where it might possibly move upward into the sprinkler system and become an obstruction to the proper operation of the sprinkler system. The plastic or thin metal vane is rolled up and inserted into the riser. Once it resumes its original shape, it forms a thin barrier that covers almost all of the interior of the riser. When a sprinkler fuses, water flowing past the vane will push the vane upward, actuating a switch. When the switch contacts close, they initiate an alarm.

Two problems that may occur occasionally with vane-type waterflow alarm initiating devices are leaks around the point of entrance into the riser and detached vanes. The laboratories that test and list such devices require that each vane have a supplementary means of attaching the vane to the device so that, even if the vane breaks off, it will be retained near its original point of connection.

Waterflow Alarms for Dry-Pipe, Preaction, and Deluge Automatic Sprinkler Systems

Automatic sprinkler systems equipped with dry-pipe valves, preaction valves, or deluge valves also must have appropriate waterflow alarms. Each of these types of valves has a passageway that is uncovered when the clapper of the valve trips into the open position. Since these three types of valves are not affected generally by surges in the water system, no retard chamber is necessary. Piping connected to the outlet from the alarm passageway of these valves may be attached to a water motor gong or to a pressure-operated switch.

Vane-type waterflow alarm initiating devices are not permitted to be used for dry-pipe valve, preaction, or deluge sprinkler systems. This prohibition comes from concern that the hydraulic shock that occurs when these valves trip would increase the possibility that the vane might become detached.

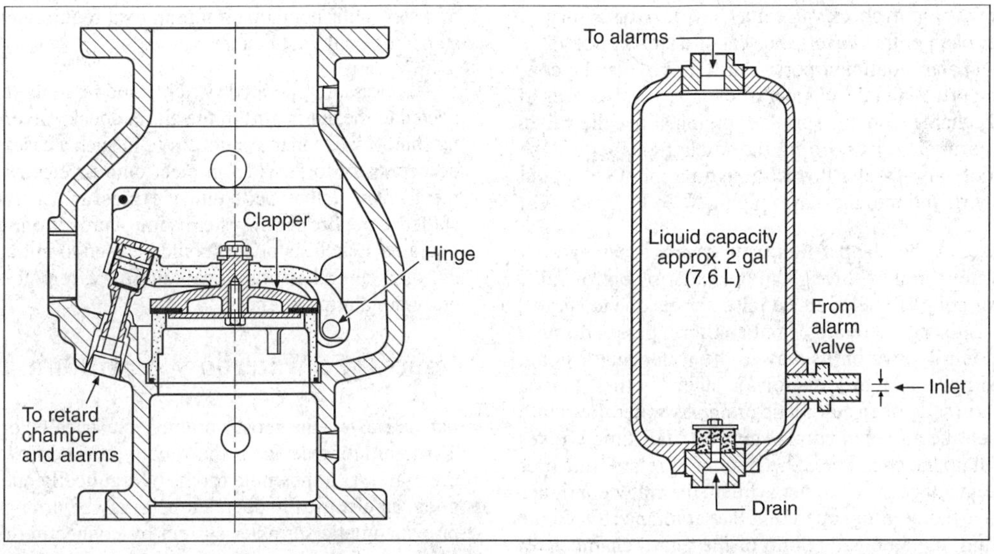

FIG. 5-4D.   A waterflow alarm valve and retard chamber. Shown are Viking Models F-1 and B-3.

Drop-in-Pressure Waterflow Alarms

An alternative arrangement designed to initiate a waterflow alarm, even if the main automatic sprinkler system control valve is shut, is the "drop-in-pressure" waterflow alarm. The supervision of all fire protective system control valves 2½-in. (63.5-mm) nominal size or larger is considered very important, especially when a fire alarm system is used to supplement or supplant standard guard service. Occasionally, it is simply not possible to obtain this valuable supervision. In such cases, it is essential that some method of detecting sprinkler system waterflow be used that will initiate an alarm even if the sprinkler control valve is shut. One method of accomplishing

this is to use a waterflow alarm initiating device that detects a drop in sprinkler system pressure as a means of detecting the discharge of water from the system.

Such devices may use an ancillary storage chamber connected to the sprinkler system through a restricted orifice. Under normal conditions, the pressure in the ancillary chamber is the same as that in the sprinkler system. When system pressure drops, the pressure

FIG. 5-4E.   A pressure-operated waterflow alarm switch.

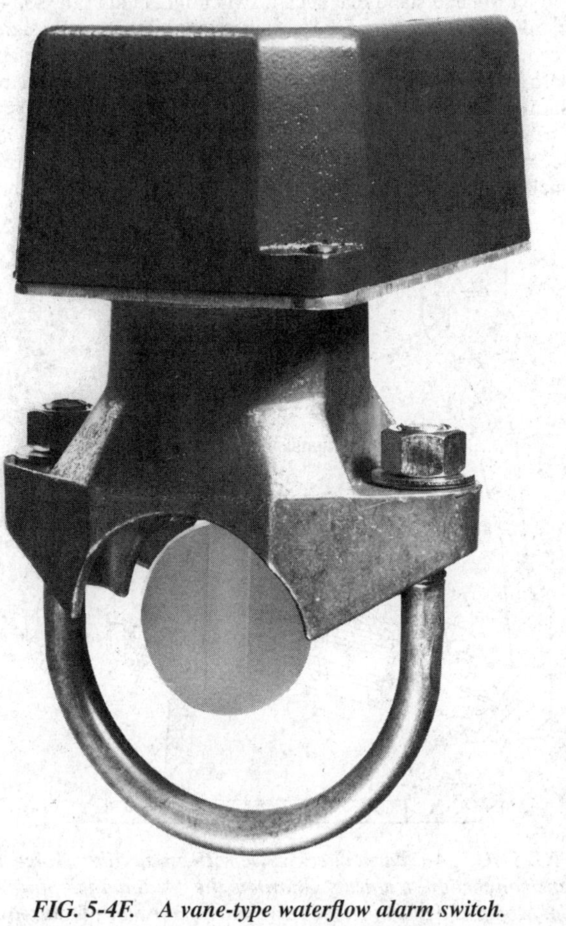

FIG. 5-4F.   A vane-type waterflow alarm switch.

stored in the chamber acts on a diaphragm to activate an alarm initiating switch.

Other such devices may employ a pressure-sensing transducer and associated electronic circuitry to detect the rate of pressure change in the system. A slow rate of change is reported as a trouble signal, while a rapid rate of change is reported as an alarm.

The most common drop-in-pressure waterflow alarm initiating device uses a pressure switch to sense the pressure in the system and, for wet-pipe systems, employs a small riser-mounted excess pressure pump. (An excess pressure pump is not needed for dry-pipe systems.) When the water pressure in a wet-pipe system or the air pressure in a dry-pipe system drops to a predetermined pressure setting, the pressure switch initiates an alarm signal. (See Figure 5-4G.)

It is possible for drop-in-pressure-type waterflow switches to be incorrectly installed on sprinkler systems supplied by fire pumps where the small excess pressure pump is not provided. For the

Pressure indicating pilot lights
Pressure connection to sprinkler system
Differential pressure switch pump control
Water discharge to sprinkler system
Excess pressure pump
Water inlet
Low pressure pump shut-off switch

FIG. 5-4G.   An excess pressure pump for installation on a wet-pipe sprinkler system. The automatic excess pressure pump maintains a pressure 29 to 47 psi (200 to 325 kPa) in excess of the water supply pressure above an alarm check valve. A continuous white pilot light, controlled by a differential switch, indicates when the pressure in the sprinkler system is more than 29 psi (200 kPa) above the supply pressure. Should the water supply fail, a low-pressure switch shuts off the current to the pump so that it will not run without water. This switch also controls a red pilot light to show when the supply pressure drops below 29 psi (200 kPa). A second pressure switch on the sprinkler system (not a part of the pump unit) is connected into the sprinkler system alarm circuit and gives an alarm if system pressure fails completely (e.g., a closed water control valve) when a sprinkler operates. A third pressure switch gives a warning by a local bell or other means in the event the pressure pump fails to maintain system pressure within the established excess pressure range. (Source: Gamewell Corp.)

switch to work in this case, the fire pump starting pressure must be lowered below the waterflow alarm switch setting. If this is not done, it is possible for the fire pump to start before the drop-in-pressure switch initiates an alarm.

Since most Authorities Having Jurisdiction will want the fire pump starting pressure be set as close as possible to the fire pump churn pressure to minimize hydraulic shock, a drop-in-pressure-type waterflow switch should not be used on wet-pipe sprinkler systems supplied by a fire pump, unless the small excess pressure pump is also installed and arranged to discharge above each alarm check valve. One excess pressure pump can supply two or more adjacent risers by connecting the discharge pump to a point above the alarm check valve through a small check valve with a hole drilled in the clapper.

Drop-in-pressure-type waterflow alarm initiating devices should be connected to the sprinkler system through an automatic stop-and-waste valve. Such a valve automatically "wastes" the pressure trapped between the valve and the device to the atmosphere. Thus, a signal is initiated whenever the valve is closed. This eliminates the possibility of tampering with the device without producing a signal.

Testing Waterflow Alarm Initiating Devices

The piping connected to the alarm outlet from alarm check valves, dry-pipe valves, preaction valves, or deluge valves usually is connected also through a normally closed valve to a point of water supply below the alarm check valve, dry-pipe valve, preaction valve, or deluge valve. Opening this test bypass valve will allow water to flow into the alarm piping, actuating the water motor gong or pressure-operated switch. Unless the dry-pipe valve, preaction valve, or deluge valve is going to be performance tested with a full actuation of the system, opening the alarm test bypass valve is the only way to test the alarms. In the case of a dry-pipe valve, care must be taken to be certain that the automatic velocity drain from the intermediate chamber of the dry-pipe valve is not stuck in the closed position and that the check valve at the alarm outlet from the dry-pipe valve intermediate chamber is not stuck in the open position. The combination of these two faults will most likely cause the dry-pipe valve to trip when the alarm test bypass valve is opened. This tripping occurs when water from the test bypass valve flows back into the intermediate chamber of the dry-pipe valve, destroying the differential between the air seat and the water seat inside the valve.

For wet-pipe automatic sprinkler systems, the only proper way to test the operation of a vane-type waterflow alarm initiating device, and the preferable way to test all other types of waterflow alarm initiating devices, is to open the valve at the inspector's test connection. This valve, installed on a 1-in. (25-mm) piece of pipe—for older installations, it is often located at the most remote point of the sprinkler system, and, for more modern installations, it is often located at the top of the riser—terminates outside the building in an orifice that can simulate the discharge from a single sprinkler having the smallest orifice used by those sprinklers installed on the system. Thus, opening the inspector's test connection will simulate the fusing of a single sprinkler.

SPRINKLER SYSTEM SUPERVISION

To ensure that an automatic sprinkler system is ready to discharge water on a hostile fire and hold the fire to its area of origin, a central station, local, remote station, or proprietary fire alarm system may be connected to supervisory initiating devices that will supervise critical operating features of the sprinkler system. Such devices may include the supervision of:

1. The position of sprinkler shutoff valves. Normally, only those valves that are 2 1/2 in. (63.5 mm) or larger are supervised. This

supervision is provided by mounting a supervisory initiating device on the valve that will initiate an off-normal signal whenever the valve is operated into a closed position, and that will initiate a restoration-to-normal signal whenever the valve is returned to a fully open position. When mounting such an initiating device where it will come in contact with the threaded shaft of the valve, care must be taken that the device is adjusted so that it will not falsely initiate a restoration signal if the shaft stops turning with the sensing portion of the device between two of the threads. This is a particularly troubling problem with larger sizes of control valves. (See Figures 5-4H and 5-4I.)

2. The air pressure in a dry-pipe sprinkler system or water pressure tank and the supervisory air pressure in a preaction sprinkler system. This is accomplished by connecting an air pressure supervisory initiating device to the system being supervised. Such a device is designed to initiate a supervisory off-normal signal whenever the air pressure climbs above or drops below a designated set-point. Conversely, the device will initiate a restoration-to-normal signal whenever the air pressure is restored to the valve at which the device is set. During installation, it is impor-

tant that provision be made to allow testing of the device by simulating its operation during abnormal conditions. One method of accomplishing this is to connect the device to the sprinkler system through a stop-and-waste valve. When operated, such a valve automatically "wastes" the pressure trapped between the valve and the device to the atmosphere. Thus, whenever the valve is closed, a signal is initiated. This eliminates the possibility of tampering with the device without producing a signal.

3. Low temperature in a building equipped with a wet-pipe automatic sprinkler system, in a closet protecting a dry-pipe valve, preaction valve, or deluge valve from freezing, or of the water inside a storage tank. This is accomplished by mounting a temperature supervisory initiating device at the location to be supervised. Such a device is designed to initiate a supervisory off-normal signal whenever the temperature drops below a designated set-point. Conversely, the device will initiate a restoration-to-normal signal whenever the temperature is restored to the value at which the device is set. (See Figure 5-4J.)

4. The level of the water in a storage tank. This is accomplished by mounting a water-level supervisory initiating device at the loca-

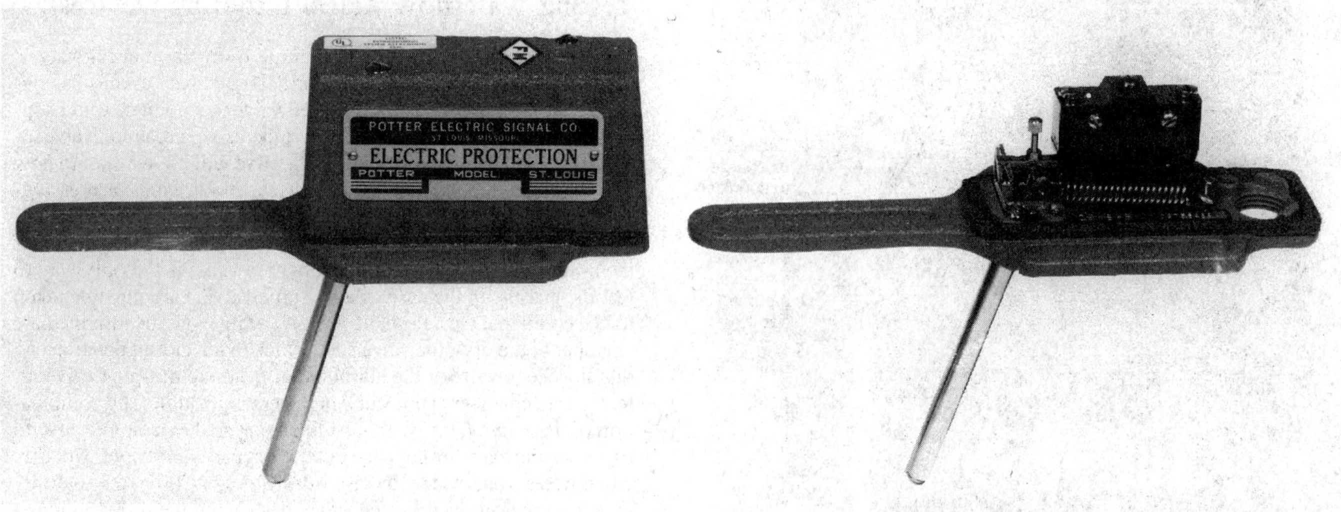

FIG. 5-4H.   A valve supervisory switch for an outside screw and yoke (OS&Y) valve.

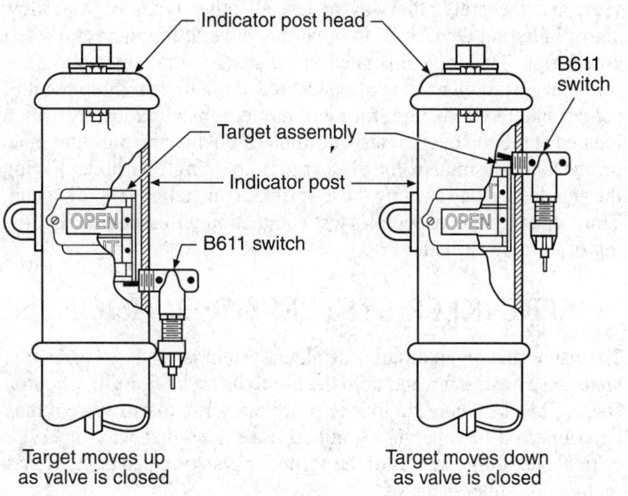

FIG. 5-4I.   A valve supervisory switch for a post-indicator valve (PIV).

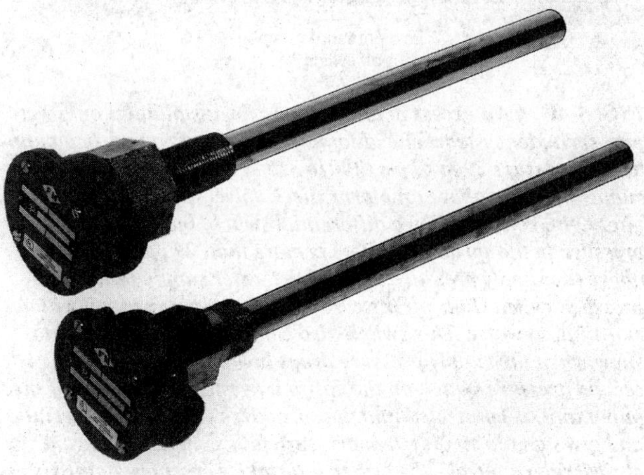

FIG. 5-4J.   Low-temperature switches for a water storage tank.

tion to be supervised. Such a device is designed to initiate a supervisory off-normal signal whenever the water level rises above or drops below a designated set-point. Conversely, the device will initiate a restoration-to-normal signal whenever the water level is restored to the value at which the device is set. During installation, it is important that provision be made to allow testing of the device by simulating its operation during abnormal conditions. (See Figure 5-4K.)

5. The operation and integrity of a fire pump. This is accomplished by connecting supervisory initiating device circuits to the "remote signal" contacts inside the fire pump controller. When an electric motor-driven fire pump operates into a running condition or when the power to the motor is interrupted, distinct supervisory signals are initiated. Similarly, when an engine-driven fire pump operates into a running condition—that is, its main switch is placed in a position other than "automatic start"—the battery charger fails, the engine experiences an abnormal condition or fails to start, or the controller fails, distinct supervisory signals are initiated. Conversely, a restoration-to-normal signal will be initiated whenever the particular feature of the fire pump that is off-normal is restored to normal.

NFPA 72 requires that three signals be identified at the fire alarm control unit: (1) alarm, (2) supervisory, and (3) trouble. In order to comply with that requirement, supervisory initiating devices must not be on the same initiating device circuit as alarm initiating devices. A common wiring configuration has been to install the waterflow alarm initiating device on the same circuit with the supervisory initiating device, e.g., an OS&Y supervisory switch. (See Figure 5-4H.) Although electrically this configuration may work, the wiring configuration with both devices on the same circuit will not provide the differentiation between a broken connection (trouble condition) and a closed valve (supervisory condition). If the wiring configuration and panel operation (such as with addressable devices) can provide the signal differentiation required by NFPA 72, then the devices can be wired on the same electrical circuit.

Supervisory Signal Initiation

In addition to outside screw and yoke (OS&Y) valve and post-indicator valve (PIV) supervision, automatic sprinkler systems and other fire suppression systems are supervised for conditions that are essential for the proper operation of the systems. Functions such as valve position, water temperature, building temperature, and high and low air pressure are required to be monitored for off-normal conditions.

Signal Initiation from Automatic Fire Suppression Systems Other Than Waterflow

There are many other types of fire suppression systems that may be found in a protected premises, such as a kitchen hood/duct suppres-

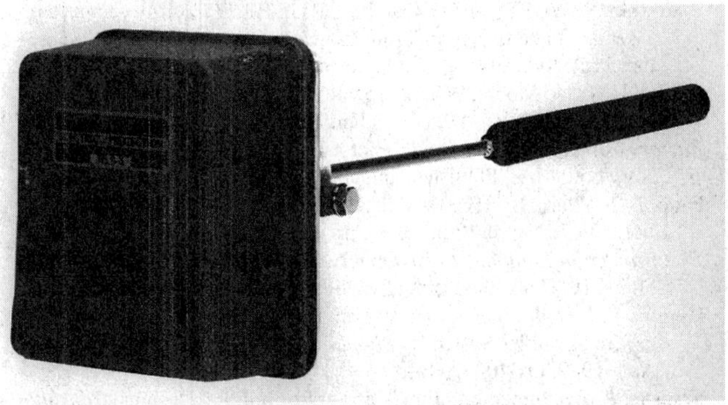

FIG. 5-4K. *A low-water-level switch for a water storage tank.*

sion system or a carbon dioxide or other gaseous suppression system. Each of these fire suppression systems must be connected to the protected premises fire alarm system to initiate an alarm. Each of these systems must also be monitored for supervision for proper operation. The supervisory devices must be listed for the intended application and installed in accordance with NFPA 72, Chapter 5.

Automatic fire pumps and special service pumps must be supervised in accordance with the requirements of NFPA 20, *Standard for the Installation of Centrifugal Fire Pumps*, and the Authority Having Jurisdiction.

Other Systems Interface

NFPA 72, Chapter 3, also provides guidance for installation requirements of fire safety control functions that will be interfaced with fire alarm systems. These fire safety control functions include fan control, door control, smoke control, etc. The four most important areas of concern when interfacing the fire alarm control unit with a fire safety function control device are: (1) the relay used to initiate control must be located within 3 ft (0.9 m) of the controlled circuit or device; (2) the relay must be listed for the intended purpose; (3) the relay must be compatible with the fire alarm control unit; and (4) the wiring between the fire alarm control unit and the relay must be monitored for integrity, unless the control device is connected in a fail-safe manner.

Elevator Recall

Fire alarm systems are interfaced to elevator control systems to activate the elevator recall sequence. The recall sequence and initiation requirements are found in Chapter 3 of NFPA 72. In most jurisdictions, only the elevator lobby and machine room smoke detectors are required to initiate recall of the elevators. The standard that originally provided the smoke detector placement requirements and requires the elevators to be recalled is ANSI/ASME A17.1, *Safety Code for Elevators and Escalators* (hereinafter referred to as ANSI/ASME A17.1). The ANSI/ASME A17.1 committee has now developed a liaison relationship with the NFPA 72, Chapter 3, committee, and all detection and interface requirements are contained in NFPA 72, Chapter 3.

Elevator Shutdown

When the elevator machine room is protected with automatic sprinklers, ANSI/ASME A17.1 requires that the power to the elevators must be shut down prior to or upon discharge of water from the sprinklers. The power shutdown can be accomplished in either of two ways. When heat detectors are used, they must be installed within 2 ft (0.6 m) of each sprinkler, have both a lower temperature rating and a higher sensitivity than the sprinkler head, and be installed in accordance with NFPA 72, Chapter 5. When pressure or waterflow switches are used to shut down the elevator power, they must not have a time delay associated with their operation. In either case, it is assumed that elevator recall will have already taken place through the operation of the smoke detectors required for the elevator recall function.

Interconnected Fire Alarm Control Units

There are numerous occasions where it is necessary or required by design to interconnect fire alarm control units. This function may occur during an addition to a building, such as a new wing to a hospital, or during a renovation to an existing building. In the past, there was no guidance as to the proper way to interconnect different manufacturers' control equipment. What all too often happened in the field was makeshift interconnections that were improper, unreliable, and unsupervised. NFPA 72 requires that the interconnected fire alarm control units be connected by listed methods and the interconnection must be monitored for integrity. In addition to monitoring the inter-connected system for alarm conditions, the interconnected control must be monitored for conditions that are essential for the proper operation of the interconnected system.

When an owner or designer suggests that two different manufacturers' control units be interconnected, the Authority Having Jurisdiction would be wise to ask each manufacturer for a letter indicating the approved interconnection method or an approved drawing showing the approved interconnection method. If the interconnection is being made using relay configurations, the relays used must be listed and compatible with the control units being interconnected.

Door Release Service

When smoke and fire doors are held in the open position to allow free movement in corridors of buildings, it is required that the doors be released when a fire alarm system is activated. These hold-open devices are commonly magnetic units that are normally powered from a 24- or 120-Vac supply that is interconnected through a relay in the smoke detector or in the fire alarm control unit. One of the common mistakes that is found when these units are installed is the perception that the door hold-open devices be connected to the standby power for the fire alarm system. This is not required by NFPA 72, because the units are supposed to operate in a fail-safe fashion. When power is lost, the hold-open units deactivate and the doors close.

Door Unlocking Devices

Based on the danger of fire deaths occurring in smoke-filled stairways where occupants are trapped after discovering that they have entered an unsafe area and cannot return to the corridor, NFPA 101®, Life Safety Code®, requires all stairway exit "locked doors" be unlocked when an alarm occurs in the building. Once again, the proper way to interconnect these unlocking devices is so that when power is lost, they unlock. This is a fail-safe operation and when the unlocking devices are wired in this fashion, the interconnecting wiring does not require monitoring for integrity.

Computer-Controlled Building Systems

The computerization of building control systems has led to the integration of these systems to the fire alarm system control unit. Because the fire safety control function devices must be listed for compatibility with the fire alarm system control unit, so as not to interfere with the fire alarm system's operation, the system designer and the installer must be careful not to interconnect computerized devices together that have not been listed for compatibility. One of the common areas that presents this problem is the interconnection of the fire alarm system to the heating, ventilating, and air-conditioning (HVAC) system direct digital control (DDC) devices used to control fire/smoke dampers. This is done instead of connecting the fire alarm system directly to the damper control. Due to the compatibility requirement and, more logically, due to non-fire alarm system technician/programmers working on the interface, the DDC type of interconnections should be approached very carefully.

Non-Fire Alarm Interfaced Equipment

There are times when an owner may wish to interconnect a process monitoring system or some other non-fire alarm-related system to the fire alarm system. Caution should be exercised in this area, and the authority having jurisdiction should be consulted. If the system being interfaced is monitoring something that will affect the life safety of the occupants or the protection of the property, then it may be required to have that system initiate an alarm condition on the fire alarm system. Otherwise, these non-fire signals must be considered supervisory in nature.

Testing Interfaced Systems

Whether the interfaced system is a security system or a special hazard suppression system, care must be taken when performing acceptance and periodic testing of the fire alarm system. The technicians performing the tests must be totally familiar with all of the systems interfaced to the fire alarm system or must have other technicians present for the test who are familiar with the interfaced systems.

Interfaced systems equipment connections must be tested by operating or simulating the operation of the equipment being supervised. Signals required to be transmitted must be verified at the control panel. The test frequency for the interfaced equipment is governed by the applicable NFPA standard(s) for the equipment being supervised by the fire alarm system.

Summary

There are many fire protection systems, fire safety control functions, and process monitoring systems that may be interfaced with a building's fire alarm system. The purpose of this section is to make the reader aware of the need to address the interface requirements early in the design stage to ensure the reliable operation of both the fire alarm system and the systems being interfaced to the fire alarm system. Ensuring compatibility and correct operation during the acceptance tests is encouraged and is important for reliable operation.

BIBLIOGRAPHY

NFPA Codes, Standards, and Recommended Practices

Reference to the following NFPA codes, standards, and recommended practices will provide further information on the interfacing of fire alarm systems. (See the latest version of *The NFPA Catalog* for availability of current editions of the following documents.)

NFPA 13, *Standard for the Installation of Sprinkler Systems*
NFPA 20, *Standard for the Installation of Centrifugal Fire Pumps*
NFPA 70, *National Electrical Code®*
NFPA 72, *National Fire Alarm Code*
NFPA 101®, *Life Safety Code®*

Reference

ANSI/ASME A17.1, *Safety Code for Elevators and Escalators*, Rules 211.3 Through 211.8, American National Standards Institute, New York, 1993.

Additional Reading

Bryant, P., "British Standard Code of Practice For the Design, Installation and Servicing of Fire Detection and Alarm Systems in Buildings, BS 5839 PART 1:1988," Third International Symposium on Fire Protection of Buildings, May 10–12, 1990, Eger, Hungary, 1990, pp. 129–134.
Bukowski, R. W., P.E., and O'Laughlin, R. J., P.E., *Fire Alarm Signaling Systems*, 2nd ed., National Fire Protection Association and Society of Fire Protection Engineers, Quincy and Boston, MA, 1994.
Cholin, J. M. and Moore, W. D., "How Reliable Is Your Fire Alarm System," *NFPA Journal*, Vol. 89, No. 1, Jan./Feb. 1995, pp. 49–53.
Factory Mutual Approval Guide, Factory Mutual Research Corporation, Norwood, MA. (Published annually.)
Faup, J. J., "Alarm Systems Ensure Safe Elevator Recall," *NFPA Journal*, Vol. 87, No. 3, May/June 1993, pp. 28, 104.
Fire Protection Equipment List, Underwriters Laboratories Inc., Northbrook, IL. (Published annually, with bimonthly supplements.)
Higgins, P. J., and Ballanco, J. A., "Backflow Protection for Water Supply to Fire Sprinkler Systems," *Plumbing Engineer*, Vol. 17, No. 6, July–Aug. 1989, pp. 40–42, 46–47.
Transue, R. E., "Integrating Fire Alarms into Building Design," *Consulting-Specifying Engineer*, Vol. 14, No. 7, Dec. 1993, pp. 36–38, 40.

FIRE ALARM SYSTEMS INSPECTION, TESTING, AND MAINTENANCE

John M. Cholin

With each passing year, improvement in the understanding of fire and its behavior enables designers to better design fire alarm systems to achieve specific objectives and levels of performance rather than compliance with a minimum prescriptive standard. Yet no performance-based or objective-oriented design is complete without an explicit quantitative assessment of the level of confidence one can have that the fire alarm system will perform as intended. All computational methods that have been developed to assess this parameter are predicated upon an inspection, testing, and maintenance program at specified regular and predictable times throughout the life of the system. Consequently, without a thorough understanding of the impact the inspection, testing, and maintenance program has on the mission effectiveness of the fire alarm system, one cannot utilize a performance-based, objective-oriented design approach.

This chapter first develops the concepts of reliability analysis and prediction that form the foundation upon which all inspection, testing, and maintenance programs should be based, then reviews the types of electronic components from which fire alarm systems are constructed. These two background areas form the basis for inspection, testing, and maintenance requirements for fire alarm systems, components, and scheduling. Further, design concepts are applied to the household fire alarm system. Finally, this chapter addresses criticality of inspection, testing, and maintenance on the mission-effectiveness of the system.

This chapter is specifically limited to inspection, testing, and maintenance of fire alarm systems. These are the electronic and electrical systems that are designed: (1) to sense a fire, transmit that information to a control unit of some type, and activate personnel warnings, whether for persons occupying the site or at a remote location; and (2) for remedial response, such as the discharge of extinguishing systems and activation of devices designed to enhance the compartmentation of the fire.

This chapter does not cover maintenance of specific remedial response systems. Inspection, testing, and maintenance of these systems is addressed in: NFPA 11, *Standard for Low-Expansion Foam;* NFPA 12, *Standard on Carbon Dioxide Extinguishing Systems;* NFPA 12A, *Standard on Halon 1301 Fire Extinguishing Systems;* NFPA 13, *Standard for the Installation of Sprinkler Systems;* NFPA 15, *Standard for Water Spray Fixed Systems for Fire Protection;* NFPA 16, *Standard for the Installation of Deluge Foam-Water Sprinkler and Foam-Water Spray Systems;* NFPA 17, *Standard for*

Dry Chemical Extinguishing Systems; NFPA 20, Standard for the Installation of Centrifugal Fire Pumps; NFPA 25, Standard for the Inspection, Testing, and Maintenance of Water-Based Fire Protection Systems; and NFPA 2001, Standard on Clean Agent Fire Extinguishing Systems.

RELIABILITY

Murphy's Law is familiar: anything that can go wrong will go wrong. It is an inescapable fact that any "system" will suffer a failure of one of its constituent components at some moment during its design lifetime. This establishes an incontrovertible requirement that a regimen of inspection, testing, and maintenance be employed with every fire alarm system.

The driving force behind every inspection, testing, and maintenance program is the need to maximize system reliability. Reliability is essential for fire alarm systems, simply due to mission objectives for these systems. Fire alarm systems are intended to fulfill three essential objectives: (1) ensure life safety, (2) conserve property, and (3) ensure continuity of the mission of the site. The life safety objective is achieved through timely warning of occupants. This is usually the highest priority. Mitigation of property damage is achieved through the timely activation of automatic fire extinguishing systems and transmission of fire alarm signals to the fire service. This is generally the second priority. After the first two objectives are met, the third objective, i.e., continuity of the mission of the site, becomes attainable. In many cases, the mission objective places greater demands on the fire alarm system, as this objective cannot be attained if the system is unstable; nor can it be attained if the system allows too large a fire to develop prior to the activation of fire extinguishing systems. These three objectives can only be achieved when the fire alarm system operates reliably.

The reliability of a system includes both the ability to: (1) detect and correctly respond to every occurrence of fire and (2) not render a fire alarm indication except when the legitimate fire alarm stimulus actually occurs. For a system to be deemed reliable, alarm signals will occur if, and only if, there is a fire. For the purposes of this chapter, reliability is defined as the measure of the certainty that the system will provide the appropriate response to the conditions that occur, as they occur, during the defined lifetime of the system.

Experience has shown that reliability of a system is the result of reliability contributions from four system elements: (1) design, (2) equipment, (3) installation, and (4) maintenance. Each of these elements contributes to the overall reliability of the system. Often

---

John M. Cholin, P.E., is an independent fire protection consultant and engineer with J.M. Cholin Consultants, Inc., Oakland, NJ.

some "problems" are designed into the system. For example, a smoke detector located above a normally occurring source of aerosols that are not indicative of a fire is a problem that is designed into the system. This represents a reduction of the system reliability that occurred during the design process. Other examples of design problems are: rate-of-rise heat detectors in an area where rapid and wide fluctuations in temperature are expected, control panels mounted on surfaces that are vibrating, or smoke and heat detectors improperly mounted relative to the ceiling plane. Initial inspection and testing of the system must identify these problems that are designed into the system, so they can be corrected before the system is accepted.

Other problems are introduced during installation. These also deduct from the attainable system reliability. For example, installation-generated problems result from the use of substandard wiring, general-purpose conduit and boxes where weathertight materials are needed, insufficient torque on screw terminals, abraded wire insulation, etc. The initial inspection and test of the system should uncover these types of problems. If they are not corrected prior to acceptance, they will create problems over the operational lifetime of the installation. Furthermore, design- and installation-related problems may not always be apparent during initial inspection. These errors are most likely discovered as a result of a system failure or in the context of the regularly scheduled maintenance program.

These examples illustrate how necessary an initial inspection and test program is to uncover problems that are literally designed or installed into the system.

There is another source of failures: the equipment itself. Most of today's electronic fire alarm system devices are extremely reliable. They are mature technologies, produced under stringent quality assurance programs that are audited by the nationally recognized testing laboratories, such as Underwriters Laboratories Inc. (UL) and Factory Mutual Research Corporation (FMRC). However, all electronic components manifest a statistical failure rate that results in failures of these devices over time. These failures occur after the system has been accepted and consequently can only be uncovered through a program of regular maintenance.

The inspection, testing, and maintenance of fire alarm systems are critical to achieving the design objectives of the system. This is the final element affecting the reliability of every system. A fire alarm system without an inspection, testing, and maintenance program is every bit as incomplete as a car with only three wheels.

While inspection, testing, and maintenance schedules have traditionally been developed based upon the qualitative judgment of technical committee members, the emergence of performance-based and objective-oriented designs demand a more rigorous method of assessing the reliability of a system. Performance-based design requires that the inspection, testing, and maintenance schedules provide the reliability necessary to achieve the performance objectives of the design. If reliable performance is to be achieved from a fire alarm system, a rigorous inspection, testing, and maintenance program on a schedule that is derived from the computed reliability of the system and its components is crucial.

Reliability Prediction Techniques For Fire Alarm Systems

The reliability of a fire alarm system, like any electronic system, can be computed using a method developed from the Lusser Product Law. This concept was developed during World War II by a German rocket scientist, named Lusser, who discovered that the reliability of a system was the product of the reliabilities of the individual components. As the number of components increased, the reliability of each individual component had to be improved commensurately in order to maintain the reliability of the overall system. The Lusser Product Law is expressed mathematically by

$$P_r = P_1 \times P_2 \times P_3 \times \lambda \times P_n$$

where

P_T = probability of successful operation for the system, or reliability;

P_n = probability of successful operation of the nth component; and

n = the number of constituent components.

This equation applies to systems in which the failure of any one component will mean the failure of the system.

The probability of successful operation of a system, P_T, is not directly computable. It is computed from failure rate data for that system. Failure rate, i.e., the probability of failure, is related to the probability of successful operation by

$$\Lambda_T = 1 - P_T$$

where

Λ_T = failure rate of the total system, and

P_T = probability of successful operation for the system, or reliability.

The failure rate, Λ, of a system is computed from the sum of the failure rates of each of the components, Λ_n, from:

$$\Lambda = q_1\lambda_1 + q_2\lambda_2 + q_3\lambda_3 + \lambda + q_n\lambda_n$$

where

Λ = failure rate of the total system,

q_n = quantity of the nth component in the system, and

λ_n = failure rate of the nth component in the system.

The component failure rates are usually expressed as failures per million device operating hours (f/mdoh), while the system failure rate is expressed as failures per million operating hours (f/moh). The above equation enables one to calculate the reliability of a system based upon the failure rates of the constituent components. Each fire alarm system component, i.e., a control panel, detector, or notification appliance, has a calculable failure rate and, hence, the system failure rate can be computed. The failure rate of each of these major components is computed from the failure rates of the individual electronic components from which it is made.

Research into the reliability of systems and components has shown that, for most physical systems when the failure rate is plotted vs. time, a curve of familiar shape results.[1] This curve is termed the *bathtub curve*. The "bathtub" curve is the sum of two bell-shaped, Gaussian curves, one centered at the time the product is made and called the *turn-on curve*, and the other centered at the design lifetime and termed the *lifetime curve*.[2]

The first, the turn-on curve, shown in Figure 5-5A, describes the rate at which failures can be expected to occur from the date of manufacture, $t = 0$, forward over time, due to inherent defects in both constituent components and manufacturing technique. It illustrates the phenomenon known as "infant mortality" in the components and finished product.

The second curve, shown in Figure 5-5B, is also a Gaussian distribution, but it is centered at $t = t_{DL}$, the design lifetime. It describes the distribution of failure rates due to long-term aging.

When the failure rates from the initial turn-on and the lifetime curves are added together (having adjusted the standard deviations

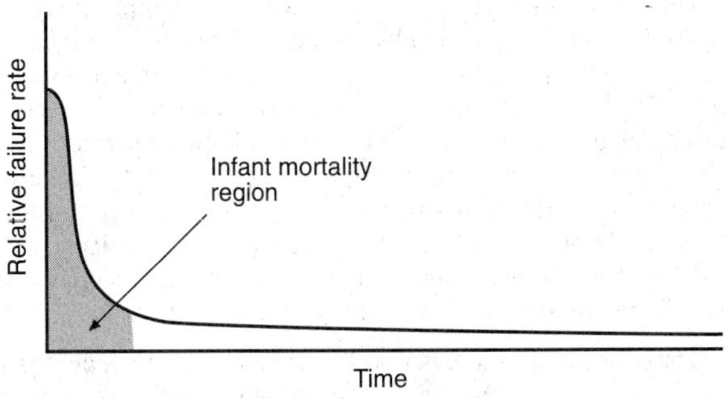

FIG. 5-5A. *The initial turn-on curve shows failure rate of a component or system at turn-on.*

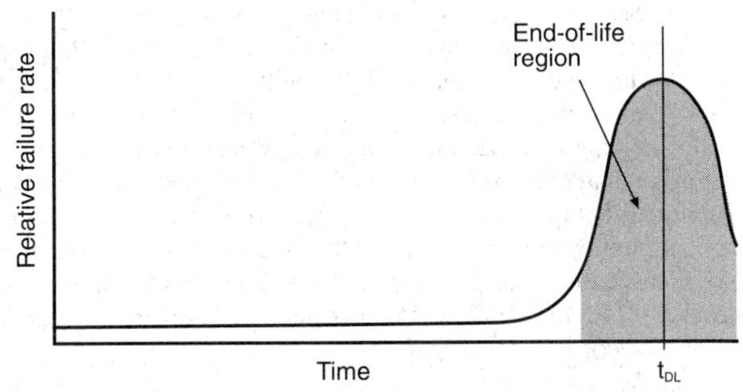

FIG. 5-5B. *The lifetime curve plots the failure rate due to long-term aging of components, and establishes the design lifetime, t_{DL}.*

for each of the distributions to match the experimentally measured failure rates for each failure mechanism), their sum yields a bathtub curve as shown in Figure 5-5C.

The bathtub curve is divided into three regions. The left-most is the "infant mortality" region. This is the initial turn-on phase where failures due to inherent defects in constituent components and manufacturing techniques usually appear. In order to ensure that these failures occur before the product is shipped to a customer, most manufacturers "burn-in" their products for time periods sufficient to allow inherently defective products to fail prior to shipment.

The righthand portion of the bathtub curve is the "end of useful life" portion. It is this portion of the curve that is used to establish the *design lifetime* of the assembly. The design lifetime of a product is that age at which the probability of operability at the end of the recommended service interval is less than 0.5, due to the acceleration in the frequency of failures resulting from long-term deterioration of constituent components. As the failure rate of the constituent

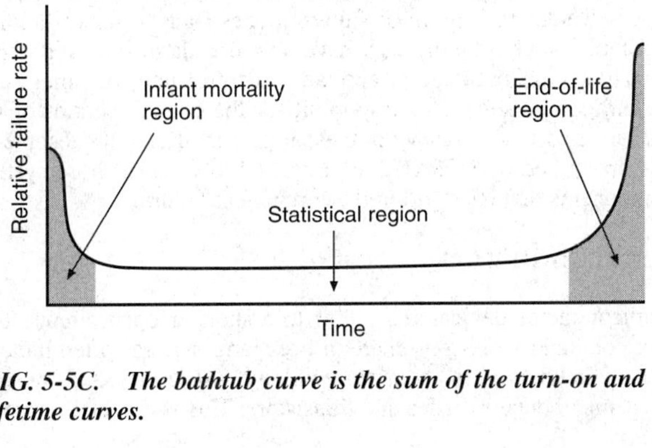

FIG. 5-5C. *The bathtub curve is the sum of the turn-on and lifetime curves.*

components continues to escalate, it becomes increasingly more difficult to maintain the operability of the assembly. However, one can correctly infer from this that the lifetime of a product can be, and often is, extended by increasing the frequency of the inspection, testing, and maintenance procedures.

The center region of the reliability curve is the "statistical failure rate region." Over this region failures are random and unpredictable. This region is essentially flat, indicating a fairly uniform failure rate per unit time over this region. While the actual failure rate may be low, it is always a finite, positive number. Regardless of the distance (time) between t_0 and t_{DL}, the failure rate in the statistical failure region will, to the limits of precision available from the actual data, approach a horizontal line indicating a constant, evenly distributed failure rate. Because this region is flat, one can use the concept of the mean time between failures (MTBF), a quantification of failure rate, to predict the length of time the system can be expected to remain operational. It is a statistical average that is valid over a population of systems, but does not provide any guidance on what will happen with any specific system at any specific time.

The failure rate of any electronic component and, hence, any electronic system can be derived from the failure rate data established in *Military Handbook: Reliability Prediction of Electronic Equipment.*[3] This handbook, known as *MIL Handbook 217*, provides experimentally derived failure rate data on electronic components commonly found in electronic assemblies, including fire alarm system components. *MIL Handbook 217* allows the engineer to compute the failure rates of the detectors, notification appliances, control panels, relay panels, interfaces, and other system components. From these failure rates the engineer can then compute the inherent failure rate (IFR) for the system. It is derived from the failure rates of the components from which the system has been assembled. It is important to keep in mind that this inherent failure rate does not include failures due to improper design or installation. Failures caused by design errors or installation errors must be added to the inherent failure rate computed from *MIL Handbook 217*.

Once the inherent failure rate has been computed, the reliability of the system can be calculated:

$$R = e^{-\Lambda t}$$

where

R = reliability of the system;

e = Napierian logarithm base, 2.71828;

Λ = inherent failure rate of the system; and

t = time period over which the system must operate.[4,5]

Once the performance of the system is defined in terms of the reliability with which it executes the performance of the design, this relation is used to determine the maintenance interval necessary to achieve that reliability level. The factor t is the interval of time between each execution of the inspection, testing, and maintenance procedures for the system. The required interval, t_R, is computed from:

$$t_R = \ln R_R /(-\Lambda)$$

where

$\ln R_R$ = Napierian log of the required reliability, R_R,

Λ = inherent failure rate of the system, and

t_R = required maintenance interval to achieve the required reliability.

An Example Analysis

To illustrate the usefulness of the above computational method, consider the fire alarm system shown in Figure 5-5D. It consists of 10 smoke detectors, a horn/strobe notification appliance, and a control panel. Its design objective is to alert the occupants of a fire, enabling them to make safe egress from the facility, with a reliability of 90 percent (0.90). For the purposes of this example only, and with the understanding that these failure rates are not derived from any specific device but are intended to be of the same order of magnitude as commercially available products, use the failure rates in Table 5-5A.

The inherent failure rate of the system as described is 43.65 f/moh:

43.65 f/moh = 0.00004365 failures per hour (f/h)

Λ = 0.3824 failures per year

Using the required interval equation

$$t_R = \ln R_R / (-\Lambda)$$

where

$\ln R_R = \ln 0.90 = -0.11$

Λ = 0.3824

t_R = −0.11/−0.3824

t_R = 0.288 yr

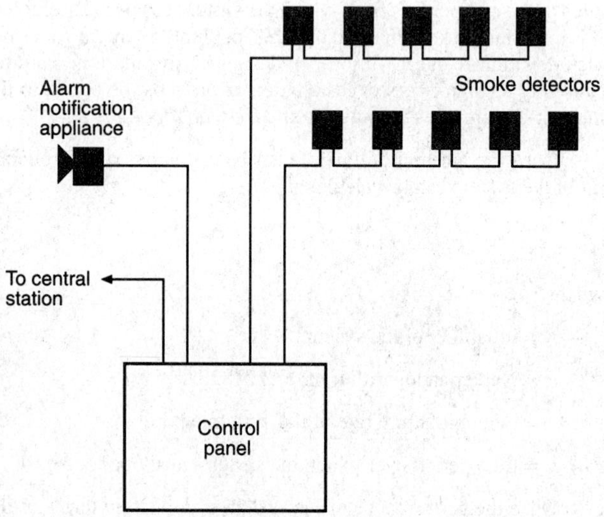

FIG. 5-5D. *A simplified example fire alarm system.*

TABLE 5-5A. *A Compilation of Computed Failure Rates Provides the Basis for the Reliability Calculation.*

System Component	Quantity (q)	Failure Rate (λ) (f/mdoh)	(q_x/λ_x) (f/moh)
Smoke detectors	10	1.00	10.00
Horn/strobe	1	2.50	2.50
Control panel	1	25.00	25.00
Screw terminals	49	0.10	4.90
Wire segments	25	0.05	1.25
Inherent failure rate, Λ			43.65

This indicates that this system, with the component failure rates used, will require a complete inspection, test, and maintenance routine every 0.288 years in order to attain a 90 percent reliability in achieving its design objectives. As performance-based, objective-oriented designs become more prevalent, reliability predictions for those systems similar to this example will be required to determine the maintenance frequency necessary to achieve the performance criteria. A reliability of 0.90 is a very high reliability, well above that of most systems. It means that there is only a 10 percent chance that there will be any form of failure during the lifetime of the system.

The concept of failure is very broad here. Failures include trouble conditions resulting from supervised conditions; spurious alarms; failure to alarm within the design sensitivity or time frame for the design fire; and failures of electronic components, even if the component failure does not cause a system performance malfunction. A reliability of 0.90 (90 percent) is much higher than most owners need. Most owners would be happy with a reliability of 0.50 (50 percent). This same system (see Figure 5-5D and Table 5-5A) could achieve a 50 percent reliability with a maintenance schedule of 1.8 years. A maintenance program predicated upon an inspection and testing regimen performed every 1.8 years will surely be less expensive than one based upon a time interval of every 0.288 years. Consequently, a balance must be achieved between the cost of the service/maintenance program and the inherent reliability of the system. Ideally, the reliability to be achieved should reflect analysis of the cost of failure, which will in turn need to distinguish among different types of failure, as listed above.

The inspection, testing, and maintenance schedules required by NFPA 72, *National Fire Alarm Code* (hereinafter referred to as NFPA 72), are prescriptive in nature and are based upon a consensus opinion of what an average system requires. They are not the result of actual reliability calculations. The facility owner/operator must keep in mind that the prescriptive requirements represent the minimum acceptance criteria, and that there are often circumstances which demand that these minimum criteria be exceeded. It is obvious from the above that the larger or more complex the system is, the more frequently it must be inspected and tested if the design performance is to be achieved.

The implications of the reliability computations cannot be overstated. The failure to maintain a system virtually assures the owner/operator of the protected site that the fire alarm system will NOT achieve the design objective.

ELECTRONIC COMPONENTS

The anticipated reliability of an electronic system, such as a fire alarm system, can be computed because the reliability of the individual electronic components has been thoroughly studied and documented, primarily under the auspices of the U.S. Department of Defense. This prior research has shown that each general type of electronic component exhibits particular and predictable failure modes. Understanding these failure modes leads to understanding the impact such a failure may have on a fire alarm system component (detector, notification appliance, control unit, or other such system component). It is important for the owner/operator or responsible party to be aware of these failure modes, as the test methods prescribed by NFPA 72 are intended to uncover those system components that have suffered a component failure.

Semiconductors

Semiconductor devices fail either to a short or open circuit. The semiconductors in use in modern fire alarm systems often include integrated circuits (IC) consisting of hundreds of thousands of simple semiconductor diodes and transistors. This is especially true of the microcomputers at the operational core of the control unit. Regardless of the complexity of the component, most semiconductor failures are catastrophic in nature, meaning that they fail completely and precipitously, often with a little puff of white smoke curling up from the circuit board.

Semiconductor failures are the result of one of three causes: (1) internal flaw, (2) overcurrent, or (3) overvoltage. Internal flaws are becoming less common as semiconductor manufacturing science continues to advance. Nevertheless, occasionally thermal ion migration or die-bond failure, typical internal flaws, cause a semiconductor to fail.

Since microcomputer memory consists of large integrated circuit arrays of transistor memory cells, there is the possibility of a memory cell failure. Often, control units have diagnostic programs that are executed during the regularly scheduled maintenance to detect memory location failures. Additionally, regular thorough functional tests provide assurance that there are no memory location failures where the executable programs are stored.

The probability of this type of event is addressed in the reliability calculations presented earlier. Given enough time, a failure of this type is likely to occur, even with the highest quality electronic equipment. The testing frequency and methods stipulated by the manufacturer and NFPA 72 are intended to identify these types of failures with a minimum elapsed time period between the occurrence of the failure and its discovery.

Overcurrent and overvoltage failures are usually associated with some other event that places an unintended stress on the system. Occasionally the event is the failure of some other electronic component subjecting adjacent components to the overvoltage or overcurrent. More frequently, overvoltages and overcurrents are the result of wiring and interconnection errors. Regardless, both overvoltages and overcurrents cause severe overheating of the semiconductor die, and the current-carrying portion of the die melts and/or vaporizes. Obviously, this type of damage is irreversible. Usually, these types of failures result in trouble signals and unwarranted activations of parts of the system, and generally do not go unnoticed. It is important to correctly identify the source of the failure quickly. Otherwise, it is likely to occur again. Electrical transients caused by lightning, power supply spikes and surges, operation of highly inductive or capacitive equipment in close proximity to the fire alarm system, and even radio frequency interference have been known to cause these types of failures. Generally, the reliability computations presented earlier do not explicitly address the contribution of these external events to the failure rate of the system.

Resistors, Capacitors, and Inductors

These components are classified as "passive" components. They are less likely to fail than semiconductor devices because their structure is much simpler. Usually, these components fail to an open circuit. This generally causes some type of inexplicable action by the system, such as a trouble signal or the unwarranted activation of some control system function. Rarely do these failures occur and on those rare occasions when they do they are immediately noticed. These failures are included in the reliability computations presented earlier.

Circuit Board Solder Bonds

A major source of electronic system failures is the solder bonds between the electronic components and the circuit board. Solder bonds are occasionally broken during shipping and installation, rendering the unit "dead on arrival." Unfortunately, broken solder bonds do not always show up immediately. Broken solder bonds can behave as either an intermittent or high-resistance connection. Both of these cause spurious operation of portions of the system or inter-mittent trouble signals. There is no guarantee that these failures will be noticed when they occur. Broken solder bonds are just as likely to allow the unit to "fail silent."

As a system ages, the lead and tin atoms in the solder begin to migrate, allowing the solder to crystalize and become brittle. Variations in temperature, vibrations, and excessive moisture also contribute to the premature failure of solder bonds. A significant percentage of the computed failure rate for a fire alarm system component is derived from the failure rate assigned to each of the hundreds of solder bonds in that component.

Transformers

Transformers are generally very reliable. Furthermore, since most transformers are used in the power supply section of a control unit, their failure will cause obvious indications, making timely repair likely.

Relays

Those switching functions that cannot be accomplished with solid-state (transistor) switches are accomplished with relays. Relays are electromechanical switches that are activated by causing current to flow through an electromagnetic coil which is situated in such a manner that the magnetic attraction developed by the coil pulls a switch from one pole to another. Often several switches are connected together so that the energization of the coil causes all of the switches (poles) to transfer at the same time.

As with any other mechanical component with moving parts, there is a possibility of a mechanical impediment to operation from dust, corrosion, or wear. There is also the possibility that, even though the electromagnet has mechanically closed the contacts, electrical current fails to flow due to contact contamination with oxide, dirt, or some other material. The continuity of the electromagnetic coil can be lost, rendering the relay inoperative.

With all of these potential failure modes, relays do not provide the level of reliability of solid-state components. Nevertheless, most fire alarm system control units are equipped with relays, because they are very useful for providing a general-purpose switching function, especially for circuits that operate at higher voltages, such as 115 V, ac.

While some relays are available in sealed versions, most relays are constructed with simple dust covers. This is usually adequate for normal indoor installations, but extra care must be exercised when testing a unit that is installed where it is exposed to a wide variation in ambient temperature, dust, excessive humidity, or corrosive aerosols. All of these environmental factors have been implicated in relay failures.

When "operation-critical" functions are performed by a fire alarm system with relays, the failure rate of the relay must be included in the reliability computation for the fire alarm control unit. The testing regimen for a fire alarm control unit must test the operability of all of the relays to verify that they are still operable.

Switches

Switches enable an operator to introduce electronic signals into the fire alarm system. The least reliable part of any switching system is the human operator. Nevertheless, once the operator is taken out of the picture, there are two general failure modes for switches: (1) failing to operate when they should and (2) operating when they should not. Since switches are usually mechanical components with moving parts, they suffer relatively high failure rates. The exception to this rule are some of the newer capacitive and membrane switch technologies where the parts do not actually move.

The first failure mode is of lesser concern, because an operator is present and the failure is immediately apparent. It is extremely rare that a fire alarm system is designed requiring that an operator activate a switch in order to secure the life-safety and property conservation objectives of the system. Usually, system switches are used to interrupt the automatic functions of the control unit. In this scenario, failure of the switch should result in a safe condition.

The second failure mode usually takes the form of intermittence and usually causes trouble signals. The listing criteria for the nationally recognized testing laboratories require that the off-normal position of any switch result in a trouble signal. This requirement grew out of the years of experience with the failures of switches.

Screw Terminals

The least reliable part of a screw terminal is the person with the wire stripper and screwdriver. However, screw terminals do occasionally vibrate loose. NFPA 72 requires that the wiring to initiating devices and notification appliances be supervised. However, wiring within the control unit enclosure is not required to be, and is often not, supervised. Consequently, the failure rate of the screw terminals must be factored into the reliability calculations.

Fuses

Fuses demonstrate one principal failure mode: they fail to an open circuit. This failure mode occurred with such regularity that it gave rise to the establishment by nationally recognized testing laboratories of a listing criterion, requiring that all fuses be supervised. The failure of any fuse should result in a trouble signal. It is required that the supervision of fuses be checked during testing.

Indicator Lamps

The invention of light-emitting diodes (LED) ended the reliability problems with indicator lamps. While it is possible for an LED to fail, it is not a common occurrence. When they do, they cease making light. LED indicators are deemed inherently reliable and are not always supervised. Their failure rate is included in the reliability prediction for the control unit.

Batteries

Batteries are the most commonly used source of secondary power for fire alarm systems. They must be capable of supporting the operation of the system for both the time period and the electrical current load required by the specific system.

It is important to understand how batteries fail in order to understand the importance of proper testing. Batteries are designed to provide a specified current for a specified time at a specified voltage. The specified current for a specified time is called capacity. Batteries undergo failure modes where they lose their capacity. Batteries can also fail in such a manner that they lose their output voltage. These two failure modes are separate and distinct, occurring independently of each other.

The loss of capacity cannot be discovered by simply measuring open terminal potential, nor does the loss of terminal potential necessarily imply a loss of capacity. When a totally "dead" battery, i.e., one that cannot provide any current at any voltage, is measured with a voltmeter under "open terminal conditions," the meter will usually show very close to nominal potential (voltage). That is because the mechanism of failure which results in the loss of capacity does not necessarily change the open terminal potential of the battery. When a battery is measured with a voltmeter and shows less than nominal open terminal potential, it may still have the ability to provide very large currents, in spite of its lower terminal voltage. Consequently, the testing regimen for battery systems must test the ability of the battery to perform its intended function: to store and supply, as needed, electrical energy. A measurement of terminal potential and whether the battery "holds charge" does not provide an adequate assessment of the battery set.

Wires

Wiring is subject to two general classes of failure: (1) breakage of the conductor and (2) loss of insulation. The continuity of wiring conductors used in initiating device circuits and notification appliance circuits is required to be supervised per NFPA 72. When fire alarm control units are used to actuate extinguishing systems, the wiring to the actuating mechanism and other critical functions should be supervised also. However, the supervision circuitry in most fire alarm system control units does not have the ability to distinguish between a normal conductor and one that is damaged. Close visual inspection is necessary on initial acceptance and whenever revisions are made.

The insulation on wiring conductors represents a different problem. Thermoplastic wire insulation behaves much like a supercooled liquid, in that it can flow over time away from a point of compression. This is particularly common when conductors are pulled taut against the edge of a metal enclosure or when pinched between an electrical box and its cover (both indicative of substandard installation practice). Under these conditions, the insulation will slowly flow away from the stress point, eventually allowing the conductor to make electrical contact with the enclosure. This creates a "ground fault."

The nationally recognized testing laboratories require that fire alarm system control units operate completely isolated from earth ground and provide ground-fault detection. These requirements ensure that a single ground fault will be detected, yet will not impair the operation of the system. It is imperative that the first ground fault be corrected before a second one occurs, as the second ground fault introduces a circuit via ground between two separate circuits in the system.

INSPECTION, TESTING, AND MAINTENANCE REQUIREMENTS FOR FIRE ALARM SIGNALING SYSTEMS AND COMPONENTS

The reliability analysis presented earlier establishes the absolute demand that fire alarm systems be maintained. NFPA 72 establishes this requirement and places the responsibility for the regular inspection, testing, and maintenance on the owner/operator of the site. NFPA 72 references the manufacturer's instructions, effectively making the manufacturer's instructions part of NFPA 72 by reference and thus enforceable by the local Authorities Having Jurisdiction.

Definitions

To discuss the inspection, testing, and maintenance of fire alarm systems one must define several terms that are not explicitly defined in NFPA 72, but are important in establishing the expected course of conduct.

An *inspection* is a close visual observation of the item being inspected without causing the item to perform its intended function. There is said to be a facility manager who insists that the inspector's nose be less than 12 in. (305 mm) from the object to qualify as an inspection. This rule differentiates between a casual glance at a device and an inspection. It works for the manager. An inspection does not require the operation of the device being inspected, but it does require that a written record be made of the observed status of the device. Is it clean? Dirty? Bent? Covered with paint? In "as installed" condition?

A *test* is a process where the device under test is provided a stimulus to determine if it is functional, and a written record of the response of the device under test is made. The term "test" does not necessarily imply that the stimulus is of a similar magnitude to that expected from the fire the unit is intended to detect. However, it does imply that the device has also been inspected while the test was conducted. Many tests require that the system be intentionally impaired to determine if the system properly responds to the impairment. *It is critical that all conditions created in the context of a test be returned to the normal state before the test is deemed completed.* The written record should include both the observed status of the unit and whether it was found to be functional.

Sensitivity measurement is the quantitative measure and recording of the stimulus necessary to achieve an alarm signal from an initiating device. This is very different from a test. A test does not imply that the stimulus is of a similar magnitude to that obtained from the design fire. A sensitivity measurement determines how large a stimulus is necessary to cause an alarm response. This measurement is to be compared to the value for the unit as shipped to quantify any change in the performance one can anticipate from the unit. Thus, the sensitivity measurement is intended to assess the ability of the detector to perform its intended function when the design fire occurs. It is understood that initiating devices are also inspected and tested when the sensitivity is measured. The written record should contain the observed status, the sensitivity measurement, and the verification of the functionality of the unit.

Bear in mind that many "intelligent/analog" detectors are actually addressable sensors where the alarm set-point is stored in a memory location at the control panel. These systems continuously measure the sensor output and compare it to the value stored in memory for that particular device. These systems often can provide a printout of the current measured values for the initiating devices and their alarm points, thus satisfying the requirements of NFPA 72. Observations of physical status and verification of operability are normally associated with measurement of sensitivity, and these must be separately performed when "intelligent/analog" detection is encountered.

An *acceptance test* of a system is the process wherein every initiating device is provided an appropriate stimulus that simulates the existence of the design fire, and the design sequence of operations of each system component in the entire system is verified and recorded in written form. This is usually achieved with a sensitivity measurement, including the functional test to initiate an alarm. Once the alarm is initiated, the technician verifies that all system functions, including interconnections to other building systems, operate as designed. It is preferable to repetitively operate all interconnections for each initiating device rather than verifying their operation once and assuming that all other initiating device alarms will cause the same operations. Often problems show up after several operational cycles that would otherwise be missed. Programming errors in some software architectures can lead to problems that can only be uncovered with complete function verifications.

A *reacceptance test* is an acceptance test performed after modifications have been made to the system. Modifications include any change, repair, adjustment to hardware or wiring, or any change in system software. The scope of a reacceptance test can be limited. All components, circuits, system operations, or software functions that are known to be affected by the change in the system must be tested, in addition to 10 percent of the initiating devices (up to a maximum of 50) known to not be directly affected by the change. A reacceptance test does not necessarily imply a complete acceptance test in all cases. However, there are times when a reacceptance test is equivalent to an acceptance test because of the nature of the modifications or repairs performed.

Documentation

Every form of evaluation of a fire alarm system must be completely documented. This includes inspections, tests, sensitivity measurements, acceptance tests, and reacceptance tests. This documentation should be retained in a secure location for future reference. The lack of complete documentation is often viewed by the courts and regulatory authorities as presumptive evidence that the required testing and inspection have not been done. A minimum compliance form for documenting the inspection and testing of a fire alarm system is provided in NFPA 72. However, many Authorities Having Jurisdiction require a more complete form as part of a test plan.

Personnel

NFPA 72 establishes qualifications criteria for those individuals undertaking the inspection and testing of fire alarm systems. The facility owner is responsible for making certain that the people performing these functions are qualified. Many fire alarm systems are connected to remote monitoring stations and automatic extinguishing systems. The personnel performing the testing of a fire alarm system must be sufficiently familiar with these interconnections to take the appropriate steps to prevent the accidental discharge of an extinguishing system or reporting of a fire alarm to the municipal fire service during the testing of the system.

Preparation and Coordination of Inspection, Testing, and Maintenance Activities

Many fire alarm systems are connected to off-premises monitoring services, fire pump controllers, special agent extinguishing systems, and building management systems. There are times when the functional testing of a fire alarm system will cause serious interruptions in site operations due to these interconnections. By the same token, it is critical to verify that the interconnections are operative and that the fire management plan for the entire site is achieved when a fire alarm signal occurs.

This presents a dilemma to the facility manager: (1) should a shutdown be endured, a major problem if the facility is the hub of an on-line computer network; or (2) should the shutdowns be bypassed, thereby failing to verify complete system operability? Usually the question is not an "all or nothing" situation, and reasoned decision must be made regarding those automatic interconnections that must be bypassed and those which should be allowed to operate. The facility manager must recognize that some interconnections are actuating extinguishing systems that represent a significant cleanup and/or recharge expense if they are actuated. Naturally, these types of systems must be disabled for the routine inspection, testing, and maintenance. However, the portion of the system that is disabled should be as small as possible. For example, many extinguishing system actuators are removable from the extinguishing agent containers, allowing all of the electrical components to be tested. This is a very desirable feature. However, since the electrical supervision cannot recognize this type of impairment, there is the danger that the system will be left with the actuators removed from the agent storage containers. Extra care must be taken to ensure that all of the actuators are properly reinstalled at the conclusion of the inspection, testing, and maintenance program.

Other interconnections, such as smoke management systems, automated process controls, and data processing systems, will interrupt the productivity of the site to the point that the cost of an unwarranted actuation is too high to allow during routine inspection, testing, and maintenance. Systems with these types of interconnections can be designed with built-in disconnect switches that are supervised when in the off-normal state. However, remember, when these functions are disabled their operability is not verified. It may be reasonable to permit the inspection, testing, and maintenance program to be performed with these types of functions disabled during normal operations time in a site that has experienced no modifications since the initial acceptance tests; however, it is not appropriate if revisions or modifications have been made. If it is critical for a function to occur in order to achieve the design objectives of a fire protection system, then it is equally critical that those functions be inspected and tested.

Finally, off-site reporting and resulting responses must be addressed. These can be effectively managed with prior notification. All impairments must be monitored. Usually, at the beginning of a service visit, the technician places a telephone call to the monitoring entity to advise them that the system is being tested and to disregard alarm and trouble signals. The monitoring entity should be under strict instructions that, when such a call is received, the following is addressed:

1. Call back the facility manager to verify that testing is about to commence.
2. Verify that an alternate procedure to report a fire is in place.
3. Require verification when testing will be completed and deem any signal occurring after that time as a real signal, unless a second call is received rescheduling the estimated completion time.
4. Require a follow-up call upon completion of the maintenance visit that confirms the system has been restored to fully automatic status.

System impairments that facilitate inspection, testing, and maintenance should not be granted casually. There have been a number of large-loss fires that became such because existing and operable fire protection systems were impaired. It is also possible that serious fires became serious only because the monitoring entity failed to report them to the public fire service, believing the system was still being tested. Fire alarm systems should be designed and serviced with these considerations in mind. Every system must be serviced, and often the service activities necessitate disabling portions of the fire alarm system or interconnections. However, the disabling should be limited to the smallest possible portion of the building fire protection assets. Furthermore, if at all possible, those portions that are directly affected by the testing should be disabled individually, therefore retaining as many building fire protection assets operative as possible.

The Test Plan

The inspection, testing, and maintenance regimen for a fire alarm system must stipulate:

1. What procedure is to be performed,
2. How it is to be performed,
3. What data are to be recorded in the documentation, and
4. How frequently the specific procedure is to be performed.

These four criteria are usually organized into a test plan. A formal test plan is often a very good investment. It establishes a formal procedure to ensure the complete inspection and testing of the system and a record of the results for future reference. Once a test plan has been developed, it is used each time the inspection and test program is performed. Consistency of the presentation of the results facilitates identifying problem areas.

NFPA 72, Chapter 7, "Inspection, Testing, and Maintenance," is organized to facilitate the development of a formal test plan. It establishes the test methods and then stipulates an inspection and testing frequency.

The test methods outline what procedures are to be performed. They are intended to be used in conjunction with the manufacturer's instructions for the specific make and model of equipment involved. Thus the manufacturer's instructions provide the specific "how to" information. The issue of documentation is addressed by NFPA 72, Chapter 7, in the section on records.

The frequency of each procedure varies, depending upon the general class of device. For example, batteries are inspected more frequently than initiating devices. The frequency of visual inspections is different (i.e., more frequent) than testing. These frequencies have been based primarily upon experience developed over the recent past, since reliability data do not yet exist for all fire alarm system components.

Supervising Station Systems

Supervising station systems monitor a number of protected premises fire alarm systems via one or more types of communication paths. The principal concerns during the inspection and testing of these systems is that each communication path is intact and operable and that the supervising station fire alarm control unit is fully operational.

The inspection, testing, and maintenance procedures for specific models of a control unit for a supervising station fire alarm control system are established by the manufacturer. Since supervising station fire alarm systems include many of the same components that make up a protected premises system, organized in a manner that fulfills the design requirements for a supervising station system, the inspection and testing requirements for these components are the same, regardless of the type of system in which they are used.

Protected Premises Systems

There are two general objectives in the testing and inspection of a protected premises fire alarm system. The first is to verify the operability of the system and identify any failures in system components that have failed to a "silent" state. Any failures identified during the course of the inspection and testing must be repaired to restore the system to a full operational state. The second is to identify changes in the occupancy that render a formerly appropriate system inappropriate, based on the hazard area to be protected. Both objectives affect proper response to a fire.

Fire Alarm Control Unit

The manufacturer for each make and model of fire alarm control unit (FACU) provides a recommended service procedure for the control unit. These procedures are included in NFPA 72 by reference. However, the fire alarm control unit is one building block in a much larger system, and the manufacturer rarely knows how the overall system is designed. Consequently, there are at least two layers of functions that must be tested. The first layer of functions are those functions that are an integral part of the FACU, irrespective of the particular system in which it is used. These functions include, e.g., supervision of relays, fuses, main power supply, batteries, wiring integrity, alarm indications, trouble indications, supervisory indications, and isolation from ground. The second layer of functions are those specific to the facility or system. These are usually in a "sequence of operations" format, establishing the response functions that occur as a result of each input (initiating or supervisory device) signal.

Functions: All functions of the system should be tested. The supervision of all initiating device circuits, all notification appliance circuits, and all signal transmission circuits must be tested. Furthermore, when the FACU is used for the activation of extinguishing systems, all circuits activating equipment whose operation is critical to successful extinguishment must also be supervised. The testing of these functions consists of creating breaks and ground faults on each circuit and verifying that the appropriate trouble signal is rendered by the FACU.

The sequence of operations must then be tested to verify the operability of all those functions the control unit is intended to implement specifically for the hazard area it protects. These functions are application specific and vary from one site to another.

The appendix of NFPA 72 provides specific tests for each type (style) of circuit. Trouble signals can be introduced on initiating device circuits by temporarily removing detectors (from their mounting bases) or wiring conductors from detectors. Similar techniques are used for notification appliance circuits. Signaling line circuits are usually interrupted at the FACU screw terminals, with a verification that the trouble signal is received at the receiving station unit. Alarm signals must be generated on each initiating device circuit or zone by causing a detector to alarm with heat, smoke, or radiant energy, as appropriate, to verify that the design alarm response occurs. The alarm response will include activation of the notification appliances, alarm indication at the control unit, activation of interconnected equipment, and signal transmission to the off-premises receiving station.

Conductors on each initiating device, notification appliance, and signaling line circuit must be temporarily grounded to verify that the ground detection circuitry and associated trouble indications are operative. The applications manual for most FACUs stipulate the amount of resistance between conductors and ground that should be detected. A fixed resistor reflecting that criterion should be used to test ground-fault detection.

Special hazards extinguishing systems employ specialized equipment to address the unique needs of the various extinguishing system equipment designs and agent limitations. Each manufacturer is required to provide an engineering manual that addresses the inspection, testing, and maintenance needs of the control unit and ancillary equipment necessary to operate the special hazards extinguishing system. These functions include cross-zone detection, matrix detection, verified detection, sequential detection, detector counting detection, abort switches, solenoid activation (release), and squib actuation (release). When these functions are implemented by the FACU, the specific inspection and testing requirements established by the manufacturer must be observed.

The functional test serves to verify that all of the system interconnects are energized when appropriate and that the system achieves the basic functions it was designed to achieve. However, this type of general "macroscopic" test does not fully evaluate all of the system components. There are specific tests to evaluate these components.

Fuses: The listing standards of the nationally recognized testing laboratories require that all fuses be supervised. However, it is possible that the supervisory circuitry could fail and not provide an indication that such a critical component was inoperative. Furthermore, it is conceivable that the fuse may have been replaced with another of the wrong value. Consequently, all fuses must be checked for supervision and correct value. Keep in mind that a fuse cannot be properly checked visually. The only way to check a fuse for continuity is with a volt-ohm meter (VOM) or multimeter.

Indicating lamps: All indicating lamps must be operated to verify that they are still operative. Modern fire alarm control panels use light-emitting diodes (LED) for indicator lamps. LEDs are much more reliable than the old incandescent lamps but can still fail. Most fire alarm control units are equipped with a lamp-test switch specifically for this purpose, while others have a key-pad code that performs the lamp-test function.

Main power supply: There are electronic component failures that could render the main power supply incapable of supporting the design current load on the system during fire conditions yet not appear under normal, quiescent conditions. The main power supply test entails temporarily disconnecting the secondary supply and applying the maximum design current load to the system while monitoring the output voltage of the system power supply. If the power supply voltage drops below that stipulated in the manufacturer's applications manual, remedial action must be taken.

Engine-driven generators: In those instances where an engine-driven generator is used as a required power source, either as a primary or a secondary supply, it must be tested according to the procedures outlined in NFPA 110, *Standard for Emergency and Standby Power Systems.*

Uninterruptable power supply (UPS): In those cases where a UPS system is dedicated to the fire alarm system, it must be tested per NFPA 111, *Standard on Stored Electrical Energy Emergency and Standby Power Systems.* Note that a UPS is used for a number of purposes, one of which is the fire alarm system. UPS is treated by NFPA 72 as being no different than public utility power, and the requirement for a secondary power supply for the fire alarm system is still enforceable.

Secondary power supply: The secondary power supply must be able to support the full current load of the fire alarm system when the primary supply is lost. Primary power is often lost during a fire. Testing of the secondary power supply entails temporarily disconnecting the primary power supply and applying the maximum design current load to the system while monitoring the output voltage of the secondary system power supply. If the secondary power supply voltage drops below that stipulated in the manufacturer's applications manual, remedial action must be taken. Additional and more specific testing is required when the secondary power supply consists of a battery set.

Batteries: The testing of a battery set consists of general tests that are performed regardless of battery chemistry and specific tests for specific types of batteries. A visual inspection is required for all types to discover any terminal corrosion, looseness in terminal connections, or electrolyte leakage. All of these are serious conditions that can cause an increase in the series resistance of the battery set, thus depriving the fire alarm system of the necessary operating potential under large current load conditions.

Batteries exhibit a number of failure modes that result in either a loss of operating potential or a loss of capacity. The exact mechanism of each class of failure mode is dependent upon the battery chemistry; nickel/cadmium batteries behave somewhat differently than lead/lead peroxide batteries.

In general, the test regimen for all batteries includes:

1. Battery charger test;
2. Discharge test, where the batteries must support a design current load;
3. Load voltage test, where the voltage is measured while the batteries support a design current load; and
4. Open circuit test, where the open-circuit potential of the battery is measured.

There is a relationship between the output potential of a battery, measured in volts, and the output capacity of the battery, measured in ampere-hours (Ah). This relationship is illustrated with a "family" of potential *vs.* discharge time curves for a number of discharge currents, as shown in Figure 5-5E. When the terminal potential reaches 85 percent of original value, the battery is generally considered effectively depleted. The discharge current multiplied by the length of time it takes to deplete the battery is its capacity (C), measured in Ah. Typically, batteries exhibit smaller capacities when the discharge rates are higher. The battery industry usually uses a 20-hr discharge rate for their specifications. A 10-Ah battery will support a 0.5-ampere load for 20 hr, and be at 85 percent of its fully charged potential after having been discharged at that rate for that time period.

The capacity of the battery is a function of the number of charge carriers stored in the unit. Unfortunately, there are chemical reactions that occur within a battery that "lock up" charge carriers into precipitates, thus removing them from the electro-chemical reservoir of available charge carriers. This results in the battery becoming depleted far sooner than expected. It is analogous to having a water tank that has become filled with rocks. While the water level is still the same (analogous to voltage), there is less water available (analogous to capacity) from the tank when needed.

The lead/acid battery chemistry is the most commonly used type in fire alarm system service. These batteries are typically found as unsealed lead/acid batteries, unsealed calcium grid (lead-calcium) batteries, sealed lead/acid batteries, and gelled electrolyte lead/acid batteries. All of these batteries use a solution of sulfuric acid in water as the electrolyte. These various types all work on the same principle, but require slightly different charging potentials and maintenance. It is imperative that the battery charger be properly matched to the battery, consistent with the listing.

There are two symptoms indicating a deterioration of the battery: (1) loss of capacity (Ah) and (2) loss of potential (voltage). These symptoms can usually be traced to the stresses inherent in recharging and maintaining charge. When the battery is being charged, lead sulfate is converted to lead and lead dioxide. As charging continues, bubbles of hydrogen and oxygen gas are formed from the electrolysis of the water in the electrolyte solution. In the sealed and gelled electrolyte batteries, the hydrogen and oxygen are contained within the battery, to a large degree, where they should recombine. In the unsealed type, these gases are allowed to escape. The loss of hydrogen and oxygen represents a loss of water from the electrolyte. This increases the concentration of the acid in the electrolyte, upsetting the chemical balance within the battery. As a re-

sult, a lead sulfate and lead sulfite precipitate begins to form, "locking up" the lead in an insoluble and, hence, unusable form. This process is called sulfation. Since the capacity of the battery is directly proportional to the quantity of lead available, a sulfated battery cannot support the Ah current load it could before the sulfation occurred.

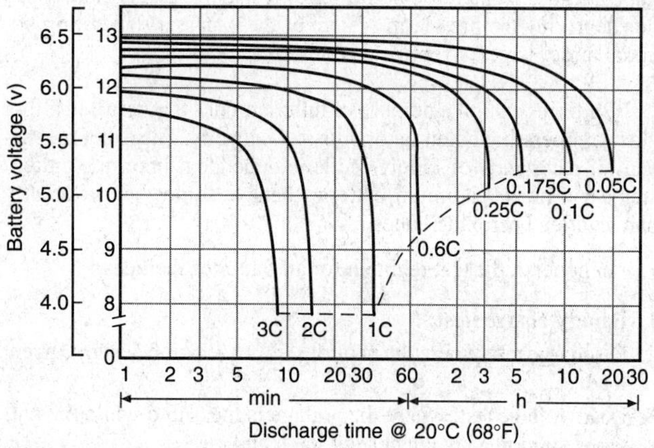

With unsealed lead acid batteries the solution to this problem is simple. The acid concentration is determined by measuring the specific gravity of the electrolyte. As water is lost, the specific gravity of the electrolyte increases. When the specific gravity reaches the upper allowable limit, reagent-grade distilled water is added, restoring the acid concentration to the design value. The measurement of electrolyte specific gravity is done with a hydrometer, similar to that shown in Figure 5-5F.

The sealed lead/acid and gelled electrolyte lead/acid batteries have no provision for returning the electrolyte to the design specific gravity range. These batteries are designed to minimize the loss of hydrogen and oxygen, and include design features to offset the effect of the slow leakage that often accompanies continuous float charge. Eventually, the loss of water allows sulfation to occur, with the resultant loss of capacity. If the recharging current is properly matched to the batteries, these types of batteries usually provide reliable service over a time period generally between 3 and 5 years, after which they must be replaced. Figure 5-5G shows a typical capacity *vs.* time curve for a sealed or gelled electrolyte lead/acid battery.

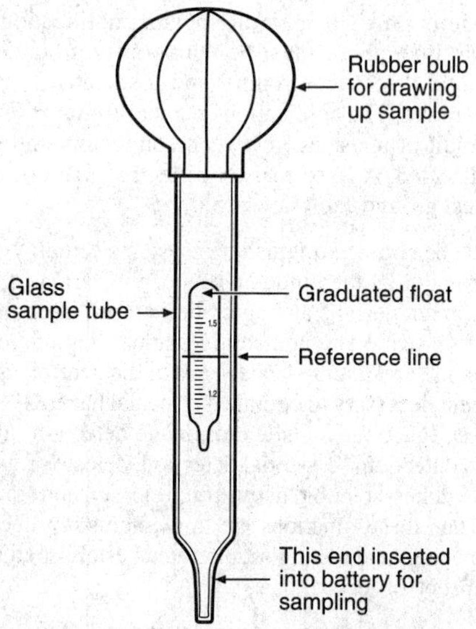

FIG. 5-5F. *A hydrometer is used to measure specific gravity of a liquid. Always use one that has a float and a graduated scale.*

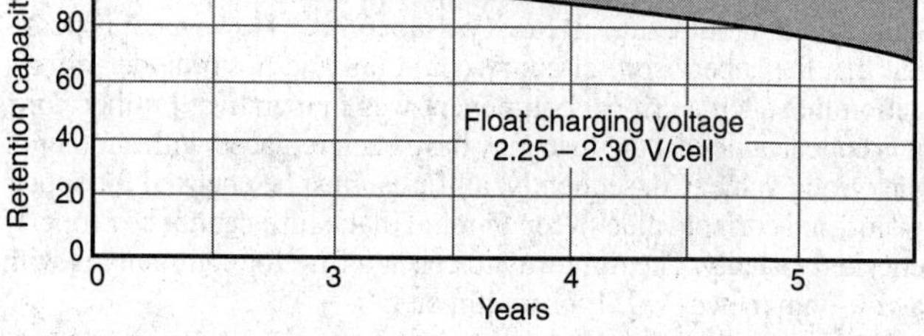

FIG. 5-5E. *The capacity (C) of a battery is a function of discharge rate, time, and voltage.*

FIG. 5-5G. *The capacity vs. time curve of a sealed or gelled electrolyte lead/acid battery.*

The symptom of a battery failure, i.e., loss of potential, comes from either the shorting of a cell or internal deterioration that causes an increase in the internal resistance of the battery. Both of these failure mechanisms can result in a battery set that cannot provide sufficient potential (voltage) for the fire alarm system to operate as intended.

Shorted cells usually result from the recharging and maintenance of charge over a period of time. Batteries are constructed by connecting a number of primary cells in series to attain the intended output potential. For example, the lead/acid cell has a nominal terminal potential of 2.2 V. To construct a nominal 12-V battery, 6 cells are connected in series to provide 13.2 V. During the course of recharging the battery, lead ions move between the lead-dioxide plate and the lead plate. Occasionally a needle-like crystal forms. As it continues to grow, it eventually extends from one plate to an adjacent plate, making an electrically conductive contact between the two. This shorts out the cell, and results in a battery with a terminal potential 2.2 V lower than originally designed.

When a cell becomes shorted, two serious problems arise. The first is that the battery charger, sensing an inordinately low battery potential, forces a greater charging current into the battery. The increased current overcharges the remaining cells, boiling off the water in the electrolyte, ultimately causing catastrophic failure. Second, if the fire alarm system loses primary power supply while connected to a battery with a shorted cell, it will not receive an adequate operating potential and will very likely not be able to support the design output load.

Batteries can also lose the ability to maintain output potential due to increases in internal resistance within the battery. During the discharge and recharge of the battery, lead ions migrate through the electrolyte between the plates within the battery. Eventually, electrical resistance within portions of the plate, interconnecting conductor bars, and electrolyte begins increasing. This causes the internal resistance of the battery to increase. This condition only shows up when the battery is required to supply current to a load (the fire alarm system). The larger the current load placed upon the battery, the larger the drop in voltage due to internal resistance and hence, the easier it is to detect.

The likelihood of a shorted cell failure and increased resistance failures increases as the length of time under recharging conditions increases. Hence, older batteries are more likely to suffer these failures as are overcharged batteries. This fact underscores the importance of using the battery that has been listed for use with the battery charger, and of performing the maintenance as prescribed by NFPA 72.

Nickel/cadmium battery secondary power supplies are usually encountered in connection with fire alarm systems only as sealed batteries. Sealed nickel/cadmium batteries manifest similar failure modes to the lead/acid units. However, there is an additional failure mechanism associated with some nickel/cadmium batteries that is unique to that chemistry.

Fire alarm systems are "float-charge" applications. These are applications where the battery is maintained at full charge by a battery charger over long periods of time and, occasionally, are partially discharged, only to be restored to full charge when the primary power is restored. Some nickel/cadmium batteries exhibit a "memory effect," where the usable capacity of the battery declines to that quantity that has been recurrently used.[6] Thus, a 20-Ah battery discharged several times by 2 or 3 Ah will lose the ability to supply more than the 2 or 3 Ah. Nickel/cadmium batteries are much better adapted to cycle charge applications where the unit is completely discharged prior to recharging. This is an additional loss of capacity failure mode that must be addressed in the test program.

Because the failure modes of batteries are so complex, battery secondary power supplies represent a challenge for any testing program. NFPA 72 provides specific tests for each type of battery. Each type requires a charger test, where the recharge potential of the battery charger is measured with a multimeter to verify that the recharge potential is within acceptable limits. Since most battery failure mechanisms are related to recharging and overcharging, the charger test is essential.

For each type of battery there is a load test. A current load is placed on the batteries and the battery potential is measured. This test is intended to identify batteries that have excessive internal resistances. Consequently, the larger the load, the better. In practice, the ideal load is one that is equivalent to the fire alarm system in full alarm with all notification appliances operating, the emergency voice evacuation system operating, and extinguishing system solenoid valves energized. This load can be simulated with a bank of power resistors, as shown in Figure 5-5H.

Where the battery is being used for a public reporting system, the load tests must be modified and the potential between each side of the battery and earth-ground must be included. The standard multimeter cannot be used for testing public reporting system power supplies due to the possible erroneous readings caused by voltage differentials in the ground potential. Figure 5-5I shows how to modify a standard multimeter for these battery tests.

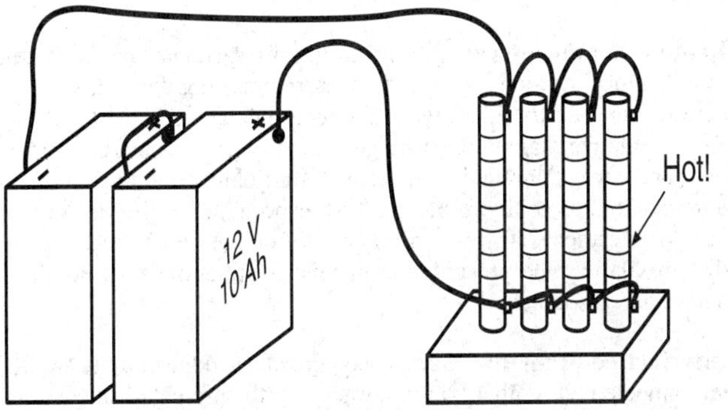

FIG. 5-5H. *A bank of power resistors can be used to simulate the system load for the battery load test. (1 24-Ω, 100-Watt resistor allows 1.0A to flow at 24 V. Current increases for parallel-connected resistors.)*

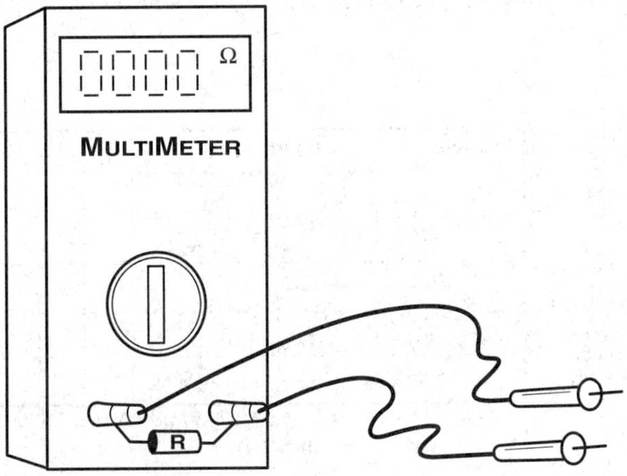

FIG. 5-5I. *A standard 10-MΩ input impedance meter can be converted to read battery potentials for a public reporting system power supply, with the addition of a properly sized ballast resistor.*

$$R = 100 \ \Omega/\text{V} \times \text{battery voltage}$$

For a 12-V battery

$$100 \ \Omega/\text{V} + 12 \ \text{V} = 1,200 \ \Omega$$

It is important to note that none of these tests actually test for battery capacity. The load test is used to infer that no serious loss of capacity has occurred. This inference is supported by the observation that loss of capacity is almost always accompanied by an increase in internal resistance. A better inference can be drawn from a timed discharge/recharge test. In this test, a fraction of the battery capacity is discharged and then the battery is returned to the battery charger, while the current is monitored over time. A curve similar to Figure 5-5J results. If the capacity returned to the battery, measured in Ah, is no more than 110 percent of the capacity removed, then the battery has retained its design efficiency and one can be more certain that no loss of capacity has occurred. The only sure way to completely verify design battery capacity is a full, measured discharge of the battery set at the design discharge rates with subsequent monitored recharge.[6]

Transient suppressors: Since fire alarm systems must maintain isolation from ground, and monitor that isolation, they are vulnerable to damage from the electromagnetic pulse (EMP) that occurs during a lightning strike. Transient suppressors are designed to absorb a finite quantity of energy before they overheat and fail, melting to a short or vaporizing to an open. These devices must be tested with a multimeter to verify that they are still intact.

Remote annunciators: Many fire alarm systems include a remote annunciator to provide information regarding the status of the control unit and the identity of alarm signals at a remote location. Representative signals must be generated at the fire alarm control unit, and correct annunciation verified. Remember that most remote annunciators serve to provide the first responding fire service members with critical information relating to the nature and location of the fire alarm. The presentation of this information must be clear and understandable.

Interface equipment: Commonly, fire alarm control units are interconnected with other control systems in the protected site. These interfaces must be inspected and tested in a manner consistent with the requirements established by the manufacturer and the expected operation of the interfaced systems.

Off-premises signal transmitters: Fire alarm systems often transmit signals to monitoring stations that are off the immediate premises. The transmitter and the signaling line circuit from the protected premises to the supervising station must be inspected and tested as part of the inspection and testing program for the protected premises fire alarm system.

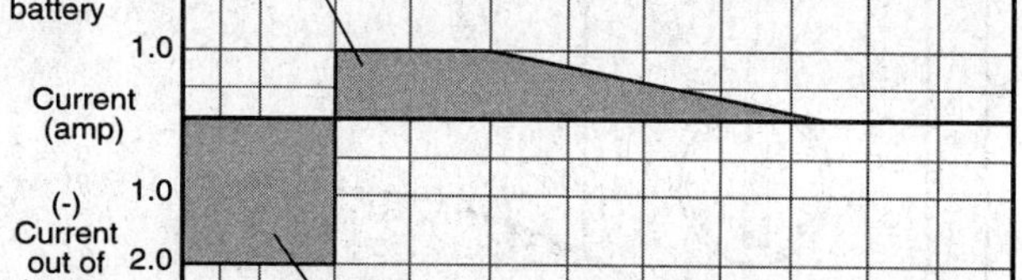

The oldest type of off-premises signaling is the McCulloh system. The fire alarm control unit terminals must be regularly inspected for corrosion, and the function must be tested for successful signal transmission, in spite of each of the following four fault conditions:

1. Open circuit in one conductor,
2. Ground fault on one conductor,
3. Short circuit between the two signal conductors, and/or
4. A concurrent open circuit and ground fault.

Currently, digital alarm communicator systems are the most commonly provided method for communication to a remote supervising station. These systems consist of a digital alarm communicator transmitter (DACT) at the protected premises and a digital alarm communicator receiver (DACR) at the supervising station. A pair of telephone lines routed through the local telephone switching center is used as the communication path. Since it is impossible to supervise the entire signal path from protected premises to supervising station, two "numbers" (lines) are used: (1) primary and (2) secondary.

Digital alarm communicator transmitters are tested by verifying that they transmit the appropriate signal to the remote supervising station receiver. When an alarm signal is created in the fire alarm system, it must be verified that the DACT seizes the line and transmits the alarm signal to the supervising station. Similarly, the line seizure and trouble signal transmission must result from a trouble signal on the fire alarm control unit. When either of the signaling lines (numbers) is disconnected, the DACT must transmit a trouble signal on the alternate line within 4 min.

Systems similar to the digital alarm communicator are now available using radio communication. These are referred to as digital alarm radio transmitters and receivers (DART and DARR, respectively). These systems must also have provisions for secondary communications paths and must be tested in a similar manner as the DACT/DACR system.

Emergency voice communication equipment: The inspection of the emergency voice communication equipment includes the remote phone jacks, handsets, phone hooks, control panel interface, and amplifiers. These components must be inspected for damage or deterioration due to ambient conditions. The testing includes a verification of the operability of both the primary and backup amplifiers, as well as the switching between them. A verification that the call-in silence, off-hook indication, and the operability of the communication from remote jacks to the base station must be achieved in the testing. Each phone set must be tested and determined to be operable. Finally, system operability with a simulated fire condition communications load, involving the concurrent operation of at least five phone sets, should be conducted to assess the voice quality and clarity of the system.

Guard tour monitoring equipment: When guard tour monitoring equipment is integrated into the fire alarm control unit, it must be added to the inspection and testing schedule for the system. The inspection and testing procedures for this equipment prescribed by the manufacturer should be followed.

Metallic conductors: Metallic conductors installed in a building are subject to damage and gradual deterioration. The inspection and testing of a fire alarm system must include an evaluation of the wiring. Stray potentials can be introduced into the fire alarm system from a number of sources. These are most often related to leakage currents and electromagnetic interference from high-voltage ac conductors that have been installed in close proximity to the fire

FIG. 5-5J. A partial discharge/recharge integration can be used to infer battery capacity.

alarm system conductors. However, sources of dc leakage currents have also been encountered. The multimeter is used to measure the conductor-to-conductor and conductor-to-ground potentials, both ac and dc, for each fire alarm system circuit. Where the potential differs from that stipulated in the manufacturer's installation manual by more than 1.0 V (ac or dc), the source of the potential must be identified and the problem corrected.

Fire alarm system conductors can also become shorted. There are many circumstances where a shorted circuit will not result in a "trouble" signal at the control panel, because the control unit circuitry does not monitor for short-circuit conditions. Consequently, fire alarm circuits should also be tested with the multimeter to verify the proper impedance on circuits that are not connected to normally closed contacts of other intentionally closed circuits.

Finally, damage to conductors can cause an increase in conductor resistance that will prevent successful operation of the devices served by the circuit. One method to test for this condition is to short the circuit at the end, and measure the loop resistance to verify it is within the limits established by the equipment manufacturer. If this method is elected, particular care must be exercised to ensure that no shorting devices can accidentally be allowed to remain in place at the conclusion of the testing. An alternative method requires that detailed records be kept of the loop resistance as installed, including: (1) the end-of-line supervisory component and (2) comparing subsequent tests to the as-installed values to identify circuits where the resistance has changed. This method may not be available when testing addressable detector fire alarm systems.

Initiating Devices

Initiating devices sense a fire or a change in the state of some other monitored condition that has an effect on the fire safety of the protected site. These initiating devices include, listed in the same order as in NFPA 72:

1. Heat detectors,
2. Smoke detectors,
3. Radiant energy sensing detectors,
4. Sprinkler system waterflow alarms,
5. Extinguishing system operation initiating devices,
6. Manually activated alarm initiating devices (manual fire alarm boxes), and
7. Supervisory signal initiating devices.

Each of these devices must be tested to verify that its ability to sense the condition for which it was designed has not deteriorated. The specific test is dependent upon the type of device and how it senses the condition it is intended to monitor.

Heat detectors: A common source of failure for heat detection devices is a layer of insulating material between the detector and the fire which retards or prevents response. This is usually found in the form of paint or some other coating. Physical damage has also been found to be a source of problems, causing both spurious alarms as well as failure to alarm.

There is a wide variety of heat sensor designs in common use. NFPA 72 categorizes heat detectors into various classes based upon their general operating characteristics and physical form. Each type has unique features and, hence, specific requirements regarding the inspection, testing, and maintenance of detectors of that type.

The method of testing is established by the manufacturer of the detector, and it is reviewed by nationally recognized testing laboratories as part of the listing evaluation. Consequently, it must be followed meticulously for the results to be credible.

Most methods involve the use of a heat source to simulate the flow of heated air and smoke beneath the ceiling across the detector surface. NFPA 72 requires that restorable, spot-type, and line-type heat detectors be tested by heating them in the manner required by the manufacturer and verifying an alarm within 1 min. Nonrestorable, spot-type, and line-type detectors, such as those that utilize an eutectic alloy of specific melting point, are not tested until they are 15 years old, at which time they are either all replaced or a sample is removed and tested by a nationally recognized testing laboratory to substantiate their continued operability. The activation of nonrestorable heat detectors is simulated by shorting the circuit conductors for verification of the circuit operability.

Smoke detectors: As with heat detectors, there is a wide variety of smoke detectors utilizing a number of different design concepts. Since smoke detectors exist as spot-type, line-type, and air-sampling-type devices, specific methods must be employed that are appropriate to the detector type. However, all inspection, testing, and maintenance methods relating to smoke detectors are designed to ensure three general criteria:

1. That smoke can enter the sensing "chamber" of the detector,
2. That the detector attains the alarm state at the smoke concentration for which the detector was listed, and
3. The detector alarm signal is received and processed by the fire alarm system control panel.

Spot-type smoke detectors are available using the ionization technology and the photoelectric technology. Most ionization detectors suffer from the accumulation of dust within the sensing chamber over time, which often interferes with the flow of electrical current through the sensing chamber in a manner similar to the effect of smoke. This usually renders the affected detector more sensitive and, hence, more prone to spurious alarms. The same effect is seen in the photoelectric detectors. The accumulation of dust increases the internal reflectance within the sensing chamber, making it more sensitive and prone to render a spurious alarm. A sensitivity measurement is critical in order to identify units that have suffered a change in sensitivity due to dirt accumulation which is not apparent from an external inspection.

Spot-type detectors are also subject to clogging from airborne dust or, worse yet, paint. The application of paint to the exterior of a spot-type smoke detector permanently changes the smoke entry characteristics of a detector and is prohibited by NFPA 72. Detectors must be carefully inspected to ensure that smoke entry is uninhibited. Room air-handling equipment can create airflows of such magnitude that it is impossible for the smoke from the design fire to reach the detectors. Consequently, an "in place" test is required to ensure the continued ability for smoke to reach the detector. Usually, an aerosol can of surrogate smoke or some other method outlined in the manufacturer's instructions is used to simulate the flow of the ceiling jet and access to the detector to verify this criterion.

Line-type smoke detectors are usually of the projected beam, light-obscuration type which use the protected room as a sampling chamber. These detectors respond when the optical transmission of light across the room is reduced due to the presence of smoke. The accumulation of dust and dirt on the optical surfaces of these units can cause an increase in detector sensitivity. Consequently, they must also be inspected. Since the alarm threshold of the projected beam-type detectors must be adjustable to accommodate different room dimensions, the manufacturers of this type of detector provide a test method that was evaluated during the listing evaluation. The sensitivity of the unit must be measurable. Often, neutral density filters or test screens are placed between the emitter and receiver units of the projected beam detector to simulate the optical effect of a smoke accumulation in the protected space.

Air-sampling smoke detectors use a set of sampling tubes and an air pump to convey air from the protected space to the detector unit. Detailed inspection and testing methods for these units are stipulated by the manufacturers. The test methods must verify that all of the air-sampling ports and tubes are operative at their design flow rates and that the detector unit, including the sampling pump, is operating within the parameters established by the listing. This implies a sensitivity measurement similar to that required of all of the other detector types.

Smoke detectors are also often used to monitor the air in heating, ventilating, and air-conditioning (HVAC) ducts. Generally, spot-type detectors are used for this type of service, installed in specially listed housings equipped with sampling tubes that allow air to be sampled from the duct. The inspection, testing, and maintenance for these units includes the inspection and testing of the spot-type detector within the housing and the testing of the airflow through the sampling tubes, similar to that required for air-sampling-type units.

Traditionally, smoke detectors were designed with fixed alarm thresholds. However, recently "analog/addressable" detectors have been introduced where the alarm threshold for each detector is a numerical value stored in the fire alarm control panel memory. The control panel sequentially addresses each detector, and the unit responds with a voltage or current that is the analog of stimulus its sensing element is sensing at that moment. The analog voltage or current is compared to the value in memory, and the "alarm" decision is made in the control panel. The sensitivity of this type of detector is determined by the difference between the sensed analog value and the stored threshold value. This type of detector and control panel design minimizes the need for individual detector sensitivity measurements, but does not change the need for regular inspections and maintenance.

Radiant energy sensing detectors: This category includes flame detectors and spark/ember detectors. These detectors depend upon the optical clarity of their lenses and windows, as well as a clear, unobstructed line of sight to the fire. The inspection of these detectors must verify these two critical parameters. The inspection must include a review of the location of the detector and any opaque objects that may have been introduced into the hazard area since the last inspection to verify that the ability of the detector to respond to the design fire has not been impaired.

A close inspection of the detector window/lens must be made to determine if any opaque materials have adhered to the optical surfaces, reducing the sensitivity of the unit. Keep in mind that many materials are transparent in the visible portion of the spectrum but opaque in either the ultraviolet or infrared portions of the spectrum, where the detectors may operate. Many radiant energy sensing detectors are available with a lens clarity monitoring feature. This feature is often associated with flame detectors utilizing ultraviolet sensing as all or part of the detection method, since very small levels of contamination can have a very significant effect on the effective sensitivity of the unit.

Window clarity monitoring is not a sensitivity measurement. Sensitivity measurement for radiant energy sensing detectors requires a means by which a radiant emitter, listed for the purpose, is used to provide emissions of the specific wavelengths and intensities representative of the fire the unit is designed to detect. The response of the detector is measured to determine whether a variation in the calibration has occurred. (See Figure 5-5K and 5-5L.)

The measurement of the sensitivity of a radiant energy-sensing detector, whether a flame detector or a spark/ember detector, is a required part of the inspection and testing regimen required by NFPA 72. Sensitivity measurements are used to identify which detectors have suffered a loss of sensitivity due to aging of internal electronic components or deterioration of the optical clarity of the detector window or lens. A common source of performance deterioration in

FIG. 5-5K. These spark detectors are installed with test lights across the duct which can be used for both the lens clarity testing and sensitivity measurement. (Source: Photonic Solutions, Inc.)

radiant energy sensing detectors is loss of clarity of the detector's optical surfaces. Particular attention must be paid to this fact when maintaining these detectors.

Sprinkler system waterflow alarms: Sprinkler system waterflow alarm initiation devices are often the only form of automatic fire detection in the protected site which can render a fire alarm signal during those periods when the building is unoccupied. The flow of water through the sprinkler system is an indication that a serious condition exists. Consequently, failure to render a signal when waterflow occurs is a doubly serious concern. This establishes the background for the aggressive inspection, testing, and maintenance schedules for these devices.

Sprinkler system waterflow alarm initiating devices are tested by operating the valve controlling flow through the inspector's test connection on the sprinkler system. If the waterflow initiating device is monitoring dry-pipe, pre-action, or deluge systems, the alarm bypass connection, specifically installed for this purpose, is used for testing of waterflow initiating devices. Operation of the appropriate test connection causes a flow equivalent to the smallest orifice sprinkler head installed in the system. The waterflow alarm initiating device must transmit a signal to the fire alarm control panel within 90 sec of the start of the waterflow. The delay between the time that flow begins and the time the alarm is transmitted to the

FIG. 5-5L.   This flame detector is equipped with a built-in sensitivity checking and test lamp that performs both functions.

fire alarm control panel is necessary to accommodate the oscillatory flows that often occur in large sprinkler systems and, as a result, surges in the public water supply pressure.

Extinguishing system operation initiating devices: These initiating devices usually take the form of discharge switches specifically designed for the particular model of extinguishing system involved. Some of these discharge switches have a means to simulate the pressure of the extinguishing agent and, thus, test the operability of the initiating device. Others do not, and the testing can only be performed by shorting the contacts at their wiring terminations. In either case, each extinguishing system discharge initiating device must be tested as specified by the manufacturer and appropriate system functions verified.

Manually activated alarm initiating devices: Manually activated alarm initiating devices, commonly called manual stations, have traditionally consisted of a mechanical means to cause an electrical switch to transfer from open to closed, thereby shorting the conductors of the fire alarm initiating device circuit. Two aspects of these devices should be inspected and tested. First, the mechanical portion of the unit should be inspected and the operability should be verified. Recent legislation establishes maximum allowable forces for the operation of manual stations. Any unit that appears to have suffered corrosion, damage, or some other potentially adverse effect should be tested to verify that the operation force is within the limits for the make and model device in question. This is especially important in hazard areas where the walls have been painted subsequent to the installation of the manual station. Second, the manually activated alarm initiating devices should be activated to verify that the internal switch contact is electrically operative. Furthermore, many fire alarm systems activate different responses due to the operation of manual stations in different areas of the building or facility. The proper sequence of operation must be verified for each area.

Public-accessible fire-alarm boxes, commonly known as street boxes, are addressed in NFPA 72. These devices exist in a wide variety of types, connected to a variety of municipal signaling systems. The maintenance of these units is a municipal function, and the appropriate means of inspection and testing is established by the municipality.

Supervisory signal initiating devices: Supervisory signal initiating devices monitor the status of critical fire protection assets. These include sprinkler system control valve status, water pressure status,

water level, air pressure, and temperature. It is imperative to recognize that the failure mode and probability of the electrical supervisory initiating device is not related to the probability of failure of the monitored condition. Any coincidence in the frequency and scheduling of the inspection and testing of a supervisory signal initiating device with the supervised device is just that, a coincidence. Control valves do not fail on the same timetable as valve status supervisory signal initiating devices. The failures of valves are due to the compression welding of bronze components, tuberculation of the piping, galvanic corrosion, and the polymerization of seat grease. Failure of status switches is related to the corrosion of electronic contacts, and the fatigue of copper wiring conductors and mounting brackets. The inspection and testing of the supervisory signal initiating devices is necessary to ensure that these devices are providing an accurate indication of the status of the supervised equipment and a reliable indication of the change in status of the equipment if and when it occurs.

Inspection and testing of the supervisory signal initiating devices usually involves a close inspection of each device to verify that the ambient conditions have not caused any deterioration in the device, followed by a temporary simulation of a change in status to verify that the supervisory signal initiating device transmits the indication of the change in status to the fire alarm control unit. Once the receipt of the change-in-status signal at the fire alarm system control unit has been verified, the simulated change in status is restored to normal and the receipt of the return-to-normal signal at the control unit is verified. Transmission of a change in status to the fire alarm system control unit should occur without any undue delay. If delays are encountered, the cause should be identified and corrected.

Notification Appliances

The principal objective of every fire alarm system is to ensure timely notification to all building occupants, enabling them to make safe egress from the fire location; therefore, the inspection and testing of the alarm notification appliances is crucial.

The three general types of notification appliances referenced in NFPA 72 are: (1) audible notification appliances, (2) visible notification appliances, and (3) speakers for emergency voice communication. Each type of notification appliance has a test method that verifies that the unit is providing the notification signal at a level consistent with the design objective.

Audible notification appliances: These include bells, horns, sirens, chimes, speakers, buzzers, and any other device designed to make a sound other than human speech. They all have a means of converting electrical energy into sound, often involving the making and breaking of an electrical switch contact connected to a bell clapper or a diaphragm and an electromagnet, thus producing a mechanical oscillation and sound. Corrosion, metal fatigue, paint clogging, mechanical damage, electrical contact erosion, and electrical conductor fatigue all can cause a deterioration in the quantity of sound a device produces. Each audible notification appliance must be inspected for signs of deterioration.

The audibility must also be tested to verify that the audibility criteria for the hazard area are still being achieved by the system. Audibility is verified through the measurement of the sound pressure produced by the audible notification appliance. Sound pressure is measured in units of decibels, dBA; the A signifies the A-scale which is designed to weigh sound pressure *vs.* frequency in a manner that approximates human hearing. The sound pressure measurement is made with a sound meter. Only sound meters that comply with ANSI S1.42, *Design Response of Weighting Networks for Acoustical Measurements*,[8] should be used for the assessment of fire alarm systems. Care must also be taken that the sound pressure meter is used as directed by the operating manual to minimize measurement errors. (See Figure 5-5M.)

There are two criteria that must be met by the audible notification appliances. The first is the minimum compliance audibility (usually 85 dBA or 15 dBA above average ambient, unless the average ambient is equal to or above 115 dBA). Since sound intensity usually varies inversely with the distance from the audible appliance, the sound pressure readings are taken at a location farthest from the audible appliance. Generally, if the minimum compliance sound level is achieved at that location, it can be safely assumed that closer locations will also be above the minimum compliance sound level. However, "usually" does not mean *always*. Occasionally buildings have unexpected acoustics that affect the sound intensities at various locations within the building. Keep this in mind when assessing audibility, and be prepared to make additional measurements where audibility seems lower than "normal."

The second criterion is the maximum allowed sound pressure level at the closest occupiable location to the audible appliance. This sound level cannot exceed 130 dBA without subjecting the occupant to potential hearing injury. System component failures rarely result in higher sound output from audible notification appliances. However, changes in the occupancy, such as installation of catwalks, stairs, mezzanines, etc., can render a formerly unoccupiable location occupiable. This is consistent with the inspection of all other system components, in that the role of the inspection includes identifying changes in the protected space that may have an impact on the fire alarm system.

Speakers: Emergency voice communication speakers are also audible notification appliances, and they are required to achieve sound levels equivalent to audible notification appliances (85 dBA or 15 dBA above average ambient, unless the average ambient is equal to or above 115 dBA). The inspection should identify any form of damage, e.g., mechanical deformation, corrosion, metal fatiguing, and paint clogging. The testing is similar to that for audible notification appliances. The sound level is measured with an ANSI S1.42 compliant sound level meter.[7] The minimum compliance sound level must be verified throughout the hazard area, and the test must verify that the maximum allowable sound level is not exceeded at any occupiable location.

An additional test criterion that is not relevant to other audible notification appliances is included when evaluating emergency voice communication speakers. This criterion is clarity and intelligibility of the voice communications. In the testing of emergency voice communication speakers, the clarity and understandability of the speech message must be assessed. There is no quantitative

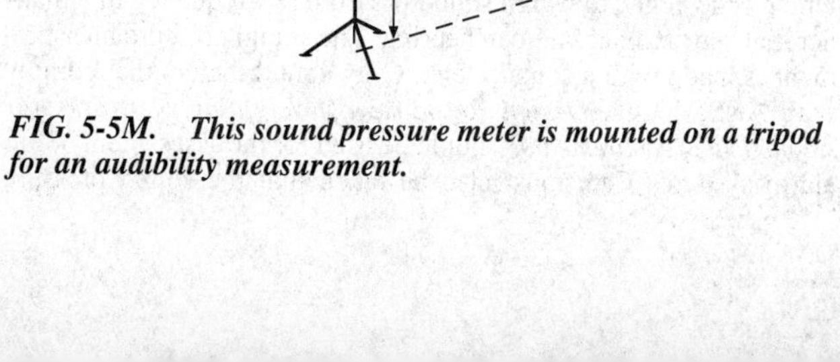

measure or standard on speech intelligibility. The Authority Having Jurisdiction is the judge of whether the communications are sufficiently clear.

Visible notification appliances: Visible notification appliances are inspected and tested to verify that they are still capable of achieving the design objectives of the fire alarm system. A light meter analogous to the sound-level meter is not in common usage as of this writing. This limits the inspection and testing to those procedures recommended by the manufacturer of the specific visible notification appliance used.

During the inspection and operational testing of the visible notification appliances, particular attention must be paid to any furnishings, e.g., shelves, ornamental plants, lamps, banners, sculptures, etc., or added construction features, e.g., stairs, catwalks, ladders, lighting fixtures, and partitions, that interfere with a clear line of sight to the visible notification appliances.

INSPECTION, TESTING, AND MAINTENANCE SCHEDULING FOR FIRE ALARM SYSTEMS AND COMPONENTS

The requirements regarding the inspection, testing, and maintenance of fire alarm signaling systems must also be scheduled. Since different components have different failure rates, those with the higher failure rates require more frequent inspections and testing if the overall system reliability is to be kept at an acceptable level. Inspection, testing, and maintenance interval recommendations come from both the predicted reliability derived from failure rate analyses and from experience data derived over time. Manufacturers provide specific schedules for the service requirements for their products, and these must be observed where they establish a more frequent regimen of inspection, testing, and maintenance.

The objectives of an inspection program are to identify cases where: (1) the hazard area has changed, rendering the existing fire alarm system inappropriate for the hazard area, and (2) some episode of physical abuse has occurred which has damaged the component, rendering it incapable of performing in a manner consistent with the system design. The frequency of these types of conditions is a function of the hazard area, not the inherent reliability of the system components. Consequently, some sites will require more frequent inspections due to the unique nature of the particular site.

NFPA 72 provides the required visual inspection frequencies, reproduced here as Table 5-5B.

Testing of fire alarm system components occurs on a schedule that is less frequent than the inspections. The testing uncovers failures that are not apparent from a visual inspection and are generally attributable to the combined effect of the ambient conditions and the inherent failure rates of the components. There are applications where the testing frequency established in NFPA 72 as a minimum compliance criterion is not adequate to achieve the required level of reliability. The effective reliability of such a fire alarm system can be improved by implementing a more frequent testing program.

NFPA 72 provides the required testing frequencies, reproduced here as Table 5-5C.

HOUSEHOLD FIRE WARNING EQUIPMENT

Household fire alarm systems have been implemented in a number of different ways. More recent design concepts correct the perfor-

FIG. 5-5M. This sound pressure meter is mounted on a tripod for an audibility measurement.

5.0 ft (1.53 m)

TABLE 5-5B. *Visual Inspection Schedules for Fire Alarm System Components*

Component	Init./Reaccpt.	Monthly	Quarterly	Semiann.	Ann.
1. Control Equipment: Fire Alarm Systems Monitored for Alarm, Supervisory, Trouble Signals					
a. Fuses	X				X
b. Interfaced Equipment	X				X
c. Lamps and LEDs	X				X
d. Primary (Main) Power Supply	X				X
2. Control Equipment: Fire Alarm Systems Unmonitored for Alarm, Supervisory, Trouble Signals					
a. Fuses	X (weekly)				
b. Interfaced Equipment	X (weekly)				
c. Lamps and LEDs	X (weekly)				
d. Primary (Main) Power Supply	X (weekly)				
3. Batteries					
a. Lead-Acid	X	X			
b. Nickel-Cadmium	X			X	
c. Primary (Dry Cell)	X	X			
d. Sealed Lead-Acid	X			X	
4. Transient Suppressors	X			X	
5. Control Panel Trouble Signals	X			X	
6. Fiber-Optic Cable Connections	X				X
7. Emergency Voice/Alarm Communications Equipment	X			X	
8. Remote Annunciators	X			X	
9. Initiating Devices					
a. Air Sampling	X			X	
b. Duct Detectors	X			X	
c. Electromechanical Releasing Devices	X			X	
d. Fire Extinguishing System(s) or Suppression System(s) Switches	X			X	
e. Fire Alarm Boxes	X			X	
f. Heat Detectors	X			X	
g. Radiant Energy Fire Detectors	X		X		
h. Smoke Detectors	X			X	
i. Supervisory Signal Devices	X		X		
j. Waterflow Devices	X		X		
10. Guard's Tour Equipment	X			X	
11. Interface Equipment	X			X	
12. Alarm Notification Appliances—Supervised	X			X	
13. Supervising Station Fire Alarm Systems—Transmitters					
a. DACT	X			X	
b. DART	X			X	
c. McCulloh	X			X	
d. RAT	X			X	
14. Special Procedures	X			X	
15. Supervising Station Fire Alarm Systems—Receivers					
a. DACR*	X	X			
b. DARR*	X			X	
c. McCulloh Systems*	X			X	
d. Two-Way RF Multiplex*	X			X	
e. RASSR*	X			X	
f. RARS*	X			X	
g. Private Microwave*	X			X	

* Reports of automatic signal receipt shall be verified daily.

mance shortcomings of older design philosophies and reflect the improved performance of modern equipment.

NFPA 72 requires the monthly inspection and testing, per the manufacturer's instructions, of all smoke detectors in a household fire alarm application, regardless of whether those detectors are single station units, interconnected multiple-station units, or detectors integrated into a complete household fire alarm system. This monthly testing should include a verification that all audible and visible notification appliances are operative. Furthermore, NFPA 72 requires that household fire alarm systems, i.e., systems consisting of detectors, notification appliances, and a control panel, be inspected and tested at least every three years by a qualified alarm technician with a procedure in compliance with the manufacturer's instructions.

TABLE 5-5C. *The Testing Frequencies for Fire Alarm System Components*

Component	Init./Reaccpt.	Monthly	Quarterly	Semiann.	Ann.
1. Control Equipment: Fire Alarm Systems Monitored for Alarm, Supervisory, Trouble Signals					
a. Functions	X				X
b. Fuses	X				X
c. Interfaced Equipment	X				X
d. Lamps and LEDs	X				X
e. Primary (Main) Power Supply	X				X
f. Transponders	X				X
2. Control Equipment: Fire Alarm Systems Unmonitored for Alarm, Supervisory, Trouble Signals					
a. Functions	X		X		
b. Fuses	X		X		
c. Interfaced Equipment	X		X		
d. Lamps and LEDs	X		X		
e. Primary (Main) Power Supply	X		X		
f. Transponders	X		X		
3. Engine-Driven Generator	X (weekly)				
4. Batteries—Central Station Facilities					
a. Lead-Acid Type					
1. Charger Test	X				X
(Replace battery as needed.)					
2. Discharge Test (30 min)	X	X			
3. Load Voltage Test	X	X			
4. Specific Gravity	X			X	
b. Nickel-Cadmium Type					
1. Charger Test	X		X		
(Replace battery as needed.)					
2. Discharge Test (30 min)	X				X
3. Load Voltage Test	X				X
c. Sealed Lead-Acid Type	X	X			
1. Charger Test		X	X		
(Replace battery as needed.)					
2. Discharge Test (30 min)	X	X			
3. Load Voltage Test	X	X			
5. Batteries—Fire Alarm Systems					
a. Lead-Acid Type					
1. Charger Test	X				X
(Replace battery as needed.)					
2. Discharge Test (30 min)	X			X	
3. Load Voltage Test	X			X	
4. Specific Gravity	X			X	
b. Nickel-Cadmium Type					
1. Charger Test	X				X
(Replace battery as needed.)					
2. Discharge Test (30 min)	X				X
3. Load Voltage Test	X			X	
c. Primary Type (Dry Cell)					
1. Load Voltage Test	X	X			
d. Sealed Lead-Acid Type					
1. Charger Test	X				X
(Replace battery every 4 years.)					
2. Discharge Test (30 min)	X				X
3. Load Voltage Test	X			X	
6. Batteries—Public Fire Alarm Reporting Systems	X (daily)				
a. Lead-Acid Type					
1. Charger Test	X				X
(Replace battery as needed.)					
2. Discharge Test (2 hr)	X		X		
3. Load Voltage Test	X		X		
4. Specific Gravity	X			X	
b. Nickel-Cadmium Type					
1. Charger Test	X				X
(Replace battery as needed.)					
2. Discharge Test (2 hr)	X				X
3. Load Voltage Test	X		X		

TABLE 5-5C. *(continued)*

Component	Init./Reaccpt.	Monthly	Quarterly	Semiann.	Ann.
c. Sealed Lead-Acid Type					
1. Charger Test	X				X
(Replace battery as needed.)					
2. Discharge Test (2 hr)	X				X
3. Load Voltage Test	X		X		
7. Fiber-Optic Cable Power	X				X
8. Control Unit Trouble Signals	X				X
9. Conductors/Metallic	X				
10. Conductors/Nonmetallic	X				
11. Emergency Voice/Alarm Communications Equipment	X				X
12. Retransmission Equipment	X				
13. Remote Annunciators	X				X
14. Initiating Devices					
a. Duct Detectors	X				X
b. Electromechanical Releasing Device	X				X
c. Fire Extinguishing System(s) or Suppression System(s) Switches	X				X
d. Fire-Gas and Other Detectors	X				X
e. Heat Detectors	X				X
f. Fire Alarm Boxes	X				X
g. Radiant Energy Fire Detectors	X			X	
h. All Smoke Detectors—Functional	X				X
i. Smoke Detectors—Sensitivity					
j. Supervisory Signal Devices (except value tamper switches)	X		X	X	
k. Waterflow Devices	X			X	
l. Value Tamper Switches					
15. Guard's Tour Equipment	X				X
16. Interface Equipment	X				X
17. Special Hazard Equipment	X				X
18. Alarm Notification Appliances					
a. Audible Devices	X				X
b. Speakers	X				X
c. Visible Devices	X				X
19. Off-Premises Transmission Equipment	X		X		
20. Supervising Station Fire Alarm Systems—Transmitters					
a. DACT	X				X
b. DART	X				X
c. McCulloh	X				X
d. RAT	X				X
21. Special Procedures	X				X
22. Supervising Station Fire Alarm Systems—Receivers					
a. DACR	X	X			
b. DARR	X	X			
c. McCulloh Systems	X	X			
d. Two-Way RF Multiplex	X	X			
e. RASSR	X	X			
f. RARSR	X	X			
g. Private Microwave	X	X			

UL 217, *Standard for Safety Single and Multiple Station Smoke Detectors*, allows a maximum failure rate of 4.0 f/mdoh for the detector circuitry.[8] This failure rate can be expressed as 1.0 failures per 250,000 device operating hours, 1.0 failure per 250,000 hr / 8760 hr/ yr = 28.5 years, or 0.035 failures per year. The battery for the smoke detector can only operate the unit for 1.5 years. Consequently, its failure rate is 0.667 failures per year. The failure rate for the household fire alarm is the failure rate per detector and battery combination, multiplied by the number of units in the household. Assuming 6 detectors in the average house and a failure rate of

$1 - (1 - .667)(1 - .035) = .679$ one obtains $6 \times 0.679 = 4.074$ failures per year for Λ. Using the equation developed earlier, i.e.,

$$R = e^{-\Lambda t}$$

where

R = reliability of the system;

e = Napierian logarithm base, 2.71828;

Λ = inherent failure rate of the system; and

t = time period between performing recommended inspection and testing, $^{1}/_{12th}$ of a year or 0.0833 year

$R = e^{-4.074 \times 0.0833}$

= 0.712 or 71.2 percent.

Thus, generally available residential smoke detectors when installed, inspected, and tested monthly per the manufacturer's instructions and NFPA 72 can be expected to provide a functional reliability of a nominal 71.2 percent.

CRITICALITY OF INSPECTION, TESTING, AND MAINTENANCE ON THE MISSION-EFFECTIVENESS OF THE SYSTEM

Electronic systems are, at this time, the only systems on which solid reliability data exist in the quantity and quality that allows one to develop a predictive computational method to assess various design alternatives. All physical systems have failure modes, and each system component demonstrates failure at a finite rate over the majority of its operational lifetime. Nonelectrical fire alarm systems are not necessarily more reliable than electronic systems. It is possible to compute the reliability of electrical systems only because good failure rate data exists.

The one concept to be gained from reliability analysis and prediction computations is that all systems are destined to fail, given enough time. It is a probability so high as to approach a *certainty*. A performance-based design cannot be deemed complete until it includes a quantitative assessment of the likelihood that the system will perform as designed. This assessment is dependent upon the service and maintenance interval for the system. With long time intervals between executions of the inspection and testing regimen, the probability increases that the system will be operating with some type of unnoticed impairment. Obviously, the time during which the system is partially and perhaps even completely impaired increases with the increase in the time interval between each exercise of the required inspection and testing. Consequently, no performance- or objective-oriented design is complete without a preventive maintenance program that includes inspection and testing, exercised on an interval that is predicated upon the size and complexity of the system itself.

Consider the example fire alarm system presented earlier. A system consisting of 10 smoke detectors, a notification appliance, and a control panel produced a predicted failure rate of 0.3824 failures per year. Table 5-5D shows the effect of varying inspection and testing intervals on the reliability of this system.

The reliability of the system and, hence, the needed maintenance interval are dependent upon the size and complexity of the system. Table 5-5E demonstrates the impact that increasing system size has on the needed inspection and testing interval in order to maintain a uniform design reliability of 71.2 percent (0.712).

Clearly, as the size or complexity of the system increases, the frequency of the inspection and testing program must be increased to retain a constant reliability. If the inspection and testing frequency is held constant, then the reliability of the system decreases each time the size and complexity of the system are increased.

Tables 5-5C, 5-5D, and 5-5E demonstrate the importance of the inspection, testing, and maintenance program. A system cannot be expected to achieve the design objectives if the inspection, testing, and maintenance program does not reflect the reliability objectives of the design. The failure to maintain a system in a manner

TABLE 5-5D. System Reliability Dependent upon the Inspection and Testing Interval System (Actual values based upon the reliability data used in the example.)

Inspection and Testing Interval	Reliability
Quarterly	0.904
Semi-Annually	0.826
Annually	0.682
Bi-Annually	0.465

TABLE 5-5E. Inspection and Testing Interval Necessary to Maintain a 71.2 Percent Reliability as System Size Is Increased (Values based upon the reliability data used in the example.)

System Size	Λ	Inspection and Testing Interval
10 Detectors, 1 Notification Appliance	0.3824	0.93 years
20 Detectors, 2 Notification Appliances	0.5278	0.68 years
40 Detectors, 4 Notification Appliances	0.8554	0.42 years
80 Detectors, 8 Notification Appliances	1.486	0.24 years

consistent with the design assures the owner/operator of the protected site that the fire alarm system will NOT achieve the design objective.

BIBLIOGRAPHY

References Cited

1. Institute of Electrical and Electronics Engineers, Inc., *Reliability: An IEEE Spectrum Compendium*, IEEE, Piscataway, NJ, 1981.
2. Moore, W. D., P.E., and Cholin, J. M., P.E., "Reliability Analysis and Prediction Techniques for Fire Protective Signaling Systems," 1995 NFPA Annual Meeting, published by the authors, 1995.
3. U.S. Department of Defense, *Military Handbook: Reliability Prediction of Electronic Equipment*, MIL-HDBK-217F, U.S. Government Printing Office, 1990.
4. Klote, J. H., D. Sc., P.E., and Milke, J. A., Ph.D., P.E., *Design of Smoke Management Systems*, American Society of Heating, Refrigerating, and Air-Conditioning Engineers, Inc., Atlanta, GA, 1992.
5. Schifiliti, R. P., *et al.,* "Design of Detection Systems," *The SFPE Handbook of Fire Protection Engineering*, 2nd ed., National Fire Protection Association, Quincy, MA, 1995.
6. General Electric, *Nickel-Cadmium Battery Application Engineering Handbook*, GET-3148A, General Electric Company Battery Business Dept., Gainesville, FL, 1975.
7. ANSI S1.42, *Design Response of Weighting Networks for Acoustical Measurements*, American National Standards Institute, New York, 1986.
8. UL 217, *Standard for Safety Single and Multiple Station Smoke Detectors*, Underwriters Laboratories Inc., Northbrook, IL, 1993.

NFPA Codes, Standards, and Recommended Practices

Reference to the following NFPA codes, standards, and recommended practices will provide further information on fire alarm system inspection, discussed in this chapter. (See the latest version of *The NFPA Catalog* for availability of current editions of the following documents.)

NFPA 11, *Standard for Low-Expansion Foam*
NFPA 12, *Standard on Carbon Dioxide Extinguishing Systems*
NFPA 12A, *Standard on Halon 1301 Fire Extinguishing Systems*
NFPA 13, *Standard for the Installation of Sprinkler Systems*

NFPA 15, *Standard for Water Spray Fixed Systems for Fire Protection*

NFPA 16, *Standard for the Installation of Deluge Foam-Water Sprinkler and Foam-Water Spray Systems*

NFPA 17, *Standard for Dry Chemical Extinguishing Systems*

NFPA 20, *Standard for the Installation of Centrifugal Fire Pumps*

NFPA 25, *Standard for the Inspection, Testing, and Maintenance of Water-Based Fire Protection Systems*

NFPA 72, *National Fire Alarm Code*

NFPA 110, *Standard for Emergency and Standby Power Systems*

NFPA 111, *Standard on Stored Electrical Energy Emergency and Standby Power Systems*

NFPA 2001, *Standard on Clean Agent Fire Extinguishing Systems*

Additional Reading

Bukowski, R. W., P.E., and O'Laughlin, R. J., P.E., *Fire Alarm Signaling Systems,* National Fire Protection Association, Quincy, MA, 1994.

Carson, W. G., P.E., and Klinker, R. L., P.E., *Fire Protection Systems Inspection, Testing, and Maintenance Manual,* 2nd ed., National Fire Protection Association, Quincy, MA, 1993.

Hall, John R., "Probability Concepts," *The SFPE Handbook of Fire Protection Engineering,* 2e. Philip DiNenno, ed., National Fire Protection Association, Quincy, MA, 1995.

Hall, John R., "Statistics," *The SFPE Handbook of Fire Protection Engineering,* 2e. Philip DiNenno, ed., National Fire Protection Association, Quincy, MA, 1995.

Hamming, R. W., *The Art of Probability for Scientists and Engineers*, Addison-Wesley, Redwood City, CA, 1991.

Moore, W. D., P.E., *National Fire Alarm Code Handbook*, National Fire Protection Association, Quincy, MA, 1994.

Roe, B. P., *Probability and Statistics in Experimental Physics*, Springer-Verlag, New York, 1992.

Western Electric, *Statistical Quality Control*, Western Electric Co., Inc., Indianapolis, IN, 1956.

HOUSEHOLD FIRE WARNING EQUIPMENT

Revised by Walter F. Schuchard

Why are household warning systems needed? Because America's unusually high fire death rate is dominated by fires in U.S. homes, and household warning systems, specifically smoke detectors, provide a rare combination of effectiveness and affordability in reducing the risk of death in home fires. The dominance of U.S. homes in fire deaths is demonstrated each year. In 1995, 79 percent of all fire deaths occurred in U.S. homes and 91 percent of fire deaths in any type of building. Analyses by NFPA indicate that, when a fire occurs, detectors cut the chances of dying nearly in half. The potential impact of detectors is far greater still because the 50 percent reduction reflects all the current gaps, such as households that fail to provide enough detectors or proper locations or working batteries.

Why do smoke detectors have such great performance and even greater potential? Actual fire tests in residential occupancies have shown that measurable amounts of smoke have preceded measurable amounts of heat in almost all cases. In the other cases, the smoke and heat appeared almost simultaneously.

This chapter will discuss the reasons household fire warning systems are needed, the history of residential fire warning systems, research into system effectiveness, the importance of having a system in the home and how to make the best use of it, and the importance of an effective escape plan. The types of systems and detectors will also be covered, along with guidelines for installing and testing household fire warning equipment.

Although heat detectors for residential use have been available since 1921, field tests have shown that they are not as effective as smoke detectors in detecting fires in the home.

Residential smoke detectors were relatively expensive when they began to appear in the marketplace in the late 1960s. In 1970, however, the introduction of a battery-operated smoke detector, along with several new line-powered smoke detectors, initiated a period of public acceptance of single-station residential smoke detectors. By 1995, 93 percent of U.S. homes had at least one smoke detector.

RESIDENTIAL FIRE DETECTOR RESEARCH

Current performance requirements and installation practices for residential fire detectors are the result of knowledge and experience

gained from several test programs. The first major test programs using heat and smoke detectors were "Operation School Burning" and "Operation School Burning No. 2," conducted by the Los Angeles (CA) Fire Department in 1959 and 1961, respectively. These tests had a major impact on fire protection and illustrated two significant points: (1) the tests proved that smoke detectors could provide a high level of life safety due to their fast response to a fire, and (2) detectable quantities of smoke usually preceded detectable levels of heat.

The next significant test series was conducted by the Bloomington (MN) Fire Department in May 1969. Although all detectors employed in the Bloomington tests were system-connected units, the results of these tests—such as time of detector response, order of response as a function of detector and fire type, and conditions at time of response—correlated well with subsequent tests.

In 1974, NFPA members adopted a major change to NFPA 74, *Household Fire Warning Equipment* (hereinafter referred to as NFPA 74). Prior to this, NFPA 74 required that, in addition to a smoke detector, a home be protected with a heat detector in every major room of the dwelling. The 1974 edition recognized two facts: (1) the mandatory total system approach was impractical for the majority of the population, and (2) heat detectors were not the most effective detectors for the home. Therefore, an NFPA committee developed a matrix of four "levels of protection" from which the homeowner could choose. The four levels of protection concept was finally adopted by the NFPA membership, but not without a fight. The opposition (principally some fire chiefs and fire marshals) felt that they could not recommend anything less than complete protection, and that effectiveness of Level 4, which provided for one or two smoke detectors in the home, had not been proved by test data.

Based on this latter argument, the National Bureau of Standards (NBS) [now the National Institute for Standards and Technology (NIST)] contracted with ITT Research Institute and Underwriters Laboratories Inc. (UL) to provide just such test data. The results of the first phase of this test program were published in August 1975 under the title "Detector Sensitivity and Siting Requirements for Dwellings." The study, more commonly known as the "Indiana Dunes Report," showed, in fact, that Level 4 coverage was inadequate under certain conditions. The report recommended that at least one smoke detector on each story of the residence be used as a minimum requirement. This was referred to as the "every level" smoke detection system. Based primarily on the results of this report, NFPA was revised again in 1978 by dropping the four levels of protection and adopting a minimum required system of one smoke detector on

Walter F. Schuchard is president of Fire Detection Consulting Service in Hingham, MA.

each story (floor level) of a home. Also, because of the slow response of heat detectors when compared to smoke detectors, NFPA 74 no longer permitted the use of heat detectors as the prime method of fire detection.

Shortly after publication of the first phase of the Indiana Dunes Report, the Minneapolis Fire Department conducted a similar independent study on smoke detector performance. The Minneapolis and Dunes tests were very similar in that each was conducted in a two-story house with basement, with detectors installed on each floor. Likewise, the conclusions of the Minneapolis and Dunes tests were essentially identical; thus, the Minneapolis tests provided an independent verification of many of the conclusions of the Dunes tests.

In 1993, NFPA 74 was combined with the other fire alarm system standards and now appears as Chapter 2 of NFPA 72, *National Fire Alarm Code* (hereinafter referred to as NFPA 72). The definitions formerly found in NFPA 74 are now in Chapter 1 of NFPA 72, and the fire alarm system test requirements have been moved to Chapter 7 of the code.

In the 1993 conversion of NFPA 74, the requirement for a smoke detector in each sleeping room in new construction was introduced in NFPA 72. This change specifically addressed the protection of the occupant of the sleeping room. Detection of smoke from a fire within the sleeping room by a detector outside the sleeping room, especially with the door closed, has been a problem.

Chapter 1 of the 1996 edition of NFPA 72 introduced several terms that directly affect household fire warning equipment. "Single station alarm," "multiple station alarm," "smoke alarm," and "heat alarm" are detectors combined with a notification appliance and an integral power source or can be directly connected to household power. These alarms not only detect, but notify.

In addition, the 1996 edition of NFPA 72 requires both primary (main) and secondary (standby) power for single- and multiple-station alarms in new construction. In existing construction a single source of power, either ac or monitored battery, is still acceptable.

ELEMENTS OF RESIDENTIAL FIRE PROTECTION

The many elements of fire safety and fire protection can be grouped into six elements that need to work together in an integrated system for best results: (1) fire prevention, (2) prevention of unusually rapid early growth or spread, (3) automatic detection, (4) automatic suppression, (5) compartmentation or containment, and (6) evacuation or escape. Each is important individually, and all are interdependent in any successful residential fire protection scheme. From the perspective of household early warning systems, these six elements need to be examined in terms of the essential steps that make the automatic detection element successful and the other fire elements that can render detector protection unsuccessful. Taking them in order, one may focus on prevention of ignition and of rapid early growth or spread under a generic title of minimizing fire hazards.

Also, although automatic residential sprinkler systems are gaining acceptance for new home construction and are mandated in several jurisdictions the general consensus among fire safety professionals is to also recommend automatic smoke detection in accordance with NFPA 72, Chapter 2, "Household Fire Warning Equipment," providing a minimum level of life safety protection.

Minimizing Fire Hazards

It is always easier to prevent a fire before it occurs than to detect it once it has occurred; therefore, minimizing fire hazards in and around the home is the first key element to minimizing hazards. This includes care in the selection and use of all potential heat sources (e.g., tobacco products, heating and cooking equipment, the electrical system, appliances, lighting appliances) and fuel sources (e.g., furnishings and contents, interior finishes). Preventing rapid early growth or spread implies special attention to removing all unessential flammable liquids and other highly combustible items from the home. Further, it should also mean strict adherence to and enforcement of building and fire codes in the construction of residential occupancies, use of only low flame-spread materials for interior finish, and use of low-flammability or fire-retardant-treated materials in residential furnishings. Proper compartmentation is also essential to slow the fire as it moves from zone to zone, extending the time to hazardous exposure for most occupants. It must be realized that, if a fire develops rapidly enough, residential smoke detectors may not provide enough time for the occupants of a dwelling to escape before their exit routes are cut off. Therefore, it is highly dangerous for people to be careless, thinking smoke detectors will protect them from their carelessness.

Installation of Smoke Detectors

Full effectiveness for household warning systems means the right number of the right kind of detectors, located in the right places and operating properly. As required by NFPA 72, Chapter 2, "Household Fire Warning Equipment," the acceptable minimum for existing construction consists of one smoke detector outside each sleeping area, one on each habitable story, and one in the basement of the home. Additionally, for new construction, a smoke detector is required in each sleeping room. It is also recommended that the minimum package be supplemented by additional smoke detectors in other areas as needed. For example, a smoke detector could be located inside each bedroom (required in new construction), in living and family rooms, and smoke or heat detectors could be located in the kitchen, attached garage, attic, and furnace room. Each additional detector could provide more time for the family to escape. When more than one smoke detector is installed, the detectors should be arranged so all of the alarm appliances sound when any one device detects smoke (required in new construction). This can be done either by installing multiple-station smoke alarms or a complete system with smoke detectors, panel, and separate notification appliances.

Equally as important as the installation is proper care and maintenance of the detectors. Each residential smoke alarm comes with an owner's booklet describing the necessary maintenance procedures. A smoke detector with a dead or missing battery will be useless in a fire. In fact, analysis of fire reports at NFPA has shown roughly one-third of U.S. home smoke alarms are not working when fires occur, typically because of dead or missing batteries. This is a major obstacle to achievement of full success with home detectors. Because of the battery problem, NFPA 72, Chapter 2, "Household Fire Warning Equipment," mandates use of both primary ac and secondary battery power. It would follow that, even in existing households, smoke alarms should also be powered by ac with standby battery.

Escape Planning

The last key element in residential fire protection is the development and practice of a family escape plan. Remember, residential smoke alarms can do no more than warn occupants of a fire in the home. Detectors will not extinguish the fire. Also, depending upon the speed at which the fire is developing, the amount of time available for escape may be limited. It is critical that all occupants leave the home immediately and *call the fire department from a neighbor's house*. Many deaths have been reported in fires where people have unwisely taken extra time, for example, to get dressed, gather valuables, or look for pets. Also, it is important that everyone have an alternate way out of each room in case the primary exit is blocked by fire. There must be a prearranged outside meeting place so every-one will know that the entire family has escaped the fire. Deaths have resulted because people reluctant to climb out of windows were overcome by smoke, and, equally tragic, many have died when they reentered a burning home to look for people who had, in fact, already escaped.

Information on how to formulate a family escape plan is contained in the homeowner's booklet provided with each residential smoke detector and by NFPA in its "E.D.I.T.H." (Exit Drills In The Home) materials.

TYPES OF SYSTEMS

Residential fire detection may range in size from one single-station smoke alarm in an apartment, manufactured home, or recreational vehicle to a centrally wired system containing numerous detectors and separate alarm notification appliances. The fire detection system may be combined with a burglar alarm system and an emergency medical alert system. It may be connected to an alarm monitoring company by leased telephone lines or by a digital alarm communication transmitter. The number of devices and the complexity of the system are determined, on the low end, by the minimum acceptable for a given dwelling, and, on the high end, by the extent of the occupant's needs and finances.

Single- and Multiple-Station Smoke Alarms

A single-station alarm (Figures 5-6A and 5-6B) is a self-contained device that consists of a detecting element, electrical control components, an alarm-sounding appliance, and provision for connection to a separate power supply source or an integral power source, or both. In new construction, the smoke alarm is mounted to an outlet box and directly connected to unswitched house power.

A direct-wired type, when provided with one or more additional wires for interconnection with other similar smoke alarms, is referred to as a multiple-station smoke alarm. (See Figure 5-6C.) When interconnected according to the instructions provided, an alarm from any one of the interconnected smoke alarms activates the alarm in all the detectors. This is important because the detector near the bedrooms should sound when a smoke alarm located in another area senses smoke. The number of smoke alarms that are permitted to be interconnected is limited to 12 by NFPA 72, Chapter 2. Additional multiple-station alarms, other than smoke alarms, are permitted with limitations as specified in Chapter 2 of NFPA 72.

NFPA 72, Chapter 2, "Household Fire Warning Equipment," now requires that low-power wireless systems comply with all of the requirements, except one, given in NFPA 72, Chapter 3, "Protected Premises Fire Alarm Systems." The exception deletes the requirement for a distinctive supervisory signal when a detector/transmitter is removed. Wireless systems are comprised of smoke detectors with built-in low-power radio transmitters that are similar to, yet far more sophisticated than, garage-door-opener transmitters. System operability, including battery condition, is verified by periodic transmission via repeaters, when necessary, to a central receiver. Both repeaters and the central receiver operate from ac power with standby battery, thus providing a more reliable system.

Single- and multiple-station smoke alarms are also used in residential applications other than single-family homes. When these detectors are used in commercial applications, such as apartments, condominiums, hotels, and motels, Chapters 2, 5, and 7 of NFPA 72 must be followed. The single- and multiple-station smoke alarms are not considered "system"-connected smoke detectors, and therefore cannot be connected to a fire alarm control unit.

Household Fire Warning System

In addition to single- and multiple-station alarms, there are household fire warning systems similar in makeup and operation to the local fire alarm systems described in NFPA 72, Chapter 3, "Protected Premises Fire Alarm Systems." Such a system typically consists of a household system control unit that derives main power from the ac house wiring, and usually contains a rechargeable standby battery capable of operating the system for at least 24 hr. (See Figure 5-6D.) In addition, this system uses smoke detectors of the system-

FIG. 5-6A. Single-station photoelectric smoke alarm. (Sources: Gentex Corp., left, and First Alert/BRK Brands, Inc., right)

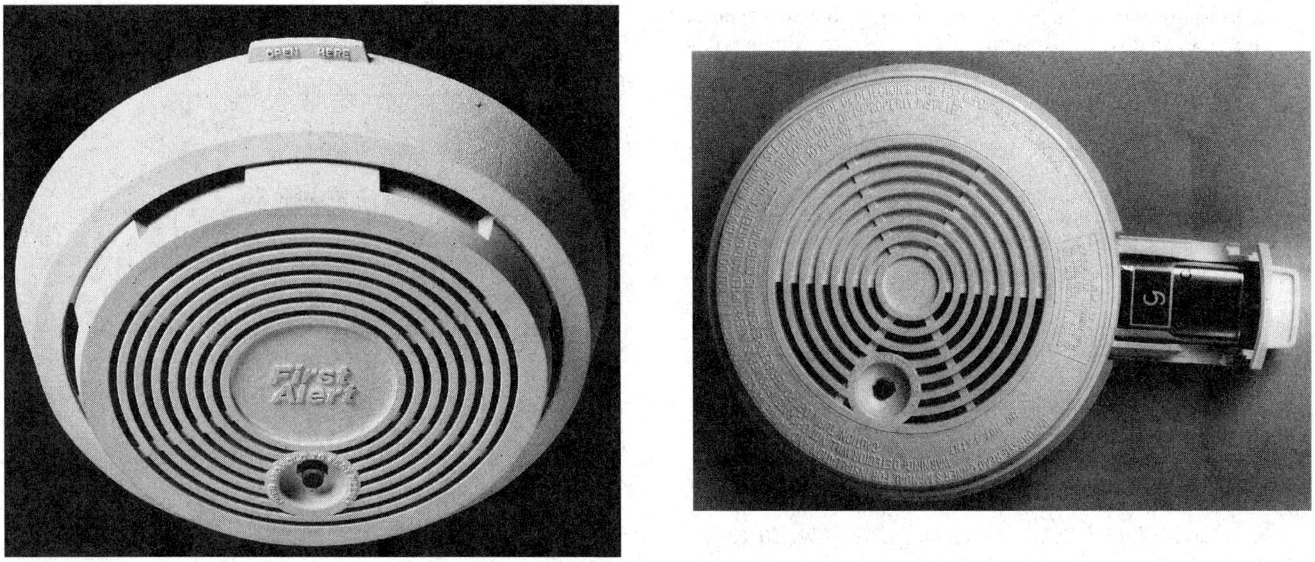

FIG. 5-6B. *Single-station ionization smoke alarm (Left) (Source: First Alert/BRK Brands, Inc.); Single-station ionization smoke alarm with dry-cell battery backup (Right) (Source: First Alert/BRK Brands, Inc.).*

connected type. The fire warning system also could use heat detectors and manual fire alarm boxes and has separate alarm notification appliances, such as bells, horns, or electronic sirens. The control unit also might have provision for connection to an automatic digital dialer, central station, or television cable to transmit the alarm to a point beyond the household. Where the system is connected to a supervising station, NFPA 72, Chapter 2, permits the station to contact the residence for verification of the alarm condition and, where acceptable assurance is provided within 90 seconds that the fire service is not needed, retransmission of the alarm is not required.

The most popular method for such a connection is digital communication. Where a digital alarm transmitter (DACT) is included, NFPA 72 permits: (1) operation with only one telephone line and (2) programming to call only one digital alarm receiver (DACR) number.

Combination Systems

Both wired and wireless household systems often include a burglar alarm in addition to fire alarm functions. This could make the system

more cost effective. In combined systems, the fire alarm signal must take precedence over the burglar alarm signal. The audible alarms are required to be distinctive so the homeowner can immediately discern the difference between a fire and a burglary. In addition, a system that provides for transmission of alarms to a point beyond the household by means of a digital dialer will sometimes have a medical alert feature that can be used to summon medical help.

DETECTION SYSTEM COMPONENTS

Smoke Detectors

A single-station smoke alarm is a self-contained fire alarm device that consists of an assembly of electronic components, including a

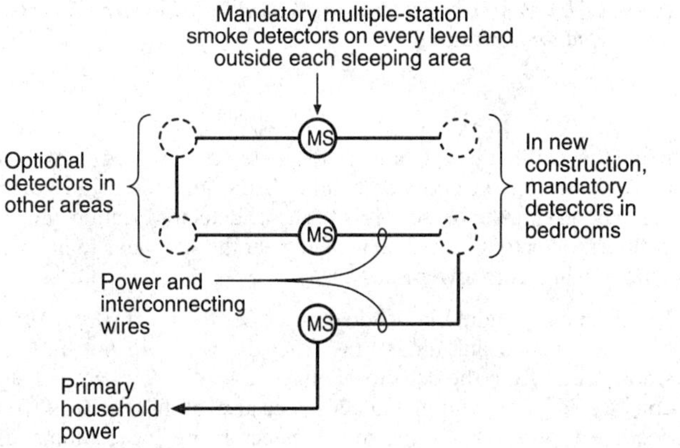

FIG. 5-6C. *Typical household multiple-station smoke alarm system.*

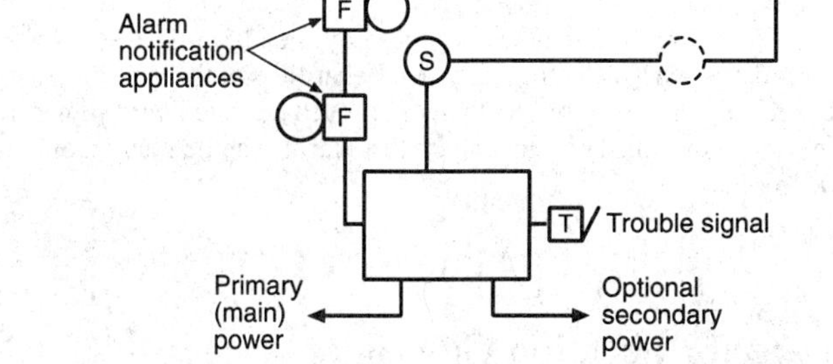

FIG. 5-6D. *Typical household fire alarm system, with separate control panel.*

smoke sensing chamber and an alarm sounding appliance, and is powered from an unswitched source of house power, with or without a standby battery, or, for use in existing construction only, by integral batteries. A standby battery is not required for ac-powered detectors in existing construction and in certain other situations but their use is encouraged. Single- and multiple-station alarms may be either of the ionization or photoelectric type.

Tests have shown that listed smoke detectors of either the ionization or photoelectric type should provide adequate warning to the occupants for most residential fires. It might be noted that, as a class, ionization detectors respond slightly faster to open flaming fires than photoelectric detectors. Conversely, the photoelectric detectors respond faster to smoldering fires. Since residential occupancies experience both flaming and smoldering fires, an extra margin of occupant safety might be achieved if the home is equipped with one detector of each type. However, both types of smoke detectors are subjected to, and must pass, the same test fires for listing by a testing laboratory. Typical test fires are described in UL 217, *Standard for Safety Single- and Multiple-Station Smoke Detectors*, and ANSI/UL 268, *Standard for Safety Smoke Detectors for Fire Protective Signaling Systems*.

It is well known that compliance with strong smoke detector laws offers the potential of significant reduction in loss of life and property from fire. For uniformity, laws adopted statewide are preferable to local ordinances. Most states and communities now require residential units from single-family dwellings to hotels and motels to be equipped with smoke detectors.

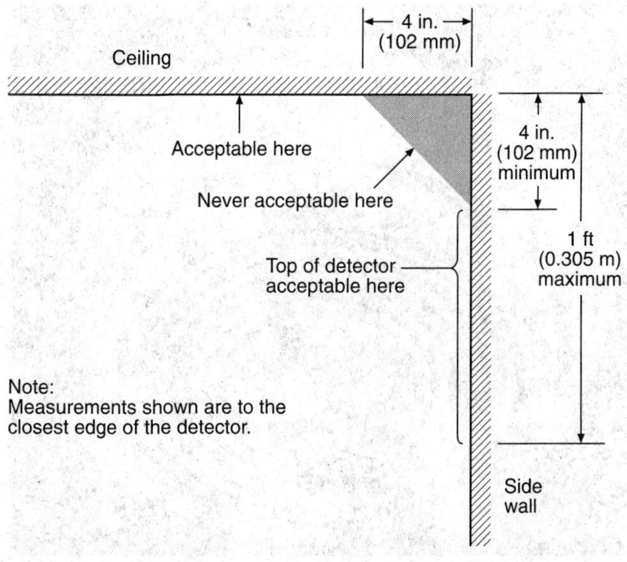

Alarm Notification Appliances

A single-station smoke alarm sounds an audible alarm at the device itself. The sound level output must be at least 85 dBA at a distance of 10 ft (3.05 m). This should be sufficient to alert most sleeping people as long as the detector is located on the same floor and near those who are sleeping. A general rule of thumb is that a device should provide at least 15 dBA above the maximum ambient sound level at the ear of the sleeping person. (For comparison, conversational speech has been shown to register 65 dBA and traffic on an "average" street corner 75 dBA). In addition to background noise, the stage of sleep affects the volume necessary to wake a person.

Long distances between the detector and the bedroom, closed bedroom doors, a noisy appliance in the bedroom such as an air conditioner or humidifier, or impaired hearing could prevent an occupant from being awakened by the detector. In such cases, an alarm should sound in the bedroom and detectors should be interconnected so at least one member of the household will be awakened.

A wired or wireless household system normally has an alarm-notification appliance circuit for connection of various appliances, such as fire alarm bells, horns, or electronic sirens. These devices can produce higher sound levels and can be located independently of the detectors. They can be clearly audible throughout the house or even outside, where neighbors might hear the alarm.

The sound frequencies produced by the alarm horn should be considered when selecting a device. Regardless of the type of detector installed in the home, each alarm should be tested with someone in each bedroom. This can ensure that alarms can be heard from all bedrooms.

INSTALLATION PRACTICES

Detector Location Guidelines

Detectors should be located on the ceiling at least 4 in. (100 mm) from the wall; or on the side wall, 4 to 12 in. (100 to 300 mm)

FIG. 5-6E. *Proper mounting of detectors.*

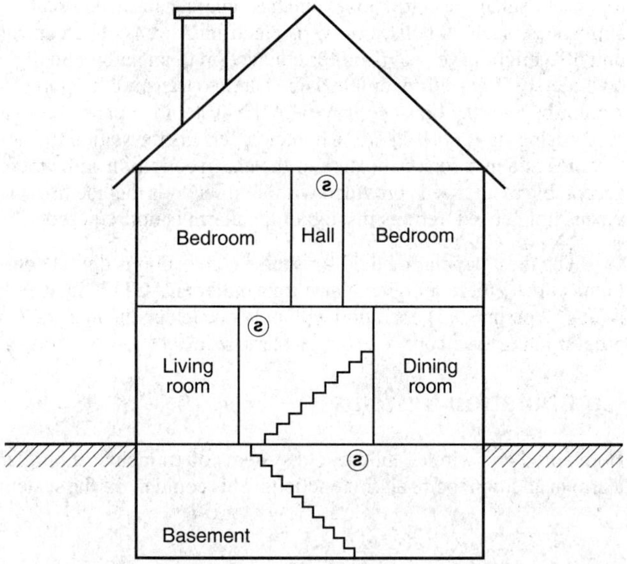

FIG. 5-6F. *A smoke detector (indicated by backward "S") shall be located on each story.*

from the ceiling to the top of the detector. (See Figures 5-6E through 5-6H.) On floors containing bedrooms, the ceiling of the hallway serving the bedrooms is the usual detector location. On other floors, detectors should be located near the stairways to intercept smoke rising from lower floors.

Detectors should be located no closer than 3 ft (0.9 m) from heating vents so that air issuing from the vent will not blow the smoke away from the detector. Some smoke detectors are not suitable for location within kitchens, because of false alarms from cooking vapors. Also, some smoke detectors are not recommended for garages, where automobile exhaust might cause alarms, or for attics or other unheated spaces where extremes of temperature or humidity might affect their operation. Before a smoke detector is in-

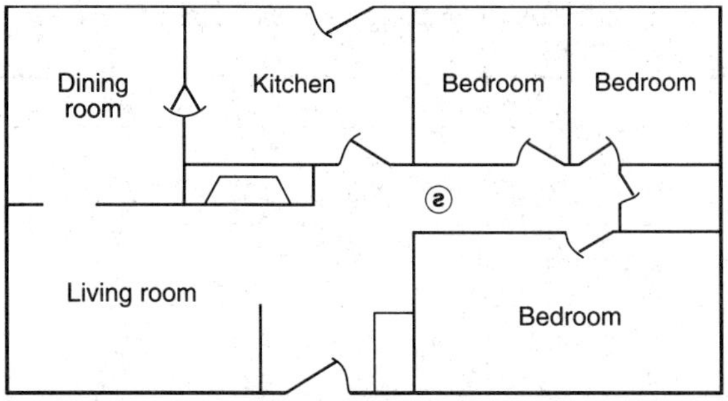

FIG. 5-6G. A smoke detector (indicated by backward "S") shall be located between the sleeping area and the rest of the family living unit.

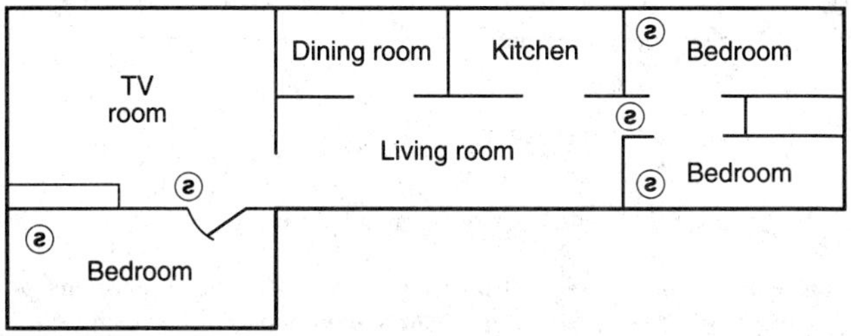

FIG. 5-6H. In family living units with more than one sleeping area, a smoke detector (indicated by backward "S") should be provided to protect each sleeping area in addition to detectors required in bedrooms.

FIG. 5-6I. Combination photoelectric smoke detector/strobe light. (Source: Gentex Corp.)

stalled in any of these locations, its specifications should be checked to ensure it is appropriate for the intended area.

Another consideration is the temperature of the mounting surface. If a smoke detector is installed on a surface that can be even a few degrees warmer or cooler than the room temperature, its response may be delayed by the cold or warm air layer at the mounting surface. Smoke detectors should not be installed on uninsulated exterior walls, ceilings below uninsulated attics, or ceilings containing radiant heating coils. If the detector is colder than the air temperature and if the relative humidity is high, condensation of water vapor may occur in the detector and either produce a false alarm or put the detector to "sleep." Condensation, dust, and insects are three of the most common things that lead to detector malfunction.

Special Considerations for the Handicapped

In general, households with handicapped or elderly occupants need a higher level of protection to provide additional escape time. If the handicapped person could not escape without assistance, provisions should be made for someone to provide help. Detectors that signal to remote receivers could be used, for instance, to sound the alarm in a neighbor's home. For hearing-impaired occupants, some detectors can be connected to bed vibrators, flashing strobes (see NFPA 72, Chapter 6, "Notification Appliances for Fire Alarm Systems"), or fans to provide tactile or visual as well as aural stimulation. (See Figure 5-6I.)

MAINTENANCE

Household fire-warning equipment should be maintained in accordance with the recommendations of the equipment manufacturer. In general, this means little more than keeping the equipment clean and free of dust, and replacing the batteries when needed. Detectors should never be disconnected due to nuisance alarms. If a detector is removed from its mounting bracket to stop a nuisance alarm caused by cooking, place the detector in an obvious location as a reminder to remount it. Most such nuisance alarms can be corrected either by moving the detector to another location, farther from the cooking source, or by replacing it with a smoke detector that is not as sensitive to cooking smoke. Also, unless the detector has a built-in sensitivity adjustment, an individual should never tamper with the factory-set sensitivity of a detector.

Testing

The importance of testing smoke detectors cannot be emphasized too strongly. Smoke detectors are basically electronic devices, and, since all devices will fail at some time, the failure rate projection of a smoke detector can be calculated. From these projections, the life expectancy or the average failure rate of the smoke detector can be determined.

To reduce the period of time that a smoke detector could be out of service because of failure, the detector that has failed can best be identified by periodic testing and then replaced. The following example is merely illustrative and is not indicative of currently available units. Table 5-6A shows that over a 30-year period, if a detector with a calculated failure rate of 3.5 failures per million hours is not tested, the probability of it failing is 60 percent. If so, the owner could be unprotected for a period of almost 900 weeks, or almost 17 $^{1}/_{2}$ years.

TABLE 5-6A.   *Probability of Failure and Average Unprotected Time for Typical Ranges of Detector Failure Rates and Test Intervals for Various Service Life Periods*

Service Period	1 Year				10 Years				20 Years				30 Years			
Failures per million hours	2.0	3.0	3.5	4.0	2.0	3.0	3.5	4.0	2.0	3.0	3.5	4.0	2.0	3.0	3.5	4.0
Unprotected time (weeks)																
Test interval																
1 week	2.5	2.5	2.5	2.5	2.7	2.8	2.9	3.0	3.0	3.2	3.3	3.5	3.2	3.6	3.8	4.0
2 weeks	3.0	3.0	3.0	3.1	3.3	3.4	3.5	3.6	3.6	3.9	4.0	4.2	3.9	4.3	4.6	4.9
1 month	4.2	4.2	4.2	4.2	4.5	4.7	4.8	5.0	4.9	5.4	5.6	5.8	5.4	6.0	6.4	6.7
3 months	8.6	8.6	8.7	8.7	9.3	9.7	9.9	10.1	10.1	10.9	11.4	11.9	10.9	12.3	13.0	13.8
6 months	15.2	15.2	15.3	15.3	16.4	17.1	17.5	17.9	17.8	19.3	20.1	21.0	19.3	21.7	23.0	24.4
1 year	28.2	28.6	28.6	28.7	30.6	32.0	32.7	33.5	33.3	36.1	37.6	39.3	36.1	40.6	43.0	45.6
No test	28.2	28.6	28.6	28.7	268	271	273	276	550	565	573	581	848	881	898	914
Probability of failure during service period (%)	1.7	2.6	3.0	3.4	16.1	23.1	26.4	29.6	29.6	40.1	45.8	50.4	40.1	54.4	60.1	65.0

Notice that over the same 30-year period, with a test only once a year, the unprotected time would fall from almost 900 weeks to approximately 43 weeks. Referring to Figure 5-6J, notice that, if the frequency of testing is increased from one year to one month, the out-of-service time would drop from 43 weeks to about 6 weeks. All of these figures have a built-in factor of two weeks, so a defective detector can be removed and a replacement can be obtained and installed. This dramatically shows how important it is to test a smoke detector frequently.

Guidelines for testing smoke detectors can be found in NFPA 72, Chapter 2, "Household Fire Warning Equipment," and Chapter 7, "Inspection, Testing, and Maintenance", and in the manufacturers' instructions.

In summary, fire prevention in the home involves good planning, good housekeeping, installation of an adequate fire warning system, and ongoing maintenance and testing of the system.

FIG. 5-6J. A plot of unprotected time vs. service life with different test intervals. The typical case of 3.5 failures per million hours is used, with the expected percent failure also indicated.

BIBLIOGRAPHY

NFPA Codes, Standards, and Recommended Practices

Reference to the following NFPA codes, standards, and recommended practices will provide further information on household warning systems discussed in this chapter. (See the latest version of *The NFPA Catalog* for availability of current editions of the following document.)

NFPA 72, *National Fire Alarm Code*

Additional Reading

ANSI/UL 268, *Standard for Safety Smoke Detectors for Fire Protective Signaling Systems*, Underwriters Laboratories Inc., Northbrook, IL, 1994.

Bendall, D., "Domestic Smoke Alarms: A Guide for Specifiers," *Fire Prevention*, No. 281, July/Aug. 1995, pp. 32–34.

Bills, Robert G., Jr., "The Response of Smoke Detectors to Smoldering-Started Fires in a Hotel Occupancy," FMRC J.I. 0Q0R4.RA, Factory Mutual Research Corporation, Norwood, MA.

Breen, D. E., "Toward More Reliable Residential Smoke Detection Systems," *Journal of Fire Protection Engineering*, Vol. 2, No. 1, 1990, pp. 1–10.

Brien, D. E., "Toward More Reliable Residential Smoke Detection Systems," *Journal of Fire Protection Engineering*, Vol. 2, No. 1, 1990, pp. 1–10.

Bruck, D. and Horasan, M., "Non-arousal and Non-action of Normal Sleepers in Response to a Smoke Detector Alarm," *Fire Safety Journal*, Vol. 25, No. 2, 1995, pp. 125–140.

Bukowski, R. W., "Studies Assess Performance of Residential Detectors," *NFPA Journal*, Vol. 87, No. 1, Jan./Feb. 1993, pp. 48–54.

Bukowski, R. W., "Notification Appliances for the Hearing Impaired," VHS Video Tape, National Institute of Standards and Technology, Gaithersburg, MD, Video, Federal Fire Forum, Current Issues in Fire Protection, June 15, 1992, Gaithersburg, MD, 1992.

Cholin, J. M. and Moore, W. D., "How Reliable Is Your Fire Alarm System," *NFPA Journal*, Vol. 89, No. 1, Jan./Feb. 1995, pp. 49–53.

"Components of Automatic Fire Alarm Systems for Residential Premises. Part 1. Specification for Self-Contained Smoke Alarms and Point-Type Smoke Detectors, British Standards Institution, London, England, BS 5446:Part 1:1990, Part 1, 1990, 26 pages.

Craig, J., "Smoke Alarms Are Not Enough," *Fire Prevention*, No. 233, Oct. 1990, pp. 22–23.

DeVoss, F., "Bringing Alarms to Light—Signaling for the Hearing Impaired," Underwriters Labs., Inc., Northbrook, IL, *UL Lab Data*, Vol. 20, No. 1, 1990, pp. 4–7.

"Domestic Smoke Alarms," *Fire Prevention*, No. 283, Oct. 1995, p. 12.

"Focus on Fire Detection: Ionization Smoke Detectors—The High-Voltage Issues," *Fire Prevention*, No. 249, May 1992, pp. 35–36.

"Fire Incident Study National Smoke Detector Project," Jan. 1995, U.S. Consumer Product Safety Commission, Washington, DC.

Fowler, J., "Reducing False Alarms," *Fire*, Vol. 88, No. 1090, April 1996, pp. 27–28.

Hall, J. R., "The Latest Statistics on U.S. Home Smoke Detectors," *Fire Journal*, Vol. 83, No. 1, 1989, pp. 39–41.

Hall, J. R., Jr., *U.S. Experience with Smoke Detectors*, National Fire Protection Association, Quincy, MA, annual.

Hall, J. R., Jr., "U.S. Experience With Smoke Detectors and Other Fire Detectors: Who Has Them? How Well Do They Work? When Don't They Work?" National Fire Protection Association, Quincy, MA, June 1995.

Hansen, A. T. and Platts, R. E., "Costs and Benefits of Smoke Alarms in Canadian Houses," Scanada Consultants, Ltd., Ontario, Canada, Canada Mortgage and Housing Corp., Ontario, Canada, Phase 2, March 1, 1990.

Hygge, S., "Smoke Detectors in Apartments and One-Family Houses: A Comparison Between the Maintenance, Care, and Performance of Free and Purchased Smoke Detectors," *Fire Safety Journal*, Vol. 15, No. 3, 1989, pp. 195–210.

Hodge, S. G., "Fire Alarm Safety Improvement," Westinghouse Hanford Co., Richland, WA, WHC-SD-631-ATP-006, October 13, 1994.

Ingrassia, E., "Fire-Alarm Systems: Seven Factors Influencing Selection," *Consulting-Specifying Engineer*, Vol. 12, No. 1, 60–62, July 1992, pp. 60–62.

Jahr, F., "Focus on Fire Detection: Detector Success for Norway," *Fire Prevention*, No. 249, May 1992, pp. 37–38.

Johnson, P. F., and Brown, S. K., "Smoke Detection of Smoldering Fires in a Typical Melbourne Dwelling," *Fire Technology*, Vol. 22, No. 4, 1986, pp. 295–310.

LeCoque, P. G., and Harris, K., "State by State . . . An Update of Residential Smoke Detector Legislation," *Fire Journal*, Vol. 84, No. 1, 1990, p. 40.

Lines, P. E., "Domestic Fire Detection BS 5839 Part 6: Some Questions and Answers," *Fire Safety Engineering*, Vol. 3, No. 2, April 1996, pp. 24–25.

March, J. H., "Domestic Fire Safety—A Problem in Hong Kong," *Fire Protection*, Vol. 17, No. 3, Sept. 1990, pp. 6–12.

Marriott, M., "Reliability and Effectiveness of Domestic Smoke Alarms," *Fire Engineers Journal*, Vol. 55, No. 177, June 1995, pp. 10–13.

Marriott, M., "Reliability of Domestic Smoke Alarms," *Fire Research News*, Vol. 13, Spring 1991, p. 8.

Moyers, R. E., "Smoke Alarms and Residential Sprinklers: Costs and Benefits," Summary Report, Canada Mortgage and Housing Corp., Ontario, Canada, ARL-MR-65, Apr. 1991.

Neily, M. L., Smith, C. L. and Shapiro, J. I., "Residential Smoke Detector Performance in the United States," Consumer Product Safety Commission, Washington, DC, European Consumer Safety Association Conference, November 22, 1993, Amsterdam, Netherlands, 1993, pp. 1–16.

Nober, E. H., Well, A. and Moss, S., "Alarms for the Hearing-Impaired," *Fire Prevention*, No. 233, Oct. 1990, pp. 28–31.

Nober, E. H., Well, A. D. and Moss, S., "Does Light Work As Well As Sound? Smoke Alarms for the Hearing-Impaired," *Fire Journal*, Vol. 84, No. 1, Jan./Feb. 1990, pp. 26–28, 30.

Salamone, R., "Retrofitting High-Rise Dorms With Alarm and Detection Systems," *Fire Journal*, Vol. 84, No. 1, Jan./Feb. 1990, pp. 37–39.

Scheidweiler, A. and Guttinger, H., "Reducing False Alarms From Smoke Detectors," *Fire International*, Vol. 14, No. 125, Oct./Nov. 1990, pp. 77–78.

Smith, P., "Smoke Alarms Fitted on Ground Floors May Not Be Loud Enough for Sleepers," *Fire*, Vol. 85, No. 1045, July 1992, p. 21.

"Smoke Detector Operability Survey Report on Findings," Oct. 1994, U.S. Consumer Product Safety Commission, Washington, DC.

Stiles, B., "'Inspector Detector' Promotes Proper Smoke Detector Care," *Fire Chief*, Vol. 36, No. 6, June 1992, p. 48.

Sultan, M. A. and Halliwell, R. E., "Optimum Location for Fire Alarms in Apartment Buildings," *Fire Technology*, Vol. 26, No. 4, Nov. 1990, pp. 342–356.

Todd, C., "Domestic Fire Detection: The New Code BS 5839:Part 6," *Fire Safety Journal*, Vol. 2, No. 6, Dec. 1995, pp. 21–24.

Tucker, R. W., "Early Detection in Jails—While Minimizing Unwanted Alarms," *SFPE Bulletin*, Sept./Oct. 1990, pp. 6–8.

UL 217, *Standard for Safety Single- and Multiple-Station Smoke Detectors*, Underwriters Laboratories Inc., Northbrook, IL, 1993.

Waterman, T. E., "Fire Detector Response Versus Available Escape Time in Residences," ITT Research Institute, Chicago, IL.

"Will Your Smoke Detectors Operate?," *Fire Protection*, Vol. 17, No. 4, Dec. 1990, p. 7.

Willey, A. E., "Reducing Home Fire Deaths in the 1990s," *Fire Journal*, Vol. 81, No. 5, 1987, pp. 12–13.

FIRE PROTECTION SURVEILLANCE AND FIRE GUARD SERVICES

by Dean K. Wilson

Effective fire protection at a commercial or industrial facility depends on the degree to which the management of the facility is committed to providing an appropriate level of fire loss prevention and fire control. Once the hazards of the facility have been identified and evaluated and built-in fire protection systems have been provided to offset the impact of those hazards, the protection features must be consistently managed. A fire alarm system may be used as a management tool to oversee fire protection surveillance. The fire alarm system can monitor the availability of other fire protection systems. It may also monitor the activities of guards making fire patrol tours.

The fire alarm system should monitor all fire alarm initiating devices. Examples are: manual fire alarm boxes; heat, smoke, radiant energy, fire-gas, or other fire detection devices; and the discharge of automatic sprinkler systems and other fire suppression or extinguishing systems, such as dry chemical, foam, foam-water, water mist, carbon dioxide, and other gaseous agents.

The fire alarm system should monitor pertinent supervisory initiating devices for both "off-normal" and "restoration to normal" conditions. These include:

- Any automatic sprinkler control valve $2^1/_2$ in. (63.5 mm) or larger
- Dry-pipe or pre-action valve high and low air pressure
- Fire protection water tank or reservoir level
- Fire protection water-pressure tank high and low air pressure
- Electric-driven fire pump running, power failure, and phase reversal
- Diesel-engine-driven fire pump running, controller switch in a position other than "automatic," engine or controller failure, and failure of the battery charger
- Steam-driven fire pump running and steam availability
- Low public water system pressure when fire protection systems are fed from a dead-end public water main, or when the public water system is deemed unreliable

In addition, for those geographic areas subject to freezing winter temperatures, the fire alarm system should monitor low temperature for:

- Buildings protected by wet-pipe automatic sprinklers
- Dry-pipe, deluge, or pre-action valve closets
- Fire-pump rooms
- Fire protection water tanks.

For those facilities where fire guard service is included as a part of the overall fire protection surveillance, the protected premises fire alarm system can also monitor the patrol tours made by the fire guards.

In order to ensure immediate attention to signals generated by the protected premises fire alarm system, to ensure that management is promptly notified of the nature of these signals, and to keep proper records of signals received, the fire alarm system should be connected to a central station, proprietary, remote station, or combination auxiliary and remote station fire alarm system. This will ensure that signals receive attention at all times.

FIRE GUARD SERVICES

Containment of fire to its area of origin is made possible by several factors, one of the most important of which is early fire detection.

During normal business hours, most areas of most facilities are occupied. By their presence, the occupants provide fire protection surveillance because they are able to detect a fire when it occurs. A carelessly discarded cigarette, for example, may ignite the contents of a wastebasket; that fire may spread until the entire building is involved. However, someone who discovers the fire while it is still in its incipient stage may be able to extinguish it with a portable fire extinguisher. In some cases, the occupants can also detect conditions, such as a malfunctioning machine, that might lead to a fire.

Similarly, most companies have come to realize that increased security surveillance is vital in guarding against fires of incendiary origin.

Fire guard services generally serve three purposes in protecting a property against fire loss: (1) to protect the property at times when the management is not present; (2) to facilitate and control the movement of persons into, out of, and within the property; and (3) to carry out procedures for the orderly conduct of some operations on the property. Guards may be facility employees or employees of outside firms established to provide these services on a contract basis. The duties of these individuals may be supplemented or, in some cases, replaced in part by various approved fire and security protective signaling systems.

Dean K. Wilson, P.E., is director of loss prevention training for Industrial Risk Insurers, Hartford, CT. He is chair of the NFPA Technical Correlating Committee on the *National Fire Alarm Code* and publisher and co-editor of the fire alarm newsletter, *The Moore-Wilson Signaling Report*.

GUARD SERVICE DIRECTION

In conjunction with fire and explosion protective systems and various other management programs for loss prevention and control, a comprehensive program of fire protection and security surveillance provides a means of: (1) continually monitoring the facility for conditions that might lead to a fire or explosion, (2) promptly notifying the public fire department or private fire brigade that a fire or explosion has occurred, and (3) effectively inhibiting unauthorized access to the facility. Thus, management must determine how this surveillance can be best achieved.

In concept, guards should patrol thoroughly all unoccupied areas of a facility often enough to be able to detect conditions likely to result in a fire or other emergency. These patrols should be supervised in some way so management can verify that the guard is indeed making the patrol tours.

The highly protected risk (HPR) property insurance carriers generally recommend that the first patrol tour begin within one-half hour of vacancy and that subsequent patrol tours be made every hour at night and every other hour (bihourly) during the day. Each tour should not exceed 45 min in actual length, allowing the guard 15 min of rest.

Management should develop a written surveillance plan for both fire protection and security in the facility. To do this, the property manager should determine which areas of the facility are unoccupied during both working and nonworking hours. The property manager should also designate a management representative who will be responsible for overseeing the surveillance program. It is very important that management retain supervisory control of surveillance to maintain program integrity. This representative should review surveillance reports daily and evaluate changes in the facility that might require modification of the surveillance plan.

The facility's fire loss prevention and control manager should be consulted during the establishment of procedures to be followed by fire guard service personnel. Procedures and instructions to guards should be geared to the specific actions required. General instructions or superficial training are of little value. Meaningful, specific instructions cannot be prepared, however, without an investment of both time and thought by the management of the property.

Management should establish a clear line of succession in event of absences. Even when there are only two guards employed, one should be designated as the leader.

Supervision of guards provided under contract with outside firms should be through the designated representatives of the company providing the guard service. In its contract or supplementary documents, that company should be given details as specific and complete as possible regarding the services expected.

When guards are assigned to patrol a property, management is responsible for providing them with adequate equipment and information with which to safeguard their own health and safety. Potential situations that must be considered include:

1. Sudden illness or injury while the guard is alone at the property.
2. The possibility of a guard being overpowered by an intruder.
3. The development of a situation requiring management decisions.

The first situation can be addressed by equipping guards with a means of prompt, efficient communication, such as portable two-way radios, which would allow them to summon aid.

In the second situation, an intruder may prevent the guard from using any communications equipment. It then becomes necessary to have a system or procedure whereby the guard's failure to transmit a signal or meet a predetermined schedule will be investigated promptly.

Third, situations that require management decisions may arise at times when the property is attended by guards only. Such decisions may involve conditions ranging from the unscheduled arrival of merchandise or supplies to anonymous bomb threats. Guards must have instructions and the means for contacting management personnel.

COMMUNICATIONS EQUIPMENT

Guards should be provided with facilities for communication within and outside the property. A control center should provide a point with which guards may communicate, and the center should have communications facilities to points outside the property. Such a center is needed even when there is very limited guard service. For example, in a facility with only one or two guards, this control center might simply be a room with a telephone. Even with a complete fire alarm and security system covering all areas of the property, a control center on the property is necessary.

If an emergency voice/alarm communication fire alarm system is provided, guards may communicate with the building fire command station by using the fire warden telephones.

Where the equipment for fire guard communications, including guards on watch patrol, requires that signals from guards be monitored, the control center should be provided with an operator. Additional operators needed to effect a 24-hr operator service should be provided at the control center, according to the character of guard service provided. For some services, runners, or guards who can be dispatched to investigate signals, should also be provided.

A directory of names and telephone numbers (including any other information to assist in making emergency calls to the outside) should be kept at the control center in a visible index or other quickly accessible form. This directory should give information about the public fire and police departments, key management personnel, and other outside agencies that may have to be contacted in an emergency.

GUARD PATROL TOUR SUPERVISION

Off-premises guard supervisory service designed to report the performance of a guard may be provided by central station, proprietary, or remote station fire alarm systems. Portable watchclocks may also be used to supervise the guard if centralized, immediate supervision is not required.

Tour supervisory systems generally consist of guard patrol tour supervisory initiating devices of some type and a means of recording signals initiated from these devices. Two types of supervisory service systems are: (1) compulsory and (2) noncompulsory.

Compulsory Tour Supervisory Systems

This system involves one or more "active" guard patrol tour supervisory initiating devices that are installed in conjunction with several "passive" guard patrol tour supervisory initiating devices. These devices are arranged so that the guard must progress in a fixed route from one station to another. Ordinarily, the guard carries some type of device called a "key." The initiating devices are arranged so that the passive devices condition the key so that it may operate the next device, whether that device is active or passive. In addition to conditioning the key for the next device, active initiating devices also either transmit a signal to an on- or off-premises central recording location, or cause a signal to be recorded inside the key which indicates the date and time that the active initiating device was operated.

This system is particularly useful when, because of the nature of the facility, it is essential that the guard follow a carefully prescribed route. One disadvantage to this system from a security standpoint is that the tour may not be varied. This is sometimes felt undesirable because a potential intruder who is observing the movement of the guard might be able to readily discern a pattern to the guard's movement.

In many compulsory tour systems, the guard initiates a starting signal at the beginning of a round and initiates a finishing or completion signal at the end of the round. All of the other stations in between would be passive initiating devices.

A fixed schedule of starting and finishing times is arranged in advance when a compulsory tour is monitored by an off-premises facility. If the guard fails to either begin or end a round within the prescribed times, upon the expiration of a "grace" period of usually no more than 10 or 15 min in length, the off-premises monitoring facility will initiate action. Normally, an attempt will be made to contact the guard by telephone. Upon failure to reach the guard, the off-premises monitoring facility will normally dispatch the police department and notify the management of the facility that the guard has failed to respond.

When a compulsory tour system is used in conjunction with a proprietary fire alarm system, the process works much the same, except that, since the receiving location is normally located on the protected premises, the operator at the receiving location will be more readily able to determine why the guard has not begun or ended his/her round.

Noncompulsory Tour Supervisory Systems

In this system, all of the initiating devices are active. The guard progresses from station to station with each initiating device indicating the date and time of the guard operating the station. By following the progression of signals, the off-premises or on-premises monitoring station can follow the progress of the guard during the course of the round. In some cases, combination manual fire alarm boxes/guard patrol tour supervisory initiating stations are grouped in the same physical housing. These stations are arranged so that they may be actuated by any person to indicate a fire, but may also be operated by a guard to indicate the progress of his/her tour.

Delinquency Reporting Systems

Some compulsory or noncompulsory tour supervisory systems are arranged so that, rather than reporting every time the guard operates an active initiating device, the system only transmits a signal when the guard has failed to operate the device within a prescribed time period. Such delinquency reporting systems are utilized when it is desired to reduce the amount of signal traffic associated with the guard patrol tour supervisory system. Many remote station fire alarm systems that supervise guard patrol tours are arranged in this fashion.

Telephone, Radio, and Other Voice Reporting Systems

Occasionally, a guard patrol tour is supervised by means of the guard reporting from a specific location by means of a voice reporting system. Probably the most common system utilizes a telephone system equipped with calling subset identification. When the guard initiates voice communication over such a system, the system indicates and records the specific location from which the guard is calling. Other systems employing means of voice communication, including radio systems, can also be utilized. In some radio systems, a tour supervisory initiating device must be inserted into the radio

before the guard establishes patrol tour contact. This device transmits a digital code that gives a specific location, thus indicating that the guard is at the prescribed tour station.

Portable Watchclocks

In many cases, management determines that it is not necessary to provide immediate supervision of the guard patrol tours, and elects to provide retrospective guard patrol tour supervision by means of a stationary or portable watchclock. (See Figure 5-7A.) The records from these devices must be reviewed the following day to determine that the guard has indeed made the prescribed tours and operated all of the tour supervisory initiating devices. While this method lacks the immediacy of the other tour supervisory methods, it is sufficiently cost effective that it is the most common tour supervisory method employed. Traditionally, a spring-wound clock mechanism—either at a single location in the case of a stationary watchclock, or in a portable case that the guard can carry from location to location—is used to drive some type of recording medium. The most common recording medium is a paper disk or a paper tape. In this case, the tour supervisory initiating devices are embossing keys that must be inserted into the portable watchclock and turned, making their impression on the paper tape. The tour supervisory initiating devices for the stationary clock are usually some type of electric switch or, in at least one design, a magneto that must be operated by the guard, causing the stationary clock to record the specific location.

The portable watchclock can be an effective supervisory tool if the control of access to the clock is carefully limited so that the

FIG. 5-7A. *Front view (top), interior view (bottom) of a portable watchclock. This clock provides an embossed record tape, made directly from type on the recording key. The tape has 24 ruled segments and is synchronized with the clock mechanism. If a guard fails to punch in, the omission is indicated by a prominent white space on the tape. This clock indicates when it was opened. (Source: Detex Corp.)*

guard cannot have access to the internal mechanism of the watch-clock. This means that someone in management should actually open the clock and retrieve the record each day. To safeguard the premature opening of the clock, a tell-tale device places a mark on the paper disk or tape whenever the clock is opened.

Even when a guard service company provides the guard patrol tour service, it is advisable to have management supervise the retrieval of records from the watchclock to prevent any collusion between upper level guards, such as sergeants or lieutenants, and the guards themselves.

FIRE GUARD SERVICE FUNCTIONS

A sufficient number of guards should be provided to accomplish the needed services. If guards are assigned to other part-time duties in addition to their regular guard services, these duties should be chosen so they will not interfere with regular guard duties.

Fire guard service can facilitate and control the movement of persons within a property when the number of persons in the property requires such a service. Duties in this category include:

1. Preventing the entry of unauthorized persons who might set a fire or do damage to the facility.
2. Controlling the activities of people authorized to be on the property, but who may not be aware of procedures established for the prevention of fire.
3. Controlling pedestrian and vehicular traffic during exit drills, and evacuating the property or parts of it during emergencies.
4. Controlling gates and vehicular traffic to facilitate access to the property by the public fire department, members of a private fire brigade, and off-duty management personnel in case of fire or emergency.

Guard service should be established to carry out certain procedures for the orderly conduct of the operations on the property, including procedures for fire loss prevention and control, both by personnel associated with the property and outside contractors. Duties in this category include:

1. Promptly discovering a fire and calling the public fire department (also the fire brigade of the property, where there is such a brigade).
2. Operating equipment provided for fire control and extinguishment *after* giving the alarm and before the response of other persons to the alarm.
3. Monitoring signals of fire alarm systems.
4. Patrolling routes chosen by management to ensure surveillance of all the property at appropriate intervals.
5. Starting up and shutting down certain equipment when no other personnel are provided for this purpose.
6. Checking permits for "hotwork," including cutting and welding, and standing by to operate fire extinguishing equipment where necessary.
7. Detecting conditions likely to cause a fire, such as leaks, spills, and faulty equipment.
8. Detecting conditions likely to reduce the effectiveness with which a fire may be controlled, such as portable fire extinguishers not in place, sprinkler valves not open, and water supplies impaired.
9. Performing operations to ensure that fire equipment will function effectively. These may include testing automatic sprinkler and other fixed fire protection systems, including fire pumps, and other equipment related to these systems and assisting in the maintenance of this equipment; checking portable fire extinguishers and fire hose and assisting in pressure tests and maintenance service on these items; testing fire alarm systems; and checking equipment provided on any fire apparatus and carrying

out the periodic tests and maintenance operations required for the apparatus.

PATROL ROUTES AND ROUNDS

Each route to be patrolled should be laid out by the responsible manager. The guard responsible for each route should be given instructions as to all details of the route, and what is expected in covering it. The route should be laid out to prevent shortcuts so that the guard is required to pass through the entire patrol area. A reasonable rest period between rounds must be provided.

Guards should make rounds at intervals determined for the particular situation by management. When operations in the property are normally suspended, rounds should be made hourly, unless management is willing to accept rounds at less frequent intervals. When there are special conditions, such as the presence of exceptional hazards or when protection is impaired, management should institute as many additional rounds as necessary to meet the fire safety requirements of such conditions.

The first round of a patrol should begin as soon as possible after the end of activities of the preceding work shift. The guards should be instructed to make a thorough inspection of all buildings or spaces on the route during their first round. Their instructions should cover the following:

1. Outside doors and gates should be closed and locked. Windows, skylights, fire doors, and fire shutters should be closed.
2. All oily waste, rags, paint residue, rubbish, and similar items should be removed from buildings or placed in approved containers.
3. All fire apparatus should be in place and not obstructed.
4. Aisles should be clear.
5. Motors or machines carelessly left running should be shut off and reported.
6. All offices, conference rooms, and smoking areas should be checked for carelessly discarded smoking materials.
7. All gas and electric heaters, coal and oil stoves, and other heating devices on the premises should be checked.
8. All hazardous manufacturing processes should be left in a safe condition. The temperature of dryers, annealing furnaces, and similar equipment that continue to operate during the night, holidays, and weekends should be noted on all rounds.
9. Hazardous materials, such as gasoline, rubber cement, and other flammable and highly volatile combustibles, should be kept in proper containers or removed from buildings.
10. All sprinkler valves should be open, with gages indicating proper pressures. If not open, the fact should be reported immediately.
11. During cold weather, all rooms should be checked to determine if they are heated properly.
12. All water faucets and air valves found leaking should be closed. If leaks cannot be stopped, the condition should be reported.
13. Particular attention should be given to new construction or alterations that may be in progress.

SELECTION OF GUARDS

Management should require individuals considered for fire guard service to satisfactorily pass a character investigation. This investigation should attempt to evaluate the individual's reliability, self-control, and potential loyalty to the employer. Applicants for a guard position should be required to be fingerprinted and to give particulars of any police records. The local police should be furnished with this information, should corroborate it, and should ask for checks by other police agencies. The fingerprint data should be cleared with state, national, or appropriate international agencies that maintain clearing facilities for police records. All applicants for a position as a fire or security guard or patrol person should be required to state any military service record and to submit evidence of such service that may assist in an evaluation of the individual's suitability for guard service.

Contracts for fire guard service should include a provision that the company furnishing guard service will replace any of its employees who, in the judgment of the company purchasing the service, are not qualified.

Management should be satisfied that individuals considered for fire guard service are mentally alert and have good powers of observation, intelligence, and judgment. Investigation should attempt to evaluate the individual's personality and temperament. Such an evaluation is more realistic than arbitrarily testing education or intelligence, or setting an age limit. Very young people may not qualify because they have not acquired a sense of responsibility or judgment. Very old people may have impaired alertness. Individuals should be sought who are known to be "clear-headed" in an emergency.

Guards should be required to pass an annual written examination dealing with information about the property protected and procedures for fire loss prevention with which they are expected to be familiar.

Management should require that individuals considered for fire guard service pass an examination to determine whether they are physically able to perform the guard duties to which they will be assigned. Guards should also be required to pass an annual physical examination. The guards do not need to be athletes, but they should not have a heart condition or other physical ailment that might work to their disadvantage in moments of stress.

TRAINING

Management should establish a continuing training program for its guards. Its scope should be established by the manager or by a fire loss prevention and control manager acting for the manager. Courses for guards and fire fighters, available through training programs of vocational training agencies, schools, universities, and other training and educational agencies, should be considered in any training program for guards.

Management should require guards to have completed at least elementary courses of instruction in the use of portable fire equipment and emergency first aid. The time spent in such preliminary training should be a minimum of two working days in each subject. During service, guards should be given not less than the equivalent of two full working days per year of training to increase their knowledge and experience in the use of portable fire extinguishers, first aid, and other useful training. Guards should be required as part of their training to participate in appropriate meetings devoted to pre-fire planning with operating personnel. Guards can help accomplish this by working together with the public fire department when it establishes its plan for the premises.

Management should require guards to know the location of portable fire extinguishers, hand hoses, standpipes and hydrants, valves controlling sprinkler systems, inside riser valves, post-indicator valves, and sectional valves in the property's own water system. They also should know how to start fire pumps. Guards also need to know the location and purpose of valves controlling water for purposes other than for fire protection, and valves controlling steam, gas, and other services. Management should require guards to know the locations of dangerous machinery or materials and inform them of hazardous manufacturing processes, especially those continuing during the night, holidays, or weekends.

EVALUATING FIRE GUARD SERVICE

There are several items to be considered when evaluating fire guard service.

Tour Records

Management's representative should ensure that:

1. All unoccupied areas of the facility are included in each tour.
2. All key stations or tour supervisory transmitters in each tour have been recorded clearly in a regular hourly pattern at night and in a bihourly pattern during the day.
3. Tours last no longer than 45 min, allowing for a rest period of at least 15 min each hour.
4. Tours begin within one-half hour of the time the area becomes unoccupied and continue to within one-half hour of the resumption of occupancy.
5. The "tell-tale" of a portable or stationary watchclock is recording each time the clock is opened. Look for indications that the clock has been opened more than once a day or at unusual times. This might indicate that unauthorized persons have access to a clock key.

Inspecting Tour Supervisory System Initiating Devices

Key stations, tour supervisory transmitters, or intermediate stations should be inspected once a month to ensure they are firmly attached and sealed with a "tamper" seal and have not been relocated or removed. The key should be checked for damage. If there is evidence of tampering, the key stations, tour supervisory transmitters, or intermediate stations should be checked more often and suitable action taken.

Guards

Management should expect, and is entitled to receive, guard service of the highest quality. Guards must be conscientious in the performance of their duties, note and report all infractions of company regulations, and closely follow the orders given to them.

1. Guards hold positions of trust, which require individuals who are physically able, mentally alert, and morally responsible. Therefore, their physical and emotional stability should be evaluated by appropriate management personnel.
2. Guards must be sufficiently intelligent, calm in the face of an emergency, mature with sound judgment, and possess the physical stamina required for the job.
3. A sufficient number of guards should be provided to maintain proper surveillance in the facility. It is undesirable for guards to be assigned part-time duties unrelated to surveillance. If they are so assigned, however, these duties must not interfere with surveillance responsibilities.
4. Guards should receive full support of management in the performance of their duties.
5. When the guards are facility employees, management should establish the scope of the service and provide necessary training and supervision for the guards.
6. If a contract guard service is used, management should not assume that it will be adequate. Rather, management should prepare detailed specifications and investigate the ability of prospective contractors to meet these specifications. When the contract has been met, management should make sure that its intent is being carried out.
7. Initial and continued training of guards should be by a formal, comprehensive, written program covering all applicable protection procedures. Each guard must be:

- Acquainted with the general nature of the facility's operations, and possess specific knowledge of any hazardous operations

- Familiar with all of the facility's manual and automatic fire protection equipment. They should be especially aware of the location of all sprinkler valves and know which area each controls. Guards should periodically accompany the person making fire protection equipment inspections, in order to gain a working knowledge of facility protection features and hazards

- Familiar with the location and operation of manual fire alarm boxes and other means of transmitting fire alarms. Such means should be provided throughout the facility so guards can easily report a fire

- Taught to notify the fire department *before* taking any other action

- Taught how to admit public fire apparatus to the property and how to direct fire department officers to the location of the fire

- Taught to properly notify company officials when an emergency occurs or when potential trouble is observed

- Taught to maintain a shift log and to prepare reports to management of observations made and actions taken during tours

8. Guard service should be integrated into the overall pre-emergency planning program.
9. General and special instructions and other data required by the guards should be written down and kept up-to-date.

While making regular tours throughout the facility, guards must be alert for all emergencies, paying special attention to known hazardous areas. Guards are in a position not only to detect and correct unsafe conditions that might develop into or contribute to a serious fire, but also to discover an incipient fire. Therefore, the guards should be familiar with the fundamentals of fire control and with the proper use of all available extinguishing equipment.

The importance of notifying the fire department *before* taking any other action should be stressed in guard training. Guards should report any situation that may endanger the facility, e.g., an exposing fire in adjacent properties. Any unusual condition that the guards cannot correct without assistance must be reported immediately to the proper official so the situation can be remedied without undue delay. Such situations include the interruption of sprinkler service or the failure of heating equipment.

In addition to carefully following written pre-emergency plans, guards must be resourceful and capable of applying common sense to any unusual conditions, e.g., "natural" hazard threats, they may encounter. If the facility is being subjected to freezing weather, for example, the guard should be alert to those areas of the facility where protective systems or process equipment might be vulnerable to freezing damage. Such areas should be checked frequently to ensure the heating system is properly maintaining the temperature necessary to avoid damage. If a thermometer is unavailable, a resourceful guard might set out a small container of very cold water to observe whether the area is approaching a dangerously low temperature. See NFPA 601, *Standard for Security Services in Fire Loss Prevention*.

BIBLIOGRAPHY

NFPA Codes, Standards, and Recommended Practices

Reference to the following NFPA codes, standards, and recommended practices will provide further information on fire alarm systems as a management tool of fire protection surveillance and fire guard services

discussed in this chapter. (See the latest version of *The NFPA Catalog* for availability of current editions of the following documents.)

NFPA 72, *National Fire Alarm Code*
NFPA 601, *Standard on Guard Service in Fire Loss Prevention*

Additional Reading

Barnett, P., "Voice Alarm Systems: Loudspeakers—Fire, Flame or Fool Proof?" *Fire Surveyor*, Vol. 21, No. 6, Dec. 1992, pp. 7–9.

Barnett, P., "Merging of Expertise in Fire Alarm and Evacuation Systems Welcomed," *Fire*, Vol. 85, No. 1050, Dec. 1992, p. 18.

Brannigan, F. L., "Preplanning Building Hazards," *Fire Engineering*, Vol. 148, No. 1, Jan. 1995, p. 111.

Bryant, P., "British Standard Code of Practice for the Design, Installation and Servicing of Fire Detection and Alarm Systems in Buildings, BS 5839 PART 1:1988," Third International Symposium on Fire Protection of Buildings, May 10–12, 1990, Eger, Hungary, 1990, pp. 129–134.

Cote, A. E., ed., *Industrial Fire Hazards Handbook*, 3rd ed., National Fire Protection Association, Quincy, MA, 1990.

Faulkner, G., "Fire Alarm Systems Advancement—Meeting the Challenges . . . And More," *Fire Engineers Journal*, Vol. 51, No. 160, Mar. 1991, pp. 9–10, 18.

"Fire Surveillance and Security for Electronic Equipment," *Cerberus Alarm*, No. 112, September 1991, pp. 1–5.

Ingrassia, E., "Fire-Alarm Systems: Seven Factors Influencing Selection," *Consulting-Specifying Engineer*, Vol. 12, No. 1, 60–62, July 1992, pp. 60–62.

Korslund, S. M., "Technical Evaluation of Equipment Maintenance on Fire Alarm Detection, Suppression, and Signaling Systems on the Hanford Site," Westinghouse Hanford Co., Richland, WA, WHC-SC-GN-TEEM-10001, Rev. 0, 1994.

Letts, J., "Fire Detection and Alarm Systems—The Future," *Fire Surveyor*, Vol. 20, No. 6, Dec. 1991, pp. 48–50.

Meacham, B. J., "The Use of Artificial Intelligence Techniques for Signal Discrimination in Fire Detection Systems," *Journal of Fire Protection Engineering*, Vol. 6, No. 3, 1994, pp. 125–136.

Moulton, G. A., ed., *NFPA Inspection Manual*, 7th ed., National Fire Protection Association, Quincy, MA, 1994.

O'Hara, T., "Balancing Security and Fire Safety," Department of State, Washington, DC, Video; Federal Fire Forum. Performance-Based Design Issues of Concern to the A/E Community, Nov. 6, 1995, Gaithersburg, MD, 1995.

Overview Manual, 3rd ed., Industrial Risk Insurers, Hartford, CT, 1989.

Pigott, B., "Computer Controlled Fire Alarms," *Building Services*, Vol. 13, No. 4, April 1991, pp. 55–56.

Rosato, C., "Radio Revolution in Central Alarm Stations," *Fire Prevention*, No. 245, Dec. 1991, pp. 30–31.

Sanderford, H. B., Jr., "Spread Spectrum Radio Alarms Give Hardwired Performance," *Fire Journal*, Vol. 84, No. 1, Jan./Feb. 1990, pp. 57–59.

"Security Equipment and Systems. Secuirty Precautions," Fire Protection Association, London, England, SEC 1, Sept. 1990.

So, A. T. P. and Chan, W. L., "A Computer-Vision-Based and Fuzzy-Logic-Aided Security and Fire-Detection System," *Fire Technology*, Vol. 30, No. 3, 1994, pp. 341–356.

Stone, A., "Developments in Fire Alarm Control Equipment," *Fire Safety Engineering*, Vol. 1, No. 5, Oct. 1994, pp. 14–16.

Todd, C., "Voice Alarms and Evacuation," *Fire Prevention*, No. 274, Nov. 1994, pp. 21–25.

Transue, R. E., "Integrating Fire Alarms into Building Design," *Consulting-Specifying Engineer*, Vol. 14, No. 7, Dec. 1993, pp. 36–38, 40.

GAS AND VAPOR DETECTION SYSTEMS AND MONITORS

John M. Cholin

There are two concerns that are the driving force behind the use of gas and vapor detection systems and monitors. First, the existence of flammable gases or vapors represents: (1) an elevation in the potential for a fire, either in and of itself as the concentration reaches the critical value sufficient to support a flame; or (2) an indication of a leak, implying an immediate potential of loss of containment of a large quantity of fuel. Second, there is a life-safety concern whenever breathing air is contaminated with gases or vapors that are not normal constituents of air. While one could argue the theoretical position that this second concern is not strictly within the purview of the fire protection engineer, in the context of actual applications it is very rare that these two concerns can be rationally separated from each other. Consequently, the fire protection specialist must have a working knowledge of gas and vapor detection systems and monitors both for flammable and toxic materials.

As of this writing, no NFPA standard exists that explicitly establishes requirements for the design, installation, and maintenance of gas and vapor detection systems, nor a standard for gas monitors. However, there are a number of NFPA standards that make reference to gas and vapor detection systems in their requirements. For example, NFPA 86, *Standard for Ovens and Furnaces* (hereinafter referred to as NFPA 86), governs the fire alarm features of drying ovens such as those where painted parts are cured. NFPA 86 allows reduced airflow rates on the condition that a system for monitoring the concentration of combustible vapor in the ventilation air is provided. Since the fire safety of the curing oven is predicated upon the vapor monitoring system, that system becomes a critical part of the fire protection for the site. Failure of the flammable vapor detection system to perform its intended function, whether due to flawed system design, installation error, inadequate maintenance or component failure, could result in a catastrophic fire loss. This dependence upon flammable gas and vapor detection systems in order to maintain the required level of fire safety argues for the application of sound fire protection engineering in the selection, design, installation, and maintenance of these systems.

This chapter addresses the detection of flammable, combustible, or toxic gases and vapors that could lead to a deflagration or injury to an occupant of the hazard area but are not the result of an uncontrolled or uncontained fire. The term *fire-gas* was used in the past in an effort to differentiate the gaseous component of smoke from the particulate component of smoke and is dealt with in other chapters. Fire-gas implies the existence of a fire. This chapter presumes a potential of fire or toxic assault.

TERMINOLOGY

The terminology used in the gas and vapor detection community may not be familiar to many fire protection specialists.

Gas: A gas is matter in which the average thermal kinetic energy of the constituent molecules at standard temperature and pressure (STP) exceeds the cohesive forces between them, and consequently the matter expands to occupy the entire containment volume. STP is 77°F (25°C) and 760 mm(Hg) absolute pressure or 1 atm. It is necessary to stipulate temperature and pressure in this definition, because any normally gaseous compound can be compressed and cooled to the point where it will exist in the liquid state.

The gases that are routinely addressed in the context of gas detection, whether flammable gas or toxic gas, include: carbon monoxide, carbon dioxide, sulfur dioxide, hydrogen sulfide, hydrogen, methane, ethane, propane, butane, acetylene, ethylene, propylene, silane, germane, diborane, phosphine, arsine, allane, fluorine, and chlorine.

Vapor: Matter in the gas phase which normally exists at STP as a liquid. This term is not essential. However, it is used too frequently not to be addressed in the context of this chapter. The term is commonly used because of the inference it provides regarding the source of the gaseous matter. Vapors are the result of the evaporation of a liquid and, consequently, are normally associated with a liquid source. There is an equilibrium state at the interface above a pool of liquid and the gas above it. As liquid molecules attain excess thermal energy, they escape the cohesive forces of adjacent molecules and leave the liquid as a gas. When gas molecules lose thermal energy, they condense back into the liquid phase. When vapors are dispersed in air, they behave as gases until a temperature is reached where they begin to condense. The principal distinction between the compounds normally thought of as gases and those as vapors is the temperature and pressure at which they condense.

Combustible/flammable gases and vapors: Combustible/flammable gases and vapors are those gases and vapors that burn in air when the concentration of the gas or vapor is within the range of concentration where combustion can occur. The distinction between combustible and flammable is established in NFPA 30, *Flammable and Combustible Liquids Code*. This distinction is predicated upon

---

John M. Cholin, P.E., is an independent fire protection consultant and engineer with J.M. Cholin Consultants, Inc., Oakland, NJ.

the flash point of the liquid. The flash point is the temperature at which the vapor pressure of the liquid is sufficiently high to produce a concentration of gaseous fuel in the air above the liquid in excess of the lower flammable limit (LFL) and, hence, sufficient to support a flame. This distinction has little relevance in the context of detection of molecular species dispersed in the ambient air, except to imply possible origin. Unfortunately, the community of manufacturers and users of gas detection systems and instruments come from an instrumentation background where the term "combustible" has been adopted. In an effort to provide consistency with the terminology and concepts presented in this handbook, this chapter will use the term *flammable gas* to include all chemical species dispersed in the gas phase in ambient air, regardless of source or chemical composition, that will burn in air.

Concentration of flammable gas is typically measured in percent of the lower flammable limit (% LFL). If the lower flammable limit of a given gas is 4.0 percent, a measurement of 25% LFL means that there is 1 percent of that gas by volume in the measured sample of air.

Lower flammable limit (LFL): The lowest concentration of a flammable gas in air which will support a flame. Below the lower flammable limit there is insufficient fuel per unit of air volume to sustain a flame through the volume. Therefore, when a source of ignition is introduced, the resulting flame will remain at the source of ignition. Once the concentration exceeds the LFL, a flame front will propagate through the air volume. Note that the term lower explosive limit (LEL) was formerly used for this concentration of flammable gas, and is occasionally still encountered in older literature.

Upper flammable limit (UFL): The highest concentration of a flammable gas in air which will sustain a flame. As the concentration of flammable gas increases, the oxygen-containing air is displaced. Eventually, the concentration of flammable gas is so high that there is insufficient oxygen remaining to sustain a flame when an ignition source is introduced. Note that the term upper explosive limit (UEL) was formerly used for this concentration of flammable gas, and is occasionally still encountered in older literature.

Toxic gases and vapors: Toxic gases and vapors are molecular species dispersed in air in the gas phase that have an adverse effect on living organisms, as defined by the Occupational Safety and Health Administration (OSHA). The concentration of toxic gas is usually measured and displayed as parts per million (ppm) concentration. To convert from percent to parts per million, multiply by the conversion factor of 10,000 ppm/%. The commonly monitored toxic gases include carbon monoxide, carbon dioxide, hydrogen sulfide, sulfur dioxide, methane, silane, arsine, phosphine, chlorine, and fluorine. Most flammable gases are also toxic at some concentration, but not all toxic gases are flammable.

Oxygen concentration: The measure of the portion of the air within a space that is oxygen. This is an important life-safety and fire protection parameter. The depletion of the oxygen in a gas volume reduces the ability of life to survive in that volume and also reduces the ability of that gas volume to support combustion. The nominal oxygen concentration of air is 23.2 percent oxygen, by weight, and 21.0 percent oxygen by volume. The difference in the numerical value of the oxygen concentration by weight *vs.* by volume comes from the different densities of the constituent gases in the mixture called "air."

Fixed gas/vapor detection system: A fixed gas/vapor detection system is a permanently installed system usually consisting of at least one sensor, transmitter, and display unit. (See Figure 5-8A.) The sensor is located in the area where the gas or vapor may occur. It produces a minute current or voltage that is proportional to the

concentration of the gas. The sensor is connected via wiring to a transmitter. Generally the distance between the sensor and transmitter is quite limited. The transmitter linearizes the signal from the sensor, corrects it for zero and span, and converts it into an electrical format that can be transmitted a long distance without interference. The most popular format is the "4 to 20 milliampere" format. The transmitter is connected to the display unit via a second set of wires. The maximum wiring distance between the transmitter and the display unit can often be as far as several thousand ft (m). The display unit expresses the output of the transmitter in a format that is understandable to the operator. (See Figures 5-8B and 5-8C.) Some displays utilize analog meters, while others provide bar-graph and numeric displays. Most display units are equipped with output relays that transfer when the measured concentration of gas exceeds pre-established threshold concentrations. This provides for different

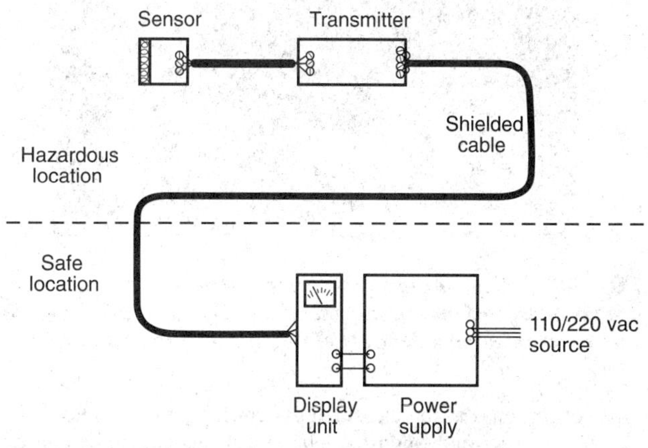

FIG. 5-8A. *A fixed gas detection system generally consists of at least one sensor, transmitter, and display unit.*

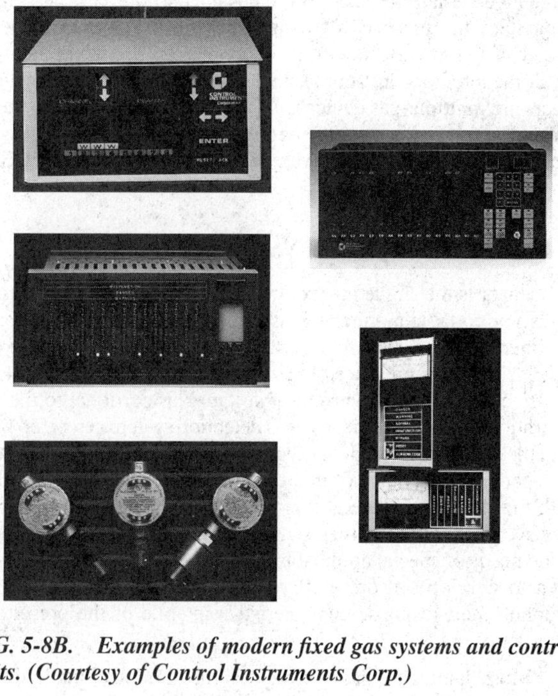

FIG. 5-8B. *Examples of modern fixed gas systems and control units. (Courtesy of Control Instruments Corp.)*

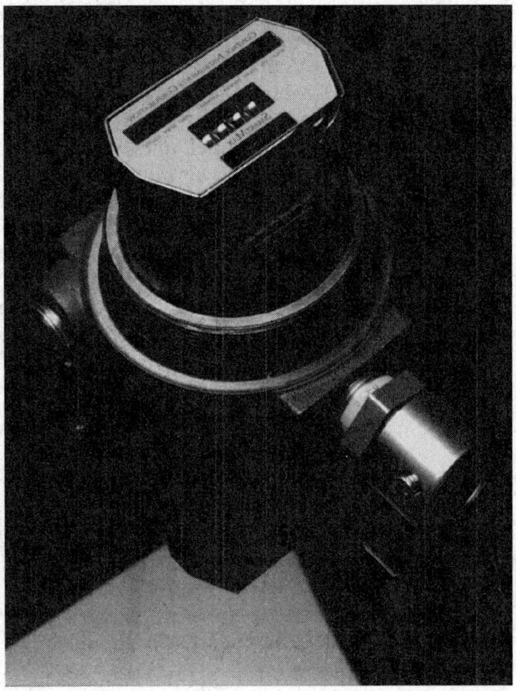

FIG. 5-8C. *Some fixed gas detection systems use a sensor and display integrated into a common unit, shown here with the enclosure cover removed. (Courtesy of Control Instruments Corp.)*

types of response as the measured gas concentration increases. For example, a warning response might be activated when the concentration reaches 20% LFL, and inerting or shutdown activated when the concentration reaches 50% LFL.

Portable gas/vapor monitoring instrument: Portable gas/vapor monitoring instruments are hand-carried portable instruments, utilizing no external source of power, designed to warn the person carrying it of the presence of dangerous levels of gas or vapor or dangerously low concentrations of oxygen in the ambient breathing air. Some monitors are simple single gas-warning devices, whereas others are multiple-gas devices. All are designed to measure the gas concentration and warn the user if the concentration begins approaching the threshold limit value (TLV), as established by OSHA, for that particular gas.

SENSORS

The sensor is the single most critical part of any fixed gas detection system or portable monitoring instrument. The sensor is the portion of either the fixed system or portable instrument that actually does the sensing of the presence of gas in the air. The sensor generates an electrical signal, either current or voltage, proportional to the concentration of gas in the air. The gas detection system can be no better than the sensors used and, consequently, the development of sensors has accounted for much of the research and development effort. As with fire detection, there are a limited number of sensor technologies in common use and then some specialized technologies tailored to the needs of special applications. In general, the design focus has been to develop sensors that are specific to a given gas and that maintain their response characteristics in spite of the presence of other gases.

Most flammable-gas sensors designed to detect the hydrocarbon gases and vapors from hydrocarbon liquids are based upon the detection of the heat of combustion of the flammable gas, using re-

sistive-thermal devices (RTDs) and the Wheatstone bridge circuit. An RTD is an electronic component whose resistance varies with temperature. RTDs include metal oxide resistors, resistance wires, and thermistors. These devices can all be wired into a circuit called the Wheatstone bridge, shown in Figure 5-8D.

The resistors shown as the sample and reference legs of the bridge circuit can be any type of RTD. The sample RTD is coated with a material that acts as a catalyst of the combustion of hydrocarbons. Alloys of both palladium and platinum are often used for this purpose. The reference RTD is enclosed in a manner that prevents contact with the ambient, gas-containing air. Electric current is passed through the bridge circuit, causing all of the legs of the bridge to become warm. When flammable gas impinges upon the sample RTD, it oxidizes, releasing heat from the combustion process occurring on the surface of the RTD. This raises the temperature of the sample RTD, and causes the electrical resistance of the sample RTD to change compared to the reference RTD, which remains at normal operating temperature. The change in the sample RTD resistance causes a potential (voltage) difference across the "voltmeter," which is proportional to the concentration of flammable gas.

Theoretically, the heat derived from the combustion of a gas on the surface of a catalyst-coated RTD is dependent upon both the concentration and the heat of combustion of the flammable gas. However, the heats of combustion of most chemical compounds encountered in the context of gas detection are very close in value. This allows the heat of combustion term to be considered nominally constant, especially when the gas concentrations are at or below the LFL. This approximation is not perfect. Carbon disulfide vapor yields readings that are lower than the actual concentration under these conditions, while natural gas yields readings that tend to be higher than the actual concentrations. These discrepancies increase as the concentrations increase. However, this is easily addressed with specific calibrations where these gases are present. For most

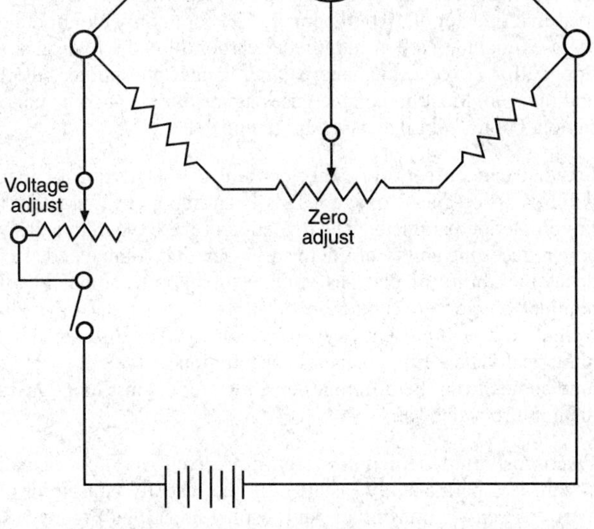

FIG. 5-8D. *The basic Wheatstone bridge circuit as used in many gas detection systems.*

purposes, where the sought-after condition is a hazardous gas-free atmosphere, the differences between various gases disappear as the actual concentration approaches zero.

Catalytic Beads

One of the more popular types of RTD is the catalyst-coated thermistor, called the *catalytic bead.* Thermistors are made from rare-earth ceramics that have very high-resistance temperature coefficients. Many have a temperature coefficient that is negative, meaning that the resistance goes down with increasing temperature.

Air mixtures containing concentrations of flammable gas much lower than the LFL, approaching zero, can "burn" on the surface of a heated catalytic material, yielding heat in direct proportion to the concentration of the flammable gas. This is called *surface* or *catalytic* combustion. Some RTD designs depend upon the flow of electric current through the thermistor bead to generate the requisite operating temperature of the catalyst. Others use a separate heat resister integrated into the bead. Obviously, the combustion of the flammable gas on the surface of the catalyst could conceivably cause ignition once the concentration exceeded the LFL. Consequently, sensors are contained within a flame arrester of either wire mesh or sintered metal, as shown in Figure 5-8E, that prevents the propagation of combustion outside the sensor. These flame arresters are generally suitable for conventional gas-air mixtures. However, if the atmosphere is oxygen enriched to concentrations greater than 21 percent oxygen by volume, or if the partial pressure of oxygen exceeds [21.27 kPa or 160 mm(Hg)] special arresters may be necessary.

Most sensors use a catalyst that is derived from the catalytic properties of platinum, palladium, and nickel. Careful control over the alloying of the catalysts and the operating temperature enables the manufacturer to make sensors that are specific for a particular group of gases. Catalytic beads have a very small mass; therefore, they respond rapidly to the heating produced by the combustion of flammable gas on their surfaces. The speed of response of a sensor is an important factor when choosing the best sensor for a given hazard.

Catalytic bead flammable-gas monitoring instruments are designed specifically for measuring combustibles in air. The oxygen in the air is necessary for the combustion of the flammable gas and the generation of the heat that the unit senses to infer the presence of flammable gas. These instruments give a reasonably accurate indication of the presence of combustible gas, even when the oxygen concentration has been substantially reduced by oxidation, biological action, or displacement. Generally, one should expect the accu-

racy of a catalytic-bead sensor to suffer as the ambient oxygen concentration decreases to less than 10 percent. Also note that inerting agents, such as Halon 1301 or similar clean agents, interfere with the combustion reaction on the catalyst surface, rendering measurements from these devices unreliable when inerting agents are present.

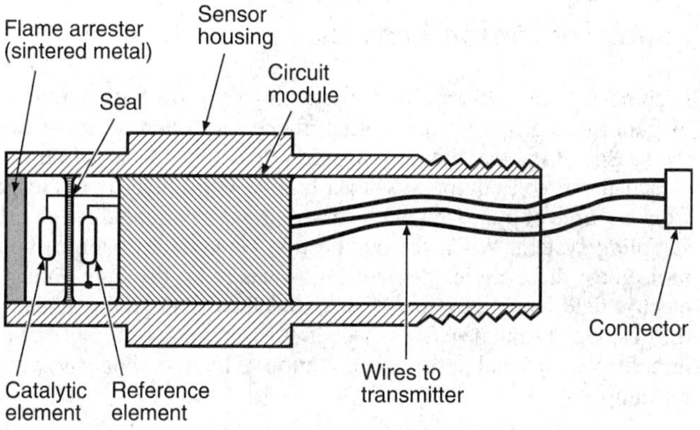

Catalytic bead sensors do suffer from "poisoning" to varying degrees. Poisoning occurs when a noncombustible material impinges upon the bead and only partially burns, leaving a residue on the surface of the bead. The coated surface no longer oxidizes the flammable gas in the air, and the sensor ceases responding to the presence of flammable gas in the air. While considerable advance has been made in reducing the tendency to become poisoned, this is a limitation of catalytic sensors that necessitates regular calibration of sensors to adjust for the progressive loss of sensing ability over time. Regular field calibration is necessary to adjust the zero and span of the display unit as compensation for drift and aging of the catalytic bead.

Many catalytic bead sensors are supplied with calibration curves showing the response of the particular unit to combustible gases other than that on which the device was calibrated. Calibration curves for one particular model of combustible gas indicator are shown in Figure 5-8F. Since all manufacturers do not use the same gas for base factory calibrations, the curves for one instrument cannot be used with curves of another type.

Calibration curves for field reference					
Gas or vapor	LEL % by volume	Curve no.	Gas or vapor	LEL % by volume	Curve no.
Acetone	2.5	5	Hydrogen	4.0	1
Acetylene	2.3	3	Methyl alcohol	6.7	2
Benzene	1.4	5	Methyl ethyl ketone	1.8	6
Carbon disulfide	1.0	10	Natural gas	4.8	3
Carbon monoxide	12.5	1	Octane	1.0	9
Ethyl acetate	2.2	7	Pentane	1.4	5
Ethyl ether	1.7	7	Propane	2.2	4
Gasoline	1.3	8	Toluene	1.3	4
Hexane	1.2	7	Xylene	1.0	7

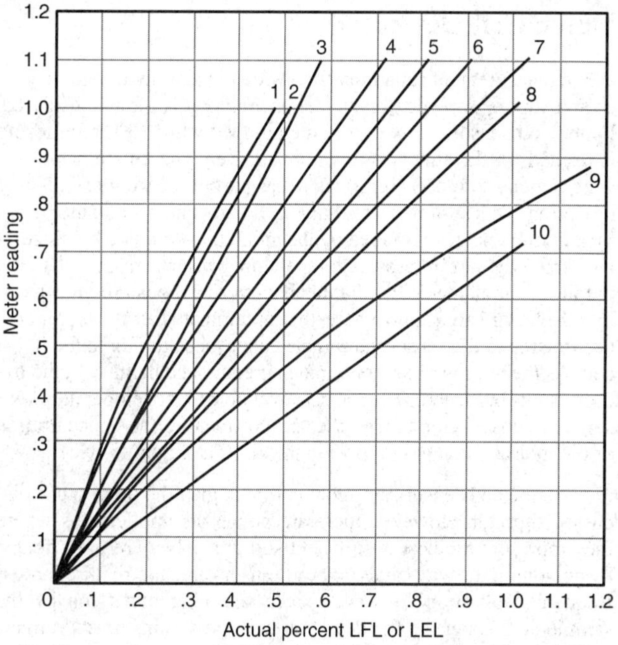

FIG. 5-8E. *A flammable-gas sensor using catalytic bead technology.*

FIG. 5-8F. *Calibration curves for a catalytic bead sensor. Note that, despite wide variations at higher concentrations, the error becomes negligible as the zero concentration is approached.*

Semiconductor (MOS) Devices

Metal oxide semiconductors (MOS) can be used to detect a number of gases. Metal oxide semiconductor devices can be fabricated with precisely controlled band-gap energies. When a gas with the right chemical structure impinges upon such a device while under voltage bias, the current through the device changes, indicating the presence of the gas. The semiconductor response can be adjusted with semiconductor doping and surface treatments, maximizing response to the desired gas. However, they will generally respond to any "reducing" gas, and care must be used to ascertain whether an unanticipated gas is causing the response of the sensor. Often, semiconductor sensors are also operated at elevated temperatures to improve the response specificity of the unit.

Electrochemical Devices

Electrochemical cells are primarily used for the detection of the toxic gases. The electrochemical cell is designed somewhat like a battery in which air can diffuse into the electrolyte of the cell. If the air contains the toxic gas for which the cell has been designed, an electrochemical reaction takes place, producing a minute voltage within the cell. This voltage causes a current to flow through the sensor electronics. The current is proportional to the concentration of the gas in the air. This enables the concentration to be measured by measuring the current from the cell. The diffusion path and the electrolyte can be varied to limit the response of the sensor to a particular gas.

Electrochemical cells are the predominant method for sensing carbon monoxide, chlorine, nitrogen oxide, sulfur dioxide, and hydrogen sulfide. Electrochemical cells are also used for oxygen-level monitoring by the same means as is used for the toxic gases. While the semiconductor and catalytic bead sensors depend upon a level of oxygen in the air in order to provide accurate results, electrochemical cells do not. Furthermore, they usually provide high levels of specificity relative to other technologies.

Flame Cell Devices

When measurement capability for a wide range of hydrocarbon gases and vapors is needed, flame cell technology is often considered. Flame cell sensors are very often employed when LFL monitoring is needed for the atmosphere within an oven where flammable solvent vehicles (vapors) are driven off as part of a curing or finishing operation. This system uses a flame cell that contains a flame fueled by a flammable gas provided by the unit, and ventilated by air sampled from the hazard area. A temperature sensor measures the heat output of the flame. When flammable hydrocarbons are introduced into the hazard area atmosphere, the contaminated air flows from the hazard area to the sensor via sample conveyance tubing to the flame cell. As the hydrocarbon-containing air enters the flame cell, the hydrocarbons burn, adding to the heat available to the temperature sensor. The sensor temperature rises proportional to the concentration of the hydrocarbon gas (vapor) in the air. (See Figure 5-8G.)

Flame cell sensors are more complex than the other types discussed thus far. However, there are numerous applications where they represent the best cost/benefit solution. They are not affected by poisoning the way catalytic bead sensors are, and require far less frequent calibration. However, since sampling tubes convey the sample atmosphere to the detection unit, provisions must be made to ensure that there is no condensation of flammable or combustible vapors on the interior surfaces of the sample conveyance tubing, as this would create serious errors in the flammable gas concentration measurements. (See Figure 5-8H.)

FIG. 5-8G. *A flame cell sensor (Left); the flame cell sensor with the enclosure open, showing the internals (right). (Courtesy of Control Instruments Corp.)*

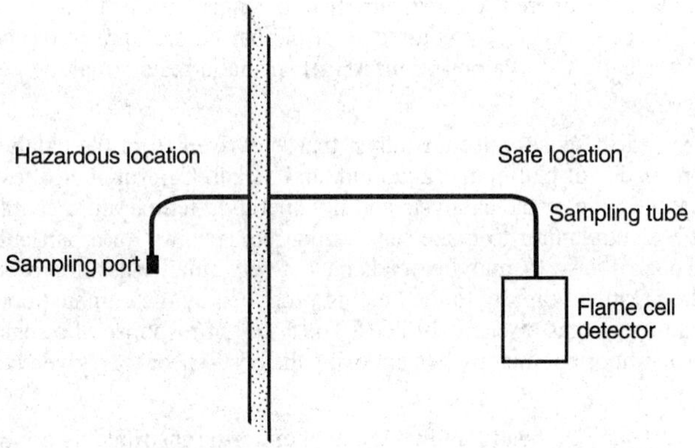

FIG. 5-8H. *The flame cell sensor is a sampling-type device, connected to the hazardous area via a sampling tube.*

Flame Ionization Devices

In some applications a means for detecting very low concentrations of flammable gas is needed. Often, flame ionization detectors are the sensor of choice. Flame ionization detectors consist of an enclosed flame cell equipped with a high-voltage field. It is fueled from a source of hydrogen, with combustion air coming through the sampling system. When the combustion air includes hydrocarbon molecules, these molecules are pulled apart by the high-voltage electric field, producing ions. These ions flow through the flame to the positive (+) and negative (−) electrodes, creating a current that is directly proportional to the concentration of hydrocarbon gas in the air sample.

Flame ionization sensors, like the flame cell sensors, are more complex than the other spot-type sensors. However, there are numerous applications where they represent the best cost/benefit solution. They are able to measure much smaller concentrations than the other types of gas sensors. They are not affected by poisoning the way catalytic bead sensors are, and require far less frequent calibration. Flame ionization sensors are installed like the flame cell devices, with a sampling tube that conveys a sample of the hazardous atmosphere to the sensor. Consequently, provisions must be made to ensure that there is no condensation of flammable or combustible vapors on the interior surfaces of the sample conveyance tubing, as this would create serious errors in the flammable gas concentration measurements. (See Figure 5-8I.)

Nondispersive Infrared (NDIR) Sensors

It is a well-known fact that gaseous molecules absorb infrared radiation at specific and characteristic wavelengths. This is analogous to the infrared emission detection that is the basis for many of the advanced technology flame detectors in use today. The correlation between specific molecular species and specific wavelengths in the infrared spectrum has recently made available infrared gas detectors. Like smoke detectors, infrared gas detectors are available either as spot devices or as open-path devices, analogous to the projected beam smoke detector.

FIG. 5-8I.   A flame ionization sensor. (Courtesy of Control Instruments Corp.)

Open-path gas detectors operate with one or more infrared emitters and one or more receivers. These units compare the optical absorbance of the air in the optical path at two wavelengths, one at which the gas absorbs and one which is a reference wavelength where at which the gas does not absorb. The detection system electronics compare the intensities at the two wavelengths. If gas exists along the beam path, it will absorb the sample wavelength while allowing the reference wavelength to pass unattenuated. The difference in received intensities is proportional to the concentration of the gas and the percentage of the path length which contains that percentage. This value is transmitted via the common 4-20 mA analog signal to the display unit. It is displayed as LFL-meters, the product of the percent LFL and the distance. (See Figure 5-8J.)

Spot-type IR gas sensors operate on a similar principle, but the optical path is contained within the detector enclosure. This eliminates the path length as a variable, and they display concentration in simple percent of LFL. The spot-type sensor often contains an integral 4-20 mA transmitter that sends the analog signal to the display unit. (See Figure 5-8K.)

Infrared gas detectors are reputed to require far less maintenance and calibration than the catalytic bead, electrochemical, and semiconductor devices.

FIXED GAS DETECTION SYSTEMS

Throughout modern industry there exist enclosed areas where gaseous contaminants can accumulate. Whether these gases are flammable, toxic, or both flammable and toxic, they represent a serious threat to personnel, the facility, and the operations housed within the facility. Reliable monitoring of known gas hazards is critical if losses are to be prevented.

Where known, permanent gas hazards exist, a fixed gas detection system is necessary. These hazards are commonly encountered in public utility electric generating stations; publicly owned water treatment works; petroleum exploration, production, refining, and distribution facilities; pharmaceutical, chemical, and petrochemical production, storage, and distribution facilities; metallurgical facilities; semiconductor fabrication facilities; distilleries; paint and varnish manufacturing facilities; and mines.

Concurrent with the detection of the presence and concentration of both flammable and toxic gases, there is the need to measure the oxygen concentration in the atmosphere in many applications. Measurement of the oxygen concentration is important as a means

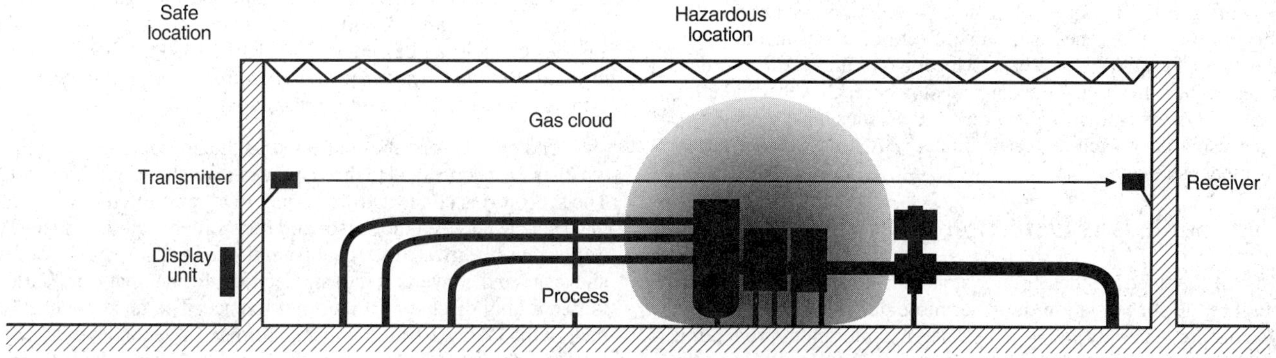

FIG. 5-8J.   The open-path NDIR gas detector covers a large area. However, the reading is the product of the distance and the concentration over that distance.

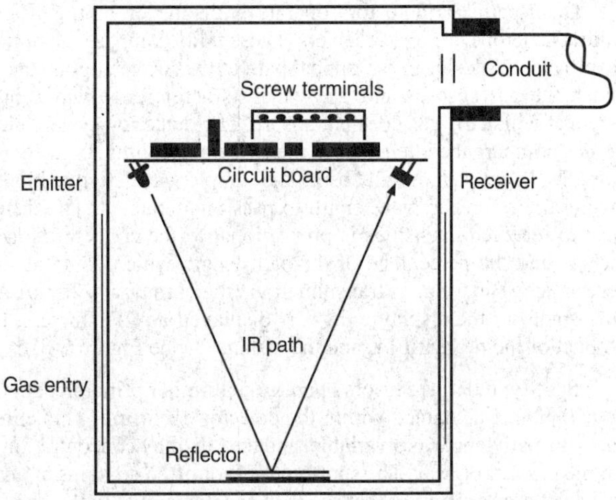

FIG. 5-8K.   *The spot-type IR gas sensor.*

TABLE 5-8A.   *Specific Densities of Commonly Encountered Gases and Vapors*

Standard Conditions of Temperature and Pressure		
Gas/Vapor	Molecular Formula	Specific Density, Air = 1.000
Air	—	1.000
Hydrogen	H_2	0.70
Carbon Monoxide	CO	0.967
Carbon Dioxide	CO_2	1.528
Methane	CH_4	0.554
Acetylene	C_2H_2	0.911
Ethane	C_2H_6	1.049
Hydrogen Sulfide	H_2S	0.9109
Phosphine	PH_3	1.184
Propane	C_3H_8	1.55
Silane	SiH_4	1.114
Sulfur Dioxide	SO_2	2.264
Acetone	$(CH_3)_2CO$	2.0
Benzene	C_6H_6	2.8
Carbon Disulfide	CS_2	2.6
Cyclohexane	C_6H_{12}	2.9
Ethanol (Ethyl Alcohol)	CH_3CH_2OH	1.6
Ethyl Ether	$CH_3Ch_2OCH_2CH_3$	2.6
Gasoline	mixture	3.4
Mineral Spirits	mixture	3.9
Naphtha (VM&P)	mixture	<1.0
Toluene	$C_6H_5CH_3$	3.1

* Note: See NFPA 86.

of evaluating the effectiveness of inerting processes, of protecting personnel from the effects of oxygen deficiency, and of determining that sufficient oxygen is present in the atmosphere to ensure reliable concentration readings from combustible gas detectors.

It is important to note that one cannot use the spacing criteria for smoke detectors in NFPA 72, *National Fire Alarm Code* (hereinafter referred to as NFPA 72), as a guide for the location and spacing of gas detectors. The spacing and location criteria in NFPA 72 for smoke detectors are based upon fire plume dynamics in which the heat released by the combustion of the fuel serves as the driving force behind the development of a fire plume and a ceiling jet. No such energy source exists in the context of a gas leak. When a contaminant gas leaks into the surrounding atmosphere, it mixes with the air, changing its density. In a mixture of gases, the density of the mixture is a function of the constituent gases and the portion of the mixture that they represent. Under still air conditions where there is no ambient air movement, higher concentrations are usually observed at higher locations within the enclosing space when lighter-than-air gases are involved. Conversely, higher concentrations are usually observed at lower locations within the enclosing space when heavier-than-air gases are involved. One can use the specific density of the gas in question to aid in predicting whether the gas in question will tend to rise or sink when a leak occurs. (See Table 5-8A.)

The data in Table 5-8A must be used carefully. It is often necessary to take into account the refrigeration effect that occurs when a gas at high-pressure leaks into a region of much lower pressure. The volumetric expansion of the leaking gas causes it to cool the air into which it is flowing. The cooled air is no longer at STP conditions of 77°F (25°C) and 760 mm(Hg) absolute pressure or 1 atm. This cooling causes the air/gas mixture to contract, increasing in density. As the mixture moves from the locus of the leak it continues to absorb heat from the ambient, and slowly reaches thermal equilibrium with the ambient. These thermodynamics must be observed when using data, such as that in Table 5-8A, to develop a sensor placement strategy.

Flammable Gas Detection

Most of the fixed gas detection systems in service are designed to measure the concentration of one or more flammable gases. These systems are generally found in petroleum exploration, production, refining, and distribution facilities; power generation plants; water treatment plants; landfills; chemical and pharmaceutical facilities; and numerous industrial processes where flammable solvent (vehi-

cle) vapors are driven off in ovens. Generally, these systems are designed to respond to a broad range of hydrocarbon gases and vapors, such as methane, ethane, propane, butane, ethylene, propylene, acetylene, and similar gases, as well as the vapors from the evaporation of hydrocarbon liquids, such as gasoline, acetone, ethyl ether, methylethyl ketone, methylisobutyl ketone, carbon disulfide, turpentine, and lacquer thinner, to list a few.

The predominant sensor for flammable gas detection in general, normally occupied spaces is the catalytic bead sensor. They are designed to comply with the performance criteria outlined in ANSI/ISA S12.13, *Performance Requirements, Combustible Gas Detectors*.[3] Sensors are situated where there is the probability that gas will accumulate when a leak occurs. When the gas to be detected is methane, natural gas, or other lighter-than-air gases, sensors are generally located above the gas-containing equipment. When propane, butane, and most vapors are involved, the sensors are located low in the air space of the hazard area. Flammable gas sensors must be placed to take advantage of normally occurring air currents and the differences in the density of air and the flammable gas(es) in question. The installation, operation, and maintenance of flammable gas detectors is addressed in ISA RP12.13, Part II-87, *Recommended Practice: Installation, Operation, and Maintenance of Combustible Gas Detection Instruments*.[4]

The gas sensors are connected to the appropriate transmitter, which is then connected to the display unit, as shown in Figure 5-8L. The sensor(s) is connected via wiring to a transmitter(s). Generally the distance between the sensor and transmitter is quite limited. The transmitter linearizes the signal from the sensor, corrects it for zero and span, and converts it into an electrical format that can be transmitted a long distance without interference. The most popular format is the "4 to 20 milliampere" format. A second set of wires connects the transmitter to the display unit. The maximum wiring distance between the transmitter and the display usually can be as far as several thousand ft (m).

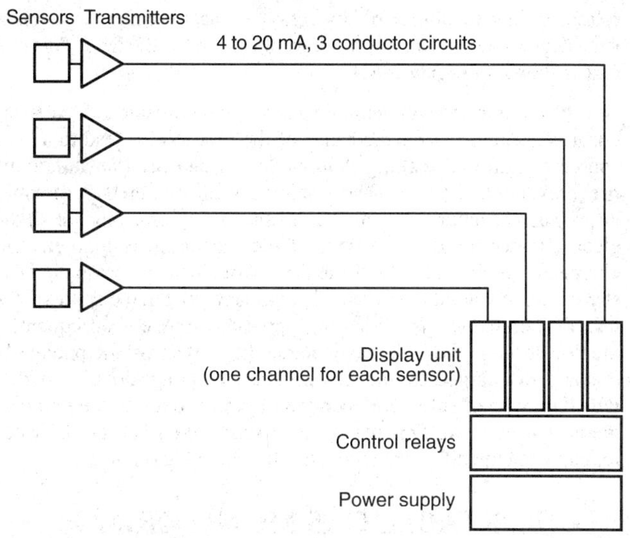

FIG. 5-8L. *The wiring architecture for the traditional and currently most likely encountered flammable gas detection system.*

Large numbers of "points" are generally wired to a centrally located display unit where the status of the facility can be monitored by facility staff. The unit displays the output of the transmitter. Some display units utilize analog meters, while others provide bargraph and numeric displays.

Recently, addressable gas detection systems have been introduced that enable the system to be wired in a manner similar to the traditional smoke detector circuit. The transmitter converts the analog data from the sensor into a binary format, and transmits it serially along with its identity to the display unit. The display unit presents the analog data in the customary fashion. (See Figure 5-8M.)

Most display units are equipped with output relays that transfer when the measured concentration of gas exceeds pre-established threshold concentrations. This provides for different types of response as the measured gas concentration increases. For example, a warning response might be activated when the concentration reaches 20% LFL, and inerting or shutdown activated when the concentration reaches 50% LFL.

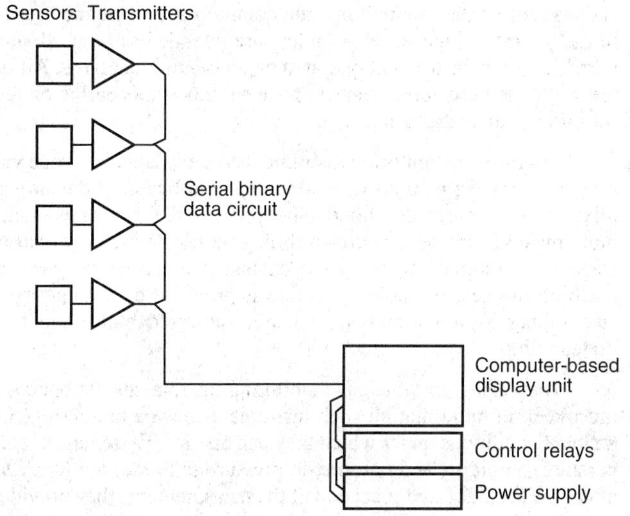

FIG. 5-8M. *The architecture of the addressable flammable gas detection system.*

ANSI/ISA S12.13[3] requires that the flammable (combustible) gas detection system supervise for:

1. The failure of the unit power supply;
2. The loss of continuity of the sensor element (catalytic bead);
3. The continuity of all conductors to the detector head (the sensor and transmitter); and
4. Down-scale reading, below zero, equivalent to a concentration of more than 5 ppm.

This is not entirely consistent with the supervisory requirements of NFPA 72. Furthermore, ANSI ISA RP 12.13 does not address supervision of any of the connections that are intended to initiate any type of automatic response to a gas concentration above threshold levels. This inconsistency is one of several system design issues that come into play when integrating or interconnecting flammable gas detection systems with fire protective systems. NFPA 72 includes specific requirements for the power supplies, electrical supervision, isolation from ground, programming security, and supervision of output circuits that might not have been observed by the manufacturer of the gas detection system. The activation of inerting, warning, and ventilation systems in response to high-level gas indication are all fire-protective actions that should be achieved with inherently reliable, supervised designs, as described in NFPA 72. Some Authorities Having Jurisdiction might apply the requirements of NFPA 72 to flammable gas detection systems when the flammable gas detection is integrated into the facility fire protection systems.

Toxic Gas Detection

The detection of many toxic gases is necessary in the same environments where flammable gases must be monitored. Most frequently, hydrogen sulfide gas is commonly encountered in petroleum exploration and production involving sour crude and/or sour gas. Consequently, many gas detection systems in place in the petroleum industries include both flammable gas and toxic gas detection for hydrogen sulfide, sulfur dioxide, and fluorine in the same system.

The architecture of most fixed gas detection systems lends itself to this reality. The systems are designed with transmitters and display units that can accommodate a number of interchangeable sensors, specifically designed for a particular gas. Combined flammable and toxic gas detection systems are commonly encountered.

The semiconductor and electrochemical sensors are most commonly used for the detection of the toxic gases. The measurements are expressed in ppm (parts per million) and generally the threshold limit value (TLV) established by the Occupational Safety and Health Administration (OSHA) for the gas in question is placed in the upper half of the display range.

ANSI/ISA S12.15, PT I-90, *Performance Requirements for Hydrogen Sulfide Detection Instruments (10-100 ppm),*[3] requires that the toxic gas detection system supervise for:

1. The continuity of all conductors to the detector head (the sensor and transmitter);
2. The failure of the unit power supply;
3. The loss of continuity of a circuit protection device;
4. A down-scale reading, below zero, equivalent to a concentration of more than 5 ppm.

This is not entirely consistent with the supervisory requirements of NFPA 72, which establishes requirements on the equipment design that are not normally assumed to be relevant when a manufacturer outside the fire protection community undertakes the design of a product. This becomes an important issue when integration of gas detection and fire alarm systems is contemplated.

Calibration and Maintenance

Both flammable and toxic gas detection systems require regular calibration and maintenance if reliable measurements are expected.[4,6] Each system has a specific set of calibration and maintenance criteria that are incorporated as part of the listing investigation by the nationally recognized testing laboratory. These procedures must be carefully and meticulously followed. In general, most systems using catalytic bead, semiconductor, and electrochemical sensors require calibration on a monthly schedule. The flame cell and flame ionization detectors usually require less frequent calibration.

The calibration of a gas detector, whether flammable or toxic, involves adjusting the electrical output of the transmitter to accurately reflect the atmosphere at the sensor. The first step is the zero adjustment. While the sensor is under normal ambient conditions with no gas present, the output of the transmitter is adjusted to 4.0 mA output to the display unit. The sensor is exposed to *test gas*. Test gas is a mixture of air and the gas to be detected, stored at elevated pressure in a specially designed cylinder with a sensor test fitting attached. The mixture is certified by the manufacturer as being accurate to within 1.0% LFL or 1.0 ppm, whichever is appropriate. Since there is an obvious hazard if the test gas concentration is at 100% LFL or 100% TLV, the test gas concentration is usually at 50% LFL or TLV, again whichever is appropriate. Exposure to the test gas is used to adjust the span output of the transmitter to the appropriate value for the test gas and sensor in question. In this way the electronics of the transmitter are used to correct for any non-linearities and drift in the sensor. The display unit merely reads the current passing through the 4-20 mA circuit to the transmitter and displays it. Consequently, there is usually no calibration necessary at the display unit.

Eventually, the zero and span adjustments on the transmitter are insufficient to achieve both a zero and a correct span adjustment. This indicates that the sensor has aged to the point where it must be replaced. The rate of aging is very application specific and can occur without any apparent warning. Consequently, regular calibration is the only means to be confident that the gas detection system is operative and displaying an accurate measurement of the actual gas concentrations in the hazard areas.

Integration with Fire Alarm Systems

There are inconsistencies between the NFPA 72 requirements for fire alarm signaling systems and the ISA standards for gas detection instruments. These inconsistencies become relevant when the conditions in the hazard area demand an integrated approach. Obviously, the presence of a flammable gas in an area sufficiently hazardous to warrant a fire alarm system represents an aspect of that hazard area that warrants immediate remedial action. Whether that remedial action consists of merely the activation of a notification appliance or whether it requires the activation of purge fans, process shutdown, deployment of an inerting agent, or some other response, these actions must be designed into the system in a manner that provides a high degree of intrinsic reliability and supervision if they are to represent a meaningful response. Typically, these responses are executed by a fire alarm system control unit that has been listed for releasing device service, and the flammable gas detection system serves as an initiating device in that system.

Since many gas detection systems use earth-ground as a circuit common, a practice that is forbidden in the context of fire alarm systems, there is often a fundamental power supply incompatibility between the gas detection control unit and the fire alarm system control unit. Consequently, the interface between the gas detection and the fire alarm system is often achieved through dry relay contacts. This is not the ideal means of signal transmission due to the relatively low reliability of dry relay contacts compared to solid-state devices. However, it is often what is found in the field. This circuit is shown in Figure 5-8N.

It is important to understand that when flammable gas leaks occur it is not unusual for the cause of the leak to also lead to a heat source of ignition, making flammable gas detection an important fire prevention strategy. The commonly encountered flammable gases, such as natural gas, methane, ethane, and propane, are all dielectric materials. Being dielectrics they accumulate static electric charges when the velocity of the flow varies widely across the flow stream. Such velocity differentials regularly exist in the context of a leak of flammable gas from a high-pressure source. Consequently, the flow through the leak may generate large electrostatic potentials that are discharged as the gas escapes, creating electric sparks, which may be of sufficient energy and power to ignite the flammable gas/air mixture. This reality underscores the importance of detecting and expeditiously correcting flammable gas leaks.

PORTABLE GAS MONITORING INSTRUMENTS

Wherever there is the reasonable possibility of an accidental release of either flammable or toxic gas, there is the need to equip personnel with a portable instrument to warn them of such a hazard upon entering. Both flammable and toxic gases are encountered in diverse locations throughout industry, e.g., as a result of some form of accidental or opportunistic event, such as leakage from piping systems (both above and below ground) into the surrounding soil; from spillage or the decomposition of natural organic material in the soil; or sanitary landfills. Regardless of the source, the possibility and probability that a dangerous concentration of a flammable or toxic gas might exist in an area within a facility establishes the need for portable gas monitors. (See Figure 5-8O.)

There are portable gas monitors designed for a wide range of performance objectives. It is important that the performance objectives for the gas monitoring instrument be fully understood before a unit is selected. Some applications require oxygen monitoring to ensure that there is sufficient oxygen for safe occupation of the space. Many applications require a monitor for only one specific gas, whereas other applications represent a number of concerns justifying the use of a multifunction monitor. Areas that contain concentrations of flammable gas above the LFL will necessitate an explosionproof or intrinsically safe instrument. Finally, some applications require that the unit operate continuously while the user is in the hazardous space, while others are intended to be used as a portable survey instrument operated on an intermittent basis. All of these intended use considerations have an important bearing on the selection of an instrument.

Portable gas monitoring instruments are available in a wide variety of forms. Some units depend upon the diffusion of the air/gas mixture into the sensor. Others have provisions for a forced sampling means. Samples are drawn through a hose or tube, which is frequently equipped with a rigid extension or probe for ease in reaching inaccessible points. Suction is provided by a hand-operated rubber aspirator bulb or a battery-powered built-in motor-driven pump.

When using portable gas monitoring instruments the operator must keep in mind that all such instruments have a finite range of ambient conditions under which they can operate. Extremes in temperature, humidity, and barometric pressure (altitude) are likely to affect the accuracy and precision of the measurements they provide. Sampling steamy atmospheres is not recommended, since condensation of water within the instrument is likely to cause errors. Sampling from ovens or dryers operating at temperatures much higher

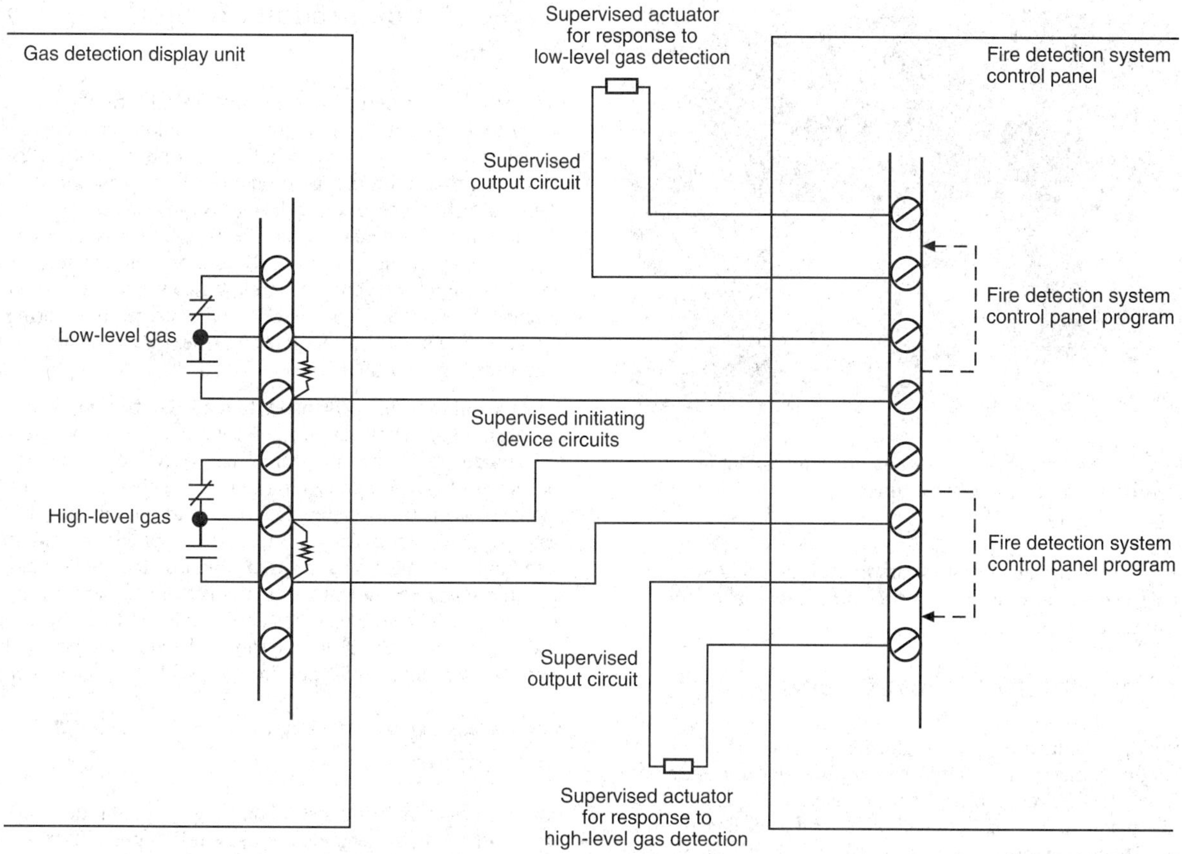

FIG. 5-8N. A commonly encountered interface between the gas detection and the fire alarm system used to activate automatic responses to the gas detection level alarms.

FIG. 5-8O. A portable gas monitor can be used to survey sites for gas leakage that do not normally warrant a fixed system.

than instrument ambient should also be avoided. Unless the solvents in question have flash points below instrument ambient temperature, vapors might condense within the sampling tubing or measuring instrument, again causing errors.

Finally, if the portable gas monitoring instrument is to be used in areas where ignitable concentrations of flammable gas might exist, it should be designed and listed for use in hazardous locations as defined in Article 500 of NFPA 70, *National Electrical Code®*. The manufacturer's label and instructions indicate the type of approval granted for the particular instrument.

Flammable (Combustible) Gas Monitoring Instruments

Flammable gas monitoring instruments generally use a catalytic bead or semiconductor sensor. The output from the sensor is directly displayed with a number of different techniques, ranging from an analog meter to digital readout. Most also include some form of audible alarm set at a level considered immediately hazardous. As a rule, flammable gas instruments must be set to read "zero" in uncontaminated air each time they are used. Like the fixed systems, the gas concentration is displayed in % LFL. Generally, flammable gas monitoring instruments respond to all flammable gases or mixtures of flammable gases, irrespective of their chemical composition. (See Figure 5-8P.) This is usually desirable in a portable monitor, as it is unnecessary to know the exact identity of the flammable gas or gases prior to being able to assess the fire risk of the area being surveyed. Where it is necessary to selectively measure the concentration of some specific flammable gas, especially while in the

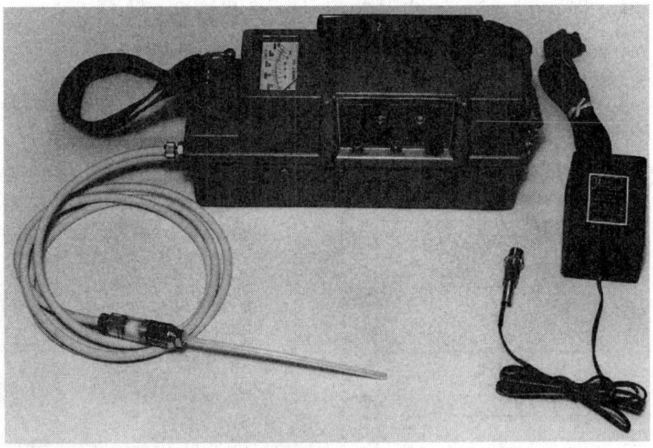

FIG. 5-8P. A portable multiple-range gas monitoring instrument. (Source: Heath Consultants, Inc.)

presence of other combustibles, infrared absorption and gas chromotographic techniques on samples transported to a laboratory by some means must be used.

Toxic Gas Monitoring Instruments

Similar to the flammable gas monitoring instruments, toxic gas monitoring instruments are often necessary in order to ensure personnel safety. Hydrogen sulfide monitors usually employ electrochemical sensors as do the sulfur dioxide, chlorine, and fluorine monitors. The carbon monoxide monitors are available using either the semiconductor sensor or the electrochemical sensor technologies. (See Figure 5-8Q.) Other techniques that involve electrical detection of gas-specific chemical reactions are also employed. These include chemical stain, electro-conductance, and membrane permeability methods. These units generally display the concentration in parts per million (ppm) and include an audible signal when the concentration reaches a threshold value above which the threat to life and health is unacceptable.

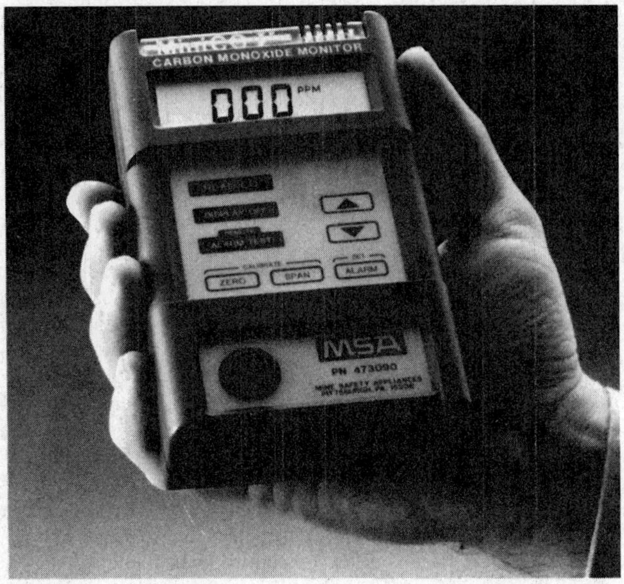

Oxygen Concentration Monitoring Instruments

Oxygen concentration monitoring instruments are often critical life-safety tools. Generally they utilize an electrochemical sensor. The partial pressure of the oxygen in the atmosphere controls the rate at which it diffuses through a porous membrane and into the electrochemical cell. Once in the cell, the oxygen enters into an electrochemical reaction, generating an electric current that is electronically scaled to display the oxygen concentration. The oxygen concentration is usually displayed as a simple "% oxygen," usually on a scale graduated from 0 to 25 percent. Portable oxygen monitoring instruments usually include an audible alarm set for a level around 18 percent. (See Figure 5-8R.)

Except for setting the meter to read 20.8 percent while aspirating fresh air at the elevation (atmospheric pressure) where tests are to be made, no other calibration is necessary for accurate reading in the range of oxygen concentrations in which it is permissible for people to work. (Current regulations require a minimum of 19.5 percent oxygen, otherwise air must be supplied or self-contained breathing equipment must be used.) Enclosed spaces should also be tested to ensure sufficient oxygen is present to permit reliable gas measurements. Where readings are intended to be significant in the low range, i.e., with inerting systems, the zero setting can be confirmed by sampling an oxygen-free gas, such as propane or nitrogen.

Calibration of Portable Gas Monitoring Instruments

Portable gas monitoring instruments require calibration on a regular basis if their measurements are to be relied upon. The procedures for calibration vary from instrument to instrument. However, as with the fixed detection systems, most employ the use of a test gas. The "zero" of the unit is adjusted. Then it is subjected to test gas containing a known, certified concentration of the gas it is designed to measure. The span adjustment is then employed to achieve a measurement reading consistent with the known gas concentration.

The frequency at which units should be calibrated is determined by the authorities responsible for the life safety of the facility staff. Some authorities recommend calibration be verified daily, before use. Such a calibration check is accomplished by passing test gas through the instrument.

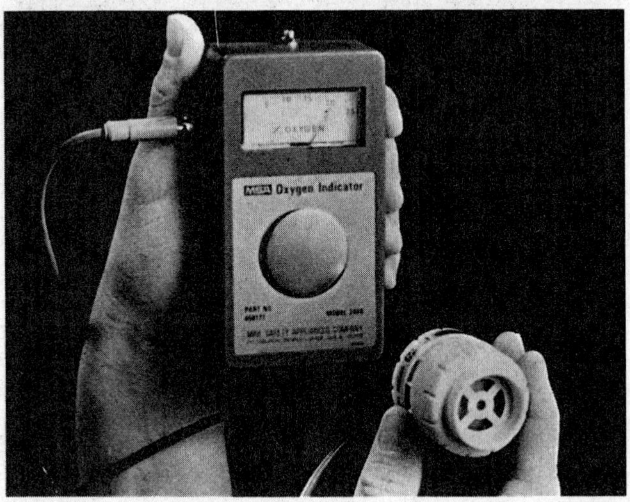

FIG. 5-8Q. This electrochemical carbon monoxide monitoring instrument includes an indicator and alarm.

FIG. 5-8R. An individual oxygen indicator for remote sampling. (Source: Mine Safety Appliances Co.)

SUMMARY

This chapter is by no means a complete, authoritative presentation on gas detection. It is an introduction to this topic. The reader must be cautioned that the gas detection industry has developed over the years as part of the instrumentation industry, serving the needs of the chemical process and petroleum industries. Generally, the language is different from that used in the fire protection industry. Furthermore, the standards that have evolved over the years relating to gas detection do not address the concerns of the fire protection community in the customary manner. For example, consider that the 4-20 mA data transmission circuit is not recognized by NFPA 72.

As fire alarm systems become increasingly interactive with the facilities and production processes they protect, conflicts are expected to emerge in the requirements of various standards. These conflicts can only be resolved when fire protection professionals understand the basics of this technology and the objectives the standards were written to ensure. As fire alarm systems are more thoroughly interfaced with the process control systems, questions of reliability, operability during impairment, and maintainability must be addressed in a manner that satisfies the needs of both fire protection and rigorous process control.

BIBLIOGRAPHY

References Cited

1. Braker, W. T., and Mossman, A. L., *The Matheson Gas Data Book,* 6th ed., Matheson Division of Searle Medical Products, Inc., Lyndhurst, NJ, 1980.
2. Heisler, S. I., P.E., *The Wiley Engineer's Desk Reference,* Wiley, New York, 1984.
3. ANSI/ISA S12.13, *Performance Requirements, Combustible Gas Detectors,* Instrument Society of America, Research Triangle Park, NC, 1986.
4. ISA RP12.13, PT II-87, *Recommended Practice: Installation, Operation, and Maintenance of Combustible Gas Detection Instruments,* Instrument Society of America, Research Triangle Park, NC, 1987.
5. ANSI/ISA S12.15, PT I-90, *Performance Requirements for Hydrogen Sulfide Detection Instruments (10-100 ppm),* Instrument Society of America, Research Triangle Park, NC, 1990.
6. ISA RP12.15, PT II-90, *Recommended Practice: Installation, Operation, and Maintenance of Hydrogen Sulfide Detection Instruments*, Instrument Society of America, Research Triangle Park, NC, 1990.

NFPA Codes, Standards, and Recommended Practices

Reference to the following NFPA codes, standards, and recommended practices will provide further information on gas and vapor detection discussed in this chapter. (See the latest version of *The NFPA Catalog* for availability of current editions of the following documents.)

NFPA 30, *Flammable and Combustible Liquids Code*
NFPA 70, *National Electrical Code*®
NFPA 72, *National Fire Alarm Code*
NFPA 86, *Standard for Ovens and Furnaces*
NFPA 325, *Guide to Fire Hazard Properties of Flammable Liquids, Gases, and Volatile Solids*

Additional Reading

Air Sampling Instruments, American Conference of Governmental Industrial Hygienists, Cincinnati, OH, latest ed.
ANSI Standard 117, *Safety Requirements for Confined Spaces,* 1-1989.
Bonn, R. J. C., "Modern Methods of Designing Fire and Gas Detection Systems," Third International Conference on Management and Engineering of Fire Safety and Loss Prevention: Onshore and Offshore, February 18–20, 1991, Aberdeen, Scotland, Elsevier Applied Science, New York, 1991, pp. 253–266.
"British Standard Methods of Test for Gases Evolved During Combustion of Electric Cables. Part 1. Method for Determination of Amount of Halogen Acid Gas Evolved During Combustion of Polymeric Materials Taken From Cables," British Standards Institution, England, BS 6425:Part 1:1990, 1990.
"Carbon Monoxide Detectors Alerts to a Deadly Gas," *Consumer Reports*, Vol. 59, No. 5, May 1994, pp. 334–336.
"Carbon Monoxide Gas Detectors for Marine Use. Standard for Safety," Underwriters Laboratories, Inc., Northbrook, IL, UL 1524, 1st Edition, February 20, 1992.
"Continuous Gas Detection Instruments Help Automate Remediation Systems," *Industrial Fire World*, Vol. 9, No. 5, Sept./Oct. 1994, pp. 41–42, 44.
Davidson, R. S., and Peacock, S. J., "Smoke, Fire, and Gas Detection at British Gas Installations," *Communication* of the 52nd Autumn Meeting of the Institution of Gas Engineers, Communication 1298, Institution of Gas Engineers, 1986.
Dean, K., "Gas Detection: Who Needs It?" *Fire Prevention*, No. 268, Apr. 1994, pp. 34–36.
Hagglund, L., "Laboratorietest och utvardering av gasindikeringsinstrument. [Laboratory Test and Evaluation of Instruments for Detection and Measurement of Combustible Gases.]," National Defence Research Institute, Stockholm, Sweden, FOA Report C40327-4.6, ISSN 0347-2124, Mar. 1994.
Hokstad, P., *et al.*, "Reliability Model for Optimization of Test Schemes for Fire and Gas Detectors," *Reliability Engineering and System Safety*, Vol. 47, No. 1, 1995, pp. 15–25.
Jackson, M. A. and Robins, I., "Gas Sensing for Fire Detection: Measurements of CO, CO_2, H_2, O_2, and Smoke Density in European Standard Fire Tests," *Fire Safety Journal*, Vol. 22, No. 2, 1994, pp. 181–205.
Kohl, D., Kelleter, J. and Petig, H., "Detection of Smoldering Fires by Gas Emission," *Proceedings* of the Tenth International Conference on Automatic Fire Detection "AUBE '95," [Internationale Konferenz uber Automatischen Brandentdeckung.] April 4–6, 1995, Duisburg, Germany, 1995, pp. 223–239.
Lion, D. R., "Smoke Alarms and CO Detectors," *Fire Fighting in Canada*, Vol. 40, No. 3, Apr. 1996, pp. 61–62.
MacDonald, S. J., Roundbehler, D. P. and Young, R., "Analysis of hydrazines From Environmental Air Samples by 'In Situ' Derivatization and Gas Chromatography With Chemiluminescence Detection," Thermedics, Inc., Woburn, MA, National Aeronautics and Space Administration, Kennedy Space Center, FL, CPIA Publication 543; Joint Army-Navy-NASA-Air Force (JANNAF) Safety and Environmental Protection Subcommittee Meeting, June 4, 1989–June 21, 1990, Livermore, CA, 1990, pp. 335–346.
Montagna, F. C., "Responding to CO Detector Activations," *Fire Engineering*, Vol. 149, No. 1, Jan. 1996, pp. 99–102, 104, 106, 108–110.
Moye, C. H., "Development of a First Approximation of a Design Method for Combustible Gas Detection Systems," Worcester Polytechnic Inst., MA, Thesis, May 1991.
Murakami, T., Takeda, M. and Tanaka, Y., "Analyzed Gases of Burning Products Collected at Fire Scenes," Japanese Association of Fire Science and Engineering, Fire Research Annual Conference, May 17–18, 1990, pp. 1–4.
Patty, P., "Carbon Monoxide Detectors Protection Against the Silent Killer," Underwriters Laboratories, Inc., Northbrook, IL, *UL Lab Data*, Vol. 1, No. 3, Fall 1995, pp. 9–14.
Sinha, A. K., Rajwar, D. P., and Ghosh, A. K., "Underground Fire Hazard Detection Through Carbon Monoxide Monitors," *Journal of Mines, Metals, and Fuels*, Vol. 35, No. 8, Aug. 1987, pp. 365–367.
Srivastava, S. C., *et al.*, "Development of the Electrochemical Cell for Carbon Monoxide Measurement," *Journal of Mines, Metals, and Fuels*, Vol. 35, No. 8, Aug. 1987, pp. 184–385.
"Tests on Gases Evolved During Combustion of Electric Cables. Part 2. Determination of Degree of Acidity of Gases Evolved During the Combustion of Materials Taken From Electric Cables by Measuring pH and Conductivity," International Electrotechnical Commission.

SUPPRESSION

Built-in fire safety measures and common-sense fire safety techniques provide a meaningful approach to combating fire. This section focuses on built-in fire safety measures fire suppression systems that use water and other suppressing agents.

Sooner or later, someone or something has to extinguish the fire. There are few fires that self-extinguish with no specific outside intervention. Each year millions of unwanted fires continue spreading until an occupant or an on-site fire suppression system extinguishes them, or worse, until a fire department or fire brigade extinguishes them.

Section 6 provides details on extinguishing agents, suppression systems, and the related design aspects of fixed and nonfixed fire suppression systems.

As stated in Chapter 1, "Water and Water Additives for Fire Fighting," water is the most common extinguishing agent. Chapter 1 describes the physical properties of water and the various means by which those properties make fire extinguishment possible. Chapter 1 also addresses the use of additives to enhance the effectiveness of water in extinguishment, and discusses situations where water is not the extinguishing agent of choice or when it must be applied with unusual care (e.g., to avoid hazards of electric shock).

Chapter 2, "Water Storage Facilities and Suction Supplies," describes specialized storage tanks and devices for obtaining water from surface sources. Chapter 3, "Water Distribution Systems," discusses public systems for water distribution. Chapter 4, "Fire Pumps," provides details on the different types of fire pumps used in fixed installations. Chapter 5, "Water Supply Requirements for Fire Protection," addresses the quantity of water supply required and the elements of the systems used. Chapter 6 "Hydraulics," reviews the physics of water and other media in motion through orifices, nozzles, and pipes. Finally, Chapter 7, "Test of Water Supplies," provides useful information needed to evaluate water supplies.

The different types of sprinkler systems and other water suppression systems are covered in Chapters 8 through 16. Chapter 8, "Theory of Automatic Sprinkler Performance," addresses mathematical concepts relevant to sprinkler design and effectiveness, including some of the more recently developed concepts like the Response Time Index (RTI). Chapter 9, "Automatic Sprinklers," discusses the many types of sprinkler hardware. Chapter 10, "Automatic Sprinkler Systems," provides an overview of basic sprinkler system design and layout. Chapter 11, "Water Supplies for Sprinkler Systems," discusses what types of supplies are acceptable and how the supply can be evaluated. Water spray system layout and design is covered in Chapter 12, "Water Spray Protection." Chapter 13 "Fast Response Sprinkler Technology," provides detailed discussion on the characteristics of the thermal element properties of sprinkler activation devices.

Two relatively recent water-based technologies are discussed Chapters 14 and 15. Chapter 14, "Ultra-High-Speed Suppression Systems for Explosive Hazards," presents information on the use of systems to contain explosions. Chapter 15, "Water Mist Suppression Systems," discusses the application of this innovative technology to control and suppress fires.

Chapter 16, "Standpipe and Hose Systems," Provides additional information on another type of water-based system that is primarily for use by the fire service. Chapter 17, "Care and Maintenance of Water-Based Extinguishing Systems," provides general information on inspection, testing, and maintenance of extinguishing systems that use water.

Nonwater-based extinguishing systems are addressed in Chapters 18, through 22. Chapter 18, "Halogenated Agents and Systems," is an older system technology that uses halons such as Halon 1301 or Halon 1211. The new alternative agents that are replacing the fire protection halons are addressed in Chapter19, "Direct Halon Replacement Agents and Systems." Chapter 20, "Carbon Dioxide and Application Systems," is on carbon dioxide systems, while dry chemical systems are addressed by Chapter 21, "Dry Chemical Agents and Application Systems." Chapter 22, "Foam Extinguishing Agents and Systems," provides information on the technologies that use foam as their extinguishing agent.

Portable extinguishers also perform an important role in suppression. Chapter 23, "Selection, Operation, Distribution, and Maintenance of Fire Extinguishers," addresses the most essential elements encountered with portable fire extinguishers. Chapter 24, "The Role of Extinguishers in Fire Protection," provides an overview of these units.

In closing this section on suppression, two chapters focus on exotic hazards and extinguishing agents. Chapter 25, "Special Systems and Extinguishing Techniques," and Chapter 26, "Combustible Metal Extinguishing Agents and Application Techniques," provide the details on systems that protect hazards that are particularly challenging.

—Robert E. Solomon
—Casey C. Grant
December 1996

WATER AND WATER ADDITIVES FOR FIRE FIGHTING

Andrew M. Wahl

Water is the most widely used and common extinguishing agent, however, it is not the proper extinguishing agent for all types of fires. This chapter discusses the properties of water as an extinguishing agent, including its advantages and limitations.

The principles of fire extinguishment are discussed in Section 1, Chapter 5, "Chemistry and Physics of Fire," and in Section 4, Chapter 14, "Explosion Prevention and Protection." The systems and devices utilized for the transportation and application of water as a fire extinguishing agent are addressed in the other chapters in this section of the handbook. For the use of water as a suitable extinguishing agent on specific materials, such as chemicals, flammable liquids, gases, and metals, see Section 4, Chapter 6, "Flammable and Combustible Liquids," and Section 4, Chapter 16, "Metals." Section 1, Chapter 9, "Environmental Issues in Fire Protection," can also be useful.

INTRODUCTION

Water is the most widely used and available fire extinguishing agent for many reasons: it is inexpensive, abundant, and effective in fire suppression; water is transportable and can be pumped from a source to the fire; and it is available from domestic water distribution systems (fire hydrants), streams, wells, ponds, lakes, and swimming pools.

Water, a very effective agent for controlling and extinguishing combustion, is the most abundant and available material on the earth's surface. It comes in three states: (1) liquid (water), (2) vapor (steam), and (3) solid (ice).

Life safety considerations must be weighed when choosing an effective fire extinguishing agent. Water as an agent is safe, nontoxic, noncorrosive, and stable. Water (H_2O) remains stable when applied to a fire and does not break down into its basic elements of hydrogen (H) and oxygen (O), both of which would encourage fire growth.

Water can be applied as an extinguishing agent with building occupants in compartments, unlike some gaseous extinguishing agents, which may cause asphyxiation or adverse side effects.

Andrew M. Wahl, P.E., is a senior professional fire protection engineering consultant with Gage-Babcock & Associates, Inc., Fairfax, VA, and serves as a member of the NFPA Foam-Water Sprinkler Committee.

PROPERTIES OF WATER

The physical properties that allow water to be an effective extinguishing agent are:

1. At ordinary temperatures, water exists as a stable liquid. Water viscosity in the temperature range of 34 to 210°F (1 to 99°C) remains consistent, which allows it to be transported and pumped.

2. Water has a high density, which allows it to be discharged from and projected from nozzles, etc. Water's surface tension allows it to exist from small droplets to a solid stream.

3. The latent heat of fusion is the amount of energy required to change the state of water from solid (ice) at 32°F (0°C) to a liquid. Water absorbs 143.4 Btu per lb (333.2 kJ/kg) in this process.

4. The specific heat of water is 1.0 Btu per lb (4.186 kJ/kg K). Therefore, raising the temperature of 1 lb (0.45 kg) of water 180°F (100°C) [from 32°F (0°C) to 212°F (100°C)] requires 180 Btu.

5. Water is effective as a cooling agent because of its high latent heat of evaporation (changing water from a liquid to vapor), which is 970.3 Btu per lb (2260 kJ/kg) as described in Section 1, Chapter 5, "Chemistry and Physics of Fire."

6. Water expands in its conversion from a liquid state to a vapor state to 1600 to 1700 times the liquid volume. One gallon (3.8 L) of liquid [which occupies 0.1337 cu ft (0.004 m³)] produces over 223 cu ft (6.3 m³) of steam. Therefore, it follows that 1 gal (3.8 L) of water at room temperature applied to a fire and converted to steam (complete conversion) will have the following results. (For SI units: °F = °C × $^9/_5$ + 32; 1 Btu = 1.055 kJ; 1 lb = 0.45 kg; 1 gal = 3.785 L; 1 cu ft = 0.0283 m³.)

Heat required to raise temperature of water to boiling:

212°F – 68°F (room temperature) = 144 Δ°F

144 Δ°F × 1 Btu/lb × 8.33 lb (weight of 1 gal) = 1200 Btu

Heat required to change water from a liquid to a vapor:

970.3 Btu/lb × 8.33 lb (weight of 1 gal) = 8083 Btu

Total heat absorbed is 1200 + 8083 = 9283 Btu/gal of water

Therefore, a fire department hose stream discharging at 100 gpm would then absorb 928,300 Btu per min in a complete conversion. The same fire department hose stream will create 22,300 cu ft of steam in a complete conversion.

EXTINGUISHING PROPERTIES

Water is a very effective extinguishing agent because of its ability to cool, remove or displace the oxygen supply, and separate/dilute the fuel source. See the discussion on the principles of fire in Section 1, Chapter 5, "Chemistry and Physics of Fire."

Extinguishing by Cooling

Water principally extinguishes fires by cooling the fuel surface. The role of cooling of a fire must not be overlooked as it is the predominant method by which fires are extinguished.

Water is effective as a cooling agent because of its high latent heat of evaporation (2.4 kJ/g at 25°C). Water introduced to a fire promotes heat loss in the heat transfer action from the fire to the water. When the heat loss exceeds the fire heat gain, the fuel surface will begin to cool until the flame can no longer exist at the surface. Water is an effective coolant for solid fuel surfaces.

Water cools solid fuels by the reduction of the radiant heat flux of the flame to the fuel surface, and reduces the rate of pyrolysis of the fuel. This includes the cooling effects from water droplets and steam.

When the heat absorption rate of the water spray approaches the total heat release rate of the fire, fire control begins. When the heat absorption rate of the water spray exceeds the heat release rate of the fire, fire suppression and, ultimately, fire extinguishment are achieved. Other factors to consider in fire control and extinguishment include heat losses through openings, and heat losses to walls, ceilings, and floors.

The amount of water required to extinguish a fire depends on the heat output [Btu (kJ)] of the fire. How quickly a fire is extinguished depends on how the water is applied, how much is applied, and in what form water is applied. To achieve extinguishment by cooling, it is best to apply water so that the maximum amount of heat will be absorbed. Water absorbs the most heat when it is converted into steam, and it will be converted into steam more easily from smaller droplets (spray) than from a solid stream.

The droplet size in application is important. A water droplet approaching the fire can evaporate in the fire plume, thus only cooling the fire plume but not effectively cooling the fuel surface. The smaller the droplet, the higher the speed with which water extracts heat from the fire and fire gases, therefore using a lower volume of water. It is not necessary to absorb all heat given off from a fire at a burning rate; absorbing 30 to 60 percent may be enough to extinguish the fire.[1]

Calculations show that the optimum diameter of a water droplet is in the range of 0.01 to 0.04 in. (0.3 to 1.0 mm), and that the best results are obtained when the droplets are fairly uniform in size. (See Figure 6-1A.) Current discharge devices are not capable of producing completely uniform droplets, although many discharge devices spray droplets that are nearly uniform over a broad range of pressures. Actual delivered density (ADD) is the actual water application rate to a fire at the surface of the burning fuel. The amount of water reaching a fire is affected by the fire burning rate and the associated fire's upward plume velocity. The water droplets must overcome the effects of the upward momentum of the fire plume, air currents, etc., in order to reach the burning fuel and be effective. For example, droplets, once produced at a sprinkler, are subject to high temperatures that evaporate the small droplets at the ceiling. If the small droplets survive the ceiling temperature, they may not have the mass or momentum to be able to penetrate the fire plume.

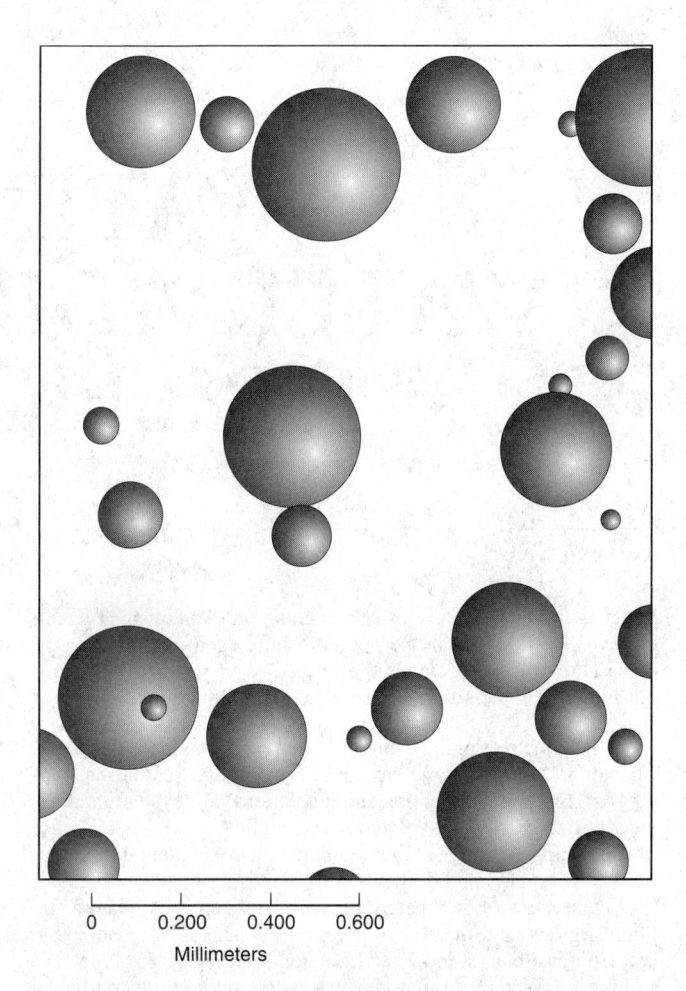

0   0.200   0.400   0.600
Millimeters

FIG. 6-1A. Diameters of typical droplets of water spray. (Not to scale.) (Source: National Board of Fire Underwriters, Research Report No. 10)

Extinguishing by Smothering

When water is applied to a fire, steam is formed. The dilution of the air (oxygen) supply around the fuel sources provides suppression by a smothering action. Suppression by this method is more effective if the steam and water droplets are confined around the fuel source. The steam and water droplets also continue to extinguish a fire by cooling as the water droplets continue to evaporate around the heated area of a fire.

Fires in ordinary combustibles are normally extinguished by the cooling effect of water—not by the smothering effect created by steam. Water mist systems, which may be used as an alternative to sprinkler systems or certain gaseous extinguishing systems, have been found to be effective in controlling *and* extinguishing fires by cooling and smothering, as discussed in Section 6, Chapter 15, "Water Mist Suppression Systems."

Water may be used to smother a burning flammable liquid when the liquid has a flash point above 100°F (37.8°C), a specific gravity greater than 1.0, and is not water soluble. To achieve this most effectively, a foam concentrate is added to the water to form a foam-water solution. The foam-water solution must then be applied gently to the surface of the flammable liquid.

In cases where oxygen is produced while a burning material decomposes, smothering by any agent is not possible.

Extinguishment by Emulsification

An emulsion is formed when immiscible liquids are agitated together and one of the liquids is dispersed throughout the others. Extinguishment by this process can be achieved by applying water to certain viscous flammable liquids, since the effect of cooling the surfaces of such liquids prevents the release of flammable vapors. With some viscous liquids (such as No. 6 fuel oil), the emulsification is a "froth" which retards the release of flammable vapors. Care must be used on liquids of appreciable depth, however, because frothing may spread the burning liquids over the sides of the container. A relatively strong, coarse water spray is normally used for emulsification. A solid stream of water should be avoided, as it will cause violent frothing.

Extinguishment by Dilution

Fires in water-soluble, flammable materials may, in some instances, be extinguished by dilution. The percentage of dilution necessary varies greatly, as will the volume of water and the time necessary for extinguishment. For example, dilution can be used successfully in a fire involving an ethyl or methyl alcohol spill if it is possible to get an adequate mixture of water and alcohol. Dilution is not a common practice where tanks are involved. The danger of overflow because of the amount of water required, and the danger of frothing should the mixture become heated to the boiling point of water, make this form of extinguishment seldom practical.

OPACITY AND REFLECTIVITY

Tests conducted at Underwriters Laboratories Inc. (UL) using water spray to provide exposure protection to a sheet metal surface from a gasoline fire indicate that, when the spray is applied as a thin film of water over the sheet metal, the temperature of the metal was contained within limits that protected the metal from significant damage. This was not true, however, when the water spray was adjusted so that it did not touch the sheet metal, but it did provide a water curtain between the metal and the fire. In this latter case, the temperature of the metal was three to four times greater than when the water flowed over the metal. These tests indicate that, because of its lack of opacity, water does not prevent the passage of radiant heat very well. The principal value of water used to protect exposures is from the cooling obtained by evaporation of a water film on the exposed surfaces.

NFPA 13, *Standard for the Installation of Sprinkler Systems*, calls for outside sprinklers for protection against exposure fires to be positioned so that the water will thoroughly wet exposed glass windows and run down over the window sash and the glass, wetting the entire window as much as possible. Similar recommendations require that as much of the cornice as possible be wetted when cornice sprinklers are installed. These recommendations reflect the experimental evidence.

In England, tests measured the transmission of radiant heat through water sprays from two types of nozzles.[2] The tests proved that the transmission of this kind of heat depends mostly on nozzle design, and that, with certain nozzles, a water curtain of low transmission could be produced for water flows comparable to those of sprinkler installations. Fire fighters have, of course, used water cur-

tains in situations where it was too hot or dangerous for them to remain exposed to flames and heat.

WATER AS AN EXTINGUISHING AGENT

It is recognized that fires involving various fuels and materials react differently to various extinguishing agents, and that water *is not* the best agent for all types of fuels.

Water is not suitable for all types of fires. There are four typical classifications of fire: (1) Class A—ordinary combustibles, (2) Class B—flammable and combustible liquids, (3) Class C—electrical, and (4) Class D—combustible metals. Water is most effective on Class A, i.e., ordinary combustible, fires. Water is not appropriate or the most desirable agent of choice for Class B, i.e., flammable and combustible liquids, and Class D, i.e., combustible metals, fires. Using water for Class C, i.e., electrical, fires must be considered carefully before the water is applied. Water may not be an acceptable agent where collateral water damage is deemed unacceptable. The most common application involves Class A fires where water is well suited.

A good reference to consult for recommendations on use of water on specific materials where problems could be encountered is the NFPA *Fire Protection Guide to Hazardous Materials*.[3]

Water applied to fires by water mist fire suppression systems has been found to be effective with fires involving flammable liquids and live electrical equipment. Water mist systems have the potential to protect shipboard engine enclosures, aircraft interiors, valuable artwork, and computer equipment. Water mist is especially effective in confined spaces when a detection system operates the water mist fire suppression system. Direction on the use of water mist as a suppression system is given in NFPA 750, *Standard on Water Mist Fire Protection Systems*.

Water can be used in some instances for fires involving chemicals and combustible metals where the quantity of water can overcome some adverse chemical reactions and extinguish the fire.

The following possible personnel hazards must be considered when choosing water as the extinguishing agent: (1) exposure and/or inhalation of steam in all fires, (2) electrocution or shock injury in electrical fires, and (3) adverse or explosive reactions in chemical and combustible metal fires.

Wetting Agents[4]

Wetting agents are compounds that are added to water to change some or all of its characteristics, e.g., surface tension or viscosity. Experience as well as testing indicates that the addition of a proper wetting agent to water will, when properly applied, increase the extinguishing efficiency of that water, with respect to quantity used and time saved. The value of such a factor may well become important, especially in rural areas where adequate quantities of water are not always available for fire fighting.

Certain types of fires, such as those in baled cotton, stacked hay, some rubber compounds, and some flammable liquids that do not ordinarily respond to treatment with water, may be extinguished when a proper wetting agent is used. This may be attributed to an increase in the penetrating, spreading, and emulsifying capability of water due to lowered surface tension. This decrease in surface tension can be described as a disruption of the forces holding the surface film of water together, permitting it to flow and spread uniformly over solid surfaces. The treated water acquires the ability to penetrate into small openings and recesses that nontreated water would flow over, by the simple bridging action of the surface film. These solutions exhibit not only penetrating and spreading qualities, but also increased absorptive speed and superior adhesion to solid surfaces.

Some wetting agents have foaming characteristics when mixed with water and air. The foam retains the wetting and penetrating characteristics of the wetting agent and provides an efficient smothering action for the extinguishment of all Class A and some Class B fires. It also provides a fluid insulation for protection against fire exposure. The foam produced in this manner has the additional advantage of breakdown at approximately 175°F (79.4°C), then returns to its original liquid state and retains the penetrating and wetting qualities.

There are numerous chemicals that fulfill the primary function of a wetting agent, which is to lower the surface tension of water. However, very few of these chemicals are suited to fire protection, because their application is complicated by hazard considerations, such as toxicity, corrosive action on equipment, and stability in water.

Wet water is water plus a wetting agent. Wet water has the same limitations as water on fires in chemicals that react with water. The use of wet water on flammable and combustible liquid fires is not common. Wet water should not be used on flammable or combustible liquid fires if the liquids are water soluble, such as the alcohols, glycols, and some ketones.

Because of its conductivity, the application of wet water solutions on live electrical equipment requires the same precautions as application of water on Class C (electrical) fires. Wet water applied by spray or fog might be used with caution due to its penetrating characteristics. Wet water may have more harmful effects on motors, transformers, and similar equipment than plain water. Any electrical equipment that has been penetrated by wet water should be thoroughly flushed and cleansed before it is returned to service.

Wetting agents may be either premixed with water or added to the water through suitable proportioning equipment at the time the water is being used. Mixing wetting agents from different manufacturers, or mixing a wetting agent with mechanical or chemical foam concentrates, is not recommended. Distinction must be made between wetting agents and wetting-agent foams and other detergent-type foams (high-expansion foam) and film-forming foam agents. Guidance for the use of wetting agents is given in NFPA 18, *Standard on Wetting Agents.*

WATER AND FLAMMABLE AND COMBUSTIBLE LIQUIDS—CLASS B FIRES

Special care must be exercised when using water as an extinguishing agent with Class B, i.e., flammable and combustible liquid, fires. Water density can create problems with flammable and combustible liquids. When the liquid is lighter (specific gravity less than 1.0) than water, the water can submerge beneath the liquid and can cause the flammable or combustible liquid to flow from its containment, spreading the fire. In crude oil, water can submerge to the bottom of the container and cause the crude oil to "boilover." (See Section 4, Chapter 6, "Flammable and Combustible Liquids.") Water is immiscible with hydrocarbon fuels, and does not provide an effective coating of the fuel surface, and does not dilute the flammable or combustible mixture below the flammable limits. Water is an effective fire suppressing agent when combined with fire-fighting foam.

Heavy fuel oil, lubricating oil, asphalt, and other high-flash-point liquids do not produce flammable vapors unless heated. Once ignited, the heat of the fire will cause pyrolysis for continued burning. If water in spray form is applied to the surface of such high-flash-point burning liquids, cooling will slow down the rate of pyrolysis enough that it may extinguish the fire. If water is applied to high-flash-point burning liquids by means of a coarse spray, extinguishment may be obtained by emulsification.

Water, combined with a fire-fighting foam concentrate, can be a very effective extinguishing agent for Class B, i.e., flammable and combustible liquid, fires.

Fire-fighting foam is: (1) an aggregate of air-filled bubbles forming from aqueous solutions (water and foam concentrate) and (2) lower in density than flammable liquids. It is used principally to form a cohesive floating blanket on flammable and combustible liquids, and prevents or extinguishes fire by excluding air and cooling the fuel. It also prevents reignition by suppressing formation of flammable vapors. (See Section 6, Chapter 22, "Foam Extinguishing Agents and Systems.")

Fire-fighting foams consist of a combination of water, foam concentrate, and air. Low-expansion-type foams typically utilize a ratio of 3 percent concentrate to 97 percent water for a foam-water solution, or a ratio of 6 percent concentrate to 94 percent water for a foam-water solution. Water remains the principal component in fire-fighting foams.

The ability of water without additives (foaming agents) to extinguish a fire is limited to low-flash-point flammable liquids, such as Class I flammable liquids [flash points below 100°F (37.8°C)], as defined in NFPA 30, *Flammable and Combustible Liquids Code.* Sprinklers and water spray are usually effective in extinguishing fires in combustible liquids with flash points of 200°F (93.3°C) and higher, in flammable liquids with a specific gravity greater than 1.0, and in water-soluble liquids. Water can be effective with high-flash-point hydrocarbon fires when introduced as a high-velocity spray causing penetration of the droplets and cooling of the surface layer.[5] If water does not evaporate and cool the fuel surface, it can accumulate and submerge and can displace the hydrocarbon. Control of fire, but not extinguishment, is possible in low-flash-point [under 200°F (93.3°C)] flammable liquids. Any water that reaches the surface of a burning, low-flash-point flammable liquid in a tank may sink and cause the tank to overflow. In the case of a spill fire, the water may cause the fire to spread. Special handling of certain types of water spray nozzles can result in extinguishment of fires in these liquids or, at a minimum, effective fire control.

Water can be used as an effective cooling agent in Class B fires; it can also: (1) protect against flame-exposure of the storage container and (2) protect exposures as a cooling agent.

See Section 4, Chapter 6, "Flammable and Combustible Liquids," for characteristics and fire preventive methods for Class B.

WATER AND LIVE ELECTRICAL EQUIPMENT—CLASS C FIRES

Water in its natural state contains impurities that make it conductive. If water is applied to fires involving live electrical equipment, a continuous circuit might be formed that would conduct electricity back to the user and cause a shock, especially if there are high voltages or potentials. Foam-type extinguishing agents are also conductive. The amount of current, rather than the voltage, determines the extent of the shock. Conductivity of water when used on live electrical equipment depends on several variables:

1. The voltage and amount of current flowing.
2. The "breakup" of the stream as a result of the nozzle design, the pressures used, and the wind conditions. This breakup influences the conductivity of the stream because the air spaces formed

between the droplets interrupt the electrical path to ground. Water spray nozzles and combination straight-stream spray nozzles (in the spray position) provide for effective dispersion of the water droplets. The hazards of these are less than those of solid streams of water.

3. The purity of the water and the relative resistivity of the water.
4. The length and cross-sectional area of the water stream.
5. The resistance to ground through a person's body as influenced by location (whether on wet ground or not), skin moisture, the amount of current the body can endure, the length of exposure to the current, and other factors, such as protective clothing.
6. The resistance to ground through the hose.

Conductivity and Hazard of Shock

There is some danger to fire fighters directing streams of water onto wires of less than 600 V to ground from a distance likely to be encountered under ordinary fire-fighting conditions. It is more dangerous if fire fighters, standing either in puddles of water or on moist surfaces, come into contact with live electrical equipment. In such cases, the fire fighters' bodies complete an electrical circuit, and the current from the electrical equipment relayed through their bodies is more readily grounded than if it were conveyed through dry, nonconductive surfaces. Rubber boots often contain enough carbon black to permit the passage of current through the body, and do not provide reliable protection.

Research conducted by UL on electric fences indicates that there are differences in the electric current to which individuals may be safely subjected, and that the maximum continuous (uninterrupted) current to which an individual may be safely subjected is 5 mA (milliamperes) ac applied on the surface of the body.[6]

Impurities in water (mostly the mineral content) also affect its conductivity. Tests of the resistivity of public water supplies in Indiana showed results ranging from 710 to 5400 ohms per cm^3; the lowest values were found in supplies from deep wells. The resistivity of deep well supplies ranged from 1000 to 2000 ohms per cm^3; and the resistivity of river waters was about 4000 ohms per cm^3. In tests conducted by the Commonwealth Edison Company in cooperation with the Chicago Fire Department, the resistivity of Chicago River water ranged from 1671 to 2393 ohms per cm^3.[7] At the time the tests were made, the normal hydrant water in the Chicago area had a resistivity of about 3800 ohms per cm^3.

Safe Distances from Live Equipment

From time to time, authorities have tried to determine safe distances between nozzles and live electrical equipment. The bibliography at the end of this chapter cites more papers on this subject.

The conductivity of water streams varies according to the type of equipment from which they are expelled, such as: (1) handheld or manually supported solid-stream nozzles, (2) handheld water spray (water fog) nozzles, (3) fixed water spray systems for fire protection services, and (4) plain water and water solution portable fire extinguishers.

Data available on the minimum safe distances between manually supported solid-stream hose lines and live electrical equipment carrying voltages higher than 600 V are not entirely consistent, because the results of different tests vary. These variations may be attributed to the different testing methods used, the variances in the purposes of the tests, the limitations of the tests caused by the physical circumstances and available equipment, and the fact that the same voltages were not used in all of the tests.

The American Insurance Services Group (AISG) (formerly AIA) suggests the distances given in Table 6-1A between solid-stream nozzles (dispensing fresh water) and electrical conductors or equipment carrying voltages higher than 600 V to be the minimum for safety. The AISG data are based on tests run in the 1930s. The voltages given in this data do not correspond to today's commonly encountered ranges, nor does the data extend to the higher ranges now widely used.[8] The AISG cautions that, for streams from contaminated or salt water, or where additives have been introduced into the water, it is not possible to give simple rules due to the variances in the conductivity of the water.

Some limited tests made in 1958 by the Hydroelectric Power Commission of Ontario, in cooperation with the Office of the Fire Marshal of Ontario, Canada,[9] resulted in recommendations for minimum safe distances from live electrical equipment for a $5/8$-in. (16-mm) solid-stream nozzle. (See Table 6-1B.) The report recommends that solid streams greater than $5/8$ in. (16 mm) should not be used near live electrical equipment, but the tests were limited to a maximum stream distance of 30 ft (9.1 m). Larger nozzles might produce sufficient stream dispersion over longer distances, which would permit their use.

The results of tests made in 1934 for the Fire Brigade of Paris, France, present perhaps the most comprehensive guide.[10] (See Table 6-1C.) The distances are based on preventing the transmission

TABLE 6-1A.   *Distance Between Solid-Stream Nozzles (Fresh Water) and Electrical Conductors*

		Diameter of Nozzle Orifice			
		$1^1/_8$ in. (29 mm)		$1^1/_2$ in. (38 mm)	
	Voltage Between Conductors	Minimum Safe Distance			
Voltage to Ground		ft	m	ft	m
635	1,100	6	1.8	9	2.7
1,270	2,200	11	3.6	16	4.9
1,905	3,300	15	4.6	22	6.7
3,175	5,500	18	5.5	27	8.2
4,215	6,600	19	5.8	29	8.8
6,350	11,000	20	6.1	30	9.1
12,700	22,000	25	7.6	33	10.1
19,050	33,000	30	9.1	40	12.2

TABLE 6-1B.   *Limit of Safe Approach to Live Electrical Equipment*

| | | $5/8$-in. (16-mm) Solid-Stream Nozzle* | |
| | Voltage Between Conductors | Minimum Safe Distance | |
Voltage to Ground		ft	m
2,400	4,160	15	4.6
4,800	8,320	20	6.1
7,200	12,500	20	6.1
8,000	13,800	20	6.1
14,400	24,900	25	7.6
16,000	27,600	25	7.6
25,000	44,000	30	9.1
66,000	115,000	30	9.1
130,000	230,000	30	9.1

*Nozzle pressure 100 psi (690 kPa) water resistance 600 ohms per cu in.

TABLE 6-1C. *Minimum Safe Distances Between Hose Nozzles and Live Electrical Equipment Recommended for the Paris, France, Fire Brigade[10]*

Voltage to Ground	Voltage Between Conductors	Diameter of Nozzle Orifice					
		1/4 in. (6 mm)		3/4 in. (19 mm)		1 1/4 in. (32 mm)	
				Safe Distance			
		ft	m	ft	m	ft	m
115	230	1.6	0.50	3.3	1.00	6.6	2.00
460	480	2.5	0.75	9.8	3.00	16.4	5.00
3,000	5,195	6.6	2.00	16.4	5.00	32.8	10.00
6,000	10,395	8.2	2.50	19.7	6.00	39.4	12.00
12,000	20,785	9.8	3.00	21.4	6.50	49.2	15.00
60,000	103,820	14.8	4.50	39.4	12.00	72.2	22.00
150,000	259,800	19.7	6.00	49.2	15.00	82.0	25.00

of a 1-mA current to a fire fighter in contact with a nozzle or hose. The tests covered only voltages to ground ranging from 115 to 150,000 V, and the groupings of the voltages do not correspond to current standard U.S. voltages. The maximum size of the nozzle used is also nonstandard in the United States.

The preceding information indicates that definite shock hazards exist unless adequate distances are maintained, and these distances can only be estimated from the available data. It is difficult for fire fighters in the field to know precisely what electrical potentials exist in any given situation. For this reason, and those given below, it is best, whenever possible, to use water spray streams rather than solid streams.

As noted previously, water spray reduces the conductivity hazard. The design of the nozzle and the characteristics of the spray determine the amount of leakage current that can actually flow in the stream, and each nozzle should be tested to determine precisely the characteristics it possesses. Tests on various commercial water spray nozzles indicate that a minimum distance of 4 ft (1.2 m) should be maintained for voltages to ground up to about 10 kV. This distance is actually no greater than is sensible to prevent personnel from getting dangerously close to live electrical equipment. Distances should be increased when attacking fires involving live electrical equipment operating above this voltage. Figure 6-1B shows the results of four researchers, as analyzed by the U.K. Fire Offices' Committee, Joint Fire Research Organization.[11]

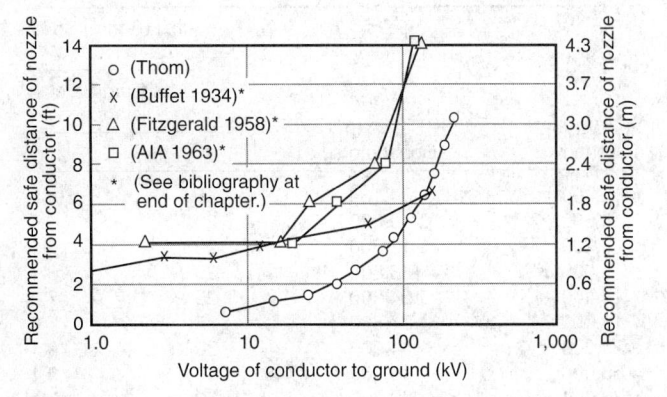

FIG. 6-1B. *Variation of safe distance with conductor voltage for spray nozzles.*

The Toledo Edison Company conducted tests in which water was discharged onto a screen at a potential to ground of 80,500 V (equivalent to a system or line voltage of 138 kV phase-to-phase). As a consequence, the Edison Electric Institute in 1967 adopted the following safety rules (the distances in these rules will limit leakage currents to less than 1 mA):

1. Using all handheld water spray nozzles, the minimum approach distance is 10 ft (3 m).
2. Using handheld, 1 1/2-in. (38-mm) straight (solid) stream nozzles, the minimum approach distance is 20 ft (6 m).
3. Using handheld, 2 1/2-in. (64-mm) straight (solid) stream nozzles, the minimum approach distance is 30 ft (9 m).

When combination straight-stream spray nozzles are used on live electrical equipment, fire fighters should be sure they have the desired spray pattern before applying the stream. The use of spray nozzles on "applicators" increases the possibility of accidental contact between the nozzle and live electrical equipment, and most authorities recommend that they not be used.

Clearance from Fixed Water Spray Systems

Fixed water spray systems are used extensively to protect high-value and/or critical electrical equipment, such as transformers, oil switches, and motors. These systems are designed to provide effective fire control, extinguishment, prevention, or exposure protection. NFPA 15, *Standard for Water Spray Fixed Systems for Fire Protection*, gives installation recommendations for such systems, and includes a table of recommended clearances between water spray equipment and unenclosed or uninsulated live electrical components at other-than-ground potential. (See Table 6-1D.) Modern practice is to coordinate the required clearance with the electrical design. The basic insulation level (BIL) values of the equipment are used as the basis, although the clearance between uninsulated live parts of the equipment and any portion of the water spray system should not be less than the minimum clearances provided elsewhere for electrical system insulation on any individual component (the minimum unshielded straight line distance from the exposed electrical parts to nearby grounded objects). The BIL [expressed in kilovolts (kV)] is the crest value of the full wave impulse test. (See Section 6, Chapter 12, "Water Spray Protection.")

Portable Extinguishers and Hazard of Shock

Water-based or water-solution portable fire extinguishers are not recommended for use on live electrical equipment, i.e., Class C

TABLE 6-1D. *Clearance from Water Spray Equipment to Live Uninsulated Electrical Components*

Nominal System Voltage (kV)	Maximum System Voltage (kV)	Design BIL (kV)	Minimum* Clearance (in.)	Minimum* Clearance (mm)
To 13.8	14.5	110	7	178
23	24.3	150	10	254
34.5	36.5	200	13	330
46	48.3	250	17	432
69	72.5	350	25	635
115	121	550	42	1067
138	145	650	50	1270
161	169	750	58	1473
230	242	900	76	1930
		1050	84	2134
345	362	1050	84	2134
		1300	104	2642
500	550	1500	124	3150
		1800	144	3658
765	800	2050	167	4242

* For voltages up to 161 kV, the clearances are taken from NFPA 70, *National Electrical Code®*. Voltages above the clearances are taken from Table 124 of ANSI C2, *National Electrical Safety Code*.

Note: BIL values are expressed as kilovolts (kV), the number being the crest value of the full wave impulse test that the electrical equipment is designed to withstand. For BIL values that are not listed in the table, clearances can be found by interpolation.

Source: NFPA 15, *Standard for Water Spray Fixed Systems for Fire Protection*.

fires. NFPA 10, *Standard for Portable Fire Extinguishers*, recommends that extinguishers specifically tested for use on Class C fires be used for such fires. When electrical equipment is de-energized, water-based extinguishing agents and extinguishers for Class A or B fires may be used safely. Conductivity tests of portable fire extinguishers containing water indicate that soda-acid (now discontinued), loaded-stream (now discontinued), foam, and antifreeze solution extinguishers, all of which produce solid streams and have a short range, are particularly hazardous. One test involving an antifreeze solution extinguisher showed a current of 157 ma for a potential of 1 kV with a $^3/_{32}$-in. (2.4-mm) stream at a distance of 1 ft (0.3 m). To reduce the current to below 1 ma, it would be necessary to use the device from a distance at which the stream is dispersed; it is generally agreed that for such extinguishers the minimum distance should be 4 ft (1.2 m) for voltages up to 1 kV. Extinguishers containing plain water might, in theory, be used at shorter distances, but it is best to maintain a 4-ft (1.2-m) distance if it is necessary to use this kind of extinguisher.

Water on Electrical, Electronic, and Computer Equipment

Automatic sprinkler protection and water spray fixed systems are valuable for fire control, even where electrical, electronic, or computer equipment may be present. There should be little concern about the possibility of shock, or of the water causing excessive damage to the equipment. Experience has proved that, if a fire activates sprinklers, the sprinklers, if properly installed and maintained, provide effective protection with virtually no hazard to personnel and with no measurable increase in damage to the equipment, as compared with the damage done by heat, flame, smoke, and the manual hose streams.[12]

NFPA 75, *Standard for the Protection of Electronic Computer/Data Processing Equipment*, recognizes the importance of automatic sprinkler protection in computer rooms. It recommends that, where sprinklers are installed to protect electronic computer equipment, the power to the equipment should be disconnected prior to the application of water.

WATER AND COMBUSTIBLE METALS— CLASS D FIRES

The reaction between water and combustible metals ranges widely from minor reactions to an explosive impact. In some instances, water can be used to overcome some adverse chemical reactions if a large quantity is applied. As a rule, water should not be used on fires involving combustible metals, e.g., magnesium, titanium, metallic sodium, and hafnium, or on metals that are combustible under certain conditions, e.g., calcium, zinc, and aluminum. However, water can be utilized to protect exposures as a cooling agent.

USE OF WATER ON SPECIAL HAZARDS

While water is generally a universal extinguishing agent, there are certain prohibitions and precautions that must be observed when it is applied manually on some burning materials that either react chemically or explosively upon contact with water. In other instances, the mechanical action of applying water must be carefully monitored to avoid creating conditions that intensify the hazard rather than controlling it.

The following paragraphs give guidance on using water on different materials that can present problems if water is used arbitrarily as an extinguishing agent. See Section 4, "Materials, Products, and Environments," for specific information on chemicals and gases.

Chemical Hazards

As a general rule, water should not be used on materials such as alkalies, anhydrides, carbides, hydrides, nitrates, peroxides (organic and inorganic), to list a few. These chemicals react with water and may release oxygen, flammable gases, and heat. When wet, certain materials, such as unslaked lime, will heat spontaneously over a period of time if heat cannot be dissipated due to storage conditions.

Radioactive Metals

Water should not be used continuously on radioactive metals. The requirements for fire protection for radioactive metals are generally consistent with their nonradioactive counterparts (for all practical purposes, radioactivity does not influence, nor is it influenced by, the fire properties of a metal). Control of contaminated runoff water is a complicating factor in using water on radioactive metals.

Flammable Gases

In gas-fire emergencies, water is generally used for the control of heat from the fire while efforts are made to shut off, or stop, the flow of escaping gas. Water in the form of spray applied from hose lines or monitor nozzles, or by fixed water spray systems, is commonly used for dispersement or dilution of concentrations of flammable gases.

WATER ADDITIVES

Freezing Temperatures and Antifreeze Additives

Because water freezes at 32°F (0°C), its use as an extinguishing agent is limited in climates or situations where freezing temperatures are encountered. There are several methods commonly used to prevent problems of freezing. These include using dry-pipe sprinkler systems instead of wet-pipe systems, circulation or heating tank water supplies held for fire protection purposes, adding freezing-point depressants to the water, or a combination of these.

The water-soluble freezing-point depressant most widely used in fire equipment is calcium chloride with a corrosion-inhibitor additive. Calcium chloride solutions are not used when fire protection systems are supplied by public water connections.

Sodium chloride (common salt) is unsatisfactory because of its limited ability to depress the freezing point of water and its highly corrosive attribute.

Antifreeze Additives for Sprinkler Systems

Chemically pure glycerine (U.S. Pharmacopoeia 96.5 percent grade) or pure propylene glycol can be used to depress the freezing point of water in portions of water suppression systems connected to public water supplies, if authorized by local health or water authorities. Diethylene glycol, ethylene glycol, or calcium chloride, as well as glycerine or propylene glycol, can be used for the same purposes where public water is not connected to the system. Both ethylene and diethylene glycol are poisonous and cannot be permitted to contaminate drinking water.

Antifreeze Additives for Water-Type Extinguishers

Alkali-metal salt solutions provide protection against freezing temperatures as low as –40°F (–40°C). Only the antifreeze solutions specified by the manufacturers should be used. Glycol solutions should not be used in extinguishers because the amount required to protect against freezing would be high, i.e., a 52.5 percent solution of ethylene glycol would be needed to depress the freezing point to –40°F (–40°C). Such amounts will alter the effectiveness of the extinguisher and may also lead to complications if the water should "boil off" and leave a strong concentration of the glycol which, under certain conditions, could be ignited.

Efforts have been made to develop other additives that can be mixed with water to lower the freezing point to –65°F (–54°C). This work has been largely stimulated by increased activity in extremely low-temperature areas, with most of the research being done by the U.S. armed forces. To date, formulas of lithium chloride, lithium chloride-calcium chloride, and lithium chloride-anhydrous sodium chromate have been used successfully. No commercial use of these solutions is known; however, the U.S. Naval Research Laboratory has developed a lithium chloride solution for fire extinguishers exposed to low temperatures of –65°F (–54°C).

Additives to Modify Flow Characteristics

Friction loss in the fire hose is always an obstacle for fire fighters. The longer the hose or the more water pumped through it, the greater the pressure loss. With a smooth fire hose lining, most of the pressure loss is the result of friction between particles of water generated by the turbulence in the flowing stream. When flow is either smooth or laminar, the friction loss tends to be very low with a slow stream of water. However, the amount of water delivered under laminar flow is generally too low for fire fighting. Fire fighting requires high-velocity streams that generate turbulence which, in turn, results in friction between water particles. This friction accounts for about 90 percent of the pressure loss in fire hose. The friction between the flowing water and the interior hose wall accounts for only 5 to 10 percent of the loss.

Until 1948 it was generally believed that not much could be done to reduce friction loss. At the time, trace quantities of certain polymers were found to reduce friction loss of turbulent streams. Most researchers report that linear polymers (i.e., polymers that form a single straight-line chemical chain with no branches) are the most effective in reducing turbulent frictional losses. Of these, the poly-chain (polyethylene oxide) is the most effective. Friction-reducing efficiency is a direct function of polymer linearity.

Poly-chain synthetics are nontoxic, have no effects on plants or marine life, and will degrade in sunlight. A poly-chain synthetic is a long linear chain, high-molecular-weight polymer, and is 2 to 3 times more effective as a friction reduction agent than other materials tested to date. It is an odorless, opaque, white slurry that weighs 9.1 lb per gal (1.1 kg/L) and must be kept within 0 to 120°F (–17 to 49°C). When it is injected into the hose stream, it dissolves completely and does not separate. It is compatible with all fire-fighting equipment and is useful in both fresh and salt water. One gallon (2.6 L) of additive treats 6000 gal (22 710 L) of water and achieves at least a 40 percent greater water delivery.

Tests conducted by the New York City Fire Department and Union Carbide Corporation found that a $1^1/_2$-in. (38-mm) hose with an additive delivered 250 gpm (946 L/min), or as much as a $2^1/_2$-in. (64-mm) hose without the additive. With the additive, a $2^1/_2$-in. (64-mm) hose was able to deliver more water than a 3-in. (76-mm) hose, and nearly as much water as a $3^1/_2$-in. (89-mm) hose. These tests also showed that the additive nearly doubled the nozzle pressure. The stream's reach was increased by nearly 30 percent, and the stream was more coherent.

Additives to Increase Water's Viscosity

The relatively low viscosity of water makes it tend to run off solid fuel surfaces quickly, and limits the ability to blanket a fire by forming a barrier at the surface. Additives make the use of water more effective on certain types of fires. Most applications of viscous water have been directed at fighting forest fires.

Viscous water has had several thickening agents added to it. In proper proportions, viscous water has the following advantages over nontreated water: ability to cling and adhere to the fuel surface; provides a continuous coating over the fuel surface; provides a layer thicker than water; absorbs heat proportional to the amount of water present; projects farther when discharged from a nozzle; and resists movement due to wind and air currents.

Viscous water has the following disadvantages over water: it does not penetrate the fuel as well as water, it creates a higher friction loss in fire hose or piping, increases the water droplet size, makes surfaces slippery and difficult to walk on, and requires mixing prior to use.

CARE IN USE OF WATER AS AN EXTINGUISHING AGENT

There are numerous factors that must be considered before using water as an extinguishing agent, as this chapter has already discussed. It must be considered if the use of water will cause the fire to grow or spread, whether there will be a chemical reaction causing harm, or if live electrical equipment can endanger personnel.

Fire fighters must also consider water runoff when using water-based fire extinguishing agents. Water can carry pollutants from a fire to a water supply or cause groundwater contamination. When using water to extinguish hazardous materials fires, e.g., pesticides and flammable liquids, some of the fuel will be carried off by the water vapor (steam), some of the water may be absorbed by the fuel, and some fuel may be carried by the runoff water.

See Section 1, Chapter 9, "Environmental Issues in Fire Protection," for precautions in the effects of using water and water-based extinguishing agents.

BIBLIOGRAPHY

References Cited

1. Mawhinney, J. R., Dlugogorski, B. Z., and Kim, A. K., "A Closer Look at the Fire Extinguishing Properties of Water Mist," *Fire Safety Science—Proceedings* of the Fourth International Symposium at the Ottawa Congress Centre, Ottawa, Ontario, Canada, Kashiwagi, T., Ph.D., ed., International Association of Fire Safety Science, 1994, p. 51.
2. Heselden, A. J. M., and Hinkley, P. L., "Measurement of the Transmission of Radiation Through Water Sprays," Fire Research Note No. 520, 1963, Dept. of Scientific and Industrial Research and Fire Offices' Committee, Joint Fire Research Organization, Boreham Wood, Herts, England.
3. NFPA, *Fire Protection Guide to Hazardous Materials*, 11th ed., National Fire Protection Association, Quincy, MA, 1994.
4. NFPA, *Standard on Wetting Agents*, National Fire Protection Association, Quincy, MA, 1995.
5. Drysdale, D., *An Introduction to Fire Dynamics*, 1st ed., John Wiley and Sons, New York, 1985, p. 224.
6. UL, "Electric Shock as It Pertains to the Electric Fence," Bulletin of Research No. 14, 1939, Underwriters Laboratories Inc., Northbrook, IL.
7. Commonwealth Edison Co., "Conductivity of Electricity Through Various Sizes and Types of Fire Streams," 1947. (An engineering report of tests conducted in cooperation with the Chicago Fire Department.)
8. AIA, "Fire Streams and Electrical Circuits," Special Interest Bulletin No. 91, 1963, American Insurance Association, New York.
9. Fitzgerald, G. W. N., "Fire Fighting Near Live Electrical Apparatus," Research Division Report No. 58–160, 1958, Hydro-Electric Power Commission of Ontario, Canada.
10. Buffet, "Peut-on Employer les Lances d'Incendie sur des Conducteurs Electriques?" 1934.
11. O'Dogherty, M. J., "The Shock Hazard Associated with the Extinction of Fires Involving Electrical Equipment," Fire Research Technical Paper No. 13, 1965, Ministry of Technology and Fire Offices' Committee, Joint Fire Research Organization. (Published by Her Majesty's Stationery Office.)
12. Keigher, D. J., "Water and Electronics Can Mix," *Fire Journal*, Vol. 62, No. 6, 1968, pp. 68–72.

Reference

Friedman, R., *Principles of Fire Protection Chemistry*, 2nd ed., National Fire Protection Association, Quincy, MA, 1989.

NFPA Codes, Standards, and Recommended Practices

Reference to the following NFPA codes, standards, and recommended practices will provide further information on water and water additives for fire fighting discussed in this chapter. (See the latest version of *The NFPA Catalog* for availability of current editions of the following documents.)

NFPA 10, *Standard for Portable Fire Extinguishers*
NFPA 11, *Standard for Low-Expansion Foam*
NFPA 11A, *Standard for Medium- and High-Expansion Foam Systems*
NFPA 13, *Standard for the Installation of Sprinkler Systems*
NFPA 15, *Standard for Water Spray Fixed Systems for Fire Protection*
NFPA 16, *Standard for the Installation of Deluge Foam-Water Sprinkler and Foam-Water Spray Systems*
NFPA 16A, *Standard for the Installation of Closed-Head Foam-Water Sprinkler Systems*
NFPA 18, *Standard on Wetting Agents*
NFPA 30, *Flammable and Combustible Liquids Code*
NFPA 70, *National Electrical Code®*
NFPA 75, *Standard for the Protection of Electronic Computer/Data Processing Equipment*
NFPA 750, *Standard on Water Mist Fire Protection Systems*

Additional Reading

ANSI C2, *National Electrical Safety Code*, American National Standards Institute, New York, 1993.
Atreya, A., "Extinguishment of Combustible Porous Solids by Water Droplets," Annual Progress Report, Michigan State Univ., East Lansing, National Institute of Standards and Technology, Gaithersburg, MD, Sept. 1990.
Back, G. G., "Water Mist as a Total Flooding Agent," *Proceedings* of the Technical Symposium on Halon Alternatives, June 27–28, 1994, Knoxville, TN, 1994, pp. 65–70.
Chow, W. K. and Shek, L. C., "Physical Properties of a Sprinkler Water Spray," *Fire and Materials*, Vol. 17, No. 6, Nov./Dec. 1993, pp. 279–292.
Clark, W. E., "Fighting Fire With Water: Using Water Wisely," *Fire Engineering*, Vol. 148, No. 9, Sept. 1995, pp. 33–34, 38, 40.
Coppalle, A., Nedelka, D., Bauer, B., "Fire Protection: Water Curtains," *Fire Safety Journal*, Vol. 20, No. 3, 1993, pp. 241–255.
Cummins, M., "Natural Water *vs.* Polarized for Fighting Fire," *American Fire Journal*, Vol. 47, No. 8, Aug. 1995, p. 27.
Everest, J., "Water Sprays Take Shape," *Fire Prevention*, No. 239, May 1991, pp. 11–13.
Fons, W. L., "Wet Water for Forest Fire Suppression," *Foam Applications for Wildland and Urban Fire Management*, Vol. 7, No. 1, May 1995, pp. 8–12; Southern California Association of Foresters and Fire Wardens, Annual Meeting, San Bernardino, California, April 7, 1950 and California Rural Fire Association, Annual Meeting, Merced, California, April 21, 1950, 1995.
Grosshandler, W. L., Lowe, D. L., Notarianni, K. A., and Rinkinen, W. J., "Protection of Data Processing Equipment with Fine Water Mist," NISTIR 5514, Oct. 1994, National Institute of Technology, Gaithersburg, MD.
Grosshandler, W. L., Lowe, D. L., Notarianni, K. A., and Rinkinen, W. J., "Suppression within a Simulated Computer Cabinet Using an External Water Spray," NISTIR 5499, Sept. 1994, National Institute of Technology, Gaithersburg, MD.
Hills, A. T., Simpson, T., and Smith, D. P., "Water Mist Fire Protection Systems for Telecommunication Switch Gear and Other Electronic Facilities," Fire and Safety International Research, Slough, England, NISTIR 5207, June 1993; National Institute of Standards and Technology, *Proceedings* of the Water Mist Fire Suppression Workshop, March 1–2, 1993, Gaithersburg, MD, 1993, pp. 123–142.
Jackman, L. A., Glockling, J. L. D., and Nolan, P. F., "Water Sprays: Characteristics and Effectiveness," *Proceedings* of the 1993 Halon Alternatives Technical Working Conference, May 11–13, 1993, Albuquerque, NM, 1993, pp. 263–273.
Kim, A. K., Mawhinney, J.R., and Su, J.Z., "Water-Mist System Can Replace Halon for Use on Electrical Equipment," *Canadian Consulting Engineer*, 1996, pp. 30, 32.
Mawhinney, J. R., "Water Mist Suppression Systems May Solve an Array of Fire Protection Problems," *NFPA Fire Journal*, Vol. 88, No. 3, May/June 1994, pp. 46–57.
Olsson, S. and Ryderman, A., "Extinguishment of Oil Spray Fires With Water: Experimental Procedures and Test Data," Swedish National Testing and Research Institute, Boras, Sweden, SP REPORT 1990:32, CIB W14/92/07 (S), 1990.

Scheffey, J. L. and Williams, F. W., "Extinguishment of Fires Using Low Flow Water Hose Streams. Part 1," *Fire Technology*, Vol. 27, No. 2, May 1991, pp. 128–144.

Scheffey, J. L. and Williams, F. W., "Extinguishment of Fires Using Low Flow Water Hose Streams. Part 2," *Fire Technology*, Vol. 27, No. 4, Nov. 1991, pp. 291–320.

Takahashi, S., "Extinguishment of Plastics Fires with Plain Water and Wet Water," *Fire Safety Journal,* Vol. 22, No. 2, 1994, pp. 169–179.

Tinker, S., and diMarzo, M., "Water Droplet Evaporation from Radiantly Heated Solids," NIST-GCR-95-665, Dec. 1994, National Institute of Technology, Gaithersburg, MD.

You, H. Z., Kung, H. C., and Han, Z., *Spray Cooling in Room Fires*, National Bureau of Standards, Washington, DC, 1986.

Wighus, R., "Extinguishment of Enclosed Gas Fires with Water Spray," *Proceedings* of the Third International Symposium on Fire Safety Science.

WATER STORAGE FACILITIES AND SUCTION SUPPLIES

Revised by William E. Wilcox

In a broad sense, water storage facilities and suction supplies include all bodies of water available as sources of supply, whether they are contained by constructed or natural barriers. Elevated or ground-level storage tanks of metal, concrete, wood, or earth lined with rubberized fabric are examples of constructed storage facilities; rivers, ponds, and harbors are examples of natural storage facilities.

Open bodies of water, such as reservoirs created by the damming of streams, are sometimes used in private fire protection to supplement public water supplies or furnish the primary source of water for fire protection if public supplies are insufficient in volume or pressure, or both, or if they lack dependability. Such arrangements are, however, feasible only in special situations. A common method is to use elevated gravity tanks, or ground-level suction tanks with fire pumps. Pressure tanks, with their limited capacity, may be used where storage requirements are relatively small.

This chapter contains information on the design, installation, and maintenance of constructed water storage facilities and the ways in which naturally occurring surface and ground-water supplies can be used for fire protection. Of particular interest are the provisions described for preventing water in storage from freezing during cold weather.

Water distribution systems and fire pumps (depending on the nature of the storage facilities), both integral parts of water supplies for fire protection systems, are covered in Chapters 3 and 4, respectively, in this section. Water-based extinguishing systems are covered in other chapters of this section.

STORAGE TANKS (GRAVITY AND SUCTION)

At the outset, it is well to recognize that, with the advent of hydraulically designed sprinkler systems, the use of elevated tanks for fire protection has declined; however, the use of ground-level suction tanks, combined with fire pumps, has increased. Nevertheless, there are many elevated gravity tanks still in service used solely for fire protection service, and they require high standards of maintenance to continue their reliability as primary sources of water for extinguishing systems.

William E. Wilcox, P.E., is a senior engineer in the Standards Division at Factory Mutual Research Corporation, Norwood, MA. He is a member of NFPA's Technical Committees on Sprinkler Systems, Fire Pumps, and on Inspection, Testing, and Maintenance of Water-Based Systems.

It is best if tanks for fire protection are not used for any other purpose. Tanks used for other purposes must usually be refilled frequently, and they become settling basins that collect large accumulations of sediment. When water is drawn from the tank, the sediment also is drawn into the yard mains or extinguishing system and may block the system.

If the tank is wood and is refilled frequently, the alternate drying and wetting of the lumber may appreciably shorten the life of the tank; with a steel tank, more frequent painting is required, which means not only greater expense, but more time out of service.

Another important consideration involving dual-purpose tanks is the water available at the time of the fire. Such tanks will seldom be full, because domestic and industrial consumption constantly draws it down. In addition, the normal water level may drop lower and lower if the industry grows. If a fire occurs several years after the tank is installed, sufficient water at sufficient pressure might not be available. To overcome this problem, the outlet for the domestic and industrial supply should be set at a water level above the water dedicated for fire protection service.

Location

A gravity tank supported on an independent steel tower with foundations placed in the ground rather than on a building is the best arrangement. The tank should be built so that it will not be subject to fire exposure from adjacent buildings or yard storage. If lack of yard room makes this impossible, the exposed steelwork should be suitably protected by fire-resistant construction or coverings. The steel protection, when necessary, should include steelwork within 20 ft (6 m) of combustible buildings or openings from which fire might issue.

If the tank or supporting trestle is to be placed on the walls of a new building, the building should be designed and built to carry the maximum loads.

Suction tanks should be located so as to minimize yard piping. The pump house is generally placed close to the tanks to minimize suction piping. The tanks should not be put where they will be exposed to fires in combustible construction, or fires issuing from windows or yard storage. (See Figure 6-2A.)

Design for Earthquake Resistance

Storage tanks can be designed to resist earthquakes. In an earthquake, the motion of Earth sets up a rocking action and a sloshing of water. These motions may produce stresses beyond those provided

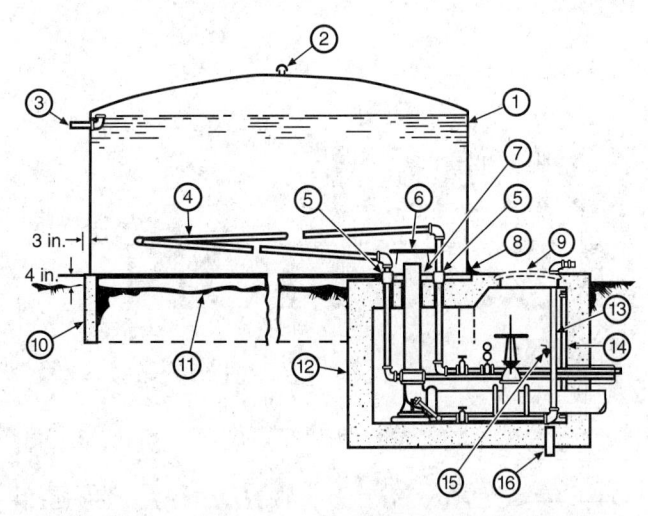

FIG. 6-2A. A ground-level suction tank showing the discharge pipe connected to the bottom of the tank in a valve pit. Nomenclature: (1) pump suction tank, (2) screened vent, (3) stub overflow pipe, (4) steam coil for heating, (5) extra-heavy couplings welded to tank bottom, (6) vortex plate, (7) watertight lead slip joint, (8) flashing around tank, (9) manhole with cover, (10) concrete ring wall, (11) sand or concrete pad (depending on soil condition), (12) valve pit, (13) drain pipe, (14) ladder, (15) drain cock, and (16) valve pit drain. (For SI units: 1 in. = 25.4 mm.) (Source: Factory Mutual Research Corp.)

for in a design that allows only for ordinary dead, live, and wind loads.

In areas where earthquakes are likely, the tanks, including tower and foundation, should meet the requirements for resisting earthquake damage by complying with the earthquake design provisions of the American Water Works Association's *Welded Steel Tanks for Water Storage*[1] or *Factory-Coated Bolted Steel Tanks for Water Storage.*[2]

There are three earthquake design categories: (1) cross-braced, (2) pedestal-type for elevated tanks, and (3) ground-supported for suction tanks. Cross-braced elevated tanks rely on the cross-bracing system to resist lateral forces. Pedestal tanks rely on the cantilever action of the shaft to resist earthquake forces. Suction tanks are affected less by earthquakes than elevated tanks; however, the tank itself needs to be designed to prevent elephant-footing, and the foundation needs to be designed to resist overturning the tank. The maximum load in the foundation is dependent on the column or shaft reactions of elevated tanks and whether suction tanks are anchored. Pipe connections, especially between suction tanks and the pumps, should be designed for sufficient flexibility.

Tank Capacities

Today it is usually uneconomical to install a gravity tank big enough and tall enough that it can be connected directly into the fire protection system and furnish adequate capacity and pressure for both hose streams from hydrants and water-based fire protection systems.

Years ago, because of limited capacity and pressure requirements, a gravity tank holding at least 30,000 gal (114 m³) with its bottom at least 75 ft (22.9 m) above the ground could adequately supply both hose lines from hydrants and the extinguishing systems. Storage tank selection is still determined by the capacity and pressure required for both hose streams and sprinklers for the possible duration of a fire, but, because of the increased requirements

of today, the suction tank and pump is more economical than a gravity tank or gravity tank and booster pump.

Gravity and suction tanks are generally erected in standard sizes. (See Tables 6-2A and 6-2B.) The capacity required is determined by the intended use of the tank and is specified in the number of gallons (cubic meters) available from the tank (1 gal = 0.00379 m³).

TABLE 6-2A. *Standard Sizes of Gravity Tanks*

Steel Tanks		Wood Tanks		Standard Height	
gal	m³*	gal	m³*	ft	m*
30,000	115	30,000	114	75	22.9
40,000	150	40,000	150	100	30.5
50,000	190	50,000	190	125	38.1
60,000	230	60,000	230	150	45.7
75,000	290	75,000	290		
100,000	380	100,000	380		
150,000	570				
200,000	760				
300,000	1100				
400,000	1480				
500,000	2000				

*Figures rounded off as approximations from nominal customary American tank sizes.

TABLE 6-2B. *Common Sizes of Steel Pump Suction Tanks*

gal	m³*	gal	m³*
50,000	190	250,000	950
75,000	290	300,000	1100
100,000	380	400,000	1500
125,000	475	500,000	2000
150,000	575	750,000	3000
200,000	750	1,000,000	4000

*Figures rounded off as approximations from nominal customary American tank sizes.

Steel gravity tanks with suspended bottoms are usually erected on towers with four columns for capacities from 50,000 to 200,000 gal (190 to 750 m³) inclusive, six columns for capacities from 200,000 to 300,000 gal (750 to 1100 m³) inclusive, and eight columns for tanks over 300,000 gal (1100 m³).

Construction of Tanks

New gravity tanks are usually built of steel and are supported by steel towers. Reinforced concrete towers are sometimes used, and tanks can also be placed directly on top of the structures they supply. In a few cases, concrete has also been used for the tank shells themselves. Typical gravity tanks are shown in Figures 6-2B and 6-2C.

Tanks should be designed and installed according to NFPA 22, *Standard for Water Tanks for Private Fire Protection* (hereinafter referred to as NFPA 22), which gives full requirements for construction materials, loads, unit stresses, details of design, foundations, accessories, and workmanship. Welding of towers should conform to code requirements for welding in building construction.[3]

Steel for tanks and towers should conform to the specifications in NFPA 22. Chief among these specifications are the American Water Works Association standards for steel tanks,[1,2] which give the thickness of steel plates and the welding and bolting practices that

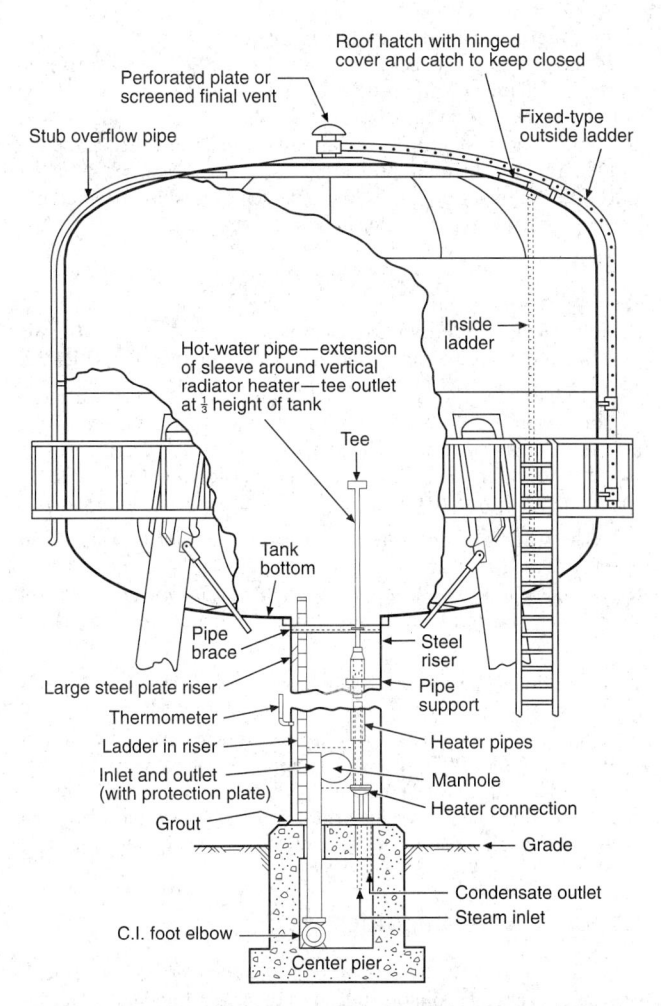

FIG. 6-2B.   A typical tower-supported double ellipsoidal tank.

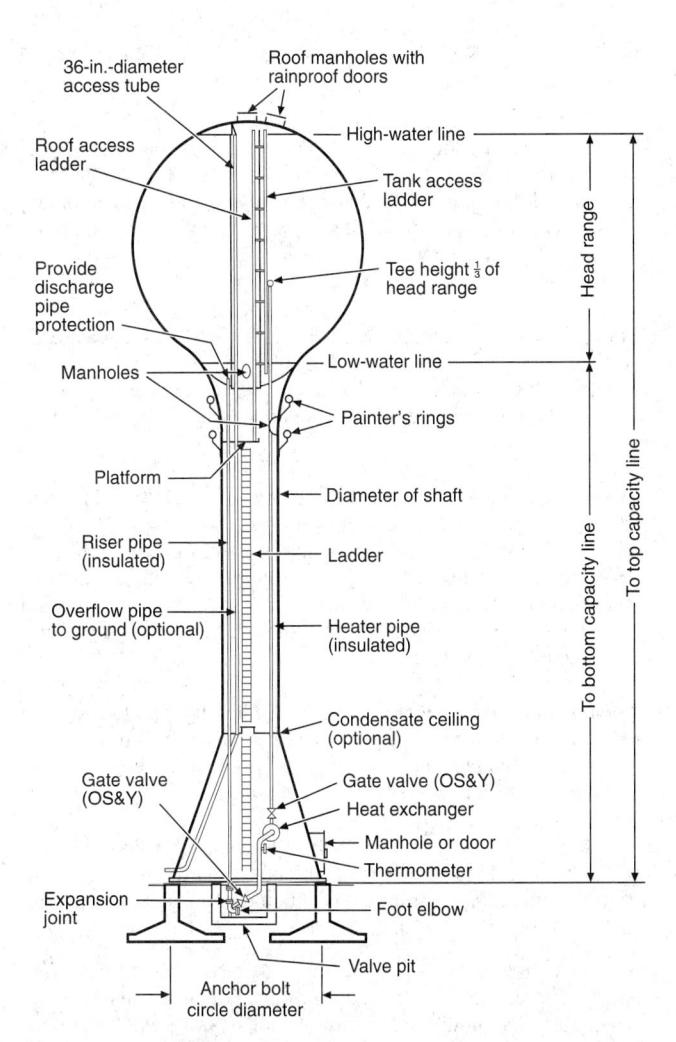

FIG. 6-2C.   A typical pedestal tank. The higher the tank, the greater the water pressure at the base of the tank. (For SI units: 1 in. = 25.4 mm.)

should be followed. Other standards referenced in NFPA 22 cover steel shapes, plate materials, bolts, anchor bolts and rods, forgings, castings, reinforcing steel, and filler material for welding.

NFPA 22 also gives information on wood tanks, including types and dimensions of wood suitable for use, processes that should be used, and proper hoop materials and design, as well as unit loads and unit stresses for steel tanks and towers, and working stresses for timber for wood tanks.

Steel tanks and towers should be riveted, welded, or bolted if factory-coated. Unfinished bolts should only be used in: (1) field connections of nonadjustable tension members carrying wind stress, and (2) field connections of compression members and grillages in towers supporting tanks of 30,000-gal (114 m³) or less capacity.

During assembly and erection, plates should be bolted firmly together before riveting. Drift pins should not be used to bring parts together or to enlarge unfair holes.

No waste material, such as boards, roofing, paint cans, etc., should be left in the tank or in the space at the top of the tank after its completion, because it may get into the water and obstruct the piping.

Tanks should be put in service promptly after completion. Wood tanks may be damaged by shrinkage if left empty.

TANK AND TOWER FOUNDATIONS

The following are principles of good foundations for suction tanks and gravity tank towers.

Foundations in the Ground

Material: Foundations should be built of concrete with compressive strength not less than 3000 psi (20.69 mPa). The cement and aggregates, and the mixing and placing of the concrete, should conform to the current American Concrete Institute requirements for reinforced concrete.[4] Concrete work should conform to all requirements of ACI 301.[5]

Pump suction tanks should be set upon crushed stone, sand, or concrete foundations. If the soil is good, at least 4 in. (100 mm) of crushed stone or sand is suggested. The material should then be laid on moistened and compacted gravel after removing unsuitable soil. Special consideration (outlined below) must be given if soil is poor.

A concrete ring wall at least 2½ ft (760 mm) deep and 10 in. (250 mm) thick should surround the tank foundation. This ring normally projects 6 in. (150 mm) above grade and includes hoop tension reinforcing steel equal to 25 percent of the cross-sectional area, or as required. If the ring wall is outside of the tank shell, asphalt flashing should be installed between the tank and the ring wall at ground level.

For poor soil, an 8-in. (200-mm) reinforced concrete slab with a concrete ring wall directly beneath the tank sides and extended below the frost line is advised. For riveted tanks, a $1^1/_2$-in. (38-mm) layer of a dry mixture of sand and cement should be placed on top of the concrete slab. For tanks of welded construction, no sand cushion is needed on the concrete slab. Piles can be used in addition to the reinforced concrete slab if the soil is very poor.

Form: The tops of foundations should be level and at least 6 in. (150 mm) above ground. The bottoms of foundation piers for towers should be located below the frost line, and, in the case of piers, at least 4 ft (1.2 m) below grade, resting on thoroughly tamped soil or rock.

Piers: Pier foundations may be of any suitable shape and may be either plain or reinforced concrete. If they support a tower, their center of gravity should lie in the continued center of gravity line of the tower column or else be designed for eccentricity. The height of piers should not be less than the mean width. The top surface should extend at least 3 in. (75 mm) beyond the bearing plates on all sides and is generally chamfered at the edges.

Anchorage: The weight of piers should be sufficient to resist the maximum net uplift that occurs when the wind blows from any direction on the empty tank. The weight of dirt directly above the base of the pier may be included in the calculations.

Anchor bolts should be arranged to engage a weight at least equal to the net uplift with the tank empty and the wind blowing from any direction. Their lower ends should be hooked or fitted with anchor plates.

Anchor bolts should be accurately located with sufficient free length of thread to fully engage their nuts. Expansion bolts are not acceptable. The minimum size of anchor bolts should be $1^1/_2$ in. (38 mm).

Grouting and flashing: Bearing or base plates should have complete bearing on the foundation or be laid on cement grout to secure a complete bearing. The stressed portion of anchor bolts should not be exposed except where necessary. If the stressed portions of anchor bolts must be exposed, they should be encased in cement mortar to protect against corrosion unless they are accessible for complete cleaning and painting. If structural shapes, plates, and bolts enter or are supported by masonry or concrete, the joint between the metal and masonry or concrete should be flashed with asphalt. (This does not refer to base plates under columns.)

Soil-bearing pressure: In order to find the proper depth of the foundation, the soil-bearing pressure must be determined by analyzing the subsurface and checking the foundations of other structures in the area. Test borings should be made by or under the supervision of an experienced soils engineer or soils testing laboratory. The borings should be made deep enough to determine the adequacy of the support, which is usually a minimum of 20 to 30 ft (6 to 9 m).

Foundations: The foundations should be designed to carry the maximum loads without excessive settlement. If wood piles are used above permanent low ground-water level, they should be protected as specified by the American Wood Preservers Association.[6]

Foundations should not be constructed over buried pipes or immediately adjacent to existing or former deep excavations unless the foundation bases go below the excavation.

Center pier: In addition to the weight of the water in a large plate riser, the weight of the column of water directly above the riser in the tank, and the weight of the steel plate, the center pier should be considered as supporting a hollow cylinder of water in the tank. If the hemispherical or ellipsoidal bottom is rigidly attached to the top of the large riser by a flat horizontal diaphragm plate, the radius of the hollow cylinder of water should be determined as specified under large risers discussed later in this chapter.

TANK TOWERS

Steel is generally used for the construction of tank towers. Details on the specific types of steel used in tanks can be found in NFPA 22.

Both live loads and dead loads should be considered when designing towers. The dead load is an estimate of the weight of the structure and all its fittings. The live load is considered as the weight of the water when the tank is filled and overflowing. Consideration should also be given to temporary loading, such as ice and snow.

The weight of the water in the riser need not be considered when figuring loads unless the riser is suspended from the tank bottom. If the riser is used to support the tank bottom, the entire weight supported by the riser should be considered, including the weight of the water.

Additional factors to consider are wind, balcony, and earthquake loads. Wind loads are based on 30 psi (207 kPa) for vertical plain surfaces and 18 psi (124 kPa) over the vertical projection of cylindrical surfaces. These loads are applied to the center of gravity of the projected areas. The balcony load consists of the weight of the balcony railing and ladder material and, if exposed, the weight of snow. Earthquake loading must be designed specifically for local conditions.

Ladders And Balconies

All tanks should have ladders on both the outside and inside, with convenient passage from one ladder to the other. Ladders are for inspection and maintenance of the tank's interior and exterior surfaces; they should be constructed of materials compatible to those of the tank and tower, and they should be easily accessible. Ladders more than 20 ft (6 m) long should be equipped with a cage or other safety device designed to protect the climber.

Balconies and walkways are recommended for towers over 20-ft (6-m) high. Normally, balconies should not be less than 24-in. (610-mm) wide and walkways not less than 18-in. (457-mm) wide. Railings should be at least 42-in. (1067-mm) high. Balconies and walkways should also be made of materials compatible with the tank and tower materials.

Tower Construction

During field erection, the columns of towers should be built on thin metal wedges driven to equal resistance so that all columns will be loaded equally after the structure is completed. The spaces beneath the base plates and the anchor holes should be completely filled with cement mortar.

Sections of structural members for towers should be symmetrical and built of standard structural shapes or tubular sections. Structural shapes should be designed with open sections so that all corrosible surfaces that will be exposed to air or moisture can be painted. Tubular sections of columns and struts should be airtight.

TANK HEATING EQUIPMENT

Adequate heating of tank equipment ranks next to structural design in importance. An ice plug in a riser pipe may make the tank water unavailable in case of fire and may break the pipe. Ice in or on tank structures has been the direct cause of collapse in several cases. The heating system must therefore be reliable and allow convenient and economical operation to 50°F (10°C). However, overheating can seriously damage wood tanks and the paint on steel tanks and should be avoided.

Determination of Heater Capacity

To prevent freezing in any part of a tank during the coldest weather, the heating system must be of such capacity that the temperature of the coldest water in the tank or riser, or both, will be maintained at or above 42°F (5.6°C). The coldest weather temperature used to determine the need of heating should be based upon the lowest mean temperature for one day, as shown on the isothermal map, Figure 6-2D.

Tables 6-2C and 6-2D show the heat losses from tanks of common sizes exposed to various atmospheric temperatures. Table 6-2D is an example of the method used to determine required heating capacities for various types and sizes of tanks.

QUESTION: What heater capacity would be needed for a 75,000-gal (284-m³) steel tank in Duluth, MN?

ANSWER: A heater capable of delivering 659,000 Btu per hr (193,000 W) under field conditions. Answer formed by interpolating from Figure 6-2D. The lowest mean temperature for one day in Duluth is –28°F (–33°C), and by interpolating from Table 6-2D, it is found that the heat loss at –28°F (–33°C) for a 75,000-gal (284-m³) steel tank is approximately 659,000 Btu per hr (193,000 W).

Selection of Heating Method

The selection of a heating method depends chiefly upon a tank's height, construction material, size, and shape, and on the lowest temperature of exposure. The recommended methods of heating are covered in detail in NFPA 22.

There are three basic methods of heating tank water: (1) gravity circulation of hot water, (2) steam coils inside tanks, and (3) direct discharge of steam into the water.

Gravity Circulation of Hot Water

Heating by gravity circulation is dependable and economical if correctly planned. Cold water received through a connection from the discharge pipe or from near the bottom of a suction tank or standpipe is heated and rises through a separate hot water pipe into the

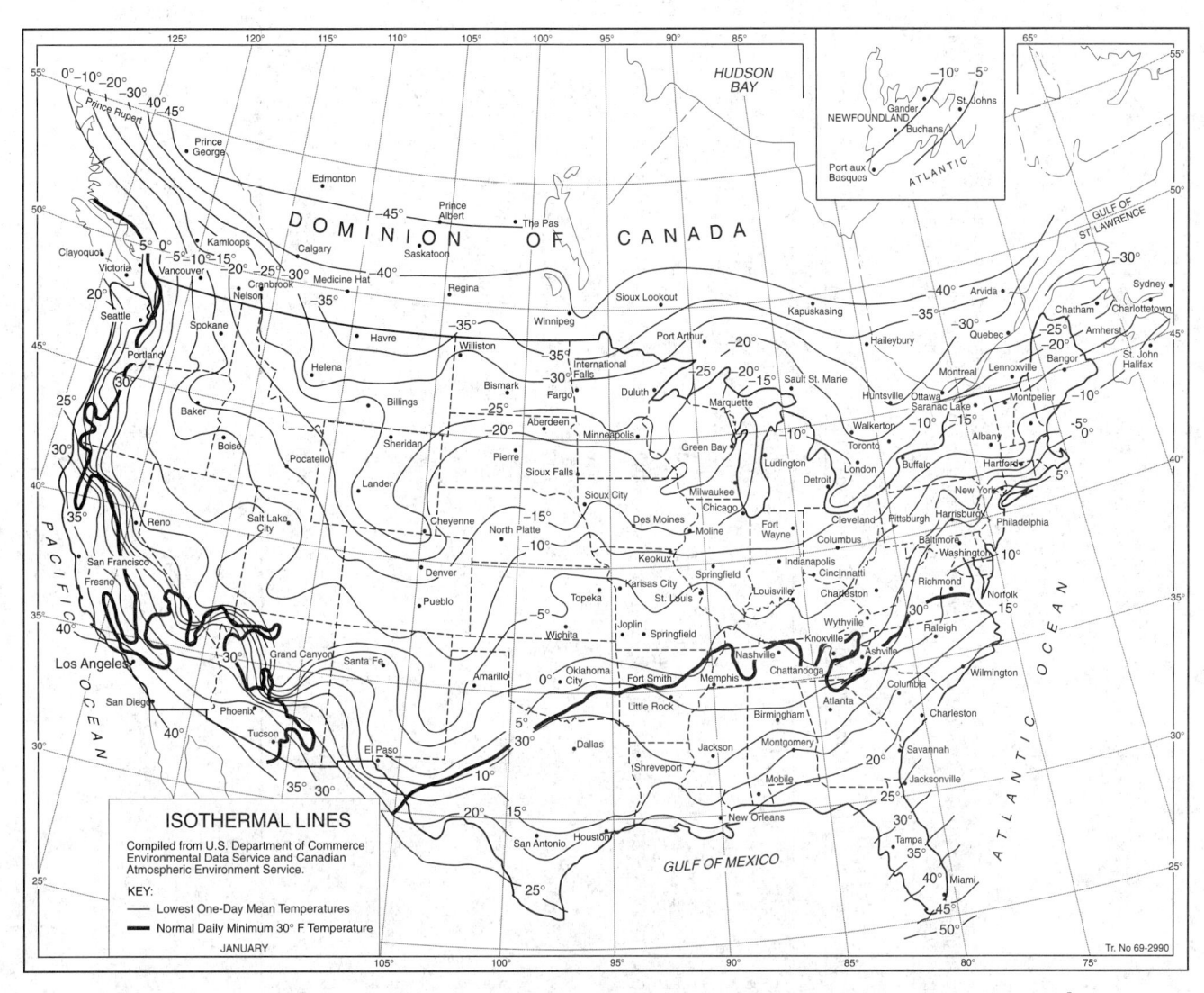

FIG. 6-2D.   Isothermal lines—lowest one-day temperature mean temperature. [For SI units: (°C = (°F – 32) × ⁵/₉.]

TABLE 6-2C. *Heat Loss from Standpipes and Steel Suction Tanks*

Thousands of British thermal units lost per hour when the temperature of the coldest water is 42°F (5.6°C). To determine capacity of heater needed, find the minimum mean atmospheric temperature for one day from the isothermal map, Figure 6-2D, and note the corresponding heat loss below.

TANK CAPACITIES—Thousands of Gallons

Temperature		100,000 gal	378 m³	150,000 gal	567 m³	200,000 gal	757 m³	300,000 gal	1135 m³	400,000 gal	1514 m³	500,000 gal	1892 m³	750,000 gal	2838 m³	1,000,000 gal
°F	°C	Btu/hr	W	Btu/hr	W	Btu/hr	W	Btu/hr	W	Btu/hr	W	Btu/hr	W	Btu/hr	W	Btu/hr
35	2	85	25	114	33	135	40	175	51	206	60	238	70	312	91	380
30	−1	121	35	162	47	193	57	248	73	294	86	340	100	445	130	542
25	−4	161	47	216	63	257	75	330	97	393	115	453	133	594	174	722
20	−7	202	59	271	79	323	95	414	121	493	144	568	166	745	218	907
15	−9	245	72	329	96	391	115	502	147	597	175	689	202	904	265	1099
10	−12	290	85	389	114	463	136	595	174	707	207	816	239	1071	314	1032
5	−15	337	99	452	132	539	158	691	202	822	241	949	278	1244	364	1514
0	−18	388	114	521	153	620	182	796	233	947	277	1093	320	1434	420	1744
−5	−21	441	129	592	173	705	207	905	265	1076	315	1241	364	1628	477	1987
−10	−23	498	146	669	196	797	234	1023	300	1216	356	1403	411	1841	539	2239
−15	−26	557	163	748	219	891	261	1143	335	1360	398	1569	460	2058	603	2503
−20	−29	619	181	830	243	989	290	1270	372	1510	442	1742	510	2286	670	2787
−25	−32	685	201	920	270	1096	321	1406	412	1673	490	1930	565	2532	742	3080
−30	−34	752	220	1010	296	1203	352	1545	453	1837	538	2119	621	2781	815	3383
−35	−37	825	242	1108	325	1320	387	1694	496	2015	590	2325	681	3050	894	3710
−40	−40	898	263	1206	353	1437	421	1844	540	2193	643	2531	742	3320	973	4039
−50	−46	1059	310	1422	417	1694	496	2175	637	2586	758	2984	874	3915	1147	4762
−60	−51	1229	360	1651	484	1966	576	2524	740	3002	880	3463	1015	4544	1331	5528

TABLE 6-2D. *Heat Loss from Elevated Tanks*

Thousands of British thermal units lost per hour when the temperature of the coldest water is 42°F (5.6°C). To determine capacity of heater needed, find the minimum mean atmospheric temperature for one day from the isothermal map, Figure 6-2D, and note the corresponding heat loss below.

Wood Tanks

Atmospheric Temperature °F	°C	10,000 gal (38 m³) Btu/hr	W	15,000 gal (57 m³) Btu/hr	W	20,000 gal (76 m³) Btu/hr	W	25,000 gal (95 m³) Btu/hr	W	30,000 gal (114 m³) Btu/hr	W	40,000 gal (151 m³) Bru/hr	W	50,000 gal (189 m³) Btu/hr	W	75,000 gal (284 m³) Btu/hr	W	100,000 gal (378 m³) Btu/hr	W
35	−2	8	2	10	3	11	3	13	4	14	4	19	6	21	6	28	8	33	10
30	−1	11	3	14	4	16	5	19	6	21	6	27	8	31	9	40	12	49	14
25	−4	15	4	20	6	21	6	25	7	28	8	36	11	42	12	54	16	65	19
20	−7	19	6	25	7	27	8	32	9	35	10	46	13	54	16	69	20	83	24
15	−9	24	7	31	9	34	10	39	11	44	13	57	17	66	19	85	25	102	30
10	−12	28	8	36	11	40	12	46	13	51	15	68	20	78	23	100	29	121	35
5	−15	33	10	43	13	47	14	54	16	60	18	78	23	92	27	117	34	142	42
0	−18	38	11	49	14	53	16	62	18	69	20	90	26	106	31	135	40	164	48
−5	−21	43	13	56	16	61	18	71	21	79	23	103	30	120	35	154	45	187	55
−10	−23	49	14	63	18	69	20	80	23	89	26	116	34	136	40	174	51	211	62
−15	−26	54	16	71	21	77	23	89	26	100	29	130	36	153	45	195	57	236	69
−20	−29	61	18	79	23	86	25	99	29	111	33	145	42	169	50	217	64	262	77
−25	−32	68	20	87	25	95	28	110	32	123	36	160	47	188	55	240	70	291	85
−30	−34	74	22	96	28	104	30	121	35	135	40	176	52	206	60	264	77	319	93
−35	−37	81	24	105	31	115	34	133	39	148	43	193	57	226	66	289	85	350	103
−40	−40	88	26	114	33	125	37	144	42	162	47	210	62	246	72	317	93	382	112
−50	−46	104	30	135	40	147	43	170	50	190	56	246	72	290	85	372	109	450	132
−60	−51	122	36	157	46	171	50	197	58	222	65	266	78	307	90	407	119	490	144

Steel Tanks

Atmospheric Temperature °F	°C	30,000 gal (114 m³) Btu/hr	W	40,000 gal (151 m³) Btu/hr	W	50,000 gal (189 m³) Btu/hr	W	75,000 gal (284 m³) Btu/hr	W	100,000 gal (284 m³) Btu/hr	W	150,000 gal (568 m³) Bru/hr	W	200,000 gal (757 m³) Btu/hr	W	250,000 gal (946 m³) Btu/hr	W	See Note Below Btu/hr	W
35	17	43	13	51	15	59	17	77	23	92	27	120	35	145	42	168	49	69	20
30	12	62	18	72	21	83	24	110	32	132	39	171	50	207	61	242	71	192	56
25	7	82	24	96	28	111	33	146	43	175	51	228	67	275	81	323	96	340	100
20	2	103	30	120	35	139	41	183	54	220	64	287	84	346	101	405	119	506	148
15	−3	125	37	146	43	169	50	222	65	267	78	347	102	419	123	491	144	692	203
10	−8	147	43	172	50	200	59	263	77	316	93	411	120	496	145	582	171	893	262
5	−13	171	51	200	59	233	68	306	90	367	108	478	140	577	169	676	198	1092	320
0	−18	197	58	231	68	268	79	352	103	423	124	551	161	664	195	779	228	1309	384
−5	−23	224	66	262	77	304	89	400	117	480	141	626	183	755	221	884	259	1536	450
−10	−28	253	74	296	87	344	101	452	132	543	159	707	207	853	250	1000	293	1771	519
−15	−33	283	83	331	97	384	113	506	148	607	178	790	231	954	280	1118	328	2020	592
−20	−38	314	92	368	108	427	125	562	165	674	197	878	257	1059	310	1241	364	2291	671
−25	−43	348	102	407	119	473	139	622	182	747	219	972	285	1173	344	1375	403	2568	752
−30	−48	382	112	447	131	519	152	683	200	820	240	1068	313	1288	377	1510	442	2860	838
−35	−53	419	123	490	144	569	167	749	219	900	264	1171	343	1413	414	1656	485	3174	930
−40	−58	456	134	534	156	620	182	816	239	979	287	1275	374	1538	451	1803	528	3494	102
−50	−68	538	158	629	184	731	214	962	282	1154	338	1503	440	1814	532	2126	623	4186	122
−60	−78	624	183	730	214	848	248	1116	327	1340	393	1745	511	2105	617	2467	723	4936	144

Note: For each lineal foot of uninsulated riser 4 ft (1.2 m) in diameter, add the number of Btu's or W's in this column to the total heat loss at the different temperatures of the various tank capacities.

tank. Steam, coal burning, or oil heaters are ordinarily used; gas heaters or electric heaters are also satisfactory.

A steam heater ordinarily consists of a cast-iron or steel shell through which water circulates by gravity around steam tubes or coils of brass or copper. It should be located in a valve pit, heater house, or in a nearby building at or near the base of the tank. When the tank is over a building, the steam heater should be located in the top story.

Heater thermometer: The convenience of a gravity system is that it permits discovery of the temperature of the coldest water in the system. A recording thermometer should be placed in the cold-water return pipe near the heater; it should be checked frequently to make sure that the temperature does not fall below 42°F (5.6°C). To fail to observe the temperature is to risk freezing the equipment. [Water has its maximum density at 39.2°F (4°C). When the temperature of the water falls below 39.2°F (4°C), there is an inversion, and the warmer water settles to the bottom of the tank while the colder water rises. Therefore, if the circulation heater is to be effective, sufficient heat must be provided so that the temperature of the coldest water will be above 42°F (5.6°C).]

Water circulating pipes: The size of pipe used for the circulation of heating water is given in Table 6-2E. Pipe should be either copper water tubing or brass (85 percent copper) throughout. The hot water discharges into the tank through a tee fitting at the end of the hot-water pipe at a height about one-third up from the bottom of the tank. The return pipe connects into the discharge pipe at a point that guarantees circulation throughout the portion of the discharge pipe subject to freezing. A typical arrangement of a circulating heater and piping in a valve pit is shown in Figure 6-2E.

Steam Coils Inside Tanks

Steam coils inside the tank do not permit convenient observation of water temperatures and have other faults that make them unsuited

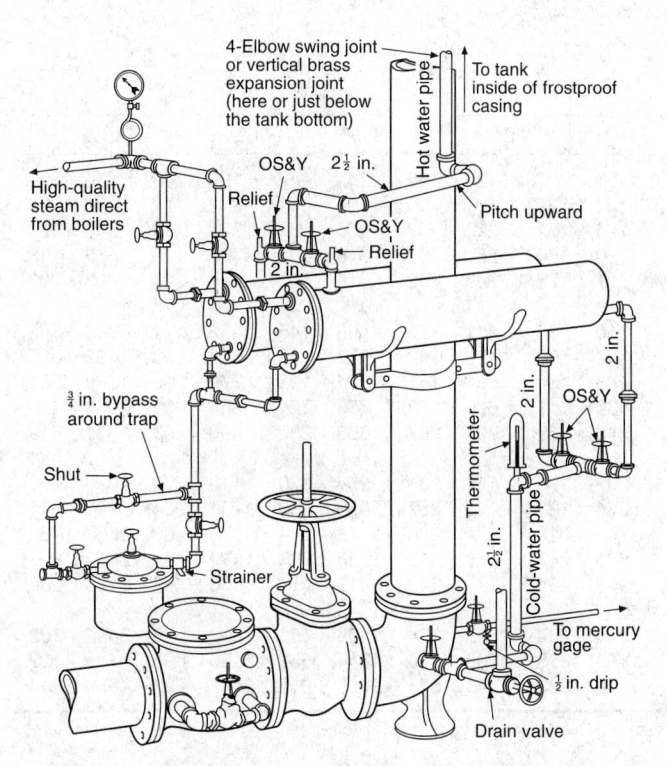

FIG. 6-2E.   The piping arrangement for a multi-unit, steam-heated water heater for a gravity circulation system.

for heating elevated tanks except in areas of the South where only intermittent heating is necessary. This method may, however, be used for heating suction tanks and standpipes with flat bottoms supported near the ground level if coils are submerged continuously. The coil consists of at least 1¼-in. (32-mm) brass or copper pipe, pitched to drain, supplied with steam at not less than 10-psi (69-kPa) pressure through a pipe of sufficient size to furnish the required quantity of steam from a reliable source.

Direct Discharge of Steam

Steam from a reliable supply is blown directly into the tank water through a pipe that enters the tank through the bottom, extends to above the maximum water level, and then returns to a point 3 or 4 ft (0.9 or 1.2 m) below the normal fire service level. An air vent and check valve in the pipe above the surface of the water keeps the water from siphoning back down the steam line. This method is employed where the lowest mean temperature for one day is 5°F (–15°C) or above.

Solar Heating of Elevated Steel Tanks

With the increase in fossil fuel prices, work was done in Canada on the feasibility of solar heating of elevated steel tanks.[7] The basic technique is to insulate the entire tank, leaving a window on the south side through the insulation. The tank shell in the opening is painted flat black to absorb heat from the sun and the opening is double-glazed.

Retrofit installations of solar heating systems on existing tanks involves special precautions in the attachment of components to the tank, such as the use of adhesives. The costs would be relatively high and the payback period long.

For new tanks, low-cost welding techniques can be used for attachment of collector components to the tank. With substantially all energy being supplied by the sun, a backup heating system can be a low-cost electrical system. Compared to a conventional tank with a fossil-fuel-fired boiler heating system, a solar-heated tank with an electrical backup heating system is likely to have a lower capital cost.

The preliminary findings indicate that solar-heated water tanks would seem appropriate for new installations.

TANK EQUIPMENT AND ACCESSORIES

A complete installation of a gravity or suction tank includes the necessary pipe connections, valve enclosures, and, where appropriate for the particular tank, such features as a frostproof casing for the riser.

Valve Pits

A pit 7 ft (2.1 m) deep and 6 ft by 9 ft (1.8 by 2.7 m) inside is usually large enough to house the necessary valves, tank heaters, and other fittings. Details of the construction of the pit and its arrangement, including clearance around equipment, waterproofing, manhole and ladder, and drainage, should conform to the recommendations given for valve pits. (See Figure 6-2F.)

Valve Enclosures

The tank heater and other fittings are sometimes installed in an enclosure above grade. In such a case, the indicating valve and the check valve are generally placed in the horizontal pipe below the frost line in a small pit heated sufficiently to maintain a temperature of at least 40°F (4.4°C) during the most severe weather.

The enclosure may be of concrete, brick, cement plaster on metal lath, or any other noncombustible material with suitable heat-

TABLE 6-2E. ***Sizes of Circulating Pipes Required for Elevated Steel Tanks***

Minimum One-Day Mean Temp		TANK CAPACITY																			
		gal 15,000	m³ 568	gal 20,000	m³ 757	gal 25,000	m³ 946	gal 30,000	m³ 1136	gal 40,000	m³ 1514	gal 50,000	m³ 1893	gal 60,000	m³ 2271	gal 75,000	m³ 2839	gal 100,000	m³ 3785	gal 150,000	m³ 5678
°F	°C	in.	mm	in.	mm	in.	mm	in.	mm	in.	mm	in.	mm	in.	mm	in.	mm	in.	mm	in.	mm
10	−12	2	51	2	51	2	51	2	51	2	51	2	51	2	51	2	51	2	51	2½	64
5	−15	2	51	2	51	2	51	2	51	2	51	2	51	2	51	2	51	2	51	2½	64
0	−18	2	51	2	51	2	51	2	51	2	51	2	51	2	51	2	51	2½	64	2½	64
−5	−21	2	51	2	51	2	51	2	51	2	51	2	51	2	51	2	51	2½	64	2½	64
−10	−23	2	51	2	51	2	51	2	51	2	51	2	51	2	51	2½	64	2½	64	2½	64
−15	−26	2	51	2	51	2	51	2	51	2	51	2	51	2½	64	2½	64	2½	64	3	76
−20	−29	2	51	2	51	2	51	2	51	2	51	2½	64	2½	64	2½	64	2½	64	3	76
−25	−32	2	51	2	51	2	51	2	51	2½	64	2½	64	2½	64	2½	64	3	76	3	76
−30	−34	2	51	2	51	2	51	2	51	2½	64	2½	64	2½	64	2½	64	3	76	3	76
−35	−37	2	51	2	51	2	51	2½	64	2½	64	2½	64	2½	64	3	76	3	76	3	76
−40	−40	2	51	2	51	2	51	2½	64	2½	64	2½	64	2½	64	3	76	3	76	3	76

insulating properties. The roof should be strong enough to support the frostproof casing and other loads without excessive deflection.

Frostproof Casings

Except in the case of a large steel riser, a frostproof casing around all exposed tank piping is necessary in localities where the lowest mean atmospheric temperature for one day as shown by the isothermal map (Figure 6-2D) is 20°F (–6.7°C) or lower. Tank piping subjected to temperatures below freezing within unheated buildings must also be adequately protected. Noncombustible frostproof casings should be used where there is danger of serious fire exposure.

Large Risers

Large steel-plate riser pipes 3 ft (0.9 m) or more in diameter, without frostproof casings, are often desirable. The fire hazard and upkeep of the frost casing are thereby avoided, the expansion joint in the discharge pipe is eliminated, and it is not necessary to have a walkway to reach the valves. When the tank is on an independent tower, a concrete valve pit at the base of the discharge pipe and the pier are usually built as a single unit to support the riser. (See Figure 6-2B.) On the other hand, the larger valve pit at the base of the riser makes the initial cost higher than it is for equipment with small risers.

Water-Level Indicators

A water-level gage of suitable design should be provided. A mercury gage is the most reliable water-level indicator for tanks. A de-

pendable, closed circuit, high- and low-water electrical alarm is a suitable substitute for the mercury gage in certain installations. The mercury gage is normally installed in a heated room, such as a boiler room, engine room, or office, where it will be readily accessible.

The gage should be accurately installed so that, when the tank is filled to the level of the overflow, the mercury level is opposite the "FULL" mark on the gage board. The procedures for installing and testing mercury gages are given in detail in NFPA 22 and in NFPA 25, *Standard for the Inspection, Testing, and Maintenance of Water-Based Fire Protection Systems.*

Overflow Pipes

The overflow pipe located at the top capacity or high-water line of the tank should not be less than 3 in. (76 mm) in diameter. If dripping water or a small accumulation of ice is not objectionable, the overflow may pass through the side of the tank near the top and extend not more than 4 ft (1.2 m), with a slight downward pitch to discharge beyond the balcony and away from the ladders.

When a stub pipe is undesirable, the overflow pipe can be extended down through the tank bottom and inside the frostproof casing or steel-plate riser, discharging through the casing near the ground or roof level. The section of the pipe inside the tank should be brass except on tanks with steel-plate risers, when overflow pipes $3\frac{1}{2}$ in. (89 mm) or larger may be of extra-heavy wrought iron or of flanged cast-iron pipe.

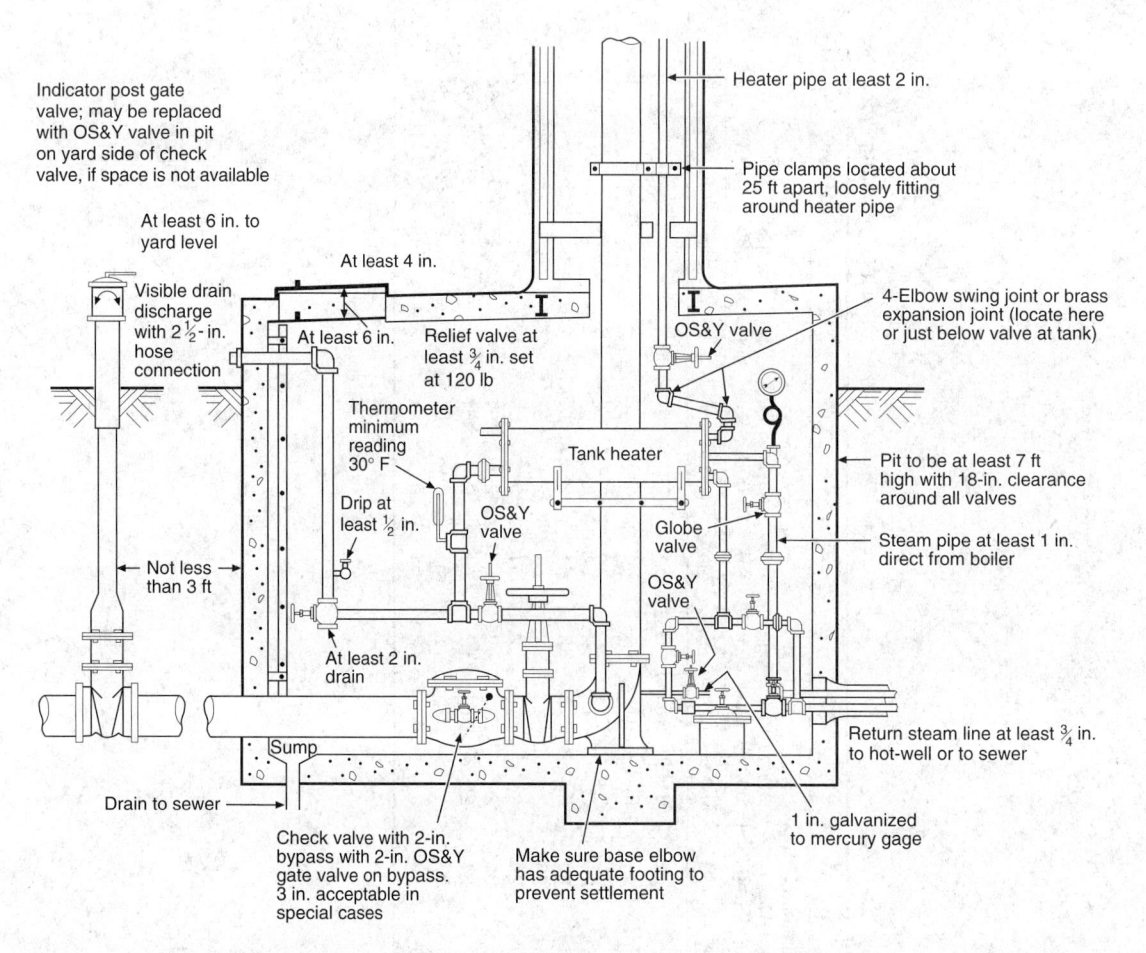

FIG. 6-2F. Valve pit and tank connections at base of riser. (For SI units: 1 in. = 25.4 mm; 1 ft = 0.305 m; 1 lb = 0.454 kg.)

Cathodic Protection

Cathodic protection can be used instead of painting to prevent interior corrosion on surfaces that are wetted by the tank's contents. Other surfaces inside the tank should be painted.

Internal corrosion is caused by galvanic current flowing from numerous anodic areas of the tank shell through the water to adjacent cathodic areas. Cathodic protection counteracts this process by passing sufficient direct current from an outside source (anodes suspended in the water) to the tank shell to keep the whole of the interior wetted surfaces of the tank at a negative potential. Low-voltage direct current is supplied from a rectifier. Frequent and regular checking of the ammeter and voltmeter readings is advisable to determine whether the electrical system is functioning.

Aluminum anodes, which are common, require renewal annually. In unheated tanks, ice may damage the anodes in winter, and early servicing in the spring is necessary. Broken parts of the anodes should be removed from the tank.

For heated tanks, anodes of fine platinum wire (which is not deteriorated by the electric current) are sometimes used, but they cost much more than aluminum anodes.

EMBANKMENT-SUPPORTED FABRIC SUCTION TANKS

Embankment-supported fabric (ESF) tanks can be used as suction tanks for fire protection. NFPA 22 contains details on construction, installation, and maintenance of ESF tanks.

ESF tanks are available in 20,000- and 50,000-gal (75- and 190-m^3) sizes, and in 100,000-gal (378.5-m^3) increments up to 1 million gal (3,785 m^3). The tank is usually composed of a reservoir liner with an integral flexible roof and is designed to be supported by earth underneath and on all four sides. The material of some ESF tanks is a nylon fabric coated with an elastomer compounded to provide resistance to abrasion and weather. Support is provided by a specifically prepared excavation and earthen berm.

Preparation of the installation site is critical to the reliability of an ESF tank. Usually, a shallow excavation the size of the bottom of the tank is made. The excavated earth is then graded to form the berm, or embankment, for the upper sides of the tank. The exterior of the berm should be graded to allow for rain and snow drainage and pipe connections, and the interior graded to match the contour of the tank, including rounded corners. ESF tanks may be placed

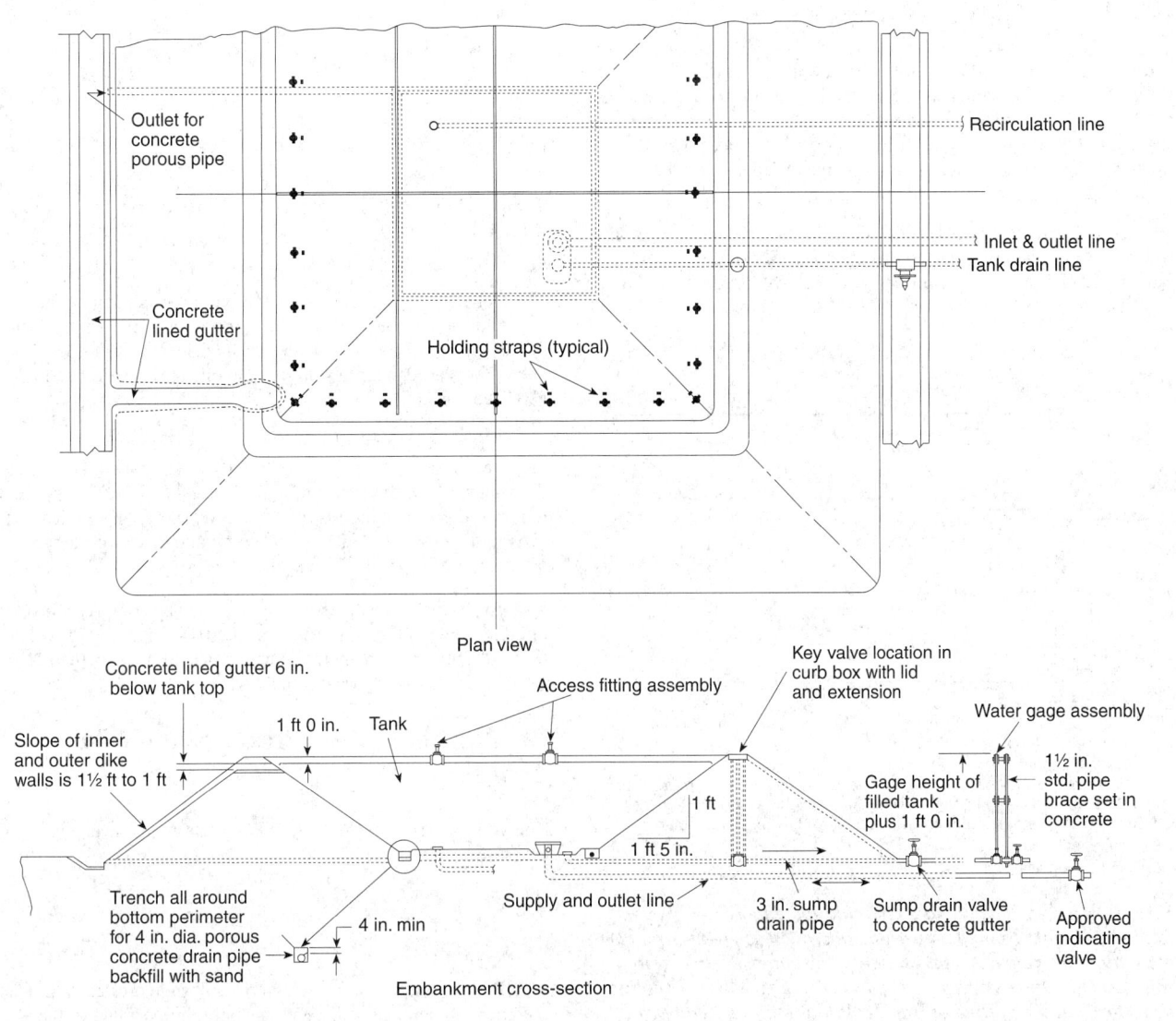

FIG. 6-2G. *Installation details of a typical embankment-supported fabric (ESF) tank, including fittings. (For SI units 1 in. = 25.4 mm; 1 ft = 0.305 m)*

underground with the top of the tank at grade level, or they may be placed above ground where the earthen berm provides the entire support.

When the excavation meets the shape requirements of the tank and all sharp objects have been carefully removed from the floor of the tank excavation, a 6-in. (150-mm) layer of sand or clean soil is spread over a 3-in. (75-mm) underlayment of pea gravel, which provides a firm base with good drainage. The tank is then placed in the foundation, and all connections are made. (Tanks are shipped to the site fully constructed.) After the tank is filled, a coating is applied to the exposed surface to protect it from the atmosphere. (See Figure 6-2G for a typical ESF installation.)

As with other types of storage tanks, water temperature in ESF tanks should be maintained at not less than 42°F (5.6°C). An acceptable method of providing heat is a water recirculation system with a heat exchanger. (See Figure 6-2H.) When the ambient water temperature drops below 42°F (5.6°C), a thermostat activates a pump that draws water from the tank through an inlet-outlet fitting and pumps the heated water back into the tank through a recirculation fitting located in the bottom of the tank diagonally opposite the inlet-outlet fitting. Heat losses in ESF tanks are given in Table 6-2F.

PRESSURE TANKS

Pressure tanks are used for limited private fire protection services, such as sprinkler systems, standpipe and hose systems, and water spray systems. They are sometimes used with fire pumps and gravity tanks that are located at as high an elevation as possible for quicker discharge. A typical pressure tank installation is shown in Figure 6-2I.

Tank capacity is considered to be the total contents, both air and water, not including dished ends. For light-hazard occupancies only, the tank may have a minimum capacity of 3000 gal (11 m³),

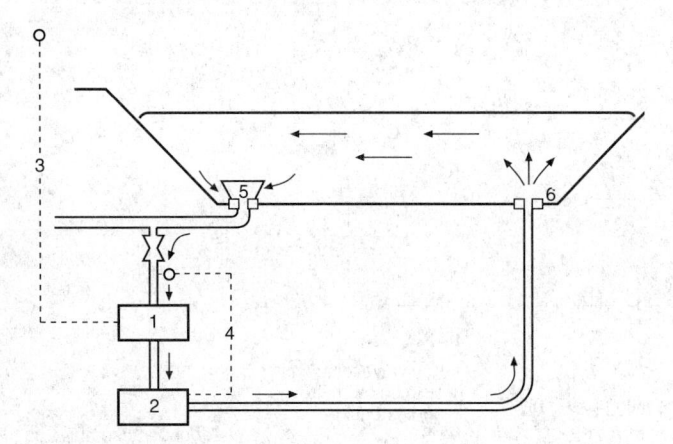

1. Recirculation pump
2. Heat exchanger
3. Unit sensing atmosphere temperature starts pump and water recirculation enabling heat stored in ground to transfer into water at higher rate.
4. Unit sensing water temperature starts heat exchanger when required.
5. Inlet-outlet fitting
6. Recirculation fitting

FIG. 6-2H. A schematic of a recirculation and heating system for an ESF tank. Nomenclature and operation sequence are: (1) recirculation pump, (2) heat exchanger, (3) unit sensing atmosphere temperature starts pump and water recirculation, enabling heat stored in the ground to transfer into water at a higher rate, (4) unit sensing water temperature starts heat exchanger when required, (5) inlet-outlet fitting, and (6) recirculation fitting.

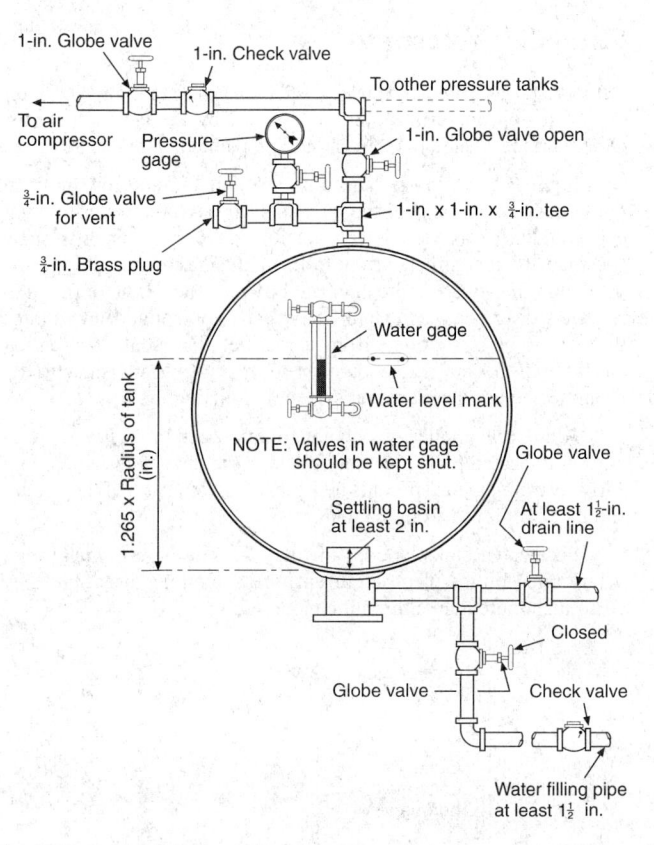

FIG. 6-2I. A typical pressure tank installation. (For SI units: 1 in. = 25.4 mm.)

of which 2000 gal (7.5 m³) is water, but the tank capacity ordinarily should be at least 4500 gal (17 m³) for ordinary-hazard occupancies. Tanks for this service generally are not over 9000-gal (34-m³) capacity. For larger supplies, more than one tank would be employed. (See Table 6-2G.)

The tank is normally kept two-thirds full of water, and a gage pressure of at least 75 psig (517 kPa) is maintained. As the last of the water leaves the pressure tank, the residual pressure shown on the gage should not be less than zero, and should give at least 15 psi (103 kPa) pressure at the highest automatic discharge device under the main roof of the building.

Air for pressure tanks may be supplied by compressors capable of delivering not less than 16 cu ft/min (0.045 m³/min) of free air for tanks of 7500-gal (28-m³) total capacity, and not less than 20 cu ft/min (0.057 m³/min) for larger sizes.

Relationship of Air Pressure and Volume in Tanks

The volume of air in the tank at any time varies inversely with the pressure:

$$\frac{P_1}{P_2} = \frac{V_2}{V_1}$$

Pressures are absolute pressures, not gage pressures, as used in the above general formula. Let us apply this formula to the special condition that exists in a tank used to supply water for sprinklers. Except where there is danger of air lock, it is desirable to have a

TABLE 6-2F. *Heat Loss from Embankment-Supported, Rubberized Fabric Suction Tanks*

Thousands of British thermal units lost per hour when the temperature of the coldest water is 42°F (5.6°C). To determine capacity of heater needed, find the minimum mean atmospheric temperature for one day from the isothermal map, Figure 6-2F, and note the corresponding heat loss below.

Atmospheric Temperature		Heat Loss Tank Radiating Surface		Tank Capacities															
				100,000 gal (378 m³)		200,000 gal (757 m³)		300,000 gal (1136 m³)		400,000 gal (1514 m³)		500,000 gal 1893 m³		600,000 gal (227 m³)		800,000 gal (3407 m3)		1,000,000 gal (3785 m3)	
				2746 sq ft (255 m²)		4409 sq ft (410 m²)		6037 sq ft (561 m²)		7604 sq ft (706 m²)		9139 sq ft (849 m²)		10,630 sq ft (988 m²)		13,572 sq ft (1261 m²)		16,435 sq ft (1527 m²)	
				Loss Per Hour—Thousands															
				Exposed Tank Surface															
°F	°C	Btu	k/m²	Btu	W	Btu	W	Btu	W	Btu	W	Btu	W	Btu	W	Btu	W	Btu	W
35	–2	22	252	61	693	98	1113	134	1522	168	1908	202	2294	235	2669	300	3407	363	4122
30	–1	29	324	78	886	126	1431	173	1965	217	2464	261	2964	304	3452	389	4417	470	5337
25	–4	35	399	96	1090	155	1760	212	2407	266	3021	320	3634	372	4224	476	5405	576	6541
20	–7	42	472	114	1295	183	2078	251	2850	315	3577	379	4304	441	5008	564	6405	682	7745
15	–9	48	545	132	1499	212	2407	290	3293	364	4134	438	4974	510	5792	652	7404	789	8960
10	–12	55	619	149	1692	241	2737	329	3736	413	4690	497	5644	579	6675	740	8403	896	10175
5	–15	61	693	167	1896	269	3055	369	4190	463	5258	557	6325	648	7359	828	9403	1003	11390
0	–18	68	767	185	2101	298	3384	408	4633	512	5814	616	6995	717	8142	916	10402	1109	12594
–5	–21	74	839	203	2305	326	3702	447	5076	561	6371	675	7665	786	8926	1004	11401	1216	13809
–10	–23	80	913	220	2498	355	4031	486	5519	610	6927	734	8335	855	9709	1092	12401	1322	15013
–15	–26	87	986	238	2703	384	4361	525	5962	659	7484	793	9005	924	10493	1180	13400	1429	16228
–20	–29	93	1060	256	2907	412	4679	564	6405	708	8040	852	9675	992	11265	1268	14399	1536	17443
–25	–32	100	1134	273	3100	441	5008	604	6859	758	8508	912	10357	1061	12049	1356	15399	1642	18647
–30	–34	106	1206	291	3205	469	5326	643	7302	807	9164	971	11027	1130	12832	1444	16398	1749	19862
–40	–40	119	1355	327	3713	526	5973	721	8188	905	10277	1089	12367	1268	14399	1620	18397	1962	22809
–46	–46	132	1498	362	4111	584	6632	799	9073	1003	11390	1207	13707	1406	15967	1796	20395	2175	24699
–60	–51	145	1648	397	4508	641	7279	878	9971	1102	12514	1326	15058	1544	17534	1972	22394	2389	27129

Note: Heat loss for a given capacity with a different radiating surface than shown above is obtained by multiplying the radiating surface by the tabulated heat loss per sq ft (m²) for the atmospheric temperature involved. The minimum radiation surface area is the wetted surface exposed to atmosphere. No heat loss is figured for tank bottoms resting on grade.

gage pressure of 15 psi (103 kPa) remaining in the tank when the last water leaves. Thus,

P_2 = residual pressure required + atmospheric pressure,

= 15 psi + 15 psi

= 30 psi

Since at this condition the tank is full of air,

$$V_2 = 1$$

If the tank is at or above the top line of sprinklers, the tank will be normally kept two-thirds full of water, so,

$$V_1 - \frac{1}{3}$$

Therefore,

$$P_1 = P_2 \times \frac{V_2}{V_1} = 30 \times \frac{1}{\frac{1}{3}} = 30 \times 3 = 90 \text{ psi}$$

The corresponding gage pressure would be 90 psi (710 kPa), minus the atmospheric pressure, 15 psi (103 kPa), or 75 psi (517 kPa).

If the tank is below the top line of sprinklers, P_2, the required residual pressure, would be increased 0.434 psi for every foot (0.912 kPa/m) of elevation represented in the distance between the base of the tank and the highest sprinkler. If we call this valve H, we have

$$P_1 = P_2 \times \frac{V_2}{V_1} = (30 + 0.434H)\frac{V_2}{V_1}$$

Gage pressure to be carried in the tank would be $P_1 - 15$, or

$$(30 + 0.434H)\frac{V_2}{V_1} - 15$$

For a tank $\frac{1}{3}$ full of air $\frac{V_2}{V_1} = 3$, and gage pressure would be

$$3(30 + 0.434H - 15) = 75 + 1.30H$$

For a tank $\frac{1}{2}$ full of air $\frac{V_2}{V_1} = 2$, and gage pressure would be

$$2(30 + 0.434H) - 15 = 45 + 0.87H$$

Air Lock

A condition known as air lock can occur when a pressure tank and a gravity tank are connected to a sprinkler system through a common riser. Air lock occurs if the gravity water pressure at the gravity tank check valve is less than the air pressure trapped in the pressure tank and the common riser by a column of water in the sprinkler system, after water has been drained from the pressure tank. For instance, if the pressure tank is kept two-thirds full of water with an air pressure of 75 psi (517 kPa), and a sprinkler opens 35 ft (10.5 m) or more above the point where connections from both tanks enter the common riser supplying the sprinkler system, the pressure tank drains, leaving an air pressure of 15 psi (103 kPa) balanced by a column of water of equal pressure [35 ft (10.5 m) head] in the sprinkler system. The gravity tank check valve is held closed unless the water

pressure from the gravity tank is more than 15 psi (103 kPa) [35 ft (10.5 m) head].

Air lock can be prevented by increasing the volume of water and decreasing the air pressure in the pressure tank so that little or no air pressure remains after water has been exhausted. For example, if the pressure tank is kept four-fifths full of water, with an air pressure of 60 psi (414 kPa), the air pressure remaining in the tank after water has been drained is zero, and the gravity tank check valve opens as soon as the pressure at that point from the pressure tank drops below the static head from the gravity tank.

Air lock may be conveniently prevented in new equipment by connecting the gravity tank and pressure tank discharge pipes together 40 ft (12 m) or more below the bottom of the gravity tank. (See Figure 6-2J.)

Construction of Pressure Tanks

Standard pressure tanks are constructed in accordance with *ASME Boiler and Pressure Vessel Code*,[8] with modifications given in NFPA 22. Important modifications are a minimum hydrostatic test at 150 psi (1034 kPa) and a test for tightness.

Tanks should be located in substantial noncombustible housings unless they are within a heated room in a building. They should be large enough to provide free access to all connections, fittings, and manholes, with at least 3 ft (0.9 m) of clearance around the valves and gages and at least 18 in. (451 mm) around the rest of the tank. The distance between the floor and any part of a tank should be at least 3 ft (0.9 m).

The interiors of pressure tanks are inspected at three-year intervals to determine if corrosion is taking place and if repainting or repairing is needed. When necessary, tanks should be scraped, wire brushed, and repainted with an approved metal-protective paint. Relief valves should be tested at least once each month.

Provisions should be made to drain each tank independently of all other tanks and have the sprinkler system drained by a pipe not less than $1\frac{1}{2}$ in. (38 mm) in diameter.

The filling supply or pump should be reliable and capable of replenishing the water to be maintained in the tank against the normal tank pressure in not more than 4 hr.

NATURAL AND CONSTRUCTED SUCTION FACILITIES

The advantages of piped systems make them the first choice for a water supply for fire protection, but sometimes other water may be more convenient or less expensive. Examples are suburbs or farms and resorts that are not near a community with a piped water system for fire protection. In many of these, however, water is available for fire protection in the ocean, lakes, ponds, rivers, and streams. Cisterns or tanks can be used at locations distant from natural water sources. Water that is provided for other uses in fairly large quantities, such as the stock tanks on a farm, can be used for fire protection if the tanks are kept full. Conversely, small ponds in parks and

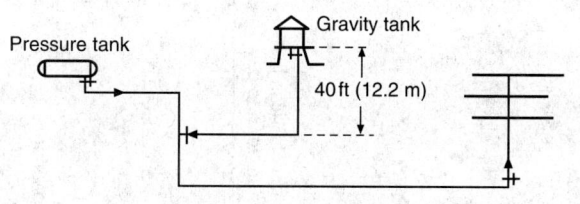

FIG. 6-2J. Gravity tank and pressure tank showing how risers are connected to avoid air lock. (For SI units: 1 ft = 0.305 m.)

TABLE 6-2G. *Typical Dimensions of Horizontal Pressure Tanks of Standard Sizes*

Approx. Gross Capacity		Approx. Net Cap 2/3 Full		Inside Diameter		Inside Length		Approx. Wt. of Water 2/3 Full	
gal	L	gal	L	in.	m	ft	m	lbs	kg
3,000	11355	2,000	7570	60	1.5	20.2	6.2	16,670	7568
3,000	11355	2,000	7570	66	1.7	17.0	5.2	16,670	7568
3,000	11355	2,000	7570	72	1.8	14.2	4.3	16,670	7568
4,500	17033	3,000	11355	66	1.7	25.4	7.7	25,000	11350
4,500	17033	3,000	11355	72	1.8	21.3	6.5	25,000	11350
4,500	17033	3,000	11355	78	2.0	18.2	5.5	25,000	11350
6,000	22710	4,000	15140	72	1.8	28.2	8.6	33,340	15136
6,000	22710	4,000	15140	78	2.0	24.2	7.4	33,340	15136
6,000	22710	4,000	15140	84	2.1	21.0	6.4	33,340	15136
7,500	28388	5,000	18925	78	2.0	30.3	9.2	41,670	18918
7,500	28388	5,000	18925	84	2.1	26.2	8.0	41,670	18918
7,500	28388	5,000	18925	90	2.3	22.7	6.9	41,670	18918
9,000	34065	6,000	22710	84	2.1	31.4	9.6	50,000	22700
9,000	34065	6,000	22710	90	2.3	27.3	8.3	50,000	22700
9,000	34065	6,000	22710	96	2.4	24.0	7.3	50,000	22700

Note: 4,500 gal (17 m³) gross capacity is the minimum ordinarily accepted for pressure tanks for automatic sprinkler systems. In the above table, the length of the tank has been figured as though the ends were flat instead of dished. The actual length of tanks of the above capacities and diameters will therefore be slightly greater.

swimming pools are not particularly good sources of emergency water because this competes with their regular purposes.

For water sources such as those mentioned, the general term "suction supplies" is used to distinguish these supplies from the conventional public water systems for fire protection. Static suction supplies must, in general, be used where they are found. While there have been instances where water in suburbs, farms, and forests was used to provide effective fire streams at very long distances, laying hose lines more than 750 to 1000 ft (225 to 300 m) seldom makes it possible to provide good protection because of the time required to lay the long lines and the greater difficulty in operating them.

Ground Tanks and Cisterns

If no hydrant on a water system is available, one way to ensure the amount of water required for fire fighting is to store it in underground tanks or cisterns at locations where hydrants would normally be installed. The fire department pumper can then get water from a connection on the tank or by dropping a suction hose into the tank. Such tanks and cisterns are filled from domestic water sources too small for fire fighting. The water does not have to be changed so frequently that it presents difficulties.

Static water supplies are of obvious value in suburban, farm, and forest areas, but they are equally important for emergency conditions in cities. Before the present public water systems were developed, some cities provided underground reservoirs or cisterns for fire protection. When water was needed for fire fighting, a pumper suction hose was simply dropped through a manhole. Some of these cisterns are still used in San Francisco and Chicago. Even though San Francisco has a regular water system and a separate system of high-pressure mains, it has been considered prudent, because of the threat of earthquake, to keep over 100 cisterns in service, most of which hold 75,000 gal (284 m³). Some of the oldest cisterns have brick walls. Newer ones are reinforced concrete.

Rivers and Ponds

Rivers and ponds are important auxiliary water sources, but merely because the water can be seen does not mean that it can be used for fire fighting. These sources of water must have suitable approaches so that fire apparatus can get near the water without becoming mired. A cleared space and some fill in many cases will provide access. These sources must be studied to determine that the water is available all year round and to select the best possible approach. They should also be recorded and mapped and given periodic attention in much the same manner that hydrants are maintained.

At such locations, it is usually necessary to provide a basin or cistern into which water can flow and the pumper suction strainer can be placed. There should be a screen or weir so that large objects will not interfere with the suction strainer. In some places, a permanent suction pipe and strainer can be installed. The details of such an arrangement are the same as those for a fire pump operating under a lift. The suction pipe should be laid with enough cover so that it will not freeze. It can be connected at the top to a hydrant with one or more pumper suction connections.

Small Streams and Brooks

Where water supply comes from a brook or stream, it is usually best to drain the water into a sump rather than attempt to use it straight from the flowing stream. If there is a marshy bank or shore, it may be best to go out into the pond or stream to a point where water can be collected in a sump. From the sump in the pond or stream, a pipe should be laid for draining to another sump, the latter located near where the fire apparatus will stand.

In some sumps, and particularly in open water that freezes over in winter, covers may not be feasible at the point where the sump is provided for pumper suction. At such locations, an anchored plug of wood can be floated where the pumper suction should be taken. Such a plug should have a minimum diameter larger than the pumper suction hose and be tapered so that, when frozen in ice, it can be driven down and out of position with a sledge hammer.

In some parts of the country, streams do not flow continuously. A covered cistern of generous capacity may be filled when water flows. Thus, the water may be stored for a longer period than it would keep in an open reservoir.

Water from streams and ponds is more important in areas without public water systems than in a city. There are small groups of farm buildings and small communities where even a brook, if dammed, and with a proper pumper approach and a suction basin, can provide good protection at little cost.

For projects in which a stream will be dammed to form static water, experienced local engineers should be consulted. Details of the dam construction and decisions about the amount of water to be impounded depend upon many factors and may require a comprehensive engineering survey. Its effect on users of the stream and flood conditions also have to be considered.

Harbors and Rivers

The principal problem with using harbors and rivers as suction supplies is that their water levels change. In harbors, ocean tide levels vary continuously throughout any given 24-hr period. River water levels are seasonal, but the problem is generally the same.

Where there is little fluctuation in the level of a natural water source, a simple pier or platform may be all that is needed to make the water available. If a pumper can be driven onto a sturdy pier, it can go to work when its suction line is dropped over the side. But where the water level varies, special arrangements are needed. If the water is always deep enough for drafting, a hinged ramp leading down to a large float is feasible. But more common is a sloping beach down which the water recedes too far to be accessible at all times. Reservoirs that can be filled at high tide are one solution to this problem.

Where a bridge passes over tidewater, water can be obtained by permanently installing a deep-well pump with a discharge pipe to a point convenient for fire department pumper use. The most practical way to use tidewater supplies is to provide pumps on boats, supplementary to any fireboats in service. Relatively large quantities of water for fire fighting can be obtained from river and harbor sources because large-capacity pumps are normally provided on fireboats, and where there are no fireboats, large pumps can be installed on a boat or barge and pressed into use to supply water for fire protection. To use the water on land, it is necessary to provide a place at which a fireboat or a barge with pumps can tie up and discharge into tanks or a special pipe system with hydrants from which land-based pumping equipment can take the water.

Wells

Wells have become increasingly popular for both domestic use and fire supply for industrial sites, shopping centers, etc., located beyond the reach of a piped water supply. Before a well is constructed, a thorough examination of the groundwater must be made. This examination generally includes an aquifer performance analysis and a review of the history of nearby wells. The water in the ground must be of sufficient capacity and dependability, and of reasonably good quality.

A vertical shaft turbine pump is used to draw the water from wells into the fire system or into a storage tank.

BIBLIOGRAPHY

References Cited

1. AWWA D100-84 (AWS D5.2-84), *Welded Steel Tanks for Water Storage*, American Water Works Association, Denver, CO.

2. AWWA D103-87, *Factory-Coated Bolted Steel Tanks for Water Storage*, American Water Works Association, Denver, CO.

3. AWS D1.1-94, *Structural Welding Code, Steel*, American Welding Society, Inc., Miami, FL, 1994.

4. ACI 318M-1989, *Building Code Requirements for Reinforced Concrete*, American Concrete Institute, Detroit, MI.

5. ACI 301-89, *Specifications for Structural Concrete for Buildings*, American Concrete Institute, Detroit, MI.

6. *Standard Specifications of the American Wood Preservers Association*, American Wood Preservers Association, Washington, DC, 1983.

7. *Testing and Monitoring Program for Solar-Heated Water Storage Tanks*, Prepared for the National Research Council of Canada, Ottawa, Ontario, 1984.

8. *Boiler and Pressure Vessel Code*, American Society of Mechanical Engineers, New York, 1992.

NFPA Codes, Standards, and Recommended Practices

Reference to the following NFPA codes, standards, and recommended practices will provide further information on water storage facilities and suction supplies discussed in this chapter. (See the latest version of *The NFPA Catalog* for availability of current editions of the following documents.)

NFPA 13, *Standard for the Installation of Sprinkler Systems*

NFPA 22, *Standard for Water Tanks for Private Fire Protection*

NFPA 25, *Standard for the Inspection, Testing, and Maintenance of Water-Based Fire Protection Systems*

Additional Reading

Bussen, R., "Anschlub von Feuerloschanlagen an Trinkwassernetze. [Connection of Fire Extinguishing Systems to Drinking Water Supply Systems.]," *VFDB*, Vol. 44, No. 1, Feb. 1995, pp. 2–4.

Clark, W. E., "Fighting Fire with Water: Using Water Wisely," *Fire Engineering*, Vol. 148, No. 9, Sept. 1995, pp. 33–34, 38, 40.

"Earthquake Protection for Water-Based Fire Protection Systems," *Loss Prevention Data 2-8*, Factory Mutual Research Corp., Norwood, MA.

"Embankment-Supported Fabric Tanks," *Loss Prevention Data 3-4*, Factory Mutual Research Corp., Norwood, MA.

"Isothermal Map for Europe," *Loss Prevention Data 3-2S*, Factory Mutual Research Corp., Norwood, MA.

"Lined Earth Reservoirs for Fire Protection," *Loss Prevention Data 3-6*, Factory Mutual Research Corp., Norwood, MA.

"Water Tanks for Fire Protection," *Loss Prevention Data 3-2*, Factory Mutual Research Corp., Norwood, MA.

WATER DISTRIBUTION SYSTEMS

Revised by Gerald R. Schultz

Most public waterworks systems of any size are designed to provide water for fire protection and normal consumption demands. Many water distribution systems on private property providing water for fire protection also supply water for sanitary purposes. In some large private plant installations, extensive water distribution systems exist primarily to provide process water for elements of manufacturing or purification and additionally provide water for fire protection. In other private plants, large or small, water distribution systems exist for the sole purpose of providing fire protection.

This chapter includes information on components that make up a system for the distribution of water from sources of supply to specific areas of use for the purpose of providing fire protection, and, in dual systems, information on simultaneously supplying water for normal consumption demands. Specifically, subjects covered are: the sources of supply, the distribution systems, the rules for laying pipe, and the types of pipe and equipment used in the systems for control and distribution purposes. The principles are the same whether the distribution system is owned by a municipality, is owned by a public utility, or is privately owned to provide water to a single property.

Water supply requirements, i.e., the amount of water necessary for adequate fire control, are covered in Section 6, Chapter 5, "Water Supply Requirements for Fire Protection"; and test procedures to measure the adequacy of water supply for fire protection can be found in Section 6, Chapter 7, "Test of Water Supplies."

A list of NFPA standards and recommended practices having application to water distribution systems can be found in the bibliography. The bibliography also includes standards and manuals of the American Water Works Association (AWWA), Underwriters Laboratories Inc.(UL), and Underwriters Laboratories of Canada (ULC) that refer to water distribution systems and their components. Many of these standards are also approved by the American National Standards Institute (ANSI).

SEPARATE AND DUAL DISTRIBUTION SYSTEMS

A distribution system can be either: (1) separate, serving fire protection only, or (2) a dual-type system, where it serves both fire protec-

Gerald R. Schultz, P.E., is a principal of Gage-Babcock & Associates, Oak Brook, IL. He is a member of the NFPA Technical Committee on Automatic Sprinklers.

tion and another purpose (e.g., process, domestic). Whenever alternatives exist, there are both advantages and disadvantages to a completely separate distribution system dedicated to fire protection.

Among the advantages are:

1. Those responsible for fire protection have complete control over the system.

2. The system is designed properly to meet all fire demands.

3. The system is not subject to reduced supply when increases in population increase the use of water for normal demands or when processes using increased amounts of water are experienced.

4. There is little danger of introducing a nonpotable water supply for fire protection into a potable supply through faulty cross connections.

One of the disadvantages, however, is that pumps, prime movers, and other machinery and equipment are usually in a "nonoperating" mode. This equipment must be started, controlled, and operated successfully during a fire emergency.

A dual-type distribution system (or combined systems) has an advantage in that pumping machinery is in operation almost continuously, and in many cases equipment is duplicated. Further, in combined systems, a malfunction surfaces quickly, as customers will likely be without water. In addition, public water systems have trained personnel readily available and not just for emergency operations.

In systems providing for normal consumption demands and for fire protection, that portion of the system extending into private property may be required to be isolated from the public portion of the system. This is usually done by the installation of listed or approved backflow prevention devices to provide protection from the possibility of contamination of the potable source. Although simple gravity check valves will accomplish the same goal, state or local municipalities may require these additional devices.

Systems on private property that provide water for process use should be made available to fire systems through the automatic opening of normally closed valves or other cross ties. Process water systems usually have substantial capacities and could be of significant value in augmenting the adequacy and reliability of fire protection systems.

FUNCTION OF SUPPLY AND DISTRIBUTION SYSTEMS

It is generally accepted that water systems are subdivided into two divisions: (1) supply systems and (2) distribution systems. However, in small water systems there may be no way of differentiating between supply and distribution systems since the functions of both may be carried out in a single element of the system.

Supply System

The supply portion of a water system is usually that part of the system where the source or sources of supply are found. It also includes the storage and transmission of that supply through large conduits and aqueducts, and, in some cases, includes the arterial feeders extending to the distribution system.

Distribution System

The distribution system is that portion of the waterworks that actually delivers water to the individual consumer connections and to which fire hydrants are attached.

SOURCES OF SUPPLY

Sources of supply have two major divisions: (1) groundwater supplies and (2) surface water supplies.

Groundwater Supplies

Of the total quantity of fresh water available for use in the world, by far the large majority is available from groundwater supplies. Groundwater supply is that water that percolates into the ground from precipitation and is stored in underground strata. The underground stratum holding water can be defined as an aquifer. The free surface of water in the underground stratum is referred to as the water table. The height of the water table varies throughout the year, depending upon variations in precipitation, water movement in the aquifer, and water withdrawn from the aquifer through springs or wells.

There are locations, now greatly diminished in number because of usage, where water is stored in the underground strata under a positive head. When this aquifer is penetrated, water will rise to an elevation higher than that in the aquifer, resulting in free flow at the surface.

The amount of water in the underground strata and its availability depends upon the nature of the material of the strata. For example, the porosity of natural sand and gravel may exceed 40 percent, while the porosity of igneous rock is approximately 1 percent.

While porosity is important, perviousness, i.e., the ability of a substance to allow water to flow through it, is of greater importance insofar as the movement of water through the strata to drilled or dug wells is dependent upon the perviousness of the soil. For example, some clays have a porosity in excess of 40 percent but are impervious to water flow.

The supply available from underground aquifers to wells is influenced, among other factors, by the precipitation falling on the area of recharge and the amount of water drawn from all of the wells that draw from the same aquifer. When the aquifer is of substantial size, static water levels usually delay the effects of drought and also delay the effects of increased precipitation. There are many different types of wells that may be constructed; however, high-capacity wells that are used for municipal water supplies are usually drilled and are equipped with deep-well turbine or submersible pumps.

NFPA 20, *Standard for the Installation of Centrifugal Fire Pumps*, permits the installation of vertical turbine pumps in properly developed and tested wells under conditions described in the standard.

Surface Supplies

Surface supplies consist of rivers, lakes, streams, and impounded supplies. As with underground supplies, the availability and reliability of the supply is dependent upon precipitation falling within the drainage area or watershed of the supply. Usually, however, surface supplies respond more quickly to diminished precipitation or drought. Water levels may vary substantially between wet and dry periods. At certain times of the year, there may be excessive runoff from the watershed, resulting in high water levels at intake structures and low lift pumping stations. At other times, under drought conditions, water levels in surface supplies could be so low as to require the installation of pumps in a reservoir discharging into normally gravity-supplied intake structures. Large surface reservoirs supplied by substantial watersheds or runoff areas are reliable sources, provided water consumption demands do not increase beyond the recharge capabilities of the watershed.

There are several factors that may affect the operation of the intake structures of surface supplies, particularly in colder climates. For example, ice formation is a hazard. There are several kinds of ice that may form and affect the functioning or threaten the stability of intake structures or cribs. Anchor ice, which forms on submerged dark metallic pipe, fittings, or gratings, can restrict flow into the intakes. Surface ice and surface ice forming ice dams may impose considerable thrust upon intake structures and cribs. Frazil ice, i.e., displaced anchor ice, or ice that has formed about small suspended particles, may clog intake parts.

When anchor or frazil ice begins to clog an intake, a relatively small change in temperature will free the obstruction. There are some intakes designed with a grid which is in essence a heating coil. In such cases, all that is required is the activation of the circuit which in turn will heat the intake element sufficiently to keep it free from ice. This method has been found economical in a number of locations, since it requires small amounts of electric power and is used only a relatively few times per year. When the intake is designed so that there is a continuous temperature recording, the imminence of the formation of anchor or frazil ice can be predicted.

A river can also be a reliable source of supply if the flow rates during drought periods are not seriously affected. A river can be susceptible to ice hazard, scouring of the bottom, changing of the channel, and silting. Before a river intake is constructed, a careful study must be made of the stream bottom, the degree of scour, and the extent of formation of surface ice; and the likelihood of the formation of ice jams must be established. An intake can easily be destroyed by an ice jam, or the entire flow of a river may be stopped by ice. Provision must be made in the design of the intake to ensure that it can withstand the forces that will act upon it during times of flood, heavy silting, or ice conditions.

GRAVITY AND PUMPING SYSTEMS

There are two basic types of water distribution systems: (1) gravity systems and (2) direct pumping systems. Most water systems are a combination of the two types.

Gravity Systems

A true gravity system is one which delivers a supply from the source directly to the distribution system without the use of pumping equipment. This type of system is usually ideal for a fire supply, provided pressures are adequate to supply fire demands and normal consumption rates. A gravity system is extremely reliable because the supply is not dependent upon the operation of mechanical equipment; however, the reliability of a well-designed and safeguarded pumping system can be developed to the extent that no distinction is made between gravity and pumping systems.

Pumping Systems

When water cannot be obtained at an elevation sufficient to provide working pressures from the elevation head, it is necessary to provide pumps on the system. These pumps are normally located at the source of supply and are used to develop the pressure needed to overcome friction loss in the supply system and to provide satisfactory working pressures in the distribution system. Public systems sometimes have water treatment facilities associated with the pumping station. Water treatment facilities will affect flow rates and quantities of water available to the distribution system. Many times, limiting features of supply are due to some element of water treatment. Therefore, it is imperative that the effects of water treatment on the availability of supply be thoroughly understood and considered. (See Section 9, Chapter 25, "Protection of Wastewater Treatment Plants.")

Combination Systems

Often associated with pumping systems are elevated water distribution storage facilities. These provide for storage of water during times of least demand and then supply water during times of peak demand.

Storage can be located so that pumps directly supply the elevated storage facility, and water flows to the distribution system from the storage facility. Elevated storage can also be provided at a remote location within the distribution system, and water can be pumped directly into the distribution system, with any excess automatically dumping into the elevated storage facility. The more water that can be maintained in elevated storage, the more reliable a system can be considered because water flowing from an elevated storage facility is a gravity system. Any failure of pumping equipment will not prevent this water from being available for fire protection purposes.

Distribution storage may also consist of large water tanks located at surface elevations equal to or even somewhat lower than the areas of the distribution system they serve. These tanks are filled during periods of relatively low consumption in the system. When demand rates are high, pumps deliver water from these storage facilities to the distribution system. Duplication of pumping equipment and proper design and operation can improve the reliability of these facilities to approach that provided by elevated storage alone.

SUPPLY CONDUITS, AQUEDUCTS, AND PIPELINES

Two terms sometimes used to describe the conveyor of water from the source of supply to the distribution system are "conduit" and "aqueduct." A conduit is an enclosed pipe capable of withstanding internal water pressures, while an aqueduct is either a closed tube or an open trench, canal, or channel in which water flows but which has no pressure on the side or bottom except that caused by the weight of the water. Aqueducts are not usually designed to withstand internal pressure other than atmospheric.

Pipelines

Pipelines are designed to withstand pressure and to distribute water to the point of use. Three classes of pipelines, or distribution mains, in a large system are:

1. Primary feeders consisting of large pipes with relatively wide spacing. They convey large quantities of water to various points of the system for local distribution to the smaller mains.
2. Secondary feeders forming a network of pipes of intermediate size. They reinforce the distribution grid within the various panels of the primary feeder system and aid the concentration of the required fire flow at any point.
3. Distributors consisting of a grid-configured arrangement of small mains. They serve the individual fire hydrants and blocks of consumers.

In order to provide for reliability, two or more primary feeders should extend by separate routes from the source of supply to the main districts of the city. Similarly, secondary feeders should be arranged as much as possible in loops to give two directions of supply to any point. This practice increases the capacity of the supply at any given point and ensures that a break in a feeder main will not completely cut off the supply.

Secondary feeders should generally be installed not over 3000 ft (914 m) apart in built-up areas.

Where water systems are divided into pressure zones, water can be transferred from one zone to another by operating valves or by using fire department pumpers to pump from the hydrants in one zone to hydrants in the other. The same sort of thing can be done between the water systems of adjoining communities or between a private system and the public system. This method of supporting a part of the water distribution network was used successfully by the Los Angeles Fire Department following the 1994 Northridge earthquake. However, great care must be taken to prevent damage from occurring by subjecting parts of the system to excessive pressures and possible contamination. Usually this is not a good practice and should not be attempted without advance planning, adequate controls, and specific written approval from health authorities.

The Size of Pipe

Pipe less than 6 in.* (150 mm) in diameter is not recommended for fire service, and 6-in. (150-mm) pipes should only be used when looped in a grid where no leg is greater than 600 ft (183 m) in length. Pipe with a diameter less than 6 in. (150 mm) can be used when it is used as the direct supply line to a sprinkler system and when hydraulic calculations prove that the smaller pipe will be effective. In congested districts, it is recommended that distributors should be not less than 8 in. (200 mm) in diameter and interconnected within every 600 ft (183 m). On principal streets and for all long lines, the distributors should be 12 in. (300 mm) or larger.

The cost of a line of pipe includes such factors as trenching (sometimes with piling), laying the pipe, backfilling, and testing. All of these factors are present regardless of the size of pipe used. To this is added the cost of the pipe delivered on the job. It is usually good practice, therefore, to install pipe for fire protection that is one or more sizes larger than the bare minimum might require. Increasing the pipe diameter only one size will often nearly double the possible flow. The figures in Table 6-3A show the relative capacity of pipe obtained by increasing sizes above 6 in. (150 mm).

In designing a system, it is also important to consider the probable development of the area under consideration; to plan, in a general way at least, protection for its ultimate development; then to install that part of the system for which there is immediate need.

---

*Nominal pipe sizes are not directly convertible into metric sizes. The following metric equivalents for common pipe sizes (rounded off for convenience in use) may be useful as points of reference: 4 in. = 100 mm; 5 in. = 125 mm; 6 in. = 150 mm; 8 in. = 200 mm; 10 in. = 250 mm; 12 in. = 300 mm; 16 in. = 400 mm; and 20 in. = 500 mm.

TABLE 6-3A. *Comparison of Pipe Capacity*

Size of Pipe		
in.	mm	Relative Capacity
6	150	1.0
8	200	2.1
10	250	3.8
12	300	6.2
14	350	9.3
16	400	13.2

The actual size of pipe needed is based on the rate of water flow required (domestic consumption plus fire flow) and the hydraulic gradient in the area.

Another way to show advantages of larger pipe is to indicate the better performance of four sizes by comparing their friction loss characteristics under a simple range of fire flow demands. (See Table 6-3B.)

Arrangements of Pipe Systems

Pipe systems should be arranged in loops wherever possible. This allows hydrants and other connections to be fed from at least two directions and greatly increases the possible delivery of water without excessive friction loss.

In private water systems where pipelines supply hydrants only and where there are adequate pressures for good streams at the hydrants, it is general practice to use 6-in. (150-mm) pipe to supply two-outlet hydrants, and 8-in. (200-mm) or larger pipe under the following conditions: dead-end mains, if more than one hydrant is to be supplied or the distance is more than 500 ft (152 m); looped mains, if two hydrants are to be supplied on a loop with over 1500 ft (457 m) of pipe; if three hydrants are to be supplied on a loop with over 1000 ft (305 m) of pipe; or if four or more hydrants are to be supplied.

Where pressures are low, or where three-outlet or four-outlet hydrants only are installed, pipes should be larger. However, the design and layout of the pipe system must provide for the delivery of the maximum fire flow rates to all locations coincident with maximum daily consumption demands.

Internal Condition of Pipe Systems

In the course of time, the internal cross-section of unlined cast-iron pipe may be reduced or its interior surface roughened because of tuberculation, incrustation, or sedimentation. Tuberculation is caused by internal pipe corrosion and a build-up of rust. Incrustations may be due to: (1) tubercular growth; (2) the depositing of chemical constituents normally in solution in the water; or (3) growth of biological or living organisms, such as zebra mussels or asiatic clams.

Sedimentation can be due to the accumulation of mud, clay, leaves, or vegetable decay. Additionally, foreign matter other than sediment can adversely affect flow capacity.

The existence of serious trouble can generally be detected by careful flushing tests. Flushing the system will remove ordinary sediment. Operation of valves will sometimes show presence of sediment or corrosion. Local water conditions are taken into account in establishing a regular procedure of flushing and testing. Pipes can be cleaned by the use of a scraper (hydraulic pig) or rotating auger. The cleaning device can be pulled through the pipe by a cable or forced through by water pressure.

When pipe has been cleaned in this manner, the rate of tuberculation following cleaning will usually be very rapid, quickly resulting in decreased carrying capacities. The addition of cement lining will retard or prevent further deterioration of capacity. Therefore, a cost analysis should be made to determine if the pipe should be lined or cleaned periodically throughout its life.

TYPES OF PIPES

Underground pipe and fittings for fire protection should be suitable for the working pressures and conditions under which the pipe is to be installed, and in accordance with American Water Works Association (AWWA) specifications. Pipe is mostly installed without blocks in flat-bottomed trenches with tamped backfill. The required earth cover over pipe varies depending on pipe size, soil type, and geographical location. Generally, because of frost penetration in the northern states, a minimum cover of 4 to 5 ft (1.2 to 1.5 m) is required for large-diameter mains and as much as 7 or 8 ft (2.1 to 2.4 m) for small-diameter mains. Classes of pipe for working pressures above 150 psi (1034 kPa) are often used if heavier and thicker pipe is desired, e.g., in unstable or corrosive soils or in locations that would be difficult to reach should leaks or breaks occur. Flexible construction is advantageous in difficult situations, such as under railroad tracks, in areas with heavy industrial machinery, in earthquake areas, or where steep slopes or unstable soil conditions are encountered.

Pipe and fittings used for underground fire service mains should also be listed by a recognized testing laboratory.

The titles and availability of design and installation standards and manuals for the various types of pipe and fittings discussed in the following paragraphs can be found in the bibliography at the end of this chapter.

Asbestos Cement Pipe

Asbestos cement pipe is particularly well adapted for locations where ferrous pipe materials without special protective linings or coverings would be attacked by corrosive soil conditions or by electrolysis. Where asbestos cement pipe must be buried in highly acid

TABLE 6-3B. *Friction Loss in Cast-Iron Pipe (Nominal Size)*
Hazen-Williams Coefficient $C = 100$

Flow		6 in.* (150 mm)		8 in.* (200 mm)		10 in.* (250 mm)		12 in.* (300 mm)	
gpm	L/min	psi	kPa	psi	kPa	psi	kPa	psi	kPa
500	1892	12.9	89	3.1	21	1.1	7.6	.4	02.7
1000	3785	46.4	320	11.1	76	3.9	26.9	1.6	11.0
1500	5678	98.0	676	23.6	163	8.2	56.5	3.4	23.4

*Nominal pipe sizes.

or alkaline soils, coatings can be provided that will protect the pipe from soil conditions.

The usual joint for asbestos cement pipe is an asbestos cement sleeve into which a specially shaped rubber gasket is inserted in a circumferential groove near each end of the sleeve. (See Figure 6-3A.) Cast-iron fittings may be used in asbestos cement pipelines.

Environmental concerns (personal health) exist when working with asbestos cement pipe. Care must be taken in the handling and preparation of the pipe. It should be noted that such pipe poses no threat to the water supply itself.

Polyvinyl Chloride (PVC) Pipe

PVC pressure pipe, manufactured in accordance with AWWA C900, is acceptable for fire service. PVC pipe often is specified in situations where severe corrosion problems are anticipated. PVC pipe is not subject to electrolytic corrosion, nor to tuberculation from corrosion byproducts.

Because of its flexibility and high-impact strength, PVC pipe manufactured to AWWA C900 specifications is not susceptible to beam, sheer, or impact failure in unstable soil conditions. The product is considered particularly suitable where shifting or heaving soils, live loading, or earthquake shock is anticipated.

AWWA C900 specifies PVC pipe with three pressure classes—100, 150, and 200. The pressure class designations define the maximum working pressure ratings for the pipe. The product is available in two series of outside diameters—cast-iron pipe (CI) and steel pipe (IPS). AWWA C900 PVC pipe is available in nominal diameters from 4 in. through 12 in. (approximately 100 to 300 mm).

PVC pipe used in potable water distribution and fire protection systems is commonly provided with gasketed (push-on) joints. Either integral bell gasketed joints or gasketed couplers that satisfy the requirements of ASTM D3139 may be specified. When installing PVC pipe with gasketed joints, thrust restraint or blocking should be provided as necessary to prevent movement of pipe or appurtenances in response to thrust.

Cast-Iron Pipe

Table 6-3C gives dimensions and weights for cast-iron pipes of thickness class 22, which is designed for standard laying conditions. The complete AWWA specifications have additional data on pipe sizes, weights, pressure classes, and thickness classes for other laying conditions. Pipe should be selected on the basis of maximum working pressure, with consideration for transient pressures and the laying condition. Cast-iron pipe is seldom used in new installations and has limited availability. Ductile iron has largely replaced it.

There are several acceptable types of joints. The most common are single gasket push-on and standard mechanical. Poured bell-and-spigot joints are seldom used now. All these joints depend on friction between the parts and surrounding earth fill to prevent separation. (See Figures 6-3B, 6-3C, and 6-3D.)

Push-on joints: The push-on joint is made up by seating a circular rubber gasket of special cross section in the valve and then forcing the spigot end of the pipe past the gasket to the bottom of the valve socket. No packing or caulking is required.

Standard mechanical joints: These are joints in which a rubber ring gasket is held in place by a follower ring bolted to the bell. A mechanical joint provides for a limited amount of flexibility; a ball-and-socket joint of similar design provides a little more. Because of

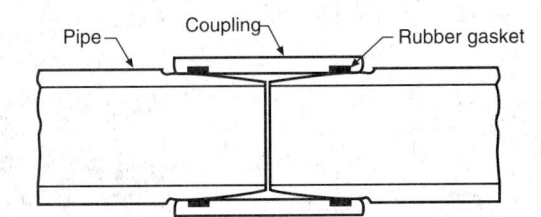

FIG. 6-3A. Coupling for asbestos cement pipes.

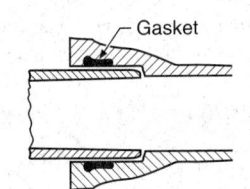

FIG. 6-3B. A push-on joint.

TABLE 6-3C. *Standard Dimensions and Weights of Cast-Iron Pipe**

| Nominal Size | | Outside Diameter | | Class 150 (150 psi or 1034 kPa) | | | | Class 200 (200 psi or 1379 kPa) | | | | Class 250 (250 psi or 1723 kPa) | | | |
| | | | | Thickness | | Weight per 18 ft (5.5 m) Laying Length† | | Thickness | | Weight per 18 ft (5.5 m) Laying Length† | | Thickness | | Weight per 18 ft (5.5 m) Laying Length † | |
in.	mm#	in.	mm	in.	mm	lb	kg	in.	mm	lb	kg	in.	mm	lb	kg
4	100	4.80	122	0.35	8.89	290	131	0.35	8.89	290	131	0.35	8.89	290	131
6	150	6.90	175	0.38	9.65	460	209	0.38	9.65	460	209	0.38	9.65	460	209
8	200	9.05	230	0.41	10.41	655	297	0.41	10.41	655	297	0.41	10.41	655	297
10	250	11.10	282	0.44	11.18	870	395	0.44	11.18	870	395	0.44	11.18	870	395
12	300	13.20	335	0.48	12.19	1,125	510	0.48	12.19	1,125	510	0.52	13.20	1,215	551
14	350	15.30	388	0.51	12.95	1,410	639	0.55	13.97	1,510	685	0.59	14.99	1,610	730
16	400	17.40	442	0.54	13.71	1,700	771	0.58	14.73	1,815	823	0.63	16.00	1,960	889
18	460	19.50	495	0.58	14.73	2,050	930	0.63	16.00	2,210	1002	0.68	17.27	2,370	1075
20	500	21.60	548	0.62	15.75	2,430	1,102	0.67	17.02	2,610	1184	0.72	18.28	2,785	1263
24	610	25.80	655	0.73	18.54	3,405	1,540	0.79	20.07	3,665	1662	0.79	20.07	3,665	1662

* ANSI/AWWA C150/A21.50-1991. Based on standard laying conditions [5 ft (1.5 m) cover, flat-bottomed trench and tamped backfill].
† Includes bell.
# Rounded off for convenience.

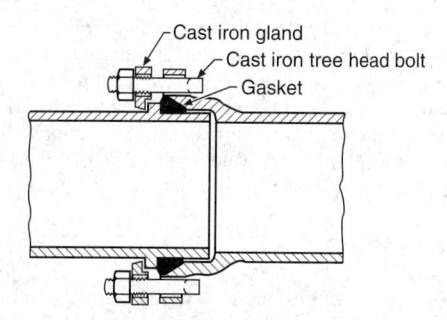

FIG. 6-3C. *A standardized mechanical joint using anchoring fittings.*

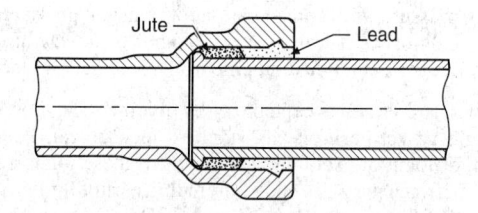

FIG. 6-3D. *A poured bell-and-spigot lead joint.*

the flexibility, pipe with these joints is often selected for lines across bridges or in unstable soil.

Poured bell-and-spigot joints: These joints have a ring packing of jute or other material and are caulked with lead. Special joint compounds are available that require no caulking.

When using bell-and-spigot pipe and fittings, changes in grade or direction should never be made by shifting the pipe in the joints. This would result in uneven packing and caulking, and such joints are likely to leak. Only a very slight variation from normal is tolerable.

Ductile-Iron Pipe

Ductile iron has the corrosion resistance of cast iron and approaches the strength and ductility of steel. Ductile iron is now used in place of cast iron. Table 6-3D gives the minimum available thicknesses for ductile-iron pipe 4 to 24 in. (100 to 610 mm) in diameter. Ductile-iron pipe is made with push-on and mechanical joints.

Cast-iron fittings are used with ductile-iron pipe. Coating or cathodic protection is not needed, except where extremely corrosive conditions exist. Cement-lined pipe is recommended for all new or replacement installations of cast-iron or ductile-iron pipe to offset the corrosive action of water. Portland cement is used extensively for lining. Coal-tar enamel linings are available but less common. A large percentage of all cast-iron pipe is cement lined at the foundry.

Steel Pipe

Steel pipe of suitable wall thickness and manufacture, when lined and coated, can be used for fire protection service, both for underground mains and for supply lines in tunnels and buildings. Because of its high tensile strength, steel pipe is particularly suitable where it may be exposed to shock or to impact from railroad tracks, highways, drop-forge equipment, etc. The greater strength of steel is also advantageous in unstable soil or on steep slopes. Approximate dimensions and weights are given in Table 6-3E.

Steel pipe joints are welded, made with flanges or mechanical couplings. (See Figure 6-3E.) Expansion joints may be needed in long runs of pipe in tunnels. Hangers and supports should conform to good engineering practice and applicable standards.

Reinforced Concrete Pipe

Several designs of pipe of concrete and steel 24 in. (610 mm) and larger in diameter are available. Concrete pipe is often used for long supply and arterial feeders, but it is not normally used in distribution systems. A "nonprestressed" design is a steel cylinder with one or two steel-cage reinforcements encased in concrete. A "modified prestressed" design is a steel cylinder with a spirally wound steel rod reinforcement prestressed to provide a slight initial tension in the cylinder and concrete lining. "Prestressed" designs consist of a

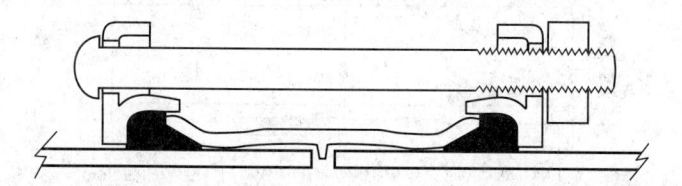

FIG. 6-3E. *A steel mechanical coupling for plain-end steel pipe.*

TABLE 6-3D. *Standard Dimensions and Weights of Ductile-Iron Pipe**

Nominal Size		Outside Diameter		Working Pressure		Thickness		Weight per 18 ft (5.5 m) Laying Length‡	
in.	mm	in.	mm#	psi	kPa	in. †	mm	lb	kg
4	100	4.80	122	350	2413	0.29	7.37	240	109
6	150	6.90	175	350	2413	0.31	7.87	375	170
8	200	9.05	230	350	2413	0.33	8.38	530	241
10	250	11.10	282	350	2413	0.35	8.89	695	316
12	300	13.20	335	350	2413	0.37	9.40	875	398
14	350	15.30	388	350	2413	0.36	9.14	1,000	454
16	400	17.40	442	350	2413	0.37	9.40	1,175	534
18	460	19.50	495	300	2068	0.38	9.65	1,360	618
20	500	21.60	548	250	1723	0.39	9.91	1,555	706
24	610	25.80	655	250	1723	0.41	10.41	1,965	892

*ANSI/AWWA C151/A21.51-1991. Based on standard laying conditions [5 ft (1.5 m) cover, flat-bottomed trench and tamped backfill].
† Minimum distance.
‡ Includes bell.
# Rounded off for convenience.

TABLE 6-3E.   *Recommended Minimum Dimensions and Weights of Steel Pipe for Fire Protection Mains*

Nominal Diameter		Outside Diameter		For Welded Joints				For Flexible Couplings or Threaded Joints			
				Minimum Wall Thickness		Weight per ft*		Minimum Wall Thickness		Weight per ft*	
in.	mm†	in.	mm	in.	mm	lb	kg	in.	mm	lb	kg
6	150	6.625	168	0.188	4.77	12.9	5.85	0.219	5.56	15.0	6.8
8	200	8.625	219	0.188	4.77	16.9	7.67	0.239	6.07	21.4	9.7
10	250	10.750	273	0.188	4.77	21.2	9.61	0.250	6.35	28.0	12.7
12	300	12.750	324	0.188	4.77	25.1	11.38	0.281	7.14	37.0	16.78
14	350	14.000	355	0.239	6.07	35.1	15.92	0.281	7.14	41.2	18.69
16	400	16.000	406	0.250	6.35	42.0	19.05	0.312	7.92	52.4	23.77

* 1 ft equals 305 mm.
† Rounded off for convenience.

concrete-lined cylinder or steel cylinder helically wrapped under tension with a high tensile strength wire. External coatings are cement mortar. Details of a reinforced concrete pipe joint are shown in Figure 6-3F.

Fittings

Fittings used should be appropriate for the same range of working pressures as the pipe with which they are used. A single class of cast-iron fittings for sizes 3 to 12 in. (75 to 300 mm) is now used generally. These are for working pressures of 250 psi (1723 kPa). Cast-iron fittings (bends and tees) are used with asbestos cement pipe. They have bells designed to be used with the asbestos cement gasket-type joint.

Cast-iron and steel fittings for underground water lines for fire protection are listed by testing laboratories.

Pipe Corrosion

Water is corrosive to cast-iron, ductile-iron, and steel pipe, and to fittings. The initial rate of corrosion for steel pipe may be more rapid than for cast- or ductile-iron, but after several years of exposure, there is little difference.

External corrosion of buried iron and steel pipe is the direct result of complicated electrochemical reactions. Soil containing metallic salts, acids, or other substances in combination with moisture causes iron ions to separate from the pipe. The mass of the metal at the pipe surface is diminished, and the pipe becomes pitted or corroded. Iron or steel pipe should not be installed under coal piles, in cinderfill, or wherever acids, alkalies, pickling liquors, etc., can penetrate the soil.

Stray electric currents from external sources may reach and follow buried pipelines to locations where the resistance to ground

is less than that of the pipeline. Ionization then occurs at points where the current leaves the pipe, producing an effect similar to that of soil corrosion.

When stray electric currents are suspected, the extent and origin should be determined by professional ground surveys. If the stray currents cannot be eliminated or diverted, the pipe, if not yet seriously corroded, can be protected by bonding all the joints and by providing direct low-resistance metallic ground connections.

Cathodic methods are widely used for the external protection of iron and steel water mains. Cathodic protection is a technique of imposing direct electric current from a galvanic anode to the buried pipeline. In many instances, cathodic protection is more economical than coating and wrapping. Cast-iron or ductile-iron pipe used in water systems should be cement-lined in accordance with AWWA standards.

A smooth lining is necessary to minimize loss of carrying capacity. Buried piping needs a coating to protect against soil corrosion. An outside coating may be applied in the field if desired, but it is practical only on large jobs. Exposed piping should be painted or otherwise protected as required by atmospheric conditions. Nuts and bolts of buried joint assemblies should be heavily coated. Any damage resulting to lining or coating should be thoroughly repaired.

RULES FOR LAYING PIPE

The depth of cover to provide protection against freezing will vary from about $2^{1}/_{2}$ ft (0.76 m) in the southern states to about 10 ft (3.05 m) in northern Canada. Because there is normally no circulation of water in fire protection mains, they require greater depth of covering than do public mains. The $2^{1}/_{2}$-ft (0.76-m) minimum should always be maintained to prevent mechanical damage. Depth of covering should be measured from top of pipe to ground level, and consideration should always be given to future or final grade and nature of soil. A greater depth is required in a loose, gravelly soil (or in rock) than in compact or clay soil. A safe rule to follow is to bury the top of the pipe not less than 1 ft (0.3 m) below the lowest frostline for the locality.

Placing pipes over raceways or near embankment walls should be avoided. Keep mains back a sufficient distance from the banks of streams or raceways to avoid any danger of freezing through the side of the bank. Where mains are laid in raceways or shallow streams, care should be taken that there will be running water over the pipe during any time when frost is expected; a safer method is to bury the main 1 ft (0.3 m) or more under the bed of the waterway.

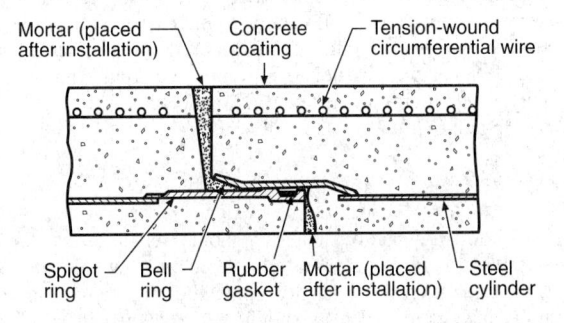

FIG. 6-3F.   A reinforced concrete pipe joint.

Protection Against Pipe Breakage

In general, it is advisable to avoid running pipe under buildings. Where mains necessarily pass under a building, the foundation walls can be arched over the pipe. Pipes passing under building walls with ground floors at grade are buried to the same depth as outdoors. Pipes under basement floors below grade may require less depth, but in no case should the cover be less than $2^1/_2$ ft (0.76 m).

Any pipe that passes through a wall or foundation must be protected from fracture. This is done by keeping a clear 1- to 3-in. (25- to 75-mm) annular space around the pipe and sealing it with a waterproof material, such as coal tar or asphalt. (See Figure 6-3G.)

Special care is necessary in running pipes under railroad tracks, highways, large piles of iron, and under buildings housing heavy machinery that could subject the buried pipes to shock or vibration. Where subject to such breakage, pipes should be run in a covered pipe trench, be sleeved, or be otherwise properly guarded. While flanged cast-iron pipe with metallic gaskets is sometimes used under buildings, it is not recommended because of its lack of flexibility, higher cost, and almost impossible access.

Frost loading is also recognized as a substantial factor that must be considered in pipe installation. Frost loading may exceed earth and live loads for which the pipe was designed. Other loads must also be considered, such as those which result from seismic shifts and tremors, unequal settlement and expansion, and subsidence of specific clay soils with changes in moisture content.

Care in Laying Pipe

Pipes should be clean inside when put in trenches, and open ends should be plugged when work is stopped to prevent stones or dirt from entering.

Pipes should be supported throughout their length and not by the bell ends only. Superior support is obtained where the bottom of the trench is shaped to fit the pipe. If the ground is soft or of a quicksand nature, special provisions must be made for supporting pipe. For ordinary conditions of soft ground, longitudinal wood stringers with cross ties will give good results. A reinforced concrete mat 3 or 4 in. (75 to 100 mm) thick in the bottom of the trench can also be used. In extreme cases, the stringers and crossties or concrete mat

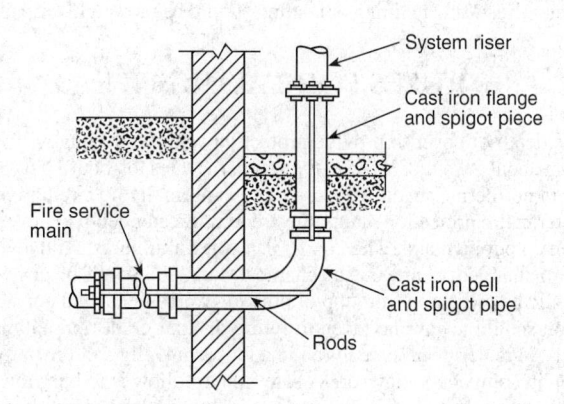

FIG. 6-3G. *Pipe is commonly run through a foundation to feed a sprinkler or standpipe riser. The anchorage of the horizontal run is determined by soil conditions. In some cases, as in earthquake areas, considerable flexibility is desirable in this run and in the first two joints outside the building. Often the run is brought into the building with the first joint of a flexible type, particularly if the pipe openings through foundation and floors are sealed with grouted concrete instead of mastic. The anchorages shown would be the usual ones.*

may have to be supported on piles. The most important aspect of laying pipe, though, is to follow the manufacturer's instructions for trench preparation, maximum deflection, and method of joining.

Pipe Anchorage

Most conventional pipe joints are not designed to resist unusual forces tending to pull them apart. Joints are expected to be kept in place by the soil in which the pipe is buried. It is necessary to determine, by tests in case of doubt, that a particular soil will actually do this. In unsatisfactory soils, trenches may have to be excavated below the final pipe level and filled to the pipe grade with soil suitable to give the pipe an even bearing throughout its length.

Forces acting on pipe laid in the ground that must be considered, among others, are: (1) internal static pressure of the water, (2) water hammer and pressure surge, (3) load from the backfill, and (4) load and impact from passing trucks and other vehicles. Static pressure and backfill loads are always present. Water hammer loads are considered separately from impact loads of passing vehicles on the theory that simultaneous action of the two would be a remote possibility. The thickness designated for the various pressure classes of pipe tends to reflect possible water hammer loads. The magnitude of water hammer depends upon the modulus of elasticity of the piping material, pipe wall thickness, pipe diameter, flow velocity, and the rate of change in velocity. Anchorages are needed to take care of additional loads due to water hammer and other forces. Trenches, which have to be unusually wide or deep, impose additional loads which may affect the choice of pipe thickness for the particular installation.

To prevent joints in cast-iron pipes from coming apart, they should be securely braced or clamped unless anchoring fittings or locked joints are used. Typical methods of anchoring joints at elbows, tees, and bends, and plugs at blanked openings are shown in Figures 6-3Hthrough 6-3O. Clamps and rods and other steel fittings at anchors should be protected against corrosion by a thick covering of asphalt.

Thrust Blocks

It is also necessary to consider the loads imposed by water moving in the pipe. This is why at bends, tees, and pipe ends, and wherever the piping changes direction, the pipe assembly must actually bear on a surface that will resist the loads imposed. Thrust blocks are used to keep the pipe assemblies in place. Guidance on sizing thrust blocks can be found in NFPA 24, *Standard for the Installation of Private Fire Service Mains and Their Appurtenances* (hereinafter referred to as NFPA 24).

The usual thrust block is made of concrete and is placed between the fitting and the trench wall. Thrust blocks may be tied to foundations where these are of sufficient size to provide a bearing.

A thrust block under a hydrant or valve to prevent upward movement in newly laid soil would require rods bent over the bells to hold the valve or hydrant to the block. Blocks under hydrants should be located so as not to prevent thehydrant from draining properly when drains are installed. Thrust blocks are shown in Figures 6-3P and 6-3Q.

Testing

All pipelines, of whatever material, should be subjected to hydrostatic tests, either by sections as completed or as a whole after completion. Such testing is usually done after the trench has been partially backfilled. All tests should be conducted in accordance with AWWA standards and NFPA 24.

FIG. 6-3H.   *Strap for bend anchor to tee.*

FIG. 6-3I.   *Anchor rods for bell-and-spigot pipe. When distance between bells is less than 12 ft (3.7 m), the next length of pipe is also anchored.*

FIG. 6-3J.   *Clamps for bell-and-spigot and short-body fittings.*

Washer

FIG. 6-3K.   *A plug for the bell end of a pipe.*

Strap

FIG. 6-3L.   *A general form of an anchor for indicator post valves. A mechanical joint pipe is illustrated.*

FIG. 6-3M.   *Hydrants are connected with a double-spigot connection anchored as illustrated. Anchor fittings and locked joints are often more convenient alternatives for rodded anchors. When the hydrant does not have lugs, the anchor rods are arranged as indicated by the dotted lines.*

FIG. 6-3N.   *An anchor at a spigot end of a tee fitting.*

FIG. 6-3O.   *An anchor at a bell end of a tee fitting. If the distance between joints is less than 12 ft (3.7 m), rods also should be run to a clamp on the next bell.*

Bearing area

Concrete thrust block

FIG. 6-3P.   *A thrust block at a one-quarter bend.*

Bearing area

Bearing area

Concrete thrust blocks

FIG. 6-3Q.   *A thrust block at a tee and plug.*

Backfilling

Earth should be well tamped under and around pipes (and puddled where possible) to prevent settlement or lateral movement, and should contain no ashes, cinders, or other corrosive materials. If the ground in which the pipe is laid is partly or wholly cinderfill, care should be taken that about a foot of cinder-free dirt is put in the trench below the pipe and no dirt containing cinders is used in backfilling around the pipe. Cinders stimulate galvanic actions which may cause the pipe to fail in a relatively short period. There are similar considerations in occupancies where salt or industrial wastes, e.g., from cattle sheds, packing operations, and chemical plants, might seep into the ground.

Rocks should not be rolled into trenches nor allowed to drop on pipes. In trenches cut through rock, backfilling is normally entirely of earth. In any case, earth should be used under and around pipe and at least 2 ft (0.6 m) above it.

Flushing

After a system of underground pipe for fire protection has been completed and before it is permanently filled with water, the entire system should be thoroughly flushed out under pressure through hydrants or other outlets. Mains supplying sprinkler systems should be flushed at a rate capable of achieving a flow of 10 ft per sec (3.1 m/s), or at the minimum demand rate of the system, whichever is greater. (See Table 6-3F.)

TABLE 6-3F. *Flow Required to Produce a Velocity of 10 ft per sec (3 m/s) in Pipes*

Pipe Size (in.)	Flow Rate (gpm)	(L/min)
4	390	1476
6	880	3331
8	1560	5905
10	2440	9235
12	3520	13323

NFPA 13, *Standard for the Installation of Sprinkler Systems*, contains information on the requirements for flushing underground connectors to ensure their reliability.

During flushing, the valves to any inside sprinkler or standpipe equipment should be closed to avoid washing stones or other debris into the interior portions of the system. Branches from the outside system should be flushed before they are connected to sprinkler or standpipe risers.

HYDRANTS

The most important features of good hydrants for fire protection are:

1. Nominal diameter of bottom valve opening at least 4 in. (100 mm) for two 2$^1/_2$-in. (64-mm) or larger outlets, 5 in. (125 mm) for three 2$^1/_2$-in. (64-mm) or larger outlets, and 6 in. (150 mm) for four 2$^1/_2$-in. (64-mm) or larger outlet hydrants. Hydrants having a bottom valve less than 4 in. (100 mm), or outlets less than two 2$^1/_2$-in. (64-mm), are usually not listed by testing laboratories. The connection between a water main and a hydrant should not be less than 6 in. (150 mm) in diameter.

2. The net area of the hydrant barrel and foot piece at the smallest part is not less than 120 percent of that of the net opening of the main valve.

3. A liberal-sized waterway and low friction loss. With the hydrant discharging 250 gpm (946 L/min) through each of two 2$^1/_2$-in. (64-mm) hose outlets, the total head loss at the hydrant should not exceed 2 psi (13.8 kPa). For a hydrant with a 4$^1/_2$-in. (114 mm) outlet, at a discharge rate of 1000 gpm (3785 L/min), the maximum permissible head loss is 5 psi (34.5 kPa). For hydrants designed or intended to deliver more than 1000 gpm (3785 L/min), the maximum permissible head loss should not exceed 5 psi (34.5 kPa) when discharging at the intended flow rate.

4. A positive-operating, corrosion-resistant drain or drip valve.

5. A uniform-sized pentagonal operating nut measuring 1$^1/_2$ in. (38 mm) from point to flat at the base and 1$^7/_{16}$ in. (36 mm) at the top. The faces should be tapered uniformly, and the height of the nut should not be less than 1 in. (25.4 mm).

Hydrant bonnets, barrels, and foot pieces are generally made of cast iron with internal working parts of bronze. Valve facings should be of a suitable, yielding material, such as rubber or a composition material. Hydrants are available with various configurations of outlets.

Types of Hydrants

There are two types of fire hydrants in general use today. The most common is the base valve (dry barrel) in which the valve controlling the water is located below the frost line between the foot piece and the barrel of the hydrant. (See Figure 6-3R.) The barrel on this type

FIG. 6-3R. A base valve, or "dry barrel," hydrant with nomenclature identified. When installed, the valve is below the frost line. This type of hydrant is also known as a "frostproof" hydrant. (Source: Mueller Company)

hydrant is normally dry with water being admitted only when there is a need. A drain valve at the base of the barrel is open when the main valve is closed, allowing residual water in the barrel to drain out. This type of hydrant is used whenever there is a chance the temperature will go below freezing, because the valve and water supply are installed below the frost line.

The other type of hydrant is the wet barrel (California) type, which is sometimes used where the temperature remains above freezing. These hydrants usually have a compression valve at each outlet, but they may have another valve in the bonnet that controls the water flow to all outlets. (See Figure 6-3S.)

Fire hydrants are covered by standards of the American Water Works Association (AWWA C502-94 and AWWA C503-88) and Underwriters Laboratories Inc. (UL 246), all listed in the bibliography at the end of this chapter.

Location

Hydrant spacing is usually determined by fire flow demand established on the basis of the type, size, occupancy, and exposure of structures. At present, there is no universally accepted method for establishing fire flows for other than for fixed extinguishing systems. Insurance Services Office (ISO) has a procedure that was developed for insurance rate-making purposes only. The procedure has been in effect for many years and is of value in arriving at hydrant spacing recommendations. Its use should be based upon sound engineering judgment that addresses the specific field situations that are encountered.

As a general rule, however, spacing should not exceed 800 ft (245 m) between hydrants. In closely built areas, 500 ft (150 m) or less between hydrants is more realistic. Hydrants should be located as close to a street intersection as possible, with intermediate hydrants along the street to meet the area requirements.

Where hydrants are located on a private water system and hose lines are intended to be used directly from the hydrants, they should be so located as to keep hose lines short, preferably not over 250 ft (75 m). At a minimum, there should be enough hydrants to make two streams available at every part of the interior of each building not covered by standpipe system protection. They should also provide hose stream protection for exterior parts of each building using

only the lengths of hose normally attached to the hydrants. It is desirable to have a sufficient number of hydrants to concentrate the required fire flow about any important building with no hose line length exceeding 500 ft (150 m).

For average conditions, hydrants normally are placed about 40 ft (12.2 m) from the buildings to be accessible during a fire event. When that is impossible, they are set where the chance of injury by falling walls or debris is small and where fire fighters are not likely to be driven away by smoke or heat. In crowded industrial yards, hydrants usually can be placed beside low buildings, near substantial stair towers, or at corners formed by masonry walls that are not likely to fall.

Hydrants that must be located in areas subject to heavy traffic need protection against damage from collision. The parking lots of shopping centers and mill yards are good examples.

Setting of Hydrants

Hydrants should be set plumb with the centerline of their outlets about 18 in. (0.46 m) above the ground and 12 in. (0.30 m) above the floor in hose houses. When hydrants are installed before grading is completed, the final grade line and accessibility should be considered. Most hydrants have a grade line indicated on the barrel of the hydrant.

Drainage is necessary for hydrants equipped with drain ports, and can be provided by excavating a pit about 2 ft (0.61 m) in diameter and 2 ft (0.61 m) deep below the base of the hydrant, and filling it compactly with coarse gravel or stones placed around the bowl of the hydrant to a level of 6 in. (150 mm) above the waste opening. If the drip valve of the hydrant is below groundwater level, it may be plugged to exclude groundwater. In that case, water remaining in the hydrant after use should be pumped out to prevent freezing.

The bowl of a hydrant should be secured to the next preceding bell with anchoring or locking joints or rods and clamps, or securely anchored by means of concrete backing. (See Figure 6-3M.)

Maintenance and Testing

Private hydrants should be tested and maintained in accordance with NFPA 25, *Standard for the Inspection, Testing, and Maintenance of Water-Based Fire Protection Systems* (hereinafter referred to as NFPA 25), which recommends that private hydrants be visually inspected monthly, lubricated twice yearly, and opened and closed yearly to ensure proper water flow and drainage.

Well-designed and properly installed hydrants present a minimum of maintenance difficulties. The dry barrel hydrant, for example, has a small drain near the base of the barrel arranged to permit water to drain out when the main valve is shut. When the main valve is opened several turns, this drain is closed. If this drain is working properly and the main valve is tight, the difficulty of water freezing in the barrel is avoided. Occasionally situations are found where ground drainage is unsatisfactory or where groundwater may stand at dangerous levels. In those cases, drains may be closed entirely and hydrant barrels pumped out periodically.

The use of salt or salt solutions to prevent freezing is not recommended because of their corrosive effect and limited usefulness. If antifreeze is used in hydrant barrels, its use must be confined to hydrants that are not part of a system supplying water for domestic consumption.

Ethylene glycol is extremely toxic, with as little as 0.1 mg/L ingested for a period of a week being fatal. This substance should not be used. Propylene glycol is not as toxic and may be used to prevent freezing but with proper precautions and in accordance with local health regulations.

FIG. 6-3S.   A wet barrel, or "California," hydrant used where freezing is not encountered. There is a compression valve at each hydrant outlet. (Source: Mueller Company)

Labels in figure: Stem "O" ring; Stem thread cleaning slots; Stem; Stuffing box; Stuffing box "O" ring; Barrel "O" ring; Upper barrel; Valve carrier; Seat washer retainer; Seat ring retaining pin; Slotted nut; Seat ring; Cap gasket; Cap; Cotter pin; Cap; Chain; Seat ring "O" ring; Seat washer

Suggestions for detecting freezing in hydrants include:

1. Sounding by striking the hand over an open outlet. Water or ice shortens the length of the "organ tube" and raises the pitch.
2. Turning the hydrant stem. If solidly frozen, the stem will not turn. If only slightly bound by ice, placing a hydrant wrench on the nut and tapping smartly may release the stem. Blows should be moderate to prevent breaking the valve rod.
3. Lowering a weight on a stout string into the hydrant. It may strike ice or come up wet, showing water in the barrel.

Probably the most satisfactory method of thawing a hydrant is by means of a steam hose. A thawing device in which steam may be rapidly produced should be standard equipment for fire departments in cold-weather climates. The steam hose is introduced into the hydrant through an outlet and pushed down, thawing as it goes.

A major item of periodic maintenance is a check for leaks in: (1) the main valve when the hydrant is closed, (2) the drip valve when the main valve is open but outlets capped, and (3) the mains near the hydrant. Stethoscope-like listening devices are available to make these checks. Maintenance routines provide for an operating test, repair of leaks, and pumping out of the hydrants where necessary. Threads of the outlet, caps, and the valve stem should be lubricated with graphite. Hydrants should be kept painted, but care should be taken to avoid accumulations of paint that might prevent easy removal of caps or operation of valve stems.

Uniform Marking of Fire Hydrants

Color coding of hydrants is based on water flow available from them. Color coding is of substantial value to water and fire departments. A test of an individual hydrant does not give as complete and satisfactory results as group testing, but such a test is a start in the right direction. The colors signify the capacity of the individual hydrant as tested, not group hydrant effect.

NFPA 291, *Recommended Practice for Fire Flow Testing and Marking of Hydrants*, recommends that hydrants be classified as shown in Table 6-3G.

TABLE 6-3G. *Hydrant Classification*

Class	Flow	Color of Bonnets and Nozzle Caps
AA	1500 gpm or greater	Light Blue
A	1000 to 1499 gpm	Green
B	500 to 999 gpm	Orange
C	Less than 500 gpm	Red

For SI units: 1 gpm = 3.785 L/min.

Capacities are rated by flow measurement and tests of individual hydrants during a period of ordinary water consumption. Rating is based upon the flow rate available at 20 psi (138 kPa) residual pressure.

The capacity-indicating color scheme provides simplicity and consistency with colors used in signal work for safety, danger, and intermediate conditions. (See Figure 6-3T.)

Within private enclosures, marking of private hydrants is at the discretion of the owner. Private hydrants on public streets are normally painted red to distinguish them from public hydrants.

Location markers for flush hydrants carry the same color background for class indication, with such data stenciled or painted on them as may be necessary.

CONTROL VALVES FOR WATER DISTRIBUTION SYSTEMS

Water distribution systems require valves at strategic locations and intervals to control flow as circumstances dictate. Nonindicating gate valves, butterfly valves, check valves, and pressure-reducing or altitude valves are the types of controlling valves used in supply systems.

Required features for gate valves, check valves, and indicator posts listed by a testing laboratory include:

For Gate Valves:

1. Stems of bronze with a minimum tensile strength of 32,000 psi (220 640 kPa).
2. Stuffing boxes with good-sized packing space.
3. Bronze-lined gland and bonnet opening, and a valve capable of being repacked under pressure.
4. Yokes bolted on bonnets in valves larger than 4 in. (100 mm).

For Check Valves:

1. Large clearances between moving parts and valve body.
2. A clapper that moves entirely out of the waterway.
3. Bronze-to-bronze bearings having large wearing surfaces.
4. Valves of the straightaway type, and iron body valves having a waterway equal to the area of the pipe.

For Indicator Posts:

1. A uniform flange to suit all sizes of gate valves.
2. Interchangeable operating stems.
3. Adjustable and interchangeable target plates.
4. Uniform-sized square operating nut measuring 1 in. (25.4 mm) square by 1 in. (25.4 mm) high.

Gate Valves

Gate valves of the nonindicating type are provided in distribution systems to allow segments to be shut off for repairs or extensions without reducing protection over a wide area. Such valves are normally a nonrising-stem type that require a key wrench to operate. A valve box is located over the valve to keep dirt from the valve and to provide a convenient access point for the valve wrench to the valve nut. (See Figure 6-3U.)

A complete record should be made for each valve in the system, including date installed, make, size, direction of opening, number of turns to open, any maintenance performed, and location by triangulation from fixed bounds, preferably under the control of the

FIG. 6-3T. Shaded areas are color coded to show the hydrant's flow capacity.

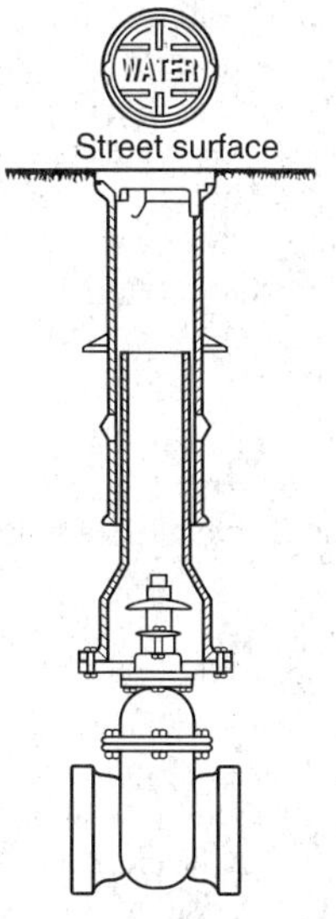

FIG. 6-3U.   A non-indicating-type gate valve for underground installation.

Parts List	
No.	Description
1	Cap
2	Operating stem
3	Operating stem oil hole screw
4	Operating wrench
5	Retaining ring
6	Target plate screw & nut
7	Target plate (shut)
8	Target
9	Indicator post staple
10	Target plate (open)
11	Indicator post
12	Extension rod (specify length)
13	Extension rod coupling
14	Coupling pin
15	Window glass
16	Window frame
17	Window frame screw
18	Cap bolt & nut
19	Set screw
20	Sleeve bonnet

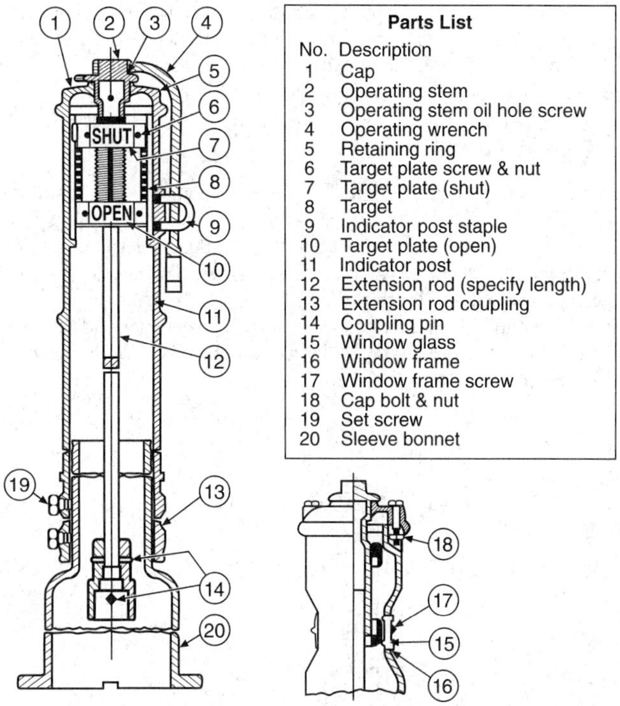

FIG. 6-3V.   A standard post indicator valve. Nearly all manufacturers of valves furnish standard valves, and their catalogs should be consulted for details of each. (Source: Henry Pratt Company)

water system management. Permanent valve records should be stored in a secure place, and copies suitably indexed should be available in all water supply vehicles that respond to emergency calls. Copies should also be available to each fire station.

Good practice dictates that valves are provided so that no single accident, break, or repair will necessitate shutting down a length of pipe greater than 500 ft (150 m) in high-density districts or greater than 800 ft (245 m) in other sections, and so that flow may be maintained through other arterial mains.

PRIVATE FIRE SYSTEMS

This part of the chapter covers the valving arrangement on a private fire protection water distribution system from the point where it connects with its supply.

Indicating Valves

The first valve in the fire line on private property should be a valve of the indicating type. It may be one of three kinds: (1) underground gate valve with indicator post attached, (2) underground butterfly indicating type with post, or (3) outside screw and yoke (OS&Y) gate valve in a pit.

The type of underground gate valve commonly used in domestic and industrial water lines (which requires a key wrench to operate) should be avoided in fire lines. The difficulties of finding such valves, lifting cover plates, and locating a key wrench that will fit are detriments to its use. If used, underground gate valves should open counter-clockwise and have the same size nut on all valve stems. The valve locations should be clearly marked on nearby buildings.

Indicator posts should have a metal plate showing what they control. Painting building names or numbers on them is advisable. The proper direction to turn for opening should be shown. Posts are often locked in the open position to prevent tampering. Lock shackles should, however, be brittle so that they can be removed readily if they are frozen and the keys will not work. Posts should have a handle, wheel, or wrench attached. A typical post indicator valve is shown in Figure 6-3V. Figure 6-3W shows a butterfly post-indicator valve.

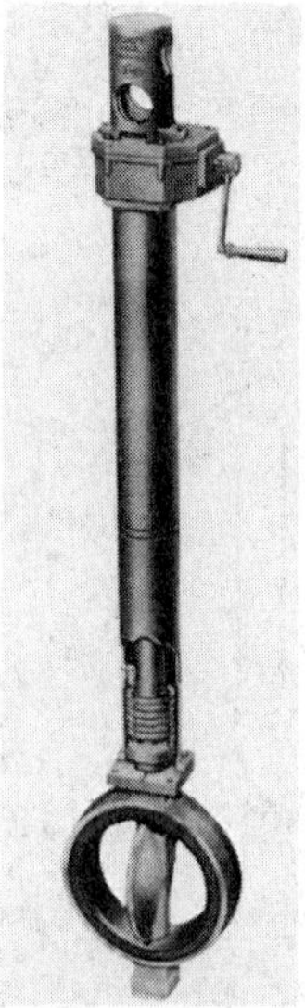

FIG. 6-3W.   A butterfly post-indicator valve. (Source: Henry Pratt Company)

OS&Y gate valves, if underground, should be in pits. Important junction points often warrant the use of pits to cover several valves. Indicating valves are used on all shut-offs. They are universally used on inside piping because they show the position of the gates at a glance. They are commonly supervised in the "open" position through the use of an electrical means (preferred) or with a chain and padlock. Where OS&Y gate valves are located in pits, a metal sleeve with the upper end closed may be slipped over the valve stem to keep off dirt. An OS&Y gate valve is shown in Figure 6-3X.

Check Valves

A check valve is installed next to the main control valve inside the property. The check valve must be installed in a pit unless it is located inside a building, even when the main control valve is a conventional buried valve with an indicating post. The check valve also needs to be accessible, so the size of the pit, the arrangement of its manhole and ladder, and the size of working space around the valve or valves in the pit should be large enough for convenient access. (See Figure 6-3Y.)

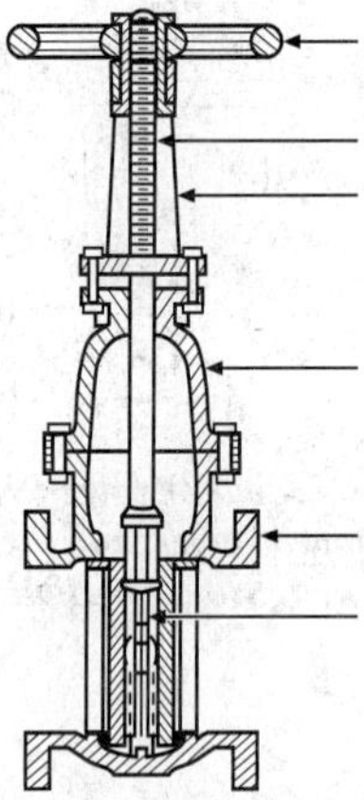

FIG. 6-3X.   A typical outside screw and yoke (OS&Y) valve in the closed position. The spindle on the valve stem indicates whether the valve is open or shut.

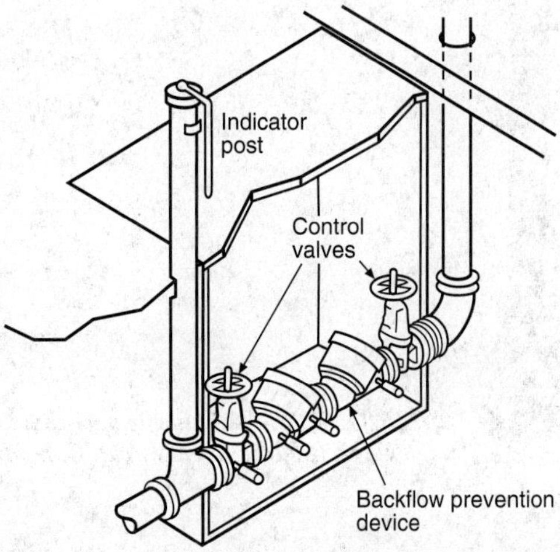

FIG. 6-3Y.   A valve pit is usually provided for a check valve or a backflow device (as shown) and for OS&Y, gate valves, meters, and other equipment on underground mains. Clearance should be 18 in. (0.46 m) around all valves and equipment in the pit. Pits are as deep as necessary to conform to the location of the pipe in which the valve is located.

A check valve allows water to flow from the public system to the private system but not from the private system into the public system. It also permits water at higher pressures in the private system than the normal pressures in the connecting public water mains in case of fire. Higher pressures can be provided in a private system through its own fire pumps, where these are provided, or by fire department pumpers when they pump through the fire department connections on automatic sprinkler systems and standpipe systems in buildings or on a private yard system.

Check valves are also installed at the base of gravity tanks and in the fire department pumper connections. The check valves at these points ensure flow of water in one direction only.

Air Gap Connections

Connections between potable and nonpotable water systems should be avoided. Where one water system must be separated positively from another, an air gap is provided so that backflow cannot occur. An example of an air gap is where water from one system is supplied through a pipe that discharges into a "break" tank above the level of the water in the tank. Such an arrangement does not allow the receiving water system to have the advantage of the pressures available in the supplying system. With a connection through a check valve, practically all of the pressure in the supplying system may be available.

Additional information can be found on cross connections in AWWA M14, *Recommended Practice for Backflow Prevention and Cross-Connection Control.*

Double Check Valves

For other situations, a double check-valve assembly may be considered. (Double check valves are required for certain situations by the regulations of some health departments.) Check valves in double installations are made with bronze working parts and rubber facings so that they will have tight seats. For each connection, two of these valves are installed in series, thus increasing the efficiency because the probability that both will leak at the same time is extremely remote. The two check valves are installed between control valves in a pit where they are readily accessible for examination. In addition, the valves are provided with pressure gages and test cocks so arranged that the tightness of each check valve may be verified in a few minutes. Figure 6-3Z shows a typical double check valve installation in a pit. A double check valve assembly is shown in Figure 6-3AA.

When the two check valves are bolted together, an 18-in. (0.46-m) space is normally provided between clappers, which is sufficient to prevent any ordinary material found in a pipeline from holding both clappers open at the same time. To further improve the efficiency of the equipment, a filling piece of pipe or spacer from 3 to 5 ft (0.9 to 1.5 m) long installed between the check valves is advised where space is available.

Certain manufacturers have developed a double check valve assembly where the 18-in. (0.46-m) space is not provided. In these devices, the design attempts to minimize the possibility of simultaneous failure.

Reduced-Pressure Backflow Preventers

Figure 6-3BB shows a reduced-pressure backflow preventer assembly [they may also be referred to as RPZ (reduced pressure zone devices)] that is intended to maintain pressure between two check valves at less than the supply pressure in the pipeline. (Figure 6-3CC shows a cutaway view of one type of RPZ assembly.) The assembly

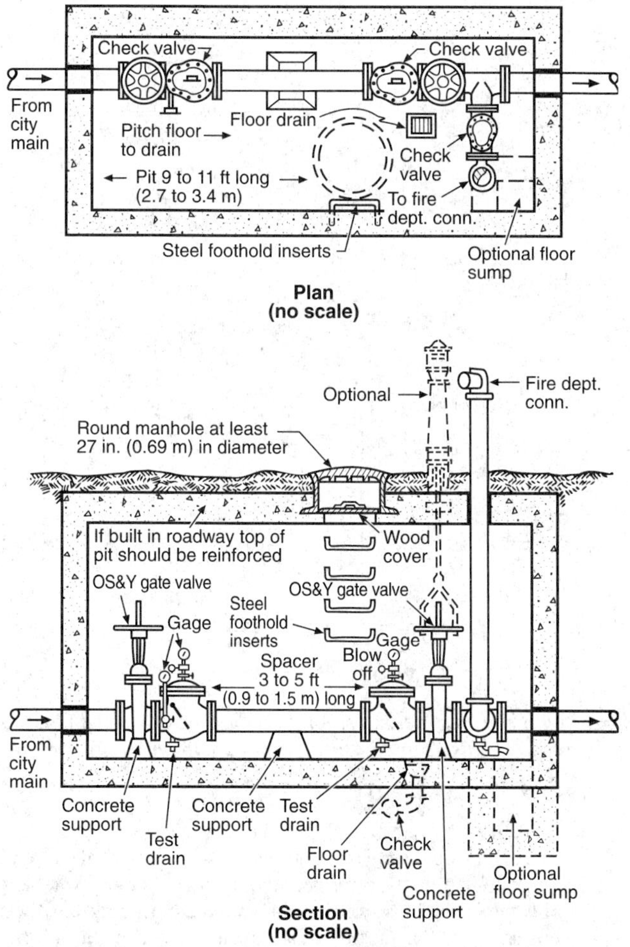

Plan
(no scale)

Section
(no scale)

FIG. 6-3Z.   *Plan and section views showing the arrangement of double check valves in a connection for fire protection service from a public water system. Indicating valves of the butterfly-type designed for fire service may be used in place of the gate valves illustrated.*

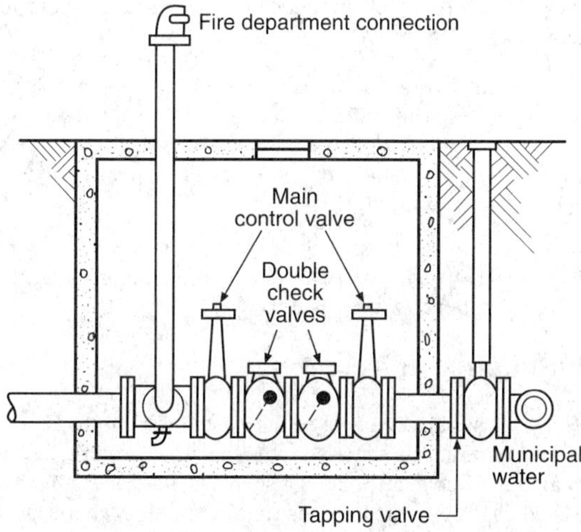

consists of two independently acting approved check valves interspaced by an automatically operated pressure differential relief valve. In case of leakage at either check valve, the relief valve operates to maintain the pressure between the check valves at less than supply pressure. There is some loss of head [generally 5 to 25 psi (34.5 to 172 kPa)] in such backflow preventers, but there may be situations where sacrificing some pressure is preferable to sacrificing all of it, as would be necessary with an air gap separation. In recent years, many municipalities have mandated the use of reduced-pressure backflow preventers on a retroactive basis. When these are installed, a careful evaluation of the impact of the loss of head should be undertaken to ensure the adequacy of the water supply. This is critical on automatic sprinkler systems.

Location of Valves in Private Fire Service Systems

Opinions vary on how many valves should be used in a system of underground mains. Making sure all sectional control valves are open is probably more critical than avoiding the problem of too few sectional valves. Nevertheless, the modern tendency is to make fairly liberal use of valves. A few well-established principles are shown in Figure 6-3DD. They include:

1. A city supply check valve (and meter, if required) located between indicating valves so it can be repaired without affecting city or plant systems.

FIG. 6-3BB.   *A reduced-pressure backflow preventer assembly for installation between two gate valves. (Source: Hersey Products, Inc.)*

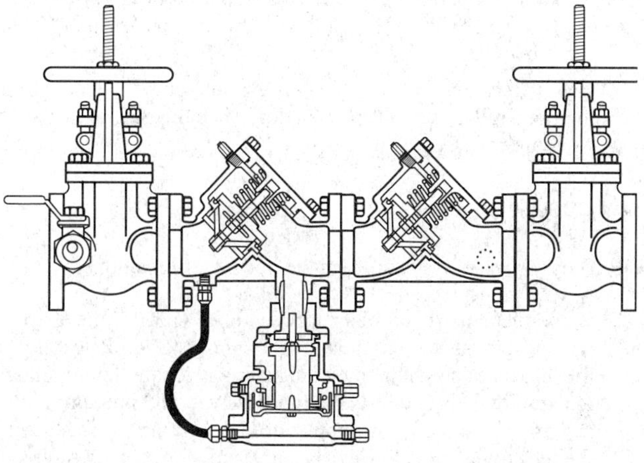

FIG. 6-3AA.   *A double check valve assembly.*

FIG. 6-3CC.   *A cutaway view of one type of reduced-pressure backflow preventer.*

FIG. 6-3DD. *Water piping for fire protection of an industrial site. Typical details shown are connections to public mains and supplies for a private fire pump, looped water mains, sectional control valves (lettered), and hydrants. (For SI units: 1 in. = 25.4 mm.)*

2. A pump check valve located between pump and indicating valves so that the latter can be used to shut off the connection to the system when making check valve and pump repairs.

3. Three sectional valves (two, G and H, to take care of present loop and one, J, for a short branch supplying a small detached building) in addition to the main water supply valve. (See detail of pit 1.) The branch will ultimately be part of a second loop. There should be a loop valve on each side of every valuable water supply to permit cutting off a part of the loop without cutting the water supply off altogether. Best practice requires that post indicators be attached to valves in pits, with the posts cemented into the concrete tops.

4. Sectional control valves (indicator posts C, E, and F) can cut the loop into four sections (in connection with valves H and G). In large or complicated underground systems, it is recommended that indicator posts controlling risers to sprinklers or standpipes be painted a different color from sectional control valves. Generally, no more than six hydrants or indicator posts should be located between section valves.

5. Gate valves must be provided on hydrant laterals to isolate the hydrant in the event it malfunctions, is damaged, or when repairs are necessary.

Records

Records for water systems should include the following:

1. A complete map and satisfactory sectional views of the distribution system made on a suitable scale, showing all the features of the distribution system, including size and location of pipelines; location of gate, check, and altitude valves, and pressure reducers; and hydrants. There should also be sectional maps showing the location of all gate valves by triangulation to fixed reference points. Copies of these maps should be distributed to the personnel who would be working on the system.

In the event of a break in the underground system or the abandonment of a flowing hydrant during a fire, it is essential that water be shut off promptly to protect the supply to other hydrants or sprinkler connections and to avoid flooding. Operating valves correctly and rapidly at the time of a fire, especially at night and under adverse weather conditions, requires accurate knowledge of the system.

2. Complete records should be kept for each fire hydrant in the system. They should show the date of installation; location, number, and size of outlets; size of the hydrant valve; size and location of the gate valve between the hydrant and street connection; the size of pipe to which it is connected, including the length of the branch connection; and inspection dates. The records should note if the hydrant is equipped with drains, if the drains have been plugged or omitted, and if there has been any problem with drainage or with groundwater entering the hydrant barrel. Records of all maintenance performed should also be kept.

Permanent hydrant records should be stored in a safe place with copies indexed for field use located in all vehicles responding to an emergency. Also, copies should be available to each fire station providing protection in the area served by the water distribution system.

Copies of the distribution map and hydrant and valve records should be used by the fire department in preplanning and developing a familiarity with the water system.

Maintenance and Testing

Check valves should be checked and tested for tightness on a periodic basis, particularly where double check valves are used to prevent backflow between potable water and yard systems. The test drains and gages provide a means for determining where a leak may exist. Even though found tight when tested periodically, check valves should be thoroughly cleaned once a year, so they will remain tight. Valves should be tested in accordance with the requirements of NFPA 25.

Where there are several sets of check valves in private fire service main connections from public mains, only one set of check valves should be overhauled and cleaned at a time, the others being left in service.

Gate valves should be operated periodically to ensure that they are in proper working condition and that gate boxes are not silted up or tarred over.

METERS FOR FIRE CONNECTIONS

Fire flow meters are devices capable of measuring small and large flows with a minimum loss of head for heavy demands. They are offered in two types: (1) detector check-valve meters that detect only small rates of flow and (2) full registration meters that measure the entire flow throughout the line in which they are installed. Meters of types other than the fire-flow type are unsatisfactory for fire protection water supplies.

Detector Check Valves

These devices consist of a check valve with a weighted clapper in the main passage and a meter in a bypass around the check. In normal operation, the smaller flows pass through the meter in the bypass and are accurately registered. Meters may be furnished up to 3 in. (76 mm) in size to serve specific needs. For heavy flows, the check valve opens, and a free unmetered waterway is provided. Beyond the point where the weighted check valve lifts, the bypass meter registers only a small part of the flow. In many situations, detector check valves give the waterworks assurance about the proper use of water. Figure 6-3EE shows a detector check valve.

CONNECTION BETWEEN PUBLIC AND PRIVATE WATER

Connections from public water systems for fire protection are used for the purpose of providing water supply to the following:

1. Automatic sprinkler systems.
2. Standpipes for fire department or occupant use.
3. Open sprinklers.
4. Yard systems with private hydrants.
5. Fire pumps.
6. Private storage reservoirs or tanks for fire protection.
7. Water spray systems.
8. Foam-water sprinkler systems.
9. Water mist systems.

A fundamental principle is that, within the property, piping systems from which water is used for fire protection purposes must be independent of systems supplying domestic and industrial service. Special conditions where this is not the case are rare and may be dealt with individually.

In public water systems, except in some unusual situations, economics favor a single water system for combined fire use and domestic service. In private properties, it is economically practicable to have separate systems; the advantages of a definite, dependable supply, free from interruptions of service inevitable in a domestic system, strongly favor the established practice. However, it should be remembered that a failure in any portion of the public system supplying the private and separate fire system may result in reduced flows or no flow into the private system.

Occasionally, a property will have two sources of water supply for fire protection: one from the public supply; the other from a private source, many times nonpotable. When this happens, adequate measures must be followed in accordance with local health department regulations to prevent the public water supply from becoming contaminated. Means of achieving this have been discussed previously in this chapter. Some communities do not allow any direct connection, and, if two supplies are required or desired, separate systems must be maintained with an air gap or backflow prevention device between the two systems.

Charges for Connections to Private Fire Protection Systems

The question of annual standby charges for connections from a public water system to fire protection systems may be controversial, largely because the nature of automatic sprinkler systems is not always understood by public officials. Whatever charges are made should be based on actual cost to the utility and not on the possible value of the installation to the customer. Annual charges for connections to fire protection systems are often established in water rate schedules for the sole purpose of obtaining additional revenue.

Although many waterworks, both public and private, make no standby charge for connections for fire protection, there may be no objections to charges that fairly reflect actual cost to the waterworks of the connection itself and the expense of servicing it, including necessary pits, valves, and meters. There could be costs to the waterworks for amortization of costs of its part of the connection and for its maintenance and inspection. However, it is usual to charge for the installation when it is made; and if repairs, maintenance, and inspection are charged at the time when work is done, there is little other cost involved.

Since the amount of water used for fire fighting overall is usually very small when compared to normal consumption, it is almost universal practice to make no charge to property owners for the water actually used in the extinguishment of fire or for authorized hydrant flow testing.

FIG. 6-3EE.   *A detector check valve assembly. (Source: Febco)*

Control of Connections During Fires

The safe location of control valves for connections for fire protection is important so that they may be shut off promptly after the fire has been extinguished, or to conserve water and pressure if pipes have been broken.

Sprinkler systems have outside indicating valves where yard space is available. This is the preferable arrangement. In many city properties having no yard space, valves are installed in a stair tower or other sufficiently cut off area. Sometimes the control valves are located outside the area protected as, for example, on the other side of a fire wall. In a relatively few instances, dependence must be placed on the accessibility of the valve on the connection to the street main.

Sprinkler control valves on private property are adequate to serve their intended use and allow prompt and safe control of fire protection systems.

Control of Water Waste

There are a number of conditions under which flow of water from private fire protection equipment other than for fire or testing purposes may occur. These are:

1. Leakage.
2. Wrongful use, such as wetting down roofs, watering lawns, or waste through ignorance.
3. Emergency use.
4. Theft.

Fire protection equipment should be inspected several times a year to ensure that the systems will be kept free from material leakage or unauthorized connections. Maintenance rules enforced by inspection authorities tend to discourage wrongful uses of these systems. These are further enforced in many places by fire department inspections, which are made to ensure that valves controlling extinguishing systems and other fire supplies are kept open and that hose and other equipment are not used for any purpose other than that intended. Occasional inspections by waterworks inspectors are desirable.

In cases where inspections alone are not a sufficient safeguard against water waste, a plan involving the securing of valves and outlets would be effective. In such cases, all sprinkler drains, hydrants, and hose outlets are sealed by the water utility which may require:

1. A report from the owner when a seal is broken.
2. A notice from the user before testing through outlets on hydrants or sprinkler drains.
3. Payment of a nominal charge for resealing.

As a further safeguard, indemnity to the water utility may be secured by requiring the user to file a bond or deposit to be forfeited in the event of willful or persistent violation of the rules, or abuse of the fire service furnished by the water utility.

The simplest cases of control of water waste to deal with are fire connections in single buildings such as those serving:

1. One or more standpipes for fire department or occupant use.
2. Automatic extinguishing systems only.
3. Combined automatic sprinkler systems and standpipes.

In connections of these types, any regulation on the part of a water utility (for the ostensible purpose of preventing water waste) that involves an excessive annual charge or a costly meter installation may impose an expense to the property owner so large as to discourage the installation of extinguishing systems or standpipes. The

additional expense may be the reason why a system is not installed in a given occupancy.

Ordinarily inspections, or at least a valve securing procedure, will give adequate protection to the utility. In extreme cases, special action may be taken by a utility against any individual violator. In adopting any procedure to apply to all users of private fire protection, the relative unimportance of a very small amount of water improperly used, as against the greater public good in the extensive installation of automatic sprinklers and other private fire equipment, should be considered.

Fire protection connections that serve several buildings with extinguishing systems and standpipes, or, in addition, a yard system with private hydrants, fire pump, and tank supplies, may introduce a slightly more difficult problem of water waste control. In general, inspection or sealing procedures will suffice, and both may be appropriate in some cases. Where meters are employed on fire connections, the extent to which they reduce pressures and obstruct or reduce flows must be determined.

BIBLIOGRAPHY

NFPA Codes, Standards, and Recommended Practices

Reference to the following NFPA codes, standards, and recommended practices will provide further information on water distribution systems discussed in this chapter. (See the latest version of *The NFPA Catalog* for availability of current editions of the following documents.)

NFPA 13, *Standard for the Installation of Sprinkler Systems*
NFPA 20, *Standard for the Installation of Centrifugal Fire Pumps*
NFPA 24, *Standard for the Installation of Private Fire Service Mains and Their Appurtenances*
NFPA 25, *Standard for the Inspection, Testing, and Maintenance of Water-Based Fire Protection Systems*
NFPA 291, *Recommended Practice for Fire Flow Testing and Marking of Hydrants*

Other Codes, Standards, and Manuals

American Society for Testing and Materials

ASTM, D3139, *Standard Specification for Joints for Plastic Pressure Pipes Using Flexible Elastomatic Seals, American Society for Testing and Materials*, Philadelphia, PA.

American Water Works Association

The following are published by American Water Works Association, 6666 W. Quincy Ave., Denver, CO 80235:
AWWA C200-86, *Standard for Steel Water Pipe 6 in. and Larger.*
AWWA C203-86, *Standard for Coal–Tar Protective Coatings and Linings for Steel Water Pipelines—Enamel and Tape—Hot Applied.*
AWWA C205-89, *Standard for Cement-Mortar Protective Lining and Coating for Steel Water Pipe—4 in. and Larger—Shop Applied.*
AWWA C206-91, *Standard for Field Welding of Steel Water Pipe.*
AWWA C207-94, *Standard for Steel Pipe Flanges for Waterworks Service —Sizes 4 in. Through 144 in. (100 mm Through 3,600 mm).*
AWWA C300-89, *Standard for Reinforced Concrete Pressure Pipe—Steel Cylinder Type, for Water and Other Liquids.*
AWWA C301-84, *Standard for Prestressed Concrete Pressure Pipe—Steel Cylinder Type, for Water and Other Liquids.*
AWWA C302-87, *Standard for Reinforced Concrete Pressure Pipe—Noncylinder Type, for Water and Other Liquids.*
AWWA C303-87, *Standard for Reinforced Concrete Pressure Pipe—Steel Cylinder Type, Pretensioned, for Water and Other Liquids.*
AWWA C400-80/(R86), *Standard for Asbestos Cement Distribution Pipe 4 in. Through 16 in. NPS for Water and Other Liquids.*
AWWA C401-83/(R86), *Standard Practice for the Selection of Asbestos Cement Distribution Pipe; 4 in. Through 16 in. for Water and Other Liquids.*

AWWA C500-93, *Standard for Gate Valves 3 in. through 48 in. NPS for Water and Sewage Systems.*

AWWA C502-94, *Standard for Dry Barrel Fire Hydrants.*

AWWA C503-88, *Standard for Wet Barrel Fire Hydrants.*

AWWA C510-92, *Standard for Double Check Valve Backflow-Prevention Assembly.*

AWWA C511-92, *Standard for Reduced-Pressure Principle Backflow-Prevention Assembly.*

AWWA C600-93, *Standard for Installation of Ductile-Iron Water Mains and Appurtenances.*

AWWA C603-90, *Standard for Installation of Asbestos Cement Pressure Pipe.*

AWWA C703-86, *Standard for Cold Water Meters—Fire Service Type.*

AWWA C900-89, *Standard for Polyvinyl Chloride (PVC) Pressure Pipe, 4 in. Through 12 in. for Water Distribution.*

AWWA C104/A21.4-90, *American National Standard for Cement-Mortar Lining for Ductile-Iron Pipe and Fittings for Water.*

AWWA C110/A21.10-93, *American National Standard for Ductile-Iron and Gray-Iron Fittings, 3 in. Through 48 in., for Water and Other Liquids.*

AWWA C111/A21.11-90, *American National Standard for Rubber Gasket Joints for Ductile-Iron and Gray-Iron Pressure Pipe and Fittings.*

AWWA C150/A21.50-91, *American National Standard for the Thickness Design of Ductile-Iron Pipe.*

AWWA C151/A21.51-91, *American National Standard for Ductile-Iron Pipe, Centrifugally Cast in Metal Molds or Sand-Lined Molds, for Water or Other Liquids.*

AWWA M6 (30006)-86, *Water Meters—Selection, Installation, Testing, and Maintenance.*

AWWA M9 (30009)-95, *Concrete Pressure Pipe.*

AWWA M11 (30011)-89, *Steel Water Pipe—Design and Installation.*

AWWA M14 (30014)-90, *Recommended Practice for Backflow Prevention and Cross-Connection Control.*

AWWA M17 (30017)-88, *Installation, Field Testing, and Maintenance of Fire Hydrants.*

AWWA M23 (30023)-80, *PVC Pipe.*

Underwriters Laboratories of Canada

The following are published by Underwriters Laboratories of Canada, 7 Crouse Road, Scarborough, Ontario:

ULC C-312, *Check Valves for Fire Protection Service.*

ULC C-789, *Indicator Posts for Fire Protection.*

ULC S-520, *Fire Hydrants.*

Underwriters Laboratories Inc.

The following are published by Underwriters Laboratories Inc., 333 Pfingsten Road, Northbrook, IL 60062:

UL 107, *Asbestos Cement Pipe and Couplings*

UL 246, *Hydrants for Fire Protection Service*

UL 262, *Gate Valves for Fire Protection Service*

UL 312, *Check Valves for Fire Protection Service*

UL 385, *Play Pipes for Water Supply Testing in Fire Protection Service*

UL 789, *Indicator Posts for Fire Protection Service*

UL 194, *Gasketed Joints for Ductile-Iron and Gray-Iron Pressure Pipe and Fittings for Fire Protection Service*

UL 753, *Alarm Accessories for Automatic Water Supply Control Valves for Fire Protection Service*

Additional Reading

Bussen, R., "Anschlub von Feuerloschanlagen an Trinkwassernetze. [Connection of Fire Extinguishing Systems to Drinking Water Supply Systems.]," *VFDB*, Vol. 44, No. 1, Feb. 1995, pp. 2–4.

Climent, J. V. J., "Risks in the Distribution of Potable Water. [Riesgos en la distribucion de agua potable.]," FREMAP, Spain, *Mapfre Seguridad* Vol. 14, No. 53, 1994, pp. 83–85.

Cooper, K., "Water Supply Problems at Uppark House Fire," *Fire*, Vol. 87, No. 1074, Dec. 1994, pp. 13–14, 16, 18.

"Environmental Assessment: Installation and Operation of the Plantwide Fire Protection Systems and Related Domestic Water Supply Systems," Department of Energy, Aiken, SC, DOE/EA-0476, Dec. 1991.

Fredericks, A. A., "Tactical Use of Fire Hydrants," *Fire Engineering*, Vol. 149, No. 6, June 1996, pp. 61–68, 70, 72.

Gamble, C., "Pre-Planning Rural Water Supplies," *Fire Chief*, Vol. 39, No. 3, Mar. 1995, pp. 68, 70, 72.

Haley, J., "Dry Hydrants," *Fire Fighting in Canada*, Vol. 37, No. 9, Nov. 1993, pp. 6, 55.

Hart, F., "Backflow Preventers," VHS Video Tape, Worcester Polytechnic Inst., MA, Federal Fire Forum, Providing Fire Safety for Equipment Use in Buildings, [This Forum Will Also be a Part of the BFRL Building Technology Symposia Series.], June 6, 1994, Gaithersburg, MD, 1994.

Hiraga, S., "Investigation of Antifreezing Method of a Hydrant," [ABSTRACT ONLY], Report of Fire Research Institute of Japan, No. 79, 1995, p. 85.

Malone, G. and Hession, M., "Failure of Water Supplies Hampered Firefighting at Cork Chemical Plant," *Fire*, Vol. 86, No. 1062, Dec. 1993, pp. 23–24.

Nayyar, M.L., *Piping Handbook*, 6th ed., McGraw-Hill, Inc., New York, 1992.

Vance, M., *Drinking Water Standards: A Bibliography*, Vance Bibliographies, Monticello, IL, 1989.

vanZyl, J. E., "Water Supplies in the RSA Not Good Enough!," *Fire Protection*, Vol. 20, No. 3, Sept. 1993, pp. 12–15.

FIRE PUMPS

Revised by John D. Jensen

This chapter covers the operating principles of stationary pumps used for fire protection, the methods of driving and controlling them, and the testing and maintenance procedures that should be followed to keep them in top operating condition.

Fire pumps are used to provide or enhance the water supply pressure available from public mains, gravity tanks, reservoirs, or other sources. The first modern fire pumps were the wheel-and-crank reciprocating type, belt-driven from mill machinery. If plant operations were stopped during a fire, the pump could not operate. At best, these pumps were inadequate.

Better water supplies became necessary as automatic sprinkler systems became more common, and mill pumps were replaced by rotary, or displacement, pumps driven by friction drive from horizontal water wheels supplying power to the plant. As steam supplanted water power, the reciprocating steam pump was adopted for fire protection. For many years the Underwriter duplex, double-acting, direct steam–driven unit was universally accepted as the "standard" fire pump.

Today, the centrifugal fire pump is standard. (See Figures 6-4A and 6-4B.) Its compactness, reliability, easy maintenance and hydraulic characteristics, and the variety of available drivers—electric motors, steam turbines, and diesel engines—have made the Underwriter pump obsolete, although not entirely extinct.

The NFPA standard on fire pumps is NFPA 20, *Standard for the Installation of Centrifugal Fire Pumps* (hereinafter referred to NFPA 20). Other NFPA documents that contain information on fire pumps include NFPA 11, *Standard for Low-Expansion Foam*; NFPA 11A, *Standard for Medium- and High-Expansion Foam Systems*; NFPA 13, *Standard for the Installation of Sprinkler Systems*; NFPA 13D, *Standard for the Installation of Sprinkler Systems in One- and Two-Family Dwellings and Manufactured Homes*; NFPA 13E, *Guide for Fire Department Operations in Properties Protected by Sprinkler and Standpipe Systems*; NFPA 13R, *Standard for the Installation of Sprinkler Systems in Residential Occupancies up to and Including Four Stories in Height*; NFPA 14, *Standard for the Installation of Standpipe and Hose Systems*; NFPA 15, *Standard for Water Spray Fixed Systems for Fire Protection*; NFPA 16, *Standard for the Installation of Deluge Foam-Water Sprinkler and Foam-Water Spray Systems*; NFPA 16A, *Standard for the Installation of*

Closed-Head Foam-Water Sprinkler Systems; NFPA 22, *Standard for Water Tanks for Private Fire Protection*; NFPA 24, *Standard for the Installation of Private Fire Service Mains and Their Appurtenances*; and NFPA 25, *Standard for the Inspection, Testing, and Maintenance of Water-Based Fire Protection Systems* (hereinafter referred to as NFPA 25).

THE CENTRIFUGAL FIRE PUMP

An outstanding feature of a horizontal or vertical centrifugal pump is the relation of discharge to pressure at constant speed, that, as the pressure head is increased, the discharge is reduced. With displacement pumps, however, the rated capacity can be maintained against any head if the power is adequate to operate the pump at rated speed and if the pump, fittings, and piping can withstand the pressure.

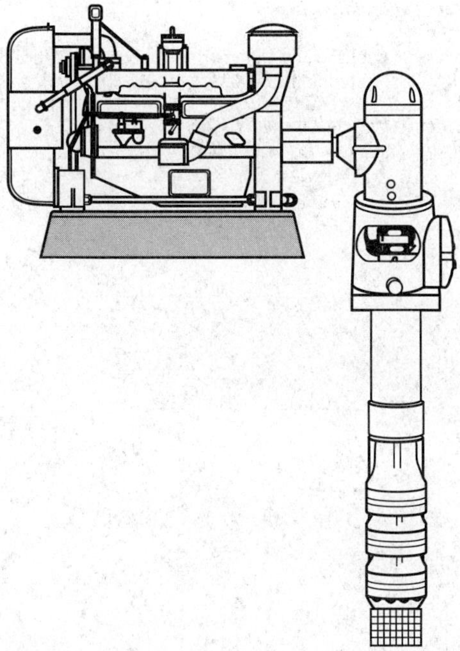

FIG. 6-4A. *A vertical turbine fire pump with a right-angle gear and engine drive. (Source: Peerless Pump)*

John D. Jensen, P. E., is president of John D. Jensen Fire Protection Consultants in Idaho Falls, ID. He is a member of NFPA's Technical Committee on Fire Pumps.

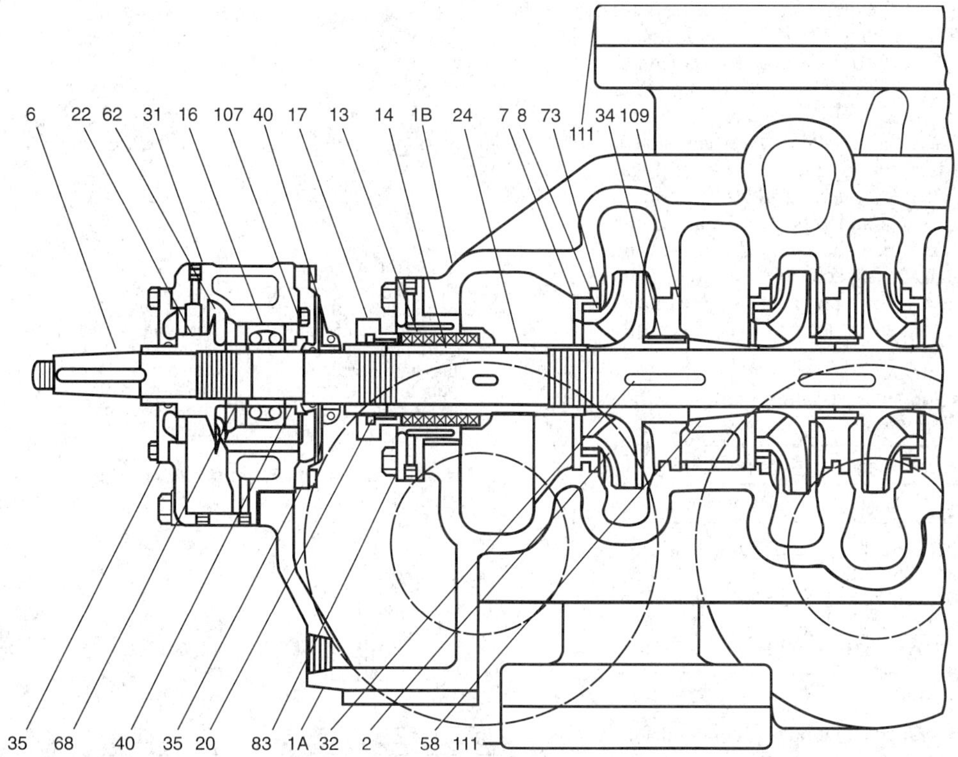

1A	Casing, Lower Half	17	Gland	37 Cover, Bearing, Outboard	83 Stuffing Box
1B	Casing, Upper Half	18	Bearing, Outboard	40 Deflector	107 Shield, Oil Retaining
2	Impeller	20	Nut, Shaft Sleeve	45 Cover, Oil, Bearing Cap	109 Diaphragm, Interstage
5	Diffuser	22	Locknut, Bearing	56 Disc or Drum, Balancing	111 Crossover, Interstage
6	Shaft, Pump	24	Nut, Impeller	58 Sleeve, Interstage	113 Bushing, Interstage Diaphragm
7	Ring, Casing	31	Housing, Bearing, Inboard	62 Thrower (Oil or Grease)	115 Ring, Balancing
8	Ring, Impeller	32	Key, Impeller	63 Bushing, Stuffing Box	117 Bushing, Pressure Reducing
13	Packing	33	Housing, Bearing, Outboard	68 Collar, Shaft	119 Coupling, Oil Pump
14	Sleeve, Shaft	34	Sleeve, Impeller Hub	72 Collar, Thrust	121 Pump, Oil
16	Bearing, Inboard	35	Cover, Bearing, Inboard	73 Gasket	

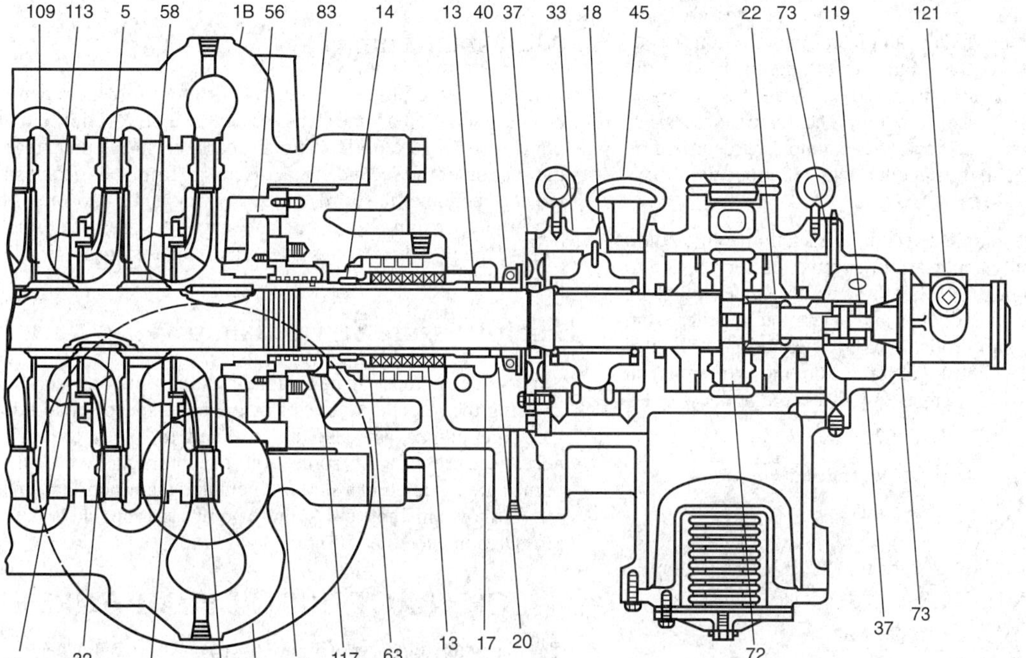

FIG. 6-4B. *Cross-section of a typical multistage centrifugal pump. The numbers shown on this drawing do not necessarily represent standard part numbers in use by any manufacturer. (Source: Hydraulics Institute)*

Listed horizontal and vertical fire pumps are available with rated capacities from 25 to 5000 gpm (95 to 18 925 L/min). Pressure ratings range from 40 to 394 psi (276 to 2758 kPa) for horizontal pumps and 26 to 510 psi (517 to 3448 kPa) for vertical turbine pumps. Listed centrifugal fire pump designs include horizontal end-suction, vertical in-line, split case (horizontal and vertical shaft), and vertical turbine types. Vertical turbine pumps are centrifugal pumps with one or more impellers discharging into one or more bowls and a vertical educator or column pipe that connects the bowls to the discharge head on which the pump driver may be mounted. It is anticipated that larger-capacity fire pumps will be listed in the future. (See Table 6-4A which provides pressure ranges and flow capacities for currently available pumps.)

TABLE 6-4A. *Pump Types and Their Pressure and Capacity Ranges*

Pump Type	Pressure Range		Capacity Range	
	psi	kPa	gpm	L/s
Horizontal end-suction	40-186	276-1282	25-750	1.6-31.5
Vertical In-line	40-186	276-1282	25-750	1.6-31.5
Split case horizontal and vertical	40-294	276-2027	150-5000	9.5-315.4
Vertical turbine	25-510	179-3516	250-5000	15.8-315.4

The "size" of a horizontal centrifugal pump is generally the diameter of the discharge outlet. However, it is sometimes indicated by both suction and discharge pipe flange diameters. The size of a vertical turbine pump is the diameter of the pump bowl. (See Figure 6-4A.)

PRINCIPLES OF OPERATION

The two major components of a centrifugal pump are a disc called the impeller and the casing in which it rotates. (See Figure 6-4C.) It operates by converting kinetic energy to velocity and pressure energy. Power from the driver, which is an electric motor, a diesel engine, or a steam turbine, is transmitted to the pump through the shaft, rotating the impeller at high speed. The way that energy is converted varies with the type of pump. The major classes are known as radial flow and mixed flow. These pumps are identified by the direction of flow through the impeller, with reference to the axis of rotation. (See Figure 6-4D.)

The horizontal shaft, single-stage, double suction volute pump is the type most commonly applied to fire protection service and to commercial use. (See Figure 6-4E.) In these pumps, waterflow from the suction inlet in the casing divides and enters the impeller from each side through an opening called the "eye." Rotation of the impeller drives the water by centrifugal force from the eye to the rim and through the casing volute to the pump discharge outlet. The ki-

Radial Flow
Pressure is developed principally by the action of centrifugal force. Liquid normally enters the impeller at the hub and flows radially to the periphery.

Mixed Flow
Pressure is developed partly by centrifugal force and partly by the lift of the vanes on the liquid. The flow enters axially and discharges in an axial and radial direction.

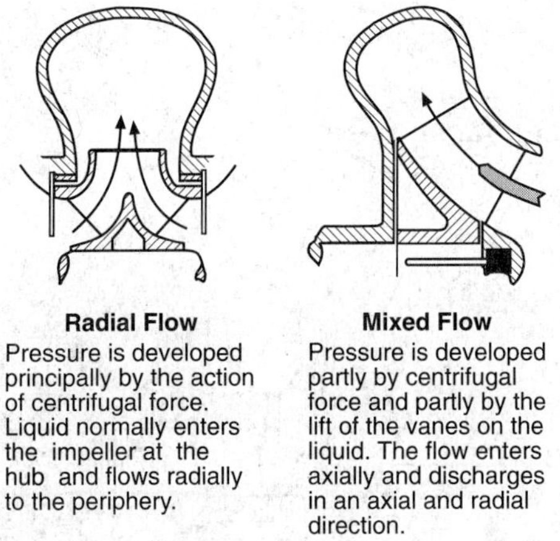

FIG. 6-4D. The two major classes of fire pumps.

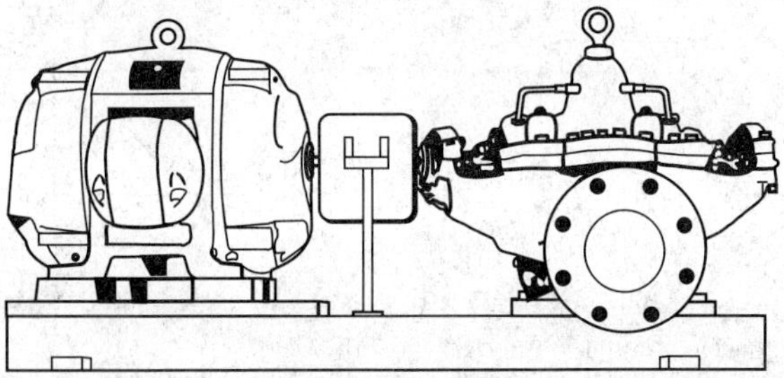

FIG. 6-4E. A horizontal shaft, single-stage centrifugal pump, with cutaway view of the pump. (Source: Peerless Pump)

netic energy acquired by the water in its passage through the impeller is converted to pressure energy by gradual reduction of velocity in the volute.

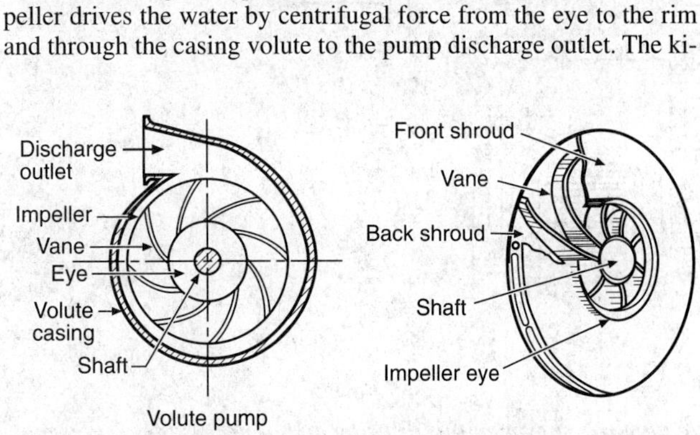

Multistage Pumps

In order to give high pressure, two or more impellers and casings can be assembled on one shaft as a single unit, forming a multistage pump. (See Figure 6-4B.) The discharge from the first stage enters the suction of the second stage, discharge from the second stage enters the suction of the third, and so on. The pump capacity is the rating in gal per min (L/min) of one stage; the pressure rating is the sum of the pressure ratings of the individual stages, minus a small head loss.

High-Pressure Service Pumps

Single-stage pumps can be designed for high-pressure service by increasing the impeller diameter or the rated speed. In addition, single-stage pumps can also be installed in series to achieve the necessary pressure. In some high-rise buildings, vertical turbine pumps are installed in a subbasement and take suction from a sump pit. As shown in Table 6-4A, these pumps are capable of developing pressures in excess of 500 psi (3448 kPa).

CHARACTERISTIC PUMP CURVES

The characteristic curves (Figure 6-4F) of a horizontal, centrifugal, or a vertical turbine-type pump are:

1. Total head *vs.* discharge (ft of head or psi of pressure *vs.* gal per min).
2. Brake horsepower *vs.* discharge.

FIG. 6-4C. A volute casing and impeller.

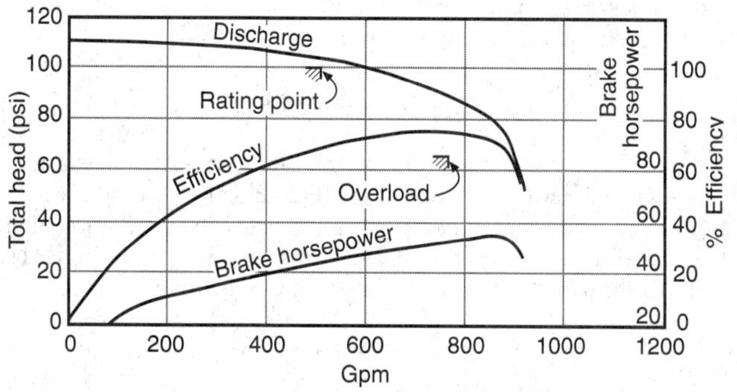

FIG. 6-4F. Typical fire pump characteristic curves. Illustrated curves are for a 500-gpm, 100-psi, 2000-rpm, diesel engine–drive pump with a 14-in. impeller and a maximum suction lift of 6.4 ft. Note that the shutoff point is 110 psi, the maximum brake horsepower is 55, and the maximum efficiency is 75 percent at overload capacity. Head is 90 psi over the 65 percent minimum required. (For SI units: 1 in. = 25.4 mm; 1 ft = 0.305 m; 1 psi = 6.89 kPa; 1 gpm = 0.38 L/min)

3. Efficiency vs. discharge $\left(\dfrac{\text{Water hp}}{\text{Input hp}} \text{ vs. gpm}\right)$

(See Figure 6-4F for SI units.)

These curves assume that the pump is operated at a constant speed equal to its rated rpm (revolutions per minute). In actual service, however, the speed of the driver may vary with changes in the flow capacity.

The flow and pressure ratings of commercial pumps are usually established on the basis of maximum efficiency and desired speed. Impellers can be designed for flat, medium, or steep head-discharge characteristics, as required for various uses. Figure 6-4G illustrates how the head-discharge curve is affected by the diameter of the eye, the width of the impeller, the number of vanes, and the shape or angle of the vanes.

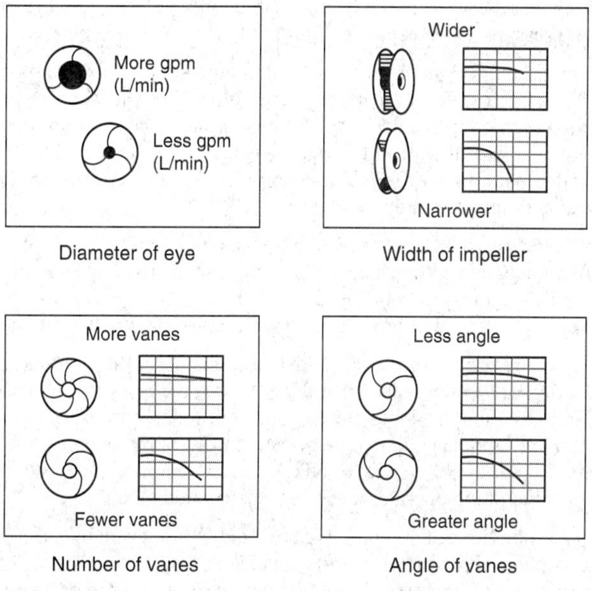

FIG. 6-4G. Effect of impeller design on head-discharge curves for fire pumps.

TOTAL HEAD

The total head of a pump is the energy imparted to the liquid as it passes through the pump. It may be expressed in various units of pressure, but for fire protection it is generally given in pounds per square inch (psi) or kilopascals (kPa), or in feet (ft) or meters (m) of liquid measured vertically. The total head is calculated by subtracting the energy in the incoming liquid from the energy in the discharging liquid. The total head (H) of a pump is calculated by the formula:

$$H = h_d + h_{vd} - h_s - h_{Vs}$$

where

H = total head, ft (m);
h_d = discharge head, ft (m);
h_{vd} = $\dfrac{Vd^2}{2g}$ = discharge velocity head, ft (m);
h_s = suction head, ft (m);
h_{Vs} = $\dfrac{Vv^2}{2g}$ = suction velocity head, ft (m);
V = average velocity, ft/sec (m/s); and
g = acceleration due to gravity, 32.2 ft/sec² (9.81 m/s²).

For a horizontal split case pump, the individual heads (h_v) are measured at the pump discharge nozzle flange and at the suction flange. (See Figure 6-4H.) The heads are read from pressure gages attached to the pump flanges. The velocity head must be calculated for the volume of liquid passing through the flanges. One expression of this relationship is:

$$h_V = \frac{V^2}{2g}$$

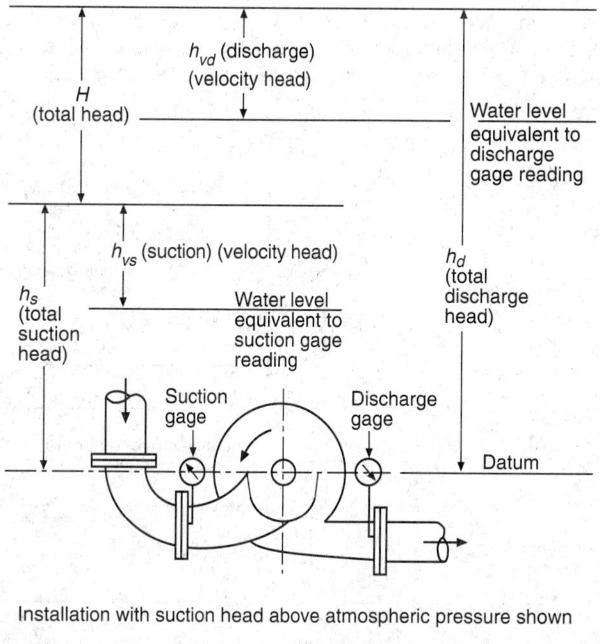

Installation with suction head above atmospheric pressure shown

FIG. 6-4H. Typical head of horizontal shaft centrifugal fire pumps.

If the flanges have the same diameters, there will be no difference between the incoming and outgoing velocity, and the calculation can be omitted.

For a vertical turbine pump, the discharge head is theoretically read at the discharge flange of the pump. Since this flange usually is inaccessible for gage readings, a gage is used at the discharge fitting at the top of the pump column pipe. (See Figure 6-4I.) The discharge pressure at the pump discharge flange therefore equals the pressure at the gage above, plus the pressure effect of the vertical distance between the two points, plus the friction loss between the two points. In most cases, the friction loss is so small that it may be disregarded.

The suction head is the vertical distance from water level to the pump discharge flange. The velocity head of the incoming liquid is assumed to be zero. Hence, the formula would now take the following form:

$$H = h_d + h_{Vd} - h_s$$
$$= \left(h_{gd} + L\right) + h_{Vd} - h_s$$

where

h_{gd} = discharge gage reading, ft (m);
L = gage pump flange, ft (m); and
h_s = water level when pumping to pump flange, ft (m).

However, $L - h_s = h =$ the vertical distance between the discharge gage and water level.

Therefore the formula becomes:

$$H = h_{gd} + h_{Vd} + h$$

Hydraulic and power losses within the pump due to turbulence, disc friction, and shock are represented by the efficiency rating.

Total head at rated capacity is used to establish the rated head of a pump. Actually, the rated head is the amount of energy given to the water.

The total head of a vertical turbine-type pump also may be defined as the vertical water-to-water dimension of the system in which the pump operates. However, there is a difference in the method of measuring total head. As shown in Figure 6-4I, it is the sum of the vertical distance between water level in the well or pit,

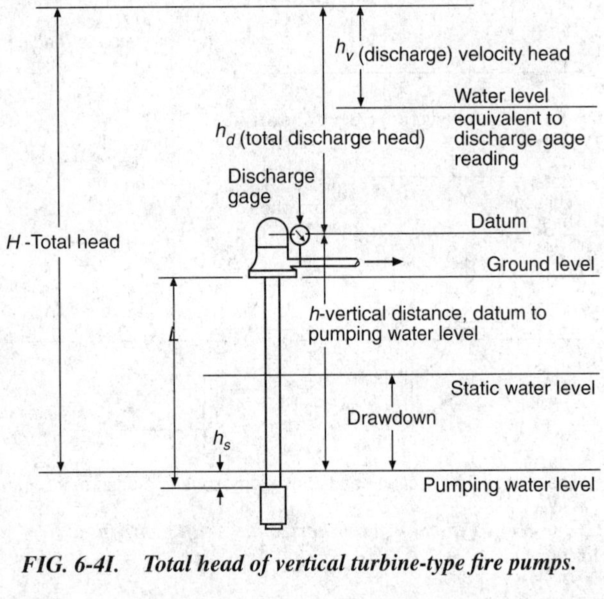

FIG. 6-4I.   Total head of vertical turbine-type fire pumps.

the discharge head indicated by the gage on the pump outlet, and the velocity head at the gage connection.

SPECIFIC SPEED (N_s)

Specific speed is a number relating the head, capacity, and speed of a centrifugal pump for design purposes. Actually, specific speed is the revolutions per minute (rpm) of a geometrically similar impeller that will discharge 1 gpm (3.8 L/min) at 1 ft (0.3 m) total head. The formula for calculating specific speed of a centrifugal pump is:

$$N_s = \frac{\text{rpm} \times \text{gpm}^{1/2}}{H^{3/4}}$$

where

N_3 = specific speed number, and

H = head in ft (m).

When values of head, speed, and capacity in the formula correspond to pump performance at optimum efficiency, the specific speed may be used as one measure of pump performance. Impellers designed for high heads usually have low specific speeds, and impellers designed for low heads have high specific speeds.

A pump of low specific speed will operate satisfactorily with greater suction lift than a pump of the same head and capacity with a higher specific speed. Specific speed is a useful guide for determining maximum suction lift or minimum suction head.

Fire pumps normally are not allowed to take suction under lift. However, some existing installations may be encountered when a centrifugal pump is subject to lift conditions exceeding 15 ft (4.5 m), and it may be necessary to provide a larger pump at less speed. With low lift or positive head on the suction, a smaller pump operating at greater speed may be used. Abnormally high suction lifts may seriously reduce pump capacity and efficiency or cause excessive vibration and cavitation.

NET POSITIVE SUCTION HEAD

Net positive suction head (NPSH) is the pressure head that causes liquid to flow through the suction pipe and fittings into the eye of a pump impeller. The pump itself has no ability to "lift," and the suction pressure depends on the nature of the supply.

Although not permitted, a pump being supplied from a pond, stream, open well, or uncovered reservoir, where the water level is below the pump, would have a suction head equal to atmospheric pressure minus the lift. If the water level is above the pump, as from a water main, penstock, or aboveground tank, the suction head is atmospheric pressure plus static pressure.

Pressure readings at the inlet flange of a pump operating under lift are negative with respect to the gage, but positive when referred to absolute pressure—hence the expression, "net positive suction head." Absolute pressure is gage pressure plus barometric pressure.

There are two kinds of NPSH to consider. Pump NPSH is a function of the pump design and varies with the capacity and speed of any one pump and with the designs of different pumps. Curves of NPSH *vs.* gpm usually can be obtained from pump manufacturers. (See Figure 6-4J.) Available NPSH is a function of the system in which the pump operates and can be calculated readily.

When the water source is above the pump, available NPSH = atmospheric pressure (ft or m) + static head on suction (ft or m) − friction and fitting losses in suction piping (ft or m) − vapor pressure of liquid (ft or m). Note that the vapor pressure of water at 90°F (32°C) is 1.6 ft (0.48 m).

For any pump installation, the available system NPSH must be equal to or greater than the pump NPSH at the desired operating conditions.

CAVITATION

Cavitation is a complex phenomenon that can take place in pumps or other hydraulic equipment. As water flows through the suction pipe of a centrifugal pump and enters the eye of the impeller, the velocity increases and pressure decreases. If the pressure falls below the vapor pressure corresponding to the temperature of the water, pockets of vapor will form. When the vapor pockets in the flowing water reach a region of higher pressure, they collapse with a hammer effect, causing noise and vibration. Tests have shown that the extremely high instantaneous pressures that may be developed in this manner may pit various parts of the pump casing and impeller. Conditions may be mild or severe, and mild cavitation may occur without much noise. Severe cavitation can cause reduced efficiency and, ultimately, failure of the pump if it is not corrected.

AFFINITY LAWS

The mathematical relationships among head, capacity, brake horsepower, and impeller diameter are called affinity laws. Law 1 assumes constant impeller diameter with change of speed. Law 2 assumes constant speed with change in diameter of the impeller. These laws are expressed by proportion, as follows:

Law 1:

$$\frac{Q_1}{Q_2} = \frac{N_1}{N_2} \quad \frac{H_1}{H_2} = \frac{N_1^2}{N_2^2} \quad \frac{bhp_1}{bhp_2} = \frac{N_1^3}{N_2^3}$$

Law 2:

$$\frac{Q_1}{Q_2} = \frac{D_1}{D_2} \quad \frac{H_1}{H_2} = \frac{D_1^2}{D_2^2} \quad \frac{bhp_1}{bhp_2} = \frac{D_1^3}{D_2^3}$$

The nomenclature for the relationship is

Q = capacity;

H = head;

N = speed;

D = impeller diameter; and

bhp = brake horsepower.

Thus

Q_1 = gpm (L/min) at N_1 or D_1 Q_2 = gpm (L/min) at N_2 or D_2
H_1 = head in ft (m) at N_1 or D_1 H_2 = head in ft (m) at N_2 or D_2
bhp_1 = brake horsepower (kW) bhp_2 = brake horsepower (kW)
at N_1 or D_1 at N_2 or D_2

Law 1 applies to common types of pumps, including horizontal centrifugal pumps and vertical turbine pumps. Law 2 applies to centrifugal pumps with reasonably close agreement between calculated and tested performance. Generally, pumps with low specific speeds show closer agreement than pumps with high specific speeds.

The affinity laws should be applied when proposed changes in a fire pump installation would increase the speed or significantly raise the pressure of the suction supply. Greater speed would increase the power demand, and high discharge pressure might be undesirable. In some instances, it is possible to trim the impeller. This should not be done without the approval of the pump manufacturer.

FIRE PUMP APPROVAL AND LISTING

NFPA standards for design and installation of various fire protection systems require the use of approved equipment, or listed equipment, or both, including fire pumps, for installations requiring them. Under the testing and listing approval system, the manufacturer is responsible for providing a listed or approved pump that will perform satisfactorily when installed in conformance with NFPA 20. Contractors or others are responsible for installing the driver-pump combination in accordance with the provisions of NFPA 20, while it is the customer's obligation to provide adequate data about the pump driver, power supply, water supply, location, and so on.

Fire pumps are designed to provide maximum reliability and specific net head-discharge characteristics. Except for periodic inspections and tests, fire pumps are idle most of the time. Pumps for commercial use, on the other hand, are chosen for maximum efficiency and economy.

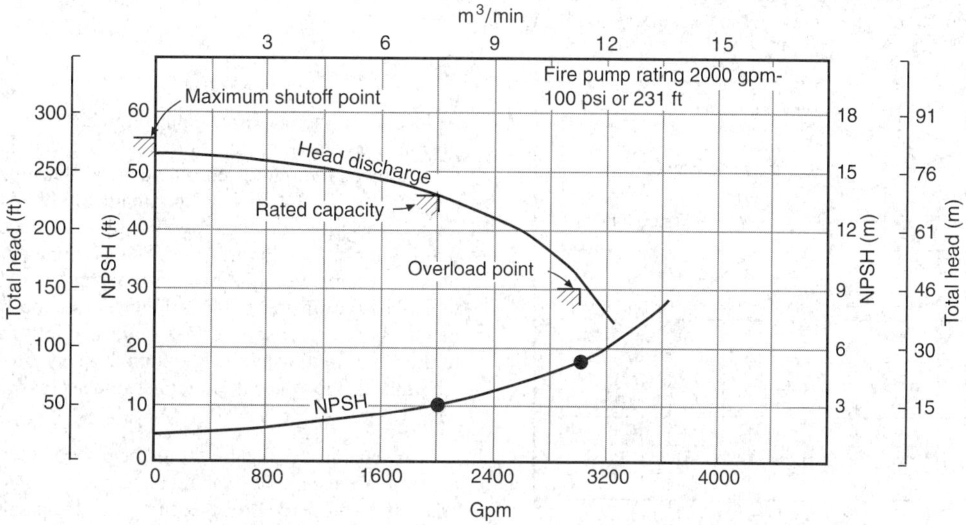

FIG. 6-4J. *NPSH curve of a typical fire pump. Note that the required NPSH at 2000 gpm is 10 ft and 18 ft at 3000 gpm. (For SI units: 1 gpm = 0.38 L/min; 1 ft = 0.30 m.)*

To obtain a listing of a new pump, the manufacturer submits plans and specifications to a testing agency for review and comment. After any revisions or corrections have been agreed to, arrangements are made for representatives of the testing agency involved to witness the required approval tests at the manufacturer's plant.

If the results are satisfactory, the new pumps are listed in the usual manner or with any restrictions considered desirable. It is the duty of the manufacturer to shop-test every unit sold and to furnish certified curves of head, efficiency, and brake horsepower *vs.* discharge. The NPSH curve of the pump should be provided upon request.

Many of the listed fire pumps used today are top-quality commercial units. These pumps are upgraded when necessary, trimmed, and fitted to meet all the approval requirements for fire protection.

STANDARD HEAD DISCHARGE CURVES

The shape of the standard head discharge curve of a fire pump is determined by three limiting points, as follows.

Shutoff

With the pump operating at rated speed and no flow, the total head of a horizontal centrifugal pump, vertical turbine pump, or an end-suction pump at shutoff must not exceed 100 to 140 percent of the rated head at 100 percent capacity.

The shutoff point represents the maximum allowable total head pressure. Otherwise, the pump would have a rising or convex characteristic curve. Such pumps are not listed. With a convex curve, there could be two flow points for one pressure.

Rating

The curve should pass through or above the point of rated capacity and head. (See Figure 6-4K.)

Overload

At 150 percent of rated capacity, the total head should not be less than 65 percent of the rated total head. Here, also, the curve should pass through or above the overload point. Most modern fire pumps have curves with a small margin above the theoretical overload, and

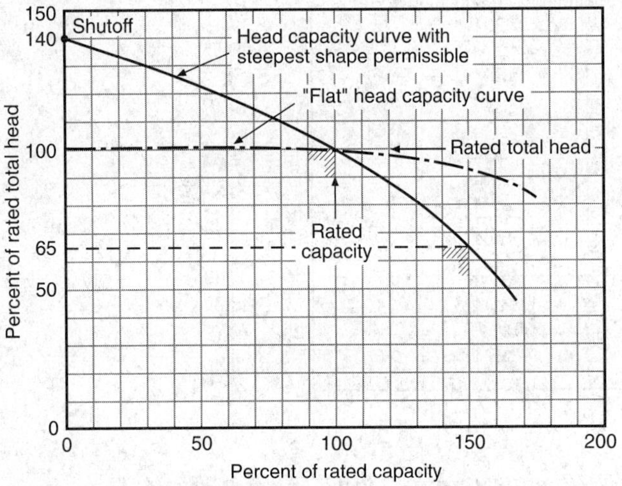

FIG. 6-4K. Pump characteristic curves.

some models have a cavitation or "break" point in the curve just beyond overload.

HORIZONTAL-SHAFT CENTRIFUGAL FIRE PUMPS

Horizontal-shaft centrifugal fire pumps are required to be installed to operate under positive suction head. If the water supply is such that suction lift cannot be avoided, vertical turbine fire pumps should be installed. NFPA 20 no longer allows the use of horizontal centrifugal fire pumps taking suction under lift for new installations.

Types of Pump

Horizontal centrifugal fire pumps are of the split case or end suction type. (See Figures 6-4E and 6-4L.) The end suction type is manufactured to ANSI specifications[1] for centrifugal pumps. In general, there are no limits on the capacities of split case fire pumps, but the maximum capacity of any listed fire pump is currently 5000 gpm (18,925 L/min).

Suction Facilities

When streams, ponds, and other open bodies of water are used, properly screened intakes should be provided to prevent fish, eels, and foreign material from entering the pump and the fire protection system. Some older pumps may have a foot valve such as the one shown in Figure 6-4M. However, pumps taking suction under lift are not presently allowed by NFPA 20.

Use of nonpotable water should be avoided when fire pumps discharge into a system that is also connected to public mains or to other potable supplies. Otherwise, there will be cross connections, which are restricted in some locations by either health or water utility authorities, or both, in most states and provinces. In those cases where cross connections cannot be avoided, suitable backflow prevention devices acceptable to health and water authorities should be provided to prevent contamination of potable water supplies. Such devices should be installed on the discharge side of the pump. Aboveground covered tanks filled with potable water are recommended for supplying fire pumps.

The volume of suction storage should be sufficient to supply the pump at overload rate for the required duration of the water demand.

Break Tanks

At locations where a direct connection between a public water supply and a private fire protection system is prohibited, either for public health or hydraulic reasons, a break tank installation may be desirable. A break tank is an automatically filled tank that provides a suction supply for a fire pump without a direct connection to a public supply. (See Figure 6-4N.) This is done by an actual physical break or gap between the public supply and a private protection system. Water from the public supply enters the break tank from a height above the tank's overflow outlet and falls freely to the surface of the water in the tank. A fire pump is needed to take suction from the tank, as the water is no longer under pressure from the public supply.

A separate break tank should be provided for each fire pump in an installation, and they should be large enough to supply pump operation at 150 percent of the pump rated capacity for the required duration. The flow into the tank from the public supply is controlled by an automatic fill mechanism, and, since the water in the tank is considered potable, the top of the tank should be closed.

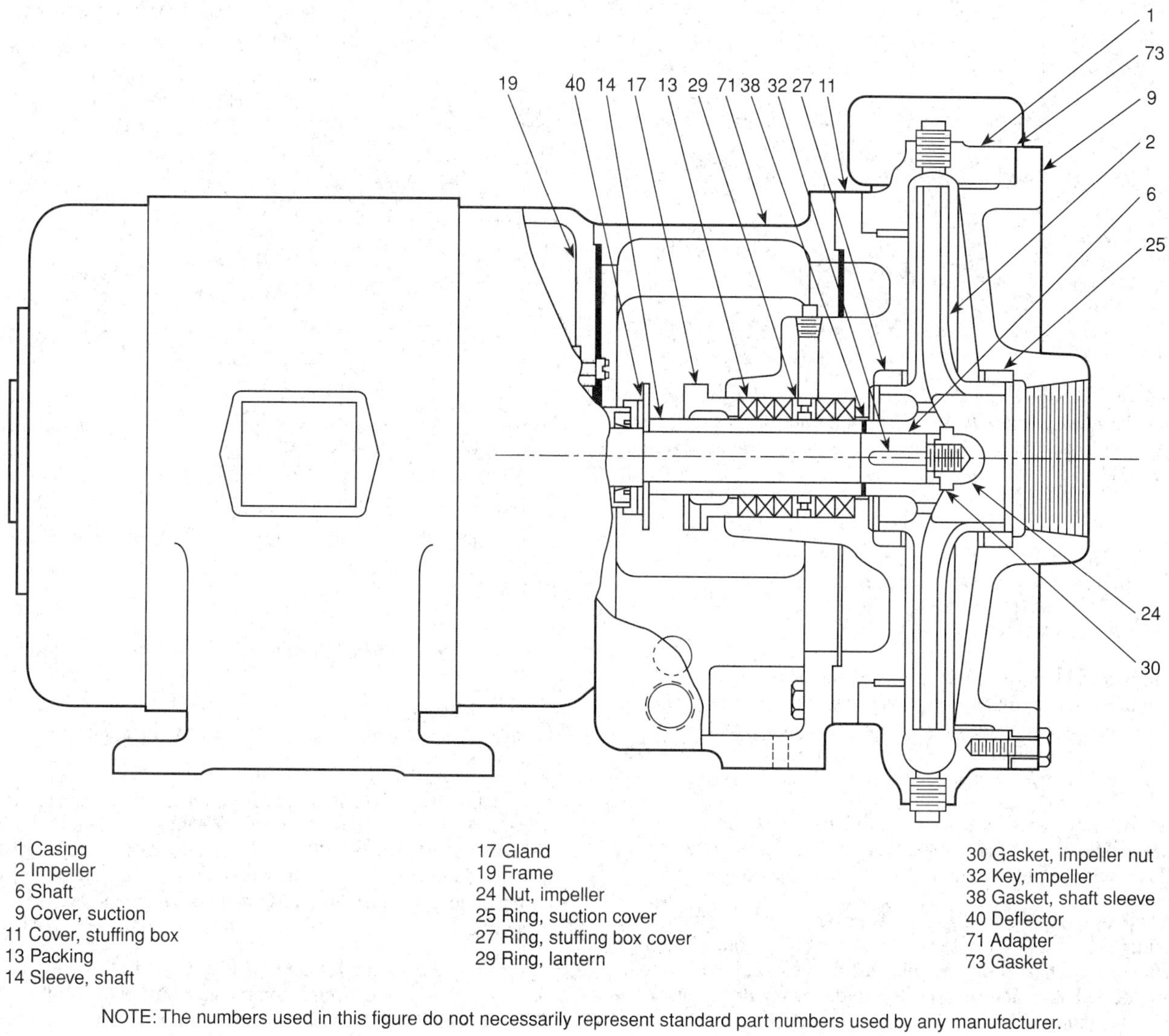

1 Casing	17 Gland	30 Gasket, impeller nut
2 Impeller	19 Frame	32 Key, impeller
6 Shaft	24 Nut, impeller	38 Gasket, shaft sleeve
9 Cover, suction	25 Ring, suction cover	40 Deflector
11 Cover, stuffing box	27 Ring, stuffing box cover	71 Adapter
13 Packing	29 Ring, lantern	73 Gasket
14 Sleeve, shaft		

NOTE: The numbers used in this figure do not necessarily represent standard part numbers used by any manufacturer.

FIG. 6-4L. *An overhung impeller, close-coupled, single-stage, end suction fire pump. (Source: Hydraulics Institute)*

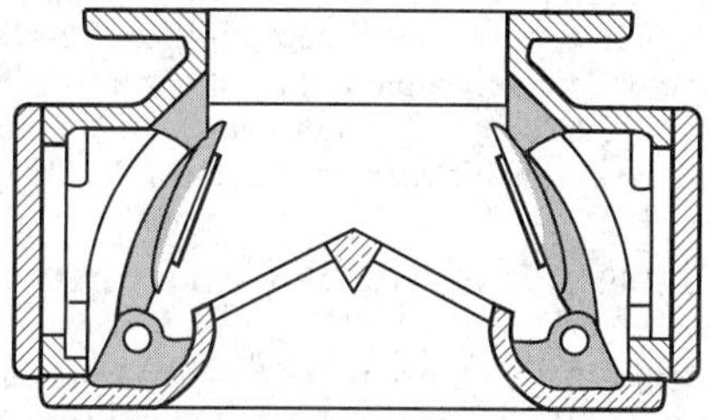

FIG. 6-4M. *A typical foot valve shown in the open position. (Not allowed for new installations.)*

Break tanks are less reliable than a good, full-sized suction tank, as the automatic fill mechanisms could fail. There is no NFPA standard for the design and installation of break tanks, although recommendations have been made on them by an insurance engineering service.[2]

Booster Pumps

These are fire pumps taking suction from public water mains or industrial water systems. (In a mechanical sense, all pumps are booster pumps.) As a prelude to installation, the available fire flow is obtained by conducting flow tests. The pump is sized by first conducting a flow test of the water supply and plotting the test on a water-supply graph having pressure *vs.* volume (gpm), semi-log $N^{1.85}$ paper. (For SI units: 1 gpm = 3.785 L/min.) The required water demand is then plotted to determine what pressure is required. The manufacturer's characteristic curves are researched to find one that adds the required pressure at a volume range of 90 percent of rated capacity to 130 percent of the rated pump capacity. The pump should be sized so that, when pumping is at 150 percent of rated (overload) capacity, the pump does not lower the water supply pressure to a gage pressure less than 0, or a safe point that may be determined by the local health authorities. Full overload capacity of the pump, plus the probable flow drawn from area hydrants by the fire department, are calculated, and they should not drop the gage pressure in the water mains below 0 or that allowed by the public health authorities. In some cases this pressure may be as high as 20 psi

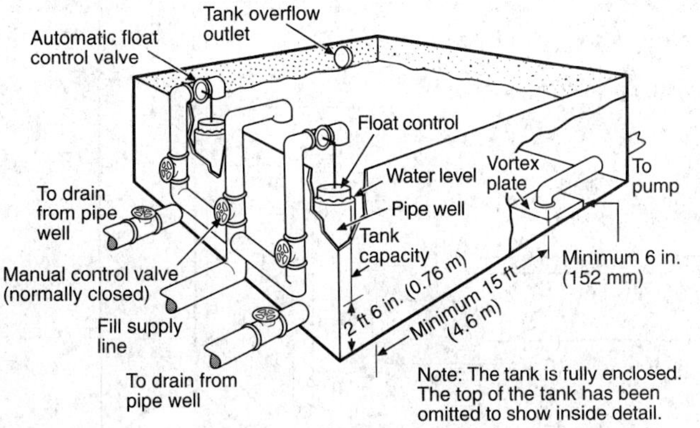

FIG. 6-4N. *A break tank used in connection with a fire pump when a direct connection between a public water supply and a private fire protection system is prohibited. (Source: Factory Mutual Research Corp.)*

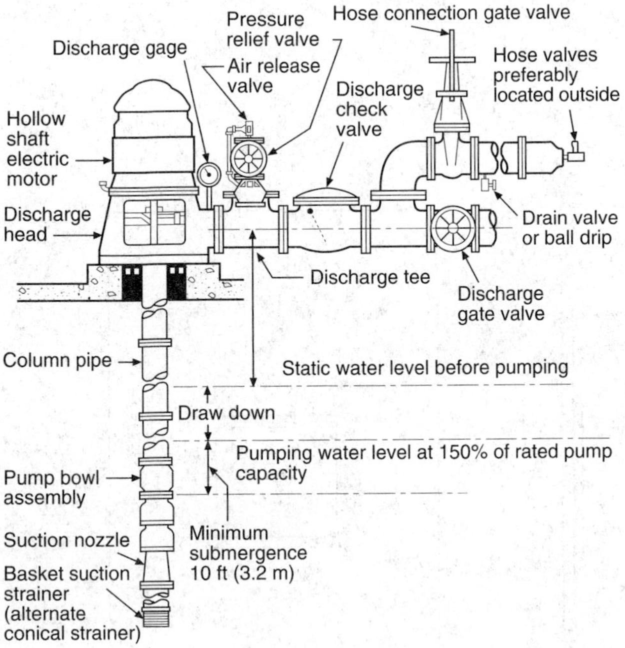

(138 kPa). Head rating of the pump should be sufficient to meet all pipe friction in the connection plus pressure demand.

Pump Accessories

Auxiliary devices have an important bearing on the complete functioning of a pump as a fire protection water supply, and their provision or omission should never be decided solely on the basis of cost. NFPA 20 gives detailed information concerning their installation. The following items are worthy of special consideration.

Relief valves: These are required on the pump discharge line when the operation of the pump can result in excess pressure. That would exceed the pressure rating of the fire protection system.

Hose valves: Approved $2^1/_2$-in. (64-mm) hose valves can be used in testing pumps and for hose stream(s) fire protection. The valves are attached to a header or manifold outside the pump room, or otherwise located to avoid water damage to the pump, driver, and controller. The number of valves needed depends on the pump capacity. The number of $2^1/_2$-in. (64-mm) outlets provided is based on a 250 gpm (946 L/min) flow per outlet.

Automatic air release valves: These are installed on the top of the pump(s) casing and arranged for automatic or remote-control operation. The purpose is to release trapped air in the casing and to minimize pump cavitation. An umbrella cock may be adequate for a pump that can be started only manually by an operator in the pump room, but an automatic air release is required on any pump with a casing that is normally full of water.

Circulation relief valves: These are installed on pumps that may be started automatically or by remote control. Their function is to open at slightly above rated pressure, when there is little or no discharge, so that sufficient water is discharged to prevent the pump from overheating. These valves are not needed on diesel engine–driven pumps where cooling water is taken from the pump discharge.

VERTICAL TURBINE TYPE FIRE PUMPS

Vertical turbine pumps were originally designed to pump water from wells. As fire pumps, they are recommended in instances where horizontal pumps would operate with suction lift. An outstanding feature of vertical pumps is their ability to operate without priming. Vertical pumps may be used to pump from streams, ponds,

FIG. 6-4O. *Vertical shaft turbine-type pump installation.*

wet pits, wells, underground water storage tanks, cisterns, and so on, as well as in booster service.

Suction from wells is acceptable if the adequacy and reliability of the well has been established and the entire installation is built in conformance with NFPA 20. The maximum pump depth is 200 ft (60 m) and the water level in the well is required to cover the pumping bowls to a depth required by the pump manufacturer's recommendations.

If the yield from a reliable well is too small to supply a standard fire pump, low-capacity, commercial well pumps can be used to fill conventional ground-level tanks or reservoirs for the fire pump supply.

A vertical fire pump assembly consists of a discharge head or right-angle gear drive, a pump column pipe and discharge fitting, an open or enclosed drive shaft, a bowl assembly containing the impellers, and a suction strainer. (See Figures 6-4O and 6-4P.) The principle of operation is comparable to that of a multistage horizontal centrifugal pump. Except for shutoff pressure, the characteristic curve is the same as it is for horizontal pumps. (See Figure 6-4K.)

Vertical Turbine Barrel or Can Pump

Sometimes vertical-type multistage pumps are installed in a casing called a "barrel" or "can" for high-pressure installations. (See Figure 6-4Q.)

Vertical pumps have the same standard capacity ratings that horizontal fire pumps have. Pressure ratings are standardized and vary by the number of bowls and diameter of the bowls. By changing the number of stages, or the impeller diameters, or both, the pump manufacturer can provide a specific total head at rated speed. Pressures are available up to 520 psi (3585 kPa).

Vertical or horizontal mounted electric motors can be used, as well as hollow shaft motors. Diesel engines or steam turbines can be used by means of right-angle gear drives.

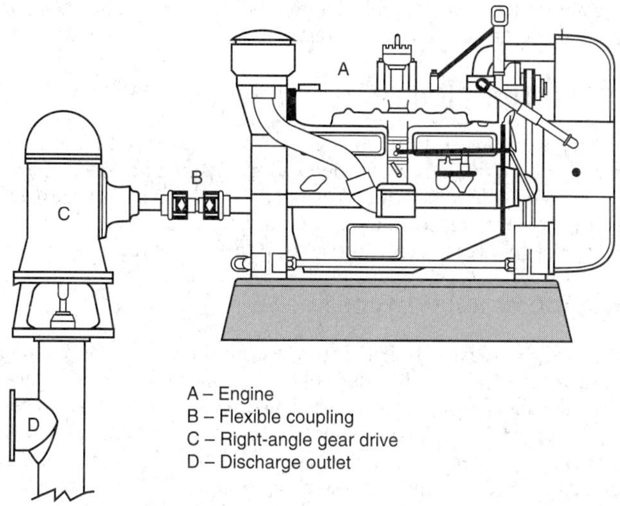

A – Engine
B – Flexible coupling
C – Right-angle gear drive
D – Discharge outlet

FIG. 6-4P. Diesel engine–driven vertical fire pump.

FIRE PUMP CAPACITY AND HEAD RATING

The capacity and pressure ratings of fire pumps must be adequate to meet flow and pressure demands consistent with water supply requirements for the property in question. Fire pumps are designed to provide their rated capacity with a safety factor built in (150 percent of rated capacity at 65 percent of rated pressure) to provide some protection in case of greater-than-expected demand at the time of a fire. The following examples show one method of how rated capacity and pressure can be determined by using the standard fire pump curve for a typical manufacturer's pump characteristic curve. (See Figure 6-4K.)

Example No. 1
Horizontal centrifugal pump. The estimated water demand for sprinklers and hose streams is 1300 gpm at 90 psi. The suction supply is a ground level storage tank, and the minimum inlet gage pressure is 0 psi at maximum flow.

PROBLEM: Determine the required rated capacity and pressure of the pump.

SOLUTION:

1. Meet the demand of 1300 gpm with the capacity of the pump, which is 135 percent of rated capacity;
2. Thus, 1300 ÷ 135 percent = 963 gpm. The nearest standard pump rating is 1000 gpm;
3. Therefore, 1300 gpm demand would be 130 percent of capacity;
4. From a manufacturer's pump characteristic curve, it is determined that, at 130 percent capacity, the total pressure is 75 percent of rated pressure;
5. Under conditions of operation, the total pressure equals the discharge pressure (90 psi) plus suction pressure (0 psi);
6. Therefore, net pressure at 1300 gpm equals 90 + 0 = 90 psi, and rated pressure at 1000 gpm = 90 ÷ 75 percent = 120 psi.

ANSWER: Pump rating should be not less than 1000 gpm at 120 psi.

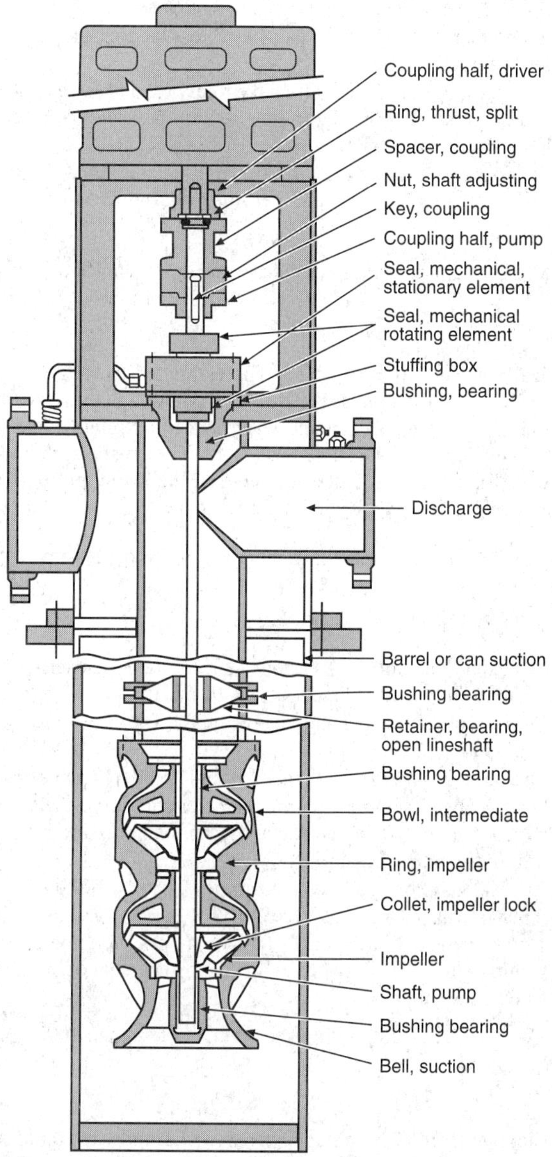

Coupling half, driver
Ring, thrust, split
Spacer, coupling
Nut, shaft adjusting
Key, coupling
Coupling half, pump
Seal, mechanical, stationary element
Seal, mechanical rotating element
Stuffing box
Bushing, bearing
Discharge
Barrel or can suction
Bushing bearing
Retainer, bearing, open lineshaft
Bushing bearing
Bowl, intermediate
Ring, impeller
Collet, impeller lock
Impeller
Shaft, pump
Bushing bearing
Bell, suction

FIG. 6-4Q. A turbine-type, vertical, multistage, barrel or can pump.

Example No. 2
Horizontal centrifugal pump (SI units). The estimated water demand for a sprinkler system is 2000 L/min at a pressure of 400 kPa. The suction supply is a well with a lift of 5 m from water surface to the pump (not allowed for new installations).

PROBLEM: Determine the required rated capacity and pressure of the pump.

SOLUTION:

1. Meet the demand of 2000 L/min with the overload capacity of the pump which is 135 percent of rated capacity;
2. Thus, 2000 ÷ 135 percent = 1481 L/min. The nearest standard pump rating is approximately 1500 L/min;
3. Therefore, 2000 L/min demand would be 2000/1500 × 100 percent = 133 percent of rated capacity;

4. From the manufacturer's pump characteristic curve, it is determined that, at 133 percent capacity, the total pressure is 78 percent of rated pressure;
5. Under conditions of operation, the total pressure equals the discharge pressure (400 kPa) minus suction pressure (5 m head = 5 × 9.81 kPa = 50 kPa);
6. Therefore, net pressure at 2000 L/min equals 400 + 50 = 450 kPa, and rated pressure at 1500 L/min = 450 ÷ 78 percent = 580 kPa.

ANSWER: Pump rating should be not less than 1500 L/min at 580 kPa.

Example No. 3
Vertical turbine pump in a well. The estimated water demand at ground level is 1100 gpm at 100 psi. Tests and weather records show that the aquifer, or underground water source, is reliable and adequate at all seasons. The static level is 45 ft below the surface. The draw down, or vertical distance, between the static and pumping water levels is 40 ft at 1100 gpm pumping rate.

PROBLEM: Determine the rated capacity and pressure of the pump.

SOLUTION:

1. Meet the demand of 1100 gpm with the rated capacity of the pump, which is 100 percent of rated capacity;
2. Thus, 1100 ÷ 100 percent = 1100 gpm. The nearest standard pump rating is approximately 1000 gpm;
3. Therefore, the 1100 gpm demand would be 110 percent capacity;
4. From a manufacturer's pump characteristic curve, it is determined that, at 110 percent capacity, the total pressure is 96 percent of the rated pressure;
5. At 1100 gpm, pressure demand at the surface = 100 psi;
6. Since the distance to water level at 1100 gpm pumping rate is 45 + 40 + 10ft (suction) = 95 ft × 0.434 = 41 psi, pumping pressure would be 100 + 41 = 141 psi, which is 136 percent of rated pressure;

ANSWER: The pump rating should be not less than 1000 gpm at 147 psi.

Example No. 4
Booster pump on public water connection. A sprinklered building in a city has an estimated sprinkler demand of 750 gpm at 60 psi. Based on fire flow tests from nearby street hydrants, 750 gpm at 27 psi is available for sprinklers at the inlet flange of the pump. Allowance has already been made for hose streams.

PROBLEM: Determine the rated capacity and pressure of the pump.

SOLUTION:

1. Meet the demand of 750 gpm with the rated capacity of the pump, which is 100 percent of rated capacity;
2. Thus, 750 ÷ 100 percent = 750 gpm, a standard pump rating;
3. The total pressure at 100 percent capacity is 100 percent rated net pressure;
4. With positive head suction supply, the net pressure equals discharge pressure minus suction pressure; thus, at 750 gpm flow, net pressure equals 60 – 27 = 33 psi;
5. Therefore, 33 ÷ 100 percent = 33 psi.

ANSWER: The pump rating should be 750 gpm at 33 psi.

NFPA 20 recommends that the pump not be used at over 140 percent capacity. This leaves a small reserve as pump curves drop off sharply after 150 percent. Generally, a one size larger pump is recommended than the calculations indicate. When the pump is used above its rated capacity, the available pressure is reduced.

HORSEPOWER OF FIRE PUMPS

Before matching a driver to a pump, it is necessary to know the maximum brake horsepower demand of the pump at rated speed. This can be determined directly from the horsepower curve provided by the pump manufacturer. Typical fire pumps reach maximum brake horsepower between 140 and 170 percent of rated capacity.

Engine Horsepower

Engines specifically designed for use with fire pumps are rated by measuring the horsepower developed with all accessories in operation and then making some allowance for wear and tear. When equipped for fire pump driver service, an allowance of not less than 10 percent greater than the maximum brake horsepower is required by the pump under any conditions of pump load. Typical bare engines and usable horsepower curves are shown in Figure 6-4R.

The engine manufacturer's test curves are based on barometric pressure of 29.61 in. (752 mm) Hg, which approximates 300 ft (90 m) above sea level and 77°F (25°C). The usable horsepower of a fire pump engine should be reduced for each 1000-ft (300-m) rise in altitude above 300 ft (90 m) by 3 percent for a diesel engine and 1 percent for each 10°F (5.6°C) rise above 77°F (25°C).

If the curves are not available, horsepower can be calculated, by the formula (in U.S. customary units):

$$bhp = \frac{5.83QP}{10,000 \times E} \text{ or } bhp = \frac{QP}{1710 \times E}$$

where

bhp = brake horsepower,

Q = gallons per minute,

P = total head (psi) or net pressure, and

$E = \text{efficiency} = \dfrac{\text{water horsepower}}{\text{input horsepower}}$

The efficiency at maximum brake horsepower is usually 60 to 75 percent.

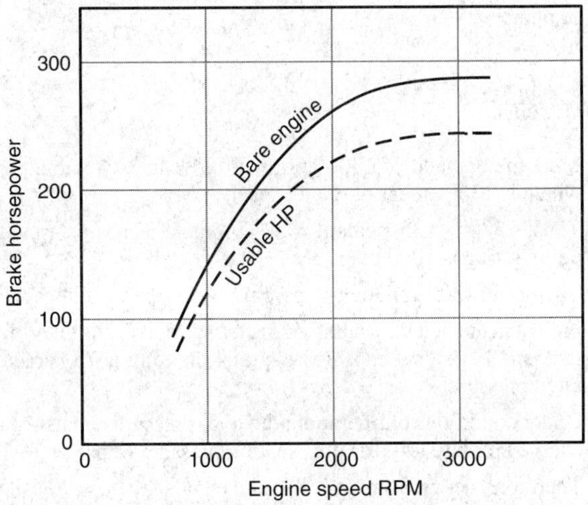

FIG. 6-4R. Typical engine horsepower curves. (For SI units: 1 bph = 745.70 W.)

EXAMPLE: Find by formula the minimum horsepower needed to drive a 1000-gpm, 100-psi, 1760-rpm horizontal centrifugal fire pump.

SOLUTION:

1. Assume 65 percent efficiency at 160 percent capacity;
2. From a standard pump curve, the pressure is 55 percent at 160 percent capacity or 55 psi at 1600 gpm;
3. By formula,

$$bhp = \frac{5.83 \times 1600 \times 55}{10,000 \times 0.65} = 79$$

ANSWER: Not less than 79 usable brake horsepower.

IN SI UNITS:

$$Output\ power = \frac{0.167 Q_m P_m}{10,000\ E}$$

where

kW = power output (kilowatts),

Q_m = liters per minute,

P_m = total head (kPa) or net pressure, and

E = efficiency = $\dfrac{water\ power\ output}{input\ power}$

EXAMPLE (SI units):   Find by formula the minimum power output needed to drive a 4000-L/min, 700-kPa, 1760 = rpm horizontal centrifugal fire pump.

SOLUTION:

1. Assume 65 percent efficiency at 160 percent capacity;
2. From a standard pump curve, the pressure is 55 percent at 160 percent capacity or 385 kPa at 6400 L/min;
3. By formula,

$$power\ output = \frac{0.167 \times 6400 \times 385}{10,000 \times 0.65}$$
$$= 63\ kW$$

ANSWER: Not less than 63 kW output.

FIRE PUMP DRIVERS

Power for driving fire pumps is selected on the basis of reliability, adequacy, safety, and economy. The reliability of utility electric power may be judged by the record of outages and by review of the power sources and distribution layout of the system in question.

Some public utilities in metropolitan areas operate steam distribution systems. When high-pressure steam is available, it is practical to use steam turbine-driven fire pumps.

Some remote industrial plants generate their own electricity using steam or hydro power, or both. Utility electric power also is used.

Diesel engines have the advantage of not being dependent upon outside sources of power.

Electric Motors

Electric motors for driving fire pumps are required to be listed for fire pump service. These motors are designed in accordance with specifications of National Electrical Manufacturers Association

(NEMA) or Electrical Manufacturers Association of Canada (EMAC). All electrical equipment and wiring in a fire pump installation should comply with NFPA 70, *National Electrical Code®*, except as modified by NFPA 20.

Fire pumps that are driven by electric motors are designed to withstand rather severe ranges of operating conditions. As a result, the power supply for these pumps is designed to accommodate large electrical loads, and pump configurations that will sacrifice the pump motor. Such design is contrary to the design of most other electrical motors and results in a high level of pump reliability.

The installing contractor is responsible for providing a motor of sufficient capacity to avoid overloading beyond the limit of the service factor at maximum brake horsepower and rated speed. The service factor is a numerical value, and it depends on the type of motor—open, splashproof, or totally enclosed—and on the resistance of the insulation on the windings to heat and breakdown.

When the service factor exceeds 1.0, the excessive amount is stamped on the nameplate along with the voltage and the full-load ampere rating. For example, a 75-hp (56 kW) motor, with a 1.15 service factor, could safely meet a demand of $75 \times 1.15 = 86.25$ bhp (64 kW).

Another use for the service factor is to determine the maximum allowable ampere demand. For example, with a 40-amp full-load rating and a 1.12 service factor, the maximum ammeter reading should not exceed $40 \times 1.12 = 45$ amps. Note also that, for a given voltage, horsepower is proportional to amperes.

Only motors wound for 208 V should be used on 208-V services.

Direct-current motors for pumps are either of the stabilized shunt type or of the cumulative compound-wound type. The speed of the motor with no load at operating temperature must not exceed the speed of the motor under full load at operating temperature by more than 10 percent.

The most commonly used alternating current motors are of the squirrel-cage induction type. They usually are equipped with across-the-line starting equipment unless their starting characteristics would be objectionable to the company furnishing the power. In the latter case, primary resistance starting may be employed, or a wound-rotor-type of motor with appropriate starting equipment may be substituted. When squirrel-cage motors are used, the voltage drop must not be so great as to prevent the motor from starting; that is, it should not be more than 10 percent below normal voltage at moment of start. While the motor is running at rated pump capacity, pressure, and speed, the line voltage should not drop more than 5 percent below motor nameplate voltage.

This type of motor should have normal starting and breakdown torque. The locked rotor currents for motors of various horsepower ratings at 460 V are specified in NFPA 20.

When an electric-driven pump is used as a part of a high-rise building fire protection system, it is required to have a backup power supply. Thus, the reliability of pumps used in this environment is especially important.

If power supplies to the pump are not reliable, the use of a generator, in conjunction with a permanent power supply, or a diesel engine drive, may be necessary. Installation of the generator must be done in accordance with NFPA 20, NFPA 37, *Standard for the Installation and Use of Stationary Combustion Engines and Gas Turbines* (hereinafter referred to as NFPA 37), and NFPA 110, *Standard for Emergency and Standby Power Systems*. A transfer switch is required to be installed for the pump and is necessary to switch the pump energy source from the permanent supply to the emergency source. This transfer switch may be provided as an integral part of the fire pump controller, or it may be installed as a separate piece of equipment. No matter which method is selected, it must be installed such that it does not reduce the degree of reliability for the entire system.

Electric Motor Controllers

The fire pump controller used for an electric-driven pump is a critical component to ensure the successful operation of the pump. Such controllers are equipped with a variety of internal components to achieve this level of reliability. These components may include elements such as circuit breakers, disconnecting means, timers, and similar devices.

Motor controllers are available for alternating-current fire pump motors operating at standard voltages up to 2300 V. A controller is a complete, assembled unit, wired and tested, and ready for service by connecting to the power supply and the proper motor terminals. Detailed specifications are described in NFPA 20.

Controllers are built for combined manual and automatic operation. They also are available for either squirrel-cage or wound-rotor motors and for one-phase to three-phase power. Across-the-line starting is recommended for most applications. Controllers for reduced voltage starting are also available. Reduced voltage starting is normally used when electric-driven pumps are served by engine generator sets.

The circuit breaker of a fire pump controller permits normal starting without tripping and provides stalled rotor and instantaneous short circuit protection. The interrupting capacity of the breaker should be adequate for the circuit in which it is located, but not less than 15,000 amps in any case.

Automatic or manual controllers not designed for locked rotor conditions, known as limited-service controllers, are available for across-the-line squirrel-cage motors of 30 hp (22 kW) or less driving fire pumps.

Steam Turbine

When adequate and reliable steam supplies are available, turbine-driven fire pumps are acceptable. Only well-built machines of good design with industrial records of proven reliability are used. Special arrangements are needed for automatic operation. Speed rating should not exceed 3600 rpm, because this is the maximum speed of listed fire pumps. Detailed requirements for steam supply, speed governors, and controls are contained in NFPA 20. Steam turbine engines are not listed for fire pump service.

Diesel Engines

Engines that are powered by diesel fuel are used for fire pump service. Engines powered by gasoline, natural gas, or LP-Gas are not recognized by NFPA 20.

In addition to NFPA 20, reference should be made to NFPA 37 and NFPA 31, *Standard for the Installation of Oil-Burning Equipment.* Installations should be made in conformance to local codes.

When unsatisfactory conditions, such as pump running, failure to start, high cooling water temperature, and low oil pressure, are required to be indicated by supervisory signals, not by shutting down the engine. The intent is to keep the pump operating just as long as possible. The importance of constant supervision for automatic pumps is obvious.

Cooling systems: An adequate cooling system is vital to the reliable operation of a diesel engine. A closed pipe system with a heat exchanger or a thermally insulated manifold is the preferred cooling arrangement for a fire pump recognized in NFPA 20. Only clean or potable water should be circulated through the engine block. Raw water is piped from the fire pump, discarded through the heat exchanger tubes to free discharge in a visible location, such as the cone of the fire pump relief valve. On some engines, the manifolds, oil coolers, and other parts are equipped with water jackets, as recommended by the engine manufacturer. (See Figure 6-4S.) Most engines require a raw water flow of 15 to 30 gpm (57 to 114 L/min).

Another method of providing cooling to the diesel engine is a closed loop radiator. The radiator can be remotely installed outside of the pump room or on the engine platform. When the radiator is mounted on the engine skid, it should be arranged to pull the air across the engine and out through the radiator. The radiator is sized, according to engine manufacturer's recommendations, to remove the engine-radiated heat and to provide the internal engine cooling requirements.

Fuel tanks: The storage tank for the diesel fuel is sized to contain at least an 8-hr supply. A larger capacity can be provided if facilities for prompt refilling are not available. Tank capacity is estimated by allowing 1 pt of diesel fuel per horsepower per hour, or 1 gal/hp. (For SI units: 1 pt = 0.473 L; and 1 gal = 3.785 L.) (See diagrams of typical fuel systems in NFPA 20.)

The fuel storage tank is required to be installed inside the fire pump room. This is because all listed diesel engines require fuel to be available by gravity, thereby eliminating the fuel pump on the engine. This also allows for easy inspection and maintenance of the tank.

Battery chargers: The energy source needed for automatic starting of a diesel-driven fire pump is dependent upon a sufficient capacity from the battery system. The chargers used to satisfy this requirement are specifically listed for fire pump service, and consist of rectifiers, transformers, and relays.

Batteries are provided in two banks, with alternate starting cycles to increase the system reliability.

FIG. 6-4S. *A typical heat-exchanger cooling system for an automatically controlled engine-driven fire pump. Raw water from the fire pump enters the system through the strainer (1), which prevents sediment from entering the system, and the pressure regulator (2), which protects the heat exchanger from excessive pressure. The solenoid valve (3) is required with automatic control of the engine. Valve (4), normally closed, may be used to bypass the regulator and solenoid valve. The exhaust manifold (5) may be cooled by the clean water circulating system.*

Engine Controllers

Controllers are used for automatic operation of engine-driven fire pumps. The specifications for construction, location, and methods of actuation of engine controllers are the same as for electric motor controllers. Automatic controllers are equipped with manual starting and stopping switches.

Audible supervisory devices are provided to indicate low oil pressure in the lubrication systems, high engine-jacket water temperature, failure of the engine to start automatically, and shutdown from overspeed.

There are additional features for controllers as given in NFPA 20. A weekly program timer is provided. This device is arranged to start the unit automatically once a week and to run for a predetermined number of minutes (30 min). A recording pressure gage records this performance.

Controllers are operated on low-voltage direct-current power from the engine's batteries. The program timer, battery charger, or other auxiliary devices not essential to pump control are powered by the regular 120-V alternating current power supply at the property.

Automatic Pump Control

Most fire pump installations are arranged for automatic operation, preferably with automatic start and manual shutdown. The choice between manual and automatic shutdown depends on the specific conditions involved in a pump's installation and use. Horizontal centrifugal pumps under automatic control must always operate under a head to avoid the need for priming.

Each controller is equipped with a separate pressure switch and sensing line that actuate the pump unit when pressure in the water system piping drops to a preset level. Unless the normal water supply static pressure is higher than the pump starting pressure, an automatic jockey (pressure maintenance) pump must be provided to maintain pressure in the system at the higher level.

Actuation of a pump by water flow instead of by pressure drop is desirable for certain installations, such as those in which the opening of a moderate number of sprinklers would not drop the system pressure enough to move the pressure switch; high-hazard occupancies in which a fire would demand fire pump service without delay; combined fire protection and plant service systems where a pressure maintenance pump would be impractical; and occupancies in which pressure fluctuates so much that a stable cut-in pressure could not be obtained.

The wiring system of a pump controller includes terminals for connection of a relay to an external alarm circuit from a sprinkler, deluge, or special fire protection system. To secure reliable pump actuation, external circuits should be installed in conformance with one of the following NFPA standards, depending upon the nature of the signaling system:

NFPA 70, *National Electrical Code®*, Article 695.

NFPA 72, *National Fire Alarm Code*.

Circuits for remote automatic starting of fire pumps should be powered from the controller power.

FIELD ACCEPTANCE TESTS

After a new fire pump has been installed, it is required to make a performance test. Defects and faults are discovered, and steps taken to remedy them. These tests enable the purchaser to determine that the contract has been properly concluded. They also demonstrate the need for future maintenance tests.

Details of the acceptance tests are given in NFPA 20. The test demonstrates the adequacy of the pump and the ability of the pump to deliver water in accordance with its head-capacity curve. The prime mover is operated under various conditions and its performance recorded. There are provisions for electric motors, steam turbines, and diesel engines. Repeated operation of the controlling equipment is required to ensure that full operation of the unit will result from either manual or automatic operation of the controller.

Flow tests are conducted to develop the pressure-discharge characteristic curve of the pump. The procedure followed is to run the pump at three to four different flows, including shutoff (no water flowing). The rate in gallons per minute (L/min) is determined with a Pitot gage at the nozzles [preferably standard 30-in. (762-mm) long Underwriter playpipes] attached to hose lines from an outside hose-valve header. The discharge is varied by changing the number of lines, the size of nozzle tips, or both; or by using an installed waterflow meter with flow to an open source. Flow is varied by opening or closing the valve on the meter line.

The nozzles may be attached directly to outdoor headers, without hose lines, if water damage can be avoided. Disposal of test water is often a problem, and the length of hose lines depends on the drainage facilities available and the exposure to property and people.

For every flow, pressure readings are taken at the suction gage and the discharge gage. The revolutions per minute are also measured using a revolution counter or a tachometer, if available.

The net pressures are calculated from the pump gage readings, and the flows in gallons per minute corresponding to the Pitot readings are obtained from discharge tables. The flow required for coolant purposes for diesel engine–driven pumps can be added to the measured flow. However, it is usually difficult to accurately measure this flow, so an estimate based on the manufacturer's recommendations may have to be used. In many cases, this flow is small enough to be considered negligible.

With the growing problems of waste water disposal, many pump installations are equipped with waterflow meters for the acceptance test and the periodic service tests. The meters should be installed in conformance with NFPA 20, in order to function properly and not interfere with the operation of the pump.

Vertical turbine fire pumps are tested in the same manner as horizontal pumps, except that there is no suction gage. The pumping water level should be recorded at each test point.

Figure 6-4T presents data obtained by a typical field acceptance test of a horizontal 1500-gpm, 100-psi, 1760-rpm (5678 L/min, 689 kPa, 1760 rpm) diesel engine-driven centrifugal pump. The net pressure and total flows are calculated from the observed data and plotted. (See Figure 6-4U.) The curve best fitting the plotted points is then drawn (Curve A). In this installation, the engine governor appeared to be out of adjustment, restricting the average speed to 1689 rpm, whereas the rated speed was 1760 rpm.

Since the pump was tested at less than rated speed, the observed net pressures and flows were converted to what they would have been at the rated speed of 1760 rpm. Curve B is the characteristic curve at rated conditions. Although the rating point was barely reached, the overload point exceeded the minimum by a good margin. With the engine adjusted to operate at full speed, the pump performance would be acceptable. The following is the conversion calculation procedure that was followed:

Flow is directly proportional to rpm; net pressure is proportional to rpm.[2]

Acceptance Test for a 1500-GPM, 100-PSI, 1760 rpm Fire Pump

			Hose	Streams				Corrected to 1760 rpm	
pm	Discharge (psi)	Suction (psi)	Net (psi)*	No.	Size (in.)	Pitot Pressure (psi)	gpm*	Total*	gpm*	Net (psi)*
1700	125	+16	109	0	–	–	–	0	0	118
1695	120	+18	102	1	1¾	70	742	742	772	110
1690	110	+16	94	2	1¾	60, 60	687, 687	1374	1420	101
1686	95	+17	79	3	1¾	55, 55, 55	657, 657 657	1971	2060	85
1675	85	+16	69	4	1¾	35, 37, 48, 48	525, 540 614, 614	2293	2410	76

* Calculated from observed data.

FIG. 6-4T. An example of a log for a fire pump acceptance test

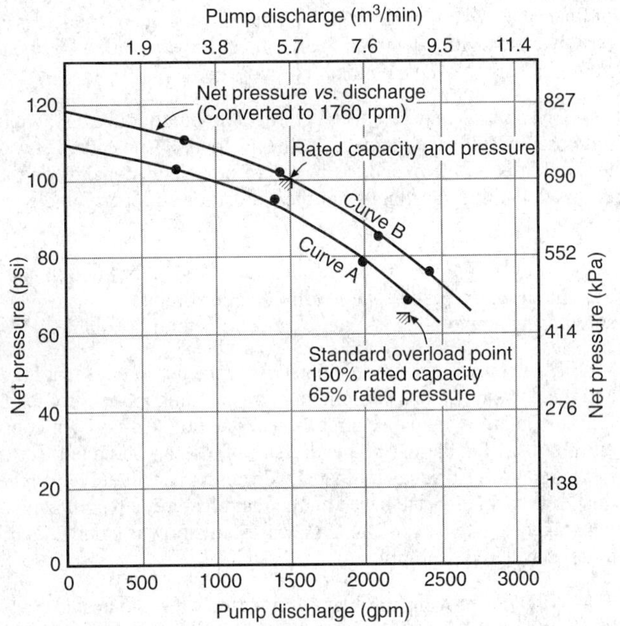

FIG. 6-4U. Head capacity curves plotted from data compiled in a fire pump acceptance test. (See Figure 6-4T for the data used to plot these curves.)

EXAMPLE: Test flow, 1971 gpm at 1686 rpm

$$\text{Flow at 1760 rpm} = 1971\left(\frac{1760}{1686}\right) = 2060 \text{ gpm}$$

Net pressure for 1971 gpm at 1686 rpm is 78 psi

$$\text{Net pressure for 2060 gpm at 1760 rpm is } 78\left(\frac{1760}{1686}\right)^2 = 85 \text{ psi}$$

Similar calculations can be carried out in SI units using the affinity laws mentioned previously in this chapter.

LOCATION AND HOUSING OF CENTRIFUGAL PUMPS

Fire pumps are preferably housed in buildings of fire-resistant or noncombustible construction. Even when the climate is so mild that there is no danger of freezing, sufficient enclosure is needed to protect against dirt, corrosion, and tampering. Structural separation of the pump room from other parts of a property is now a requirement.

Pump rooms and power facilities should be as free as possible from exposure to fire, explosion, flood, and windstorm damage.

Light, heat, ventilation, and floor drainage should be provided for pump rooms. A dry location above grade is preferred. For a diesel engine driven unit, heat, ventilation, and above grade location are essential.

Pump rooms should be large enough to facilitate easy access to all equipment and devices for inspection, testing, and maintenance.

Fire pumps are located as close as possible to those areas where protection is most important. In some large properties, it may be necessary to have water supplies at more than one point to obtain the most favorable distribution system.

PUMP TESTS

A fire pump must be tested annually to ensure that the pump, driver, suction, and power supply function properly and to correct faults that may be revealed. The hydraulic performance of the pump is measured by a flow test, using a flow meter or hose and nozzles connected to the pump header or yard hydrants. Three points on the standard curve are checked: (1) shutoff, (2) overload (150 percent of rated capacity or more), and (3) at capacity rating.

Automatic operation is tested by lowering the system pressure, giving due consideration to the layout of the fire protection system, i.e., pressure drop or water flow actuation, jockey pump, and so on.

The water level of wells, ponds, and reservoirs and the condition of suction screens and intakes, above- and below ground tanks, and so on, should be carefully inspected.

The history of power outages, low water, and failure of any kind that involves pump, driver, or associated equipment also should be investigated and gage records from engine controllers recorded for future evaluation.

PUMP OPERATION AND MAINTENANCE

A fire pump can be depended upon to work in an emergency only if it is properly operated and maintained. (See NFPA 25 for requirements for inspection, testing, and maintenance of these systems.) It is required that an inspection and testing program that includes written instructions and recordkeeping be implemented. It is further required to have someone at the property at all times who is designated and instructed to operate the pump and its driver. A short (30-min diesel engine; 10-min electric motor) test is required to be made each week by discharging water from some convenient outlet or recirculating through a flow meter.

When a fire alarm occurs or a supervisory signal indicates an automatic fire pump is operating, the person responsible for the fire pump should proceed to its location immediately. The pump should be allowed to run until the emergency is over, when it may be shut down manually. During this and every other operating period, the equipment should be checked carefully to see that it is performing properly.

To minimize too frequent starting and stopping, an electric motor controller has a timer that keeps the motor running for at least 1 min for each 10-hp motor rating; not less than 10 min is required. With electric motors, it is necessary to allow the electrical windings to cool down after starting, to minimize damage to the windings.

It is preferable with all pump drivers to permit the unit to run until it is shut down manually.

When there is more than one automatic fire pump, the controllers are arranged to operate the pumps in a predetermined sequence. The diesel engine is usually first, followed by the electric motor or second diesel engine. Control of the pump from one or more remote pushbuttons, which will start but not stop the pump, may be provided, if desired. If there is deluge valve control of an open discharge device system, the pump may be started by a drop out relay in a closed circuit.

The cooling and lubrication of a centrifugal fire pump is so dependent upon water that the pump must never be run unless the pump casing is full of water. Close attention should be given to the bearings and stuffing boxes during the first few minutes of running to see that they do not heat up and need no adjustment. When water reaches the water seal, a small leak at the stuffing box glands is desirable. The suction inlet and discharge outlet pressure gages should be read weekly to see that the inlet is not obstructed by a choked screen or partially closed valve.

With a vertical shaft turbine-type fire pump, the water level can be observed if suction is from a visible supply. If the pump takes suction from a well, water level testing equipment must be used. The groundwater level at the pump should be checked at the weekly test interval during the year and the draw down should be determined during the annual 150 percent capacity test. These tests should indicate any important change in the groundwater supply.

The direction of rotation of the pump and the speed of operation should be checked each week.

Power Supply Maintenance

The source of electric power for the pump should be checked weekly. With an electric motor drive this means current supply for the motor and its auxiliary equipment. For steam turbine drive, it means the steam supply up to the control valve and the absence of condensate from supply, turbine, and exhaust. If the pump is driven by a diesel engine, there must be adequate fuel for 8 hr of operation. The batteries must be fully charged.

The starting equipment must be test-operated and its functioning carefully checked. Any evidence of a drop in voltage (−15 percent) to an electric motor or a drop in steam pressure to a turbine must be investigated.

With a diesel engine drive, the crankcase oil must be replenished or renewed as needed, the oil filter and air cleaner given the necessary attention, the automatic battery charging equipment checked, and the specific gravity of battery electrolyte determined at least once a month.

RECIPROCATING STEAM FIRE PUMPS

Although no new reciprocating steam fire pumps are known to have been installed in recent years, a few may still be in service. These fire pumps are commonly referred to as Underwriter steam fire pumps.

The general features of a direct-acting duplex steam pump are shown in Figure 6-4V. The size of the steam and exhaust ports of this type of pump is larger than it is in the general-purpose steam pump, thus permitting higher speeds. The steam supply pipe should be an independent line run from boiler to pump in such a way that it will not be damaged by fire or other hazards.

Automatic Control

Automatic control can be provided by a pressure governor to regulate steam supply to the pump in accordance with the water pressure on the pump discharge. For successful operation, it is nearly always necessary to provide a small, automatically controlled jockey pump to maintain system pressure, control supply leakage, and avoid continuous operation of the large pump.

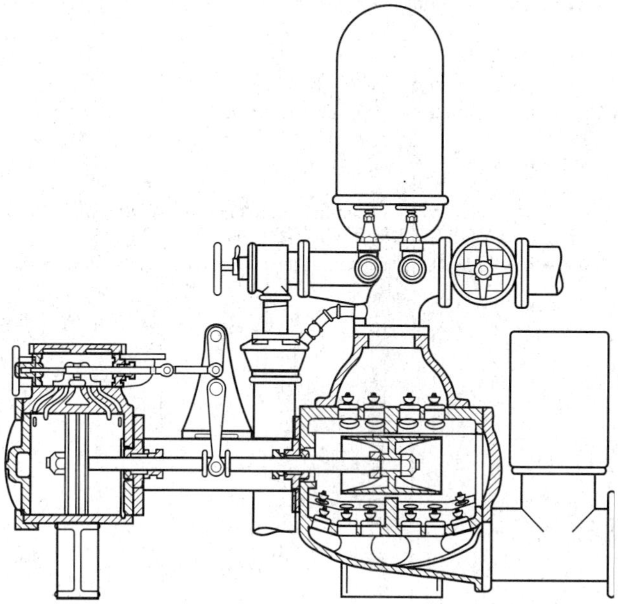

FIG. 6-4V.   Sectional view of an approved duplex steam pump.

BIBLIOGRAPHY

References Cited

1. ANSI B73.1, *Specification for Horizontal End-Suction Centrifugal Pumps for Chemical Process,* American National Standards Institute, New York, 1991.
2. "Break Tanks," Loss Prevention Data 3-251, Factory Mutual Research Corp., Norwood, MA, May 1983.

NFPA Codes, Standards, and Recommended Practices

Reference to the following NFPA codes, standards, and recommended practices will provide further information on fire pumps discussed in this chapter. (See the latest version of *The NFPA Catalog* for availability of current editions of the following documents.)

NFPA 11, *Standard for Low-Expansion Foam*
NFPA 11A, *Standard for Medium- and High-Expansion Foam Systems*
NFPA 13, *Standard for the Installation of Sprinkler Systems*
NFPA 13D, *Standard for the Installation of Sprinkler Systems in One- and Two-Family Dwellings and Manufactured Homes*
NFPA 13E, *Guide for Fire Department Operations in Properties Protected by Sprinkler and Standpipe Systems*
NFPA 13R, *Standard for the Installation of Sprinkler Systems in Residential Occupancies up to and Including Four Stories in Height*
NFPA 14, *Standard for the Installation of Standpipe and Hose Systems*
NFPA 15, *Standard for Water Spray Fixed Systems for Fire Protection*
NFPA 16, *Standard for the Installation of Deluge Foam-Water Sprinkler and Foam-Water Spray Systems*
NFPA 16A, *Standard for the Installation of Closed-Head Foam-Water Sprinkler Systems*
NFPA 20, *Standard for the Installation of Centrifugal Fire Pumps*
NFPA 22, *Standard for Water Tanks for Private Fire Protection*
NFPA 24, *Standard for the Installation of Private Fire Service Mains and Their Appurtenances*
NFPA 25, *Standard for the Inspection, Testing, and Maintenance of Water-Based Fire Protection Systems*
NFPA 31, *Standard for the Installation of Oil-Burning Equipment*
NFPA 37, *Standard for the Installation and Use of Stationary Combustion Engines and Gas Turbines*
NFPA 54, *National Fuel Gas Code*

NFPA 58, *Standard for the Storage and Handling of Liquefied Petroleum Gases*

NFPA 70, *National Electrical Code*®

NFPA 72, *National Fire Alarm Code*

NFPA 110, *Standard for Emergency and Standby Power Systems*

Additional Reading

Brown, T. C., "Fire Pumps: Ensuring Performance," *Consulting Specifying Engineer,* Vol. 16, No. 1, July 1994, pp. 64–65.

Cameron Hydraulic Data, 17th ed., Ingersoll Rand Co., New York, 1988.

Damon, W. A., "Testing Building Fire Pumps. Part 2," *Fire Engineering,* Vol. 145, No. 5, May 1992, pp. 59–60, 62, 64–65.

Harvey, B., "Fire Pumps: Problems and Solutions. Part 5. Frequently Encountered Problems," *Sprinkler Age,* Vol. 12, No. 10, Oct. 1993, pp. 20–21, 24–25, 27.

Isman, K. E., "Fire Pump Sensing Lines," *Sprinkler Quarterly,* No. 89, Winter 1994, p. 10.

Titus, J. J., "Hydraulics," *The SFPE Handbook of Fire Protection Engineering,* 2nd ed., DiNenno, P.J., *et al.,* eds., National Fire Protection Association, Quincy, MA, 1995.

WATER SUPPLY REQUIREMENTS FOR FIRE PROTECTION

Revised by Kenneth W. Linder

The amount of water needed and the economics of supplying it are basic questions underlying the planning of water supply systems for fire suppression.

This chapter gives information on planning for public water supply systems and outlines the methods used to evaluate public supplies for their adequacy and reliability. The chapter does not cover the specifics of water supply requirements for automatic extinguishing systems within individual properties. They are covered in the following chapters of Section 6: Chapter 11, "Water Supplies for Sprinkler Systems"; Chapter 12, "Water Spray Protection"; Chapter 13, "Fast Response Sprinkler Technology"; Chapter 15, "Water Mist Suppression Systems"; and Chapter 16, "Standpipe and Hose Systems."

Components of a water supply system are covered in Section 6, Chapter 3, "Water Distribution Systems." Testing procedures to determine adequacy of flows can be found in Section 6, Chapter 7, "Test of Water Supplies."

FACTORS AFFECTING PUBLIC WATER SUPPLY DESIGN

Most public water supply systems serving a substantial number of customers are, and should be, designed for a dual purpose: (1) to supply water for normal domestic demands, such as drinking and sanitary purposes, as well as for processing and industrial uses, and (2) to provide water to fire hydrants for emergency use by fire departments and to supply fixed fire protection systems, such as automatic sprinklers, foam systems, and standpipe systems.

The variety of hazards encountered in most communities and the need to plan for future growth must be considered when evaluating these objectives. In some cities, there may be a heavy industrial demand. At the same time, demands for industrial use and lawn sprinkling may affect the required capacity of the system. The adequacy of a public water system for fire protection cannot be taken for granted, and the other demands must be determined to estimate their effects on the capacity of the system.

Public water systems also must be reliable. Since the occurrence of a fire should not affect domestic demands, the water system must be designed to meet the simultaneous demand rates for both purposes. In addition, portions of the pumping or distribution sys-

tem may be out of service due to breakdown or scheduled maintenance, so the consequences of these interruptions must be examined for acceptability when evaluating system reliability.

DEMANDS FOR DOMESTIC PURPOSES

To determine the domestic demands on a public water supply, it is necessary to focus on variations in water consumption with respect to the time of year, the day of the week, and even the hour of the day. Obviously, as more water is used in a given system for normal consumption, less remains for fire protection. Normal consumption demands are usually expressed in the following terms:

1. Average daily demand, or the average of the total amount of water used each day during a 1-year period.
2. Maximum daily demand, or the maximum total amount of water used during any 24-hr period in a 3-year period. Unusual situations that might have caused an excessive use of water, such as refilling a reservoir after cleaning, should not be considered when determining this figure.
3. Peak hourly demand, or the maximum amount of water used in any given hour of a day.

A joint report of committees from the American Society of Civil Engineers, the American Water Works Association (AWWA), and others suggested that the maximum general service demand on a waterworks system be taken as the peak hourly demand during a test year.[1] This, the report noted, was the only figure that can be fairly compared with the maximum fire flow requirement.

The maximum daily demand can be estimated as 1.5 times the average daily consumption if the actual maximum is not known. The peak hourly rate usually varies from two to four times the normal hourly rate. The effect these varying consumption rates have on the ability of the system to deliver required fire flows varies with the system design. Both maximum daily consumption and peak hourly consumption should be considered to ensure that water supplies and pressures do not reach dangerously low levels during these periods and that adequate water will be available if there is a fire.

WATER REQUIREMENTS FOR FIRE FIGHTING

The water requirements for fire fighting include the rate of flow, the residual pressure required at that flow, and the total quantity required. The American Water Works Association defines the required fire flow as "the rate of water flow, at a residual pressure of 20 psi

Kenneth W. Linder is assistant director of research, Industrial Risk Insurers, Hartford, CT.

and for a specified duration that is necessary to control a major fire in a specific structure."[2]

Calculating Fire Flow Rates

The required flow rate for properties protected by automatic sprinklers is based on the sprinkler system design as required by NFPA 13, *Standard for the Installation of Sprinkler Systems* (hereinafter referred to as NFPA 13). (Also see NFPA 13R, *Standard for the Installation of Sprinkler Systems in Residential Occupancies up to and Including Four Stories in Height*, and NFPA 13D, *Standard for the Installation of Sprinkler Systems in One- and Two-Family Dwellings and Manufactured Homes*.) The flow required is that of the sprinkler system plus the anticipated hose stream, or manual fire-fighting requirements.

There are several methods currently used to calculate required water flow rates for non-sprinklered properties. These include:

- The Insurance Services Office (ISO) method
- The Iowa State University (ISU) method
- The Illinois Institute of Technology Research Institute method.

Insurance Services Office (ISO) method: One of the most comprehensive and widely recommended methods for estimating fire flow requirements is found in the Insurance Services Office's *Fire Suppression Rating Schedule*.[3] It provides guidance for estimating fire flow requirements for specific structures and was designed for insurance rating purposes. The flows determined by this method are generally considered a good estimate, and, as a result, the ISO method has received widespread use. The ISO method considers building construction, occupancy, adjacent exposed buildings, and communication paths between buildings.

The basic formula in the schedule is:

$$NFF_i = (C_i)(O_i)(X + P)_i$$

where NFF_i is needed fire flow (NFF) in gal per min (L/min); C_i is a construction factor that depends upon the construction of the structure under consideration; O_i is an occupancy factor that depends upon the combustibility of the occupancy; and $(X + P)_i$ is an exposure factor that depends upon the extent of exposure from and to adjacent structures.

The subscripts in the formula indicate that, where portions of a building have differing characteristics, a factor can be calculated for each section and multiplied by the percentage it represents of the effective area to obtain a weighted factor. The weighted C_i factor should not be less than the individual factor required for any individual section.

Construction factor. The construction factor, C_i, is calculated from the following formula:

$$C_i = 18F\sqrt{A_i}$$

where

F = coefficient related to the class of construction,

= 1.5 for construction class 1 (frame),

= 1.0 for construction class 2 (joisted masonry),

= 0.8 for construction class 3 (noncombustible) or construction class 4 (masonry, noncombustible),

= 0.6 for construction class 5 (modified fire-resistant) or construction class 6 (fire-resistive), and

A_i = effective building area.

The effective building area is the total square foot (m²) area of the largest floor plus:

- Construction classes 1 through 4—50 percent of all other floors.
- Construction classes 5 and 6—25 percent of the area not exceeding the two other largest floors when all of the vertical openings have at least 1¹/₂-hr-rated protection, or 50 percent of the area not exceeding eight other floors when vertical openings are unprotected or have less than 1¹/₂-hr protection.

The value of C_i should not be less than 500 gpm (1,893 L/min) nor greater than 8,000 gpm (30,280 L/min) for construction classes 1 and 2, and 6,000 gpm (22,710 L/min) for construction classes 3, 4, 5, and 6, or any single, one-story building, regardless of construction.

Occupancy factor. The occupancy factor, O_i, reflects the combustibility of the occupancy on the needed fire flow and is determined from the occupancy combustibility class. Occupancy factors can be found in Table 6-5A. Typical occupancies and their classification can be found in Table 6-5B.

TABLE 6-5A. Occupancy Factors

Occupancy Combustibility Class	Occupancy Factor (O_i)
C-1 (Noncombustible)	0.75
C-2 (Limited-combustible)	0.85
C-3 (Combustible)	1.00
C-4 (Free burning)	1.15
C-5 (Rapid burning)	1.25

TABLE 6-5B. Occupancy Classification

C-1 (Noncombustible)
Steel or concrete products storage, unpackaged

C-2 (Limited-combustible)
Apartments
Ceramics manufacturing
Churches
Concrete or cider products manufacturing
Courthouses
Dormitories
Hospitals
Hotels
Metal products fabrication
Metals inductries (primary)
Motels
Offices
Parking garages
Schools

C-3 (Combustible)
Amusement park buildings, including arcades and game rooms
Automobile sales and services
Baking and confectionery
Dairy processing
Department stores
Discount stores
Food and beverage—sales, service, or storage
General merchandise—sales or storage
Hardware, including electrical fixtures and supplies
Leather processing
Motion picture theaters
Pharmaceutical retail sales and storage
Repair or service shops *continued*

TABLE 6-5B. *Occupancy Classification*

Soft-drink bottling
Supermarkets
Tobacco processing
Unoccupied buildings

C-4 (Free burning)
Aircraft hangers, with or without servicing/repair
Apparel manufacturing
Auditoriums
Breweries
Building material sales and storage
Cotton gins
Food processing
Freight depots, terminals
Furniture—new or second-hand
Metal coating or finishing
Paper and paper product sales and storage
Paper products manufacturing
Printing shops and allied industries
Rubber products manufacturing
Theaters, other than motion picture
Warehouses
Wood product sales and storage
Woodworking industries

C-5 (Rapid burning)
Cereal or flour mills
Chemical manufacturing
Chemical sales and storage
Cleaning and dyeing material sales and storage
Distilleries
Meat or poultry processing
Paint sales and storage
Plastic or plastic product sales and storage
Plastic products manufacturing
Rag sales and storage
Textile manufacturing
Textile products fabrication, except clothes
Upholstering shops
Waste and reclaimed materials sales and storage

Exposure and communication $[(X + P)_i]$ factors. The exposure and communication factors are determined as:

$$(X + P)_i = 1 + \sum_{i=1}^{n} X_i + P_i$$

where

n = number of sides of subject building, and

$(X + P)_i$ has a maximum value of 1.75.

The exposure factor, X_i, reflects the need for additional water to reduce the exposure to adjacent buildings. It depends on the separation distance, the construction of the exposed building wall, and a length-height value—that is, the length of the exposed wall in feet (m) multiplied by the height in stories. Values can be obtained from Table 6-5C.

The communication factor, P_i, reflects the potential fire spread through open or enclosed communicating passageways between buildings and is taken from Table 6-5D. Where there is more than

one connection, only the one with the largest factor is used. Where there are no openings, $P_i = 0$.

Needed fire flow (NFF). The needed fire flow is calculated from the formula given previously and from the above factors. The NFF calculated from the formula should be rounded to the nearest 250 gpm (946 L/min) for flows under 2,500 gpm (9463 L/min) and to the nearest 500 gpm (1893 L/min) for larger flows and then adjusted by the following:

- For buildings with a wood roof, add 500 gpm (1893 L/min).

- The needed flow should not exceed 12,000 gpm (45,420 L/min) nor be less than 500 gpm (1893 L/min). The practical reason for these figures is that manual fire-fighting methods using hose streams and heavy stream appliances are not likely to need a larger supply, considering the general arrangement of buildings and the availability of hydrants.

- For habitational buildings, use the calculated NFF up to 3,500 gpm (13,248 L/min) maximum.

- For groupings of one-family and small two-family dwellings not more than two stories high, the required fire flow given in Table 6-5E may be used.

Iowa State University (ISU) method: The Iowa State University method is another common method used to determine water flow rates for fire fighting. It uses a more theoretical approach and is based on the amount of water needed to deplete the oxygen in a confined area when the water is vaporized into steam by the heat of the fire. Tests conducted by the university indicate that a fire is best controlled if the amount of water necessary to deplete the oxygen is applied within 30 sec.

The required flow in gpm is given as:

$$\text{Required flow} = \frac{V}{100}$$

where V is the enclosed volume in cubic feet. (For SI units: 1 gpm = 3.785 L/min; 1 cu ft = 0.0283 m^3.)

This method is unique, in that it does not consider the occupancy hazard, only the volume of the building to be filled with steam.

Due to inefficiencies in applying water, some experts feel that the rate should be 2 to 4 gal per 100 cu ft (7.6 to 15 L/2.8m^3) of building volume, rather than the 1 gal per 100 cu ft (3.785 L/2.8m^3) in the formula. Other variations include changing the value in the denominator based on the occupancy hazard.

This formula has been used for approximately 30 years and is extremely simple to apply. For most buildings, the entire volume of the structure should be used, including the volume of basements, attics, crawl spaces, and other concealed spaces. For groups of buildings, the largest flow rate should be used.

Illinois Institute of Technology Research Institute method: The Illinois Institute of Technology method was based on a survey of 134 fires in the Chicago area. The results of the survey were used with regression analysis to develop fire flow formulas based on building area. The fire flow rate is based on one of the following formulas:

Flow for residential occupancies = $9 \times 10^{-5}A^2 + 50 \times 10^{-2}A$

Flow for other occupancies = $-1.3 \times 10^{-5}A^2 + 42 \times 10^{-2}A$

where A is the area of the fire in square feet. (For SI units: 1 sq ft = 0.0929 m^2.)

TABLE 6-5C. *Factors for Exposure, X_i*

Construction of Facing Wall of Subject Building	Distance (ft) to the Exposed Building	Length-Height of Facing Wall of Exposed Building		Construction of Facing Wall of Exposed Building Classes		
				2, 4, 5, and 6		
			1 and 3	Unprotected Openings	Semi-Protected Openings (wired glass or outside open sprinklers)	Blank Wall
Frame, Metal or Masonry with Openings	0-10	1-100	0.22	0.21	0.16	0
		101-200	0.23	0.22	0.17	0
		201-300	0.24	0.23	0.18	0
		301-400	0.25	0.24	0.19	0
		Over 400	0.25	0.25	0.20	0
	11-30	1-100	0.17	0.15	0.11	0
		101-200	0.18	0.16	0.12	0
		201-300	0.19	0.18	0.14	0
		301-400	0.20	0.19	0.15	0
		Over 400	0.20	0.19	0.15	0
	31-60	1-100	0.12	0.10	0.07	0
		101-200	0.13	0.11	0.08	0
		201-300	0.14	0.13	0.10	0
		301-400	0.15	0.14	0.11	0
		Over 400	0.15	0.15	0.12	0
	61-100	1-100	0.08	0.06	0.04	0
		101-200	0.08	0.07	0.05	0
		201-300	0.09	0.08	0.06	0
		301-400	0.10	0.09	0.07	0
		Over 400	0.10	0.10	0.08	0

Blank Masonry Wall   Facing wall of the exposed building is higher than subject building.

Use the above table, EXCEPT use only the length-height of facing wall of the exposed building ABOVE the height of the facing wall of the subject building. Buildings five stories or over in height consider as five stories.

When the height of the facing wall of the exposed building is the same or lower than the height of the facing wall of the subject building, $X_i = 0$.

Source: Insurance Services Office © 1980.
For SI units: 1 ft = 0.305 m.

Other Flow Considerations

Regardless of the method used to determine the flow rate, the required fire flow should be available simultaneously with consumption at the maximum daily rate.

When evaluating the flow required for general public protection, both the AWWA and the ISO suggest that 3,500 gpm (13,248 L/min) is the upper limit to be provided and that large facilities or those with severe hazards that need flow rates up to 12,000 gpm (45,420 L/min) be individually analyzed to determine the required flow rate.

There are fires in which quantities of water in excess of the required fire flow are used. Water supplies of 50,000 gpm (189,250 L/min) or greater have been used in fire suppression, but designing systems capable of delivering flows of that magnitude is not cost-effective or practical.

Fire Water Duration

The number of hours during which the required fire flow should be available varies from 2 to 10 hr, as indicated in Table 6-5F. It should be noted that many water authorities place an upper limit of 2 to 4 hr on fire water supply duration due to economics.

Evaluating System Capacity

The capacity of a water system is determined by the total amount of water it must furnish. This is the sum of water required for domestic or industrial uses and water required for the fire service. In small towns, the requirements for fire protection almost always exceed other requirements.

The AWWA recommends that the rate used be either the peak hourly rate or the maximum daily rate plus fire flow, whichever is larger. In most large cities, the peak hourly rate exceeds the maximum daily consumption rate plus fire flow and is therefore the controlling factor in the supply system design. In smaller communities, however, the reverse is true and the maximum daily consumption rate plus fire flow is the controlling factor. For many years, water consumption has been increasing in most municipalities, resulting in increased peak hourly rates. Consequently, there has been an increase in the number of municipalities in which the peak hourly rate controls designs of the supply system. However, there is no guarantee that a fire will not occur at the peak hour, and some recommend that the capacity of the system be sufficient to meet the peak hourly rate plus the fire flow rate.

Fire flows are a very important consideration in all areas served by the distribution system and, in many instances, they govern the size of pipe used in these locations. In all systems, the supply should be sufficient to provide for automatic sprinklers and other automatic

TABLE 6-5D.   *Factors for Communications, P_i*

Description of Protection of Passageway Openings	Fire-Resistive, Noncombustible, or Slow-Burning Communications				Communications with Combustible Construction					
	Open	Enclosed			Open			Enclosed		
	Any Length	10 ft or Less	11 ft to 20 ft	21 ft to 50 ft +	10 ft or Less	11 ft to 20 ft	21 ft to 50 ft +	10 ft or Less	11 ft to 20 ft	21 ft to 50 ft +
Unprotected	0	+ +	0.30	0.20	0.30	0.20	0.10	+ +	+ +	0.30
Single Class A Fire Door at One End of Passageway	0	0.20	0.10	0	0.20	0.15	0	0.30	0.20	0.10
Single Class B Fire Door at One End of Passageway	0	0.30	0.20	0.10	0.25	0.20	0.10	0.35	0.25	0.15
Single Class A Fire Door at Each End or Double Class A Fire Doors at One End of Passageway	0	0	0	0	0	0	0	0	0	0
Single Class B Fire Door at Each End or Double Class B Fire Doors at One End of Passageway	0	0.10	0.05	0	0	0	0	0.15	0.10	0

+ For over 50 ft, $P_i = 0$.
+ + For unprotected passageways of this length, consider the two buildings as a single fire division.
Note: When a party wall has communicating openings protected by a single automatic or self-closing Class B fire door, it qualifies as a division wall* for reduction of area. Where communications are protected by a recognized water curtain, the value of P_i is 0.
For SI units: 1 ft = 0.305 m.

fire protection systems, in addition to the other demand rates imposed upon the system. For example, many smaller cities and large towns restrict lawn watering in summer months to specified periods, usually 2 to 4 hr in the evening. The demand rates imposed by lawn watering are excessive in many water supply systems, depleting storage facilities and reducing pressure throughout the system for many hours. In such situations, there would be little or no water available for fire protection systems, particularly at higher elevations.

Pressure Characteristics of Systems

The pressures for which systems are normally designed are the result of practical attempts to provide adequate pressures for both domestic consumption and for fire protection. If either demands special ranges of pressure, materials and design methods are available that will allow almost any desired range.

For example, San Francisco has a separate system, designated the "high-pressure system," that is under the control of the fire department. All of the pipe is heavy cast-iron, tar-coated and lined, and it is tested on installation and repair to 450 psi (3103 kPa). Two steam-operated pump stations can pump water from San Francisco Bay into the system, and 20,000 gpm (75,700 L/min) at 250 psi (1724 kPa) can be delivered to most of the principal mercantile district. San Francisco provided this system primarily because an earthquake might put the regular public water system out of service. A few other cities have similar high-pressure systems.

Modern fire department pumpers make heavy streams and high pressures available from ordinary water systems where adequate

TABLE 6-5E.   *Fire Flows for Groups of Dwellings*

Exposure Distances		Suggested Required Fire Flow	
ft	m	gpm	L/min
Over 100	30.5	500	1893
31 to 100	9.5–30.5	750–1,000	2839–3785
11 to 30	3.4–9.2	1,000	3785
10 or less	3.1 or less	1,500	5678

TABLE 6-5F.   *Duration of Required Fire Flow*

Required Fire Flow					Required Fire Flow				
gpm	L/min	Million gallons per day	Million liters per day	Duration hours	gpm	L/min	Million gallons per day	Million liters per day	Duration hours
1,000	3785	1.44	5.45	2	4,500	17,033	6.48	24.53	4
1,250	4731	1.80	6.81	2	5,000	18,925	7.20	27.25	5
1,500	5678	2.16	8.18	2	5,500	20,818	7.92	29.99	5
1,750	6624	2.52	9.54	2	6,000	22,710	8.64	32.71	6
2,000	7570	2.88	10.90	2	7,000	26,495	10.08	38.16	7
2,250	8516	3.24	12.26	2	8,000	30,280	11.52	43.61	8
2,500	9463	3.60	13.63	2	9,000	34,065	12.96	49.06	9
3,000	11,355	4.32	16.35	3	10,000	37,850	14.40	54.51	10
3,500	13,248	5.04	19.08	3	11,000	41,635	15.84	59.96	10
4,000	15,140	5.76	21.80	4	12,000	45,420	17.28	65.41	10

volume is provided. Cities that formerly had separate systems of fire mains operating at so-called high pressures now generally keep them at normal public water pressures. The second system still provides an advantage, though, even if it is not at high pressure, because it is still available.

Public water systems reflect a compromise on the question of pressures. Pressures in the range of 65 to 80 psi (448 to 552 kPa) are common. This range, which is adequate for ordinary consumption in buildings up to about ten stories, will provide a good supply of water for automatic sprinkler systems in buildings of about four stories, in which occupancies are classified as "ordinary." Where pressures of this order are provided, it is reasonably easy to compensate for local fluctuations in draft.

Because of the increased cost of energy, greater care is being given to the water pressure that systems should provide. A reduction in water pressure will substantially reduce pumping costs. Before a general reduction is made, however, a study should be conducted as to the effects it could have on sprinklers and other fixed fire protection systems. If a reduction in pressure is planned, it is imperative that the system remain capable of meeting anticipated demand rates, or steps must be taken to lower the demand rates to a point where they will be within the capability of the system.

A minimum residual pressure of least 20 psi (138 kPa) should be maintained at hydrants delivering the required fire flow. Pumpers can be operated, but with difficulty, where hydrant pressures are less. Where hydrants are well distributed and of the proper size and type so that friction losses in the hydrant and suction line will not be excessive, it may be possible to set 10 psi (69 kPa) as the minimum pressure. Sufficient hydrant pressure should be maintained to prevent a negative pressure from developing in the street mains, which might cause back siphonage of polluted water from some interconnected source.

Pressures in a public water system may be considered excessive as they approach 150 psi (1034 kPa). As pressures increase, they tend to cause leaks in domestic plumbing, and special attention is required to restrain pipelines in the ground. Pipe and fittings used in ordinary public water systems are designed for maximum working pressures of 150 psi (1034 kPa), but it is not good practice to operate with pressures that high. Pressure-reducing valves can be used in sections of a system where variations in topography result in excessive pressures. Individual water services to buildings may require pressure-reducing valves to keep the pressure on domestic piping at safe levels.

Systems for Higher Elevations

When water must be supplied to high elevations, a separate water distribution system is usually provided for the elevated section so that reasonable pressures are maintained. In such cases, the elevated area should be provided with its own water storage facility, and pumps may be provided to boost the water from other parts of the system.

Likewise, the upper stories of a high-rise building are sometimes provided with dedicated risers (express risers) to deliver water to these upper floors. High-rise structures are normally divided into a number of pressure zones, and zones of more than 12 stories tend to get outside the normal pressure ranges. In any case, each pressure zone must be provided with water in the amounts needed for sprinkler system and hose stream use, and each system is usually supplied by a series of pumps and tanks arranged so that each zone is fed from the zone below. Care should be taken to ensure that pumps will be able to operate even during power failures.

For information and guidance concerning water supplies for high-rise structures, see NFPA 13; NFPA 14, *Standard for the Installation of Standpipe and Hose Systems*; and NFPA 20, *Standard for the Installation of Centrifugal Fire Pumps*.

ADEQUACY AND RELIABILITY OF SUPPLY

The adequacy of any given water supply system can be determined by engineering estimates. The source (e.g., storage facilities and distribution system) must be sufficient to furnish all the water that combined fire and domestic needs may call for at any one time. The arrangement of the supply and pumping facilities may render the supply inadequate or affect its reliability.

Pumping systems commonly are arranged so that one set of pumps takes suction from wells or from a river, lake, or other body of water. If the water does not have to be filtered, the pumps may discharge directly into the distribution system. Where filtration or other treatment is necessary, pumps take suction from the primary or raw water source and discharge to sedimentation basins or other facilities and then to filter beds. After the water is processed, it flows to clear-water reservoirs from which a second set of pumps takes suction and discharges the water directly into the supply system. Unfortunately, failure of any part of the equipment may affect the entire system.

When assessing the reliability of the supply works, one should evaluate: minimum yield; the frequency and duration of droughts; the condition of intakes; the possibility of earthquakes, floods, and forest fires; ice formations; silting up or shifting of river channels; and the absence of guards or watchmen, where needed, to protect the facility from physical injury. Reservoirs out of service for cleaning and the interdependence of parts of waterworks also affect reliability. The condition, arrangement, and dependability of individual units of plant equipment, such as pumps, engines, generators, electric motors, fuel supply, electric transmission facilities, and similar items, are also factors. Pumping stations of combustible construction may be destroyed by fire unless they are protected by automatic sprinkler systems.

Duplication of pumping units and storage facilities, and arrangement of mains and distributors so that water may be supplied to any area from more than one direction, are measures that can ensure continuous operation. The importance of duplicate facilities is shown by the frequency of their use. Many utilities design their systems so that the peak hourly rate and required fire flow rate can be supplied when any pump or section of the distribution system is out of service.

The amount of water needed to control and extinguish a fire in a given property currently cannot be established in precise terms. Better fire experience data bases should make it possible to tailor fire flows more specifically to conditions that might be expected at the time of a fire. Better analysis may indicate a need to increase fire flow beyond what is presently required, or it may result in a water system design based upon a balance between the risk involved and the economics of maintaining the water system.

A detailed discussion of all the factors to be considered in designing a water supply system is beyond the scope of this handbook. An overview of the subject can be found in AWWA M24, *Dual Water Systems*, and AWWA M31, *Distribution System Requirements for Fire Protection*.

THE ROLE OF CODES AND ORDINANCES

Fire prevention codes can effectively limit hazards and ignition sources within buildings which, in turn, limits the number and size of fires by controlling combustibles in a fire area. A good building code further reduces the chance for a serious fire by requiring construction materials and building assemblies that will contain a developing fire to a given area. Codes alone can reduce considerably the amount of water needed for fire fighting. Zoning ordinances that establish distances between properties can be effective in controlling exposure situations.

THE ROLE OF FIRE DETECTION AND EXTINGUISHING SYSTEMS

The increased use of automatic extinguishing systems, whether they use water or some other agent, will affect the quantities of water required. Until more widespread use is made of early warning systems and automatic extinguishing systems, however, it will not be possible to equate the effect of these systems with required fire flow. Water supply requirements are just one factor in a system that determines what the potential for a fire is, how extensive that fire will be, and what measures will be needed to suppress it. Someday, research will measure all these factors and allow us to establish fire flows on the basis of thoroughly researched and documented principles.

HISTORY

Historically, water supply systems for cities and towns were developed primarily to provide drinking water and water for sanitary purposes, rather than for fire protection. However, it was found that large cities that required substantial amounts of water for domestic purposes usually had enough water to provide a useful supply for fire-fighting purposes.

This led to inquiries in the late 19th century into the additional cost of waterworks that could provide water for fire fighting, as well as other uses. A number of distinguished engineers associated with individual waterworks examined the problem and presented their findings in technical papers at engineering society meetings. Papers by J. Herbert Shedd,[4] J. T. Fanning,[5] and Emil Kuichling[6] give the details of discussions in which standards for American and Canadian waterworks were first developed. (See Table 6-5G.)

TABLE 6-5G.   Estimates of Fire Flow

Population, Thousands	Number of Fire Streams Required Simultaneously				
	Shedd 1889[4]	Fanning 1892[5]	Freeman 1892[7]	Kuichling 1897[6]	NBFU 1911[8]
1			2–3	3	4
4		7		6	8
5	5		4–8	6	9
10	7	10	6–12	9	12
20	10		8–15	12	17
40	14		12–18	18	24
50		14		20	26
60	17		15–22	22	28
100	22	18	20–30	28	36
150		25		34	44
180	30			38	48
200			30–50	40	48

Number of Hose Streams

In these discussions, the starting point for computing the cost of water for fire protection was to estimate the number of hose streams that a fire department might need for fire fighting. This was usually estimated using as a basis the central portion of a city where the largest buildings were located and where there was the greatest

building congestion. The number of streams was found to be related, in a very rough way, to the population. Shedd's proposal,[4] which was the first, used hose streams discharging 200 gpm (757 L/min). He suggested that a community of 5,000 would need about five such streams and that the needs of other cities could be graduated up to 30 streams in a city of 180,000. Fanning[5] proposed streams requiring about 54 psi (372 kPa) pressure. His figures were similar to Shedd's,[4] beginning at seven streams for a community of 4,000 and going up to 25 streams for a city of 150,000.

Kuichling[6] suggested a formula in which the number of streams required would be the square root of the population in thousands multiplied by 2.8. There were arithmatical differences as to how these estimates worked out for individual cities, but they were all of the same general order. (See Table 6-5G.) Most important, they provided a basis from which waterworks designers could estimate costs.

The most important paper produced during this time was John R. Freeman's "The Arrangement of Hydrants and Water Pipes for the Protection of a City Against Fire," published in 1892.[7] Freeman had done the fundamental work on water flow through hose and nozzles, so he was able to pin down the definition of a standard fire stream to one with a discharge of 250 gpm (946 L/min) at 40 to 50 psi (276 to 345 kPa) pressure. He said that the relationships suggested by Shedd[4] and Fanning[5] between population and the number of streams required were of the right order, but he did not think the needs of individual cities could be quite so definitely ascertained. He suggested two to three streams as a minimum for a population of 1000, graduated up to 30 to 50 streams at 200,000. (See Table 6-5G.) Most significantly, he warned that, "Ten streams, or as large a proportion thereof as the financial consideration will permit, may be recommended for a compact group of large, valuable buildings, irrespective of a small population."[7]

ENGINEERING: DISTRIBUTION NETWORK, HYDRANT SPACING, STORAGE

Freeman[7] noted a fundamental difference between systems designed to supply ordinary water needs and those for fire protection: Fire draft required concentration of the water, whereas domestic draft was a matter of distribution.

Freeman[7] asserted that, if a water system was to supply fire protection needs, the distribution system should be designed to concentrate the needed amounts of water. Small pipes were sufficient for distribution, but larger ones were needed to concentrate supply to fire streams. He suggested 6-in. (152-mm) diameter pipe as the minimum for residential districts and noted that 8-in. (203-mm) pipe was adequate only if it formed part of a network of distribution pipes whose intersections were not far apart.

Another important point Freeman[7] made was that hydrants should be placed where they could concentrate streams at specific blocks or groups of buildings, rather than placed arbitrarily a certain number of feet (m) apart on the street mains. His work on hose streams showed how long hose lines reduced the water that can be delivered promptly to a fire. He therefore suggested a working rule for hydrant spacing of 250 ft (76 m) between hydrants in compact mercantile and manufacturing districts, and 400 ft (122 m) to 500 ft (153 m) in residential districts. These working rules can still be used as guides for good design.

Freeman[7] further insisted that fire supply should be in addition to maximum domestic consumption, and laid the foundation for the eventual recognition of this principle. He also calculated how much water should be stored in standpipes or in elevated reservoirs. He figured that flow for all of the hose streams required should be supplied from a reliable source, such as an elevated storage reservoir, for at least 6 hr when the system was also furnishing maximum demands for domestic and other uses and he calculated that, to supply the combined fire and domestic needs in a system provided with reliable pump capacity, a 1-hr supply in a standpipe or elevated reservoir would be acceptable.

The Insurance Grading Schedule

As early as 1889, the National Board of Fire Underwriters (NBFU) began to make fire protection surveys of municipalities. This work was intensified in 1904 after a conflagration in Baltimore, MD. Today, the ISO surveys communities in all but a few states. The ISO survey, called the *Fire Suppression Rating Schedule*,[3] was discussed in detail earlier in this chapter.

The current ISO procedure for establishing fire flows is different from the original NBFU formula, which was commonly used for many years as a guide to determine the fire flow required in downtown business districts of municipalities. (See Table 6-5G.) The formula gave the fire flow, G, in gpm as a function of the population, P, in thousands. (For SI units: 1 gpm = 3.785 L/min.)

$$G = 1020\sqrt{P}\left(1 - 0.01\sqrt{P}\right)$$

In making fire protection surveys, engineers from NBFU and insurance bureaus estimated the fire flow requirements in sections of municipalities outside the downtown business district.

As cities became more decentralized, the formula based on population became less reliable as a guide for the fire flow needed downtown. In addition, it became apparent that a guide to engineering judgment was needed for the other sections of the cities. In 1948, a paper by A. C. Hutson,[9] assistant chief engineer at NBFU, provided some specific suggestions for estimating fire flow requirements in these sections, based on the type of construction and the area of a building. This eventually led to the current ISO rating system.

BIBLIOGRAPHY

References Cited

1. ASCE, "Fundamental Considerations in Rates and Rate Structures for Water and Sewage Works: A Joint Report of Committees of the American Society of Civil Engineers and the Section of Municipal Law of the American Bar Association and of Representatives of the American Water Works Association, National Association of Railroad and Utilities Commissioners, Municipal Finance Officers Association, Federation of Sewage Works Association, American Public Works Association, and Investment Bankers Association of America," *ASCE Bulletin No. 2*, American Society of Civil Engineers, New York, 1951.
2. AWWA M31, *Distribution System Requirements for Fire Protection*, American Water Works Association, Denver, CO, 1989.
3. *Fire Suppression Rating Schedule*, Insurance Services Office, New York, 1980.
4. Shedd, J. H., *Discussion* on a paper by William B. Sherman, "Ratio of Pumping Capacity to Maximum Consumption," *Journal of New England Water Works Association*, Vol. 3, 1889, p. 113.
5. Fanning, J. T., "Distribution Mains and the Fire Service," *Proceedings* of the American Water Works Association, Vol. 12, 1892, p. 61.
6. Kuichling, E., "The Financial Management of Water Works," *Transactions* of the American Society of Civil Engineers, Vol. 38, 1897, p. 16.
7. Freeman, J. R., "The Arrangement of Hydrants and Water Pipes for the Protection of a City Against Fire," *Journal of the New England Water Works Association*, Vol. 7, 1892, p. 49.
8. Metcalf, L., Kuichling, E., and Hawley, W. C., "Some Fundamental Considerations in the Determination of a Reasonable Return for Public Fire Hydrant Service," *Proceedings* of the American Water Works Association, Vol. 31, 1911, p. 55.
9. Hutson, A. C., "Water Works Requirements for Fire Protection," *Journal of the American Water Works Association*, Vol. 40, No. 9, 1948, p. 936. [Also reprinted in Special Interest Bulletin No. 266, National Board of Fire Underwriters (now American Insurance Service Group), New York.]

Other Reference

Davis, L. W., *Rural Firefighting Operations*, International Society of Fire Service Instructors, Ashland, MA, 1986.

NFPA Codes, Standards, and Recommended Practices

Reference to the following NFPA codes, standards, and recommended practices will provide further information on water supply requirements for fire protection discussed in this chapter. (See the latest version of *The NFPA Catalog* for availability of current editions of the following documents.)

NFPA 13, *Standard for the Installation of Sprinkler Systems*
NFPA 13D, *Standard for the Installation of Sprinkler Systems in One- and Two-Family Dwellings and Manufactured Homes*
NFPA 13R, *Standard for the Installation of Sprinkler Systems in Residential Occupancies up to and Including Four Stories in Height*
NFPA 14, *Standard for the Installation of Standpipe and Hose Systems*
NFPA 20, *Standard for the Installation of Centrifugal Fire Pumps*
NFPA 24, *Standard for the Installation of Private Fire Service Mains and Their Appurtenances*
NFPA 1231, *Standard on Water Supplies for Suburban and Rural Fire Fighting*

Additional Reading

"AFSA TAC Voices Praise, Concern for Limited Water Supply Sprinkler System," *Sprinkler Age*, Vol. 11, No. 5, May 1992, pp. 19–20.
Bill, R. G., Jr. and Hill, E. E., Jr., "Sprinkler Protection of Manufactured Homes With Sloped Ceilings Using Prototype Limited Water Supply Sprinklers," *Fire Technology*, Vol. 31, No. 1, First Quarter 1995, pp. 17–43.
Bill, R. G., Jr. and Kung, H. C., "Limited-Water-Supply Sprinklers for Manufactured (Mobile) Homes," *Fire Technology*, Vol. 29, No. 3, Third Quarter 1993, pp. 203–205.
Bussen, R., "Anschlub von Feuerloschanlagen an Trinkwassernetze. [Connection of Fire Extinguishing Systems to Drinking Water Supply Systems.]," *VFDB*, Vol. 44, No. 1, Feb. 1995, pp. 2–4.
"Cabin Water Sprays for Fire Suppression: Design Considerations and Safety Benefit Analysis Based on Past Accidents," Civil Aviation Authority, London, England, CAA Paper 93010, Aug. 1993.
Carey, B., "Reduction in the Quantity of Stored Water for MHSS Systems," Underwriters Laboratories, Inc., Northbrook, IL, National Institute of Standards and Technology, *Proceedings* of the Demonstration of Limited Water Supply Fire Suppression Technologies. Proceedings. March 12, 1996, Gaithersburg, MD, 1996, pp. 1–24.
Clark, W. E., "Fighting Fire With Water: Using Water Wisely," *Fire Engineering*, Vol. 148, No. 9, Sept. 1995, pp. 33–34, 38, 40.
Cooper, K., "Water Supply Problems at Uppark House Fire," *Fire*, Vol. 87, No. 1074, Dec. 1994, pp. 13–14, 16, 18.
Costigan, E. J., "Rural Water Supply," *Fire Engineering*, Vol. 142, No. 3, 1989, pp. 93–95.
Damon, W. A., "Testing Pressure-Regulating Devices in Water-Based Fire Protection Systems," *Fire Engineering*, Vol. 146, No. 11, Nov. 1993, pp. 51, 55–57.
Eckman, W. F., "Writing Specs for Mobile Water Supply Apparatus," *Fire Engineering*, Vol. 143, No. 11, Nov. 1990, pp. 40–44, 46–48.
"Environmental Assessment: Installation and Operation of the Plantwide Fire Protection Systems and Related Domestic Water Supply Systems," Department of Energy, Aiken, SC, DOE/EA-0476, December 1991.
Gamble, C. C., Jr., "Training the Water Supply Officer," *Fire Chief*, Vol. 37, No. 10, Oct. 1993, pp. 43–45.
Gamble, C., "Pre-Planning Rural Water Supplies," *Fire Chief*, Vol. 39, No. 3, Mar. 1995, pp. 68, 70, 72.
Hussey, J., "Water Supply Systems and Devices Used in Vapor Mitigation," *Industrial Fire Safety*, Vol. 2, No. 2, Mar./Apr. 1993, pp. 33–37.

Keith, G. and Garner, D., "Sprinkler System Water Supply Analysis," *Fire Engineering*, Vol. 144, No. 12, Dec. 1991, pp. 67–68, 70–71.

Laughlin, J. W., and Williams, C., eds., *Water Supplies for Fire Protection*, 4th ed., IFSTA, Oklahoma State University, Stillwater, OK, 1988.

"Limited Water Supply Fire Suppression Technologies Demonstrated at NIST," *U.S. Fire Sprinkler Reporter*, Vol. 10, No. 10, Apr. 1996, pp. 1, 3–10.

Loikkanen, P. and Tuomisaari, M., "Suppression of Compartment Fires With a Small Amount of Water. Part 1," VTT-Technical Research Center of Finland, Espoo, Research Report PAL2192/92, Dec. 30, 1992.

Madrzykowski, D., "Evaluation of Limited Water Supply Fire Suppression Technologies," Progress Report, National Institute of Standards and Technology, Gaithersburg, MD; National Institute of Standards and Technology, *Proceedings* of the Demonstration of Limited Water Supply Fire Suppression Technologies, March 12, 1996, Gaithersburg, MD, 1996, pp. 1–14.

Malone, G. and Hession, M., "Failure of Water Supplies Hampered Firefighting at Cork Chemical Plant," *Fire*, Vol. 86, No. 1062, Dec. 1993, pp. 23–24.

Mawhinney, J. R., "Engineering Criteria for Water Mist Fire Suppression Systems," National Research Council of Canada, Ottawa, Ontario, NISTIR 5207, June 1993; National Institute of Standards and Technology, *Proceedings* of the Water Mist Fire Suppression Workshop, March 1–2, 1993, Gaithersburg, MD, 1993, pp. 37–73.

Mawhinney, J. R. and Hadjisophocleous, G. V., "Role of Fire Dynamics in Design of Water Mist Fire Suppression Systems," National Research Council of Canada, Ottawa, Ontario, ISBN 0-9516320-9-4; *Proceedings* of the 7th International Interflam Conference, Interflam '96, March 26–28, 1996, Cambridge, England, Interscience Communications Ltd., London, England, 1996, pp. 415–424.

Mawhinney, J. R., "Design of Water Mist Fire Suppression Systems for Shipboard Enclosures," National Research Council of Canada, Ottawa, Ontario, SP Report 1994:03; *Proceedings* of the International Conference on Water Mist Fire Suppression Systems, November 4–5, 1993, Boras, Sweden, 1993, pp. 16–44.

McPhee, R. A., "Questions About Limited-Water-Supply Sprinklers for Manufactured (Mobile) Homes," *Fire Technology*, Vol. 30, No. 1, First Quarter 1994, pp. 179–182.

Milke, J. A. and Bryan, J. L., "Development of a Technique to Assess the Adequacy of the Municipal Water Supply for a Residential Sprinkler System," Maryland Univ., College Park, National Institute of Standards and Technology, Gaithersburg, MD, Fire Administration, Emmitsburg, MD, NIST-GCR-91-600, Nov. 1991.

Notarianni, K. A., "Water Mist Fire Suppression Workshop Summary," *SFPE Bulletin*, Summer 1993, pp. 8–9.

Notarianni, K. A. and Jason, N. H., "Water Mist Fire Suppression Workshop," National Institute of Standards and Technology, Gaithersburg, MD, NISTIR 5207, June 1993; National Institute of Standards and Technology, *Proceedings* of the Water Mist Fire Suppression Workshop, March 1–2, 1993, Gaithersburg, MD, 1993, 156 pages.

Scheffey, J. L. and Williams, F. W., "Extinguishment of Fires Using Low Flow Water Hose Streams. Part 1," *Fire Technology*, Vol. 27, No. 2, May 1991, pp. 128–144.

Scheffey, J. L. and Williams, F. W., "Extinguishment of Fires Using Low Flow Water Hose Streams. Part 2," *Fire Technology*, Vol. 27, No. 4, Nov. 1991, pp. 291–320.

Shaw, H., "Limited Water Supply Sprinklers," VHS Video Tape, Consultant, Federal Fire Forum, Developments in Fire Protection, November 6, 1992, Gaithersburg, MD, 1992.

Stephens, J., "Improving Sprinkler Fire Control With Lower Water Pressures," *Fire Prevention*, No. 257, Mar. 1993, pp. 32–36.

Stroup, D. W. and Evans, D. D., "Suppression of Post-Flashover Compartment Fires Using Manually Applied Water Sprays," National Institute of Standards and Technology, Gaithersburg, MD, Brandforsk, Stockholm, Sweden, General Services Administration, Washington, DC, NISTIR 4625, July 1991.

Tammisaari, M., "Huoneistopalon sammuttaminen pienella vesimaaralla. [Extinguishing Apartment Fires With Small Quantities of Water.]," VTT-Paloteknikan Laboratorio, Espoo, *Palontorjuntateknikka*, Vol. 4, 1991, pp. 24–27.

Tammisaari, M., "Huoneistopalon sammuttaminen pienella vesimaaralla. Osa 2: Suihkuputkien vertailukokeet. [Extinguishing Apartment Fires With Small Quantities of Water. Part 2: Nozzle Comparisons," VTT-Paloteknikan Laboratorio, Espoo, *Palontorjuntatekniikka*, Vol. 1, 1994, pp. 10–15.

Tuomisaari, M., "Suppression of Compartment Fires With a Small Amount of Water," VTT-Technical Research Center of Finland, Espoo, SP Report 1994:03; *Proceedings* of the International Conference on Water Mist Fire Suppression Systems, November 4–5, 1993, Boras Sweden, 1993, pp. 167–170.

vanZyl, J. E., "Water Supplies in the RSA Not Good Enough!" *Fire Protection*, Vol. 20, No. 3, Sept. 1993, pp. 12–15.

Yang, J. C., Boyer, C. I. and Grosshandler, W. L., "Minimum Mass Flux Requirements to Suppress Burning Surfaces With Water Sprays," National Institute of Standards and Technology, Gaithersburg, MD, NISTIR 5795, April 1996.

HYDRAULICS

Revised by Kenneth W. Linder

Fire protection hydraulics, a category of fluid mechanics, involves the flow of water through pipes and orifices, such as hydrant outlets, hose nozzles, or sprinklers. This chapter describes the physical properties of water that are pertinent to hydraulic calculations and the formulas used to calculate flow and pressure loss in fire protection systems.

HYDRAULIC PROPERTIES OF WATER

As used in this handbook, water refers to fresh water unless otherwise specified. All calculations are made in U.S. gallons (gal), unless otherwise indicated. One U.S. gal equals 3.78 L. An Imperial gal equals 1.20 U.S. gal (4.54 L).

Material Properties

Density: Density, ρ, is defined as mass per unit volume;

$$\rho = \left(\frac{\text{mass}}{\text{volume}}\right)$$

The density of water, as with many other liquids, varies with temperature. The maximum density of water occurs at 39.2°F (4.0°C) and is 62.43 lbm (pound mass) per cu ft [1000 kg/m³ (kilograms per cubic meter)] in vacuum or 62.35 lbm per cu ft (998.7 kg/m³) in air. For most hydraulic calculations, an approximate value of 62.4 lbm per cu ft (1000 kg/m³) is usually used. Average seawater has a density of 64.1 lbm per cu ft (1030 kg/m³) at 39.2°F (4.0°C).

Specific weight: The specific weight of a material, w, is defined as $w = \rho g$, where g is the acceleration due to gravity. The specific weight is usually measured in lbf (pound force) per cu ft in U.S. custom units and kgf (kilogram force)/m³ in SI units and is:

$$w = \rho g = 62.4 \frac{\text{lbm}}{\text{ft}^3} \times 32.2 \frac{\text{ft}}{\text{sec}^2} \times 1 \frac{\text{lbf sec}^2}{32.2 \text{ lbm ft}}$$

$$= 62.4 \frac{\text{lbf}}{\text{ft}^3} \left(1000 \text{ kgf per m}^3\right)$$

A common, though incorrect, practice is to use the terms pound mass (lbm) and pound force (lbf) interchangeably since one

lbm has a weight of one lbf under standard gravity. [A similar situation exists in SI units with kg (kilogram) and kgf (kilogram force).] In this chapter, pound (lb) means pound force (lbf) and kilogram (kg) means kilogram force (kgf), as is common in engineering practice.

One cu ft (0.028 m³) equals 7.48 U.S. gal. Assuming the specific weight of water is 62.4 lb per cu ft (1000 kg/m³), one gal of water therefore weighs 62.4 lb per cu ft ÷ 7.48 gal per cu ft, or 8.34 lb (3.78 kg).

Viscosity: Viscosity is a measurement of a fluid's resistance to flow, and is usually measured in pound sec per square foot (lb sec/ft²) in U.S. custom units or newton sec per square meter (N sec/m²) in SI units. Viscosity, like density, varies with temperature. At 32°F (0.0°C), water has an absolute viscosity, μ, of 3.746×10^{-5} lb sec/ft² (1.793×10^{-5} N sec/m²). In hydraulic problems, viscosity is often divided by density. This relative viscosity, called kinematic viscosity, υ, is defined as:

$$\upsilon = \frac{\mu}{\rho}$$

While viscosity is an important factor in fluid flows, most fire protection hydraulics applications assume water at ambient conditions, and the empirical formulas normally used to calculate losses do not take into account changes in viscosity.

Pressure

Pressure, p, is the unit that measures the force, caused by compression, per unit area in a fluid. In fire protection hydraulics, pressure is normally measured in pounds per square inch (psi) or in kilopascals (kPa), as indicated by a pressure gage, or as head, h, in feet (ft) or meters (m) of water. Pressure is also commonly measured as a head of mercury.

For water flow in pipes, the total pressure, p_t, is the sum of normal pressure, p_n, and velocity pressure, p_υ.

$$p_t = p_n + p_\upsilon$$

Normal pressure: Net or normal pressure is the pressure exerted against the side of a pipe or container by the liquid in the pipe or container with or without flow. Without flow, this pressure is called "static pressure" or "static head." With flow, this pressure is called "residual" pressure.

Kenneth W. Linder is assistant director of research for Industrial Risk Insurers, Hartford, CT.

The pressure exerted by a column of water is related to its specific weight. Expressed slightly differently, the specific weight is:

$$\frac{62.4 \text{ lb}}{\text{ft}^3} = \frac{62.4 \text{ lb}}{\text{ft}^2 ft} \times \frac{1 \text{ ft}^2}{144 \text{ in.}^2} = 0.433 \text{ psi per ft}$$

It follows that pressure and static head are related by the following formulas:

$$p = wh = 0.433h$$

$$h = \frac{p}{w} = \frac{p}{0.433} = 2.31p$$

For SI units, the weight of a 1-m column of water equals a force of 9.81 kPa or:

$$p = 9.81h$$

$$h = \frac{1}{9.81} = 0.102p$$

A 1-in. (25.4-mm) head of mercury generates a pressure of 0.491 psi (3.39 kPa), and is equivalent to a 1.135-ft (0.3456-m) head of water.

Normal atmospheric pressure is taken as 14.7 psi (101.4 kPa), equivalent to a head of water of 33.95 ft (10.35 m) and a head of mercury of 29.9 in. (760 mm).

Velocity head or velocity pressure: The velocity, v, produced in a mass of water by pressure acting upon it is the same as if the mass was to fall freely, starting from rest, through a distance equivalent to the pressure head in feet. This relationship is represented by Torricelli's equation:

$$v = \sqrt{2gh}$$

where

v = the velocity produced in ft/sec (m/sec), g = the acceleration due to gravity or 32.2 ft/sec² (9.81 m/sec²), and h = the head in ft (m) producing the velocity.

Just as a static head can be converted into a velocity head, velocity head can be converted to an equivalent static pressure head. This relation is:

$$h_v(\text{velocity head}) = \frac{v^2}{2g}(\text{ft})$$

Because

$$p_v = 0.433h_v$$

the velocity pressure can be expressed as:

$$p_v = 0.433\frac{v^2}{2g}(\text{psi})$$

In SI units:

$$p_{vm}(\text{kPa}) = 9.81\frac{v^2}{2g}$$

for v in m/sec.

Values of velocity pressure for different rates of flow in various pipe sizes are shown in Figure 6-6A. (For SI units see Figure 6-6B.) Velocity head or velocity pressure may be calculated by formulas involving velocity and pipe diameter:

$$h_v = \frac{v^2}{64.4} \text{ or } p_v = \frac{0.433v^2}{64.4} = \frac{v^2}{149}$$

In SI units:

$$h_{vm} = 0.0151v^2 \text{ and } p_{vm} = 0.5v^2$$

A convenient equation for calculating velocity in feet per second (fps) from the rate of flow can be developed from the principle of conservation of mass. For a steady-state one-dimensional flow with average velocity v, this principle can be stated as

$$Q = av$$

It follows that:

$$v = \frac{Q}{a}$$

where

v = the average velocity in ft/sec, Q = the flow in cu ft/sec, and a = the cross-sectional area of the pipe in sq ft.

For a pipe with flow in gpm and diameter in in., the velocity is:

$$v(\text{fps}) = \frac{Q \text{ (gal/min)}}{60 \text{ sec/min} \times 7.48 \text{ gal/ft}^3} \div \frac{\pi d^2 (\text{in})^2}{4 \times 144 \text{ in}^2 / \text{ft}^2}$$

$$= \frac{Q \times 4 \times 144}{60 \times 7.48 \times \pi d^2} = \frac{0.4085 \times Q}{d^2}$$

It follows that h_v and p_v are:

$$h_v = \frac{v^2}{2g} = \frac{(0.4085Q)^2}{(d^2)^2} \div 64.4 = \frac{Q^2}{d^4}\frac{(0.4085)^2}{64.4} = \frac{Q^2}{386d^2}$$

$$p_v = \frac{Q^2}{d^4} \times \frac{0.433}{386} = \frac{Q^2}{891^4}$$

In SI units, the formula for velocity pressure is expressed as:

$$p_{vm} = 225\frac{Q_m^2}{d_m^4}$$

where

p_{vm} = velocity pressure, kPa,

Q_m = flow, L/min, and

d_m = inside diameter, mm.

EXAMPLE 1:

Find the velocity pressure in 1-in. Schedule 40 pipe with 36 gpm flowing. The actual diameter of the pipe is 1.049 in.

SOLUTION:

$$p_v = \frac{Q^2}{891(d)^4} = \frac{36^2}{891(1.049)^4} = 1.20 \text{ psi}$$

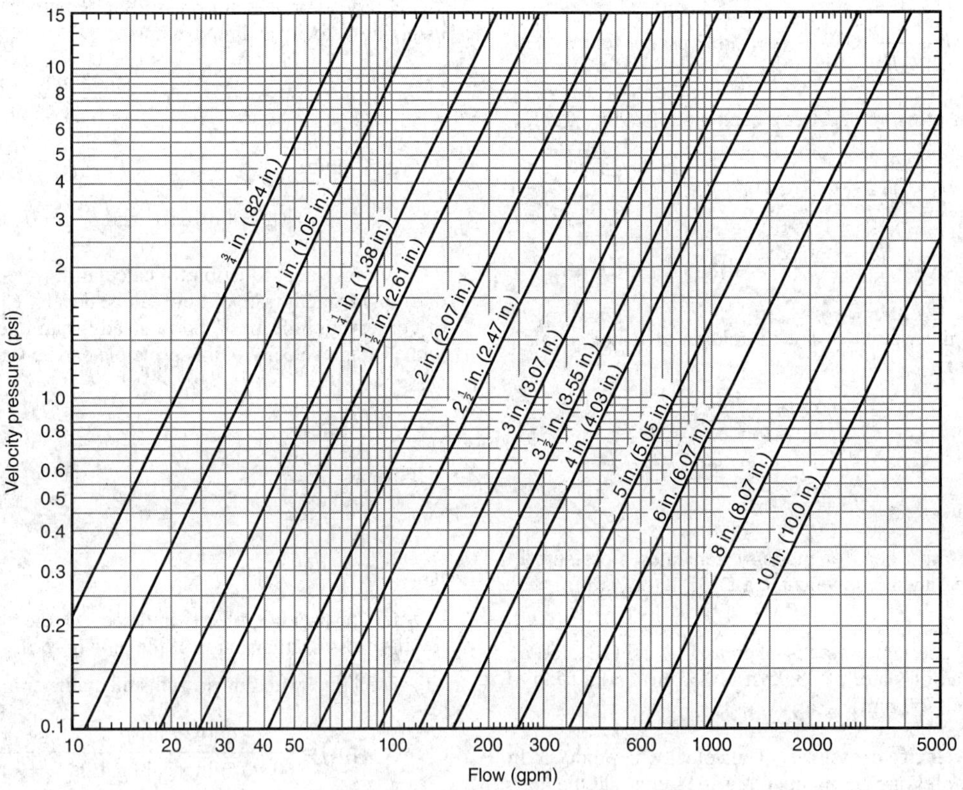

FIG. 6-6A.   *Graph for the determination of velocity pressure.*

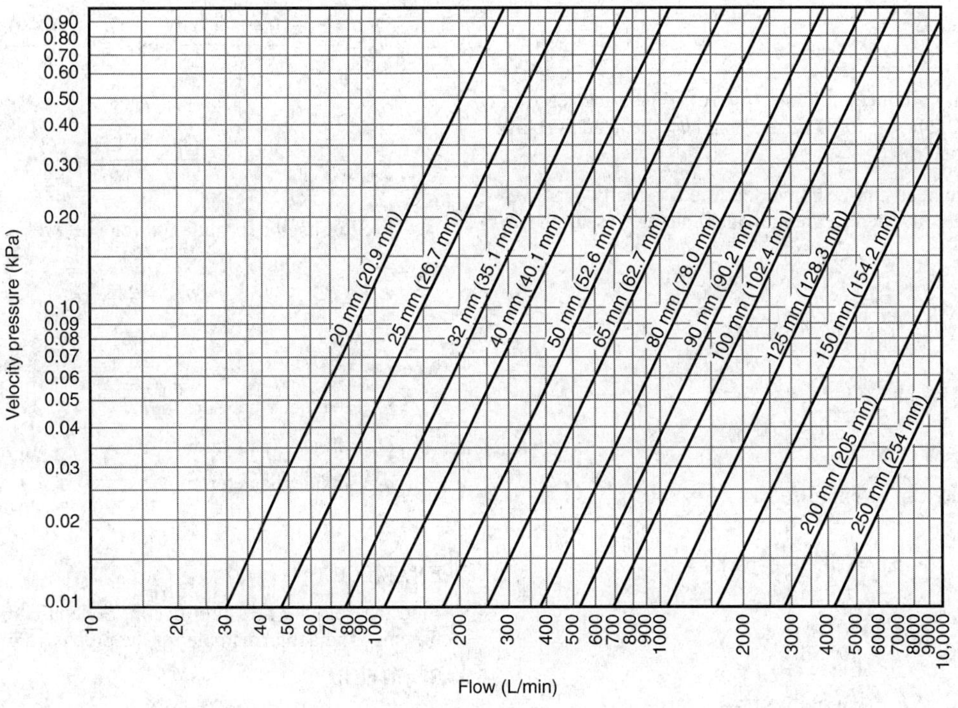

FIG. 6-6B.   *Graph for the determination of velocity pressure (SI units).*

EXAMPLE 2:

Find the velocity pressure in a 25-mm (actual diameter) pipe with 100 L/m flowing.

SOLUTION:

$$p_{vm} = 225 \frac{Q_m^2}{(d_m)^4} = \frac{225 \times (100)^2}{(25)^4} = 5.8 \text{ kPa}$$

Total head: At any point within a piping system that contains water in motion, there is a pressure head, h_p (normal pressure head), acting perpendicular to the pipe wall independent of velocity, and a velocity head, h_v, acting parallel to the wall but exerting no pressure against the wall. Therefore, the total head, $H = h_p + h_v$, expressed as pressure (psi) instead of feet is:

$$p_t = 0.433 h_p + 0.433 \frac{v^2}{2g}$$

In SI units, total head expressed in kPa is:

$$p_{tm} = 9.81 h_{pm} + 9.81 \frac{v_m^2}{2g}$$

where

p_{tm} = total pressure (kPa),

h_{pm} = head (m), and

v_m = velocity (m/s).

Pressure Sources

The sources of pressure head commonly found in fire protection hydraulic systems include the following.

Gravity (elevated tanks, reservoirs, standpipes): Head is the elevation of the water supply surface above the point under consideration, measured directly in ft (m) or converted from a pressure gage reading.

Pumping: Head is the combination of pump discharge pressure and any difference in elevation between the pump discharge gage and the point under consideration.

Pneumatic (pressure tanks): Head is the tank air pressure combined with any difference in elevation in tank water surface and the point under consideration.

Combination: Any combination of the above pressure sources.

BERNOULLI'S THEOREM

Bernoulli's theorem expresses the physical law of conservation of energy applied to problems of incompressible fluid flow. The theorem can be defined as follows: "In steady flow without friction, the sum of the velocity head, pressure head, and elevation head is constant for any incompressible fluid particle throughout its course." In other words, the total pressure (head) is the same at all locations within the system.

[Note that, in Bernoulli's theorem, all of the individual head terms, i.e., velocity head, pressure head, elevation head, and lost head, are expressed in ft (m). When using velocities in fps (m/sec) and gage pressure in psi (kPa), they must be converted to ft (m), or all of the terms expressed as pressures.]

Real systems are not frictionless, however, and in practice, losses due to pipe friction and other factors are accounted for. Expressed mathematically, Bernoulli's theorem, when applied to locations "A" and "B," is:

$$\frac{v_A^2}{2g} + \frac{p_A}{w} + z_A = \frac{v_B^2}{2g} + \frac{p_B}{w} + z_B + h_{AB}$$

where

v (v_m) = velocity in fps (m/s),

g (g_m) = acceleration of gravity [32.2 fps (9.81 m/sec²)],

p (p_m) = pressure [lb/sq ft (kPa)]

z (z_m) = elevation head [(distance above assumed datum), in ft (m)], and

w (w_m) = specific weight of the fluid in lb/cu ft (64.4 pcf or 9.81 kN/m³ for water).

$\dfrac{v^2}{2_g} \left(\dfrac{v_m^2}{2_g} \right) =$ velocity head in ft (m)

$\dfrac{p}{\omega} \left(\dfrac{p_m}{\omega_m} \right) =$ pressure head in ft (m)

$h_{ab} (h_{abm}) =$ lost head between location "A" and location "B" in ft (m)

Application of Bernoulli's Theorem

Consider a reservoir and a pipeline discharging water to atmosphere at "B." (See Figure 6-6C.) Assuming datum through "B," Bernoulli's theorem applied from the water surface at "A" to the outlet at "B" is:

$$\frac{v_A^2}{2g} + \frac{p_A}{w} + z_A = \frac{v_B^2}{2g} + \frac{p_B}{w} + z_B + h_{AB}$$

The velocity at "A" is practically zero, because the tank is very large, and the gage pressure is zero because only atmospheric pressure works on the water surface. At "A," the elevation is z_A measured in ft (m) above the datum.

At "B," the elevation above the datum is zero; the gage pressure is zero, since the water is discharging to atmosphere; and only velocity pressure is available as the water leaves the outlet. (A gage at a right angle to the emerging stream would register zero pressure).

Therefore:

$$0 + 0 + z_A = \frac{v_B^2}{2g} + 0 + 0 + h_{AB}$$

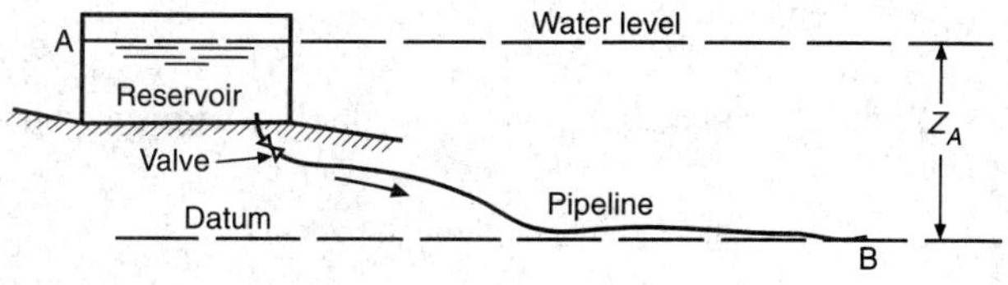

FIG. 6-6C. Graphic respresentation of the application of Bernoulli's theorem to a reservoir and pipelines.

or:

$$\frac{v_B^2}{2g} = z_A - h_{AB}$$

Lost head, h_{AB}, is the sum of: (1) hydraulic losses at the reservoir where water enters the pipeline, at the valve, and at the discharge outlet, plus (2) the friction loss in the pipeline. The values of the components producing lost head can be estimated, as discussed later in the chapter.

As another example, calculate the head loss across 1,000 ft of 8-in. pipe with 750 gpm flowing from a $2^1/_2$-in. hydrant outlet at "B" and a residual pressure at hydrant "A" of 40 psi. With no flow, hydrant "A" has a 60 psi static pressure, and hydrant "B" has an 80 psi static pressure. Assume a datum through hydrant "B." Again, Bernoulli's theorem applied from point "A" to point "B" applies:

$$\frac{v_A^2}{2g} + \frac{p_A}{w} + z_A = \frac{v_B^2}{2g} + \frac{p_B}{w} + z_B + h_{AB}$$

Rearranging:

$$h_{AB} = \frac{v_A^2}{2g} + \frac{p_A}{w} + z_A - \frac{v_B^2}{2g} - \frac{p_B}{w} - z_B$$

Substituting:

$$v_A = \frac{Q}{a} = \frac{Q}{\frac{\pi d^2}{4 \times 144}}$$

$$= \frac{750 \text{ gpm}}{\frac{7.48 \text{ gal/ft}^2 \times 60 \text{ sec/min}}{\frac{3.1416 \times (8 \text{ in.})^2}{4 \times 144 \text{ in.}^2/\text{ft}^2}}} = 4.79 \text{ fps}$$

$$\frac{v_A^2}{2g} = \frac{(4.79 \text{ fps})^2}{64.4 \text{ fps/sec}} = 0.36 \text{ ft} = 0.4 \text{ ft}$$

$$\frac{p_a}{w} = \frac{40 \text{ psi}}{62.4 \text{ pcf}} = 144 \text{ in.}^2/\text{ft}^2 = 92.3 \text{ ft}$$

$$z_A = 80 \text{ psi} - 60 \text{ psi} \times 2.31 \text{ ft/psi} = 46.2 \text{ ft}$$

$$v_B = \frac{Q}{a} = \frac{750 \text{ gpm}}{\frac{7.48 \text{ gal/ft}^2 \times 60 \text{ sec/min}}{\frac{3.1416 \times (2.5)^2}{4 \times 144 \text{ in.}^2/\text{ft}^2}}} = 49.02 \text{ fps}$$

$$\frac{v_B^2}{2g} = \frac{(49.02 \text{ fps})^2}{64.4 \text{ fps}} = 37.3 \text{ ft}$$

$$\frac{p_B}{w} = 0.$$

There is no normal pressure, as the flow discharges to atmosphere.

$z_B = 0$, since the datum goes through hydrant "B"

Thus

$$h_{AB} = 0.4 + 92.3 + 46.2 - 37.3 - 0 - 0 = 101.6 \text{ ft}$$

One further problem is expressed in SI units. Water is pumped via a pipeline, up a 5.0-m incline from "A" to "B." The pipeline at "A" has an inside diameter of 80 mm and a static pressure of 300 kPa. If the pipeline has changed in diameter to 70 mm at "B" and there is frictional head loss over the length of the pipe (h_{AB}) of 12 m, determine the residual pressure at "B" for a flow rate of 4200 L/min.

The solution is expressed as:

$$\frac{v_A^2}{2g} + \frac{p_A}{w} + z_A = \frac{v_B^2}{2g} + \frac{p_B}{w} + z_B + h_{AB}$$

Rearranging for the static at "B":

$$\frac{p_B}{w} = \frac{v_A^2}{2g} + \frac{p_A}{w} + z_A - \frac{v_B^2}{2g} - z_B - h_{AB}$$

Substituting:

$$v_a = \frac{Q}{a_a} = \frac{4200 \text{ L/min}}{\frac{60 \text{ sec/min} \times 1000 \text{ L/m}^3}{\frac{\pi (80 \text{ mm})^2}{4 \times 10^6 \text{ mm}^2/\text{m}^2}}}$$

$$\frac{v_A^2}{2g} = \frac{(13.9 \text{ m/s})^2}{2 \times 9.81 \text{ m/s}^2} = 9.8 \text{ m}$$

$$\frac{p_A}{w} = \frac{300 \text{ kPa}}{9.81 \text{ kPa/m}} = 30.6 \text{ m}$$

$$z_A = 0 \quad (\text{datum through "A"})$$

$$v_B = \frac{Q}{a_B} = \frac{4200 \text{ L/min}}{\frac{60 \text{ sec/min} \times 1000 \text{ L/m}^3}{\frac{\pi (70 \text{ mm})^2}{4 \times 10^6 \text{ mm}^2/\text{m}^2}}} = 18.2 \text{ m/sec}$$

$$\frac{v_B^2}{2g} = \frac{(18.2 \text{ m/s})^2}{2 \times 9.81 \text{ m/s}^2} = 16.9 \text{ m}$$

$$z_B = 5.0 \text{ m}$$

$$h_{AB} = 12 \text{ m} \quad (\text{friction loss})$$

Then

$$\frac{p_A}{w} = 9.8 + 30.6 + 0 - 16.9 - 5 - 12 = 6.5 \text{ m}$$

In pressure terms

$$6.5 \text{ m} \times 9.81 \frac{\text{kPa}}{\text{m}} = 64 \text{ kPa}$$

i.e., the residual pressure at "B" is 64 kPa.

FLOW OF WATER THROUGH ORIFICES

As a liquid leaves a pipe, conduit, or container through an orifice and discharges to atmosphere, the normal pressure is converted to velocity pressure. The rate of flow though an orifice can be expressed in terms of velocity and cross-sectional area of the stream, the basic relations being $Q = av$, where Q = rate of flow in cu ft/sec (m³/s); a = area of cross section in sq ft (m²); and v = velocity at the cross section in ft/sec (m/sec). (See Table 6-6A.) From the previous discussion in this chapter on velocity head, it is known that

$$Q = a\sqrt{2gh}$$

and h (ft) = 2.307p (psi). It follows that, with the orifice diameter in inches, Q in gal/min, and h in psi, would equal:

$$Q = 60 \times 7.4805 \times \frac{\pi d^2}{4 \times 144} \sqrt{64.4 \times 2.30 p_\upsilon}$$

$$Q = 29.8 d^2 \sqrt{p_\upsilon}$$

In SI units, the flow formula is expressed as:

$$Q_m = 0.0666 d_m^2 \sqrt{p_{\upsilon m}}$$

where

Q_m = flow rate (L/m),

d_m = inside diameter (mm), and

$p_{\upsilon m}$ = velocity pressure (kPa).

The above equations (and tables derived therefrom) assume that: (1) the jet is a solid stream the full size of the discharge orifice, and (2) 100 percent of the available total head is converted to velocity head, which is uniform over the cross section. This is a theoretical situation only, however, as these two conditions are not totally attainable, as the following discussion will show.

Coefficients of Flow

In actual flow from nozzles or orifices, the velocity, considered to be average velocity across the entire cross section of the stream, is somewhat less than the velocity calculated from the head. The reduction is due to friction of the water against the nozzle or orifice and turbulence within the nozzle, and is accounted for by a coefficient of velocity, designated c_υ. Values of c_υ are determined by laboratory tests. With well-designed nozzles, the coefficient of velocity is nearly constant and is approximately equal to 0.98.

Some nozzles are designed so that the actual cross-sectional area of the stream is somewhat less than the cross-sectional area of the orifice. This difference is accounted for by the coefficient of contraction, designated c_c. Coefficients of contraction vary greatly with the design and quality of the orifice or nozzle. For a sharp-edged orifice, the value of c_c is about 0.62.

The coefficients of velocity and contraction are usually combined as a single coefficient of discharge, designated c_d:

$$c_d = c_\upsilon \times c_c$$

The basic flow equation can now be written as:

$$Q = 29.8 c_d d^2 \sqrt{p_\upsilon}$$

In SI units the formula is:

$$Q = 0.0666 c_d d_m^2 \sqrt{p_{\upsilon m}}$$

The coefficient of discharge, c_d, is defined as the ratio of the actual discharge to the theoretical discharge. For any specific orifice or nozzle, values of c_d are determined by standard test procedures using this definition. The actual rate of flow is measured by calibrated meters, or "weigh tanks." The theoretical flow is calculated using c_d = 1.0, the carefully measured orifice or nozzle diameter, and the measured velocity pressure in the flow equation.

Standard orifice: An orifice with a sharp entrance edge, shown as form (1) in Figure 6-6D, is known as a standard orifice, and is commonly used to measure water flow. As water leaves the orifice, it contracts to form a jet whose cross-sectional area is less than that of the orifice. The contraction is complete at the plane a' which is located a distance from the plane of the orifice equal to approximately half the diameter of the jet. (See Figure 6-6E.)

The quantity flowing is obviously the same at the orifice a as at the contracted section a', so the quantity flowing could be obtained by measuring the velocity and area at either of these planes. Expressed in a formula, where Q is cu ft/sec (m³/sec), υ is velocity in ft/sec (m/sec), and a is the area in sq ft (m²):

$$Q = \upsilon a = \upsilon' a'$$

The coefficient of discharge for a standard orifice is the product of the coefficient of velocity and the coefficient of contraction, or $c = 0.98 \times 0.62 = 0.61$.

Other orifices: The hydraulic characteristics of good solid-stream nozzles are consistent within a wide range of flow conditions. The velocity at the surface of the stream of most such nozzles is reduced slightly by friction against the orifice or nozzle. A coefficient of velocity of 0.97 is usually applied to nozzles of fire stream sizes to account for this friction.

Coefficients of discharge are available for the flow through hydrants, hose nozzles, automatic sprinklers, and other common fire protection discharge outlets. Representative values for the coefficients of

(1)   (2)   (3)   (4)

FIG. 6-6D. *Orifices of various shapes. If the shape of the orifice is changed so as to decrease the contraction, its capacity will be increased. Form 1 in the illustration is a standard orifice having a sharp edge on the approach side. Form 2, when in a thin plate, gives the same stream characteristics as Form 1. Form 3 is the reverse of 1. In Form 4, the edge is rounded to conform to the shape of the stream. The coefficients of discharge of 3 and 4 are greater than those of the standard orifices, approaching a value of 1.0 in the case of 4.*

FIG. 6-6E. *Flow through a standard orifice.*

TABLE 6-6A. Theoretical Flow Through Circular Orifices—gpm (L/min)

This table is computed from the formula $Q = 29.84 cd^2 \sqrt{p}$ $\left(Q_m = 0.0666 cd^2 m \sqrt{P_{vm}}\right)$ with $c = 1.00$. The theoretical discharge of seawater, as from fireboat nozzles, may be found by subtracting 1 percent from the figures in the following table, or from the formula $Q = 29.84 cd^2 \sqrt{p}$ $\left(Q_m = 0.065 cd^2 m \sqrt{P_{vm}}\right)$.

When pressures are read with a Pitot gage at a nozzle, the nozzle discharge in most cases will correspond to the values in the table within a range of 1 to 3 percent for nozzles up to 1³/₈ in. (35 mm) in diameter. For larger diameter nozzles, the principles discussed in "The Nozzle Method of Measuring Flow" in this chapter of the handbook apply. Appropriate coefficients should be applied where it is read from a hydrant outlet. Where more accurate results are required, a coefficient appropriate to the particular nozzle must be selected and applied to the figures of the table.

The discharge from circular openings of sizes other than those in the table may readily be computed by applying the principle that quantity discharged under a given head varies as the square of the diameter of the opening.

Orifice Diameter in. (mm)

Pressure psi (kPa)	Velocity ft/sec (m/sec)	3/8 (9.53)	1/2 (12.7)	5/8 (15.9)	3/4 (19.1)	7/8 (22.2)	1 (25.4)	1-1/8 (28.6)	1-1/4 (31.8)	1-1/2 (38.1)	1-3/4 (44.5)	2 (50.8)	2-1/4 (57.2)	2-3/8 (60.3)	2-1/2 (63.5)	2-5/8 (66.7)	2-3/4 (69.9)	3 (76.2)	3-1/4 (82.6)	3-1/2 (88.0)	3-3/4 (95.3)	4 (102)	4-1/2 (114)
1 (6.90)	12.2 (3.72)	4.20 (15.9)	7.46 (28.2)	11.7 (44.1)	16.8 (63.5)	22.8 (86.5)	29.8 (112.9)	37.8 (143)	46.6 (176)	67.1 (254)	91.4 (346)	119 (452)	151 (572)	168 (637)	187 (706)	206 (778)	226 (854)	269 (1020)	315 (1190)	366 (1380)	420 (1590)	477 (1810)	604 (2290)
2 (13.8)	17.2 (5.30)	5.93 (22.5)	10.6 (39.9)	16.5 (62.4)	23.7 (89.8)	32.3 (122)	42.2 (159.7)	53.4 (202)	65.9 (250)	95.0 (359)	129 (489)	169 (639)	214 (809)	238 (901)	264 (998)	291 (1100)	319 (1210)	380 (1440)	446 (1690)	517 (1960)	593 (2250)	675 (2560)	855 (3230)
3 (20.7)	21.1 (6.44)	7.27 (27.5)	12.9 (48.9)	20.2 (76.4)	29.1 (110)	39.6 (150)	51.7 (195.6)	65.4 (248)	80.8 (306)	116 (440)	158 (599)	207 (783)	262 (900)	292 (1100)	323 (1220)	356 (1350)	391 (1480)	465 (1760)	546 (2070)	633 (2400)	727 (2750)	827 (3130)	1050 (3960)
4 (27.6)	24.4 (7.43)	8.39 (31.8)	14.9 (56.5)	23.3 (88.2)	33.6 (127)	45.7 (173)	59.7 (225.9)	75.5 (286)	93.3 (353)	134 (508)	183 (692)	239 (904)	302 (1140)	337 (1270)	373 (1410)	411 (1560)	451 (1710)	537 (2030)	630 (2390)	731 (2770)	839 (3180)	955 (3610)	1210 (4570)
5 (34.5)	27.3 (8.31)	9.38 (35.5)	16.7 (63.1)	26.1 (98.7)	37.5 (142)	51.1 (193)	66.7 (252.6)	84.4 (320)	104 (395)	150 (568)	204 (773)	267 (1010)	338 (1280)	376 (1420)	417 (1580)	460 (1740)	505 (1910)	601 (2270)	705 (2670)	817 (3090)	938 (3550)	1070 (4040)	1350 (5110)
6 (41.4)	29.9 (9.11)	10.3 (38.9)	18.3 (69.2)	28.6 (108)	41.1 (156)	56.0 (212)	73.1 (276.7)	92.5 (350)	114 (432)	164 (622)	224 (847)	292 (1110)	370 (c1400)	412 (1560)	457 (1730)	504 (1910)	553 (2090)	658 (2490)	772 (2920)	895 (3390)	1030 (3890)	1170 (4430)	1480 (5600)
7 (48.3)	32.3 (9.83)	11.1 (42.0)	19.7 (74.7)	30.8 (117)	44.4 (168)	60.4 (229)	78.9 (298.8)	99.9 (378)	123 (467)	178 (672)	242 (915)	316 (1200)	400 (1510)	445 (1690)	493 (1870)	544 (2060)	597 (2260)	711 (2690)	834 (3160)	967 (3660)	1110 (4200)	1260 (4780)	1600 (6050)
8 (55.2)	34.5 (10.5)	11.9 (44.9)	21.1 (79.9)	33.0 (125)	47.5 (180)	64.6 (245)	84.4 (319.5)	107 (404)	132 (499)	190 (719)	258 (978)	338 (1280)	427 (1620)	476 (1800)	528 (2000)	582 (2200)	638 (2420)	760 (2880)	891 (3370)	1030 (3910)	1190 (4490)	1350 (5110)	1710 (6470)
9 (62.1)	36.6 (11.2)	12.6 (47.6)	22.4 (84.7)	35.0 (132)	50.4 (191)	68.5 (259)	89.5 (338.8)	113 (429)	140 (529)	201 (762)	274 (1040)	358 (1360)	453 (1720)	505 (1910)	560 (2120)	617 (2330)	677 (2560)	806 (3050)	946 (3580)	1100 (4150)	1260 (4760)	1430 (5420)	1810 (6860)
10 (69.0)	38.6 (11.8)	13.3 (50.2)	23.6 (89.3)	36.9 (140)	53.1 (201)	72.2 (273)	94.4 (357.2)	119 (452)	147 (558)	212 (804)	289 (1090)	377 (1430)	478 (1810)	532 (2010)	590 (2230)	650 (2460)	714 (2700)	849 (3210)	997 (3770)	1160 (4380)	1330 (5020)	1510 (5710)	1910 (7230)
11 (75.9)	40.4 (12.3)	13.9 (52.7)	24.7 (93.6)	38.7 (146)	55.7 (211)	75.8 (287)	99.0 (374.6)	125 (474)	155 (585)	223 (843)	303 (1150)	396 (1500)	501 (1900)	558 (2110)	619 (2340)	682 (2580)	748 (2830)	891 (3370)	1050 (3960)	1210 (4590)	1390 (5270)	1580 (5990)	2000 (7590)
12 (82.8)	42.2 (12.0)	14.5 (55.0)	25.8 (97.8)	40.4 (153)	58.1 (220)	79.1 (300)	103 (391.3)	131 (495)	162 (611)	233 (880)	317 (1200)	413 (1570)	523 (1980)	583 (2210)	646 (2450)	712 (2700)	782 (2960)	930 (3520)	1090 (4130)	1270 (4790)	1450 (5500)	1650 (6260)	2090 (7920)
13 (89.7)	44.0 (13.4)	15.1 (57.3)	26.9 (102)	42.0 (159)	60.5 (229)	82.4 (312)	108 (407.2)	136 (515)	168 (636)	242 (916)	329 (1250)	430 (1630)	545 (2060)	607 (2300)	672 (2550)	741 (2810)	814 (3080)	968 (3670)	1140 (4300)	1320 (4990)	1510 (5730)	1720 (6520)	2180 (8250)
14 (96.6)	45.6 (13.9)	15.7 (59.4)	27.9 (106)	43.6 (165)	62.8 (238)	85.5 (324)	112 (422.6)	141 (535)	174 (660)	251 (951)	342 (1290)	447 (1690)	565 (2140)	630 (2380)	698 (2640)	769 (2910)	844 (3200)	1000 (3800)	1180 (4460)	1370 (5180)	1570 (5940)	1790 (6760)	2260 (8560)
15 (103)	47.2 (14.4)	16.3 (61.5)	28.9 (109)	45.1 (171)	65.0 (246)	88.5 (335)	116 (437.4)	146 (554)	181 (683)	260 (984)	354 (1340)	462 (1750)	585 (2210)	652 (2470)	722 (2730)	796 (3010)	874 (3310)	1040 (3940)	1220 (4620)	1420 (5360)	1630 (6150)	1850 (7000)	2340 (8860)
16 (110)	48.8 (14.9)	16.8 (63.5)	29.8 (113)	46.6 (176)	67.1 (254)	91.4 (346)	119 (451.8)	151 (572)	187 (706)	269 (1020)	366 (1380)	477 (1810)	604 (2290)	673 (2550)	746 (2820)	822 (3110)	903 (3420)	1070 (4070)	1260 (4770)	1460 (5530)	1680 (6350)	1910 (7230)	2420 (9150)
17 (117)	50.3 (15.3)	17.3 (65.5)	30.8 (116)	48.1 (182)	69.2 (262)	94.2 (357)	123 (465.7)	156 (589)	192 (728)	277 (1050)	377 (1430)	492 (1860)	623 (2360)	694 (2630)	769 (2910)	848 (3210)	930 (3520)	1110 (4190)	1300 (4920)	1510 (5700)	1730 (6550)	1970 (7450)	2490 (9430)
18 (124)	51.7 (15.8)	17.8 (67/4)	31.7 (120)	49.5 (187)	71.2 (270)	96.9 (367)	127 (479.2)	160 (606)	198 (749)	285 (1080)	388 (1470)	506 (1920)	641 (2430)	714 (2700)	791 (2990)	872 (3300)	957 (3620)	1140 (4310)	1340 (5060)	1550 (5870)	1780 (6740)	2030 (7670)	2560 (9700)

TABLE 6-6A. Theoretical Flow Through Circular Orifices—gpm (L/min) (Continued)

pressure psi (kpa)	Velocity ft/sec (m/sec)	3/8 (9.53)	1/2 (12.7)	5/8 (15.9)	3/4 (19.1)	7/8 (22.2)	1 (25.4)	1-1/8 (28.6)	1-1/4 (31.8)	1-1/2 (38.1)	1-3/4 (44.5)	2 (50.8)	2-1/4 (57.2)	2-3/8 (60.3)	2-1/2 (63.5)	2-5/8 (66.7)	2-3/4 (69.9)	3 (76.2)	3-1/4 (82.6)	3-1/2 (88.0)	3-3/4 (95.3)	4 (102)	4-1/2 (114)
19 (131)	53.2 (16.2)	18.3 (69.2)	32.5 (123)	50.8 (192)	73.2 (277)	99.6 (377)	130 (492.3)	165 (623)	203 (769)	293 (1110)	398 (1510)	520 (1970)	658 (2490)	734 (2780)	813 (3080)	896 (3390)	984 (3720)	1170 (4430)	1370 (5200)	1590 (6030)	1830 (6920)	2080 (7880)	2630 (9970)
20 (138)	54.5 (16.6)	18.8 (71.0)	33.4 (126)	52.1 (197)	75.1 (284)	102 (387)	133 (505.1)	169 (639)	209 (789)	300 (1140)	409 (1550)	534 (2020)	676 (2560)	753 (2850)	834 (3160)	920 (3480)	1010 (3820)	1200 (4550)	1410 (5340)	1630 (6190)	1880 (7100)	2140 (8080)	2700 (10200)
21 (145)	55.9 (17.0)	19.2 (72.8)	34.2 (129)	53.4 (202)	76.9 (291)	105 (396)	137 (517.6)	173 (655)	214 (809)	308 (1160)	419 (1590)	547 (2070)	692 (2620)	771 (2920)	855 (3230)	942 (3570)	1030 (3910)	1230 (4660)	1440 (5470)	1680 (6340)	1920 (7280)	2190 (8280)	2770 (10500)
22 (152)	57.2 (17.4)	19.7 (74.5)	35.0 (132)	54.7 (207)	78.7 (298)	107 (406)	140 (529.8)	177 (670)	219 (828)	315 (1190)	429 (1620)	560 (2120)	709 (2680)	789 (2990)	875 (3310)	964 (3650)	1060 (4010)	1260 (4770)	1480 (5600)	1710 (6490)	1970 (7450)	2240 (8480)	2830 (10700)
23 (159)	58.5 (17.8)	20.1 (76.2)	35.8 (135)	55.9 (212)	80.5 (305)	110 (415)	143 (541.7)	181 (686)	224 (846)	322 (1220)	438 (1660)	572 (2170)	724 (2740)	807 (3060)	894 (3390)	986 (3730)	1080 (4100)	1290 (4870)	1510 (5720)	1750 (6640)	2010 (7620)	2290 (8670)	2900 (11000)
24 (166)	59.7 (18.2)	20.6 (77.8)	36.5 (138)	57.1 (216)	82.2 (311)	112 (424)	146 (553.3)	185 (700)	228 (865)	329 (1240)	448 (1690)	585 (2210)	740 (2800)	825 (3120)	914 (3460)	1010 (3810)	1110 (4180)	1320 (4980)	1540 (5840)	1790 (6780)	2060 (7780)	2340 (8850)	2960 (11200)
25 (172)	61.0 (18.6)	21.0 (79.4)	37.3 (141)	58.3 (221)	83.9 (318)	114 (423)	149 (564.7)	189 (715)	233 (882)	336 (1270)	457 (1730)	597 (2260)	755 (2860)	842 (3190)	933 (3530)	1030 (3890)	1130 (4270)	1340 (5080)	1580 (5960)	1830 (6920)	2100 (7940)	2390 (9040)	3020 (11400)
26 (179)	62.2 (19.0)	21.4 (81.0)	38.0 (144)	59.4 (225)	85.6 (324)	116 (441)	152 (575.0)	193 (729)	238 (900)	342 (1300)	466 (1760)	609 (2300)	770 (2920)	858 (3250)	951 (3600)	1050 (3970)	1150 (4360)	1370 (5180)	1610 (6080)	1860 (7050)	2140 (8100)	2430 (9210)	3080 (11700)
27 (186)	63.4 (19.3)	21.8 (82.5)	38.8 (147)	60.6 (229)	87.2 (330)	119 (449)	155 (586.9)	196 (743)	242 (917)	349 (1320)	475 (1800)	620 (2350)	785 (2970)	875 (3310)	969 (3670)	1070 (4040)	1170 (4440)	1400 (5280)	1640 (6200)	1900 (7190)	2180 (8250)	2480 (9390)	3140 (11900)
28 (193)	64.5 (19.7)	22.2 (84.0)	39.5 (149)	61.7 (233)	88.8 (336)	121 (458)	158 (597.6)	200 (756)	247 (934)	355 (1340)	484 (1830)	632 (2390)	799 (3030)	891 (3370)	987 (3740)	1090 (4120)	1190 (4520)	1420 (5380)	1670 (6310)	1930 (7320)	2220 (8400)	2530 (9560)	3200 (12100)
29 (200)	65.7 (20.0)	22.6 (85.5)	40.2 (152)	62.8 (238)	90.4 (342)	123 (466)	161 (608.2)	203 (770)	251 (950)	362 (1370)	492 (1860)	643 (2430)	814 (3080)	906 (3430)	1000 (3800)	1110 (4190)	1220 (4600)	1450 (5470)	1700 (6420)	1970 (7450)	2260 (8550)	2570 (9730)	3250 (12300)
30 (207)	66.8 (20.4)	23.0 (87.0)	40.9 (155)	63.8 (242)	91.9 (348)	125 (474)	163 (618.6)	207 (783)	255 (967)	368 (1390)	501 (1890)	654 (2470)	827 (3130)	922 (3490)	1020 (3870)	1130 (4260)	1240 (4680)	1470 (5570)	1730 (6530)	2000 (7580)	2300 (8700)	2620 (9900)	3310 (12500)
31 (214)	67.9 (20.7)	23.4 (88.4)	41.5 (157)	64.9 (246)	93.5 (354)	127 (481)	166 (628.8)	210 (796)	260 (983)	374 (1410)	509 (1930)	665 (2520)	841 (3180)	937 (3550)	1040 (3930)	1140 (4330)	1260 (4760)	1500 (5660)	1750 (6640)	2040 (7700)	2340 (8840)	2660 (10100)	3360 (12700)
32 (221)	69.0 (21.0)	23.7 (89.8)	42.2 (160)	65.9 (250)	95.0 (359)	129 (489)	169 (638.9)	214 (809)	264 (998)	380 (1440)	517 (1960)	675 (2560)	855 (3230)	952 (3600)	1060 (3990)	1160 (4400)	1280 (4830)	1520 (5750)	1780 (6750)	2070 (7830)	2370 (8980)	2700 (10200)	3420 (12900)
33 (228)	70.1 (21.4)	24.1 (91.2)	42.9 (162)	67.0 (253)	96.4 (365)	131 (497)	171 (648.8)	217 (821)	268 (1010)	386 (1460)	525 (1990)	686 (2600)	868 (3280)	967 (3660)	1070 (4060)	1180 (4470)	1300 (4910)	1540 (5840)	1810 (6850)	2100 (7950)	2410 (9120)	2740 (10400)	3470 (13100)
34 (234)	71.1 (21.7)	24.5 (92.6)	43.5 (165)	68.0 (257)	97.9 (370)	133 (504)	174 (658.6)	220 (834)	272 (1030)	391 (1480)	533 (2020)	696 (2630)	881 (3330)	981 (3710)	1090 (4120)	1200 (4540)	1320 (4980)	1570 (5930)	1840 (6960)	2130 (8070)	2450 (9260)	2780 (10500)	3520 (13300)
35 (241)	72.1 (22.0)	24.8 (94.0)	44.1 (167)	69.0 (261)	99.3 (376)	135 (512)	177 (668.2)	223 (846)	276 (1040)	397 (1500)	541 (2050)	706 (2670)	894 (3380)	996 (3770)	1100 (4180)	1220 (4600)	1340 (5050)	1590 (6010)	1860 (7060)	2160 (8190)	2480 (9400)	2820 (10700)	3570 (13500)
36 (248)	73.2 (22.3)	25.2 (95.3)	44.8 (169)	69.9 (265)	101 (381)	137 (519)	179 (677.7)	227 (858)	280 (1060)	403 (1520)	548 (2080)	716 (2710)	906 (3430)	1010 (3820)	1120 (4240)	1230 (4670)	1350 (5120)	1610 (6100)	1890 (7160)	2190 (8300)	2520 (9530)	2860 (10800)	3630 (13700)

TABLE 6-6A. Theoretical Flow Through Circular Orifices—gpm (L/min) (Continued)

Pressure psi (kPa)	Velocity ft/sec (m/sec)	3/8 (9.53)	1/2 (12.7)	5/8 (15.9)	3/4 (19.1)	7/8 (22.2)	1 (25.4)	1-1/8 (28.6)	1-1/4 (31.8)	1-1/2 (38.1)	1-34 (44.5)	2 (50.8)	2-1/4 (57.2)	2-3/8 (60.3)	2-1/2 (63.5)	2-5/8 (66.7)	2-3/4 (69.9)	3 (76.2)	3-1/4 (82.6)	3-1/2 (88.9)	3-3/4 (95.3)	4 (102)	4-1/2 (114)
37 (255)	74.2 (22.6)	25.5 (96.6)	45.4 (172)	70.9 (268)	102 (386)	139 (526)	182 (687)	230 (870)	284 (1070)	408 (1550)	556 (2100)	726 (2750)	919 (3480)	1020 (3880)	1130 (4290)	1250 (4730)	1370 (5200)	1630 (6180)	1920 (7260)	2220 (8420)	2550 (9660)	2900 (11000)	3680 (13900)
38 (262)	75.2 (22.9)	25.9 (97.9)	46.0 (174)	71.9 (272)	103 (392)	141 (533)	184 (696.2)	233 (881)	287 (1090)	414 (1570)	563 (2130)	736 (2780)	931 (3520)	1040 (3930)	1150 (4350)	1270 (4800)	1390 (5270)	1660 (6270)	1940 (7350)	2250 (8530)	2590 (9790)	2940 (11100)	3720 (14100)
39 (269)	76.2 (23.2)	26.2 (99.2)	46.6 (176)	72.8 (276)	105 (397)	143 (540)	186 (705.3)	236 (893)	291 (1100)	419 (1590)	571 (2160)	745 (2820)	943 (3570)	1050 (3980)	1160 (4410)	1280 (4860)	1410 (5330)	1680 (6350)	1970 (7450)	2280 (8640)	2620 (9920)	2980 (11300)	3770 (14300)
40 (276)	77.1 (23.5)	26.5 (100)	47.2 (179)	73.7 (279)	106 (402)	144 (547)	189 (714.3)	239 (904)	295 (1120)	425 (1610)	578 (2190)	755 (2860)	955 (3620)	1060 (4030)	1180 (4460)	1300 (4920)	1430 (5400)	1700 (6430)	1990 (7550)	2310 (8750)	2650 (10000)	3020 (11400)	3820 (14500)
41 (283)	78.1 (23.8)	26.9 (102)	47.8 (181)	74.6 (282)	107 (407)	146 (554)	191 (723.2)	242 (915)	299 (1130)	430 (1630)	585 (2210)	764 (2890)	967 (3660)	1080 (4080)	1190 (4520)	1320 (4980)	1440 (5470)	1720 (6510)	2020 (7640)	2340 (8860)	2690 (10200)	3060 (11600)	3870 (14600)
42 (290)	79.0 (24.1)	27.2 (103)	48.3 (183)	75.5 (286)	109 (412)	148 (560)	193 (732)	245 (926)	302 (1140)	435 (1650)	592 (2240)	774 (2930)	979 (3710)	1090 (4130)	1210 (4570)	1330 (5040)	1460 (5540)	1740 (6590)	2040 (7730)	2370 (8970)	2720 (10300)	3090 (11700)	3920 (14800)
43 (297)	80.0 (24.4)	27.5 (104)	48.9 (185)	76.4 (289)	110 (417)	150 (567)	196 (740.6)	248 (937)	306 (1160)	440 (1670)	599 (2270)	783 (2960)	991 (3750)	1100 (4180)	1220 (4630)	1350 (5100)	1480 (5600)	1760 (6670)	2070 (7820)	2400 (9070)	2750 (10400)	3130 (11900)	3960 (15000)
44 (303)	80.9 (24.7)	27.8 (105)	49.5 (187)	77.3 (293)	111 (421)	152 (574)	198 (749.2)	251 (948)	309 (1170)	445 (1690)	606 (2290)	792 (3000)	1000 (3790)	1120 (4230)	1240 (4680)	1360 (5160)	1500 (5670)	1780 (6740)	2090 (7910)	2420 (9180)	2780 (10500)	3170 (12000)	4010 (15200)
45 (310)	81.8 (24.9)	28.1 (107)	50.0 (189)	78.2 (296)	113 (426)	153 (580)	200 (757.7)	253 (959)	313 (1180)	450 (1700)	613 (2320)	801 (3030)	1010 (3840)	1130 (4270)	1250 (4740)	1380 (5220)	1510 (5730)	1800 (6820)	2110 (8000)	2450 (9280)	2810 (10700)	3200 (12100)	4050 (15300)
46 (317)	82.7 (25.2)	28.5 (108)	50.6 (192)	79.1 (299)	114 (431)	155 (586)	202 (766)	256 (970)	316 (1200)	455 (1720)	620 (2350)	810 (3060)	1020 (3880)	1140 (4320)	1260 (4790)	1390 (5280)	1530 (5790)	1820 (6890)	2140 (8090)	2480 (9380)	2850 (10800)	3240 (12300)	4100 (15500)
47 (324)	83.6 (25.5)	28.8 (109)	51.1 (194)	79.9 (302)	115 (436)	157 (593)	205 (774.3)	259 (980)	320 (1210)	460 (1740)	627 (2370)	818 (3100)	1040 (3920)	1150 (4370)	1280 (4840)	1410 (5340)	1550 (5860)	1840 (6970)	2160 (8180)	2510 (9490)	2880 (10900)	3270 (12400)	4140 (15700)
48 (331)	84.5 (25.8)	29.1 (110)	51.7 (196)	80.8 (306)	116 (440)	158 (599)	207 (782.5)	262 (990)	323 (1220)	465 (1760)	633 (2400)	827 (3130)	1050 (3960)	1170 (4410)	1290 (4890)	1420 (5390)	1560 (5920)	1860 (7040)	2180 (8270)	2530 (9590)	2910 (11000)	3310 (12500)	4190 (15800)
49 (338)	85.4 (26.0)	29.4 (111)	52.2 (198)	81.6 (309)	117 (445)	160 (605)	209 (790.6)	264 (1000)	326 (1240)	470 (1780)	640 (2420)	836 (3160)	1060 (4000)	1180 (4460)	1310 (4940)	1440 (5450)	1580 (5980)	1880 (7120)	2210 (8350)	2560 (9680)	2940 (11100)	3340 (12600)	4230 (16000)
50 (345)	86.2 (26.3)	29.7 (112)	52.8 (200)	82.4 (312)	119 (449)	162 (611)	211 (798.6)	267 (1010)	330 (1250)	475 (1800)	646 (2450)	844 (3190)	1070 (4040)	1190 (4500)	1320 (4990)	1450 (5500)	1600 (6040)	1900 (7190)	2230 (8440)	2580 (9780)	2970 (11200)	3380 (12800)	4270 (16200)
52 (359)	87.9 (26.8)	30.3 (115)	53.8 (204)	84.1 (318)	121 (458)	165 (624)	215 (814.5)	272 (1030)	336 (1270)	484 (1830)	659 (2490)	861 (3260)	1090 (4120)	1210 (4590)	1340 (5090)	1480 (5610)	1630 (6160)	1940 (7330)	2270 (8600)	2640 (9980)	3030 (11500)	3440 (13000)	4360 (16500)
54 (372)	89.6 (27.3)	30.8 (117)	54.8 (207)	85.7 (324)	123 (467)	168 (635)	219 (830)	278 (1050)	343 (1300)	493 (1870)	672 (2540)	877 (3320)	1110 (4200)	1240 (4680)	1370 (5190)	1510 (5720)	1660 (6280)	1970 (7470)	2320 (8770)	2690 (10200)	3080 (11700)	3510 (13300)	4440 (16800)
56 (386)	91.3 (27.8)	31.4 (119)	55.8 (211)	87.2 (330)	126 (475)	171 (647)	223 (845.2)	283 (1070)	349 (1320)	502 (1900)	684 (2590)	893 (3380)	1130 (4280)	1260 (4770)	1400 (5280)	1540 (5820)	1690 (6390)	2010 (7610)	2360 (8930)	2740 (10400)	3140 (11900)	3570 (13500)	4520 (17100)
58 (400)	92.9 (28.3)	32.0 (121)	56.8 (215)	88.8 (336)	128 (484)	174 (659)	227 (860.2)	288 (1090)	355 (1340)	511 (1940)	696 (2630)	909 (3440)	1150 (4350)	1280 (4850)	1420 (5380)	1570 (5930)	1720 (6500)	2050 (7740)	2400 (9090)	2780 (10500)	3200 (12100)	3640 (13800)	4600 (17400)

TABLE 6-6A. Theoretical Flow Through Circular Orifices—gpm (L/min) (Continued)

Pressure psi (kPa)	Velocity ft/sec (m/sec)	3/8 (9.53)	1/2 (12.7)	5/8 (15.9)	3/4 (19.1)	7/8 (22.2)	1 (25.4)	1-1/8 (28.6)	1-1/4 (31.8)	1-1/2 (38.1)	1-34 (44.5)	2 (50.8)	2-1/4 (57.2)	2-3/8 (60.3)	2-1/2 (63.5)	2-5/8 (66.7)	2-3/4 (69.9)	3 (76.2)	3-1/4 (82.6)	3-1/2 (88.9)	3-3/4 (95.3)	4 (102)	4-1/2 (114)
60 (414)	94.5 (28.8)	32.5 (123)	57.8 (219)	90.3 (342)	130 (492)	177 (670)	231 (874.9)	293 (1110)	361 (1370)	520 (1970)	708 (2680)	925 (3500)	1170 (4430)	1300 (4930)	1440 (5470)	1590 (6030)	1750 (6620)	2080 (7870)	2440 (9240)	2830 (10700)	3250 (12300)	3700 (14000)	4680 (17700)
62 (428)	96.0 (29.3)	33.0 (125)	58.7 (222)	91.8 (347)	132 (500)	180 (681)	235 (889.3)	297 (1130)	367 (1390)	529 (2000)	720 (2720)	940 (3560)	1190 (4500)	1330 (5020)	1470 (5560)	1620 (6130)	1780 (6730)	2110 (8000)	2480 (9390)	2880 (10900)	3300 (12500)	3760 (14200)	4760 (18000)
64 (441)	97.6 (29.7)	33.6 (127)	59.7 (226)	93.3 (353)	134 (508)	183 (692)	239 (903.6)	302 (1140)	373 (1410)	537 (2030)	731 (2770)	955 (3610)	1210 (4570)	1350 (5100)	1490 (5650)	1640 (6230)	1810 (6830)	2150 (8130)	2520 (9540)	2920 (11100)	3360 (12700)	3820 (14500)	4830 (18300)
66 (455)	99.1 (30.2)	34.1 (129)	60.6 (229)	94.7 (358)	136 (516)	186 (703)	242 (917.6)	307 (1160)	379 (1430)	545 (2060)	742 (2810)	970 (3670)	1230 (4650)	1370 (5180)	1520 (5730)	1670 (6320)	1830 (6940)	2180 (8260)	2560 (9690)	2970 (11200)	3410 (12900)	3880 (14700)	4910 (18600)
68 (469)	101 (30.7)	34.6 (131)	61.5 (233)	96.1 (364)	138 (524)	188 (713)	246 (931.4)	311 (1180)	384 (1460)	554 (2100)	754 (2850)	984 (3730)	1250 (4720)	1390 (5250)	1540 (5820)	1700 (6420)	1860 (7040)	2210 (8380)	2600 (9840)	3010 (11400)	3460 (13100)	3940 (14900)	4980 (18900)
70 (483)	102 (31.1)	35.1 (133)	62.4 (236)	97.5 (369)	140 (532)	191 (723)	250 (945)	316 (1200)	390 (1480)	562 (2130)	765 (2890)	999 (3780)	1260 (4780)	1410 (5330)	1560 (5910)	1720 (6510)	1890 (7150)	2250 (8500)	2640 (9980)	3060 (11600)	3510 (13300)	3990 (15100)	5060 (19100)
72 (497)	103 (31.5)	35.6 (135)	63.3 (240)	98.9 (374)	142 (539)	194 (734)	253 (958.4)	320 (1210)	396 (1500)	570 (2160)	775 (2930)	1010 (3830)	1280 (4850)	1430 (5410)	1580 (5990)	1740 (6600)	1910 (7250)	2280 (8630)	2670 (10100)	3100 (11700)	3560 (13500)	4050 (15300)	5130 (19400)
74 (510)	105 (32.0)	36.1 (137)	64.2 (243)	100 (380)	144 (547)	197 (744)	257 (971.6)	325 (1230)	401 (1520)	578 (2190)	786 (2980)	1030 (3890)	1300 (4920)	1450 (5480)	1600 (6070)	1770 (6690)	1940 (7350)	2310 (8740)	2710 (10300)	3140 (11900)	3610 (13700)	4110 (15500)	5200 (19700)
76 (524)	106 (32.4)	36.6 (138)	65.0 (246)	102 (385)	146 (554)	199 (754)	260 (984.6)	329 (1250)	406 (1540)	585 (2220)	797 (3020)	1040 (3940)	1320 (4980)	1470 (5550)	1630 (6150)	1790 (6780)	1970 (7450)	2340 (8860)	2750 (10400)	3190 (12100)	3660 (13800)	4160 (15800)	5270 (19900)
78 (538)	108 (32.8)	37.1 (140)	65.9 (249)	103 (390)	148 (561)	202 (764)	264 (997.5)	334 (1260)	412 (1560)	593 (2240)	807 (3050)	1050 (3990)	1330 (5050)	1490 (5630)	1650 (6230)	1820 (6870)	1990 (7540)	2370 (8980)	2780 (10500)	3230 (12200)	3710 (14000)	4220 (16000)	5340 (20200)
80 (552)	109 (33.2)	37.5 (142)	66.7 (253)	104 (395)	150 (568)	204 (773)	267 (1010)	338 (1280)	417 (1580)	601 (2270)	817 (3090)	1070 (4040)	1350 (5110)	1510 (5700)	1670 (6310)	1840 (6960)	2020 (7640)	2400 (9090)	2820 (10700)	3270 (12400)	3750 (14200)	4270 (16200)	5400 (20500)
82 (566)	110 (33.7)	38.0 (144)	67.6 (256)	106 (400)	152 (575)	207 (783)	270 (1020)	342 (1290)	422 (1600)	608 (2300)	828 (3130)	1080 (4090)	1370 (5180)	1520 (5770)	1690 (6390)	1860 (7050)	2040 (7730)	2430 (9200)	2850 (10800)	3310 (12500)	3800 (14400)	4320 (16400)	5470 (20700)
84 (579)	112 (34.1)	38.5 (146)	68.4 (259)	107 (404)	154 (582)	209 (793)	273 (1040)	346 (1310)	427 (1620)	615 (2330)	838 (3170)	1090 (4140)	1380 (5240)	1540 (5840)	1710 (6470)	1880 (7130)	2070 (7830)	2460 (9320)	2890 (10900)	3350 (12700)	3850 (14600)	4380 (16600)	5540 (21000)
86 (593)	113 (34.5)	38.9 (147)	69.2 (262)	108 (409)	156 (589)	212 (802)	277 (1050)	350 (1330)	432 (1640)	623 (2360)	847 (3210)	1110 (4190)	1400 (5300)	1560 (5910)	1730 (6550)	1910 (7220)	2090 (7920)	2490 (9430)	2920 (11100)	3390 (12800)	3890 (14700)	4430 (16800)	5600 (21200)
88 (607)	114 (34.9)	39.4 (149)	70.0 (265)	109 (414)	157 (596)	214 (811)	280 (1060)	354 (1340)	437 (1660)	630 (2380)	857 (3240)	1120 (4240)	1420 (5360)	1580 (5980)	1750 (6620)	1930 (7300)	2120 (8010)	2520 (9540)	2960 (11200)	3430 (13000)	3940 (14900)	4480 (17000)	5670 (21500)
90 (621)	116 (35.3)	39.8 (151)	70.8 (268)	111 (419)	159 (603)	217 (820)	283 (1070)	358 (1360)	442 (1670)	637 (2410)	867 (3280)	1130 (4290)	1430 (5420)	1600 (6040)	1770 (6700)	1950 (7380)	2140 (8100)	2550 (9640)	2990 (11300)	3470 (13100)	3980 (15100)	4530 (17100)	5730 (21700)
92 (634)	117 (35.7)	40.2 (152)	71.6 (271)	112 (423)	161 (609)	219 (829)	286 (1080)	362 (1370)	447 (1690)	644 (2440)	877 (3320)	1140 (4330)	1450 (5480)	1610 (6110)	1790 (6770)	1970 (7460)	2160 (8190)	2580 (9750)	3020 (11400)	3510 (13300)	4020 (15200)	4580 (17300)	5800 (21900)
94 (648)	118 (36.0)	40.7 (154)	72.3 (274)	113 (428)	163 (616)	222 (838)	289 (1100)	366 (1390)	452 (1710)	651 (2460)	886 (3350)	1160 (4380)	1460 (5540)	1630 (6180)	1810 (6840)	1990 (7550)	2190 (8280)	2600 (9860)	3060 (11600)	3540 (13400)	4070 (15400)	4630 (17500)	5860 (22200)

TABLE 6-6A. Theoretical Flow Through Circular Orifices—gpm (L/min) (Continued)

Pressure psi (kPa)	Velocity ft/sec (m/sec)	3/8 (9.53)	1/2 (12.7)	5/8 (15.9)	3/4 (19.1)	7/8 (22.2)	1 (25.4)	1-1/8 (28.6)	1-1/4 (31.8)	1-1/2 (38.1)	1-3/4 (44.5)	2 (50.8)	2-1/4 (57.2)	2-3/8 (60.3)	2-1/2 (63.5)	2-5/8 (66.7)	2-3/4 (69.9)	3 (76.2)	3-1/4 (82.6)	3-1/2 (88.0)	3-3/4 (95.3)	4 (102)	4-1/2 (114)
96 (662)	119 (36.4)	41.1 (156)	73.1 (277)	114 (432)	164 (622)	224 (847)	292 (1110)	370 (1400)	457 (1730)	658 (2490)	895 (3390)	1170 (4430)	1480 (5600)	1650 (6240)	1830 (6920)	2010 (7630)	2210 (8370)	2630 (9960)	3090 (11700)	3580 (13600)	4110 (15600)	4680 (17700)	5920 (22400)
98 (676)	121 (36.8)	41.5 (157)	73.9 (280)	115 (437)	166 (629)	226 (856)	295 (1120)	374 (1420)	462 (1750)	665 (2520)	905 (3420)	1180 (4470)	1500 (5660)	1670 (6310)	1850 (6990)	2040 (7700)	2230 (8460)	2660 (10100)	3120 (11800)	3620 (13700)	4150 (15700)	4730 (17900)	5980 (22600)
100 (690)	122 (37.2)	42.0 (159)	74.6 (282)	117 (441)	168 (635)	228 (865)	298 (1130)	378 (1430)	466 (1760)	671 (2540)	914 (3640)	1190 (4520)	1510 (5720)	1680 (6370)	1870 (7060)	2060 (7780)	2260 (8540)	2690 (10200)	3150 (11900)	3660 (13800)	4200 (15900)	4770 (18100)	6040 (22900)
105 (724)	125 (38.1)	43.0 (163)	76.4 (289)	119 (452)	172 (651)	234 (886)	306 (1160)	387 (1460)	478 (1810)	688 (2600)	936 (3540)	1220 (4630)	1550 (5860)	1720 (6530)	1910 (7230)	2110 (7970)	2310 (8750)	2750 (10400)	3230 (12200)	3750 (14200)	4300 (16300)	4890 (18500)	6190 (23400)
110 (759)	128 (39.0)	44.0 (167)	78.2 (296)	122 (463)	176 (666)	240 (907)	313 (1180)	396 (1500)	489 (1850)	704 (2670)	958 (3630)	1250 (4740)	1580 (6000)	1770 (6680)	1960 (7400)	2160 (8160)	2370 (8960)	2820 (10700)	3310 (12500)	3830 (14500)	4400 (16700)	5010 (19000)	6340 (24000)
115 (793)	131 (39.9)	45.0 (170)	80.0 (303)	125 (473)	180 (681)	245 (927)	320 (1210)	405 (1530)	500 (1890)	720 (2730)	980 (3710)	1280 (4840)	1620 (6130)	1800 (6830)	2000 (7570)	2200 (8350)	2420 (9160)	2880 (10900)	3380 (12800)	3920 (14800)	4500 (17000)	5120 (19400)	6480 (24500)
120 (828)	134 (40.7)	46.0 (174)	81.7 (309)	128 (483)	184 (696)	250 (947)	327 (1240)	414 (1570)	511 (1930)	735 (2780)	1000 (3790)	1310 (4950)	1650 (6260)	1840 (6980)	2040 (7730)	2250 (8530)	2470 (9360)	2940 (11100)	3450 (13100)	4000 (15200)	4600 (17400)	5230 (19800)	6620 (25100)
125 (862)	136 (41.6)	46.9 (178)	83.4 (316)	130 (493)	188 (710)	255 (967)	334 (1260)	422 (1600)	521 (1970)	751 (2840)	1020 (3870)	1330 (5050)	1690 (6390)	1880 (7120)	2090 (7890)	2300 (8700)	2520 (9550)	3000 (11400)	3520 (13300)	4090 (15500)	4690 (17800)	5340 (20200)	6760 (25600)
130 (897)	139 (42.4)	47.8 (181)	85.1 (322)	133 (503)	191 (724)	260 (986)	340 (1290)	431 (1630)	532 (2010)	766 (2900)	1040 (3940)	1360 (5150)	1720 (6520)	1920 (7260)	2130 (8050)	2340 (8870)	2570 (9740)	3060 (11600)	3590 (13600)	4170 (15800)	4780 (18100)	5440 (20600)	6890 (26100)
135 (931)	142 (43.2)	48.8 (185)	86.7 (328)	135 (513)	195 (738)	265 (1000)	347 (1310)	439 (1660)	542 (2050)	780 (2950)	1060 (4020)	1390 (5250)	1760 (6640)	1960 (7400)	2170 (8200)	2390 (9040)	2620 (9920)	3120 (11800)	3660 (13900)	4250 (16100)	4880 (18500)	5550 (21000)	7020 (26600)
140 (966)	144 (44.0)	49.7 (188)	88.3 (334)	138 (522)	199 (752)	270 (1020)	353 (1340)	447 (1690)	552 (2090)	794 (3010)	1080 (4090)	1410 (5350)	1790 (6770)	1990 (7540)	2210 (8350)	2430 (9210)	2670 (10100)	3180 (12000)	3730 (14100)	4330 (16400)	4970 (18800)	5650 (21400)	7150 (27100)
145 (1000)	147 (44.8)	50.5 (191)	89.8 (340)	140 (531)	202 (765)	275 (1040)	359 (1360)	455 (1720)	561 (2130)	808 (3060)	1100 (4170)	1440 (5440)	1820 (6890)	2030 (7670)	2250 (8500)	2480 (9370)	2720 (10300)	3230 (12200)	3800 (14400)	4400 (16700)	5050 (19100)	5750 (21800)	7280 (27500)
150 (1030)	149 (45.5)	51.4 (195)	91.4 (346)	143 (540)	206 (778)	280 (1060)	365 (1380)	463 (1750)	571 (2160)	822 (3110)	1120 (4240)	1460 (5530)	1850 (7000)	2060 (7800)	2280 (8650)	2520 (9530)	2760 (10500)	3290 (12400)	3860 (14600)	4480 (16900)	5140 (19500)	5850 (22100)	7400 (28000)

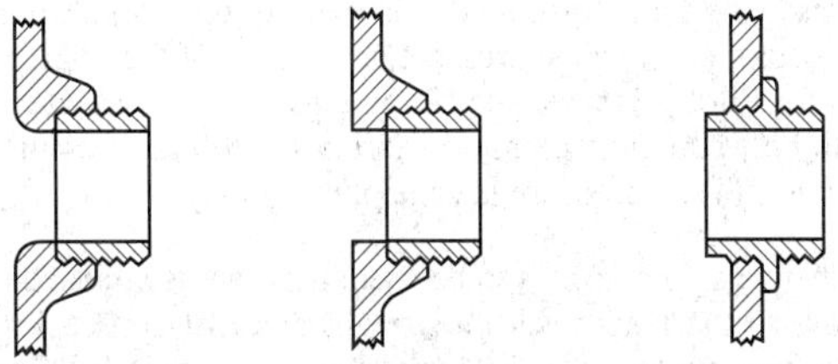

FIG. 6-6F.   Three general types of hydrant outlets and their coefficients of discharge.

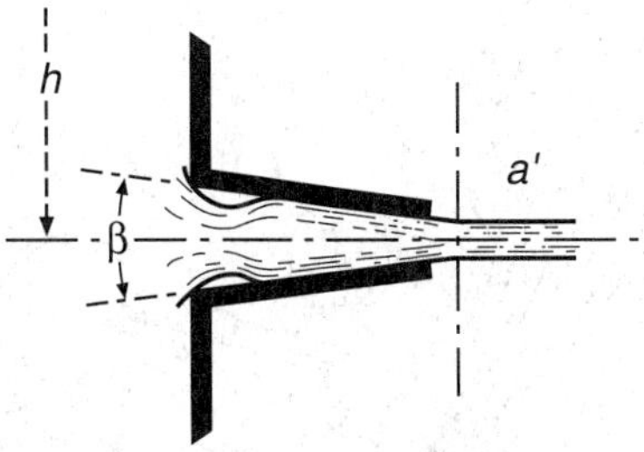

FIG. 6-6H.   Flow in short conical converging tube.

discharge are given in Table 6-6B. Again, these coefficients only apply when there is flow through the full orifice or nozzle opening with a reasonably uniform velocity profile.

TABLE 6-6B.   *Typical Discharge Coefficients of Solid Stream Nozzle*

Spray sprinkler, average (nominal $1/2$-in. dia.)	0.75
Spray sprinkler, average (nominal $17/32$-in. dia.)	0.95
Large-drop sprinkler (0.64 in. dia.)	0.90
Standard orifice (sharp edge)	0.62
Smooth-bore nozzles, general	0.96–0.98
Underwriter playpipes or equal	0.97
Deluge or monitor nozzles	0.997
Open pipe, burred opening	0.80
Open pipe, smooth, well rounded	0.90
Hydrant butt, smooth and well-rounded outlet, flowing full*	0.90
Hydrant butt, square and sharp at hydrant barrel*	0.80
Hydrant butt, outlet square, projecting into barrel*	0.70

*See Figure 6-6F.

Flow in short tubes: A tube attached to an orifice is known as a standard short tube if it is $2^1/_2$ to 3 times longer than the diameter of the orifice, and its diameter is the same as the orifice. A shorter tube will not flow full, and friction losses in a longer tube will affect results when used as a measuring device, hence the specified length limit.

The characteristics of a standard short tube and a short conical converging tube are shown in Figures 6-6G and 6-6H, respectively. The principles of flow in orifices apply, but with different coefficients. With the conical tube, the coefficients c_v and c_c vary with the angle β. When β is 0 degrees, the converging tube becomes a cylindrical tube, with $c_c = 1$ and $c_v = 0.82$; c_d is then 0.82. As the angle β

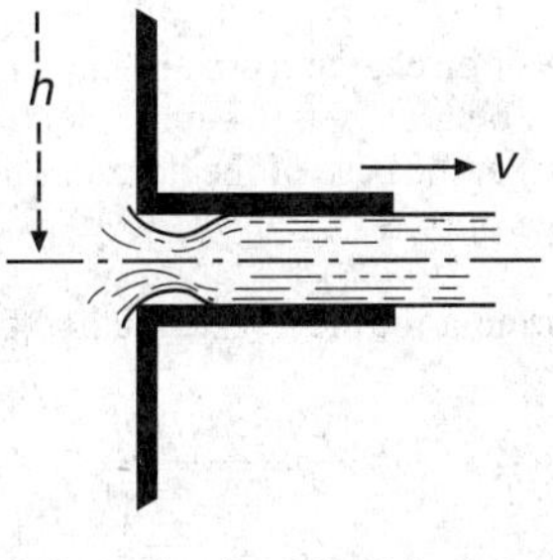

increases, the coefficient of contraction (c_c) develops, and the coefficient of velocity (c_v) increases, approaching the 0.98 value for a sharp-edge orifice. Relations are such that the coefficient of discharge attains a maximum value of 0.94, with a β angle of about 13 degrees.

FLOW MEASUREMENT

Pitot Tube Method of Measuring Flow

The most commonly used method of measuring flow in an open stream discharging from an orifice, nozzle, or open pipe is by direct measurement of the velocity head that produces the flow. This measurement process makes use of the well-known Pitot tube and pressure gage combination of which representative forms are shown in Section 6, Chapter 7, "Test of Water Supplies."

When the small opening [usually not over $1/_{16}$ in. (1.6 mm) in diameter] is inserted into the center of a stream at the point of maximum contraction, with the opening directly in the line of flow, the gage will indicate the total head at that location. With the stream open to the atmosphere, there will be no pressure head, so the indicated reading will be velocity head alone, and thus the velocity of the stream can be calculated directly. As a result, velocity pressure is sometimes referred to as Pitot pressure.

If the area of the cross-section of the stream at the location of the velocity measurement location is known, the quantity flowing can be determined from the relation, $Q = a\upsilon = 29.8c_d d^2 \sqrt{p}$ or in SI units, $Q_m = 0.0666c_d d_m^2 \sqrt{p_{\upsilon m}}$, previously derived. In practice, discharge tables are usually used to determine the flow from hydrants and nozzles. (See Table 6-6A.)

A typical Pitot tube as used in the measuring of the flow from a fire stream nozzle is shown in Figure 6-6I. For the usual forms of orifices and nozzles, the coefficient of discharge (c_d) is accurately known, so that $a = c_d \times$ actual discharge opening. For example, with a sharp-edge orifice the area of the stream may be determined from the actual diameter of the orifice opening and the use of the 0.62 coefficient of discharge, as outlined in the previous section on flow-through orifices.

When measuring flow from a straight stream fire nozzle, the use of the Pitot tube method only holds with reasonable accuracy for tip sizes up to $1^3/_8$ in. (35 mm) supplied from $2^1/_2$-in. (64-mm) hose. Above that, the error rate increases beyond acceptable limits, as the assumptions of uniform velocity and full flow become less valid. An exception is the Underwriters playpipe which maintains a uniform coefficient over a wide range of flows and pressures for tip sizes of $1^1/_8$ or $1^3/_4$ in. (29 or 45 mm).

The Pitot tube method is also commonly used to measure the flow discharging from the outlets of fire hydrants to determine the water supply available for fire protection. Unlike the flow entering a nozzle on the end of a pipe or hose line, the flow through large

FIG. 6-6G.   Flow in cylindrical short tube.

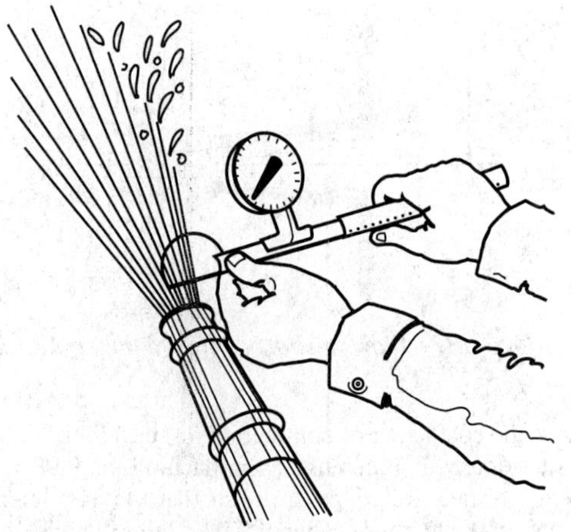

*FIG. 6-6I. **Taking nozzle pressure with a Pitot tube.***

hydrant outlets, or through smaller hydrant outlets at high velocities, has neither a uniform velocity profile nor full flow, since the additional turbulence generated by the flow passing through the hydrant has not dissipated.

In cases such as these, the flow conditions must be changed so that the assumptions needed for the Pitot tube method are valid, or an alternate method, such as the one described next under "The Nozzle Method of Measuring Flow," should be used. If flow is from an open hydrant outlet, hoses and nozzles, a reducer, or a short tube can often be connected to the outlet to improve the flow characteristics.

The Nozzle Method of Measuring Flow

The rate of discharge can also be calculated from the gage pressure at the base of the nozzle. The flow formula for using base pressure is:

$$Q = \frac{29.8cd^2\sqrt{p_1}}{\sqrt{1 - c^2\left(\frac{d}{D}\right)^4}}$$

where

Q = flow in gallons per minute;

c = coefficient of discharge;

d = diameter of outlet, in inches;

p_1 = gage pressure at base of nozzle, in psi; and

D = inside diameter of fitting to which gage is attached, in inches.

For SI units, the formula is expressed as:

$$Q_m = \frac{0.0666cd_m^2\sqrt{p_{1m}}}{\sqrt{1 - c^2\left(\frac{d_m}{D_m}\right)^4}}$$

where

Q_m = flow in liters per minute;

c = coefficient of discharge;

d_m = diameter of outlet, in mm;

p_{1m} = gage pressure at base of nozzle, in kPa; and

D_m = inside diameter of fitting to which gage is attached, in mm.

This is the same formula that is used for discharge from an orifice, except that: (1) gage pressure at the base of the nozzle is substituted for Pitot pressure, and (2) a factor is added that represents the ratio between gage pressure (normal) and total pressure at the nozzle base. (Total pressure is gage plus velocity pressure.)

When base pressure is to be used, the gage is attached to a fitting close to the nozzle with a straight piece of approach pipe or hose to eliminate turbulence or unstable flow conditions. To obtain greater accuracy than provided by a simple fitting, a piezometer fitting may be used. With this device, the gage is connected to an annular tube or channel having a number of small holes drilled into the waterway around the circumference. The mean or resultant static pressure indicated by the gage is p_1 in the formula above.

Although accurate and convenient for fixed test arrangements, the measurement of pressure at the base of a nozzle is not practical for usual hose stream operations. However, because a Pitot gage is useless with spray-type nozzles, or other devices producing special types of discharge, the base pressure method is necessary.

Discharge Calculations

The most common method of estimating nozzle or orifice discharge is to use Table 6-6A. The flow from the table, corresponding to the measured Pitot pressure and orifice diameter, is multiplied by the discharge coefficient. (See Table 6-6B.)

EXAMPLE 1:

A Pitot reading of 20 psi was measured on an open $2^1/_2$-in. smooth and well-rounded hydrant butt. From Table 6-6B, the discharge coefficient is 0.90. The theoretical flow for a 20-psi velocity head is 834 gpm. The discharge is:

$Q = 834 \times 0.90 = 751$ gpm. (Or subtract 10 percent from 834 and get 751 gpm.)

EXAMPLE 2:

A Pitot reading was 200 kPa at a 38.1-mm pipe with a square sharp opening (discharge coefficient = 0.80). From Table 6-6A, 200 kPa gives a theoretical flow of 1370 L/min. The actual flow would be 0.8 × 1370 = 1100 L/min.

Discharge curves, such as those shown in Figure 6-6J, are available for many nozzles and are sufficiently accurate for most fire flow calculations. These usually incorporate the specific discharge coefficients for the nozzle involved, but occasionally plot theoretical flow. Nozzle discharge can also be determined by the standard formulas previously discussed. (See "Flow of Water Through Orifices" and "The Nozzle Method of Measuring Flow," in this chapter).

EXAMPLE 3:

Calculate the rate of discharge from a 2-in. (51-mm) nozzle with a pressure measured by a $2^1/_2$-in. (64-mm) piezometer ring and gage of 80 psi (552 kPa) at the base of the nozzle. The nozzle has a coefficient of discharge of 0.99.

Using the formula for the nozzle method of measuring flow:

$$Q = \frac{29.8cd^2\sqrt{p_1}}{\sqrt{1 - c^2\left(\frac{d}{D}\right)^4}}$$

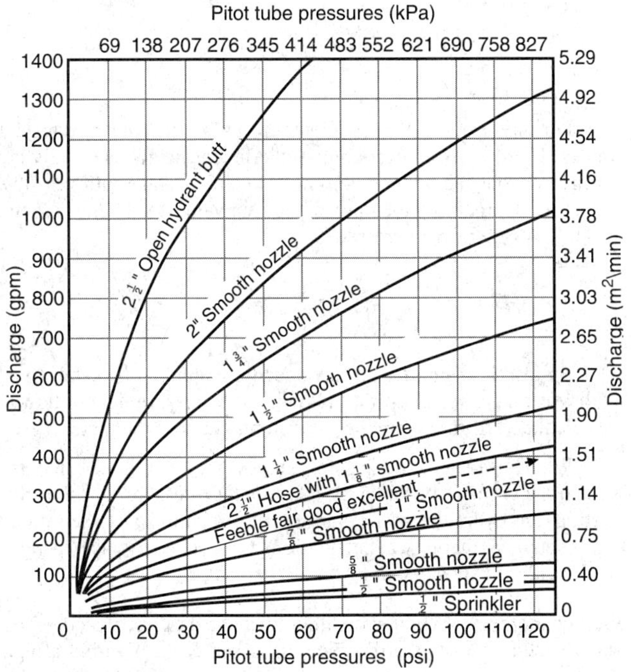

FIG. 6-6J. Relative discharge curves. (Source: Chemical Engineers' Handbook, McGraw-Hill.) (For SI units: 1 in. = 25.4 mm.)

TABLE 6-6C. *Values of k for Various Discharge Orifices*

Type of Orifice	Nominal Diameter		k Factor	
	(in.)	(mm)	gpm/p$^{1/2}$	Lpm/ Pm$^{1/2}$
Sprinkler	1/4	7	1.3–1.5	1.9–2.2
Sprinkler	5/16	8	1.8–2.0	2.6–2.9
Sprinkler	3/8	10	2.6–2.9	3.7–4.2
Sprinkler	7/16	11	4.0–4.4	5.8–6.3
Sprinkler	1/2	13	5.3–5.8	7.6–8.4
Sprinkler	17/32	14	7.4–8.2	10.6–11.8
Sprinkler	5/8	16	11.0–11.5	15.9–16.6
Sprinkler	3/4	19	13.5–14.5	19.5–20.1
Nozzle	1/2	13	7.2	10.3
Nozzle	7/8	22	22.2	32.0
Nozzle	1	25	29.1	41.9
Nozzle	1 1/16	27	32.8	47.2
Nozzle	1 1/8	29	36.8	53.0
Nozzle	1 3/16	30	41.0	59.0
Nozzle	1 1/4	32	45.4	65.4
Nozzle	1 5/16	33	50.1	72.1
Nozzle	1 3/8	35	54.9	79.1
Nozzle (c = 0.97 for all nozzles)	1 7/16	37	60.0	86.4
Nozzle	1 1/2	38	65.4	94.2
Nozzle	1 9/16	40	70.9	102.0
Nozzle	1 5/8	41	76.8	110.6
Nozzle	1 11/16	43	82.8	119.2
Nozzle	1 3/4	44	89.0	128.2
Nozzle	1 13/16	46	95.5	137.5
Nozzle	1 7/8	48	102.0	146.9
Nozzle	2 15/16	49	109.0	157.0
Nozzle	2	51	116.0	167.0
Hydrant butt (c = 0.90)	2	51	107.4	154.7
Hydrant butt (c = 0.90)	2	57	135.9	195.7
Hydrant butt (c = 0.90)	2	64	167.8	241.6

$$Q = \frac{29.8 \times 0.99 \times (2)^2 \sqrt{80}}{\sqrt{1 - (0.99)^2 \left(\frac{2}{2.5}\right)^4}}$$

$$Q = 1366 \text{ gpm } (5170 \text{ L/min}).$$

To simplify calculations for a specific orifice or nozzle, the constants in the flow formula can be combined, reducing the formula to:

$$Q = k\sqrt{p}$$

where k combines the constants 29.8 (0.0666 in SI units), c_d, and d^2. Table 6-6C lists k factors of some common discharge orifices used for fire protection. For SI units, use the k_m values.

Because

$$k = \frac{Q}{\sqrt{p}}$$

the k values of spray nozzles can be calculated from data in testing laboratory listings of nozzles. For some nozzles, the flow rate at 25 and 125 psi (170 and 162 kPa) pressure is given, from which the k factor can be calculated.

EXAMPLE 4:

A certain fire service spray nozzle is rated for 83 gpm at 50 psi. Therefore,

$$k = \frac{83}{\sqrt{50}} = \frac{83}{7.07} = 11.7$$

At 25 psi pressure, the discharge would be

$$11.7\sqrt{25} = 11.7 \times 5 = 58.5 \text{ gpm}$$

EXAMPLE 5:

Determine the discharge from a 51-mm hydrant butt (c_d = 90) at a pressure of 350 kPa.

$$k_m = 0.0666cd_m^2 = 0.0666 \times 0.90 \times 51^2 = 156$$

Thus,

$$Q_m = k_m\sqrt{p_m} = 156\sqrt{350} = 2918 \text{ L/min}$$

Alternatively, from Table 6-6C, k_m = 154.7. Thus,

$$Q_m = k_m\sqrt{p_m} = 154.7\sqrt{350} = 2894 \text{ L/min}$$

Flow Meters

When it is not convenient to discharge water to the atmosphere, meters are used to determine flow.

Venturi tube: The Venturi principle has a number of applications in fire protection. The Venturi tube is essentially a tapered constriction in a pipe. In the constricted part, the velocity must be greater than in the straight tube, and the pressure is correspondingly less, in accordance with Bernoulli's theorem. If the increase in velocity through the restricted portion is sufficient, the pressure at that point will be less than atmospheric, and a suction will be created at any opening into the side of the tube. The Venturi tube is illustrated in Figure 6-6K. The diverging portion of a Venturi tube serves only to restore the system pressure with a minimum of friction loss.

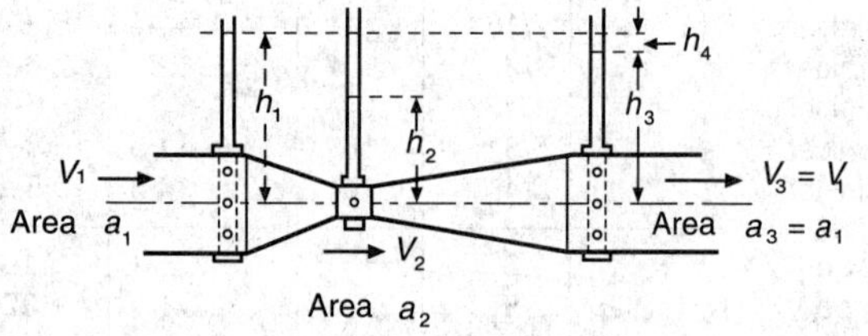

FIG. 6-6K.   The Venturi tube.

Venturi meter: The Venturi principle as applied in the Venturi meter for the measurement of flow in closed pipelines under pressure is as follows.

With no elevation difference along the line of flow, Bernoulli's theorem becomes:

$$\frac{v_1^2}{2g} + \frac{p_1}{w} + 0 = \frac{v_2^2}{2g} + \frac{p_2}{w} + 0$$

In Figure 6-6K,

$$\frac{p_1}{w}$$

is represented by h_1, and

$$\frac{p_2}{w}$$

by h_2, which are used in the following equations.

The quantity of liquid passing through all portions of the Venturi meter must be the same. Therefore,

$$Q = a_1 v_1 = a_2 v_2 \text{ or } v_1 = \frac{Q}{a_1} \text{ or } v_2 = \frac{Q}{a_2}$$

Substituting in Bernoulli's theorem,

$$\frac{\left(\frac{Q}{a_1}\right)^2}{2g} + h_1 = \frac{\left(\frac{Q}{a_2}\right)^2}{2g} + h_2$$

$$\frac{\left(\frac{Q}{a_2}\right)^2}{2g} - \frac{\left(\frac{Q}{a_1}\right)^2}{2g} = h_1 - h_2$$

$$Q^2\left(\frac{1}{a_2^2} - \frac{1}{a_1^2}\right) = 2g(h_1 - h_2)$$

$$Q^2\left(\frac{a_1^2 - a_2^2}{a_1^2 a_2^2}\right) = 2g(h_1 - h_2)$$

$$Q = \frac{a_1 a_2}{\sqrt{a_1^2 - a_2^2}}\sqrt{2g(h_1 - h_2)}$$

For any specific Venturi meter, a_1 and a_2 are known constant values. There is also a friction loss coefficient, which is usually determined by test and which does not remain constant with very low velocities. Combining the known constant values, the Venturi meter formula is generally expressed as:

$$Q = k\sqrt{h_1 - h_2}, \text{ or } Q = k\sqrt{\frac{p_1}{w} - \frac{p_2}{w}}$$

By test, a value of k for any specific meter can be established with reasonable accuracy, thus allowing the measurement of flow to be calculated from the pressure differential across the meter.

When used as a device for inducing gas or liquid into the stream, as is made possible by the reduced pressure, in the throat section, the hydraulic performance will not be in strict accordance with the above theoretical calculations because energy is expended on the induced substance.

Orifice meters: When water flows through a thin sharp-edged orifice (such as the standard orifice discussed earlier) within a pipe, the flow diameter contracts the same way it does when the flow discharges to atmosphere, and then increases back to the full pipeline diameter. The pressure at the pipe wall is reduced due to this change in flow, and is related to the flow rate. (See Figure 6-6L.)

Bernoulli's equation can be applied to an orifice meter in the same way as it is applied to the Venturi meter. The resultant k factor or discharge coefficient depends upon several factors, including the ratio of orifice/pipe diameter and the location of the pressure taps upstream and downstream of the orifice.

ASME flow nozzle: Flow can also be measured by an ASME flow nozzle, as shown in Figure 6-6M. The pressure taps for this device are normally one pipe diameter upstream of the nozzle and one-half pipe diameter downstream from the nozzle inlet. Combining known constant values, the Venturi meter formula is generally expressed as:

$$Q = k\sqrt{h_1 - h_2}, \text{ or } Q = k\sqrt{\frac{p_1}{w} - \frac{p_2}{w}}$$

FLOW OF WATER IN PIPES

The theory of liquid flow in pipes involves the same continuity principles used in the previous discussions. These include continuity of energy (Bernoulli's theorem with friction) and continuity of flow.

When water flows through a pipe, there is always a drop in pressure. Theoretically, the lost head between two points is caused by: (1) friction between the moving water and the pipe wall and (2) friction between water particles, including that produced by turbulence when flow changes direction or when a rapid increase or decrease in velocity takes place, such as at abrupt changes in pipe diameter. A change in velocity results in some conversion of velocity head to pressure head, or vice versa.

At low velocity in a smooth pipe, very little turbulence is produced, and the flow is called "laminar" or "streamline" flow. With this condition, all particles of water move along the pipe in definite paths, which are essentially straight lines, in concentric layers. Friction loss occurs due to shear stress, mainly in a thin boundary layer at the pipe wall, and also between adjacent stream layers. The friction loss is small compared to that of turbulent flow.

The flow within either a smooth or a rough pipe remains laminar until the velocity reaches what is called critical velocity. At this point, there is a range of unstable flow which is neither laminar nor completely turbulent. This is called the transition zone.

As the flow continues to increase, it becomes turbulent. In turbulent flow the fluid moves in an eddying mass, and, at any point, the individual water particles move rapidly in a random manner rather than in a straight line.

Reynolds demonstrated that, for any fluid, the critical point at which the flow changes from laminar to turbulent could be predicted. In circular pipes, the critical point occurs when the dimensionless parameter $dv\rho/\mu$ (called the Reynolds number) is approximately 2000. The transition to complete turbulence is complete for Reynolds numbers exceeding 4000.

Most fire protection systems and water distribution mains function under turbulent flow conditions, and friction losses within the pipe itself account for most of the lost head. Other losses are usually considered together and are called "minor losses" or "losses in fittings."

Friction (Head) Loss Flow Formulas

Experimental data have established that frictional resistance in a pipe is:

1. Independent of pressure in the pipe.
2. Proportional to the amount and character of the flow.
3. Variable with the velocity of the flow (nearly proportional to the second power of the velocity for velocities above the critical; if velocity is below critical, resistance varies with the first power).

The Chezy formula: One of the best-known and oldest expressions relating velocity to friction loss in piping is known as the Chezy formula:

$$v = c\sqrt{rs}$$

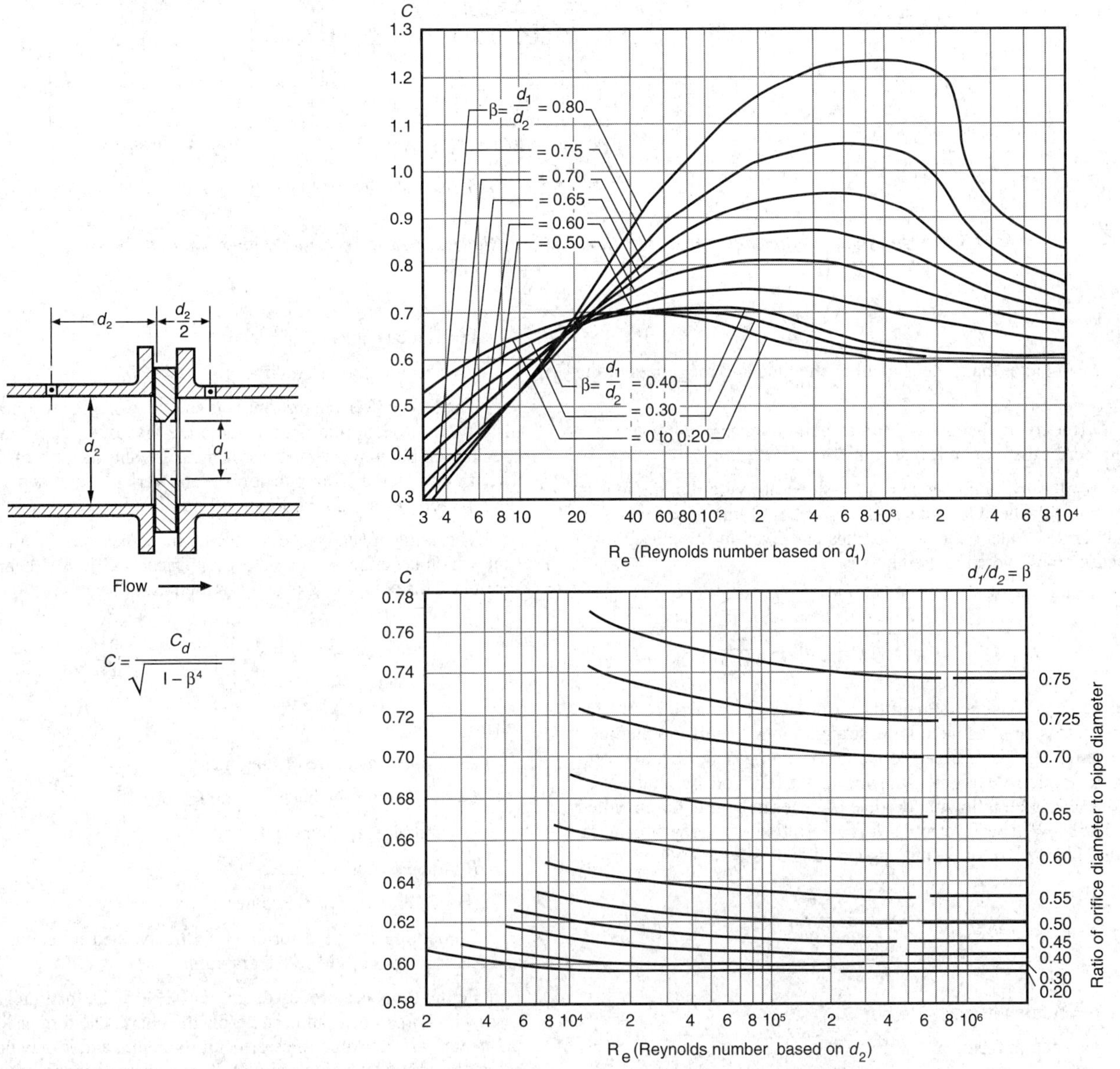

$$C = \frac{C_d}{\sqrt{1 - \beta^4}}$$

FIG. 6-6L. *Flow coefficient C for square-edged orifice meters. (Source: Crane Technical Paper No. 410, 1976)*

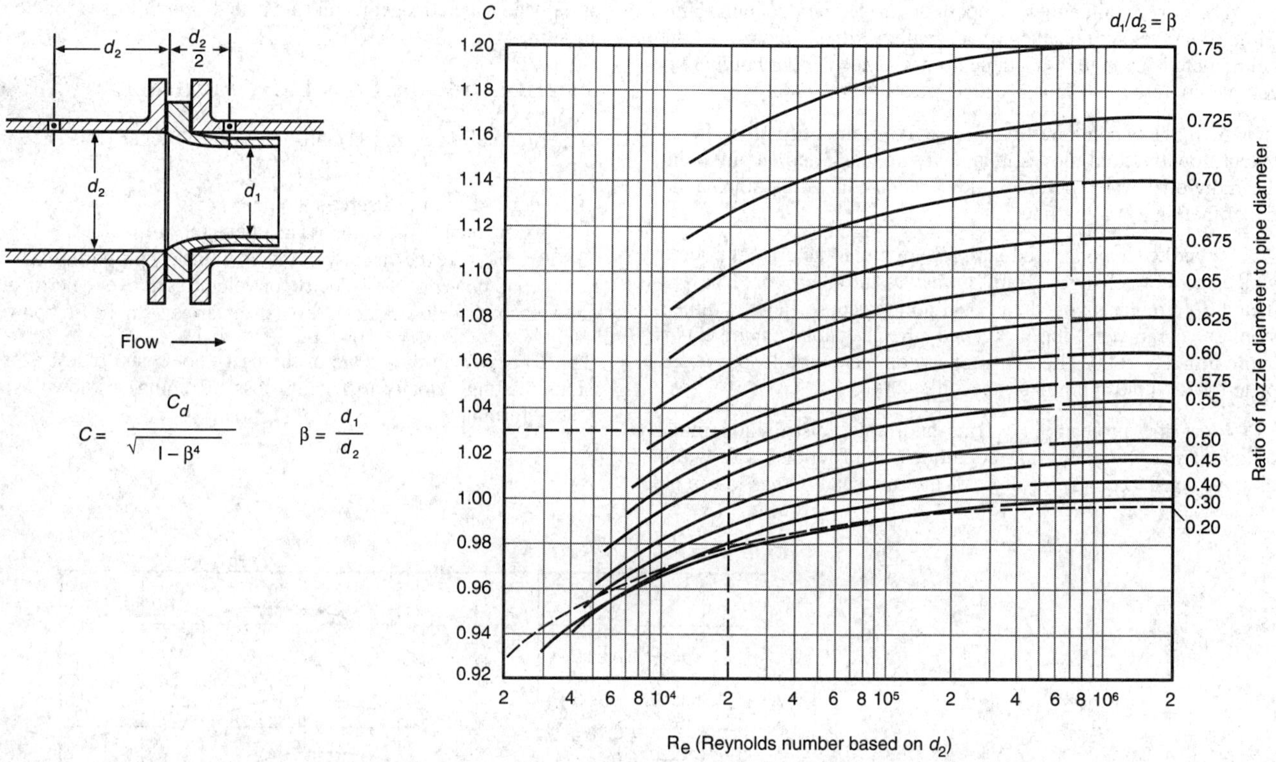

$$C = \frac{C_d}{\sqrt{1 - \beta^4}} \qquad \beta = \frac{d_1}{d_2}$$

FIG. 6-6M.   *Flow coefficient C for ASME flow nozzles. (Source: Crane Technical Paper No. 410, 1976)*

where

  c = a factor that is dependent on the kind and roughness of the pipe;

  r (the hydraulic radius) = area/circumference = $d/4$, where d = diameter of the pipe in ft (m); and

  s = the hydraulic slope = h/l = slope of hydraulic gradient in which h is the head loss in length of pipe l in ft (m). (See Section 6, Chapter 2, "Water Storage Facilities and Suction Supplies," for a discussion of hydraulic gradients.)

  Therefore:

$$v = c\sqrt{d/4 \times h/l} \quad \text{or} \quad h = \frac{4lv^2}{c^2 d}$$

The Darcy-Weisbach formula: Another classical formula for friction loss, applicable to long, straight pipes of uniform diameter and roughness, is ascribed to Darcy, Manning, Fanning, and others. In modern textbooks, the formula is derived by analysis of forces acting on a flowing particle of water in a pipe. Often called the Darcy-Weisbach formula, it is a variation of Chezy's formula, with a friction factor f replacing c, and expressed as follows:

$$h = f \frac{l}{d} \frac{v^2}{2g}$$

where

  h = friction head,

  f = friction factor,

  l = length of pipe,

  d = diameter of pipe,

  v = velocity, and

  g = acceleration of gravity.

  The Darcy-Weisbach formula is suitable for all Newtonian fluids. (A Newtonian fluid is one where the viscosity is constant at a specific temperature, regardless of pressure and the rate of shear.) The friction factor f is dimensionless and variable, and depends on the pipe roughness and the Reynolds number.

  The value of f can be computed by the Colebrook-White equation, which is neither a completely empirical nor rigidly theoretical formula. This equation is usually written as:

$$\frac{1}{\sqrt{f}} = -2 \log_{10}\left[\frac{\varepsilon}{3.7D} + \frac{2.51}{R\sqrt{f}} \right]$$

where

  ε = a linear measure of roughness,

  f = the Darcy-Weisbach friction factor,

  D = the pipe diameter in ft, and

  R = the Reynolds number.

  (For SI units: 1 ft = 0.305 m.)

  Computing f by the formula can be avoided by using tables and charts known as "Moody" diagrams.

  Figure 6-6N is a Moody diagram (© Hydraulic Institute, 1954) from which f can be read directly off the chart. The dimensionless parameter, ε/D, is sometimes difficult to obtain, and it may be necessary to assume a value for ε/D, based on experience and judgment. The roughness factor of new pipe usually can be provided by the manufacturer.

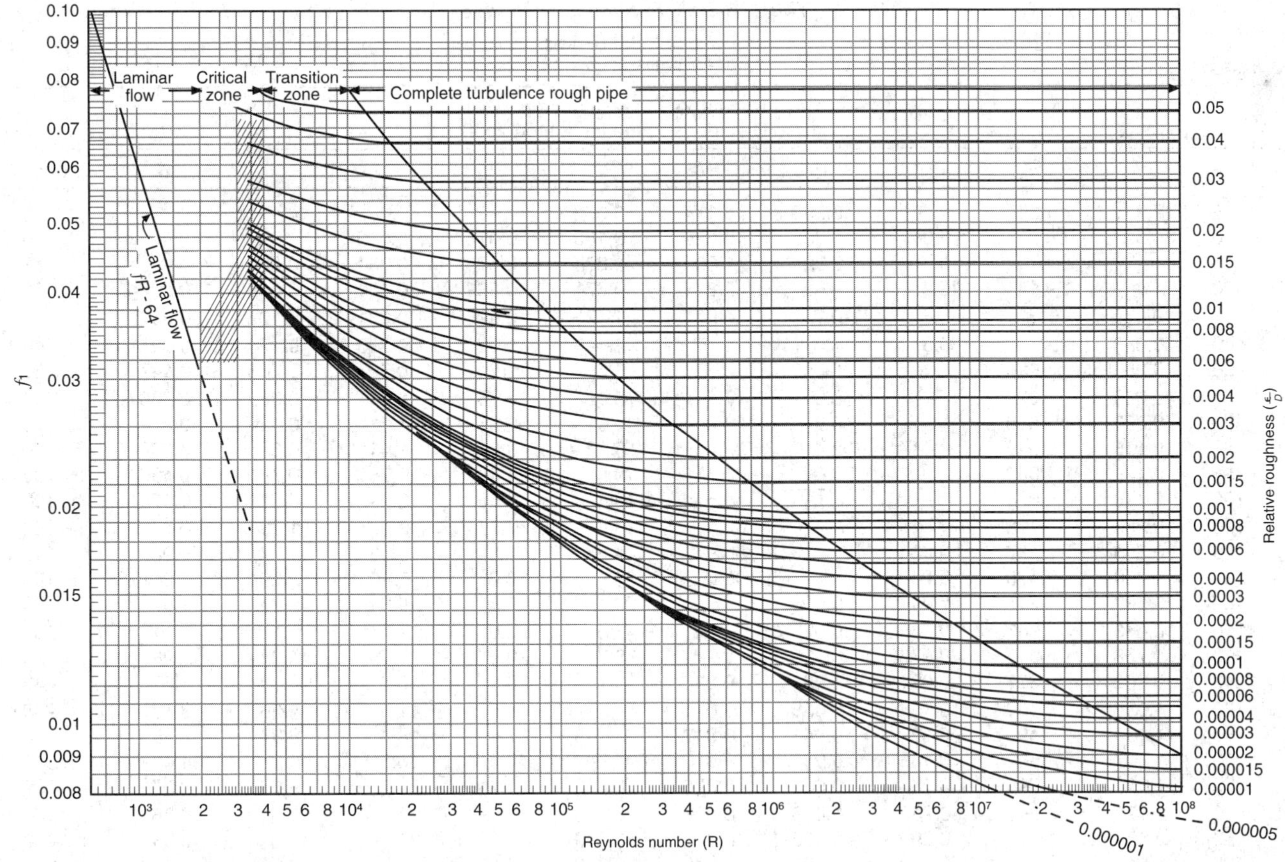

FIG. 6-6N. *Moody Diagram for friction in pipe. Values for friction factor are on the vertical scale, at left. (See Friction Manual Hydraulic Institute.)*

The Hazen-Williams formula: The friction-flow formulas commonly used in fire protection hydraulics have been developed by experiment and experience. These formulas (which are variations of the Chezy formula) are usually exponential in the form $v = Cr^x s^y$, where v is velocity, c the coefficient of friction, r the hydraulic radius (area divided by circumference), and s the hydraulic slope (loss of head divided by length). The most popular exponential formula is the Hazen-Williams, its basic form being $v = 1.31cr^{0.63}s^{0.54}$. The friction coefficients in formulas of this type are constant for a specific roughness of pipe and are independent of velocity, and thus the accuracy of these formulas is variable. However, the fixed values generally assumed for viscosity and density are considered adequate for most fire protection hydraulic work.

The basic form of the Hazen-Williams formula ($v = 1.31Cr^{0.63}s^{0.54}$) is not practical for ordinary fire protection flow calculations. Given pressure loss in psi and flow in gpm, the formula becomes:

$$p = \frac{4.52Q^{1.85}}{C^{1.85}d^{4.87}}$$

in which p is the pressure loss in psi per ft of pipe; Q is the rate of flow in gpm; and d is the inside diameter of the pipe in in. (Numbers to the 1.85 power are given in Table 6-6D.)

In SI units the formula is:

$$p_m = 6.06 \times \frac{Q_m^{1.85}}{c^{1.85}d_m^{4.87}} \times 10^7$$

where

p_m = pressure loss (kPa) per m of pipe,

Q_m = rate of flow (L/min), and

d_m = inside diameter (mm).

Friction loss calculations: The solutions of many fire protection problems involving pipe flow and friction do not require direct calculation using formulas, since tables and charts are readily available. However, in using the simplifying charts and tables, great care must be taken to identify the C value (coefficient of friction) on which the chart or table is based. Where the type or condition of a pipe necessitates the use of a different C value, the friction loss from the table must be multiplied by a conversion factor to obtain the correct results for the desired C value.

By way of illustration, Table 6-6D gives values of p when $C = 100$ for standard pipe sizes from $1/2$ in. to 30 in. in diameter. For values of C other than 100, the tabular value losses are multiplied by the corresponding factor in Table 6-6E. Where a different type of

TABLE 6-6D. *Friction Loss in Pipe*
lb/in.² per 100 ft of Pipe
Hazen-Williams C = 100*

Actual Diameter of Pipe 1/2 through 3 1/2 in.†‡
Nominal Diameter of Pipe for 4 through 30 in.‡

gpm	1/2	3/4	1	1 1/4	1 1/2	2	2 1/2	3	3 1/2	4	gpm
5	17.9	4.55	1.40	0.369	0.174	0.052	—	—	—	—	5
10	64.5	16.4	5.06	1.33	0.629	0.186	0.078	0.030	—	—	10
15		34.7	10.7	2.82	1.33	0.394	0.166	0.064	0.028	—	15
20	*5*	59.1	18.2	4.89	2.27	0.671	0.282	0.109	0.048	0.027	20
30	0.019	*6*	38.6	10.2	4.80	1.42	0.598	0.231	0.102	0.057	30
40	0.033	—	65.8	17.3	8.17	2.42	1.02	0.393	0.174	0.097	40
50	0.050	.020	*8*	26.2	12.3	3.66	1.54	0.593	0.263	0.147	50
60	0.069	.029	—	36.6	17.3	5.12	2.16	0.831	0.369	0.206	60
70	0.092	.038	—	48.7	23.0	6.81	2.87	1.11	0.490	0.274	70
80	0.118	.049	—	62.4	29.4	8.72	3.67	1.41	0.628	0.350	80
90	0.147	.060	—	77.6	36.6	10.8	4.56	1.76	0.781	0.435	90
100	0.178	.074	—	*10*	44.5	13.1	5.55	2.14	0.949	0.529	100
120	0.250	.103	—	—	62.3	18.5	7.77	3.00	1.33	0.741	120
140	0.333	.137	0.034	—	82.9	24.6	10.3	3.98	1.77	0.986	140
160	0.426	.175	0.043	—	106.0	31.4	13.2	5.10	2.26	1.26	160
180	0.529	.218	0.054	0.018	*12*	39.1	16.5	6.34	2.81	1.57	180
200	0.643	.265	0.065	0.022	—	47.5	20.0	7.71	3.42	1.91	200
220	0.768	.316	0.078	0.026	—	56.7	23.9	9.19	4.08	2.28	220
240	0.902	.371	0.091	0.031	0.013	*14*	28.0	10.8	4.79	2.67	240
260	1.05	.430	0.106	0.036	0.015	—	32.5	12.5	5.56	3.10	260
280	1.20	.493	0.122	0.041	0.017	—	37.3	14.4	6.37	3.55	280
300	1.36	.562	0.138	0.047	0.019	—	42.3	16.3	7.24	4.04	300
350	1.81	.746	0.184	0.062	0.026	0.012	*16*	21.7	9.63	5.37	350
400	2.32	.955	0.235	0.079	0.033	0.015	—	27.8	12.3	6.88	400
450	2.88	1.19	0.292	0.099	0.041	0.019	—	34.6	15.3	8.55	450
500	3.51	1.44	0.353	0.120	0.049	0.023	0.012	42.0	18.6	10.4	500
550	4.18	1.72	0.424	0.143	0.059	0.028	0.015	50.1	22.2	12.4	550
600	4.91	2.02	0.498	0.168	0.069	0.033	0.017	58.8	26.1	14.6	600
650	5.70	2.34	0.577	0.195	0.080	0.038	0.020	68.2	30.3	16.9	650
700	6.53	2.69	0.662	0.223	0.092	0.043	0.023	*18*	34.7	19.4	700
750	7.42	3.05	0.752	0.254	0.104	0.049	0.026	—	39.4	22.0	750
800	8.36	3.44	0.848	0.286	0.118	0.056	0.029	—	44.5	24.8	800
850	9.35	3.85	0.948	0.320	0.132	0.062	0.032	—	49.7	27.7	850
900	10.4	4.28	1.05	0.356	0.146	0.069	0.036	—	*20*	30.8	900
950	11.5	4.73	1.17	0.393	0.162	0.076	0.040	—	—	34.1	950
1,000	12.6	5.20	1.28	0.432	0.178	0.084	0.044	—	—	37.5	1,000
1,250	19.1	7.85	1.94	0.653	0.269	0.127	0.066	—	—	*24*	1,250
1,500	*30*	11.0	2.71 (18.7)	0.914	0.376	0.178	0.093	—	—	—	1,500
1,750	—	—	3.61	1.22	0.501	0.236	0.123	—	—	—	1,750
2,000	.007	—	4.62	1.56	0.641	0.303	0.158	0.089	0.053	0.022	2,000
2,250	.009	—	—	1.94	0.797	0.376	0.196	0.111	0.066	0.027	2,250
2,500	.011	—	—	2.35	0.969	0.457	0.239	0.134	0.081	0.033	2,500
2,750	.013	—	—	2.81	1.16	0.545	0.285	0.160	0.096	0.040	2,750
3,000	.016	—	—	3.30	1.36	0.641	0.334	0.188	0.113	0.046	3,000
4,000	.027	—	—	—	2.31	1.09	0.569	0.321	0.192	0.079	4,000
5,000	.040	—	—	—	3.49	1.65	0.860	0.485	0.290	0.119	5,000

* To convert friction loss at C = 100 or other values of C, see Table 6-6E.
† Schedule 40 pipe sizes 1/2 through 3 1/2 in. steel pipe.
‡ SI units: 1 psi = 6.895 kPa; 1 gpm = 0.378 L/min; 1 in. = 25.4 mm.
Note: Actual inside diameter for sizes 1/2 in. through 3 1/2 in. is given for greater accuracy as these sizes include sprinkler branch lines and the smaller sizes of cross mains. For sizes 4 in. and greater, the nominal diameters were used as a fairly safe average for the diameters of various types of underground pipes as follows: cast-iron unlined and Enameline, greater than nominal; cast iron cement lines and Class 200 asbestos cement, less than nominal; Class 150 asbestos-cement sizes 6 and 8 in. less than nominal, and other sizes even nominal. (A 0.10 variation is true for Class 150 cement lined only—see ASHD FT-9 through 45 for actual IDs.)

This table will be useful in approximating friction loss in flow through existing underground piping where the type, inside diameter, and condition are frequently unknown. However, in such cases, a flow test is recommended.

When the type, inside diameter, and condition are known, and in designing new systems for all sizes and types of pipes, the friction loss tables should be used. Friction tables based on Hazen-Williams formula are published in *Automatic Sprinkler Hydraulic Data* by "Automatic" Sprinkler Corporation of America, and tables based on Darcy-Weisbach formula are published in *Standards of the Hydraulic Institute*.

TABLE 6-6E. *Conversion Factors for Friction Loss in Pipe for Values of Coefficient Other Than 100*

C	Factor	C	Factor	C	Factor
150	0.472	110	0.838	70	1.93
145	0.503	105	0.914	65	2.22
140	0.537	100	1.00	60	2.57
135	0.574	95	1.10	55	3.02
130	0.615	90	1.22	50	3.61
125	0.662	85	1.35	45	4.38
120	0.714	80	1.51	40	5.48
115	0.772	75	1.70	35	6.97

pipe is used, the friction loss from the table can be corrected using the formula:

$$\Delta \rho_a = \Delta \rho_{40}\left(\frac{d_{40}}{d_a}\right)^{4.87}$$

where

Δp_a = actual friction loss,

Δp_{40} = friction loss in Schedule 40 pipe,

d_{40} = internal diameter of Schedule 40 pipe, and

d_a = internal diameter of actual pipe.

EXAMPLE:

Determine the friction loss with 700 gpm, flowing in 700 ft of 8-in. cast-iron pipe having a C value of 80.

SOLUTION:

From Table 6-6D, the loss for 700 gpm per 100 ft of 8-in. pipe with C = 100 is 0.662 psi. From Table 6-6E, the factor for C = 80 is 1.51. Because friction loss is directly proportional to the length of pipe, multiply 0.662 × 7 × 1.5 = 6.95 psi (answer). Since other than Schedule 40 steel pipe is used, this value must be adjusted using the previously discussed formula.

$$\Delta \rho_a = \Delta \rho_{40}\left(\frac{d_{40}}{d_a}\right)^{4.87}$$

$$= 6.95\left(\frac{8.071}{8.23}\right)^{4.87}$$

$$= 6.32 \text{ psi}$$

The Hazen-Williams diagram: Figure 6-6O is a graphical representation of Table 6-6D, except that it is based upon a Hazen-Williams coefficient of C = 120 rather than C = 100, as used in the table. It is limited in scope to pipes not over 10 in. in diameter, and due to the reduced scale is less accurate than the table.

Figure 6-6P gives data in SI units for pipes up to 10 in. (254 mm) in diameter, again based on a C = 120.

FIG. 6-6O. *Friction loss in Schedule 40 steel pipe, Hazen-Williams C = 120. For other C values use: Value of C: 80 100 120 130 140 150; multiplying factor: 2.12 1.40 1.00 0.86 0.75 0.66.*

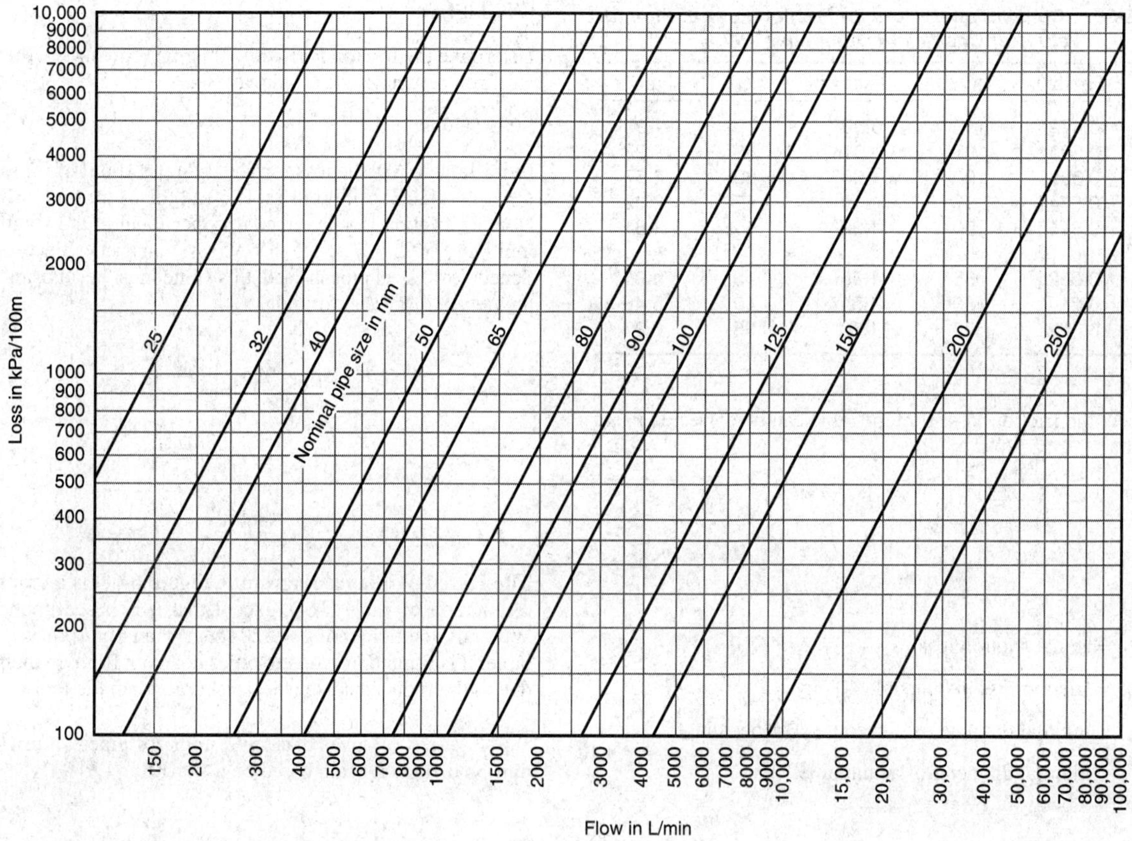

FIG. 6-6P. *Friction loss (SI units) in Schedule 40 steel pipe, Hazen-Williams C=120. For other C values use: Value of C 80 100 120 130 140 150; multiplying factor 2.12 1.40 1.00 0.86 0.75 0.66.*

EXAMPLE 1:

What is the friction loss in 300 ft (90 m) of 8-in. cement-lined cast-iron pipe, at 1500 gpm (5678 L/min)?

SOLUTION:

From the intersection of the 1500 gpm vertical line with the sloping 8-in. pipe diameter line in Figure 6-6O, read left horizontally for a loss of head value, which is found to be 0.019 psi per foot. For 300 ft, the loss would be 300×0.019 psi or 5.7 psi. The probable C value of cement-lined pipe is 140 (Table 6-6G), and the conversion factor is 0.75 (Figure 6-6O). Therefore, the friction loss is $5.7 \times 0.75 = 4.3$ psi. Since other than Schedule 40 steel pipe is used, this value must be adjusted using the previously discussed formula.

$$\Delta \rho_a = \Delta \rho_{40} \left(\frac{d_{40}}{d_a} \right)^{4.87}$$

$$= 4.3 \left(\frac{8.071}{7.98} \right)^{4.87}$$

$$= 4.54 \text{ psi}$$

EXAMPLE 2:

What is the friction loss in 400 m of 150 mm 30-year-old unlined cast-iron pipe ($C = 80$) for a flow of 10,000 L/min?

SOLUTION:

From the intersection of the 10,000 L/min vertical line with the sloping 150 mm pipe diameter line in Figure 6-6P, read left horizontally for a loss of head value, which is 480 kPa per 100 m. For 400 m, the

loss would be 4×480 kPa or 1920 kPa, based on $C = 120$. For cast-iron pipe, $C = 80$ and the conversion factor is 2.12. (See Figure 6-6F). Therefore, the friction loss is $1920 \times 2.12 = 4070$ kPa.

Equivalent pipes: Problems involving piped water supplies and fire protection systems occasionally require substitution of one pipe for another. The term "equivalent pipe" usually means a pipe having the same friction loss as the pipe for which it is being substituted. The formula for using the friction loss table is applicable here.

EXAMPLE:

What length of 8-in. pipe ($C = 110$) is equivalent to 700 ft of 6-in. pipe ($C = 85$)?

SOLUTION:

$$N_1 \times \frac{L_1}{100} \times T_1 = N_2 \times \frac{L_2}{100} \times T_2$$

where

$N = C$ factor conversion for each pipe,

L = length of each pipe, and

T = friction loss from Table 6-6D.

Assume a rate of flow (say 1000 gpm) and substitute known values:

$$1.35 \times \frac{700}{100} \times 5.20 = 0.838 \times \frac{L_2}{100} \times 1.28$$

Solving for L_2:

$$\frac{1.35 \times 700 \times 5.20}{0.838 \times 1.28} = 4581 \text{ ft}$$

which can be rounded off to 4600 ft.

Minor Losses

While the friction loss within the pipe normally accounts for most of the lost head, head losses also occur when the flow in a pipe changes direction, the pipe size changes, or a valve or other fitting is encountered. These losses are typically referred to as minor losses even though they can be significant for some fittings, such as swing check valves or backflow preventers, that are commonly found in fire protection systems.

The amount of minor losses from fittings can be found in many references and are often expressed in various forms. The most common are as an equivalent length (l/d), a resistance coefficient (k), or a flow coefficient (C_v).

Equivalent length: For most fire protection calculations, friction loss is obtained by using the equivalent length method from a table such as Table 6-6F, which expresses the friction loss of the fitting as an "equivalent pipe length" having the same friction loss as the fitting. This length is then added to the length of the pipe to which the fittings are connected to obtain the total friction loss of pipe and fittings.

EXAMPLE:

Calculate total friction loss for 600 L/min flow through 200 m of nominal 2-in. (50-mm) pipe ($C = 120$) which incorporates three 90-degree elbows, one tee junction (90-degree turn), and two butterfly valves.

SOLUTION:

Using Table 6-6F for equivalent lengths of fittings ($C = 120$):

Fittings or Pipe	Number	Equivalent Length per Fittings (m)	Total Equivalent Length (m)
Std. elbow	3	1.5	4.5
T-junction	1	3.1	3.1
Butterfly	2	1.8	3.6
Pipe	1	200.0	200.0
		Total equivalent pipe length =	211.2

TABLE 6-6F.   *Equivalent Pipe Length Chart*

	Fittings and Valves Expressed in Equivalent ft (m) of Pipe						
	3/4 in. (20 mm)	1 in. (25 mm)	1 1/4 in. (32 mm)	1 1/2 in. (40 mm)	2 in. (50 mm)	2 1/2 in. (50 mm)	3 in. (80 mm)
45-degree Elbow	1 (0.3)	1 (0.3)	1 (0.3)	2 (0.6)	2 (0.6)	3 (0.9)	3 (0.9)
90-degree Standard Elbow	2 (0.6)	2 (0.6)	3 (0.9)	4 (1.2)	5 (1.5)	6 (1.8)	7 (2.1)
90-degree Long-Turn Elbow	1 (0.3)	2 (0.6)	2 (0.6)	2 (0.6)	3 (0.9)	4 (1.2)	5 (1.5)
Tee or Cross (Flow Turned 90°)	4 (1.2)	5 (1.5)	6 (1.8)	8 (2.4)	10 (3.1)	12 (3.7)	15 (4.6)
Gate Valve	—	—	—	—	1 (0.3)	1 (0.3)	1 (0.3)
Butterfly Valve	—	—	—	—	6 (1.8)	7 (2.1)	10 (3.1)
Swing Check*	4 (1.2)	5 (1.5)	7 (2.1)	9 (2.7)	11 (3.4)	14 (4.3)	16 (4.9)

	Fittings and Valves Expressed in Equivalent ft (m) of Pipe						
	3 1/2 in. (90 mm)	4 in. (100 mm)	5 in. (125 mm)	6 in. (150 mm)	8 in. (200 mm)	10 in. (250 mm)	12 in. (300 mm)
45-degree Elbow	3 (0.9)	4 (1.2)	5 (1.5)	7 (2.1)	9 (2.7)	11 (3.4)	13 (4.0)
90-degree Standard Elbow	8 (2.4)	10 (3.1)	12 (3.7)	14 (4.3)	18 (5.5)	22 (6.7)	27 (8.2)
90-degree Long-Turn Elbow	5 (1.5)	6 (1.8)	8 (2.4)	9 (2.7)	13 (4.0)	16 (4.9)	18 (5.5)
Tee or Cross (Flow Turned 90 degrees)	17 (5.2)	20 (6.1)	25 (7.6)	30 (9.2)	35 (10.7)	50 (15.3)	60 (18.3)
Gate Valve	1 (0.3)	2 (0.6)	2 (0.6)	3 (0.9)	4 (1.2)	5 (1.5)	6 (1.8)
Butterfly Valve	—	12 (3.7)	9 (2.7)	10 (3.1)	12 (3.7)	19 (5.8)	21 (6.4)
Swing Check*	19 (5.8)	22 (6.7)	27 (8.2)	32 (9.8)	45 (13.7)	55 (16.8)	65 (19.8)

Use with Hazen-Williams $C = 120$ only. For other values of C, the figures in this table should be multiplied by the factors below:

Value of C	80	100	120	130	140	150
Multiplying factor	0.472	0.713	1.00	1.16	1.32	1.51

(This is based upon the friction loss through the fitting being independent of the C factor applicable to the piping.)

Specific friction loss values or equivalent pipe lengths for alarm valves, dry pipe valves, deluge valves, strainers, and other devices or fittings should be made available to the Authority Having Jurisdiction.

* Due to the variations in design of swing check valves, the pipe equivalents shown in this table should be considered average.

Note: Use the equivalent ft (m) value for the "standard elbow" on any abrupt 90-degree turn, such as the screw-type pattern. Use the equivalent ft (m) value for the "long-turn elbow" on any sweeping 90-degree turn, such as a flanged, welded, or mechanical joint elbow type.

TABLE 6-6G. *Guide for Estimating Hazen-Williams C*

Kind of Pipe	Value of C		
	1*	2†	3‡
Cast iron, unlined:			
10 years old	110	90	75
15 years old	100	75	65
20 years old	90	65	55
30 years old	80	55	45
50 years old	70	50	40
Cast iron, unlined, new	120		
Cast iron, cement-lined	140		
Cast iron, bitumastic enamel-lined	140		
Average steel, new	140		
Riveted steel, new	110		
Asbestos-cement	140		
Reinforced concrete	140		
Plastic	150		

* Water mildly corrosive. Use same values for fire-protection mains having no mill-use or domestic draft.
† Water moderately corrosive.
‡ Water severely corrosive.
Note: C values chosen for design of piping systems should be based on applicable NFPA standards or the Authority Having Jurisdiction.

From Figure 6-6P, the total friction loss of 600 L/min flow in 50 mm (2 in. nominal) pipe is 500 kPa per 100 m of pipe. Because friction loss is directly proportional to length of pipe, total loss is given by:

$$\frac{211.2}{100} \times 500 = 1056 \text{ kPa}$$

The loss due to fittings is given by

$$\frac{11.2}{100} \times 500 = 56 \text{ kPa}$$

or 5.3 percent of the total friction loss.

Resistance coefficients: Resistance coefficients are sometimes used to express the head loss in a fitting as a function of velocity according to the relationship:

$$h_f = k\frac{v^2}{2g}$$

The equivalent length and the resistance coefficient methods are related. Using the basic Darcy-Weisbach formula:

$$h = f\frac{1}{d}\ \frac{v^2}{2g} = k\frac{v^2}{2g}$$

or:

$$k = f\frac{1}{d}$$

Most entrance losses are calculated using resistance coefficients. Figure 6-6Q provides resistance coefficients for some common entrance conditions and fittings.

Flow coefficients: Where the loss in a fitting is defined by a flow coefficient C_v, the coefficient is defined as the flow of water that will produce a known friction loss (usually 1 psi) through the fitting. The relationship is usually shown as:

$$Q = C_v\sqrt{h}$$

Substituting p/w for h, and solving for p:

$$p = 0.433\frac{Q^2}{C_v^2}\text{ psi}$$

The flow coefficient can be shown to be related to the resistance coefficient k by:

$$C_v = \frac{29.9d^2}{\sqrt{k}}$$

WATER HAMMER

Water hammer is an effect of the pressure rise (surge) that accompanies a sudden change in the velocity of water flowing in a pipe. When deceleration of velocity is rapid or completely stopped, the kinetic energy of the moving water column is absorbed temporarily by elastic deformation of the pipe and by the compressibility of water. A pressure wave is then formed that is reflected back and forth within the pipe.

Pressure surges may be initiated by the closing of a valve, stopping of a pump, or by the sudden development of abnormal water demand when a water main breaks. Occasionally, the operation of automatic control valves in sprinkler systems may result in the reversal of flow and a buildup of high pressure in the fire protection system.

Consideration of water hammer and transient pressure surges is traditionally based on the elastic wave theories of Joukowski and Allievi. Chapter 4-2 of *The SFPE Handbook of Fire Protection Engineering* provides an alternative discussion based on the elastic wave theories of Zhukovsky.

The force of water hammer is sometimes sufficient to rupture pipes, fittings, or hose lines. Theoretically, the resultant forces could be infinite if the system were totally inelastic.

The elasticity of hose tends to reduce the danger from water hammer, but the sudden closing of shutoff nozzles on long hose lines may cause a pressure rise sufficient to rupture the hose. Tests conducted by the New York City Fire Department indicated that pressure surges from closing of nozzles approximated twice the hydrant pressure. These tests point to the advantage of operating nozzle valves slowly.

Discharge lines from pumps are subject to water hammer caused by water column separation. This may occur when a pump suddenly stops (due to power failure, manual shutdown, etc.), or if the discharge valve is closed too quickly with the pump operating. Separation takes place somewhere downstream, especially at a summit, or where the downward slope of the pipe increases sharply. When forward movement becomes exhausted, the flow reverses direction and closes the gap.

When a pump is located at an elevation above the system outlet, a vacuum breaker in the line may provide effective control. When there is static head on a pump at discharge, it is practically impossible to completely eliminate a water column reversal. Surge suppressors or special-type vacuum breakers designed to bypass a portion of the reverse flow water column around the check valve or control valve may be effective.

FRICTION LOSS COEFFICIENTS

Head loss = $h = k \dfrac{v^2}{2g}$

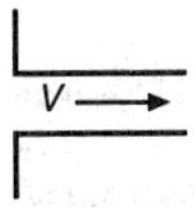

Sharp-edged inlet

$V \longrightarrow$ $k = 0.5$

Inward projecting pipe

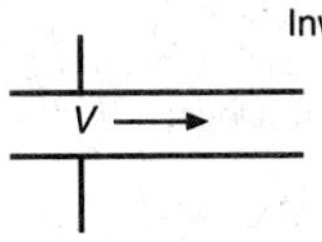

$V \longrightarrow$ $k = 0.8 - 1.0$

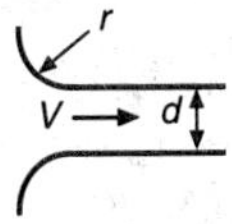

Rounded inlet

$V \longrightarrow d$

r/d	0.05	0.1	0.2	0.3	0.4
k	0.20-0.25	0.09-0.17	0.08	0.05	0.04

SUDDEN CONTRACTION

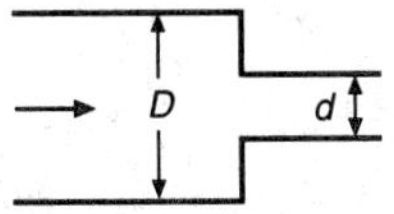

$\longrightarrow D$ d

D/d	1.1	1.5	2.0	3.0	10
k	.15	0.28-0.30	0.36-0.40	0.42-0.50	0.50

GRADUAL CONTRACTION

Note: Use k with V of small pipe section

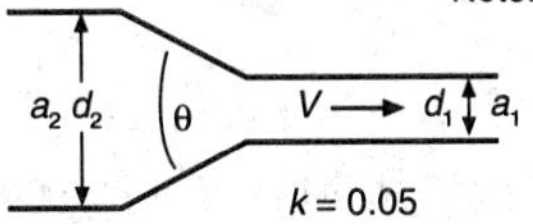

$a_2\ d_2$ (θ $V \longrightarrow d_1\ a_1$ NOTE: Use k with V from large pipe

$k = 0.05$ $\beta = \dfrac{d_1}{d_2}$

IF $\theta \leqslant 45$ degrees

$$k = \frac{0.8 \sin \frac{\theta}{2} (1 - \beta^2)}{\beta^4}$$

$\theta > 45$ degrees

$$k = \frac{0.5 (1 - \beta^2) \sqrt{\sin \frac{\theta}{2}}}{\beta^4}$$

SUDDEN ENLARGEMENT

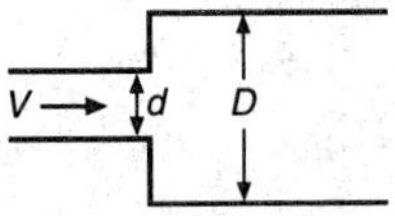

$V \longrightarrow d$ D

d/D	0.1	0.2	0.3	0.4	0.5	0.6	0.7	0.8	0.9
k	0.98	0.92	0.83	0.71	0.56	0.41	0.28	0.13	0.04

GRADUAL ENLARGEMENT

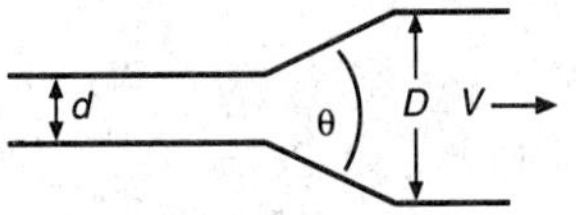

d θ D $V \longrightarrow$ NOTE: Use k with V in large pipe

$\beta = d/D$

If $\theta \leqslant 45$ degrees

$$k = \frac{2.6 \sin \frac{\theta}{2} (1 - \beta^2)}{\beta^4}$$

If $\theta > 45$ degrees $k = \dfrac{(1 - \beta^2)^2}{\beta^4}$

FIG. 6-6Q. Friction loss coefficients.

The restarting of a fire pump too quickly after a tripout may cause excessive surging, and installations subject to intermittent operation should be protected by time-delay relays. Simple relief valves are considered useless, because their operation is too slow to counteract the speed of the pressure rise. In fire protection systems, a jockey pump is used to maintain a high pressure on the system and reduce the water hammer causing surge when the fire pump starts.

The principal factors contributing to water column separation are: (1) rate of flow stoppage, either by rapid closing of a valve, or the fast deceleration of a pump; (2) length of pipe system (this determines the time that pressure continues to fall before positive-pressure waves returning from the far end of the line counteract the initial pressure drop); (3) the normal operating pressure at critical points, such as the crests of hills; and (4) the velocity of the water just before pump stoppage or valve closure occurs; the greater the velocity the larger the size of the void, reverse flow velocity, and the final pressure rise.

Elastic Wave Theory

The basic concepts of elastic wave theory (EWT) are:

1. The magnitude of the pressure rise is proportional to the fluid velocity destroyed and to the velocity of the pressure wave.
2. The pressure rise is independent of the length and profile of the pipe.
3. The velocity of the pressure wave is the same as the velocity of sound through water.

The theoretical pressure rise when flow is stopped instantly may be calculated by the formula:

$$\Delta p = \frac{0.433_{av}}{g}$$

where

Δp = pressure rise in psi,

a = velocity of the pressure wave (tps),

v = water flow velocity (fps), and

g = acceleration of gravity (fps/sec²).

In SI units, the formula is:

$$\Delta p_m = \frac{9.81 a_m v_m}{g_m}$$

where

Δp_m = pressure rise (kPa),

a_m = velocity of the pressure wave (m/sec),

v_m = water flow velocity (m/s), and

g_m = acceleration of gravity (m/s²).

In practice, the calculated Δp may be reduced, allowing for valve closure characteristics, and friction loss in the pipe. Usually, this is a matter of judgment and experience. For the pipe sizes used in fire protection systems, an allowance of 100 to 125 psi (690 to 860 kPa) is suggested.

The pressure rise, Δp, is at maximum when the flow is stopped in a time equal to or less than the critical time of the pipe, which is the time required for the pressure wave to travel from the point of closure to the end of the pipe and return. The formula for critical time is:

$$T = \frac{2L}{a}$$

where

L = length of pipe.

The wave velocity, a, is:

$$a = \frac{12}{\sqrt{\dfrac{w}{g} \times \left(\dfrac{1}{k} + \dfrac{d}{E \times e}\right)}}$$

where

w = weight of water (lb/ft³),

g = acceleration of (ft/sec²),

k = bulk modulus of compressibility of water (psi),

E = Young's modulus of elasticity of pipe wall material (psi),

e = thickness of the pipe wall (in.), and

d = inside diameter of the pipe (in.).

To avoid calculating a, use the chart in Figure 6-6R.

The calculated pressure rise in a 6-in. cast-iron pipe is about 60 psi per foot (1357 kPa/m) of arrested velocity.

The water hammer potential of distribution systems, especially those with automatic pumps, should be examined, and practical steps taken to reduce the probability of destructive pressure surges. It should be noted, however, that because of the conditions under which dedicated fire protection distribution systems are designed (automatic operation of pumps with check valves in their discharge lines), the potential for water hammer cannot be completely eliminated.

To prevent water hammer, valves and hydrants should be maintained in good condition, and operated carefully to prevent an abrupt reduction in flow; and remote-controlled power-operated valves should be carefully timed to prevent them from closing too fast (never less than 5 sec).

BIBLIOGRAPHY

References

AWWA M11, *Steel Pipe—A Guide for Design and Installation*, American Water Works Association, Denver, CO, 1989.
Baumeister, T., Avallone, E. A., Baumeister, T., III, eds., *Marks' Standard Handbook for Mechanical Engineers*, 8th ed., McGraw-Hill, New York, 1978.
Casey, James F., ed., *Fire Service Hydraulics*, 2nd ed., Dun-Donnelley, New York, 1970.
Crocker, S., and King, R. C., eds., *Piping Handbook*, McGraw-Hill, New York, 1967.
DeNevers, N., *Fluid Mechanics*, Addison-Wesley, Reading, 1970.
DiNenno, P. J., ed., *SFPE Handbook of Fire Protection Engineering*, 2nd ed., National Fire Protection Association, Quincy, 1995.
Flow of Fluids Through Valves, Fittings, and Pipe, *Crane Technical Paper No. 410*, Crane Co., Chicago, 1976.
Hydraulic Institute, *Engineering Data Book*, Hydraulic Institute, Cleveland, OH, 1970.
Idelchik, I. E., ed., *Handbook of Hydraulic Resistance*, 2nd ed., Hemisphere, Washington, 1986.
Jeppson, R. W., *Analysis of Flow in Pipe Networks*, Butterworth, Boston, 1976.
Parmakian, J., *Water Hammer Analysis*, Dover, New York, 1963.
Perry, R. H., and Chilton, C. H., *Chemical Engineers Handbook*, 5th Ed., McGraw-Hill, New York, 1973.
Wass, H. S., *Sprinkler Hydraulics*, IRM Insurance, White Plains, NY, 1983.

Additional Reading

Allan, H. A., "Engineering Approach to Sprinkler Design. Session 5-Current Realities-Case Studies," CSIRO Division of Building Construction and Engineering, North Ryde New South Wales, Australia, CSIRO International Fire Safety Engineering Conference, Concept and the Tools. Presented for FORUM for International Cooperation on Fire Research, Supported by SFPE, NFPA, AFPA, AAFA, October 18–20, 1992, Sydney, Australia, 1992, pp. 1–6.
Ashfield, A., "Gridded Sprinkler Systems," *Fire Surveyor*, Vol. 19, No. 3, June 1990, pp. 20–27.
Ashfield, A., "Sprinkler Calculations: Computer Programs," *Fire Surveyor*, Vol. 22, No. 1, Feb. 1993, pp. 14–15.
Ashfield, A., "Computer Calculations for Determining Design of Sprinkler Installations," *Fire*, Vol. 83, No. 1022, Sept. 1990, pp. 26, 28–29.
Bollinger, B., "CAD Design for Sprinklers . . . The Year 2000," *Sprinkler Age*, Vol. 13, No. 3, Mar. 1994, pp. 12, 14.

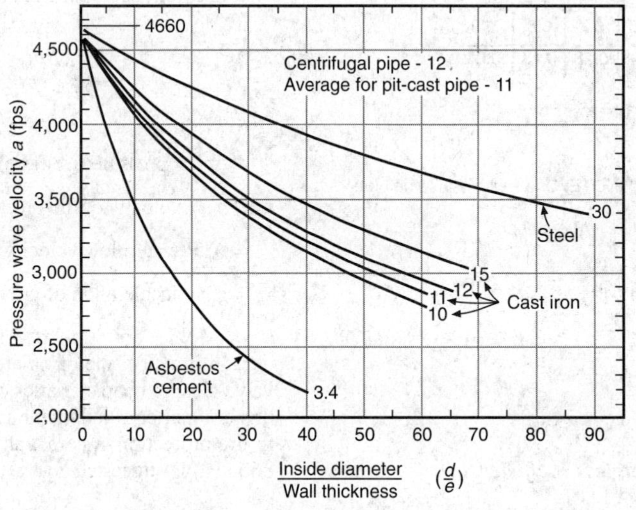

FIG. 6-6R. *Surge wave velocity chart for water. The figures on the curves represent values of E in millions of pounds per square inch (bulk modulus of elasticity).*

Brock, P. D., "Critical Considerations in the Hydraulic Design of Automatic Sprinklers," *FPD Today*, Vol. 2, No. 1, Jan./Feb. 1990, pp. 15, 21.

Brown, P., "Pre-Designed Fire Sprinkler Systems," *Sprinkler Age*, Vol. 14, No. 2, Feb. 1995, p. 16.

Buckley, G., "Short Note on Sprinkler System Hydraulic Calculations Using Foam," *Fire Engineers Journal*, Vol. 54, No. 172, Mar. 1994, p. 28.

Caulfield, H. J., "Hydraulics," *Fire Engineering*, Vol. 138, No. 6, 1987, p. 56.

Chan, T. S., "Measurements of Water Density and Drop Size Distributions of Selected ESFR Sprinklers," *Journal of Fire Protection Engineering*, Vol. 6, No. 2, 1994, pp. 79–97.

Chow, W. K. and Shek, L. C., "Physical Properties of a Sprinkler Water Spray," *Fire and Materials*, Vol. 17, No. 6, Nov./Dec. 1993, pp. 279–292.

Chow, W. K. and Wong, V. M. K., "Water Penetration Ratio of a Sprinkler Water Spray," First Asian Conference on Fire Science and Technology (ACFST), October 9–13, 1992, Hefei, China, International Academic Publishers, China, 1992, pp. 467–474.

Edwards, C. C., "Fireground Hydraulics in Your Head," *Fire Engineering*, Vol. 142, No. 3, 1989, pp. 79–80, 82–83.

Fleming, R. P., "Reducing Design Areas With Quick Response Sprinklers," *Sprinkler Quarterly*, No. 93, Winter 1995, pp. 22, 24–26.

Huggins, R., "ESFR System Design: Basic Parameters for Sprinkler Layout," *Sprinkler Age*, Vol. 15, No. 5, Mar. 1996, pp. 22.

Kandola, B. S., "Introduction to Mechanics of Fluids," *SFPE Handbook of Fire Protection Engineering*, 2nd ed., National Fire Protection Association, Quincy, MA, 1995.

Keith, G. and Garner, D., "Sprinkler System Water Supply Analysis," *Fire Engineering*, Vol. 144, No. 12, Dec. 1991, pp. 67–68, 70–71.

Koffel, W. E., "Designing Warehouse Sprinkler Systems," *NFPA Journal*, Vol. 87, No. 4, July/Aug. 1993, pp. 16, 90.

Koffel, W. E., "Improving Sprinkler Performance by Design," *NFPA Journal*, Vol. 87, No. 6, Nov./Dec. 1993, pp. 12, 20.

Lee, Y. C., "Application of *Hypercard* for the Design of Sprinkler Installation," Hong Kong Polytechnic, Hong Kong, CIB 92; Conseil International du Batiment, Congres Mondial du Batiment, World Building Congress CIB 92, *Proceedings [Compte-rendu]*, Tome 2, Session 5B, Computers as a Design Tool, May 18–22, 1992, Montreal, Canada, 1992, pp. 629–630.

Linder, K. W., "*THE* Sprinkler Program, V. 1.2, Software Review," *Journal of Fire Protection Engineering*, Vol. 5, No. 1, Jan./Mar. 1993, pp. 11–16.

Linder, K. W., "Does Sprinkler Design Match the Use of the Building? Part 1. The Role of the Insurance Carrier," *Sprinkler Age*, Vol. 9, No. 4, Apr. 1990, pp. 20–21.

Linder, K. W., "Does Sprinkler Design Match the Use of the Building? Part 2. The Role of the Insurance Carrier," *Sprinkler Age*, Vol. 9, No. 5, May 1990, pp. 28–29.

Linder, K. W., "Does Sprinkler Design Match the Use of the Building? Part 3. The Role of the Insurance Carrier," *Sprinkler Age*, Vol. 9, No. 6, June 1990, pp. 21–22.

Melley, B. W., "Comparison of Several Computer Hydraulics Programs for the IBM PC and Compatibles," *Fire Technology*, Vol. 25, No. 4, 1989, pp. 336–363.

Titus, J. J., "Hydraulics," *The SFPE Handbook of Fire Protection Engineering*, 2nd ed., National Fire Protection Association, Quincy, MA, 1995.

Milke, J. A.; Bryan, J. L., "Development of a Technique to Assess the Adequacy of the Municipal Water Supply for a Residential Sprinkler System," Maryland Univ., College Park, National Institute of Standards and Technology, Gaithersburg, MD, Fire Administration, Emmitsburg, MD, NIST-GCR-91-600, Nov. 1991.

Mowrer, F. W., "'THE' Sprinkler Program," *Fire Technology*, Vol. 29, No. 1, Feb. (First Quarter) 1993, pp. 60–68.

Ryan, M. W., *et al.*, "Effect of Pressure on Leakage of Automatic Sprinklers," *Sprinkler Age*, Vol. 12, No. 5, May 1993, pp. 16, 18, 20–24.

Schulte, R. C., "Reviewing Sprinkler System Hydraulic Calculations Generated by Computers," *Building Official and Code Administrator*, Vol. 24, No. 3, May/June 1990, pp. 19–23.

Smith, W. W., "Computerizing Fire Sprinkler Hydraulic Calculations," *Consulting/Specifying Engineer*, Vol. 5, No. 4, Apr. 1989, p. 49; *Journal of Applied Fire Science*, Vol. 1, No. 2, 1990–1991, pp. 153–161.

Smith, W. W., "Integrating Hydraulic Sprinkler Calculations and CAD," *Fire Journal*, Vol. 84, No. 2, Mar./Apr. 1990, pp. 56–57.

"Sprinkler Costs Offset by Alternative Design Options," *Sprinkler Age*, Vol. 15, No. 1, Jan. 1996, p. 14.

Stephens, J., "Improving Sprinkler Fire Control With Lower Water Pressures," *Fire Prevention*, No. 257, Mar. 1993, pp. 32–36.

Wood, M. S., "Full Hydraulic Calculations for Sprinkler Systems," *Fire Surveyor*, Vol. 17, No. 4, Aug. 1988, pp. 64–73.

TEST OF WATER SUPPLIES

Revised by Gerald R. Schultz

An existing water supply system is one of the most important factors in public or private protection. Fire departments and fire protection engineers, as well as those responsible for the design, operation, and maintenance of water systems, are concerned with two aspects of the water supply: (1) its adequacy and (2) its reliability.

Adequacy in the case of a water system supplying water for normal consumption and for fire protection can be defined as having the capability of simultaneously supplying water for maximum normal consumption demands and water that may be needed to combat and extinguish a major fire within the area served by the water system.

Reliability of a water system can be defined as having the capability of supplying the maximum daily consumption plus a required fire demand, even in the event of a malfunction or the outage of important system components, such as a pipe line break, valve failure, power failure, or pump outage.

This chapter explains the objectives of water supply testing, test procedures, and graphical solutions to test and flow problems.

The hydraulics principles underlying waterflow and pump testing can be found in Section 6, Chapter 4, "Fire Pumps," and Section 6, Chapter 6, "Hydraulics." Information on water supply requirements and systems can be found in Section 6, Chapter 3, "Water Distribution Systems," and Section 6, Chapter 5, "Water Supply Requirements for Fire Protection."

TEST OBJECTIVES

Hydrant flow tests conducted on public water systems are usually made to determine the rate at which water is available at specific locations within the distribution system to determine points of connection for pipe line extensions, for booster pump applications, to verify the accuracy of distribution models, and for a variety of other purposes.

Of particular interest to fire departments and insurance companies is the rate and quantity at which water is available to concentrated high-value areas, such as shopping centers, industrial parks, institutions, and built-up residential areas.

Private fire systems are usually tested for water flow and integrity annually. This includes testing the operating performance of fire

pumps, elevated tanks, municipal connections, other water sources, and the overall adequacy of the system.

Flow testing of public water systems should not be conducted without the consent and assistance of waterworks personnel, and possibly the local police and fire departments. Fire or water departments will usually aid in the test work; police may be needed to control vehicular and pedestrian traffic.

Hydrant and flow tests are made so infrequently and the results obtained are so important that it is imperative for anyone charged with the responsibility for operation, management, or maintenance of a water supply system to be aware of how tests are conducted and to be able to interpret the results correctly.

Hydrant flow tests should not be attempted until all the operational characteristics of a water system are known. Results may differ substantially depending upon the operation of pumping equipment, water levels in the system's storage facilities, rates of consumption, and points of demand in the system. Even though it is possible to conduct accurate tests within acceptable tolerances, the results obtained will often vary from day to day and even at different periods during the same day because of the many variables involved.

TEST EQUIPMENT

In order to conduct hydrant flow tests, the following equipment should be provided, calibrated in customary American units or appropriate SI units. (For SI units: 1 in. = 25.4 mm; 1 lb = 0.454 kg; 1 psi = 6.895 kPa.)

1. A 6-in. steel ruler with $^1/_{16}$-in. divisions.
2. A Pitot tube, together with a test-pressure gage, suitable for the pressures to be expected, with $^1/_2$-lb graduations (usually a 60-psi gage is satisfactory). (See Figure 6-7A.)
3. A $2^1/_2$-in. hydrant cap with fittings, together with a test-pressure gage suitable for the pressures to be expected, with 1-lb graduations (usually a 200-psi gage is satisfactory). (See Figure 6-7B.)
4. A convenient form to record test data and to include a sketch of the location at which the test was conducted showing hydrants and any other salient features.

A Pitot-tube-and-gage combination is indispensable in conducting flow tests from hydrants and nozzles. The small opening at the end of the tube, not over $^1/_{16}$ in. (1.6 mm) in diameter, is inserted in the center of the stream, with the opening in a direct line with the flow and at a distance in front of the opening of one half the diame-

Gerald R. Schultz, P.E., is a principal of Gage-Babcock & Associates, Oak Brook, IL. He is a member of the NFPA Technical Committee on Automatic Sprinklers.

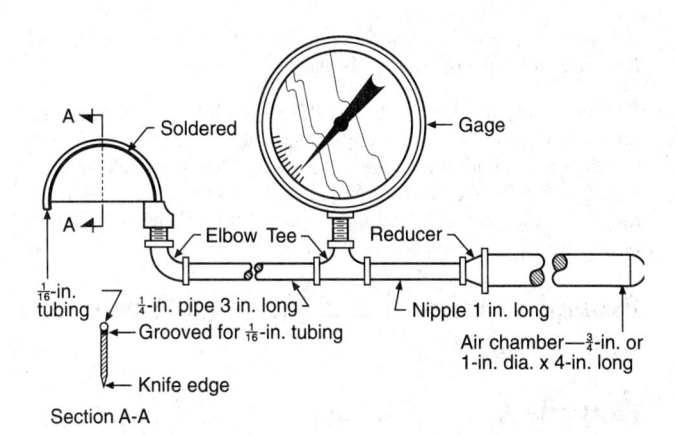

FIG. 6-7A.   A typical Pitot tube assembly. (For SI units: 1 in. = 25.4 mm.)

ter of the opening. Velocity pressure is registered on the gage attached to the tube.

Note in Figure 6-7B the petcock blow-off. Its operation will permit air from the hydrant barrel to be vented when the hydrant is opened and will allow air to re-enter when the hydrant is closed. Failure to open the petcock during hydrant closing may subject the pressure gage to a partial vacuum which may introduce errors in future gage readings. Also note the 3/4-in. (19-mm) garden-hose thread on the 2½-in. (64-mm) hydrant cap. Removal of the cap from the remainder of the fitting allows the gage assembly to be attached to a house silcock connection.

Test-quality gages should be used in accordance with ASME B40.1, *Gauges—Pressure Indicating Dial Type—Elastic Element*, grade AA. The use of good quality test gages produces results that are considered reasonably accurate within the scope of the test procedure. However, care should be taken to protect the gages from rough handling. They should be tested periodically by means of a dead-weight tester throughout the range of operation. Calibration sheets should be kept for each gage and correction factors affixed to the back of each gage before a test series is begun. All test data should be recorded in a neat, systematic fashion, along with any system operating conditions that might affect the test results.

CONDUCTING FLOW TESTS

Hydrant flow tests are fairly simple to conduct and the results usually are easily interpreted. Once the objective of the test has been determined, it is only necessary to discharge water at a known rate from one or more hydrants. The pressure drop the discharge produces is observed simultaneously at a second hydrant or at other points of connection to the system that is supplying the flowing hydrants.

The usual procedure for conducting a flow test on a water system is to flow a sufficient number of hydrants to determine the capacity of the system in the area tested. Pressures observed without test hydrants flowing are called "static pressures"; those obtained with hydrants flowing are "residual pressures." One hydrant is chosen for observing the static and residual pressures, preferably located in the center of the group or where the best average pressure conditions might be expected. (See Figure 6-7C.) Avoid using a flowing hydrant for this purpose.

Once flow and pressure have stabilized, Pitot readings are taken simultaneously on the flowing hydrants. Residual pressure at the nonflow hydrant should be recorded at the same time. After the flowing hydrants have been slowly shut off, the static pressure should be observed and compared with the initial reading.

Pitot readings of less than 10 psi (69 kPa) or over 30 psi (207 kPa) at any open hydrant should be avoided. To keep within these pressure limits, the rate of flow can be controlled by throttling the hydrant or opening a second outlet, or both. However, the flow hydrants should be opened sufficiently so that hydrant drains are closed. Water continuously discharged through the drains tends to erode the soil from the base of the hydrant. Avoid using the larger pumper connection on a hydrant for testing unless flow and pressure are strong enough to produce a full stream. When pumper outlets are used, a proper coefficient of discharge must be determined, based upon the extent to which the orifice is completely filled with water. On occasion, it may be desirable to obtain an average velocity pressure by moving the Pitot tube through the entire vertical dimension of the orifice.

Figure 6-7D is the record of a typical flow test. Hydrant No. 1 was the gaging point, and hydrant No. 2 was the point where flow tests were actually made. Figure 6-7E is a graphical plot of the results of the tests.

Tests of this kind show the available flow over and above water consumption that occurs during the test. To evaluate the adequacy and reliability of a system properly, consideration is given to the sources of supply, the levels of water in distribution storage, and the overall operating condition of the system. The hour of the day, the day of the week, the month of the year, the weather, or an impairment made necessary by construction or system expansion are all factors that may affect the results of flow tests. Fire flow may

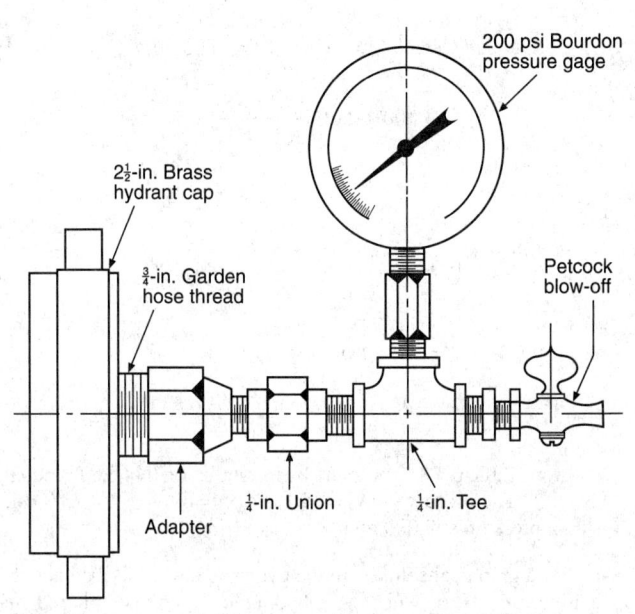

FIG. 6-7B.   A typical hydrant cap and gage. (For SI units: 1 in. = 25.4 mm; 1 psi = 6.895 kPa.)

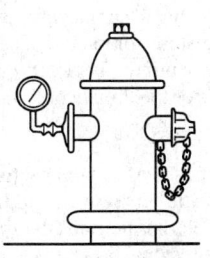

FIG. 6-7C.   Gage on nonflowing hydrant to measure pressure.

Location: Adams St. between Cox St. and Baker St.						
Hydrants: 2½-in. square shape assumed C = 0.80						
Pressure	Pitot pressure				gpm	Total gpm
Hyd 1	Hyd 2	Hyd 3	Hyd 4			
72	–	–	–		0	0
62	18	–	–		633	633
50	10 10				472 472	944

Gage 103 at hydrant No. 1 – gage 79 for pitot readings

FIG. 6-7D. *Log of a public waterflow test. (For SI units: 1 in. = 25.4 mm; 1 gpm = 3.785 L/min.)*

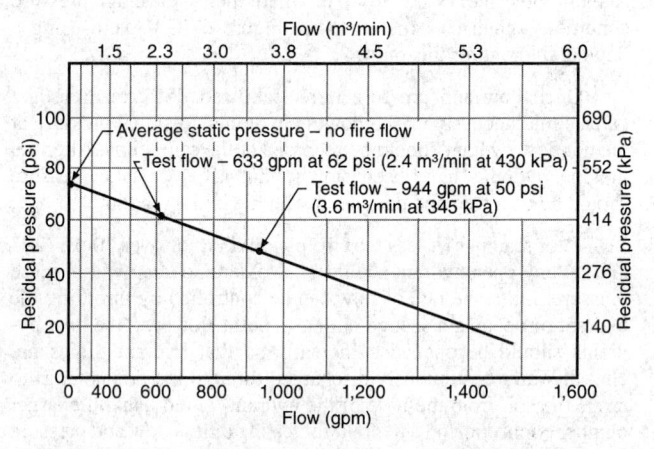

FIG. 6-7E. *Curve plotted from flow test data given in Figure 6-7D.*

be adequate one time and inadequate another because of variations in consumption or in the mode of operation of the system.

Sometimes unsuspected faults other than insufficient supplies are disclosed by hydrant flow tests. Valves may be partly or entirely shut; sometimes, they are found broken or with bent valve stems. Valve boxes may be filled with muck or sand, or even paved over with concrete or asphalt cement. Silt, stones, fish, and other foreign material may be found in pipe systems or hydrants, effectively reducing the available water supply. Many or all of these faults may be due to lack of maintenance and operational control.

It is customary to report the results of hydrant flow tests conducted in public systems in gallons per minute (L/min) available at 20-psi (138-kPa) residual pressure, which is the minimum recommended residual pressure at hydrants for fire engine use. The observed flow and pressures can be converted to any desired residual pressure or flow rate by a simple proportion derived from the Hazen-Williams pipe flow formula. This formula shows that the flow rate in gallons per minute is directly proportional to the 0.54 power of the head loss or the drop in pressure from static to residual observed during the test.

Head loss, which is mostly pipe friction, is the difference between the observed static pressure, S, and the residual pressure, R. The gage value of S seldom indicates the true static (no flow) pressure; actually, it is the residual pressure for the normal pipe flow occurring before and during the test flow. Observed residual pressure

is a function of total flow or $Q \pm DQ$, where Q is the normal rate of flow and DQ is the measured discharge during the test.

For many fire flow tests, the difference between the true and observed values of S is not usually significant. When meter readings or other consumption data are not available for the period of time during which flow tests were conducted, rates of water consumption may be estimated using records of consumption on preceding days or by comparing similar water systems.

Examples of Calculating Flow and Pressure Conversion

Example A

A 2500-gpm test flow from a group of street hydrants dropped the pressure from 69 to 44 psi. Then, (1) what flow would be available at 20 psi? and (2) what would the residual pressure be if flow were increased to 3000 gpm?

SOLUTION 1: Since the flow rate, Q, is directly proportional to the 0.54 power of the head (friction) loss, by proportion

$$Q_2 = \frac{(S-R_2)^{0.54}}{Q_1(S-R_1)^{0.54}} \quad \text{and} \quad Q_2 = Q_1 \frac{(S-R_2)^{0.54}}{(S-R_1)^{0.54}}$$

Substituting the known values:

$$S - R_2 = 69 - 20 = 49 \text{ psi}$$

From Table 6-7A, $49^{0.54} = 8.18$

$$S - R_1 = 69 - 44 = 25 \text{ psi}$$

From Table 6-7A, $25^{0.54} = 5.69$

$$Q_2 = 2,500 \frac{8.18}{5.69} = 2,500 \times 1.438 = 3,595 \text{ gpm (answer)}.$$

SOLUTION 2: Calculate R_2 when $Q_2 = 3,000$ gpm.

$$3,000 = 2,500 \frac{(69-R_2)^{0.54}}{(69-44)^{0.54}}$$

or

$$(69-R_2)^{0.54} = \frac{3,000}{2,500}(69-44)^{0.54}$$

Since $(69-44)^{0.54} = 5.69$, then

$$(69-R_2)^{0.54} = 1.20 \times 5.69 = 6.83$$

From Table 6-7A, it is found by interpolation that 6.83 is equal to $35.1^{0.54}$. Therefore, $69 - R_2 = 35.1$ and $R_2 = 69 - 35.1 = 33.9$ psi, residual pressure at 3,000 gpm.

For a graphical solution to the two problems described above, see Figure 6-7H(a), which is a hydraulic flow curve plotted on semiexponential paper. Note the minor differences due to rounding or graphical interpretation.

TABLE 6-7A.   Numbers to 0.54 Power

h	$h^{0.54}$	h	$h^{0.54}$	h	$h^{0.54}$	h	$h^{0.54}$	h	$h^{0.54}$
1	1.00	36	6.93	71	9.99	106	12.41	141	14.47
2	1.45	37	7.03	72	10.07	107	12.47	142	14.53
3	1.81	38	7.13	73	10.14	108	12.53	143	14.58
4	2.11	39	7.23	74	10.22	109	12.60	144	14.64
5	2.39	40	7.33	75	10.29	119	12.66	145	14.69
6	2.63	41	7.43	76	10.37	111	12.72	146	14.75
7	2.86	42	7.53	77	10.44	112	12.78	147	14.80
8	3.07	43	7.62	78	10.51	113	12.84	148	14.86
9	3.28	44	7.72	79	10.59	114	12.90	149	14.91
10	3.47	45	7.81	80	10.66	115	12.96	150	14.97
11	3.65	46	7.91	81	10.73	116	13.03	151	15.02
12	3.83	47	8.00	82	10.80	117	13.09	152	15.07
13	4.00	48	8.09	83	10.87	118	13.15	153	15.13
14	4.16	49	8.18	84	10.94	119	13.21	154	15.18
15	4.32	50	8.27	85	11.01	120	13.27	155	15.23
16	4.48	51	8.36	86	11.08	121	13.33	156	15.29
17	4.62	52	8.44	87	11.15	122	13.39	157	15.34
18	4.76	53	8.53	88	11.22	123	13.44	158	15.39
19	4.90	54	8.62	89	11.29	124	13.50	159	15.44
20	5.04	55	8.71	90	11.36	125	13.56	160	15.50
21	5.18	56	8.79	91	11.43	126	13.62	161	15.55
22	5.31	57	8.88	92	11.49	127	13.68	162	15.60
23	5.44	58	8.96	93	11.56	128	13.74	163	15.65
24	5.56	59	9.04	94	11.63	129	13.80	164	15.70
25	5.69	60	9.12	95	11.69	130	13.85	165	15.76
26	5.81	61	9.21	96	11.76	131	13.91	166	15.81
27	5.93	62	9.29	97	11.83	132	13.97	167	15.86
28	6.05	63	9.37	98	11.89	133	14.02	168	15.91
29	6.16	64	9.45	99	11.96	134	14.08	169	15.96
30	6.28	65	9.53	100	12.02	135	14.14	170	16.01
31	6.39	66	9.61	101	12.09	136	14.19	171	16.06
32	6.50	67	9.69	102	12.15	137	14.25	172	16.11
33	6.61	68	9.76	103	12.22	138	14.31	173	16.16
34	6.71	69	9.84	104	12.28	139	14.36	174	16.21
35	6.82	70	9.92	105	12.34	140	14.42	175	16.26

Example B (SI units)

A 5000 L/min test flow from a group of street hydrants dropped the pressure from 500 to 350 kPa. What flow would be available at 140 kPa (approximately 20 psi)?

SOLUTION: Again, flow rate, Q, is directly proportional to the 0.54 power of the head (friction) loss, by proportion:

$$Q_2 = \frac{(S-R_2)^{0.54}}{Q_1(S-R_1)^{0.54}} \quad \text{and} \quad Q_2 = Q_1\frac{(S-R_2)^{0.54}}{(S-R_1)}$$

Substituting known values:

$$S - R_2 = 500 - 140 = 360 \text{ kPa and } 360^{0.54} = 24.01$$

$$S - R_1 = 500 - 350 = 150 \text{ kPa and } 150^{0.54} = 14.97$$

$$Q_2 = 5000 \times \frac{24.01}{14.97} = 8019 \text{ L/min.}$$

See Figure 6-7H(b) for a graphical solution on semiexponential paper to the problem in SI units described in Example B. Note the minor differences due to rounding or graphical interpretation.

Flow Test of Public Main at Plant Site

A common procedure for using street hydrants to test the water supply for a connection to an industrial plant follows, using an example to describe the step-by-step testing procedures. The sprinkler system in the example is connected to the dead-ended Adams Street main. (See Figure 6-7F.) Therefore, the entire supply comes from the 10-in. (254-mm) main in Baker Street. There are no yard hydrants. The test was conducted as follows:

Step No. 1:   A gage was attached to No. 1 hydrant with a cap. The hydrant was opened, and the static pressure [72 psi (496 kPa)] was recorded. (See Figure 6-7D.) No. 1 hydrant was chosen for the gaging point. The gage could have been located on the sprinkler riser in the building, but the hydrant location probably was more convenient.

Step No. 2:   The caps were removed from No. 2 hydrant. The diameter of the outlets was measured, and the outlets were found to be square and sharp. (See Figure 6-6F.) After one cap was replaced, the hydrant was opened, and a Pitot reading of 18 psi (124 kPa) was made and recorded. Before the flow was shut off, the residual pressure at No. 1 hydrant was recorded as 62 psi (427 kPa).

Step No. 3:   The second butt at No. 2 hydrant was opened, and 10-psi (69-kPa) Pitot readings were noted on both streams. At No. 1 hydrant, the residual pressure was now 50 psi (345 kPa). It is always desirable to obtain data for at least two rates of flow, one of which should be as large as facilities, conditions, and time will allow.

Step No. 4:   No. 2 hydrant was shut down slowly and carefully, and the caps were replaced. The static pressure at No. 1 hydrant was read and found to be 72 psi (496 kPa), as before the test. The hydrant was then shut down, the gage cap removed, and the regular cap replaced.

Computing the flow: The rate of discharge is best determined by using the table of theoretical discharge (Table 6-6A) and by applying a suitable coefficient from Table 6-6B. In this instance, the hydrant outlets justified the use of $C = 0.80$ (square and sharp hydrant outlets in Figure 6-6F). Table 6-6A shows that, when the Pitot gage reading of velocity pressure from a $2^1/_2$-in. (64-mm) orifice is 18 psi (124 kPa), the flow will be 791 gpm (2990 L/min). The actual flow then would be 791×0.8 or 633 gpm (2990×0.8 or 2392 L/min) at a residual pressure of 62 psi (427 kPa). With both outlets flowing, the Pitot reading on each was 10 psi (69 kPa), corresponding to 590

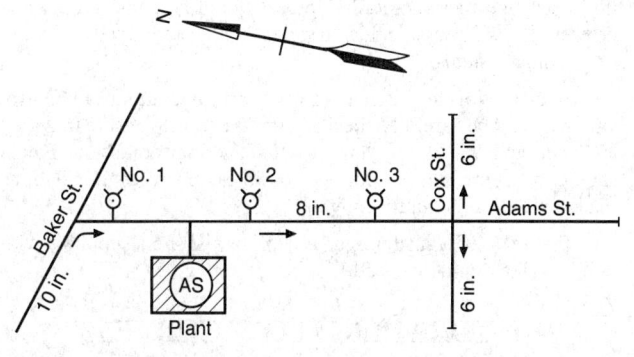

FIG. 6-7F.   Data and sketch of a flow test of a public main at a plant site. (For SI units: 1 in. = 25.4 mm.)

gpm (2233 L/min). The total actual flow was $2 \times 590 \times 0.8 = 944$ gpm ($2 \times 2233 \times 0.8 = 3573$ L/min) at a residual pressure of 50 psi (345 kPa). It is now possible to calculate a flow at any residual pressure. It should be remembered, however, that this flow test procedure establishes the available flows to the point of connection of No. 1 hydrant to Adams Street and *not* to the point of connection of the sprinkler supply line to Adams St.

There may be times when a hydrant's open outlet cannot be flowed because of the damage it will do. In that case, additional or alternate hydrants should be flowed. When water supplies are very weak and the available water does not give a good Pitot reading with an open hydrant outlet flowing, a nozzle with a diameter less than the hydrant outlet may be attached directly to the hydrant outlet. Be sure to measure accurately the diameter of the nozzle used. When this is done the coefficient of the nozzle is used.

Annual Tests

Insurance companies usually require annual tests of private water supplies. When there are multiple sources, each source should be tested separately and in combination to determine if the total supply is sufficient to meet the maximum required fire flow.

Many faults are disclosed by flow tests. Among them are:

1. Valves partly or wholly closed, or inoperative.
2. Stones, silt, and other foreign material in the mains.
3. Tuberculated mains causing high friction loss.
4. Empty or partly filled gravity tanks.
5. Check valves leaking or installed backward.
6. Mains smaller than indicated on plans.
7. Broken meters or clogged strainers.
8. Existence of meters and valves not previously known.
9. Inoperative hydrants.

Hydrants and gaging points are carefully chosen. Tests should be conducted so that the available flow and pressure at high-value or hazardous areas can be determined readily. Make sure that water from the test streams does not cause flooding or property damage. Do not use flowing hydrants for pressure readings: Loss of head in hydrant and connection is not determined and is difficult to estimate.

Care should be taken not to drop the residual pressure in a public water supply system to less than 20 psi (140 kPa), or the minimum pressures that are established by municipal authorities.

Claims of a health hazard being created by a sprinkler system have resulted in the installation of backflow prevention devices on many systems despite the lack of compelling or convincing evidence that such fears are justified.

Backflow prevention devices will introduce additional hydraulic losses, and these additional losses should be considered when determining the available flow to sprinkler and other fixed-water extinguishing systems.

Because flow testing may involve valve operations, all control valves should be carefully checked after the tests to ensure that they are open and that the system is left in normal condition. Faults should be corrected as soon as possible, and recommendations made for desirable improvements.

Figure 6-7G illustrates how a private system supplied with public water would be tested.

HYDRAULIC FLOW CURVES

Many problems involving water tests and flow in pipes can be solved readily by graphs plotted on semiexponential, semilogarith-

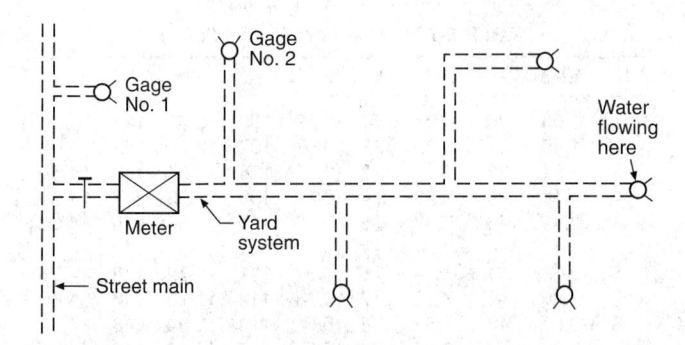

FIG. 6-7G. Method of testing city water supply to private system. Gage No. 1 shows street pressure and should, for gallons discharged, be consistent with data previously secured by street tests, if such tests have been made. Gage No. 2 on branch line will show flowing pressures in main at the point where the branch takes off. The difference in pressure between Gage Nos. 1 and 2 equals the loss in meter, pipe, and fittings. Friction losses to be expected in fittings and pipes can be estimated from Tables 6-6D and 6-6F. Fire-flow meter friction loss can be checked to see if it is unusually high by obtaining the manufacturer's literature.

mic, or linear cross-section paper. Semiexponential paper, commonly called $N^{1.85}$ or "hydraulic" paper, is recommended because a master sheet can be drawn easily and copies easily made by duplicating equipment.

The problems can be solved equally well in U.S. custom units or SI units, although only problem examples in U.S. custom units will be presented in the remainder of this chapter. The Factory Mutual Research Corporation has a data sheet with examples worked graphically in both sets of units. (See bibliography.)

The design of $N^{1.85}$ paper is based on the Hazen-Williams pressure-drop relation: Head loss, which is static pressure less residual pressure, is proportional to the 1.85 power of flow. This relationship holds equally for head loss and flow calculations in SI units. Table 6-7B gives values of unit flows from 1 to 20 gpm in $\frac{1}{2}$-gal/min increments raised to the 1.85 power (see Table 6-7C), and the corresponding scale values when the distance from 0 to the 1-gpm point is 0.05 in. For plotting pressure, the vertical spacing is linear.

The unit values on either vertical or horizontal scales may be multiplied or divided by any constant that will best fit the problem, and the graph paper is suited equally to U.S. customary and SI units.

See Figure 6-7H(a) for an example of a flow curve in U.S. custom units plotted on $N^{1.85}$ semiexponential scale. Figure 6-7H(b) is a graphical solution in SI units.

FLOW IN LOOP SYSTEMS

It is sometimes necessary to estimate the friction loss and flow characteristics of loops or parallel pipe systems. Such problems are readily solved graphically by procedures based on the principle that pressure drop through a simple loop is the same in every leg; this is true regardless of size, condition, and length of pipe. The method described is not applicable to grid or network systems, although it is sometimes possible to treat a grid as a loop system by making certain assumptions. Grids can be solved by network analyzers or computers, or they can be calculated by the Hardy Cross method and other regression formulas. Because these matters are outside the scope of this handbook, the words of John R. Freeman might well be recalled: "A day of (flow) testing is worth a week of calculating."

TABLE 6-7B.   *Data for Making $N^{1.85}$ Hydraulic Graph Paper*

1	2	3	1	2	3
gpm N	$N^{1.85}$	Scale Value, inches from 0	gpm N	$N^{1.85}$	Scale Value, inches from 0
1	1.00	0.05	10.5	77.48	3.87
1.5	2.12	0.11	11	84.44	4.22
2	3.60	0.18	11.5	91.68	4.59
2.5	5.45	0.27	12	99.19	4.96
3	7.63	0.38	12.5	107.0	5.35
3.5	10.15	0.50	13	115.0	6.75
4	13.00	0.65	13.5	123.3	6.16
4.5	16.16	0.81	14	131.9	6.59
5	19.64	0.98	14.5	140.8	7.04
5.5	23.42	1.17	15	149.9	7.49
6	27.52	1.37	15.5	159.3	7.96
6.5	31.90	1.60	16	168.9	8.44
7	36.60	1.83	16.5	178.8	8.95
7.5	41.58	2.08	17	189.0	9.45
8	46.85	2.34	17.5	199.4	9.96
8.5	52.41	2.62	18	210.0	10.50
9	58.26	2.91	18.5	221.0	11.05
9.5	64.39	3.22	19	232.1	11.60
10	70.80	3.54	19.5	243.5	12.18
			20	255.2	12.76

Note:
Column 1. gpm or rate of flow in other units.
Column 2. $N^{1.85}$ from Table 6-7C or by interpolation from that table.
Column 3. Column 2 multiplied by distance in inches from $N = 0$ to $N = 1$ (0.05 in.).
Examples:
Scale value for $N = 7$ is $36.60 \times 0.05 \times 1.83$ in.
Scale value for $N = 16$ is $168.9 \times 0.05 = 8.44$ in.
If scale value of $N = 1$ is 0.08, what is scale for $N = 9$?
Answer: $58.26 \times 0.08 = 4.66$ in.

TABLE 6-7C.   *Numbers to 1.85 Power*

N	$N^{1.85}$	N	$N^{1.85}$	N	$N^{1.85}$
5	19.64	39	877.9	230	23,400
6	27.52	40	920.1	240	25,320
7	36.60	41	963	250	27,300
8	46.85	42	1,007	260	29,360
9	58.26	43	1,052	270	31,480
10	70.80	44	1,097	280	32,910
11	84.44	45	1,144	290	34,310
12	99.19	46	1,192	300	38,250
13	115.0	47	1,240	350	50,880
14	131.9	48	1,289	400	65,150
15	149.9	49	1,339	450	80,990
16	168.9	50	1,390	500	98,440
17	189.0	55	1,658	550	117,400
18	210.0	60	1,948	600	137,900
19	232.1	65	2,259	650	159,900
20	255.2	70	2,591	700	183,400
21	279.3	75	2,944	750	208,400
22	304.4	80	3,317	800	234,800
23	330.5	85	3,710	850	262,700
24	357.6	90	4,124	900	291,900
25	385.7	95	4,558	950	322,700
26	414.7	100	5,012	1,000	354,800
27	444.7	110	5,979	1,200	497,200
28	475.6	120	7,022	1,400	661,100
29	507.5	130	8,144	1,600	846,400
30	540.4	140	9,339	1,800	1,053,000
31	574.1	150	10,610	2,000	1,279,000
32	608.9	160	11,960	2,200	1,526,000
33	644.5	170	13,370	2,400	1,792,000
34	681.2	180	14,870	2,600	2,079,000
35	718.7	190	16,440	2,800	2,384,000
36	757.1	200	18,070	3,000	2,708,000
37	794.6	210	19,770	4,000	4,611,000
38	836.7	220	21,550	5,000	6,968,000

Note:
Figures in this table are for use with the Hazen-Williams formula.
Determining the 1.85 power of numbers by proportional parts between the figures given (linear interpolation) will produce results with error of less than 2 percent.

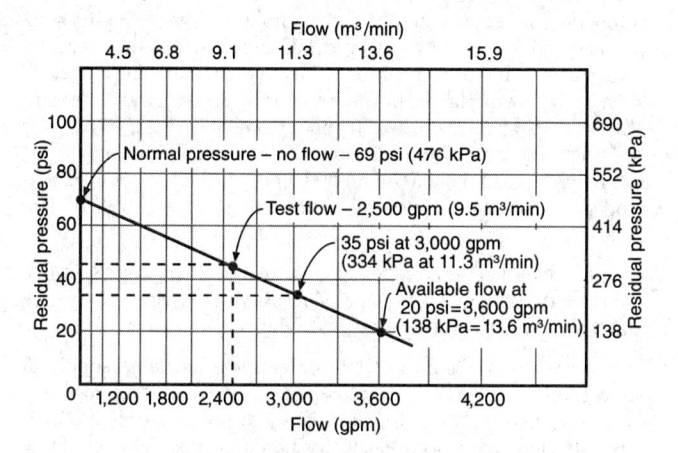

FIG. 6-7H(a).   *An example of a flow curve plotted on $N^{1.85}$ semiexponential paper. The illustration is a graphical plot of data involved in Example A of a flow test of a public water system described earlier in this chapter.*

Figure 6-7I illustrates a graphical solution of a simple loop having two legs, one 800 ft of 8-in. pipe and the other 1200 ft of 6-in. pipe. The Hazen-Williams coefficient is $C = 100$ (unlined cast-iron pipe about 15 to 20 years old and in good condition). Using Table 6-6D, the calculations would be as follows:

Step No. 1:   Friction loss in the 8-in. pipe was calculated for an assumed flow of 1500 gpm as follows: $1 \times 8 \times 2.71 = 21.6$ psi in 800 ft of 8-in. pipe. Friction loss of 21.6 was plotted as Point "a" on the vertical 1500 gpm line in Figure 6-7I. A straight line was drawn to connect Point "a" with 0, forming the Loss Curve "A" for the 8-in. leg.

Step No. 2:   Assuming a flow of 600 gpm, follow the same procedure as followed in Step No. 1: $1 \times 12 \times 2.02 = 24.2$ psi loss in 1200 ft of 6-in. pipe. Friction loss of 24.2 was plotted as Point "b" on the vertical 600-gpm line and connected to 0, thus forming Loss Curve "B" for the 6-in. pipe.

Step No. 3:   The flows corresponding to a convenient pressure of 10 psi on each curve (Points "c" and "d") were added together, and Point "e" plotted at 1400 gpm. This represents 400 gpm at Point "d" + 1000 gpm at Point "c." The straight line connecting Point "e" with 0 is the Curve C for the whole loop.

GRAPHICAL DETERMINATION OF YIELD FROM COMBINED SUPPLIES

Determination of the yield from combined water supplies made by actual testing is not always possible. A supply may be impaired, as when a gravity tank is emptied for repairs or painting, or it may be

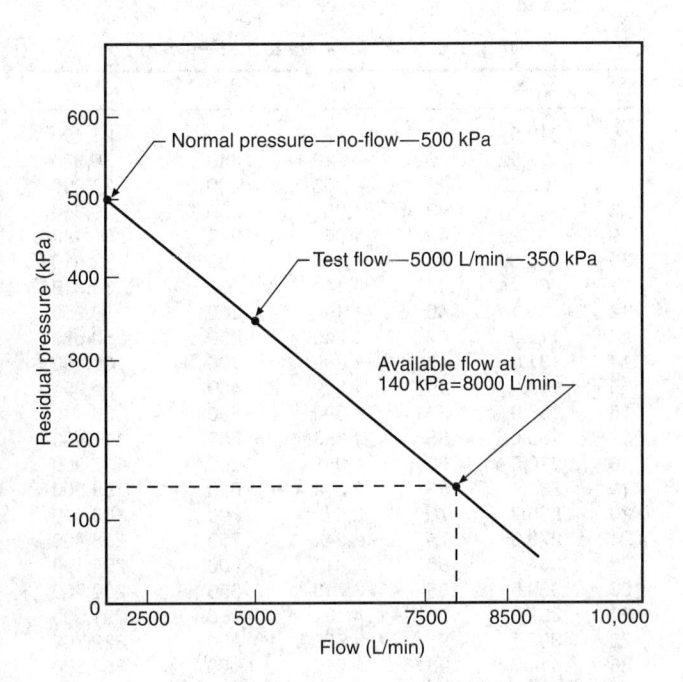

FIG. 6-7H(b). *A graphical solution to the flow problem (Example B described earlier), expressed in SI units on $N^{1.85}$ semiexponential paper.*

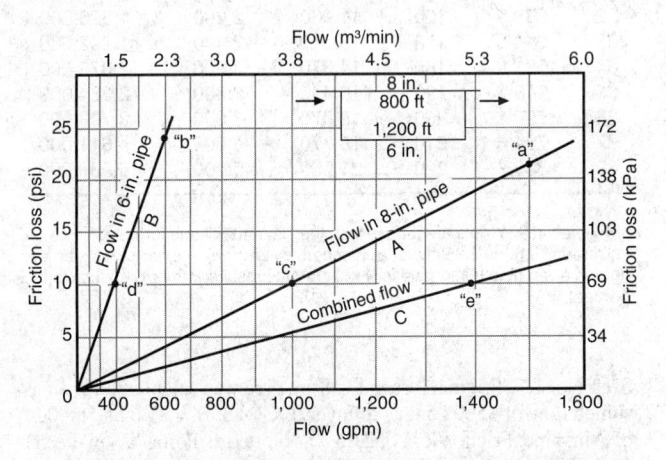

FIG. 6-7I. *Graphical demonstration of flow in loop piping. The assumed friction coefficient is C = 100. Curve A represents flow in 800 ft of 8-in. pipe; Curve B represents flow in 1200 ft of 6-in. pipe. (For SI units: 1 in. = 25.4 mm; 1 ft = 0.305 m.)*

necessary to estimate the yield when a new supply is to be added to a one-source system.

Figure 6-7J shows how to develop the combined yield curve from a private gravity tank and a public water connection. Each source had been tested separately, but a combined test was impossible. The test flows had been taken at the yard hydrant, and the pressures on the nearby sprinkler riser.

The curves were plotted from the test data as follows: 1260 gpm flowing from the tank (Curve A) reduced pressure from 65 to 40 psi; 900 gpm flowing from the public supply (Curve B) reduced pressure from 90 to 43 psi.

All the water would come from the public main until the residual pressure dropped to the static pressure of the tank—in this case,

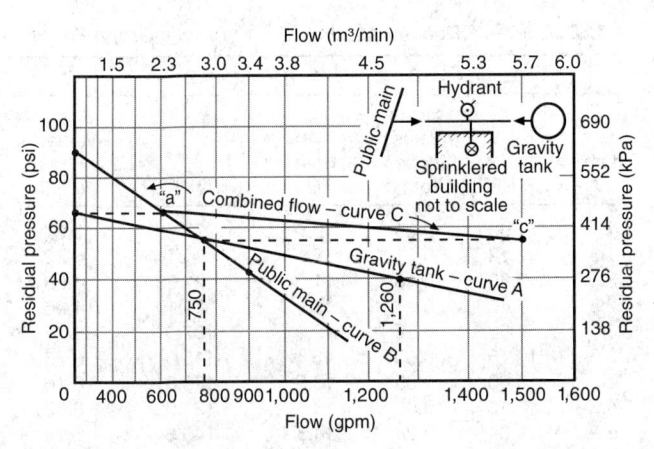

FIG. 6-7J. *Graph of a combined water supply from a public system and a private gravity tank.*

to 65 psi at approximately 600 gpm plotted as Point "a." From then on, water would come from both sources.

The curves cross at 750 gpm at 55 psi, plotted as Point "b." With 750 gpm from each, the total flow at 55 psi would be 1500 gpm, plotted as Point "c." Connecting Points "a" and "c" with a straight line produces Curve C, which is the desired combined curve.

ANALYZING TEST DATA

A major purpose of flow testing is to determine whether the available water supply can meet the water demand required for acceptable protection. Plotting test data on $N^{1.85}$ hydraulic graph paper offers a simple and convenient method of analyzing water supply and demand. (See "Hydraulic Flow Curves," in this chapter.)

As an example of analyzing data, assume that flow tests have been made on a water supply to a sprinklered building and that the following data have been recorded: no flow at 72 psi; 633 gpm at 62 psi; and 944 gpm at 50 psi. Curve A in Figure 6-7K was drawn by fitting a straight line to the three points. By extending the curve, residual pressures at larger flows can be read off directly. For example, the residual pressure at 1200 gpm would have been 40 psi. It is not necessary, nor should it become common practice, to extrapolate much beyond the maximum flow point, because tests from networks often produce supply curves that flatten out as the rate of flow increases.

The fact that all three test points fix a practically straight line indicates that the observed static of 72 psi was close to a true figure and that the normal rate of flow was relatively small.

Curve B in Figure 6-7K represents the available supply at street level to the sprinkler system over and above a 500-gpm allowance for probable hose stream use by the fire department. Obviously, if more water were taken for hose streams, there would be less remaining for the sprinklers.

Curve B was developed by subtracting 500 gpm from Curve A at various pressures. The point at zero flow was obtained by moving horizontally from the intersection of Curve A and the vertical 500-gpm line. (See Figure 6-7K.) The next point was found by plotting 444 gpm (944–500) at 50 psi, which was the residual pressure for the 944-gpm test. Other points were plotted in the same way.

Curves developed by this method are approximations at best, because the rate of flow is dependent on the 0.5 power of discharge pressure (see Section 6, Chapter 6, "Hydraulics") and the 0.54 power

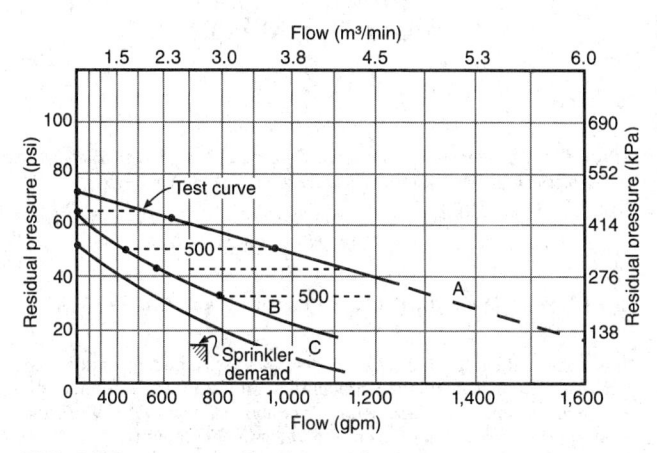

FIG. 6-7K. *An example of a graphic water supply evaluation. Curve A is the test flow; Curve B is the test flow less 500 gpm for hose streams; and Curve C is Curve B corrected for elevation (30-ft elevation in this example).*

of head loss (see Table 6-7A). However, this manner of evaluating the water supply to sprinkler systems is practical and acceptable.

Curve C in Figure 6-7K is actually Curve B corrected for a 30-ft (9.1-m) elevation difference between the residual pressure at street level and the top line of sprinklers. Therefore, every point of Curve C is 13 psi (89.6 kPa) below the corresponding point of Curve B. Friction loss in pipe and fittings between the city main and the top of the sprinkler riser in the building was not considered, but the losses could have been estimated with certain details assumed, as follows (no meter or fire department connection to sprinklers or backflow prevention device):

System Component	Equivalent Length (ft) (see Section 6, Chapter 6)
Eighty ft of 6-in. pipe, $C = 120$ (riser and connection)	80.0
One $8 \times 8 \times 6$-in. tee (connection to main)	21.5
Two 6-in. standard elbows (top and bottom of riser)	20.0
Two 6-in. gate valves (waterworks valve and sprinkler valve)	04.0
One 6-in. check valve	23.0
	Total 148.5 ft

It is assumed in the foregoing example that the sprinkler system in the building requires a flow of 750 gpm at the point of supply to the sprinklers, while residual pressure at the top line of sprinklers does not drop below 15 psi. Curve C of Figure 6-7K shows that 750 gpm would be available at approximately 20 psi residual pressure.

HYDRAULIC GRADIENT

A hydraulic gradient is a profile of residual pressure whose function is to present graphically the flow characteristics of a pipe line. The

hydraulic gradient is an important factor in the design of water supply conduits and trunk mains, and is useful for investigating the condition of a public or private main when tests produce less than the expected flow.

The relationships between pressure and elevation in a pipe having uniform flow are shown in Figure 6-7L. The axioms accompanying the diagram should be noted carefully.

Hydraulic gradient tests of a private fire protection system usually involve shorter runs of pipe than tests of public mains. To reduce the number of tests, pipes should be chosen that are typical of the age and condition of the system. Relatively heavy flows should be induced through the test section to obtain a maximum pressure drop, thereby minimizing the effect of fluctuating pressure or inaccurate gage readings.

In public water supply systems, residual pressures should not be lowered to a value less than that required by municipal authorities.

The data obtained from a hydraulic gradient are readily applied to calculating the C values (internal pipe roughness coefficient) of the pipes tested. The head loss in valves and fittings, if any, should be deducted from the observed pressure drop before calculating C or the value obtained would be too low. (See Section 6, Chapter 6, "Hydraulics.") In municipal water supply systems, however, this deduction is not made, as tested pipe lengths are relatively long and losses due to fittings are considered to be minor. Obviously, consideration of the C value is not part of gradient development; it is a convenient and widely used measure in determining the condition of the interior of the pipe.

If there are more than two gaging points, an attempt should be made to take simultaneous readings, but satisfactory results can usually be obtained (if rates of consumption are relatively low and steady) by moving the gage progressively from hydrant to hydrant while the test flow is maintained in the pipe. True static pressure obtained under no-flow conditions would plot as a horizontal line.

In a municipal system with normal flow, the observed static pressures are actually residual pressures and would trace the normal

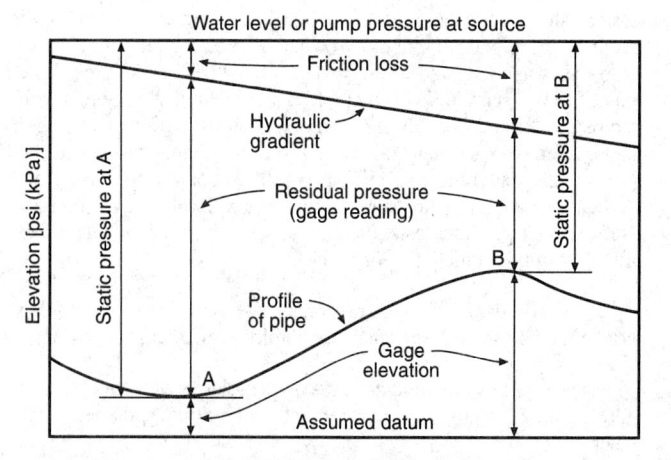

FIG. 6-7L. *Principle of a hydraulic gradient. The following axioms apply: Static-pressure readings measure distance below source. The higher the pipe elevation, the lower the static pressure. Static pressure plus gage elevation [expressed in psi (kPa)] is constant for all points along pipe. Static pressure minus residual pressure equals total friction loss from source to point of measurement. Residual pressure plus gage elevation equals hydraulic-gradient elevation. Friction loss is independent of elevation.*

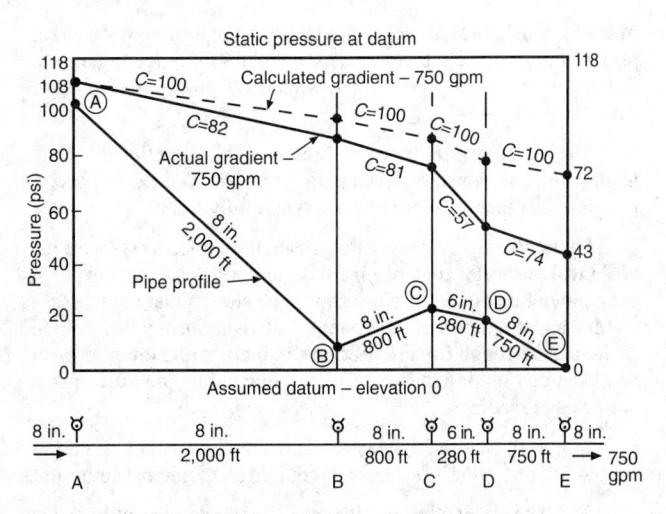

FIG. 6-7M. *Hydraulic gradient and pipe profile. (For SI units: 1 in. = 25.4 mm; 1 ft = 0.305 m; 1 psi = 6.895 kPa; 1 gpm = 3.785 L/min.)*

gradient. Therefore, efforts should be made to find the true static pressure. If the system is supplied by gravity, it may be possible to determine the static by elevation differences, topographic maps, or other survey data. If it is necessary to depend on the static readings, it would be desirable to take readings between 1:00 a.m. and 3:30 a.m., when normal use is at a minimum and observed static pressures are closer to the true levels.

Static pressure on fire protection mains usually can be obtained readily because there is little or no normal flow except in properties with combined fire and industrial systems. With street mains, it is often possible to close a valve below the downstream end of the section being tested, thereby reducing the normal flow temporarily.

It is usually desirable to plot the profile of the pipe being tested, along with the gradient. When improvements and changes are in order, it is desirable to plot a calculated gradient for comparison with the one tested. Both gradients must be based on the same rate of flow. Figure 6-7M is a graph of a gradient test, together with a pipe profile and a calculated gradient. A uniform flow of 750 gpm is assumed. If a gradient falls below the pipe line, pressure in the pipe is less than atmospheric. This condition could impair the flow and cause dangerous pressure surges. When C values less than 80 are found by gradient tests, the pipe usually should be cleaned and lined by standard methods, or replaced.

Regardless of C value, however, small-size pipe causing a steep gradient slope should be replaced with pipe with a larger diameter.

The choice of method depends on the relative costs of cleaning and lining, as opposed to replacement, and on other practical considerations.

Table 6-7D shows the data for Figure 6-7M and explains the calculation. Table 6-7E covers the calculated gradient. Table 6-7F explains the calculation of C values.

BIBLIOGRAPHY

References

ASME B40.1, *Gauges—Pressure Indicating Dial Type—Elastic Element*, American Society of Mechanical Engineers, New York 1991.

"Hydraulics of Fire Protection Systems," *Loss Prevention Data Sheet 3-0*, Factory Mutual Research Corp., Norwood, MA, 1977.

NFPA Codes, Standards, and Recommended Practices

Reference to the following NFPA codes, standards, and recommended practices will provide further information on test of water supplies discussed in this chapter. (See the latest version of *The NFPA Catalog* for availability of current editions of the following documents.)

NFPA 24, *Standard for the Installation of Private Fire Service Mains and Their Appurtenances*

NFPA 25, *Standard for the Inspection, Testing, and Maintenance of Water-Based Extinguishing Systems*

NFPA 291, *Recommended Practice for Fire Flow Testing and Marking of Hydrants*

Additional Reading

ANSI B40.1, *Gauges—Pressure-Indicating Dial-Type—Elastic Element*, American National Standards Institute, New York, 1991.

Brock, P. D., *Fire Protection Hydraulics and Water Supply Analysis*, Fire Protection Publications, Oklahoma State University, Stillwater, OK, 1990.

Carson, W. G., and Klinker, R. J., *Fire Protection Systems, Inspection, Test & Maintenance Manual*, 2nd ed., National Fire Protection Association, Quincy, MA, 1993.

Damon, W. A., "Testing Pressure-Regulating Devices in Water-Based Fire Protection Systems," *Fire Engineering*, Vol. 146, No. 11, Nov. 1993, pp. 51, 55–57.

Gamble, C. C., Jr., "Training the Water Supply Officer," *Fire Chief*, Vol. 37, No. 10, Oct. 1993, pp. 43–45.

Gamble, C., "Pre-Planning Rural Water Supplies," *Fire Chief*, Vol. 39, No. 3, Mar. 1995, pp. 68, 70, 72.

Keith, G. and Garner, D., "Sprinkler System Water Supply Analysis," *Fire Engineering*, Vol. 144, No. 12, Dec. 1991, pp. 67–68, 70–71.

Madrzykowski, D., "Evaluation of Limited Water Supply Fire Suppression Technologies," Progress Report, National Institute of Standards and Technology, Gaithersburg, MD; National Institute of Standards and Technology, *Proceedings* of the Demonstration of Limited Water Supply Fire Suppression Technologies, March 12, 1996, Gaithersburg, MD, 1996, pp. 1–14.

Milke, J. A. and Bryan, J. L., "Development of a Technique to Assess the Adequacy of the Municipal Water Supply for a Residential Sprinkler System," Maryland Univ., College Park, National Institute of Standards and Technology, Gaithersburg, MD, Fire Administration, Emmitsburg, MD, NIST-GCR-91-600, Nov. 1991.

Stroup, D. W. and Evans, D. D., "Suppression of Post-Flashover Compartment Fires Using Manually Applied Water Sprays," National Institute of Standards and Technology, Gaithersburg, MD, Brandforsk, Stockholm, Sweden, General Services Administration, Washington, DC, NISTIR 4625, July 1991.

TABLE 6-7D. *Data for Hydraulic Gradient*

(1)	(2)	(3)	(4)	(5)	(6)	(7)	(8)	(9)
			Gage Pressure					
Gage Location	Actual Pipe Length (ft)	Pipe Diameter (in.)	Static (psi)	Residual (psi)	Total Loss (psi)	Loss Between Stations (psi)	Gage Elevation Above Datum (psi)	Gradient Elevation (psi)
A	—	—	18	8	10	—	100	108
B	A – B = 2000	8	110	78	32	A – B = 22	8	86
C	B – C = 800	8	95	54	41	B – C = 9	23	77
D	C – D = 280	6	100	35	65	C – D = 24	18	53
E	D – E = 750	8	118	43	75	D – E = 10	0	43

Note:
Explanation: Columns 1–5: Data from actual gradient test.
Column 6: Static pressure – residual pressure = total loss.
Column 7: Difference in total loss, station to station.
Column 8: Gage elevation above datum = static pressure at datum (118 psi in this example) minus observed static at each location. Thus, A = 118 – 18 = 100; B = 118 – 100 = 8; C = 118 – 95 = 23; D = 118 – 100 = 18; and E = 118 –118 = 0.
Column 9: Gradient elevation = gage elevation + residual pressure.
Thus, A = 100 + 8 = 108; B = 8 + 78 = 86; C = 23 + 54 = 77; D = 18 + 53; E = 0 + 43 = 43.
For SI units: 1 in. = 25.4 mm; 1 ft = 0.305 m; 1 psi = 6.895 kPa.

TABLE 6-7E. *Data for Calculated Hydraulic Gradient*
(Rate of Flow – 750 gpm – *C* = 100)

(1)	(2)	(3)	(4)	(5)	(6)	(7)	(8)	(9)
Gage Location	Actual Pipe Length (ft)	Pipe Diameter (in.)	Gage Elevation Above Datum (psi)	Static Pressure (psi)	Calculated Loss From Station to Station (psi)	Total Loss (psi)	Residual Pressure (psi)	Gradient Elevation (psi)
A	—	—	100	18	10	10	8	108
B	A – B = 2000	8	8	110	15	25	85	93
C	B – C = 800	8	23	95	6	31	64	87
D	C – D = 280	6	18	100	9	40	60	78
E	D – E = 750	8	0	118	6	46	72	72

Note:
Columns 1–5: Data taken from Columns 1, 2, 3, 8, and 4 of Table 5-8D, respectively.
Column 6: Calculated pressure loss from station to station. (See Section 5, Chapter 2 for friction loss calculations). The losses are:

$$\text{A to B} = 1 \times \frac{2000}{100} \times 0.752 = 15$$

$$\text{B to C} = 1 \times \frac{800}{100} \times 0.752 = 6$$

$$\text{C to D} = 1 \times \frac{280}{100} \times 3.05 = 9$$

$$\text{D to E} = 1 \times \frac{750}{100} \times 0.752 = 6$$

Column 7: Total loss (accumulative loss at each location): Thus:
A = 10 psi
B = Loss A to B + Loss A = 15 + 10 = 25
C = Loss B to C + Loss B = 6 + 25 = 31
D = Loss C to D + Loss C = 9 + 31 = 40
F = Loss D to E + Loss D = 6 + 40 = 46.
Column 8: Residual pressure: equals the static pressure (Column 5) minus the total loss (Column 7) at each location.
Column 9: Gradient elevation is gage elevation (Column 4) plus the residual pressure (Column 8).
For SI units: 1 in. = 25.4 mm; 1 ft = 0.305 m; 1 psi = 6.895 kPa; 1 gpm = 3.785 L/min.

TABLE 6-7F.   *Calculations for C Values*
(Rate of Flow—750 gpm)

(1)	(2)	(3)	(4)	(5)	(6)	(7)
Gage Location	Actual Pipe Length (ft)	Pipe Diameter (in.)	Loss From Station to Station (psi)	Loss per 100 ft (psi)	Factor	C
A	—	—	—	—	—	—
B	A – B = 2000	8	A – B = 22	1.10	1.46	82
C	B – C = 800	8	B – C = 9	1.125	1.49	81
D	C – D = 280	6	C – D = 24	8.6	2.82	57
E	D – E = 750	8	D – E = 10	1.35	1.76	74

Note:
Columns 1–4: Data taken from Columns 1, 2, 3, and 7 in Table 6-7D, respectively.
Column 5: Pressure loss from station to station (Column 4) divided by the actual pipe length (Column 2) multiplied by 100.
Column 6: The actual pressure loss by test for each 100 ft of pipe (Column 5) divided by friction loss for 100 ft of pipe with C=100 (see Table 6-6D).
Column 7: Values for C interpolated from table. (Approximate linear interpolation is acceptable.) Example: For Station C, the factor is 1.49. Find C:

C	Factor	Factor
85	1.35	1.35
?		1.49
80	1.51	
5	14	16

$$\frac{14}{16} \times 5 = 4.4$$

85 – 4.4 = 80.6 = 81 (answer).

THEORY OF AUTOMATIC SPRINKLER PERFORMANCE

Russell P. Fleming

Most fire sprinkler systems are designed so that each individual sprinkler can react to heat from a fire, operating to distribute water over the source of that heat. The complex nature of the interaction of fire and sprinklers historically has prevented an analytical approach to sprinkler system design, but today there is a growing appreciation for measuring the performance characteristics of sprinklers and for using those characteristics in new combinations to achieve specific fire protection goals. The characteristics include thermal response, water distribution, and fire suppression and control capability.

This chapter discusses the response, distribution, and fire extinguishing characteristics of automatic sprinkler systems, and the mechanisms by which it is believed that those characteristics work together for successful sprinkler system performance.

General information on fire sprinklers can be found in Section 6, Chapter 9, "Automatic Sprinklers," and Section 6, Chapter 10, "Automatic Sprinkler Systems." Additional information on sprinkler response and sensitivity ranges appears in Section 6, Chapter 13, "Fast-Response Sprinkler Technology." Related information also is contained in Section 1, Chapter 5, "Chemistry and Physics of Fire"; Section 6, Chapter 1, "Water and Water Additives for Fire Fighting"; Section 6, Chapter 6, "Hydraulics"; and Section 6, Chapter 12, "Water Spray Protection."

THERMAL RESPONSE OF SPRINKLERS

While it is possible to equip sprinkler systems with elaborate detection systems intended to recognize other fire products or signatures, it is heat detection that serves as the basis of sprinkler system response. Therefore, a basic understanding of heat production and movement in a fire is an important first step in an analysis of the way fire sprinkler systems work.

Convective Heat Flow in Fires

Since combustion is an exothermic process, burning fuels produce heat. Even smoldering fires produce heat, although the level of heat can be so low as to be ineffective in actuating thermal detectors. However, smoldering fires generally develop to a flaming stage before producing untenable room conditions. Once the fire reaches the

flaming stage, the heat release rate of the fire will grow in a manner related to the nature of the fuel and its arrangement, the ventilation conditions, and other factors.

Heat is released from a fire in several forms: radiation, conduction, and convection. It has been determined that convective heat transfer is most important in activating sprinklers.[1] Convective heat transfer involves heat transfer through a circulating medium, which, in the case of fire sprinklers, is the room air. The air heated by the fire rises in a plume, entraining other room air as it rises. When the plume hits the ceiling, it generally splits to produce a ceiling gas jet. (See Figure 6-8A.) The thickness of this ceiling jet flow is approximately 5 to 12 percent of the height of the ceiling above the fire source, with the maximum temperature and velocity occurring 1 percent of the distance from the ceiling to the fire source.[2] The heat-sensing elements of sprinklers within this ceiling jet are then heated by conduction of the heat from the air.

Quantifying Sprinkler Sensitivity

Tremendous effort has been made in recent years to quantify the thermal sensitivity of sprinklers and other heat-actuated detectors. This effort is aimed at making it possible to use analytical methods to predict the time of sprinkler response.

FIG. 6-8A. The plume-ceiling interaction.

Russell P. Fleming, P.E., is vice president of engineering for the National Fire Sprinkler Association, Inc., Patterson, NY. He has served on nine different NFPA Technical Committees, including the Committee on Automatic Sprinklers.

Several terms have come into use as part of this process. One is the time constant τ (tau), which is a measure of the thermal sensitivity of a body and is defined as:

$$\tau = mc/h_c a$$

where

m = mass of the body,

c = specific heat of the body,

h_c = convective heat transfer coefficient, and

a = area of the body exposed to gas flow.

The time constant has units of seconds but cannot be determined readily for any specific body due to difficulties in estimating the convective heat transfer coefficient. This term varies with the velocity of the gases that pass by the body. As a result, the time constant is measured using a plunge test apparatus, first developed for use with sprinklers in 1976.[3] This apparatus generally consists of a circulating air oven with a known air temperature and velocity. (See Figure 6-8B.) The air temperature is set well above the nominal operating temperature of the sprinkler. At time $t = 0$, a sprinkler is "plunged" into the heated air. The amount of time it takes for the sprinkler to operate is recorded and is presumed to be the time necessary for the sprinkler operating mechanism to move from room temperature to the nominal operating temperature.

The use of the plunge test permits determination of the time constant, because it provides two data points with regard to the sensitivity of a sprinkler. It is known that the sprinkler was at ambient room temperature, T_a, at $t = 0$, and it is known that the sprinkler reached its nominal operating temperature, T_L at time $t = t_{act}$, the time of sprinkler activation. Based on the known shape of the curve by which a body takes on or loses heat to its environment, the time constant can be determined. Under plunge test conditions involving a constant environmental temperature, the time constant is effectively the amount of time that would be needed for the body to move 62.8 percent of the way to the temperature of its heated environment. (See Figure 6-8C.) In an environment in which the temperature is constantly increasing, the time constant is the amount of time by which the body lags behind its environment after some initial period of time equal to approximately four times the time constant. (See Figure 6-8D.)

Due to the fact that the time constant varies with the velocity at which it is measured, it is not a particularly useful term. Factory Mutual Research Corporation (FMRC) researchers therefore devel-

oped a new term, "response time index" (RTI), as a measure of sprinkler sensitivity independent of velocity.[4] FMRC observed that the convective heat transfer coefficient of blunt bodies in cross-flows was roughly proportional to the square root of the velocity of those flows. By multiplying the time constant by the square root of its corresponding velocity, FMRC could eliminate the effects of the convective heat transfer coefficient:

$$\text{RTI} = \tau \, u^{1/2} = \text{a constant}$$

It should be noted that, depending on the units of velocity used, the actual units of RTI are either $m^{1/2}\text{sec}^{1/2}$ or $ft^{1/2}\text{sec}^{1/2}$. U.S. customary RTI values are 1.81 times the equivalent metric RTI values.

Laboratory determination of the RTIs of various devices has permitted the development of computer programs that predict the operating time of heat detectors and sprinklers based on the heat release history of a fire in conjunction with the ceiling height and radial distance of the detector or sprinkler from the fire.[5] These programs must be used with caution, since they represent only a rough correlation to actual conditions. It should also be noted that they are not appropriate for use with sprinklers following the start of water discharge.

The models of sprinkler sensitivity have recently been modified to account for conduction losses from the sprinkler operating mechanism to the sprinkler frame, fittings, and water in the adjacent piping. These new complexities are not taken into account by the original computer models.

The importance of conduction losses was established in 1986 sprinkler industry test work comparing various sprinklers.[6] These tests showed that the apparent RTI of some sprinklers increased at low velocities, particularly for sprinklers with high conductivity be-

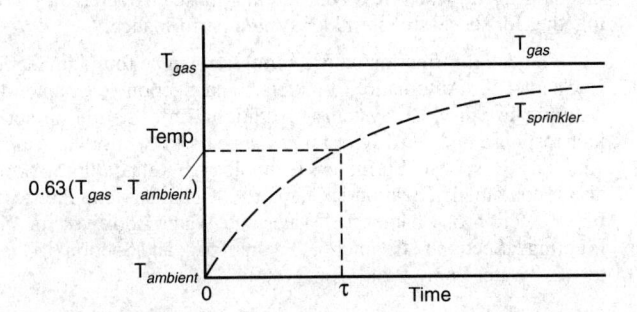

FIG. 6-8C. *Representation of time constant for constant temperature condition.*

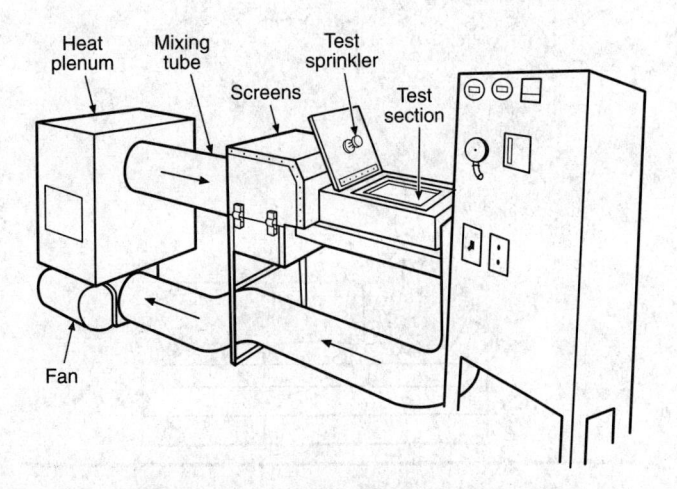

FIG. 6-8B. *Sprinkler plunge test apparatus.*

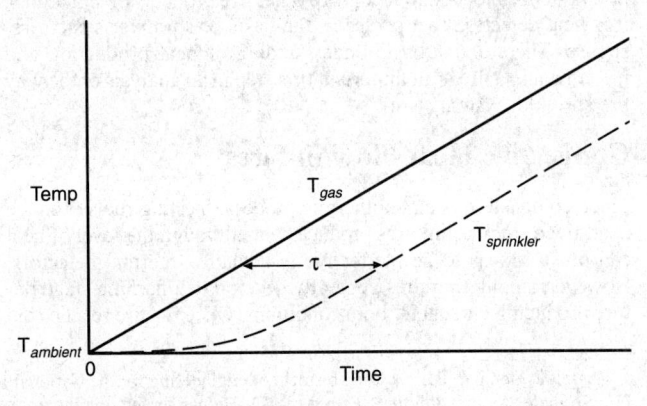

FIG. 6-8D. *Representation of time constant for constantly increasing temperature condition.*

tween the sprinkler operating mechanism and the sprinkler body. (See Figure 6-8E.)

While it might seem beneficial to seek to control the conduction effect to maintain a constant RTI, the tests also demonstrated that the phenomenon could be helpful in preventing excessive sprinkler operations, since velocities tend to decrease with distance from the fire.

Following those tests, the FMRC researchers reviewed the original thermal response model and RTI concept, and the plunge test devised for measuring RTI. They then introduced a modified model that incorporated a conductive heat loss factor.[7]

Additional testing using the plunge test apparatus yields a value for C, the conductivity factor. For purposes of using computer models to estimate the time of sprinkler activation, a value of "virtual RTI" can be calculated as:

$$RTI_v = RTI/\left(1 + C/u^{1/2}\right)$$

The virtual RTI can be used successfully whenever the gas velocity is fairly constant, or does not change rapidly with time. With the exception of small turbulent fluctuations, this is usually the case with actual fire situations.

The revised model assumes that the sprinkler fitting is essentially at ambient temperature and that the conductive heat loss rate is proportional to the temperature rise of the sprinkler operating element. Experimental programs have demonstrated the accuracy of the revised model.

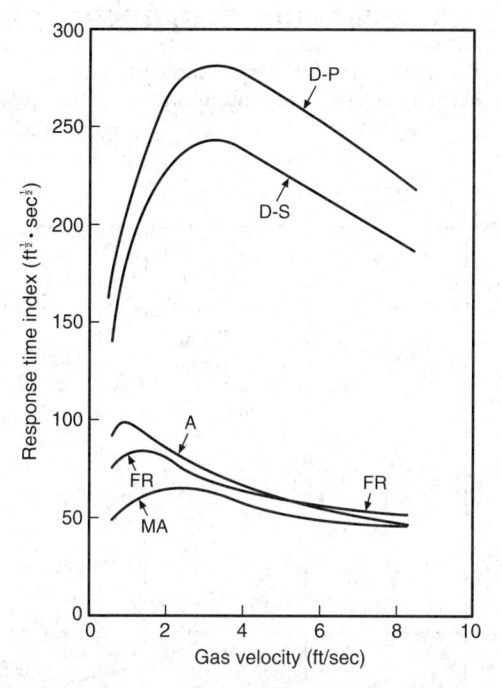

FIG. 6-8E.   Response time indexes as a function of velocity.[6] Sprinklers D-P and D-S represent specific models of standard pendent and sidewall sprinklers, A and MA represent quick-response flush-type pendent and sidewall sprinklers, and FR represents a specific quick-response sprinkler model with conventional frame arms.

WATER DISTRIBUTION AND SPRAY COOLING CHARACTERISTICS OF SPRINKLERS

Relatively little can be accomplished from a design engineering standpoint with regard to the specific spray patterns produced by sprinklers. The distribution pattern of most types of sprinklers is tested only for overall coverage under specific geometric conditions. At present, there is no method of predicting the actual amount of water that will be delivered to a specific unit of floor area under actual fire conditions, especially since spray patterns vary with water discharge pressures.

Figure 6-8F demonstrates the variability of the spray pattern of a typical $1/2$-in. (12.7-mm) pendent spray sprinkler when discharged under nonfire conditions at selected pressures. A 7-psi (48.3-kPa) operating pressure is considered minimum by NFPA 13, *Standard for the Installation of Sprinkler Systems* (hereinafter referred to as NFPA 13). At that minimum pressure, the extent of the spray pattern roughly approximates that of the maximum light-hazard spacing permitted by NFPA 13, when the sprinkler is located 8 ft (2.4 m) above the floor. The spray pattern enlarges as the operating pressure is increased to about 70 psi (483 kPa), then begins to contract at higher pressures, becoming more elliptical in shape at the upper end of the allowable pressure range. Within the wetting area, the water application rates vary considerably, such that the concept of an average density over the coverage area of a sprinkler provides only the roughest approximation of the amount of water that might actually be reaching a particular unit of floor area.

Some studies have been made of individual droplet size and motion. For geometrically similar sprinklers, the median droplet diameter in the sprinkler spray has been found to be inversely proportional to the $1/3$ power of water pressure and directly proportional to the $2/3$ power of sprinkler orifice diameter,[8] such that:

$$d_m \propto D^{2/3}/p^{1/3} \propto D^2/Q^{2/3}$$

where

d_m = median droplet diameter,

D = orifice diameter,

p = pressure, and

Q = rate of water flow.

Total droplet surface area is proportional to the total water discharge rate, divided by the median droplet diameter

$$A_s \propto Q \setminus d_m$$

where

A_s = total droplet surface area.

Combining these relationships, it can be seen that:

$$A_s \propto \left(Q^3 p D^{-2}\right)^{1/3}$$

The heat absorption rate of a sprinkler spray can be expected to depend on the total surface area of the water droplets, A_s, and the temperature of the ceiling gas layer, in excess of the droplet temperature. The total cooling ability of a sprinkler also depends on the depth of the ceiling jet or smoke layer through which the droplets pass.

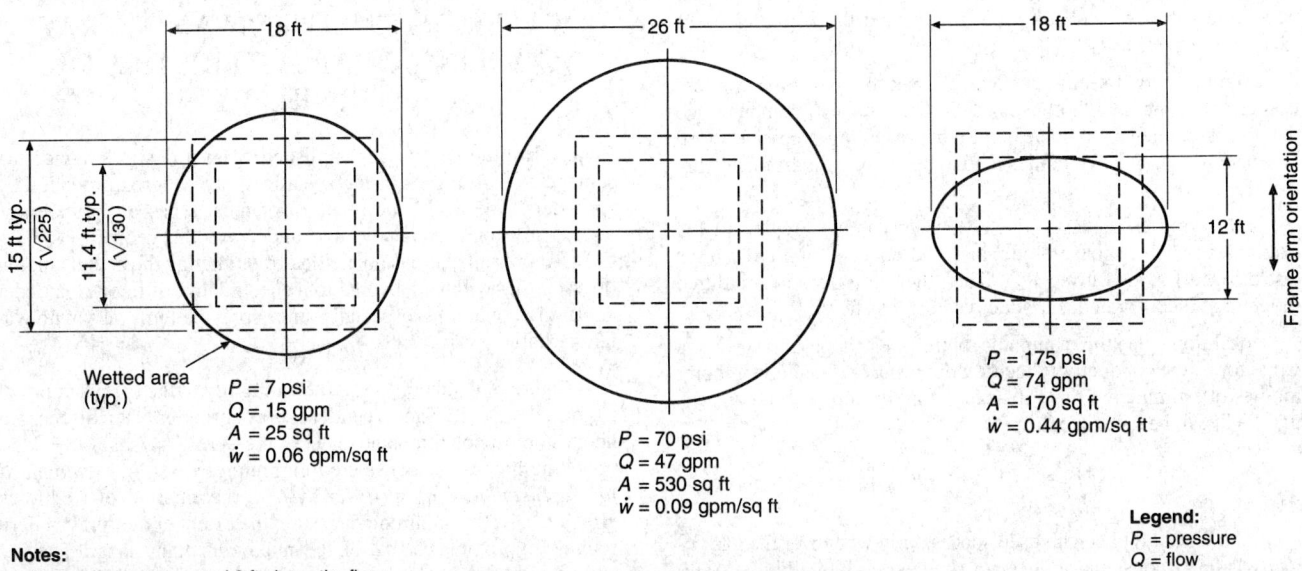

Notes:
1. The sprinkler is mounted 8 ft above the floor.
2. The smaller dashed square indicates the tributary area of the sprinkler at maximum uniform ordinary hazard spacing, per NFPA 13.
3. The larger dashed square indicates the tributary area of the sprinkler at maximum uniform light hazard spacing, per NFPA 13.

FIG. 6-8F. Spray pattern (floor level) vs. discharge pressure. (For SI units: 1 ft = 0.305 m; 1 ft² = 0.0929 m²; 1 psi = 6.875 kPa; 1 gpm = 3.785 L/min.)

INTERACTION OF SPRINKLER DISTRIBUTION AND BURNING FUEL

Sprinklers can be effective against fires in a number of ways. One of the most efficient, in terms of use of water, is simply through the cooling effects of the water spray. The production of a mist of fine water droplets can cause significant cooling, which reduces the radiative feedback to the fire below that which is needed to sustain combustion. In addition, the evaporation of water droplets can produce steam with a volume more than 1,700 times that of the water and can deprive the fire of needed oxygen. This combination of "spray cooling" and "smothering," while efficient, has its limitations. It generally works well only if the fire is contained within an unventilated enclosure.[9]

The greatest efficiency is in the smallest enclosures, since almost all water droplets can be evaporated by contact with enclosing surfaces, and smothering is enhanced.[10]

In 1955, Nash and Rasbush[10] reviewed extensive tests of manual water extinguishment of fully developed room fires and suggested that there is a minimum "critical rate" of water application below which the fire cannot be extinguished. With increased rate of water application above the critical rate, the time needed to control or extinguish the fire falls rapidly. With further increases in the rate of water application, the time to extinguish diminishes more slowly. (See Figure 6-8G.) Under the best of conditions, fires in model rooms could be extinguished with scanned jets or sprays with about 1 (imperial) gal per 1000 ft³ (4.5 L/28.4 m³) of room volume. It was found that a fixed spray with a medium degree of ventilation could extinguish the fire with about 2 (imperial) gal per 1000 ft³ (9 L/28 m³) of room volume.

Automatic sprinklers, of course, are expected to operate before total room involvement. To some extent, this makes their job more difficult, since the sprinkler spray is not readily evaporated to steam.

In a large, open or well-ventilated area, a strong fire can create updrafts which sweep away the small droplets, making them inef-

fective. For this reason, traditional sprinkler system design is based on the idea of distributing a variety of droplet sizes over the burning fuel area to maintain relatively low ceiling temperatures while controlling or extinguishing the fire.

FIRE CONTROL BY SPRINKLERS

The traditional method by which sprinkler systems control fires currently is being termed the "fire control" approach. This method an-

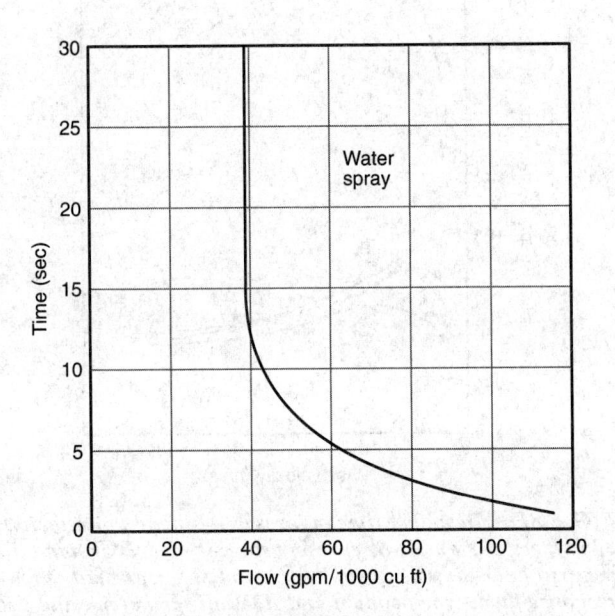

FIG. 6-8G. Time to extinguish fire in model room at various flows.[10] (For SI units: 1 ft³ = 0.0283 m³.)

ticipates that a certain number of sprinklers will be opened surrounding the fire area. While the sprinklers immediately over the fire may not be able to actually extinguish the fire, they will work with other open sprinklers to cool the atmosphere and to prevent sprinklers outside the general vicinity of the fire from operating. In the meantime, the open sprinklers outside the immediate fire area also can be expected to pre-wet adjacent combustibles, helping to prevent the spread of fire.

Figure 6-8H represents the fire control condition that can be achieved by sprinklers. In this steady-state condition, two separate energy balances must take place. At the level of the fuel, the water reaching the fire must be capable of reducing the burning rate to the point at which, in combination with the pre-wetting of adjacent combustibles that has taken place, the fire will not spread to additional fuel. Simultaneously, at the ceiling level, the cooling effect of the water spray from open sprinklers must be sufficient to absorb the heat of the fire plume so as to prevent additional sprinklers from operating and to maintain temperatures below those which would result in structural damage to the building.

In the fire control condition, the area over which sprinklers open generally exceeds the maximum area of the fire. The appendix of NFPA 231C, *Standard for Rack Storage of Materials* (hereinafter referred to as NFPA 231C), provides typical examples of relative sizes of damaged areas of the test array and sprinkler operating areas. (See Table 6-8A.)

TABLE 6-8A.   *Sprinkler Operation in Test Array*

Density gpm/sq ft		Fire Damage in Test Array		Sprinkler Operation (165°F); Area sq ft
		%	sq ft	
0.30	(Ceiling only)	22	395	4500–4800
0.375	(Ceiling only)	17	306	1800
0.45	(Ceiling only)	9	162	700
0.20	(Ceiling only)	28–36	504–648	13,100–14,000
0.20	(Sprinklers at ceiling and in racks)	8	144	4100
0.30	(Sprinklers at ceiling and in racks)	7	126	700

For SI units: 1 ft = 0.305 m; °C = $\frac{5}{9}$ (°F – 32); 1 gpm/sq ft = 40.746 (L/min) × m².

With sprinklers operating in the fire control mode, the central fuel package is eventually depleted, and the fire is extinguished. For fire control by sprinklers, the heat release rate of the fire over time can be characterized, as shown in Figure 6-8I.

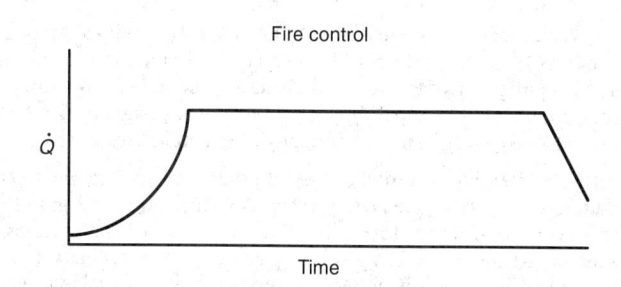

FIG. 6-8H.   *Fire control by sprinklers (conceptual).*

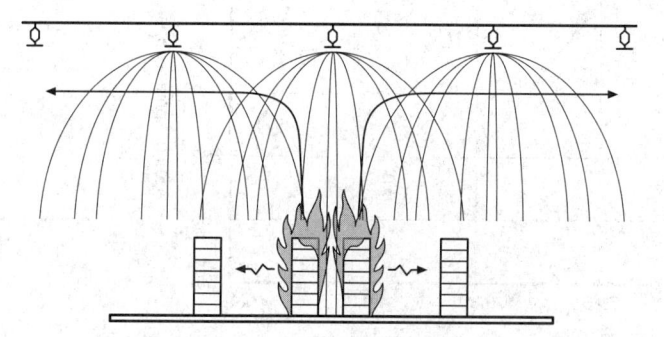

FIG. 6-8I.   *Representation of fire control by sprinklers; heat release rate ($\dot{Q}$) vs. time.*

Various studies have been made of the water application rate needed to extinguish or control fires in fully ventilated areas, both on simple surfaces and in complex fuel arrangements.

The critical rate for wood appears to be between 0.0022 and 0.0044 gpm/ft² [0.09 and 0.18 (L/min/m²)], increasing to a degree with preburn.[11] For some plastic surfaces, the critical rate appears to be in the range of 0.02 to 0.065 gpm/ft² [0.8 to 2.65 (L/min/m²)], higher with external radiation.[12] These values are per unit of burning surface area, not floor area, and are determined under laboratory conditions.

For realistic fuel arrangements, the unit surface area of burning combustibles can be many multiples of the unit floor area protected by sprinklers. British fire tests have indicated that, for midsize wood cribs, water applied through impinging jet nozzles at rates less than 0.275 gpm/ft² [11.2 (L/min)/m²] of unit surface area had little effect on the actual burning zone.[13] The tests also indicated that application rates as low as 0.0625 gpm/ft² [2.56 (L/min)/m²] of unit surface area were found to be sufficient to limit flame spread within the crib.

Area/Density Curves

The area/density curves that form the basis of standard sprinkler system design can be considered to reflect historical experience with the fire control concept. Although most fires are extinguished or contained by only a few sprinklers, the more troublesome deep-seated fires historically have opened more sprinklers and justifiably have influenced the standards writers. System design criteria must address a reasonably "worst-case" situation.

The first edition of NFPA 13, published in 1896, contained simple pipe schedules which limited the maximum number of sprinklers that could be fed through each size of pipe. Combined with a minimum supply pressure of 25 psi (3.6 kPa) at the highest line of sprinklers, the pipe schedule was expected to adequately provide water to at least 15 sprinklers.

The pipe schedule for sprinkler systems was altered many times, and the concept of different pipe schedules for various "hazards" was instituted with the development of "Class B" (later to become light-hazard) criteria in 1930 and the development of an extra-hazard pipe schedule in 1940.

The use of hydraulic calculation methods for fire sprinkler systems actually started in 1929, with the first formal compilation of Hazen-Williams friction loss tables and other hydraulic data in 1931.[14] NFPA 13 first addressed hydraulic calculations in its 1955 edition, but the use of such calculations was limited essentially to deluge system and water spray applications until the inclusion of area/density criteria in the 1972 edition of the standard. NFPA 231C, the first standard developed based on fire test data, was adopted in 1971, and 1972 saw the insertion of area/density curves into NFPA 231, *Standard for General Storage*, although basic criteria had appeared as early as 1965.

The area/density curves in NFPA 13 were modified significantly in 1974 and again in 1991 as part of an overall reorganization and simplification of the standard. As developed in 1974, the curves reflect the hydraulic capability of pipe schedule designs which had proved satisfactory in nearly a century of experience.

The merging of ordinary-hazard groups 2 and 3 in the 1991 edition of NFPA 13 created a system design point of 0.2 gpm/ft² [8.18 (L/min)/m²] over 1500 ft² (135 m²) for a hazard that could essentially be considered the processing and display of ordinary combustibles. This relates well to the historical development and use of the standard sprinkler system. The 1991 area/density curves are shown in Figure 6-8J.

The area/density curves have traditionally provided some flexibility in system pipe sizing. Under provisions of NFPA 13, meeting any point on the appropriate curve is acceptable. This permits the use of higher densities over smaller areas, or lower densities over larger areas. Higher densities with smaller areas will generally result in larger branch-line piping, but lower main sizes and lower overall water supply requirements. For this reason, a high-density small-area point is often selected as the most economical.

Modifications to the area/density curves have traditionally been made to reflect special conditions. The use of high-temperature sprinklers, where permitted for fast-developing fires, is expected to help limit the area over which sprinklers operate. The delay in water delivery resulting from the use of a dry-pipe system is expected to open additional sprinklers prior to the onset of fire control, necessitating a larger sprinkler design area.

Beginning with the 1996 edition of NFPA 13, a decrease in design area was permitted with the use of quick-response sprinklers in wet-pipe systems in light and ordinary hazard occupancies based on ceiling height. While the faster actuation of quick-response sprinklers is expected to more quickly establish the conditions for fire control, the real basis for the area reduction is the recognition that ceiling height plays a major role in the fire control performance of sprinklers. Higher ceilings permit fires to grow larger prior to actuation of sprinklers. As far back as 1967, it was found that the size of the fire at the time of sprinkler operation increased with the square of ceiling height when the fire was located directly beneath a sprinkler, and linearly with ceiling height when the fire was offset 5 ft (1.5 m) horizontally from the sprinkler.[13] The 1996 edition of NFPA

13 also introduced a corresponding design area increase for ceilings with slopes exceeding 2 in 12, based on modeling studies that showed operating patterns of sprinklers could be irregular under such conditions, causing larger numbers of sprinklers to operate.[15]

Compartmentation and Fire Control

It has been observed in a number of test programs that walls tend to assist the sprinkler system in the control mode by limiting the number of sprinklers that can operate. If a fire takes place near a wall, the total number of sprinklers that operate will tend to be reduced by the fact that presumed "outer ring" sprinklers are not available on at least one side.

This will tend to conserve water for the operating sprinklers, resulting in higher application rates and earlier achievement of the fire control energy balances. Carried to an extreme, it is the benefit of walls that is recognized in the "room design method" permitted by NFPA 13, as an alternate to the area/density method.

However, one might ask how the same amount of heat can be absorbed by fewer operating sprinklers. In other words, if those fringe sprinklers are not available to operate and absorb some of the heat from the fire plume, doesn't that same heat flow back across the room in the other direction and simply open more sprinklers on the opposite side?

Two factors argue against this. One is that, unless the fire is almost exactly along the wall, it will entrain the same amount of air as a centered fire, producing the same amount of ceiling jet gases. However, the wall will prevent it from breaking into an axisymmetric ceiling jet, and the resulting ceiling jet or smoke layer will be thicker. The thicker smoke layer actually will permit more cooling to take place from the discharge of each sprinkler, since the path of each droplet through the layer is longer.

The other factor is that the spray cooling ability of each sprinkler is enhanced at higher pressures. Cooling depends on droplet surface area, which is increased with flow and with the production of smaller-diameter droplets, both of which result from higher pressures. With fewer sprinklers operating in the vicinity of the fire, higher water supply pressures are available and more cooling takes place.

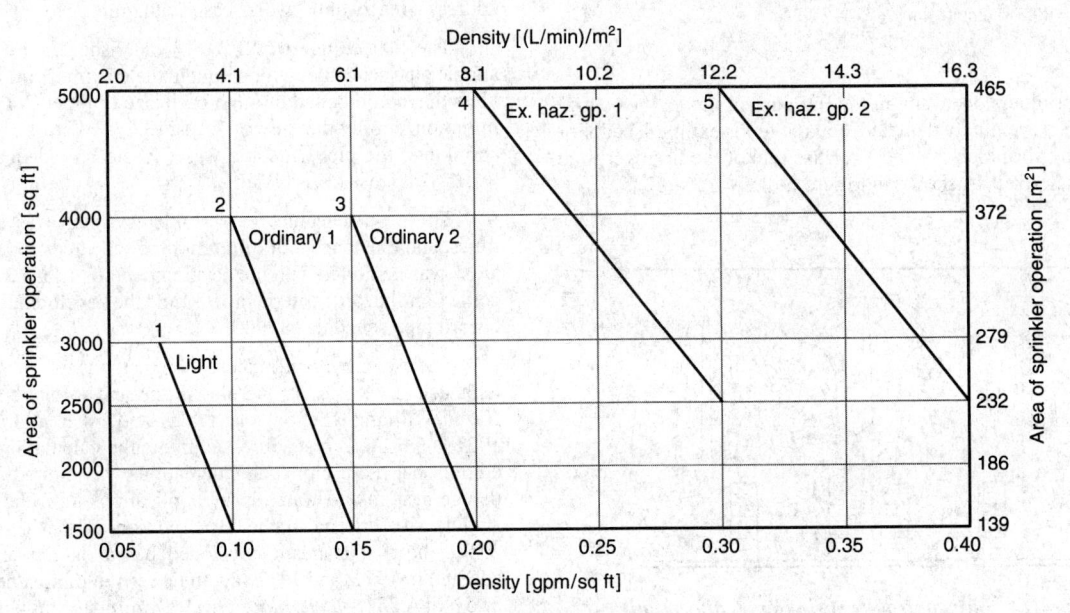

FIG. 6-8J. Area/density curves.

FIRE SUPPRESSION BY SPRINKLERS

Recent years have seen the development of new types of sprinklers that purport not only to control a fire but to actively suppress the fire. To achieve fire suppression, water must be delivered by sprinklers to the burning fuel surface in sufficient quantity to disrupt the combustion process, knocking down the rate of heat release and preventing fire regrowth. (See Figure 6-8K.) If suppression is achieved at an early stage of the fire, only the sprinklers immediately over the fire area are expected to operate.

For this reason, the term "early suppression" has been developed. Today the term is being used both as the name for a concept and for a particular type of sprinkler. The concept is that fast response of sprinklers can produce an advantage in a fire if the response is accompanied by an effective discharge density—that is, a sprinkler spray capable of fighting its way down through the fire plume in sufficient quantities to suppress the burning fuel package.

The first sprinkler to include the term in its name, the early suppression fast-response (ESFR) sprinkler, was developed to apply the concept to a particular class of fire: protection of high-rack storage of FMRC's standardized plastic commodity, using ceiling sprinklers only.

As part of the ESFR development program, FMRC researchers created new terms to investigate and to define the early suppression phenomenon: RDD and ADD.

Required Delivered Density

Required delivered density (RDD) is the minimum rate of water application which, if delivered to the top of the fuel package, is capable of providing early suppression. For the purposes of FMRC's ESFR program, early suppression was, in turn, defined as a sustained knockdown of the fire sufficient to prevent the operation of sprinklers beyond the initial ring.

Specialized test apparatus was developed by FMRC researchers to measure RDD. The apparatus consists of a series of perforated pipes suspended over the burning fuel package that are capable of spreading a fairly uniform water distribution directly over the top of the fuel, such that the water does not have to fight its way down through the fire plume. (See Figure 6-8L.) The pipes and fuel are all positioned below the fire products collector, a large calorimeter which collects all fire gases and determines the total rate of heat release of the fire as a function of time. The water application is initiated at a time simulating the operation of sprinklers, based on an assumed sprinkler location, sensitivity, and temperature rating. The heat-release history generated while the water fights the fire can then be used to predict the operation of sprinklers beyond the first ring, based on their presumed location with respect to the fire.

Figure 6-8M shows the results of several RDD tests for a particular fuel package. Suppression can be considered to occur when

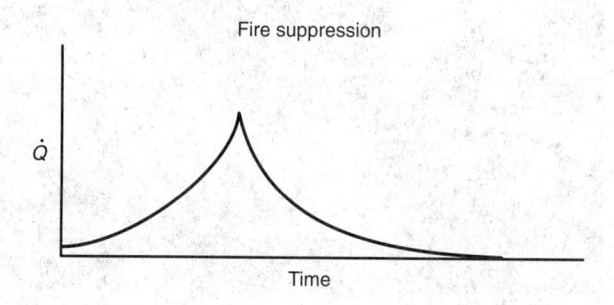

FIG. 6-8L. *RDD test apparatus. (© Factory Mutual Research Corp. Reprinted with permission.)*

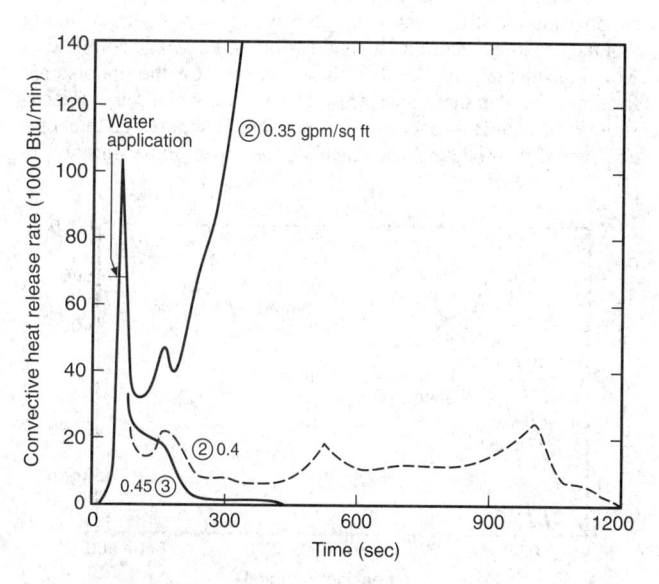

FIG. 6-8M. *RDD test results using three different water application rates. (© Factory Mutual Research Corp. Reprinted with permission.) (For SI units: 1 gpm = 3.785 L/min; 1 ft² = 0.0929 m²; 1 Btu = 1.055 kJ.)*

FIG. 6-8K. *Representation of fire suppression by sprinklers; heat release rate ($\dot{Q}$) vs. time.*

the heat-release rate of the fire is quickly knocked down and is prevented from resurging.

With regard to the standard plastic commodity stored in racks, FMRC confirmed that a relationship existed between RDD and height of storage:

Standard Plastic Commodity—RDD Determination

Storage Height in Racks (Fast response sprinklers—30-ft racks)

3 tiers (15 ft)	0.35 gpm/ft^2
4 tiers (20 ft)	0.45 gpm/ft^2
5 tiers (25 ft)	0.65 gpm/ft^2

(For SI units: 1 ft = 0.305 m; 1 ft^2 = 0.0929 m^2; 1 gpm = 3.785 L/min.)

Under the sponsorship of the National Fire Protection Research Foundation, there has been an attempt to extend the early suppression concept to sprinkler protection of ordinary hazards.

Prior freeburn testing had been performed to determine the heat release histories of selected fuel packages. Among the fuel packages tested were the upholstered furniture corner scenario used during the development of the residential sprinkler and a 6.7-ft (2.0-m) high storage configuration of the standardized plastic commodity in corrugated cardboard cartons that was used in the ESFR sprinkler program. Nineteen RDD tests were conducted by Underwriters Laboratories Inc. under the foundation's sponsorship to compare the relative suppressibilities of the two different fuel packages.[16,17]

In the freeburn testing, both these fuel packages were found to produce a total heat release history that could be approximated by an "ultra-fast t-squared" fire, with a maximum heat release rate of about 5 MW. This refers to a standardized curve in which heat release grows proportionally to the square of time. (See Figure 6-8N.)

The specific water application rates needed to provide early suppression are still under study. However, it was observed clearly that the densities required for suppression of the plastic commodity were considerably higher than those required for the upholstered furniture for the same size fires. The tests conclusively demonstrated that some fuel packages are more easily suppressed than others, even if their heat release rate histories appear to be similar.

Actual Delivered Density

Actual delivered density (ADD) is the actual rate of water application that a particular configuration of flowing sprinklers is capable of delivering to the top of a fuel package, depending on the strength of the upward fire plume.

The ADD measuring apparatus consists of a series of collection pans situated over a gas burner. (See Figure 6-8O.) The 5.4 ft^2 (0.5 m^2) pans simulate the top of a storage array and measure the rate at which each such area receives water from any particular configuration of flowing sprinklers. The gas burner is used to produce upward fire plume forces through the simulated flue spaces between the pans. In this manner, the ability of the flowing sprinkler(s) to deliver water to the fuel package can be checked against a wide range of fire strength conditions, simulating different points in the fire history. Typically, ADD decreases as the fire gets larger.

Since the water spray distribution of a sprinkler is not uniform, the collection pans collect different quantities of water during an ADD test. The development of ADD values from test results is accomplished using a probability distribution developed by, and named for, W. Weibull of Sweden. It was first applied to the distribution of liquid droplet sizes in 1951.[18] Using the Weibull distribution, the pan collections of each sprinkler spray acting through a fire plume of a given strength are fitted to a probability curve with a characteristic or average ADD (Φ) and a uniformity value (b). The higher the uniformity value, the more likely it is that a characteristic ADD (Φ) higher than the RDD will suppress the fire, since the needed water is being distributed in usable fashion.

The uniformity factor affects the basic shape of the probability distribution curve. Figures 6-8P and 6-8Q indicate the relative effects of changing the scale parameter Φ and the uniformity parameter b on the frequency distribution of the statistical model. The variable x represents any given measurement. Typically, the ADD

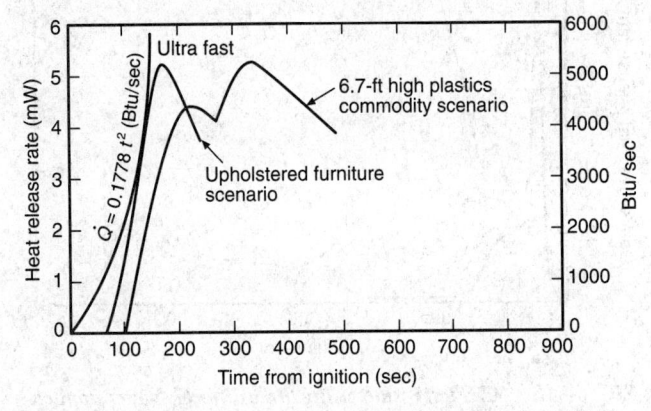

FIG. 6-8N. Comparison of fire growth rates for selected fuel packages and the "ultra-fast" t^2 fire. (For SI units: 1 ft = 0.305 m; 1 Btu = 0.055 kJ.)

FIG. 6-8O. ADD measuring apparatus for rack storage. (Factory Mutual Research Corp. Reprinted with permission.)

uniformity values range from about 1 to 7. A uniformity value of infinity would represent a perfectly uniform distribution.

The probability concept can also be applied to RDD data. A uniformity value b can be determined despite the fact the RDD is measured in a test using the special water applicator intended to provide fairly uniform distribution. A "characteristic" RDD can be calculated as:

$$\Phi_{RDD} = \frac{RDD \text{ (proposed)}}{(1/b)!}$$

The probability distribution concept can also be used to rank the relative suppression capability of sprinklers. If ADD measurements are taken for three prototype sprinklers under the same fire, spacing, and flow conditions, and values of characteristic ADD and uniformity are found for each, the probability that each sprinkler will meet or exceed the characteristic RDD at any selected point on the fuel surface can be calculated from the relationship:

$$P\left(ADD \geq \Phi_{RDD}\right) = \exp\left[-\left(\Phi_{RDD}/\Phi_{ADD}\right)b^{ADD}\right]$$

FUTURE DEVELOPMENTS

The preceding discussion shows that the present understanding of the theory of fire sprinkler performance is incomplete. For some aspects of sprinkler system performance, such as thermal response

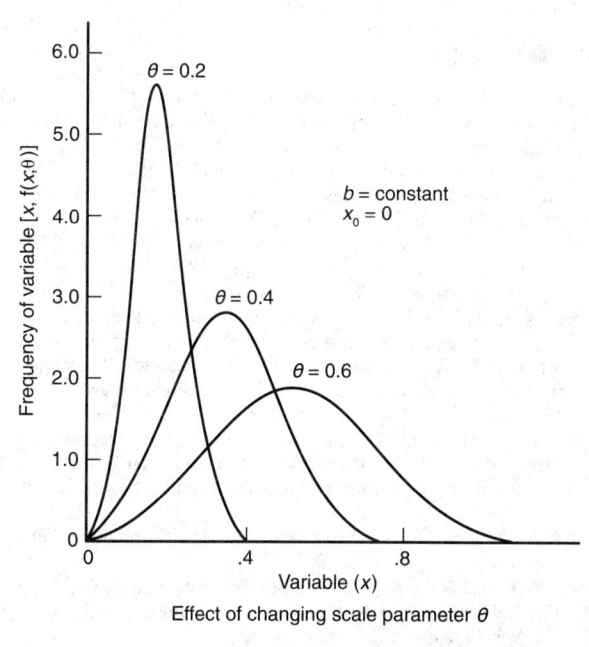
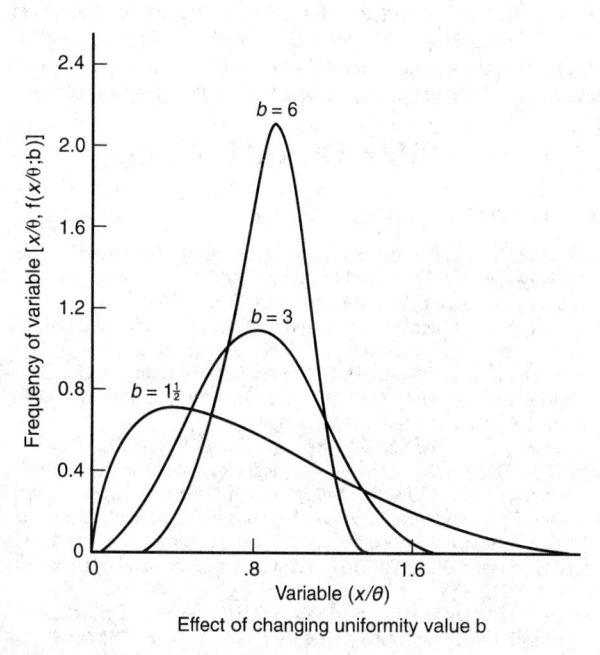

FIG. 6-8P.   Effect of changing scale parameter θ (left) and uniformity value b (right).

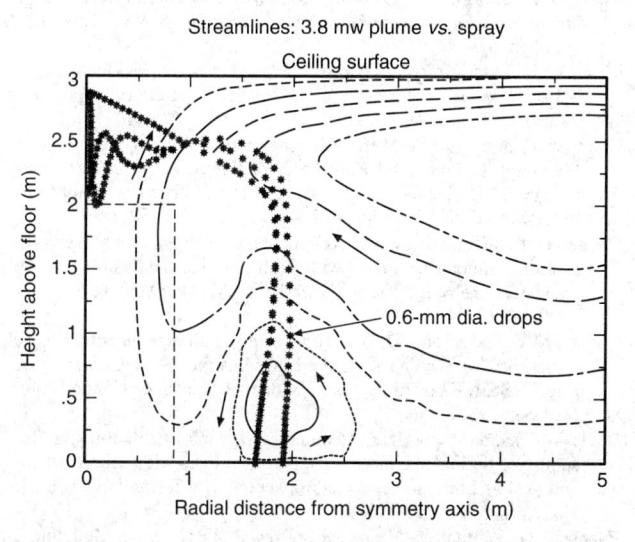
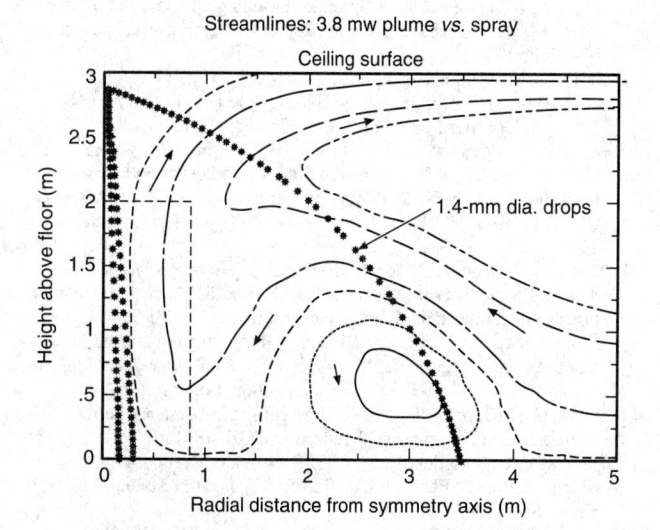

FIG. 6-8Q.   These computer-generated figures demonstrate the superior ability of larger water droplets to penetrate the fire plume when a large fire is located directly below a sprinkler. At left, the 0.6-mm diameter drops are swept away by the upward plume velocities, while, at right, the 1.4-mm diameter drops are able to follow their trajectories.[19] (For U.S. units: 1 in. = 25.4 mm; 1 ft = 0.305 m.)

and the hydraulics of water delivery, research has led to a good understanding of the mechanisms at work. On the other hand, the science of understanding how the water is distributed remains very weak. Almost all the advances made in mathematical modeling of sprinkler performance in the past 20 years deal only with the time up to which the sprinkler starts to put water on the fire.

Research in the 1980s at FMRC, funded partly through the National Institute of Standards and Technology (NIST), has been aimed at using computers to analyze the complex interaction between water droplet sprays and the gas flows induced by a building fire.[19] The results were promising, but the work was based on very simplified assumptions, with the presumed fire located directly under a single sprinkler producing droplets of uniform size in a few specific directions. (See Figure 6-8Q.) Current efforts are underway at NIST to study the interaction of sprinklers, vents, and draft curtains. As part of this effort, a sprinkler/spray interaction model is expected to be developed to work in conjunction with existing fire-induced fluid flow simulation techniques. Much more work will be needed before this research can help build a better sprinkler.

BIBLIOGRAPHY

References Cited

1. Heskestad, G., and Smith, H. F., "Investigation of a New Sprinkler Sensitivity Approval Test: The Plunge Test," FMRC No. 22485, Factory Mutual Research Corp., Norwood, MA, Dec. 1976.
2. Alpert, R. L., "Calculation of Response Time of Ceiling-Mounted Fire Detectors," *Fire Technology*, Vol. 8, No. 3, Aug. 1972.
3. Heskestad, G., and Smith, H. F., "Plunge Test for Determination of Sprinkler Sensitivity," J.I. 3A1E2.RR, Dec. 1980, Factory Mutual Research Corp., Norwood, MA, Dec. 1980.
4. Heskestad, G., "The Sprinkler Response Time Index (RTI)," *Paper (RC81-TP-3)* presented at the Technical Conference on Residential Sprinkler Systems, Factory Mutual Research Corp., Apr. 28–29, 1981.
5. Evans, D. D., and Stroup, D. W., "Methods to Calculate the Response Time of Heat and Smoke Detectors Installed Below Large Unobstructed Ceilings," NBSIR 85-3167, National Bureau of Standards, Gaithersburg, MD, 1985.
6. Pepi, J. S., "Design Characteristics of Quick-Response Sprinklers," Grinnell Fire Protection Systems Co., Providence, RI, May 1986.
7. Heskestad, G., and Bill, R. G., "Conduction Heat-Loss Effects on Thermal Response of Automatic Sprinklers," FMRC J.I. ONOJ5.RU, Factory Mutual Research Corp., Norwood, MA, Sept. 1987.
8. Dundas, P. H., "Optimization of Sprinkler Fire Protection, Progress Report No. 10, The Scaling of Sprinkler Discharge: Prediction of Drop Size," FMRC No. 18792, Factory Mutual Research Corp., Norwood, MA, June 1974.
9. Braidech, M. M., and Neale, J. A., "The Mechanism of Extinguishment of Fire by Finely Divided Water," NBFU Research Report No. 10, 1955, Underwriter Laboratories Inc., Chicago, IL.
10. Nash, P., and Rasbash, D. J., "The Use of Water in Fire Fighting," F.R. Note No. 202/1955, Fire Research Station, Borehamwood, England.
11. Heskestad, G., "The Role of Water in Suppression of Fire: A Review," CIB/W14/80-35, Factory Mutual Research Corp., Norwood, MA, 1980.
12. Magee, R. S., and Reitz, R. D., "Extinguishment of Radiation-Augmented Plastic Fires by Water Sprays," Fifteenth Symposium (International) on Combustion, The Combustion Institute, 1975.
13. O'Dogherty, M. J., Nash, P., and Young, R. A., "A Study of the Performance of Automatic Sprinkler Systems," *Fire Research Technical Paper No. 17*, Joint Fire Research Organization, London, 1967.
14. Wood, C. M., *Hydraulic Data for Fire Protection Systems*, Automatic Sprinkler Corp. of America, Youngstown, OH, 1961.
15. Heskestad, G., "Model Study of ESFR Sprinkler Response under Sloped Ceilings," FMRC J.I. 0N0E3.RU(2), Factory Mutual Research Corp., Norwood, MA, Nov. 1988.
16. Carey, W., *Group 2 RDD Test Program*, National Fire Protection Research Foundation, Quincy, MA, 1989.
17. Budnick, E., and Fleming, R., "How Quick-Response Sprinklers Perform," *Fire Journal*, Vol. 83, No. 5, Sept./Oct. 1989.
18. Goodfellow, D. G., "A Statistical Model for Analysis of Sprinkler Water-Spray Distributions," Factory Mutual Research Corp., Norwood, MA, Aug. 1985.
19. Alpert, R. L., and Delichatsios, M. M., "Calculated Interaction of Water Droplet Sprays with Fire Plumes in Compartments," NBS-GCR 86-520, National Bureau of Standards, Gaithersburg, MD, Dec. 1986.

NFPA Codes, Standards, and Recommended Practices

Reference to the following NFPA codes, standards, and recommended practices will provide further information on the theory of automatic sprinkler extinguishment discussed in this chapter. (See the latest version of *The NFPA Catalog* for availability of current editions of the following documents.)

NFPA 13, *Standard for the Installation of Sprinkler Systems*
NFPA 231, *Standard for General Storage*
NFPA 231C, *Standard for Rack Storage of Materials*

Additional Reading

Allan, H., "Using Heat Release to Optimize the Design of Combined Sprinkler/Vent Systems," Commonwealth Scientific and Industrial Research Organisation, New South Wales, Australia, CIB W14/93/2 (J); *Proceedings* of the First Japan Symposium on Heat Release and Fire Hazard, Session 2, Fire Modeling as a Tool for Reaction to Fire Evaluation, May 10–11, 1993, Tsukuba, Japan, 1993, pp. II/13–18.

Alpert, R. L., "Modeling of Sprinkler Spray Suppression of Fires," First International Conference on Fire Suppression Research, May 5–8, 1992, Stockholm, Sweden, 1992, pp. 245–260.

Atreya, A., Crompton, T., and Suh, J., "Experimental and Theoretical Study of Mechanisms of Fire Suppression by Water," Michigan Univ., Ann Arbor, NISTIR 5499, Sept. 1994; National Institute of Standards and Technology, Annual Conference on Fire Research: Book of Abstracts, October 17–20, 1994, Gaithersburg, MD, 1994, pp. 67–68.

Bill, R. G. and Heskestad, G., "Comments on 'Thermal Response of Sprinkler Heads in Hot Air Stream With Speed Less Than 1 M/S'," Letters to the Editors, *Fire Science and Technology*, Vol. 13, No. 1/2, 1993, p. 89.

Bill, R. G., "Thermal Sensitivity Limits of Residential Sprinklers," *Fire Safety Journal*, Vol. 21, No. 2, 1993, pp. 131–152.

Chow, W. K. and Fong, N. K., "Numerical Simulation on Cooling of the Fire-Induced Air Flow by Sprinkler Water Sprays," *Fire Safety Journal*, Vol. 17, No. 4, 1991, pp. 263–290.

Chow, W. K. and Ho, P. L., "Thermal Responses of Sprinkler Heads in Hot Air Stream with Speed Less Than 1 ms," *Fire Science and Technology*, Vol. 12, No. 1, 1992, pp. 7–12.

Chow, W. K. and Ho, P. L., "Thermal Responses of Sprinkler Heads," *Building Services Engineering Research and Technology*, Vol. 11, No. 2, 1990, pp. 37–47.

Chow, W. K. and Shek, L. C., "Physical Properties of a Sprinkler Water Spray," *Fire and Materials*, Vol. 17, No. 6, Nov./Dec. 1993, pp. 279–292.

Chow, W. K. and Wong, V. M. K., "Water Penetration Ratio of a Sprinkler Water Spray," First Asian Conference on Fire Science and Technology (ACFST), October 9–13, 1992, Hefei, China, International Academic Publishers, China, 1992, pp. 467–474.

Cooper, L. Y., "Estimating the Environment and the Response of Sprinkler Links in Compartment Fires with Draft Curtains and Fusible Link-Actuated Ceiling Vents: Theory," *Fire Safety Journal*, Vol. 16, No. 2, 1990, pp. 137–163.

Cooper, L. Y., and Stroup, D. W., "Test Results and Procedures for the Response of Near Ceiling Sprinkler Links in a Full-Scale Compartment Fire," NBSIR 87-3633, National Bureau of Standards, Gaithersburg, MD, Sept. 1987.

Davis, W. D. and Cooper, L. Y., "Computer Model for Estimating the Response of Sprinkler Links to Compartment Fires with Draft Curtains and Fusible Link-Actuated Ceiling Vents," *Fire Technology*, Vol. 27, No. 2, May 1991, pp. 113–127.

Evans, D. D., Walton, W. D., and McCaffrey, B., "Progress in Modeling Sprinkler Performance," *9th Joint Panel Meeting of the UJNR Panel on Fire Research and Safety*, NBSIR 88-3753, National Institute of Standards and Technology, Gaithersburg, MD, Apr. 1988, pp. 325–335.

Heskestad, G., and Bill, R. G., "Quantification of Thermal Responsiveness of Automatic Sprinklers Including Conduction Effects," *Fire Safety Journal*, Vol. 14, Nos. 1 & 2, July 1988, pp. 113–125.

Theobald, C. R., "FRS Ramp Test for the Thermal Sensitivity of Sprinklers," *Journal of Fire Protection Engineering*, Vol. 1, No. 1, 1989, pp. 23–34.

Theobald, C. R., and Westley, S. A., "Factors Affecting the Sensitivity of Sprinklers," *Fire Surveyors*, Vol. 17, No. 3, June 1988, pp. 5–11.

Thorne, P. F., Theobald, C. R., and Melinek, S. J., "Thermal Performance of Sprinkler Heads," *Fire Safety Journal*, Vol. 14, Nos. 1 & 2, July 1988, pp. 89–99.

Gandhi, P. D., "Temperature and Velocity Correlations in Room Fires for Estimating Sprinkler Actuation Times," *Fire Technology*, Vol. 31, No. 2, May 1995, pp. 137–157.

Gupta, A. K., "Estimating Sprinkler Actuation Time in Compartments," *Proceedings* of the First International Conference on Fire Science and Engineering, ASIAFLAM '95, March 15–16, 1995, Kowloon, Hong Kong, 1995, pp. 219–230.

Heskestad, G., "Sprinkler/Hot Layer Interaction," Factory Mutual Research Corp., Norwood, MA, National Institute of Standards and Technology, Gaithersburg, MD, American Architectural Manufacturers, Des Plaines, IL, NIST-GCR-91-590, May 1991.

Holmstedt, G., "Extinction Mechanisms of Water Mist," Lund Univ., Sweden, SP Report 1994:03; Swedish National Testing and Research Institute, *Proceedings* of the International Conference on Water Mist Fire Suppression Systems, November 4–5, 1993, Boras, Sweden, 1993, pp. 89–100.

Ingason, H., "Thermal Response Models for Glass Bulb Sprinklers: An Experimental and Theoretical Analysis," BRANDFORSK Project 618–911, Swedish National Testing and Research Institute, Boras, Sweden, SP REPORT 1992:12, CIB W14/92/18 (S), 1992.

Jackman, L. A., Nolan, P. F., and Morgan, H. P., "Characterization of Water Drops From Sprinkler Sprays," First International Conference on Fire Suppression Research, May 5–8, 1992, Stockholm, Sweden, 1992, pp. 159–184.

Ndubizu, C., Motevalli, V. and Tatem, P., "Study of the Suppression Mechanisms of Small Flames by Water Mist," Naval Research Laboratory, Washington, DC, *Proceedings* of the 1995 Fall Technical Meeting of the Combustion Institute/Eastern States Section on Chemical and Physical Processes in Combustion, October 16–18, 1995, Worcester, MA, 1995, pp. 451–454.

Shanley, J. H. and Budnick, E. K., "Study of the Utility of Heat Collectors in Reducing the Response Time of Automatic Fire Sprinklers Located in Production Modules of Building 707," Hughes Associates, Inc., Wheaton, MD, EG&G, Golden, CO, RFP-4874, Jan. 1990.

Takahashi, S., "Heat Transfer Mechanisms in the Extinguishment of Charcoal Fire with Water," Japanese Association of Fire Science and Engineering, Fire Research Annual Conference, May 17–18, 1990, pp. 127–130.

Wang, M., "Thoretische Grundlagen der Loschwirkung und der Effizienz von Sprinklern. [Theoretical Foundation of the Extinguishment and Efficiency of Sprinklers.]," University of Kaiserslautern, Shanghai, China, Thesis D386, 1994.

AUTOMATIC SPRINKLERS

Revised by Kenneth E. Isman

Automatic sprinklers are thermosensitive devices designed to react at predetermined temperatures by automatically releasing a stream of water and distributing it in specified patterns and quantities over designated areas. The automatic distribution of water is intended to extinguish a fire or to prevent its spread in the event that the initial fire is out of range of the sprinklers or is of a type that cannot be extinguished by water discharged from sprinklers. The water is fed to the sprinklers through a system of piping, ordinarily overhead, with the sprinklers placed at intervals along the pipes.

This chapter covers the operating principles, including operating temperatures, of the various types of automatic sprinklers currently available and describes their construction in some detail. Sprinklers for special service conditions are also covered.

The National Fire Protection Association (NFPA) does not have standards that cover the manufacture of sprinklers. However, it does insist through requirements in its many fire protection standards containing sprinkler protection requirements that only sprinklers listed by reputable product evaluation organizations be used in sprinkler systems meeting the requirements of NFPA 13, *Standard for the Installation of Sprinkler Systems* (hereinafter referred to as NFPA 13). Listing of sprinklers in a national organization's listing service is an indication that the sprinklers have been scrutinized for their reliability and compliance with the organization's test criteria for sprinklers.

Since they were introduced in the latter part of the 19th century, the performance and reliability of automatic sprinklers have been improved continually through experience and the efforts of manufacturers and testing organizations.

In 1952 and 1953, a new sprinkler was developed with a more efficient umbrella-shaped spray pattern. This "spray sprinkler" became the standard sprinkler in 1958 for use in conformance with NFPA 13; and sprinklers of the older design became known as "old-style" or "conventional" sprinklers. Redesign of the deflector was the principal feature of the new standard sprinklers.

Due to the development of many new types of sprinklers during the 1980s, it became questionable whether any single type of sprinkler could be considered standard in the future. As such, the term "spray sprinkler" was brought back into use with the 1991 edition of NFPA 13. This also conforms better to internationally accepted terminology.

OPERATING PRINCIPLES OF AUTOMATIC SPRINKLERS

In order to appreciate the ruggedness, mechanical simplicity, reliability in operation, and freedom from premature operation of an automatic sprinkler, a familiarity with the basic principles of its design, construction, and operation is necessary.

Operating Elements

Under normal conditions, the discharge of water from an automatic sprinkler is restrained by a cap or valve held tightly against the orifice by a system of levers and links or other releasing devices pressing down on the cap and anchored firmly by struts on the sprinkler.

Fusible sprinklers: A common fusible-style automatic sprinkler operates when a metal alloy of predetermined melting point fuses. Various combinations of levers, struts, and links or other soldered members are used to reduce the force acting upon the solder so that the sprinkler will be held closed with the smallest practical amount of metal and solder. This minimizes the time of operation by reducing the mass of fusible metal to be heated.

The solders used with automatic sprinklers are alloys of optimum fusibility composed principally of tin, lead, cadmium, and bismuth; all have sharply defined melting points. Alloys of two or more metals may have a melting point that is lower than that of the individual metal with the lower melting point. The mixture of two or more metals that gives the lowest melting point possible is called an "eutectic" alloy.

Bulb sprinklers: A second style of operating element utilizes a frangible bulb. (See Figure 6-9A.) The small bulb, usually of glass, contains a liquid that does not completely fill the bulb, leaving a small air bubble trapped in it. As heat expands the liquid, the bubble is compressed and finally absorbed by the liquid. Once the bubble disappears, the pressure rises substantially, and the bulb shatters, releasing the valve cap. The exact operating temperature is regulated by adjusting the amount of liquid and the size of the bubble when the bulb is sealed.

Kenneth E. Isman, P.E., is director of engineering standards for the National Fire Sprinkler Association, Patterson, NY. He currently serves on eleven different NFPA technical committees, including the Committee on Automatic Sprinklers and the Committee on Inspection, Testing, and Maintenance of Water-Based Extinguishing Systems.

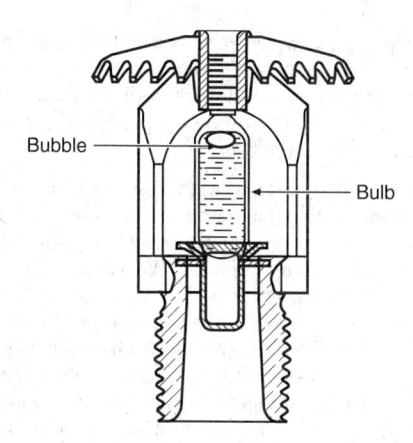

FIG. 6-9A. Grinnell Quartzoid, Issue D.

FIG. 6-9C. Representative arrangement of a solder-type link-and-lever automatic sprinkler.

Other thermosensitive elements: Other styles of thermosensitive operating elements employed to provide automatic discharge include bimetallic discs, fusible alloy pellets, and chemical pellets. (See Figure 6-9B.)

Sprinkler Dynamics

Figure 6-9C shows how the mechanical force exists in a solder-type link-and-lever-style automatic sprinkler. The construction shown is diagrammatic and does not represent any particular sprinkler.

The mechanical force normally exerted on the top of the cap or valve is many times that developed by the water pressure below, so that the possibility of leakage, even from water hammer or exceptionally high water pressure, is practically eliminated. The mechanical force in a link-and-lever sprinkler is produced by tension in the sprinkler frame, usually created by tightening the screw that holds the deflector down against the toggle joint formed by the levers. This pressure is applied against the valve or cap, but the line of force is not direct. The eccentricity of the loading permits a leveraged reduction of the force, first by the toggle effect of the two levers, and second by the mechanism of the link parts. The force resisted by the solder is made relatively low because solder of the composition needed to give the desired operating temperatures is subject to cold flow under high stress. The sprinkler frame or other parts usually possess a degree of elasticity to provide the energy that produces a positive, sharp release of the operating parts.

Sprinklers illustrated in Figure 6-9D use modifications of the common link-and-lever construction, some of which employ solder under compression, tension, and shear.

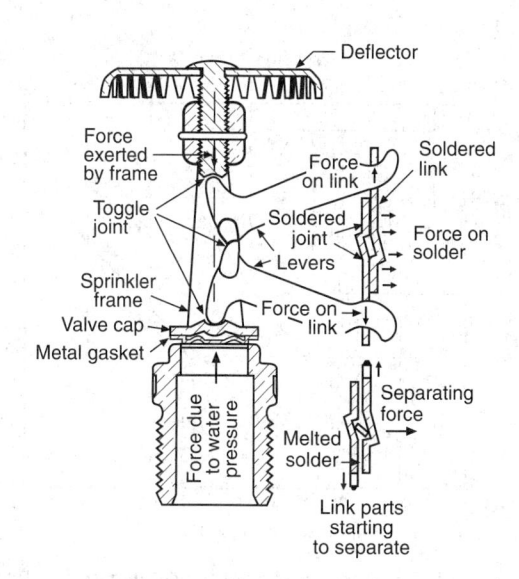

FIG. 6-9D. Some modifications of the common link-and-lever construction. At left is Grinnell F950 Duraspeed; at center is Reliable Model G; and at right is Firematic Type S.

To ensure that cold flow will not be a problem, the laboratories that test and list sprinklers use statistical methods to simulate long-term loading of heat-responsive elements. Statistical methods are also employed to ensure that the crush strength of glass bulbs is sufficiently higher than the frame loads that will be applied to the bulbs.

Deflector Design

Attached to the frame of the sprinkler is a deflector or distributor against which the stream of water is directed and converted into a spray designed to cover or protect a certain area. The amount of water discharged depends upon the flowing water pressure and the size of the sprinkler orifice. A flowing pressure of 7 psi (48 kPa) is generally considered a minimum for development of a reasonable spray pattern. At this pressure, a sprinkler having a nominal $1/2$-in. (12.7-mm) orifice will discharge 15 gpm (58 L/min). At the same 7-psi (48-kPa) pressure, a nominal $17/32$-in. (13.5-mm) orifice sprinkler will discharge 21 gpm (79 L/min). (See Figure 6-9E for the quantity of water discharged at various water pressures.)

In order to have even the minimum flowing pressure at sprinklers that are remote from the point of water supply, especially when

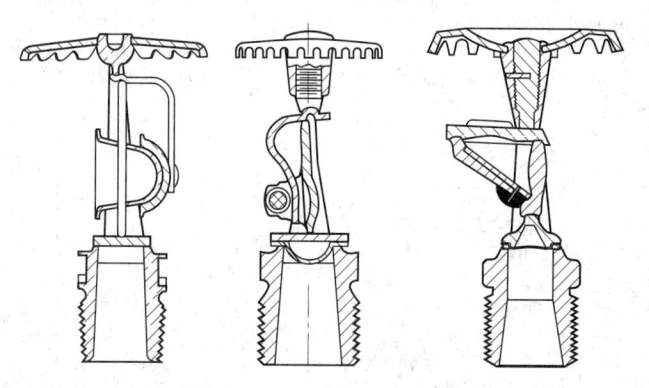

FIG. 6-9B. Central "Flow Control" (on-off) sprinkler.

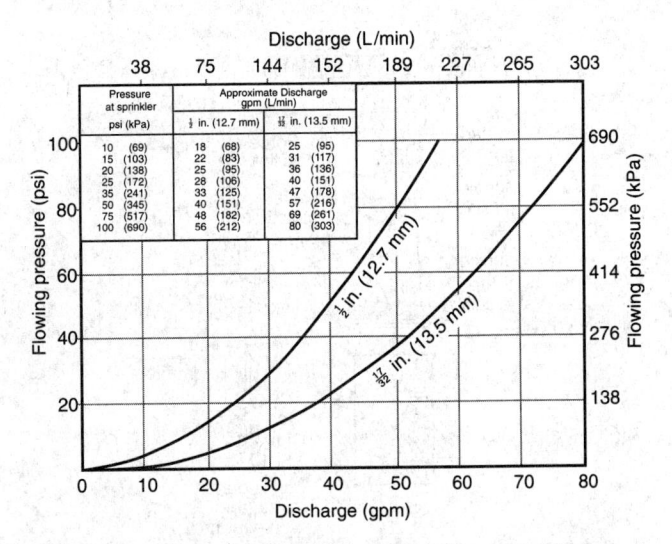

Pressure at sprinkler		Approximate Discharge gpm (L/min)	
psi	(kPa)	½ in. (12.7 mm)	17/32 in. (13.5 mm)
10	(69)	18 (68)	25 (95)
15	(103)	22 (83)	31 (117)
20	(138)	25 (95)	36 (136)
25	(172)	28 (106)	40 (151)
35	(241)	33 (125)	47 (178)
50	(345)	40 (151)	57 (216)
75	(517)	48 (182)	69 (261)
100	(690)	56 (212)	80 (303)

FIG. 6-9E. *Water discharge rates of typical nominal ¹/₂-in. (12.7-mm) and ¹⁷/₃₂-in. (13.5-mm) orifice automatic sprinklers.*

a number of sprinklers are operating simultaneously, water supply pressures in the range of 30 to 100 psi (316 to 690 kPa) are customarily provided. Hydraulically calculated systems are designed around the normally available water supply volume and pressure.

The distribution of water from a sprinkler is not expected to be axisymmetric, particularly since sprinkler frame arms obstruct the spray pattern and the serrated edges of the deflector create a "fingering" effect. A contour plot of the flow pattern of a pendent spray sprinkler is shown in Figure 6-9F.[1]

Because the distribution of water from sprinklers is a complex phenomenon, the testing laboratories have fairly broad requirements in this area. For upright and pendent spray sprinklers, for example, Underwriters Laboratories Inc. (UL) requires a turntable test to "fingerprint" the spray pattern. For a ¹/₂-in. (12.7-mm) orifice sprinkler flowing 15 gpm (58 L/min), or a large ¹⁷/₃₂-in. (13.5-mm) orifice sprinkler flowing 21 gpm (79 L/min), water distribution is not permitted outside a 16-ft (4.9-m) diameter circle in a horizontal plane 4 ft (1.2 m) below the sprinkler. A separate "16 pan" distribution test checks the ability of four flowing sprinklers to cover the area between them with their overlapping spray patterns. Factory Mutual Research Corporation (FMRC) applies an additional six-sprinkler flow test, which looks at the area covered between two adjacent sprinklers.

TEMPERATURE RATINGS OF AUTOMATIC SPRINKLERS

Automatic sprinklers have various temperature ratings that are based on standardized tests in which a sprinkler is immersed in a liquid and the temperature of the liquid is raised very slowly until the sprinkler operates. (See Table 6-9A.)

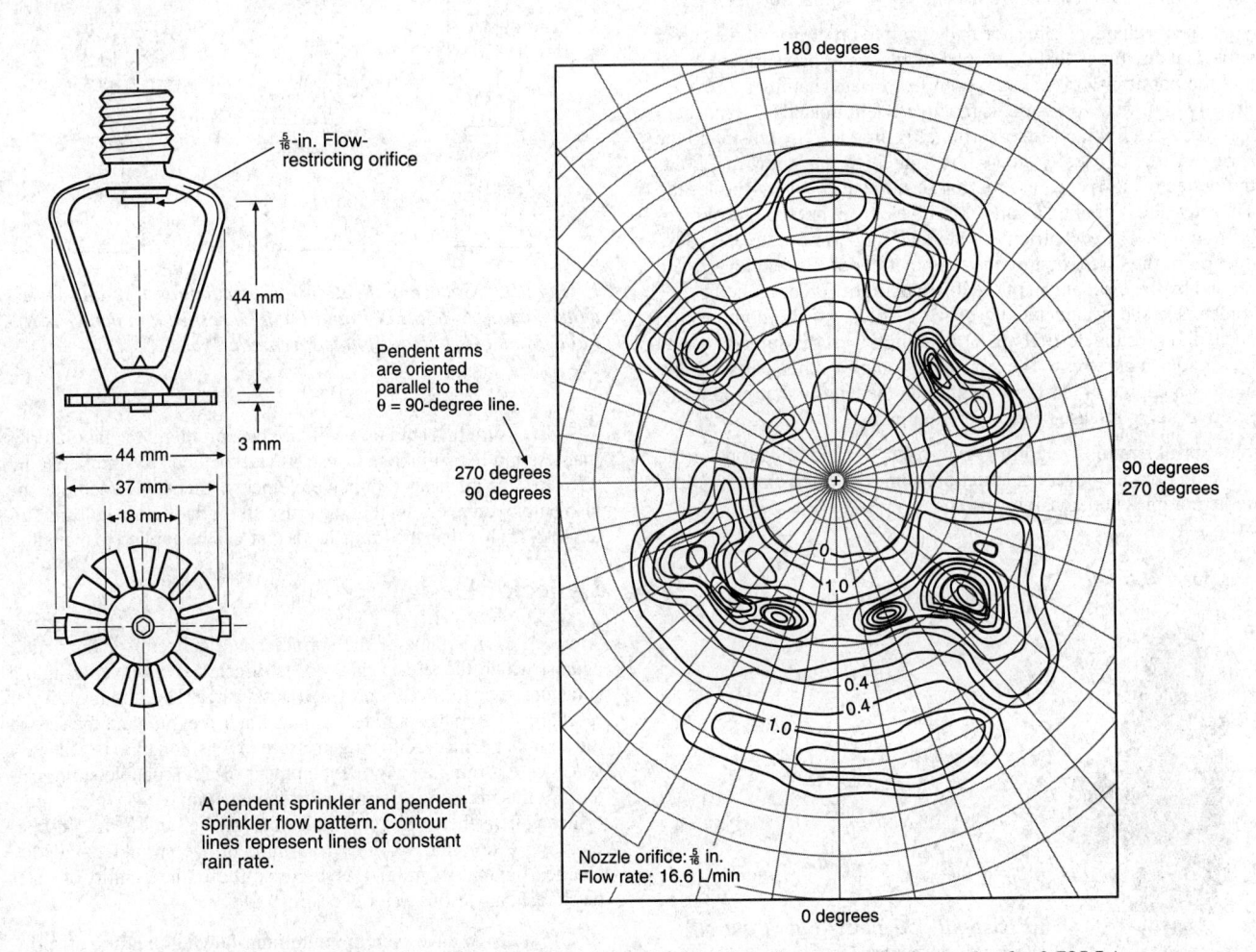

FIG. 6-9F. *Pendent spray sprinkler and flow contours.[1] (For SI units: 1 in. = 25.4 mm; 1 gal = 3.785 L.)*

TABLE 6-9A.   *Temperature Ratings, Classifications, and Color Codings*

Max. Ceiling Temp.		Temperature Rating		Temperature Classification	Color Code	Glass Bulb Colors
°F	°C	°F	°C			
100	38	135 to 170	57 to 77	Ordinary	Uncolored or black	Orange or red
150	66	175 to 225	79 to 107	Intermediate	White	Yellow or green
225	107	250 to 300	121 to 149	High	Blue	Blue
300	149	325 to 375	163 to 191	Extra high	Red	Purple
375	191	400 to 475	204 to 246	Very extra high	Green	Black
475	246	500 to 575	260 to 302	Ultra high	Orange	Black
625	329	650	343	Ultra high	Orange	Black

* Source: NFPA 13, *Standard for the Installation of Sprinkler Systems.*

The temperature rating of all fusible-element-style automatic sprinklers is stamped upon the soldered link. For bulb sprinklers, the temperature rating must be stamped or cast on some visible part. Color codes are also used for glass bulbs and for frame arms of fusible-element sprinklers.

The recommended maximum room temperature is restricted for both bulbs and fusible-element sprinklers. This is because solder begins to lose its strength somewhat below its actual melting point. Premature operation of a solder sprinkler usually depends on the extent to which the normal room temperature is exceeded, the duration of the excessive temperature, and the load on the operating parts of the sprinkler. While glass bulb sprinklers do not lose strength at temperatures close to their operating temperatures, using them at such temperatures can result in continuous loss and reforming of the air bubble, which creates stresses on the bulb.

The general rule of not using sprinklers of ordinary [135 to 170°F (57 to 77°C)] temperature rating where temperatures exceed 100°F (38°C) is necessary to provide a margin of safety. General practices regarding the use of automatic sprinklers of higher than the ordinary rating are given in Tables 6-9B and 6-9C. Sprinklers of ordinary temperature rating can be used inside buildings and other places where they are not subject to direct sun rays, except in monitors and in blind attics without ventilation, under metal or tile roofs, near or above heat sources, or in confined spaces where normal temperatures may be exceeded.

When there is doubt as to maximum temperatures at sprinkler locations, maximum reading thermometers should be used and the temperature determined under conditions that would show the highest readings to be expected.

Sprinklers with high temperature ratings are often used in place of ordinary-rated sprinklers in situations where a high rate of heat release can be anticipated. The sprinklers rated for higher temperatures may have the advantage of reducing the number of sprinklers that would otherwise operate outside the fire area.

Automatic sprinklers may require a longer time period to operate when exposed to a slowly developing fire as opposed to the

TABLE 6-9B.   *Temperature Ratings of Sprinklers Based on Distance from Heat Sources* *

Type of Heat Condition	Ordinary Degree Rating	Intermediate Degree Rating	High Degree Rating
1. Heating ducts (a) Above	More than 2 ft 6 in.	2 ft 6 in. or less	—
(b) Side and below	More than 1 ft 0 in.	1 ft 0 in. or less	—
(c) Diffuser	Any distance except as shown under Intermediate Degree Rating column	*Downward Discharge:* Cylinder with 1 ft 0 in. radius from edge, extending 1 ft 0 in. below and 2 ft 6 in. above *Horizontal Discharge:* Semi-cylinder with 2 ft 6 in. radius in direction of flow, extending 1 ft 0 in. below and 2 ft 6 in. above	—
2. Unit heater (a) Horizontal discharge	—	*Discharge Side:* 7 ft 0 in. to 20 ft 0 in. radius pie-shaped cylinder [see Figure 4-3.1.3.2] extending 7 ft 0 in. above and 2 ft 0 in. below heater; also 7 ft 0 in. radius cylinder more than 7 ft 0 in. above unit heater	7 ft 0 in. radius cylinder extending 7 ft 0 in. above and 2 ft 0 in. below unit heater
(b) Vertical downward discharge	—	7 ft 0 in. radius cylinder extending upward from an elevation 7 ft 0 in. above unit heater	7 ft 0 in. radius cylinder extending from the top of the unit heater to an elevation 7 ft 0 in. above unit heater
3. Steam mains (uncovered)			
(a) Above	More than 2 ft 6 in.	2 ft 6 in. or less	—
(b) Side and below	More than 1 ft 0 in.	1 ft 0 in. or less	—
(c) Blowoff valve	More than 7 ft 0 in.	—	7 ft 0 in. or less

For SI units: 1 in. = 25.4 mm; 1 ft = 0.305 m.
* Source: NFPA 13, *Standard for the Installation of Sprinkler Systems.*

TABLE 6-9C. *Ratings of Sprinklers in Specified Locations**

Location	Ordinary Degree Rating	Intermediate Degree Rating	High Degree Rating
Skylights	—	Glass or plastic	—
Attics	Ventilated	Unventilated	—
Peaked roof: Metal or thin boards, concealed or not concealed, insulated or uninsulated	Ventilated	Unventilated	—
Flat roof: Metal, not concealed, insulated or uninsulated	Ventilated or unventilated	Note: For uninsulated roof, climate and occupancy may necessitate intermediate sprinklers. Check on job.	—
Flat roof: Metal, concealed, insulated or uninsulated	Ventilated	Unventilated	—
Show windows	Ventilated	Unventilated	—

Note: A check of job condition by means of thermometers may be necessary.
* Source: NFPA 13, *Standard for the Installation of Sprinkler Systems.*

high heat release of a rapidly developing fire. They are designed to operate quickly enough to control a fire and prevent its spread.

The speed of operation depends on the physical properties of the sprinkler's mechanism. The time involved to operate depends, among other factors, upon the shape, size, and mass of the thermosensitive mechanism, the temperature differential between the surrounding atmosphere and the operating temperature of the sprinkler, and the velocity of the heated fire gases past the sprinkler operating element.

In cases where extreme speed of operation is necessary owing to the likelihood of a rapidly developing and spreading fire as, for example, in explosives manufacturing, the practice is to use deluge systems incorporating open sprinklers. These systems can be activated by heat detectors, special detection systems using infrared or ultraviolet light detection, smoke detection, or other means to open a valve and quickly allow water to enter the system and discharge through the open sprinklers. In cases where extremely fast response is required, the sprinkler piping is pre-filled with water up to the open sprinklers, which are equipped with blow-off plugs or caps. This is sometimes referred to as a "hot" deluge system.

STANDARD SPRAY SPRINKLERS

Standard spray sprinklers are generally similar in appearance to old-style sprinklers and use the same style frame and linkage or other release mechanism. The essential difference is in the deflector; seemingly minor differences in deflector design make major differences in discharge characteristics. Several representative standard spray sprinklers are illustrated in Figure 6-9G.

On the assumption that discharging water against the ceiling was essential to fire extinguishment, early research on automatic sprinklers was largely concerned with securing reasonably uniform distribution of water over the area protected by one sprinkler and with wetting the ceiling. Later research showed that more effective extinguishment and a larger area of coverage could be secured by directing all the water downward and horizontally. Research further showed that, with this pattern, discharge is effective even in controlling fires on the ceiling above the sprinklers because of the improved cooling effect of the spray, better high-level water distribution, and decreased exposure to the ceiling because of more effective direct discharge of water on burning materials below.[2]

The design of the deflector causes the solid stream of water issuing from the orifice of a standard sprinkler to break up to form an umbrella-shaped spray. The pattern is roughly that of a half-sphere filled with spray.

Standard spray sprinklers are made for installation in an upright or pendent position and must be installed in the position for which they are designed. (See Figure 6-9H.) It is customary to replace old-style sprinklers with spray sprinklers in existing installations, although NFPA 13 permits replacing old-style sprinklers with similar devices. Some manufacturers, however, have discontinued producing old-style sprinklers.

Double Deflectors

Several manufacturers developed a double deflector as their first design for use as standard spray upright sprinklers. A typical double

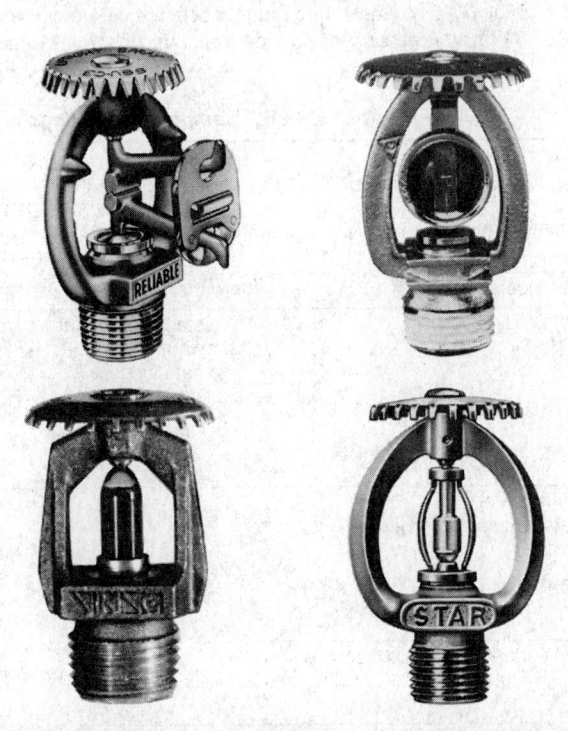

FIG. 6-9G. Standard spray sprinklers showing the various arrangements of releasing mechanisms. Clockwise from upper left is a fusible link-and-lever style (Reliable Issue C); a perforated heat-collector style, which is a variation of the link-and-lever style (Grinnell F950 Duraspeed); a center-strut style with solder under compression (Chemetron Stargard Model H); and a frangible-bulb style (Viking Model M).

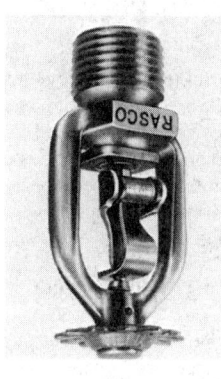

FIG. 6-9H.   A listed sprinkler showing upright (right) and pendent (left) models of the same issue; note the difference in design of the deflectors. The sprinklers shown are Reliable Model G.

deflector is shown in Figure 6-9I. Such sprinklers are no longer manufactured but, if previously "listed," are as acceptable as the currently approved upright spray sprinklers with a one-piece deflector.

WATER DISCHARGE

The general patterns of water discharge from the old-style and from the standard spray sprinklers are shown in Figures 6-9J and 6-9K, respectively.

The water distribution characteristics of sidewall automatic sprinklers, picker trunk sprinklers, window sprinklers, and cornice sprinklers are described later in this chapter.

Experimentation, engineering judgment, and experience determined that, for pipe schedule systems, a favorable rate of water discharge from an automatic sprinkler would be that of a $1/2$-in. (12.7-mm) diameter orifice. Often, this is not the case with hydraulically designed systems. Therefore, sprinklers of various orifice sizes are used. Standard automatic sprinklers have a nominal $1/2$-in. (12.7-mm) orifice that may be of the ring-nozzle or tapered-nozzle style. (See Figure 6-9L.)

The rate of water discharge from a sprinkler follows hydraulic laws and depends upon the size of the orifice or nozzle and on the water pressure. Approximate rates of discharge at different pressures may be obtained from the plotted curve or from the table given in Figure 6-9E.

Sprinklers discharging water through smaller or larger orifices are discussed later in this chapter. With similar forms of sprinkler

FIG. 6-9J.   Principal distribution pattern of water from old-style/conventional sprinklers.

FIG. 6-9K.   Principal distribution pattern of water from standard spray sprinklers.

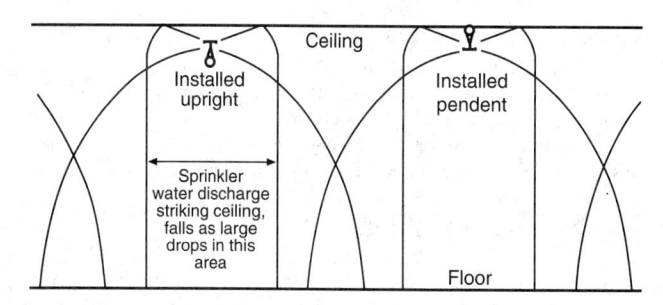

FIG. 6-9L.   Typical sprinklers, showing a straight discharge orifice (ring style) at left and tapered-nozzle orifice at right.

nozzles and at the same water pressure, the discharge from these styles of sprinklers is approximately proportional to the nominal size of the orifice.

To obtain acceptance or approval of their sprinklers, manufacturers submit them to fire-testing organizations. After extensive tests and verification of the manufacturer's ability to manufacture the product properly, sprinklers found satisfactory are listed. Acceptance of a sprinkler by inspection departments or other regulatory agencies is based on such a listing.

Standard spray sprinklers are designed to be installed and operated in their proper position—that is, upright or pendent, as indicated by a stamping upon the deflector bearing the appropriate word or the letters "SSU" (spray sprinkler upright) or "SSP" (spray sprinkler pendent).

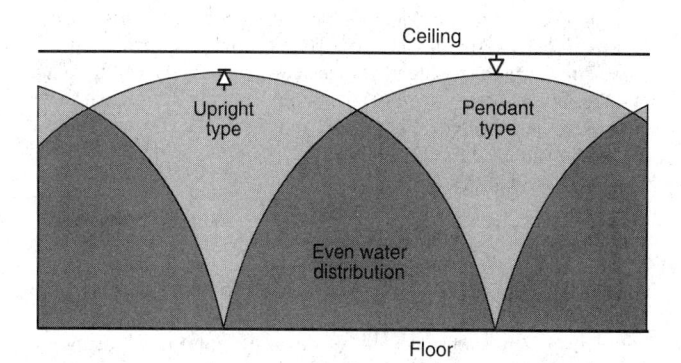

FIG. 6-9I.   Standard upright spray sprinkler with double deflector.

Recessed Sprinklers

A recessed sprinkler is a type of ceiling sprinkler in which part or most of the body of the sprinkler, other than the part that connects to the piping, is mounted in a recessed housing. Its operation is similar to that of a standard pendent sprinkler. (See Figure 6-9M.)

Flush-Type Sprinklers

A typical flush-style sprinkler is shown in Figure 6-9N. The special design of this type of ceiling sprinkler allows a minimum projection of the working parts of the sprinkler below the ceiling in which it is installed without adversely affecting the heat sensitivity or the pattern of water distribution.

In an effort to provide sprinkler protection in low-risk occupancies where aesthetics are important, sprinklers have been designed that attractively blend with the ceiling. Only the ceiling plate and thermosensitive assembly are visible from the floor when these sprinklers are installed. When a fire occurs and the thermosensitive element operates, the deflector drops to a position below the ceiling, and water discharge commences.

Concealed Sprinklers

A concealed sprinkler is a type of ceiling sprinkler whose entire body, including the operating mechanism, is above its concealing cover plate. When a fire occurs, the cover plate drops, exposing the thermosensitive assembly. The subsequent operation of the thermosensitive assembly initiates discharge. (See Figure 6-9O.)

Ornamental Sprinklers

Ornamental sprinklers are automatic sprinklers that have been decorated by plating or enameling to give desired surface finishes. Ornamentation or special decorative design must not affect the operation or the water distribution unfavorably. Listed styles of ornamental sprinklers are for pendent installation in accordance with NFPA 13. (See Figure 6-9P.) Such decorative coatings can only be applied by the manufacturer of the device.

Dry Pendent and Dry Upright Sprinklers

Dry pendent and dry upright sprinklers are used to provide sprinkler protection in unheated areas, such as freezers, where individual sprinklers are supplied from a drop or riser pipe from a wet-pipe system outside the unheated area. To prevent damage to the system, the drop/riser must remain free of water. Dry pendent sprinklers also are used on dry-pipe systems where it is desirable to install pendent sprinklers.

A seal is provided at the entrance of the dry sprinkler to prevent water from entering until the sprinkler operates. The heat-sensitive operating mechanisms are adaptations of those used with automatic standard sprinklers. See Figure 6-9Q for an example of a dry pendent automatic sprinkler.

SPRINKLERS FOR SPECIAL SERVICE CONDITIONS

The sprinkler system using listed upright or pendent spray sprinklers is adaptable to a wide variety of conditions. However, there are

FIG. 6-9M. A Central Model H recessed sprinkler.

FIG. 6-9N. A flush-style ceiling sprinkler; the diagram at right shows the sprinkler after operation. This style of sprinkler is used where appearance is considered to be of prime importance (Reliable Model B).

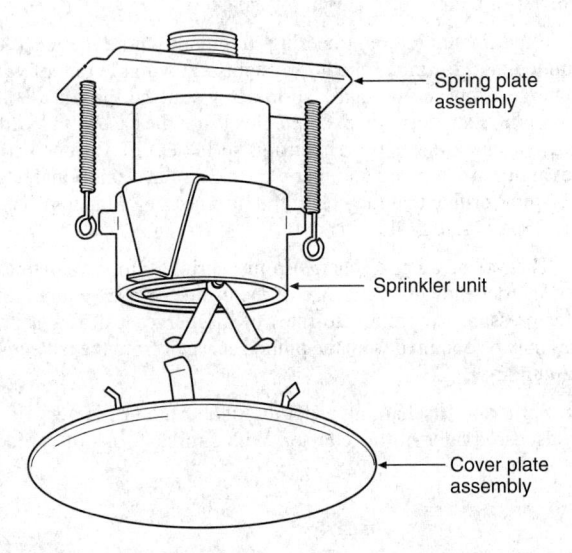

Spring plate assembly

Sprinkler unit

Cover plate assembly

FIG. 6-9O. Concealed ceiling sprinkler. Cover plate drops away when heat is applied to bottom side of plate (Stargard Model G).

FIG. 6-9P. Prussag SFH glass-bulb-style decorative sprinkler.

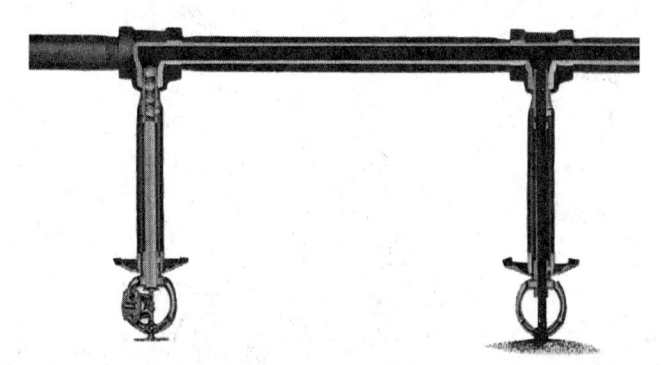

FIG. 6-9Q. *Representative dry pendent automatic sprinkler. When the ambient temperature rises beyond the operating temperature of the soldered link, the solder melts and the link plates separate on roller key. Levers held in place by a deflector screw are released, and the fixed tension of the frame, acting as a spring, ejects the levers and link parts clear of the sprinkler. Since the inner tube, which also serves as a discharge orifice, is no longer held in place by the levers, it moves to a predetermined position. With the support of the inner tube eliminated, the elements forming the watertight seal at the piping inlet pass through the inner tube and away from the sprinkler, allowing water to flow through the unobstructed waterway and strike the deflector, which distributes it in a spray pattern comparable to that of a ¹/₂-in. (12.7-mm) standard sprinkler. Shown is Reliable's Model C Dry Pendent. Other manufacturers have similar arrangements of proprietary equipment for dry pendent automatic systems.*

situations for which special styles of sprinklers and special sprinkler arrangements are suited. In some cases, such as with sidewall sprinklers, special patterns of water distribution are necessary; in others, unusual ambient temperatures or corrosive atmospheres call for special design or construction features.

Residential Sprinklers

Residential sprinklers are sprinklers that have been specifically listed for use in residential occupancies. These fast-response sprinklers have special low-mass fusible links or bulbs that make the time of temperature actuation much less than that of a sprinkler with a conventional operating element.

Residential sprinklers also have different discharge characteristics than spray sprinklers. They are required to throw water within 28 in. (711 mm) of the ceiling. This high-wall wetting pattern, along with the faster response, helps the residential sprinkler control or suppress typical residential fires with flows much lower than standard spray sprinklers.

Because of the life safety aspects of fast response, these sprinklers can be used in a variety of residential occupancies. Where a conventional system installed according to the provisions of NFPA 13 might not be as practical, such as in a single-family dwelling or a low-rise residential property, NFPA 13D, *Standard for the Installation of Sprinkler Systems in One- and Two-Family Dwellings and Manufactured Homes*, and NFPA 13R, *Standard for the Installation of Sprinkler Systems in Residential Occupancies up to and Including Four Stories in Height*, may be used. These standards have been developed specifically around the life-safety capabilities of the residential sprinkler. A representative residential sprinkler is shown in Figure 6-9R. (Residential sprinkler installations are discussed in detail in Section 6, Chapter 13, "Fast-Response Sprinkler Technology.")

FIG. 6-9R. *A representative residential sprinkler.*

Large Drop Sprinklers

Large drop sprinklers are special sprinklers with a K factor between 11.0 and 11.5. [K factors for standard ¹/₂-in. (12.7-mm) sprinklers range from 5.3 to 5.8.] The deflector of a large drop sprinkler is specially designed and, combined with the greater discharge, produces large drops of such size and velocity as to enable the spray to penetrate strong updrafts generated by high-challenge fires. The special requirements for large drop sprinklers are found throughout NFPA 13, and are organized by subject category. These special requirements supersede the density/area requirements used for standard spray sprinklers. A representative large drop sprinkler is shown in Figure 6-9S.

On/Off Sprinklers

An on/off sprinkler cycles on and off as needed. The model shown in Figure 6-9T operates as a normal fusible element sprinkler, but with a supplemental activation system based on a pilot valve principle. Under normal conditions, the pilot valve is held closed by the bimetallic snap disc. Water in the diaphragm chamber holds the diaphragm closed. When the snap disc is heated to the rated temperature, it opens the pilot valve and releases the water from the diaphragm chamber. This allows the diaphragm to open and water to flow from the sprinkler. When the snap disc cools to approximately 100°F (37.8°C), it closes the pilot valve. Water enters the diaphragm chamber, and the difference in pressure closes the valve. The sprinkler is ready for repeated operation should the snap disc again be heated to the rated temperature, although the sprinkler is intended to be replaced after a fire. Both a pendent and a recessed model are available.

Sprinklers for Corrosive Conditions

Measures have been developed to protect automatic sprinklers from corrosive conditions, and testing organizations have studied the value

FIG. 6-9S. *A representative large drop sprinkler.*

Installation diagram
Flow control – FC pendent

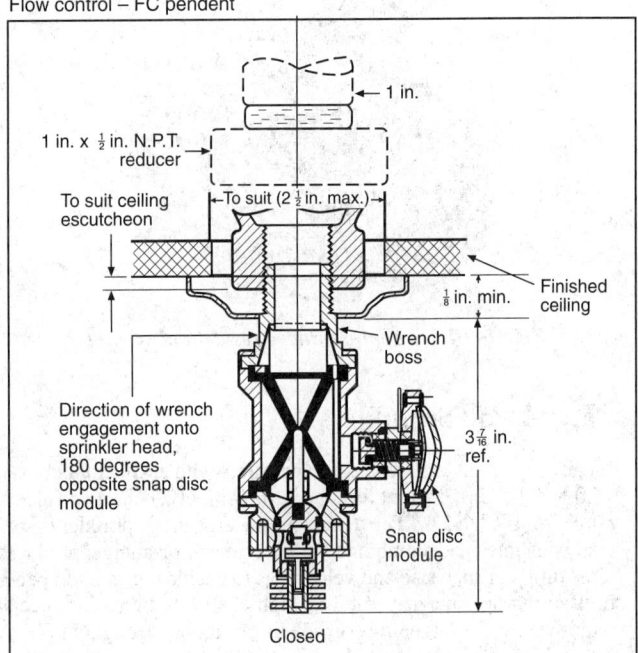

Closed

Cross section
Flow control – FC

Activated

FIG. 6-9T. *An on/off sprinkler showing its operating elements. In a fire, the melting of the fusible element opens the main waterway, and the bimetallic snap disc responds by opening the pilot valve, which relieves pressure from the diaphragm chamber. (For SI units: 1 in. = 25.4 mm.) (Source: Central Sprinkler Corp.)*

of each method. A complete covering of wax that has a melting point slightly below the temperature at which the sprinkler operates is the protective coating most commonly used. A lead coating for the body of the sprinkler and the levers in combination with wax for protecting fusible elements is also common. Traditional coatings include wax only (Figure 6-9U), asphalt only, lead only, wax over lead, and asphalt over lead. In recent years, sprinklers have been listed with enamel, polyester, and Teflon® coatings. Stainless steel sprinklers are available from some manufacturers. Other coatings are also available and should be selected on the basis of the expected environment.

Whatever the protective measures taken, they must not delay the operation of the sprinkler, nor interfere with the release of operating parts, nor significantly alter the pattern of water distribution. As with decorative sprinklers, these coatings may also only be applied by the manufacturer.

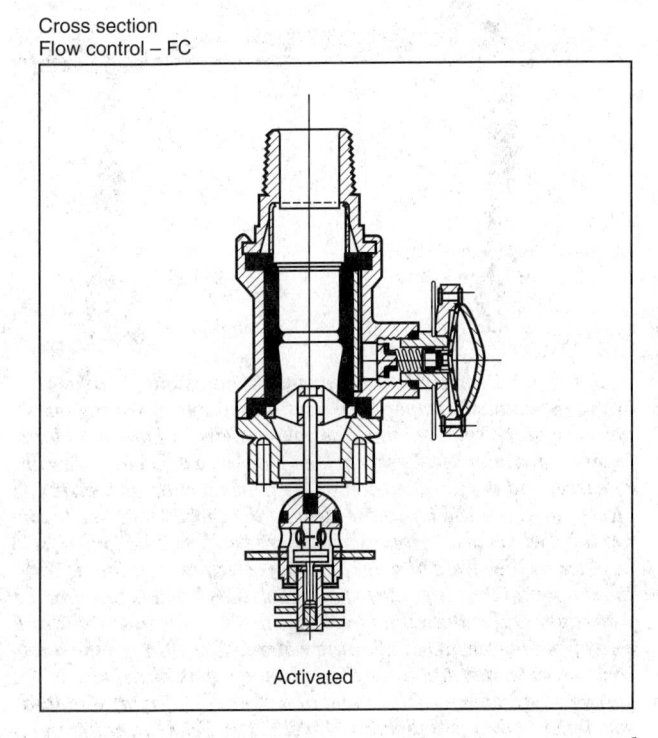

FIG. 6-9U. *A wax-coated upright sprinkler for corrosive atmospheres. (Source: "Automatic" Sprinkler Corporation of America)*

Sidewall Sprinklers

Sidewall sprinklers have the components of standard sprinklers except for a special deflector, which discharges most of the water toward one side in a pattern somewhat resembling one-quarter of a sphere. A small proportion of the discharge wets the wall behind the sprinkler. Located and mounted either vertically or horizontally along the junction between a ceiling and sidewall, sidewall sprinklers provide protection adequate for light-hazard occupancies, such as hotel lobbies, dining rooms, executive offices, and other areas where the usual sprinkler pipes would be objectionable in appearance. Some sidewall sprinklers have been tested and listed for use in ordinary-hazard occupancies.

The directional character of the discharge from sidewall sprinklers makes them applicable to occasional special protection problems. They may be installed to give discharge in any desired direction.

Figure 6-9V shows typical sidewall sprinklers. Selection has been made to show the different shapes of a variety of deflectors. Many vertically mounted sidewall sprinklers may be installed in either the pendent or upright position.

Extended coverage sidewall sprinklers: These are special sidewall sprinklers, used in the horizontal position, that have larger areas of coverage than those allowed for standard sidewall sprinklers. They may be used in light-hazard occupancies, particularly in hotels and similar occupancies where a sprinkler system can be installed in an existing building without having piping exposed in living areas, which could be objectionable aesthetically.

The water pressure required to obtain the greater coverage is specified in the listings for the sprinklers and is greater than that required for standard sidewall sprinklers. Installation requirements are also a part of the listing.

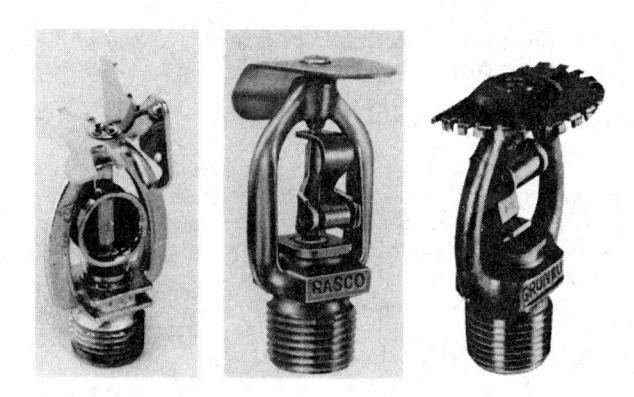

FIG. 6-9V. A representative selection of listed sidewall sprinklers showing various shapes of deflectors.

An extended coverage sidewall sprinkler with a conventional link release is shown in Figure 6-9W.

Open Sprinklers

Automatic sprinklers, from which the valve cap and heat-responsive elements have been omitted, are used in deluge sprinkler systems where the water supply is controlled by an automatic water control valve actuated independently of the automatic sprinklers. (See Figure 6-9X.) The water distribution pattern and the density of the discharge are designed to be appropriate for the hazard to be protected.

Small- and Large-Orifice Sprinklers

Automatic sprinklers having a water discharge rate greater or less than that of a standard $1/_2$-in. (12.7-mm) orifice sprinkler operating

at the same pressure have characteristics desirable in protecting certain occupancies. High-discharge, large-orifice sprinklers operating at discharge densities of 0.45 gpm/sq ft [18 (L/min)/m^2] and greater provide a quantity discharge not available from standard nominal $1/_2$-in. (12.7-mm) orifice sprinklers. The pattern of the water discharge from small- and large-orifice sprinklers is similar to that of the standard $1/_2$-in. (12.7-mm) sprinkler.

Whenever sprinklers are replaced, care must be taken to replace them with the proper styles. Replacement of sprinklers with others of the correct orifice size is critical in hydraulically calculated systems.

$1/_4$-in. through $7/_{16}$-in.-orifice automatic sprinklers: These sprinklers have been listed with orifices ranging from $1/_4$ through $7/_{16}$ in. (6.4 through 11.1 mm) and k factors ranging from 1.3 to 4.4. $1/_4$-in. through $7/_{16}$-in.-orifice sprinklers are identified by a pintle extending above the deflector and by the size stamped on the base of the sprinkler. (See Figure 6-9Y.)

$17/_{32}$-in.-orifice automatic sprinklers: These sprinklers have a discharge rate of 140 percent of that of the standard nominal $1/_2$-in. (12.7-mm) sprinkler operating at the same pressure with K factors between 7.4 and 8.2. (See Figure 6-9Z.) $17/_{32}$-in.-orifice sprinklers intended for use in new sprinkler installations are identified by a $3/_4$-in. (19.1-mm) iron pipe thread connection. Those intended for use in existing installations are manufactured with a $1/_2$-in. (12.7-mm) iron pipe thread connection and an identifying pintle on the deflector and with size marking on the sprinkler frame.

FIG. 6-9W. A representative extended coverage sidewall sprinkler.

FIG. 6-9Y. A representative $7/_{16}$-in.-orifice sprinkler in the pendent position. Note the pintle extending from deflector.

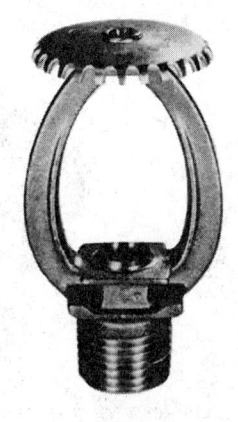

FIG. 6-9X. A standard upright sprinkler with its operating elements removed.

FIG. 6-9Z. A representative $17/_{32}$-in.-orifice sprinkler. (Source: Viking Corp.)

⁵/₈ in.-extra large-orifice automatic sprinklers: These sprinklers have been listed with a K factor of 11.0 to 11.5, and discharge 200 percent of the standard nominal ¹/₂-in. (12.7-mm) sprinkler operating at the same pressure. Manufactured with either a ¹/₂-in. (12.7-mm) or ³/₄-in. (19.1 mm) iron pipe thread connection, they are identified with a pintle.

While NFPA 13 does not limit the use of ⁵/₈-in.-orifice sprinklers, other NFPA standards do. For instance, in order to use ⁵/₈-in.-orifice sprinklers to protect general storage in accordance with NFPA 231, *Standard for General Storage*, or rack storage in accordance with NFPA 231C, *Standard for Rack Storage of Materials*, the sprinkler must be listed for use in storage occupancies and can only be used at a minimum of 10 psi (69 KPa).

³/₄ in.-very-extra-large-orifice (VELO) automatic sprinklers: Sprinklers with a K factor of 13.5 to 14.5 are defined in this category. Originally designated for use with ESFR sprinklers, ³/₄ in.-orifice sprinklers are now available that are intended for use with traditional, standard spray design options to discharge 250 percent of the standard nominal ¹/₂-in sprinkler operating at the same pressure.

The discharge characteristics of various large- and small-orifice sprinklers are given in Table 6-9D.

Picker Trunk Sprinklers

Picker trunk automatic sprinklers have a small, smooth deflector that aids in reducing collections of lint and fiber on the sprinklers when they are placed inside ducts or enclosures where moving air carries such foreign materials in suspension. Freedom from obstruction and a general breakup of the water stream are of more importance than any specific pattern of distribution. (See Figure 6-9AA.)

Intermediate-Level Sprinklers

Intermediate-level sprinklers, also referred to as in-rack sprinklers, have large discs (shields) designed to protect the thermosensitive assembly from the spray of sprinklers suspended at higher levels. Without the protective shields, the impinging water could cool the thermosensitive element and retard sprinkler operation. Both upright and pendent intermediate-level sprinklers are listed by testing organizations. (See Figure 6-9BB.)

TABLE 6-9D. *Sprinkler Discharge Characteristics Identification*

Nominal Orifice Size		K Factor*	Percent of Nominal ¹/₂ in. Discharge	Thread Type	Pintle	Nominal Orifice Size Marked on Frame
(in.)	(mm)					
¹/₄	6.4	1.3–1.5	25	¹/₂ in. NPT	Yes	Yes
⁵/₁₆	8.0	1.8–2.0	33.3	¹/₂ in. NPT	Yes	Yes
³/₈	9.5	2.6–2.9	50	¹/₂ in. NPT	Yes	Yes
⁷/₁₆	11.0	4.0–4.4	75	¹/₂ in. NPT	Yes	Yes
¹/₂	12.7	5.3–5.8	100	¹/₂ in. NPT	No	No
¹⁷/₃₂	13.5	7.4–8.2	140	³/₄ in. NPT or ¹/₂ in. NPT	No / Yes	No / Yes
⁵/₈	15.9	11.0–11.5	200	¹/₂ in. NPT or ³/₄ in. NPT	Yes / Yes	Yes / Yes
³/₄	19.0	13.5–14.5	250	³/₄ in. NPT	Yes	Yes

* K factor is the constant in the formula $Q = K\sqrt{p}$
  where Q = Flow in gpm
  p = Pressure in psi

For SI units: $Q_m = K_m\sqrt{P_m}$
  where Q_m = Flow in L/min
  P_m = Pressure in bars
  $K_m = 14\ K$

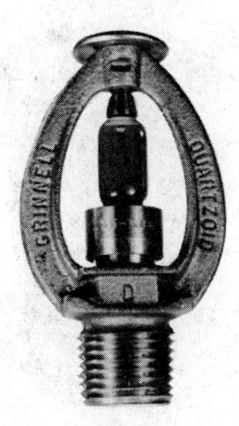

FIG. 6-9AA. A picker trunk automatic sprinkler. (Source: Grinnell)

FIG. 6-9BB. Intermediate-level (rack storage) sprinklers showing integral shields that protect operating elements from the discharge of sprinklers installed at higher levels. (Star Model LD)

BIBLIOGRAPHY

References Cited

1. Wendt, B., and Prahl, J. M., "Discharge Distribution Performance for an Axisymmetric Model of a Fire Sprinkler Head," NBS-GCR-86-517, Oct. 1986, Case Western Reserve University, sponsored by National Bureau of Standards, Gaithersburg, MD.
2. Thompson, N. J., *Fire Behavior and Sprinklers*, National Fire Protection Association, Quincy, MA, 1964.

NFPA Codes, Standards, and Recommended Practices

Reference to the following NFPA codes, standards, and recommended practices will provide further information on automatic sprinklers discussed in this chapter. (See the latest version of *The NFPA Catalog* for availability of current editions of the following documents.)

NFPA 13, *Standard for the Installation of Sprinkler Systems*
NFPA 13D, *Standard for the Installation of Sprinkler Systems in One- and Two-Family Dwellings and Manufactured Homes*
NFPA 13R, *Standard for the Installation of Sprinkler Systems in Residential Occupancies up to and Including Four Stories in Height*
NFPA 25, *Standard for the Inspection, Testing, and Maintenance of Water-Based Fire Protection Systems*
NFPA 231, *Standard for General Storage*
NFPA 231C, *Standard for Rack Storage of Materials*

Additional Reading

"An Overview of Sprinkler Technology," *Fire Journal*, Vol. 83, No. 6, Nov.–Dec. 1989, p. 48.
Beitel, J. J., "Quick-Response Sprinkler Tests, Group 2 Performance Tests," National Fire Protection Research Foundation, Quincy, MA, Dec. 1988.
Bill, R. G., and Kung, H. C., "Evaluation of an Extended Coverage Sidewall Sprinkler and Smoke Detector in a Hotel Occupancy," *Journal of Fire Protection Engineering*, Vol. 1, No. 3, 1989.
Bryan, J. L., "Automatic Sprinklers," *Automatic Sprinkler and Standpipe Systems*, 2nd ed., National Fire Protection Association, Quincy, MA, 1990, pp. 193–269.
Budnick, E. K., and Fleming, R. P., "Developing an Early Suppression Design Procedure for Quick-Response Sprinklers," *Fire Journal*, Vol. 83, No. 6, Nov.–Dec. 1989, p. 41.
Budnick, E. K.,and Fleming, R. P., "How Quick-Response Sprinklers Perform and What It Means for Their Application," *Fire Journal*, Vol. 83, No. 5, Sept.–Oct. 1989, p. 48.
Carey, W. M., "Report of the Quick-Response Sprinkler Project, Group 2 Performance Tests," Mar. 1988, National Fire Protection Research Foundation, Quincy, MA.

Cooper, L. Y., and Stroup, D. W., "Test Results and Procedures for the Response of Near Ceiling Sprinkler Links in a Full-Scale Compartment Fire," NBSIR 87-3633, Sept. 1987, National Bureau of Standards, Gaithersburg, MD.
Cote, A. E., "QRS: Do You Understand the Technology?" *Fire Journal*, Vol. 81, No. 1, Jan. 1987, p. 27.
Evans, D. D., Walton, W. D., and McCaffrey, B., "Progress in Modeling Sprinkler Performance," *9th Joint Panel Meeting* of the UJNR Panel on Fire Research and Safety, NBSIR 88-3753, National Institute of Standards and Technology, Gaithersburg, MD, Apr. 1988, pp. 325–335.
Factory Mutual Approval Guide, Factory Mutual Research Corporation, Norwood, MA (published annually).
Fire Protection Equipment List, Underwriters Laboratories Inc., Northbrook, IL (published annually).
Guenther, P., "Sprinklers—A German Viewpoint," *Fire Prevention*, No. 200, June 1987, pp. 23–24.
Heskestad, G., and Bill, R. G., "Quantification of Thermal Responsiveness of Automatic Sprinklers Including Conduction Effects," *Fire Safety Journal*, Vol. 14, Nos. 1 & 2, July 1988, pp. 113–125.
Kirson, D., "Automatic Sprinklers in Exhaust Systems," *Heating, Piping and Air Conditioning*, Vol. 61, No. 1, Jan. 1989, pp. 163–166.
Lougheed, G. D., and Richardson, J. K., "Sprinklers in Combustible Concealed Spaces," *Journal of Fire Protection Engineering*, Vol. 1, No. 2, 1989, pp. 49–62.
Pepi, J. S., "Design Characteristics of Quick-Response Sprinklers," TR 86-3, 1986, Society of Fire Protection Engineers, Boston, MA.
Solomon, R. E., ed., *Automatic Sprinkler Systems Handbook*, 5th ed., National Fire Protection Association, Quincy, MA, 1994.
Standard for Automatic Sprinklers for Fire Protection Service, 8th ed., Underwriters Laboratories Inc., Northbrook, IL, 1990.
Theobald, C. R., "FRS Ramp Test for the Thermal Sensitivity of Sprinklers," *Journal of Fire Protection Engineering*, Vol. 1, No. 1, 1989, pp. 23–34.
Theobald, C. R., and Westley, S. A., "Factors Affecting the Sensitivity of Sprinklers," *Fire Surveyors*, Vol. 17, No. 3, June 1988, pp. 5–11.
Thorne, P. F., Theobald, C. R., and Melinek, S. J., "Thermal Performance of Sprinkler Heads," *Fire Safety Journal*, Vol. 14, Nos. 1 & 2, July 1988, pp. 89–99.
Unoki, J., "Fire Extinguishing Time by Sprinkler," *Proceedings* of the First International Symposium on Fire Safety Science, Hemisphere, Washington DC, 1986, pp. 1187–1196.
Vincent, B. G., Stavrianidis, P., and Kung, H. C., "Large-Scale Fire Testing of Fast-Response Sprinklers and Conventional Response Sprinklers in a Fire-Control-Mode Scenario," Factory Mutual Research Corp., Norwood, MA, June 1989.
Walton, W. D., and Budnick, E. K., "Quick-Response Sprinklers in Office Configurations: Fire Test Results," NBSIR 88-3695, 1988, National Institute of Standards and Technology, Gaithersburg, MD.
Yao, C., "FMRC Sprinkler Research," *9th Joint Panel Meeting* of the UJNR Panel on Fire Research and Safety, NBSIR 88-3753, Apr. 1988, National Institute of Standards and Technology, Gaithersburg, MD, pp. 314–324.

AUTOMATIC SPRINKLER SYSTEMS

Revised by Robert E. Solomon

The first document published by NFPA, in 1896, provided the initial set of standardized rules for automatic sprinkler systems. The document was awkwardly titled, *Rules and Regulations of the National Board of Fire Underwriters for Sprinkler Equipment's, Automatic and Open Systems as Recommended by the National Fire Protection Association.* This NFPA document evolved to be NFPA 13, *Standard for the Installation of Sprinkler Systems* (hereinafter referred to as NFPA 13).

The fundamental principles and assumptions about sprinkler systems have remained fairly consistent in the last 100 yr. The modern-day version of NFPA 13 is an expansion of those early principles. In addition, NFPA 13 represents the adaptation of precise sprinkler criteria that allows a number of options to the designers and installers. These options involve the types of valves, the materials for piping systems, and the design approach that may be used to establish the system flow rates, among other design considerations.

This chapter provides an introduction and overview to the fundamentals of automatic sprinklers systems and how the current edition of NFPA 13 has evolved to govern a number of situations. The benefits of sprinklers are discussed, followed by a review of current rules and practices that are associated with these systems.

Several other chapters in this handbook directly or indirectly relate to sprinkler system design and performance. For example, see the following: Section 6, Chapter 9, "Automatic Sprinklers"; Section 6, Chapter 11, "Water Supplies for Sprinkler Systems"; Section 6, Chapter 13, "Fast-Response Sprinkler Technology"; and Section 6, Chapter 18, "Care and Maintenance of Water-Based Extinguishing Systems."

While sprinkler systems are perhaps the oldest and most established of the water-based fire protection systems, they are not the only system that utilizes water as the fire control or fire suppression mechanism. Related technologies include water spray systems (see Section 6, Chapter 12, "Water Spray Protection"), foam-water sprinkler systems (see Section 6, Chapter 22, "Foam Extinguishing Agents and Systems") and water mist systems (See Section 6, Chapter 15, "Water Mist Suppression Systems").

Properly designed and installed systems are a part of the fundamental fire protection envelope that is so integral to modern building design and construction. Compartmentation, detection,

and suppression are the key features that oftentimes allow buildings of a certain height or area to be constructed. Proper understanding of what sprinklers can and cannot do is an important element for the Authority Having Jurisdiction, the design professional, and the system installer.

DEVELOPMENT OF AUTOMATIC SPRINKLERS

In lieu of defining what a sprinkler is, NFPA 13 provides a series of definitions for various types of sprinklers. A sprinkler is basically a device that is designed to discharge water over a certain area, is only activated when a fire generates a sufficient quantity of heat, and will control or suppress the fire once it has activated.

The water supply for the sprinkler is provided through a set of specially sized pipes or tubes. A number of control valves are typically present to allow the water supply to be manually shut off to the pipe network and the sprinklers.

Early "sprinkler" systems involved the use of steel pipe with drilled holes or perforations provided along the length of pipe. This type of system often involved the use of a manually operated water supply. If the fire occurred when the building was not occupied, the system would be of little or no benefit. The piping for these systems was often corroded, the holes were often plugged, and the water discharge from the pipe could generously be described as poor.

A derivative of the perforated pipe involved attachment of an open sprinkler device to the pipe. This improved the distribution pattern, but the other limitations of manual intervention and corrosion were the same.

The concept for the use of a heat-actuated device dates back to approximately 1860. It was not until 1875, however, that this concept was incorporated into a device with the advent of the Parmelee sprinkler. This sprinkler is given credit as the first "automatic" sprinkler. While the device was crude when compared to current designs, it established a specific water distribution pattern that was effective and efficient at controlling a fire. Figure 6-10A shows the discharge pattern from a Parmelee sprinkler.

Other variations soon followed and each one had its own, unique shape or form. Among the first group of devices that were available when the 1896 edition of NFPA 13 was published were: Neracher (Figure 6-10B), National Manufacturing Company (Figure 6-10C), New Grinnell (Figure 6-10D), and Kane (Figure 6-10E).

Robert E. Solomon, P.E., is chief building fire protection engineer in the NFPA Building Fire Protection and Life Safety Department, Quincy, MA.

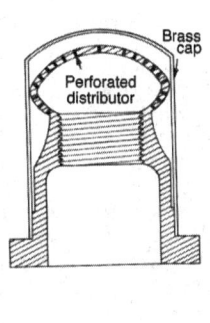

FIG. 6-10A. *An early automatic sprinkler. Water is shown discharging from a Parmelee No. 3 upright sprinkler, which was first used in 1875. It consisted of a brass cap soldered over a perforated distributor and was designed to screw onto a nipple. A cross-sectional view of the sprinkler is at right.*

FIG. 6-10C. *National Manufacturing Company sprinkler.*

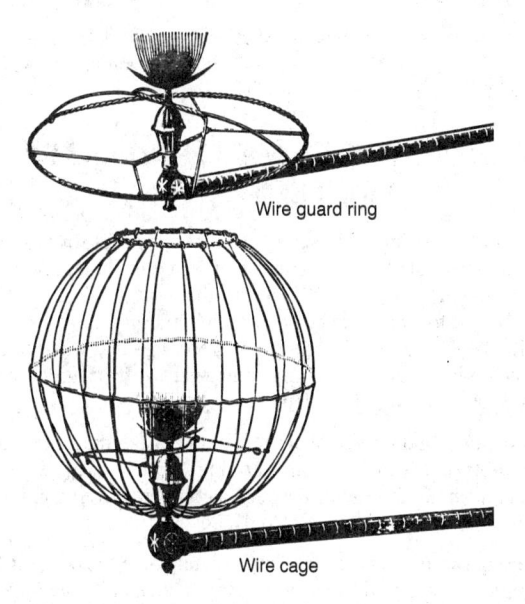

FIG. 6-10B. *Neracher sprinkler.*

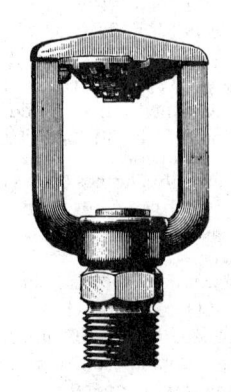

FIG. 6-10D. *New Grinnell sprinkler.*

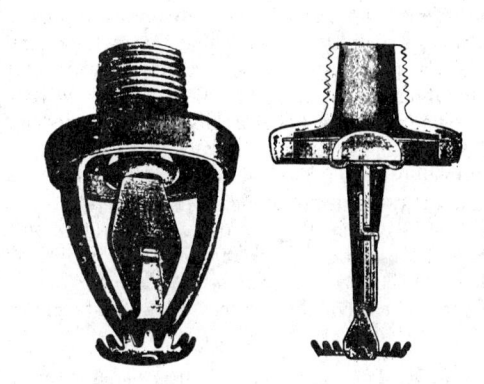

FIG. 6-10E. *Kane sprinkler.*

The advent of these devices, which were automatic in nature and did not require manual operation of the water supply, provided a catalyst for greater acceptance and use of sprinklers.

VALUE OF AUTOMATIC SPRINKLER PROTECTION

While the insurance industry can be given much credit for promulgating and expanding the use of sprinkler systems, the accelerated industrial growth of the United States in the early 1900s is just as likely a catalyst for the acceptance of sprinklers. Initial system use was justified by the acute reduction in property loss that was evident in fires involving sprinklers when compared to those fire events where no sprinklers were installed.

Warehouses and large industrial manufacturing facilities were typically protected with automatic sprinkler systems. This distinction often meant the difference between a company staying in business *vs.* not being able to recover from a devastating loss.

Although sprinklers were not immediately recognized for their life safety features, a trend was evident that indicated reductions in fire-related injuries and death when sprinklers were present. Large-area, high-rise, high-value, and dense population structures soon became primary environments where sprinkler systems would be of benefit.

The life safety benefit of sprinklers was intuitive since operation of sprinklers would sound an alarm, thereby warning the building occupants of a dangerous condition. More importantly, the discharge of water would reduce the fire size, control the fire, and diminish the threat to the occupants. Early system designs could typically provide protection to building occupants immediately outside of the fire area. Today's technology includes the residential sprinkler, which is designed to maintain a tenable environment for occupants in the room of fire origin. Section 6, Chapter 13, "Fast Response Sprinkler Technology" provides information on the development of the residential sprinkler. This device is quite versatile, as it is intended to be used in everything from a single-family dwelling to a high-rise residential occupancy, such as a hotel or condominium.

NFPA *101*®, *Life Safety Code*® (hereinafter referred to as NFPA *101*), recognizes the attributes of sprinkler protection for the types of occupancies that are governed by NFPA *101*. Extended travel distances are permitted, reductions in rated wall assemblies are permitted, and interior finish materials may be used with higher flame-spread ratings when sprinklers are present. These features are especially important for existing buildings and configurations. NFPA *101* also will mandate sprinklers in all buildings that are defined as high-rise buildings, as well as in those buildings where evacuation of the occupants is impractical. This latter category includes health care facilities and nursing homes with nonambulatory residents.

Up until 1970, NFPA maintained a data base on the performance and effectiveness of sprinklers. Unfortunately, collection of this type of information was discontinued when the data became skewed toward those instances where large numbers of sprinklers tended to operate or where the dollar losses were quite large. Such data appeared to indicate that sprinklers were not as effective as one would think. While substantial dollar losses can occur in fires in sprinklered buildings, such fires tend to involve partial or misapplied sprinkler systems or properties with large dollar values in very small areas.

Some limited data can be found concerning the number of sprinklers that tend to operate in certain fires. A small number of sprinklers tend to control most of the fires. Table 6-10A provides a

summary of how many sprinklers operated, based upon 2,860 fires, all of which occurred in buildings protected with sprinklers.

TABLE 6-10A. *Sprinkler Operation 1978–1987*

Number Operating	Fires	Cumulative Number	Percent of Total	Cumulative Percent
1	797	797	28	28
2	520	1317	18	46
3	291	1608	10	56
4	221	1829	8	64
5	147	1976	5	69
6	146	2122	5	74
7	74	2196	3	77
8	77	2273	3	80
9	44	2317	1	81
10	55	2372	2	83
11–15	172	2544	6	89
16–20	100	2644	3	92
21–25	49	2693	2	94
26+	167	2860	6	100

Source: Factory Mutual Research Corporation.

NFPA's Fire Analysis and Research Division has produced a report on the topic of sprinklers. It is entitled *U.S. Experience with Sprinklers: Who Has Them? How Well Do They Work?* This report, updated annually, discusses a number of benefits of automatic sprinklers. It also provides several comparative situations for fires in similar structures—some of which are provided with sprinklers and some of which are not provided with sprinklers.

One of the most compelling topics discussed in the report is the life safety benefit that is derived from sprinkler systems. The following statements are taken directly from the 1995 report:

NFPA has no record of a fire killing more than two people in a completely sprinklered building where the system was properly operating, except in an explosion or flash fire or where industrial fire brigade members or employees were killed during fire suppression operations.

A further breakdown of this statement is directed at specific occupancies, as follows:

NFPA has no record of a fire killing more than two people in a completely sprinklered public assembly, educational, institutional, or residential building where the system was properly operating.

When single fatalities do occur in sprinklered buildings, the victims tend to be in close proximity to the fire, even intimate with ignition. In other cases, the victim may have been incapacitated or otherwise incapable of self-preservation.

The "Fire Watch" column in *NFPA Journal* usually contains one or two sprinkler stories. In addition, the information collection process through the National Fire Incident Reporting System (NFIRS), NFPA's Fire Incident Data Organization (FIDO), and insurance summaries all indicate that sprinklers are an effective means to reduce fire fatalities and damage.

Sprinkler protection mandates will continue to come in the form of building code regulations, insurance company requirements, and local and state laws. There is every reason to believe that sprinkler systems will continue to enjoy a high rate of success, as long as systems are properly designed, installed, and maintained.

ECONOMIES OF SPRINKLER SYSTEM INSTALLATION

A number of devastating fires have been investigated by NFPA over the years. In most of these fires, the lack of automatic sprinklers has been a key factor in explaining the accelerated growth and spread of many of these fires.

On a smaller scale, sprinklers are estimated to reduce the average loss per fire by one-half to two-thirds. These losses are tangible and can be measured in real dollars. Shortly after these losses are determined, the more challenging task of establishing the business interruption losses, the economic impact on the employees, and even an attempt to determine if the business can withstand a major fire must be completed.

Another element, i.e., pending lawsuits, is also a measurable value that often exceeds the other combined direct and indirect losses, especially when there has been a loss of life. These items, while typically not a consideration to most building owners or business owners, can devastate a company, the employees, and ultimately the community if the business cannot recover.

Automatic sprinkler systems have the capability to intercede at an early stage in the fire and minimize the impact of a fire. In addition, many systems are provided with a water-flow alarm, which will automatically notify the fire department about operation of the sprinkler system. This will allow for final extinguishment and overhaul by the fire department.

Unlike the TV portrayal of sprinkler systems, where all sprinklers in the building may appear to operate at once, only those sprinklers that are installed near the source of the fire will operate. As previously shown in Table 6-10A, a relatively small number of sprinklers tend to operate in most fire scenarios. In reality, there is considerably less water damage associated with sprinkler system activation than when sprinklers are not present.

Anxiety about water damage and inadvertent system discharge are real concerns of building owners. In reality, neither of these concerns have ever been a true problem. Reports in some media accounts of fires have focused on this element, thus reinforcing the public misperception. For instance, some news accounts have been selectively prepared, indicating that the fire damage was limited to, say, $1,000 while the collateral water damage from sprinklers was perhaps $1,500. No mention is made about the total value of the contents and the building; no mention is made about no one being injured; and no mention is made of the fatalities that did not occur.

Inadvertent discharge of water from system piping, valves, or from a sprinkler is rare but not unheard of. System components are carefully controlled by NFPA 13 to help ensure the integrity of the system. In addition, NFPA 13 also governs a number of methods and techniques that will help to protect against such problems. Quality control of installation practices and procedures is also an important matter. Special materials and methods are specified for use in corrosive atmospheres, areas subject to freezing, and for systems installed in areas requiring seismic protection.

There are those rare occasions when everything is done in accordance with NFPA 13, but some malfunction (resulting in an inadvertent discharge of water) still takes place. Factory Mutual Research Corporation (FMRC) has estimated that 1 in 16,000,000 installed sprinklers will fail to operate for no apparent reason. In addition, FMRC estimates that 1 in 2,500,000 installed systems will fault for no apparent reason and result in an unwanted discharge of water. These items are the exception and clearly not the rule, thus there are no extensive studies available on this topic.

Incentives for Sprinklers

Just as investing in real estate can provide one with an economic return, installing a sprinkler system will have additional benefits beyond what has been described. Property owners and managers begin to realize a savings on insurance as soon as a system is complete. Sprinkler system costs can be amortized over the life of the building, as can the annual savings associated with a reduced insurance premium.

Mandatory laws for installing sprinklers on a retroactive basis are frequently allowed to be done in phases over some specified time period. This permits building owners to coordinate the installation with some other building upgrade.

In addition, most building codes will permit certain construction modifications when systems are installed in new construction. Installing sprinklers in existing buildings can also offset other fire protection deficiencies. Increases to building heights and areas are permitted, reductions to hourly partitions may be permitted, and reductions in physical separation between adjacent structures may be permitted. All of these changes can quickly add up to offset the cost of the sprinkler system. Finally, there is the value of the life safety protection afforded by sprinklers and the security that sprinklers will provide to the building occupants. While difficult to put into dollar terms, this is the greatest tangible benefit of all.

STANDARDS FOR INSTALLING SPRINKLERS

NFPA 13 is the fundamental document that governs the design and installation criteria for these specialized fire protection systems. NFPA 13 is a standard, thus it provides the necessary requirements and guidance with respect to the specifics of "how" to design, lay out, and install a system. It does not tell when a system is needed—that is a function of NFPA 101 or a building code.

The scope of NFPA 13 notes that the document will provide the minimum requirements for the system, including water supply, types of materials, and components, and the design criteria associated with the system. The standard is organized in a manner that follows a logical approach, from the system concept through to the completion of the system acceptance. The outline of the standard is as follows:

Chapter 1. General Information
Chapter 2. System Components and Hardware
Chapter 3. System Requirements
Chapter 4. Installation Requirements
Chapter 5. Design Approaches
Chapter 6. Plans and Calculations
Chapter 7. Water Supplies
Chapter 8. System Acceptance
Chapter 9. Marine Applications
Chapter 10. System Maintenance

Each chapter contains specific pieces of information that are interconnected and, when combined, will result in an effective and functional system. The basis of NFPA 13 presumes a number of items in order to keep the levels of protection at an elevated state, yet keep the system costs reasonable. For instance, the standard presumes that only one fire will occur in a building at a given time—it does not assume multiple ignition sources. It does not attempt to establish the probability or likelihood of fire occurring in any given area—it assumes that an ignition could be present in any part of the structure. As a result of these assumptions, NFPA 13 presupposes that a limited number of sprinklers will operate.

The fundamental basis of the system design focuses on the contents of the building and the fire hazard associated with those contents. For example, a building that is of concrete construction is not given any special preference with respect to the selection of the system density when compared to a structure that is constructed of wood framing members. The use and, hence, the hazard of the contents of that building must be evaluated when embarking on a system design selection and approach. In other words, the contents of the building are expected to behave in the same manner no matter what type of construction materials are utilized.

DEFINITIONS AND DESCRIPTIONS

A number of variations exist that must be considered with respect to everything from the types of systems, the anticipated performance of the system, the types of sprinklers, and the specific design criteria that can be used. Some of the basic definitions are included in NFPA 13. Some of the more important definitions are paraphrased, as follows.

A *sprinkler system* is defined as a combination of underground and overhead piping that is installed throughout the building. The piping within the building is used to carry the water supply throughout the structure. The sprinklers are then connected to this part of the system. Sprinklers are usually activated by heat from the fire, and only those devices in the immediate vicinity of the fire will operate. Figure 6-10F shows an overview of the various elements that can be involved in a total system package.

System Types

While the definition of a sprinkler system is fairly basic, there are four basic types of sprinkler systems: (1) wet pipe, (2) dry pipe, (3) preaction, and (4) deluge. Some of these systems are more common than others, and the selection of any specific type is usually determined by the designer, the environment, or the end user.

Wet-pipe system: This system is the most common, easiest to design, and simplest to maintain. These systems contain water under pressure at all times and utilize a series of closed sprinklers. Once a fire occurs and produces enough heat to operate one or more sprinklers, the water will discharge immediately from any of the open sprinklers. Wet-pipe systems should be the first choice of designers and installers, and should only be used when the temperature of the protected area is maintained at or above 40°F (4°C).

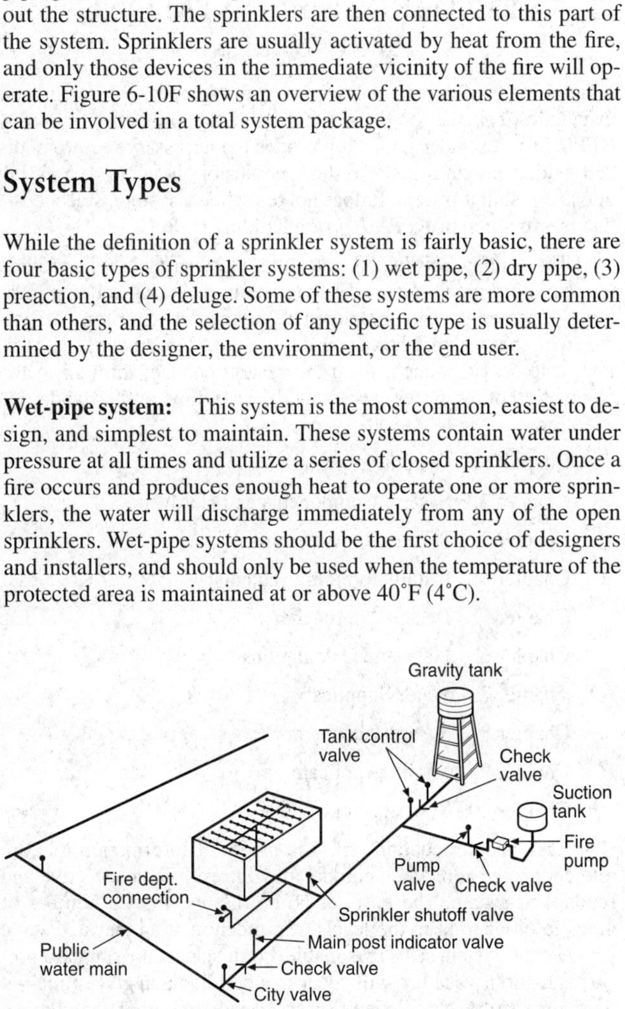

FIG. 6-10F. A typical sprinkler installation showing all common water supplies, outdoor hydrants, and underground piping.

This system is typically found in office buildings, stores, manufacturing facilities, hotels, and health care facilities. (See Figure 6-10G.)

Dry-pipe system: These systems are found in environments where the temperature is maintained below 40°F (4°C). The system piping contains air under pressure [40 psi (2.7 bar) maximum] under normal circumstances. A dry-pipe valve is used to hold back the water supply and to serve as the water/air interface. This valve acts on a pressure differential principle, the surface area of the valve face on the air side being greater than the surface area on the water side. Figure 6-10H shows this principle.

When a fire occurs and enough heat is generated, one or more sprinklers will operate, the system air pressure will then escape through the open sprinklers, drop to a predetermined level, and allow the dry pipe valve to open. Once the valve opens, the water supply will be admitted into the system piping, fill the pipe network, and water will discharge from any sprinklers that have operated.

These systems are more complex, require a reliable air supply source and involve specific design limitations such as the volume of pipe that can be governed by one dry-pipe valve, and special adjustments that are necessary for the anticipated area of operation.

Dry-pipe systems can be found in buildings that are not maintained at the 40°F (4°C) limit, such as outside canopies and structures, and cold-storage warehouses. See Figure 6-10I.

Preaction system: The piping for these systems is typically provided with some minimal quantity of air pressure, thus the pipe network has no water in it under normal circumstances. The water supply is held back by means of a preaction valve. The system is equipped with a supplemental detection system. Operation of the detection system allows the preaction valve to automatically open and admit water into the pipe network. Water will not discharge from the system until a fire has generated a sufficient quantity of heat to cause operation of one or more sprinklers. In essence, the system appears as a wet-pipe system once the preaction valve operates. [See Figure 6-10J (top).]

The small amount of air, which is maintained in the pipe, is used to monitor the integrity of the pipe. If the pipe develops a leak, the air pressure will drop and an alarm will sound indicating a low air-pressure condition exists within the pipe. The preaction valve stays in its normal position until the detection system is activated.

Another type of preaction system is commonly referred to as a double-interlock preaction system. This system has characteristics as previously described for preaction systems and characteristics of a dry-pipe system. In order to admit water into this type of system, the detection system must operate *and* the fire must generate a sufficient quantity of heat to cause operation of one or more sprinklers, thereby allowing a loss of pressure.

Preaction systems are typically found in environments that house computer equipment or communications equipment, museums, and other facilities where inadvertent water discharge is of major concern to the end user. The double-interlock system is most common in deep-freeze facilities where accidental valve operation may result in freezing of the pipe in a matter of minutes. [See Figure 6-10J (bottom).]

Deluge systems: Rapidly growing and spreading fires are most effectively protected with this type of system. Deluge systems, as the name implies, are intended to deliver large quantities of water over a large area in a relatively short period of time. The sprinklers that are used in a deluge system have their operating elements removed. These open sprinklers are attached to branch-line piping in the same manner as other types of sprinklers.

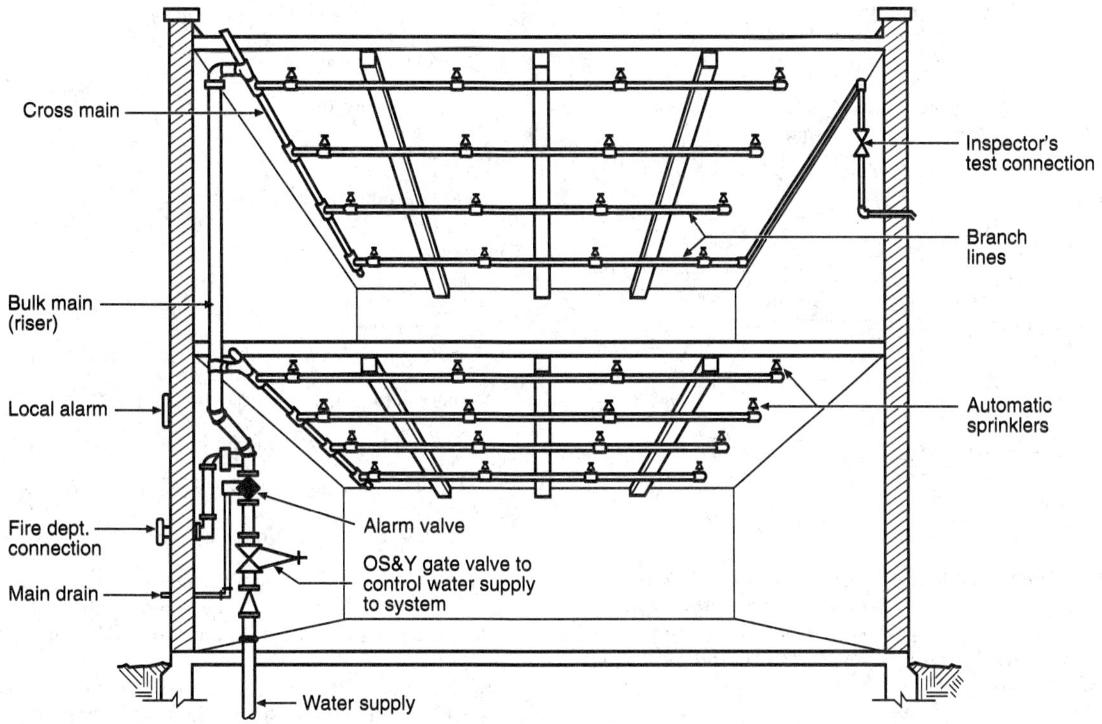

FIG. 6-10G. *A wet-pipe sprinkler system is under water pressure at all times so that water will be discharged immediately when an automatic sprinkler operates. The automatic alarm valve shown causes a warning signal to sound when water flows through the sprinkler piping.*

A deluge valve is used to control the system water supply. The sprinkler system pipe is at atmospheric pressure, since the open sprinklers are attached to it. The system water supply is maintained to the system side of the deluge valve. In a similar manner to the pre-action system, a supplemental detection system is provided throughout the same area as the sprinklers. Upon activation of this detection system the deluge valve is electrically opened, thereby admitting water into the pipe network. As the water reaches each sprinkler in the system, it immediately discharges from the open sprinkler.

The nature of this system makes it appropriate for facilities that contain combustible or flammable liquids. In addition, this system is used for situations in which thermal damage is likely to occur in a relatively short period of time. Deluge systems are required for use in most types of aircraft hangars. (See Figure 6-10K.)

Several variations to each one of these four basic systems may be found. Antifreeze systems are basically wet-pipe systems with a certain percentage of antifreeze concentrate added in to depress the freezing point. This type of system can be used to protect small areas, such as may be found at outside loading docks or exterior canopies. NFPA 13 specifies select types of antifreeze concentrate and percentages.

Design Objectives

Today's sprinkler design criteria allow the designer to select a system for fire control or fire suppression. Fire control is defined as limiting the size of the fire by decreasing the rate of heat release and pre-wetting adjacent combustibles. If fire suppression is desired, then a system selection could be made to accomplish this goal. Suppression of the fire is evidenced by a sharp reduction in the heat-release rate and application of large quantities of water through the

fire plume to the surface of the burning fuel. Figure 6-10L shows a simplified representation of these phenomena.

Beyond this, the designer is sometimes faced with the choice of using residential sprinklers. These devices are intended to maintain a tenable environment in the room of fire origin. The use of these devices is strictly limited to residential-type hazards. In the last 5 years, building codes and NFPA *101* have mandated the use of residential sprinklers or quick-response sprinklers in apartments, hotels, and patient care rooms in health care environments.

In order to achieve the design objective, NFPA 13 defines three fundamental hazard levels: (1) light hazard, (2) ordinary hazard, and (3) extra hazard. These are defined in a subjective manner and generally focus on the heat-release rate and the quantity of combustible fuel. The ordinary hazard and extra-hazard categories are both subdivided into Group 1 and Group 2 categories. Larger quantities of fuel are typically present in the respective Group 2 categories.

As previously stated, the properties of the contents in the building rather than the materials of construction of the building determine the classification. An extensive list of examples of various buildings is placed in the appendix of NFPA 13. Defining the hazard is thus the first step in selecting and designing the system.

Some examples of occupancies in each of these classifications is as follows:

Light hazard:

Office buildings

Schools

Residential occupancies

Public assembly

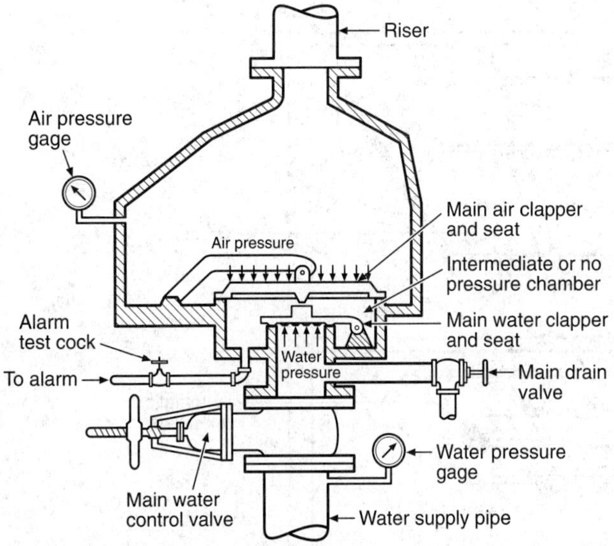

FIG. 6-10H. *The principle of a dry-pipe system is illustrated by this simplified drawing of a dry-pipe valve. Compressed air in the sprinkler system holds the dry valve closed, preventing water from entering the sprinkler piping until the air pressure has dropped below a predetermined point.*

Ordinary hazard:

Canneries (Group 1)

Electronic plants (Group 1)

Restaurant service areas (Group 1)

Dry cleaners (Group 2)

Libraries (Group 2)

Repair garages (Group 2)

Wood product assembly (Group 2)

Extra hazard:

Combustible fluid use area (Group 1)

Printing (inks with a flash point below 100°F (38°C) (Group 1)

Upholstering with plastic foams (Group 1)

Flammable liquids spraying (Group 2)

Manufactured home assembly (Group 2)

Plastics processing (Group 2)

Beyond this basic grouping, special circumstances must be considered for building uses that contain substantially larger quantities of materials than what is contemplated by these descriptions. A warehouse with 35 ft (10.7 m) rack storage is outside the protection features of NFPA 13. NFPA 231C, *Standard for Rack Storage of Materials,* must be used to determine an appropriate set of design criteria. The same is true of a flammable liquids storage facility. NFPA 30, *Flammable and Combustible Liquids Code*, must be used to properly design the sprinkler system. These other codes and standards defer to NFPA 13 for many of the specific procedures and general installation rules. The specific density and system areas of operation are regulated by these other documents.

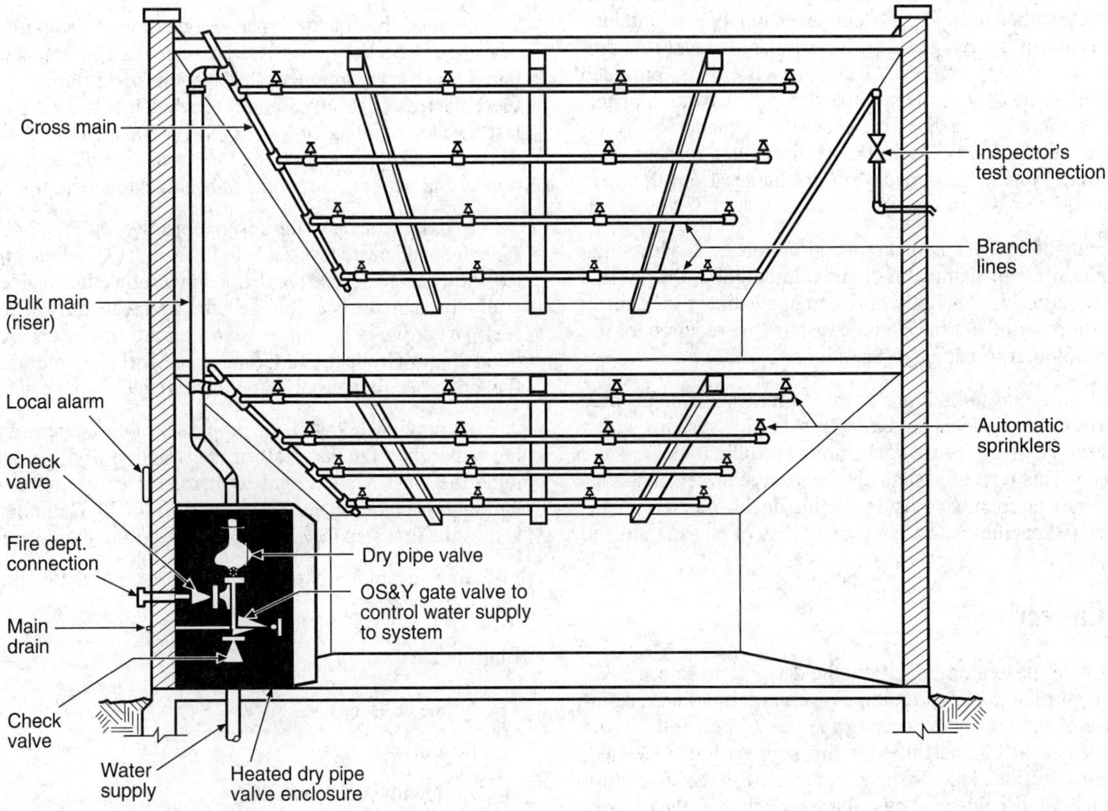

FIG. 6-10I. *Dry-pipe system for unheated properties. The system riser, cross mains, and branch lines are maintained at some air pressure.*

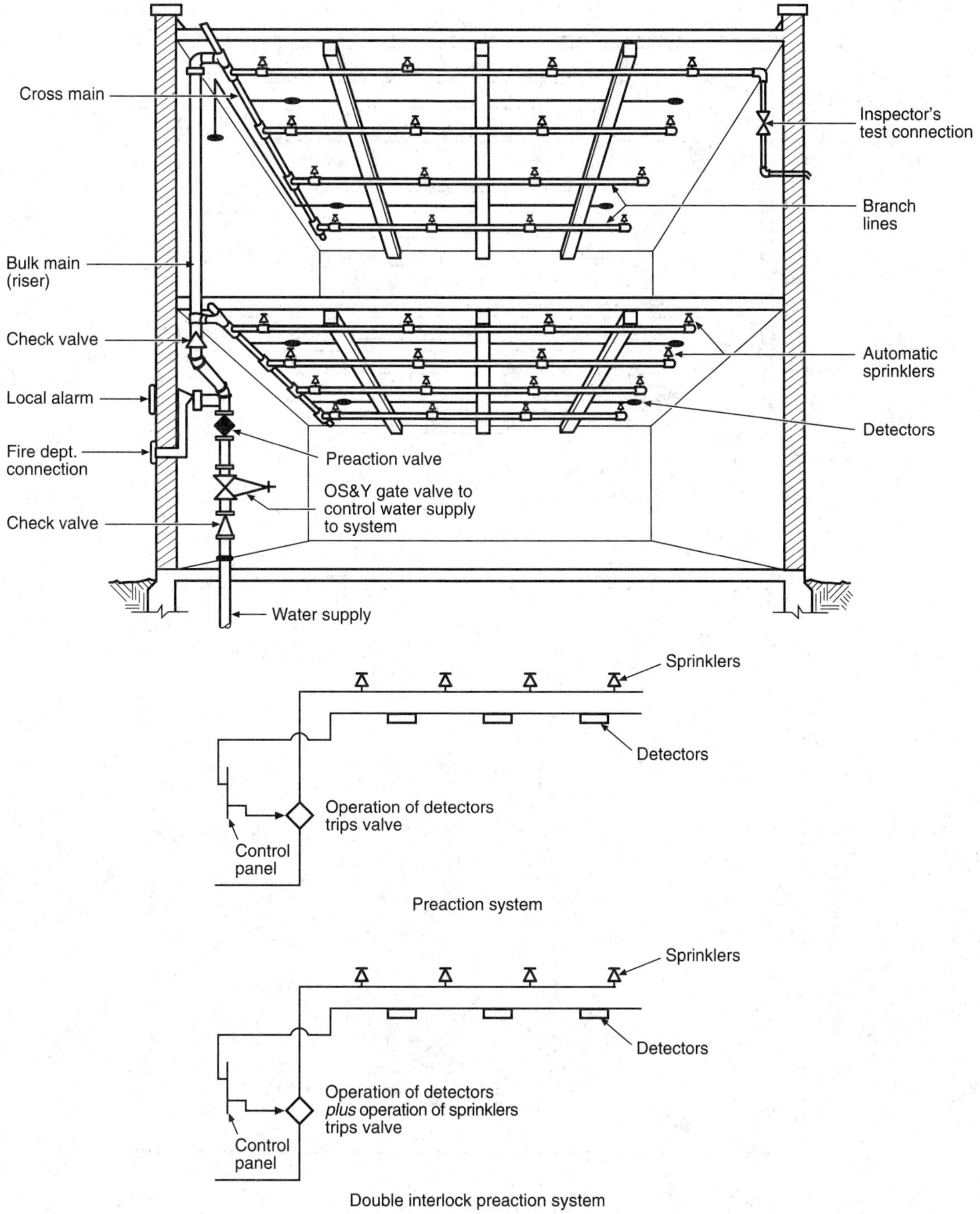

Cross main

Inspector's
test connection

Branch
lines

Bulk main
(riser)

Check valve

Automatic
sprinklers

Local alarm

Detectors

Fire dept.
connection

Preaction valve

OS&Y gate valve to
control water supply
to system

Check valve

Water supply

Sprinklers

Detectors

Operation of detectors
trips valve

Control
panel

Preaction system

Sprinklers

Detectors

Operation of detectors
plus operation of sprinklers
trips valve

Control
panel

Double interlock preaction system

FIG. 6-10J.   (Top) A typical preaction sprinkler system; (Bottom) comparison of operating features for the types of preaction systems.

Construction Types

Two basic types of construction are recognized by NFPA 13: (1) obstructed and (2) unobstructed. These are important to know when deciding on a number of installation features, including the permitted area of coverage of a sprinkler and the deflector distance below the ceiling. Identification of these features has been an integral part of NFPA 13 since the first edition was prepared in 1896. Special derivative or supplemental construction types are described in the appendix of NFPA 13.

Obstructed construction is described as being one in which "members impede heat flow or water distribution in a manner that materially affects the ability of the sprinklers to control or suppress a fire." In other words, the depth, spacing, and openness of the structural members will impact the flow of heat to the automatic sprinkler. In addition, once the sprinkler has operated, these same features may disrupt the spray pattern of the operating sprinkler.

Unobstructed construction is, as expected, the opposite of the previously discussed term. This definition describes conditions in

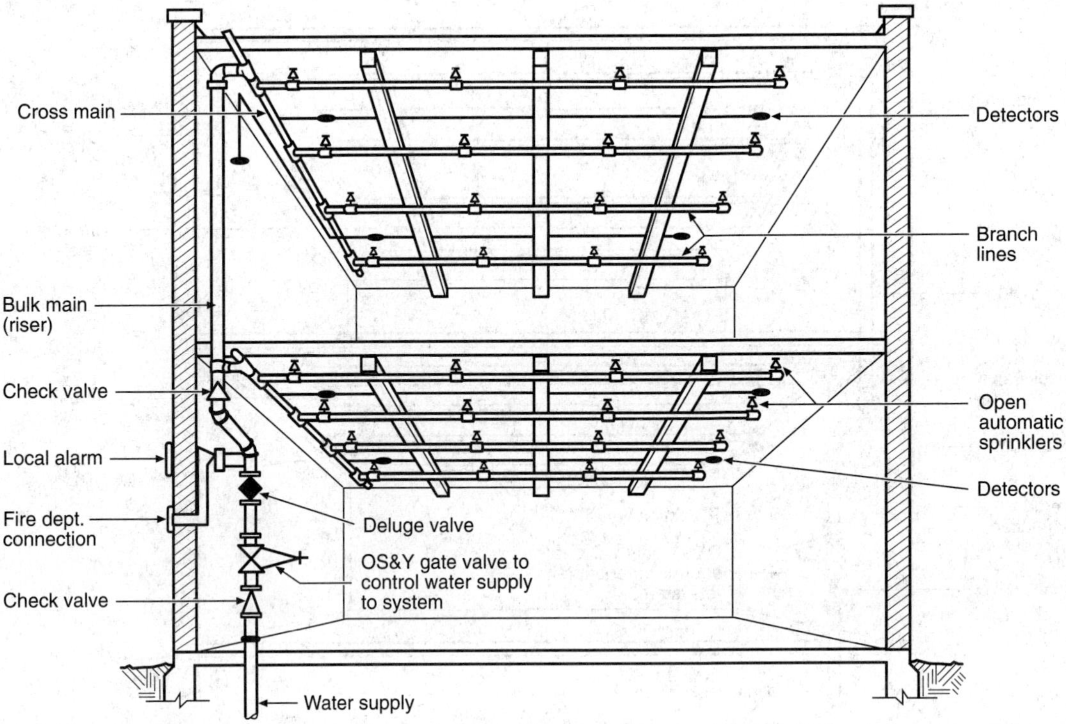

FIG. 6-10K.   Deluge system.

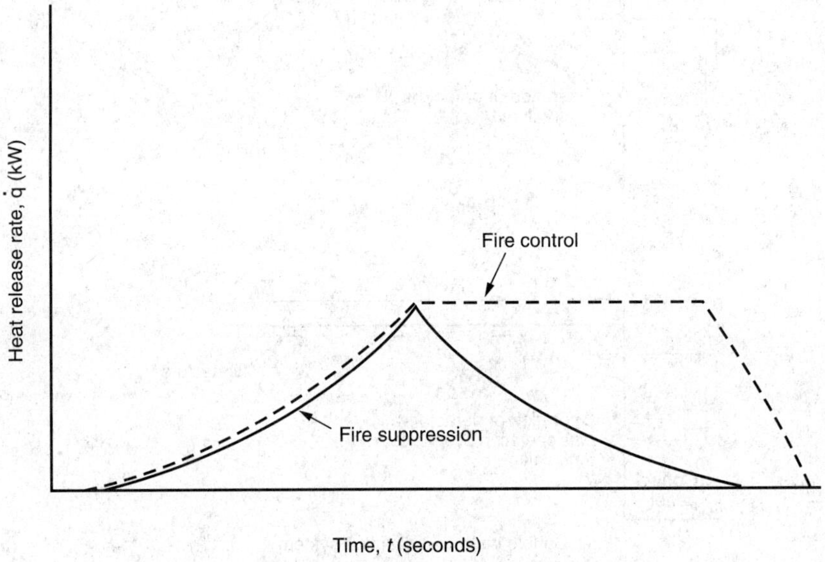

FIG. 6-10L.   Simplified fire control/fire suppression analogy.

which the structural members must be configured or spaced in order to fit into this category. These include:

1. Openings in the member must be at least 70 percent of the cross-sectional area.

2. Depth of the member does not exceed the least dimension of the openings.

3. Any case where the members are spaced more than 7 ft 6 in. (2.3 m) on center.

Figure 6-10M illustrates an example of these general construction definitions.

SYSTEM COMPONENTS AND MATERIALS

Quality control of automatic sprinkler systems is policed in several ways. One method to ensure the installation of a reliable system is to restrict the types of materials that are permitted for use. As a gen-

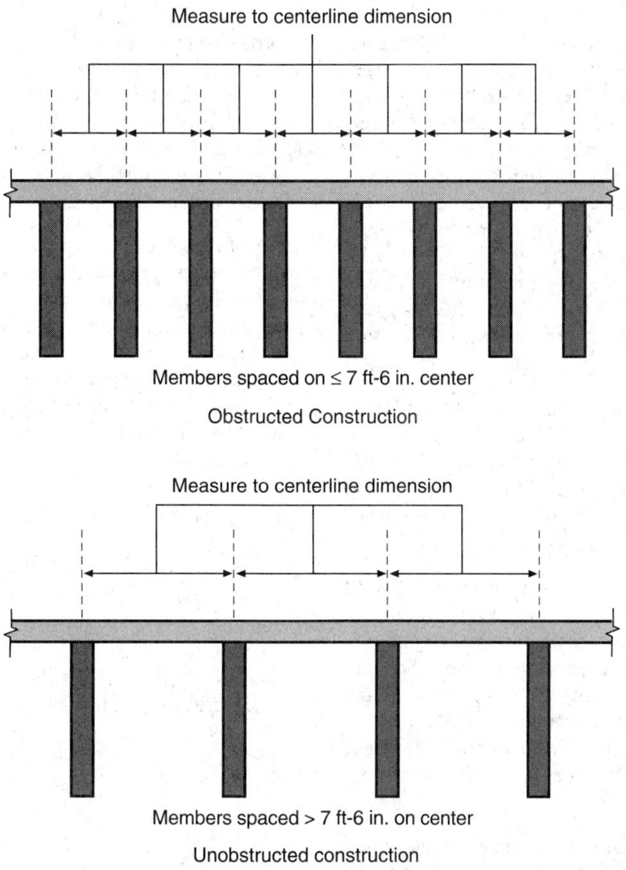

FIG. 6-10M. *(Top) Obstructed and (Bottom) unobstructed construction.*

eral rule, many materials and components that are deemed acceptable for use in sprinkler systems are required to be listed. This applies to sprinklers, valves, and pipe supports, as an example. A review of some of the acceptable materials is discussed in this section.

Sprinklers

Sprinklers must be listed, and only new sprinklers are permitted for use. Assurances must be made that the proper sprinkler has been se-

lected for the intended hazard. Sprinklers with certain orifice sizes are restricted for use in some systems. Table 6-10B catalogs sprinklers according to certain physical characteristics. One of these characteristics is the K-factor. Sprinklers with a "small" orifice are restricted for use in wet-pipe systems for light-hazard occupancies only. Additional restrictions include that the system must be hydraulically calculated and, for systems using sprinklers with an orifice of less than $3/8$ in. (9.5 mm), that a listed strainer be installed on the system.

TABLE 6-10B. *Sprinkler Discharge Characteristics Identification*

Nominal Orifice Size			Percent of Nominal $1/2$ in.			Nominal Orifice Size Marked
in.	mm	K Factor[1]	Discharge	Thread Type	Pintle	on Frame
$1/4$	6.4	1.3–1.5	25	$1/2$ in. NPT	Yes	Yes
$5/16$	8.0	1.8–2.0	33.3	$1/2$ in. NPT	Yes	Yes
$3/8$	9.5	2.6–2.9	50	$1/2$ in. NPT	Yes	Yes
$7/16$	11.0	4.0–4.4	75	$1/2$ in. NPT	Yes	Yes
$1/2$	12.7	5.3–5.8	00	$1/2$ in. NPT	No	No
$17/32$	13.5	7.4–8.2	140	$3/4$ in. NPT or $1/2$ in. NPT	No / Yes	No / Yes
$5/8$	15.9	11.0–11.5	200	$1/2$ in. NPT or $3/4$ in. NPT	Yes / Yes	Yes / Yes
$3/4$	19.0	13.5–14.5	250	$3/4$ in. NPT	Yes	Yes

[1]K factor is the constant in formula $Q = K\sqrt{p}$ where Q = flow in gpm, and p = pressure in psi.

For SI units: $Q_m = \sqrt{P_m}$ where Q_m = flow in L/min, and P_m = pressure in bars and $K_m = 14 K$.

In general, sprinklers have been compared to the $^1/_2$-in. (12.7-mm) or standard orifice sprinkler, thus Table 6-10B recognizes sprinklers with orifices that are either smaller than, or larger than, the $^1/_2$-in. (12.7 mm) orifice sprinkler. Sprinklers must also satisfy other criteria in this table, such as the allowable thread type, presence of a pintle, and a requirement to mark the orifice size on the frame. Manufacturers and the laboratories that list sprinklers are bound to the limits of the characteristics that are noted in Table 6-10B. As an example, a sprinkler with a nominal orifice size of $^3/_4$ in. (19.1 mm) (very extra large) could not be listed with a $^1/_2$-in. (12.7-mm) NPT thread. Restrictions on thread type and diameters are imposed on this type of sprinkler, as a recognition of the poor hydraulic features associated with attempting to move large quantities of water through a smaller opening.

Large-drop and ESFR sprinklers are also identified in Table 6-10B. These sprinklers are not really identified as being a particular "orifice type," but they do have certain restrictions imposed on them. It has been proposed for the 1996 edition of NFPA 13 to remove these devices from the table, since they are really a type of extra-large or very-extra-large orifice sprinkler.

Another feature of the sprinkler that is governed by NFPA 13 is the temperature rating. Table 6-10C shows the categories that must be considered. The first item states the maximum ambient temperature to which the sprinkler is expected to be exposed. The majority of buildings would fall into the ordinary-temperature rating classification. Sprinklers with higher temperature ratings may be necessary in buildings that are not provided with air conditioning or in those buildings that may be used for some type of manufacturing process involving a heat-producing process.

While these general rules are appropriate, there are additional situations where higher temperature ratings are necessary. The installation rules in NFPA 13 govern these situations and include locations such as where sprinklers would be installed under skylights in a manner where the sprinkler would be directly exposed to the rays of the sun. Other localized conditions must be evaluated on a case-by-case basis to determine if a higher temperature rating is in order.

As noted in Table 6-10C, sprinklers are color coded to make identification of the temperature classification easier from a distance. Eutectic metal or solder-type sprinklers use the color scheme in the next to last column, while glass-bulb sprinklers use the color scheme in the last column. The liquid in a glass-bulb sprinkler is basically a low-boiling-point alcohol. An air bubble within the glass bulb is volume controlled to achieve a certain temperature rating. The larger the bubble, the higher the operating temperature. The liquid is then dyed the appropriate color to equate with the color code system in the table. Selection of the correct temperature rating for the environment is critical to ensure that sprinklers will not prematurely operate just due to exposure to excess localized heat from some benign source.

While sprinklers serve an important purpose in the overall fire protection plan for a structure, they can be manufactured with special coatings which are either functional or aesthetic. Sprinklers that are intended to be installed in corrosive atmospheres can be provided with a lead or wax coating to protect the sprinkler from corrosive effects. Some types of sprinklers may also be provided with special finish coatings to allow them to blend into architectural features, such as ceiling and wall finish materials or woodwork.

In both the case of listed corrosion-resistant sprinklers and sprinklers with special decorative finishes, such finishes may only be applied by the manufacturer. Under no circumstance can sprinklers be "field coated" with any type of material. Sprinklers that have been coated or painted must be replaced. Such coatings may impair the spray pattern of the sprinkler or they may react with some functional part of the sprinkler assembly, such as the operating element or the Teflon® seal, which is used to seat the orifice cap on the orifice.

Pipe and Tube

A number of materials are deemed acceptable for use in sprinkler systems. Steel, copper, and nonmetallic pipe materials are all acceptable for use in sprinkler systems. These pipe materials must meet certain pipe manufacturing standards, certain listing requirements, or both. Limiting the types of pipe that are acceptable for use will increase reliability of not only the system installation, but it will also ensure the integrity of the system piping during a fire event. The methods and materials used to join the pipe are also controlled by NFPA 13.

Steel pipe: Steel pipe has been a viable option for use in sprinkler systems since the first edition of NFPA 13, published in 1896. A number of American Society for Testing and Materials (ASTM) standards are listed in NFPA 13 for use in sprinkler systems. Steel pipe that is identified as being schedule 10 or greater need only comply with the pipe standard which appears in the table. Table 6-10D provides a list of acceptable steel piping materials that satisfy this basic criteria. Steel pipe can be used in any environment within the confines of NFPA 13.

Copper tube: Copper tube was first discussed as an optional material for sprinkler systems in 1954. Concern over the failure of solder materials or the brazing materials used to join the copper tube was expressed as being a potential failure point of the system during a fire. Joint research by the listing laboratories and the Copper Development Association addressed those concerns, and recommendations were made to restrict the types of brazing material and solder materials. Table 6-10E indicates the relevant standards for both copper tube and the joining materials. Copper tube systems have been installed in a large number of buildings. Some notable buildings with copper systems are the Trans-America building in San Francisco, the Monterey Aquarium, and the Library of Congress.

TABLE 6-10C. *Temperature Ratings, Classifications, and Color Codings*

Max. Ceiling Temp.		Temperature Rating		Temperature Classification	Color Code	Glass Bulb Colors
°F	°C	°F	°C			
100	38	135 to 170	57 to 77	Ordinary	Uncolored or black	Orange or red
150	66	175 to 225	79 to 107	Intermediate	White	Yellow or green
225	107	250 to 300	121 to 149	High	Blue	Blue
300	149	325 to 375	163 to 191	Extra high	Red	Purple
375	191	400 to 475	204 to 246	Very extra high	Green	Black
475	246	500 to 575	260 to 302	Ultra high	Orange	Black
625	329	650	343	Ultra high	Orange	Black

TABLE 6-10D. *Pipe or Tube Materials and Dimensions*

Materials and Dimensions	Standard
Ferrous Piping (Welded and Seamless)	
Specification for Black and Hot-Dipped Zinc Coated (Galvanized) Welded and Seamless Steel Pipe for Fire Protection Use	ASTM A795
Specification for Welded and Seamless Steel Pipe	ANSI/ ASTM A53
Wrought Steel Pipe	ANSI B36.10M
Specification for Elec.-Resistance Welded Steel Pipe	ASTM A135

TABLE 6-10E. *Pipe or Tube Materials and Dimensions*

Materials and Dimensions	Standard
Copper Tube (Drawn, Seamless)	
Specification for Seamless Copper Tube	ASTM B75
Specification for Seamless Copper Water Tube	ASTM B88
Specification for General Requirements for Wrought Seamless Copper and Copper-Alloy Tube	ASTM B251
Fluxes for Soldering Applications of Copper and Copper Alloy Tube	ASTM B813
Brazing Filler Metal (Classification BCuP-3 or BCuP-4)	AWS A5.8
Solder Metal, 95-5 (Tin-Antimony-Grade 95TA)	ASTM B32
Alloy Materials	ASTM B446
	ASTM B467

TABLE 6-10F. *Specially Listed Pipe or Tube Materials and Dimensions*

Materials and Dimensions	Standard
Nonmetallic Piping	
Specification for Special Listed Chlorinated Polyvinyl Chloride (CPVC) Pipe	ASTM F442
Specification for Special Listed Polybutylene (PB) Pipe	ASTM D3309

Nonmetallic pipe: Two types of plastic pipe currently exist on the U.S. market for use in automatic sprinkler systems. Polybutylene (PB) pipe was first listed for use in systems in NFPA 13D, *Standard for the Installation of Sprinkler Systems in One- and Two-Family Dwellings and Manufactured Homes* (hereinafter referred to as NFPA 13D), in 1984. An obvious concern with plastic pipe is its failure when exposed to a fire, and even its possible contribution to fire growth. This concern was addressed by requiring the pipe to be installed behind a protective membrane. Such a protective barrier would shield the pipe from direct heat exposure. The exposing temperature would be expected to be substantially decreased, since the operating sprinkler would be controlling the fire.

A second type of nonmetallic pipe was introduced in 1986, chlorinated polyvinyl chloride (CPVC). CPVC pipe was listed for use in systems complying with NFPA 13D and for systems complying with NFPA 13 for light-hazard occupancies. Soon thereafter, PB pipe also had its listing expanded for the identical use.

Nonmetallic piping materials must be manufactured in accordance with the piping standards shown in Table 6-10F *and* they must be specifically listed for use in automatic sprinkler systems. The listing criteria includes a number of conditions and restrictions:

1. Limited for use in systems that comply with NFPA 13D; NFPA 13R, *Standard for the Installation of Sprinkler Systems in Residential Occupancies up to and Including Four Stories in Height* (hereinafter referred to as NFPA 13R); and NFPA 13 (light-hazard occupancies only).
2. Limited for use in wet-pipe systems only.
3. Must be installed behind a thermal barrier. [Note: CPVC pipe may be installed exposed if used with listed residential sprinklers that are installed in accordance with their listing or if used with listed quick response (QR), where the QR sprinklers are installed within 8 in. (203 mm) of the ceiling.]
4. Must be joined with listed fittings or materials. (Example: PB may be joined with listed heat-fusion tool. CPVC may be joined with listed two-step or one-step solvent cement system.)
5. Not permitted to be installed in concealed combustible spaces that require sprinkler protection by NFPA 13, NFPA 13D, or NFPA 13R.

NFPA 13 requires that all of the manufacturer's installation instructions be followed, in addition to the design and installation rules of NFPA 13.

Specially listed steel pipe: In the same manner that NFPA 13 previously allowed manufacturers to produce pipe with alternative materials besides steel and copper, that identical provision also permits the use of other types of steel pipe that have a schedule thickness of less than schedule 10. These materials are typically referred to as thin-wall steel pipe, and may have wall thicknesses that replicate schedule 5. In most cases, these pipe materials must still be produced in accordance with one of the standards identified in Table 6-10D and they must be specifically listed for use in automatic sprinkler systems.

The listing evaluation includes an analysis of items, such as the hydraulic characteristics, corrosion resistance properties, and beam strength (for determining hanger spacing for piping support), and methods of joining the pipe.

Pipe Hangers

The philosophy for supporting system piping is basically unchanged in the last 100 years, i.e., prevent the pipe from coming down during normal conditions and during times when the system is operating. This is accomplished with the use of components that are specified in NFPA 13 for dimensional criteria and that are used as a part of a listed assembly of hanger components.

Listed hangers are evaluated to prove that they can support five times the weight of water-filled pipe plus 250 lb (113 kg). This becomes the basis for manufacturers to design and evaluate their product. This also forms the basis for a system designer to evaluate alternative hanger materials or shapes for unusual conditions or situations. The fastener tables in Chapter 2 of NFPA 13 consider this criteria as well.

Figure 6-10N shows typical hangers for use in sprinkler systems. The type of hanger used is usually a function of the type of construction (wood, steel, concrete), the spacing of the structural members, and the ability of the structural system to support the added load of the system pipe. With respect to this last point, the building structure must be capable of supporting the weight of the water-filled pipe plus 250 lb (113 kg). This is similar to the hanger criteria, but instead of a five-time rule, it is a one-time rule.

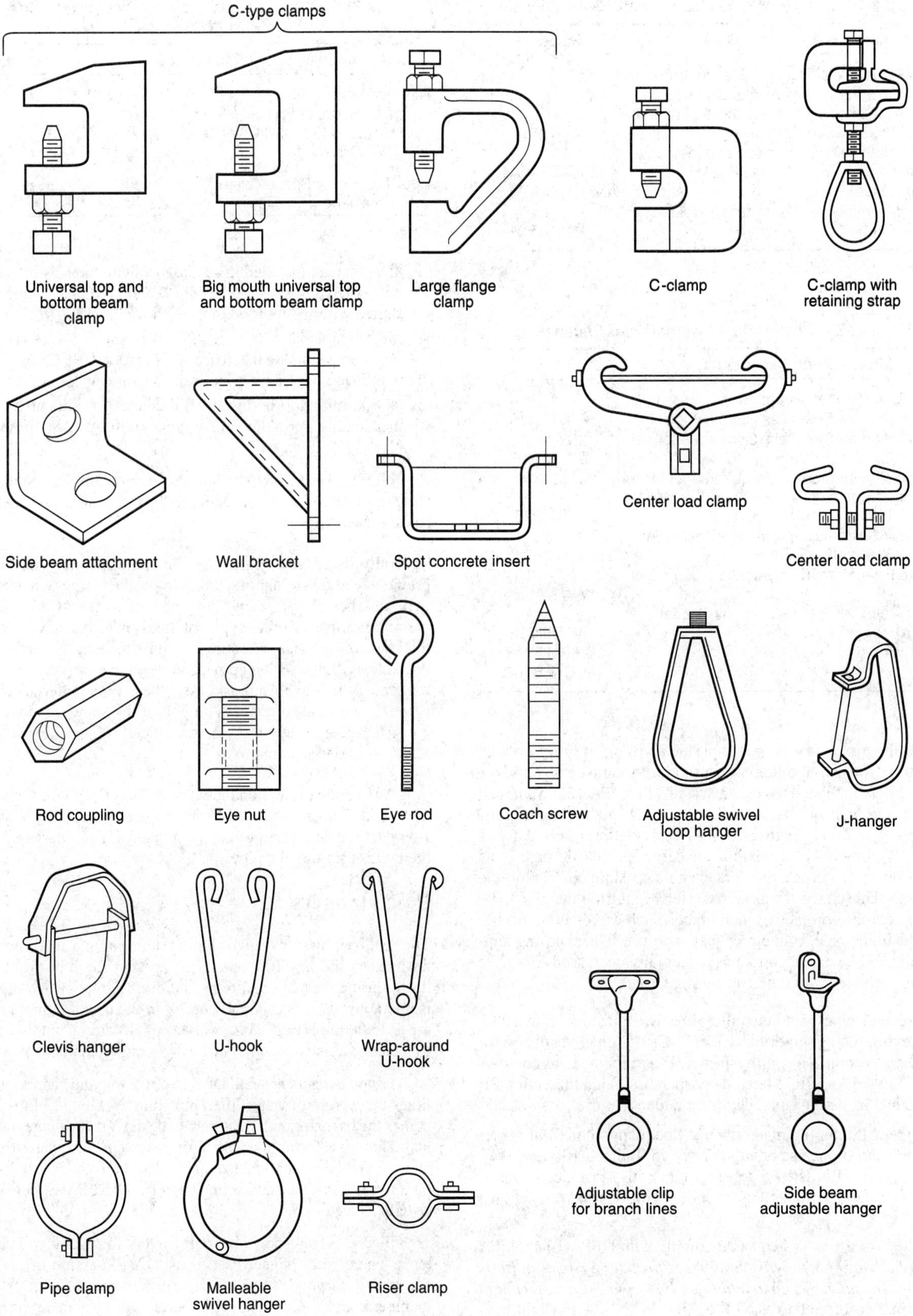

C-type clamps

Universal top and bottom beam clamp

Big mouth universal top and bottom beam clamp

Large flange clamp

C-clamp

C-clamp with retaining strap

Side beam attachment

Wall bracket

Spot concrete insert

Center load clamp

Center load clamp

Rod coupling

Eye nut

Eye rod

Coach screw

Adjustable swivel loop hanger

J-hanger

Clevis hanger

U-hook

Wrap-around U-hook

Adjustable clip for branch lines

Side beam adjustable hanger

Pipe clamp

Malleable swivel hanger

Riser clamp

FIG. 6-10N. *Hanger types.*

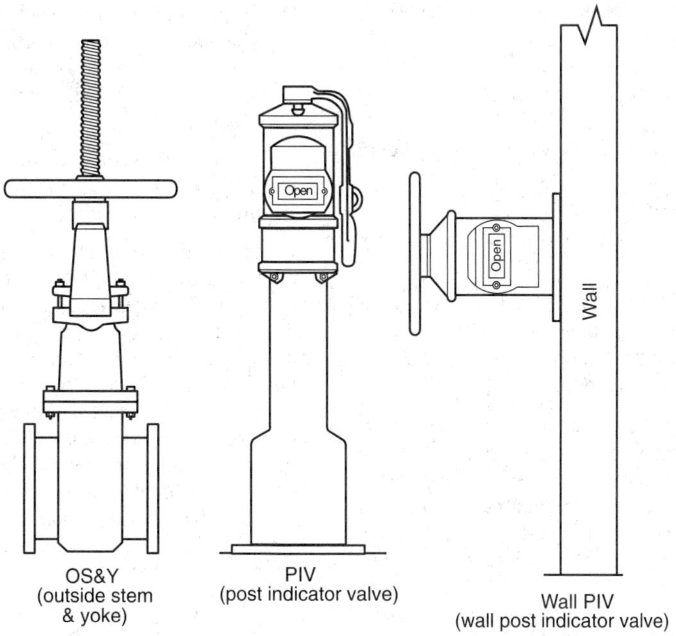

FIG. 6-10O. *Sprinkler system indicating control valves.*

The type of structural system and the diameter of pipe also drive the type and number of required fasteners. For example, a U-hook installed in conjunction with a solid wood joist is permitted to utilize drive screws, as long as the pipe diameter is not over 2 in. (51 mm). U-hooks installed with pipe sizes over 2 in. (51 mm) in diameter must use lag screws. In both cases, NFPA 13 controls the type of fastener, the length of the fastener, and the diameter of the fastener.

System Valves

Automatic sprinkler systems are required to have at least one valve installed to allow for the system to be shut down. As a matter of course, sprinkler systems should never be shut down except when system modifications are being accomplished or during the time following a fire to allow for replacement of any sprinklers that operated. The types of valves used in a sprinkler system vary with the location of the valve in the water supply system, or portion thereof.

Valves may be of the isolation type to shut the system off, they may be a one-way or check valve to prevent flow of water in one direction, or they may be a pressure-regulating valve to permit the pressure entering part of the system to be maintained at a certain level. In all cases, the valves must be listed for use in fire protection systems, they must be supervised in their normally open position, and they must be identified as to the area or segment of the system that they control.

Listed control valves are evaluated for characteristics, such as the maximum working pressure to which they can be exposed, the hydraulic profile or equivalent pipe lengths that result, and the means of supervising equipment that they may use.

Indicating features: The purpose of an indicating feature on a valve is to allow a method to quickly determine if the valve is in the open or closed position. One of the more common methods for accomplishing this is the use of a listed outside stem and yoke (OS&Y)

valve. The rising stem feature of this valve is the indicator if the valve is open or closed. The stem rises as the valve is set to the open position. The opposite end of the stem is equipped with a gate valve that raises out of the pipe network as the valve wheel is operated. OS&Y valves are oftentimes installed at, or near, the system riser.

A post-indicator valve (PIV) is frequently installed outside of the structure. The indicating feature on this device is a viewing window with a sliding metal plate. As the valve is operated with the special wrench, the stem, which is connected to the gate device on the end of the stem (the stem is internal to the post on a PIV), also repositions the sign. When the valve is in the open position, the sign labeled "OPEN" will be evident in the small viewing window on the side of the PIV housing. Likewise, when the valve is operated to the closed position, the sign labeled "CLOSED" will be evident in the viewing window.

Listed indicating butterfly valves (IBV) do not have the same type of physical features of either the OS&Y valve or the PIV valve, thus it must use an indicating feature that is substantially different from either of the first two valves. The indicating feature with an IBV is usually a directional arrow, mounted on the housing of the valve. This feature points in the "position" of the valve. For example, when the valve is open, the arrow is aligned with the flow direction of the water. When the valve is in the closed position, the arrow points in a direction perpendicular to the pipe, indicating that water cannot travel beyond the valve. Figure 6-10O shows commonly used control valves.

Supervisory features: Another requirement of NFPA 13 is to provide some means to ensure that the valve remains in its normal position. Supervision of system valves dates back to the 1896 edition of NFPA 13, in which Rule 60 stated, "All gate valves in supply pipes to sprinklers to be of indicator pattern, and to be strapped open with riveted leather straps passing around the riser and spoke of the wheel, or secured open with a seal." This basic method remains as one of four options in 1996. The methods for supervising valves include:

1. Electrically supervised in a manner to allow a signal to be heard at a central station facility.
2. Electrically supervised in a manner to allow a signal to be heard at a constantly attended point at the protected facility.
3. Locking valves in the open position.
4. Sealing valves in the open position, with a weekly surveillance program.

It is important to remember that, while NFPA 13 allows any of these options, other codes and standards may require a specific type of valve supervision, depending on the type of building or the type of building fire alarm system that is provided.

In addition to those valves that are used to control the system supplies or a portion of the system supply, other types of valves are used to test the system or drain the system pipe. While such valves are not required to meet the same level of reliability as control valves, they must still be installed to facilitate testing of the system, as well as to allow for the system to be drained during a period of maintenance. The main drain or test valve is located at the system riser. Its purpose is two-fold. First, it may be used as a central point to drain water from the entire system. Its second purpose is to conduct the main drain test. This test is done when the system is commissioned, in order to allow for a set of baseline values to be established. The points of interest are: (1) the static pressure and (2) the residual pressure. Figure 6-10P shows the common arrangement of the main drain valve.

Another type of valve used is the inspector's test valve, or the alarm test connection. This valve is usually installed at some point out in the system on the end of a branch line. When installed on a wet-pipe system, the main purpose of this device is to test the operation of the water-flow device. When installed on a dry-pipe system, its purpose is two-fold. First, it is used to establish the trip time of the dry-pipe valve and the transit time of the water to the connection. Second, it establishes the time for the water-flow device to operate and the actual operation of the alarm device.

INSTALLATION RULES FOR AUTOMATIC SPRINKLER SYSTEMS

The proper installation rules for automatic sprinkler systems are summed up in three basic rules:

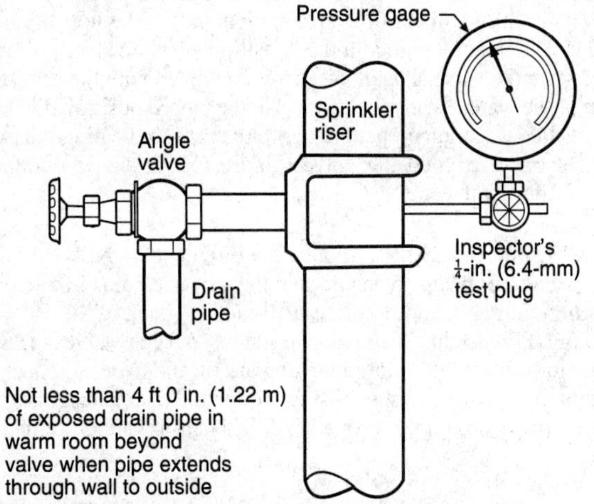

FIG. 6-10P. Drain connection for system riser.

1. Sprinklers are to be installed throughout the premises.
2. The maximum allowed protection area per sprinkler is not to be exceeded.
3. Sprinklers are positioned to allow for timely operation and distribution.

When viewed individually, each item is a stand-alone issue that requires a number of details to be completed in order to achieve the requirements for a proper installation.

Installing sprinklers throughout the premises is necessary in order to allow for a high level of protection to be provided. The rules in NFPA 13 are intended to achieve a high level of property protection as one of its goals; thus, this first rule is necessary to ensure that this will be satisfied. Codes which may have rules that would not require sprinklers in certain areas of the building (small closets, bathrooms) recognize that the threat to life may be minimal from fires that originate in such spaces. NFPA 13 recognizes a similar concept in some of these areas, such that fire in these areas does not often constitute a threat to the structure. Due to the scope of the rules contained in NFPA 13, property conservation is of paramount importance, thus very few exceptions to the placement of sprinklers can be found in NFPA 13.

The areas of coverage afforded by each sprinkler is predicated on the basis that sprinklers are evaluated on their performance to cover relatively small areas [100 sq ft (9.3 m^2) in the case of an ESFR sprinkler], as well as to cover larger areas [400 sq ft (37.2 m^2) in the case of an extended coverage sprinkler]. The limit for which a sprinkler has been listed for use must be strictly adhered to. Attempts to exceed the value by which a sprinkler is listed may result in the failure of the sprinkler's ability to properly control the fire. Strict limits are imposed on the sprinkler to prevent misuse.

The final point in the installation philosophy is to position the sprinkler in such a manner as to minimize the impact that various types of obstructions can have on the discharge pattern of the sprinkler. This third provision can be the most difficult to manage. A number of rules have been developed to allow for better control of the operating time of the sprinkler and to minimize the impact of obstructions that may be in the same areas as the sprinklers. A discussion on each one of these tenets follows.

Installing sprinklers throughout the premises: Since there is no method to predict where a fire is likely to start in a structure, the basic rules of NFPA 13 require sprinklers to be installed in most areas of the building. In general, a fire that originates in an unsprinklered area is expected to overpower the capabilities of a system, which may be installed in a nearby or adjacent space. There have been instances where sprinklers did actually contain or extinguish a fire that originated in an unsprinklered space. The most notable of these fires was the One Meridian Plaza fire in Philadelphia in 1991. In this incident, an uncontrolled fire spread between eight stories in a high-rise office building. Upon reaching the 30th floor, which had been equipped with sprinklers, a total of nine sprinklers operated to completely stop the further spread of the fire.

There are several exceptions within NFPA 13 that allow sprinklers to be omitted from certain areas of the building. Such spaces include bathrooms in residential-type occupancies that are not over 55 sq ft (5.1 m^2) in area and are composed of noncombustible or limited-combustible materials. Small closets in hotel occupancies are also not required to be protected with sprinklers. In both instances, it is posited that the fuel loading in such spaces is minimal and that fires which originate in such spaces do not normally become threatening to either the occupants or the structure.

Two recently added exceptions to sprinkler protection center on electrical equipment rooms and elevator shafts. Electrical equipment rooms that (1) use dry-type equipment, (2) are used for no other purpose, and (3) are constructed of 2-hr construction are not required to be protected with automatic sprinklers. In this instance, strict controls are imposed on the use of the room and the materials of construction.

Passenger elevator shafts that use noncombustible construction and utilize elevator cabs constructed in accordance with ASME A17.1, *Safety Code for Elevators and Escalators*, are not required to be provided with a sprinkler at the top of the shaft. Sprinklers are not required at the bottom of similar shafts, unless the elevator lift mechanism utilizes hydraulic fluid. It is important to note that such exceptions do not extend to the elevator equipment or machine rooms.

Concealed spaces, such as those found above a ceiling within the building, do not require sprinklers, provided the space is constructed of noncombustible materials. Conversely, concealed spaces that are formed by combustible construction or materials are required to be protected with sprinklers. There are a number of exceptions to this general rule, however. Exceptions to sprinkler protection center on two basic ideas: (1) that physical limitations within a space may prevent a meaningful arrangement of sprinklers to be installed, and (2) that the nature and arrangement of the combustible materials may have some inherent fire resistance. The 1996 edition of NFPA 13 contains ten exceptions that address the specific instances in which sprinklers can be omitted from combustible spaces.

Areas of Coverage

A number of limits are imposed on the size of the area that a particular type of sprinkler can protect. In arriving at what area of coverage is permitted, a number of variables are considered. These include: the type of sprinkler being considered, the nature of the hazard being protected, the type of construction contemplated, and the spacing of the construction members. Tables 6-10G through 6-10L

TABLE 6-10G. *Protection Areas and Maximum Spacing (SSU/SSP)*

Construction Type	Light Hazard Protection Area ft²	Light Hazard Spacing (max.) ft	Ordinary Hazard Protection Area, ft²	Ordinary Hazard Spacing (max.), ft	Extra Hazard Protection Area, ft²	Extra Hazard Spacing (max.), ft	High-Piled Storage Protection Area, ft²	High-Piled Storage Spacing (max.), ft
Noncombustible obstructed and unobstructed and combustible unobstructed	225	15	130	15	100	12	100	12
Combustible obstructed	168	15	130	15	100	12	100	12

For SI units: 1 ft² = 0.0929 m²; 1 ft = 0.305 m.

TABLE 6-10H. *Protection Areas and Maximum Spacing (Standard Sidewall Spray Sprinkler)*

	Light Hazard Combustible Finish	Light Hazard Noncombustible or Limited-Combustible Finish	Ordinary Hazard Combustible Finish	Ordinary Hazard Noncombustible or Limited-Combustible Finish
Maximum distance along the wall (S)	14 ft	14 ft	10 ft	10 ft
Maximum room width (L)	12 ft	14 ft	10 ft	10 ft
Maximum protection area	120 ft²	196 ft²	80 ft²	100 ft²

For SI units: 1 ft = 0.305 m; 1 sq ft = 0.0929 m².

TABLE 6-10I. *Protection Areas and Maximum Spacing (Extended Coverage Upright and Pendent Spray Sprinklers)*

Construction Type	Light Hazard Protection Area (ft²)	Light Hazard Spacing (ft)	Ordinary Hazard Protection Area (ft²)	Ordinary Hazard Spacing (ft)	Extra Hazard Protection Area (ft²)	Extra Hazard Spacing (ft)	High Pile Storage Protection Area (ft²)	High Pile Storage Spacing (ft)
Unobstructed	400	20	400	20	196	14	196	14
	324	18	324	18	144	12	144	12
	256	16	256	16				
			196	14				
			144	12				
Obstructed noncombustible (when specifically listed for such use)	400	20	400	20	196	14	196	14
	324	18	324	18	144	12	144	12
	256	16	256	16				
			196	14				
			144	12				
Obstructed combustible	NA	NA	NA	NA	NA	NA	NA	NA

For SI units: 1 ft² = 0.0929 m²; 1 ft = 0.305 m.
NA, not applicable

TABLE 6-10J. *Protection Area and Maximum Spacing for Extended Coverage Sidewall Sprinklers*

Construction Type	Light Hazard Protection Area (ft²)	Light Hazard Spacing (ft)	Ordinary Hazard Protection Area (ft²)	Ordinary Hazard Spacing (ft)
Unobstructed, smooth, flat	400	28	400	24

TABLE 6-10K. *Protection Areas and Maximum Spacing for Large-Drop Sprinklers*

Construction Type	Protection Area (ft²)	Maximum Spacing (ft)
Noncombustible unobstructed	130	12
Noncombustible obstructed	130	12
Combustible unobstructed	130	12
Combustible obstructed	100	10

For SI units: 1 ft = 0.305 m; 1 ft² = 0.0929 m².

give a summary of the maximum areas of coverage that are permitted for different types of sprinklers, hazards, and building construction features.

The area of coverage of a sprinkler is established through a set of principles that are based upon an evaluation of the relationship between the distance between sprinklers located on the branch line and the distance between sprinklers on adjacent branch lines. The first value is defined as "S" (along the branch line), while the second value is defined as "L" (between adjacent branch lines). In the same manner that maximum limits are imposed on the overall area of coverage, limits are also placed on these distances. Once again, the type of sprinkler, the hazard to be protected, and the characteristics of the construction are factored into these limits. Tables 6-10G through L show the permitted values for "S" and "L."

The area of coverage is then determined by multiplying the L and S values together. It should be noted that sprinklers installed on the ends of a branch line near a wall and those sprinklers installed on the branch line that runs closest to a wall receive special treatment and may alter the selected values for S and L. (See NFPA 13 for a further discussion on this issue.) Even a slight deviation in these limits may result in the sprinkler system not being able to adequately control the fire.

Extended coverage (EC) sprinklers use a slightly different method of determining their area of coverage. In essence, the area of coverage is identical to whatever the sprinkler is listed to. Pendent-

mounted extended coverage sprinklers are listed with equal-sided even-numbered dimensions. An EC sprinkler that is listed for 16 × 16 ft (4.9 × 4.9 m) has an area of coverage of 256 sq ft (23.8 m²). These values become fixed with respect to system design. For example, even if this type of sprinkler is used in an area or room with dimensions of 15 × 15 ft (4.6 × 4.6 m), it must still be designed for the listed spacing of 16 × 16 ft (4.9 × 4.9 m), or 256 sq ft (23.8 m²).

Sensitivity and Distribution Issues

The final position of the sprinkler with respect to the ceiling and any nearby construction features, such as lighting fixtures, construction elements, and the like, can impact the performance of the sprinkler in two distinct ways. First, if the sprinkler is located in a manner so that it will not be exposed to the heat from the fire, the subsequent delay in operation of the sprinkler may result in the fire being too large to be properly controlled by the sprinkler. In order to reduce the chance of this happening, sprinkler system design standards will control the positioning of the sprinkler with respect to walls and ceilings.

The second area of concern lies within the distribution pattern of the sprinkler. In other words, once the sprinkler operates, is it likely that the water that discharges from the sprinkler will be able to reach the protected object(s)? There is virtually an unlimited number of conditions and scenarios that can impact the distribution pattern of the sprinkler. The question then becomes: which items can impact this aspect of the sprinkler, and to what extent is some of this impact tolerable?

The current requirements of NFPA 13 that deal with the ensurance of timely operation of the sprinkler are specific to the type of sprinkler involved. While a set of specific ranges can be found for each type of sprinkler, as an example, the principles are the same: (1) control the distance down from the ceiling on which the sprinkler is located, and (2) determine if a particular construction feature will somehow impede the heat generated by the fire from reaching the sprinkler. This latter item is a prime consideration for the construction types previously discussed.

The final position of the sprinkler beneath a ceiling is predicated on the fact that a layer of heat that banks down from the ceiling surface is actually responsible for the operation of the sprinkler. In essence, heat from the fire rises to the ceiling surface where it may spread across the surface. As the layer collects or builds, the thermally sensitive components of the sprinklers will eventually activate, allowing for the release of the water. It is for this reason that NFPA 13 limits the distance that the sprinkler can be positioned below the ceiling. Without this type of control, a measurable time delay is likely. This concern is also factored into the type of construction feature permitted for a particular type of sprinkler. In general, sprinklers with very focused performance characteristics,

TABLE 6-10L. *Protection Areas and Maximum Spacing of ESFR Sprinklers*

Construction Type	ESFR Sprinkler up to 30 ft in Height Protection Area (ft²)	ESFR Sprinkler up to 30 ft in Height Spacing (ft)	ESFR Sprinkler up to 40 ft in Height Protection Area (ft²)	ESFR Sprinkler up to 40 ft in Height Spacing (ft)
Noncombustible unobstructed	100	12	100	10
Noncombustible obstructed	100	12	100	10
Combustible unobstructed	100	12	100	10
Combustible obstructed	NA	NA	NA	NA

For SI units: 1 ft = 0.305 m; 1 ft² = 0.0929 m².
NA, not applicable

such as suppression (ESFR) or life safety (residential), will have the most restrictive set of rules for allowable distances below a ceiling. Table 6-10M shows a summary of these rules for the various types of sprinklers.

TABLE 6-10M. *Summary of Maximum Allowable Deflector Distances below the Ceiling for Various Sprinklers**

Sprinkler Type	Maximum Distance
SSU/SSP	12 in.
Standard sidewall	6 in.
EC upright and pendent	12 in.
EC sidewall	6 in.
Large drop	8 in.
ESFR	14 in.

*Greater distances are permitted based on construction type, special listings, or both.

The subject of obstructions that affect the discharge pattern of the sprinkler is not an easy one. A building with a completely smooth, flat horizontal ceiling that utilizes recessed lighting fixtures, or flush-mounted heating, ventilating, and air-conditioning (HVAC) exhaust grilles, and that otherwise has no projections from the ceiling is an ideal situation that most designers would prefer to encounter. This ideal configuration is the exception.

The many combinations of architectural features, lighting components, structural members, and related building systems create the opportunity for a system's performance to be impaired or otherwise not up to its full ability to control or suppress a fire. Since the first edition of NFPA 13 was published in 1896, many rules and requirements have been provided in NFPA 13 to try to address the issue. Both the 1991 edition of NFPA 13, in which a complete reorganization and substantial rewrite was completed, and the 1996 edition contain major changes to the section on obstructions to discharge of the sprinkler. It is important to note that the fundamental philosophy has not changed in these last 100 years, but the level of detail of better defining what type, size, and shape of obstruction is important and how close that obstruction has to be before it becomes a concern.

The latest edition of NFPA 13 has segregated a section for each type of sprinkler: upright and pendent spray; sidewall spray; upright (which does not yet exist) and pendent-extended coverage spray; sidewall-extended coverage spray; large drop; and ESFR. The user can now find a thorough set of guiding principles and requirements for the type of obstruction and the particular rules for how such an obstruction is managed. Three specific types of obstruction, or perhaps more appropriately, three specific regions of influence created by a potential obstruction, have been developed. In general, all three regions are a concern for all types of sprinklers, but the rules for managing such obstructions will differ depending on the type of sprinkler that one is contemplating.

The first region of obstruction is that which falls within 18 in. (457 mm) of the sprinkler. As one may suspect, 18 in. (457 mm) is not a random number but rather the distance established in 1952 for the clearance below the newly developed spray sprinkler. It was determined that this amount of space was necessary to allow the spray pattern to develop in order to provide an effective distribution pattern. In general, any object (structural element, light fixture, or component) as small as $^1/_2$ in. (12.7 mm) within this zone is potentially damaging to the effectiveness of the spray pattern. A comparison of the obstruction rules in the 1989 edition of NFPA 13 indicated that, in very broad-based terms, the obstruction rules established in 1952

for horizontal separation between the sprinkler and some structural element resulted in the allowable blockage of a segment of the spray pattern of approximately 11.5 degrees.

This view of what the obstruction rules did and did not do became the basis for the rewrite of this section in the 1991 standard and the refinement of same in the 1996 standard. It is by no means perfect, but it is much more specific than ". . . sprinkler spray patterns shall not be obstructed."

The second region of influence is that area which is more than 18 in. (457 mm) below the sprinkler but not more than 48 in. (1.22 m) below the sprinkler. Once again, the 48-in. (1.22-m) figure is not random. It forms the basis for the point at which the spray pattern has evolved to its full diameter. (Considered to be approximately 16 ft (4.88 m) for an upright or pendent spray sprinkler). Objects in this region that have the potential for blocking or otherwise impacting the discharge of the sprinkler have a set of rules. Intuitively, obstruction in this region can usually be somewhat larger than in the first region discussed. In other words, the farther removed the object is from the sprinkler, the larger its dimensional criteria can be.

The third region of influence can best be characterized as being outside of the 48-in. (1.22-m) region and containing any feature that prevents the spray pattern from reaching the hazard. This is the traditional "4-ft (1.22-m) rule," which has also existed since the first edition of NFPA 13. In this case, the shielding caused by the obstruction is considered to be severe enough that supplemental sprinklers are intended to be installed below such obstructions. The general exception to this rule is for those conditions in which the obstruction occurs at the wall, such as with a soffit. In this case, the concern must be analyzed on a case-by-case basis [even for a soffit that has a projection as small as 6 in. (152.4 mm)] to determine if any additional protection measures are necessary.

Figure 6-10Q is extracted from the 1996 edition of NFPA 13 to illustrate the three regions of concern when determining the potential impact of obstructions to the sprinkler.

It is important to remember the purpose of these rules, i.e., to minimize the impact of any obstructions that are in close proximity to the sprinkler as opposed to not allowing any obstructions to be near the sprinkler. In essence, these rules will help the designer to manage the obstructions. Very few (if any) buildings are designed with completely flat ceilings with no obstructions of any type.

It is also important to remember that certain types of sprinklers are more "obstruction sensitive," in that they have little or no tolerance for objects that encroach into the spray pattern discharge space. Large-drop and ESFR sprinklers both fit into this category.

Support of Pipe

Earlier discussion touched on what types of components are necessary to properly support the system piping. The application of these components is a concern associated with the installation of the system, thus additional information is necessary. For example, the allowable distance permitted between the hangers for a certain type or diameter of pipe must be known to ensure that the system is properly supported. NFPA 13 provides requirements for the support of system risers, feed mains, cross mains, and branch-line piping. Each type of piping segment must be treated somewhat differently and, depending on the orientation of the pipe, select types of hangers are permitted.

The installation rules for the hangers typically only involve determining two items: (1) ascertaining the maximum allowable distance between two hangers and (2) determining that the structural point of attachment of the hanger is capable of supporting the weight of water-filled pipe plus 250 lb (113.4 kg). Beyond this there

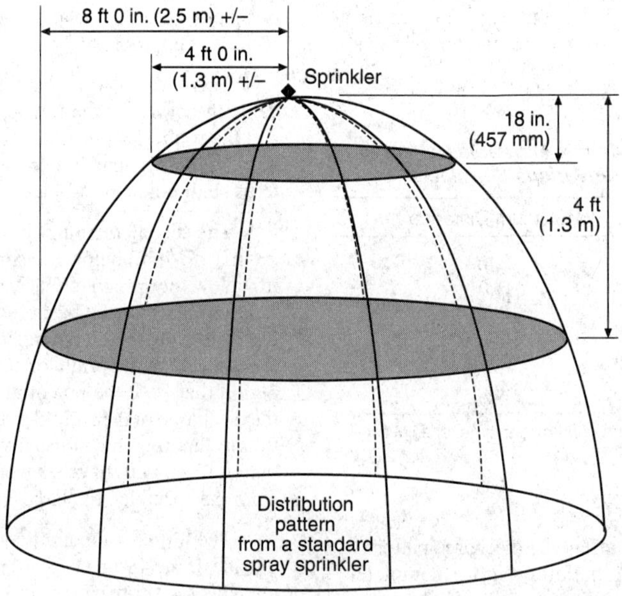

FIG. 6-10Q. *Typical distribution pattern from a standard spray sprinkler.*

are only a few special circumstances or conditions in which deviations are made to these two items.

One of the conditions noted involves a case where an armover in excess of 24 in. (610 mm) is used on a system. A supplemental hanger is required in this case. If a system has an operating pressure in excess of 100 psi (689.5 kPa), special hangers are required for use with pendent sprinklers. Such hangers are intended to restrict the vertical movement of the system piping and the sprinkler. Such movement may result in the sprinkler deflector relocating above the ceiling and subsequently being impaired by the ceiling.

Table 6-10N shows the basic spacing rules for listed hangers for the types of pipe that are permitted for use with automatic sprinkler systems. Note that the smaller the pipe material is in terms of diameter, the more closely spaced the hangers are required to be. This seems counter intuitive, however the beam strength of the smaller diameter pipe is less than that of larger diameter pipe, thus more hangers are required so that the pipe does not sag between adjacent points of support.

Earthquake Bracing of Pipe

The 1947 edition of NFPA 13 first mentioned the need to consider the potential impact of an earthquake on the performance of a sprinkler system. Since that time, a number of requirements have been included in NFPA 13 to specifically address this issue. The methods,

materials, and concepts for accomplishing the proper bracing of the piping system are contained in NFPA 13. The building code in effect in the jurisdiction must be consulted to determine *if* such protection is required. If so, then the details on accomplishing this can be found in NFPA 13.

Protecting the sprinkler system pipe from a measurable seismic event is predicated on three concepts (1) providing clearance for the sprinkler system pipe at or near immovable objects, such as structural walls and between floors; (2) utilizing listed flexible couplings at critical junction points within the sprinkler system; and (3) installing longitudinal and lateral bracing at pipe length segments to allow for pipe movement in conjunction with the building's own structural system. The combination of all three approaches will improve the chance that the system piping will remain intact for any seismic event in which the building survives.

An annular clearance around the pipe in the vicinity of a wall allows the pipe to move in any direction without binding at the penetration and possibly being damaged. Clearance is typically provided at any through penetration of the pipe in a wall or a floor. Such clearance becomes important, since the piping system is not only supported by the floor or ceiling above it but it is also braced from the same point. In a seismic event, movement of walls and floors will not necessarily be in the same direction at the same time. Clearance becomes an important consideration in order to prevent damage occurring to the pipe.

TABLE 6-10N. *Maximum Distance between Hangers (ft – in.)*

Nominal Pipe Size (in.)	3/4	1	1 1/4	1 1/2	2	2 1/2	3	3 1/2	4	5	6	8
Steel pipe except threaded lightwall	NA	12–0	12–0	15–0	15–0	15–0	15–0	15–0	15–0	15-0	15-0	15-0
Threaded lightwall steel pipe	NA	12–0	12–0	12–0	12–0	12–0	12–0	NA	NA	NA	NA	NA
Copper tube	8–0	8–0	10–0	10–0	12–0	12–0	12–0	15–0	15–0	15–0	15–0	15–0
CPVC	5–6	6–0	6–6	7–0	8–0	9–0	10–0	NA	NA	NA	NA	NA
Polybutylene (IPS)	NA	3–9	4–7	5–0	5–11	NA	NA	NA	NA	NA	NA	NA
Polybutylene (CTS)	2–11	3–4	3–11	4–5	5–5	NA	NA	NA	NA	NA	NA	NA

For SI units: 1 in. = 25.4 mm; 1 ft = 0.305 m.
Note: (IPS) Iron Pipe Size, (CTS) Copper Tube Size; NA, not available.

Listed flexible couplings, as defined by NFPA 13 are couplings that permit at least 1 degree of movement in any direction. This type of movement allows the motion of the pipe to be dissipated at each flexible coupling. The movement of the pipe in this case is more or less the result of a movement of a floor (by which the pipe is supported) or a wall in which the pipe may be passing through.

Bracing of the pipe is intended to allow the pipe to move with the building's own structural system. It is also used to arrest and contain proximate accelerations of the pipe from an external source.

Longitudinal bracing is provided to minimize/control this type of movement along the length of the pipe. Such bracing is typically installed in a parallel direction to the system pipe and may not be spaced at more than an 80-ft (24.4-m) interval.

Lateral bracing is provided to minimize/control this type of movement in a side-to-side manner. Lateral bracing is therefore installed in a perpendicular direction to the system pipe. It is limited to a maximum spacing of 40 ft. (12.2 m).

Both types of bracing are required on system feed mains and cross mains. Branch-line piping is normally excluded from these requirements, unless it has a diameter of $2^1/_2$ in. (63.5 mm) or greater. The allowable shapes and strength provisions of NFPA 13 have a built-in factor of safety that includes a weight contribution from a "typical" branch line. The bracing provisions for these larger diameter branch lines are required, since this diameter is beyond what was originally considered by NFPA 13.

Branch lines, however, do not completely escape from all of the conditions for seismic criteria. Even smaller diameter [less than $2^1/_2$ (63.5 mm)] branch-line piping must be restrained in some manner when located in proximity to building finish materials. There have been situations where sprinklers have sheared at a finished ceiling, resulting in unwanted discharge of water. In order to better contain this problem, branch lines do, as a minimum, have to be provided with some additional restraint so as to minimize this problem. Such restraint can be provided by the use of lateral bracing, U-type hooks, or splayed seismic bracing wire. This last option has been used more frequently since it was first recognized in 1991.

A new concept in seismic bracing, never before permitted by NFPA 13 until the 1996 edition was released, involves the use of tension-only bracing systems. Such systems utilize a version of aircraft cabling and at least, for the time being, will primarily be considered as a specially listed system. The 1996 edition of NFPA 13 has added a number of items to be considered by the listing agencies when evaluating these tension-only systems. The items to be considered include: (1) prestretching requirements for the cable; (2) corrosion resistance of the cables; and (3) evaluation of the break strength of the cable.

DESIGN CONSIDERATIONS FOR AUTOMATIC SPRINKLER SYSTEMS

The basic design element associated with an automatic sprinkler system is centered on proving that the water supply source has both an adequate pressure, flow rate, and volume to support the system. Some of the installation criteria will influence this analysis. For example, depending upon how the construction was categorized (obstructed or unobstructed), the area of coverage selected for a particular sprinkler and the presence of any concealed combustible spaces will all factor into the minimum requirements on the selected water supply.

At present, two general design methods exist in NFPA 13: (1) the pipe schedule method and (2) the hydraulic design method. The pipe schedule method dates back to the late 1800s. It basically uses a table that allows a predetermined number of sprinklers to be sup-

plied from a certain diameter pipe. NFPA 13 contains such schedules for both copper tube and steel pipe. The use of this particular design method is severely restricted by NFPA 13, almost to the point where it is impractical to consider in most instances.

In addition to the schedule or table that is provided for a light-hazard or ordinary-hazard occupancy (pipe schedule design is no longer permitted for extra hazard), a second table is used to specify the minimum flow and pressure at the base of the system riser. The flow is a minimum and, in the case of a light-hazard occupancy, that flow must be made available at the base of the system riser at a pressure of 15 psi (103.4 kPa) plus the pressure that is necessary to reach the highest sprinkler. In the case of a light-hazard occupancy that has its highest sprinkler positioned 38 ft (11.2 m) above the base of the riser, the minimum flow would have to be available at a pressure of 31.5 psi (217.2 kPa). Table 6-10O indicates that not less than 500 gpm (1,892.5 L/min) must be provided at this point. Such flows are not typical of most municipal water supplies.

TABLE 6-10O. *Water Supply Requirements for Pipe Schedule Sprinkler Systems*

Occupancy Classification	Minimum Residual Pressure Required	Acceptable Flow at Base of Riser	Duration in Minutes
Light hazard	15 psi	500–750 gpm	30–60
Ordinary hazard	20 psi	850–1500 gpm	60–90

For SI units: 1 gpm = 3.785 L/min; 1 psi = 0.0689 bar.

The more accurate method of determining the demand for a sprinkler system is through the use of the hydraulic calculation procedure. This method is more accurate and it provides the designer and the building owner with a more correct, cost-effective system. This method offers a number of other advantages in that it is not limited to use with a particular type of pipe, a particular type of sprinkler, or a particular type of occupancy hazard. These are all limitations associated with the pipe schedule approach.

The process by which the calculations are completed is based upon an evaluation of the sprinkler system from the hydraulically most remote point. This is that portion of the system which is expected to produce the greatest demand on and tax the system to its greatest extent. It sometimes, but not always, coincides with the physically most remote point in the system, i.e., the point farthest away from the base of the system riser. The concept here is that, if one can satisfy the most remote point hydraulically, then one can also satisfy any point which is hydraulically closer to the system water supply.

In determining this most remote point, a design area or system area of operation is determined. The size and shape of this area is considered to be a rectangle with a minimum area of 1,500 sq ft (139.4 m^2) for light- and ordinary-hazard occupancies and 2,500 sq ft (232.3 m^2) for extra-hazard occupancies. The selected area is then matched to its corresponding density. Figure 6-10R shows the "area-density" curve as found in NFPA 13. The designer need only satisfy any single point on this curve in order to produce an acceptable design.

There is no mandated way of picking a point from this figure; however, it is more advantageous to select the smallest corresponding area of operation (vertical axis) and highest density (horizontal axis). Other points that move higher on the curve are acceptable, but only result in the system requiring more water overall. Interestingly,

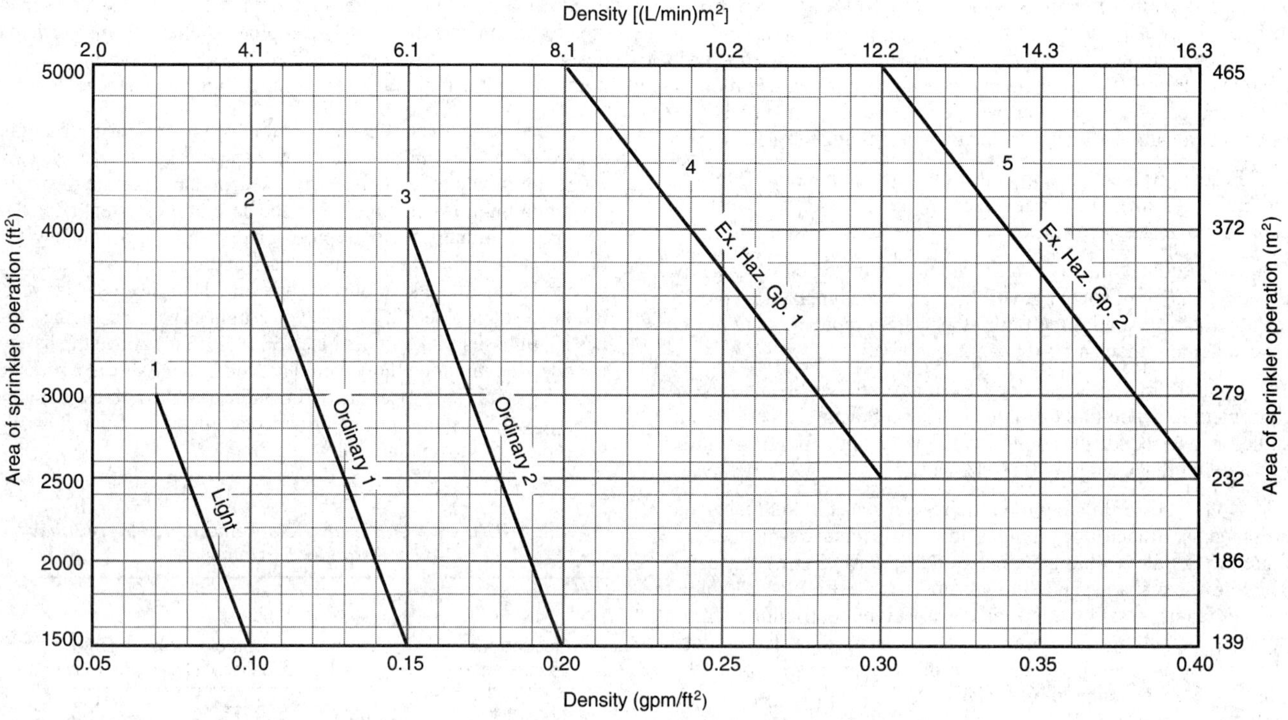

FIG 6-10R. Area-density curve.

less water is made available to any given area on the floor when such points are selected.

Up until this point, the only characteristic needed is the occupancy hazard that one wishes to protect. Once the point is selected one must now be aware of other influencing factors. One of these involves the protection (or lack thereof) of concealed combustible spaces. In some instances, the design area must be increased from 1,500 to 3,000 sq ft (139.4 to 278.7 m²). Another factor is the area of coverage for each sprinkler and the spacing that is provided between sprinklers on each branch line (dimension S, which was previously described).

This type of information will become critical since the area of system operation is to be converted to a number of sprinklers expected to operate. First, the number of sprinklers expected to operate is the result of the system area of operation (in units of sq ft) divided by the area of coverage for a sprinkler (in units of sq ft per sprinkler). (For SI units: 1 sq ft = 0.0929 m².) A sprinkler that covers a 200-sq ft (18.6-m²) area in a light-hazard occupancy, and in which a system area of operation was selected to be 1,500 sq ft (139.4 m²), could also be said to have a system area of operation of 1,500 sq ft (139.4 m²) divided by 200-sq ft (18.6-m²) sprinkler or 7.5 sprinklers. In this process, any fractional amount is always rounded up to the next whole number. In this case, the 1,500 sq ft (139.4 m²) hydraulically most remote area may also be described as being an 8-sprinkler hydraulically most remote area. This translates to the need to provide enough water to these 8 sprinklers in order to control a fire.

The shape of the rectangular area is required to be 1.2 times longer in the direction of the system branch lines. This value is also translated into the area being 1.2 "sprinklers" longer in the direction of the branch lines. The formula determining this is shown in NFPA 13 as:

$$\frac{1.2\sqrt{A}}{S}$$

The result of this makes the hydraulically most remote area a certain number of sprinklers in length. Taking this example a step further and stating that the spacing of sprinklers on the branch line is 10 ft, then the length of the remote area becomes:

$$\frac{1.2\sqrt{1,500}}{10} = 4.6 \text{ sprinklers}$$

(For SI units: 1 ft = 0.305 m.)

As noted before, always round the calculated value up to the nearest whole number. In this case, 5 sprinklers is the correct value. In this example, the remote area would consist of 8 sprinklers total; 5 on the first branch line in the remote area and 3 on the second branch line in the remote area. Once this is established, the calculations can begin by multiplying the area of coverage of the most remote sprinkler by the selected density. This gives the flow from this sprinkler. The pressure for that first sprinkler is then determined by the relationship between the sprinkler orifice (K factor), the flow (Q), and the pressure (p). This flow is then carried upstream to the next sprinkler. The pressure at this sprinkler can then be determined. This pressure is merely the calculated pressure for the first-flowing sprinkler plus the pressure needed to overcome the losses due to friction for the flow to reach that sprinkler. Once again, the relationship between flow (Q), sprinkler characteristic (K), and pressure (p) is used. The K factor is known and the pressure is known, thus the flow from that sprinkler can be determined.

The total system flow at this point is simply the additive discharge from these first two sprinklers. This value is now used to calculate up to the next sprinkler, and the process is repeated as previously described.

The method of conducting this hydraulic calculation is through the use of the Hazen-Williams formula. This method, although developed in the 1920s, is still used for many types of fire protection calculations in the 1990s. It is an empirically derived process that can be completed with the formula or through the use

TABLE 6-10P.   *Friction Loss—PSI per Lineal Foot of Pipe, American Standard Weight Black Steel Pipe*

GPM	1 in. (1.049)	1¼ in. (1.380)	1½ in. (1.610)	2 in. (2.067)	2½ in. (2.469)	3 in. (3.068)	3½ in. (3.548)
1	.0005	.0001	.0001				
2	.0018	.0005	.0002	.0001			
3	.0039	.0010	.0005	.0001	.0001		
4	.0066	.0017	.0008	.0002	.0001		
5	.0100	.0026	.0012	.0004	.0002	.0001	
6	.0140	.0037	.0017	.0005	.0002	.0001	
7	.0187	.0049	.0023	.0007	.0003	.0001	
8	.0239	.0063	.0030	.0009	.0004	.0001	.0001
9	.0297	.0078	.0037	.0011	.0005	.0002	.0001
10	.0361	.0095	.0045	.0013	.0006	.0002	.0001
11	.0431	.0113	.0053	.0016	.0007	.0002	.0001
12	.0506	.0133	.0063	.0019	.0008	.0003	.0001
13	.0587	.0154	.0073	.0022	.0009	.0003	.0002
14	.0673	.0177	.0084	.0025	.0010	.0004	.0002
15	.0764	.0201	.0095	.0028	.0012	.0004	.0002
16	.0861	.0226	.0107	.0032	.0013	.0005	.0002
17	.0963	.0253	.0120	.0035	.0015	.0005	.0003
18	.107	.0282	.0133	.0039	.0017	.0006	.0003
19	.118	.0311	.0147	.0044	.0018	.0006	.0003
20	.130	.0342	.0162	.0048	.0020	.0007	.0003
21	.142	.0375	.0177	.0052	.0022	.0008	.0004
22	.155	.0408	.0193	.0057	.0024	.0008	.0004
23	.169	.0443	.0209	.0062	.0026	.0009	.0004
24	.182	.0480	.0226	.0067	.0028	.0010	.0005
25	.197	.0517	.0244	.0072	.0030	.0011	.0005
26	.211	.0556	.0262	.0078	.0033	.0011	.0006
27	.227	.0596	.0281	.0083	.0035	.0012	.0006
28	.243	.0638	.0301	.0089	.0038	.0013	.0006
29	.259	.0681	.0321	.0095	.0040	.0014	.0007
30	.276	.0725	.0342	.0101	.0043	.0015	.0007
31	.293	.0770	.0363	.0108	.0045	.0016	.0008
32	.310	.0817	.0385	.0114	.0048	.0017	.0008
33	.329	.0864	.0408	.0121	.0051	.0018	.0009
34	.347	.0913	.0431	.0128	.0054	.0019	.0009
35	.366	.0964	.0455	.0135	.0057	.0020	.0010
36	.386	.102	.0479	.0142	.0060	.0021	.0010
37	.406	.107	.0504	.0149	.0063	.0022	.0011
38	.427	.112	.0530	.0157	.0066	.0023	.0011
39	.448	.118	.0556	.0165	.0069	.0024	.0012
40	.469	.123	.0582	.0172	.0073	.0025	.0012
41	.491	.129	.0610	.0181	.0076	.0026	.0013
42	.513	.135	.0637	.0189	.0079	.0028	.0014
43	.536	.141	.0666	.0197	.0083	.0029	.0014
44	.560	.147	.0695	.0206	.0087	.0030	.0015
45	.583	.153	.0724	.0214	.0090	.0031	.0015
46	.608	.160		.0223	.0094	.0033	.0016
47	.632	.166	.0785	.0232	.0098	.0034	.0017
48		.173	.0816	.0242	.0102	.0035	.0017
49	.683	.180	.0848	.0251	.0106	.0037	.0018
50	.709	.186	.0880	.0261	.0110	.0038	.0019

of a set of tables that give a range of values for different types of pipe, C factors, and flows. Table 6-10P shows a portion of one such table. Section 6, Chapter 6, "Hydraulics," provides a detailed discussion of some of the other considerations involving hydraulic calculation procedures.

Special Design Features

Certain types of sprinklers involve design parameters that do not evolve around the density or system area concept. These include residential, ESFR, and large-drop sprinklers.

TABLE 6-10Q. *Large-Drop Sprinkler Data Pressure and Number of Design Sprinklers Required for Various Hazards for Large-Drop Sprinklers*

Hazard	Type of System	Minimum Operating Pressure,[1] psi (bar)			Hose Stream Demand, gal/min (dm³/min)	Water Supply Duration, hr
		25 (1.7)	50 (3.4)	75 (5.2)		
		Number Design Sprinklers				
Palletized[2] Storage						
Class I, II, and III commodities up to 25 ft (7.6 m) with maximum 10 ft (3.0 m) clearance to ceiling	Wet	15	Note 4	Note 4	500 (1900)	2
	Dry	25	Note 4	Note 4		
Class IV commodities up to 20 ft (6.1 m) with maximum 10 ft (3.0 m) clearance to ceiling	Wet	20	15	Note 4		
	Dry	Does not apply	Does not apply	Does not apply	500 (1900)	2
Unexpanded plastics up to 20 ft (6.1 m) with maximum 10 ft (3.0 m) clearance to ceiling	Wet	25	15	Note 4		
	Dry	Does not apply	Does not apply	Does not apply	500 (1900)	2
Expanded plastics commodities up to 18 ft (5.5 m) with maximum 8 ft (2.4 m) clearance to ceiling	Wet	Does not apply	15	Note 4	500 (1900)	2
	Dry	Does not apply	Does not apply	Does not apply		
Idle wood pallets up to 20 ft (6.1 m) with maximum 10 ft (3.0 m) clearance to ceiling	Wet	15	Note 4	Note 4		
	Dry	25	Note 4	Note 4	500 (1900)	1½
Solid Piled[2] Storage						
Class I, II, and III commodities up to 20 ft (6.1 m) with maximum 10 ft (3.0 m) clearance to ceiling	Wet	15	Note 4	Note 4		
	Dry	25	Note 4	Note 4	500 (1900)	1½
Class IV commodities and unexpanded plastics up to 20 ft (6.1 m) with maximum 10 ft (3.0 m) clearance to ceiling	Wet	Does not apply	15	Note 4		
	Dry	Does not apply	Does not apply	Does not apply	500 (1900)	1½
Double-Row Rack Storage[3] with Minimum 5.5 ft (1.7 m) Aisle Width and Multiple-Row Rack Storage with Minimum 8.0 ft (2.5 m) Aisle Width						
Class I and II commodities up to 25 ft (7.6 m) with maximum 5 ft (1.5 m) clearance to ceiling	Wet	20	Note 4	Note 4		
	Dry	30	Note 4	Note 4	500 (1900)	1½
Class I and II commodities up to 30 ft (9.2 m) with maximum 5 ft (1.5 m) clearance to ceiling	Wet	20 plus one level of in-rack sprinklers[5]	Note 4	Note 4		
	Dry	30 plus one level of in-rack sprinklers[5]	Note 4	Note 4	500 (1900)	1½
Class I, II, and III commodities up to 20 ft (6.1 m) with maximum 10 ft (3.0 m) clearance to ceiling	Wet	15	Note 4	Note 4		
	Dry	25	Note 4	Note 4	500 (1900)	1½
Class I, II, and III commodities up to 25 ft (7.6 m) with maximum 10 ft (3.0 m) clearance to ceiling	Wet	15 plus one level of in-rack sprinklers[5]	Note 4	Note 4		1½
	Dry	25 plus one level of in-rack sprinklers[5]	Note 4	Note 4	500 (1900)	

In the case of residential sprinklers when used in accordance with NFPA 13, the design requirement is to simply provide calculations to prove that the single sprinkler flow can be satisfied for the hydraulically most remote sprinkler, and that the multiple sprinkler flow can be satisfied for the hydraulically most remote four residential sprinklers.

TABLE 6-10Q. *Continued*

Hazard	Type of System	Minimum Operating Pressure,[1] psi (bar)			Hose Stream Demand, gal/min (dm³/min)	Water Supply Duration, hr
		25 (1.7)	50 (3.4)	75 (5.2)		
		Number Design Sprinklers				
Class IV commodities up to 20 ft (6.1 m) with maximum 10 ft (3.0 m) clearance to ceiling	Wet	Does not apply	20	15		
	Dry	Does not apply	Does not apply	Does not apply	500 (1900)	2
Class IV commodities up to 25 ft (7.6 m) with maximum 10 ft clearance to ceiling	Wet	Does not apply	20 plus one level of in-rack sprinklers[5]	15 plus one level of in-rack sprinklers[5]		
	Dry	Does not apply	Does not apply	Does not apply	500 (1900)	2
Unexpanded plastics up to 20 ft (6.1 m) with maximum 10 ft (3.0 m) clearance to ceiling	Wet	Does not apply	30	20	500 (1900)	2
	Dry	Does not apply	Does not apply	Does not apply		
Unexpanded plastics up to 25 ft (7.6 m) with maximum 10 ft (3.0 m) clearance to ceiling	Wet	Does not apply	30 plus one level of in-rack sprinklers[4]	20 plus one level of in-rack sprinklers[4]	500 (1900)	2
	Dry	Does not apply	Does not apply	Does not apply		
Class IV commodities and unexpanded plastics up to 20 ft (6.1 m) with maximum 5 ft (1.5 m) clearance to ceiling	Wet	Does not apply	15	Note 4	500 (1900)	2
	Dry	Does not apply	Does not apply	Does not apply		
Class IV commodities and unexpanded plastics up to 25 ft (7.6 m) with maximum 5 ft (1.5 m) clearance to ceiling	Wet	Does not apply	15 plus one level of in-rack sprinklers[4]	Note 4	500 (1900)	2
	Dry	Does not apply	Does not apply			
On-end Storage of Roll Paper[2]						
Heavyweight paper in closed array, banded in open array, or banded or unbanded in a standard array, up to 26 ft (7.9 m) with maximum 34 ft (10.4 m) clearance to ceiling	Wet	Does not apply	15	Note 4		
	Dry	Does not apply	Does not apply	Does not apply	0 (Note 6)	4 (Note 6)
Any grade of paper, except lightweight paper with stacks in closed array, or banded or unbanded in a standard array, up to 20 ft (6.1 m) with maximum 10 ft (3.0 m) clearance to ceiling	Wet	Does not apply	15	Note 4		
	Dry	Does not apply	25	Note 4	0 (Note 6)	4 (Note 6)
Medium weight paper completely wrapped (sides and ends) in one or more layers of heavyweight paper, or lightweight paper in two or more layers of heavyweight paper, with closed array, banded in open array, or unbanded in a standard array, up to 26 ft (7.9 m) with maximum 34 ft (10.4 m) clearance to ceiling	Wet	Does not apply	15	Note 4	Note 6	Note 6
	Dry	Does not apply	Does not apply	Does not apply	Does not apply	Does not apply

The ESFR sprinkler has a design approach that is even simpler. This basically involves being able to supply adequate pressure and flow to the hydraulically most demanding 12 ESFR sprinklers. The "shape" of the ESFR's most remote area is a configuration that is four sprinklers along the branch line and three branch lines wide. The minimum operating pressure of an ESFR sprinkler is never less than 50 psi (344.8 kPa), thus the estimated flow can be readily determined from such a system.

Large-drop sprinkler design is similar, but the actual number of design sprinklers is not always the same. The hazard being protected, the available minimum operating pressure, and the configuration (height and rack *vs.* pile) of the storage will determine the number of large-drop sprinklers to be used. (See Table 6-10Q.)

Another special rule involves system designs that incorporate extended-coverage sprinklers. In order to maintain some level of safety, NFPA 13 mandates that no fewer than five EC sprinklers be

TABLE 6-10Q. *Continued*

Hazard	Type of System	Minimum Operating Pressure,[1] psi (bar)			Hose Stream Demand, gal/min (dm³/min)	Water Supply Duration, hr
		25 (1.7)	50 (3.4)	75 (5.2)		
		Number Design Sprinklers				

Record Storage

Hazard	Type of System	25 (1.7)	50 (3.4)	75 (5.2)	Hose Stream	Water Supply
Paper records and/or computer tapes in multitier steel shelving up to 5 ft (1.5 m) in width and with aisles 30 in. (76 cm) or wider, without catwalks in the aisles, up to 15 ft (4.6 m) with maximum 5 ft (1.5 m) clearance to ceiling	Wet	15	Note 4	Note 4	500 (1900)	1½
	Dry	25	Note 4	Note 4		
Same as above, but with catwalks of expanded metal or metal grid with minimum 50 percent open area in the aisles	Wet	Does not apply	15	Note 4	500 (1900)	1½
	Dry	Does not apply	15	Note 4		

For SI units: 1 ft = 0.305 m; 1 psi = 0.0689 bar.
Notes:
1. Open wood joist construction. Fully firestop each joist channel to its full depth at intervals not exceeding 20 ft (6.2 m). In unfirestopped open wood joist construction, or if firestops are installed at intervals exceeding 20 ft (6.1 m), increase the minimum operating pressures of Table 6-10R by 40 percent.
2. See NFPA 231, *Standard for General Storage.*
3. With rack storage, use conventional wood pallets only; no slave pallets.
4. The high pressure may be used, but the required number of design sprinklers may not be reduced from that required for the lower pressure.
5. Install in-rack sprinklers in accordance with NFPA 231C, *Standard for Rack Storage of Materials.*
6. Hose stream demands and water supply durations may vary for roll paper storage depending on local conditions. See NFPA 231F, *Standard for the Storage of Roll Paper.*

included in the design of a system. Without this rule, a system that used an extended-coverage sprinkler listed for 400 sq ft could be designed with only four sprinklers (1,500 sq ft ÷ 400 sq ft = 3.75 or 4 sprinklers). (For SI units: 1 sq ft = 0.0929 m².)

A new provision in the 1996 edition of NFPA 13 permits a reduction in the system area of operation when QR sprinklers are used. This provision is limited to those situations in which the occupancy is light hazard and in which certain ceiling configurations are present. Figure 6-10S shows the allowable reduction. It is limited to smooth, flat, horizontal ceilings that are no higher than 20 ft (6.1 m). This special rule was developed as the result of a series of tests done in 1988, which compared the number of QR sprinklers operating to the number of standard response sprinklers operating under similar fire scenarios.

A final point on system design involves fire department operations. Thus far, this section has dealt with those items necessary to consider for correct operation of the system. An integral part of the system is the final extinguishment (when necessary) by the fire department. In order to ensure that the water and pressure that is necessary for the sprinkler system is not usurped by the fire department operation, an allowance for hose streams is considered. This will basically warrant that the fire department will have some quantity of water available for final suppression operations without adversely impacting those automatic features associated with the system water supply. NFPA 13E, *Guide for Fire Department Operations in Properties Protected by Sprinkler and Standpipe Systems,* provides guidance on the fire department interaction with the sprinkler system.

PLANS, CALCULATIONS, AND SYSTEM ACCEPTANCE

Implementation of the items discussed thus far does not occur by accident. Knowledgeable designers must be able to present a design package to the Authority Having Jurisdiction (AHJ) for review and/ or approval prior to the commencement of any work. The purpose

of such plans is to ascertain that the appropriate system elements have been considered and incorporated. Detailed plans should include items such as the layout of the pipe, the location of the sprinklers, the location or spacing of supports, the location of the system

Note: $y = \frac{-3x}{2} + 55$

For ceiling height ≥ 10 ft and ≤ 20 ft, $y = \frac{-3x}{2} + 55$;

For ceiling height < 10 ft, $y = 40$;

For ceiling height > 20, $y = 0$.

For SI units: 1 ft = 0.31 m.

FIG. 6-10S. Allowable reduction in system area when "quick response" sprinklers are used.

riser, the location of any water supply control valves, and the location of the system water supply source. This represents a fraction of the 43 items that are mandated by NFPA 13.

Hydraulic calculations should be detailed and comprehensive. A summary sheet showing the design area of operation, the design density, the occupancy hazard, the hose stream allowance, the water supply information, and the total system demand should also be included. Detailed calculation sheets that document the proper sequence of sprinkler demand points must also be provided. These calculations must originate in the hydraulically most demanding portion of the system and reflect what happens between that point and the water supply source.

The system acceptance or checkout test is, to some extent, a formality, but a very important formality. It is used to verify that items previously submitted on the contract plans and reviewed by the AHJ were indeed provided, installed, incorporated, and located as indicated. Test provisions may include items such as a hydrostatic test of the system piping. This test is used to verify the quality of the pipe and fitting installation. In addition to being hydrostatically tested, dry-pipe systems must also undergo a loss of air pressure test for 24 hr to make sure that air is not being lost to faulty installation of fittings or pipe.

Upon completion of these items (as well as several other items not mentioned here), the contractors material and test certificate is signed by the installing contractor, the AHJ, and the building owner (or the designated representative). This certificate becomes the basis for final acceptance and commissioning of the system. It is more than just another form, in that it often becomes a determining factor in allowing the certificate of occupancy to be issued for the structure.

Once this step is completed, it is now incumbent on the building owner to follow up with periodic inspection, testing, and maintenance of the system. Future changes to the building or the contents may have an impact on the sprinkler system. Careful consideration should be given to the impact that any such items would have on the sprinkler system.

Innocent-appearing changes to a building's architectural features may create a severe obstruction to the sprinkler system and may result in less-than-effective performance. Caution in such changes should always be considered.

MARINE APPLICATIONS

Another new item contained in the 1996 edition of NFPA 13 is the addition of a chapter on the rules needed to properly install a sprinkler system on a marine vessel. Section 9, Chapter 37, "Marine," discusses the development of an entirely new NFPA document on fire protection of merchant vessels.

The U.S. Coast Guard has provided, since 1991, a circular on merchant vessel sprinkler systems. This circular basically requires the use of a NFPA 13-type sprinkler system on certain categories of vessels. The prevalence of gaming-type excursion vessels on both inland and coastal routes has been a motivating factor in this aspect of sprinkler protection.

Sprinkler systems are not exclusively limited to these vessel types. The span of marine vessels that will require systems in the near future includes not only excursion boats for both gaming and pleasure, but also cruise ships and merchant vessels used to haul various commodities. The Safety of Life at Sea (SOLAS) regulations, promulgated by the International Maritime Organization (IMO), also require sprinkler systems.

The newly created marine chapter in NFPA 13 deals with those aspects of a ship that are unique to that type of operating environment. In general, all of the current rules of NFPA 13 will still be ap-

plicable, except those that are amended or modified by this particular chapter. Many design considerations are similar between a ship and building, but slight differences in the terminology should be considered. These include references to decks instead of floors, shore connections instead of fire department connections, and sprinkler pumps instead of fire pumps. The terms and connotations are similar but the arrangement and importance of each will differ.

One aspect of the design of a sprinkler system on a merchant vessel that is underway, as compared to a building, is that the ship must be entirely self-reliant. That is, occupants cannot be readily evacuated; the fire department cannot be summoned to assist with manual fire suppression, and a maintenance contractor cannot be called in at a moments notice to repair a system fault. Because of these factors, additional redundant features will be found on the ship, which otherwise would not be considered in the design of a land-based structure.

Other conditions that must be considered include the potential for the heaving and rolling that a ship may experience and the resultant impact that would have on the sprinkler system piping. This problem can be dealt with by incorporating some of the seismic design features that are discussed in NFPA 13.

The exposure of certain pipe materials and sprinklers to corrosive atmospheres is also a concern. The use of nonmetallic pipe and the special corrosion-protection measures that are discussed in NFPA 13 are applicable to this situation.

Perhaps the most crucial design consideration is the arrangement and reliability of the water supply. In most circumstances, a nearly unlimited supply is readily available to a merchant vessel, but how it is arranged and supplied to the ship's sprinkler system is of paramount importance. A number of water supplies (never less than two) are deemed acceptable, and must be arranged to ensure that a high level of reliability is obtained.

A pressure tank is considered to be an acceptable method to satisfy part of the water supply requirements. This source must be charged with a quantity of fresh water, which will supply the system for a period of at least 1 min, in the case of a wet pipe system. The supply for a dry-pipe system should be adequate to allow for complete filling of the pipe volume plus a quantity of water to permit 1 min of system discharge.

A sprinkler pump is one of the supply sources that must always be provided. The sprinkler pump has characteristics that are similar to a fire pump, and it must be specifically listed for marine application. The power supply for a sprinkler pump must be arranged to minimize the chances of a fire event affecting the starting or operation of the pump. This is done by obtaining the power supply for the pump from the main switchboard on the ship. A backup source of power from a generator must also be provided.

In addition to these arrangements, the water supply for the onboard sprinkler system must also be interconnected with the ship's fire main system. The fire main system is the equivalent of an underground distribution system that is also used to supply strategically located hose outlets on the ship. Figure 6-10T shows one basic configuration between a pressure tank, a fire pump, and the ship's fire main.

The fire department connection depicted in Figure 6-10T should typically be of the same style and size of those used on fixed structures. Such connections may be supplied by land-based fire department pumpers. In other cases, the source of supplying this connection may be from a fire boat when the vessel is in or near a port area. The companion International Shore Connection (ISC) is a universal water supply hook-up for vessels operating at the time of a fire outside of their home port. As shown in Figure 6-10T, both types of connections may be used to supplement the other automatic sources of supply for the sprinkler system.

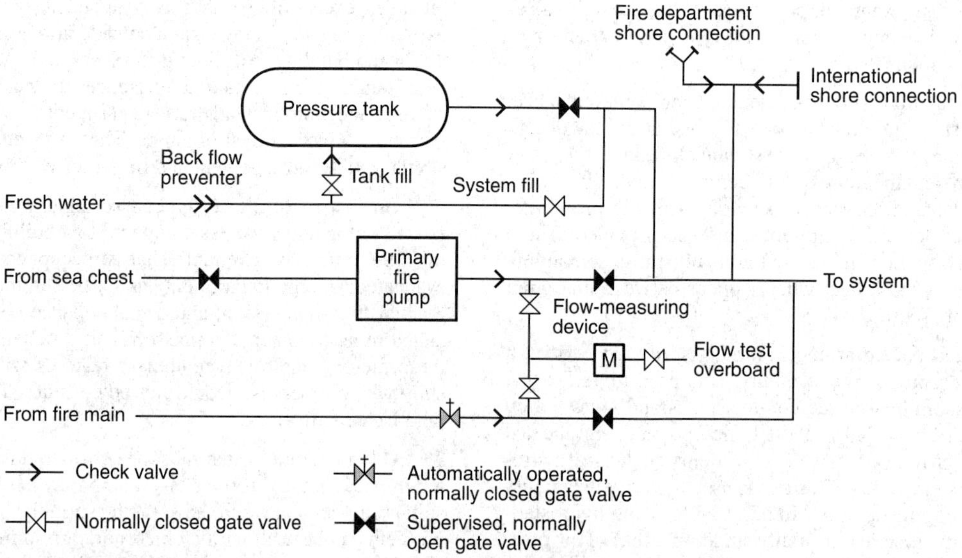

FIG. 6-10T. Abbreviated example of a water supply with fire pump backup for marine applications.

Another feature associated with this unique operating environment is the need to carry additional spare parts to facilitate basic repairs while the vessel is underway. Certain merchant vessels and cruise ships may be as much as four days away from the nearest port capable of making repairs to correct system impairments.

OTHER CONSIDERATIONS

The use of automatic sprinkler systems is still one of the most effective methods of protecting occupants of structures as well as the contents and the structure itself. Sprinkler systems are uniquely qualified to be installed in everything from manufactured homes, high-rise buildings, warehouse facilities, manufacturing facilities and more recently, merchant vessels. Advances continue in the refinement of piping materials, sprinklers, fittings, and hangers. Advances in the technology, when properly applied, will allow sprinkler systems to contribute to at least another 100 years of success. More widespread acceptance of these systems by the general public is an important consideration and should not be overlooked when contemplating future technology or adoption of sprinkler system requirements for state and local ordinances.

BIBLIOGRAPHY

NFPA Codes, Standards, and Recommended Practices

Reference to the following NFPA codes, standards, and recommended practices will provide further information on automatic sprinkler systems discussed in this chapter. (See the latest version of *The NFPA Catalog* for availability of current editions of the following documents.)

NFPA 13, *Standard for the Installation of Sprinkler Systems*

NFPA 13D, *Standard for the Installation of Sprinkler Systems in One- and Two-Family Dwellings and Manufactured Homes*

NFPA 13E, *Guide for Fire Department Operations in Properties Protected by Sprinkler and Standpipe Systems*

NFPA 13R, *Standard for the Installation of Sprinkler Systems in Residential Occupancies up to and Including Four Stories in Height*

NFPA 30, *Flammable and Combustible Liquids Code*

NFPA 101®, *Life Safety Code®*

NFPA 231, *Standard for General Storage*

NFPA 231C, *Standard for Rack Storage of Materials*

Additional Reading

"1991 Updated: Legislation and Codes Affecting the Fire Sprinkler Industry," *Sprinkler Age*, Vol. 10, No. 11, Nov. 1991, pp. 12–15, 17.

"AFSA TAC Voices Praise, Concern for Limited Water Supply Sprinkler System," *Sprinkler Age,* Vol. 11, No. 5, May 1992, pp. 19–20.

Allan, H., "Using Heat Release to Optimize the Design of Combined Sprinkler/Vent Systems," Commonwealth Scientific and Industrial Research Organisation, New South Wales, Australia, CIB W14/93/2 (J); *Proceedings* of the First Japan Symposium on Heat Release and Fire Hazard, Session 2, Fire Modeling as a Tool for Reaction to Fire Evaluation, May 10–11, 1993, Tsukuba, Japan, 1993, pp. II/13–18.

Alpert, R. L., "Modeling of Sprinkler Spray Suppression of Fires," First International Conference on Fire Suppression Research, May 5–8, 1992, Stockholm, Sweden, 1992, pp. 245–260.

Arvidson, M. and Isaksson, S., "Equivalency Sprinkler Fire Tests," Swedish National Testing and Research Inst., Boras, Sweden, SP Report 1995:19, Nordtest Project 1152-94, ISBN 91-7848-548-7, 1995, 125 pages.

"Attic Sprinklers (TM): The Next Step in New Technology," *Sprinkler Age*, Vol. 14, No. 3, Mar. 1995, pp. 16–17.

"Automatic Fire Sprinkler System Performance," *Fire Engineering*, Vol. 147, No. 8, Aug. 1994, pp. 167–168.

Automatic Sprinkler System Performance and Reliability in United States Department of Energy Facilities," DOE-EP-0052, 1980, U.S. Department of Energy, Washington, DC, June 1981.

Ballast, D. K., *Sprinkler Use in Buildings for Fire Protection: Recent Periodical Literature*, Vance Bibliographies, Monticello, IL, 1988.

Bathurst, D., "Sprinkler Retrofit in Buildings With Asbestos Fireproofing," VHS Video Tape, General Services Administration, Washington, DC, Video; Federal Fire Forum, Current Issues in Fire Protection, June 15, 1992, Gaithersburg, MD, 1992.

Beitel, J. J., "National Quick Response Sprinkler Research Project: Group 1 Retrofit Test Series. [Group 1 Warehouse Retrofit Test Series.]," Southwest Research Inst., San Antonio, TX, National Fire Protection Research Foundation, Quincy, MA, Group 1, June 1990, 50 pages.

Bill, R. G., *et al.*, "Development of a Limited-Water-Supply Sprinkler for Mobile, Manufactured Homes," Twelfth Joint Panel Meeting of the U.S./Japan Government Cooperative Program on Natural Resources (UJNR) on Fire Research and Safety, October 27–November 2, 1992, Tsukuba, Japan, Building Research Inst., Ibaraki, Japan, 1992, pp. 121–130.

Bill, R. G., Jr. and Heskestad, G., "Comments on 'Thermal Response of Sprinkler Heads in Hot Air Stream With Speed Less Than 1 M/S'," Letters to the Editors, *Fire Science and Technology*, Vol. 13, No. 1/2, 1993, p. 89.

Bill, R. G., Jr. and Hill, E. E., Jr., "Sprinkler Protection of Manufactured Homes With Sloped Ceilings Using Prototype Limited Water Supply

Sprinklers," *Fire Technology*, Vol. 31, No. 1, First Quarter 1995, pp. 17–43.

Bill, R. G., Jr. and Kung, H. C., "Limited-Water-Supply Sprinklers for Manufactured (Mobile) Homes," *Fire Technology*, Vol. 29, No. 3, Third Quarter 1993, pp. 203–225.

Bill, R. G., Jr., *et al.*, "Predicting the Suppression Capability of Quick Response Sprinklers in a Light Hazard Scenario. Part 1. Fire Growth and Required Delivered Density (RDD) Measurements," *Journal of Fire Protection Engineers*, Vol. 3, No. 3, July/Aug./Sept. 1991, pp. 81–93.

Bill, R. G., Jr., *et al.*, "Predicting the Suppression Capability of Quick Response Sprinklers in a Light Hazard Scenario. Part 2. Actual Delivered Density (ADD) Measurements and Full-Scale Fire Tests," *Journal of Fire Protection Engineers*, Vol. 3, No. 3, July/Aug./Sept. 1991, pp. 95–107.

Braga, A. C., "Avoiding Ice Plugs in Sprinkler Systems Protecting Freezers," *Sprinkler Age*, Vol. 14, No. 1, Jan. 1995, pp. 11–12, 14.

Brock, P. D., "Critical Considerations in the Hydraulic Design of Automatic Sprinklers," *FPD Today*, Vol. 2, No. 1, Jan./Feb. 1990, pp. 15, 21.

Brown, I., "'Green Shoots' Identified in the UK Market for Automatic Sprinklers," *Fire*, Vol. 86, No. 1061, Nov. 1993, p. 45.

Bryan, J. L., *Automatic Sprinkler and Standpipe Systems*, 2nd ed., National Fire Protection Association, Quincy, MA, 1990.

Budnick, E. K. and Fleming, R. P., "How Quick Response Sprinklers Perform and What it Means for Their Applications," *Fire Journal*, Vol. 83, No. 5, Sept./Oct. 1989, p. 48.

Bychowski, J. A., "Automatic Sprinklers: Simple, Successful Fire-Suppression Devices," *Consulting-Specifying Engineer*, Vol. 16, No. 5, Nov. 1994, pp. 71–72, 74, 76, 78, 80.

Chan, T. S., "Measurements of Water Density and Drop Size Distributions of Selected ESFR Sprinklers," *Journal of Fire Protection Engineering*, Vol. 6, No. 2, 1994, pp. 79–97.

Cheng, Y., "The Development of the ESFR Sprinkler System," *Fire Safety Journal*, Vol. 14, Nos. 1 and 2, 1988, pp. 65–73.

Chow, W. K., "Sprinklers in Atrium Buildings," *Hong Kong Engineer*, Vol. 16, No. 9, Sept. 1988, pp. 41, 43.

"Comparative Study on the Performance of the U. S. Standard Spray Sprinkler and the European Conventional Style Sprinkler. Current Research," *Fire Technology*, Vol. 27, No. 4, Nov. 1991, pp. 350–355.

Cote, A. E., "Fifty Editions of NFPA 13," *Fire Journal*, Vol. 77, No. 3, May 1983, pp. 36–38, 124–126.

Cote, A. E., "Final Report of Field Test of a Retrofit Sprinkler System," National Fire Protection Research Foundation, Washington, DC, Feb. 1983.

Coughlin, P., "Residential Fire Sprinklers on Trial," *Operation Life Safety Newsletter*, Vol. 10, No. 1, Jan. 1995, pp. 1–2.

Damon, W. A., "Preventing Pitfalls in Fire Protection Systems," *Heating/Piping/Air Conditioning*, April 1985, pp. 61–64, 69–71.

Davis, W. D. and Cooper, L. Y., "Computer Model for Estimating the Response of Sprinkler Links to Compartment Fires With Draft Curtains and Fusible Link-Actuated Ceiling Vents," *Fire Technology*, Vol. 27, No. 2, May 1991, pp. 113–127.

Davis, W. D., Forney, G. P. and Bukowski, R. W., "Simulating the Effect of Sloped Beamed Ceilings on Detector and Sprinkler Response," National Institute of Standards and Technology, Gaithersburg, MD, NISTIR 5499, Sept. 1994; National Institute of Standards and Technology, Annual Conference on Fire Research: Book of Abstracts, October 17–20, 1994, Gaithersburg, MD, 1994, pp. 147–148.

"ESFR Sprinklers: Your Way to Fight Aerosol Fires," *Sentinel*, Second Quarter 1991, p. 6.

Estepp, M. H., "Strategies to Implement One- and Two Family Dwellings Sprinkler Initiatives," Prince George's County Fire Dept., MD, International Association of Fire Chiefs (IAFC), Fire-Rescue International Conference Proceedings, Annual Conference, 1993, August 28–September 1, 1993, Dallas, TX, 1993, pp. 171–177.

Factory Mutual Approval Guide, Factory Mutual Engineering Corporation, Norwood, MA, published annually.

Fire Protection Equipment Directory, Underwriters Laboratories Inc., Northbrook, IL. Published annually in January, with one supplement in July.

Feld, J. M., "Developing a Performance Based Sprinkler System Standard," *Proceedings* of the SFPE Engineering Seminars on Performance-Based Fire Safety Engineering, November 15–17, 1993, Phoenix, AZ, SFPE, Boston, MA, 1993, pp. 31–38.

Fleming R. P., "A Closer Look at the NFPA Residential Sprinkler Standards," *Fire Journal*, Vol. 82, No. 2, March–April 1988, p. 51.

Fleming, R. P., "Automatic Sprinkler System Calculations," *The SFPE Handbook of Fire protection Engineering*, NFPA, Quincy, MA., 1994.

Fleming, R., "Bracing for Earthquake," *Sprinkler Quarterly*, No. 49, Spring 1984, p. 43.

Fleming, R. P., "Current Fire Sprinkler Research in the U. S. Current Research," *Fire Technology*, Vol. 28, No. 2, May 1992, pp. 174–176.

Fleming, R. P., "Current Fire Sprinkler Research," *Sprinkler Quarterly*, No. 77, Winter 1991, pp. 22–23.

Fleming, R. P., "Few Controversies at Residential Sprinkler Forum," *Sprinkler Quarterly*, No. 87, Summer 1994, pp. 28, 30.

Fleming, R. P., "Reducing Design Areas With Quick Response Sprinklers," *Sprinkler Quarterly*, No. 93, Winter 1995, pp. 22, 24–26.

Fleming, R. P., "Residential Sprinkler Plan Review Guide," National Fire Sprinkler Assoc., Patterson, NY, Review Guide, 1990.

Fleming, R. P., "Sprinkler System Performance in the Northridge Earthquake," *Sprinkler Quarterly*, No. 86, Spring 1994, pp. 20–22.

Franchuk, D. M. and Waggoner, W. W., "Engineers, Sprinklers, and the Authority Having Jurisdiction: Working Together to Benefit Fire Fighters and the Public," *NFPA Journal*, Vol. 89, No. 2, Mar./Apr. 1995, pp. 85–89.

Gaiser, K. E., "Challenge: Sprinklers in Hise-Rises, Manufactured Homes, One- and Two Family Dwellings...Bull or No Bull,"Georegetown City Fire Dept., SC, International Association of Fire Chiefs (IAFC), Fire-Rescue International Conference Proceedings, Annual Conference, 1993, August 28-September 1, 1993, Dallas, TX, 1993, pp. 161–170.

Hall, J. R., Jr., "U.S. Experience With Sprinklers: Who Has Them? How Well Do They Work?" National Fire Protection Assoc., Quincy, MA, June 1995,.

Hamer, A. and Beyler, C., "Modeling the Activation of Dry Pipe Sprinklers in High Rack Storage," *Proceedings* of the International Conference on Fire Research and Engineering, September 10–15, 1995, Orlando, FL, SFPE, Boston, MA, 1995, pp. 65–70.

Hankins, J., "ESFR Sprinklers: The Facts You Should Know," *Fire Safety Journal*, Vol. 2, No. 6, Dec. 1995, pp. 8–9

Hart, S., "City of Los Angeles Civil Disturbance: Fire Sprinkler Performance," *Journal of Applied Fire Science*, Vol. 2, No. 2, 1992–1993, pp. 133–155.

Hodge, B. K., "Analysis and Design Program for Series Piping Systems," *Fire Technology*, Vol. 21, No. 4, 1985, pp. 310–319.

Hoffmann, N. Galea, E. R., and Markatos, N. C., "Mathematical Modeling of Fire Sprinkler Systems," *Applied Mathematical Modeling*, Vol. 13, No. 5, May 1989, pp. 415–424.

Huggins, R., "ESFR System Design: Basic Parameters for Sprinkler Layout," *Sprinkler Age*, Vol. 15, No. 5, Mar. 1996, p. 22.

Isman, K. E., "Fire Sprinklers Help Building Owners Comply With the Americans With Disabilities Act," *Sprinkler Quarterly*, No. 80, Fall 1992, pp. 15, 17.

Isman, K. E., "Sprinkler Protection for General Storage Occupancies. Part 2," *Sprinkler Quarterly*, No. 94, Spring 1996, pp. 10, 12.

Isman, K. E., "New/Advanced Sprinkler Technology," *Proceedings* of the Technical Symposium on Halon Alternatives, June 27–28, 1994, Knoxville, TN, 1994, pp. 53–56.

Isman, K. E., "Sprinkler Protection for General Storage Occupancies," *Sprinkler Quarterly*, No. 93, Winter 1995, pp. 6, 8, 10.

Isner, M. S., "Hospital Fire, Sprinkler Success, Weymouth, Massachusetts, January 24, 1993. Summary," Fire Investigation Report, National Fire Protection Association, Quincy, MA, 1993.

Isner, M. S., "Nursing Home Fire, Sprinkler Success, Ashland, Kentucky, June 2, 1993. Summary," Fire Investigation Report, National Fire Protection Association, Quincy, MA, Fire Investigation Report, 1993.

Isner, M. S., "Sprinklers Prevent Tragedy in Two Health Care Facility Fires," *NFPA Journal*, Vol. 87, No. 5, Sept./Oct. 1993, pp. 49–52, 54–55.

Kim, A., "Protection of Glazing in Fire Separations by Sprinklers," Sixth International Fire Conference, Interflam '93, Fire Safety, March 30–April 1, 1993, Oxford, England, Interscience Communications Ltd., London, England, 1993, pp. 83–93.

Kim, A. K. and Lougheed, G. D., "Protection of Glazing Systems With Dedicated Sprinklers," *Journal of Fire Protection Engineers*, Vol. 2, No. 2, 1990, pp. 49–59.

Koffel, W. E., "Improving Sprinkler Performance by Design," *NFPA Journal*, Vol. 87, No. 6, Nov./Dec. 1993, pp. 12, 20.

Kung, H. C., "Evolution and Future of Sprinkler Technology," *Sprinkler Age*, Vol. 7, No. 5, Jan. 1993, pp. 14, 16.

Lake, J. D., "Egremont Sprinkler Demonstration Drives Point Home," *Sprinkler Quarterly*, No. 71, Summer 1990, pp. 12–14.

Laverick, G., "New Specific Application Sprinkler Listings," *Sprinkler Age*, Vol. 14, No. 3, Mar. 1995, pp. 10, 12–14.

Linder, K. W., "Does Sprinkler Design Match the Use of the Building? Part 1. The Role of the Insurance Carrier," *Sprinkler Age*, Vol. 9, No. 4, Apr. 1990, pp. 20–21.

Linder, K. W., "Does Sprinkler Design Match the Use of the Building? Part 2. The Role of the Insurance Carrier," *Sprinkler Age*, Vol. 9, No. 5, May 1990, pp. 28–29.

Linder, K. W., "Does Sprinkler Design Match the Use of the Building? Part 3. The Role of the Insurance Carrier," *Sprinkler Age*, Vol. 9, No. 6, June 1990, pp. 21–22.

Lougheed, G. and Mawhinney, J., "Automatic Sprinkler Protection for Compact Mobile Shelving," *Fire Research News*, No. 67, Winter 1993, pp. 1–3.

Lougheed, G. D., Mawhinney, J. R. and O'Neill, J., "Full-Scale Fire Tests and the Development of Design Criteria for Sprinkler Protection of Mobile Shelving Units," *Fire Technology*, Vol. 30, No. 1, First Quarter 1994, pp. 98–132.

Madrzykowski, D., "Evaluation of Sprinkler Activation Prediction Methods," *Proceedings* of the First International Conference on Fire Science and Engineering, ASIAFLAM '95, March 15–16, 1995, Kowloon, Hong Kong, 1995, pp. 211–218.

Mawhinney, J. R., "Effects of Automatic Sprinkler Protection on a Smoke Control System," *Fire Research News*, No. 68, Spring 1993, pp. 1–3; *Journal of Applied Fire Science*, Vol. 3, No. 1, 1993/1994, pp. 43–48, 1993/1994; *Canadian Fire Alarm Association Newsletter*, Sept. 1993, pp. 7–11

Mawhinney, J. R. and Tamura, G. T., "Effect of Automatic Sprinkler Protection on Smoke Control Systems," *ASHRAE Transactions*, Vol. 100, No. 1, 1994.

Marryatt, H. W., *Fire—A Century of Automatic Sprinkler Performance in Australia and New Zealand*, 1886-1986, Australian Fire Protection Associations Melbourne, Australia, 1988. (Available from NFPA.)

Marryatt, H. W., "Suppression and Extinguishment of Fire-98 Years of Experience with Automatic Sprinkler Systems in Australia," *SFPE Bulletin*, Vol. 85, No. 1, Jan. 1985, pp. 19–23.

Milke, J. A.; Bryan, J. L., "Development of a Technique to Assess the Adequacy of the Municipal Water Supply for a Residential Sprinkler System," Maryland Univ., College Park, National Institute of Standards and Technology, Gaithersburg, MD, Fire Administration, Emmitsburg, MD, NIST-GCR-91-600, Nov. 1991.

Miller, G. A., "Residential Fire Loss Trends: Sprinklers Make the Difference," Sprinklers in Residential and Commercial Buildings, August 19–20, 1990, Charleston, SC, National Conference States on Building Codes and Standards, 1990, pp. 155–172.

Moore, J., "Why Sprinkler Systems Fail," *Fire Engineering*, Vol. 142, No. 5, 1989, pp. 36–40.

Morris, J., "Life Safety and Sprinkler Protection," *Fire Prevention*, No. 209, May 1988, pp. 18–21.

Nash, P. and Young, R. A., *Automatic Sprinklers for Fire Protection*, Victor Green, London, 1982.

"National Quick Response Sprinkler Research Project: Group 2 Performance Tests. Phase 1. RDD Tests. [Group 2 Performance Tests: RDD Test Results.]," Underwriters Labs., Inc., Northbrook, IL, National Fire Protection Research Foundation, Quincy, MA, Group 2, Phase 1, 1991.

"National Quick Response Sprinkler Research Project: Preliminary Report of the Group 2 Performance Tests. Phase 2. ADD Tests. [Group 2 Performance Tests: ADD Test Results.]," Underwriters Labs., Inc., Northbrook, IL, National Fire Protection Research Foundation, Quincy, MA, Group 2, Phase 2, 1991.

Notarianni, K. A., "Modeling High Bay Sprinkler Response," VHS Video Tape, National Institute of Standards and Technology, Gaithersburg, MD, Video; Federal Fire Forum, Current Issues in Fire Protection, June 15, 1992, Gaithersburg, MD, 1992.

Notarianni, K. A., "Predicting the Response of Sprinklers and Detectors in Large Spaces," Extended Abstracts of the SFPE Engineering Seminars on Large Fires: Causes and Consequences, November 16–18, 1992, Dallas, TX, Society of Fire Protection Engineers, Boston, MA, 1992, pp. 23–28.

Notarianni, K. A. and Davis, W. D., "Use of Computer Models to Predict the Response of Sprinklers and Detectors in Large Spaces," National Institute of Standards and Technology, Gaithersburg, MD, Society of Fire Protection Engineers, Worcester Polytechnic Institute, *Proceedings* of Computer Applications in Fire Protection, June 28–29, 1993, Worcester, MA, 1993, pp. 27–33.

O'Rourke, G., "Polybutylene Piping for Automatic Sprinkler Systems," *Heating/Piping/Air Conditioning*, April 1985, pp. 54–58.

Pennel, G., "Computer Based Sprinkler Plans, " *Fire Safety Journal*, Vol. 9, No. 2, 1985, pp. 165–170.

"Pocket Guide to Automatic Sprinklers," Factory Mutual Engineering Corp., Norwood, MA, 1990.

Prince George's County Fire Department, NAHB Research Center, Inc., "Express Residential Fire Sprinkler Design Guide," *Prince George's County Implementation Version 1.0*, Prince George's County Fire Dept., MD, NAHB Research Center, Inc., Upper Marlboro, MD, May 1994.

Puchovsky, M. T., "Entering the Next 100 Years of Standardized Sprinkler System Technology," *Sprinkler Age*, Vol. 15, No. 3, Mar. 1996, pp. 10, 12, 14–15, 18–19.

Richardson, J. K., "An Assessment of the Performance of Automatic Sprinkler Systems," *Fire Technology*, Vol. 19, No. 4, Nov. 1983, pp. 275–279.

Rousseau, P., "ESFR Sprinklers: Listing Limitations Should Not be Overlooked," *Sprinkler Age*, Vol. 15, No. 3, Mar. 1996, p. 30.

Smith, W. W., "Computerized Sprinkler Calculations—The Advantages Are Too Great to Ignore," *Fire Journal*, Vol. 83, No. 3, May–June 1989, p. 67.

Sanders, R. E., "Challenge: Sprinklers in High Rises," Louisville Fire Dept., KY, International Association of Fire Chiefs (IAFC), Fire-Rescue International Conference Proceedings, Annual Conference, 1993, August 28–September 1, 1993, Dallas, TX, 1993, pp. 179–184.

Schmitt, R. F., "Priorities: Sprinkler Systems, Fire Safety and Cost," Sprinklers in Residential and Commercial Buildings, August 19–20, 1990, Charleston, SC, National Conference States on Building Codes and Standards, 1990, pp. 111–124.

Schultz, G. R., "Keeping Up With the Latest in Sprinkler Technology," *NFPA Journal*, Vol. 88, No. 3, May/June 1994, pp. 38–42.

Shaw, H., "Limited Water Supply Sprinklers," VHS Video Tape, Federal Fire Forum. Developments in Fire Protection, November 6, 1992, Gaithersburg, MD, 1992.

Smith, W. W., "Integrating Hydraulic Sprinkler Calculations and CAD," *Fire Journal*, Vol. 84, No. 2, March–April 1989, p. 56.

Smoluk, G. R., "What Makes Plastic the Choice in, of All Things, Fire Sprinkler Systems?" *Modern Plastics*, Vol. 66, No. 7, July 1989, pp. 55–56.

Solomon, R., ed., *Automatic Sprinkler Systems Handbook*, 6th ed., National Fire Protection Association, Quincy, MA, 1994.

Spees, P. L. and Wright, J. R., "Cycling Waterflow in Sprinkler Systems," *Fire Engineering*, Vol. 142, No. 4 (Apr. 1989).

Tomes, W. J. and Golinveaux, J., "Fire Test Performance of Extra Large Orifice Sprinklers for Rack Storage of Group A Plastics in Warehouse-Type Retail Occupancies," *Proceedings* of the International Conference on Fire Research and Engineering, September 10–15, 1995, Orlando, FL, SFPE, Boston, MA, 1995, pp. 415–420.

Tomes, W., "Model Building Codes—Direct Influence on Future Sprinkler Requirements," *Sprinkler Age*, Vol. 9, No. 1, Jan. 1990, p. 58.

Troup, J. M. A., Tomes, W. J. and Golinveaux, J., "Full-Scale Fire Test Performance of Extra Large Orifice Sprinklers for Protection of Rack Storage of Group A Plastic Commodities in Warehouse-Type Retail Occupancies," Factory Mutual Research Corp., Norwood, MA, Tomes, Van Rickley and Associates, San Diego, CA, Central Sprinkler Co., Lansdale, PA, ISBN 0-9516320-9-4; *Proceedings* of the Seventh International Interflam Conference, Interflam '96, March 26–28, 1996, Cambridge, England, Interscience Communications Ltd., London, England, 1996, pp. 405–413.

Tuten, H. A., "Computer-Assisted Sprinkler Design," *Plumbing Engineer*, Vol. 15, No. 4, May 1987, pp. 22–29.

Vincent, B. G. and Kung, H. C., "Comparison of European Conventional and U.S. Spray Sprinklers," *Journal of Fire Protection Engineering*, Vol. 5, No. 1, Jan./Mar. 1993, pp. 17–28.

Watson, F., "Trip Times for Dry Pipe Valves in Grid Systems Theoretical Analysis," *Sprinkler Quarterly*, March 1982.

Yao, C., "The ESFR Sprinkler System—A New Approach to High-Challenge Storage Protection," *Fire Journal*, Vol. 79. No. 2, March 1985, pp. 30–33, 70–72.

Yao, C., "Overview of FMRC's Sprinkler Technology Research," First International Conference on Fire Suppression Research, May 5–8, 1992, Stockholm, Sweden, 1992, pp. 57–82.

WATER SUPPLIES FOR SPRINKLER SYSTEMS

Revised by Wayne M. Martin

Every automatic sprinkler system must have at least one automatic water supply of adequate pressure, capacity, and reliability. An "automatic" supply is one that is not dependent on any manual operation, such as making connections, operating valves, or starting pumps, to supply water at the time of a fire. Both the rate of flow and the duration of that flow must be considered.

This chapter identifies the types of water supplies that are acceptable for sprinkler systems; the variables that must be considered, including the hazards of occupancies, in evaluating requirements for both sprinklers and supplementary hose line supplies; and the fundamentals of hydraulically designing sprinkler systems to match sprinkler piping sizes with the characteristics of available water supplies.

Other chapters in this handbook of interest in understanding the complexities of providing adequate water supplies for sprinkler systems are those chapters in Section 6 covering water supplies for fire protection. Of particular interest are Chapter 2, "Water Storage Facilities and Suction Supplies"; Chapter 4, "Fire Pumps"; Chapter 6, "Hydraulics," which discusses the fundamentals of water flow through pipes and orifices; and Chapter 12, "Water Spray Protection," which is of value because there is no definitive line between sprinkler systems and water spray systems, and many of the water supply requirements pertaining to sprinkler systems apply equally to water spray systems.

TYPES OF SUPPLIES

Sprinkler systems may be supplied with water from one source or a combination of sources, such as street mains, gravity tanks, reservoirs, fire pumps, pressure tanks, rivers, lakes, and wells.

In theory, a single water supply would seem to be all that is necessary for satisfactory protection. However, a single supply may be temporarily out of service; it may be disabled at the time of a fire or before a fire is completely extinguished; or its pressure or capacity may be below normal during an emergency. Therefore, a secondary supply may be necessary depending on the strength and reliability of the primary supply; the value and importance of the property; the area, height, and construction of the building; the occupancy; and the outside exposures.

Wayne M. Martin, P.E., is the senior fire protection engineer and technical section commander for the City of Los Angeles Fire Department. He is a member of NFPA's Technical Committee on Automatic Sprinklers.

Connections to Public Waterworks Systems

A connection from a reliable public waterworks system of adequate capacity and pressure is the preferred single or primary supply for automatic sprinkler systems. In determining its adequacy, one must consider not only the normal capacity and pressure of the system, but also the probable minimum pressures and flows available at unfavorable times, such as the summer months; periods of heavy demand on the system, such as those created by large industrial uses; or periods of impairment caused by flood or by ice conditions in winter. These conditions are typically referred to as seasonal lows.

Daily demands will fluctuate as well. Consider the early morning hours when the population is preparing for work or school. Demands created by domestic plumbing system use will be greater than at other times of the day.

The size and arrangement of street mains and feeders from public water supplies are also important. Connections from large mains fed two ways or from two mains on a grid system may provide an excellent supply. Street mains less than 6 in. (152 mm) in diameter are usually considered inadequate and unreliable. Feeds from dead-end mains are also undesirable. Water meters, if required by the water supply authority, should be of types listed for the fire service. Flow and pressure tests under varying conditions of demand are generally necessary to determine the amount of public water available for fire protection.

Cross-Connections Between Public and Private Supplies

Where a secondary supply is needed to supplement the public water supply, public and private fire service mains can be connected to feed into a single fire protection system. When the private fire service main is from a nonpotable source and it is connected to the public main, the systems are said to be cross-connected. In some localities, cross-connections may be prohibited or closely regulated by health authorities.

Federal regulations approved in 1986 require states to provide quality water when it is intended for public consumption. States and municipal governments have taken various steps beyond the federal mandate to protect the potable water supply. Double check valves and reduced pressure zone (RPZ) backflow prevention devices may be required when the sprinkler system will be supplied by a potable water source.

These devices, which are intended to protect the public water supply from potential contamination, are often-times costly and have a negative impact on the available water supply and pressures to the sprinkler system. Although no compelling documentation that any actual problems exist with respect to sprinkler systems and potable water exists, many state health and environmental agencies have been successful in mandating such devices to protect the public from some perceived threat. NFPA 24, *Standard for the Installation of Private Fire Service Mains and Their Appurtenances*, provides a discussion on these devices.

Gravity Tanks

Gravity tanks of adequate capacity and elevation make a good primary supply and may be acceptable as a single supply. Details of the construction, heating, and maintenance of gravity tanks are given in NFPA 22, *Standard for Water Tanks for Private Fire Protection* (hereinafter referred to as NFPA 22). In determining tank size and elevation, consideration should be given to the number of sprinklers expected to operate, the duration of operation, the arrangement of underground supply piping, and the presence of standpipes, hydrants, and fire department connections.

Suction Tanks

With the advent of hydraulically designed sprinkler systems, the use of gravity tanks for fire protection has declined. The use of fire pumps combined with suction tanks has increased. These tanks are typically constructed of steel or concrete. NFPA 22 should be consulted for details.

Fire Pumps

A fire pump with a reliable source of power and a reliable suction water supply is a desirable piece of equipment. Fire pumps are used to a great extent because of the hydraulic advantages of having a water supply available at high pressure. With ample water, a fire pump can maintain a high pressure over a long period of time and may be a necessary part of some installations requiring greater water pressure than would be available otherwise. For details of power sources, pump construction, installation, and methods of control and operation, consult NFPA 20, *Standard for the Installation of Centrifugal Fire Pumps*.

Most fire pumps today are started automatically. Automatic control of fire pumps is usually needed where an immediate water demand is necessary, as with a deluge system. Automatic fire pumps must have their suction under a positive pressure to avoid the delays and uncertainties of priming.

Most fire pumps are powered by electric motors or diesel engines. Where a reliable source of power is available at all times, an electrically driven installation may be the most desirable. Where the power supply is questionable, a diesel-driven fire pump would be preferred. In some critical installations, such as hospitals and high-rise buildings, a diesel-driven emergency power generator may be used to supply secondary power to the electric motor. The use of a diesel-driven fire pump in a critical installation would eliminate the need for much of the capacity of the emergency generator.

The automatic control of electrically driven centrifugal pumps must be arranged to prevent frequent, repeated starting of the motor, either by initiating continuous running until stopped manually or by using a timing device that stops the motor automatically only after a predetermined period of operation.

Pressure Tanks

Pressure tanks have several possible uses in automatic sprinkler protection. An important limitation is the small volume of water that can be stored in such tanks.

In situations where an adequate volume of water can be supplied by a public or private source but where the pressure is not sufficient to serve a sprinkler system directly, the pressure tank gives a good starting pressure for the first sprinklers that operate. In tall buildings where the public water pressure is too low for effective water distribution from the highest sprinklers, pressure tanks may be used to supply such sprinklers during the time required for an automatically operated fire pump to begin supplying water.

Each proposed use of pressure tanks calls for special consideration and analysis of water capacity, location, and arrangement of the connection to the sprinkler system. Each installation must usually have specific approval. Details on the construction and installation of pressure tanks are given in NFPA 22.

Fire Department Connections

Under fire conditions that cause a considerable number of sprinklers to operate, public water or tank supplies may not provide water at sufficient pressure for effective sprinkler discharge and distribution. In addition, the pressure in many public water supplies to sprinkler systems may be reduced materially by hose streams from hydrants. In such cases, a connection through which the public fire department can pump water into the sprinkler system provides an important secondary supply. Fire department connections are therefore a standard part of sprinkler systems.

Fire department connections must be located and arranged so that hose lines can be readily and conveniently attached without interference from nearby objects. Each fire department connection must be properly identified with a sign. Each connection must be fitted with a check valve, but not with a gate valve, so that the connection will not be shut off inadvertently. There should be a proper drain, as well as a drip device between the check valve and the outside hose coupling. Figures 6-11A and 6-11B show the main features of a fire department connection. Other details of installation and pipe size are given in NFPA 13, *Standard for the Installation of Sprinkler Systems* (hereinafter referred to as NFPA 13).

Where a wet-pipe sprinkler system has a single riser, the fire department connection should be attached to the system side of the controlling valve. For a dry-pipe system, the fire department connection should be between the system control valve and the dry-pipe valve. This makes it possible to pump water into the system even if

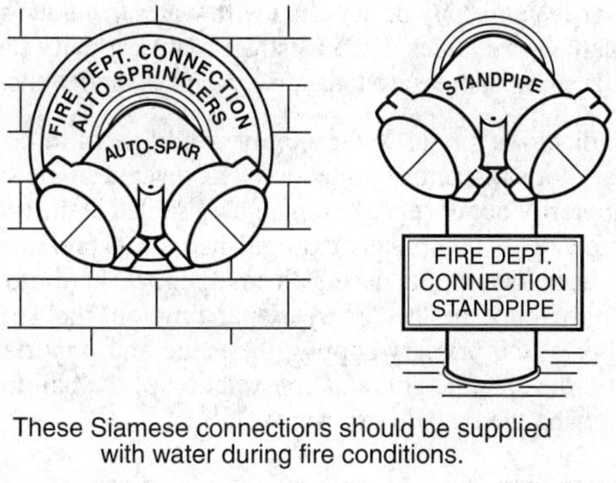

These Siamese connections should be supplied with water during fire conditions.

FIG. 6-11A. Typical Siamese connections for sprinkler systems and standpipes. A check valve allows the use of a single hose line.

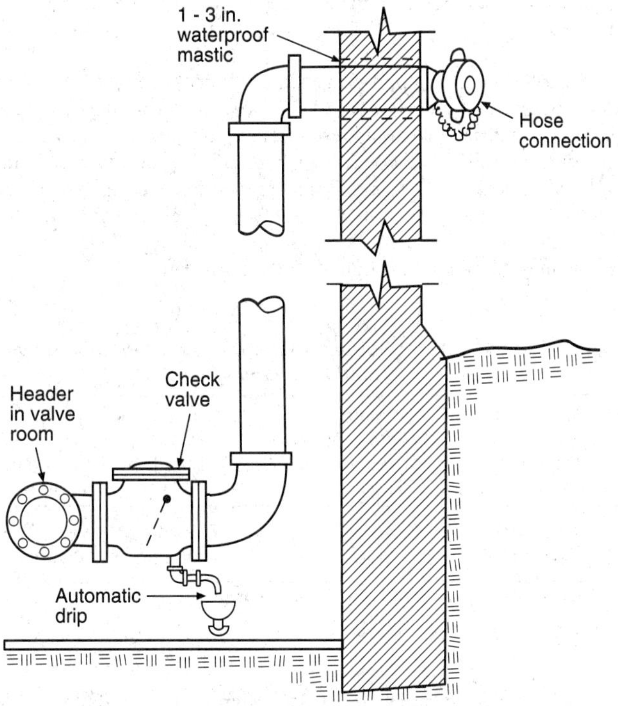

FIG. 6-11B. Typical fire department connection. (For SI units: 1 in. = 25.4 mm.)

the valve is closed. If there are two or more sprinkler system risers, each with its own separate connection to a public main, each system must have its own fire department connection. If more than one riser is connected to a yard system, the fire department connection should feed into the yard system on the supply side of all riser shutoff valves, and there must be a check valve in all other water supply connections into the yard system to prevent backflow and loss of water supplied through the fire department connection. If one riser is shut off, the fire department connection can still supply all other risers. If two or more sprinkler system risers are fed by one sprinkler connection, one fire department connection can supply both risers.

In an emergency, a fire department can pump water from public hydrants or other sources of water into a sprinkler system through hose connected to a yard hydrant or other hose connection using a double female hose coupling, if other supply connections have a check valve or a gate valve that can be closed.

INFLUENCE OF VARIOUS FACTORS ON WATER SUPPLY NEED

Determining the water supply requirement for most sprinkler systems is not always easy because of the many variables involved. If a water source that can supply all the sprinklers is available, there is no problem, but such a water supply is seldom practical except in the case of small systems. The water supply requirement for any sprinkler system is directly related to the number of sprinklers expected to operate, which is determined by the hazard classification and the anticipated use of outside and inside hose stream demands. This, in turn, depends on so many other variables and uncertain factors that no exact mathematical solution is possible.

Records kept by NFPA before 1970 show that the absence of an adequate water supply accounts for a significant share of the comparatively few fires in sprinklered buildings where sprinkler performance was unsatisfactory. Thus, water supply is a significant

concern, particularly with large sprinkler systems and with systems protecting greater-than-ordinary hazards.

Establishing the water supply requirement for any particular sprinkler system requires good engineering judgment based on all the factors relating to sprinkler control. Where the cooling effect from the water discharged by sprinklers is greater than the heat liberated by the fire, the sprinklers can gain control. When the reverse occurs, as when a water supply is overtaxed or system design is inadequate, the sprinklers cannot control the fire and the sprinkler system may ultimately fail. Where all conditions are favorable, the fire should be controlled by the operation of only a small number of sprinklers. Because conditions vary with different classes of occupancy, areas, and types of buildings, the number of sprinklers expected to operate to control a fire may range up to the total number in the design area, and the water supply should be provided accordingly.

The primary factors affecting the number of sprinklers that might open in a fire, and which therefore must be considered in determining the water supply requirement, include the following.

Hazard of Occupancy (Including Flash Fire Hazard and Potential Rate of Heat Release)

This is an important factor and one requiring experienced judgment to evaluate. Where a flash fire hazard is present, it is usually necessary to provide enough water to operate all of the sprinklers in any individual fire area.

Initial Water Pressure

At a pressure of 15 psi (103 kPa), a standard $1/2$-in. (12.7-mm) sprinkler will discharge 22 gpm (83 L/min), or an average of approximately 0.17 gal per sq ft per min [6.8 (L/min)/m^2] over an area of 130 sq ft (12 m^2). At 30 psi (207 kPa), the discharge is 31 gpm (125 L/min), and at 50 psi (345 kPa), 40 gpm (155 L/min). At higher pressures, the discharge is correspondingly greater. With a greater discharge, there is a better chance of fire control from a small number of sprinklers and less need for large volumes of water to supply a large number of sprinklers.

Obstructions to Distribution of Water from Sprinklers

With obstructions, such as high-piled stocks, pallets, racks, and shelving, there is less likelihood that fire will be controlled in its initial stages, and a greater chance that a large number of sprinklers needing large water supplies will open.

High Ceilings and Air Current Conditions

With ceilings of unusual height, there is a greater chance that natural air movement will carry heat away from the sprinklers immediately over a fire, resulting not only in a delay in the application of water, but also in the opening of sprinklers remote from the fire's place of origin. More water is usually needed under such conditions. The same situation exists wherever there are drafts, such as in areas open on the sides to the outside where winds can divert heat from sprinklers over the fire.

Unprotected Vertical Openings

Sprinkler systems in multistory buildings are usually designed on the assumption that fire will be controlled on the floor of origin. Where there are unprotected openings up through which heat and fire may spread, it may be expected that more sprinklers will open, particularly in the case of a fire originating near the vertical opening.

In the case of high combustibility, the interconnected floors may need to be considered as one fire area. This means more water and larger pipe sizes in risers and supply lines.

Wet-Pipe *vs.* Dry-Pipe Systems

Because of the delay due to exhausting air from dry-pipe systems, more sprinklers are expected to open on dry-pipe systems than on wet-pipe systems. This may call for greater water supplies. Design areas are increased by 30 percent to account for this delay.

Size of Undivided Areas

A large undivided area has a greater number of sprinklers, a possibility of a greater maximum number of sprinklers operating. This may result in a greater water demand for systems designed with the pipe schedule method.

Floor-Mounted and Ceiling-Mounted Obstructions and Concealed Spaces

Columns, beams, girders, bar joists, light fixtures, soffits, and HVAC ductwork can obstruct water distribution to the point where additional sprinklers may be necessary to account for the obstruction, the result being more sprinklers in the design area. Floor-mounted obstructions, such as privacy curtains, free-standing partitions, and room dividers, require similar special treatment.

Concealed combustible spaces also have an impact on sprinkler design areas. When they are unprotected, the sprinkler design area is usually doubled to account for the uncontrolled spread of fire in such areas.

Extent of Coverage and Exposures

Any fire in an unsprinklered space extending to an area with automatic sprinklers places an abnormal demand on the sprinkler system and requires increased water supplies so that the system may function effectively.

The preceding factors must be considered individually and collectively, and it is not feasible to derive any general formula or simple method of arriving at water supply requirements. However, certain general statements on this subject may be made. One is that any situation may be protected effectively with much less water where the water is applied automatically instead of manually. Another is that it is good practice to provide more water at higher pressure than will probably be needed to extinguish any fire. Hose streams may be used to supplement sprinklers, even when it is not necessary, and an ample supply of water provides a margin of safety.

Regardless of the size of the building, NFPA 13 requires a specific design area that relates to the occupancy fire hazard rather than the size of the building. The idea behind this concept is to provide a water supply that is adequate to cover this design area for a given period. The sprinklers in the most remote portion (hydraulically remote areas) of this area are used as the basis of design. The capability of supplying sprinklers in this area automatically means that sprinklers that are physically closer to the water supply source can be supplied, as well.

WATER SUPPLY REQUIREMENTS FOR SPRINKLER SYSTEMS

Notwithstanding the general problems involved in arriving at water supply requirements, the fire hazard represented by different building occupancies has made it possible to establish requirements for water supplies for sprinkler systems. The occupancy factor is the primary consideration, with latitude allowed for the contributing factors.

Selection of the appropriate water supply guide table in NFPA 13 is based on the design approach for sprinkler piping. The pipe schedule approach is the oldest version, and first appeared in the standard in 1896. The concept of hydraulically designed systems was advanced in 1969 and is the predominate design approach method used today. Minimum water supply requirements for the pipe schedule design method, as shown in Table 6-11A, is based on occupancy classification (light or ordinary hazard).

TABLE 6-11A. *Water Supply Requirements for Pipe Schedule Sprinkler Systems*

Occupancy Classification	Minimum Residual Pressure Required	Acceptable Flow at Base of Riser	Duration in Minutes
Light Hazard	15 psi	500–750 gpm	30–60
Ordinary Hazard	20 psi	850–1500 gpm	60–90

For SI units: 1 gpm = 3.785 L/min; 1 psi = 0.0689 bar.

The total water supply required is determined from Table 6-11A, and the water for hose streams is added if required. If stored water is used, the total water flow required would have to be multiplied by the duration of flow to determine whether the capacity of the stored supply is adequate, a factor that would not be pertinent for a reliable municipal water source.

The guide should be used only with experienced judgment, but it can serve for all cases qualifying as the light-hazard and ordinary-hazard occupancy classifications that constitute the larger percentage of sprinkler installations. Other classifications, such as extra-hazard occupancy and high-piled storage, usually involve more complex factors requiring special consideration and must use the hydraulic calculation design method.

Where fire pumps contribute to the water supply, pumps of standard size used with adequate rates of discharge are required. A suction supply for a pump must preferably be large enough for continuous operation.

Where pressure tanks furnish the water supply, the tank installation should comply with the appropriate provisions of NFPA 22. Where a combination of different water supplies is provided in the interest of reliability, it is good practice to have the rate of supply from each source at least equal to the minimum requirement for the system.

Light-Hazard Occupancies

Examples of light-hazard occupancies are apartment buildings, dormitories, office buildings, the seating areas of restaurants, and hospitals. In these occupancies, the potential rate of heat release is low, areas are usually subdivided, and a small number of sprinklers normally should control any fire. For hydraulically designed systems, 150 gpm (568 L/min) for sprinkler operation and an additional 100 gpm (378 L/min) for hose stream demand will generally make a good starting point for system demand. A total of 500 to 750 gpm (1893 to 2839 L/min) for pipe schedule systems may be necessary for the same hazard, thereby making this a less desirable design approach.

Ordinary-Hazard Occupancies

Group 1: This ordinary-hazard classification includes occupancies, such as in garages, bakeries, laundries, and canneries, in which the combustibility of contents is moderate, stockpiles of combustibles do not exceed 8 ft (2.4 m), and the rate of heat release is greater than that for the light-hazard classification. Ordinary-hazard Group 1 occupancies will require a minimum of 225 gpm (851 L/min) for sprinkler operation, and an additional 250 gpm (946 L/min) to support exterior hose streams. If a pipe schedule approach is selected, a minimum flow of 850 gpm (3217 L/min) is necessary.

Group 2: This ordinary-hazard classification includes occupancies, such as clothing factories, mercantiles, pharmaceutical manufacturing, certain woodworking occupancies, and shoe factories. With this group, the combustibility of contents, storage heights, and obstruction features are moderate to high, stockpiles do not exceed 12 ft (3.7 m), and the rate of heat release is greater than Group 1. Ordinary-hazard Group 2 occupancies will require a minimum flow of 300 gpm (1125 L/min) for sprinkler operation, plus an additional 250 gpm (946 L/min) to operate the exterior hose streams. If design is based upon the pipe schedule method, a minimum total flow of 850 gpm (3217 L/min) is necessary.

Extra-Hazard Occupancies

Extra-hazard occupancies consist of properties where quantity and combustibility of contents is very high. Flammable and combustible liquids, dust, lint, or other materials may be present, introducing the probability of rapidly developing fires with high rates of heat release. Closely spaced sprinklers and larger pipe sizes are expected. Extra-hazard occupancies are divided into two groups:

1. Extra-hazard occupancies (Group 1) include those that may produce severe fires where small amounts of flammable or combustible liquids are present. These include such occupancies as die casting, plywood and particleboard manufacturing, printing [using inks with below 100°F (37.8°C) flash points], rubber manufacturing operations, saw mills, textile operations, and plastic foam manufacturing.
2. Extra-hazard occupancies (Group 2) include those that may produce severe fires where moderate to substantial amounts of flammable or combustible liquids are present, or where shielding of combustibles is extensive. These include such occupancy hazards as asphalt saturating, flammable liquids spraying, flow coating, open oil quenching, solvent cleaning, and varnish and paint dipping.

In any treatment of hazards by general groups of occupancy, it must be noted that individual properties differ markedly and that buildings of the same nominal occupancy classification may show widely different individual hazards that should be considered in any determination of water supply.

WATER SUPPLY REQUIREMENTS FOR HOSE STREAM PROTECTION

The values given in Table 6-11A include hose stream requirements. In considering water requirements for hose streams, one should realize that, if sprinklers perform effectively, little hose stream assistance is required. Although this is generally the case, a realistic view must be taken of possible contingencies and of the amount of water that might be needed for hose stream protection under adverse conditions. Hose streams may also be necessary to accomplish final extinguishment of the fire. However, it should be noted that current sprinkler system technology is intended to control a fire, not necessarily to extinguish it.

In evaluating hose stream requirements, one should consider several possibilities, such as the amount of water necessary for final extinguishment and cleanup operations, or the steps that may have to be taken in the event that sprinklers are retarding fire spread but are not fully effective in gaining control of the fire and extinguishing it.

HYDRAULICALLY DESIGNED SPRINKLER SYSTEMS

Planning new water supplies or evaluating existing supplies for sprinkler systems requires information about the hydraulic behavior of sprinkler piping systems.

Hydraulic Calculations

A hydraulically designed sprinkler system is one in which pipe sizes are selected on a pressure loss basis to provide a prescribed density in gallons per minute per square foot [(L/min)/m²], or a prescribed minimum discharge pressure or flow per sprinkler, distributed with a reasonable degree of uniformity over a specified area. This permits the selection of pipe sizes in accordance with the characteristics of the water supply available. The stipulated design density and area of application will vary with occupancy hazard. This is the verbatim definition of a hydraulically designed system, taken from NFPA 13. This definition sets the basic premise for the preferred method of sprinkler system design.

Table 6-11B and Figure 6-11C are used to determine density, area of sprinkler operation, and water supply requirements for hydraulically designed sprinkler systems. Systems must be calculated to satisfy a single point on the appropriate design curve, and interior piping must be based on this design point. It is not necessary to meet all points on the selected curve. In some cases where a large water supply is available, it might be advantageous, because of the elimination of a fire pump, to design the system using the top portion of the curve. In other cases, it might be better to design with smaller areas of application and higher densities, because these tend to limit fire size, even though a fire pump is required.

TABLE 6-11B. *Hose Stream Demand and Water Supply Duration Requirements*

Hazard Classification	Inside Hose (gpm)	Total Combined Inside and Outside Hose (gpm)	Duration in Minutes
Light	0, 50, or 100	100	30
Ordinary	0, 50, or 100	250	60–90
Extra Hazard	0, 50, or 100	500	90–120

For SI units: 1 gpm = 3.785 L/min.

When selecting a point on the curve in Figure 6-11C, it should be noted that more total water is necessary in terms of flow and capacity as one moves up the curve. As an example, for a light-hazard system, a 1500-sq-ft (139-m²) area requires a density of 0.1 gpm/sq ft [4.1 (L/min)/m²], or a flow of 150 gpm (568 L/min). If a 2500-sq-ft (232-m²) area is selected, the density is only 0.08 gpm/sq ft [3.2 (L/min)/m²], but the total flow is increased to 200 gpm (757 L/min).

The same hazard occupancy classifications that apply to pipe schedule sprinkler systems apply to hydraulically designed sprinkler systems. However, the recommended water supply figures are somewhat lower due to the greater efficiency of a calculated system.

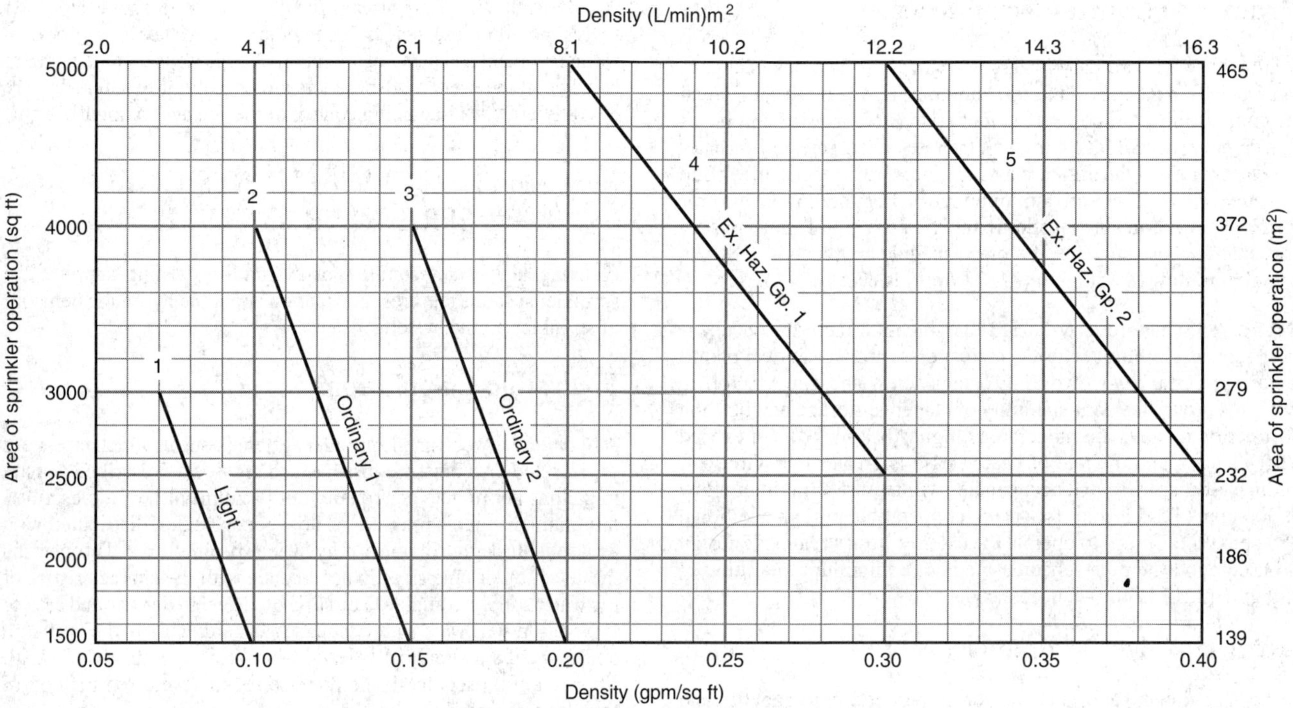

FIG. 6-11C.   Design curves for hydraulically calculated sprinkler system.

The water allowances for inside and outside hose streams may be combined and added to the system requirement at the system connection to the underground main. The total water requirement must be calculated through the underground main to the point of supply.

For deluge and water spray systems with open orifices, calculations are essential. Automatic sprinkler systems protecting high-piled storage require a specific water density for fire control. Because the hydraulic design approach results in superior design, it always should be selected as the preferred method.

Hydraulic calculation programs: Since NFPA 13 first permitted the use of hydraulic calculations, several computer programs have been developed that can assist the designer with this particular task. These programs are not required to be listed nor are they governed by NFPA 13. They must, as a minimum, be capable of providing the information required for any hydraulic calculation procedure. This information includes, but is not limited to, the hydraulic reference points, the flow and pressure at specific points, the pipe diameter, and, for a grid system, a printout showing the flow direction and quantities.

The capabilities and limitations of computer programs are as wide-ranging as the cost differences among them. Before buying one of these programs, the user should be aware of the type of sprinkler system design addressed. *A Comparison of Several Computer Hydraulics Programs for the IBM PC and Compatibles*[1] is considered the "consumer's report" for hydraulic software programs, and it should be consulted.

Flow Calculation Methods

Methods of making flow calculations for sprinkler systems are given in NFPA 13; NFPA 15, *Standard for Water Spray Fixed Systems for Fire Protection; Sprinkler Hydraulics*[2]; *Hydraulic Design and Estimation of Fire Sprinkler Systems*[3]; *Automatic Sprinkler Hy-*

draulic Data[4], "Hydraulics of Fire Protection Systems"[5]; and "Water Flow Characteristics of Sprinkler Systems."[6]

The design area for flow calculations is the most hydraulically demanding area for the sprinkler system. The area should have a dimension parallel to the branch line of 1.2 times the square root of the area of anticipated sprinkler operation. The 1.2 times the square root of the area is added to ensure that the long side of the rectangle would be along the branch line. In grid systems, the hydraulically most demanding area must be verified using at least two additional sets of calculations. This technique is referred to as "peaking," and it is merely a check to prove that the most demanding area has been calculated.

Each sprinkler in the design area must discharge at a flow rate at least equal to the stipulated minimum water application rate or density. Begin calculations at the sprinkler hydraulically farthest from the supply connection. In common system configurations, this will be the end sprinkler on the end branch line. The minimum operating pressure for any sprinkler must not be less than 7 psi (48 kPa), or a higher minimum pressure as specified in the laboratory listing. Special sprinklers, such as extended-coverage, early suppression fast response (ESFR), in-rack, and large drop, will require operating pressures well in excess of this 7-psi (48 kPa) minimum.

The Most Remote Sprinkler

Assuming a minimum pressure of 10 psi (69 kPa) at the most remote sprinkler and a discharge coefficient of 0.75 for a standard $1/_2$-in. (12.7 mm) orifice sprinkler, there is a discharge of 17.7 gpm (67 L/min) calculated from the formula

$$Q = 29.8cd^2\sqrt{P}$$

used in calculating flows through orifices and short tubes. For a given sprinkler, the product of these numbers in the equation— $29.8\ cd^2$—becomes a constant and is denoted as the sprinkler K-factor. In the case of a $1/_2$-in. (12.7-mm) orifice sprinkler, the constant is 5.6 and is derived by multiplying the following numbers: $29.8 \times 0.75 \times (^1/_2)^2 = 5.6$. A simplied equation may now be used to describe flow through the sprinkler orifice. The new equation is

$$Q = K\sqrt{P}$$

Velocity pressure is not a factor at the most remote sprinkler, but it may be considered at all the other sprinklers in the example that follows. Some organizations ignore velocity pressure in their calculations. The error introduced is on the safe side. NFPA 15, *Standard for Water Spray Fixed Systems for Fire Protection*, recommends considering velocity only when it constitutes more than 5 percent of the total pressure.

Example for Second Sprinkler from the End

Assuming sprinklers 10 ft (3.05 m) apart on branch lines, with the end section of pipe 1 in. (25.4 mm) nominal diameter, the friction loss at 17.7 gpm (67 L/min) flow with a Hazen-Williams formula coefficient of 120 (value for black steel pipe) will be 1.0 psi (7 kPa).

The total pressure at the second sprinkler will be 10.0 + 1.0 psi (69 + 7 kPa) = 11.0 psi (76 kPa). Velocity pressure, if used in the procedure, is calculated from the formula

$$P_v = 0.001123 \frac{Q^2}{D^2}$$

where

P_v = velocity pressure in psi,

Q = flow in gpm, and

D = inside pipe diameter in inches.

The velocity pressure relates to the system design by:

$$P_n = P_t - P_v$$

where

P_n = normal pressure,

P_t = total pressure, and

P_v = velocity pressure.

Of this, velocity pressure based on a flow of 17.7 gpm (67 L/min) will be 0.3 psi (2 kPa). The normal pressure (pressure acting perpendicular to the pipe wall) acting on the second sprinkler is the total pressure of 11.0 psi (76 kPa), less the velocity pressure of 0.3 psi (2 kPa), which equals 10.7 psi (74 kPa). On all sprinklers except the end sprinkler, only normal pressure is considered as acting on the sprinklers. The discharge from the second sprinkler, at a pressure of 10.7 psi (74 kPa), will be 18.3 gpm (69 L/min).

The pipe between the second and third sprinkler, also 1 in. (25.4 mm) in diameter, 10 ft (3.05 m) long, and with a flow of 17.7 gpm (67 L/min) + 18.3 gpm (69 L/min) = 36.0 gpm (136 L/min), will have a friction loss of 3.8 psi (26 kPa) and a velocity pressure of 1.2 psi (8 kPa). Normal pressure at the third sprinkler is 10.7 psi (74 kPa) + 3.8 psi (26 kPa) – 1.2 psi (8 kPa), or 13.3 psi (92 kPa).

Other Sprinklers on a Branch Line

Figure 6-11D can be used to establish the velocity pressure correction needed for various pipe sizes and flows. This figure may be used for any pipe type, but it is based on the use of schedule 40 pipe. The difference in velocity pressure is miniscule as it relates to schedule of pipe.

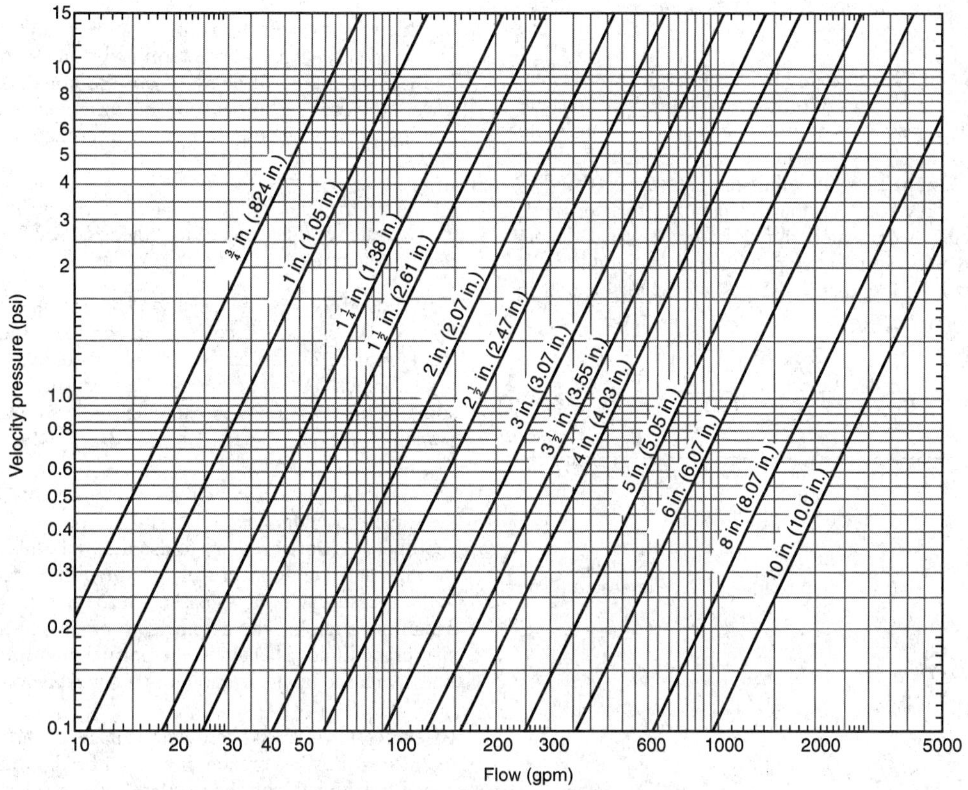

FIG. 6-11D. Graph for the determination of velocity pressure.

Figure 6-11E is an example of the velocity pressure component, determined as follows: The flow between the last sprinkler and the next to last sprinkler on the branch line was determined to be 17.7 gpm (67 L/min), and the pipe size is 1 in. (25.4 mm). The graph in Figure 6-11D is now used to locate a flow of 17.7 gpm (67 L/min) on the x-axis, going vertically from this point until the 1-in. (25.4-mm) line is intercepted. The corresponding pressure is taken from the y-axis, and is determined to be 0.3 psi (2.1 kPa). This procedure can now be repeated for each combination of pipe sizes and flows.

Up to this point, velocity pressure has been based on flow downstream from the sprinkler being considered; this has been confirmed by tests.[6] It has also been shown by those tests that, beyond the second sprinkler, velocity pressure should be figured from the flow on the upstream side of the sprinkler being considered. This is done by trial and error, assuming a flow from the sprinkler, calculating the velocity pressure from the total flow, determining a normal pressure, and calculating a flow from the normal pressure. If the calculated flow is not reasonably close to the assumed flow, assume a different flow and repeat the procedure until the two are close.

For example, assume a flow from the third sprinkler of 19.0 gpm (72 L/min) and assume that the pipe between the third and fourth sprinkler is $1^1/_4$ in. (32 mm). Total flow is 36.0 gpm (136 L/min) + 19.0 gpm (72 L/min) = 55.0 gpm (208 L/min). Velocity pressure is 0.9 psi (6 kPa), and normal pressure at the third sprinkler is therefore 15.7 psi (108 kPa) – 0.9 psi (6 kPa), or 14.8 psi (102 kPa). Corrected flow then becomes 21.6 gpm (82 L/min), which is not close enough to the 19.0 gpm (72 L/min) assumed. Try an assumed flow of 21.4 gpm (81 L/min). Velocity pressure at 57.4 gpm (217 L/min) is 1.0 psi (7 kPa); normal pressure is 14.7 psi (101 kPa), and the new corrected flow is 21.5 gpm (81 L/min). Total flow at the third sprinkler then becomes 36.0 gpm (136 L/min) + 21.5 gpm (81.4 L/min) = 57.5 gpm (217.6 L/min). The calculating procedure for the other sprinklers on the branch line is the same as it is for the third sprinkler.

If this set of conditions is compared to pipe schedule design, it will be seen that the 15-psi (103-kPa) minimum riser pressure has been exceeded, unless, as is quite probable, the pressure with 57.5-gpm (217.6-L/min) flow is substantially higher than that with a 500-gpm (1893-L/min) flow. Whether the pressure with 57.5-gpm (217.6-L/min) flow is higher than 15 psi (103 kPa) depends on the characteristics of the water supply. In any case, it appears that, with not many more sprinklers open, the pressure at the most remote sprinkler will be less than the 10 psi (69 kPa) selected in this example.

	Flow (gpm)	Friction loss (psi)	Velocity pressure (psi)	Normal pressure (psi)
① ②	17.7	1	0.3	11
② ③	36	3.8	1.2	13.3
③ ④	57.5	3.9	1.0	15.0

FIG. 6-11E. Example of velocity pressure.

Branch Lines, Cross Mains, Risers, and Fittings

Cross main pressure at the branch line connection: This is the normal pressure at the nearest open sprinkler increased by the friction loss and the velocity pressure in the intervening pipe. If the branch line is fed through a tee and nipple, additional friction loss allowances must be made, although the friction loss in nipples less than 6 in. (152 mm) long is customarily neglected.

Two branches in one line of sprinklers: These may have the same or different numbers of sprinklers. The pressure at the entrance to the two branches will always be the same. The computations starting at the end sprinklers will be duplicated for the number of open sprinklers.

After the discharge from any number of sprinklers on a branch line has been computed and the pressure needed to produce the flow has been determined, the entire branch line can be considered to have the discharge characteristics of a single orifice, and the discharge constant K in the formula $Q = K\sqrt{P}$ can be determined, P being the net pressure where flows are taken from tees in the cross main.

Branches on opposite sides of a cross main: These branches may have different numbers of sprinklers open, in which case the cross-main pressure must be the higher of the two computed values. This increases the discharge from the branch giving the lower computed pressure, and the actual discharge must be calculated for the higher pressure using the equation

$$\frac{Q_1}{Q_2} = \sqrt{\frac{P_1}{P_2}}$$

in which P_2 is taken as the higher pressure, Q_2 the corresponding increased discharge to be determined, and P_1 and Q_1 the pressure and corresponding discharge from the branch requiring only the lower pressure. This proportion and its associated concept is important in that only one pressure can be present at a particular junction point.

After the appropriate increased discharge has been determined, the two rates of flow can be combined and K calculated for the combined branches.

When sprinklers on the second branch line are assumed to have opened, starting at the cross-main sprinkler, the opened sprinkler most remote from the cross main is considered as the end sprinkler in the branch line computation, the next opened is the second, and so on, regardless of nonoperating sprinklers on the outer end of the branch.

Cross-main pressures: Cross-main pressures are calculated with the same procedure used for sprinklers on a single branch line, except that it is not necessary to use the trial-and-error procedure for the third and additional branch lines since the effect of change in velocity pressure with flows passing through tees in the cross main is usually negligible. The net pressure producing the flow in successive branch lines is taken as the normal pressure at the end branch line, increased by the friction loss in the pipe between the branches.

Riser pressure: Riser pressure is taken as the normal pressure at the nearest flowing branch, increased by the total friction loss between this branch and the riser and by the velocity pressure in the cross main at the riser connection.

Friction loss in fittings: This is generally included in calculations only when the fitting involves a change in the direction of flow. An exception to this is the fitting immediately preceding the sprinkler. Friction loss in control, gate, and check valves and in strainers, meters, backflow devices, and similar devices is always included. The friction loss in piping between the source of supply and the opened sprinklers obviously must be included in all calculations.

Differences in elevation must be allowed for on the basis that each foot (meter) of height represents 0.434 psi (9.82 kPa/m). In multistory buildings, this may be a substantial factor. Feed mains, cross mains, and branch lines within the same system may be looped or gridded to divide the total water flowing to the design area.

Grid systems: Many grid systems are currently being installed. The use of grid systems means that every sprinkler is fed from several directions, with a resulting reduction in pipe sizing. The hydraulic calculations required for a grid system are complex and almost impossible without the use of computers.

Sprinkler System Waterflow Curves

To avoid repetition of laborious computation of water flows and pressures when such information is needed in cases involving standard sprinkler, spray, or open-device systems, it is possible to prepare diagrams or waterflow curves from which riser pressures and corresponding total sprinkler flows may be determined for different numbers of opened sprinklers. One such series of curves, developed by Factory Mutual Research Corporation, and the piping arrangement and assumed pattern of opened sprinklers are shown in Figure 6-11F. Programmable, handheld calculators are also useful, fast, and accurate.

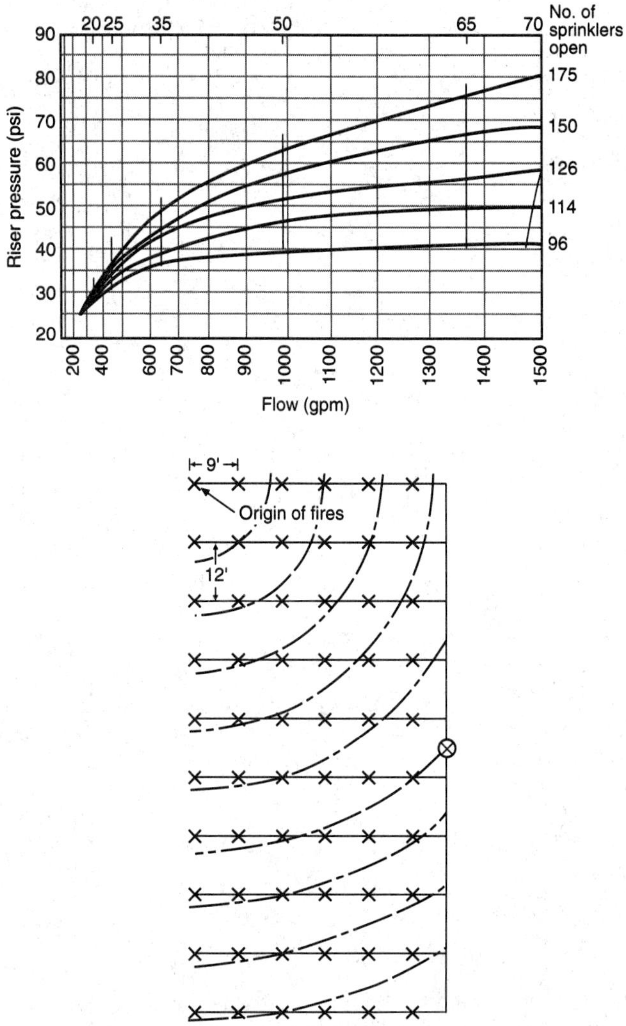

BIBLIOGRAPHY

References Cited

1. A Comparison of Several Computer Hydraulics Programs for the IBM PC and Compatibles, Society of Fire Protection Engineers, Boston, MA.
2. Wass, H. E., *Sprinkler Hydraulics*, IRM Insurance, White Plains, NY, 1983.
3. Crowley, J. E., *Hydraulic Design and Estimation of Fire Sprinkler Systems*, Crowley Design Group, Inc., King of Prussia, PA, 1982.
4. Wood, C. M., *Automatic Sprinkler Hydraulic Data*, "Automatic" Sprinkler Corporation of America, Cleveland, OH, 1980.
5. "Hydraulics of Fire Protection Systems," Loss Prevention Data Sheet 3-0, Factory Mutual Research Corp., Norwood, MA.
6. Nickerson, M. H., "Water Flow Characteristics of Sprinkler Systems," *Proceedings* of the Fifty-eighth Annual Meeting, National Fire Protection Association, May 17–21, 1954, Washington, DC, pp. 140–152.

NFPA Codes, Standards, and Recommended Practices

Reference to the following NFPA codes, standards, and recommended practices will provide further information on water supplies for sprinkler systems discussed in this chapter. (See the latest version of *The NFPA Catalog* for availability of current editions of the following documents.)

NFPA 13, *Standard for the Installation of Sprinkler Systems*
NFPA 15, *Standard for Water Spray Fixed Systems for Fire Protection*
NFPA 20, *Standard for the Installation of Centrifugal Fire Pumps*
NFPA 22, *Standard for Water Tanks for Private Fire Protection*
NFPA 24, *Standard for the Installation of Private Fire Service Mains and Their Appurtenances*
NFPA 231, *Standard for General Storage*
NFPA 231C, *Standard for Rack Storage of Materials*

FIG. 6-11F. *A flow curve for a side-central feed to sprinklers on a system having six sprinklers on each branch line is shown on the above graph. Below is the pattern of sprinklers opening on a side-central feed system. (Source: Factory Mutual Research Corp.)*

NFPA 231D, *Standard for Storage of Rubber Tires*
NFPA 231E, *Recommended Practice for the Storage of Baled Cotton*
NFPA 231F, *Standard for the Storage of Roll Paper*
NFPA 291, *Recommended Practice for Fire Flow Testing and Marking of Hydrants*

Additional Reading

Brock, P. D., "Critical Considerations in the Hydraulic Design of Automatic Sprinklers," *FPD Today*, Vol. 2, No. 1, Jan./Feb. 1990, pp. 15, 21.
Bryan, J. L., *Automatic Sprinkler and Standpipe Systems*, Chapter 4, 2nd ed., National Fire Protection Association, Quincy, MA, 1990.
Carson, W. G., and Klinger, R. L., *Fire Protection Systems: Inspection, Testing, and Maintenance Manual,* 2nd ed., National Fire Protection Association, Quincy, MA, 1993.
Hasegawa, H. K., and Lambert, H.E., "Reliability Study on the Lawrence Livermore National Laboratory Water-Supply System," *Proceedings* of the First International Symposium on Fire Safety Science, Hemisphere, 1986, pp. 611–620.
Keith, G. and Garner, D., "Sprinkler System Water Supply Analysis," *Fire Engineering*, Vol. 144, No. 12, December 1991, pp. 67–68, 70–71.
Milke, J. A., and Bryan, J. L., "Development of a Technique to Assess the Adequacy of the Municipal Water Supply for a Residential Sprinkler

System," Maryland Univ., College Park, National Institute of Standards and Technology, Gaithersburg, MD, Fire Administration, Emmitsburg, MD, NIST-GCR-91-600, Nov. 1991.

NFPA, *Automatic Sprinkler Systems Handbook*, 7th ed., National Fire Protection Association, Quincy, MA, 1996.

The SFPE Handbook of Fire Protection Engineering, 2nd ed., DiNenno, P. J., *et al.*, eds. National Fire Protection Association, Quincy, MA, 1995.

Schulte, R. C., "Reviewing Sprinkler System Hydraulic Calculations Generated by Computers," *Building Official and Code Administrator*, Vol. 24, No. 3, May/June 1990, pp. 19–23.

WATER SPRAY PROTECTION

Revised by Christopher L. Vollman

The term "water spray" refers to water that is discharged from specially designed nozzles or devices to produce a predetermined pattern, particle size, velocity, and density. The use of these designations cannot be taken as an indication of any specific discharge pattern or spray characteristics of the nozzles. A new NFPA standard is being developed, i.e., NFPA 750, *Standard on Water Mist Fire Protection Systems*, to address the use of very fine water droplets from spray nozzles. (See Section 6, Chapter 15, "Water Mist Suppression Systems.") The primary distinction between a water spray and a water mist system is that of specific coverage *vs.* general area coverage. Water spray systems have typically been provided to protect a specific piece of equipment with surface coverage.

Water spray for fire protection has been called water fog and various trade-name designations.

The discharge from water spray nozzles differs, generally, from the discharge from sprinklers. The pattern of the water spray discharged from spray nozzles onto a surface may be elliptical or circular, and the cross-section of the projected discharge is conical. The water spray is forcefully directed onto the object or surface being protected. The pattern of spray nozzle discharge must carry water spray over the distance between the nozzle and the target, compensate for wind and draft conditions, and effectively hit the surface to be protected. The required discharge density in gal per min per sq ft [(L/min)/m^2)] and complete coverage of the area to be protected are also essential elements.

Unlike most automatic sprinkler systems, water spray systems are of the deluge type. Spray nozzles are not equipped with fusible operating elements.

Spray nozzles may be equipped with internal pilot-pressure-controlled valves or with frangible or removable plugs or caps that will be displaced when the water spray deluge system is actuated. Pilot-pressure-controlled valves, plugs, or caps prevent corrosive atmospheres from entering the piping system. They may be used to retain water in the piping system for the purpose of speeding its delivery from the spray nozzles.

This chapter covers the uses and applications of water spray systems for fire suppression, control, and extinguishment and describes the components of spray systems and the specialized uses of the systems. Because of the similarities between sprinkler systems and water spray systems, their water supply requirements, some of the equipment used in the systems, and the hydraulic calculations for determining water supplies, other chapters in Section 6 should be consulted. These include Chapter 1, "Water and Water Additives for Fire Fighting," including the extinguishing properties of water and safe clearances from electrical equipment; Chapter 10, "Automatic Sprinkler Systems"; and Chapter 11, "Water Supplies for Sprinkler Systems."

Section 5, Chapter 2, "Automatic Fire Detectors," including releasing devices that can be used with water spray systems, and Section 9, Chapter 20, "Materials-Handling Equipment," including protection of conveyor openings, should also be consulted.

NFPA 15, *Standard for Water Spray Fixed Systems for Fire Protection* (hereinafter referred to as NFPA 15), is the standard applicable to water spray systems and should be consulted on details of system design and installation not covered in NFPA 13, *Standard for the Installation of Sprinkler Systems*.

USES FOR WATER SPRAY PROTECTION

Generally, water spray can be used effectively to extinguish a fire, control a fire, protect exposures, and/or prevent a fire.

Extinguishment

Water spray extinguishes a fire by cooling it, smothering it with the steam produced, emulsifying or diluting some flammable liquids, or by combining these factors.

Controlled Burning

With its consequent limitation of fire spread, controlled burning may be applied if the burning combustibles cannot be extinguished by water spray, or if extinguishment is not desirable.

Exposure Protection

Exposures are protected by applying water spray directly to the exposed structures or equipment to remove or reduce the heat transferred to them from the exposing fire. Water spray curtains mounted at a distance from the exposed surface are less effective than direct application.

Christopher L. Vollman, P.E., C.S.P. senior consulting engineer for Rolf Jensen and Associates, Inc., based in Houston, TX. He is chairman of the NFPA Technical Committee on Water Spray Fixed Systems.

Prevention of Fire

It is sometimes possible to use water spray to dissolve, dilute, disperse, or cool flammable or combustible materials before they are ignited by an exposing ignition source.

APPLICATION OF WATER SPRAY SYSTEMS

Water spray can be used to protect the following types of materials or equipment:

1. Ordinary combustible materials, such as paper, wood, and textiles, particularly to extinguish fires in such materials rather than control them.
2. Electrical equipment installations, such as transformers, oil switches, and rotating electrical machinery.
3. Flammable gases and liquids, particularly to control fires in these materials and to extinguish certain types of fires involving combustible liquids.
4. Flammable liquid and gas tanks, processing equipment, and structures, as protection against exposure fires.
5. Open cable trays and runs containing electrical cables or tubing.

Fixed water spray systems are designed specifically to provide optimum control, extinguishment, or exposure protection for special fire protection problems. Limitations to the use of water spray that should be recognized involve the nature of the equipment to be protected, the physical and chemical properties of the materials involved, and the environment of the hazard.

FIXED WATER SPRAY SYSTEMS

A water spray system is a special fixed pipe system connected to a reliable supply of fire protection water, and equipped with water spray nozzles for specific water discharge and distribution over the surface or area to be protected. The piping system is connected to a water supply through a deluge valve that can be actuated both automatically and manually to initiate the flow of water.

Automatic system actuation valves for spray systems can be actuated electrically by the operation of automatic detection equipment, such as heat detectors, relay circuits, and gas detectors, or mechanically by hydraulic or pneumatic systems, depending upon the operating mode of the individual valves. Generally, each manufacturer of system actuation valves, most of which can do dual service in deluge systems, provides its own particular combination of system actuation valve, releasing mechanism, detection system, and supervisory service.

Application of Systems

Fixed water spray systems are most commonly used to protect equipment from exposure fires in flammable liquid and gas tankage, piping, and equipment; in electrical equipment, such as transformers, oil switches, rotating machinery, and cable trays; in structural supports; and in conveyor systems and the openings in firewalls and floors through which they pass. The type of water spray required for any particular hazard will depend on the nature of the hazard and the purpose for which the protection is provided.

A water spray installation being tested at a group of liquefied petroleum gas tanks is shown in Figure 6-12A. The spray system shown is designed to give complete surface wetting with a specified water density, taking into consideration nozzle types, sizes, and spacing; the influence of wind and drafts; the probability of water rundown; the prevention of the formation of difficult-to-wet deposits of soot or carbon on surfaces; the overlap of water discharge pat-

FIG. 6-12A. Water spray protection for LP-Gas tanks. The spray keeps the tanks cool in case of fire, prevents the liquid contents from boiling away, and protects the tank shells against rupture due to localized high-temperature flame impingement.

terns onto the surfaces; and the ability of the water supply to furnish adequate pressure to all of the nozzles. Ordinarily, it is neither desired nor expected that water spray extinguish escaping liquefied petroleum gas. However, the cooling effect of the water on the tanks may reduce and control the rate of burning and reduce the severity of the exposure until the gas supply to the fire is exhausted or can be shut off.

Water spray protection for a group of oil-filled electric transformers is shown under test in Figure 6-12B. Due to the relatively high flash point and boiling point of transformer oil, transformer fires can be extinguished quickly by properly designed water spray systems. However, the metal surfaces of the transformer case and its structural supports must be protected from radiant heat while the burning transformer oil is being extinguished. Care must be taken to

FIG. 6-12B. A water spray system for oil-filled electric power transformers. A thick layer of crushed stone and subsurface drainage is provided around the base of the transformer installation to prevent the possibility that burning oil might flow beyond the ground area protected by the spray.

locate nozzles, piping, and supports the prescribed distances from electrically charged parts of the transformer installation, and direct application of water spray to electrically charged terminals or insulating bushings should be avoided. Electricity to the transformer should be shut off automatically before water spray is applied.

Design of Systems

The practical location of the piping and nozzles with respect to the surface to which the spray is to be applied or to the zone in which the spray is to be effective is determined largely by the physical arrangement and protection needs of the installation requiring protection. Once the criteria are established, the size of the nozzles to be used, the angle of the nozzle discharge cone, and the water pressure needed can be determined.

The first factor to determine is the water density required to extinguish the fire or to absorb the expected heat from exposure or combustion. When this has been determined, a nozzle may be selected that will provide that density at a velocity adequate to overcome air currents and to carry the spray to the equipment to be protected. Each nozzle chosen must have the proper angle of discharge to cover the area to be protected by the nozzle.

Determining the proper density needed for extinguishment requires considerable engineering judgment and, in the case of flammable or combustible liquids, depends on such characteristics of the fuel as vapor pressure, flash point, viscosity, water solubility, and specific gravity. The density varies between 0.2 gpm per sq ft and 0.5 gpm per sq ft [8.1 (L/min)/m^2 to 20.4 (L/min)/m^2] of protected surface.

For exposure protection of vessels, a density of 0.25 gpm per sq ft [10.2 (L/min)/m^2] should provide sufficient cooling to limit an exposure fire's heat input through the vessel walls to 6,000 Btu per hr per sq ft (18 930 W/m^2). The water density required for exposure protection of structural supports and miscellaneous equipment, such as cable trays and runs, pipe racks, transformers, and belt conveyors, varies from 0.10 gpm per sq ft [4.1 (L/min)/m^2] to 0.3 gpm per sq ft [12.2 (L/min)/m^2] of exposed surface area. NFPA 15 should be consulted for more details on required distribution densities.

Once the type of nozzle has been selected and the location and spacing to give the desired area coverage have been determined, hydraulic calculations are made to establish the appropriate pipe sizes and water supply requirements.

When water spray is to be used to protect oil-filled electrical equipment, such as transformers and large switch gear, special care must be taken to provide safe electrical clearances. Special fixed water spray nozzles have been developed to provide adequate spray density and range to accommodate wind, along with a simplified piping arrangement that is spaced safely from energized electrical parts.

Size of Water Spray Systems

Many factors govern the size of a water spray system, including the nature of the hazard or combustibles involved, the amount and type of equipment to be protected, the adequacy of other protection, and the size of the area which could be involved in a single fire. The size of the system needed may be minimized by taking advantage of possible subdivision by firewalls; by limiting the potential spread of flammable liquids with dikes, curbs, or special drainage; by installing water curtains or heat curtains; or by using a combination of these features.

The water demand will be heavy because water spray systems must perform as deluge-type systems with all nozzles open and because high densities of water are needed. Each hazard should be protected by its own water spray system to limit the cumulative demand on the water supply.

NFPA 15 advises that the size of a single water spray system be limited only by the available water supply so that the designed discharge rate will be calculated at the minimum pressures for which the nozzles are effective. Experience has shown that, in most installations, a single system should not exceed a design discharge rate of 3,000 gpm (11 356 L/min). Separate hazard areas should be protected by separate systems.

Water Supplies

Fixed spray systems usually are supplied from one or more of the following:

1. Connections from a reliable waterworks system of adequate capacity and pressure.
2. Automatic fire pumps with reliable power and a water supply of adequate capacity and reliability.
3. An elevated gravity tank of adequate capacity and elevation.

The capacity of pressure tanks generally is not adequate to supply water spray systems. However, they may be acceptable as water supplies to small systems whose water and pressure requirements do not exceed the tank's capabilities.

Water Demand Rate

The water supply must be adequate to supply the operating water spray system(s) with the required gpm (L/min) at effective pressure. Water spray systems adjacent to the hazard initially protected may require additional water. The water supply should be able to supply hose streams simultaneously. The total required water supply pressure and flow rates should be considered when the system is designed. The duration of the discharge required will vary according to the nature of the hazard, the purpose for which the system is designed, and other factors which can be evaluated only for each installation.

Water demand is specified in terms of the density of a uniformly distributed spray measured in gal per min per sq ft [(L/min)/m^2] of area protected. The discharge rate per unit of area depends on whether the spray system is installed to extinguish a fire, to control a fire, or to protect an exposure, and on the characteristics of the materials involved.

Pipe Sizes

Pipe sizes must be calculated for each system so that the water at the spray nozzles will have adequate pressure, to provide the necessary flow and spray pattern. A procedure for making the hydraulic calculations for a fixed pipe water spray system is given in the Appendix to NFPA 15.

Selection and Use of Spray Nozzles

The selection of spray nozzles should take into consideration such factors as the character of the hazard to be protected, the purpose of the system, and the possibility of severe winds or drafts.

High-velocity spray nozzles, generally used in piped installations, discharge in the form of a spray-filled cone, while low-velocity spray nozzles usually deliver a much finer spray in the form of a spray-filled spheroid or cone.

Nozzles of one type cannot be substituted for nozzles of another type when a water spray system is being repaired or nozzles are being replaced without taking into account the possibility of seriously affecting the effectiveness of the system. In general, the velocity and distribution of size of water droplets affects the "reach" or range and the area of coverage of the water spray pattern.

Some spray nozzles produce spray by giving the water streams a rotary motion in spiral passages inside the nozzles. These rotating streams are mixed internally with a center stream to project a solid cone of water spray from the nozzle. These nozzles are known as the internally impinging type. Sectional views of spray nozzles having internal spiral water passages are shown in Figure 6-12C.

Another type of water spray nozzle uses the deflector principle of a sprinkler. The water-discharge orifice projects a solid, cylindrical stream of water onto a deflector that alters the stream into a conical distribution of water spray. This type of water spray nozzle is shown in Figure 6-12D.

One characteristically different type of water spray nozzle discharges water along the axis of a spiral of diminishing inside diameter. This spiral continuously peels off a thin layer of water from the surface of the cone, and this thin layer of water breaks into spray as it leaves the spiral. (See Figure 6-12E.)

Strainers

Strainers are ordinarily required in the supply lines of fixed piping spray systems to keep the nozzles from clogging. (See Figure 6-12F.) The baskets of those selected should have a mesh small enough to protect the smallest water passages in the nozzles used.

Water spray nozzles with very small water passages may have their own internal strainer, as well as a supply-line strainer to remove larger foreign material.

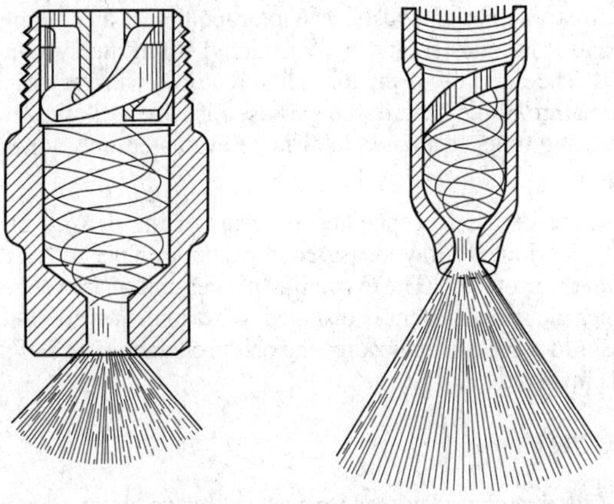

FIG. 6-12E.   A spiral-type water spray nozzle.

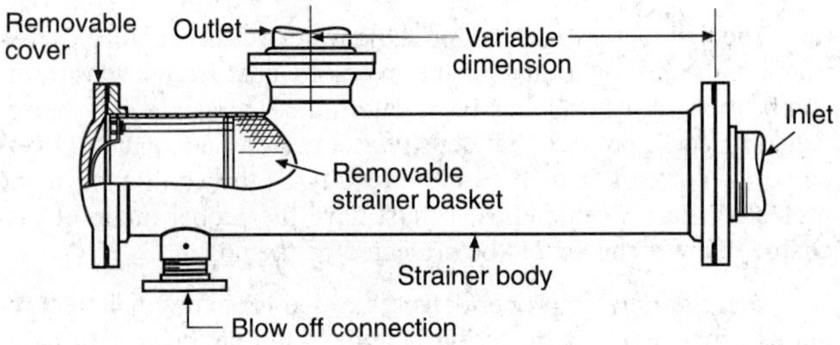

FIG. 6-12F.   A Grinnell Model A strainer having a basket-type screen of corrosion-resistant metal. It is available in 3-in., 4-in., 6-in., 8-in., and 10-in. sizes. (For SI units: 1 in. = 25.4 mm.)

Drainage

Fixed water spray systems discharge large quantities of water. Special drainage and disposal facilities are important to control the spread of flammable or hazardous liquids to areas adjacent to the fire. Pitched floors, curbs, trenches, dikes, and sumps should be provided to direct the mixture of flammable or hazardous liquids and water runoff. These should be connected to a facility that separates flammable and hazardous liquids from the water. These liquids should be impounded in controlled areas until they can be disposed of safely.

Maintenance

It is important that fixed water spray systems be inspected and maintained on a regular basis. Chapter 7 of NFPA 25, *Standard for the Inspection, Testing, and Maintenance of Water-Based Fire Protection Systems*, provides specific requirements for periodic inspection, testing, and maintenance of water spray systems. Included in a scheduled inspection program are such items as strainers, piping, control valves, system actuation valves, heat-actuated devices, and the spray nozzles, particularly those equipped with strainers. Flow tests are conducted frequently to ensure satisfactory operation. After the tests have been completed, it is necessary to clean all strainers and to check all valves to be sure the system is in normal operating condition.

Nozzles with blowoff caps or plugs are provided to protect the interior of the piping and the nozzles located in areas containing paint vapor, corrosive vapors, dusts, or other foreign matter. Other blowoff caps or plugs are attached to the nozzles to allow them to retain water in the piping system, and thus reduce the time between the operation of the deluge valve and the discharge of water spray from the nozzles. (See Figure 6-12G.)

FIG. 6-12C.   Water spray nozzles having internal spiral water passages.

FIG. 6-12D.   Water spray nozzles using the deflector principle of the standard automatic sprinkler.

FIG. 6-12G.   A Grinnell Mulsifyre nozzle with blowoff cap.

Regular inspection and maintenance should confirm that caps and plugs are in place.

SPECIALIZED SYSTEMS UTILIZING WATER

Specialized systems using water spray have been developed to fill particular fire control needs in the explosives industry. These systems are referred to as ultra-high-speed water deluge spray systems. (See Section 6, Chapter 14, "Ultra High-Speed Suppression Systems for Explosive Hazards," for additional information on these systems.)

BIBLIOGRAPHY

NFPA Codes, Standards, and Recommended Practices

Reference to the following NFPA codes, standards, and recommended practices will provide further information on water spray protection discussed in this chapter. (See the latest version of *The NFPA Catalog* for availability of current editions of the following documents.)

NFPA 13, *Standard for the Installation of Sprinkler Systems*

NFPA 14, *Standard for the Installation of Standpipe and Hose Systems*

NFPA 15, *Standard for Water Spray Fixed Systems for Fire Protection (Note especially Appendix A-4-5.2.)*

NFPA 18, *Standard on Wetting Agents*

NFPA 20, *Standard for the Installation of Centrifugal Fire Pumps*

NFPA 22, *Standard for Water Tanks for Private Fire Protection*

NFPA 24, *Standard for the Installation of Private Fire Service Mains and Their Appurtenances*

NFPA 25, *Standard for the Inspection, Testing, and Maintenance of Water-Based Fire Protection Systems*

NFPA 30, *Flammable and Combustible Liquids Code*

NFPA 69, *Standard on Explosion Prevention Systems*

NFPA 70, *National Electrical Code®*

NFPA 72, *National Fire Alarm Code*

NFPA 80A, *Recommended Practice for Protection of Buildings from Exterior Fire Exposures*

NFPA 321, *Standard on Basic Classification of Flammable and Combustible Liquids*

NFPA 325, *Guide to Fire Hazard Properties of Flammable Liquids, Gases, and Volatile Solids*

Additional Reading

Acton, M. R., Sutton, P., and Wickens, M. J., "Investigation of the Mitigation of Gas Cloud Explosions by Water Sprays," *Gas Engineering and Management*, Vol. 31, No. 7–8, July/Aug. 1991, pp. 166–172; Institution of Chemical Engineers, Piper Alpha: Lessons for Life-Cycle Safe-ty Management Symposium, IChemE Symposium Series No. 122, Sept. 26–27, 1990, London, England, 1991.

Atreya, A., Crompton, T., and Suh, J., "Experimental and Theoretical Study of Mechanisms of Fire Suppression by Water," Univ. of Michigan, Ann Arbor, NISTIR 5499, Sept. 1994.; National Institute of Standards and Technology, Annual Conference on Fire Research: Book of Abstracts, October 17–20, 1994, Gaithersburg, MD, 1994, pp. 67–68.

Brenton, J. R., Thomas, G. O., and Al-Hassan, T., "Small-Scale Studies of Water Spray Dynamics During Explosion Mitigation Tests," *Proceedings* of the Institution of Chemical Engineers Symposium Series, Hazards XII: European Advances in Process Safety Conference, No. 134, 1994, Rugby, England, Institution of Chemical Engineers, Rugby, England, 1994, pp. 393–403.

Chow, W. K., and Shek, L. C., "Physical Properties of a Sprinkler Water Spray," *Fire and Materials*, Vol. 17, No. 6, Nov./Dec. 1993, pp. 279–292.

Chow, W. K., and Wong, V. M. K., "Water Penetration Ratio of a Sprinkler Water Spray," First Asian Conference on Fire Science and Technology (ACFST), October 9–13, 1992, Hefei, China, International Academic Publishers, China, 1992, pp. 467–474.

Evans, D. D., "Overview of Fire Suppression with Water," *Chemical and Physical Processes in Combustion*, 1988 Technical Meeting, Clearwater Beach, 1988.

Everest, J., "Water Sprays Take Shape," *Fire Prevention*, No. 239, May 1991, pp. 11–13.

Jackman, L. A., Nolan, P. F., and Morgan, H. P., "Characterization of Water Drops From Sprinkler Sprays," Swedish Fire Research Board and National Institute of Standards and Technology, First International Conference on Fire Suppression Research, May 5–8, 1992, Stockholm, Sweden, 1992, pp. 159–184.

Jones, A., and Thomas, G. O., "Mitigation of Small-Scale Hydrocarbon-Air Explosions by Water Sprays," *Transactions Institution of Chemical Engineers*, Vol. 70, No. A, 1992, pp. 197–199.

Jones, A., and Thomas, G. O., "Action of Water Sprays on Fires and Explosions: A Review of Experimental Work," *Process Safety Environment*, Vol. 71, No. 1, 1993, pp. 41–49.

Kokkala, M., Andsten, T., and Bjorkman, J., "Extinguishment of Liquid Fires with Sprinklers and Water Sprays: Description of Tests," *Valtion Teknillinen Tutkimuskeskus*, No. 593, 1989, Technical Research Institute of Finland, Espo, Finland.

Kokkala, M., "Extinguishment of Liquid Fires with Sprinklers and Water Sprays: Analysis of the Test Results," VTT-Technical Research Center of Finland, Espoo, VTT Research Reports 696, Project PAL9002, Aug. 1990.

Kokkala, M. A., "Fixed Water Sprays Against Open Liquid Pool Fires," Swedish Fire Research Board and National Institute of Standards and Technology, First International Conference on Fire Suppression Research, May 5–8, 1992, Stockholm, Sweden, 1992, pp. 129–158.

Lev, Y., "Cooling Sprays for Hot Surfaces," *Fire Prevention*, No. 222, Sept. 1989, pp. 42–47.

Lev, Y., and Strachan, D. C., "Dry Testing of Water Spray Systems," *Journal of Fire Sciences*, Vol. 7, No. 1, Jan.-Feb. 1989, pp. 3–21.

Milke, J. A., Evans, D. D., and Hayes, W., "Water Spray Suppression of Fully Developed Wood Crib Fires in a Compartment," NBSIR 88-37453, June 1988, National Institute of Standards and Technology, Gaithersburg, MD.

Nishro, G., and Machida, S., "Pool Fires under Atmosphere and Ventilation in Steady-state Burning," *Fire Technology*, Vol. 23, No. 2, 1987, pp. 146–155.

Qiao, Y. M., and Chandra, S., "Evaporative Cooling Enhancement by Addition of a Surfactant to Water Drops on a Hot Surface," Toronto Univ., Ontario, Canada, HTD-Vol. 304; *Proceedings* of the Thirtieth National Heat Transfer Conference, 1995, Combustion and Fire Research, Heat Transfer in High Heat-Flux Systems, Volume 2, HTD-Vol. 304, Aug. 6–8, 1995, Portland, OR, 1995, pp. 63–71.

Rasbash, D. J., "Extinction of Fire with Plain Water: A Review," *Proceedings* of the First International Symposium on Fire Safety Science, Hemisphere, Washington DC, 1986, pp. 1145–1163.

Schnieder, M.E., and Kent, L.A., "Measurements of Gas Velocities and Temperatures in a Large Open Pool Fire," *Fire Technology*, Vol. 25, No. 1, Feb. 1989, pp. 51–80.

Schneider, V., and Hofmann, J., "Feldmodell-Simulation von Kohlenwasserstoff-Raumbranden und Spruhnebel-Loschversuchen. [Field-Model

Simulation of Hydrocarbon Room Fires and Fire Fighting Tests With Water Spray.]," *VFDB*, Feb. 1993, pp. 67–73.

Schoen, W., and Droste, B., "Investigations of Water Spraying Systems for LPG Storage Tanks by Full-Scale Fire Tests," *Journal of Hazardous Materials*, Vol. 20, Dec. 1988, pp. 73–82.

Schultz, N., "Fire Protection for Cable Trays in Petrochemical Facilities," *Plant/Operations Progress*, Vol. 5, No. 1, 1986, pp. 35–39.

Shokri, M.Y., *Predicting Thermal Radiation from Large Open Pool Fires*, MS Thesis, Worcester Polytechnic Institute, Worcester, MA, 1987.

Stroup, D. W., and Evans, D. D., "Suppression of Post-Flashover Compartment Fires Using Manually Applied Water Sprays," National Institute of Standards and Technology, Gaithersburg, MD,

Taylor, C. D., and Furno, A. L., "Evaluation of High-Pressure Front-Mounted Water Jets for Frictional-Ignition Suppression," Report No. 9237, 1989, United States Bureau of Mines, Washington, DC.

Wighus, R., "Active Fire Protection: Extinguishment of Enclosed Gas Fires with Water Sprays," Norwegian Fire Research Lab., Trondheim, Norway, STF25 A91028, June 1989.

Wighus, R., "Extinguishment of Enclosed Gas Fires with Water Spray," *Proceedings* of the Third International Symposium on FireSafety Science, Elsevier Applied Science, New York, 1991, pp. 997–1006.

Yang, J. C., Boyer, C. I., and Grosshandler, W. L., "Minimum Mass Flux Requirements to Suppress Burning Surfaces with Water Sprays," National Institute of Standards and Technology, Gaithersburg, MD, NISTIR 5795, April 1996.

Yow, H. Z., Kung, H. C., and Han, Z., *Spray Cooling in Room Fires*, National Bureau of Standards, Gaithersburg, MD, 1986.

FAST-RESPONSE SPRINKLER TECHNOLOGY

Arthur E. Cote and Russell P. Fleming

The ability to measure and control sprinkler sensitivity was the single most significant development in sprinkler technology in the 1980s. This ability has led to the development of new fast-response sprinkler technology that is demonstrating improved fire protection for both life and property.

This chapter covers the development of residential and other fast-response sprinklers, first for the residential environment and later for other applications. It also discusses the specific types of fast-response sprinklers currently available and some of the incentives that have been considered to encourage the widespread use of residential sprinklers.

General information on fire sprinklers can be found in and Section 6, Chapter 9, "Automatic Sprinklers," and Section 6, Chapter 10, "Automatic Sprinkler Systems." A discussion of the theory of fire sprinkler performance is in Section 6, Chapter 8, "Theory of Automatic Sprinkler Performance."

MEASURING SPRINKLER SENSITIVITY

Much of the original work in the area of measuring sprinkler sensitivity was done at Factory Mutual Research Corporation (FMRC) under the sponsorship of the United States Fire Administration (USFA) during the development of the residential sprinkler. [1,2] Important contributing research was also performed at the British Fire Research Station and the National Institute of Standards and Technology (NIST). [3,4,5,6]

The decade of progress in this area climaxed late in 1990, when agreement was reached within the working group on sprinkler and water spray equipment of the International Standards Organization (ISO) for a standardized approach to sprinkler sensitivity requirements and testing. The agreement, included in ISO 6182/1, "Requirements and Methods of Test for Sprinklers," uses a combination of sprinkler test procedures developed by laboratories in the United States and Europe and establishes the three ranges of sprinkler sensitivity characteristics shown in Figure 6-13A.

These ranges of sensitivity are based both on the response time index (RTI) of the device and on its conductivity (C). RTI is a measure of pure thermal sensitivity, which indicates how fast the sprinkler can absorb heat from its surroundings sufficient to cause activation. The conductivity factor is important in measuring how much of the heat picked up from the surrounding air will be lost to the sprinkler fittings and waterway.

Figure 6-13A shows three broad ranges of sprinkler sensitivity: standard, special, and fast response. Traditional sprinkler hardware falls into the standard-response category. The fast-response category is being used for new types of sprinklers for which fast response is considered important. The special-response category is being used in some countries for special types of sprinklers that may be installed in conformance with appropriate national installation standards. In the United States, this includes some of the extended coverage sprinklers.

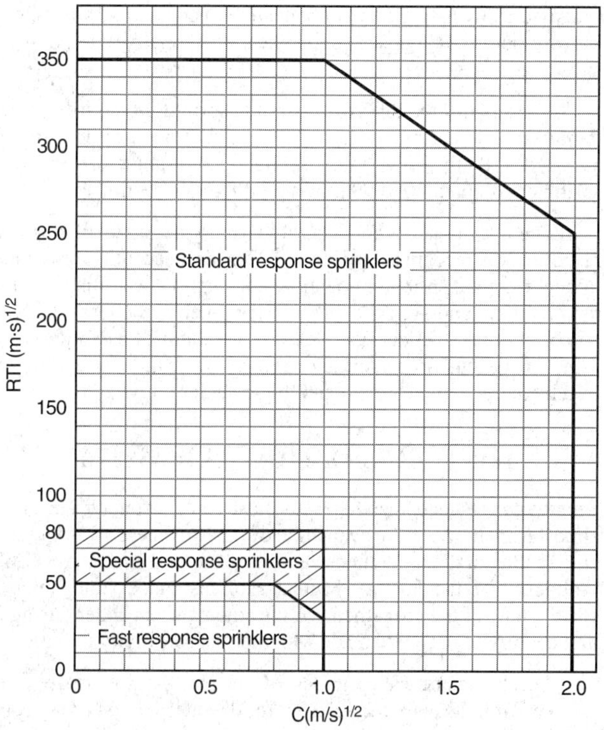

FIG. 6-13A. *International sprinkler sensitivity ranges, RTI vs. conductivity. (For U.S. conversion: 1 ft = 0.305 m.)*

Arthur E. Cote, P.E., is senior vice president, operations, of the NFPA.
Russell P. Fleming, P.E., is vice president of engineering of the National Fire Sprinkler Association.

Sprinkler response time as a function of the temperature rating of the operating element is well understood, that is, a 165°F (74°C) rated sprinkler would operate when its temperature reaches 165°F (74°C), plus or minus a few degrees. Because of the "thermal lag" of the link or bulb mass, however, the air temperature may be as high as 1000°F (538°C) before the element operates. The smaller mass of the operating element of a fast-response sprinkler permits it to follow a temperature rise in the surrounding air more rapidly, resulting in faster operation. The actual sensitivity requirements of the first fast-response sprinklers, intended as residential sprinklers, were arrived at somewhat by trial and error during developmental test work. To measure sensitivity, FMRC researchers first applied the concept of the "tau" factor and later developed the RTI.

Both the tau factor and RTI refer to the performance of a sprinkler or its operating element in a standardized air oven tunnel test. The test is known as a "plunge" test because a sprinkler at room temperature is plunged into a heated air stream of known constant temperature and velocity.[1,2] In the plunge test, the tau factor is the time at which the excess temperature of the sensing element of the sprinkler is approximately 63 percent of the excess gas temperature. In other words, the tau factor is the time at which the temperature of the sprinkler thermal element has risen 63 percent of the way to the higher temperature of the heated air. The smaller the tau factor, the faster the sprinkler sensing element heats up and operates.

The tau factor is independent of the air temperature used in the plunge test, but is inversely proportional to the square root of the air velocity. During the early development of the residential sprinkler, a tau factor of 21 sec was considered to indicate the needed level of sensitivity, but this was associated with the specific velocity of 5 ft/sec (1.52 m/s) used in the FMRC plunge test. Since the tau factor changes with the velocity of heated air moving past the sprinkler, it is a fairly inconvenient measure of sprinkler sensitivity.

The RTI has replaced the tau factor as the measure of sensitivity and is determined simply by multiplying the tau factor by the square root of the air velocity at which it is found. The RTI is therefore practically independent of both air temperature and air velocity. Comparisons of RTI give a good indication of relative sprinkler sensitivity.

The smaller the RTI, the faster the sprinkler operation. Standard response sprinklers have RTIs in the range of 180 to 650 $\text{sec}^{1/2}\text{ft}^{1/2}$ (100 to 350 $\text{s}^{1/2}/\text{m}^{1/2}$), while the RTI range for residential sprinklers is about 50 to 90 $\text{sec}^{1/2}\text{ft}^{1/2}$ (28 to 50 $\text{s}^{1/2}/\text{m}^{1/2}$).

The need to add a conductivity term to the model of sprinkler response was recognized in 1986.[7,8] This term accounts for the loss of heat from the sprinkler operating element to the sprinkler frame, its mounting, and even the water in the pipe. These losses can become significant under low-velocity conditions, particularly for some of the flush-type sprinkler designs with little insulation between the operating element and the sprinkler body.

HISTORICAL DEVELOPMENT

Sensitivity was recognized as a key attribute of automatic sprinklers almost from the start. An evaluation of "sensitiveness" was included among the first comprehensive tests ever performed by C.J.H. Woodbury of the Factory Mutual Fire Insurance Companies in 1884. Many of the refinements to the automatic sprinkler during its first century of use were made with thermal sensitivity in mind.

Yet the standard sprinkler of the 1970s was really little changed from its ancestors. Except for the introduction of the spray sprinkler in the early 1950s to improve the sprinkler discharge pattern, sprinkler design, including the sensitivity of release mechanisms, remained about the same. Sprinklers were essentially devices intended to protect property, although their effectiveness in providing life safety from fires was outstanding.

Development of NFPA 13D

In 1973, in response to recommendations in the Presidential Commission report, "America Burning," the NFPA Committee on Automatic Sprinklers appointed a Subcommittee on Residential and Light-Hazard Occupancies to prepare a residential sprinkler standard. In 1975, the first edition of NFPA 13D, *Standard for the Installation of Sprinkler Systems in One- and Two-Family Dwellings and Manufactured Homes* (hereinafter referred to as NFPA 13D), was published, based on expert judgment and the best information available at that time. (Note the first several editions of NFPA 13D used the term "Mobile Home" in the title).

The purpose of the standard was "to provide a sprinkler system that will aid in the detection and control of dwelling fires and thus provide improved protection against injury, life loss, and property damage." The standard required the use of an NFPA 13, *Standard for the Installation of Sprinkler Systems* (hereinafter referred to as NFPA 13), light-hazard water application density of 0.10 gpm/ft² [4 mm/min], but it permitted other concessions. The water supply could be based on the area of the largest room or 25 gpm (95 L/min), whichever was less; the total water supply required was to be only 250 gal (946 L); and sprinkler spacing of 256 sq ft (23.7 m²) was permitted, even though NFPA 13 allowed only 225 sq ft (20.9 m²). Further, NFPA 13D permitted sprinklers to be omitted from certain areas where the incidence of life loss from fires was shown statistically to be low. NFPA 13 had always required complete sprinkler protection in order to safeguard property properly. In departing from this ideal, the 1975 edition of NFPA 13D became the first attempt at a "life safety" sprinkler standard. In spite of these concessions, actual installations based on this standard were rare, primarily due to cost.

Residential Sprinkler Research

Beginning in 1976, the National Fire Prevention and Control Administration (NFPCA), which was later renamed the U.S. Fire Administration (USFA), acted on its mandate to reduce the nation's fire losses and funded research programs focusing on the residential fire problem in general and residential sprinkler protection in particular. The NFPCA/USFA programs included studies to assess the impact sprinklers would have on reducing deaths and injuries in residential fires.[9] Other studies evaluated the design, installation, practical usage, and water acceptance factors that would have an impact on achieving reliable and acceptable systems[10]; the minimum water discharge rates and automatic sprinkler flow required, and response sensitivity and design criteria[11-13]; and full-scale tests of prototype residential sprinkler systems.[14-18]

The research showed that a more sensitive sprinkler was needed to respond faster to both smoldering and fast-developing residential fires for two reasons. First, fires had to be controlled quickly in order to prevent the development of lethal conditions in small residential compartments. In addition, fires had to be attacked while still small if they were to be controlled with the water supplies typically available in residences, i.e., 20 to 30 gpm (76 to 114 L/min), and if low costs were to be achieved.

Full-scale tests conducted by FMRC resulted in the development of a prototype fast-response sprinkler that could control or suppress typical residential fires with the operation of not more than two sprinklers. It could also operate fast enough to maintain survivable conditions within the room of fire origin.[14] Survivable conditions were established as follows:

1. Maximum gas temperature at eye level of 200°F (93°C).

2. Maximum ceiling surface temperature of 500°F (260°C).
3. Maximum carbon monoxide concentration of 1,500 parts per million.

Thus, the sprinkler concept expanded from the traditional role of property protection to include life safety. Full-scale field tests were then conducted in Los Angeles to establish system design parameters using the new prototype fast-response "residential sprinkler."[15-18]

The data from these tests were studied by the National Fire Protection Association Technical Committee on Automatic Sprinklers and used to establish the criteria for the 1980 edition of NFPA 13D.

Revised NFPA 13D Design Requirements

The design criteria in the 1980 edition of NFPA 13D included for the first time the requirement that all sprinklers be "listed residential sprinklers." Figure 6-13B shows the first two listed residential sprinklers.

Other initial basic design requirements in the revamped NFPA 13D were as follows.

Performance criteria: To prevent flashover in the room of fire origin, when sprinklered, and to improve the chance for occupants to escape or be evacuated.

Design criteria:

1. Only listed residential sprinklers to be used.
2. Minimum 18 gpm (68 L/min) to any single operating sprinkler and 13 gpm (49 L/min) to all operating sprinklers in the design area up to a maximum of two sprinklers.
3. Maximum area protected by a single sprinkler of 144 sq ft (13.4 m²).
4. Maximum distance between sprinklers of 12 ft (3.7 m).
5. Minimum distance between sprinklers of 8 ft (2.4 m).
6. Maximum distance from a sprinkler to a wall or partition of 6 ft (1.8 m).

Application rates, design areas, areas of coverage, and minimum design pressures other than those specified above were permitted to be used with special sprinklers listed for such special residential installation conditions.

Sprinkler coverage: Sprinklers to be installed in all areas with the following exceptions.

Exception No. 1:   Sprinklers allowed to be omitted from bathrooms no larger than 55 sq ft (5.1 m²).

FIG. 6-13B.   The first two listed residential sprinklers. Shown are the Grinnell Model F954 (left) and Central "Omega" Model R-1 (right) pendent sprinklers.

Exception No. 2:   Sprinklers allowed to be omitted from closets where the least dimension does not exceed 3 ft (0.9 m), the area does not exceed 24 sq ft (2.2 m²), and the walls and ceiling are surfaced with noncombustible materials.

Exception No. 3:   Sprinklers allowed to be omitted from open-attached porches, garages, carports, and similar structures.

Exception No. 4:   Sprinklers allowed to be omitted from attics and crawl spaces that are not used or intended for living purposes or storage.

Exception No. 5:   Sprinklers allowed to be omitted from entrance foyers that are not the only means of egress.

TYPES OF FAST-RESPONSE SPRINKLERS

Today, fast-response sprinkler technology has been incorporated into several different types of sprinklers, each with unique attributes and applications.

Residential Sprinklers

In the sixteen years following the development of the residential sprinkler, special listings involving expanded protection areas and reduced flows proliferated to the point that the original flow and spacing criteria have become all but obsolete. Residential sprinklers are now listed for coverage areas up to 400 sq ft (37.2 m²) per sprinkler. Table 6-13A contains a 1995 compilation of residential sprinklers listed by Underwriters Laboratories Inc. (UL), along with their pressure and flow requirements.

It is important to recognize that, in addition to their fast-response characteristics, residential sprinklers have a special water distribution pattern. Because the effective control of residential fires often depends on a single sprinkler in the room of fire origin, the distribution of residential sprinklers is required to be more uniform than that of standard spray sprinklers, which in large areas can rely upon the overlapping patterns of several sprinklers to make up for voids. Additionally, residential sprinklers are required to protect sofas, drapes, and similar furnishings at the periphery of the room. In their discharge patterns, therefore, the sprinklers must not only be capable of delivering water to the walls of their assigned areas, but high enough up on the walls to prevent the fire from getting "above" the sprinklers. The water delivered close to the ceiling not only protects the portion of the wall close to the ceiling, but also enhances the capacity of the spray to cool gases at the ceiling level, thus reducing the likelihood of excessive sprinkler openings.

Because of their differences, residential sprinklers are not listed by product evaluation organizations under the same product standards as standard sprinklers. Underwriters Laboratories Inc., for example, has developed UL 1626 for residential sprinklers, and FMRC has published its Approval Standard FM 2030 for residential sprinklers. Both of these standards include a plunge test with specific sensitivity requirements and a distribution test that checks the spray pattern in the vertical plane, as well as the horizontal plane. The product standards for standard spray sprinklers contain neither test.

Both UL 1626 and FM 2030 also include a fire test that is intended to simulate a residential fire in the corner of a room containing combustible materials representative of a living room environment. The fire test arrangements for UL 1626 are shown in Figure 6-13C.

The UL 1626 test procedure contains a time-temperature curve that is reproducible in actual tests. The curve parallels the time-temperature relationship developed during the tests of the prototype residential sprinkler in Los Angeles that was part of the ongoing residential sprinkler research effort.[15, 17, 18]

TABLE 6-13A. *1996 Residential Sprinkler Listings*

The following updated roster of residential sprinkler listings of NFSA member manufacturers was prepared with data provided from the manufacturers. All temperatures are in degrees Fahrenheit, spacings in feet, flows in gallons per minute (gpm) and pressure in pounds per square inch (psi). Second (multiple) sprinkler flows have been increased in some cases to accommodate the new NFPA 13/13R/13D requirement for 7 psi minimum pressures.

Model	Style	K	Temp °F	Maximum Spacing (ft)	1-spr Flow (gpm)	(Pres) (psi)	Multiple Sprinkler Flow (gpm)	(Pres) (psi)	Non-Standard Notes
Badger Fire Protection Systems - Charlottesville, Virginia (804-973-4361)									
HR-1	Pendent	4.3	155	12×12	14	(10.6)	11.5	(7.1)	
				14×14	14	(10.6)	11.5	(7.1)	
				16×16	16	(13.8)	12	(7.8)	
				18×18	18	(17.5)	16	(13.8)	
HR-1	Recessed pendent	4.3	155	12×12	14	(10.6)	11.5	(7.1)	
				14×14	15	(12.2)	12	(7.8)	
				16×16	16	(13.8)	12	(7.8)	
HR	Pendent or recessed	4.2	155	14×14	18	(18.0)	13	(10.0)	
				16×16	21	(25.0)	15	(13.0)	
HR	Horizontal sidewall	5.5	155	14×14	25	(21.0)	18	(11.0)	
				14W×16L	30	(30.0)	21	(15.0)	
HR	Recessed sidewall	5.5	155	14×14	25	(21.0)	18	(11.0)	
Central Sprinkler Corporation - Lansdale, Pennsylvania (800-523-6512)									
R-1M	Flush pendent	3.9	160	12×12	10.5	(7.2)	10.5	(7.2)	
				14×14	10.5	(7.2)	10.5	(7.2)	
				16×16	14	(12.9)	11	(8.0)	
				18×18	14	(12.9)	12	(9.5)	
				20×20	16	(16.8)	16	(16.8)	
EC-20A	Flush pendent	5.6	160	12×12	19	(11.5)	16	(8.1)	
				14×14	25	(19.9)	18	(10.3)	
				16×16	30	(28.7)	21	(14.1)	
				18×18	32	(32.7)	29	(26.8)	
HEC-12 RES	Flush sidewall	5.6	160	12×12	26	(21.6)	18	(10.3)	
				14×14	27	(23.2)	21	(14.1)	
GBR	Pendent	4.3	155	14×14	14	(10.6)	11.5	(7.1)	
				16×16	16	(13.8)	12	(7.8)	
				18×18	19	(19.5)	14	(10.6)	
GBR	Recessed pendent	4.3	155	14×14	14	(10.6)	11.5	(7.1)	
				16×16	16	(13.8)	12	(7.8)	
				18×18	19	(19.5)	16	(13.8)	
GBR-2	Pendent or recessed	4.3	155	12×12	13	(9.1)	11.5	(7.1)	
				14×14	18	(17.5)	14	(10.6)	
				16×16	18	(17.5)	14	(10.6)	
				18×18	20	(21.6)	16	(13.8)	
				20×20	20	(21.6)	16	(13.8)	

TABLE 6-13A. *(Continued)*

Model	Style	K	Temp °F	Maximum Spacing (ft)	1-spr Flow (gpm)	(Pres) (psi)	Multiple Sprinkler Flow (gpm)	(Pres) (psi)	Non-Standard Notes
GBR	Concealed pendent	4.3	155	12×12	13	(9.1)	11.5	(7.1)	
				14×14	16	(13.8)	13.5	(9.8)	
				16×16	21	(23.8)	15	(12.2)	
GBRS/W	Horizontal sidewall or recessed	5.4	155	12×12	20	(13.7)	18	(11.1)	
				16W×18L	22	(16.6)	20	(13.7)	
				16W×20L	30	(30.9)	25	(21.4)	
ROC	Concealed pendent	4.2	165	12×12	18	(18.4)	13	(9.6)	
				14×14	18	(18.4)	13	(9.6)	
				16×16	18	(18.4)	13	(9.6)	
				18×18	24	(32.7)	17	(16.4)	
				20×20	24	(32.7)	17	(16.4)	
GBR-LF	Sidewall or recessed sidewall	3.0	155&175	12×12	14	(16.0)	11	(9.9)	Deflector distance 4–12 in. below ceiling for spacings 14×14 and smaller
				14×14	14	(16.0)	12	(11.8)	
				16×16	16	(20.9)	13	(13.8)	
GBR-LF	Pendent or recessed pendent	3.0	155	12×12	10	(11.1)	8	(7.1)	
				14×14	10	(11.1)	8	(7.1)	
				16×16	10	(11.1)	8	(7.1)	
				18×18	14	(21.8)	10.5	(12.3)	
				20×20	16	(28.4)	13.5	(20.3)	
GBR-LF	Pendent or recessed pendent	3.0	175	12×12	10	(11.1)	8	(7.1)	
				14×14	12	(16.0)	8.5	(8.0)	
				16×16	12	(16.0)	8.5	(8.0)	
				18×18	15	(25.0)	10.5	(12.3)	
				20×20	16	(28.4)	13.5	(20.3)	
GBR-LP	Pendent "extended"	3.0	155&175	12×12	11	(13.4)	8	(7.1)	Deflector distance extended 4–8 in. below ceiling
				14×14	12	(16.0)	9	(9.0)	
				16×16	14	(21.8)	10.5	(12.3)	
				18×18	14	(21.8)	12	(16.0)	
				20×20	16	(28.4)	15	(25.0)	

Globe Fire Sprinkler - Standish, Michigan (800-248-0278)

Model	Style	K	Temp °F	Maximum Spacing (ft)	1-spr Flow (gpm)	(Pres) (psi)	Multiple Sprinkler Flow (gpm)	(Pres) (psi)	Non-Standard Notes
J/JN	Pendent	4.2	155	12×12	13	(9.6)	11.5	(7.5)	
				14×14	20	(22.7)	14	(11.1)	
				16×16	20	(22.7)	14	(11.1)	
				20×20	19	(20.5)	16	(14.5)	
J/JN	Recessed pendent	4.2	155	12×12	13	(9.6)	11.5	(7.5)	
				14×14	18	(18.4)	13	(9.6)	
				16×16	18	(18.4)	13	(9.6)	
				18×18	18	(18.4)	14	(11.1)	
				20×20	19	(20.5)	16	(14.5)	

TABLE 6-13A. *(Continued)*

Model	Style	K	Temp °F	Maximum Spacing (ft)	1-spr Flow (gpm)	(Pres) (psi)	Multiple Sprinkler Flow (gpm)	(Pres) (psi)	Non-Standard Notes
J/JN	Concealed pendent	4.2	155	12×12	13	(9.6)	11.5	(7.5)	
				14×14	17	(16.4)	13	(9.6)	
				16×16	22	(27.4)	16	(14.5)	
J/JN	Horizontal sidewall or recessed sidewall	4.2	155	12×12	19	(20.5)	14	(11.1)	
				14×14	21	(25.0)	16	(14.5)	
				16×16	24	(32.7)	19	(20.5)	
	Horizontal sidewall or recessed sidewall	5.6	155	12×12	20	(12.8)	15	(7.2)	
				14×14	23	(16.9)	17	(9.2)	
				16×16	24	(18.4)	20	(12.8)	
JRES2/ JNRES2	Horizontal sidewall or recessed sidewall	5.6	155	12×12	18	(10.3)	15	(7.2)	
				16W×18L	30	(28.7)	22	(15.4)	
				16W×20L	35	(39.1)	26	(21.6)	

Grinnell Corporation - Exeter, New Hampshire (603-778-9200)

Model	Style	K	Temp °F	Maximum Spacing (ft)	1-spr Flow (gpm)	(Pres) (psi)	Multiple Sprinkler Flow (gpm)	(Pres) (psi)	Non-Standard Notes
F983	Pendent	4.4	155	12×12	12	(7.4)	12	(7.4)	
				14×14	15	(11.6)	12	(7.4)	
				16×16	15	(11.6)	12	(7.4)	
				18×18	8	(16.7)	14	(10.1)	
				20×20	20	(20.7)	17	(14.9)	
			175	12×12	12	(7.4)	12	(7.4)	
				14×14	16	(13.2)	12	(7.4)	
F973	Recessed pendent	4.4	155	12×12	12	(7.4)	12	(7.4)	
				14×14	15	(11.6)	12	(7.4)	
				16×16	15	(11.6)	12	(7.4)	
				18×18	18	(16.7)	14	(10.1)	
				20×20	20	(20.7)	17	(14.9)	
F990	Flush pendent	4.2	160	12×12	12	(8.2)	11.5	(7.5)	
				14×14	14	(11.1)	11.5	(7.5)	
				16×16	16	(14.5)	12	(8.2)	
				18×18	18	(18.4)	14	(11.1)	
				20×20	22	(27.4)	18	(18.4)	
F992	Flush pendent	5.6	160	12×12	17	(9.2)	15	(7.2)	
				14×14	20	(12.8)	17	(9.2)	
				16×16	22	(15.4)	19	(11.5)	
				18×18	31	(30.6)	24	(18.4)	
F978	Concealed pendent	4.2	155	12×12	16	(14.5)	12	(8.2)	
				14×14	16	(14.5)	12	(8.2)	

TABLE 6-13A. *(Continued)*

Model	Style	K	Temp °F	Maximum Spacing (ft)	1-spr Flow (gpm)	(Pres) (psi)	Multiple Sprinkler Flow (gpm)	(Pres) (psi)	Non-Standard Notes
				16×16	16	(14.5)	12	(8.2)	
				18×18	24	(32.7)	17	(16.4)	
				20×20	25	(35.4)	18	(18.4)	
FR-1	Horizontal sidewall	5.6	165	12×12	21	(14.1)	17	(9.2)	
				16W×20L	40	(51.0)	37	(43.7)	
		8.1	140&165	12×12	26	(10.3)	24	(8.8)	
				16W×20L	50	(38.1)	45	(30.9)	
F984	Horizontal sidewall	4.4	55	12×12	13	(8.7)	12	(7.4)	
				14×14	21	(22.8)	15	(11.6)	
				16×16	21	(22.8)	15	(11.6)	
F974	Recessed horizontal sidewall	4.4	155	12×12	13	(8.7)	12	(7.4)	
				14×14	21	(22.8)	15	(11.6)	
				16×16	22	(22.8)	15	(11.6)	
F993	Flush sidewall	5.6	160	12×12	22	(15.4)	15.5	(7.7)	
				14×14	30	(28.7)	21	(14.1)	

The Reliable Automatic Sprinkler Company—Mount Vernon, New York (800-431-1588)

Model	Style	K	Temp °F	Maximum Spacing (ft)	1-spr Flow (gpm)	(Pres) (psi)	Multiple Sprinkler Flow (gpm)	(Pres) (psi)	Non-Standard Notes
A	Pendent	4.15	165	14×14	18	(18.8)	13	(9.8)	
				16×16	25	(36.3)	18	(18.8)	
		5.5	165	12×12	21	(14.6)	15	(7.4)	
				14×14	24	(19.0)	17	(9.6)	
A	Recessed pendent	4.15	165	14×14	19	(21.0)	13.5	(10.6)	
				16×16	25	(36.3)	18	(18.8)	
		5.5	165	12×12	21	(14.6)	15	(7.4)	
				14×14	24	(19.0)	17	(9.6)	
A	Horizontal sidewall	4.15	165	12×12	21	(25.6)	15	(13.1)	
				14×14	25	(36.3)	18	(18.8)	
A/FZ	Recessed sidewall	4.15	165	12×12	21	(25.6)	15	(13.1)	
				14×14	25	(36.3)	18	(18.8)	
ZX-RES	Flush pendent	4.1	165	14×14	18	(19.3)	13	(10.1)	
				16×16	24	(34.3)	17	(17.2)	
		5.4	165	14×14	26	(23.2)	18	(11.1)	
				16×16	30	(30.9)	21	(15.1)	
				18×18	32.5	(36.2)	23	(18.1)	
ZX RES/2	Flush pendent	4.1	165	12×12	12	(8.6)	11	(7.2)	
				14×14	13	(10.1)	11	(7.2)	
				16×16	14	(11.7)	12	(8.6)	
				18×18	16	(15.2)	14	(11.7)	
				20×20	18	(19.3)	16	(15.2)	

TABLE 6-13A. *(Continued)*

Model	Style	K	Temp °F	Maximum Spacing (ft)	1-spr Flow (gpm)	(Pres) (psi)	Multiple Sprinkler Flow (gpm)	(Pres) (psi)	Non-Standard Notes
ZX-RES	Horizontal sidewall	5.4	165	14×14	24	(19.8)	17	(10.0)	
F1/RES	Pendent	5.6	155	14×14	20	(12.8)	15	(7.2)	
				16×16	20	(12.8)	20	(12.8)	
				18×18	28	(25.0)	22	(15.4)	
F1/RES	Horizontal sidewall	5.6	155	12×12	25	(19.9)	18	(10.3)	
				14×14	25	(19.9)	18	(10.3)	
				16×16	30	(28.7)	21	(14.1)	
				16×18	34	(36.9)	24	(18.4)	
				16×20	37	(43.7)	26	(21.6)	
F1/RES/2	Pendent and recessed pendent*	4.2	155	12×12	12	(8.2)	11	(7.0)	
				14×14	13	(9.6)	11	(7.0)	
				16×16	15	(12.8)	12	(8.2)	
				18×18	17	(16.4)	15	(12.8)	
				20×20	19	(20.5)	17	(16.4)	
F1/RES /F2	Recessed pendent	5.6	155	14×14	20	(12.8)	15	(7.2)	
				16×16	20	(12.8)	20	(12.8)	
F1/RES /F2	Recessed horizontal sidewall	5.6	155	12×12	25	(19.9)	18	(10.3)	
				14×14	25	(19.9)	18	(10.3)	
F1/RES/3	Pendent and recessed pendent **	3.9	155&175	12×12	10.5	(7.2)	10.5	(7.2)	
			155&175	14×14	12	(9.5)	10.5	(7.2)	9-ft minimum
			155&175	16×16	12	(9.5)	10.5	(7.2)	spacing between
			155	18×18	15	(14.8)	10.5	(7.2)	sprinklers
			155	20×20	18	(21.3)	18	(21.3)	
F1/RES/3	Concealed pendent	3.9	155	12×12	11	(8.0)	10.5	(7.2)	
				14×14	12	(9.5)	10.5	(7.2)	
				16×16	15	(14.8)	12	(9.5)	
				18×18	16	(16.8)	13	(11.1)	
				20×20	18	(21.3)	14	(12.9)	
F1/RES/3	Horizontal sidewall	3.9	155	12×12	13	(11.1)	11	(7.9)	
				14×14	14	(12.9)	12	(9.5)	
				16×16	17	(19.0)	14	(12.9)	
				16×18	18	(21.3)	16.5	(17.9)	
				16×20	25	(41.1)	19.5	(25.0)	
				18×18	22	(31.8)	21	(29.0)	
				12×12	13	(11.1)	13	(11.1)	Deflector 6–12 in.
				14×14	17	(19.0)	16	(16.8)	distance
				16×16	20	(26.3)	17	(19.0)	below ceiling

TABLE 6-13A. *(Continued)*

Model	Style	K	Temp °F	Maximum Spacing (ft)	1-spr Flow (gpm)	(Pres) (psi)	Multiple Sprinkler Flow (gpm)	(Pres) (psi)	Non-Standard Notes
			175	12×12	14	(12.9)	11	(7.9)	
				14×14	14	(12.9)	12	(9.5)	
				16×16	17	(19.0)	14	(12.9)	
				16×18	19	(23.7)	16.5	(17.9)	
				16×20	25	(41.1)	20	(26.3)	
				12×12	13	(11.1)	13	(11.1)	Deflector
				14×14	17	(19.0)	16	(16.8)	distance 6–12 in.
				16×16	20	(26.3)	17	(19.0)	below ceiling
F1/RES/3 /F2	Recessed horizontal sidewall	3.9	155	12×12	13	(11.1)	11	(7.9)	
				14×14	14	(12.9)	12	(9.5)	
				16×16	17	(19.0)	14	(12.9)	
				16×18	18	(21.3)	16.5	(17.9)	
				16×20	25	(41.1)	19.5	(25.0)	
				18×18	22	(31.8)	21	(29.0)	
				12×12	13	(11.1)	13	(11.1)	
				14×14	17	(19.0)	16	(16.8)	
				16×16	20	(26.3)	17	(19.0)	
			175	12×12	14	(12.9)	11	(7.9)	
				14×14	14	(12.9)	12	(9.5)	
				16×16	17	(19.0)	14	(12.9)	
				16×18	19	(23.7)	16.5	(17.9)	
				16×20	25	(41.1)	20	(26.3)	
				12×12	13	(11.1)	13	(11.1)	Deflector
				14×14	17	(19.0)	16	(16.8)	distance 6–12 in.
				16×16	20	(26.3)	17	(19.0)	below ceiling
F4/RES	Concealed	5.5	135	12×12	22	(16.0)	20	(13.2)	
				14×14	27	(24.1)	22.5	(16.7)	
				16×16	30	(29.8)	23	(17.5)	

Star Sprinkler Corporation—Milwaukee, Wisconsin (800-558-5236; in Wisconsin 414-570-5000)

Model	Style	K	Temp °F	Maximum Spacing (ft)	1-spr Flow (gpm)	(Pres) (psi)	Multiple Sprinkler Flow (gpm)	(Pres) (psi)	Non-Standard Notes
LD-2	Pendent or recessed	4.5	165	12×12	18	(16.0)	13	(8.3)	
				14×14	20	(19.8)	14	(9.7)	
				16×16	20	(19.8)	15	(11.1)	
LD-2	Horizontal sidewall or recessed sidewall	4.5	165	12×12	17	(14.3)	12	(7.1)	
				14×14	21	(21.8)	15	(11.1)	
				16×16	25	(30.9)	18	(16.0)	
		4.8	165	12×12	17	(12.5)	13	(7.3)	
				16W×20L	38	(62.7)	27	(31.6)	
SG-R	Pendent	4.3	155	12×12	16	(13.8)	11.5	(7.1)	
				14×14	16	(13.8)	11.5	(7.1)	
				16×16	17	(15.6)	13	(9.1)	
				18×18	19	(19.5)	14	(10.6)	
				20×20	20	(21.6)	17	(15.6)	

TABLE 6-13A. *(Continued)*

Model	Style	K	Temp °F	Maximum Spacing (ft)	1-spr Flow (gpm)	(Pres) (psi)	Multiple Sprinkler Flow (gpm)	(Pres) (psi)	Non-Standard Notes
SG-R	Recessed	4.3	155	12×12	13	(9.1)	11.5	(7.1)	
				14×14	16	(13.8)	11.5	(7.1)	
				16×16	17	(15.6)	13	(9.1)	
				18×18	17	(15.6)	13	(9.1)	
				20×20	21	(23.9)	17	(15.6)	
SG-RES	Horizontal sidewall or recessed sidewall	4.3	155	12×12	15	(12.2)	12	(7.8)	
				12×12	17	(15.6)	12	(7.8)	6–12 in.
				14×14	16	(13.8)	14	(10.6)	
				14×14	19	(19.5)	15	(12.2)	6–12 in.
				16×16	18	(17.5)	14	(10.6)	
				16W×18L	21	(23.9)	19	(19.5)	
		5.8	155	12×12	18	(9.6)	15.5	(7.1)	
				12×12	17	(8.6)	15.5	(7.1)	6–12 in.
				16W×18L	28	(23.3)	23	(15.7)	
				16W×20L	29	(25.0)	26	(20.1)	
Q	Concealed	4.3	155	12×12	13	(9.1)	11.5	(7.5)	
				14×14	17	(15.6)	13	(9.1)	
				16×16	20	(21.6)	15	(12.2)	

Viking Corporation—Hastings, Michigan (616-945-9501)

Model	Style	K	Temp °F	Maximum Spacing (ft)	1-spr Flow (gpm)	(Pres) (psi)	Multiple Sprinkler Flow (gpm)	(Pres) (psi)	Non-Standard Notes
M	Pendent or recessed†	4.2	155&175	12×12	15	(12.8)	11.5	(7.0)	
				14×14	18	(18.4)	13	(9.6)	
				16×16	24	(32.7)	17	(16.4)	
M-2	Pendent or recessed†	5.5	155&175	12×12	23	(17.5)	16	(8.5)	
				14×14	24	(19.0)	17	(9.6)	
M	Horizontal sidewall or recessed sidewall†	5.5	155&175	12×12	21	(14.6)	17	(9.6)	
				14×14	25	(20.7)	18	(10.7)	
M	Horizontal sidewall	5.5	155	16×16	27	(24.1)	23	(17.5)	
				16×16	28	(25.9)	23	(17.5)	6–10 in.
				16×18	28	(25.9)	23	(17.5)	4–10 in.
				16×20	37	(45.2)	30	(29.7)	
				16×20	38	(47.7)	31	(31.7)	6–10 in.
				18×16	29	(27.8)	25	(20.6)	4–10 in.
				18×18	31	(31.7)	26	(22.3)	4–10 in.
H	Flush pendent	4.2	165	12×12	14	(11.1)	11.5	(7.5)	
				16×16	17	(16.4)	12	(8.2)	
				18×18	20	(22.7)	14	(11.1)	
				20×20	29	(47.7)	21	(25.0)	

TABLE 6-13A. *(Continued)*

Model	Style	K	Temp °F	Maximum Spacing (ft)	1-spr Flow (gpm)	(Pres) (psi)	Multiple Sprinkler Flow (gpm)	(Pres) (psi)	Non-Standard Notes
		5.5	165	12×12	18	(10.7)	14.5	(7.0)	
				16×16	22	(16.0)	16	(8.5)	
				18×18	22	(16.0)	18	(10.7)	
				20×20	30	(29.8)	21	(14.6)	
H-2, H-3	Flush pendent	3.8	165	12×12	10	(7.0)	10	(7.0)	
				14×14	11	(8.4)	10	(7.0)	
				16×16	13	(11.7)	11	(8.4)	
				18×18	16	(17.7)	13	(11.7)	
				20×20	19	(25.0)	17	(20.0)	
H-3	Flush pendent	3.8	165	16×16	16	(17.7)	12	(10.0)	Listed for sloped ceilings to pitch of 6 in 12
A-1	Concealed	4.2	155	14×14	18	(18.4)	13	(9.6)	
				16×16	25	(35.4)	19	(20.5)	
		5.5	155	14×14	24	(19.0)	17	(9.6)	
M-4	Pendent or recessed pendent ††	4.3	155	12×12	11.5	(7.1)	11.5	(7.1)	
				14×14	13	(9.1)	11.5	(7.1)	
				16×16	13	(9.1)	11.5	(7.1)	
				18×18	16	(13.8)	13.5	(9.9)	
				20×20	19	(19.5)	17	(15.6)	
M-4	Horizontal sidewall or recessed sidewall	4.3	155	12×12	19	(19.5)	15	(12.2)	6–12 in. below sloped ceilings to pitch of 4 in 12
				14×14	22	(26.2)	17	(15.6)	
				16×16	22	(26.2)	17	(15.6)	
M-5	Horizontal sidewall or recessed sidewall	5.5	155	12×12	21	(14.6)	15	(7.4)	6–12 in. below sloped ceilings to pitch of 6 in 12
				14×14	24	(19.0)	21	(14.6)	
				16×16	24	(19.0)	21	(14.6)	

* Recessed pendent is model F1/RES/2/F2
** Recessed pendent is model F1/RES/3/F2
† F-1 escutcheon for recessed version
†† E-1 escutcheon for recessed version

In summary, to meet the UL 1626 test criteria, residential sprinklers, installed in a fire test enclosure with an 8-ft (2.4-m) ceiling are required to control a fire for 10 min with the following limits:

1. The maximum gas or air temperature adjacent to the sprinkler—3 in. (76.2 mm) below the ceiling and 8 in. (203 mm) horizontally away from the sprinkler—must not exceed 600°F (316°C).

2. The maximum temperature—5 ft, 3 in. (1.6 m) above the floor and half the room length away from each wall—must be less than 200°F (93°C) during the entire test. This temperature must not exceed 130°F (54°C) for more than a 2-min period.

3. The maximum temperature $1/4$ in. (6.3 mm) behind the finished surface of the ceiling material directly above the test fire must not exceed 500°F (260°C).

4. No more than two residential sprinklers in the test enclosure can operate.

The test enclosure is built to be twice the size of the proposed maximum spacing of the residential sprinkler. In a typical test, the walls of one corner of the test enclosure are covered with combustible cellulosic acoustical panels $1/8$-in. (3.2-mm) thick. Two combustible structures with urethane foam, representing stuffed furniture, are positioned at the same corner. (See Figure 6-13C.) The enclosure is kept at an initial ambient temperature of 80°F (27°C)±

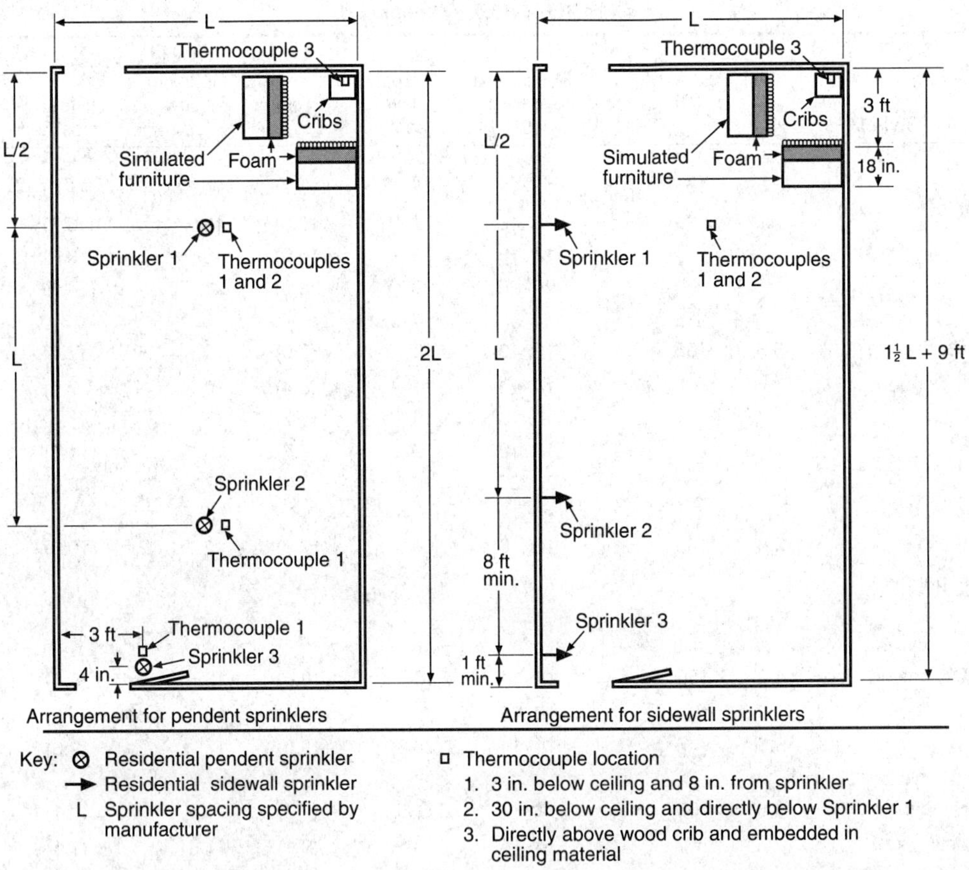

FIG. 6-13C. Fire test arrangements for UL 1626 for pendent sprinklers (left) and sidewall sprinklers (right). The simulated furniture consists of 3 in. (76 mm) thick uncovered urethane foam cushions 30 in. (762 mm) high by 30 in. (762 mm) wide. The foam has a density of 1.25 lb/sq ft (6.1 kg/m³) and is attached to a wood frame. The walls of the test room are covered with 4 by 8 ft × ¹/₈-in. (1.2 × 2.4 m × 3.2 mm) decorative plywood paneling (flame spread rating 200) attached to wood furring strips. The fire source consists of 12 × 12 × 12 in. (0.3 × 0.3 × 0.3 m) wood cribs weighing 12 to 13 lb (5.4 to 5.9 kg). The ceiling of the test room is 8 ft (2.4 m) high and covered with 2 × 4 ft × ¹/₂ in. (6.1 × 1.2 m × 12.7 mm) thick acoustical panels attached to wood furring strips. (For SI units: 1 in = 25.4 mm; 1 ft = 0.305 m.)

5°F (3°C), and it is ventilated through two door openings on opposite walls. The fire source is a wood crib of 12 to 13 lb (5.4 to 5.9 kg) mass, measuring approximately 12 by 12 by 12 in. (0.3 by 0.3 by 0.3 m).

The wood crib is ignited by 8 oz. (24 mL) of N-heptane in a pan directly below the crib. Forty sec after ignition, the excelsior [¹/₄ lb (0.11 kg)] on the floor next to the simulated furniture is ignited.

The fire test is conducted for 10 min after the ignition of the wood crib. The water flow to the first sprinkler that operates and the total water flow when the second sprinkler operates are specified as part of the listing limitations for the sprinklers in the test. The total water flow for two sprinklers must be a minimum of 1.2 times the minimum flow for a single sprinkler.

The water distribution test requirements are based on the distribution pattern of the prototype residential sprinkler used in the Los Angeles test fires.[18] The distribution requirements involve collections in both the horizontal and vertical planes.

All residential sprinklers in the test must discharge water at the flow rate specified by the manufacturer for a 10-min period simulating one sprinkler operating and two sprinklers operating. The quantity of water collected on both the horizontal and vertical surfaces is measured and recorded.

Sprinklers being tested are required to discharge a minimum of 0.02 gpm per sq ft [0.8(L/min)/m²] at both flow rates over the entire design area, plus an area 2 ft (0.61 m) beyond the listed coverage area in each direction (only three directions for horizontal sidewall-type sprinklers). They must also wet the walls of the test enclosure to a height not less than 28 in. (711 mm) below the ceiling with one sprinkler operating and not less than 36 in. (914 mm) with two sprinklers operating. Each wall surrounding the coverage area is required to be wetted with a minimum of 5 percent of the sprinkler application densities at both flow rates.

Since 1985, the use of residential sprinklers has also been permitted under some conditions in accordance with NFPA 13. Essentially, NFPA 13 allows residential sprinklers in dwelling units located in any occupancy, provided they are installed in conformance with the requirements of their listing and the positioning requirements of NFPA 13D. A dwelling unit is defined as one or more rooms arranged for the use of one or more individuals living together, as in a single housekeeping unit normally having cooking, living, sanitary, and sleeping facilities. Dwelling units include hotel rooms, dormitory rooms, sleeping rooms in nursing homes, and similar living units. Occupancies encompassing dwelling units include apartment buildings, board and care facilities, dormitories, condominiums, lodging and rooming houses, and other multiple-family dwellings.

For NFPA 13 applications involving residential sprinklers in dwelling units, the design area is required to consist of the four most hydraulically demanding sprinklers. (See Figure 6-13D).

Other areas, such as attics, basements, or other types of occupancies outside of dwelling units but within the same structure, are required to be protected in accordance with regular provisions of NFPA 13, including the appropriate water supply requirements. The decision as to which areas are to be protected with sprinklers is also regulated in accordance with the normal provisions of NFPA 13. This means, for example, that combustible concealed spaces generally require sprinklers. Although the four-sprinkler design area can be used in the dwelling units when protected with residential sprinklers, any sprinklers installed within such concealed spaces would have to use a different design approach.

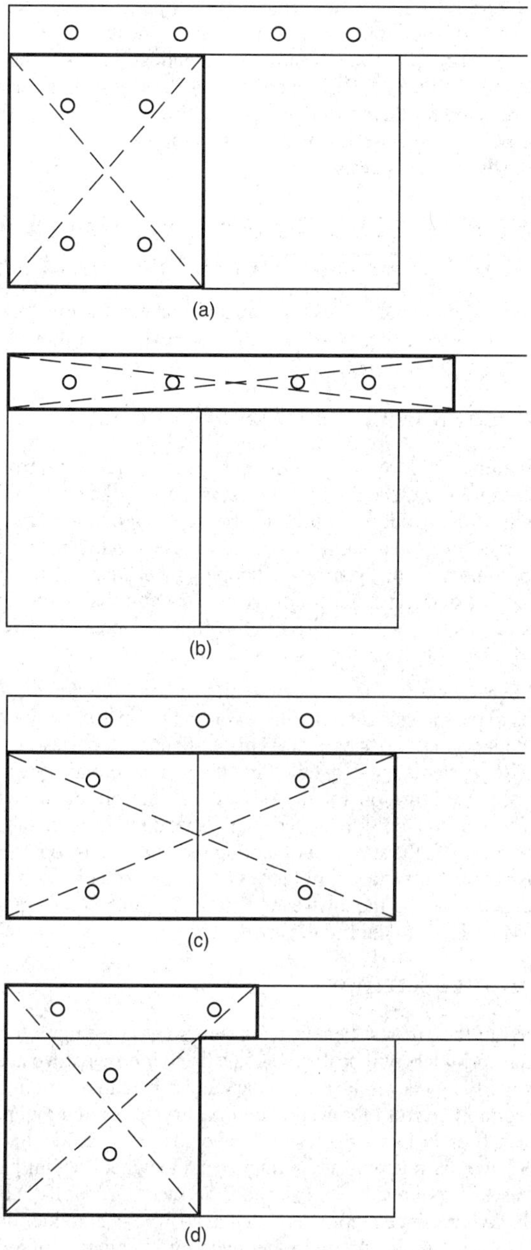

FIG. 6-13D.   *Examples of design areas for dwelling units within NFPA 13 occupancies.*

Residential sprinklers installed in systems designed to NFPA 13 requirements are spaced and positioned in accordance with their residential listings, not with the spacing requirements of NFPA 13. The water demands for the residential sprinklers are the same as in NFPA 13 applications, except that the multiple sprinkler flow requirement is extended to four sprinklers rather than the two stipulated for one- and two-family dwellings and manufactured homes in NFPA 13D. The more liberal piping, component, hanger location, and water supply duration allowances of NFPA 13D are not permitted in these systems.

Beginning in 1996, NFPA 13 requires residential sprinklers or quick-response sprinklers in residential areas.

In 1989, a new standard was developed to bridge the gap between NFPA 13 and NFPA 13D. The new standard is NFPA 13R, *Standard for the Installation of Sprinkler Systems in Residential Occupancies up to and Including Four Stories in Height* (hereinafter referred to as NFPA 13R).

Like NFPA 13D, NFPA 13R is oriented toward economical life safety protection. Sprinklers are omitted from building areas that have been found to have a low incidence of fatal fires originating in them, including combustible concealed spaces, small bathrooms and closets, and attached porches. As with NFPA 13D, residential sprinklers are required throughout dwelling units, with some minor exceptions. A four-sprinkler design area is required unless the largest compartment contains fewer sprinklers.

In recognition of the greater risk associated with multifamily occupancies, however, NFPA 13R is more conservative than NFPA 13D, in some areas. Requirements for plans, hydraulic calculations, and system acceptance certificates parallel those of NFPA 13. Unlike NFPA 13D, NFPA 13R requires a consideration of the likelihood that simultaneous domestic flows might occur through combined service piping. In addition, pumps and other key equipment are required to be listed.

In NFPA 13R systems, areas outside dwelling units can be protected with standard spray sprinklers, using NFPA 13 design criteria.

Quick-Response Sprinklers

Quick-response sprinklers are designed to be used in hotels, motels, offices, and other buildings where faster sprinkler operation could enhance life safety. UL first listed these sprinklers as quick-response sprinklers based on performance comparable to a standard response sprinkler equipped with a quick-response electronic squib actuator, a device marketed by one sprinkler manufacturer in the 1970s.

Quick-response sprinklers, which are essentially standard spray sprinklers equipped with fast-response operating mechanisms, were first introduced in 1983. They were developed by manufacturers in two different ways. One was to replace the actuating mechanism of standard spray sprinklers with the more sensitive heat-responsive element used in residential sprinklers. This sprinkler was then resubmitted to a testing laboratory for testing and listing under the product standard for standard spray sprinklers. The other way was to simply submit a residential sprinkler to a testing laboratory for testing and listing under the provisions of a product standard for standard spray sprinklers.

It was not until 1988 that full-scale testing showed that quick-response sprinklers could be used effectively with the traditional area/density design approaches found in sprinkler installation standards.[19,20] In "control-mode" testing of a 6-ft, 8-in. (2.0-m) arrangement of a standard plastic commodity under a 20-ft (6.1-m) ceiling, multiple tests involving sprinklers identical in all respects except sensitivity indicated that on average fewer quick-response sprinklers activated to control the fire than standard response sprinklers.

Current NFPA standards require the use of quick-response sprinklers throughout buildings of light-hazard occupancy. They are also permitted for use in ordinary hazard occupancies but have not yet been recognized for use in extra-hazard occupancies or for use as ceiling sprinklers in occupancies involving high-piled storage. Beginning with the 1995 edition of NFPA 231C, *Standard for Rack Storage of Materials*, quick-response sprinklers were permitted for use as in-rack sprinklers. Research is currently underway to determine if quick-response sprinklers can be used successfully in dry systems as well as wet systems.

Early Suppression Fast-Response Sprinklers

The early suppression fast-response (ESFR) sprinklers were first envisioned by the Factory Mutual Research Corporation.[21,22] Working with FMRC, the National Fire Protection Research Foundation funded full-scale testing to prove that fast-response sprinklers made suppression of a high-challenge fire feasible.[23]

Like its predecessor, the large-drop sprinkler, the ESFR sprinkler is intended to prevail against the strong fire plumes associated with fires in industrial and warehouse occupancies. However, its delivery of water to the burning fuel surfaces is accomplished by means of spray momentum rather than the gravitational effects of large droplets.

Operating at a minimum pressure of 50 psi (7.25 kPa) so as to produce a minimum flow per sprinkler of 100 gpm (380 L/min), the original ESFR sprinkler could protect challenges up to and including 25 ft (7.6 m) of plastics storage under a 30-ft (9.1-m) ceiling, without the use of in-rack sprinklers. Subsequent research and development has expanded the use of the ESFR sprinkler to protect commodities up to 35 ft (10.6 m) high under a 40-ft (12.2-m) ceiling without the use of in-rack sprinklers, and has also extended the application of the sprinkler for the protection of aerosols and other special hazards.

"Quick-Response, High-Challenge" Large-Drop Sprinklers

The "quick-response, high-challenge" sprinkler resulted from one manufacturer's insertion of a fast-response link into an approved large-drop sprinkler. The large-drop sprinkler concept, pioneered during the 1970s by FMRC, was based on the theory that larger spray droplets would be more capable of achieving gravitational penetration of a strong upward fire plume.

Product approval and installation standards were developed, and specialized system design criteria found its way into NFPA standards beginning with the 1985 edition of NFPA 13. The large-drop sprinkler has a characteristic large deflector and is an upright sprinkler with a 0.64-in. (16-mm) diameter orifice.

The concept of combining the penetration capabilities of the large-drop sprinkler with fast response was basic to the development of the ESFR sprinkler program. In tests of the large-drop sprinklers with and without fast-response links, the addition of fast response to the penetration abilities of the sprinkler provided superior protection against a high-challenge fire.

This particular sprinkler, however, later referred to as the "quick-response, high-challenge" sprinkler, fell by the wayside during the screening for the ESFR prototype sprinkler, due to its inability to provide sufficient water directly under the sprinkler. As an upright sprinkler, it could not handle the scenario in which a fire is started directly under a single sprinkler with 5 to 15 ft (1.5 to 4.6 m) of clearance below a 30-ft (9.1-m) ceiling. The sprinkler's water

penetration capability was considered insufficient, due to an opening of the spray umbrella. The ESFR prototype eventually selected was a pendent with a larger orifice.

Nevertheless, this sprinkler had already established its worth in the eyes of many. A variation of that sprinkler was listed by UL as a "specific application" sprinkler, permitted to protect commodities up to 20 ft (6.1 m) high under a 25-ft (7.6-m) ceiling with ceiling sprinklers only. Beginning with the 1994 edition of NFPA 13, the NFPA Committee on Automatic Sprinklers expanded the allowable types of ESFR sprinklers to permit such sprinklers to fall within one category of ESFR sprinklers.

Quick-Response, Early Suppression Sprinklers

Quick-response, early suppression (QRES) sprinklers are considered a future area of product development within NFPA 13. With the QRES sprinkler, the same principles used in the development of the ESFR sprinkler could be applied to a smaller-orifice sprinkler, which could be used in a wide variety of business, mercantile, public assembly, educational, and other applications. The proposed NFPA definition of QRES sprinklers is "fast-response sprinklers that are listed for their ability to provide fire suppression of specific hazards." It is expected that the final development of such sprinklers is still several years away.

INCENTIVES TO MORE WIDESPREAD USE OF RESIDENTIAL SPRINKLERS

There are certain incentives that can stimulate interest in residential sprinklers. These incentives are discussed in the following paragraphs.

Reduction in Government Spending

Reduction in all forms of government spending, resulting from public pressure to reduce property taxes, is a prime factor in the future growth of the residential sprinkler concept. Many fire departments are forced to protect larger areas and more subdivisions with the same number of or even fewer people since financial restrictions hamper a fire department's ability to grow with the community. As a result, alternates to traditional fire-fighting techniques must be found. One of them is the use of residential sprinklers.

San Clemente, CA, was the first community in the United States to pass a residential sprinkler ordinance in 1980 as part of the fire department's master plan. This ordinance requires automatic sprinkler systems to be installed in all new residential construction. The prime motivation for the passage of this ordinance was San Clemente's cutbacks in government spending brought about by Proposition 13, the state's tax-capping measure. Many communities across the country face similar situations. Automatic sprinklers in residences may be the answer to fewer fire fighters and longer response times from the fire department.

Insurance Savings

Although the greatest benefit from widespread installation of residential sprinklers will be the lives saved and injuries prevented, lower property losses will be a secondary and substantial benefit. An ad-hoc committee from the insurance industry sponsored a number of the test fires in Los Angeles and concluded that residential sprinklers have the potential for reducing homeowners' claim payment expenses.[24] As a result, the Insurance Services Office (ISO) Personal Lines Committee recommended that a 15-percent reduction in the homeowner's policy premium be given for installation of an NFPA 13D residential sprinkler system. While this would not pay for the system over a short period of time, as is the case in many commercial installations, the continuing increases in the cost of insuring a single-family home make this a significant incentive nonetheless.

Real Estate Tax Reductions

In 1981, the State of Alaska enacted into law a significant piece of legislation that has a dramatic impact on the installation of sprinkler systems throughout that state. The law provides that 2 percent of the assessed value of any structure is exempt from taxation if the structure is protected with a fire protection system. The word "structure" is significant in the law, since it also applies to homes. In effect, if a home were assessed at $100,000 for purposes of taxation, the assessed value would be computed at $98,000, provided that it contained a fire protection system.

Buyers' Attitude

While one study indicated that 30 percent of the people interviewed perceived no need for a residential sprinkler system,[10] a 1980 survey published by the National Association of Home Builders on luxury features that buyers want in a new home showed that 14.3 percent indicated "fire suppression systems" as a choice. For potential buyers with incomes over $50,000, the percentage rose to over 20 percent.

Zoning

Greater land use may be possible with zoning changes that would permit fully sprinklered residences to be built on smaller parcels of land. The assumption is that the space between houses will not be as important from a fire protection standpoint if an entire street or neighborhood is fully sprinklered. One could argue, however, that if the sprinkler system fails, the resultant fire involving a number of residences could be much greater.

Sprinkler Legislation

In addition to the San Clemente ordinance, a number of other California communities have passed residential sprinkler legislation, including Orange County and Los Angeles County. By 1993, more than 4 million Californians lived in communities in which residential sprinklers were mandated in all new homes.[25]

Since 1982, Greenburgh, NY, and several surrounding communities have enacted sprinkler ordinances that require the installation of automatic sprinklers in virtually all new construction, including all new multiple- and one- and two-family dwellings. A similar law went into effect in Prince Georges County, MD, in 1992.

The State of Florida in 1983 passed a law requiring that all public lodging and time-share units three stories or more high in the state be sprinklered. It also required that all existing units be sprinklered by 1988.

In 1983, the City of Honolulu, HI, adopted legislation that required all new and existing high-rise hotels, which are those more than 75 ft (23 m) above grade, to be sprinklered.

In the late 1980s, additional jurisdictions, including Atlanta, GA, the State of Connecticut, and the Commonwealth of Massachusetts, acted to retroactively require the installation of sprinkler systems in high-rise residential buildings.

In 1990, the federal government enacted the Hotel and Motel Fire Safety Act, which contains strong incentives for complete sprinkler protection of hotels, since only hotels with satisfactory levels of fire protection are eligible for federal employee travel.

The Federal Fire Safety Act of 1992 requires automatic sprinkler systems or an equivalent level of safety in all federally assisted housing four or more stories in height, as well as in office buildings owned or leased for more than 25 federal employees.

Perhaps the most significant legislation promoting the use of sprinkler systems, however, is the 1990 Americans with Disabilities Act. In 1991, the U.S. Department of Justice published criteria that became mandatory for places of public accommodation and commercial facilities designed for first occupation after January 26, 1993. Alterations to existing buildings must also comply. A key feature of the criteria is the need for areas of refuge. Floors of buildings that do not have direct access to the exterior at grade must provide areas of rescue assistance, except those buildings that have a supervised automatic sprinkler system.

Construction Incentives

Many Authorities Having Jurisdiction have used building code modifications as an incentive to install sprinklers. Cobb County, GA, was one of the first communities to amend its Buildings and Construction Code include to such an approach for multifamily structures equipped with residential sprinkler systems.

While these construction alterations can be a major incentive to install residential sprinklers, the disaster potential must always be considered if a fire, for whatever reason, should overpower the sprinkler system. This is especially true if the system is designed with the minimal water supplies required by NFPA 13D.

The City of Dallas, TX, adopted a building code that requires all new buildings or those undergoing major renovation, having an area greater than 7,500 sq ft (697 m^2), to have automatic sprinklers. At the same time, this building code encourages the installation of sprinkler systems by allowing design options that may allow different levels of "passive" fire protection features in exchange for "active" automatic sprinkler alternatives.

Performance

Since 1983, Operation Life Safety, a project of the International Association of Fire Chiefs, has been collecting incident reports of residential sprinkler activations in systems installed in accordance with NFPA 13D or NFPA 13R as of July 31, 1995. A total of 551 incidents have been reported. Of this total, 35.2 percent of the fires that activated sprinklers were reported to have originated in kitchens, and 14.9 percent in bedrooms, with 20.3 percent listed as "not reported." Table 6-13B shows the reported activations by room of origin. Figure 6-13E shows activations of residential sprinklers by occupancy, as well as the number of sprinkler activations per fire.[26]

TABLE 6-13B. *Room of Origin of Residential Sprinkler Activation*

Kitchen	35.2%
Bedroom	14.9%
Living room	7.8%
Closet	3.6%
Laundry room	2.7%
Basement	2.2%
Bathroom	1.5%
Storeroom	2.4%
Garage	4.4%
Others	5.1%
Not reported	20.3%

Note: Data as of July 31, 1995.

Code Requirements

Beginning with the 1991 edition, NFPA *101*®, *Life Safety Code*®, required the use of quick-response or residential sprinklers in new

FIG. 6-13E. Residential sprinkler activations.

health care occupancies in smoke compartments containing patient sleeping rooms. This generally means all patient rooms and their adjacent corridors. Beginning with the 1994 edition, quick-response or residential sprinklers were also required in all new hotel and dormitory guest rooms and guest room suites.

Because of these and other incentives the use of residential and quick-response sprinklers nearly tripled in the United States between 1987 and 1994, even though the total number of sprinklers installed increased only slightly. There is growing recognition of the enhanced ability of fast-response sprinklers to protect life and property from fires.

BIBLIOGRAPHY

References Cited

1. Heskestad, G., and Smith, H., "Investigation of a New Sprinkler Sensitivity Approval Test: The Plunge Test," FMRC No. 22485, Dec., 1976, Factory Mutual Research Corp., Norwood, MA.
2. Heskestad, G., and Smith, H., "Plunge Test for Determination of Sprinkler Sensitivity," FMRC J.I.3A1E2.RR, 1980, Factory Mutual Research Corp., Norwood, MA.
3. Pickard, R. W., Hird, D., and Nash, P., "The Thermal Testing of Heat-Sensitive Fire Detectors," F.R. Note No. 247, Jan. 1957, Fire Research Station, Borehamwood, Herts, U.K.
4. Evans, D. D., and Madrzykowski, D., "Characterizing the Thermal Response of Fusible-Link Sprinklers," NBSIR 81–2329, Aug. 1981, National Bureau of Standards, Washington, DC.
5. Theobald, C. R., "FRS Ramp Test for Thermal Sensitivity of Sprinklers," *Journal of Fire Protection Engineering*, Vol. 1, No. 1, Jan.–Mar. 1989, p. 23.
6. Beever, P. F., "Estimating the Response of Thermal Detectors," *Journal of Fire Protection Engineering*, Vol. 2, No. 1, Jan.-Mar. 1990, p. 11.
7. Pepi, J. S., "Design Characteristics of Quick-Response Sprinklers," Grinnell Fire Protection Systems Co., Providence, RI, May 1986.
8. Heskestad, G., and Bill, R. G., "Conduction Heat-Loss Effects on Thermal Response of Automatic Sprinklers," FMRC J.I. OMOJ5.RU, Sept. 1987, Factory Mutual Research Corp., Norwood, MA.
9. Halpin, B. M., Dinan, J. J., and Deters, O. J., "Assessment of the Potential Impact of Fire Protection Systems on Actual Fire Incidents," Johns Hopkins University—Applied Physics Laboratory (JHU/APL), Laurel, MD, Oct. 1978.
10. Yurkonis, P., "Study to Establish the Existing Automatic Fire Suppression Technology for Use in Residential Occupancies," Rolf Jensen & Associates, Inc., Deerfield, IL., Dec. 1980.
11. Kung, H. C., Haines, D., and Green, R., Jr., "Development of Low-Cost Residential Sprinkler Protection," Factory Mutual Research Corp., Norwood, MA, Feb. 1978.
12. Henderson, N. C., Riegel, P. S., Patton, R. M., and Larcomb, D. B., "Investigation of Low-Cost Residential Sprinkler Systems," Battelle Columbus Laboratories, Columbus, OH, June 1978.
13. Clark, G., "Performance Specifications for Low-Cost Residential Sprinkler System," Factory Mutual Research Corp., Norwood, MA, Jan. 1978.
14. Kung, H. C., Spaulding, R. D., and Hill, E. E., Jr., "Sprinkler Performance in Residential Fire Tests," Factory Mutual Research Corp., Norwood, MA, Dec. 1980.
15. Cote, A. E., and Moore, D., "Field Test and Evaluation of Residential Sprinkler Systems," Los Angeles Test Series (A report for the NFPA 13D Subcommittee), National Fire Protection Association, Quincy, MA, Apr. 1980.
16. Moore, D., "Data Summary of the North Carolina Test Series of USFA Grant 79027 Field Test and Evaluation of Residential Sprinkler Systems, (a report for the NFPA 13D Subcommittee)," National Fire Protection Association, Quincy, MA, Sept. 1980.
17. Kung, H. C., Spaulding, R. D., Hill, E. E., Jr., and Symonds, A. P., "Technical Report, Field Evaluation of Residential Prototype Sprinkler Los Angeles Fire Test Program," Feb. 1982, Factory Mutual Research Corp., Norwood, MA.
18. Cote, A. E., "Final Report on Field Test and Evaluation of Residential Sprinkler Systems," July 1982, National Fire Protection Association, Quincy, MA.
19. Vincent, B. G., Stavrianidis, P., and Kung, H. C., "Large-Scale Fire Testing of Fast-Response Sprinklers and Conventional Response Sprinklers in a Fire-Control-Mode Scenario," FMRC J.I. OQ2P6.RA,070(A), June 1989, Factory Mutual Research Corp., sponsored by National Fire Sprinkler Association.
20. Budnick, E., and Fleming, R., "Developing an Early Suppression Design Procedure For Quick-Response Sprinklers," *Fire Journal*, Vol. 83, No. 6, Nov. 1989.
21. Yao, C., and Marsh, W., "Early Suppression—Fast Response: A Revolution in Sprinkler Technology," *Fire Journal*, Vol. 78, No. 1, Jan. 1984.
22. Yao, C., "ESFR Sprinkler System—A New Approach to High-Challenge Storage Protection," *Fire Journal*, Vol. 79, No. 2, Mar. 1985.
23. Chicarello, P., Troup, J., and Dean, R., "Large-Scale Fire Test Evaluation of Early Suppression Fast-Response (ESFR) Automatic Sprinklers," FMRC J.I. OM2R5.RR/OMJ7.RR, May 1986, Factory Mutual Research Corp., sponsored by National Fire Protection Research Foundation.
24. Jackson, R. J., "Report of 1980 Property Loss Comparison Fires," 1980, FEMA/USFA, Washington, DC.
25. Hart, S., "Executive Summary Report on the 1993 Fire Sprinkler Ordinance Survey," May 31, 1993, Fire Sprinkler Advisory Board of Southern California, Cerritos, CA.
26. "Residential Sprinkler Activations—Annual Report," *Operation Life Safety Newsletter*, International Association of Fire Chiefs, Vol. 10, No. 9–10, Sept.–Oct., 1995.

NFPA Codes, Standards, and Recommended Practices

Reference to the following NFPA codes, standards, and recommended practices will provide further information on fast-response sprinkler technology discussed in this chapter. (See the latest version of *The NFPA Catalog* for availability of current editions of the following documents.)

NFPA 13, *Standard for the Installation of Sprinkler Systems*
NFPA 13D, *Standard for the Installation of Sprinkler Systems in One- and Two-Family Dwellings and Manufactured Homes*
NFPA 13R, *Standard for the Installation of Sprinkler Systems in Residential Occupancies up to and Including Four Stories in Height*
NFPA 101®, *Life Safety Code*®
NFPA 231C, *Standard for Rack Storage of Materials*

Additional Reading

Anderson, N., "Sprinklers in the Home—A Life Saver," *Fire International*, Vol. 12, No. 108, Dec.–Jan. 1988, pp. 54–55.
Beitel, J. J., "Quick-Response Sprinkler Tests, Group 2 Performance Tests," National Fire Protection Research Foundation, Quincy, MA, Dec. 1988.
Benjamin/Clarke Associates, "Operation San Francisco Smoke/Sprinkler Test Technical Report," International Association of Fire Chiefs, Washington, DC, Apr. 1984.

Beitel, J. J., "Quick-Response Sprinkler Tests, Group 2 Performance Tests," National Fire Protection Research Foundation, Quincy, MA, Dec. 1988.

Budnick, E. K., "Estimating Effectiveness of State-of-the-Art Detectors and Automatic Sprinklers on Life Safety in Residential Occupancies," NB-SIR 84-2819, Jan. 1984, National Bureau of Standards, Washington, DC.

Carey, W. M., "Report of the Quick-Response Sprinkler Project, Group 2 Performance Tests," National Fire Protection Research Foundation, Quincy, MA, Mar. 1988.

Casaccio, E. K., "ESFR Installation Guidelines Set," *Building Standards*, Vol. 57, No. 2, Mar.–Apr. 1988, pp. 20–21.

Chopra, I. S., "Subject 1626: Residential Automatic Sprinklers," *Lab Data*, Underwriters Laboratories, Northbrook, IL, Vol. 15, No. 2, 1984.

Coleman, R. F., "Sprinklers in the Home—The Evolution of an Idea," *Fire International*, Vol. 11, No. 102, 1987, pp. 43, 46–47.

Cote, A. E., "Final Report on Field Test of a Retrofit Sprinkler System," National Fire Protection Research Foundation, Washington, DC, Feb. 1983.

Cote, A. E., "QRS: Do You Understand the Technology?" *Fire Journal*, Vol. 81, No. 1, Jan. 1987, p. 27.

Cote, A. E., "Behind the Technology of Quick-Response Sprinklers," *Consulting Specifying Engineer*, Vol. 1, No. 5, 1987, pp. 68–71.

Cote, A. E., "Use of Quick-Response Sprinklers in Residential Occupancies—What You Should Know," *Sprinkler Quarterly*, Spring 1985.

Cote, A. E., "Highlights of a Field Test of a Retrofit Sprinkler System," *Fire Journal*, Vol. 77, No. 3, May 1983.

Cote, A. E., "Update on Residential Sprinkler Protection," *Fire Journal*, Vol. 77, No. 6, Nov. 1983, p. 69.

Cote, A. E., "Field Test and Evaluation of Residential Sprinkler Systems, Part I," *Fire Technology*, Vol. 19, No. 4, Nov. 1983, pp. 221-232.

Cote, A. E., "Field Test and Evaluation of Residential Sprinkler Systems," Part II, *Fire Technology*, Vol. 20, No. 1, Feb. 1984, pp. 48-58.

Cote, A. E., "Field Test and Evaluation of Residential Sprinkler Systems, Part III," *Fire Technology*, Vol. 120, No. 2, May 1984, pp. 41–46.

"Development of an Experimental Prototype Low-Cost Electronic Sensor/Actuator for a Residential Automatic Sprinkler Head," Battelle Columbus Laboratories, Columbus, OH, 1982.

Evans, D. D., "Calculating Sprinkler Activation Time in Compartments," National Bureau of Standards, Washington, DC, 1984.

Fleming, R. P., "Applications/Limitations of QRS Technology," *Fire Safety Journal*, Vol. 14, Nos. 1&2, July 1988, pp. 75–88.

Fleming, R. P., "Quick-Response Sprinklers—A State-of-the-Art Technical Report," National Fire Protection Research Corporation, Quincy, MA, 1985.

Fleming, R. P., "The Special Listings," *Sprinkler Quarterly*, No. 52, Spring 1985.

Fleming, R. P., "Understanding the New Sprinklers," *Sprinkler Quarterly*, National Fire Sprinkler Association, No. 47, Fall 1983.

FM 2030, *Approval Standard for Residential Sprinklers*, Factory Mutual Research Corp., Norwood, MA, 1983.

Foehl, J. M., "In Quest of an Economical Automatic Fire Suppression System for Multi-Family Residential Complexes," *Fire Journal*, Vol. 70, No. 1, Jan. 1976, pp. 48-57.

Fuller, S. K., and Ruegg, R. T., "Economics of Fast-Response Residential Sprinkler Systems," *Fire Journal*, Vol. 79, No. 3, May–June 1985, pp. 18–22, 115.

Goodfellow, D. G., "Apparatus and Procedure for Measuring the Actual Delivered Density (ADD) of Sprinklers in Simulated Rack Storage Fires," Factory Mutual Research Foundation, Norwood, MA, Jan. 1985.

Hall, J.R., Jr., "The U.S. Experience with Sprinklers: Who Has Them? How Well Do They Work?" *NFPA Journal*, Nov./Dec. 1993, p. 44.

Hammerman, D. M., "Project Home—A Pilot Program for a Low-Cost Residential Sprinkler System," *Fire Journal*, Vol. 75, No. 2, Mar. 1981, pp. 66–69.

Harmathy, T. Z., "On the Economics of the Mandatory Sprinklering of Dwellings," *Fire Technology*, Vol. 24, No. 3, 1988, pp. 245–261.

Isner, M. S., "Successful Residential Sprinkler Activation," *Fire Command*, Vol. 53, No. 1, Jan. 1986, pp. 22–27.

Kung, H. C., Spaulding, R. D., and Hill, E. E., "Sprinkler Performance in Living-Room Fire Tests: Effect of Link Sensitivity and Room Size," Factory Mutual Research Corporation, July 1983.

Kung, H. C., "Low-Cost Residential Sprinkler Protection," *Fire Technology*, Vol. 12, No. 2, May 1976, pp. 85.

Miller, C., "Early Suppression-Fast Response Sprinklers: A New Level of Industrial Fire Protection," *National Engineer*, Vol. 93, No. 9, Sept. 1989, pp. 18–19.

Moore, D. A., "Field Test and Evaluation of Residential Sprinkler Systems," *Fire Journal*, Vol. 74, No. 6, Nov. 1980.

O'Neil, J. G., "Fast-Response Sprinklers in Patient Room Fires," *Fire Technology*, Vol. 17, No. 4, Nov. 1981.

O'Rourke, G. W., "Automatic Sprinkler Systems: What's New?" *Heating/Piping/AirCondition*, Vol. 55, No. 4, April 1983, pp. 37–41.

"Outline of Proposed Investigation for Residential Automatic Sprinklers for Fire Protection Service," Subject 1626, Underwriters Laboratories, Oct. 27, 1980.

Pepi, J. S., "Design Characteristics of Quick-Response Sprinklers," TR 86-3, 1986, Society of Fire Protection Engineers, Boston, MA.

Pepi, J. S., "Concept and Development of the Residential and Fast-Response Sprinklers," *Sprinkler Quarterly*, No. 52, Spring 1985.

Proceedings of the Fourth Conference on Low-Cost Residential Sprinkler Systems, Sept. 26–27, 1979, Federal Emergency Management Agency/U.S. Fire Administration, Aug. 1980.

Proceedings of the Third Conference on Low-Cost Residential Sprinklers, November 29–30, 1977, National Fire Prevention and Control Administration, June 1978.

Ruegg, R. T., and Fuller, S. K., "A Benefit-Cost Model of Residential Fire Sprinkler Systems," NBS Technical Note 1203, Nov. 1984, National Bureau of Standards, Washington, DC,.

Shaw, H., "Fast-Response Sprinklers—The Answer to the Toxicity Problem," *Sprinkler Quarterly*, No. 47, Fall 1983.

Teague, P. E., "Residential Sprinklers: An Idea Whose Time Has Almost Come," *Fire Journal*, Vol. 82, No. 5, Sept.–Oct., 1988, p. 46.

UL 1626, *Standard for Safety Residential Sprinklers for Fire-Protection Service*, Underwriters Laboratories Inc., Northbrook, IL, 1988.

Viniello, J. A., "Residential and Quick-Response Fire Sprinklers—You Need to Know the Difference!" *Sprinkler Quarterly*, No. 52, Spring 1985.

Walton, W. D., and Budnick, E. K., "Quick-Response Sprinklers in Office Configurations: Fire Test Results," NBSIR 88-3695, 1988, National Institute of Standards and Technology, Gaithersburg, MD.

Weber, R. A., and Smith, D. T., "Evaluation of Residential Sprinklers for Use in Military Housing," Technical Report, Sept. 1987, U.S. Army Corps of Engineers, Champaign, IL., p. 28.

"What Residential Sprinklers Can Do," *Fire Journal*, Vol. 82, No. 5, Sept.–Oct. 1985, p. 60.

Yoa, C., "Development of the ESFR Sprinkler System," *Fire Safety Journal*, Vol. 14, Nos. 1&2, July 1988, pp. 73–75.

Yao, C., "Early Fire Suppression with Fast-Response Sprinklers," *Sprinkler Quarterly*, Winter 1983–84, pp. 19–23.

ULTRA-HIGH-SPEED SUPPRESSION SYSTEMS FOR EXPLOSIVE HAZARDS

Robert M. Gagnon

Ultra-high-speed suppression systems for explosive hazards have come to symbolize many things to different people. To fire protection engineers, they are the cutting edge of today's fire protection technology. To some contractors, they may be too difficult to install. To the general public, they may connote an air of mystery and foreboding. Clients may consider them to be "temperamental" or hard to maintain. It is this diversity of impressions that has necessitated further developments in ultra-high-speed fire suppression system technology.

Two fundamental types of suppression systems exist that are available to fire protection professionals for the control or extinguishment of explosive hazards: (1) ultra-high-speed water spray systems, which are discussed in significant detail in this chapter, and (2) explosion suppression systems, which are covered briefly in this chapter, and are well detailed in Section 4, Chapter 14, "Explosion Prevention and Protection."

Since the ultra-high-speed water spray system is essentially a highly sophisticated water spray system, a review of the fundamentals of water spray system design may be of considerable assistance. See Section 6, Chapter 12, "Water Spray Protection."

Further information is available in Section 1, Chapter 6, "Explosions"; Section 4, Chapter 12, "Explosives and Blasting Agents"; Section 5, Chapter 2, "Automatic Fire Detectors"; Section 6, Chapter 25, "Special Systems and Extinguishing Techniques"; and Section 7, Chapter 7, "Venting Practices."

THE ULTRA-HIGH-SPEED WATER SPRAY SYSTEM

Definition and Overview

An ultra-high-speed water spray system is generally defined as one that can commence the initiation of discharge of water to a volatile commodity within 100 milliseconds of detection, measured from presentation of an energy source to the detector, to flow of water from the nozzle being tested. This level of response is generally re-

quired where materials of very high flame spread and heat release, such as propellants and pyrotechnic materials, are involved.

The NFPA Technical Committee on Water Spray Fixed Systems has determined that a line of demarcation must exist between ultra-high-speed water spray systems, covered in NFPA 15, *Standard for Water Spray Fixed Systems for Fire Protection* (hereinafter referred to as NFPA 15), and explosion suppression systems, covered in NFPA 69, *Standard on Explosion Prevention Systems* (hereinafter referred to as NFPA 69). Explosion suppression systems are to be designed in accordance with NFPA 69 for enclosed vessels or other devices or conveyances where overpressurization or rupture is the primary concern. Ultra-high-speed water spray systems are predominantly designed in accordance with NFPA 15 for rooms or open, unconfined areas where rupture resulting from overpressurization is not of major concern. Tightly sealed rooms, and enclosed, or partially enclosed, process equipment are examples of applications that may be appropriate for the design of either ultra-high-speed water spray systems or explosion suppression systems, depending upon the configuration of the commodity and its associated manufacturing process. Explosion suppression systems are not normally used in ordinance operations, because their response time is slower than ultra-high-speed water spray systems for the volume of the space requiring protection.

Both ultra-high-speed water spray systems and explosion suppression systems can be satisfactorily designed for the suppression of deflagrations, or fires where the expanding flame front travels at a velocity less than the speed of sound. Neither system has been found to be effective for applications where a detonation, or a fire where the expanding flame front travels at a velocity greater than the speed of sound, could occur. Test data or field testing of the commodity may be necessary before selecting a suppression system for a particular explosive hazard.

Examples of facilities where ultra-high-speed water spray systems are advantageous include rocket fuel manufacturing or processing; solid propellant manufacturing or handling; ammunition manufacturing; pyrotechnics manufacturing; maintenance/renovation/demilitarization of ordinance items containing pyrotechnic materials; and the manufacture or handling of other volatile solids, chemicals, dusts, and gases. An example of a situation where an ultra-high-speed suppression system is not likely to be effective would be the protection of stored munitions whose manufacture has been completed.

Certain types of munitions, such as C-4 explosives, may promote a fire scenario that transitions from a deflagration to a detona-

Robert M. Gagnon, P.E., S.E.T., M.S.F.P.E., is president of Gagnon Engineering, a fire protection consulting firm in Ellicott City, MD. He is a member of NFPA's committees on Water Spray Fixed Systems, Foam-Water Sprinkler Systems, Water Tanks, Private Water Supply Piping Systems, and Water Cooling Towers. Mr. Gagnon also serves as a lecturer at the University of Maryland Department of Fire Protection Engineering and as an adjunct faculty member at Montgomery College in Rockville, MD.

tion. An ultra-high-speed water spray system has been found to be effective for the suppression of such fires when the expansion of the flame front has been halted in its incipient stages.[1]

For military applications, it has become customary to use a nine-hazard classification system developed by the United Nations Organization (UNO) to promote safe manufacture, storage, and transport of explosives. Class 1 contains the majority of energetic materials that are ordinarily protected by ultra-high-speed water spray systems. Class 1 is subdivided into six divisions, with Divisions 3 and 4 providing the most opportunity for high-speed suppression:

Class 1, Division 1:	Mass detonation. Includes explosives, ammunition, and liquid propellants that are too energetic to be protected by an ultra-high-speed water spray system.
Class 1, Division 2:	Explosion with fragments. Includes ammunition and explosives that may be too energetic to be protected by an ultra-high speed water spray system.
Class 1, Division 3:	Mass fire. Includes ammunition and explosives that may be extinguished during manufacture if suppressed at an early stage. The majority of ultra-high-speed water spray systems protect commodities within this division.
Class 1, Division 4:	Moderate fire. Includes ammunition and explosives that can be extinguished by an ultra-high-speed water spray system.
Class 1, Division 5:	Blasting agents.
Class 1, Division 6:	Extremely sensitive ammunition.

Reaction Time

Reaction time is defined as the total time required from initial fire event to the presence of water at the nozzle, and can be broken down into eleven segments for clarity as to what is included in reaction time and what is currently not being considered during the testing of ultra-high-speed water spray systems.[2]

Time Segment 1:	The energetic material is subject to excessive energy input.
Time Segment 2:	Deflagration begins.
Time Segment 3:	The deflagration develops to a point that puts it in the detector's field of vision.
Time Segment 4:	The detector begins to react to the fire.
Time Segment 5:	The detector determines that there is sufficient light energy to be considered a fire.
Time Segment 6:	The detector sends out a fire signal.
Time Segment 7:	A high-speed module on the control panel receives the signal from the detector and activates the squib or solenoid-actuated valve.
Time Segment 8:	Mechanical components associated with the ultra-high-speed water spray system go into motion.
Time Segment 9:	Water leaves the nozzle and travels toward the target.
Time Segment 10:	Water spray impinges upon the target.
Time Segment 11:	Water flow is maintained to achieve cooling, dispersion, or extinguishment.

The 100-millisecond maximum response time used by NFPA 15 considers the period from detector reaction to flow at nozzles, or the time elapsed from segments 4 through 9.

Since different detectors have differing reaction times, it is imperative that a detector be chosen with the shortest reaction time available, that the detectors be placed as close as possible to the target, and that no retards or delays be present.

Of great concern to fire protection engineers is the travel time for the water to hit the target, segment 10. To minimize water travel time, nozzles are to be placed as close as physically or practically possible to the target. It is highly recommended that the water travel time be considered in the total response time where feasible. High-speed photographic equipment, or other sophisticated time measurement tools, would be needed to consider water travel time as a component of the total response time.

General Detection Concerns

To achieve the aforementioned rapid response time, optical flame detection for open areas, or pressure detection for areas subject to overpressurization, is necessary. Examples of optical flame detection devices include ultraviolet or infrared detectors placed in close proximity to the hazard. Such detectors instantaneously detect flashes of light associated with the wavelengths normally found in combustion. Pressure detectors sense a buildup of explosive pressure in an enclosure; they are very effective, but generally provide slower actuation times than optical flame detectors.

The signal emitted by a detector is received by a high-speed module on the control panel. To enhance actuation response times, it is recommended that the control panel be specifically designed for ultra-high-speed systems. It is the panel and its associated detection system that makes an ultra-high-speed suppression system unique from other types of water-delivery fire protection systems. The proper choice and design of supervisory equipment is essential to the successful functioning of the system.

NFPA 70, *National Electrical Code®*, must be followed closely for all detection system wiring and connections. Circuits between sensing devices and their controllers must be continuous with no splices, and must be individually shielded.

General System Concerns

The piping methods of agent delivery to the hazard are as varied as the hazards they protect. The nozzles should be in very close proximity to the hazard, with water in the piping and primed up to the nozzle, prepared for quick delivery of water to the hazard. One type of ultra-high-speed water spray system, the squib-actuated system, involves loosely capped nozzles, and/or a rupture disc at each nozzle, a preprimed piping system, and a deluge valve with an explosive squib. (See Figure 6-14A.) Another type, the solenoid-actuated system, involves pilot-actuated nozzles, with a solenoid valve or system of solenoid valves, installed on the pilot piping, which actuate upon receiving a signal from the detection system. (See Figure 6-14B.) A third type, possessing explosive squibs at each nozzle, is a variation of the squib-actuated system. Its distinguishing features, i.e., prepriming and squib actuation, will be covered in detail in this chapter.

Two general design approaches, point protection and area protection, will apply to ultra-high-speed water spray system design. Point protection is the application of concentrated high-speed water spray onto a likely point of ignition, and is accomplished by providing two or more nozzles at positions as close as physically possible to the point of ignition. (See Figure 6-14C.) Area protection is the application of high-speed water spray over the area of a room or over the surface area of an object. (See Figure 6-14D.) In both cases, the nozzles must be as close to the intended target as possible to minimize the time for the water to travel from the nozzle to the target. Some hazards may be conducive to using the attributes of both

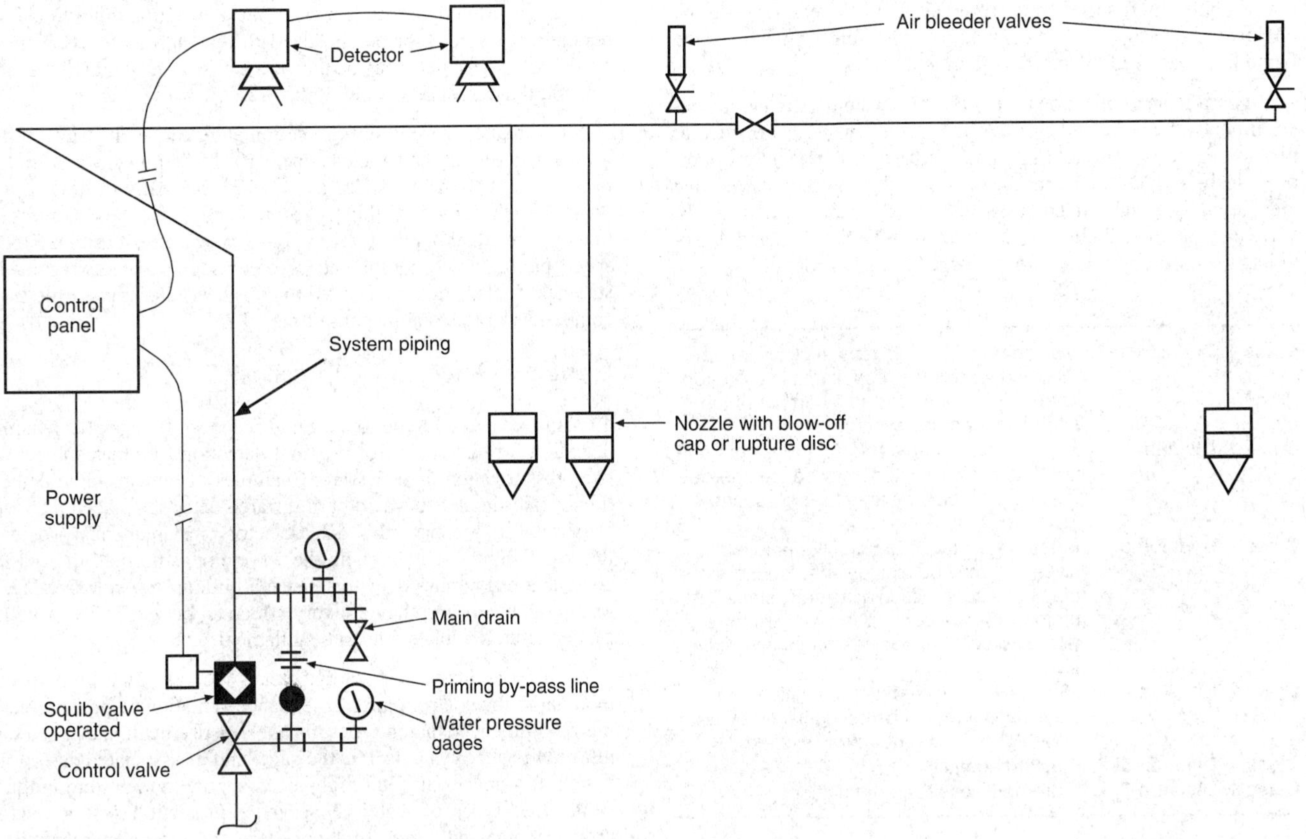

FIG. 6-14A. Squib-actuated system.

point protection and area protection. This hybrid system is referred to as a combination system. (See Figure 6-14E.)

The process of detection, transfer of signal to the control panel, interpretation of the signal, transmission of signal to the valve, and delivery of water to the hazard happens before one's senses are aware of a problem. It is not unusual for an installer to receive complaints of false trips after the installation of an ultra-high-speed water spray system. These occurrences will include real fires, undetectable by human senses. It is recommended that a diary of events be kept before and after system installation for analysis. (See Figure 6-14O.)

The first high-speed systems that protected munitions and propellant facilities were primarily preprimed deluge systems that used standard deluge valves and pneumatic detection. Response times for such systems were in the vicinity of one second. This type of system can still be used when it is desired for standard water spray deluge systems to have an enhanced response time.

Improvements and innovations were required to protect processes involving materials that require faster response times than the preprimed deluge system. The currently available ultra-high-speed water spray systems are covered in the sections that follow.

Selecting a Detection System

In order to provide the best possible results in protecting munitions and other hazards requiring fast-response technology, it is recommended that a four-step method be used to select a detection system:[3]

1. Choose a detector capable of reacting to the wavelengths that would be encountered in a fire involving the hazardous material in question. For example, an ultraviolet (UV) detector will respond to a hydrogen fire, but an infrared (IR) detector operating in the 4.4 micron range will not.

2. Survey the hazard area, the process, and the adjoining areas to determine what sources of false signals may exist. For example, if arc welding is expected to occur within the view of a detector, UV detectors must either be shielded from view or disabled.

3. Determine what elements could interfere with the optimum operation of the detector. Dust or ice could severely hamper the detector's ability to function in accordance with its design parameters, and will determine a detector's features or placement. Detectors with air shields to automatically clear dust from the detector lens are available.

4. Ascertain the speed with which the detector is required to respond. UV detectors are recognized as having the fastest available response time.

Two basic categories of flame detection systems are used in conjunction with ultra-high-speed water spray systems. One type, UV detectors, responds to radiation in the range of 1,850 to 2,850 angstroms (0.185 to 0.285) microns. Detection ranges are graphically shown on a radiation spectrum chart. (See Figure 6-14F.) The other type of flame detector is the IR detector.

Note how the radiation spectrum chart shows an intersection between the IR detector sensitivity (between 0.8 and 1.0 microns) and the graph of solar radiation that reaches the earth. One can surmise by this intersection that an IR detector would be receptive to and would be adversely affected by sunlight radiation.

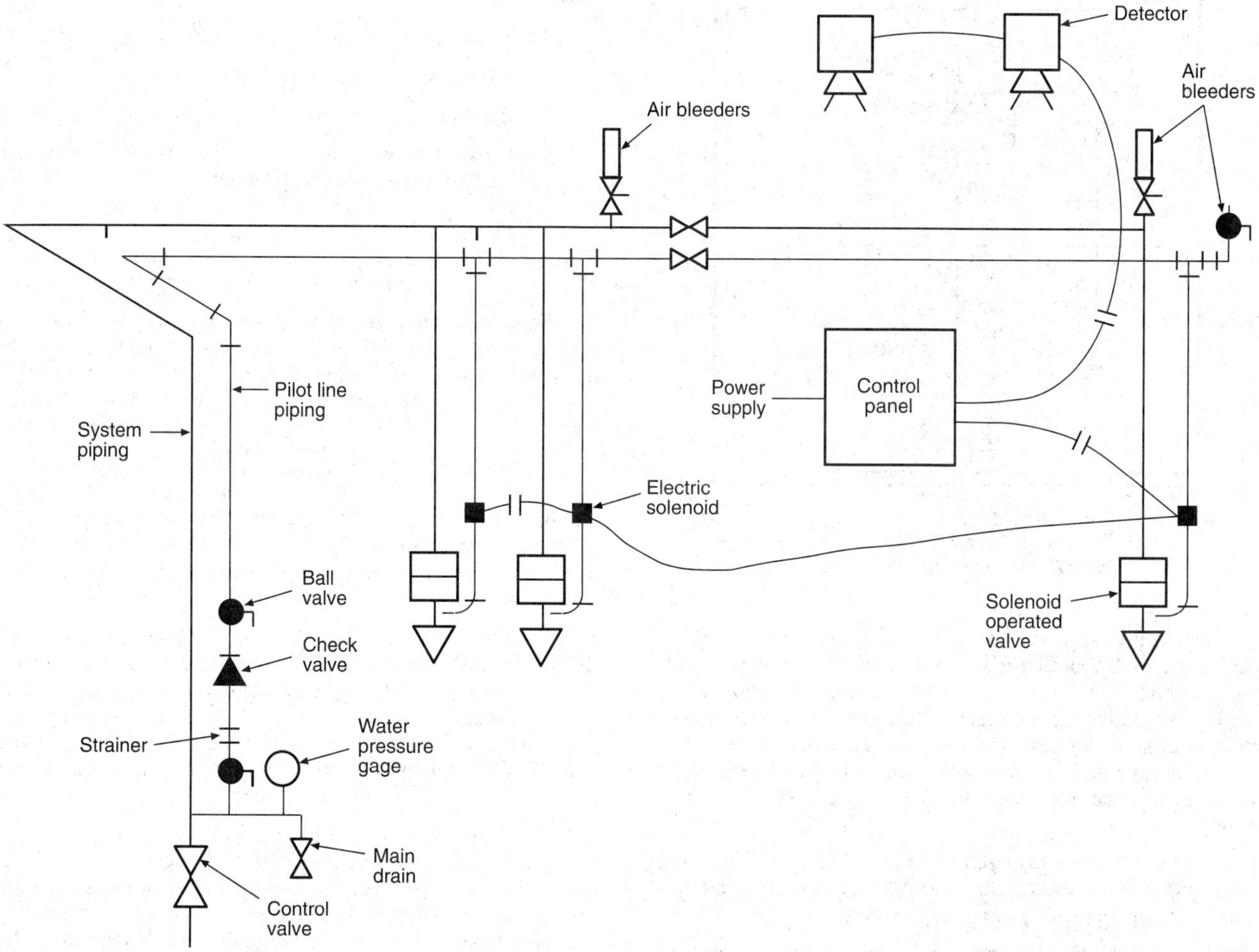

FIG. 6-14B.   Solenoid-actuated system.

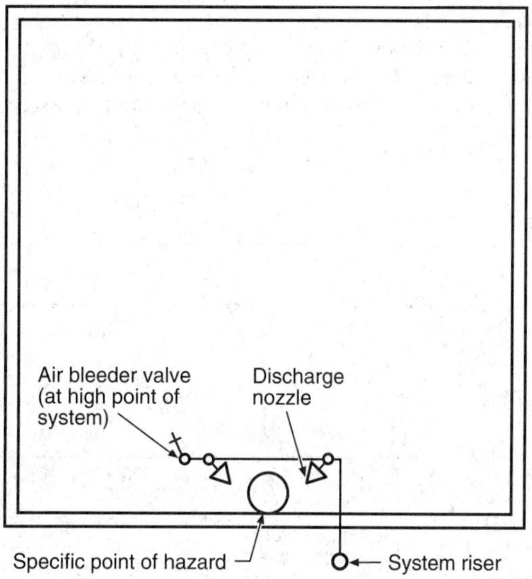

FIG. 6-14C.   Local point application design.

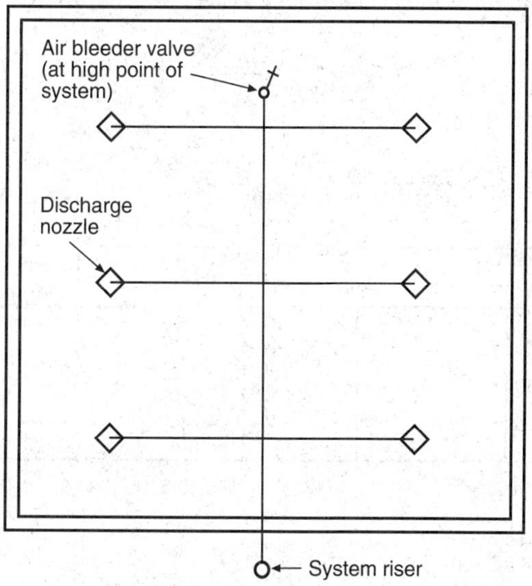

FIG. 6-14D.   Area application design.

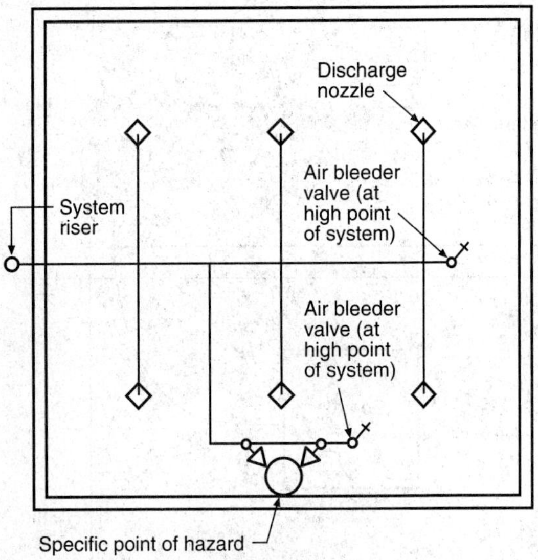

FIG. 6-14E. *Combination application.*

This sensitivity will strongly affect the manner in which IR detectors are used. UV detectors are not affected by sunlight, according to the graph, and can be located without regard to the sun factor. High-intensity lights and high-temperature bodies will affect IR detectors, with no effect upon UV detectors. However, arc welding, lightning, X-rays, and gamma rays will affect UV detectors, with a lessened effect upon IR detectors that have been properly aimed and shielded.

Since the two types of detectors, i.e., UV and IR, are searching for differing types of radiation and are affected by differing interfering media, their applications are generally quite different.

UV detectors, being insensitive to sunlight, would be well suited for surveillance of an entire process area, or a process requiring a wide angle of vision, while IR detectors would be better suited for close surveillance of a specific contact point within a processing machine, or supervision of the interior of an enclosed vessel. (See Figure 6-14G.)

UV detectors are subject to the inverse square laws of optics, where detector sensitivity varies inversely with the distance between the detector and the energy source:

$$S = \frac{(K)pe^{cd}}{d^2}$$

where

S = radiation power reaching the detector,

d = lineal distance from the fire to the detector,

K = constant,

p = radiation power emitted by the fire, which is a function of the fuel characteristics and fire size,

e = base logarithm, and

c = extinction coefficient of air, or the radiation energy absorbed or diffused by air, a function of humidity, dust, etc.

From this famous law of physics, one can see that doubling the sight distance of the detector will result in a dramatic reduction in the radiation received by the detector. Conversely, reducing the sight distance greatly increases the radiation received by the detector.

The wisest use of UV detectors is to provide area detection of the entire process by overlapping the area detection system with additional detectors that closely view critical points in the process. Reliability increases dramatically when critical areas are viewed by an overlapping cone of vision from different detectors in the same detection zone.

Because of their sensitivity to sunlight, IR detectors are best located in enclosed, partially enclosed, or shielded spaces. While IR detectors also obey the inverse square law, the hazard of receiving unwanted sunlight signals negatively affects their performance at longer distances. Shields are available to restrict the cone of vision of an IR detector to prevent unwanted actuation, and downward positioning reduces dust buildup on the lens.

The actuation or release apparatus for an ultra-high-speed suppression system will consist of detectors, a controller capable of monitoring and supervising up to eight detectors, a relay module to alert personnel and provide remote annunciation capability, and a detonator module that activates the squibs or solenoids that release an ultra-high-speed deluge system. (See Figures 6-14H and 6-14I.) The modular control components are installed in a mounting cage, and are served by a voltage converter that transforms ac building current into dc system current.

THE MUNITIONS MANUFACTURING PROCESS

The munitions manufacturing process is one of the most prominent applications for ultra-high-speed water spray systems, and is among the most hazardous fire protection challenges that a system designer or contractor can encounter. A delay of only a few milliseconds can

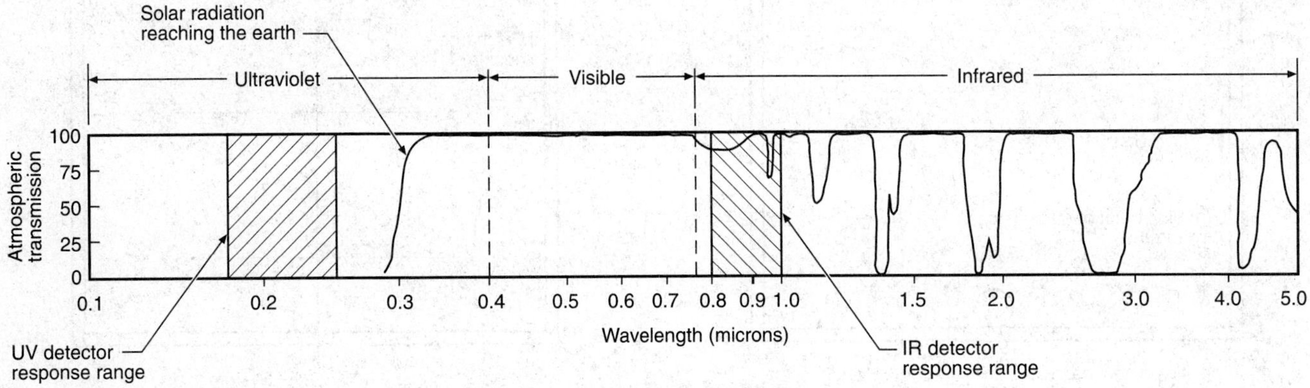

FIG 6-14F. *Radiation spectrum chart.*

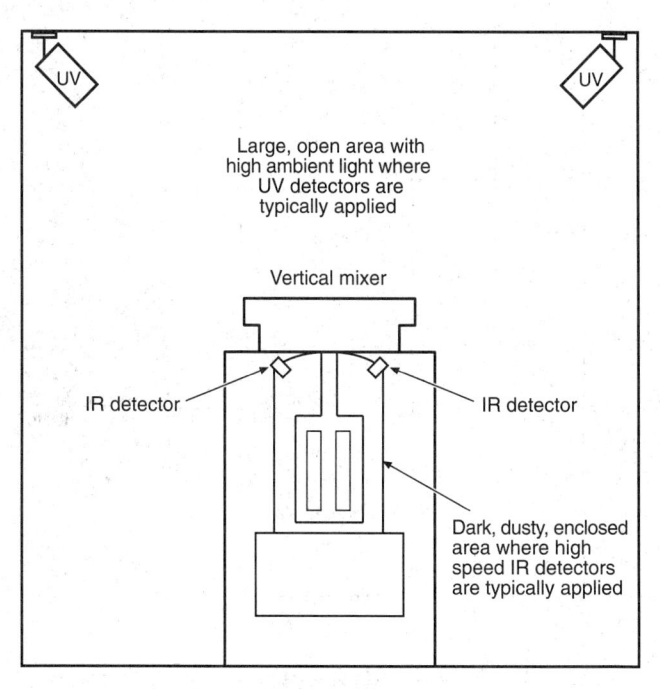

FIG. 6-14G.   UV and IR detectors.

mean disaster. The stakes are frighteningly high, and munitions manufacturing personnel are routinely required to perform operations involving significant risk. The U.S. Army Armament, Munitions, and Chemical Command and its Army ammunition plants have played a leading role in the development and improvement of ultra-high-speed water spray systems.

Ultra-high-speed water spray systems have been successfully employed with many types of munitions, including propellants, diazodinitropherol (DDNP), trinitrotoluene (TNT), tetranitromethylaniline (TETRYL), tetranitrocellulose-type explosives (RDX), C-4, black powder, pyrotechnics, fireworks, magnesium Teflon™ flares, and smoke-generating devices. Each commodity involves differing manufacturing processes.

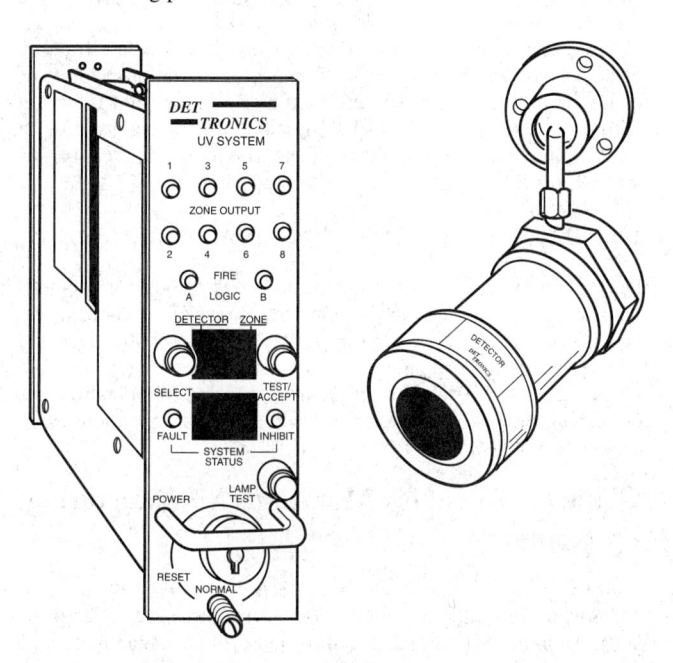

FIG. 6-14H.   Detectors and controllers.

In general, the fourteen processes[1,4] that describe most munitions manufacturing operations include:

1. Weighing munitions,
2. Pressing munitions,
3. Pelletizing,
4. Propellant loading,
5. Melting,
6. Extrusion,
7. Mixing,
8. Blending,
9. Screening,
10. Sawing,
11. Granulating,
12. Drying,
13. Pouring, and
14. Machining.

Each munitions manufacturing process requires individual consideration and careful engineering to fully analyze the process and determine the proper and appropriate protection measures. In each case, the specifying engineer or designer must determine the objective of the system.

The three objectives that are to be considered in ultra-high-speed suppression system design are: (1) life safety and personnel protection, (2) equipment protection and facility protection, and (3) total extinguishment or propagation cessation. The location and extent of protection and detection will vary with the size of the operation, location of critical friction points, and the presence and location of the operating crew. The choice of objective or combination of objectives must be the result of a thorough risk analysis of the hazard and of the financial and life safety effects of a catastrophic event in a processing area.

A strong consideration to be given in the choice of general area protection *vs.* specific point protection is the mobility of the equipment. In many cases munitions processing machinery can be moved from one location to another, making protection at specific contact points impractical or impossible. It is recommended that a combination of area protection and point protection be provided wherever feasible, and that protection for personnel be provided in all cases.

Weighing Munitions

If the munition is a powder, a special situation will arise as a result of the dust created by the operation. Static charges could result in ignition of the dust. The detection system should be as close to the hazard as possible, while maintaining the ability to observe the pouring operation. Nozzles should be placed in such a way as to wet down the entire operation, including the operator, with overlapping sprays from opposing angles and with as heavy a density as is possible. Extra nozzles may be required at critical points in the process. Solid weighing processes will involve, in general, area detection and area nozzle placement. By protecting the entire room, the pathways between operations will be supervised and protected by heavy-density water spray.

Pressing and Pelletizing Munitions

A major risk in pressing and pelletizing operations is one of ignition of the munition at the contact points of the operation, and propagation of flame to the bulk storage of the munition. Detection and protection must be located as close to the points of contact as possible. Normally, the operator's arms will be protected by spray applied to the contact points. Strong consideration should also be given to add nozzles for the protection of the entire machine and the operator.

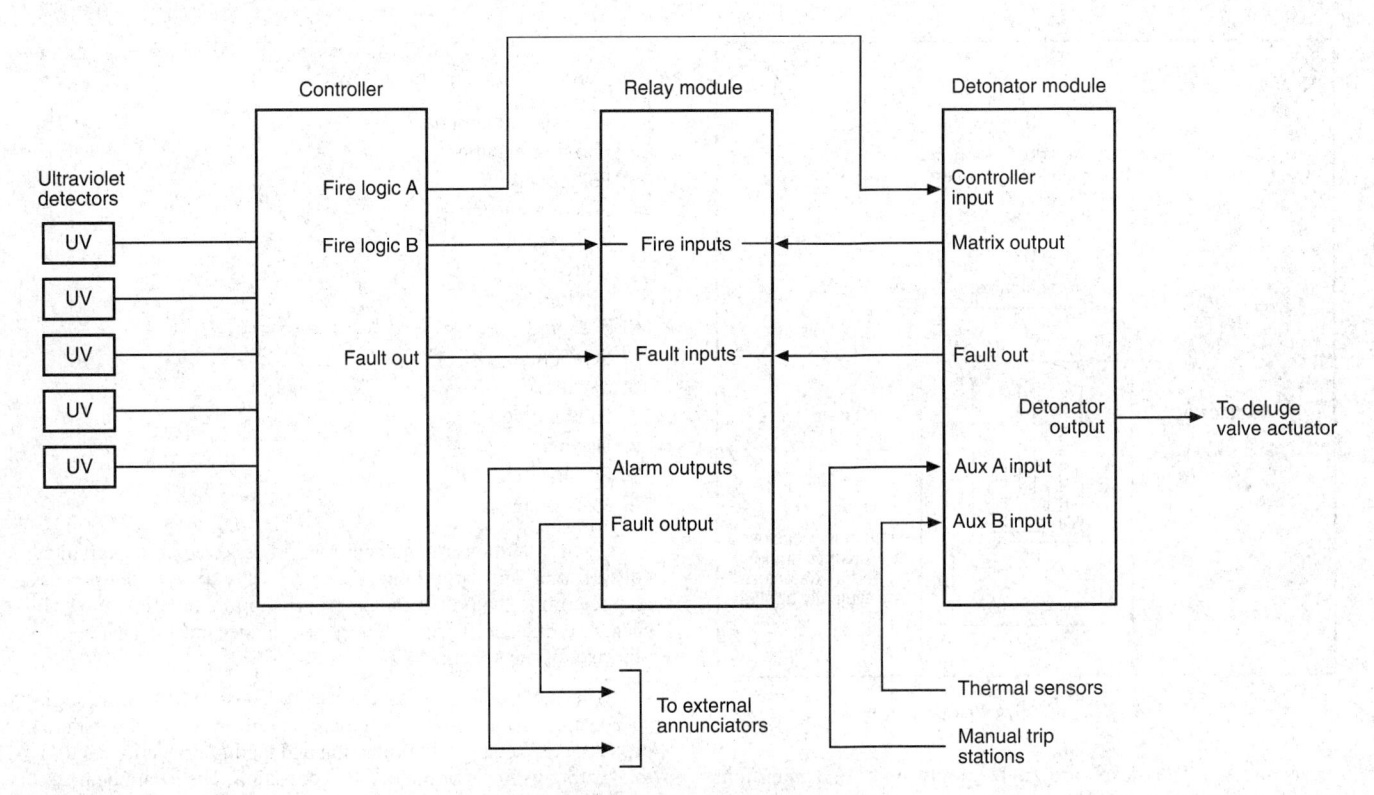

FIG. 6-14I. UV and IR schematics.

Propellant Loading

The fire protection objective in explosive propellant loading is to halt the flame front caused by an ignition of propellant, therefore preventing propagation from the loading area to the main hopper, and the resulting deflagration.

Area protection and detection is recommended for the loading area, the storage area, and the travel routes between the munitions operation, the storage area, and the loading area. Extra protection is recommended for critical friction or contact points. UV detection is best suited for this hazard.

Melting Munitions

Kettles are used to melt munitions, with steam or hot water as the predominant energy source. One or more detectors, placed as close to the kettle as possible, and nozzles directed over the exterior of the kettle, especially at the opening where materials are added, will be required. The proximity of detection and nozzles to the kettle is critical to successful fire protection during a melting operation. UV detection is recommended.

Extrusion of Munitions

Extrusion involves the process of pressing the propellant or pyrotechnic material through an opening or dye to form it into a desired shape. The point of interest here is the opening through which the munition exits. Detection and nozzle placement close to this point should prevent any ignited munition from spreading to the bulk storage area of the machine. Personnel and machine protection can also be considered.

Mixing and Blending Munitions

Such operations can take place in an open vat or in a closed container. The open mixing or blending process may be approached in a manner similar to the melting process. If the container is closed, the detector and nozzle should penetrate the enclosure, with appropriate precautions taken to avoid dust loading or obscuration of the detector and nozzle.

Screening, Sawing, and Granulating Munitions

Common hazards to be considered in these processes are friction, dust, and sparks that can occur during these operations. Points of contact must be completely observed by detection, and protected with ultra-high-speed water spray. Locating the nozzle as close to the point of contact as possible will enhance timely delivery of water to the critical point most likely to support ignition.

Drying, Pouring, and Machining Munitions

Drying operations are usually done in ovens or heating cabinets, with area nozzle protection over the surface area of the cabinets, especially where the operator opens the cabinet to remove or add items to be dried. Pouring requires point nozzle protection, and area nozzle protection is also often required. Machining operations require point nozzle protection similar to extrusion operations, with area coverage also often required.

Choice of Agent for Munitions Manufacturing Protection

Water is the agent of choice in the protection of explosive munitions hazards, for its ability to cool the combustible, insulate the exposed surfaces, dilute a hazardous material, and cool the expanding flame front.

Munitions and propellants differ from most fire protection hazards in that they contain a built-in supply of their own oxidizer, mixed in with the fuel. The oxidizer is needed to ensure rapid burning of the material upon demand in tightly enclosed areas or oxygen-deficient atmospheres, such as rocket motor combustion chambers or artillery gun tubes. Most oxidizer additives in the munitions industry fall under the general classification of nitrates and chlorates, such as potassium nitrate and potassium chlorate.

In the choice of an agent for munitions manufacturing, the presence of this built-in oxygen supply effectively eliminates from consideration the commonly available smothering or blanketing agents, such as carbon dioxide. Halon typically fails to stop the flame front in munitions fires. It begins to chemically break down and so fails to provide the cooling necessary to inhibit propagation.

It must therefore be said that the primary objective in munitions protection is cooling, not unlike most hazards encountered by the fire protection engineer. By introducing a cooling agent in the early stages of a potential deflagration, one prevents the feedback of heat energy from spreading to unburnt fuel, thus effectively limiting, then ceasing combustion. Failure to apply water in the very early stages of munitions combustion could result in burrowing, or flame propagation within a pile of munitions mixture covered by extinguished and water-soaked munitions material.[1]

Also to be considered are the combustion gases produced by burning munitions. There exists in munitions fires a point of no return, where the production of combustion gases creates a barrier sufficient to prevent water spray from reaching the fuel. This explains why sophisticated mechanical and electrical fire protection equipment, designed for response in a few milliseconds, is essential for the protection of munitions hazards.

Retardation of ultra-high-speed fire suppression system response time from 100 milliseconds to 1 second or more creates a scenario whereby the process reaches a point where the expanding flame front gets out of control, negating the effectiveness of the suppression system, and resulting in the loss of the process.

Flame Detection Devices

Flame detection is the detection method of choice for ultra-high-speed deluge systems. Two types of flame detectors in common usage are UV detectors and IR detectors, both consisting of a photoelectric cell and a photoconductive cell.

UV detectors are designed to detect radiation falling within the ultraviolet wavelengths below 4000 angstroms (0.4 microns). Fires emitting radiation with wavelengths of this nature generally have high-intensity flames associated with diffusion flame combustion. Close proximity of the detector to the source of the combustion is critical to operation within its intended range of sensitivity.

Shielding or screening may be necessary to isolate the UV detector from unwanted signals, such as lightning and welding arcs. It should be noted that a UV detector is a flame detector, not a smoke detector. An anticipated fire that would be preceded by or accompanied by significant quantities of thick or dark smoke may not be amenable for UV detection, since the smoke could prevent or delay the operation of a UV detector.

A limitation to the use of UV detectors is the fogging or loading of the lens that can occur when these detectors are used in dusty or moist atmospheres, adversely affecting their reaction time. Unless special arrangements are provided, such as air shields to divert dust away from the lens, or a tightly controlled 24-hr maintenance and supervision program, UV detectors may be unsuitable for some dusty or greasy environments.

IR detectors are amenable for use at distances up to 50 ft (15.3 m), and respond to infrared radiation involving wavelengths from 8500 to 12,000 angstroms (0.8 to 1.2 microns). Using IR detectors at extended distances is generally not recommended for ultra-high-speed suppression systems for explosive hazards. There is, however, research currently underway relative to the possible use of some of the newer IR detectors that show great promise as a possible detection source for ultra-high-speed water spray systems for specialized applications.

Some common heat sources that can be sources of false alarms for IR detectors include automobile headlights, sunlight, room lighting, heaters, lighted tobacco products, and matches. IR detectors may not be able to discriminate between a common heat source and a combustion scenario.

Most IR detectors have a built-in timing unit that can delay actuation by as much as 1 to 30 sec. This length of delay is usually unsuitable for the detection of munitions fires. If IR detectors are used, the distance between hazard and detector must be kept to a minimum, effectively screening out most sources of unwanted signals, and the time delay must be eliminated or minimized. A special ultra-high-speed IR detector is available with the delay feature eliminated.

Most UV and IR detectors are listed by Underwriters Laboratories, Inc. (UL), and approved by Factory Mutual Research Corporation (FMRC) for fire protection use. While they are approved for both indoor and outdoor use, it is recommended that they not be used outdoors for ultra-high-speed detection, so that unwanted signals can be minimized. Sources of flickering light, such as lights mounted on operating or vibrating machinery, may be perceived as being a fire by the detector, and should be screened or eliminated.

Combination UV and IR devices are commercially available. While such devices are less prone to provide false alarms, they are more prone to delay system actuation, since the detector must see both wavelengths before notifying the control panel.

SQUIB-ACTUATED ULTRA-HIGH-SPEED WATER SPRAY SYSTEM

The squib-actuated system is a preprimed deluge system with a squib-operated deluge valve, a system of piping preprimed with water, specially capped nozzles and/or rupture discs, and an ultra-high-speed flame detection system. Developed in the early 1960s, it was the first commercially available ultra-high-speed water spray system, and continues to be in active use today.

It is recommended that this system be used for hazards such as ammunition loading, explosive powder drying operations, high-energy fuel processing, self-oxidizing fuel processing and storage, silane containment cells, explosive dust collecting and conveying operations, roll mills, and casting pits. Other hazards requiring enhanced reaction times may be suitable for the squib-actuated ultra-high-speed water spray system.

Preprimed Piping System

The preprimed piping system is supplied by a priming bypass valve that provides a water head of up to a maximum of 12 psi to the blow-off caps installed to hold back the priming water. Water pressure required to remove the blow-off caps will vary with the make and model of cap used. A wire retains the blow-off cap to the nozzle for ease of retrieval and to prevent it from falling into material or devices being protected. It is recommended that the nozzles be as close as possible to the hazard being protected, and directed toward the critical points of the hazard. (See Figure 6-14J.)

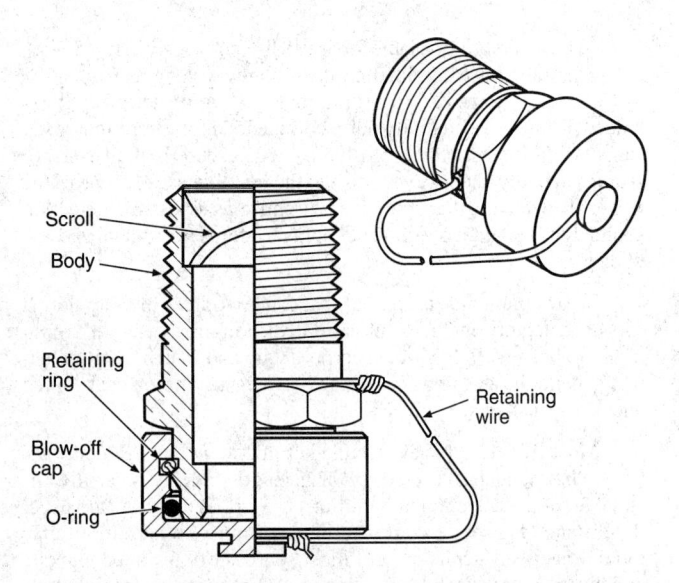

FIG. 6-14J. Nozzles and blow-off caps.

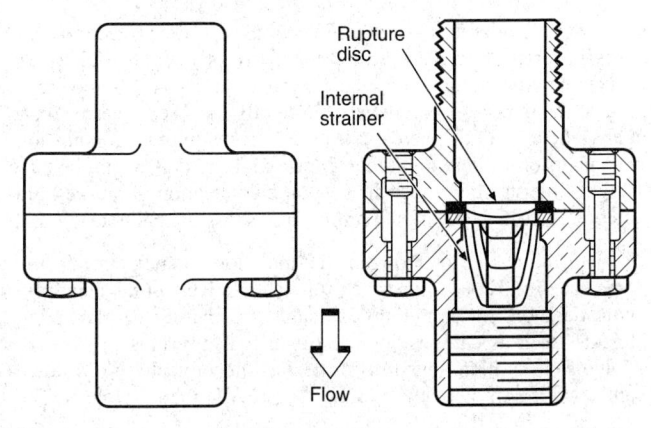

FIG. 6-14K. Rupture discs.

Sprinklers may be used if there is a possibility of a fire occurring in other materials that could result in a fire propagating to the munitions, such as a fire starting in combustible packaging material and spreading to a munitions item.

Allowing the water to be preprimed in the piping has a strong advantage over standard evacuated piping deluge systems. The time it would have taken for the water to travel from the deluge valve to the nozzle has been eliminated. This can be a significant difference, especially for systems requiring the deluge valve to be remote from the hazard.

The preprimed aspect of this system requires that the hazard area and valve riser be heated. Heat tracing of the piping is not recommended because loss of power, improper tracing installation, cracking or crimping of the tracing wire, or other potential failures of the tracing would damage and compromise the ability of the system to function when required.

The priming static pressure that can be maintained on the blow-off caps for a squib-actuated system, without danger of accidental leakage, is 12 psi (82.74 kPa) or less, depending upon the cap manufacturer. Rupture discs or other devices capable of maintaining a minimum system static pressure of 50 psi (344.75 kPa) are strongly recommended. The system must be hydraulically designed to ensure a minimum residual pressure of 50 psi (344.75 kPa) at each nozzle, with sufficient static pressure to allow for automatic removal of all blow-off caps or rupture of all rupture discs upon system actuation. (See Figure 6-14K.)

The Squib-Actuated Valve

The squib-actuated valve is an electrically actuated valve with two explosive primers, wired in parallel. The valve is intended to be installed vertically, but will also function, with increased response times, in the horizontal position. The primers consist of .004 oz (0.12 g) of lead styphnate charge, and actuate upon the receipt of an electric signal from the detection system.

The primers are designed so that the force of only one primer is needed to actuate the system. Firing of the primers shears a rod holding the clapper latch, allowing the plunger to rise, resulting in the rapid movement of the clapper and rapid pressurization of the priming water as water begins to flow through the valve. The blow-

off caps are ejected or rupture discs broken once their inlet pressure exceeds their respective release pressure.

Although squib-actuated valves are rated for use in Class 2 locations, it is recommended that the valve not be located directly adjacent to a hazardous process, and in a nonhazardous area, if physically possible. (See Figures 6-14L and 6-14M.)

Squib-actuated valves are available for use with 2-in. (51-mm) and 2¹/₂-in. (64-mm) risers, and releases are available for 4-in. (102-mm) and 6-in. (152-mm) deluge valves. Increased system size has a very strong effect on the increasing of response time. A 500-gal (1893-L) water-holding capacity is recommended as the maximum system size, and best results would be obtained by subdividing the systems such that the number of nozzles served by one squib-actuated valve is between two and six nozzles.

The Detection System

The detection system for a squib-actuated system is ordinarily a flame detection system. These photo-conductive cells monitor the hazard and can detect small amounts of radiant energy emitted by a fire in the hazard area. Such detectors are monitoring light within certain predetermined wavelengths normally associated with fire. Should the photo-conductive cells receive more light than was specified to be ambient, the cells' resistance will change, sending a signal to an amplifier in the control panel. This amplifier increases the signal's strength and sends it to the primer for actuation. Signal transmission time is normally two or three milliseconds.

Normal actuation times for a squib-actuated system can vary with piping configuration, size of the system, distance from the valve to the nozzles, water pressure, and hazard conditions. Response times in the 100-millisecond range are possible for very small systems with risers very close to the hazard, and with high water supply pressures.

Resetting a Squib-Actuated System

After operation, the two primers and primer holders are replaced, and a new break rod installed. Nozzles are then recapped or plugged, or rupture discs replaced. Breakglasses must be replaced if the system was operated manually. The system is then reprimed by slowly opening the priming bypass valve and bleeding the air from bleeder valves located at high points in the piping system. Care must be taken to purge all air to ensure complete priming of the system. The control panel is then reset and returned to service. It is recommended that spare primers, holders, rods, blow-off caps, rupture discs, and breakglasses be kept in a convenient nearby location for ease of resetting.

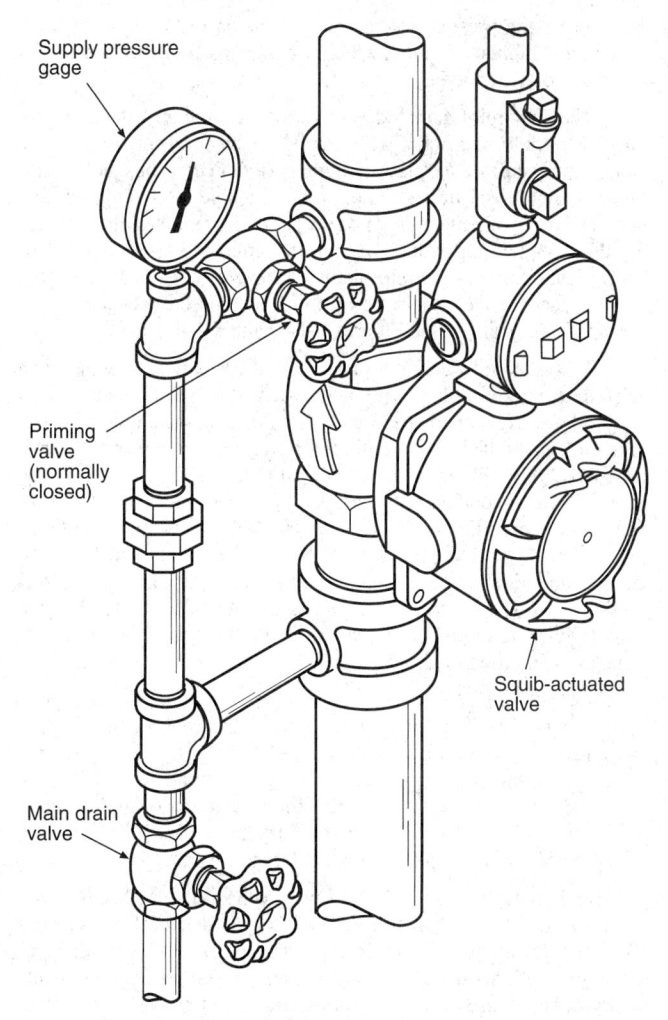

FIG 6-14L.   *Squib-actuated valve.*

It is essential that the system be operated at least once a year to ensure that all parts are functioning properly and ready for use.

System Actuation Time

The piping in a preprimed deluge system must be sloped, with air bleed valves placed at the piping high points, to facilitate the removal of trapped air bubbles in the piping, which are known to have the effect of increasing response time.

Another major factor influencing actuation time is the system water supply pressure. When all other variables are held constant, the actuation time decreases with increasing water supply pressure. For example, if one could double the supply pressure from 50 to 100 psi (344.75 to 689.5 kPa), the actuation time would decrease by about 20 to 25 percent. Actuation times assume that the priming water is at rest at the nozzle orifices. The system pressure and the resulting enhanced actuation times therefore result from inertia effects and are effectively independent of pipe friction, head elevation, and nozzle orifice coefficients.

Other factors influencing the effectiveness of a squib-actuated system are number of nozzles on the system, orifice diameter of the nozzles, length of run from deluge valve to the nozzles, number of fittings on the supply and branch pipes, cap or disc blow-off pressure, and cross-sectional area of the supply pipe.

Actuation time for squib-actuated systems is shown in Figure 6-14N. The valve unlatching time and cap blow-off time are strong functions of system pressure. The graph also shows that 4- and 6-in. (102- and 152-mm) squib-actuated valves take three milliseconds longer to unlatch than the 2- and 2$\frac{1}{2}$-in. (51- and 64-mm) valves.

Most U.S. Army ammunition plants maintain static pressure for ultra-high-speed water spray systems in the range of 50 to 75 psi (344.75 to 517.13 kPa), usually supplied by elevated gravity tanks, to enhance actuation time.

Squib-Actuated Ultra-High-Speed Water Spray System Design

1. It is recommended that the squib-actuated valves be installed in the vertical position, even though a horizontal position is possible. Installation in the horizontal position could trap air, negatively affecting actuation time.
2. Care should be taken to make the pipe routing from squib-actuated valve to the nozzles as direct as possible, with as few fittings and bends as possible.
3. Slope all piping up to vent valves at piping high points, to facilitate the elimination of air from the system. The pipe should be sloped as radically as possible, especially for smaller pipe sizes.
4. For point protection, position the nozzles into counter-opposed groups of no fewer than two nozzles for efficient delivery of water to a likely source of ignition. Counter-opposition of the nozzles involves nozzles aimed from differing directions at a common point.
5. Ensure the highest possible water supply static pressure, enhancing system pressure with a surge (pressure) tank if city or plant supply is insufficient. A minimum of 50 psi (344.75 kPa) static is recommended.

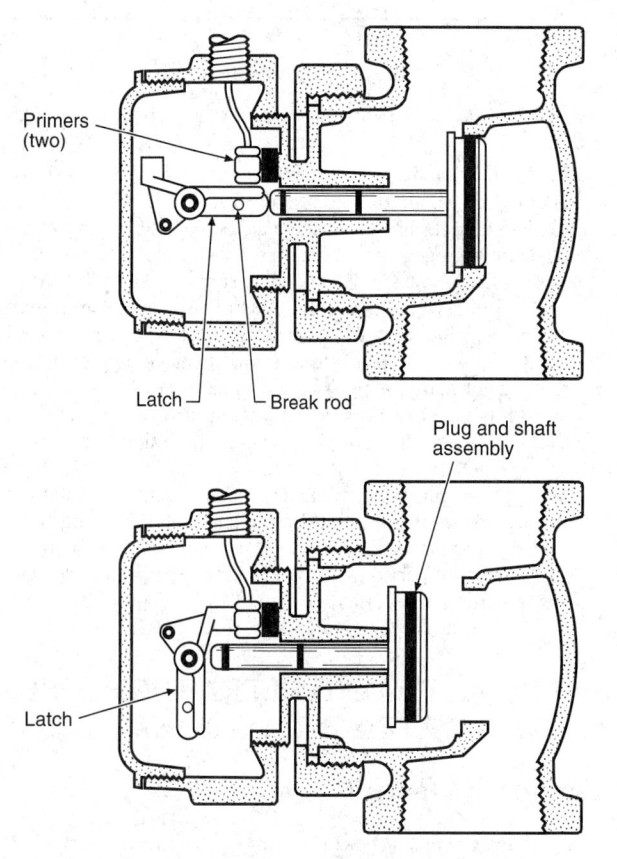

FIG. 6-14M.   *Squib-actuated valve unlatching.*

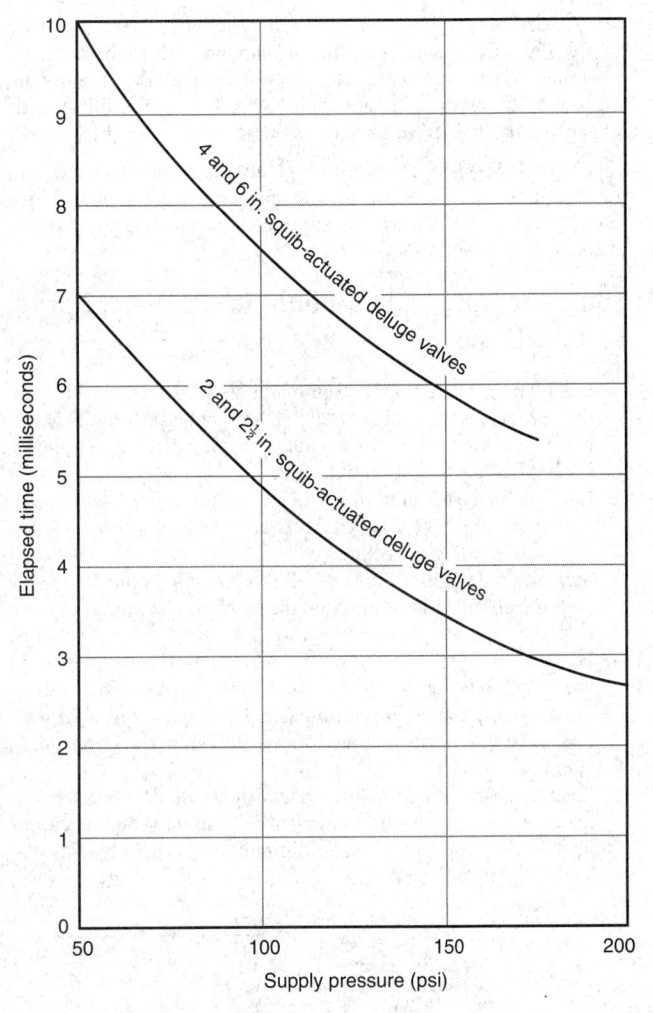

FIG. 6-14N. Valve-unlatching time.

6. Perform hydraulic calculations of the system to verify pipe sizes and ensure minimum nozzle discharge pressures of 50 psi (344.75 kPa). Densities must be calculated based upon water that actually impinges upon the hazard area. Unless the nozzles are wisely placed, the spray that impinges upon areas of little consequence could be considerable, reducing the actual delivered density on the hazard area. Also, care must be taken to overlap sprays as much as possible to supplement the lighter densities that are normally found along the outside perimeters of the conical spray patterns. For spot protection, it is recommended that the concept of average density [raw gal (L) available at nozzle divided by total sq ft (m²) of floor area covered] be replaced by experienced engineering evaluation of actual nozzle distribution to the hazard area in question. Each nozzle should discharge no less than 25 gpm (95 L/min) per nozzle for point protection, nor less than 0.50 gpm per sq ft (20.37 L/min × m²) for area protection. Some commodities, such as certain pyrotechnics, may require densities as high as 3.0 gpm per sq. ft. (122.24 L/min × m²).

PILOT-ACTUATED ULTRA-HIGH-SPEED WATER SPRAY SYSTEM

Early Development

The first pilot-actuated nozzle was introduced in the early 1960s. This early system has as its basis similar types of mechanical components as the present system. Each revision to the early concept was in the interest of speed, which became possible with the advent of more sophisticated electronic technology.

The first pilot-actuated nozzles had an integral deflector attached to the lower body, with a male adapter for connection of a water-spray nozzle. Today, the integral deflector has been augmented by the addition of other bodies capable of accommodating a wide variety of water spray nozzles, allowing engineering flexibility. Each nozzle individually functions similar to a wet-pilot-actuated deluge valve. The pilot-actuated nozzle was once called the "aerospace" nozzle, as a result of its use in hyperbaric chambers for the space program.[5] (See Figures 6-14O and 6-14P.)

The standard detection arrangement for early systems was a system of pilot piping with a heat-actuated pilot release. This method was developed specifically for, and became commonly employed with, naval ship munitions magazine protection systems. The naval requirement for the early pilot-actuated system response was 150 milliseconds, with an operation time, measured from release of the pilot line to delivery of water, in the 10- to 20-millisecond range. The time that it took for the heat-actuated device to actuate and commence evacuation of the pilot line was not included in this time. The heat-actuated means of detection eventually gave way to ultraviolet electronic detection, with solenoids on the pilot line, as the method of choice, once the electronic detection and control technology became available.

What made the pilot-actuated system different was its positive priming at fire main pressure, its poppet-valve nozzle and pilot system, and simultaneous opening of all nozzles connected to one pilot line. The basic system in its earliest form had certain similarities to the pneumatically actuated deluge system, and also bears a strong resemblance to a wet-pilot-actuated deluge system.

Pneumatic heat-actuated and eutectic heat-actuated devices are significantly slower than an electronic UV-actuated release system. With the development of more sophisticated circuit boards, microprocessors, high-speed release modules, solenoid improvements and other response time innovations, improvements in system actuation time were possible.

Current Arrangement

A fixed pilot-actuated ultra-high-speed water spray system consists of a primed discharge system with pilot-actuated nozzles; a hydraulic wet pilot system, supervised and discharged by high-speed solenoid valves; a UV or IR flame detection system; and a central control panel with standby batteries and charger. (See Figures 6-14Q and 6-14R.)

The Discharge System

While the pilot-actuated nozzle has been available for many years, recent developments in control modules and panels make the pilot-actuated system more amenable to hazards requiring extremely fast response times. The pilot-actuated nozzle consists of a two-piece body, with ¹/₂-in. (12.7-mm) male National Pipe Thread (NPT) threaded connection for attachment to the deluge piping system, a ¹/₄-in. (6.4-mm) female inlet for connection to the pilot piping system, and an outlet for the insertion of water spray nozzles with ³/₄-in. (19.1-mm), 1-in. (25.4-mm), or 1¹/₄-in. (31.8-mm) male NPT connections.

The pilot-actuated nozzle is connected to two systems (i.e., discharge and pilot) of piping. The discharge system, normally filled with water to the nozzle, delivers the water for distribution to the hazard. The pilot system applies pressure onto a poppet valve in such a way that water in the discharge system, in its standby state, is prevented from flowing through the nozzle. The nozzles are lo-

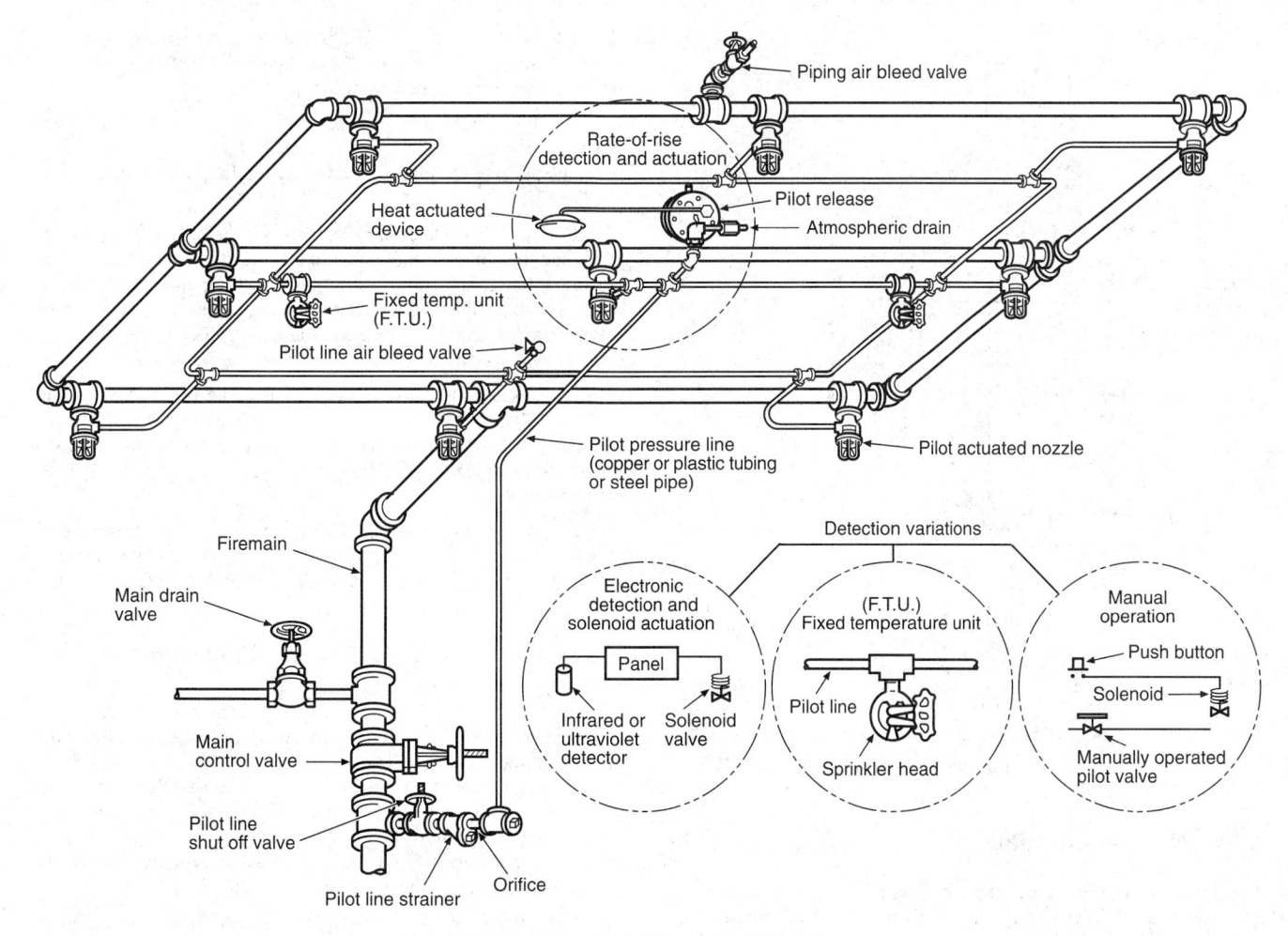

FIG. 6-14O.   Early pilot-actuated system.

cated as close to the hazard as possible. Release of pilot pressure allows the deluge system to overcome the pressure differential, forcing the poppet valve up and initiating instantaneous discharge.

Pilot-actuated nozzles are capable of accepting inserts that allow custom designing of water application to the specific hazard in question. Discharge piping can be gridded when an area protection concept is employed. A strainer is required on the main riser of the discharge system to minimize debris that may accumulate between the poppet and the poppet seat, and may cause leakage from the nozzle and possible system actuation. Sloping of the discharge system to air vents will serve to purge air from the discharge system, and reduce system discharge time. Piping with a diameter of $1^1/_2$-in. (38-mm) or less must be sloped a minimum of 1 in. (25.4 mm) per 10 ft (3.1 m) of pipe. Piping with a diameter of 2 in. (51 mm) and larger must be sloped no less than $^3/_4$-in. (19.1-mm) per 10 ft (3.1 m) of pipe.

The Pilot System

The pilot system is monitored by normally closed two-way solenoid valves spaced throughout the system. Ideally, a solenoid valve should be provided for every pilot-actuated nozzle, but in certain cases one solenoid valve could control up to three nozzles. Engineering judgement must be exercised in determining a solenoid-to-nozzle ratio, since the response time increases as the distance between the pilot-actuated nozzle and the solenoid increases. Solenoids should be equipped with an enclosure rated NEMA type 4

(watertight), NEMA type 7 (hazardous locations Class I, Group C or D), and NEMA type 9 (E, F, or G hazardous locations, Class II, Group E, F, or G). Detectors and detector housings should be rated for Class I and Class II locations. (See Figure 6-14T.)

It is essential that every effort be made to minimize air pockets in the pilot line, with piping radically sloped to air vents placed at all high points, facilitating the bleeding of air from the pilot system, and enhancing the speed of reaction. The pilot line slope should be equal to or greater than the discharge piping slope, but should not be less than 1 in. (25.4 mm) per 10 ft (3.1 m) of pipe. For best results, keep branchlines and their associated pilot lines very short, and slope them as radically as possible.

Air in the pilot line is of paramount concern because of the smaller cross-sectional area of the pilot line, as compared to the calculated pipe sizes for the discharge piping system. An automatic air vent is a feasible option. (See Figure 6-14U.) To protect the solenoid orifices, a strainer must be installed on the pilot line feed.

The two systems of piping, i.e., pilot line and discharge, are simultaneously pressurized by an excess pressure pump, or some other pressure maintenance device, with static system pressures normally ranging from 170 to 180 psi (1172.2 to 1241.1 kPa), with a relief valve and pressure-limiting switch ensuring this range. A less desirable alternate to an excess pressure pump is an elevated gravity tank with a minimum static pressure of 75 psi (517.13 kPa). A specially designed orifice union in the pilot line maintains the pressure ratio necessary to keep the pilot-actuated nozzles closed.

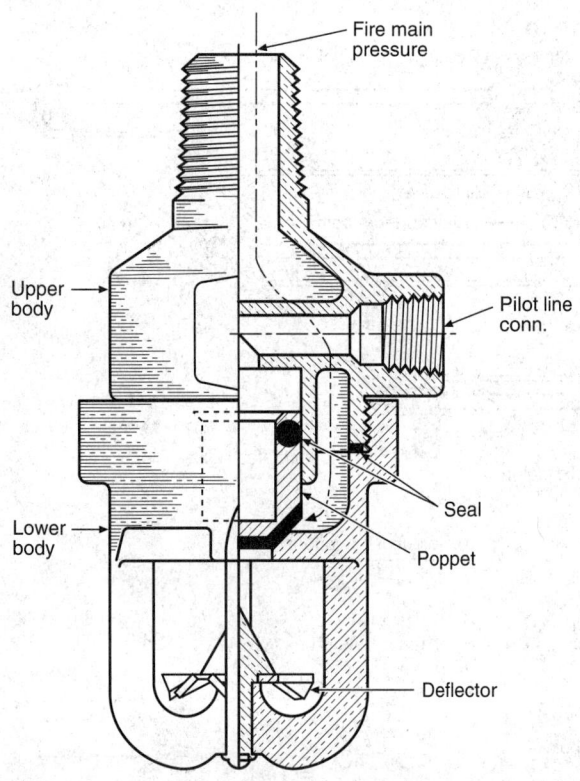

FIG. 6-14P. *Early pilot-actuated nozzle.*

The Detection System

UV detectors currently used on pilot-actuated systems comprise an extra-sensitive UV detection tube housed in a stainless steel, brass, or aluminum explosion proof housing. The detector is mounted on a swivel base to give maximum field adjustment, up to an angle of 240 degrees. Most UV and IR detectors have a field of vision between 80 and 90 degrees.

Flame detectors are installed as close to the points of anticipated hazard as practicable, and angled and spaced so that all areas where the volatile substance might be found are observed by the detectors. Placement of detectors on an ultra-high-speed water spray system requires experienced judgment and thorough field inspection.

The Control Panel

The control panel, with its sensitive high-speed modules, is kept in a protected area remote from the hazard, usually an area classified safe for non-explosionproof equipment. Where it is necessary or required to install the panel in the hazard area, the control panel must be installed in a rated enclosure. The detector, module, and control panel system are specifically designed and customized for the hazard in question. The primary concern of each component is speed, and, as a result, the telemetry system is considerably faster than the system of the early 1960s. The control system is designed to provide supervision over all remote electronic devices.

The control panel should be comprised of a controller, high-speed modules, control relays, a lamp test push button, trouble silence switch, panel reset button, ac power fail light, water flow lamp, tamper switch fault lamps, alarm lights, and indicators for special functions for pumps and tanks, with terminals for other functions, if required.

The controller powers and monitors the ultraviolet detectors and their signals, and notifies the high-speed modules of a trouble or fire signal from the detectors.

The high-speed module receives a signal from the controller and will operate the high-speed solenoids upon receipt of a fire signal, thus purging the pilot system of hydraulic pressure, and opening the nozzles.

Control relays supervise gate valves, pressure switches, alarm bells, and the control panel operation. Contacts can be provided for customer functions, such as shutting down equipment. An automatic reset capability option allows the high-speed module to cease water distribution after a predetermined period of time (usually 2 min), provided that the detectors no longer view a hazard.

Buttons and indicators annunciate the control functions of the panel and display the status of the pilot-actuated system. The control panel should be equipped with a capability for resetting the solenoids automatically from the panel, eliminating lengthy system shutdowns.

A battery charger recharges the batteries when needed, and rectifies the 120-V ac input power to 24-V dc for system use. Upon loss of primary 120-V ac power, the charger will transfer to the secondary 24-V dc battery supply.

Sequence of Operation for a Pilot-Actuated System

The flame detectors constantly view the hazard areas, searching for changes in light intensity. The status is reported to the controller for analysis. Light intensities are screened by the detectors' ability to recognize common signals, such as house lighting and sunlight. Physical screening may be necessary to prevent detector signals from such intermittent light sources as welding torches, lightning, or sun reflecting from the windshield of a car.

Upon detection of a valid signal, the controller, constantly interpreting the system status, receives the signal. The signal is interpreted and relayed to the appropriate high-speed module. Indicator lights and indicating devices commence operation, and a signal is sent simultaneously to all the high-speed solenoids on the system.

The normally closed solenoids open, allowing water to flow through the solenoids, thus purging the pilot system of water and relieving the pilot pressure. The pressure drop allows the poppet valves in the pilot-actuated nozzles to move, allowing water to flow through the nozzle and onto the hazard. The pilot-actuated nozzle insert directs the water flow into a wide-, medium-, or narrow-angle spray. Minimum static pressure at each nozzle should be no less than 50 psi (344.75 kPa), with 180 psi (1241.1 kPa) strongly recommended.

Factors Influencing the Speed of a Pilot-Actuated System

There are many variables that can enhance the speed of a pilot-actuated system, and all are considered in each engineered layout.

Trapped air in the pilot line can be a significant hindrance to the hydraulic evacuation of the pilot line. To prevent this, the pilot line is sloped as radically as possible, with air-release vents, either manual or automatic, at all high points of the pilot system. It is a common practice to bleed air from the system after it is reset, and repeat the process at the start of the next working day to ensure that all air is removed from the system.

Excess system volumetric capacity of a pilot-actuated system is a substantial contributor to increasing the response time of the system. It is important to maintain a maximum system volumetric

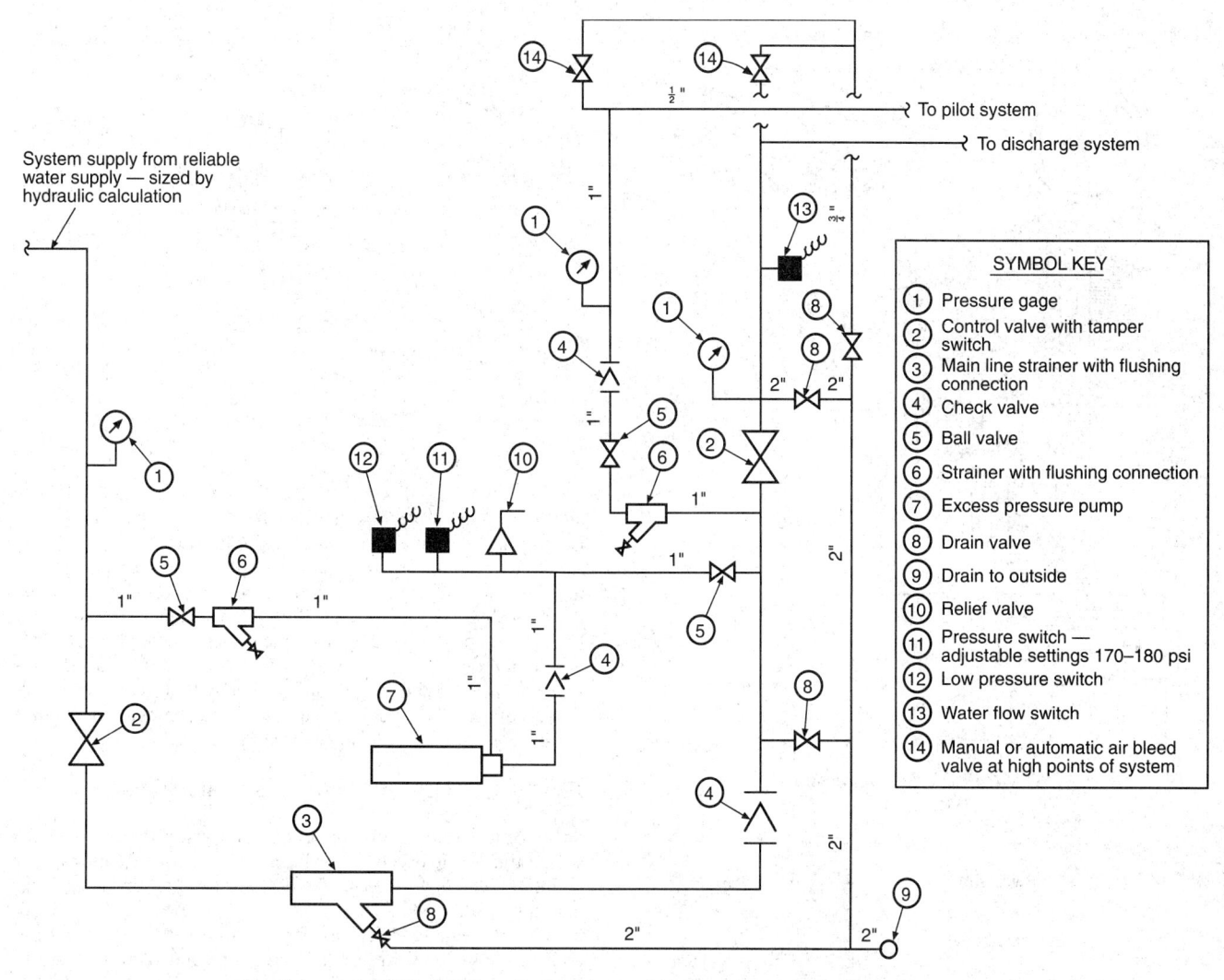

FIG. 6-14Q. *Control riser solenoid operated system.*

capacity of less than 500 gal (1892.5L), and to endeavor to subdivide the systems such that the maximum number of nozzles on a single system is about 10. It is also advantageous to minimize the number of fittings and changes of water direction, and to minimize the length of the feed main to the system.

The distance from the solenoid valve on the pilot line to the pilot connection of the pilot-actuated nozzle can be a major factor in the speed of the evacuation of the pilot line. The ideal situation would be to place a solenoid on the pilot line adjacent to each nozzle. This would release the pilot pressure to the nozzle almost immediately, and evacuation of the entire pilot line would not become a factor. When large systems are necessary, it may be electronically impractical to provide a ratio of one solenoid per nozzle.

The diameter of the pilot line can be a factor in pilot evacuation. For short runs of pilot piping, $1/2$-in. (12.7-mm) galvanized steel pipe is sufficient, with a very short run of $3/8$-in. (9.5-mm) internal diameter copper piping for the nozzle pilot connections. Longer runs of either steel or copper piping may require consideration of larger internal cross-sectional diameters.

Pilot pressure can be a major factor in response time. An excess pressure pump should be provided for each system or group of systems to maintain 170 to 180 psi (1172.2 to 1241.1 kPa) static pressure on the system piping and pilot line. This high pressure at

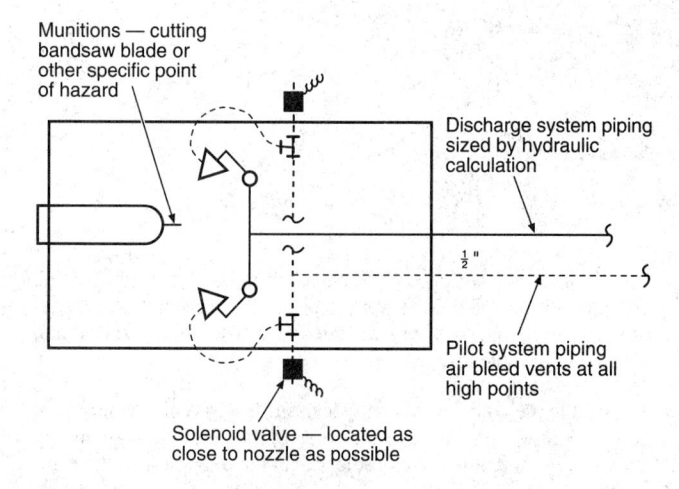

FIG. 6-14R. *Plan view.*

each nozzle provides the impetus for rapid evacuation of pilot pressure and rapid delivery of water to the hazard. A less desirable alternative is the use of an elevated gravity tank that would provide a minimum of 75 psi (517.13 kPa) static pressure on the system.

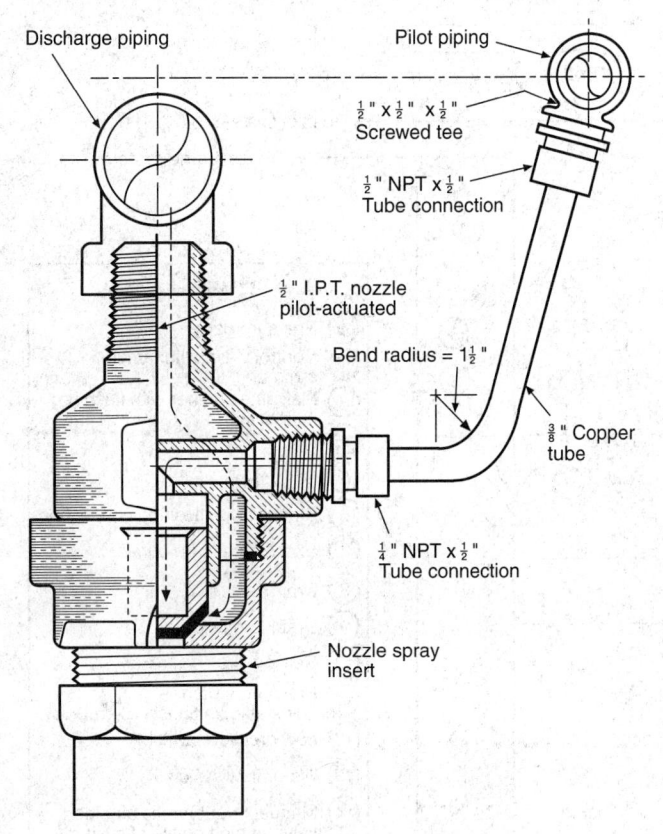

FIG. 6-14S. *Pilot connection.*

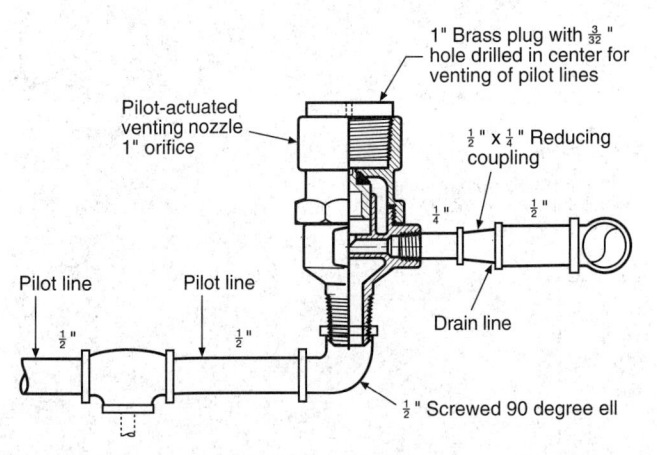

FIG. 6-14T. *Solenoid valve.*

To increase the speed of solenoid actuation, the initial voltage provided to the solenoids from the panel, upon receipt of a signal from the detectors, should provide delivery of a high-voltage signal, enhancing the solenoid operating time.

While the voltage eventually levels off to 24-V dc, which is the normal control system operating voltage, the initial burst of energy may damage solenoids rated for less than the maximum voltage encountered.

An orifice restriction device, either an orifice union or a drilled ball valve, is placed in the pilot line at the riser to maintain a high pressure in the pilot system. Once the solenoids open, residual pressure drop in the pilot line is minimized by this restrictor, increasing the speed of pilot evacuation.

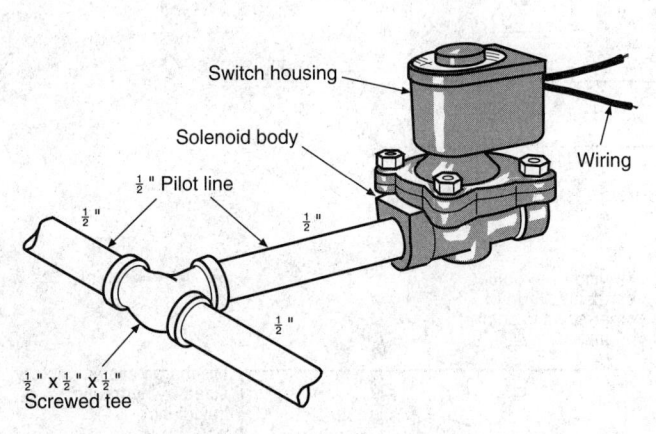

FIG. 6-14U. *Automatic pilot vent.*

No less than 25 gpm (95 L/min) per nozzle should be designed for point protection, nor less than 0.50 gpm per sq ft (20.37 L/min × m²) of exposed area for area protection. Point protection should be accomplished with no fewer than two counter-opposed nozzles per point of likely ignition, and area protection should be accomplished with a nozzle spacing of between 30 and 60 sq ft (2.8 and 5.6 m²), depending upon the configuration of the hazardous commodity or process. Counter-opposition of the nozzles involves nozzles aimed from differing directions at a common point. Some commodities, such as certain pyrotechnics, may require densities as high as 3.0 gpm per sq ft (122.24 L/min × m²).

Response Time for Pilot-Actuated Systems

Specifications written for pilot-actuated systems mandate a response time not to exceed 100 milliseconds, measured from presentation of an energy source to the detector to water discharge from the most distant pilot-actuated nozzle. Whenever possible, travel time for the water to hit its intended target should be included in the total response time. Research is currently underway that would evaluate and further define this time span.[1,6]

Method of Timing the Response of a Pilot-Actuated System

The following procedure can be used to perform a functional time test of a pilot-actuated system:

1. Install a pressure switch on the nozzle being tested. Each nozzle on the system must be tested to ensure that the solenoids are functioning properly, and that operation speeds are being maintained at each nozzle.
2. Slowly fill the system piping.
3. Bleed all air from the pilot system through air vent valves. Opening these valves too quickly may operate the system.
4. Close the air vent valves and record the static system pressure at the riser.
5. Connect the timer to the nozzle pressure switch and to a switch on the control panel. High-speed photography is also an effective timing device.
6. Take three consecutive readings for each system.
7. Record the static pressure at the riser before and after each test.
8. A UV light source will be used to activate a UV detector for the system being tested. UV, IR, and UV/IR testing lamps for tripping ultra-high-speed water spray systems are commercially available.
9. An electrical impulse from the detectors will start the timers.

10. Two timers should be used in order to ensure accuracy.
11. The timers will be stopped by the water discharge reaching the pressure switch.
12. Record the test results, including residual pressure at the riser.
13. Reset the system at the control panel.

Other Applications for Pilot-Actuated Systems

Although munitions provide the "lion's share" of current pilot-actuated system applications, this system has been used, with a preset timer to limit discharge, to protect valuable commodities such as mail conveyors and chutes, ovens, and antique book storage facilities. Where enhanced system response is desired, but where a 100-millisecond response time is unnecessary, the system can be modified to resemble the early pilot-actuated system, with thermal, pneumatic, or electric detection and standard control modules.

PORTABLE ULTRA-HIGH-SPEED WATER SPRAY SYSTEMS

Portable ultra-high-speed water spray systems have been created to supply cost-efficient, mobile, short-term protection for U.S. Army ammunition plants and other facilities requiring enhanced actuation times for exposed energetic materials.[7] This system is a portable, self-contained fire detection and suppression system capable of delivering water to the nozzles within 50 to 100 milliseconds from time of detection, and can protect an area up to 100 sq ft (9.3 m^2). The system features UV detection with a high-speed controller, a solenoid-actuated deluge valve, and a portable pressurized water storage tank that is capable of being temporarily cross-connected to a building water supply, if available.

EXPLOSION SUPPRESSION SYSTEMS

The first explosion suppression systems were introduced by the British in the late 1940s for protection of aircraft, and were later extended to industrial use. Explosion suppression systems consist of a cylinder filled with a suppressing agent, a detection system, and a system of control circuitry. They can ordinarily be visually differentiated from ultra-high-speed water spray systems, by their lack of system piping, by the existence of uniquely shaped agent containers, and by the wide variety of agents, including water, that can be employed with such systems. Halon, a commonly used agent for explosion suppression systems, has been scheduled for elimination by the Montréal Protocol.[8] A clean agent or halon substitute, in accordance with NFPA 2001, *Standard on Clean Agent Fire Extinguishing Systems*, may be required for a new explosion suppression system. Explosion suppression systems are designed in accordance with NFPA 69 to protect enclosed vessels or containers from rupture due to overpressurization resulting from a deflagration.

EXPLOSION VENTING

An enhancement to the effectiveness of an ultra-high-speed water spray system or an explosion suppression system is explosion venting. These passive vents are sometimes used in lieu of an ultra-high-speed water spray system or an explosion suppression system. Venting, however, will relieve pressure without any active effort to control the continued burning of the fuel in the container. Care must be taken to avoid placing explosion venting louvers in a location where an operator or passerby could be injured by an explosion relief blast. NFPA 68, *Guide for Venting of Deflagrations*, should be used when designing an explosion venting system.

An alternative to a suppression system could be to provide no protection of any kind. In a facility with an enclosed and separated or isolated automatic process that is normally unoccupied, where vessels are manufactured with sufficient strength to withstand overpressurization resulting from a deflagration, or where there is enough process redundancy to avoid process stoppage, a firm may consider the option of no protection. Repair or replacement of a process area, the spreading of fire or blast damage to adjacent areas with a corresponding result of possible deaths or injuries, unfavorable publicity, and loss of income could prove, in almost all cases, to be significantly more costly than an ultra-high-speed suppression system or an explosion suppression system.

CONCLUSION

A major difference exists between ultra-high-speed water spray systems and explosion suppression systems. Even though both systems suppress explosive hazards, ultra-high-speed water spray systems are used predominantly for open machine protection or room area protection, while ultra-high-speed explosion suppression systems protect enclosed vessels. Although system names are sometimes used interchangeably, this chapter helps clarify correct terminology.

There is relatively little information available on practical testing of various munitions manufacturing explosive levels for differing munitions commodities and processes. However, years of successful field performance and experience, and system time and reliability testing demonstrate the effectiveness of ultra-high-speed water spray systems, with respect to the suppression of munitions explosive hazards. Artificial retardation of the normal actuation times of ultra-high-speed water spray systems means the difference between extinguishment and total loss of control in the explosive process.

BIBLIOGRAPHY

References Cited

1. Loyd, R. A., "Ultra-High-Speed Deluge Systems for Ordinance Operations," *Proceedings*, Federal Fire Forum, Nov. 3, 1989.
2. Goedeke, A.D., and Fadorsen, G.A., "Evaluation of State-of-the-Art High-Speed Deluge Systems Presently in Service at Various U.S. Ammunition Plants," Document WL-TR-93-3510, Sept. 1993, Wright Laboratory, Wright-Patterson AFB, Dayton, OH.
3. Detector Electronics Corporation, "Fire Detection System Selection Guide," Form 92-1002-01, Dector Electronics Corp, Minneapolis, MN. Undated.
4. Fadorsen, G. A. "Ultra-High-Speed Fire Suppression for Explosives Facilities," American Chemical Society Symposium for Design Considerations for Toxic Chemical and Explosives Facilities, R. A. Scott and L.J. Doemeny, eds., 1987.
5. Ault, W. E., and Carter, D. I., "The Influence of Hyperbaric Chamber Pressure on Water-Spray Patterns," *Fire Journal*, Nov. 1967, pp. 48–49.
6. Loyd, R.A., "Design and Installation of Ultra-High-Speed Deluge Systems," *Proceedings*—24th Department of Defense Explosives Safety Seminar, Aug. 27–30, 1990.
7. Loyd, R. A., "Portable Ultra-High-Speed Deluge Systems," *Proceedings*—23rd Department of Defense Explosives Safety Seminar, Aug. 9–11, 1988.
8. Grant, C. C., "Halons and the Ozone Layer: An Overview," *Fire Journal*, Sep./Oct., 1989, pp. 59–81.

NFPA Codes, Standards, and Recommended Practices

Reference to the following NFPA codes, standards, and recommended practices will provide further information on ultra-high-speed suppression systems for explosive hazards discussed in this chapter. (See the latest version of *The NFPA Catalog* for availability of current editions of the following documents.)

NFPA 13, *Standard for the Installation of Sprinkler Systems*
NFPA 14, *Standard for the Installation of Standpipe and Hose Systems*
NFPA 15, *Standard for Water Spray Fixed Systems for Fire Protection*
NFPA 18, *Standard on Wetting Agents*

NFPA 20, *Standard for the Installation of Centrifugal Fire Pumps*
NFPA 22, *Standard for Water Tanks for Private Fire Protection*
NFPA 24, *Standard for the Installation of Private Fire Service Mains and Their Appurtenances*
NFPA 25, *Standard for the Inspection, Testing, and Maintenance of Water-Based Fire Protection Systems*
NFPA 30, *Flammable and Combustible Liquids Code*
NFPA 61, *Standard for the Prevention of Fires and Dust Explosions in Agricultural and Food Products Facilities*
NFPA 68, *Guide for Venting of Deflagrations*
NFPA 69, *Standard on Explosion Prevention Systems*
NFPA 70, *National Electrical Code®*
NFPA 72, *National Fire Alarm Code*
NFPA 77, *Recommended Practice on Static Electricity*
NFPA 325, *Guide to Fire Hazard Properties of Flammable Liquids, Gases, and Volatile Solids*
NFPA 495, *Explosive Materials Code*
NFPA 2001, *Standard on Clean Agent Fire Extinguishing Systems*

Additional Reading

American National Standards Institute Publications. Available from ANSI, 11 W. 42nd Street, New York, NY 10036.
B16.1-89, *Cast Iron Pipe Flanges and Flanged Fittings.*
B16.3-92, *Malleable Iron Threaded Fittings.*
B18.2.1-81, *Square and Hex Bolts and Screws Inch Series.*
B18.2.2-87, *Square and Hex Nuts (Inch Series).*
American Society for Testing and Materials Publication. Available from, ASTM 1916 Race St., Philadelphia, PA 19103.
E1226-94, *Standard Test Method for Pressure and Rate of Pressure Rise for Combustible Dusts.*

Acton, M. R., Sutton, P. and Wickens, M. J., "Investigation of the Mitigation of Gas Cloud Explosions by Water Sprays," *Gas Engineering and Management*, Vol. 31, No. 7–8, July/Aug. 1991, pp. 166–172; Institution of Chemical Engineers, Piper Alpha: Lessons for Life-Cycle Safety Management Symposium, IChemE Symposium Series No. 122, Sept. 26–27, 1990, London, England, 1991.
"Automatic" Sprinkler Corporation of America, *Data Book and Engineering Guide*, ASCOA, Cleveland, OH, 1989.
Bonney, M. J., "Suppressing Dust Explosions," *Fire*, Vol. 88, No. 1086, Dec. 1995, p. 40.
Bragg, K., "Parker Reactive Explosion Suppression System (PRESS) Proof-of-Concept Demonstration," Final Report, March 1989–Sept. 1990, Flight Dynamics Directorate, Wright-Patterson AFB, OH, JTCG/AS Central Office, Washington, DC, WRDC-TR-90-3064, May 1992.
Brenton, J. R., Thomas, G. O., and Al-Hassan, T., "Small-Scale Studies of Water Spray Dynamics During Explosion Mitigation Tests," *Proceedings* of the Institution of Chemical Engineers Symposium Series, Hazards XII: European Advances in Process Safety Conference, No. 134, 1994, Rugby, England, Institution of Chemical Engineers, Rugby, England, 1994, pp. 393–403.
Bryan, J. L., *Automatic Sprinkler and Standpipe Systems*, 2nd ed., National Fire Protection Association, Quincy, MA, 1990, Chapters 9 and 11.
Bryan, J. L., 1993. *Fire Suppression and Detection Systems*, 3rd ed., Macmillan Publishing Co., Inc., New York, 1993, Chapter 8, pp. 303–320.
Butz, J. R., and French, P., "Application of Fine Water Mists to Hydrogen Deflagrations," *Proceedings* of the 1993 Halon Alternatives Technical Working Conference, May 11–13, 1993, Albuquerque, NM, 1993, pp. 345–355.
Chatrathi, Dr. K., "Explosion Testing," *Safety and Technology News*, Vol. 3, Iss. 1, 1989. (Published by Fike Corporation.)
Cortese, R. A., and Sapko, M. J., "Flame-Powered Trigger Device for Activating Explosion Suppression Barrier," Bureau of Mines, Pittsburg, PA, RI 9365, 1991.
Creighton, J. R., "Ignition Temperature of Solid Explosives Exposed to a Fire," *PSAM-II Proceedings*, An International Conference Devoted to the Advancement of System-Based Methods for the Design and Operation of Technological Systems and Processes, Vol. 2, Sessions 37–72, Mar. 20–25, 1994, San Diego, CA, 1994, pp. 037/5–10.
Cote, A. E., and Bugbee, P., *Principles of Fire Protection*, National Fire Protection Association, Quincy, MA, 1988, pp. 59–63.
Department of the Army, "AMC Safety Manual, C-1," AMC-R-385-1, Mar. 16, 1990, U.S. Dept. of the Army, Washington, DC.

Department of Defense, "Military Handbook—Fire Protection for Facilities—Engineering, Design, and Construction," MIL-HDBK-1008B, Jan. 15, 1994, Dept. of Defense, Washington, DC.
Department of Defense, "Ammunition and Explosives Safety Standards," Document DOD 6055.9-STD, July 1984, Dept. of Defense, Washington, DC.
"Explosives Incidents Report, 1993," Bureau of Alcohol, Tobacco and Firearms, Washington, DC, ATF P 3320.4, July 1994.
"Explosives," Department of the Interior, Washington, DC, Annual Review, 1994.
Factory Mutual Research Corp., *Factory Mutual Approval Guide*, , Publications Section, FMRC, Norwood, MA. 1996.
"Fire and Gas Detector Product Catalogue," Form 52-1002-01, Detector Electronics Corporation, Minneapolis, MN, Undated.
Fitzwater, T., "Introduction to Explosives," *American Fire Journal*, Vol. 45, No. 4, April 1993, pp. 12–17.
Fitzwater, T., "Introduction to Explosives. Part 2. High Explosives," *American Fire Journal*, Vol. 46, No. 3, March 1994, pp. 12–16.
Fitzwater, T., "Introduction to Explosives. Part 3. Slurries, Water Gels, Emulsions, Binary Explosives and Detonating Cord," *American Fire Journal*, Vol. 47, No. 11, Nov. 1995, pp. 12–14.
Gagnon, R. M., *Design of Water-Based Fire Protection Systems*, Delmar Publishers, P.O. Box 15015, Albany, NY 11212.
Gagnon, R. M., "Ultra-High-Speed Deluge Fire Suppression Systems," *Fire Protection Contractor*, June 1993.
Grinnell Fire Protection Systems Company, *Data Book and Systems Engineering Guide*, GFPSC, Exeter, NH, 1989.
Johnson, M. A., "The Technical Aspects of Flammability," *Aerosol Age*, Nov. 1987.
Jones, A., and Thomas, G. O., "Mitigation of Small-Scale Hydrocarbon-Air Explosions by Water Sprays," *Transactions Institution of Chemical Engineers*, Vol. 70, No. A, 1992, pp. 197–199.
Jones, A., and Thomas, G. O., "Action of Water Sprays on Fires and Explosions: A Review of Experimental Work," *Process Safety Environment*, Vol. 71, No. 1, 1993, pp. 41–49.
Jones, J. C., "Model for Critical Behavior in a Liquid Explosive Experiencing Dual Ambient Temperatures," *Journal of Fire Sciences*, Vol. 13, No. 5, Sept./Oct. 1995, pp. 399–406.
Jones, J. C. and Moore, A., "Model for Critical Behavior in a Liquid Explosive Experiencing Dual Ambient Temperatures. Part 2. Temperature Trajectories and Times to Explosion," *Journal of Fire Sciences*, Vol. 13, No. 5, Sept./Oct. 1995, pp. 407–413.
Kruse, D.H., "Typical Configurations of High-Speed Deluge Systems for Various LAP Equipment," *Proceedings*—Annual Meeting of the Load, Assemble, and Packaging Section, American Defense Preparedness Association, Colorado Springs, CO, Mar. 8–10, 1988.
Lloyd, R. A., "Ultra-High-Speed Deluge Systems," *Proceedings*—22nd Department of Defense Explosives Safety Seminar, Aug. 26–28, 1986.
Pickett, M., "Explosives: What Firefighters Must Know. Part 1. Why Firefighters Should Know About Explosives and Hot To Identify Common Commercial Types," *Firehouse*, Vol. 20, No. 9, Sept. 1995, pp. 108, 110, 111.
"Munitions Capabilities Brochure," Form 92-1004, Detector Electronics Corporation, Minneapolis, MN, undated.
"Munitions Manufacturing," Form 91-1005, 6-87, Detector Electronics Corporation, Minneapolis, MN, undated.
Senecal, J. A., "Detection and Suppression of Process Dust Deflagrations: An Overview with Examples," ASME Symposium, Chicago, IL, Nov. 28, 1988.
Senecal, J. A., "Explosion Protection in Occupied Spaces: The Status of Suppression and Inertion Using Halon and Its Descendants," 1993 International CFC and Halon Alternatives Conference, Stratospheric Ozone Protection for the 90's, Oct. 20–22, 1993, Washington, DC, 1993, pp. 767–772.
Serena III, J. M., "Hazard Division 1.3 Passive Structural System Design Guide," HNDED-CS-93-7, Rev. 1, Nov. 1, 1994, Huntsville Division U.S. Army Corps of Engineers, Huntsville, AL.
Tamanini, F., "Explosion Suppression for Industrial Applications," Factory Mutual Research Corp., Norwood, MA, NISTIR 5766, Nov. 1995; National Institute of Standards and Technology, Solid Propellant Gas Generators: Proceedings of the 1995 Workshop, June 28–29, 1995, Gaithersburg, MD, 1995, pp. 208–217.
UL Fire Protection Equipment Directory, Underwriters Laboratories Inc., Northbrook, IL, 1996.

United States Government Specifications WW-F-406b, WW-P-406c, and WW-P-521g, Government Printing Office, Washington, DC.

Technical Reports on Ultra-High-Speed Water Spray Systems

These reports can be ordered from the Defense Technical Information Center, Cameron Station, Alexandria, VA, 22314. Phone (703)-274-6733. Documents that are approved for public release are noted (APR). Documents that contain confidential or proprietary information, and whose distribution is limited to government agencies only are noted (DL).

"Design of a Deluge System to Extinguish Lead Azide Fires," No. AD-E400-204, Aug. 1978. (APR)

"Evaluation of Pyrotechnic Fire Suppression Systems for Six Pyrotechnic Compositions," No. AD-E401-306, Mar. 1985. (APR)

"Engineering Guide for Fire Protection and Detection Systems at Army Ammunition Plants," (Vol. I: Selection and Design), No. AD-E400-531, Dec. 1980. (APR)

"Engineering Guide for Fire Protection and Detection Systems at Army Ammunition Plants," (Vol. II: Testing and Inspection), No. AD-E400-874, Dec. 1982. (DL)

"On-Site Survey and Analysis of Pyrotechnic Mixer Bays," No. AD-E401-141, Feb. 1984. (APR)

"Feasibility Study to Develop a Water Deluge System for Conveyor Lines Transporting High Explosives," Tech. Report. No. 4889, Aug. 1975. (APR)

" Development of a Water Deluge System to Extinguish M-1 Propellant Fires," No. AD-E400-217, Sept. 1978. (APR)

"Design of a Water Deluge System to Extinguish M-1 Propellant Fires in Closed Conveyors," No. AD-E400-216, Sept. 1978. (APR)

"Fire Suppression System Safety Evaluation," No. AD-E401-083 Dec. 1983. (APR)

"Dynamic Model of a Water Deluge System for Propellant Fires," No. AD-E400-315, May 1979. (APR)

"Deluge Systems in Army Ammunition Plants," prepared by Science Applications, Inc., for the U.S. Army Munitions Production Base Modernization Agency, June 30, 1981.

"Minutes of the Rapid Action Fire Protection Systems Seminar," U.S. Army Armament, Munitions, and Chemical Command, Oct. 23–24, 1984.

"Water Deluge Fire Protection System for Conveyor Lines Transporting High Explosives," No. AD-E400-034, Dec. 1977. (APR)

"Evaluation of an Improved Fire Suppression System for Pyrotechnic Mixes," No. AD-E401-569, Sept. 1986. (DL)

"Analysis of Mixer Bay Designs for Pyrotechnic Operations," No. AD-E401-602, Nov. 1986. (APR)

"Calculating Sprinkler Actuation Time in Compartments," Center for Fire Research, National Institute for Standards and Technology, Mar. 1984.

"Guidelines for a Thermochemical Kinetics Computer Program," Los Alamos National Laboratory, LA-10361-MS, May 1895.

"Mathematical and Numerical Methods Used in Thermal Hazards Evaluation," Naval Weapons Center China Lake, NWC-TP-6510, Dec. 1986.

"Technical Report on the Testing of Ultraviolet-Actuated Deluge Systems Utilizing Solid-State Controllers and Detonator-Actuated Valves to Extinguish Black Powder Fires," Day and Zimmerman, Lone Star Division, Nov. 1986

WATER MIST FIRE SUPPRESSION SYSTEMS

Jack R. Mawhinney and Robert Solomon

Interest in the use of very fine water sprays (water mist) in fire suppression systems has been intense for the last 10 years. The economic force behind this interest has been the need to develop water-based fire suppression systems for applications where water was not previously considered practical and as a potential replacement for Halon 1301. Innovative fire suppression systems have been conceptualized, tested against specific fire scenarios, and installed as working systems, primarily in marine applications. Recently, system concepts that originated in Europe have been evaluated to North American standards and have received the "approval" of fire equipment listing agencies, such as Factory Mutual Research Corporation (FMRC). In 1993, the National Fire Protection Association (NFPA) began work on a design and installation standard, NFPA 750, *Standard on Water Mist Fire Protection Systems* (hereinafter referred to as NFPA 750), to consolidate design concepts emerging from the research and development work.

This chapter describes the principles involved in the design of water mist fire suppression systems. It starts by describing the objectives and details of several applications of water mist systems for which acceptable performance has been confirmed. Included is an example of an unsuccessful trial, which was nonetheless instructive in that it clarified limits of performance. From the examples, it is possible to identify generalizable objectives and principles of operation. These principles involve the extinguishing mechanisms, the influence of fuel type and compartment ventilation on effectiveness, spray characteristics, and spray generation methods.

Having reviewed the general principles involved in the operation of water mist systems, this chapter proceeds to give details on the components, calculation methods, and methods of activation and control that are involved in their design. It must be noted, however, that an understanding of the "first principles" of operation is not yet sufficient to confidently predict the outcome of a particular new fire scenario. There is sufficient uncertainty that full-scale fire testing is justified to confirm performance.

EXAMPLES OF APPLICATIONS OF WATER MIST

Before discussing the details and general principles of water mist systems, it is instructive to review several scenarios where water mist has been applied, both successfully and unsuccessfully. The following examples represent the "proving ground" for water mist fire suppression technology.

Aircraft Passenger Compartments

Starting in 1985, the Civil Aviation Administration (CAA) in the U.K., and the Federal Aviation Administration (FAA) in the U.S., with the involvement of Transport Canada, Airbus, Boeing, and other aviation industry agencies, developed a fine water spray fire control system for aircraft passenger compartments.[1] The objective of this international research and development effort was to find a way to decrease the number of fatalities in air crashes that involve fuel spill fires, i.e., the post-crash fire scenario. The Manchester, U.K., air crash of 1984 was the particular incident that sparked this interest. In that incident, the fatalities were caused not by the crash but by the fire from spilled fuel that engulfed the fuselage and spread into the passenger compartment. Aircraft are not at present equipped with fixed active fire protection systems to protect against a post-crash fire.

The aircraft cabin water spray system was not intended to suppress the external liquid fuel pool fire. The performance objective was to prevent spread of fire into the cabin and to cool the hot gases that enter the cabin, in order to extend the length of time that conditions are survivable. The target was to provide an extra 3 min of survivable conditions, which is deemed to be enough to allow all passengers to escape the aircraft through an unexposed exit. For narrow-bodied aircraft, the final optimized design of a water spray system divided the fuselage into 2.4-m long zones of spray nozzles, each zone activated by a single thermal detector. Nozzles in each zone were mounted on three lines, one along the ceiling of the center aisle and one under each of the luggage racks. A successful arrangement was also developed for wide-bodied aircraft. The passenger cabin water spray system achieved a 7-min extension of survivability, using a 3-min discharge of approximately 91 L (24 U.S. gal) of water.[2]

The testing program verified that the aircraft passenger compartment water mist system met or exceeded the researchers' objectives for the post-crash fire scenario. Having determined the details of the design, a cost-benefit study was conducted. The study con-

Jack R. Mawhinney, P.Eng. is a senior engineer at Hughes Associates, Inc. based in Orleans, Ontario. At the time this chapter was written he was senior research officer at the Institute for Research in Construction, National Research Council Canada. Robert Solomon, P.E., is chief building fire protection engineer in the Building Fire Protection and Life Safety Department at the National Fire Protection Association.

cluded that the cost of outfitting a fleet of aircraft with passenger compartment water mist systems would be too high, relative to the number of lives that would be saved, to justify mandatory requirements for their installation. As a result, both industry and regulatory partners in the aviation industry decided not to make it mandatory to install such systems on existing or new aircraft. Instead, focus has shifted to the problem of developing water mist systems for on-board cargo compartment fires during flight.

The aircraft passenger cabin water spray system can be characterized as follows:

Safety objective: Save lives by protecting an aircraft passenger compartment from a post-crash, external fire threat, in order to extend the survivable evacuation time.

Mechanisms: Blocking convective and radiant heat transfer, cooling of hot fire gases, and wetting of fuels (plastic and cloth).

System features: Zoned piping, automatic thermal activation of each zone, 3-min supply of stored water in pressurized cylinders, single discharge, limited duration of protection adequate for a well-defined performance objective.

Industry acceptance: Developed through collaboration between aviation industry and regulatory authorities, but not made mandatory on the basis of an unfavorable cost-benefit study.

Marine Machinery Spaces

Among the marine machinery compartment fires of greatest concern are those involving diesel fuel and lubricating or hydraulic oil spills, ignited by hot engine parts, overheated bearings, or electrical arcing.[3,4] Pool or spray hydrocarbon fires are fast developing, very hot, and have the potential to cripple or destroy a ship. International shipping safety regulations, developed by the International Maritime Organization (IMO) and as written in the rules of SOLAS (Safety of Life at Sea), have for many years required that machinery spaces on large vessels be equipped with either CO2 or halon fire suppression systems. With the phase-out of halon, and increased awareness of the safety hazard that CO2 represents, there is strong interest in finding an alternative fire suppression agent for machinery compartments on ships. Further incentive for improved suppression systems for machinery spaces comes from IMO regulation that will require all ships capable of carrying more than 36 passengers to be equipped with automatic sprinkler systems by 1997. Marine engineers dislike the stability problems created by the weight of large-diameter sprinkler piping in the superstructure of a ship and have concerns about the potential to flood compartments and alter ship stability. Marine engineers are therefore interested in water mist fire suppression systems that promise to use small-diameter piping and deliver minimum quantities of water.

The potential to use finely divided water sprays to suppress hydrocarbon pool and spray fires has been known for some time, at least since the 1950s.[5,6] Since 1990, a number of companies have developed water mist systems for marine machinery compartments. The designs include a detection system, control panel, water storage, filtration and pressurization equipment, and specially designed nozzles. Some marine Authorities Having Jurisdiction have accepted water mist systems for machinery spaces, based on witnessing successful tests in mock-up machinery compartments, usually of limited size, and with little variation in test parameters. These demonstrations generally involved enclosed, ventilation-limited diesel fuel fires, where the fire was large, relative to the compartment size. Such fires are the easiest type of fire to extinguish with water mist. Other Authorities Having Jurisdiction wanted to know more about the limits of performance of water mist systems before

accepting the systems. To that end, various research projects have been conducted to explain the mechanisms of extinguishment, and limits of performance, and to develop system design criteria.[7,8,9] The IMO Fire Protection Subcommittee has drafted a test protocol for water mist systems that are intended to replace sprinkler systems in marine machinery compartments. The protocol[9] requires that the water mist system extinguish shielded diesel pool fires and spray fires, under fully ventilated conditions. These conditions are much more challenging to water mist than the early demonstration tests in which ventilation-limited fires were extinguished. Research in this area is continuing.

The marine machinery compartment water mist system can be characterized as follows:

Safety objective: Extinguish a compartment fire involving spilled or sprayed diesel fuel or hydraulic oils, and provide cooling of hot gases, using a limited amount of water. Fire sizes between 1 and 20 MW. Extinguishment in 1 to 3 min. Replace halon or CO2, and in some cases AFFF foam, which are the traditional agents for Class B fires in machinery compartments.

Mechanisms: Convective cooling of flame, hot gases, and fuel; steam displacement of oxygen and dilution of air/fuel vapor mixtures; blocking convective and radiant heat transfer to surfaces of unburned fuels and surrounding objects. The ability to ensure optimum conditions for any of the extinguishing mechanisms decreases as compartment size, ventilation, fuel complexity, and number of obstructions increases.

System features: Deluge systems with open nozzles in grid at ceiling and sometimes at intermediate levels. Nozzles may be: high- or low-pressure jet, twin-fluid, or impingement types. May be combined with other suppression systems (e.g., bilge foam systems) where control is beyond the capability of mist alone. Activated by a separate detection system. Water supply may be salt or fresh water. Maritime regulations (IMO) require 30-min water supply.

Industry acceptance: The U.S. Coast Guard, U.S. Navy, and U.S. Army continue to conduct tests to the IMO draft protocol. They are reported to be ready to accept water mist systems for marine machinery rooms for applications within their own jurisdictions.[10] Internationally, other maritime Authorities Having Jurisdiction have already accepted the systems as replacements for halon or CO_2.

Turbine Enclosures

Turbine enclosures are a type of marine machinery space. Unlike general machinery spaces, which may contain many fire sources, the source of fire in a turbine enclosure is limited to the turbine and its fuel lines. Large diesel-driven turbines are standard equipment in off-shore drilling platforms. Fuel line breaks may lead to spray fires, or 3-dimensional spill fires with pool fires in a bilge area under the turbine. The hot, rotating turbine blades are susceptible to serious damage if cooled rapidly. For that reason, it was in the past considered dangerous to apply water to turbine fires, at least until the machine had stopped rotating. The turbines are enclosed in ventilation-controlled compartments. Halon 1301 total flooding systems have been the preferred method of providing automatic fire suppression for turbine enclosures.

Wighus et al.[4] confirmed the ability of a water mist fire suppression system to extinguish fires in a turbine enclosure. Subsequent studies were done in collaboration with turbine manufacturers, to address concerns about thermal shock to the turbine blades. Several systems were installed in turbine enclosures on Norwegian offshore platforms.[11] by the Ginge Kerr Company of Denmark. The Ginge Kerr system, tested by Wighus et al., incorporated a twin-fluid (air/

water) misting nozzle first developed by British Petroleum (BP) Ventures in Great Britain.

Securiplex Technologies, Inc., of Dorval, Quebec, Canada, holds the patents on the BP Ventures twin-fluid (air/water) nozzles that were installed by Ginge Kerr on the Norwegian drilling platforms. Securiplex submitted its system design to FMRC in the U.S., to obtain an FMRC approval listing. Through full-scale fire testing, FMRC has confirmed the arrangement of nozzles needed to control all likely fire scenarios, and it has tested the reliability of the combined detection, activation, and cycling features of the system. FMRC was satisfied that application of short, timed sprays reduces the risk of thermal shock of water sprays on hot turbine blades. The first FMRC approval listing of the Securiplex system includes a specific geometry and nozzle arrangement for a turbine enclosure up to 80 m³ in volume. The nozzle locations and operating conditions are fixed to the geometry of the tested arrangement. For that reason, it is in essence a "pre-engineered system" approval. At the time of this writing, several companies are testing designs for turbine enclosures up to 260 m³.

The turbine enclosure water mist system can be characterized as follows:

Safety objective: Rapidly extinguish a fire involving spilled or sprayed diesel fuel on or under an operating diesel turbine, using a limited amount of water, and without damaging the turbine blades through thermal shock. Provide cooling of hot gases in the compartment. Replace Halon 1301 or CO_2.

Mechanisms: Convective cooling of flame, hot gases, and fuel; steam displacement of oxygen and dilution of air/fuel vapor mixtures; blocking convective and radiant heat transfer to surfaces of unburned fuels and other objects; cooling hot metal surfaces.

System features: Deluge systems with open nozzles in a pre-tested pattern at the ceiling, around all sides, and under a turbine. In the Securiplex system, nozzles are twin-fluid (air/water), so both water and compressed air are stored. A separate fire detection system and control panel are used for activation. All systems are tested to ensure that the rate of cooling of a metal surface does not exceed a certain value. A large pre-engineered fire suppression system.

Industry acceptance: Norwegian petroleum industry platform operators and regulators accept water mist as an alternative to halon in turbine enclosures. FMRC has approved the Securiplex system for compartments up to 80 m³. New tests are being run for 260 m³ compartments. Approvals are based on full-scale validation of all elements of the system, i.e., detection, actuation, delivery, nozzle placement, and extinguishment.

Crews' Quarters and Passenger Cabins

The new maritime regulations that will require installation of sprinkler systems on passenger-carrying ships have stimulated interest in developing water-based suppression systems that require less water than standard sprinklers, can use smaller diameter piping, and be compatible with the water supplies, detection, and control systems on a ship. To that end, considerable effort is going into developing water mist systems as equivalent to sprinkler systems.

A test protocol has been proposed by IMO to evaluate the use of water mist in crews quarters and ship corridors.[12] Testing to date has focused on small cabins, i.e., 12 m² for crews quarters, and 25 m² for luxury cabins. Unlike the other water mist systems, which were intended for application to Class B (hydrocarbon fuel) fires, the crews' quarters fire scenario involves ordinary, or Class A, combustibles. Whereas large droplets are a disadvantage for hydrocarbon fires, they are beneficial to the extinguishment of Class A combustibles, by causing wetting of unburned fuel. Mist intended for Class A combustibles can have more large droplets than sprays intended for liquid fuel fires. The need for a water-based suppression system that weighs less than a standard sprinkler system has led to inclusion of sprays with drop sizes up to 1000 microns within the definition of water mist in NFPA 750.

Grinnell Wormald of the U.S., and GW Sprinkler of Denmark, at the time of this writing, are developing thermally activated mist-sprinklers that are still undergoing evaluation to the IMO draft test protocol. The systems are reported to extinguish fires in foam furnishings, to prevent flashover in the compartment, and to prevent fire spread into the corridor. In the 12 m² cabin scenario, two nozzles control the fire: one installed in the room and one in the corridor outside the room, which activates automatically when the nozzle in the room activates due to heat. In the 25 m² luxury cabin, three nozzles are involved: two in the cabin and one in the corridor outside the door.

The ship's crew quarters and luxury cabin water mist system can be characterized as follows:

Safety objective: Save lives by controlling or extinguishing fires in Class A combustibles, in living quarters on ships. Prevent spread of fire from the compartment into the corridor, or through conductive bulkheads to adjacent compartments.

Mechanisms: Wetting of wood, paper, cloth, and plastic fuels in furnishings and wall coverings. Some steam displacement of oxygen; convective cooling of flame, hot gases, and fuel; blocking convective and radiant heat transfer to surfaces of unburned fuels and other objects, thereby preventing flashover.

System features: Similar to wet-pipe sprinkler systems, with normally closed, thermally activated nozzles, one or two per compartment. The activation by heat of one nozzle may trigger the simultaneous activation of one or two other nozzles in a pre-determined group. The nozzles have smaller orifices, operate at higher pressures, and produce a higher percentage of finer drops than standard sprinklers. Nozzle designs include both impingement and high-pressure jet types.

Industry acceptance: The IMO test protocol for evaluation of water mist systems as equivalent to sprinkler systems for crews' quarters and corridors on ships will form the test standard for confirmation of performance of sprinkler-equivalent systems in this application. Maritime authorities around the world will likely accept systems that are shown to meet the pass criteria for the IMO tests, if conducted in internationally recognized testing laboratories.

"Hush House"—Jet Engine Test Facility

A number of tests have been carried out based on the expectation that, if a fire can be totally surrounded by water mist, it will be extinguished by self-entrainment of mist into the fire. The tests invariably confirmed instead that fires will burn quite well in spite of being engulfed in mist. An experiment that demonstrates this was carried out by the U.S. Air Force in an evaluation of water mist for a jet engine test facility. The U.S. Air Force tests jet aircraft engines in insulated hangars, called "hush houses," that are designed to muffle the sound of the jet engines, and pass the exhaust of high-velocity jets through ventilation ducts out of the building. Makeup air is provided to match the exhaust flow rate. The tests were conducted in a Quonset™-type hangar with ceiling height at the peak of approximately 6 m. The test objective was to determine whether a system using high-pressure nozzles mounted on the ceiling of the hush house could extinguish a combined JP-4 jet-fuel spray fire and pool fire on an engine, where it would be shielded from the spray by the wing of the aircraft.

It is difficult to extinguish well-ventilated, shielded fires with water mist, particularly so in the absence of an enclosure. It was expected, then, that a fire located under the wing of the aircraft in a large-volume compartment would pose a major challenge to water mist applied from ceiling nozzles. To overcome the shielding effect of the wing, it was proposed to use high-velocity sprays that (it was hypothesized) would strike the floor with sufficient momentum to push mist laterally into the spray/pool fire and cause extinguishment. A mock-up of a jet engine was constructed, on which the spray/spill pool fire scenario with JP-4 jet fuel was simulated. For the hush-house tests, no wing shielding was constructed; i.e., the engine mock-up was in the open, not shielded under the wing of an aircraft.

The largeness of the space prevented the most effective extinguishing mechanisms of water mist from being brought into play. Flame cooling was not uniform, because the random nature of reflected spray does not impart sufficient momentum for fine spray to penetrate to the inner core of the flame. Persistent flame reignited local areas that may have been momentarily extinguished. The spray was not strong enough to push the water vapor or oxygen-depleted fire gases onto the fuel surface, where it would *dilute* the vapor/air mixture below the flammable region. Fuel *wetting* was of little value on a liquid fuel. Fuel *cooling* was ineffective because the flash point of the JP-4 jet fuel was below the temperature of the water. Well-ventilated fires in low-flash point liquids can be extinguished by water mist, but only when the spray momentum and flux density are sufficient to push most of the water vapor formed back onto the fuel surface, or, for spray fires, into the flame jet.[13]

The characteristics of the system tested in the hush-house trial can be summarized as follows:

Safety objective: Extinguish well-ventilated jet fuel spray, spill, and pool fires, completely shielded from line-of-sight spray impact, in a high-ceiling open area, using only nozzles located at ceiling level, before serious fire damage can be done to an aircraft.

Mechanisms: Spray reflected from the floor, and moved about by the turbulence, would transport enough spray under the wing obstructions to extinguish spray and pool fires through flame cooling and oxygen displacement.

System features: Total flooding of a compartment by water mist from a single row of ceiling-mounted nozzles. The test system used high-pressure nozzles with high spray velocity, intended to overcome the height and distance to the seat of the fire.

Analysis of failure: The downward-directed spray from a 6 m height lost velocity before reaching floor level. Most of the mass flux was lost as the spray struck the floor. The portion of the spray that was not deposited on the floor was scattered in random directions. The velocity of the reduced concentration of deflected spray was too random to extinguish a strong spill fire. The shielding created by a wing would have exacerbated this problem.

User acceptance: The U.S. Air Force did not accept water mist as a viable replacement for Halon 1301 in the hush-house application.

SUMMARY OF WATER MIST SYSTEM FEATURES

From the above descriptions, it is evident that there are wide variations in the objectives, fire scenarios, system details, and levels of acceptance for water mist fire suppression systems. These can be summarized as follows.

Objectives

1. Protect lives and property in a compartment from external fire threat;
2. Control fire temperatures in a compartment to prevent flashover and fire spread to adjacent compartments;
3. Extinguish Class A and Class B fires;
4. Use water to provide protection where there was no previous protection (aircraft passenger compartments);
5. Use water where halon or CO_2 was previously used (Class B fires in machinery rooms) and
6. Provide freedom from sprinkler system design constraints (e.g., allow use of smaller pipe sizes), with equivalent level of safety.

Fire Scenarios

1. Large external pool fire threatening to spread into a compartment;
2. Liquid hydrocarbon fuels with high- and low-flash point liquids, in spray, spill, and pool fires, outdoors and in compartments;
3. Class A combustibles (furniture, bedding, books on fixed shelving) in small compartments; and
4. Machinery compartments, under-ventilated liquid fuel fires; and
5. Large machinery compartments with obstructions, well-ventilated conditions, and small to large fire sizes.

System Details

1. Low-pressure to high-pressure systems: 5 bar to 272 bar (75 psi to 4000 psi);
2. Single- and twin-fluid systems;
3. Normally open, nozzles, deluge-type systems, with separate detection system for activation;
4. Normally closed, thermally activated nozzles, singly or in groups;
5. Continuous discharge; and
6. Repeated short discharges.

Levels of Acceptance

1. IMO full-scale fire test protocols are in place. Systems that meet the fire performance criteria in the IMO test protocols are acceptable to national and international maritime authorities. The test protocols address suppression effectiveness only. The system components, such as pumps, valves, and detection/activation equipment must meet existing SOLAS requirements.
2. FMRC approval of a pre-engineered system for turbine enclosures in a compartment of fixed maximum dimensions. The performance evaluation includes not only the suppression effectiveness, but also the performance of the detection system, effects of water on the equipment being protected, the performance of the detection/actuation equipment, flow duration, and sensitivity to partial dysfunction.
3. The components of water mist systems, including nozzles, pumps, actuators, gas cylinders, pressure control equipment, detection system panel, control valves, etc., must be listed by Underwriters Laboratories Inc. (UL) or FMRC for use in water mist systems. The listing approval may not extend to the performance of the whole system. An assembly of listed components will not necessarily extinguish fire in a given fire scenario.
4. NFPA 750 specifies materials and components to be used in water mist systems. It does not provide a generic design approach based on first principles, but refers to internationally recognized test protocols to confirm the system will meet its objectives. NFPA 750 describes what must be taken into account in full-scale fire test evaluations of water mist systems, but defers to testing agencies to conduct those tests. If tests conducted by a

credible laboratory demonstrate that the proposed design will meet the performance objectives, then it is expected that Authorities Having Jurisdiction will accept the design.

Having reviewed the types of water mist systems that have achieved a level of recognition, the fundamental principles involved in extinguishing fires with water mist can be addressed. Ultimately, an understanding of the "first principles" governing the performance of water mist systems will allow the designer to adapt the range of misting technologies to the range of fire scenarios, to meet a variety of system objectives.

GENERAL PRINCIPLES OF WATER MIST SYSTEMS

This section describes the general principles of water mist fire suppression systems technology and includes extinguishing mechanisms, spray characteristics, generic design elements, and methods of generating water mist. Extinguishing mechanisms are discussed first, as they are the basis upon which the importance of specific spray characteristics can be appreciated. Technical details relating to spray characteristics are presented next. General principles governing the performance of water mist systems are outlined. Methods of generating water mist are also described.

Extinguishing Mechanisms

A basic description of how water mist extinguishes fires was provided as early as 1955 by Braidech[5] and confirmed by Rasbash et al.[6] To quote Braidech:

> The extinguishing action of sprays of finely divided water applied to commonly encountered fires appears to be due predominantly to dilution of the air (oxygen) supply in the zone of burning with vapor (steam) resulting from evaporation of water droplets in the heated area surrounding the fire. The cooling effects of the water may also be important factors in extinguishment in many cases. In order to obtain extinguishment, the water droplets comprising the spray must be relatively small, and the amount of water applied must be adequate in relation to the specific fire.

Research conducted over four decades later has not altered the accuracy of this description. The recent work has, however, suggested other processes that are involved and identified the mechanisms that are most important for design of effective fire suppression systems.

Figures 6-15A and 6-15B illustrate the mechanisms involved in extinguishing fire with mist. The mechanisms that act together to extinguish fire can be described as three primary and two secondary mechanisms.[14] The primary mechanisms are: (1) heat extraction, (2) oxygen displacement, and (3) blocking of radiant heat. Wighus[3] and Hanauska and Back[8] described the mechanisms as "gas-phase cooling and steam inerting," making no reference to radiant heat attenuation. Although all three mechanisms are involved to some degree in every extinguishment, in tests conducted at the National Research Council, some fires were extinguished predominantly through heat extraction (cooling) and others predominantly through displacement of oxygen. The difference depends on whether the fire was poorly or well-ventilated, and the properties of the fuel. The benefit of radiation attenuation in a compartment is evident in reduced thermal feedback to burning and unburned fuel surfaces. In the National Research Council of Canada (NRCC) experiments, however, radiation attenuation could not be identified as the dominant mechanism of extinguishment.

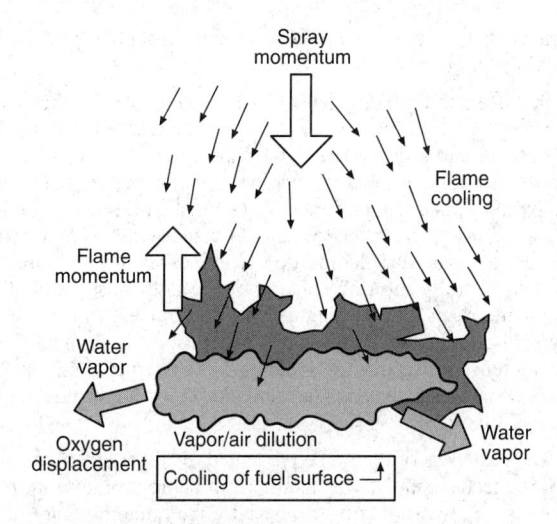

FIG. 6-15A. *Interaction of mist with pool fire flame, showing the extinguishing mechanisms.*

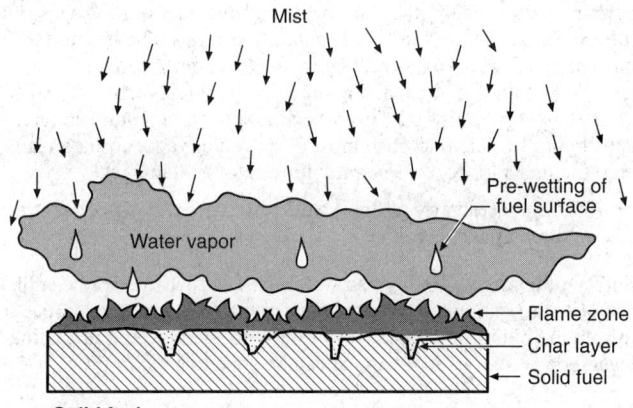

Solid fuel:
1. Low flame, harder to cool,
2. Char burns at lower O_2 concentration, and
3. Fuel wetting.

FIG. 6-15B. *Flame on a solid fuel with charring and mist.*

There are two secondary mechanisms that play a role in extinguishment, but it is difficult to quantify their importance. These are: (1) vapor/air dilution and (2) kinetic effects. The dilution of vapor/air mixtures by the mixing of water vapor and entrained air in the flammable vapor zone above a fuel surface can be deduced to have some effect on pool fires. Also, the velocity of the flame front in a flammable gas mixture may be inhibited (or accelerated) by the presence of small water droplets dispersed in a volume of flame, and this phenomenon may play a role in the extinguishment of spray fires and in inhibiting deflagrations.[15]

For engineering design of reliable water mist fire suppression systems, it may be sufficient to understand the three dominant mechanisms, i.e., heat extraction, oxygen displacement, and radiation attenuation. For scientific purposes, computer modelers need to consider all extinguishing mechanisms in order to develop algorithms to simulate the extinction of fire by mist. An ideal computer model would account for flame and fuel cooling, oxygen reduction, concentration changes in vapor/air mixtures at the fuel surface, reduced radiant feedback from flame to fuel surface, and possible kinetic effects caused by the interaction of mist with a moving flame front. Computational fluid dynamics (CFD) models to that level of detail are, at this time, in their early stages of development.[16]

The mechanisms of extinguishment are described in detail in the following sections. Table 6-15A summarizes the points to be discussed.

Heat extraction (cooling):   When water is applied to a fire, heat is absorbed in three areas: (1) from the hot gases and flames, (2) from the fuel, and (3) from the objects and surfaces in the vicinity of the fire. Cooling of the fuel and nearby objects contributes to reducing fire spread, but it does not necessarily require fine drop sizes. In fact, wetting and cooling of solid fuels is easier to achieve by using larger droplets (> 400 microns) and delivering more water than can usually be delivered as water mist.

Compared with coarser sprays, finely divided water sprays enhance the speed at which the spray extracts heat from the hot gases and flame. Reducing the drop size increases the surface area of the water mass and thereby increases the rate of heat transfer. The conversion of water droplets to steam absorbs heat. If sufficient heat is withdrawn, the gas-phase temperature of the flame can be dropped below that necessary to sustain the combustion reaction, and flame will be extinguished. Theoretical considerations suggest that the combustion reaction in a diffusion flame will cease if the flame temperature drops below approximately 1600 K (1327°C).[17]

As for what constitutes "sufficient" heat absorption, various authors note that it is not necessary to absorb all of the heat given off by the fire at the rate of burning. Absorbing 30 to 60 percent may be enough to cause burning to stop.[3] An estimate of the minimum evaporation rate to extinguish a fire of a given heat release rate (HRR) is of little practical value, however, because extinguishment usually involves more than just flame cooling. The concurrent effect of oxygen reduction could mean the fire could be extinguished with only a fraction of the theoretical minimum required for flame cooling.

For liquid fuel fires, the evaporation of mist cools the flame, which, in turn, reduces the radiant heat flux to the surface of the fuel, resulting in a reduction in the evolution of flammable vapors. The combination of reduced flame temperature and reduced evolution of vapors results in a reduced burning rate and, in some cases, complete extinction. Fires in liquid fuels with flash points (FP) above normal ambient temperatures, e.g., diesel fuel (FP ≈ 60°C), can be extinguished with relative ease by flame cooling and reduced radiation to the surface. Fires in liquid fuels with flash points below normal ambient temperatures, e.g., heptane (C_7H_{16}, FP ≈ –4°C), are much harder to extinguish by cooling alone because temperatures cannot be reduced enough to reduce the vapor/air mixture above the surface of the fuel to below the lean flammability limit.[18] Any hot object or fragment of flame can cause reignition.

The cooling of flames above solid fuels also reduces the radiant heat flux to the fuel surface and the rate of pyrolysis of the fuel. As illustrated in Figure 6-15B, however, with charring substances, the combustion reaction occurs within the carbon-rich porous zone that forms on the fuel surface.[17] Cooling of the diffusion flame above an established char zone may not suppress radiant feedback within the crevices of the char. Either water mist must be applied early in the fire, before a deep char zone has developed, or water droplets must penetrate the char zone and reach the actual interface between burning and unburned fuel. On the other hand, the flame height and plume velocities for glowing combustion are relatively low, so that the larger drop sizes in a spray may not be evaporated and will wet the fuel surface.

Extinguishment of solid fuels such as Class A combustibles, therefore, depends on the geometry of the fuel arrangement and the depth of the char layer. Exposed surfaces facing the source of the spray might be extinguished, while shielded surfaces (such as within a wood crib) will be almost impossible to extinguish by cooling alone.

Oxygen displacement: Braidech et al.[5] concluded correctly in 1955 that the suppression effect of water mist "appears to be due predominantly to dilution of the air (oxygen) supply in the zone of burning with vapor (steam)." That is to say, oxygen displacement appears to play a stronger role than flame cooling. Water droplets expand approximately 1900-fold upon evaporation (at 95°C, 1 atm

TABLE 6-15A.   *Mechanisms of Extinguishment by Water Mist and Their Application*

Mechanism	Principle of Application
Primary	
Heat extraction	The drop size distribution, momentum and mass flow rate delivered to the fire, after losses to interior surfaces and obstructions, must be sufficient to absorb a critical percentage of the heat released by the fire.
Oxygen displacement	Design to: 1. Enclose fire to contain evaporated water, or 2. Use nozzle dynamics to force water vapor into the base of the fire.
Radiant heat attenuation: 1. To unburned surfaces, and 2. To burning surfaces.	Mist must: 1. Surround the fire, and 2. Penetrate the flame.
Secondary	
Vapor/air dilution: 1. By water vapor, and 2. By entrained air.	Significant for liquid fuel pool or spray fires. Must have enclosure or control of dynamic spray properties to distribute diluent over the fuel surface. Nozzle design may influence air entrainment, hence dilution. Difficult to predict or control
Kinetic effects: 1. Reduce flame velocity, and 2. Accelerate combustion reactions.	Applies to deflagration control by reducing velocity of the flame front, hence explosion overpressure. Unpredictable: mist may suppress or invigorate combustion

pressure). If evaporation occurs rapidly, the water vapor displaces the air in the vicinity of the drop. Injection of a finely divided water spray into a hot compartment results in rapid evaporation, expansion, and in displacement of the air in the compartment by steam. If the amount of oxygen available for combustion is reduced below a critical level, the fire burns inefficiently and will be easier to extinguish by cooling.

Unfortunately, the dilution of O_2 by water vapor in a suppression scenario is limited by the temperature. The injection of mist cools the air below the temperature at which it can hold enough vapor to significantly reduce the O_2 concentration.

The Braidech et al.[5] study, as with most studies of the extinguishment of fires with water mist, concentrates on hydrocarbon pool fires, with a few tests on wood cribs. Recent tests by Wighus et al.,[4] Mawhinney,[19] and Hanauska and Back[8] on diesel and heptane pool and spray fires confirmed that oxygen displacement ("steam inerting") was the dominant mechanism by which water mist extinguished the flames, both in compartments and in open-area pool fires. The compartmentation allowed the water vapor concentration to build up so that hidden and shielded fires could be extinguished. Hanauska and Back[8] used a very fine spray with very low momentum and flow rate to extinguish heptane pool and spray fires, after 30 to 50 sec of mist application, in a 28 m^3 compartment. Fires in such enclosures can be extinguished with about one-tenth the mass flux of water needed for pool fires in open areas.[3,4] In open-area fire tests at NRCC, extinguishment of pool fires could not be achieved unless the spray was oriented and applied with enough force, push water vapor created in the outer regions of a flame onto the fuel surface. The open-area fires required high-momentum sprays, properly oriented. This suggests that the water vapor concentration at the fuel surface can be increased either by confining the fire in an enclosure or by using spray momentum to push the water vapor against the surface. Of course, the additional mixing created by the higher momentum spray also increased the gas-phase cooling. It is not easy to separate the effects of cooling and oxygen displacement, other than to note that, in the NRCC plenum tests,[20] turbulent mixing alone did not bring about extinction if there was no component of the spray velocity directed toward the fuel surface.

The minimum amount of free oxygen needed to support combustion varies with the type of fuel. In general, hydrocarbon gases and vapors cease burning at oxygen concentrations below 13 percent, while charring solid fuels may burn with oxygen concentrations as low as 7 percent.[17] This explains why it is easier to extinguish hydrocarbon pool fires (diesel and heptane) than wood crib fires; the reduction in oxygen needed to extinguish a hydrocarbon flame is easy to obtain, compared to that needed to arrest combustion in burning wood.

The effects of oxygen displacement also explain why it is easier to extinguish "large" fires than "small" fires in a compartment. "Large" and "small" are defined loosely in terms of whether the fire will affect the average temperature and oxygen concentration in the compartment within the activation time of the water mist system. A "large" fire releases more heat into a compartment in the early stages than a "small" fire, so that more heat is available to evaporate the fine water droplets. A "small" fire will have a continuous source of fresh combustion air at normal oxygen concentrations. A "large" fire will reduce the ambient oxygen concentration to the point that combustion efficiency will already be reduced, prior to introducing the water mist. With the combined effects of vitiated combustion air, plus further dilution by water vapor, "large" fires can be extinguished with lower flux densities than "small" fires.

The first commercially developed water mist fire suppression systems were accepted (by European maritime authorities) on the basis of their ability to suppress machinery room pool and spray fires involving hydrocarbon fuels, usually diesel. The dramatic extinction events shown in those tests represented the best case for a water mist fire suppression system, i.e., relatively high flash point fuel, much available heat due to large size of fire, and confinement of the water vapor by an enclosure. To be useful in a compartment that cannot tolerate the damage that could be done by a "large" fire, however, careful thought must be given as to how to achieve heat extraction and oxygen displacement with small fires early in their development.

Blocking radiant heat: Another mechanism of extinguishment, i.e., blocking radiant heat, plays a role in stopping the fire from spreading to unignited fuel surfaces and reduces the vaporization or pyrolysis rate at the fuel surface. On a "macro-scale," radiation attenuation provided by water mist protects objects and personnel in a space from radiant heat damage, whether or not extinguishment occurs.

Theoretical work on radiation attenuation by water sprays[22,23] indicates that the attenuation of radiation depends on drop diameter and mass density of the droplets. As the concentration of drops with diameters smaller than 50 microns increases, the degree of attenuation of radiant heat increases.[22] This is one reason why water mists, having high concentrations of very fine drops, prove to be very effective at reducing radiant heat transfer.

One question that arises is how much of the reduction in radiation to surfaces is due to the reduced size of the flame, and what proportion is due to actual attenuation of radiant heat due to the presence of mist. Theoretical considerations suggest that mist or steam that enters the space between a flame and the fuel surface will reduce the radiant heat flux to that surface.[15] This is difficult to measure, but it is hypothesized that radiant heat from the flame could be absorbed by unevaporated droplets and steam, which reradiate at the lower temperature of the droplet or vapor. With some liquids, such as diesel fuel (FP ≈ 60°C), this reduction in the heat flux to the fuel surface may reduce the vaporization rate of the liquid fuel and contribute to the extinguishment. The effect would be expected to be less important for liquids with flash points below normal ambient temperatures, such as heptane (FP ≈ -4°C).

Radiant heat attenuation between flame and fuel surface is important for computer modeling of fire suppression using computational fluid dynamics. In developing a CFD model of fire suppression using water mist, An et al.[21] calculated the heat release rate of a liquid pool fire, taking into account the coefficient of absorption of water vapor in each control volume. A reduction of radiant heat transfer to the fuel pan was predicted to measurably reduce the vaporization rate, hence the heat release rate of the burning vapors. More experiments are needed to compare measurements of heat flux in this zone, to predictions of radiant heat attenuation obtained using the CFD models.

Dilution of vapor/air mixture: Air and water vapor entrained in a water spray may dilute the vapor/air mixture to below the lean flammability limit. With diesel fuels (FP ≈ 60°C), cooling of the flame reduces the thermal energy to the fuel surface, which reduces the rate of evaporation. Coupled with a dilution of the vapors by the addition of entrained air, the vapor/air concentration falls below the lean flammability limit. In contrast, it is much harder to reduce a heptane/air mixture to below its lean flammability limit, because of the low flash point temperature (≈ -4°C) and high vapor pressure of heptane. Dilution of pyrolyzed vapors emitted from solid fuels may also contribute to extinction. This is referred to as a secondary mechanism, because it is difficult to see how dilution alone could result in extinguishment. It requires uniform mixing of mist and entrained air throughout the space between the flame and the fuel surface, to dilute all of the vapor/air mixture within the vaporization region. Mixing at fuel surfaces is often turbulent and nonuniform, so it is likely that there will always be some region of the vapor/air cloud that is in the flammable range.

For purposes of designing suppression systems, it is not yet feasible to quantify the relationship among the flammability limits of different fuels, the fuel vaporization rates, spray evaporation rate, and the mass flow rates of water mist and entrained air. A knowledge of dilution effects can, however, be used to explain differences in the effectiveness of water mist on different fuels, as for example, diesel and heptane. It should eventually be possible to calculate dilution effects on gas concentrations using computer field models based on computational fluid dynamics.

Kinetic effects of mist on flames: A liquid pool fire is sometimes intensified by the application of water spray. A "flare-up" often occurs in the first second of contact with the water mist, and it is evident in some fire tests that the burning rate is increased for longer periods. The flare-up at the instant of application of water spray on liquid fuel fires is familiar to fire fighters, who are trained not to be startled by the event. In many cases, the flare-up is followed by a quick knockdown and extinguishment of the flames. If the spray dynamics are insufficient to bring about extinguishment, the fire will continue to burn violently in spite of the mist.

The momentary intensification of the fire has been attributed by some authors to the effect of droplets striking the fuel surface and causing splashing and increased vaporization rate. Kokkala[24] referred to a "flame ball" which occurred when a sprinkler spray hit a hot, high boiling point liquid and explosively vaporized. The same phenomenon was noted by Mawhinney,[20] during suppression tests in which a fine water mist was applied from below a diesel pan fire in a direction concurrent with the fire plume, such that very few droplets could reach the surface of the diesel fuel. The visual evidence, supported by thermocouple readings, was that the flame within the plenum became violently turbulent and appeared to intensify during the time of application of the mist. One possible explanation for the observed intensification is that the turbulence acted to increase the burning rate of the fuel.

Mawhinney *et al.*,[14] at NRCC, measured the heat release rate of open-area heptane and diesel pan fires during the application of water mist. Figure 6-15C shows the measurements of heat release rate during the suppression of a diesel pool fire and a heptane pool fire, respectively, using the same downward-oriented water spray, in both cases. The HRR curves under suppressed conditions are plotted with the curves for unsuppressed fires in the same pans. In the case of diesel fuel, there is a brief "spike" in the HRR, followed by complete extinguishment within 60 sec once the system is activated. For the heptane fire, a brief "spike" also occurred, followed by a partial suppression of the fire, and then by a continued increase in the burning rate of the heptane. The water spray failed to extinguish the fire, and turbulent flaming continued until the fuel was consumed. Over a portion of the burning time, the HRR of the fire with mist exceeded the HRR for the unsuppressed fire, although, as expected, the overall burning time was reduced by the interval of intensified combustion. It is possible that the combustion process benefited from the additional air brought into the flame zone by the turbulent spray.

Jones and Thomas[15] report similar intensification of the rate of combustion in the use of water mists to quench gaseous explosions. The mist suppresses the maximum overpressure in an enclosed gaseous ignition, but the peak pressure may be attained sooner than for an unsuppressed explosion. Furthermore, the authors[15] report that it is "never immediately obvious whether application of a water spray will quell or invigorate an explosion." The conflicting influences of cooling, inerting, dilution, and enhanced turbulence and fuel mixing

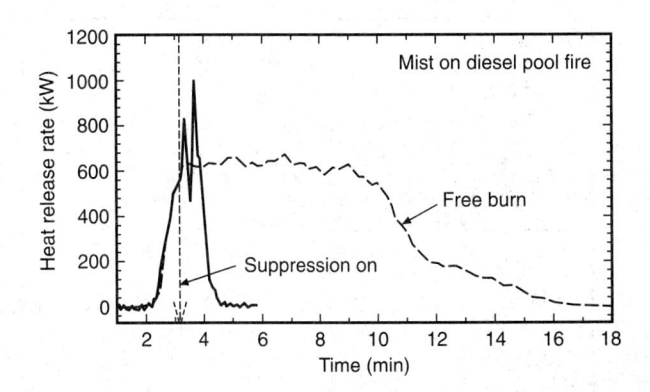

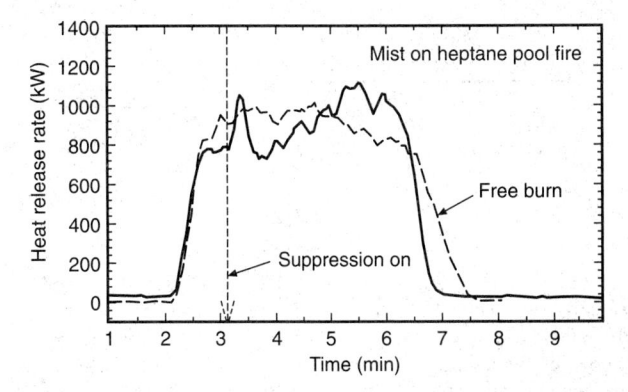

FIG. 6-15C. Plots showing momentary and sustained increases in the heat release rate of diesel and heptane pool fires with application of water mist.

lead to a degree of unpredictability in the effects of water mist on gas-phase burning.

There is reason to be concerned that a water mist system that is unable to extinguish a liquid fuel pool or spray fire could instead momentarily increase the heat release rate of the fire. Further research is needed to investigate the conditions under which flare-up or flame invigoration occurs.

Enclosure effects: Enclosure effects enhance the performance of water mist systems. The enhanced performance can be attributed to restricted ventilation and heat entrapment. A fire in a "small" compartment can reduce the average oxygen concentration in the compartment by several percentage points within the first few minutes of growth. Thus, a fire that is large enough to affect the average oxygen concentration in a compartment could be considered to be "poorly ventilated." Fires burn less efficiently at lowered oxygen concentration, which makes them easier to suppress.

Figure 6-15D illustrates how enclosure effects contribute to fire suppression with water mist. Heat from the fire trapped in the compartment evaporates the finest portion of the mist, so that the expanding water vapor pushes the air out of the compartment. Then, oxygen-depleted, hot fire gases at the ceiling of the compartment are cooled by the mist and pushed down to floor level, mixing water vapor, oxygen-depleted air, and water drops with the fire. The combined effects of reduced combustion efficiency and flame cooling usually result in extinguishment. With enclosure effects, it is possible to extinguish even shielded fires with low-momentum sprays in heavily obstructed compartments. Where enclosure effects can be relied upon, the flux density required for extinguishment can be as much as 10 times lower than that required for unconfined, well-ventilated fires.[3,4]

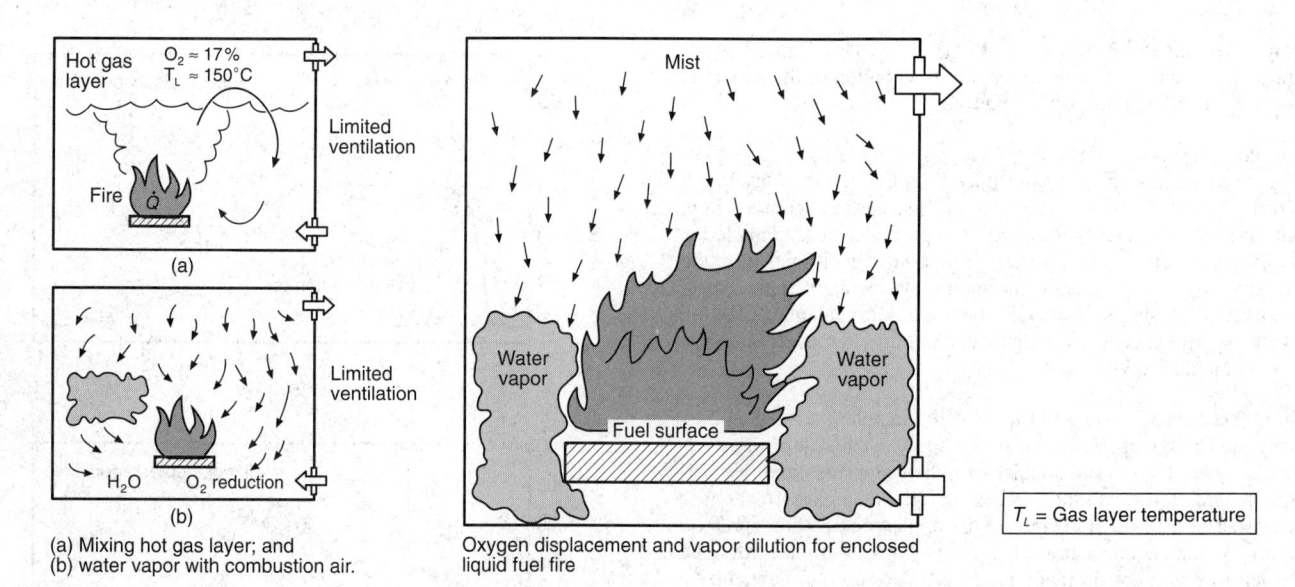

(a) Mixing hot gas layer; and
(b) water vapor with combustion air.

Oxygen displacement and vapor dilution for enclosed liquid fuel fire

T_L = Gas layer temperature

FIG. 6-15D. *Illustration of enclosure effects.*

Of the many water mist suppression experiments reported in recent years, all comment on the difficulty of extinguishing "small" fires and the ease of extinguishing "large" fires. This can be explained by the fact that the small flame has little effect on the average oxygen concentration in the compartment, and the small amount of water vapor that is produced is carried away from the fuel-flame interface. In the case of the well-ventilated or unconfined fire, water mist must have strong enough momentum to push water droplets into the flame and to force the water vapor that is created in the flame zone against the fuel surface. Thus, nozzle and mist dynamics must compensate for the lack of confinement of heat and water vapor.

Time to extinguishment and flow duration: The time needed to extinguish a fire with water mist varies with the fuel type, compartment geometry, and the drop size distribution, momentum, and application rate of the water mist. There have been many demonstrations in which water mist extinguishes liquid fuel pool fires in 10 to 20 sec, using small amounts of water. Figure 6-15C, for example, shows a 600 kW, unenclosed diesel pool fire that was extinguished in less than 60 sec by water mist. Such successes occur in circumstances in which all suppression properties act together, i.e., oxygen displacement, flame cooling, radiant heat blocking, and vapor dilution.

Rapid extinguishment is not characteristic of all water mist fire suppression systems, however. There are circumstances in which it can take several minutes to extinguish a fire, without it being considered a "failure." One example is where flame cooling cannot extinguish the fire immediately, but, after several minutes, the buildup of water vapor in the compartment (an enclosure effect) combines with flame cooling to bring about extinguishment. With combustibles that form a char layer, e.g., fabric, wood, plastics, or electrical wiring, the fire may not be extinguished immediately. Fuel wetting begins to reduce the size of the fire and it is eventually extinguished. As long as the mist limits the heat release rate, causes the fire to progressively diminish in intensity, and cools the compartment, the level of fire control can be considered adequate.

There are reasons to flow water from a system for a longer period of time than the minimum extinguishing time. First, extinguishment times vary, even under controlled test conditions; and conditions in "real" compartments cannot be closely controlled.

Second, with liquid fuel fires, reignition may occur if hot surfaces are not sufficiently cooled, or if there is a remnant of flame in a shielded area. In principle, the flow should be sustained until it is confirmed that all fire has been extinguished and reignition will not occur. Such confirmation is difficult to obtain in real fire conditions, however. Practically, it is tempting to assign an arbitrary minimum flow duration, based on conservative assumptions about the probability of fire being extinguished. NFPA 750 recommends that the water supply be sized based on a 30-min minimum flow duration. An exception is provided for "pre-engineered" systems, for which the minimum duration must be sufficient for two complete discharges. The length of one "complete discharge" is determined by the full-scale testing upon which the listing was based. A second exception allows a fire protection engineer to use "standard methods of fire hazard analysis" to determine the design duration.

Spray Characteristics

The term "water mist" implies a very fine water spray that remains suspended in air for an extended period of time. The term reflects one of the qualities of a spray, i.e., the drop sizes are "small," relative to rain, or to sprinkler sprays, for example. This one characteristic of drop size is not the only characteristic of a spray that must be controlled in order to generate an effective fire suppression medium. There are three other characteristics that determine its effectiveness as an extinguishing agent: (1) the density of the spray (mass of suspended water per unit volume of space); (2) the velocity with which it is delivered to the seat of the fire; and (3) the quality of the water itself (which may contain dissolved additives to enhance suppression effectiveness). The following discussion includes all four characteristics of a water mist for fire suppression purposes:

1. Drop size distribution,
2. Flux density,
3. Spray momentum, and
4. Additives.

Drop size distribution: It is difficult to discuss "water mist" without measuring and analyzing its most notable feature, the fact that it is made up of "finely divided water drops." At the time of this writing, except for computer modeling, the drop size distribution of the spray cannot be used explicitly as a variable in the design of a water mist system. That stage awaits the results of further research work. The value in knowing the drop size distribution (and other characteristics) of a water spray lies in the fact that it relates directly to system performance: how spray is affected by obstructions, the dynamics of the spray interaction with the fire, and how extinguishment occurs. For computer modeling of the dynamics of suppression with water mist, a quantitative measure of the drop size characteristics of the spray is essential.

The term "drop size distribution" refers to the range of drop sizes contained in a representative sample of a spray or mist. There is a distribution of small and large drops, which varies with location in the spray as well as with time. For a continuous discharge, the distribution of drop sizes changes with distance from the source as drops collide, evaporate, or hit surfaces and fall out. For a short duration discharge, the distribution measured at a point in space changes with time as the larger drops pass through quickly, leaving increasingly finer drops, which move at lower velocity.

There are a number of ways to present data about drop size distributions of sprays.[25,26] It is customary in some fields to refer to the size of particles in a spray by a single drop size parameter, such as a "Sauter Mean Diameter (SMD), or volumetric Median Diameter (VMD)." Such terms reveal little about the *range* of drop sizes in a spray, however. As is evident in Point A in Figure 6-15E, two sprays with the same Volumetric median diameter (Dv0.5) may have distinctly different ranges in drop size. It is important to know about the presence of larger diameter drops in a spray, to avoid splashing on the surface of a liquid fuel, for example, or perhaps to know about the potential to achieve fuel wetting.

NFPA 750 has adopted the "cumulative percent volume" (CPV) *vs.* "diameter" curve to represent the distribution of drop sizes in a water mist. Reasons for this choice are that: (1) the cumulative percent volume plot visually reveals the range of sizes, and (2) the "volume" distribution converts readily to "mass" distribution, which is the most relevant term for analyzing heat transfer and evaporation rates using computer modeling. Figure 6-15E illustrates cumulative percent volume curves, measured 3 ft (0.9 m) from the nozzle, for several nozzles used in water mist, water spray, and sprinkler systems, all at pressures below 175 psi (11.9 bar).

Commercial software conforming to ASTM E799, *Standard Practice for Determining Data Criteria and Processing for Liquid Drop Size Analysis*, for particle size measurement,[25] provides output in several formats, one of which is the cumulative percent volume *vs.* diameter curve. For comparison with other sprays, the "S" shape of the curve can be roughly described using only three representative diameters: Dv0.1, Dv0.5, and Dv0.9. These parameters appear as standard output from a particle size analyzer conforming to ASTM E799. For a spray with a Dv0.9 of 400 microns, it means that "90 percent of the volume of a sample of the spray, at the sampling location, is contained in drops with diameters less than 400 microns." The cumulative percent volume distribution *vs.* diameter curves can also be expressed as a mathematical function, which can be used in computer modeling of water spray/fire plume interactions.[26]

The drop size distribution must be measured at a point that is statistically representative of the spray under the conditions of interest. For fire suppression, the point of interest is where the spray interacts with the flame or fuel, which may be several feet (m) away from a nozzle. NFPA 750 recommends that the drop size distribution be measured at several locations in the spray cone, on a plane 3.28 ft (1 m) from the nozzle, where the spray is expected to have the average properties of the spray elsewhere in the compartment. At that distance, the cone is at its maximum diameter, the entrained air has acquired a velocity, and the velocities of large- and small-diameter drops approach equality.

A single measurement in the region of interest would not capture variations of coarser or finer sprays within the cone. Therefore measurements should be taken at several points in the spray. The method described in the Appendix of NFPA 750 is illustrated in Figure 6-15F(a). It is proposed to take 9 drop size measurements in a single quadrant of a horizontal plane through the spray cone, 3.28 ft

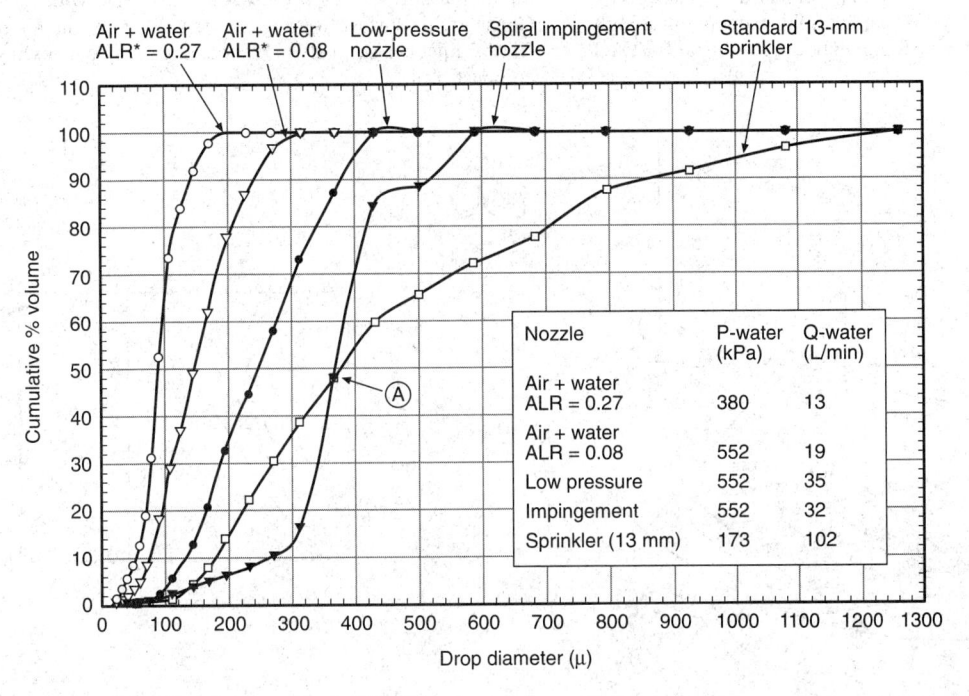

FIG. 6-15E. Comparison of drop size distribution curves for mist nozzles.

(1 m) below the nozzle. All nine cumulative percent volume curves should be plotted to show an "envelope" of curves [Figure 6-15F (b)]. However, neither the finest (to the left), nor the coarsest (to the right), cumulative percent volume curve is uniquely representative of the "true" drop size distribution of the spray. All 9 curves should be combined into a single, representative drop-size distribution curve, using statistically acceptable techniques. One approach is to "weight" the curves according to the magnitude of the flux densities measured at each of the 9 locations. Thus, the CPV curve for the area of highest flux density would be deemed to be representative of a greater proportion of the spray than the curve for a region of lower flux density. A second, possibly better, approach is to select the locations of measurement of the spray so that each reading represents an equal area of the spray plane, then to calculate an arithmetic average. The procedures for obtaining meaningful drop size distributions for water mist nozzles are still under development.

Early research with fine water sprays was motivated by interest in using "small" drop sizes to enhance evaporation and heat extraction, using small volumes of water.[5,6] The research focused primarily on liquid fuel pool fires, for which the desired benefits are achievable with sprays that have most of the mass contained in drops of diameters of 400 microns or less.[20] Recent interest in using water mist for applications involving solid fuels, however, suggests that sprays with drop sizes greater than 400 microns need to be included. Drop sizes larger than 400 microns provide for fuel wetting, an important mechanism in extinguishing fires in solid fuels and charring substances. For that reason, the definition of "water mist" in NFPA 750 includes sprays for which 99 percent of the volume of the spray is contained in droplets less than 1000 microns in diameter (Dv0.99 < 1000μ). Water mist systems, therefore, are applicable to a wide range of fire scenarios, including both liquid and solid fuels.

Classification of water sprays by drop size distribution: The definition of "water mist" in NFPA 750 is so broad that some important differences in the qualities of sprays are disguised. It includes, for example, all water sprays used in NFPA 15, *Standard for Water Spray Fixed Systems for Fire Protection* (hereinafter referred to as NFPA 15), sprays produced by standard sprinklers operating at high pressure, and light mists suitable for greenhouse misting and HVAC humidification systems.

The drop size distribution that will be most effective in one fire scenario will not necessarily be the best for a second scenario. There is no "one size fits all." The type of fuel (solid or liquid), the compartment dimensions, the velocity at which the mist is delivered, and the performance objective (e.g., temperature control *vs.* extinguishment) collectively determine the most appropriate drop size distribution for a given application.

As a means of allowing distinctions to be made among "coarser" and "finer" sprays across the 1000-micron spectrum of the NFPA 750 definition of water mist, this chapter adopts a classification system that subdivides "water mist" into Class 1, 2, or 3 mists, according to their representative cumulative percent volume (CPV) *vs.* diameter distribution curve. The boundaries of the three classifications of water mist are shown in Figure 6-15G and described in the following paragraphs.

A spray for which the Dv0.9 ≤ 200 microns is a Class 1 mist (^{1}Dv0.9 ≤ 200). This category includes the "finest" water mists. Usually the fineness is obtained at the expense of mass flow rate, spray velocity, and energy required to produce the spray. Thus, generation of a Class 1 mist requires significant input of energy to produce useful quantities. Because drag effects have a greater influence on small (low mass) particles than on large particles, it is more difficult to deliver a Class 1 mist at high velocity than a coarser mist. A Class 1 mist would be suitable for situations where enclosure effects reduce the need for momentum, and where fuel wetting is not critical to performance, as, for example, on liquid fuel fires and spray fires in enclosed spaces. Class 1 mists can be generated by twin-fluid nozzles; high-pressure, pressure-jet nozzles; some small impingement nozzles; and by flashing of super-heated water.

Again with reference to Figure 6-15G, the CPV curve for a Class 2 mist falls in the range 200 < (^{2}Dv0.9) ≤ 400 microns. Such sprays can be generated by low- and intermediate-pressure, pressure-jet nozzles; twin-fluid nozzles; and some types of impingement nozzles. Due to the presence of larger drops, it is easier to achieve higher mass flow rates with Class 2 mists than with Class 1 mists. Considerable surface wetting occurs with sprays in this range, so a Class 2 mist is also likely to be effective on fires involving ordinary combustibles.

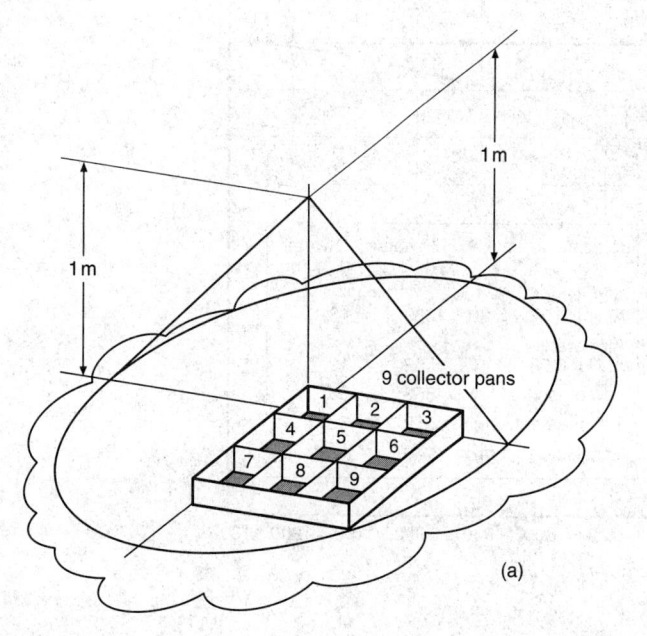

(a)

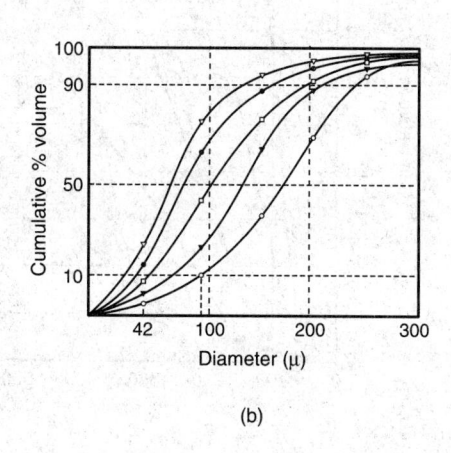

(b)

FIG. 6-15F. *Recommended method for obtaining a representative measurement of the drop size distribution produced by a water mist nozzle.*

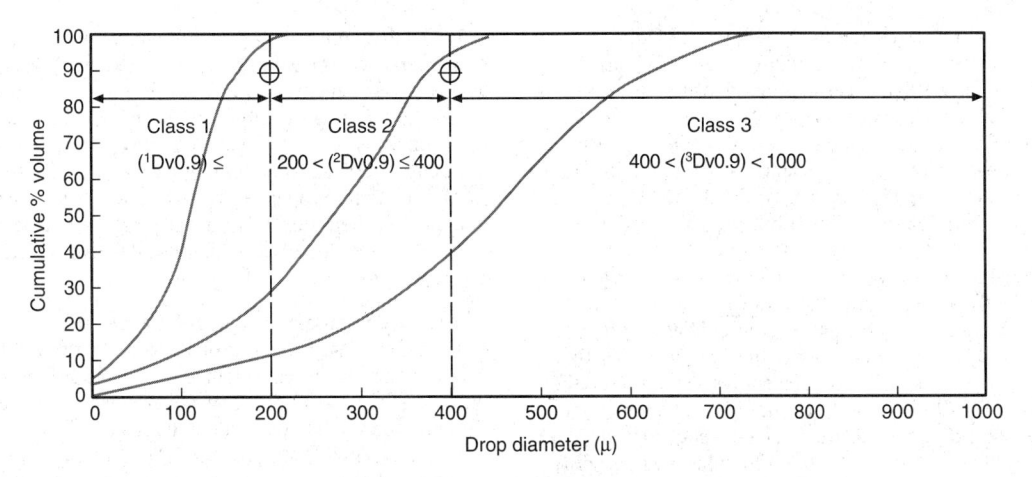

FIG. 6-15G.   *Classification of drop size distributions—Classes 1, 2, and 3.*

A Class 3 mist has a CPV curve that lies in the range $400 < (^3Dv0.9) \leq 1000$ microns. Such sprays are typically generated by impingement nozzles of various sorts, including low- and intermediate-pressure, small-orifice sprinklers; and fire hose fog-nozzles. High mass flow rates are possible. Class 3 mists generated by impingement nozzles have been tested effectively in the IMO equivalent sprinkler system for crews' quarters and corridors on ships.[12]

The relationship between drop size distribution and extinguishing capacity of a water mist is complex. In general, Class 1 and Class 2 mists are successful at extinguishing liquid fuel pool fires and spray fires without agitation of liquid pool surfaces. Given an appropriate geometry, Class 3 sprays are reported to have extinguished pool fires, however.[24] Also, in general, it is difficult to extinguish Class A combustibles with Class 1 sprays, which may not achieve the fuel wetting necessary to penetrate the char layer. Class A fires can be extinguished with Class 1 mists, however, particularly if the velocity is high, the burning is surficial, or enclosure effects enhance the degree of oxygen reduction. This evidence confirms that drop size distribution alone does not determine the ability of a spray to extinguish a given fire. Factors, such as fuel properties, enclosure effects, spray flux density, and spray velocity (momentum), are all involved in determining whether a fire will be extinguished.

Flux density: The ability of water mist to extinguish a fire depends only partly on having the appropriate drop size and spray velocity. It also requires that the mass of water spray that interacts with the fire be sufficient to absorb a critical portion of the heat given off by the fire. Spray flux density is, therefore, an important characteristic of water mist for fire suppression systems. It can be expressed in *volume units* of gpm/cu ft (L/min/m³), although it is more practical to measure it as *area-units* of gpm/sq ft (L/min/m²).

Expression of spray density in volume units is useful for the limited case of a total-flooding system in a compartment with no obstructions, and nozzles with uniform distribution within their spray cones. With few exceptions, the flux density distribution of mist within a single spray cone is non-homogeneous. When mixed with overlapping spray cones, air movement caused by the fire plume and the nozzles themselves, and the scrubbing effects of obstructions, the end result is that the volume concentration of spray at any point in the compartment is unpredictable. It is, therefore, difficult to establish a meaningful correlation between volume concentration of mist and extinguishing effectiveness.

On the other hand, the average flux density in area units of gpm/sq ft (L/min/m2) can be measured easily using a grid of collection cups to measure the volume of liquid collected per unit of area

per unit time. It is possible to relate success or failure to extinguish a fire to the flux density measured at that point (under nonfire conditions). It is, therefore, more useful to express flux density in area units, as a design variable, than to express it in volume units. Both representations are poor approximations of the actual flux density during extinguishment, however. That number is likely to vary dynamically with fire conditions and spray velocity, and, therefore, cannot be used as a unique design parameter.

The density distributions from spray nozzles are far from uniform. Figure 6-15H shows an example of the flux density distribution on a plane 9.9 ft (3.0 m) below two SSC 3/4 7G5 nozzles, spaced 6.6 ft (2.0 m) apart, discharging 9.2 gpm (35 L/min) each.[27] These nozzles have been proposed for a "waterfog" system in a hydraulic equipment compartment on British (Royal Navy) submarines. At 100 psi (6.8 bar), the nozzles produce a Class 2 spray. The area "covered" by the spray cone at the level of collection was about 5 ft (1.5 m) square, for an area of 25 sq ft (2.25 m²). The "nominal flux density" for this array is calculated as the nozzle

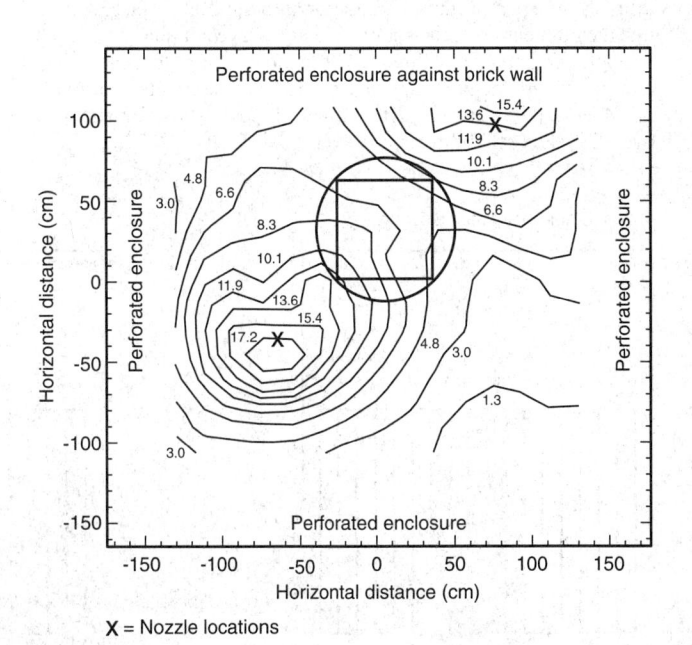

X = Nozzle locations

FIG. 6-15H.   *Actual flux density measured 9.9 ft (3.0 m) below two SSC 3/4 7G5 nozzles, spaced 6.6 ft (2.0 m) apart, discharging at a nominal flux density of 0.37 gpm/sq ft (15.1 L/min/m²).*

discharge divided by the coverage area of the spray cone at the plane of interest. In this case, the nominal flux density was 0.37 gpm/sq ft (15.1 L/min/m²). The measured flux density ranged from 0.42 gpm/sq ft (17.2 L/min/m²) directly below the nozzle, to 0.07 gpm/sq ft (3 L/min/m²) at the outer edge; and about 0.16 gpm/sq ft (6.6 L/min/m²) midway between the two nozzles. Thus, within a distance of less than 3.3 ft (1 m), the flux density distribution from these two nozzles ranged from 0.2 to 1.1 times the nominal flux density.

The extinguishing effectiveness of these two nozzles on heptane and diesel fuel pan fires varied with fire location.[27] Directly beneath the nozzles, where both flux density and spray direction (momentum) were sufficient to overpower the fire plume, all fires were extinguished in less than 10 sec. Placed midway between the nozzles, the diesel fires took longer to extinguish, and some heptane fires were not extinguished. Midway between nozzles, the overlapping spray cones create a zone of turbulence in which spray goes in all directions. The flux density of 0.16 gpm/sq ft (6.6 L/min/m²) would likely have been sufficient, had the spray momentum been better aimed against the fire plume. The strong heptane fire plume could rise through the mist and deflect the rest of the spray cone.[16]

These results illustrate that extinguishing effectiveness of water mist can only be related to actual, not nominal, flux density. Many nozzles concentrate a high percentage of the water spray into the center of the cone area. This occurs because a negative pressure develops in the inner core of a hollow-cone spray, which draws droplets from the outer fringes into the center. Further distortion of actual from nominal flux densities occurs because of obstructions. Obstructions "scrub" water mist from suspension, drastically reducing the flux density. Obstructions also deflect the spray randomly, reducing its momentum, rendering it less effective.

Spray momentum: The difference between success and failure of a fire suppression test can often be attributed to variations in spray *momentum*. Three factors, i.e., (1) the spray velocity, (2) its direction relative to the fire plume, and (3) the mass of the water droplets transported into the flame or onto the fuel surface, constitute the momentum of the spray. Velocity is a vector quantity, which has both magnitude and direction. The more control that can be exercised over momentum, the greater will be the ability to control total water requirements, time to extinguishment, water damage, and overall system reliability.

The extinguishing capacity of water mist is the result of a complex interaction of fuel properties, enclosure effects, drop size distribution, flux density, and spray momentum. Of course, the velocity cannot be too great, or liquid fuels could be displaced, worsening fire conditions. The efficiency of a spray generation method is going to be highest when the maximum amount of the spray has the desired directional properties relative to the fire plume. Control over momentum, then, depends not only on the nozzle characteristics, but also on the degree to which nozzles can be strategically positioned around the fire.

One very important aspect of the selection of a mist generation method is the degree of control it allows over the "directionality" of the spray produced. If oriented strategically, narrow-cone angle sprays direct more of their flow toward the fire source than wide-angle spray cones. This was demonstrated in a series of tests conducted by NIST[28] and at NRCC, involving fires in computer cabinets.[29] The fire was located between two parallel printed circuit boards in an array of circuit boards, as illustrated in Figure 6-15I. The fires were extinguished only when the spray direction was close to parallel to the circuit boards. A single, wide-spray-angle nozzle mounted in an open space above the circuit boards could only extinguish fires in a narrow band on either side of the vertical axis of the cone, in which the spray could enter the narrow spaces between the circuit boards.[29] As the horizontal component of the spray increased toward the edges of the array, the spray could not penetrate the space between the circuit boards. Increasing the nozzle pressure and mass flow rate did not widen the effective zone: it simply increased its depth. When the wide-angle nozzle was replaced with a series of closely spaced, narrow spray-angle nozzles, the effective zone was widened to include the entire cabinet. By converting from a "fan-shaped" distribution of spray to a linear distribution of parallel sprays, it became possible to extinguish fires between any of the circuit boards with very little water.

The experiment described illustrates that controlling the "directionality" of the spray can be more important than controlling the drop size distribution or mass flow rate. Drop size distributions can vary by 100 microns (Dv0.9) or more, and be apparently equally effective. Most water supply systems deliver more water than is needed for the minimum performance of a single nozzle, so there is no penalty for being over-designed on flow rate. There is little to be gained by designing to narrow tolerances on those parameters. On the other hand, small modifications to the direction of application of the spray can mean the difference between success or failure to extinguish.

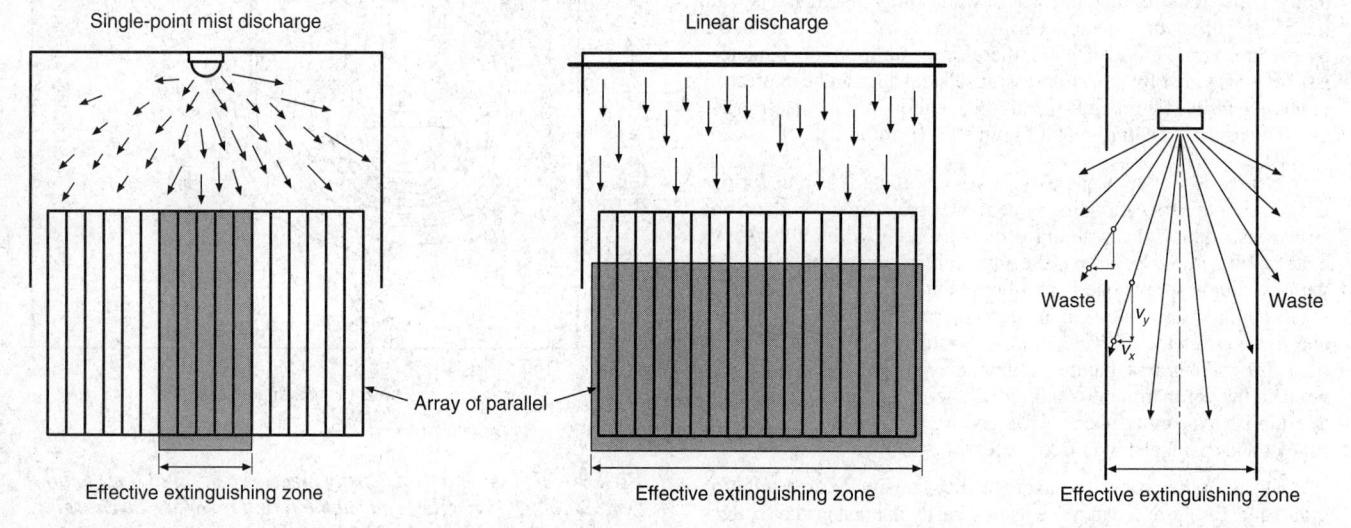

FIG. 6-15I. The effect of controlling spray directionality or momentum in extinguishing fires in printed circuit boards.[29]

Additives: The chemical nature of the water supplied to the nozzles influences the performance of water mist as a fire suppressant. Experiments at NRCC demonstrated that the extinguishing effectiveness of a water mist was enhanced by addition of sodium chloride to the water.[20] Mist made with "seawater" (2.5 percent by weight sodium chloride solution) extinguished diesel fuel pan fires at lower flux densities, and 40 to 50 percent faster, than mist made with fresh water. In addition, the seawater mist was less influenced by obstructions than fresh-water mist. One possible explanation for the improved performance is that dissolved alkali salts crystallize in the flame zone as drops evaporate. The salt crystals are small enough to be in the optimum effectiveness range of dry chemical extinguishing agents,[30] many of which are alkali salts. This hypothesis suggests that other alkali salts, which are better dry chemical fire suppressants than sodium chloride, could significantly improve the extinguishing effectiveness of water mist. Various products are already sold on the basis of their ability to enhance suppression effectiveness of water sprays.

Fire laboratories that have conducted water mist suppression testing report that addition of a low percentage of a film-forming agent (e.g., 0.3 percent AFFF) greatly improves the effectiveness of water mist on hydrocarbon pool fires. This is particularly useful for suppressing spill fires in bilge areas that are sheltered from the water spray.

The chemicals that could be added to the water supply for a water mist system include antifreeze, film-forming agents, Class A or B foaming agents, surfactants to increase penetration of water into solid fuels, fire-retardant salts, biocides to prevent algae growth in stored water, etc. The benefit provided by the additive may be offset by the introduction of secondary problems, e.g., such as increased corrosivity or increased toxicity. The tolerance of equipment for increased corrosivity must be evaluated by the property owner/insurer on the basis of degree of corrosivity and the acceptable level of loss. Toxicity concerns, on the other hand, invoke government regulation. At the time of the aviation industry studies on aircraft passenger cabin water mist systems,[2] the U.S. Surgeon General expressed concerns about the effects of inhalation of fine water spray on people. Those concerns were raised also by the U.S. Environmental Protection Agency (EPA) when water mist was being evaluated under the Significant New Alternatives Policy Program (SNAP) as a halon replacement. In response to the EPA concerns, the Halon Alternatives Research Committee (HARC), with the participation of the NFPA 750 committee, convened a health panel of medical experts to review data on this potential hazard.

The HARC Health Panel Report[31] concluded that water mist supplied by water, equivalent in quality to potable water (i.e., containing no additives other than those used in drinking water), or by "natural seawater," poses no health risk to persons exposed to it, either under circumstances of a fire, or due to nonfire actuation. The EPA found the panel's findings were credible and adopted its conclusions as the basis to a ruling. The resultant U.S. Federal Register regulation,[32] therefore, recognizes that water mist systems supplied by potable water or natural seawater are "acceptable without restriction," as both a Halon 1301 substitute in occupied spaces, and as a streaming agent to replace Halon 1211.

Water mist systems with additives are of interest for applications such as the engine compartments of a variety of vehicles and in machinery spaces. The EPA has accepted the use of water mist combined with a small percentage of a surfactant (water mist/Surfactant Blend A) for such applications in unoccupied areas.[32] Similarly, other additives, such as antifreeze, foaming agents, and fire-retardant salts, will be limited to unoccupied spaces unless the proponent provides EPA with the results of a medical peer review panel study addressing the toxicity question.

GENERIC DESIGN PRINCIPLES OF WATER MIST SYSTEMS

It is not yet possible to design a water mist system entirely from generic, or "first," principles. The 1996 edition of NFPA 750, for example, requires that the performance of proposed designs be verified by full-scale fire testing conducted by internationally recognized fire testing laboratories. Unfortunately, the expense of a "design by testing" approach inhibits the development of water mist systems. Ultimately, the designer would like to rely on engineering knowledge about how the different influencing factors relate, in order to select system features most appropriate for the particular hazard and control objectives. Such an approach constitutes "design by first principles."

The present level of understanding of the interaction of the water mist with fire is sufficient to permit a fire protection engineer to evaluate the potential for its application in a given situation. As has already been stated, the relationship between fuel type, fire scenario, spray drop size, spray momentum, flux density, and performance is complex. The controlling variables can be combined in different proportions to achieve the same objective. Designing a cost-effective fire suppression system using those relationships constitutes a multivariable design problem. Figure 6-15J is a decision tree that presents the multiple variables that affect the ability of a

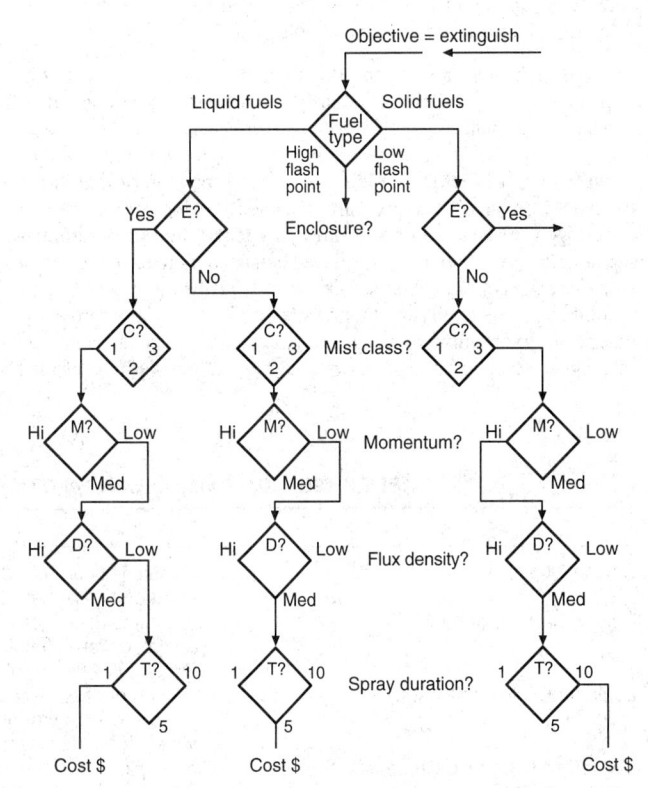

Water Mist System Design Decision Process
Objective = Extinguish Branch

Legend:
E = Enclosure conditions.
C = Class of mist; 1, 2, or 3.
M = Momentum; high, medium, or low.
D = Flux density; high, medium, or low
T = Duration time; e.g., 1, 5, or 10 min.

FIG. 6-15J. *Conceptual decision tree for a generic approach to design of a water mist system.*[33]

water mist system to achieve the fire safety objective.[33] It leads the designer conceptually through consecutive steps, such that each choice affects the options on the next decision. As each condition or characteristic is selected or decided, the subsequent choices change. Examples of considerations invoked at each decision or selection point are described in Table 6-15B.

METHODS OF GENERATING WATER MIST

General

Methods to generate water mist range from simple to elaborate. Some of the more elaborate methods include high-speed spinning discs, ultra-sonic vibrations, flashing and re-condensation of super-heated liquids, rapid release of dissolved gases, and explosives. For fire suppression purposes, the choice of method is narrowed by the facts that the mass flow rates and velocities needed to be effective are beyond the capacity of some methods. The methods that are practical for fire suppression systems are those that break up water into drops smaller than 1,000 microns, at mass flow rates and spray velocities appropriate for full-scale fire scenarios, which can be protected against plugging and which reliably sustain that flow as long as is required. The economics of providing the stored energy, or installing special pumps and piping systems, are also determinants in selection of a mist-generating technology for a given application.

Water mist systems technology is currently in a state of constant innovation. This is particularly evident in the mist generation concepts that continue to emerge. In general, water mist nozzle designs involve one of three basic principles: (1) impingement of a jet of water on a deflector (impingement), (2) expulsion of a high-velocity jet from an orifice (pressure jet), and (3) using compressed air or nitrogen to shear water into fine spray (twin-fluid or air-atomizing nozzles). All three methods have been used by manufacturers of spray nozzles for many years. The innovations being made by manufacturers interested in fire suppression are intended to improve efficiency or to optimize some spray characteristics such as mass flow rate, spray velocity, drop size, or cone shape. New methods of mist generation operate on different principles or combinations of the three basic principles. Examples of new methods will be described briefly under the heading of "Innovations in Mist Generation."

There is a common misconception that all water mist systems require "high" pressures. One source of this impression is the fact that some mist-generating technologies, which appeared early in the development period for fire suppression, require pressures near 1,000 psi (68 bar), and even as high as 3,000 psi (204 bar). Water mist-generating technologies have long been available that work at low pressures, however. Most impingement nozzles operate at much lower pressures, certainly below 500 psi (34 bar), and more often in the range of 150 psi (10 bar) or less. Twin-fluid nozzles typically operate at 60 psi (4 bar) liquid pressure, with 75 psi (5 bar) compressed gas pressure. In some cases, high pressure is an advantage in that the energy converts to high spray momentum (which enhances suppression effectiveness), or allows delivery of the water through small-diameter piping. The advantages must be weighed against the cost of providing the energy for high operating pressures, however. There is economic incentive not to have pressures higher than are necessary to produce an effective water mist fire suppression system.

NFPA 750 distinguishes among the pressure regimes at which mist-generating technologies operate by introducing terms for low-, intermediate-, and high-pressure systems. "Low-pressure" systems operate at pressures of 175 psi (12 bar), or less; "intermediate pressure" systems involve pressures greater than 175 psi (12 bar), but less than 500 psi (34 bar); and "high-pressure" systems operate at pressures of 500 psi (34 bar) or greater. The operating range of *low-pressure systems* is similar to standard fire protection systems, such as sprinklers and standpipes. The requirements for pipe, fittings, valves, pumps, or hydraulic calculations can be met by conventional materials and installation practices. Pipe, fittings, and valves for *intermediate-pressure systems* are also commonly available, although there may be no "listed" centrifugal fire pumps that produce pressures in the range of 500 psi. Therefore, there are some aspects of *intermediate-pressure systems* that lie outside of standard fire protection engineering experience. Special problems are invoked with *high-pressure systems*, for pipe, fittings, pumps, valves, pressure regulators, cylinders, and tanks. The technology of *high-pressure systems*, although new to fire protection engineering practice, is

TABLE 6-15B. *Decisions Involved in Designing a Water Mist System from First Principles*

Decision	Considerations
Decide Objective	Extinguish or control? Control temperatures, control fire spread, control smoke. The choice of mist characteristics and spray duration could vary greatly, depending on the objective.
Determine Type of Fuel	Solid or liquid, high or low flash point, the degree of combustibility determines extinguishing mechanism; governs (choice of) drop size, flux density, and momentum; and suggests limits on spacing and nozzle orientation.
Evaluate Enclosure Conditions	Enclosed or open, well ventilated, poorly ventilated. Full compartment or local application. If enclosure effects can be relied upon, flux density and spray momentum can be lower than otherwise. If not, nozzles must be located differently, momentum increased.
Determine Drop Size Characteristics (from listing)	Class 1, 2, or 3 drop size distribution; may be pre-determined by choice of mist-generating method.
Determine Momentum Characteristics (from listing)	High to low; may be pre-determined by choice of mist-generating method; restrictions on distance from nozzles to fuel; type of fuel (liquid or solid). May define type of nozzles to use.
Decide Minimum Flux Density, Average Density, Peak Density	High to low; uniform distribution or irregular; group activation *vs.* automatic nozzles. This determines selection of nozzle cone angle, the water flow rate, the overall water storage requirements, and anticipated problems with water damage.
Decide Spray Duration	Long, short, cycled, Limited by cost of water/air supplies. Concerns about water damage, extinguishing mechanisms, re-ignition, volume of water storage, and weight of system.
Choose System Activation Concept	Determine type of nozzles, e.g., individually thermally activated (automatic); separate detection system and control panel; difficulties with zoning; choice of signal for activation, etc.
Calculate System Costs and Address Reliability Concerns	Based on choice of mist generation technology; minimum flux density and duration; volume of storage; degree of control over ventilation, etc.

well developed in other industries, such as the hydraulics and off-shore petroleum drilling industries.

The *method* of generating mist does not uniquely determine suppression capability. Matching the spray characteristics of drop size distribution, flux density and spray momentum, to the fire hazard, is more important. The choice of mist-generating method does influence factors, such as the cost-effectiveness and reliability of a system.

Water mist produced by groups of nozzles differs from the spray produced by a single nozzle. The turbulent interaction of overlapping spray cones changes the drop size distribution. The characteristics of the individual nozzle spray cones, combined with the spacing between nozzles, determines the flux density distribution in the compartment. The coverage area and spacing of nozzles must conform to the manufacturers' installation guidelines provided with the listing information. It is the intent of NFPA 750 that nozzle listing information be based on full-scale fire suppression tests. It is further intended that the conditions of application be identical to the conditions of the tests conducted to obtain the nozzle listing.

Table 6-15C provides a matrix for organizing spray generation concepts according to the mist-generating principle and operating pressure regime. Some, not all, manufacturers that produce nozzles representative of the type indicated, are named. Details are provided in the following discussion.

Impingement nozzles: Impingement nozzles operate on the principle of a solid jet of water striking a deflector and breaking up into small drops. They include standard sprinklers, nozzles used in "traditional" (NFPA 15) water spray and deluge systems, and nozzles used for a myriad of industrial applications. The jet velocity and the shape of the impingement surface determine the cone angle, drop size distribution, flux uniformity, and spray momentum. Figure 6-15K shows several types of impingement-style nozzles used in water mist systems.

Impingement nozzles operating at low pressures are simple in design and therefore cost less to construct than nozzles that require more precise machining. In 1992, NRCC[34] measured the spray characteristics of several commercially available impingement nozzles, including sprinkler-type deflector plates, spiral deflectors, and pin-type deflectors, operating at pressures less than 175 psi (12 bar). (See Figure 6-15K.) These nozzles produced Class 2 and 3 sprays, with cone angles between 60 degrees and 120 degrees. The size and shape of the deflector determined the cone angle. The tests revealed that impingement nozzles have a limited axial spray projection distance, because of the energy lost as the jet strikes the deflector. A slight tilt in the deflector plate will skew the spray distribution in one direction. Deflector supports can cause shadow effects, and irregularities in the impingement surface can magnify to large irregularities in the spray distribution. These sprays controlled or extinguished fires, not only with Class A combustibles, but also in

TABLE 6-15C. *Methods of Generating Water Mist for Fire Suppression Systems*

System Type	Nozzle Type		
	Impingement	Pressure Jet (swirl type)	Twin-Fluid
Low Pressure Up to 175 psi (12 bar)	Large-diameter orifice plus deflector: standard sprinklers, spiral type (BFN, SSC), pin type (BFN, SSC), AquaMist (Grinnell), others	Small-orifice, swirl chamber, multijet assemblies: SSC, BFN, many varieties	Air inlet plus water inlet plus internal chamber: STI, Ginge Kerr, BP, SSC, BFN, others.
Intermediate Pressure >175 to <500 psi (> 12 to < 34 bar)	Small-diameter orifice plus plate-type deflector: AquaMist (Grinnell), GW Sprinkler (Denmark)	SSC, BFN, Tyco, MicroDrop	
High Pressure > 500 to 4000 psi (>34 to 272 bar)		Marioff Hy-fog, Semco, Unitor, Reliable-Baumac, others	
Innovations			
Flashing Jets Low pressure, High temperature 175°C		Jet of super-heated water plus simple orifice: dust explosion suppression (Ireland)	Super-heated water plus vapor injection into internal chamber (experimental)
Impinging Jets Intermediate pressure	Group of pressure jets oriented to collide in high-velocity zone, causing mechanical breakup (Kidde-Fenwall)		
Nucleated Agent High pressure		Pressure-jet with compressed nitrogen pre-mixed in water (Marioff)	
Impulse Spray High Velocity		Gas-generator: explosive expulsion of water through orifice or slot; pressure generated by gas-generator devices used for air bags.	IFEX 3000 Impulse - (handheld): trigger release of high-pressure gas into water-filled chamber
Linear Sprays To provide maximum control over spray direction and momentum		Individual closely spaced nozzles producing narrow cone jets, which direct mist in a preferred direction, rather than in a 360° cone	

NOTES: SSC = Spraying Systems Company; BFN = Bete Fog Nozzle Co.; STI = Securiplex Technologies, Inc.

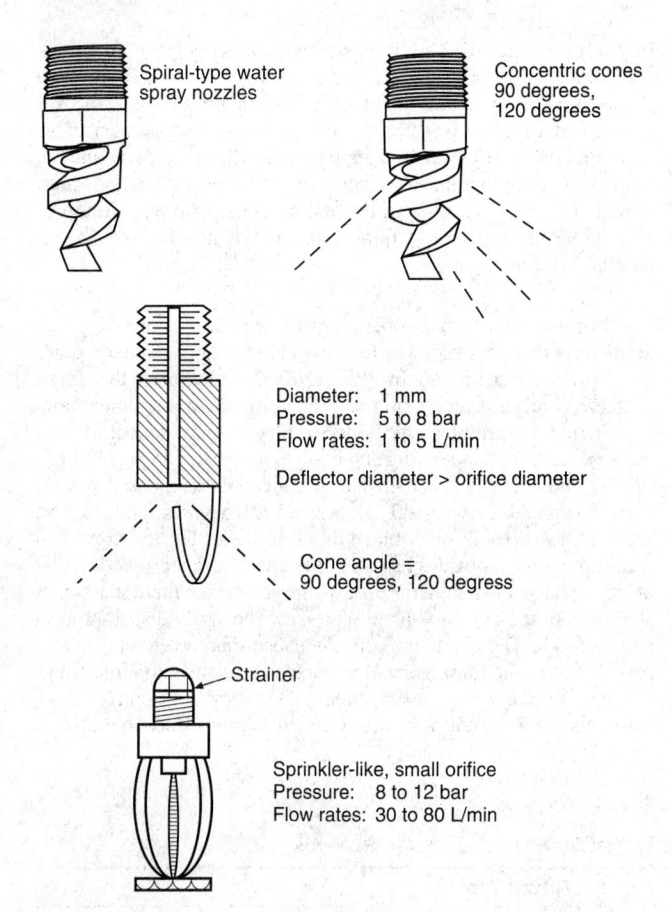

Spiral-type water spray nozzles

Concentric cones 90 degrees, 120 degrees

Diameter: 1 mm
Pressure: 5 to 8 bar
Flow rates: 1 to 5 L/min

Deflector diameter > orifice diameter

Cone angle = 90 degrees, 120 degress

Strainer

Sprinkler-like, small orifice
Pressure: 8 to 12 bar
Flow rates: 30 to 80 L/min

FIG. 6-15K. Several types of impingement nozzles for fine sprays.

liquid fuel fires in enclosures, where enclosure effects made spray momentum less critical. Overall, the reduced forward spray momentum and irregular flux distribution, which were typical of the impingement nozzles tested, were limitations on their potential use in a water mist system for large machinery spaces. Nevertheless, several sprinkler companies are manufacturing an impingement-type, low-pressure water mist nozzle for use in ship cabins and crew areas, where they have been shown to be effective against Class A fires (ordinary combustibles).

The relationship between nozzle pressure and discharge rate for impingement nozzles is reliably calculated using the expression familiar to sprinkler system designers: $q = k\sqrt{P}$ where q is the discharge rate, P is the gage pressure at the orifice, and k is a coefficient characterizing the nozzle.

Pressure jet nozzles: Pressure jet nozzles work by pushing a jet of water through a relatively small-diameter orifice at a high velocity. As the jet leaves the orifice, complex physical events occur, involving the fluid viscosity, and the jet diameter and velocity relative to the air.[26] The break-up of the thin jet of water into droplets begins within a few millimeters of the orifice. Figure 6-15L illustrates several pressure jet nozzles. The disintegration of the jet into droplets can be enhanced by inserting swirl devices in a chamber inside the nozzle, to impart a rotation to the jet as it emerges from the orifice. If water is pushed through a narrow slot instead of a circular orifice, the jet takes the form of a sheet, which becomes unstable and disintegrates into droplets. In a study conducted at NRCC on a selection of commercially available pressure jet nozzles used for water mist,[34] orifice diameters ranged from 0.008 in. to 0.125 in. (0.2 mm

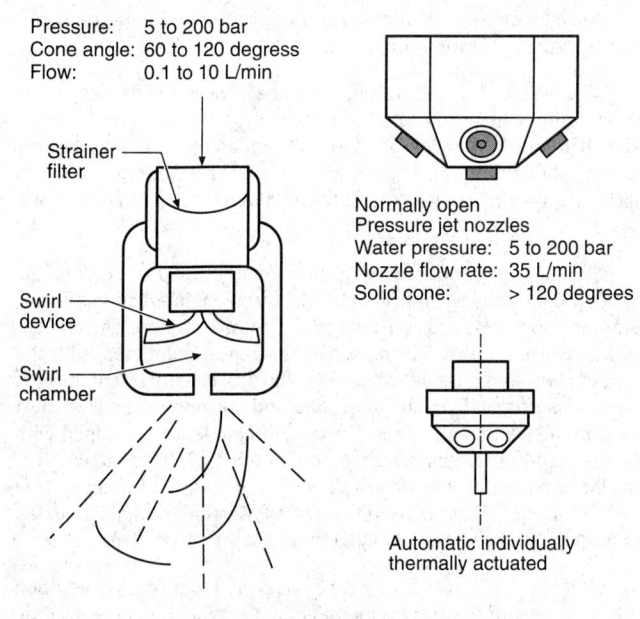

Pressure: 5 to 200 bar
Cone angle: 60 to 120 degress
Flow: 0.1 to 10 L/min

Strainer filter

Swirl device

Swirl chamber

Normally open
Pressure jet nozzles
Water pressure: 5 to 200 bar
Nozzle flow rate: 35 L/min
Solid cone: > 120 degrees

Automatic individually thermally actuated

FIG. 6-15L. Pressure jet nozzles.

to 3 mm). The mass flow rates from the nozzles ranged from about 0.25 gpm (1 L/min) for single-orifice nozzles, to 12 gpm(45 L/min) for a multiorifice assembly. Spray cone angles ranged from less than 20 degrees to 150 degrees.

Individual pressure jet orifices can be assembled in groups on a single base, or at intervals along a pipe (linear arrangement). Where several pressure jets are assembled in close proximity on a single base, the combined effect of all jets entrains air into the composite spray cone. The diameter of the fully developed composite cone is not a simple function of initial jet velocity and direction. The trajectory of drops from the individual jets is influenced by the pressure regime within the spray cone, which depends on the manner in which air is drawn toward the nozzle and entrained into the cone. With some nozzles, the spray cone may not become fully developed if free air movement to the back of the nozzle is restricted.

Pressure jet nozzles operate over a wide range of pressures, from 75 psi to 4000 psi (12 bar to 272 bar). The small diameter of the orifice, internal details, e.g., the shape of the swirl chamber, and the presence of filters on each orifice dictate that pressure jet nozzles require higher operating pressures than impingement nozzles. The drop size distribution typically becomes finer as pressure increases. Often there is an upper limit, at which further increase in pressure has little effect on the drop size distribution. There may be incremental increase in mass flow rate or momentum, but with no significant improvement in the coverage area of the spray cone. NFPA 750 assumes that the listing agency will provide information about the operating pressure range needed to produce the desired spray qualities.

The relationship between pressure and flow from multi-orifice pressure jet nozzles may not follow the simple $q = k\sqrt{P}$ formula used for single-orifice nozzles, such as sprinklers. It is possible that external factors, such as the mutual influence of adjacent jets in a multijet assembly, and air entrainment into the composite spray cone, will affect the discharge rate. Again, the listing information for the nozzle must provide information on how to calculate discharge rates as a function of pressure from pressure jet nozzles.

The Royal Navy in the U.K. investigated the use of a commercially available pressure jet nozzle (SSC 3/4 7G5) in a low-pressure "water fog" system on submarines. At 70 psi (4.8 bar), this nozzle produces a Class 2 spray and discharges approximately 10 gpm (38 L/min) with a cone angle of about 150 degrees. The Marioff Hy-fog company of Finland has pioneered the development of high-pressure [4000 psi (272 bar)] systems using pressure jet nozzles with flow rates ranging between 5 and 10 gpm (19 and 38 L/min) and higher. A very low flow rate [0.25 gpm (1 L/min)] high-pressure jet nozzle [1000 psi (68 bar)], adapted from greenhouse misting systems, is under development by the Reliable Automatic Sprinkler Company. This nozzle has a very low flow rate and momentum, but can be installed in groups or at intervals along a tube. The low flow rate per nozzle allows the use of small-diameter tube.

Pressure jets can be aimed to collide with other jets, causing a physical disintegration of impinging streams. Kidde-Fenwal has developed a nozzle, based on this principle of impinging pressure jets, that operates in an intermediate pressure range. Further innovations in the design of pressure jet nozzles for fire suppression applications are expected.

Twin-fluid nozzles: Twin-fluid, also called air-atomizing nozzles generate water mist by injecting water and a compressed gaseous atomizing medium into a chamber in such a way that the compressed gas shears the water into fine droplets, and discharges the resulting mist through one or more orifices which dictate the shape of the spray cone. Twin-fluid nozzles allow for efficient control over drop-size distribution, cone angle, spray momentum and discharge rate, at water and compressed gas pressures between 50 and 100 psi (3 and 7 bar) (low-pressure regime). They are available with cone angles between 20 and 120°C. Nozzle manufacturers have produced twin-fluid nozzles for many years for applications such as agricultural spraying, building humidification systems, and paint spraying. Innovation in twin-fluid nozzle design by British Petroleum (BP) Ventures in the 1980s introduced an efficient nozzle for fire suppression use. The BP design is licensed to Securiplex Technologies, Inc., of Canada, and is used in the Securiplex Fire-Scope 2000 Fine Water Spray System, which has been listed by FMRC for turbine enclosure water mist systems. Figure 6-15M illustrates two designs for twin-fluid nozzles for generating water mist. Other types of twin-fluid water mist nozzle designs are under development.

The reliability of twin-fluid nozzles is well established by their widespread use in industrial spray systems. Relatively large-diameter orifices (>2 mm) are less likely to plug than smaller orifices in pressure jet nozzles. Both water and atomizing medium lines operate in the low-pressure range, so that commonly available pipe fittings and valves can be used. Their primary disadvantage for use in fire suppression systems, which require installation of many nozzles, is the cost of installing two supply lines to each nozzle, plus storing a sufficient quantity of the pressurized atomizing media. The pneumatic controls to regulate the balance of air and water pressures introduce additional complexity.

The relationship between water discharge and atomizing medium (air) discharge from a twin-fluid nozzle cannot be characterized by the $q = k\sqrt{P}$ expression used for simple orifice nozzles. Because the air and water lines are joined (inside the nozzle), the pressure in one line works against the pressure in the other line. As air pressure increases, the air flow increases, which, in turn, forces a reduction in the water flow and modifies the water system pressure. Conversely, if the water pressure is increased, it reduces the airflow—if too high, water will flow through the nozzle into the air line. If the atomizing medium is restricted too much, the mist will not form, and the drop size distribution will be very poor.

Figure 6-15N shows a plot of atomizing medium and water flow rates for a commercial twin-fluid nozzle. The term "air to liquid ratio" (ALR) is used to describe the balance of atomizing medium and liquid demand across the nozzle. To adjust a system in the field, and

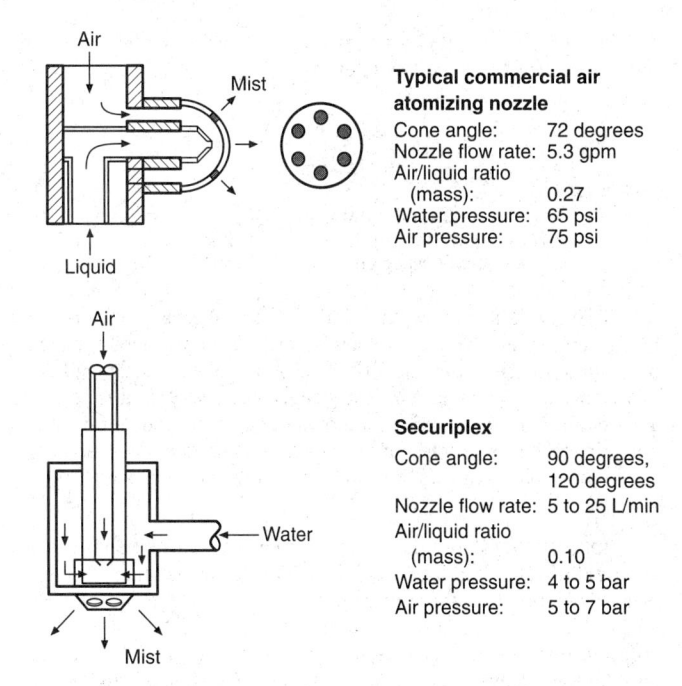

Typical commercial air atomizing nozzle
Cone angle:	72 degrees
Nozzle flow rate:	5.3 gpm
Air/liquid ratio (mass):	0.27
Water pressure:	65 psi
Air pressure:	75 psi

Securiplex
Cone angle:	90 degrees, 120 degrees
Nozzle flow rate:	5 to 25 L/min
Air/liquid ratio (mass):	0.10
Water pressure:	4 to 5 bar
Air pressure:	5 to 7 bar

FIG. 6-15M. *Two designs for air-atomizing nozzles (twin-fluid).*

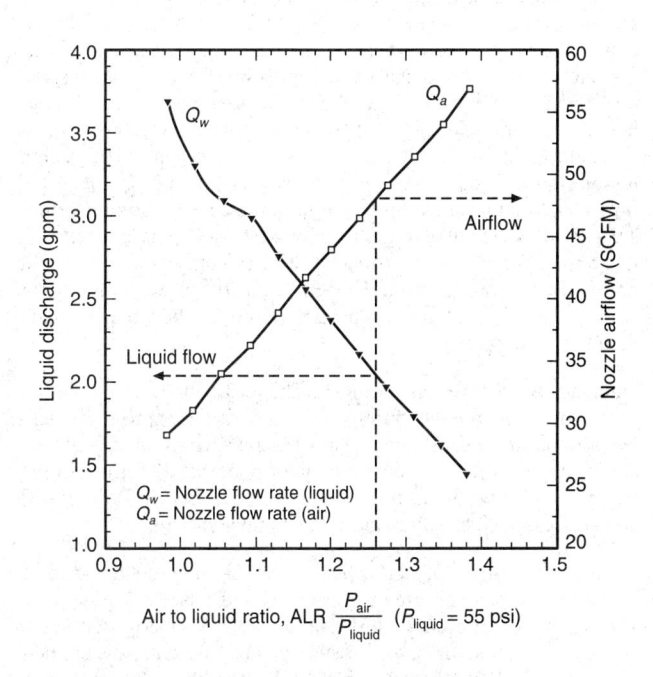

FIG. 6-15N. *Air and water flow rates in twin-fluid nozzles.*

to determine air and water discharge rates from manufacturers' data sheets, it is convenient to use the air-to-liquid *pressure* ratio, ALR_p.

$$ALR_p = \frac{P_a}{P_w}$$

where,
ALR_p = Air-to-liquid pressure ratio;
P_a = Air pressure, gage, psi (bar);
P_w = Water pressure, gage, psi (bar).

For comparing the efficiencies of different nozzles, and performing pneumatic system calculations to determine pipe sizes, pressure regulator settings, and the total quantity of compressed gas required for a system, it is better to work with air-to-liquid ratio expressed in terms of mass flow rates, ALR_m.

$$ALR_m = \frac{M_a}{M_w}$$

where,
ALR_m = Air-to-liquid mass ratio;
M_a = Mass of air per unit time, lb/min (kg/s); and
M_w = Mass of water per unit time, lb/min (kg/s).

The nozzle manufacturer should identify the range of air-to-liquid ratios (pressure and mass) over which the nozzle achieves the optimum spray characteristics. The total compressed-gas demand of many nozzles in an open-nozzle system can be costly, so it is desirable to minimize the demand for atomizing medium. An "efficient" operating point for a twin-fluid nozzle is one that produces the desired spray characteristics at as low an ALR_m, as possible, e.g., between 0.05 and 0.10.

Table 6-15D shows the results of measurements taken by the author at the National Research Council of Canada on a commercially available, low pressure twin-fluid nozzle, operating at $P_w =$ 55 psi (3.7 bar), and three different air pressures. Increased air pressure resulted in a reduction in the Dv0.5 and Dv0.9 values, but all three sprays remained a Class 1 mist (^{1}Dv0.9 ≤ 200μ). For fire suppression purposes, the benefit of the additional "fineness" was not measurable. Reducing the air pressure reduced the air demand from 56.1 to 34.8 SCFM, and increased the water flow rate by a factor of 2, from 1.9 to 4.1 gpm (7.2 to 15.5 L/min). The increased water flow rate translates to increased flux density. Testing showed that flux density had a greater effect on suppression effectiveness than the change of a few tens of microns in the fineness of the spray. The reduced demand for atomizing medium was a benefit, because it reduced the cost of the system. The design lesson, then, is that, in many circumstances, one can reduce the flow of atomizing medium to a minimum, tolerate some overdesign on the water flow, but still have adequate spray characteristics and suppression performance. One circumstance that might not fit that approach would be an aircraft system, where, for weight reasons, it is more important to reduce the water demand than the air demand.

Innovations in mist generation: The term "innovation" is used here to describe modifications that are intended to improve efficiency, or to optimize some spray characteristic such as mass flow rate, spray velocity, drop size distribution, or cone shape, for use in fire suppression applications. Several of these are shown in Table 6-15C and are described briefly in the following paragraphs.

Flashing of super-heated water. The "flashing of super-heated liquid" phenomenon is usually associated with gaseous fire suppression agents, but it occurs with water, as well. A superheated liquid is one that is maintained, by pressure, in its liquid state, at a temperature above its boiling point. When released suddenly from the pressurized container, the liquid partly "flashes" to vapor, and partly disintegrates into spray by the shearing effect of rapidly expanding

bubbles.[36,37] The pressure in the closed container is determined by the vapor pressure of the liquid at the given temperature. For water, standard steam tables indicate that, at 348°F (176°C), the gage pressure inside the container will be 118 psi (8 bar). This self-generated pressure is enough to discharge the water from the cylinder. The storage cylinder can be further pressurized by addition of compressed nitrogen, if necessary.

Once released to atmospheric pressure through a simple orifice, super-heated water boils instantaneously. Tiny gas bubbles form, expand, then "explode" in a powerful release of energy that fractures still-liquid water to small droplets. The violence of the bubbles coming out of the liquid can be loosely compared to an "explosion in reverse." The rapid release of energy propels the mist throughout the compartment. The percentage of the water that flashes to vapor can be estimated from standard steam tables and ideally could be as high as 14 percent at 348°F (176°C). However, the flashed water vapor promptly cools to much lower temperatures, and condenses, forming an opaque fog of aerosol-sized, (20 micron) water drops. These ultra-fine drops are the result of condensation, not the physical break-up of liquid water by mechanical energy. Within the cloud of white fog, the portion of the liquid that does not flash to vapor is distributed as a "rainout" of Class 1 mist (^{1}Dv0.9 ≤ 200 microns).[29,38,39]

In open areas, where ambient air can entrain freely with the spray, the energy released in the flashing process is sufficient to drop the temperature of the spray to a safe 35°C, within 4 in. (102 mm) of the orifice. If discharged into a small space, such as a switchgear cabinet, however, temperatures can rise to the point where scalding could occur.

The rapid generation of mist by the flashing process has been demonstrated to be effective in quenching dust explosions.[39] For hydrocarbon pool fires in enclosures, tests conducted at NRCC[29] showed that the spray was no more effective than other Class 1 sprays. Although the ultra-fine particles were distributed widely throughout the compartment, the zone of effective extinguishment coincided with the distribution region of the Class 1 spray within the cloud. The ultra-fine fraction contained insufficient mass of water to extinguish a hydrocarbon pool or a fire in bundled cables.

Impinging jets. Impinging jets combine the principles of pressure jets and impingement nozzles. Kidde-Fenwall has developed a nozzle that orients a group of pressure jets so that they collide obliquely several centimeters from the nozzle tip. By eliminating the fixed impingement surface, "shadow effects" that leave gaps in the distribution of spray are eliminated. The nozzle operates in the intermediate pressure range.

Nucleated, high-pressure spray. The Marioff Hi-fog Company of Finland has pioneered many innovations in high-pressure water mist technology. One Marioff innovation involves injecting nitrogen into the water line, so that nitrogen and water mix thoroughly before reaching the nozzle. It is reported that, when discharged from the Marioff pressure-jet nozzle, the resulting spray has properties that enhance its fire suppression effectiveness.

TABLE 6-15D. *The Effect of Varying Air Pressure on the Discharge Characteristics of a Twin-Fluid Nozzle*

	P_w (psi)	P_a (psi)	Q_w (gpm)	M_w (lb/min.)	Q_{air} (SCFM)	M_{air} (lb/min.)	ALR_m (mass)	Dv0.5* (Micron)	Dv0.9* (Micron)	Class of Mist
Condition 1	55	76	1.9	15.8	56.1	4.29	0.27	60	140	1
Condition 2	55	65	3.2	26.5	44.2	3.38	0.13	75	160	1
Condition 3	55	55	4.1	34.0	34.8	2.66	0.08	90	160	1

*Drop size distribution measured by forward light-scattering laser instrument (Malvern), on spray centerline, 26 in. (0.66 m) below vertically oriented nozzle.

Impulse spray. The IFEX GmbH company of Germany has developed the IFEX 3000 Impulse spray generation method. It is based on the explosively rapid injection of air into a cylindrical, open-ended chamber that has been pre-filled with water. A "slug" of water is formed and is propelled for several yards (meters) from the end of the discharge tube. Largely due to the power of the delivery, i.e., the high impact velocity of the spray, extinguishment of Class A combustibles can be achieved using much less water than would be required using conventional hose streams. Its suitability for use in a fixed water mist fire suppression system has not yet been explored.

Linear sprays. Water mist nozzles typically discharge water in a 360° cone, sending mist out in all directions. In order to minimize the number of nozzles needed to protect a space, the usual design approach is to try to fill as large a volume as possible from a few point sources. Some fuel arrays, however, are in linear geometries, such that the mist will be most effective if discharged in one direction—water discharged in other directions is wasted. Examples of linear fuel arrangements include fixed library shelving, cable trays, and arrays of circuitboards in switchgear frames. For these applications, control over the direction of spray discharge can be obtained by using linear sprays.

A linear spray is generated by installing narrow cone spray jets at close spacing along the length of a pipe. Spray jets can be customized as to discharge velocity, discharge rate, and drop size distributions, according to the needs of the scenario.

Automatic, Nonautomatic, and Hybrid Nozzles

NFPA 750 introduces the terms *automatic, nonautomatic,* and *hybrid nozzles* to distinguish between three types of nozzles that are already on the market. Automatic nozzles are normally closed but open automatically when exposed to a fire. Nonautomatic nozzles are normally open and deliver water in groups when a valve controlling water entry to the piping is activated. Hybrid nozzles are normally closed but can be opened individually, automatically by local conditions, or in groups, remotely by an external action.

Automatic nozzles operate independently of other nozzles by means of a detection/activation device built into each nozzle. The sprinkler-like, low-pressure impingement nozzles, used in shipboard crews' quarters and corridor systems, fall into this category. A heat-activated link or bulb is built into the nozzle which, when exposed to heat, releases and allows discharge from that single nozzle, in exactly the same manner as a sprinkler.

Flammable liquid spill and pool fires can seldom be extinguished by a single nozzle operating on one edge of the fire. The machinery room fire scenario requires that mist be generated from an array of nozzles operating simultaneously in all parts of the fire zone. The term *nonautomatic nozzle* describes open-orifice nozzles that operate as a grouping, similar to a deluge system or traditional NFPA 15 water spray system. Water flows to these nozzles only when an independent system indicates there is a fire, and the detection system control panel actuates a valve controlling water flow to the piping system. The design of the independent actuation system, including selection of heat, smoke, or other fire signatures as the condition requiring activation, becomes an integral part of the water mist system design. The overall system operates "automatically," meaning without manual intervention, but the nozzles themselves do not automatically respond to the presence of fire.

Water mist produced by groups of nozzles differs from the spray produced by a single nozzle. The spacing between nozzles determines the uniformity of the spray flux density distribution, as well as the velocity patterns and mixing dynamics. The required spacing and orientation of nozzles must be described in the manufacturers; design manual. That information must be based on the results of full-scale fire suppression tests.

Hybrid nozzles operate using a combination of normally-closed nozzles with automatic activation of groups of nozzles. These nozzles contain built-in activation devices that can be automatically or remotely activated by a detection/control system. The Marioff Company has experimented with a high-pressure nozzle with a thermally activated release device, with a secondary device that allows other nozzles to be activated. Once one nozzle is activated by heat, it automatically causes the activation of immediately adjacent nozzles as a mini-group. Nozzles not sent a release signal from the first nozzle remain closed, unless heat spreads to other areas. Hybrid nozzles can be electrically released in some cases; one idea uses an electric signal to cause a heating element, wrapped around the glass-bulb thermal sensor on the nozzle, to heat the glass bulb and break it, as if it had been exposed to hot gases. Trials have been done using a hydraulic pilot line connection between pairs of nozzles, so that activation of the first nozzle in a compartment will automatically release a second nozzle located outside the door in the corridor. Other design concepts incorporate a thermally activated "group control valve" that opens to direct water to pre-determined arrays of normally open nozzles.

It is the need for activation of groups of nozzles, rather than individual nozzles, that is driving the need for innovative control concepts for water mist systems. At this early stage in water mist systems development, proposed solutions to system activation problems arise as fast as new situations are encountered.

Reliability Considerations

Fire protection engineers prefer to minimize the number of potential modes of failure of automatic fire protection systems. All of the mist-generating methods described above depart in some way from the "keep it simple" philosophy of traditional fire suppression system design. The mist-generating technologies involve smaller orifices, smaller diameter pipe, higher pressures, higher power requirements, and more complex valving and pressure-control arrangements than were considered feasible or reliable in sprinkler or water spray systems design. The incentive to accept higher complexity in water mist systems design is the ability to use water in place of other, environmentally harmful, fire suppression agents, where water was never used before. However, the potential benefits will only be realizable if the reliability of the innovative systems can approach existing fire suppression system performance expectations. Steps must be taken to counter the additional potential failure modes for these systems. Table 6-15E illustrates some of the areas of concern.

Choice of Mist Generation Method

It is not possible to say that one mist-generation method is better than another, with respect to fire suppression. High-pressure nozzles are an advantage in some circumstances, because they impart momentum and have an acceptable discharge rate and drop size distribution. The cost of the higher discharge energy can be traded off against increased nozzle spacing, a lower flux density, and hence, lower water demand. Equivalent suppression performance could be achieved at lower pressures by another method, perhaps with a higher water demand, but at lower total cost and higher reliability. Ultimately, the "best" choice of mist-generation method will be determined by: the cost of providing the pumping capacity, the water and energy storage to meet the minimum duration of discharge, filtration requirements, compressed gas storage, and the confidence of owners and Authorities Having Jurisdiction about overall reliability. Unfortunately, manufacturers of particular equipment are committed to their particular technology. They are motivated by that

TABLE 6-15E. *Potential Failure Concerns and Suggested Countermeasures to Improve Reliability in Water Mist Systems*

Potential Failure Concern	Examples of Countermeasures to Improve Reliability
Plugging of small orifices in nozzles, and small-diameter passages in hydraulic and pneumatic valves	Adopt filtration technologies and standards from hydraulics industry. Set water quality standards.
Start-up of high-pressure, electrically driven, positive-displacement pumps	Add requirements to NFPA 20 for positive displacement pump controllers and power supplies.
Use of solenoid valves in system controls	Provide fail-safe valves, emergency power supplies, hydraulic over-ride. Adopt standards from process control technologies.
Limited storage volume and rapid depletion of stored-pressure cylinders	Require backup arrangements, secondary sources.
Failure of pressure-regulating valves in twin-fluid systems	Adopt standards from process control technologies in other industries. Upgrade testing for listing procedures for pressure regulating valves.
Failure of pilot lines and pneumatic, hydraulic, or electrical interconnections between hybrid nozzles	Dictate installation details in installation standard.

preference to apply their particular technology to every potential application. This situation places a great deal of responsibility on the fire protection engineer to understand the pros and cons of all available technologies, in order to select one that is appropriate for the application in question.

DESIGN CALCULATIONS FOR WATER MIST SYSTEMS

Calculation Fundamentals

The calculation procedures for water mist systems range from fundamental methods for single-fluid, low-pressure systems to somewhat more complex methods for twin-fluid and high-pressure systems.

As for any engineered hydraulic system, the designer must select pipe sizes, confirm flows and pressures at discharge points, and determine total flow and pressure requirements for the system in order to select pumps and size storage reservoirs. Such calculations are done routinely for sprinkler systems and water spray systems. Water mist systems introduce problems that are not commonly encountered in calculations for sprinkler systems, however. For example, high-pressure systems may involve positive displacement pumps, or stored pressure reservoirs with continuously declining pressure during the discharge. Twin-fluid systems involve the use of pressure-regulating and solenoid valves on both liquid and compressed gas lines. All types of systems may incorporate long runs of small-diameter piping (smaller than sprinkler pipe). Where small diameter piping may be used, flow velocities can exceed what is usually encountered in sprinkler or water spray systems. Many of the differences from standard sprinkler system design require a change in calculation procedures. This section reviews some of the special calculation problems encountered in design of water mist systems.

Calculations for Low-Pressure, Single-Fluid Systems

NFPA 750 defines a low-pressure water mist system as one that operates at 175 psi (11.9 bar) or less. The operating range of low-pressure systems is similar to standard fire protection systems, such as sprinklers (see NFPA 13, *Standard for the Installation of Sprinkler Systems;* standpipes (see NFPA 14, *Standard for the Installation of Standpipe and Hose Systems*); and water spray systems (See NFPA 15, *Standard for Water Spray Fixed Systems*). The requirements for pipe, fittings, valves, pumps, and hydraulic calculations, then, can be met by conventional materials and design practices. It is expected that piping materials will conform to NFPA 13, which allows the use of black steel piping and drawn copper tubing. It is also expect-

ed that pipes will be sized approximately in accordance with NFPA 13, which implies that velocities in the piping will be in the same range as in sprinkler piping. Based on this assumed similarity with sprinkler piping, NFPA 750 accepts the use of the Hazen-Williams equation to calculate pressure losses in piping for low-pressure water mist systems. (See Table 6-15F.)

TABLE 6-15F. *Hazen-Williams Equation for Calculation of Pressure Loss in Low Pressure Piping*

US Units	SI Units
$h_f = 4.52 \dfrac{Q^{1.85}}{C^{1.85}} \dfrac{L}{d^{4.87}}$	$h_m = 6.05 \dfrac{Q_m^{1.85}}{C^{1.85}} \dfrac{L_m}{d_m^{4.87}} 10^5$
h_f = friction loss, psi	h_m = friction loss, bars
Q = flow rate, gpm	Q_m = flow rate, L/min
L = pipe length, ft	L_m = pipe length, m
C = friction coefficient	C = friction loss coefficient
d = actual inside diameter, in.	d_m = actual inside diameter, mm

Not all low-pressure water mist system piping will necessarily be similar to sprinkler piping, however. The designer may choose to use small-diameter pipe in order to reduce system weight, or to trade-off high friction losses in order to be able to install piping in a restricted space, such as in an aircraft cargo compartment. The use of small-diameter pipe will put the velocities higher than is "normal" in sprinkler systems, which introduces the probability that the Hazen-Williams equation will not be accurate. Values of the friction loss coefficient, C, used in the Hazen-Williams equation, are accurate only if the flow velocity is close to that at which the value of C was measured.[40] It is a matter of judgment as to what velocity is "too high" for the Hazen-Williams equation. American Water Works Association (AWWA) data, quoted in reference 40, lists C factors measured at a velocity of 3 ft/sec (0.9 m/s), yet it is standard practice in sprinkler calculations to accept velocities in sprinkler piping between 10 and 30 ft/sec (3.05 and 9.1 m/s). Similarly, the tables of equivalent lengths for fittings and valves, used by sprinkler system designers, are based on fittings and valve types typical of sprinkler systems. Water mist systems may incorporate different types of fittings and valves, for which the Hazen-Williams based equivalent length values will be incorrect.

It is also important to note that the Hazen-Williams equation contains no terms that account for the temperature, hence, density and viscosity of the liquid. It assumes that the water contains no additives and is close to 60°F (15.6°C). If viscosity or water temperature depart significantly from typical sprinkler system water supply conditions, consideration should be given to using the Darcy-Weisbach equation regardless of the pressure regime or flow velocities. (See Table 6-15G.)

Calculations for Intermediate- and High-Pressure, Single-Fluid Systems

Table 6-15G shows a summary of the Darcy-Weisbach equation and the related equations for use in intermediate and high pressure systems.

The Darcy-Weisbach (D-W) equation (Equation 1) allows for input of actual fluid properties of temperature and viscosity, actual velocity and turbulence conditions, and pipe roughness factors. It should be used instead of the Hazen-Williams equation for intermediate- and high-pressure systems, and for low-pressure systems where additives may alter the density and viscosity of the water.

Water mist system calculations can use the following form of the Darcy-Weisbach equation:

$$\Delta p = \frac{0.000216 f L \rho Q^2}{d^5}$$

In this form, the elements are defined as:

Δp = pressure drop (psi)
f = friction factor (dimensionless)
L = length of pipe (ft)
ρ = weight density of fluid (lbs/ft³)
Q = volumetric flow rate (gpm)
d = internal diameter of the pipe or tube (in.)

Most of the parameters in Equation 1 will be known by the designer, such as the diameters and lengths of pipe (including equivalent lengths for fittings) and the density of the fluid. Density may vary according to water source (seawater or potable water) and the possible presence of additives, such as wetting agents, fire retardant salts, or antifreeze. The flow rates will be determined by the nozzle characteristics, such as the "k" factor. The unknown value 'f' (friction factor) must be derived through a three-step process as follows:

(a) Equation 2 in Table 6-15G is used to calculate the Reynolds number (Re);
(b) Equation 3 in Table 6-15G is used to calculate the relative roughness factor (ε/d); and
(c) the Moody diagram (Figure 6-15O) is used to obtain a value for the friction factor (f).

Tables 6-15H and 6-15I provide values of water density (ρ), dynamic viscosity (μ), and absolute roughness (ε) for use in solving Equations 2 and 3. Once the Reynolds number and the relative roughness are determined, the values are located in the Moody diagram (see Figure 6-15O). The intersection point for the Re and ε/d values is located on the diagram. A horizontal line is projected from the intersection point to the left axis of the Moody diagram, which provides a value for the friction factor (f).

The value of "f" and other known parameters are then plugged in to Equation 1 to calculate a total pressure loss over each length of pipe. The value of 'f' changes as flow quantities accumulate along the pipe length, and as pipe diameters change across the system. From this point on the hydraulic calculation procedures are comparable to standard sprinkler calculations — discharges and pressure losses are accumulated over the entire piping network, to provide an estimate of the total water flow rate and pressure required at the water source (pump). It can be expected that computerized calculation procedures based on the Darcy-Weisbach equations will be developed in the near future.

Fire protection systems designers are familiar with the use of "equivalent lengths" of pipe to represent fittings and valves, for use with the Hazen-Williams equation. The equivalent length tables available in many NFPA standards are based on the Hazen-Williams equation and should be checked before being used with the Darcy-Weisbach equation. The head losses incurred across valves and fittings may be calculated using resistance coefficients (K), equivalent lengths (L/D), and flow coefficients, (C_v). Valve manufacturers often provide C_v values with data sheets on their valves. Practical engineering guidance on calculating head losses through valves and fittings using the Darcy-Weisbach equation is available in reference 41.

Calculations for Declining Pressure Systems

Systems that utilize stored energy involving compressed gas will operate in one of two ways. Low-pressure systems may operate through a pressure regulator, to modify flow from a high-pressure source to a lower, constant pressure for the system piping. Discharge after the source pressure drops below the design point is not considered part of the design flow. Calculations for such systems are the same as for any constant pressure system.

Other types of stored-energy systems, particularly high-pressure systems, operate on the principle of declining pressure, hence declining flow rate, throughout the discharge period. Special calculation procedures are required to ensure that nozzle pressures, flow rates, and discharge duration are adequate to produce the desired

TABLE 6-15G. *Darcy-Weisbach and Associated Equations for Pressure Loss in Intermediate- and High-Pressure Systems*

U.S. Units	SI Units	
Darcy-Weisbach Equation $$\Delta p = \frac{0.000216 f L \rho Q^2}{d^5}$$	$$\Delta p_m = 2.252 \frac{f L_m \rho_m Q_m^2}{d_m^5}$$	(1)
Reynolds Number $$\mathrm{Re} = 50.6 \frac{Q \rho}{d \mu}$$	$$\mathrm{Re} = 21.22 \frac{Q_m \rho_m}{d_m \mu_m}$$	(2)
Relative Roughness Relative roughness $= \dfrac{\varepsilon}{D}$	Relative roughness $= \dfrac{\varepsilon_m}{D_m}$	(3)

Δp	= friction loss, psi gage	Δp_m	= friction loss, bar
L	= length of pipe, ft	L	= length of pipe, m
f	= friction factor, psi/ft	f	= friction factor, bar/m
Q	= flow, gpm	Q	= flow, L/min
d	= internal pipe diameter, in.	d_m	= internal pipe diameter, mm
D	= internal pipe diameter, ft	ε	= pipe wall roughness, mm
ε	= pipe wall roughness, ft	ρ	= weight density of fluid, kg/m³
ρ	= weight density of fluid, lb/ft³	μ	= absolute (dynamic) viscosity, centipoise (cP)
μ	= absolute (dynamic) viscosity, centipoise (cP)		

TABLE 6-15H. *Approximate Values of* μ, *Absolute (dynamic) Viscosity, and* ρ *for Clean Water, over the Temperature Range 40°F to 100°F (4.4°C to 37.8°C)*

Temperature (°F)	Temperature (°C)	Weight Density of Water (lbs/ft³) ρ	Weight Density of Water (kg/m³) ρm	Absolute (dynamic) Viscosity, μ (centipoise)
40	4.4	62.42	999.9	1.5
50	10.0	62.38	999.7	1.3
60	15.6	62.34	998.8	1.1
70	21.1	62.27	998.0	0.95
80	26.7	62.19	996.6	0.85
90	32.2	62.11	995.4	0.74
100	37.8	62.00	993.6	0.66

TABLE 6-15I. *Recommended Values of Absolute Roughness or Effective Height of Pipe Wall Irregularities, for Use in Relative Roughness Equation*

Pipe Material (new)	Design Value of ε (ft)	Design Value of ε (mm)
Copper, copper nickel, drawn tubing	0.000 005	0.0015
Stainless steel, commercial steel, wrought iron	0.00015	0.045
Galvanized iron	0.0005	0.15

spray characteristics. The designer is interested in knowing the length of time each nozzle will discharge in the pressure range needed for producing the design spray characteristics, the settings for cylinder pressure and pressure regulators, and how much water must be stored for one complete discharge.

As a first approximation of the total flow from a declining-pressure system, the designer could simply multiply the average discharge rate per nozzle by the duration of the interval of declining pressure, and multiply by the number of nozzles. Adding an arbitrary percentage as a "safety factor," the result would likely be close enough to ensure that water storage capacity and discharge duration were adequate for the hazard. A more thorough calculation procedure is needed, however, to provide a degree of control over the design of the water supply, storage capacity, system piping, spray characteristics and discharge duration.

Figure 6-15P and the following text illustrate a procedure for determining total water demand for a declining-pressure system. It is first necessary to estimate the "pressure *vs.* time" condition at each nozzle location. (It is admittedly not easy to determine such curves. It may be possible to adapt hydraulic calculation software used for designing gaseous agent piping systems to handle single-phase flow to multiple outlets.) As a start, the pressure *vs.* time curve at each nozzle can be estimated as a straight line using the maximum and minimum desired nozzle pressures for effective spray generation, as indicated by the listing information, and the desired duration of flow.

The pressure *vs.* time plot for each nozzle should be expressed numerically as a function - "F": $P = F\{t\}$. Also required is an expression for the "discharge *vs.* pressure" relationship for the nozzle, $Q = F\{P\}$, which will be provided by the nozzle manufacturer as part of the listing information. It is coupled with the "pressure *vs.* time" equation, $(P = F\{t\})$, to obtain an expression for "discharge *vs.* time" $(Q = F(P\{t\}))$ specific to each nozzle location. Calculating the area under the plotted line (i.e., integrating the equation for $Q = F(P\{t\})$), determines the total volume of water discharged from each nozzle over the discharge interval. The "total" includes the portion discharged when the nozzle pressure is high enough to gen-

erate "effective" spray, followed by an amount discharged after the nozzle pressure drops below the effective range. The total volume of water that must be stored for one discharge event is then the sum of the individual nozzle discharges. A conservative design would require storage of enough water to supply both the "effective" and "ineffective" spray quantities. A more aggressive design might choose to store only enough water to supply the spray during the "effective" range, perhaps relying on a low-pressure switch setting to shutoff the control valve to stop water flow when pressure drops below the effective range.

Hydraulic calculations are then done to determine the total friction losses in the system piping. In Figure 6-15P, the maximum flow for an assumed 20 nozzles is estimated as twenty times the peak flow from the most remote nozzle (Nozzle 1), increased by 10 percent for the inevitable "overage." Using the Darcy-Weisbach equation, the pressure losses in system piping were calculated for the peak flow rate, and then for the lowest effective flow rate. The D-W equation does not plot to a straight line on the $N = 1.85$ hydraulic grid used in conventional sprinkler calculations, so the calculation must be done for enough conditions to define the curve. Plotted on an arithmetic grid, these points generate the characteristic head loss curve for the piping, also known as the "System Head Curve." It is this curve that shows the total flow and pressure that must be available at the supply point, to deliver the desired flow to the 20 nozzles in the example, at any time during the discharge. The cylinder pressure is treated in the same way as "elevation head" in the example, i.e., as a step-increase to the position of the SHC. As the cylinder pressure declines, the SHC moves down, while the flow point moves to the left.

Finally, the designer must confirm that the pressurized cylinder (or bank of cylinders), connected via discharge valves and header, is able to provide the maximum pressure and flow rate at the peak and minimum design conditions. Information on the discharge characteristics of the water supply assembly should be provided as part of the listing of equipment for use in water mist fire suppression systems.

The suggested calculation procedure for a declining-pressure system becomes increasingly complex as the number of nozzles on the system increases. It is expected that manufacturers of declining-pressure water mist systems will provide guidance on procedures to determine discharge rates and total water demand for their systems. The responsibility to validate those procedures will rest with the listing agencies.

Calculations for Twin-Fluid Systems

Twin-fluid systems require sources of pressurized water and compressed gas, in order for the nozzles to produce spray as required for fire suppression. The two "fluids" are delivered to the nozzles through two independent piping systems. The designer must calculate the pressure losses in each piping system in order to ensure that nozzles perform as intended, and to determine the minimum storage

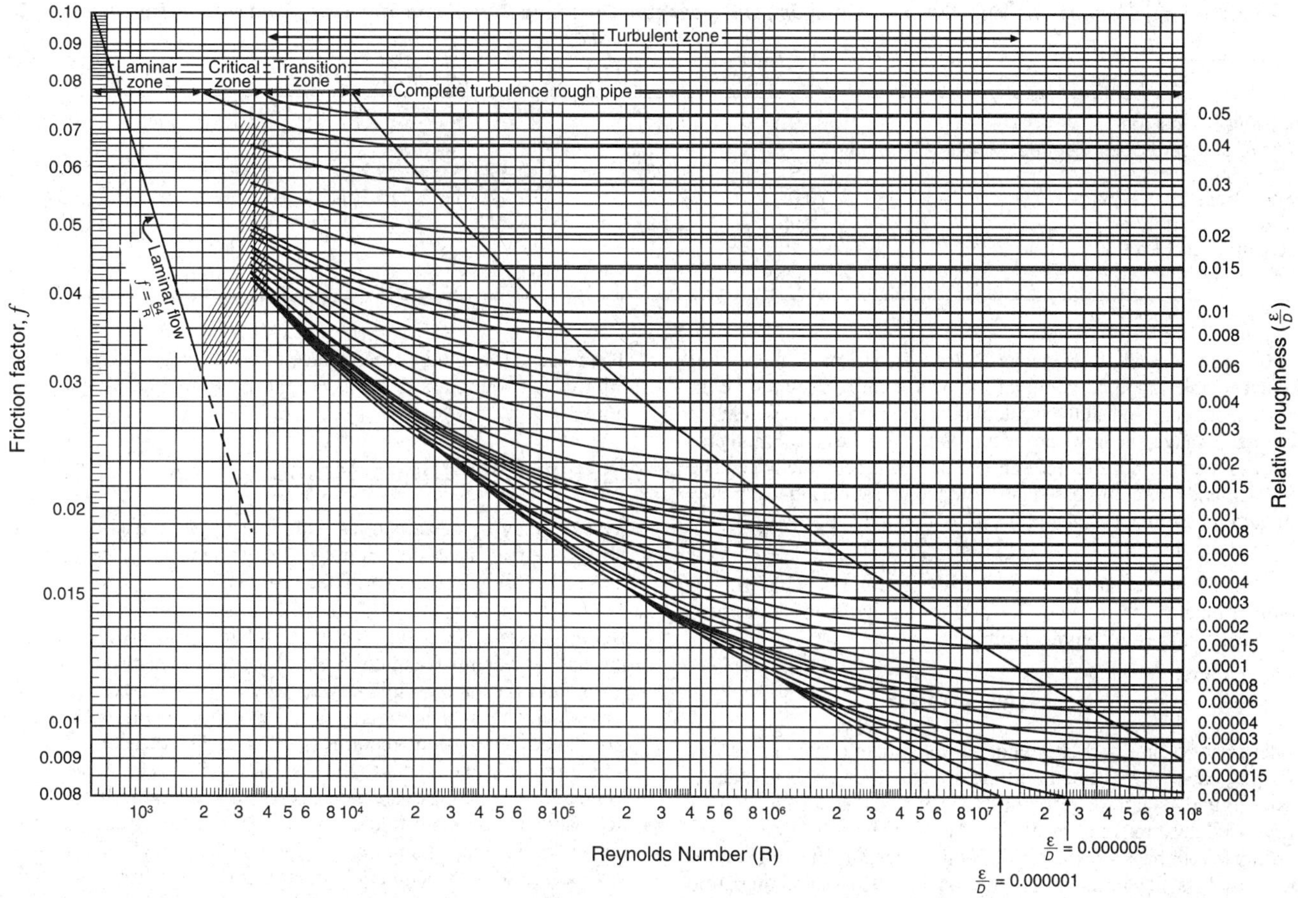

FIG. 6-15O.   Moody diagram.

quantities and pressure for each agent. The quantities of water and of compressed gas ("air") discharged at each nozzle depend on the Air-to-Liquid Ratio (ALR) characteristics of the nozzle. If the "air" pressure is too high, it could prevent water from entering the nozzle. If the water pressure is too high, not enough air will enter the nozzle and the spray quality will deteriorate. Twin fluid systems, therefore, require special calculation procedures in order to ensure adequate nozzle performance, select pipe sizes, and determine agent storage requirements.

There is a simple (but costly) way to ensure that the water pressure and atomizing medium pressure at each nozzle stay within acceptable ranges - that is to select generously large diameter pipes for both lines so that the friction losses between discharge points in each system will be negligible. Each piping system acts as a manifold, in which pressures are equal at all points along its length. This approach has been used on twin-fluid water spray systems for applications other than fire suppression. It is generally not practical where there is a need to minimize the cost and weight of the system, however. It is preferable to streamline the piping systems to the smallest diameters that still can provide for proper performance of all nozzles. To achieve that, there is no way to avoid doing both hydraulic and pneumatic calculations.

One approach to the problem is to perform both the hydraulic and the pneumatic calculations simultaneously, checking to ensure that the gauge pressures, and pressure differences between liquid and compressed gas, at each nozzle, fall within a specified range. The listing information for a twin-fluid nozzle could, for example, identify the maximum and minimum operating pressures for both liquid and atomizing media, as shown in Table 10. The designer's task is to select preliminary pipe sizes on both piping systems, then to work methodically through two sets of calculations, comparing the conditions at each nozzle. The H-W equation can be used for the low pressure, water distribution piping. The calculations for "compressible flow" (compressed gas line) should follow standard engineering practice as described in reference [46]. At any point, if any of the pressure/flow conditions in Table 10 exceed or fall short of the desired operating point, an adjustment to the piping system should be made. After several iterations, it is possible to arrive at a reasonable estimate (± 15%) of total system demand.

The key step in the calculation procedure involves confirming that the ALR_m, that is, the mass ratio of the water discharge and the atomizing media discharge, at each nozzle, is within an acceptable range. That is,

$$ALR_m(min) < ALR_m(actual) < ALR_m(max)$$

Figure 6-15Q illustrates one model for this calculation procedure. Values for the water pressure and flow rate and air pressure and flow rate, at the most "remote" nozzle, are set based on the manufacturer's recommended values. Flow rates must be converted from volumetric to mass units, e.g., gpm → lb/min (L/min → kg/min), and SCFM → lb/min (m³/min → kg/min). The ALR_m value (ALR1 in Figure 6-15Q) is calculated. Then, for the given flow rates through the pipes supplying the most remote nozzle, the total pressure losses in each leg are calculated using the appropriate equations, and added to the beginning pressures. The "new" pressures at the upstream nozzle (Pw_2 and Pa_2 in Figure 6-15Q) are used, with the manufacturer's nozzle data, such as shown in Figure 6-15N, to determine the new flow rates, Qw_2 and Qa_2. These values are con-

TABLE 6-15J.   Example Showing Minimum, Optimum, and Maximum Operating Conditions for a Twin-Fluid Nozzle, Nominal Rating 1.32 gpm (5 L/min)

Conditions at Nozzle	Minimum		Optimum		Maximum	
Air Pressure, psi (bar)	77.0	(5.24)	82.0	(5.58)	87.0	(5.92)
Airflow, Standard Cubic Feet per Minute (m³/min)	9.8	(0.28)	10.4	(0.29)	15.6	(0.44)
Water Pressure, psi (bar)	70.0	(4.76)	75.0	(5.10)	80.0	(5.44)
Waterflow, gpm (L/min)	0.95	(3.60)	1.28	(4.83)	1.82	(6.89)
Pressure Ratio: Pa/Pw	1.10		1.09		1.09	
Mass Ratio: Ma/Mw	0.093		0.074		0.077	

verted to mass units, to calculate $ALRm_2$. If $ALRm_2$ is not within the acceptable range, then one of the pipe diameters should be increased, or decreased, and the calculation repeated. If it is within the acceptable range, the flows are accumulated and the calculation proceeds to the next nozzle, until all nozzles have been included. The total demands for compressed gas flow and pressure, and water flow rate and pressure are then compared to the total capacity of the respective supplies.

It is expected that manufacturers of twin fluid systems will develop computerized methods for performing hydraulic and pneumatic calculations in combined piping systems

SYSTEM TYPES

Water mist systems currently under development come in a variety of types, each with distinctly different features. In order to make it easier to understand the different types of systems, NFPA 750 has introduced descriptive terms which classify water mist systems according to four basic characteristics of their design. The four parameters used as the basis of description are: 1) system operation method, 2) operating pressure, 3) application, and 4) nozzle type.

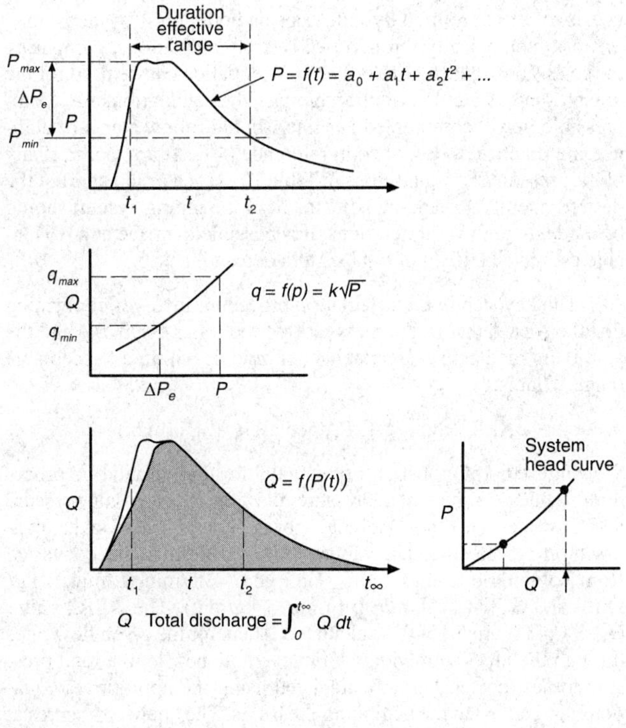

The reasons for and the significance of classifying systems according to these four classifications are presented in the following sections.

System Operation Method

Systems may be described as deluge, dry-pipe, preaction, or wet pipe systems. These descriptions are fundamentally the same as those used for automatic sprinkler systems.

Deluge systems: Deluge systems utilize open, nozzles spaced throughout a compartment or around an object to be protected. Water is held out of the piping system by a special valve (a *deluge* valve), which can be automatically released on a signal from a control panel. The system requires a separate fire detection system, which provides the signal to the control panel to release the water. Manual release features are always provided as well. When the deluge valve opens, water fills the piping system and discharges from all water mist nozzles simultaneously. The deluge system operation is widely used for machinery compartments, and other applications where it is necessary to completely surround a fire with mist.

Wet-pipe systems: A wet-pipe system consists of an array of closed nozzles on a piping network that is filled with water. Each nozzle is equipped with a thermal sensing element, like a sprinkler, which opens the nozzle when exposed to fire (hence, an *automatic*

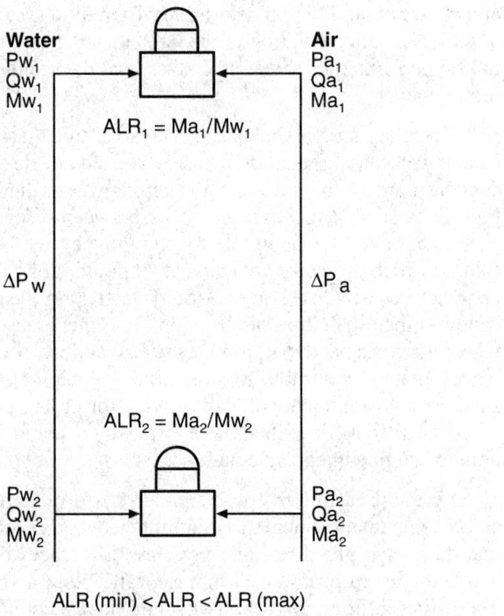

FIG. 6-15P.   A procedure for determining total discharge from a nozzle in a declining-pressure water mist system.

FIG. 6-15Q.   Simplified Air to Liquid Ratio Model.

nozzle). As nozzles are opened by heat, water discharges immediately. This sprinkler–system–like configuration is suitable for applications in which there may be only three or four nozzles in a compartment, or where it is not necessary to completely surround the hazard with mist. Nozzles that are "downstream" in the thermal plume must have enough energy to project spray through the fire plume to the seat of the fire.

Dry-pipe systems: Dry-pipe systems utilize normally-closed, thermally-actuated nozzles on a closed piping system. The piping system is pressurized with compressed gas, which holds down the seat on a differential dry-pipe valve to prevent water from entering the piping system. Like the wet-pipe system, closed nozzles can be individually thermally actuated when exposed to fire. As nozzles are opened by heat, air is released and the air pressure in the piping drops to the point at which the valve will open. Water enters the piping system and discharges only from the nozzles which have been opened by heat.

Preaction systems: A preaction system is similar to a dry-pipe system, in that it utilizes closed nozzles on a piping system that contains no water. Water is prevented from entering the piping network by means of a valve which can be opened automatically by a control panel. A supplementary detection system is required to signal that there is a fire in the compartment, and to open the valve. Water then enters and fills the piping, at which point the system operation is identical to a wet-pipe system. Water is not discharged until the heat is sufficient to actuate a nozzle.

Operating Pressure

A second characteristic that differentiates system types is the range of operating pressure. Three categories have been proposed by NFPA 750: (1) low-pressure, (2) intermediate-pressure, and (3) high-pressure systems.

Low-pressure systems: A water mist system where the distribution piping is exposed to pressures of 175 psi (12.1 bar) or less.

Intermediate-pressure systems: A water mist system where the distribution system piping is exposed to pressures greater than 175 psi (12.1 bar) but less than 500 psi (34.5 bar).

High-pressure systems: A water mist system where the distribution system piping is exposed to pressures of 500 psi (34.5 bar) or greater.

The working pressure regime affects the system cost and performance in a number of ways. It affects the qualities of the spray, such as drop size distribution, mass flow rate, and spray momentum. Also, increased pressure combined with higher flow rates generally requires more pumping energy, which impacts upon system cost. However, it is not the case that high-pressure systems are necessarily "better" or that low-pressure systems are necessarily "cheaper". For example, the cost of high-pressure pumping may be offset by advantages in reduced piping diameters or increased nozzle spacing. Manufacturers tend to adopt a method of spray generation, based on their access to the technology, which operates in one or the other pressure regimes. The "choice" of pressure regime then may occur at the time an end user chooses between proposals from competing manufacturers.

Application

Three different types of applications are described in NFPA 750, to relate the type of protection to the hazard. These are (1) local, (2) total compartment, and (3) zoned applications.

Local application: This configuration is typically used to protect a specific hazard or object. An example may be the protection of a piece of equipment in a large room or compartment. The system would be designed to discharge water mist directly onto the object. Figure 6-15R depicts a local application system.

Total compartment application: This type of system provides protection to all fire hazards and all areas in a compartment. The open nozzles are positioned in a grid so that water mist discharges approximately uniformly throughout the entire volume. Figure 6-15S illustrates a Total Compartment Application System.

Zoned application: This type of system is configured to discharge mist from portions of a larger system, as required to control fire in a specific part of a compartment. It would be considered, for example, in circumstances where the water demand for a total compartment system, i.e., a deluge system, would be beyond the capability of the water supply or pump capacity. Zoning the water mist piping network, however, requires the installation of a detection system capable of pinpointing the exact location of the fire in the compartment, and a piping control system capable of opening or closing specific valves in the distribution network. Figure 6-15T illustrates essential elements of a zoned application system.

One alternative to complex fire detection and valving arrangements to achieve the objectives of zoning, involves using thermally activated nozzles (*automatic* nozzles) so that only the nozzles near the fire open. That configuration may have disadvantages, however, with respect to the ability to control certain types of fires, and to maximize the suppression efficiency of water mist.

Nozzle Type

The nozzle type constitutes the fourth parameter for classifying water mist systems. Nozzles have been described as "open" or "closed" in previous text. Open nozzles have also been referred to as "nonautomatic" nozzles because they play no role in whether water is released. That is determined by the supplemental detection system. However, "closed" nozzles are designed with thermal sensing elements or other features which allow them to open, individually and "automatically", in the presence of fire. Thus, NFPA 750 has adopted the terms (1) nonautomatic, (2) automatic, and (3) hybrid, to describe the operation principle of types of nozzles.

Nonautomatic nozzles: These are open orifices. Water discharges from all nozzles when the system is turned on by an independent detection system.

Automatic nozzles: These nozzles operate independently of other nozzles. They may consist of an orifice held closed by a thermal sensing-and-release mechanism (like a sprinkler). They have also been referred to in some literature as "individually thermally actuated" nozzles.

Hybrid nozzles: The term "hybrid" denotes a normally closed nozzle that combines the individual thermal actuation principle of an automatic nozzle, with a feature that allows it to be opened on a signal from a control panel. The concept arises from the interest of water mist systems designers to operate groups of nozzles to surround a fire, rather than rely on a "one-by-one" response to heat. To give an example, a hybrid nozzle closest to the fire would activate thermally, due to the heat of the fire. At the same time it would send a signal of its condition to a control panel. The control panel would then send an electrical signal to activate a pre-determined pattern of adjacent nozzles, without waiting for them to be opened thermally by the fire. The Marioff Hy-fog Company of Finland has experimented with hybrid nozzle designs, utilizing both electrical and hydraulic principles.

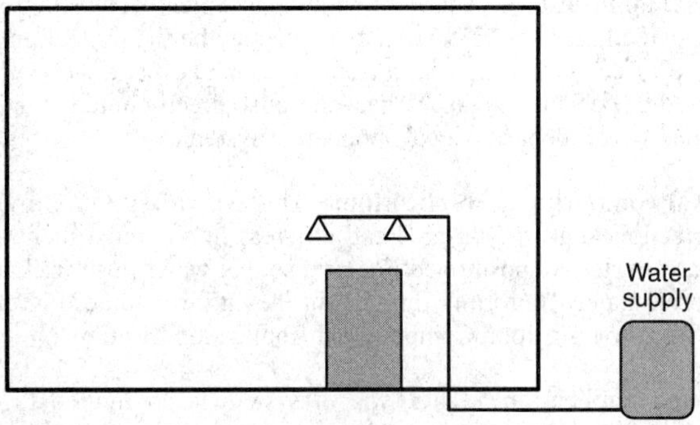

FIG. 6-15R. Local application systems.

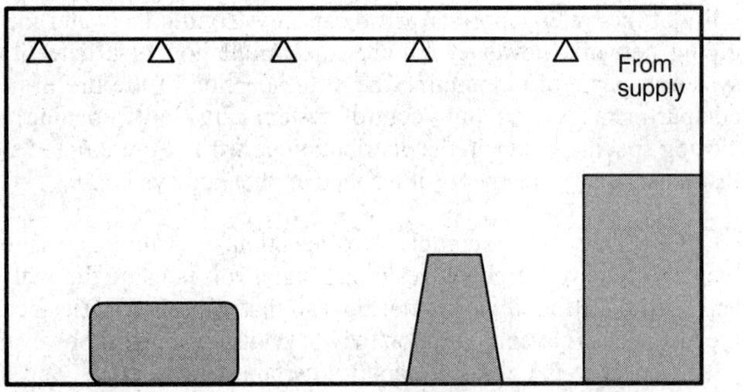

FIG. 6-15S. Total Compartment application.

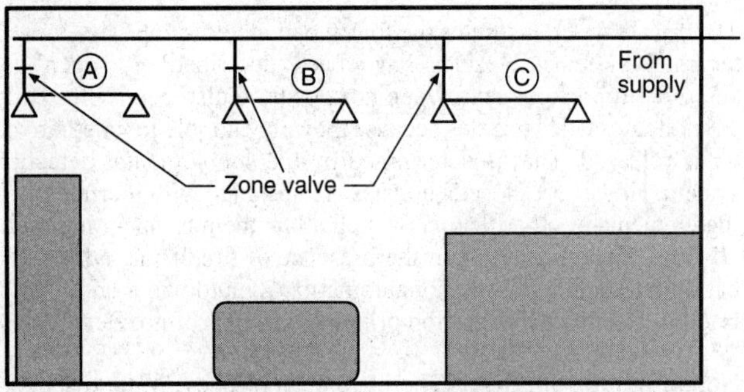

FIG. 6-15T. Zoned application.

SYSTEM COMPONENTS

As with all fire protection systems, water mist systems require their own unique set of hardware features, components, and materials to increase the level of reliability and performance. Basic components include, but are not limited to pipe, fittings, valves, hangers, gas and water containers, nozzles, strainers, pumps, and detection system components.

The quality of system components and the overall reliability of water mist systems can be ensured by using materials that comply with internationally recognized standards, codes or material specifications. This is addressed in NFPA 750 by requirements for most components of systems to be "listed" for specific water mist system use. The term "listed" as used by NFPA implies that the equipment or material meets identified standards or has been tested and found suitable for the specified purpose. The following sections provide information on quality control aspects of key components of water mist systems.

Gas and Water Containers

Single-fluid systems are permitted to utilize any number of sources for their water supply. The same is true for twin-fluid systems, with respect to both their water supply and atomizing media supply. One such option for each system is to derive the fluid source from a captured supply, such as a storage container.

Pressurized containers of smaller volumes that could be categorized as shipping containers must be designed to meet the requirements of the U.S. Department of Transportation (DOT) for 3A or 3AA-1800 seamless steel cylinders. Companion standards such as those from Transport Canada, are also considered as meeting this requirement.

Containers used to store a supply of water or atomizing medium must be equipped with a nameplate that is to include:

1. *Water containers*
   - Type of liquid (including any additives) in the container
   - Water volume
   - Pressure level to be maintained
2. *Atomizing media*
   - Type of gas
   - Weight of gas
   - Weight of container
   - Gas volume
   - Pressure level to be maintained

In lieu of providing individual information on each container, it is acceptable to have a placard in the vicinity that includes the necessary information.

Pipe and Tube

Two primary considerations when selecting this part of the system hardware are: (1) the potential pressure to which the pipe will be exposed to during system operation and (2) the potential for corrosion.

The working pressure of the pipe and associated fittings is important, since operating pressures vary. Low pressure systems require the use of pipe and fittings that can sustain a working pressure of 175 psi (12 bar). Likewise, intermediate-pressure systems must be able to accommodate working pressures of up to just below 500 (34. bar) while a high pressure system must be able to accommodate pressures well beyond 500 psi (34.5 bar).

Pipe and tube materials should have a corrosion resistance that is equal to or better than the types of materials shown in Table 6-15K. These materials are capable of sustaining a range of operating pressures. In addition, they have low corrosion rate properties. Resistance to corrosion is important for water mist systems, since the small-diameter orifices that are used are prone to clogging if a number of precautions are not taken. (See "Water Supplies," later in this chapter.)

The standards listed in Table 6-15K are based on low-pressure systems. Systems that must operate at pressures above 175 psi (12 bar) and do not meet the criteria of materials in Table must satisfy ANSI B31.1, *Power Piping*[42]. While ANSI B 31.1 is not commonly used in fire protection systems, it forms the basis for a variety of fluid piping systems that operate at high pressures.

Support of the pipe network involves general principles, such as the use of listed hangers, spacing of hangers to adequately support the pipe, fittings, and attachment of the hangers to a part of the structure that can support the load. Hangers must be capable of sustaining a static load, as well as any dynamic loads during system op-

TABLE 6-15K. *Pipe or Tube Standards*

Materials and Dimensions	Standard
Copper Tube (Drawn, Seamless)	
Standard Specification for Solder Metal [95-5 (Tin-Antimony-Grade 95TA)]	ASTM B 32
Standard Specification for Seamless Copper Tube	ASTM B 75
Standard Specification for Seamless Copper Water Tube	ASTM B 88
Standard Specification for General Requirements for Wrought Seamless Copper and Copper-Alloy Tube	ASTM B 251
Standard Specification for Liquid and Paste Fluxes for Soldering Applications of Copper and Copper-Alloy Tubes	ASTM B 813
Standard Specification for Filler Metals for Brazing and Braze Welding (Classification BCuP-3 or BCuP-4)	ASTM A5.8
Stainless Steel	
Standard Specification for Seamless and Welded Austenitic Stainless Steel Tubing for General Service	ASTM A 269
Standard Specification for Seamless and Welded Austenitic Stainless Steel Tubing (Small-Diameter) for General Service	ASTM A 632
Standard Specification for Welded, Unannealed Austenitic Stainless Steel Tubular Products	ASTM A 778
Standard Specification for Seamless and Welded Ferritic/Austenitic Stainless Steel Tubing for General Service	ASTM A 789/A 789M

eration. Both types of loads can be evaluated on a case-by-case basis, although it is more likely that the listing criteria for pipe supports will consider both types of loads.

Water Mist Nozzles

Quality control for the design of water mist nozzles is achieved by requirements in NFPA 750 for nozzles to be "listed." In either Underwriters Laboratories Inc. (UL) or Factory Mutual Research Corporation (FMRC) will be involved in listing evaluations for water mist nozzles.

The draft standard "Proposed First Edition of the Standard for Water Mist Nozzles for Fire Protection Service, UL 2167, January 1996" describes the steps to be taken to achieve a water mist nozzle listing. The manufacturer must first demonstrate that fire test objectives are achieved, by passing the full-scale fire tests described in the standard. Once the fire suppression performance conditions are established, the nozzles must undergo a series of evaluations to measure flow characteristics, water distribution patterns, drop size distributions and velocity, corrosion resistance, leak resistance, response to heat exposure, and more. Nozzles that "pass" all of the tests receive a "listing" from the agency. This listing will specify the limitations for the use of the nozzle in fire scenarios similar to the fire tests conducted.

Pumps

Pumps used to deliver the necessary flow and pressure to system pipe networks will be of the centrifugal type or the positive displacement type. The reliability of centrifugal fire pumps, pump drivers and controllers are well regulated by NFPA 20. Centrifugal pumps are suitable for use in low-pressure and possibly intermediate-pressure systems. They are not, however, able to generate water pressures in the high-pressure regime (> 500 psi [34.5 bar]). High-pressure water mist systems utilize positive displacement pumps ("PD" pumps). At present, NFPA 20 does not address PD pumps, so it has been necessary to rely on the standards of industries more familiar with them.

Positive displacement pumps utilize pistons, such that water is allowed into a chamber, then forced out of it at high pressure by the force of a piston. As a result, the discharge *vs.* pressure (Q-P) curve for a PD pump is quite different than the Q-P curve for a centrifugal pump. This feature of PD pumps will require pressure release and flow by-pass arrangements that are not addressed in NFPA 20. They should be evaluated as part of the listing for the pump. It is expected that the innovations associated with water mist systems will require changes to several NFPA standards.

System features, such as the electrical power supply connections, location and sizing of circuit breakers, and supervision, are no different than conventional fire pump installations, and NFPA 20 should be followed.

Detection, Actuation, Alarm, and Control Equipment

System type will influence many of the options available to the designer for the equipment that can be used for this part of the system design.

Systems that use nonautomatic and hybrid nozzles will, in general, require the use of a control configuration that is multifaceted. The detection system must be able to detect fire; the control panel must then operate one or more water supply control valves; and, in some instances, operate one or more zone control valves.

Detection systems must be designed and installed in accordance with NFPA 72 (CAN/ULC S524-M86 and S529-M87). The releasing device must also comply with these codes when appropriate. The detection and control equipment used to activate a preaction system or a deluge system have a number of components that all must work at the time of the fire. In addition to increasing the reliability of the system, evaluations and performance tests evaluate system controls to ensure that inadvertent activation is either eliminated or minimized to the extent possible.

In the same manner that pumps require a specific set of features to be supervised, supervision of certain detection and control features is also necessary. System alarm, trouble signals, or both are necessary for critical parts of the electrical control system. These alarms include the following:

1. System activation (alarm),
2. Power failure (supervisory),
3. Valve position (supervisory),
4. Detector operation (alarm), and
5. Detector status (supervisory).

Some systems may use pneumatic lines to activate system valves. These control lines should be protected against mechanical damage as well as fire damage.

In the unlikely event that all of the automatic control equipment fails, for whatever reason, water mist systems should also be equipped with a manual means of operation. This method of system activation is also useful for those situations where the fire is discovered by people in the vicinity of the fire, prior to the fire being detected by the detection system. The manual means of activation should be independent of the automatic features, to the extent possible, to allow for the system to operate in the event of a catastrophic control failure. Manual releasing stations should be provided at the main system control valves or at other conspicuous areas within or near the protected hazard.

Miscellaneous Components

The discussion so far has centered on those components that are different than those used in similar types of water-based fire protection systems or special hazard systems.

Water mist systems may use a variety of valves that control everything from the discharge of the water to the discharge of the atomizing medium. Valves should be selected for their particular function and should be capable of sustaining the system pressure to which they will be subjected. All valves should be labeled as to the portion of the system that they control, and they should be supervised in the normal (open or closed) position.

Some of these valves require manual operation while other valves are automatically operated upon activation of the detection system or a manual release mechanism. Valves in this category are identified as control/activation valves. Discharge of water through the mist nozzle may require the operation of two or more of these valves.

Pressure-regulating devices will be found on any part of a system that may subject the valve, pipe, fitting, or nozzle to an operating pressure higher than the working pressure of the device. They may also be prevalent on twin-fluid systems in order to control the volume and pressure from the atomizing medium source. A precise balance is necessary between the water supply and the atomizing medium supply at the nozzle and pressure-regulating devices are useful to maintain this balance.

Check valves may be necessary in a number of locations on water mist systems. Check valves should be installed:

1. At the point between the system piping and the system water supply.
2. On twin-fluid systems, in the pipe network, to prevent atomizing medium from entering the water supply pipe and to prevent the water supply from entering the atomizing medium.

Pressure gages are another common component associated with water mist systems. Gages are necessary to monitor key points within the system to verify that proper water pressure or atomizing medium pressure is available. These locations include the following:

1. Pressurized side of all water and atomizing media connections,
2. Pressurized side of all system control valves, and
3. Pressurized storage containers.

In general, the gages are used to verify system status as part of a preventive maintenance program. gages should be selected such that they can withstand at least two times the working pressure associated with the particular system.

SYSTEM WATER SUPPLY AND ATOMIZING MEDIUM CONSIDERATIONS

Reliability of the system water supply and atomizing medium supply (if applicable) is crucial to ensure that the system will deliver the necessary flow and pressure to the system nozzles. A number of items must be considered to ascertain this level of reliability. The types of system components that have been mentioned have favorable corrosion-resistance properties. This characteristic will minimize the likelihood of system obstructions that may create blockages within the water mist nozzle.

Water Supplies

Other measures include provisions for having an automatic supply for both the water source and the atomizing medium source. The sources should be situated close to the hazard that they protect, in order to minimize the transit time of the fluid (s) between the source and the hazard. Once in operation, these supplies must at least meet the minimum flow rate and duration to achieve the desired level of fire protection. Consideration should be given to the number of systems as well as the number of water mist nozzles expected to operate. The type of system application that is selected, i.e., local, total compartment, or zoned is one variable in this equation. The type of nozzles selected, automatic or nonautomatic, will also have a bearing on these flow rates.

In general, the supplies should be available for 30 min. In some circumstances, this can be reduced depending on the system trait, such as a pre-engineered system which may be permitted for use in specific size compartments to protect a specific type of fire hazard. In other situations, an analysis of the entire scenario may indicate that a reduction in the duration is possible. It should be noted that the 30-min recommendation represents a minimum duration. Some hazards may require a longer duration depending on the extent of the fuel load, geometry, and objective of the system.

Reserve supplies may sometimes be desirable to allow for continuous protection following the operation of the system. While this is not crucial for land-based systems, which can usually receive service in a reasonable period of time, it is critical for marine vessel applications, in that vessels may be several days from a port of call or their final destination. Continuity of protection is an important consideration in this environment.

Water Quality

Quality control issues, with respect to the system water supply, begin with the need to obtain a source of water that is relatively free from particulate matter or items large enough to block nozzle orifices. This may best be handled by using a potable source for the water supply. Potable water supply sources must meet numerous criteria for clarity and content; thus they would make a good source.

Nonpotable sources may also be acceptable depending on their characteristics with respect to foreign object content. Screens and filters can be used to remove most of the particulate matter. The concern with the content of the water supply does not stop at the source of the supply, however. A number of protection measures are needed within the confines of the supply piping network, as well. These measures include the use of high-quality water, strainers, filters, or a combination of all three to achieve a high level of quality control. While these measures may appear extreme when compared to the steps taken to protect sprinkler system piping or traditional water spray systems, they are necessary to achieve the goal of providing reliable performance of water mist systems. The selection of protection measure is a function of the diameter of the smallest portion of the water-way in the system nozzle.

Special measures are required to protect water mist nozzles from plugging. These measures include the use of filters or strainers at each connection to the water supply, filters or strainers at each nozzle, and in certain cases the use of demineralized water. This latter criterion applies only when the smallest pathway in the system, or nozzle orifices, are under 50 microns.

Atomizing Media

Twin-fluid systems require a supply of atomizing medium in addition to a reliable water supply. Air and nitrogen are the most commonly used compressed gases for water mist systems. The source of compressed gas must have the same level of reliability as the water supply. Two options are: the used of dedicated compressed gas cylinders manifold in banks; and, the use of "plant air" supplies in facilities that have the necessary capacity. If used, plant air supplies must be available on a continuous basis, not only while the plant is operating. Also, NFPA 750 limits moisture content to 25 ppm: plant air supplies may not always meet this quality standard. An air compressor that is listed for fire protection service may also be used as an acceptable supply of compressed gas. The unit should be connected to a backup power supply.

Compressed gas supplies must be monitored for high and low pressure, and protected against physical or mechanical damage.

SYSTEM MAINTENANCE CONSIDERATIONS

A number of acceptance and commissioning tests must be performed once the installation of a system is completed. Measures specified in Chapter 9 of the first edition of NFPA 750 include flushing of water supply connections, internal cleansing of piping and tubing, hydrostatic testing and actuation/control verification. Following commissioning, however, the system should be inspected at regular intervals to ensure that it remains in operating condition. As with other fire protection systems, the owner is responsible for system maintenance.

Chapter 10 of NFPA specifies both inspection frequencies and testing frequencies. Inspections are largely visual, and can be conducted (frequently) by a conscientious representative of the owner. Testing is done less frequently, and requires actual hands-on operation of the system: it should be performed by an experienced technician. Many of the recommended inspection and testing steps were compiled from recommendations for other fire protection systems. Some, however, are unique to water mist systems, as they involve equipment that is non-standard for fire protection systems. Examples include the use of high-pressure, positive displacement pumps; electrically actuated sectional control valves; pressure-regulating valves, and so on. It is expected that, as experience with water mist systems grows, the inspection/testing requirements will change, and eventually be transferred to NFPA 25.

BIBLIOGRAPHY

References Cited

1. Civil Aviation Authority, "International Cabin Water Spray Research Management Group: Conclusions of Research Programme," CAA Paper 93012, 1993.
2. Hill, R. G., Marker, T. R. and Sarkos, C. P, "Evaluation and Optimization of an On-Board Water Spray Fire Suppression System in Aircraft," *Proceedings* of Water Mist Fire Suppression Workshop, NIST, Gaithersburg, MD, 1993.
3. Wighus, R., "Extinguishment of Enclosed Gas Fires with Water Spray," *Proceedings* of the Third International Symposium on Fire Safety Science, The University of Edinburgh, Edinburgh, Scotland, 1991.
4. Wighus, R., Aune, P., Drangsholt, G., and Stensaas, J. P., "Fine Water Spray System—Extinguishing Tests in Medium- and Full-Scale Turbine Hood," *Proceedings* of the International Water Mist Conference, Swedish Testing Institute, Borås, Sweden, Nov. 5-7, 1993.
5. Braidech, M. M., Neale, J. A., Matson, A. F., and Dufour, R. E., "The Mechanism of Extinguishment of Fire by Finely Divided Water," Underwriters Laboratories Inc. for the National Board of Fire Underwriters, New York, 1955.
6. Rasbash, D. J., Rogowski, Z. W., and Stark, G. W. V., "Mechanisms of Extinction of Liquid Fuel Fires with Water Sprays," *Combustion and Flame*, 4, 1960, pp. 223-234.
7. Mawhinney, J. R., "Engineering Criteria for Water Mist Fire Suppression Systems," *Proceedings* of Water Mist Fire Suppression Workshop, NIST, Gaithersburg, MD, 1993.
8. Hanauska, C. P., and Back, G. G., "Halons: Alternative Fire Protection Systems, An Overview of Water Mist Fire Suppression Systems Technology," Hughes Associates, Inc., Columbia, MD, 1993.
9. International Maritime Organization, "Draft Test Method for Equivalent Sprinkler System for Class A Machinery Compartments (up to 500 m3) Volume," IMO Fire Test Procedures Correspondence Group, submittal for the IMO Fire Protection Subcommittee, July 1995.
10. Personal correspondence, Darwin, R., NAVSEA, Crystal City, VA, July 1995.
11. Berner, F., and Vemmestad, J., "Development of Fine Water Spray Technology for Application on BP Norwegian Continental Shelf Offshore Installations ULA and GYDA," in *Proceedings*: Halon Alternatives Conference, Phoenix, AZ, Nov. 1993.
12. International Maritime Organization, "Draft Test Protocol for Equivalent Sprinkler System for Cabin and Corridor Fires," July 1995.
13. Hansen, *et al.*, "Tyndal Airforce Base Hush-House Mist Trials," test report provided to the NFPA 750 Water Mist Committee, Jan. 25, 1995.
14. Mawhinney, J. R., Dlugogorski, B. Z., and Kim, A. K., "A Closer Look at the Extinguishing Properties of Water Mist," in *Proceedings*: International Association for Fire Safety Science (IAFSS) Conference, Ottawa, Canada, June 13-17, 1994.
15. Jones, A., and Thomas, G. O., "The Action of Water Sprays on Fires and Explosions," *Transactions* of the Institution of Chemical Engineers, 71: Part B, 41-49, 1993.
16. Hadjisophocleous, G. V., Kim, A. K., and Knill, K., "Physical and Numerical Modeling of the Interaction Between Watersprays and a Fire Plume," National Research Council, Canada, 1994.
17. Drysdale, D., *An Introduction to Fire Dynamics*, 1st ed., pp. 222-225, John Wiley and Sons, New York, 1985.
18. Kanury, A. M., "Ignition of Liquid Fuels," *The SFPE Handbook of Fire Protection Engineering*, 2nd. ed., DiNenno, P.J., *et al.*, National Fire Protection Association, Quincy, MA, 1995.
19. Mawhinney, J. R., "Water Mist Fire Suppression Systems for Marine Applications: A Case Study," *Proceedings* of IMAS 94: Fire Safety on Ships—Developments into the 21st Century, Institute of Marine Engineers, London, UK, May 1994.
20. Mawhinney, J. R., "Characteristics of Water Mist for Fire Suppression in Enclosures," *Proceedings* of the Halon Alternatives Technical Working Conference, New Mexico Engineering Research Institute (NMERI), Albuquerque, NM, 1993.
21. An, M. W., Sousa, A. C. M., and Venart, J. E. S., "Modeling of a Shipboard Waterfog Fire Suppression System," University of New Brunswick, Report No. 991-1042-2045, 1994, Fredericton, NB.
22. Coppalle, A., Nedelka, D., and Bauer, B., "Fire Protection: Water Curtains," *Fire Safety Journal*, 20, 1993, pp. 241-255.
23. Ravigururajan, T. E., and Beltrav, M. P., "A Model for Attenuation of Fire Radiation Through Water Droplets," *Fire Safety Journal*, 15: 2, 1989, pp. 171-181.
24. Kokkala, M. A., "Extinction of Liquid Pool Fires with Sprinklers and Water Sprays," Valtion Teknillinen Tutkimuskeskus, Statens Teniska Forskningscentral (Technical Research Centre of Finland), Espoo, Finland, 1989.
25. ASTM E799, *Standard Practice for Determining Data Criteria and Processing for Liquid Drop Size Analysis*, American Society for Testing and Materials, W. Conshohocken, PA, 1992.
26. Lefebvre, A., *Atomization and Sprays*, Hemisphere Publishing Corporation, New York, 1989.

27. Dlugogorski, B. Z., Mawhinney, J. R., Kim, A. K., and Crampton, G., unpublished laboratory data, National Fire Laboratory, National Research Council Canada, Ottawa, 1994.
28. Grosshandler, W., Lowe, D., Notarianni, K., and Rinkinen, W., "Protection of Data Processing Equipment with Fine Water Sprays," NISTIR 5514, 1994, National Institute of Standards and Technology, Gaithersburg, MD.
29. Mawhinney, J. R., Taber, B., and Su, J. Z., "The Fire Extinguishing Capability of Mists Generated by Flashing of Super-Heated Water," *Proceedings*: American Institute of Chemical Engineers, 1995 Summer National Meeting, Boston, MA, July 30 to Aug. 2, 1995.
30. Ewing, C. T., Faith, F. R., Romans, J. B., and Hughes, J. T., "Extinguishment Properties of Dry Chemicals: Extinction Weights for Small Diffusion Pan Fires and Additional Evidences for Flame Extinguishment by Thermal Mechanisms," *Journal of Fire Protection Engineering*, Vol. 4 , 1992, pp. 35-52.
31. Cortina, T. A., ed., "Water Mist Fire Suppression System Health Hazard Evaluation: Response to Questions Posed by the US Environmental Protection Agency," Halon Alternatives Research Corporation, Washington, DC, August 1995.
32. U.S. Environmental Protection Agency, "Protection of Stratospheric Ozone; Acceptable Substitutes for the Significant New Alternatives Policy (SNAP) Program," 40 CFR Part 82, U.S. Federal Register, Vol. 60, No. 145, Friday, July 28, 1995, p. 38731.
33. Mawhinney, J. R., Hadjisophocleous, G. V., "The Role of Fire Dynamics in Design of Water Mist Fire Suppression Systems" in Proceedings: Interflam '96, Cambridge, UK, March 26-28, 1996 (in press).
34. Mawhinney, J. R., and Carpenter, D. W., "Waterfog Fire Suppression System Project Full-Scale Fire Tests Summary Report," for Department of National Defense, Canada, Navy (DMEE 4-2), Client Report CR-A4007.4, Mar. 1993, National Research Council Canada, Ottawa.
35. YARD Report YR 3693, "Project Hulvul: Water-fog Evaluation Trials," British Ministry of Defense, Navy, 1988.
36. Brown, R., and York, J. L., "Sprays Formed by Flashing Liquid Jets," *American Institute of Chemical Engineers Journal*, Vol. 8, No. 21, May 1962, pp. 149-153, 1962.
37. Sallet, D. W., Rod, S. R., Palmer, M. E., Nastoll, W., and Guhler, M., "Discharge of Flashing Liquid Through Tubes," *Proceedings*: Second Multi-Phase Flow and Heat Transfer Symposium—Workshop, Miami Beach, FL, 1979.
38. Keary, J., and McGovern, F., "Characterization of a Heated Water Extinguishing System: Report Prepared for M. O'Connell," Department of Experimental Physics, University College Galway, Galway, Ireland, May 1994.
39. O'Connell, M., Personal Correspondence, Ballineen, County Cork, Ireland, 1994.
40. Titus, J. J. "Hydraulics*," SFPE Handbook of Fire Protection Engineering*, 2nd ed., DiNenno, P. J., *et al.*, eds., National Fire Protection Association, Boston, MA, 1995.
41. Crane Co., "Flow of Fluids Through Valves, Fittings, and Pipe," Technical Paper 410M, Crane Valves, St. Laurent, Quebec, Canada, 1986.
42. ANSI B31-1, *Power Piping*, American National Standards Institute, New York, 1995.
43. Underwriters Laboratories, Inc., "Proposed First Edition of the Standard for Water Mist Nozzles for Fire Protection Service, UL 2167, January 1996"

NFPA Codes, Standards, and Recommended Practices

Reference to the following NFPA codes, standards, and recommended practices will provide further information on water mist fire suppression systems discussed in this chapter. (See the latest version of *The NFPA Catalog* for availability of current editions of the following documents.)

NFPA 13, *Standard for the Installation of Sprinkler Systems*
NFPA 14, *Standard for the Installation of Standpipe Systems*
NFPA 15, *Standard for Water Spray Fixed Systems for Fire Protection*
NFPA 20, *Standard for the Installation of Centrifugal Fire Pumps*
NFPA 72, *National Fire Alarm Code*
NFPA 750, *Standard on the Installation of Water Mist Fire Protection Systems*

Additional Reading

Arvidson, M., "Efficiency of Different Water Mist Systems in a Ship Cabin," Swedish National Testing and Research Inst., Boras, Sweden, SP Report 1994:03; *Proceedings* of the International Conference on Water Mist Fire Suppression Systems, November 4–5, 1993, Boras, Sweden, 1993, pp. 45–62.
Atreya, A., Crompton, T., and Suh, J., "Experimental and Theoretical Study of Mechanisms of Fire Suppression by Water," Michigan Univ., Ann Arbor, NISTIR 5499, September 1994.; National Institute of Standards and Technology, Annual Conference on Fire Research: Book of Abstracts, Oct. 17–20, 1994, Gaithersburg, MD, 1994, pp. 67–68.
Back, G. G., "1995 Progress Report: Water Mist Fire Suppression System Technologies," *SFPE Bulletin*, Fall 1995, pp. 11–18.
Back, G. G., "Experimental Evaluation of Water Mist Fire Suppression System Technologies Applied to Flammable Liquid Storeroom Applications," *Proceedings* of the National Institute of Standards and Technology (NIST) and Society of Fire Protection Engineers (SFPE) International Conference on Fire Research and Engineering, September 10-15, 1995, Orlando, FL, SFPE, Boston, MA, 1995, pp. 127–132.
Back, G. G., "Water Mist as a Total Flooding Agent," *Proceedings* of the Technical Symposium on Halon Alternatives, June 27–28, 1994, Knoxville, TN, 1994, pp. 65–70.
Back, G. G., Darwin, R. L., and Leonard, J. T., "Full Scale Tests of Water Mist Fire Suppression Systems for Navy Shipboard Machinery Spaces," Hughes Associates, Inc., Baltimore, MD, Naval sea Systems Command, Washington, DC, Naval Research Laboratory, Washington, DC, ISBN 0-9516320-9-4; *Proceedings* of the 7th International Interflam Conference, Interflam '96, March 26–28, 1996, Cambridge, England, Interscience Communications Ltd., London, England, 1996, pp. 435–444.
Back, G. G., *et al.* "Evaluation of Water Mist Fire Extinguishing Systems for Flammable Liquid Storeroom Applications on U.S. Army Watercraft," Hughes Associates, Inc., Baltimore, MD, Naval Research Laboratory, Washington, DC, Department of Transportation, Cambridge, MA, NRL/MR/6180-95-7795, November 30, 1995.
Back, G. G., *et al.,* "Full Scale Tests of Water Mist Fire Suppression Systems for Navy Shipboard Machinery Spaces. Part 2. Obstructed Spaces," September 1993–December 1994, Hughes Associates, Inc., Baltimore, MD, Naval Sea Systems Command, Washington, DC, Naval Research Laboratory, Washington, DC, Naval Sea Systems Command, Washington, DC, NRL/MR/6180-96-7831, March 8, 1996.
Back, G. G., *et al.,* "Full Scale Tests of Water Mist Fire Suppression Systems for Navy Shipboard Machinery Spaces. Part 1. Unobstructed Spaces," September 1993–December 1994, Hughes Associates, Inc., Baltimore, MD, Naval Sea Systems Command, Washington, DC, Naval Research Laboratory, Washington, DC, Naval Sea Systems Command, Washington, DC, NRL/MR/6180-96-7830, March 8, 1996.
Beyler, C. L., *et al.*, "Review of Water Mist Technology for Fire Suppression," Naval Research Laboratory, Washington, DC, Office of Naval Research, Arlington, VA, NRL/MR/6189-94-7624, Sep. 30, 1994.
Bill, R. G., Jr., "Water Mist in Residential Occupancies," Factory Mutual Research Corp., Norwood, MA, National Institute of Standards and Technology, *Proceedings* of the Demonstration of Limited Water Supply Fire Suppression Technologies, March 12, 1996, Gaithersburg, MD, 1996, pp. 1–19.
Budnick, E. K., "USFA Residential Water Mist Test Program (Initial Feasibility)," Hughes Associates, Inc., Baltimore, MD, National Institute of Standards and Technology, *Proceedings* of the Demonstration of Limited Water Supply Fire Suppression Technologies, March 12, 1996, Gaithersburg, MD, 1996, pp. 1–48.
Butz, J. R., and French, P., "Application of Fine Water Mists to Hydrogen Deflagrations," *Proceedings* of the 1993 Halon Alternatives Technical Working Conference, May 11–13, 1993, Albuquerque, NM, 1993, pp. 345–355.
Cousin, C. S., "Developments in the Use of Water Fog," Fire Research Station, Borehamwood, England, FM260, PD 58/93, Feb. 1993.
Gameiro, V. M., "Fine Water Spray Fire Extinguishing System," *Proceedings* of the Technical Symposium on Halon Alternatives, June 27–28, 1994, Knoxville, TN, 1994, pp. 71–84.
Gameiro, V. M., and Girard, P., "Fine Water Spray Fire Suppression Alternative to Halon 1301 in Machinery Spaces," Securiplex Technologies Inc., Dorval, Canada, U. S. Environmental Protection Agency, Environment Canada, and United National Environmental Program, Inter-

national CFC and Halon Alternatives Conference, 1993, Stratospheric Ozone, Protection for the 90's, October 20–22, 1993, Washington, DC, 1993, pp. 830–839.

Gameiro, V. M., "Fine Water Spray Fire Suppression Alternative to Halon 1301 in Gas Turbine Enclosures," *Proceedings* of the 1993 Halon Alternatives Technical Working Conference, May 11–13, 1993, Albuquerque, NM, 1993, pp. 317–344.

Grosshandler, W. L., *et al.*, "Protection of Data Processing Equipment With Fine Water Sprays," Annual Report, September 1993–September 1994, National Institute of Standards and Technology, Gaithersburg, MD, Federal Emergency Management Agency, Emmitsburg, MD, NISTIR 5514, October 1994.

Hanauska, C. P., and Back, G. G., "Halons: Alternative Fire Protection Systems—An Overview of Water Mist Fire Suppression System Technology," Hughes Associates, Inc., Columbia, MD, U. S. Environmental Protection Agency, Environment Canada, and United National Environmental Program, International CFC and Halon Alternatives Conference, 1993, Stratospheric Ozone, Protection for the 90's, October 20–22, 1993, Washington, DC, 1993, pp. 820–829.

Hills, A. T., Simpson, T., and Smith, D. P., "Water Mist Fire Protection Systems for Telecommunication Switch Gear and Other Electronic Facilities," Fire and Safety International Research, Slough, England, NISTIR 5207, June 1993; National Institute of Standards and Technology, *Proceedings* of the Water Mist Fire Suppression Workshop, March 1–2, 1993, Gaithersburg, MD, 1993, pp. 123–142.

Holmstedt, G., "Extinction Mechanisms of Water Mist," Lund Univ., Sweden, SP Report 1994:03; *Proceedings* of the International Conference on Water Mist Fire Suppression Systems, November 4–5, 1993, Boras, Sweden, 1993, pp. 89–100.

Jacobsen, S. E., "Approval of Water Mist Systems on Ships. Consideration of Equivalency to Sprinkler and Water Spray Systems," Det Norske Veritas Classification AS, Norway, SP Report 1994:03; *Proceedings* of the International Conference on Water Mist Fire Suppression Systems, November 4–5, 1993, Boras Sweden, 1993, pp. 171–175.

Kim, A. K., Dlugogorski, B. Z., and Mawhinney, J. R., "Effect of Foam Additives on the Fire Suppression Efficiency of Water Mist," National Research Council of Canada, Ottawa, Ontario, NRCC 37902, IRC Paper 3748; *Proceedings* of the Halon Options Technical Working Conference, May 3–5, 1994, Albuquerque, NM, 1994, pp. 347–358.

Ladouceur, H. D., Fleming, J. W., and Brown, E. F., "Fire Suppression by Water Mist: A Numerical Study," *Proceedings* of the 1995 Fall Technical Meeting of the Combustion Institute/Eastern States Section on Chemical and Physical Processes in Combustion, October 16–18, 1995, Worcester, MA, 1995, pp. 455–458.

Leonard, J. T., Darwin, R. L., and Back, G. G., "Full Scale Tests of Water Mist Fire Suppression Systems for Machinery Spaces," *Proceedings* of the National Institute of Standards and Technology (NIST) and Society of Fire Protection Engineers (SFPE) International Conference on Fire Research and Engineering, September 10–15, 1995, Orlando, FL, SFPE, Boston, MA, 1995, pp. 109–114.

Log, T., "Radiant Heat Attenuation in Fine Water Sprays," Stord/Haugesund College, Norway, ISBN 0-9516320-9-4; *Proceedings* of the 7th International Interflam Conference, Interflam '96, March 26–28, 1996, Cambridge, England, Interscience Communications Ltd., London, England, 1996, pp. 425–434.

Marttila, P., "Water Mist in Total Flooding Applications," *Proceedings* of the 1993 Halon Alternatives Technical Working Conference, May 11–13, 1993, Albuquerque, NM, 1993, pp. 309–316.

Mawhinney, J. R., "Water Mist Suppression Systems May Solve an Array of Fire Protection Problems," *NFPA Journal*, Vol. 88, No. 3, May/June 1994, pp. 46–50, 52–54, 56–57.

Mawhinney, J. R., "Water-Mist Fire Suppression Systems for the Telecommunication and Utility Industries," *Fire Research News*, No. 74, Fall 1994, pp. 1–3.

Mawhinney, J. R., and Hadjisophocleous, G. V., "Role of Fire Dynamics in Design of Water Mist Fire Suppression Systems," National Research Council of Canada, Ottawa, Ontario, ISBN 0-9516320-9-4; *Proceedings* of the 7th International Interflam Conference, Interflam '96, March 26–28, 1996, Cambridge, England, Interscience Communications Ltd., London, England, 1996, pp. 415–424.

Mawhinney, J. R., "Characteristics of Water Mists for Fire Suppression in Enclosures," *Proceedings* of the 1993 Halon Alternatives Technical Working Conference, May 11–13, 1993, Albuquerque, NM, 1993, pp. 291–302.

Mawhinney, J. R., "Design of Water Mist Fire Suppression Systems for Shipboard Enclosures," National Research Council of Canada, Ottawa, Ontario, SP Report 1994:03; *Proceedings* of the International Conference on Water Mist Fire Suppression Systems, November 4–5, 1993, Boras, Sweden, 1993, pp. 16–44.

Mawhinney, J. R., "Engineering Criteria for Water Mist Fire Suppression Systems," National Research Council of Canada, Ottawa, Ontario, NISTIR 5207, June 1993; National Institute of Standards and Technology, *Proceedings* of the Water Mist Fire Suppression Workshop, March 1–2, 1993, Gaithersburg, MD, 1993, pp. 37–73.

Milke, J. A., and Gerschefski, C. E., "Overview of Water Mist Research for Library Applications," *Proceedings* of the National Institute of Standards and Technology (NIST) and Society of Fire Protection Engineers (SFPE) International Conference on Fire Research and Engineering, September 10-15, 1995, Orlando, FL, SFPE, Boston, MA, 1995, pp. 133–138.

Ndubizu, C., Motevalli, V., and Tatem, P., "Study of the Suppression Mechanisms of Small Flames by Water Mist," *Proceedings* of the 1995 Fall Technical Meeting of Combustion Institute/Eastern States Section on Chemical and Physical Processes in Combustion, October 16–18, 1995, Worcester, MA, 1995, pp. 451–454.

Notarianni, K. A., "Water Mist Fire Suppression Workshop Summary," *SFPE Bulletin*, Summer 1993, pp. 8–9.

Notarianni, K. A., and Jason, N. H., "Water Mist Fire Suppression Workshop," National Institute of Standards and Technology, Gaithersburg, MD, NISTIR 5207, June 1993; National Institute of Standards and Technology, *Proceedings* of the Water Mist Fire Suppression Workshop, March 1–2, 1993, Gaithersburg, MD, 1993.

Notarianni, K. A., "Water Mist Fire Suppression Systems," National Institute of Standards and Technology, Gaithersburg, MD, Society of Fire Protection Engineers and PLC Education Foundation, *Proceedings* of the Technical Symposium on Halon Alternatives, June 27–28, 1994, Knoxville, TN, 1994, pp. 57–64.

Papavergos, P. G., "Fine Water Sprays for Fire Protection: A Halon Replacement Option," B. P. Research, Sunbury-on-Thames, England, 1990, 13 pages; *Proceedings* of the 1991 Halon Alternatives Technical Working Conference, April 30-May 1, 1991, Albuquerque, NM, 1990, pp. 206–217.

Parsons, M. L., and Swartwood, K. S., "Investigation Into Halon 1301 Replacements: A Water-Based System for the Airline Lavatory Application," *Proceedings* of the 1993 Halon Alternatives Technical Working Conference, May 11-13, 1993, Albuquerque, NM, 1993, pp. 611–620.

Pepi, J. S., "Fire Test Status Report on the Evaluation of the AquaMist Fixed Water Mist Deluge System in Ventilated Marine Machinery Spaces," *Proceedings* of the Technical Symposium on Halon Alternatives, June 27–28, 1994, Knoxville, TN, 1994, pp. 85–102.

Ransden, N., "Water Mist: a Status Update," *Fire Prevention*, No. 287, March 1996, pp. 16–21.

Ryderman, A., "Development of Standards and Test Methods for Water Mist Systems," Swedish National Testing and Research Inst., Boras, Sweden, SP Report 1994:03; *Proceedings* of the International Conference on Water Mist Fire Suppression Systems, November 4–5, 1993, Boras Sweden, 1993, pp. 8–15.

Schatz, H., "Brand und Loschversuche mit Wassernebel an Pappkartons und Kasten aus Polypropylen im Hochregal. [Fire- and Extinguishing Tests With Water Fog on Cardboard Boxes and Boxes of Polyethylene in High Rack Storage.]," *VFDB*, Vol. 45, No. 1, Feb. 1996, pp. 31–36.

Simpson, T., and Smith, D. P., "Fully Integrated Water Mist Fire Suppression System for Telecommunications and Other Electronics Cabinets," Fire and Safety International, UK, SP Report 1994:03; *Proceedings* of the International Conference on Water Mist Fire Suppression Systems, Nov. 4–5, 1993, Boras Sweden, 1993, pp. 153–166.

Smith, D. P., "Water Mist: Fire Suppression Systems," *Fire Safety Engineering*, Vol. 2, No. 2, April 1995, pp. 10–15.

Sprakel, K. J., "Water Fog: An Alternative to Fighting Fires With Plain Sprinklers, Foam, Powder or Gas?" KAMAT-Pumpen GmbH and Co KG, ISBN 0-9527398-3-6; Institution of Fire Engineers; University of Sunderland, Fire Research Station, CIB W14, Tyne and Wear Metropolitan Fire Brigade, Fire Safety by Design, Conference *Proceedings*, Volume 3, Research Papers, July 10–12, 1995, UK, 1995, pp. 277–280.

Tuomisaari, M., "Vesisumu halonien korvaajana? [Water Fog as a Replacement for Halons?]," *Palontorjuntatekniikka*, Vol. 1, 1994, pp. 6–8.

Turner, A. R. F., "Water Mist in Marine Applications," Marioff Hi-Fog Oy, Vantaa, Finland, NISTIR 5207, June 1993; National Institute of Standards and Technology, *Proceedings* of the Water Mist Fire Suppression Workshop, March 1–2, 1993, Gaithersburg, MD, 1993, pp. 105–122.

"Water Fog in Western Europe," *Fire Engineering*, Vol. 148, No. 9, September 1995, p. 52.

"Water Mist Fire Protection in Telecommunications Facilities," International Symposium on Fire Protection for the Telecommunications Industry, May 14–15, 1992, New Orleans, LA, 1992, pp. 1–12.

"Water Replaces Halon in Firefighting Technology," *Petroleum Review*, Vol. 46, No. 549, Oct. 1992, p. 508.

Wighus, R.,, *et al.*, "Full Scale Mist Experiments," Norwegian Fire Research Lab., Trondheim, Norway, SP Report 1994: 03; Proceedings of the International Conference on Water Mist Fire Suppression Systems, Nov. 4–5, 1993, Boras, Sweden, 1993, pp. 101–152.

STANDPIPE AND HOSE SYSTEMS

Jeffrey M. Shapiro

Standpipe systems are fixed piping systems and associated equipment that provides a means to transport water from a reliable water supply to designated areas of buildings where hoses can be deployed for fire fighting. Such systems are typically provided in tall and large-area buildings.

By eliminating the need for long and cumbersome hose lays from fire apparatus to a fire, standpipe systems can significantly improve the efficiency of manual fire-fighting operations. Even in buildings that are protected by automatic sprinklers, standpipe systems play an essential role in building fire safety by serving as a backup for, and complement to, sprinklers.

The U.S. standard regulating the design, installation, and testing of standpipe systems is NFPA 14, *Standard for the Installation of Standpipe and Hose Systems* (hereinafter referred to as NFPA 14). Note that NFPA 14 does not cover where standpipe systems are required. Required installations are set forth in code documents, such as NFPA *101®*, *Life Safety Code®*; NFPA 1, *Fire Prevention Code*; and model building codes or they may be established by insurance regulations.

NFPA 13, *Standard for the Installation of Sprinkler Systems*, is a useful reference when considering hydraulically designed systems or combined sprinkler and standpipe systems. Other NFPA standards and recommended practices that may be helpful when dealing with standpipe systems are included in the bibliography at the end of this chapter.

For other information relevant to standpipe systems, see Section 6, Chapter 2, "Water Storage Facilities and Suction Supplies"; Section 6, Chapter 4, "Fire Pumps"; and Section 6, Chapter 17, "Care and Maintenance of Water-Based Extinguishing Systems."

STANDPIPE SYSTEM DESIGN CONSIDERATIONS

All standpipe systems share the common purpose of delivering water for manual fire fighting. However, system designs used to accomplish this purpose will vary. While one system may involve a simple pipe network for conveying water from fire department apparatus to inside hose connections, another may involve a fully automatic water supply and preconnected hoses.

The design of a standpipe system is generally dictated by NFPA 14 but local, state, and model building codes and insurance standards may also have an influence. The requirements among these codes and standards may vary, so it is essential that an individual interested in standpipe system design requirements determine the applicable codes and standards adopted by the regulating entity.

Because the 1996 edition of NFPA 14 is recognized as the U.S. design standard for standpipe systems, the following discussion on design considerations focuses on the regulatory requirements in NFPA 14. In addition, model code requirements and engineering and historical perspectives are also discussed.

The Design Process

The design process begins by determining the intended use of a system, i.e., whether it is for full-scale fire fighting, first-aid fire fighting, or both. These three uses correspond with the three categories of standpipe systems, commonly referred to as "classes" of systems. Most aspects of system design, such as the water supply, layout, and system components, are affected or dictated by the class of system. Other aspects, such as whether piping is required to be wet or dry, may involve other factors, such as building height and area.

Classes of Systems

The three classes of standpipe systems are designated as Class I, Class II, and Class III. Some sources refer to a fourth class of system called a "combined" system. In fact, combined systems are simply Class I or Class III standpipe systems that also supply water to a sprinkler system. Note that sprinkler systems with occasional hose connections are not considered to be combined systems.

The design of a combined system is similar to any other Class I or Class III system, except that the water supply and pipe sizes may be different, based on the sprinkler system.

Class I systems: Class I systems provide $2\frac{1}{2}$-in. (64-mm) hose connections at designated locations in a building for full-scale fire fighting. These systems are generally intended for use by fire departments.

Class I systems are generally required in both sprinklered and nonsprinklered buildings that are more than three stories high, because of the excessive time and personnel required to lay hoses from

Jeffrey M. Shapiro, P.E., is president of International Code Consultants, a consulting engineering firm specializing in the development and application of building and fire codes based in Austin, TX. He serves as secretary to the NFPA Committee on Standpipes.

fire apparatus to floors above the third story. Class I systems are commonly required in malls for similar reasons.

Class II systems: Class II systems provide $1^1/_2$-in. (38-mm) hose connections at designated locations in a building for first-aid fire fighting. These systems are generally intended for use by fire brigades and perhaps building occupants before the fire department arrives. They are occasionally provided for use by fire departments, as well. With Class II systems, a hose, a nozzle, and a hose rack are typically installed on each hose connection.

Class II systems are often required in large unsprinklered buildings. They may also be required to protect special hazard areas, such as exhibit halls and stages.

Members of the fire protection community have historically disagreed about the desirability of having standpipe systems available for occupant use. Concerns focus on the ability of untrained occupants to safely use a 100-ft (30.5-m) long hose flowing up to 100 gpm (378 L/min) and on the wisdom of encouraging occupants to fight a fire instead of evacuating. These concerns have led to a trend of reducing the requirements for, or eliminating the installation of, standpipe systems with preconnected hoses. Consequently, the use of Class II systems is declining.

Class III systems: Class III systems combine the features of Class I and Class II systems. They are provided for both full-scale and first-aid fire fighting. These systems are generally intended for use by fire departments, fire brigades, and perhaps building occupants. Because of their multiple uses, Class III systems are provided with both Class I and Class II hose connections. This is sometimes accomplished by using $2^1/_2$-in. (64-mm) hose valves with easily removable $2^1/_2$-in. (64-mm) to $1^1/_2$-in. (38-mm) adapters that are permanently chained to hose connections.

Class III systems are sometimes used when both Class I and Class II systems are required and no useful purpose would be served by installing separate systems. Like Class II systems, the use of Class III systems is declining because of concerns about the safety and effectiveness of untrained occupants fighting fires.

Types of Systems

In addition to being subdivided into classes, standpipe systems are further designated into categories, referred to as "types." Standpipe classes delineate the intended use of a system, i.e., fire department use or occupant use. Standpipe types delineate the basic characteristics of system design, i.e., whether the piping will be filled with water or not (wet or dry), and whether the water supply for fire fighting will be automatically available or not (automatic, semiautomatic, or manual).

Beginning with the 1993 edition of NFPA 14, standpipe system types were completely redefined. The result was the creation of five new categories, as follows:

1. *Automatic-wet systems* have piping that is filled with water and have a water supply capable of supplying the fire-fighting water demand that is automatically available.
2. *Automatic-dry systems* have piping that is normally filled with pressurized air; systems are arranged, through the use of devices such as a dry-pipe valve, to automatically admit water into system piping from a water supply capable of supplying the fire-fighting water demand when a hose valve is opened.
3. *Semiautomatic-dry systems* have piping that is normally filled with air; systems are arranged through the use of devices, such as a deluge valve, to admit water into system piping from a water supply capable of supplying the fire-fighting water demand

when a remote actuation device located at a hose station is operated.
4. *Manual-dry systems* have piping that is normally filled with air; there is no permanent water supply connected to such systems, and a fire department connection must be used to manually supply fire-fighting water demands.
5. *Manual-wet systems* have piping that is normally filled with water from a domestic water-fill connection for the purpose of ensuring detection of leaks and maintaining system integrity; there is no permanent water supply capable of supplying fire-fighting water demands, and a fire department connection must be used to supply the system manually.

In summary, wet systems have water-filled piping, and dry systems do not. Automatic systems provide a water supply for fire fighting by simply opening a hose valve. Semiautomatic systems are connected to a water supply for fire fighting but require activation of a device at a hose valve in addition to opening the valve to get water. Manual systems must have the water supply for fire fighting provided through a fire department connection.

NFPA 14 specifies several design limitations based on the type of system to be used. For example, manual standpipe systems are not allowed to be used in high-rise buildings. This is partly to provide an increased level of safety in such buildings and partly in recognition of the fact that a sprinkler system will be required by building codes, almost surely necessitating a water supply pump. Once a pump is called for, it is considered reasonable to require that it be designed to supply the standpipe system demand, given the marginal additional cost.

Manual systems are also prohibited for Class II and Class III systems, because systems with preconnected hoses obviously need an available water supply to be of benefit. In addition, dry systems are not allowed for Class II or Class III systems because of the risk to users that would be created by delaying the availability of water. Also, dry systems are only allowed in areas subject to freezing, because wet systems have a higher degree of integrity. Dry standpipes have been known to fail in the past because of open valves or severe leaks that went undetected until the system was charged with water. The NFPA 14 committee considered this risk of failure avoidable for most cases by simply requiring system piping to be filled with water, unless a risk of freezing is present.

Water Supply Requirements

Once the class and type of system to be installed have been identified, the required water supply can be determined. There are three aspects of water supplies that are always of concern: (1) the minimum flow rate required, (2) the minimum pressure required, (3) and the type of supply to be used. In addition, the maximum pressure available may also be of concern.

Minimum flow rate and pressure (system demand): The minimum required flow and pressure are together referred to as "system demand." The system demand for a standpipe system is a function of several variables:

1. The required flow rate and pressure at the hydraulically most remote hose connections.
2. Required flow rates for additional standpipes.
3. The water supply requirements for automatic sprinklers sharing common piping.
4. Required flow duration.
5. Pressure losses from friction and changes in elevation.

Flow rate, pressure, and duration requirements for standpipe systems are determined by the class of system. According to the 1996 edition of NFPA 14, Class I and Class III systems are typically required to deliver 500 gpm (1,893 L/min) for the first standpipe, plus 250 gpm (946 L/min) for each additional standpipe, up to a maximum of 1,250 gpm (4731 L/min). The maximum flow is reduced to 1,000 gpm (3,785 L/m) for combined systems to grant some benefit for the sprinkler protection. NFPA 14 also requires the water supply to be capable of providing the minimum flow rate for 30 min. Figure 6-16A illustrates these design flow rate requirements for Class I and Class III systems.

NFPA 14 also requires that a water supply for Class I and Class III systems be able to deliver a residual pressure of 100 psi (690 kPa) at the outlet of the topmost hose connection of each standpipe. This pressure must be available while flowing 250 gpm (946 L/min) from each of the two topmost hose connections of the hydraulically most remote standpipe, plus 250 gpm (946 L/min) from each additional standpipe, up to the required maximum. Note that, NFPA 14 allows a reduction in the minimum pressure requirement, if desired by the Authority Having Jurisdiction, based on fire-fighting tactics. Specifically, this exception is provided to accommodate fire departments committed exclusively to using smooth-bore nozzles in fire-fighting operations. If smooth-bore nozzles will be the only type to ever be used by the fire department, lower outlet pressures may be adequate.

For Class II systems, NFPA 14 requires that the water supply be capable of delivering 100 gpm (379 L/min) for 30 min. It must also be strong enough to maintain a residual pressure of 65 psi (448 kPa) at the outlet of the hydraulically most remote hose connection with 100 gpm (379 L/min) flowing. See Table 6-16A for a summary of requirements.

In addition to minimum pressure and flow requirements, there are maximum pressure limits for hose connections as well. For a discussion on maximum pressure limits, see the subsection "Maximum Pressure and Zoning in Tall Buildings" later in this chapter.

Also, note that for manual systems (described previously), the system demand can be supplied exclusively through the fire department connection.

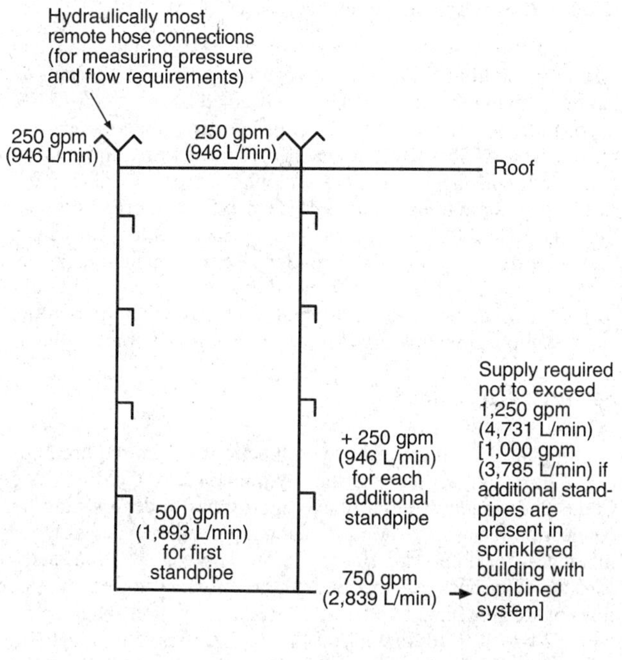

FIG. 6-16A. *Basic flow rate requirements for Class I and Class III systems.*

History on outlet pressures and fire-fighting nozzles: NFPA 14 was modified in the 1993 edition to increase the minimum outlet pressure for Class I and Class III outlets from 65 psi (448 kPa) to 100 psi (690 kPa) because of questions raised regarding the adequacy of a 65-psi (448-kPa) minimum design pressure on automatic standpipes. From a performance perspective, the NFPA 14 committee agreed that the minimum pressure at any hose connection should be adequate to properly operate a nozzle while overcoming friction loss in a hose. From a design perspective, the establishment of a minimum design pressure for hose connections has proven difficult because nozzle operating pressures and friction losses in hoses vary, depending on the equipment used by each local fire department.

The concern over nozzle pressure became an issue when fog nozzles were introduced for fire fighting in the mid-1900s. Before using fog nozzles, fire fighters used solid stream nozzles, which operated well at fairly low inlet pressures. As the use of fog nozzles became common, the fire protection community moved to increase the minimum pressure required at the hydraulically most remote hose connections on standpipe systems to accommodate the greater nozzle pressure required by fog nozzles. In response, the NFPA Committee on Standpipe and Hose Systems originally recommended a pressure of 65 psi (448 kPa), which was widely implemented. In establishing this pressure, the committee had tried to minimize the pressure required for standpipe system water supplies and assumed that nozzle designers and testing laboratories would establish criteria to ensure that fog nozzles would operate satisfactorily with a 65-psi (448-kPa) pressure at hose connections. Unfortunately, this assumption never materialized, and many questioned the 65-psi (448-kPa) minimum pressure requirement.

Until such time as a nationally recognized standard is developed to establish a minimum inlet pressure at which fog nozzles must satisfactorily perform, the minimum pressure required at standpipe hose connections must be chosen conservatively. This is because fog nozzles, unlike solid stream nozzles, vary significantly with respect to the minimum inlet pressure required to achieve satisfactory streams. Though some fire service fog nozzles operate satisfactorily with as little as 70 psi (483 kPa), others have discharge patterns that quickly deteriorate as the nozzle pressure drops below 100 psi (690 kPa). As a result of this inconsistency, the only pressure that will ensure satisfactory nozzle performance in all cases is the pressure at which nozzles are tested and rated, 100 psi (690 kPa). This is the residual test pressure for fog nozzles established in NFPA 1964, *Standard for Spray Nozzles (Shutoff and Tip)*. Also, it is interesting to note that some fire departments now specify nozzles that are rated at lower pressures.

To deliver a pressure of 100 psi (690 kPa) at the nozzle, the pressure at the standpipe hose connection must exceed 100 psi (690 kPa) by a value great enough to accommodate friction loss in the hose connecting the nozzle to the outlet. At a minimum, a design should provide an additional 25 psi (172 kPa) for this purpose. This is adequate to compensate for friction loss in 200 ft (61 m) of 2$\frac{1}{2}$-in. (64-mm) rubber-lined hose flowing 250 gpm (946 L/min), in 200 ft (61 m) of 2-in. (51-mm) rubber-lined hose flowing 125 gpm (473 L/min), and in 150 ft (46 m) of 1 3/4-in. (45-mm) rubber-lined hose flowing about 100 gpm (379 L/min).

Given the controversy surrounding outlet pressures, NFPA 14 reassessed this issue in the 1993 edition and increased the minimum pressure requirement for standpipe hose connections. NFPA 14 now specifies a minimum design outlet pressure of 100 psi (690 kPa). The basis of this figure is an assumption that a fire pump will typically be installed to supply an automatic standpipe and that a pump designed to deliver 100 psi (690 kPa) at the most remote outlet at full system demand will deliver 120 psi (827 kPa) or more at lower flows. This takes advantage of excess pressure at the low-flow end of a fire pump curve. (See NFPA 20, *Standard for the Installation of Centrifugal Fire Pumps.*)

The 120 to 130 Psi (827 to 896 kPa) initial pressure is adequate to accommodate a 100 psi (690 kPa) nozzle inlet pressure plus friction loss in the connecting hose for the first one or two hose lines deployed. It is assumed that, if fire-fighting operations grow beyond this level, fire apparatus will be available to supply additional pressure to meet fire nozzle demands. Alternatively, smooth-bore nozzles could be deployed. If these assumptions are not valid for a particular installation, consideration should be given to exceeding the minimum pressure specified by NFPA 14 to accommodate fire-fighting needs.

Although the 100-psi (690-kPa) minimum design pressure now specified by NFPA 14 may seem excessive when compared to the 65-psi (448-kPa) pressure specified in the past it is consistent with established firefighting practices. Fire departments for decades have used 100 psi (690 kPa) as the target nozzle pressure when pumping hoses supplying fog nozzles. In fact, fire hoses supplying fog nozzles are routinely pumped by fire apparatus at 150 psi (1,034 kPa) or more in ground-level fire-fighting operations.

For hose connections on Class II or Class III systems with preconnected hoses installed, nozzles with operating pressures in the 50-psi (345-kPa) range should be specified, recognizing that this may require special nozzle testing. By purchasing low-pressure nozzles, the minimum pressure at the hydraulically most remote hose connection with preconnected hoses can be limited to 65 psi (448 kPa).

Type and number of water supplies: Manual standpipe systems may be supplied exclusively by a fire department connection (FDC), with a reliable and adequate water source accessible nearby.

Automatic and semiautomatic standpipe systems require a minimum of one preconnected water supply that is capable of supplying the standpipe system's hydraulic demand for the minimum required duration. Several types of water supplies may be acceptable. These include:

1. Public waterworks systems, with a booster pump when higher pressures are required
2. Fire pumps connected to a reliable fixed water source
3. Pressure tanks
4. Gravity tanks

In addition to the primary water supply on an automatic or semiautomatic standpipe system, one or more FDCs with a reliable water source accessible nearby are required.

Special water supply requirements: In some cases, special water supply requirements may be applicable. For example, buildings in active seismic areas are often required by local regulations to use in-house tanks and pumps to supply standpipe systems rather than relying on public water supplies, which are susceptible to earthquake-induced failure.

Water supply for high-rise buildings: For high-rise buildings, two remotely located FDCs are required for each zone within the pumping range of fire apparatus in addition to the automatic water supply. Two FDCs reduce the possibility of having falling glass cut all supply hoses and interrupt the secondary water supply during a fire.

Very high buildings with floors located above the pumping range of fire department apparatus require a permanent secondary water source instead of FDCs because failure of the primary supply would otherwise render a building completely unprotected. If the primary water supply uses a pump, the backup arrangement could consist of a second pump, driver, and controller, or, perhaps, high-level water storage with additional pumping capability. The point where a system is beyond the capability of fire department apparatus depends on the apparatus' pump and the residual pressure available from the water source. A permanent secondary water source is needed, if the following is "true":

$$AP + RP \leq SD$$

where

AP = apparatus pressure boost available from a fire department pumper measured at the system demand flow rate;

RP = residual pressure available from the public or private water supply measured at the system demand flow rate; and

SD = system demand pressure at the system demand flow rate, measured at the pumper truck outlet supplying the hose to the FDC.

If the system demand exceeds the sum of the fire apparatus pressure boost and the residual supply pressure, fire apparatus will not be able to adequately supply upper portions of the standpipe system, and a secondary water supply is necessary. As a rule, buildings exceeding 300 ft (91.5 m) in height should be evaluated to determine whether the pumping range of the fire department apparatus has been exceeded.

Water supply for mid-rise buildings: An issue that is often raised in the design of standpipe systems for mid-rise buildings is whether the sprinkler demand, the standpipe demand, or both, must be supplied by the automatic water supply for a sprinklered building. The issue can have a major cost impact on mid-rise buildings, because the sprinkler demand can often be supplied by the public water supply without the use of a pump, while the standpipe demand almost always requires a pump.

In the 1993 edition of NFPA 14, an effort was made to provide further guidance of this issue. Although the solution is not clearly stated in the document, the intent is clear.

NFPA 14 now allows a Class I standpipe system to be a manual system in buildings that are not high-rise buildings (defined as a building having the floor level of an occupied floor located 75 ft (23 m) or more above the lowest level of fire department access). Remember that a Class I system that is combined with a sprinkler system is still considered a Class I standpipe system. By allowing the standpipe system to be a manual system, NFPA 14 does not require the water supply connected for supplying sprinklers to meet the standpipe demand. Only the sprinkler system demand must be met. The combined system, from a standpipe perspective, is merely a manual-wet standpipe system which, like any other such system, can be supplied exclusively by the fire department connection.

System Layout

Number and location of hose connections: The required number of hose connections is primarily a function of building design. Codes and standards employ two approaches to determining hose connection locations. The first method, termed the "actual length" method, is used only for $1^{1}/_{2}$-in (38-mm) hose connections and locates connections such that enough are provided to reach all portions of the area served with a 100-ft (30.5-m) hose that has a nozzle with a 20- to 30-ft (6- to 9-m) reach. The 100 ft (30.5 m) length limit is required for ease of handling. Distances should be measured around partitions and obstructions for hose length, and along unobstructed straight-line paths for nozzle reach.

Since standpipe systems with preconnected hoses are usually intended for extinguishing incipient fires before the fire department arrives, hose stations should be located in central areas, such as corridors, where they are clearly visible and readily accessible.

The second method, termed the "exit location" method, is used for $2^1/_2$-in. (64-mm) hose connections and locates connections based upon a building's exiting system. Using this method, hose connections are located in exit stairs, at horizontal exits, and in exit passageways. Since exits must be reasonably distributed within buildings to provide adequate egress, hose connections are considered to be adequately distributed when located at the points of egress from areas served. Figure 6-16B illustrates the exit location method of locating hose connections.

In prior editions of NFPA 14, Class I hose connections were located using the actual length method. In the 1993 and subsequent editions, the method specified is the exit location method. This switch was made because the actual length method encouraged locating hose connections in open-floor areas to meet distance requirements. It is now considered preferable to locate $2^1/_2$-in. (64-mm) hose connections only where they will be protected from a fire, such as within protected stair enclosures or on opposite sides of horizontal exits. This allows fire fighters to hook up to a system in relative safety and to have a charged hose available before entering a fire area.

Some have argued that extending hoses from the inside of an exit enclosure violates enclosure integrity and places occupants at risk. However, the alternative is an unreasonable expectation of fire fighters to enter an area filled with smoke and heated gases to search for a hose connection, placing them in danger unnecessarily. In addition, time lost while searching would prolong an uncontrolled fire, potentially posing a much greater threat to occupants.

NFPA 14 does allow for additional $2^1/_2$-in. (64-mm) hose connections to be required by the fire department when exit travel distances exceed 150 ft (46 m) in nonsprinklered buildings and 200 ft (61 m) in sprinklered buildings. This special option is provided because exit travel distance for occupants going out effectively equates to the maximum hose lay distance for fire fighters going in. If additional connections outside of exit enclosures are considered necessary by the fire department, they should only be located in areas protected by fire-resistive construction so that fire fighters will have a protected environment in which they can connect hoses.

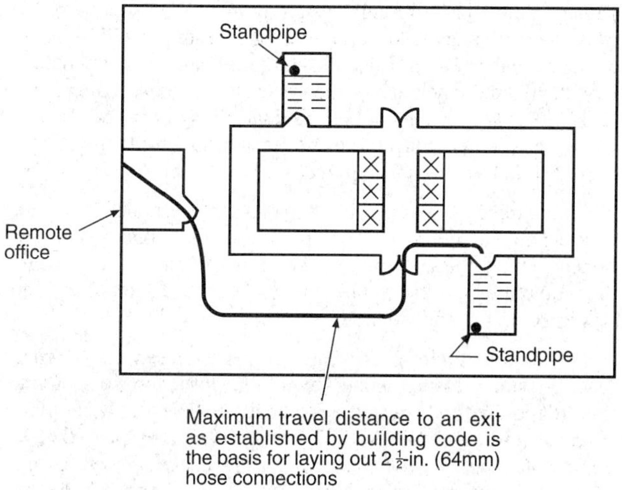

FIG. 6-16B. *Exit location method for locating hose connections.*

One other aspect of locating $2^1/_2$-in. (64-mm) hose connections that is somewhat new and worthy of discussion is the positioning of connections within stair enclosures. In the past, hose connections were located at floor levels within stair enclosures. We now know that this positioning actually hampered fire-fighting efforts. Fire departments typically use connections at the floor below the fire floor to leave the doorway to the fire floor uncongested and give fire fighters a safer location for connecting hoses. Unfortunately, connecting hoses an entire story below the fire floor uses up valuable hose length and obstructs the adjacent doorway. The solution now included in NFPA 14 is to place connections at intermediate floor landings. This allows hoses to be connected somewhat below the fire floor while minimizing the additional hose length required to do so. It also avoids problems with having hose connections located adjacent to doors and areas of wheelchair refuge within stairwells.

In addition to the normal distribution, additional hose connections are sometimes required for protection of special hazards. When there is a risk of exposure from a fire involving an adjacent building, hose connections can be located near openings in exterior walls, at grade along the perimeter (wall hydrants), and on the roof for exposure protection. When large combustible rooftop structures, such as cooling towers, are present, multiple-outlet rooftop hose connections are sometimes provided, as well. This practice is becoming less frequent, due to the increased use of noncombustible construction materials for structures on roofs.

Rooftop hose connections are also required for system testing, since water is easily disposed of from most roofs. An alternative to providing rooftop hose connections is to provide additional connections at the top landings of stair enclosures having stairway access to the roof. From there, hoses can be easily extended to the roof when necessary, and the need to make roof penetrations and provide freeze protection for piping is avoided.

Sizing pipe: There are two methods of sizing standpipe system piping: (1) the pipe schedule method, and (2) the hydraulic method.

With the pipe schedule method, pipe sizes are prescribed by a schedule, in NFPA 14, based upon the class of system and the height of the standpipe. According to the pipe schedule, Class I and Class III standpipes not exceeding 100 ft (30.5 m) in height must be a minimum of 4-in. (102-mm) nominal pipe size, and standpipes over 100-ft (30.5-m) high must be a minimum of 6-in. (152-mm) nominal pipe size [the top 100 ft (30.5 m) may be 4 in. (102 mm)]. NFPA 14 also specifies that standpipes serving both sprinklers and hose connections must be a minimum of 6-in. (152-mm) nominal pipe size, regardless of height, for pipe schedule systems. Figure 6-16C illustrates the pipe schedule sizing method for a Class I system.

For Class II systems, the pipe schedule specifies that standpipes less than 50 ft (15 m) in height must be a minimum of 2-in. (51-mm) nominal pipe size, and standpipes more than 50-ft (15-m) must be not less than $2^1/_2$-in. (64-mm) nominal pipe size. When the pipe schedule design option is used, the water supply must be capable of delivering a residual pressure of 100 psi (690 kPa) at rated flow at the highest $2^1/_2$-in (64-mm) hose connection, and 65 psi (448 kPa) at the highest $1^1/_2$-in. (38-mm) hose connection. This residual pressure need not consider friction loss. Only elevation loss is of concern for pipe schedule systems.

With the hydraulic design method, pipe sizes are calculated so that a system will deliver the minimum required pressure and flow. Pipe sizes must be large enough to allow each water supply, including fire department apparatus, to deliver the required flow at the hydraulically most remote hose connection.

When performing a hydraulic design, the hydraulic characteristics of each water supply must be known. The procedure for determining the hydraulic characteristics of permanent water supplies,

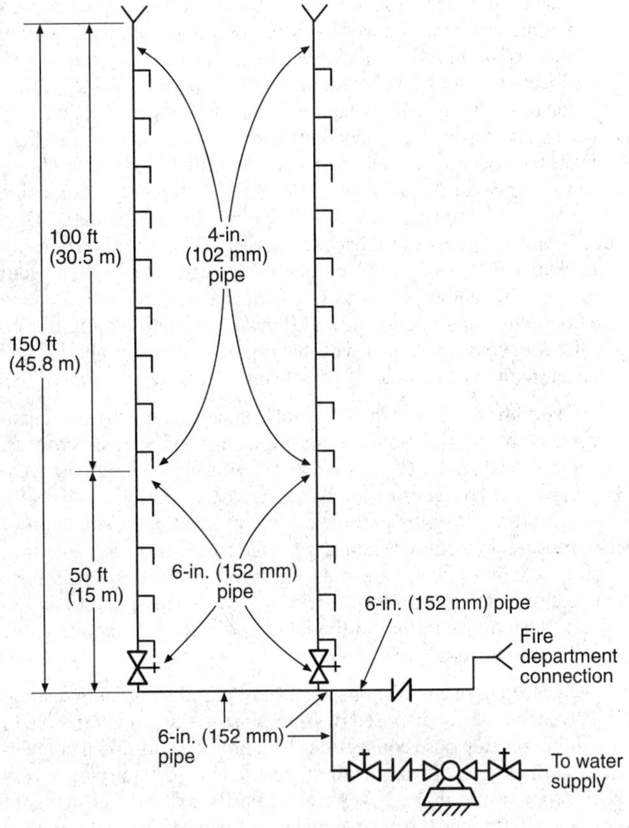

FIG. 6-16C.   *Pipe schedule sizing method for a Class I system.*

such as pumps, is fairly straightforward and is described in NFPA 20 and in other chapters of this handbook. The procedure for determining the hydraulic characteristics of fire apparatus supplying a standpipe system are similar. Lacking better information about local fire apparatus, a conservative design would accommodate a 1,000-gpm (3,785-L/min) fire department pumper performing at the level of design specifications set forth in NFPA 1901, *Standard for Automotive Fire Apparatus* (hereinafter referred to as NFPA 1901). NFPA 1901 specifies that fire department pumpers must be able to achieve three pressure/flow combinations. These are 100 percent of rated capacity at 150-psi (1,034-kPa) net pump pressure, 70 percent of rated capacity at 200-psi (1,379-kPa) net pump pressure, and 50 percent of rated capacity at 250-psi (1,724-kPa) net pump pressure. Therefore, a 1,000-gpm (3,785-L/min) pumper can be expected to deliver no less than 1,000 gpm (3,785 L/min) at 150 psi (1,034 kPa), 700 gpm (2,650 L/min) at 200 psi (1,379 kPa), and 500 gpm (1,893 L/min) at 250 psi (1,724 kPa). Residual supply pressure on the suction side of a pump from a municipal or other pressurized water supply can also be added.

To perform a hydraulic design, one should determine the minimum required pressure and flow at the hydraulically most remote hose connection and calculate this demand back through system piping to each water supply, accumulating losses for friction and elevation and adding flows for additional standpipes and sprinklers at each point where such standpipes or sprinklers connect to the hydraulic design path. When considering fire apparatus as a water supply, flows must be calculated from system piping through the fire department connection and back through connecting hoses to the pump. If the pressure available at each supply source exceeds a standpipe system's pressure demand at the designated flow, the de-

sign is acceptable. Otherwise, the piping design or the water supply must be adjusted.

Maximum pressure and zoning in tall buildings: In tall buildings, standpipe systems are often divided into subsystems to limit system pressure by controlling the maximum height of a water column. These subsystems are called zones. The goal of zoning is to limit the pressure that can be developed in system piping and at hose connections to reduce the need for high-pressure fittings and pressure-reducing valves.

Beginning with the 1993 edition, a performance approach to system zoning was adopted by NFPA 14 to replace the prior fixed height limitation for system zones. The new approach limits zone heights by limiting the maximum pressure to which a system component can be subjected without creating a safety problem or a risk of failure. There are three main criteria affecting the establishment of maximum system pressures that were considered:

1. The maximum pressure rating of system components
2. The maximum pressure that should be permitted at hose connections with and without preconnected hoses
3. The maximum pressure that should be permitted at sprinkler system connections on combined sprinkler and standpipe systems

The maximum pressure in a standpipe system is always limited commensurate with the pressure ratings of system components. Fittings, valves and pressure regulating devices are usually the weakest links among system components. Fittings generally have a maximum pressure rating of 175 psi (1,207 kPa); however, listed fittings are commonly available with pressure ratings of up to 350 psi (2,413 kPa). Limited valves and pressure regulating devices are also available with 350 psi (2,413 kPa) ratings. Since 350 psi (2,413 kPa) is the maximum pressure rating for commonly available and cost-effective equipment, NFPA 14 selected this pressure as the maximum pressure allowed within a system. This became the performance basis for establishing the maximum allowable zone height. However, zone heights must be further limited if equipment with pressure ratings less than 350 psi (2,413 kPa) is used.

Although the maximum zone height corresponding to a particular maximum system pressure can be calculated, zone height limits will be determined by trial and error in most cases. To determine if a single zone can be used, a designer can lay out a system with one zone and calculate the required pressures and flows from hydraulic demand points back through system piping to the supply to determine the system demand. Assuming that the supply will be a pump, a single zone can be justified if a pump is available that would satisfy the hydraulic demand of the system without exceeding 350 psi (2,413 kPa) at churn, including residual supply pressure. If a pump can not be found to meet these criteria, adjustments to the design or subdivision into zones is required.

Standpipe system pressures in the 350 psi (2,413 kPa) range pose a danger to users and a risk of failure of sprinkler system components on combined systems. Therefore, pressure control devices are required when system pressure could exceed safety limits hose or sprinkler at connections.

The safety limits at hose connections are based on two criteria: (1) hose burst pressure, and (2) handling. Hoses for fire department use are required to be tested periodically to 250 psi (1,724 kPa) or more, according to NFPA 1962, *Standard for the Care, Use, and Service Testing of Fire Hose, Including Couplings and Nozzles* (hereinafter referred to as NFPA 1962). Preconnected standpipe hoses installed on new systems are required to be tested periodically to 150 psi (1,034 kPa) or more, according to NFPA 1961, *Standard for Fire Hose.* Though a hose may not fail up to these pressures, experience indicates that 175 psi (1,207 kPa) is about the maximum pressure on a hose that can be satisfactorily maneuvered by fire fighters and that 100 psi (690 kPa) is the maximum pressure that an untrained user should be expected to handle. Therefore, pressure regulating devices, are required to limit static and residual pressures to 175 psi (1,207 kPa) at hose connections without pre-connected hoses and to limit residual pressure to 100 psi (690 kPa) at hose connections with preconnected hoses.

The established pressure limit for sprinkler connections is also 175 psi (1,207 kPa). This is the maximum pressure for which sprinklers are listed, also serves as a safety factor to prevent leakage. When higher pressure is expected at a sprinkler connection, both pressure regulating and pressure relief valves are required. To allow a buffer between the pressure settings for the valves, regulating valves should be set at 165 psi (1,138 kPa) and relief valves should be set at 175 psi (1,207 kPa).

Table 6-16A summarizes the limits discussed above. Figures 6-16D and 6-16E illustrate sample multiple-zone systems. Note that, on multiple-zone systems, fire pumps at the same level may be used in series. In addition, two separate supply risers are always required for high zones supplied by pumps at lower levels so that a water supply can be maintained when one riser is out of service. Separate fire department connections are also required for each zone.

Interconnection of piping: One last item to remember when laying out a system is that standpipes must be interconnected at the bottom, or at the top when gravity tanks are used. Interconnection of standpipes ensures that water supplies can service all portions of a system.

TABLE 6-16A. *Pressure Limits for Standpipe System Outlets*

Outlet Use	Minimum Residual Pressure[a]	Pressure Regulating Device Required When Pressure Exceeds[b]
1¹/₂–in (38 mm) Hose Connection with Pre-connected hose	65 psi (448 kPa)	100 psi (690 kPa)[d]
1¹/₂–2¹/₂–in (38–64 mm) Hose Connection without Pre-connected Hose	100 psi (690 kPa)	175 psi (1207 kPa)[c]
Sprinkler System Connection	N/A	175 psi (1207 kPa)[e]

[a]Measured at the hydraulically most remote hose connection while flowing at a rate specified by NFPA 14.

[b]Pressures in system piping should not exceed the rating of any system component.

[c]This is a residual pressure limit. Static pressure cannot exceed 175 psi (1,207 kPa).

[d]This setting will accommodate 200 ft (61 m) of hose with approximate flows as follows: 250 gpm (947 L/min) or more from 2- (51 mm) or 2¹/₂-in. (64 mm) hose, 150 gpm (568 L/min) from 1³/₄-in. (45 mm) hose, and 125 gpm (473 L/min) from 1¹/₂-in (38 mm) hose. A setting of 165 psi (1138 kPa) can be used when installed on a combined sprinkler/hose connection on combined systems while achieving acceptable flows and nozzles pressure.

[e]Set pressure regulating device at 165 psi (1138 kPa). Note that the pressure relief valves are required by NFPA 13, *Installation of Sprinkler Systems. Set at 175 psi (1207 kPa).*

N/A = Not applicable.

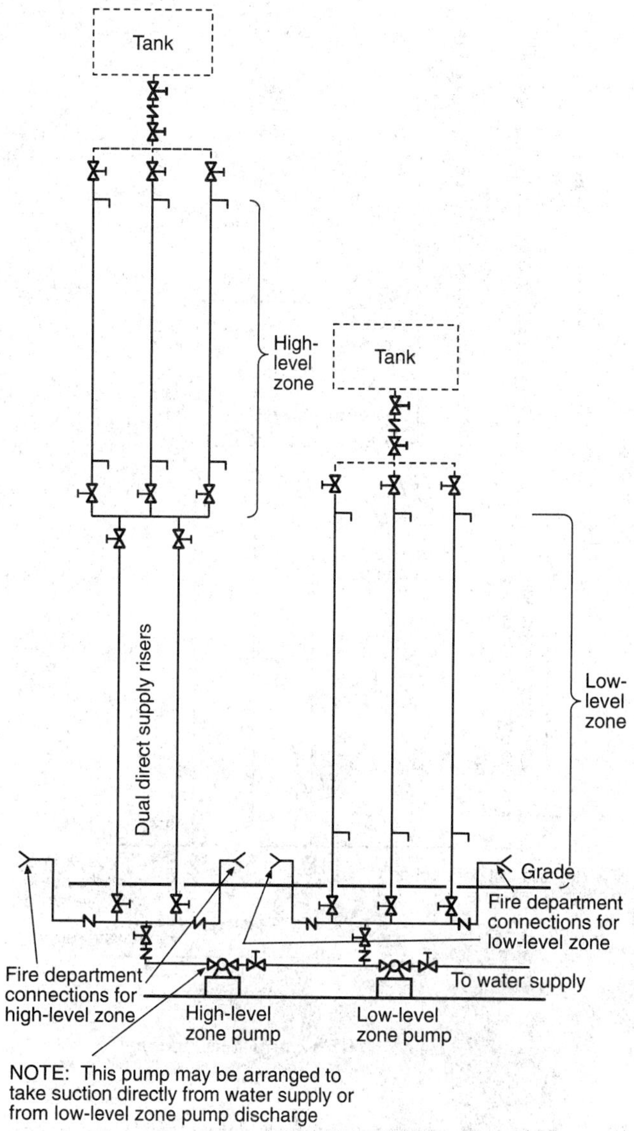

NOTE: This pump may be arranged to take suction directly from water supply or from low-level zone pump discharge

FIG. 6-16D. With low-level pumps.

System Components

The selection of standpipe system components is determined largely by the class of system and the system layout. Particular attention should be given to ensuring that components are rated for pressures to which they may be subjected.

Piping and fittings: Steel pipes assembled with welded joints, screwed fittings, flanged fittings, rubber-gasketed fittings, or a combination of these are the most common configurations of materials used for standpipe system piping and fittings. Ductile-iron pipe and copper tubing, the latter assembled with brazed joints, also are used occasionally. Piping should be capable of withstanding the maximum pressures that can be developed in a system.

Fittings are required to be rated to withstand either 175 psi (1,207 kPa) or maximum system pressures, whichever is greater. When pressures are expected to exceed 175 psi (1,207 kPa), extra-heavy fittings or fittings listed for greater pressures should be used for pressures up to 300 psi (2,069 kPa), and specially designed fittings should be used for pressures exceeding 300 psi (2,069 kPa).

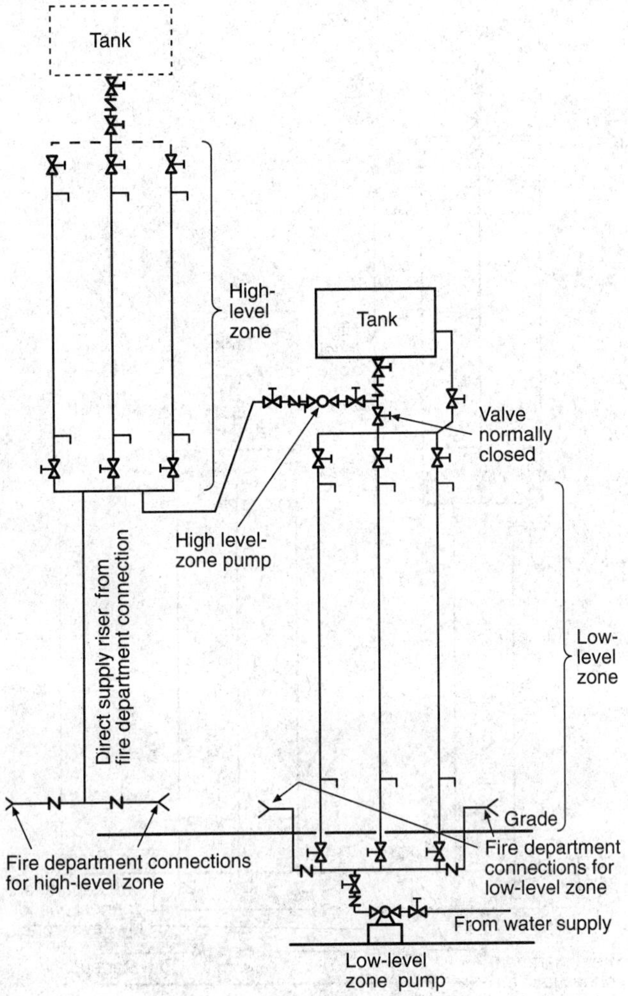

FIG. 6-16E. *With a high-level pump.*

Hose, hose racks, nozzles, and hose cabinets: When preconnected hoses are required to be installed on Class II or Class III systems, lined hose must be used. Unlined linen hose has been phased out, because it is subject to rapid deterioration under moist or wet conditions. Usually, the size of choice is $1^1/_2$-in. (38-mm) hose. However, the use of listed hoses smaller than $1^1/_2$ in. (38 mm) is permitted by NFPA 14, in light-hazard occupancies when approved by the Authority Having Jurisdiction. The allowance for smaller hose mounted on a reel recognizes that an untrained person might be more capable of using such equipment in the event of a fire. Similar equipment has been used in Europe for many years.

Preconnected hoses on Class II and Class III standpipes are generally limited to 100 ft (30.5 m) in length to minimize the difficulty that untrained users may have advancing a hose and to minimize kinking. Hoses are always required to be kept on compatible racks and should always be positioned in a readily accessible location within convenient reach of a person standing on the floor. Hoses should also be clearly visible and located in a place not likely to be obstructed. Where hoses are kept in cabinets or closets, the doors should have a glass panel or some other means to allow easy identification.

Preconnected hoses are required to be equipped with approved $^3/_8$- or $^1/_2$-in. (10- or 13-mm) open nozzles or combination spray/solid pattern nozzles. It is preferable to use nozzles that have oper-

ating pressures in the 50-psi (345-kPa) range. This minimizes the pressure demand on the system and helps to ensure that hoses can be handled by occupants. Since nozzle shutoff valves complicate the operation of a hose, they are considered undesirable unless the users are trained.

For typical hose rack and valve connections, see Figure 6-16F. A typical hose cabinet is shown in Figure 6-16G.

Valves and pressure regulating devices: Several different types of valves may be used as components of standpipe systems. These include gate valves, check valves, and hose valves. Like piping and fittings, valves and pressure regulating devices must be capable of withstanding the maximum pressure that can be developed in a system. The following discussion highlights the most common valves and pressure control devices encountered.

Starting at permanent water supplies, indicating gate valves are required at each permanent water source to allow isolation of any water source for servicing. Note that gate valves are not permitted to be installed between a fire department connection and the standpipe system that it serves. This ensures the fire department's ability to pump into a system.

Next, each permanent water source should be protected by a check valve to prevent backflow. Check valves should also be installed between fire department connections and wet standpipe systems that they serve. These valves allow piping that is connected to fire department connections to normally be kept dry, when provided with automatic drain valves at low points. Keeping the FDC dry protects it from freezing and prevents the possibility of having FDC caps "thread-lock" if the internal clapper valves leak.

Moving farther into a system, gate valves should be installed at feed main connections to each standpipe to allow standpipes to be serviced independently without impairing protection to an entire building.

Drain valves also should be provided to allow individual standpipes, as well as an entire system, to be drained for servicing. Finally, each outlet from a standpipe system must terminate at a hose valve or at a sprinkler control valve on combined systems. When excessive pressures are expected at outlets, pressure-regulating devices are required.

Pressure-regulating devices are used to limit system outlet pressures. Pressure-regulating devices used in standpipe systems can be categorized into three basic types: (1) pressure restricting, (2) pressure reducing, and (3) pressure control. Pressure-restricting devices reduce pressures in flowing conditions only. They do not compensate for changes in input pressure to maintain a constant discharge pressure, and they do not control pressure in static conditions. For these reasons, their use is limited to $1^1/_2$-inch (38-mm) hose connections for limiting residual pressure to 100 psi (690 kPa) where the static pressure does not exceed 175 psi (1,207 kPa). One example of a pressure-restricting device is an orifice plate. An orifice plate is a metal disk with a restricting orifice. Orifice plates may be inserted into connections to standpipe systems to reduce residual pressure on the downstream side. Though relatively inexpensive, orifice plates are generally undesirable because they do not maintain a steady discharge pressure, and because they can damage hoses if the pressure jet from an orifice strikes the inner wall of a hose. If used, orifice plates should always be installed at hose couplings to allow for easy removal.

The preferred devices for regulating excessive pressures are pressure control and pressure-reducing valves. These valves are designed to actively regulate outlet pressures in both static and flowing conditions, and they are required for outlets of systems where static pressure could exceed 175 psi (1,207 kPa). Pressure control and pres-

FIG. 6-16F.   Typical hose racks and valve connections. (Photos courtesy of Potter-Roemer, Inc.)

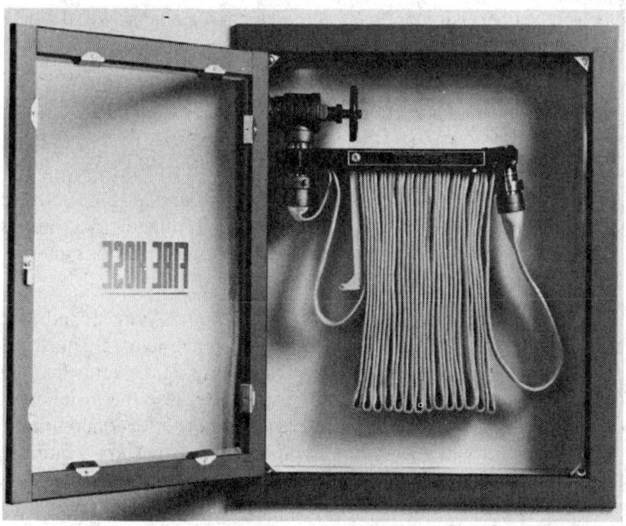

FIG. 6-16G.   Typical hose cabinet.

sure-reducing valves are preferred over pressure-restricting devices because hoses and sprinkler systems connected to standpipes are exposed to both static and flowing conditions; and uncontrolled static pressures can cause equipment failure and pose a danger to users.

Pressure control and pressure-reducing valves employ a mechanism that compensates for variations in inlet pressures by balancing water pressure in an internal chamber or chambers, typically against a spring. Pressure control valves, which are a type of pressure reducing valve that is pilot operated, are considered by many to be the most reliable method of controlling pressure. Figure 6-16H illustrates one example of a pressure control valve. In the example shown, the outlet pressure is a function of the setting on the regulating spring. Often, the outlet pressure is not field-adjustable, and valves are preset at the factory for use at a specific floor level in a building. In such cases, it is critical to ensure that valves are installed at the intended location.

Unfortunately, pressure-regulating devices, particularly pressure-reducing valves, have a history of high failure rates. As a result, standpipe systems having such devices are now required to be designed such that each device may be tested, and routine testing must be conducted. To accommodate the need for testing pressure-regulating devices, standpipe systems using such devices must have a dedicated drainage system with connections at every other floor and a central means for measuring flow. Figure 6-16I shows one possible

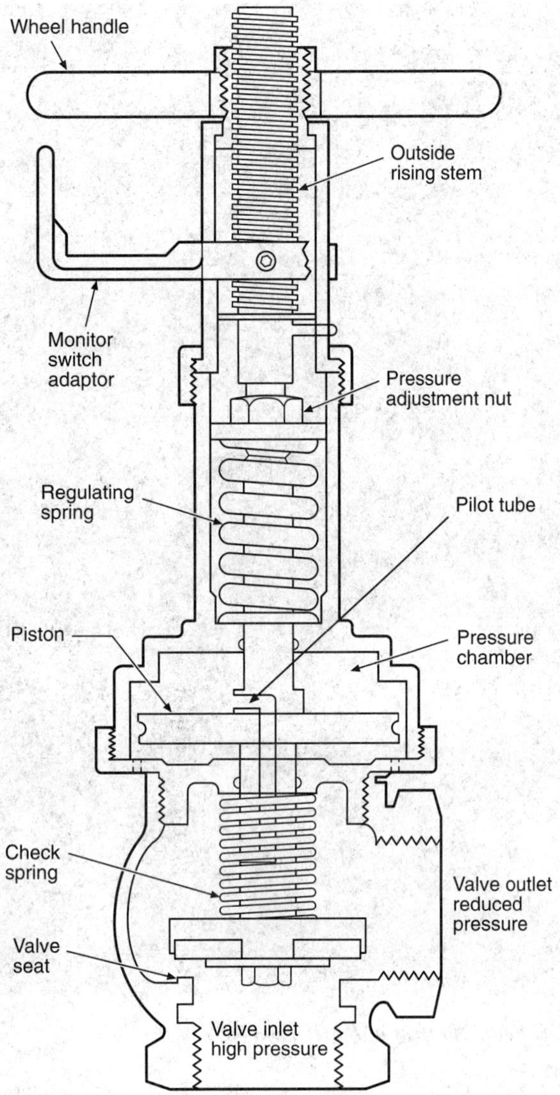

Wheel handle

Outside rising stem

Monitor switch adaptor

Pressure adjustment nut

Regulating spring

Pilot tube

Piston

Pressure chamber

Check spring

Valve outlet reduced pressure

Valve seat

Valve inlet high pressure

FIG. 6-16H. Cutaway of a typical pressure control valve.

design of a floor valve assembly for such an arrangement. The case shown assumes a standpipe pressure in excess of 175 psi (1,207 kPa) and a combined sprinkler and standpipe system with a directly connected drainage riser for pressure-regulating device testing. As an option, the drainage riser may simply have a hose connection at every other floor where a short piece of hose can be connected to the two closest pressure-regulating devices for testing.

Where system pressures are extreme, pressure relief valves, which reduce pressures by flowing water into a drain, may be provided as a backup to pressure-regulating valves. Pressure relief valves are commonly available in two types: (1) spring operated, and (2) pilot operated. Pilot-operated valves are preferred, since they respond more quickly to changes in system pressures; however, these valves require increased maintenance to prevent excessive accumulation of debris in strainers.

Fire department connections: (See additional discussion under "Type and number of water supplies" earlier in this chapter) Fire department connections (FDC) are required for all Class I and Class III systems. On manual standpipe systems, FDCs serve as the sole

water supply. On automatic and semiautomatic standpipe systems, they serve as a reliable auxiliary water supply.

Fire department connections should be located where they are accessible to fire department vehicles and close to a fire hydrant. They should not be obstructed by shrubbery, vehicles, fences, or similar obstructions. For high-rise buildings, two or more connections, located remotely from each other, are required for each zone. This redundancy helps to ensure the fire department's ability to provide an uninterrupted water supply during a major fire. It is required because falling glass from broken windows can cut supply hoses and prevent personnel from connecting new hoses to the affected FDC.

As a rule, most fire department connections are provided with one $2^1/_2$-in. (64-mm) inlet for each 250 gpm (946 L/min) of design flow rate. Some fire departments require large-diameter inlets that are designed to connect to special high-pressure, large-diameter hoses. Even when large-diameter hoses are used, a second inlet is needed to allow for a backup hose that could be charged if the primary hose ruptures or separates from a coupling.

Another important issue that should not be overlooked is the size of the pipe connecting a fire department connection to system mains. This pipe should be sized to deliver the system demand, as discussed previously under "Sizing pipe." It is preferable to use 4-in. (102-mm) or larger piping in all cases to avoid obstruction by debris.

Figure 6-16J illustrates a typical arrangement for a fire department connection on a wet standpipe system.

System monitoring and supervision devices: To ensure the reliability and performance of standpipe systems, provisions must be made for monitoring system pressure and supervising control valves. Gages should be provided to monitor at each water supply and at the top of each standpipe so that system pressure can be monitored.

Valve supervision can be accomplished by electrical or mechanical means. Supervision is generally required for water supply isolation valves, standpipe isolation valves, and similar valves that are capable of interrupting the water supply to large portions of a system. Supervision is necessary to ensure that the valves remain open at all times.

Occasionally, waterflow alarms are also specified for monitoring a system. This is usually limited to systems having preconnected hoses where immediate notification of waterflow is desired. Also, electronic supervision of hose cabinet doors may be provided where vandalism is a concern.

STANDPIPE SYSTEM INSTALLATION CONSIDERATIONS

Although a proper design establishes the groundwork for a reliable and usable standpipe system, the installation ultimately determines how well a system will perform. A few key considerations are discussed below.

Since most standpipe systems are hydraulically designed, it is essential that the installation conform to design drawings. This is especially true for combined standpipe and sprinkler systems that are based on hydraulically calculated designs. Because the installation of additional pipe or fittings will have an adverse effect on hydraulic calculations, as will reductions in pipe sizes, plans should indicate that all such changes should be reported to the designer for approval. Plans should also caution installers of systems with preset pressure-regulating valves to ensure that the valves are installed at the proper floor levels.

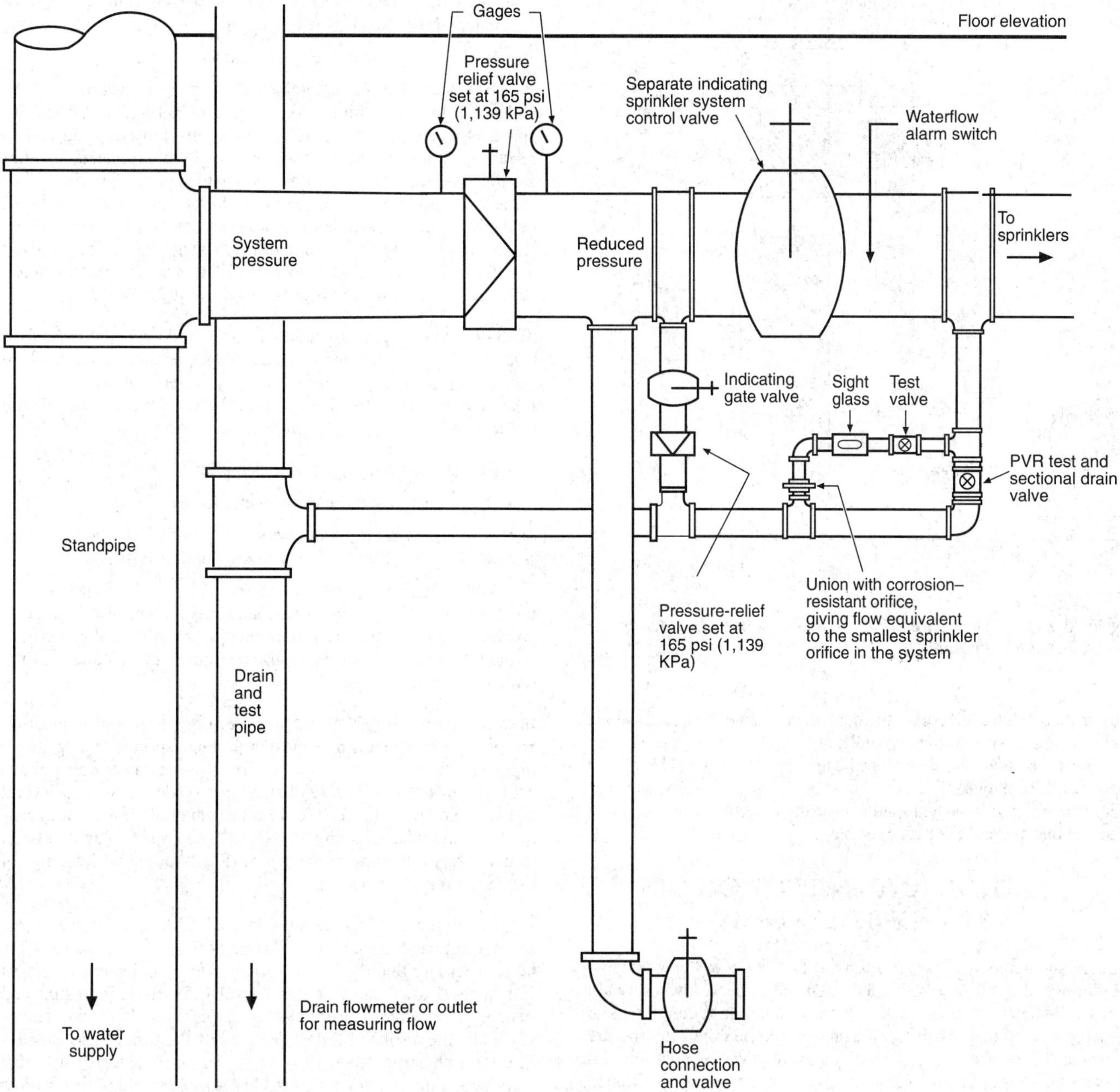

FIG. 6-16I.   *One possible arrangement for valving at a floor connection for a combined system. (Not to scale)*

To ensure operability and maintainability, it is preferable to label piping and indicate the direction of water flow. Valve locations must also be clearly identified and accessible. Particular attention should be given to providing access space around hose connections, recognizing that fire fighters will be using large gloves when connecting hoses and operating valves. Twelve inches (0.3 m) of clear space from the valve handles and hose outlets to walls or other obstructions is needed.

The key concerns for locating fire department connections are accessibility and visibility. Fire department connections should be clear of shrubbery, parking for vehicles, and other obstructions that could hinder their use. In addition, signs should be provided, preferably reflective, to indicate whether a connection serves stand-pipes, sprinklers, or both, and to indicate the area served by the connection.

During construction of a building that will have a standpipe system, it is important that the system be made operable before work progresses to a point where deployment of hoses from fire apparatus no longer provides an efficient means of fire fighting. Local codes usually dictate the point at which standpipes must be made operable during construction.

The need for standpipes during construction has been demonstrated in many fires. Even noncombustible structures often represent extreme fire hazards during construction because combustible shoring and a variety of ignition sources may be present. To provide protection during construction, the permanent system piping is typically

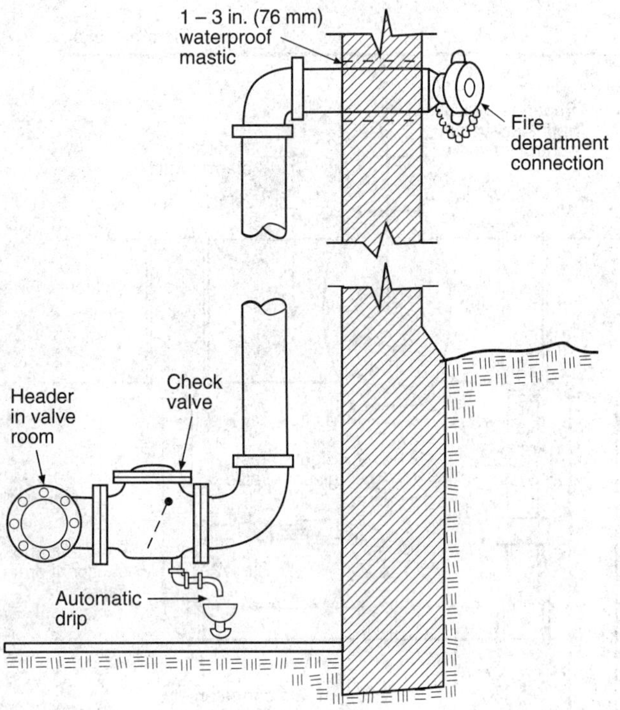

FIG. 6-16J. Typical fire department connection for a standpipe with water-filled piping.

used, along with a temporary fire department connection. At least one hose connection should be provided at each floor level, and all hose connections should be kept capped with an easily removable cap and protected from injury. Systems should be extended to keep pace with construction progress, and routine inspection should be scheduled to ensure that systems are kept operable.

TESTING AND INSPECTION OF STANDPIPE SYSTEMS

Since standpipe systems are essential fire safety systems in large buildings, they must be routinely tested and inspected which verify that a system performs to the original installation requirements to ensure operational readiness. Testing should be conducted before system acceptance and at intervals adequate to verify reliability. Inspection, testing, and maintenance of standpipe systems is regulated by NFPA 25, *Standard for the Inspection, Testing, and Maintenance of Water-Based Fire Protection Systems.* Full system tests which verify that a system performs to the original installation requirements are called for on five-year intervals. System components are required to be inspected and tested on a more frequent basis.

Pre-Inspection: Prior to conducting a complete performance test, a visual inspection should be conducted to verify that:

- Water supplies have been properly maintained.

- Valves to water supplies and standpipes are open and supervised by a chain and lock, seal, or electrical device.

- Piping is securely joined at connections. If piping will be enclosed in a concealed space, it should be left exposed until initial acceptance testing is completed.

- Hose connections are accessible, valve wheels are present, and hose threads on all are protected by caps. On dry systems, en-

sure that valves are closed. Hose threads on connections must match those used by the jurisdiction.

- Hose stations are properly maintained.

Fire hose must be maintained in the proper position on racks and be in good condition. Each hose should be removed and inspected for cuts, abrasions, faulty gaskets, and loose couplings, then re-racked, at an interval not exceeding five years immediately after the purchase date and every three years thereafter. Periodic hydrostatic testing of lined hose is required. Unlined linen hose should not be hydrostatically tested because it is susceptible to rot and mildew if left even slightly damp. In addition, nozzles should be removed and examined for foreign objects. Specific inspection and maintenance instructions for fire hose are set forth in NFPA 1962.

Testing: Once a pre-inspection is completed hydrostatic and flow tests should be conducted. Pressure gages should be checked on each standpipe during testing to verify the interconnection of standpipes. Dry systems should be air-tested before they are charged with water. Flow tests should be conducted to verify:

- Operation of automatic water supplies.

- Operation of each fire department connection.

- Proper flow rates, static pressures, and residual pressures for each standpipe and for pressure regulating valves.

When standpipes are flow tested by flowing water at the roof level, a flow diffuser is recommended at the point of discharge to control the water stream and intercept any debris that may be ejected. Particular attention should be paid to the adequacy of the drainage route for water.

Pressure Regulating Valves: Flow testing of pressure-regulating valves is probably the single most important aspect of testing standpipe systems with these devices. Pressure-regulating valves have been implicated as the cause of standpipe system failures in several catastrophic fires. The need for routine testing of these devices cannot be overemphasized. In fact, NFPA even issued an alert bulletin to this effect following a major fire. (See "Additional Reading" at the end of this chapter).

When flow-testing pressure-regulating valves, it is preferable to test each valve through a full range of flows and pressures. This helps to ensure that all devices perform to specifications. Using a full range of flows also exercises the valve. The need for testing and maintenance of pressure-regulating devices has also been demonstrated by the results of tests conducted by the City of Los Angeles Fire Department between 1984 and 1986. In these tests, 413 pressure-regulating valves were tested in a dozen buildings. Test results indicated that more than 75 percent of the valves tested required recalibration, repair, or replacement.

If a drain riser is not available for testing purposes, possible solutions include: (1) draining one of the standpipes and flowing from one standpipe to the other using a hoseline between stairwells, (2) laying a hoseline in the stairwell of the standpipe being tested for use as a drain, and (3) removal of valves on every other floor, alternating between standpipes, so that each floor has at least one valve installed; then bench testing the valves.

FIRE DEPARTMENT OPERATIONS USING STANDPIPE SYSTEMS

Preplanning: For standpipe systems to be effective, firefighters must be trained in standpipe operations, and they should pre-plan systems that they will use. As part of the pre-plan, firefighters should become familiar with the locations of fire hydrants, fire department connections, water supply valves, pumps, tanks, and hose connections. The presence of orifice plates in hose connections or pressure-reducing or pressure control valves also should be noted. Fire departments also should develop standard operating procedures that assign personnel and apparatus responsibilities for using standpipe systems.

Equipment: To facilitate the use of standpipe systems, "standpipe packs" should be carried on responding apparatus. The packs should contain the equipment necessary to use standpipe system hose connections. Standpipe packs are typically carried on carts or in bags and contain between 100 and 200 ft (30.5 and 61 m) of hose, a nozzle, and miscellaneous tools and wrenches. A gated wye may also be carried to allow a second hose to be connected to a single hose connection. The length of hose carried should be based upon local practices regarding which hose connections are expected to be used (at the fire floor or the floor below) and local building code requirements for locating hose connections. Probably the most versatile hose arrangement is one that includes $1^3/_4$- or 2-in. (45- or 51-mm) hose and a high-volume nozzle capable of flowing 200 gpm (757 L/min) or more. Such an arrangement can provide a flow capability comparable to a $2^1/_2$-in. (64-mm) hose with a selectable flow fog nozzle when adequate pressure is available, but it is significantly easier to handle. A fog nozzle that produces an effective pattern at operating pressures less than 100 psi (690 kPa) should be used to take maximum advantage of the pressure available.

In addition to standpipe packs, fire apparatus should also carry coiled or packed sections of $2^1/_2$-in. (64-mm) hose for cases where large volumes of water are needed and the pressure available from a standpipe is limited. Such cases may occur in very tall buildings that have floor levels beyond the range of fire apparatus or where pressure regulating valves have been improperly adjusted.

Operations: It is common for the first fire-fighting crew on the scene of an incident in a building with a standpipe system to take a standpipe pack into the building when investigating the call. Meanwhile, an engine should connect to a fire department connection to supply the system. More than one supply hose should be connected to a system if a fire is discovered to prevent the water supply from being interrupted if a supply hose breaks. It is also preferable for additional apparatus to connect to a second FDC, during major fires when more than one FDC is present. This reduces the risk of interruption of the water supply if glass breakage on upper floors makes conditions at some parts of a building's perimeter unsafe, perhaps cutting supply hoses.

When fire fighters determine that hoses are needed, they should select a hose connection located in a protected stairway enclosure, as opposed to a connection located in a corridor or open area (when such connections are present) since it is desirable to have a charged hose before entering a floor if smoke or flame is detected. In multiple-story buildings, a hose connection located at the intermediate landing below or one floor below a fire floor is often preferred over a connection at a fire floor because it provides additional space to spread a hose before charging it, and minimizes congestion at the fire floor entrance. It should be noted that using a hose connection on a lower level will create a need for additional hose to reach all areas on a fire floor. Fire fighters should also be aware that, depending on desired flows, they may have to remove pressure-restricting devices or to adjust pressure-regulating valves if they are field adjustable.

The pressure to which a standpipe system should be charged varies from one fire department to another based upon local preferences. Some departments use recommendations in NFPA 13E, *Guide for Fire Department Operations in Properties Protected by Sprinkler and Standpipe Systems*, which suggest that the pressure at the FDC should be initiated at 100 psi (690 kPa) when solid-pattern

nozzles will be used at elevations of 100 ft (30.5 m) or less, and at 150 psi (1,034 kPa) when fog nozzles will be used at elevations of 100 ft (30.5 m) or less. For elevations above 100 ft (30.5 m), 5 psi (34.5 kPa) should be added for each floor level above 100 ft (30.5 m). Other departments simply pump 150 psi (1,034 kPa) plus 5 psi (34.5 kPa) for each floor level above grade. The latter method is more commonly used and will deliver adequate pressure for fog nozzles in all cases.

OUTSIDE HOSE SYSTEMS

Outside hose systems are not standpipe systems. Rather, they are appendages to a water supply. However, outside hose systems and standpipe systems have similar purposes. They both provide an efficient means for manually applying water to fires. Though not mandated by model building or fire codes, outside hose systems may be requested by insurance authorities or by owners as an additional safety measure to minimize potential fire losses.

Outside hose systems are typically found at industrial sites, and since they usually include equipment for delivering high-volume fire streams, their installation or use is not appropriate where there is not a trained fire brigade. Systems will usually consist of yard-type fire hydrants, strategically placed around buildings, adjacent to which fire hose and fire-fighting accessories are stored in hose houses. Rooftop hose houses or hose stations may also be provided when rooftop or exposure hazards are present.

Design

The design of an outside hose system should be tailored to its intended uses. These uses may include deployment by a trained fire brigade until the fire department arrives, serving as backup protection for sprinkler systems when such systems are not in service, and serving as a backup water supply for impaired sprinkler systems by providing a ready means of supporting a fire department connection. NFPA 24, *Standard for the Installation of Private Fire Service Mains and Their Appurtenances* (hereinafter referred to as NFPA 24), establishes the most commonly used design standards.

Fire hydrant spacing and placement of hose houses should be adequate to allow fire hoses to reach the desired locations while minimizing the distance that hose must be advanced. Selection of fire hydrant locations and hose lengths should consider the path of travel, as well as any obstructions between the hydrants and the buildings or areas that they protect. When interior standpipe systems are present, it should not be necessary to provide coverage using outside hoses to all portions of a protected building's interior. When rooftop hazards are present, a means of advancing hoses to the roof, such as fixed exterior ladders, hoisting ropes, and hose straps, should be provided unless there are rooftop hose houses or hose stations.

When designing outside hose systems, the needs of the fire department for on-site fire hydrants must also be considered. For large buildings, on-site hydrants may be required to serve as a water source for fire department apparatus, in addition to supplying hose systems. In such cases, underground mains must be designed for the required flow, including demands for fire apparatus, hoses connected directly to hydrants, sprinkler systems, and domestic service. In addition, fire hydrants must be placed in locations required by the local fire code.

Equipment

NFPA 24 sets forth nationally recognized standards for fire hydrants, construction of hose houses, and equipment to be maintained in hose houses. The following is an overview of some of the major considerations for outside hose system equipment.

FIG. 6-16K. Typical hose houses. (Photo courtesy of Potter-Roemer, Inc.)

FIG. 6-16L. Typical rooftop hose station. (Photo courtesy of Potter-Roemer, Inc.)

Fire hydrants should have outlets for use by fire brigades and fire department apparatus as necessary. Hose threads and lugs on fire hydrants must always match those used by the local fire department. Also, it is preferable to equip each outlet with an independent valve to avoid the need to shut a fire hydrant when connecting additional hoses.

The principal features considered in hose house construction are ventilation and protection against weather. Ventilation can be obtained by using a slatted floor and slatted shelves and by allowing space for ventilation under roof overhangs. Vents should be arranged so that rain cannot enter the structure. Hose houses should be of substantial construction and large enough to contain and provide operating space for all equipment. Where it is necessary to lock hose houses, special locks with brittle shackles or latches behind break-glass panels can be used to provide ready access in an emergency.

Hoses for outside hose systems may be preconnected and stored in hose houses, or kept on hose carts that can deliver hose to a fire scene, or both. Preconnected hoses are often preferred to min-

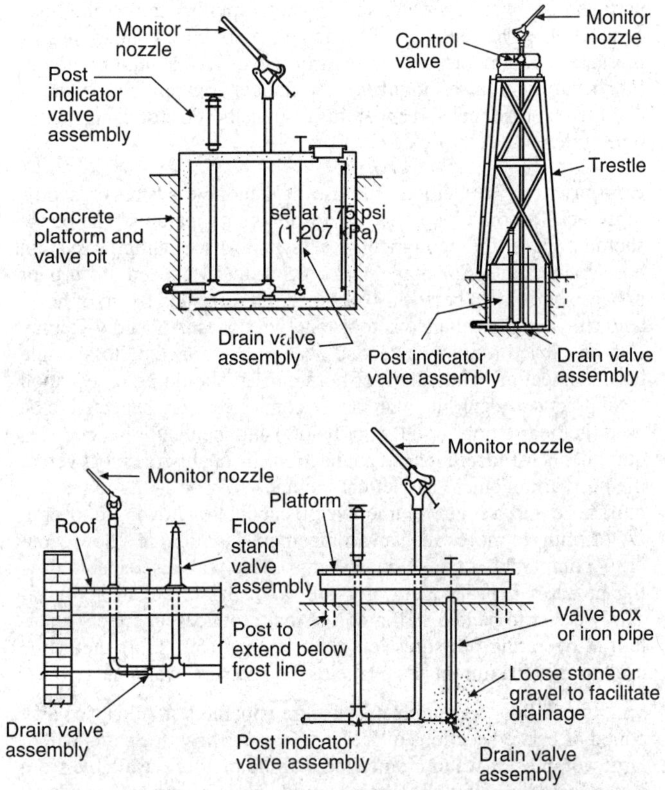

FIG. 6-16M. Various arrangements of monitor nozzles.

imize the time necessary to begin fire fighting. The length of hose required will depend on hydrant locations. For hose lays greater than 150 ft (46 m), hose carts or other vehicles should be considered. One-and-three-quarter-in. (45-mm) and 2-in. (51-mm) hoses with $1^1/_2$-in. (38-mm) couplings are preferred sizes, providing high-flow characteristics with minimal friction loss and relatively easy maneuvering. Other possible arrangements include 100 to 150 ft (30.5 to 46 m) of $2^1/_2$-in. (64-mm) hose with a gated wye at the end, connecting to an additional length of smaller hose. The wye allows for subsequent connection of an additional hose closer to a fire.

Figures 6-16K and 6-16L are examples of hose houses and a rooftop hose station, respectively, that are commercially available.

Accessories

Accessory equipment in a hose house should include hydrant wrenches, nozzles, spare hose washers, gated wyes, reducing couplings, spanner wrenches, fire axes, and other tools, as necessary. When selecting nozzles, the particular fire hazard present should be considered. For example, spray nozzles are preferred for flammable liquid and electrical fires, and solid-stream nozzles may be preferred for lumber storage.

Hose Carts

Hose carts are useful where a hose supply is needed to supplement that stored in hose houses. This may occur where long travel distances are anticipated for deployment, such as in large factories or warehouses; and where there are outlying buildings or yard storage areas beyond easy reach of hose houses. Carts may be motorized, hand-pulled, or modular devices stored for pickup and transportation by a lift truck. They should be stored in a central location, where possible.

Monitor Nozzles

Where large amounts of combustible materials, such as log piles, lumber piles, flammable or combustible liquids, or railway cars, are stored in yards, it may be necessary to provide a means of quickly delivering large volumes of water to control a fire. This can be best accomplished by installing permanent monitor nozzles around such storage at grade level, on special trestles, or on the roofs of buildings, and by having portable monitor nozzles available for use. Figure 6-16M illustrates various arrangements of monitor nozzles.

BIBLIOGRAPHY

NFPA Codes, Standards, and Recommended Practices

Reference to the following NFPA codes, standards, and recommended practices will provide further information on standpipe and hose systems discussed in this chapter. (See the latest version of *The NFPA Catalog* for availability of current editions of the following documents.)

NFPA 1, *Fire Prevention Code*
NFPA 13, *Standard for the Installation of Sprinkler Systems*
NFPA 13E, *Guide for Fire Department Operations in Properties Protected by Sprinkler and Standpipe Systems*
NFPA 14, *Standard for the Installation of Standpipe and Hose Systems*
NFPA 20, *Standard for the Installation of Centrifugal Fire Pumps*
NFPA 22, *Standard for Water Tanks for Private Fire Protection*
NFPA 24, *Standard for the Installation of Private Fire Service Mains and Their Appurtenances*
NFPA 25, *Standard for the Inspection, Testing, and Maintenance of Water-Based Fire Protection Systems*
NFPA 72, *National Fire Alarm Code*
NFPA 101®, *Life Safety Code*®
NFPA 1901, *Standard for Automotive Fire Apparatus*
NFPA 1961, *Standard for Fire Hose*
NFPA 1962, *Standard for the Care, Use, and Service Testing of Fire Hose Including Couplings and Nozzles*
NFPA 1963, *Standard for Fire Hose Connections*
NFPA 1964, *Standard for Spray Nozzles (Shutoff and Tip)*

Additional Reading

Bond, H., "Water for Fire Fighting in High-Rise Buildings," *Fire Technology,* Vol. 2, No. 2, pp. 159–163.
Bryan, J. L., *Automatic Sprinkler and Standpipe Systems*, 2nd ed., National Fire Protection Association, Quincy, MA, 1990.

Carson, W. G., and Klinker, R. L., *Fire Protection Systems: Inspection, Test and Maintenance Manual*, 2nd ed., National Fire Protection Association, Quincy, MA, 1992.
Chapman, E. F., "Standpipe Operations," *Fire Engineering*, Vol. 142, No. 1, 1989, pp. 47–49.
City of Fort Worth, Texas, *Fire Department Standard Operating Procedure for Standpipe Testing.*
Factory Mutual Approval Guide, Factory Mutual Research Corporation, Norwood, MA (published annually).
Fire Protection Equipment List, Underwriters Laboratories Inc., Northbrook, IL (published annually with bimonthly supplements).
Fornell, D. P., "Standpipe Operations," *Fire Engineering,* Vol. 144, No. 8, August 1991, pp. 71–74, 76, 78–84.
Fredericks, A. A., "Standpipe System Operations: Engine Company Basics," *Fire Engineering*, Vol. 149, No. 2, February 1996, pp. 33–42.
Hammack, J. M., "Combined Sprinkler System and Standpipes," *Fire Journal*, Vol. 65, No. 5, pp. 68–69
Jensen, R., ed., *Fire Protection for the Design Professional*, Cahners, Boston, 1975, pp. 66–71.
Johnson, B. P., "Additives for Hose Reel Systems: Trials of Foam on Tire Fires," Home Office, London, England, Publication 5/91, SC/88 42/1087/1, 1992, 63 pages.
Kimball, W. Y., "Can Pumpers Supply Standpipes in High-Rise Buildings," *Fire Journal*, Vol. 59, No. 5, Sept. 1965.
Loss Prevention Data Sheet 3-10, Factory Mutual Research Corporation, Norwood, MA.
Loss Prevention Data Sheet 3-11, Factory Mutual Research Corporation, Norwood, MA.
Lyons, P. R., "Dry Standpipe Survey in Los Angeles," *Fire Journal*, Vol. 63, No. 3, May 1969, pp. 63–66.
National Building Code, Chapter 9, Building Officials and Code Administrators International, Inc., Country Club Hills, IL, 1996.
NFPA Alert Bulletin No. 91-3, "Pressure-Regulating Devices in Standpipe Systems," May 1991.
Nolan, J. W., "How to Approach Standpipe Design," *Actual Specifying Engineer*, Vol. 24, No. 7, July 1970, pp. 85–89.
Private Fire Protection and Detection, Fire Protection Publications, Oklahoma State University, 1979, pp. 111–114.
Solomon, R. E., "Pressure-Regulating Devices in Standpipe Systems," *NFPA Journal*, Vol. 85, No. 5, September/October 1991, pp. 70–71, 88–89.
Standard Building Code, Chapter 9, Southern Building Code Congress International, Inc., Birmingham, AL, 1997.
Uniform Building Code and Uniform Building Code Standards, Chapter 9, International Conference of Building Officials, Whittier, CA, 1997.

CARE AND MAINTENANCE OF WATER-BASED EXTINGUISHING SYSTEMS

Revised by James M. Fantauzzi

In the continuing evolution of active fire protection, the advent of NFPA 25, *Standard for the Inspection, Testing, and Maintenance of Water-Based Fire Protection Systems* (hereinafter referred to as NFPA 25), marks a new step in the advancement of water-based fire protection. The inclusion of NFPA 25 as a standard and not just a recommended practice was a major accomplishment and recognizes some of the problems that those active in the fire protection community are now facing.

As existing, active systems continue to age, the reliability of these systems will diminish if the maintenance procedures are not practiced on a regularly scheduled basis, as is required by NFPA so as to maintain the integrity of these systems.

The design of and installation practices for these water-based systems in the past were based on property salvage and the economic impact of their proper operation. The major support for the property-saving aspect comes from the insurance companies. Now that the building codes also require active fire protection systems to be installed for their life-saving purposes, the life safety aspect of water-based fire protection systems has moved into the forefront, and property conservation has taken a back seat to governmental regulation of their installation.

Another factor is the current technological advances and tendencies toward proprietary and special application types of active fire protection equipment. The pace of these advances is a potential time bomb, due to misapplication of these products caused by faulty design; installation; lack of flexibility for future change in occupancy reorganizations; and lack of monitoring of the influence of other building factors, such as modifications of fire doors; walls; soffits; heating, ventilating, and air-conditioning (HVAC); and other building components.

The last major concern is the perceived minimization of safety factors for the installation of materials and proper application of methods. These changes may shorten the useful life expectancy of these new systems, and the effective operation of these systems becomes more dependent on proper maintenance procedures.

IMPORTANCE OF EXTINGUISHING SYSTEM MAINTENANCE

Care and maintenance includes more than just the inspection and testing of system devices and equipment. For effective protection, the proper relation between hazard and protection must be balanced. This calls for careful consideration of the fire hazards to be protected. Required as well are the decisions regarding the adequacy of the protection for the varying conditions of the property. Further, a proper inspection must ascertain whether the originally installed system is adequate for the current fire hazards being protected.

Water-based extinguishing systems employing standard devices and installed in accordance with established rules are sturdy and durable, and require minimum expenditure for maintenance. However, like other types of equipment, they may suffer deterioration or impairment through neglect or from certain conditions of service.

An inspection determines the operational condition of the equipment and is directly tied to the knowledge of the maintenance personnel. Three considerations involving an inspection are:

1. An organized, methodical procedure for determining the operating conditions of devices and equipment.
2. Assessment of the qualifications of personnel caring for the system.
3. Regular inspection for detection of conditions that require maintenance, repair, or remedy.

To maintain the proper level of protection for a water-based extinguishing system, there are four areas of concern:

1. Adhere to a regular inspection schedule as determined by the needs of the system.
2. Execute any special investigations or tests that can assess the performance of the devices and equipment.
3. Take care to repair or keep devices and equipment in dependable operating order.
4. Ensure that all personnel involved in maintenance are able to properly and correctly execute the procedures for the performance and usage of the equipment.

Inspection and maintenance functions are closely related and often the two areas overlap. Inspections often reveal areas that are in need of maintenance. Similarly, the first step in a proper maintenance schedule involves an inspection of the system's operations.

James M. Fantauzzi is president of North East Fire Protection Systems, Inc., located in Burnt Hills, NY, and is a member of the NFPA Technical Committee on Inspection, Testing, and Maintenance of Water Based Fire Protection Systems.

Ultimately, management is responsible for inspection, testing, and maintenance of their active fire protection system.

RESPONSIBILITY FOR MAINTENANCE

Many of the troubles experienced with extinguishing systems are due to their passive nature. They are designed for emergency conditions. They do not operate on a daily basis. The establishment of appropriate inspection, testing, and maintenance procedures is the responsibility of the building owner or designated representative.

These procedures can be established independently by management or in conjunction with its insurance company, otherwise guidelines established by NFPA 25 should be followed. Any acceptable procedure must be dictated by NFPA 25 or equivalent codes.

Inspection procedures may be affected by frequency as well as seasonal changes. The following are examples of inspections governed by seasonal effects.

Spring inspection: When the danger of freezing temperatures [32°F (0°C)] has passed, spring inspections may be performed. Pertinent operations are:

1. Open cold-weather valves (it should be noted that these are prohibited for use on new systems),
2. Inspect, test, and reset dry-pipe valves,
3. Test water motor gongs,
4. Conduct waterflow tests, and
5. Test fire pump.

Fall inspection: At the approach of freezing weather, the system inspector(s) must:

1. Close cold-weather valves and drain pipes exposed to freezing temperatures (drain valves on the exposed piping are left slightly open), and test the specific gravity of the solution in an antifreeze sprinkler system;
2. Check dry-pipe valves to make sure that the systems are holding air properly and that the pressure switches and water motor alarms are in order, check drains at the low points of the dry piping to ensure that they are properly clear of water, and check heating provisions for dry-pipe valve enclosures;
3. Examine gravity tanks to verify that they are protected against freezing, and check that the employed heating system is in operating condition;
4. Check the condition of fire pump reservoirs and suction intakes from other water sources; and
5. Carefully inspect buildings to make sure that cold air will not enter or unduly expose sprinkler piping to freezing.

TYPES OF SPRINKLER SYSTEM INSPECTIONS

In addition to the indispensable inspection procedures followed by the property owner, other inspection services are available.

Insurance Inspections

In insured properties, extinguishing systems may be given special attention by the insurance carrier. Routine testing of sprinkler systems and devices at regular intervals is a service extended by some insurance companies. This service can be of common interest to both the insurance carrier and the property owner. By these routine tests, the equipment can be shown to be in good operating condition or in need of repair or replacement. Since the tests are made at the owner's responsibility and risk, it behooves the property owner to intelligently cooperate in conducting the tests.

Fire Department Inspections

Inspections are made of extinguishing system equipment by many fire departments at varying intervals. This type of inspection is principally done to make sure that valves are fully open and allows the fire department to familiarize itself with the system. Inspections are customarily made by the fire department in the district where the property is located.

Sprinkler Contractor Services

Standardized sprinkler equipment inspection and maintenance services are offered by sprinkler manufacturers, sprinkler contractors, and inspection service companies. This service provides periodic examinations and reports, and is of value to the property owner not only for a regular checkup of the condition of sprinkler system and components, but also for the valuable instruction that can be given to employees in the process. In addition to sprinkler devices and equipment, a contract can also cover other items, such as tanks and fire pumps that play an important role in the fire protection of property.

Central Station Supervisory Service

Central station supervision of sprinkler alarm and control devices provided under contract is a valuable aid to any maintenance program. The reporting of each incident involving waterflow or valve closure or other supervised events keeps a constant check on the condition of the equipment and encourages care on the part of the plant fire organization as well as the building owner.

All of the above inspection procedures should follow the prescribed guidelines in NFPA 25 and NFPA 72, *National Fire Alarm Code.* Appendix B of the 1995 edition of NFPA 25 contains the prescribed frequency table and generally accepted inspection and testing forms. See Figures 6-17A and 6-17B for sample forms. Table 6-17A shows the various system components which are to be inspected, tested, or maintained and the frequency needed to ensure reliability.

GENERAL MAINTENANCE OF SPRINKLERS AND SPRINKLER PIPING

Automatic sprinklers, when installed where they are not subjected to abnormal conditions, such as loading, corrosion, high temperatures, or mechanical abuse, will yield continued satisfactory performance for a great many years. Some sprinklers accepted by insurance organizations when first installed proved reliable after years of service. Sprinklers listed by leading laboratories since 1900 have generally given satisfactory service even though superseded by new or improved sprinklers. NFPA 25 recommends testing representative samples of sprinklers 50 years or older at 10-year intervals. There are three exceptions to this rule:

1. Sprinklers manufactured prior to 1920 should be replaced.
2. Fast-response sprinklers that have been in service for 20 years or more should be removed and tested. They should be retested every 10 years thereafter.
3. Representative samples of solder-type sprinklers with temperature ratings of 325°F (163°C) or greater that are exposed on a semicontinuous to continuous basis to a maximum allowable ambient temperature condition should be tested for operation at 5-year intervals by a testing laboratory or the sprinkler manufacturer.

For more information, see Chapter 2 of NFPA 25.

Form for Inspection, Testing and Maintenance of Fire Sprinkler Systems

Information on this form covers the minimum requirements of **NFPA 25-1995** for fire sprinkler systems connected to distribution systems without supplemental tanks or fire pumps. Separate forms are available to inspect, test and maintain fire pumps and water tanks. Additional forms are also available for standpipe and hose systems, private fire service mains, water spray fixed systems and foam-water sprinkler systems. More frequent inspection, testing and maintenance may be necessary depending on the conditions of the occupancy and the water supply.

Owner: _____

Owner's Address: _____

Property Being Inspected: _____

Property Address: _____

Date of Inspection: _____ All responses refer to the current inspection performed on this date.

This inspection is (*check one*): ❑ Daily ❑ Weekly ❑ Monthly ❑ Quarterly ❑ Semiannual ❑ Annual ❑ Third Year ❑ Fifth Year

Note: All questions are to be answered *Yes*, *No*, or *Not Applicable*. All "No" answers are to be explained in the comments portion of this form.

Part I - Owner's Section
A. Is the building occupied? ❑ Yes ❑ No ❑ N/A
B. Has the occupancy classification and hazard of contents remained the same since the last inspection? ❑ Yes ❑ No ❑ N/A
C. Are all fire protection systems in service? ❑ Yes ❑ No ❑ N/A
D. Has the system remained in service without modification since the last inspection? ❑ Yes ❑ No ❑ N/A
E. Was the system free of actuations of devices or alarms since the last inspection? ❑ Yes ❑ No ❑ N/A

Owner or Representative (print name) ___ Signature and Date ___

Part II - Inspector's Section
A. Inspections
1. Daily, or weekly if low temperature alarms are installed
Enclosures around dry-pipe, preaction or deluge valves maintaining a minimum of 40^0F? ❑ Yes ❑ No ❑ N/A
2. Weekly Inspection Item
Relief port on reduced pressure backflow prevention assemblies free of continuous dischare? ❑ Yes ❑ No ❑ N/A
3. Weekly inspection items which can be performed monthly if the items are electrically supervised or secured with locks
A. Gauges on dry, preaction and deluge systems in good condition and showing normal air and water pressure? ❑ Yes ❑ No ❑ N/A
B. Control valves and isolation valves on backflow prevention devices:
　1. In correct (open or closed) position? ❑ Yes ❑ No ❑ N/A
　2. Sealed, locked or supervised and accessible? ❑ Yes ❑ No ❑ N/A
4. Monthly Inspection Items
A. Preaction and Deluge Valves:
　1. Free from physical damage? ❑ Yes ❑ No ❑ N/A
　2. Trim valves in appropriate (open or closed) position and no leakage from valve seat? ❑ Yes ❑ No ❑ N/A
　3. Electrical components in service? ❑ Yes ❑ No ❑ N/A
B. Dry-Pipe Valves:
　1. Free from physical damage? ❑ Yes ❑ No ❑ N/A
　2. Trim valves in appropriate (open or closed) position? ❑ Yes ❑ No ❑ N/A
　3. No leakage from intermediate chamber? ❑ Yes ❑ No ❑ N/A
C. Sprinkler wrench with spare sprinklers? ❑ Yes ❑ No ❑ N/A
D. Gauges on wet-pipe system in good condition and showing normal water supply pressure? ❑ Yes ❑ No ❑ N/A
E. Alarm Valves:
　1. Gauges show normal supply water pressure? ❑ Yes ❑ No ❑ N/A
　2. Free from physical damage? ❑ Yes ❑ No ❑ N/A
　3. Valves in correct (open or closed) position? ❑ Yes ❑ No ❑ N/A
　4. No leakage from retarding chamber or drains? ❑ Yes ❑ No ❑ N/A
5. Quarterly Inspection Items
A. Sprinkler Pressure Regulating Control Valves:
　1. In open position and not leaking? ❑ Yes ❑ No ❑ N/A
　2. Maintaining downstream pressure per design criteria? ❑ Yes ❑ No ❑ N/A
　3. In good condition with handwheels not broken? ❑ Yes ❑ No ❑ N/A

5. Quarterly Inspection Items (Continued)
B. Fire Department Connections:
　1. Visible and accessible? ❑ Yes ❑ No ❑ N/A
　2. Couplings and swivels not damaged and rotate smoothly? ❑ Yes ❑ No ❑ N/A
　3. Plugs or caps in place and undamaged? ❑ Yes ❑ No ❑ N/A
　4. Gaskets in place and in good condition? ❑ Yes ❑ No ❑ N/A
　5. Identification sign(s) in place? ❑ Yes ❑ No ❑ N/A
　6. Check valve is not leaking? ❑ Yes ❑ No ❑ N/A
　7. Automatic drain valve in place and operating properly? ❑ Yes ❑ No ❑ N/A

(Note: If plugs or caps are not in place, inspect the interior for obstructions and verify that the valve clapper is operational over its full range.)
C. Alarm devices free from physical damage? ❑ Yes ❑ No ❑ N/A
D. Hydraulic nameplate, if provided, securely attached to riser and legible? ❑ Yes ❑ No ❑ N/A
6. Annual Inspection Items
A. Proper number and type of spare sprinklers? ❑ Yes ❑ No ❑ N/A
B. Visible sprinklers:
　1. Free of corrosion? ❑ Yes ❑ No ❑ N/A
　2. Free of obstructions to spray patterns? ❑ Yes ❑ No ❑ N/A
　3. Free of foreign materials including paint? ❑ Yes ❑ No ❑ N/A
　4. Free of physical damage? ❑ Yes ❑ No ❑ N/A
C. Visible pipe:
　1. In good condition? ❑ Yes ❑ No ❑ N/A
　2. Free of mechanical damage and not leaking? ❑ Yes ❑ No ❑ N/A
　3. No external corrosion? ❑ Yes ❑ No ❑ N/A
　4. Properly aligned? ❑ Yes ❑ No ❑ N/A
　5. No external loads? ❑ Yes ❑ No ❑ N/A
D. Visible pipe hangers and seismic braces not damaged or loose? ❑ Yes ❑ No ❑ N/A
E. Must be done before cold weather
　1. Adequate heat in areas with wet piping? ❑ Yes ❑ No ❑ N/A
　2. Low temperature alarms in dry-pipe, preaction and deluge valve enclosures functioning? ❑ Yes ❑ No ❑ N/A
　3. Interior of pipe in preaction and dry-pipe systems which passes through freezers free of ice blockage? ❑ Yes ❑ No ❑ N/A
7. Annual, or every fifth year for valves which can be reset without opening:
Interior of dry-pipe, preaction and deluge valves passed internal inspection? ❑ Yes ❑ No ❑ N/A
8. Fifth Year Inspection Items
A. Alarm valves and their associated strainers, filters and restriction orifices passed internal inspection? ❑ Yes ❑ No ❑ N/A
B. Check valves internally inspected and all parts operate properly, move freely and are in good condition? ❑ Yes ❑ No ❑ N/A
C. Strainers, filters, restricted orifices and diaphragm chambers on dry-pipe, preaction and deluge valves passed internal inspection? ❑ Yes ❑ No ❑ N/A

FIG. 6-17A. Sample form for inspection, testing, and maintenance of fire sprinkler system.

B. Testing
The following tests are to be performed at the noted intervals. Report any failures on Part III of this form.

1. Quarterly Tests

A. Sprinkler system main drain test:
   1. Record Static Pressure _____ psi and Residual Pressure _____ psi. Was flow observed? ❑ Yes ❑ No ❑ N/A
   2. Are results comparable to previous test? ❑ Yes ❑ No ❑ N/A

B. Waterflow alarm devices passed tests? ❑ Yes ❑ No ❑ N/A
   1. Inspectors test connection opened? (wet-pipe when not in freezing weather) ❑ Yes ❑ No ❑ N/A
   2. Bypass connection opened? (wet-pipe systems in freezing weather, dry-pipe, preaction, or deluge). ❑ Yes ❑ No ❑ N/A
   3. Alarms actuated and flow observed? ❑ Yes ❑ No ❑ N/A

C. Control Valves (except OS&Y and gear-operated indicating butterfly valves) opened until spring or torsion is felt in the rod, then closed back one-quarter turn? ❑ Yes ❑ No ❑ N/A

D. Dry-pipe and preaction systems:
   1. Priming water level correct? ❑ Yes ❑ No ❑ N/A
   2. Low air pressure signal passed test? ❑ Yes ❑ No ❑ N/A

E. Quick opening devices passed test? ❑ Yes ❑ No ❑ N/A

F. Valve supervisory switches indicate movement? ❑ Yes ❑ No ❑ N/A

2. Annual Tests

A. Are all sprinklers in service dated 1920 or later? ❑ Yes ❑ No ❑ N/A

B. Fast Response sprinklers in service for less than 20 years? *If "no" test sample now and every 10 years.* ❑ Yes ❑ No ❑ N/A

C. Standard sprinklers less than 50 years old? ❑ Yes ❑ No ❑ N/A
   If "no" has a sample been tested within 10 years? ❑ Yes ❑ No ❑ N/A
   If "no" test sample now and every 10 years.

D. Specific gravity of antifreeze correct? ❑ Yes ❑ No ❑ N/A

E. All control valves operated through full range and returned to normal position? ❑ Yes ❑ No ❑ N/A

F. Low temperature alarms in dry-pipe, preaction and deluge valve enclosures passed test? ❑ Yes ❑ No ❑ N/A

G. Preaction and deluge valve full flow trip test: (except deluge valves where water can't be discharged) *(Test all systems together which will operate simultaneously.)*
   1. Water discharge from all nozzles unimpeded? ❑ Yes ❑ No ❑ N/A
   2. Pressure reading at hydraulically most remote nozzle _____ psi.
   3. Residual pressure reading at valve _____ psi. Was flow observed? ❑ Yes ❑ No ❑ N/A
   4. Are above readings comparable to design values? ❑ Yes ❑ No ❑ N/A
   5. Manual activation devices passed test? ❑ Yes ❑ No ❑ N/A
   6. Automatic air pressure maintenance devices passed test? ❑ Yes ❑ No ❑ N/A

H. Dry-pipe valve partial flow trip test:
   1. Record initial air pressure _____ psi and water pressure _____ psi.
   2. Record tripping air pressure _____ psi and tripping time ___ (sec.).
   3. Are above results comparable to previous tests? ❑ Yes ❑ No ❑ N/A

I. Automatic air maintenance devices on dry-pipe and preaction systems passed test? ❑ Yes ❑ No ❑ N/A

J. Backflow devices passed backflow test? ❑ Yes ❑ No ❑ N/A

K. Backflow devices passed full flow test? ❑ Yes ❑ No ❑ N/A

L. All sprinkler pressure regulating control valves passed full flow test? ❑ Yes ❑ No ❑ N/A

3. Dry-pipe full flow trip test to be done every third year:

1. Record initial air pressure _____ psi and water pressure _____ psi.
2. Record tripping air pressure _____ psi and tripping time ___ (sec.).
3. Was water delivered to inspectors test connection? ❑ Yes ❑ No ❑ N/A
4. Are above results comparable to previous tests? ❑ Yes ❑ No ❑ N/A

4. Tests to be done every fifth year.

A. Extra High, Very Extra High and Ultra High Temperature sprinklers tested? ❑ Yes ❑ No ❑ N/A

B. Gauges checked against calibrated gauge or replaced? ❑ Yes ❑ No ❑ N/A

C. Maintenance
1. Regular Maintenance Items

A. If sprinklers have been replaced, were they proper replacements? ❑ Yes ❑ No ❑ N/A

B. Air leaks in dry-pipe system resulting in air pressure loss more than 10 psi/week repaired? ❑ Yes ❑ No ❑ N/A

C. Dry-pipe systems being maintained in dry condition? ❑ Yes ❑ No ❑ N/A

D. If any of the following were discovered, was an obstruction investigation conducted and the system flushed? ❑ Yes ❑ No ❑ N/A
Explain reason(s) and obstruction investigation findings in Part III
1. Defective intake screen for pumps taking suction from open sources.
2. Obstructive material discharged during waterflow tests.
3. Foreign materials found in dry-pipe valves, check valves or pumps.
4. Heavy discoloration of water during drain test or plugging of inspectors test connection.
5. Plugging of sprinklers found during activation or alteration.
6. Plugging found in piping dismantled during alterations.
7. Failure to flush yard piping or surrounding public mains following new installation or repairs.
8. Record of broken mains in the vicinity.
9. Abnormally frequent false-tripping of dry-pipe valves.
10. System is returned to service after an extended period out of service (greater than one year).
11. There is reason to believe the system contains sodium silicate or its derivatives.

2. Annual Maintenance Items

A. Operating stem of all OS&Y valves lubricated, completely closed, and reopened? ❑ Yes ❑ No ❑ N/A   B. Interior of dry-pipe, preaction and deluge valves cleaned? ❑ Yes ❑ No ❑ N/A

C. Low points drained in dry-pipe, preaction and deluge systems prior to the onset of freezing weather? ❑ Yes ❑ No ❑ N/A

D. Sprinklers and spray nozzles protecting commercial cooking equipment and ventilating systems replaced except for bulb-type which show no signs of grease buildup? ❑ Yes ❑ No ❑ N/A

Part III - Comments
(Any "No" answers, test failures or other problems found with the sprinkler system must be explained here.)

_____
_____
_____
_____
_____
_____
_____
_____
_____

Part IV - Inspector's Information

Inspector: _____ Company: _____
Company's Address: _____
I state that the information on this form is correct at the time and place of my inspection, and that all equipment tested at this time was left in operational condition upon completion of this inspection except as noted in Part III above.

Signature of Inspector _____ Date: ____

FIG. 6-17A. Continued

Weekly Report of Inspection
of Water Based Fire Protection Systems
ALL QUESTIONS ARE TO BE FULLY ANSWERED AND ALL BLANKS TO BE FILLED

This form is being offered to assist in the performance and recording of the results of Weekly Scheduled Inspection Tasks of the various types of Fire Sprinkler Systems and component parts as listed below.

Inspecting Firm:

Name of Property:

Inspector Name: Date:

Page   of

Wet Sprinkler and Standpipe Systems:

A-1.1 Spkr. Supply Gauge   psi
A-1.2 Spkr. System Gauge   psi
A-1.3 StPipe Supply Gauge psi
A-1.4 StPipe System Gauge   psi
A-1.5 StPipe (Top Flr) Gauge   psi

		Y	N/A	N
A-2.0	System In Service On Inspection			
A-2.1	Spkr. Control Valves Sealed Open			
A-2.2	StPipe Control Valves Sealed Open:			
A-3.1	Trim Piping leak tight:			
A-4.1	Backflow Asmb. Valves Sealed Open:			
A-5.1	Control Valves accessible:			
A-8.1	Signage/Identification Tags in Place:			
A-9.1	**ALARM PANEL CLEAR**			
A-9.2	**SYSTEMS LEFT IN SERVICE**			
A-10.1	**COMMENTS:**			

Dry Pipe Sprinkler System:

B-1.1 Air pressure Gauge: psi
B-1.2 Accelerator Gauge: psi
B-1.3 Water pressure Gauge: psi

		Y	N/A	N
B-2.0	System In Service On Inspection			
B-2.1	Compressor Operational:			
*B-2.2	Oil level Full:			
B-3.1	Control Valve Sealed Open:			
B-3.2	Control Valves accessible:			
B-3.3	Alarm Test valve closed:			
B-3.4	Alarm Line Valve Open:			
B-4.1	Intermediate Chamber leak tight:			
B-4.2	Low point drum drips drained:			
	(As frequently as needed)			
B-5.1	Valve Enclosure Secured:			
*B-5.2	Low Temperature Alarm Operational:			
B-5.3	Heater Operational:			
B-8.1	Signage/Identification Tags in Place:			
B-9.1	**ALARM PANEL CLEAR**			
B-9.2	**SYSTEM LEFT IN SERVICE**			
B-10.1	**COMMENTS:**			

* The Inspection tasks noted with an asterisk (*) are required to be performed on a Monthly frequency schedule; however, due to varying conditions that may exist on any individual project, it is suggested that these tasks be performed on a Weekly frequency schedule.

Fire Dept. Connection:

		Y	N/A	N
*C-1.1	Caps or Plugs on FDC:			
*C-1.2	Swivel rotation nonbinding:			
*C-2.1	FDC location plainly visible:			
*C-2.2	FDC easily accessible:			
*C-2.3	FDC Identification Plate in place:			
*C-3.1	Ball Drip drain leak tight:			
*C-4.1	Wall Hydrant plainly visible:			
*C-4.2	Wall Hydrant easily accessible:			
*C-4.3	Wall Hydrant Identification Plate			
	in place:			
*C-10.1	COMMENTS:			

Sprinkler Heads:

		Y	N/A	N
*D-1.1	Extra Heads in Spare head cabinet:			
*D-1.2	Heads appear of proper temperature:			
*D-1.3	Head Wrench for each type of Head:			
*D-2.1	Head in Cooler appears free of ice, corrosion:			
*D-2.2	Head appears free of leakage or damage:			
*D-2.3	Head appears free of paint:			
*D-2.4	Heads appear free of non-approved coverings:			
*D-3.1	Standard Head less than 50 year:			
*D-3.2	Residential Head less than 20 year:			
*D-3.3	Fast Response Heads 20 year:			
*D-3.4	High Temperature Heads 5 year:			
*D-10.1	COMMENTS:			

(All "NO" answers to be fully explained.)

INSPECTOR'S INITIAL _____   OWNER/DESIGNATED REP. INITIAL _____   DATE _____

(AFSA Form 94-105A)

Page 1 of 2

FIG. 6-17B. *Sample weekly report form for inspection of water-based fire protection systems.*

Weekly Report of Inspection of Water Based Fire Protection Systems ...*continued*

Preaction and Deluge System:

E-1.1 Air pressure Gauge: psi

E-1.2 Water pressure Gauge: psi

	Y	N/A	N
E-2.0 Systems In Service On Inspection			
E-2.1 Control Valves Sealed Open:			
E-2.2 Control Valves accessible:			
E-3.1 Enclosure Secured:			
E-3.2 Heater Operational:			
E-3.3 Low Temperature Alarm Operational:			
E-4.1 Trim valves normally open:			
E-4.2 Alarm Test valve closed:			
E-4.3 Trim Piping leak tight:			
E-5.1 Low point drum drips drained:			
(As frequently as needed)			
E-8.1 Signage/Identification Tags in Place:			
E-9.1 ALARM PANEL CLEAR			
E-9.2 SYSTEM LEFT IN SERVICE			

E-10.1 COMMENTS:

_____

_____

Other

_____

_____

_____

_____

_____

_____

_____

_____

_____

_____

_____

_____

_____

_____

Fire Pump:

F-1.1 Suction pressure Gauge: psi

F-1.2 Discharge pressure Gauge: psi

	Y	N/A	N
F-2.0 Pump In Service On Inspection			
F-2.1 Control Valves Sealed Open:			
F-2.2 Control Valves accessible:			
F-3.1 Pump enclosure secured:			
F-3.2 Pump enclosure heated (40° F):			
F-3.3 Adequately Lighted:			
F-4.1 Weekly run test: (No water flow)			
F-4.2 Shaft seals dripping water properly:			
F-4.3 Casing relief valve free of damage:			
F-4.4 Pressure relief valve free of damage:			
F-5.1 Jockey Pump Operational:			
F-6.1 Bearings and Valves Lubricated:			
F-6.2 Valves, fittings, pipe leak free:			
F-7.1 Controllers power "ON":			
F-7.2 Controllers set on "AUTO":			
F-8.1 Hose Header control valve closed:			
F-9.1 Diesel tank ⅔ full:			
F-9.2 Oil level full:			
F-9.3 Water level full:			
F-9.4 Water hose condition good:			
F-9.5 Water jacket heater working:			
F-9.6 Antifreeze protection adequate:			
F-9.7 Cooling line Strainer appears clear of debris:			
F-9.8 Water Jacket piping leak tight:			
F-9.9 Batteries fully charged:			
F-9.10 Battery charger appears operating properly:			
F-9.11 Battery terminals clean:			
F-9.12 Battery state of charge checked:			
F-9.13 Solenoid valve appears operating correctly:			
F-10.1 Condensate drain clear:			
F-10.2 Louvers, in-take duct clean:			
F-14.1 Signage/Identification Tags in Place:			
F-15.1 ALARM PANEL CLEAR			
F-15.2 SYSTEM LEFT IN SERVICE			

F-16.1 COMMENTS:

_____

_____

_____

(All "NO" answers to be fully explained.)

INSPECTOR'S INITIAL _____   OWNER/DESIGNATED REP. INITIAL _____   DATE _____

(AFSA Form 94-105A)

Page 2 of 2

FIG. 6-17B.   Continued

TABLE 6-17A. *Summary of Sprinkler System Inspection, Testing, and Maintenance*

Item	Activity	Frequency
Gages (dry, preaction deluge systems)	Inspection	Weekly/monthly
Control valves	Inspection	Weekly/monthly
Alarm devices	Inspection	Quarterly
Gages (wet-pipe systems)	Inspection	Monthly
Hydraulic nameplate	Inspection	Quarterly
Buildings	Inspection	Annually (prior to freezing weather)
Hanger/seismic bracing	Inspection	Annually
Pipe and fittings	Inspection	Annually
Sprinklers	Inspection	Annually
Spare sprinklers	Inspection	Annually
Fire department connections	Inspection	(see Table 6-17B)
Valves (all types)	Inspection	
Alarm devices	Test	Quarterly
Main drain	Test	Quarterly
Antifreeze solution	Test	Annually
Gages	Test	5 years
Sprinklers— extra-high temp.	Test	5 years
Sprinklers— fast response	Test	At 20 years and every 10 years thereafter
Sprinklers	Test	At 50 years and every 10 years thereafter
Valves (all types)	Maintenance	Annually or as needed
Obstruction investigation	Maintenance	5 years or as needed

Where sprinklers are subject to "loading" (the accumulation of foreign material), or corrosion, even to a minimal extent, their condition should be monitored carefully and frequently. Should the condition of the sprinklers show signs of loading, a representative minimum sample of four sprinklers or 1 percent of the total number of sprinklers per individual model, whichever is greater, should be removed and sent for testing to an approved laboratory. The results of such testing will be an indicator of the condition of similar sprinklers remaining in the property. When removing the sprinklers, take care to ensure that no damage results and that they are well packed to avoid further damage in transit.

To prevent mechanical injury and distortion when installing or removing sprinklers to be cleaned and reinstalled, utilize the proper wrench for the sprinkler provided by the manufacturer.

Accumulation of Foreign Materials on Sprinklers

In many classes of properties, conditions exist that cause an accumulation of foreign material on automatic sprinklers. This accumulation is commonly known as loading. The operation of the sprinkler may be retarded or prevented by the loading effect.

Deposits on sprinklers tend to retard their operation as a result of the heat insulating effect of the loading material. If the deposit is hard, it may physically retard or prevent the sprinkler from operating. The best practice is to replace loaded sprinklers with new sprinklers, rather than to attempt to clean the sprinkler. Attempts at cleaning, particularly in instances where deposits are hard, are likely to damage the sprinkler, thus rendering it inoperative or possibly causing leakage.

Deposits of light dust on sprinklers, such as may be found in woodworking plants and grain elevators, are less serious than hard deposits. Dust may be expected to partially delay the operation of sprinklers, but ordinarily will not prevent the eventual discharge of water. Dust deposits can be blown or brushed off. Blowing by compressed air should not be undertaken where it can create a dust explosion or ignition hazard. When using a brush, make sure that it is soft to avoid possible injury to sprinkler parts. If these two methods prove unsatisfactory, then a damp, soft cloth may be used to clean the sprinkler. Water-solution cleaning liquids of caustic or acid type are likely to be destructive to sprinklers and should not be used for cleaning. Hot solutions of any kind should not be used either.

Sprinklers are sometimes protected when ceilings or sprinkler piping are being painted by temporarily placing small, lightweight paper or plastic bags over them, and securing them with rubber bands. Bags, however, are likely to delay the operation of the sprinklers and should be removed immediately after the painting is completed. Should paint cover one of the sprinklers, take out the sprinkler and replace it. There is no known method of adequately removing paint under the water cap or on the fusible link. Sprinklers that have been painted by anyone other than the manufacturer must be replaced with new sprinklers.

Sprinklers in spray booths present a special problem. To help minimize this problem, one acceptable solution is to conduct the spray process in a downward manner to avoid the sprinkler. Cleaning will be necessary; therefore, install the sprinklers in an easily accessible location. To facilitate the washing or wiping off of deposits, use a light coating of a mild, soft soap. Unless cleaning is done very carefully, deposits are likely to accumulate to such an extent as to interfere seriously with sprinkler operation. The use of paper, polyethylene, or cellophane bags to protect sprinklers in spray booths is fairly common and is permitted by NFPA 25 when the thickness of such bags does not exceed 0.0003 in.

Corrosion of Automatic Sprinklers

Corrosive conditions are likely to make automatic sprinklers inoperative or to retard the speed of their operation. They can also seriously impede the waterways of spray nozzles. Corrosive vapors may seriously affect not only the heat-actuated element and the water-retaining mechanism of an automatic sprinkler, but also may be severe enough to weaken or destroy other portions of the sprinkler. In most instances, such corrosive action is slow and chronic. Thus, the process must be watched vigilantly. Illustrations of some typical corroded sprinklers are shown in Figure 6-17C.

Some types of sprinklers are less susceptible than others to corrosive conditions. Nonferrous metal is used for some sprinkler parts, but special protective coatings are still necessary for all types of sprinklers which are exposed to extreme corrosive conditions. Approved corrosion-resistant or special coated sprinklers are needed in locations where chemicals, excessive moisture, or corrosive vapors exist.

Protection of Pipe Against External Corrosion

Under some conditions, corrosive vapors may cause rapid deterioration of steel pipe and hangers. Proper protection can prevent the frequent replacement of the steel pipe and hangers. Cast-iron fittings are generally not affected by corrosive vapors

In severe corrosive conditions where protective methods are not wholly satisfactory, copper or special alloy, non-corrosive pipe will yield the best results. Galvanized steel may offer the most economi-

FIG. 6-17C.   Examples of corroded automatic sprinklers.

cal method of obtaining a reasonably long life for the piping system. Galvanized piping systems have met with success in chemical plants, saltworks, or similar properties where severe corrosion exists. Stainless steel and copper piping have also been used in some cases.

Corrosion of existing equipment becomes a maintenance problem. The options available are: (1) replacement of the components or (2) corrective measures, such as an application of a recognized protective coating.

Advance Preparations Before Shutoff

When work requires systems to be shut off, the work should be planned for a time when the least amount of hazard exists. In industrial plants, about three times as many fires occur during operations than during idle periods. Therefore, work that requires system shutoff should be performed during weekends or other idle periods. Special guard services may be required to aid in the detection of any fire that might develop while the systems are shut down.

Sectional valves, rather than main valves, should be used to reduce to a minimum the number of systems removed from service and to take advantage of multiple water supplies. All personnel, materials, and tools should be on hand before the protection is impaired.

If underground mains are involved, a tapping machine should be used when possible to avoid shutting off the water. If mains are to be opened, wood or other plugs or caps and clamps to close the ends of pipes should be available and ready to install. Emergency measures should be taken to maintain the maximum possible water supply. One possibility is to make a temporary hose connection from a hydrant that is still in service to the system riser(s). Other options include the use of a domestic or industrial supply to the riser(s). These connections are usually made to the nominal 2-in. (51-mm) drain with the drain valve and hydrant valve left open. Such arrangements are shown in Figure 6-17D. Adapters for connecting $2^1/_2$-in. (64-mm) hose to sprinkler systems should be kept on hand.

Definite procedures should be followed for closing valves, notifying the fire department and insurance companies, and making waterflow tests after the work is completed. Such impairment procedures will help to ensure that systems are out of service for a minimal period of time and that complete system protection is restored.

Emergency Measures for Maintaining Protection During Repairs or Alterations

Fire records show the seriousness of having extinguishing systems shut off when a fire starts. There is a danger if the water supply is shut off for extensions or alterations to piping, repairs due to accidental damage to sprinklers and spray nozzles, replacement of sprinklers after a fire, or maintenance or replacement of sprinklers and other system devices. When, of necessity, protection is inter-

FIG. 6-17D.   If it is necessary to shut off the water supply to a sprinkler system when repairs or extensions are to be made, a temporary supply can be provided by a hose line, keeping all or part of the system in service.

rupted, every effort must be made to limit the extent and duration of the interruption. A cardinal rule is to notify the fire department whenever an impairment exists so they will not place false reliance on the systems. Insurance companies also may request that building owners advise them when there is an interruption in protection, so that alternate means of protection can be arranged if necessary or desirable.

CARE AND MAINTENANCE OF SPECIFIC COMPONENTS

The following items of care and maintenance are treated in relation to routine inspection, test, and maintenance programs. They are, however, independent of the organization performing the inspection and maintenance functions. These are in-house responsibilities and should be left in the care of trained and competent individuals.

Public and Private Water Supply Equipment

The inspection and testing of public water supplies and of private water-supply equipment, such as tanks and fire pumps, is covered elsewhere in Section 6. See Table 6-17B for a summary of valve and fire department connection inspection, testing, and maintenance.

TABLE 6-17B. *Summary of Valve and Fire Department Connection Inspection, Testing, and Maintenance*

Item	Activity	Frequency
Control Valve		
Sealed	Inspection	Weekly
Locked	Inspection	Monthly
Tamper Switch	Inspection	Monthly
Alarm valve		
Exterior	Inspection	Monthly
Interior	Inspection	5 years
Strainers, filters, orifices	Inspection	5 years
Check valve		
Interior	Inspection	5 years
Preaction/Deluge Valve		
Enclosure (during cold weather)	Inspection	Daily/weekly
Exterior	Inspection	Monthly
Interior	Inspection	Annually/5 years
Strainers, filters, orifices	Inspection	5 years
Dry-pipe valves/quick-opening devices		
Enclosure (during cold weather)	Inspection	Daily/weekly
Exterior	Inspection	Monthly
Interior	Inspection	Annually
Strainers, filters, orifices	Inspection	5 years
Pressure regulating and relief valves		
Sprinkler systems	Inspection	Quarterly
Hose connection	Inspection	Quarterly
Hose rack	Inspection	Quarterly
Fire pump		
Casing relief valve	Inspection	Weekly
Pressure relief valve	Inspection	Weekly
Backflow prevention assemblies		
Reduced pressure	Inspection	Weekly/monthly
Reduced pressure detector	Inspection	Weekly/monthly
Fire department connections	Inspection	Quarterly
Main drain	Test	Quarterly
Waterflow alarm	Test	Quarterly
Control valve		
Position	Test	Quarterly
Operation	Test	Annually
Supervisory	Test	Quarterly
Preaction/deluge valve		
Priming water	Test	Quarterly
Low-air-pressure alarms	Test	Quarterly
Full flow	Test	Annually
Dry pipe valves/quick-opening devices		
Priming water	Test	Quarterly
Low-air-pressure alarm	Test	Quarterly
Quick-opening devices	Test	Quarterly
Trip test	Test	Annually
Full flow trip test	Test	3 years
Pressure regulating and relief valves		
Sprinkler system	Test	Annually
Circulation relief	Test	Annually
Pressure relief valve	Test	Annually
Hose connection	Test	5 years
Hose rack	Test	5 years
Backflow prevention Assemblies	Test	Annually
Control valve	Maintenance	Annually
Preaction/deluge valve	Maintenance	Annually
Dry pipe valve/quick-opening device	Maintenance	Annually

Source: NFPA 25.

Control Valves and Meters

Inspection and supervision of valves in fire protection systems are necessary since closed valves are the leading cause of unsatisfactory performance, when it occurs. The following paragraphs summarize some of the more salient points of valve inspection and supervision.

Service valves: Service valves at private fire system connections to public systems are usually under the control of the water department and are seldom operated. Their condition is usually indicated adequately by waterflow tests made from the private protection system, as covered later in this chapter.

Meters: Meters in public water system connections are also generally under the control of the water department, but are sometimes located in pits on the protected property.

Check valves: Check valves in public water connections, where needed to prevent backflow into the public systems, are usually a part of the protection system for which the property owner is responsible. Tightness of check valves should be determined periodically by proper tests, depending on water supplies.

Check valves and any corresponding control valves should be properly arranged and located in accordance with the appropriate NFPA standards for sprinkler systems, water spray protection, or standpipe and hose systems.

Control valves: All control valves should be readily accessible and unobstructed so they can be closed promptly and examined to see that they are open and in good operative condition, turn easily, and do not leak.

Pits for gate valves and check valves should be kept reasonably dry and clean, so that valves can be tested, examined, and maintained in good condition. Manhole covers should be kept clear of snow and ice. See Section 6, Chapter 3, "Water Distribution Systems," for an example of a well-arranged check valve pit with access for inspection.

Each control valve should be numbered or otherwise identified at the valve and cataloged, giving location, use, or portions of the system controlled, on a plant fire inspection form. A plan showing valve locations should be posted at a central point known to plant and public fire officials.

Each valve should have a sign showing what it controls, with the legend "Must Be Open at All Times," or other proper wording. Underground valve sites should be shown by location and by distance markings on nearby buildings and on accurate plans of the property.

All control valves should be sealed or locked open, unless there is a central station supervisory service. Wrenches should be kept at post indicator valves or at locations where they are readily accessible.

For larger facilities or plants, there should be someone on premises at all times who knows the use and location of all the control and drain valves. This includes the person on watch or any others who may be on duty at night.

Sprinklers and Sprinkler Piping

In the inspection of the sprinkler system, the following eight categories should be considered:

1. *Absence of sprinklers.*   Observe whether there is any room or building from which sprinklers have been omitted. The following should be used as a basic checklist:

   - Basement
   - Lofts
   - Show windows
   - Concealed spaces
   - Towers
   - Under stairwells
   - Under skylights
   - Inside elevator shafts
   - Vertical shafts
   - Small enclosures, such as those for drying and heating
   - Closets, unless open on top

2. *Improper location of sprinklers.*   Observe whether there are sprinklers under the following:

   - Air ducts over 4 ft (1.3 m) in width
   - Shelves
   - Benches and tables
   - Overhead storage racks
   - Platforms or similar obstructions

3. *Check the impact of adding pipe to the system with hydraulic calculations.*   Check the hydraulic calculations to ensure that the current water pressure will suffice. The calculations should be reviewed with the additions so that distribution piping is not overtaxed for the sprinklers it must supply.

4. *Check for proper clearance for the sprinklers.*   There are a few areas to review for clearance of sprinklers:

   - Sprinklers cannot be obstructed by high-piled stock, walls, or partitions that might prevent free and proper water distribution.
   - There must be a clear space of 18 in. (457 mm) below the deflectors of the sprinklers. This value is 36 in. for large drop and ESFR sprinklers.
   - Installation guidelines of special sprinklers have not been violated. This may include installation under certain construction types.
   - Refer to NFPA 13, *Standard for the Installation of Sprinkler Systems* (hereinafter referred to as NFPA 13), for spacing recommendations for various occupancy classes, types of sprinklers, and construction factors.

5. *Proper position of deflectors.*   The distance of deflectors from the ceiling or the bottom of beams or joists should conform to NFPA 13.

6. *Proper pitch of dry-pipe systems.*   Note whether all pipes in a dry-pipe system have the proper pitch. This feature is of special importance, because water remaining in pockets or low points is likely to freeze and cripple the system.

7. *Proper support of piping.*   Are any hangers loose or pipes not properly supported? Observe whether sprinkler piping is used to support stock or clothing. Sprinkler piping and hangars should never be used for such purposes.

8. *Proper sprinkler installation.*   There are five areas to check for the proper installation of sprinklers:

   - Sprinklers must be installed in the positions for which they are intended, in compliance with the rules for special application, whether they be pendent, upright, or sidewall types.
   - Note the type and design of the sprinklers, the year of their manufacture, and the date of their installation.
   - Are all sprinklers of the proper temperature rating? Ordinary-degree sprinklers should be substituted for high-degree sprinklers when the latter are unnecessary. Wherever the temperature

around the sprinkler exceeds 100°F (37.8°C), substitute those sprinklers with intermediate-degree sprinklers.

- Are any sprinklers loaded or corroded? If so, remove immediately for testing and replace the questionable samples. In areas where corrosion is common, use sprinklers with the proper coating to decrease the likelihood of future corrosion.

- Sprinklers with coatings of paint, whitewashing, bronzing, or excessive deposits or incrustations should be replaced immediately with new sprinklers.

A supply of extra sprinklers should be kept in a spare sprinkler cabinet so that any sprinklers that have operated or been damaged in any way can be replaced promptly. These sprinklers should correspond in type, style, and in temperature ratings to the sprinklers in the property. The cabinet should be situated in a cool, easily accessible location. The number of sprinklers stocked for replacement purposes is governed by the size and number of systems. Other considerations include: (1) the location of the protected property relative to the source of supply for replacement sprinklers and (2) the number of sprinklers likely to be opened in the event of a fire.

The stock of spare sprinklers, under normal conditions, should be as follows:

System Size	Number of Spares
Up to 300 sprinklers	6 sprinklers
From 300 to 1,000 sprinklers	12 sprinklers
More than 1,000 sprinklers	24 sprinklers

For systems aboard vessels or in isolated locations, a greater number of sprinklers should be carried to permit restoring equipment to service promptly after a fire.

A sprinkler wrench should be kept in the sprinkler cabinet to be used for the removal and installation of sprinklers. This wrench should always be used for installing new sprinklers in order to minimize the chance of damaging the sprinkler.

System Waterflow Tests

NFPA 13 calls for a water supply test of pipe and pressure gages to be provided at locations that will permit flowing tests to be made to determine whether water supplies and connections are in order. A 2-in. (51-mm) drain at the sprinkler riser may suffice as a water supply test pipe. The valve of the water supply test pipe must be installed so that the valve may be opened wide for a sufficient time to ensure a proper test, without causing water damage. A similar main drain valve in a water spray system can serve the same purpose.

There should be provision for checking the system's gage and its pressure recordings with an inspector's test gage.

The pressure on the sprinkler system side of the alarm or check valves may be higher than that of the water supply, because any momentary high pressure on the supply will be transmitted to the system and retained by the check valves. This excess pressure is relieved when a waterflow test on the system side of such valves is made.

The drop in pressure below the normal static pressure with the 2-in. (51-mm) main drain pipe wide open can be noted, and general flow conditions evaluated. For instance, if the normal pressure is 50 psi (345 kPa), and if previous tests have shown a pressure drop to 45 psi (310 kPa) when the main drain valve is opened, it will be apparent that, if at a subsequent inspection the pressure drops to 35 psi (241 kPa) or lower, there is some obstruction to the flow, which should be investigated. There may also be a defect that should be located and remedied.

At each inspection, an individual flow test should be made for each water supply and each connection from a supply. This can be accomplished by closing all water supplies temporarily, except the one under test. Too much emphasis cannot be placed on the value of flow tests to determine whether there is any obstruction to full flow. It is mandatory that all water supplies be returned to service immediately after testing.

If there is a waterflow supervisory service, tests should not be made without first notifying the fire department, central station, or other alarm receiving facilities.

There should be a minimal 1-in. (25.4-mm) test pipe having a test orifice of equivalent size at the remote point of each sprinkler system. This should be operated at each inspection of a wet system to make sure that there is free flow at good pressure and to test the waterflow alarm. In dry-pipe systems, this test pipe can also be used to trip the dry-pipe valve and determine water delivery time.

Dry-Pipe Sprinkler Systems

The best practice is to keep dry-pipe systems in the dry condition throughout the year. When water flows into sprinkler piping, rust scaling and other foreign materials tend to be carried into the smaller pipes. Accumulation and obstruction can occur due to the water flow. Annual tripping of the dry-pipe valve to make the system wet may accelerate this condition.

When systems are wet, with the dry-pipe valve tripped, the waterflow alarm feature of the dry valve would have to be taken out of service.

Any old, obsolete, unapproved dry-pipe, or mechanically damaged valves should be replaced with the proper approved listed device.

Instruction charts are provided for the maintenance of dry-pipe valves by the sprinkler manufacturing company and should be posted at or near the valves.

All dry-pipe valves should be numbered and listed on inspection report forms.

The air pressure on each dry-pipe system should be checked at least once a week. A daily check is recommended for the first week after a dry-pipe valve is reset. If pressure is lost rapidly, requiring frequent replenishment of the air pressure, the piping system should be carefully inspected and made tight. Air pressure higher than that called for by the valve manufacturer's instructions should be avoided as this may damage the gasketing on the valve or severely delay the operation of the valves.

Make sure that the priming water is maintained at the proper level above the dry-pipe valve.

Slight freezing of a dry-pipe valve may cause it to be inoperative. It is extremely important to make sure that adequate provisions are made for heating the valve enclosure or the room in which it is located, and that the heating equipment is safe and in proper working order.

Dry-pipe system piping should be thoroughly drained at all drum drips or low point drains before freezing weather and kept clear of water during the winter. The freezing of a small amount of water in the piping may cause rupture of the piping or sprinklers and valve disfunction. Make sure that all low-point drains of the system are kept free of water and that the automatic drip or drain is clear and free to operate.

Dry-pipe valves should be examined externally at frequent intervals to detect evidence of deterioration.

Thoroughly clean and reset each dry-pipe valve yearly during the warm weather. On such occasions, the valve body should be thoroughly washed out with warm water. Make sure that all small ports and piping leading to the alarm connections are free from obstructions.

Operating tests of dry-pipe valves and their quick-opening devices should be conducted annually. Such tests may be combined with the annual cleaning and resetting. Operating tests should also be conducted when any necessary repairs, such as renewal of rubber gaskets, or the adjustment of gages, alarm devices, connecting piping, and quick-opening devices, are performed.

When dry-pipe valves are tripped for testing purposes, the procedure should be such that only the minimum flow of water needed to trip the valve is admitted to the sprinkler riser. This can usually be done by opening the control valve only two to six turns and by closing it immediately after the dry-pipe valve trips.

Operating tests and servicing of dry-pipe valves, including quick-opening devices, should be conducted either by a sprinkler installing company or other fully qualified personnel.

No grease or other sealing material should be used on seats of dry-pipe valves in an effort to stop leaks. Force should not be used to make dry-pipe valves tight.

Dry-pipe valves should carry a tag or card showing the date on which the valve was last tripped and the name of the organization making the test. Such tags are usually available from the installing company or the insurance company.

Quick-Opening Devices

The operation of quick-opening devices usually can be tested either with or without operating the dry-pipe valve. The manufacturer's instructions for testing and resetting the valve must be followed.

If the device does not operate properly during a test, it can, if necessary, be removed and the sprinkler system kept in operation. Repair parts can be ordered from the manufacturer. The device can also be sent to the manufacturer for repair, adjustment, or replacement.

Pressure Gages

Water pressure and air pressure gages should be tested for accuracy whenever the dry-pipe valve is cleaned and reset, and at other times if the pressure indication is questionable.

Waterflow Alarm Devices

Waterflow alarm devices should be tested at quarterly scheduled inspections of fire protection equipment. Electric alarms should be tested by the bypass test valve at dry-pipe valves and alarm valves. Actual waterflow alarm tests should be made when cold weather does not make the outdoor discharge of water from test or drain valves objectionable. Water motor gongs should never be tested in freezing weather.

Keep the small valve or cocks controlling the water supply to alarm devices sealed in open position.

Deluge and Preaction Systems

Complete charts are furnished by installing companies showing in detail the proper method of operating and testing thermostatically controlled systems. Only competent persons, fully instructed with respect to the details and operation of such systems, should be employed in their repair and adjustment. It is highly advisable for the owner to arrange with the installing company for the regular periodic inspection, testing, and maintenance of the equipment.

The automatic valves controlling the flow of water into these systems operate through the effect of fire temperatures on actuating devices. Ordinarily, when it is necessary to repair the actuating system of a preaction system, as distinguished from the piping system itself, the piping system can be run "wet," and automatic sprinkler protection is thus maintained on a preaction system, provided there is no danger of freezing.

Sprinkler System Supervisory Systems

Central station, proprietary, remote station, fire alarm, and supervisory services for sprinkler systems interpret sprinkler system waterflow signals as fire alarms and immediately initiate occupant notification and fire department response. Signals of other types, such as valve supervision and pressure or temperature supervision, do not serve as fire alarms, but initiate appropriate courses of action as prearranged between the supervisory service and the property owner. It is extremely important that the headquarters of any of these services be notified and any special arrangements made in advance of any inspection or tests of equipment that will cause a signal to be transmitted.

Fire Department Connections

Inspect each fire department connection regularly to make sure caps are in place, threads are in good condition, the ball drip or drain is in order, and the check valve is not leaking. A hydrostatic test should be conducted periodically on old fire department connection piping to ensure that it will withstand the required pressure.

Open Sprinkler Equipment

Outside or open sprinkler equipment should be tested once each year during warm weather. These tests should be made preferably in conjunction with the Authority Having Jurisdiction and, if desired, the fire department.

Before making operational tests, care should be exercised to make sure that all windows and doors through which water might enter are tightly closed. Proper precautions must be taken to prevent damage from discharge or accumulation of water to sidewalks, streets, pedestrian ways, or adjoining buildings.

Determine by test whether the sprinklers and system piping are in good condition and free from plugging. Any piping or sprinklers found clogged should be removed, cleaned, and replaced at once.

OBSTRUCTIONS IN PIPING

Obstructions in system piping or yard mains will reduce or cut off the flow of water from all or part of the system. This will compromise the expected performance of the protection. Obstructed piping is a well recognized cause of unsatisfactory sprinkler performance. One source of obstruction is foreign material in the water supply system. Another is foreign material originating from the interior of the pipe itself. Figure 6-17E graphically shows an example of an obstruction in system piping.

Foreign material, such as sand, gravel, stones, and pieces of wood, may enter the underground mains as a result of the carelessness of workers when laying or repairing the mains. Sand, silt, or wood chips may also enter the mains through inadequately protected fire pump suction inlets. Sometimes when gravity tanks are cleaned, foreign material enters the drop pipes unless preventive care measures are taken.

Keeping foreign material out of piping for water-spray systems is particularly critical. If the waterways in discharge nozzles are less than $3/_8$ in. (9.5 mm) in size or if it is suspected that the water is likely to contain obstructive material, then a strainer is required in the pipeline supplying the system. The small nozzles are prone to clogging if waterborne debris reaches them. Thus, it is imperative that strainers are maintained in good condition. Strainers used in fire

FIG. 6-17E. A stone lodged in the tee of a dry-pipe system's cross main. Good maintenance programs enable discovery of such obstructions before they can become a problem. (Source: Factory Mutual Research Corp.)

FIG. 6-17F. Silt buildup in this branch sprinkler line has totally obstructed the waterway. (Source: Factory Mutual Research Corp.)

protection systems should be approved and listed by a recognized testing laboratory for fire protection service.

Scale, corrosion, and incrustations can also form in the piping. For example, in dry-pipe sprinkler systems the condensation of moisture in the air supply may result in hard scale forming along the bottom of piping. Deep well water and water containing natural salts tend to corrode pipe interiors. Therefore, the interior of some wet pipe systems may be found obstructed by rust and corrosion. Obstructing material is carried into piping when systems operate or are refilled after draining. It may plug one or more fittings or obstruct a number of sprinklers at the end of a system. Pipe diameters may also be reduced by pipe incrustation where the water contains lime or magnesia.

If the water is badly discolored during waterflow tests, or if gravel or stones are discharged with the water, foreign material is probably present in the piping. Finding foreign material in a fire pump suggests that there may be similar material down stream in the system piping. Finding foreign material when resetting dry-pipe valves or examining check valves indicates a possible serious impairment of the entire fire protection system.

Obstruction to the flow of water from sprinklers usually comprises of concentrations of the lighter materials, such as silt, sand, and small pebbles, piling up in the ends of the cross mains and in the nearby branch lines, with the heaviest solids collecting near the system risers. (See Figure 6-17F.) There are some cases where the foreign material deposits extend so far back into the system from the ends of the cross mains that complete cleaning of the entire system is necessary. More often, however, a preliminary investigation will disclose that, while foreign material may cause complete stoppage of waterflow from sprinklers on branch lines connected to the ends of cross mains, there may be no blockage at any point on branch lines farther back in the system.

Conditions Indicating Possible Obstructions in Piping

Evidence of possible obstruction in piping is given by any one of the following:

1. Plugging of test connections, or discharge of dirty or colored water.
2. Discharge of foreign material during routine water tests.
3. Foreign material in dry-pipe valves, check valves, or fire pumps.
4. Obstructing material found in piping which is dismantled for building alterations.
5. Defects found in fire pump suction screens when the water supply source is from open bodies of water.
6. Repairs made to public water mains in vicinity.
7. Underground piping not flushed before connection to systems.
8. Plugged sprinklers or spray nozzles at time of fire.
9. Old equipment, especially in dry-pipe systems.

Investigation of Severity and Extent of Obstruction

An investigation of conditions must be made when evidence of foreign materials in systems is found or when obstructions are suspected. Where needed, cleaning or flushing procedures subsequently must be carried out to remove obstructing materials. Inspection and flushing methods outlined in this chapter are treated more completely in NFPA 25.

Flushing Feed Mains

Existing yard piping suspected of containing obstructing material should be thoroughly flushed through hydrants at dead ends of the system, blow-off valves, or groups of riser drains, by opening simultaneously as many outlets as possible and allowing the water to run until clean. If the water is supplied from more than one direction or through a looped system, divisional valves should be closed to produce a high-velocity flow through each single line.

It is usually desirable to learn the nature and extent of foreign material that may be in the piping. This can best be done by fastening burlap bags to the flowing outlets and examining the effluent remaining in the bag.

The connections to system risers should next be flushed through drain valves. If the foreign material is too large to pass through a main drain valve, some ingenuity must be used to procure a larger opening. Fire department connections can be used by removing the check valve. Flanged check valves or check valve covers, or flanged fittings can be removed and a Siamese fitting can be substituted to facilitate such flushing.

Flushing System Piping

In choosing the critical points for examination, consideration should be given to the arrangement of the piping and the probable pattern of the flow of water when a system is rapidly filled (as in a dry-pipe system), or when a large flow of water from the system occurs due to operation of many sprinklers on a wet-pipe system.

After selection of the area in which the heaviest deposit of foreign material ordinarily would be expected, the investigation should take into consideration how much of the system's equipment in that area might be made ineffective by the obstructing material at the time of the fire. It is important that the preliminary investigation sufficiently indicates the extent of cleaning that may be required.

Flow Tests

Flow tests at critical points in a sprinkler system provide a positive means of determining whether there is foreign material in the system that may cause stoppage of flow from any of the sprinklers, and also of indicating how much of the equipment is affected. To achieve this, valved hose connections are installed at the ends of two, three, or four branch lines on one side of and at the end of the affected cross main in the selected area. Flows are then taken simultaneously from these points and examined.

If stoppage occurs at any of the branch lines tested simultaneously, similar tests may be necessary at points upstream on the system toward the system riser. Tests may next be extended to other floors to help make a judgment on the extent of cleaning operations that may be necessary for removing all foreign material that might cause stoppage.

Foreign deposits in sprinkler systems tend to be accumulated toward the ends of the cross mains and in the ends of branch lines. Of course, there will be exceptions to this general statement. For example, where highly corrosive material has been introduced into the system, severe corrosion and scaling of the interior surfaces of the piping will be evident uniformly throughout the system.

Visual Examination

The visual inspection of a system is a critical starting point of any flushing program. This first step will not tell the entire story of the condition of the piping, however it can aid the trained person in an accurate determination of the piping condition. The visual examination encompasses the inspection of the interior and exterior of the piping, the condition of said piping, and the tuberculation in said piping.

The initial inspection should start by removing the end sprinklers at the upper and lower most portion of the system. Take extra care when removing any pendent sprinklers to check for sediment clogging the orifice.

The removal of the flushing connection is an easy method to determine the condition of the cross main.

Should these steps in the visual inspection arouse any suspicion and an internal inspection becomes necessary, the following should be used as a guideline.

1. Approximately 10 percent of the system should be examined when an internal inspection is recommended.
2. Remove
   a. Three to four endlines,
   b. Riser nipples, and
   c. Elbows and tee feed for the cross main.
3. Examine any areas where the water velocity decreases rapidly when draining the system piping.

If the internal inspection reveals substantial tuberculation and piping deterioration, a determination must be made by cost efficiency analysis. The options available are flushing, whether it be by the hydropneumatic method or just with water, or the partial to total replacement of the affected area of the piping.

The expense of flushing can be high when labor costs and materials are coupled. Even though the system is flushed and back on line, such piping may have a shorter life expectancy than a new system.

EXTENT OF CLEANING REQUIRED

When preliminary tests show evidence of obstruction in sprinkler system piping, the next step is to undertake a cleaning program to remove the obstructing material. There are three degrees of cleaning: (1) complete cleaning, (2) limited cleaning, and (3) flushing extremities only.

Generally, when stoppage occurs only in the branch lines at extremities of the system, the "flushing extremities" only procedure can be used to clean the system. If stoppage occurs at several test points, but indications are that free flows are available on the other lines back toward the riser, a limited cleaning procedure may be followed. If the stoppage is more extensive, then a complete cleaning procedure should be undertaken.

Before any cleaning is undertaken, however, written specifications or a flushing program covering the entire proposed operations should be understood by all concerned. After completion of work, the sprinkler contractor or other party performing the cleaning operations should furnish a written statement to those concerned indicating that all operations have been completed in accordance with the approved procedure, that the specified examinations and tests have been made, and that the system has been left in normal condition.

Complete Cleaning

There are some cases where the foreign material is of such character and extends so far back into the system from the ends of the cross mains that complete cleaning of the entire system will be necessary. In such cases, it is of prime importance that a plan of the piping system be prepared and be marked numerically to show the proper sequence of the cleaning operations, the size and length of hose to be used, the points where the hose is attached, and the number of flows to be made at each point.

Limited Cleaning

Where points of impairment can be identified as occurring within restricted areas rather than throughout the system, apply the limited cleaning procedure. If these points exist in any appreciable amount but only in the ends of cross mains and in the branch lines connected at or near those points, not all branch lines need to be cleaned. The cleaning procedure may be limited to the branch lines at or near these points and to the cross mains on each floor.

Limited cleaning operations serve to minimize the time during which the sprinkler system is impaired. They lessen the disturbance to normal activities within the premises where the cleaning operations are under way. They materially lower the cost of the work as compared with the cost of more extensive cleaning. Since only a fraction of the total number of branch lines is being cleaned directly under the limited procedure, periodic flow tests should be made as the cleaning progresses, in order to determine the effectiveness of the work. The limited procedure takes into account the purging effect in branch lines not actually flushed, caused by high-velocity water and airflow along the cross main.

Cleaning System Extremities Only

This cleaning method should be used only in cases in which obstruction of dry-pipe sprinkler system piping has been found to occur solely at the system extremities, and it relates to every floor protected, regardless of the number of floors. Extreme care should be exercised before this cleaning method is employed. Adequate examinations and flow tests are made to make certain that only the sprinklers located at the extremities of the system are subject to plugging. It should not be used where the deposits extend back into the cross main beyond the second branch line from the end of the cross main.

CLEANING METHODS

There are two methods, in general, used for cleaning sprinkler systems: (1) water flushing and (2) hydropneumatic. Convenience is a key factor that may make the hydropneumatic system preferable to the water flushing method. With the former, there is no open discharge of water outside the building; there is less disturbance of plant operations; frequent fillings and drainings of the sprinkler system are unnecessary; there is considerably less pipe work; and a complete operation is performed at a given flush point.

Both the water flushing and hydropneumatic methods are appropriate for complete cleaning and limited cleaning procedures. Good practice is first to flush the underground piping and then to clean all private sources of water supply that might cause silting in the underground main. Both of these methods are thoroughly covered in NFPA 25.

Branch Line Testers

If it is anticipated that a sprinkler system may need periodic testing for obstruction or repeated flushing, a branch line tester may be installed permanently as a nippled fitting at the end of branch lines or between the last sprinkler and the end fitting on the branch lines. The tester has a normally closed and capped outlet connection for hose. Turning a wrench fitting opens a passage to the outlet which has a 1-in. (25.4-mm) hose coupling thread. It is not necessary to remove the sprinkler when this type of device is used.

SPRINKLER LEAKAGE

The term "sprinkler leakage" is ordinarily intended to cover all leakage or discharge of water from the sprinkler system, except in case of fire. The system includes sprinklers, piping, tanks, and other water supplies used for fire protection.

The danger of accidental water discharge is often exaggerated, because a sprinkler system is rugged. The pipe, sprinklers, and fittings are made to endure far greater water pressures than are met with in practice. The supports are strong also, and the system is so installed that it should remain intact even under severe mechanical strain, including any anticipated earthquake shock or movement.

There is no reasonable ground for omission of sprinklers from areas in hospitals and other institutions occupied by patients. While it is true that accidental discharge of cold water from sprinklers might, for example, have an adverse effect on certain patients in a hospital, the probability of loss of life is greater if sprinklers are not installed than from an accidental discharge from the sprinklers. Conditions that might result in accidental discharge of water from sprinklers are highly unlikely where ordinary hospital care would

preclude the existence of either freezing or excessively high temperatures as to cause the opening of sprinklers, except in the case of fire.

Automatic sprinkler protection for property particularly vulnerable to water damage, such as computers, finely made complicated mechanical equipment, works of art, or valuable books, should not be omitted due to fear of premature operation of sprinklers if reasonable precautions are taken against freezing, overheating, and mechanical injury. There are also cases where special types of systems, such as single and double interlock, preaction systems are available for use.

Some of the more important conditions responsible for sprinkler leakage and water loss are explained in the following paragraphs.

Mechanical Injury

Automatic sprinklers are designed to withstand at least 500 psi (3448 kPa) pressure without injury or leakage. If properly installed, there is little danger of the sprinkler breaking apart, unless it is physically damaged in some way. All of the listed sprinklers are very durable and will withstand considerable abuse. The frames can be bent somewhat, without opening the sprinkler.

When sprinklers are located in areas where they may suffer mechanical injury, they can be protected by commercially available metal guards. (See Figure 6-17G.)

The use of lift trucks or tractors in handling materials indoors must be carefully planned and supervised to avoid hitting sprinklers and sprinkler piping. Precautions should be taken where fast-response sprinklers are installed to reduce the chances of mechanical and environmental damage.

Improper Installation and Maintenance

Improper installation includes lack of proper supports for risers and feed mains, lack of proper tie rods where inside piping is connected to cast-iron fittings, and defective pipe hangers or supports.

Breaks or leaks sometimes occur in underground work due to improper laying of underground mains or poor quality control in making joints. Sometimes breaks occur in properly installed systems due to undue stresses on the piping system, such as settling of a building or foundation or lack of clearance around pipes where they pass through a foundation, deflection of joints, or a combination of stress and deflection.

Every piping system should be frequently inspected and properly maintained. The condition of hangers and supports should be visually checked to determine whether they are in good order, loose, or deteriorating.

FIG. 6-17G. *Metal guard protection against mechanical injury*

Sprinkler piping should never be used as a support for ladders, stock, or other materials. Any piping system under constant internal pressure should receive proper care and not be subjected to mechanical abuse.

Freezing

The freezing of water in pipes is a common cause of trouble and the remedy in most cases is self-evident. Where this is a local condition and not unduly severe, ice in the sprinkler or fittings may rupture the sprinkler without damage to the pipe or fittings. When the ice melts, water is discharged. In other cases, the pipe or fittings may burst, causing more serious damage and more costly repairs.

In some types of sprinklers, a moderate amount of freezing may not break the sprinkler open, but may cause it to leak at the seat after the ice has melted.

Some of the conditions that cause trouble from freezing are:

1. Insufficient heat for severe weather conditions in certain portions, such as concealed spaces under roofs or blind attics, large open doorways for trucks or railway cars, entryways, or space under buildings.
2. Windows left open during severely cold weather.
3. Insufficient heat due to a shortage of heating fuel or power failure.
4. Branch lines of dry-pipe systems not correctly pitched to drain.
5. Feed mains or piping in unheated areas properly protected to prevent freezing.
6. Freezing of tank risers due to lack of heat or lack of adequate protection; or obstructed circulating pipes of heating systems.
7. Freezing of water in underground mains not buried deep enough.

Monitoring dry-pipe systems for low air pressure becomes cost effective, because the time between low air signal and tripping of the dry-pipe valve is usually sufficient to circumvent most, if not all, water damage.

Overheating

Overheating is another cause of accidental discharge from sprinklers and is the term used when an automatic sprinkler operates as a result of abnormally high temperatures, without the presence of fire. This condition is usually caused by hot manufacturing processes, artificial heating, or lack of ventilation. It may result from a sudden increase in temperature which operates the thermal response element of the sprinkler, or from longer exposure to temperatures sufficiently high to cause gradual weakening of the element. In the solder-type of sprinkler, the latter condition, sometimes termed "cold flow," is indicated by partial separation of the soldered members.

In practice, there are a number of conditions that cause unsafe temperatures, such as:

1. Climatic conditions exceeding 100°F (41.3°C).
2. Ordinary temperature sprinklers located in close proximity to steam mains, unit heaters, or heating ducts. Temperatures higher than anticipated may occur after installation of sprinklers. Changes in heating methods, installation of unit heaters, changes in manufacturing processes or equipment, after sprinklers are installed, frequently occur.
3. Changes in temperatures due to changes in occupancy or processes which cause higher temperatures than contemplated. These changes may be general or in isolated portions of a building.
4. In some cases, a fire that opens a number of sprinklers may weaken other sprinklers in that vicinity. This may not be appar-

ent. Change all sprinklers in an area of fire influence and not just those that have opened.
5. Low-mass fusible linkage of quick-response sprinklers will enhance the negative influence as mentioned above. The installation and maintenance of this device requires great care.

NFPA 13 and NFPA 25 provides some information on sprinkler temperature ratings near unit heaters. Unit heaters should never be used without a careful study of conditions, and, in most cases, sprinklers near unit heaters must be replaced with a higher temperature rating in order to avoid accidental operation.

When conditions occur that require sprinklers higher than ordinary-degree rating, the maximum temperature at the sprinkler level must be determined and sprinklers of the correct temperature rating installed if trouble is to be avoided.

REDUCING WATER DAMAGE

It is evident that, if leakage occurs when there is no one in the building and no proper means of alarm notification to outsiders exists, the leakage may continue for a considerable amount of time. This can result in serious damage to property, even with a small amount of water from just one leaking sprinkler.

While good, or "standard," guard service is of value, it may not be adequate unless there is a sprinkler alarm device, which will notify the guard if water is flowing. Every sprinkler system should have a waterflow alarm device and properly located water motor gong to notify the proper personnel in case of fire or discharge of water due to inadvertent system discharge.

Large losses from sprinkler leakage can result from lack of, or failure of, isolated sprinkler devices or the lack of waterflow alarms. Losses can also be attributed to the failure to promptly notify some authorized person(s) who will investigate immediately and shut off the water if there is no fire. In some cases, water may run for many hours, undetected.

Sprinkler supervisory service is of great value in the prompt discovery of system operation from a fire event. In addition, it is also useful to detect leakage from the system, and in securing appropriate action to minimize the water loss.

BIBLIOGRAPHY

NFPA Codes, Standards, and Recommended Practices

Reference to the following NFPA codes, standards, and recommended practices will provide further information on the care and maintenance of water-based extinguishing systems discussed in this chapter. (See the latest version of *The NFPA Catalog* for availability of current editions of the following documents.)

NFPA 13, *Standard for the Installation of Sprinkler Systems*
NFPA 14, *Standard for the Installation of Standpipe and Hose Systems*
NFPA 15, *Standard for Water Spray Fixed Systems for Fire Protection*
NFPA 24, *Standard for the Installation of Private Fire Service Mains and Their Appurtenances*
NFPA 25, *Standard for the Inspection, Testing, and Maintenance of Water-Based Fire Protection Systems*
NFPA 72, *National Fire Alarm Code*
NFPA 409, *Standard on Aircraft Hangars*

Additional Reading

Boyd, H. D., and Locurto, C. A., "Reliability and Maintainability for Fire Protection Systems," *Proceedings* of the First International Symposium on Fire Safety Science, Hemisphere, Washington, DC, 1986, pp. 963–970.
Bryan, J. L., *Automatic Sprinkler and Standpipe Systems*, 2nd ed., National Fire Protection Association, Quincy, MA, 1990.

Carson, W. G., and Klinker, R. L., *Fire Protection Systems: Inspection, Test, and Maintenance Manual*, 2nd ed., National Fire Protection Association, Quincy, MA, 1993.

Damon, W. A., "Testing Pressure-Regulating Devices in Water-Based Fire Protection Systems," *Fire Engineering*, Vol. 146, No. 11, Nov. 1993, pp. 51, 55–57.

Ebling, S., and Laverick, G., "Maintaining Standards for Sprinkler System Effectiveness," *Sprinkler Age*, Vol. 21, No. 2, Feb. 1992, pp. 10, 12, 14.

"Environmental Assessment: Installation and Operation of the Plantwide Fire Protection Systems and Related Domestic Water Supply Systems," Department of Energy, Aiken, SC, DOE/EA-0476, Dec. 1991.

Factory Mutual Research Corporation, *Loss Prevention Data Sheets, Nos. 2-11 through 2-95* (various titles); information on the care and maintenance of sprinkler system components (dry-pipe valves, quick-opening devices, deluge and preaction system valves, etc.), by manufacturers' names and model numbers (various dates).

Keith, G., and Garner, D., "Sprinkler System Water Supply Analysis," *Fire Engineering*, Vol. 144, No. 12, Dec. 1991, pp. 67–68, 70–71.

Leyton, S. M., "Preparing for NFPA 25: Write Your Own Headlines on the Issue of Fire Sprinkler System Maintenance," *Industrial Fire World*, Vol. 9, No. 5, Sept./Oct. 1994, pp. 9, 12.

Ryan, M. W., *et al.*, "Effect of Pressure on Leakage of Automatic Sprinklers," *Sprinkler Age*, Vol. 12, No. 5, May 1993, pp. 16, 18, 20–24.

Santavuori, A., "Automaattisten sammutuslaitteistojen tarkastustoiminta. [Inspection of Sprinkler Systems.]," *Palontorjuntatekniika*, March 1992, pp. 16–17.

HALOGENATED AGENTS AND SYSTEMS

Revised by Gary M. Taylor

Halogenated extinguishing agents are hydrocarbons in which one or more hydrogen atoms have been replaced by atoms from the halogen series: fluorine, chlorine, bromine, or iodine. This substitution confers not only nonflammability but flame extinguishment properties to many of the resulting compounds. Halogenated agents are used both in portable fire extinguishers and in extinguishing systems. Streaming agents, such as Halon 1211 and 2402, have most often been used in manually applied fire equipment and local application-type fixed systems. Halon 1301 has most often been used in total flooding-type fixed systems.

Halons 1301, 2402, and 1211 have been identified as stratospheric ozone-depletion agents. (See Figure 6-18A.) The Montreal Protocol on Substances that Deplete the Ozone Layer, an international agreement, requires a complete phaseout of the production of these halons by January 1, 2000, except to the extent necessary to satisfy essential uses, for which no adequate alternatives are available. As of the end of 1996, essential uses in commercial aviation, in military applications, and as an inerting agent for the protection of flammable liquids and gases have been supplied using halons recovered from less critical use applications, such as conventional computer rooms. The Montreal Protocol establishes minimum compliance requirements; individual nations may require more stringent controls on the production, use, or disposal of ozone-depleting substances, including the halons.

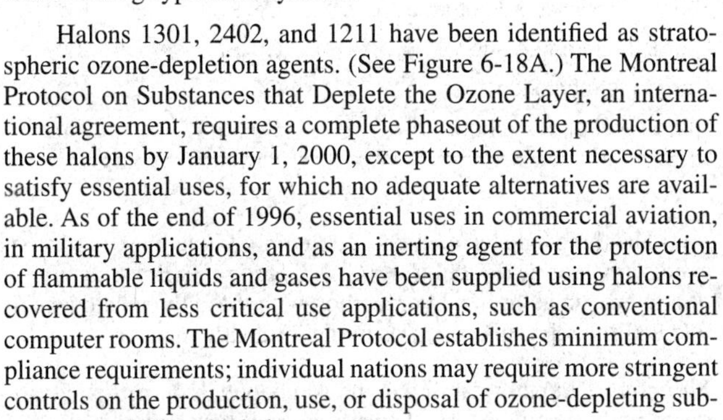

FIG. 6-18A. *Chemical mechanisms of Halon 1301 and ozone.*

Gary M. Taylor is a fire protection consultant employed by Taylor/Wagner, Inc., a fire protection consulting firm located in Willowdale, Ontario, Canada.

The use and development of halogenated fire-extinguishing agents have evolved over many years. The first member of this family of chemicals was carbon tetrachloride (Halon 104). Use as a fire extinguishant probably occurred before 1900, and by 1910 portable fire extinguishers, tested by independent agencies, had appeared. The growing popularity of the automobile, and other uses of internal combustion engines, signaled an increasing need for fire extinguishants suitable for use on flammable liquids fires.

By 1917, there were discussions regarding the possible effects that carbon tetrachloride could have on the human system. In 1919 the first recorded deaths due to carbon tetrachloride use occurred. Two men working on the construction of a submarine were killed. One man's clothing had caught fire, and the other man extinguished the fire with a carbon tetrachloride agent fire extinguisher. Both were overcome by the fumes and later died. During the 1920s the discussions regarding the toxicity of carbon tetrachloride continued, with particular attention to the possibility that freezing point depressants and impurities were contributing factors.

Methyl bromide (Halon 1001) gained popularity after it was discovered in the late 1920s. Due to its high toxicity, it was never popular for use in portable extinguishers, although it was used in British and German aircraft and ships during World War II. During World War II, Germany developed bromochloromethane (Halon 1011) to replace methyl bromide. In 1947 a report by Underwriters Laboratories Inc. (UL) showed that the toxicity of carbon tetrachloride (Halon 104) and bromochloromethane (Halon 1011) were comparable, but bromochloromethane was a more efficient fire-extinguishing agent.

In the post–World War II era, the addition of stearate to sodium bicarbonate-based dry chemicals provided improved flow and moisture repellency characteristics to sodium bicarbonate-based dry chemicals. This, in turn, encouraged the use of portable dry chemical fire extinguishers as an alternative to vaporizing liquid extinguishers that used early halons as extinguishants.

By the 1950s the era of the early halons (Halons 104, 1001, and 1011) was ending. Increased popularity of dry chemicals had decreased the need for widespread use of these early halons, and growing concerns with their toxic effects resulted in their "official" death by the 1960s.

In 1947, the Purdue Research Foundation conducted a systematic evaluation of more than 60 new candidate extinguishing agents. Simultaneously, the U.S. Army Corps of Engineers undertook toxicological studies of these same compounds. As a result, four halons were selected for further study: dibromodifluoromethane (Halon 1202), bromochlorodifluoromethane (Halon

1211), bromotrifluoromethane (Halon 1301), and dibromotetrafluoroethane (Halon 2402). Testing showed that Halon 1202 was the most effective fire extinguishant, but it was also the most toxic. Halon 1301 ranked second in fire extinguishing effectiveness and least toxic. As a direct result of this program, a portable fire extinguisher employing Halon 1301 was developed for use by the U.S. Army, primarily for use inside armored personnel carriers and tanks. The U.S. Air Force selected Halon 1202 for military aircraft engine protection, and the U.S. Federal Aviation Administration approved the use of Halon 1301 for commercial aircraft engine fire protection.

Because of this basic research, some general conclusions were drawn, based on the contribution of the various halogens to resultant characteristics of the halon as a useful fire extinguishant. These are outlined in Table 6-18A.

TABLE 6-18A. *Halogen Characteristics*

Characteristic/Halogen	Fluorine	Chlorine	Bromine
Stability of Compound	Enhances	—	—
Toxicity	Reduces	Enhances	Enhances
Boiling Point	Reduces	Enhances	Enhances
Thermal Stability	Enhances	Reduces	Reduces
Extinguishing Effectiveness	—	Enhances	Enhances

In the 1960s attention began to focus on the use of Halon 1301 as a total flooding extinguishant for the protection of computer rooms. In the past 20 years, Halon 1301 has grown in usage as an agent for use in fixed fire protection systems, primarily for the protection of vital electronics facilities, such as computer rooms and communications equipment rooms. Other significant applications for Halon 1301 systems include: museums, shipboard machinery spaces, and pipeline pumping stations.

In the United States, early Halon 1301 total flooding systems were installed using carbon dioxide equipment and technology. An example of such a system, installed in the Winterthur Museum, was described.[1]

In 1966, NFPA organized a Technical Committee on Halogenated Fire Extinguishing Agent Systems to develop standards covering installation, maintenance, and use of such systems. NFPA 12A, *Standard on Halon 1301 Fire Extinguishing Systems* (hereinafter referred to as NFPA 12A), and NFPA 12B, *Standard on Halon 1211 Fire Extinguishing Systems* (hereinafter referred to as NFPA 12B) (discontinued), were approved in the early 1970s. NFPA 12CT was a tentative standard for Halon 2402; however, the standard was never adopted as a full standard due to lack of acceptance of this agent by users of NFPA standards. Halon 2402 has been employed extensively within the Russian federation as both a total flooding system agent and a streaming agent for use in portable fire equipment. Because of its rather recent recognition, there is limited experience with Halon 2402 in the U.S., and, therefore, it cannot be treated as extensively as the other agents in this handbook.

During 1966, attention began to focus on the use of Halon 1301 to protect computer rooms and electronic data processing (EDP) equipment. In 1972, following extensive testing by several major companies on the effects of Halon 1301 decomposition products on electronic equipment,[2] the NFPA Committee on Electronic Computer/Data Processing Equipment recognized Halon 1301 total flooding systems as suitable for protection of electronic computer/data processing equipment. With suitable precautions, such as time delays, Halon 1211 total flooding systems have been used in Europe.

CHEMICAL COMPOSITION AND CLASSIFICATION

The halogenated extinguishing agents are currently known simply as halons, and the halon system for naming the halogenated hydrocarbons was devised by the U.S. Army Corps of Engineers.[3] This simplified system of nomenclature describes the chemical composition of the materials without the use of chemical names or possibly confusing abbreviations (e.g., "BT" for bromotrifluoromethane and "DB" for dibromodifluoromethane). Examples of this system are shown in Table 6-18B. The first digit of the number represents the number of carbon atoms in the compound molecule; the second digit, the number of fluorine atoms; the third digit, the number of chlorine atoms; the fourth digit, the number of bromine atoms; and the fifth digit, the number of iodine atoms (if any). If the fifth digit is a zero, it is not expressed; bromotrifluoromethane ($BrCF_3$), for example, is referred to as Halon 1301 (not 13010), although its chemical formula shows one carbon atom, three fluorine atoms, no chlorine atoms, one bromine atom, and no iodine atoms.

TABLE 6-18B. *Sample Halon Numbers for Various Halogenated Fire Extinguishing Agents*

Chemical Name	Formula	Halon No.
Methyl bromide	CH_3Br	1001
Methyl iodide	CH_3I	10001
Bromochloromethane	CH_2BrCl	1011
Dibromodifluoromethane	CF_2Br_2	1202
Bromochlorodifluoromethane	CF_2BrCl	1211
Bromotrifluoromethane	CF_3Br	1301
Carbon tetrachloride	CCl_4	104
Dibromotetrafluoroethane	$C_2F_4Br_2$	2402

The three halogen elements commonly found in extinguishing agents are fluorine (F), chlorine (Cl), and bromine (Br). Substitution of a hydrogen atom in a hydrocarbon with these three halogens influences the relevant properties in the following manner.

Fluorine: Imparts stability to the compound, reduces toxicity, reduces boiling point, increases thermal stability.

Chlorine: Imparts fire extinguishing effectiveness, increases boiling point, increases toxicity, reduces thermal stability.

Bromine: Same effects as chlorine, but to a greater degree.

Thus, compounds containing combinations of fluorine, chlorine, and bromine can possess varying degrees of extinguishing effectiveness, chemical and thermal stability, volatility, and toxicity.

The halogen is linked to the carbon atom by a "covalent" chemical bond, which means that, unlike inorganic halogen compounds, e.g., common table salt (sodium chloride), there is no tendency to ionize or become electrically conductive in the presence of water. The presence of a fluorine atom in the molecule increases the C-Cl and C-Br strength and improves the overall chemical stability of the compound.

Because they are either gases or liquids that rapidly vaporize in fire, halons leave no corrosive or abrasive residue after use. They are nonconductors of electricity and have high liquid densities that permit use of compact storage containers. The areas of major halon use are for the protection of electrical and electronic equipment, petroleum production facilities, engine compartments (ships, military vehicles, and aircraft), and other areas where rapid extinguishment is important, or where damage to equipment or materials or cleanup after use must be minimized.

PROPERTIES OF HALOGENATED AGENTS

Some of the physical properties of the common halogenated fire extinguishing agents referred to in this chapter are shown in Table 6-18C. More details on bromotrifluoromethane (Halon 1301) and bromochlorodifluoromethane (Halon 1211) are given in the following paragraphs.

Physical Properties

Halon 1301 is a gas at 70°F (21°C), with a vapor pressure of 199 psig (1372 kPa). Although this pressure would adequately expel the material, it decreases rapidly to 56 psig (386 kPa) at 0°F (–17°C) and to 17.2 psig (119 kPa) at –40°F (–40°C). It is, therefore, usual to increase the pressure on portable extinguishers or systems with nitrogen either to 360 psig (2482 kPa) or to 600 psig (4137 kPa), which ensures adequate performance at all temperatures. The high volatility of Halon 1301 has led to its use in total flooding systems, where rapid vaporization is an asset.

Halon 1211 is also a gas at 70°F (21°C), with a vapor pressure of 22 psig (152 kPa) and a boiling point of 25°F (–4°C). Its relatively high boiling point allows it to be projected as a liquid stream, thus enabling portable extinguishers and local application systems to have a greater range than possible with other gaseous materials. For portable extinguishers, the normal superpressurization with nitrogen is approximately 100 psig at 70°F (689 kPa at 21°C), which permits the use of inexpensive hardware. Halon 2402, with a boiling point of 117°F (47°C), is a liquid at room temperature.

Corrosion and Other Effects on Materials

The early nonfluorinated halogenated agents had significant corrosion problems. The fluorinated agents currently in use are chemically more stable, and neither Halon 1301, Halon 1211, nor Halon 2402 has any significant corrosive action on the commonly used construction metals, unless free water is present. (Free water is defined as the presence of a separate water phase in the liquid halon. When present in a small quantity, free water can provide a site for concentrating acid impurities into a corrosive liquid. It has been determined that dissolved water is not a problem.)

The effect of these agents on plastics and elastomers depends upon the precise grade or composition of the plastics and elastomers. In general, normally satisfactory effects are obtained with rigid plastics, such as polytetrafluoroethylene, nylon, and acetal copolymers. Elastomers of particular suitability are neoprene, Buna N®, and "Viton®" fluoroelastomer rubbers. More precise guidelines should be obtained from the manufacturers of the materials. Halon 1301 generally has considerably less effect on plastic and elastomeric materials than does Halon 1211 or Halon 2402.

EXTINGUISHING CHARACTERISTICS

The extinguishing mechanism of the halogenated agents is not clearly understood. However, a chemical reaction undoubtedly occurs that interferes with the combustion processes. The agents act by removing the active chemical species involved in the flame chain reactions (a process known as "chain breaking"). While all the halogens are active in this way, bromine is much more effective than chlorine or fluorine.

A number of possible mechanisms by which bromine inhibits combustion have been proposed. The most active species in hydrocarbon combustion are oxygen and hydrogen atoms, and hydroxyl radicals (O, H$^\bullet$, and OH$^\bullet$). Under fire conditions, the halogenated agent—e.g., bromotrifluoromethane (Halon 1301)—will release a bromine atom:

$$CBrF_3 \rightarrow CF_3{}^\bullet + Br^\bullet$$

The Br atom can react with a hydrocarbon molecule to form hydrogen bromide, HBr:

$$R = H + Br^\bullet \rightarrow R^\bullet = HBr$$

The HBr then reacts with an active hydrogen or hydroxyl radical, releasing the bromine atom for further inhibition reactions:

$$OH^\bullet + Hbr \rightarrow H_2O + Br^\bullet$$

In this way, chain carriers are removed from the system while the inhibiting HBr is continuously regenerated.

Another "ionic" theory supposes that the uninhibited combustion process includes a step in which oxygen ions are formed by the

TABLE 6-18C. *Some Physical Properties of the Common Halogenated Fire Extinguishing Agents*

Agent	Chemical Formula	Halon No.	Type of Agent	Approx. Boiling Point °F†	Approx. Freezing Point °F†	Specific Gravity of Liquid at 68°F† (Water = 1)	Approx. Critical Temp. °F†	Estimated Pressure psig‡ At 130°F†	At Critical Temp.	Latent Heat of Vaporization cal/g Water = 540 cal/g CO$_2$ = 138 cal/g
Carbon tetrachloride	CCl$_4$	104	Liquid	170	–8	1.595	—	—	—	46
Methyl bromide	CH$_3$Br	1001	Liquid	40	–135	1.73	—	—	—	62
Bromochloromethane	CH$_2$BrCl	1011	Liquid	151	–124	1.93	—	—	—	—
Dibromodifluoromethane	CF$_2$Br$_2$	1202	Liquid	76	–223	2.28	389	23	585	29
Bromochlorodifluoromethane	CF$_2$BrCl	1211	Liquefied Gas*	25	–257	1.83	309	75	580	32
Bromotrifluoromethane	CF$_3$Br	1301	Liquefied Gas	–72	–270	1.57	153	435	560	28
Dibromotetrafluoroethane	C$_2$F$_4$Br$_2$	2402	Liquid	117	–167	2.17	—	3.8	—	25

*May be kept as a liquid at reduced temperatures.
† $\frac{5}{9}$ (°F – 32) = °C.
‡ 1 psig = 6.895 kPa.

capture of electrons that come from ionization of hydrocarbon molecules. Since bromine atoms have a much higher cross section for the capture of slow electrons than has oxygen, the bromine inhibits the reaction by removing the electrons that are needed for activation of the oxygen.[4]

Fire Extinguishing Effectiveness

The effectiveness of the different halons as extinguishing agents has been the subject of many studies. The initial results of some of the early studies are contained in the October 1954 issue of the *NFPA Quarterly*.[3] It is now recognized that comparison of effectiveness of agents depends upon whether portable extinguishers or fixed systems are being considered and, particularly in tests on portable units, whether the agents were evaluated with hardware of optimum nozzle and value design. Therefore, the following information should be used only as a general guide, because other considerations, such as range of the unit (which varies on the gas or liquid characteristic of the individual halon and the total weight and volume of the equipment), may be of equal importance, depending upon use.

Table 6-18D gives the approximate pounds of agent required per unit of Class B rating obtained with small portable extinguishers currently listed with Underwriters Laboratories Inc. (UL), or approved by Factory Mutual Research Corporation (FMRC).

TABLE 6-18D. *Relative Effectiveness of Halogenated Agents on Small UL Class B Fires*

Agent	Halon No.	Weight of Agent (lb)*	UL B:C	lb per Unit B Rating*
Bromotrifluoromethane	1301	4.0	5	0.8
Bromochlorodifluoromethane	1211	2.5	5	0.5
Halon mixture	1301/1211	0.9	1	0.9
Carbon dioxide	—	5.0	5	1.0

*For SI units: 1 lb = 0.36 kg.

In total flooding systems, the effectiveness of the halogenated agents on flammable liquid and vapor fires is quite dramatic. Rapid and complete extinguishment is obtained with low concentrations of agent. On a worldwide basis, the development of systems using Halon 1301 and Halon 1211, together with the development of realistic test methods that have been applied to both these agents and recognized by NFPA, has shown that, in total flooding application for flame extinguishment or inerting, Halon 1301 requires an average of 10 percent less material on a gas volume basis than does Halon 1211 for any given fuel. Table 6-18E shows a comparison of flame extinguishment values included in NFPA 12A, and formerly in NFPA 12B. It is generally recognized that, on a weight-of-agent basis, both agents are approximately two and one-half times more effective than carbon dioxide.

The flame extinguishing ability of Halon 2402 vapors is very good and quite similar to that for Halon 1301 and Halon 1211. However, because it is a liquid at room temperature and is intended mainly for local application situations, its performance is not directly comparable to the data obtained for Halon 1301 and Halon 1211, with total flooding in mind.

The effectiveness of halogenated agents on Class A fires is less predictable. It depends to a large extent upon the specific burning material, its configuration, and how early in the combustion cycle the agent is applied. Most plastics behave as flammable liquids—they can be extinguished rapidly and completely with 4 to 6 percent

TABLE 6-18E. *Comparison of Flame Extinguishment Values for Halon 1301 and Halon 1211*

Fuel	Halon 1301	Halon 1211
Methane	3.1	3.5
Propane	4.3	4.8
n-Heptane	4.1	4.1
Ethylene	6.8	7.2
Benzene	3.3	2.9
Ethanol	3.8	4.2
Acetone	3.3	3.6

Columns under: Average Percent by Volume of Agent in Air Required for Flame Extinguishment

Note: Design flame extinguishment concentrations for Halon 1301 and 1211 total flooding systems are calculated from the tested value by adding a 20 percent safety factor. However, they are never less than 5 percent.

concentrations of Halon 1211 or Halon 1301. Other materials, particularly cellulosic products, can, in certain forms, develop deep-seated fires in addition to flaming combustion. The flaming portion of such fires can be extinguished with low 4 to 6 percent concentrations of agent, but the deep-seated portion may continue glowing under some circumstances.[5] Even so, the deep-seated fire will be controlled in that its rate of burning and consequent heat release will be reduced. Considerably higher agent concentrations (18 percent to 30 percent) are required to achieve complete extinguishment, and these levels are seldom economical to apply. However, the concept of controlling deep-seated fires with halogenated agents has been accepted in the respective NFPA standards.

TOXIC AND IRRITANT EFFECTS

The toxicology of Halon 1301, Halon 1211, and Halon 2402 has been studied extensively in both animals and humans. As a result, safety guidelines for these agents have been written.

Early animal studies by the U.S. Army Chemical Center, summarized in Table 6-18F, determined the approximate lethal concentration (ALC) for 15-min exposures to Halon 1301, Halon 1211, and Halon 2402 to be 83 percent, 32 percent, and 13 percent by volume, respectively. The ALC for decomposed vapors of these agents is also shown.

Animals exposed to halon concentrations below lethal levels exhibit two distinct types of toxic effects: (1) central nervous system changes, such as tremors, convulsions, lethargy, and unconsciousness at high airborne concentrations (above 30 percent by volume for Halon 1301, and 10 percent for Halon 1211); and (2) cardiovascular effects, including hypotension, decreased heart rate, and occasional cardiac arrhythmia (lack of rhythm in the heartbeat). Effects are transitory and disappear rapidly after exposure.

The inhalation of many halocarbons and hydrocarbons can make the heart abnormally sensitive to elevated adrenaline levels, resulting in cardiac arrhythmia and possibly death. This phenomenon has been referred to as cardiac sensitization. The halon compounds can also sensitize the heart. In the case of Halon 1301, this occurs only at high exposure levels. For example, in standard cardiac sensitization screening studies in dogs using 5-min exposures and large doses of injected adrenaline, the threshold for sensitization is in the 7.5 to 10 percent range.[6] The sensitization level for Halon 1211 is 1 to 2 percent and for Halon 2402 is 1,000 to 2,500 ppm.[7,8]

Human exposures to both Halon 1301 and Halon 1211 have shown that Halon 1301 concentrations up to 7 percent by volume and Halon 1211 concentrations below 2 to 3 percent by volume have little noticeable effect on the subject. At Halon 1301 concentrations

TABLE 6-18F. *Approximate Lethal Concentrations for 15-min Exposure to Vapors of Various Fire Extinguishing Agents**

Agent	Formula	Halon No.	Approximate Lethal Concentration in Parts per Million	
			Natural Vapor	Decomposed Vapor
Bromotrifluoromethane	CF_3Br	1301	83,200	14,000†
Bromochlorodifluoromethane	CF_2BrCl	1211	324,000	7,650
Carbon dioxide	CO_2	—	658,000	658,000
Dibromodifluoromethane	CF_2Br_2	1202	54,000	1,850
Bromochloromethane	CH_2BrCl	1011	65,000	4,000
Dibromotetrafluoroethane	$C_2F_4Br_2$	2402	126,000	1,600‡
Carbon tetrachloride	CCl_4	104	28,000	300
Methyl bromide	CH_3Br	1001	5,900	9,600

*Based on tests with white rats by the Medical Laboratories, U.S. Army Chemical Center.

† Subsequent tests by Kettering Laboratory of the University of Cincinnati (unpublished data) with a commercial Halon 1301 of improved quality indicated that the lethal concentration of decomposed vapor is at least 20,000 ppm. Other tests have given the ALC (approximate lethal concentration) value of Halon 1301 decomposition products as low as 2,500 ppm. The variance is based on differing analytical procedures.

‡ This figure does not agree with manufacturer's data on its product.

between 7 and 10 percent and Halon 1211 concentrations between 3 and 4 percent, subjects experienced dizziness and tingling of the extremities, indicating mild anesthesia. At Halon 1301 concentrations above 10 percent and Halon 1211 concentrations above 4 to 5 percent, the dizziness becomes pronounced, the subjects feel as if they will lose consciousness (although none have), and physical and mental dexterity is reduced.[9,10,11] In human subjects exposed to 7 percent Halon 1301 for periods up to 30 min, the effects that appeared within the first 5 to 10 min of exposure remained constant throughout the remainder of the exposure, then disappeared quickly after exposures were stopped.[12] When used as intended, no significant adverse health effects have been reported from the use of Halon 1301 or Halon 1211 as a fire extinguishant since their introduction into the marketplace 30 years ago.

Halon 1301 has been tested for effects on pregnant rats during the critical stages of gestation. Results showed this agent was not embryotoxic (causing growth retardation or death of the fetus) or teratogenic. In addition, Halon 1301 was not mutagenic when tested in the salmonella/microsome assay (Ames test).

A mutagen is a chemical that causes changes in the DNA structure of a cell. A standardized test, called the Ames test or assay, performed on a special strain of salmonella bacteria, provides a preliminary indication that a chemical may be a mutagen and possibly a carcinogen in humans. A high proportion of known carcinogens do exhibit mutagenic activity in the Ames assay. Halon 1301 concentrations up to 40 percent by volume did not exhibit mutagenic activity in this test.[13]

A teratogen is a chemical that causes permanent structural or functional alteration of the normal processes of fetal development. Pregnant female rats were exposed (by inhalation) to 5 percent Halon 1301 by volume. No evidence of teratogenic or embryotoxic effects was seen in their offspring.[14]

From the extensive medical data available, exposure guidelines have been produced for use of Halon 1301, Halon 1211, and Halon 2402. (See Table 6-18G.)

NFPA 12A permits Halon 1301 design concentration up to 10 percent in normally occupied areas, and up to 15 percent in areas not normally occupied. Because the required extinguishing concentration of Halon 1211 is near or above its limit for safe exposures, Halon 1211 systems were not recognized in NFPA 12B for use in normally occupied areas.

NFPA 12CT set acceptable concentrations for human exposure to Halon 2402 vapors at 0.05 to 0.10 percent by volume. Additionally, the standard recognized only outdoor local application systems

TABLE 6-18G. *Permitted Exposure Times to Halon 1301, Halon 1211, and Halon 2402*

	Concentration Percent by Volume	Permitted Time of Exposure
Halon 1301	Up to 7	15 min
	7–10	1 min
	10–15	30 sec
	Above 15	Prevent exposure
Halon 1211	Up to 4	5 min
	4–5	1 min
	Above 5	Prevent exposure
Halon 2402	0.05	10 min
	0.10	1 min

that may be outdoors with appropriate personnel safeguards, or indoors in unoccupied areas.

DECOMPOSITION PRODUCTS OF HALONS

Consideration of life safety during use of halogenated agents also must include the effects of breakdown products, which have a relatively higher toxicity to humans. Decomposition of halogenated agents takes place on exposure to flame, or to surface temperatures above approximately 900°F (482°C). In the presence of available hydrogen (from water vapor or the combustion process itself), the main decomposition products of Halon 1301 are hydrogen fluoride (HF), hydrogen bromide (HBr), and free bromine (Br_2). Although small amounts of carbonyl halides (COF_2, $COBr_2$) were reported in early tests, more recent studies have failed to confirm the presence of these compounds. The decomposition products of Halon 1211 and Halon 2402 are similar, but in the case of Halon 1211 include hydrogen chloride (HCl) and free chlorine (Cl_2), as well.

The approximate lethal concentrations (ALC) for 15-min exposures to some of these compounds are given in Column 2 of Table 6-18H. Column 3 gives the concentrations of these materials that have been quoted as "dangerous" for short exposures.[15]

Even in minute concentrations of only a few parts per million, the decomposition products of the halogenated agents have characteristically sharp, acrid odors. This characteristic provides a built-in warning system for the agent, and at the same time creates a noxious, irritating atmosphere for those who must enter the hazard area following a fire. It also serves as a warning that other potentially toxic products of combustion (such as carbon monoxide) will be present.

TABLE 6-18H. *Approximate Lethal Concentrations for Predominant Halon 1301 and Halon 1211 Decomposition Products*

Compound	ALC for 15-min Exposure (ppm by Volume in Air)	Dangerous Concentrations (ppm by Volume in Air)*
Hydrogen fluoride, HF	2500	50–250
Hydrogen bromide, HBr	4750	—
Hydrogen chloride, HCl	—	—
Bromine, Br_2	550	—
Chlorine, Cl_2	—	50
Carbonyl fluoride, COF_2	1500	—
Carbonyl chloride, $COCl_2$	100–150	—
Carbonyl bromide, $COBr_2$	—	—

*From reference 15.

It is necessary to establish the concentrations likely to be encountered when extinguishing fires with halogenated agents and to relate them not only to the absolute toxicity of the agent, but also to the toxic effects of the normal products of combustion.

The expected concentration of decomposition products depends upon many factors (i.e., size of room, size of fire, presence of large quantities of hot surfaces, and elapsed time before extinguishment). Test results, which may help to put the situation into perspective, have been made for both Halon 1301 and Halon 1211 in total flooding and portable applications. As an example of total flooding, Halon 1301 was used to extinguish 0.1-sq ft, 1.0-sq ft, and 10-sq ft (0.009 m^2, 0.09 m^2, and 0.9 m^2) n-Heptane fires in a 1695-cu-ft (48 m^3) test enclosure, at agent discharge times of 15, 10, and 5 sec. The results are shown in Table 6-18I, relating the resulting decomposition products to the ratio of fuel area to enclosure volume and flame extinguishment time.[16] It is obviously advantageous to attack the fire at an early stage while it is still small, and to release the extinguishing agent as rapidly as practical.

TABLE 6-18I. *Halon 1301 Decomposition Produced by n-Heptane Fires**

Enclosure volume: 1695 cu ft (48 m^3)
Halon 1301 concentration: 4% by volume

Fuel Surface Area (Area per Unit Enclosure Volume ft²/ft³ or m²/m³)	Flame Extinction Time (sec)	Total Decomposition Products† (ppm)
0.06	11.5–15.4	4.5–5.6
0.06	7.1–7.6	2.8–4.2
0.06	4.0–4.8	3.3–4.5
0.60	20–37	94–289
0.60	11.5–13.5	64–284
0.60	4.7–6.7	11.5–169
6.00	20–22	2252–2304
6.00	13.0–16.3	1292–1590
6.00	5.2–10.0	358–778

*Edmonds, 1972.
†Sum of HF, HBr, and Br_2.

For a similar example in portable extinguishers, bromochlorodifluoromethane (Halon 1211) was used in 3-lb (1.36-kg) porta-

ble units to extinguish a 2.5-sq-ft (0.23-m^2) n-Heptane fire in a 2500-cu-ft (70.8-m^3) room. The results of two tests, one normal extinguishment in 1 sec, and the second deliberately extended extinguishment in 10 sec, are given in Table 6-18J. In the case of normal extinguishment, the levels of breakdown products are below the limits allowed by Occupational Safety and Health Administration (OSHA) regulations for 8-hr continuous exposure.[17]

TABLE 6-18J. *Concentrations of Breakdown Products Obtained from Extinguishing 2.5 sq ft (0.23 m²) n-Heptane Fires with Halon 1211 Portable Extinguishers*

Concentration of Breakdown Products—ppm by Volume

Compounds	Test I Normal Extinction in 1 sec	Test II Prolonged Extinction in 10 sec
HCl + HBr	2	50
HF	0.5	10
Cl_2 + Br_2	Not detected*	2.5
$COCl_2$	Not detected†	Not detected†

* Limit of detection 0.1 ppm.
† Limit of detection 0.25 ppm.

The extensive studies on the toxicological effects of halogenated agents indicate that little, if any, risk is attached to the use of Halon 1301 or Halon 1211 when used in accordance with provisions of NFPA standards governing use of these agents.

CLEAN EXTINGUISHING AGENT RESEARCH

This section presents an overview of alternative agent development and is largely extracted from the "Halon Fire Extinguishing Agents Technical Options Report" of August 11, 1989.[18] The report was prepared pursuant to Article (6) of the Montreal Protocol on Substances that Deplete the Ozone Layer, under the auspices of the United Nations Environment Program.

Approaches to Alternative Agents

The halon/ozone issue is not a single problem, and there is no single solution. Multiple alternative agents, varying according to application, are the most likely outcome. The development of one or two general-purpose "drop-in" replacements having all significant characteristics equal to those of the existing halons may not be a realistic research goal. On the other hand, alternative low or zero ozone depletion potential (ODP) clean agents for specific applications are realistic objectives.

Five requirements make halons difficult to replace: (1) cleanliness, (2) low ODP, (3) low toxicity, (4) effectiveness, and (5) low global warming potential (GWP, the calculated ability of a volatile compound to affect global climate through absorption and emission of infrared radiation). There are, of course, other considerations for replacement agents; cost, storage stability, and compatibility with engineering materials are among these.

Commonly used halon fire extinguishants can be separated into two groups according to their application: Halons 1211 and 2402 in one group, and Halon 1301 in the second. Different types of chemicals will likely be needed for replacement of each group. Halons 1211 and 2402 are usually applied by streaming (directed discharge from a nozzle positioned remote from the fire). Such agents are used in localized applications and are often applied manually. Halon 1301 is most often used in total flooding applications (filling of an enclosed volume with sufficient agent to suppress combustion). Both clean streaming agents and clean total flooding agents are needed. Some very specialized applications (e.g., fire protection systems for aircraft engines) may be best served by agents that combine characteristics of both streaming and total flooding agents. Other applications may be difficult to categorize exactly. Nevertheless, it is useful to consider and contrast streaming and total flooding agents.

Ideally both alternative streaming agents and alternative total flooding agents must be clean and must have zero or very low ODPs. Cleanliness is the primary reason for the use of halon agents in most applications, and acceptable ODP is the primary driving force for development of new agents. However, there are important differences in certain other requirements for the two agent types. A streaming agent requires good deliverability; the material must not be too gaseous. Toxicity requirements are less stringent for streaming agents, since personnel are not usually exposed to high agent concentrations. In contrast, a total flooding agent must be more gaseous to encourage dispersion and to avoid stratification, and a very low toxicity is essential for application in normally inhabited spaces.

The steps needed to determine alternative agents have been vigorously debated. Some favor a broad-based effort with a foundation and/or parallel effort of basic research. This type of program begins with consideration of all possible families of chemicals, and screens these to end up with a set of final agents. Others favor a targeted approach, directing their effort toward those chemical families considered to offer the most promise as alternative agents.

Existing Efforts to Develop Alternative Agents

Most of the work to date on alternative agents has emphasized low-ODP halocarbons, due to cleanliness, fire suppression capabilities, and relatively well-developed commercial processes for manufacture. Four major problems exist:

1. No limits on "acceptable" ODP values have been established. In fact, some doubt exists in the mind of many as to whether any ODP value greater than zero is acceptable. (Controversy about the reliability of ODP calculations is beyond the scope of this chapter.) While many consider very low ODP halocarbons to be a solution to the problem, rather than the source of the problem, this is not universally accepted. Unfortunately, even if the international environmental and political community were to adopt values for acceptable ODPs, there is no guarantee that these limits would not change.
2. The concern about global warming is increasing. GWPs are calculated from atmospheric models similar to those used for ODP. GWP is becoming of increasing concern, and many of the candidates being evaluated as alternative clean fire-fighting agents have significant GWPs. Particularly bothersome about this is that one group of halocarbons having zero ODPs, i.e., perfluorocarbons, have very high GWPs.
3. Atmospheric lifetime is also a concern. Long atmospheric lifetime is a contributing factor to GWP, in some cases; however, a greater concern may be accumulation in the atmosphere, with unknown consequences.
4. In assessing the commercial viability of new, clean extinguishing agents, chemical manufacturers must consider trends regarding acceptability, toxicity, and environmental evaluations. It is likely that future environmental requirements may either be more stringent or limit production quantities. While it is difficult to fault such an approach by chemical manufacturers, this cautious attitude significantly increases the time required to introduce new halon replacements.

Research and development programs for halon alternatives have been established by a number of chlorofluorocarbon (CFC) and halon producers. Some nonindustrial development projects are also underway. Some promising candidates are now undergoing testing.

The Halon Alternatives Research Corporation (HARC) has been established in the United States to encourage and facilitate research and development activities. This consortium is a group of government and industry leaders who are supporting cooperative efforts for information exchange and complementary research programs. These efforts are intended to support the scientific basis for development and commercialization of clean, safe, and reliable fire extinguishing agents.

Conclusions: Work to date indicates that the development of general-purpose, direct replacements having attributes equal to those of the present halons may be unrealistic in the short term. On the other hand, clean alternative agents with zero or very low ODPs for specific uses are now available for most applications, if tradeoffs in fire extinguishment capability, toxicity, and/or other characteristics are acceptable. In particular, such agents could be available in the short term for selected manually applied equipment and local application systems. Toxicity considerations make total flooding agents for normally inhabited areas much more problematic. Streaming agents are inherently easier to develop, since physical properties that give improved streaming performance can partially offset poorer inherent flame-suppression capabilities, and toxicity requirements are not as stringent for an agent delivered by streaming as they are for a total flooding agent.

It should be pointed out that fully halogenated chlorofluorocarbons have significant fire extinguishing capabilities. Furthermore, a number of these materials have well-defined, low toxicities. Since these CFCs have lower ODPs than do the present halon fire extinguishing agents, substitution of CFCs for halons in selected fire protection applications could reduce the threat to the ozone layer. On the other hand, these materials are regulated under the Montreal Protocol and are in a different category than are the halons. This makes substitution a highly sensitive issue and unlikely to be widely supported. Certainly, such substitutions could not be regarded as a long-term solution.

Uncertainties about future ODP, GWP, atmospheric lifetime, and toxicity requirements are of great concern in considering the development and commercial viability of halon alternatives.

APPLICATION SYSTEMS

A system differs from a portable or mobile appliance primarily in that the agent discharge stream is not directed by a person. The discharge stream or pattern is usually determined in advance, as is either the quantity of agent or discharge rate, or both, and the number and types of nozzles provided. A system consists of a supply of agent, a means for releasing or propelling the agent from its container, and one or more discharge nozzles to apply the agent into the hazard or directly onto the burning object. A system may also contain other elements, such as one or more detectors, remote and local alarms, a piping network, mechanical and electrical interlocks to close fire doors and to shut down ventilation, directional control valves, installed reserve agent supplies, etc. The extent of the auxiliary functions of a system is usually dependent upon the nature of the hazard, in keeping with the desires and resources of the user.

A system is usually a fixed or stationary apparatus, but some portable or mobile systems have been designed. A portable or mobile system may be moved from one hazard to a similar one, but it is placed in a stationary position while it is in service. In this regard, all systems may be considered to be "fixed," but not necessarily permanently placed.

Halogenated agent systems are broadly classified by their method of applying agent to the hazard. The two main types recognized in NFPA 12A, and formerly in NFPA 12B, are: (1) total flooding and (2) local application systems. Another category, termed "specialized systems," includes those systems that are designed to protect special or unique hazards, and which have been tested and approved under these specific conditions.

Total Flooding Systems

These systems protect enclosed, or at least partially enclosed, hazards. A sufficient quantity of extinguishing agent is discharged into the enclosure to provide a uniform fire extinguishing concentration of agent throughout the entire enclosure. Examples of total flooding systems using halogenated agents are found in computer rooms, electrical switchgear rooms, magnetic tape storage vaults, and electronic control rooms; storage areas for artwork, books, and stamps; aerosol filling rooms; machinery spaces in ships; cargo areas in large transport aircraft; processing and storage areas for paints, solvents, and other flammable liquids; etc. Halon 1301, by virtue of its lower toxicity, higher volatility, and lower molecular weight, offers particular advantages for use in total flooding systems. Halon 1211 total flooding systems, with suitable safeguards, are used outside the United States.

Total flooding systems may be further distinguished by their method of design or installation. An engineered system is custom designed for a particular hazard, using components that are approved or listed only for their broad performance characteristics. Components may be arranged into an almost unlimited variety of configurations. (See Figure 6-18B.)

In pre-engineered systems, the number of components and configurations are determined in advance and included in the description of the system's approval or listing. While the degree of pre-engineering can differ from one system to another, the following limits of components and configurations must be considered:

1. Maximum number of cylinders per manifold.
2. Maximum and minimum size and length of piping.
3. Maximum and minimum size and number of elbows, tees, and discharge nozzles.
4. Container volume, fill density, and level of nitrogen superpressurization.

Modular systems consist of single containers connected to discharge nozzles, with minimal piping. A group of container-nozzle assemblies may be distributed throughout the protected area and interconnected electrically. Modular systems permit a more attractive and less expensive initial installation, but often result in higher maintenance costs. Figure 6-18C illustrates the installation of a typical modular system.

Central storage systems locate all agent containers in one centralized location. The agent can be stored in multiple tanks connected by a manifold arrangement (Figure 6-18D) or in a single bulk storage tank. A piping network distributes the agent to the various discharge nozzles. Ease of maintenance is the primary advantage of central storage systems.

Local Application Systems

As the name implies, these systems discharge extinguishing agent in such a manner that the burning object is surrounded locally by a high concentration of agent to extinguish the fire. In local application systems, neither the quantity of agent nor the type or arrangement of discharge nozzles is sufficient to achieve total flooding of the enclosure containing the object. Often, too, a local application system is required because the enclosure itself may not be suitable to provide total flooding. Examples of areas protected by local application are printing presses, dip and quench tanks, spray booths, oil-filled electric transformers, vapor vents, etc. (See Figure 6-18E.)

Because of its lower volatility, Halon 1211 is well suited for local application systems. The lower volatility, plus a high liquid density, permits the agent to be sprayed as a liquid and thus propelled into the fire zone to a greater extent than is possible with other gaseous agents. Halon 2402 also is well suited for local application systems, but its major usage has been outside the United States.

The material relating to local application systems, in NFPA 12B, was largely theoretical and intended for use by equipment manufacturers and testing laboratories in designing and evaluating

FIG. 6-18C. A modular Halon 1301 installation in a computer room. The spherical container holds 105 lb (47.6 kg) of Halon 1301 and is mounted above a false ceiling. Only a short length of pipe protrudes into the room, and it connects to two discharge nozzles. The mounting bracket of the container is visible at upper right, and braided cable in the center is the electrical connector to the release valve on the container outlet. Six identical container assemblies are installed in this room; they are interconnected to discharge simultaneously.

1. Automatic fire detectors installed both in room proper and in underfloor area.
2. Control panel connected between fire detectors and cylinder release valves.
3. Storage containers for room proper and underfloor area.
4. Discharge nozzles installed both in room proper and in underfloor area.
5. Control panel might also sound alarms, close doors, and shut off power to the area.

FIG. 6-18B. A total flooding system installed in a room with a raised floor. (Source: The Ansul Company)

FIG. 6-18D.   A Halon 1301 centralized storage system installed in the control room of an electrical generating plant. The large manifold has two 90-lb (40.8-kg) cylinders and discharges into the room proper. The smaller manifolds, each composed of two 56-lb (25.4-kg) cylinders, protect two individual underfloor areas. Control panels are visible at upper center and upper right. Cylinder valves are activated pneumatically through small hose connected to the top center of the valve. The large hose connected to the side of the valve is a Halon 1301 outlet.

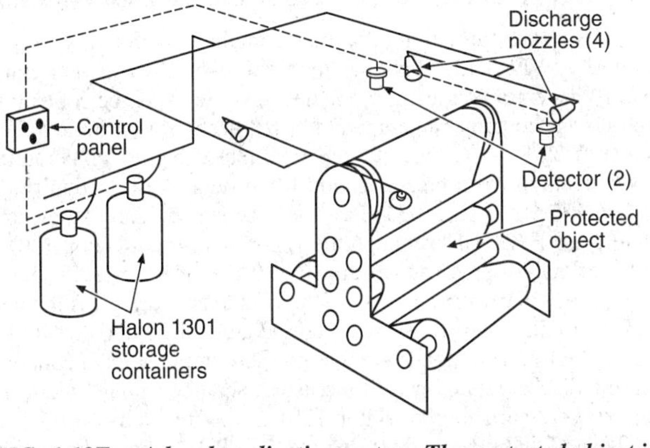

FIG. 6-18E.   A local application system. The protected object is not enclosed; therefore, discharge nozzles and rate of application must be capable of enveloping the object. The agent supply must be sufficient to maintain flow for the required time of protection, usually several minutes. Nozzle design is critical and must be determined by extensive testing.

components for local application systems. The component of overriding importance was the discharge nozzle, and its performance characteristics must be known with a high degree of reliability.

Specialized Systems

These systems, using both Halon 1301 and Halon 1211, are widely used throughout the world. Systems to protect aircraft engine nacelles, racing cars, military vehicles, emergency generator motors, etc., all fall into this category. The distinguishing characteristic of a specialized system is that it can be applied only to the specific hazard for which it was designed and tested. For example, a system designed to protect the jet engines of a Boeing 757 aircraft cannot be applied, per se, to an M-1 battle tank. For each such application, the system is developed by a comprehensive test program.

Systems protecting racing cars are somewhat less specialized, although they properly fit into this category. Sanctioning bodies, such as the Federation International de l'Automobile, the United States Automobile Club, and the National Hot Rod Association, have formulated general rules governing these systems. One study showed a wide variety of sizes and performance characteristics of commercially available Halon 1301 systems available in the U.S.[19] Virtually none of these systems is listed or approved by testing laboratories. In spite of this, they have been credited with saving a large number of lives in the racing field.

Another type of specialized system is a "thermatic" unit. This unit consists of a container connected directly to a thermally actuated release device, which is similar to an automatic sprinkler. When heat from the fire activates the valve, the extinguishing agent is released to attack the fire. Sizes of thermatic units range from a $1^1/_2$-lb (0.68-kg) Halon 1301 unit to protect engine compartments of small pleasure boats to 150-lb (68-kg) units containing Halon 1301 and Halon 1211 designed for total flooding of small rooms.

DESIGN CONSIDERATIONS

This section highlights many of the aspects and requirements contained in NFPA 12A and formerly in NFPA 12B.

Uses and Limitations

Halogenated agent systems are generally considered useful for the following types of hazards:

1. Where a clean agent is required.
2. Where live electrical or electronic circuits exist.
3. For flammable liquids or flammable gases.
4. For surface-burning flammable solids, such as thermoplastics.
5. Where the hazard contains a process or objects of high value, and where use of other extinguishing agents could cause extensive damage or downtime.
6. Where the area is normally occupied by personnel (use of Halon 1301 only).
7. Where availability of water, or space for systems using other agents, is limited.

There are several types of flammable materials on which halogenated agents are ineffective. These are:

1. Fuels that contain their own oxidizing agent, such as gunpowder, rocket propellants, cellulose nitrate, organic peroxides, etc.
2. Reactive metals, such as sodium, potassium, NaK eutectic alloy, magnesium, titanium, and zirconium.
3. Metal hydrides, such as lithium hydride.
4. Chemicals capable of autothermal decomposition, such as organic peroxides and hydrazine.

In the first category—in which the compound contains its own oxygen supply, often built into the fuel molecule—the halogenated agent is unable to penetrate into the reaction zone quickly enough to put out the fire. The oxidizer is in too close physical proximity to the fuel to permit interaction with the extinguishing agent. Often the only effective agent for such fuels is a water deluge to dilute the fuel and attempt to remove heat ahead of the combustion front. In the second category, the reactive metals and metal hydrides are too reactive at flame temperatures for the halogenated agent to operate effectively. Also, the flame chemistry of metal fires is quite different from that of hydrocarbon fires.[20]

A more commonly encountered limitation of the capabilities of an agent is its limited effectiveness on deep-seated Class A fires at concentrations below 10 percent by volume. When the concept of control is applied to using halogenated agent systems on Class A fires, rapid response by outside help is necessary. Otherwise, a re-flash potential will exist once the extinguishing concentration is dissipated.

Safety

While both Halon 1301 and Halon 1211 have low vapor toxicity, there are hazards in exposing personnel to high concentrations of either agent (above 10 percent Halon 1301, or above 4 percent Halon 1211). Further, the inhalation hazard produced by the fire itself, such as heat, smoke, oxygen depletion, and toxic combustion and decomposition products, may be substantial. Several general safety precautions are listed in the appendix of NFPA 12A and were listed in the appendices of NFPA 12B and NFPA 12CT.

Detection and Actuation

Halogenated agent systems generally have operation requirements similar to systems using other types of agents. However, there are strong recommendations in NFPA 12A and were in NFPA 12B for the use of automatically actuated systems. The primary reason behind these recommendations is to limit the size and severity of fire with which the system must deal, thus minimizing decomposition of the agent during extinguishment. In a Class A fire, automatic actuation, coupled with sensitive detectors, often can prevent the fire from becoming deep seated.

The detector portion of the system is therefore of paramount importance in system design. The detector must be sensitive enough to the fuel in question to rapidly respond to the fire at an early stage. But too great a sensitivity may produce false actuations, thus imposing an economic burden on the owner, causing unnecessary outage of fire protection, and creating a lack of confidence in the system. Designers have successfully combined sensitivity and reliability by utilizing multiple detectors connected in a double circuit or "cross-zoned" mode of operation. (See Figure 6-18F.) Actuation of either zone alone activates local and remote alarms, but does not discharge the extinguishing agent. Actuation of both zones simultaneously causes the agent to discharge. This arrangement prevents a false signal by an individual detector from discharging the entire system. The two zones may contain the same or different types of detectors. Ionization and photoelectric detectors are commonly used in computer and EDP areas.

In another arrangement, appropriate for generally attended locations, one zone of ionization detection provides early warning, and one zone of rate-compensated thermal detection releases the extinguishing agent.[21]

Particular attention must be given to locating detectors within a hazard. The amount and location of combustibles, the listed sensitivity spacing relationship of the detectors considered, and ventilation characteristics of the hazard are all important factors.

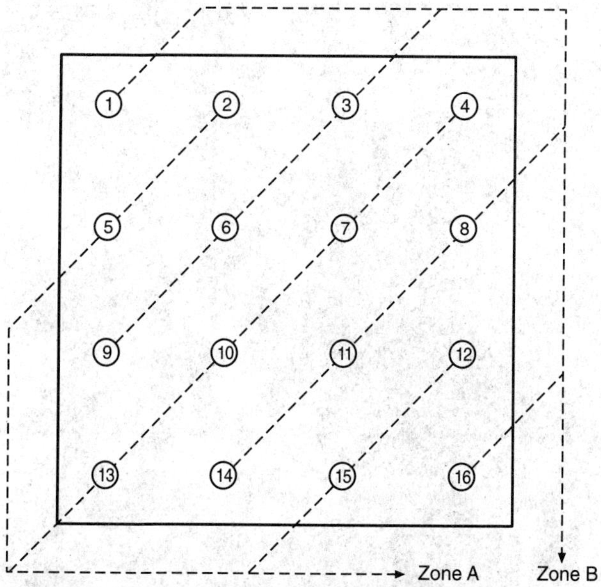

FIG. 6-18F. *A cross-zoned detector circuit. Sixteen detectors in a 4-by-4 array are connected so that adjacent detectors are on different circuits, or "zones." For example, consider detector No. 7, connected to Zone A. Adjacent detectors, Nos. 3, 6, 8, and 11, are connected to Zone B.*

Agent Supply

The relatively high cost of agents and the specialized nature of systems using them dictate that a specific supply of agent be provided to protect against a given hazard or set of hazards. Conventionally, the agent is contained in one or more pressurized vessels, which are installed near or within the protected area. Steel pressure cylinders or spheres containing from a few pounds to several hundred pounds of agent (1 lb = 0.454 kg) are commonly used. Larger "bulk" tanks containing several tons (1 ton = 907 kg) of agent are used in Europe and Canada and have been proposed in the United States as well.

Extra pressurization of storage containers with nitrogen is required by NFPA 12A, and was required by NFPA 12B. For Halon 1211, this extra pressurization is necessary to expel the contents of the storage container at temperatures below the atmospheric boiling point of 25°F (–4°C) and at a practical rate at higher temperatures. Two levels of extra pressurization for systems were recognized in NFPA 12B: 150 psig (1034 kPa), and 360 psig (2482 kPa) total pressure at 70°F (21°C). Although the vapor pressure of Halon 1301 at all reasonable temperatures above –30°F (–34°C) is sufficient to completely expel the contents of the container, extra pressurization is used to flatten the pressure *vs.* temperature characteristics during storage, provide a higher average pressure in the storage container during discharge, and permit the use of smaller piping. Two levels of extra pressurization of Halon 1301 are permitted by NFPA 12A: 360 psig (2482 kPa), and 600 psig (4137 kPa) total pressure at 70°F (21°C).

The fill density of a container is defined as the weight of agent divided by the internal volume of the container. NFPA 12A, and formerly NFPA 12B, permits maximum fill densities of 70 lb per cu ft (1121 kg/m^3) for Halon 1301, and 102 lb per cu ft (1634 kg/m^3) for Halon 1211. These limits are set by Department of Transportation (DOT) regulations for shipping containers, which require that the container not become liquid-full at temperatures below 130°F (54°C). The maximum fill densities permitted in the standards coincide with the liquid densities of the two agents at 130°F (54°C) in the presence of nitrogen at the highest level of extra pressurization, again allowing for a 5 percent maximum error in filling the container. Most manufacturers utilize a variable fill density in their design to obtain maximum economy in the system. There is no lower limit on permissible fill densities in NFPA 12A, and was none in NFPA 12B.

Flow Characteristics

The flow of Halon 1301 and Halon 1211 through pipe and tubing has been determined experimentally. For Halon 1301, the flow is two-phase liquid-vapor. Although the agent exists as a single-phase liquid in the storage container, it begins to vaporize as soon as its pressure drops during flow. By the time the agent reaches the discharge nozzle, this has substantially reduced agent density. Therefore, classical single-phase flow equations of the Moody or Fanning type cannot be used to predict the frictional pressure losses during flow of Halon 1301.

There are two methods in NFPA 12A to estimate flow characteristics. One is a simplified chart method to be used for balanced piping systems only. (A balanced system is one in which the actual pipeline lengths, the equivalent pipeline lengths, and the flow rates to each nozzle are equal.) The other is a rigorous two-phase equation that may be used either for balanced or unbalanced piping systems. The chart method can be applied to complex balanced systems using hand calculations, but even relatively simple unbalanced systems calculated by the two-phase equation require a computer solution. Most U.S. manufacturers of Halon 1301 systems have developed computer programs to help perform these flow calculations.

Because the solubility of nitrogen in the liquid phase of Halon 1211 is only one-half as great as for Halon 1301, the flow properties of Halon 1211 were treated as single-phase liquid flow in NFPA 12B.

Total Flooding Systems

The great majority of Halon 1301 and Halon 1211 systems throughout the world are of the total flooding type. As described earlier, a total flooding system is one that develops a uniform extinguishing concentration of agent throughout an enclosure, such as a room. This system is capable of extinguishing a fire within the enclosure, regardless of the location of the fire. Wickham[22] outlines the main design elements of a total flooding system, as follows:

1. Define the hazard, which includes the dimensions and configuration of the enclosure, the maximum and minimum net volumes, the fuels involved, the expected temperature range in the hazard area, ventilation and unclosable openings, and occupancy status.
2. Establish a minimum design concentration, based upon the fuels involved.
3. Calculate the minimum agent quantity, based upon the minimum design concentration, the maximum net volume, the minimum expected temperature of the enclosure, and compensation for losses from ventilation and unclosable openings.
4. Calculate the maximum possible concentration that could occur if the hazard is at the conditions of minimum net volume and maximum temperature. This concentration must not be greater than that permitted by NFPA 12A, or was permitted by NFPA 12B, for the occupancy status of the hazard.
5. Select the agent storage container, based upon the design quantity of the agent and the sizes of standard containers available from the equipment manufacturers.
6. Determine the minimum agent flow rate by dividing the design agent quantity by the maximum permissible discharge time.

This time is currently set by NFPA 12A, and was set by NFPA 12B, at 10 sec, but is subject to some discretion.
7. Determine the size of piping, considering the location of the agent storage containers and locations of the discharge nozzles. This step must be performed concurrently with the following Step 8.
8. Determine the number, size, and locations of discharge nozzles. Discharge rate and nozzle area coverage data must be obtained from laboratory listings or from design manuals of the equipment manufacturers.

Jensen regards the first of the preceding elements as the single most important.[23] Certainly it is the most important one in which the end user can usually participate. The remainder of the system is generally designed by the equipment manufacturer.

In establishing the minimum design concentration (Step 2 above), one must consider the conditions under which the fuels are used. For flammable liquids or vapors where an explosion is unlikely, a flame extinguishment concentration may be used. If an explosion potential is considered likely, a higher inerting concentration must be used. An inerting concentration is sufficient to prevent combustion of any fuel-air ratio that might occur in the enclosure. It, therefore, must be applied upon development of a dangerous condition, and not after ignition has occurred. Minimum design concentrations for both flame extinguishment and inertion for some flammable liquids and gases are included in NFPA 12A, and were included in NFPA 12B. Guidelines for determining which level should be used are/were also given in these standards, respectively.

For combustible solids or Class A fires, the possibility and consequences of developing a deep-seated fire must be considered. Rapid detection and prompt application of agent also can help prevent a Class A fire from becoming deep seated. Remote alarms to summon outside assistance should be an integral part of a fire protection system.

Enclosures with fresh air make-up ducts, and exhaust ducts to the outside must have a means of self-closure at the time of agent discharge. It is not necessary to shut down closed-loop ventilation systems in which all the exhaust air is returned to the room. In fact, continued operation of such systems will improve agent distribution and reduce the rate at which agent is lost through unclosable openings. Losses through unclosable openings are more insidious, in that they often are undetected in a casual inspection of a plan of the enclosure. Because of the high density of the vapors of halogenated agents, there is a definite tendency of agent-air mixtures to find their way out of openings, particularly ones in the lower portion of the enclosure. As this mixture leaves the enclosure, it is replaced by fresh air that enters and collects near the top of the enclosure. This often gives the impression that the agent is separating itself from the air, which it is not. Also, the tendency frequently is to compensate for such losses by providing a higher concentration initially. This, of course, simply aggravates the situation by increasing the rate at which the loss occurs. An extended agent discharge or provision for mechanical mixing during the soaking period are the only effective ways to overcome a serious loss of agent through unclosable openings. Fortunately, since the fire hazard in most applications is located in the lower half of the room, some agent losses in this manner can be tolerated.

Local Application Systems

Design methods for local application systems for both Halon 1301 and Halon 1211 have not yet become well established. The information currently appearing in the NFPA halon standards on local application systems is useful only as a guide to the equipment manufacturer or testing laboratory.

The principle of local application differs greatly from total flooding. Local application is similar to a portable fire extinguisher, in that the discharge nozzle is directed at the surface or object on which the fire is anticipated. For this reason, the single most important element in a local application system is the discharge nozzle. The discharge velocity and rate must be sufficient to penetrate the flames and produce extinguishment, but not so great as to cause splashing of fuel and thus increase the fire hazard. The performance of the discharge nozzle must be determined in advance by laboratory testing; the location, position, and orientation of nozzles in the installation must be in strict accordance with these listings. To date, there is no listed or approved equipment for local application systems for either Halon 1301 or Halon 1211 in the United States.

TESTING AND MAINTENANCE SYSTEMS

Following installation, a system must first be tested to ensure that it will perform in accordance with its design, and thereafter it must be maintained to ensure its continued performance at some later time. Although minimum requirements for testing and maintenance of halogenated agent systems are given in the NFPA halon standards, some elaboration on both topics seems in order.

Testing of Halogenated Agent Systems

While the performance of individual components comprising the system has been determined by laboratory testing and listings, certain tests must be made on the completed installation to ensure that it has been installed properly and that, as a whole, the system will perform according to design. This process begins with the review of plans, hydraulic calculations, and specifications that have been submitted before work begins. Where field conditions necessitate a material change from the original submission, the change should be reviewed.

The complete, installed system should be tested by qualified personnel to meet the approval of the Authority Having Jurisdiction. This testing should include all mechanical and electrical portions of the system. Specifically, the following should be checked.

Hazard configuration and containment capability: The configuration of the protected space should be carefully compared to the original hazard specification. Halon total flooding systems are designed to establish a concentration of halon by volume. Should the volume of the hazard change, two significant problems could result: (1) where the hazard volume has increased, the design quantity could be insufficient to provide an extinguishing concentration, and (2) where the hazard volume has decreased, personnel exposure to unacceptably high halon concentrations during discharge could result.

It is important that the enclosure be capable of containing the halon/air mixture that results from discharge for an acceptable period of time. In order to ensure this capability, the enclosure should be thoroughly inspected for sources of leakage. Air circulation systems should shut down automatically upon halon system actuation, and dampers should automatically close. Access doors to protected enclosures should be equipped with self-closing devices. Door latches should be adequate to hold the doors closed during pressure variations that may be experienced during halon discharge.

The enclosure should be tested to determine actual leakage. A method for conducting a test of enclosure integrity, using a door fan, is outlined in Appendix B of NFPA 12A. A more thorough explanation of this methodology is also outlined in the "Enclosure Integrity Procedure for Halon 1301 Total Flooding Fire Suppression Systems,"[24] available from the National Fire Protection Research Foundation. Basically, the door fan procedure involves using a fan, mounted inside a panel that fits into a door opening to pressurize or de-pressurize the enclosure. Readings are then taken to determine the rate of leakage of air to or from the enclosure being tested. Calculations are then performed to predict the worst-case loss of a halon/air mixture from the protected enclosure. This test can be repeated on an annual basis to ensure the ongoing viability of the integrity of the enclosure.

Agent containers: All agent containers should be properly located as shown on the drawings. The weight of halon should be verified by weighing, or calculated using liquid fill level as the basis of the calculation. All containers should be mounted in brackets and securely fastened.

Piping: The piping system should be inspected to ensure that it has been installed in accordance with the submitted hydraulic calculations and plans, and that components provided are in agreement with the submissions. This should include inspection of the means of pipe size reduction and attitudes of tees. Piping joints, discharge nozzles, and piping supports should be inspected to ensure that they are securely fastened to prevent leakage or hazardous movement during system discharge. Especially ensure that nozzle discharge is not at head height of personnel in the normal work area, and also that it will not impinge on any loose objects or shelves, cabinet tops, or similar surfaces where loose objects could be present and become missiles.

After a thorough visual inspection, the piping system should be subjected to a closed-circuit pneumatic pressure test to ensure tightness. Care should be taken in conducting this test, and the area should be evacuated of personnel, as a rupture of the piping system could have serious consequences.

Detection and controls: Automatic detection devices, such as heat and/or smoke detectors, should be checked for proper type and location. It is particularly important that detectors not be located near obstructions, supply air vents, humidifier discharge, or other equipment or conditions that would adversely affect their ability to respond to heat, smoke, or products of combustion emitted by a fire.

Manual stations used to cause actuation of the halon system should be securely installed, readily accessible, accurately identified as to their purpose, and properly protected to prevent damage or accidental operation. Manual stations should be located at egress points from the hazard to ensure that personnel safety is not compromised.

Where abort switches are provided, they should be of the "dead-man" type, requiring constant manual pressure to operate. Manual stations must always override the operation of abort switches.

The control unit must be installed securely in a readily accessible location. Power to the control unit should be supplied from a dedicated circuit. Where circuit breakers are used, they should be protected to prevent accidental shutoff.

All wiring should be properly installed in accordance with local codes. During installation, the field circuitry should be measured for ground-fault and short-circuit conditions. This should be accomplished before electronic detection devices are connected, as this test could result in damage to these devices.

All auxiliary devices, such as audible alarm devices, graphic displays, air-handling shutdown, power shutdown, etc., should be checked for proper function.

Functional test: A complete, nondestructive test of all system functions should be undertaken before acceptance. This should include operation of all detection devices; tests of all required sequences, including ability to actuate discharge valves; and tests of auxiliary operations, including shutdown functions and remote alarm transmission.

A "puff test" should also be considered. This involves use of compressed air or carbon dioxide in lieu of halon in the system containers for the purposes of the test only. Care must be taken to ensure that pressure safety requirements of the agent containers are not exceeded. The purpose of the puff test is to ensure valve operation against a pressure equivalent to the normal agent container pressure, and to check for continuous and obstruction-free piping. Blow-off indicators should be placed over nozzles before conducting this test. Absence of the devices at the conclusion of the test indicates that test gas reached the nozzle.

Full-discharge tests using Halon 1301 should be avoided. Halon 1301 is a potent ozone depletion agent, and should not be released to the atmosphere, except to extinguish fire. Should there be a special need to consider a discharge test as necessary, a substitute test gas should be used. The most current edition of NFPA 12A should be consulted when considering the use of a test gas.

The functional test is a good opportunity to provide training for operating personnel or others who normally work in the protected area.

Upon successful completion of all tests and correction of any deficiencies to auxiliary functions, the system should be restored to, or initially placed in, full-operating condition.

Maintenance of Halogenated Agent Systems

NFPA 12A specifies, and NFPA 12B specified, certain items that must be checked at semiannual and annual intervals. Semiannually, the system should be visually inspected for evidence of corrosion or other damage, and the storage containers checked for loss of agent. This latter item involves a twofold check. First, the pressure corrected for temperature should be measured to ensure no loss of pressurizing gas, and second, each container must be weighed to determine loss of agent. Neither check alone is sufficient; both are required. Liquid level indicators are sometimes used to indicate the quantity of agent, rather than weighing, which is often cumbersome and time-consuming.

At least annually, the operational characteristics of the system should be retested by knowledgeable and qualified personnel. It is of particular importance that all necessary measures be taken to ensure that inadvertent halon release does not occur as a result of the service procedures. Consideration should be given for the inclusion of a full functional test and test of enclosure integrity. Reports should be filed with the owner of the system. Needless to say, all impediments found in these inspections must be corrected promptly.

AT THE END OF USEFUL LIFE—RECYCLE

Halons are potent ozone-depletion substances and should never be released to the atmosphere, except to extinguish actual fires. When the halon system is no longer required to provide protection for a particular hazard, the halon should be made available for recycling. Recycled halon can then be provided to protect other important facilities. Recycling halon is important—it is a wise use of an important fire protection resource and it negates the need to produce equivalent quantities of an environmentally harmful substance.

Decommissioning Halon Systems

Halon cylinders and valves contain a liquified compressed gas, and halon cylinder valves are designed to accommodate the high flow rates required to complete discharge in less than 10 sec. As a result,

halon cylinders have the potential to cause serious damage, injury, or death. Before a halon cylinder is removed from the system cylinder retaining brackets, the cylinder valve must be made safe and an anti-recoil device placed in the discharge opening. Several incidences of damage, injury, and fatality have occured from failure to take these precautions. The manufacturer of the equipment should be contacted for instructions and necessary parts to accomplish decommissioning in a safe manner.

BIBLIOGRAPHY

References Cited

1. Dowling, J., and Ford, C., "Halon 1301 Total Flooding System for Winterthur Museum," *Fire Journal*, Vol. 63, No. 6, Nov. 1969.
2. Ford, C., "Halon 1301 Computer Fire Test Program—Interim Report," 1972, E. I. duPont de Nemours & Co., Inc., Wilmington, DE.
3. "The Halogenated Extinguishing Agents," *NFPA Quarterly*, Vol. 48, No. 8, Part 3, Oct. 1954.
4. Hough, R., "Determination of a Standard Extinguishing Agent for Airborne Fixed Systems," WADD Technical Report 60-552, 1960, Wright Air Development Division, Wright-Patterson Air Force Base, OH.
5. Ford, C., "Overview of Halon 1301 Systems," *Symposium* on the Mechanism of Halogenated Extinguishing Agents, ACS Symposia Series, 1962.
6. duPont, duPont Haskell Laboratory, unpublished data, 1969.
7. duPont, duPont Haskell Laboratory, unpublished data, 1970.
8. Beck, P. S., Clark, D. G., and Tinston, D. J., "The Pharmacologic Actions of Bromochlorodifluoromethane (BCF)," *Toxicology and Applied Pharmacology*, pp. 24, 20–29, 1973.
9. Hine, "Clinical Toxicologic Studies on Freon® FE 1301," Report No. 1, The Hine Laboratories, Inc., San Francisco, CA, 1968. (Unpublished.)
10. duPont, duPont Haskell Laboratory, unpublished data, 1966.
11. duPont, duPont Haskell Laboratory, unpublished data, 1966.
12. Stewart, R. D., Newton, P. E., Wu, A., Hake, C. L., and Krivanek, N. D., "Human Exposure to Halon 1301," Medical College of Wisconsin, Milwaukee, WI, 1978. (Unpublished)
13. duPont, duPont Haskell Laboratory, unpublished data, 1976.
14. duPont, duPont Haskell Laboratory, unpublished data, 1978.
15. Sax, N., *Dangerous Properties of Industrial Materials*, 6th ed., Van Nostrand Reinhold, New York, 1984.
16. Ford, C., "Extinguishment of Surface and Deep-Seated Fires with Halon 1301," *An Appraisal of Halogenated Fire Extinguishing Agents*, National Academy of Sciences, Washington, DC, 1972.
17. *Guide to OSHA Fire Protection Regulations*, Vol. 1, 2nd ed., National Fire Protection Association, Quincy, MA, pp. 22140–22142, 1972.
18. "Halon Fire Extinguishing Agents Technical Options Report," Aug. 11, 1989, United Nations Environment Program, New York.
19. Curry, T. H., "Halon 1301 Protects Racing Cars," *Fire Journal*, Vol. 67, No. 2, Mar. 1973.
20. Wharry, D., and Hirst, R., *Fire Technology: Chemistry and Combustion*, Institution of Fire Engineers, Leicester, UK, 1974.
21. Grabowski, G. J., "Fire Detection and Actuation Devices for Halon Extinguishing Systems," *An Appraisal of Halogenated Fire Extinguishing Agents*, National Academy of Sciences, Washington, DC, 1972.
22. Wickham, R. T., "Engineering and Economic Aspects of Halon Extinguishing Equipment," *An Appraisal of Halogenated Fire Extinguishing Agents*, National Academy of Sciences, Washington, DC, 1972.
23. Jensen, R., "Halogenated Extinguishing Agent Systems," *Fire Journal*, Vol. 66, No. 3, May 1972, pp. 37–39.
24. Grant, C., "Enclosure Integrity Procedure for Halon 1301 Total Flooding Fire Suppression Systems," National Fire Protection Research Foundation, Quincy, MA, 1989.

References

DiNenno, P. J., and Budnick, E. K., *Halon 1301 Discharge Testing—A Technical Analysis*, National Fire Protection Research Foundation, Quincy, MA, 1988.
"Halon Fire Extinguishing Agents Technical Options Report," Aug. 1989, United Nations Environment Program, Nairobi, Kenya.

Strasiak, R., "The Development of Bromochloromethane (CB)," WADC Technical Report 53-279, 1954, Wright Air Development Center, OH.

NFPA Codes, Standards, and Recommended Practices

Reference to the following NFPA codes, standards, and recommended practices will provide further information on halogenated agents and systems discussed in this chapter. (See the latest version of *The NFPA Catalog* for availability of current editions of the following documents.)

NFPA 10, *Standard for Portable Fire Extinguishers*
NFPA 12A, *Standard on Halon 1301 Fire Extinguishing Systems*
NFPA 75, *Standard for the Protection of Electronic Computer/Data Processing Equipment*

Additional Reading

Adcock, J. R., *et al.*, "Fluorinated Ethers: A New Family of Halons?" *Proceedings* of the 1991 Halon Alternatives Technical Working Conference, April 30–May 1, 1991, Albuquerque, NM, 1991, pp. 83–96.

Anderson, S. O., "Halons and the Stratospheric Ozone Issue," *Fire Journal*, Vol. 81, No. 3, May 1987, p. 56.

Anderson, S. O., *et al.*, "Halons, Stratospheric Ozone and the U.S. Air Force," *Military Engineer*, Vol. 80, No. 523, Aug. 1988, pp. 485–492.

Anderson, S. O. and Mauzerall, D. L., "Halon Phase-Out in the USA," *Fire Protection*, Vol. 17, No. 4, December 1990, pp. 23–25.

Ashmore, F., "How Hazardous Is Halon 1301?" *Fire Prevention*, No. 209, May 1988, pp. 34–36.

Bailey, B., "Halons: The Situation in a Nutshell," *Fire Fighting in Canada*, Vol. 38, No. 3, April 1994, pp. 32–33.

Baker, R. W., *et al.* "Membrane Vapor Separation Systems for the Recovery of CFCs, Halons and Their Alternatives," *Proceedings* of the 1992 International CFC and Halon Alternatives Conference on Stratospheric Ozone Protection for the 90's, September 29–October 1, 1992, Washington, DC, 1992, pp. 869–877.

Boyce, M. G., "Halon Systems and the COSHH Regulations," *Fire Surveyor*, Vol. 19, No. 3, June 1990, pp. 14–15.

Brabson, G. D., Tapscott, R. E. and Morehouse, E. T., "Initial Fire Suppression Reactions of Halons. Phase 1. Development of Experimental Approach," Final Report, March 1988–June 1988, New Mexico Univ., Albuquerque, Air Force Engineering and Services Center, Tnydall AFB, FL, ESL-TR-89-50, WA3-67(3.36), September 1991, 67 pages.

Brashear, W. T., "QSAR Evaluation of Halon 1301 (CF3Br) and CF3I," Interim Report, September-November 1993, Health Design Inc., Rochester, NY, Armstrong Laboratory, Wright-Patterson AFB, OH, January 1994, 21 pages.

Carhart, H. W., *et al.*, "Discharge System Tests of Halon 1301 Test Gas Simulants," Interim Report, June 1988–September 1988, Naval Research Lab., Washington, DC, Hughes Associates Inc., Wheaton, MD, Naval Sea Systems Command, Washington, DC, NRL Memorandum Report 6898, September 19, 1991, 115 pages.

Catchpole, D. V., "Halon Recycling corporation: A U.S. Approach to Halon Bank Management," 1993 International CFC and Halon Alternatives Conference on Stratospheric Ozone Protection for the 90's. October 20–22, 1993, Washington, DC, 1993, pp. 773–779.

Chines, S. A., "Halon Design Discharge Testing—An Authority Having Jurisdiction Point of View," *Seminar Paper* for Fire Protection Halons and the Environment, NFPA Annual Meeting, Cincinnati, OH, May 1987.

"Chubb Plans to Phase Out Portable Halons," *Fire Prevention*, Vol. 229, May 1990, p. 5.

Clay, J. T., "Fundamental Studies on the Chemical Mechanism of Flame Extinguishment by Halons," *Proceedings* of the 1991 Halon Alternatives Technical Working Conference, April 30–May 1, 1991, Albuquerque, NM, 1991, pp. 97–108.

Cohn, B. M., "Laboratory Halon Systems: Rehab, Replace or Remove?" *Journal of Applied Fire Science*, Vol. 5, No. 1, 1995/1996, pp. 45–54.

Connell, E., "Essential Uses of Halon," VHS Video Tape, Department of Energy, Washington, DC, Federal Fire Forum, Developments in Fire Protection, November 6, 1992, Gaithersburg, MD, 1992.

Crews, G., "Halon 1211 Fire Protection Conservation and Recycling in the U.S. Marine Corps," *Proceedings* of the 1992 International CFC and Halon Alternatives Conference on Stratospheric Ozone Protection for

the 90's, September 29–October 1, 1992, Washington, DC, 1992, pp. 691–693.

Crews, G. W., "Halon 1211 Conservation on a Military Base," 1993 International CFC and Halon Alternatives Conference on Stratospheric Ozone Protection for the 90's. October 20–22, 1993, Washington, DC, 1993, pp. 792–794.

Davey, P. and Greig, A., "Halon Issue is in the Bank," *Fire Protection*, Vol. 21, No. 1, March 1994, pp. 6–7.

Daws, S., "Halon Users Face Difficult Dilemma Over 'Essential Uses' Definition," *Fire*, Vol. 83, No. 1027, January 1991, pp. 6, 8.

DiNenno, P., and Budnick, E., "Halon 1301 Discharge Testing: A Technical Analysis," U.S. Environmental Protection Agency and National Fire Protection Research Foundation, Quincy, MA, 1988.

DiNenno, P., and Taylor, G. M., "Best and Essential Halon Use: A Methodology," National Fire Protection Research Foundation, Quincy, MA, Dec. 1989.

DiNenno, P. J., *et al.*, "Evaluation of Halon 1301 Test Gas Simulants: Enclosure Leakage," *Fire Technology*, Vol. 25, No. 1, Feb. 1989, pp. 24–40.

DiNenno, P. J., *et al.*, "Halon 1301 Test Gas Simulants," *Naval Engineers Journal*, Vol. 103, January 1991, pp. 59–69.

DiNenno, P., *et al.*, "Hydraulic Performance Tests of Halon 1301 Test Gas Simulants," *Fire Technology*, Vol. 26, No. 2, May 1990, pp. 121–140.

DiNenno, P. J., *et al.*, "Shipboard Tests of Halon 1301 Test Gas Simulants," February 1989-September 1989, Naval Research Lab., Washington, DC, NRL Memorandum Report 6708, August 22, 1990, 113 pages.

Eckholm, W. A., "Testing One of the World's Biggest Halon Systems. . . Without Using Halon," *Fire Journal*, Vol. 83, No. 4, July 1989, p. 71.

"Extinguishing a Threat to the Atmosphere: Halon 1211 Recovery/Recharge Equipment," *UL Lab Data*, Vol. 22, No. 2, 1992, pp. 4–7.

Floden, J. R., Scheil, G. and Klamm, S., "Final Firefighter Exposure Results: A Comparison of Halon 1211, HCFC 123 and PFH," *Proceedings* of the 1993 Halon Alternatives Technical Working Conference, May 11–13, 1993, Albuquerque, NM, 1993, pp. 645–658.

Franchuk, D., "Halon Update," *FACTs*, Vol. 4, No. 1, Winter 1995, pp. 1–3.

Gann, R. G., "Potentiating the Search for Alternatives to Halons 1301 and 1211," CFR Video Seminar, National Institute of Standards and Technology, Gaithersburg, MD, Video, November 13, 1990.

Gann, R. G., *et al.*, "Preliminary Screening Procedures and Criteria for Replacements for Halons 1211 and 1301," National Institute of Standards and Technology, Gaithersburg, MD, Air Force Engineering Lab., Tyndall AFB, FL NIST TN 1278, August 1990.

Goodall, K., "Halon Fire Protection: Are Discharge Tests Necessary," *Fire Surveyor*, Vol. 17, No. 4, Aug. 1988, pp. 57–62.

Grant, C. C., "Computer-Aided Halon 1301 Piping Calculations," *Fire Safety Journal*, Vol. 9, No. 2, 1986, pp. 171–180.

Grant, C. C., "Controlling Fire Protection Halon Emissions," *Fire Technology*, Vol. 24, No. 1, 1988.

Grant, C. C., "Enclosure Integrity Procedure," National Fire Protection Research Foundation, Quincy, MA, Jan. 1989.

Grant, C. C., "Fire Protection Halons and the Environment: An Update Symposium," *Fire Technology*, Vol. 24, No. 1, Feb. 1988, p. 63.

Grant, C. C., "Halon and Beyond: Developing New Alternatives," *NFPA Journal*, Vol. 88, No. 6, November/December 1994, pp. 40–43, 45–47, 49–54.

Grant, C. C., "Halons and the Ozone Layer: An Overview," *Fire Journal*, Vol. 83, No. 5, Sept. 1989, p. 58.

Grant, C. C., "Halon Design Calculations," *The SFPE Handbook of Fire Protection Engineering*, 2nd ed., DiNenno, P. J., *et al.*, eds., National Fire Protection Association, Quincy, MA, 1995.

Grant, C. C., "Life Beyond Halon," *Fire Journal*, Vol. 84, No. 3, May/June 1990, pp. 52–54, 56, 58–59.

"Halon in the 90s? Use of Halons for Firefighting Will Continue," *Burning Issues*, Vol. 10, No. 2, Summer 1988.

"Halon Update: New EPA Rules Define Acceptable Halon Substitutes," *Consulting Specifying Engineer*, Vol. 16, No. 1, July 1994, p. 53.

"Halonisammuttimien tarkastus, huolto ja kayttotarkoituksen muuttaminen. [Inspection, Maintenance and Alteration of Halon Extinguishers.]," *Palontorjuntatekniikka*, February 1993, p. 16.

"Halons Still the Best," *New Scientist*, Vol. 134, No. 1824, June 6, 1992, p. 19.

"Halons To Be Phased Out," *Fire Protection*, Vol. 17, No. 3, September 1990, pp. 20–21.

Hanauska, C. P., "Coping With the Halon 1301 Production Phase-Out," *SFPE Bulletin*, September/October 1993, pp. 9–12.

Harrington, J. L., "Halon Phaseout Speeds Up," *NFPA Fire Journal*, Vol. 87, No. 2, March/April 1993, pp. 38–42.

Hebert, J., "Essential Uses of Fire Protection Halons," *Fire Systems*, Vol. 5, No. 2, 1991, pp. 20-32; Halon Alternatives Technical Working Conference Proceedings, April 30–May 1, 1991, Albuquere, NM, 1991, pp. 245–259.

Hillaert, J. A. and Trench, R. C., "Recovery of Halon 1301 Superpressurized With Nitrogen: A Technology Review," *Proceedings* of the 1992 International CFC and Halon Alternatives Conference on Stratospheric Ozone Protection for the 90's, September 29–October 1, 1992, Washington, DC, 1992, pp. 673–679.

Hoke, S. H. and Herud, C., "Real-Time Analysis of Halon Degradation Products," *Proceedings* of the 1993 Halon Alternatives Technical Working Conference, May 11–13, 1993, Albuquerque, NM, 1993, pp. 185–203.

Holtham, P. D., "Australia's Answer to Reducing the Use of Halons," *Fire Engineers Journal*, Vol. 50, No. 156, March 1990, p. 18.

Hommstedt, G., Andersson, P and Andersson, J., "Investigation of Scale Effects on Halon and Halon Alternatives Regarding Flame Extinguishing, Inerting Concentration and Thermal Decomposition Products," *Proceedings* of the Fourth International Symposium on Fire Safety Science, Int'l. Assoc. for Fire Safety Science, Boston, MA, 1994, pp. 853–864.

Hough, E., "Halons Use: The Big Question. . . But Is There a Real Answer?" *Fire International*, No. 138, February 1993, pp. 47–49.

Humphrey, B. J., Smith, B. R. and Skaggs, S. R., "Toxicity of Halon 2402," Final Report, August 1985–September 1986, New Mexico Engineering Research Inst., Albuquerque, NM, Air Force Engineering and Services Center, Tyndall AFB, FL, ESL-TR-88-59; WA3-82(3.18), September 1990, 42 pages.

Huston, P. O., "Halon 1211 Recycling Initiatives in ULI and NFPA Standards," 1993 International CFC and Halon Alternatives Conference on Stratospheric Ozone Protection for the 90's. October 20–22, 1993, Washington, DC, 1993, pp. 790–791.

Huston, P. O., "Modular Concept of Halon 1301 Recycling," 1993 International CFC and Halon Alternatives Conference on Stratospheric Ozone Protection for the 90's. October 20–22, 1993, Washington, DC, 1993, pp. 795–799.

Jacobson, E., Bar, I. and Rosenwaks, S., "Neutralizing Halogenated Hydrocarbons (CFC and HALONS)," *Proceedings* of the 1993 Halon Alternatives Technical Working Conference, May 11–13, 1993, Albuquerque, NM, 1993, pp. 583–596.

Jennings, P., "Insufficient Enclosure Sealing Can Compromise Halon's Effect," *Fire*, Vol. 84, No. 1037, November 1991, pp. 26, 32.

Kibert, C. J., "Halon 1211/1301 Replacement Program Status: U.S. Air Force Ground Based Fire Suppression Applications," *Proceedings* of the 1993 Halon Alternatives Technical Working Conference, May 11–13, 1993, Albuquerque, NM, 1993, pp. 519–527.

Kiljunen, M., "Attempts to Decrease Use of Halons," *Palontorjuntatekniikka*, January 1991, p. 17.

Laakkonen, E., "Ratkaisu haloniongelmann? [Solution to the Halon Problem?]," Riskienhallintaosasto Pohjota-yhtiot, *Palontorjuntateknikka*, Vol. 3, 1990, pp. 32–33.

Lavid, M., *et al.*, "Practical Conversions of CFCs and Halons by Selective Photothermal Hydrodehalogention (PTH)," Final Report, May 1991-November 1991, M. L. Energia, Inc., Princeton, NJ, Civil Engineering Laboratory, Tyndall AFB, FL, CEL-TR-92-03; MLE-113091, May 1992, 94 pages. [AUTHORIZED TO U.S. GOVERNMENT AGENCIES ONLY]

Lawson, D., "Halon Protected Rooms: Air Leakage Testing, Identification and Sealing," *Fire Surveyor*, Vol. 20, No. 2, April 1991, pp. 26–30.

Leonard, J., "Training Simulant for Halon 1211 Portable Extinguishers," Naval Research Laboratory, Washington, DC, NRL/MR/6180-94-7615, September 8, 1994, 35 pages.

"Life After Halon," *Fire International*, No. 145, Winter 1994, p. 35.

"Living Without Halons and CFCs," *Record*, Vol. 69, No. 4, July/August 1992, pp. 3–8.

Maier, B., "HAL: A Complete System for Halon 1301," *Proceedings* of the 1992 International CFC and Halon Alternatives Conference on Stratospheric Ozone Protection for the 90's, September 29–October 1, 1992, Washington, DC, 1992, pp. 671–672.

Mallard, W. G., "Property Estimation of Halons," *Proceedings* of the 1991 Halon Alternatives Technical Working Conference, April 30–May 1, 1991, Albuquerque, NM, 1991, pp. 269–276.

Mauzerall, D. L., "Protecting the Ozone Layer: Phasing Out Halon by 2000," *Fire Journal*, Vol. 84, No. 5, September/October 1990, pp. 22–24, 26–31.

Maynard, M. J., "NASA's Approach to Halon Issues," *Proceedings* of the 1992 International CFC and Halon Alternatives Conference on Stratospheric Ozone Protection for the 90's, September 29–October 1, 1992, Washington, DC, 1992, pp. 615–622.

Merrick, D., "Methodology for the Analysis of Special Hazard Halon 1301 Systems and Replacement Alternatives," *Proceedings* of the 1993 Halon Alternatives Technical Working Conference, May 11–13, 1993, Albuquerque, NM, 1993, pp. 465–472.

Metchis, K., "Regulation of Halons and Halon Substitutes: An Update," *Proceedings* of the 1993 Halon Alternatives Technical Working Conference, May 11–13, 1993, Albuquerque, NM, 1993, pp. 11–30.

Mitchell, M. D., "Feasibility of Systematic Recycling of Aircraft Halon Extinguishing Agents," Final Report, Walter Kidde Aerospace, Inc., Wilson, NC, Federal Aviation Administration, Atlantic City International Airport, NJ, DOT/FAA/CT-91-21, December 1991, 224 pages.

"Montrealin poltakirja maarittelee otsonin-kayton rajoitukset maailmanlaajuisesti. [Current Legislative Restrictions on the Use of Halons.]," *Palontorjuntatekniikka*, February 1993, p. 15.

Moore, T. A. and McCarson, T. D., "Halon and CFC Destruction, Recovery, Recycling, and Reclamation Technologies," *Proceedings* of the 1991 Halon Alternatives Technical Working Conference, April 30–May 1, 1991, Albuquerque, NM, 1991, pp. 109–121.

Morehouse, T., "Halon Update," VHS Video Tape, Department of the Air Force, Washington, DC, Federal Fire Forum, Developments in Fire Protection, November 6, 1992, Gaithersburg, MD, 1992.

Mulhaupt, R., "Beyond Fire Protection: The Halons and the Ozone Layer," *Fire Protection*, Vol. 17, No. 2, June 1990, pp. 24–26, 28–29.

Pallavicini, M. R., "Information Technology Protection After Halon," *Proceedings* of the First International Conference on Fire Science and Engineering, ASIAFLAM '95, March 15–16, 1995, Kowloon, Hong Kong, 1995, pp. 583–585.

Parsons, M., "Replacements for Halon 1301 in Total Flooding Fire Suppression Systems," *Firehouse*, Vol. 18, No. 2, December 1993, pp. 28–29.

Parsons, M. and Meserve, W., "Halon 1301 Recycling System," *Proceedings* of the 1992 International CFC and Halon Alternatives Conference on Stratospheric Ozone Protection for the 90's, September 29–October 1, 1992, Washington, DC, 1992, pp. 657–666.

Parsons, M. L., "Halon 1211: Its Current and Future Status," *Industrial Fire World*, Vol. 9, No. 4, July/August 1994, pp. 20–29.

Pignato, J. A., "How to Have the Effect of Halon: Without Using It," *Fire*, Vol. 87, No. 1072, October 1994, pp. 27–28, 32.

Pinto, F., "Halons: 1211 Recycling," *Proceedings* of the 1992 International CFC and Halon Alternatives Conference on Stratospheric Ozone Protection for the 90's, September 29–October 1, 1992, Washington, DC, 1992, p. 818.

Pitts, W. M., *et al.*, "Construction of an Exploratory List of Chemicals to Initiate the Search for Halon Alternatives," National Institute of Standards and Technology, Gaithersburg, MD, Air Force Engineering Lab., Tyndall AFB, FL NIST TN 1279, August 1990.

"Plumbers Pull Plug on Halon Extinguishers," *Journal of the Institution of Engineers*, Vol. 61, No. 19, Oct. 6, 1989, p. 21.

"Post-Halon Computer Protection," *Record*, Vol. 70, No. 5, Nov./Dec., pp. 3–9.

"Protection for Computers Without Using Halons," *Fire*, Vol. 87, No. 1072, October 1994, pp. 35–36.

"Questions and Answers on Halons and Their Substitutes," *Sentinel*, Vol. 50, No. 4, Fourth Quarter 1993, pp. 12–15.

"Relationship Between Halons, Chlorofluorocarbons and the Ozone Layer," *Building Official and Code Administrator*, Vol. 26, No. 5, Sept./Oct. 1992, pp. 29–33.

"Report on the Halon Technical Options Committee," *Meeting* of the NFPA Technical Committee on the Halogenated Fire Extinguishing Systems, Washington, DC, May 1989.

Riley, J. F., "Fire Protection in the '90s Without Halon 1211 and Halon 1301," *Fire Systems*, Vol. 5, No. 2, 1991, pp. 12–19; Halon Alternatives Technical Working Conference Proceedings, April 30–May 1, 1991, Albuquere, NM, 1991, pp. 1–15.

Rozman, T., "Fire Protection for Floating Roof Tanks With Halon 1211 Automatic Extinguishing System," *Industrial Fire World*, Vol. 4, No. 6, Dec. 1989/Jan. 1990, pp. 4–5, 25, 27.

Sanderson, L. E., "Halon 1211 Recycling and Conservation Strategies for Developing Countries Working Together 'What We Can Accomplish Today'," *Proceedings* of the 1992 International CFC and Halon Alternatives Conference on Stratospheric Ozone Protection for the 90's, Sept. 29-Oct. 1, 1992, Washington, DC, 1992, pp. 685–690.

Seaton, M., "Halon: Searching for Solutions," *NFPA Journal*, Vol. 89, No. 6, November/December 1995, pp. 46–47, 49–50, 53.

Senecal, J. A., "Explosion Protection in Occupied Spaces: The Status of Suppression and Inertion Using Halon and Its Descendants," 1993 International CFC and Halon Alternatives Conference on Stratospheric Ozone Protection for the 90's. Oct. 20–22, 1993, Washington, DC, 1993, pp. 767–772.

Sheinson, R. S., *et al.*, "CF3I: Halon 1301 Total Flooding Fire Extinguishment Comparison," Naval Research Laboratory, Washington, DC, GEO-CENTERS, Inc., Fort Washington, MD, Hughes Associates, Inc., Columbia, MD, Naval Sea Systems Command, Washington, DC, NIS-TIR 5499, Sept. 1994; National Institute of Standards and Technology, Annual Conference on Fire Research: Book of Abstracts, October 17-20, 1994, Gaithersburg, MD, 1994, pp. 59–60.

Sheinson, R. S., *et al.,* "From Halon 1301 to a Replacement Agent: Integrated Development of a Fire Protection System," Naval Research Laboratory, Washington, DC, ISBN 0-9516320-9-4; *Proceedings* of the Seventh International Interflam Conference, Interflam '96, March 26–28, 1996, Cambridge, England, Interscience Communications Ltd., London, England, 1996, pp. 459–466.

Simpson, K., "Practical Guide to Halon Phase-Out," *Fire Safety Journal*, Vol. 2, No. 6, Dec. 1995, pp. 13–14.

Sirola, J., "Halonien kayton rajoittaminen paloviranomaisten nakokulmasta. [Restrictions on Halons From the Point of View of the Fire Official.]," *Palontorjuntatekniikka*, February 1993, p. 14.

"Special Hazards Protection—Manufacturers and Equipment," *Fire Journal*, Vol. 81, No. 4, July–Aug. 1987, p. 43.

Steciak, J. and Zalosh, R. G., "Reliability Methodology Applied to Halon 1301 Extinguishing Systems in Computer Rooms," Factory Mutual Research Corp., Norwood, MA, Worcester Polytechnic Inst., MA, ASTM STP 1150; American Society for Testing and Materials, Fire Hazard and Fire Risk Assessment, Sponsored by ASTM Committee E-5 on Fire Standards, ASTM STP 1150, Dec. 3, 1990, San Antonio, TX, ASTM, Philadelphia, PA, 1990, pp. 161–182.

Taflinger, N., "Aircraft Halon Use and Management in the Air Force," *Proceedings* of the 1993 Halon Alternatives Technical Working Conference, May 11–13, 1993, Albuquerque, NM, 1993, pp. 225–238.

"Taking Steps to Reduce Halon Emissions," *Fire Prevention*, No. 224, Nov. 1989, pp. 16–17.

Tapscott, R., "Banking Halon After Its Phase-Out," *Fire Protection*, Vol. 19, No. 4, Dec. 1992, pp. 16–17.

Tapscott, R., "Chemical Options to Halon for Aircraft Use," Final Report, Federal Aviation Administration, Atlantic City International Airport, NJ, DOT/FAA/CT-95/9, ACD-240, February 1995, 62 pages.

Taylor, G., "Achieving the Best Use of Halons," *Fire Journal*, Vol. 81, No. 3, May–June 1987, p. 69.

Tenhunen, H., "Halonien kaytto sammutusaineena. [Halons in Fire Fighting.]," *Palontorjuntatekniikka*, February 1993, pp. 12–13.

Tsang, W., Miziolek, A. W., Finnerty, A. E., "Kinetic Aspects of Flame Inhibition by Halons and Halon Alternative Compounds," *Proceedings* of the 1993 Halon Alternatives Technical Working Conference, May 11–13, 1993, Albuquerque, NM, 1993, pp. 247–255.

Ulmer, P. E., "Halon 1301 Use in Oil and Gas Production Facilities: Alaska's North Slope," *Proceedings* of the 1991 Halon Alternatives Technical Working Conference, April 30–May 1, 1991, Albuquerque, NM, 1991, pp. 240–244.

VanTiggelen, P. J. and Thill, L., "Halon 1301 and Gaseous Detonations," *Proceedings* of the 1993 Halon Alternatives Technical Working Conference, May 11–13, 1993, Albuquerque, NM, 1993, pp. 481–491.

Varanka, V., "Taydellista halonin korviketta ei viela ole valmiina. [No Perfect Substitute for Halon.]," *Palontorjuntatekniikka*, Vol. 4, 1994, pp. 16–18.

Verkonik, D. P., "ASTM Emergency Standard, ES-24, Specification for Halon 1301, Bromotrifluoromethane (CF3Br)," *Proceedings* of the 1993 Halon Alternatives Technical Working Conference, May 11–13, 1993, Albuquerque, NM, 1993, pp. 191–203.

Wackerlig, H. U., "Less Halon at Equal Safety Level," *Proceedings* of the 1992 International CFC and Halon Alternatives Conference on Stratospheric Ozone Protection for the 90's, September 29–October 1, 1992, Washington, DC, 1992, pp. 623–629.

Walters, E. A., *et al.,* "Initial Fire Suppression Reactions of Halons. Phase 3. Molecular Beam Experiments Involving Halon Clusters," Final Report, June 1988–April 1989, New Mexico Univ., Albuquerque, Engineering and Services Lab., Tyndall AFB, FL, ESL-TR-89-50, SS 2.10(1), September 1990.

Walters, E. A., *et al.,* "Initial Fire Suppression Reactions of Halons. Phase 2. Verification of Experimental Approach and Initial Studies," Final Report, June 1988–April 1989, New Mexico Univ., Albuquerque, Engineering and Services Lab., Tyndall AFB, FL, ESL-TR-89-50, SS 2.08(2), September 1990.

Waters, S., "Alternative Test Agent for Halon Systems," *Fire Systems*, Vol. 2, 1987.

Warner-Howard, D. and Horsman, G., "Safety Without Halon," *Fire Prevention*, No. 277, March 1995, pp. 31–33.

Whiteley, R. A., "Fan Testing for Halon Protected Enclosures," *Fire Surveyor*, Vol. 20, No. 2, April 1991, pp. 23–25.

Whiteley, R. A., "Integrity Testing for Halon—The State of the Art," *Fire Surveyor*, Vol. 18, No. 5, Oct. 1989, pp. 17–20.

Willey, E. A., "Summary of International Conference on Fire Protection Halons and the Environment," National Fire Protection Association, Quincy, MA, Aug. 1988.

Womeldorf, C. A., "Recommendations for Further Testing: Simulants for Halon 1301: Phase 1," Letter Report, National Institute of Standards and Technology, Gaithersburg, MD, Naval Air Warfare Center, Lakehurst, NY, 1994.

Womeldorf, C. A., "Experimental Results and Recommendations for Halon 1301 Simulant. Phase 2," Letter Report, National Institute of Standards and Technology, Gaithersburg, MD, Naval Air Warfare Center, Lakehurst, NY, Jan. 20, 1995.

Womeldorf, C. A., Yang, J. C. and Grosshandler, W. L., "Halon 1301 Surrogates for Engine Nacelle Fire Suppression System Certification," National Institute of Standards and Technology, Gaithersburg, MD, Naval Air Warfare Center, Lakehurst, NY, NISTIR 5499, September 1994; National Institute of Standards and Technology, Annual Conference on Fire Research: Book of Abstracts, Oct. 17–20, 1994, Gaithersburg, MD, 1994, pp. 9–10.

Yang, J. C., Breuel, B. D. and Grosshandler, W. L., "Solubilities of Nitrogen and Freon-23 in Alternative Halon Replacement Agents," *Proceedings* of the 1993 Halon Alternatives Technical Working Conference, May 11–13, 1993, Albuquerque, NM, 1993, pp. 107–114.

Yang, J. C., *et al.,* "Storage and Discharge Characteristics of Halon Alternatives," *Proceedings* of the 1995 International CFC and Halon Alternatives Conference and Exhibition on Stratospheric Ozone Protection for the 90's, Oct. 21–23, 1995, Washington, DC, 1995, pp. 594–603.

Young, R., "Effective Alternatives to Halon 1301," *Fire Prevention*, No. 289, May 1996, pp. 16–18.

Young, R., "Evaluating Halon 1301 Alternatives," *Fire Prevention*, No. 267, March 1994, pp. 34–37.

Young, R., "Gaseous Fire Extinguishing Agents as Halon 1301 Alternatives," *Fire Prevention*, No. 277, March 1995, pp. 24–27.

Young, R., "Is There Life After Halon?" *Fire Prevention*, No. 234, Nov. 1990, pp. 18–21.

Young, R., "Halons: The Current Position," *Fire Prevention*, No. 252, Sept. 1992, pp. 16, 18.

Ziemba, J., "Halon Retrofit Technologies: Gazing Into the Crystal Ball," *Fire Systems*, Vol. 5, No. 2, 1991, pp. 4–11.

Zlochower, I. A. and Hertzberg, M., "Inerting of Methane-Air Mixtures by Halon 1301 (CF3Br) and Halon Substitutes," *Proceedings* of the 1991 Halon Alternatives Technical Working Conference, April 30–May 1, 1991, Albuquerque, NM, 1991, pp. 122–130.

Zoi, D., "Halon 1211 Recycling Seminars for Developing Countries," *Proceedings* of the 1992 International CFC and Halon Alternatives Conference on Stratospheric Ozone Protection for the 90's, Sept. 29–Oct. 1, 1992, Washington, DC, 1992, pp. 681–683.

DIRECT HALON REPLACEMENT AGENTS AND SYSTEMS

Philip J. DiNenno

The regulation of Halon 1301 under the Montreal Protocol and its amendments culminated in the phaseout of production of halons in the developed countries on December 31, 1993. This regulation engendered tremendous research and development efforts across the world in a search for replacements and alternatives. There are currently over 12 commercialized total flooding, clean agent alternatives to Halon 1301, and development continues on others. In addition to clean agent total flooding gaseous alternatives, new technologies, e.g., water mist and fine solid particulate, are being introduced. This chapter focuses on total flooding clean agent halon replacements.

Table 6-19A is a summary of the most important halocarbon and inert gas extinguishing agents developed to date. It gives the chemical name, trade name, American Society of Heating, Refrigeration and Air-Conditioning Engineers (ASHRAE) designation for halocarbons, and chemical formula.

The primary design and installation standard used in North America is NFPA 2001, *Standard on Clean Agent Fire Extinguishing Systems* (hereinafter referred to as NFPA 2001). The first edition of the standard was published in 1994. The second and current edition was issued on February 2, 1996. International Standards Organization (ISO) efforts are underway, and a draft ISO standard is being prepared by ISO TC21/SC5/WG8.

Clean fire suppression agents are defined as fire extinguishants that vaporize readily and leave no residue.[1] Clean agent halon replacements fall into two broad categories: (1) halocarbon compounds and (2) inert gases and mixtures. Halocarbon replacements include compounds containing carbon, hydrogen, bromine, chlorine, fluorine, and iodine. They are grouped into five categories: (1) hydrobromofluorocarbons (HBFC), (2) hydrofluorocarbons (HFC), (3) hydrochlorofluorocarbons (HCFC), (4) perfluorocarbons (FC or PFC), and (5) fluoroiodocarbons (FIC).

While the characteristics of halocarbon clean agents vary widely, they share several common attributes:

1. All are electrically nonconductive.
2. All are clean agents; they vaporize readily and leave no residue.
3. All are liquefied gases or display analogous behavior (e.g., compressible liquid).

Philip J. DiNenno, P.E., is vice president of Hughes Associates, Inc., a fire protection engineering research and development firm. He has been actively involved in the testing and development of halon replacement chemicals and alternative fire suppression technologies.

4. All can be stored and discharged from typical Halon 1301 hardware [with the possible exception of HFC-23, which more closely resembles 600 psig (40 bar) superpressurized halon systems].
5. All (except HFC-23) use nitrogen superpressurization in most applications for discharge purposes.
6. All are less efficient fire extinguishants than Halon 1301, in terms of storage volume and agent weight; the use of most of these agents requires increased storage capacity.
7. All are total flooding gases after discharge; many require additional care relative to nozzle design and mixing.
8. All produce more decomposition products (primary HF) than Halon 1301, given similar fire type, fire size, and discharge time.
9. All are more expensive at present than Halon 1301 on a weight (mass) basis.

Inert gas alternatives include nitrogen and argon, and blends of these. One inert gas replacement has a small fraction of carbon dioxide. Carbon dioxide is not an inert gas, because it is physiologically active and fatal at low concentrations (approximately 9 percent). Inert gas clean agents are stored as pressurized gases and, hence, require substantially greater storage volume. They are electrically non-conductive, form stable mixtures in air, and leave no residue.

EXTINGUISHING MECHANISMS

Halocarbon clean agents extinguish fires by a combination of chemical and physical mechanisms, depending upon the compound. Chemical suppression mechanisms of HBFC and HFIC compounds are similar to Halon 1301; i.e., the Br and I species scavenge flame radicals, thereby interrupting the chemical chain reaction. Other replacement compounds suppress fires primarily by extracting heat from the flame reaction zone, reducing the flame temperature below that which is necessary to maintain sufficiently high reaction rates by a combination of heat of vaporization, heat capacity, and the energy absorbed by the decomposition of the agent.

Burgess *et al.*[2] performed kinetic simulations on a range of fluorinated compounds (CF_4, CF_3H, CF_2H_2, CF_3-CF_3, CF_3-CF_2H, and CF_3-CFH_2) for chemical kinetic fire suppression effects of fluorine compounds. Of the agents examined, only CF_3H was seen to have any chemical effect on flame suppression. They concluded that the differences in flame suppression properties can be explained by differences in heat capacity and relative amounts of heat release during agent combustion. Battin-LeClerc *et al.*[3] indicate that, while clearly some noncatalytic flame inhibition may occur with FC and HFC

TABLE 6-19A. *Commercialized Halon Replacement Nomenclature*

Chemical Name	Trade Name	Designation	Formula
Perfluorobutane	CEA-410	FC-3-1-10	C_4F_{10}
Heptafluoropropane	FM-200	HFC-227ea	C_3F_7H
Trifluoromethane	FE-13	HFC-23	CHF_3
Chlorotetrafluoroethane	FE-24	HCFC-124	C_2HCIF_4
Pentafluoroethane	FE-25	HFC-125	C_2HF_5
Dichlorotrifluoroethane (4.75%) Chlorodifluoromethane (82%) Chlorotetrafluroethane (9.5%) Isopropenyl-1-methylcyclohexene (3.75%)	NAF-SIII	HCFC Blend A	$CHCl_2CF_3$ $CHClF_2$ $CHClFCF_3$
Trifluoroiodide	Triodide	FIC-131I	CF_3I
$N_2/Ar/CO_2$	Inergen	IG-541	N_2 (52%) Ar (40%) CO_2 (8%)
N_2/Ar	Argonite	IG-55	N_2 (50%) Ar (50%)
Argon	Argon	IG-01	Ar (100%)

compounds, this effect is expected to be limited. Richter *et al.*[4] draw similar conclusions, owing to the inability of fluorine atoms to be continuously recycled in the flame, as occurs with bromine.

Oxygen depletion also plays an important role in reducing flame temperature. The energy absorbed in decomposing the agent by breaking fluorine and chlorine bonds is quite important, particularly with respect to decomposition production formation. There is undoubtedly some degree of "chemical" suppression action in flame radical combustion with halogens, but it is considered to be of minor importance, since it is not catalytic (e.g., one F radical combines with one H flame radical).

The lack of significant chemical reaction inhibition in the flame zone by HCFC, HFC, and FC compounds results in higher extinguishing concentrations, relative to Halon 1301. The relative importance of the energy sink, represented by breaking halogen species bonds, results in higher levels of agent decomposition, relative to Halon 1301.

Inert gas agents suppress flames by reducing the flame temperature below thresholds necessary to maintain combustion reactions. This is done by reducing the oxygen concentration and by raising the heat capacity of the atmosphere supporting the flame. For exam-

ple, the addition of a sufficient quantity of nitrogen to reduce the oxygen concentration below 12 percent (in air) will extinguish flaming fires. The agent concentration required (and, hence, the minimum oxygen level) is a function of the heat capacity of the inert gas added. Hence, there are differences in minimum extinguishing concentration between inert gases.

The differences between argon and nitrogen are clearly shown in Table 6-19B, with nitrogen extinguishing concentrations in the range of 30 percent (~14.6 percent O_2), and argon extinguishing concentrations of 41 percent (12.3 percent O_2). Inert gas blends with more nitrogen than argon will have lower minimum extinguishing concentrations. When one considers nitrogen as the primary inert gas in examining oxygen concentrations at extinction in fire experiments, the minimum oxygen concentration at extinction can be as low as 12 percent, or a 42.5 percent "agent" concentration.[5,6]

FLAME SUPPRESSION EFFECTIVENESS

Flame suppression effectiveness of total flooding halon replacement agents has been evaluated in a number of ways. The predominant small-scale test method for establishing flame extinguishing con-

TABLE 6-19B. *n-Heptane Cup Burner Extinguishing Values from Various Investigators (from NFPA 2001* except as noted)*

Reference	Nitrogen	FC-3-1-10	HFC-227ea	HFC-23	Halon 1301	CO_2†	Argon†
Sheinson	30.0†	5.2	6.6	12	3.1	21.0	41.0
3M		5.9	—	—	3.9		
NMERI		5.0	6.3	12.6	2.9		38.0
Senecal, Fenwal		5.5	5.8	12 (13)	3 (3.5)		
Robin			5.9	12.7	3.5		
NIST	32.0	5.3	6.2	12	3.2		
Hamins *et al.*‡						23.0	41.0
Hirst and Booth§						20.5	

* See reference 1.
† See reference 7.
‡ See reference 8.
§ See reference 9.

centrations for liquid and gaseous fuels is the ICI cup burner or variations thereof.

Figure 6-19A is a schematic of the ICI cup burner. A small laminar flame is established above a "cup" of fuel surrounded by a cylindrical chimney. An air/agent mixture flows up the chimney surrounding the flame. The minimum concentration of agent (in air) at which the flame is extinguished is the minimum extinguishing concentration (MEC). There are many variations of the basic device as used by different laboratories; these variations include cup and chimney diameter, different mixing and measuring methods, chimney height, and agent/air mixture velocity past the flame.[10] Table 6-19B gives some indication of the variation in cup burner extinguishing concentration for a range of extinguishment concentrations for a range of agents with n-Heptane as the fuel.

Despite this wide variation in test methods, the minimum extinguishing concentrations measured from these different devices are reasonably close.

The cup burner can also be used to establish MEC for inert gases; Ansul obtained a value of 2.91 percent for IG-541. This is in contrast to concentrations of 32, 41, and 23 percent by volume measured by National Institute of Standards and Technology (NIST) for nitrogen, argon, and carbon dioxide, respectively, the components of Inergen.

NIST has conducted investigations on a wide range of halon replacement chemicals for aviation use. In order to give a wider perspective on the type and range of chemicals being evaluated for fire suppression use, Table 6-19C from reference 8 is included. This table gives cup burner n-Heptane flame extinction data for a wide range of potential halon replacements.

Table 6-19D presents cup burner MEC for a range of fuels and agents taken from various sources. Where multiple values of the MEC were found, they are given. These data should not be used for design purposes without ensuring that the concentrations are consistent with system manufacturer requirements and third-party approvals.

In addition to the cup burner apparatus, researchers at NIST have utilized an opposed flow diffusion flame (OFDF) apparatus to rank halon replacements for fire-extinguishing effectiveness. The OFDF burner is commonly used for combustion research. It has many advantages as a research tool for fundamental combustion studies. Its primary advantage is in the ability to relate the results to fundamental predictions of flame structure and conditions at flame extinction. The oxidizer (and suppressant) stream is forced down onto the fuel surface, exhaust gases are drawn down through an annulus or jacket around the fuel cup, and a flat flame is established.

The OFDF burner can vary the turbulence intensity or strain rate of the flame. For most applications of clean agent fire suppression, this is not a major concern; however, in specialized

TABLE 6-19C. *Agent Fraction in the Oxidizer Stream at Extinction of n-Heptane Cup Burner Flames (from Reference 8)*

Agent Type	Agent	Mass Percent	Volume Percent
Inert	N_2	31	32
	CO_2	31	23
	He	6.0	31
	Ar	38	41
Nitrogen containing	NF_3	*	*
Silicon containing	SiF_4	36	13
Sodium containing	$NaHCO_3$ (10-20μm)	3.0	†
Hydrofluorocarbons	CF_3H	25	12
	CF_2H_2	‡	‡
	CF_2H_2/C_2HF_5	30	15
	CH_2FCF_3	29	10
	CHF_2CF_3	29	8.7
	$CF_3CH_2CF_3$	27	6.5
	C_3HF_7	28	6.2
Fluorocarbons	CF_4	37	16
	C_2F_6	30	8.1
	C_3F_6	29	7.3
	C_3F_8	30	6.3
	C_4F_8	32	6.3
	C_4F_{10}	32	5.3
Chlorine containing	CHF_2Cl	28	12
	$CHCl_2F$	32	11
	CH_3CF_2Cl	‡	‡
	$CF_2=CHCl$	‡	‡
	$CF_2=CFCl$	31	10
	$CHFClCF_3$	26	7.0
Bromine containing	CF_3Br	14	3.1
	CF_2Br_2 (I)	16	2.6
	CH_2BrCF_3 (I)	17	3.5
	$CH_2=CHBr$	‡	‡
	$CF_2=CFBr$	27	6.3
	$CF_2=CHBr$	24	6.0
Iodine containing	CF_3I	18	3.2

* Acted as an oxidizer, promoted flame stability.
† Solid powder not expressed in volume percent.
‡ Agent observed to be flammable.

TABLE 6-19D. *Cup Burner Minimum Extinguishing Concentrations*

Fuel	Cup Burner Extinguishment Concentration (% by Volume)								
	HFC-227ea*	FC-3-1-10	HFC-23	HCFC Blend A	N₂	FIC-131I	IG-55	IG-01†‡	IG-541
Acetone	6.8	5.5§	11.8‖	9.5#		**		30	31.1††
Acetonitrile	3.7			7.0#				15	26.7
Aviation Gasoline	6.7			11.4#				29	35.8
Avtur (Jet A)									36.2
n-Butanol	7.1			12.2#				32	37.2
n-Butyl Acetate	6.6			9.8					
Cyclohexane hex-ane				9.9#					
				10.9#					
hydrogen		5.6§		21.0#				33	34.4
Cyclopentanone	6.7							32	42/1
Diesel No. 2	6.7			9.6		3.3		25	
Diethyl Ether									34.2
Ethane									29.5
Ethanol	8.1	6.8§		11.0		3.0		32	35.0
Ethyl Acetate	5.6			10.6		3.0		31	32.7
Ethylene Glycol	7.8			11.4		2.4		27	42.1
Gasoline (unleaded)	6.5			9.3		3.6		30	35.8
n-Heptane	6.5–6.6‡‡	5.3–5.5‡‡	10.29–16.8‖	>11***	33–33.6‡‡	3.0	35 (est.)†††	41†††	29.1
	6.0§§	5.2‖‖	12.0‖‖	12.6##	32§§			38‡	
	5.8–6.6##	5.0##	12.7–12.9‡‡	9.9#	30†††				
			12.6##						
Hydraulic Fluid	5.8	4.3–4.5§§		9.6#	22–26§§	2.3		19	
i-Propanol	7.3	6.2		10.8				28	
Iso-octane		5.5		9.8#				27.0	
Isopropyl Alcohol			11.7‖						
JP-4	6.6			10.2		3.3		31	
JP-5	6.0§§	4.8§§		9.0	27§§	3.2		30	
	6.6‖‖								
Methane	6.2			13.7		2.0		29	15.4
Methanol	10.0	9.4§	21.2‖	15.1		3.8		38	44.2
Methyl Ethyl Ketone	6.7					4.4		35	35.8
Methyl Isobutyl Ketone	6.6			9.4		2.9			32.3
Morpholine	7.3								
Natural Gas				12.4#					
Nitromethane								36	
Octane									35.8
Pentane									37.2
Petroleum Ether									35.0
Propane	6.3	6.0§§		12.6	32.5§§	3.0		35.0	32.3
Pyrolidine	7.0			10.1		2.8			
Tetrahydrofuran	7.2			12.0					
Toluene	5.8		9–9.3‖	7.0				27	31.1
Turbo Hydraulic Oil 2380	5.1					2.1			
Vinyl Acetate									34.4
Xylene	5.3			8.7		5.5		24	

* See reference 11.
† Average value of 10 tests, n-Heptane standard deviation 1.6.
‡ See reference 12.
§ See reference 13.
‖ See reference 14.
See reference 15.
** See reference 16.
†† See reference 17.
‡‡ See reference 18.
§§ See reference 8.
‖‖ See reference 19.
See reference 20.
*** See reference 21.
††† See reference 7.

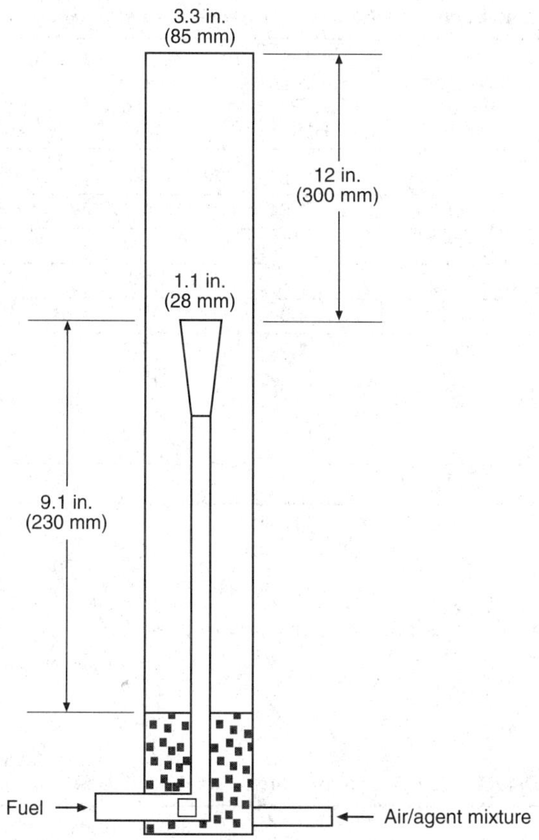

FIG. 6-19A.   *Schematic of ICI cup burner apparatus.*[10]

Figure 6-19C, from reference 8, shows the relationship between MEC for cup and OFDF burners. As expected, co-flowing cup burner concentration is quite similar to the OFDF concentration at a low strain rate (25 s⁻¹), typical of natural fires. In all cases, the MEC of an agent is much lower for high strain rate flames. This further reinforces the value of the cup burner (and OFDF) apparatus for evaluation of minimum extinguishing concentration.

One of the important aspects of strain rate relates to full-scale testing of gaseous fire suppressants and the need to ensure that nozzle-generated turbulence does not cause enhanced fire extinguishment. In real-scale testing, this generally requires shielding or baffling of flames. Results of full-scale tests where flames were not adequately baffled, particularly where fire extinguishment occurs prior to the end of agent discharge, should be further evaluated.

Tables 6-19E and 6-19F summarize the fire suppression capability of halon replacement agents, relative to Halon 1301 for halocarbon and inert gas agents, respectively. Agent weight and storage volume equivalents for Halon 1301 are given as 1. All comparisons are based on 120 percent of the agent manufacturer's recommended MEC, based on the cup burner value for n-Heptane. The design values used are provided by the manufacturer. Design values for HCFC Blend A and argon are problematic, because these values are below the MEC as measured by the cup burner.

Note that all clean agent alternatives require at least 60 percent more agent by weight and storage volume. This is a consequence of the elimination of bromine in the compounds and subsequent level of catalytic recombination of flame radicals. These data should be taken as representative values, as there are variations among hardware manufacturers.

The storage volume equivalents are based on the maximum fill density permitted in a storage cylinder with a pressure rating as recommended by the manufacturer. The approximate 10:1 storage volume requirement for inert gases is a consequence of the inability to liquefy these gases at ambient temperature. The storage volume requirement for inert gases is reduced with higher pressure storage cylinders.

The storage volume equivalent does not translate directly to a required area or volume for storage cylinders. The relative "footprint" of these storage volume equivalents will vary with the volume of the space protected and the maximum storage cylinder size offered by a manufacturer for a particular gas. In general, the floor

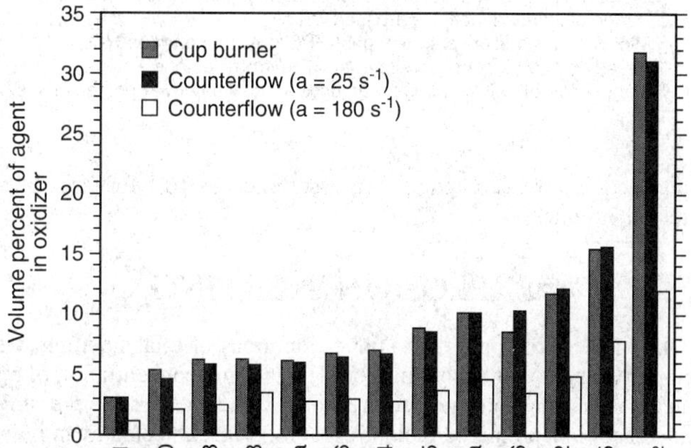

applications, such as engine nacelles with high fuel and oxidizer flow rates or in high-pressure spray or jet fires, the strain rate will substantially impact the minimum condition for extinguishment. Figure 6-19B is a sample plot from reference 8, showing the variation of the mole fraction of extinguishing agent *vs.* the strain rate at extinction for n-heptane fuels for a range of suppressants. For typical natural fires, the strain rate is approximately 25 s⁻¹. At high strain rates, the flame is extinguished at lower agent concentrations. In most total flooding applications, the minimum agent concentration should be measured at low strain rates.

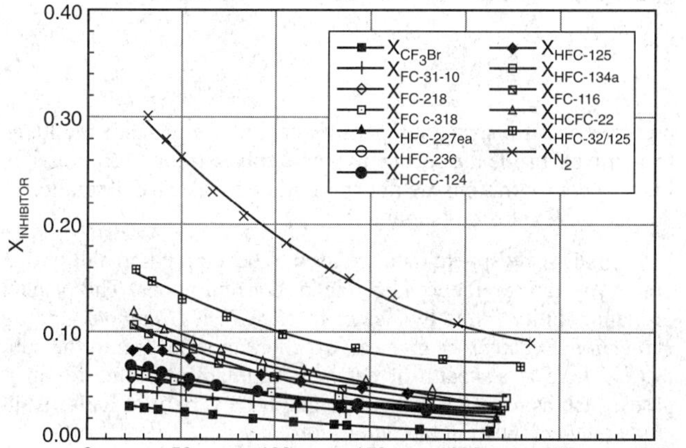

FIG. 6-19B.   *Mole fraction of various suppressants as a function of strain rate at extinction for n-Heptane.*[8]

FIG. 6-19C.   *Comparison of n-Heptane extinction results for the cup burner and OFDF apparatus at two strain rates.*[8]

TABLE 6-19E. *Weight and Storage Volume Equivalent Data for New Technology Halocarbon Gaseous Alternatives*

Trade Name	Designation	Formula	BP °C*	Cup Burner % V/V†	Min. Design Conc. % V/V	Ratio Agent Mass Req'd to H1301‡	Ratio Agent Storage Vol. Req'd to H1301§	Storage Pressure psi 20 °C
Halon 1301	Halon 1301	CF_3Br	−58	2.9–3.9	5‖	1	1	360
CEA-410	FC-3-1-10	C_4F_{10}	−2	5.0–5.9	6#**	1.9	1.7	360
FM-200	HFC-227ea	C_3F_7H	−16.4	5.8–6.6	7#**	1.7	1.6	360
FE-13	HFC-23	CHF3	−82.1	12–13	16#	1.7	2.2	609
FE-25	HFC-125	C_2HF_5	−48.5	8.1–9.4	10.9#	1.9	2.3	166
NAF-SIII	HCFC Blend A	HCFC–22 82% HCFC–123 4.75% HCFC–124 9.5% Organic 3.75%		>11%	8.6††	1.1	1.4	360
Triodide	FIC-131I	CF_3I	−22.5	2.7–3.2	5.0	~1	~1	

Notes:
* For U.S. custom unit: °F = °C × ⁹/₅ + 32.
† Range of independently established values.
‡ Ratio of halon design concentration 20.6 lb/cu ft at 70°C to new agent.
§ Ratio of halon storage volume required at minimum design concentration (max. fill density 70 lb/cu ft) to new agent.
‖ Minimum design concentration per NFPA 12A, *Standard Halon 1301 Fire Extinguishing Systems;* cup burner value approximately 3 percent.
Based on 120 percent of cup burner value for n-Heptane.
** Based on 120 percent of cup burner value, verified by listing/approval tests.
†† Based on listing/approval tests.

TABLE 6-19F. *Agent Weight and Storage Volume Equivalent Data for New Technology Inert Gas Alternatives*

Trade Name	Designation	Formula		Cup Burner % V/V (n-Heptane)	Min. Design Conc. % V/V	Ratio Agent Mass Req'd to H1301*	Ratio Agent Storage Vol. Req'd to H1301†	Storage Pressure psi (bar) 20°C
Halon 1301	Halon 1301	CF_3Br		2.9–3.9‡	5§	1	1	360 (25)
Inergen	IG-541	N_2 A CO_2	52% 40% 8%	29.1‡	35‖	2.0	10.0#	2180 (150)
Argonite	IG-55	N_2 A	50% 50%	30#	~36‖	2.0	10.0#	2220 (153)
Argon	IG-01	A	100%	30#	~36‖	2.0	8.0#	2370 (163)

NOTES:
* Ratio of halon design concentration 20.6 lb/cu ft at 70°C to new agent.
† Ratio of halon storage volume required at minimum design concentration (max. fill density 70 lb/cu ft) to new agent.
‡ Range of independently established values.
§ Minimum design concentration per NFPA 12A, *Standard on Halon 1301 Fire Extinguishing Systems;* cup burner value approximately 3 percent.
‖ Based on 120 percent of cup burner value for n-Heptane.
Manufacturers' values. MEC for argon is approximately 41 percent. (See reference 8.)

area required for storage of inert gases exceeds 10:1 for large protected volumes.

EXPLOSION INERTING

One of the most important application areas of total flooding fire suppressants is explosion inertion. The inerting concentration of an agent is the concentration required to prevent unacceptable pressure increases in a premixed fuel/air/agent mixture subjected to an ignition source. Inertion concentrations are typically measured in small laboratory-scale spheres, with an electric spark initiator.

The measured inerting concentration of an agent is dependent on the details of the test apparatus used, particularly the ignition source strength and "allowable" pressure rise. The "allowable"

pressure rise is a surrogate measurement of the distance the flame front travels inside the constant volume sphere prior to suppression. Inerting concentrations are not appropriate for use in deflagration or detonation (explosion) suppression.

Small-scale sphere data are used to develop flammability diagrams for various fuel/oxidizer/agent concentrations. The chapter on flammability limits by Beyler in *The SFPE Handbook of Fire Protection Engineering* gives an excellent introduction to the subject.[22] There is a wealth of data on flammability limits for inert gases, such as nitrogen and argon, and for a variety of fuels, available in the combustion literature.

Table 6-19G provides inerting concentration data for several agents and fuels, taken from small-scale inertion spheres. There are some substantial differences in results. Heinonen[23] has identified both ignition source type and strength as important variables with differences of approximately 40 percent for Halon 1301 inerting concentrations reported. Figure 6-19D shows flammability diagrams derived from small-scale inertion data as presented in Table 6-19E, along with points taken from a large-scale 795 cu ft (22.5 m^3) explosion vessel. While the small- to large-scale agreement is reasonable, there are scale effects.

TABLE 6-19G. *Explosion Inerting Concentrations, Small-Scale Inertion Sphere*

Agent	Inerting Concentration (% by Volume) of Fuel			
	Propane	Methane	i-Butane	Pentane
FC-3-1-10	10.3* 9.5†	~7.8†	—	
HFC-227ea	12.0† 11.6‡	8.0†	11.3*	11.6‡
HFC-23	20.2* 19.8†	20.2* 14.0†	—	
IG-541	49.0§	43.0§	—	
HCFC Blend A	18.0†	13.3†		
CO$_2$	28-36.9‖	23-29.6‖		
N$_2$	42-49.4‖	37-41.7‖		
Argon	61.7‖	44-55.4‖		

Notes:
* See reference 25.
† See reference 26.
‡ See reference 11.
§ See reference 27.
‖ See reference 28.

Explosion suppression systems employ rapid delivery of agent following very early detection of an ignition. Such systems employ significantly higher agent quantities (than flame suppression or inertion) delivered at higher rates. The total agent delivery time is on the order of 100 milliseconds.

Explosion suppression systems must be specifically designed for a particular application. There are no generic design requirements or standards currently available for such systems.

Senecal *et al.*[29,30] report on explosion suppression testing for occupied armored fighting vehicles and aerosol filling rooms. Results were obtained on premixed fuel droplet sprays. In contrast to flame suppression or inerting, suppression or deflagration requires significantly more agent. The aerosol filling room tests employed 44.1 lb (20 kg) of HFC-227ea, FC-3-1-10, HFC-236fa, and 22 lb (10 kg) of water in a 2827-cu-ft (80-m^3) test room to suppress a 3.2-oz. (90-g) propane release in a simulated aerosol fill station. Suppression of the propane-air deflagration was achieved, and the maximum flame front extension was approximately 4 ft (1.22 m). Suppression tests of heated diesel fuel droplet cloud deflagrations were conducted in simulated armored fighting vehicle crew compartments.

Table 6-19H summarizes typical data for flame suppression, inertion, and deflagration suppression concentrations. Note these values are for comparison purposes only. They should not be used in any way for design purposes.

Suppression of detonations requires substantially higher agent concentrations. An excellent discussion is given in reference 8.

TOXICITY

Introduction[31]

Toxicology tests are performed based on: (1) the duration and frequency of the exposure, and (2) determining specific biological effects. The studies examining duration and frequency are divided into four categories:[32]

1. Acute—less than 24-hr exposure,
2. Subacute—1 month or less exposure,
3. Subchronic—1- to 3-month exposure, and
4. Chronic—more than 3-month exposure.

The specific biological effects examined are many, but for the purposes of halocarbon fire-fighting agents, the following are the most common:

1. Inhalation toxicity,
2. Developmental toxicity,
3. Reproductive toxicity,
4. Genetic toxicity,
5. Cardiotoxicity, and
6. Central nervous system (CNS) effects.

The separation into categories is based on experience that acute exposure toxicity and chronic exposure toxicity can be significantly different. For example, the acute (single) exposure toxicity for organic solvents is typically central nervous system disturbances that are reversible once the exposure ends. Chronic (repeated, long-term exposure) toxicity for organic solvents can include liver damage and cancer, which are not considered reversible. Each chemical may cause different specific acute and chronic toxicity problems that need to be addressed. These include cancer, birth defects, liver toxicity, heart toxicity, etc.[32] Different chemicals within the same family may cause different effects or similar effects over a wide range of concentrations. Lastly, the route of exposure is examined. For gaseous fire-fighting agents, inhalation studies are required.

A series of tests is required, beginning with the acute studies. This is done to build on the results from previous shorter term studies. One important reason is to establish the correct exposure concentrations for testing. An improper dosage could cause the invalidation of a long-term test halfway or more through the study, i.e., too many test specimens may die from an excessive test dosage. An equally important reason is to preclude running expensive, longer term chronic studies when the shorter term studies do not suggest any need. The shorter term studies can be used as screens to give indications, but not direct results, for the type of toxicity problems that longer term tests may need to address. The toxicology series is not a "cookbook" process. Professional toxicologists must evaluate the results through each step of the process and make educated decisions on the next phase of testing.

A minimum set of tests required by the U.S. Environmental Protection Agency (EPA) under the Significant New Alternatives Policy (SNAP) program would likely include:[33]

1. Acute toxicity "range finder," e.g., limit or LC$_{50}$ test;
2. Cardiac sensitivization test;
3. Developmental and reproductive toxicity;
4. 4- or 13-week subchronic test;
5. Genetic toxicity screening test (such as Ames test); and
6. Thermal decomposition product test.

As noted, acute toxicity studies are usually the first to be done. These tests provide information on acute lethality. For inhalation this is called the LC$_{50}$, and is the concentration at which 50 percent of the animals die within 14 days of the exposure. This is usually a single 4-hr exposure, and generally performed with mice and rats.

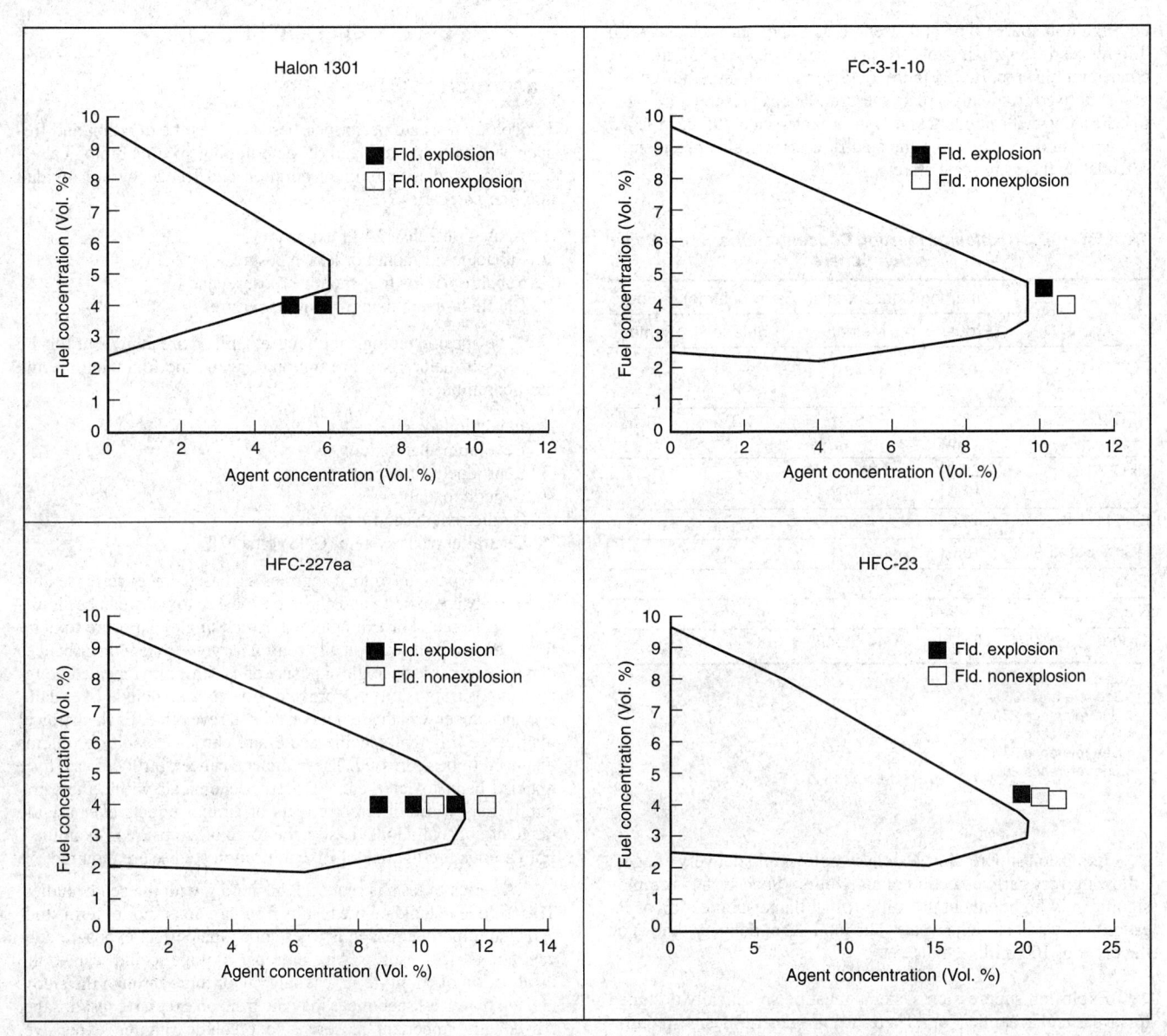

FIG. 6-19D. *Small-scale flammability diagrams for propane and several replacement agents. Squares denote explosion/no explosion points in large-scale 792 cu ft (22.5 m³) chamber.*[24]

TABLE 6-19H. *Comparison of Concentrations for Flame Extinguishment, Inerting, and Deflagration Suppression*

	Volume %		
Agent	Typical Value Flame Suppression	Inerting Concentration in Propane	Diesel Fuel Droplet Deflagration Suppression
Halon 1301	3	6–7	12
FC-3-1-10	5.5	10.3	8
HFC-227ea	5.8	~12	11
HFC-23	12	20.2	—
IG-541	29	49.0	—

Subacute studies are the lowest tier of the repeated exposure tests. They are used: (1) to discover specific toxicity problems above the acute effects, and (2) for refining the proper doses for follow-on tests. Subacute studies generally run 14 days.

Subchronic studies are used: to (1) establish final dosage data to be used in chronic studies and (2) identify further the specific toxicity concerns of long-term exposures. These studies are usually performed for 90 days on one or two species.

Chronic studies are generally performed the same as the subchronic tests, except that the exposure is increased beyond 3 months, and up to $2^1/_2$ years. Often chronic studies will incorporate carcinogenicity (cancer) tests to preclude running separate tests.

Developmental and reproductive toxicity studies are designed to look at the adverse effects on the reproductive systems and on offspring. Developmental tests examine the effects on a developing organism any time during its life span, as the result of repeated exposure by either parent before conception or during pregnancy, and postnatally until puberty. Reproductive toxicology is a multigenerational study concerned with the adverse effects to the reproductive system that may be passed on to the first generation and second generation offspring by either parent. The specific study of defects caused between conception and birth is called teratology.[32]

Genetic toxicity studies examine mutangenic potential, i.e., the capability of a chemical to cause changes in the genes of a cell. These studies are also effective as screens and tests for the potential to cause cancer. Two of the most common tests used for halocarbon agents are: (1) the Ames test and (2) the micronucleus test. The Ames test examines the ability of the chemical to cause changes or mutations in the genes of a cell. The micronucleus test studies look at chromosome aberrations. Both screens are used as indicators of the potential to cause cancer. Results of these screens are used to decide whether longer term cancer studies, other genetic toxicity studies, or developmental and reproductive studies are required.[32]

Beyond the above tests, specific toxicology concerns may warrant requiring other, less standard, toxicity studies. This is the case for halocarbons where they have been found to depress the heart rate. In the presence of adrenaline, the heart can experience mild conditions of an irregular heartbeat to severe conditions of a full heart attack. This effect was first seen with chloroform when it was widely used as an anesthetic. The test developed to assess this type of cardiotoxicity is called "cardiac sensitization."

The data gathered from these tests on nonhuman species is only an indication of the toxicity for humans. Toxicologists use these studies to do risk assessments to predict the toxicity of a chemical or specific mixture of chemicals in a particular exposure scenario. The acute tests are used to establish limits for single or very infrequent exposures. For example, acute toxicity is used in determining the effects that can be expected when a fire suppression system is discharged. Subchronic and chronic studies are used to establish limits for repeated exposures for workers who are subjected to the halocarbon on a more frequent basis, such as production plant workers and maintenance personnel who fill/refill cylinders.

The EPA requires, as a minimum for halocarbon fire suppression agents, cardiac sensitization and developmental toxicology studies to set acute exposure limits. To set worker/manufacturer exposure limits, the EPA uses, as a minimum, genetic toxicology, 90-day subchronic and developmental toxicology studies.[34]

While all halocarbon agents are tested for long-term health hazards, the primary endpoint is acute, or short-term, exposure. The primary acute toxicity effects of the halocarbon agents described in this chapter are anesthesia and cardiac sensitization. For inert gases, the primary physiological concern is reduced oxygen concentration.

Halocarbon Agents

Cardiac sensitization is the primary short-term "acute" toxicity problem for fire suppression applications. Cardiac sensitization is a term describing the sudden onset of cardiac arrhythmias in the presence of a concentration of an agent, caused by sensitization of the heart to epinephrine. The presence of epinephrine is critical to the onset of arrhythmias. This is important in fire protection applications, due to the increased production of epinephrine by the body under stress.

The toxicity endpoints used to describe cardiotoxicity and allowable exposure levels are no observed adverse effect level (NOAEL) and the lowest observed adverse effect level (LOAEL). The NOAEL is the highest concentration of an agent at which no "marked" or adverse effect occurred. The LOAEL is the lowest concentration at which an adverse effect was measured.

The procedures used to evaluate cardiac sensitization vary somewhat. The procedure involves intravenous dosing of male beagles with epinephrine for 5 minutes. Continuous inhalation exposure to the agent follows for five minutes. Following this inhalation exposure, the dog is dosed again with epinephrine and monitored for five minutes to determine the effect of the agent and epinephrine. The protocol is performed at higher doses until an effect occurs.

Effects are monitored by electrocardiograph (ECG) measurements. An adverse effect is generally considered to be the appearance of five or more arrhythmias or ventricular fibrillation. The data from these tests are evaluated by medical experts, and the appropriate NOAEL and LOAEL values are reported by the EPA under the SNAP program.

There is no direct correlation between the experimental results from dogs to humans. It is generally accepted due to the combination of the high doses of epinephrine in the tests and the similarity in cardiovascular function between dogs and humans that the results can be applied to humans. It is believed that the results of these tests are conservative with respect to direct application of the results to humans in typical fire suppression applications.

In addition to the short-term chronic exposure limits of interest in fire suppression system design, the EPA evaluates longer term inhalation data for these compounds. Table 6-19I summarizes NOAEL, LOAEL, and LC_{50} values. Note that the LC_{50} values greatly exceed the NOAEL.

The use of halocarbon agents in occupied areas is generally subject to the constraint that the design concentration must be less than the NOAEL. In the 1996 edition of NFPA 2001, an exception allows the use of total flooding gases up to LOAEL for Class B hazards where a predischarge alarm and time delay are provided and where acceptable to the Authority Having Jurisdiction. The rationale for the exception is based on: (1) the conservative nature of NOAEL/LOAEL, (2) consistency with EPA guidelines, and (3) positive benefit of higher agent concentrations in certain flammable/combustible liquid hazards. While it is recommended that all systems employ predischarge alarms and that personnel evacuate prior to system actuation, it is understood that inadvertent discharges and short-term exposures will occur, hence the limitation. It is expected that emergency exposures for up to several minutes at or below the NOAEL and possibly the LOAEL are reasonably safe. In no case should systems be designed or installed where intentional exposure of any duration is anticipated.

It has been proposed by the EPA that agents be permitted for use at concentrations up to the LOAEL where evacuation will occur in less than 60 seconds. This has not been widely integrated into design standards to date due to the vagaries of accidental exposure conditions. The EPA continues work on methods that will permit correlation between the agent concentration and the expected exposure time.

Based on the limitation that the design concentration must be below the NOAEL in most cases, it can be seen from Table 6-19I that three agents are acceptable for use in normally occupied areas for flame extinguishant purposes. These are HFC-227ea, HFC-23, and FC-3-1-10.

TABLE 6-19I. *Toxicity Data for Halocarbon Clean Agent Fire Suppressants*

Trade Name	Designation	Formula	NOAEL % V/V*	LOAEL % V/V*	LC_{50} or ALC†‡
CEA-410	FC-3-1-10	C_4F_{10}	40	>40§	>80%
FM-200	HFC-227ea	C_3F_7H	9.0	>10.5	>80%
FE-13	HFC-23	CHF_3	30	>50§	>65%
FE-24	HCFC-124	C_2HCIF_4	1	2.5	23–29
FE-25	HFC-125	CH_2HF_5	7.5	10.0	>70%
NAF-SIII	HCFC Blend A	HCFC-22 82% HCFC-123 4.75% HCFC-124 9.5% Organic 3.75%	10	>10	64%
Triodide	FIC-131I	CF_3I	0.2	—	—

Notes:
* From EPA SNAP documents.
† ALC = Allowable Lethal Concentration
‡ From NFPA 2001 (see reference 1).
§ Maximum concentration before oxygen depletion concerns.

Inert Gas Agents

Inert gas agents are effectively physiologically inert. The primary physiological problem with these agents is the reduced oxygen concentration caused by the high agent design concentrations. One inert gas blend employs a low concentration of CO_2 (which is not physiologically inert) in order to counter the effects of the reduced oxygen concentration. The mechanism of this effect is discussed in reference 35.

The EPA has determined that inert gases can be used in concentrations up to 43 percent in normally occupied areas subject to similar limits placed on halocarbon agents when used at the NOAEL. This results in a residual oxygen concentration of 12 percent. The analog to the LOAEL is 52 percent agent concentration, with a 10 percent residual oxygen concentration. NFPA 2001 requirements mirror the EPA limitations at 43 percent agent concentration, with the exemption permitting use up to 52 percent concentration for Class B hazards. There appears to be a growing consensus among independent toxicologists that inert gases could be safely used up to 52 percent concentration for a 5-minute exposure time.

ENVIRONMENTAL FACTORS[36]

Two main environmental impacts need to be considered with respect to halocarbon extinguishing agents: (1) ozone depletion and (2) global warming. Another environmental factor, atmospheric lifetime, can be considered in the discussion of each of these environmental impacts. It is also a "stand-alone" environmental concern that needs to be addressed. International, national, state, and local governments currently regulate halocarbon fire-fighting agents based on their effect on ozone depletion. However, the U.S. EPA also takes into account atmospheric lifetimes and global warming potentials in the implementation of its SNAP program under Section 612 of the Clean Air Act, as amended in 1990. Although no specific national environmental regulations cover global warming and atmospheric lifetimes, the consideration of these issues in the EPA's implementation of existing regulations makes these issues of concern.

Ozone Depletion

Ozone (O_3) is a naturally occurring gas that is found in the atmosphere. Most naturally occurring ozone is created by the reaction of O_2 with ultraviolet (UV) light coming from the sun. This is illustrated below:

$$O_2 + UV \longrightarrow O + O$$

$$O + O_2 \longrightarrow O_3$$

The UV light causes this reaction to occur. Ozone is also created in nature when the high energy output from lightning initiates a similar reaction.

Ozone is also formed by a number of manmade sources, mainly as a result of air pollution. When carbon monoxide, methane, and other hydrocarbons meet nitrogen oxide (e.g., from car exhaust) and ordinary sunlight, ozone is produced.[37] It is also produced by laser printers and electrical motors, and is responsible for the pungent odor often associated with the devices. Manmade ozone is often called "smog" and is a considerable health risk.

With respect to the environment, an important difference exists between naturally occurring and manmade sources of ozone. Manmade ozone is produced in the region of our atmosphere called the troposphere. This is the portion that starts at the earth's surface and extends 3.7 to 10.6 mi. (6 to 17 km) above the surface. (The thickness of this region is different at the poles *vs.* the equator.) Naturally occurring ozone is produced primarily in the stratosphere. The stratosphere is the region directly above the troposphere and extends to approximately 31 mi. (50 km) above the earth's surface.[38] The ozone produced in the stratosphere by the reaction described above is what is known as the "ozone layer" and serves a very important purpose. Ozone absorbs the UV-b portion of the total UV in sunlight, and prevents it from entering further into the atmosphere. This has two net effects: (1) it lowers the amount of harmful UV-b that plants and animals are exposed to and (2) it prevents more ozone from being created naturally in the troposphere.

Recently, the U.S. has begun to give "ozone alerts," typically as part of daily weather reports. The ozone referred to by these alerts is tropospheric ozone and does not refer, in any way, to stratospheric ozone or ozone depletion. There are, however, concerns with tropospheric ozone. These concerns will be addressed along with the discussion of global warming.

Halons, HCFCs, and other halocarbons containing chlorine or bromine have been shown to cause the destruction of stratospheric ozone. It is not a measure of the exact amount of ozone destroyed. Instead, it is the relative amount of ozone destroyed as compared with an arbitrary standard. The standard chosen is CFC-11, which has been assigned an ozone depletion potential (ODP) of 1. ODP of all other halocarbons relates to their relative effect on the destruction of ozone, as compared with CFC-11. Halon 1301 has an ODP of 13, meaning it will destroy 13 times as much ozone as CFC-11 on a pound-for-pound (kg for kg) basis. A compound having an ODP of 0.1 would have 10 percent of the relative ozone-depleting effects of CFC-11. All ODP values are based on mass (weight) and not on moles (numbers of molecules).

An alternative method for measuring the relative ODP of a compound has recently been suggested. The standard ODP defined above considers the overall atmospheric lifetime, but does not take into account the stratospheric portion of the lifetime. It not only matters that the halocarbon gets to the stratosphere, but that it breaks down to liberate chlorine and bromine atoms. A short stratospheric lifetime means that the ozone depletion depicted by its ODP must occur over a shorter time than for one with a long stratospheric lifetime. In the short term, more ozone depletion than predicted occurs. A time-dependent ozone depletion potential has been suggested to remedy this problem by looking at the ozone depleted over a given time period, or horizon, and relating it to the net effect of CFC-11 over the same time horizon. As the time horizon approaches infinity, the time-dependent ODP values approach the standard values.[39] The idea of time-dependent ODPs is new and has not come into wide use, as of yet.

Atmospheric Lifetimes

When one thinks of a chemical species lifetime, the term half-life is often used. This is most common in the nuclear field where calculations are made to determine how long it takes a species to decay to $1/2$ its original concentration. Atmospheric lifetime values used in the determination of ozone depletion and global warming potentials are *not* half-lives. They are $1/e$ lifetimes, sometimes called e-folding lifetimes.

It has been determined that greenhouse gases break down in the atmosphere according to the following equation:

$$C = C_0 e^{-kt}$$

where

C = concentration at time, t;

C_0 = initial concentration at time $t = 0$; and

k = an experimentally determined rate constant (units = 1/time).

A mathematical manipulation can be made to express this equation as a function of more readily quantifiable terms. This is accomplished by defining the atmospheric lifetime, L, as an e-folding lifetime, or the time it takes for the ratio of $C:C_0$ to be equal to $1/e$. (e is the base of the natural logarithm system and has a numerical value of approximately 2.718.) The resulting equation is as follows:

$$C = C_0 e^{-t/L}$$

Although the term "half-life" is most commonly used when describing the decay of a species concentration, the use of the e-folding lifetime allows for quantification of concentration in terms of a compound's atmospheric lifetime, as previously shown. A half-life refers to the time it takes for half a given amount of compound to be broken down in the atmosphere, thus the ratio $C:C_0$ is equal to $1/2$. Using an e-folding lifetime, the ratio is equal to $1/e$ or 0.368. After one lifetime, the concentration will be equal to 0.368 times its original value. After two lifetimes, the concentration will be 0.135 or $(0.368)^2$ its original value. After three lifetimes, the concentration will be 0.0498 or $(0.368)^3$ times its original value, etc.

An important factor when using the equation above to solve for concentration as a function of time is that, no matter what value is used for time, the concentration never equals zero. Some portion will always exist in the atmosphere. When the atmospheric lifetime is small, the concentration over hundreds of years may become negligible. When the atmospheric lifetime is very large, on the order of thousands of tens of thousands of years, the concentration in the atmosphere may not be negligible.

The environmental concern is one of "what if?". The halocarbons were believed to be "safe" for many years after their release into the atmosphere began. It was not until many years later that they were linked to ozone depletion. There is a concern that these other halocarbons, or other compounds, may cause other environmental impacts, that yet are unknown. If the atmosphere is filled with these chemicals and they exist in appreciable amounts for hundreds, thousands, and millions of years, what then? Might the damage we do to our environment be more than humans can cope with? Nature can probably overcome this impact, but in tens of millions of years. This is a short time frame for nature, but not for man. Current implementation of the EPA SNAP program places restrictions on perfluorocarbons that have very large or "outlying" atmospheric lifetimes, as compared with the rest of the halocarbons. This restriction is not based on any known or anticipated environmental problem. It is a response to the "what if " concerns raised above.

Global Warming Potential

A basic understanding about the earth's climate and atmosphere is needed to understand what global warming potential (GWP) is and how to attempt its measurement.

The atmospheric region closest to Earth, the troposphere, represents approximately 81 percent of the Earth's atmosphere. It is made up mainly of nitrogen (N_2) and oxygen (O_2), representing 99 percent of its composition. The remaining 1 percent is made up of mostly argon, with smaller amounts of water, 0.2 percent; and CO_2, 0.03 percent [300 parts per million by volume (ppmv)]. The atmosphere also contains several compounds in very small amounts, known as "trace gases." These are methane, nitrous oxide, carbon monoxide, hydrogen, ozone, and halocarbons, which are so dilute that they are measured in parts per billion by volume (ppbv).[38,40]

Climate and weather are not the same. Weather is the short-term, day-to-day conditions of temperature, humidity, rainfall, winds, etc. Climate is the long-term condition averaged over many years. Based on the climate of a region, particular weather is expected or predicted. Many factors determine a region's climate. These are a complex interaction of both natural and manmade activities. The biggest factor on the overall behavior of the Earth's climate is the sun. The sun's energy hits Earth more at the tropics and equator than at the poles. Driven by diffusion principles, Earth distributes heat away from the tropics, which leads to the major wind and ocean currents.[38]

Estimates of the average temperature of Earth's atmosphere have been made, assuming two theoretical conditions. The first assumes no atmosphere at all, and the second assumes an atmosphere made up solely of N_2 and O_2 (normally representing 99 percent of our atmosphere). In both cases, models have predicted that Earth's surface would be some 91°F (33°C) cooler than it is.[40] The temperature difference is believed to be a direct result of the very small amounts of trace gases, water vapor, and CO_2. These act very much like the glass on a greenhouse that lets the sunlight in but helps to prevent the heat from escaping; hence, the terms "greenhouse effect" and "greenhouse gases." A greenhouse gas is defined as any gas that absorbs infrared (IR) radiation.

It should not be surprising that certain constituents in the air can cause it to stay warmer. For example, it is common knowledge that the temperature on a hot, humid night does not drop as much as on a less humid night. During the winter, a clear night will have a larger drop in temperature than a cloudy night. In both cases, the difference is due to the water vapor present in the air. In fact, water is the largest contributor to the natural greenhouse effect, followed second by CO_2.[38]

Climate change and global warming are not synonymous. Climate change includes cooling and warming of the atmosphere. Global warming only deals with the aspects of climate change that result in warming of the atmosphere.

It is estimated that about one-third of the total solar radiation is reflected off Earth's atmosphere. Most of the remaining two-thirds passes through the atmosphere and is absorbed by Earth's surface, causing it to warm.[41] Earth cools itself by releasing heat, or IR. In order for a balance to be maintained, the solar radiation coming in must equal the radiation going out. If more radiation entered the atmosphere than left, Earth would be constantly getting warmer. The opposite would be true if more radiation left Earth than came in, i.e., the Earth would be constantly getting cooler.

O_2 and N_2 are transparent to IR, meaning that these compounds easily let IR radiation or heat pass through. Greenhouse gases are not transparent to IR, they absorb it. Their ability to absorb IR is finite, however, and some of what they absorb is re-radiated out of the atmosphere. The IR re-radiated is what is important. For a given concentration of greenhouse gases, it depends on two factors: (1) how much IR comes in and (2) the temperature of the gas. Given the premise that the atmosphere wants to establish an equilibrium, then the IR from Earth's surface and the amount re-radiated from these gases want to be equal. Because the IR coming from Earth's surface does not change, the only remaining variable to affect the equilibrium is temperature. The atmosphere will heat until the IR balance is achieved. As stated previously, this has been predicted as a 91°F (33°C) rise in the temperature of the atmosphere.[40]

On a day-to-day basis, the premise that the solar radiation is equal to the sum of IR radiated from Earth's surface and the IR re-radiated from the greenhouse gases is not necessarily true. Daily, weekly, monthly, and yearly changes in weather are always occurring. Nevertheless, over the years, the averages are very close to equal and have resulted in a stable climate. To understand the greenhouse effect, scientists studying the climate have developed the concept of "radiative forcing." They define radiative forcing as anything that will cause the energy balance at the top of the troposphere to no longer be in balance. Put another way, it is when radiation in and the radiation out of the top of the troposphere are not equal. This unbalanced radiation forces the climate to adapt, hence the term "radiative forcing." The theory is that the climate will change to equalize the incoming and outgoing radiation. Any condition that results in a positive radiative forcing will cause a rise in the average temperature and, therefore, is responsible for global warming. Any condition that results in a negative radiative forcing will cause a drop in the average temperature and, therefore, is responsible for global cooling.[41]

To estimate the amount of global warming expected from a release of a particular greenhouse gas, a scale was developed by the International Panel on Climate Change (IPCC), based on the idea of radiative forcing. This scale is called the global warming potential (GWP) and relates to positive radiative forcing that will cause the atmosphere to heat up. The GWP is the cumulative amount of radiative forcing between the present and some future time caused by a unit mass (weight) of a compound, as compared with the same unit mass (weight) of an arbitrary standard. CO_2 is the most common reference, and 20-year, 100-year, and 500-year time periods, or horizons,

are the most common time references cited in the literature.[41] The choice of time horizons is a policy issue and not a technical one.[42]

Since the GWP is a cumulative effect, summed year by year over a given time horizon, the quantity present in the atmosphere must be known year by year over that horizon. The atmospheric lifetime is used to perform these calculations. GWPs are also affected by the specific IR-absorbing capabilities of the chemical. Analogous to ODPs, GWPs are not exact numbers showing the precise effect on global warming. A 100-year horizon GWP of 6200 for Halon 1301 means that 1 lb (0.454 kg) of Halon 1301 will cause as much global warming as 6200 lb (2812 kg) of CO_2. A 500-year horizon GWP for methylene chloride of 0.3 means that 1 lb (0.454 kg) of methylene chloride will cause the same warming as 0.3 lb (0.14 kg) of CO_2. (See Table 6-19J.)

GWPs are used to determine the future global warming contribution of a substance *over a given time* by multiplying the weight of the greenhouse gas by a specific time-horizon-GWP. The resultant number can be compared with others to decide which will have the least (or greatest) impact *over that time horizon*. Emitting a large quantity of a small GWP greenhouse gas may cause less global warming than a small release of a very large GWP greenhouse gas over a certain time horizon. Different time horizons may lead to different results.

The choice of which time horizon to use is a policy issue and not a technical one. A standard measure to predict future global warming commitments was developed as a tool to help policy makers and regulators in making decisions of which greenhouse gases would provide the biggest "bang for the buck." Currently, there is no standard in choosing a time horizon. In the future, however, this may change. It is also possible that different time horizons will be used for different greenhouse gases.

Environmental Regulation of Halon Replacements

The evaluation of clean agent fire suppressants includes a consideration of environmental factors. International, national, and local government regulations control the use of any alternatives in this regard. The primary environmental consideration is ODP. This is a measure of a chemical's ability to deplete stratospheric ozone with CFC-11 as a basis with an ODP of 1. All chemicals with a non-zero ODP are subject to phaseout under the Montreal Protocol and its amendments. Table 6-19K summarizes environmental impact data for halocarbon alternatives. Note that FC and HFC compounds have zero ozone depletion potential. The HCFC compounds have quite low ODP. Other HCFC compounds are widely used as CFC replacements for refrigerants.

In the United States, the U.S. EPA has the responsibility for regulating the use of halon replacements. Through the SNAP program, the EPA may prohibit or restrict the use of halon replacements on the basis of certain environmental factors or toxicity. The current status of EPA regulation of halon replacements is given in Table 6-19L. There are two basic types of restrictions given to date: (1) restrictions on use in occupied *vs.* unoccupied areas, and (2) use restrictions (e.g., HCFC-22, HCFC-124, CF_3I) related to high atmospheric lifetime (e.g., C_4F_{10}, C_3F_8). The EPA periodically updates this list as new agents are proposed or as new data becomes available. It is relatively common to have NOAEL/LOAEL values updated, usually increased, as additional data becomes available. This may change the acceptability of an agent for use in occupied areas.

TABLE 6-19J.   *Radioactive Forcing of Climate Change: Global Warming Potentials, Referenced to the Absolute GWP for CO_2.*
(The typical uncertainty is ±35% relative to the CO_2 reference.)

Species	Chemical Formula	Lifetime (yr)	Global Warming Potential (Time Horizon)		
			20 yr	100 yr	500 yr
Methane*	CH_4	14.5±2.5†	62	24.5	7.5
Nitrous oxide	N_2O	120	290	320	180
CFCs					
CFC-11	$CFCl_3$	50±5	5000	4000	1400
CFC-12	CF_2Cl_2	102	7900	8500	4200
CFC-13	$CClF_3$	640	8100	11700	13600
CFC-113	$C_2F_3Cl_3$	85	5000	5000	2300
CFC-114	$C_2F_4Cl_2$	300	6900	9300	8300
CFC-115	C_2F_5Cl	1700	6200	9300	13000
HCFCs, etc.					
HCFC-22	CF_2HCl	13.3	4300	1700	520
HCFC-123	$C_2F_3HCl_2$	1.4	300	93	29
HCFC-124	C_2F_4HCl	5.9	1500	480	150
HCFC-141b	$C_2FH_3Cl_2$	9.4	1800	630	200
HCFC-142b	$C_2F_2H_3Cl$	19.5	4200	2000	630
HCFC-225ca	$C_3F_3HCl_2$	2.5	550	170	52
HCFC-225Cb	$C_3F_5HCl_2$	6.6	1700	530	170
Carbon tetrachloride	CCl_4	42	2000	1400	500
Methylchloroform	CH_3CCl_3	5.4±0.6	360	110	35
Bromocarbons					
H-1301	CF_3Br	65	6200	5600	2200
Other					
HFC-23	CHF_3	250	9200	12100	9900
HFC-32	CH_2F_2	6	1800	580	180
HFC-43-10	$C_3H_2F_{10}$	20.8	3300	1600	520
HFC-125	C_2HF_5	36.0	4800	3200	1100
HFC-134	CHF_2CHF_2	11.9	3100	1200	370
HFC-134a	CH_2FCF_3	14	3300	1300	420
HFC-152a	$C_2H_4F_2$	1.5	460	40	44
HFC-143	CHF_2CH_2F	3.5	950	290	90
HFC-143a	CF_3CH_3	55	5200	4400	1600
HFC-227ea	C_3F_7H	41	4500	3300	1100
HFC-236fa	$C_3H_2F_6$	250	6100	8000	6600
HFC-245ca	$C_3H_3F_5$	7	1900	610	190
Chloroform	$CHCl_3$	0.55	15	5	1
Methylene chloride	CH_2Cl_2	0.41	28	9	3
Sulphur hexafluoride	SF_6	3200	6500	24900	36500
Perfluoromethane	CF_4	50000	4100	6300	9800
Perfluoroethane	C_2F_6	10000	8200	12500	19100
Perfluorocyclobutane	C_4F_8	3200	6000	9100	13300
Perfluorohexane	C_6F_{14}	3200	4500	6800	9900

* The methane GWP includes the direct effect and those indirect effects due to the production of tropospheric ozone and stratospheric water vapor. The indirect effect due to the production of CO_2 is not included.
† For methane, the adjustment time is given, rather than the lifetime.

TABLE 6-19K. *Environmental Factors for Halocarbon Clean Agents*

Trade Name	Designation	Formula	ODP	GWP (100 yr)	Atmospheric Lifetime (yr)
Halon 1301	Halon 1301	CF_3Br	16	5800	100
CEA-410	FC-3-1-10	C_4F_{10}	0	5500	2600
FM-200	HFC-227ea	C_3F_7H	0	2050	31
FE-13	HFC-23	CHF_3	0	9000	280
FE-24	HCFC-124	C_2HClF_4	0.022	440	7
FE-25	HFC-125	C_2HF_5	0	3400	41
NAF-SIII	HCFC Blend A	HCFC-22 82% HCFC-123 4.75% HCFC-124 9.5% Organic 3.75%	0.05	1600	16
Triodide	FIC-131I	CF_3I	<0.2	0	—

TABLE 6-19L. *Summary of EPA Total Flooding*

Agent	Decision	Conditions	Comments
HCFC-22 (3/94)	Acceptable	NOAEL = 2.5% LOAEL = 5.0%	MEC = 13.9% Unlikely to be used in occupied areas*†§¶
HCFC-124 (3/94)	Acceptable	NOAEL = 1.0% LOAEL = 2.5%	MEC = 8.4% Unlikely to be used in occupied areas*†§¶
HCFC Blend A (3/94)	Acceptable	NOAEL = 10.0% LOAEL = 10.0%	Agent should be recovered and recycled*†§¶
HFC-23 (3/94)	Acceptable	NOAEL = 30.0% LOAEL = 7.50%	MEC = 14.4%*†§¶
HFC-125 (3/94)	Acceptable	NOAEL = 7.5% LOAEL = 10.0%	MEC = 11.3%, unlikely to be used in occupied areas*†§¶
HFC-134a (3/94)	Acceptable	NOAEL = 4.0% LOAEL = 8.0%	MEC = 12.6%, unlikely to be used in occupied areas*†§¶
HFC-227ea	Acceptable	NOAEL = 8.0% LOAEL = 10.5%	MEC = 7.0% LOAEL probably greater than 10.0% *†§¶
FC-3-1-10	Acceptable where other alternatives are not technically feasible, due to performance or safety requirements: 1. due to their physical or chemical properties, or 2. where human exposure to the extinguishing agents may approach unacceptable cardiosensitivity levels or result in other unacceptable health effects under normal operating conditions.	NOAEL = 30.0%	The comparative design concentration based on cup burner values is approximately 8.8%. Users must observe the limitations on PFC acceptability by making reasonable efforts to undertake the following measures: 1. conduct an evaluation of foreseeable end use, 2. determine that human exposure to the other alternative extinguishing agents may approach or result in cardiosensitization or other unacceptable toxicity effects under normal operating conditions, and 3. determine that the physical or chemical properties or other technical constraints of the other available agents preclude their use. Documentation of such measures must be available for review upon request. The principal environmental characteristic of concern for PFCs is that they have high GWPs and long atmospheric lifetimes. Actual contributions to global warming depend upon the quantities of PFCs emitted. For additional guidance regarding applications in which PFCs may be appropriate, users should consult the description of potential users which is included in the rulemaking.*†§¶
IG-541 (3/94)	Acceptable	Design concentration must result in at least 10% oxygen, and no more than 5% carbon dioxide.	Individuals can remain in 10–12% oxygen atmosphere for 30–40 min.*†

TABLE 6-19L.   *(Continued)*

Agent	Decision	Conditions	Comments
C$_3$F8 (6/95)	Acceptable where other alternatives are not technically feasible, due to performance or safety requirements: 1. due to their physical or chemical properties, or 2. where human exposure to the extinguishing agents may approach unacceptable cardiosensitivity levels or result in other unacceptable health effects under normal operating conditions.	NOAEL = 30.0%	The comparative design concentration based on cup burner values is approximately 8.8%. Users must observe the limitations on PFC acceptability by making reasonable efforts to undertake the following measures: 1. conduct an evaluation of foreseeable end use, 2. determine that human exposure to the other alternative extinguishing agents may approach or result in cardiosensitization or other unacceptable toxicity effects under normal operating conditions, and 3. determine that the physical or chemical properties or other technical constraints of the other available agents preclude their use. Documentation of such measures must be available for review upon request. The principal environmental characteristic of concern for PFCs is that they have high GWPs and long atmospheric lifetimes. Actual contributions to global warming depend upon the quantities of PFCs emitted. For additional guidance regarding applications in which PFCs may be appropriate, users should consult the description of potential users which is included in the rulemaking.*†§¶
Powdered Aerosol A (3/94)	Acceptable in unoccupied areas		For use in occupied areas, additional decomposition product and health effect data are required.
Powdered Aerosol B (3/94)	Acceptable in unoccupied areas		Agency review for occupied areas incomplete.
FIC-131 I (6/95)	Acceptable in normally unoccupied areas.	EPA requires that any employee who could possibly be in the area must be able to escape within 30 sec. The employer must ensure that no unprotected employees enter the area during agent discharge.	Manufacturer has not applied for listing for use in normally occupied areas. Preliminary cardiosensitization data indicate that this agent would not be suitable for use in normally occupied areas. EPA is awaiting results of ODP calculations.*†§¶
Gelled halocarbon/dry chemical suspension (6/95)	Acceptable in normally unoccupied areas.	EPA requires that any employee who could possibly be in the area must be able to escape within 30 sec. The employer must ensure that no unprotected employees enter the area during agent discharge.	The manufacturer's SNAP application requested listing for use in unoccupied areas only.†
Inert gas/powdered aerosol blend (6/95)	Acceptable as a Halon 1301 substitute in normally unoccupied areas.	In areas where personnel could possibly be present, as in a cargo area, EPA requires that the employer must provide a predischarge employee alarm, capable of being perceived above ambient light or noise levels for alerting employees before system discharge. The predischarge alarm must provide employees time to safely exit the discharge area prior to system discharge.	The manufacturer's SNAP application requested listing for use in unoccupied areas only.†
Sulphur hexafluoride (6/5)	Acceptable as a discharge test agent in military uses and in civilian aircraft uses only.		This agent has an atmospheric lifetime greater than 1,000 years, with an estimate 100-year, 500-year, and 1,000-year GWP of 16,100, 26,100, and 32,803, respectively. Users should limit testing only to that which is essential to meet safety or performance requirements. This agent is only used to test new Halon 1301 systems.

TABLE 6-19L. *(Continued)*

Agent	Decision	Conditions	Comments
HFC-32 (6/95)	Unacceptable		Agent is flammable.
Water mist using potable or natural seawater	Acceptable		
Water mist surfactant Blend A	Acceptable		
Water mist surfactant Blend A	Acceptable in normally unoccupied areas.		

NOTES:
* Must conform with OSHA 29, CFR 1910, Subpart L, Section 1910.160 of the U.S. Code.
† Per OSHA requirements, protective gear (SCBA) must be available in the event personnel must reenter the area.
§ Discharge testing should be strictly limited only to that which is essential to meet safety or performance requirements.
¶ The agent should be recovered from the fire protection system in conjunction with testing or servicing, and recycled for later use or destroyed.

Fire Suppression and Explosion Protection Pending Substitutes

HFC Blend A	Pending receipt of further data requested by EPA.
IG-55 (formerly inert gas Blend B)	Proposed acceptable (forthcoming).
IG-01 (formerly inert gas Blend C) (7/95)	Proposed acceptable (forthcoming).
Water mist surfactant Blend A	Pending peer review for use in normally occupied areas.
Water mist systems with additives (7/95)	Must be individually submitted to EPA and reviewed on a case-by-case basis. No submissions have been received to date.

THERMOPHYSICAL PROPERTIES

Tables 6-19M and 6-19N give thermophysical properties of clean agent replacements from NFPA 2001,[1] in U.S. custom and SI units, respectively. Table 6-19O, extracted from reference 43, gives independent data and estimates for some thermophysical properties. Additional thermophysical and transport property data can be found in reference 11 for FM-200 and reference 43 for a range of halocarbon alternatives.

Representative isometric diagrams for various clean agents are given in Figures 6-19E, 6-19F, and 6-19G for HFC-227ea, HFC-23, and IG-541, respectively. The features of a typical isometric diagram for a liquified halocarbon, superpressurized to 360 psig, can be seen in Figure 6-19E. The importance of an isometric diagram is that it determines the maximum fill density for an agent in a cylinder with a fixed pressure rating. The basic rule is that the cylinder must not become liquid full at 130°F (54°C) (for U.S. DOT) and/or the pressure developed at 130°F (54°C) must not exceed $5/4$ cylinder design pressure. The pressure developed is a function of the agent, the superpressurization level, and the temperature. For 360 psig (at 70°F) cylinders, pressure is limited to $5/4$ of the design working pressure of 500 psig at 130°F. For HFC-227ea, as seen in Figure 6-19E, this yields a maximum fill density of 72 lb/cu ft. The 360 psig pressurization level is not based on characteristics of halon replacement halocarbons, but is a vestige of standard Halon 1301 systems.

Figure 6-19F is an isometric diagram for HFC-23. HFC-23 is not pressurized with nitrogen. At a maximum fill density of 54 lb/cu ft, it has a vapor pressure of 609 psi at 70°F. Above its critical point of 78.6°F, there is no distinction between liquid and gas, with no liquid/gas separation in the cylinder. Further, compression does not result in formation of a liquid phase. Since it is not possible for the cylinder to be "liquid" full at 130°F, the pressure developed at a maximum fill density of 54 lb/cu ft must be less than $5/4$ of the container design level working pressure of 1800 psi or 2250 psig, as shown in the graph. Below the critical temperature, the agent will exist as a liquid with a vapor or gas phase above it.

Figure 6-19G is a typical isometric diagram for inert gases exhibiting ideal gas law behavior at the pressures and temperatures of interest. Since these are compressed gases only, there is no fill density variation, only a linear relationship between temperature and pressure.

USE CONSIDERATIONS FOR CLEAN AGENT SYSTEMS

The primary advantages of total flooding clean agent systems are: (1) the ability to extinguish shielded, obstructed, or three-dimensional fires in complex geometries; (2) the ability, through the use of appropriate detector-based actuation, to extinguish fires at a very early stage, well before direct or indirect fire/smoke damage occurs; and (3) they cause no collateral damage due to agent discharge. These three technical attributes drive much of the usage of these agents in both flammable and combustible liquid hazards (e.g., shipboard engine rooms, pump rooms, etc.) and electronics equipment areas.

There are, however, technical disadvantages with respect to the use of clean agents:

1. They require a reasonably intact enclosure with the doors closed and external ventilation secured prior to discharge.
2. These agents provide virtually no cooling. This is an important consideration where large fires with long preburn times are expected.
3. Electrical power to wire/cable and equipment must be secured, or sufficiently high agent concentrations and hold times must be provided to ensure extinguishment at the expected electrical power density in the cable.
4. All elements of the detection, control, actuation, release, and distribution system must perform as designed. This places increased reliance on both acceptance testing and post-installation maintenance and test.

TABLE 6-19M. *Thermophysical Properties of Clean Halocarbon Agents (U.S. Custom Units)*

	Units	FC-3-1-10	HCFC Blend A	HCFC-124	HFC-125	HFC-227ea	HFC-23	IG-541	IG-55	IG-01
Molecular Weight	N/A	238.03	92.90	136.5	120.2	170.03	70.01	34.0	33.95	39.9
Boiling Point @ 760 mm Hg	°F	28.4	−37.0	12.2	−55.3	2.6	−115.7	−320	−310.2	−302.6
Freezing Point	°F	−198.8	≤−161.0	−326.0	−153	−204	−247.4	−109	−327.5	−308.9
Critical Temperature	°F	235.8	256.0	252.0	158.8	215.0	78.6	—	−210.5	−188.1
Critical Pressure	psia	337	964	524.5	521	422	701	—	602	711
Critical Volume	ft³/lbm	0.0250	0.0280	0.0283	0.0281	0.0258	0.0305	N/A	N/A	N/A
Critical Density	bm/ft³	39.3	36.00	35.28	35.68	38.76	32.78	N/A	N/A	N/A
Specific Heat, Liquid @ 77°F	Btu/lb-°F	0.25	0.30	0.270	0.301	0.2831	0.370	N/A	N/A	N/A
Specific Heat, Vapor @ Constant Pressure (1 atm) and 77°F	Btu/lb-°F	0.192	0.16	0.177	0.191	0.1932	0.176	0.195	0.187	0.125
Heat of Vaporization at Boiling Point @ 25°C	Btu/lb	41.4	97	83.2	70.8	57.0	103.0	94.7	77.8	70.1
Thermal Conductivity of Liquid @ 77°F	Btu/h ft°F	0.0310	0.052	0.0417	0.0376	0.040	0.0450	N/A	N/A	N/A
Viscosity, Liquid @ 77°F	lb/ft hr	0.783	0.508	0.723	0.351	0.443	0.201	N/A	N/A	N/A
Relative Dielectric Strength @ 1 atm @ 734 mm Hg 77°F (N_2 = 1.0)	N/A	5.25	1.32	1.55	0.955 @ 70°F	2.00	1.04	1.03	1.01	1.01
Solubility of Water in Agent @ 70°F	N/A	0.001% by weight	0.12% by weight	0.07% by weight @ 77°F	0.07% by weight @ 77°F	0.06% by weight	500 ppm @ 50°F (10°C)	0.015%	0.006%	0.006%
Vapor Pressure @ 77°F	psi	42.0	1.37	56	199	66.4	686.0	2207	N/A	N/A

The decision to use total flooding clean agents is a relatively complex risk-management decision. There are several recommendations on the use of these systems in the context of an overall fire protection program. Some common-sense recommendations are given below.

1. In general, it does not make sense to protect a hazard area with a total flooding gas system within a building in which all other areas are not protected. Consider the case of a sensitive electronics facility within an office building. The probability of a fire occurring outside the space is at least equal to, and probably greater than, a fire occurring inside the space. Given the cost of protecting the electronics space, the cost/benefit of providing alternative, e.g., automatic sprinkler protection, in the remainder of the building is much lower. For a fixed investment, it makes more sense to protect areas outside the electronics area first. As a general rule, never protect an area with a clean agent total flooding system unless the remainder of the building is adequately protected. Notable exceptions to this rule are cases where the only or predominant fire hazard is in the space to be protected with a clean agent.

2. In facilities where the room or enclosure geometry changes frequently, sufficient administrative control of such modifications is required to ensure a clean agent total flooding system is not rendered inoperable or useless by changes in partitioning, mod-

ifications to detection equipment, etc. While the same can be said for other types of fire protection, total flooding gas systems are particularly vulnerable to enclosure volume changes, enclosure integrity, and the integrity of the originally designed and installed detection, actuation, and control system.

3. Wherever possible, individual fire suppression equipment item failures should be anticipated and accounted for in the design. Examples include the use of higher design concentration to account for floodable volume changes or leakage changes, the addition of an extra cylinder to account for valve actuation failure, and provision of very early warning sampling detectors to permit human intervention prior to discharge.

4. Given the cost of halon replacement clean agent systems, it is expected that one significant use area will be in critically important sensitive electronics facilities. Substantial additional risk reduction at very high benefit/cost ratios may be realized by protecting these facilities with *both* clean agent total flooding systems and, at a relatively low incremental cost, standard wet-pipe automatic sprinklers. Since both systems are designed to perform at different levels, the provision of an additional wet-pipe sprinkler system can be viewed as a back-up system to the total flooding gas system. While one expects higher direct and indirect damage levels in sensitive electronics, due to the difference in actuation time of sprinklers, they are relatively inexpensive

TABLE 6-19N. *Physical Properties of Clean Halocarbon Agents (SI Units)*

	Units	FC-3-1-10	HCFC Blend A	HCFC-124	HFC-125	HFC-227ea	HFC-23	IG-541	IG-55	IG-01
Molecular Weight	N/A	238.03	92.90	136.5	120.2	170.03	70.01	34.0	33.95	39.9
Boiling Point @ 760 mm Hg	°C	–2.0	–38.3	–11.0	–48.5	–16.4	–82.1	–196		
Freezing Point	°C	–128.2	≤–107.2	–198.9	–102.8	–131	–155.2	–78.5		
Critical Temperature	°C	113.2	124.4	122.2	66.0	101.7	25.9	—		
Critical Pressure	kPa	2323	6647	3614	3595	2912	4836	—		
Critical Volume	cc/mole	371	162	241.6	210	274	133	N/A	N/A	N/A
Critical Density	kg/m³	629	577	565	571	621	525	N/A	N/A	N/A
Specific Heat, Liquid @ 25°C	kJ/kg°C	1.047	1.256	1.13	1.260	1.184	1.549	N/A	N/A	N/A
Specific Heat, Vapor @ Constant Pressure (1 atm) and 25°C	kJ/kg°C	0.804	0.67	0.741	0.800	0.8082	0.737	0.574		
Heat of Vaporization at Boiling Point @ 25°C	kJ/kg°C	96.3	225.6	194	164.7	132.6	239.6	220		
Thermal Conductivity of Liquid @ 25°C	W/m°C	0.0537	0.900	0.0722	0.0651	0.069	0.0779	N/A	N/A	N/A
Viscosity, Liquid @ 25°C	centipoise	0.324	0.21	0.299	0.145	0.184	0.083	N/A	N/A	N/A
Relative Dielectric Strength @ 1 atm @ 734 mm Hg 25°C (N_2 =1.0)	N/A	5.25	1.32	1.55	0.955 @70°F	2.00	1.04	1.03	1.01	1.01
Solubility of Water in Agent @ 21°C	N/A	0.001% by weight	0.12% by weight	0.07% by weight @ 77°F	0.07% by weight @ 77°F	0.06% by weight	500 ppm @ 50°F (10°C)	0.015%	0.006%	0.006%
Vapor Pressure @ 25°C	kPa	289.6	948	386	1371	457.7	4730	15200	N/A	N/A

TABLE 6-19O. *Selected Thermophysical Properties of Agents (extracted from reference 43)*

	FC-3-1-10	HCFC-227ea	HCFC-124	HFC-125	HFC-23
Boiling Point @ 0.101 MPa (°C)	–2.0	–16.4	–13.2	–48.6	–82.1
Crtical Temperature (°C)	113.2	101.7	122.5	66.3	25.6
Critical Pressure (Mpa)	2.32	2.9	3.65	3.62	4.82
Vapor Pressure @ 25°C (MPa)	0.27	0.47	0.38	1.38	4.69
Liquid Density at 25°C (kg/m³)	1497	1395	1357	1190	0.685
Liquid Heat Capacity @ Boiling Point (kJ/kg K)	0.951	1.074	1.080	1.107	1.269
Liquid Heat Capacity at 25°C (kJ/kg K)	1.017	1.177	1.111	1.358	—
Latent Heat of Vaporization at Boiling Point (kJ/kg)	96	131	162	160	240

and a reliable means of preventing excessively large losses in the event of the primary system (i.e., clean agent system) failure. For a typical 8-ft (2.4-m) high compartment, the additional cost of providing a simple wet-pipe sprinkler system as a secondary system can be as low as 5 percent of the installed clean agent system cost.

Clean agent total flooding systems can offer unequaled performance for the very early extinguishment of fires, which is critical in certain electronics facilities and the suppression of three-dimensional flammable liquid fires. However, the system must function as designed through the life of the facility. In the case of these total flooding gases, the system includes not only the detection, alarm, and suppression system, but the hazard enclosure, doors, dampers, ventilation/fans, and electrical power isolation. The need to control and maintain this "system" cannot be overstated.

CLEAN AGENT SYSTEM DESIGN

The basic process of designing a total flooding gaseous extinguishing system is outlined below:

1. Determine the design concentration,
2. Determine the total agent quantity,

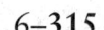

FIG. 6-19E. *Isometric diagram of HFC-227ea, pressurized to 360 psig with N_2 at 70°F.[30] For SI units: 1 lb per cu ft.*

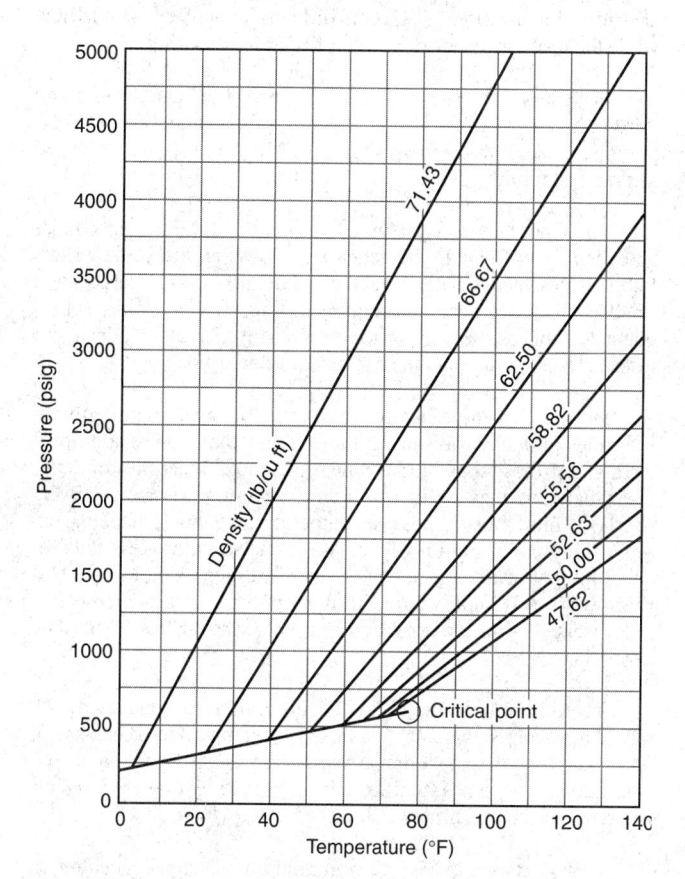

FIG. 6-19F. *Isometric diagram of HFC-23.[1] For SI units: 1 lb per cu ft.*

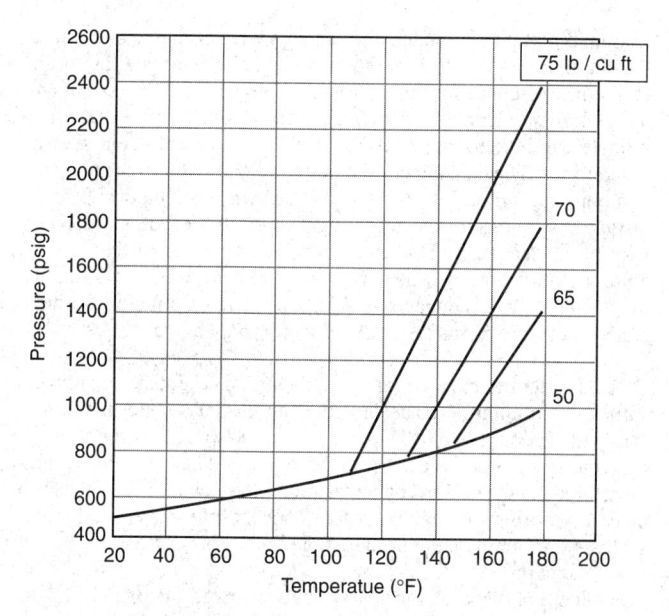

FIG. 6-19G. *Isometric diagram of inert gases and blends treated as ideal gases pressurized to 2175 psig at 70°F.[1]*

5. Design piping network and select nozzles to deliver required concentration at required discharge time to ensure mixing,

6. Evaluate compartment over/underpressurization and provide venting if required, and

7. Establish minimum agent hold requirements and evaluate compartments for leakage.

These attributes apply only to the mechanical design of the system.

The detection and actuation system is a critical and integral part of a clean agent system design. The detection system should be designed to actuate the system, with appropriate pre-discharge alarms, before unacceptable thermal or nonthermal damage occurs. This is particularly important, relative to the thermal decomposition where products of halocarbon clean agents are a concern. Section 5, Chapter 2, "Automatic Fire Detectors," provides engineering methods and calculation procedures for this purpose.

In addition to the detection, actuation, and alarm system, the enclosure itself is critical in the design of any total flooding gaseous suppression system. The most important considerations are that the enclosure's integrity is sufficient to: (1) prevent preferential agent loss during discharge and (2) prevent excessive agent/air mixture loss after discharge to ensure adequate hold time.

As a general rule, all openings, notably doors and ventilation fans and/or openings, must be secured prior to discharge in conjunction with the detection and alarm system. System installation in rooms with unenclosable openings should not be attempted with these agents, unless sufficient test data are available to ensure

3. Establish the maximum discharge time,

4. Select piping material and thickness consistent with pressure rating requirements,

adequate concentrations. Some enclosures, such as very tightly sealed low electromagnetic field (EMF) emission electronics spaces, require additional care to avoid compartment damage, due to over or underpressurization during agent discharge.

Design Concentrations

Design concentrations for various agents and fuel combinations are generally determined by a combination of small-scale testing, large-scale testing, independent laboratory approval of hardware, and addition of design safety factors.

Historically, minimum design concentrations for Halon 1301 were set by the cup burner extinguishing concentration plus a 20 percent safety factor. A minimum Halon 1301 design concentration of 5 percent was also established for all applications. For n-heptane, the cup burner value was approximately 3 percent; with a 20 percent safety factor, a design concentration of 3.6 percent is obtained. At the minimum design concentration set by NFPA 12A of 5 percent, a 66 percent safety factor was achieved. For fuels with cup burner extinguishing concentrations greater than 4.2 percent, the safety factor remained at 20 percent.

The basic requirement for determining the design concentration of clean agents in NFPA 2001 is twofold. First, the minimum extinguishing concentration as determined by the cup burner must be established. After this minimum is established by the *system* manufacturer, full-scale third-party approval testing is conducted using the manufacturer's hardware on n-heptane, wood crib, and selected flammable liquids. These tests are performed at the cup burner minimum extinguishing concentration, not the design concentration. Further, they are conducted with flooding factors lower than utilized in design. Hence, the minimum set by the cup burner or equipment manufacturer, whichever is higher, is tested in full scale as part of the approval/listing process for the agent/system combination. Often, hardware manufacturers will establish a minimum concentration greater than the cup burner value to account for nozzle efficiency.

Reliance on the cup burner to set a minimum value is supported by the observation that no full-scale test conducted at the cup burner concentration plus 80 percent safety factor has failed to extinguish a flammable liquid fire in an enclosure. It is critical to recognize that, in real system design, the agent and the hardware delivering it, particularly the nozzle, are not independent. They must be taken as a system and designed and installed per the listing/approval guidelines and limitations.

There has been some full-scale test work that indicates that the 20 percent safety factor may be insufficient for certain applications, depending on the performance required of the system. Sheinson noted significant improvement in extinguishing time performance with a safety factor of 40 percent.[44] Brockway noted similar results, with no performance improvement beyond a safety factor of 40 percent.[45]

It has also been noted by several investigators that higher safety factors result in lower thermal decomposition products.[44,45,46] None of the above referenced investigations utilized listed or approved hardware for the specific agents tested; as in most cases, the tests were performed before such hardware was available.

There was an exception in the first edition of NFPA 2001 to the general rule that a minimum extinguishing concentration be established by the cup burner method. It was alleged that reliable cup burner data were not available for HCFC Blend A due to two facts: (1) the agent was a blend and (2) one of the blend components heats at a low vapor pressure. In the case of this agent, a minimum extinguishing concentration of 7.2 percent and, hence, a design concentration of 8.4 percent was established through limited full-scale

testing. Since at the time insufficient data were available to evaluate the claim, the exception that requires full-scale testing at minimum extinguishing concentration consistent with UL 1058, *Halogenated agent Extinguishing Systems Units*,[65] was invoked. Since that time, reliable cup burner data were obtained for the blend from several laboratories. The data are consistent with MEC values for the blend components, primarily HCFC-22. Furthermore, some full-scale testing has indicated that the design concentration of 8.6 percent may be inadequate.[47] This exception was removed in the second edition of NFPA 2001. The current requirement for Class B fires is that the cup burner concentration for the worst-case fuel in the protected enclosure forms the MEC basis for that hazard.

It is important to recognize that the use of n-heptane cup burner values as a minimum for design concentration has no technical basis. With Halon 1301, n-Heptane was a reasonable "worst case" fuel, but this is not the case with the replacements. It is important, therefore, that the design concentration be based on the "worst case" or fuel requiring the highest design concentration when designing systems for flammable or combustible liquid risks.

Moore and MacGregor[48] have demonstrated that the use of n-heptane as a "standard" or reference flammable liquid fuel for cup burner evaluation of fire suppressants is not justified. The extinguishing concentration required for n-Heptane relative to other fuels varies substantially by agent type, e.g., for one agent n-heptane is relatively easy to extinguish (low MEC), and difficult for another (high MEC), while the result is reversed for other fuels such as diesel.

For Class A fires, NFPA 2001 requires full-scale testing and third-party approval for evaluating design concentration on solid polymeric materials. In many cases, the MEC for n-Heptane is used as practical minimum.

There has been no systematic evaluation of these agents under so-called "deep-seated" fire scenarios. However, the Underwriters Laboratories Inc. (UL) and Factory Mutual Research Corporation (FMRC) listing procedures require testing on wood cribs subsequent to long preburn times (approximately 5 min). Under these tests, surface oxidation and char reactions do occur.

Design concentrations for fire scenarios involving long preburn times in thick arrays of cellulosic fuels may require additional testing. For most applications where incidental quantities of cellulosic materials may be involved and preburn times are relatively short (< 5 min) time frames (i.e., automatic actuation), the flame extinguishing concentrations for Class A fuels that are less than, or equal to, that of n-heptane can be used. Surface oxidation or charring reactions do not occur with most polymers; hence, so-called "deep-seated" fires are not a concern where most Class A fuels are involved.

There have been thousands of full-scale tests conducted with halon replacement agents. Table 6-19P summarizes a subset of these data for a range of compartment sizes, agent types, agent concentrations, fuel types, and fire sizes. For all hydrocarbon agents listed, the discharge time was between 5 and 15 sec.

IG-541 is used at 37.5 percent minimum design concentration where Class A materials are involved. IG-55 may have slightly higher minimum design concentrations; IG-01 (argon) should have substantially higher MEC.

The minimum design concentration is a function of the fuel, the agent, and the delivery system. Design concentrations for specific hazards must be determined in accordance with the system manufacturer's approval or listing.

TABLE 6-19P. *Summary of Fire Extinguishment Test Results*

Compartment Size	Agent	Design Concentration (%)	Fuel/Fire Type	Extinguishment Time(s)	Reference
2000 ft³ (56.6 m³)	HFC-23	11.5 – 20.9	(5 approx. 500 kW) heptane/ pool spray	4 – 40	*
2000 ft³ (56.6 m³)	HFC-227ea	8.0 –10.8	(5 approx. 500 kW) heptane/ pool spray	7 –26	*
2000 ft³ (56.6 m³)	FC-3-1-10	5.7 –7.3	(5 approx. 500 kW) heptane/ pool spray	7–24	*
840 m³	HFC-23	16–18.0	~3–8 MW heptane and diesel spray/pool/wood crib	4 –12	†
840 m³	HFC-227ea	9.2–10	~3–8MW heptane and diesel spray/pool/wood crib	4 –28	†
560 m³	HFC-23	16.0	1 –5.5 MW heptane/diesel spray/pool	<2 –5	‡
560 m³	HFC-227ea	7.0	heptane/diesel spray/pool	<2 –17	‡
560 m³	HCFC Blend A	8.6	heptane/diesel spray/pool	30 –>90	‡
560 m³	FC-3-1-10	6.0	heptane/diesel spray/pool	<2 – 22	‡
41.8 ft³ (1.2 m³)	HFC-23	10.0 – 13.6	heptane pans	2 –24	§
41.8 ft³ (1.2 m³)	HFC-227ea	5.8 – 9.2	heptane pans	3.2 –22	§
41.8 ft³ (1.2 m³)	FC-3-1-10	4.8 – 6.6	heptane pans	3.2–21	§
41.8 ft³ (1.2 m³)	HCFC Blend A	10.9 –14	heptane pans	5–25	§
1027 ft³ (29.1 m³)	HFC-23	12.5 – 12.8	50 – 250 kW heptane pans	9.4 –10.4	§
1027 ft³ (29.1 m³)	HFC-227ea	7.3 – 9.4	50 – 250 kW heptane pans	5.1 – 9	§
1027 ft³ (29.1 m³)	FC-3-1-10	5.4 –7.1	50 – 250 kW heptane pans	6.1 – 10.1	§
2562 ft³ (72.5 m³)	HFC-227ea	7.0	shredded paper	8.2 – 25	‖
2562 ft³ (72.5 m³)	HFC-227ea	7.0	printed circuit board	2 –13.3	‖
2562 ft³ (72.5 m³)	HFC-227ea	7.0	PVC cable/heptane		
2562 ft³ (72.5 m³)	HFC-227ea	7.0	35 kW plastic magnetic-tape reels (baffled)	5.7 – 32.0	‖

NOTES:
* See reference 19.
† See reference 49.
‡ See reference 50.
§ See reference 51.
‖ See reference 52.

Safety Factors

The 20 percent safety factor requirement in NFPA 2001 used for developing design concentrations should be considered a minimum. It can reasonably be expected to account for minor errors arising from the design and installation process, such as errors in room volume calculations, cylinder filling/agent weight, and cup burner MEC values. Consideration should be given to increasing the safety factor and, hence, the minimum design concentration where any of the following conditions apply:

1. Where the nozzle is installed at or near its maximum listed ceiling height or area coverage;

2. For the protection of combustible or flammable liquid hazards where limiting quantity of thermal decomposition products (for halocarbon agents) and/or rapid (< 15 sec) extinguishment times are required;

3. Where a large number of highly imbalanced flow splits occurs in a piping network protecting large and small enclosures simultaneously;

4. Where the floodable room volume is highly variable (e.g., record storage), unless the maximum floodable volume is used;

5. Where unenclosable openings exist or pressure-relief vents are provided and located such that excessive agent loss during discharge will occur;

6. Where excessive post-discharge agent leakage is expected.

Increasing the safety factor or design concentration is not warranted in many applications and may not substantially improve the reliability or performance of the system. For example, increasing the safety factor will not improve system reliability if doors are not closed. The design safety factor is not a panacea and cannot be used in lieu of other system design requirements.

Agent Quantity

Once the design concentration is established, the quantity of agent necessary to achieve that concentration is determined. The quantity of halocarbon agent necessary is determined by the following equation:

$$w = \frac{V}{S}\left(\frac{C}{100 - C}\right)$$

where

w = specific weight of agent required,

S = specific volume (cu ft/lb or m^3/kg), and is determined by

$$S = k_1 + k_2(T)$$

where

T = minimum ambient temperature of the protected space,

k_1 and k_2 are constants,

V = net volume of protected space, and

C = design concentration (%).

Values for k_1 and k_2 used in Equation 2 are given in Table 6-19Q.

The flooding factor in Equation 1 [$C/(100 - C)$] implies that the agent/air mixture "lost" during discharge is well mixed and has an agent concentration of C. This formula makes no assumption regarding leakage of the enclosure. During UL/FMRC approval testing, the agent is evaluated with a flooding factor of ($C/100$), essentially assuming that losses during discharge are 100 percent air.

For inert gases, the following formula is used:

$$X = 2.303\frac{V}{S}\log\left(\frac{100}{100 - C}\right)V_S$$

where

X = volume of inert gas required at 70°F (21°C),

V_S = specific volume at 70°F (21°C),

TABLE 6-19Q. *Specific Volume Constants*

Agent	°F		°C	
	k_1	k_2	k_1	k_2
FC-3-1-10	1.409	0.0031	0.0941	0.0003
HCFC Blend A	3.612	0.0079	0.2413	0.00088
HCFC-124	2.352	0.0057	0.1578	0.0006
HFC-125	2.724	0.0063	0.1701	0.0007
HFC-227ea	1.885	0.0046	0.1269	0.0005
HFC-23	4.731	0.0107	0.2954	0.0012
IG-541	9.7261	0.0211	0.649	0.00237
IG-01	8.514	0.0185	0.5685	0.00208
IG-55	10.0116	0.0217	—	—

V = net protected hazard volume, and

S = specific volume at ambient temperature in protected volume (from Equation 2) = $k_1 + k_2(T)$.

The flooding factor used here [log ($100/(100 - C)$)] is derived assuming that leakage from the compartment during discharge occurs with a varying concentration of agent from zero to C, from beginning to the end of discharge. It is identical to the expression used in CO_2 system design. It assumes that the displaced atmosphere is freely vented from the enclosure.

Discharge Time

The maximum discharge time permitted for halocarbon clean agent systems is 10 sec. This discharge time is taken to be the point where all liquid agent has cleared the nozzle. The total discharge time will be longer as agent vapor and nitrogen is expelled from the system.

The 10-sec discharge time limitation for halocarbon agents is designed to aid four objectives:

1. Provide high flow rates through nozzles to ensure adequate mixing of agent with air inside the enclosure;
2. Provide sufficient velocity through pipes to ensure homogeneous flow of liquid and vapor;
3. Limit the formation of agent thermal decomposition products; and
4. Minimize direct and indirect fire damage, particularly in fast-developing fire scenarios.

The most important of these objectives relative to discharge time is the minimization of agent thermal decomposition product formation. Items 1 and 2 alone are determined by the piping system design.

The discharge time requirement for inert gases is currently 60 sec.[1] Longer discharge times are typically used for these systems in Europe. The primary reasons to constrain the discharge time of inert gas agents which form no decomposition products are to limit the direct and indirect fire damage and to minimize the length of time that the fire burns in a depleted oxygen atmosphere. As more information is developed on the effect of discharge time on inert gas agent performance, this 60-sec limit may be increased for certain applications.

In some applications, such as flammable liquid hazards and explosion inerting, it is necessary to discharge the agent quickly to minimize direct fire damage or to ensure that the agent concentration is achieved prior to the lower explosive limit (LEL) being reached.

Thermal Decomposition Products

All of the halocarbon replacement agents form higher levels of decomposition products than Halon 1301 under similar conditions. For a given fuel, the primary variables determining the level of decomposition products are: (1) the size of the fire at the time of discharge, (2) the time required to reach an extinguishing concentration in the compartment, and (3) the agent design concentration.

The dependence of thermal decomposition product formation on discharge time and fire size has been extensively evaluated.[44–47,51,53–55] Figure 6-19H is a plot of peak HF concentration as a function of fire size to room volume ratio. The data encompass room scales of 1.2 m^3 to 972 m^3. The 526 m^3 results are from USCG testing; the 972 m^3 results are based on NRL testing.

These fires include diesel and n-Heptane pool and spray fires. The design concentrations in all cases, except HCFC Blend A (at 8.6 percent), are at least 20 percent above the cup burner value. For fires where the extinguishment times were greater than 17 sec, the extin-

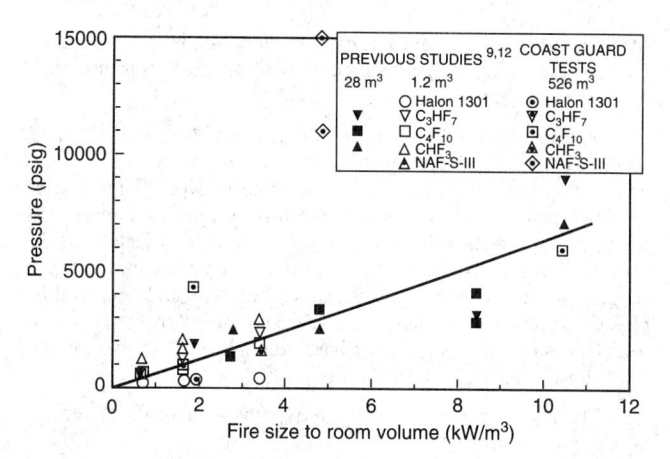

FIG. 6-19H. *Maximum HF concentration resulting from extinguishment of test fires with nominal 10-sec discharge times with data from U.S. Coast Guard Research and Development Center testing[47] and from Naval Research Laboratory testing.[51] (For U.S. Custom units: 1 cu ft = 0.0283 m³.)*

guishment time is noted in braces. Note that excessively high extinguishment times (> 60 sec), generally an indication of inadequate agent concentrations, yield qualitatively higher HF concentrations. In addition, Halon 1301 will yield bromine or bromine acids as well as HF. Likewise, HCFC Blend A will produce HCl in addition to HF.

The quantity of HF formed is approximately three to eight times higher for all halocarbon replacements relative to Halon 1301. There may be differences between the various HFC/HCFC compounds tested, but it is not clear from these data whether: (1) such differences occur, (2) are the differences attributable to agent mixing and distribution, or (3) attributable to locally high velocities or concentrations of agent from the nozzle. In all of the data reported, the fire source, n-heptane pans of varying sizes, was baffled to prevent direct interaction with the agent jet.

While these results are based on flammable and combustible liquid fires, solid polymeric fires of similar size will generally produce lower HF concentrations. Figure 6-19I reports 10-min average HF concentration data for the 1.2 m³ and 29 m³ enclosures, as opposed to the peak values shown in Figure 6-19H. (For U.S. units: 1 cu ft = 0.0283 m³.) These 10-min average values show HF concentrations 50 percent (or less) of the peak values.

Figure 6-19J shows the dependence of the quantity of decomposition products formed on the discharge time. Increased discharge time will also result in longer extinguishment times. The relationship is approximately linear, with a doubling in discharge time resulting in a 100 percent increase in HF production. Clearly, increasing the amount of time the flame is exposed to a subextinguishing concentration of agent will increase the amount of agent decomposition expected.

Sheinson[56] has shown that increasing the design concentration of HFC-227ea from 8 percent (approximately 1.2 × cup burner) to 9.8 percent (1.8 × cup burner) results in a decrease in peak HF concentration from 8000 ppm to 2500 ppm for real-scale large combustible liquid fires. He also found HF concentrations 4 to 5.5 times higher than the peak HF concentration of Halon 1301 under similar conditions. He did not report HBr or bromine formation in these tests.

Although other thermal decomposition products have been identified in some cases, it appears that HF is the primary decomposition product of interest relative to human safety and equipment damage.

One of the characteristics of HF, like HCl, is that it is an irritant gas, detectable at very low concentrations; however, there is a very large difference between the human detection threshold, the severe sensory irritant threshold, and the ALC (approximately two and three orders of magnitude, respectively).

Fire sizes necessary to generate short-term lethal concentrations of HF in an enclosure (on the order of 1000 ppm) are large enough in some cases that the hazard posed to personnel in the protected space during a discharge in a fire incident is primarily from the fire and its effects rather than the secondary impact of agent decomposition products. This, however, should be verified for a particular application under a range of fire scenarios using engineering methods discussed by Hanauska et al.[57,58]

The impact of decomposition products on electronic equipment is another area of concern. There is not sufficient data at present to predict the effects of a given HF exposure scenario on all electronic equipment. Several evaluations of the impact of HF on electronic equipment have been performed relative to the decomposition of Halon 1301, where decomposition products include HF and HBr. One of the more notable evaluations was a NASA study where the space shuttle Orbiter electronics were exposed to 700, 7000, and 70,000 ppm HF and HBr[57]. In these tests, exposures up to 700 ppm HF and HBr caused no failures. At 7000 ppm, severe

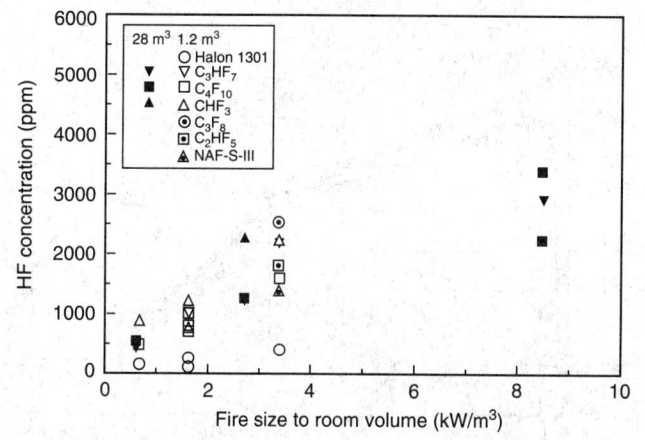

FIG. 6-19I. *Average HF concentration resulting from extinguishment of n-Heptane fires with nominal 15-sec total discharge time.[51] (For U.S. Custom units: 1 cu ft = 0.0283 m³.)*

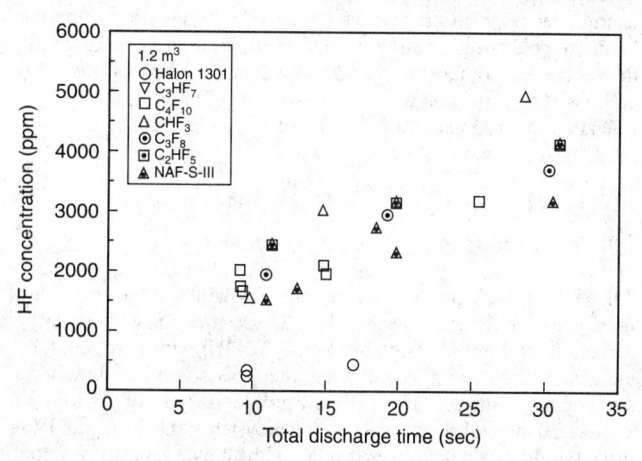

FIG. 6-19J. *Maximum HF concentration resulting from extinguishment of 4.0 kW n-Heptane fires.[51] (For U.S. Custom units: 1 cu ft = 0.0283 m³.)*

corrosion was noted; there were some operating failures at this level.

Dumayas[59] exposed IBM-PC-compatible multifunction cards to environments produced by a range of fire sizes as part of an evaluation program on halon alternatives.[59] He found no loss of function of these boards following a 15-min exposure to a postfire extinguishment atmosphere up to 5000 ppm HF, with unconditioned samples stored at ambient humidity and temperature conditions for up to 30 days. Forssell et al.[60] exposed multifunction boards for 30 min in the postfire extinguishment environment; no failures were reported up to 90 days posttest. HF concentrations up to 550 ppm were evaluated.

While no generic rule or statement can be made at this time, it appears that short-term damage (< 90 days) resulting in electronic equipment malfunction is not likely for exposures of between 500 and 1000 ppm HF for up to 30 min. This, however, is dependent on the characteristics of the equipment exposed, post-exposure treatment, exposure to other combustion products, and relative humidity. Important equipment characteristics include its location in the space, existence of equipment enclosures, and the sensitivity of the equipment to damage.

All HCFC, PFC, and HFC clean agents form more decomposition products than Halon 1301, given similar fire sizes and discharge times. The primary variable controlling the quantity of decomposition products is the size of the fire at the time of agent discharge. Through evaluation of the fire size at the time of system actuation, the potential hazard posed can be adequately managed.

Hanauska et al.[57,58] have indicated that the degree of decomposition products in electronic applications of agents can be safely managed where automatic detection and actuation is provided. Full-scale testing with typical Class A fuel packages in conjunction with typical detection system installation[60] has shown that the level of decomposition products is acceptable in typical computer/electronics spaces. For installation in hazard areas where very rapidly developing large fires are likely, the degree of thermal decomposition formation should be evaluated in the context of the hazard posed by the fire and the performance of alternative fire protection systems.

For large flammable or combustible liquid fire hazards utilizing manual system operation, the concentration of thermal decomposition products may be high enough to preclude entry into the space without special precautions. The use of an inexpensive water spray, wash-down cooling system in conjunction with a gaseous suppressant will provide substantial cooling, a reduction in the total decomposition products generated, and a reduction in the post-discharge decomposition product concentration due to scrubbing. While the enhanced cooling would benefit all total flooding gases, the reduced decomposition product potential applies only to halocarbons. The provision of such a system would permit more rapid re-entry into the space and may be of particular benefit for shipboard engine room protection.

Hydraulic and System Discharge Characteristics

All of the halocarbon replacement agents will exhibit two-phase flow behavior. Since all, except HFC-23, are used in cylinders pressurized to 360 or 600 psig (2482 or 4137 kPa) with nitrogen, they are also multiple component flows. Inert gases and blends will be single-phase compressible fluid flow problems. Most of the following discussion will not apply to inert gas systems; the testing and approval of flow calculations is similar, with emphasis placed on flow parameters relevant to compressible gas flow, ignoring phase separation issues such as the orientation. As in the case of engineered Halon 1301 systems, all flow calculation procedures used must be listed or approved by a third-party laboratory and within the limitations of the flow calculation method determined during the engineered system approval process.

The characteristics that differentiate two-phase pipe flow from incompressible fluid (e.g., water) pipe flow is the existence of gas and liquid phases simultaneously in the pipe network. This coupled with the relatively short flow times results in significant challenges to correctly predicting the flow. Among the important factors are the change in density of the fluid with pressure, the release of nitrogen in the cylinder and pipe as the fluid pressure and temperature change, differences in agent mass delivered caused by the flow time imbalances between nozzles, and preferential distribution of phases (and subsequently agent mass) at tee splits.

The need for accurate flow predictions is driven by three design requirements:

1. Control of agent discharge time;
2. Maintenance of adequate nozzle flow and pressure to ensure agent distribution and mixing at the listed coverage area; and
3. Delivery of adequate, but not excessive, agent quantities to different rooms within the same protected area, when such rooms are flooded simultaneously.

In addition, agent flow rate and thermodynamic state properties are necessary for estimating compartment pressurization levels during agent discharge.

For pre-engineered systems, limits on discharge time and nozzle pressure are built into the limits on piping system geometry. Agent distribution is handled by constraining preengineered systems to balanced flow conditions (i.e., the same agent mass is distributed from each nozzle). For adequate design of preengineered systems, accurate methods for predicting these elements are required.

Figure 6-19K is an idealized plot of cylinder and nozzle pressure during discharge. The discharge process can be divided into five sections, as shown schematically in Figures 6-19L through 6-19O.

Throughout the discharge process the amount of agent vapor and liquid as well as dissolved and gaseous nitrogen varies. As the pressure decreases in the cylinder and piping system, more agent is vaporized, and nitrogen is released from solution in the agent. The formation of additional vapor and nitrogen bubbles lowers the average density of the fluid. The rapid vaporization of agent is more

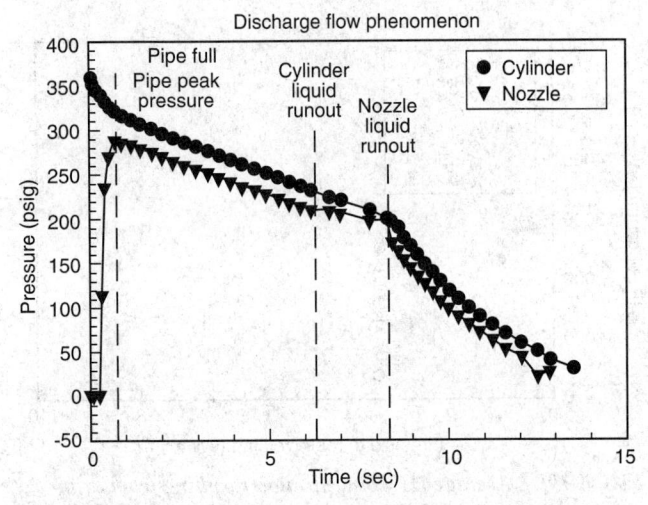

FIG. 6-19K. *Idealized cylinder and nozzle pressure time curves for halocarbon agents. (For SI units: 1 psi = 6.895 kPa.)*

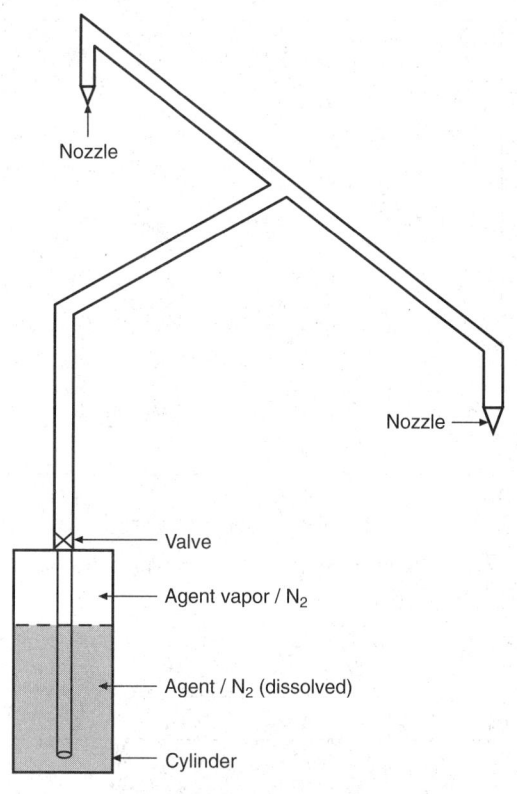

Nozzle

Nozzle

Valve

Agent vapor / N_2

Agent / N_2 (dissolved)

Cylinder

FIG. 6-19L.   Initial conditions.

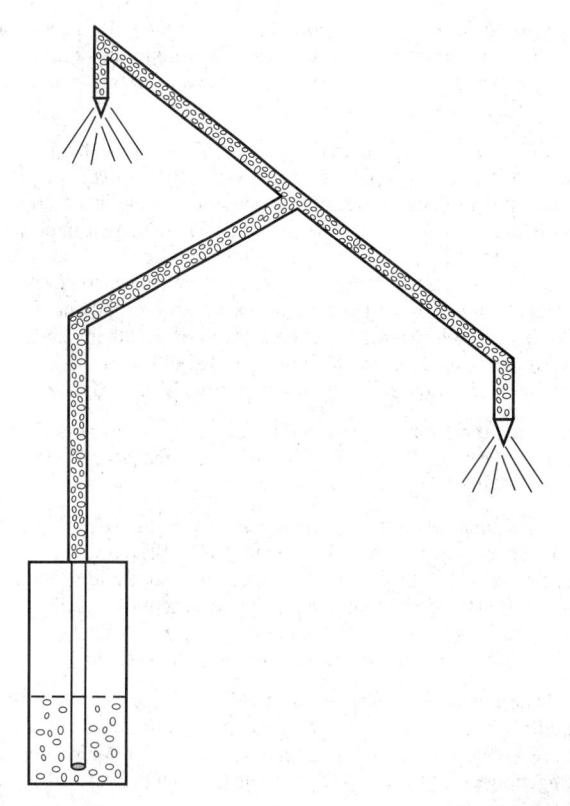

FIG. 6-19N.   Quasi-steady flow, liquid throughout network.

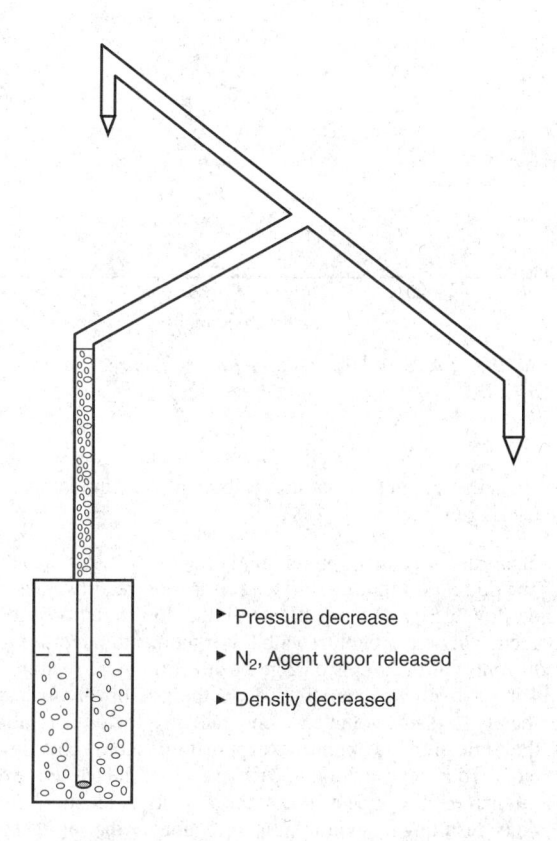

▸ Pressure decrease

▸ N_2, Agent vapor released

▸ Density decreased

FIG. 6-19M.   Valve open, pipe filling.

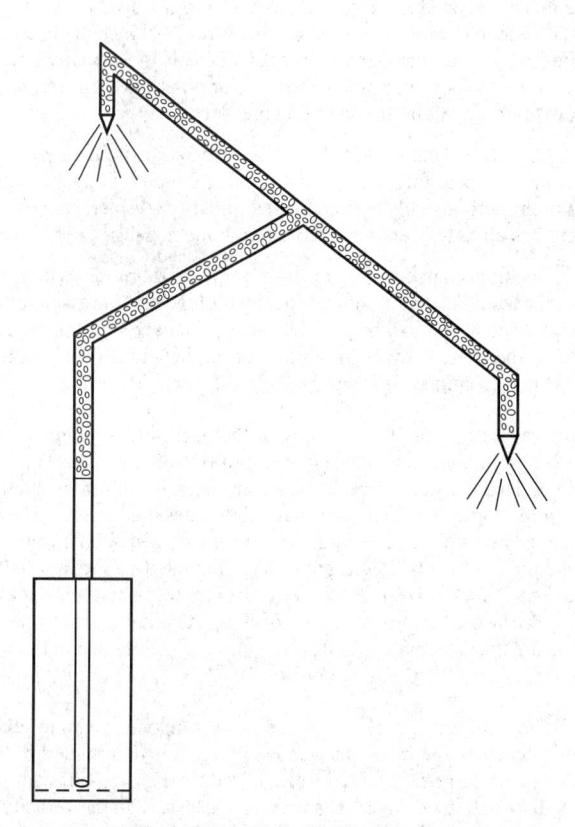

FIG. 6-19O.   Cylinder liquid runout.

pronounced in low boiling point/high vapor pressure agents. The fluid temperature also varies with time and along the length of the piping network, impacting the degree of agent vaporization and nitrogen release as well as liquid agent density.

The first step in the discharge process is filling the pipe with agent. (See Figure 6-19M). The rate at which this occurs is driven by the speed of the agent interface moving through the network. This is driven by either the sonic velocity at the agent interface or the discharge of displaced air through the nozzle. This phase determines the time at which the agent discharges from each nozzle. For systems with high degrees of imbalance in terms of flow path length or large pipe volume differences between nozzles, there can be significant delay in agent reaching one nozzle before another. This has a dramatic effect on the distribution of mass from each nozzle.

Once the agent reaches the nozzle and is compressed in the pipeline, the so-called nozzle peak pressure is reached. At this point, agent is discharging from each nozzle.

The next step in the discharge process is the so-called quasi-steady agent flow regime. (See Figure 6-19N.) This is generally the longest portion of the discharge, particularly for systems with low pipe volume to agent volume ratios. This point of the discharge process is the basis for the simplified pressure drop calculations embodied in NFPA 12A for balanced Halon 1301 systems.

The next milestone during the discharge process is cylinder liquid runout, where no liquid agent remains in the cylinder. (See Figure 6-19O.) At this point, an interface between liquid agent and nitrogen/agent vapor forms and travels through the network.

When the trailing liquid/vapor interface reaches the first nozzle, nozzle liquid runout occurs. This is important in two ways. First, this liquid runout occurs at different times during the discharge for each nozzle and can significantly impact the quantity of agent flowing from any given nozzle. Second, it is possible in many circumstances to discharge sufficient vapor/gas mixture from the first nozzle to reduce the pressure in the piping below that necessary to flow the remaining agent in the network. This is especially important for low vapor pressure agents and for nozzle designs requiring relatively high minimum operating pressures.

Once all of the nozzles have been cleared of liquid agent, the system is discharging a combination of nitrogen and agent vapor. This regime is usually ignored, since most (> 95 percent) of the agent has already been delivered through the nozzles.

The importance of the pipe filling and nozzle runout with these alternatives is relatively more critical with low vapor pressure alternatives, due to the inability of the agent to deliver significant pressure to the system by boiling, and the higher fluid densities that occur in the piping relative to Halon 1301.

Flow regime: If the flow velocity of the agent in the piping is not high enough, the flow may separate into two distinct phases in the piping. This causes severe problems at tee splits and in evaluating pressure drops. Therefore, minimum flow rates that ensure a homogeneous mixture of agent liquid and vapor/nitrogen bubbles must be maintained. Various flow regimes are illustrated in Figures 6-19P and 6-19Q for horizontal and vertical pipe, respectively.[61] One of the objectives of approval testing of flow calculation procedures is to ensure that homogeneous flow regimes are maintained in the piping throughout the discharge process.

Flow division at tees: For a single component, single-phase flow condition, the flow split at a tee junction is determined by the flow rate of the nozzles downstream of the tee. For two-phase fluids, flow distribution occurs at tees which are sensitive to the velocity of the flow along each branch of the tee, the orientation of the tee, the

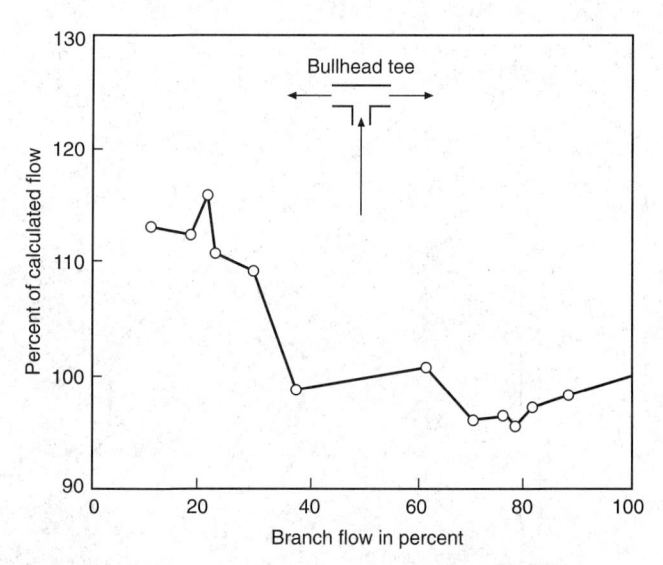

FIG. 6-19P.   *Bullhead tee flow split corrections for Halon 1301.*[61]

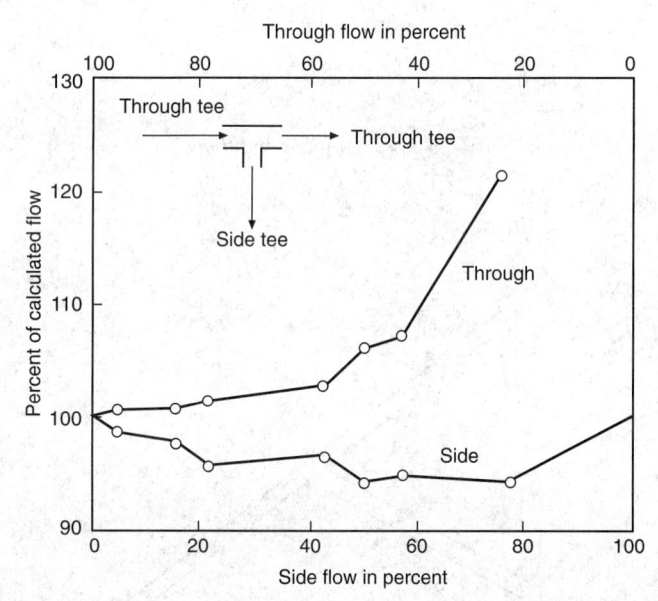

FIG. 6-10Q.   *Side and through tee flow split corrections for Halon 1301.*[61]

pressure at the tee, and the phase distribution of the fluid (gas or liquid) entering the tee.

The primary cause of preferential flow splits at tees is the inertia of the liquid *vs.* vapor/gas phase. This is most readily envisioned for side-flow tees, where one branch of the flow is required to turn 90 percent. Gas/vapor bubbles with lower momentum relative to the liquid agent will make this change of direction more readily. This results in relatively less mass flow down the side-flow branch at approximately the same volumetric flow rate or velocity. For bullhead tees, the same problem applies, except that it is more subtle involving velocity differences through each branch of the tee. For evenly split (50 percent/50 percent) flows, the velocity is identical in both directions, resulting in no flow split correction; as the split becomes greater the velocity differences are greater, and inertial effects of the gas/vapor relative to the liquid cause significant redistribution of mass through each branch of the tee.

The dependence was understood for Halon 1301 and described in detail by Williamson.[62] Similar processes occur in all two-phase flows, including air/water, steam/water, and refrigerant flows. In the context of clean agent system design calculations, this flow distribution is dealt with using empirical factors that redistribute the flow relative to the pure pressure-driven flow distribution that would occur without preferential phase distribution at tees.

Figures 6-19P and 6-19Q illustrate these correction factors for Halon 1301 flows in bullhead and side-flow tees, respectively.[62] All of the halocarbon agent flow predictions require similar treatment. Side-flow tees and bullhead tees require independent empirical correction factors. One of the most important limitations to any flow calculation procedure is the maximum flow split allowed for each type of tee. For a bullhead tee, as one moves further away from 50 percent/50 percent splits, the correction factor becomes greater, and at some point usually in the range of 80 percent/20 percent, it becomes so large that the prediction becomes unreliable. For allowable side-flow splits, ranges between $^{75}/_{25}$ percent and $^{90}/_{10}$ percent are typical. This correction of flow splits at tees is one reason that final approval of engineered system designs should be constrained to calculation methods that have undergone testing within the range of flow splits required.

Pressure drop due to friction loss: The pressure drop caused by friction in the pipeline is calculated differently for two-phase fluids. The presence of agent vapor and gas affects the pressure drop per unit length of pipe. There are numerous methods for dealing with two-phase fluid pressure drop.[63,64] Those typically used for fire suppression agent calculations involve either correcting the pressure drop estimated for single-phase fluid as a function of liquid to vapor/gas volume fraction, or empirical correlations of the pressure drop to average fluid density. Figure 6-19R illustrates the dependence of pressure drop on liquid volume fraction. In all cases, for purposes of design of fire protection systems, the pressure drop is calculated on the basis of a homogeneous flow assumption where changes in the liquid fraction are seen as density changes in the assumed homogeneous fluid.

Effect of system parameters on flow behavior: The variation of important system flow characteristics (discharge time, nozzle pressure, and flow rate) with cylinder and system parameters is a source of some confusion. Table 6-19R presents a simple summary of the effects of system parameters and illustrates some of the important variations.

Testing and approval of design methods: The approval or listing of a clean agent flow calculation procedure including inert gases is part of the approval granted for engineered systems. Since some aspects of two-phase flow calculations are empirically based (e.g., flow regime, pressure drop, flow splits) and all calculation procedures have some bounds on their validity, testing is performed to

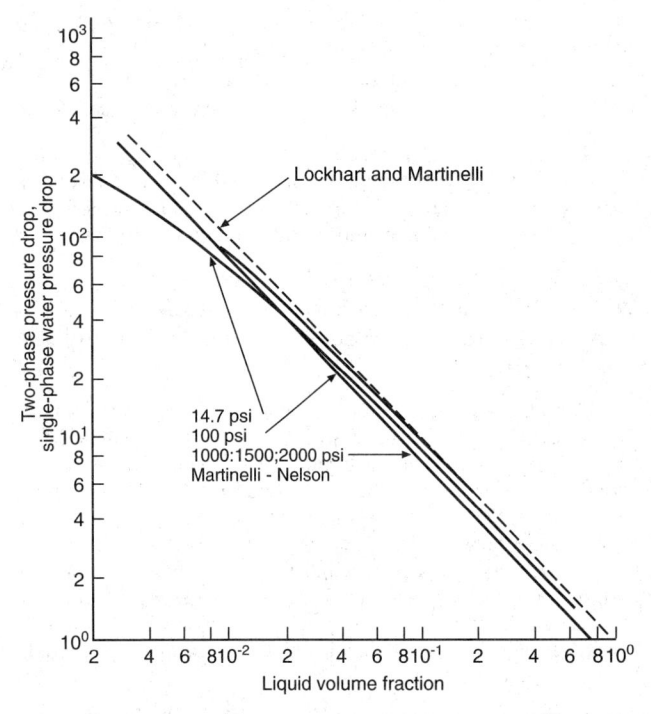

FIG. 6-19R. *Pressure drop vs. liquid volume fraction.[63] (For SI units: 1 psi = 6.895 kPa.)*

verify the predictions and establish the limits of the calculation procedure. These limitations are crucial in helping to ensure that system designs do not exceed verified limits of calculation.

One of the most rigorous approval procedures used in verifying design methods is outlined by Underwriters Laboratories Inc.[65] UL 1058, *Halogenated Agent Extinguishing Systems Units*, was used for evaluating engineered Halon 1301 systems, but the same approach is taken for all clean agent alternatives. Design method limitations are described by ten parameters, and tests are required to verify the accuracy of the calculation procedure at all of these limits. Some of these parameters will not apply to inert gas systems.

1. Percent of agent in piping (maximum);
2. Minimum and maximum discharge times;
3. Minimum pipeline flow rates;
4. Variance of piping volume to each nozzle;
5. Maximum variance of nozzle pressures within a piping arrangement;
6. Maximum ratio of nozzle diameter to inlet pipe diameter;
7. Arrangement most likely to exhibit vapor time-imbalance condition at nozzle;

TABLE 6-19R. *Effects of System Parameters*

Parameters (holding all other variables constant)	Variation	Average Cylinder Pressure	Nozzle Pressure	Discharge Time
Cylinder fill density	increase	decrease	decrease	increase
Pipe volume	increase	decrease	decrease	increase
Pipe length	increase	—	decrease	increase
Nozzle area	increase	decrease	decrease	decrease
Valve equivalent length	increase	—	decrease	increase

8. All types of tee splits, including through tees, bullhead tees, and tee orientation (vertical *vs.* horizontal);
9. Minimum and maximum container fill density; and
10. Minimum and maximum flow split for each type of tee.

These parameters are related to the important attributes of the agent discharge process previously discussed. Full-scale testing is performed to evaluate the performance of the design method. The limits on flow calculation method performance are as follows:

1. Actual *vs.* predicted discharge time ±1 sec;
2. Actual *vs.* predicted nozzle pressure ±10 percent; and
3. Actual *vs.* predicted mass flow through a nozzle, −5 + 10 percent.

The accuracy limits required for mass flow rate through a nozzle, particularly the −5 percent allowable variation, are quite challenging given the complex nature of the problem. This allowable variation points to an area where increased agent quantity (or safety factor) may be appropriate. For networks where a small fraction of the total agent flow is directed to a simultaneously protected but segregated enclosure *and* a large number of tees or flow splits are involved, the provision of additional agent to provide an allowance for the mass flow accuracy through greater than 5 or 6 tees should be considered. This is generally not a problem where all of the agent is discharged into the same enclosure.

Testing in conjunction with a particular manufacturer's hardware is important. Two critical hardware-dependent verifications are: (1) ensuring that pressure drop through a particular valve assembly is calculated properly and (2) the nozzle orifice discharge coefficient evaluation.

Several generic flow calculation routines have been developed.[51,66–69] Of these, two are directed at single-nozzle systems with very short discharge times,[68] or relatively simple balanced networks.[67] It is not recommended that any generic calculation procedure be used for final design purposes, unless it has been tested with the specific hardware to be installed and the system is within the limitations derived by tests.

In order to preserve a 10-sec discharge time, the mass flow rate of these clean agents must be higher than Halon 1301. The increased density of some of the alternative agents in the piping, caused by lower vapor pressures and nitrogen solubility differences, may result in high enough mass flow rates to retrofit existing Halon 1301 systems. While agent cylinders and nozzles will require changeout, it is often possible to preserve the existing Halon 1301 pipe network. This often requires the use of lower fill density cylinders to increase the average system pressure throughout the discharge time. Any such retrofit using existing Halon 1301 piping must be carefully evaluated with respect to hydraulic performance, with particular care given to preserving minimum required nozzle pressures and flow divisions at tees.

Nozzle Area Coverage and Height Limitations

One of the most important requirements of a gaseous total flooding fire suppression system is the ability of the system to deliver a uniform concentration of agent throughout the protected enclosure. The nozzle design and minimum nozzle pressure are critical in ensuring this distribution of agent. The performance of the nozzle is evaluated by full-scale approval testing, such as UL 1058.[65] The basic testing performed to evaluate nozzles is as follows:

1. Establish minimum nozzle pressure and maximum nozzle height by ensuring extinguishment of n-heptane fires located throughout a space with a height equal to the maximum allowable, at the minimum allowable nozzle pressure; and

2. Establish maximum nozzle coverage area by extinguishing tests in a plenum at the minimum height [generally less than 1.6 ft (0.5 m)] at the maximum nozzle coverage area [on the order of 1076 sq ft (100 m²)] and minimum nozzle operating pressure.

There are substantial differences between hardware manufacturers relative to minimum nozzle pressure, maximum ceiling height, and maximum average coverage. All nozzle orientations should be evaluated. In general, maximum nozzle heights are on the order of 13 to 16 ft (4 to 5 m), nozzles area coverage on the order of 97 to 108 sq ft (9 to 10 m²), and minimum nozzle pressures between 3 and 6 bar. It is critical to ensure that the nozzle spacing, height, and minimum pressure limits are not exceeded for a particular manufacturer's hardware in a specific design.

The flow, mixing, and distribution of an agent from a nozzle into an enclosure can be predicted theoretically for relatively simple nozzle designs, using sophisticated computer models.[67] Further development of such methods for complex nozzle designs and compartment geometries may eventually form the basis of a design procedure. At present, however, the primary means of ensuring adequate nozzle performance is the hardware approval process and real-scale testing.

Since many of the halocarbon replacements have lower vapor pressures than Halon 1301, there is often a much higher percentage of liquid at the nozzle. This makes the task of vaporizing and mixing the agent in the compartment more difficult. In general, nozzle designs used for Halon 1301 systems are not adequate for the halocarbon replacement agents. Due to the increased liquid fraction at the nozzle, it is critical to ensure that no unenclosed openings exist along the trajectory of the nozzle orifices. This may result in significant preferential loss of agent through these openings. This further emphasizes the need for third-party approval testing of nozzle performance. In any retrofit situation, the nozzles will need to be replaced even if the piping is adequately sized to deliver adequate agent flow rates.

Compartment Pressurization

The rapid discharge of agent into a compartment will cause rapid changes in the compartment pressure. Depending on the agent and rate of discharge, the initial pressure change may be negative. Figure 6-19S is a plot of compartment pressure *vs.* time for the discharge of HFC-227ea into a 989- cu-ft (28-m³) room with a 56-sq-in. (360 cm²) leakage area.[51] Immediately after discharge, the pressure in the

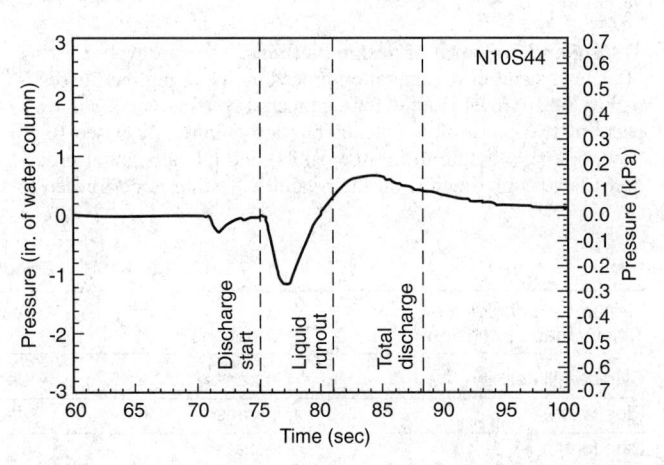

FIG. 6-19S. *Pressure measured in 989-cu-ft (28-m³) enclosure during HFC-227ea discharge with nominal 15-sec discharge time and 10-in. (254-mm) pan of n-Heptane.*[51]

compartment drops below ambient to a minimum of –0.3 kPa; at approximately 1.5 sec after discharge began, the pressure then begins to increase to a maximum of approximately 0.14 kPa after nozzle liquid runout. Similar results were obtained for FC-3-1-10. HFC-23 discharge exhibited much higher compartment overpressurization without the marked initial negative pressure. The maximum overpressure for HFC-227ea and FC-3-1-10 discharge was similar to that of Halon 1301. Inert gas agents will, in general, result in positive pressure throughout the discharge.

As the halocarbon agent is discharged into the space, it vaporizes rapidly, cooling the compartment and lowering the pressure. As the agent/air mixture gains heat from the walls or other objects in the space, the pressure recovers and, as additional agent is added, the pressure increases over ambient as mass is added to the compartment.

The expected maximum and minimum compartment pressure during discharge will be a function of:

1. Thermodynamic state of the agent at the nozzle (determined by the agent and piping arrangement),
2. Nozzle design,
3. Compartment volume and wall surface area,
4. Size of fire,
5. Initial conditions in space,
6. Leakage area from compartment, and
7. Agent flow rate.

For inert gases, significant compartment overpressurization can occur during discharge, unless adequate free vent area is provided. Calculation of required open area for venting is a part of the design manual for IG-541 systems.[70]

No generalized design procedure for calculating under/overpressurization has been established. Forssell and DiNenno[51] have developed a procedure for estimating the compartment pressure as a function of agent, agent flow rate, agent thermodynamic state at the nozzle, compartment volume, and surface area and leakage area. For inert gases utilizing relatively high agent concentrations, the required vent area calculation used for CO_2 systems is often used.

One area of concern with respect to vents is the possibility of preferential leakage of agent (as opposed to leakage of an agent/air mixture). This can occur where an uncloseable opening or vent opening during discharge is located where a jet (from the nozzle) with a high agent concentration directly impinges on the opening. In this case, the possibility of higher than expected (in the flooding factor assumption) agent losses during discharge may occur. This is particularly important for small rooms or areas with high wall area/room volume ratios and is probably more a problem for liquified halocarbon agents. Direct impingement by horizontal projection nozzles is most problematic where vents are located high on the walls.

Note that this preferential agent loss will not occur with the distributed leakage normally found in a room. A simple fix is to baffle any unenclosable opening or vent opening so that direct impingement by a nozzle jet on the opening is avoided. Ductwork attached to the relief vent discharge will also help prevent excessive agent loss.

Agent Hold Time and Leakage

Traditionally, total flooding gas systems were required to maintain a minimum concentration for a specified time period (10-20 min) after discharge. The minimum required hold time was a function of:

1. Soak time required for deep-seated Class A fuels;
2. Response time of emergency personnel; and
3. Time required to prevent reflash due to presence of hot surfaces and other reignition sources, particularly in flammable and combustible liquid applications.

Currently, there is no specified minimum hold or soak time for clean agents. The variables described above will vary among installations, and there is no significant data base on the performance of these agents on "deep-seated" fires other than wood cribs. The designer will be required to specify the minimum soak time consistent with the requirements of the hazard being protected. A practical minimum of 10 min is generally recommended, with some applications requiring substantially longer hold times.

The ability of a compartment to maintain adequate agent concentration is a function of the leakage of the compartment. Historically, this was done with Halon 1301 through the use of discharge tests. Discharge testing for this purpose was rendered unnecessary by the introduction of door fan pressurization leakage tests. Appendix B of NFPA 2001 describes a complete procedure for evaluating agent hold time as a function of compartment leakage measured by the door fan pressurization method.

The only difference between alternative agents and Halon 1301, in this regard, is the density of the agent/air mixture that is the driving force for leakage in quiescent environments. The mixture density can be estimated as follows:[1]

where

ρ_m = clean agent/air mixture density (kg/m³),

ρ_a = air density (1.202 kg/m³),

C = clean agent concentration (percent), and

V_d = agent vapor density at 21˚C (kg/m³).

Agent vapor densities at 70˚F (21˚C) are given below.

FC-3-1-10	0.615 lb/ft³ (9.85 kg/m³)
HBFC-22B1	0.346 lb/ft³ (5.54 kg/m³)
HCFC Blend A	0.240 lb/ft³ (3.84 kg/m³)
HFC-124	0.364 lb/ft³ (5.83 kg/m³)
HFC-125	0.316 lb/ft³ (5.06 kg/m³)
HFC-227ea	0.453 lb/ft³ (7.26 kg/m³)
HFC-23	0.182 lb/ft³ (2.915 kg/m³)
IG-541	0.089 lb/ft³ (1.43 kg/m³)
IG-55	0.088 lb/ft³ (1.41 kg/m³)
IG-01	0.106 lb/ft³ (1.70 kg/m³)
Halon 1301	0.392 lb/ft³ (6.283 kg/m³)

All agents, except inert gases, have higher mixture densities than Halon 1301 at 5 percent when used at their design concentrations. This will require slightly more leaktight enclosures to maintain the same hold time.

Note that IG-541 and IG-55 have densities very close to air. This will result in excellent concentration retention or, conversely, relatively high allowable leakage areas, minimizing problems associated with leak integrity testing and required hold time.

In general, the allowable leakage area calculated from the enclosure integrity test procedure given in Appendix B of NFPA 2001[1] will be very conservative. The actual hold time can be expected to be quite longer than the calculated hold time. This conservatism arises from the necessary, conservative simplifying assumption that the total equivalent leakage area measured by the door fan is evenly distributed at the top and bottom of the enclosure being tested.

SUMMARY

A wide range of inert gas and halocarbon total flooding clean agents has been introduced over the past several years. More will be commercialized in the near future. The use of an agent must be consistent with applicable environmental regulations. The selection of an agent is driven by its fire performance characteristics; agent and system space and weight concerns; toxicity, particularly for use in occupied areas; and the availability of approved system hardware.

The design of clean agent systems must be carefully done in accordance with third-party listing and approval limitations on both agent and hardware. Given the relative lack of experience with systems employing these new agents, particular care in design, installation, inspection, testing, and maintenance is warranted. Design and installation standards, such as NFPA 2001[1] and the associated hardware approval standards, form the minimum requirements for these new technologies.

As generalized design methods and more detailed requirements evolve, the ability to design and install systems on a performance basis will increase. A critical part of the installation process is post-installation inspection and testing. NFPA 2001[1] contains requirements for the approval and post-installation inspection and test of clean agent systems. Bearing in mind the relative complexity of these systems and the importance of the detection system and enclosure integrity, post-installation inspection and testing should be rigorously performed.

BIBLIOGRAPHY

References Cited

1. NFPA 2001, *Standard on Clean Agent Fire Extinguishing Systems, 1996*, 2nd ed., National Fire Protection Association, Quincy, MA, 1996.
2. Burgess, D., *et al.*, "Kinetics of Fluorine-Inhibited Hydrocarbon Flames," *Proceedings, Halon Options Technical Working Conference 1994*, Albuquerque, NM, May 3–5, 1994, pp. 489–500.
3. Battin-LeClerc, F., *et al.*, "The Chemical Inhibiting Effect of Some Fluorocarbons and Hydrofluorocarbons Proposed as Substitutes for Halons," *Halon Replacements: Technology and Science*, Miziolek and Tsang, eds., American Chemical Society, 1995, pp. 289–303.
4. Richter, H., *et al.*, "Flame Inhibition of Current Fire Extinguishers and of Potential Substitutes," *Halon Replacements: Technology and Science*, Miziolek and Tsang, eds., American Chemical Society, 1995, pp. 304–320.
5. Brown, R., and Bailey, J. L., "Effect of Pressure and Oxygen Concentration on Pan Fires in an Enclosed Space," NRL Ltr Rpt Ser 6180-25, Feb. 20, 1992.
6. Morehart, J. H., Zukowski, E. E., and Kubota, T., "Characteristics of Large Diffusion Flames Burning in a Vitiated Atmosphere," *Fire Safety Science—Proceedings of the Third International Symposium*, Elsevier Science Publishers, New York, 1991, pp. 575–583.
7. Sheinson, R. S., *et al.*, "The Physical and Chemical Action of Fire Suppressants," *Fire Safety Journal*, 15, 1989, pp. 437–450.
8. Hamins, A., *et al.*, "Flame Suppression Effectiveness," *Evaluation of Alternative In-Flight Fire Suppressants for Full-Scale Testing in Simulated Aircraft Engine Nacelles and Dry Bays*, Grosshandler, *et al.*, eds., NIST SP 861, National Institute of Standards and Technology, Gaithersburg, MD, Apr. 1994.
9. Hirst, R., and Booth, K., "Measurement of Flame Extinguishing Concentrations," *Fire Technology*, 13, 1977.
10. Moore, T. A., "Cup Burner Analysis," *Halon Substitute Program Review*, Albuquerque, NM, May 14, 1993.
11. Robin, M. L., "Properties and Performance of FM-200TM," *Proceedings* of the Halon Options Technical Working Conference 1994, Albuquerque, NM, May 3–5, 1994, pp. 531–542.
12. Moore, T. A., *et al.*, "Cup Burner Testing of Argon," NMERI 95/35/32630, Oct. 1995, New Mexico Engineering Research Institute, Albuquerque, NM.
13. Ferreira, M. J., *et al.*, "Thermal Decomposition Product Results Utilizing PFC-410," *Proceedings* of the Halon Options Technical Working Conference 1992, Albuquerque, NM, May 13, 1992.
14. Senecal, J., "Cup Burner Extinguishing Concentrations, Fe-13," CRC Technical Note 405, Rev. 1, Fenwal Safety Systems, Ashland, MA, Mar. 17, 1995.
15. MacGregor, L., and Moore, T., "Comparison of Extinguishing Values for Various Fuel/Agent Combinations," presented at 1995 International CFC and Halon Alternatives Conference, Washington, DC, Oct. 23–25, 1995.
16. Moore, T.A., *et al.*, "The Development of CF$_3$I as a Halon Replacement," New Mexico Engineering Research Institute, Albuquerque, NM, Nov. 1994.
17. Fidler, T., NFPA 2001, 1995 Fall Meeting Report.
18. Saito, N., *et al.*, "Improvement on Reproducibility of Flame Extinguishing Concentration Measured by Cup Burner Method," *Proceedings* of Halon Options Technical Working Conference, May 9–11, 1995, Albuquerque, NM.
19. Sheinson, R., *et al.*, "Halon 1301 Total Flooding Fire Testing, Intermediate Scale," *Proceedings* of the Halon Options Technical Working Conference 1994, Albuquerque, NM, May 3–5, 1994, pp. 43–53.
20. Moore, T. A., *et al.*, "Intermediate Scale (645 ft3) Fire Suppression Evaluation of NFPA 2001 Agents," *Proceedings* of the Halon Options Technical Working Conference 1993, Albuquerque, NM, May 11–13, 1993, pp. 115–127.
21. Sheinson, R. S., and Maranghides, A., "Halon Replacement n-Heptane Cup Burner Values," Naval Research Laboratory Letter Report 6180/0222, Apr. 1995.
22. Beyler, C. L., "Flammability Limits of Premixed and Diffusion Flames," *The SFPE Handbook of Fire Protection Engineering*, 2nd ed., DiNenno, P.J., *et al.*, eds., National Fire Protection Association, Quincy, MA, 1995.
23. Heinonen, E. W., "The Effect of Ignition Source and Strength on Sphere Inertion Results," *Proceedings* of the Halon Options Technical Working Conference 1993, Albuquerque, NM, May 11–13, 1993, pp. 565–576.
24. Moore, T.A., "Large-Scale Inertion Evaluation of NFPA 2001 Agents," *Proceedings* of the 1993 International CFC and Halon Alternatives Conference, Washington, DC, Oct. 20–22, 1993.
25. Senecal, J. A., "Agent Inerting Concentrations for Fuel-Air Systems," Fenwal Safety Systems CRC Technical Note No. 361, May 27, 1992.
26. Heinonen, E., "Laboratory-Scale Inertion Results," *Halon Substitutes Program Review*, CGET/NMERI, Albuquerque, NM, May 14, 1993.
27. Tamanini, F., "Determination of Inerting Requirements for Methane/Air and Propane/Air Mixtures by an Ansul Inerting Mixture of Argon, Carbon Dioxide, and Nitrogen," Factory Mutual Research Corp., Norwood, MA, Aug. 24, 1992.
28. Saito, N., *et al.*, "Flammability Peak Concentrations of Halon Replacements and Their Function as Fire Suppressants," *Halon Replacements: Technology and Science*, Miziolek and Tsang, eds., American Chemical Society, 1995, pp. 243–257.
29. Senecal, J. A., "Explosion Protection in Occupied Spaces: The Status of Suppression and Inertion Using Halon and Its Descendants," *Proceedings* of the 1993 International CFC and Halon Alternatives Conference, Washington, DC, Oct. 20–22, 1993, pp. 767–772.
30. Senecal, J. A., Ball, D. N., and Chattaway, A., "Explosion Suppression in Occupied Spaces," *Proceedings* of the Halon Options Technical Working Conference 1994, Albuquerque, NM, May 3–5, 1994, pp. 79–86.
31. Verdonik, D., "Introduction to Fire Suppression Agent Toxicology," Hughes Associates, Inc., Baltimore, MD, 1995.
32. *Casarett and Doull's Toxicology: The Basic Science of Poisons*, 4th ed., Amdur, M. O., Ph.D., Doul, J., Ph.D., M.D., and Klaassen, C. D., Ph.D., Pergamon Press, Inc., Maxwell House, Fairview Park, Elmsford, New York, 1991.
33. Skaggs, S. R., Moore, T. A., and Tapscott, R. E., "Toxicological Properties of Halon Substitutes," *Halon Replacements: Technology and Science*, Miziolek and Tsang, eds., American Chemical Society, 1995, pp. 99–109.
34. Rubenstein, R., "Halon Alternatives in Health Effects Assessment," *Proceedings* of the Halon Options Technical Working Conference 1995, Albuquerque, NM, May 9–11, 1995.
35. "Research Basis for Improvement of Human Tolerance to Hypoxic Atmospheres in Fire Prevention and Extinguishment," EBRDC Report 10.30.92, Oct. 1992, Environmental Biomedical Research Data Center, Institute for Environmental Medicine, University of Pennsylvania, Philadelphia, PA.
36. Verdonik, D., "Understanding the Environmental Impact of Halocarbon Fire Suppression Agents," Baltimore, MD, 1995.
37. *Reporting on Climate Change: Understanding the Science*, National Safety Council of the Environmental Health Center, Washington, DC, Nov. 1994.

38. *A Matter of Degrees: A Primer on Global Warming*, Ministry of Supply and Services, Ottawa, Canada, 1993.

39. Solomon, S., and Abritton, L. L., "Time-Dependent Ozone Depletion Potentials for Short- and Long-Term Forecasts," *Nature*, 33, p. 357, 1992.

40. Schlesinger, M. E., "Greenhouse Policy," *Research and Exploration—National Geographic Society*, Vol. 9, No. 2, Spring 1993.

41. *Climate Change 1994 Radiative Forcing of Climate Change and an Evaluation of the IPC 1992 Emission Scenarios*, Intergovernmental Panel on Climate Change, Cambridge, Cambridge University Press, London, 1994.

42. TELCOM with Dr. Tapscott.

43. Yang, J. C., and Bruel, B. D., "Thermophysical Properties of Alternative Agents," *Evaluation of Alternative In-Flight Fire Suppressants for Full-Scale Testing in Simulated Aircraft Engine Nacelles and Dry Bays*, NIST SP 861, National Institute of Standards and Technology, Gaithersburg, MD, Apr. 1994.

44. Sheinson, R. S., et al., "Halon 1301 Replacement Total Flooding Fire Testing, Intermediate Scale," *Proceedings* of Halon Options Technical Working Conference 1994, Albuquerque, NM, May 3–5, 1994, pp. 43–53.

45. Brockway, J. C., "Recent Findings on Thermal Decomposition Products of Clean Extinguishing Agents," 3M Report presented to NFPA 2001 Committee, Ft. Lauderdale, FL, Sept. 19–22, 1994.

46. Moore, T. A., et al., "Intermediate Scale (645 ft3) Fire Suppression Evaluation of NFPA 2001 Agents," *Proceedings* of Halon Options Technical Working Conference 1993, Albuquerque, NM, May 11–13, 1993, pp. 115–128.

47. Back, G. G., et al., "Full-Scale Machinery Space Testing of Gaseous Halon Alternatives," USCG R&D Center, Groton, CT, Sept. 1994.

48. Moore, T., and MacGregor, L., private communication, 1995.

49. Sheinson, R., et al., "Large-Scale (840 m3) HFC Total Flooding Extinguishment Results," *Proceedings* of the Halon Options Technical Working Conference 1995, Albuquerque, NM, May 9–11, 1995, pp. 637–648.

50. Hansen, R., et al., "USCG Full-scale Shipboard Testing of Gaseous Agents," *1994 International CFC and Halon Alternatives Conference Proceedings*, Washington, DC, Oct. 24–25, 1994, pp. 386–394.

51. Forssell, E. W., and DiNenno, P. J., "Evaluation of Alternative Agents for Use in Total Flooding Fire Protection Systems," Contract NAS 10-1181, National Aeronautics and Space Administration, John F. Kennedy Space Center, FL, Oct. 1994.

52. Forssell, E., et al., "Hazard Assessment of Thermal Decomposition Products of FM-200TM in Electronics and Data Processing Facilities," Hughes Associates, Inc., Baltimore, MD, Jan. 6, 1995.

53. DiNenno, P. J. et al., "Thermal Decomposition Testing of Halon Alternatives," *Proceedings* of the Halon Alternatives Technical Working Conference 1993, Albuquerque, NM, May 1993.

54. Purser, D. A., "Toxicity Assessment of Combustion Products," *SFPE Handbook of Fire Protection Engineering*, 2nd ed., DiNenno, et al., eds., National Fire Protection Association, Quincy, MA, 1995.

55. Dierdorf, D. S. et al., "Decomposition Product Analysis During Intermediate-Scale (645 ft3) Testing of NFPA 2001 Agents," *Proceedings* of the Halon Alternatives Technical Working Conference 1993, Albuquerque, NM, May 1993.

56. Sheinson, R., "The US Navy Halon Total Flooding Replacement Program: Laboratory through Full Scale," *Halon Replacements: Technology and Science*, Miziolek and Tsang, eds., American Chemical Society, 1995, pp. 175–189.

57. Hanauska, C. P., "Hazard Assessment of HFC Decomposition Products," presented at 1994 International CFC and Halon Alternatives Conference, Washington, DC, Oct. 1994.

58. Hanauska, C. P., et al., "Hazard Assessment of Thermal Decomposition Products of Halon Alternatives," *Proceedings* of the Halon Alternatives Technical Working Conference 1993, Albuquerque, NM, May 1993.

59. Dumayas, W. A., "Effect of HF Exposure on PC Multifunction Cards," Senior Research Project, Department of Fire Protection Engineering, University of Maryland, College Park, MD, 1992.

60. Forssell, E. F., et al., "Draft Report: Performance of FM-200 on Typical Class A Computer Room Fuel Packages," Oct. 1994, Hughes Associates, Inc., Columbia, MD.

61. Barnea, D., and Taitel, Y., "Flow Pattern Transition in Two-Phase Gas-Liquid Flows," *Encyclopedia of Fluid Mechanics*, Vol. 3, Cheremisinoff, N.P., ed., Gulf Publishing Company, Houston, TX, 1986.

62. Williamson, H. V., "Halon 1301 Flow in Pipelines," *Fire Technology*, Vol. 13, No. 1, 1976, pp. 18–32.

63. Chisholm, D., "Predicting Two-Phase Flow Pressure Drop," *Encyclopedia of Fluid Mechanics*, Vol. 3, Cheremisinoff, N.P., ed., Gulf Publishing Company, Houston, TX, 1986.

64. Hsu, Y. Y., and Graham, R. W., *Transport Processes in Boiling and Two-Phase Systems*, Hemisphere Publishing Corporation, Washington, DC, 1976.

65. Underwriters Laboratories, Inc., UL 1058, *Halogenated Agent Extinguishing Systems Units*, Underwriters Laboratories, Inc., Northbrook, IL, 1995.

66. DiNenno, P. J., et al., "Modeling the Flow Properties and Discharges of Halon Replacement Agents," *Proceedings* of the Halon Options Technical Working Conference 1994, Albuquerque, NM, May 1994.

67. Bird, E. B., et al., "Development of Computer Model to Predict the Transient Discharge Characteristics of Halon Alternatives," *Proceedings* of the Halon Options Technical Working Conference 1994, Albuquerque, NM, May 1994.

68. Cleary, T. G., et al., "Flow of Alternative Agents in Piping," *Proceedings* of the Halon Options Technical Working Conference 1994, Albuquerque, NM, May 1994.

69. Pitts, W. M., et al., "Fluid Dynamics of Agent Discharge," *Evaluation of Alternative In-Flight Fire Suppressants for Full-Scale Testing in Simulated Aircraft Engine Nacelles and Dry Bays,* Grosshandler, et al., eds., NIST SP 681, National Institute of Standards and Technology, Gaithersburg, MD, Apr. 1994.

70. Ansul Co., "Inergen System Design Installation and Maintenance Manual," Ansul Co., July 1994.

NFPA Codes, Standards, and Recommended Practices

Reference to the following NFPA codes, standards, and recommended practices will provide further information on halon replacement discussed in this chapter. (See the latest version of *The NFPA Catalog* for availability of current editions of the following documents.)

NFPA 12A, *Standard on Halon 1301 Fire Extinguishing Systems*
NFPA 2001, *Standard on Clean Agent Fire Extinguishing Systems*

Additional Reading

"3M Announces Halon Replacements," *Sprinkler Age*, Vol. 13, No. 2, February 1994, pp. 24, 26.

Abe, T., "Study for the Development of New Halon Alternatives Supported by Environmental Agency," *Proceedings* of the 1991 Halon Alternatives Technical Working Conference, April 30–May 1, 1991, Albuquerque, NM, 1991, pp. 307–311.

Adcock, J. R., et al., "Fluorinated Ethers: A New Family of Halons?" *Proceedings* of the 1991 Halon Alternatives Technical Working Conference, April 30-May 1, 1991, Albuquerque, NM, 1991, pp. 83–96.

Ahola, H., "High Sensitivity Smoke Detectors for the Fire Protection of Computer Rooms: An Alternative to Halons," Third International Symposium on Fire Protection of Buildings, May 10–12, 1990, Eger, Hungary, 1990, pp. 153–162.

Andreev, V. A., et al., "Replacement of Halon in Fire Extinguishing Systems," *Proceedings* of the 1993 Halon Alternatives Technical Working Conference, May 11–13, 1993, Albuquerque, NM, 1993, pp. 409–412.

Baker, R. W., et al. "Membrane Vapor Separation Systems for the Recovery of CFCs, Halons and Their Alternatives," *Proceedings* of the 1992 International CFC and Halon Alternatives Conference on Stratospheric Ozone Protection for the 90's, September 29–October 1, 1992, Washington, DC, 1992, pp. 869–877.

Bannister, W. W., et al., "Non-Volatile Precursors to Alternative Halon Fire Extinguishing Agents With Reduced Global Environmental Impacts," 1993 International CFC and Halon Alternatives Conference on Stratospheric Ozone Protection for the 90's. October 20–22, 1993, Washington, DC, 1993, pp. 761–766.

Bannister, W. W., et al., "Recent Advances in Development of Non-Volatile Precursors [NBPs] to Alternative Halon Fire Extinguishing Agents With Reduced Global Environmental Impacts," *Proceedings* of the 1993 Halon Alternatives Technical Working Conference, May 11–13, 1993, Albuquerque, NM, 1993, pp. 605–610.

Barondes, C. D., "Design Tool for the Evaluation of Replacements for Halon in Aircraft Dry Bay Fire Suppression Systems, Air Force Institute of Technology, Wright-Patterson AFB, OH, AFIT/GSE/ENY/92-D-02, December 1992, 299 pages. [AUTHORIZED TO U.S. GOVERNMENT AGENCIES ONLY]

Bennett, M., "Halon Replacement for Aviation Systems," *Proceedings* of the 1992 International CFC and Halon Alternatives Conference on Stratospheric Ozone Protection for the 90's, September 29–October 1, 1992, Washington, DC, 1992, pp. 667–670.

Burkhart, D. J., "Selecting Halon Alternatives: No Cause for Alarm," *Consulting-Specifying Engineer*, Vol. 15, No. 5, May 1994, pp. 54–56, 58.

Chorba, J., "Clean Agents for Critical Fire Protection Applications. Halon Alternatives," *Sprinkler Age*, Vol. 14, No. 2, February 1995, p. 15.

Cohn, B. M., "Laboratory Halon Systems: Rehab, Replace or Remove?" *Journal of Applied Fire Science*, Vol. 5, No. 1, 1995/1996, pp. 45–54.

"Construction of an Exploratory List of Chemicals to Initiate the Search for Halon Alternatives: Executive Summary, Background and Scope," *Fire Systems*, Vol. 5, No. 2, 1991, pp. 91–99.

Cortina, T., "Halon Alternatives Become the Norm in New Applications," *Consulting-Specifying Engineer*, Vol. 19, No. 5, May 1996, pp. 52–53.

Cortina, T., "Halon Alternatives Gain Industry Confidence," *Consulting-Specifying Engineer*, Vol. 17, No. 5, May 1995, pp. 54–56.

Cousin, C. S., "Potential of Fine Water Sprays as Halon Replacements for Fires in Enclosures," Fire Research Station, Borehamwood, England, SP Report 1994:03; *Proceedings* of the International Conference on Water Mist Fire Suppression Systems, November 4–5, 1993, Boras, Sweden, 1993, pp. 63–74.

DiNenno, P. J., "Perspective on Halon Replacements and Alternatives. Keynote," *Proceedings* of the 1993 Halon Alternatives Technical Working Conference, May 11–13, 1993, Albuquerque, NM, 1993, pp. ix–xxxi.

DiNenno, P. J. and Hanauska, C. P., "Design Issues for Halon Replacement Systems," *Proceedings* of the Technical Symposium on Halon Alternatives,. June 27–28, 1994, Knoxville, TN, 1994, pp. 9–15.

DiNenno, P. J., *et al.*, "Evaluation of Alternative Agents for Halon 1301 in Total Flooding Fire Suppression Systems: Thermal Decomposition Product Testing," *Proceedings* of the 1993 Halon Alternatives Technical Working Conference, May 11–13, 1993, Albuquerque, NM, 1993, pp. 161–184.

Eaton, H. G., *et al.*, "Evaluation of Halon Replacement Extinguishants in a Fire Environment: Gas Sampling and Analysis," *Proceedings* of the 1992 International CFC and Halon Alternatives Conference on Stratospheric Ozone Protection for the 90's, September 29–October 1, 1992, Washington, DC, 1992, pp. 633–635.

Edwards, R. J., "Propane Inerting Concentrations of Two Halon Replacement Gases," Worcester Polytechnic Inst., MA, Thesis, December 1993, 118 pages

Fernandez, R. E., "Du Pont's Halon Alternatives Program. Technical Progress," *Proceedings* of the 1991 Halon Alternatives Technical Working Conference, April 30–May 1, 1991, Albuquerque, NM, 1991, pp. 39–70.

Filipczak, R. A., "Relative Extinguishment Effectiveness and Agent Decomposition Products of Halon Alternative Agents," *Proceedings* of the 1993 Halon Alternatives Technical Working Conference, May 11–13, 1993, Albuquerque, NM, 1993, pp. 149–159.

"Fine Water Spray 'Is Alternative to Halon'," *Fire*, Vol. 85, No. 1050, December 1992, p. 20.

Floden, J. R., "Halon Replacement Program in U. S. Air Force," First International Conference on Fire Suppression Research, May 5–8, 1992, Stockholm, Sweden, 1992, pp. 319–325.

Floden, J. R., Scheil, G. and Klamm, S., "Final Firefighter Exposure Results: A Comparison of Halon 1211, HCFC 123 and PFH," *Proceedings* of the 1993 Halon Alternatives Technical Working Conference, May 11–13, 1993, Albuquerque, NM, 1993, pp. 645–658.

Forssell, E. W., *et al.*, "Predicting the Two-Phase Flow Properties of Halon Replacement Agents," *SFPE Bulletin*, Summer 1994, pp. 10–16.

Gameiro, V. M., "Fine Water Spray Fire Suppression Alternative to Halon 1301 in Gas Turbine Enclosures," *Proceedings* of the 1993 Halon Alternatives Technical Working Conference, May 11–13, 1993, Albuquerque, NM, 1993, pp. 317–344.

Gameiro, V. M. and Girard, P., "Fine Water Spray Fire Suppression Alternative to Halon 1301 in Machinery," 1993 International CFC and Halon Alternatives Conference on Stratospheric Ozone Protection for the 90's. October 20–22, 1993, Washington, DC, 1993, pp. 830–839.

Gann, R. G., "Halon Substitute Program," VHS Video Tape, National Institute of Standards and Technology, Gaithersburg, MD, National Institute of Standards and Technology, Federal Fire Forum on Technology Update, November 1, 1990, Gaithersburg, MD, 1990.

Gann, R. G., "Potentiating the Search for Alternatives to Halons 1301 and 1211," CFR Video Seminar, National Institute of Standards and Technology, Gaithersburg, MD, Video, November 13, 1990.

Gann, R. G., "Replacement for the Halogenated Fire Suppressants: A Research Strategy and Plan," National Institute of Standards and Technology, Gaithersburg, MD, NISTIR 4449; Eleventh Joint Panel Meeting of the U. S./Japan Government Cooperative Program on Natural Resources (UJNR) on Fire Research and Safety, October 19-24, 1989, Berkeley, CA, 1990, pp. 45–53.

Gann, R. G., *et al.*, "Agent/System Compatibility for Halon 1301 Aviation Replacement," 1993 International CFC and Halon Alternatives Conference on Stratospheric Ozone Protection for the 90's. October 20–22, 1993, Washington, DC, 1993, pp. 753–760.

Gann, R. G., *et al.*, "Preliminary Screening Procedures and Criteria for Replacements for Halons 1211 and 1301," National Institute of Standards and Technology, Gaithersburg, MD, Air Force Engineering Lab., Tyndall AFB, FL NIST TN 1278, August 1990, 311 pages.

Goodall, K., "New Fire Extinguishants—A Search for Halons' Replacements," *Fire Surveyor*, Vol. 20, No. 6, December 1991, pp. 52–55.

Grant, C. C., "Fire Protection Alternatives for High Value Halon Applications," Third International Symposium on Fire Protection of Buildings, May 10–12, 1990, Eger, Hungary, 1990, pp. 163–172.

Grant, C. C., "Halon and Beyond: Developing New Alternatives," *NFPA Journal*, Vol. 88, No. 6, November/December 1994, pp. 40–43, 45–47, 49–54.

Grant, C. C., "Life Beyond Halon," *Fire Journal*, Vol. 84, No. 3, May/June 1990, pp. 52–54, 56, 58–59.

Greig, A., "Alternative Technology Options for Replacing Fire-Fighting Halons," *Fire Protection*, Vol. 21, No. 2, June 1994, pp. 7–10.

Grosshandler, W. L. and Hamins, A., "Flame Suppression Effectiveness of Halon Alternatives," National Institute of Standards and Technology, Gaithersburg, MD, Air Force Materials Lab., Wright-Patterson AFB, OH, Naval Air Systems Command, Washington, DC, Army Aviation and Troop Command FAA Technical Center, Atlantic City, NJ, NISTIR 5499, September 1994; National Institute of Standards and Technology, Annual Conference on Fire Research: Book of Abstracts, October 17–20, 1994, Gaithersburg, MD, 1994, pp. 3–4.

Grosshandler, W. L., *et al.*, "Agent Screening for Halon 1301 Aviation Replacement," 1993 International CFC and Halon Alternatives Conference on Stratospheric Ozone Protection for the 90's. October 20–22, 1993, Washington, DC, 1993, pp. 744–752.

Grosshandler, W. L., *et al.*, "Assessing Halon Alternatives for Aircraft Engine Nacelle Fire Suppression," *Journal of Heat Transfer*, Vol. 117, May 1995, pp. 489–494.

Grzyull, L. R. and Parrish, C. F., "Innovative Approach for the Screening and Development of Halon Alternatives," *Proceedings* of the 1992 International CFC and Halon Alternatives Conference on Stratospheric Ozone Protection for the 90's, September 29–October 1, 1992, Washington, DC, 1992, pp. 647–656.

Guglielmi, E., "NAF S-III and NAF P-II: Environmentally Safer Alternatives to Halon 1301 and Halon 1211," [ABSTRACT ONLY], 1993 International CFC and Halon Alternatives Conference on Stratospheric Ozone Protection for the 90's. October 20–22, 1993, Washington, DC, 1993, p. 742.

"Halon alternatives—What Is Acceptable?" *Fire Protection*, Vol. 22, No. 1, March 1995, pp. 10, 12–13.

"Halon Update: New EPA Rules Define Acceptable Halon Substitutes," *Consulting Specifying Engineer*, Vol. 16, No. 1, July 1994, p. 53.

"Halons Still the Best," *New Scientist*, Vol. 134, No. 1824, June 6, 1992, p. 19.

"Halons: The Search for Alternatives," *Fire Surveyor*, Vol. 20, No. 1, February 1991, pp. 5–8.

Hanauska, C., "Perfluorocarbons as Halon Replacement Fire Extinguishing Agents," *Proceedings* of the 1991 Halon Alternatives Technical Working Conference, April 30–May 1, 1991, Albuquerque, NM, 1991, pp. 299–300.

Hanauska, C. P. and Back, G. G., "Halons: Alternative Fire Protection Systems—An Overview of Water Mist Fire Suppression System Technology," 1993 International CFC and Halon Alternatives Conference on Stratospheric Ozone Protection for the 90's. October 20–22, 1993, Washington, DC, 1993, pp. 820–829.

Hanauska, C. P., Forssell, E. W. and DiNenno, P. J., "Hazard Assessment of Thermal Decomposition Products of Halon Alternatives," *Proceedings* of the 1993 Halon Alternatives Technical Working Conference, May 11–13, 1993, Albuquerque, NM, 1993, pp. 577–582.

Herbert, J. P., "Developing Alternatives to Halon," *Sprinkler Age*, Vol. 12, No. 5, May 1993, pp. 10, 12, 14–15.

Hofmeister, C. and Zalosh, R., "Heptane Inerting concentrations of Candidate Halon 1301 Replacement Agents," Worcester Polytechnic Inst., MA, NISTIR 5499, September 1994; National Institute of Standards and Technology, Annual Conference on Fire Research: Book of Abstracts, October 17–20, 1994, Gaithersburg, MD, 1994, pp. 25–26.

Hommstedt, G., Andersson, P and Andersson, J., "Investigation of Scale Effects on Halon and Halon Alternatives Regarding Flame Extinguishing, Inerting Concentration and Thermal Decomposition Products," *Proceedings* of the Fourth International Symposium on Fire Safety Science, Intl. Assoc. for Fire Safety Science, Boston, MA, 1994, pp. 853–864.

Ingason, H., Persson, H. and Ryderman, A., "Foam Sprinklers as a Replacement for Halon in Engine Rooms," Swedish National Testing and Research Institute, Boras, Sweden, SP REPORT 1992:37, 1992, 45 pages.

Inoue, Y., *et al.*, "Evaluation of Fire Suppression Efficiency and Practical Applicability of Halon Replacements," *Report of Fire Research Institute of Japan*, No. 79, 1995, pp. 1–7.

"IRI Outlines Position on Halon Replacements," *Sentinel*, Vol. 50, No. 4, Fourth Quarter 1993, pp. 10–11.

Kibert, C. J., "Combustion Toxicology Testing of Halon Replacement Agents," Fire Safety '91 Conference, Volume 4, November 4–7, 1991, St. Petersburg, FL, Am. Technologies Int'l., Pinellas Park, FL, 1991, pp. 227–237.

Kibert, C. J., "Halon 1211/1301 Replacement Program Status: U.S. Air Force Ground Based Fire Suppression Applications," *Proceedings* of the 1993 Halon Alternatives Technical Working Conference, May 11–13, 1993, Albuquerque, NM, 1993, pp. 519–527.

Larsen, E. R., "Halons: Comments on the Development of Replacement Agents," *Journal of Fire Sciences*, Vol. 13, No. 3, May/June 1995, pp. 171–176.

Maranghides, A., *et al.*, "Effects of Fire Size on Extinguishment Time and Acid Byproducts During Real Scale Halon Replacement Testing," *Proceedings* of the 1995 Fall Technical Meeting of the Combustion Institute/Eastern States Section on Chemical and Physical Processes in Combustion, October 16–18, 1995, Worcester, MA, 1995, pp. 195–198.

McDougal, J. N., Dodd, D. E. and Skaggs, S. R., "Recommending Exposure Limits for Halon Replacements," *Proceedings* of the 1993 Halon Alternatives Technical Working Conference, May 11–13, 1993, Albuquerque, NM, 1993, pp. 63–73.

McKay, R. S., "Fine Water Spray: A Halon Replacement Option," *Fire Surveyor*, Vol. 22, No. 3, June 1993, pp. 16–20.

Mehta, H, K., "Aviation Industry Halon Replacement and Management Plan," 1993 International CFC and Halon Alternatives Conference on Stratospheric Ozone Protection for the 90's. October 20–22, 1993, Washington, DC, 1993, pp. 780–789.

Merrick, D., "Methodology for the Analysis of Special Hazard Halon 1301 Systems and Replacement Alternatives," *Proceedings* of the 1993 Halon Alternatives Technical Working Conference, May 11–13, 1993, Albuquerque, NM, 1993, pp. 465–472.

Metchis, K., "Regulation of Halons and Halon Substitutes: An Update," *Proceedings* of the 1993 Halon Alternatives Technical Working Conference, May 11–13, 1993, Albuquerque, NM, 1993, pp. 11–30.

Moore, D. W., "FE-13 Now Playing. Halon Alternatives," *Sprinkler Age*, Vol. 14, No. 2, February 1995, pp. 13–14.

Mossel, J., "ICI's Alternative Halon Program," *Proceedings* of the 1991 Halon Alternatives Technical Working Conference, April 30–May 1, 1991, Albuquerque, NM, 1991, pp. 295–298.

Nelson, D. F. and Bannister, W. W., "Non-Volatile Precursors to Alternative Halon Fire Extinguishing Agents With Reduced Global Environmental Impacts," *Proceedings* of the 1992 International CFC and Halon Alternatives Conference on Stratospheric Ozone Protection for the 90's, September 29–October 1, 1992, Washington, DC, 1992, pp. 637–646.

Nimitz, J. and Landford, L., "Ultimate Halon Replacements Are In Sight," *Proceedings* of the 1993 Halon Alternatives Technical Working Conference, May 11–13, 1993, Albuquerque, NM, 1993, pp. 597–604.

Nimitz, J. and Lankford, L., "Fluoroiodocarbons as Halon Replacements," 1993 International CFC and Halon Alternatives Conference on Strato-

spheric Ozone Protection for the 90's. October 20–22, 1993, Washington, DC, 1993, pp. 810–819.

Nimitz, J. S., *et al.*, "Halocarbons as Halon Replacements: Technology Review and Initiation," Final Report, December 1988-May 1989, New Mexico Univ., Albuquerque, Air Force Engineering and Services Center, Tnydall AFB, FL, ESL-TR-90-38; SS 2.03(1), March 1991, 138 pages.

Papavergos, P. G., "Fine Water Sprays for Fire Protection: A Halon Replacement Option," B. P. Research, Sunbury-on-Thames, England, 1990, 13 pages.

Parsons, M., "Replacements for Halon 1301 in Total Flooding Fire Suppression Systems," *Firehouse*, Vol. 18, No. 2, December 1993, pp. 28–29.

Parsons, M. L. and Swartwood, K. S., "Investigation Into Halon 1301 Replacements: A Water-Based System for the Airline Lavatory Application," *Proceedings* of the 1993 Halon Alternatives Technical Working Conference, May 11–13, 1993, Albuquerque, NM, 1993, pp. 611–620.

Pelton, D. A., "INERGEN Fire Extinguishing Agent: The Environment-Friendly Replacement for Halon 1301," *Sprinkler Age*, Vol. 14, No. 2, February 1995, pp. 11–12.

Pitts, W. M., *et al.*, "Construction of an Exploratory List of Chemicals to Initiate the Search for Halon Alternatives," National Institute of Standards and Technology, Gaithersburg, MD, Air Force Engineering Lab., Tyndall AFB, FL NIST TN 1279, August 1990, 206 pages.

Pitts, W. M., *et al.*, "Dynamics of the Release of Alternate Halon Replacement Agents From Pressurized Bottles," *Proceedings* of the 1993 Halon Alternatives Technical Working Conference, May 11–13, 1993, Albuquerque, NM, 1993, pp. 75–82.

Povey, N., "Searching for Suitable Halon Replacements on Passenger-Carrying Aircraft," *Fire Prevention*, No. 270, June 1994, pp. 26–28.

"Preliminary Screening Procedures and Criteria for Replacements for Halons 1211 and 1301: Executive Summary and Background," *Fire Systems*, Vol. 5, No. 2, 1991, pp. 84–90.

"Questions and Answers on Halons and Their Substitutes," *Sentinel*, Vol. 50, No. 4, Fourth Quarter 1993, pp. 12–15.

Register, W. D., "FM-200 (HFC-227ea) as a Halon 1301 Replacement," *Proceedings* of the Technical Symposium on Halon Alternatives,. June 27–28, 1994, Knoxville, TN, 1994, pp. 43–51.

Register, W. D., "FM-200: An Optimum Replacement for Halon 1301," *Sprinkler Age*, Vol. 14, No. 2, February 1995, pp. 10–11.

Riley, J. F., "Intermediate Scale Determination of Extinguishment Concentration and Measurement of Products of Decomposition of Selected Alternatives to Halon 1301," *Proceedings* of the 1991 Halon Alternatives Technical Working Conference, April 30–May 1, 1991, Albuquerque, NM, 1991, pp. 131–141.

Robin, M. L., "Evaluation of Halon Alternatives," *Proceedings* of the 1991 Halon Alternatives Technical Working Conference, April 30–May 1, 1991, Albuquerque, NM, 1991, pp. 16–38.

Rochefort, M. A., Dees, B. R. and Risinger, C. W., "Halon 1211 Replacement Agent Evaluation: Perfluorohexane and Halotron I," Final Report, Applied Research Associates, Inc., Tyndall AFB, FL, WL-TR-93-3520, November 1993, 29 pages.

Rubenstein, R. and Bellin, J. S., "Short Duration High-Level Exposure to Halon Substitutes: Potential Cardiovascular Effects," *Proceedings* of the 1993 Halon Alternatives Technical Working Conference, May 11–13, 1993, Albuquerque, NM, 1993, pp. 37–55.

Sakei, R., *et al.*, "Flame Extinguishing Concentrations of Halon Replacements for Flammable Liquids," *Report of Fire Research Institute of Japan*, No. 80, September 1995, pp. 36–42.

Sarkos, C. P., "Status of Halon Replacement Agent Evaluation in Commercial Aircraft and Airport Applications," Fire and Materials," Third International Conference and Exhibition on Fire and Materials, October 27–28, 1994, Arlington, VA, 1994, pp. 189–197.

Sarkos, C. P., "Status of Halon Replacement Evaluation in Commercial Aircraft and Airport Applications," *Proceedings* of the Twentieth International Conference on Fire Safety, Volume 20, January 9–13, 1995, Millbrae, CA, Product Safety Corp., Sunnyvale, CA, 1995, pp. 72–81.

Senecal, J. A., "Halon Replacement Chemicals: Perspectives on the Alternatives," *Fire Technology*, Vol. 28, No. 4, November 1992, pp. 332–344; International Symposium on Fire Protection for the Telecommunications Industry, May 14–15, 1992, New Orleans, LA, 1992, pp. 1–14.

Sheinson, R. S., "Halon Alternatives Extinguishment Pathways," *Proceedings* of the 1991 Halon Alternatives Technical Working Conference, April 30–May 1, 1991, Albuquerque, NM, 1991, pp. 71–82.

Sheinson, R. S., et. al., "CF3I: Halon 1301 Total Flooding Fire Extinguishment Comparison," Naval Research Laboratory, Washington, DC, GEO-CENTERS, Inc., Fort Washington, MD, Hughes Associates, Inc., Columbia, MD, Naval Sea Systems Command, Washington, DC, NISTIR 5499, September 1994; National Institute of Standards and Technology, Annual Conference on Fire Research: Book of Abstracts, October 17–20, 1994, Gaithersburg, MD, 1994, pp. 59–60.

Sheinson, R. S., et al., "Conducting Shipboard Total Flooding Fire Testing With Halon 1301 Replacements," Naval Research Laboratory, Washington, DC, GEO-CENTERS, Inc., Fort Washington, MD, NAVSEA 03G2, Crystal City, VA, University of South Alabama, Mobile, MPR Associates, Inc., Alexandria, VA, Hughes Associates, Inc., Columbia, MD, Naval Sea Systems Command, Washington, DC, NISTIR 5499, September 1994; National Institute of Standards and Technology, Annual Conference on Fire Research: Book of Abstracts, October 17–20, 1994, Gaithersburg, MD, 1994, pp. 61–62.

Sheinson, R. S., et al., "From Halon 1301 to a Replacement Agent: Integrated Development of a Fire Protection System," Naval Research Laboratory, Washington, DC, ISBN 0-9516320-9-4; Proceedings of the Seventh International Interflam Conference, Interflam '96, March 26–28, 1996, Cambridge, England, Interscience Communications Ltd., London, England, 1996, pp. 459–466.

Sheinson, R. S., et al., "Halon Replacement Considerations for Shipboard Machinery Spaces," Proceedings of the Twentieth International Conference on Fire Safety, Volume 20, January 9–13, 1995, Millbrae, CA, Product Safety Corp., Sunnyvale, CA, 1995, pp. 112–119.

Sheinson, R. S., et al., "Quest for Halon Replacements," Sixth International Fire Conference on Fire Safety, Interflam '93, March 30–April 1, 1993, Oxford, England, Interscience Communications Ltd., London, England, 1993, pp. 739–746.

Sheinson, R. S., et al., "Shipboard Total Flooding Fire Testing With Halon 1301 Replacements: Test Results," Naval Research Laboratory, Washington, DC, GEO-CENTERS, Inc., Fort Washington, MD, Worcester Polytechnic Inst., MA, Hughes Associates, Inc., Columbia, MD, Naval Sea Systems Command, Washington, DC, NISTIR 5499, September 1994; National Institute of Standards and Technology, Annual Conference on Fire Research: Book of Abstracts, October 17–20, 1994, Gaithersburg, MD, 1994, pp. 63–64.

Skaggs, S. R., "Second Generation Halon Replacements," Proceedings of the 1993 Halon Alternatives Technical Working Conference, May 11–13, 1993, Albuquerque, NM, 1993, pp. 239–246.

Skaggs, S. R. and Moore, T., "Toxicity and Physiology of Halon Alternatives," Proceedings of the Technical Symposium on Halon Alternatives, June 27–28, 1994, Knoxville, TN, 1994, pp. 33–36.

Smith, B. D., "On Demand, Non-Halon, Fire Extinguishing Systems," Department of the Navy, Arlington, VA, Serial 083,403, Navy Case No. 75238, June 23, 1993, 26 pages.

Smith, W. D., et al., "Efficacy Model for the Assessment of Alternative Halon Candidates in a Total Flooding Fire Suppression Discharge System," Proceedings of the 1992 International CFC and Halon Alternatives Conference on Stratospheric Ozone Protection for the 90's, September 29–October 1, 1992, Washington, DC, 1992, p. 779.

Smith, W. D., et al., "Halogen Acid Production Evaluation in an Intermediate-Scale Halon 1301 Replacement Suppressant Test," Sixth International Fire Conference on Fire Safety, Interflam '93, March 30–April 1, 1993, Oxford, England, Interscience Communications Ltd., London, England, 1993, pp. 757–764.

Tapscott, R. E., "Halon Substitutes: An Overview," Proceedings of the 1993 Halon Alternatives Technical Working Conference, May 11–13, 1993, Albuquerque, NM, 1993, pp. 1–10.

Tapscott, R. E., "Second-Generation Replacements for Halon," First International Conference on Fire Suppression Research, May 5–8, 1992, Stockholm, Sweden, 1992, pp. 327–335.

Tapscott, R. E., et al., "Next-Generation Fire Extinguishing Agent. Phase 5. Initiation of Halon Replacement Development," Final Report, July 1988-December 1988, New Mexico Univ., Albuquerque, Air Force Engineering and Services Center, Tyndall AFB, FL, ESL-TR-87-03, NMERI-SS-2.06(2), September 1990, 92 pages.

Tsang, W., Miziolek, A. W., Finnerty, A. E., "Kinetic Aspects of Flame Inhibition by Halons and Halon Alternative Compounds," Proceedings of the 1993 Halon Alternatives Technical Working Conference, May 11–13, 1993, Albuquerque, NM, 1993, pp. 247–255.

Tuomisaari, M., "Vesisumu halonien korvaajana? [Water Fog as a Replacement for Halons?]," Palontorjuntatekniikka, Vol. 1, 1994, pp. 6–8.

Varanka, V., "Taydellista halonin korviketta ei viela ole valmiina. [No Perfect Substitute for Halon.]," Palontorjuntatekniikka, Vol. 4, 1994, pp. 16–18.

Walters, E. A. and Molina, M. J., "Global Environmental Characteristics of Second-Generation Halon Replacements," Proceedings of the 1993 Halon Alternatives Technical Working Conference, May 11–13, 1993, Albuquerque, NM, 1993, pp. 31–36.

"Water Replaces Halon in Firefighting Technology," Petroleum Review, Vol. 46, No. 549, October 1992, p. 508.

Welch, J., "Which Halon Alternative?" Fire Prevention, No. 285, December 1995, pp. 34–36.

"What Are the Safe Alternatives to Halon?" Fire Prevention, Vol. 227, March 1990, p. 5.

Whiteley, R. A., "Halon Alternatives and Replacements," Third International Conference on Management and Engineering of Fire Safety and Loss Prevention: Onshore and Offshore, February 18–20, 1991, Aberdeen, Scotland, Elsevier Applied Science, New York, 1991, pp. 341–351.

Yang, J. C., Breuel, B. D. and Grosshandler, W. L., "Solubilities of Nitrogen and Freon-23 in Alternative Halon Replacement Agents," Proceedings of the 1993 Halon Alternatives Technical Working Conference, May 11–13, 1993, Albuquerque, NM, 1993, pp. 107–114.

Yang, J. C., et al., "Storage and Discharge Characteristics of Halon Alternatives," Proceedings of the 1995 International CFC and Halon Alternatives Conference and Exhibition on Stratospheric Ozone Protection for the 90's, October 21–23, 1995, Washington, DC, 1995, pp. 594–603.

Young, R., "Effective Alternatives to Halon 1301," Fire Prevention, No. 289, May 1996, pp. 16–18.

Young, R., "Evaluating Halon 1301 Alternatives," Fire Prevention, No. 267, March 1994, pp. 34–37.

Young, R., "Gaseous Fire Extinguishing Agents as Halon 1301 Alternatives," Fire Prevention, No. 277, March 1995, pp. 24–27.

Ziemba, J. "Halon Retrofit Technologies: Gazing Into the Crystal Ball," Fire Systems, Vol. 5, No. 2, 1991, pp. 4–11.

Zlochower, I. A. and Hertzberg, M., "Inerting of Methane-Air Mixtures by Halon 1301 (CF3Br) and Halon Substitutes," Proceedings of the 1991 Halon Alternatives Technical Working Conference, April 30–May 1, 1991, Albuquerque, NM, 1991, pp. 122–130.

CARBON DIOXIDE AND APPLICATION SYSTEMS

Revised by Thomas J. Wysocki

Carbon dioxide (CO_2) has been in widespread use for a long time and so has safely extinguished more fire than any other gaseous fire extinguishing agent. In the process, carbon dioxide systems have saved lives, maintained livelihoods, and prevented property damage. The benefits of carbon dioxide systems, properly applied, are enormous.

To gain the benefits of CO_2 extinguishing systems while minimizing risk to people, serious attention must be given to personnel safety in the design, installation, and maintenance of CO_2 systems. Training of personnel is essential.

This chapter will discuss the properties of CO_2 and its proper use in fixed fire extinguishing systems; its limitations as a fire extinguishing agent; life safety considerations; methods of application; system components; design considerations; and requirements for testing, maintenance, and training. NFPA 12, *Standard on Carbon Dioxide Extinguishing Systems* (hereinafter referred to as NFPA 12), provides guidance to those responsible for the purchase, design, installation, testing, inspection, operation, and maintenance of carbon dioxide systems.

Other chapters in the handbook that reference CO_2 systems include Section 3, Chapter 25, "Storage and Handling of Records"; Section 6, Chapter 14, "Ultra-High-Speed Suppression Systems for Explosive Hazards"; and Section 9, Chapter 37, "Marine."

USAGE

Carbon dioxide has been used for many years to extinguish flammable liquid fires; gas fires; fires involving electrically energized equipment; and, to a lesser extent, fires in ordinary combustibles, such as paper, cloth, and other cellulosic materials. CO_2 will suppress fire effectively in most combustible material. Exceptions are a few active metals and metal hydrides, and materials, such as cellulose nitrate, that contain available oxygen. Further, practical limitations of CO_2 are related to the physiological effects of CO_2 and to restrictions imposed by the hazard itself.

PROPERTIES OF CARBON DIOXIDE

Carbon dioxide has a number of properties that make it a desirable fire extinguishing agent. It is noncombustible, it does not react with

most substances, and it provides its own pressure for discharge from the storage container. Since carbon dioxide is a gas, it can penetrate and spread to all parts of a fire area. As a gas or as a finely divided solid called "snow" or "dry ice," it will not conduct electricity and, therefore, can be used on energized electrical equipment. It leaves no residue, thus eliminating cleanup of the agent itself.

Thermodynamic Properties

At room temperature and pressure, carbon dioxide is a gas. It is easily liquefied by compressing and cooling, and, with further compressing and cooling, it can be converted to a solid. The effect of temperature changes on compressed carbon dioxide in a closed container is shown in Figure 6-20A.

On the part of the curve between $-69.9°F$ ($-57°C$) and the critical temperature of $87.8°F$ ($31°C$), carbon dioxide in a closed container may be a gas or a liquid. The pressure is related to the temperature, as long as both vapor (gaseous) and liquid states are present. As the temperature and pressure increase, the density of the vapor phase increases while the density of the liquid phase decreases. At $87.8°F$ ($31°C$), the density of the vapor becomes equal to the density of the liquid, and the clear demarcation between the two phases disappears. Above the critical temperature, high-pressure carbon dioxide exists only in a gaseous form.

When the temperature is reduced to $-69.9°F$ ($-57°C$) at 75 psia (517 kPa), carbon dioxide may be present in vapor, liquid, and solid forms in equilibrium with each other. Hence, the term "triple point" to describe this condition. Below the triple point, only vapor and solid phases can exist. Thus, when liquid carbon dioxide is discharged to atmospheric pressure, a portion instantly flashes to vapor while the remainder is cooled by evaporation and converted to finely divided snow, or dry ice, at a temperature near $-110°F$ ($-79°C$). The proportion of CO_2 converted to dry ice depends upon the temperature of the stored liquid. Approximately 46 percent of the liquid stored at $0°F$ ($-18°C$) will be converted to dry ice, compared to approximately 25 percent for liquid stored at $70°F$ ($21°C$).

Storage

Liquid carbon dioxide may be stored in high-pressure cylinders with storage temperature varying with ambient temperature or in low-pressure refrigerated containers designed to maintain a storage temperature near $0°F$ ($-18°C$). Freezing (solidifying) in storage is not a problem. However, any substantial reduction in storage temperature—and corresponding storage pressure—could reduce the

Thomas J. Wysocki is a consultant with Guardian Services, Inc., Frankfort, IL. He is active on several NFPA technical committees, including those covering carbon dioxide and halon fire-extinguishing systems.

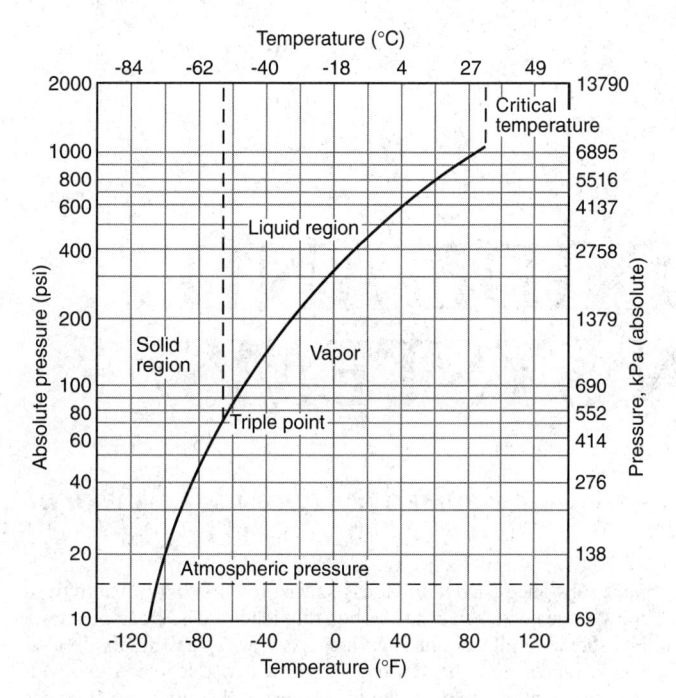

FIG. 6-20A. *Effect of pressure and temperature change on the physical state of carbon dioxide. Above the critical temperature of 87.8°F (31°C), irrespective of pressure, it is entirely gas. Between 87.8°F (31°C) and the temperature of the triple point, which is –69.9°F (–57°C), in a closed container, it is part liquid and part gas. Below the triple point, it is either a solid or a gas, depending upon the pressure and temperature.*

discharge flow rate below acceptable design limits. High-pressure systems normally are designed to operate properly with storage temperatures ranging from 32 to 120°F (0 to 49°C). Low-pressure systems, which operate normally at 0°F (–18°C), would not be affected unless the ambient temperature surrounding the storage container dropped to –10°F (–23°C) or lower for a prolonged time. Extreme climatic environments may require special design considerations to ensure proper operation. For low storage temperatures, high-pressure cylinders may require special "winterization" treatment, while low-pressure storage units may require heaters with circulation pumps.

Discharge Properties

A typical discharge of liquid carbon dioxide has a white, cloudy appearance due to finely divided dry ice particles carried along with the flash vapor. Because of the low temperature, some water vapor from the atmosphere will condense, creating additional fog that will persist for a time after the dry ice particles have settled out or sublimed.

Static Electricity

The dry ice particles produced during a discharge of carbon dioxide can carry a charge of static electricity. Static charge can also build up on ungrounded discharge nozzles. To prevent shock hazard to personnel or unwanted static discharge in a potentially explosive atmosphere, all discharge nozzles must be grounded. This is particularly important in the case of nozzles and playpipes used in hand hose line systems.

Vapor Density

Carbon dioxide gas has a density of one and one-half times the density of air at the same temperature. The cold discharge has a much greater density, which accounts for its ability to replace air above burning surfaces and maintain a smothering atmosphere when used in local application systems. When carbon dioxide is used for total flooding, the resulting mixture of CO_2 and air will be more dense than the ambient atmosphere.

Physiological Effects

Carbon dioxide is normally present in the atmosphere at a concentration of approximately 0.03 percent. It is present in humans and animals as a normal byproduct of cellular respiration. In the human body, carbon dioxide acts as a regulator of breathing, thus ensuring an adequate supply of oxygen to the system. Up to a point, an increase in carbon dioxide in the blood causes an increase in breathing rate. The maximum increase in respiration occurs when breathing 6 to 7 percent CO_2 in air. Higher concentrations of CO_2 slow down breathing. Finally, with 25 to 30 percent CO_2 in air, a narcotic effect takes over and stops breathing almost immediately—even with a sufficient supply of oxygen in the air. Reduced oxygen supplies will cause a very much lower concentration of carbon dioxide to suppress breathing and cause death from asphyxiation. The exact concentration of carbon dioxide in air that will cause a decrease in respiration varies from person to person and is not constant even in the same person from time to time.

Six to 7 percent CO_2 is considered the threshold level at which harmful effects become noticeable in human beings. At concentrations above 9 percent, most people lose consciousness within a short time. Since the minimum concentrations of CO_2 in air used to extinguish fire far exceed 9 percent, adequate safety precautions must be designed into every carbon dioxide fire extinguishing system.

The dry ice that is produced during a discharge can produce "burns," due to the extreme low temperature. Personnel should be warned not to handle any residual snow after a discharge.

EXTINGUISHING PROPERTIES OF CO_2

The primary mechanism by which carbon dioxide extinguishes fire is oxygen reduction (smothering). The cooling effect of carbon dioxide is relatively small but does make some contribution to fire extinguishment, particularly when carbon dioxide is applied directly to the burning material.

Extinguishment by Smothering

In any fire, heat is generated by rapid oxidation of a combustible material. Some of this heat raises the unburned fuel to its ignition temperature, while a large part of the heat is lost by radiation and convection, especially in the case of surface burning materials. If the atmosphere that supplies oxygen to the fire is diluted with carbon dioxide vapor, the rate of heat generation is reduced until it is below the rate of heat loss. When the fuel is cooled below its ignition temperature, the fire dies out and is extinguished completely.

The minimum concentration of carbon dioxide needed to extinguish surface burning materials, such as liquid fuels, can be determined accurately, since the rate of heat loss by radiation and convection is reasonably constant. Table 6-20A lists the minimum concentrations of CO_2 for some common liquid and gaseous fuels as determined by the U.S. Bureau of Mines.[1] The theoretical minimum CO_2 concentration is the actual concentration of CO_2 required to extinguish and prevent fire in a given fuel. The minimum design concentration is 20 percent more than the theoretical minimum CO_2 concentration, but never is less than 34 percent (per

NFPA 12). It is difficult to obtain similar data for solid materials because the rate of heat loss by radiation and convection can vary widely, depending upon shielding effects caused by the physical arrangement of the burning material. Design concentrations for hazards containing solid fuels have been determined from testing and experience. NFPA 12 gives design concentrations for a number of such hazards.

TABLE 6-20A. *Minimum Carbon Dioxide Concentrations for Extinguishment*

Material	Theoretical Min. CO_2 Concentration (%)	Minimum Design CO_2 Concentration (%)
Acetylene	55	66
Acetone	27*	34
Aviation Gas Grades 115/145	30	36
Benzol, Benzene	31	37
Butadiene	34	41
Butane	28	34
Butane - I	31	37
Carbon Disulfide	60	72
Carbon Monoxide	53	64
Coal Gas or Natural Gas	31*	37
Cyclopropane	31	37
Diethyl Ether	33	40
Dimethyl Ether	33	40
Dowtherm	38*	46
Ethane	33	40
Ethyl Alcohol	36	43
Ethyl Ether	38*	46
Ethylene	41	49
Ethylene Dichloride	21	34
Ethylene Oxide	44	53
Gasoline	28	34
Hexane	29	35
Higher Paraffin Hydrocarbons $C_nH_{2m} + 2m - 5$	28	34
Hydrogen	62	75
Hydrogen Sulfide	30	36
Isobutane	30*	36
Isobutylene	26	34
Isobutyl Formate	26	34
JP-4	30	36
Kerosene	28	34
Methane	25	34
Methyl Acetate	29	35
Methyl Alcohol	33	40
Methyl Butene-I	30	36
Methyl Ethyl Ketone	32	40
Methyl Formate	32	39
Pentane	29	35
Propane	30	36
Propylene	30	36
Quench, Lubricating Oils	28	34

NOTE: The theoretical minimum extinguishing concentrations in air for the above materials were obtained from a compilation of Bureau of Mines Limits of Flammability of Gases and Vapors (*Bulletins* 503 and 627). Those marked with * were calculated from accepted residual oxygen values.

Extinguishment by Cooling

Although the temperatures involved in a carbon dioxide discharge may approach –110°F (–79°C), the cooling capacity of the carbon dioxide is quite small compared to an equal weight of water. The latent heat of one pound (1 lb = 0.4536 kg) of liquid CO_2 is about 120 Btu (126 kJ) from low-pressure storage and 64 Btu (67.5 kJ) from storage at 70°F (21°C). The cooling effect is most apparent when the agent is discharged directly on the burning material by "local application." A massive application quickly covering the entire surface area smothers the fire and helps cool the fuel.

LIMITATIONS OF CO_2 AS AN EXTINGUISHING AGENT

The use of carbon dioxide on general Class A fires is limited by its relatively low cooling capacity, compared to that of water, and by enclosures incapable of retaining an extinguishing atmosphere. True surface burning fires are extinguished easily because natural cooling takes place quickly. On the other hand, if the fire penetrates below the surface of the fuel and the mass of fuel provides a layer of thermal insulation that slows down the rate of heat loss (generally referred to as "deep-seated burning"), a higher concentration of carbon dioxide and a much longer holding time are needed for complete extinguishment.

Oxygen-Containing Materials and Reactive Chemicals

Carbon dioxide is not an effective extinguishing agent for fires involving chemicals that contain their own oxygen supply, such as cellulose nitrate. Fires involving reactive metals, such as sodium, potassium, magnesium, titanium, zirconium, and the metal hydrides, cannot be extinguished by carbon dioxide because the metals and hydrides decompose CO_2.

LIFE SAFETY CONSIDERATIONS

To gain the benefits of CO_2 extinguishing systems while minimizing risk to people, serious attention must be given to personnel safety. Personnel safety must be considered in the design, installation, maintenance, and use of carbon dioxide systems. NFPA 12 discusses personnel safety at length.

General Safety Guidelines

Total flooding carbon dioxide should not be used in normally occupied spaces unless arrangements can be made to ensure evacuation before discharge. The same restriction applies to spaces that are not normally occupied but in which personnel may be present for maintenance or other purposes. It may be difficult to ensure evacuation if the space is large or if egress is in any way impeded by obstacles or complicated passageways. Escape is even more difficult after the discharge starts, because noise, greatly reduced visibility, and the physiological effects of the carbon dioxide concentration may confuse the occupants.

Consideration also should be given to the possibility of large volumes of carbon dioxide vapor leaking or flowing into unprotected lower levels, such as cellars, tunnels, or pits.

Periodic maintenance is required to ensure proper function of these systems. Those doing maintenance must be thoroughly trained in the functions of the system. Whenever maintenance involves work on system controls, discharge valves, or other devices that could possibly initiate CO_2 discharge, personnel must be evacuated from the protected space.

System Design Features/Operating Procedures for Life Safety

Audible and visual pre-discharge alarms are required for all systems. Personnel who will enter an area protected by a carbon dioxide system must be trained as to the correct response to these alarms.

Every precaution must be taken to ensure that personnel in a protected area are evacuated before the area is flooded with carbon dioxide. This is the primary purpose of a pre-discharge alarm, which must provide a suitable time delay before the CO_2 discharge.

Odorizers, such as oil of wintergreen, may be added to the carbon dioxide discharge to provide an olfactory indication of the presence of carbon dioxide. Odorizers are particularly useful to indicate that carbon dioxide has leaked from the protected area to adjoining areas.

Manual controls for the system must be located to avoid confusion, and they must be clearly labeled with safe operating procedures.

System "lockouts" should be provided to prevent accidental or deliberate discharge when persons unfamiliar with the system and its operation are present in the protected space. Lockouts are also needed when personnel egress from a space is hampered, e.g., when persons are climbing ladders, working on scaffolds, or entering confined spaces. System lockouts should be mechanical and designed to prevent both automatic and manual discharge. The emergency manual release required by NFPA 12 should *not* override the lockout. The lockout should be supervised to give an indication to those responsible for the system when the system is locked out. An adequate fire watch should be established to deal with any fire that might occur during a system lockout.

Self-contained breathing apparatus (SCBA) should be provided. This apparatus should be used by trained personnel to effect rescue of anyone who might be trapped in a carbon dioxide discharge.

Consideration should be given to any possibility that substantial volumes of carbon dioxide vapor could leak or flow into adjacent low-level areas, such as cellars, pits, or narrow hallways; i.e., spaces not included in the protected area. If this potential exists, alarms and warning signs should be extended to include these areas. Arrangements for safely ventilating all CO_2-flooded spaces following a discharge must be made.

METHODS OF APPLICATION

Two basic methods are used to apply carbon dioxide in extinguishing fires. One method is to discharge a sufficient amount of the agent into an enclosure to create an extinguishing atmosphere throughout the enclosed area. This is called "total flooding." The second method is to discharge the agent directly onto the burning material without relying on an enclosure to retain the carbon dioxide. This is called "local application."

Total Flooding

In total flooding systems, carbon dioxide is applied through nozzles designed and located to develop a uniform concentration of CO_2 in all parts of an enclosure. (See Figure 6-20B.) Calculation of the quantity of carbon dioxide required to achieve an extinguishing atmosphere is based upon the volume of the room and the concentration of CO_2 required for the combustible materials in it.

The integrity of the enclosure is a very important part of total flooding, particularly if deep-seated fire potential exists in the hazard. If the room is tight, especially on the sides and bottom, the CO_2

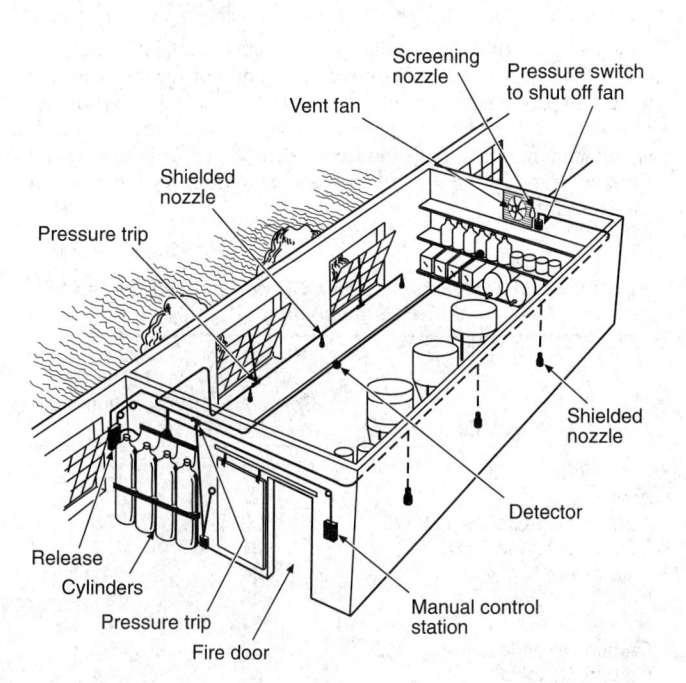

FIG. 6-20B. *Diagram of a high-pressure carbon dioxide system for total flooding.*

extinguishing atmosphere can be retained for a long time to ensure complete control of the fire. If there are openings on the sides and bottom, however, the heavier mixture of carbon dioxide and air may leak out of the room rapidly. If the extinguishing atmosphere is lost too rapidly, glowing embers may remain and cause reignition when air reaches the fire zone. Therefore, it is important to close all openings to minimize leakage or to compensate for the openings by discharging additional carbon dioxide. Because of the relative weight of carbon dioxide, an opening in the ceiling helps to relieve internal air pressure during the discharge, with very little effect on leakage rate after the discharge.

Extended Discharge

An extended discharge of CO_2 is used when an enclosure is not tight enough to retain an extinguishing concentration as long as is needed. The extended discharge normally is at a reduced rate, following a high initial rate used to develop the extinguishing concentration in a reasonably short time. The reduced rate of discharge should be a function of the leakage rate, which can be calculated on the basis of leakage area, or of the flow rate through ventilating ducts that cannot be shut down.

Extended discharge is particularly applicable to enclosed rotating electrical equipment, such as generators, where it is difficult to prevent leakage until rotation stops. Extended discharge can be applied to ordinary total flooding systems, as well as to local application systems where a small hot spot may require prolonged cooling.

Local Application

In local application systems, carbon dioxide is discharged directly on the burning surfaces through nozzles designed for this purpose. The intent is to cover all combustible areas with nozzles located so they will extinguish all flames as quickly as possible. Any adjacent area to which fuel may spread also must be covered, because any residual fire could cause reignition after the CO_2 discharge ends. (See Figure 6-20C.)

FIG. 6-20C. A local application carbon dioxide extinguishing system that has activated. (Source: The Ansul Company)

Local application discharge nozzles usually are designed for relatively low velocity to avoid splashing and air entrainment. Automatic detection is a necessity to provide fast response and minimize heat buildup. Although not essential, an enclosure would help retain carbon dioxide in the fire area. Local application of CO_2 can also be used for fast fire knockdown in an enclosure where final total flooding can provide absolute assurance that extinguishment will be complete.

Hand Hose Lines

Carbon dioxide systems can consist of hand hose lines permanently connected by means of fixed piping to a fixed supply of CO_2. Such systems frequently are provided for manual protection of small localized hazards. Although not a substitute for a fixed system, a hose line may be used to supplement a fixed system where the hazard is accessible for manual fire fighting, as well as to supplement portable equipment.

There must be a sufficient carbon dioxide supply to permit the use of the hand hose line nozzle for at least 1 min. Nozzles should be connected to a good electrical ground to permit static electrical charges produced by the carbon dioxide discharge to drain off. The wire-braided hose properly coupled to the discharge piping can provide a path to ground. Figure 6-20D shows a hand hose line system.

Standpipe Systems and Mobile Supply

Total flooding, local application, and hand hose line systems without a permanently connected carbon dioxide supply are known as gas standpipe systems. They are supplied by containers of carbon dioxide mounted on mobile units that can be moved and quickly coupled to a standpipe, in case of fire.

Gas standpipe systems can be used as a supplement to complete fixed fire protection systems or as the only protection of hazards under certain circumstances. Mobile supplies may be equipped with hand hose lines for the protection of scattered hazards. If the

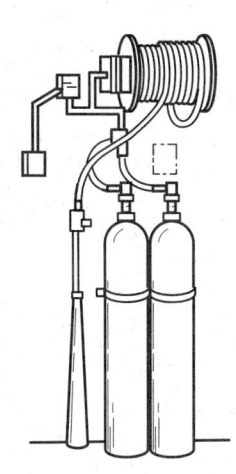

FIG. 6-20D.   Carbon dioxide hand hose extinguishing system with hose mounted on a reel. This system offers flexibility for attacking fires. (Source: Walter Kidde and Company, Inc.)

possible fire damage that might occur while the mobile supply is moved to the scene and deployed is unacceptable, a fixed automatic system should be used.

Special Applications

Carbon dioxide systems are sometimes used for inerting or purging. Carbon dioxide vapor is typically used in these applications. NFPA 69, *Standard on Explosion Prevention Systems*, gives guidance on these applications. Fire suppression in coal storage silos also can use carbon dioxide vapor. Low-pressure storage with an in-line vaporizer can conveniently supply the large quantities of carbon dioxide vapor often required for these applications.

COMPONENTS OF CARBON DIOXIDE SYSTEMS

The main components of a carbon dioxide system are the carbon dioxide supply, the discharge nozzles, and the piping system. These components, along with control valves and other operating devices, dispense the carbon dioxide and provide effective fire extinguishment.

Carbon Dioxide Storage

The CO_2 supply may be stored in high- or low-pressure containers. Because of the differences in pressure, system design is influenced by the storage method.

High-pressure storage: High-pressure containers, usually cylinders, are designed to store liquid carbon dioxide at atmospheric temperature. Since the maximum pressure in the cylinder or other container is affected by the ambient temperature, it is important that the container be designed to withstand the maximum expected pressure. Figure 6-20E shows the change in gage pressure with temperature.

Storage cylinders are designed, tested, and filled to U.S. Department of Transportation (DOT) specifications. The maximum filling density permitted is equal to 68 percent of the weight of water that the container can hold at 60°F (16°C). Standard cylinder capacities are 5, 10, 15, 20, 25, 35, 50, 75, and 100 lb (2.3, 4.5, 6.8, 9.1, 11.3, 15.9, 22.7, 34, and 45.4 kg) of carbon dioxide. Fire extinguishing cylinders are fitted with an internal dip tube so that liquid will

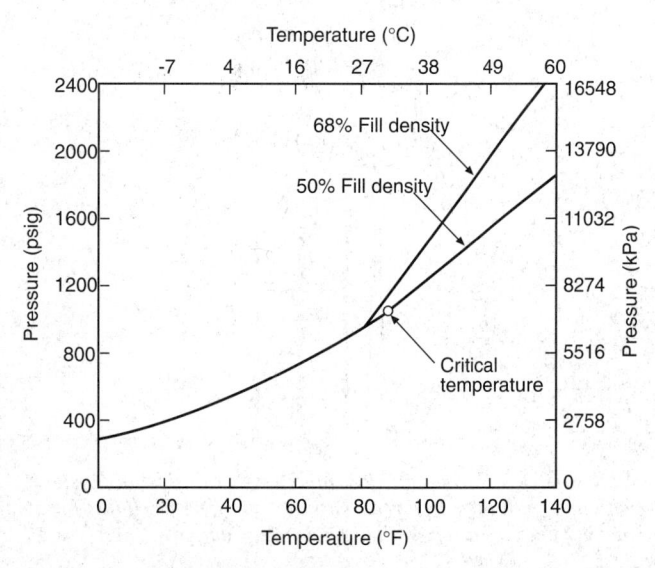

FIG. 6-20E. Variation in pressure of carbon dioxide with change in temperature, showing the effect of filling density in high-pressure storage.

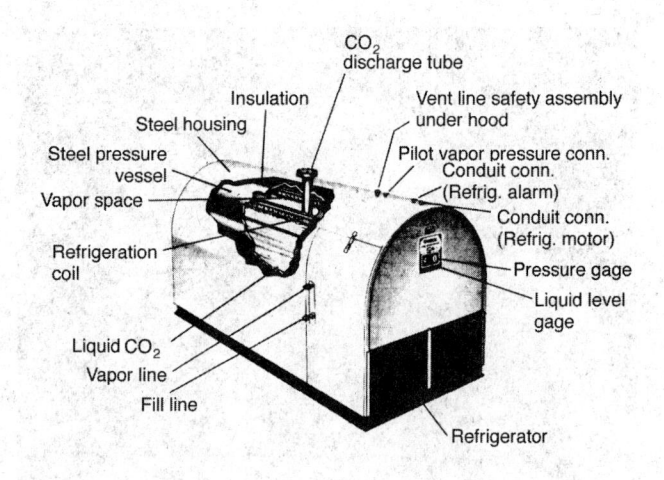

FIG. 6-20F. A low-pressure storage unit. The carbon dioxide is maintained at 0°F (–18°C) by refrigeration, and the pressure is 300 psi (2069 kPa). (Source: Cardox Division of Chemetron Corporation)

be discharged from the bottom when the cylinder is upright and the valve is opened.

Abnormally low storage temperatures adversely affect the rate of discharge. For this reason, NFPA 12 does not permit storage temperatures below 0°F (–18°C) for total flooding systems or below 32°F (0°C) for local application systems. Lower storage temperatures may be permitted only if the design includes special features to compensate for the reduced pressure. Techniques for "winterizing" cylinders are also available. Proper use of "winterizing" techniques can permit use of high-pressure storage at temperatures as low as –60°F (–51°C).

Low-pressure storage: Low-pressure storage containers are pressure vessels with a design working pressure of at least 325 psi (2240 kPa). These containers are maintained at a temperature of approximately 0°F (–18°C) by use of insulation and mechanical refrigeration. At this temperature, the pressure is approximately 300 psi (2069 kPa). A compressor, controlled by a pressure switch in the tank, circulates refrigerant through coils near the tank top. Tank pressure is controlled by condensation of carbon dioxide vapor by the coils. In the event of refrigeration failure, pressure-relief valves bleed off some of the vapor to keep the pressure within safe limits. This permits some of the liquid to evaporate, creating a self-refrigerating effect that reduces the pressure in the tank. Figure 6-20F shows a typical low-pressure container. Heaters are required if ambient temperatures will cool the tank contents below approximately –10°F (–23°C) [250 (1724 kPa) pressure level].

With low-pressure storage, it is common practice to protect multiple hazards from one central storage unit. The quantity of carbon dioxide discharged on a particular hazard is controlled by opening and closing the discharge valves in a preset timed sequence. Central storage units may have capacities ranging from less than one ton to a hundred tons or more (1 ton is approximately 907 kg). For large hazards, the distance between hazard and storage may be several hundred feet (100 ft is approximately 30.5 m).

Piping

Piping systems, normally empty, convey carbon dioxide from the storage container to open nozzles where there is a fire. Since the proper rate of flow is a critical requirement for fire extinguishment, it is important that the piping be designed and installed accurately. Minimum pressure in the pipeline must be kept well above the triple point pressure of 75 psia (5.2 bars). If the pressure of the flowing carbon dioxide falls below the triple point pressure, dry ice will form in the pipe and block orifices in the discharge nozzles, thus stopping the flow of carbon dioxide. NFPA 12 limits design nozzle pressures to a minimum of 300 psia (20.7 bars) for high-pressure storage and a minimum of 150 psia (10.3 bars) for low-pressure storage.

Carbon dioxide drawn from the bottom of the storage container enters the piping as a liquid. Friction causes loss in pressure. As pressure drops, the liquid boils, resulting in a mixture of liquid and vapor in the piping. The vapor increases in volume as the mixture passes through the piping, with a further drop in pressure. Thus, the flow is two-phase, a mixture of liquid and gas, a fact that pressure drop calculations must take into account.

NFPA 12 covers the calculation of CO_2 flow in some detail and provides pertinent equations and data tables. Although charts and tables are available for manual calculation of system piping, the use of available computer programs speeds and simplifies the design of piping systems.

The piping must be adequately supported to prevent movement during the discharge, and provision must be made for its contraction and expansion. Because liquid carbon dioxide is a refrigerant, it will substantially reduce the pipe temperature during discharge. Low-pressure liquid, in particular, starts at 0°F (–18°C) and may reach temperatures as low as –50°F (–46°C) in the piping before the discharge ends.

Valves and Operating Devices

Valves for controlling the discharge of carbon dioxide must withstand the maximum operating pressure, be absolutely bubbletight when closed, and be capable of both manual and automatic operation. Valves and allied devices, such as timers and pressure switches, must be listed or approved for use in CO_2 systems.

Discharge Nozzles

Many nozzle types are available for fire extinguishment applications. Nozzles used in total flooding simply may be orifices producing high-velocity jet streams, or they may be partially shielded to achieve reduced velocity or a specific discharge pattern. High-velocity types provide substantial mixing to ensure uniform concentration of carbon dioxide throughout an enclosure. Low-velocity types have a tendency to create high concentrations in the lower levels, which may be desirable under certain conditions.

Nozzle types used in local application systems normally are designed for relatively low discharge velocity. This helps to avoid splashing of liquid fuels and minimizes turbulence and air entrainment. All such nozzles must be tested for fire extinguishing characteristics and listed or approved by a testing laboratory on the basis of test performance.

Figure 6-20G shows a typical performance curve for a nozzle designed for overhead mounting. The rate of discharge used is based on the height of the nozzle above the hazard or the fuel surface and is given in listing cards or approvals. (Testing is based on the maximum rate that will not splash liquid fuels.) The maximum fire area that the nozzle will extinguish also is based on the height above the hazard surface, using the design flow rate.

Performance information enables the system designer to select and properly locate the number of nozzles necessary to cover the entire fire area of the hazard. It also establishes the design flow rate that must be used on the hazard. The effective design flow rate must continue for a minimum of 30 sec.

Figures 6-20H and 6-20I show typical performance curves for nozzles designed for mounting on the sides of open tanks containing flammable liquids. In such cases, the design flow rates are based on the area of coverage for the tankside nozzles or on the width of the hazard for linear nozzles mounted to extend the full length of one side. In each case, the design flow rate must be in the shaded area of the curves.

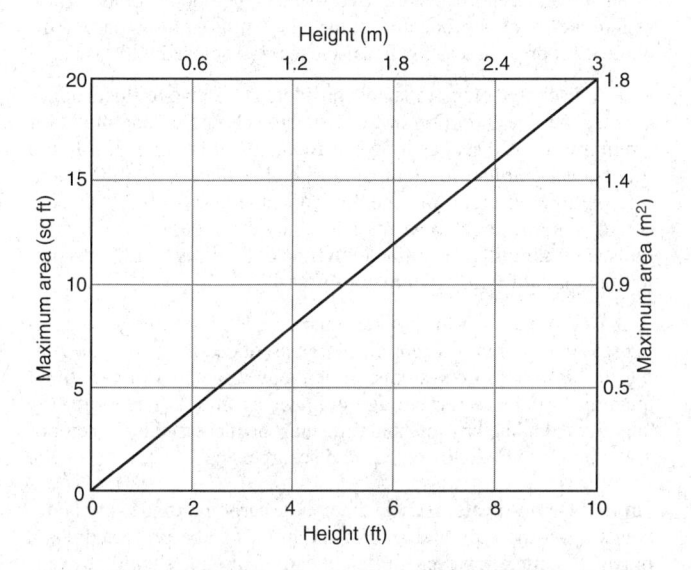

FIG. 6-20H.   *A typical listing or approval curve of a tankside carbon dioxide nozzle showing the flow rate vs. area coverage.*

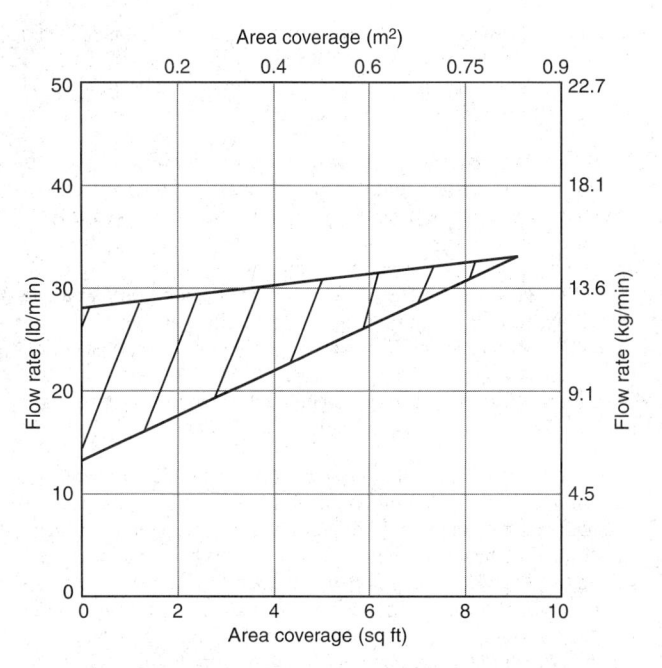

FIG. 6-20I.   *A typical listing or approval curve of a linear carbon dioxide nozzle showing flow rate vs. hazard width.*

CO_2 SYSTEM DESIGN CONSIDERATIONS

To organize basic components into a CO_2 fire protection system, a number of factors must be considered. These include the quantity of stored carbon dioxide, the method of actuation, the location of pre-discharge alarms, ventilation shutdown, pressure venting, and anything else involved in ensuring safe, prompt, and effective fire extinguishment and personnel safety.

Quantity Requirements

The quantity of carbon dioxide required for extinguishment depends upon the type of fire, the type of extinguishing system, and

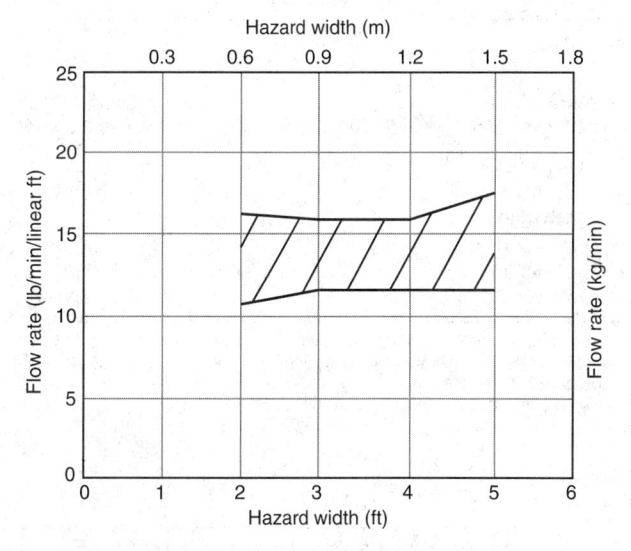

FIG. 6-20G.   *A listing or approval curve of a typical carbon dioxide nozzle showing maximum area vs. height or distance from liquid surface.*

conditions in the fire area. Where continuous protection is required, the quantity of carbon dioxide on hand should be at least twice that needed for extinguishment. Detailed methods of determining the quantity required for extinguishment are covered in NFPA 12.

Total flooding systems: The design concentration for a given enclosure should be sufficient to extinguish fires in all the fuels that are present in the hazard. The minimum concentration used in total flooding systems is 34 percent carbon dioxide by volume. Minimum design concentrations for various liquids and gases are given in Table 6-20A. NFPA 12 requires a 50 percent concentration for electrical wiring hazards, including small electrical machines; 65 percent for bulk paper and fur storage vaults; and 75 percent for dust collectors. These are specific hazards for which there is a background of test experience. Other materials should be tested to determine minimum CO_2 concentrations and holding time.

The quantity of carbon dioxide must be sufficient to achieve a minimum design concentration and to hold it until the fire is extinguished. A series of specific flooding factors has been established for surface fire hazards. These factors include an allowance for distributed leakage due to cracks around doors, porosity of the walls, and other small openings based on room size. The factors are greater for smaller rooms, because the anticipated leakage would be greater relative to volume. (See Table 6-20B and Figure 6-20J.)

TABLE 6-20B. *Flooding Factors to Achieve 34 percent Design Concentration*

Volume of Space (cu ft)	Volume Factor		Calculated Quantity Not Less than (lb)
	(cu ft/lb CO_2)	(lb CO_2/cu ft)	
Up to 140	14	0.072	—
141–500	15	0.067	10
501–1600	16	0.063	35
1601–4500	18	0.056	100
4501–50,000	20	0.050	250
Over 50,000	22	0.046	2500

Notes: For hazards requiring a 34 percent concentration of CO_2, the basic quantity of CO_2 is calculated by multiplying the volume of the enclosure by the appropriate flooding factor from Figure 6-20A.
For SI units: 1 cu ft = 0.0283 m³; 1 lb = 0.454 kg.

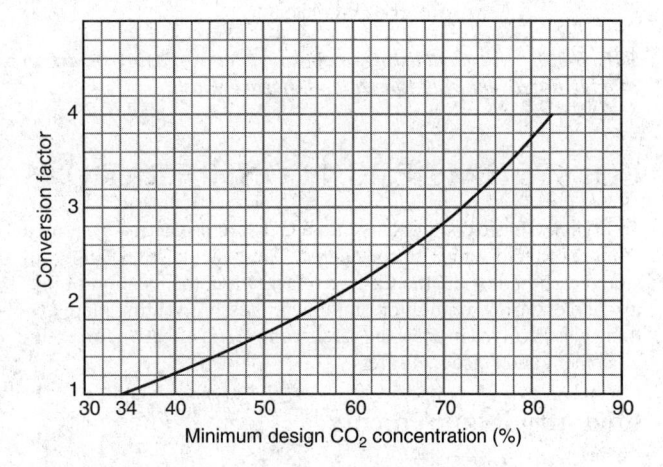

FIG. 6-20J. Minimum design CO_2 concentration conversion factors.

Note: If a hazard requires a design concentration greater than 34 percent, the basic quantity of CO_2 needed would be calculated by multiplying the appropriate flooding factor from Table 6-20B by the conversion factor determined from Figure 6-20J.

If the enclosure has obvious openings that cannot be closed, or ventilating systems that cannot be shut off in the event of a fire, additional CO_2 must be added to the "basic quantity" calculated using the flooding factor and material conversion factor.

Surface burning fires, such as flammable liquid fires, are normally extinguished during a 1-min carbon dioxide discharge. Leakage compensation must be in addition to the basic quantity.

Deep-seated fires require higher concentrations and much longer holding times. The rate of discharge must be high enough to develop a concentration of 30 percent in not more than 2 min, and the final design concentration must be achieved in not more than 7 min. Enclosures for deep-seated fires must be relatively tight, or it quickly becomes uneconomical to maintain the CO_2 design concentration. The basic quantity of CO_2 needed for deep-seated fire hazards is calculated using flooding factors given in NFPA 12.

Extended discharge systems: The most efficient way to compensate for leakage where the concentration must be maintained for a substantial time is to use an extended discharge at a reduced rate equal to the rate of leakage. The quantity of carbon dioxide required is based on the rate of discharge multiplied by the time that the concentration must be maintained. This is in addition to the quantity required to develop the design concentration.

Local application systems: These systems may be designed using either "rate-by-area" or "rate-by-volume" methods. When rate-by-area is used, the hazard is considered as a number of equivalent surface areas, each protected by a single nozzle. The rate of discharge from each nozzle is determined from listing information, and the total discharge rate for the system is the sum of the design rates required for each nozzle.

Where the hazard consists of three-dimensional irregular objects that cannot be reduced easily to equivalent surface areas, the rate-by-volume method may be used. In this case, the discharge rate of the system is based on the volume of an assumed enclosure entirely surrounding the hazard. Because this method is not as precise as the rate-by-area procedure, it usually results in more carbon dioxide discharge, unless the hazard is nearly enclosed with walls.

A discharge of vapor is not considered effective for local application. The liquid portion of the discharge should be continued for a minimum of 30 sec, or longer if required for cooling. If a liquid fuel has an autoignition temperature higher than its boiling point, e.g., paraffin wax and some cooking oils, the mass of liquid can become "superheated" above its autoignition temperature. For such fuels, a minimum local application time of 3 min is required to permit the mass of liquid fuel to cool.

Additional carbon dioxide must be added to storage to compensate for any initial vapor discharge produced in cooling the piping. In high-pressure systems, only about 70 percent of the stored quantity will leave the storage cylinder as liquid. Therefore, the quantity calculated by rate and time must be increased by a factor of 1.4 to help ensure a 30-sec liquid discharge.

Hand hose line systems: In theory, the carbon dioxide supply depends upon the type and size of the hazard to be protected. As a practical matter, however, sufficient carbon dioxide should be available to counteract possible waste by inexperienced users. In any case, there should be sufficient carbon dioxide to permit operation of the system for at least 1 min.

Venting Requirements

When liquid carbon dioxide is discharged into a closed room, the atmosphere will shrink initially due to the sudden refrigerating effect. At a later flooding stage, the combined volume of the carbon dioxide and air will become greater than the initial room volume. Thus, while the first result will be to create a vacuum, or draw in air, the final result must be to increase the pressure or to exhaust the excess volume through vent openings. Actual air temperatures are greatly reduced during the discharge, but they soon return to normal as heat is absorbed from solid surfaces in the room.

Experience has shown that most ordinary rooms have sufficient leakage through cracks around doors and windows and through general porosity to prevent noticeable vacuum or pressure buildup. From the viewpoint of efficient total flooding, it is better to have major vent openings in or near the ceiling rather than near the floor. In rooms that may be tightly sealed, a safe vent area for relatively low-strength structures can be estimated on the basis of calculated discharge flow rate:

$$X = \frac{Q}{1.3\sqrt{p}}$$

where:

X = vent area in sq in.,

Q = flow rate in lb/min, and

p = allowable strength of the enclosure in lb/sq ft.

For SI units:

$$X_m = \frac{23.9 Q_m}{\sqrt{p_m}}$$

where:

X_m = vent area in mm^2,

Q_m = flow rate in kg/min, and

p_m = allowable strength of the enclosure in kg/m^2.

More information on room venting can be found in NFPA 12.

System Control

Total flooding and local-application carbon dioxide systems normally should be designed to operate automatically. The detection device may be any of the listed or approved devices that are actuated by heat, smoke, flame, flammable vapors, or other abnormal process conditions that could lead to a fire or explosion. Automatically operated systems are required to have an independent means of manual actuation.

Modern carbon dioxide systems usually have an electric control panel. The control panel supervises the interconnections among all components necessary for control of the system and life safety. Connections among the control device and detectors, manual electric release stations, system valve actuators, and alarms are typically supervised. System lockouts must also be monitored. Standby power sources also are incorporated.

When a detector or manual station is activated, the control panel operates the pre-discharge alarms to warn personnel of the impending discharge. The panel also can control shutdown of fuel lines, closure of dampers, and shutdown of ventilating and process equipment. After a suitable time delay and pre-discharge alarm, the control panel operates the system discharge valves.

Some older carbon dioxide systems used timer-driven cams to operate switches that controlled the various discharge functions. Totally pneumatic systems using pneumatic detection and control were also common.

Whether supplemental to automatic actuation or the sole means of placing a system in operation, manual controls must be easy to operate, accessible in case of fire, and located close to the valves they control. Remote manual controls usually are located near an exit for a total flooding installation or near a hazardous area for local application installations. Instructions for safe operation of the manual controls must be prominently displayed at the control.

Marine Systems

Systems used on offshore platforms and marine vessels are subject to different regulations and design procedures. Quantity calculations, piping design, nozzle sizing, and system control requirements will vary from those discussed in this chapter. The guidance of the Authority Having Jurisdiction over the vessel must be followed for such systems.

TESTING AND MAINTAINING CO$_2$ SYSTEMS

An elaborate fire protection system could be worthless unless it is properly tested to demonstrate adequate performance and maintained in good working condition.

Acceptance Testing

No newly installed system should be accepted until it has been inspected properly and tested to prove performance in accordance with design specifications. The inspection should ascertain whether the system is installed in accordance with approved plans. The piping is particularly important, because any changes might affect the flow rates that were calculated on the basis of the pipe sizes shown on the plans.

NFPA 12 requires a "full discharge test" of all newly installed systems. The final proof of performance is to discharge the system and, for total flooding hazards, to measure the concentrations of CO$_2$ and the holding time in the fire zone. Local application systems cannot be subjected to concentration testing. Their effectiveness is judged by observation during the test discharge. NFPA 12 gives the minimum requirements for approval of new installations.

Inspections

To maintain a system in proper operating condition, it is essential to conduct periodic inspections. At least once a month, the system must be inspected to verify its operational status. The status of the carbon dioxide supply should be checked at least weekly for low-pressure supplies, at six-month intervals for high-pressure cylinders, and immediately after the system has operated. The entire system should be inspected annually. In particular, inspectors should look for changes in the scope of the hazard, changes in the condition of any enclosure, and other changes that could affect the adequacy of the fire protection system.

Maintenance

All operating devices should be tested annually. Although much of this testing can be accomplished without discharging the system, a partial discharge may be necessary in some cases. Any required maintenance found by testing should be performed without delay.

Training

Personnel must be made aware of the importance of the fire protection as well as the potential effects of exposure to high concentrations of CO_2. Training should include how the system extinguishes fire, physiological effects of exposure to CO_2, how the system is actuated, proper response to alarms (including a detailed evacuation plan), and proper safety precautions that are to be observed before manually actuating the system. New personnel should be trained before they are permitted to enter spaces that could be affected by a CO_2 discharge or areas where CO_2 system controls are located. Periodic "refresher" training courses should be mandatory.

Persons who are not normally associated with the protected hazard and adjoining spaces, but who are given temporary assignment there, must also be briefed on the safety issues related to the fire protection system. This applies likewise to outsiders, such as construction and maintenance people.

BIBLIOGRAPHY

Reference Cited

1. Coward, H. W., and Jones, G. W., "Limits of Flammability of Gases and Vapors," *Bulletin* 503, 1952, USDI Bureau of Mines, Washington, DC.

NFPA Codes, Standards, and Recommended Practices

Reference to the following NFPA codes, standards, and recommended practices will provide further information on carbon dioxide and application systems discussed in this chapter. (See the latest version of *The NFPA Catalog* for availability of current editions of the following documents.)

NFPA 12, *Standard on Carbon Dioxide Extinguishing Systems*
NFPA 69, *Standard on Explosion Prevention Systems*

Additional Reading

Carbon Dioxide Fire Extinguishing Systems, Walter Kidde and Co., Wake Forest, NC, 1981, rev. Aug. 1984.

Carbon Dioxide Safety Manual, Chemetron Fire Systems, University Park, IL, 1993.

Cromer, P. B., "Carbon Dioxide Fire Extinguishing Systems," *Fire Engineering*, Vol. 147, No. 6, June 1994, pp. 68, 70, 72, 76, 78.

Fire Extinguishing Systems Design Manual: Carbon Dioxide, Walter Kidde and Co., Wake Forest, NC, 1982.

Hansen, B., and Campbell, D., "Responders Use CO_2 on Difficult Shipboard Fire," *NFPA Journal*, Vol. 86, No. 6, Nov./Dec. 1992, pp. 48–52, 54–55.

Hoare, J., "Carbon Dioxide Fire Extinguishing Systems," *Fire Safety*, Vol. 3, No. 1, Feb. 1996, pp. 15–17.

Noronha, J. A., and Schiffhauer, E. J., "Explicit Equations for Two-Phase Carbon Dioxide Flow," *Fire Technology*, Vol. 10, No. 2, May 1974, pp. 101–109.

Schaenman, P., "Major Ship Fire Extinguished by CO_2, Seattle, Washington, Sept. 16, 1991," USFA Fire Investigation Technical Report Series, TriData Corp., Arlington, VA, Federal Emergency Management Agency, Washington, DC, Report 058, 1992.

"Special Hazards Protection—Manufacturers and Equipment," *Fire Journal*, Vol. 81, No. 4, July–Aug. 1987, p. 43.

Wolverton, C. D., Jr., Rynasko, J. D., and McLain, W. H., "Localized Fire Extinguishing System (LFES): Low Pressure Carbon Dioxide Agent Tests (Horn and Projection Nozzles) with Large and Small Machinery Mockups," Final Report, Coast Guard, Groton, CT, Department of Transportation, Washington, DC, CG-D-10-92, Dec. 1991.

Zalosh, R., and Hung, C. W., "Carbon Dioxide Discharge Test Modelling," *Proceedings* of the Fourth International Symposium on Fire Safety Science, Intl. Assoc. for Fire Safety Science, Boston, MA, 1994, pp. 889–900.

DRY CHEMICAL AGENTS AND APPLICATION SYSTEMS

Revised by David R. Hague

Dry chemical is a powder mixture that is used as a fire extinguishing agent. It is intended for application by means of portable extinguishers, hand hose line systems, or fixed systems. Borax- and sodium bicarbonate-based dry chemical were the first such agents developed. Sodium bicarbonate became the standard because of its greater effectiveness as a fire extinguishing agent. About 1960, sodium bicarbonate-based dry chemical was modified to render it compatible with protein-based low-expansion foams to permit a dual agent attack. Multipurpose (monoammonium phosphate base) and "Purple K" (potassium bicarbonate base) dry chemicals then were developed for fire extinguishing use. Shortly thereafter, "Super K" (potassium chloride base) was developed to equal "Purple K" in effectiveness. In the late 1960s, the British developed urea-potassium-bicarbonate-based dry chemical. Presently, there are five basic varieties of dry chemical extinguishing agents.

This chapter contains information on the properties of dry chemicals affecting their use as extinguishing agents, their uses and limitations, methods of storage and handling, and quality control. Portable dry chemical extinguishers are discussed in Section 6 of this handbook. Dry powder extinguishing agents for combustible metal fires are discussed in Chapter 26 of this section, "Combustible Metal Agents and Application Techniques"; and fire and explosion control are discussed in Section 1, Chapter 8, "Theory of Fire Extinguishment."

The terms "regular dry chemical" and "ordinary dry chemical" generally refer to powders that are listed for use on Class B and Class C fires.[1] "Multipurpose dry chemical" refers to powders that are listed for use on Class A, B, and C fires. The terms "regular dry chemical," "ordinary dry chemical," and "multipurpose dry chemical" should not be confused with "dry powder" or "dry compound," which are used to identify powdered extinguishing agents developed primarily for use on combustible metal fires.[2]

Dry chemical is exceptionally efficient in extinguishing fires in flammable liquids. It can also be used on fires involving some types of electrical equipment. Regular dry chemical has certain limited applications in extinguishment of flash surface fires with ordinary combustibles, but the chemical requires water to put out deep-seated smoldering fires. Multipurpose dry chemical can be used on fires in flammable liquids, fires involving energized electrical equipment, and fires in ordinary combustible materials. Multipurpose dry chemical seldom needs the help of water to completely extinguish fires in Class A materials.

PHYSICAL PROPERTIES OF DRY CHEMICALS

The principal base chemicals used in the production of currently available dry chemical extinguishing agents are sodium bicarbonate, potassium bicarbonate, potassium chloride, urea-potassium bicarbonate, and monoammonium phosphate. Various additives are mixed with these base materials to improve their storage, flow, and water repellency characteristics. The most commonly used additives are metallic stearates, tricalcium phosphate, or silicones, which coat the particles of dry chemical to make them free-flowing and resistant to the caking effects of moisture and vibration.

Stability

Dry chemical is stable at both low and normal temperatures. However, since some of the additives may melt and cause sticking at higher temperatures, an upper storage temperature limit of 120°F (49°C) is recommended for dry chemical. [Temperatures up to 150°F (66°C) may be acceptable for very short durations.] At fire temperatures, the active ingredients either disassociate or decompose while performing their function in fire extinguishment. Of extreme importance is the danger caused by indiscriminate mixing of the various dry chemicals. For example, mixing multipurpose (monoammonium base) dry chemical, which is acidic, with an alkaline dry chemical (most of the other dry chemicals) will result in an undesirable reaction that releases free carbon dioxide gas and causes caking. Extinguisher shells have been known to explode because of this phenomenon. Dry chemical agents may also react violently with other chemicals.

Toxicity

Generally dry chemical extinguishing agents are considered to be nontoxic and noncarcinogenic. No chronic medical conditions have been reported from long-term exposure. The acute effects of exposure are irritation to mucous membranes, with the possibility of chemical burns of the skin, eyes, and mucous membranes lining the respiratory system. Moist skin may enhance this effect. In all cases, self-contained breathing apparatus (SCBA) should be worn to protect against smoke, fumes, and the dust created by fixed system discharge. Consult the appropriate material safety data sheets (MSDS) for detailed information, instructions, and special precautions.[3]

David R. Hague is a fire protection specialist on the staff of NFPA.

Particle Size

Particles of dry chemical range in size from less than 10 microns up to 75 microns (1 micron = 0.000029 in.). Particle size has a definite effect on extinguishing efficiency, and careful quality control is necessary to prevent particles from exceeding upper and lower limits of this performance range. Each dry chemical has a unique size limiting factor, below which the particles completely decompose and vaporize, above which the particles do not completely decompose or vaporize.[4] The best results are obtained by a heterogeneous mixture with a median particle size of 20 to 25 microns. Particle size also has an effect on the flow characteristics of dry chemical. Solid particles suspended in a gas stream present higher pressure losses than gas alone. Coarse powders cause excessive surging, low flow rates, and will require larger expellant gas quantities. Fine powder will produce similar results, but not to the same degree. Particle size is also responsible for the aerodynamic drag phenomenon (ADP). The momentum gained by larger particles will transport smaller particles to penetrate the updraft of a flame. Ordinarily, smaller particles would decompose or vaporize prior to penetration.[5] Underwriters Laboratories Inc. (UL) and Factory Mutual Research Corporation (FMRC) listings and ratings are based upon the specified use of the particular type of dry chemical set forth by the equipment manufacturer.

EXTINGUISHING PROPERTIES

Fire tests on flammable liquids have shown potassium bicarbonate-based dry chemical to be more effective than sodium bicarbonate-based dry chemical in extinguishment. Similarly, monoammonium phosphate has been found equal to or better than sodium bicarbonate in extinguishment effectiveness.[6] The effectiveness of potassium chloride is about equivalent to potassium bicarbonate, and urea-potassium bicarbonate exhibits the greatest effectiveness of all the dry chemicals tested.

When introduced directly to the fire area, dry chemical causes extinguishment almost at once. Smothering, cooling, and radiation shielding contribute to the extinguishing efficiency of dry chemical, but studies suggest that a chain-breaking reaction in the flame is the principal cause of extinguishment.[7]

Smothering Action

For many years it was widely held that regular dry chemical extinguishing properties relied primarily on the smothering action of the carbon dioxide released when sodium bicarbonate was heated by fire. The carbon dioxide does undoubtedly contribute to the effectiveness of dry chemical, as does the like volume of water vapor released when dry chemical is heated. However, tests have generally disproved the belief that these gases are a major factor in extinguishment.

When multipurpose dry chemical is discharged into burning ordinary combustibles, the decomposed monoammonium phosphate leaves a sticky residue (metaphosphoric acid) on the burning material. This residue seals glowing material from oxygen, thus helping to extinguish the fire and prevent reignition.

Cooling Action

It cannot be substantiated that the cooling action of dry chemical is an important reason for its ability to promptly extinguish fires. The heat energy required to decompose dry chemicals plays an undeniable role in contributing toward their individual extinguishing abilities, but the effect, per se, is minor. To be effective, any dry chemical must be heat sensitive and, as such, absorbs heat in order to become chemically active.[8]

Radiation Shielding

Discharge of dry chemical produces a cloud of powder between the flame and the fuel; this cloud shields the fuel from some of the heat radiated by the flame. Tests to evaluate this factor concluded that the shielding factor is of some significance.[8]

Chain-Breaking Reaction

The preceding extinguishing actions, when combined, exert a minimal effect. The rapidity of extinguishment is due to the interference of the dry chemical particles with the propagation of the combustion chain reaction, which reduces the concentration of "free" radicals present within the flames. To accomplish this, the dry chemical must become thermally decomposed.

The discharge of dry chemical into the flames prevents reactive particles from coming together and continuing the combustion chain reaction. The explanation is referred to as the chain-breaking mechanism of extinguishment.[9,10]

SAPONIFICATION

For special applications, such as kitchen range, hood, duct, and fryer fire protection, the extinguishing mechanism for dry as well as wet chemical is based upon the process of saponification. Saponification is the process of chemically converting the fatty acid contained in the cooking medium to soap, or foam, and it accomplishes extinguishment by forming a surface coating.[11] Saponification value is a measure of the amount of potassium hydroxide [milligrams] consumed by the saponification of 1 gram of fat. The most common types of cooking media, i.e., animal fats (or lard), vegetable oils, and peanut oils, have similar saponification values. Other fats, such as cocoa, have substantially higher saponification values and are difficult to extinguish.

The soap produced by saponification is readily broken down by exposure to heat. As a result, since dry chemicals do not offer a substantial cooling effect, the cooking medium may reflash after a brief period of time.

USES AND LIMITATIONS

Dry chemical is primarily used to extinguish flammable liquid fires. Because it is electrically nonconductive, it can also be used on flammable liquid fires involving live electrical equipment. Regular dry chemical extinguishers have been tested and found suitable for use on flammable liquid and electrical fires (Class B and C fires) by fire equipment testing laboratories.

Due to the rapidity with which dry chemical extinguishes flame, dry chemical is used on surface fires involving ordinary combustible materials (Class A fires). There are several areas in the textile industry, notably opener-picker rooms and carding rooms in cotton mills, where regular dry chemical has been used effectively. However, wherever regular dry chemical is provided for use on surface-type Class A fires, it should be supplemented by water spray for extinguishing smoldering embers or in case the fire gets beneath the surface. In some baled cotton storage areas, the tops of bales can be covered with regular dry chemical to prevent surface spread should fire break out. This preventive measure does not eliminate the need for automatic sprinkler protection in such areas. Since multipurpose dry chemical becomes sticky when heated, it is not recommended for textile card rooms or other locations where removal of the residue from fine machine parts may be difficult.

Dry chemical does not produce a lasting inert atmosphere above the surface of a flammable liquid; consequently, its use will not result in permanent extinguishment if reignition sources, such as hot metal surfaces or persistent electrical arcing, are present.

Dry chemical should not be used in installations where relays and delicate electrical contacts are located (e.g., in telephone exchanges and computer equipment rooms), since the insulating properties of dry chemical might render such equipment inoperative. Because some dry chemicals are slightly corrosive, they should be removed from all undamaged surfaces as soon as possible after fire extinguishment.

Regular dry chemical will not extinguish fires that penetrate beneath the surface, or fires in materials that supply their own oxygen for combustion. Dry chemical may be incompatible with mechanical (air) foam unless the dry chemical has been specially prepared to be reasonably foam compatible.

Specifications have been established by fire equipment testing laboratories to ensure the positive and consistent performance of dry chemical as an extinguishing agent. These specifications control moisture content, water repellency, electrical resistivity, storage at elevated temperatures, flow capability, caking resistivity, and abrasive action. The discharge characteristics of the device in which the dry chemical is to be used are also evaluated. Extinguishing effectiveness is determined by performance tests of the application to standard fires under conditions recommended by the manufacturer.

Though dry chemical had been used for many years in fire extinguishers, it was not until 1954 that the first dry chemical extinguishing system was tested and listed by a testing laboratory. In 1952 the NFPA Committee on Dry Chemical Extinguishing Systems was established, and in 1957 the first edition of NFPA 17, *Standard for Dry Chemical Extinguishing Systems*, was adopted.

DRY CHEMICAL EXTINGUISHING SYSTEMS

Dry chemical extinguishing systems can be used in those situations where quick extinguishment is desired and where reignition sources are not present. Dry chemical systems are used primarily for flammable liquid fire hazards, such as dip tanks, flammable liquid storage rooms, and areas where flammable liquid spills may occur. Systems have been designed for kitchen range hoods, ducts, and associated rangetop hazards, such as deep-fat fryers. However, recent concerns have been raised regarding dry chemicals' ability to prevent reflashing of a fire in this application. At present, dry chemical systems have experienced difficulty in meeting the criteria of the latest test standards for deep-fat fryer applications. Where it is necessary to extinguish a flammable liquid or gas fire being fed by fuel under pressure, dry chemical hand hose line systems can be used, followed by closure of fuel shutoff valves.

Since dry chemical is electrically nonconductive, extinguishing systems using this agent can be used on electrical equipment that is subject to flammable liquid fires, such as oil-filled transformers and oil-filled circuit breakers. Dry chemical system protection is not recommended, however, for delicate electrical equipment, such as telephone switchboards and electronic computers. Such equipment is subject to damage by dry chemical deposit and, because of the insulating properties of the dry chemical, may require excessive cleaning to restore operation.

Hand hose line systems containing regular or ordinary dry chemical have been used to a limited extent for quick-spreading surface fires on ordinary combustible material. In such applications, the dry chemical system only stops or prevents a rapid surface spread, and must be supplemented by a water-type extinguishing device to put out deep-seated smoldering fires. Fixed systems containing multipurpose dry chemical are available and are suitable for

the protection of ordinary combustibles, provided the dry chemical can reach all burning surfaces.

METHODS OF APPLICATION

The two basic types of dry chemical systems are referred to as: (1) fixed systems and (2) hand hose line systems. Other methods of applying dry chemical are by portable and wheeled-type extinguishers.

Fixed Systems

Fixed dry chemical systems consist of a supply of dry chemical, an expellant gas, an actuating method, fixed piping, and nozzles through which the dry chemical can be discharged into the hazard area. Fixed dry chemical systems are of two types: (1) total flooding and (2) local application.

In total flooding applications, a predetermined amount of dry chemical is discharged through fixed piping and nozzles into an enclosed space or enclosure around the hazard. (See Figure 6-21A.) Total flooding systems are applicable only when the hazard is totally enclosed or when all openings surrounding a hazard can be closed automatically, before or simultaneously when the system is discharged. In some instances, additional dry chemical can be included in the system design to compensate for unclosable openings. Total flooding can be used only where no reignition is anticipated, since the extinguishing action is transient.

Local application differs from total flooding in that the nozzles are arranged to discharge directly onto the fire or burning surface. Local application is practical in those situations where the hazard can be isolated from other hazards so that fire will not spread beyond the area protected, and where the entire hazard can be protected. The principal use of local application systems is to protect open tanks of flammable liquids. As with total flooding systems, local application is ineffective unless extinguishment can be immediate and no reignition sources are present.

Hand Hose Line Systems

Hand hose line systems consist of a supply of dry chemical and expellant gas with one or more hand hose lines to deliver the dry chemical to the fire. (See Figure 6-21B.) The hose stations are connected

FIG. 6-21A.   *This 15- by 24-ft (4.6- by 7.3-m) flammable liquids storage building is protected by a total flooding dry chemical system. Upon actuation of the system, by a heat detector, nitrogen is discharged into the 150-lb (68-kg) storage container and dry chemical is expelled through eight nozzles in the roof.*

FIG. 6-21B. *A dry chemical hand hose line system consisting of an expellant gas assembly, dry chemical storage tank assembly, and a discharge assembly. Dry chemical systems of this type range from 125 to 2,000 lb (57 to 907 kg); the 2,000-lb (907-kg) system shown has four nitrogen cylinders and two $^3/_4$-in. (19-mm) hoses with shutoff nozzles. (Source: The Ansul Company)*

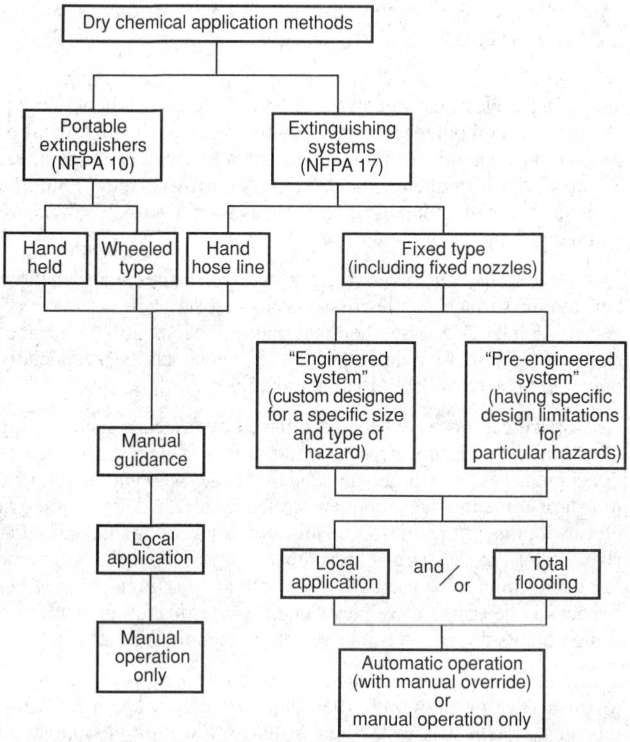

to the dry chemical container either directly or indirectly by means of intermediate piping. They can provide a large quantity of extinguishing agent for quick knockdown and extinguishment of relatively large fires, such as might be experienced at gasoline loading racks, flammable liquid storage areas, diesel and gas turbine locomotives, and aircraft hangars.

DESIGN OF DRY CHEMICAL SYSTEMS

Usually, dry chemical systems consist of dry chemical and expellant gas storage tanks, piping and/or hose to carry the agent to the fire area, nozzles to ensure proper distribution of the agent into the fire or area to be protected, and automatic and/or manual actuating mechanisms. (See Figure 6-21C.)

Dry chemical systems are called either engineered or pre-engineered, depending upon how the quantity of dry chemical, rate of flow, size and length of piping, and number and size of fittings are determined. An engineered system is one in which individual calculation and design is needed to determine the flow rate, nozzle pressures, pipe sizes, quantity of dry chemical, and the number, types, and placement of nozzles for the hazard being protected. A pre-engineered system, sometimes called a package system, is one in which the size of the system (i.e., the quantity of dry chemical, pipe sizes, maximum and minimum pipe lengths, number of fittings, and number and types of nozzles) is predetermined by fire tests for specific sizes and types of hazards. Installation within these limits of hazard and system design ensures adequate flow rate, nozzle pressure, and pattern coverage without individual calculation. (See Figure 6-21D.)

Pre-engineered systems are frequently used for kitchen range and hood fire protection, including deep-fat fryers. Only alkaline dry chemicals can be used in these cases (sodium bicarbonate, potassium bicarbonate, etc.) in order to saponify fats and oils. Multipurpose dry chemical must never be used. (See Figure 6-21E.)

FIG. 6-21C. *Methods of dry chemical application.*

FIG. 6-21D. *A double 30-lb (13.6-kg) stored-pressure dry chemical system with pneumatic release for automatic actuation. Each cylinder contains 30 lb (13.6 kg) of dry chemical pressurized to 350 psi (2413 kPa) with nitrogen.*

Storage of Dry Chemical and Expellant

Dry chemical is stored in a pressure container, usually of welded steel construction, either under atmospheric pressure until the sys-

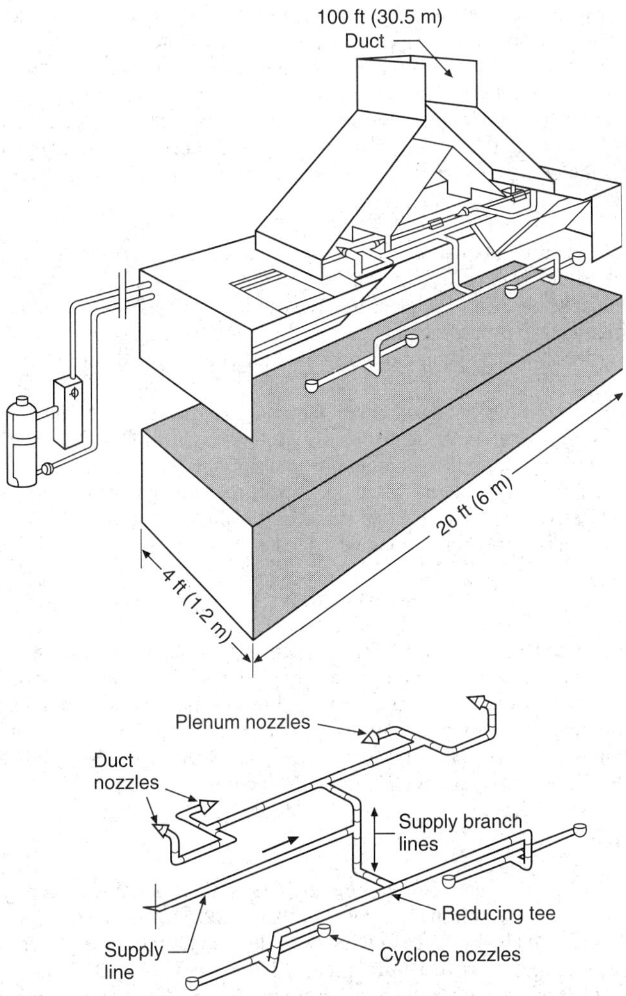

100 ft (30.5 m)
Duct

20 ft (6 m)

4 ft (1.2 m)

Plenum nozzles

Duct nozzles

Supply branch lines

Reducing tee

Supply line

Cyclone nozzles

FIG. 6-21E.   A typical single 30-lb (13.6-kg) cartridge-operated dry chemical system with fusible links for automatic operation; it is pre-engineered for kitchen range, hood, duct, and fryer fire protection. (Source: The Ansul Company)

tem is actuated or under the pressure of the internally stored expellant gas.

Containers in which dry chemical is stored separately under atmospheric pressure are equipped with an expellant gas inlet, a moisture-sealed fill opening, and a dry chemical outlet. The gas inlet leads to an internal gas tube arrangement constructed so that, when it flows into the tank, it agitates and permeates the powder, making it fluidlike. The dry chemical outlet is provided with a rupture disc or valve to permit buildup of proper operating pressure in the tank before the dry chemical can start to flow. The expellant gas assembly consists of a pressure storage vessel together with necessary valves, pressure regulators, and piping to deliver the expellant gas to the dry chemical storage tank at the correct pressure and rate of flow. (See Figure 6-21A.) The expellant gas is usually dry nitrogen; however, dry air or other gases may be used. The volume and storage pressure of the expellant are dictated by the gas used and the requirements of the system. Containers in which dry chemical and the expellant gas are stored together are equipped with a moisture-sealed fill opening, a valve with integral discharge outlet and expellant gas charging inlet, and a pressure gage. (See Figure 6-21E.)

It is desirable to locate the dry chemical expellant gas assemblies as near as practicable to the hazard to be protected. An area in

which temperatures stay between –40 and +120°F (–40 and +49°C) is desirable to maintain the quality of the dry chemical.

Dry chemical is commercially available in various sized packages of 10 lb (4.5 kg) or more in weight, or in metal drums, plastic-lined paperboard containers, or plastic bags. Whatever the container, it should be kept tightly closed and stored in a dry location to prevent absorption of moisture. Storage in a dry location is also essential for extinguisher recharging. Once the powder has lost its free-flowing characteristic, it should be discarded.

System Actuation

Actuation of fixed systems is initiated by automatic mechanisms that incorporate sensing devices located in the hazard area and automatic, mechanical, or electrical releases that initiate the flow of dry chemical, actuate alarms, and shut down process equipment. In systems with separate dry chemical and expellant gas containers, the flow of dry chemical is started by releasing the expellant gas, which pressurizes the dry chemical chamber to the point where the rupture disc in the dry chemical outlet operates. The dry chemical is then carried by the expellant gas through the distribution system to the hazard. In stored-pressure systems, however, the flow of dry chemical is started by merely opening the valve on the dry chemical chamber. An easily accessible device for manual operation is required for all automatically operated systems.

When automatically actuated systems are used, consideration should be given to the reduced visibility and temporary breathing difficulty sometimes caused by quick discharge of large amounts of dry chemical in a restricted space. In all cases where there is a possibility that personnel may be stationed in such locations, suitable alarms and safeguards should be incorporated in the system to ensure adequate warning and prompt evacuation.

Operation of a hand hose line system requires two or, at the most, three steps. *Step one*: pressurizing the dry chemical chamber by opening the expellant gas valve (if the dry chemical and expellant gas containers are separate), or opening the main discharge valve if dry chemical is under stored pressure. *Step two*: operating the nozzle at the end of the hose line. *Step three*: if multiple hose stations are supplied by the same dry chemical supply, a distribution valve must be opened to direct the flow to the particular hose station to be used. It is extremely important to play out the hose fully (without kinks) to ensure the proper chemical discharge rate.

Fixed piping systems normally blow themselves clear. However, in hand hose line systems where the operator exercises control over the amount of dry chemical discharged, it is vital that all pipes and hose be independently blown clear before recharging to prevent blockage.

Distribution System

For fixed systems, the provisions for conveying the dry chemical to the hazard area and discharging it properly consist of piping or, for hand hose line systems, piping and hose lines or hose lines alone. Nozzles designed to emit wide, flat, or round streams in desired patterns are available to meet specific hazard requirements. Adjustable nozzles permitting an operator to vary the range and shape of the discharge are also available for use with hand hose line systems.

The piping and valving for dry chemical systems are of special design because dry chemical, while tending to behave as a fluid, has distinct flow characteristics.

Control of dry chemical flow starts at the dry chemical storage tank. In systems of separate dry chemical and expellant gas containers, expellant gas must be admitted to the tank in such a way as to properly fluidize the dry chemical while the pressure builds up equally throughout the entire volume of the tank before the dry chemical is released from the tank. Should the pressure increase too rapidly above the dry chemical, the powder will not be properly fluidized and, upon release, the pressure drop in the piping or hose line will be excessive and result in a rate of dry chemical flow too low for proper extinguishing effectiveness.

If the expellant gas is permitted to channel to the outlet before the top of the dry chemical storage tank is properly pressurized, insufficient dry chemical will be carried by the flowing stream of gas, and the fire extinguishing effectiveness of the system again will be greatly reduced.

After release from the storage tank, dry chemical is carried through the piping at high velocities by the expellant gas and is thrown to the wall of the piping by centrifugal force whenever the direction of flow is sharply altered, as by an elbow. Should an elbow be directly connected to a tee and lie in the same plane as the tee, it is obvious that the two branches of the tee will carry appreciably different proportions of gas and dry chemical. This condition is overcome by installing all tees in planes perpendicular to the planes of adjacent elbows, by allowing sufficient length of straight pipe between tees and elbows, or by inserting special venturi devices between tees and elbows to ensure proper redistribution of dry chemical in the stream of gas before the mixture enters the tee.

Another critical factor in dry chemical distribution system design is the pressure drop through various lengths of pipe, hose, or fittings. Consequently, the length and size of piping and hose, and the number and size of fittings, must be selected to provide the required rate of discharge and nozzle distribution. In pre-engineered systems, this is already ensured by virtue of the limitations established for piping and fittings. In engineered systems, this selection must be made on the basis of actual calculation of pressure drop and subsequent flow rate.

For engineered systems, pressure drop data have been obtained at various rates of flow for various sizes of pipe and fittings. In one manufacturer's system, the pressure drop through a standard elbow is equivalent to the pressure drop through 20 ft (6 m) of straight pipe of the same size.[12] Pressure drop through a tee is equivalent to that of 45 ft (14 m) of straight pipe of the same size.

Nozzles

The selection of the proper types and sizes of nozzles for fixed systems is necessary to obtain proper coverage of the hazard.

Nozzles for hand hose line systems may be of either a one- or two-position type. One-position nozzles provide a modified straight stream, while two-position nozzles provide a straight stream or a fan. Since the straight stream has a longer reach than the fan discharge, it is considered better for initial attack. The shorter, wider fan-shaped stream is usually preferred for a close-range attack to complete the extinguishment. However, the present state of the art seems to prefer the use of the modified straight stream only. It is more effective and less confusing in the hands of an experienced operator.

Quantity and Rate of Application of Dry Chemical

The quantity of dry chemical and rate of flow must be sufficient to create a fire extinguishing concentration throughout all parts of the enclosure protected by a total flooding system, or over the specific fire area protected by a local application system. It must be realized that the fire extinguishing concentration accomplished throughout an enclosure (total flooding) can occur only during discharge. Fol-

lowing discharge, dry chemical rapidly settles out, and diminishes. This is in contrast to gaseous diffusion, where concentrations persist following discharge, such as in carbon dioxide and halon systems. Minimum rate of flow is critical because dry chemical will not extinguish a fire if applied too slowly. In local application systems, application at too high a rate may result in uneven discharge, or complete discharge of dry chemical before extinguishment is accomplished.

Optimum quantities of dry chemical and application rates have been determined by experiment for engineered total flooding and local application systems. For pre-engineered systems, the maximum sizes of hazards that can be protected by the various sizes of systems within specific piping limitations have been determined.

From tests with one manufacturer's engineered total flooding system, it has been determined that the weight in lbs (kg) of dry chemical required is obtained by multiplying the number of cu ft (m^3) in the net volume of the open space to be flooded by 0.0385 (in SI units 0.616). Net volume is the gross volume less the volume of machinery or other permanently located bulky objects.

The minimum flow rate in lbs per sec (kg/sec) is determined by multiplying the net volume in cu ft (m^3) by 0.00125 (in SI units 0.020). These volume factors apply only to this manufacturer's equipment and are cited merely to illustrate a procedure used.[12] Quantities and flow rates determined in this manner are contingent upon a nozzle arrangement that provides even distribution throughout the volume. The rates also assume that devices will be installed to close automatically all doors, windows, ventilators, or other openings through which dry chemical could escape from the enclosure.

In one manufacturer's pre-engineered total flooding system, a 1000 cu ft (28 m^3) space not longer than 20 ft (6 m) can be protected by a 30-lb (14-kg) system with 4 nozzles and no more than 90 ft of $^3/_4$-in. pipe (27 m of 19-mm pipe), 13 elbows, 3 tees, and 3 venturi devices.[12] As with engineered systems, successful extinguishment is contingent upon proper nozzle location, shutdown of ventilation, and the closing of all openings, such as doors and windows, in addition to closure of fuel valves.

Although simultaneous shutdown of ventilation with the actuation of a total flooding system is the normal procedure, one exception exists. This exception is the special case of kitchen range hood and duct protection wherein pre-engineered systems are specifically designed to provide extinguishment, regardless of whether the ventilation is operating or not. Actually, this latter type of system is a combination of both total flooding and local application.

Quantities of dry chemical and flow rates for engineered local application systems are obtained from graphs plotted from tests using different flammable liquid surface areas, weights of dry chemical, and rates of application. These graphs are available from the manufacturer. For pre-engineered local application systems, the limitations of hazard size, system size, and piping arrangement are established by test and used in the same manner as for total flooding systems.

The minimum recommended quantity of dry chemical for hand hose line systems is sufficient to permit use of the system for 30 sec. Capacities of hand hose line systems range from 125 to 2,000 lb (57 to 900 kg), and as many as 4 hose lines can be operated from a single system. As with fixed systems, a minimum flow rate must be maintained to prevent surging or interruption of flow. Minimum rates are determined by the equipment manufacturer and depend upon the equipment used.

INSPECTION, TESTING, AND MAINTENANCE PROCEDURES

All dry chemical systems should be inspected by the owner on a monthly basis. Systems subjected to process deposits (such as paint and dust) and corrosive conditions may require more frequent inspection of components. This inspection or quick check should include: a check to ensure that the system is in its proper location, manual pull stations are unobstructed, tamper indicators and seals are intact, no obvious physical damage is evident, pressure gages are within the operating range, and nozzle blowoff caps (where provided) are intact. This inspection should be recorded in report form and made available for review by the local Authority Having Jurisdiction. Any deficiencies should be corrected immediately.

On a semiannual basis, routine maintenance in addition to a more detailed inspection should be performed. During this inspection, the hazard should be reviewed to determine if it has been expanded or otherwise modified to extend beyond the capabilities of the system. A thorough examination of all system components is necessary. Each component should be inspected for wear, contaminant loading, corrosion, or any other condition that may render the component or system inoperable. The dry chemical agent is normally inspected at this time for evidence of caking. If caking is present, the dry chemical must be discarded and the system recharged in accordance with the manufacturer's instructions. In stored pressure systems, the dry chemical must be checked every six years.

The distribution piping should be checked for obstructions during this procedure. This can be accomplished by several means. If the system is small and accessible, the piping can be dismantled and inspected visually. If this is not practical, then purging with nitrogen or compressed air may be done. A full or partial discharge test may also be performed to verify the condition of the piping; however, all of the agent must be removed from the piping following the test to prevent caking and subsequent blockage of the pipe.

Periodically, a discharge test may be required during the performance of routine maintenance procedures. In this case a "bag" test may be performed. During a bag test, the discharge nozzles are covered with a collection bag (a burlap bag or pillow case) to retain discharged dry chemical. Following discharge of the system, the chemical should be weighed to determine the quantity of chemical contained. This can be compared to the amount required by the manufacturer's listed maintenance manual. Insufficient amounts may indicate possible piping obstructions.

The amount of expellant gas should be checked semiannually to ensure that there is sufficient gas to provide an effective discharge, and to automatically clean out the piping after the dry chemical has dissipated. In systems with separate expellant gas containers, this is accomplished by checking the pressure (if nitrogen) or weight (if carbon dioxide) against the manufacturer's recommended minimums. In stored-pressure systems, the pressure gage is checked to see that it is in the operable range.

Most dry chemicals are alkaline (sodium and potassium bicarbonate and potassium carbonate). One is neutral (potassium chloride), and the multipurpose type (monoammonium phosphate) is acidic. Inadvertent mixing of different base dry chemicals can initiate undesirable reactions, generating carbon dioxide gas and double decomposition, resulting in equipment failure or loss of discharge capability.

During the periodic inspections, the nozzles should be examined to see that they are properly aimed, free of obstruction, and in good operating condition. Nozzles in kitchen range hood and duct protection systems must be fitted with grease seals that are tight, yet easily blown clear. The system actuating devices, such as fusible links, pneumatic heat detectors, or electric thermostats, should also be checked to ensure that they are not loaded with residues, or otherwise impaired. Freedom of movement should be verified for cable and pulley mechanisms.

The inspection and maintenance of hand hose line systems will vary with the location and climate conditions. Equipment located in extremely hot or humid areas will require more frequent checking because the heat can cause cylinder pressure to increase and possibly cause leakage. Inspection of hand hose line systems consists primarily of checking the pressure of the expelling gas container, or the pressure of the unit itself (if it is of the pressurized type). It is also advisable to inspect all hose lines and nozzles to be sure that they are unobstructed and in good operating condition.

Systems containing dry chemical chambers of less than 150-lb (69-kg) nominal capacity must be hydrostatically tested every 12 years. The hydrostatic test should be applied to the auxiliary pressure containers, valve assemblies, hoses, fittings (not including field piping), check valves, directional valves, manifolds, and hose nozzles, in addition to the dry chemical chambers. The dry chemical removed from the chambers should be discarded. Care must be exercised to ensure that all equipment tested is thoroughly dried prior to recharging. If there is no automatic, connected reserve supply, alternate fire protection should be provided.

BIBLIOGRAPHY

References Cited

1. Underwriters Laboratories Inc., *Fire Protection Equipment List*, Northbrook, IL. (Published annually with bimonthly supplement.)
2. Factory Mutual Research Corp., *Factory Mutual Approval Guide*, Factory Mutual Research Corporation, Norwood, MA. (Published annually.)
3. Genium Publishing Corp., "Material Safety Data Sheets Collection," Aug. 1989, Schenectady, NY.
4. Ewing, C. T., *et al.*, "Extinguishing Class B Fires with Dry Chemicals: Scaling Studies," *Fire Technology*, Vol. 31, No. 1, First Quarter 1995, pp. 17–43.
5. Ewing, C. T., *et al*, "Extinguishing Class A Fires with Multipurpose Chemicals," *Fire Technology*, Vol. 31, No. 3, Aug. 1995, pp. 195–210.
6. Guise, A. B., "Potassium Bicarbonate-Based Dry Chemical," *NFPA Quarterly*, Vol. 56, No. 1, 1962, pp. 21–27.
7. Haessler, W. M., *The Extinguishment of Fire*, revised ed., National Fire Protection Association, Quincy, MA, 1974.
8. McCamy, C. S., Shoub, H., and Lee, T. G., "Fire Extinguishment by Means of Dry Powder," Sixth *Symposium* on Combustion, 1956, The Combustion Institute, Van Nostrand Reinhold, New York, pp. 795–801.
9. Guise, A. B., "The Chemical Aspects of Fire Extinguishment," *NFPA Quarterly*, Vol. 53, No. 4, 1960, pp. 330–336.
10. Haessler, W. M., "Fire and Its Extinguishment," *NFPA Quarterly*, Vol. 56, No. 1, 1962, pp. 89–96.
11. Kirk-Othmer, *Encyclopedia of Chemical Technology*, 2nd ed., Interscience Publishers, New York, 1965.
12. Guise, A. B., and Lindlof, J. A., "A Dry Chemical Extinguishing System," *NFPA Quarterly*, Vol. 49, No. 1, 1955, pp. 52–60.

NFPA Codes, Standards, and Recommended Practices

Reference to the following NFPA codes, standards, and recommended practices will provide further information on dry chemical agents and application systems discussed in this chapter. (See the latest version of *The NFPA Catalog* for availability of current editions of the following documents.)

NFPA 10, *Standard for Portable Fire Extinguishers*
NFPA 17, *Standard for Dry Chemical Extinguishing Systems*

Additional Reading

Carson, W. G., and Klinker, R. L., *Fire Protection Systems—Inspection, Test & Maintenance Manual*, 2nd ed., National Fire Protection Association, Quincy, MA, 1994.

Ewing, C. T., "Flame Extinguishment Properties of Dry Chemicals," *Fire Technology*, Vol. 25, No. 2, May 1989, pp. 134–149.

Ewing, C. T., *et al.*, "Flame Extinguishment Properties of Dry Chemicals: Extinction Weights for Small Diffusion Pan Fires and Additional Evidence for Flame Extinguishment by Thermal Mechanisms," *Journal of Fire Protection Engineering*, Vol. 4, No. 2, 1992, pp. 35–52.

Fire Control Engineering Co., *Dry Chemical Fire Extinguishing Characteristics*, expanded 2nd ed., Ft. Worth, TX.

Jensen, R. H., "Compatibility of Mechanical Foam and Dry Chemical," *NFPA Quarterly*, Vol. 57, No. 3, Jan. 1964, pp. 296–303.

Kaminski, E. J., "Practical Aspects of Wet and Dry Chemical Extinguishing System Inspection and Acceptance," *Building Official and Code Administrator*, Vol. 29, No. 4, July/Aug. 1995, pp. 34–37.

Meldrum, D. N., "Combined Use of Foam and Dry Chemical," *NFPA Quarterly*, Vol. 56, No. 1, July 1962, pp. 28–34.

"Special Hazards Protection—Manufacturers and Equipment," *Fire Journal*, Vol. 81, No. 4, 1987, p. 43.

Standard for Dry Chemical Fire Extinguishers, 7th ed., Underwriters Laboratories Inc., Northbrook, IL, 1984.

Tuve, R. L., "Light Water and Potassium Bicarbonate Dry Chemical—A New Two-Agent Extinguishing System," *NFPA Quarterly*, Vol. 58, No. 1, July 1964, pp. 64–69.

Underwriters Laboratories Inc., "The Compatibility Relationship Between Mechanical Foam and Dry Chemical Fire Extinguishing Agents," *UL Bulletin of Research No. 54*, July 1963, Northbrook, IL.

Wesson, H. R., "Studies of the Effects of Particle Size on the Flow Characteristics of Dry Chemical," *Fire Technology*, Vol. 8, No. 3, Aug. 1972, pp. 173–180.

Woolhouse, R. A., and Sayers, D. R., "Monex Compared with Other Potassium-Based Dry Chemicals," *Fire Journal*, Vol. 671, No. 1, Jan. 1973, pp. 85–88.

Xiujuan, G., "Combined Agent (Dry Chemical-Foam) Fire Fighting Model Test," First Asian Conference on Fire Science and Technology (ACFST), October 9–13, 1992, Hefei, China, International Academic Publishers, China, 1992, pp. 360–365.

FOAM EXTINGUISHING AGENTS AND SYSTEMS

Revised by Joseph L. Scheffey

Fire-fighting foam is an aggregate of gas-filled bubbles formed from aqueous solutions of specially formulated concentrated liquid foaming agents. Normally, the gas used is air, but in certain applications may be an inert gas. Since foam is lighter than the aqueous solutions from which it is formed, and lighter than flammable liquids, it floats on flammable or combustible liquids, producing an air-excluding, cooling, continuous layer of vapor-sealing, water-bearing material that halts or prevents combustion.

Foam is produced by mixing a foam concentrate with water at the appropriate concentration, and then aerating and agitating the solution to form the bubble structure. Some foams are thick and viscous and form tough, heat-resistant blankets over burning liquid surfaces and vertical areas; other foams are thinner and spread more rapidly. Some foams are capable of producing a vapor-sealing film of surface-active water solution on a liquid surface. Some, such as medium- or high-expansion foam, are meant to be used as large volumes of wet gas cells for inundating surfaces and filling cavities.

Foams are defined by their expansion ratio, which is the ratio of final foam volume to original foam solution volume before adding air. They are arbitrarily subdivided into three ranges: (1) low-expansion foam—expansion up to 20:1; (2) medium-expansion foam—expansion 20 to 200:1; and (3) high-expansion foam—expansion 200 to 1,000:1.

Foams are also defined according to their effectiveness on hydrocarbon fuels, water-miscible fuels, or both.

There are various methods of generating and applying foams. This chapter covers the basic characteristics of various foaming agents and the methods for producing fire-fighting foams, as well as specific applications of equipment and systems useful for different types of hazards.

USES AND LIMITATIONS OF FIRE FIGHTING FOAMS

Low-expansion foam is used principally to extinguish burning flammable or combustible liquid spill or tank fires by application to develop a cooling, coherent blanket. Foam is the only permanent extinguishing agent used for fires of this type. Its application allows fire fighters to extinguish fires progressively. A foam blanket covering a tank's liquid surface can prevent vapor transmission for some time, depending upon the stability and depth of the foam. Fuel spills are quickly rendered safe by foam blanketing. The blanket may be removed after a suitable period of time; often, it has no detrimental effect on the product with which it comes into contact.

Foams may be used to diminish or halt the generation of flammable vapors from nonburning liquids or solids and may be used to fill cavities or enclosures where toxic or flammable gases may collect.

Foam is of great importance where aircraft are fueled and operated. Sudden, large fuel spills resulting from aircraft accidents or malfunction require rapid foam application. Hangar fire protection is best accomplished by properly designed foam systems.

Increasingly, warehouses and buildings storing large quantities of combustible and flammable liquids are protected by foam-water sprinkler systems. The protection required is a function of the type and quantity of liquid stored, building height, and storage configuration.

Foams of the medium- or high-expansion type (20 to 1,000 times) may be used to fill enclosures such as basement room areas or holds of ships where fires are difficult or impossible to reach. Here, foams act to halt convection and access to air for combustion. Their water content also cools and diminishes oxygen by steam displacement. Foams of this type (with expansion ratios of 400 to 500) may be used to control liquefied natural gas (LNG) spill fires and help disperse the resulting vapor cloud.

Many foams are generated from solutions with very low surface tension and penetration characteristics. Foams of this type are useful where Class A combustible materials are present. In such instances, the water solution draining from the foam cools and wets the solid combustible.

Foam breaks down and vaporizes its water content under attack by heat and flame. It therefore must be applied to a burning liquid surface in sufficient volume and rate to compensate for this loss, with an additional amount applied to guarantee a residual foam layer over the extinguished liquid. Foam is unstable and may be broken down easily by a physical or mechanical force, such as a water hose stream. Certain chemical vapors or fluids may also destroy foam quickly. When certain other extinguishing agents are used in conjunction with foam, severe breakdown of the foam may occur. Turbulent air or violently uprising combustion gases from fires may divert foam from the burning area.

The mechanisms of foam fire extinguishment on two-dimensional hydrocarbon fuel fires have not been fully developed. Recent theoretical and preliminary experimental work has attempted to

---

Joseph L. Scheffey, P. E., is director of Fire Protection Research and Development at Hughes Associates, Inc., Baltimore, MD.

correlate foam extinguishment with fundamental fluid spread mechanisms. More research is required to develop foam suppression theory.

Foam solutions are conductive and therefore not recommended for use on electrical fires. If foam is used, a spray is less conductive than a straight stream. However, because foam is cohesive and contains materials that allow water to conduct electricity, foam spray is more conductive than water spray.

Engineering design requirements and recommended application methods must be followed for successful use of foams. These requirements can be found in NFPA 11, *Standard for Low-Expansion Foam* (hereinafter referred to as NFPA 11); NFPA 11A, *Standard for Medium- and High-Expansion Foam Systems* (hereinafter referred to as NFPA 11A); NFPA 11C, *Standard for Mobile Foam Apparatus* (hereinafter referred to as NFPA 11C); NFPA 16, *Standard for the Installation of Deluge Foam-Water Sprinkler and Foam-Water Spray Systems* (hereinafter referred to as NFPA 16); NFPA 16A, *Standard for the Installation of Closed-Head Foam-Water Sprinkler Systems* (hereinafter referred to as NFPA 16A); and NFPA 403, *Standard for Aircraft Rescue and Fire Fighting Services at Airports* (hereinafter referred to as NFPA 403). Additionally, NFPA codes and standards, such as NFPA 30, *Flammable and Combustible Liquids Code* (hereinafter referred to as NFPA 30), and NFPA 409, *Standard on Aircraft Hangars* (hereinafter referred to as NFPA 409), provide application rate and installation details on special hazard systems.

In general, the following criteria for the hazardous liquid must be met for a foam to be fully effective:

1. The liquid must be below its boiling point at the ambient conditions of temperature and pressure.
2. Care must be taken in application of foam to liquids with a bulk temperature higher than 212°F (100°C). At these fuel temperatures and above, foam forms an emulsion of steam, air, and fuel. This may produce a fourfold increase in volume when applied to a tank fire, with dangerous frothing or slopover of the burning liquid.
3. The liquid must not be unduly destructive to the foam used, or the foam must not be highly soluble in the liquid to be protected.
4. The liquid must not be water-reactive.
5. The fire must be a horizontal surface fire. Three-dimensional (falling fuel) or pressure fires cannot be extinguished by foam unless the hazard has a relatively high flashpoint and can be cooled to extinguishment by the water in the foam.

TYPES OF FOAM

There are a number of types of foaming agents available, known as foam concentrates, some of which are designed for specific applications. Some are suitable for extinguishing all types of flammable liquids, including water-soluble and foam-destructive liquids. Descriptions of the common types of foam follow.

Aqueous Film-Forming Foam Agents (AFFF)

Aqueous film-forming foam agents are composed of synthetically produced materials that form air-foams similar to those produced by the protein-based materials. In addition, these foaming agents are capable of forming water solution films on the surface of flammable hydrocarbon liquids, hence the term "aqueous film-forming foam" (AFFF). AFFF concentrates are available for proportioning to a final concentration of 1 percent, 3 percent, or 6 percent by volume, with either fresh water or seawater.

The air-foams generated from AFFF solutions possess low viscosity, have fast spreading and leveling characteristics, and, like other foams, act as surface barriers to exclude air and halt fuel vaporization. These foams also develop a continuous aqueous layer of solution under the foam, maintaining a floating film on hydrocarbon fuel surfaces to help suppress combustible vapors and cool the fuel substrate. This film, which can also spread over fuel surfaces not fully covered with foam, is self-healing following mechanical disruption and continues to spread as long as there remains a reservoir of nearby foam. Film effectiveness may be reduced on hot surfaces and aromatic hydrocarbons. To ensure fire extinction, an AFFF blanket should cover the fuel surface entirely, as with other types of foam.

AFFF fluidity and film strength on kerosene makes it particularly suitable for jet aircraft fuel spill fire fighting.

AFFF concentrates contain fluorinated, long-chain synthetic hydrocarbons with particular surface active properties.

Because of the extremely low surface tension of the solutions draining from AFFF, these foams may be useful under mixed-class fire situations (Classes A and B), where deep penetration of water is needed in addition to the surface spreading action of foam itself.

Foam-generating devices yielding stable, homogeneous foams are not necessarily needed in the employment of AFFF. Less sophisticated foaming devices, such as water spray nozzles and sprinklers (unlike with most other foaming agents), may be used because of the inherent rapid and easy foaming capability of AFFF solutions. However, such foams drain relatively rapidly, and they may provide less burnback resistance compared to protein-based foams. AFFF also may be used, without compatibility problems, in conjunction with dry chemical agents. Although AFFF concentrates must not be mixed with other types of foam concentrates, foams made from them do not break down other types of foams in fire-fighting operations. The normal-use temperature range for these agents is 35 to 120°F (1.7 to 49°C).

Fluoroprotein (FP) Foaming Agents

The concentrates utilized for generating fluoroprotein foams are similar in composition to protein foam concentrates, but, in addition to protein polymers, they contain fluorinated surface active agents that confer a "fuel shedding" property to the foam generated. This makes them particularly effective for fire-fighting conditions where the foam becomes coated with fuel, such as in the method of subsurface injection of foam for tank fire fighting, and nozzle or monitor foam applications where the foam may often be plunged into the fuel. Fluoroprotein foams are very effective for in-depth crude petroleum or other hydrocarbon fuel fires because of this fuel shedding property. In addition, these foams demonstrate better compatibility with dry chemical agents than do the regular protein-type foams. They also possess superior vapor securing and burnback resistance characteristics. Fluoroprotein-type concentrates are available for proportioning to a final concentration of either 3 percent or 6 percent by volume, using either fresh water or seawater. They are nontoxic and biodegradable after dilution. The normal-use temperature range for these agents is 20 to 120°F (−7 to 49°C).

Film-Forming Fluoroprotein (FFFP) Agents

Film-forming fluoroprotein (FFFP) agents are composed of protein together with film-forming fluorinated surface-active agents, which make them capable of forming water solution films on the surface of most flammable hydrocarbons and of conferring a fuel-shedding property to the foam generated.

Air-foams generated from FFFP solutions have fast-spreading and leveling characteristics and, just as other foams, act as surface barriers to exclude air and prevent vaporization. Like AFFF, they generate a self-healing, continuous floating film on hydrocarbon fuel surfaces which helps suppress combustible vapors. However, to ensure fire extinction, an FFFP blanket, as with other types of foam, should cover the entire fuel surface.

Because of the rapid and easy foaming capability of FFFP solutions, water spray devices may be used in many situations. However, the foams produced drain rapidly and may have less burnback resistance compared to protein-based foams.

Film-forming fluoroprotein-type concentrates are available for proportioning to a final concentration of either 3 or 6 percent by volume, using either fresh water or seawater. They may be used in conjunction with dry chemical agents without compatibility problems.

Protein (P) Foaming Agents

Protein-type air-foams utilize aqueous liquid concentrates proportioned with water for their generation. These concentrates contain high-molecular-weight natural proteinaceous polymers derived from a chemical digestion and hydrolysis of natural protein solids. The polymers give elasticity, mechanical strength, and water retention capability to foams generated from them. The concentrates also contain dissolved polyvalent metallic salts which aid the protein polymers in their bubble-strengthening capability when the foam is exposed to heat and flame. Protein-type concentrates are available for proportioning to a final concentration of either 3 percent or 6 percent by volume, using either fresh water or seawater. In general, these concentrates produce dense, viscous foams of high stability, high heat resistance, and good resistance to burnback, but they are less resistant to breakdown by fuel saturation than are AFFF and fluoroprotein foams. They are nontoxic and biodegradable after dilution. The normal-use ambient temperature range for these concentrates is 20 to 120°F (–7 to 49°C).

Low-Temperature Foaming Agents

This type of foam concentrate is protected for storage and use at low temperature by the inclusion of freezing-point depressants. Low-temperature foaming agents may be used at ambient temperatures as low as –20°F (–29°C). They are available for use at either 3 percent or 6 percent by volume concentration in either fresh water or seawater, and may be of the AFFF or protein-based type.

Alcohol-Type Foaming Agents (AR)

Air-foams generated from ordinary agents are subject to rapid breakdown and loss of effectiveness when they are used on fires that involve fuels that are water soluble, water miscible, or of a "polar solvent" type. Examples of this type of fuel are alcohols, enamel and lacquer thinners, methyl ethyl ketone, acetone, isopropyl ether, acrylonitrile, ethyl and butyl acetate, and the amines and anhydrides. Even small amounts of these substances mixed with common hydrocarbon fuels, such as gasohol, may cause rapid breakdown of ordinary fire-fighting foams.

Therefore, certain special foaming agents, called alcohol-type concentrates, have been developed. These alcohol-resistant concentrates are proprietary compositions of several types, some containing a protein, fluoroprotein, or an aqueous film-forming foam concentrate base. The most common are usually described as polymeric alcohol-resistant AFFF concentrates which produce foams suitable for application to spill or in-depth fires of either hydrocarbon or water-miscible flammable liquids by any foam-generating device. They exhibit AFFF characteristics on hydrocarbons and produce a floating gel-like mass for foam buildup on water-miscible fuels. Agents of this type have no transit time limitations. Normal-use temperatures for any of the alcohol-type agents are 35 to 120°F (1.7 to 49°C).

Medium- and High-Expansion Foaming Agents (SYNDET)

Medium- and high-expansion foams are agents for control and extinguishment of Class A and some Class B fires, and are particularly suited as a flooding agent for use in confined spaces. The foam is an aggregation of bubbles, mechanically generated by aspiration or a blower-fan, which forces air or some other gas through a net or screen that is wetted by an aqueous solution of surface-active foaming agents. Under proper conditions, fire-fighting foams of expansions from 20:1 up to 1,000:1 can be generated.

Medium- or high-expansion foam is a unique vehicle for transporting wet foam masses to inaccessible places, for total flooding of confined spaces, and for volumetric displacement of vapor, heat, and smoke. Tests have shown that when used under certain circumstances in conjunction with water from automatic sprinklers, the foam will provide more positive control and extinguishment than either extinguishing agent by itself. Optimum efficiency in any one type of hazard is dependent upon the rate of application and the foam expansion and stability.

Liquid concentrates for producing medium- and high-expansion foams consist of synthetic hydrocarbon surfactants of a type that will foam copiously with a small input of turbulent action. They are generally used in approximately 2 percent proportion in water solution. Medium-expansion foam may also be generated from solutions of fluoroprotein, protein, or AFFF concentrates at 3 or 6 percent proportion.

Medium- or high-expansion foams are particularly suited for indoor fires in confined spaces. Their use outdoors may be limited because of the effects of weather. These foams have several effects on fires:

1. When generated in sufficient volume, they can prevent the air necessary for continued combustion from reaching the fire.
2. When forced into the heat of a fire, the water in the foam is converted to steam, which reduces the oxygen concentration by diluting the air.
3. The conversion of the water to steam absorbs heat from the burning fuel. Any hot object exposed to the foam will continue breaking down the foam, converting the water to steam, and being further cooled.
4. Because of its relatively low surface tension, foam solution not converted to steam will tend to penetrate Class A materials. However, deep-seated fires may require overhaul.
5. When accumulated in depth, high-expansion foam can provide an insulating barrier for protection of exposed materials or structures not involved in a fire, thereby preventing fire spread.

Research has shown that using air from inside a burning building to generate high-expansion foam has an adverse effect on the volume and stability of the foam produced. Combustion and pyrolysis products, when they react chemically with the foaming agent, can reduce the volume of foam produced and increase the drainage rate. The high temperature of the air breaks down the foam as it is generated. Physical disruption also takes place, apparently caused by vapor and solid particles from the combustion process. These factors that cause foam breakdown may be compensated for by higher rates of foam generation.

Entry to a foam-filled passage must not be attempted without use of self-contained breathing apparatus. The foam mass also reduces vision and hearing, and life lines must be used for personnel entering the foam.

Foam of approximately 500:1 expansion can be successfully used for control of fires and reduction of vaporization from LNG spills. As water slowly drains from the foam in small amounts, it forms a thin ice layer which floats on the LNG and supports the high-expansion foam blanket. Fixed and portable high-expansion foam facilities of this type have been provided for LNG storage and manufacturing plants.

Other Synthetic Hydrocarbon Surfactant Foaming Agents

There are many synthetically produced surface active compounds that foam copiously in water solution. When these are formulated properly, they may be used as wetting agents or as fire-fighting foams and employed in much the same manner as other types of foam. [See NFPA 18, *Standard on Wetting Agents* (hereinafter referred to as NFPA 18).]

Hydrocarbon surfactant foam liquid concentrates are employed in 1 to 6 percent proportions in water. When these solutions are used in conventional foam-making devices, the resulting air-foam possesses low viscosity and spreads quickly over liquid surfaces. Its fire-fighting characteristics depend upon: (1) the volume of the foam layer on the burning surface, which halts access to air and controls combustible vapor production; and (2) the minor cooling effect of the water in the foam, which becomes available due to a relatively rapid breakdown of the foam mass. This water solution does not possess film-forming characteristics on the flammable liquid surface, although, under some conditions, it may produce a temporary water emulsion due to its wetting agent or "detergent-type" properties. Because of the low surface tension and wetting properties of the water solutions of these foams, they also may be used as extinguishing agents for Class A fires, although they were primarily developed with medium- or high-expansion foam in mind.

Synthetic hydrocarbon surfactant foams are generally less stable than other fire-fighting foams. Their water solution content drains away rapidly, leaving a bubble mass which is highly vulnerable to fuel heat or mechanical disruption. Usually these foams must be applied at higher rates than other fire-fighting foams to achieve extinction. Many formulations of this type of foam concentrate break down other foams if used simultaneously or sequentially.

Caution should be used when using wetting agents described in NFPA 18 for Class B situations. The test criteria used to classify the wetting agent for use on Class B fires may be significantly different than that used to classify AFFF, fluoroprotein, and protein foams. Fire suppression of Class A, B, and D fires using wetting agents is the subject of ongoing investigations.

Foams for Vapor Suppression

Fire-fighting foams may be used to suppress vapors from unignited flammable liquids for which the foams are proven extinguishing agents. To ensure good vapor suppression, the foam must be applied gently, so as to not mix fuel with the foam, and must cover the entire fuel surface. Care should be taken not to agitate the fuel during foam application, because static spark ignition of volatile hydrocarbons can result from plunging and turbulence from a foam or water stream.

There are flammable, toxic, or otherwise hazardous liquids on which fire-fighting foams are unstable and are therefore unsuitable for use either for fire fighting or vapor suppression. Special vapor-mitigating foams have been developed that are relatively stable on many hazardous liquids; one type is an alcohol-resistant fire-fighting foam of the low-expansion type, which can be additionally stabilized to enhance its effective foam life and stability on toxic, flammable, or corrosive liquids. Other types are medium-expansion foams especially formulated to be stable on either acidic or alkaline hazards. These medium-expansion foams are not effective fire-fighting foams.

In general, vapor-mitigating foams should be applied to spills of toxic liquids only by trained personnel wearing suitable protective gear. These foams require special engineering consideration, and, to date, no standards have been developed for their use. The manufacturers should be consulted for use, limitations, and application data.

Chemical Foam Agents and Powders

These foam-producing materials have become obsolete because of the superior economics and ease of handling of the liquid foam-forming concentrates previously discussed. Chemical foam is formed from the chemical reaction in aqueous solution between aluminum sulfate ("A," acidic) and sodium bicarbonate ("B," basic) which also contains proteinaceous foam stabilizers. Foam is formed by the generation of carbon dioxide gas trapped in the bubbles of the foaming solution.

GUIDELINES FOR FIRE PROTECTION WITH FOAMS

The following general rules apply to the application and use of ordinary air-foams:

1. The more gently the foam is applied, the more rapid the extinguishment and the lower the total amount of agent required.
2. Successful use of foam is also dependent upon the rate at which it is applied. Application rates are described in terms of the amount by volume of foam solution reaching the fuel surface (in terms of total area) every minute. If the foam has an expansion rate of 8:1, an application rate of 0.1 gpm per sq ft [4.1 (L/min)m^2] will provide 0.8 gal per sq ft (32.8 L/m^2) of finished foam every minute. Increasing the foam application rate over the recommended minimum will generally reduce the time required for extinguishment. However, little time advantage is gained if application rates are increased more than three times the minimum recommended. If application rates are less than the minimum, extinguishment time will be prolonged or may not be accomplished at all. If application rates are so low that the rate of foam loss by heat or fuel attack equals or exceeds the rate of foam application, the fire will not be controlled or extinguished.
3. The minimum recommended application rate is the rate found by test to be the most practical in terms of speed of control and amount of agent required. The general curve in Figure 6-22A illustrates the rate-time relationship for foam application to a hazard. The curve may be displaced right or left depending upon fuel, method of application, and type of foam concentrate; hence, the need for carefully engineered systems based on the appropriate standard and actual test information.
4. In general, air-foams will be more stable when they are generated with water at ambient temperature. Preferred water temperatures range from 35 to 80°F (1.7 to 27°C). Either fresh water or seawater may be used. Water containing known foam contaminants, such as detergents, oil residues, or certain corrosion inhibitors, may adversely affect foam quality.
5. Foams are adversely affected by air containing certain combustion products. While the effect is minor with ordinary air-foam and ordinary hydrocarbon fuels, it is desirable to locate fixed foam-makers on the sides of, rather than directly over, the hazard.
6. Recommended pressure ranges should be observed for all foam-making devices. Foam quality will deteriorate if these limits (high and low) are exceeded.
7. Many air-foams are adversely affected by contact with vaporizing liquid extinguishing agents, their vapors, and by some dry chemical agents.

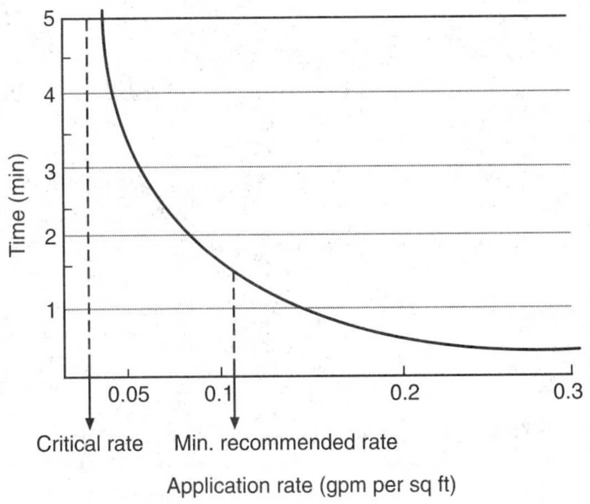

FIG. 6-22A.   *General relationship of foam application rate to time of application necessary for extinction. [1 gpm per sq ft equals 40.746 (L/min)/m².]*

FOAM-GENERATING METHODS

The process of producing and applying fire-fighting air-foams to hazards requires three separate operations, each of which consumes energy. They are: (1) the proportioning process, (2) the foam generation phase, and (3) the distribution method. A flow diagram illustrating the relationship of the three operations is given in Figure 6-22B.

In general practice, air-foam generation and distribution occur nearly simultaneously within the same device. There are also many types of proportioning. In certain portable devices, all three functions are combined into a single device. The design and performance requirements of foam systems dictate the choice of types of proportioning, generating, and distributing equipment for the protection of specific hazards.

Foam Concentrate Proportioners

It is very important that foam concentrate be proportioned accurately into the water stream. Proportioning equipment, foam concentrate, and discharge equipment must be matched to produce the proper solution concentration at system design operating pressures. If proportioning is low, the foam will be relatively weak and unstable; if too high, the foam may be stiff and concentrate will be wasted, thus reducing effective system operating time.

So that a predetermined volume of liquid foam concentrate may be mixed with a water stream to form a foam solution of fixed concentration, the following two general methods are used:

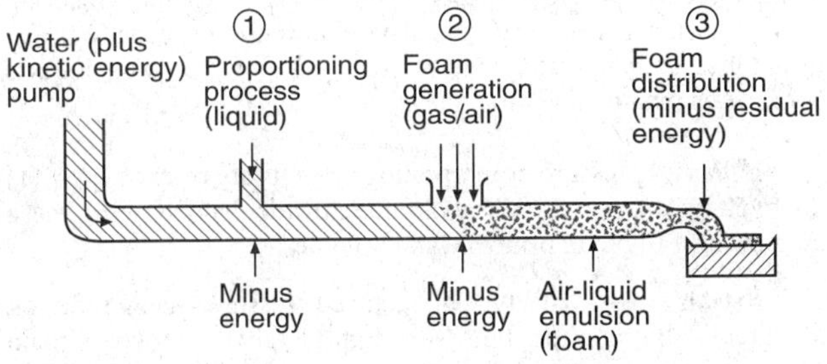

1. Methods that utilize the pressure energy of the water stream by venturi action and orifices to induct concentrate.
2. Methods that utilize external pumps or pressure heads to inject concentrate into the water stream at a fixed ratio to flow.

Figures 6-22C and 6-22D illustrate the general principles of the two different proportioning methods. Specific system designs of proportioning equipment are given below.

The nozzle eductor: This type of foam concentrate proportioner is of simple design and is widely used in portable foam-making nozzles where foam concentrate is available in 5-gal (19-L) pails or drums as shown in Figure 6-22E. Incorporating a modified venturi within the foam-making nozzle section, the nozzle eductor drafts concentrate from a portable container through a pickup tube. Using a properly sized orifice or pipe section at the low-pressure cavity, the concentrate is mixed in proper proportion to the fixed flow and operating pressure of the nozzle, and foam generation proceeds.

In-line eductor: This type of proportioner educts or drafts foam concentrate from a container or tank by venturi action, utilizing the operating pressure of the hose water stream on which it is installed, and injecting concentrate into that flow of water. Its correct operation is very sensitive to water flow and pressure. Changes in either of these factors from those for which the eductor was designed will result in incorrect proportioning. Distances of more than 6 ft (1.8 m) elevation from the eductor to the lowest liquid level of the foam concentrate container also may result in incorrect proportioning. Foam-generation devices and maximum downstream lengths of hose recommended for each eductor must be carefully adhered to. This proportioning device may be used in the hose line leading to the foam-generation device. An eductor of this type may also be installed at the foam concentrate tank in a fixed system or at the pump discharge of a mobile pumper.

Some designs of this proportioning device incorporate metering valves in the foam concentrate intake line so various volume percentages of concentrate in the water stream may be obtained. A check valve is usually placed in this intake so that water cannot flow back into the foam concentrate container if a blockage occurs or a valve is closed in the downstream hose. The use of this eductor requires an allowance for a pressure loss of approximately 30 percent in the system or layout.

Around-the-pump proportioner: This type of proportioner also operates on the venturi principle, except that this proportioner must be situated at the pump and connected to both suction and pressure sides. Its advantage is that pressure recovery of the venturi action is attained, and pump delivery pressures to the foam-making device or devices downstream require no compensation for pressure loss except for that in the layout hose length. The net delivery volume of

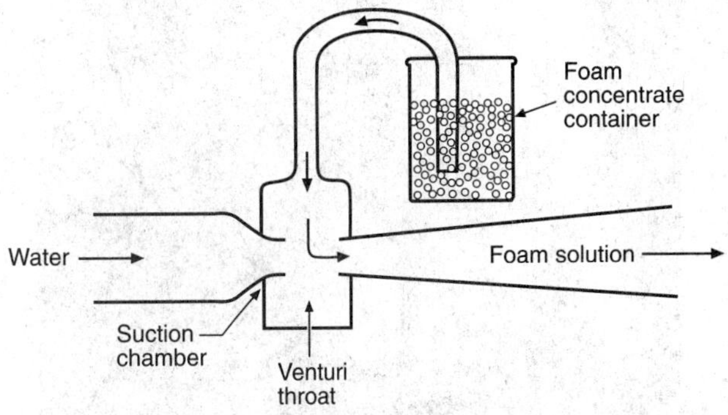

FIG. 6-22B.   *A diagram illustrating the steps in air-foam generation.*

FIG. 6-22C.   *Venturi induction (in-line) proportioner.*

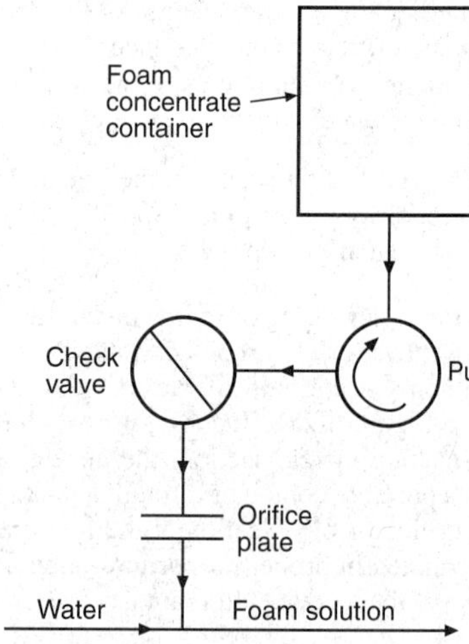

FIG. 6-22D. Foam concentrate pump proportioner.

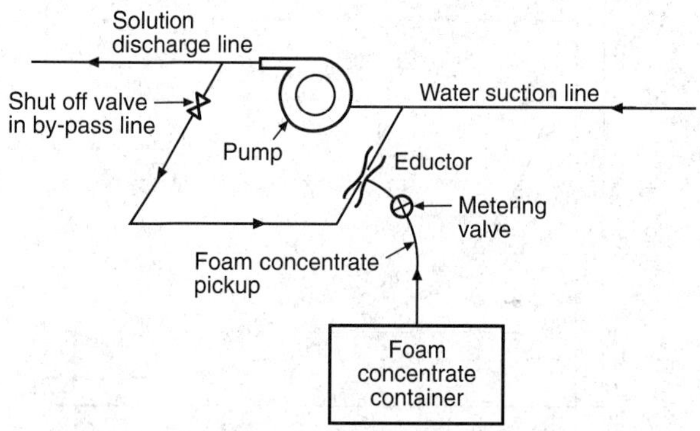

FIG. 6-22F. An around-the-pump proportioner.

the pump will be decreased by 10 to 40 gpm (38 to 150 L/min) in this method of proportioning. (See Figure 6-22F.) The small portion of the pump discharge flows through a bypass line to the suction side of the pump. A venturi eductor in this line produces a negative pressure on the foam concentrate pickup line from the foam concentrate container. Foam concentrate is led to the eductor, where it mixes with water and is delivered to the suction side of the pump.

Multiple foam-makers may be supplied with foam solution by this type of proportioner when it is supplied with a multiported metering valve designed for the required flows.

Pressure proportioning tank: This type of proportioner may consist of one tank or of two tanks separately connected to the water and foam solution lines. The tank or tanks in the system may each be fitted with a flexible diaphragm or bladder to separate the "driving" water from foam concentrate. Tanks may rely simply on differences in density of the two liquids to retard mixing during operation (without diaphragm or bladder), when proportioning protein or fluoroprotein concentrate.

FIG. 6-22E. Air-foam nozzle with built-in eductor.

The principle of this device is simple. A small amount of the flowing water volumetrically displaces foam concentrate into the main water stream. The design pressure of the vessel must be above the maximum static water pressure encountered in the system.

Water is allowed to enter the foam tank from the main stream with as little friction loss as possible. An orifice meters liquid in the tank into the low pressure area. Advantages of this system are its low pressure drop, automatic proportioning over range of flows and pressures, and its freedom from external power. Its disadvantages are a long refill time (since it is a "batch" method, the tank must be drained of its driving water and the tank or the bladder refilled with concentrate), and an economic limit on size.

Coupled water motor-pump proportioner: This proportioner consists of two positive displacement rotary pumps mounted on a common shaft. Water delivered to the larger pump (motor) causes it to drive the smaller pump, which is used to draft concentrate from a container and deliver it (at line pressure) to the water discharge line from the larger pump. By proportioning the sizes of the two pumps, the correct volume of concentrate is delivered to the water stream.

This proportioner is manufactured in only two sizes for water throughputs of 60 to 180 gpm (227 to 680 L/min) and 200 to 1000 gpm (757 to 3800 L/min). Both sizes are designed for proportioning foam concentrates recommended for use at 6 percent concentration. A pressure drop of 25 to 30 percent in the water stream supplied to the device is required for its operation.

Balanced pressure proportioner: The use of a separate pump for pressurizing foam concentrate to be delivered in the correct proportions to a flowing water stream offers the greatest advantage for reliable and accurate operation of a foam concentrate proportioning system which must function at varying rates of waterflow or pressure during its use. (See Figure 6-22G for a fixed design and Figure 6-22H for a foam truck-mounted design.) Many systems require manual attention to flow meters, duplex gages, or other measuring devices during system operation. These designs have been replaced by automatically operated proportioning systems that do not require continuous manual attention.

Balanced pressure proportioning systems are of two types: (1) the bypass system and (2) the variable flow demand system. These systems are simple in principle and reliable.

Bypass system. This type of balanced pressure system relies on a hydraulically monitored pressure control valve in a bypass foam concentrate line from a pump back to the concentrate supply tank, a correctly sized venturi-type proportioning controller in the water line for correct and automatic proportioning over wide ranges of

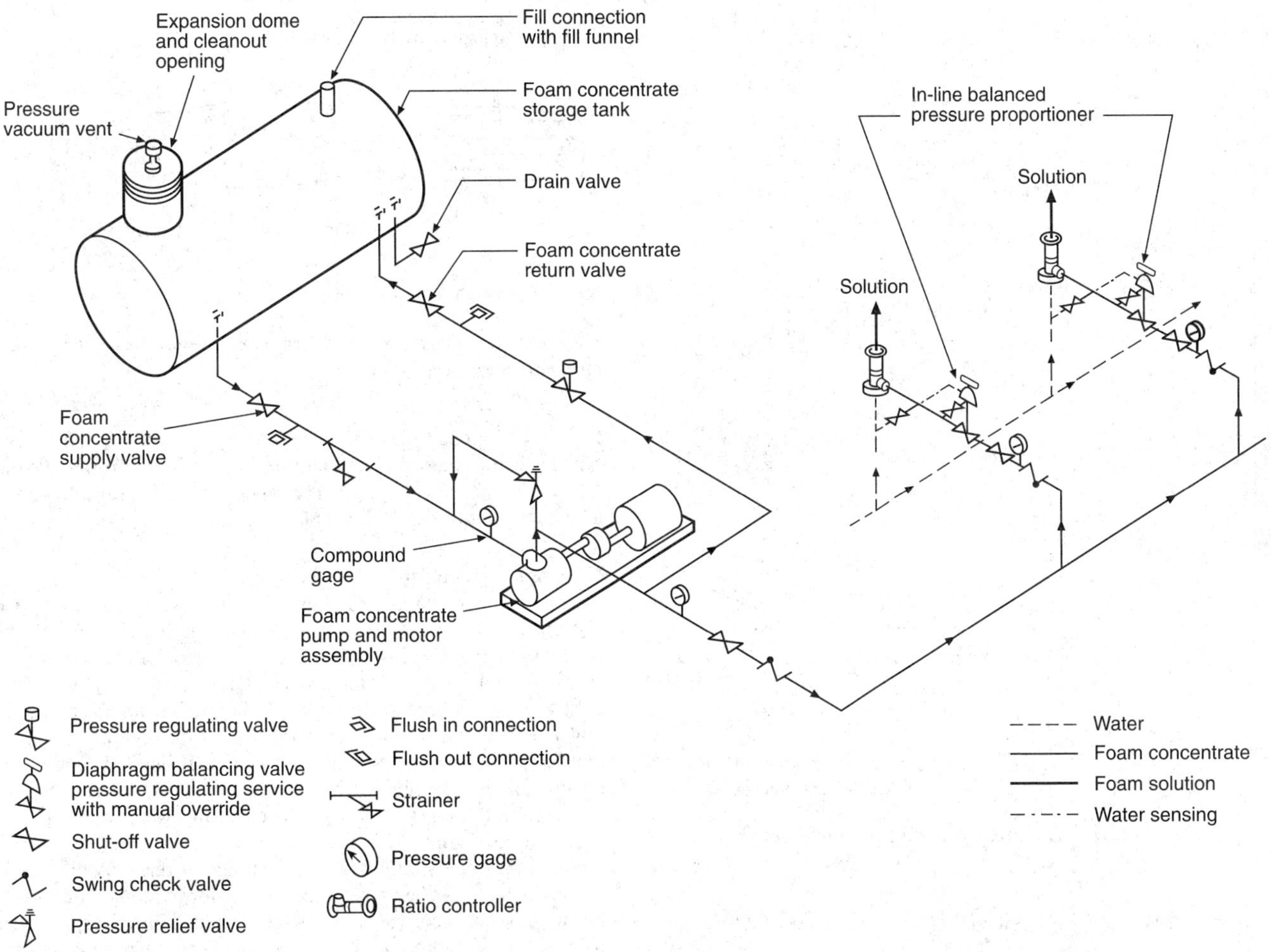

FIG. 6-22G. Balanced pressure proportioning with multiple injection points (metered proportioning).

flow, and a fixed or variable metering orifice between the pump and the proportioning controller.

Variable flow demand system. This type of balanced pressure system utilizes a variable speed mechanism to drive a pump delivering foam concentrate to the system. Only a correctly sized venturi-type proportioning controller in the water line (for correct and automatic proportioning over wide ranges of flow) and a fixed or variable metering orifice between the pump and proportioning controller are used. The variable drive mechanism is controlled by a feedback system which can be designed to operate hydraulically or electronically. Once activated, the pump output is monitored automatically so foam concentrate flow and pressure are always proportional to system demand for foam solution. (See Figure 6-22I.) The variable flow demand system may be fixed or truck-mounted.

Variable orifice, variable flow demand proportioner: This type of balanced pressure system utilizes a specially designed, flow-sensitive, moveable piston section in the water supply that controls a variable orifice in the foam concentrate line. A pump supplies concentrate at monitored pressures to the metering orifice, which changes in size proportionally to the system demand for foam solution. The device is especially designed for large-capacity systems. Its design provides accurate proportioning over ranges of flow of water of approximately 2.3 to 1.

Premixed foam solution: This method of proportioning is a batch type of mixing of concentrate with water, usually in a container that can be pressurized. The measured volume of concentrate is poured into a measured volume of water to yield a foam solution of the recommended strength; e.g., for a 3 percent solution in a container that holds 100 gal (378.5 L) of liquid, 97 gal (367 L) of water are poured into it and 3 gal (11.5 L) of foam concentrate are mixed with it to give a solution of 3 percent by volume. The final solution mixture is then educted from a tank to a pump or placed in a pressurizing vessel.

Some foam concentrates are not suitable for storage in premix solution. Even those that are formulated for this use may show gradual loss of effectiveness. Manufacturers' recommendations should be followed carefully.

Portable Aspirating Equipment and Systems

Because of the difficulties associated with pumping or transporting generated foam in pipes or hoses and the familiarity of fire fighters with nozzles, the earliest designs of air-foam generators were devised to be used in much the same manner as water nozzles. They incorporated a crude venturi design whereby a jet or jets of foam solution enter an open contracted portion of a large-diameter foam tube. This action lowers the atmospheric pressure surrounding the jets, and air is drawn or aspirated into the throat of the tube. Downstream of the contracted portion of the tube, a high turbulence and mixing of air and foam solution occurs. This turbulence may be increased by internal turbulence-accelerating devices, such as screens

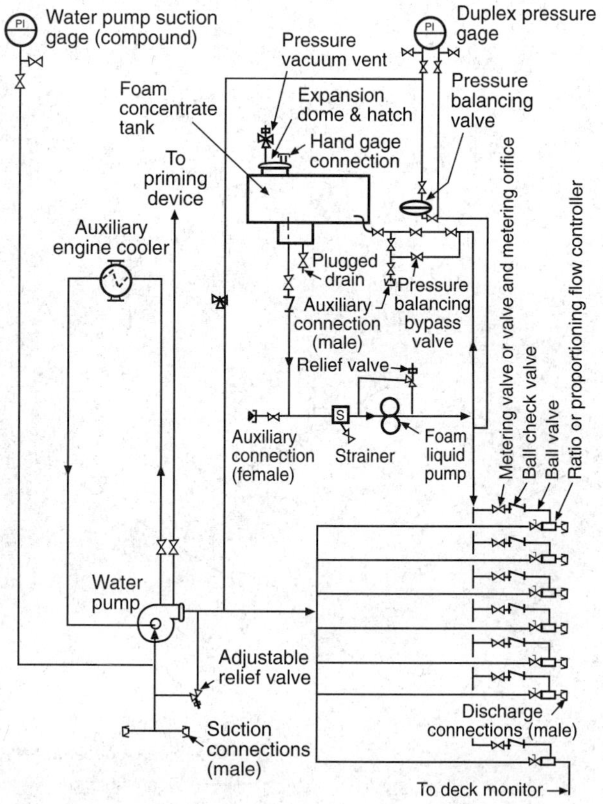

FIG. 6-22H. *Balanced pressure proportioning.*

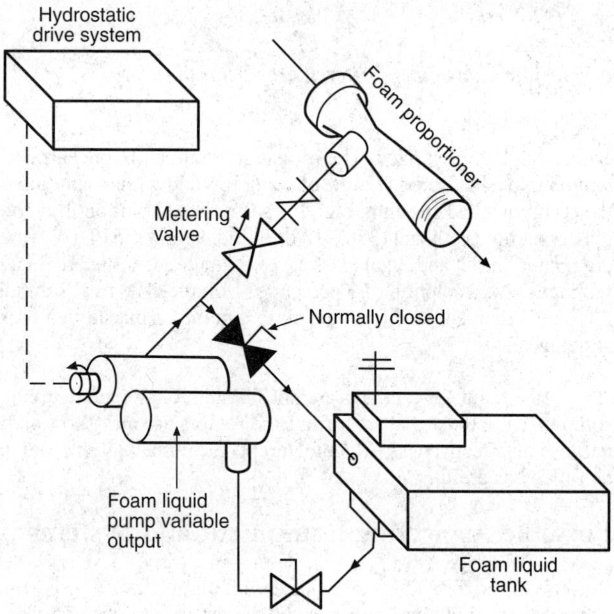

FIG. 6-22I. *Balanced pressure proportioning demand system.*

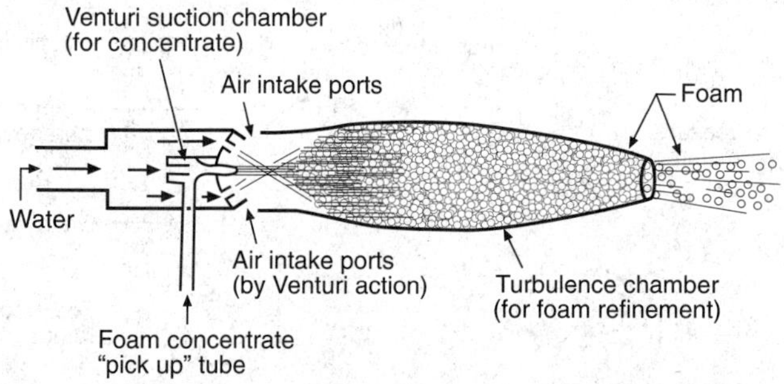

FIG. 6-22J. *Cross-section of an aspirating foam-maker with a concentrate pickup tube.*

foams with greatly differing characteristics. However, all types of nozzles incorporating foam solution jets leading into free air-mixing cavities, followed by discharge apertures of one kind or another, use the aspirating action for making foam.

Hose line foam nozzle: This is the most widely used portable, air-aspirating foam device for flammable liquid fire fighting. It is manufactured in a variety of capacities up to approximately 350 gpm (1325 L/min). (See Figure 6-22K.) Supplied with foam solution from a proportioner or by means of a pickup tube, it is used for combating fires resulting from spills of flammable liquid, or fires in tanks or fuel pits. To provide a variety of foam stream patterns that may be needed for the extinguishment operation, these nozzles often contain built-in devices that allow continuous foam pattern variation from a solid, straight stream to an inverted, filled umbrella shape. (See Figure 6-22L.) As with water streams, foam is employed in a solid, straight stream for range or reach. A flat or wide "bushy" shape is used for gentle "snowstorm" application on the burning fuel surface, and a very wide, circular "bushy" shape is used for radiance shielding of the operator during fire extinguishment or penetration into the fire area.

Another type of hose line foam nozzle is especially designed for quick, portable, one-person use during emergency operations from airport crash-rescue vehicles. Customarily called a handline foam nozzle, it is equipped with a foam pattern changing device, and some types are supplied with a valve control for converting the water or solution stream into a water spray for cooling purposes.

Foam and foam-water monitors: In large-scale fuel-fire firefighting operations, it may become necessary to position a foam-making nozzle with a high discharge rate at an advantageous position for continuous application of foam to one point or over an area. Devices for this purpose are available in a variety of types, e.g., as part of mobile truck apparatus, on a trailer, wheeled, or permanently mounted. The foam pattern change is accomplished by moving a deflector into the foam stream or by opening or closing the "jaws" at the exit end of the large tube.

High back-pressure foam-maker: Certain circumstances necessitate that foam be generated and supplied under pressure for transmission in pipes under a definite pressure head. The subsurface foam injection method for fuel tank fire extinguishment requires this type of foam-maker. The high back-pressure foam-maker, or "forcing foam-maker," is a venturi device that is carefully designed to make foam by air aspiration, and to supply it under pressure at a carefully selected ratio (from 2:1 to 4:1) of air-to-foam solution. Approximately 20 to 40 percent of the inlet pressure is recoverable. In use, this foam device is usually brought to the fixed piping foam inlet, installed, and then supplied with foam solution by a portable

or baffles. The kinetic energy of the fluid contributes to this mixing action so that a useful, stable foam exits the tube at a relatively low pressure. (See Figure 6-22J.)

The basic design principles of this method of producing air-foam have been changed in many ways to yield, for many purposes,

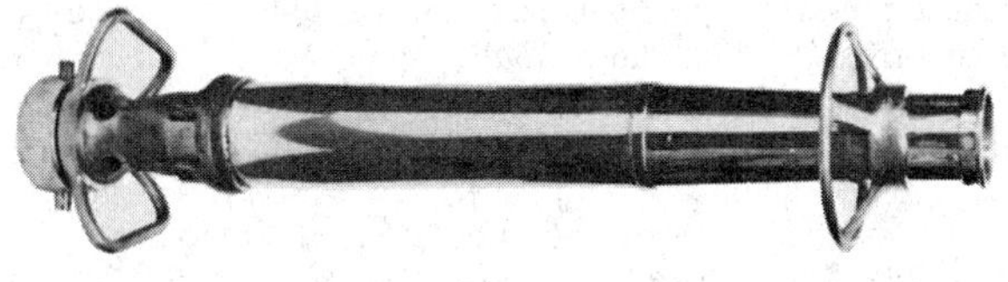

FIG. 6-22K.   Hose line foam nozzle.

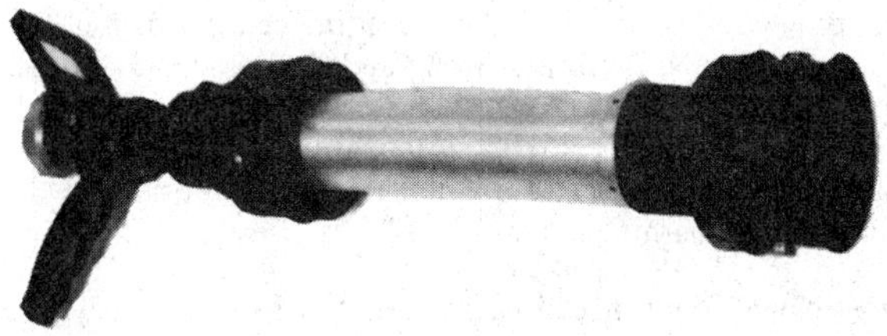

FIG. 6-22L.   Adjustable hose line foam nozzle. (Source: Chubb National Foam.)

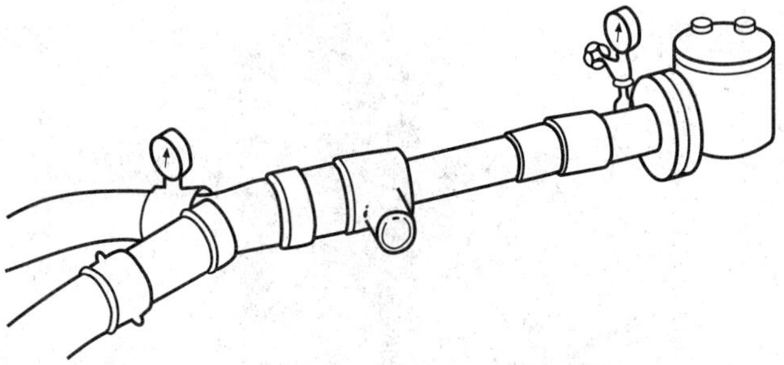

FIG. 6-22M.   A "high back-pressure-type" (or "forcing"-type) foam-maker.

or mobile pumper, or it may be permanently installed as part of a fixed system. (See Figure 6-22M.)

Medium- and high-expansion foam-generating devices: There are two principal methods used for the generation of these types of fire-fighting foams. One method utilizes a modified venturi action with air aspiration flow, while the other requires use of a blower and screen to form the finished foam. The latter system produces high-expansion foam containing sufficient residual kinetic energy to enable it to be forced through large tubes and passageways.

Figure 6-22N illustrates the operating principles of high-expansion foam-generating devices.

Water fog or spray nozzles: Several types of adjustable water fog or spray nozzles for portable use provide an acceptable fire-fighting foam of adequate characteristics when supplied with certain foam concentrates. The most universal design of these is shown in Figure 6-22O. An adjustable water spray monitor nozzle designed for this purpose is shown in Figure 6-22P.

These portable water spray nozzles are used with aqueous film forming foam (AFFF) solutions for combatting flammable liquid tank and spill fires and are used in this manner on crash-rescue vehicles. The foam resulting from the discharge of AFFF solution devices that do not aspirate air is generally fast-draining and does not impart the same degree of burnback resistance as the foam produced from AFFF agents when foam-generating devices of an air-aspirating type are used.

Foam-Generating Equipment

Where flammable liquid fire protection is required for permanently installed hazards, such as fuel storage tanks or dip tanks containing flammable or combustible liquids, air-foam-generating and distributing devices are installed integrally with the hazard. These fixed devices, which are piped to a source of foam solution, may be arranged for manual control or automatic activation by fire detectors in the event of fire.

Open dip tank, quench tank system: This system consists of a small, aspirating foam-maker supplied by a water line and foam concentrate educted to the foam-maker. Foam discharges into a

mixing box, which also acts as a surface distributor for gentle foam application.

Foam chambers for large fuel storage tanks: Fire protection of large outdoor fuel storage tanks requires that several foam chambers with foam-makers be installed at equally spaced positions slightly below the curb angle on the top periphery of the tanks. These chambers are connected to lines on the ground that supply foam solution to each foam-maker simultaneously in case of ignition of the flammable contents of the tank. Frangible seals at the discharge outlet of the foam chamber prevent vapor from entering the foam piping. These seals are designed to burst when foam pressure is applied. A screen for the air inlet to the aspirating foam-maker prevents clogging from foreign matter, such as bird nesting material. A universal or swing pipe joint is installed at ground level in the foam solution inlet pipe to prevent fracturing of the supply piping if an explosion precedes a tank fire. (See Figure 6-22Q.)

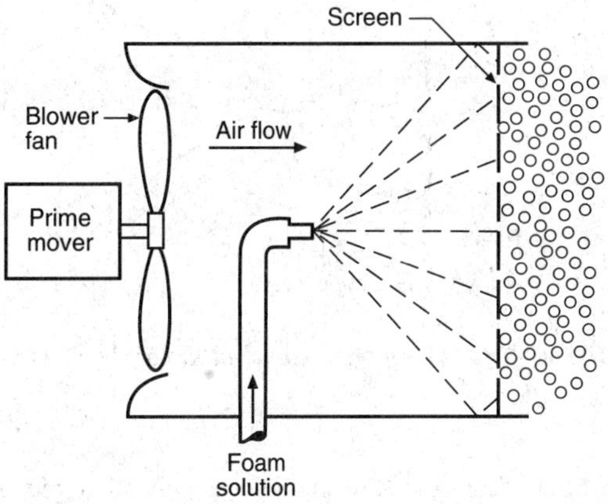

FIG. 6-22N.   A fan-blower-type high-expansion foam-generator.

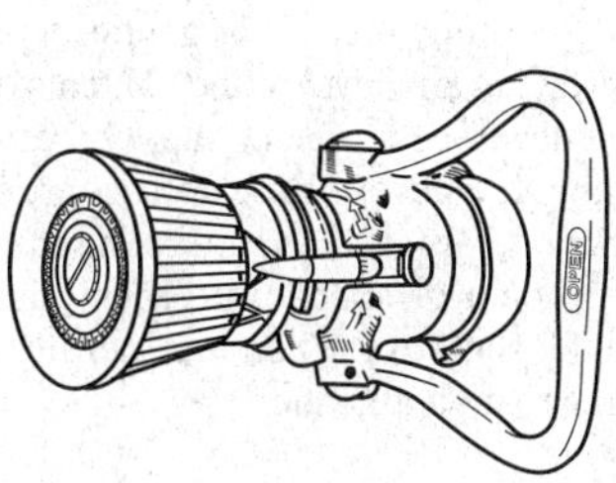

FIG. 6-22O.   Variable-pattern water fog nozzle foam-maker for use with AFFF.

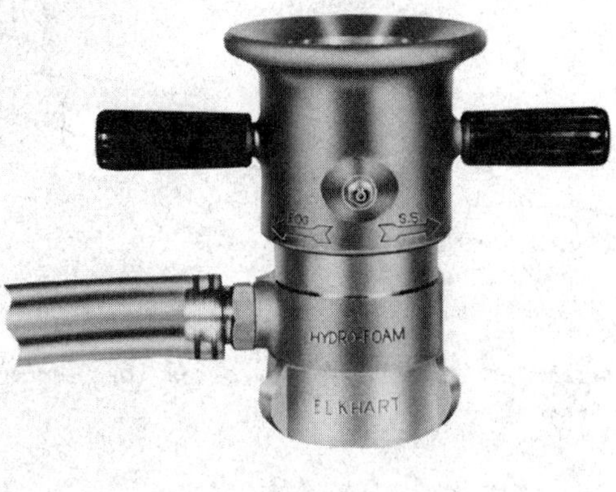

FIG. 6-22P. Adjustable 350 to 500 gpm (1,325 to 1,893 L/min) water spray foam monitor nozzle. (Source: Elkhart Brass Mfg. Co.)

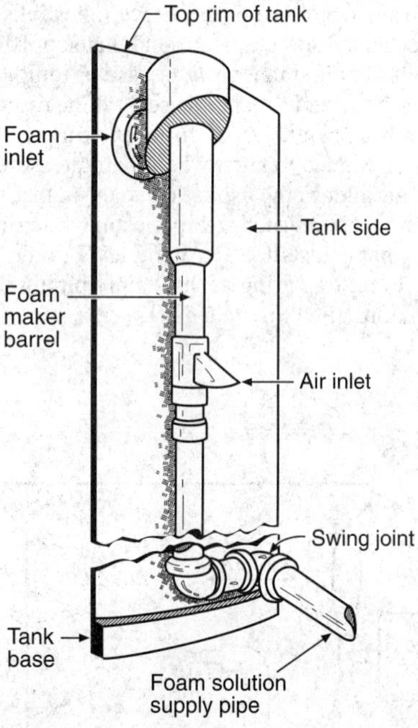

FIG. 6-22Q. Air-foam chamber at top of storage tank.

Internal tank foam distributing devices: A prime requirement for efficient fuel tank extinguishment by topside foam devices has always been that the foam must be applied to the burning surface without undue plunging into the fuel, or allowing the foam to become coated with burning fuel. This gentle application of foam must be accomplished at any level of the contents of the tank. Many devices have been developed to gently apply foam from one point, regardless of burning fuel level. These devices are listed as "Type I" foam-discharge outlets for tanks and are required for some alcohol-type foams. When foam discharge into a tank is deflected to run down the inside tank shell to the burning fuel surface, it is called a "Type II" outlet for foam application.

Central "foam house" distributing systems: These systems, as shown in Figure 6-22R, consist of an enclosure housing a foam concentrate supply tank and a proportioning device of an automatic or

balanced pressure type. Foam solution is supplied under adequate pressure from this foam house to the piping system, and controlled by appropriate valves so that the foam chambers with foam-makers on the burning tank receive foam solution.

Semifixed systems of similar design are more frequently used with mobile foam concentrate supply from foam trucks. The truck proportions and pumps foam solution to the pipe laterals feeding the foam-makers from a safe location outside the dike.

Intermediate back-pressure system: Although this system of tank protection is similar to the preceding central foam house distributing systems, it utilizes strong and well-braced foam-delivery pipes on the side of the tank, which act as supports to prevent buckling of the tank from heat. This system also uses a foam truck to proportion and pump foam solution to the piping outside the dike or firewall of the burning tank.

Subsurface foam-injection systems: The problems inherent with the application of foam from above the burning surface (topside) are sometimes difficult to combat. Problems may consist of explosion or fire damage to the foam-makers or tankside piping; forceful upward fire-drafted air currents that prevent the falling foam from reaching the burning surface; hazard to workers attempting to erect portable foam distributing devices near the burning tank; or inability of foam applied from the periphery of a large tank [greater than about 200 ft (61 m) in diameter] to flow and form a complete center seal during fire attack.

The obvious solution to these problems is to apply foam from the underside of the fire, causing it to come up through the contents of the tank. The subsurface foam-injection system accomplishes this by injecting foam under the pressure of the head of fuel in the tank, using the high back-pressure foam-maker referred to previously in this chapter.

Entry of foam may be provided at several points at the base of the tank (base injection), or it may be accomplished by means of the product line. Where large tanks are involved, a branched pipe foam distributor may be installed on or slightly above the floor of the tank above any expected water level and connected to a central foam-injection point outside the tank.

Mobile foam concentrate proportioning equipment is used with these systems, pumping from a protected position outside the tank dike. (See Figure 6-22S.)

Subsurface application of foam is recommended only for protection of cone-roof tanks containing hydrocarbon fuels. It is not recommended for use on floating-roof tanks or for protection of polar or water-miscible fuels.

These topside and subsurface injection systems require careful design. Their minimum installation requirements are found in NFPA 11.

Portable foam devices for tank protection: Mobile foam monitors of high-capacity discharge (foam cannon) may be used to direct a stream of foam over the open top rim of a burning tank so foam will fall into the burning area. These devices waste foam because of crosswinds, fire updrafts, and inability to place the equipment in an advantageous position. A foam-application rate 60 percent higher therefore must be included in the design; details are contained in NFPA 11. Nozzles of this type also are used to extinguish fires in the overflow space inside the dike surrounding the tank. Care is needed in directing the foam stream from such devices, as shown in Figure 6-22T.

In recent years, mobile foam monitors with capacities up to 4,000 gpm (15,144 L/min) have been developed and used successfully for extinguishment of hydrocarbon fires in tanks up to 195 ft

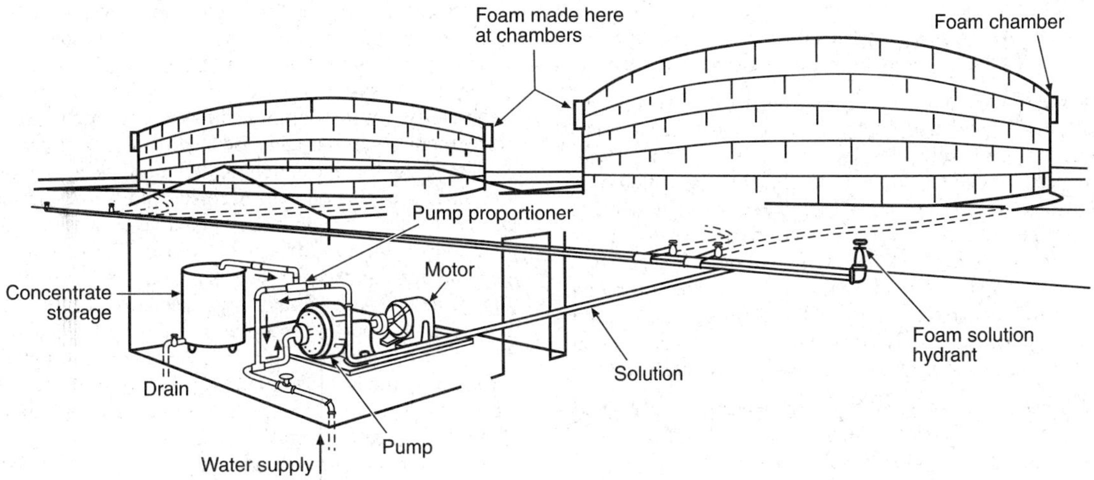

FIG. 6-22R.   Schematic arrangement of air-foam protection for storage tanks.

(60 m) in diameter. Careful planning is required to ensure availability of adequate water flow and pressure, foam concentrate, proportioning equipment, trained personnel, and safe accessibility to the hazard when relying upon such devices.

Other considerations for fuel storage tanks: The floating-roof storage tank has a good chance of freedom from fire; consequently, fixed foam systems are usually not required for their protection. Under certain circumstances, however, there may be a need for a foam-flooding system to flood the rim area.

As indicated by its name, the "covered" floating-roof tank is totally enclosed above the floating roof with a properly vented steel roof. Usually Class IB flammable liquids, such as gasolines and crudes, are stored in "open top" or covered floating-roof tanks. During storage operations, there is no vapor space between the bottom of the floating roof and the stored product surface. However, during periods of initial fill, a flammable vapor space will exist until the floating roof is buoyant. In the case of covered floating-roof tanks, the space between the fixed and the floating roof will be within the flammable range during periods of initial fill and longer, depending upon atmospheric temperatures, wind, and vapor pressure of the stored product. Class II and III combustible liquids are usually stored in fixed-roof tanks.

In an open-top floating-roof tank, a rim of fire is all that can be expected. Rim fires may be caused by atmospheric disturbances, such as lightning, but usually do not occur if the floating roof is bonded properly to the shell as specified in NFPA 780, *Standard for the Installation of Lightning Protection Systems.* Rim fires due to atmospheric disturbances do not occur in covered floating-roof tanks because of the "Faraday effect" of the fixed roof. Rim fires may occur in either type of floating-roof tank due to a serious exposure fire.

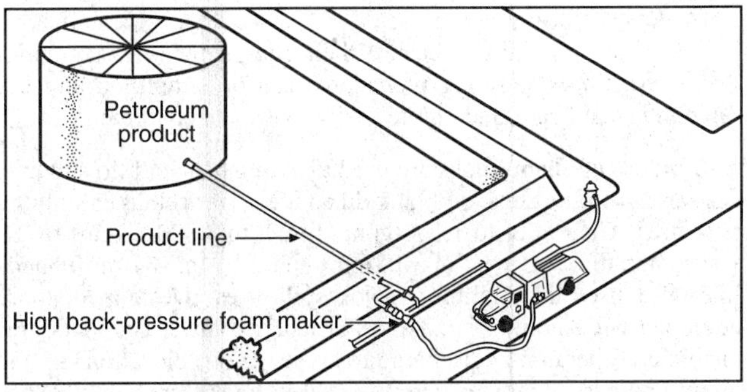

Fixed foam fire protection for the open top floating-roof consists of aspirating-type foam-makers installed in such a way that when supplied with foam solution, their foam discharge floods the annular area covered by the seal around the tank periphery. A metal foam dam may be needed to restrict the foam to this area. Further details of such construction are contained in the Appendix to NFPA 11.

When foam protection is desired for covered floating-roof tanks, a foam system similar to those described for fixed-roof tanks may be provided. Subsurface injection systems are not recommended for open top or covered floating-roof tanks because of the possibility of tilted or sunken roofs resulting in improper foam distribution.

Fixed foam systems employing foam spray nozzles or monitors are used to protect large oil-water separators, pump areas, and

FIG. 6-22S.   Semifixed subsurface foam installation.

FIG. 6-22T.   Portable (trailer) foam monitor (cannon). (Source: Angus Fire Armour Corp.)

oil piping manifolds. Fixed foam systems are not generally used to protect tank diked areas. Large wheeled monitors or trailer-mounted foam cannons, as shown in Figure 6-22T, are used for this purpose.

Certain hazardous areas, e.g., aircraft hangars, may require additional foam-making and distributing devices, such as foam monitors mounted near floor level. Often, these devices are provided with an automatic oscillator so that foam is continually distributed over the floor to extinguish burning spilled fuel under obstacles, such as aircraft wings. These are typically combined with air-aspirating or non-air-aspirating sprinklers installed at the roof of the hangar. Application rates for non-air-aspirating AFFF sprinklers are lower than air-aspirating sprinklers, in recognition of the fire plume penetration capability of AFFF discharged from ordinary sprinklers. NFPA 409 provides design details for overhead and low-level foam systems.

Fixed foam systems are used to protect petroleum piers and wharves where products and petroleum crude are handled. Foam monitors of 1,000 gpm (3,800 L/min) and greater capacity are mounted on towers and remotely operated to cover both the pier and the tanker deck in the pipe manifold area. Below deck, foam spray systems or oscillating monitors are installed to keep the pier tenable in the event of a spill fire floating on the water surface under the pier.

Fixed systems consisting of automatically operated combinations of foam spray systems and foam monitors are often installed to protect chemical processing plants. Alcohol-resistant foams are usually required. In these designs, where there may be a high risk, process vessels, pumps, and piping often are all included within the foam distribution pattern for overall protection. The system can be automatically activated by heat or fire detectors. See NFPA 11 for design of foam spray systems.

Foam-water sprinkler systems: In areas where flammable and combustible liquids are processed, stored, or handled, a water discharge may be ineffective for controlling or extinguishing fires. The foam-making sprinklers (aspirating-type) and deluge or spray nozzles using AFFF foams have successfully replaced water sprinkler nozzles for such systems so that fires in these occupancies may be controlled and property safeguarded.

When supplied with foam solution, sprinkler system piping grids provided with foam-water nozzles generate air-foam in essentially the same water sprinkler pattern as when water is discharged from the same nozzle. This dual capability affords the system Class A and B extinguishment ability. Design requirements can be found in NFPA 16 and NFPA 16A.

Obviously, fixed sprinkler systems using these nozzles require that foam concentrate tanks, proportioners, and suitable pumps be provided to supply the system with foam solution or water. Detection devices may also be used to activate the system, or the system may be activated manually. Closed-head sprinklers may also be used, and are now recognized in NFPA 30.

DELIVERING FOAM FROM VEHICLES

The majority of the mobile fire protection vehicles using foam consist of airport crash-rescue trucks and industrial foam trucks used at oil refineries and petrochemical plants. Many of these vehicles are also equipped to discharge dry chemical in combination with foam. These have combined, or "twinned," agent equipment, and are described later in this chapter.

Crash-Rescue Trucks

Mobile foam trucks, developed by military and municipal authorities in the United States and other countries, are large, custom-designed vehicles with oversize running gear which allows them to travel over all types of terrain. The trucks carry their own water supply as well as foam concentrate. Automatic balanced pressure proportioning is usually provided, although in some designs an around-the-pump proportioning system is used. The trucks are equipped with separate engine-driven water pumps and foam concentrate pumps so that the turret monitor nozzles and roadway foam nozzles can be discharged while the truck is in motion. Recent designs accomplish this using one or two engines for all necessary power requirements. Crash-rescue trucks are equipped with large-capacity, adjustable foam monitors of 500- to 1,500-gpm (1,900 to 5,700 L/min) discharge, depending upon truck size. The foam monitors are usually installed on top of the cab with remotely operated controls. The trucks are also equipped with portable handlines. Although limited, their supply of foam concentrate and water is sufficient for fire fighters to form a passageway for access to a burning aircraft for rescue. Their primary purpose is rescue of people and not necessarily a total extinguishment of fire. Aircraft crash-rescue foam vehicles are specially designed and require special attention to detail in their fabrication and performance requirements. NFPA 414, *Standard for Aircraft Rescue and Fire Fighting Vehicles*, provides standards for such vehicles.

Industrial Foam Truck Design

Foam trucks are manufactured by vendors who make a specialty of this design and by some vendors of fire department pumpers. Generally, using NFPA 11C and NFPA 1901, *Standard for Automotive Fire Apparatus*, as a design basis, these trucks are fabricated on a suitable commercial or custom truck chassis. In the United States, they are available with water pumps in sizes from 750 to 2,000 gpm (2839 to 7570 L/min), using gasoline or diesel engine drives.

Balanced pressure proportioning is provided for maximum flexibility and simplification. A positive displacement-type foam concentrate pump takes suction from a 500- to 1,000-gal (1,900- to 3,800-L) foam concentrate storage tank. The foam concentrate pump is driven by a power takeoff or hydraulic motor. The water pump is driven by a transfer gear from the main drive shaft behind the transmission. A metering valve at each truck outlet can vary foam proportioning from 3 to 6 percent.

In addition to the foam concentrate, each foam truck usually is equipped with a 500- to 1,000-gpm (1,900- to 3,800-L/min) monitor and carries fire hose, foam nozzles and adjustable water spray nozzles, hose adapters, and other accessories. The maximum truck height is limited to permit passage beneath overhead obstructions. When the truck is equipped with a large-capacity nozzle on a telescopic or articulated boom, a second power takeoff is used to power the boom hydraulic system.

Foam trucks for special hazard application, instead of fixed foam systems, are popular for refinery and petrochemical plant use. Their advantages are:

1. Capability of discharging their maximum capacity at any hazard in the plant, rather than only to the limited areas covered by a fixed system.
2. Improved reliability, because their equipment is easily maintained; thus, operating procedures can be simplified and fire fighters can be trained more easily to use the equipment.

Industrial foam trucks are used to extinguish spill fires in process areas, piping runs, and tank diked areas, as well as in fighting tank fires. Their 500- to 1,000-gpm (1,900- to 3,800-L/min) foam-water monitors and 1 1/2-in. (38-mm) and 2 1/2-in. (64-mm) handlines are used for fighting major spill fires. Designs that are equipped with monitor nozzles on articulated or telescopic booms enable the operator to discharge foam at various elevations in and around process equipment. In some cases, a tank fire can be extinguished with this equipment. Figure 6-22U illustrates a typical in-

FIG. 6-22U.   A typical industrial foam truck. (Source: Chubb National Foam)

FIG. 6-22V.   Foam truck with telescopic boom and nozzle with ladder and basket. (Source: Chubb National Foam)

FIG. 6-22W.   A twinned nozzle applicator for combined agent use. The AFFF nozzle is in operator's left hand, and the dry chemical nozzle is in the right hand.

FIG. 6-22X.   Industrial twinned nozzle applicator for combined agent use. The AFFF nozzle is below, and the dry chemical nozzle is above. (Source: The Ansul Co.)

dustrial foam truck with monitor. Figure 6-22V illustrates a typical foam truck with a telescopic boom, ladder, and basket. The foam trucks with booms may or may not be equipped with a ladder and/or basket.

COMBINED AGENT OR "TWINNED" EQUIPMENT

The capability of dry chemical agent (especially potassium-salt-types) for very fast flame control and three-dimensional flowing fuel extinguishment is well documented. However, dry chemicals have no cooling or vapor-suppressing capability, and reflash can occur. With dry-chemical-compatible foam concentrates, it is possible to apply a coating of vapor-securing foam to a burning fuel surface that has been freshly extinguished by the chemical action of dry chemical discharges.

The logical extension of these developments has been incorporated into portable and mobile devices with dual-trigger valve nozzles and monitor nozzles discharging AFFF and dry chemical. Figure 6-22W illustrates a type of twinned pistol grip trigger valve dual-nozzle device previously used by the U.S. Navy for dual hose line use. Figure 6-22X shows an improved industrial design used on offshore oil rigs. The use of this combined agent attack allows three-dimensional as well as spill fire extinguishment of flammable liquid, with both speed and freedom from reignition. The fire-fighting vehicle shown in Figure 6-22Y is for aircraft crash-rescue purposes and for extinguishment on freeways or in plants, using combined agents.

In use, the twinned hose from the hose reel supplies dry chemical and AFFF to a dual nozzle for one-person operation. Similarly, the turret nozzles at the front of the vehicle discharge dry chemical from the upper nozzle and AFFF from the lower. With a sweeping motion, the flames are controlled with the dry chemical, quickly followed by a vapor-securing, cooling layer of foam to prevent reflash. This rapid action halts flame radiation and makes advancement over the fire area safe and cool.

A design especially for petroleum refinery plant protection involves the combined agent concept in a triple-use arrangement consisting of a 2,000-lb (908-kg) dry chemical tank, a 200-gal (757-L) AFFF premixed solution tank, nitrogen bottles for pressurization of both tanks, a 500-gal (1,900-L) foam concentrate tank, a 1,500-gpm (5,700-L/min) water pump, and balanced pressure proportioning up to 2,000 gpm (7,600 L/min). Two monitors mounted behind the truck cab are twinned for operation as a unit and consist of a 50-lb-per-sec (23-kg/sec) dry chemical turret nozzle and a 500-gpm (1,900-L/min) AFFF or foam nozzle as shown in Figure 6-22Z. A

FIG. 6-22Y.   A combined-agent vehicle. (Source: Emergency One, Inc., and Feecon Corp.)

FIG. 6-22Z.   Triple-agent truck. (Source: Chubb National Foam)

different configuration uses three monitors mounted behind the truck cab—twinned 50-lb-per-sec (23-kg/sec) dry chemical and 180-gpm (680-L/min) AFFF nozzles, and one adjacent 1,000-gpm (3,800-L/min) adjustable foam monitor. In addition, the truck carries two hose reels, each with 100 ft (30.5 m) of twinned hose, and hose nozzles for dry chemical and AFFF arranged for twinned operation. The hose reels are equipped with an electric motor rewind. Foam nozzles, adjustable water fog nozzles, fire hose, and other accessories are carried on the truck.

The dry chemical turret nozzle will extinguish three-dimensional or pressure fires at a distance of 100 ft (30.5 m) and spill fires at a distance of approximately 150 ft (46 m). In addition, for long-range operation with lengthy duration, the 1,000-gpm (3,800 L/min) foam-water monitor has a range of more than 200 ft (61 m). The truck is capable of fighting fires in fixed-roof oil storage tanks with a maximum diameter of 160 ft (49 m). The proportioning system is designed to permit an external tank, such as a tank truck or trailer, to be used for foam concentrate supply without interrupting foam solution proportioning.

Twinned agent hose reels are often used on offshore oil platforms for equipment protection and in naval shipboard engine rooms. These systems have suitably sized potassium bicarbonate

(Purple K) dry chemical containers that are pressurized by nitrogen when put into operation, and AFFF central pumped systems using balanced pressure proportioning. Similar twinned agent skid-mounted systems are used in refineries and petrochemical plants.

MEDIUM- AND HIGH-EXPANSION FOAM-GENERATING EQUIPMENT AND SYSTEMS

Medium- or high-expansion foam is an aggregation of bubbles resulting from the mechanical expansion of a foam solution by air or by other gases, with expansion ratios in the range of from 20:1 to approximately 1,000:1. There are three types of systems: (1) total flooding, (2) local application, and (3) portable.

The foam generators for these systems are of two types: (1) aspirator and (2) blower. The portable or fixed aspirator-type device utilizes jet streams of water-foam solution, entraining suitable amounts of air to produce foam. With the blower-type, foam solution is discharged onto a screen through which an air stream, developed by a fan or blower, is passing. As the air passes through the screens wetted with the foam solution, large masses of bubbles or foam are formed. The blower may be powered by a hydraulic or water motor, compressed air or gas, an electric motor, or an internal combustion engine.

System Design and Use

Detailed design information on medium- and high-expansion foam systems can be found in NFPA 11A.

Basically, medium- and high-expansion foam systems are used to control or extinguish fires involving surface fires in flammable and combustible liquids and solids, and deep-seated fires involving solid materials subject to smoldering. Three-dimensional fires in flammable liquids (falling or flowing under pressure) with flash points below 100°F (38°C) generally cannot be extinguished with this technique, although they may be kept under control. Key factors to consider in determining the design adequacy of medium- and high-expansion foam systems are:

1. The quality and adequacy of water supply, the adequacy of supply of foam liquid concentrate, and the source of the air supply.
2. Suitability of the generator and foam delivery system (piping, fittings, valves, ducts).
3. Needed submergence volume and time as influenced by the space being protected, the nature of the hazards involved, leakage of the foam, and similar factors.
4. If outdoors, the effect of wind, since some of the foam may be dissipated by air currents.

High-expansion foam may be useful in controlling LNG fires and unignited spills by forming an ice layer on the liquid and by helping disperse the vapor cloud.

Total Flooding Systems

A total flooding system may be used where there is an enclosure surrounding the hazard being protected which will permit the required amount of medium- or high-expansion foam to be built up to extinguish or control the fire. Examples of such enclosures are rooms, vaults, pits, and basement areas. Even an entire building may be so protected where the foam generators are of sufficient capacity and steps are taken to ensure effective distribution and retention of the foam. Since the efficiency of the system depends upon the development and maintenance of a suitable quantity of foam within the particular enclosure, leakage of the foam from the enclosure must be avoided. Thus, it is important that doorways or windows be designed to close automatically, with consideration being given to the evacuation of personnel. High-level venting is required for the air that is displaced by the foam.

For adequate protection, sufficient high-expansion foam must be discharged at a rate to fill the space to an effective depth above the hazard before an unacceptable degree of damage occurs. The depth of the foam above the hazard will vary, depending upon the type of materials creating the hazard. Generally, the minimum depth above the hazard should be 2 ft (0.6 m). The time allowed to cover the hazard will likewise vary depending upon the type of material involved, construction features, and whether the enclosure also has an automatic sprinkler system or similar protection. Consideration has to be given to the disintegration of the foam by the heat of the fire; by normal foam shrinkage; by leakage around doors and windows, and through unclosable openings; and by the effects of sprinkler discharge where sprinkler protection is provided.

Local Application Systems

Local application systems can be used where total flooding systems may be impractical or unnecessary. Such hazards may be indoors or outdoors, where air currents are not likely to be severe. These systems are best adapted to the protection of flammable or combustible liquids in dip tanks and associated drainboards, and for pits and trenches. Medium- or high-expansion foam may be used for extinguishing spill fires where it is feasible to apply the foam from fixed or portable nozzles at adequate rates of discharge to develop a foam blanket to achieve complete extinguishment.

Precautions to Be Observed with Medium- or High-Expansion Foam

Medium- or high-expansion foam is normally nontoxic to persons who may be trapped in a space filled with it, since the air entrained in the foam is generally not contaminated. However, because of the foam bubbles, some difficulty may be experienced in breathing, and breathing discomfort will increase with reduction of foam expansion. Air is usually taken from a clean source, but, where products of combustion are introduced, the foam quality may be substandard because of the contaminants present and the temperature of the heated air. Entering a foam-filled space should be avoided unless adequate precautions are taken, because loss of vision and disorientation introduce life and injury hazards. A coarse water spray may be used to "cut" a path in the foam. Personnel should wear self-contained breathing apparatus and employ a lifeline when entering an area filled with high-expansion foam.

FOAM EQUIPMENT TESTING AND SURVEILLANCE

The continued effective emergency performance of foam equipment depends entirely upon fully adequate maintenance procedures, with periodic testing where possible. The many variations in system design and equipment applications for hazards requiring foam make it impossible to establish anything other than general procedures for periodic inspection. Because air-foam concentrates, foam equipment, and fire protection systems are subject to change, variation, and even malfunction over long unattended spans of time, they must be kept adequate for the purpose for which they are designed.

Foam Concentrate Surveillance

Because all air-foam concentrates are water solutions of organic and inorganic chemicals of one sort or another, they must be carefully observed for changes in constitution and characteristics. They must be stored in shipping containers and in storage tanks according to the manufacturer's recommendations. Exposure to extreme heat, cold, or contamination, or mixing with other materials must be avoided. Sedimentation or precipitate formation in containers or tanks of concentrate should be carefully checked periodically. The manufacturer or the manufacturer's representative is best qualified to test and determine the extent of reliability of foam concentrates under questionable conditions of deterioration. Foam concentrates should be tested at least annually.

Equipment Testing

The performance of all foam equipment under emergency conditions is guaranteed best by the initial acceptance test and periodic inspection of the equipment. Original design specifications of fixed installations should include provision for periodic testing of proportioners, pumps, and other ancillary equipment, without the trial distribution of foam to the hazard being protected.

Problems of corrosion, clogged orifices, sticky valves, and electric circuitry malfunction may be detected by suitable means without full system activation.

In the absence of an actual fire, and without full system activation, the complete testing of foam equipment performance for fire protection adequacy may be accomplished by various means. In the absence of an actual fire, complete testing may be accomplished by several physical test methods. However, because these tests are similar in nature to "fingerprint" tests, their results can only be interpreted by comparison with similar tests conducted at the installation acceptance of the equipment, or as provided for under the performance guarantee of the manufacturer.

Foam equipment performance can be tested for the following physical characteristics of foam:

1. The dimensions of the discharge pattern or patterns of the device or system.
2. The percent concentration of foam concentrate in the finished foam solution.
3. The degree of expansion of the finished foam.
4. The rate at which water drains from the foam, e.g., 25-percent drainage time. This correlates with its viscosity or rate of spreading over fuel surfaces.
5. Film-forming capacity of the foam concentrate.

The preceding tests require specially designed test equipment and standardized techniques. They require qualified operators and may or may not be carried out easily in the field. Detailed information concerning the equipment required, methods used, and some interpretations of the test results can be found in the Appendices of NFPA 11 and NFPA 412, *Standard for Evaluating Aircraft Rescue and Fire Fighting Foam Equipment*, which includes tests for foam discharge patterns, AFFF-type foams, and protein-based foams.

The test for the concentration of the foam concentrate is very important because it indicates the efficiency of operation of the concentrate proportioning device in the system. It is performed easily in the field with a hand refractometer and volume-measuring device or portable electrical conductivity meter.

ENVIRONMENTAL ISSUES

The use of fire-fighting foams is essential for the control of flammable liquid fire threats. The ability of foam to rapidly extinguish flammable liquid fires has contributed to life safety and property conservation. However, with the increasing global environmental awareness, fire-fighting foams are being scrutinized for their potential environmental impact. The primary concerns are fish toxicity, biodegradability, treatability in wastewater treatment plants, and nutrient loading.

Discharge Scenarios

In order to assess the impact of foam on the environment, likely discharge scenarios should be considered. These scenarios include the following:

1. Uncontrolled fire situations;
2. Potential hazardous situations (e.g., unignited fuel spills);
3. Fire-fighting training evolutions; and
4. Fixed or mobile vehicle suppression system discharge:
   (a) intentional system test, or
   (b) inadvertent discharge.

Under many fire-fighting situations, control of foam and water/fuel effluent will be difficult. Control and extinguishment are usually of paramount importance, and rapid control by foam can substantially reduce the potential total effluent from an incident. Where possible, blocking of sewer drains, along with the use of portable dikes or booms, can be used to implement mitigation strategies.

Training scenarios provide greater opportunities to control foam flow, particularly by means of collection. New training techniques, using simulated pool fires and simulated foams, are available.

System tests can also be performed, to a certain extent, with the use of simulants. Alternatively, small amounts of foam can be flowed and contained for proper disposal in accordance with local regulations and manufacturer's recommendations. Inadvertent system discharges may be handled by preplanned, engineered containment strategies or by portable containment devices, where containment is deemed appropriate.

Environmental Properties and Concerns

Foam solutions generally have a high biological oxygen demand (BOD), i.e., they extract high levels of oxygen to break down. This is an issue both in the natural environment and where foam is discharged to wastewater treatment plants.

Foaming can occur in natural waterways and in treatment plants. In wastewater plants, foaming can cause bubbles to trap activated sludge, upsetting the normal operation of the plant. Before discharge of foam to a wastewater treatment plant, the plant operator should be contacted to discuss the type of foam, estimated BOD, total volume, and time over which the foam may be discharged to the plant. This should be done, where possible, as part of the emergency planning process.

In sufficient concentrations, foam may affect aquatic life. Where sufficiently diluted, foam solutions may not have an impact on aquatic life; however, localized concentrations in streams or ponds may create levels that are toxic to fish. Under emergency conditions, attempts should be made to prevent discharge to waterways (recognizing the primary concern of controlling the fire). Under controlled discharge conditions, prevention of foam discharge to natural waterways is appropriate.

Fluorochemical surfactants in film-forming foams resist biodegradation. Because of this persistence, foaming at very low concentrations may occur. The fluorochemicals may also contribute to toxicity.

The U.S. Environmental Protection Agency (EPA) requires reporting of certain releases of chemicals in broadly defined categories. The glycol ether category of chemicals was placed on the list of hazardous air pollutants, under the 1990 Clean Air Act Amendments. These chemicals are used as solvents in certain fire-fighting foams. While the EPA has assigned no reportable quantity for any glycol ethers at this time, these chemicals still are designated as a hazardous substances, under the Comprehensive Environmental Response Compensation and Liability Act (CERCLA). Research and development is underway to identify replacement solvents for glycol ethers used in fire-fighting foam that are more environmentally friendly.

Disposal Alternatives

The uncontrolled release of foam solutions to the environment should be avoided. Alternative disposal options include the following:

1. Discharge to a wastewater treatment plant, with or without pretreatment;
2. Discharge to the environment after pretreatment;
3. Solar evaporation; and
4. Transportation to a wastewater treatment plant or hazardous waste facility.

When foam is collected, any fuel from a fire should be separated. Dilution with water may be required before discharging foam to a wastewater treatment plant. Defoamers or other pretreatments may also be required. Any discharge to a wastewater treatment plant should be coordinated in advance with the treatment facility personnel. Foam manufacturers can advise on proper disposal techniques for specific circumstances.

BIBLIOGRAPHY

NFPA Codes, Standards, and Recommended Practices

Reference to the following NFPA codes, standards, and recommended practices will provide further information on foam extinguishing agents and systems discussed in this chapter. (See the latest version of *The NFPA Catalog* for availability of current editions of the following documents.)

NFPA 11, *Standard for Low-Expansion Foam*
NFPA 11A, *Standard for Medium- and High-Expansion Foam Systems*
NFPA 11C, *Standard for Mobile Foam Apparatus*
NFPA 16, *Standard for the Installation of Deluge Foam-Water Sprinkler and Foam-Water Spray Systems*
NFPA 16A, *Standard for the Installation of Closed-Head Foam-Water Sprinkler Systems*
NFPA 18, *Standard on Wetting Agents*
NFPA 30, *Flammable and Combustible Liquids Code*
NFPA 403, *Standard for Aircraft Rescue and Fire Fighting Services at Airports*
NFPA 409, *Standard on Aircraft Hangars*
NFPA 412, *Standard for Evaluating Aircraft Rescue and Fire Fighting Foam Equipment*
NFPA 414, *Standard for Aircraft Rescue and Fire Fighting Vehicles*
NFPA 780, *Standard for the Installation of Lightning Protection Systems*
NFPA 1901, *Standard for Automotive Fire Apparatus*

Additional Reading

A Position on Foam-Water Sprinkler Research, Foam-Water Research Coalition, Washington, DC, 1989.
Andrews, R. C., Jr., "Environmental Impact of Firefighting Foam," *Industrial Fire Safety*, Vol. 1, No. 1, November/December 1992, pp. 26–31.
Andsten, T., "CEN TC 191/WG/3: Sammutusvaahtojen testausmenetelmat. [CEN TC 181/WG 3: Testing Methods for Extinguishing Foams.]," *Palontorjuntatekniikka*, Vol. 2, 1994, pp. 25–26.
Arata, M. J., Jr., "Foam Supply: The Shuttle Approach," *Industrial Fire Safety*, Vol. 2, No. 2, March/April 1993, pp. 44–45.
Aubert, J. H., Kraynik, A. M., and Rand, P. B., "Aqueous Foams," *Scientific American*, Vol. 19, No. 1, 1988, pp. 74–82.
Bailey, B., "Foam: A Fire Fighting Primer for the Industrial Sector," *Fire Fighting in Canada*, Vol. 38, No. 2, March 1994, pp. 8–9.
Banerjee, S. C. and Acharya, A. K., "High-Expansion Foams—Its Use in Mine Fires," *Journal of Mines, Metals, and Fuels*, Vol. 34, No. 10, 1986, pp. 448–451.

Blankenship, P. L., "Class A Foam for 1993 Wildland Fires," *American Fire Journal*, Vol. 45, No. 6, June 1993, pp. 17–18.

Blankenship, P. L., "Foam: Should We Use It?" *American Fire Journal*, Vol. 43, No. 7, July 1991, pp. 45–47, 49.

Bobert, M., "Skumvatskor-Koppling mellan viskositetsdata och rorstromning. [Foam Concentrates: Correlation Between Viscosity Properties and Pipeflow.]," Swedish National Testing and Research Inst., Boras, Sweden, SP Report 1995:10, ISBN 91-7848-539-8, ISSN 0284-5172, 1995.

Brackin, M., *et al.*, "Pros and Cons of Class A Foam," *Fire Engineering*, Vol. 145, No. 7, July 1992, pp. 59–60, 62, 64–66.

Bradley, M., "Heavy Foam Cover," *Fire Engineering*, Vol. 140, No. 6, 1987, p. 52.

Briggs, A. A., and Webb, J. S., "Gasoline Fires and Foams," *Fire Technology*, Vol. 24, No. 1, Feb. 1988, pp. 48–58.

Brittain, J., "FFFP or AFFF Fire Foam? One View From the 'Other Side of the Fence'," *Fire*, Vol. 85, No. 1050, December 1992, p. 31.

Brittain, J., "Foam Versus the Environment," *Fire Prevention*, No. 259, May 1993, pp. 24–27.

Buckley, G., "Short Note on Sprinkler System Hydraulic Calculations Using Foam," *Fire Engineers Journal*, Vol. 54, No. 172, March 1994, p. 28.

Budnick, E. K., *et al.*, "Preliminary Evaluation of High Expansion Foam Systems for Shipboard Applications. March 1991–October 1991," Hughes Associates, Inc., Baltimore, MD, Naval Research Laboratory, Washington, DC, Naval Sea Systems Command, Washington, DC, NRL/MR/6180-95-7785, Nov. 27, 1995.

Carey, W. M., "Knockdown, Exposure and Retention Tests. National Class A Foam Research Project Technical Report. Phase 1," Underwriters Laboratories, Inc., Northbrook, IL, National Fire Protection Research Foundation, Quincy, MA, Technical Report, December 1993.

Carey, W. M., "Structural Fire Fighting: Room Burn Tests. National Class A Foam Research Project Technical Report. Phase 2," Underwriters Laboratories, Inc., Northbrook, IL, National Fire Protection Research Foundation, Quincy, MA, Technical Report, December 1994.

Carlson, G. P., "What's All the Talk About Class A Foam?" *Fire Engineering*, Vol. 144, No. 10, October 1991, pp. 10,12.

Carringer, R., "Class 'A' Foam as Part of Suppression Strategy for Tire Fires," *Foam Applications for Wildland and Urban Fire Management*, Vol. 7, No. 2, September 1995, pp. 10–12.

Carringer, R., "Class A Foam," *American Fire Journal*, Vol. 46, No. 2, February 1995, pp. 20–21.

Carringer, R., "Everything You Wanted to Know About "Class A" Foam (But Didn't Know Who to Ask)," *Firefighter's News*, February/March 1991, pp. 32–34.

Cash, T., "Fighting Fires With Foam. Part 1," *Fire Safety Engineering*, Vol. 2, No. 2, April 1995, pp. 25–27.

Cash, T., "Fighting Fires With Foam. Part 3," *Fire Safety Engineering*, Vol. 2, No. 5, October 1995, pp. 10–16.

Cash, T., "Foam Types," *Fire Fighting in Canada*, Vol. 40, No. 2, March 1996, pp. 14, 32.

Cash, T., "Fighting Fires With Foam. Part 2," *Fire Safety Engineering*, Vol. 2, No. 4, August 1995, pp. 9–14.

Clark, A., "Class A Foam Extinguishers Tire Fire," *Fire Engineering*, Vol. 147, No. 3, March 1994, pp. 73–74.

Clarke, M. "Fire Fighting Foams," *Fire Surveyor*, Vol. 21, No. 6, December 1992, pp. 11–13.

Colletti, D. J., "Class A Foam for Structure Firefighting," *Fire Engineering*, Vol. 145, No. 7, July 1992, pp. 47–48, 50–52, 53–57.

Colletti, D. J., "Class A Foam: An Emerging Technology. Compressed-Air Foam Mechanics," *Fire Engineering*, Vol. 147, No. 3, March 1994, pp. 61–64, 66.

Colletti, D. J., "Class A Foams," *Firefighter's News*, February/March 1991, pp. 30–31.

Colletti, D. J., "Class A. Foam: Q&A," *Firehouse*, Vol. 18, No. 3, March 1993, pp. 54, 56–57.

Colletti, D. J., "Quantifying the Effects of Class A Foam in Structure Firefighting: The Salem Tests," *Fire Engineering*, Vol. 146, No. 2, February 1993, pp. 41–42, 44.

Correia, R. C., Jr., "Foam-Water Sprinkler Systems: What the Firefighter Needs to Know," *Fire Engineering*, Vol. 146, No. 10, October 1993, pp. 53–56.

Cowan, G., "Class A Foam," *Fire Fighting in Canada*, Vol. 40, No. 2, March 1996, pp. 16–17, 36.

Cowan, G., "Wave of Future is Class 'A' Foam," *Fire Fighting in Canada*, Vol. 39, No. 2, March 1995, pp. 8–9, 18–19.

Cox, R., "Class A Foam Extinguishes Coal Fire," *Foam Applications for Wildland and Urban Fire Management*, Vol. 5, No. 1, 1993, pp. 14–16.

Dahlberg, M., "Foam Spread Experiments on a Water Surface," Swedish National Testing and Research Institute, Boras, Sweden, SP REPORT 1994:16, 1994.

Darley, P. C., "Use of Class 'A' Foam and Compressed Air Foam Systems (CAFS) in Firefighting," *Foam Applications for Wildland and Urban Fire Management*, Vol. 7, No. 2, September 1995, pp. 15–22.

Darwin, R. L., Ottman, R. E., Norman, E.C., Gott, J. E., and Hanauska, C. P., "Foam and the Environment: A Delicate Balance," *NFPA Fire Journal*, Vol. 89, No. 3, May/June 1995, pp. 67–73.

Davis, L., "Class A Foams—New Technology Brings Class A Foams From Wildland to Structural Firefighting," *Firehouse*, Vol. 16, No. 4, April 1991, pp. 49–50, 75.

Duncan, S., "Foam on the (Firing) Range - Class A Foam in Department of the Army," *Foam Applications for Wildland and Urban Fire Management*, Vol. 7, No. 1, May 1995, p. 14.

Dutton, D., "Los Angeles County fire Department Class 'A' Foam Operations," *Foam Applications for Wildland and Urban Fire Management*, Vol. 7, No. 2, September 1995, pp. 1, 3.

Edwards, C.B., "Getting Started with Class A Foam Solution," *Fire Engineering*, Vol. 149, No. 2, February 1996, pp. 56–60.

Evans, J. L., "Fire Fighting Foams," *Fire Surveyors*, Vol. 17, No. 5, Oct. 1988, pp. 5–10.

"Fire Extinguishing Agent, Aqueous Film Forming Foam (AFFF) Liquid Concentrate, Six Percent for Fresh and Sea Water," *Military Specification MIL-F-24385 (NAVY)*, Amendment 8, Naval Ship Engineering Center, Department of the Navy, Hyattsville, MD.

"Fire Fighting Foams," *Industrial Fire World*, Feb. 1987, pp. 20–23.

"Fixed Foam Systems," *Sprinkler Age*, Vol. 11, No. 8, August 1992, pp. 16–18.

Foster, J., "Use of Foam Against Large-Scale Petroleum Fires Involving Lead-Free Petrol," *Fire Research News*, No. 14, Autumn 1992, pp. 11–13.

Foster, J. A. and Johnson, B. P., "Use of Firefighting Foam in the UK Fire Service," First International Conference on Fire Suppression Research, May 5–8, 1992, Stockholm, Sweden, 1992, pp. 291–299.

Fox, P., "Tank Fire: A Case Study History," *Industrial Fire World*, May/June 1986, pp. 9–12.

Fox, P. G., "Use of Foam on aircraft Interior Fires," *Industrial Fire Safety*, Vol. 2, No. 5, Sept./Oct. 1993, pp. 56–57.

Frank, J. A., and Woodworth, S. P., "Foam System Considerations for Conventional Fire Apparatus," *Fire Engineering*, Vol. 141, No. 10, 1988, pp. 60–67.

Franklin, S. E., "Foam Applications for Prescribed Fire," *American Fire Journal*, Vol. 42, No. 2, Feb. 1990, pp. 40–43.

Gainey, T., "B-Plant Canyon Fire Foam System," Westinghouse Hanford Co., Richland, WA, WHC-SC-WM-OTP-172, Jan. 11, 1995.

Gainey, T., "New Supply for Canyon Fire Foam System," Westinghouse Hanford Co., Richland, WA, WHC-SC-WM-ATP-117, Dec. 1, 1994.

Gillespie, P., "Protecting Petrochemical Storage Tanks with Fire-fighting Foam," *Fire Journal*, Vol. 83, No. 3, May/June 1989, p. 44.

"Guide for Fighting Fires in and Around Petroleum Storage Tanks," *API Publication 2021*, American Petroleum Institute, 3rd ed., 1991.

Hanauska, C. P., Scheffey, J. L., Roby, R. J., and Gottuk, D. T., "Improved Formulations of Firefighting Agents for Hydrocarbon Fuel Fires," SBIR Phase I Final Report for the U.S. Air Force, Jan. 1994, Hughes Associates, Inc., Baltimore, MD.

Harker, M., "Foam in Sprinkler Systems," *Fire Surveyor*, Vol. 19, No. 4, Aug. 1990, pp. 4–6.

Hayashi, K. and Hiraga, S., "Pressure Loss of Fire Fighting Foam in Pressurized Transportation," Japanese Association of Fire Science and Engineering, Fire Research Annual Conference, May 17–18, 1990, pp. 147–150.

"Helicopter Video and Fast Foam Supply Aid Refinery Battle," *Fire International*, No. 124, August/September 1990, p. 23.

Hickey, H. E., "Foam System Calculations," *SFPE Handbook of Fire Protection Engineering*, 2nd ed., National Fire Protection Association, Quincy, MA, 1995.

Hird, D., "Fire-Fighting Foams for Flammable Liquid Fires," *Fire Prevention*, No. 202, Sept. 1987, p. 20.

Holtham, P. D., "Saponification Foam Process Proves Its Effectiveness," *Fire Engineers Journal*, Vol. 48, No. 152, Mar. 1989, p. 14.

Hoshino, M., "Extinguishing Performance of Fractionated Keratin Hydrolysates and of Various Petroleum-Fire Fighting Foams: Extinguishment of Gasoline Fires by the Plugging Application," *Bulletin of Japanese Association of Fire Science and Engineering*, Vol. 42, No. 1, 1994, pp. 13–19.

Hoshino, M. and Hayashi, K., "Extinguishing Abilities of Fire-Fighting Foams for Petroleum Tank Fires," *Bulletin of Japanese Association of Fire Science and Engineering*, Vol. 39, No. 2, 1990, pp. 41–49, 1990; *Report of Fire Research Institute of Japan*, No. 71, March 1991, pp. 53–62.

Howells, P., "Electrostatic Hazards of Foam Blanketing Operations," *Industrial Fire Safety*, Vol. 2, No. 4, July/August 1993, pp. 18, 21–24.

Ingason, H., Persson, H. and Ryderman, A., "Foam Sprinklers as a Replacement for Halon in Engine Rooms," Swedish National Testing and Research Institute, Boras, Sweden, SP REPORT 1992:37, 1992.

"International Foam-Water Sprinkler Research Project. Task 4. Palletized Storage Fire Tests 1 Through 13," Technical Report, Underwriters Laboratories, Inc., Northbrook, IL, National Fire Protection Research Foundation, Quincy, MA, Feb. 1992.

"International Foam-Water Sprinkler Research Project. Task 5. Rack Storage Fire Tests," Underwriters Laboratories, Inc., Northbrook, IL, National Fire Protection Research Foundation, Quincy, MA, Task 5, Nov. 1992.

Johnson, B., "Comparison of Various Foams When Used Against Large-Scale Petrol Fires," *Fire Research News*, No. 15, Spring 1993, pp. 5–6.

Johnson, B., "Fire Fighting Trials of Foam Additives on Wooden Crib Fires," *Fire Research News*, Vol. 13, Spring 1991, pp. 4–5.

Johnson, B. P., "Additives for Hose Reel Systems: Trials of Foam on Tire Fires," Home Office, London, England, Publication 5/91, SC/88 42/1087/1, 1992.

Josephson, R. A. and Shafer, P. R., "Foam Systems for Highway Tunnels," *Industrial Fire Safety*, Vol. 1, No. 1, November/December 1992, pp. 43–44, 46–48, 50.

Kerr, A., "Storing Foam Concentrates," *Fire Engineers Journal*, Vol. 50, No. 156, March 1990, pp. 20–21.

Kim, A. K. and Dlugogorski, B. Z., "Effective Fixed Foam System Using Compressed Air," *Proceedings* of the National Institute of Standards and Technology (NIST) and Society of Fire Protection Engineers (SFPE) International Conference on Fire Research and Engineering, September 10–15, 1995, Orlando, FL, SFPE, Boston, MA, 1995, pp. 71–76.

Kim, A. K., Dlugogorski, B. Z. and Mawhinney, J. R., "Effect of Foam Additives on the Fire Suppression Efficiency of Water Mist," National Research Council of Canada, Ottawa, Ontario, NRCC 37902, IRC Paper 3748; *Proceedings* of the Halon Options Technical Working Conference, May 3–5, 1994, Albuquerque, NM, 1994, pp. 347–358.

Lawlor, M., "Class A Foam Utilization by a Career Department," *Fire Fighting in Canada*, Vol. 38, No. 2, March 1994, p. 18.

Liebson, J., "Change Your Policies—Then Adopt Class A Foam," *American Fire Journal*, Vol. 48, No. 3, March 1996, p. 6.

Liebson, J., "Class A Foam in the Structural Fire Service," *Foam Applications for Wildland and Urban Fire Management*, Vol. 7, No. 1, May 1995, pp. 24–25.

McKenzie, D., "Venturi Foam Proportioning System," *Foam Applications for Wildland and Urban Fire Management*, Vol. 3, No. 1, 1990, pp. 4–7.

McKenzie, D. W., "Compressed Air Foam System Powered by Allison World Transmission," *Foam Applications for Wildland and Urban Fire Management*, Vol. 7, No. 2, September 1995, pp. 13–15.

McKenzie, D. W., "Compressed Air Foam Systems for Use in Wildland Fire Applications," Forest Service, San Dimas, CA, Report 9251 1203-SDTDC, September 1992.

Nash, P., and Yount, R. A., *Automatic Sprinkler Systems for Fire Protection*, 1st ed., Ch. 15, Victor Green Publications Ltd., London, UK, 1978.

National Fire Protection Association, "Speakers Papers', International Conference on Aviation Fire Protection," Interlaken, Switzerland, Sept. 1987.

Noll, G., "Evaluating the Use of Fire Fighting Foams," *Fire Engineering*, Vol. 139, No. 2, 1986, p. 44.

Omans, L. P., "Fighting Flammable Liquid Fires: A Primer. Part 1. The Family of Foams," *Fire Engineering*, Vol. 146, No. 1, January 1993, pp. 50–54, 57–61.

Patterson, T., "Joshua Tree National Park Converts to Compressed Air Foam System (CAFS)," *Foam Applications for Wildland and Urban Fire Management*, Vol. 7, No. 2, September 1995, pp. 4–6.

Pekkala, P., "Kalvovaahdon kaytto sprinklerilaitteistoissa. [Using Film-Forming Foam in Sprinkler Systems.]," *Palontorjuntatekniika*, March 1992, pp. 18–21.

Person, H., "Fire Extinguishing Foams Resistance Against Heat Radiation," First International Conference on Fire Suppression Research, May 5–8, 1992, Stockholm, Sweden, 1992, pp. 359–376.

Persson, B. and Dahlberg, M., "Simple Model of Foam Spreading on Liquid Surfaces," Swedish National Testing and Research Institute, Boras, Sweden, SP REPORT 1994:27, 1994.

Persson, H. and Wetterlund, H., "Slackforsok med alkoholbestandiga skum. [Extinguishing Tests With Alcohol Resistant Foam.]," Swedish National Testing and Research Inst., Boras, Sweden, SP Report 1994:35, 1994.

Persson, H., "Basutrustning fur skumslackning—Forsoksresultat och rekommendationer som underlag for dimensionering och utforande. [Recommendations for a Mobile Foam Protection System for Transportation Crashes.]," (Abstract in English), Swedish National Testing and Research Institute, Boras, Sweden, SP Report 1990:36, 1990.

Persson, H., "Fire Extinguishing Foams—Resistance Against Heat Radiation," Brandforsk Project 609-903, SR Report 1992:54, 1992, Swedish National Testing and Research Institute, Sweden.

Persson, H., and Dahlberg, M., "A Simple Method for Predicting Foam Spread over Liquids," Fire Safety Science—*Proceedings* of the Fourth International Symposium, International Association for Fire Safety Science, Ottawa, 1994, pp. 265–276.

Peterson, H. F., "Suppression Capability of Foam on Runways," *NFPA Aviation Bulletin* No. 250, National Fire Protection Association, Quincy, MA.

Rhein, R. A., "Experimental Study of the Use of Liquid Argon and Argon-Filled Aqueous Foams for the Extinction of Lithium Fires," *Fire Technology*, Vol. 28, No. 4, November 1992, pp. 290–316.

Rochna, R., "Compressed Air Foam Systems," *Fire Systems*, Vol. 5, No. 2, 1991, pp. 42–49.

Rochna, R. R., "Class A High Expansion Foam," *Foam Applications for Wildland and Urban Fire Management*, Vol. 7, No. 1, May 1995, pp. 15–18.

Rochna, R. R., "Foam on the Range," *Fire Chief*, Vol. 38, No. 6, June 1994, pp. 34-38.

Rochna, R. R., "Hi-Ex Foam Has It Covered in the Wildland," *American Fire Journal*, Vol. 46, No. 2, February 1995, pp. 12–15.

Rochna, R., "Compressed Air Foam: What It Is!" *American Fire Journal*, Vol. 43, No. 8, August 1991, pp. 28–31, 34.

Ruth, R. A., "Specifying Industrial Foam Pumps," *Industrial Fire World*, Vol. 9, No. 2, March/April 1994, pp. 6–9, 11–14.

Scheffey, J. L., "Foam Agents and AFFF System Design Considerations," *The SFPE Handbook of Fire Protection Engineering*, 2nd ed., Quincy, MA, 1995.

Scheffey, J. L., Darwin, R. L., Leonard, J. T., Fulper, C. R., Ouellette, R. J., and Siegmann, C. W., "A Comparative Analysis of Film-forming Fluoroprotein Foam (FFFP) and Aqueous Film-forming Foam (AFFF) for Aircraft Rescue and Fire Fighting Services, Report 2108-A01-90 for the NFPA Aviation Committee, June 1990, Hughes Associates, Inc., Baltimore, MD.

Scheffey, J. L., and Leonard, J. T., "AFFF Protection for Weapons Staging Areas," *Fire Safety Journal*, Vol. 14, Nos. 1&2, 1988, pp. 47–63.

Scheffey, J. L., Wright, J., and Sarkos, C., "Analysis of Test Criteria for Specifying Foam Firefighting Agents for Aircraft Rescue and Firefighting," FAA Technical Report, DOT/FAA/CT-94/04, Aug. 1994, FAA Technical Center, Atlantic City, NJ.

Scheffey, J. L., Darwin, R. L. and Leonard, J. T., "Evaluating Firefighting Foams for Aviation Fire Protection," *Fire Technology*, Vol. 31, No. 3, 3rd Quarter, Aug. 1995, pp. 224–243.

Schlobohm, P., "Language of Foam," *Foam Applications for Wildland and Urban Fire Management*, Vol. 3, No. 1, 1990, pp. 1–4.

Schlobohm, P., "More Class A Foam on the Shelf," *Foam Applications for Wildland and Urban Fire Management*, Vol. 7, No. 1, May 1995, pp. 26–27.

Sharma, T. P., *et al.*, "Experiments on Extinction of Liquid Hydrocarbon Fires by a Foam Technique," *Proceedings* of the Fourth International Symposium on Fire Safety Science, July 13–17, 1994, Int'l. Assoc. for Fire Safety Science, Boston, MA, 1994, pp. 865–876.

Silensky, P., "Foams Applied in Chemical Spill," *Fire Engineering*, Vol. 121, No. 6, 1988, pp. 97–99.

"Special Hazards Protection—Manufacturers and Equipment," *Fire Journal*, Vol. 81, No. 4, 1987, p. 43.

Szonyi, S. and Cambon, A., "Influence of Water-Soluble Polymers/New Fluorochemical Surfactants Interaction According to Extinguishing Efficiency of Multipurpose Foam Compounds," *Fire Safety Journal*, Vol. 16, No. 5, 1990, pp. 353–365.

Szonyi, S. and Cambon, A., "New Multipurpose AFFF/MP Extinguishing foam With a Rheological Newtonian Behavior," *Fire Technology*, Vol. 28, No. 2, May 1992, pp. 123–133.

Tabar, D. C., "Foam-Water Sprinkler Protection for Flammable Liquids," *Plant/Operations Progress*, Vol. 8, No. 4, Oct. 1989, pp. 218–224.

Thomas, M., "Trials of a Compressed Air Foam System," *Fire Research News*, Vol. 17, Summer 1994, pp. 12–14.

Timms, G., and Hagger, P., "Foam Concentration Measurement Techniques," *Fire Technology*, Vol. 26, No. 1, Feb. 1990, pp. 41–50.

UL 162, *Foam Equipment and Liquid Concentrates*, 7th ed., Underwriters Laboratories Inc., Northbrook, IL, Mar. 30, 1994.

Walters, A., "Avon's New Foam Transport System Has Many Other Potential Applications," *Fire*, Vol. 85, No. 1054, April 1993, pp. 6, 12.

White, D., "Storage Tank Fire Protection," *Industrial Fire World*, May/June 1986, pp. 20–24.

Whiteley, B., "Foaming Sprinklers and Flammable Liquid Fires," *Fire Prevention*, No. 262, Sept. 1993, pp. 32–35.

Wieder, M., "Applying Class B Fire Fighting Foams," *Speaking of Fire*, Vol. 2, No. 2, Summer 1995, pp. 18–19.

Wieder, M., "Foam Concentrates," *Speaking of Fire*, Spring 1995, pp. 18–19.

Willson, M., "Fixed Foam Systems: Why Do We Need Them?" *Fire Surveyor*, Vol. 22, No. 6, Dec. 1993, pp. 21–25.

Wilson, M., "WASP and Foam-Water Sprinkler Systems," *Fire Safety Engineering*, Vol. 1, No. 5, Oct. 1994, pp. 22–24.

Xiujuan, G., "Combined Agent (Dry Chemical-Foam) Fire Fighting Model Test," First Asian Conference on Fire Science and Technology (ACFST), October 9–13, 1992, Hefei, China, International Academic Publishers, China, 1992, pp. 360–365.

FIRE EXTINGUISHER USE AND MAINTENANCE

Revised by Mark T. Conroy

Before choosing an extinguisher, it is important to know the nature of the fuels present, who will use the extinguisher, the physical environment in which the extinguisher will be placed, and whether any chemicals present in the area will react adversely with an extinguishing agent. When choosing from among various extinguishers, one should consider whether it is effective on the specific hazards present, whether it is easy to operate, and what maintenance and upkeep it requires.

This chapter specifically discusses these important fire extinguisher concerns within the following topics: matching extinguishers to hazards, available personnel, physical environment, health and operational safety, operation and use, obsolete extinguishers, distribution, inspections, and maintenance.

Related information can be found elsewhere in this section.

MATCHING EXTINGUISHERS TO HAZARDS

By far the most important consideration when selecting extinguishers is the nature of the area to be protected. NFPA 10, *Standard for Portable Fire Extinguishers* (hereinafter referred to as NFPA 10), classifies fires as either Class A, Class B, Class C, or Class D, according to the fuel involved. Extinguishers are classified for use on one or more of these types of fires.

In addition, the relative hazard in a building varies according to its fire load, or the amount of combustibles it contains. NFPA 10 establishes three types of hazards. A light, or low, hazard exists when there are few combustibles. Light-hazard occupancies include offices, churches, schoolrooms, assembly halls, and so on. An ordinary, or moderate, hazard exists when the combustibles present are substantial but ordinary in nature or where there are small quantities of combustibles capable of rapid fire growth. Examples of ordinary-hazard occupancies are mercantile storage and display areas, auto showrooms, and parking garages. Some offices, schools, and so on may contain a sufficient amount of combustible materials to be classified as ordinary hazard. An extra, or high, hazard exists in areas where there are substantial quantities of on-site

combustibles whose nature or configuration could readily support rapid fire growth and large fire size, such as woodworking areas, aircraft servicing areas, and warehouses with high-piled combustibles.

Hazard type is an important factor when selecting extinguishers. For example, a $2^1/_2$-gal (9.5-L) water extinguisher may only be suitable in low- and moderate-hazard areas. In a high-hazard area, an extinguisher with a minimum rating of 4-A is needed.

Class-A-rated extinguishers are used most often for ordinary building protection. Among the agents classified for Class A use are water, loaded stream, aqueous film-forming foam (AFFF), film-forming fluoroprotein foam (FFFP), multi-purpose (ammonium-phosphate-base) dry chemical, and halogenated types. Due to the environmental problems associated with halogenated agents, however, they are no longer being produced and their use is being discouraged except in specific circumstances.

Class-A-rated extinguishers are not the only type needed for building protection. In most areas of a restaurant, for example, the principal combustibles are wood, paper, and fabrics. In the kitchen, however, the hazard of greatest concern is cooking grease, which requires a Class-B-rated extinguisher. In hospitals, there is a general need for Class-A-rated extinguishers in rooms, corridors, and offices, but Class-B:C-rated extinguishers should be placed in laboratories, kitchens, generator rooms, and areas where anesthetics are stored or used. In short, the extinguishers in any one area should correspond to the hazards of that area.

Class B fires are fires in flammable liquids. Class-B-rated extinguishing agents include carbon dioxide, dry chemicals, AFFF, FFFP, and halogenated types. There are three general types of flammable liquids fires: (1) fires in liquids of appreciable depth [deeper than $1/_4$ in. (6.4 mm)], such as those in industrial dip tanks and quench tanks; (2) spill fires or running fires in liquids of a depth of $1/_4$ in. (6.4 mm) or less; and (3) pressurized flammable liquid or gas fires from damaged vessels or product lines. Portable fire extinguishers should not be relied upon when the surface area of an indoor open tank is in excess of 10 sq ft (1 m^2); fires in such areas give off so much heat and smoke that it is dangerous for anyone to remain in the area.

Pressurized flammable liquids and pressurized gas fires present special hazards. Only dry chemical type extinguishers have proved to be effective. In addition, special nozzles and higher rates of agent application are often required. Extinguishers for these

---

Mark T. Conroy is Senior Fire Protection Engineer in NFPA's General Engineering Division. He serves as staff liaison to NFPA's Technical Committee on Portable Fire Extinguishers.

types of fires should therefore be selected on the basis of manufacturers' recommendations. *No* attempt should be made to extinguish these fires unless there is reasonable certainty that the source of fuel can be shut off promptly.

Class-C-rated extinguishers contain extinguishing agents that are electrically nonconductive and therefore are preferred for fires in charged electrical equipment. They should be chosen based on the materials used in constructing the electrical equipment, the degree of agent contamination that can be tolerated, and the types of other combustibles in the area. Of these, the first is particularly important. For example, a power panel will contain more Class A insulating materials than an oil-filled transformer, which will contain mostly Class B material. The agents classified for Class C use include carbon dioxide, dry chemicals, and the halogenated types.

Class D fires are fires in combustible metals. These require special extinguishing agents, generally known as dry powder.

Once an area has been analyzed for hazards, appropriate extinguishers can be chosen. For the Class A fire, there are three basic types of extinguishers that can be used: water-based extinguishers, multi-purpose (ammonium-phosphate-base) dry chemical extinguishers, or halogenated agent extinguishers. For Class B fires, there are carbon dioxide, dry chemical, halogenated agent, and AFFF extinguishers. (Some small extinguishers are rated less than 5-B and, as indicated by their omission from Table 6-23A, may not be permit-

TABLE 6-23A.   *Characteristics of Extinguishers*

Extinguishing Agent	Method of Operation	Capacity*	Horizontal Range of Stream†	Approximate Time of Discharge	Protection Required Below 40°F (4°C)	UL or ULC Classifications‡
Water/Antifreeze	Stored-Pressure or Cartridge	2½ gal	30 to 40 ft	1 min	Yes	2-A
	Pump	2½ gal	30 to 40 ft	1 min	Yes	2-A
	Pump	4 gal	30 to 40 ft	2 min	Yes	3-A
	Pump	5 gal	30 to 40 ft	2 to 3 min	Yes	4-A
Water (Wetting Agent)	Stored-Pressure	1½ gal	20 ft	30 sec	Yes	2-A
	Carbon Dioxide Cylinder	25 gal (wheeled)	35 ft	1½ min	Yes	10-A
	Carbon Dioxide Cylinder	45 gal (wheeled)	35 ft	2 min	Yes	30-A
	Carbon Dioxide Cylinder	60 gal (wheeled)	35 ft	2½ min	Yes	40-A
Loaded Stream	Stored-Pressure or Cartridge	2½ gal	30	1 min	No	2 to 3-A:1-B
	Carbon Dioxide Cylinder	33 gal (wheeled)	50 ft	3 min	No	20-A
AFFF, FFFP	Stored-Pressure	2½ gal	20 to 25 ft	50 sec	Yes	3-A: 20 to 40-B
	Nitrogen Cylinder	33 gal	30 ft	1 min	Yes	20-A: 160-B
Carbon Dioxide	Self-Expelling§	2½ to 5 lb	3 to 8 ft	8 to 30 sec	No	1 to 5-B:C
		10 to 15 lb	3 to 8 ft	8 to 30 sec	No	2 to 10-B:C
		20 lb	3 to 8 ft	10 to 30 sec	No	10-B:C
		50 to 100 lb (wheeled)	3 to 10 ft	10 to 30 sec	No	10 to 20-B:C
Dry Chemical (Sodium Bicarbonate)	Stored-Pressure	1 to 2½ lb	5 to 8 ft	8 to 12 sec	No	2 to 10-B:C
	Cartridge or Stored-Pressure	2¾ to 5 lb	5 to 20 ft	8 to 20 sec	No	5 to 20-B:C
	Cartridge or Stored-Pressure	6 to 30 lb	5 to 20 ft	10 to 25 sec	No	10 to 160-B:C
	Cartridge or Stored-Pressure	50 lb (wheeled)	20 ft	25 to 35 sec	No	160-B:C
	Nitrogen Cylinder or Stored-Pressure	75 to 350 lb (wheeled)	15 to 45 ft	20 to 105 sec	No	40 to 320-B:C
Dry Chemical (Potassium Bicarbonate)	Cartridge or Stored-Pressure	2 to 5 lb	5 to 12 ft	8 to 10 sec	No	5 to 20-B:C
	Cartridge or Stored-Pressure	5½ to 10 lb	5 to 12 ft	8 to 20 sec	No	10 to 80-B:C
	Cartridge or Stored-Pressure	16 to 30 lb	10 to 20 ft	8 to 25 sec	No	40 to 120-B:C
	Cartridge or Stored-Pressure	48 to 50 lb (wheeled)	20 ft	30 to 35 sec	No	120 to 160-B:C
	Nitrogen Cylinder or Stored-Pressure	125 to 315 lb (wheeled)	15 to 45 ft	30 to 80 sec	No	80 to 640-B:C

TABLE 6-23A. *Characteristics of Extinguishers (Continued)*

Extinguishing Agent	Method of Operation	Capacity*	Horizontal Range of Stream†	Approximate Time of Discharge	Protection Required Below 40°F (4°C)	UL or ULC Classifications‡
Dry Chemical (Potassium Chloride)	Cartridge or Stored-Pressure	2 to 5 lb	5 to 8 ft	8 to 10 sec	No	5 to 10-B:C
	Cartridge or Stored-Pressure	5 to 9 lb	8 to 12 ft	10 to 15 sec	No	20 to 40-B:C
	Cartridge or Stored-Pressure	9½ to 20 lb	10 to 15 ft	15 to 20 sec	No	40 to 60-B:C
	Cartridge or Stored-Pressure	19½ to 30 lb	5 to 20 ft	10 to 25 sec	No	60 to 80-B:C
	Cartridge or Stored-Pressure	125 to 200 lb (wheeled)	15 to 45 ft	30 to 40 sec	No	160-B:C
Dry Chemical (Ammonium Phosphate)	Stored-Pressure	1 to 5 lb	5 to 12 ft	8 to 10 sec	No	1 to 2-All and 2 to 10-B:C
	Stored-Pressure or Cartridge	2½ to 9 lb	5 to 12 ft	8 to 15 sec	No	1 to 4-A and 10 to 40-B:C
	Stored-Pressure or Cartridge	9 to 17 lb	5 to 20 ft	10 to 25 sec	No	2 to 20-A and 10 to 80-B:C
	Stored-Pressure or Cartridge	17 to 30 lb	5 to 20 ft	10 to 25 sec	No	3 to 20-A and 30 to 120-B:C
	Stored-Pressure or Cartridge	45 to 50 lb (wheeled)	20 ft	25 to 35 sec	No	20 to 30-A and 80 to 160-B:C
	Nitrogen Cylinder or Stored-Pressure	110 to 315 lb (wheeled)	15 to 45 ft	30 to 60 sec	No	20 to 40-A and 60 to 320-B:C
Dry Chemical (Foam Compatible)	Cartridge or Stored-Pressure	4¾ to 9 lb	5 to 20 ft	8 to 10 sec	No	10 to 20-B:C
	Cartridge or Stored-Pressure	9 to 27 lb	5 to 20 ft	10 to 25 sec	No	20 to 30-B:C
	Cartridge or Stored-Pressure	18 to 30 lb	5 to 20 ft	10 to 25 sec	No	40 to 60-B:C
	Nitrogen Cylinder or Stored-Pressure	150 to 350 lb (wheeled)	15 to 45 ft	20 to 150 sec	No	80 to 240-B:C
Dry Chemical (Potassium Bicarbonate Urea Based)	Stored-Pressure	5 to 11 lb	11 to 22 ft	13 to 18 sec	No	40 to 80-B:C
	Stored-Pressure	9 to 23 lb	15 to 30 ft	17 to 33 sec	No	60 to 160-B:C
		175 lb (wheeled)	70 ft	62 sec	No	480-B:C
Halon 1301 (Bromotrifluoromethane)	Stored-Pressure	2½ lb	4 to 6 ft	8 to 10 sec	No	2-B:C
Halon 1211 (Bromochlorodifluoromethane)	Stored-Pressure	0.9 to 2 lb	6 to 10 ft	8 to 10 sec	No	1 to 2-B:C
		2 to 3 lb	6 to 10 ft	8 to 10 sec	No	5-B:C
		5½ to 9 lb	9 to 15 ft	8 to 15 sec	No	1-A:10-B:C
		13 to 22 lb	14 to 16 ft	10 to 18 sec	No	2 to 4-A and 20 to 80-B:C
		50 lb	35 ft	30 sec	No	1 to 2-B:C
		150 lb	20 to 30 ft	30 to 44 sec	No	30-A:160-B:C
Halon 1211/1301 (Bromochlorodifluoromethane/ Bromotrifluoromethane) Mixtures	Self-Expelling Stored-Pressure	0.9 to 5 lb	3 to 12 ft	8 to 10 sec	No	1 to 10-B:C
		9 to 20 lb	10 to 18 ft	10 to 22 sec		1-A: 10-B:C to 4-A: 80-B:C

NOTES: *1 gal = 3.78 L; 1 lb = 0.45 kg.

† 1 ft = 0.30 m.

‡ UL and ULC ratings checked as of December 9, 1983. Readers concerned with subsequent ratings should review the pertinent "lists" and "supplements" issued by these laboratories: Underwriters Laboratories Inc., 333 Pfingsten Road, Northbrook, IL 60062, or Underwriters' Laboratories of Canada, 7 Crouse Road, Scarborough, Ont., Canada M1R 3A9.

§ Carbon dioxide extinguishers with metallic horns do not carry a "C" classification.

‖ Some small extinguishers containing ammonium phosphate-base dry chemical do not carry an "A" classification.

NB: Vaporizing liquid extinguishers (carbon tetrachloride or chlorobromomethane base) are not recognized in NFPA 10, *Standard for Portable Fire Extinguishers.*

NB: Soda acid and inverting-type foam extinguishers are no longer acceptable for use. These extinguishers can be extremely dangerous to use due to the explosion hazard.

ted as a required extinguisher.) For Class C fires there are carbon dioxide, dry chemical, or halogenated agent extinguishers. The various types of extinguishing agents for Class D combustible metal fires are discussed separately. (See Section 4, Chapter 16, "Metals.")

AVAILABLE PERSONNEL—EASE OF USE

Before extinguishers are selected, consideration should be given to who will use them. Evaluation should include the potential user's physical abilities, reactions under stress, and previous training. The more choices the user must make in an emergency, the greater the chance for error. Many firms have standardized their extinguishers so that employees need only learn one set of instructions. In companies employing trained fire brigades, there may be more variation.

An individual's emotional reaction to a fire will be greatly influenced by his or her familiarity with the extinguisher, experience in using it or observing its use, and self-confidence. Training is, therefore, very important. Many companies have employees practice with extinguishers when the units are scheduled for recharging. Employees should be adequately trained so they will not jeopardize their own safety.

Sometimes, the size or weight of an extinguisher is important. Some extinguisher models have lightweight shells or use agents with more extinguishing capacity per unit of weight. The most common fire extinguisher holds $2^1/_2$ gal (9.5 L) of water and weighs about 30 lb (13.6 kg). The weight of many extinguishers varies. For example, one manufacturer offers two dry chemical extinguishers with a rating of 40-B:C. The sodium bicarbonate-based model has a 20-lb (9.1-kg) capacity and weighs 27 lb (12.3 kg), but the potassium bicarbonate-based model has a 9-lb (41-kg) capacity and weighs only 14 lb (6.4 kg).

PHYSICAL ENVIRONMENT

Still another factor that influences extinguisher selection is the area in which the extinguishers will be placed. Is the extinguisher affected by extreme temperatures, for example? ANSI standards require evaluation of water-based extinguishers at temperatures between 40 and 120°F (4 and 49°C) and all other types between –40 and 120°F (–40 and 49°C). If extinguishers are installed in areas subject to higher or lower temperatures, they should be listed for those areas or put in an enclosure where the proper temperature is maintained. Some stored-pressure extinguishers for Class B fires use nitrogen rather than carbon dioxide as the pressurizing force. For units that must perform in temperatures as low as –65°F (–54°C), nitrogen pressurization is used.

Other conditions that may affect extinguisher performance are direct sunlight, snow, rain, airborne debris, and corrosive fumes. If extinguishers must be placed outdoors, installing them in cabinets and sheltered areas or placing a protective cover over them will help prevent damage or premature deterioration. (See Figures 6-23A and 6-23B.)

Corrosives can cause extinguishers to fail, and extinguishers capable of withstanding corrosive conditions are listed separately. For example, extinguishers that may be used in salt-air atmospheres appear under the heading "Marine Type, USCG." Where corrosive fumes exist at industrial sites, special analysis should be made before extinguishers are chosen.

Extinguishers may also be affected by the vibration found in places such as a drop forge or hammer mill, trains, vehicles, and power boats. In such cases, they should be of rugged design, mounted securely, and inspected at frequent intervals.

HEALTH AND OPERATIONAL SAFETY CONSIDERATIONS

Potential health hazards should always be considered when selecting extinguishers. Manufacturers normally provide prominent labels of caution on extinguishers that could produce toxic vapors or

FIG. 6-23A.   An extinguisher mounted on an outside wall, covered with plastic to shield it from substances that would interfere with operation. (Source: The Ansul Company)

decomposition vapors. Sometimes, however, the danger resides not in the extinguisher but in the area in which it will be used. Useful safeguards include posting warning signs at the entries to confined areas, providing extinguishers with long-range nozzles, installing special ventilation, or supplying employees in the area with self-contained breathing apparatus.

All water-based extinguishers are rated only for Class A fires, except AFFF and FFFP models, which are also used on Class B fires. If the nonfoam/water types are used on Class B fires, the fire may flare up, spread, or injure the operator in some way. If water-based extinguishers are used on fires in or near live electrical equipment, the water stream may transmit an electric shock to the operator.

Although carbon dioxide is not itself toxic, it will not support life when used in concentrations high enough to extinguish a fire. If a carbon dioxide extinguisher is used in an unventilated area, it dilutes the oxygen supply, and those remaining in the area may become unconscious or even die if they do not receive oxygen. In addition, the thick cloud of condensing water vapor that forms upon discharge may cause persons to become disoriented.

Older carbon dioxide extinguishers may have metal horns that transmit a shock to the user when the horns touch live electrical

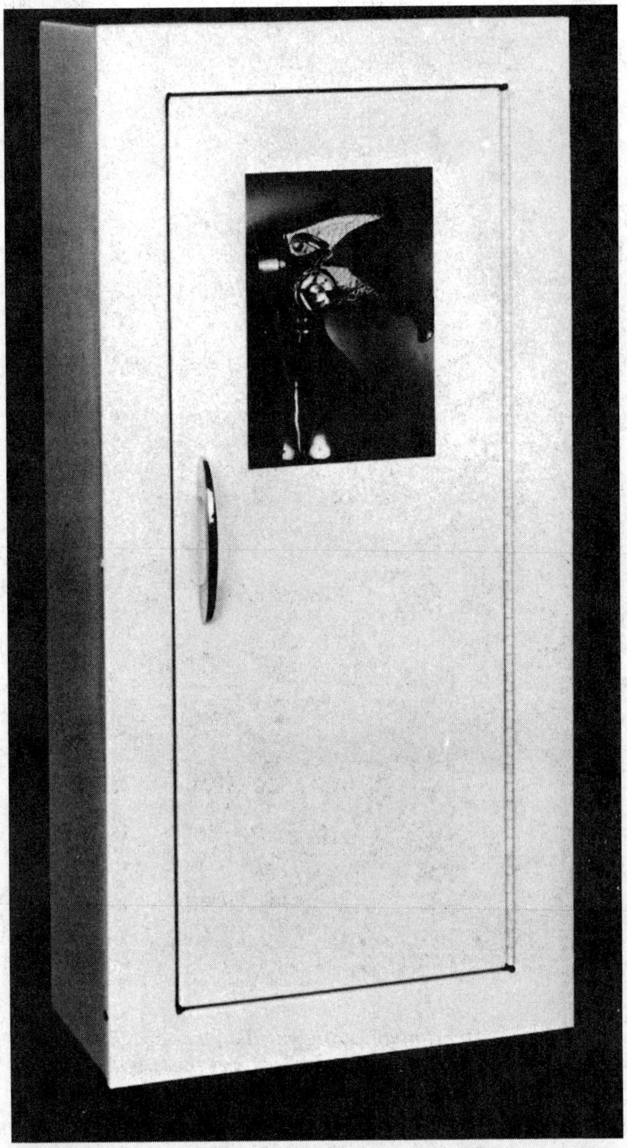

FIG. 6-23B. A surface-mounted extinguisher cabinet with a hinged door. (Source: Larsen's Manufacturing Company)

and thoroughly. Finally, the initial discharge of agent from an extinguisher has considerable force. If it is aimed at close range at a small flammable liquid or grease fire, it may cause the fire to spread extensively before it can be brought under control.

Extinguishers containing halogenated agents—i.e., Halon 1301, Halon 1211, and Halon 1301/1211 mixtures—present some degree of toxicity under normal operating conditions. However, the decomposition products of these agents can be hazardous. When using these extinguishers in unventilated places, such as small rooms, closets, motor vehicles, or other confined spaces, it is best to avoid breathing the vapors or the gases produced by thermal decomposition.

Class D fires involving burning metal chips can be scattered if the full force of a dry powder extinguisher is used at close range. To avoid spreading the fire, the discharge lever should be squeezed slowly at a safe distance.

Virtually every fire produces toxic products of combustion, and some burning materials create highly toxic gases. Until the fire has been extinguished and the area well ventilated, it is important either to stay out of the area or to wear protective breathing apparatus.

OPERATION AND USE

Whether an extinguisher is effective often depends on who is using it. One person may be able to extinguish a fire that someone else, using the same equipment, may not. Many extinguishers discharge their entire contents in 8 to 15 sec, leaving little time for experimentation. Occasionally, improper use of an extinguisher injures the operator and delays extinguishment.

There are several kinds of extinguishers. Because differences exist among extinguishers, it is imperative that people be trained to use them properly. Ideally, this includes the general public. In any case, fire fighters and others responsible for fire protection, such as industrial fire brigades, should be trained thoroughly in the operation and use of extinguishers.

Currently, listed fire extinguishers are classified into five major groups, based on the extinguishing medium each contains. They are: (1) water-type, (2) carbon dioxide, (3) halogenated agent, (4) dry chemical, and (5) foam. Dry powder for Class D fires receives special testing and listing based on specific combustible metals. Information about each of these can be found in the following subsections and in Table 6-23A.

In addition, extinguishers that are no longer manufactured may still be found in some areas. Some information on them is provided under the heading "Obsolete Extinguishers."

Water-Based Extinguishers

Water-based extinguishing agents include water, antifreeze, loaded stream, wetting agent, soda-acid, and foam. All except AFFF and FFFP are for use on Class A fires only. Soda-acid and inverting-type foam extinguishers are no longer available and are discussed later. Antifreeze, loaded stream, AFFF, and wetting agent all use water as a base to which chemicals are added to improve the extinguisher's performance. Both antifreeze and loaded stream models are specially treated to withstand low temperatures; the additive in loaded stream extinguishers is an alkali metal-salt solution. In wetting agents, a material is added to reduce the surface tension of the water so that it will spread and penetrate better.

Originally, water-based extinguishers were of three basic designs: (1) stored-pressure, (2) pump tank, and (3) inverting. In 1969, however, the manufacture of all inverting extinguishers was discontinued. Consequently, the soda-acid and foam agents used exclusively in inverting extinguishers also became obsolete. The two designs that remain, stored-pressure and pump tank, are discussed later.

equipment. These metal horns should be replaced with nonmetallic horns. Occasionally, the operator may receive a shock even when no contact has been made; these result from built-up static electricity and are generally more annoying than hazardous. They are usually caused by the turbulence of the agent discharging through a damaged or defective pickup tube, which can be replaced easily.

Dry chemical extinguishers are not considered toxic, but their discharge can be irritating if breathed for a long time. Monoammonium phosphate is the most irritating, followed by potassium-based agents. Sodium bicarbonate is the least irritating. If dry chemicals are discharged in a confined area, they may reduce visibility and cause disorientation. Because dry chemical agents are nonconductors, deposits left on electrical contacts can reduce or negate the ability of the contacts to conduct. They may also clog air conditioning and air cleaning filters if discharged nearby.

Multi-purpose dry chemical (monoammonium-phosphate-base) is acidic and, if mixed with even a small amount of water, will corrode some metals unless all agent residues are removed promptly

If ordinary, or nonfoam, water-based extinguishers are used on flammable liquids or electrical fires, they may spread the fire, or injure the operator, or both. After activating an extinguisher, the operator should point the stream at the base of the flames and work from side to side or around the fire. When flames are high, the range of the stream [about 30 ft (9 m)] should be used to best advantage. As the flames diminish, it is possible to move closer and change the solid stream to a spray by holding a fingertip over the end of the nozzle. A spray stream is more effective on burning embers. Deep-seated smoldering or glowing areas should be wetted thoroughly. If necessary, a hand tool should be used to poke apart burning materials to do this.

Stored-pressure: For the capacities, ratings, discharge times, stream range, and temperature requirements of the various stored-pressure water-based extinguishers available, see Table 6-23A. The most common stored-pressure water-based extinguisher contains $2^1/_2$ gal (9.5 L) and weighs about 30 lb (13.6 kg). It can be operated intermittently, is rechargeable, and has a comparatively long stream range and discharge time. It consists of a single chamber that contains both the agent and expellant gas. The cap, or head assembly, includes a siphon tube, a combination carrying handle/operating lever, a discharge valve, an air pressure valve and gage, a discharge hose, and a nozzle. (See Figure 6-23C.) The extinguisher is pressurized with air or inert gas by means of an automobile-tire-type valve. Charging pressures range from 90 to 125 psi (620 to 862 kPa). A "fill mark" is stamped on many older stored-pressure water-based extinguishers about 6 in. (152.4 mm) from the top; when refilling, this level should never be exceeded. Some models use a special tube device to prevent overfilling.

In most models, the operating lever is locked by a ring pin that prevents accidental discharge. To activate the extinguisher, set it on the ground. Hold the combination handle loosely in one hand and pull out the ring pin or release a small latch with the other hand. Then grab the hose in one hand and squeeze the discharge lever with the other.

Pump tanks: Two different types of pump tanks are available: one is cylindrical, and the other is made to be used as a backpack. The sizes, ratings, discharge times, stream ranges, and weather requirements of the various pump tanks are given in Table 6-23A. The cylindrical model has carrying handles either attached to the container or built into the pump handle, and the water is discharged by a built-in, hand-operated vertical piston pump with an attached short rubber hose. The pump is double-acting and discharges water on both the up and the down strokes. (See Figure 6-23D.) The cylindrical models are available with copper, steel, or plastic shells.

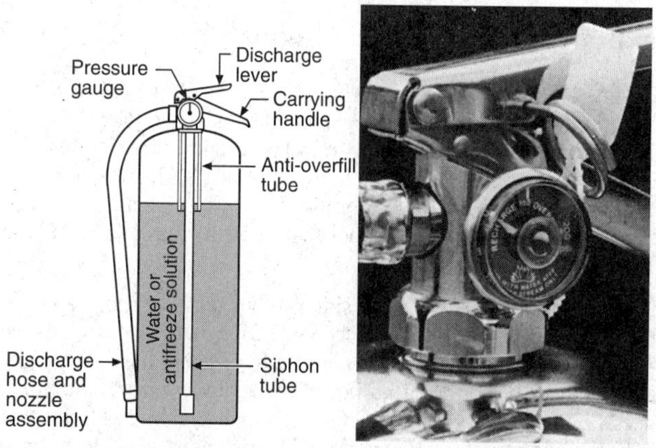

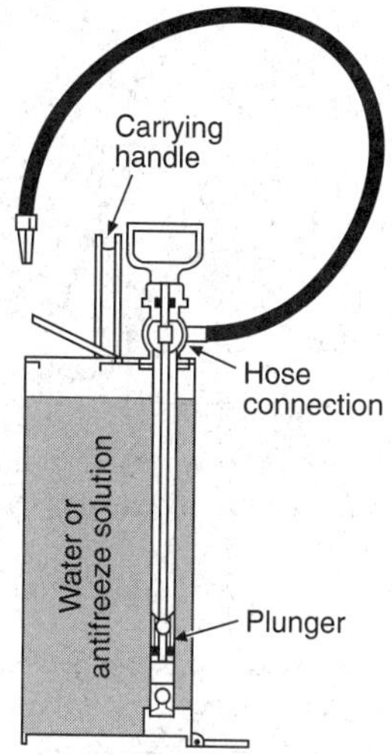

FIG. 6-23D.   *A pump tank fire extinguisher.*

To operate a cylindrical pump tank, place it on the ground and put one foot on the extension bracket at the base to steady the unit. To force water through the hose, pump the handle up and down. Cylindrical pump tanks have two characteristics that may be slight disadvantages: (1) to move the extinguisher, the operator must stop pumping, and (2) the force, range, and duration of the stream are dependent, in part, on the operator.

Cylindrical pump tanks may be filled with either plain water or antifreeze charges recommended by the manufacturer. Common salt or other freezing depressants may corrode the extinguisher or damage the pump assembly. Copper- or plastic-shell models do not corrode as easily as steel and are recommended for use with antifreeze.

Backpack pump tanks are used chiefly for fighting outdoor brush and wildland fires. As the name implies, they are carried on the operator's back. The most common backpack pump tank holds 5 gal (19 L) and weighs about 50 lb (23 kg) when full. (See Figures 6-23E and 6-23F.) Although listed, backpack pump tanks do not have a designated rating. Generally, the tank is filled with plain water, though antifreeze agents, wetting agents, and other special water-based agents may be used. The tank may be made of fiberglass, stainless steel, galvanized steel, or brass, or it may be a flexible water bag. Some models have a large opening with a tight-fitting filter that allows speedy refilling and prevents foreign matter from entering and clogging the pump. This design also permits refilling from sources such as ponds, lakes, and streams.

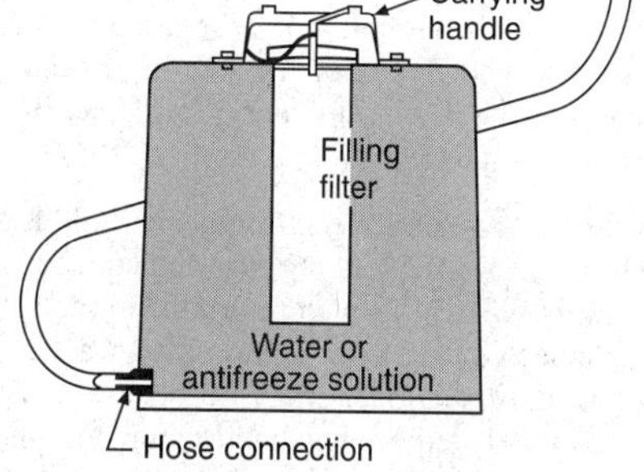

FIG. 6-23C.   **A stored-pressure water extinguisher with close-up of cap assembly. (Source: Badger-Powhatan)**

FIG. 6-23E.   **A backpack pump tank fire extinguisher.**

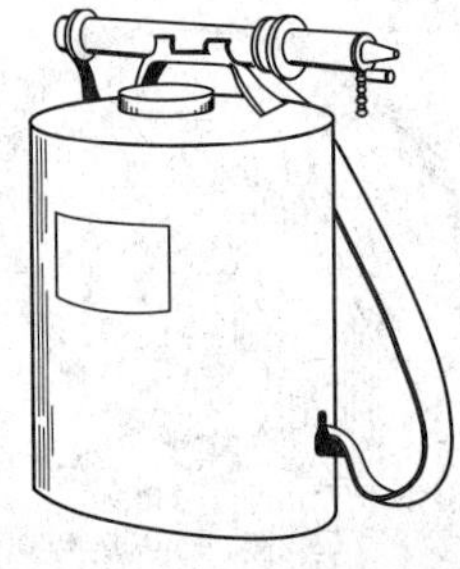

FIG. 6-23F. Backpack pump tank extinguisher with 5-gal (19-L) tank made of fiberglass. Trombone-type pump action is at nozzle.

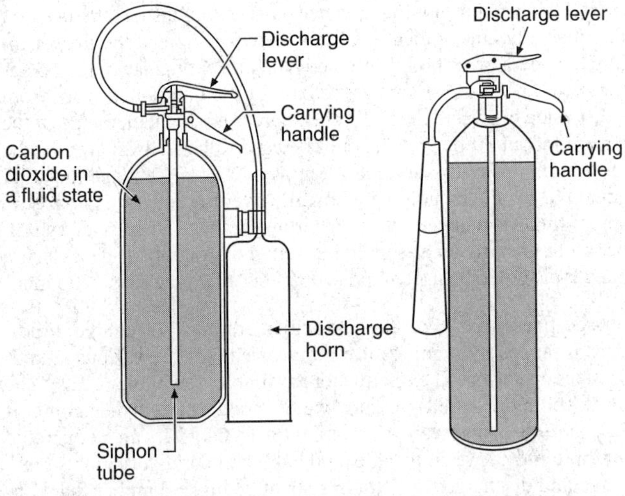

The most common backpack pump tank has a trombone-type, double-acting piston pump connected to the tank by a short length of rubber hose. To discharge the device, the operator holds the pump in both hands and moves the piston back and forth. Other models have compression pumps mounted on the right side of the tank. In these, the expellant pressure is built up with about ten strokes of the handle and then maintained by slow, easy pumping. The left-hand controls discharge by means of a lever-operated shutoff nozzle at the end of the hose.

Carbon Dioxide Extinguishers

Carbon dioxide (CO_2) is a compressed gas agent. Though rated for use on Class B and C fires, it is often effectively used on small Class A fires.

Carbon dioxide prevents combustion by displacing the oxygen in the air surrounding a fire and by cooling the fuel. Its principal advantage is that it does not leave a residue, a consideration that may be important in laboratories and in areas in which food is prepared or electronic equipment is present. The rapid expansion and pressure reduction of the pressurized liquid carbon dioxide in the cylinder results in the cooling and refrigeration effect. However, carbon dioxide extinguishers have a relatively short range because the agent is expelled in the form of a gas/snow cloud; they are also affected by wind or drafts. If a carbon dioxide extinguisher is used in a confined or unventilated area, precautions should be taken so that people are not overcome from lack of oxygen. For the various sizes, stream ranges, discharge times, temperature requirements, and ratings of carbon dioxide extinguishers available, see Table 6-23A.

In all carbon dioxide extinguishers, the agent is retained as a liquid at 800 to 900 psi (5516 to 6205 kPa) at temperatures below 88°F (31°C), and it is self-expelling. The extinguisher design consists of a pressure cylinder, or shell; a siphon tube and valve for releasing the agent; and a discharge horn or horn-hose combination. The siphon tube extends from the valve almost to the bottom of the shell, so that normally only liquid CO_2 reaches the discharge horn until about 80 percent of the content has been released. The remaining 20 percent enters the siphon tube as a gas. The rapid expansion from liquid to gas when most of the CO_2 is discharged converts about 30 percent of the liquid into a very cold "snow" or "dry ice," which then sublimes into a gas.

To operate a carbon dioxide extinguisher, hold it upright by its carrying handle, remove the locking pin, and squeeze the operating lever. (See Figure 6-23G.) In smaller portable models, the discharge horn often is connected to the valve assembly by a metal tube/swing joint connector, and the models may be operated with one hand. Larger portable models require two hands, and their discharge horns are attached to a length of hose. Wheeled extinguishers have a cylinder valve with a locking ring pin, a long hose of 15 to 40 ft (4 m

FIG. 6-23G. Carbon dioxide fire extinguisher.

to 12 m), and a projector that consists of a horn, a long handle, and a control valve. Once the cylinder valve has been opened, the operator controls the discharge with the valve on the projector handle.

Care should be taken not to touch the discharge horn during operation, because it can become extremely cold. If carbon dioxide extinguishers are used in subzero temperatures, the valve must remain open at all times or the discharge may become blocked unless a special low-temperature charge has been added.

Because carbon dioxide extinguishers have a limited range and are affected by draft or wind, they should be applied as close to the base of a fire as possible. Agent should be applied even after the flames have been extinguished to allow time for cooling and to prevent reflash. Due to the relatively short discharge range, carbon dioxide extinguishers are of limited effectiveness on large Class B fires. For flammable liquid fires, the usual method is to begin at the near edge and sweep from side to side toward the back of the fire. However, there is another method called "overhead application." In this method, the discharge horn is pointed downward at an angle of about 45 degrees toward the center of the burning area. Usually, the horn is not moved, and the agent spreads out in all directions. The side-to-side sweeping method may give better results on spill fires, while the overhead method may be best for confined fires.

For fires involving electrical equipment, the discharge should be directed at the source of the flames. It is important to deenergize the equipment as soon as possible to prevent possible reignition.

Halogenated Agent Extinguishers

In general, bromochlorodifluoromethane, or Halon 1211, is similar to carbon dioxide in that it is a "clean agent." (See Figure 6-23H.) Though intended for use mostly on Class B and C fires, Halon 1211 is effective on Class A fires. Sizes of 9-lb (4.1-kg) capacity and greater have Class A ratings. For listed ratings, capacities, discharge times, stream range, and weather requirements for Halon 1211, see Table 6-23A.

Halon 1211 leaves no residue, is virtually noncorrosive and nonabrasive, and is at least twice as effective on Class B fires as carbon dioxide when compared on a weight of agent basis. It also has about twice the range of carbon dioxide. Like carbon dioxide, it needs no cold weather protection. Though Halon 1211 does not have the cooling effect that carbon dioxide has, it is not as severely affected by wind as carbon dioxide, although strong air currents may disperse the agent too rapidly. When Halon 1211 is used on a

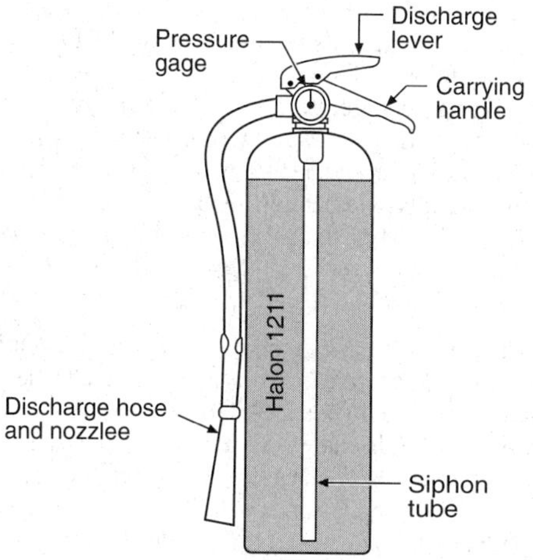

FIG. 6-23H. Halon 1211 stored-pressure fire extinguisher.

fire, the decomposition products include hydrogen chloride, hydrogen fluoride, hydrogen bromide, and traces of free halogens. Normally, only small quantities of these chemicals are formed, and, as a warning of their presence, they give off an acrid odor. Although Halon 1211 is retained under pressure in a liquid state [40 psi at 70°F (276 kPa at 21°C)], a booster charge of nitrogen is added to ensure proper operation.

Several manufacturers have produced Halon 1301 extinguishers during the past 30 years in a $2^1/_2$-lb (1.1-kg) size. (See Figure 6-23I.) These extinguishers only achieve the B:C rating and have a significantly less toxicity hazard. Because of its low vapor pressure, the discharge stream of Halon 1301 is principally an invisible gas, which may complicate effective application.

Halogenated agent extinguishers are also available containing a blend of Halon 1211 and Halon 1301. Halogenated agent extinguishers should be operated and applied to fires in the same general way that carbon dioxide extinguishers are.

Again, due to environmental problems associated with halogenated agents, their use is being discouraged except in specific circumstances.

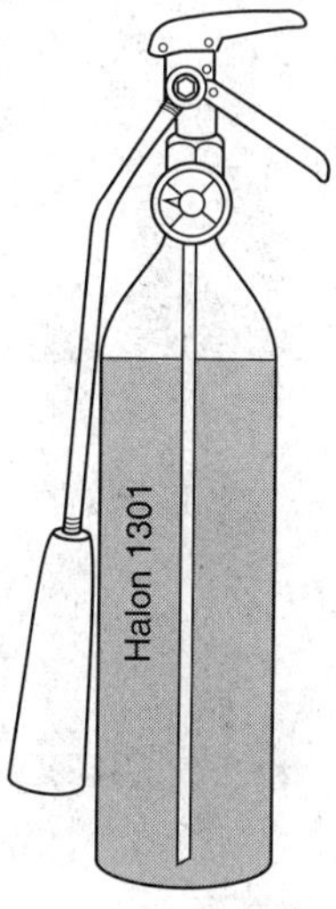

FIG. 6-23I. Halon 1301 fire extinguisher.

Dry Chemical Extinguishers

There are two basic kinds of dry chemical agents: (1) ordinary dry chemicals, which are rated B:C and include sodium bicarbonate, potassium bicarbonate (Purple K), urea potassium bicarbonate (Monnex®), and potassium chloride-based agents; and (2) multipurpose dry chemicals, which achieve the A:B:C rating. An ammonium-phosphate-based agent is the only multipurpose dry chemical currently manufactured. There are also two basic designs of dry chemical extinguishers: one uses a separate pressurized cartridge to expel the agent, and the other pressurizes the agent chamber for the same purpose. The stored-pressure, or rechargeable type, is the most widely used. It is best suited to areas where infrequent use is anticipated and where skilled personnel with professional recharge equipment are available. By contrast, the cartridge-operated type can be refilled quickly in remote locations without special equipment.

The size, rating, and method of operation of any particular extinguisher depends both on its design and on the agent. Table 6-23A gives specific information on the capacities, ratings, discharge times, and stream ranges of the dry chemical extinguishers now available.

In areas where Class A:B:C protection is needed, water-based extinguishers may be omitted if multipurpose dry chemical extinguishers are installed. However, they are not replacements for water-based extinguishers under all conditions.

When choosing a dry chemical extinguisher for Class C protection, it is important to remember that potassium chloride is more corrosive than the other dry chemicals and that multipurpose agent (ammonium phosphate) also is corrosive and will be more difficult to remove because it hardens when it cools.

The various dry chemical extinguishers possess certain common characteristics. Dry chemical extinguishers can be discharged while being carried, and all can be briefly stopped from discharging momentarily to conserve the agent. Dry chemicals also can be used simultaneously with water, either straight stream or spray. Special long-range nozzles may be necessary for some very hazardous areas, such as those with pressurized gases and flammable liquids. Dry chemicals are dispersed less easily by wind than either carbon dioxide or halogenated agents, though dry chemicals are most effective during windy conditions if the wind is at the operator's back.

If dry chemicals are used on wet, energized electrical equipment, such as rain-soaked utility poles, high-voltage switch gear, and transformers, electrical leakage problems may be aggravated. The combination of dry chemical with moisture provides an electrical path that can keep insulation from being effective. In such cases, all traces of dry chemical should be removed after the fire has been extinguished.

Although all dry chemicals are treated for water repellancy, they may harden if exposed to water. It is, therefore, important to avoid exposing them to any moisture during storage, handling, and recharging. Manufacturers' specifications must be followed carefully.

Ordinary dry chemicals are rated for use on Class B:C fires, but they may be used on Class A fires to knock down flames rapidly until something more suitable can be obtained. When used on flammable liquid fires, the stream should be directed at the base of the flame. Attack the fire near the edge, and move the discharge toward the back of the fire while sweeping the nozzle rapidly from side to side. Do not direct the initial discharge directly at the burning surface at close range [less than 5 to 8 ft (1.5 to 2.4 m)], because the high velocity of the stream may splash or scatter the burning material.

Multi-purpose dry chemical extinguishers should be used in exactly the same manner as ordinary dry chemical agents on Class B fires. On Class A fires, the multipurpose agent softens and sticks to hot surfaces. In this way, it forms a coating that smothers and isolates the fuel from the fire. The agent itself has little cooling effect and cannot penetrate below the burning surface. For this reason, it may be hard to extinguish deep-seated fires unless the agent is discharged below the surface or the material is broken apart and spread out.

Cartridge-operated: Cartridge-type dry chemical extinguishers have a chamber in which the agent is kept at atmospheric pressure; the chamber has a large opening at the top through which the extinguisher may be filled. Attached to the side of the extinguisher, on most models, is a small cartridge of propellant gas, either carbon dioxide or nitrogen, threaded into a puncture valve and gas tube assembly. The agent is discharged through a hose attached to the bottom edge of the shell, and the discharge is controlled by a squeeze-grip nozzle at the end of the hose. (See Figure 6-23J.) Cartridge-operated extinguishers should be recharged once they have been pressurized. Even if no agent has been released, the propellant gas can leak away in several hours.

To activate a cartridge-operated extinguisher, hold it upright or place it on the ground. Remove the nozzle from its holder, and hold it in one hand while first pushing down the puncture lever before squeezing the nozzle. Pushing the puncture lever releases the propellant from the cartridge and pressurizes the agent cylinder. On some models, it may be necessary to remove a locking pin before pushing the puncture lever. This operation requires two hands, one to hold the device and the other to release and direct the discharge. Do not stand directly over the extinguisher when pressurizing it.

Stored-pressure: Dry chemical extinguishers with this expellant method are available in both rechargeable and disposable shell models. Most disposable models have a disposable, factory-sealed cylinder containing agent and propellant gas that is threaded into the valve and nozzle assembly. Some small models are designed so that the entire device can be discarded after use. Once this type of device has been used, it should be replaced, even if only a small amount of agent has been discharged, because the propellant gas will leak, leaving no pressure to discharge to agent.

Rechargeable stored-pressure extinguishers come in two types. In both, the agent and the propellant gas are in the extinguisher shell. When the extinguisher is activated, the agent is forced up the siphon tube, where its release is controlled by the operator. One type of extinguisher has a threaded valve assembly and combination carrying handle/operating lever that screws into an opening on the top of the shell. To activate it, release the locking device and squeeze the operating lever. (See Figure 6-23K.) On smaller sizes of this extinguisher, only one hand is needed to operate the device because the nozzle is part of the valve assembly. Larger sizes require two hands for operation, one to squeeze the lever and carry the device, and the other to direct the agent discharge from the hose.

The other rechargeable stored-pressure dry chemical extinguisher (pre-1975) has a release lever and a hose attached to the cap assembly that covers the fill opening. The hose has a squeeze-grip nozzle for controlling the discharge. To activate it, pull the ring pin and push down the release lever. The agent will travel down the hose to the nozzle, where its discharge can be controlled by squeezing the nozzle. Two hands are needed to operate this extinguisher, one to carry the extinguisher, and the other to release and aim the discharge stream.

Dry Powder Extinguishers

Dry powder extinguishers are intended for use on Class D fires, which involve combustible metals. The agent, extinguisher, and method of application should be chosen according to manufacturers' recommendations. The agent may be applied to the fire from an extinguisher or with a scoop and shovel, depending on the type of both agent and metal. In any case, the agent should be applied so that it covers the fire and provides a smothering blanket. More agent may be necessary on hot spots. Care should be taken not to scatter the burning material, which should be left undisturbed until it has cooled.

If there is a fire in finely divided combustible metal or combustible metal alloy scrap that is wet, or if such a fire is wetted with water or water-soluble machine lubricants, the metal may burn rapidly and violently. Such fires may become so hot that it is impossible to get close enough to apply extinguishing agent. If burning metal is on a combustible surface, it should be covered with dry powder. Then a 1- or 2-in. (25- or 50-mm) layer of powder should be spread out nearby, and the burning metal shoveled into it. (See Section 6, Chapter 26, "Combustible Metal Agents and Application Techniques.")

One cartridge-operated portable [30 lb (13.6 kg)] dry powder extinguisher currently is available. (See Figure 6-23L.) Wheeled models are available in 150- and 350-lb (68- and 159-kg) sizes. The extinguishing agent is sodium chloride with additives that render it free-flowing. Thermoplastic material also is added to bind the sodium chlo-

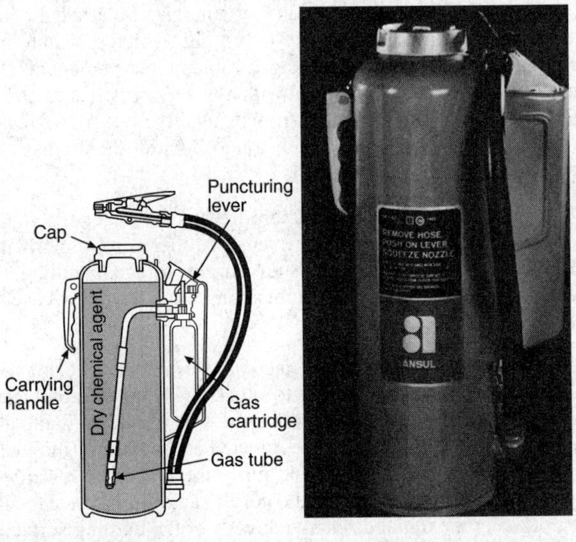

FIG. 6-23J. *A cartridge-operated dry chemical fire extinguisher. (Source: The Ansul Company)*

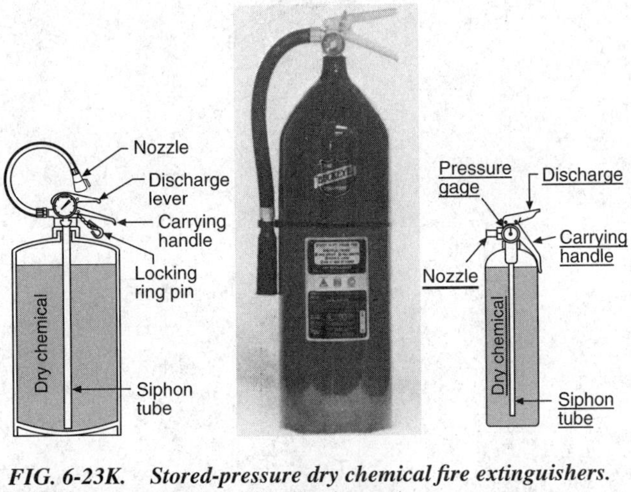

Puncturing lever
Cap
Dry chemical agent
Carrying handle
Gas cartridge
Gas tube

FIG. 6-23K. *Stored-pressure dry chemical fire extinguishers. (Photo source: Buckeye Fire Equipment)*

Nozzle
Discharge lever
Carrying handle
Locking ring pin
Dry chemical
Siphon tube

Pressure gage
Discharge
Carrying handle
Nozzle
Dry chemical
Siphon tube

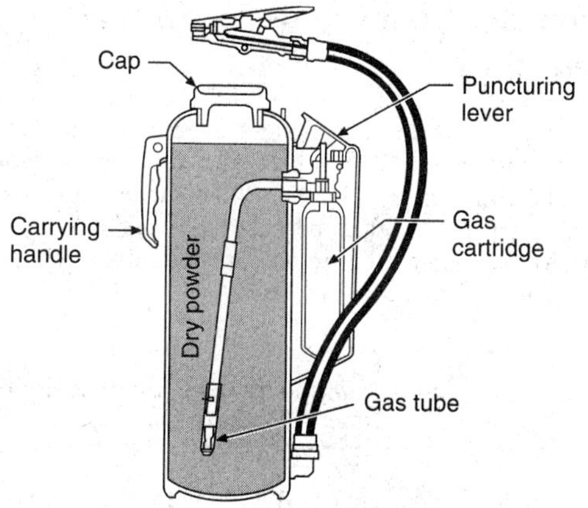

FIG. 6-23L. Dry powder extinguisher.

ride particles into a solid crust when applied to a fire. With the nozzle fully open, the portable model has a range of 6 to 8 ft (1.8 to 2.4 m).

The method of agent application depends on the type, quantity, and physical form of the burning metal. If the fire is very hot, discharge should begin at the maximum range with the nozzle fully open. Once control is established, the nozzle should be partly closed to produce a soft, heavy flow so that complete coverage can be accomplished at close range. More complete details should be obtained from the manufacturer.

In bulk form, dry powder agents are available in 40- or 50-lb (18- or 23-kg) pails and 350-lb (159-kg) drums. The two most common agents are sodium chloride and G-1 powder. The latter consists of graded, granular graphite to which compounds containing phosphorous have been added. The sodium chloride can be used in an extinguisher or applied by hand. The G-1 agent must be applied by hand. When G-1 is applied to a metal fire, the heat of the fire causes the phosphorous compounds to generate vapors that blanket the fire and prevent air from reaching the burning metal. The graphite, being a good conductor of heat, cools the metal.

Aqueous Film-Forming Foam (AFFF) and Film-Forming Fluoroprotein Foam (FFFP) Extinguishers

These extinguishing agents contain aqueous film-forming surfactant. When added to water, AFFF and FFFP form solutions that create a foam when discharged through an aspirating nozzle. On Class A fires, these agents both cool and penetrate to reduce temperatures below the ignition level. On Class B fires, they act as a barrier to exclude oxygen from the surface of the fuel.

AFFF and FFFP currently are available in $2^1/_2$-gal (9.5-L) stored-pressure models rated at 3-A:20-B. They discharge their contents in approximately 50 sec and have a range of 20 to 25 ft (6 to 8 m). They should be installed only in areas not subject to temperatures below 40°F (4°C). There are also 33-gal (125.4-L) unit with 20A:160-B ratings.

These types of extinguishers closely resemble the stored-pressure water extinguisher, except for their special aspirating nozzles. (See Figure 6-23M.) On flammable liquid fires of appreciable depth, the best results are obtained when the discharge is played against the inside of the back wall of the vat or tank just above the burning surface, which should permit the natural spread of the foam back over the burning liquid. If this is not possible, the operator should stand far enough away from the fire to allow the foam to fall lightly on the burning surface instead of splashing it into the burning liquid. Where

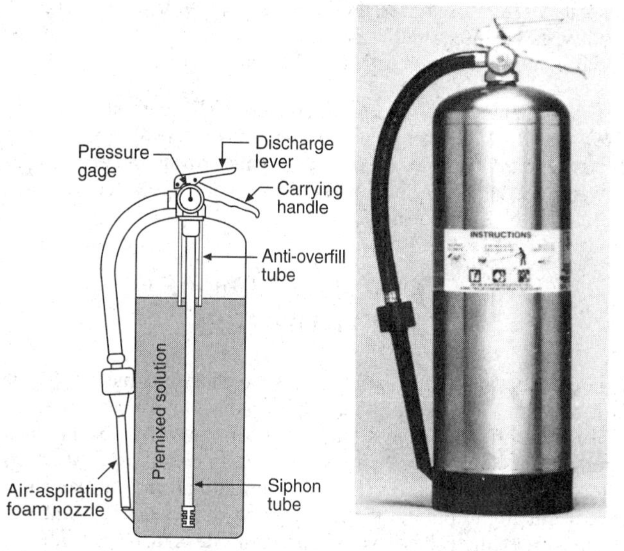

FIG. 6-23M. Aqueous film-forming foam (AFFF) extinguishers. (Photo source: Amerex)

possible, the operator should walk around the fire while directing the stream to get maximum coverage during discharge. For flammable liquid spill fires, the foam may be flowed over a burning surface by bouncing it off the floor just in front of the burning area. The operator should never turn his or her back on the fire, as unburned fuel may reignite. For fires in ordinary combustibles, the foam may be used to coat the burning surface directly. Foam is not effective on flammable liquids and gases escaping under pressure, nor is it suitable for fires involving ethers, alcohols, esters, acetone, lacquer thinners, carbon disulphide, and other flammable liquids that either break down or penetrate the foam blanket.

OBSOLETE EXTINGUISHERS

In 1969, the manufacture and testing of all inverting-type extinguishers—soda-acid, foam, and cartridge-operated water and loaded stream—was halted in the United States. It was not only the difficult and unorthodox "upside-down" method of actuation that brought about the discontinuance of these extinguishers. Of greater importance was the fact that, after 10 to 15 years, many inverting types tended to fail the minimum test pressure requirements of NFPA 10 (original factory test pressure). When pressurized, the failure record, including hydrostatic test failures, was alarmingly high. Since the inverting types are not normally pressurized, potential failures generally are not evident until the time of operation or hydrostatic testing.

When inverted, normal operating pressures are about 100 psi (690 kPa). Should the discharge elbow or hose become blocked, however, pressures in excess of 300 psi (2069 kPa) may occur. Container failures have been known to injure the operator seriously.

NFPA 10 was revised in 1978, and stated that extinguishers with copper or brass shells joined by soft solder or rivets could no longer be hydrostatically tested, which would result in their removal from use within five years of the latest hydrostatic test date. Reliability and safety of this type of construction could not be determined by standard hydrostatic test procedures.

In 1978, NFPA 10 called for the removal from service, by or before 1983, of all riveted and soft-solder copper and brass shell extinguishers of the inverting type. The one exception is stainless-steel shells and brass shells that are brazed, as opposed to soft-soldered.

Also in 1978, stored-pressure water and/or antifreeze fiberglass shell types were required to be removed from service immediately because they had been recalled by the manufacturer.

Vaporizing liquid extinguishers with CCl_4 (carbon tetrachloride) and CBM (chlorobromomethane) became obsolete in the late 1960s. The toxic properties of these agents expose the operator to unwarranted health hazards, and their use was supplanted by safer, more effective extinguishers.

DISTRIBUTION OF FIRE EXTINGUISHERS

No matter how carefully extinguishers are chosen to match the hazards of an area and the ability of the people who will use them, they will not be effective unless they are readily available. Sometimes extinguishers are kept at hand, as in welding operations, but, more often, one has to travel from the fire to the extinguisher and back before beginning to put out the fire. In such cases, travel distance to the nearest extinguisher is very important. Travel distance is the actual distance around partitions, through doorways and aisles, and so on that someone must walk to reach the extinguisher.

When placing extinguishers, select locations that provide uniform distribution, easily accessible and relatively free from temporary blockage, near normal paths of travel, near exits and entrances, free from the potential of physical damage, and readily available.

Mounting Extinguishers

If an extinguisher falls, it may injure someone or be so damaged that it must be replaced. Most extinguishers are mounted on walls or columns by securely fastened hangers so that they are supported adequately. Some extinguishers are mounted in cabinets or wall recesses. The operating instructions should always face outward, and the extinguisher should be placed so that it can be removed easily. Cabinets should be kept clean, dry, and ventilated if installed in outdoor locations.

Where extinguishers may become dislodged, brackets specifically designed to cope with this problem are available. In areas where they are subject to physical damage, such as warehouse aisles, protection from impact is important. In large open areas, such as aircraft hangars, extinguishers may be mounted on moveable pedestals or wheeled carts. In order to maintain some pattern of distribution and specify intended placement, locations should be marked on the floor.

NFPA 10 specifies floor clearance and mounting heights, based on extinguisher weight, as follows:

1. Extinguishers with a gross weight not exceeding 40 lb (18 kg) should be installed so that the top of the extinguisher is not more than 5 ft (1.5 m) above the floor.
2. Extinguishers with a gross weight greater than 40 lb (18 kg), except wheeled types, should be installed so that the top of the extinguisher is not more than $3^1/_2$ ft (1 m) above the floor.
3. In no case can the clearance between the bottom of the extinguisher and the floor be less than 4 in. (100 mm).

When extinguishers are mounted on industrial trucks, vehicles, boats, aircraft, trains, and so on, special mounting brackets, available from the manufacturer, should be used. It is also important that the extinguisher be located at a safe distance from the hazard so that it will not become involved in the fire.

Extinguisher Distribution for Class A Combustibles

Table 6-23B is a guide to determining the minimum number and rating of extinguishers for Class A fires needed in any particular area. Sometimes extinguishers with ratings higher than those indicated in Table 6-23A may be necessary because of process hazards, building configuration, and so on, but in no case should the recommended maximum travel distance be exceeded.

TABLE 6-23B. *Fire Extinguisher Size and Placement for Class A Hazards*

	Light (Low) Hazard Occupancy	Ordinary (Moderate) Hazard Occupancy	Extra (High) Hazard Occupancy
Minimum rated single extinguisher	2-A*	2-A*	4-A†
Maximum floor area per unit of A	3,000 sq ft	1,500 sq ft	1,000 sq ft
Maximum floor area for extinguisher	11,250 sq ft‡	11,250 sq ft‡	11,250 sq ft‡
Maximum travel distance to extinguisher	75 ft	75 ft	75 ft

* Up to two water-type extinguishers each with 1-A rating can be used to fulfill the requirements of one 2-A-rated extinguisher for light (low) hazard occupancies.
† Two $2^1/_2$-gal (9.5-L) water-type extinguishers can be used to fulfill the requirements of one 4-A-rated extinguisher.
‡ See Appendix E-3-3 of NFPA 10, *Standard for Portable Fire Extinguishers.*
For SI units: 1 ft = 0.305 m; 1 sq ft = 0.0929 m^2

The first step when calculating how many Class A extinguishers are needed is to determine whether an occupancy is light-, ordinary-, or extra-hazard, according to NFPA 10. Next, the extinguisher rating should be matched with occupancy hazard to determine the maximum area that an extinguisher can protect. Table 6-23B also specifies the maximum travel distance, or actual walking distance, allowed. For Class A extinguishers, it is 75 ft (23 m). For example, each $2^1/_2$-gal (9.5-L) stored-pressure water extinguisher rated 2-A will protect an area of 3000 sq ft (279 m^2) in an ordinary-hazard occupancy, but only 2000 sq ft (186 m^2) in an extra-hazard occupancy.

The figure of 11,250 sq ft (1044 m^2) in Table 6-23B is used instead of 12,000 sq ft (1115 m^2), which would appear to be a normal progressive increment. This is because the largest square inside a circle with a 75-ft (23-m) radius would be 106×106 ft (32×32 m), or 11,250 sq ft (1044 m^2). Because buildings are usually rectangular, this is the largest open area one can have and still comply with the 75-ft (23-m) travel distance rule. (See Figure 6-23N for an illustration of this concept.)

The area that can be protected by one fire extinguisher with a given A rating is shown in Table 6-23C. These values are determined by multiplying the maximum floor area per unit of A shown in Table 6-23B by the various A ratings until a value of 11,250 sq ft (1044 m^2) is exceeded.

NFPA 10 also provides that up to one-half of the complement of extinguishers for Class A fires, as specified in Table 6-23B, may be replaced by uniformly spaced small-hose [$1^1/_2$ in. (38 mm)] stations. However, the hose stations and extinguishers should be located so that the hose stations do not replace more than every other extinguisher.

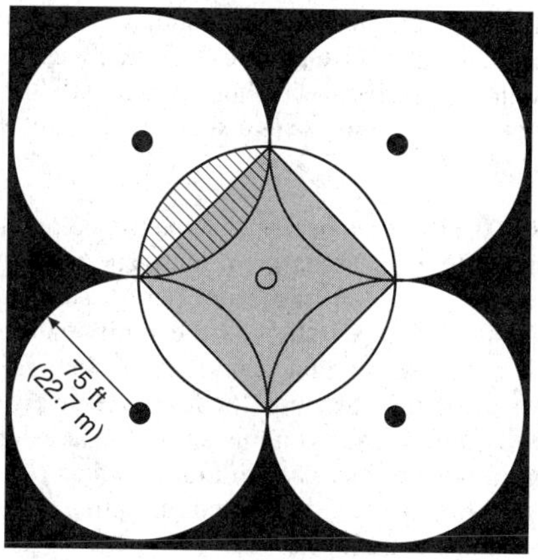

FIG. 6-23N.   The dotted square shows the maximum area [11,250 sq ft (1044 m²)] that an extinguisher can protect within the limits of a 75-ft (23-m) radius.

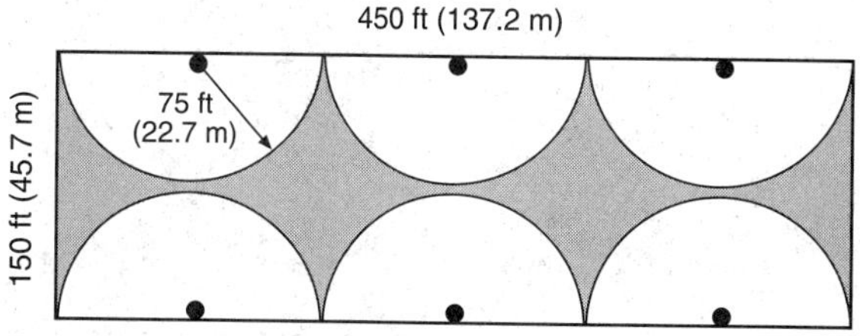

FIG. 6-23O.   A diagrammatic representation of extinguishers located along the outside walls of a 150- × 450-ft (46- × 137-m) building. The large dots represent extinguishers. The shaded areas indicate "voids" that are farther than 75 ft (23 m) from the nearest extinguisher.

TABLE 6-23C.   Maximum Area to Be Protected per Extinguisher (sq ft)

Class A Rating Shown on Extinguisher	Light (Low) Hazard Occupancy	Ordinary (Moderate) Hazard Occupancy	Extra (High) Hazard Occupancy
1A	—	—	—
2A	6,000	3,000	—
3A	9,000	4,500	—
4A	11,250	6,000	4,000
6A	11,250	9,000	6,000
10A	11,250	11,250	10,000
20A	11,250	11,250	11,250
30A	11,250	11,250	11,250
40A	11,250	11,250	11,250

For SI units: 1 sq ft = 0.0929 m².
Note: 11,250 sq ft (1044 m²) is considered a practical limit.

The following examples illustrate the number and placement of extinguishers according to occupancy hazard and extinguisher rating. The sample building is 150 by 450 ft (46 by 137 m), which gives a floor area of 67,500 sq ft (6270 m²). Although several different ways of placing extinguishers are given, a number of other locations could have been used with similar results.

The first example demonstrates placement at the maximum protection area limits [11,250 sq ft (1044 m²)] allowed per extinguisher in NFPA 10 for each class of occupancy. Installing extinguishers with higher ratings will not affect distribution or placement.

EXAMPLE 1:

$$\frac{67,500}{11,250} = 6$$

4-A   Extinguishers for light-hazard occupancy
10-A   Extinguishers for ordinary-hazard occupancy
20-A   Extinguishers for extra-hazard occupancy

This placement along outside walls would not be acceptable because it clearly violates the travel distance rule. (See Figure 6-23O.) Instead, relocation and/or additional extinguishers are needed.

Example 2 is for fire extinguishers having ratings that correspond to protection areas of 6000 sq ft (557 m²). Example 3 is for extinguishers having the minimum ratings permitted by Table 6-23B

with corresponding minimum protection areas. As the number of lower-rated extinguishers increases, meeting the travel distance requirement generally becomes less of a problem.

EXAMPLE 2:

$$\frac{67,500}{6,000} = 12$$

2-A   Extinguishers for light-hazard occupancy
4-A   Extinguishers for ordinary-hazard occupancy
6-A   Extinguishers for extra-hazard occupancy

EXAMPLE 3:

$$\frac{67,500}{6,000} = 12$$ 2-A   Extinguishers for light-hazard occupancy

$$\frac{67,500}{3,000} = 23$$ 2-A   Extinguishers for ordinary-hazard occupancy

$$\frac{67,500}{4,000} = 17$$ 4-A   Extinguishers for extra-hazard occupancy

Extinguishers could be mounted on exterior walls or, as shown in Figure 6-23P, on building columns or interior walls, and conform to both distribution and travel distance rules.

The arrangement illustrated in Figure 6-23Q shows extinguishers grouped together on building columns or interior walls in a manner that still conforms to distribution and travel distance rules.

Extinguisher Distribution for Class B Combustibles

Class B fire hazards fall into two distinct categories. The first includes liquids ¼-in. (6.4-mm) deep or less, and the other includes liquids deeper than ¼ in. (6.4 mm).

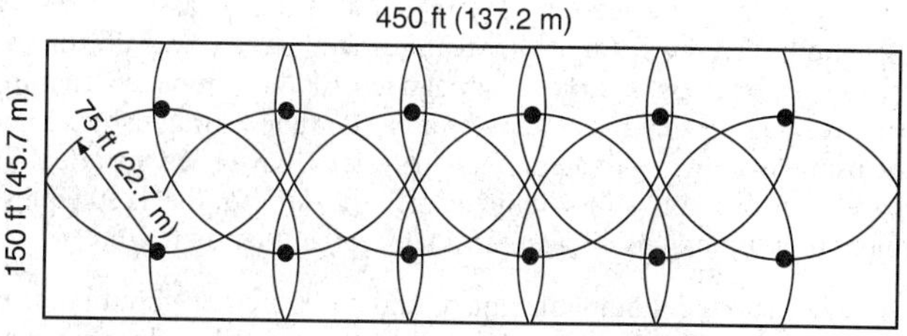

FIG. 6-23P.   Requirements for both travel distance and extinguisher distribution are met in this configuration of 12 extinguishers mounted on building columns or interior walls.

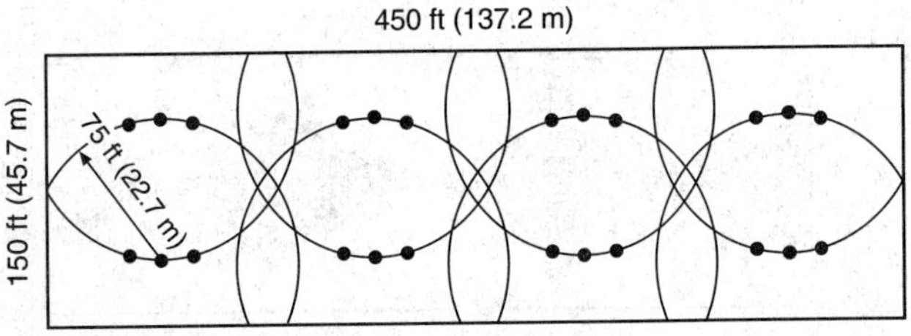

FIG. 6-23Q.   Extinguishers grouped together.

In areas where liquids do not reach an appreciable depth, extinguishers should be provided according to Table 6-23D. The reason the basic maximum travel distance to Class B extinguishers is 50 ft (15 m), as opposed to 75 ft (23 m) for Class A extinguishers, is that flammable liquids fires reach their maximum intensity almost immediately, and thus the extinguisher must be nearer to hand. With lower-rated extinguishers, the travel distance is reduced to 30 ft (9 m).

Even though Table 6-23D specifies maximum travel distances to extinguishers for Class B fires, judgment should be used when actually placing the devices. The closer the extinguisher to the hazard, the better—up to the point at which a fire might damage or obstruct access to the device.

TABLE 6-23D.   *Fire Extinguisher Size and Placement for Class B Hazard Excluding Protection of Deep-Layer Flammable Liquid Tanks*

Type of Hazard	Basic Minimum Extinguisher Rating	Maximum Travel Distance to Extinguishers, ft (m)
Low	5-B	30 (9)
	10-B	50 (15)
Moderate	10-B	30 (9)
	20-B	50 (15)
High	40-B	30 (9)
	80-B	50 (15)

When flammable liquids reach an appreciable depth, the extinguisher's rating number, except for foam types, should be at least twice the number of square feet (m²) of surface area of the largest tank in the area, assuming that other requirements are met. The travel distances specified by Table 6-23D should also be used to locate extinguishers for protection of spot hazards. Sometimes, one extinguisher can be installed to provide protection against several different hazards, provided that travel distances are not exceeded.

Where there are open process tanks of flammable liquids with surface areas greater than 10 sq ft (1 m²), complete dependence should not be placed on fire extinguishers. It is often advisable to consider installing fixed fire protection systems for these tanks because of the potential size and intensity of the fire which could occur. If a fixed protection system suitable for Class B fires is installed, then the extinguisher requirement may be waived for the hazard, or hazards, it protects, though not for the complete structure or any other special hazards in the area. Portable extinguishers may be useful even when fixed protection systems have been installed, in case a burning tank spills liquid outside the range of the fixed equipment or a fire begins adjacent to a tank rather than in it.

Pressurized flammable liquids and gases are not stored in open containers, and it is not possible to select extinguishers for them according to the square foot (m²) requirements. Instead, extinguishers with special nozzles and rates of agent application should be chosen. The manufacturers of these specialized extinguishers usually recommend what equipment they are suited for. In general, however, no attempt to extinguish pressurized fuel fires should be made unless the source of fuel can be shut off promptly. As before, the travel distances to portable extinguishers should not exceed those specified by Table 6-23D.

Because Class B fires become so intense so quickly, the flow rate and the duration of discharge of an agent are very important. For these reasons, NFPA 10 does not permit substituting two or more extinguishers of lower rating for the minimum ratings given in Table 6-23D, except for certain foam extinguishers. The exception permits the substitution because foam progressively "secures" a fire, and the foam from one extinguisher is effective during the time it takes to bring other units into operation. With carbon dioxide and dry chemical extinguishers, however, four extinguishers rated 5-B are not equal to one extinguisher rated 20-B; the larger extinguisher has a higher flow rate in pounds per second (kg/s) and a longer continuous discharge time than the smaller models.

Wheeled extinguishers with ratings from 20-B to 480-B are available. They are designed chiefly for outdoor fire fighting and should be used only by trained employees. They should be distributed according to the 50-ft (15-m) travel distance rules of Table 6-23D, though longer travel distances may be authorized by the Authority Having Jurisdiction.

Figure 6-23R provides a sample problem for protection by portable extinguishers.

Extinguisher Distribution for Class C Fires

Extinguishers for Class C fires are required wherever there is live electrical equipment. This sort of extinguisher contains a nonconducting agent, usually carbon dioxide, dry chemical, or a halogenated agent.

Once the power to live electrical equipment is cut off, the fire becomes a Class A or Class B fire, depending on the nature of the burning electrical equipment and the burning material in the vicinity. Extinguishers for Class C fires should be selected according to the size of the electrical equipment; the configuration of the electrical equipment, particularly the enclosures of units, which influence agent distribution; and the range of the extinguisher's stream. At large installations of electrical equipment where continuity of power is critical, fixed fire protection is desirable. But even when fixed fire protection is present, it is recommended that some Class C extinguishers be provided to handle incipient fires.

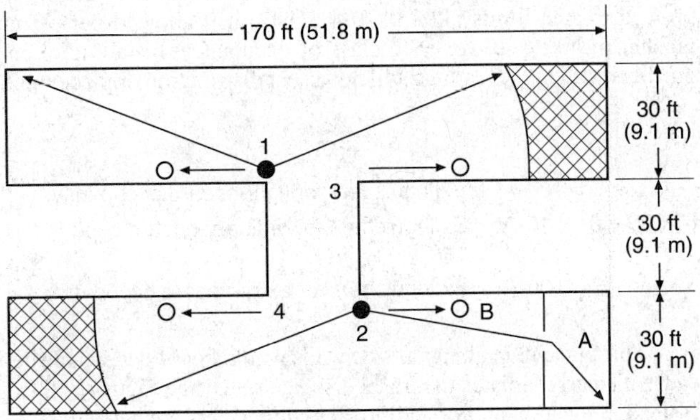

FIG. 6-23R.   Sample problem for protection by portable extinguishers.

Class D Extinguisher Distribution

It is particularly important that the proper extinguishers be available for Class D fires. Because the properties of combustible metals differ, even an agent for Class D fires may be hazardous if used on the wrong metal. Agents should be chosen carefully according to the manufacturers' recommendations. The amount of agent needed is normally figured according to the surface area of the metal, plus the shape and form of the metal, which could contribute to the severity of the fire and cause "bake-off" of the agent. For example, fires in magnesium filings are more difficult to put out than fires in magnesium scrap, and more agent is needed for magnesium filings. The maximum travel distance to all extinguishers for Class D fires is 75 ft (23 m). (See Section 6, Chapter 26, "Combustible Metal Agents and Application Techniques," for additional information.)

INSPECTION AND MAINTENANCE OF FIRE EXTINGUISHERS

Once a fire extinguisher has been purchased, it becomes the responsibility of the purchaser or an assigned agent to maintain the device. Adequate maintenance consists of periodically inspecting each extinguisher, recharging each extinguisher following discharge, and performing periodic hydrostatic tests or as needed.

A fire equipment servicing agency is considered the most reliable way for the general public to maintain extinguishers, but large industries often train employees to handle this maintenance themselves. Many jurisdictions require service technicians to be licensed.

Inspection

An inspection is a quick check that visually determines whether the fire extinguisher is properly placed and will operate. Its purpose is to give reasonable assurance that the extinguisher is fully charged and will function effectively if needed. An inspection should determine that the extinguisher is in its designated place; is conspicuous; is not blocked in any way; has not been activated and partially or completely emptied; has not been tampered with; and has not sustained any obvious physical damage or been subjected to an environment that could interfere with its operation, such as corrosive fumes. If the extinguisher is equipped with a pressure gage and/or tamper indicators, the inspection should show conditions to be satisfactory. In addition, the maintenance tag can be checked to determine the date of the last thorough maintenance check.

In order to be effective, inspections must be frequent, regular, and thorough. In a small building with few extinguishers, the manager, property owner, or some designated person could check extinguishers at the beginning of each workday. If a building is large enough to employ security guards or a surveillance service, extinguishers should be inspected at least once during each 8-hr shift. In industrial plants, the plant fire brigade or fire inspector often inspects extinguishers either daily, weekly, or monthly. The maximum time period between inspections is 30 days.

An individual evaluation must be made of each property to determine how frequent inspections should be made. If a particular operation is fire-prone or is crucial to the use of the property, more frequent inspections should be made. For example, if all the products manufactured in a particular industrial plant had to be painted, a fire in the painting room could be disastrous, so inspections should be very frequent.

It is important to consider the nature of the hazards present which influence the potential use of the extinguisher: the exposure of the extinguisher to tampering, vandalism, and malicious mischief; extraordinary weather conditions; the likelihood of accidental damage to the extinguisher; and the possibility of visual or physical obstructions to the accessibility of extinguishers.

Maintenance

Maintenance, as distinguished from inspection, means a complete and thorough examination of each extinguisher. A maintenance check involves examining all its parts, cleaning and replacing any defective parts, and reassembling, recharging, and, where appropriate, pressurizing the extinguisher.

Maintenance checks sometimes reveal the need for special testing of extinguisher shells or other components. For example, they may reveal that the extinguisher container should be hydrostatically tested or even replaced.

Maintenance should be performed periodically, but at least once every 12 months, after each use, or when an inspection shows that the need is obvious. If an inspection reveals evidence of serious damage by corrosion, for example, the extinguisher should be subjected to a thorough maintenance check, even though it may have undergone one recently. Similarly, if an inspection shows evidence of tampering, agent leakage, or physical damage, a complete maintenance check should be initiated. NFPA 10 contains specific details related to maintenance.

Tags and tamper seals: For many years, extinguisher tags have been used as a convenient means for recording maintenance checks. For routine maintenance, a tied-on tag or pressure-sensitive label is used to record the date and the inspector's initials. Newer technology using bar code readers also can be used. Tamper seals also should be used. Lead and wire seals were used commonly until plastic seals were introduced in 1972. As long as the seal remains intact, one can be reasonably sure that the extinguisher has not been used. However, it is important to note that a stored-pressure extinguisher can develop a leak and lose its pressure even though the tamper indicator remains intact or the pressure gage reads normal. During annual maintenance, the tamper seal is removed by operating the pull pin or locking device. This is done to ensure that they operate freely. A new tamper seal is then installed.

Maintenance Operations

In any maintenance test of a portable fire extinguisher, there are three basic items that need to be checked: (1) the mechanical parts of the device—that is, the extinguisher shell and other component parts; (2) the amount and condition of the extinguishing agent; and (3) the condition of the means for expelling the agent.

A record of the date of purchase, as well as maintenance dates, should be kept for each extinguisher. A separate record is also desirable and should include the maintenance date and the name of the person or agency performing the maintenance; the date last recharged and the name of the person or agency performing the recharge; the hydrostatic retest date and the name of the person or agency performing the hydrostatic test; description of dents remaining after passing a hydrostatic test; and the date of the six-year maintenance for certain stored-pressure dry chemical and halogenated agent extinguishers.

Individuals who own extinguishers often neglect them because there is no planned periodic follow-up program. It is recommended that owners become familiar with their extinguishers so that they can detect telltale signs which may suggest the need for maintenance. An alternative is to have the dealer from whom the extinguisher was purchased establish an annual follow-up maintenance program.

On properties where extinguishers are maintained by the occupant, a supply of recharging materials should be kept on hand. When recharging extinguishers other than those with plain water, use only those recharging materials specified on the extinguisher's nameplate. Other recharging materials may impair the efficiency of the extinguisher or cause it to malfunction and injure the operator. Special precautions are necessary for certain types of extinguishers, and they are discussed elsewhere in this chapter.

NFPA 10 has an appendix with a checklist of items that require maintenance. The checklist is divided into two sections. One section of the checklist is arranged to pinpoint the mechanical parts common to most extinguishers. The other section addresses agent and expelling means.

Water-Type Extinguishers

Inspection and maintenance for stored-pressure models is too often thought of as quite simple. Unless the extinguisher has been used or a routine inspection reveals a defect, maintenance is sometimes left until the five-year hydrostatic test is performed. This practice has resulted in many failures and malfunctions. The principal items that need to be checked during an inspection are a loose, worn, or damaged hose; a plugged nozzle; a dented shell; a damaged indicator gage; and a damaged or jammed ring pin. Normal maintenance service, as specified in NFPA 10, should be performed annually.

Mechanical pump extinguishers, such as pump tanks and backpacks, are easy to inspect and maintain. The tanks containing the water or antifreeze solution must be checked to ensure they are in good condition, have not been weakened by corrosive action, and do not contain sediment that might block the stream during the pumping operation. If the device has not been used in the last 12 months, the liquid should be removed from the tank and the tank should be flushed, and filled with new liquid. Pumps may require lubrication from time to time. They should be checked annually to ensure that the plungers are in proper condition, that seals and washers have not deteriorated, and that hoses are not clogged. Antifreeze solutions may need to be checked to ensure that they can withstand the temperatures in the area and that they are not contaminated.

Dry Chemical Extinguishers

Dry chemical extinguishers should be inspected monthly, and they should also undergo normal annual maintenance. The quantity of agent for a cartridge-operated model can be checked by weighing or by removing the fill cap and checking it visually. The gas cartridge also may be checked by weighing. On stored-pressure models, the pressure gage will indicate if adequate pressure is maintained, and the agent quantity can be checked by weighing.

Dry chemical extinguishers should be refilled promptly after use, even if they have been discharged only partially. When refilling, extreme caution must be taken to ensure that no water, moisture, or foreign material enters the cylinder. Even though dry chemical agents are treated for moisture repellency, they can eventually harden if moisture is present. When these extinguishers are hydrostatically tested, they must be dried thoroughly so that no trace of water or moisture remains. Before new agent is added, all of the unused agent should be removed by dumping or vacuuming, and any residue should be removed from the hose.

The type of dry chemical used in recharging must be the one recommended by the manufacturer. For example, an extinguisher containing a Class B:C dry chemical (bicarbonate-base or potassium chloride) should not be refilled with a Class A:B:C agent (monoammonium-phosphate-base). Mixing different dry chemical agents can result in malfunction, or damage to the extinguisher, or both. The bicarbonate-based agent is chemically alkaline and will react with the acidic ammonium phosphate. This reaction is aggravated by exposure to heat or the presence of moisture in amounts as small as 0.1 percent. One result is caking of the agent; another possible result is the internal corrosion of the extinguisher. Under certain conditions, the reaction will cause excess pressure to build up within the shell, which may damage or rupture the extinguisher.

Substituting another manufacturer's dry chemical of the same type is not recommended unless it has the same chemical and physical characteristics, or has been tested and found to give equivalent performance. The problems involved in substitution include:

1. An altered flow rate of the dry chemical, which could influence extinguishing efficiency.
2. The amount of agent that could be placed in the extinguisher, which could influence discharge time.
3. Incompatibility of one dry chemical to another, including chemical reactions, caking, and jamming of hose or nozzles. The flow rate can change with the particle size and density of the agent. A discharge nozzle designed to dispense one dry chemical efficiently might well be less efficient for another. An extinguisher designed to hold a given weight of one formula may not hold the same amount of another. Dip tubes, hoses, or valves could become plugged.

Recharge operations for the cartridge-operated models are relatively easy. The rubber sleeves on the gas tube should be checked for cracking or overstretching, and the gasket and gasket seats on the shell and cap should be wiped clean. The cap should be screwed on handtight. When replacing the cartridge, care should be taken that the threads are not dirty, cross-threaded, or otherwise damaged. For stored-pressure models, the entire valve body, stem, and "O" ring should be carefully wiped clean and blown out with dry nitrogen to get rid of all traces of agent residue which might cause a pressure leak.

Nitrogen gas used for pressurizing should be of the standard industrial grade, with a dew point of –60°F (–51.1°C) or lower so that it is free of moisture. Once the extinguisher has been pressurized, it should be given a leak test or allowed to stand for about 12 hr and then checked for leaks before it is returned to service.

Some stored-pressure extinguishers are factory-sealed and are not rechargeable. No effort should be made to recharge these units in the field. Normally, the shell of the extinguisher is discarded following use; only the valve and nozzle assembly are retained. Refill units should be obtained from the manufacturer, and detailed instructions for replacement are included on the extinguisher nameplate.

The general inspection and maintenance procedures for dry powder (Class D) extinguishers are the same as for cartridge-operated dry chemical extinguishers.

Carbon Dioxide Extinguishers

Weighing is the only way to determine whether carbon dioxide extinguishers are fully charged. They should be weighed at least semi-annually for loss of weight and inspected for deterioration and/or physical damage. The full and empty weights are stamped on the valve assembly. Any carbon dioxide extinguisher that has a weight loss of 10 percent or more should be recharged and tested for leaks. Recharging is generally done by an extinguisher servicing company. Should on-site recharging be desirable, however, the manufacturer should be contacted for assistance in setting up a recharging station.

Recharging extinguishers with carbon dioxide from dry-ice converters is no longer permitted due to the high potential for moisture in the carbon dioxide. The vapor phase of carbon dioxide must be not less than 99.5 percent carbon dioxide. The water content of the liquid phase must be not more than 0.01 percent by weight [–30°F (–34.4°C) dew point]. The oil content of the carbon dioxide must not exceed 10 ppm by weight.

Specific recommendations are given for such recharging in NFPA 10. The required source of carbon dioxide is a low-pressure carbon dioxide supply [300 psi at 0°F (206.8 kPa at −17.8°C)], either directly from the source or through dry cylinders used as intermediaries. It cannot be emphasized too strongly that internal moisture can seriously corrode an extinguisher, and that moisture is most easily introduced into a cylinder either during recharging or following a hydrostatic test.

Halogenated Agent Extinguishers

The general inspection and maintenance procedures for halogenated agent extinguishers are similar to the requirements for other extinguishers. Bromochlorodifluoromethane (Halon 1211) extinguishers are physically similar to stored-pressure dry chemical extinguishers. The specific requirements of each manufacturer should be followed.

HYDROSTATIC TESTING OF EXTINGUISHERS

The purpose of hydrostatic testing of portable fire extinguishers that are subject to internal pressures is to protect against unexpected, in-service failure. Such a failure may be due to undetected internal corrosion caused by moisture in the extinguisher; external corrosion caused by atmospheric humidity or corrosive vapors; damage caused by rough handling, which may or may not be obvious by external inspection; repeated pressurizations; manufacturing flaws in the construction of the extinguisher; improper assembly of valves or safety relief discs; or exposure of the extinguisher to abnormal heat, as during a fire.

NFPA 10 requires that hydrostatic pressure tests be performed on extinguishers according to Table 6-23E. The first hydrostatic retest may be conducted between the fifth and sixth year for those with a designated test interval of five years. However, carbon dioxide types require a strict five-year interval according to federal law. Hydrostatic tests also should be conducted immediately upon discovery of any mechanical injury or corrosion to the extinguisher shell.

In the United States, various U.S. Department of Transportation (DOT) rules apply to specific types of cylinders used for fire extinguishers. Title 49 of the *Code of Federal Regulations* calls, with some exceptions, for the hydrostatic retesting every five years of compressed-gas cylinders offered for interstate transportation in a charged condition. Paragraph 173.34(e) calls for periodic retesting

of DOT compressed-gas cylinders whether or not the cylinders are shipped interstate.

A great many different DOT specification cylinders are used for dry chemical extinguishers, including Types DOT-4B and DOT-4BA. The requirements for these types are peculiar to them.

The U.S. Coast Guard Regulations Title 46, *Code of Federal Regulations*, Part 147, Paragraph 147.65, specifies that compressed-gas cylinders brought aboard a vessel must have been tested hydrostatically within five years. After that, a cylinder continuously installed on board a vessel as part of the vessel's equipment for a period exceeding five years must be removed from the vessel after twelve years have elapsed from the date of previous test and marking. Its contents must be discharged, and the cylinder must be retested and remarked. The Board of Transport Commissioners of Canada has published similar rules covering these same subjects.

The preferred method of hydrostatic testing of compressed-gas cylinders is the water-jacketed, volumetric expansion method recommended in NFPA 10. The guide to follow in conducting this type of test is Pamphlet C-1, "Methods for Hydrostatic Testing of Compressed Gas Cylinders," published by the Compressed Gas Association, Inc. (CGA). For visually evaluating the condition of cylinders made to DOT specifications, CGA's Pamphlet C-6, "Standards for Visual Inspection of Steel Compressed Gas Cylinders," is helpful.

Procedures for testing extinguishers other than compressed-gas extinguishers or affixed compressed-gas cylinders are detailed in NFPA 10. Table 6-23F gives the basic data on the hydrostatic test pressure requirements. Figures 6-23S and 6-23T illustrate equipment used in hydrostatic testing work.

TABLE 6-23F. *Hydrostatic Test Pressure Requirements*

Carbon dioxide extinguishers Carbon dioxide and nitrogen cylinders (used with wheeled extinguishers)	$5/3$ service pressure stamped on cylinder
Carbon dioxide extinguishers with cylinder specification ICC3	3000 psi (20 685 kPa)
All stored-pressure and bromo-chlorodifluoromethane (1211)	Factory test pressure not to exceed 3 times the service pressure
Carbon dioxide hose assemblies	1,250 psi (8619 kPa)
Dry chemical and dry powder hose assemblies	300 psi (2068 kPa) or at service pressure, whichever is higher

Note: The factory test pressure is the pressure at which the shell was tested at time of manufacture. This pressure is shown on the nameplate.
The service pressure is the normal operating pressure as indicated on the gage and nameplate.

It is not necessary to hydrostatically test certain extinguishers, such as pump tanks, backpacks, and similar devices. Factory-sealed, nonrefillable, disposable fire extinguishers cannot be hydrostatically tested. When such extinguishers are damaged, they must be replaced.

Proper hydrostatic testing requires competent personnel and suitable testing equipment and facilities. The preparation for testing and the removal of all traces of water and moisture with special drying equipment after testing are very important and reinforce the recommendation that only those experienced in this type of work should undertake such servicing. The DOT requires that agencies testing carbon dioxide extinguishers hydrostatically be certified by the DOT.

TABLE 6-23E. *Hydrostatic Test Interval for Extinguishers*

Extinguisher Type	Test Interval (Years)
Stored-pressure water loaded stream and/or anti-freeze	5
Wetting agent	5
AFF (aqueous film-forming foam)	5
FFP (film-forming fluoroprotein foam)	5
Dry chemical with stainless-steel shells	5
Carbon dioxide	5
Dry chemical, stored-pressure, with mild steel shells, brazed brass shells, or aluminum shells	12
Dry chemical, cartridge or cylinder operated, with mild steel shells	12
Halogenated agents	12
Dry powder, stored pressure, cartridge- or cylinder-operated, with mild steel shells	12

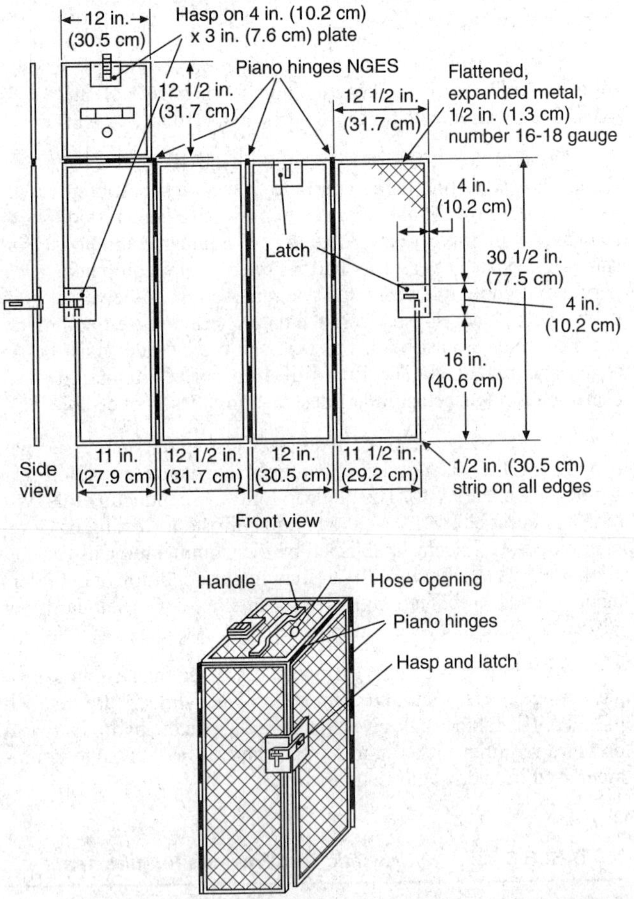

FIG. 6-23S. *A low-pressure portable hydrostatic test cage used for hydrostatic tests of low-pressure cylinders. Cage is not used for hydrostatic testing of high-pressure cylinders. Cage should not be anchored to floor during testing. Such cages can be made by any metal fabricator. (For SI units: 1 in. = 25.4 mm.)*

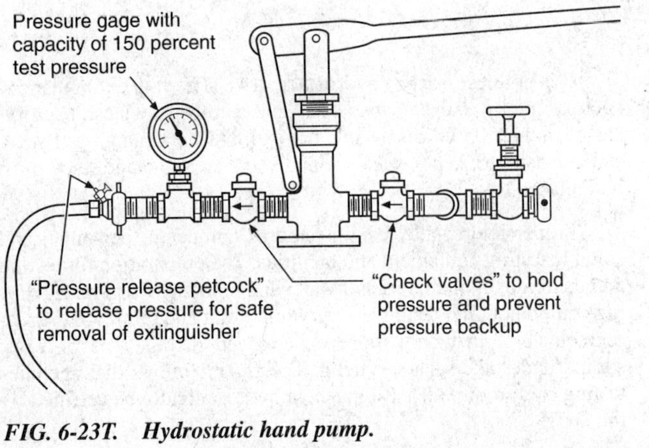

FIG. 6-23T. *Hydrostatic hand pump.*

Extinguisher hoses of certain types also need to be hydrostatically tested, and the details of these tests can be found in NFPA 10.

Because hydrostatic test results are of major importance, they must be recorded on the extinguisher. For high-pressure cylinders and cartridges passing a hydrostatic test, the month and year is stamped into the cylinder, in accordance with CGA Pamphlet C-1, or the Canadian Transport Commission regulations. It is important that the recording, or stamping, be placed only on the shoulder, top head, neck, or footing (when so provided) of the cylinder. For low-pressure cylinder extinguishers, the test information should be recorded on metal or some equally durable material. The label should be affixed by a heatless process to the shell and should be self-destructive if removal is attempted. (See Figure 6-23U.) The label must include the following information:

1. The month and year the test was performed, indicated by a perforation, such as by a hand punch.
2. The test pressure used.
3. The name or initials of the person performing the test, and the name and the agency performing the test.

Above all, remember the following: Allow only competent personnel with proper equipment and facilities to do the testing. Do not use air or gas for pressure testing because a violent rupture may occur in cases of failure. Place any extinguisher undergoing a test in a protective cage before applying the test pressures. Remove all traces of moisture from dry chemical, carbon dioxide, halogenated agent, and dry powder extinguishers before refilling after each test; a heated air stream not exceeding 150°F (65°C) is recommended for such drying. Destroy any extinguisher shell that fails a hydrostatic test—do not attempt to repair it.

Destroy any cylinder or shell if it has one or more of the following conditions. Do not hydrostatically test it:

1. Repairs by soldering, welding, brazing, or use of patching compounds. For welding or brazing on mild steel shells, consult the manufacturer of the extinguisher.
2. Damaged cylinder or shell threads.
3. Pitting caused by corrosion, even under the removable nameplate band assemblies.
4. Burns from a fire.
5. Calcium chloride agent in a stainless steel extinguisher.

EXTINGUISHER MAINTENANCE SERVICES

Fire extinguisher maintenance, particularly hydrostatic testing, is a specialized activity and should be performed by competent, dependable people. Fire extinguishers are provided to protect life and property; this means that there should be no doubt as to their reliability or safe use in an emergency. Extinguisher owners are thus urged to

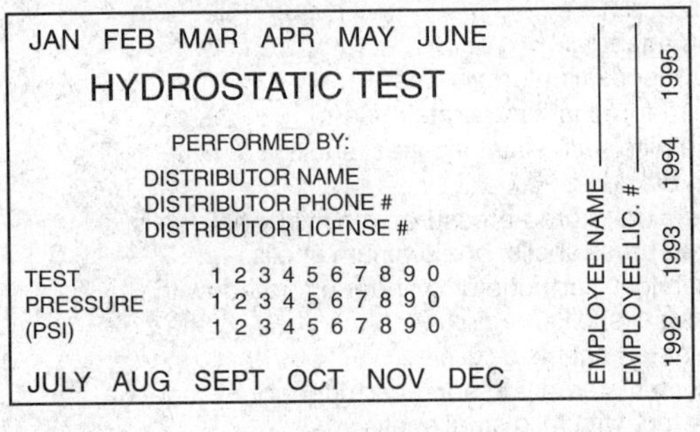

FIG. 6-23U. *A typical hydrostatic test label.*

seek out the services of reliable fire extinguisher maintenance firms. Such firms should be able to show proof that they are competent, that their facilities are adequate, and that they are licensed or registered in accordance with any local or federal law.

BIBLIOGRAPHY

NFPA Codes, Standards, and Recommended Practices

Reference to the following NFPA codes, standards, and recommended practices will provide further information on fire extinguisher use and maintenance discussed in this chapter. (See the latest version of *The NFPA Catalog* for availability of current editions of the following documents.)

NFPA 10, *Standard for Portable Fire Extinguishers*

Additional Reading/Video

Andsten, T., "CEN TC 70: Kasisammuttimien nykyiset laatuvaatimukset. [CEN TC 70: Current Quality Standards for Portable Extinguishers.]," *Palontorjuntatekniikka*, Vol. 2, 1994, pp. 24–25.

CGA C-6 "Standards for Visual Inspection of Steel Compressed Gas Cylinders," 7th ed., 1993, Compresssed Gas Association Inc., Arlington, VA.

Clark, A., "Class A Foam Extinguishers Tire Fire," *Fire Engineering*, Vol. 147, No. 3, March 1994, pp. 73–74.

Dalton, J., "Quality Evaluation: Navy Fleet-Returned Fire Extinguisher Cartridges, P/N 894846," Final Report, Naval Surface Warfare Center, Indian Head, MD, Naval Air Systems Command, Washington, DC, IHTR 1532, Oct. 1, 1992. [AUTHORIZED TO U.S. GOVERNMENT AGENCIES ONLY]

Dalton, J., "Surveillance: Army Service-Returned Fire Extinguisher Cartridge, P/N 897899 (DODIC MT20)," Final Report, Naval Surface Warfare Center, Indian Head, MD, Army Armament, Munitions, and Chemical Command, Rock Island, IL, NSWC IHTR 1533, May 22, 1992. [AUTHORIZED TO U.S. GOVERNMENT AGENCIES ONLY]

Dickerson, L. and Blake, D., "Effectiveness of Hand-Held Fire Extinguishers on Cargo Container Fire," Technical Note, Federal Aviation Administration, Atlantic City International Airport, NJ, Federal Aviation Administration, Atlantic City International Airport, NJ, DOT/FAA/CT-TN92/42, Feb. 1993.

"Extinguishers Should Never Malfunction," *Fire Protection*, Vol. 19, No. 3, September 1992, p. 12.

Fighting Fires with Portable Extinguishers, Video, National Fire Protection Association, Quincy, MA, 1986.

"Fire Extinguisher and Booster Hose. Standard for Safety," Underwriters Laboratories, Inc., Northbrook, IL, UL 92, 10th Edition, July 27, 1993, 17 pages.

Fire Extinguishers: Fight or Flight? Video, National Fire Protection Association, Quincy, MA, 1988.

"Fire Extinguishers Rating the Flame Fighters," *Consumer Reports*, Vol. 59, No. 5, May 1994, pp. 340–343.

"Getting a Handle on Fire With Portable Fire Extinguishers," Factory Mutual Research Corp., Norwood, MA, P8411, 1994.

"Halonisammuttimien tarkastus, huolto ja kayttotarkoituksen muuttaminen. [Inspection, Maintenance and Alteration of Halon Extinguishers.]," *Palontorjuntatekniikka*, Feb. 1993, p. 16.

Hock, B. H., "Set Up for Repairs," *Fire Command*, Vol. 56, No. 4, Apr. 1989, pp. 39–40.

"How Useful Are Portable Fire Extinguishers?" *Fire Prevention*, Vol. 227, Mar. 1990, pp. 16–17.

Leonard, J., "Training Simulant for Halon 1211 Portable Extinguishers," Naval Research Laboratory, Washington, DC, NRL/MR/6180-94-7615, Sept. 8, 1994.

Leonard, J., *et al.,* "Use of Copper Powder Extinguishers on Lithium Fires," Naval Research Laboratory, Washington, DC, Hughes Associates, Inc., Columbia, MD, Naval Sea Systems Command, Washington, DC, NRL/MR/6180-94-7490, Funding Numbers 61-M259-X3, Jul. 8, 1994.

Nash, P., "Portable Extinguishers: Standards, Design and Use," *Fire Surveyor*, Vol. 20, No. 3, June 1991, pp. 23–26.

"PASS: Fighting Fire With Portable Fire Extinguishers," Factory Mutual Research Corp., Norwood, MA, VHS Tape, P9409, 1994.

"Portable Fire Extinguishers. Fire Protection Equipment," Fire Protection Association, London, England, PE 4, Sept. 1990.

Pretious, T., "Methyl Bromide Poisoning Attributed to Old Fire Extinguisher," *Fire Research News*, No. 14, Autumn 1992, pp. 13–14.

Riley, J. F., and Kraus, M. R., "Selecting and Using Wheeled Fire Extinguishers: Training, Inspection, and Maintenance," *Plant Engineering*, Vol. 40, No. 22, 1986, pp. 58–60.

Salka, J., Jr., "2 1/2-Gallon Extinguisher," *Fire Engineering*, Vol. 147, No. 6, June 1994, pp. 22, 24.

Toledo, R. L., "Quality Evaluation: Navy Fleet-Returned Fire Extinguisher Cartridges, NAVAIR 850AS137 (DODICS MF73, MF74, and MF75)," Final Report, Naval Surface Warfare Center, Indian Head, MD, Naval Air Systems Command, Washington, DC, IHTR 1564, Sept. 28, 1992. [AUTHORIZED TO U.S. GOVERNMENT AGENCIES ONLY]

"Water Fire Extinguisher," *Fire Prevention*, No. 283, October 1995, p. 34.

THE ROLE OF EXTINGUISHERS IN FIRE PROTECTION

Marshall E. Petersen

Virtually all fires are small at first and might be extinguished easily if the proper type and amount of extinguishing agent were applied promptly. Portable fire extinguishers are designed for this purpose, but their successful use depends on the following conditions:

1. The extinguisher must be located properly and be in good working order.
2. The extinguisher must be the proper type for the fire that occurs.
3. The fire must be discovered while still small enough for the extinguisher to be effective.
4. The fire must be discovered by a person ready, willing, and able to use the extinguisher.

Fire extinguishers give the first opportunity to control or extinguish unfriendly fires and should be installed regardless of other fire control measures. However, an alarm should be given and the fire department notified as soon as a fire is discovered. Notification should never be delayed in the hope that the use of an extinguisher will be sufficient to control the fire.

This chapter extensively discusses the role of portable fire extinguishers. It begins by giving historical background on the development of extinguishers and then discusses their reliability and design safety. Finally, current extinguisher identification and fire extinguishment in the home are detailed.

For further information on allied topics, see Section 6, Chapter 23, "Fire Extinguisher Use and Maintenance."

HISTORICAL BACKGROUND

The first real portable fire extinguishers were developed in the late 1800s. They contained glass bottles of acid, which was released into a soda solution to produce a mixture with sufficient gas pressure to expel the solution. Cartridge-operated water extinguishers of the inverting type were introduced in the late 1920s. In 1928, a nonfreeze alkali metal-salt solution called "loaded stream" was developed for use in cartridge-operated extinguishers. Stored-pressure water extinguishers were developed in 1959, and, over the next ten years, they gradually replaced cartridge-operated models. In 1969, the manufacture of all inverting extinguishers was discontinued in the United States, and these extinguishers are no longer listed or approved by testing laboratories.

Marshall E. Petersen, P.E., is president of Marshall Petersen & Co., Mount Prospect, IL.

The first foam extinguisher was developed in 1917, and it looked and worked much like the soda-acid extinguisher. The use of foam extinguishers steadily increased over the years until, during the 1950s, dry chemical extinguishers gained widespread acceptance.

In 1976, aqueous film-forming foam (AFFF) extinguishers were introduced. These proved to be a direct replacement for the inverting foam type. The latest addition to water-type extinguishers came in 1988 with the introduction of film-forming fluoroprotein foam (FFFP).

NFPA 10, *Standard for Portable Fire Extinguishers* (hereinafter referred to as NFPA 10), no longer recognizes inverting types of extinguishers—that is, soda-acid, foam, and cartridge-operated models. Extinguishers of this type should be permanently removed from service and replaced with currently available models of the proper type and size.

Vaporizing Liquids

Carbon tetrachloride (CCl_4) was one of the first chemicals used in portable fire extinguishers in 1908. However, it was subsequently found that its vapors were toxic, and, when used on a fire, this substance could produce highly toxic hydrogen chloride and phosgene. Slightly less toxic chlorobromomethane (CH_2ClBr) was produced after World War II, when the term "vaporizing liquid" was first used to designate these extinguishers. However, some federal agencies banned vaporizing-liquid extinguishers in the 1950s because they were poisonous. By the mid-1960s, many states, cities, and industrial firms also had banned them. In the late 1960s, listings of vaporizing liquids by testing laboratories were discontinued.

Halogenated Agents

Although vaporizing liquids proved unacceptable, less toxic halogenated hydrocarbon chemicals were used in the form of compressed or liquefied gases. Bromotrifluoromethane, or Halon 1301, was first introduced in 1954 in a high-vapor-pressure compressed-gas extinguisher for use on fires in flammable liquids and live electrical equipment. A medium-vapor-pressure extinguisher using bromochlorodifluoromethane, or Halon 1211, became available in 1973. In 1974, extensive testing was begun on extinguishers containing low-vapor-pressure dibromotetrafluoromethane, or Halon 2402, which is a liquid at room temperature. Its use in portable extinguishers has proved to be impractical due to high costs and potential toxicity problems.

Halon is a chlorofluorocarbon (CFC). CFC emissions have been identified as a major contributor to the decay of the ozone layer in Earth's atmosphere. The Montreal Protocol, signed in 1987, is designed to limit ozone-depleting substances and has set production limits on CFCs, including Halons 1211 and 1301.

Under the Clean Air Act, the United States has banned the production of Halon 1211 and Halon 1301 beginning January 1, 1994, in compliance with the Montreal Protocol. It is legal, however, in the United States to purchase recycled halons for the purpose of recharging fire extinguishers. Currently, there are no federal laws prohibiting halon emissions, but the United States Environmental Protection Agency (EPA), NFPA, the National Association of Fire Equipment Distributors (NAFED), and the Fire Equipment Manufacturers Association (FEMA) all strongly discourage unnecessary emissions of halons for training purposes or discharges during testing, and also recommend precaution measures against accidental discharges.

It is evident that the future of halon as an extinguishing agent is in question, and alternate methods for fire extinguishment should be considered.

Carbon Dioxide

The first carbon dioxide extinguishers were produced during World War I, and, during World War II, they were the leading extinguisher for flammable liquids fires. By 1950, however, dry chemical agents had frequently replaced them as the preferred agent for flammable liquids fires. A further decline in the use of carbon dioxide extinguishers occurred as the popularity of halogenated agent extinguishers increased. With the phase-out of halon extinguishers, CO_2 use may increase.

Dry Chemicals

Although the extinguishing ability of sodium bicarbonate was recognized as early as the late 1800s, it was not until 1928 that an effective cartridge-operated dry chemical extinguisher was developed. Research and development produced an improved, finely granulated agent in 1943 and a further improved model in 1947.

As the use of flammable liquids increased, so did the development of effective dry chemical agents. In 1959, a potassium-bicarbonate-based agent called "Purple K," about twice as effective as the ordinary sodium-bicarbonate-based agent, was introduced.

In 1961, manufacturers introduced a new kind of agent called "multipurpose (ABC) dry chemical." It had the added advantage of being approximately 50 percent more effective than ordinary dry chemical on flammable liquid and electrical fires, and also was capable of extinguishing fires in ordinary combustibles. Originally, diammonium phosphate was used because it was less expensive, but monoammonium phosphate soon became preferred because it is considerably less hygroscopic.

In 1968, an agent with a potassium chloride base, known as "Super K," was introduced. It was approximately 80 percent more effective than ordinary dry chemical, but it was more corrosive and hygroscopic than potassium bicarbonate. A urea-potassium-bicarbonate-based agent (potassium carbonate), called "Monnex™," was developed in Europe in 1967 and brought to the United States in 1970. It has been judged at least $2^{1}/_{2}$ times as effective as ordinary dry chemical.

Dry Powder

The increased use of combustible metals, such as magnesium, sodium, and lithium, established the need for special agents to extinguish fires involving them. The designation "dry powder" was specifically chosen to indicate an agent's suitability for use on Class D, or combustible metal fires, while the term "dry chemical" was reserved for agents effective on Class A:B:C or B:C fires. In 1950, dry powder extinguishers using sodium chloride as a base agent were first marketed.

Wet Chemical

Recently, a wet chemical extinguisher, which contains a special potassium-based agent that was developed for use with restaurant fire suppression systems, was tested and listed. These wet chemical extinguishers provide an increased fire-fighting capacity and cooling effect to combat fires in modern restaurants that utilize more efficient and hotter cooking appliances. A greater fire challenge may also be encountered with the use of unsaturated cooking oils that burn at a higher temperature.

Wheeled Extinguishers

Over the years, most of the agents contained in hand-portable extinguishers have also been available in wheeled units. They have been typically utilized in high-hazard applications to increase the fire suppression capability of the operator. An increase in utilization of wheeled extinguishers may be realized as technology in manufacturing and warehouse operations increases and manpower requirements diminish. The most common models currently available are carbon dioxide, in sizes of 50 to 100 lb (22.7 to 45.4 kg), and dry chemical, in sizes of 50 to 350 lb (22.7 to 158.7 kg). Other wheeled extinguishers with dry powder or AFFF are also available. Specific information on these extinguishers is contained in NFPA 10.

RELIABILITY AND DESIGN SAFETY OF FIRE EXTINGUISHERS

Portable fire extinguishers may remain idle for many years, but they must always be capable of functioning at any time with maximum efficiency and without hazard to the user. Although most extinguishers are pressure vessels, they may rupture if not properly designed, constructed, and maintained. The initial responsibility for safe extinguishers belongs to the manufacturer, who is subject to the design standards, testing, inspection, labeling procedures, and follow-up program of responsible fire testing laboratories. But the owner is responsible for maintaining an extinguisher once it is in service, and may retain the services of a qualified service company.

NFPA 10

NFPA 10 contains requirements for the selection, installation, inspection, maintenance, and testing of extinguishers. It has been widely adopted by property owners, sales/servicing agencies, and enforcing officials.

The first edition of this standard was published in 1921 from data assembled by the NFPA Committee on Field Practices, the predecessor of the present committee. This standard is reviewed continuously, and revised editions are published every few years. Starting with the 1978 edition, more detailed information, with specific examples, was added to give guidance on selection and distribution.

Extinguisher Testing, Labeling, and Inspection

In North America, manufacturers have traditionally submitted their extinguishers to Underwriters Laboratories Inc. (UL), Underwriters Laboratories of Canada (ULC), and Factory Mutual Research Corporation (FMRC) for evaluation. Both UL and ULC have their own standards for construction, testing, and rating of extinguishers.

To assist purchasers of extinguishers in the United States and Canada, UL, ULC, and FMRC authorize manufacturers to affix a label to extinguishers that have been constructed and evaluated according to their standards. (See Figure 6-24A.) UL and ULC also require periodic examinations and tests of samples of extinguishers taken from a manufacturer's current production and/or stock. NFPA standards call for the use of listed or approved extinguishers, and most responsible authorities and property owners rely on this requirement.

Typically, extinguishers manufactured in the United States have both been listed by UL and approved by FMRC. The combined effect of the standards of these two testing agencies resulted in a quality and reliability consistent with the objectives of NFPA 10.

In recent years, other testing organizations have begun to evaluate portable fire extinguishers. In an effort to establish uniformity and maintain the high levels of quality and reliability that have been achieved, the NFPA Extinguisher Committee incorporated specific standards that must be used by any organization evaluating extinguisher products. In addition to a specific standard for each type of fire extinguisher, the American National Standards Institute (ANSI) also publishes a standard that requires a "third-party certification program." The following U.S. and Canadian standards are included in the requirements of NFPA 10.

Fire Test Standards*

ANSI/UL 711, *Standard for Safety Rating and Fire Testing of Fire Extinguishers;* CAN 4-S508-M83.

Performance Standards*

Carbon dioxide (CO_2) types: ANSI/UL 154, *Standard for Safety Carbon-Dioxide Fire Extinguishers;* CAN 4-S503-M83.

FIG. 6-24A. *Close-up of an extinguisher label, showing approvals from both Underwriters Laboratories Inc. (UL) (top) and Factory Mutual Research Corp. (FMRC) (bottom right).*

---

*ANSI (American National Standards Institute), UL (Underwriters Laboratories Inc.), ULC (Underwriters Laboratories of Canada), and CAN (Standards Council of Canada).

Dry chemical types: ANSI/UL 299, *Standard for Safety Dry Chemical Fire Extinguishers;* ULC-S504.

Water types: ANSI/UL 626, *Standard for Safety 2$^1/_2$-Gallon Stored-Pressure Water-Type Fire Extinguishers;* CAN 4-S507-M83.

Halogenated agent types: ANSI/UL 1093, *Standard for Safety Halogenated Agent Fire Extinguishers;* ULC-S512.

Foam types: *Standard for Safety Foam Fire Extinguishers;* ANSI/UL 8.

Factory follow-up on third-party certified portable fire extinguishers: ANSI/UL 1803, *Standard for Safety Factory Follow-up on Third-Party Certified Portable Fire Extinguishers.*

According to NFPA 10, extinguishers manufactured after January 1, 1986, must have the identification of the listing and labeling organization, the fire test, and the performance standard that the extinguisher meets or exceeds clearly marked on each extinguisher.

Substandard Extinguishers

Extinguishers are designed for use in emergencies, and it is vital that they operate effectively. Only extinguishers that are now tested in accordance with ANSI standards should be purchased, because otherwise it is difficult to know whether an extinguisher is reliable and effective. An inadequate extinguisher gives its owner a false sense of security.

Some common types of untested extinguishers are aerosol cans, "glass bulb grenades," and dry chemical "shaker units." Their instructions may claim that they can be used on all types of fires, both indoors and outdoors. However, they are unlikely to list the exact composition and amount of extinguishing agent. Even if unlisted extinguishers do contain recognized extinguishing agents, one cannot assume that they are as safe and effective as those tested in accordance with ANSI standards.

RELATION OF EXTINGUISHERS TO CLASSES OF FIRES

Because various extinguishing agents are not uniformly effective on different fires, NFPA 10 classifies fires into the following four types:

Class A: Fires in ordinary combustible materials, such as wood, cloth, paper, rubber, and many plastics, that require the heat-absorbing, cooling effects of water or water solutions, the coating effects of certain dry chemicals that retard combustion, or the interruption of the combustion chain reaction by dry chemical or halogenated agents.

Class B: Fires in flammable or combustible liquids, flammable gases, greases, and similar materials that must be put out by excluding air (oxygen), by inhibiting the release of combustible vapors with AFFF or FFFP agents, or by interrupting the combustion chain reaction.

Class C: Fires in live electrical equipment. The operator's safety requires the use of electrically nonconductive extinguishing agents, such as dry chemical or halon. When electrical equipment is deenergized, extinguishers for Class A or B fires may be used.

Class D: Fires in certain combustible metals, such as magnesium, titanium, zirconium, sodium, and potassium, that require a heat-absorbing extinguishing medium that does not react with the burning metals.

Some portable extinguishers will put out only one class of fire, and some are suitable for two or three, but none is suitable for all four. Most extinguishers are labeled so that users may quickly identify the class of fire for which they may be used. This classification is contained in NFPA 10, and it gives the applicable picture symbol or symbols. Color coding also is used. (See Figures 6-24B and 6-24C.) The NFPA classification is required on the extinguisher label in accordance with ANSI standards.

A rating number also is used on the labels of extinguishers for Class A and Class B fires. The rating number gives the relative extinguishing effectiveness of the extinguisher. For example, an extinguisher rated 4-A, 20-B:C indicates that it should extinguish approximately twice as much Class A fire as a 2-A-rated extinguisher, that it should extinguish considerably more Class B fire than a 1-B-rated extinguisher, and that it is suitable for use on energized electrical equipment.

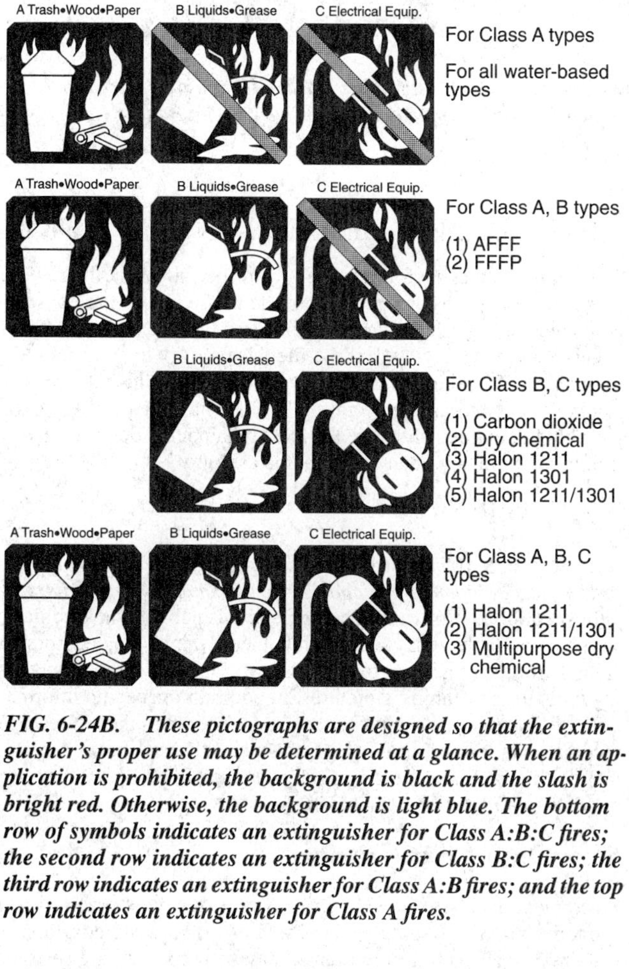

FIG. 6-24B. *These pictographs are designed so that the extinguisher's proper use may be determined at a glance. When an application is prohibited, the background is black and the slash is bright red. Otherwise, the background is light blue. The bottom row of symbols indicates an extinguisher for Class A:B:C fires; the second row indicates an extinguisher for Class B:C fires; the third row indicates an extinguisher for Class A:B fires; and the top row indicates an extinguisher for Class A fires.*

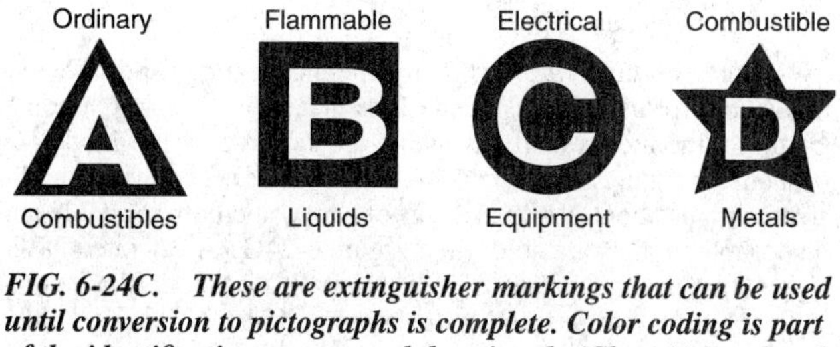

FIG. 6-24C. *These are extinguisher markings that can be used until conversion to pictographs is complete. Color coding is part of the identification system, and the triangle (Class A) is colored green, the square (Class B) red, the circle (Class C) blue, and the five-pointed star (Class D) yellow.*

Extinguishers rated for Class B fires have number ratings, based on the relative quantity of burning flammable liquid [2 in. (51 mm) of n-heptane] in a flat square pan that can be extinguished during a laboratory test. Again, an extinguisher rated 20-B can extinguish much more fire than one rated 5-B.

No rating numbers are used for extinguishers labeled for Class C fires. Since electrical equipment has either ordinary combustibles or flammable liquids, or both, as part of its construction, an extinguisher for Class C fires should be chosen according to the nature of the combustibles in the immediate area.

Extinguishers for Class D fires contain different dry powders that are effective on fires in different kinds of combustible metals. An extinguisher for magnesium fires may not work on a sodium fire, or at least not with the same effectiveness. For that reason, general numerical ratings are not used. Instead, each extinguisher for Class D fires has a nameplate detailing the type of metal in which the particular agent will extinguish a fire.

Extinguishers that are effective on more than one class of fire have multiple "letter" and "number-letter" classifications and ratings. Fractional ratings are not recognized in NFPA 10 and are not included in the requirements of the ANSI standards.

The most recently recommended marking system combines pictographs of both uses and non-uses, which could be hazardous to the operator, on a single label. (See Figure 6-24B.) Letter-shaped symbol markings, as shown in Figure 6-24C, are recommended for use until full conversion to the newer pictographs is completed.

Sometimes extinguishers produced by different manufacturers or even by the same manufacturer have the same quantity of the same extinguishing agent, but different ratings. In such cases, the difference in performance is usually due to differences of design, including rates of discharge, nozzle design, discharge patterns, and so on.

IDENTIFICATION OF EXTINGUISHERS

Clearly marked extinguishers and extinguisher locations are of the utmost importance. In an emergency, it is essential that extinguishers be located quickly and used while the fire is still small.

The rating class and number of an extinguisher should be visible so that the proper unit can be selected for the particular fire at hand. Manufacturers are required to provide permanently attached markings that describe not only the type and rating but also the way the unit is operated. ANSI standards specify general requirements for the form of the operating instructions that must be placed on the front of the extinguisher. (See Figures 6-24D and 6-24E.) Maintenance and other information can only be included on a rear label.

A stenciled inventory number on extinguishers and mountings also may aid in inventory and maintenance record-keeping. Computerized record keeping and the use of bar coding are also available for inventory, inspections, maintenance, and other services performed.

If the pictograph marking system is used, the operating instructions must be visible from the front of the extinguisher as it hangs. If class and rating markings are applied to wall panels in the vicinity of extinguishers, they should be easily legible at a distance of 15 ft (4.5 m).

The locations of wall- or column-mounted extinguishers can easily be marked by painting a red rectangle or band about 8 to 10 ft (2.4 to 3.0 m) above them. (See Figure 6-24F.) It is also good practice to paint the background on which it is mounted red.

If separate extinguishers for different classes of fire are mounted at the same location, extra attention should be given to distinguishing between them. Where extinguishers may easily be

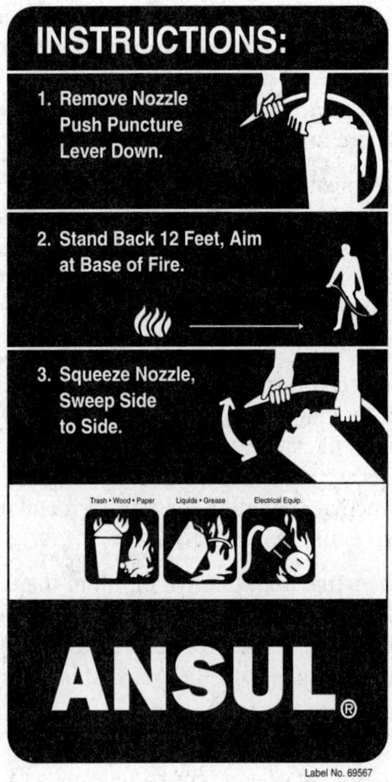

FIG. 6-24D. Example of the type of instruction/identification label required by ANSI standards for cartridge-operated dry chemical extinguishers. (Source: ANSUL)

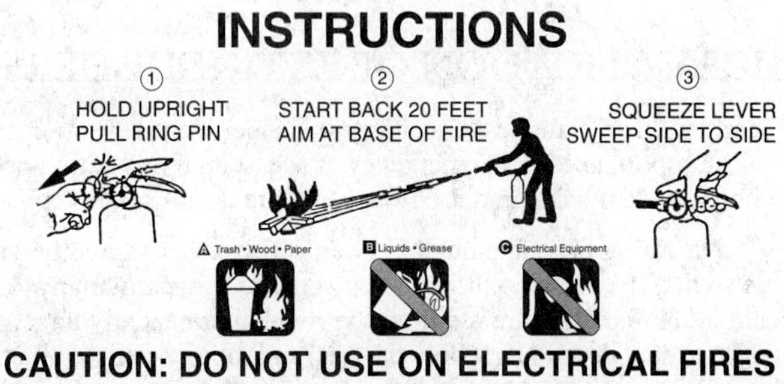

FIG. 6-24E. Example of the type of instruction/identification label required by ANSI standards for stored-pressure water extinguishers. (Source: AMEREX)

blocked by storage or equipment, barriers or floor markings can be employed. If the problem persists, it may be more practical to find another location.

Extinguishers installed in recessed cabinets or wall recesses generally are more difficult to find unless clearly marked. Where possible, the cabinet frame or wall recess should be painted red. When such installations are in long halls or corridors, however, they may still be very difficult to find unless a sign is mounted perpendicular to the cabinet wall at a height of about 8 ft (2.4 m).

In many public and commercial buildings, extinguisher locations are camouflaged for aesthetic reasons or installed in hidden areas, such as closets or stairwells. Unless one knows where to look or happens to see a small sign on a door, these extinguishers are of little value in an emergency.

FIG. 6-24F. An extinguisher marker. It should be placed so that it is clearly visible from a distance. (Source: The Badger Company)

FIRE EXTINGUISHMENT IN THE HOME

The cardinal rules to follow when a fire occurs in the home are: make sure that everyone gets out of the house before the exits become blocked by heat or smoke; be certain that the fire department is called immediately; and, if the fire is small, try to control or extinguish it. To achieve the last, every home should have readily available, easy-to-use fire-fighting equipment. Listed or approved portable fire extinguishers are the most reliable first defense against fire. In addition, a garden hose may be useful.

NFPA 10R

In 1992, NFPA 10R, *Recommended Practice for Portable Fire Extinguishing Equipment in Family Dwellings and Living Units*, was published for the first time. The scope of this publication specifically covers portable extinguishers and other extinguishing equipment suitable for the protection of one- and two-family dwellings and living units in multifamily structures. In addition to specific information on inspection, maintenance, and servicing, there are also recommendations on applications for special hazards related to residential living. Significant material contained in NFPA 10 that also pertains to NFPA 10R is included.

Selection of Extinguishers

Since most dwelling fires are Class A fires, the most practical and economical extinguishing agent is water. Although stored-pressure water extinguishers or pump tanks may be used, they are selected infrequently because they are large, unattractive, heavy, expensive, and limited to Class A fires.

Small, efficient, ABC-type dry chemical extinguishers are ideally suited for use on most small fires that may occur in the home. Of these, the most readily available are stored-pressure models in capacities ranging from 1 to 5 lb (0.5 to 2.3 kg). The most popular size contains about $2^1/_2$ lb (1.1 kg) of agent, and both refillable and disposable models are sold. (See Figure 6-24G.) A 2A-rated ABC-type extinguisher can have the advantage of a longer discharge time (12 sec) and a hose assembly for a more efficient discharge and a wider, softer pattern. Currently, these small extinguishers can be purchased with ordinary, or bicarbonate-base, agent or multipurpose agent. It is important to note that the multipurpose agent can

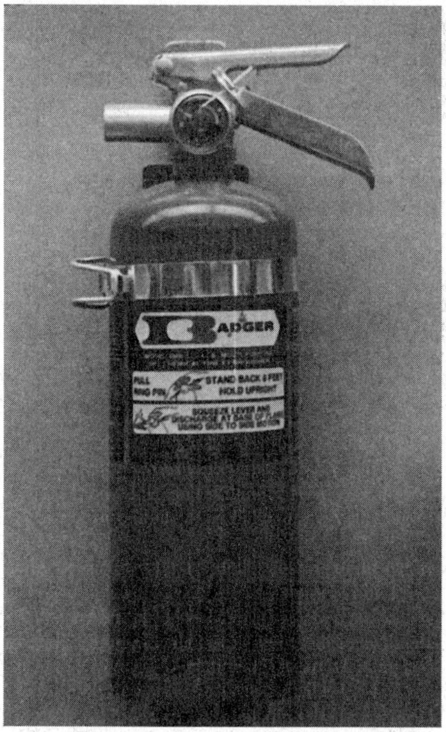

FIG. 6-24G.   A small, dry chemical fire extinguisher suitable for home protection. (Source: The Badger Company)

cause corrosion and may seriously damage stoves, heaters, appliances, and vehicles unless all the agent deposits are removed *promptly* and *thoroughly* after extinguishment. The 5-lb (2.3-kg) and larger sizes of ordinary and multipurpose dry chemical extinguishers also can be obtained in cartridge-operated models.

In general, dry chemical extinguishers will provide adequate protection for small Class B fires involving grease or cooking oils in the kitchen, paints and solvents in the basement or utility area, and oils or gasoline in the garage or outdoors. However, it is important to stand back at least 6 ft (2 m) when discharging the extinguisher, or the grease, oil, or gasoline may splash and cause the fire to spread. When multipurpose extinguishers are used on fires of ordinary combustibles, the small sizes will be of limited value. The larger sizes, which are given higher ratings by the testing laboratories, are more suitable. Multipurpose dry chemicals are surface-coating agents. Even though extinguishers of this type may rapidly extinguish the flames in combustible materials, it is important that the deep-seated burning embers, especially those in furniture cushions, pillows, or bedding, be thoroughly wetted with water.

It is also essential that, once the extinguisher has been used, it be refilled or replaced as soon as possible. Even if only a short burst of agent is released, the extinguisher will probably lose the rest of its pressure very shortly.

Possibly the best approach to home fire extinguishment is to provide fire extinguishers, such as ordinary or dry chemical, for Class B:C protection and the means for rapid flame "knock-down" of Class A fires. A water extinguisher or permanently connected hose should then be used for final extinguishment of the Class A fire. (See the subsection on garden hose in the home later in this chapter.)

Homeowners who have extensive home workshops, gasoline-powered equipment, etc., should consider installing additional, con-veniently located fire extinguishers to control any fire that might occur in such areas or equipment. A logical choice would be a dry chemical extinguisher with a capacity of 5 lb (2.3 kg) or greater.

For private automobiles, a multipurpose dry chemical extinguisher is effective on engine fires and any electrical fires that also include Class A insulating materials. For small trucks, NFPA committees have recommended extinguishers in the 5- to 20-B:C range, depending on the type of vehicle. Due to the corrosive characteristics of multipurpose dry chemical, it is important to *promptly* and *thoroughly* clean all agent deposits off metal surfaces.

Location of Extinguishers

Most dwelling fires start in the kitchen, living room, or bedroom. Thus, at least one extinguisher should be located where it can be reached quickly from both the kitchen and the living room. It is best to locate each extinguisher near the path of exit travel so that, if the fire cannot be controlled readily with the device, a quick escape can be made. In bedrooms, the extinguisher should be located in a handy closet or cabinet. In basements, the head of the basement stairs is preferred, except when a basement workshop may dictate otherwise.

It is important that the location of each extinguisher be known to each member of the family who is able to use the device. A permanently mounted bracket on which to hang the extinguisher will help to ensure that it remains in place and is not tucked away in a closet or some other place that is difficult to reach.

Use of Extinguishers

The mere presence of an extinguisher in the home is not worthwhile unless the homeowner is willing to learn how to use the device properly, instruct family members who may have to use it, and maintain and recharge it according to the manufacturer's instructions. The manufacturer of listed extinguishers is required to provide an instruction manual with each unit. This manual contains instructions and cautions for the installation, operation, inspection, and maintenance of the extinguisher. For safety and to obtain the maximum benefit of having a fire extinguisher, the manual should be studied carefully, kept in a safe place, and reviewed periodically.

It is important that homeowners understand that extinguishers of the sizes discussed here have a discharge time of only 8 to 15 sec. In actual use, no time can be wasted determining the best way to use the device. Operating instructions are given on each listed or approved extinguisher, and most of them are easily understood. When possible, families should practice using an extinguisher in a safe outdoor location. Instructions on using fire extinguishers also may be obtained from local fire department personnel.

Maintenance of Extinguishers

Maintenance requirements vary for each extinguisher, and the information is given on each device. Most small, modern extinguishers have a pressure gage that should be checked on a regular basis. To be sure that the device can be readily removed from its bracket and is in good physical condition, the extinguisher also should be lifted from its bracket occasionally and handled. It is important that the safety locking pin is in place and/or the tamper seal is not broken. If this problem is encountered, it is advisable to have the extinguisher serviced. The family should periodically review how each extinguisher should be operated and used.

Range Hood Systems

Systems designed for protection of residential kitchen range-top cooking appliances have been tested and listed by UL. (See Figure 6-24H.) The system will actuate either automatically or manually

Flames up   Gas releases   Puts out flame

FIG. 6-24H.   Residential range hood systems are designed to provide protection for fires that occur from cooking activities on the burners of gas or electric ranges. (Source: Pem Systems, Inc.)

and is designed only to protect cooking activities on top of the burners of gas or electric ranges. Actuation will shut off gas or electric power. In addition to the regular inspection, maintenance, and servicing of the agent cylinder, it is essential that the grease deposits in the hood, ducts, filters, and fusible links be cleaned at regular intervals.

BIBLIOGRAPHY

NFPA Codes, Standards, and Recommended Practices

Reference to the following NFPA codes, standards, and recommended practices will provide further information on the role of extinguishers in fire protection discussed in this chapter. (See the latest version of *The NFPA Catalog* for availability of current editions of the following documents.)

NFPA 10, *Standard for Portable Fire Extinguishers*
NFPA 10R, *Recommended Practice for Portable Fire Extinguishing Equipment in Family Dwellings and Living Units*

ANSI/UL Safety Standards [American National Standards Institute (ANSI), New York, NY; Underwriters Laboratories Inc. (UL), Northbrook, IL].

ANSI/UL 8, *Standard for Safety Foam Fire Extinguishers*
ANSI/UL 154, *Standard for Safety Carbon-Dioxide Fire Extinguishers*
ANSI/UL 299, *Standard for Safety Dry Chemical Fire Extinguishers*

ANSI/UL 626, *Standard for Safety 2 1/2-Gallon Stored-Pressure Water-Type Fire Extinguishers*
ANSI/UL 711, *Standard for Safety Rating and Fire Testing of Fire Extinguishers*
ANSI/UL 1093, *Standard for Safety Halogenated Agent Fire Extinguishers*
ANSI/UL 1803, *Standard for Safety Factory Follow-Up on Third-Party Certified Portable Fire Extinguishers*

Additional Reading

"A Computerized Study of Fire Extinguisher Effectiveness," National Association of Fire Equipment Distributors, Chicago, IL, 1979 (revised 1995), p. 11.
"Getting a Handle on Fire with Portable Fire Extinguishers," Factory Mutual Research Corp., Norwood, MA, P8411, 1994.
"How Useful Are Portable Fire Extinguishers," *Fire Prevention*, No. 227, March 1990, pp. 16–17.
James, D., "Fire Extinguishers," *Fire Prevention Handbook*, Butterworths, Boston, 1986, pp. 36–65.
"PASS: Fighting Fire with Portable Fire Extinguishers," Factory Mutual Research Corp., Norwood, MA, VHS Tape, P9409, 1994.
Selection Guide for Portable Fire Extinguishers, National Association of Fire Equipment Distributors, Chicago, IL, 1995.

SPECIAL SYSTEMS AND EXTINGUISHING TECHNIQUES

Revised by Peter F. Johnson

A range of fire extinguishing and control systems, agents, devices, and techniques developed and used with varying degrees of success do not fit into those commonly recognized categories of systems described in previous chapters of this section. Many of the systems and techniques are novel or still in the experimental stage. Often, they are directed toward special applications, such as aircraft or coal mines, for which traditional extinguishing methods are not entirely suited. This chapter describes some of these special systems and techniques by grouping them as follows:

1. Techniques based on solids and powders.
2. Systems using water or water solutions for fire control.
3. Inerting gases used for fire or explosion suppression.
4. Combined agent systems.
5. Special applications.

For a complete discussion of the more conventional extinguishing systems, refer to these other chapters in this section: Chapter 15, "Water Mist Suppression Systems"; Chapter 18, "Halogenated Agents and Systems"; Chapter 20, "Carbon Dioxide and Application Systems"; Chapter 21, "Dry Chemical Agents and Application Systems"; Chapter 22, "Foam Extinguishing Agents and Systems"; and Chapter 26, "Combustible Metal Agents and Application Techniques." For information on removal of emulsifying agents, see Section 3, Chapter 32, "Hazardous Waste Control."

SOLIDS AND POWDERS

Solid materials in powder or granular form have been used for many years to extinguish fires. Use of solids to cover and suppress fires by excluding oxygen is a common technique; rows of sand-filled buckets for fire fighting were once a familiar sight. Solids in the form of dry powder systems are still a widely used extinguishing technique. Special proprietary powders and a range of common materials, such as talc, graphite, and soda ash, are capable of suppressing various metal fires, as well.

A number of recent developments have taken place in the use of powders and other solid materials to suppress special fire hazards.

Granules for Flammable Liquid Pool Fires

Problems associated with fighting large fires of flammable liquids, such as major spills from storage tanks, have led to considerable re-search in this area. One of the more dangerous fire situations occurs when a major spill of liquefied natural gas (LNG) flows into a dike surrounding a storage tank. Conventional fire-fighting methods for large LNG fires have serious limitations, which include the following:

1. Vast quantities of dry chemical are required to guarantee total LNG fire extinguishment.
2. Because of their high water content, low-expansion foams are unsuitable for LNG fires.
3. High- and medium-expansion foams can control, but not extinguish, LNG fires, and large quantities of foam are required continually to replace foam destroyed by fire.

A novel concept, therefore, has been to use a "solid foam" consisting of $3/4$- to $1^1/_2$-in. (20- to 40-mm) cubes of cellular glass in a layer about 8 in. (200 mm) deep to cover the full extent of an LNG spill where it is contained in a storage dike.[1] The granules are of nonflammable, low-density solid material and are stored within the dike in bags. Loose-filled material (with a suitable binder on the top of the liquid) is also used. When an LNG spill occurs, the granules float to the top of the liquid and control the rate of burning. Both small-tray and large-scale tests have shown the glass granules to be most effective. The material shields the LNG surface from back radiation from the flames and reduces the evaporation of the fuel from the liquid surface. This limits the fuel burning rate and the overall size of the fire. In addition, radiation at a distance from the spill fire, which is the major contributor to fire spread, is reduced by approximately 90 to 95 percent, compared with LNG fires without the cellular glass granules.

Some practical difficulties with this approach to suppression of LNG fires have emerged, however. Lambert[2] reports that obtaining the cellular glass granules in the large volumes required is both difficult and costly. In addition, research in the U.K. has shown that, if the granules are stacked in diked areas for long periods, they deteriorate due to weathering and other means.

Carbon Microspheres for Metal Fires

Specially formulated extinguishing powders are generally used to suppress fires involving metals. Because of the reaction many metals have with water, sprinklers and the use of other water-based agents are not appropriate and, in some cases, quite dangerous.

However, many of the special agents for metal fires are at times unsatisfactory because they are corrosive, applied manually rather than by an extinguishing system, capable of clogging extinguishing nozzles, and expensive.

Peter F. Johnson is manager of Arup Fire Engineering, a fire protection consulting firm, in Melbourne, Australia.

Studies have been undertaken to examine the effectiveness of carbon microspheres or microspheroids to extinguish fires involving alkali metals, such as sodium, sodium-potassium, and lithium.[3] These microspheres are petroleum-coke-based particles with a diameter of approximately 100 to 500 microns. The particles possess high thermal conductivity, chemical inertness, and excellent flow characteristics and are capable of being directed onto fires from dry-chemical-type extinguishers and conventional nozzles.

Tests have shown that carbon microspheres compare favorably in performance to other metal extinguishing agents.[3] In particular, experiments with carbon microspheroids incorporating neutron absorbers have been effective in extinguishing fires involving nuclear fissionable materials, such as uranium metal powder.[4] The excellent flow characteristics and noncaking properties of these microspheres suggest an effective way to extinguish radioactive metal fires within the inert atmosphere glovebox enclosures used in the nuclear industry.

Ground Cover Materials—Aircraft Fuel Fires

Aircraft fuel spill fires are another significant hazard in which solid materials have been used for extinguishment. Research in the United States suggests that crushed rock graded to approximately $3/8$ in. (10 mm) can reduce fire growth and the spread of flames in hydrocarbon spill fires.[5] The suppression mechanism seems to be similar to that for cellular glass granules—that is, the solid material largely blocks back-radiation of the flames to the liquid surface, and prevents heat from being distributed to the fuel in front of the flame by convection currents. Ground rock materials do not float and must cover the fuel surface by 1 to 2 in. (25 to 50 mm) to retard flame growth effectively.

Dry Chemical Projectile

Since the 1940s, a number of interesting suppression techniques have been developed using projectiles of dry chemical that can be fired at a target fire from a distance.

One such system developed in the United States for coal mine fires used a cannon-like dispersal system. Upon an alarm from an ultraviolet detector, mono-ammonium phosphate powder was propelled from the cannon by high-pressure nitrogen to quench potential coal explosions and methane-air ignitions.[6] In reality, the devices proved somewhat impractical because of space limitations.

Dr. John L. Bryan of the University of Maryland reported on a related extinguishing technique developed some years ago that used a device known as a "gren gun." This gun consisted of cylinders that rotated in turn into the firing position. The projectiles contained $1^3/4$ lb (0.8 kg) of dry chemical that, when fired with compressed air, could reach distances of 200 ft (60 m).[7] The gren gun was designed to operate with a minimum air pressure of 150 psi (1030 kPa) from a compressed-air cylinder carried on the back of the operator. The gun itself was constructed of lightweight aluminum, weighed approximately 10 lb ($4^1/2$ kg), and was fired from the shoulder. Being portable, it was ideal as a first-attack suppression device that could place dry chemical on a target from a safe, remote distance.

Vermiculite for Flammable Liquids

Foam, dry powder, and halon systems have traditionally been used to extinguish or control flammable liquid fires. However, recent research in India[8] has examined the effectiveness of exfoliated vermiculite in gasoline and kerosene fires.

Vermiculite is a hydrated laminar mineral that is soft and flaky, and consists mainly of aluminum, iron, and magnesium silicates in proportions shown in Table 6-25A.

TABLE 6-25A. *Chemical Composition of Vermiculite*

SiO_2	39–45%
MgO	15–20%
Al_2O_3	14–20%
Fe_2O_3—FeO	6–11%
Other compounds	4–26%

When vermiculite is heated under controlled conditions between 1292 and 1832°F (700 and 1000°C), its flakes exfoliate or expand from 7 to 15 times their original volume, resulting in a highly insulating material of very low density.

Tests have shown that a layer of 63 in. (160 mm) of exfoliated vermiculite can be used to blanket or smother flammable liquid fires of gasoline and kerosene up to 54 sq ft (5 m²) in size. While comparative studies with foam have shown that 2.2 to 6.6 lb (1 to 3 kg) of foam is required per sq ft (1 sq ft = 0.09 m²) of gasoline fire extinguished, compared with 28.6 lb (13 kg) of exfoliated vermiculite, the latter material appears to have an indefinite shelf life and to be unaffected by fuel properties or environmental degradation. Further research obviously is needed to determine the practical aspects of supply, placement in the event of a fire, and performance on flammable liquid fires of larger area.

SYSTEMS UTILIZING WATER OR WATER SOLUTIONS

A number of specialized systems using water or water-based agents have been developed to fill particular fire control needs. These extinguishing agents remove heat from the combustion process or form a noncombustible emulsion with the fuel.

Ultra-High-Speed Water Spray Systems

These systems are designed to handle extremely rapid fires, such as those which can occur during the handling of solid propellants or sensitive chemicals, or in any industrial process or oxygen-enriched environment.

Wetting Agent Systems

Most wetting agent systems—as opposed to wetting agents used through manually operated hose lines from mobile tank supplies through portable proportioners—are designed to dispense wetting agent foam. The design, volume, and purpose of each installation depends upon the hazard being protected. Normally, wetting agent systems that meet the requirements of NFPA 18, *Standard on Wetting Agents*, can use standard water spray, sprinkler, or foam system equipment. However, this should be carefully checked to ensure maximum system efficiency.

The detergent action of a wetting agent solution may introduce some special problems. Wet-water foam is a mixture of wet water with air that forms a cellular structure foam. This foam breaks down into its original liquid state at temperatures below the boiling point of water. Breakdown is proportional to the heat exposure and is intended to cool sufficiently the surface on which the foam is applied. These properties improve the ability of water to protect exposed properties against heat transfer, and water requirements are thus appreciably reduced. Wet-water systems can introduce the potential of increased water damage to exposed stocks due to the high absorption ability of wet water, and can increase the load on any flooring affected because of the retention of large volumes of wet water. These factors must be considered in the design of such installations.

Viscous Water Systems

Systems for applying viscous, or "thickened," water vary in detailed components, depending upon whether the equipment is designed for application from the ground or from aircraft. Most ground applications are from conventional fire tankers, with some modifications. The NFPA report, "Chemicals for Forest Fire Fighting," lists the following three basic systems for ground applications.[9]

Injector-recirculating ground tanker system: In this system, water is drawn from the tank bottom through a centrifugal pump and bypassed through an added return line to the injector-disperser, where the liquid and dry powder combine and flow into the tank. The centrifugal pump serves as a mixing device, continuing until the correct amount of powder is added.

Demand viscous water tanker-mixer: This system consists of an auxiliary mixing tank, an auger feed mechanism, and a positive displacement rotary meter that acts as a power unit for the feed system. This system uses the tank, centrifugal pump, and discharge lines of a conventional tanker.

Slip-on chemical tanker-mixer: This system is a skid-mounted, all-metal unit for mounting onto a heavy-duty tractor trailer for off-road assignments. The major assemblies consist of a mixing tank, chemical hopper, hose reel, engine, pump, and pump priming tank. This system has only been used experimentally, employing a modified ordinary agricultural sprayer.

Conventional fire nozzles are used for viscous water or light gels. Those nozzles producing coarse droplets normally give satisfactory results, but fine spray nozzles should be avoided. Slurry-type fire-retardant chemicals need special nozzles because these chemicals quickly erode most conventional brass and aluminum nozzles. Cast aluminum inserts (a ceramic aluminum oxide) seem to be satisfactory, however, and can be manufactured inexpensively.

"Chemicals for Forest Fire Fighting"[9] gives additional information on equipment for mixing the chemicals, on pumps for handling the chemicals, on distribution systems, on storage of mixed and unmixed chemicals, and on means for judging the quality of the solutions in the field.

The NFPA report "Air Operations for Forest, Brush, and Grass Fires"[10] contains information on the use of aircraft for fire control operations. This manual covers air operations plans, airports, heliports, helistops, suitable aircraft, operating principles and procedures, and amphibious operations. Special equipment is needed to use aircraft for these fire control purposes.

Emulsifying Agents—Detergents

Various types of emulsifying agents or detergents have been used with varying degrees of success to extinguish or control flammable liquid spill fires on land and on water. These agents also have been used on unignited spills of such liquids to minimize the possible ignition hazard. In general, these agents mix the oil and water to form a noncombustible emulsion.

A major problem with the use of emulsifying agents and detergents in lakes, rivers, harbors, and so on is that the resultant emulsions often have a more severe effect on aquatic forms of life than does oil alone. However, relatively nontoxic emulsifying agents may be developed in the future.

One interesting example of the use of local-application emulsifying agent in cooking range hood protection is a system that uses an aqueous potassium carbonate solution with minor amounts of sodium dichromate added to inhibit corrosion, and ethylene glycol to prevent freezing. It is dispersed in the form of a dense liquid spray from conventional nozzles located in the range hood. (See Figure 6-25A.) The alkaline solution not only removes heat by steam generation, but undergoes a saponification reaction with burning fat or oil to produce a soap foam or emulsified layer that, in turn, produces inerting carbon dioxide. Similar aqueous potassium carbonate extinguishing systems have also been used for range hoods on naval vessels.[11]

A further extension of the use of emulsifying agents in kitchen range hoods has been to link this type of suppression system to a special thermal line detector that can be distributed around the inside of the range hood. When a fixed temperature of approximately 347°F (175°C) is exceeded, a pressure is generated within a $1/4$-in. (6-mm) diameter stainless steel line detector that automatically initiates the rapid discharge of the aqueous emulsifying agent.

INERTING GAS SUPPRESSION

For many years, a range of gases such as carbon dioxide, nitrogen, and helium has been used successfully to prevent ignition of potentially flammable mixtures or to extinguish fires, particularly those involving flammable liquids or gases. These methods are based on the knowledge that a fire can be extinguished if a sufficient volume of an inert gas is introduced into an enclosed space where the fire is burning. The concentration must be maintained in the space for an adequate period to keep the burned material from rekindling.

Traditionally, this method of extinguishment has been explained as a reduction of the fuel or oxygen concentration to the point where combustion is prevented. Modern fire dynamics show the added inerting gas acts as "thermal ballast" to reduce the temperature of the flame/vapor mixture below the limiting adiabatic flame temperature necessary to sustain combustion.[12] The gaseous extinguishing agent actually absorbs the combustion energy, and its thermal capacity is the important factor in inerting explosive or burning materials.

The thermal capacities of the more commonly used inerting agents are shown in Table 6-25B.

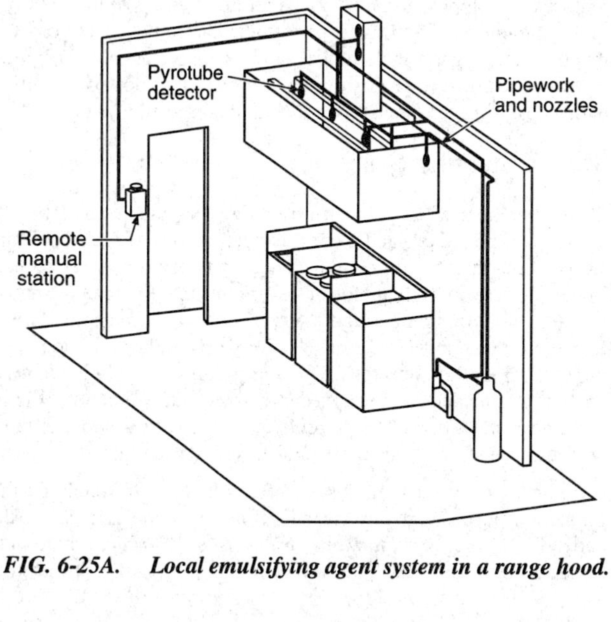

FIG. 6-25A. Local emulsifying agent system in a range hood.

TABLE 6-25B. *Thermal Capacities of the More Commonly Used Inerting Agents*

Agent	Symbol	Thermal Capacity at 340°F (727°C)	
		Btu/mol °F	J/mol-k
Carbon dioxide	CO_2	12.9	54.3
Steam	$H_2O(g)$	9.8	41.2
Nitrogen	N_2	7.8	32.7
Helium	He	5.0	20.8

Table 6-25B reveals why carbon dioxide is an important gaseous extinguishing agent and why helium, with its lower heat-absorbing capacity, is not widely used. Halon extinguishing agents have not been included in the table. While they have even higher thermal capacity than carbon dioxide, their extinguishment mechanism is more related to chemical interaction in the combustion process.

Nitrogen

Nitrogen has been used in a number of inerting applications, mostly those related to aircraft fire suppression systems.

Nitrogen has been considered for aircraft engine nacelle fires, but halons are generally far superior for this application.[13] The use of nitrogen has been directed more toward fuel tank inerting where cryogenic nitrogen in Dewar vessels is directed into the fuel tank ullage. Nitrogen up to an additional 80 percent of the gross volume to be protected is added so that a maximum oxygen concentration of 9 percent by volume is not exceeded in the atmosphere above the fuel.[14]

Nitrogen fire-fighting systems have also been considered for cargo compartment underfloor areas, wheel wells, and wing areas in both military and civilian aircraft.[13] This technique is being used on U.S. Air Force C-5A transport aircraft, and consideration is being given to using nitrogen in parked aircraft to reduce oxygen concentrations to below 10 percent by volume, thus eliminating cabin fires in unattended aircraft.

One of the problems with an inerting gas is to ensure that it is well mixed and that all parts of the volume being protected have the minimum concentration required for fire suppression. This is particularly true in enclosures where the amount of extinguishing agent required for suppression must be balanced by the potential damage to equipment and injuries to personnel that could arise from too much inerting gas. Scale-model studies of nitrogen distribution within pressurizable enclosures have proved that poor distribution patterns of nitrogen can be identified and corrected.[15]

Steam Inerting Systems

Steam may be used to smother fires in the same manner as other inert gases. It effectively reduces the concentrations of fuel vapor and oxygen and absorbs sufficient heat to stop the combustion process. The use of steam systems for fire extinguishment precedes the use of other modern smothering systems, such as carbon dioxide and foam extinguishing systems, and is rarely used today. It is clearly an impractical method, except where a large steam supply is continuously available and can be tapped effectively and efficiently when a fire emergency arises. The possible burn hazard to personnel must be considered in any steam extinguishing installation.

Steam extinguishing systems are not recommended for fire protection purposes in any current NFPA standards. However, the appendix to NFPA 86, *Standard for Ovens and Furnaces* (hereinafter referred to as NFPA 86), does offer suggestions for fire protection of steam systems which may be followed "where steam flooding is the only alternative" after automatic sprinklers or water spray systems and approved types of supplementary fire protection, such as carbon dioxide, foam, or dry chemical systems, have already been considered. To protect ovens, this standard calls for steam outlets to supply at least 8 lb (3.6 kg) of steam per minute for each 100 cu ft (2.8 m³) of oven volume. One lb (0.45 kg) of saturated steam at 212°F (100°C) and normal atmospheric pressure has a volume of 26.75 cu ft (0.76 m³). Steam outlets should be located near the bottom of the oven, but they may be located at the top, pointing downward if the oven is not more than 20 ft (6.0 m) high. Pipe sizes and the steam supply should be sufficient to deliver the required amount of steam and to maintain a minimum pressure of 15 psi (100 kPa) at the outlet. NFPA 86 recommends manual release devices for such systems and recommends that controls be arranged to close down oven outlets as far as practicable. Additional guidance includes arrangements to reduce hazards to personnel.

Steam smothering systems used to be employed to protect cargo spaces and the holds of steamships. This method is no longer recommended. Tests indicating the relative inefficiency of such systems in controlling cotton cargo fires were conducted by the U.S. Coast Guard during "Operation Phobos."[16]

One more interesting example of the use of steam inerting is the system used to prevent explosions in certain coal pulverizers in the United States.[17] In essence, the inerting system introduces steam into the pulverizing system to reduce the oxygen concentration below the flammability limit for ignition of a coal-dust atmosphere. Steam was chosen for this application rather than carbon dioxide or nitrogen because it was readily available, needed no storage system, was abundant from a reliable source, did not require tight enclosure of the inerted space, had no toxic effects, and was inexpensive. Onsite tests and plant experience have shown the system to be quite effective in preventing coal dust explosions.

Use of Combustion Gases for Fire Extinguishment

Experiments have been conducted on use of the gases of combustion to achieve extinguishment. Inert gas was generated by an experimental turbojet engine specially designed for the test.[13] The engine burned most of the oxygen in the air it used, and the hot combustion gases were cooled by vaporizing water introduced as a fine spray. The most efficient inert gas produced by the experimental appliance contained 46 percent nitrogen, 44 percent water vapor, 3 percent carbon dioxide, and 7 percent oxygen. The highest rate of delivery used was 45,000 cu ft/min (21 m³/s) when the appliance consumed 84 gpm (320 L/min) of water and 7.8 gpm (29.5 L/min) of kerosene. The emergent gas was cooled to 194 to 248°F (90 to 120°C). In this experimental work, either the inert gas alone was used or a high-expansion foam was made with the inert gas. By spraying a suitable detergent solution on a screen and blowing the inert gas through the wetted screen at a velocity not exceeding 5 ft/sec (1.5 m/s), an inert gas high-expansion foam was generated with an expansion ratio of 1000 to 1. Between 50 and 80 percent of the gas was made into foam. A flexible ducting made of terylene, with a neoprene coating to make it impermeable, was used to convey the gas or foam to the fire site.

In similar investigations,[7,18] a smaller inert gas generator was constructed that produced approximately 3.300 cu ft/min (1.6 m³/s) of the mixture, consisting of 68 percent water vapor, 28 percent nitrogen, and 4 percent carbon dioxide at a temperature of 194°F (90°C). Depending upon the fuel involved, it was reported that the inert gas concentration must reduce the oxygen level to between 9 and 12 percent for effective extinguishment. In addition, a quantity of inert gas equal to twice the building volume involved must be produced within 15 min for successful extinguishment.

Air Agitation for Oil Tank Fire Control

Fuel tanks often are protected by topside foam protection systems or by subsurface foam injection into the fuel itself. Fire in an oil storage tank may, under some conditions, be controlled or extinguished by introducing air under pressure near the bottom of the tank. The principle is founded upon the fact that flammable and combustible liquids require a temperature greater than their flash point temperature to sustain burning. If a tank of oil with a flash point above the actual temperature of the liquid ignites when the surface temperature is raised above the flash point, or the oil has become contaminated with a low-flash-point liquid, the resulting fire may be controlled or extinguished by agitating the oil mass. By agitating the mass, cool oil circulated at a proper rate decreases the temperature of burning oil at the surface by displacement and mixing to a point below its flash point, thus slowing or inhibiting further combustion.

The time element during which such control operations must be pursued actively is critical. If the oil mass heats too quickly, the cooling effect might be inadequate; and if the burning liquid is crude oil, the time before boiling occurs can be shortened considerably. This technique is not considered a standard method of combating oil tank fires. Under most circumstances where fire protection is provided for such storage tanks, foam fire-fighting equipment meeting the recommendations contained in NFPA 11, *Standard for Low-Expansion Foam*, should be employed.

COMBINED AGENT SYSTEMS

It is not unusual in fire-fighting practice to use two or more agents simultaneously or in rapid sequence, taking advantage of the best qualities of any two agents in achieving fast fire control and permanent extinguishment. Some common combinations used in manual fire control include water, usually water spray, and foam; carbon dioxide and foam; and certain dry chemicals and foam. Aircraft rescue and fire-fighting operations have long employed such combined agent methods, using water spray, carbon dioxide, or dry chemical to secure rapid flame knockdown, plus simultaneous or sequential use of foam to permanently smother the fire. During World War II, the combined use of bulk applications of carbon dioxide from mobile low-pressure carbon dioxide trucks, with simultaneous applications of foam from specialized crash trucks, led to many significant personnel rescues in aircraft fires and achieved quick fire extinguishment or control, as well. This technique, and the combined use of foam-compatible dry chemicals and foam, is still being used. While these techniques do involve the simultaneous and/or sequential use of two agents, each agent system is independent of the other, most often located on separate mobile vehicles.

Recent developments in dual-agent systems combine light water with potassium salt-based dry chemical, or mechanical (air) foam with potassium salt-based dry chemical, providing the foam is compatible with dry chemical. These are systems in the sense that they are designed to provide a balanced discharge of the agents, although the agents themselves are not premixed. The two agents are discharged through independent but twinned nozzles, both of which are independent but under the control of one fire fighter. Some experiments have entrained certain dry chemicals in the water-foam concentrate turret stream with encouraging results.

It has been shown that certain salts, such as diammonium hydrogen phosphate and calcium hydrogen carbonate, can be used effectively to reduce the quantity of water required for extinguishment by up to 73 percent. It was found that the time to extinguishment was also significantly reduced on wood crib fires and in full-scale furniture fire tests. Apart from the combined agent systems using foam, other combined agent systems using combinations of halons and water and halons with dry chemical (potassium bicarbonate) have also been tried and found to be effective.[6] This work has been extended more recently, with foams being generated using halons rather than air. A halon gas is added to the mixture of water and foam agents and distributed by traditional fire-fighting jets. Several large-scale tests have been performed using aqueous film-forming foam (AFFF) agent, together with Halon 2420 and Halon 2330, and rapid extinguishment with a relatively small amount of agent has been achieved.

SPECIAL APPLICATIONS

A series of examples from selected industries serves to illustrate how special techniques, combinations of agents, or unusual control mechanisms may be used to overcome specialized fire problems.

Bulk Sulphur Storage

Sulphur is one of industry's most widely used chemicals. It comes in the form of rock, known as brimstone, or as a powder. Solid, finely ground sulphur is both a fire and an explosion hazard. Sulphur is unusual in that it has a low melting point of approximately 239°F (115°C) and a low ignition temperature of 374 to 392°F (190 to 200°C). It burns with a pale blue flame to form mainly sulphur dioxide.

Thompson indicates that small fires in bulk sulphur can be extinguished by carefully shoveling more sulphur onto the fire, thus removing the oxygen supply.[19] For larger fires, water is the most effective extinguishing agent, but it must be applied gently in the form of a water fog to avoid raising sulphur dust and creating an explosion hazard.

Microelectronic Fabrication Facility

These fabrication facilities are exceptionally expensive to build and operate and require very extensive cleanroom provisions to maintain low particle counts to avoid product contamination. Even a small fire, then, can cause widespread damage to sensitive microchips and lead to significant business interruption losses.

Much of the manufacturing in these facilities occurs in so-called "wet benches" that handle flammable, corrosive, and toxic chemicals under strong ventilation conditions. Recent experimental work[20] suggests that a halon suppression system, initiated by fast-response infrared flame detectors and backed with fast-response automatic sprinklers, is the best combination to extinguish a fire quickly and prevent contamination of adjoining cleanroom facilities. Further work remains to be done to investigate the use of carbon dioxide or nitrogen.

Anechoic Chambers

The combination of halon and automatic sprinklers is also appropriate for anechoic chambers. These chambers, designed to be free of echoes or reverberations, use geometrically shaped foam in the form of wedges or pyramids to achieve electromagnetic radiation absorption on all internal surfaces. The foam, which is impregnated with carbon to improve its conductivity, represents a major fuel load and, if it is ignited, can lead to a large fire and great damage.

Research[21] has shown that a special telescoping sprinkler system, forming part of a deluge system to minimize accidental water damage, is the most appropriate water-based suppression system. (See Figure 6-25B.) This is supported by a fast-response halon system using cross-zoning of optical and ionization smoke detectors.

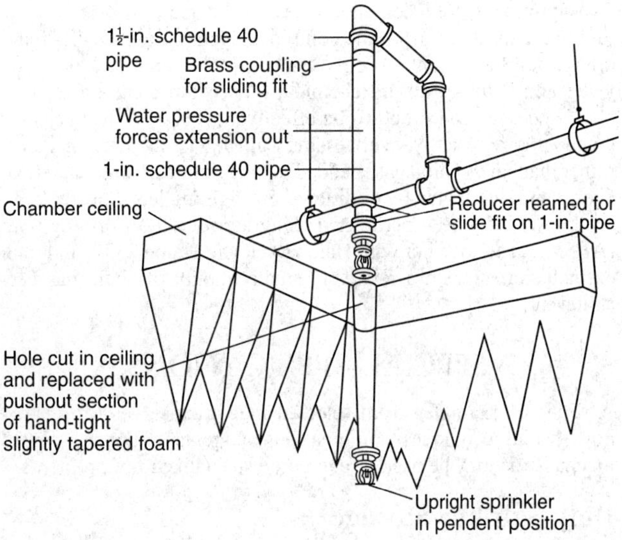

FIG. 6-25B. Telescoping drop-down sprinkler. (For SI units: 1 in. = 25.4 mm.)

Given ventilation rates of up to 700 air changes/hour, however, experimental work has shown that a VESDA™-type smoke detection system is perhaps a more appropriate sensing system for anechoic chambers. VESDA™, an acronym for "very early smoke detection apparatus," is the generic trade name for a type of smoke detector manufactured by IEI (Aust.) Pty. Ltd. This would support other work[22] that indicates that high-sensitivity-sampling smoke detection systems provide the best form of detection in high airflow situations.

Filter Plenums and Gloveboxes

In the nuclear industry, the handling of plutonium, uranium, and beryllium represents a hazard that requires special fire protection precautions. Two special risks are: (1) gloveboxes, in which nuclear materials are processed, and (2) filter plenums that house the HEPA filters used to prevent loss of nuclear material to the atmosphere.

One approach to this hazard[23] is to use special heat detectors to initiate an automatic deluge and water curtain system. Special spray nozzles are used and hot fire gases are cooled before reaching the combustible HEPA filters. The automatic deluge system is backed by a separate manual deluge system for increased reliability.

A unique feature of the heat detectors used to initiate water deluge is the special heaters surrounding the detector sensors to allow remote testing in gloveboxes. Entry to these gloveboxes, which contain highly radioactive material, is very difficult, so testing by remote activation of heating elements means that routine maintenance and assurance of operation can be checked regularly.

Submarines

In confined pressurizable spaces, such as submarines, U.S. Navy research has shown that nitrogen pressurization can be an effective fire suppressant, while still providing a habitable environment.[24]

This technique of nitrogen pressurization is based on the premise that life depends upon the partial pressure of oxygen, whereas fires depend upon percentage of oxygen in the atmosphere. If the compartment pressure is raised by adding nitrogen, a range of different fires can be extinguished.

The amount of nitrogen required for suppression depends upon a number of important parameters, including fuel type, compartment geometry, and pressurization rate. Tests have shown pressurization to 1.8 atm with nitrogen to be effective with a range of fuels. However, the weight of nitrogen to be stored is a penalty in submarines.

Plastics

Plastics are involved in many fires because they are in widespread use, and plastics fires can evolve a great deal of heat, smoke, and toxic gases. Cable fires, automobile fires, and fires involving stored plastics in warehouses can often represent severe fire challenges.

Research in Japan[25] has shown that "wet water" can be more effective than ordinary water for many fires involving plastics. "Wet water" can be made by diluting common foam agents by up to 10,000 times, depending on the foam type.

The research indicated that the greater effectiveness of "wet water" on plastics fires was due to its much greater adhesiveness to plastic surfaces. Nevertheless, fires in unretarded foam plastics proved difficult to extinguish, even with "wet water."

Future of Halon

Chlorofluorocarbons (CFCs), such as halon, have been identified as contributors to ozone depletion in Earth's atmosphere. In 1987, the Montréal Protocol was signed to limit ozone-depleting substances and to set production limits on CFCs, including Halons 1301 and 1211. As a result, the use of halons as fire extinguishing agents is being phased out. A whole range of new gaseous suppression agents and new techniques, such as water mist, is being developed for these special hazard situations.

One of the most interesting new halon replacement agents is the solid particulate aerosol fire suppressant.[26] These aerosols are usually produced by pyrotechnic action of solid compound formulations (not unlike solid rocket propellant).

When burned, the solid materials produce large numbers of aerosol particles in the 1 to 3 μ diameter range, plus some gaseous products. Like dry chemicals, these aerosols are thought to suppress fires by chemical inhibition and heat absorption, and other mechanisms.

Aerosols have been shown to suppress n-heptane fires and fires in other materials. The aerosols exhibit three-dimensional behavior and appear to offer an alternative to halons, particularly in military applications.

BIBLIOGRAPHY

References Cited

1. Lev, Y., "A Novel Method for Controlling LNG Pool Fires," *Fire Technology*, Vol. 17, No. 4, 1981, pp. 275–284.
2. Lambert, G. M. S., unpublished data, 1989.
3. McCormick, J. W., and Schmitt, C. R., "Extinguishment of Selected Metal Fires Using Carbon Microspheroids," *Fire Technology*, Vol. 10, No. 3, Aug. 1974, pp. 197–199.
4. Schmitt, C. R., "Carbon Microspheres as Extinguishing Agents for Fissionable Material Fires," *Fire Technology*, Vol. 11, No. 2, May 1975, pp. 95–98.
5. Geyer, G. B., "The Use of Ground Cover Materials to Suppress Fuel Fires at Airports," *Fire Technology*, Vol. 9, No. 4, Nov. 1973, pp. 254–262.
6. "Fire Safety Aspects of Polymeric Materials—Vol. 10, Mines and Bunkers," Publication NMAB 318-10, National Academy of Sciences, Washington, DC, 1980.

7. Bryan, J. L., *Fire Suppression and Detection Systems*, Glencoe Press, Encino, CA, 1982.

8. Sharma, T. P., Badami, G.N., Lal, B. B., and Singh, J., "A New Particulate Extinguishant for Flammable Liquid Fires," *Fire Safety Science, Proceedings of the Second International Symposium*, Hemisphere Publishing Corporation, New York, 1989, pp. 667–677.

9. NFPA Forest Committee, "Chemicals for Forest Fire Fighting," Report, 3rd ed., National Fire Protection Association, Quincy, MA, 1977.

10. NFPA Forest Committee, "Air Operations for Forest, Brush, and Grass Fires," Report, rev. ed., National Fire Protection Association, Quincy, MA, 1975.

11. Neill, R. R., and Peterson, H. B., "An Investigation of the Heating and Fire Extinguishment of Deep Frying Fat," N.R.L. Memorandum Report 2080, 1969, Naval Research Laboratory, Washington, DC.

12. Drysdale, D. D., *Fire Dynamics*, John Wiley & Sons, Chichester, UK, 1985.

13. Dyer, J. H., Marjoram, M. J., and Simmons, R. F., "The Extinction of Fires in Aircraft Jet Engines—Part III, Extinction of Fires at Low Airflows," *Fire Technology*, Vol. 13, No. 2, May 1977, pp. 126–130.

14. "Fire Safety Aspects of Polymeric Materials—Vol. 6, Aircraft: Civil and Military," Publication NMAB 318-6, Appendix C, D, National Academy of Sciences, Washington, DC, 1977.

15. Corlett, R. C., Stone, J. P., and Williams, F. W., "Scale Modeling of Inert Pressurant Distribution," *Fire Technology*, Vol. 16, No. 4, Nov. 1980, pp. 259–273.

16. *Proceedings* of the Fifty-First Annual Meeting of the National Fire Protection Association, May 26-29, 1947, National Fire Protection Association, Quincy, MA, pp. 119–123.

17. McGuire, J. H., "Large-Scale Use of Inert Gas to Extinguish Building Fires," *Research Paper 246*, National Research Council of Canada, Division of Building Research, Ottawa, Canada, 1964.

18. John, R., "New Additives to Extinguishing Water for Fire Fighting," *New Technology to Reduce Fire Losses and Costs*, Grayson, S. J., Smith, D. A., eds., Elsevier Applied Science Publishers, New York, 1986, pp. 280–287.

19. Thompson, C., "Fire Safety in Bulk Sulphur Handling Locations," *Fire Prevention*, No. 208, Apr. 1988, pp. 28–31.

20. Fisher, F. L., Williamson, R. B., Toms, G. L., and Crinnion, D. M., "Fire Protection of Flammable Work Stations in the Clean Room Environment of a Microelectronic Fabrication Facility," *Fire Technology*, Vol. 22, No. 2, May 1989, pp. 148–165.

21. Kirson, D., "Anechoic Chambers—A Unique Challenge to the Fire Protection Engineer," *Papers of IFPEI-V International Fire Protection Engineering Institute, Ottawa*, Vol. 2, Society of Fire Protection Engineers, Boston, 1989.

22. Johnson, P. F., "Very Early Smoke Detection for Computer and Telecommunications Industries," *Fire Safety Journal*, Vol. 14, Nos. 1 & 2, July 1988, pp. 13–24.

23. Campbell, D. G., "Fire Protection for Filter Plenums at a Nuclear Facility," *Papers of IFPEI-V International Fire Protection Engineering Institute, Ottawa*, Vol. 2, Society of Fire Protection Engineers, Boston, 1989.

24. Carhart, H. W., Sheinson, R. S., Tatem, P. A., and Lugar, J. R., "Fire Suppression Research in the U.S. Navy," First International Conference on Fire Suppression Research, Stockholm and Borås, Sweden, May 1992, pp. 337–357.

25. Takahashi, S., "Extinguishment of Plastic Fires with Plain Water and Wet Water," *Fire Safety Journal*, Vol. 22, 1994, pp. 167–179.

26. Kibert, C. J., and Dierdorf, D., "Solid Particulate Aerosol Fire Suppressants," *Fire Technology*, 4th Quarter, 1994, pp. 387–399.

NFPA, Codes, Standards, and Recommended Practices

Reference to the following NFPA codes, standards, and recommended practices will provide further information on special systems and extinguishing techniques discussed in this chapter. (See the latest version of *The NFPA Catalog* for availability of current editions of the following documents.)

NFPA 11, *Standard for Low-Expansion Foam*
NFPA 18, *Standard on Wetting Agents*
NFPA 86, *Standard for Ovens and Furnaces*

Additional Reading

Acton, M. R., Sutton, P. and Wickens, M. J., "Investigation of the Mitigation of Gas Cloud Explosions by Water Sprays," *Gas Engineering and Management*, Vol. 31, No. 7–8, July/Aug. 1991, pp. 166–172; Institution of Chemical Engineers, Piper Alpha: Lessons for Life-Cycle Safety Management Symposium, IChemE Symposium Series No. 122, Sept. 26–27, 1990, London, England, 1991.

Atreya, A., "Extinguishment of Combustible Porous Solids by Water Droplets," Annual Progress Report, Michigan State Univ., East Lansing, National Institute of Standards and Technology, Gaithersburg, MD, Sept. 1990.

Atreya, A., Crompton, T. and Suh, J., "Experimental and Theoretical Study of Mechanisms of Fire Suppression by Water," Univ. of Michigan, Ann Arbor, NISTIR 5499, Sept. 1994.; National Institute of Standards and Technology, Annual Conference on Fire Research: Book of Abstracts, Oct. 17–20, 1994, Gaithersburg, MD, 1994, pp. 67–68.

Banerjee, S. P., "Experience of Inert Gas Injection in Combating Fires in Indian Coal Mines," *Proceedings* of the 3rd Mine Ventilation Symposium, Society of Mining Engineers of AIME, 1987, pp. 438–443.

Bannister, W. W., Walker, J. L., and Morehouse, T., "Applications of Amine Gelling Agents in Fire Technology," *Plant/Operations Progress*, Vol. 8, No. 2, April 1989, pp. 80–81.

Bowman, J. W., and Perkins, R. P., "Preventing Fires with High-Temperature Vaporizers," *Plant/Operations Progress*, Vol. 9, No. 1, Jan. 1990, pp. 39–43.

Butz, J. R., and French, P., "Application of Fine Water Mists to Hydrogen Deflagrations," *Proceedings* of the 1993 Halon Alternatives Technical Working Conference, May 11–13, 1993, Albuquerque, NM, 1993, pp. 345–355.

Cohen, P., *et al.*, "Trial by Fire, Understanding the Design Requirements for Agents in Complex Environments," *AI Magazine*, Vol. 10, No. 3, Fall 1989, pp. 32–48.

Cortese, R. A., and Sapko, M. J., "Flame-Powered Trigger Device for Activating Explosion Suppression Barrier," Bureau of Mines, Pittsburg, PA, RI 9365, 1991.

Cummins, M., "Natural Water *vs.* Polarized for Fighting Fire," *American Fire Journal*, Vol. 47, No. 8, Aug. 1995, p. 27.

Ewing, C. T., Beyler, C. L., and Carhart, H. W., "Extinguishment of Class B Flames by Thermal Mechanisms; Principles Underlying a Comprehensive Theory; Prediction of Flame Extinguishing Effectiveness," *Journal of Fire Protection Engineering*, Vol. 6, No. 1, 1994, pp. 23–54.

Ewing, C. T., *et al.*, "Flame Extinguishment Properties of Dry Chemicals: Extinction Weights for Small Diffusion Pan Fires and Additional Evidence for Flame Extinguishment by Thermal Mechanisms," *Journal of Fire Protection Engineering*, Vol. 4, No. 2, 1992, pp. 35–52.

Garg, P. C., "Development of Nitrogen Infusion Technology for Fighting and Inhibition of Fires," *Journal of Mines, Metals and Fuels*, Vol. 35, No. 8, Aug. 1987, pp. 368-377, 394.

Jones, A., and Thomas, G. O., "Action of Water Sprays on Fires and Explosions: A Review of Experimental Work," *Process Safety Environment*, Vol. 71, No. 1, 1993, pp. 41–49.

Kim, A. K., Dlugogorski, B. Z., and Mawhinney, J. R., "Effect of Foam Additives on the Fire Suppression Efficiency of Water Mist," National Research Council of Canada, Ottawa, Ontario, NRCC 37902, IRC Paper 3748; *Proceedings* of the Halon Options Technical Working Conference, May 3-5, 1994, Albuquerque, NM, 1994, pp. 347-358.

Leonard, J., *et al.*, "Use of Copper Powder Extinguishers on Lithium Fires," Naval Research Laboratory, Washington, DC, Hughes Associates, Inc., Columbia, MD, Naval Sea Systems Command, Washington, DC, NRL/MR/6180-94-7490, Funding Numbers 61-M259-X3, July 8, 1994.

Lomai, T., and Isei, T., "Underground Fire-Fighting System by a Rapid Evaporation Method of Liquid Nitrogen," *Mining Science and Technology*, Vol. 8, No. 2, 1989, pp. 145–152.

Nyden, M. R., *et al.*, "Flame Inhibition Chemistry and the Search for Additional Fire Fighting Chemicals," National Institute of Standards and Technology, Gaithersburg, MD, NIST SP 861, April 1994.

Olsson, S., and Ryderman, A., "Extinguishment of Oil Spray Fires with Water: Experimental Procedures and Test Data," Swedish National Testing and Research Institute, Boras, Sweden, SP REPORT 1990:32, CIB W14/92/07 (S), 1990.

Ramaswamy, A., and Katiyar, P. S., "Experiences with Liquid Nitrogen in Combating Coal Fires Underground," *Journal of Mines, Metals and Fuels*, Vol. 36, No. 9, Sept. 1988, pp. 415–424.

Rhein, R. A., "Experimental Study of the Use of Liquid Argon and Argon-Filled Aqueous Foams for the Extinction of Lithium Fires," *Fire Technology*, Vol. 28, No. 4, Nov. 1992, pp. 290–316.

Sato, K., "Extinguishment of Liquefied Gas Fires by Carbon Dioxide and Liquid Nitrogen," *Bulletin of Japanese Association of Fire Science and Engineering*, Vol. 40, No. 1, 1990, pp. 29–35.

Simpson, J. M., and Gardener, N. J. L., "Graphite-Based Extinguishants for Liquid Metal Fires," *Proceedings* of the International Conference on Science and Technology of Fast Reactor Safety, Thomas, Telford, UK, 1986, pp. 603–606.

"Special Fire Risks Require Special Extinguishing Systems," *National Safety and Health News*, Vol. 134, No. 3, Sept. 1986, pp. 43-48.

Wrenn, C., "Inerting for Safety," *Plant/Operations Progress*, Vol. 5, No. 4, Oct. 1986, pp. 225–227.

Xiujuan, G., "Combined Agent (Dry Chemical-Foam) Fire Fighting Model Test," First Asian Conference on Fire Science and Technology (ACFST), October 9–13, 1992, Hefei, China, International Academic Publishers, China, 1992, pp. 360–365.

Young, R., "Evaluating Halon 1301 Alternatives," *Fire Prevention*, Vol. 267, March 1994, pp. 34–37.

Zallen, D. M., and Morehouse, T., "Fire Extinguishing Agents for Oxygen-Enriched Atmospheres," *Proceedings* of the Third International Symposium on Flammability and Sensitivity of Materials in Oxygen-Enriched Atmospheres, Vol. 3, American Society for Testing and Materials, W. Conshohocken, PA, 1988, pp. 391–412.

COMBUSTIBLE METAL EXTINGUISHING AGENTS AND APPLICATION TECHNIQUES

Revised by Robert E. Tapscott

There are a limited number of proprietary combustible metal extinguishing agents that have been approved by recognized testing or listing agencies. The majority of agents available, and most notably those developed for special metals of limited commercial use, do not carry approvals or listings. The following agents have either been listed for use or have shown effectiveness in extinguishing metal fires.

MET-L-X POWDER

MET-L-X powder, available in 50-lb (23-kg) pails, is suitable for use on fires involving magnesium, sodium, potassium, and sodium-potassium alloy (NaK). The material is listed by Underwriters Laboratories Inc. (UL) for application by hand scoop and shovel. An approximate guide for application of agent is:

Magnesium turnings/chips	5 lb agent/lb of fuel
Molten sodium or potassium	6 to 8 lb agent/lb of fuel greater than $1/2$ in. fuel depth
Molten NaK alloy	15 lb/lb of fuel

(For SI units: 1 lb = 0.4536 kg; 1 in. = 25.4 mm.)

This dry powder, with its particle size controlled for optimum extinguishing effectiveness, is composed of a sodium chloride base with additives. A thermoplastic polymer is added to bind the sodium chloride particles into a solid mass under fire conditions.

MET-L-X powder is noncombustible, and secondary fires do not result from its application to burning metal. No known health hazard results from the use of this agent. It is nonabrasive and nonconductive.

When stored in extinguishers or original containers, MET-L-X powder is not subject to decomposition or a change in physical properties. Periodic replacement of the extinguisher charge is unnecessary. Extinguishers range from 30-lb (14-kg), carbon-dioxide-cartridge-operated hand portables through 150- and 350-lb (68- and 160-kg) wheeled units to 2000-lb (910-kg) stationary or piped systems. The wheeled units and piped systems employ nitrogen as the expellant gas.

MET-L-X powder is suitable for fires in solid chunks, such as castings, because of its ability to cling to hot vertical surfaces. To control and extinguish a metal fire, the nozzle of the extinguisher is fully opened and a thin layer of agent is cautiously applied over the burning mass from a safe distance to prevent the burning metal from blowing into other areas. Once control is established, the nozzle valve is used to throttle the stream to produce a soft, heavy flow. The metal can then be covered completely and safely from close range with a heavy layer. The heat of the fire causes the powder to cake, forming a crust that excludes air and results in extinguishment.

MET-L-X extinguishers are available for fires involving magnesium, sodium (spills in depth), potassium, and sodium-potassium alloy (NaK). In addition, MET-L-X has been successfully used where zirconium, uranium, titanium, and powdered aluminum present serious hazards.

Na-X POWDER

Na-X has been developed as a low- or non-chloride-containing agent for use on sodium metal fires. Na-X has a sodium carbonate base with various additives incorporated to render the agent resistant to moisture and to increase its fluidity for use in pressurized extinguishers. The agent also contains a polymer that will soften and form a crust over an exposed surface of burning sodium metal.

Na-X is noncombustible and does not cause secondary fires when applied to burning sodium metal above temperatures ranging from 1200 to 1500°F (649 to 816°C). No known health hazard results from the use of this agent on sodium fires, and it is nonabrasive and nonconductive.

Stored in 50-lb (23-kg) pails, 30-lb (14-kg) hand portables, and 150- and 350-lb (68- and 160-kg) wheeled and stationary extinguishers, Na-X is UL-listed for fires involving sodium metal (spills in depth) at a fuel temperature of 1200°F (649°C). Na-X has been tested on sodium metal (spills in depth) at fuel temperatures as high as 1500°F (816°C). Stored in the supplier's metal pails and extinguishers, Na-X is not subject to decomposition, so periodic replacement of the agent is unnecessary.

OTHER COMBUSTIBLE METAL EXTINGUISHING AGENTS

G-1 Powder/MetalGuard™ Powder

"Pyrene" G-1 powder is composed of screened, graphitized foundry coke to which an organic phosphate has been added. A combination

Robert E. Tapscott, Ph.D., is director, Center for Global Environmental Technologies, Albuquerque, NM.

of particle sizes is used to provide good packing characteristics when applied to a metal fire. The graphite acts as a heat conductor and absorbs heat from the fire to lower the metal temperature below the ignition point, which results in extinguishment. The closely packed graphite also smothers the fire, and the organic material in the agent breaks down with heat to yield a slightly smoky gas that penetrates the spaces between the graphite particles, excluding air. The powder is nontoxic and noncombustible.

MetalGuard™ powder is identical to G-1 powder in composition and is simply a trade-name variation.

G-1 powder is effective for fires in magnesium, sodium, potassium, titanium, lithium, calcium, zirconium, hafnium, thorium, uranium, and plutonium. It has been recommended for special applications on powder fires in aluminum, zinc, and iron. The products of combustion of thorium, uranium, beryllium, and plutonium can be a health hazard, and precautions should be observed consistent with the usual procedures in combating fires in radioactive material.

G-1 powder is not commercially available at present.

Lith-X Powder

This dry powder is composed of a special graphite base with additives. The additives render it free-flowing so it can be discharged from an extinguisher. The technique used to extinguish a metal fire with this agent is the same as that used with MET-L-X. Lith-X does not cake or crust over when applied to the burning metal. It excludes air and conducts heat away from the burning mass to effect extinguishment. It does not cling to hot metal surfaces, so it is necessary to cover the burning metal completely.

Lith-X will successfully extinguish lithium fires and is suitable for the control and extinguishment of magnesium and zirconium chip fires. It will extinguish sodium spills and sodium fires in depth. It will also extinguish sodium-potassium alloy spill fires and control fires in depth.

Ternary Eutectic Chloride (TEC) Powder

Ternary eutectic chloride (TEC) powder is a mixture of potassium chloride, sodium chloride, and barium chloride and is effective in extinguishing fires in certain combustible metals. The powder tends to seal the metal, excluding air. On a hot magnesium chip fire, the method of extinguishment is air exclusion by the formation of a layer of molten salt on the metal surface. In tests reported in *Fire Technology*, TEC powder was the most effective salt for control of sodium, potassium, and sodium-potassium alloy fires.[1]

Small uranium and plutonium fires within scientific gloveboxes have been extinguished with TEC. A small plastic bag filled with powder is simply placed directly on the metal fire. The barium chloride in the mixture is toxic, but this is of no practical concern if the glovebox remains intact. In other locations, the operator should avoid inhaling the airborne powder.

Boralon

Boralon is the name given to a mixture of trimethoxyborane (TMB) and Halon 1211 (bromochlorodifluoromethane). The addition of halogenated hydrocarbons, specifically halons, reduces the problems associated with aging, low-temperature viscosity, and flammability. Although the addition of halon improves the physical characteristics of TMB, the material remains susceptible to hydrolysis; the colorless liquid hydrolyzes readily to form boric acid and methanol. Contact with moist air or other sources of water must be avoided to prevent hydrolysis.

Extinguishment involves the thermal breakdown of TMB. The typical application of boralon to metal fires results in the formation of molten boric oxide. TMB would ordinarily produce a secondary class B fire due to the excess methanol, but the presence of halon reduces this possibility. The molten boric oxide coating on the hot metal prevents contact with air. Cooling may then occur normally or through the cautious introduction of cooling agents, such as water. It is necessary to avoid fracturing the coating during this process.

Boralon has been developed and tested extensively for use on magnesium fires. It is applied from a sealed storage tank actuated by pressurizing with nitrogen. Because the agent is susceptible to hydrolysis, periodic maintenance and replacement of agent may be necessary. Due to phaseout of Halon 1211 production under international regulatory actions to protect stratospheric ozone, boralon is not expected to be available in the future.

Copper Powder

Advances in the science of alternative propulsion systems have led to the development of copper powder as a viable extinguishing agent for combustible metals. Work sponsored by the Naval Sea Systems Command was conducted to evaluate the adequacy of existing lithium fire suppression agents and to develop new agents should deficiencies exist.[2]

Copper powder was found to be superior to known lithium fire extinguishing agents in extinguishing capacity. The dry powder is of uniform particle size and extinguishes a lithium fire more quickly and efficiently than existing agents. The process of extinction is by formation of a copper-lithium alloy, which is nonreactive and forms preferentially on the surface of the molten lithium. The alloy becomes an exclusion boundary between air and the molten metal, preventing reignition and promoting cooling of the unreacted lithium.

Copper powder can be applied from hand-portable extinguishers. The nominal charge for each extinguisher is 30 lb (14 kg); it is 150 and 350 lb (68 and 160 kg) for wheeled units, as well as fixed systems. Argon is used as the propellant. The method of application is similar to that of other metal fire powders, in that the fuel surface is coated with the copper powder in an initial pass, with a throttled application following once control is achieved. Typical application densities are 8 lb (3.6 kg) of copper powder per lb of lithium for complete extinguishment of the lithium. A 40-lb (18-kg) lithium fire can be fully controlled in 30 sec and completely extinguished in 9 min. Copper powder also has demonstrated the ability to extinguish magnesium and aluminum fires.

NONPROPRIETARY COMBUSTIBLE METAL EXTINGUISHING AGENTS

Foundry Flux

In magnesium foundry operations, molten magnesium is protected from contact with air by layers of either molten or crust-type fluxes. These fluxes, which are also used as molten metal cleaning agents, consist of various amounts of potassium chloride, barium chloride, magnesium chloride, sodium chloride, and calcium fluoride. The fluxes are stored in covered steel drums. When applied to burning magnesium, they melt on the surface of the solid or molten metal, excluding air. The thin layer of protection can be provided by properly applying relatively small amounts of flux.

Fluxes are valuable in extinguishing magnesium spill fires from broken molds or leaking pots, and in controlling and extinguishing fires in heat-treating furnaces. In open fires, the flux is applied with a hand scoop or a shovel. Areas of furnaces that are difficult to reach can be coated by means of a flux-throwing device similar to those used to throw concrete onto building forms.

Although fluxes can rapidly extinguish chip fires in machine shops, they are not recommended for such use. The fluxes are hygroscopic, and the water picked up from the air combines with the salts and causes severe rusting of equipment.

Talc (Powder)

Talc, an agent that has been used industrially on magnesium fires, acts to control rather than extinguish fire. Talc acts as an insulator to retain the heat of the fire, rather than as a coolant. However, it does react with burning magnesium to provide a source of oxygen. The addition of organic matter, such as protein, to talc assists in the controlling action, but does not prevent the reaction that releases oxygen to the fire.

Graphite Powder

Graphite powder, or plumbago, has been used as an extinguishing agent for metal fires. Its action is similar to that of G-1 powder, in that the graphite acts as a coolant. Unless the powder is finely divided and closely packed over the burning metal, some air does get through to the metal, and extinguishment is not as rapid as with G-1 powder.

Sand

Dry sand often has been recommended as an agent for controlling and extinguishing metal fires. At times, it seems to be satisfactory, but hot metal, such as magnesium, usually obtains oxygen from the silicon dioxide in the sand and continues to burn under the pile. Sand is seldom completely dry, and burning metal reacting with the moisture in the sand produces steam. Under certain conditions, it may even produce an explosive metal-water reaction. By laying fine, dry sand around the perimeter of the fire, it can be used to isolate incipient aluminum dust fires.

Cast-Iron Borings

Cast-iron borings of turnings are frequently available in the same machine shops as the various combustible metals. Clean iron borings applied over a magnesium chip fire cool the hot metal and help extinguish the fire. This agent is used by some shops for handling small fires where, with normal good housekeeping, only a few combustible chips are involved. Contamination of the metal chips with iron may be an economic problem. Oxidized iron chips must be avoided to prevent possible thermite reaction with the hot metal. The iron chips must be free of moisture.

Sodium Chloride

Alkali metal fires can be extinguished by sodium chloride, which forms a protective blanket that excludes air over the metal so that the metal cools below its burning temperature. Sodium chloride is used to extinguish sodium and potassium fires. It can also be used to extinguish magnesium fires.

Soda Ash

Sodium carbonate or soda ash (not dry chemical) is recommended for extinguishing sodium and potassium fires. Its action is similar to that of sodium chloride.

Lithium Chloride

Lithium chloride is effective in extinguishing lithium metal fires. However, its use should be limited to specialized applications because the chemical is hygroscopic to a degree and the reaction between moisture and lithium may present problems.

Zirconium Silicate

This agent has been used successfully to extinguish lithium fires.

Dolomite

If zirconium or titanium in the form of dry powder ignites, neither can be extinguished easily. Control can be effected by spreading dolomite, a carbonate of calcium and magnesium, around the burning area and adding more powder until the burning pile is covered completely.

Boron Trifluoride and Boron Trichloride

Boron trifluoride and boron trichloride have both been used to control fires in heat-treating furnaces containing magnesium. The trifluoride is considerably more effective. In the case of small fires, the gases provide complete extinguishment. In the case of large fires, they control the flames and rapid burning, but the hot metal reignites on exposure to air. A combined attack of boron trifluoride gas followed by application of foundry flux completely extinguishes the fire. For details of gas application, see NFPA 480, *Standard for the Storage, Handling, and Processing of Magnesium Solids and Powders* (hereinafter referred to as NFPA 480.)

Inert Gases

In some cases, inert gases, such as argon and helium, will control zirconium fires if they can be used under conditions that will exclude air. Gas blanketing with argon has been effective in controlling lithium, sodium, and potassium fires.[1] Caution should be exercised when using the agent in confined spaces because of the danger of suffocation to personnel.

Water

If automatic sprinklers open during a fire in a shop where magnesium or other combustible metals (except alkali metals or radioactive materials) are being machined or fabricated, the large volume of water from the sprinklers normally will extinguish both Class A and magnesium fires. Where automatic sprinkler protection is provided, a deflecting shield or hood should be provided over furnaces, reactors, or other places where hot molten metal may be present. Additional information on the use of sprinkler protection in shops handling magnesium or titanium can be found in NFPA 480 and NFPA 481, *Standard for the Production, Processing, Handling, and Storage of Titanium.*

When burning metals are spattered with limited amounts of water, the hot metal extracts oxygen from the water and promotes combustion. The hydrogen released in a free state ignites readily. Since small amounts of water accelerate combustible metal fires, particularly where chips and fines are involved, use of common portable extinguishers containing water is not recommended except to control fires in adjacent Class A materials.

However, water is a good coolant and can be used on some combustible metals, under proper conditions and applications, to reduce the temperature of the burning metals below the ignition point. The following paragraphs discuss the advantages and limitations of using water on fires involving combustible metals.

Water on sodium, potassium, lithium, NaK, barium, calcium, and strontium fires: Water applied to sodium, potassium, lithium, sodium-potassium alloys (NaK), barium, and probably calcium and strontium will induce chemical reactions that will lead to fire or explosion even at room temperature. Therefore, water must *not* be used on fires involving these metals.

Water on zirconium fires: Powdered zirconium that has been wetted with water is more difficult to ignite than dry powder. Once that ignition takes place, however, wet powder burns more violently than dry powder. Powder containing about 5 to 10 percent water is considered the most dangerous. Small volumes of water should not be applied to burning zirconium, but large volumes of water can be used successfully to completely cover solid chunks or large chips of burning zirconium. For example, the metal may be drowned in a tank or barrel of water. Hose streams applied directly to burning zirconium chips may yield violent reactions.

Water on plutonium, uranium, and thorium fires: Limited amounts of water add to the intensity of a fire in natural uranium and thorium, and greatly increase the contamination cleanup required after the fire. A natural uranium scrap fire can be fought with water by personnel, wearing face shields and gloves, using long-handled shovels to put the burning scrap into a drum of water in the open. The hydrogen formed may ignite and burn off above the top of the drum. The radioactivity hazard of natural uranium is relatively low. In reality, uranium is a metal poison, although it is considerably less toxic than lead. The use of water on enriched uranium or plutonium, which are fissionable materials, is generally prohibited. If ingested or inhaled, plutonium is considerably more hazardous to humans than uranium.

Water on magnesium fires: Although water in small quantities accelerates magnesium fires, rapid application of large amounts of water is effective in extinguishing them because of the cooling effect of the water. Automatic sprinklers will extinguish a typical shop fire where the quantity of magnesium is limited. However, water should not be used on any fire involving a large number of magnesium chips when it is doubtful that there is sufficient water to handle the large area. (A few burning chips can be extinguished by dropping them into a bucket of water.) Small streams from portable extinguishers will accelerate a magnesium chip fire violently.

Burning magnesium parts, such as castings and fabricated structures, can be cooled and extinguished with coarse streams of water applied with standard hoses. A straight stream scatters the fire, but coarse drops produced by a fixed nozzle operating at a distance or produced by an adjustable nozzle flow over and cool the unburned metal. Some temporary acceleration normally takes place with this procedure, but rapid extinguishment follows if the technique is pursued. Well-advanced fires in several hundred pounds (100 lb equals 45.4 kg) of magnesium scrap have been extinguished in less than 1 min with two 1 $^1/_2$-in. (38-mm) fire hoses. Water fog, on the other hand, tends to accelerate, rather than cool, such a fire. Water should not be applied to magnesium fires where quantities of molten metal are likely to be present because the steam formation and possible metal-water reactions may be explosive.

Water on titanium fires: Water must not be used on fires in titanium fines and should be used with caution on other titanium fires. Small amounts of burning titanium other than fines can be extinguished and considerable salvage realized by quickly submerging the burning material in a large volume of water. Hose streams can be used effectively on fires outside piles of scrap, but violent reactions resulting in serious injuries can also occur if water is applied to hot or burning titanium.

MISCELLANEOUS AGENTS

Agents that have been tested for cooling capacity and ability to extinguish magnesium fires include the following: boric acid dissolved in triethylene glycol, followed by foam to extinguish the secondary fire[3]; di-isodecyl phthalate combined with chlorobromomethane[4]; and tri-cresyl phosphate, followed by foam to handle the secondary fire. To extinguish burning magnesium, a suitable powder, consisting essentially of powdered polyvinyl chloride and sodium borate, was developed.[5] The powder forms a coating on the molten metal and enables a fine water spray to be used to cool the metal to extinction. Research has been conducted at two U.S. laboratories on optimum methods of controlling zirconium and uranium fires with mixed halogen organic derivatives.[6]

Among the more recent literature on combustible metals is a report on computer modeling of titanium combustion. The report describes an extensive program of analytical and experimental investigation to establish the parameters that govern the ignition and self-sustained combustion of metals, such as titanium. The analytical model allows prediction of frictional heating and exposure to high ambient temperatures.[7] Other pertinent publications include suggestions for safely machining magnesium metal[8] and for determining the pyrophorosity of the liquid mixing of combustible metal powders. The flammability of metal charges dried after liquid and dry mixing is said to be different and is connected with a substitution of oxygen absorbed on the metal surface. From the viewpoint of fire safety, maximum diluents are suggested to minimize the possibility of fire or explosion.[9]

BIBLIOGRAPHY

References Cited

1. Rodgers, S. J., and Everson, W. A., "Extinguishment of Alkali Metal Fires," *Fire Technology*, Vol. 1, No. 2, May 1965, pp. 103–111.
2. Lee, M. E., Stepetic, T. J., Watson, J. D., and Moore, T. A., "Lithium Fire Suppression Study, Phase 2 (Medium-Scale)," *New Mexico Engineering Research Institute Report* NMERI OC 90/10, Nov. 1989, Naval Undersea Warfare Engineering Station, Keyport, WA.
3. McCutchan, R. T., "Investigation of Magnesium Fire Extinguishing Agents," *WADC Technical Report* No. 55-5, 1954, U.S. Air Force Wright Air Development Center (by Southwest Research Institute), Wright-Patterson AFB, OH.
4. Greenstein, L. M., and Richman, S. I., "A Study of Magnesium Fire Extinguishing Agents," *WADC Technical Report* No. 55-107, 1988, U.S. Air Force Wright Air Development Center (by Francis Earle Laboratories, Inc.), Wright-Patterson AFB, OH.
5. Nash, P., "A Dry Powder Extinguishing Agent for Magnesium Fires," *Institution of Fire Engineers Quarterly*, Vol. 23, No. 51, Sept. 1963, pp. 275–276.
6. Lawrence, K. D., "New Agents for the Extinguishment of Magnesium Fires," *CEEDO Technical Report* No. 78-19, 1978, Civil and Environmental Development Office, Tyndall Air Force Base, FL.
7. Glickstein, M. R., "Computer Modeling of Titanium Combustion," *Tri-Service Conference on Corrosion*, Vol. II, U.S. Air Force Wright Aeronautical Laboratories, Wright-Patterson AFB, OH, 5–7 Nov. 1980, pp. 93–122.
8. Morales, H. J., "Machining Magnesium Safely," *American Machinist*, Vol. 124, No. 10, Oct. 1980, pp. 154–155.
9. Afanasieva, L. F., Chernenko, E. V., and Rozenband, V. I., "Pyrophorosity of Liquid Mixing of the Charge for Titanium Carbide Production," *Poroshk. Metall.*, Vol. 5, May 1981, pp. 90–94.

References

Beeson, H. D., Tapscott, R. E., and Mason, B. E., "Extinguishing Agent for Magnesium Fire: Phases V and VI," *ESL-TR-86-65*, 1987, Engineering and Services Laboratory, Air Force Engineering and Services Center, Tyndall Air Force Base, FL.
Tuve, R. I., Gipe, R. L., Peterson, H. B., and Neil, R. R., "The Use of Tri-methoxyboroxine for the Extinguishment of Metal Fires," *NRL Report No. 4933*, 1957, Naval Research Laboratory, Washington, DC.
Zeratsky, E. D., "Extinguishing Agents for Combustible Metals and Special Chemicals," *Safety Maintenance*, Vol. 120, No. 2, Aug. 1960, pp. 28–32.

NFPA Codes, Standards, and Recommended Practices

Reference to the following NFPA codes, standards, and recommended practices will provide further information on combustible metal extinguishing agents and application techniques discussed in this chapter. (See the latest version of *The NFPA Catalog* for availability of current editions of the following documents.)

NFPA 480, *Standard for the Storage, Handling, and Processing of Magnesium Solids and Powders*

NFPA 481, *Standard for the Production, Processing, Handling, and Storage of Titanium*

NFPA 482, *Standard for the Production, Processing, Handling, and Storage of Zirconium*

Additional Reading

Charpenel, J., *et al.*, "Combined Sodium Fires—Experiments and Code Developments," *Proceedings* of the International Conference on Science and Technology of Fast Reactor Safety, Vol. 1, Thomas Telford, London, 1986, pp. 85–88.

Leihbacher, D., "Tactics for Large Magnesium Fires: Yonkers, New York Magnesium Incident," *Fire Engineering*, Vol. 148, No. 4, April 1995, pp. 38–40.

Leonard, J., *et al.*, "Use of Copper Powder Extinguishers on Lithium Fires," Naval Research Laboratory, Washington, DC, Hughes Associates, Inc., Columbia, MD, Naval Sea Systems Command, Washington, DC, NRL/MR/6180-94-7490, Funding Numbers 61-M259-X3, July 8, 1994.

"Lithium Fire Fighting Study 1, July 1988 through 20 December 1989," Navy Sea Systems Command, PMS 406.

Rhein, R. A., "Experimental Study of the Use of Liquid Argon and Argon-Filled Aqueous Foams for the Extinction of Lithium Fires," *Fire Technology*, Vol. 28, No. 4, Nov. 1992, pp. 290–316.

Rhein, R.A., and Carlton, C.M., "Extinction of Lithium Fires: Thermodynamic Computations and Experimental Data From Literature," *Fire Technology*, Vol. 29, No. 2, 1993, pp. 100–130.

Sachen, J. B., "Industry Practices for Magnesium Fires," *Fire Engineering*, Vol. 148, No. 4, April 1995, pp. 40, 43–46.

Sharma, T. P., Varshney, B. S., and Kumar, S., "Studies on the Burning Behavior of Metal Powder Fires and Their Extinguishment. Part 2. Magnesium Powder Heaps on Insulated and Conducting Material Beds," *Fire Safety Journal*, Vol. 21, No. 2, 1993, pp. 153–176.

Simpson, J. M., and Gardener, N. J. L., "Graphite-Based Extinguishants for Liquid Metal Fires," *Proceedings* of the International Conference on Science and Technology of Fast Reactor Safety, Thomas Telford, 1986, pp. 603–606.

"Tactics for Large Magnesium Fires," *Fire Engineering*, Vol. 148, No. 4, April 1995, pp. 38–47.

Varshney, B. S., Kumar, S. and Sharma, T. P., "Studies on the Burning Behavior of Metal Powder Fires and Their Extinguishment. Part 1. Mg, Al, Al-Mg Alloy Powder Fires on Sand Bed," *Fire Safety Journal*, Vol. 16, No. 2, 1990, pp. 93–117.

Weman, D.G., "Tactics for Large Magnesium Fires: St. Louis, Missouri, Magnesium Incident," *Fire Engineering*, Vol. 148, No. 4, April 1995, pp. 46–48.

CONFINING FIRES

The *Fire Protection Handbook* emphasizes that fire safety must be addressed through system design and analysis of all elements and their relationships to each other. Nowhere is this more essential than in the decisions that determine strength, location, and other features of barriers, surfaces, and spaces in a building. Confining fires is an important aspect of building fire protection.

Effective fire-safe design begins with analysis and decision making early in the design process. Chapter 1 provides the fundamentals of building and site layout, with particular attention given to the exterior features of buildings in terms of both fire department access and exposure hazards. Chapter 2 addresses building construction features and structural components. Basic building assemblies and systems, including columns, walls, partitions, floors, trusses, ceilings, and rooms, are described in terms of function and alternative configurations. Chapter 3, "Interior Finish," considers flame spread characteristics in confining fires and testing materials.

Chapters 4, 5, and 6 address the performance of buildings during a fire. In Chapter 4, the structural integrity of buildings during a fire is addressed, including protection of critical building features and the protection of building occupants. Test procedures and methods for calculating the design of fire conditions and relating these conditions to measurable fire performance properties of the building are also discussed. Chapter 5 addresses the actual confinement of fires in buildings, including the application of fire barriers. Chapter 6 addresses smoke movement in buildings under fire conditions and introduces the calculation procedures used to predict smoke movement both horizontally and vertically. It includes hot and cold smoke movement in buildings, smoke movement dynamics in tall buildings, and managing smoke movement through the use of active smoke removal systems. Chapter 7 addresses ventilation requirements of buildings under fire conditions to prevent buildup of smoke and remove hot gases.

Chapter 8 covers the structural elements associated with one- and two-family dwellings, while Chapter 9 discusses those aspects of ventilation systems in commercial cooking establishments that can have an effect on fire spread, containment, and suppression.

Chapter 10 delves into the problems associated with a number of special structures, such as, towers and piers.

After a fire, it is important to estimate the amount of structural damage to the building and any impact on its structural integrity. Chapter 11 addresses evaluating building fire damage in the post-fire condition.

Chapter 12 addresses fire hazards during construction, alterations, and demolition of buildings. This includes fire issues associated with temporary buildings and loss prevention during construction.

Chapter 13 covers miscellaneous building services. This chapter addresses fire spread through building services and features, such as firestopping, to prevent the spread of fire in buildings.

Chapter 14 describes the fire protection requirements for air-conditioning and ventilation systems (HVAC) in modern buildings.

Other chapters that emphasize fire-safe design of building systems include: Section 1, Chapters 1, 2, and 3; Section 4, Chapters 1 and 2; Section 8, Chapters 1 and 2; and Section 11, Chapter 4. Calculative methods are described in Section 11, Chapters 5, 6, 7, and 8 are essential to every facet of design for fire confinement and will be of interest to building designers. A more detailed treatment of these methods, and application of theory and models to building fire safety design, may be found in *The SFPE Handbook of Fire Protection Engineering*.

—Robert J. Vondrasek
December 1996

BUILDING AND SITE PLANNING FOR FIRE SAFETY

Revised by Albert M. Comly, Jr.

Effective, fire safe design begins with conscious analysis and decision making early in the design process. This broad overall approach includes consideration of both interior building functions and layout, as well as exterior site planning. This chapter will present some broad guidelines to be considered by building designers in analyzing fire protection measures that must be provided for a fire safe environment for structures and their sites. Detailed information of the various aspects of fire safe site planning and building design can be found elsewhere in this handbook, particularly in the other chapters in this section.

To effectively incorporate the building's fire defenses into the design, the fire safety objectives must first be identified. Section 1, Chapter 3, "Systems Concepts for Building Fire Safety," briefly discusses a systematic approach to assessment of fire safety objectives. Decisions must then be made regarding the means to achieve the objectives. The fire safety concepts tree in Section 1, Chapter 3, can be used to obtain an overall perspective of the fire safety system. Venting in conjunction with sprinkler protection continues as a subject of some controversy and is discussed in greater detail in Section 7, Chapter 7, "Venting Practices."

Concepts of egress design, an extremely important part of the design process, are discussed in Section 8, Chapter 2, "Concepts of Egress Design." That chapter, based on NFPA *101*®, *Life Safety Code*®, describes the elements that must be considered when designing adequate egress in buildings for life safety from fire.

BUILDING INTERIOR DESIGN CONSIDERATIONS

The architectural design of a building has a significant influence on its fire safety capabilities and performance. In addition to the design of the structure and shell of the building itself, the interior layout and spatial flow, circulation patterns, interior finish materials, and building services are all important factors in the level of fire safety in a building. The design of the building and the relationship of the building to the site or lot can enhance the efficiency of fire department operations by reducing dependence on manual suppression, if the design is executed appropriately. Consequently, fire department operations should be considered in all phases of design.

Albert M. Comly, Jr., AIA, is program manager in the Justice Planning and Design Program at Vitetta Group, Architects and Engineers, in Philadelphia, PA., and is safety officer for the Wissahickon Fire Co. of Amber, PA. He is a member of the NFPA Technical Committee on Automatic Sprinklers.

The primary concept of creating fire safety in structures today involves the philosophy that the building itself must be designed to be either self-protecting or self-controlling from fire and fire spread. This is particularly important as buildings become more complex and exhibit features that are contradictory to fire department operations, such as sealed environments for energy efficiency and security against intrusion. The design features that are used should consider the local fire-fighting resources that are available and be effected to reduce the reliance on the intervention of the fire service to ensure fire safety. A building should not be designed today that relies solely on the fire service to provide complete protection for the occupants of the building and property; it must include both active and passive building fire defenses to provide reasonable safety from the effects of fire.

Interior Layout

During a fire the occupants of the entire building are exposed to two types of danger: (1) flames and hot products of combustion, the danger of which rapidly decreases as the distance from the fire increases; and (2) smoke and toxic gases, which present a different type of hazard. The greatest number of fire deaths result from these latter products of combustion, and the danger is present at considerable distances from the location of the fire.

Even relatively minor fires can produce large quantities of smoke and gases. These products of combustion can obscure vision and irritate the eyes to the extent that visibility is reduced to zero. Additionally, these products can in surprisingly small quantities cause a person to become disoriented and get lost in a building. Even occupants familiar with their surroundings may experience great difficulty in locating means of egress. This problem is compounded for transients and occasional visitors to the building.

Architectural layout and normal circulation patterns are important elements in emergency evacuation. For example, many large office buildings are a maze of offices, storage areas, and meeting rooms. Even under normal conditions, a visitor can become confused and have difficulty exiting the building. Clearly marked emergency travel routes can enhance life safety features in all buildings.

The building height is also related to the interior layout and fire safety. Under optimum conditions, modern fire department aerial equipment reaches only to approximately the seventh floor of a building. Occupants above the seventh floor cannot be evacuated by exterior aerial equipment. Consequently, interior layout becomes more significant to occupant evacuation and protection and to fire department operations.

BUILDING DESIGN AND FIRE DEPARTMENT OPERATIONS

From the beginning of the building project, the building designer should consult fire department personnel about fire suppression operations in a building. Normally the code review and planning process will serve this purpose. Such consultation is particularly important in the design phase of a building project. In this way, the designer can incorporate features into the building that will aid, rather than hinder, fire suppression and emergency operations.

Briefly, the operations may be broadly grouped as rescue, fire control, and property conservation. The first priority of any fire-fighting operation is rescue, which is defined as the removal of trapped and injured occupants and the search of all areas to ensure that all occupants are out of the building. The larger the building, the greater the number of fire-fighting personnel that must be committed to the rescue operation. In some cases, simultaneous fire suppression operations will be necessary to complete the rescue operation. Significant design features that are necessary in larger buildings include areas of refuge for the physically disabled, fire department controls for ventilation systems, and emergency communications to critical locations in the building.

Accessibility for Fire Fighting

Building design features are important factors in the spread of fire and the ease of suppression. One of the most important design features involves access to the fire. This includes access to the exterior and interior of the building by fire fighters and their equipment.

Access to the building itself requires roads or fire lanes that are designed to withstand the weight of the heaviest apparatus owned by the municipality in which the building is located. These access lanes must be designed to allow fire equipment to get close enough to all sides of the building for rescue and fire suppression functions. These access lanes must be of sufficient width to ensure they will not be obstructed when needed. Street or road widths should not be less than 28 ft (8.5 m) curb face to curb face in low-density residential areas and should be much wider in commercial and industrial areas. Secondary access and emergency fire lanes should be readily visible and at least 16 ft (4.9 m) wide [20 ft (6.1 m) wide at buildings], with adequate turning radii at each corner or turn. Accepted standards dictate a 22-ft (6.7-m) inside turning radius and a 50-ft (15-m) outside turning radius, but larger equipment may need even greater radii.

Access to the interior of a building can be greatly hampered where large areas exist and where buildings have blank walls, false facades, solar screens, or signs covering a high percentage of exterior walls. Obstacles that prevent ventilation allow smoke to accumulate and obscure fire fighters' vision. Lack of adequate interior access can also delay or prevent fire department rescue of trapped occupants or initiation of fire suppression efforts.

Windowless buildings and basements present unique fire-fighting problems, requiring careful consideration on the part of the building designer to mitigate these problems. The lack of natural ventilation facilities, such as windows, contributes to the buildup of dense smoke and intense heat, which hampers fire-fighting operations and endangers occupants trapped in these occupancies. For this reason, automatic sprinkler protection is required for windowless buildings by most building codes, to extinguish or control a fire before it reaches life- and property-endangering intensity.

Any spaces in which adequate fire-fighting access and operations are restricted because of architectural, engineering, or functional barriers should be provided with automatic extinguishing systems. A complete automatic sprinkler system is generally the best solution to this problem. Other methods can be incorporated in appropriate design situations, such as access panels in interior walls and floors, fixed nozzles in floors with fire department connections, automatic smoke venting, and pressurization of certain building areas, although none of these items can take the place of, or be considered equivalent to, a complete automatic sprinkler system.

Ventilation

Ventilation is an important fire-fighting operation. It involves the removal of smoke, gases, and heat from building spaces. Ventilation of building spaces performs the following important functions:

1. Protection of life by removing or diverting toxic gases and smoke from locations where building occupants must find temporary refuge.
2. Improvement of the environment in the vicinity of the fire by removal of smoke and heat. This enables fire fighters to advance close to the fire and extinguish it with a minimum of time, water, and damage.
3. Control of the spread or direction of fire by setting up air currents that cause the fire to move in a desired direction. In this way, occupants or valuable property can be more readily protected.
4. Provision of a release for unburned, combustible gases before they develop a flammable mixture, thus avoiding a backdraft or smoke explosion.

The building designer should be conscious of these important fire ventilation functions and provide effective means of facilitating venting practices whenever possible. This may involve access panels, moveable windows, skylights, or other means of readily venting a structure in case of a fire emergency. Emergency controls on mechanical equipment may also be an effective means of accomplishing fire ventilation. Each building has unique features, and, consequently, a unique solution should be incorporated into each design. NFPA 204M, *Guide for Smoke and Heat Venting*, should be consulted for additional information on venting.

Water Supply and Use

Water is the principal agent used to extinguish building fires. Although other agents may occasionally be employed (e.g., carbon dioxide, dry chemical, foams and surfactants, and halons), water remains the primary extinguishing agent of the fire service. Consequently, the building designer should anticipate the needs of both the fire department and automatic extinguishing systems and provide an adequate supply of water at adequate residual pressure.

In cities water is normally supplied to the building site by mains that are part of the municipal water distribution system. Cities may not be able to supply a sufficient amount of water at required pressures to every part of the city. Consequently, water supplied to hydrants, standpipes, or sprinklers may need to be boosted by pumps located on fire department apparatus or in the buildings themselves.

Careful attention must be given to water supply, distribution, and pressure for emergency fire conditions. High-rise buildings are particularly sensitive because of the higher water pressures required due to the building height. Large-area buildings must also be given careful attention as to the water supply needs.

Fire conditions or fire loads that require operation of a large number of sprinklers or use of a large number of hose streams can reduce pressure in sprinkler and standpipe systems to the point where residual pressures in the distribution system are adversely affected. Fire department connections for sprinkler and standpipe systems, so the fire department can augment system pressure, are therefore important components of the building fire defenses. The building designer must carefully consider installation details of fire department connections to ensure they will be easily located, readily accessible, and properly marked. Locations for the connections should always be approved by the local fire department.

Water removal: Watertight floors are important in this respect. Salvage efforts can be greatly affected by the integrity of the floors. Of greater importance is the number and location of floor drains. If interior drains and/or scuppers are available, salvage teams can remove water effectively with a minimum of damage. Removal of floor-mounted toilets can also accomplish that task.

BUILDING DESIGN AND FIRE SUPPRESSION

The building designer has a significant influence on the relative ease and effectiveness of fire suppression operations. It is important not only to incorporate the needs of the fire service into the functional layout of the building and its services but also to properly plan for automatic fire detection and extinguishing capabilities.

Compartmentation

Fire-fighting efficiency decreases rapidly as the fire propagates vertically. Fires on two or more floors are extremely difficult to control and extinguish by manual fire-fighting methods. Building designs that utilize unprotected vertical openings, such as open stairwells, curtain wall construction not sealed at each floor, poke-through assemblies, and air-handling ductwork (without dampers), provide avenues for vertical fire spread. It is more difficult to isolate a space completely in modern construction than it was in pre–World War II construction, yet isolation can be accomplished if attention is paid to the details of design and construction. Tested details for all components of the modern building are now documented and readily available, as well as computer modeling programs that can demonstrate what might happen within a building without exposing that building to the effects of an actual fire.

Detection, Alarm, and Communications

Time is the most significant factor concerning fire control. The time for fire control includes several distinct segments, including fire detection, occupant notification, fire department notification, fire department response, fire situation assessment, suppression activities, and overhaul operations. In the early development stages of a fire, a matter of minutes is a significant factor both in life safety and in manual fire extinguishment capabilities. Consequently, effective fire detection becomes a vital element in fire safety. After a fire or smoke detector is activated, appropriate action must be initiated. This may involve alarm, communications, fire suppression, or a combination of these activities. An alarm may alert building occupants to the existence of a fire, but what action should then be taken? Evacuation is the normal procedure in smaller buildings. High-rise buildings, on the other hand, cannot reasonably be totally evacuated without allowing for a greater period of time. Consequently, zoned evacuation and/or safe areas of refuge should be provided within the building, and instructions should be given to building occupants concerning the appropriate course of action.

Communications are vital in occupant safety and fire control. Communications help warn occupants, provide instructions for action, alert the fire department early about the existence and location of a fire, and can assist the fire department in manual fire suppression. Too often, fire department notification has been delayed because occupants were under the mistaken impression that a local fire alarm system notified the fire department.

Automatic Fire Suppression

For more than a century, automatic sprinklers have been the most important single system for automatic control of fires in buildings. Many desirable aesthetic and functional features of buildings that might offer some concern for fire safety can be protected by the installation of a properly designed sprinkler system.

Among the advantages of automatic sprinklers is that they operate directly over a fire and are not affected by smoke, toxic gases, or reduced visibility. In addition, much less water is used because only those sprinklers fused by the heat of the fire operate, particularly if the building is compartmented. Sprinkler systems must be designed and installed to provide this level of protection. (See NFPA 13, *Standard for the Installation of Sprinkler Systems.*) Other automatic extinguishing systems (carbon dioxide, dry chemical, halon agents, high-expansion foam) also provide protection in certain portions of buildings or types of occupancies for which they are particularly suited.

Another advantage that fully sprinklered properties may offer is the reduction in number of fire department equipment access lanes, because these access lanes (necessary for rescue and fire-fighting operations) are not as necessary for fully sprinklered properties.

SITE PLANNING

The architect must utilize a given site in designing the building and adapt the functional and engineering considerations to the particular site conditions that are present. In a similar manner, the architect should consider site features in arriving at decisions on fire protection. A particular set of site characteristics may significantly influence the type of building fire defenses incorporated by the design team. Among the more significant features are traffic and transportation conditions, fire department access, and water supply. Water supply was discussed briefly earlier in this chapter. A few additional considerations are given here.

Traffic and Transportation

Time is a vital factor in fire control. The response time of the fire department can be a major component of the overall time for fire control. Fire apparatus must respond to a location via streets at all times of the day. Depending upon traffic conditions, apparatus can be seriously delayed in responding to some locations.

It may or may not be helpful to consider limited access highways for fire department response. Highways may attract motorists, thus improving traffic conditions on other city streets. On the other hand, the limited access highway may divide a community into sections. Means of travel to locations intermediate to the access points can be delayed. This can have a significant effect on response times and must be considered by the design team in selecting the appropriate fire defenses for the building. Lengthy response times may dictate greater built-in fire protection capability.

Fire Department Access

Ideal exterior accessibility occurs where a building can be approached by fire department apparatus from all sides. This, unfortunately, is not often possible. In congested areas, only the sides of buildings facing streets may be accessible. In other areas, topography or other obstacles can prevent effective use of apparatus in combatting the fire.

Some shopping centers and buildings that are located some distance from the street make the approach of apparatus difficult. If obstructions or topography prevent apparatus from being located close enough to the building for effective use, such equipment as aerial ladders, elevating platforms, and water tower apparatus are rendered useless. Valuable manpower must be expended to carry ground ladders or hand carry hose lines long distances.

The matter of access to buildings has become far more complicated in recent years. The building designer must consider this important aspect in the planning stage. Inadequate attention to site details can place the building in an unnecessarily vulnerable position. If its fire defenses are compromised by preventing adequate fire department access, more complete internal protection in the building itself must make up the difference. If fire department access is limited or nonexistent, complete automatic sprinkler protection should be provided throughout the structure.

Location of Hydrants

An adequate water supply delivered with the necessary pressure is important to control a fire. In many new developments, such as shopping centers, apartments, and housing projects, inadequate attention is given to the number and location of hydrants. There are situations that present extremely difficult fire-fighting problems where water mains are so small and hydrants are so poorly spaced that adequate water to control a fire is unavailable. The building designer must consider these conditions when planning for the fire defenses.

EXPOSURE PROTECTION

A fire in one building creates an external fire hazard to neighboring structures by exposing them to heat by radiation and convection, as well as to the danger of flying brands from the fire. Any or all of these sources of heat transfer may be sufficient to cause an ignition in the exposed structure or its contents. This part of the chapter describes a method for evaluating exposure severity, and suggests methods for protecting buildings from exterior fire exposures. These methods are aimed at protecting combustibles within, and on the exterior of, an exposed building. NFPA 80A, *Recommended Practice for Protection of Buildings from Exterior Fire Exposures*, is a source for further information on exposure protection.

Factors Influencing Severity of Exposure

When discussing protection from exposure fires, there are two conditions to be considered: (1) exposure to horizontal radiation and (2) exposure to flames issuing from the roof or top of a burning building in cases where the exposed building is greater in height than the burning building. Radiation exposure can result from an interior fire where the radiation passes through windows and other exterior wall openings. It can also result from the flames issuing from the windows of the burning building or from flames of the burning facade itself. (See Figures 7-1A, 7-1B, and 7-1C.)

A number of factors significantly influence the danger and intensity of exposing fires to neighboring exposed structures. Among the more important parameters is the severity of the exposing fire (total energy of the fire) itself. It involves both the temperatures developed within the exposing fire and the duration of burning. NFPA 80A, *Recommended Practice for Protection of Buildings from Exterior Fire Exposures*, describes three levels of exposure severity as light, moderate, or severe. The classifications are based on: (1) the average combustible load per unit of floor area and (2) the characteristics of an average flame spread rating of the interior wall and ceiling finishes. Tables 7-1A and 7-1B serve as a guide in assessing severity on the basis of these properties. The more severe of the two classifications should govern when using these tables.

FIG. 7-1A.   *The Burlington Building (right), Chicago, March 15, 1922. Fire in a group of brick, wood-joisted buildings (left) spread across the 80-ft (24-m) street to ignite the upper seven floors of the 15-story office building, which was of fire-resistive construction but lacked exposure protection for street-front windows. Wind carried hot gases against the Burlington Building. Updraft due to convection made exposure most severe above the level of exposing buildings.*

FIG. 7-1B.   *Typical ignition of wood-frame dwelling by heat radiated from an exposure fire. Distances of 10 to 20 ft (3 to 6.1 m) between frame buildings, as commonly specified by building codes, do not eliminate exposure hazard, but usually provide space for fire department operations.*

The duration of the exposing fire and the total heat produced by the fire are related to the guide in Table 7-1A. The speed of fire buildup is influenced by both the nature of the combustibles and by the combustibility of the interior walls and ceiling finish. Table 7-1B relates the flame spread classification with severity.

FIG. 7-1C.   Fire starting in Wellesley, MA (left), ignited a building in Newton (right) by radiated heat across the Charles River, a 100-ft (30-m) distance, against the wind. The river prevented access for fire department operations to protect the exposure.

TABLE 7-1A.   *Classification of Exposure Severity Based on Fire Loading**

Classification of Severity	Fire Loading in lb/sq ft (kg/m^2) of Floor Area	
Light	0–7*	(0–34)
Moderate	7–15	(34–73)
Severe	15–up	(73–up)

*Excluding any appreciable quantities of rapidly burning materials, such as certain foamed plastics, excelsior, or flammable liquids. Where these materials are found in substantial quantities, the severity should be classed as moderate or severe.

Source: NFPA 80A, *Recommended Practice for Protection of Buildings from Exterior Fire Exposures.*

TABLE 7-1B.   *Classification of Exposure Severity Based on Flame Spread of Interior Finish**

Classification of Severity	Average Flame Spread Rating of Interior Wall and Ceiling Finish
Light	0–25
Moderate	26–75
Severe	75–up

Note: Where only a portion of the exposing building has combustible interior finish (i.e., some rooms only, ceiling only, some walls only, etc.), this factor is considered in judging severity classification in as much as it reduces the average flame spread rating.

Source: NFPA 80A, *Recommended Practice for Protection of Buildings from Exterior Fire Exposures.*

Besides the temperature and duration of the exposing fire, other variables influence the severity of exposure on buildings. Some of these variables include the following:

1. Exposing fire
   a. Type of construction of exterior walls and roofs.
   b. Width of exposing fire.
   c. Height of exposing fire.
   d. Percent of openings in exposing wall area. Exterior walls that are combustible or that do not have sufficient resistance to contain the fire should be treated as having 100 percent openings.

e. Ventilation characteristics of the burning room.
f. The fuel dispersion, or surface-to-volume ratio of the fuel.
g. The size, geometry, and surface-to-volume ratio of the room involved.
h. The thermal properties, conductivity, specific heat, and density of the interior finish.
i. Whether or not the exposing building is fully sprinklered.

2. Exposed building
   a. Type of construction of exterior walls and roofs.
   b. Orientation and surface area of exposed exterior walls.
   c. Percent of openings in exterior wall area.
   d. Protection of openings.
   e. Exposure of interior finish and combustibles to the radiation, convection, and flying brands of the exposing fire.
   f. Thermal properties, conductivity, specific heat, density, and fuel dispersion of the interior finish materials and the building contents.
   g. Whether or not the exposed building is fully sprinklered.

3. Site and protection features
   a. Separation distance between exposing and exposed building.
   b. Shielding effect of intervening noncombustible construction.
   c. Wind direction and velocity.
   d. Air temperature and humidity.
   e. Accessibility for fire-fighting operations.
   f. Extent and character of fire department operations.
   g. Built-in fire protection features.

Design Guidelines for Exposure Protection

When analyzing or designing a building for exposure protection, two situations are considered. The first occurs where the exposing building is equal or greater in height than the exposed building. In this case, only the thermal radiation from walls or wall openings is considered. The second case arises when the exposing building is shorter than the exposed building.

The criteria for protection are based on the separation distance between the exposing and the exposed buildings for these two cases. Tables 7-1C and 7-1D provide guidelines for determining separation distances that should protect exposed structures from fire spread. They are based on the condition that separation distances are great enough that ignition of the exposed building or its contents is unlikely, assuming that no means of protection are installed in connection with either building.

In Table 7-1C, the parameters include severity classification (described in Tables 7-1A and 7-1B), width and height of exposing fire, and percent of openings in exposing wall area. The definitions of the terms are as follows.

Width of exposing fire (w):   The length in feet or meters of the exposing wall between interior fire separations or between exterior end walls where no fire separations exist. Fire separations, such as fire partitions or fire walls, should have sufficient resistance to contain the expected fire.

Height of exposing fire (h):   The height in feet or meters of the number of stories involved in the exposing fire, considering such factors as building construction, closure of vertical openings, and fire resistance of floors. The relevant fire separations must have sufficient fire resistance to contain the expected fire.

Percent of openings in exposing wall area: This is the percentage of the exposing wall that consists of doors, windows, or other openings within the assumed height and width of the exposing fire.

TABLE 7-1C. *Guide Numbers for Minimum Separation Distances*

Severity (Percent Openings)			Width/Height or Height/Width Ratio																
			1.0	1.3	1.6	2.0	2.5	3.2	4	5	6	8	10	13	16	20	25	32	40
Light	Moderate	Severe	Guide Number (Multiply by Lesser Dimension, Add 5 ft, to Get Building-to-Building Separation)																
20	10	5	0.36	0.40	0.44	0.46	0.48	0.49	0.50	0.51	0.51	0.51	0.51	0.51	0.51	0.51	0.51	0.51	0.51
30	15	7.5	0.60	0.66	0.73	0.79	0.84	0.88	0.90	0.92	0.93	0.94	0.94	0.95	0.95	0.95	0.95	0.95	0.95
40	20	10	0.76	0.85	0.94	1.02	1.10	1.17	1.23	1.27	1.30	1.32	1.33	1.33	1.34	1.34	1.34	1.34	1.34
50	25	12.5	0.90	1.00	1.11	1.22	1.33	1.42	1.51	1.58	1.63	1.66	1.69	1.70	1.71	1.71	1.71	1.71	1.71
60	30	15	1.02	1.14	1.26	1.39	1.52	1.64	1.76	1.85	1.93	1.99	2.03	2.05	2.07	2.08	2.08	2.08	2.08
80	40	20	1.22	1.37	1.52	1.68	1.85	2.02	2.18	2.34	2.48	2.59	2.67	2.73	2.77	2.79	2.80	2.81	2.81
100	50	25	1.39	1.56	1.74	1.93	2.13	2.34	2.55	2.76	2.95	3.12	3.26	3.36	3.43	3.48	3.51	3.52	3.53
***	60	30	1.55	1.73	1.94	2.15	2.38	2.63	2.88	3.13	3.37	3.60	3.79	3.95	4.07	4.15	4.20	4.22	4.24
***	80	40	1.82	2.04	2.28	2.54	2.82	3.12	3.44	3.77	4.11	4.43	4.74	5.01	5.24	5.41	5.52	5.60	5.64
***	100	50	2.05	2.30	2.57	2.87	3.20	3.55	3.93	4.33	4.74	5.16	5.56	5.95	6.29	6.56	6.77	6.92	7.01
***	***	60	2.26	2.54	2.84	3.17	3.54	3.93	4.36	4.82	5.30	5.80	6.30	6.78	7.23	7.63	7.94	8.18	8.34
***	***	80	2.63	2.95	3.31	3.70	4.13	4.61	5.12	5.68	6.28	6.91	7.57	8.24	8.89	9.51	10.05	10.50	10.84
***	***	100	2.96	3.32	3.72	4.16	4.65	5.19	5.78	6.43	7.13	7.88	8.67	9.50	10.33	11.15	11.91	12.59	13.15

For SI units: 1 ft = 0.305 m.
Source: NFPA 80A, *Recommended Practice for Protection of Buildings from Exterior Fire Exposures.*

TABLE 7-1D. *Separation Distance Based on Building Height*

Number of Stories Likely to Contribute to Flaming Through the Roof	Horizontal Separation Distance or Height of Protection Above Exposing Fire ft (m)
1	25 (7.6)
2	32 (9.8)
3	40 (12.2)
4	47 (14.3)

Source: NFPA 80A, *Recommended Practice for Protection of Buildings from Exterior Fire Exposures.*

Walls without the ability to withstand fire penetration for the expected duration of the fire should be treated as having 100 percent openings.

In using Table 7-1C, for example, assume that a building has a moderate severity [7 to 15 lb per sq ft (34 to 73 kg/m²) of fire load]. Assume further that the width and height of exposing fire is 100 ft (30 m) and 50 ft (15 m), respectively, and that 60 percent of the exposed wall area is open. Using a width to height ratio of 100/50 (30/15) = 2.0, a moderate severity, and 60 percent openings, a guide number of 2.15 is obtained from Table 7-1C. This indicates that the minimum separation an unprotected building should have from an exposing fire in the building described should be 2.15 multiplied by the smaller dimension plus 5 ft (1.5 m).*

Naturally, site conditions and economic constraints preclude the opportunity of providing most buildings with the calculated separation distance. When this occurs, the separation may be reduced by protecting the structure. Some of the means of protecting buildings from exposure fires, and thus reducing the desirable separation distance, are:

1. Complete automatic sprinkler protection.
2. Blank walls of noncombustible materials.

3. Barrier walls (self-supporting) between building and exposure. (See Figure 7-1D.)
4. Extension of exterior masonry walls to form parapets or wings.
5. Automatic outside water curtains for combustible walls.
6. Elimination of openings by filling them with equivalent construction.
7. Glass block panels in openings.
8. Wired glass in steel sash (fixed or automatic closing) in openings, coupled with automatic or deluge sprinklers outside over openings. (See Figure 7-1E.)
9. Automatic (rolling steel) fire shutters on openings.
10. Automatic fire doors on door openings.
11. Automatic fire dampers on wall openings.

The application of the means of protection described previously will enable the separation distance to be reduced. Guidelines for the distance reduction are given in Table 7-1E.

It should be noted that none of these measures is necessary, or even desirable, when both the exposing and exposed properties are provided with a complete fire sprinkler system.

TABLE 7-1E. *Adjustments for Reducing Separation Distances*

Means of Protection	Separation Distance Adjustment
Frame or Combustible Exterior Wall:	
1. Replace with blank fire-resistive wall (3-hr minimum)	Reduce to 0 ft (0 m)
2. Install automatic deluge water curtain over entire wall with no windows or with wired-glass windows closed by 3/4-hr protection	Reduce to 5 ft (1.5 m)

For SI units: 1 in. = 24.5 mm; 1 ft = 0.305 m.
Source: NFPA 80A, *Recommended Practice for Protection of Buildings from Exterior Fire Exposures.*

*The 5 ft (1.5 m) is added to the computed values of separation distance partly to account for the horizontal projection of flames from windows and partly to guard against the risk of ignition by direct flame impingement where small separations are involved.

TABLE 7-1E. *(Continued)*

Means of Protection	Separation Distance Adjustment
Frame or Combustible Exterior Wall:	
3. Install automatic deluge water curtain over entire wall with ordinary glass windows	Reduce by 50 percent
Noncombustible Exposed Exterior Wall (Fire Resistance Less than 3 hr):	
1. Replace wall with blank fire-resistive wall (3-hr minimum)	Reduce to 0 ft (0 m)
2. Close all wall openings with material equivalent to wall, or with $^3/_4$-hr protection and eliminate combustible projections	Reduce by 50 percent
3. Install automatic deluge water curtain over entire wall with no windows or with glass windows or with windows closed by $^3/_4$-hr protection	Reduce to 5 ft (1.5 m)
4. Install automatic deluge water curtain on all wall openings equipped with ordinary glass and on combustible projections	Reduce by 50 percent
Veneered Exposed Exterior Wall (Combustible Construction Covered by a Minimum of 4 in. of Masonry):	
1. Replace wall with blank fire-resistive wall (3-hr minimum)	Reduce to 0 ft (0 m)
2. Close all wall openings with $^3/_4$-hr protection and eliminate combustible projections	Reduce by 50 percent
3. Close all wall openings with material equivalent to wall construction and eliminate combustible projections	Reduce to 5 ft (1.5 m)

TABLE 7-1E. *(Continued)*

Means of Protection	Separation Distance Adjustment
Veneered Exposed Exterior Wall (Combustible Construction Covered by a Minimum of 4 in. of Masonry):	
4. Install automatic deluge water curtain over windows equipped with wired glass or over $^3/_4$-hr closed openings and on combustible projections	Reduce to 5 ft (1.5 m)
5. Install automatic deluge water curtain over windows equipped with ordinary glass and on combustible projections	Reduce by 50 percent
Fire-Resistive Exposed Exterior Wall (Minimum 3-hr Rating:	
1. Close all openings with material equivalent to wall or protect all wall openings with 3-hr protection	Reduce to 0 ft (0 m)
2. Protect all openings with $1^1/_2$-hr protection	Reduce by 75 percent [maximum required = 10 ft (3 m)]
3. Protect all openings with $^3/_4$-hr protection	Reduce by 50 percent [(maximum required = 20 ft (6 m)]
4. Install automatic deluge water curtain on all wall openings with wired glass or with $^3/_4$- or $1^1/_2$-hr protection	Reduce to 5 ft (1.5 m)
5. Install automatic deluge water curtain on all wall openings equipped with ordinary glass	Reduce by 50 percent

For SI units: 1 in. = 24.5 mm; 1 ft = 0.305 m.
Source: NFPA 80A, *Recommended Practice for Protection of Buildings from Exterior Fire Exposures.*

FIG. 7-1D. Above are two types of free-standing fire walls. At left is a brick fire wall reinforced with concrete abutments to make it free-standing. At right is a reinforced concrete fire wall that protected the dwelling from a fire in an adjacent lumberyard.

FIG. 7-1E. *This wired-glass window in a metal frame, though cracked by intense heat of an exposure fire, remained in place and combined with an open sprinkler to prevent fire from entering the building. Note the open sprinkler at top outside the window.*

FIG. 7-1F. *This is an illustration of an inadequate fire separation wall that failed to prevent the spread of fire to an adjacent property. Stair tower in the center of the photograph would not offer refuge to occupants fleeing fire.*

BIBLIOGRAPHY

NFPA Codes, Standards, and Recommended Practices

Reference to the following NFPA codes, standards, and recommended practices will provide further information on building and site planning for fire safety in this chapter. (See the latest version of *The NFPA Catalog* for availability of current editions of the following documents.)

NFPA 13, *Standard for the Installation of Sprinkler Systems*
NFPA 80A, *Recommended Practice for Protection of Buildings from Exterior Fire Exposures*

FIG. 7-1G. *This is an illustration of a fire separation wall that served to stop the spread of fire to an adjacent structure.*

NFPA *101®, Life Safety Code®*
NFPA 204M, *Guide for Smoke and Heat Venting*
NFPA 1231, *Standard on Water Supplies for Suburban and Rural Fire Fighting*

Additional Reading

Anchor, R. D., Malhotra, H. L., and Purkiss, J. A., eds., *Proceedings* of the International Conference on Design of Structures Against Fire, Elsevier, New York, 1986.

Brannigan, F. L., *Building Construction for the Fire Service*, National Fire Protection Association, Quincy, MA, 1982.

Breen, D. E., "Improved Life Safety Through More Reliable Smoke Detection Systems," *Proceedings* from the Conference on New Technology to Reduce Fire Losses and Costs, Elsevier, New York, 1986, pp. 227–234.

Bugbee, P., and Cote, A., *Principles of Fire Protection*, National Fire Protection Association, Quincy, MA, 1988.

Clark, W. E., *Fire Fighting Principles and Practices*, Donnelly Publishing Corporation, New York, 1974.

Egan, M. D., *Concepts in Building Firesafety*, Robert Krieger Publishing Co., Malabar, FL, 1986.

Fried, E., *Fireground Tactics*, H. M. Ginn Corporation, Chicago, IL, 1972.

Gilardi, S. A., "Planning for Fire Safety," *Construction Specifier*, Vol. 41, No. 4, Apr. 1988, pp. 33–34.

Glon, P. L., "Building Safety—A New Educational Approach," *Building Standards*, Vol. 56, No. 2, 1987, pp. 20–21.

Grosse, L., and Malven, F., "Fire Safety in Buildings," Professional Development, Monograph 4, National Council of Architectural Registration Boards, Washington, DC, Aug. 1996.

Harvey, C. S., "A Flexible Approach to Fire-Code Compliance," *Architectural Record*, Vol. 176, No. 12, Oct. 1988, pp. 130–135.

He, Y., and Beck, V., "Estimation of Neutral Plane Position in High-Rise Buildings," *Jounal of Fire Science*, Vol. 14, No. 3, May/June 1996, pp. 235–248.

Jennings, C., "High-Rise Office Building Evacuation Planning: Human Factors Versus 'Cutting Edge' Technologies," *Journal of Applied Fire Science*, Vol. 4, No. 4, 1994/1995, pp. 289–302.

Johnson, W. H., and Matthews, W. R., "Disaster Plan Simulates Plane Crash Into High-Rise Building," *NFPA Journal*, Vol. 87, No. 6, Nov./Dec. 1993, pp. 35–40, 42–43.

Law, M., "Heat Radiation from Fires and Building Separation," Technical Paper No. 5, Joint Fire Research Organization, Borehamwood, Hertfordshire, England, 1963.

Lie, T. T., "Safety Factors for the Fire Loads," *Canadian Journal of Civil Engineering*, Vol. 6, 1979, p. 617.

McGuire, J. H., "Spatial Separation of Buildings," *Fire Technology*, Vol. 1, No. 4, Nov. 1965, pp. 278–287.

Morris, J., "Life Safety and Sprinkler Protection," *Fire Prevention*, No. 209, May 1988, pp. 18–21.

Nakamura, H., "Investment Model of Fire Protection Equipment for Office Building," *Proceedings* of the First International Symposium on Fire Safety Science, Hemisphere, Washington, DC, 1986, pp. 1019–1028.

Nelson, H. E., "Jefferson National Memorial Historical Site Analysis of Impact of Fire Safety," NBSIR 84-2897, Mar. 1985, National Bureau of Standards, Gaithersburg, MD.

Ow, C. S., and Thin, C. C., "Upgrading Fire Protection in Existing Buildings," *Fire International*, Vol. 12, No. 112, Aug.–Sept. 1988, pp. 39–40.

Shields, T. J., and Silcock, G. W. H., *Buildings and Fire*, Longman Scientific and Technical, Essex CM20 2JE, England, 1987.

Shorter, G. W., *et al.*, "The St. Lawrence Burns," *NFPA Quarterly*, Vol. 53, No. 4, Apr. 1960, National Fire Protection Association, Quincy, MA, pp. 300–316.

Thomas, P. H., "The Size of Flames from Natural Fires," Ninth Symposium on Combustion, Academic Press, 1963, pp. 844–859.

Williams-Leir, G., "Approximations of Spatial Separation," *Fire Technology*, Vol. 2, No. 2, May 1966, pp. 137–145.

Wright, R. N., Rosenfeld, A. H., and Fowell, A. J., "National Planning for Construction and Building R&D," National Institute of Standards and Technology, Gaithersburg, MD, NISTIR 5759, Dec. 1995.

Wright, R. N., Rosenfeld, A. H., and Fowell, A. J., "Rationale and Preliminary Plan for Federal Research for Construction and Building," National Institute of Standards and Technology, Gaithersburg, MD, Department of Energy, Washington, DC, NISTIR 5536, Nov. 1994.

BUILDING CONSTRUCTION

Revised by Richard J. Davis

A well-established means of codifying fire protection and fire safety requirements for buildings is to classify them by types of construction, based upon the materials used for the structural elements, and the degree of fire resistance afforded by each element.

In early codes, only two classifications of construction were identified: (1) "fireproof" and (2) "nonfireproof." The term "fireproof" was replaced by the term "fire resistive," since it was recognized that no material or building is totally fireproof, and that the building contents can produce a significant fire without involving the structure. It is possible, however, to design buildings that will resist a fire without suffering serious structural damage. Appropriate fire-resistive design, balanced against anticipated fire severity, is the objective of structural fire protection requirements in modern codes.

Several distinct types of construction that use combustible framing were originally classified based on the materials—masonry or wood—used in the exterior wall construction and the type and size of the framing members, i.e., heavy timber *vs.* conventional framing. As fire-resistance ratings for construction assemblies were recognized in building codes, subclassifications of building types were added for both noncombustible and combustible types of construction, based on the degree of fire resistance provided.

Code regulations governing the size, area, and height of buildings and their allowable uses are usually predicated on the relative fire load represented by the occupancy and the construction materials used in the building.

This chapter identifies the basic types of building construction that are recognized in NFPA 220, Standard on Types of Building Construction (hereinafter referred to as NFPA 220), and in the model building codes. It also recognizes the current classification system of building construction equated with the traditional descriptive terms used to identify building types, e.g., "fire resistive," "noncombustible," "ordinary," "frame," etc., that no longer are prime references to construction types. It also addresses fire protection of building elements; the value of exterior walls, interior walls, and partitions; floor framing systems; trusses; floor/ceiling assemblies; and roof framing and coverings.

CONSTRUCTION CLASSIFICATIONS

The construction types presently identified in NFPA 220, as well as in the three model building codes, i.e., UBC (Uniform Building Code); BNBC (BOCA National Building Code); and SBC (Standard Building Code), fall into ten subtypes in NFPA 220 and the BNBC. The UBC and SBC have nine subtypes of construction. Subtypes are derived from five fundamental construction types, in almost every case: (1) fire resistive, (2) noncombustible, (3) ordinary (exterior protected), (4) heavy timber, and (5) wood frame. The SBC uses six construction types. These descriptive names are now being discontinued because they no longer define the construction types as precisely as needed. The names are helpful, however, in tracing the development of building types. It is anticipated that by the year 2000 there will be only one model building code in the United States.

Although the fire-resistive construction types differ in detail from code to code, the three model building codes and NFPA 220 all define subtypes of fire-resistive construction, using as a basis the amount of fire resistance, (e.g., 2, 3, or 4 hr) required for the building components.

Where the interior structural members and floors are of noncombustible materials with fire-resistance ratings of 1 hr or less, the type of construction is generally identified as noncombustible. Most codes subdivide the noncombustible classification into two types: (1) protected and (2) unprotected.

NFPA 220 and the three model building codes employ three broad classifications for combustible types of construction: (1) exterior protected ordinary, (2) heavy timber, and (3) wood frame. Exterior protected ordinary and wood frame each include two subtypes, i.e., protected and unprotected. They are almost identical in most respects, except for their exterior wall requirements. Heavy timber construction is unique because it is identified by detailed requirements relating mainly to the size of structural members and their connections. Properties, such as combustibility or fire resistance, are not specifically included in the requirements for heavy timber construction, except that exterior walls are required to be of noncombustible construction.

Canadian Classifications

Construction type classification in building codes is more of a convenience than a necessity. The National Building Code of Canada (NBCC) does not classify buildings in the traditional manner as do United States codes, but rather specifies fire-resistive requirements

---

Richard J. Davis, P.E., is manager of the Construction Section of the Standards Division of Factory Mutual Research Corporation (FMRC), Norwood, MA. He is a member of the NFPA Technical Committee on Building Construction.

for the structural components of a building, depending upon its occupancy and its story height and floor area. In this code, two basic types of construction, i.e., combustible and noncombustible, are recognized. These are further subdivided by the characteristics of the materials used in construction under fire conditions.

BDMC (Board for the Development of the Model Codes)

With the formation of the International Code Council (ICC) by the three model building code organizations, the Board for the Coordination of Model Codes (BCMC) has been renamed the Board for the Development of Model Codes (BDMC) with the following objectives.

- To identify and make recommendations for the elimination of conflicts among International Codes and other codes and standards.

- To identify those areas where existing conflicts cannot be resolved and recommend new changes as may be necessary.

- To evaluate and incorporate, where appropriate, new technology and concepts which pertain to the requirements being investigated.

To achieve better uniformity in building codes, the Model Codes Standardization Council (MCSC), predecessor to the Board for the Coordination of the Model Codes (BCMC), established a committee in 1972 to study the classifications and fire resistance requirements for types of construction used in the model building codes and to develop recommendations for the model building code organizations.

As a result of its comparative study, the BCMC proposed that the basic types of construction now recognized in the codes be continued but that they be reordered to some degree and be divided into two groups: (1) noncombustible and (2) combustible. It was also proposed that the identifying names for types of construction, such as "fireproof," "ordinary," "heavy timber," etc., be dropped because current design methods and architecture no longer follow the concepts in vogue when the named building types were established. The classifications proposed are shown in Table 7-2A. (These are the classifications recognized in the current edition of NFPA 220.)

The MCSC also concluded that to rationally compare the various types of construction, a notational system was needed to identify the fire resistance required for three basic elements of the building. These elements are: (1) the exterior wall, (2) the primary structural frame, and (3) the floor construction. A three-digit notation was developed, as follows:

1. *First digit*—Hourly fire resistance requirement for exterior-bearing wall fronting on a street or lot line.
2. *Second digit*—Hourly fire resistance requirement for structural frame or columns and girders supporting loads from more than one floor.

TABLE 7-2A.   Model Codes Standardization Council Recommended Types of Construction

Noncombustible		
Type I (443) Type I (332)	Type II (222) Type II (111) Type II (000)	
Combustible		
Type III (211) Type III (200)	Type IV (2HH)	Type V (111) Type V (000)

3. *Third digit*—Hourly fire resistance requirement for floor construction.

For heavy timber construction, the notation "H" and not a digit was used for the structural frame and floor construction designations.

Thus, for example, a "332" building would have 3-hr fire-resistant exterior bearing walls, a 3-hr fire-resistant structural frame, and 2-hr fire-resistant floor construction, and would correspond to the NFPA 220 Type I (332) building, the BOCA National Building Code Type IB building, the Uniform Building Code Type I FR (fire resistive) building, and the Standard Building Code Type II building.

A comparison of types of construction, based on the MCSC notational system, as found in NFPA 220 and the current editions of three model building codes, is shown in Table 7-2B.

A standard nomenclature was also developed for identifying and defining the structural elements in buildings as they relate to fire resistance. For example, it was found in reviewing various codes and fire protection standards that floor construction was referred to by such terms as "floors," "floor assemblies," "floor and ceiling assemblies," and "floor deck construction." If codes agree, for example, that "floor construction" includes the floor deck and all structural elements directly supporting the loads from the floor, as recommended by MCSC, then some misinterpretation of a code's intent would be avoided.

CLASSIFICATION OF BUILDING TYPES

Following the completion of the MCSC recommendations in 1974, the model building code organizations and NFPA adopted a number of changes to their requirements for types of construction to agree with the MCSC classifications. However, it was recognized that some conflicts among the model building codes still remained. In 1975, therefore, the BCMC established a committee to develop more detailed recommendations for types of construction.

In 1980 the committee's recommended definitions of types of construction and fire resistance requirements were finalized. The requirements are based on five basic types of construction. Two are

TABLE 7-2B.   A Comparison of Construction Types (Based on the MCSC National System)

NFPA 220	I (443)	I (332)	II (222)	II (111)	II (000)	III (211)	III (200)	IV (2HH)	V (111)	V (000)
UBC	—	I FR	II FR	II-1 hr	II N	III-1 hr	III N	IV HT	V 1-hr	V-N
BNBC	1A	1B	2A	2B	2C	3A	3B	4	5A	5B
SBC	I	II	—	IV 1-hr	IV unp	V 1-hr	V unp	III	VI 1-hr	VI unp

identified as noncombustible construction and three as combustible construction types. Table 7-2C gives the fire resistance requirements for the structural frame, interior bearing walls, floor construction, and roof construction of the five basic types of construction.

The table lists the building components that are essential to the stability of the building as a whole and comprise the "structural frame." The members of floor or roof panels that have no connection to the columns are considered part of the floor or roof construction and are not classified as a part of the structural frame.

Type I Construction

Type I construction (formerly referred to as fire resistive) is construction in which the structural members are noncombustible and have a fire resistance as specified in Table 7-2C. This classification is divided into two subtypes, i.e., Type I (443) and Type I (332). The basic difference between the subtypes is in the level of fire resistance specified for the structural frame.

For both subtypes, the required fire resistance of those portions of the structural frame and bearing walls supporting roof loads only may be reduced by 1 hr.

The fire resistance requirements for Type I (433 and 332) construction were selected because they provide reasonable fire safety for the structure for occupancies with moderate- and low-combustible contents. In occupancies with higher fire loads and hazardous uses, fire resistance may be supplemented by additional protection, usually including an automatic fire extinguishing system. Even in occupancies with moderate fire loads, such as in mercantile and in some factory industrial and storage uses, supplementary fire safety precautions are required. These include restrictions on the building size or requirements for automatic fire extinguishing equipment.

In Type I construction, only noncombustible materials are permitted for the structural elements of the building. This is an accepted regulation that appears in practically every modern building code. Obviously, if combustible structural materials were allowed in noncombustible building types, the whole concept of their allowable use (height and area) would become meaningless. However, for practical reasons, the use of some combustible materials in Type I and Type II buildings are permitted for other than structural components. Roof coverings, some types of insulating materials, and limited amounts of interior finish and flooring do not add significantly to the fire hazard or fire load if these materials are properly regulated and qualified by fire tests.

Some codes have attempted to regulate combustible materials by using two or three alternatives that allow for the acceptance of materials having relatively low fuel content and surface burning characteristics. The purpose of this definition was to recognize certain materials or nonhomogenous assemblies containing limited amounts of combustible, such as gypsum wallboard which, although covered with paper, is used as a fire-resistive material. These alternate definitions include limits on surface flame spread rating (per NFPA 255, *Standard Method of Test of Surface Burning Characteristics of Building Materials*) and on the heat content (per NFPA 259, *Standard Test Method for Potential Heat of Building Materials*)—the latter being 3500 Btu per lb (8050 J/kg) for limited-combustible material, somewhat less than one-half that of untreated wood.

Rather than complicate the definition for the accommodation of certain materials, a more fundamental approach is to define limited uses and combustibility characteristics of materials that may be acceptable in buildings of noncombustible construction. This approach is followed in the NBCC and by the BCMC committee in its recommendations for the allowable kinds and extent of use of combustible material in noncombustible buildings.

TABLE 7-2C. *Fire Resistance Requirements for Type 1 through Type V Construction (Source: NFPA 220)*

	Type I		Type II			Type III		Type IV	Type V	
	443	332	222	111	000	211	200	2HH	111	000
EXTERIOR BEARING WALLS—										
Supporting more than one floor, columns, or other bearing walls	4	3	2	1	0*	2	2	2	1	0*
Supporting one floor only	4	3	2	1	0*	2	2	2	1	0*
Supporting a roof only	4	3	1	1	0*	2	2	2	1	0*
INTERIOR BEARING WALLS—										
Supporting more than one floor, columns, or other bearing walls	4	3	2	1	0	1	0	2	1	0
Supporting one floor only	3	2	2	1	0	1	0	1	1	0
Supporting a roof only	3	2	1	1	0	1	0	1	1	0
COLUMNS—										
Supporting more than one floor, bearing walls, or other columns	4	3	2	1	0	1	0	H+	1	0
Supporting one floor only	3	2	2	1	0	1	0	H+	1	0
Supporting roofs only	3	2	1	1	0	1	0	H+	1	0
BEAMS, GIRDERS, TRUSSES, AND ARCHES										
Supporting more than one floor, bearing walls, or columns	4	3	2	1	0	1	0	H+	1	0
Supporting one floor only	3	2	2	1	0	1	0	H+	1	0
Supporting roofs only	3	2	1	1	0	1	0	H+	1	0
FLOOR CONSTRUCTION	3	2	2	1	0	1	0	H+	1	0
ROOF CONSTRUCTION	2	1 1/2	1	1	0	1	0	H+	1	0
EXTERIOR NONBEARING WALLS	0*	0*	0*	0*	0*	0*	0*	0*	0*	0*

■ Those members listed that are permitted to be of approved combustible material.
 * Requirements for fire resistance of exterior walls located in close proximity to property lines, other buildings, or exposures; the provision of spandrel wall sections and the limitation or protection of wall openings are not related to construction type. They might be specified in other standards and codes, where appropriate, and might be required in addition to the requirements of NFPA 220.
 +"H" Indicates heavy timber members; see NFPA 220.

Type II Construction

Type II construction (formerly referred to as noncombustible) is a construction type in which the structural elements are entirely of noncombustible or limited combustible materials permitted by the code and protected to have some degree of fire resistance, either 2 hr [Type II (222)], 1 hr [Type II (111)], or completely unprotected except for exterior walls in Type II (000) construction.

The fire resistance required in Type II (222 or 111) construction will afford adequate fire safety for residential, educational, institutional, business, and assembly occupancies, without supplementary restrictions. Height limits, however, are commonly prescribed for this type of construction. When used for other occupancies involving a greater fire loading, additional fire safety precautions are usually required, such as more stringent area limitations and automatic fire extinguishing equipment. In occupancies with low-combustible contents, the absence of fuel in noncombustible construction not only helps prevent the spread of fire but also reduces potential risk of a fire starting within the structure itself.

The noncombustible feature is valuable because it prevents fire from spreading through concealed spaces or involving the structure itself. Because of this attribute, a fire in a building of noncombustible construction can be controlled more readily. (See Figure 7-2A.)

Type III Construction

Type III construction (formerly referred to as exterior protected combustible or ordinary construction) is a construction type in which all or part of the interior structural elements may be of combustible materials or any other material permitted by the particular building code being applied. The exterior walls are required to be of noncombustible or limited noncombustible materials acceptable to the code and have a degree of fire resistance depending upon the horizontal separation and the fire load. Type III construction is further divided into protected and unprotected subtypes. Protected construction, Type III (211), has a 1-hr fire resistance for the floors and structural elements. Type III (200) construction has no fire resistance for the floors or structural elements. Whether or not fire resistance is provided, it is essential that all concealed spaces be properly firestopped in buildings of combustible construction. This must be done with care in all furred spaces, partitions, ceiling spaces, and attics. Codes are very specific as to the materials to be used for firestopping and the locations where firestopping is required. To be effective, firestopping must completely close off and subdivide the combustible construction into limited areas, thereby restricting the spread of fire and hot gases and allowing additional time for detection and evacuation of the building or area involved.

The 1-hr fire resistance provided in Type III (211) construction offers a measure of safety for fire fighting and evacuation before the

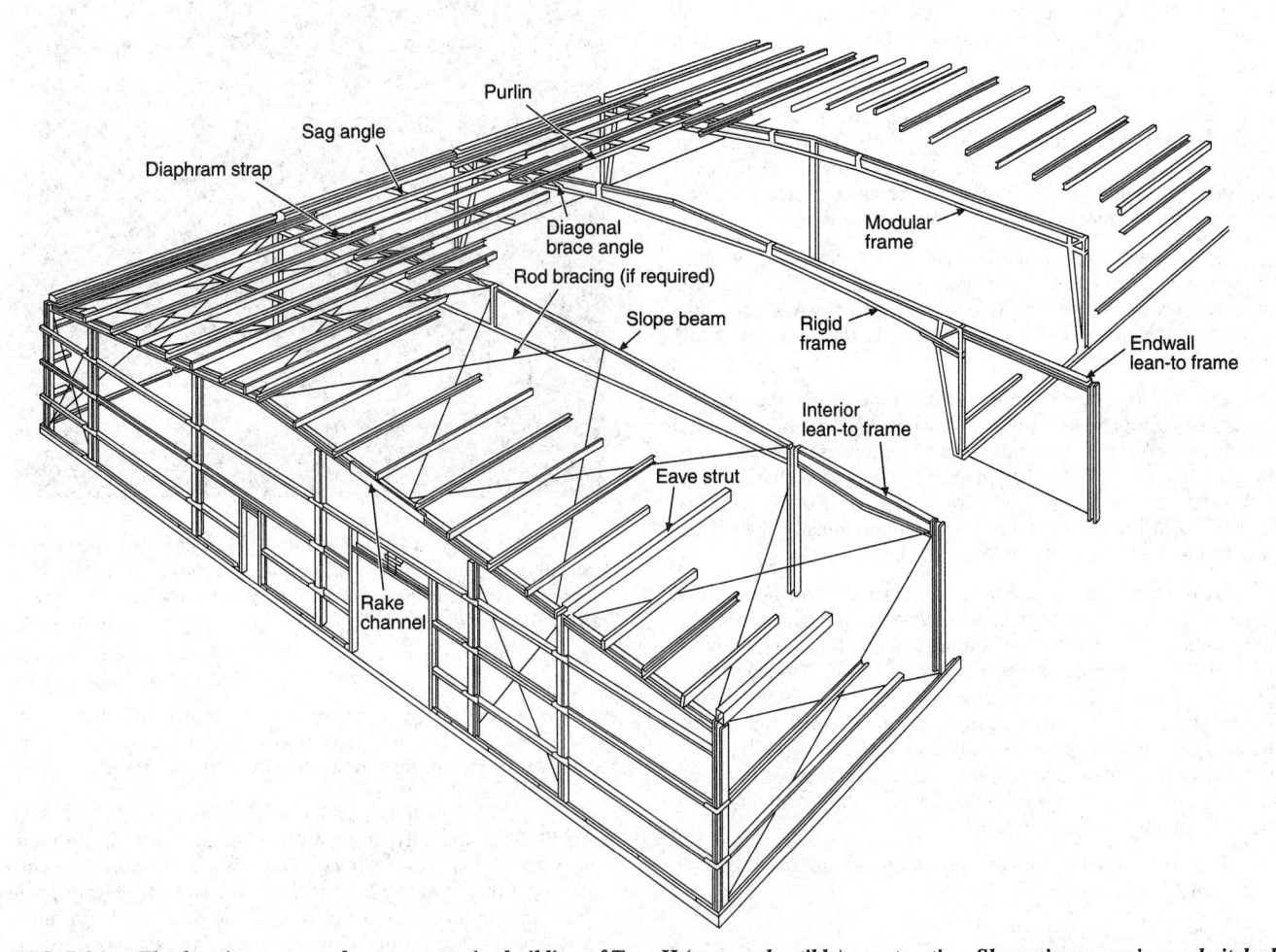

FIG. 7-2A. *The framing system of a representative building of Type II (noncombustible) construction. Shown is pre-engineered pitched roof and lean-to framing with structural elements of unprotected steel. Requirements for exterior walls of Type II construction are specified by the applicable building code or standard that is used by the Authority Having Jurisdiction.*

TABLE 7-2D. *Recommended Nominal Dimensional Requirements for BCMC Type IV (2HH) Construction*

	Supporting Floors	Supporting Roofs
Columns	8 in. × 8 in.	6 in. × 8 in.
Beams and girders	6 in. × 10 in.	4 in. × 6 in.
Arches	8 in. × 8 in.	6 in. × 8 in., 6 in. × 6 in., 4 in. × 6 in.
Trusses	8 in. × 8 in.	4 in. × 6 in.
Floors	3 in. T & G or 4 in. on edge w/1-in. flooring	
Roofs		2 in. T & G or 3 in. on edge or 1⅛ in. plywood

Note: T and G is tongue and groove.
For SI units: 1 in. = 25.4 mm.

construction itself becomes involved. It has been well established, however, that combustible parts of any fire-rated assembly will be burning actively before the end of the rated time period. For this reason, that portion of the fire load represented by combustible structure must be considered as part of the total potential fire load, whether or not the construction is protected.

Type IV Construction

Type IV construction (formerly referred to as heavy timber) is a construction type in which structural members, i.e., columns, beams, arches, floors, and roofs, are basically of unprotected wood (solid or laminated) with large cross-sectional areas. No concealed spaces are permitted in the floors and roofs or other structural members, with minor exceptions. NFPA 220 and most model building codes are specific in the minimum dimensions permitted for the various wood structural members and minimum fire-resistive ratings required for interior columns, arches, beams, girders, and trusses of materials other than wood that may be permitted as acceptable alternatives to wood members. (See Table 7-2D.)

Walls, both interior and exterior, including structural members framed into them, can be of noncombustible or limited combustible materials acceptable to the code being applied. Brick and stone were the traditional materials used in early heavy timber, or "mill" construction.

During a fire, heavy timber construction resists failure longer than a conventional wood frame structure because the structural members are larger, have a smaller surface to mass ratio, and take longer to burn. As the wood member burns, a layer of char develops which acts like insulation and slows down the rate of burning. The large wood members, therefore, can continue to carry their structural loads due to the mass of unburned wood.

Heavy timber construction is more properly considered as a building system, not just a construction type using large-size framing members. It was developed during the mid-1800s by insurance interests for the purpose of reducing fire losses in the many textile factories, paper mills, and storage buildings in the New England states. Through the intelligent use of combustible materials of sufficient mass, the absence of concealed spaces, and by paying attention to details to avoid sharp corners and ignitable projections, the chance of rapid spread of fire are lessened and the probability of serious structural damage is reduced.

Examples of heavy timber construction are shown in Figures 7-2B and 7-2C.

Type V Construction

Type V construction (formerly referred to as wood frame) is a type of construction in which the structural members are entirely of wood

FIG. 7-2B. Elements of a building of Type IV (heavy timber) construction. Note the large size of the columns and beams and the absence of concealed spaces. The exterior wall at far left is of lightweight corrugated steel.

or any other material permitted by the code being applied. (See Figure 7-2D.) Depending upon the exterior horizontal separation, the exterior walls may or may not be required to be fire resistive.

Type V construction is probably more vulnerable to fire, both internally and externally, than any other building type. Accordingly, it is essential that greater attention be given to the details of construction of this basically light wood-frame building. Firestopping in exterior and interior walls at ceiling and floor levels, in furred spaces, and other concealed spaces can retard the spread of fire and hot gases in these vulnerable areas. Type V construction is subdivided into two subtypes, i.e., Type V (111) construction, which has 1-hr fire resistance throughout, including the exterior walls; and

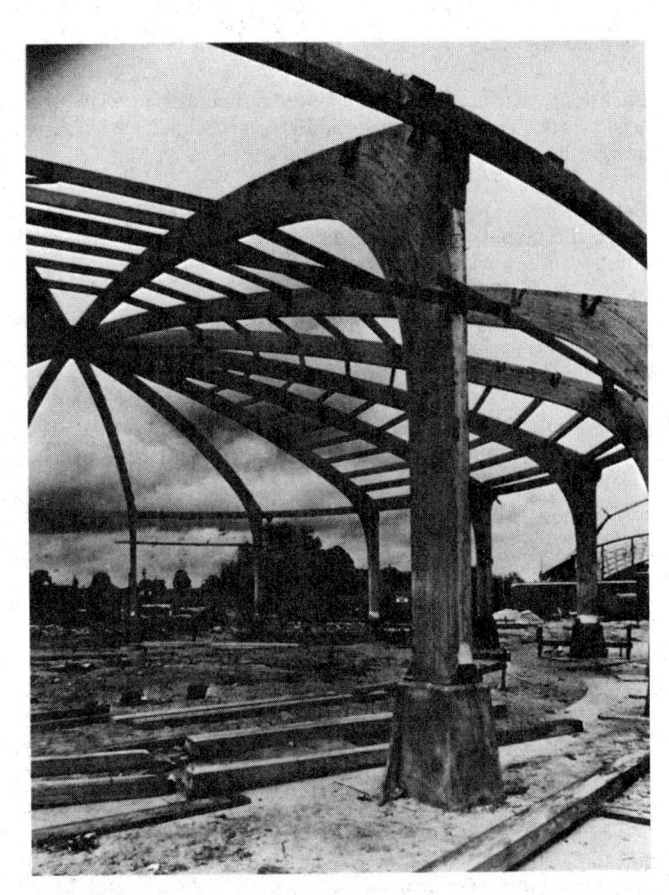

FIG. 7-2C. *A variation of Type IV (heavy timber) construction. Shown are haunched arches made of laminated wood (glue-laminated construction). Beams are anchored to the arches by steel hangers.*

Type V (000) construction, which has no fire protection or fire resistance requirements, except for the exterior walls when horizontal separation is less than 10 ft (3 m).

Mixed Types of Construction

Where two or more types of construction are used in the same building, it is generally recognized that the requirements for occupancy or height and area would apply for the least fire-resistive type of construction. However, in cases where each building type is separated by adequate fire walls or area separation walls [see NFPA 221, *Standard for Fire Walls and Fire Barrier Walls* (hereinafter referred to as NFPA 221)] having appropriate fire resistance, each portion may be considered as a separate building.

Another general limitation included in some codes prohibits construction types of lesser fire resistance to support construction types having higher required fire resistance. In the event of a fire, the risks of a major structural collapse are generally too great to permit this type of design. This limitation does not necessarily apply where construction supports nonbearing separating partitions, that provide protection for exit corridors or tenant spaces.

FIRE PROTECTION OF BUILDING ELEMENTS

Fire protection of building elements is provided for two reasons: (1) to prevent the spread of fire within or into the building during a protracted or uncontrolled fire exposure and (2) to ensure that, even under that exposure, the building frame or elements of that frame will not collapse. Such collapse or even the threat of collapse will render fire fighting measures less effective than they might be otherwise.

There are two groups of building elements: (1) load bearing and (2) nonload bearing. Load-bearing elements are those that support loads other than their own weight. Nonload-bearing elements support only their own weight. Removal of nonload-bearing elements

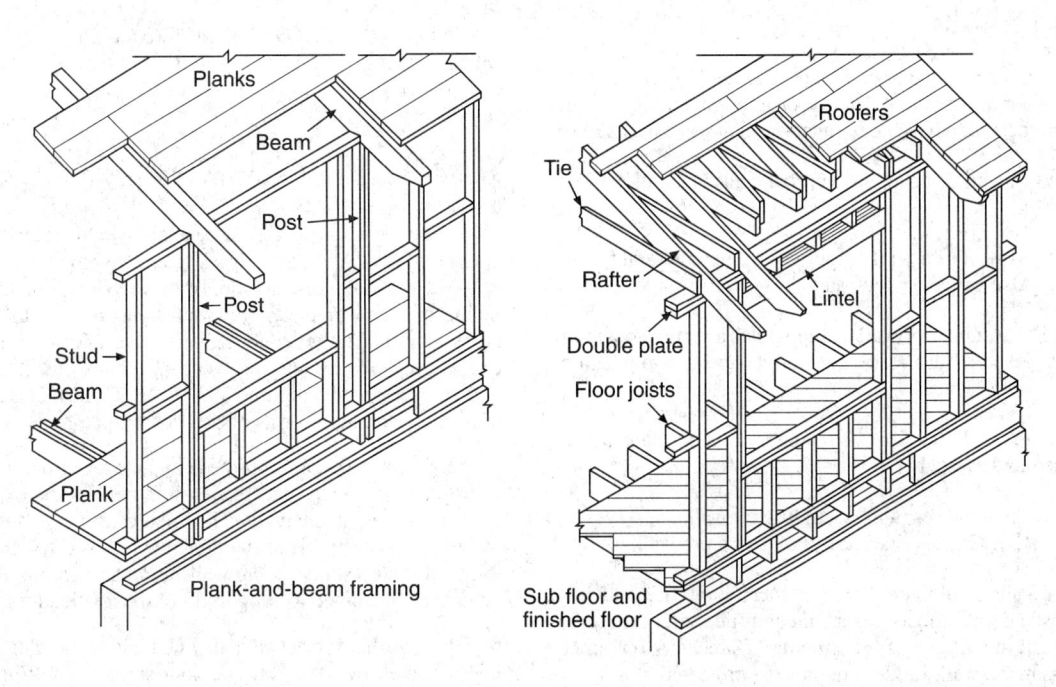

FIG. 7-2D. *Two variations on basic Type V (wood frame) construction. At left is plank-and-beam framing in which a few large members replace many small members used in typical wood framing. At right is conventional wood framing (western or platform construction). Firestopping is essential in concealed spaces.*

would have no effect on the structural behavior of the building as a whole. Table 7-2E lists the structural elements that are common to most buildings, separated into these two general categories.

TABLE 7-2E. *Structural Elements of a Building*

Load-Bearing Elements	Nonload-Bearing Elements
Compressively stressed columns	Curtain or panel walls
Walls, exterior or interior	Partitions
Flexurally stressed beams, girders, trusses floors, roofs	Ceilings

Building codes provide requirements for both structural loads and superimposed live loads. Structural failures are generally the result of application of unanticipated loadings. In a fire situation, "loads" are induced by heat which may cause thermal stresses. These stresses may be increased if the members are in any way restrained against expansion. Also, heat may cause a loss of strength, and increase in deflection, if not actual consumption of the member. While restraint in some designs may offset any loss in strength, there is the increased likelihood of collapse in the event of a protracted fire. Properly designed and constructed assemblies, having a fire resistance adequate for the exposure, will retain adequate strength during fire exposure and continue to resist collapse.

Columns

Columns serve to carry building loads to the foundation where the loads are distributed to the supporting ground or rock. The material of the column is established by the type of construction—steel or reinforced concrete if noncombustible, or wood, including heavy timber designs, if combustible. In a fire situation, interior columns are the most severely exposed of any structural members, as they may be enveloped in fire on all sides.

Functions of Walls

Building walls may serve one or more functions and they may be either bearing or nonbearing, and exterior or interior. The following brief definitions describe the general functions that walls may serve.

Bearing wall: A wall that supports vertical loads in addition to its own weight.

Curtain wall: An exterior wall supported by the structural frame of the building. Also called a panel wall.

Enclosure wall: An interior wall that separates a vertical opening for a stairway, elevator, duct space, etc., and connects two or more floors.

Exterior wall: A wall that forms a boundary to a building and is usually exposed to the weather.

Fire partition or fire barrier wall: An interior wall that serves to restrict the spread of fire, but does not qualify as a fire wall.

Fire wall: A wall of sufficient fire resistance, durability, and stability to withstand the effects of an uncontrolled fire exposure, which may result in collapse of the structural framework on either side. Openings in the wall, if allowed, must be protected.

Parapet wall: A portion of an exterior, fire, or party wall that penetrates through and extends above the roof line.

Nonbearing wall: A wall that supports only its own weight.

Partition: An interior wall, not more than one story in height, that separates two areas in the same building but is not intended to serve as a fire barrier.

Party wall: A wall that lies on a common lot line for two buildings and is common to both buildings. Most of these walls may be constructed in a wide range of materials or assemblies.

Cavity wall: A wall of two parallel wythes (vertical wall of bricks or concrete blocks) with an air space between them. The wythes are connected by metal ties.

Faced wall: A wall composed of two different masonry materials: (1) the facing wythe and (2) the backup wythe, which are bonded together to act as a single unit.

Hollow wall: A wall of two parallel wythes, with an air space between them, but without ties to hold the wythes together.

Sandwich wall: A nonbearing wall whose outer faces enclose an insulating core material.

EXTERIOR WALLS

The primary function of all exterior walls is to protect the inside of the building from the elements, such as heat and cold, water, wind, and windblown dust and dirt. In addition, some exterior walls may support floor and roof framing systems and may include portions that support girders or beams.

The fire resistance required for exterior walls is determined not only by the type of construction, but also by the distance the wall is set back from a lot line or an exposing structure. The greater that distance, the less severe the exposure and, hence, the less fire resistance that would be needed.

NFPA 80A, *Recommended Practice for Protection of Buildings from Exterior Fire Exposures,* may be used to determine acceptable combinations of separation distance and exterior wall fire resistance.

Load-Bearing Exterior Walls

Load-bearing exterior walls are generally constructed of either masonry units (such as stone, brick, concrete block, or a combination of these materials) or concrete. Brick veneer is sometimes used as a facing in wood or metal-stud wall construction. In this case, the studs support the applied loads, while the veneer provides an attractive, useful exterior surface. Veneers are also used as the exterior face of cavity walls. Figures 7-2E, 7-2F, 7-2G, 7-2H, and 7-2I illustrate some common forms of exterior wall construction.

Exterior walls of reinforced concrete are either poured in place or precast. Masonry veneers are often used as the exposed surface of reinforced concrete walls. The brick veneer is tied to the concrete by metal ties fastened to the concrete and set into the bed mortar joints of the masonry. If the walls are load bearing, the reinforced concrete is designed to support all of the applied loads.

Exterior masonry walls may be nonload bearing, i.e., supporting only their own weight. The floor and roof framing is supported by columns that transfer the loads to the foundation. Usually each story of the exterior wall is supported on spandrel beams, which frame into the columns.

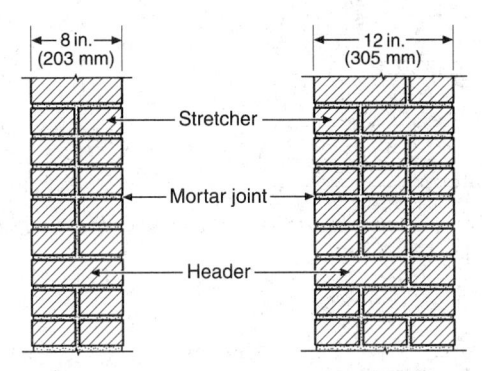

FIG. 7-2E.   Typical types of wall and partition assemblies showing an 8-in. (203-mm) brick bearing wall and a 12-in. (305-mm) brick bearing wall.

Openings in Fire Subdivisions

Some of the structural difficulties experienced with masonry walls in fires are illustrated in Figure 7-2J.

Openings, such as doors and windows, in masonry walls must have supports to carry the weight of the wall structure above the openings. These supports, called lintels, can be constructed of several materials, including steel shapes, precast concrete beams, and stone or brick arches. Figures 7-2K and 7-2L illustrate several types of lintels. Steel lintels used in fire-resistive walls should have a fire-protective covering, such as concrete, or a spray-applied cementitious or mineral coating to prevent failure of the lintel and wall section above the opening. Openings may need to be limited in size and number when the exterior wall is close to the lot line. Openings that are protected do not have fire resistance as such, but can impede the spread of fire even though they have little or no resistance to the transmission of heat.

Openings for which protection (windows, doors) does not have the insulating value as that of the fire barrier wall or fire wall may allow autoignition of combustibles located near the unexposed face of the opening. Maintaining a minimum separation distance (perpendicular to the face of the opening) equal to the maximum dimension of the opening will help prevent this. Such a situation can also hamper egress or cause personal injury in a fire.

Fire-rated glass will remain in place for short-duration fires (about $3/4$ hr to 1 hr); however, most have negligible insulating value. Special insulating fire-rated glass is available; however, it cannot be used on exterior walls in cold climates due to usage limitations of the insulating gel.

Fire doors may be of wood, steel, or steel with an insulated core of wood or mineral material. While some doors have negligible insulating value, others may have a heat transmission rating of 250°F, 450°F, and 650°F (121°C, 232°C, and 343°C). This means the doors will limit temperature rise on the unexposed side to that respective value when exposed to the standard time-temperature for 30 min. This aids egress, particularly in stairwells in multistory buildings, and provides some protection against autoignition of combustibles near the opening's unexposed side.

Curtain Walls

Most multi-story buildings consist of a protected steel or reinforced-concrete frame (columns, girders, and beams). The building envelope may be enclosed with nonbearing walls supported at each level. These walls are often identified as either curtain or panel walls.

Many materials and types of construction are used for curtain walls. Aluminum and stainless steel curtain walls are, by far, the most common, but copper and copper alloys, carbon steel, galvanized metals, porcelain enamel finish, concrete, glass, and plastics are also used. Large window areas are common in this type of wall. Figure 7-2M shows an example of contemporary curtain wall construction.

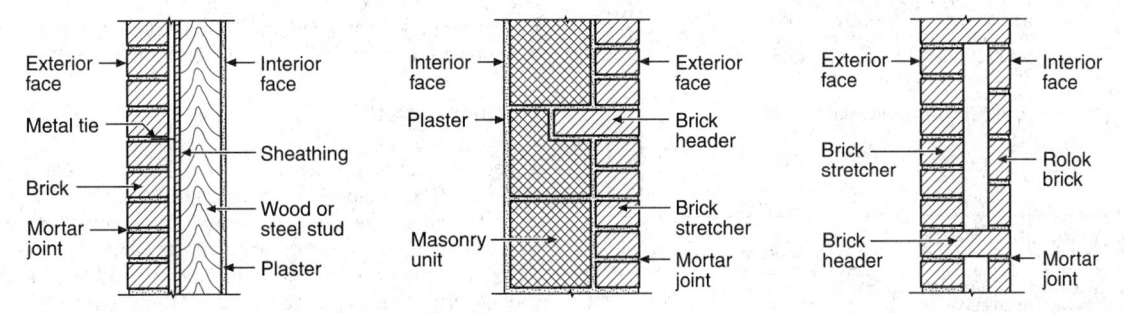

FIG. 7-2F.   Typical types of wall assemblies showing (left to right) an exterior brick veneer on wood-frame wall, a 12-in. (305-mm) exterior faced or veneered wall, and an 8-in. (203-mm) exterior hollow Rolok Bak® brick wall.

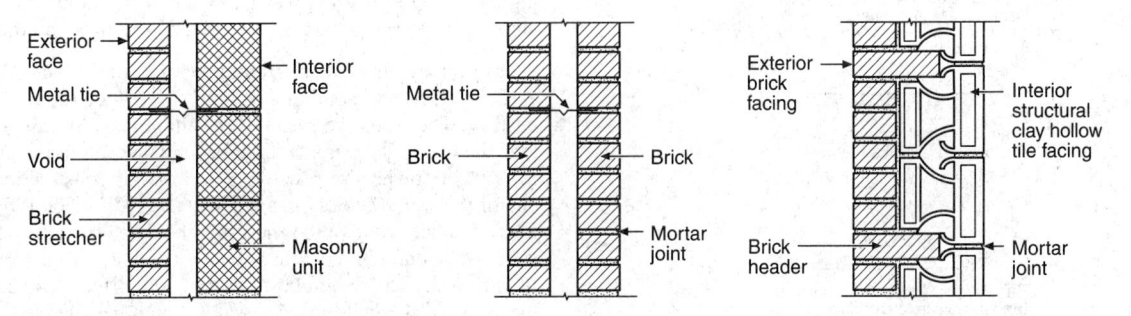

FIG. 7-2G.   Typical types of wall assemblies showing (left to right) two examples of exterior nonbearing cavity walls and a 12-in. (305-mm) exterior bearing wall.

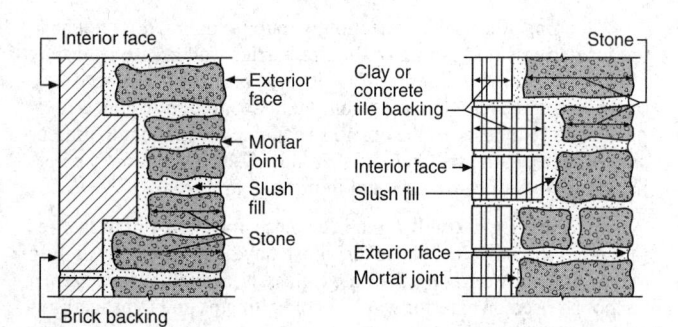

FIG. 7-2H. *Stone-faced wall assemblies, one with brick backing, the other with tile.*

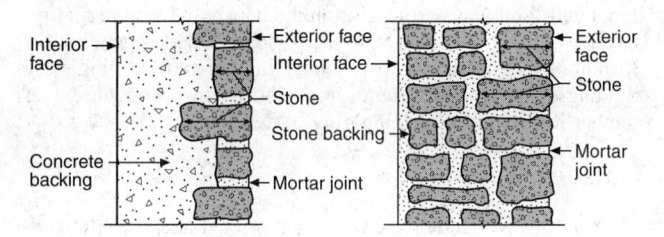

FIG. 7-2I. *At left, a wall of concrete backing and stone face; at right, a solid stone wall.*

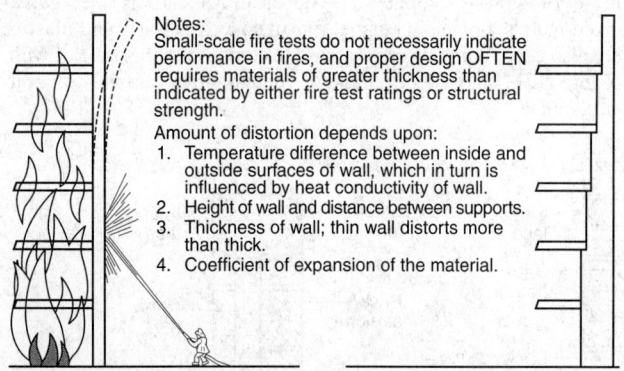

Notes:
Small-scale fire tests do not necessarily indicate performance in fires, and proper design OFTEN requires materials of greater thickness than indicated by either fire test ratings or structural strength.
Amount of distortion depends upon:
1. Temperature difference between inside and outside surfaces of wall, which in turn is influenced by heat conductivity of wall.
2. Height of wall and distance between supports.
3. Thickness of wall; thin wall distorts more than thick.
4. Coefficient of expansion of the material.

Inside of wall, heated by fire, expands. Results in wall leaning out at top and falling away from fire. Cooling by hose streams on outside of wall, while inside remains hot, may hasten collapse.

Masonry walls with extra thickness on lower floors are less likely to collapse, as the greater thickness mass results in less tipping.

Masonry structures with well-bonded cross walls at frequent intervals are not likely to collapse from heat expansion as is the case in high walls without lateral support.

Pilasters furnish another method of providing lateral support.

Panel walls between rigid supports tend to bow in toward fire.

FIG. 7-2J. *Heat expansion effects on ordinary masonry walls.*

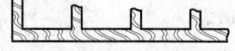

FIG. 7-2K. *Lintels in brick-faced, concrete block walls. (For SI units: 1 in. = 25.4 mm.)*

FIG. 7-2L. *Use of concrete and brick lintels in masonry walls.*

A complete curtain wall assembly may consist of a panel with finished outside and inside surfaces, insulation, and means of attachment to the building frame. However, this type is not used as frequently as is a metal or glass skin assembly reinforced by conventional construction.

Curtain walls are required to have fire-resistance ratings up to 2 hr, depending upon the separation distance between the wall and a lot line or exposing structure. Secure fastening of curtain walls is important for many reasons, including stability during fire exposure. Curtain walls are generally bolted to clips attached to the columns, spandrel, or floor slab. There is usually a space between the outer end of the floor slab and the inside face of the curtain wall. Unless adequately firestopped, this space may act as a route for vertical fire spread for the entire height of a building.

Parapet Walls

A parapet wall is an extension of the wall construction above the roof line. The parapet on a fire wall helps to prevent a fire from spreading across the roof over the fire wall. Where walls extend only a few inches above the roof, ignition of an adjoining roof may occur, depending upon wind direction and velocity and the character of the roof covering. NFPA 220 and Factory Mutual Research Corporation (FMRC), for example, specifies a parapet height of 30 in. (762 mm). Many building codes permit lesser heights, and in large sections of the country, 18-in. (457-mm) parapets are considered standard. The higher the parapet, the greater the degree of safety, but considerations of expense and appearance dictate practical compromises.

Fire may also extend around the ends of fire walls where the exterior walls of the building on both sides of the fire wall are combustible. There are two methods of minimizing this danger: (1) extend the fire wall several feet out beyond the wall of the building or (2) provide blank fire-resistive exterior wall construction at each end of the wall. (See Figures 7-2N(a) and (b) and Table 7-2F.) The fire resistance rating of the end walls should be from the outside and should be a minimum of 1 hr but not more than 2 hr less than that of the fire wall. Where combustible roofs or cornices project beyond the walls of the building, fire walls should be extended to form a break in such combustible construction, as shown in Figure 7-2O. Otherwise effective fire walls have failed, because wood platforms

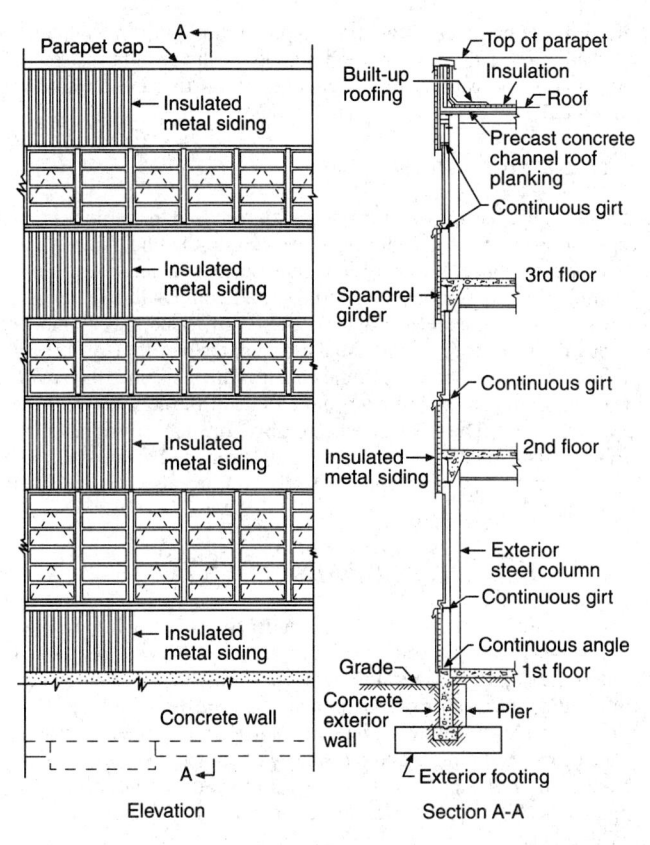

FIG. 7-2M. *Elevation of typical bay of steel frame curtain wall construction.*

TABLE 7-2F. *Length of End Wall Protection*

Height of Exposing Area [ft (m)]	Length of End Wall Protection* [ft (m)]
Up to 40 (12.2)	6 (1.8)
41 to 70 (21.3)	10 (3.1)
71 (21.6) and over	14 (4.3)

*Protection should consist of blank, fire-rated construction.

or projecting canopies along the sides of buildings have spread fire from one fire area to another.

INTERIOR WALLS AND PARTITIONS

Interior walls and partitions used for building corridors or rooms may be either bearing or nonbearing. Bearing partitions are common in the older, wall-bearing construction systems and are also employed with industrialized building systems, particularly those of precast concrete. Newer construction tends to use nonbearing partitions.

The need for open space flexibility in modern construction has led to floor and interior partition design that allows nonbearing partitions to be installed at any location. These moveable partitions are often made of steel studs and gypsum board, and they may be installed, then dismantled, and reinstalled at another location when occupancy needs change. Normally, moveable partitions extend only from the floor to the underside of the ceiling.

Interior partitions, particularly nonload-bearing partitions, can be installed with a number of other different materials. Wood stud

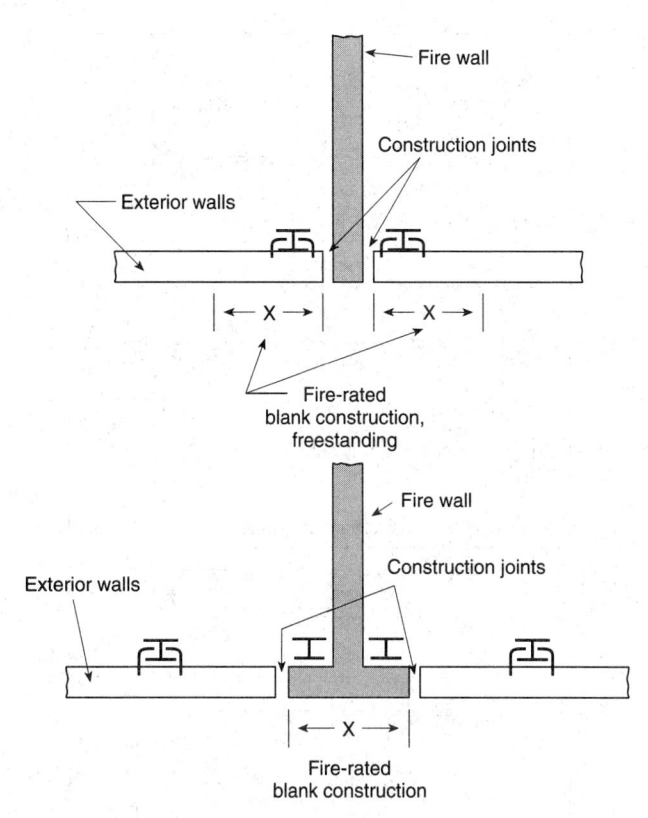

FIG. 7-2N. *End wall exposure protection: (top) end walls tied to structural framing (bottom) end walls not tied to structural framing. (See Table 7-2F.)*

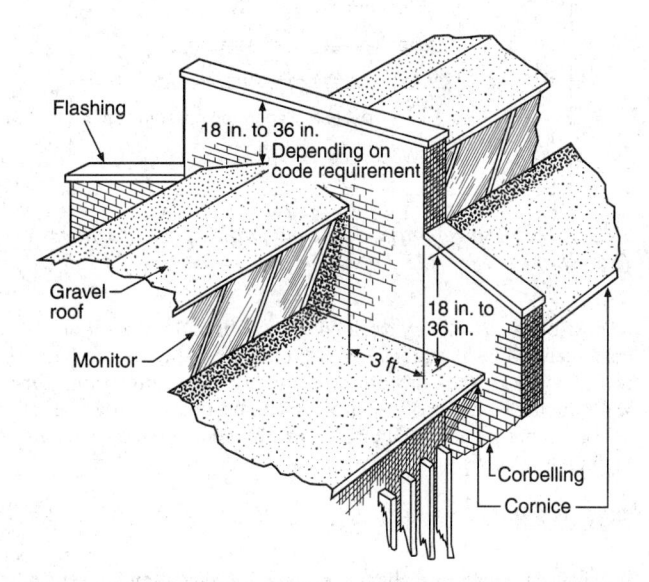

FIG. 7-2O. *Fire wall installation on building with combustible roof and monitor. (For SI units: 1 in. = 25.4 mm; 1 ft = 0.305 m.)*

and gypsum board or plaster is another very common type of partition. Partitions constructed of masonry units, such as concrete block, structural clay tile, terra cotta, and gypsum block, are or have been commonly used in a wide variety of buildings and are more permanent. Figure 7-2P shows typical partition constructions.

Properly designed interior partitions can act as barriers to the spread of fire. To protect certain areas more completely than would

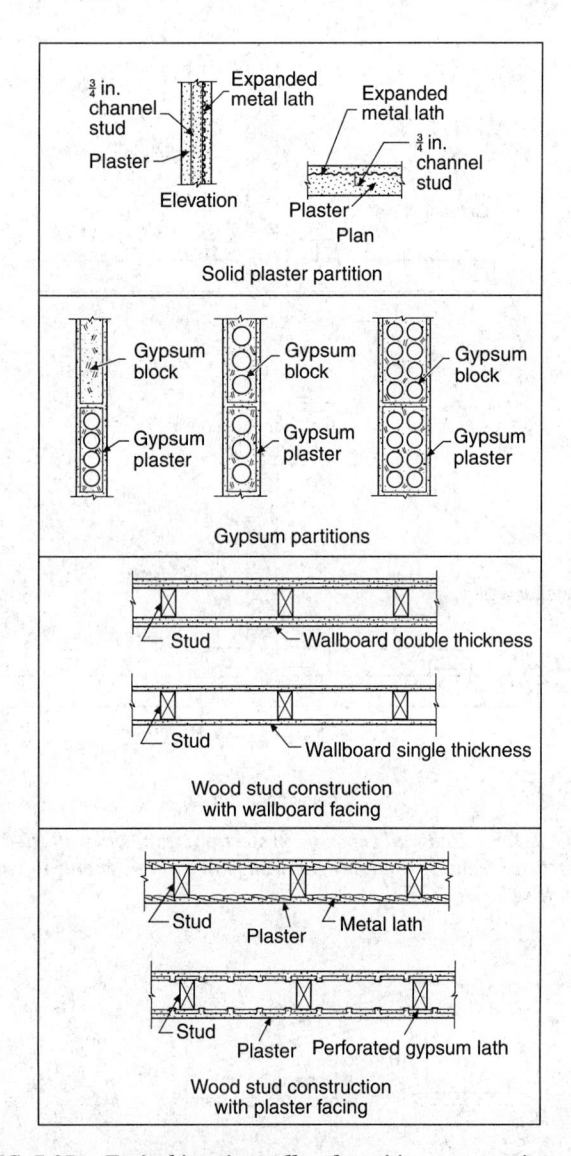

FIG. 7-2P. Typical interior wall and partition constructions. (For SI units: 1 in. = 25.4 mm.)

be possible with ordinary partitions, fire walls and fire partitions are constructed. Fire partitions may also be referred to as fire barrier walls; however, they are often incorrectly called fire walls. Fire walls are fire resistive, of superior noncombustible construction, often self-supporting, and cannot be readily modified to meet changing building needs.

Fire Walls

Fire walls are interior walls providing a fire separation between areas of the same building. They should be designed to maintain structural integrity, even in cases of complete collapse of the structure on either side of the fire wall. They are constructed of reinforced concrete, concrete block, prestressed concrete, and sometimes brick. Concrete or masonry walls generally require some type of steel reinforcement to withstand heat expansion effects. Adequate cover thickness of concrete must be provided over steel reinforcement. This is particularly important in the case of prestressed concrete walls. When designing prestressed concrete walls, refer to *Design for Fire Resistance of Precast Prestressed Concrete*.[1] Also, they may be buttressed by pilasters if of considerable height or length. In

fire-resistive buildings, structure-supported fire division walls may be used, provided the structure has equal or greater fire resistance than that of the wall. Requirements for fire walls can be found in NFPA 221.

Fire Barrier Walls

Usually, a fire barrier wall (or fire partition) subdivides a floor or an area and is erected to extend from the floor to the underside of the floor or roof above. Fire partitions may be constructed of noncombustible, limited combustible, or protected combustible materials, and should be attached to, and supported by, structural members having fire resistance at least equal to that of the partition. A fire barrier wall normally possesses somewhat less fire resistance than a fire wall and does not extend from the basement through the roof, as does a fire wall. The fire-resistance ratings for such partitions range from $1/2$ to as much as 4 hr. Requirements for fire barrier walls are less involved than those for fire walls and are also contained in NFPA 221.

FLOOR FRAMING SYSTEMS

Floor framing systems typically include not only the flooring assembly but also its supporting beams, girders, or trusses. Beams and girders are nearly always an integral part of the floor system, while trusses may serve other purposes (and hence are discussed separately). The range of different types of floor framing systems that include beams and girders and the variety of components that may be included in a given assembly is best portrayed through drawings. Figures 7-2Q through 7-2AA are typical of many designs that are currently used.

Figures 7-2Q and 7-2R are generalized framing plans for steel and reinforced-concrete floor assemblies. The steel beams shown in Figure 7-2Q support the floor assembly and are, in turn, supported by girders. Interior girders support beams on both sides. Girders are often concealed within partitions where their depth is not apparent, and they can directly support the weight of partitions in the space above. The girders frame into steel columns. Figure 7-2R illustrates

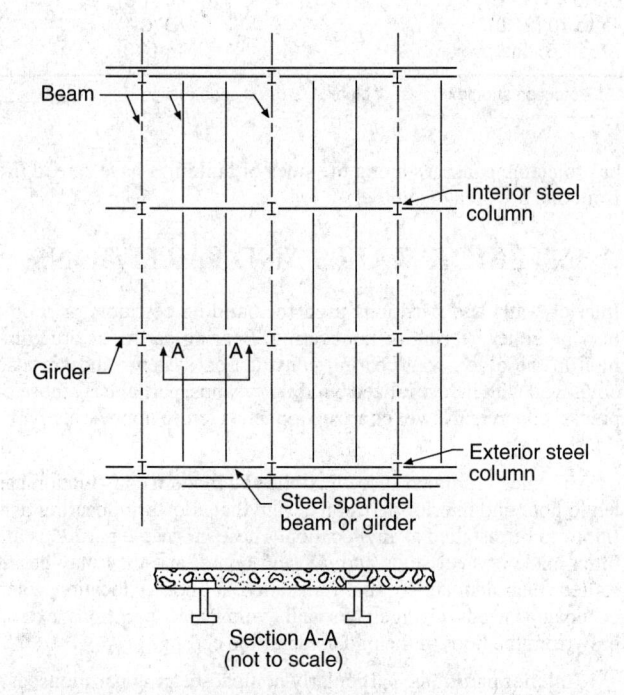

FIG. 7-2Q. Portion of floor plan of steel frame structure.

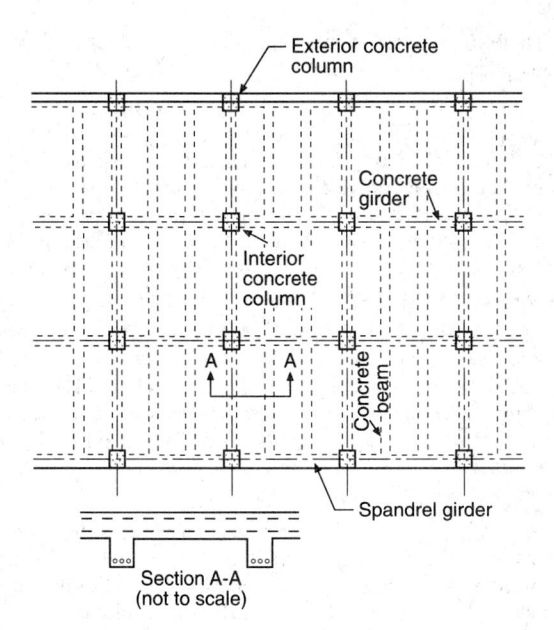

FIG. 7-2R. *Portion of floor plan of reinforced concrete beam and girder construction.*

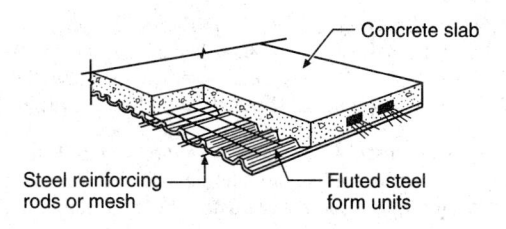

FIG. 7-2S. *Example of concrete floor construction showing composite floor assembly.*

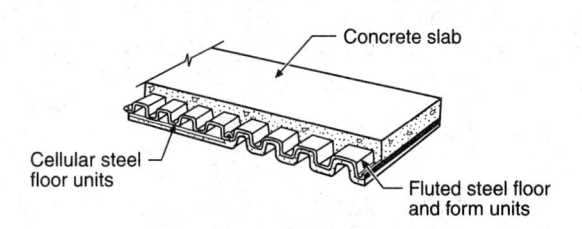

FIG. 7-2T. *Example of concrete floor construction showing light gage cellular floor panels alternated with light gage steel forms.*

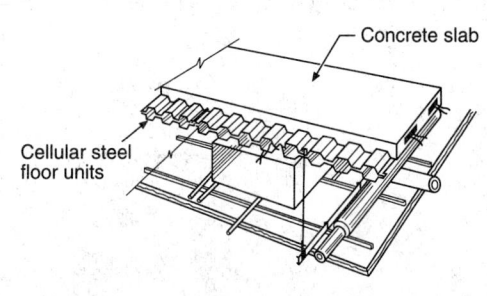

FIG. 7-2U. *Example of a concrete floor poured over light gage cellular floor panels. The cellular panels can carry all three basic building services—power, telephone, and data transmission lines.*

a slab beam girder column system cast monolithically (no joints) of reinforced concrete. The functions of the various components are the same as those described for the steel frame.

The floor slab supported by the steel beams shown in Figure 7-2Q need not be of reinforced concrete. Alternative types of deck are the composite decks shown in Figures 7-2S, 7-2T, and 7-2U. This type of floor system offers many construction advantages. The metal deck acts as a form for the concrete that remains in place after curing. In addition, the cellular panels can be used for building service raceways where desired. A feature of this floor system is that the steel deck also acts as the tensile reinforcement of the concrete floor slab. A spray-applied fire-resistive coating on the steel deck underside may be required in some cases.

Another floor deck modification is to replace the reinforced-concrete floor slab with a precast concrete system. Figure 7-2V illustrates

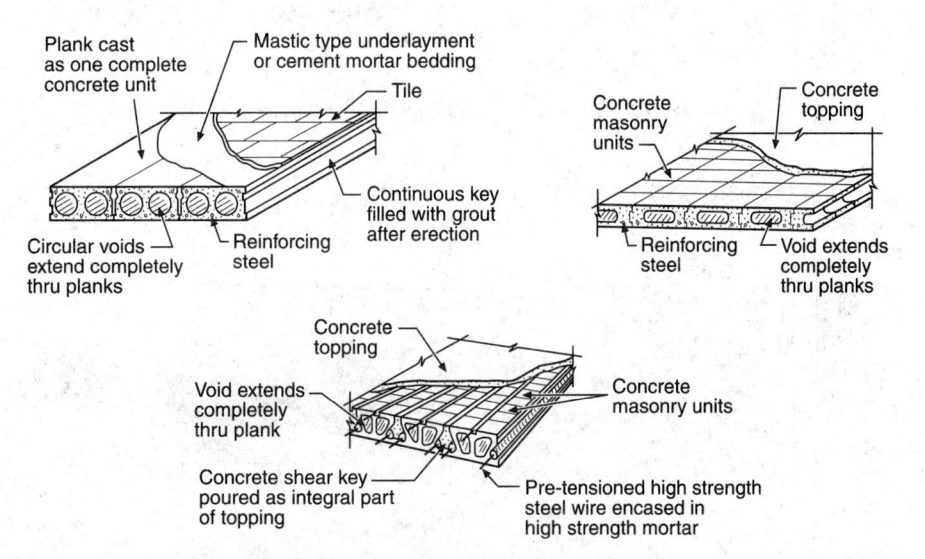

FIG. 7-2V. *Three examples of precast concrete floor slab construction.*

three common precast concrete floor slabs. The precast concrete planks generally have voids through their length which reduce the weight of the planks and act as building service raceways. The planks may be supported by either steel beams or reinforced concrete beams.

Open-web joists, which are lightweight prefabricated trusses, may replace steel beams as the intermediate flexural framing and are a very common type of construction both for floors and roofs. The spacing of floor joists is approximately 2 ft (0.61 m), and generally about 4 ft (1.22 m) for roof joists. Figure 7-2W illustrates this type of construction.

Because the spacing of the joists is so close, the floor slab can be thinner and still support the same load. Corrugated steel forms with 2 $^1/_2$ in. (64 mm) of concrete as a wearing surface is a typical design. The corrugated deck acts as a form for the concrete and con-

tributes to the support of live loads. Lightweight concrete is often used in place of normal-weight concrete to reduce the dead load.

Similarly, the reinforced concrete beams shown in Figure 7-2R may be spaced more closely together, permitting both a thinner floor slab and smaller beams. This type of construction is generally called a ribbed reinforced-concrete system or a concrete joist system. The ribbed system can be either cast in place or precast. Figure 7-2X illustrates the concrete joist floor, while Figure 7-2II illustrates two of the more common forms of ribbed precast systems. In each illustration, the ribs are the intermediate flexural framing.

The two-way flat slab is an increasingly popular floor system. Two-way reinforced-concrete slabs, used since the turn of the century, include reinforcement placed in two directions. The slab performs the functions of deck, intermediate flexural framing, and primary flexural framing. The underside of the slab may be either flat or ribbed in two directions. Figures 7-2Y and 7-2Z illustrate two different forms of this type construction.

Wood floor systems can be designed to perform the same functions as the basic model in Figure 7-2Q. If the reinforced-concrete deck were replaced by heavy wood planking 2 in. (51 mm) or more in thickness, and the steel beams, girders, and columns with large wood members with a minimum dimension of 6 in. (152 mm), the basic form would be that of mill construction. This construction, shown in Figure 7-2AA, was quite common for industrial buildings and warehouses during the last century. The general design is still used in locations where wood has aesthetic appeal.

Lightweight, wood-frame construction is usually used in buildings of limited size. Floor joists in such construction are nor-

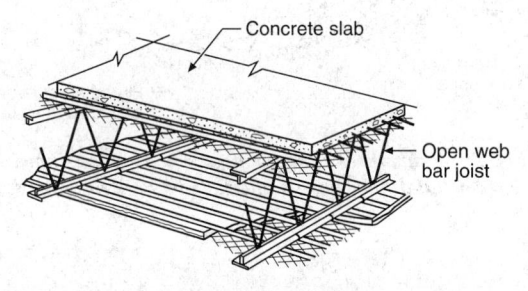

FIG. 7-2W.   Example of concrete floor construction showing open-web steel joist.

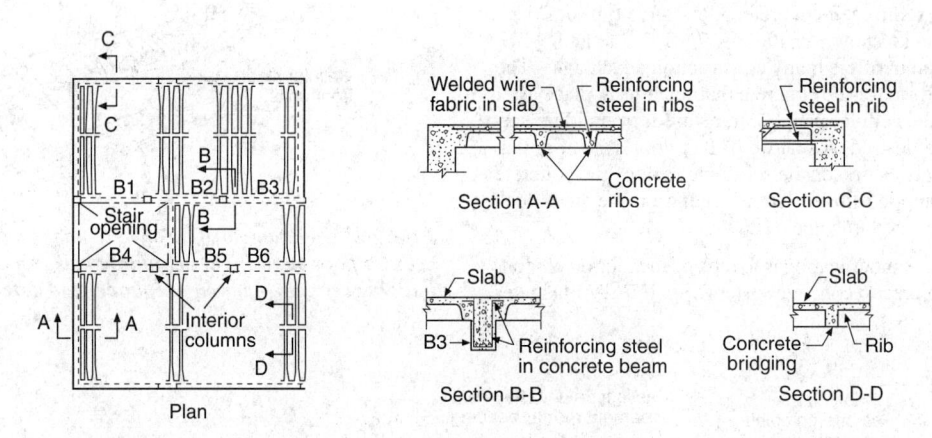

FIG. 7-2X.   Partial plan of a concrete joist floor with four sectional views.

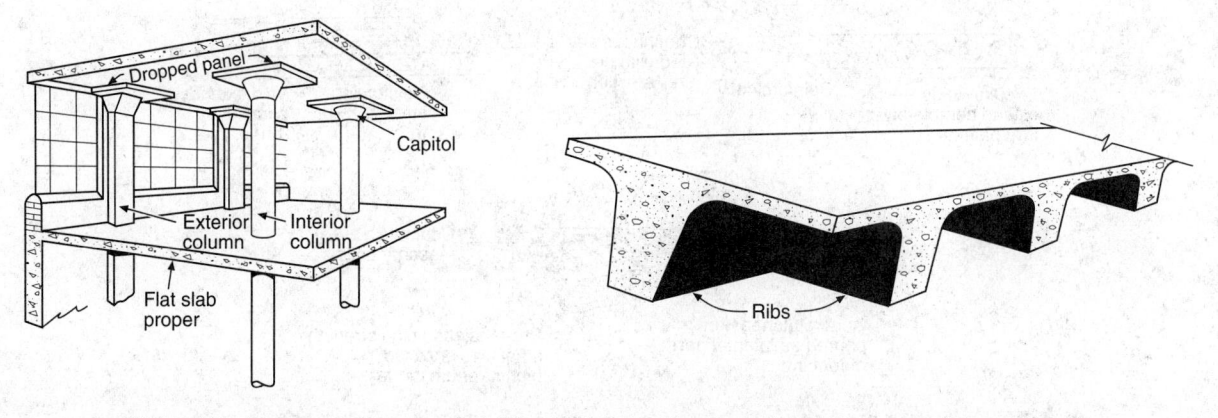

FIG. 7-2Y.   Two types of two-way flat slabs: (left) flat underside and (right) two-way ribbed underside.

mally spaced 16 in. (406 mm) on center, and the vertical supports are often 2 by 4 in. (51 by 102 mm) or 2 by 6 in. (51 by 152 mm) wall-bearing studs, again spaced 16 in. (406 mm) on center. The stud walls may extend only from floor to floor in platform framing (Figure 7-2BB), or they may be extended continuously for two or three floors, a type of framing identified as balloon construction. (See Figure 7-2CC.)

Wood-frame construction has little fire resistance because flames and hot gases can penetrate into the spaces between the joists or the studs. Firestopping must be installed at critical locations throughout the building to prevent the rapid vertical and horizontal spread of fire. Firestops are normally made of wood blocks or non-combustible material. Figures 7-2CC and 7-2DD illustrate critical locations for firestopping in balloon frame and in platform frame construction; Figure 7-2EE shows the firestopping for these constructions in greater detail.

Watertightness and Drainage of Floors

In addition to the effectiveness of floor framing systems as barriers to fire spread, there is another consideration involving floors in multi-story buildings that can have a bearing on their performance in fire situations. This consideration is the degree of watertightness of construction and drainage arrangements for the floors. Massive quantities of water are sometimes needed on fires in unsprinklered buildings of combustible construction or buildings of noncombustible construction with significant occupancy fuel loads. Without good drainage and/or waterproofing arrangements, water could pose the threat of increased water damage on floors below.

Where acceptable, one inexpensive method of drainage is to provide scuppers in the exterior walls. Floor drains connected to ample size headers or a combination of scuppers and floor drains may be used. The number of scuppers or drains is dependent upon the hazard and the amount of water likely to be used. For average conditions, the recommendations in Table 7-2G are considered good practice.

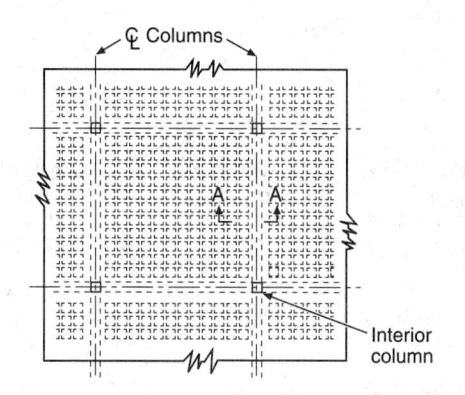

FIG. 7-2Z. *Partial typical interior two-way ribbed floor slab with section view A-A.*

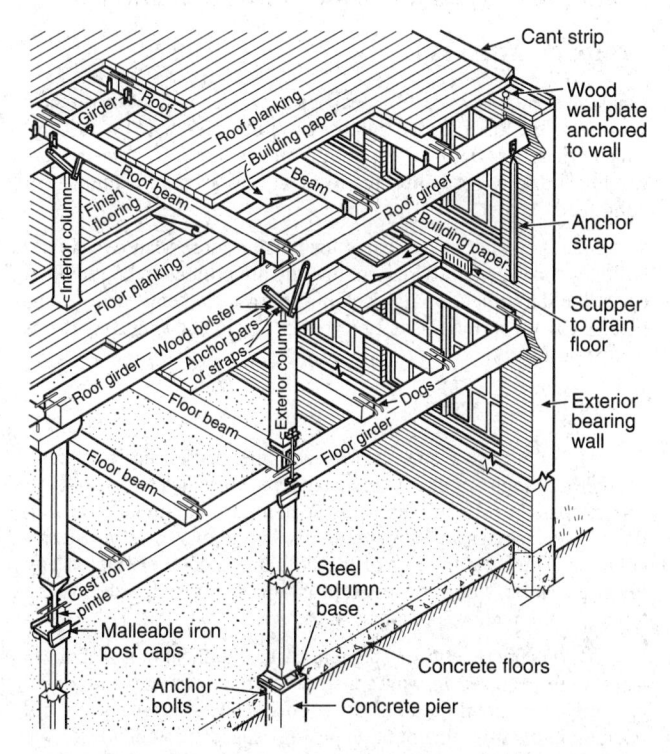

FIG. 7-2AA. *Components of a heavy timber building showing floor framing and identifying components of a type known as semimill.*

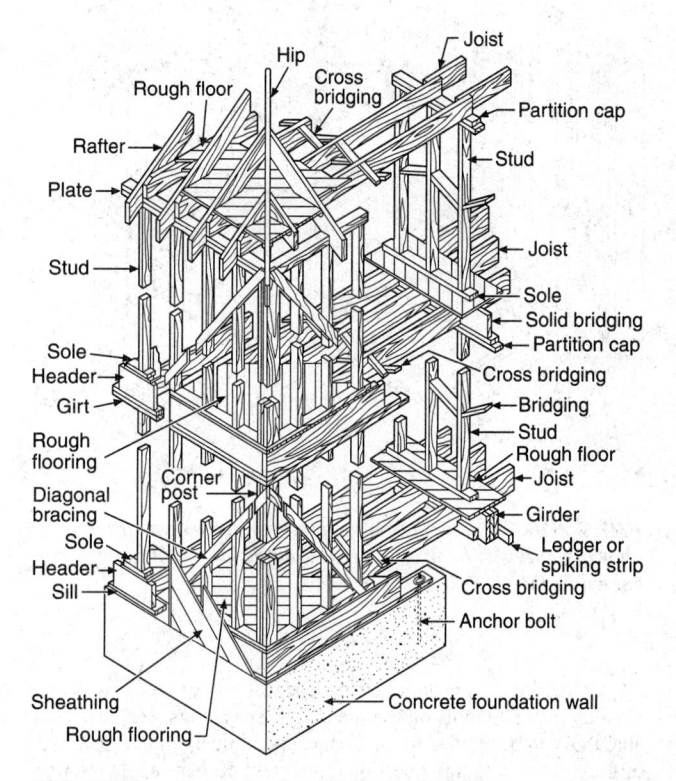

FIG. 7-2BB. *Example of wood-frame platform construction common to dwellings. Structural members are identified.*

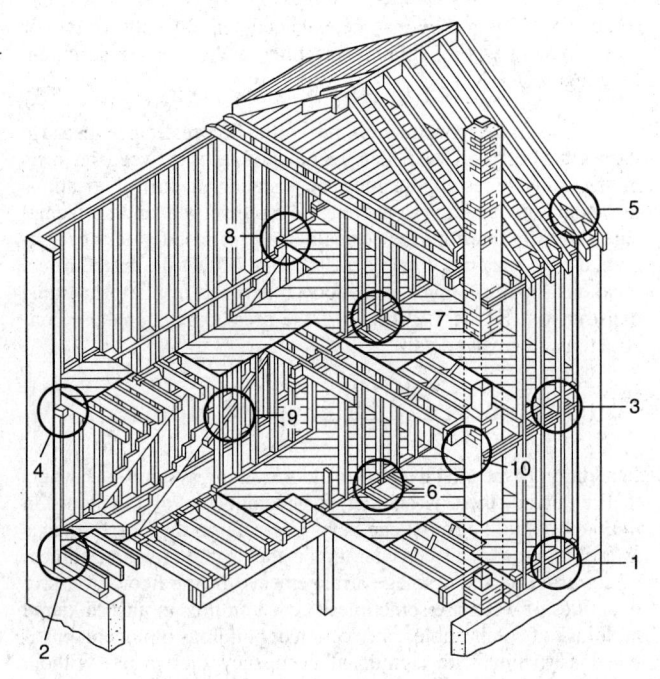

FIG. 7-2CC. *Example of wood balloon frame construction showing points to be firestopped. Expanded views of points are shown in Figure 7-2EE.*

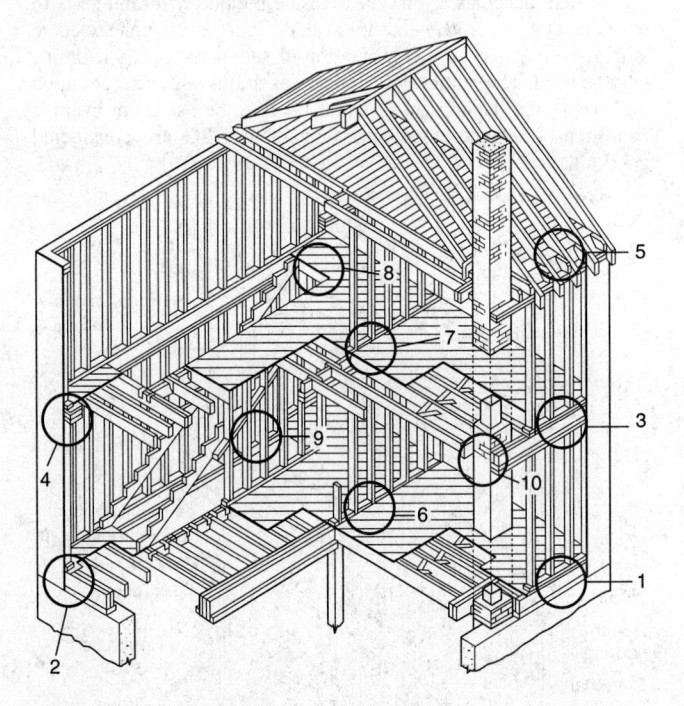

FIG. 7-2DD. *Example of wood platform frame construction showing points to be firestopped. Expanded views of points are shown in Figure 7-2EE.*

Waterproofing techniques, such as waterproof membranes, hot mastics, asphalt emulsions, cementitious surfacings, etc., are available that can be applied to the floors depending upon the basic type of floor construction in question. The intent of these applications is to provide a watertight barrier on the floor itself which extends up 4 to 6 in. (50 to 150 mm) at walls and columns. Water impounded by

TABLE 7-2G. *Sizes for Scuppers and Drains*

Floor Areas	No. of 4-in. (102-mm) Scuppers or Drains
500 sq ft (46 m²) or less	2
750 sq ft (70 m²)	3
1000 sq ft (93 m²)	4

Additional scuppers for areas over 1000 sq ft (93 m²):

For Extra Hazard Occupancies, Group 1 or 2 (quantity and combustibility of contents is very high), or floors of questionable watertightness, or contents especially subject to water damage—one scupper for each additional 500 sq ft (46 m²).

For Ordinary Hazard Group 2 Occupancies (quantity and combustibility of contents is moderate) with watertight floors—one scupper for each additional 1000 sq ft (93 m²) or fraction thereof.

For Ordinary Hazard Occupancies Group 1, or Light Hazard Occupancies (quantity and combustibility of contents is low) with strictly watertight floors—one scupper for each additional 2000 sq ft (186 m²) or fraction thereof.

these protective measures must be carried off before its weight becomes a loading threat to the building.

TRUSSES

Where large areas must be column free or where special occupancy requirements may warrant, trusses may be used for purposes other than roof support. Three such applications are identified as: (1) transfer trusses, (2) staggered trusses, and (3) interstitial trusses.

Transfer Trusses

Load transfer trusses create a clear space on a lower floor by directly supporting the loads from columns above the truss or, at times, from tension members below. This design allows large, column-free areas for auditoriums, ballrooms, etc., at any level in a multistory building.

Since transfer trusses carry loads from more than two floors, building codes generally require that the fire protection be provided by individual protection for each of the truss elements, or by complete enclosure of the truss (for its entire height and length) with envelope protection.

Staggered Trusses

The staggered truss system is primarily intended for high-rise residential buildings. Typically, in a staggered truss design, story-high trusses span the full building width at alternate column lines on each floor. These trusses are supported only on the two rows of exterior columns. Thus, the interior of the building is column free; at any given elevation, floor construction is alternately supported on the top and bottom chords of adjacent trusses.

It is characteristic of staggered truss applications that these trusses are enclosed in wall construction that separates individual apartments or hotel/motel guest rooms. There may be a control opening in the truss that permits passage space for a corridor.

Since staggered trusses are usually enclosed in walls, the entire wall assembly must have a fire resistance required for this condition.

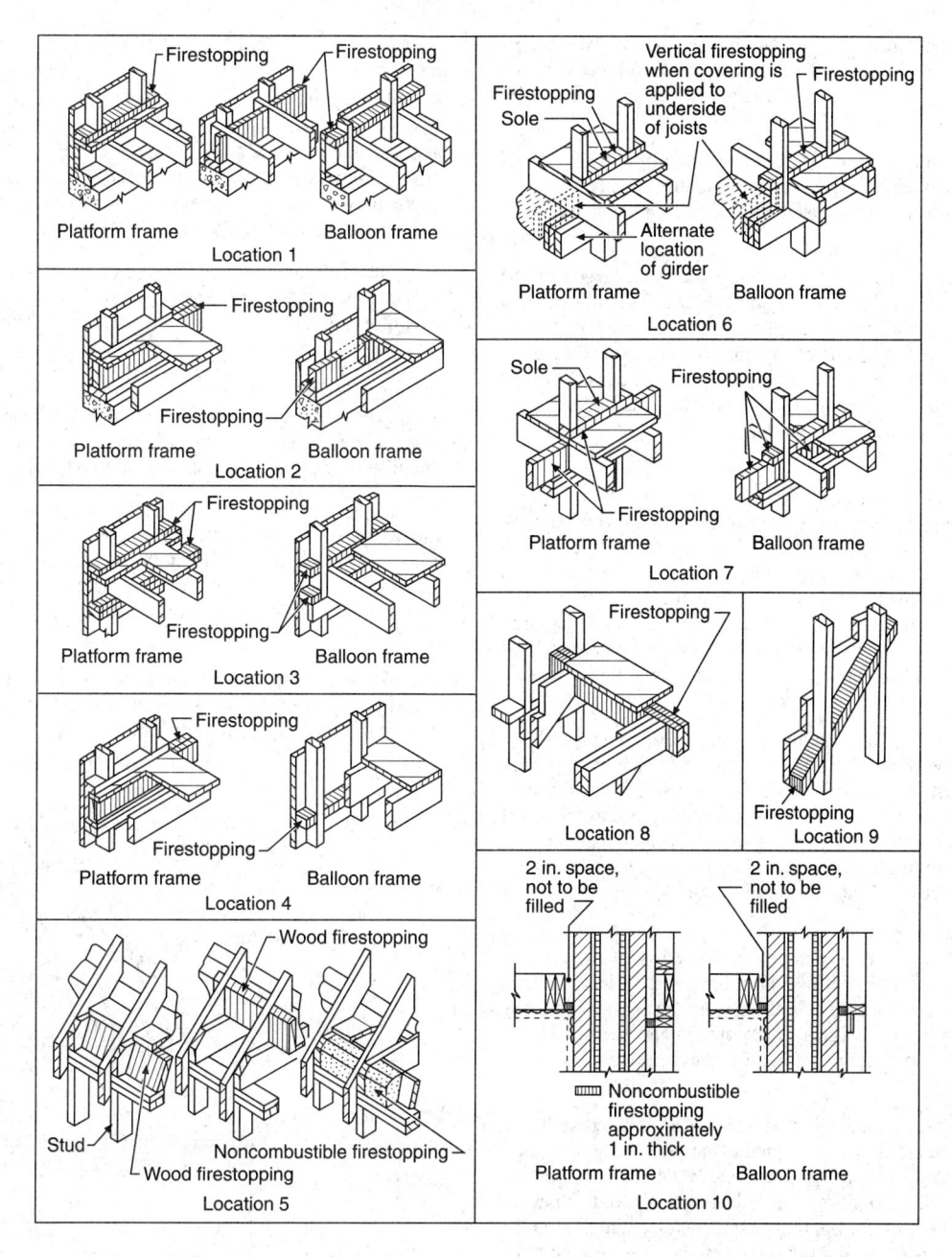

FIG. 7-2EE. *Details of the application of firestopping to platform and balloon framing. Location numbers coincide with locations circled in Figures 7-2CC and 7-2DD. (For SI units: 1 in. = 25.4 mm)*

Interstitial Trusses

The interstitial truss concept was developed primarily to solve the functional needs of hospitals. While steel trusses in interstitial framing systems are quite deep [on the order of 8 ft (2.5 m) in height], they do not extend from floor to floor as do staggered trusses. The top chords support the floor above, while the bottom chords support a suspended ceiling system and a walking surface for maintenance purposes. As such, interstitial trusses are analogous to very deep open-web steel joists.

The interstitial space thus created provides a convenient location for the complex mechanical and electrical systems that are necessary components of a modern hospital. Direct access for maintenance, renovation, and replacement of the various system components within these interfloor spaces is provided without significant interference with normal operations.

Since each interstitial truss supports only one floor, all three conventional fire protection methods for trusses are permitted by the model building codes. However, because of practical considerations, individual protection of each truss element or ceiling membrane protection are the methods most widely used.

FLOOR/CEILING ASSEMBLIES

Ceiling components are important elements in the performance of a fire-resistive floor/ceiling assembly. In the event of fire within a room, the ceiling acts as a barrier to protect the structural framing above it. The degree of protection, of course, depends upon the type of material, its installation, and its completeness. Combustible ceilings or ceilings that do not remain in place when subjected to the pressures and temperatures of a fire do not provide a significant degree of protection.

It is important to note that the membrane ceiling is one part of the floor/ceiling assembly. The entire assembly, acting together, provides the designated fire resistance. The ceiling itself has insufficient fire resistance.

Ceilings may be applied directly to, or they may be suspended from, the underside of the floor framing. Figure 7-2II illustrates a ceiling connected directly to the floor framing system, while Figure 7-2U shows a ceiling suspended from the floor slab. It is also common to apply a fire-protective material, such as a plaster or sprayed mixture, directly to the bottom surface of the floor slab. Figure 7-2FF illustrates a floor slab with plaster applied directly to the steel floor deck.

There are several types of suspended membrane ceilings in common use. One type consists of a lay-in system in which the ceiling panels are supported by an exposed metal grid. Another common type of membrane has the metal grid concealed by recesses in the edges of the ceiling panels. Suspended membrane ceilings may also consist of traditional lath and plaster or permanently secured gypsum wallboard with finished joints. Both of these latter systems can be either suspended below or attached directly to the basic subflooring system.

In the variety of arrangements currently used, the space above the ceiling may be of combustible construction and may also contain other combustible materials, such as duct insulation, wire insulation, and vapor seals. Air-handling systems may use ducts or the void space itself (plenum) as a means of supplying, exhausting, or recirculating air; but associated materials must meet stringent requirements for combustibility and smoke evolution.

The material used for the membrane ceiling is important. Gypsum plaster and lath were a common construction of early membranes. Today special mineral tile formulations are generally used for acoustical panel systems that will not warp during fire exposure. The support framing is designed to accommodate thermal expansion with a minimum of distortion, so that the ceiling panels stay in place.

If lighting fixtures and duct openings are included in membrane ceilings that are part of fire-rated assemblies, they both must be of suitable design, properly spaced, and often require shielding to reduce heat transmission into the ceiling space. Fire test performance is based on specific lighting fixtures and spacing, so indiscriminate substitution of lighting fixtures can cause premature failure. If air duct openings are installed in a membrane-type ceiling, they should be suitably protected or dampered, as determined by a fire test on the ceiling assembly that incorporates the air-handling components.

ROOF FRAMING

The design and construction of roof framing follow the general pattern for floor framing systems—both must support vertical loads and distribute these loads to walls or columns. Roof loads are usually smaller than floor loads. In addition, architectural considerations may demand longer spans than floor framing, and the shape of the roof need not be flat.

The roof covering may be supported by steel or wood joists or by a truss where a longer span is needed. Roof trusses can be made to conform to any roof shape, whether flat, pitched, or curved.

Open-web steel or wood joists are often used in flat roof construction. An open-web joist is merely a lightweight, parallel chord truss. They are usually spaced about 6 ft (1.8 m) on center. This close spacing allows the roof deck to be designed to span between the joists. Figure 7-2GG illustrates this type of roof framing.

In single-story buildings that require a large open space, rigid frames are sometimes used both as columns and roof supports. The portion spanning the roof is rigidly connected to columns on opposite sides of the building to form a single member. The frame itself may provide for a sloping roof. Figure 7-2HH shows a rigid frame structure.

Flat slab and ribbed concrete slab systems may be used for roof construction in those buildings in which that is the predominant type of framing. Other systems utilizing long-span prestressed con-

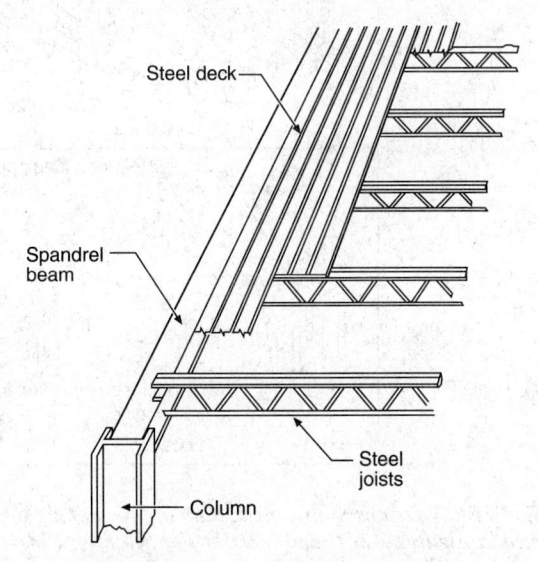

FIG. 7-2GG.   Long-span joist framing.

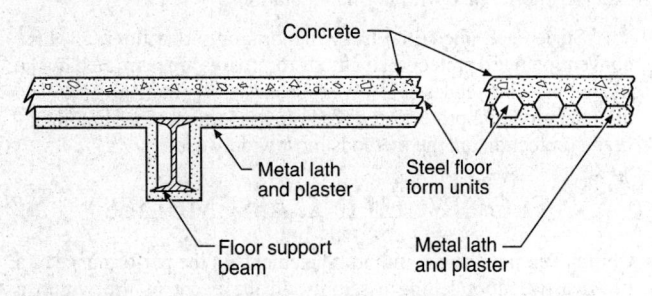

FIG. 7-2FF.   Plaster applied directly to underside of floor slab construction.

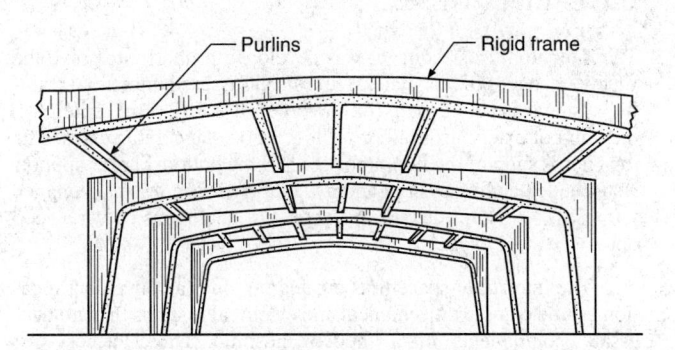

FIG. 7-2HH.   A rigid frame.

crete double-tee or channel sections are also common. These units serve both as support and as the deck. (See Figure 7-2II.)

Roof decks may also consist of precast reinforced-concrete planks fastened to the structural steel supports by galvanized metal clips, as shown in Figure 7-2JJ. These planks are manufactured in three general shapes: (1) square edge, (2) channel slab, and (3) tongue and groove. To reduce weight they may be fabricated with aerated concrete using lightweight aggregates.

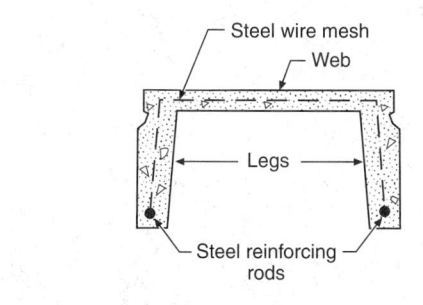

FIG. 7-2II. *Sections through channel and double-tee concrete slabs. The slab at bottom shows a metal lath and plaster ceiling attached directly to the ribs of the slab.*

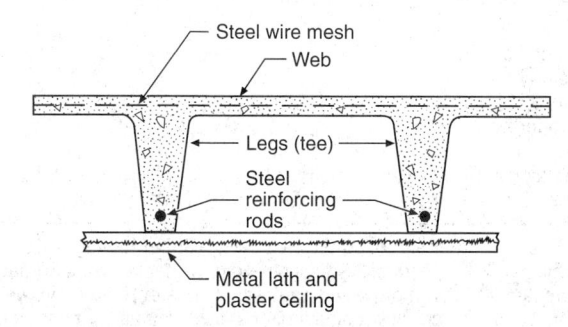

Lightweight precast concrete channel roof plank

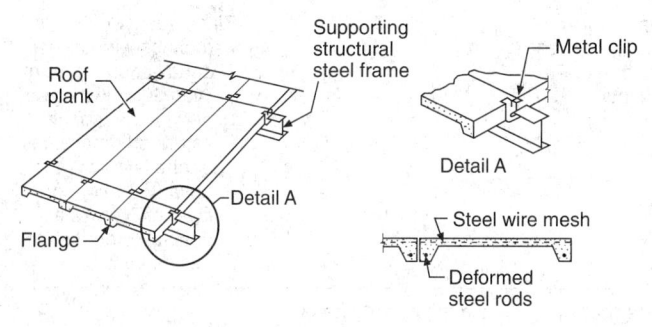

Lightweight tongue and groove concrete roof plank

FIG. 7-2JJ. *Lightweight precast concrete planks.*

ROOF COVERINGS

In resistance to ignition and burning, roof coverings range from combustible wood shingles with no fire-retardant treatment to built-up or prepared coverings that are effective against severe external fire exposure. Insofar as well-designed roof coverings can protect buildings from exposure fires, the likelihood of fire spread from one building to another can be reduced considerably.

Fire-Retardant Roof Coverings

Definitions of three classes of roof covering that include criteria for resistance to ignition have been prepared by Underwriters Laboratories Inc. (UL). The test standard for roof coverings is NFPA 256, *Standard Methods of Fire Tests of Roof Coverings* (also known as ASTM E108). The three classes are:

Class A coverings: Include roof coverings that are effective against severe fire exposures. These coverings are not readily flammable, do not carry or communicate fire, afford a fairly high degree of fire protection to the roof deck, do not slip from position, and possess no flying brand hazard.

Class B coverings: Include roof coverings that are effective against moderate fire exposures. These coverings are not readily flammable, do not readily carry or communicate fire, afford a moderate degree of fire protection to the roof deck, do not slip from position, and possess no flying brand hazard; however, they may require repairs to maintain their fire retardant properties.

Class C coverings: Include roof coverings that are effective against light fire exposure. These coverings are not readily flammable, afford at least a slight degree of fire protection to the roof decks, do not slip from position, and possess no flying brand hazard; however, they may require repairs or renewals to maintain their fire-retardant properties.

Building codes commonly require Class A or B coverings within the fire limits of cities or wherever fire-resistive construction is required. Class C roofing is appropriate for other buildings. Many cities specify Class C as the minimum standard for roofing.

Fire testing laboratories classify metal deck roof assemblies by type of construction.[3] (See Table 7-2H.)

Built-up and Prepared Coverings

Fire-retardant roof coverings fall within one of two groups. One is the built-up covering which, as the name implies, consists of several layers of materials applied or "built up" on the roof decks according to specifications. What is known as a "tar and gravel" roof is actually a built-up roof covering consisting of several layers of roofing felts and insulating panels or sheets bonded together by hot or cold cements and topped with roofing gravel. (See Figure 7-2KK.)

The second group (prepared coverings) includes fire-retardant shingles and sheet coverings, which can be applied only to roof decks capable of receiving and retaining nails and which are sloped sufficiently to permit drainage. In some instances, however, the slope must not exceed a specified maximum in order to qualify for Class A, B, or C fire-retardant rating. Asphalt organic felt shingles, i.e., "composition" roofs, are a common example of prepared roof coverings and are frequently used on dwellings. (See Figure 7-2LL.)

Prepared roof coverings also include such noncombustible coverings as brick, concrete, tile, and slate. Tables 7-2H and 7-2I list various types of prepared and built-up roof coverings, respectively, and their fire-retardant classifications. Any roof covering can be used for a less severe exposure than the one for which it is listed.

TABLE 7-2H.   *Some Typical Prepared Roof Coverings**†

Description	Minimum/ Maximum Roof Slope, in. per ft	Class A	Class B	Class C
Brick Concrete Tile Slate		Brick, $2^1/_4$ in. thick. Reinforced portland cement, 1 in. thick. Concrete or clay floor or deck tile, 1 in. thick. Flat or French-type clay or concrete tile, $^3/_8$ in. thick with $1^1/_2$ in. or more end lap and head lock, spacing body of tile $^1/_2$ in. or more above roof sheathing, with underlay of one layer of Type 15 asphalt-saturated asbestos‡ felt or one layer of Type 30 or two layers of Type 15 asphalt-saturated organic felt. Clay or concrete roof tile, Spanish or Mission pattern, $^7/_{16}$ in. thick, 3 in. end lap, same underlay as above. Slate, $^3/_{16}$ in. thick, laid American method.		Metal roofing
Metal roofing	12/unlimited	Sheet roofing of 16 oz copper or of 30-gage steel or iron protected against corrosion. Limited to noncombustible roof decks or noncombustible roof supports when no separate roof deck is provided.	Sheet roofing of 16 copper or of 30-gage steel or iron protected against corrosion or shingle-pattern roofings with underlay of one layer of Type 15 saturated asbestos-felt‡, or one layer of Type 30 or two layers of Type 15 asphalt-saturated organic felt.	Sheet roofing of 16 oz copper or of 30 gage steel or iron, protected against corrosion, or shingle-pattern roofings, either without underlay or with underlay of rosin-sized paper.
	$2^1/_4$ (standing seam) or $2^1/_2$ (lap seam)/ unlimited	Standing or lap seam metal panel with glass fiber batts below on steel purling or joints.		Zinc sheets or shingle roofings with an underlay of one layer of Type 30 or two layers of Type 15 asphalt-saturated organic-felt or one layer of 14 lb unsaturated or one layer of Type 15 asphalt-saturated asbestos felt‡.
Cement-asbestos shingles†	Exceeding 4/unlimited	Laid to provide two or more thicknesses over one layer of Type 15 asphalt-saturated asbestos felt‡.	Laid to provide one or more thicknesses over one layer of Type 15 asphalt-saturated asbestos felt‡.	

*Prepared roof coverings as classified as applied over square-edge wood sheathing of 1-in. nominal thickness, or the equivalent, unless otherwise specified. See footnotes (built-up roof coverings) to Table 7-2I. Laid in accordance with instruction sheets accompanying package. Limited to decks capable of receiving and retaining nails.

Where organic-felt is indicated, asbestos felt of equivalent weight can be substituted.

End lap means the overlapping length of the two units, one placed over the other. Head lap in shingle-type roofs is the distance a shingle in any course overlaps a shingle in the second course below it. However, with shingles laid by the Dutch-lap method, where no shingle overlaps a shingle in the second course below, the head lap is taken as the distance a shingle overlaps one in the next course below.

Prepared roofings are labeled by Underwriters' Laboratories to indicate the classification when applied in accordance with direction for application included in package.

†1 in. = 25.4 mm; 1 oz. = 29 mL; 1 ft = 0.305 m.; 1 lb = 0.454 kg.

‡Roofing materials containing asbestos are no longer used in the U.S.; however, they may still be encountered in existing construction.

TABLE 7-2H. *(Continued)*

Description	Minimum/ Maximum Roof Slope, in. per ft	Class A	Class B	Class C
Asphalt-asbestos felt sheet coverings‡	2¼/not exceeding 12	Factory-assembled sheets of 4-ply asphalt and asbestos material.	Factory-assembled sheets of 3-ply asphalt and asbestos material or sheet coverings of single thickness with a grit surface.	Single thickness smooth surfaced.
Asphalt-asbestos felt shingle coverings‡	Exceeding 4/Uunlimited		Asphalt-asbestos felt grit surfaced.	
Organic-felt (previously referred to as rag felt) sheet coverings	Exceeding 4/unlimited			Sheet coverings of asphalt organic felt either grit surfaced or aluminum surfaced.
Organic-felt (previously referred to as rag felt) shingle covering, with special coating	Sufficient to permit drainage	Mineral surfaced, two or more thicknesses.	Mineral surfaced, two or more thicknesses.	
Organic-felt (previously referred to as rag felt) shingle coverings	Sufficient to permit drainage	Mineral surfaced, two or more thicknesses.	Mineral surfaced, two or more thicknesses.	Mineral surfaced shingles, one or more thicknesses.
Asphalt glass fiber mat shingle coverings	Sufficient to permit drainage	Mineral surfaced, two or more thicknesses.	Mineral surfaced, one or more thicknesses.	Mineral surfaced shingles, one or more thicknesses.
Asphalt glass mat sheet covering	Sufficient to permit drainage			Mineral surfaced.
Fire-retardant treated red cedar wood shingles and shakes	Sufficient to permit drainage			Treated shingles or shakes, one or more thicknesses; shakes require at least one layer of Type 15 felt underlayment.

*Prepared roof coverings as classified as applied over square-edge wood sheathing of 1-in. nominal thickness, or the equivalent, unless otherwise specified. See footnotes (built-up roof coverings) to Table 7-2I. Laid in accordance with instruction sheets accompanying package. Limited to decks capable of receiving and retaining nails.

Where organic-felt is indicated, asbestos felt of equivalent weight can be substituted.

End lap means the overlapping length of the two units, one placed over the other. Head lap in shingle-type roofs is the distance a shingle in any course overlaps a shingle in the second course below it. However, with shingles laid by the Dutch-lap method, where no shingle overlaps a shingle in the second course below, the head lap is taken as the distance a shingle overlaps one in the next course below.

Prepared roofings are labeled by Underwriters' Laboratories to indicate the classification when applied in accordance with direction for application included in package.

†1 in. = 25.4 mm; 1 oz. = 29 mL; 1 ft = 0.305 m.; 1 lb = 0.454 kg.

‡Roofing materials containing asbestos are no longer used in the U.S.; however, they may still be encountered in existing construction.

In addition to the built-up roof coverings described in Table 7-2I, there are several other types consisting of organic felt, glass fiber, aluminum, steel, and tile applied in a special manner with special listed proprietary securements that are listed by testing laboratories.

Roof Deck Insulations and Vapor Barriers

Since the 1953 General Motors fire at Livonia, MI, major attention has been paid to the combustibility of insulated metal roof decks. In that fire, the asphalt mopping in the felt vapor barrier beneath the insulation on the 1,502,500 sq ft (139,582 m²), flat, steel-deck roof melted and vaporized. Asphalt vapor under pressure entered the

building through the joints in the steel deck and burned inside. As a result of this and similar fires, research was conducted to find a solution to the problem. The significant findings are summarized in the following paragraphs.

The combustibility of a roof was historically determined by the physical characteristics of the roof deck and its supporting structure, with no consideration given to the combustibility of the roof insulation and covering. Thus, combustible fiberboard insulation and asphalt-impregnated felt roof covering normally were not considered to affect the classification of a noncombustible roof deck on noncombustible supports. Large-scale fire tests, confirmed by actual fire experience, however, have shown that under certain conditions some metal deck assemblies can contribute to an interior fire.

TABLE 7-2I. *Built-up Roof Coverings**†

Description	Minimum Incline, in. per ft	Class A	Class B	Class C
Asphalt organic-felt, bonded with asphalt and surfaced with 400 lb of roofing gravel or crushed stone, or 300 lb of crushed slag per 100 sq ft of roof surface, on coating of hot mopping asphalt.	3	4 (plain) or 5 (perforated) layers of Type 15 felt. 1 layer of Type 30 felt and 2 layers of Type 15 felt. 1 layer of Type 15 felt and 2 layers of Type 15 or 30 cap or base sheets. 3 layers of Type 15 or 30 cap or base sheets. 3 layers of Type 15 felt. Limited to non-combustible decks.	4 layers of perforated Type 15 felt. 3 layers of Type 15 felt. 2 layers of Type 15 or 30 cap or base sheets.	
Tar-asbestos-felt or organic-felt bonded with tar and surfaced with 400 lb of roofing gravel or crushed stone, or 300 lb of crushed slag per 100 sq ft of roof surface on a coating of hot mopping tar.	3	4 layers of 14 lb asbestos-felt or Type 15 organic-felt. 3 layers of 14 lb asbestos-felt or Type 15 organic-felt.	3 layers of Type 14 lb asbestos-felt or Type 15 organic-felt.	
Steep tar organic-felt	5	4 layers of Type 15 tar-saturated organic-felt, bonded with steep coal-tar pitch, surfaced with 275 lb of ⁵⁄₈ in. crushed slag per 100 sq ft of roof surface on steep coal-tar pitch.		
Asphalt organic-felt, plain or perforated, bonded and surfaced with a cold application coating.	12			3 layers of Type 15 felt. 1 layer of Type 30 felt and 1 layer of Type 15 felt. 2 layers of Type 15 or 30 cap or base sheets. 2 layers of Type 15 felt and 1 layer of Type 15 or 30 cap or base sheets.

*Built-up roof coverings are classified as applied over square-edge wood sheathing of 1-in. nominal thickness, or the equivalent, unless otherwise specified. From the standpoint of relative effectiveness of the different types of wood roof sheathing, the tongue-and-groove boards and ³⁄₄-in. moisture-resistant plywood give better results in the brand and flame tests than square-edge sheathing with boards spaced about ¹⁄₄ in. apart. For classifications based on square-edge sheathing, tongue-and-groove or plywood sheathing can be substituted. Square-edge sheathing boards should be butted together as closely as possible. Reference to ¹⁄₄-in. spacing is to indicate fire test procedure intended to simulate actual conditions after shrinkage of boards due to age or other reasons.
The minimum weight of cementing material between separate layers of felt is considered to be 25 lb per 100 sq ft of roof surface.
Types 15 and 30 felts are defined as saturated felts weighing a minimum of 14 lb and 28 lb per 100 sq ft of the finished materials, respectively. Where saturated felts are referred to by weight, the weight is minimum and is expressed in pounds per 100 sq ft of the finished material.
Materials intended for built-up roof coverings are labeled by Underwriters Laboratories. The classifications indicated are of generally accepted combinations.
†1 in. = 25.4 mm; 1 ft = 0.305 m; 1 sq ft = 0.0929 m²; 1 lb = 0.454 kg.

A series of full-scale tests were conducted in 1955 at FMRC testing facilities to compare the performance of various metal roof deck constructions with respect to fire propagation along the underside of the deck. The tests were run in cooperation with the Metal Roof Deck Technical Institute (now Steel Deck Institute) and the Insulation Board Institute, and the results were published in a technical report.[3] With the insulation or vapor barriers adhered to the metal deck with complete moppings of asphalt, it was shown that the asphalt contributed a significant amount of fuel when exposed to the heat of an intensive interior fire below the deck. (See Figure 7-2MM.) The asphalt was vaporized by the heat of the fire below, and unable to escape upward through the roof covering, was forced down through the joints in the deck, when it ignited and contributed to the spread of fire under the roof deck. This type of assembly is called a Class 2 steel decks. Because of poor wind-loss experience, FMRC no longer recommends using adhesives to secure above-deck components to steel deck. However, other combinations of combustible above-deck components (such as vapor barriers, insulation, and roof covers) may create a Class 2 steel-deck assembly.

Recognized fire testing laboratories now list metal roof deck construction and components that will not contribute significantly to an interior fire and that meet certain performance standards. Such assemblies are called Class 1 steel decks.

Metal roof deck constructions use both combustible and noncombustible insulation fastened with mechanical fasteners to the upper surface of the deck. Prior to the early 1980s, asphalt was accepted as an adhesive to secure insulation to steel decks if it was ap-

FIG. 7-2KK. *Typical built-up roof coverings with hot asphalt mopping or coal tar pitch between each two layers.*

FIG. 7-2LL. *Installation of a typical prepared roof covering.*

FIG. 7-2MM. *Typical built-up roof covering with a combustible vapor barrier adhered to the roof deck and to roof insulation by a combustible adhesive.*

plied in strips (with a mechanical applicator), and if the total amount used between the deck and the insulation did not exceed 12 to 15 lb per 100 sq ft (5.4 to 6.8 kg/9.3 m²) of roof area.

There is no restriction on the use of asphalt in conventional built-up roofing used above the insulation. When a vapor barrier is used between the deck and the insulation, a listed, noncombustible, or material with limited fuel contribution is recommended. Testing laboratories list various manufacturers of acceptable vapor barriers and adhesives.

FMRC divides insulated metal deck roof assemblies into two classes. Class I construction is any insulated steel roof deck construction having a combination of above-deck components that will not contribute significantly to an interior fire and that will meet or equal FMRC performance standards. Class II construction is any insulated metal roof deck construction using a combination of above-deck components which, when exposed to fire locally, will produce

sufficient heat release to allow a self-propagating fire on the underside of an unprotected deck.

Constructions that are found acceptable are listed by FMRC.[4]

Wood Shingles

Untreated wood shingles may be readily ignited by small sparks from chimneys or exposure fires, by radiated heat, or by burning brands. Burning shingles themselves also produce brands, which may be carried by the wind to start other fires. Age and humidity increase the susceptibility of wood-shingled roofs to ignition.

Treatment of wood shingles in place with fire-retardant coatings has not proved practical to date. Ordinary flame-resistant treatments lose their effectiveness with continued exposure to the weather. Wood shingles impregnated with a fire-retardant solution are more durable. Shingles of this type have been tested and carry a Class B or C rating.

Untreated wood shingle roofs are prohibited by law in the congested sections of practically all large cities, and a very large number of cities and towns prohibit their use altogether within municipal limits.

BIBLIOGRAPHY

References Cited

1. *Design for Fire Resistance of Precast Prestressed Concrete*, 2nd ed., Prestressed Concrete Institute, Chicago, IL, 1989.
2. "Protection Against Liquid Damage," Loss Prevention Data Sheet 1-24, Factory Mutual Research Corp., Norwood, MA, 1983.
3. "Insulated Metal Roof Deck Fire Tests," Factory Mutual Research Corp., Norwood, MA, 1955.
4. "Approval Guide," Factory Mutual Research Corp., Norwood, MA, (published annually).

NFPA Codes, Standards, and Recommended Practices

Reference to the following NFPA codes, standards, and recommended practices will provide further information on building construction discussed in this chapter. (See the latest version of *The NFPA Catalog* for availability of the current editions of the following documents.)

NFPA 80, *Standard for Fire Doors and Fire Windows*
NFPA 80A, *Recommended Practice for Protection of Buildings from Exterior Fire Exposures*
NFPA 203D, *Guide on Roof Coverings and Roof Deck Constructions*
NFPA 220, *Standard on Types of Building Construction*
NFPA 221, *Standard for Fire Walls and Fire Barrier Walls*

NFPA 251, *Standard Methods of Tests of Fire Endurance of Building Construction and Materials*

NFPA 255, *Standard Method of Test of Surface Burning Characteristics of Building Materials*

NFPA 256, *Standard Methods of Fire Tests of Roof Coverings*

NFPA 258, *Standard Research Test Method for Determining Smoke Generation of Solid Materials*

NFPA 259, *Standard Test Method for Potential Heat of Building Materials*

NFPA 703, *Standard for Fire Retardant and Impregnated Wood and Fire Retardant Coatings for Building Materials*

Additional Reading

Anchor, R. D., *et al.*, *Design of Structures Against Fire*, Elsevier, New York, 1986.

ASTM E136, *Standard Test Method for Behavior of Materials in a Vertical Tube Furnace at 750°C*, W. Conshohocken, PA, 1994.

ASTM, *Standard Test Method for Surface Burning Characteristics of Building Materials*, W. Conshohocken, PA, 1995.

Brannigan, F. L., "Are Wood Trusses Good for Your Health," *Fire Engineering*, Vol. 141, No. 6, 1988, pp. 73–80.

Brannigan, F. L., "Beware the Truss," *Fire Engineering*, Vol. 138, No. 7, 1985, p. 30.

Brannigan, F. L., "Ceiling Suspended Loads," *Fire Engineering*, Vol. 14, No. 3, 1987, p. 30.

Brannigan, F. L., "Know Your Roof," *Fire Engineering*, Vol. 139, No. 9, 1986, pp. 44–51.

Brannigan, F. L., "The Ties That Bind," *Fire Engineering*, Vol. 142, No. 6, 1989, pp. 57–61.

Building Officials and Code Administrators, "Guidelines for Determining Fire Resistance Ratings of Building Elements," 1994.

Carling, O., "Fire Resistance of Joint Details in Loadbearing Timber Construction, A Literature Survey," Research Report No.18, Building Research Association of New Zealand, 1989.

Corbett, G. P., "Lightweight Wood Truss Floor Construction: A Fire Lesson," *Fire Engineering*, Vol. 141, No. 7, 1988, pp. 41–47.

Coursey, R., "Do Roof Fire Tests Assure Performance," *Plant Engineering*, Apr. 14, 1988, pp. 54–57.

Cutter, B. E., "Working Together to Solve the Wood Truss Problem," *Fire Journal*, Vol. 84, No. 1, Jan./Feb. 1990, pp. 11, 70.

Day, T., "Roofs and Fire," *Fire Prevention*, No. 201, Jul.–Aug. 1987, pp. 18–23.

Debrody, R. F., "Constructing a Durable Burn Building," *Firehouse*, Vol. 17, No. 1, Mar. 1992, pp. 34, 36, 38.

"Design of Loadbearing Light Steel Frame Walls for Fire Resistance," *Build*, Nov. 1995, pp. 28, 30.

Dunn, V., "Building Construction and Fire Spread," *Firehouse*, Vol. 21, No. 6, June 1996, pp. 16, 18–21.

Dunn, V., "Common Construction Types," *Firehouse*, Vol. 19, No. 2, February 1994, pp. 20–23.

Favro, P.C., "Fire-Retardant Treatments for Wood Roofs: A Permanent Solution," Fire Safety '91 Conference, Volume 4, Nov. 4–7, 1991, St. Petersburg, FL, Am. Technologies Intl., Pinellas Park, FL, 1991, pp. 136–145.

Fire Resistance Design Manual, 13th ed., Gypsum Association, Washington, DC.

Fire Retardant Chemicals Association, *Fire Safety Problems Leading to Current Needs and Future Opportunities*, Technomic, Lancaster, PA, 1989.

Fitzgerald, R. W., "Selection of Fire Endurance Requirements for Building Design. [Integration of Fire Science Into Building Requirements.]," *Fire Safety Journal*, Vol. 17, No. 2, 1991, pp. 159–163.

Gerlich, J. T., "Design of Loadbearing Light Steel Frame Walls for Fire Resistance. Fire Engineering Research Report," University of Canterbury, Christchurch, New Zealand, Fire Engineering Research 95/3, Aug. 1995.

Gerlich, J. T., Collier, P. C. R., and Buchanan, A. H., "Design of Light Steel-Framed Walls for Fire Resistance," *Fire and Materials*, Vol. 20, No. 2, March/April 1996, pp. 79–96.

Harlow, T. D., "New Roofs Over Old," *Fire Engineering*, Vol. 140, No. 11, 1987, p. 51.

Howell, I. D., "Examination of Various Passive Fireproofing Materials in the Laboratory," *Proceedings* from the Conference on Offshore Passive Fire Protection, Plastics and Rubber Inst., London, England, 1987, pp. 1–21.

Kirby, B. R., "Recent Developments and Applications in Structural Fire Engineering Design—A Review," *Fire Safety Journal*, Vol. 11, No. 3, 1986, pp. 141–180.

Manny, W. F., "Lightweight Wood Trusses: More to Consider," *Fire Engineering*, Vol. 144, No. 5, March 1991, pp. 62–63.

Messersmith, J. J., "Fire Resistance Rating Requirements for Exterior Bearing Walls," *Building Official and Code Administrator*, Vol. 24, No. 3, May/June 1990, pp. 34–40.

Mittendorf, J., "Joist-Rafter *Versus* Lightweight Truss," *Fire Engineering*, Vol. 141, No. 7, 1988, pp. 48–50.

Mittendorf, J. W. and Brannigan, F. L., "Timber Truss: Two Points of View," *Fire Engineering*, Vol. 144, No. 5, March 1991, pp. 51–54, 56, 58–60.

Odeen, K., "Fire Spread Along Roofs—Some Experimental Studies," *Proceedings* of the First International Symposium on Fire Safety Science, Hemisphere, 1986, pp. 829–837.

Peterson, C. E., "Construction Helps Limit High-Rise Fire," *Fire Command*, Vol. 56, No. 2, Feb. 1989, pp. 20–24.

Rogowski, B. F. W., "Investigating the Contribution to Fire Growth of Combustible Materials Used in Building Components," *Proceedings* of the Conference on New Technology to Reduce Fire Losses and Costs, Elsevier, 1986, pp. 88–105.

Smith, A., "Quick Guide to UBC Building Types,"*American Fire Journal*, Vol. 42, No. 5, May 1990, pp. 22–26.

Sutherland, R. J. M., "Design of Structures Against Fire: Some Practical Constructional Issues," *Proceedings* of the International Conference on the Design of Structures Against Fire, Elsevier, 1986, pp. 177–187.

Takeda, H. and Yung, D., "Fire Types and Probabilities in Residential Buildings," *IRC Fire Research News*, No. Issue 61, Winter 1991, pp. 1–3.

Tomann, J., "Lattianpaallysteiden palotestaus. [Fire Testing of Floorings.]," *Palontorjuntatekniikka*, Vol. 1, 1992, pp. 22–25.

UL, *Building Materials Directory and Annual Supplement*, Underwriters Laboratories Inc., Northbrook, IL. (Directory published annually in January; supplement annually in July.)

UL, *Fire Resistance Directory*, Vols. 1 and 2, Underwriters Laboratories Inc., Northbrook, IL.

Williams, R. J., "Timber—Its Role in Aesthetic Fire Protection," *Fire Surveyor*, Vol. 18, No. 5, Oct. 1989, pp. 25–27.

Willse, P. J. G. and Smith, A. C., "Fire Walls: The Last Line of Defense," *Chemical Engineering*, Vol. 100, No. 7, July 1993, pp. 110–113.

Woolhead, F., "Fire Test Report: Insulated Metal Roof Decks," *Fire Surveyor*, Vol. 19, No. 1, Feb. 1990, pp. 25–29.

INTERIOR FINISH

Revised by Donald W. Belles

The best fire protection strategy for buildings is one that limits fires to small size. In instances where fires may grow large, the growth rate must be slow enough for occupants to react and for the public fire forces to have an opportunity to respond and combat the fire manually, while it is still limited in size.

Fires should be limited in size because fire hazard is related to fire size. A material's burning rate—measured by either the rate of heat release or the rate of mass loss—is the single most important parameter affecting fire hazard, because burning rates determine fire size. Burning rates directly affect temperature rise, and they establish the quantity of soot released into a given volume, which determines visibility. Moreover, in a given volume, burning rates also determine the concentration of toxic products, which dominates the onset of toxic hazard.

Large fires in buildings always represent an acute threat to occupants, irrespective of what is burning; and the faster a fire grows to large size, the greater the threat. Any building material, furnishing, or content that causes a fire, which would otherwise stay small, to become large, has undesirable flammability properties.

Building interior finishes, principally wall and ceiling linings, can have a major impact on both fire growth and ultimate fire size. Interior wall and ceiling linings may act like a "fuse," spreading flames away from fire origin to involve other objects, causing the fire to grow to large size. Interior finishes may also provide a large, unbroken surface over which flame spreads. As the flame spreads to involve greater surface area, a fire's rate of heat release, and therefore, fire size, increases. The flames from the interior finish may release sufficient energy to cause the formation of a hot gas layer. If the layer temperature approaches 1112°F (600°C) and the layer becomes thick, descending toward the lower portion of the room, all combustible furnishings and combustible room contents may be ignited, leading to flashover or full room fire involvement. Although some involve extended periods of slow, undetected fire development, fires that cause large life loss or high property loss nearly always grow rapidly to large size, reaching flashover or full room involvement within minutes of having been small and easy to control. Combustible interior finishes, such as low-density fiberboard ceilings, wood paneling, textile wall coverings, and plastic floor coverings, have been found to be significant factors in many multiple death fires.[1-10] Interior finishes are regulated by NFPA *101*®, *Life Safety Code*®, (hereinafter referred to as NFPA *101*®), to slow initial fire growth and to reduce the likelihood of fire growing to large size.

This chapter reviews the effect of wall and ceiling finish materials on fire growth. Interior finish materials, such as wood, plaster, wallboard, acoustical tile, insulating materials, and decorative materials, are discussed. Interior floor finishes are also discussed. This chapter examines the method of application of interior finishes and the effects on flammability behavior. Fire tests used to evaluate the performance of interior finishes and interior floor finishes are also reviewed.

DEFINITION OF INTERIOR FINISH AND INTERIOR FLOOR FINISH

Most codes, such as NFPA *101*®, define interior finishes as those materials or assemblies of materials that form the exposed interior surfaces of walls and ceilings in a building. Interior floor finishes are the exposed floor surfaces of buildings and include floor coverings, such as carpets and floor tiles, that may be applied over, or in lieu of, a finished floor. The term "interior lining" is used by the international fire community to define interior finish and the role of interior surfaces in the growth of fire in the room of origin. The term "interior lining" is gaining wider acceptance in North America and may ultimately replace the term "interior finish."

There is some variability in the way interior finish regulations are applied. For example, some regulatory officials consider countertops and building cabinets as interior finish. It is sometimes difficult to differentiate between interior finishes and furnishings. Furnishings, which, in some cases, may be fastened in place, are not normally considered interior finish. For example, fixed plastic-covered restaurant "booth" seating and permanently mounted work counters and cabinets at a nursing station in a health care facility are not usually classified as interior finish. On the other hand, trim is ordinarily regulated as an interior finish. However, less rigid requirements are set for trim and other incidental finish materials (like doors), when the trim or incidental finish is less than 10 percent of the aggregate wall and ceiling area. The "10 percent rule" is based upon the trim or incidental finish being uniformly distributed and lacking sufficient continuity to allow flames to spread over any significant surface area. Concentration of the trim in one area, so that the trim forms one "fuel package," would violate the principle underlying the "10 percent rule."

Free-hanging draperies that cover most or all of a wall surface have, on occasion, been subject to the requirements for interior finish.

Donald W. Belles, P.E., is a fire protection engineer at Donald W. Belles and Associates, Inc., Madison, TN.

The normal test for interior finish combustibility, NFPA 255, *Standard Method of Test of Surface Burning Characteristics of Building Materials* (hereinafter referred to as NFPA 255), may not adequately assess the fire behavior of such a product. Free-hanging drapery materials might be more properly evaluated by NFPA 701, *Standard Methods of Fire Tests for Flame-Resistant Textiles and Films*. When applied to a solid backing, such as gypsum wallboard, drapery material should be considered as a textile wall covering and would be evaluated as an interior finish. NFPA *101®* contains special requirements for textile wall coverings. Some textile wall coverings are able to obtain good flame spread ratings (less than 25) when tested in accordance with NFPA 255, and yet these same textile wall coverings in room/corner tests sometimes propagate flame and cause fires to grow to full-room size, i.e., flashover, within minutes.

Movable walls, folding partitions, and flexible or folding door assemblies are all normally treated as interior finishes. Movable partitions, used, for example, as a portion of office furnishings or work stations, are sometimes regulated as interior finish. Office cubicle partitions and work stations are sometimes formed of textile materials mounted over various substrates. Full-scale tests at the National Institutes of Standards and Technology (NIST) involving office work stations consisting of computer terminals, 5-ft (1.5-m) high partitions of textile covering over fiberglass on metal frames, wood desks, and typical office contents revealed that fires involving such work stations may reach large size in 300 to 600 sec.[11,12] It has been industry practice to evaluate movable partitions as an assembly in the tunnel test, NFPA 255. There is a question about the NFPA 255 ratings and their relationship to the performance of movable partitions in fires.

Many local building regulations include provisions that regulate floorings and floor coverings. NFPA *101®* considers floorings and floor coverings to be "interior floor finish." Experience has shown that traditional floor coverings, such as wood, vinyl tile, and linoleum, are not likely to affect early fire growth. In most instances, there will be little gain in safety achieved by regulating traditional floor coverings. Where the Authority Having Jurisdiction determines that a floor covering or flooring material poses an unusual hazard or where NFPA *101®* specifically contains restrictions for such materials, floorings and floor coverings should comply with appropriate interior floor finish provisions. NFPA 253, *Standard Method of Test for Critical Radiant Flux of Floor Covering Systems Using a Radiant Heat Energy Source* (hereinafter referred to as NFPA 253), is used to evaluate the flammability of interior floor finishes and flooring. Independent of codes, carpet is regulated by the U.S. Consumer Product Safety Commission (CPSC). All carpet manufactured for sale in the United States must meet a federal flammability standard, FF 1-70, commonly referred to as the "pill test."

TYPES OF INTERIOR FINISH

The types of interior finish materials are numerous and include such commonly used materials as plaster, gypsum wallboard, wood, plywood paneling, fibrous ceiling tiles, plastics, and a variety of wall coverings. Collectively, these finishes serve several functions, e.g., aesthetic, acoustical, and insulating, as well as protecting against wear and abrasion.

Wall coverings that are no thicker than $1/_{28}$ in. (0.9 mm) are not generally regulated as interior finish, except where they are deemed to be a hazard by the Authority Having Jurisdiction. The exception for thin coverings, i.e., with surface burning characteristics no greater than paper, eliminates regulation of products such as gypsum wallboard, paints, wallpaper, and similar thin materials. Thin coverings, secured to a substrate, will tend to behave like the substrate. Thin coverings, such as paint on steel and paper over gypsum, will not generally spread flames or contribute to fire growth

during the developing stage of a fire. However, thicker coverings, such as multiple layers of vinyl, can, and have, contributed to rapid fire growth. For example, multiple layers of vinyl wall coverings contributed to rapid fire growth in the multi-death fires at the Holiday Inn in Cambridge, OH, in 1979,[13] and a 1989 office building fire in Atlanta.[9]

The development of certain types of cellular plastics in board, poured-in-place, and sprayed-on foam has yielded lightweight materials with exceptional thermal insulation. The incorporation of fire retardants into some cellular plastic products makes it possible for them to meet the NFPA 255-related requirements for interior finishes. As a result, cellular plastics, particularly the sprayed-on type, have been used as exposed insulation. Rapid fire spread in several widely publicized fires involving exposed cellular polyurethane and polystyrene materials resulted in recommendations for protection of such surfaces against ignition and fire spread by the use of 15-min. thermal barriers.[14,15,16] NFPA *101®* states that cellular or foamed plastic materials must not be used as interior finish unless actual fire tests substantiate, on a reasonable basis, their combustibility characteristics for the intended use. For a further discussion of room/corner testing of foam plastics, refer to the subsection entitled "The Corner Test," later in this chapter.

NFPA *101®* also states that cellular or foamed plastics may be used as trim around doors or windows or as chair rails, for example, when they do not exceed 10 percent of the wall or ceiling area, the density is not less than 20 lb/cu ft (320 kg/m³), they are no more than $1/_2$ in. (12.7 mm) thick and 4 in. (102 mm) wide, and they comply with NFPA 255 Class A or B flame spread rating without the smoke rating limitation. The exception for "plastic trim" is based upon the presumption that the plastic trim performance will be comparable or better than traditional wood products. The required Class B flame spread rating is better than the rating achieved by most wood materials. A minimum density of 20 lb/cu ft (320.4 kg/m³) is set to prohibit the use of lightweight foam plastics, which typically have densities in the 1 to 3 lb/cu ft (16 to 48 kg/m³) range. The width and thickness limits are to restrict the combustible mass. The "10 percent rule" has been previously discussed and applies to the use of plastic trim.

THE ROLE OF INTERIOR FINISH IN FIRES

Most building fires begin when decorative materials, furnishings, or waste accumulations ignite. Interior finishes are less often the first items ignited. When interior finish is first ignited, heat sorces include overheated electrical circuits, careless use of torches and direct impingement of flame from a candle or a match. Once a fire has started and intensified, however, the interior finish can become involved and contribute extensively to the spread of fire.

Interior finish affects fire hazard in four ways it can: (1) affect the rate of fire buildup to flashover conditions, (2) contribute to fire extension by flame spread over its surface, (3) add to the intensity of a fire by contributing additional fuel, and (4) produce smoke and toxic gases that can contribute to life hazard and property damage.[17] From a fire safety standpoint, the most desirable interior finish is one made of a relatively dense and noncombustible material that is a good conductor of heat, does not speed up flashover, does not add fuel to the fire, provides no path for surface flame spread, and produces little or no smoke or toxic gases. Materials that exhibit high rates of flame spread, contribute substantial quantities of fuel to a fire, or produce hazardous quantities of smoke or toxic gases are undesirable.

Combustible interior finish in an enclosure can reduce the rate of heat release necessary for flashover and can reduce the time to flashover. Flashover is caused by thermal radiation feedback from the ceiling and upper walls, which have been heated by the fire. This radiation feedback gradually heats the contents of the fire area. When all the combustibles in the space have become heated to their ignition temperatures, simultaneous ignition occurs. Interior finish plays an important role in the occurrence of flashover, i.e., an insulating interior finish that lowers heat transfer to upper walls or ceilings can reduce both the time and the rate of heat release required to reach flashover. If the finish material is combustible, it will be a source of fuel for the fire. Considering the nature of thermal radiation, the size and shape of the space in which the fire occurs becomes a critical factor.[18] The term "flameover" was coined to denote the rapid spread of flame over one or more surfaces. This reaction was observed in corridor tests conducted by the National Institutes of Standards and Technology.[19]

Full-scale experiments have been conducted to determine how location and area of a wall finish affects the spread of fire in a corridor.[20] A series of experiments were conducted in a 6-ft wide by 8-ft high by 55-ft long (1.8 × 2.4 × 16.8 m) corridor exposed to hot gases from a burner. The corridor ceiling and upper walls were exposed to temperatures of approximately 1450°F (788°C). Corridor ceilings had flame spread ratings between 0 (FS0) and 25 (FS25), while wall coverings had a flame spread rating of 90 (FS90). These experiments revealed that ceiling finish plays an important role in flame spread behavior of walls. It was determined that FS90 wall coverings could be used safely with FS25 ceilings when the wall covering was limited to the lower 4 ft (1.2 m) of wall surface. Most of the test configurations with FS90 material covering the full height of the corridor walls and FS25 ceilings resulted in rapid flame propagation and unsafe conditions. It was deduced that the FS25 ceiling increased flame spread mainly due to the fact that it was a better insulator than the FS0 ceiling. In other words, the FS25 ceiling influenced flame spread over walls by its resistance to heat transfer.

Once a room fire reaches full involvement, or flashover, heat, smoke, and combustion gases escape through openings to adjoining spaces. Ignition of combustible interior finish may become a significant factor in the spread of fire to other areas. Several full-scale fire tests have demonstrated that heat, smoke, and noxious combustion gases from burning furnishings may be sufficient to create a threat to the life safety of persons unable to evacuate the room of fire origin without any contribution from combustible interior finishes.[21,22] However, even in such a case, the life safety of occupants remote from the fire and, in addition, property damage potential may be greatly affected by the nature of the interior finishes.

METHODS OF APPLICATION

The method of applying an interior finish can have a large effect on its fire behavior. For this reason, the manufacturer's specifications for application should be strictly followed. Special attention must be paid to fastening details including the type and application of adhesives, and the number of coats and the application rate of fire-retardant coatings. Application details are equally important in repair or replacement of interior finishes. Any new material used in retrofit should meet current code requirements.

The substrate material to which surface finishes are attached plays an important role in the flame spread behavior of thin products.[23] A thin combustible finish applied to a noncombustible substrate may present little hazard. The same finish material on a combustible backing could constitute a serious hazard. In the former situation, the substrate will not ignite and will absorb heat during early fire development. In the latter case, both the surface finish and the backing material may ignite and spread flames.

The adhesive used is also a factor in the fire behavior of interior finishes. The best performance is likely when an intimate bond is maintained between a product and substrate. However, adhesives

can also contribute fuel to a fire; and too much adhesive, or the use of a highly flammable adhesive, may result in flame propagation. The failure in bond between a surface finish and a substrate can also result in poor performance. Adhesives that soften at moderate temperatures may allow wall and ceiling finishes to drop or partially peel away during the growth stage of a fire. Perhaps the "worst case" for an interior finish is where partial release or partial delamination occurs. In such a case, the material separates from the substrate but remains in position. This not only increases the susceptibility of the surface material to ignition but it also exposes the substrate material which, if combustible, adds fuel to the fire.

Some interior finishes, such as polystyrene ceiling tiles, achieve acceptable performance in an entirely different manner. They are designed to fall away early in a fire, thereby avoiding fire exposure. Satisfactory performance for such products results from evading flame during the fire growth phase. However, if products fall or delaminate early in a fire, they may interfere with persons attempting to escape. Where interior finish materials may fall or delaminate at relatively low temperatures and where the products involved have significant mass, the behavior of such products should be assessed in terms of their impact on occupants and the effect on evacuation. Some building codes specify that wall and ceiling finishes should not become detached when exposed to elevated temperatures of 175 to 300°F (79 to 149°C) for a specified time interval, usually 30 min.

Thermally thin products, such as wood paneling $1/4$ in. (6.4 mm) thick or less, will tend to spread flame more rapidly when the paneling is installed over studs or furring strips with air space than would be the case if installed over a solid substrate, such as cement board or gypsum board. As a consequence, some building regulations require that Class B and C interior finish materials, less than $1/4$ in. (6.4 mm) thick, be applied over a substrate of noncombustible material. Concern also exists about the possibility of flame spread over concealed surfaces of combustible interior finish materials where air space exists behind the finish. Model building codes[24,25] have application sections that restrict the application of interior finishes over air space in noncombustible assemblies. Interior finishes installed over an air space are restricted to products having a Class A rating on both faces. The model building codes regulate concealed spaces created by furring strips by requiring firestopping, or that the space be filled with inorganic or Class A materials.

The use of sprayed-on materials for interior finishes has proliferated. Some of these materials are coatings that are inherently resistant to fire. Other materials, however, especially the sprayed-on cellular or foam plastics used for insulation or decorative effects, may readily spread flames. The use of such products should be permitted only if the fire behavior can be substantiated under actual fire conditions.

FIRE TESTS FOR INTERIOR FINISHES

The nature of interior finish materials and the fire environment to which they may be exposed vary so widely that development of a fire test becomes highly complex. Three factors must be taken into account:

1. The start and growth of fire in a building is affected by the ignition source, space geometry, ventilation, and by the nature, amount, and location of other processes and materials in the building.
2. There are many time-varying events associated with fires, such as oxygen concentration, rate of heat release, and effects caused by protection systems.
3. Materials vary in form, composition, density, and application.

With these variables in mind, the difficulty in designing a test that will provide a basis for predicting performance under a specific fire exposure becomes obvious. Equally obvious is the impracticality of designing tests to represent all fire conditions. A test designed to represent a "typical" fire or to expose materials to one set of "standard test" conditions may not provide a reliable basis for predicting "real-life" performance of all materials tested. Thus, there is a constant search for improved test methods that will provide a range of results for a variety of fire conditions.

The use of bench-scale test data to predict product behavior through the use of models is under study. Use of data from small-scale tests, such as the cone calorimeter and lift apparatus,[26,27] to predict product behavior under a variety of circumstances, may result in improved regulation of interior finish materials in the future. The models may be able to link the actual hazard created by the use of a product in a specific situation to regulatory limits. This future form of regulation must be contrasted with the current regulation of interior finishes using a standardized procedure (NFPA 255) in which the behavior of one product is simply compared with another. At present, this may be the most practical approach for building regulation; however, one should understand that good behavior in a standardized test (like NFPA 255) does not necessarily ensure good behavior in an actual fire.

The Steiner Tunnel Test

The 25-ft (7.62-m) Steiner tunnel test, also know as ASTM E84, NFPA 255, and UL 723, was developed by A. J. Steiner at Underwriters Laboratories Inc.[28] After a series of fatal fires, the need for a method of controlling interior finishes was recognized and the use of the tunnel test proposed. The method was adopted by the American Society for Testing and Materials (ASTM) as a tentative standard in 1950 and as an official standard in 1958. NFPA adopted the test method tentatively in 1953 and officially in 1958 as NFPA 255.

Figures 7-3A and 7-3B show the general appearance and basic layout of the tunnel test apparatus. A detailed description can be found in NFPA 255. In summary, a 20-in. (508-mm) wide by 25-ft (7.6-m) long test specimen is placed face down on a ledge at the top of the furnace to form the ceiling of the test chamber. A gas burner

projects a 4½-ft (1.4-m) long flame onto the underside of the test specimen while air flows through the tunnel at a prescribed rate. Flame spread over the face of the test specimen is observed through windows in the side of the "tunnel." Flame propagation is recorded as a function of time over the 10-min test duration. The record of flame propagation vs. time is used to calculate a flame spread index. Red oak flooring—a calibration material—achieves a flame spread index of approximately 100. According to the test standard, red oak flooring spreads flame 24 ft (7.3 m) to the end of the test specimen in 5½ min ± 15 sec. The flame spread ratings of products are calculated on the basis of flame spread vs. time. The lower the flame spread index, the better the presumed performance. However, there is not necessarily a correlation between the measured flame spread index and product behavior in actual fires. NFPA 255 simply provides a basis for comparing one product with another under a standardized set of test conditions.

The reduction in light across the tunnel exhaust caused by smoke generated during the test is measured by a photometer system consisting of a lamp and photo cell. The attenuation of light by the passing smoke and particulate is used to compute a smoke-developed value. The smoke-developed value for a test specimen is determined by comparing the reduction in light caused by the smoke from red oak flooring to the reduction of light by smoke from a test specimen. Red oak is assigned a value of 100. Therefore, a product achieving a 450 smoke-developed value produces a curve of light attenuation vs. time which has 4.5 the area of the curve produced by red oak. In simplistic terms, one may think of a product having a 450 rating as one that produces 4.5 times the amount of smoke that red oak flooring generates during a 10-min tunnel test.

Products should be tested as proposed for use. For example, the use of high-temperature mortar cement to mount a thin wall covering to an inorganic reinforced cement board substrate may result in good test performance, but such an assembly is not likely to simulate actual installation practices. A product installed in a manner different from that used in testing may perform differently and, in some cases, unsatisfactorily. To illustrate, products that are to be installed over gypsum board with adhesives, such as vinyl wall covering, should be tested over gypsum board, because the paper facing can affect behavior.

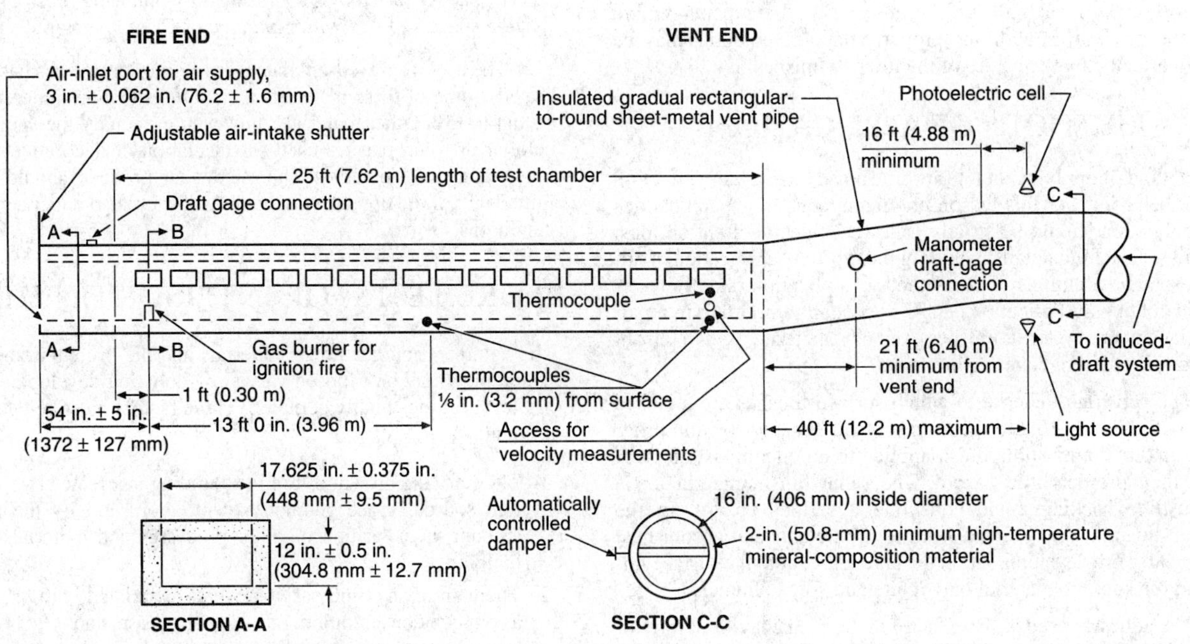

FIG. 7-3A.   A schematic diagram of the Steiner tunnel test apparatus used for the fire hazard classification of building materials.

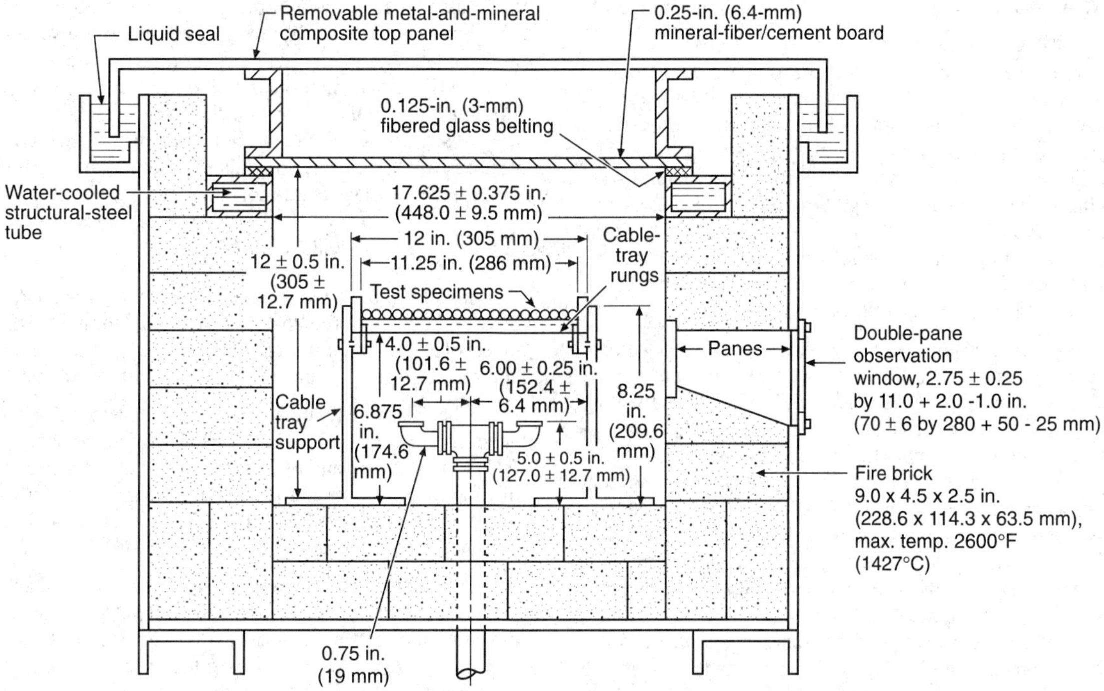

Liquid seal
Removable metal-and-mineral composite top panel
0.25-in. (6.4-mm) mineral-fiber/cement board
0.125-in. (3-mm) fibered glass belting
Water-cooled structural-steel tube
17.625 ± 0.375 in. (448.0 ± 9.5 mm)
12 in. (305 mm)
11.25 in. (286 mm)
12 ± 0.5 in. (305 ± 12.7 mm)
Test specimens
Cable-tray rungs
4.0 ± 0.5 in. (101.6 ± 12.7 mm)
6.00 ± 0.25 in. (152.4 ± 6.4 mm)
8.25 in. (209.6 mm)
Panes
Double-pane observation window, 2.75 ± 0.25 by 11.0 + 2.0 -1.0 in. (70 ± 6 by 280 + 50 - 25 mm)
Cable tray support
6.875 in. (174.6 mm)
5.0 ± 0.5 in. (127.0 ± 12.7 mm)
Fire brick 9.0 x 4.5 x 2.5 in. (228.6 x 114.3 x 63.5 mm), max. temp. 2600°F (1427°C)
0.75 in. (19 mm)

FIG. 7-3B. A cross-sectional view of the Steiner tunnel test apparatus.

The significance of adhesives has been previously addressed. Products should be tested with fastening details that are representative of conditions of actual use. NFPA *101*® requires that interior finish materials be tested as they are to be installed. For example, Paragraph 6-5.1.4 of NFPA *101*® states, "Classification of interior finish materials shall be in accordance with tests made under conditions simulating actual installations . . . ." This idea is reinforced in the test method itself. For another example, Paragraph 2-2.1 of NFPA 255 states, "The specimen shall be truly representative of the material for which test results are desired."

Specimen thickness can also affect flame spread. Thinner sections of the same material often will spread flame more rapidly than thicker sections. For example, data show that the flame spread rate remains constant for acrylic (polymethylmethacrylate) with a thickness greater than approximately $3/8$ in. (9.5 mm).[29] Decreasing thickness increases the rate of flame spread for acrylic. These findings suggest that products to be used in varying thicknesses should be tested in different thicknesses. Similarly, products that are offered in different densities should be tested in a range of densities.

Other mounting details can be important, too. As noted earlier, thermally thin products, such as wood paneling with thickness of $1/4$ in. (6.4 mm) or less, that are tested when directly applied to an inorganic-reinforced cement board may behave differently if they were installed over furring strips with air space. The installation of paneling with air space will likely result in a greater tendency for surface flame spread. Similarly, where a product is to be applied in multiple layers, it should be tested in the maximum thickness intended for use. For example, a vinyl wall covering tested by itself may perform satisfactorily. On the other hand, multiple layers of the same product applied to a wall covering may spread flame and, in recent years, multiple layers of vinyl wall covering applied to gypsum board have contributed to flame propagation in a number of fires where multiple lives were lost.

The tunnel test is not able to adequately evaluate certain types of materials. Some materials, such as foam plastics, can achieve low flame spread ratings and still spread flames readily in actual fires. In recent years, a research program involving textile wall coverings demonstrated that NFPA 255 is not always able to evaluate textile wall coverings properly.[30] Vinyl film with foam backing is another example of a product that may achieve a low NFPA 255 flame spread rating and yet may readily propagate flame in an actual fire.

One should be "suspicious" of tunnel results for thermoplastic, lightweight materials (products with low thermal inertia) and those that drip to the floor of the tunnel during testing. When questions exist about performance, products should be evaluated on a full-scale basis, using an ignition source simulating the type of fire likely to be encountered in actual use. Products should be mounted in a manner that is representative of the conditions of actual use.

Application of tunnel results: NFPA *101*® classifies interior finish materials using flame spread indices and smoke-developed values. It is recognized that considerable variability exists in tunnel data, and precise values are not likely to be obtained in replicate tests. Further, small differences in ratings (10–20 points, for example) are not likely to be discernible in terms of actual fire behavior. Accordingly, flame spread ratings are usually rounded to the nearest whole number divisible by 5. The flame spread classifications also involve a broad range of values, as follows:

Classification	Flame Spread	Smoke Developed
A	0–25	0–450
B	26–75	0–450
C	76–200	0–450

As can be noted in the above chart, interior wall and ceiling finish classifications are based mainly on flame spread ratings with a single smoke development limit value of 450, which is common to all classifications.

Flame spread ratings are to provide information on the rate of flame spread across the surface of a material. Radiant heat transfer would be the principal hazard associated with flame spread and would only affect a person who was in close proximity to the fire. As a consequence, different classes of interior finish materials are specified within a building, depending upon location. For example, exit enclosures in which travel paths are severely restricted have the most conservative flame spread ratings, since individuals in such a location would be restricted in their travel paths and, therefore, might not be able to avoid the heat from the flame.

A smoke-developed limit of 450 is used for all interior finish classifications. The smoke-developed ratings assess product behavior in terms of obscuration of exit paths and loss of visibility by smoke. Smoke interferes with vision on the basis of the accumulated soot within a given volume. The soot concentration will be dependent upon both time and the rate of smoke production. A product that has a low smoke-development value would be expected to maintain visibility in egress paths for a longer period of time than a material with a high smoke rating. The same smoke development is used for all three classifications (i.e., classes A, B, and C). The recommended 450 limit acknowledges that smoke affects visibility both in the immediate vicinity of the fire and in remote areas as well. Large buildings can be quickly filled with smoke as the result of a fire. An upper limit has been established that is applicable to interior finishes, regardless of where the materials are installed.

The smoke-developed limit of 450 was developed from research conducted by Underwriters Laboratories Inc. (UL).[31] In this research, products were burned in the NFPA 255 "tunnel test" and the smoke discharged into a 5,000 cu ft (141.5 m³) room. The room was equipped with illuminated exit signs. The time required for exit signs to be obscured was recorded and compared to the smoke-developed rating for the different materials tested. The UL report states that "materials having smoke-developed ratings above 325 showed 'good' to 'marginal' visibility in a few cases; other materials produced conditions of 'marginal' to obscuration in the 6-min period." The 450 limit has been used for many years and is thought to set a "reasonable" limit for smoke-developed ratings. The 450 smoke limit takes into account both the time required to produce the smoke and the effect of the smoke on vision distance.

There is no direct relationship between flame spread and smoke-developed values. In the UL report referenced above, for example, one material had a flame spread rating of 490 and a smoke-developed factor of 57, whereas another had a flame spread rating of 44 and a smoke-developed factor of 1387.

The smoke-developed limit from NFPA 255 is solely determined on the basis of obscuration of light. It should be recognized that smoke may act as an irritant, further affecting visibility, and the smoke may, in addition, have a debilitating effect on persons attempting to escape. The irritability and other physiological effects are not evaluated by the 450 smoke-developed limit.

Different classes of interior finish materials are specified, depending upon location within a building and depending upon the occupancy involved. Interior finish restrictions are less stringent, e.g., for a general office area, than would be the case for an exit stair enclosure or exit access corridor. Similarly, occupancies having less capable occupants have stricter interior finish requirements than occupancies used by mobile and alert occupants. For example, the interior finish restrictions for health care facilities are quite restrictive, due to the inability of hospital patients to take action for their own self-preservation.

The limits on interior wall and ceiling finish are set forth in the 1994 edition of NFPA 101® and are summarized in Table 7-3A. The table also includes requirements for interior floor finishes. As is noted in a footnote to Table 7-3A, automatic fire suppression, such as automatic sprinklers, will largely mitigate the hazard associated with interior finish materials. Accordingly, NFPA 101® allows interior finish materials to drop one classification when automatic sprinkler systems are provided. For example, if the code normally requires a Class A interior finish and automatic sprinklers are installed, then a Class B interior finish is normally allowed.

The Radiant Panel Test

Attempts have been made to develop a smaller scale test that could be used in place of the 25-ft (7.6-m) long "tunnel" test. The advantages of such a method are obvious: less space required for the equipment, smaller specimens needed, and lower costs.

The Forest Products Laboratory developed a reduced-scale version of the 25-ft (7.6-m) tunnel. The reduced-scale tunnel was 8 ft (2.4 m) in length and used flame spread observations as a function of time to determine flame spread ratings. The 8-ft (2.4-m) tunnel test was adopted by ASTM in 1968 for research and development. The method was withdrawn by ASTM in 1993.

A radiant panel apparatus was developed at NIST (formerly National Bureau of Standards). The radiant panel apparatus uses a much smaller test specimen of 6 by 18 in. (152 by 457 mm) and measures hazard in terms of both flame spread and rate of heat release after ignition. The test method was adopted as a tentative standard by ASTM in 1960 and as a full standard for research and development purposes in 1967.[32]

In the radiant panel test, ASTM E162,[32] the specimen is positioned at an angle of 30 degrees to vertical in front of a gas-fired, porous, refractory, vertical panel. The specimen slants toward the radiant panel at the top and a small pilot flame at that location ignites flammable gases released from the surface of the specimen. Air is drawn over the surface of the specimen at a controlled rate by a fan in the hood under which the equipment is located. The rate of flame travel from the top of the specimen to the bottom is noted, as are temperatures that develop in the stack. The amount of smoke generated can be measured by weighing deposits on a filter located in the stack. The test is continued for 15 min or until surface flaming reaches the lower edge of the specimen. This radiant panel apparatus is available commercially and is used in many research, commercial, and industrial laboratories. The results of tests are expressed on a scale similar to that of NFPA 255. In the radiant panel, the airflow is against or across the direction of flame travel. This provides a clear definition of the interfacial burning. With the 25-ft (7.6-m) furnace, the flame travel and airflow coincide, and, with some materials, the flame front may be several feet in front of the interfacial burning. See Figure 7-3C.

The scope of ASTM E162 limits the use of the method to research and development. ASTM E162 is not intended for regulatory use and should not be relied upon as a substitute for evaluating products in accordance with NFPA 255.

The Corner Test

Seeking a more realistic assessment of the hazard of interior finishes, several laboratories conducted room/corner tests some 30 years ago. These usually comprised an 8-ft (2.44-m) high corner construction with 2- to 4-ft (0.1- to 1.22-m) wing walls. A simulated ceiling of the same material was provided. A wood crib placed in the corner at floor level was used for ignition. The degree of flame spread and the rate and amount of smoke developed were the primary observations.

Interest in the rapidity of flame spread across the surface of what was considered low-flame-spread cellular plastics in actual fires reinforced the use of the corner test. Large-scale corner tests

TABLE 7-3A. *Life Safety Code Requirements for Interior Finish*

Occupancy	Exits	Access to Exits	Other Spaces
Assembly—New*			
Class A or B	A	A or B	A or B
Class C	A	A or B	A, B, or C
Assembly—Existing*			
Class A or B	A	A or B	A or B
Class C	A	A or B	A, B, or C
Educational—New*	A	A or B	A or B
			C on low partitions†
Educational—Existing*	A	A or B	A, B, or C
Day-Care Centers—New	A	A	A or B
	I or II	I or II	
Day-Care Centers—Existing	A or B	A or B	A or B
Day-Care Homes—New	A or B	A, or B	A, B, or C
Group Day-Care Homes—Existing	A or B	A, B, or C	A, B, or C
Family Day-Care Homes—Existing	A or B	A, B, or C	A, B, or C
Health Care—New (A.S. Mandatory)	A or B	A or B	A or B
		C lower portion of	C in small individual
		corridor wall†	rooms†
Health Care—Existing	A or B	A or B	A or B
Detention and Correctional—New	A*	A*	A, B, or C
	I	I	
Detention and Correctional—Existing	A or B*	A or B*	A, B, or C
	I or II	I or II	
Residential, Hotels, and Dormitories—New	A	A or B	A, B, or C
	I or II	I or II	
Residential, Hotels, and Dormitories—Existing	A or B	A or B	A, B, or C
	I or II‡	I or II†	
Residential, Apartment Buildings—New	A	A or B	A, B, or C
	I or II†	I or II†	
Residential, Apartment Buildings—Existing	A or B	A or B	A, B, or C
	I or II‡	I or II‡	
Residential, Board and Care§	See *Life Safety Code®*, Chapters 22 and 23		
Residential, 1- and 2-Family, Lodging or Rooming Houses	A, B, or C	A, B, or C	A, B, or C
Mercantile—New*	A or B	A or B	A or B
Mercantile—Existing Class A or B*	A or B	A or B	ceilings—A or B
			existing on walls
			—A, B, or C
Mercantile—Existing Class C*	A, B, or C	A, B, or C	A, B, or C
Office—New	A or B	A or B	A, B, or C
	I or II	I or II	
Office—Existing	A or B	A or B	A, B, or C
Industrial	A or B	A, B, or C	A, B, or C
Storage	A or B	A, B, or C	A, B, or C
Unusual Structures	A or B	A, B, or C	A, B, or C

*Interior wall and ceiling finish in corridors, exits, and any space not separated from the corridors and exits by a partition capable of retarding the passage of smoke shall meet this interior finish classification.
†Refer to NFPA *101®*.
A.S. Automatic Sprinklers.
‡Previous installed floor coverings may be continued in use, subject to the approval of the Authority Having Jurisdiction.
§See Chapter 21 of NFPA *101®* for details.
NOTES:
Class A interior wall and ceiling finish—flame spread 0–25, smoke developed 0–450.
Class B interior wall and ceiling finish—flame spread 26–75, smoke developed 0–450.
Class C interior wall and ceiling finish—flame spread 76–200, smoke developed 0–450.
Class I interior floor finish—minimum 0.43 Btu per sq ft (0.45 w/cm^2 per sec).
Class II interior floor finish—mimimum 0.19 Btu per sq ft (0.22 w/cm^2 per sec).
Automatic Sprinklers—where a complete standard system of automatic sprinklers is installed, interior finish with flame spread rating not over Class C may be used in any location where Class B is normally specified and with rating of Class B in any location where Class A is normally specified; similarly, Class II interior floor finish may be used in any location where Class I is normally specified and no critical radiant flux rating is required where Class II is normally specified.
Exposed portions of structural members complying with the requirements for heavy timber construction may be permitted.

have been conducted by Factory Mutual Research Corporation (FMRC) and Underwriters Laboratories Inc. Corners up to 25 ft (7.6 m) high with wing walls up to 50 ft (15.4 m) long have been used to determine the burning characteristics of cellular plastics and the effectiveness of protective measures, such as automatic sprinkler protection or thermal barriers.[33] Results have reinforced the premise that fire hazard cannot be fully judged on the basis of any single fire test method.[34] (See Figure 7-3D.)

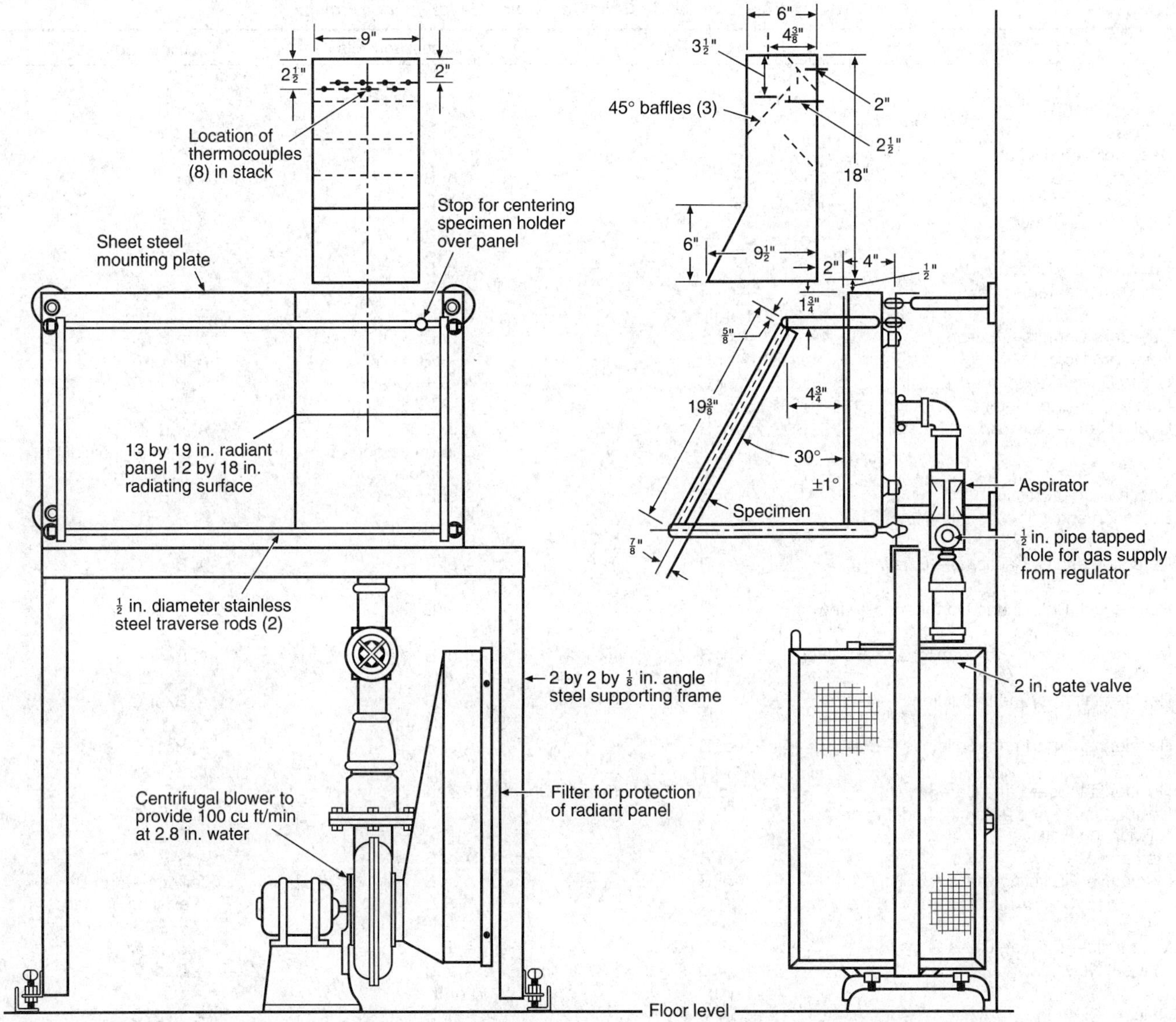

FIG. 7-3C. Diagram of the radiant panel tests. (For SI units: 1 in. = 25.4 mm; 1 cu ft/min. = 0.0283 m³/min.)

Model building codes and NFPA *101*® both specify the use of room/corner tests for judging the flammability behavior of certain interior finishes. NFPA *101*®, for example, requires the use of NFPA 265, *Standard Methods of Fire Tests for Evaluating Room Fire Growth Contribution of Textile Wall Coverings* (hereinafter referred to as NFPA 265), to evaluate the performance of textile wall coverings. In the 1980s, researchers at the University of California-Berkeley Fire Research Laboratory showed that wall coverings with NFPA 255 (ASTM E84) ratings as low as 15 sometimes performed poorly, producing flashover in room/corner tests.[35] It was determined that room/corner tests (e.g., NFPA 265) provide a better indication of the hazard presented by these particular wall coverings than do the NFPA 255 flame spread ratings.

The NFPA 265 test method consists of an 8-ft wide by 12-ft long by 8-ft high (2.44 x 3.66 x 2.44 m) room with a single door opening. (See Figure 7-3E.) Products are mounted on two walls joined at right angles to form a corner. A diffusion burner is positioned near the corner of the room and provides the required fire exposure. The smoke from the room is collected in a hood and duct

exhaust system where instrumentation is used to determine rate of heat release (through oxygen depletion) and smoke optical density. Heat flux and temperature measurements are also made within the room. In NFPA 265, the burner provides 5-min 40 kW fire exposure followed by a 10-min 150 kW exposure. The burner output and location can be adjusted during research for product development. Burner outputs up to 300 kW have been used to evaluate the performance of interior finish materials in room/corner procedures.

NFPA *101*® prohibition for foam plastics acknowledges that NFPA 255 flame spread ratings may not accurately reflect the flammability behavior of such materials. If foam or cellular plastics are to be used exposed within a building, the application should be substantiated on the basis of tests that are representative of the conditions of actual use. Room/corner tests, such as NFPA 265, have been used to evaluate foamed plastic wall coverings. NFPA 265 uses a gas diffusion burner as the fire source. At the start of a test, a burner output of 40 kW simulates a "waste basket" sized fire exposure.[30] This exposure is to determine if flames will spread away from a small fire to ignite other nearby objects. At 5 min, the burner output

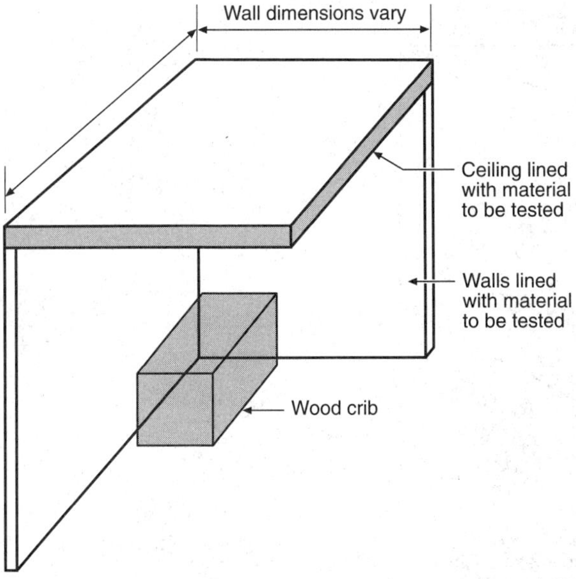

FIG. 7-3D. Corner test.

is increased to 150 kW to determine if the wall lining will release sufficient energy to cause full room involvement or flashover. At the 150 kW exposure, flames from the burner will just reach the ceiling of the 8-ft (2.44-m) high test chamber, absent any contribution by a product undergoing tests.

The "standard" fire exposure specified within NFPA 265 is believed to be appropriate for wall coverings. However, ceiling coverings may not be adequately evaluated by this procedure since the burner placement and 150 kW output (assuming no involvement of a test specimen) produces a flame that just barely reaches the ceiling. For ceiling installations, consideration could be given to using the burner placement and output detailed within the ISO 9705 room/corner test.[36] The ISO room/corner test might be employed using a 5-min, 40 kW initial exposure, followed by a 10-min, 160 kW exposure. The 160 kW exposure specified in the ISO room/corner test (absent any contribution by the products undergoing test) will result in burner flames hitting the ceiling and turning to expose the underside of the ceiling for 18–24 in. (457–610 mm). The direct flame exposure to the ceiling is considered necessary to adequately evaluate combustible ceiling finish behavior.

An updated version of the room/corner test procedure has recently been developed and is referenced in the *Uniform Building Code;* [37] it is also proposed for reference in the *Standard Building Code* [38] for the regulation of foam plastic insulated garage doors in commercial applications. The test procedure evolved from a research project sponsored by the National Association of Garage Door Manufacturers.[39] This new room/corner test method and associated regulatory limits are unique in that rate of heat release and smoke release limits have been developed that are linked to hazard. The measured pass/fail criteria eliminate the need for subjective judgments of product behavior required by the older editions of room/corner tests. The methodology employed by the new room/corner procedure could be adapted for other applications of interior finishes.

Full-scale test procedures, like room/corner tests, allow products to be evaluated as they will be installed using an ignition source representative of those likely to be encountered during actual use. Accordingly, room/corner test results are believed to be more indicative of product behavior during actual fires than are standardized tests, such as NFPA 255.

Smoke Density Chamber

A "bench test" for measuring smoke generation is available for research, development, and production quality control. Described in NFPA 258, *Standard Research Test Method for Determining Smoke Generation of Solid Materials* (hereinafter referred to as NFPA 258), it measures the total smoke generated in a chamber from test specimens up to 1 in. (25.4 mm) thick. Despite the specific intent of NFPA 258 to limit the test procedure to use as a research and development tool, some jurisdictions specify its use for regulatory purposes.

This method is known to have numerous shortcomings. The method involves a closed, static chamber, whereas smoke transfer in actual fires is dynamic. The chamber is closed and oxygen-deficient atmospheres can develop that may artificially affect the burning behavior of the products being tested. Deposits of soot on wall surfaces and optics can affect measurements. In summary, numerous questions exist as to the relevance of data from NFPA 258 to the behavior of products in actual fires.

FIRE TESTS FOR INTERIOR FLOOR FINISHES

Fire tests for interior finishes were developed primarily with wall and ceiling finishes in mind, long before carpeting became popular as an interior floor finish. Official attention was directed toward a possible hazard with soft floor coverings in 1960 and 1961 as a result of a series of small fires in the Washington, DC, area.[40] A well-publicized dwelling fire on the West Coast in 1967, followed by the Harmer House Nursing Home fire in Ohio in 1970, focused attention on carpet floor coverings, although carpets were a contributing, rather than a causative, factor in these incidents.[41,42]

In 1965, the Public Health Service of the U.S. Department of Health, Education, and Welfare issued a directive setting flame spread limits on floor coverings in federally funded hospitals, citing the 25-ft (7.6-m) tunnel test as the test method. In the 25-ft (7.6-m) tunnel test, a floor covering would be mounted upside down and on the ceiling of the apparatus. Since neither the mounting or location correctly simulated an actual installation, it is not surprising that the tunnel test was ultimately judged to be inadequate for floor coverings. Additional problems existed for carpets made with synthetic fibers of the type that melt, drip, or delaminate when exposed to elevated temperatures. As a further deficiency, carpets with separate underlays could not be effectively tested in the tunnel. The need for a test specifically designed for interior floor finishes was clearly identified.

Methanamine Pill Test

NIST (formerly National Bureau of Standards) developed the methanamine pill test as a means of preventing the use of highly flammable carpet floor coverings. The pill test was adopted in 1970 as DOC FF-1-70 (Federal Register 1970a) for carpets and DOC FF-2-70 for rugs.[43,44]

In this test, eight 9-in. (229-mm) square sections of a carpet are conditioned to drive off excess moisture, brought to room temperature in a desiccator, and tested. Each specimen, in turn, is placed on the bottom of a 1-ft (0.30-m) enclosed cube, which is open at the top, and held in place by a 9-in (229-mm) square metal plate having an 8-in. (203-mm) circular cutout. A methanamine tablet is placed at the center of the circle and lighted. The pill produces a candle-like flame for about 100 sec. A specimen fails if the flame advances to within 1 in. (25.4 mm) of the metal ring in the hold-down plate. At least seven of the eight specimens must pass the test to meet the established criteria. All carpet manufactured for sale in the United States has been required, since April 1971, to pass this test. (See Figure 7-3F).

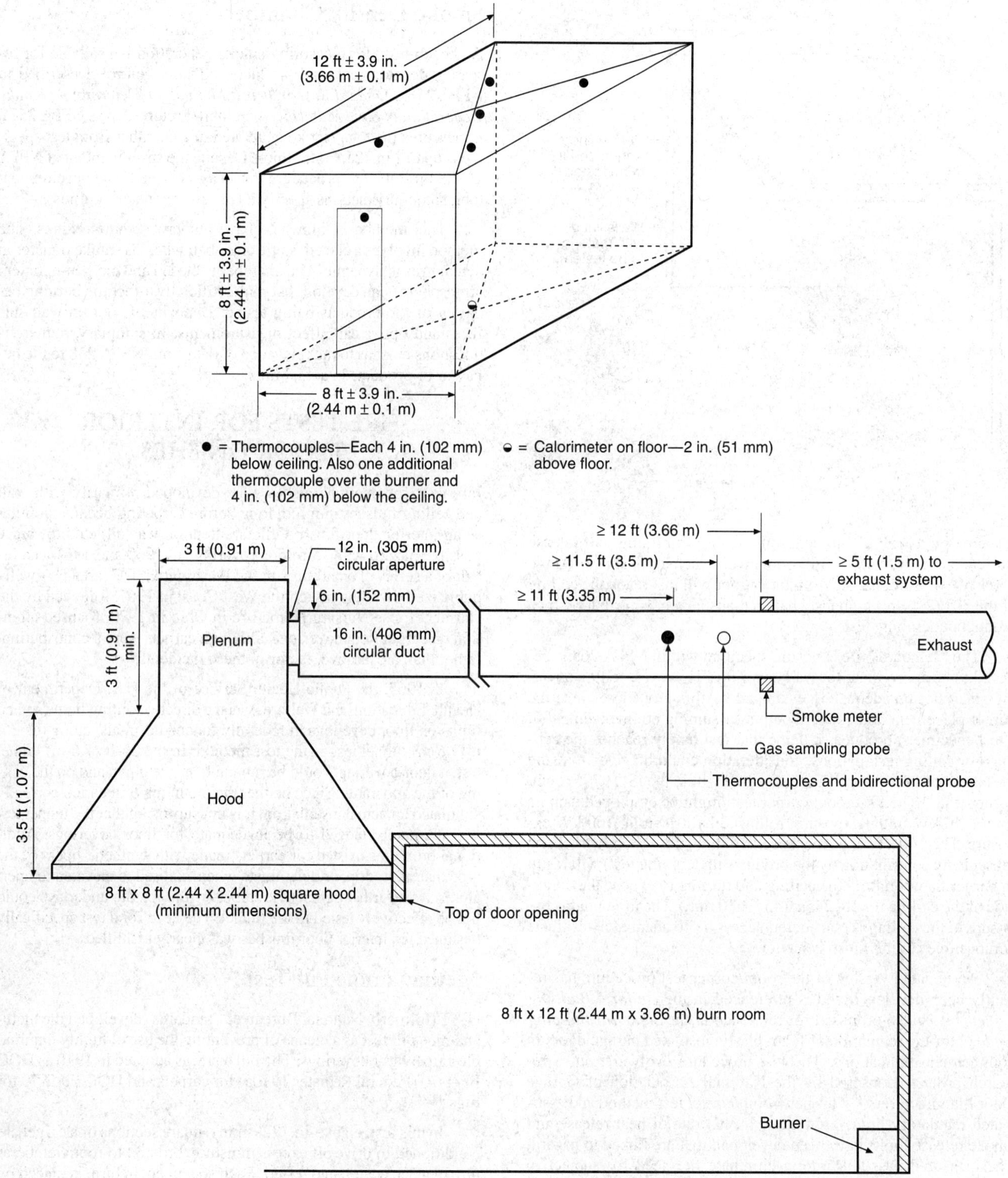

● = Thermocouples—Each 4 in. (102 mm) below ceiling. Also one additional thermocouple over the burner and 4 in. (102 mm) below the ceiling.

◒ = Calorimeter on floor—2 in. (51 mm) above floor.

FIG. 7-3E.   NFPA 265 room/corner test.

Although the "pill test" is a small-scale test, the results have large implications in terms of floor covering performance. It has been shown that the pill test provides adequate "first to ignite" protection for carpet located in rooms.[45,46] Experiments conducted at NIST[47,48] revealed that carpet and floor coverings were unlikely to propagate flame in rooms until room fires approach or exceed flashover. As a consequence, little additional safety is to be gained by regulating floor coverings within rooms beyond the pill test. These findings were used, in part, to establish the current NFPA 101® restrictions for floor coverings.

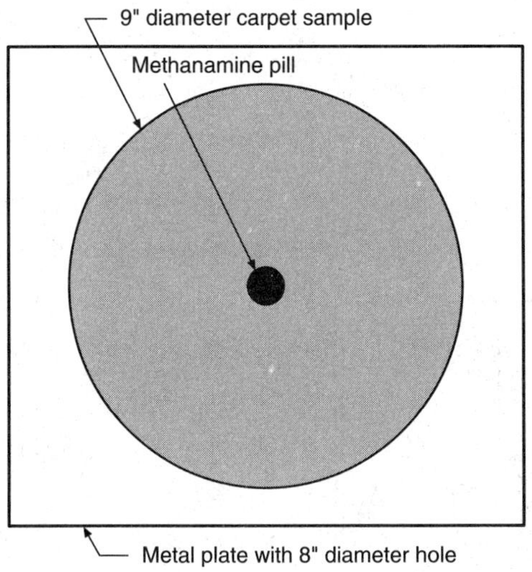

9" diameter carpet sample

Methanamine pill

Metal plate with 8" diameter hole

FIG. 7-3F.   Methanamine pill test.

Floor Covering Chamber Test

In 1969, UL started work on a new test for floor coverings with the test specimens mounted in a floor position.[49] The program was sponsored by the U.S. Public Health Service and resulted in what became known as UL 992, the floor covering chamber test. In essence, UL 992 is an 8-ft (2.44-m) version of the 25-ft (7.6-m) tunnel furnace, with the test specimen mounted on the floor of the tunnel and with appropriate modifications in heat input, burner design, airflow, and other test specifications. This test never gained widespread acceptance and is no longer used.

Critical Radiant Flux Test

Floor coverings having modest resistance to flame spread are unlikely to become involved in the early growth of a fire within a room. Regulation of floor coverings is generally where specified by the code or where a need is indicated by some unusual product. It has been shown that floor coverings will propagate flame only under the influence of a sizable exposure fire. Corridor floor coverings, for example, may spread flame when subjected to radiant heat from the hot gases discharging from the doorway of a fully fire-involved room. The fire discharges flame and hot gases into the upper level of the corridor, and this results in radiant heat energy transfer to the floor. The level of energy radiating onto the floor is a significant factor in determining whether or not progressive flaming will occur. NFPA 253 measures the minimum energy required to be impinged on floor coverings to sustain flame propagation. The minimum value required to sustain flame is termed "critical radiant flux." The higher the critical radiant flux, the better the performance. In summary, the NFPA 253 flooring radiant panel test is used for evaluating the fire performance of a floor covering in a building corridor or exit. Floor coverings in open building spaces and in rooms within buildings generally merit no further regulation, provided the floor covering is at least as resistant to flame spread as a material that will meet the federal flammability standard FF-1-70 "pill test."

NFPA *101*® specifies use of the NFPA 253 flooring radiant panel test for floor coverings in exits and corridors of certain occupancies.[50,51] The flooring radiant panel test consists of a closed chamber having a radiant panel inclined at a 30-degree angle from a 39-in. (1000-mm) long test specimen. The test specimen is exposed to radiant heat that ranges from approximately 10 kW/m^2 to 1 kW/m^2. Flame spread over the face of the floor covering specimen

is initiated by flames from a gas burner. (See Figure 7-3G.) NFPA 253 determines the minimum energy required for a floor covering to spread flame.

This test method simulates conditions that have been observed and defined in large-scale corridor experiments. Both the NFPA and the ASTM test methods indicate that the method is intended for evaluating floor coverings installed in building corridors having little or no combustible wall or ceiling finish. Corridors with combustible wall and ceiling finishes would be expected to produce a potentially greater fire hazard. Experimental results suggest that floor coverings would contribute much less to fire spread than wall or ceiling finish in the initial growth of fires.[52]

NFPA *101*® classifies interior floor finishes on the basis of critical radiant flux. Specifically, floor finishes are grouped into two classes:

Class I: A minimum critical radiant flux of 0.43 Btu per sq ft per sec (0.45 W/cm^2)

Class II: A minimum critical radiant flux of 0.19 Btu per sq ft per sec (0.22 W/cm^2)

See Table 7-3A for assignment of the two classes of floor finishes to the various occupancies identified in NFPA *101*.®

The NFPA 253 flooring radiant panel test requires test specimens to be representative of the product to be installed. Where carpet is to be loose laid over concrete, for example, it should be tested over a thick, high-density inorganic reinforced cement board that simulates concrete. Where carpet, with or without integral pad, is to be bonded to concrete, an adhesive should be used, following the manufacturer's directions as to the type and application rate.

The use of a separate underlayment (pad) will normally lower the critical radiant flux of a test specimen and, therefore, increase the tendency for flame spread. Research at NIST[53] has shown that underlayment lowers the critical radiant flux of a test specimen, due to the insulative effect of the underlayment. The NFPA 253 flooring radiant panel test allows a manufacturer of carpet to test over a "standard" pad or the actual underlayment proposed for use. Research at NIST has shown the use of a standard pad—a rubber-coated jute and animal hair or fiber pad a minimum of $^3/_8$ in. (9.5 mm) thick and weighing 50 oz/sq yd (1.86 kg/m^2)—is likely to yield lower critical radiant flux values than that likely to be produced by other underlayments in use.[53,54] If the manufacturer tests a carpet over the "standard pad," NFPA 253 allows the results of such a test to be applied to any carpet/underlayment combination. Alternatively, if the manufacturer chooses to test using the actual pad proposed for use, the results would be valid only for the actual combination tested.

Floor coverings are not generally regulated on the basis of smoke generation. Regulation of floor coverings on the basis of smoke is not considered practical or necessary. Floor coverings will not normally spread flame until after a fire grows to large size. Most fires will be so large and the smoke production so great when floor coverings become involved that the minimal benefits achieved by imposing smoke limits on floor coverings would not generally warrant such regulation. In addition, regulation of products based on smoke generation is not considered practical, because no suitable apparatus currently exists that has been shown capable of generating data that correlates with product performance in actual fires. Nevertheless, some jurisdictions establish additional criteria for permissible floor finishes based on a test for smoke generation. In such instances, the NFPA 258 test method is used. Conceived as a research and development method, this test provides a procedure for measuring the total smoke generated in a closed chamber. The

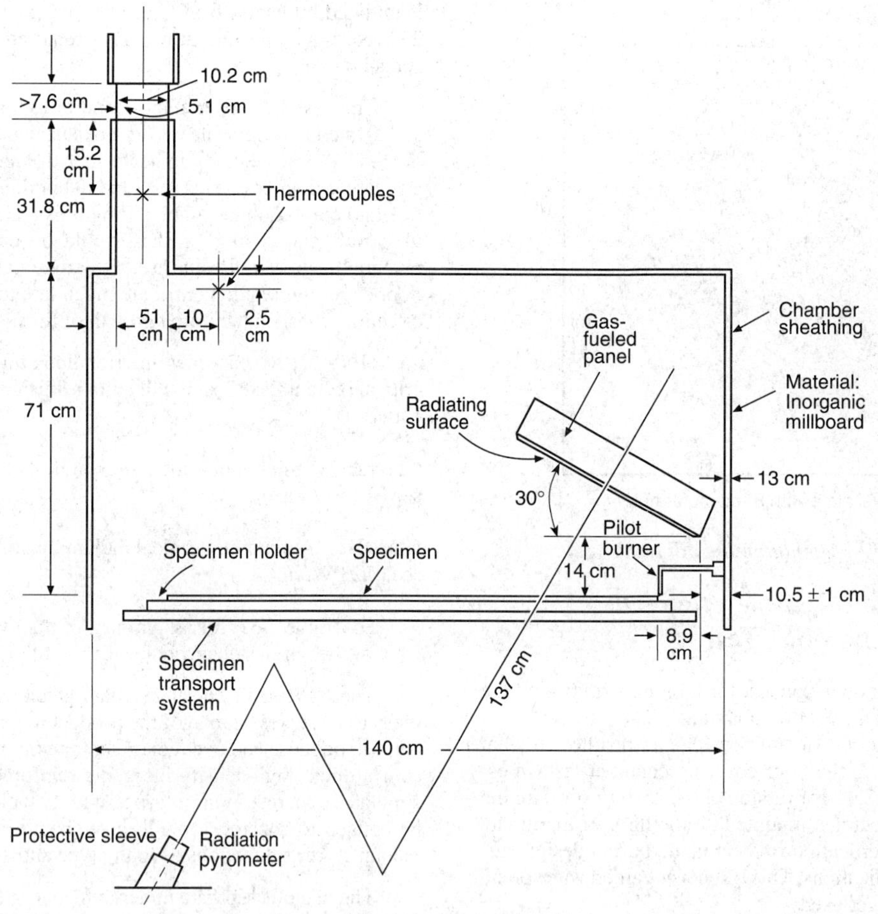

Note: in. = cm x 0.3937.

FIG. 7-3G.   *Flooring radiant panel test schematic (size elevation).*

method is known to have numerous shortcomings. Some of the method deficiencies have been previously identified.

FULL-SCALE TESTS

Many large-scale fire tests have been conducted using existing buildings slated for demolition or available for other reasons. When carefully planned, such tests can, and have, added significantly to knowledge about the growth and spread of fire in buildings.

One of the first of these tests was conducted in Great Britain in 1949, under the auspices of the Fire Research Station.[55] Two two-story dwellings were used, one with cellulose fiber insulation board lining and the other with plasterboard over fiberboard. One test in each house, both with real or simulated furnishings, was conducted with bedroom doors open. The other test was conducted with the doors closed. With both types of lining, temperatures in the living room where the fire was started became intolerable within a few minutes. As in many other tests and actual fires, the value of the closed door also was clearly demonstrated in minimizing the life hazard in bedrooms.

In 1958, the Division of Building Research of the National Research Council of Canada conducted a series of full-scale tests, known as the "St. Lawrence Burns," in six dwellings and two larger buildings.[56] Again, the one objective was to study fire spread as it would affect the life safety of the occupants of the second-floor bed-

room behind open and closed doors. All dwellings became "smoke logged" within 6 min of the ignition of the wood cribs, regardless of the type of interior finish. Temperatures and gaseous combustion products developed much faster in the dwellings with combustible finishes and in those tests in which the bedroom doors were open.

Many large buildings scheduled for demolition were partially burned under the auspices of the IIT Research Institute. In these, and in extensive room/corridor tests, much information was developed relative to fire spread, flashover phenomena, and factors affecting life safety in fires.

The Office of the State Fire Marshal in California sponsored a series of tests in unused buildings at Camp Parks in the early 1970s. The objective was to develop criteria for egress facilities, particularly corridors. The final report includes a discussion of the reaction and interaction of wall, ceiling, and floor finishes.[57] Precedent for the full-scale corridor burns was established earlier in 1959 and 1960 by the City of Los Angeles (CA) Fire Department. Final reports on the Los Angeles tests were published in two volumes by NFPA.[58,59]

Furnished room burnout and corridor tests conducted at the Forest Products Laboratories of the U.S. Department of Agriculture compared the relative importance of room furnishings *vs.* room finishes.[21] The conclusion that lethal levels of combustion products developed from burning furnishings before the wall finishes were seriously involved was supported by later work at Southwest Research Institute.[22] These and other studies recognize that, once a fire is well established, it is highly probable that all combustibles in the room will become involved. It is also recognized that early detection, containment, and extinguishment in the room of fire origin is important, not only for early evacuation from the fire area, but also for the safety of persons elsewhere in the building and for the protection of the building itself.

From 1979 to 1980, the U.S. Fire Administration sponsored a series of 75 tests to evaluate the performance of residential sprinklers in single-family and mobile homes. These tests showed the rapid buildup of heat and toxic gases that was possible with ordinary-hazard furnishings, particularly when the fire was ignited in the corner of the room or near curtains. There was a significant difference in the levels of combustion products and the times to reach those levels between fires in noncombustible interior finish and combustible interior finish.[60]

Room fire tests were commonly used in the 1980s for research purposes to evaluate interior finishes, wall coverings, and room furnishings. To help make sense of the proliferation of testing procedures, ASTM E603, *Standard Guide for Room Fire Experiments*,[61] was developed. This guide provides information on how room fire tests can be implemented and what combustion parameters are measured. Testing instrumentation may include, but is not limited to, oxygen, carbon dioxide, carbon monoxide, and other gas measurements; rate of heat release; heat flux; flame spread; temperature; air velocity profiles; mass loss rates; and smoke optical densities.

BIBLIOGRAPHY

References Cited

1. Juillerat, E. E., Jr., "The Hartford Hospital Fire," *NFPA Quarterly*, Vol. 55, No. 3, 1962, pp. 295–303.
2. Sears, A. B., Jr., "Nursing Home Fire, Marietta, Ohio," *Fire Journal*, Volume 64, No. 3, May 1970.
3. Isner, M. S., "Three Die, Eight Injured in Athletic Club Fire," *Fire Journal*, Nov./Dec. 1992.
4. Coté, R., and Timoney, T., "Boarding House Fire Causes Fifteen Deaths," *Fire Journal*, Jan. 1985.
5. "Investigation Report on the Las Vegas Hilton Hotel Fire," *Fire Journal*, Jan. 1982.
6. Bukowski, R. W., "Analysis of the Happy Land Social Club Fire with Hazard I," *Fire and Arson Investigator*, Vol. 42, No. 3, Mar. 1992, pp. 36–37.
7. Hall, J. R., Report prepared for the House Subcommittee on Science, Research, and Technology on H.R. 94, the Hotel and Motel Firesafety Act of 1989, Mar. 2, 1989, Quincy, MA, NFPA Fire Analysis and Research Division, National Fire Protection Association, Quincy, MA.
8. Klem, T., "Investigation Report on the DuPont Plaza Hotel Fire," 1987, National Fire Protection Association, Quincy, MA.
9. Isner, M. S., "Five Die in High-Rise Office Building Fire," *Fire Journal*, Vol. 84, No. 4, July/Aug. 1990, pp. 50–57, 59.
10. Watrous, L. D., "Ten Die in Convalescent Home Fire," *Fire Journal*, Vol. 66, No. 5, Sept. 1972, pp. 16–20.
11. Madrzykowski, D., and Vettori, R., "A Sprinkler Fire Suppression Algorithm for the GSA Engineering Fire Assessment System," National Institutes of Standards and Technology, Building and Research Laboratory, NISTIR 48, 33, May 1992.
12. Belles, D. W., "Designing Fire Safe Interiors," *NFPA Journal*, July/Aug. 1992.
13. Demers, D. D., "Familiar Problems Caused Ten Deaths in Hotel Fire," *Fire Journal*, Vol. 74, No. 1, Jan. 1980, pp. 52–56.
14. "Disclosure Requirements and Prohibitions Concerning the Flammability of Plastics," Federal Trade Commission, Washington, DC, Aug. 1974.
15. "Fire Safety Guidelines for Use of Rigid Urethane Foam Insulation in Building Construction," *Urethane Safety Group Bulletin*, The Society of Plastics Industry, Inc., New York, NY, May 1974.
16. SBCCI, *Standard Building Code-1994*, Section 2603.3.
17. Christian, W. J., "The Effect of Structural Characteristics on Dwelling Fire Fatalities," *Fire Journal*, Vol. 68, No. 1, Jan. 1974, pp. 22–28.
18. Waterman, T. E., "Room Flashover-Model Studies," *Fire Technology*, Vol. 8, No. 4, Nov. 1972, pp. 316–325.
19. "NBS Corridor Fire Tests: Energy and Radiation Models," NBS Technical Note 794, National Bureau of Standards, Washington, DC, Oct. 1973.
20. Waterman, T. E., and Christian, W. J., "Flame Spread in Corridors: Effect of Location and Area of Wall Finish," *Fire Journal*, July 1971.
21. Bruce, H. D., "Experimental Dwelling Room Fires," Report No. 1941, Apr. 1959, U.S. Department of Agriculture, Forest Products Laboratory, Madison, WI.
22. Pryor, A. M., "Full-Scale Fire Tests of Interior Wall Finish Assemblies," *Fire Journal*, Vol. 63, No. 2, Mar. 1969, pp. 14–20.
23. Waksman, D., and Ferguson, J. B., "Fire Tests of Building Interior Covering Systems," *Fire Technology*, Vol. 10, No. 3, Aug. 1974.
24. "BOCA National Building Code," Section 804, BOCA, 1993.
25. "Standard Building Code," Section 803.8, SBCCI, 1994.
26. ASTM E1354, *Standard Test Method for Heat and Visible Smoke Release Rates for Materials and Products Using an Oxygen Consumption Calorimeter*, American Society for Testing and Materials, W. Conshohocken, PA, 1994.
27. ASTM E1321, *Standard Test Method for Determining Material Ignition and Flame Spread Properties*, American Society for Testing and Materials, W. Conshohocken, PA, 1993.
28. Wilson, J. A., "Surface Flammability of Materials: A Survey of Test Methods and Comparison of Results," ASTM STP No. 301, American Society for Testing and Materials, W. Conshohocken, PA, 1961.
29. Drysdale, D., *Introduction to Fire Dynamics*, John Wiley and Sons, 1985, pp. 234-235.
30. Belles, D. W., Fisher, F. L., and Williamson, R. B. Ph.D., "How Well Does ASTM E84 Predict Fire Performance of Textile Wallcoverings?" *Fire Journal*, Vol. 82, No. 1, Jan./Feb. 1988.
31. Underwriters Laboratories Inc., "Study of Smoke Ratings Developed in Standard Fire Tests in Relation to Visual Observations," *Bulletin of Research*, No. 56, Apr. 1965.
32. ASTM E162, *Standard Test Method for Surface Flammability of Materials Using a Radiant Heat Energy Source*, American Society for Testing and Materials, W. Conshohocken, PA, 1994.
33. Christian, W. J., and Waterman, J. E., "Fire Behavior of Interior Finish Materials," *Fire Technology*, Vol. 6, No. 3, Aug. 1970, pp. 165–178.
34. Maroni, W. F., "Large-Scale Fire Tests of Rigid Cellular Plastic Wall and Roof Insulations," *Fire Journal*, Vol. 67, No. 6, Nov. 1973, pp. 24–30.
35. Fisher, F. L., MacCracken, W., Williamson, R. B., "Room Fire Experiments of Textile Wall Coverings, A Final Report of All Materials Tested Between March 1985 and January 1986," ES-7853, 1986, Service to Industry Report No. 86-2, Fire Research Laboratory.
36. Fire Tests—Full-scale Room Test for Surface Products, ISO 9705, International Organization for Standardization, Geneva, Switzerland.
37. 1995 Amendments, ICBO, *Uniform Building Code*, Standard 26-8, Room Fire Test Standard for Garage Doors Using Foam Plastic Insulation.
38. 1995 Amendments, SBCCI, Standard Building Code, Section 2603.3.
39. Belles, D. W., "Regulating Foam Plastic Insulated Garage Doors," *Building Standards*, ICBO, Jan.–Feb., 1995.
40. Yuill, C. H., "Floor Coverings: What Is the Hazard," *Fire Journal*, Vol. 61, No. 1, Jan. 1967, pp. 11–19.
41. "Fire in Acrylic Carpeting," *Fire Journal*, Vol. 62, No. 2, May 1968, pp. 13-14.
42. Sears, A.B., Jr., "Nursing Home Fire, Marietta, Ohio," *Fire Journal*, Vol. 64, No. 2, May 1970, pp. 5–9.
43. "Carpets and Rugs—Notice of Standard," *Federal Register*, Vol. 35, No. 74, Apr. 16, 1970.
44. "Small Carpets and Rugs—Notice of Standard," *Federal Register*, Vol. 35, No. 251, Dec. 29, 1970.
45. Benjamin, I. A., Adams, C. H., "The Flooring Radiant Panel Test and Proposed Criteria," *Fire Journal*, Vol. 70, No. 2, Mar. 1976.
46. Benjamin, I. A., and Davis, S., "Flammability Testing for Carpet," NBSIR 78-1436, Apr. 1978, Center for Fire Research, National Bureau of Standards.
47. Tu, King-Mon, and Davis, S., "Flame Spread of Carpet Systems Involved in Room Fires," NBSIR 76-1013, June 1976, Center for Fire Research, National Bureau of Standards.

48. Davis, S., Unpublished Series of Experiments, Fire Standards Research—Furnishings 4927677, Program: Fire Control-Furnishings, Center for Fire Research, NBS, Jan. 11, 1977.

49. "Standard Method of Test for Flame Propagation Classification of Flooring and Floor Covering Material," Subject 992, Underwriters Laboratories Inc., Northbrook, IL, Feb. 1971.

50. Hartzel, L. G., "Development of a Radiant Panel Test for Flooring Materials," *Journal of Fire and Flammability*, Consumer Product Flammability Supplement, Vol. 1, Dec. 1974, pp. 305–353.

51. Adams, C. H., and Davis, S., "Development of the Flooring Radiant Panel Test as a Standard Test Method," NBSIR 79-1954, Center for Fire Research, National Bureau of Standards, Mar. 1980.

52. McGuire, J. H., "The Spread of Fire in Corridors," *Fire Technology*, Vol. 4, No. 2, 1968, pp. 103–108.

53. Alderson, S., and Breden, L., "Evaluation of the Fire Performance of Carpet Underlayments," NBSIR 76-1018, Center for Fire Research, NBS, Sept. 1976.

54. Benjamin, I. A., and Adams, C. H., "Proposed Criteria for Use of the Critical Radiant Flux Test Method," NBSIR 75-950, Center for Fire Research, NBS, Dec. 1975.

55. "British Fire Tests of Fiberboard," *NFPA Quarterly*, Vol. 45, No. 3, Jan. 1952, pp. 218–224.

56. Shorter, G. W., *et al.*, "The St. Lawrence Burns," *NFPA Quarterly*, Vol. 53, No. 4, April 1960, pp. 300–316.

57. "Project Corridor: Fire and Life Safety Research," *Western Fire Journal*, North Highlands, CA, 1974.

58. "Operation School Burning: Official Report on a Series of School Fire Tests Conducted April 16 to June 30, 1959, by the Los Angeles Fire Department," National Fire Protection Association, Quincy, MA, 1959.

59. "Operation School Burning, No. 2: Official Report on a Series of Fire Tests in an Open Stairway, Multi-Story School Conducted June 30, 1960 to July 30, 1960, and February 6, 1961 by the Los Angeles Fire Department," 1961, National Fire Protection Association, Quincy, MA.

60. Cote, A. E., and Moore, D., "Field Test and Evaluation of Residential Sprinkler Systems," National Fire Protection Association, Quincy, MA, 1981.

61. ASTM E603, *Standard Guide for Room Fire Experiments*, American Society for Testing and Materials, W. Conshohocken, PA, 1995.

References

"New Fire Research Building," *Dimensions*, Vol. 58, No. 6, June 1974, pp. 123–124.

Shorter, G. W., *et al.*, "SLRP Analysis of Recommended Protection for Foamed-Plastic Wall-Ceiling Building Insulations," *Fire Journal*, Vol. 68, No. 5, Sept. 1974, pp. 51–55.

NFPA Codes, Standards, and Recommended Practices

Reference to the following NFPA codes, standards, and recommended practices will provide further information on interior finish in this chapter. (See the latest version of *The NFPA Catalog* for availability of current editions of the following documents.)

NFPA 101®, *Life Safety Code®*
NFPA 253, *Standard Method of Test for Critical Radiant Flux of Floor Covering Systems Using a Radiant Heat Energy Source*
NFPA 255, *Standard Method of Test of Surface Burning Characteristics of Building Materials*
NFPA 258, *Standard Research Test Method for Determining Smoke Generation of Solid Materials*
NFPA 265, *Standard Methods of Fire Tests for Evaluating Room Fire Growth Contribution of Textile Wall Coverings*
NFPA 701, *Standard Methods of Fire Tests for Flame-Resistant Textiles and Films*

Additional Reading

America Burning: The Report of the National Commission on Fire Prevention and Control, 1973, Superintendent of Documents, U.S. Government Printing Office, Washington, DC.

Babrauskas, V., "Estimating Room Flashover Potential," *Fire Technology*, Vol. 16, No. 2, May 1980, pp. 94–104.

Brannigan, F. L., "Flame Spread Can Be Accelerated by Interior Finishes," *Fire Engineering*, Vol. 138, No. 4, 1985, pp. 36–37.

Brehob, E. G., and Kulkarni, A. K., "Time-Dependent Mass Loss Rate Behavior of Wall Materials Under External Radiation," *Fire and Materials*, Vol. 17, No. 5, Sept./Oct. 1993, pp. 249–254.

Bukowski, R. W., *et al.*, "Fire Risk Assessment Method: Case Study 4, Interior Finish in Restaurants," National Institute of Standards and Technology, Gaithersburg, MD, National Fire Protection Assoc., Quincy, MA, Benjamin/Clarke Associates, Inc., Kensington, MD, National Fire Protection Research Foundation, Quincy, MA, NISTIR 90-4246, May 1990.

Castino, G. T., *et al.*, "Flammability Studies of Cellular Plastics and Other Building Materials Used for Interior Finish," UL Report, 1975, Underwriters Laboratories Inc., Northbrook, IL.

D'Souza, M. V., and McGuire, J. H., "ASTM E84 and the Flammability of Foamed Thermosetting Plastics," *Fire Technology*, Vol. 13, No. 2, May 1977, pp. 85–94.

Day, T., "Surface Coatings: Will Time Affect Their Performance?" *Fire Prevention*, No. 222, Sept. 1989, pp. 40–41.

Dembsey, N. A., and Williamson, R. B., "Effect of Ignition Source Exposure and Specimen Configuration on the Fire Growth Characteristics of a Combustible Interior Finish Material," *Fire Safety Journal*, Vol. 21, No. 4, 313–330, 1993.

Den Braven, K., "Radiative Ignition of Some Typical Floor Covering Materials," NBSIR 75-967, Dec. 1975, National Bureau of Standards, Gaithersburg, MD.

Denyes, W., and Quintiere, J., "Experimental and Analytical Studies of Floor Covering Flammability with a Model Corridor," *Journal of Fire and Flammability/Consumer Products Flammability Supplement*, Vol. 1, Mar. 1974, pp. 32–109.

Dietenberger, M. A., *et al.*, "Room/Corner Tests of Wall Linings With 100/300 kW Burner," *Proceedings* of the Fourth International Conference and Exhibition on Fire and Materials, Nov. 15–16, 1995, Crystal City, VA, 1995, pp. 53–62.

Fang, J. B., "Fire Buildup in a Room and the Role of Interior Finish Materials," Technical Note 879, 1975, National Bureau of Standards, Washington, DC.

Finley, G. M., "Factors Affecting the Performance of Wall Lining Materials in the ASTM E84 Test," Seventeenth International Conference on Fire Safety, Vol. 17, January 13–17, 1992, Millbrae, CA, Product Safety Corp., Sunnyvale, CA, 1992, pp. 314–335.

Fung, F. C. W., *et al.*, "The NBS Program on Corridor Fires," *Fire Journal*, Vol. 67, No. 3, May 1973, pp. 41–48.

Grexa, O., Janssens, M., and White, R., "Analysis of Cone Calorimeter Data for Modeling of the Room/Corner Tests on Wall Linings," *Proceedings* of the Fourth International Conference and Exhibition on Fire and Materials, Nov. 15–16, 1995, Crystal City, VA, 1995, pp. 63–71.

Groah, W. J., "The ASTM Tunnel Test: What It Is and How It Works," *The Building Official and Code Administrator*, Vol. 8, No. 6, June 1974, pp. 4–6.

Harkleroad, M. F., "Fire Properties Database for Textile Wall Coverings," NISTIR 89-4065, Mar. 1989, Center for Fire Research, Gaithersburg, MD.

Harmathy, T.Z., "A New Look at Compartment Fires, Part I and Part II," *Fire Technology*, Vol. 8, No. 3, Aug. 1972, pp. 196–217, 326–351

Hartzell, L. G., "Development of a Radiant Panel Text for Flooring Materials," NBSIR 74-495, May 1974, National Bureau of Standards, Gaithersburg, MD.

Hovde, P. J., "EUREFIC: A Research Programme to Develop New Fire Test Methods and Classification Criteria for Wall and Ceiling Linings," Fifth International Fire Conference Interflam '90, Fire Safety, Sept. 3–6, 1990, Canterbury, England, Interscience Communications Ltd., London, England, 1990, pp. 203–210.

Karlsson, B., "Models for Calculating Flame Spread on Wall Lining Materials and the Resulting Heat Release Rate in a Room," Lund Univ., Sweden, CIB W14/93/2 (J); Japan Symposium on Heat Release and Fire Hazard, First (1st) *Proceedings*, Session 2, Fire Modeling as a Tool for Reaction to Fire Evaluation, May 10–11, 1993, Tsukuba, Japan, 1993, pp. II/25–44.

Karlsson, B., "Models for Calculating Flame Spread on Wall Lining Materials and the Resulting Heat Release Rate is a Room," *Fire Safety Journal*, Vol. 23, No. 4, 1994, pp. 365–386.

Karlsson, B., and Magnusson, S. E., "Combustible Wall Lining Materials: Numerical Simulation of Room Fire Growth and the Outline of a Reliability Based Classification Procedure," *Proceedings* of the Third International Symposium on Fire Safety Science Safety Science, Elsevier Applied Science, New York, 1991, pp. 667–678.

Kim, C. I., and Kulkarni, A. K., "Upward Flame Spread on Vertical Walls Made of Practical, Finite-Thickness Materials," 1990 Fall Technical Meeting of the Combustion Institute/Eastern States Section on Chemical and Physical Processes in Combustion, Dec. 3–5, 1990, Orlando, FL, 1990, pp. 49/1–4.

Klitgaard, P. S., and Williamson, R. B., "The Impact of Content on Building Fires," *Journal of Fire and Flammability/Consumer Product Flammability Supplement*, Vol. 2, Mar. 1975, pp. 84–113.

Lee, B. T., "Quarter-Scale Room-Fire Tests of Interior Finishes," *Fire and Materials*, Vol. 9, No. 4, 1985, pp. 185–191.

Lieberman, P., and Bell, D., "Smoke and Fire Propagation in Compartment Spaces," *Fire Technology*, Vol. 9, No. 2, May 1973, pp. 91–100.

Magnusson, S. E., and Sundstrom, B., "Modeling of Room Fire Growth—Combustible Lining Materials," Report LUTVDG/(TVBB-3019), 1984, Lund Institute of Technology, Lund, Sweden.

Magnusson, S. E., and Thelandersson, S., "A Discussion of Compartment Fires," *Fire Technology*, Vol. 10, No. 3, Aug. 1974, pp. 228–246.

McGuire, J. H., and D'Souza, M. V., "The E162 Radiant Panel Flammability Test and Foamed Plastics," *Fire Technology*, Vol. 15, No. 2, May 1979, pp. 102–106.

Moulen, A. W., *et al.*, "The Early Fire Behavior of Combustible Wall Lining Materials," *Fire and Materials*, Vol. 4, No. 4, 1980, p. 165.

Mowrer, F. W., and Williamson, R. B., "Flame Spread Behavior of Char-Forming Wall/Ceiling Insulating Materials," *Proceedings* of the Third International Symposium on Fire Safety Science Safety Science, Elsevier Applied Science, New York, 1991, pp. 689–698.

Newman, J. S., and Tewarson, A., "Flame Spread Behavior of Char-Forming Wall/Ceiling Insulating Materials," *Proceedings* of the Third International Symposium on Fire Safety Science Safety Science, Elsevier Applied Science, New York, 1991, pp. 679–688.

Parker, W. J., "An Investigation of Fire Environment in the ASTM E84 Tunnel Test," NBS Technical Note 945, Aug. 1977, National Bureau of Standards, Gaithersburg, MD.

Pryor, A. J., "Full-Scale Fire Tests of Interior Wall Finish Assemblies," *Fire Journal*, Vol. 63, No. 2, Mar. 1979.

Quintiere, J. G., "A Characterization and Analysis of NBS Corridor Fire Experiments in Order to Evaluate the Behavior and Performance of Floor Covering Materials," NBSIR 75-691, June 1975, National Bureau of Standards, Gaithersburg, MD.

Quintiere, J. G., "Comments On 'Full-Scale/Bench-Scale Correlations of Wall and Ceiling Linings' [*Fire and Materials*, Vol. 16, 15-22, 1992] by Ulf Wickstrom and Ulf Goransson," *Fire and Materials*, Vol. 17, No. 3, 1993, pp. 149–150.

Quintiere, J. G., "Some Factors Influencing Fire Spread Over Room Linings and in the ASTM E84 Tunnel Test," *Fire and Materials*, Vol. 9, No. 2, 1985.

Quintiere, J. G., and Bromberg, K., "Calculations of Radiant Heat Flux in the Proposed Floor Covering Flame Spread Test Apparatus," NBSIR 75-706, Dec. 1975, National Bureau of Standards, Gaithersburg, MD.

Quintiere, J. G., Haynes, G., and Rhodes, B. T., "Applications of a Model to Predict Flame Spread Over Interior Finish Materials in a Compartment," *Journal of Fire Protection Engineering*, Vol. 7, No. 1, 1995, pp. 1–13.

Quintiere, J. G., Hopkins, M., and Hopkins, D., Jr., "Room-Corner Fire Prediction for Textile Wall Materials," *Proceedings* of the International Conference on Fire Research and Engineering, Sept. 10–15, 1995, Orlando, FL, SFPE, Boston, MA, 1995, pp. 272-277.

Richardson, L. R., Cornelissen, A. A., and Dubois, R. P., "Modelling Fire Performance of Wall Covering Materials," Seventeenth International Conference on Fire Safety, Vol. 17, Jan. 13–17, 1992, Millbrae, CA, Product Safety Corp., Sunnyvale, CA, 1992, pp. 276–313.

Richardson, L. R., Cornelissen, A. A., and Dubois, R. P., "Physical Modelling Fire Performance of Wall Covering Materials," *Fire and Materials*, Vol. 17, No. 6, Nov./Dec. 1993, pp. 293–303.

Schaffer, E. L., and Eickener, H. W., "Corridor Wall Linings—Effect on Fire Performance," *Fire Technology*, Vol. 1, No. 4, Nov. 1965, pp. 243–255.

Sherad, S. E., ed., *Interior Finish and Fire Spread*, National Fire Protection Association, Quincy, MA, 1977.

Thomas, P. H., "Old and New Look at Compartment Fires," *Fire Technology*, Vol. 11, No. 1, Feb. 1975, pp. 42–47.

Thomm, E. C., "Effect of Carpet Variables on the Methanamine Pill Test," *Journal of Fire and Flammability*, Vol. 4, July 1973, pp. 197–209.

Tsantaridis, L. D., "Wood Products As Wall and Ceiling Linings," Swedish National Testing and Research Institute, Boras, Sweden, ISBN 0-9516320-9-4; *Proceedings* of the Seventh International Interflam Conference, Interflam '96, March 26–28, 1996, Cambridge, England, Interscience Communications Ltd., London, England, 1996, pp. 827–834.

Wickstrom, U., and Goransson, U., "Full-Scale/Bench-Scale Correlations of Wall and Ceiling Linings," *Fire and Materials*, Vol. 16, No. 1, Jan./March 1992, pp. 15–22.

Wickstrom, U., and Goransson, U., "Full-Scale/Bench-Scale Correlations of Wall and Ceiling Linings," Chapter 13, *Heat Release in Fires*, Elsevier Applied Science, NY, 1992, pp. 461–477.

Williams-Lier, G., "The Steiner Tunnel Index as a Measure of Flame Spread Rate," *Fire Technology*, Vol. 14, No. 3, Aug. 1978, pp. 223–225.

Williamson, R. B., and Baron, F. M., "A Corner Fire Test to Simulate Residential Fires," *Journal of Fire and Flammability*, Vol. 4, Apr. 1973, pp. 99–105.

Yuill, C. H., "Flame Spread Tests in a Large Tunnel Furnace," ASTM STP No. 344, American Society for Testing Materials, W. Conshohoken, PA, 1962, pp. 3–17.

Zhang, J., Shields, T. J., and Silcock, G. W. H., "Fire Hazard Assessment of Polypropylene Wall Linings Subjected to Small Ignition Sources," Journal of Fire Sciences, Vol. 14, No. 1, Jan./Feb. 1996, pp. 67–84.

STRUCTURAL INTEGRITY DURING FIRE

Revised by Robert W. Fitzgerald

The selection of building materials and the design of the details of construction have always played an important role in building fire safety. Two of the important structural fire considerations are the ability of the structural frame to avoid collapse and the ability of the barriers to prevent ignition and resulting flame spread into adjacent spaces.

Three approaches to structural frame and barrier design for fire resistance are in use or under development today. They include:

1. Standard fire-resistance testing combined with building code requirements;

2. Analytical calculations to determine the resistance to a standard fire test exposure as a substitute for laboratory testing; and

3. Analytical structural fire engineering design methods based on real fire exposure characteristics.

The traditional method of treating the structural aspects of fire protection is through building code requirements. Codes require structural frames and barrier assemblies to be selected on their ability to pass a laboratory fire-resistance test.

Classification of buildings, with regard to fire-resistance requirements, is an approach that has the principal advantage of being easy to incorporate into the design process and to administer from a code-enforcement viewpoint. It has a major disadvantage of being unable to predict accurately the field performance of a structural frame or barrier. In other words, code compliance alone is not sufficient to ensure performance success of the barriers or structural frame in an actual fire. Studies that correlate building code requirements and fire test results with field performance have not been made.

Although the fire-resistance test can be criticized for its shortcomings, it is the only method universally accepted in building codes today. The first part of this chapter describes the fire-resistance test procedures common in building construction today. Latter parts of the chapter discuss methods of calculating fire-endurance (resistance) ratings and the behavior of building materials at elevated temperatures. The chapter concludes with a brief discussion of some of the more advanced approaches to designing for structural integrity in a fire.

Robert W. Fitzgerald, Ph.D., is professor of fire protection and civil engineering at Worcester Polytechnic Institute, Worcester, MA.

FIRE-RESISTANCE TESTING AND STRUCTURAL ASSEMBLIES

The fire endurance of the beams, girders, and columns that comprise the structural frame and the endurance of the walls, partitions, floor/ceiling assemblies, and roof/ceiling assemblies that serve as barriers to flame movement have been a historical basis for classifying buildings and rating frame and barrier capabilities. This part of the chapter will describe the history and procedure for fire testing, as well as discuss the interpretation of laboratory fire tests. In addition, structural behavior of frames and barriers is described.

History of Fire-Resistance Tests

One review of fire test methods for building constructions refers to tests on metal and masonry conducted in Germany as early as 1884–1886.[1] The first large-scale fire tests in this country are reported to have been conducted on masonry arches in Denver, CO, in 1890. These were followed by tests in New York City in 1896.

Efforts to establish an acceptable test procedure were initiated by Professor Ira H. Woolson of Columbia University and Rudolph P. Miller, chief engineer, Building Bureau, New York City. Preliminary tests by the Bureau, "necessitated by the rapid development of the skyscraper," led to the development of a test furnace (using railroad ties as fuel). This provided a means of establishing hourly ratings for floor constructions.

After the Baltimore, MD, conflagration of 1905, the American Society for Testing Materials [now the American Society for Testing and Materials (ASTM)] established a committee to standardize the test method, with Prof. Woolson as chairman and Mr. Miller as secretary. A test method for floor constructions was proposed in 1906 and adopted by ASTM in 1907. A procedure for testing wall and partition constructions was proposed in 1908 and adopted in 1909.[2]

These standards were presented to the NFPA Committee on Fire-Resistive Construction for consideration in 1914. In 1916, a joint committee composed of representatives from eleven engineering societies, including NFPA, was organized to revise and update the standard. This was done, and the revised standard test method controlled by a standard time-temperature curve was adopted subsequently by NFPA, ASTM, and the American Engineering Committee [now the American National Standards Institute (ANSI)]. The standard was adopted by NFPA as a tentative standard in 1917 and advanced to official standard status in 1918. Time-temperature curves and testing procedures developed in other countries follow the same general pattern, with minor differences.

Fire Test Procedure

Fire test procedures usually require that columns, floors, partitions, walls, and other structural elements be loaded in a manner calculated to develop, as nearly as practicable, the theoretical working stresses expected in the design. Separate test procedures are provided for load-bearing and non-load-bearing constructions and for constructions involving restrained and unrestrained beams and girders.

The Appendices to NFPA 251, *Standard Methods of Tests of Fire Endurance of Building Construction and Materials* (hereinafter referred to as NFPA 251), contains detailed specifications for test procedures, a guide to the determination of restraint required, if any, and a suggested report form. The standard specifies in detail the preparation and conditioning of the test specimens.

Acceptance criteria are specific for the construction or element tested and on predetermined conditions of test (load or no load-restrained or unrestrained). The criteria may include:

1. Failure to support load.
2. Temperature increase on the unexposed surface 250°F (121°C) above ambient.
3. Passage of heat or flame sufficient to ignite cotton waste.
4. Excess temperature (as specified) on steel members.
5. Failure under hose streams (walls and partitions).

The end restraint conditions during the fire test influence significantly the test results and, consequently, the ratings. Appendix E of NFPA 251 defines a restrained condition as one in which expansion at the supports of a load-carrying element is prevented during the test. This usually provides some unknown degree of rotational restraint at the ends of the element in addition to prestressing some parts of the assembly. An unrestrained condition is one in which the load-carrying element is free to expand and rotate at its supports.

One justification for incorporating two differing conditions of end restraint is to simulate simple (statically determinate) and continuous (statically indeterminate) constructions. Figure 7-4A illustrates simple and continuous construction. In general, continuous construction is inherently stronger than simple construction. It is possible to calculate the increase in load-carrying capacity of indeterminate structural members. However, the amount of increased strength in an indeterminate structure can be quite variable, depending on the type of construction materials used, the location of structural members within the structure, the degree of indeterminacy, the loading conditions, and the details of construction.

Test results based on the two conditions of support (restrained and unrestrained) were introduced in 1970. The degree of restraint for structural members and assemblies tested and accepted for building construction prior to 1970 is unknown.

Fire-Resistance Ratings

The fire resistance (also described as the fire endurance) is the time duration the member or assembly withstood the fire test without failure. While the actual time is recorded to the nearest integral minute, fire-resistance ratings are given in standard intervals. The usual fire-resistance ratings for all types of members, structural assemblies, doors, and windows are 15 min, 30 min, 45 min, 1 hr, 1 $^1/_2$ hr, 2 hr, 3 hr, and 4 hr. Therefore, a 1-hr rating indicates that the assembly withstood the standard test for 1 hr or longer. A 2-hr rating indicates that the assembly withstood the standard test for longer than 2 hr without failure by any one of the failure criteria listed in the fire test protocol.

The principal agencies in the United States that test assemblies of building construction materials for fire resistance are Underwriters Laboratories Inc. (UL) and the National Institute of Standards and Technology (NIST), previously the National Bureau of Standards (NBS). There are a number of other agencies that have furnaces or equipment for fire tests of various special assemblies. These include the Factory Mutual Research Corporation, Norwood, MA; the Forest Products Laboratory at Madison, WI; the Engineering Experiment Station at Ohio State University, Columbus; University of California at Berkeley; the Portland Cement Association, Skokie, IL; Armstrong Cork Co., Lancaster, PA; the Corbetta Construction Company at College Point, Queens, NY; and the National Gypsum Co., Buffalo, NY. Some of these test furnaces may not conform to the specifications described in NFPA 251. Consequently, those furnaces may be used for research purposes, rather than for rating assemblies.

In England, there are the facilities at the Testing Station of the Joint Fire Research Organization of the Fire Offices' Committee and the Department of the Environment at Borehamwood, Hertfordshire. In Canada, there are facilities at the Underwriters Laboratories of Canada in Toronto and the Building Research Division of the National Research Council of Canada in Ottawa. Throughout the world, many well-equipped fire testing laboratories are making significant contributions to the literature and to the development of international fire test standards.

The results of fire tests conducted by Underwriters Laboratories Inc. are given in the *Fire-Resistance Directory*, published annually by UL.

The results of fire tests by governmental agencies have been published in various reports and technical papers. Those issued by the National Institute of Standards and Technology may be consulted in many depository libraries, and lists of those available for purchase may be obtained on request from the NIST. The bulletins and reports of the British Ministry of Technology on the subject are

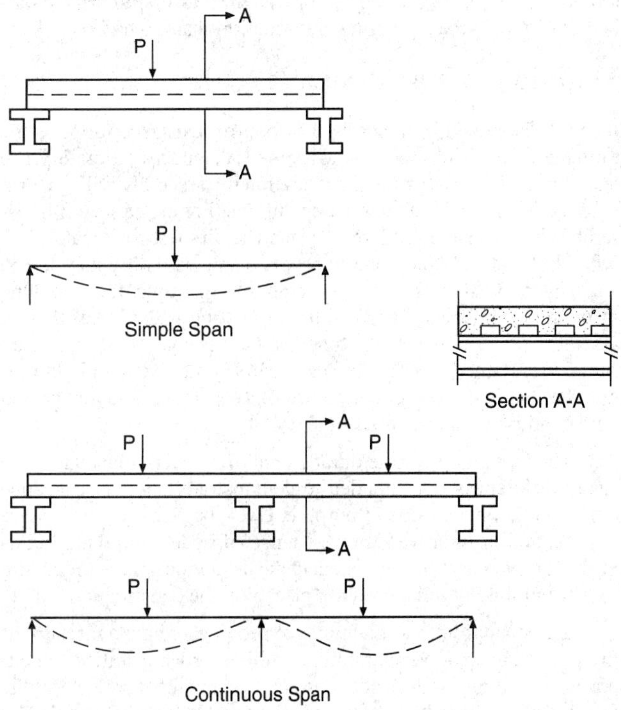

FIG. 7-4A.   *Influence of structural continuity on collapse mechanisms.*

obtainable by purchase from H.M. Stationery Office or the British Information Services, New York, NY.

The Factory Mutual Research Corporation publishes annually in its *Approval Guide* illustrations of building assemblies that have met the test requirements for various fire-resistance hourly ratings.

Various building codes, which specify fire resistance in terms of fire test results, may include tabulations of construction forms that will be accepted as meeting code requirements for specific fire-resistance ratings. Such tables are usually based upon fire test data, but some include ratings determined on a judgmental basis or estimated from limited test data.

Not listed are the many useful data sheets and summary tables published by trade associations of the building materials manufacturers. These sources are not cited because they are so numerous and because the information published either pertains to the products of a specific manufacturer or consists of tabulated ratings for a single type of construction material.

As indicated above, the number of test facilities in use today and the variety of technical papers, reports, compilations, etc., reporting on test results that are available would make it almost impossible to tabulate systematically in a single source the data on all the various assemblies that have been subjected to tests. Users of this handbook are urged to consult the fire-resistance index ratings listing and reports mentioned above for the specifics on design specifications for assemblies that have been tested and for which fire-resistance ratings have been assigned.

As a matter of information and to illustrate the scope of testing activities, tabular compilations of test results or representative assemblies will be found later in this chapter. They are included in this handbook to help give an understanding of basic requirements for test specimens and how the results of tests are recorded. In addition, the information contained in the tables can be useful in identifying, within reasonable limits, the fire-resistance capabilities of assemblies as they are found in the field and for which precise data on their fire resistance are not available or are no longer available. The tables are not presented as references for design but to familiarize handbook users with the type of test information that is available.

Variation in Test Results

It should be recognized clearly that the fire-resistance rating is the time the member or assembly withstood a standard laboratory fire test. It is not necessarily the time duration the assembly will perform without failure under actual fire conditions. For example, a 2-hr assembly can withstand failure in a standard fire test for greater than 2 hr. Under actual field conditions, that same assembly may fail in a considerably shorter or a considerably longer time duration. One of the major misconceptions in fire protection is the belief that an assembly rating indicates the time that the assembly will survive an actual fire. Rather, it is the time the assembly survived, without failure, the standard test fire. Differences between the standard test and field conditions can be substantial.

The fact that test-time duration and field performance may be vastly different is not intended to demonstrate that fire-resistance ratings do not have value or purpose. Over the years this procedure has resulted in improved fire resistance of code-complying buildings. However, one must understand the limitations of the procedure so that unthinking reliance is not placed on the construction.

The standard fire test attempts to provide a relative measure of fire performance of comparable assemblies under specified fire exposure conditions. It does not consider its suitability for use after the fire. Many effects from fire tests are observed indirectly, if at all. For example, the temperature gradient through a wall or floor slab results in internal strains and deflections. The strains may cause spal-

ling or other disruptions, and distortions may be severe enough to crack floor slabs and walls, sometimes leading to collapse. The greater the area exposed, the more serious are the results of unequal expansion. General deterioration of the test specimen is not considered, except when it has been involved in the failure of the test specimen.

While the test standard specifies the preparation and conditioning of specimens, there are many opportunities for differences between a test specimen and an actual structural element in a building. The test specimen may be superior in both materials and workmanship to those found in a building; the test specimen is usually smaller than the construction it represents. Restraints to thermal expansion may be of different magnitudes. Therefore, discretion in the interpretation of fire test results is in order.

Differences in the results of fire tests on apparently equal test specimens of building constructions arise from many factors, such as undetermined differences in the quality of materials, workmanship, moisture content, and test procedures. The standard method of test requires that materials in the specimen and the workmanship to be representative of those in actual buildings. Considerable variation in field practice can occur.

The test method permits the intensity of the fire exposure to deviate as much as 10 percent (formerly 15 percent) from that prescribed as standard and requires adjustment in the reported results to correct for such deviations only for tests of $1/2$-hr duration or greater. The effects of several of the variables encountered in fire testing procedures are found in the technical reports listed in the bibliography at the end of the chapter.

The character and proportions of aggregates and binders have an important influence on the results of fire-resistance tests. For example, gypsum plaster with lightweight aggregates, mixed with from 2 to 3 cu ft (0.057 to 0.085 m^3) to each 100 lb (45 kg) of gypsum, has been shown by test to provide fire resistance from 10 to 70 percent greater than equal thicknesses of sanded gypsum plaster. The use of such aggregate in specific constructions should be adopted only after consulting the lists of ratings based on tests in which these aggregates have been used.

Vermiculite aggregate plaster, if too wet when applied, may be subject to shrinkage cracks. Such cracks, if they develop, are likely to occur within a short time. Some expanded perlite aggregates tend to absorb atmospheric moisture and may show destructive expansion after a few years. There may be considerable variation in this effect, as perlite is a natural volcanic glass or rock from many sources and varies in composition.

Finishing lime produced from dolomite contains magnesium oxide. That which is designated as "normal" finishing hydrate may not be sufficiently hydrated, and sometimes subsequent gradual hydration results in destructive expansion. Plasters containing such lime are subject to rapid destruction in the event of fire. This difficulty is avoided with lime hydrated under high temperature and pressure and known as "special" or "autoclaved" lime.

So-called stabilized gypsum plaster containing a small percentage of the normal hydrated lime may similarly be subject to deterioration with age and, in tests, may not provide the fire resistance to be expected from such constructions.

The effects mentioned are not such as to preclude the use of these materials where long life is not a factor or where adequate guarantees are provided. They are merely illustrations of the variability in test results that can occur as a result of differences in materials.

Although standard fire tests are made on fairly large specimens representative of building constructions or assemblies, they do not necessarily produce heat-expansion effects similar to those of fires in buildings having larger wall or floor areas. It may be necessary to specify thickness of construction or lateral support, in addition to rated fire resistance, to guard against the adverse effects of heat expansion or temperature gradients.

In addition to the material variation noted, construction differences may be even more significant. For example, a test does not provide full information on performance of assemblies constructed with components or lengths other than those tested. Performance is often reduced when the spans are increased. Also, great care must be exercised in approving "or equal" clauses in construction specifications. Often, substitutions in materials are approved for components that have not been listed with the assembly.

The standard test does not incorporate information regarding the effect on fire endurance of conventional openings in the assembly, such as electrical receptacle outlets, pipe chases, etc. Poke-through construction for electrical services or piping often is not patched properly. Transfer grilles, windows, and other penetrations are often incorporated into barriers without considering their impact on the fire resistance.

The factors noted above, in addition to the many deviations to plans and specifications during and after construction, significantly affect the fire-resistance ratings of members and assemblies. Often the code calls for, and the architect may think he or she is providing, a specified fire resistance, only to have that resistance reduced because of inattention to penetrations and important details of construction.

Structural Frame Systems

The fire-resistance rating of structural framing members, such as beams, girders, and columns, may be achieved in several ways. Generally, the members are protected by encasing them in a material sufficiently inert to prevent excessive thermal penetration or by providing a membrane protection that delays thermal penetration to the members.

The general behavior of reinforced concrete, structural steel, and barrier assemblies are described in the following paragraphs. Illustrative fire-resistance ratings are shown in tabular material later in this chapter.

Reinforced Concrete Systems

Reinforced concrete construction has had a good record with regard to structural collapse. Because concrete has a low thermal conductivity and a low thermal capacity, it provides an effective cover for reinforcing steel. For example, Figure 7-4B shows the temperature

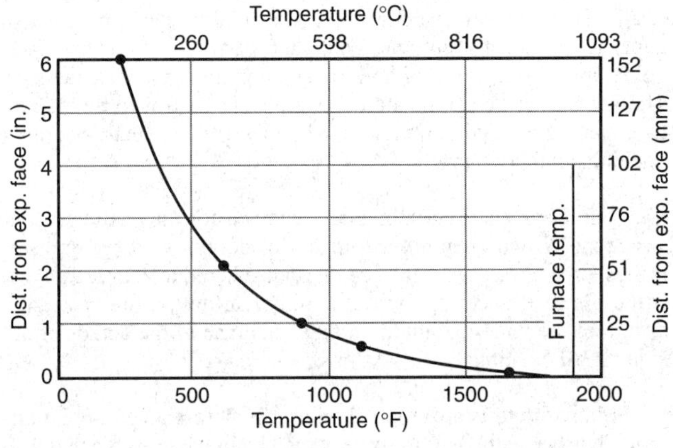

FIG. 7-4B.   Thermal gradient in a 6-in. slab after 2-hr fire exposure.[3] [For SI units: 1 in. = 25.4 mm; °C = 5/9 (°F −32).]

gradient in a 6-in. (152-mm) slab after a 2-hr fire exposure. Although undoubtedly the moisture in the concrete greatly influences the values, the significant feature is the fact that the temperatures vary considerably throughout the thickness, even after a considerable time exposure.

This feature provides some insight into one reason that reinforced-concrete systems usually perform comparatively well during fire exposure. Consider, for example, a continuous, monolithic reinforced-concrete beam or slab, as shown in Figure 7-4C. Considering the temperature gradient of Figure 7-4B, it will take some time before the tension steel at midspan is affected. Even after it reaches its yield value, the negative steel over the supports has not been seriously affected because of the insulating effect of the concrete and the moisture.

Continuous construction of this type has inherent capabilities far greater than statically determinate (simple) construction. Considerable stress redistribution can take place before collapse will occur. It takes time before excessive rotation will develop at all the necessary locations, causing structural collapse of the member. Although the member is weakened by the fire, structural stability against collapse will remain for a considerable period of time.

The level of stress in a reinforced-concrete member exposed to the elevated temperatures of a fire has a significant influence on its endurance. Table 7-4A illustrates the effects of stress level on the fire resistance of reinforced-concrete columns. The columns used for these tests were 15 by 15 in. (381 by 381 mm) containing four number-nine reinforcing bars.[4] It can be seen that the magnitude of stress during a fire causes significant reductions in capacity. This is attributed primarily to the reduction in the mechanical properties of steel and concrete at elevated temperatures.

TABLE 7-4A.   *The Influence of Stress on the Fire-Resistance of Concrete Columns*

Applied Load (% design load)	Fire Resistance (min)
150	68
100	124
75	198
50	248
30	358

Steel Construction

The popularity of steel-frame building construction is due to its high strength, ease of fabrication, and ensured uniformity of quality. Exposed structural steel, however, is vulnerable to fire damage. In order to have fire resistance, it must be protected from high temperatures encountered in fires.

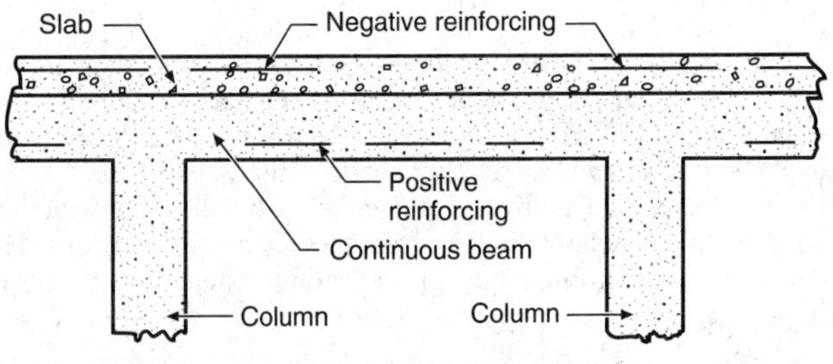

FIG. 7-4C.   *Monolithic reinforced-concrete beam and slab.*

Protection for steel beams, girders, and columns, such as encasements of concrete, clay, tile, or gypsum blocks, have been generally superseded by plastered or sprayed-on applications applied either to a furred plaster base, such as expanded metal lath, or to the surface of the member to be protected. The applications may be conventional plasters of portland or gypsum cements combined with appropriate aggregates, or one of the many combinations of mineral fibers with binders. Currently, gypsumboard encasement is a popular technique for protecting steel structural elements.

Barrier Systems

Exterior walls, interior partitions, floors, and floor/ceiling assemblies are components that define the architectural layout of rooms in a building. In the normal functional use of a building, these components are used to provide privacy, security, protection from the elements, and noise control. They also provide fire protection by delaying or preventing flames from moving from one room to an adjacent room.

The effectiveness of a barrier in preventing movement of an ignition from a fully involved room to an adjacent room depends upon several factors. One is the fire severity. The fire severity is influenced by the fuel in the space, as well as the size and location of openings, the size of the room, and the thermal properties of the wall and ceiling construction. The fire severity is the heat application to the barrier surface. This is resisted by the materials of construction of the barrier, as well as other factors such as the applied loads, the details at connections, and the dimensions. Of even greater importance are the effect of openings and penetrations in the barrier. The most common cause of fire movement from one room to an adjacent room is through unprotected openings in barriers. Often, code requirements and expensive constructions are rendered ineffective because of lack of attention to the details of opening protection.

Building codes through construction classifications identify the fire-endurance requirements of barriers. In addition, special locations, such as around vertical shafts, are required to provide specific resistances.

Fire-resistance ratings are determined by subjecting the barrier assembly to a standard fire test, as described in NFPA 251. The length of time the assembly withstands the laboratory fire without failure describes the fire endurance. Fire resistance is the endurance time rounded down to the standard durations. Both combustible and noncombustible barriers can obtain fire-resistance ratings in the standard fire test.

The cautions expressed earlier in this chapter with regard to the interpretation of the fire-resistance time durations also apply here. That is, the time durations reflect laboratory times and are not necessarily the values to be expected in a real fire. For example, one should not expect a 2-hr barrier assembly to withstand an actual building fire for 2 hr. Failure could occur earlier or later, depending upon construction details.

Floor/Ceiling Assemblies

The fire resistance of floor/ceiling assemblies is important in the prevention of flame movement vertically from floor to floor. Fire-resistance durations from the standard fire test are determined from unpenetrated assemblies. The anticipated fire resistance may be reduced greatly or destroyed completely in building construction when holes and poke-throughs are not protected adequately or when details do not conform to the tested assembly. Attention to details is important.

The method of attaching ceilings is a major factor in determining the fire resistance of floors. For example, the nailing of gypsum lath, metal lath, or gypsum wallboards to the soffits of wood joists is often critical in plaster construction. The longer thinner nails, particularly those with cement coatings, conduct less heat to char the wood surrounding them than do the common types of wire nails. Similarly, the integrity of suspended ceilings must be maintained to reduce the likelihood of premature failure on some assemblies. Details, such as attention to expansion of the supporting framework, are essential to preventing premature ceiling failures.

Self-tapping screws, made particularly for the attachment of gypsumboard, offer greater holding power and less damage to the core materials than do nails. Such screws can be used to attach wallboards to either wood or cold-rolled channels, without previous drilling.

Consideration must also be given to the character of the plaster base with respect to loosening of plaster mixes from the base on application of heat sufficient to char combustible surfaces. The use of wire or better yet, wire fabric, to reinforce the plaster mixes applied to such plaster bases ensures increased fire resistance. Clearances for longitudinal expansion of metal furring members are required to prevent damage from buckling. The tendency of certain plaster bases and plaster mixes to expand or contract with changes in atmospheric humidity should also be given consideration where resultant cracking might affect the fire resistance of structures incorporating such plaster.

Suspended ceilings with openings for air diffusers and light troughs should be designed so that such openings are not points of vulnerability to fire. Continuous construction above recesses for lighting fixtures and properly designed self-closing dampers for air ducts provide protection.

Roof/Ceiling Assemblies

Roof/ceiling assemblies are tested and rated in a manner similar to the floor/ceiling assemblies. The results are generally comparable. However, it should be noted that roof assemblies often have a given thickness of insulation in place at the test. If additional thicknesses of insulation are desired, the fire-resistance rating may be reduced.

Walls and Partitions

The fire resistance of walls and partitions will delay or prevent flames from moving horizontally from one room to an adjacent room. These assemblies are also tested in accordance with NFPA 251. The cautions expressed earlier in this section concerning the relationship between test duration and actual fire performance also apply to these constructions. Walls and partitions are particularly vulnerable to the effects of horizontal communication mechanisms. The fire resistance can be almost completely destroyed by ineffective opening protection for doors, windows, grilles, ducts, and other openings in these barriers.

The fire-resistance ratings are obtained for unpenetrated constructions. When openings are made for doors, windows, transfer grilles, pipe chases, etc., the fire resistance is reduced or destroyed. If the fire resistance of a barrier is important to the fire protection design of the building, attention must be given to the details of design for all openings.

Many fire tests are stopped arbitrarily before an end-point criterion has been attained. In some cases, the indications are that, had the test been continued, the construction would have qualified for a higher fire-resistance rating than that assigned to it.

ILLUSTRATIVE FIRE-RESISTANCE RATINGS FOR STRUCTURAL FRAME AND BARRIER CONSTRUCTIONS

Fire-resistance testing has produced a large amount of experimental data. To illustrate the effect of materials and constructions on the fire-resistance rating of assemblies and structural members, several graphs and tables have been included. These graphs and tables can be used both to give guidance on the needs to achieve fire-resistance ratings and to provide a basis for judgmental interpretations of the fire resistance of construction encountered in field evaluations.

Reinforced-Concrete Members

The details of construction and the kind of aggregates affect the fire resistance of concrete structures.

The fire-resistance ratings for reinforced concrete beams are given in Table 7-4B and for prestressed concrete girders, beams, joists, and slabs in Table 7-4C. As the cold-drawn, high-strength steel tendons used in prestressed concrete are more adversely affected by high temperatures than normal reinforcement steel, these tendons require a thicker protective cover than is required in conventional reinforced concrete. The fire resistance of reinforced concrete columns is shown in Table 7-4D.

Structural Steel Members

Fire protection generally is provided for structural steel by encasing the members in concrete, lath and plaster, gypsumboard, or sprayed fibers. A second common method of providing fire resistance is by installing a membrane barrier to delay or prevent the heat from raising the temperature of the steel to the critical temperatures. Table 7-4E gives minimum thicknesses of portland cement encasement for steel beams, while Tables 7-4F and 7-4G illustrate the influence on the fire resistance of columns for various methods of encasement. Figures 7-4D and 7-4E illustrate different methods of encasement for steel columns. Figure 7-4F provides a means to estimate the necessary encasing requirements for steel columns.

Floor/Ceiling Assemblies

Fire resistance for floor/ceiling assemblies of reinforced concrete, structural steel, and wood can be achieved in several ways. For example, Figure 7-4G illustrates the influence of floor thickness and aggregate type on the fire resistance of reinforced-concrete slabs, while Table 7-4H shows the fire resistance for several types of concrete floor constructions. Table 7-4I shows the test results for structural steel joist floor and roof constructions, and Table 7-4J provides some data on the fire resistance of steel-formed concrete floor systems. Illustrative fire resistance ratings for wood floors are provided in Tables 7-4K and 7-4L.

TABLE 7-4B. *Reinforced-Concrete Beams and Girders of Medium Size*

Concrete Grade*	Protective Cover of Reinforcement (in.)	Fire-Resistance Rating† (hr)
1	3/4	1
	1	2
	1 1/4	3
	1 1/2	4
2	3/4	1/2–1 ‡
	1	1–2 ‡
	1 1/2	2–3 ‡
	2	2–4 ‡

*For lightweight concrete having an oven-dried density of 110 pcf or less, the cover shown for Concrete Grade 1 may be reduced 25 percent.

†May be increased if some bars are better protected by being away from corners or in an upper layer or if beam is large. Should be decreased if beam is small.

‡Variable depending on spalling characteristics of aggregate. The use of mesh to hold cover in place will give ratings about as high as for concrete of Grade 1.

For SI units: 1 in. = 25.4 mm.

TABLE 7-4C. *Prestressed Concrete Girders, Beams, Joists, and Slabs (Grade I Concrete)*

Type of Unit	Condition of Restraint	Cross-Sectional Area (sq in.)†	Cover for Fire Rating Shown (in.)*			
			1 hr	2 hr	3 hr	4 hr
Girders, beams and joists	Unrestrained	40 to 150	2	2.5	—	—
		150 to 300	1.5	2.5	3.5‡	—
		Over 300	1.5	2.25	3‡	4‡
	Axially restrained	40 to 150	1.5	2	—	—
		150 to 300	1	1.5	2	—
		Over 300	1	1.5	1.5	2
Slabs, solid or covered, with flat undersurface	Unrestrained		1	1.5	2	2.5
	Biaxial restraint		0.75	1.25	1.5	2

*Cover for an individual steel tendon is measured to the nearest exposed surface. For several tendons in the same member having different concrete covers, the minimum cover may be reduced slightly. The covers shown may be reduced 25 percent for lightweight concrete having an oven-dried density of 110 pcf or less. For Grade 2 concrete, the cover may need to be increased.

†In computing the cross-sectional area of joists, the area of the flange must be added to the area of the stem, but the total width of the flange so used must not exceed three times the width of the stem.

‡Provide against spalling of the cover by means of a light, 2-in., U-shaped mesh, covered about 1 in.

NOTE: Data in this table are based on 67 standard ASTM fire tests.

For SI units: 1 in. = 25.4 mm.

TABLE 7-4D. *Fire Resistance of Reinforced-Concrete Columns*

| | Column | | | Concrete | | | Reinforcement | | | | | | |
| | Section | | Load in 1000 lbs | Aggregates | | Mix | Vertical | | | Lateral | | Concrete Cover | Fire-Resistance |
No.	(in.)	Area (sq in.)		Fine	Coarse	Cement Fine Coarse	Number	Bar Size No.	(sq in.)	Diam. (in.)	Spacing (in.)	Thickness	hr min
70	16 × 16	256	101	Fox River sand	Chicago limestone	1:2:4	4	9	4.00	1/4	12	2 1/4	8 40 (13+ —)
71	16 × 16	256	101	Long Island sand	New York trap rock	1:2:4	4	9	4.00	1/4	12	2 1/4	7 22
72	17 dia	227	107.5	Fox River sand	Chicago limestone	1:2:4	6	9	6.00	1/4	12	2 1/2	8 04 (12+ —)
73	17 dia	227	107.5	Long Island sand	New York trap rock	1:2:4	6	9	6.00	1/4	12	2 1/2	7 57
74	17 dia	227	129	Fox River sand	Chicago limestone	1:2:4	6	6	2.64	1/4	1 1/2	2 1/4	8 06 (13+ —)
75	17 dia	227	129	Long Island sand	New York trap rock	1:2:4	6	6	2.64	1/4	1 1/2	2 1/4	8 02
25	16 × 16	256	92	Pittsburgh sand	Pittsburgh gravel	1:2:4	4	8	3.16	1/4	12	1 1/2	4 — (6 30)
44	16 × 16	256	92	Long Island sand	Pure quartz gravel	1:2:4	4	8	3.16	1/4	12	1 1/2	4 — (6 —)
51	16 × 16	256	92	Pittsburgh sand	Blast furnace slag	1:2:4	4	8	3.16	1/4	12	1 1/2	4 — (8 —)
56	16 × 16	256	92	Pittsburgh sand	New Jersey trap rock	1:2:4	4	8	3.16	1/2	12	1 1/2	4 — (7 —)
7	18 dia	254	99.75	Pittsburgh sand	Pittsburgh gravel	1:2:4	8	6	3.52	1/4	12	1 1/2	5 — (6 —)
2	18 dia	254	141	Pittsburgh sand	Pittsburgh gravel	1:2:4	8	6	3.52	3/16	2	1 1/2	4 —
48	18 dia	254	141	Pittsburgh sand	Blast furnace slag	1:2:4	8	6	3.52	3/16	2	1 1/2	4 — (10+ —)
85	18 dia	254	141	Elgin, IL, sand	Elgin, IL, sand	1:2:4	8	6	3.52	3/16	2	1 1/2	4 — (12+ —)
12	18 dia	254	81	Pittsburgh sand	Pittsburgh gravel	1:2:4	None	—	—	—	—	—	4 —
33	12 dia	113	51	Pittsburgh sand	Pittsburgh gravel	1:2:4	4	5	1.24	1/4	2 1/8	1 1/2	Avg. of 2 3 —

*Column Nos. 70 to 75 tested at UL. Test No. 70 of the group was stopped at 8 hr, 40 min; others at failure or at 8 hr, and all loaded to failure at end of fire exposure (NBS Tech. Paper 184). All other columns tested at NBS Laboratory at Pittsburgh, PA. The fire endurance test of Col. No. 7 stopped at 5 hr, all others of series stopped at 4 hr. Figures in parentheses are estimates of fire resistance if tests had continued to failure (NBS Tech. Paper 272).
For SI units: 1 in. = 25.4 mm; 1 lb = 0.45 kg.

Walls and Partitions

Walls and partitions are commonly constructed of masonry, wood, or metal studs faced with fire-resistant materials. Table 7-4M gives the fire resistance of load-bearing brick or clay tile walls. Figure 7-4H shows a graph that allows the fire resistance of burned clay brick walls to be estimated. Tables 7-4N and 7-4O provide fire-resistance ratings for wood stud walls, while Figure 7-4I gives the fire resistance of stud walls faced with common construction materials and can be used to estimate quickly and conveniently. Table 7-4P and

Figure 7-4J are very useful descriptions of the fire resistance of solid, nonload-bearing partition materials.

CALCULATING FIRE ENDURANCE

The traditional approach to structural fire protection is to specify in the building code the fire-endurance ratings for identified construction classifications. Structural members and assemblies are subjected to the standard fire test (NFPA 251) to establish individual fire-endurance ratings.

TABLE 7-4E.   *Concrete Protection for Steel Beams*
(All re-entrant portions filled)

Fire-Resistance Rating		Thickness of Concrete Protection (in.)							
		Grade 1 Size of Member (flange width)				Grade 2 Size of Member (flange width)			
hr	min	2 to 3³/₄ in.	4 to 5³/₄ in.	6 to 7³/₄ in.	8 in. and over	2 to 3³/₄ in.	4 to 5³/₄ in.	6 to 7³/₄ in.	8 in. and over
4	—	4	3¹/₄	2¹/₂	2	4³/₄	3³/₄	3	2¹/₂
3	—	3¹/₂	2¹/₂	2	1¹/₂	4	3	2¹/₂	2
2	—	2¹/₂	2	1¹/₂	1	3	2¹/₂	1³/₄	1¹/₄
1	30	2	1¹/₂	1	1	2¹/₂	1³/₄	1¹/₄	1
1	—	1¹/₂	1	1	1	1³/₄	1¹/₄	1	1

NOTE: Protective concrete having thickness one-fourth of flange width or less must have steel wire reinforcement spaced not more than four times the thickness of the concrete covering the flange.
For SI units: 1 in. = 25.4 mm.

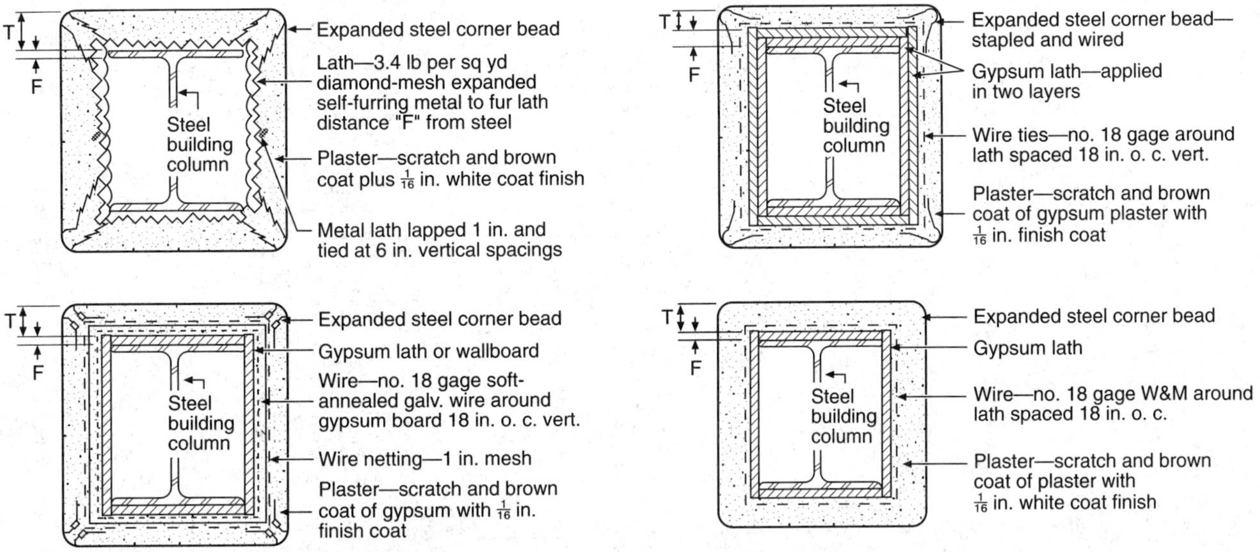

FIG. 7-4D.   *Typical steel column protection of lath and plaster assemblies. (For SI units: 1 in. = 25.4 mm; 1 lb = 0.454 kg.)*

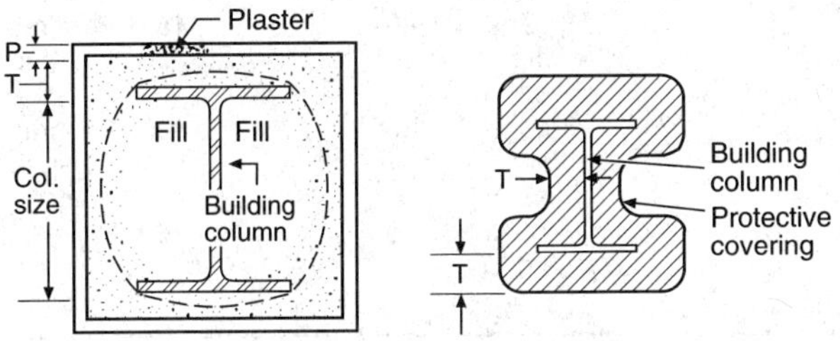

FIG. 7-4E.   *Typical steel column protections of concrete, masonry, or sprayed fibers.*

During the past three decades, a substantial amount of research has been conducted on structural behavior in fires. Computer models of compartment fire behavior, heat transfer, and structural performance at elevated temperatures have been developed. These models utilize basic research on mechanical and thermal properties of materials at elevated temperatures, as well as experimental validation of the computer models. These studies have resulted in more realistic predictions of structural behavior in fires than has been possible with the traditional code and standard fire test procedures of the past.

Calculation methods offer advantages in economy and better predictability over the traditional test approach. The calculation methods may be grouped into two categories: (1) calculation of the fire endurance that would have been obtained in the standard fire test and (2) calculation of structural or thermal performance in an actual building fire compartment.

Building code requirements specify fire endurance as reflected by the standard fire test. Consequently, it has been more productive to develop methods to calculate the fire endurance that would have been obtained in the standard fire test. Illustrative equations and graphs for this type of evaluation are given below. A brief description of procedures to calculate performance in actual building fires is given at the end of this chapter.

Equivalent Fire Endurance

Structural steel: Fire testing has been ongoing for many years, yielding a large amount of data and experience. The procedures described below reflect the type of methods that will enable equivalent fire resistance to be calculated. It should be noted that many of these calculation methods are obtained from test data. Consequently, one should be cautious when applying these methods to materials that have not been used in the tests that were the basis for the calculation

TABLE 7-4F. ***Fire Resistance of Steel Columns with Lath and Plaster Protective Coverings***

Metal Column				Protective Covering										
				Type of Covering								Fire-Resistance Rating		
				Plaster										
Type of Section	Size (in.)	Weight per lin ft (lb)	Area of Metal sq in.	Design	Type	Aggregate	Mix, Volumes	Thickness T† (in.)	Furring T† (in.)	Bond of Covering	Total Area of Materials (sq in.)	hr	min	Notes
H	6	44	13	—	Portland cement & lime	Sand	$1:1/10:2\frac{1}{2}$	$\frac{7}{8}$	1	Metal lath	40	—	45	Metal lath furred out
H	6	31	9	—	Two thicknesses of above	Sand	$1:1/10:2\frac{1}{2}$	$\frac{7}{8}+\frac{7}{8}$	1 & 1	Metal lath	80	1	30	Metal lath furred out
H	10	49	14.5	A	Gypsum-cement mixture	Lightweight*	Mill mix	$1\frac{3}{4}$	$\frac{3}{8}$	Wire fabric	125	3	25	
H	10	49	14.5	A	Gypsum	Lightweight*	Mill mix	$1\frac{7}{8}$	$\frac{5}{8}$	Metal lath	125	4	—	$\frac{1}{2}$-in. channel behind lath
H	10	49	14.5	A	Gypsum	Lightweight*	$1\frac{1}{2}:2$ $1\frac{1}{2}:3$	$1\frac{3}{4}$	$\frac{3}{8}$	Metal lath	125	4	—	Self-furring lath
H	10	49	14.5	A	Gypsum	Lightweight*	$1\frac{1}{2}:2$ $1\frac{1}{2}:3$	$1\frac{3}{8}$	$\frac{3}{8}$	Metal lath furred out	102	3	—	Self-furring lath
H	10	49	14.5	A	Gypsum	Lightweight*	$1\frac{1}{2}:2$ $1\frac{1}{2}:3$	1	$\frac{3}{8}$	Metal lath	78	2	—	Self-furring lath
H	10	49	14.5	B	Gypsum	Lightweight*	$\frac{1}{2}:3\frac{1}{2}$ $\frac{1}{2}:4$	$2\frac{1}{8}$	$\frac{1}{2}$	18-gage wire & wire fabric	145	4	—	Gypsum
H	10	49	14.5	C	Gypsum	Lightweight*	$1\frac{1}{2}:2$ $1\frac{1}{2}:3$	$1\frac{1}{2}$	1	18-gage wire & wire fabric	140	4 to 15 4 to 40		Gypsum lath
H	10	49	14.5	B	Gypsum	Lightweight*	$1\frac{1}{2}:2\frac{1}{2}$	$1\frac{1}{2}$	$\frac{1}{2}$	18-gage wire & wire fabric	110	3	40	Gypsum lath
H	10	49	14.5	D	Gypsum	Lightweight*	$1:2\frac{1}{2}$	$\frac{1}{2}$	$\frac{3}{8}$	18-gage wire ties	53	1	20	Perforated gypsum lath
H	10	49	14.5	D	Gypsum	Lightweight*	$1:2\frac{1}{2}$	$\frac{5}{8}$	$\frac{3}{8}$	18-gage wire ties	60	1	30	Perforated gypsum lath
H	10	49	14.5	D	Gypsum	Lightweight*	$1\frac{1}{2}:2$ $1\frac{1}{2}:3$	1	$\frac{3}{8}$	18-gage wire ties	80	2	15	Perforated gypsum lath
H	10	49	14.5	D	Gypsum	Lightweight*	$1\frac{1}{2}:2$ $1\frac{1}{2}:3$	$1\frac{1}{2}$	$\frac{3}{8}$	18-gage wire ties	104	2	30	Perforated gypsum lath
O	7	51	15.5	—	Portland cement & lime	Sand	$1:\frac{1}{10}:2\frac{1}{2}$	$1\frac{1}{4}$	$\frac{7}{8}$	$\frac{7}{8}$-in. rib lath	70	2	45	Metal lath on cast column

*Lightweight aggregate can be either perlite or vermiculite.
†Dimensions as shown in Figure 7-4D.
For SI units: 1 in. = 25.4 mm; 1 lb = 0.454 kg.

procedures. For example, the data for structural steel has been based on testing of A7 or A36 structural steel. In recent years, high-strength steels have become more popular. The mechanical properties of high-strength steels at both normal and elevated temperatures are different than those of A7 and A36 steel. Consequently, for fire-resistance calculations, when the term structural steel is used in this section, A7 and A36 steels are intended.

The fire endurance of structural steel beams and columns can be improved by insulating the members. Equation 1 calculates the fire endurance of steel beams and columns protected by light insulation.[5,6] The members may be protected by boxing the slope or by a contour protection. The equation is:

$$R = \left(C_1 \frac{M}{D} + C_2 \right) l \qquad (1)$$

where

R = fire endurance (min),

M = mass of the member (lb/ft),

TABLE 7-4G. Fire Resistance of Steel Columns Encased with Concrete, Masonry, or Sprayed Fibers

Steel Column			Protective Covering							Fire-Resistance Rating		
Type of Section	Size (in.)	Weight per Lin Ft (lb)	Design	Type of Covering	Thickness Outside Steel TI (in.)	Re-entrant Portion Filled	Plaster Thickness (in.)	Section Area of Solid Material (sq in.)	Bond of Covering	hr	min	Notes
H	8	34		None	0	No	0	8	—	—	10	Bare column
H	6	20	E	Siliceous gravel concrete, 1:2$\frac{1}{2}$:3$\frac{1}{2}$ mix	2	Yes	0	100	8-gage wire spiral 8-in. pitch	3	30	NBS test
Plate & angle	6	34	E	Trap rock or cinder concrete, 1:6 mix*	2	Yes	0	130	6-gage wire spiral	3	45	UL test
H	8	34	E	Limestone concrete, 1:6 mix*	2	Yes	0	144	6-gage wire spiral	6	30	UL test
H	8	34	E	Limestone concrete, 1:6 mix*	4	Yes	0	256	6-gage wire spiral	7	30	UL test
H	8	34	E	Trap rock, granite, cinders, 1:6 mix*	4	Yes	0	256	6-gage wire spiral	7	—	UL test
Plate & angle	6	34	E	Gypsum concrete†	2	Yes	$\frac{1}{2}$	114	4-in. mesh fabric	6	30	NBS test gypsum plaster
Plate & angle	6	34	E	Gypsum block	2	No	$\frac{1}{2}$	107	1-by $\frac{1}{8}$-in. O clamps	4	—	NBS test gypsum plaster
Plate & angle	6	34	E	Cinder block	3$\frac{3}{4}$	Yes	$\frac{3}{4}$	240	Block bond	7	—	NBS test gypsum plaster
H	8	34	E	Common Brick	4$\frac{1}{4}$	Yes	0	270	Brick bond	7	—	UL test
H	8	34	E	Semi-fire clay hollow tile	2	No	0	96	Wire ties	1	30	UL test
H	8	34	E	Semi-fire clay hollow tile	4	No	0	158	Wire ties	1	30	UL test
H	10	49	F	Sprayed mineral fiber‡	2$\frac{1}{4}$	Yes	—	164	No special adhesive	5	—	
H	10	49	F	Sprayed mineral fiber‡	3$\frac{3}{8}$	Yes	—	238	Special adhesive	5	—	
H	8	28	F	Sprayed mineral fiber‡	2	Yes	—	44	Special adhesive	5	—	
I	8	28	F	Sprayed asbestos fiber‡	2	Yes	—	38	No special adhesive	3	—	
	8	35	E	Sprayed asbestos fiber‡	1	Yes	—	90	No special adhesive	2	—	
	8	35	E	Sprayed asbestos fiber‡	1$\frac{3}{4}$	Yes	—	98	No special adhesive	4	—	
	8	35	E	Sprayed asbestos fiber‡	1$\frac{7}{8}$	Yes	—	120	No special adhesive	4	—	
I	8	28	F	Sprayed asbestos fiber‡	1$\frac{1}{2}$	Yes	—	28	No special adhesive	2	—	UL test
O	7.6	24		None (bare steel pipes filled with concrete)§	0	Yes	0	46	—	—	35	UL test

*Concrete mix—1 part cement to 6 parts total aggregate including sand and coarse aggregate.
†Gypsum concrete—7 parts gypsum stucco to 1 part wood shavings, by weight.
‡Mineral fibers, with bonding agent as required, sprayed on to all surfaces of column shaft to thicknesses indicated. (Thickness different on account of characteristics of fiber and binder.)
§Concrete-filled columns require vent holes to prevent explosion in the event of fire.
║Dimensions as shown in Fig. 7-4E.
For SI units: 1 in. = 25.4 mm; 1 lb = 0.454 kg.

D = heated perimeter (in.),

l = thickness of protection (in.), and

C_1, C_2 = constants that are empirically derived for the insulating units.

In SI units:

$$R = \left(0.672 C_1 \frac{M}{D} + 0.039 C_2 \right) l \qquad (1m)$$

where

R = fire endurance (min),

M = mass of the member (kg/m),

D = heated perimeter (mm),

l = thickness of protection (mm), and

C_1, C_2 = empirical constants, the same as used for U.S. customary units.

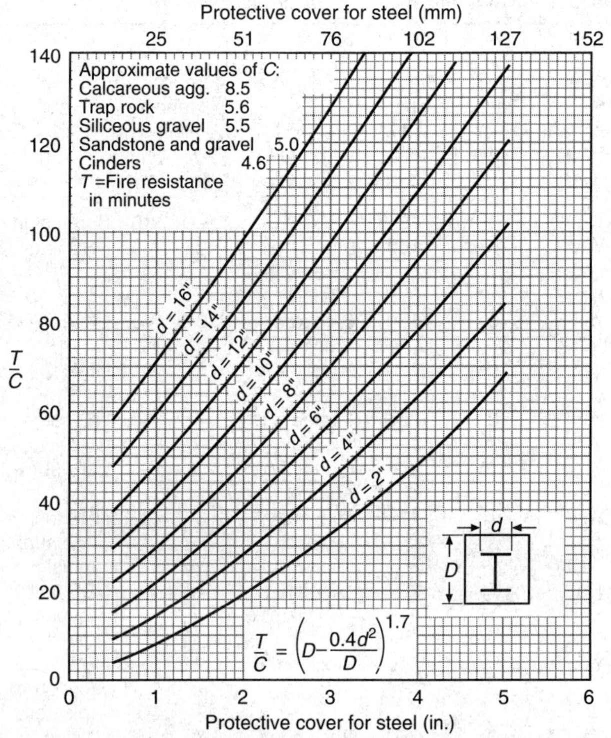
Protective cover for steel (mm)
25 51 76 102 127 152
Approximate values of C:
Calcareous agg. 8.5
Trap rock 5.6
Siliceous gravel 5.5
Sandstone and gravel 5.0
Cinders 4.6
T = Fire resistance in minutes
d = 16"
d = 14"
d = 12"
d = 10"
d = 8"
d = 6"
d = 4"
d = 2"
T/C
d
D
T/C = (D − 0.4d²/D)^1.7
Protective cover for steel (in.)

FIG. 7-4F. ***Encasing requirements of steel columns and other free-standing steel members. (For SI units: 1 in. = 25.4 mm.)***

For insulating materials, such as mineral fibers, vermiculite, and perlite, having densities, r, in the range of 20 to 50 lb per cu ft (32 to 80 kg/m^3), the factors C_1 and C_2 are:

$$C_1 = 1{,}200/r, \text{ and } C_2 = 30$$

For insulating materials of the same range of densities, but incorporating cement pastes or gypsum, the factors C_1 and C_2 are:

$$C_1 = 1{,}200/r, \text{ and } C_2 = 72$$

Heavy unprotected steel columns are capable of exhibiting some significant fire resistance. Equations 2, 3, 2m, and 3m can be used to predict the fire resistance of unprotected steel columns.[5]

$$R = 10.3\left(\frac{M}{D}\right)^{0.7} \quad \text{for } \frac{M}{D} < 10 \qquad (2)$$

$$R = 8.3\left(\frac{M}{D}\right)^{0.8} \quad \text{for } \frac{M}{D} > 10 \qquad (3)$$

where

R = fire endurance (min),

M = mass of the member (lb/ft), and

D = heated perimeter (in.).

In SI units:

$$R = 75.1\left(\frac{M}{D}\right)^{0.7} \quad \text{for } \frac{M}{D} < 171 \qquad (2m)$$

$$R = 60.5\left(\frac{M}{D}\right)^{0.7} \quad \text{for } \frac{M}{D} > 171 \qquad (3m)$$

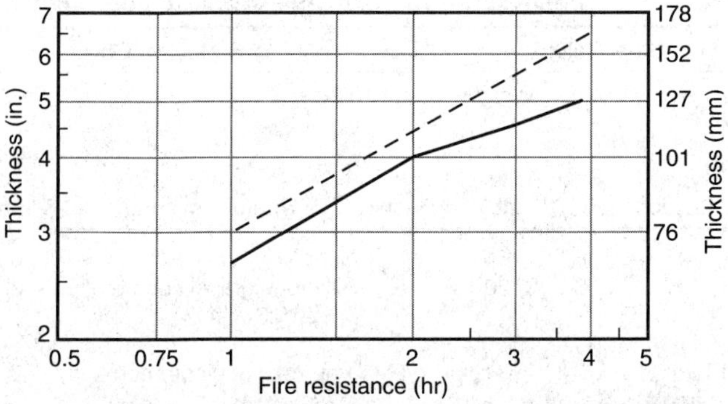
Thickness (in.)
7
6
5
4
3
2
Thickness (mm)
178
152
127
101
76
0.5 0.75 1 2 3 4 5
Fire resistance (hr)

FIG. 7-4G. ***Fire-resistance ratings of reinforced concrete floors of varying thicknesses. The dotted line represents concrete floors made with regular aggregates, and the solid line represents lightweight concretes.***

where

R = fire endurance (min),

M = mass of the member (kg/m), and

D = heated perimeter (mm).

Reinforced concrete: Thermal and mechanical properties vary for all materials with the temperature of the material. Because the heat transmission through concrete is quite slow, the variability of properties throughout the member is quite nonuniform. As a result, the behavior of reinforced concrete in fire is very complex. Nevertheless, equations have been developed to predict the fire endurance of reinforced concrete columns. Equation 4 illustrates this type of calculation.

$$R = \frac{t}{4f} - 1 \qquad (4)$$

where

R = fire endurance (hr),

t = minimum dimension of the column (in.), and

f = a factor that considers overdesign, effective length, and percent of steel (Table 7-4Q provides values for f).

In SI units:

$$R = \frac{t}{101.6f} - 1 \qquad (4m)$$

where

R = fire endurance (hr),

t = minimum dimension of the column (mm), and

f = the same factor from Table 7-4Q.

Equation 5 gives calculated fire endurance for normal-weight concrete slabs.[7,8]

$$R = 0.031\rho^{1.2}L^{1.85} \qquad (5)$$

where

R = fire resistance (hr),

ρ = concrete density (lb/cu ft), and

L = slab thickness (ft).

TABLE 7-4H. *Concrete Floor Constructions*

Materials	Fire-Resistance Rating hr	Fire-Resistance Rating min
Reinforced Concrete (Free or partly restrained, 1500–2500 psi)		
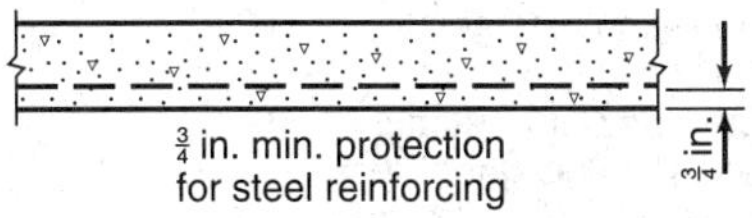¾ in. min. protection for steel reinforcing ¾ in.		
3 in. thick	—	45
4 in. thick	1	15
6 in. thick, 1-in. minimum protection for steel	2	—
Reinforced Concrete on Precast Joists		
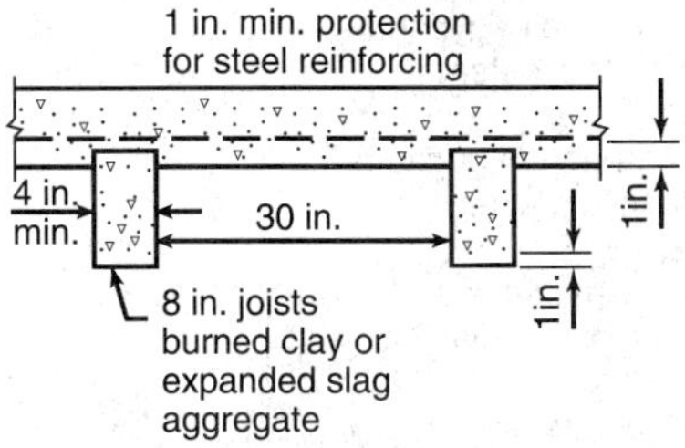1 in. min. protection for steel reinforcing 4 in. min. 30 in. 1 in. 1 in. 8 in. joists burned clay or expanded slag aggregate		
Reinforced concrete, 1:3½:4, 3 in. thick, no ceiling	45	—
Reinforced concrete, 1:3½:4, 3 in. thick, ceiling of gypsum wallboard ½-in. thick, nailed to wood, strips wired to joists	1	—
Combination of Tile and Concrete Floors		
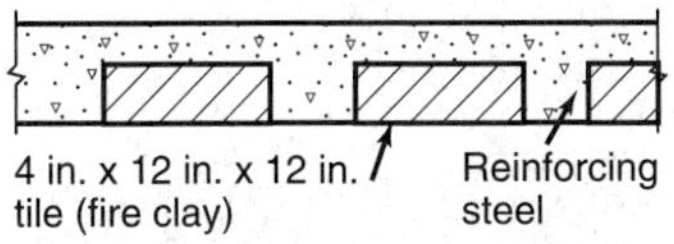4 in. x 12 in. x 12 in. tile (fire clay) Reinforcing steel		
Concrete 2 in. or 1½ in. thick, and fire clay tile 6 in. or 4 in. thick, no ceiling finish	1	—
Concrete 1½ in. thick, and tile 4 in. thick with gypsum-sand plaster ceiling finish, 1:3 mix, ⅝ in. thick	1	30
Concrete 2 in. thick, and fire clay tile 6 in. thick with gypsum plaster ceiling finish, 1:3 mix, ⅝ in. thick	2	—
Concrete 2½ in. thick, limestone aggregate 4 in. thick, expanded slag concrete tile	3	—
Reinforced Concrete Ribbed Slab		
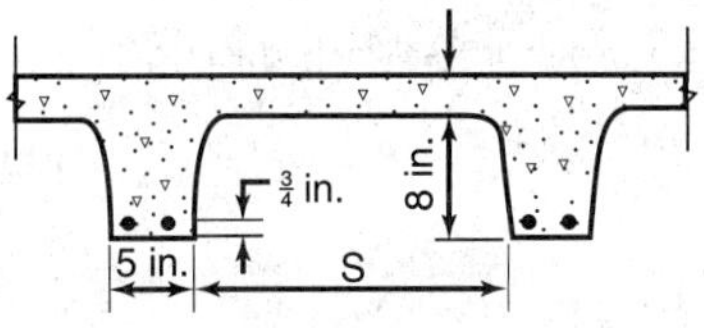¾ in. 8 in. 5 in. S		
Concrete, ribbed slab, limestone aggregate, t = 1½ in. S = 20 in.	—	20
Concrete, ribbed slab, limestone aggregate, t = 2½ in., S = 20 in.	—	45
Same with metal lath and ⅞-in. gypsum-sand plaster ceiling, 1:2, 1:3 mix	2	30
Concrete, ribbed slab, limestone aggregate, t = 3 in. S = 30 in.	1	—

For SI units: 1 in. = 25.4 mm; 1 psi = 6.89 kPa.

In SI units:

$$R = 0.031\rho^{1.2}L^{1.85} \qquad (5m)$$

where

R = fire resistance (min),

ρ = concrete density (kg/m^3), and

L = slab thickness (m).

Timber: When timber members are tested, they burn. As burning continues, a char layer is formed on the exposed surfaces. This char layer reduces the usable strength of the timber members. The time duration that elapses before the member reaches its critical failure load is the fire endurance.

The fire endurance for wood beams is calculated by determining the breaking strength of the uncharred residual cross section. The time to reach a critical depth may be calculated from the following equations.

$$t_c = (D - d)/\beta \qquad (6)$$

$$t_c = (D - d)/2\beta \qquad (7)$$

where

t_c = fire endurance time (min),

D = original depth of the beam [in. (mm)],

d = critical residual depth of the beam [in. (mm)], and

β = charring rate for wood in inches (mm). [An average charring rate for wood is $^1/_{40}$ in. (0.6 mm) per min.]

The critical residual depth, d, can be obtained from Figures 7-4K and 7-4L. The family of curves shows ratios of allowable bending strength (Y) to ultimate bending strength (a) for the beam.

STRUCTURAL MATERIALS AT ELEVATED TEMPERATURES

All structural materials used in building construction are adversely affected by the elevated temperatures caused by fire. The degree and significance of this adverse behavior depends primarily on the function of the elements and on the degree of protection afforded. The mechanical properties of strength and stiffness decrease as the temperature rises. Other adverse behavior, such as excessive expansion and accelerated creep, also develops with increasing temperatures. In general, however, the design parameters that are of concern at normal temperatures are the same parameters that are of concern at elevated temperatures.

Structural Steel

Steel is the backbone of modern building design. Whether it acts as the reinforcement for concrete or the skeleton framework for buildings, it constitutes the major load-carrying material in modern building construction. From the designer's viewpoint, steel possesses many qualities, such as high strength and good ductility, which will make it an ideal structural material. However, steel, like all other materials, is adversely affected by fire.

Steel is noncombustible and does not contribute fuel to a fire. In the past, these properties have sometimes provided a false sense of security with regard to its durability in a fire because they overshadow the fact that steel loses strength when subjected to temperatures easily attained in a fire. The relative seriousness of the problem depends on several factors, such as the function of the steel

TABLE 7-4I. *Steel Joist Floor or Roof Constructions*

Joists		Floor Slab	Thickness (in.)	Furring	Ceiling Kind	Thickness (in.)	Fire-Resistance Rating (hr)	Fire-Resistance Rating (min)
Type	Depth (in.)							
I or S*	8	T & G wood flooring on 2- by 2-in. wood strips	$^{25}/_{32}$	3.4-lb metal lath	Gypsum—sand plaster	$^3/_4$	—	45†
I or S*	8	T & G wood flooring on 2- by 2-in. wood strips	$1^5/_8$	3.4-lb metal lath	Gypsum—sand plaster	$^3/_4$	1†	—
I or S*	8	Reinforced concrete, precast concrete, or gypsum planks	2	3.4-lb metal lath	Gypsum—sand or portland cement-sand plaster	$^3/_4$	1	—
S	8	Reinforced concrete or precast gypsum tile	$2^1/_4$	3.4-lb metal lath	Gypsum—sand plaster, 1:2; 1:3 mix	$^3/_4$	2	—
S	10	Reinforced concrete or reinforced gypsum tile or planks	2	3.4-lb metal lath	Neat†† gypsum, or gypsum-vermiculite plaster, 1:2; 1:3	$1^3/_4$	2	30
S	10	Reinforced concrete	$2^1/_2$	3.4-lb metal lath	Gypsum—sand plaster	$^7/_8$	2	30
		Reinforced concrete	$2^1/_2$	3.4-lb metal lath	Gypsum-perlite of gypsum-vermiculite plaster, $1^1/_2$:2; $1^1/_2$:3	$^3/_4$	3	—
S	8	Reinforced concrete perlite, or vermiculite aggregate	$2^1/_2$	3.4-lb metal lath	Gypsum-perlite or gypsum-vermiculite plaster, $1^1/_2$:2; $1^1/_2$:3	$^3/_4$	3	—
S	10	Reinforced concrete, 1:2:4 gravel aggregate	$2^1/_2$	3-lb metal lath	Gypsum-vermiculite or gypsum-perlite plaster, $1^1/_2$:2; $1^1/_2$:3	1	4	—
S	10	Reinforced concrete, 1:2:4 gravel aggregate	2	Gypsum lath†	Gypsum-perlite or gypsum-vermiculite plaster, $1^1/_2$:$2^1/_2$	$^5/_8$	1	—
S	10	Reinforced concrete, 1:2:4 gravel aggregate	2	Gypsum & wires§	Gypsum-perlite or gypsum-vermiculite plaster, $1^1/_2$:$2^1/_2$	$^1/_2$	2	—
S	10	Reinforced concrete, 1:2:4 gravel aggregate	2	Gypsum‖	Gypsum-vermiculite or perlite plaster, $1^1/_2$:2; $1^1/_2$:3	1	4	—
S	10	Reinforced concrete, 1:2:3.4 gravel	$2^1/_2$	Gypsum & wires§	Sprayed-on mineral fiber	$^3/_4$	3	—
S	10	Reinforced concrete, 1:2.5:3.5 gravel	2	Special Z section#	Special acoustical tiles (see UL list)	$^5/_8$	2	—
S	12	Reinforced concrete, gravel aggregate	2	Nailing channels 16 in. o.c. $2^3/_4 \times ^7/_8$ in.	Type X** wallboard	$^5/_8$	1	30
S	12	Reinforced concrete, 1:3:$3^2/_3$ gravel aggregate	2	26-gage 14 in. o.c. channels	Type X** wallboard applied with No. 6 by 1-in. wallboard screws	$^5/_8$	1	30
S	10	Reinforced concrete, 1:2:4 gravel aggregate	2	25-gage nailing channels 16 in. o.c.	Gypsum wallboard applied with $1^1/_4$-in. long barbed nails $^3/_8$-in. diam. head	$^5/_8$	1	—

*I-beam or open-web type joists.
†Combustible construction.
‡All gypsum lath $^3/_8$-in. perforated type.
§Gypsum lath and No. 20-gage wires attached to nailing channels. Wires attached diagonally to reinforce and support lath and plaster.
‖ 1-in. hexagonal mesh wire fabric to reinforce plaster and hold up lath and plaster.
#Special No. 25 gage galvanized steel Z runners 12 in. o.c.
**Type X gypsum wallboard designates gypsum wallboard with a specially formulated core that provides greater fire resistance than regular gypsum wallboard of the same thickness.
††Unsanded wood-fiber plaster.
For SI units: 1 in. = 25.4 mm; 1 lb = 0.454 kg.

element, its level of stress, the support conditions, its surface area and thickness, and the temperature within the steel itself. This temperature can be quite different from the ambient temperature in the compartment.

From a structural viewpoint, the yield stress of steel is the significant parameter in establishing load-carrying capacity. The tensile stress strain diagram for A36 steel at various temperatures is shown in Figure 7-4M. It can be seen that both the yield strength and the modulus of elasticity decreases with increasing temperatures. Figure 7-4N shows the ratios of modulus of elasticity and yield

stress for A36 steel at various temperatures. As can be seen, both values decrease with an increase in temperature.

This reduction in strength, combined with the reduction in the modulus of elasticity, causes compression members to be more sensitive to higher temperatures than tensile or flexural members. Figure 7-4O shows the influence of temperature on the critical stress of compression members of A36 steel.

It should be recognized that the temperature considered for stress limitations is the temperature within the steel, and not the am-

TABLE 7-4J. *Floors of Concrete on Steel Floor and Form Units*
(Plaster or Sprayed-on Fire Protective Covering)

	Type A		Type B		Type C		
Type of Floor Unit	Thickness of Floor (in.)	Furring	Protective Covering Material	Application	Thickness (in.)	Fire-Resistance Rating hr	min
A	$5^5/_8$	None	Mineral fibers applied to floor units	Sprayed	$^1/_2$ to 2	5	—
C	$5^1/_2$	None	Same	Sprayed	$1^1/_2$	5	—
A	$5^5/_8$	None	Vermiculite or perlite acoustical plastic	Sprayed	$1^1/_{16}$ to $3^1/_{16}$	4	—
B	$4^1/_2$	None	Same	Sprayed	4	—	—
B	$4^1/_2$	See Footnote*	Gypsum-vermiculite or perlite plaster	Troweled or sprayed	$^3/_8$ to $1^5/_8$	4	—
C	4	None	Mineral fiber applied to floor units	Sprayed	$^3/_4$	3	—
A	$5^5/_8$	None	Vermiculite acoustical plastic, cellular floor units	Sprayed	$^1/_2$ to 2	2	—
B	$5^1/_4$	None	None†	—	—	2	—
B	$4^1/_4$	None	None‡	—	—	1	—
A	8	See Footnote§	Gypsum-vermiculite or perlite plaster, 100 lb gypsum to 2 cu ft for scratch coat and 3 cu ft for browncoat plaster, white finish $^1/_{16}$ in.	Troweled	$^3/_8$	4	—
A	8	See Footnote§	Same	Troweled	1	5	—
A	$6^5/_8$	See Footnote‖	Acoustical tiles, T & G edges with saw kerfs	Sheet metal clips	$^3/_4$	4	4

*Expanded metal lath tack welded or tied to bottom of corrugated steel floor units.
†Floor slab, limestone concrete, $5^1/_4$ in. thick.
‡Floor slab, limestone concrete, $4^1/_4$ in. thick.
§24-gage, 3.4-lb, $^3/_8$-in. mesh expanded metal lath suspended $2^1/_2$ to $7^1/_2$ in. below floor units.
‖ Special furring system to which acoustical tiles are clipped $10^3/_4$ in. below floor units.
For SI units: 1 in. = 25.4 mm; 1 lb = 0.454 kg; 1 cu ft = 0.0283 m³.

bient temperature. Because steel has a high thermal conductivity, it can transfer heat away from a localized heat source rather quickly. This property, in conjunction with its thermal capacity, enables steel to act as a heat sink. When the steel has an opportunity to transfer heat to cooler regions, it can take a relatively long time for a member to reach its critical value. On the other hand, an extensive fire that distributes heat simultaneously over a greater area reduces this time considerably.

Related to this thermal activity is the effect of mass and surface area of structural steel members. Heavy, thick sections have a far greater resistance to the effects of building fires than do lighter ones. Unprotected lightweight sections, such as those found in trusses and open-web joists, can collapse after 5 or 10 min of exposure.

Another property of steel that has an effect upon its performance at elevated temperatures is its coefficient of expansion. The linear coefficient of expansion of steel at temperatures up to 1100°F (600°C) is given as:

$$\alpha = 0.0000061 + 0.000000022\ \Delta t$$

where:
α = the coefficient of expansion
Δt = temperature change in degrees Fahrenheit.

In SI units:
α = the coefficient of expansion
Δt = temperature change in degrees Celsius.

$$\alpha = 6.1 \times 10^{-6} + 3.96 \times 10^{-9}\ \Delta t$$

This high coefficient affects the structure in two ways. If the ends of a structural member are axially restrained, the attempted expansion due to the heat causes thermal stresses to be induced in the member. These stresses combine with those of the normal loading, causing more rapid collapse. If the structural member is not axially restrained, the increased stresses described do not occur; instead, movement takes place. This movement causes the ends of steel columns to be moved laterally, producing an eccentrically loaded column. In other cases, walls can be moved to the point of collapse by expansion of beams. This creates an extremely hazardous condition both for the building itself and for the fire fighters.

To illustrate the magnitude of movement, consider a 50-ft (15.24-m) long steel beam that is heated uniformly over its length from 72°F to 972°F (22.2°C to 522°C). The average value for alpha is 0.0000073, and the increase in length, delta, becomes approximately,

$$\delta = \alpha\,L\,\Delta t = (7.3 \times 10^{-6})(50 \times 12)(900)$$

Thus $\delta = 3.9$ in.

In NFPA 251, alternate tests may be used to evaluate protection of steel beams and columns. These alternate tests evaluate the thermal protection of the insulating materials by measuring the temperature in the steel for the standard time-temperature exposure. In these alternate tests, failure occurs when the average temperature reaches 1000°F (538°C). Often, this temperature, or a temperature

TABLE 7-4K. *Wood Joist Floors with Plaster Ceilings (Combustible)*

	Plaster Base					Plaster					Fire-Resistance Rating	
	Thickness in inches or weight per sq yd	Nails										
Type		Size or Length (in.)	Gage	Head Diam (in.)	Spacing (in.)	Type	Aggregate	Mix	Thickness (in.)	Ceiling Thickness (in.)	hr	min
Wood lath	3/8	3d	15	11/64	1 7/8	Lime	Sand	1:4 1/2	5/8	1	—	30
Wood lath	3/8	3d	15	11/64	1 7/8	Gypsum	Sand	{1:2, 1:3}	7/8	—		35
Gypsum, perforated	3/8	1 1/8	13	5/16	3 1/2	Gypsum	Sand	1:2	1/2	7/8	—	30
Gypsum, perforated	3/8	1 1/8	13	3/8	3 1/2	Gypsum	Sand	1:2	1/2	7/8	—	45
Gypsum, perforated	3/8*	1 3/4	12	1/2	3 1/2	Gypsum	Sand	1:2	1/2	7/8	1	—
Gypsum, plain	3/8†	{5d, 8d}	{13, 12 1/2}	{5/16, 5/16}	{3 1/2, 8}	Gypsum	Vermiculite or perlite	1 1/2:2 1/2	1/2	7/8	1	40
Gypsum, perforated	3/8	4d	—	—	3 1/2	Gypsum	Vermiculite or perlite	1 1/2:2 1/2	1/2	7/8	1	—
Metal lath	3.4 lb	{6d, 11 1/4}	{11 1/2, 11}	{1/4, 3/8}	{6, 6}	Gypsum	Sand	{1:2, 1:3}	3/4	3/4	—	45
Metal lath	3.4 lb	1 1/2	11	5/16	6	Gypsum	Sand	{1:2, 1:3}	3/4	3/4	1	—
Metal lath	3.0 lb	8d	11 1/2	1/4	6‡	Gypsum Portland cement§	Sand	{1:2, 1:3}	3/4	3/4	1	15
Metal lath	3.4 lb	1 1/2	11 1/2	7/16	5		Sand	{1:2, 1:3}	3/4	3/4	1	—
Metal lath	3.4 lb	1 1/2	11	7/16	5 1/2	Gypsum	Vermiculite or perlite	{1 1/2:2, 1 1/3:3}	3/4	3/4	1	30

*3-in.-wide strips of metal lath with two 1³/₄-in. 12-gage ¹/₂-in. head barbed roofing nails in each joist covering joints of lath to reinforce plaster.
†Plaster reinforced with 1-in. hexagonal mesh wire fabric (1 lb/sq yd) nailed to joists with 8d 2³/₄-in.-long cement-coated nails spaced 8 in. on centers at each joist.
‡Additional support for metal lath by ties of 18-gage wire spaced 27 by 32 in. and nailed 2 in. up on sides of joists.
§3-lb asbestos fiber and 15-lb hydrated lime per bag of cement added.
For SI units: 1 in. = 25.4 mm; 1 lb = 0.454 kg.

of 1100°F (593°C) is perceived as the critical temperature for structural steel. While numbers such as these are useful in a conceptual sense, they should not be taken literally.

If critical temperature for structural steel is intended to describe the steel temperature at which collapse is impending, these temperatures in the range of 1000°F (538°C) or 1100°F (593°C) may be viewed as a reasonable approximation. The actual temperatures at which collapse is impending can differ from these values, depending upon the load and support conditions, dimensions, and geometry of the structural member.

Rather than focus exclusively on temperature, it is more appropriate to focus on the load-carrying capacity of a member at different temperatures. For example, one may question, what is the load-carrying capacity of the structural member when the temperature reaches x°F? The related question may then be asked, what is the

likelihood that the steel temperature will reach this value in a room fire? Modern calculation methods enable both of these questions to be addressed with a reasonable degree of confidence, although not yet as a simple, routine evaluation.

Fire Protection of Structural Steel

Because unprotected structural steel loses its strength at high temperatures, it must be protected from exposure to the heat produced by building fires. This protection, often referred to as "fireproofing," insulates the steel from the heat. The more common methods of insulating steel are encasement of the member, application of a surface treatment, or installation of a suspended ceiling as part of a floor-ceiling assembly capable of providing fire resistance. Additional methods, such as sheet steel membrane shields around members and box columns filled with liquid, have been introduced.

TABLE 7-4L.   *Wood Joist Floors with Wallboard Ceilings (Combustible)*

	Wallboard			Nails					Fire-Resistance Rating	
Type	Thickness (in.)	Core Materials	Type	Size	Gage	Length (in.)	Spacing (in.)		Hr	Min
Gypsum	$5/8$	Type "X"* special fire-retardant gypsum	Cement-coated wire	6d	13	$1^7/_8$	6		1	—
Gypsum	$1/2$	Type "X"* special fire-retardant gypsum	Cement-coated wire	5d	$13^1/_2$	$1^5/_8$	6		—	45
Gypsum	$3/8$	Type "X"* special fire-retardant gypsum	Cement-coated wire	4d	14	$1^3/_8$	6		—	30
Two thicknesses of gypsum	$1/2 + 1/2$	Gypsum	Box / wire	5d (1) / 6d (2)	14 / $10^1/_4$	$1^3/_4$ / $2^1/_2$	18 / 6		1†	—
Two thicknesses of gypsum	$3/8 + 1/2$	Gypsum	Plasterboard / cement-coated	$1^1/_2$ in. / 6d	13 / 13	$1^1/_2$ / $1^7/_8$	7 / 6		—	40
Two thicknesses of gypsum	$1/2 + 3/8$	Gypsum	Plasterboard / cement-coated	$1^1/_2$ in. / 6d	13 / 13	$1^1/_2$ / $1^7/_8$	7 / 6		—	35
Two thicknesses of gypsum‡	$3/8 + 3/8$	Gypsum	Box	$4^1/_2$d (1) / $4^1/_2$d (2)	— / —	$1^1/_2$ / $1^1/_2$	6 / 6		—	35
Gypsum	$1/2$	Gypsum	Box	$4^1/_2$d	15	$1^1/_2$	6		—	25
Gypsum§	$3/8$	Gypsum	Box	$4^1/_2$d	15	$1^1/_2$	6		—	25
None‖	—	—	—	—	—	—	—		—	14
Acoustical Tile	$5/8$	12- by 12-in. mineral fiber tiles mounted on special channels							1	—

* Type X gypsum wallboard designates gypsum wallboard with a specially formulated core that provides greater fire resistance than regular gypsum wallboard of the same thickness.
† 1-in. hexagonal mesh 20-gage wire fabric between wallboards nailed with 8d nails 8 in. o.c.
‡ NBS test on floor $4^1/_2$ by 9 ft; joints of wallboard staggered, but no tape and joint finisher.
§ NBS test; bottom of ceiling covered with 14-lb asbestos paper applied with paperhanger's paste and casein paint.
‖ NBS test on 2 specimens of open-joist floors, each $4^1/_2$ by 9 ft; fire endurance 15 min and 12 min.
NOTE: for other similar constructions, see UL Building Materials List.
For SI units: = 1 in. = 25.4 mm; 1 lb = 0.454 kg.

Over the years, encasement of the structural steel member has been a very common and a very satisfactory method of insulating steel to increase its fire resistance. In floor systems of reinforced concrete slabs supported by structural steel beams, the encasement can be placed monolithically with the floor. Figure 7-4P illustrates an old technique of encasement. The major disadvantage of this procedure is the cost, which is related both to the added formwork and concrete and to the increased weight of the supporting members due to the added dead load. To reduce the cost of encasement, systems utilizing lath and plaster or gypsumboard, or spray-on mineral fibers, have been developed, as shown in Figures 7-4Q and 7-4R.

Because of labor costs and the weight increases of encasement, surface treatments applied directly to the member have become quite popular. Sprayed-on mineral fiber coatings are widely used for protecting structural steel. While the protection is excellent if applied correctly, it can easily be scraped off the member during construction or renovations. Consequently, sprayed-on mineral fiber coatings are suspect in their effectiveness over long-term use.

Cementitious materials also have been used as sprayed-on coverings. During a fire, however, they can spall. Adhesion problems also have been experienced with this type of sprayed-on coating. Thus, effective application, complete coverage, and long-term maintenance are attributes that must be evaluated in considering sprayed-on applications.

Intumescent paints and coatings have been utilized to increase the fire endurance of structural steel. These coatings intumese, or swell, when heated, thus forming an insulation around the steel. They are primarily used for nonexposed steel subject to elevated temperatures, as prolonged exposure to flame can destroy the char coating.

Suspended ceilings consisting of lath and plaster, gypsum panels, or acoustical tile supported on a grid system as part of a floor-ceiling assembly are a popular method of fireproofing. The grid system can be suspended from wire hangers or it can be attached directly to the bottom chord of joists or to the bottom flange of beams. Sometimes ceiling tiles are either mechanically fastened or fitted into splines to prevent the pressures that occur in building fires from lifting them out of place.

The overall effectiveness of this type of barrier protection, often called membrane protection, is questionable. This is due to

TABLE 7-4M. *Load-Bearing Brick and Clay Tile Walls*

Material	Wall Thickness (in.)	Solid Content of Walls (percent)	Hollow Units		Fire-Resistance Ratings (hr)			
			Number of Cells in Wall Thickness	Thickness of Shells of Unit (in.)	Combustible Members Framed 4 in. into Wall		No Combustible Members Framed into Wall	
					No Plaster	Plaster on Two Sides	No Plaster	Plaster on Two Sides
Brick, clay, or shale	12	90 to 100	—	—	8	9	10	12
	10*	72	{ 2-in. cavity	—	2	2½	5	7
	8	90 to 100	—	—	2	2½	5	7
	4†	90 to 100	—	—	—	—	1	1½
Load-Bearing hollow tile (not partition tile)	12	45	3	0.7	2½	3½	3	6
	12‡	48	4	⅝	2½	4	5	7½
	10*	36	{ 2 + 2-in. cavity	—	—	1¼	—	4
	8	48	3 or 4	—	1	1¾	2½	3½
	8	40	2	—	¾	1½	2	3
	6†	40	2	⅝	—	—	¾	1½

* Cavity wall with metal ties across cavity.
† Nonload-bearing wall restrained on all edges.
‡ Two units, 8 by 12 by 12-in. 6-cell and 3¾ by 12 by 12-in. 3-cell tiles, in wall thickness.
For SI units: 1 in. = 25.4 mm.

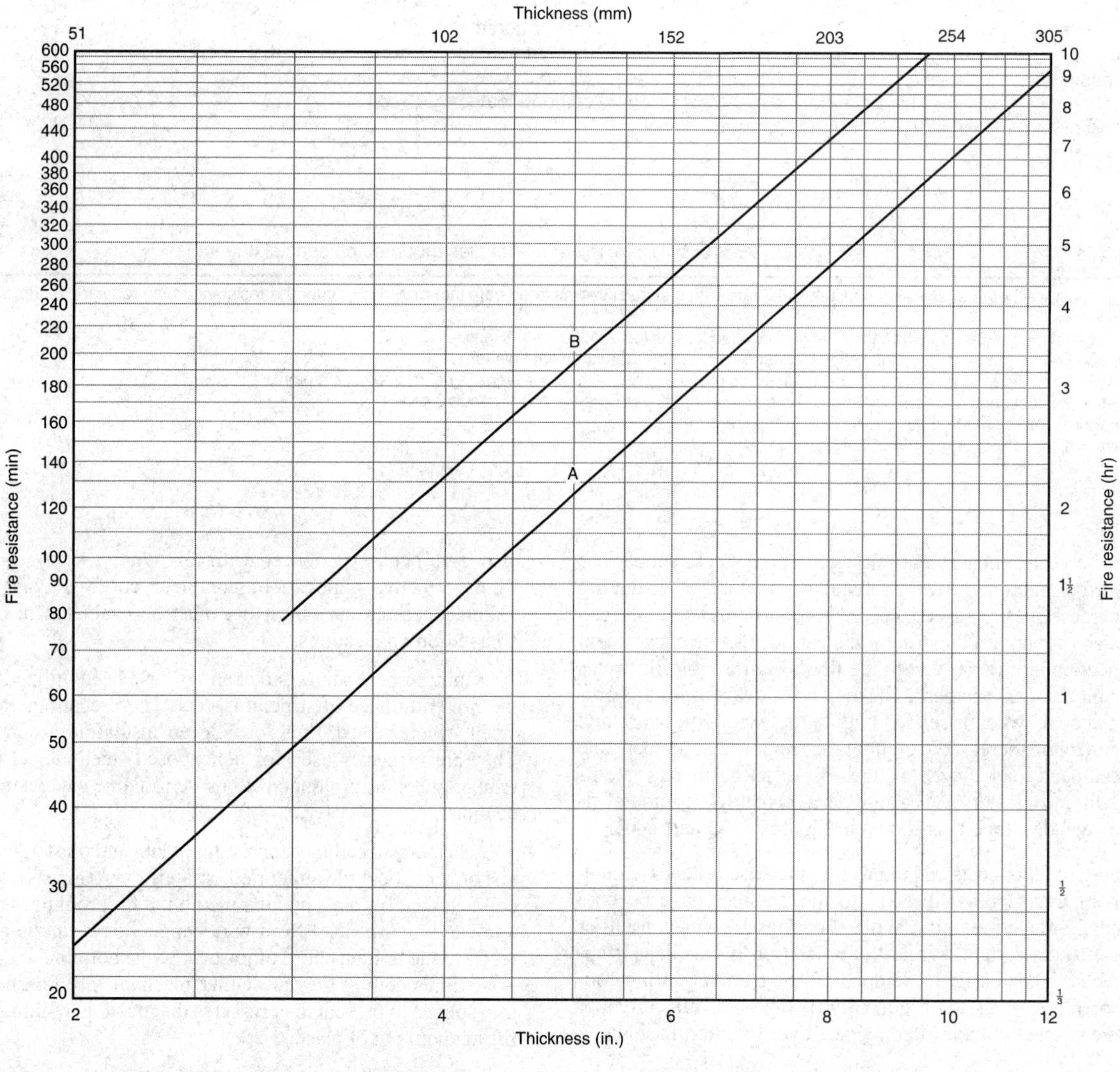

FIG. 7-4H. *Fire resistance of burned clay brick walls: A, unplastered; B, plastered both sides.*

TABLE 7-4N. *Wood Stud Walls and Partitions (Combustible)*
(Bearing and nonbearing: 2- by 4-in. studs spaced 16 in. on center, firestopped)

Material	Fire-Resistance Rating			
	Partition Hollow		Partition Filled with Mineral Wool*	
	hr	min	hr	min
Plasterless Types of Construction				
The following are applied to both sides of studs:				
Sheathing boards (tongue-and-groove) 3/4 in. thick	—	20	—	35
Gypsum wallboard, 3/8 in. thick	—	25	—	—
Gypsum wallboard, 1/2 in. thick (nonload-bearing only for mineral wool filled)	—	40	1	—
Gypsum wallboard, 3/8 in. thick, in two layers each face	1	—	—	—
Gypsum wallboard, 1/2 in. thick, in two layers each face	1	30	—	—
Gypsum wallboard, 1/2 in. thick, Type X, one layer each face	—	45	—	—
Gypsum wallboard, 5/8 in. thick, Type X, one layer each face	1	—	—	—
Gypsum wallboard, 5/8 in. thick, Type X, on fire retardant wood fibreboard, 1/2 in. thick	1	—	—	—
Fir plywood, 1/4 in. thick	—	10	—	—
Fir plywood, 3/8 in. thick	—	15	—	—
Fir plywood, 1/2 in. thick	—	20	—	—
Fir plywood, 5/8 in. thick	—	25	—	—
Cement-asbestos board, 3/16 in. thick	—	10	—	40
Cement-asbestos board, 3/16 in. thick, on gypsum wallboard, 3/8 in. thick	1	—	—	—
Cement-asbestos board, 3/16 in. thick, on gypsum wallboard, 1/2 in. thick	1	25	—	—

TABLE 7-4N. *(Continued)*

Material	Fire-Resistance Rating			
	Partition Hollow		Partition Filled with Mineral Wool*	
	hr	min	hr	min
Plaster and Lath Construction				

Material	hr	min	hr	min
The following are applied to both sides of studs:				
Gypsum-sand plaster, 1:2, 1:3, 1/2 in. thick on wood lath	—	30	1	—
Lime-sand plaster, 1:5, 1:7.5, 1/2 in. thick on wood lath	—	30	—	45
Gypsum-sand plaster, 1:2, 1:2, 1/2 in. thick on 3/8 in. perforated gypsum lath	1	—	—	—
Gypsum-sand plaster, 1:2, 1:2, 3/4 in. thick on metal lath	1	—	1	30
Gypsum lath, 1/2 in. thick, Type X, and 1/8 in. gypsum-sand plaster, each face	1	—	—	—
Gypsum wallboard, 1/2 in. thick, Type X, and 1/16 in. gypsum plaster, each face	1	—	—	—
Portland cement-lime-sand plaster, 1:1/30:2, 1:1/30:3 and asbestos-fiber plaster, 7/8 in. thick on metal lath	1	—	—	—
Gypsum-vermiculite, or perlite plaster, 100-lb gypsum to 2 1/2 cu ft aggregate, 1/2 in. thick on 3/8 in. perforated gypsum lath	1	—	—	—
Gypsum perlite plaster, 1:2, 3/4 in. thick on metal lath	1	—	—	—

Exterior Bearing Wall				

Material	hr	min	hr	min
Outside: Cement-asbestos shingles, 5/32 in. thick, on layer of asbestos felt on wood sheathing, 3/4 in. thick, on wood studs; inside: Cement-asbestos facing, 1/8 in. thick on fiberboard, 7/16 in. thick	—	30	—	—
Outside: Gypsum sheathing, 1/2 in. thick; inside 1:2 gypsum-sand plaster, 1/2 in. thick on 3/8 in. perforated gypsum lath	1	30	—	—

* Mineral wool fill requires some degree of anchorage so as to be held in place after partition facing has been burned away.
For SI units: 1 in. = 25.4 mm; 1 lb = 0.454 kg; 1 cu ft = 0.0283 m³.

experiences where a lack of control during construction resulted in improper installation procedures, such as inattention to expansion control in the suspension system. In addition, maintenance of ductwork and fixtures in the plenum area is frequently done by personnel who are not aware of the importance of the integrity of the ceiling to the fire-protection system. Removed tiles are not replaced in a manner that will ensure their integrity during a building fire. Consequently, the unprotected steel in the plenum area is exposed to fire and hot gases, which reduce its strength.

Water-filled columns have received attention for their fire-resistive capabilities. During a fire, heat is transferred away from the location of exposure by convection currents set up in the liquid. Consequently, the heat is transferred before it has an opportunity to raise the temperature of the unprotected steel to its critical value. Although the principle of operation is valid and a few tests have been performed, this type of construction has never been exposed to an actual building fire.

TABLE 7-4O. *Various Finishes Over Wood Framing, One Side (Combustible) with Exposure on Finish Side*
(See Table 7-4L for 2-side ratings.)

Material	Fire-Resistance Rating* (min)
Fiberboard, $^1/_2$ in. thick	5
Fiberboard, flameproofed, $^1/_2$ in. thick	10
Fiberboard, $^1/_2$ in. thick, with $^1/_2$ in. 1:2, 1:2 gypsum-sand plaster	15
Gypsum wallboard, $^3/_8$ in. thick	10
Gypsum wallboard, $^1/_2$ in. thick	15
Gypsum wallboard, $^5/_8$ in. thick	20
Gypsum wallboard, laminated, two $^3/_8$ in.	28
Gypsum wallboard, laminated, one $^3/_8$ in. plus one $^1/_2$ in. thick	37
Gypsum wallboard, laminated, two $^1/_2$ in. thick	47
Gypsum wallboard, laminated, two $^5/_8$ in. thick	60
Gypsum lath, plain or indented, $^3/_8$ in. thick, with $^1/_2$ in. 1:2, 1:2 gypsum-sand plaster	20
Gypsum lath, perforated, $^3/_8$ in. thick, with $^1/_2$ in. 1:2, 1:2 gypsum-sand plaster	30
Gypsum-sand plaster, 1:2, 1:3, $^1/_2$ in. thick, on wood lath	15
Lime-sand plaster, 1:5, 1:7.5, $^1/_2$ in. thick, on wood lath	15
Gypsum-sand plaster, 1:2, 1:2, $^3/_4$ in. thick, on metal lath (no paper backing)	15
Neat gypsum plaster, $^3/_4$ in. thick on metal lath (no paper backing)†	30
Neat gypsum plaster, 1 in. thick, on metal lath (no paper backing)†	35
Lime-sand plaster, 1:5, 1:7.5, $^3/_4$ in. thick, on metal lath (no paper backing)	10
Portland cement plaster, $^3/_4$ in. thick, on metal lath (no paper backing)	10
Gypsum-sand plaster, 1:2, 1:3, $^3/_4$ in. thick, on paper-backed metal lath	20

*From National Bureau of Standards BMS-92.
†Unsanded wood-fiber plaster.
For SI units: 1 in. = 25.4 mm.

FIG. 7-4I. *Fire resistance of wood or metal stud partitions faced with gypsum wallboards or gypsum plaster on metal lath: A, Type X gypsum wallboards or wood fiber gypsum plaster; B, 1:1 gypsum-sand plaster; C, 1:2 gypsum-sand plaster; and D, 1:2 and 1:3 gypsum-sand plaster.*

Structural steel members also can be protected by sheet steel membrane shields. The sheet steel holds inexpensive insulation materials in place, thus providing a greater fire resistance. In addition, polished sheet steel has been used in tests to protect spandrel girders. The shield reflects radiated heat and protects the load-carrying spandrel.

Reinforced Concrete

Concrete is often used as a protective covering for other materials, as well as a primary load-bearing structural material. Consequently, reinforced-concrete buildings give a sense of fire security. However, concrete as a material also is adversely affected by the heat of a fire. Although collapse of reinforced-concrete structures is comparatively rare, loss in strength, spalling, and other deleterious effects do occur.

When the temperature of a reinforced-concrete member is raised, the member loses strength. The amount of strength reduction is influenced by a number of factors. Among the more significant from a structural viewpoint are the type of aggregate, moisture content, type of loading, and level of stress during the fire exposure.

One of the more significant factors in determining the change in strength and the thermal characteristics of concrete is the type of aggregate. Aggregate types can vary widely in various sections of the country. Consequently, numerical values of strength properties are related to a percentage of the original strength, rather than a spe-

cific stress value. The qualitative behavior of concrete is generally accurate, however.

Lightweight concrete performs better at elevated temperatures than does normal-weight concrete. Not only does it retain more of its strength during heat buildup, it also has a lower coefficient of thermal conductivity. Concrete using aggregates or vermiculite or perlite is particularly good at protecting structural steel from a heated environment.

Figures 7-4S and 7-4T illustrate the effect of aggregates on the fire resistance of reinforced-concrete slabs. The lightweight aggregates, such as expanded shale and expanded slag, have considerably more fire endurance than do normal-weight concretes made from carbonate and siliceous aggregates.

The moisture content of concrete has a significant influence on its thermal performance. A considerable quantity of the heat energy of a fire is expended in vaporizing the absorbed and capillary moisture in concrete. In the case of horizontal members, the water vapor is driven upward and maintains a temperature at the top of the member of 212°F (100°C) until the water has been driven off. This increases the fire endurance, as it keeps the temperature on the unexposed side below that defined as the failure temperature. The voids caused by the evaporation of water contribute to shrinkage and a decrease in concrete strength.

The mechanical properties of concrete are significantly reduced at elevated temperatures. Figure 7-4U shows the effect on the compressive strength and modulus of elasticity by increasing the

TABLE 7-4P. *Solid Partitions: Nonbearing*

Materials	Fire-Resistance Rating hr	Fire-Resistance Rating min
Sheathing planks (tongue-and-groove), in 2 layers, each $^3/_4$ in. thick (1 in. nominal) and with joints staggered	—	15*
Same, with layer of 30-lb asbestos felt between planks	—	25*
Planking, pine (tongue-and-groove), 2-in. thick (nominal), wet vertically	—	12*
Wallboard, $^3/_8$-in. gypsum, full height facings on two thicknesses $^1/_2$-in. coreboard, full height, cemented with staggered vertical joints to form 1$^3/_4$-in. thick partition, external joints finish taped	1	—
Wallboard, $^5/_8$-in. gypsum, Type X‡, full height facings, cemented and nailed or screwed to ribs made of two thicknesses of $^1/_2$-in. gypsumboard 3$^1/_2$ or 6$^1/_2$ in. wide and 1- by 1$^5/_8$-in. wood runners top and bottom, external joints finish taped	1	—
Wallboard, $^1/_2$-in. gypsum, full height, nailed and cemented to 1-in. thick coreboard factory laminated from two $^1/_2$-in. thick by 24-in. wide coreboards with staggered edges, external joints staggered, butted and finish taped with joint finisher	2	—
Gypsum tile, 3 in. thick, cored	2	—
Gypsum tile, 4 in. thick, cored	2	—
Gypsum tile, 3 in. solid, no cores	3	—
Gypsum tile, 3 in. cored, $^1/_2$-in. of 1:3 gypsum-sand plaster on each side	3	—
Gypsum tile, 4 in. cored, $^1/_2$-in. of 1:3 gypsum-sand plaster on each side	4	—
Gypsum-vermiculite or perlite plaster, 1$^1/_2$:2, 1$^1/_2$:3 by vol., $^3/_4$-in. thick on each side of $^1/_2$-in. gypsum lath, vertical full height, tied to floor and ceiling runners, no studs	1	30
Gypsum-sand plaster, 1:2, 1:3 by wt, $^3/_4$ in. thick on each side of $^1/_2$-in. gypsum lath vertical full height, tied to floor and ceiling runners, no studs	1	—
Partition tile, burned clay, 4 in. thick, 1 cell in thickness	—	10†
Cinder block, 4 in. thick, solid	1†	—
Cinder block, 6 in. thick, 1 cell in thickness	1	15†
Calcareous gravel concrete tile, 4 in. thick, 65 percent solid	—	45‡
Calcareous gravel concrete tile, 8 in. thick, 55 percent solid	2	30

*Combustible.

†When plastered on both sides with $^1/_2$-in. 1:3 gypsum-sand plaster, the tile partition described has 45-min fire resistance and the cinder block and 4-in. concrete tile assemblies described have 2-hr fire resistance.

‡Type X gypsum wallboard designates gypsum wallboard with a specially formulated core that provides greater fire resistance than regular gypsum wallboard of the same thickness.

For SI units: 1 in. = 25.4 mm; 1 lb. = 0.454 kg.

temperature. There is some question about whether concrete completely regains its strength after exposure to high temperatures.

Prestressed Concrete

The factors that are significant in the behavior of prestressed concrete are similar to those that affect reinforced concrete. In addition to the influences of moisture content and aggregate, the higher

TABLE 7-4Q. *Values for f*

Overdesign factor	Effective length, kL (kL_{max} = 12 ft)	Effective length, kL 12 ft< kL < 24 ft $p \le 0.03$ $t \le 12$ in.	Effective length, kL 12 ft< kL < 24 ft All other cases
1.00	1.0	1.2	1.0
1.25	0.9	1.1	0.9
1.50	0.8	1.0	0.8

For SI units: 1 ft = 0.305 m; 1 in. = 25.4 mm.

FIG. 7-4J. Fire resistance of solid partitions of metal lath and plaster: A, wood fiber gypsum; B, 1:$^1/_2$ gypsum-sand; C, 1:1 gypsum-sand; D, 1:1$^1/_2$ gypsum-sand; E, 1:2 gypsum-sand; F, 1:2 and 1:3 gypsum-sand; G, 1:2 and 1:3 Portland cement + 0.2 lime to cement.

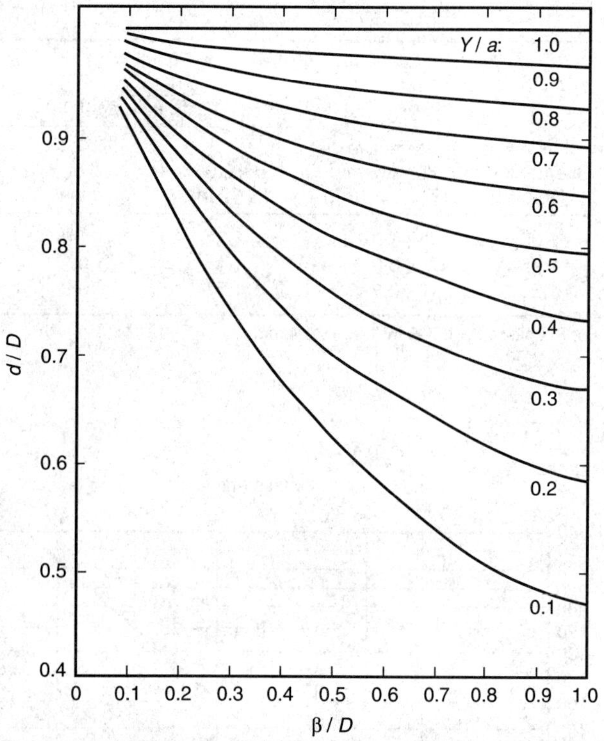

FIG. 7-4K. *Critical depth of solid timber beams of rectangular cross section exposed on three sides to fire.*[9]

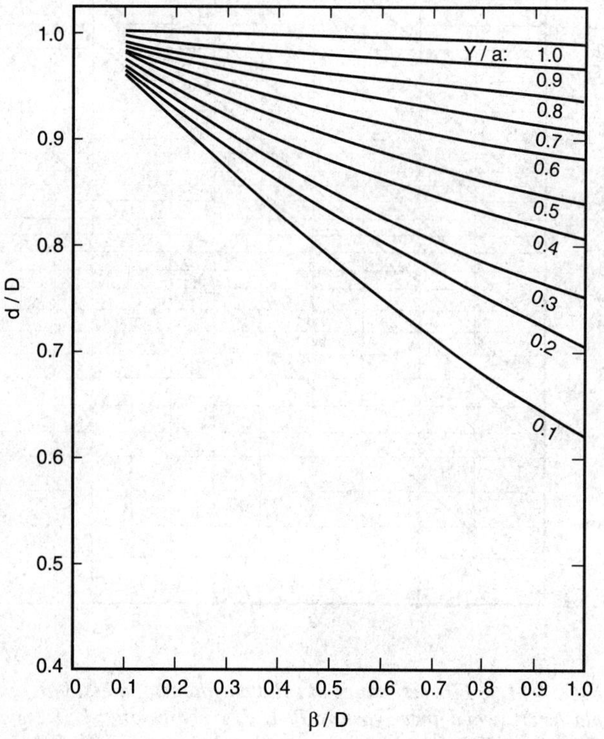

FIG. 7-4L. *Critical depth of solid timber beams of rectangular cross section exposed on four sides to fire.*[9]

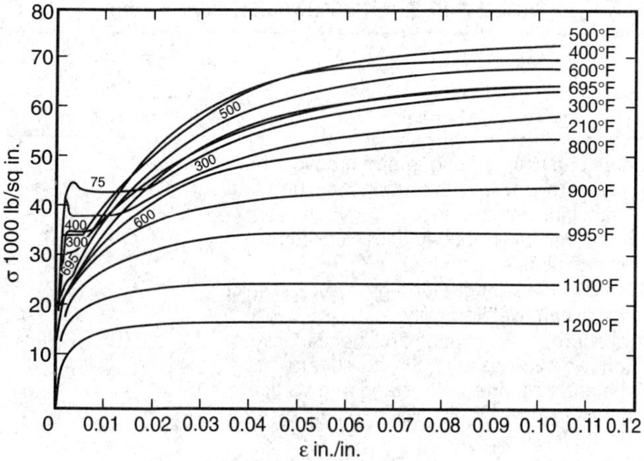

FIG. 7-4M. *Stress-strain curves for an ASTM A36 steel.*[10]

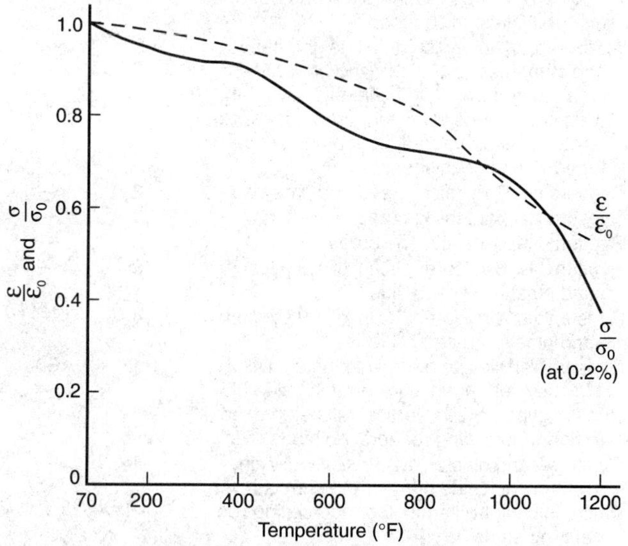

FIG. 7-4N. *Ratio of modulus and yield strength with temperature—A36 steel.*[11]

strength concrete and the function and type of steel used for prestressing are important considerations.

The concrete used for prestressed concrete is of a higher strength than that used in ordinary reinforced-concrete construction. The overall fire resistance of this concrete is somewhat better than the lower strength concrete. The heat transmission is about the same for the two systems. However, there is a somewhat greater tendency for prestressed concrete to spall, thus exposing the prestressing steel.

The greatest problem of prestressed concrete subjected to elevated temperatures involves the prestressing steel. The fact that the steel is put under high initial stress, coupled with the fact that the reinforcing wires are high-carbon, cold-drawn, rather than low-carbon, hot-rolled steel, is the root of the problem. Normal prestressing losses due to deformation and creep are considered in prestressed concrete design. At elevated temperatures these losses are accelerated. Creep in the steel increases, and the modulus of elasticity of the prestressing wires is reduced by about 20 percent when the temperature reaches 600°F (316°C). These losses reduce the carrying capacity of the member.

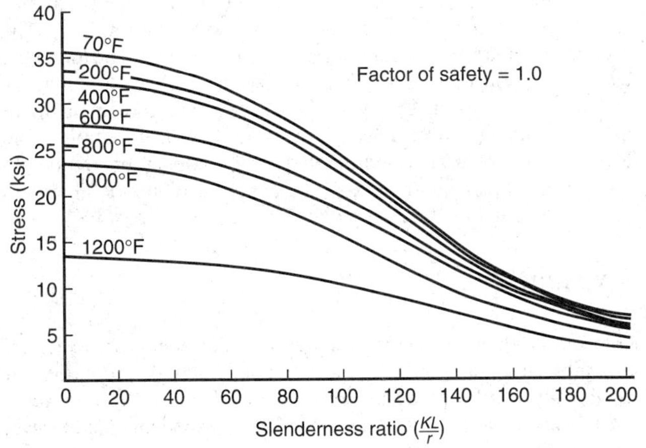

FIG. 7-4O. *Critical stress as a function of the slenderness ratio for various temperatures—A36 steel.*[11]

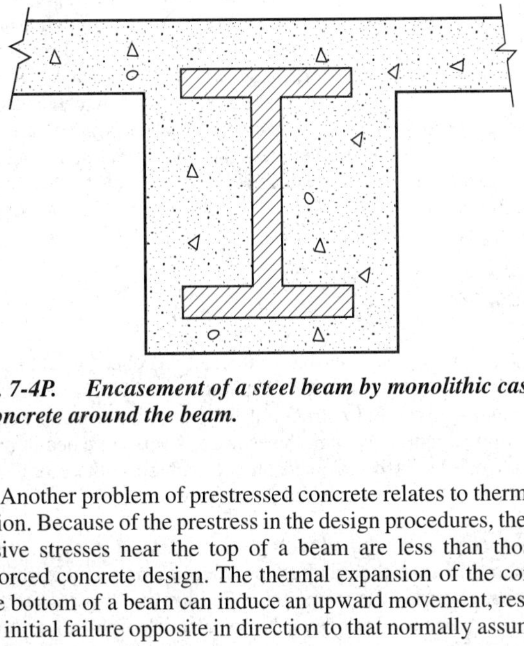

FIG. 7-4Q. *Spray-on fireproofing.*[12]

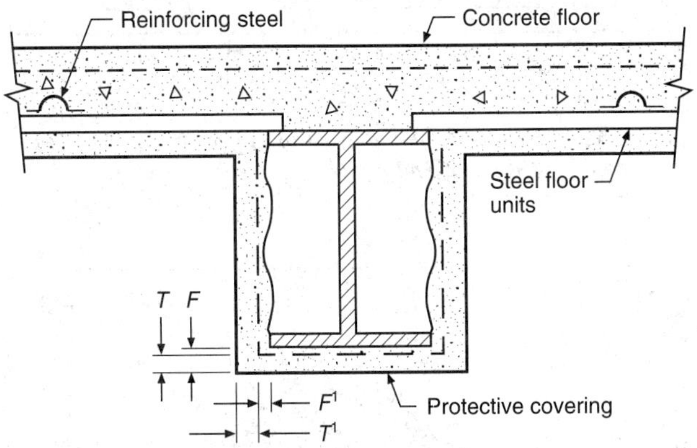

FIG. 7-4R. *Furred steel beams with noncombustible protective coverings.*

FIG. 7-4P. *Encasement of a steel beam by monolithic casting of concrete around the beam.*

Another problem of prestressed concrete relates to thermal expansion. Because of the prestress in the design procedures, the compressive stresses near the top of a beam are less than those for reinforced concrete design. The thermal expansion of the concrete at the bottom of a beam can induce an upward movement, resulting in an initial failure opposite in direction to that normally assumed in collapses.

The type of steel used for prestressing is more sensitive to elevated temperatures than the steel used in reinforced-concrete construction. Not only is its strength reduced at temperatures somewhat less than those for hot-rolled steel, but also that strength is not regained after cooling. Prestressing wires are permanently weakened when they reach a temperature of about 800°F (427°C).

Wood

Depending on its form, wood may or may not provide reasonable structural integrity in a fire. The important factors that influence fire endurance are the physical size and the moisture content of the members. Fire-retardant treatments may be used to delay ignition and retard combustion. This provides time for extinguishing procedures. However, all wood will burn.

The burning of wood produces a charcoal on the surface at an average rate of about $1/40$ in. per min (0.6 mm/min). This charcoal provides a protective coating that insulates the unburned wood and isolates it from the flame, thus retarding pyrolysis. Therefore, thicker members provide much more structural integrity over the period of fire exposure than do thin ones.

Heavy timber construction has proved to be an excellent form of construction. It maintains its integrity during a fire for a relatively long time, providing an opportunity for extinguishment. When the fire is extinguished relatively soon, much of the original strength of the members is retained and reconstruction is possible.

Glued laminated frames, arches, and beams have become increasingly popular. These members also provide reserve strength during a fire, and, if the layer is not deep, the char may be removed by sandblasting to restore the aesthetic appearance.

Wood-frame construction utilizes structural members considerably smaller than mill construction. The exposed area is greater, relative to mass, and the fire resistance considerably reduced. When this type of construction is exposed to a fire, it offers relatively little structural integrity. Therefore, protective surfaces, such as gypsum wallboard or plaster, are important to provide fire endurance.

FIRE BEHAVIOR OF OTHER BUILDING MATERIALS

Many materials other than steel, concrete, and wood are commonly used in modern building construction. They frequently make up a large volume and/or surface area of the structure. Nonbearing partitions, insulation, building services, and finish materials are all important parts of building construction. Some of the nonstructural, thermally inert materials are used as fireproofing. Others may contribute significant fuel to a potential fire.

Glass

Glass is utilized in three common ways in building construction. The most obvious is glazing for windows and doors. In this capacity

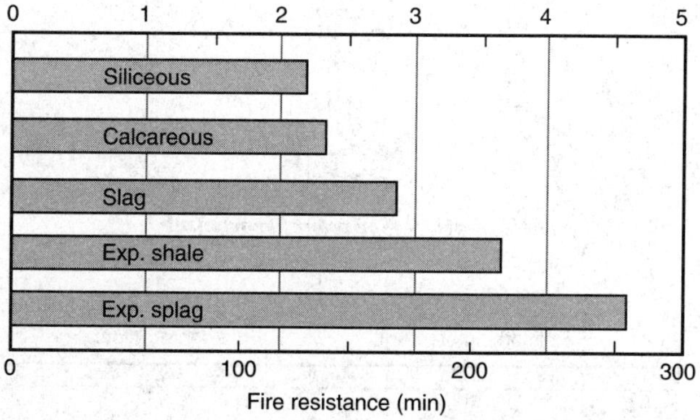

FIG. 7-4S.   *Effect of various types of aggregate on the fire endurance of 4 $^3/_4$-in. (121-mm) slabs.*[3]

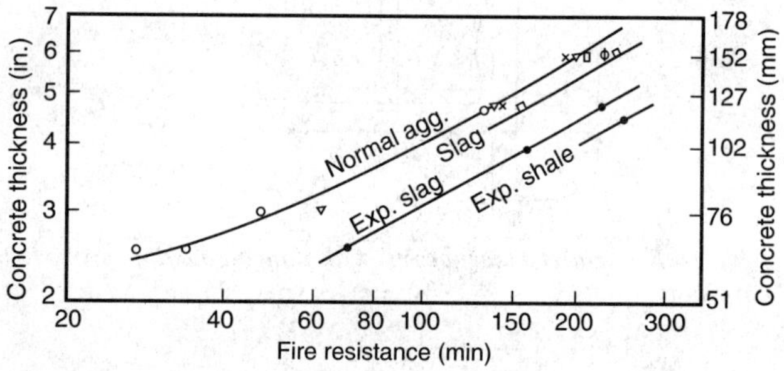

FIG. 7-4T.   *Relationship of slab thickness and type of aggregate to fire endurance.*[3]

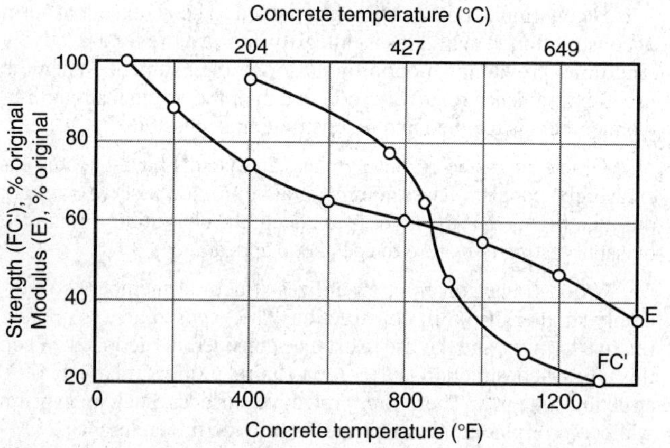

FIG. 7-4U.   *Effect of temperature on modulus of elasticity and compressive strength. The curves are taken from two differing specimens.*[3]

the glass has little resistance to fire. It quickly cracks because of the temperature difference between the surfaces. Double glazing does not provide much improvement. Wire-reinforced glass is an improvement, as it provides somewhat greater integrity in a fire if it is properly installed. However, no glazing should be relied upon to remain intact in a fire.

A second common use of glass in buildings is in fiberglass insulation. The fiberglass does not burn and is an excellent insulator. The fiberglass often is coated with a resin binder, which is combustible and can spread flames, although this occurs relatively slowly.

A third form in which glass is found in buildings is as reinforcement for fiberglass-reinforced plastic building products. The products include translucent window panels, siding, and prefabricated bathroom units. They have distinct advantages of economy and aesthetic appeal. The fiberglass acts as reinforcement for a thermo-setting resin, usually a polyester. The resin, combustible even with fire retardants incorporated in the composition, frequently comprises 50 percent or more of the material. While the fiberglass itself is noncombustible, the products are quite combustible.

Gypsum

Gypsum products, such as plaster and plaster board, are excellent fire protection materials. The gypsum has a high proportion of chemically combined water. Evaporation of this water requires a great deal of heat energy, making gypsum an excellent, inexpensive, fire-retardant building material.

Lightweight Concrete

Lightweight concrete, made with noncombustible aggregates, resists high temperatures extremely well without degradation. Vermiculite and perlite are the most common types of light concrete used for this purpose. Vermiculite is an inert aggregate made from weathered mica. The mica is crushed and roasted, which causes it to expand to form pellet-like aggregates. Perlite is a volcanic rock which is crushed and heat treated. The heat treatment causes the rock to expand in volume. During the process, water vapor is absorbed by the rock particles, and the perlite takes the form of glass-like, cellular-structured particles.

Asbestos

Asbestos is a mineral fiber that historically has been used in several forms in building construction. When it is used with a cementitious binder and sprayed onto structural members, it forms an excellent fireproofing agent. However, the asbestos fibers are a health hazard, and they can be easily scraped off the members; therefore, its use has been significantly curtailed in recent years.

Asbestos is combined with portland cement to make asbestos cement products. Although these products are noncombustible, they often shatter during the temperature buildup of a fire, thus reducing their effectiveness. However, asbestos also is combined with materials other than portland cement to form products such as asbestos insulation board and asbestos wood. These products behave quite well in fires, providing a great deal of fire resistance and protection.

Masonry

Brick, tile, and concrete masonry products behave well when subjected to the elevated temperatures of a fire. Hollow concrete blocks may crack from the heat, but they generally retain their integrity. Brick can withstand high temperatures without severe damage.

Plastics

There is a wide range of plastic products used by the building industry. They provide numerous aesthetic, physical, and economic advantages in building applications. Their major disadvantage is that all plastics are combustible. Although there are certain treatments that increase ignition temperatures or inhibit flame spread, there is no known additive that will make them noncombustible.

CONTEMPORARY TRENDS IN STRUCTURAL FIRE PROTECTION

There is an increased interest internationally in the development of more rational methods of structural fire design. These methods attempt to establish a theoretical base for the design of structural members and assemblies that will predict their actual fire performance more accurately than present methods. This section briefly addresses some of these modern trends.

Usual Structural Design Procedures for Fire Conditions

The standard fire-resistance test, combined with building code requirements that specify fire-endurance ratings for structural frames and barrier assemblies, is the usual basis for structural fire design throughout the world today. The building code identifies building-construction classifications based on materials and fire-endurance requirements for barriers and structural framing members. The code relates these construction classifications with use groups, the building height and area, and adjustments for fire-protection features. These code requirements for fire resistance are the subjective judgments of the code-writing groups.

The numbers and assemblies are laboratory tested in accordance with a standard fire test. This test is the equivalent to NFPA 251 for the jurisdiction served by the code. An acceptable design occurs when the test results exceed the code requirements.

Although this procedure is common throughout the industrial world today, its validity is subject to increasing doubts. The fire test itself is being severely criticized. In addition to the concerns identified earlier in this chapter, considerable variation can occur for the same assembly when it is tested in different laboratories. The relationship between test results and field performance is unknown. The code requirements have a weak technical base. The procedure, however, is relatively easy to administer from a regulatory viewpoint, and many other code requirements can be related to the form of construction anticipated by the fire-resistance ratings.

Analytical Methods of Design

Research has been conducted along two different but related aspects of structural fire design.[13] Both attempt to substitute rational analytical procedures for predicting structural fire behavior.

One procedure is to develop analytical methods by which the fire resistance of a member or assembly may be calculated. This method computes the fire resistance that one would expect from the standard fire test. This method uses structural characteristics of the members, thermal properties of the materials, heat-transfer characteristics of the assembly, mechanical properties of structural materials and compartment, and design service loads. This method enables the important considerations of structural restraint to be taken into account more easily than the standard fire test. This procedure is identical to the usual design procedure, except that calculated fire-resistance ratings are substituted for experimental ratings.

Another approach to which considerable research effort has been devoted is an analytical design based on real fire-exposure characteristics. There are basically two methods in this approach. The first relates the time-temperature characteristics of real fires to the structural behavior at elevated temperatures. The second method relates the properties of real fire development to an equivalent standard time-temperature curve. This method relates a theoretically equivalent time to experimental fire test results.

Design Procedure

The Swedish Institute of Steel Construction publishes a design manual, entitled "Fire Engineering Design of Steel Structures."[14] A building designer has the option of designing the structural system in accordance with the traditional code or by the procedures described in the manual.

The manual describes a rational fire-engineering design process for steel structures on the basis of performance requirements. The manual consists of four parts. The first part is a relatively detailed description of fire energy impacts and structural behavior at elevated temperatures. This part provides the foundation for the design procedures.

The second part presents the detailed design procedure. Design charts and curves which provide the numerical base are provided. Worked examples comprise the third part, and an alternate design method based on the concept of equivalent fire duration is given in the fourth.

The American Iron and Steel Institute publishes *Fire-Safe Structural Steel, A Design Guide*,[15] for exterior structural members. Exterior members, of course, cannot be tested in the standard laboratory test. The guide provides a design procedure by which interior members can be designed for fire conditions. The design guide presents step-by-step procedures to arrive at a design.

The basis for this design guide is *Design Guide for Fire Safety of Bare Exterior Structural Steel.*[16] Part One of the AISI guide presents the theory and validation of the method, while Part Two reviews the state of the art of relevant research programs conducted during the past 20 years.

The Portland Cement Association has published a procedure to analytically determine the fire resistance of prestressed concrete structures. This method is described in *PCI Design for Fire Resistance of Precast Prestressed Concrete.*[17] The method calculates fire resistance by obtaining temperatures at various locations for the various standard fire-test times, using tables and diagrams based on test results. This information is then used to compute a fire-resistance value.

BIBLIOGRAPHY

References Cited

1. Clay, W., "Standard Fire Tests," *Proceedings* of the 13th Annual Meeting of the Building Officials' Conference of America, Chicago, IL, 1927, pp. 74–88.
2. Shaub, H., "Early History of Fire Endurance Testing in the United States," *ASTM Special Technical Publication 301*, American Society for Testing and Materials, Philadelphia, PA, 1961, pp. 1–9.
3. Benjamin, I. A., "Fire Resistance of Reinforced Concrete," *Symposium on Fire Resistance of Concrete*, American Concrete Institute, Detroit, MI, 1961, pp. 29 and 31.
4. "Fire Tests of Columns Protected with Gypsum," NBS *Research Paper No. 563*, National Bureau of Standards, Washington, DC, 1961.
5. Lie, T. T., and Stanzak, W. W., "Fire Resistance of Protected Steel Columns," *Engineering Journal*, Vol. 10, No. 3, 1973.
6. *Designing Fire Protection for Steel Columns*, American Iron and Steel Institute, Washington, DC.
7. Harmathy, T. Z., "Thermal Performance of Concrete Masonry Walls and Fire," *Special Technical Publication 464*, American Society for Testing and Materials, Philadelphia, PA, 1970.
8. Allen, L. W., and Harmathy, T. Z., "Fire Endurance of Selected Concrete Masonry Units," *Journal of the ACI*, Vol. 69, 1972.
9. Schaffer, E. W., "State of Structural Timber Fire Endurance," *Wood and Fiber*, Vol. 9, No. 2, 1977.
10. Harmathy, T. Z., and Stanzak, T. T., "Elevated-Temperature Tensile and Creep Properties of Some Structural and Prestressing Steels," *ASTM*

Special Technical Publication 464, American Society for Testing and Materials, Philadelphia, PA, 1969, pp. 186–208.

11. DeFalco, F. D., "An Investigation of Modern Structural Steels at Fire Temperatures," Ph.D. Thesis, University of Connecticut, Storrs, CT, 1974.

12. Przetak, L., *Standard Details for Fire-Resistive Building Construction*, McGraw-Hill, New York, 1974.

13. Pettersson, O., "Structural Fire Protection," *Fire and Materials*, Vol. 4, No. 1, 1980.

14. Pettersson, O., and Magnusson, S. E., *et al.*, "Fire Engineering Design of Steel Structures," *Publication 50*, Swedish Institute of Steel Construction, Stockholm, Sweden, 1976.

15. *Firesafe Structural Steel, A Design Guide*, American Iron and Steel Institute, New York, 1979.

16. Law, M., *Design Guide for Fire Safety of Bare Exterior Structural Steel*, Ove Arup & Partners, London, England, 1977.

17. Gustaferro, A. H., and Martin, L. D., *PCI Design for Fire Resistance of Precast Prestressed Concrete*, Prestressed Concrete Institute, Chicago, IL, 1977.

Other Reference

Stanzak, W. W., and Lie, T. T., "Fire Resistance of Unprotected Steel Columns," *Journal of the Structural Division*, Vol. 99, No. ST5, 1973.

NFPA Codes, Standards, and Recommended Practices

Reference to the following NFPA codes, standards, and recommended practices will provide further information on structural integrity during fire discussed in this chapter. (See the latest version of *The NFPA Catalog* for availability of current editions of the following documents.)

NFPA 251, *Standard Methods of Tests of Fire Endurance of Building Construction and Materials*
NFPA 252, *Standard Methods of Fire Tests of Door Assemblies*
NFPA 257, *Standard for Fire Tests of Window Assemblies*

Additional Reading

Ahmed, G. N., Huang, C. L. D. and Fenton, D. L., "Modeling of Coated and Uncoated Concrete Walls Exposed to Fires," Kansas State Univ., Manhattan, American Institute of Aeronautics and Astronautics/American Society of Mechanical Engineers (AIAA/ASME), Heat and Mass Transfer in Fires, June 18-20, 1990, Seattle, WA, 1990, pp. 7–14.

Baker, D. J. and Xie, Y. M., "Elasto-Plastic-Creep Analysis of Restrained Steel Columns Exposed to Fire," *Proceedings* of the Fourth International Symposium on Fire Safety Science, Intl. Assoc. for Fire Safety Science, Boston, MA, 1994, pp. 1113–1124.

Barnett, J. R., "New Design Approach for Steel Structures Exposed to Fires," *Journal of Fire Protection Engineers*, Vol. 3, No. 1, 1991, pp. 1–8.

Beilicke, G., "Basic Thoughts on Structural Fire Protection of Protected Historic Buildings," *Fire Science and Technology*, Vol. 11, No. 1-2, 1991, pp. 57–59.

Brannigan, F. L., "Concrete, Fire Resistive or Non-Combustible," *Fire Fighting in Canada*, Vol. 35, No. 3, April 1991, p. 17.

Brannigan, F. L., "Safety at Building Structural Fires," *Fire Engineering*, Vol. 145, No. 2, Feb. 1992, pp. 14,16–17.

Brannigan, F. L., *Building Construction for the Fire Service*, 3rd ed., National Fire Protection Association, Quincy, MA, 1994.

Burgess, I. W., Olawale, A. O. and Plank, R. J., "Failure of Steel Columns in Fire," *Fire Safety Journal*, Vol. 18, No. 2, 1992, pp. 183–201.

Carling, O., "Fire Resistance of Joint Details in Loadbearing Timber Construction, A Liturature Survey," *Research Report No. 18*, Building Research Association of New Zealand, 1989.

Chandrasekaran, V. and Suresh, R., "Analysis of Steel columns Exposed to Fire," Third International Conference and Exhibition on Fire and Materials, October 27-28, 1994, Arlington, VA, 1994, pp. 41–50.

Chandrasekaran, V., Suresh, R. and Grubits, S., "Performance Mechanisms of Fire Walls at High Temperatures," Third International Conference and Exhibition on Fire and Materials, October 27-28, 1994, Arlington, VA, 1994, pp. 51–62.

Clifton, C., "HERA: Fire Resistance of Structural Steel," *BUILD*, February 1991, pp. 28–29.

Collier, P., "Design of Light Timber Framed Walls and Floors for Fire Resistance. Technical Recommendation," Building Research Association of New Zealand, Judgeford, BRANZ Technical Recomm. 9, Aug. 1991.

Collier, P., "Model for Predicting the Fire-Resisting Performance of Small-Scale Cavity Walls in Realistic Fires," *Fire Technology*, Vol. 32, No. 2, Second Quarter/April/May 1996, pp. 120–136.

Commission of the European Communities, "EUROCODE 2: Design of Concrete Structures. Part 1.2: Structural Fire Design," ISO/TC92/SC2 WG2 N 223, Oct. 1993.

Commission of the European Communities, "EUROCODE 3: Design of Steel Structures. Part 1.2: Structural Fire Design. Draft," ISO/TC92/SC2 WG2 N 224, CEN/TC250/SC3, prENV 1993-1-2, Aug. 1993.

Connolly, R. J., "Fire Safety Engineering and the Structural Eurocodes," *Fire Engineers Journal*, Vol. 56, No. 180, January 1996, pp. 27–30.

Connolly, R. J., "Implications of the Structural Eurocodes for Fire Safety Design in the UK," *Fire Engineers Journal*, Vol. 55, No. 9, Nov. 1995, pp. 10–12.

Cooke, G. M. E., "An Introduction to the Mechanical Properties of Structural Steel at Elevated Temperatures," *Fire Safety Journal*, Vol. 13, No. 1, 1988, pp. 45–54.

Cooke, G. M. E., "Can Integrity in Fire Resistance Tests Be Predicted?," Sixth International Fire Conference on Fire Safety, Interflam '93, March 30-April 1, 1993, Oxford, England, Interscience Communications Ltd., London, England, 1993, pp. 321–330.

Cooke, G. M. E., Virdi, K. S. and Jeyarupalingam, "Thermal Bowing of Brick Walls Exposed to Fire On One Side," Fire Research Station, Garston, England, City University, London, England, ISBN 0-9516320-9-4; *Proceedings* of the Seventh International Interflam Conference, Interflam '96, March 26-28, 1996, Cambridge, England, Interscience Communications Ltd., London, England, 1996, pp. 915–919.

Corlett, R. C. and Tuttle, M. E., "Testing Structural Composite Materials for Fire Performance," *Proceedings* of the Eighteenth International Conference on Fire Safety, Volume 18, January 11-15, 1993, Millbrae, CA, Product Safety Corp., Sunnyvale, CA, 1993, pp. 194-197.

Cutter, B. E., "Working Together to Solve the Wood Truss Problem," *Fire Journal*, Vol. 84, No. 1, January/February 1990, pp. 11, 70.

"Design of Loadbearing Light Steel Frame Walls for Fire Resistance," *Build*, Nov. 1995, pp. 28, 30.

Dickey, W. L., "Fire Endurance of Brick Walls and Veneer," *Building Standards*, Vol. 65, No. 2, March/April 1996, pp. 6–10.

Eisner, H., "Evaluating Structural Stability: Specialists Observe, Confer to Maintain Safety," *Firehouse*, Vol. 20, No. 10, Oct. 1995, pp. 67–69.

Fattal, S. G., "Strength of Partially-Grouted Masonry Shear Walls Under Lateral Loads," National Institute of Standards and Technology, Gaithersburg, MD, NISTIR 5147, June 1993.

Favro, P. C., "Fire-Retardant Treatments for Wood Roofs: A Permanent Solution," Fire Safety '91 Conference, Volume 4, November 4-7, 1991, St. Petersburg, FL, Am. Technologies Intl., Pinellas Park, FL, 1991, pp. 136–145.

Fields, B. A. and Fields, R. J., "Prediction of Elevated Temperature Deformation of Structural Steel Under Anisothermal Conditions," National Institute of Standards and Technology, Gaithersburg, MD, American Iron and Steel Institute, Washington, DC, NISTIR 4497, Jan. 1991.

Fitzgerald, R. W., "Selection of Fire Endurance Requirements for Building Design. [Integration of Fire Science Into Building Requirements.]," *Fire Safety Journal*, Vol. 17, No. 2, 1991, pp. 159–163.

Fleischmann, C., "Analytical Methods for Determining Fire Resistance of Concrete Members," *The SFPE Handbook of Fire Protection Engineering*, 2nd ed., DiNenno, P. J., ed., NFPA, Quincy, MA, 1995.

Franssen, J. M. and Dotreppe, J. C., "Fire Resistance of Columns in Steel Frames," *Fire Safety Journal*, Vol. 19, No. 2-3, 1992, pp. 159–175.

Fredlund, B., "Calculation of the Fire Resistance of Wood Based Boards and Wall Constructions," Lund Institute of Science and Technology, Sweden, LUTVDG/TVBB-3053, 1990.

Gamble, W. L., "Predicting Protected Steel Member Fire Endurance Using Spread-Sheet Programs," *Fire Technology*, Vol. 25, No. 3, 1989, pp. 256–273.

Gerlich, J. T., "Design of Loadbearing Light Steel Frame Walls for Fire Resistance. Fire Engineering Research Report," University of Canterbury, Christchurch, New Zealand, Fire Engineering Research 95/3, Aug. 1995.

Gerlich, J. T., Collier, P. C. R. and Buchanan, A. H., "Design of Light Steel-Framed Walls for Fire Resistance," *Fire and Materials*, Vol. 20, No. 2, March/April 1996, pp. 79–96.

Grant, C. C., "When the Walls Came Tumbling Down," *NFPA Journal*, Vol. 87, No. 5, Sept./Oct. 1993, pp. 94–101,103.

Harada, K. and Terai, T., "Fire Resistance of Concrete Walls," First Asian Conference on Fire Science and Technology (ACFST), October 9-13, 1992, Hefei, China, International Academic Publishers, China, 1992, pp. 215–220.

Harmathy, T. Z., Sultan, M. A., and MacLaurin, J. W., "Comparison of Severity of Exposure in ASTM E119 and ISO 834 Fire Resistance Tests," *Journal of Testing and Evaluation*, Vol. 15, No. 6, Nov. 1987, pp. 371–375.

Hasemi, Y. and Yoshida, M., "Wall Fire Modeling and Its Relevance to Building Fire Safety Design," Building Research Inst., Ibaraki, Japan, Science University of Tokyo, Japan, '93 Asian Fire Seminar, October 7-9, 1993, Tokyo, Japan, 1993, pp. 117–127.

Holmberg, S. and Anderberg, Y., "Computer Simulations and Design Method for Fire Exposed Concrete Structures," Swedish Fire Research Board, Stockholm, Sweden, Document 92050-1, FSD Project 92-50, Sept. 27, 1993.

Hosser, D., Dorn, T. and Richter, E., "Evaluation of Simplified Calculation Methods for Structural Fire Design," *Fire Safety Journal*, Vol. 22, No. 3, 1994, pp. 249–304

Howe, S., "Glass and Glazing Systems," *Fire Surveyor*, Vol. 18, No. 4, Aug. 1989, pp. 12–16.

Huttu, I., "Kantavien terasbetonirakenteiden palovauriot. [Fire Damage to Bearing Reinforced Concrete Structures.]," *Palontorjuntatekniikka*, March 1995, pp. 14–16.

"Importance of Structural Protection," *Fire*, Vol. 88, No. 1082, August 1995, p. 20.

Inha, T., "Terasrakentamisen palotekniset suojaustavat. [Fire Prevention in Steel Structures.]," *Palontorjuntateknikka*, January 1993, pp. 14–17.

Isa, D., "Fire Safety Engineering in the Structural Design in Buildings," Malaysia Fire Dept., Kuala, Lumpur, CIB W14/95/1 (J); Tokyo Fire Department, Firesafety Frontier '94, International Fire Conference and Exhibition in Tokyo, Creating a Safe Tomorrow, Oct. 18-22, 1994, Tokyo, Japan, 1994, pp. 39–45.

Jeanes, D. C., "Developing Design Concepts for Structural Fire Endurance Using Computer Models," *Proceedings* of the International Conference on the Design of Structures Against Fire, Elsevier, New York, 1986, pp. 127–153.

Jennings, P., "Testing for Integrity," *Fire Prevention*, No. 244, Nov. 1991, pp. 24–25.

Kirby, B. R., "Recent Developments and Application in Structural Fire Engineering Design," *Fire Safety Journal*, Vol. 11, No. 3, 1986, pp. 141–180.

Kodur, V. K. R. and Lie, T. T., "Fire Performance of Concrete-Filled Hollow Steel Columns," *Journal of Fire Protection Engineering*, Vol. 7, No. 3, 1995, pp. 89–98.

Kodur, V. K. R. and Lie, T. T., "Structural Fire Protection Through Concrete Filling for Hollow Steel Columns,"*Proceedings* of the International Conference on Fire Research and Engineering, Sept. 10-15, 1995, Orlando, FL, SFPE, Boston, MA, 1995, pp. 527–532.

Kruppa, J., *et. al.*, "EUROCODE 4: Design of Composite Steel and Concrete Structures. Part 1.2: Structural Fire Design. Draft," ISO/TC92/SC2 WG2 N 225, CEN/TC250/SC4 N56, prENV 1994-1-2, Nov. 22, 1993.

Kruppa, J., *et al.*, "Fire-Endurance Testing and Development in Fireproofing Technology," Chapter 12, Council on Tall Buildings and Urban Habitat, *Fire Safety in Tall Buildings. Tall Building Criteria and Loading*, Committee 8A, McGraw-Hill, Inc., Blue Ridge Summit, PA, 1992, pp. 201–227.

Lawson, M., "How Light Steel Used for Structural Purposes Can Be Made Fire Resistant," *Fire*, Vol. 86, No. 1064, Feb. 1994, pp. 27–28, 30.

Lawson, R. M., "Behavior of Steel Beam-to-Column Connections in Fire," *Structural Engineer*, Vol. 68, No. 14, July 17, 1990, pp. 263–271.

Lie, T. T. and Celikkol, B., "Method to Calculate the Fire Resistance of Circular Reinforced Concrete Columns," *ACI Materials Journal*, Vol. 88, No. 1, Jan./Feb. 1991, pp. 84–91.

Lie, T. T. and Chabot, M., "Concrete Filling: Fire Protection of Hollow Steel Columns," *Canadian Consulting Engineer*, May/June 1990, pp. 39–40.

Lie, T. T. and Chabot, M., "Experimental Studies on the Fire Resistance of Hollow Steel Columns Filled With Plain Concrete," National Research Council of Canada, Ottawa, Ontario, Internal Report 611, Jan. 1992.

Lie, T. T. and Chabot, M., "Method to Predict the Fire Resistance of Circular Concrete Filled Hollow Steel Columns," *Journal of Fire Protection Engineering*, Vol. 2, No. 4, 1990, pp. 111–126.

Lyon, D. E., *et al.*, "Evaluation of Strength Properties of Fire-Retarded Treated Wood Using the 1983 National Forest Products Association Protocol," *Forest Products Journal*, Vol. 38, No. 6, 1988, pp. 13–18.

MacIsaac, K. S., "Warning: Risk of Sudden Roof Collapse," *Fire Chief*, Vol. 39, No. 2, Feb. 1995, pp. 23–25.

Malanga, R., "Fire Endurance of Lightweight Wood Trusses in Building Construction," Technical Note, *Fire Technology*, Vol. 31, No. 1, First Quarter 1995, pp. 44–61.

Malhotra, H. L., "Specifying the Performance of Elements of Construction," FIREX, First International Conference on Fire Safety in Buildings--The Influence of the EEC Construction Products Directive, Conference Papers, June 12-14, 1990, Birmingham, England, 1990, pp. 1–9.

Manny, W. F., "Lightweight Wood Trusses: More to Consider," *Fire Engineering*, Vol. 144, No. 5, March 1991, pp. 62–63.

Melinek, S. J., "Prediction of the Fire Resistance of Insulated Steel," *Fire Safety Journal*, Vol. 14, No. 3, 1989, pp. 127–134.

Messersmith, J. J., "Fire Resistance Rating Requirements for Exterior Bearing Walls," *Building Official and Code Administrator*, Vol. 24, No. 3, May/June 1990, pp. 34–40.

"Methodologies and Available Tools for the Design/Analysis of Steel Components at Elevated Temperatures," Work Package No. FR2, Steel Construction Inst., UK, FR2, July 1991.

Meyer-Ottens, C., "On the Problem of Spalling of Concrete Structural Components Under Fire-Induced Stress," Foreign Technology Div., Wright-Patterson AFB, OH, FTD-ID(RS)T-0392-90, June 1990.

Milke, J. A., "Analytical Methods for Determining Fire Resistance of Steel Members," *The SFPE Handbook of Fire Protection Engineering*, 2nd ed., DiNenno, P. J., ed., NFPA, Quincy, MA, 1995.

Mittendorf, J. W. and Brannigan, F. L., "Timber Truss: Tow [sic] Points of View," *Fire Engineering*, Vol. 144, No. 5, March 1991, pp. 51-54,56,58–60.

Narayanan, P., "Fire Severities for Structural Fire Engineering Design," Building Research Association of New Zealand, Judgeford, Study Report 67, Dec. 1995.

O'Meagher, A. J. and Bennetts, I. D., "Modelling of Concrete Walls in Fire," *Fire Safety Journal*, Vol. 17, No. 4, 1991, pp. 315–335.

Pattersson, O., "Practical Need of Scientific Material Modes for Structural Fire Design General Review," *Fire Safety Journal*, Vol. 13, No. 1, 1988, pp. 1–8.

Pettersson, O., "Rational Structural Fire Engineering Design, Based on Simulated Real Fire Exposure," *Proceedings* of the Fourth International Symposium on Fire Safety Science, Intl. Assoc. for Fire Safety Science, Boston, MA, 1994, pp. 3–26.

Phan, L. T., Cheok, G. S. and Todd, D. R., "Strengthening Methodology for Lightly Reinforced Concrete Frames: Recommended Design Guidelines for Strengthening With Infill Walls," National Institute of Standards and Technology, Gaithersburg, MD, NISTIR 5682, July 1995.

Phan, L. T., Hendrickson, E. M. and Marshall, R. D., "Measurement of Structural Response Characteristics of Full-Scale Buildings: Analysical [sic] Modeling of the San Bruno Commercial Office Building," National Institute of Standards and Technology, Gaithersburg, MD, NISTIR 4782, March 1992.

Prendergast, E. J., "Fire Breaks Down Concrete," *Fire Engineering*, Vol. 142, No. 9, Sept. 1989, pp. 32–36.

"Proposal of Structural Design Method of Stainless Steel Buildings Structures," Building Research Inst., Tokyo, Japan, BRI Research Paper 141, September 1994.

Purkiss, J. A., "Computer Modelling of Concrete Structural Elements Exposed to Fire," Fifth International Conference on Fire Safety, Interflam '90, September 3-6, 1990, Canterbury, England, Interscience Communications, Ltd., London, England, 1990, pp. 67–75.

Purkiss, J. A., "Developments in the Fire Safety Design of Structural Steelwork," *Journal of Constructional Steel Research*, Vol. 11, No. 3, 1988, pp. 149–173.

Richardson, L. R. and Onysko, D. M., "Fire-Endurance Testing of Building Constructions Containing Wood Members," *Fire and Materials*, Vol. 19, No. 1, Jan./Feb. 1995, pp. 29–33.

Roitman, V. M., Demekhin, V. N. and Majeed, M. A. A., "Simplified Prediction Method of Real Fire Exposure as a Basis for an Analytical Structural Fire Design," *Proceedings* of the Third International Symposium on Fire Safety Science, Elsevier Applied Science, New York, 1991, pp. 751–759.

Rossiter, W. J., Jr., *et al.*, "Performance Approach to the Development of Criteria for Low-Sloped Roof Membranes," National Institute of Standards and Technology, Gaithersburg, MD, NISTIR 4638, July 1991.

Routley, J. G., "How Wood Trusses Perform During a Real Fire," *Fire Journal*, Vol. 83, No. 1, 1989, pp. 49–53.

Rudolph, K., *et al.*, "Principles for Calculation of Load-Bearing and Deformation Behaviour of Composite Structural Elements Under Fire Action," *Proceedings* of the First International Symposium on Fire Safety Science, Hemisphere, 1986, pp. 301–310.

Sakumoto, Y., *et. al.*, "Fire Resistance of Fire-Resistant Steel Columns," *Journal of Structural Engineering*, Vol. 120, No. 4, April 1994, pp. 1103–1121.

Sarvaranta, L., Elomaa, M. and Jarvela, "Study of Spalling Behavior of PAN Fiber-Reinforced Concrete by Thermal Analysis," *Fire and Materials*, Vol. 17, No. 5, Sept./Oct. 1993, pp. 225–230.

Scawthorn, C., "Structural Damage," *Fire Engineering*, Vol. 147, No. 8, August 1994, pp. 68, 71-72,74,76,78,80–82.

Schaffer, E. L., "Fire-Resistive Structural Design," *Proceedings* of the International Seminar on Wood Engineering, April 3, 6, 8, 10, 1992, Boston, MA, 1992, pp. 47–60.

Schaffer, E. L., "How Well Do Wood Trusses Really Perform During a Fire," *Fire Journal*, Vol. 88, No. 2, 1988, p. 57.

Schleich, J. B., and Mathieu, J., "Metallic Structure Subjected to Localized Fire," *Cahiers d'Informations Techniques-Revue de Metallurgie*, Vol. 86, No. 5, May 1989, pp. 429–438.

Schneider, U., "Concrete at High Temperatures—A General Review," *Fire Safety Journal*, Vol. 13, No. 1, 1988, pp. 55–68.

Sharples, J. R., Plank, R. J. and Nethercot, D. A., "Load-Temperature-Deformation Behaviour of Partially Protected Steel Columns in Fires," *Engineering Structures*, Vol. 16, No. 8, Nov. 1994, pp. 637–643.

Skowronski, W., "Load Capacity and Deflection of Fire-Resistant Steel Beams," *Fire Technology*, Vol. 26, No. 4, Nov. 1990, pp. 310–328.

Smith, C. I., "Structural Fire Engineering Design," *Structural Engineering*, Vol. 65A, No. 2, 1987, pp. 51–55.

Smith, F. P., "Concrete Spalling: Controlled Fire Tests and Review," *Fire and Arson Investigator*, Vol. 44, No. 1, Sept. 1993, pp. 43-47.

Sullivan, P. J. E. and Sharshar, R., "Performance of Concrete at Elevated Temperatures (As Measured by the Reduction in Compressive Strength)," *Fire Technology*, Vol. 28, No. 3, Aug. 1992, pp. 240–250.

Sullivan, P. J. E., Terro, M. J. and Morris, W. A., "Critical Review of Fire-Dedicated Thermal and Structural Computer Programs," *Journal of Applied Fire Science*, Vol. 3, No. 2, 1993/1994, pp. 113–135.

Talamona, D. R., Franssen, J. M. and Recho, N., "Buckling of Eccentrically Loaded Steel Columns Submitted to Fire," *Proceedings* of the International Conference on Fire Research and Engineering, Sept. 10-15, 1995, Orlando, FL, SFPE, Boston, MA, 1995, pp. 533–538.

Talamona, D., *et. al.*, "Factors Influencing the Behavior of Steel Columns Exposed to Fire," *Journal of Fire Protection Engineering*, Vol. 8, Nov. 1, 1996, pp. 31–43.

Tomann, J., "Lattianpaallysteiden palotestaus. [Fire Testing of Floorings.]," *Palontorjuntatekniikka*, Vol. 1, 1992, pp. 22–25.

Uesugi, H., "Load Bearing and Deformation Capacity of Fire Resistance Steel Tubular Columns at Elevated Temperature," *Proceedings* of the Fourth International Symposium on Fire Safety Science, Intl. Assoc. for Fire Safety Science, Boston, MA, 1994, pp. 1149–1158.

Uesugi, H., *et. al.*, "Computer Modeling of Fire Engineering Design for High Rise Steel Structure," '93 Asian Fire Seminar, Oct. 7-9, 1993, Tokyo, Japan, 1993, pp. 21–32.

Wade, C. A., "Performance of Concrete Floors Exposed to Real Fires," *Journal of Fire Protection Engineering*, Vol. 6, No. 3, 1994, pp. 113–124.

Wade, C. A., "Report on a Finite Element Program for Modelling the Structural Response of Steel Beams Exposed to Fire. Study Report," Building Research Association of New Zealand, Judgeford, Public Good Science Fund, New Zealand, Building Research Levy, New Zealand, BRANZ Study Report 55, Feb. 1994.

Wade, C., "Fire Performance of External Wall Claddings Under a Performance-Based Building Code," *Fire and Materials*, Vol. 19, No. 3, 1995, pp. 127–132.

Wade, C., "Method for Fire Engineering Design of Structural Concrete Beams and Floor Systems. Technical Recommendation," Building Research Association of New Zealand, Judgeford, BRANZ Technical Recomm. 8, Feb. 1991.

White, R. and Cramer, S., "Improving the Fire Endurance of Wood Truss Systems," *Proceedings* of the Pacific Timber Engineering Conference (PTEC) on Timber Shaping the Future, Fortitude Valley MAC, Queenland, Australia: Timber Research Development and Advisory Council, Vol. 1, July 11-15, 1994, Gold Coast, Australia, 1994, pp. 582–589.

White, R. H., "Analytical Methods for Determining Fire Resistance of Timber Members," *The SFPE Handbook of Fire Protection Engineering*, 2nd ed., DiNenno, P. J., ed., NFPA, Quincy, MA, 1995.

White, R. H., "Fire Resistance of Wood-Frame Wall Sections," Forest Service, Madison, WI, Sept. 1990.

CONFINEMENT OF FIRE IN BUILDINGS

Revised by John A. Campbell

People and property not directly exposed to a fire can be protected by confining heat and smoke to the area of origin until the fire either is extinguished or burns itself out. People can also be protected by delaying the spread of fire and smoke until the occupants can be relocated to a place of safety.

This chapter discusses barriers that are intended to limit the spread of fire. Smoke movement in buildings is discussed in Section 7, Chapter 6, "Smoke Movement in Buildings," and smoke and heat venting in Section 7, Chapter 7, "Venting Practices." The same barriers used to prevent the spread of fire will almost always be the bulwark of systems designed to prevent the spread of smoke. To be contained, a fire must be bounded by barriers that limit the transmission of heat and hot gases. Such barriers must maintain their continuity and stability under the thermal and physical forces of a fire and during the structural distortions and collapse that sometimes occur within the fire area.

The fire resistance required of a barrier depends upon its intended purpose and on the expected severity of the fire to which it may be exposed. In most cases, the temperature history of the fire provides an approximate but adequate description of fire severity. If the purpose of the barrier is to limit probable fire size independent of any fire suppression actions, the barrier must then withstand the maximum expected fire that would occur during a burnout of combustibles inside the fire zone. However, if the purpose of the barrier is to contain the fire until occupants are evacuated and the fire department gains access, the time necessary for these actions determines the time that the barrier must hold back the fire, unless the fire is expected largely to burn itself out first.

Properly designed and constructed barriers that satisfy fire-endurance test criteria are assigned a fire-resistance rating that is measured in hours or fractions thereof. Any breach or unprotected opening in a fire barrier will negate its fire-resistance rating and may indeed contribute to rapid fire spread through the barrier.

FIRE SEVERITY

The fire severity to which a barrier can be exposed is related to the intensity of a fully developed fire in the space adjacent to the fire barrier. For a room fire, a fully developed fire would be a post-flashover fire; however, in large open volumes, such as those found in many industrial plants, a fully developed fire may occur in one area without flashover of the entire volume. It is the fully developed fire that first imposes extensive thermal and physical stresses on fire barriers. The initial possibility of barrier penetration occurs prior to full development. A typical developing fire is characterized by a flame front moving over a surface and/or flames localized to a stationary source. Although such a fire does not massively attack fire barriers, it can spread through faults or openings in the barrier and cause local destruction of barrier fire resistance.

A growing fire may or may not continue to flashover or become a fully developed area wide fire. Unless it reaches these stages, it is unlikely to threaten fire-resistive barriers except through unprotected openings or serious defects in the barrier system.

Flashover of an enclosure is likely to occur if the temperature of the upper gas layer reaches approximately 1100°F (600°C). Full-scale fire experiments and energy balance analysis of room fires have shown that the temperature of this upper gas layer depends upon the heat released by the burning fire, ventilation of the enclosure, dimensions of the enclosure, and the type of material lining the enclosure.[1] These factors are also interdependent with each other to some degree. The larger the enclosure, the higher the heat release necessary to produce the upper gas temperature needed for flashover. The more heat lost through walls, doors, and windows, the larger the fire and/or longer the time needed to produce flashover. Conversely, the greater the combustibility of the enclosure lining, the smaller the fire needed to produce flashover. The combustibility of the lining, or interior finish, is particularly significant. Combustible interior finish in an enclosure can greatly reduce the rate of heat release necessary for flashover as well as reduce the preburn time before flashover. A practical difficulty in estimating the likelihood of room flashover is in determining a value for the heat-release rate possible from the room contents. Typical heat-release rates from some common furnishings have been summarized from full-scale experimental data.[2] These data can be used in calculating the likelihood of flashover in a particular room.

The intensity and duration of a fully developed fire depend upon the quantity of combustibles available, their burning rates, and the air available for combustion. Fire intensity will also be somewhat lower when walls and ceilings absorb significant amounts of energy rather than act primarily as insulation or radiation barriers.

The burning rate of a fully developed fire involving a specific material or grouping of similar materials is determined either by: (1) the fuel surface area available to participate in the combustion reaction or (2) the amount of oxygen available for combustion. These are referred to as fuel-surface-controlled and ventilation-controlled

---

John A. Campbell, P.E., is president of Triodyne Fire and Explosion Engineers, Inc., Oakbrook, IL.

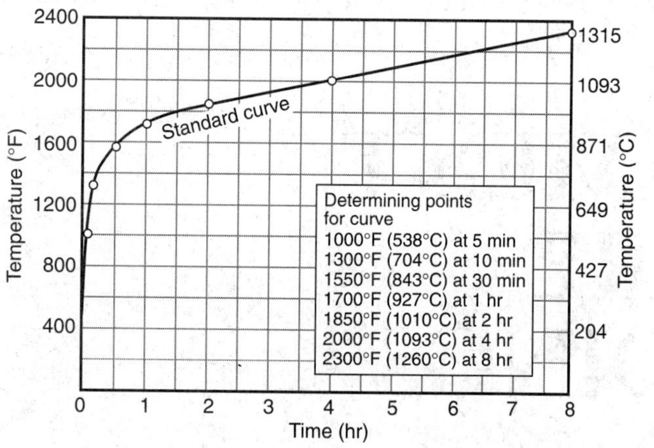

FIG. 7-5A. *The standard time-temperature curve.*

Determining points for curve
1000°F (538°C) at 5 min
1300°F (704°C) at 10 min
1550°F (843°C) at 30 min
1700°F (927°C) at 1 hr
1850°F (1010°C) at 2 hr
2000°F (1093°C) at 4 hr
2300°F (1260°C) at 8 hr

combustion, respectively. The maximum rate of heat transfer into fire separation barriers occurs at the point where the ventilation is just sufficient so that combustion is controlled at the fuel surface.[3] At higher ventilation rates, more heat is removed from the fire by the excess air. At lower ventilation rates, the combustion heat-release rate is less, and more unburned pyrolysis products and fuel particles are vented outside the fire area. Burning of unburned pyrolysis products outside fire compartment windows can increase the threat of floor-to-floor and building-to-building fire spread.

The possibility of failure of fire separation barriers can exist long after the fully developed fire begins to decay. However, in many real fire situations, this threat is mitigated by fire suppression activities. The decay in air temperature in a fire room has been reported as 27 to 36°F (15 to 20°C) per minute after fully developed fires of 10- to 15-min duration. Other data indicate a decay rate of 18°F (10°C) per minute for longer duration fires.[4] Cooling can be even slower in large debris piles. Fire-endurance testing in the United States does not simulate a fire decay period, although the decay period is simulated by some European testing and has been proposed in U.S. testing.

Poorly ventilated fires that are encountered in spaces such as basements, ship holds, or enclosed interior rooms often produce sufficient heat over a long period of time to penetrate separation barriers. These fires typically start with flaming combustion and, as the air in the space is consumed, revert to a state of mixed smoldering and glowing combustion with isolated or intermittent flaming. However, at present, there is no adequate experimental data or theoretical approach capable of reliably estimating the effect of these fires on fire-resistive barriers.

Standard Time-Temperature Curve

Fire-resistive barriers are evaluated in a testing furnace by exposure to a fire whose severity follows a time-varying temperature curve known as the standard time-temperature curve. The specified time-temperature history is tabulated in NFPA 251, *Standard Methods of Tests of Fire Endurance of Building Construction and Materials,* and illustrated in Figure 7-5A. The standard time-temperature curve was adopted by the American Society for Testing and Materials (ASTM) in 1918 and has been the basis of almost all fire-resistive testing ever since.

Following adoption of the curve, the National Institute of Standards and Technology (NIST), formerly the National Bureau of Standards (NBS), conducted a number of full-scale fire tests to determine how actual building fires compared with the temperatures represented on the curve.[5,6] The tests included two actual buildings

that were allowed to burn to destruction and a series of fires in fire-resistive test buildings containing contents representative of office, record room, and household occupancies.

The principal variable considered in these occupancy fire tests was the amount of combustible materials present, which is defined as the fire load. Although the ventilation in the test buildings was not reported, the windows were equipped with steel shutters that could be adjusted to control ventilation and maximize fire severity. The quantitative importance of ventilation on fire severity was not identified until more than twenty-five years after these tests. These tests conducted by NIST provided quantitative data on the temperature history of fires that were representative of various occupancies and fire load at that period of time. Fire load was expressed as the weight of ordinary combustibles in the room divided by the floor area of the room. Loading is the average amount of ordinary combustible material per square foot (m²) of floor area. The temperature history of the fully developed fires in three test occupancies was approximately bounded by the standard time-temperature curve.

NIST developed the concept of equivalent fire severity to define the severity of actual fires that had various temperature histories. This concept states that the area above a baseline under the time-temperature curve of a test fire, which is expressed in degree hours, is an approximate representation of the severity of a fire involving ordinary combustibles. The baseline used represents the temperature the materials can be exposed to without impairing their fire-resistive capabilities. Two fires with differing temperature histories are considered to have equivalent severity when the areas under their time-temperature curves are similar. This concept permitted comparison of any fire test data to the standard time-temperature curve by relating the area under the test curve to the area under the standard curve.

FIRE LOAD

The original concepts of fire severity and fire load are very important even though they are technically obsolete. These concepts are the basis for many of the fire-resistance requirements of building codes and for government agencies. In many cases, this original fire severity/fire load relationship was more severe than is indicated by more accurate analysis. Such results are conservative since the resultant error is on the safe side.

Analysis of NIST tests developed an approximate relationship between fire loading and an exposure to a fire severity equivalent to the standard time-temperature curve. The weight per square foot (m²) of ordinary combustibles [wood, paper, and similar materials with a heat of combustion of 7,000 to 8,000 Btu per lb (16,282 to 18,608 J/kg)] was related to hourly fire severity, as described in Table 7-5A.

The fire severity/fire load relationship was the first method developed to predict the severity of a fire that would be anticipated in various occupancies. It was used to determine resistance required of fire barriers as well as structural components. Although the technique has its limitations, the fire severity/fire load relationship still provides an approximate but conservative estimate of the probable maximum fire severity with combustibles having a high heat-release rate and when fire conditions can produce temperatures significantly higher or lower than the standard time-temperature curve.

Fire load is a measure of the maximum heat that would be released if all the combustibles in a given fire area burned. Maximum heat release is the product of the weight of each combustible multiplied by its heat of combustion. In a typical building, the fire load includes combustible contents, interior finish, floor finish, and structural elements. Fire load is commonly expressed in terms of the average fire load, which is the equivalent combustible weight divided by the fire area in square feet (m).

TABLE 7-5A. *Estimated Fire Severity for Offices and Light Commercial Occupancies*

(Data applying to fire-resistive buildings with combustible furniture and shelving)

Combustible Content Total, Including Finish, Floor, and Trim (lb/sq ft)	Heat Potential Assumed* Btu per sq ft	Equivalent Fire Severity Approximately Equivalent to That of Test under Standard Curve for the Following Periods
5	40,000	30 min
10	80,000	1 hr
15	120,000	1 1/2 hr
20	160,000	2 hr
30	240,000	3 hr
40	320,000	4 1/2 hr
50	380,000	7 hr
60	432,000	8 hr
70	500,000	9 hr

*Heat of combustion of contents taken at 8,000 Btu per lb up to 40 lb/sq ft; 7,600 Btu per lb for 50 lb; and 7,200 Btu for 60 lb and more to allow for relatively greater proportion of paper. The weights contemplated by the table are those of ordinary combustible materials, such as wood, paper, or textiles.
For SI units: 1 lb/sq ft = 4.9 kg/m²; 1 Btu/sq ft = 1.14 J/m².

Equivalent combustible weight is defined as the weight of ordinary combustibles having a heat of combustion of 8,000 Btu per lb (18,608 J/kg) that would release the same total heat as the combustibles in the space. For example, the equivalent weight of 10 lb per sq ft (48.8 kg/m²) of a plastic with a heat of combustion of 12,000 Btu per lb (27 912 J/kg) would be:

10 lb per sq ft × 12,000 Btu per lb = 120,000 Btu per sq ft

120,000 Btu per sq ft ÷ 8,000 Btu per lb ordinary combustibles = 15 lb per sq ft

Technically accurate methods for calculating the actual fire severity and fire-resistance requirements are available for many common building-occupancy-contents combinations. Technical limitations are primarily related to availability of input data rather than the analytical tools. These methods have been accepted to varying degrees under performance-based building codes adopted in some countries. Such analytical approaches have not been widely or routinely accepted by code authorities in the United States.

Occupancy Fire Load

A number of surveys have identified the fire loads found in various occupancies.[5,7,8] (See Tables 7-5B and 7-5C.)

FIG. 7-5B. Expected distributions of sample fire loads.

Data from some fire-load surveys as well as the inherent nature of combustible contents likely to be encountered suggest that the dispersion of fire load within a certain class of rooms can be approximated by either a normal or moderately skewed frequency distribution curve. (See Figure 7-5B.) The standard deviation, included in Tables 7-5B and 7-5C, can be used to determine the probability that a particular fire-load value will not be exceeded in a class of rooms. A fire load that is one standard deviation above the mean value of a normal distribution curve would represent an upper boundary for 84.13 percent of the fire loads in rooms of that class. Two standard deviations above the mean would bound 97.73 percent of the fire loads in that class of rooms, and three standard deviations, 99.86 percent of the fire loads. Thus, if a fire barrier were to be designed on the basis of two standard deviations above the mean, there would be a 97.73 percent probability that this fire load would not be exceeded in a similar room.

The above percentages are exact only if the distribution of fire loads is perfectly normal. If the distribution is more accurately defined by a moderately skewed curve, the percentages only represent close approximations.

Derated Fire Loads

Ordinary combustibles that are completely or largely enclosed in steel containers will not burn completely during a room fire and therefore will not contribute a full 8,000 Btu per lb (18,608 J/kg) to the fire load. The General Services Administration (GSA) has developed guidelines for determining a derated fire load for office buildings, which can be applied to other occupancies having similar classes of combustibles.[9] The total contents fire load is divided into three categories: (1) weight of materials completely enclosed

TABLE 7-5B. *Characteristic of Fire Loads in Office Buildings*

Room Use	Government Buildings			Private Buildings		
	No. of Rooms Sampled	Total Fire Load (lb/sq ft)		No. of Rooms Sampled	Total Fire Load (lb/sq ft)	
		Mean	Std. Dev.		Mean	Std. Dev.
General	342	7.3	4.4	479	7.7	4.3
Clerical	77	5.8	5.2	146	6.8	4.0
Lobby	15	2.6	1.4	45	5.0	4.2
Conference	39	4.2	6.1	57	5.9	4.6
File	10	17.9	11.9	20	16.2	12.9
Storage	35	11.7	19.2	77	13.2	11.7
Library	2	30.2	7.8	10	23.6	10.8

NOTES: Fire in steel enclosures. Weight of combustibles was converted to an equivalent weight of combustibles having a heat of combustion of 8,000 Btu/lb.
For SI units: 1 lb/sq ft = 4.88 kg/m².

TABLE 7-5C. *Samples of Typical Fire Loads*

Type of Room	Contents Fire Load (lb/sq ft)	Standard Deviation (lb/sq ft)
Living room	3.9	1.13
Family room	2.7	0.65
Bedroom	4.3	1.15
Dining room	3.6	1.02
Kitchen	3.2	0.77
Hospital patient room	1.2	0.36
Nursing home patient room	2.6	0.62

For SI units: 1 lb/sq ft = 4.68 kg/m^2.

in containers, such as steel desks or file cabinets, W_E; (2) weight of materials enclosed on five sides, such as in a steel bookcase, W_{PE}; and (3) weight of free combustibles, W_{FE}.

Completely enclosed combustibles will be heated by the room fire and pyrolyze; and the escaping pyrolysis products will burn. Therefore, the heat that the enclosed combustibles release depends upon the extent to which they are *pyrolyzed*. This is related to the intensity and duration of the room fire, which in turn is related to the total combustibles in the room. The extent to which the enclosed contents are derated is determined by considering the ratio of the total weight of enclosed combustibles, W_E, to the total weight of all combustibles in the room, F_T. Thus:

$$F_T = W_E + W_{PE} + W_F$$

A derating factor is assigned to W_E, as tabulated below:

Ratio W_E/F_T	Derating Factor, K
Under 0.5	0.4
0.5 to 0.8	0.2
Over 0.8	0.1

Ordinary combustibles enclosed on five sides by steel, such as in a bookcase, are derated to 75 percent of their weight.

The total derated fire load is given by:

$$F_{DR} = K \times W_E + 0.75 W_{PE} + W_F$$

The specific fire load is then computed by dividing by the floor area.

This derating procedure can be applied if other than ordinary combustibles are included in the fire load. The total weight of other combustibles must be expressed in terms of an equivalent weight of ordinary combustibles that would release the same total heat in burning. This is accomplished by multiplying the weight of the other combustibles by the ratio of their heat of combustion in Btu per lb, to 8,000 Btu per lb (18,608 J/kg), the heat of combustion of ordinary combustibles.

British Fire-Loading Studies

The British also have graded building occupancies according to hazard. Three classifications—low, moderate, and high fire loads—were defined in terms of Btu per sq ft (J/m^2), as follows.

Occupancies of low fire load: The fire load of an occupancy is described as low if it does not exceed an average of 100,000 Btu per sq ft (114,000 J/m^2) of net floor area of any compartment, nor an average of 200,000 Btu per sq ft (228,000 J/m^2) in limited isolated areas, provided that storage of combustible material necessary to the

occupancy may be allowed to a limited extent if separated from the remainder and enclosed by appropriate-grade fire-resistive construction. Examples of occupancies of normal low fire load are retail shops, factories, and workshops.

Occupancies of moderate fire load: The fire load of an occupancy is described as moderate if it exceeds an average of 100,000 Btu per sq ft (114,000 J/m^2) of net floor area of any compartment but does not exceed an average of 200,000 Btu per sq ft (228,000 J/m^2), nor an average of 400,000 Btu per sq ft (456,000 J/m^2) in limited isolated areas, provided that storage of combustible material necessary to the occupancy may be allowed to a limited extent if separated from the remainder and enclosed by fire-resistive construction of an appropriate grade. Examples of occupancies of normal moderate fire load are retail shops, factories, and workshops.

Occupancies of high fire load: The fire load of an occupancy is described as high if it exceeds an average of 200,000 Btu per sq ft (228,000 J/m^2) of net floor area but does not exceed an average of 400,000 Btu per sq ft (456,000 J/m^2) of net floor area, nor an average of 800,000 Btu per sq ft (912,000 J/m^2) in limited isolated areas. Examples of occupancies with normal high fire load are warehouses and other buildings used for the bulk storage of commodities of a recognized nonhazardous nature.

EFFECT OF VENTILATION

The burning rate of fully developed fires may depend upon the available fuel surface area or on the air available for combustion. When ample air is available, the burning rate of a fire depends upon the exposed surface area and on the properties of the combustible itself. Fully developed fires involving ordinary residential furnishings burn at a rate, R, that can be approximated by the equation:

$$R = 0.09 A_f \text{ (lb per min)}$$

where A_f = the exposed fuel surface in sq ft (m^2).[10]

For the general case, the equation becomes:

$$R = W_f A_f$$

where W_f = surface rate of burning of the fuel, lb per sq ft (kg/m^2) of exposed surface per minute.

Typical furniture found in homes and offices has a surface-area-to-weight ratio between 0.55 and 0.90 lb per sq ft (0.113 and 0.185 kg/m^2).[11] Loose combustibles, cellular material, hanging clothes, and similar materials have much higher surface-to-weight ratios and are associated with more rapidly developing fires.

When a fire cannot get sufficient air to maintain the burning rate associated with fuel-surface-controlled combustion, it will burn at a ventilation-controlled rate. The burning rate for fully developed fires in ordinary combustibles to which air is supplied through broken windows or a doorway is approximated by:

$$R = 0.62 A_o H_o^{1/2} \text{ (lb per min)}$$

where A_o = the opening area in sq ft (m^2), and H_o = the height of the opening in ft (m).[12]

Airflow into the fire, through a single opening, has been found to be approximately equal to:

$$\dot{m} = 5.06 \, C_D \, A_o H_o^{1/2} \text{(lb per min)}$$

where C_D = the orifice coefficient, normally about 0.5 to 0.7.[13]

Where there are multiple openings or a window wall, the flow is less than predicted by the area, and a C_D of about 0.35 is recommended.

The maximum fire intensity, from the standpoint of rate of heat input into barriers, occurs at a ventilation rate just sufficient to sustain a surface burning fire. This can be expressed as:

$$W_f A_f = 5.06 C_D W_a A_o H_o^{1/2}$$

where W_a = the weight of fuel each lb (kg) of air will burn completely. Typical values of W_a are presented in Table 7-5D.[13]

TABLE 7-5D. *Fuel Consumed by One Pound of Air at a Stoichiometric Fuel/Air Ratio*

Combustible	lb fuel/lb air	g fuel/kg air
Wood	0.175	175
Polyethylene	0.069	69
Polystryene	0.076	76
Polyurethane	0.135	135
Benzene	0.076	76
Methane	0.060	60

The preceding relationships assume that wind is not a factor in ventilation. In addition, if the ventilating opening is an interior door, such as from a corridor, adequate combustion air must be available for the fire.

Considerable ventilating area is required for a fully developed fire to burn at a fuel-surface-controlled rate. For example, over one-fourth of the wall area would have to be open in a 20- × 20-ft (6.1- × 6.1-m) room with an 8-ft (2.4-m) ceiling and an exposed combustible surface of 800 sq ft (74.3 m²) of ordinary combustibles. Most building fires will be ventilation controlled during the period when containment is critical from the initial well-developed phase until the start of decay. A fully developed fire can change from ventilation-controlled to fuel-surface-controlled as ventilation changes (such as when windows break out). Fires in modern buildings that have large glass windows may burn at ventilation-controlled rates.

FIRE MODELING

A number of mathematical models have been developed to calculate the temperature history of fully developed room fires as a function of ventilation, fuel, and the walls and ceiling of the room.[3,4,11,13-18] Fire models are increasingly being used in design as well as the evaluation of hazards and reconstruction of incidents. An extensive and expanding data base is available that provides heat release rate histories for combustible contents and finishes as well as the physical and thermal properties of construction materials. The most widely used models in the United States are classified as zone models and exemplified by the CFAST component of HAZARD.[16] Zone models are available with varying degrees of complexity and can be run on modern personal computers. A typical assumption of zone models is the filling of the fire compartment from the top down with hot gases of a uniform temperature. Field models do not make such assumptions but use computational fluid dynamics; this requires numerical solution of the differential equations describing the principles of conservation of mass, energy, momentum, and species. Field models are considerably more complex and expensive to run than zone models and are not practical to run on even state-of-the-art personal computers.

Calculations also can be made that yield the temperature rise on the unexposed wall and ceiling surfaces. Such calculations are meaningful only on the types of barriers that have extensive fire test and actual fire experience, because temperature rise is only one of the ways a fire barrier can fail. Mathematically modeling other failures can be difficult with many common constructions. In determining fire severity, this technique represents an improvement over the single-parameter fire load concept; however, it also is an approximation and has definite limitations.

All fire models have limitations and require some assumptions just as do all other engineering models used to design buildings and building systems. For example, the assumption that the temperature is uniform in the fire room is associated with the less complex post-flashover fire models. The turbulence associated with a post-flashover fire is assumed to mix the hot gases so that no appreciable temperature gradients exist. In fire models, this assumption is used to calculate mass airflow into the room, heat transfer from the room to walls and ceiling, and pyrolysis of the fuel. This assumption is an approximation, because temperature differences of between 250 and 500°F (139 and 278°C) have been measured within a room during post-flashover fires; larger differences could be encountered in larger compartments.[19-21]

The maximum intensity of post-flashover room fires occurs when the ventilation is just sufficient to permit fuel-surface-controlled combustion.[3] At either higher or lower ventilation levels, the maximum fire temperature is lower.[4] At higher ventilation levels, the fire duration will be shorter; at lower levels, it will be longer. The ventilation necessary to sustain fuel-surface-limited combustion will depend upon the combustibles in the room, their exposed surface area, surface burning rate, and the amount of air each lb (kg) of fuel requires for combustion. More ventilation would be needed to sustain fuel-surface-controlled combustion if a room contains blocks of foam rubber than if it contains an equivalent fire load of ordinary furnishings. The effect ventilation area has on the temperature of a fully developed room fire is qualitatively shown in Figure 7-5C.

The majority of experimental fires, which form the basis for the previously described relationships, have emphasized one-room fires, fuel consisting of either ordinary household and business furnishings or wood cribs, fuel loads under 10 lb per sq ft (48.8 kg/m²), and at least a level of ventilation provided by an open door or large window representing 10 to 15 percent of the room area. The results of these fires have shown the following:

1. Fire room temperatures may rise much faster than the standard time-temperature curve, remain above the curve for a period of time, and then drop below it.
2. Within a normal ventilation range of about 10 to 15 percent of room area on the low end and 25 to 30 percent on the high end, the maximum fire temperature depends primarily upon fire load, which, in turn, is related to the exposed fuel surface area.
3. Maximum fire temperatures involving ordinary furnishings generally range between 1,400 and 1,800°F (760 to 982°C). However,

FIG. 7-5C. Typical fully developed fire temperature-ventilation curve.

ordinary combustible materials with a very high exposed fuel surface area of 8 sq ft per sq ft (0.74 m²/m²) of floor area can result in maximum fire room temperatures over 2,000°F (1,093°C).

Temperature histories of some typical small room fires in which ventilation was provided by an open corridor door are illustrated in Figure 7-5D. Higher temperatures associated with the large fuel surface areas of wood cribs are shown in Figure 7-5E; the fuel surface area was about 1,000 sq ft (93 m²) for the 10 lb per sq ft (48.8 kg/m²) fire loading.

FIRE SPREAD

Fire spread rarely occurs by heat transfer through, or structural failure of, walls and floor-ceiling assemblies. The common mode of fire spread in a compartmented building is through open doors, unenclosed stairways and shafts, unprotected penetrations of fire barriers, and non-fire-stopped combustible concealed spaces. Even in buildings of combustible construction, the common gypsumboard or lath-on-plaster protecting wood stud walls or wood joist floors provides 25 to 30 min of resistance to a fully developed fire, as determined by a standard fire test.[23] When such barriers are properly constructed and maintained and have protected openings, they will normally contain fires of maximum expected severity in light-hazard occupancies. However, no fire barrier will reliably protect against fire spread if it is not properly constructed and maintained and openings in the barrier are not protected.

Although typical wall and floor-ceiling constructions generally have adequate inherent fire resistance in light-hazard occupancies, there are practical advantages in requiring rated fire-resistive assemblies in new construction. Materials and construction methods required for fire-rated assemblies provide quality controls not found in typical construction.

Fire can spread horizontally and vertically beyond the room or area of origin and through compartments or spaces that do not contain combustibles. Heated unburned pyrolysis products from the fire will mix with fresh air and burn as they flow outward. This results in extended flame movement under noncombustible ceilings, up exterior walls, and through noncombustible vertical openings. This is a common way fire spreads down corridors and up open stairways and shafts. Similar phenomena have been observed in flame spread underneath the ceiling in industrial and storage buildings. Combus-

tible interior finish in corridors and vertical openings, which by itself may be incapable of propagating flames, will be heated and may produce pyrolysis products. These products add to those from the main fire and increase the intensity and length of flames.

Once a room fire becomes fully developed, fire can spread very quickly to adjacent rooms in the absence of any fire separation barriers. When a building contains similar size rooms, the total volume of the rooms involved in fully developed fires tends to increase at an exponential rate.[24] This is illustrated in Figure 7-5F, which shows the form of percentage increase in total fire volume, based on the initial room volume as a function of time after the initial room flashed over.

Floor-to-floor fire spread along the outside of a building is not a common occurrence, but it can and does occur, such as in the Hilton Hotel Fire in Las Vegas, NV, in 1980. Glass panes normally break when heated to 550 to 600°F (288 to 316°C), although some breakage may occur at lower temperatures.[25] The hazard is greater in high-rise buildings because of fire-fighting difficulties and because there are more floors to which fire can spread. Wide spandrels and projections, such as balconies, will reduce the risk of, although not necessarily prevent, external fire spread.

Whether a barrier will or will not contain a fire can be considered as a probability and reliability problem that parallels reliability problems encountered in the operation of electronic, electrical, and mechanical equipment. Using this analogy, barrier failures can be considered to fall into one of three categories: (1) early failures, (2) random failures, and (3) degradation failures. Early failures would be the result of occurrences such as a door not closing or an opening not being protected. Random failures would be the result of faults in materials or construction, a fire of unexpected severity, or design weaknesses not identified by standard testing methods. Random failure probabilities typically follow an exponential law and would predominate until barrier thermal degradation became a factor. Degradation failures would be similar to wearout failures encountered in electronic, electrical, and mechanical equipment. The probability of

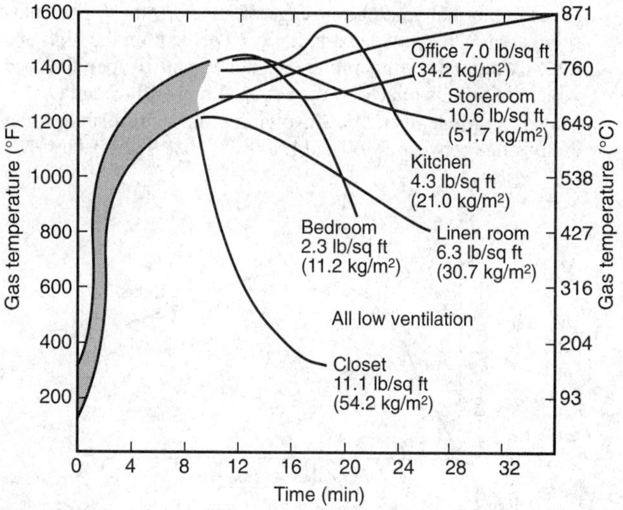

FIG. 7-5D. Gas temperature at ceiling of burning room.[21]

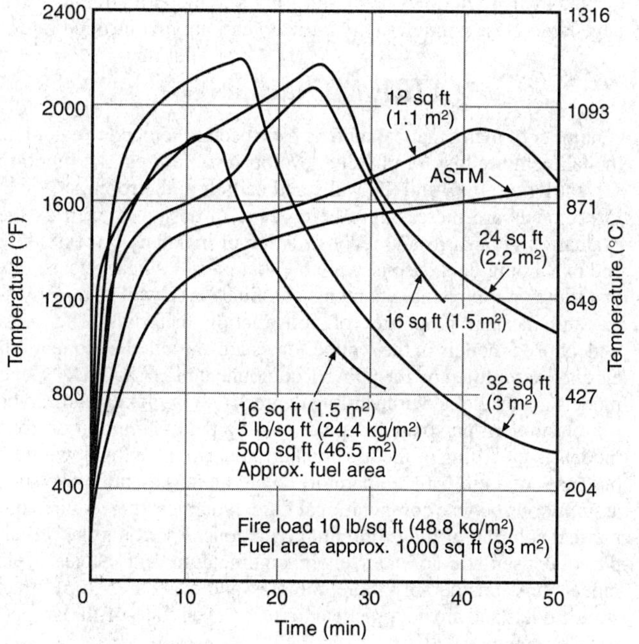

FIG. 7-5E. Effect of window area on average fire room temperature.[22]

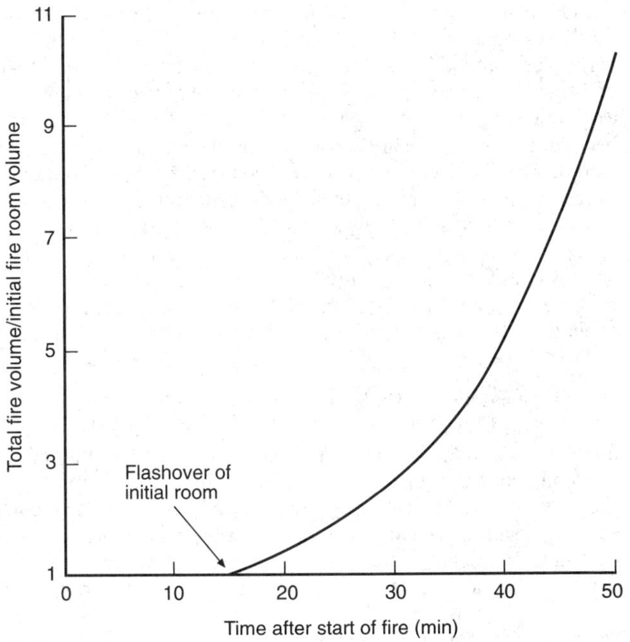

FIG. 7-5F. Exponential fire volume increase for a room flash-over time of 15 min.

a wearout failure typically follows a normal or Gaussian distribution law, and the probability of barrier degradation failures should be similarly distributed. This technique of considering the three categories of barrier failure probability permits a more realistic comparison to be made between compartmentation and other protective measures. It is also apparent from this analogy that a barrier having a certain fire-resistive rating is not necessarily twice as reliable as one having one-half that rating.

Concealed Spaces

Buildings may contain a wide variety of concealed spaces behind walls, above suspended ceilings, in utility chases, behind soffits, under computer room floors, in attics, and elsewhere. The spaces range in size from $15/_8$-in. (41.3-mm) stud spaces to even full floor height interstitial spaces. Fires in concealed spaces burn out of sight of building occupants, and detection is frequently delayed. Manual fighting of concealed-space fires can be very difficult because of limited accessibility and inherent venting problems. Vertical concealed spaces also can act as flues to spread fire and hot gases.

Fire development and spread in combustible concealed spaces has long been a cause of large losses in fires. The fire may originate in, or spread into, a combustible concealed space and burn for a long time, spreading throughout the concealed volume. The disastrous Beverly Hills Supper Club and MGM Grand Hotel fires both originated in combustible concealed spaces.[26,27] Fires in horizontal concealed spaces, such as cock lofts, attics, interstitial spaces, and soffits, have spread undetected until they either burned through the floor or roof above or caused parts of the ceiling below to drop. At that time, a fresh air supply, preheated combustible materials, and hot pyrolysis products combine to produce a rapidly developing fire, which may break out of the concealed space over a wide area. Vertical fire spread inside walls and chases can spread fire between floors and into attic spaces where it mushrooms out horizontally. The rapid vertical spread of fire was quantitatively demonstrated during several full-scale fire tests in wood-frame housing in the Bushwick section of Brooklyn, NY. In one test, temperatures in ex-

cess of 1,500°F (816°C) were recorded at the top of a 32-ft (9.75-m) shaft one minute after a fire was started at the bottom.[28]

Flame spread through horizontal noncombustible concealed spaces is rarely a hazard unless the space contains a significant amount of combustible materials, such as communications or electrical cable trays, plastic pipe, cellular foam insulation, and insulation with a combustible vapor barrier. Isolated combustible items are not likely to propagate fire horizontally; however, they may when grouped together. For example, grouped electrical cables in trays may spread a fire, although any fire spread will be slow and not likely to extend very far from the ignition source; one exception is foam-insulated coaxial cable.[29] This does not mean that such combustibles will not assist in fire spread when they are preheated by hot fire gases from another source. Movement of flame through a noncombustible ceiling plenum space was reported as contributing to the horizontal spread of fire in One New York Plaza.[30] The interstitial space contained a number of electrical communications cables and exposed foam plastic insulation on the inside of the exterior walls. Insulation installed with a combustible vapor barrier exposed can spread fire even if its fuel contribution is very small.

Electrical cables that are difficult to ignite are available and being used in nuclear power plants. In tests, they did not propagate fire horizontally, except in a multiple-stacked tray configuration.[31] Tests have also shown that use of fire-retardant coatings can be effective in inhibiting horizontal fire spread; however, there were major performance differences among the products tested.[32] Although some combustibles do not pose a serious threat of horizontal fire spread in noncombustible concealed spaces, they are likely to pose a serious threat of spread in vertical concealed spaces.

Vertical Openings

Unprotected vertical openings have been responsible for many large loss-of-life fires. These openings can act as a chimney—smoke, hot gases, and burning pyrolysis products flow up under stack effect and then mushroom out horizontally. Because of their function, openings such as elevator shafts, stairways, laundry chutes, and air shafts cannot be cut off at floor levels. Vertical openings are enclosed with fire-rated walls to prevent fire spread and, in the case of vertical exitways, to provide a safe path of egress for occupants and access for fire department personnel. Vertical interior exitways are often the only avenue of escape to the outside or to adjacent floors for building occupants. NFPA 101®, *Life Safety Code®* (hereinafter referred to as NFPA 101®) provides a comprehensive treatment of exits in general and for various types of occupancies involving both new and existing buildings. Detailed requirements are given for enclosing walls, doors, stairs, ramps, escalators, and other components of an exit system.

Shafts containing power, communications cable, combustible ducts or pipes, and other combustible materials should normally be enclosed and often firestopped at each floor. The reliability of the protection of combustible material penetrations through enclosure walls will normally be lower than for noncombustibles. If fire enters a vertical shaft containing combustible materials, it can spread rapidly upward, generate considerable smoke, and possibly spread to other floors. An eleventh-floor storeroom fire in the World Trade Center, New York City, spread into a telephone closet through a louvered door.[33] The closets on each floor were connected through an unfirestopped hole in the slab to the closet on the floor above, up to the forty-first story of the structure. The fire ignited electrical wiring and plywood mounting boards and then spread up and down grouped cables that penetrated each floor. The fire spread up to the sixteenth and down to the tenth floor before it could be controlled.

Room or Suite Compartmentation

The goals of compartmentation in confining a fire to the room or suite of rooms of origin are generally the following:

1. Segregate a space that has a higher fire hazard than the surrounding area. This is commonly found around rooms or suites used for storage, trash, flammable liquids, furnaces, laboratories, maintenance, and painting.
2. Minimize risk of loss to an occupant of one space as a result of a fire in space controlled by another. This is common in apartments, office suites, motel/hotel rooms, row houses, and other multitenant buildings.

An added benefit is that compartmentation limits the size of the fire, which limits the amount of smoke that will be generated and facilitates fire suppression.

Properly designed and installed compartmentation has been successful in limiting many fires to the unit of origin. In a number of high-rise fires, occupants who were unable to escape safely through corridors were able to "sit out" a major fire in an apartment adjacent to theirs.[34] A common cause of "failure" of this level of compartmentation has been a door that did not have a closer or was left open.

When a single room is compartmented, there may be insufficient ventilation for a fire to develop fully unless a window is open or happens to break. Occasionally such fires will completely self-extinguish.

Compartmentation is also used to protect locations of high value or critical operations from a fire in the surrounding area. Computer rooms, some control rooms, vaults, records, and high-value storage are often protected from a fire involving the rest of the building.

Protection of Corridors

Fire-resistive corridor partitions may be required: (1) to protect the means of access to an exit for sufficient time to permit occupant evacuation or (2) to provide a protected means of access to the fire department personnel. In addition, when used in conjunction with room-to-room or suite-to-suite fire-resistive separation, the corridor wall forms an element in preventing fire spread.

Corridor partitions typically are required to have a 1-hr fire resistance rating, and codes may require the partition to extend from the floor to the underside of the floor above. Either automatic or self-closers are generally required on doors opening into the corridor. Since smoke entering the corridor is a significant problem, as the corridor provides access to the exits, automatic door closers, if used, are generally actuated by smoke detectors. Additional information on corridor doors can be found later in this chapter where protection for openings is discussed.

Building Separation

Walls separating two buildings or dividing a building into separate fire areas limit the maximum probable property loss and also can divide one building into what is legally two or more buildings to comply with the height and area limits of building codes. Stability becomes a very important consideration in this class of wall. Although all fire separation walls must maintain their stability under fire exposure for their rated duration, walls separating buildings or dividing a building into two fire areas must maintain stability during a complete burnout on either side, which is often accompanied by structural collapse of the burned out side.

If at least one of the buildings has a fire-resistive frame—such as reinforced concrete or protected steel—panel walls built in along

a column line can provide fire separation. The frame must have sufficient resistance to withstand a fire of the maximum expected severity for that building. Care must be taken so that the floor construction will not adversely affect stability of the wall under fire exposure. In buildings of Type V (wood frame), Type IV (heavy timber), or Type III (ordinary) construction, the combustible framing members can be tied into the fire wall so that they are self-releasing in case of collapse. The wall then remains supported by the framing on the unexposed side. The use of self-releasing framing is considered to have been generally successful in combustible construction but has questionable reliability in steel-framed buildings. Caution must be exercised in the design of these fire walls because some methods of framing combustible members into the wall will lower its fire resistance.

Freestanding fire walls are most commonly used in one- or two-story industrial and storage buildings of unprotected steel-frame construction. The maximum height is generally limited by economic considerations. A freestanding fire wall is entirely self-supporting under vertical loads and is not directly connected to the building framing. The adjacent building framing can provide support under horizontal loading. During a fire, the wall must be capable of acting as a cantilever to withstand all horizontal forces directed toward the fire side. Freestanding fire walls should be designed to withstand at least a 5 lb per sq ft (20 kg/m^2) uniform lateral loading without any support from the framing on either side.[35]

Tension stress in mortar joints limits the height of unreinforced 12-in. (0.3-m) concrete block walls to 15 ft (4.6 m). Higher walls either require massive construction, which is rarely practical, or reinforcement. A common reinforcement technique is installation of vertical steel bars, anchored to the foundation, within hollow masonry units which are then filled with concrete. Reinforced pilasters can also be built into the wall at periodic points along the perimeter. A typical freestanding fire wall is illustrated in Figure 7-5G.

Freestanding fire walls have very definite limitations under horizontal forces, which may preclude or restrict their use. They are particularly vulnerable to collapse under seismic loads and wind forces received by exterior walls. Additional support must be provided if the walls are temporarily used as an exterior wall or become

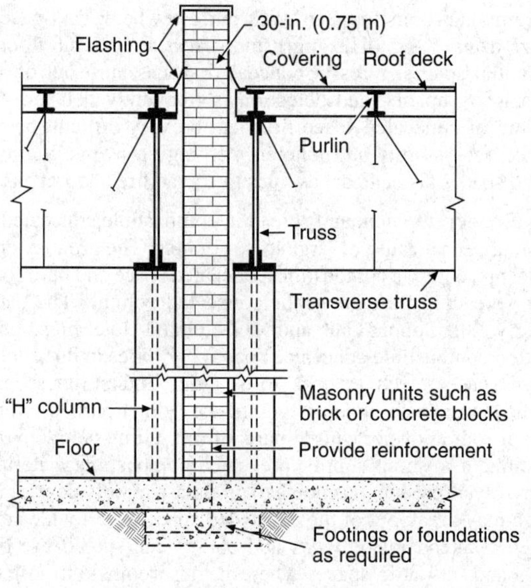

FIG. 7-5G. *Typical freestanding fire wall, used at expansion joints or at joints between buildings. (Source: Factory Mutual Research Corp.)*

an exterior wall as a result of a fire. Piping, conveyors, and similar items whose collapse in a fire will exert a horizontal pull should not pass through freestanding fire walls.

Fire walls in steel-frame buildings may be tied into the building structure so the forces of collapse on one side are resisted by the framing on the other side. The wall may be built along a column line and the columns and horizontal framing tied into the wall. The columns and framing in the wall line must be protected so that they have the same fire resistance as the wall. The walls have to be located in the framing so that the strength of framing on one side will be capable of withstanding the pull of collapsing framing on the other side. The construction of a typical tied fire wall is shown in Figure 7-5H.

Fire separation between buildings can also be provided by two blank exterior walls of adequate fire resistance and located very close together. Each wall is tied to the frame in its building; collapse of the frame pulls the wall in, leaving the other wall to resist the fire.

Pipes, conveyors, cable trays, and similar through-penetrations of fire walls can present special problems in seismic areas. The need for mobility may conflict with opening protection criteria. The complexity of the problem depends upon whether the systems are to be designed to protect against one hazard at a time (fire or earthquake) or whether the simultaneous or consecutive occurrence of earthquake and fire should be considered in protection.

Fire Plumes above Roofs

Fire walls must extend through and above a combustible roof to reduce the risk of fire spread over the top of the wall. Codes and stan-

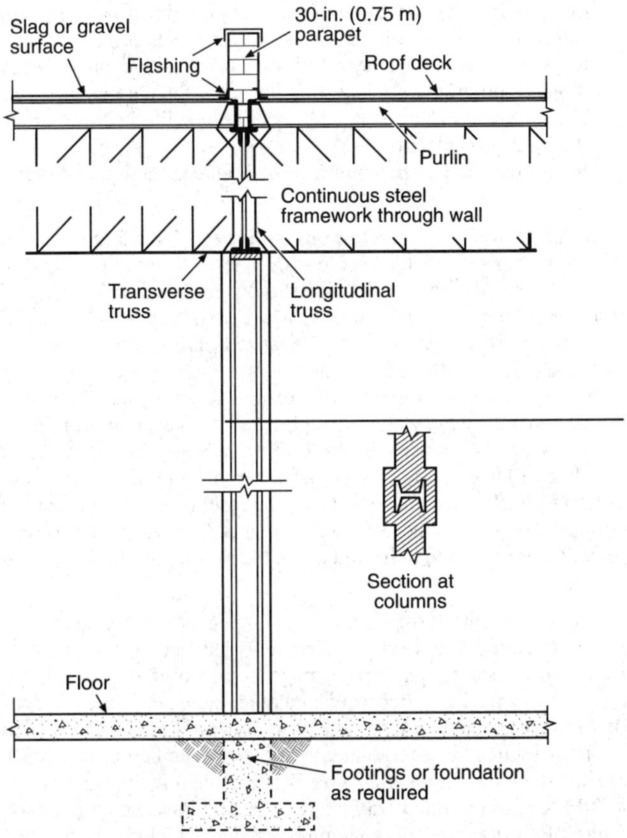

FIG. 7-5H.   Typical tied fire wall, used with continuous building framework. (Source: Factory Mutual Research Corp.)

dards typically call for fire walls to have a parapet 18 to 36 in. (0.46 to 0.9 m) above a combustible roof. The parapet minimizes the risk of direct flame spread over the fire wall and reduces the risk of fire spread by radiant heat from flames above the roof. The risk of direct flame spread over an 18- to 36-in.-high (0.46- to 0.9-m) parapet is very low. However, risk of fire spread by radiant heat depends upon the combustibility of the roof surface, the height of the parapet, the dimensions of the fire plume through the roof of the fire zone, the wind velocity, and, if the roof covering is resistive to ignition by radiant heat, its insulating properties and the duration of the exposing fire. If a roof surface is combustible, the risk of fire spread by radiant heat past a parapet can be quite high.

The threat of fire spread by radiant heat over a parapet can be calculated using the thermal radiation values identified in Appendix A of NFPA 80A, *Recommended Practice for Protection of Buildings from Exterior Fire Exposures,* and the plume dimensions.[36] Standard heat transfer procedures can then be applied, using the formula:

$$I = I_o \Phi$$

where

  I_o = radiating intensity of the fire plume,

  Φ = configuration factor or view factor, and

  I = unit incident radiation on the exposed surface.

Instead of using the referenced radiant intensities, the average temperature of the plume flames can be estimated and the radiation calculated using the relation:

$$I = \sigma T^4 \Phi$$

where

  σ = Stefefan-Boltzman constant, and

  T = abolute temperature.

This assumes the flame plume is sufficiently thick so that it radiates as a black body.

These formulae can be used with either U.S. custom engineering or SI units, as long as the units are consistent.

The configuration factor for plume-roof relationships can be calculated using the formula found in a cited reference.[37] The effect of wind velocity, which deflects the fire plume, must be accounted for in calculation of the configuration factor. At high winds, a fire plume can be deflected so it is almost horizontal above a burning roof. However, in high winds, a fire plume also tends to break up and is expected to be cooled by excess air. The angle of the fire plume selected is a matter of professional judgment; use of 45 degrees is considered reasonable by some engineers.

If the roof surface is combustible, the value of the incident thermal radiation calculated will indicate whether ignition of the roof is expected. If the surface is noncombustible or resistive to ignition by radiant heat, it is then necessary to calculate the temperature of the combustible subsurface material to which heat will be transferred from the surface. This is an unsteady-state heat transfer problem readily solved by standard numerical procedures.

A simple example of the calculation of incident thermal radiation can be used to illustrate the method of analysis. For the example, consider a 12-ft-high (3.7-m) and 60-ft-wide (18.3-m) frame building with a wood-surfaced roof. The building is subdivided by a masonry fire wall with a 3-ft (0.91-m) parapet. (See Figure 7-5I.) The expected height of flames above the roof would be about 1.4 times the height of the building, or approximately 17 ft (5.2 m).[37] Assuming a flame temperature of 1400°F (760°C), the thermal radiation from the fire plume would be about 5.8 Btu per sq ft per sec (6.6 W/m²). In a moderate wind, a fire plume deflection of 45 degrees is possible. The maximum configuration factor would be about 0.3 and the incident radiation on the wood roof surface about 1.75 Btu per sq ft per sec (2 W/m²).

At an incident radiation of 1.1 Btu per sq ft per sec (1.25 W/m²), a wood roof can be sufficiently heated so that a small burning brand will ignite it;[38] this is referred to as pilot ignition. Since in the example the incident thermal radiation is well above that required for pilot ignition, flame spread over this fire wall is to be expected if the wind deflects the fire plume toward the exposed roof.

Fire-Resistive Wall Requirements for Structural Stability

Fire-rated wall requirements may be intended to ensure structural stability rather than to provide a fire barrier. In many cases, for example, hourly rated exterior bearing walls are permitted to have unlimited unprotected window areas; there, it is obvious that the rating refers to structural stability rather than provision of a fire barrier. Required fire ratings of interior bearing walls may or may not be intended to provide a fire barrier. Interior bearing walls with an hourly rating for fire separation do require protection of all openings; however, interior bearing walls that have an hourly rating only for structural stability purposes would not need opening protection. Caution must be exercised in selecting the proper hourly rating for interior bearing walls intended for structural stability only. A barrier separation wall is exposed to fire on only one side, where an interior bearing wall that would be providing only structural stability could be exposed to fire on both sides. The same hourly rating wall will not provide the same protection in both cases.

Fire-resistive walls of a high hourly rating are occasionally improperly specified to separate areas for protection against an explosion hazard rather than a long-duration fire. The fire resistance of a wall has no relationship to its ability to withstand explosion forces. A wall with a 1-hr rating may withstand explosion overpressures better than a 4-hr wall; no conventional wall will remain in place, unless the explosion pressure is limited by venting or suppression.

PROTECTION OF OPENINGS

The protection for openings in a fire-rated wall generally has less resistance to fire spread than is provided by the wall. This is allowable

FIG. 7-51. *A fire plume exposing a wood roof over a parapeted fire wall.*

because: (1) easily ignited combustibles are not normally found against doors and other opening protection constructions, and (2) fire suppression forces can usually extinguish small localized fires that may spread past the wall. However, because openings do reduce the effectiveness of a fire barrier, fire doors generally cannot occupy more than 25 percent of the perimeter of any fire separation wall.

Factors that reduce the fire resistance of opening protection as compared to resistance of walls include the following:

1. There is no maximum temperature rise limitation on the unexposed sides of most types of fire doors. The unexposed side of steel doors may actually glow red during fire testing. When there is a limit to, or listing of, a temperature rise on a fire door, it applies only to the temperature rise during the initial 30 min of the fire test. (Temperature rise limits are often specified in building codes for doors on exit stair enclosures.) Fire doors also buckle during the fire tests and are permitted to move out of their frames in a direction perpendicular to the plane of the door as much as 2⅞ in. (72 mm). Some flaming is also permitted on unexposed sides of fire doors during the test procedures.

2. Requirements for the protection of duct penetrations place no limit on the temperature rise on the duct itself on the unexposed side of the wall.

3. Piping, cable trays, and conduit can penetrate fire walls and, although the opening must be firestopped, the temperature rise on the metal penetration itself would normally be above the 250°F (139°C) average, or 325°F (161°C) single-point temperature-rise limits required of the wall. However, some codes require use of through-penetration firestopping systems that have a temperature rise limitation on the penetrating item also.

4. The standard fire test furnace is operated at a negative pressure so that air flows in around protected openings rather than outward. In a real fire there may be a higher pressure on the fire side that is more likely to force flame through indirect paths or firestopped openings. (Fire doors are tested under certain international standards in a test furnace at a positive pressure. A research program on this testing has just been completed in the United States. The results are under review by the appropriate standards organization.)

5. Vision panels may transmit heat by both conduction and radiation.

Thermally induced distortions of between 4 and 5 in. (102 and 127 mm) were observed on some foreign tests of swinging steel fire doors.[39] Considerably smaller distortions were noted with wood and insulated metal swinging doors. However, wood doors were noted as being more likely to fail by localized carbonation at hinges, locks, and knobs. Rolling steel fire doors generally do not deform appreciably during a fire. Large distortions of some rated fire doors, both wood and metal, were also reported in "Project Corridor," a series of fire tests conducted by the California State Fire Marshal in a simulated light-hazard occupancy.[40] In some countries, extensible bolts and latches are added to swinging steel fire doors so that the door is held in the frame at five or six points. Distortions of this type of door were reported to be under 1 in. (2.54 mm) during a standard fire test.[39]

However, the most common failure mode of fire doors observed in actual fires is not closing. This has occurred because of lack of maintenance; physical damage to the closer, door, guides or tracks; blockage in the doorway; and other faults. Reliable fire door performance cannot be ensured, unless doorways are kept clean and doors maintained in operating condition, as specified in NFPA 80, *Standard for Fire Doors and Fire Windows* (hereinafter referred to as NFPA 80), and the manufacturer's instructions. Some common faults often observed on swinging fire doors include inoperative latches and improperly adjusted closing devices. Overhead and sliding doors that are not used regularly are particularly susceptible to track and guide damage. The closing mechanism on overhead doors is a more complicated and less visible mechanism than the closing mechanism on most other fire doors. Inspection, testing, and proper preventive maintenance are always essential.

Fire doors, windows, and shutters are the most widely used and accepted means of protecting openings in fire-resistive walls. Suitability of these closures is determined through tests by recognized testing laboratories, and doors not tested cannot be relied upon for effective protection. Fire door assemblies for the protection of openings depend upon the use of labeled fire doors and frames (where frames are required) and listed or labeled hardware. Hardware for fire doors is referred to as "builder's hardware," which includes fire exit hardware and fire door hardware.

Doors must be installed in a properly constructed wall and floor so that fire cannot spread below, around, or above the door. Special noncombustible sills are required for door openings in buildings with combustible floors, except for openings protected by 20- or 30-min doors. However, combustible floor coverings may extend through openings protected by fire doors rated at $1^1/_2$ hr or less if the cover has a critical radiant flux of 0.22 Watts/mm^2 or greater. Critical radiant flux is to be determined per NFPA 253, *Standard Method of Test for Critical Radiant Flux of Floor Covering Systems Using a Radiant Heat Energy Source.* In general, the critical radiant flux for vinyl sheet and flooring, wool carpet, and some synthetic carpets will exceed 0.22 Watts/mm^2. Combustible floor coverings may not extend through openings required to be protected by 3-hr or higher-rated doors.

Classification of Openings, Doors, and Windows

Fire doors are now rated according to the duration of fire exposure and, optionally, the temperature rise on the unexposed surface after 30 min, when tested as per NFPA 252, *Standard Methods of Fire Tests of Door Assemblies* (hereinafter referred to as NFPA 252). Doors are rated as 4 hr, 3 hr, $1^1/_2$ hr, $^3/_4$ hr, 30 min, and 20 min. Temperature rises are listed at 250, 450, and 650°F (121, 232, and 343°C); absence of a temperature rating indicates a rise of over 650°F (343°C) on the unexposed surface of the door after 30 min of testing. There is no temperature rise limitation for most types of fire doors. Fire doors tested under NFPA 252 are exposed to a standard time-temperature curve and a hose stream test. The hose stream test is not always required by the Authority Having Jurisdiction for 20- and 30-min doors. Those doors that are fire tested without a hose stream test indicate this fact on the label. The categories in which fire doors are classified are also identified on the label; these include: access, bullet-resisting, chute, curtain, dumbwaiter, freight elevator, rolling steel, service counter, sliding, special-purpose, and swinging-type doors.

Existing fire doors and windows may be classified by an hourly rating designation, an alphabetical letter designation, or a combination of these. At one time the alphabetical letter designation commonly was used. Appendix E of NFPA 80 refers to openings as A, B, C, D, and E, in accordance with the character and location of the wall. The alphabetical classification of the opening does not apply to the closure; however, in actual practice, the distinction between opening classification and door classification was rarely maintained. A 3-hr door for use in a Class A opening was commonly called a Class A door.

Building codes and NFPA 80 commonly specify a door by its fire-protection rating instead of by the classification of opening to be protected. Fire doors are now available with $^1/_2$ hr and $^1/_3$ hr ratings, which do not fall within the specified protection criteria for the classes of opening. The following paragraphs summarize current application of doors, windows, and shutters for openings in fire-resistive walls.

Four- and three-hour fire doors: Openings in walls separating buildings or dividing a building into different fire areas are generally protected by 4- or 3-hr fire doors. Many codes only require 3-hr fire doors in separation walls required to have a fire resistance of 3 hr or more. However, some authorities require two 3-hr fire doors, one on each side of the opening, whenever a wall is required to have a fire resistance of 4 hr. Four-hour fire door listings are relatively recent and their availability may not be reflected in all codes.

One- and one-half-hour fire doors: Openings in 2-hr enclosures of vertical openings in buildings are protected by $1^1/_2$-hr fire doors. Many codes also permit the use of $1^1/_2$-hr fire doors to protect openings in walls separating buildings or dividing buildings into different fire areas when the wall is only required to have a fire resistance of 2 hr. One and one-half-hour fire doors are used in walls required to have a 2-hr fire resistance to enclose hazardous areas.

One and one-half-hour fire doors and shutters: Openings in exterior walls that can be subjected to severe fire exposure from outside the building are protected by $1^1/_2$-hr fire doors and shutters.

One-hour fire doors: Openings in 1-hr enclosures of vertical openings in buildings, such as stairs and shafts, are protected by 1-hr fire doors.

Three-quarter-hour fire doors: Openings in room partitions and walls around some hazardous areas are protected by $^3/_4$-hr fire doors. Sometimes they are also permitted in partitions that subdivide floors of a building. Although $^3/_4$-hr fire-rated doors are for use in 1-hr corridor partitions, many codes permit installation of other types of doors, such as $^1/_2$- or $^1/_3$-hr fire-rated or $1^3/_4$-in. (445-mm) solid-bonded wood-core doors. Either automatic or self-closers are generally required on corridor doors. Since smoke control is also necessary to protect a corridor as an access to an exit, automatic closers are generally actuated by smoke detectors.

Three-quarter-hour fire doors and shutters: Openings in exterior walls subject to a moderate or light fire exposure from outside the building are protected by $^3/_4$-hr fire doors and shutters.

Three-quarter-hour fire windows: Openings in corridor or room partitions or in exterior walls subject to a moderate or light external fire exposure can be protected by $^3/_4$-hr fire windows.

One-half-hour (30-min) and one-third-hour (20-min) fire doors: Doors with these ratings are for use where smoke control is a primary consideration. They are also used for the protection of openings in partitions between a habitable room and a corridor when the wall is constructed to have a fire-resistance rating of not more than 1 hr or across corridors where a smoke partition is required. Some codes permit the use of $1^3/_4$ in. (445 mm) solid-bonded wood-core doors in corridor, room, and smoke-stop partitions that are required to have a fire resistance of not more than 1 hr. These doors cannot be used for corridor doors to hazardous areas or to protect openings in enclosures of vertical openings in buildings.

Types of Doors

There are several types of construction for fire doors. They are:

Composite doors: These are of the flush design and consist of a manufactured core material with chemically impregnated wood-edge banding and untreated wood-face veneers or laminated plastic faces, or surrounded by and encased in steel.

Hollow metal doors: These are of formed steel of the flush and paneled designs of 20-gage or heavier steel.

Metal-clad doors: These are of flush and paneled design consisting of metal-covered wood cores or stiles and rails and insulated panels covered with steel of 24 gage or lighter.

Sheet-metal doors: These are of formed 22-gage or lighter steel and of the corrugated, flush, and paneled designs.

Rolling steel doors: These are of the interlocking steel-slat design or plate-steel construction.

Tin-clad doors: These are of two- or three-ply wood-core construction, covered with 30-gage galvanized steel or terneplate [maximum size 14 × 20 in. (0.35 × 0.50 m)]; or 24-gage galvanized steel sheets not more than 48 in. (1.2 m) wide.

Curtain-type doors: These consist of interlocking steel blades or a continuous formed spring steel curtain in a steel frame.

Wood-core doors: These doors consist of wood, hardboard, or plastic-faced sheets bonded to a wood block or wood particleboard core material with untreated wood edges.

Unrated Door Protection

The fire resistance of existing doors can be improved by coating the doors with an intumescent paint, although no formal fire-resistance rating for such a door has been established. This method is not commonly accepted in the United States, although it has been used in England. Intumescent paint is listed for reducing flame spread, not for improving fire resistance. However, tests conducted by NBS showed that a single layer of intumescent paint, 7 to 8 mil (0.18 to 0.20 mm) thick, significantly delayed char penetration into a soft wood joist.[41] Ordinary wood-panel doors, in which both door and frame were coated with two layers of intumescent paint, withstood fully developed room fires similarly to $1^3/_4$-in. (44.5-mm) thick solid-core wood doors in simulated nursing home room fires.[19] In another test, a wood-panel door was protected with gypsumboard on the thin panel and intumescent paint on the remainder of the wood. This upgraded door withstood a standard fire test, modified to simulate the positive pressure of a room fire, longer than a 20-min-rated fire door.[42] However, there is no fire-protection rating assigned to any upgraded fire doors; it is entirely up to the Authority Having Jurisdiction whether or not this is acceptable for improving the fire safety in existing buildings.

Glazing for Fire Doors, Windows, and Walls

Glazing materials are available for use in fire-rated doors, windows, and walls; they are used as vision panels in fire doors and as windows in fire-rated corridor walls. In addition, they are frequently used in smoke-stop barriers and to enclose open stairways in older buildings.

Until recently, wired glass was the only transparent material that could be used in fire-rated doors and windows. It is a glass sheet containing an imbedded net of steel that helps distribute heat, lowers thermal stresses, and increases the strength of the assembly. When exposed to a fire, the glass starts cracking after a minute or two but does not open up. The glass begins to weaken when heated to about 1,470°F (799°C) and gradually starts to soften. At about 1,600°F (871°C) it deforms so badly that it will drop out.[39] Its normal 45-min fire-resistance rating corresponds to 1,638°F (892°C) on the standard time-temperature curve. In actual fires, the endurance of wired glass is temperature-dependent; if it is never heated above the temperature at which it weakens, it could remain in place for an extended period of time. Currently, wired glass is the most economical fire-protection-rated glazing material.

Clear fire-protection-rated glazing has been developed in recent years as an alternative to wired glass and to provide a higher hourly fire rating than had been possible with wired glass. These products include fire-protection-rated glass and transparent ceramics as well as fire-resistance-rated glass. The clear glazing provides more architectural freedom to achieve openness while still meeting building codes' often increased compartmentation requirements. Clear fire-protection-rated glazing materials are available that also can provide security protection.

Glazing materials for use in fire-rated windows, doors, frames, and sidelights are tested in accordance with NFPA 257, *Standard on Fire Test for Window and Glass Block Assemblies.* This test includes exposure to a standard time-temperature curve fire on one side for the rated period of time plus the hose stream test. There also are fire-resistant glazing materials that are suitable for use in specific wall and partition assemblies. These are tested in accordance with NFPA 263, *Standard Method of Test for Heat and Visible Smoke Release Rates for Materials and Products.* This test includes the same temperature-rise limitation on the unexposed surface of the glazing material as on the wall.

Vision panels are permitted in fire doors rated at 3 hr or less. However, vision panels in 3-hr fire doors and fire doors with a $1^1/_2$-hr rating used in protecting severe exterior exposures must be tested as a component of the door assembly in accordance with NFPA 252. These vision panels are limited to 100 sq in. (0.065 m²) or less. Wired glass may not be used in 3-hr fire doors or $1^1/_2$-hr doors used for severe exposure protection. Wired glass vision panels used in $1^1/_2$-hr and 1-hr fire-rated doors is limited to 100 sq in. (0.065 m²). In $3/_4$-hr-rated doors and windows and 20- and 30-min doors, wired glass is limited to 1,296 sq in. (0.84 m²) per light with a maximum dimension of 54 in. (1.37 m). In all applications, the maximum exposed area and area per light for glazing material other than wired glass is listed individually for each product.

Glazing materials used in fire doors, windows, sidelights, and commercial lights must also meet applicable safety standards if these are subject to human impact. These current standards include the U.S. Consumer Product Safety Commission (CPSC) *Safety Standard for Architectural Glazing*, and the less stringent, in terms of impact resistance, ANSI Z79.1, *Safety Performance Specifications and Methods of Test for Safety Glazing Materials Used in Buildings.* Wired glass is specifically exempt from the CPSC standard but not the ANSI standard.

Glazing materials in fire doors, windows, and frames are commonly installed on the job site and not provided by the manufacturer of the fire door. The door manufacturer provides grooves in the door for the glazing, and the glazing manufacturer or distributor supplies the glazing in the necessary light sizes. NFPA 80 requires each light to have a visible listing mark when installed. This may be the only way to identify whether the glazing installed is clear fire-protection-rated material or ordinary commercial glass. Installation and caulking requirements are very specific and specified in the individual listings. Grooves and installation methods and materials used for wired glass will generally not be suitable for use with other glazing materials.

Vision panels and windows differ from a solid wall primarily in the transmission of radiant heat. Conductive heat transmission is also increased, but high conductive heat transmission is also found indoors, door and window hardware, and frames. The intensity of radiant heat, I, to which a combustible or person would be exposed through a wired-glass panel from a fully developed room fire, is given by:

$$I = k \, \Phi \, I_o$$

where k is the transmissivity of the window, Φ is the configuration factor, and I_o is the radiant intensity of the exposing fire. This equation may be used for U.S. custom or SI units, as long as their use is consistent.

A radiant intensity of 5 Btu per sq ft per sec (5.7 W/m²) has been measured for a room fire involving ordinary residential furnishings with average ventilation, such as an open door.[43] Measurements of well-ventilated fires in light-hazard occupancies indicate a peak radiant intensity of 7.4 Btu per sq ft per sec (8.4 W/m²).[28] Transmissivity of glass windows is reported to range from 0.4 to 0.6.[38] These values of transmissivity are reasonable for the early period of fire exposure; as the glazing is exposed to the fire for a period of time its transmissivity will be reduced. The configuration factor, Φ, can be determined from any specific situation for graphs found in most heat-transfer texts.

If wood on the unexposed side of the wired glass is exposed to a radiant intensity above 3 Btu per sq ft per sec (3.4 W/m²), autoignition may occur. Most synthetic materials will require a higher level of radiant heat intensity for autoignition.

When wired-glass panels are installed in walls on a dead-end corridor or other location in which building occupants have no alternate escape path, the heat radiated through the glass may prevent escape. If persons moving past the glass absorb sufficient heat to cause severe pain, their ability to escape could be impaired. The human threshold of severe pain for exposure to radiant heat depends upon both the intensity and the time over which it is received. Experimental data that has defined this limit[44] can be approximated by the equation:

$$Q_{mas} = 2.31^{0.25} t \leq 45 \text{ sec}$$

where Q_{max} is the total heat absorbed by the skin in Btu per sq ft (W/m²) over t sec at the threshold of severe pain. As a person moves past the wired-glass panel, the view factor, Φ, will change. The absorbed heat can be calculated by determining Φ for a number of points along the person's path, and using a stepwise integration over the time required to move past each point. The normal walking speed used in exit calculations is 3.5 ft per sec (1.07 m/s).[9] If the total heat absorbed exceeds the limits in the equation, the escape of occupants past the wired-glass panels may be impaired.

Firestopping

The term *firestopping* is used to describe both barriers to restrict the spread of fire in concealed spaces and materials used to fill gaps around penetrations in walls and ceilings.

In wood-frame construction, the normal firestop inside walls and ceiling spaces is a 2-in.-thick (50.8-mm) (nominal) piece of lumber. However, for other applications, firestopping should normally be noncombustible and have a sufficiently high melting point so that it will remain in place under fire exposure. Gypsumboard, sheet metal, plaster, brick, cement grout, mineral fiber insulation, asbestos cement board, and ceramic fiberboards are all examples of commonly used firestopping. Firestop protection for penetrations of fire-rated walls and floors by noncombustible pipe, tube, and conduit has commonly been provided by filling the annular space around the penetrant with a noncombustible material, such as cement grout, mineral wool, cement plaster, etc. This type of firestopping is regarded as having performed effectively in actual fires. However, lack of firestopping[27,30] or improperly protected openings, such as the combustible foam plastic used at Brown's Ferry Nuclear Plant,[45] have been responsible for a number of serious fire losses.

Considerable attention and research has been directed toward wall and floor penetrations by plastic drain, waste, and vent (DWV) pipes.[46,47,48,49] Full-scale tests have shown that the integrity of a 2-hr fire-resistive chase enclosure could be maintained by encasing the pipe in an 18-in. (0.46-m) steel sleeve and penetrating the chase at a 45-degree downward angle. These results were applicable to 3-in. (76-mm) laterals of both DVC and ABS DWV pipe. Unless otherwise tested and approved, chases should be firestopped at each floor. In many buildings, much of the plumbing is installed in wall cavities. Use of plastic DWV pipe inside fire-resistive wall assemblies has been extensively tested, and a number of methods of protection that will not compromise the fire resistance or create a risk of vertical fire spread have been identified.

Cable-tray penetrations present another challenge; the size of the penetrant is large and contains a mixture of combustible and noncombustible material. Research and development programs were sponsored by the Nuclear Regulatory Commission (NRC) on firestops for use in nuclear power plants. The results of these programs, plus work by suppliers, led to new firestopping products.

Recent editions of some building codes now require approved and tested firestops wherever conduit, pipe, cables, or other utilities pass through fire-rated—and sometimes nonfire-rated—walls, floors, and ceilings. These firestops can be qualified under either ASTM E814, *Standard Test Method for Fire Tests of Through-Penetration Fire Stops* (hereinafter referred to as ASTM E814), or under a fire exposure corresponding to ASTM E119, *Standard Test Method for Fire Tests of Building Construction and Materials* (hereinafter referred to as ASTM E119). The test furnace must be operated at a specified positive pressure, such as 0.01 or 0.03 in. of water column. Currently, neither of these tests evaluates smoke spread.

There are two types of tested firestop protection: (1) through-penetration firestop devices and (2) through-penetration firestop systems. The firestop devices are manufactured products used primarily for protecting cable and poke-through penetrations. Through-penetration firestop systems are field-erected constructions for protecting penetrations of floors or walls by conduit, pipe, duct, cable, cable trays, etc. The basic standard used to investigate these firestops is UL 1479, *Standard for Safety Fire Tests of Through-Penetration Firestops*, which is based on ASTM E814.

ASTM E814 was developed for evaluating firestops that are installed to prevent spread of fire through an opening where cable trays, pipe, conduits, ducts, etc., pass through a wall or floor. This standard was first published in 1981. The ASTM E814 test has many similarities to ASTM E119, but it can be conducted on a much smaller scale and at a documented positive pressure. In addition, all test items are subject to a hose stream test similar to that required for some constructions tested under ASTM E119. ASTM E119 requires testing of at least a 100-sq-ft (9-m²) sample of a wall and a 180-sq-ft (16.2-m²) sample of a floor. A wall or floor sample less than 9 sq ft (0.81 m²) can be used in ASTM E814. The ASTM E814 furnace is heated to follow the temperature history of ASTM E119. There is no standard "aging" test for the new firestopping materials; one proprietary product recently was recalled because its reliability apparently deteriorated with age.

Two types of hourly time ratings are established for a tested firestop: F and T. The conditions for these ratings are broader. The F rating is a measure of the ability of the firestop system to maintain its physical integrity under the fire exposure and hose stream tests and to prevent occurrence or passage of flame on the unexposed surface. The T rating is intended to measure whether combustible materials on the unexposed side of a fire barrier would be ignited by conduction of heat through the firestop material or along the penetrating conduit or pipe. The same maximum temperature limit is used in the ASTM E814 test as in the ASTM E119 test. The hourly F or T rating is the duration of the fire test that the through-penetration firestop system passed.

The T rating achieved can be less than the F rating; there is nothing similar to this in the ASTM E119 test. If the temperature limit is reached before the completion of a successful test for an F rating, the T rating will be based on the time at which the temperature limit was reached. However, the F rating assigned would be the total hourly duration of the successful test. Building codes that require firestop systems tested under ASTM E814 differ as to if and when a T rating is necessary.

There are many types of listed through-penetration firestop system constructions, ranging from relatively simple to complex arrangements. A typical through-penetration firestop system requires installation of both a form and a fill material. The form material is generally a ceramic fiber mat; a hole is cut in it the size of the penetrating conduit (or pipe) and the form is slipped over the conduit into the opening in the wall or floor. The form material serves to place and restrain the fill material that is installed in the annular space around the conduit to a specified thickness. Fill materials include one- and two-part foams, silicone elastomers, proprietary caulking and grouting materials, and specific proprietary cementitious mixtures.

Some through-penetration firestop systems also require use of an additional filling material, such as ceramic fibers, and some only require a single material, such as a proprietary caulk or grout. There are a number of through-penetration firestop systems that are much more complex and require additional features, such as sleeves, special supports to retain the system in the opening, and insulation on the conduit or pipe extending out from the surface of the floor or wall.

Some of these systems permit penetrating items to be easily removed or replaced. This facilitates future additions of pipe, tube, or cable through the penetration opening. However, restoration of the integrity of the original firestop may or may not be possible under the conditions of the system's listing by a testing laboratory. Any additions or changes that require removal and restoration of a through-penetration firestop system have to be tightly controlled so that: (1) the restoration is performed in an approved manner, (2) the allowable number of penetrating items is not exceeded, (3) only permitted penetrants are installed, and (4) adequate clearances are maintained among penetrating items and between penetrating items and the sides of the opening.

Some designs for curtain walls, particularly in high-rise buildings, offer the danger of vertical spread of fire behind the curtain walls for the entire height of the building. This can occur when the panels of a curtain wall do not rest tightly for their full length on a floor slab. Rather, there is a continuous relieving angle attachment between the panels and the floor slab, creating vertical flue spaces between the panels and the end of the floor slab. Where firestopping is not inherent in the design, it is necessary to fill this space with appropriate firestopping materials. This arrangement is generally less satisfactory because of the tendency for both expansion and contraction of the exterior wall and the likelihood of maintenance operations causing unplanned openings through and around the firestopping.

Atria

Atria, which are becoming more common in many new buildings, depart from the traditional compartmentation concept of fire containment by enclosing all vertical openings. Alternative arrangements must be provided in atrium buildings to achieve the same level of protection found in buildings using the traditional compart-

mentation concept. The occupancy should be one in which a rapidly spreading fire would not be encountered, and all occupants should be expected to be aware of any hazard before they are endangered, and in time to move to a place of safety if necessary.

Fire protection in atrium buildings is generally provided by a combination of methods, such as compartmentation, ventilation, automatic suppression, and contents and material control. NFPA 101® requires atria in new construction to be separated from adjacent spaces by fire barriers having at least a 1-hr fire-resistance rating. However, as many as three levels of a building may open directly to the atrium without the need for a fire-resistive barrier. As an alternative where 1-hr barriers are normally required, glass walls may be used, provided there are sprinklers spaced 6 ft (1.8 m) apart not more than 1 ft (0.3 m) from the glass along both sides of the glass wall. The sprinklers must be located to ensure that the glass surface is wet when the sprinklers operate. Sprinklers are not required on the atrium side of the glass wall if there is no walkway or other floor area on that side above the main floor level.

Large-volume atria will dilute fire gases and, with proper ventilation, direct their flow safely to the outside. Small atria may not have the capacity to dilute fire gases, although if they are properly designed and ventilated, fire gases can be channeled to the outside and the risk of horizontal fire spread minimized.

BIBLIOGRAPHY

References Cited

1. McCaffrey, B. J., *et al.*, "Estimating Room Temperatures and the Likelihood of Flashover Using Fire Data Correlations," *Fire Technology*, Vol. 17, No. 2, 1981, pp. 98–119.
2. Quintiere, J. G., "A Simple Correlation for Predicting Temperature in a Room Fire," NBSIR 832712, 1983, U.S. Department of Commerce, National Bureau of Standards, Gaithersburg, MD.
3. Berry, D. L., and Minor, E. E., "Nuclear Power Plant Fire Protection—Fire Hazards Analysis (Subsystems Study Task 4)," Report NUREG/CR-0654, SAND 79-0324, 1979, Sandia National Laboratories, Albuquerque, NM.
4. Lie, T. T., "Characteristic Temperature Curves for Various Fire Severities," *Fire Technology*, Vol. 10, No. 4, 1974, pp. 315–326.
5. Ingberg, S. H., "Fire Tests of Office Occupancies," *NFPA Quarterly*, Vol. 20, No. 3, 1927, pp. 243–252.
6. Ingberg, S. H., "Tests of the Severity of Building Fires," *NFPA Quarterly*, Vol. 22, No. 1, 1928, pp. 43–61.
7. Culver, C. G., "Characteristics of Fire Loads in Office Buildings," *Fire Technology*, Vol. 14, No. 1, 1978, pp. 51–60.
8. Campbell, J. A., "Fire Safety Systems Analysis Data Base," reported at NFPA Fall Meeting, Montreal, Quebec, Canada, 1978.
9. "Handbook, Building Firesafety Criteria," PBS P5920.9, Change 6, 1977, General Services Administration, Washington, DC.
10. Waterman, T. E., *et al.*, "Prediction of Fire Damage to Installations and Built-up Areas from Nuclear Weapons," Final Report-Phase III, Experimental Studies, Appendices A-G, Contract No. DCA-8, 1964, National Military Command System Support Center, Washington, DC.
11. Harmanthy, T. Z., "Design of Buildings for Fire Safety—Part I," *Fire Technology*, Vol. 12, No. 2, 1976, pp. 95–108.
12. Heselden, A. J. M., *et al.*, "Burning Rate of Ventilation-Controlled Fires in Compartments," *Fire Technology*, Vol. 6, No. 2, 1970, pp. 123–125.
13. Babrauskas, V., and Williamson, R. B., "Post-Flashover Compartment Fires," Report UCB FRG 75-1, 1975, Fire Research Group, University of California, Berkeley, CA.
14. Kawagoe, K., "Estimation of Fire Temperature-Time Curve in Rooms," *Building Research Paper No. 29*, Ministry of Construction, Japanese Government, Tokyo, Japan, 1967.
15. Miller, H. E., and Emmons, H. W., "Documentation for the Fifth Harvard Computer Fire Code," Home Fire Project Technology Report No. 45, 1981, Harvard University, Cambridge, MA.
16. Bukowski, R. B., Peacock, R. D., Jones, W. W., and Forney, C. L., "Technical Reference Guide for the HAZARD I Fire Hazard Assess-

ment Method," *Handbook 146,* Vol. II, National Institute of Standards Technology, Gaithersburg, MD, 1989.

17. Davis, W. D., and Cooper, L. Y., "Estimating the Environment and the Response of Sprinkler Links in Compartment Fires with Draft Curtains and Fusible-Link-Actuated Ceiling Vents, Part II: User Guide for the Computer Code LAVENT," NISTIR 89–4122, 1989, National Institute of Standards and Technology, Gaithersburg, MD.

18. Mitler, H. E., and Rockett, J. A., "User's Guide to FIRST, A Comprehensive Single-Room Fire Model," NISTIR 89-4122, Sept. 1987, National Institute of Standards and Technology, Gaithersburg, MD.

19. "Fire Tests in a Nursing Home Patient Room," Report HEW, Contract HSA 105-74-116, 1975, American Health Care Association, Washington, DC.

20. Zinn, B. T., *et al.,* "Fire Spread and Smoke Control in High-Rise Fires," *Fire Technology,* Vol. 10, No. 1, 1974, pp. 35–53.

21. Christian, W. J., and Waterman, T. E., "Characteristics of Full-Scale Fires in Various Occupancies," *Fire Technology,* Vol. 7, No. 3, pp. 204–218.

22. "Fire Severity at the Exterior of a Burning Building," Report 67NK5227A, 1975, American Iron and Steel Institute, Washington, DC.

23. "Fire Resistance Ratings of Less Than One Hour," National Board of Fire Underwriters, New York, 1956.

24. Waterman, T. E., "Experimental Structural Fires," Report J6269, Contract DAHC 20-72-C-0290, DCPA Work Unit 2562B, 1974, IIT Research Institute, Chicago, IL.

25. Roytman, M. Y., *Principles of Fire Safety Standards for Building Construction,* Construction Literature Publishing House, Moscow, USSR, 1969; TT-71-58002, National Bureau of Standards, Amerind Publishing Co., PVT Ltd., New Delhi, India, 1975.

26. Best, R. L., "Tragedy in Kentucky," *Fire Journal,* Vol. 72, No. 1, Jan. 1978, pp. 18–35.

27. Best, R. L., and Demers, D., "Fire at the MGM Grand," *Fire Journal,* Vol. 76, No. 1. Jan. 1982, pp. 19–37.

28. DeCicco, P. R., "What to Do with Existing Row-Frame Residential Buildings," *Fire Journal,* Vol. 70, No. 6, 1976, pp. 23–31.

29. Riches, W. M., *Full-Scale Horizontal Cable Tray Fire Tests,* Fermi National Accelerator Laboratory, Batavia, IL, 1988.

30. Powers, W. R., "One New York Plaza Fire," New York Board of Fire Underwriters, New York, 1970.

31. Kamerus, L. J., "A Preliminary Report on Fire Protection Research Program (July 6, 1977 Test)," SAND 77-1424, 1977, Sandia National Laboratories, Albuquerque, NM.

32. Kamerus, L. J., "A Preliminary Report on Fire Protection Research Program Fire Barriers and Fire-Retardant Coating Tests," NUREG CR-0381, SAND 78-1456, 1978 Sandia National Laboratories, Albuquerque, NM.

33. Lathrop, J. K., "World Trade Center Fire," *Fire Journal,* Vol. 69, No. 4, July 1975, pp. 19–21.

34. Watrous, L. D., "Fire in a High-Rise Apartment Building, Hawthorne House, Chicago," *Fire Journal,* Vol. 63, No. 3, May 1969, pp. 5–11.

35. "MFL Fire Walls: Barriers against Destruction," *Factory Mutual Record,* Factory Mutual Research Corp., Norwood, MA, 1982, p. 13.

36. Pingree, D., "Looking at Fire Hazards: The Height of Flames above a Roof," *Fire Journal,* Vol. 62, No. 3, May 1968, pp. 24–27, 32.

37. Hamilton, D. C., and Morgan, W. R., "Radiant-Interchange Configuration Factors," *Technical Note 2836,* National Advisory Committee for Aeronautics, 1952.

38. Law, M., "Heat Radiation from Fires and Building Separation," *Technical Paper No. 5,* Joint Fire Research Organization, Her Majesty's Stationery Office, London, England, 1963.

39. Bushev, V. P., *et al., Fire Resistance of Buildings,* Construction Literature Publishers, Moscow, USSR, 1970; IT 73-52030, National Bureau of Standards, Amerind Publishing Co., PVT Ltd., New Delhi, India, 1978.

40. Reagan, R., *et al.,* "Project Corridor, Fire and Life Safety Research," *Western Fire Journal,* Bellflower, CA, 1974.

41. "A Compendium of Fire Testing," Feedback, Operation Breakthrough, Vol. 5, HUD-PDR-28-3, 1976, U.S. Department of Housing and Urban Development, Washington, DC.

42. Fisher, F. L., *et al.,* "A Study of Potential Post-Flashover Fires in Wheeler Hall and the Results from a Full-Scale Fire Test of a Modified Wheeler Hall Door Assembly," Report UCX 77-3, 1977, Fire Research Laboratory, University of California, Berkeley, CA.

43. Battelle Columbus Laboratories, "Space Age Contribution to Residential Fire Safety," *Fire Journal,* Vol. 68, No. 2, Mar. 1974, pp. 18–25.

44. Beuttner, K., "Effects of Extreme Heat on Man, III: Surface Temperature, Pain, and Heat Conductivity of Living Skin in Experiments with Radiant Heat," Project 21-26-002, Report No. 3, USAF School of Aviation Medicine, 1951.

45. Hanauer, S. H., Collins, H. E., Levine, S., Minners, W., Moore, V. A., Panciera, V. W., and Sayfirt, K. V., "Recommendations Related to Brown's Ferry Fire," Report by Special Review Group, NUREG-0050, Feb. 1976, U.S. Nuclear Regulatory Commission, Washington, DC.

46. Atwood, P. C., "Penetration of Fire Partitions by Plastic Pipe," *Fire Technology,* Vol. 16, No. 1, 1980, pp. 37–62.

47. Benjamin, I. A., and Parker, W. J., "Fire Spread Potential of ABS Plastic Plumbing," *Fire Technology,* Vol. 8, No. 2, 1972, pp. 104–119.

48. McGuire, J. H., "Penetration of Fire Partitions by Plastic DWV Pipe," *Fire Technology,* Vol. 9, No. 1, 1973, pp. 5–14.

49. Williamson, R. B., "Installing ABS and PVC Drain, Waste, and Vent Systems in Fire-Resistant Buildings," *Fire Journal,* Vol. 73, No. 2, Mar. 1979, pp. 36–45.

NFPA Codes, Standards, and Recommended Practices

Reference to the following NFPA codes, standards, and recommended practices will provide further information on the methods for confinement of fire in buildings discussed in this chapter. (See the latest version of *The NFPA Catalog* for availability of current editions of the following documents.)

NFPA 80, *Standard for Fire Doors and Fire Windows*

NFPA 80A, *Recommended Practice for Protection of Buildings from Exterior Fire Exposures*

NFPA 101®, *Life Safety Code*®

NFPA 251, *Standard Methods of Tests of Fire Endurance of Building Construction and Materials*

NFPA 252, *Standard Methods of Fire Tests of Door Assemblies*

NFPA 253, *Standard Method of Test for Critical Radiant Flux of Floor Covering Systems Using a Radiant Heat Energy Source*

NFPA 257, *Standard on Fire Test for Window and Glass Block Assemblies*

NFPA 263, *Standard Method of Test for Heat and Visible Smoke Release Rates for Materials and Products*

References

Alpert, R. L., and Ward, E. J., "Evaluation of Unsprinklered Fire Hazards," *Fire Safety Journal,* 1982, pp. 7, 127.

Babrauskas, V., and Walton, W. D., "A Simplified Characterization for Upholstered Furniture Heat Release Rates," *Fire Safety Journal,* Vol. 11, 1986, pp. 181–192.

Babrauskas, V., "Fire Tests and Hazard Analysis of Upholstered Chairs," *Fire Journal,* Vol. 74, No. 2, Mar. 1980, pp. 35–39.

Cooper, L. Y., "Estimating the Environment and the Response of Sprinkler Links in Compartment Fires with Draft Curtains and Fusible-Link-Actuated Ceiling Vents—Part I: Theory," NBSIR 88-3734, Apr. 1988, National Bureau of Standards, Gaithersburg, MD.

Davis, S., "Assessment of Fire Hazards from Furniture," International Fire, Security, and Safety Conference, London, England, Apr. 24–28, 1978.

Harmathy, T. Z., "A New Look at Compartment Fires—Parts I and II," *Fire Technology,* Vol. 8, Nos. 3 and 4, 1972.

Lathrop, J. K., *et al.,* "In Osceola a Matter of Contents—Hospital Fire Kills Seven," *Fire Journal,* Vol. 69, No. 3, 1975, pp. 20–29.

Lee, B. T., "Heat Release Rate Characteristics of Some Combustible Fuel Sources in Nuclear Power Plants," NBSIR 85-3195, 1985, National Bureau of Standards, Washington, DC.

Quintiere, J. G., "Growth of Fire in Building Compartments," *Fire Standards and Safety,* ASTM STP 614, Robertson, A. F., ed., American Society for Testing and Materials, 1977, pp. 131–167.

Sharry, J. A., "Another Pennsylvania Nursing Home Fire, Wayne, Pennsylvania," *Fire Journal,* Vol. 68, No. 3, May 1974, pp. 11–14.

Vodvarka, F. J., and Waterman, T. E., "Fire Behavior, Ignition to Flashover," Final Report, Task 2-C, OCD-PS-64-50, OCD Work Unit 2536-C, 1965, IIT Research Institute, Chicago, IL.

Additional Reading

"Firestops: For Life Safety and Property Protection," *NFPA Journal*, Supplement 1996, p. 51.

Beason, D., "Fire Endurance of Sprinklered Glass Walls," *Fire Journal*, Vol. 80, No. 4, July 1986, p. 43.

Belles, D. W., and Beitel, J. J., "Preventing Vertical Fire Spread in Aluminum Curtain Walls," *Building Standards*, Vol. 57, No. 1, 1988, pp. 10–16.

Birmingham, J. A., "Firestops Protect Property—and Lives," *Fire Journal*, Vol. 80, No. 2, Mar. 1986, p. 41.

Bordenaro, M., "Firestopping Options Expanded," *Building Design and Construction*, Vol. 34, No. 2, February 1993, pp. 44–46.

Brady, B., "Closing the Doors on Fire," *Fire Prevention*, No. 220, June 1989, pp. 20–24.

Choi, K. K., "Fire Stops for Plastic Pipe," *Fire Technology*, Vol. 23, No. 4, 1987, pp. 267–279.

Chow, W. K., "Smoke Movement and Design of Smoke Control in Atrium Buildings," *International Journal for Housing Science and Its Applications*, Vol. 13, No. 4, 1989, pp. 307–322.

Chow, W. K., "Sprinklers in Atrium Buildings," *Hong Kong Engineer*, Vol. 16, No. 9, Sept. 1988, pp. 41, 43.

Degenkolb, J. G., "Atriums," *BOCA Magazine*, Vol. 17, No. 6, Dec. 1983.

Degenkolb, J. G., "Field Test of an Atrium Exhaust System," *Building Standards*, Vol. 57, No. 2, 1988, pp. 18–19.

Emmons, H. W., "Progress in Fire Modeling," Harvard Univ., Cambridge, MA, Home Fire Proj. Tech. Rpt. 81, May 1, 1990, 10 pages; Business Communications Co., Inc. (BCC), Recent Advances in Flame Retardancy of Polymeric Materials--Materials, Applications, Industry Developments, Markets, Special Conference, May 15-17, 1990, Stamford, CT, 1990, pp. 1–9.

Fan, W., "Some New Aspects of Computer Modeling in Fire Science," *Fire Safety Science*, Vol. 1, No. 1, September 1992, pp. 15-24; First Asian Conference on Fire Science and Technology (ACFST), October 9-13, 1992, Hefei, China, International Academic Publishers, China, 1992, pp. 39–49.

Finnimore, B., "Need for Atria Fire Codes," *Fire Prevention*, No. 205, Dec. 1987, pp. 30–33.

Fire Resistance Design Manual, 12th ed., Gypsum Association, Evanston, IL, 1988.

Ford, N. J., "Safety: Smoke and Fire Curtains," *H&V Engineer*, Vol. 16, No. 683, pp. 7–8.

Gewain, R. G., *et. al.*, "Evalution of Duct Opening Protection in Two-Hour Fire Walls and Partitions," *Fire Technology*, Vol. 27, No. 3, August 1991, pp. 204-218.

Gewain, R. G., *et. al.*, "Protection of Duct Openings in Two-Hour Fire Resistant Walls and Partitions," Hughes Associates, Inc., Wheaton, MD, EG&G Rocky Flats, Golden, CO, 1990, 7 pages.

Grisanzio, J. A., "Changing Terminology for a Growing Firestop Industry," *NFPA Journal*, Vol. 90, No. 2, March/April 1966, pp. 79–81.

Hadjisophocleous, G. V. and Cacambouras, M., "Computer Modeling of Compartment Fires," *Journal of Fire Protection Engineering*, Vol. 5, No. 2, April/May/June 1993, pp. 39–52.

Harmathy, T. Z., and Oleszkiewicz, I., "Fire Drainage System," *Fire Technology*, Vol. 23, No. 1, 1987, pp. 26–48.

Irwin, W. T., "Duct Penetrations of One-Hour Fire Resistance Rated Non-load-bearing Walls/Partitions," *ASHRAE Transactions*, Vol. 92, 1986, pp. 556–565.

Janssens, M., "Mathematical Modeling of Compartment Fires," Abstracts of the First U. S. Symposium on Heat Release and Fire Hazard, December 1991, San Diego, CA, 1991, pp. 45–47.

Lorigheed, G. D., and Richardson, J. K., "Sprinklers in Combustible Concealed Spaces," *Journal of Fire Protection Engineering*, Vol. 1, No. 2, 1989, pp. 49–62.

Martin, S. B., "Perspective on Fire Modeling," Sixteenth International Conference on Fire Safety, Volume 16, January 14-18, 1991, Millbrae, CA, Product Safety Corp., Sunnyvale, CA, 1991, pp. 370–376.

Mitler, H. E., "Mathematical Modeling of Enclosure Fires," National Institute of Standards and Technology, Gaithersburg, MD, NISTIR 90-4294, May 1991, 47 pages.

Morehart, J. L., "Sprinklers in the NIH Atrium: How Did They React during the Fire Last May?" *Fire Journal*, Vol. 83, No. 1, Jan.–Feb. 1989, p. 57.

Nelson, H. E., "Introduction to Computer Fire Modeling," National Institute of Standards and Technology, Gaithersburg, MD, National Institute of Standards and Technology, Federal Fire Forum on Computer Programs for Fire Protection, November 7, 1991, Gaithersburg, MD, 1991.

Parry, L., "Fire Resisting Timber Doorsets," *Fire Surveyor*, Vol. 18, No. 6, Dec. 1989, pp. 18–21.

Read, R. E. H., "External Fire Spread: Building Separation and Boundary Distances," Fire Research Station, Borehamwood, England, BRE BR 187, 1991, 50 pages.

Richardson, J. K., and Oleszkiewicz, I., "Fire Tests on Window Assemblies Protected by Automatic Sprinklers," *Fire Technology*, Vol. 23, No. 2, 1987, pp. 115–132.

Sather, L., "Firestopping Can Make a Critical Difference in Protecting Human Life and Preventing Property Damage," *Industrial Fire World*, Vol. 8, No. 6, November/December 1993, 18–19.

Schwartz, J. K., Jensen, R. H., and Antell, J., "Role of Dampers in a Total Fire Protection System Analysis," *ASHRAE Transactions*, Vol. 92, 1986, pp. 566–576.

Semple, J. B., and Bruce, D. O., "Review of Current Fire Damper Requirements in NFPA Standard 90A and Model Codes," *ASHRAE Transactions*, Vol. 92, 1986, pp. 541–549.

Sinclair, J., "Selecting the Correct Door Control," *Fire Prevention*, No. 205, Dec. 1987, pp. 38–41.

Sumathipala, K., "Spatial Separations and Unprotected Openings: Future Directions," *Fire Research News*, No. 70, Fall 1993, pp. 1-3.

Waters, R. A., "Stansted Terminal Building and Early Atrium Studies," *Journal of Fire Protection Engineering*, Vol. 1, No. 2, 1989, pp. 63–76.

Watts, J. M., Jr., "Future of Fire Modeling. Editorial," *Fire Technology*, Vol. 27, No. 2, May 1991, pp. 97–98.

SMOKE MOVEMENT IN BUILDINGS

Revised by John H. Klote and Harold E. "Bud" Nelson

Smoke and fire gases, inherent in all unwanted fires, are dangerous products of combustion that have critical influences on life safety, property protection, and fire suppression practices in buildings. In some fires, the volume of smoke is so great that it may fill an entire building and obscure visibility at the street level to such an extent that it is difficult to identify the fire-involved building. In other incidents, the volume of smoke generated may be considerably less, although the danger to life is not necessarily diminished because of the presence of other airborne products of combustion.

This chapter gives information on the techniques used to evaluate the physical characteristics of smoke movement through both short and tall buildings as a basis for designing smoke-control systems. It also covers the approaches that can be used to test the effectiveness of designed smoke-control systems in the absence of actual performance tests involving test fires.

For more information on controlling the hazards of smoke, see the following chapters in this section: Chapter 7, "Venting Practices," and Chapter 14, "Airconditioning and Ventilating Systems." Also see Section 11, Chapter 10, "Simplified Fire Growth Calculations."

This chapter provides general background, a discussion of relationships, and selected equations useful in understanding smoke movement and smoke management in buildings. Of necessity, the information is not sufficient for detailed design analysis, but design information is available from a number of sources. The 1992 book by Klote and Milke,[1] *Design of Smoke Management Systems*, provides a consolidation and systematic presentation of data and calculations necessary for the design of systems to manage smoke movement. Specific design information is provided in that publication for pressurized stairwells, pressurized elevators, zoned smoke control, and smoke management in large spaces including atria and shopping malls. The smoke-control chapter of the 1995 *SFPE Handbook of Fire Protection Engineering*[2] summarizes much of the general information from Klote and Milke. NFPA 92A, *Recommended Practice for Smoke-Control Systems* (hereinafter referred to

as NFPA 92A), was first published in 1988 and provides additional recommendations for stairwell pressurization systems and zoned smoke-control systems, including suggested levels of pressurization for such systems in sprinklered and unsprinklered buildings. NFPA 92B, *Guide for Smoke Management Systems in Malls, Atria, and Large Areas* (hereinafter referred to as NFPA 92B), was first published in 1991, and is a technical guide for the design of smoke management systems in shopping malls, atria, and other large-volume spaces.

CLASSIFICATION OF SMOKE ZONES

As a fire burns, it:

1. Generates heat.
2. Changes major portions of the burning material or fuel from its original chemical composition to one or more other compounds, such as carbon dioxide, carbon monoxide, water, and/or other compounds.
3. Often, due to less than 100 percent combustion efficiency, transports a portion of the fuel as soot or other material that may or may not have undergone a chemical change.

A major portion of the heat generated as a fuel burns remains in the mass of products liberated by the fire. This mass expands, is lighter than the surrounding air, and rises as a plume. The rising plume is turbulent and, because of this, entrains large quantities of air from the surrounding atmosphere into the rising gases. This entrainment:

1. Increases the total mass and volume of the plume.
2. Cools the plume by mixing the cool entrained air with the rising hot gases. Normally, the rising plume is hotter at its center and cooler toward the edges where cooler air is entrained.
3. Dilutes the concentration of fire products in the plume.

Smoke, as discussed in this chapter, is therefore defined as a mixture of hot vapors and gases produced by the combustion process along with unburned decomposition and condensation matter and the quantity of air that is entrained or otherwise mixed into the mass.

For the purposes of describing smoke movement in buildings, the treatment of smoke movement is divided into two general areas: (1) the hot smoke zone and (2) the cool smoke zone.

John H. Klote, D.Sc., P.E., is research engineer at the Building and Fire Research Laboratory, NIST. He is a member of the SFPE, the NFPA Air Conditioning Committee, and the NFPA Smoke Management Systems Committee. Dr. Klote is also a member of the ASHRAE Fire and Smoke Control Committee. Harold E. Nelson, P.E., is a senior research fire protection engineer with Hughes Associates, Inc., Baltimore, MD. He is past president of the SFPE, and chairman of the NFPA Smoke Management Systems Committee.

Hot Smoke Zone

This zone includes those areas in a building where the temperature of the smoke is high enough so that the natural buoyancy of the body of smoke tends to lift the smoke toward the ceiling while clean, or at least less polluted, air is drawn in through the lower portion of the space. Normally, this condition exists in the room of fire origin. Depending upon the level of energy produced by the fire and the size of connecting openings, such as open doors, hot smoke zones can readily exist in adjacent rooms or corridors. Industrial and warehouse smoke and heat venting, atria smoke removal, and the movement of smoke in corridors open to spaces that have flashed over, all involve a hot smoke zone where the smoke is lifted and driven by the buoyant forces produced by the fire.

Cool Smoke Zone

This zone includes those areas in a building where mixing and other forms of heat transfer have reduced the effect of the driving force of the fire to the point at which buoyant lift in the smoke body is a minor factor. In these areas, the movement of smoke is primarily controlled by other forces, such as wind and stack effects, and the mechanical heat, ventilating, air conditioning, or other air-movement systems. In these areas, the movement of smoke is essentially the same as the movement of any other pollutant.

SMOKE MOVEMENT IN THE HOT SMOKE ZONE

The volume of combustion products entrained in a rising plume in the hot smoke zone is relatively small, compared with the volume of air in the total mixture. Consequently, the smoke produced by a fire will approximate the volume of air drawn into the rising plume. Figure 7-6A illustrates the process.

In situations in which the height of the plume, as measured from the top of the fire to the level of the smoke layer, is more than about twice the height of the solid body of flame, it is reasonable to estimate the amount of smoke using developed formulas.[3,4]

In general, the equations given in this chapter for conditions in the hot smoke zone should be used where the fire is small compared to the height of the space involved. For locations where this is not true, approaches such as those contained in Section 7, Chapter 7, "Venting Practices"; Section 11, Chapter 5, "Deterministic Computer Fire Models"; and Section 11, Chapter 10, "Simplified Fire Growth Calculations" are more appropriate.

The following equation is based on research conducted at Factory Mutual Research Corporation (FMRC) and is the equation used

for smoke production in NFPA 92B. The amount of smoke generated can be estimated as

$$\dot{m} = 0.071 K^{2/3} Q^{1/3} Z^{5/3} + 0.0018 Q_c$$

where

$\dot{m}$ = mass flow in plume at height z, kg/s;

Q_c = convective heat release rate of fire, kW;

z = height above top of fuel, m; and

k = wall factor (see Figure 7-6B).

The above equation is the same as the corresponding equation in NFPA 92B for the value of $k = 1$.

The expression also includes a series of assumptions, the most important of which are:

1. The tip of the flame is a significant distance below the bottom of the smoke layer. The formula, while useful, is much less accurate in spaces with a low ceiling relative to the height of the fire involved.
2. The fire bed itself covers an area whose length and width are reasonably approximate to each other. The original formula is based on the assumption of a circular fire. The degree of error in the formula increases as the relationship of length to width increases.
3. The ceiling is sufficiently high so that a correction for the virtual origin of the fire is unnecessary. This is true where the fire is small compared to the height of the space involved, as is the case for small fires in rooms or for design applications involving atria or other large-volume spaces.

Flame Height

A reasonable estimate of the visible flame height[5] can be obtained from the expression:

$$z_f = 0.166(Q/k)^{0.4}$$

where

z_f = mean flame height, m;

Q = heat release of the fire, kW; and

k = wall factor (see Figure 7-6B).

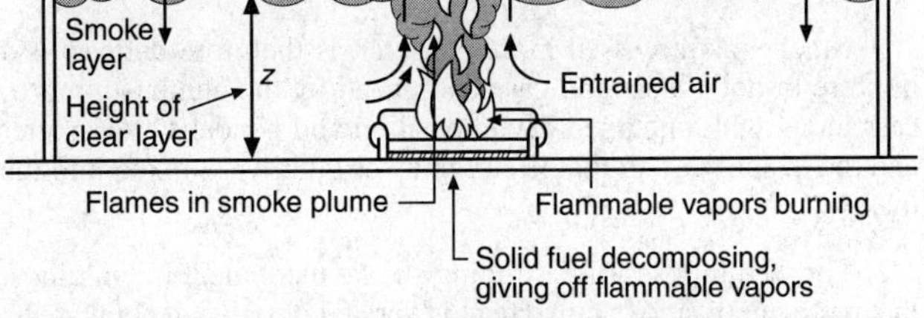

FIG. 7-6A. *The production of smoke from a fire.*

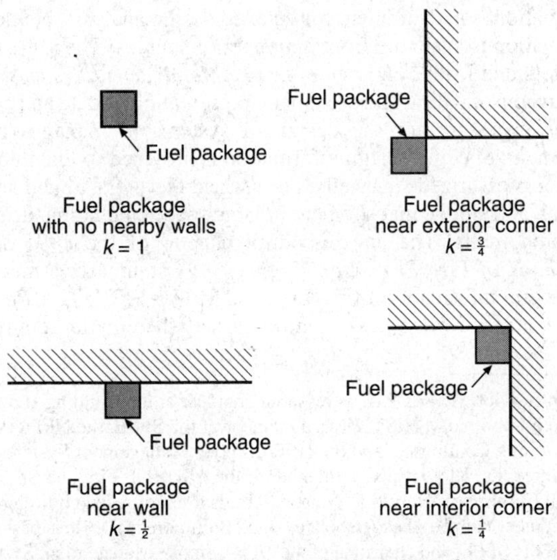

FIG. 7-6B. *Wall factors for fuel package locations.*

The above equation is the same as the corresponding equation in NFPA 92B for the value of $k = 1$. The convective portion of the heat release, rate, Q_c, can be expressed as

$$Q_c = \xi Q$$

where ξ is the convective fraction of heat release. The convective fraction depends on the heat conduction through the fuel and the radiative heat transfer of the flames, but a value of 0.7 is often used for ξ. The results of this equation for a convective fraction of 0.7 are shown graphically in Figure 7-6C.

Average Plume Temperature

Detailed engineering equations for fire plumes have been presented.[6] However, the average temperature of a fire plume is

$$\Delta T = \frac{Q_c}{\dot{m}C_p}$$

where

ΔT = average temperature increase above room temperature, °C;

$\dot{m}$ = mass flow in plume at height z, kg/s;

Q_c = convective heat release rate of fire, kW; and

C_p = specific heat of plume gases, 1.00 kJ/kg °C.

The average temperature difference should not be confused with the centerline plume temperature, which is hotter. The mass flow can be estimated by the plume equations already presented. These plume equations are for strongly buoyant plumes. For small increases over room temperature, errors due to low buoyancy could be significant. This topic needs further study, and, in the absence of better data, it is recommended that the plume equations not be used when the average temperature increase is small [less than 4°F (2°C)]. The average temperature rise of a plume for a fuel package with no nearby walls is shown in Figue 7-D.

Volumetric Plume Flow

The volumetric flow rate of a plume is

$$V = \frac{\dot{m}\left(T_p + 273\right)}{353}$$

where

V = volumetric flow rate of plume at height z, m³/s;

$\dot{m}$ = mass flow in plume at height z, kg/s; and

T_p = average temperature of plume gases at height z, °C.

$$T_p = \Delta T + \text{ room temperature}$$

In custom units, this equation is

$$v = 1.51\dot{m}\left(T_p + 460\right)$$

where

V = volumetric flow rate of plume at height z, cu ft/min;

$\dot{m}$ = mass flow in plume at height z, lb/s; and

T_p = average temperature of plume gases at height z, °F.

$$T_p = \Delta T + \text{ room temperature}$$

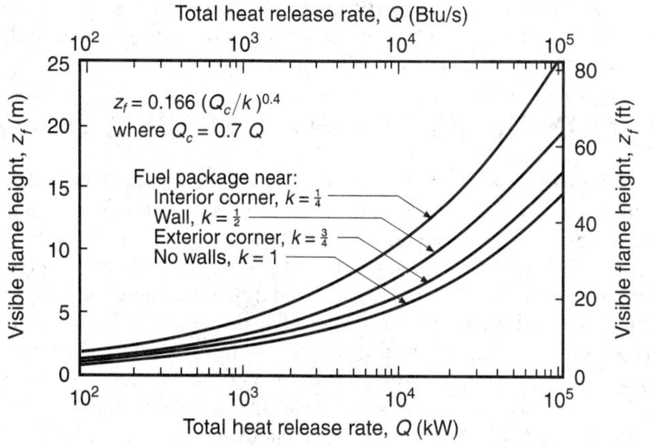

FIG. 7-6C.   Flame height vs. fire heat release rate.

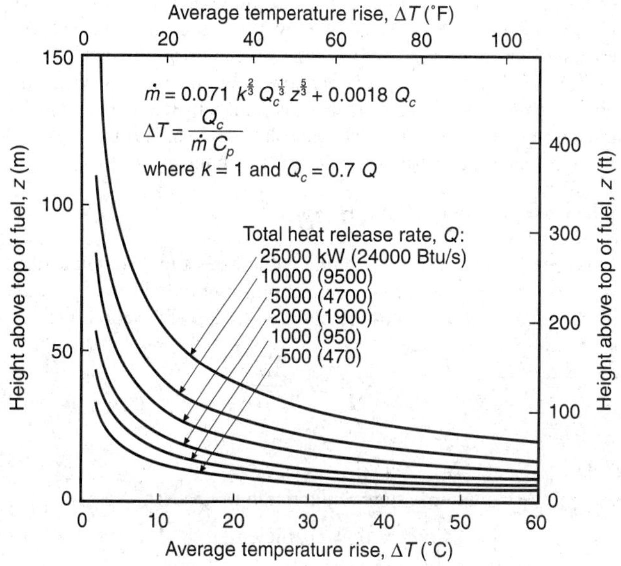

FIG. 7-6D.   Average plume temperature rise.

SMOKE MOVEMENT IN THE COLD SMOKE ZONES

As smoke is transmitted from the area of fire origin, it is cooled by entrainment of air; by the transfer of heat from the smoke body to building materials, primarily those in the walls and ceilings; and, as the smoke cools, by radiant energy losses. When smoke from a fire flows through a relatively small crack, the entrainment of cool air on the unexposed side tends to cool the smoke very quickly. When the leakage is through larger openings, there may be less entrainment relative to the mass of smoke movement at such junctures and, therefore, cooling will be slower. Once the smoke has cooled to a significant degree, however, it is transported in the same manner as any other pollutant, and the primary moving forces are those presented by the stack effect, the wind effect, and mechanical air movement systems.

When hot smoke is transported from one area of a building to another through a confined passageway, such as a duct, shaft, or stairwell, there will be little or no cooling due to entrainment. In such cases, cooling will be limited to heat lost by conduction from the moving smoke to the shaft material. Often, this loss is modest, and hot smoke can be transported significant distances with only minor cooling by such confined passageways.

PRINCIPLES OF SMOKE MOVEMENT

Smoke can behave very differently in tall buildings than in low buildings. In the lower buildings, the influences of the fire, such as heat, convective movement, and fire pressures, can be the major factors that cause smoke movement. Smoke removal and venting practices reflect this behavior. In tall buildings, these same factors are complicated by the stack effect, which is the vertical natural air movement through the building caused by the differences in temperatures and densities between the inside and outside air. This stack effect can become an important factor in smoke movement and in building design features used to combat that movement.

The predominant factors that cause smoke movement in tall buildings are the stack effect, the influence of external wind forces, and the forced air movement within the building. The following text describes the theoretical natural air movement, which is affected by the first two factors. Forced air movement caused by the building air-handling equipment is presented in Section 7, Chapter 7, and Section 7, Chapter 13, of this handbook, but it should be noted that air movement can be influenced significantly by the mechanical systems of the building. Many design solutions to the problem of tenability use emergency operation of the mechanical systems.

Flow Through Openings

For a crack, gap, or other opening with a pressure difference across it, a flow will result from the higher pressure to the lower pressure. The *orifice equation* is commonly used to describe such flow:

$$V = CA\sqrt{\frac{2\Delta P}{\rho}}$$

where

V = volumetric flow rate through the path, m³/s;

C = dimensionless flow coefficient;

A = flow area (also called leakage area), m²;

ΔP = pressure difference across path, Pa; and

ρ = density gas in path, kg/m³.

In the context of flows through gaps around doors and through construction cracks, the coefficient is generally in the range of 0.6 to 0.7. For standard air density of $\rho = 1.20$ kg/m³ (0.075 lb/cu ft) and for $C = 0.65$, the flow equation above can be expressed as

$$V = 0.839A\sqrt{\Delta P}$$

where

V = volumetric flow rate through the path, m³/s;

A = flow area (also called leakage area), m²; and

ΔP = pressure difference across path, Pa.

In custom units, this equation is

$$V = 2610A\sqrt{\Delta P}$$

where

V = volumetric flow rate through the path, cu ft/min;

A = flow area (also called leakage area), sq ft; and

ΔP = pressure difference across path, in. of water.

Stack Effect

Under normal conditions, the stack effect can account for a major part of the natural air movement in buildings. During a fire, the stack effect is often responsible for the wide distribution of smoke and toxic gases in high-rise buildings.

The stack effect is characterized by a strong draft from the ground floor to the roof of a tall building. The magnitude of this stack effect is a function of the building height, the air-tightness of the exterior walls, the air leakage between floors of the building, and the temperature difference between the inside and outside of the building.

To illustrate the principle of stack effect, consider the schematic of a box with a single opening near the bottom and another near the top, as shown in Figure 7-6E. The theoretical natural draft between the two openings is caused by the difference in weight of the column of air within the box and that of a corresponding column of air of equal dimensions outside the box. The magnitude of the theoretical natural draft may be computed using the following formula:

$$\Delta P = 2.96 H B_{o\rho}\left(\frac{1}{T_o} - \frac{1}{T_1}\right)$$

where

ΔP = theoretical pressure difference, in. of water;

H = vertical distance between the inlet and the outlet, ft;

B_o = barometric pressure, in. of mercury;

T_o = temperature of outside air; and

ρ = density of air at 0°F and 1 atmosphere pressure, lb/cu ft.

Assuming values of $B_o = 29.9$ in., and $\rho = 0.0862$ lb/cu ft, this expression reduces to

$$\Delta P = 7.63 H\left(\frac{1}{T_o} - \frac{1}{T_i}\right)$$

Vertical air movement in a building is caused by this natural draft or stack effect. The magnitude of the stack effect depends on the difference between the inside and outside temperatures and on the vertical distance between openings. If the inside and outside

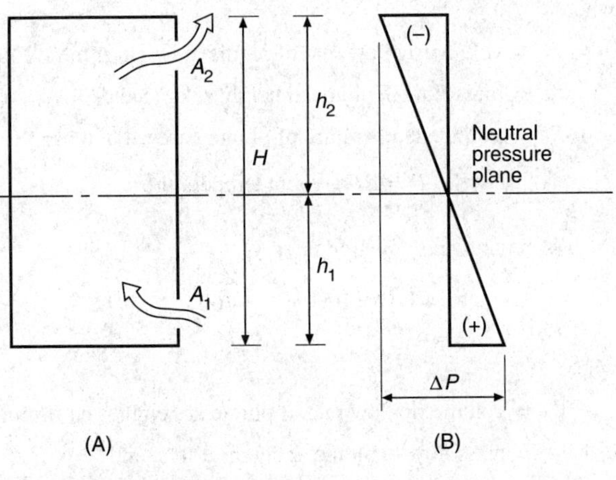

FIG. 7-6E.   Air movements caused by pressure (A) and location of neutral pressure plane (B) in a structure without horizontal barriers and with the two openings shown.

temperatures are equal ($T_i = T_o$), no natural air movement takes place. When $T_o \leq T_i$, the air moves vertically upward, with the lower opening acting as the inlet and the upper opening as the outlet. A reverse stack effect occurs when $T_o \geq T_i$. In this case, the upper opening is the inlet and the lower opening becomes the outlet.

Part (B) of Figure 7-6E illustrates the pressures that cause these movements. If it is assumed in this figure that $T_o \leq T_i$, the exterior pressure will be greater than the interior pressure at the lower opening. This is a positive pressure which forces outside air into the building at that location. The outside pressure at the upper opening is lower than the inside pressure, which creates a negative pressure at that location which forces the inside air outside. The pressure distribution between these two locations is assumed to be linear.

If an opening were present in the exterior wall in a region of positive pressure, air would flow into the building. An opening in a region of negative pressure would cause air to flow out of the building. The neutral pressure plane indicates where inside and outside pressures are equal. If there were an opening at this level, air would move neither inward nor outward. The location of the neutral pressure plane in a structure without horizontal barriers and with the two openings shown in Figure 7-6E can be determined from the following relationship:

$$\frac{h_1}{h_2} = \frac{A_2^2 T_o}{A_1^2 T_i}$$

where

h_1 and h_2 represent the distances from the neutral pressure plane to the lower and upper openings, respectively;

A_1 and A_2 represent the cross-sectional areas of the lower and upper openings, respectively; and

T_i and T_o represent the absolute temperatures of the air inside and outside the building, respectively.

The magnitude of the pressures created by the stack effect are described by the equation

$$d_i = 7.63H\left(\frac{1}{T_o} - \frac{1}{T_i}\right).$$

Examination of Figure 7-6F illustrates the significant differences between tall and short buildings with regard to air movement

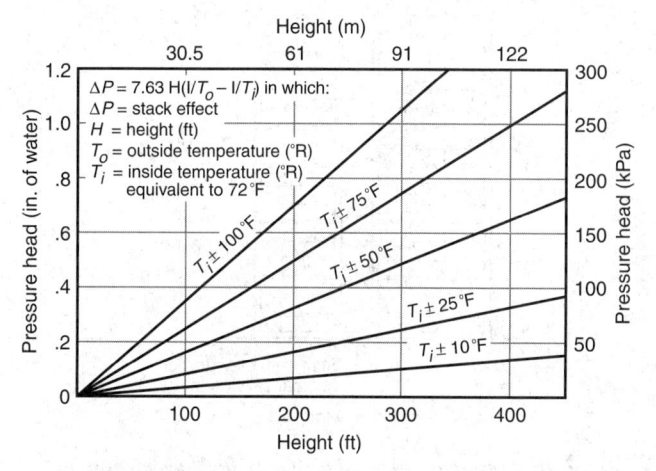

FIG. 7-6F. *Stack effect due to height and temperature difference* [$°C = (°F - 32) \times 5/9$].

by stack effect. For example, assume that a fire develops a pressure of 0.06 in. of water (25 Pa) in a compartment. Assume further that the outside temperature is 50°F (10°C) lower than the inside temperature and that the fire occurs at the same level as the lower opening. The curve $T_i \pm 50°F$ (10°C) indicates that, if the upper outlet were approximately 40 ft (12 m) above the fire, the inlet stack pressure would balance the pressure caused by the fire. A building taller that 40 ft (12 m) would create a greater stack pressure, and, theoretically, the outside air would move into the building.

Influence of Floors and Partitions

The theoretical draft described by Figure 7-6E and the final reduced equation are modified in real buildings by the presence of floors and partitions. These barriers impede the free movement of air, although a significant flow can take place through openings in the assemblies.

The magnitude and location of the leakage areas in a building naturally vary with the building's function and type of construction. The National Research Council of Canada conducted studies of airtightness for major separations on four buildings ranging from 9 to 44 stories high. The measurements were used for computer modeling of the air movement for a simulated 20-story building with a floor plan dimension of 120 by 120 ft (36 by 36 m) and a floor to floor height of 12 ft (3.6 m).[7] The data from the National Research Council of Canada has been established as a table of calculation.[7] The data are given in Table 7-6A.

TABLE 7-6A. *Typical Leakage Areas for Walls and Floors of Commercial Buildings*

Construction Element	Wall Tightness	Area Ratio A/A_w*
Exterior building walls (includes construction cracks, cracks around windows and doors)	Tight†	0.70×10^{-4}
	Average†	0.21×10^{-3}
	Loose†	0.42×10^{-3}
	Very Loose‡	0.13×10^{-2}
Stairwell walls (includes construction cracks but not cracks around windows or doors)	Tight§	0.14×10^{-4}
	Average§	0.11×10^{-3}
	Loose§	0.35×10^{-3}
Elevator shaft walls (includes construction cracks but not cracks around doors)	Tight§	0.18×10^{-3}
	Average§	0.84×10^{-3}
	Loose§	0.18×10^{-2}
		A/A_F
Floors (includes construction cracks and cracks around penetrations)	Average#	0.52×10^{-4}

*A = leakage area; A_w = wall area; and A_F = floor area.
†Tamura and Shaw 1976.
‡Tamura and Wilson 1966.
§Tamura and Shaw 1976.
#Tamura and Shaw 1978.

These leakage areas are sufficient to allow a substantial air movement throughout the building. Most of the air will flow into vertical shafts, such as stairwells and elevator shafts. Some will flow vertically from floor to floor through the minor openings in the floor-ceiling assembly. This floor-to-floor movement is always caused by a pressure differential between the floors.

Part (A) of Figure 7-6G illustrates the pressure difference characteristics of a building in which stack action causes air movement.

The slopes of the pressure lines represent differences between any two regions at the same height. Airflow from one region to another will always be in the direction of the region whose pressure curve is more to the left. This is illustrated by the airflow directions represented by the arrows in Part (B) of Figure 7-6G.

Wind Effects

Wind action is another important feature in the movement of smoke. Again, tall and short buildings behave somewhat differently in this regard. Figure 7-6H illustrates the air pressure distribution along the four sides and the roof of a building. The plan view of the pressures shows that the windward wall is subjected to an inward pressure, while the leeward wall and the two side walls have an outward pressure, or suction. The flat roof has an upward pressure, with the maximum amount occurring at the windward edge.

These pressures are caused by the movement of a mass of air around and over the structure. A short, wide building will cause the major volume of air to move over the roof, with correspondingly less air movement around the sides. A tall, narrow building, on the other hand, will cause the major volume of air to follow the path of least resistance around the building, with less movement over the top. The velocities of these movements are the primary cause of the amount and directions of the pressures on the building.

Wind velocities and direction vary over any face of a building. The most important effects are:

1. *Wind velocity.* The higher the wind velocity, the greater the effects of the following two influences.
2. *Ground effect.* Unless influenced by unusual arrangements of structures or terrain, the friction and turbulence that occur as air moves over the ground results in the lowest velocity at ground level and increases with increases in height.
3. *Structures.* Buildings and other man-made or natural features, such as trees, can produce localized effects that can increase, decrease, or alter the direction of wind forces.

The effect of wind pressures and suctions modifies the natural air movement within a building. For example, the negative pressure on the roof of a tall building can have an aspirating effect on a vertical shaft opened at the roof level. This can cause the observed draft to exceed the theoretical draft shown in Figure 7-6I.

Horizontal pressures and suctions cause the neutral planes in exterior walls to move. Positive wind pressure would tend to raise the neutral pressure plane, while negative pressure will lower it. Figure 7-6I illustrates the influence of wind action on air movement in a building.

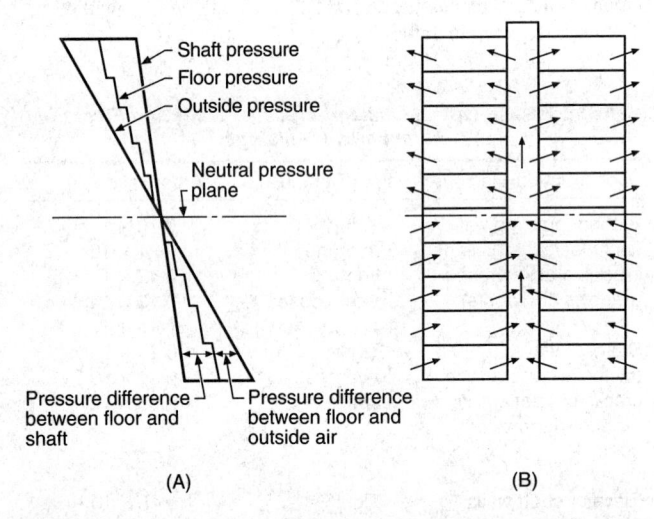

FIG. 7-6G. *The pressure difference characteristics of a building in which stack action causes air movement.*

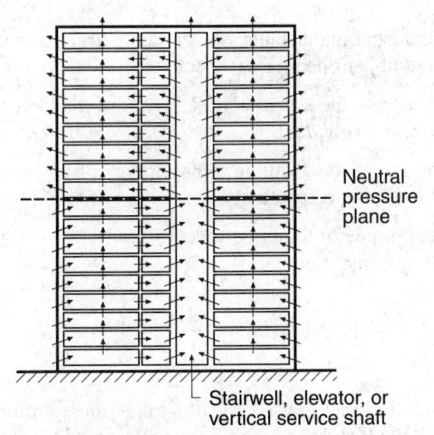

FIG. 7-6H. *The air pressure distribution along the four sides and the roof of a building.*

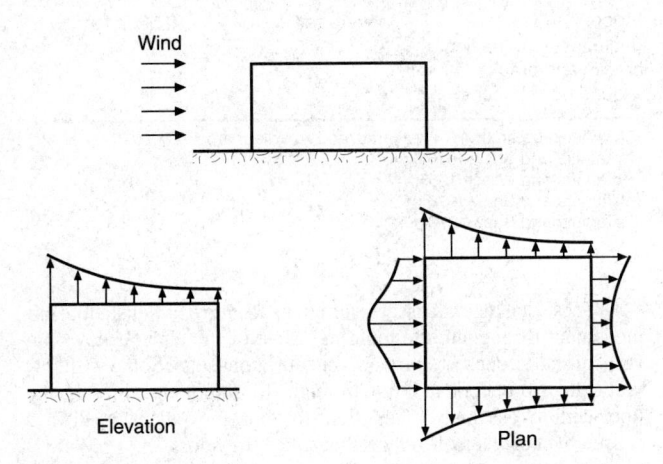

Negligible wind

Significant wind

FIG. 7-6I. *Influence of wind action on air movement in a building. Note how the neutral pressure plane changes location throughout the building in the presence of significant wind.*

SMOKE MANAGEMENT

The term "smoke management," as used in this section, includes all methods that can be used alone or in combination to modify smoke movement for the benefit of occupants or fire fighters or to reduce property damage. The mechanisms of compartmentation, dilution, airflow, pressurization, and buoyancy are used by themselves or in combination to manage smoke conditions in fires. These mechanisms are discussed below.

Compartmentation

Barriers with sufficient fire endurance to remain effective throughout a fire exposure have a long history of providing protection against fire spread. In such fire compartmentation, the walls, partitions, floors, doors, and other barriers provide some level of smoke protection to spaces remote from the fire. This section discusses the use of passive compartmentation, while the use of compartmentation in conjunction with pressurization is discussed later. Many codes, such as NFPA *101®*, *Life Safety Code®*, provide specific criteria for the construction of smoke barriers, including doors and smoke dampers in these barriers. The extent to which smoke leaks through such barriers depends on the size and shape of the leakage paths in the barriers and on the pressure differences across the paths.

There is no formalized analytical method for determining the rate of smoke leakage through barriers and the resulting levels of hazard in areas to be protected. However, emerging fire and smoke transport models can address the smoke leakage through barriers. A first-order approximation of the leakage can be made using the equation for flow through an opening, typical leakage areas listed in Table 7-6A, estimates of the dimensions of paths such as gaps around doors, and the procedures for estimating effective flow areas. More accurate calculations await better data and improved calculation procedures. Full appraisal of the impact of such leakage requires knowledge of the smoke toxicity or an assumed design value of acceptable smoke concentration in protected spaces. A formalized approach to smoke compartmentation should include development of appropriate methods of acceptance testing and routine testing. More effort is needed to increase understanding of the passive capabilities of barriers in order to maximize the usefulness of this oldest and most fundamental method of smoke management.

Dilution

Dilution of smoke is sometimes referred to as smoke purging, smoke removal, smoke exhaust, or smoke extraction. Dilution can be used to maintain an acceptable smoke concentration in a compartment subject to smoke infiltration from an adjacent space. This can be effective if the rate of smoke leakage is small compared to either the total volume of the safeguarded space or the rate of purging air supplied to and removed from the space. Dilution also can be beneficial to the fire service for removing smoke after a fire has been extinguished. Sometimes, when doors are opened, smoke will flow into areas intended to be protected. Ideally, doors will only be open for short periods during evacuation. Smoke that has entered spaces remote from the fire can be purged by supplying outside air to dilute the smoke.

Some people have unrealistic expectations about what dilution can accomplish in the fire space. There is no theoretical or experimental evidence that using a building's heating, ventilation, and air conditioning (HVAC) system for smoke dilution will result in any significant improvement in tenable conditions within the fire space. HVAC systems promote a considerable degree of air mixing within the spaces they serve. Because of this and the fact that building fires can produce very large quantities of smoke, dilution of smoke by an HVAC system in the fire space will not result in any practical improvement in the tenable conditions of that space. Thus, smoke-purging systems intended to improve hazard conditions within a fire space or in spaces connected to a fire space by large openings should not be used.

The following is a simple analysis of smoke dilution for spaces in which there is no fire. At time zero ($t = 0$), a compartment is contaminated with some concentration of smoke, and no further smoke flows into or is generated within the compartment. In addition, the contaminant is considered uniformly distributed throughout the space. The concentration of contaminant in the space can be expressed as:

$$a = \frac{1}{t}\ \log_e\left(\frac{C_o}{C}\right)$$

$$t = \frac{1}{a}\ \log_e\left(\frac{C_o}{C}\right)$$

where

C_o = initial concentration of contaminant;

C = concentration of contaminant at time t;

a = dilution rate in number of air changes per min;

t = time after smoke stops entering space or time after which smoke production has stopped, in min; and

e = constant, approximately 2.178.

The concentrations C_o and C must be expressed in the same units, and they can be any units appropriate for the particular contaminant being considered. McGuire, Tamura, and Wilson[8] evaluated the maximum levels of smoke obscuration from a number of fire tests and a number of proposed criteria for tolerable levels of smoke obscuration. Based on this evaluation, they stated that the maximum levels of smoke obscuration are greater by a factor of 100 than those relating to the limit of tolerance. Thus, they indicate that an area can be considered "reasonably safe" with respect to smoke obscuration if its atmosphere will not be contaminated to an extent greater than 1 percent by the atmosphere prevailing in the immediate fire area. It is obvious that such dilution would also reduce the concentrations of toxic smoke components. Toxicity is a more complicated problem, and no parallel statement has been made regarding dilution needed to obtain a safe atmosphere with respect to toxic gases.

In reality, it is impossible to ensure that the concentration of the contaminant is uniform throughout the compartment. Because of the buoyancy, it is likely that higher concentrations would tend to be near the ceiling. Therefore, an exhaust inlet located near the ceiling and a supply outlet located near the floor would probably dilute smoke even faster than indicated by the above equations. Caution should be exercised in locating the supply and exhaust points to prevent the supply air from blowing into the exhaust inlet and thus short circuiting the dilution operation.

EXAMPLE: Smoke purging after the fire is extinguished.

1. After the fire department puts out a fire, the smoke must be cleared quickly so that an inspection can be made to determine if the fire is completely out. If the smoke HVAC system is capable of a dilution rate of six air changes per hr, how long will it take to reduce the smoke concentration to 1 percent of the initial value?

The dilution rate, a, is 0.1 changes per min, and C_o/C is 100.

$$t = \frac{1}{0.1}\ \log_e(100) = 46 \text{ min to purge smoke}$$
$$\text{to 1 percent of initial value}$$

Considering the fire department's desire to inspect the area quickly, such a long purging time will probably be excessive.

2. If the fire department wants the space to be purged in 10 min, what dilution rate is needed?

The dilution time, t, is 10 min, and C_o/C is 100.

$$a = \frac{1}{10} \log_e(100) = 0.46 \text{ changes per min}$$
$$(28 \text{ changes per hr})$$

Pressurization

Systems using pressurization produced by mechanical fans are referred to as smoke-control systems in NFPA 92A. Pressurization results in airflows of high velocity in the small gaps around closed doors and in construction cracks, thereby preventing smoke backflows through these openings. The pressurization systems most commonly used are pressurized stairwells and zoned smoke control. Elevator smoke control is less common. Klote and Milke[1] present the public domain computer program ASCOS for analysis of smoke-control systems that use pressurization. Another public domain program, called CONTAM,[9] has extended capabilities for smoke-control analysis and runs more efficiently.

Many pressurized stairwells are designed and built with the goal of providing a tenable environment within the escape route in the event of a building fire. It is obvious that a pressurized stairwell can meet its objectives, even if a small amount of smoke infiltrates the stairwell. The three major design concerns with pressurized stairwells are:

1. Nonuniform pressure differences that occur over the stairwell height.
2. Large pressure fluctuations caused by doors being opened and closed.
3. The location of supply air inlets and fans.

At first, it might appear that the pressure differences from the stairwell to the building would be essentially the same over the height of the stairwell. Unfortunately, this is not the case. For a building without vertical leakage through floors or shafts other than the stairwell, the pressure profile is linear. Of course, this leakage characteristic is not representative of many buildings. However, this case is useful because it has been analytically solved, and it represents a worst case. The analysis has been addressed. It is a worst-case scenario in that its minimum pressure difference is less than that for other, more realistic leakage configurations and its maximum pressure difference is greater than that for other leakage configurations. Computer analysis can be performed to include the effects of more complicated building leakage arrangements.

When a door is opened in a pressurized stairwell, the pressure difference across the remaining closed doors can drop dramatically. The two classes of design concepts that have been used to deal with this problem are over-pressure relief and feedback control. An over-pressure relief system that has gained attention as being simple and cost-effective is the "Canadian System." The essential features of this system are that air is supplied by one or more fans at relatively constant flow rates, and the ground-floor exterior stairwell door opens automatically when the system activates. This system eliminates the source of the most severe pressure fluctuations—the opening and closing of the exterior door.

There is concern about locating supply air inlets near the exterior ground-floor doors of the stairwell. If a supply inlet is located near this door, it is possible that much of the supply air will flow directly through the exterior doorway when it is opened, thus effectively reducing stairwell pressurization. It is believed that locating inlets only one floor away from exterior doors eliminates this potential.

In the late 1960s, the concept of the "pressure sandwich" evolved. This consisted of exhausting the fire floor and pressurizing surrounding floors to limit smoke movement to the fire floor. The pressure sandwich concept has evolved into today's zoned smoke-control systems. According to the concept of zoned smoke control, a building can be divided into a number of smoke zones, each separated from the others by partitions and floors. A smoke-control zone can consist of one floor or more than one floor, or a floor can consist of more than one smoke zone. In the event of fire, pressure differences and airflows produced by mechanical fans can be used to restrict smoke spread to the zone in which the fire began, or the smoke zone. The concentration of smoke in this zone may render it untenable. Accordingly, in zoned smoke-control systems, building occupants should evacuate the zone in which the fire occurs as soon as possible after the fire has been detected.

Airflow

Airflow has been used extensively to manage smoke from fires in subway, railroad, and highway tunnels. Large flow rates of air are needed to control smoke flow, and these flow rates can supply additional oxygen to the fire. Because of the need for complex controls, airflow is not used as extensively in buildings. The control problem consists of having very small flows when a door is closed, and then having those flows increase significantly when that door opens.

Thomas[10] determined that airflow in a corridor in which there is a fire can almost totally prevent smoke from flowing upstream of the fire. As illustrated in Figure 7-6H, the smoke forms a surface that slopes into the direction of the oncoming airflow. Molecular diffusion is believed to result in the transfer of trace amounts of smoke, producing no hazard upstream, just the odor of smoke. There is a minimum velocity below which smoke will flow upstream, and Thomas[10] developed the following empirical relation for this critical velocity. This relation, evaluated at air density of 0.081 lb/cu ft and temperature of 81°F (27°C) is:

$$V_k = 5.68\left(\frac{E}{W}\right)^{1/3}$$

where

V_k = critical air velocity to prevent smoke backflow, ft/min;

E = energy release rate into corridor, Btu/hr; and

W = corridor width, ft.

This relation can be used when the fire is located in the corridor or when the smoke enters the corridor through an open doorway, an air transfer grille, or some other opening. The critical velocities calculated are approximate, because only an approximate value of k was used. However, the critical velocities from this relation are indicative of the kind of air velocities required to prevent smoke backflow from fires of different sizes.

EXAMPLE: Rough estimates of airflow for a doorway.

1. Thomas[10] indicated that his relationship for critical velocity can be used to obtain a rough estimate for doorways. A room fully involved in fire could have an energy release rate on the order of 8×10^6 Btu/hr. What estimate of critical velocity is obtained from the Thomas[10] equation for a door 3 ft (0.9 m) wide?

$$V_k = 5.68\left(8 \times 10^6/3\right)^{1/3} = 800 \text{ ft/min}$$

If the door has an area of 20 sq ft, this would amount to a flow of 1600 cu ft/min.

2. Consideration of a smaller fire, such as a wastebasket fire, may be appropriate for many situations. What flow rate does the Thomas[10] relation indicate is needed to prevent backflow for the above door? A wastebasket fire has an energy release rate near 0.5×10^6 Btu/hr.

$$V_k = 5.68\left(0.5 \times 10^6 / 3\right)^{1/3} = 300 \text{ ft/min}$$

For a door area of 20 sq ft, this would amount to a flow of 6000 cu ft/min.

Buoyancy in Large Spaces

Buoyancy of hot combustion gases is employed in both fan-powered and nonpowered smoke management systems for large-volume spaces. The spaces where such systems are employed include atria, arcades, covered shopping malls, sports arenas and exhibition halls. In general, these buoyancy systems are used for spaces with floor to ceiling heights of at least 33 ft (10 m). The following are approaches that can be used to manage smoke in large spaces.

1. *Smoke filling:* This approach consists of allowing smoke to fill the large-volume space while occupants evacuate the atrium. This approach applies only to spaces where the smoke filling time is sufficient for both decision making and evacuation. Evacuation time can be estimated by people movement analysis.[11,12] Smoke filling time can be estimated by either computer fire models or by the filling time equations in NFPA 92B.
2. *Unsteady clear height with upper layer exhaust:* This approach consists of exhausting smoke from the atrium top at a rate such that occupants will have sufficient time for decision making and evacuation. This approach requires an analysis of people movement and a fire model analysis of smoke filling.
3. *Steady clear height with upper layer exhaust:* This approach consists of exhausting smoke from the top of the atrium in order to achieve a steady clear height for a steady fire. (See Figure 7-6J.) Design analysis of this system is based on the fact that the mass flow of smoke entering the upper smoke layer equals that of the exhaust. For a fuel package away from walls, the exhaust airflow rates are shown in Figure 7-6K.

Computer fire models include the Harvard Code,[13] ASET,[14] ASET-B,[15] the BRI Model,[16] FIREFORM,[17] CCFM,[18] and CFAST.[19] The University of Maryland has made modifications to CCFM, specifically for atrium smoke management design.[20] De-

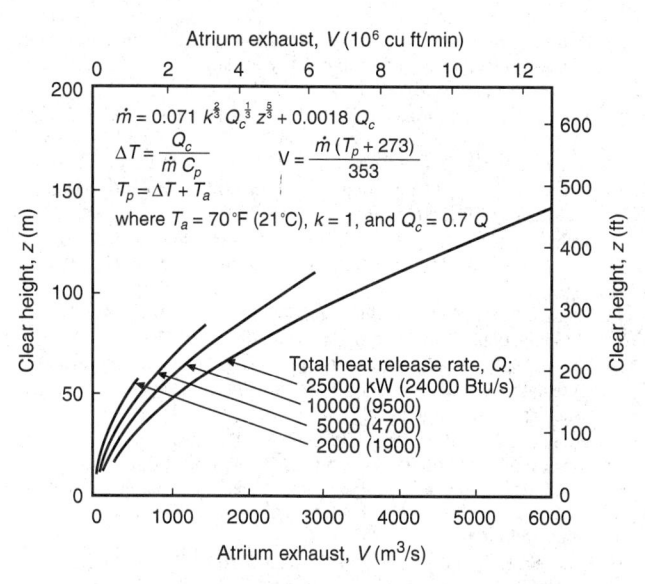

FIG. 7-6J. *Atrium smoke exhaust to maintain a smoke-free clear height.*

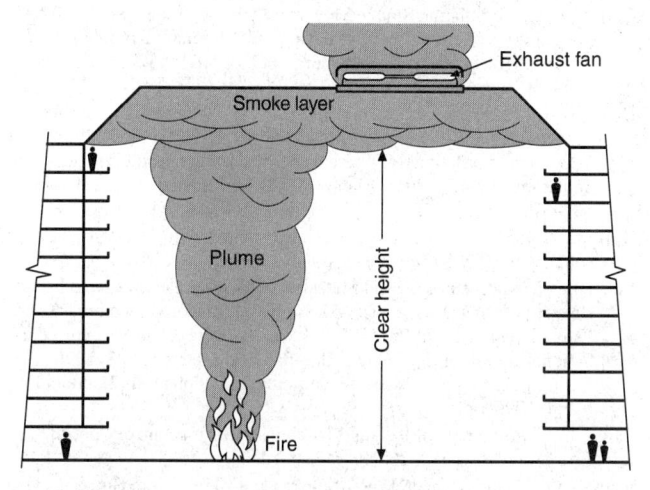

FIG. 7-6K. *Atrium exhaust needed to maintain a clear height for various heat release rates.*

scriptions of zone models are provided by Bukowski,[21] Friedman,[22] Jones,[23] Mitler and Rockett,[24] Mitler,[25] and Quintiere.[26] Klote[27] provides an overview of atrium smoke management and a public domain computer program, entitled "Atrium Smoke Management Engineering Tools" (ASMET).

BIBLIOGRAPHY

References Cited

1. Klote, J. H., and Milke, J. M., *Design of Smoke-Management Systems*, American Society of Heating, Refrigerating, and Air-Conditioning Engineers, Atlanta, GA, 1992.
2. Klote, J. H., "Smoke Control," *The SFPE Handbook of Fire Protection Engineering*, 2nd ed., DiNenno, P. J., ed. National Fire Protection Association, Quincy, MA, 1995.
3. Thomas, P. H., *et al.*, "Investigations into the Flow of Hot Gases in Roof Venting," *Fire Research Technical Paper No. 7*, Joint Fire Research Organization, London, England, 1963.
4. Butcher, E. G., and Parnell, A. C., *Smoke Control in Fire Safety Design*, E. and F. N. Spon, London, England, 1979.
5. Heskestad, G., "Fire Plumes," *The SFPE Handbook of Fire Protection Engineering*, 2nd ed., DiNenno, P. J., ed., National Fire Protection Association, Quincy, MA, 1995.
6. Heskestad, G., "Engineering Relations for Fire Plumes," *SFPE Technology Report 82-8*, Society of Fire Protection Engineers, Boston, MA, 1982.
7. Tamura, G. T., "Computer Analysis of Smoke Movement in Tall Buildings," *Annual Meeting*, American Society of Heating, Refrigerating, and Air Conditioning Engineers, June 1969.
8. McGuire, J. H., Tamura, G. T., and Wilson, A. G., "Factors in Controlling Smoke in High Buildings," *Symposium* on Fire Hazards in Buildings, ASHRAE Semiannual Meeting in San Francisco, CA, 1970, pp. 8–13.
9. Walton, W. D., "CONTAM93 User Manual," NISTIR 5385, 1994, National Institute of Standards and Technology, Gaithersburg, MD.
10. Thomas, P. H., "Movement of Smoke in Horizontal Corridors Against an Airflow," *Institute of Fire Engineers Quarterly*, Vol. 30, No. 77, 1970, pp. 45–53.
11. Nelson, H. E., and MacLennan, H. A., "Emergency Movement," *The SFPE Handbook of Fire Protection Engineering*, 2nd ed., DiNenno, P. J., ed., National Fire Protection Association, Quincy, MA, 1995.

12. Pauls, J., "Movement of People," *SFPE Handbook of Fire Protection Engineering*, 2nd ed., DiNenno, P. J., ed., National Fire Protection Association, Quincy, MA, 1995.

13. Mitler, H. E., and Emmons, H. W., Documentation for CFC V, the fifth Harvard Computer Code, Home Fire Project Tech. Rep. #45, Harvard University, Cambridge, MA, 1981.

14. Cooper, L. Y., "ASET: A Computer Program for Calculating Available Safe Egress Time," *Fire Safety Journal*, Vol. 9, pp. 29–45, 1985.

15. Walton, W. D., "ASET-B: A Room Fire Program for Personal Computers," NBSIR 85-3144-1, 1985, National Bureau of Standards, Washington, DC.

16. Tanaka, T., "A Model of Multiroom Fire Spread," NBSIR 83-2718, 1983, National Bureau of Standards, Washington, DC.

17. Nelson, H. E., "FIREFORM: A Computerized Collection of Convenient Fire Safety Computations," NBSIR 88-3308, 1986, National Bureau of Standards, Washington, DC.

18. Cooper, L. Y., and Forney, G. P., "Fire in a Room with a Hole: A Prototype Application of the Consolidated Compartment Fire Model (CCFM) Computer Code," presented at the 1987 Combined Meetings of Eastern Section of Combustion Institute and NBS Annual Conference on Fire Research, 1987.

19. Peacock, R. D., Forney, G. P., Reneke, P., Portier, R., and Jones, W. W., "CFAST: The Consolidated Model of Fire Growth and Smoke Transport," NIST Technical Note 1299, 1993, National Institute of Standards and Technology, Gaithersburg, MD.

20. Milke, J. A., and Mower, F. W., "Computer-Aided Design for Smoke Management in Atria and Covered Malls," ASHRAE Transactions, Vol. 100, Part 2, 1994.

21. Bukowski, R. W., "Fire Models, the Future Is Now!" *NFPA Journal*, No. 85, Vol. 2, Mar./Apr., 1991, pp. 60–69.

22. Friedman, R., "An International Survey of Computer Models for Fire and Smoke," *Journal of Fire Protection Engineering*, Vol. 4, No. 3, 1992, pp. 81–92.

23. Jones, W. W., "A Review of Compartment Fire Models," NBSIR 83-2684, 1983, National Bureau of Standards, Washington, DC.

24. Mitler, H. E., and Rockett, J. A., "How Accurate Is Mathematical Fire Modeling?" NBSIR 86-3459, 1986, National Bureau of Standards, Washington, DC.

25. Mitler, H. E., "Zone Modeling of Forced Ventilation Fires," *Combust. Sci. Tech.*, Vol. 39, 1984, pp. 83–106.

26. Quintiere, J. G., "Fundamentals of Enclosure Fire 'Zone' Models," *Journal of Fire Protection Engineering*, Vol. 1, No. 3, 1989, pp. 99–119.

27. Klote, J. H., "Method of Predicting Smoke Movement in Atria with Application to Smoke Management," NISTIR 5516, 1994, National Institute of Standards and Technology, Gaithersburg, MD.

References

Cooper, L. Y., "The Development of Hazardous Conditions in Enclosures with Growing Fires," NBSIR 82-2622, 1982, National Bureau of Standards, Washington, DC.

Tamura, G. T., and Shaw, C. Y., "Studies on Exterior Wall Air Tightness and Air Infiltration of Tall Buildings," *Transactions*, American Society of Heating, Refrigerating, and Air Conditioning Engineers, Vol. 82, Part 1, 1976, pp. 122–134.

Tamura, G. T., and Shaw, C. Y., "Air Leakage Data for the Design of Elevator and Stair Shaft Pressurization Systems," *Transactions*, American Society of Heating, Refrigerating, and Air Conditioning Engineers, Vol. 83, Part II, 1976, pp. 179–190.

Tamura, G. T., and Shaw, C. Y., "Experimental Studies of Mechanical Venting for Smoke Control in Tall Office Buildings," *Transactions*, American Society of Heating, Refrigerating and Air Conditioning Engineers, Vol. 86, Part I, 1978, pp. 54–71.

Tamura, G. T., and Wilson, A. G., "Pressure Differences for a Nine-Story Building as a Result of Chimney Effect and Ventilation System Operations," *Transactions*, American Society of Heating, Refrigerating, and Air Conditioning Engineers, Vol. 72, Part I, 1966, pp. 180–189.

Tamura, G. T., and Wilson, A. G., "Pressure Differences Caused by Chimney Effect in Three-Story-High Buildings," *Transactions*, American Society of Heating, Refrigerating, and Air Conditioning Engineers, Vol. 73, Part II, 1967.

NFPA Codes, Standards, and Recommended Practices

Reference to the following NFPA codes, standards, and recommended practices will provide further information on smoke movement in buildings discussed in this chapter. (See the latest version of *The NFPA Catalog* for availability of current editions of the following documents.)

NFPA 92A, *Recommended Practice for Smoke Control Systems*

NFPA 92B, *Guide for Smoke Management Systems in Malls, Atria, and Large Areas*

NFPA 101®, *Life Safety Code*®

NFPA 204M, *Guide for Smoke and Heat Venting*

NFPA 258, *Standard Research Test Method for Determining Smoke Generation of Solid Materials*

Additional Reading

Baum, H. R., McGrattan, K. B., and Rehm, R. G., "Large Eddy Simulations of Smoke Movement in Three Dimensions," National Institute of Standards and Technology, Gaithersburg, MD, ISBN 0-9516320-9-4; *Proceedings* of the Seventh International Interflam Conference, Interflam '96, March 26–28, 1996, Cambridge, England, Interscience Communications Ltd., London, England, 1996, pp. 189–198.

Bodart, X. E., Curtat, M. R., and Fromy, P. C., "Iteration of Smoke Movement Simulation Processes for an Optimized Specification of Ventilation Parameters, With a View to Control Fire-Induced Smoke Flows in Atrium Buildings," *Proceedings* of the Fourth International Symposium on Fire Safety Science, Intl. Assoc. for Fire Safety Science, Boston, MA, 1994, pp. 983–993.

Chow, W. K., and Lo, A. C. W., "Scale Modelling Studies on Atrium Smoke Movement and the Smoke Filling Process," *Journal of Fire Protection Engineering*, Vol. 7, No. 3, 1995, pp. 55–64.

Chow, W. K., and Siu, W. M., "Visualization of Smoke Movement in Scale Models of Atriums," *Journal of Applied Fire Science*, Vol. 3, No. 2, 1993/1994, pp. 93–111.

Chow, W. K., "Smoke Movement and Design of Smoke Control in Atrium Buildings," *International Journal for Housing Science and Its Application*, Vol. 13, No. 4, 1989, pp. 307–322.

Cole, J., "Smoke Movement in Single-Story Buildings," *Fire Surveyor*, Vol. 18, No. 1, Feb. 1989, pp. 25–32.

Cooper, L. Y., "Compartment Fire-Generated Environment and Smoke Filling," *The SFPE Handbook of Fire Protection Engineering*, 2nd ed., DiNenno, P. J., ed., National Fire Protection Association, Quincy, MA, 1995.

Cooper, L. Y., "Overview of a Theory for Simulating Smoke Movement Through Long Vertical Shafts in Zone-Type Fire Models," National Institute of Standards and Technology, Gaithersburg, MD, NISTIR 5499, September 1994; National Institute of Standards and Technology, Annual Conference on Fire Research: Book of Abstracts, October 17–20, 1994, Gaithersburg, MD, 1994, pp. 93–94.

Cooper, L. Y., "Simulating Smoke Movement Through Long Vertical Shafts in Zone-Type Compartment Fire Models," National Institute of Standards and Technology, Gaithersburg, MD, NISTIR 5526, November 1994, 30 pages.

Davis, W. D., Notarianni, K. A., and Tapper, P. Z., "Modelling of Smoke Movement and Detector Performance in High Bay Spaces," *Proceedings* of the International Conference on Fire Research and Engineering, September 10–15, 1995, Orlando, FL, SFPE, Boston, MA, 1995, pp. 307–311.

DeLuga, G. F., "Meeting Control Needs in Smoke Control Systems," *Consulting/Specifying Engineer*, 1989, pp. 32–39.

Fukutani, H., Matsushita, T., and Matsumoto, M., "Spread of Smoke Front Below a Ceiling: Analysis of Axisymmetric Smoke Movement," Kobe Univ., Japan, ISBN 0-9516320-9-4; *Proceedings* of the Seventh International Interflam Conference, Interflam '96, March 26-28, 1996, Cambridge, England, Interscience Communications Ltd., London, England, 1996, pp. 199–208.

Gross, D., "Estimating Air Leakage Through Doors for Smoke Control," *Fire Technology*, Vol. 26, No. 1, Feb. 1990, pp. 75–81.

Hansell, G. O., and Morgan, H. P., "Smoke Control in Atrium Buildings Using Depressurization," PD 66/88, 1988, Fire Research Station, Borehamwood, Hertfordshire, UK.

Hansell, G. O., and Morgan, H. P., "Design Approaches for Smoke Control in Atrium Buildings," BR 258, 1994, Fire Research Station, Borehamwood, Hertfordshire, UK.

Heselden, A. J. M., and Baldwin, R., "The Movement and Control of Smoke and Escape Routes in Buildings," *BRECP*, Building Research Establishment, Borehamwood, Hertfordshire, UK.

Heskestad, G., "Note on Maximum Rise of Fire Plumes in Temperature-Stratified Ambients," *Fire Safety Journal*, Vol. 15, No. 4, 1989, pp. 271–276.

Heskestad, G., "Smoke Movement and Venting," *Fire Safety Journal*, Vol. 11, Nos. 1 & 2, 1986, pp. 77–83.

Heskestad, G., and Hill, J. P., "Propagation of Fire Smoke in a Corridor," *Proceedings* of the 1987 ASME-JSME Conference, Vol. 1, ASME, 1987.

Hokugo, A., Yung, D., and Hadjisophocleous, G. V., "Experiments to Validate the NRCC Smoke Movement Model for Fire Risk-Cost Assessment," *Proceedings* of the Fourth International Symposium on Fire Safety Science, Intl. Assoc. for Fire Safety Science, Boston, MA, 1994, pp. 805–816.

Huo, R. and Li, C., "Some Investigations of Smoke Movement Through a Horizontal Ceiling Vent in Enclosure Fires," *Proceedings* of the First International Conference on Fire Science and Engineering, ASIAFLAM '95, March 15–16, 1995, Kowloon, Hong Kong, 1995, pp. 311–321.

Isner, M. S., "Smoky Fire Kills Four in New York High-Rise," *Fire Journal*, Vol. 82, No. 5, 1988.

Jones, W. W., "Modeling Smoke Movement Through Compartmented Structures," *Journal of Fire Sciences*, Vol. 11, No. 2, March/April 1993, pp. 172-183; Thirty-ninth Sagamore Army Materials Research Conference, September 16-17, 1992, Plymouth, MA, 1992; Twelfth Joint Panel Meeting of the U.S./Japan Government Cooperative Program on Natural Resources (UJNR) on Fire Research and Safety, October 27–November 2, 1992, Tsukuba, Japan, Building Research Inst., Ibaraki, Japan, Fire Research Inst., Tokyo, Japan, 1992, pp. 34–41.

Jones, W. W., and Forney, G. P., "Improvement in Predicting Smoke Movement in Compartmented Structures," *Fire Safety Journal*, Vol. 21, No. 4, 1993, pp. 269–297.

Jones, W. W., and Forney, G. P., "Modeling Smoke Movement Through Compartmented Structures," National Institute of Standards and Technology, Gaithersburg, MD, NISTIR 4872, July 1992, 34 pages; *Proceedings* of the 1991 Fall Technical Meeting of the Combustion Institute/Eastern States Section on Chemical and Physical Processes in Combustion, October 14–16, 1991, Ithaca, NY, 1991, pp. 88/1–4.

Jones, W. W., Matsushita, T., and Baum, H. R., "Smoke Movement in Corridors: Adding the Horizontal Momentum Equation to a Zone Model," Twelfth Joint Panel Meeting of the U. S./Japan Government Cooperative Program on Natural Resources (UJNR) on Fire Research and Safety, October 27–November 2, 1992, Tsukuba, Japan, Building Research Inst., Ibaraki, Japan, 1992, pp. 42–54.

Klote, J. H., "A Method for Calculation of Elevator Evacuation Time," *Journal of Fire Protection Engineering*, Vol. 5, No. 3, 1993, pp. 83–96.

Klote, J. H., "Design of Smoke Control Systems for Areas of Refuge," ASHRAE *Transactions*, Vol. 99, Part 2, pp. 793–807, American Society of Heating, Refrigerating, and Air Conditioning Engineers, Atlanta, GA.

Klote, J. H., "Fire and Smoke Control: An Historical Perspective," *ASHRAE Journal*, Vol. 36, No. 7, 1994, pp. 46–50.

Klote, J. H., "A General Routine for Analysis of Stack Effect," NISTIR 4588, 1991, National Institute of Standards and Technology, Gaithersburg, MD.

Klote, J. H., "Considerations of Stack Effect in Building Fires," NISTIR 89-4035, 1989, Center for Fire Research, National Institute of Standards and Technology, Gaithersburg, MD.

Klote, J. H., "Fire Experiments of Zoned Smoke Control at the Plaza Hotel in Washington," DC, ASHRAE *Transactions*, Vol. 96, Part 2, 1990.

Klote, J. H., "An Overview of Smoke Control Technology," NBSIR 87-3626, Sept. 1987, Center for Fire Research, Gaithersburg, MD.

Klote, J. H., "Computer Modeling for Smoke Control Design," *Fire Safety Journal*, Vol. 9, No. 2, 1986, pp. 181–188.

Klote, J. H., "Fire Safety Inspection and Testing of Air-Moving Systems," NBSIR 87-3660, 1987, National Bureau of Standards, Gaithersburg, MD.

Klote, J. H., "An Analysis of the Influence of Piston Effect on Elevator Smoke Control," NBSIR 88-3751, 1988, National Bureau of Standards, Gaithersburg, MD.

Klote, J. H., and Tamura, G. T., "Design of Elevator Smoke Control Systems for Fire Evacuation," ASHRAE *Transactions*, Vol. 97, Part 2, 1991, pp. 634–642.

Klote, J. H., and Tamura, G. T., "Elevator Piston Effect and the Smoke Problem," *Fire Safety Journal*, Vol. 11, No. 3, 1986, pp. 227–233.

Klote, J. H., and Tamura, G. T., "Smoke Control and Fire Evacuation by Elevators," ASHRAE *Transactions*, Vol. 92, Part 1a, 1986, pp. 231–245.

Klote, J. H., and Tamura, G. T., "Experiments of Piston Effect on Elevator Smoke Control," ASHRAE *Transactions*, Vol. 93, Part 2a, 1987, pp. 2217–2228.

Klote, J. H., and Tamura, G. T., "Smoke Control and Fire Evacuation by Elevators," ASHRAE *Transactions*, Vol. 92, Part 1a, 1986, pp. 231–245.

Ling, W. C. T., and Williamson, R. B., "Use of Probabilistic Networks for Analysis of Smoke Spread and Egress of People in Buildings," *Proceedings* of the First International Symposium for Fire Safety Science, Hemisphere, NY, 1986, pp. 953–962.

Marchant, R., "Sandwich Pressurization Systems for Smoke Control," *ASHRAE Journal*, Nov., 1992.

Marshall, N. R., "The Behavior of Hot Gases Flowing within a Staircase," *Fire Safety Journal*, Vol. 9, No. 3, 1987, pp. 245–255.

Marshall, N. R., "Air Entrainment into Smoke and Hot Gases in Open Shafts," *Fire Safety Journal*, Vol. 10, No. 1, 1986, pp. 37–46.

Matsushita, T., Fukai, H., and Terai, T., "Calculation of Smoke Movement in Building in Case of a Fire," *Proceedings* of the First International Symposium for Fire Safety Science, Hemisphere, NY, 1986, pp. 1123–1132.

Matsushita, T., and Klote, J. H., "Smoke Movement in a Corridor—Hybrid Model, Simple Model and Comparison with Experiments," NISTIR 4982, National Institute of Standards and Technology, Gaithersburg, MD.

Morgan, H. P., "Smoke Control in Shopping Malls and Atria," Fire and Safety in Buildings, *Symposium* in Hong Kong, 1991, Hong Kong Polytechnic.

Morgan, H. P., and Gardner, J. P., "Design Principles for Smoke Ventilation in Enclosed Shopping Centres," BR 186, 1990, Fire Research Station, Borehamwood, Hertfordshire, UK.

Morgan, H. P., and Hansell, G. O., "Atrium Buildings: Calculating Smoke Flows in Atria for Smoke-Control Design," *Fire Safety Journal*, Vol. 12, 1987, pp. 9–36.

Morgan, H. P., and Marshall, N. R., "Depth of Void-Edge Screens in Shopping Malls," *Fire Engineers Journal*, Vol. 48, No. 152, 1989, pp. 7–9.

Nagaoka, T., Tsujimoto, M., and Takenouchi, T., "Scaling Law of Smoke Movement in a Closed Space With Openings. Part 2," Japanese Association of Fire Science and Engineering, Fire Research Annual Conference, May 17–18, 1990, pp. 17–20.

Nelson, H. E., "Performance-Based Smoke Movement Design," Hughes Associates, Inc., Baltimore, MD, Video; Federal Fire Forum, Performance-Based Design Issues of Concern to the A/E Community, November 6, 1995, Gaithersburg, MD, 1995.

Notarianni, K. A., and Davis, W. D., "The Use of Computer Models to Predict Temperature and Smoke Movement in High-Bay Spaces," NISTIR 5304, 1993, National Institute of Standards and Technology, Gaithersburg, MD.

Said, M. N. A., "A Review of Smoke Control Models," *ASHRAE Journal*, Vol. 30, No. 4, 1988, p. 36.

Savilonis, B., and Richards, R., "Survey and Evaluation of Existing Smoke Movement Models," *Fire Safety Journal*, Vol. 13, Nos. 2 & 3, 1988, pp. 87–98.

Sugawa, O., *et al.*, "Full-Scale Test of Smoke Leakage from Doors of a High-rise Apartment," *Proceedings* of the First International Symposium on Fire Safety Science, Hemisphere, 1986, pp. 891–900.

Tamura, G. T., and Klote, J. H., "Experimental Fire Tower Studies on Elevator Pressurization Systems for Smoke Control," ASHRAE *Transactions*, Vol. 93, Part 2, 1987, pp. 2235–2257.

Tamura, G. T., and Klote, J. H., "Experimental Fire Tower Studies on Adverse Pressures Caused by Stack and Wind Action: Studies on Smoke Movement and Control," ASTM International *Symposium* on Characterization and Toxicity of Smoke, Dec. 5, 1988, Phoenix, AZ.

Tamura, G. T., and Klote, J. H., "Experimental Fire Tower Studies on Mechanical Pressurization to Control Smoke Movement Caused by Fire

Pressures," *Proceedings* of the 2nd International Symposium on Fire Safety Science, Tokyo, Japan, 1987.

Tamura, G. T., and Klote, J. H., "Experimental Fire Tower Studies on Elevators: Pressurization Systems for Smoke Control," *Elevator World*, Vol. 37, No. 6, 1989, pp. 80–89.

Tamura, G. T., and Klote, J. H., "Experimental Fire Tower Studies on Elevator Pressurization Systems for Smoke Control," ASHRAE *Transactions,* Vol. 93, Part II, 1987, pp. 2235–2257.

Tamura, G. T., *Smoke Movement and Control in High-rise Buildings,* NFPA, Quincy, MA, 1994.

Tamura, G. T., "Stair Pressurization Systems for Smoke Control: Design Considerations," ASHRAE *Transactions*, Vol. 95, Part 2, 1989, pp. 184–192.

Thomas, P. H., "Designing Stair Pressurization Systems," *Fire Safety Journal*, Vol. 12, No. 3, 1987, pp. 191–204.

Tsujimoto, M., Takenouchi, T., and Uehara, S., "Scaling Law of Smoke Movement in Atrium," Nagoya Univ., Japan, Takenaka Corp., Tokyo, Japan, NISTIR 4449; Eleventh Joint Panel Meeting of the U. S./Japan Government Cooperative Program on Natural Resources (UJNR) on Fire Research and Safety, October 19–24, 1989, Berkeley, CA, 1990, pp. 181–192.

Yao, J. and Fan, W., "Application of Radiation Model in Numerical Simulation of Smoke Movement," *Fire Safety Science*, Vol. 4, No. 1, March 1995, pp. 26–33.

Zhang, H., *et al.*, "Experimental Study of Salt Water Simulation of Smoke Movement in a Confined Space," *Fire Safety Science*, Vol. 3, No. 2, September 1994, pp. 48–57.

Zukoski, E. E., "Scaling Rules for Smoke Movement in Corridors. Annual Report," California Institute of Technology, Pasadena, National Institute of Standards and Technology, Gaithersburg, MD, Annual Report, June 1990.

VENTING PRACTICES

Gunnar Heskestad

The importance of the buoyant properties of heat and smoke was realized at an early date, as evidenced by an NFPA standard adopted in 1903 that called for smoke vents over theater stages and in the ceilings of theater auditoriums. However, until the advent of effective artificial lighting, buildings were generally small enough so that windows provided adequate smoke venting, other than venting of fire through the roof.

The main body of this chapter covers venting practices as they would be applied to nonsprinklered buildings, following the recommendations of NFPA 204M, *Guide for Smoke and Heat Venting* (hereinafter referred to as NFPA 204M). An addendum at the conclusion of the chapter (in which the several different approaches are incorporated) discusses the current state of the technology involving the sprinkler/vent issue.

HISTORICAL BACKGROUND

Large undivided floor areas present extremely difficult fire-fighting problems, since fire fighters must enter these areas to fight fires in central portions of the building. If fire fighters are unable to enter because of the accumulation of heat and smoke, fire-fighting efforts may be reduced to ineffective application of hose streams to perimeter areas while fire consumes the interior. (See Figure 7-7A.)

Great impetus to the subject of smoke and heat venting was provided by the General Motors fire at Livonia, MI, in 1953, when fire spread horizontally under an unvented metal roof, with 34 acres (137,600 m²) of undivided area. Fire protection engineers were in general agreement that this fire could have been greatly reduced if there had been effective roof venting. The General Motors fire led to a new approach to the subject by the NFPA Committee on Building Construction, which prepared NFPA 204M, adopted by NFPA in May 1961. In 1968, the venting guide was expanded to include a new section on inspection and maintenance.

A reconfirmation action failed in 1975, as concerns had surfaced over use of NFPA 204M in conjunction with automatically sprinklered buildings. Because of this controversy, work on a revision of NFPA 204M continued at a slow pace, but concluded with the 1982 edition.

The 1982 edition of NFPA 204M distinguished between unsprinklered and sprinklered buildings. The unsprinklered part was considered to represent a major advance in engineered smoke and heat venting. The sprinklered part, limited to a single chapter (Chapter 6), did not recognize that venting is necessarily desirable in a sprinklered building and did not offer a design basis, pending the resolution of some technical questions.

No changes were made for the 1985 edition of NFPA 204M, except for a limited revision of Chapter 6 dealing with venting of sprinklered buildings. Reference was made to new test data on deploying vents in a test building with automatic sprinklers. However, the new data did not permit conclusions to be developed whether sprinkler control was impaired or enhanced by the presence of automatic roof vents of typical spacing and area.

Changes for the 1991 edition of NFPA 204M again focused on Chapter 6. While the two previous editions had stated that a generalized design basis had not been developed and was not available for using sprinklers and vents together for hazard control, the 1991 edition stated that such a design basis "has not been universally recognized" and did not offer one. Furthermore, words were added to the effect that the concern for combining sprinklers and vents relates to "occupancies that present a high challenge to sprinkler systems." As in previous editions, "the designer is encouraged to use the available tools and data referenced in the document for solving problems peculiar to a particular type of hazard control."

The technical questions that remain unresolved in application of venting to sprinklered buildings are related to: (1) the effects of sprinkler discharge on venting effectiveness and (2) the effects of fresh air introduced into the building on the burning process and the water demand of the sprinkler system. Sprinkler discharge will cool the fire gases, perhaps to the extent that the vent discharge is reduced, hence reducing venting effectiveness. Additionally, the sprinkler sprays will entrain ambient smoke and air, transporting smoke to the floor level and possibly drawing gas from the vent exhaust wherever nozzles are located close to vents, further reducing venting effectiveness. Unless the building is very large, fresh air flowing into the building and replacing smoke escaping through the vents will increase the oxygen concentration in the fire space. The increased oxygen concentrations may cause a more vigorous fire than would otherwise exist, with the possible outcome of an increased number of operating sprinklers, in some cases overtaxing the water supply. Against these possible adverse effects must be balanced the likelihood that some improvement in visibility for fire fighting will result in many cases, and the possibility that the number of operating sprinklers may sometimes be reduced because of

Gunnar Heskestad, Ph.D., is principal research scientist in the Research Division, Factory Mutual Research Corporation, Norwood, MA.

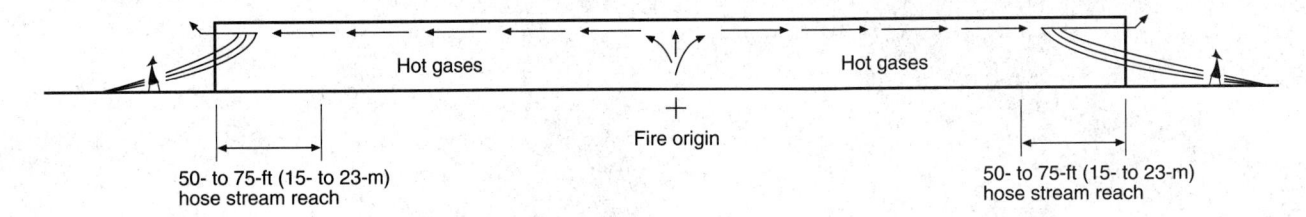

50- to 75-ft (15- to 23-m) hose stream reach

Hot gases

Hot gases

+ Fire origin

50- to 75-ft (15- to 23-m) hose stream reach

FIG. 7-7A. *Behavior of hot gases under a flat-roofed building.*

the cooling effect of vent flows. There are, as yet, no universally accepted conclusions from either fire experience or research.

Essential features specified in NFPA 204M include roof vents operated by fusible links, with curtain boards to confine heat and prevent lateral fire spread, as well as openings for fresh air makeup at low levels. (See Figure 7-7B.) Two general classes of fires are considered: (1) limited growth fires, which are not expected to grow past a predictable maximum size, and (2) continuous growth fires, which can be expected to grow indefinitely until intervention by fire fighters. Mechanical exhaust at ceiling apertures is treated as an option to natural ventilation at roof vents. The provisions of NFPA 204M may be applied to one-story buildings or to the top story of multiple-story buildings. With mechanical exhaust, lower stories of multiple-story buildings can also be vented.

INDUSTRIAL BUILDING VENTILATION

Although any opening in a roof will reduce some heat and smoke, building designers and fire protection engineers cannot rely on casual inclusion of skylights, windows, or monitors as adequate means of venting. Standards now exist (from Factory Mutual Research Corporation, Underwriters Laboratories Inc., etc.) that include design criteria and test procedures for unit vents and also call for simulated fire tests as well as engineering analysis.

Unit Vents

Unit vents are relatively small, usually 16 to 100 sq ft (1.49 to 9.29 m²) in area. Automatic unit vents are of two types, based on manner of operation—fusible link or drop-out plastic. (See Figure 7-7C.) The former type consists of a metal housing with lids that depend upon a temperature-rated fusible link to trip the lid mechanism, but alternative modes of operation (by heat, smoke, etc.) are possible. The latter type depends upon a temperature-sensitive, transparent or translucent, thermoplastic dome that deforms from its setting and falls out of the roof. Vents designed for manual operation are constructed with metal lids to resist elevated fire temperatures and may be opened from the floor with wires and cables. These units can also be modified for automatic operation.

Vents may be a single unit, in which the entire unit opens fully with a single sensor, or multiple units in rows, clusters, groups, or other arrays that will satisfy the venting requirements for the specific hazard.

Mechanical roof ventilators actuated by fire conditions are possible alternatives to roof vents based on natural ventilation, especially in lower stories of multiple-story buildings. These ventilators must be capable of functioning under expected high-temperature fire conditions.

Curtain Boards

In large-area buildings, unless vented areas are subdivided by means of walls or partitions, curtain boards are important for prompt and positive actuation of vents because they bank up heat in the curtained area. They also limit the spread of heat and smoke beneath the ceiling during the design duration of the venting system. Curtain boards can be made from any substantial noncombustible material that will resist the passage of smoke.

The depth of curtain boards would normally be selected to correspond with the design depth of the smoke layer. However, the depth should not be less than 20 percent of the ceiling height to prevent spillage of smoke under the curtain where depth and ceiling height are referenced to the center of the lowest roof vent. It would rarely be desirable to have the curtain extend below 10 ft (3 m) from the floor. Around special hazards, the curtain should preferably extend down to this limit. (See Figure 7-7D.)

The distance between curtain boards should not exceed eight times the ceiling height to ensure that vents remote from the fire will be effective within the curtained compartment. Smaller curtained areas may be desirable where occupancies are particularly vulnerable to damage. However, it is important that the distance between curtain boards not be less than two times the ceiling height, unless the curtain boards extend down to a depth of at least 40 percent of the ceiling height. The increased curtain depth is needed for these

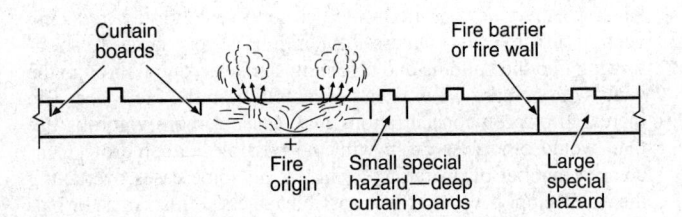

Curtain boards

Fire barrier or fire wall

+ Fire origin

Small special hazard—deep curtain boards

Large special hazard

FIG. 7-7B. *Behavior of combustion products under vented and curtained roof.*

FIG. 7-7C. *Building with roof vents.*

FIG. 7-7D. Deep curtain board around a special hazard. Equipped with proper venting, a noncombustible curtain board extending down from a ceiling around special hazards will prevent smoke and heat from mushrooming throughout the facility. A curtain board for a heat-treating department is shown.

small curtain spacings to prevent significant spillage of fire gases underneath the curtain because of the close proximity of the fire.

Dimensioning and Spacing of Vents

When the area of an individual vent becomes sufficiently large, there is a possibility that a core of clear air from beneath the smoke layer will be included in the exhaust from the vent, which reduces the effectiveness of the vent. To prevent inclusion of clear air, the area of a unit vent or cluster of vents should not exceed $2d^2$, where d is the design depth of the smoke layer or curtain board. In the case of a row of unit vents or a monitor vent, the width of the row or monitor should not exceed d.

A large number of small vents on close spacing is preferable to a small number of large vents on wide spacing. This ensures early operation of the first vents in a fire, reducing the likelihood of initial smoke excursions beyond the curtained area above the fire. In no case should the vent spacing in a rectangular matrix exceed $2H$, where H is the ceiling height measured from the floor to ceiling for flat roofs, and from the floor to the center of the vent for sloped roofs. Alternatively, in a nonrectangular vent matrix or plan, the distance between any point on the floor and the nearest vent should not exceed $2.8H$ (the diagonal of a square whose side is $2H$).

The total vent area for each curtained compartment under the ceiling depends upon the severity of the expected fire and is discussed later in this chapter.

Fresh Air Makeup

Openings must be provided at or near floor level to allow the introduction of makeup air. This is critical in today's tightly constructed and insulated buildings. The total area of these openings must normally be at least as great as the installed vent area for each curtained compartment; otherwise, the inlets will effectively throttle the vent flow. If doors and windows below the designed smoke layer cannot meet the total required inlet area, special air inlet provisions are necessary.

It is essential that a dependable means be provided for admitting inlet air within approximately one minute after the first vent opens. If prompt air inlet is not provided, the entire building may quickly fill with smoke, which will clear only slowly at lower levels (in the design clear layer) in response to the fully developed vent flow after the air inlets have been actuated.

Fire Characterization*

Each curtained compartment, or the ceiling area of buildings requiring no curtain boards, must be furnished with a total installed vent area (or exhaust capacity in case of mechanical ventilation) sufficient to vent fires of the expected severity. In addition to the expected fire severity, the installed vent area (or exhaust capacity) will depend upon the depth of the curtain boards or design depth of the smoke layer. Furthermore, unless the occupancy or hazard is such that the expected fire will peak or level off at a predictable maximum size, the installed vent area (or exhaust capacity) will also depend upon the minimum clear visibility design time as measured from the time the vents first operate.

The fire severity is expressed differently according to the class of the fire, which for purposes of calculating vent areas is classified as either a limited-growth or a continuous-growth fire.

Limited-growth fires: These are fires that are not expected to grow past a predictable maximum size in terms of heat-release rate, expressed in Btu per sec or Watt. Special-hazard fires can be assigned to this class, as can fires in occupancies with concentrations of combustibles separated by sufficiently wide aisles. The minimum aisle width to prevent lateral spread (by radiation), W_{min},[1] can be estimated from an equation for radiant flux from a fire and a (conservatively low) value for the ignition flux of most materials (1.8 Btu per sq ft per sec or 20.4 kW/m^2):

$$W_{min} \text{ (ft)} = 0.14 \left[Q \text{ (Btu/sec)} \right]^{1/2} \qquad (1)$$

For example, if the predicted maximum fire size is 30,000 Btu/sec (31.6 MW), the aisle width should be at least 24 ft (7.3 m). Table 7-7A contains examples of heat-release data, expressed as Btu per sq ft per sec (W/m^2) of floor area. To obtain heat release in Btu per sec (Watt), a number in the table is multiplied by the floor area underneath the combustible.

Continuous-growth fires: These are fires that can be expected to grow indefinitely until intervention by fire fighters. (See Figure 7-7E.) Starting after an incubation period, the heat-release rate of these fires grows continuously, proportional to the square of time. The growth time of a given fire is defined as the interval of time between the effective ignition time and the time when the fire reaches an intermediate energy release rate of 1000 Btu per sec (1054 kW). Any reference heat-release rate could have been chosen with a corresponding change in the growth time. The rate 1000 Btu per sec (1054 kW) was chosen for convenience. Table 7-7B contains examples of growth times for continuous-growth fires.

---

*The theory and application of venting outlined in this chapter is presented in U.S. custom units. These units reflect the system used in the original research and are consistent with the design recommendations and background materials contained in NFPA 204M. For those more familiar with SI units, quantities such as curtain board depth (in meters) and heat release rate (in kW) should be converted to the appropriate conventional units (ft and Btu/sec, respectively), using the conversion factors given at the end of this footnote and at various points throughout the text. After conversion, design data should be inserted in unit-dependent formulas to calculate vent areas and mass flow rates. As in NFPA 204M, 1 ft = 0.305 m; 1 Btu/sec = 1.054 kW.

TABLE 7-7A. *Limited Growth Fires*

Heat-release rate per unit floor area of fully involved combustibles, assuming 100 percent combustion efficiency. (PE = polyethylene; PS = polystyrene; PVC = polyvinyl chloride; PP = polypropylene; PU = polyurethane; FRP = fiberglass-reinforced polyester)

		Btu/sec/ft² of Floor Area
1.	Wood pallets, stack 1¹/₂ ft high (6–12% moisture)	120
2.	Wood pallets, stacked 5 ft high (6–12% moisture)	340
3.	Wood pallets, stacked 10 ft high (6–12% moisture)	600
4.	Wood pallets, stacked 15 ft high (6–12% moisture)	900
5.	Mail bags, filled, stored 5 ft high	35
6.	Cartons, compartmented, stacked 15 ft high	200
7.	PE letter trays, filled, stacked 5 ft high on cart	750
8.	PE trash barrels in cartons, stacked 15 ft high	260
9.	FRP shower stalls in cartons, stacked 15 ft high	110
10.	PE bottles packed in Item 6	550
11.	PE bottles in cartons, stacked 15 ft high	170
12.	PU insulation board, rigid foam, stacked 15 ft high	170
13.	PS jars packed in Item 6	1300
14.	PS tubs nested in cartons, stacked 14 ft high	450
15.	PS toy parts in cartons, stacked 15 ft high	180
16.	PS insulation board, rigid foam, stacked 14 ft high	280
17.	PVC bottles packed in Item 6	300
18.	PP tubs packed in Item 6	380
19.	PP and PE film in rolls, stacked 14 ft high	540
20.	Methyl alcohol	65
21.	Gasoline	200
22.	Kerosene	200
23.	Diesel oil	170

For SI units: 1 ft = 0.31 m; 1 sq ft = 0.093 m²; 1 Btu = 1.054 kW.

TABLE 7-7B. *Continuous Growth Fires*

Growth times of developing fires in various combustibles, assuming 100 percent combustion efficiency. (Pe = polyester; PE = polyethylene; PS = polystyrene; PVC = polyvinyl chloride; PP = polypropylene; PU = polyurethane; FRP = fiberglass-reinforced polyester)

		Growth Time (sec)
1.	Wood pallets, stack 1¹/₂ ft high (6–12% moisture)	160–320
2.	Wood pallets, stacked 5 ft high (6–12% moisture)	95–190
3.	Wood pallets, stacked 10 ft high (6–12% moisture)	80–120
4.	Wood pallets, stacked 16 ft high (6–12% moisture)	75–120
5.	Mail bags, filled, stored 5 ft high	190
6.	Cartons, compartmented, stacked 15 ft high	60
7.	Paper, vertical rolls, stacked 20 ft high	17–27
8.	Cotton (also Pe, Pe/Cot Acrylic/Nylon/Pe), garments in 12 ft high rack	22–43
9.	"Ordinary combustibles" rack storage, 15–30 ft high	40–270
10.	Paper products, densely packed in cartons, rack storage, 20 ft high	470
11.	PE letter trays, filled, stacked 5 ft high on cart	190
12.	PE trash barrels in cartons, stacked 15 ft high	55
13.	FRP shower stalls in cartons, stacked 15 ft high	85
14.	PE bottles packed in Item 6	85
15.	PE bottles in cartons, stacked 15 ft high	75
16.	PE pallets, stacked 3 ft high	150
17.	PE pallets, stacked 6–8 ft high	32–56
18.	PU mattress, single, horizontal	120
19.	PU insulation board, rigid foam, stacked 15 ft high	8
20.	PS jars packed in Item 6	55
21.	PS tubs nested in cartons, stacked 14 ft high	110
22.	PS toy parts in cartons, stacked 15 ft high	120
23.	PS insulation board, rigid foam, stacked 14 ft high	7
24.	PVC bottles packed in Item 6	9
25.	PP tubs packed in Item 6	10
26.	PP and PE film in rolls, stacked 14 ft high	40
27.	Distilled spirits in barrels, stacked 20 ft high	25–40

For SI unit: 1 ft = 0.305 m.

INSTALLED VENT AREA

Limited-Growth Fires

Using theory for air entrainment in fire plumes and gas flow through roof vents, required vent areas can be calculated. The results are presented in Figure 7-7F as recommended vent area per curtained compartment (in sq ft) *vs.* expected maximum heat-release rate (in Btu per sec) of the combustibles underneath the curtained compartment. Several ceiling heights in the range 15 ft to 60 ft (4.6 m to 18.3 m) are represented. The figure pertains to a curtain depth (or design smoke layer depth) that is 20 percent of the ceiling height. It is assumed that the fire is located on the floor, a conservative assumption since the lower the fire source, the greater the air entrainment rate from the clear layer and the required vent area. Moreover, an aerodynamic discharge coefficient of 0.6 has been assumed, which is normal for commercial unit vents. If the discharge coefficient is different from 0.6, the recommended vent areas need to be multiplied by the ratio of 0.6 to the actual discharge coefficient.

For each ceiling height in Figure 7-7F, the respective curve begins at a heat-release rate where vents operated by fusible links rated at 100°F (37.8°C) to 220°F (104°C) above ambient are first expected to operate promptly. Furthermore, for each ceiling height, the respective curve terminates near a heat-release rate, $Q_{feasible}$, beyond which the feasibility of roof venting, using the present approach, might be questioned. $Q_{feasible}$ can be estimated from the equation:

$$Q_{feasible}(Btu/sec) = 1130(H - d)^{5/2} \qquad (2)$$

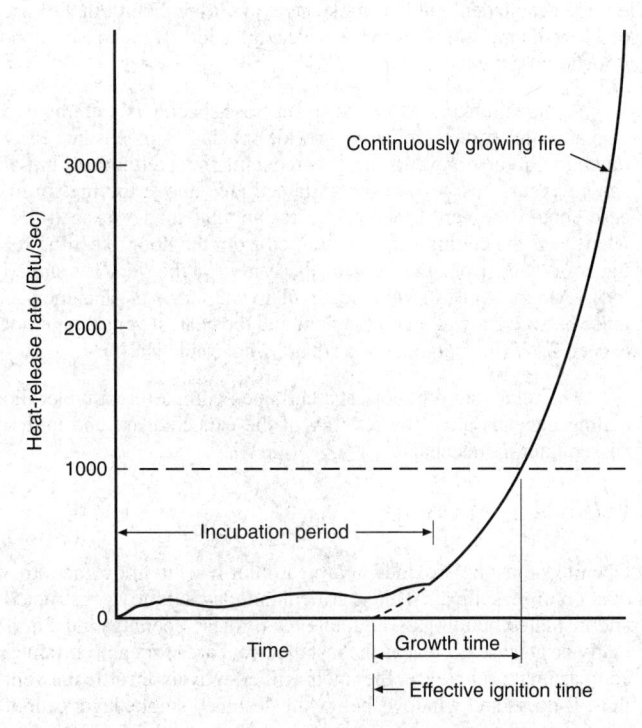

FIG. 7-7E. *Conceptual illustration of continuous fire growth.*

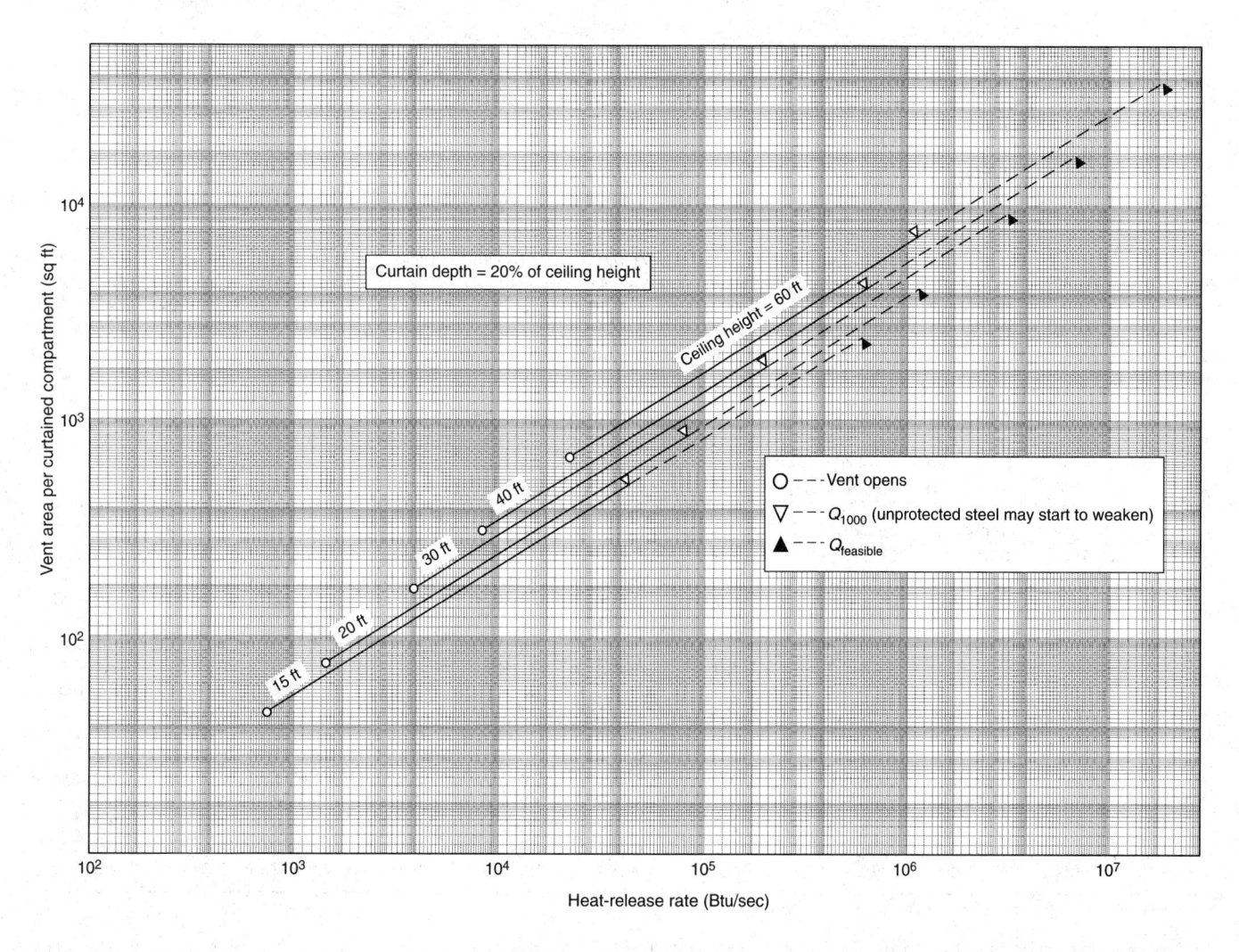

FIG. 7-7F. *Limited fire growth; recommended vent areas for each curtained compartment for various maximum heat-release rates (Btu per sec or kW/sec). (For SI units: 1 ft = 0.305 m; 1 sq ft = 0.0929 m².)*

where H is the ceiling height and d is the curtain depth ($0.2H$ in this case). At $Q_{feasible}$, the fire may become controlled by the ventilation rate allowed by the roof vents. Ventilation-controlled fires do not support a clear layer. Venting at heat-release rates greater than $Q_{feasible}$ to maintain a clear layer would require larger vent areas than indicated in Figure 7-7F.

Along the dashed segments of the curves in Figure 7-7F, gas temperatures in excess of 1000°F (538°C) may be reached; unprotected structural steel may begin to lose strength and the possibility of flashover exists within the curtained area. The lowest rate of heat release at which this may occur, Q_{1000}, has been estimated from the equation

$$Q_{1000}(\text{Btu/sec}) = 69(H - d)^{5/2} \qquad (3)$$

where, in the case of Figure 7-7F, $d = 0.2H$.

For curtain depths greater than 20 percent of the ceiling height, the vent areas read from Figure 7-7F may be multiplied by the factors indicated in Table 7-7C. New values for $Q_{feasible}$ and Q_{1000} can be calculated using Equations 2 and 3, respectively, by inserting the appropriate values of H and d.

TABLE 7-7C. *Multiplication Factors for Vent Areas in Figure 7-7F for Curtain Depths Other Than 20 Percent of Ceiling Height*

Curtain Depth in Percent of Ceiling Height	Multiplication Factor
30	0.71
40	0.53
50	0.40
60	0.29
70	0.20
80	0.13

Continuous-Growth Fires

Required vent areas for each curtained compartment for fires in this class depend upon ceiling height (H), growth time, spacing of curtain boards (S_c), vent spacing, and means of vent actuation. They also depend upon the desired minimum clear visibility time from the time the first vents operate. The minimum clear visibility design time will facilitate such activities as locating the fire, appraising the severity and extent of the fire, evacuating the building, and making

TABLE 7-7D. *Vent Areas for Curtained Compartments*

Vent area (in 1000 sq ft) per curtained compartment for heat-responsive device-operated vents with various curtain board spacings (S_c) and minimum clear-visibility design times (5, 10, or 15 min).

Ceiling Height, H	Growth Time (sec)	$S_c = 2 \times H$			$S_c = 4 \times H$			$S_c = 8 \times H$		
		5 min	10 min	15 min	5 min	10 min	15 min	5 min	10 min	15 min
15 ft	20				(1.9–2.1)			(2.2–2.5)		
	40	0.87–0.97			(0.98–1.1)	(1.8–2.0)		(1.1–1.4)	(2.0–2.3)	
	80	(0.44–0.52)	(0.82–0.90)		(0.52–0.64)	(0.90–1.0)	(1.3–1.4)	(0.64–0.83)	(1.0–1.2)	(1.5–1.7)
	150	0.25–0.32	(0.43–0.50)	(0.63–0.70)	0.31–0.41	(0.50–0.60)	(.70–.80)	(0.41–0.58)	(.60–.77)	(0.81–0.98)
	300	0.15–0.20	0.23–0.29	0.32–0.38	0.20–0.28	0.28–0.36	0.37–0.46	0.27–0.41	0.36–0.51	(0.46–0.61)
	600	0.10–0.14	0.13–0.18	0.17–0.23	0.13–0.21	0.17–0.25	0.21–0.29	0.20–0.34	0.24–0.38	0.29–0.43
20 ft	20	2.2–7.3			(2.5–2.7)			(2.7–3.2)		
	40	(1.1–1.2)	2.1–2.2		(1.2–1.5)	(2.2–2.5)	(3.3–3.6)	(1.5–1.8)	(2.5–2.9)	(3.6–4.0)
	80	0.56–0.68	(1.0–1.1)	1.5–1.6	0.68–0.86	(1.1–1.3)	(1.6–1.8)	(0.87–1.2)	(1.3–1.6)	(1.8–2.1)
	150	0.34–0.43	0.55–0.65	(0.77–0.88)	0.43–0.58	0.64–0.80	(0.88–1.0)	0.58–0.83	(0.80–1.1)	(1.0–1.3)
	300	0.21–0.29	0.30–0.39	0.40–0.50	0.28–0.41	0.38–0.52	0.48–0.63	0.40–0.65	0.51–0.76	0.62–0.88
	600	0.14–0.22	0.19–0.27	0.23–0.32	0.20–0.33	0.25–0.38	0.30–0.44	0.32–0.56	0.37–0.61	0.42–0.67
30 ft	20	(2.9–3.2)	5.6–6.0		(3.2–3.8)	(6.0–6.6)		(3.8–4.6)	(6.7–7.6)	
	40	1.5–1.8	(2.7–3.0)	4.0–4.3	(1.8–2.2)	(3.0–3.4)	(4.3–4.8)	(2.2–2.9)	(3.5–4.2)	(4.8–5.6)
	80	0.82–1.0	1.4–1.6	(2.0–2.2)	1.0–1.4	1.6–2.0	(2.2–2.6)	1.4–2.0	(2.0–2.6)	(2.6–3.2)
	150	0.51–0.71	0.78–0.99	1.1–1.3	0.68–0.99	0.96–1.3	1.3–1.6	0.97–1.5	1.3–1.8	(1.6–2.2)
	300	0.34–0.53	0.47–0.66	0.59–0.80	0.48–0.77	0.61–0.91	0.74–1.1	0.74–1.3	0.88–1.4	1.0–1.6
	600	0.26–0.44	0.32–0.50	0.38–0.57	0.38–0.67	0.44–0.74	0.50–0.81	0.62–1.2	0.69–1.2	0.76–1.3
40 ft	20	(3.6–4.1)	6.7–7.3	10–11	(4.1–4.9)	(7.4–8.2)	(11–12)	(5.0–6.3)	(8.3–9.7)	(12–13)
	40	1.9–2.3	(3.3–3.8)	(4.8–5.3)	2.3–3.0	(3.8–4.5)	(5.3–6.1)	(3.0–4.1)	(4.5–5.6)	(6.1–7.3)
	80	1.1–1.5	1.7–2.1	2.4–2.8	1.4–2.0	2.1–2.7	2.8–3.4	2.0–3.0	2.7–3.7	(3.4–4.5)
	150	0.72–1.1	1.0–1.4	1.4–1.8	0.98–1.5	1.3–1.9	1.7–2.2	1.5–2.4	1.8–2.8	2.2–3.2
	300	0.51–0.84	0.66–1.0	0.81–1.2	0.74–1.3	0.89–1.4	1.1–1.6	1.2–2.2	1.4–2.3	1.5–2.5
	600	0.41–0.74	0.48–0.81	0.55–0.89	0.61–1.1	0.69–1.2	0.77–1.3	1.1–2.0	1.1–2.1	1.2–2.2
60 ft	20	4.9–5.9	(8.9–10)	(13–14)	5.9–7.4	(10–12)	(14–16)	(7.5–10)	(12–14)	(16–19)
	40	2.8–3.6	4.6–5.5	6.5–7.4	3.5–4.8	5.4–6.7	(7.3–8.8)	4.8–7.0	(6.7–9.1)	(8.7–11)
	80	1.7–2.5	2.5–3.4	3.4–4.3	2.3–3.5	3.2–4.4	4.1–5.4	3.4–5.6	4.3–6.6	5.2–7.6
	150	1.2–2.0	1.6–2.4	2.1–2.9	1.7–2.9	2.2–3.4	2.6–3.9	2.8–4.9	3.2–5.4	3.7–5.9
	300	0.95–1.7	1.2–1.9	1.3–2.1	1.4–2.6	1.6–2.8	1.8–3.1	2.4–4.6	2.6–4.8	2.9–5.1
	600	0.82–1.6	0.92–1.7	1.0–1.8	1.3–2.4	1.4–2.6	1.5–2.7	2.2–4.4	2.3–4.5	2.5–4.6

For SI units: 1 ft = 0.3048 m; 1 sq ft = 0.093 m².
NOTES:
1. Vents are assumed to be spaced at one-half of the curtain board spacing.
2. Curtain depth assumed at 20 percent of ceiling height.
3. Each entry is the vent-area range (in 1000 sq ft) associated with heat-responsive devices rated between 100°F and 220°F (37.8°C and 104°C) above ambient temperature.
4. No entries: Heat-release rates greater than $Q_{feasible}$.
5. Entries boxed-in: Not possible, but needed for curtain depths greater than 20 percent.
6. Entries in parentheses: Correspond to levels of heat release greater than Q_{1000}.

an informed decision on deployment of personnel and equipment to be used for fire fighting.

Table 7-7D lists recommended vent areas per curtained compartment for: (1) the minimum recommended curtain depth of 20 percent of ceiling height and (2) vents spaced at no more than one-half of the curtain board spacing. For other than square curtains, the spacing S_c is interpreted as the largest spacing defined by the curtained area. The tabulated areas are approximate, pertaining to vents that are operated by heat-responsive devices of an average thermal inertia, or response-time index (RTI), of 520 (ft • sec)$^{1/2}$[287 (m • sec)$^{1/2}$] and rated between 100°F (37.8°C) and 220°F (104°C) above ambient temperature.[2] Each entry in the table gives the range of vent areas in 1000 sq ft (92.9 m²) associated with the selected range of temperature ratings. Entries boxed-in are not possible (since the vent areas exceed the largest possible curtained area of $S_c \times S_c$); however, these entries may be needed for curtain depths greater than 20 percent of ceiling height as treated in a later paragraph. Where values are not given, heat-release rates are greater than $Q_{feasible}$, as

given in Equation 2; the vent area associated with $Q_{feasible}$ is $A_{feasible}$, which has been calculated from:

$$A_{feasible}\left(ft^2\right) = 8.5(H-d)^{5/2}/d^{1/2} \tag{4}$$

where, in the case of Table 7-7D, $d = 0.2H$. Entries in parentheses in the table correspond to levels of heat release greater than Q_{1000}, as given in Equation 3; the vent area associated with Q_{1000} is A_{1000}, which has been calculated from:

$$A_{1000}\left(ft^2\right) = 1.6(H-d)^{5/2}/d^{1/2} \tag{5}$$

where, in the case of Table 7-7D, $d = 0.2H$.

To illustrate the use of Table 7-7D, consider an installation with heat-responsive devices rated approximately 100°F (37.8°C) above ambient, a ceiling height of 20 ft (6.1 m), a growth time of approximately 150 sec, a (square) curtain spacing of 80 ft or 24.4 m

$(S_c = 4 \times H)$, a curtain depth of 4 ft or 1.2 m ($^4/_{20}$ = 20 percent of ceiling height), and a minimum clear visibility time of 10 min. In Table 7-7D, the lower limit 100°F (37.8°C) of the appropriate entry indicates a vent area per curtained compartment of $0.64 \times 1000 = 640$ sq ft (59.5 m²) for this case.

The recommended vent area for each curtained compartment is reduced if larger curtain depths than minimum (20 percent of ceiling height) are installed. The reduced areas are calculated by multiplying the values listed in Table 7-7D by the appropriate multiplication factor listed in Table 7-7C. To determine if Notes 4 or 6 from Table 7-7D apply to the newly derived values for vent area, the value of A_{feasible} associated with Q_{feasible} is calculated from Equation 4, and the value of A_{1000} associated with Q_{1000} is calculated from Equation 5. Note 4 applies if a vent area is larger than A_{feasible}, and Note 6 applies if a vent area is larger than A_{1000}.

Vent areas for each curtained compartment should be distributed among individual vents within the constraints discussed in a subsection, "Dimensioning and Spacing of Vents," of this chapter. In some cases, the calculated number of vents may be so large that the vent spacing will be considerably smaller than the design spacing for vents assumed in Table 7-7D ($^1/_2 S_c$). The closer vent spacing implies earlier operation of the first vents than is the case for the designs of Table 7-7D. Earlier operation, such as that provided by an auxiliary fire detection system, would provide extra clear visibility times beyond the values of 5, 10, and 15 min selected in Table 7-7D. The extra time available with vents spaced at less than $^1/_2 S_c$ may be considered to represent a safety factor.

Selection of Design Basis

The vent area in a curtained compartment need not exceed the vent area recommended for the largest limited-growth fire predicted for combustibles beneath the curtained area. Using sufficiently small concentrations of combustibles and minimum aisle widths according to Equation 1, it may be possible to satisfy the venting needs using smaller vent areas than required by a continuous-growth fire vent design. For example, in the illustration discussed previously for the use of Table 7-7D, the continuous-growth fire vent design for a growth time of 150 sec led to a required vent area for each curtained compartment of 640 sq ft (59.5 m²). A limited-growth fire design for a 50 sq ft (4.7 m²) floor area concentration at 200 Btu per sec per sq ft (2271 kW/m²) (if that is the heat-release rate per unit floor area of the combustible considered) involves a total heat-release rate per fuel concentration of 10,000 Btu per sec (10540 kW) and a minimum aisle width of 14 ft (4.3 m), according to Equation 1. For the relevant vent parameters of a ceiling height of 20 ft (16.1 m), curtain depth at 20 percent of ceiling height, and heat-release rate of 10,000 Btu per sec (10,540 kW), Figure 7-7F indicates a vent area per curtained compartment of 250 sq ft (23.2 m²), which is considerably smaller than the 640 sq ft (59.5 m²) vent area according to the design for continuous-growth fire. Of course, the combustible and floor area concentrations must be carefully controlled when limited-growth fire vent designs are selected. Furthermore, designs incorporating heat-release rates greater than Q_{1000} are risky, since the fire may flashover to all the combustibles under the curtained area. Such designs should be based on the potential heat-release rate of all the combustibles under the curtained area and should not be attempted at all if the resulting heat-release rate approaches Q_{feasible}. For continuous-growth fires, designs giving vent areas greater than A_{1000} cannot be recommended because of the flashover potential.

Mechanical Ventilators

For mechanical venting systems capable of functioning under the expected fire exposure, recommended exhaust capacities per curtained compartments are obtained by simple conversion from the

TABLE 7-7E.   Conversion from Vent Area of Gravity Vents to Equivalent Mechanical Exhaust Capacity

Curtain Depth (ft)	Mechanical Exhaust Capacity per Unit Area of Gravity Vent (SCFM*/sq ft)
6	354
8	409
10	457
12	501
16	578
20	647
24	708

*SCFM = Standard cubic ft per min (standard temperature and pressure).
For SI units: 1 ft = 0.305 m; 1 sq ft = 0.093 m².

recommended vent areas per curtained compartment. The conversion depends upon the depth of the curtain board or the design depth of the smoke layer, according to Table 7-7E.

ELEMENTS OF VENTING THEORY

To stop the descent of a smoke layer, the mass flow rate of hot gas out of the vents, $\dot{m}_v$, must match the mass injection rate of gas by the fire plume at the interface with the smoke layer, $\dot{m}_p$. (See Figure 7-7G.) Therefore, expressions are needed for the vent flow and for the plume flow.

Vent Flow

The vent flow can be calculated from the following equation:[3,4]

$$\dot{m}_v = \left(2\rho_o^2 g\right)^{1/2} \left[\frac{T_o(T - T_o)}{T^2}\right]^{1/2} A_v d^{1/2} \qquad (6)$$

where $\dot{m}_v$ is the mass flow rate through the vent (lb per sec or kg/sec); ρ_o is the density of the ambient air (lb per cu ft or kg/m³); g is the acceleration of gravity (ft/sec² or m/sec²); T_o is the ambient temperature (°R or K); T is the smoke layer temperature (°R or K); A_v is the aerodynamic vent area (sq ft or m²); and d is the depth of the smoke layer (ft or m), generally interpreted as the height of the center of the vent above the smoke interface.

The aerodynamic vent area, A_v, is always smaller than the geometric vent area. For simple apertures, A_v may be taken as 0.6 times the geometric through-flow area.

The temperature function $\left[\dfrac{T_o(T - T_o)}{T^2}\right]^{1/2}$

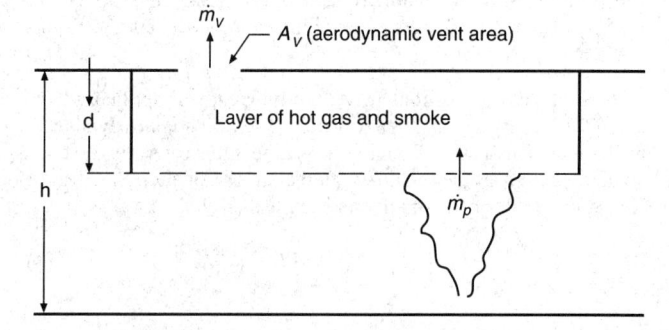

FIG. 7-7G.   A schematic of a venting system.

can be approximated at 0.50 for smoke layer temperatures in the range of 275 to 1400°F (135 to 760°C), which covers most vent applications. Then Equation 6 becomes simply

$$\dot{m}_v = \left(\rho_o{}^2 g/2\right)^{1/2} A_v d^{1/2} \qquad (7)$$

To prevent air from the clear layer below from being entrained in the vent exhaust, it has been proposed that the area of the vent not exceed $2d^2$.[5] If the vent is long and narrow, an analogous requirement would be that the width of the vent not exceed d.

Plume Flow

The mass flow rate injected into the smoke layer by the fire plume is practically equal to the air entrainment rate of the plume from the clear layer; the contribution of the fire source itself is small in comparison. Before the 1982 revision of NFPA 204M an entrainment theory[6] was already in existence. However, that theory was considered somewhat unwieldy for the format of NFPA 204M and an alternative, more adaptable, and probably more accurate theory was developed in the Appendix of 204M.

The plume flow, $\dot{m}_p$, from the new theory is calculated according to either of two equations, depending upon whether the convective heat-release rate of the fire is larger or smaller than a critical value, Q_c, defined as (fire assumed to be located on floor of building):

$$Q_c \ (\text{Btu/sec}) = 11.3(H-d)^{1/2} \qquad (8)$$

where H and d are in ft.

The plume flow equations are:

For $Q \le Q_c$

$$\dot{m}_p \ (\text{lb/sec}) = 0.022 Q^{1/3}(H-d)^{5/3}$$

$$\left[1 + 0.19 Q^{2/3}/(H-d)^{5/3}\right] \qquad (9)$$

For $Q \ge Q_c$

$$\dot{m}_p \ (\text{lb/sec}) = 0.097(H-d)^{5/2}\left(Q/Q_c\right)^{3/5} \qquad (10)$$

where Q and Q_c are in Btu/sec, and H and d are in ft.

Required Vent Area

For $Q \le Q_c$, where Q_c is given by Equation 8, the expression for $\dot{m}_p$ in Equation 9 is set equal to the expression for $\dot{m}_v$ in Equation 7 and solved for A_v, the aerodynamic vent area. For $Q > Q_c$, the expression for $\dot{m}_v$ in Equation 10 is set equal to the expression for m_v in Equation 7 and solved for A_v.

The resulting relations for A_v can be greatly simplified whenever Q/Q_c is larger than approximately 0.2, which is nearly always the case in vent design. Then the two expressions for A_v can be consolidated into a single relation with the aid of Equation 8, valid for heat-release rates both smaller and larger than Q_c:

$$A_v \ \left(\text{ft}^2\right) = 0.075 \ Q^{3/5} \frac{H-d}{d^{1/2}} \qquad (11)$$

where Q is in Btu per sec, and H and d are in ft.

For fires growing with the second power of time (continuous-growth fires), Equation 11 can be written:

$$A_v \ \left(\text{ft}^2\right) = 4.8\left[(t_d + t_r)/t_g\right]^{6/5}(H-d)/d^{1/2} \qquad (12)$$

where H and d are in ft; t_d is the detection time, or the time the first vents operate; t_r is the additional, clear visibility time; and t_g is the growth time defined in the subsection "Fire Characterization."

These are the basic relations used for calculating vent areas in the 1982 edition of NFPA 204M with adjustments for a discharge coefficient of 0.6 (aerodynamic vent areas, A_v, divided by 0.6). In Equation 12, the detection time, t_d, was taken as the time of operation of the first vent in a square matrix (farthest possible vent from the fire location). The vents were assumed to be actuated by heat-responsive devices of various temperature ratings above the ambient temperature. Link actuation times (t_d) were calculated from a thermal response equation for heat-responsive devices derived previously,[7] together with generalized data on gas temperatures and velocities under extensive flat ceilings.[8,9] The RTI value of the heat-responsive device was taken as 520 (ft•sec)$^{1/2}$ [287(m•sec)$^{1/2}$], which is considered to be a conservatively high value for heat-responsive devices listed by testing laboratories.

Limiting Heat-Release Rates

Q_{1000} was defined in conjunction with Equation 3 as the heat-release rate at which gas temperatures in excess of 1000°F (538°C) may be reached with possible weakening of structural steel and potentials for flashover within the curtained areas. The expression for Q_{1000} was derived first using Equation 10, with substitution for Q_c from Equation 8, to obtain $\dot{m}_p$ expressed as:

$$\dot{m}_p \ (\text{lb/sec}) = 0.0226(H-d)Q^{3/5} \qquad (13)$$

where Q is in Btu/sec, and H and d are in ft.

The average temperature rise in the smoke layer, ΔT_s (°F), follows from:

$$\dot{m}_p c_p \Delta T = Q \qquad (14)$$

where c_p (Btu/lb/°F or kJ/kg/k) is the specific heat of air and where it has been assumed that no heat is lost to the ceiling and walls. With substitution of Equation 13 for $\dot{m}_p$, c_p = Btu/lb/°F(1.00kJ/kg/k) and ΔT = 1000°F (538°C) in Equation 14, the resulting expression can be solved for Q to give Equation 3. Of course, the result is quite conservative, since a significant heat loss may occur to the ceiling and the walls of the curtained compartment.

Q_{feasible} was described in conjunction with Equation 2 as a heat-release rate beyond which the feasibility of roof venting, using the present approach, might be questioned because of the onset of ventilation control of the burning process. Experience has indicated that ventilation-controlled fires do not support a clear layer as, for example, in the case of the typical, flashed-over room fire. According to experiments on wood crib fires in enclosures, vented through doors or windows,[10] ventilation control appears to set in when the so-called "ventilation parameter," $A_w\sqrt{h_w}$, in ratio to the mass burning rate, R, is less than 266 ft$^{5/2}$ per lb per sec), where A_w is the window (or door) area, and h_w is the window (or door) height. Generalized to any combustible, the limiting ratio might be expressed as $A_w\sqrt{h_w}$, in ratio to the stoichiometric air requirement associated with the mass burning rate, Rr, where r is the stoichiometric mass ratio, air to combustible (assumed at 5.5 for wood). Now, the ventilation parameter, $A_w\sqrt{h_w}$, is proportioned to the mass flow rate of air through a window or door opening, $\dot{m}_v$.[11]

$$\dot{m}_v \ (\text{lb/sec}) = 0.064 A_w \sqrt{h_w} \qquad (15)$$

where A_w is in sq ft, and h_w is in ft.

The limiting condition for ventilation control can, therefore, be expressed in terms of the mass venting rate of air in ratio to the stoichiometric air requirement, $\dot{m}_v R_r$, which turns out to be approximately 3. Using the fact that Rr can be written $Q/(H_c/r)$, where Q is the heat-release rate and H_c is the heat of combustion, the limiting mass rate ratio can be converted to a limiting heat-release rate:

$$Q = \frac{1}{3} \dot{m}_v (H_c/r) \qquad (16)$$

In application to roof vents, $\dot{m}_v$ can be expressed as in Equation 7. Furthermore, H_c/r is quite constant among common combustibles. Substituting Equation 7 and taking $H_c/r = 1328$ Btu per lb (3086 kJ/kg), Equation 16 becomes:

$$Q(\text{Btu/sec}) = 133 A_v d^{1/2} \qquad (17)$$

where A_v is in sq ft, and d is in ft.

In the venting approach used in NFPA 204M, the vent area A_v is given by Equation 11, which, divided by 0.6 to convert to geometric vent area and then substituted in Equation 17, gives a relation that can be solved for Q. The result is the expression for the limiting heat-release rate, Q_{feasible}, in Equation 2.

BIBLIOGRAPHY

References Cited

1. Alpert, R. L., and Ward, E. J., "Evaluation of Unsprinklered Fire Hazards," *Fire Safety Journal*, Vol. 7, 1984, pp. 127–143.
2. Heskestad, G., and Smith, H. F., "Plunge Test for Determination of Sprinkler Sensitivity," Technical Report FMRC 3A1E2.RR, 1980, Factory Mutual Research Corporation, Norwood, MA.
3. Thomas, P. H., *et al.*, "Investigations into the Flow of Hot Gases in Roof Venting," Fire Research Technical Paper No. 7, 1963, Department of Scientific and Industrial Research and Fire Offices' Committee, Joint Fire Research Organization, Her Majesty's Stationery Office, London, England.
4. Thomas, P. H., and Hinkley, P. L., "Design of Roof-Venting Systems for Single-Story Buildings," Fire Research Technical Paper No. 10, 1964, Department of Scientific and Industrial Research and Fire Offices' Committee, Joint Fire Research Organization, Her Majesty's Stationery Office, London, England.
5. Thomas, P. H., and Hinkley, P. L., "Design of Roof-Venting Systems," Department of Scientific and Industrial Research and Fire Offices' Committee, Joint Fire Research Organization, Her Majesty's Stationery Office, London, England.
6. Thomas, P. H., *et al.*, "Investigations into the Flow of Hot Gases," Department of Scientific and Industrial Research and Fire Offices' Committee, Joint Fire Research Organization, Her Majesty's Stationery Office, London, England.
7. Heskestad, G., and Smith, H. F., "Investigation of a New Sprinkler Sensitivity Approval Test: The Plunge Test," Technical Report FMRC 22485, 1976, Factory Mutual Research Corporation, Norwood, MA.
8. Heskestad, G., "Similarity Relations for the Initial Convective Flow Generated by Fire," Paper No. 72-WA/HT-17, 1972, American Society of Mechanical Engineers, New York.
9. Heskestad, G. and Delichatsios, M. A., "The Initial Convective Flow in Fire," Seventeenth Symposium (International) on Combustion, 1979, The Combustion Institute, Pittsburgh, PA, pp. 1113–1123.
10. Croce, P. A., "Modeling of Vented Enclosure Fires, Part I: Quasi-Steady Wood-Crib Source Fire," Technical Report FMRC 7A0R5.GU, 1978, Factory Mutual Research Corporation, Norwood, MA.
11. Harmathy, T. Z., "Ventilation of Fully-Developed Compartment Fires," *Combustion and Flame*, Vol. 37, 1980, pp. 25–39.

NFPA Codes, Standards, and Recommended Practices

Reference to the following NFPA codes, standards, and recommended practices will provide further information on venting practices discussed in this chapter. (See the latest version of *The NFPA Catalog* for availability of current editions of the following documents.)

NFPA 204M, *Guide for Smoke and Heat Venting*

Additional Reading

Chow, W. K., and Fong, N. K., "Numerical Studies on the Interaction Between a Sprinkler Water Spray and a Natural Venting Inside a Building," Hong Kong Polytechnic, Hong Kong, HTD-Vol. 141; American Institute of Aeronautics and Astronautics/American Society of Mechanical Engineers (AIAA/ASME) Thermophysics and Heat Transfer Conference, Heat and Mass Transfer in Fires, HTD-Vol. 141, June 18–20, 1990, Seattle, WA, Am. Soc. of Mechanical Engineers, New York, 1990, pp. 93–100.
Courtier, G. A. C., and Wild, J. A., "Development of Axial Flow Fans for the Venting of Hot Fire Smoke," *Fire Technology*, Vol. 27, No. 3, Aug. 1991, pp. 250–265.
Davis, W. D., and Cooper, L. Y., "Estimating the Environment and the Response of Sprinkler Links in Compartment Fires with Draft Curtains and Fusible-Link-Activated Ceiling Vents, Part I," NBSIR 88-3734, Apr. 1988, Center for Fire Research, Gaithersburg, MD.
Davis, W. D., and Cooper, L. Y., "Estimating the Environment and Response of Sprinkler Links in Compartment Fires with Draft Curtains and Fusible-Link-Activated Ceiling Vents, Part 2: User Guide for the Computer Code LAVENT," NISTIR 89-4122, Aug. 1989, National Institutes for Standards and Technology, Gaithersburg, MD.
Degenkolb, J. G., "Field Test of an Atrium Exhaust System," *Building Standards*, Vol. 57, No. 2, 1988, pp. 18–19.
Edwards, J., "Theatre Fires—The Importance of Smoke Ventilation," *Fire Prevention*, No. 198, Apr. 1987, pp. 37–38.
Edwards, J., "Vents and Sprinklers—Controversy Resolved," *Fire Prevention*, No. 211, July/Aug. 1988, pp. 39–42.
Heskestad, G., "Smoke Movement and Venting," *Fire Safety Journal*, Vol. 11, Nos. 1 and 2, 1986, pp. 77–83.
Hansell, G., "Problems of Effective Venting in Warehouses," *Fire Prevention*, No. 262, Sept. 1993, pp. 40–42.
Harada, K., *et al.*, "Study on Design Methodologies of Smoke Venting Systems. Part 2. Design of Air Supply Openings," *Bulletin of Japanese Association of Fire Science and Engineering*, Vol. 41, No. 2, 1993, pp. 29–35.
Hinkley, P. L., "Case for Combining Venting and Sprinkler Systems," *Fire Engineers Journal*, Vol. 47, No. 145, June 1987, pp. 17–22.
Hinkley, P. L., "The Effect of Smoke Venting on the Operation of Sprinklers Subsequent to the First," *Fire Safety Journal*, Vol. 14, No. 4, 1989, pp. 221–240.
Hinkley, P. L., *et al.*, "Large-Scale Experiments with Roof-Vents and Sprinklers. Part 2. The Operation of Sprinklers and the Effect of Venting with Growing Fires," *Fire Science and Technology*, Vol. 13, No. 1/2, 1993, pp. 43–59.
Hinkley, P. L., "Rates of Production of Hot Gases in Roof Venting Experiments," *Fire Safety Journal*, Vol. 11, No. 1, 1986, pp. 57–65.
Hinkley, P. L., "Smoke and Heat Venting," *SFPE Handbook of Fire Protection Engineering*, 2nd ed., National Fire Protection Association, Quincy, MA, 1995.
Hinkley, P. L., "Sprinkler Operation and the Effect of Venting: Studies Using a Zone Model," Fire Research Station, Borehamwood, England, BRE BR 213, 1992.
Holmstedt, G., "Sprinkler and Fire Venting Interaction. Literature Survey on Modelling Approaches and Experiments Available and Recommendations for Further Studies," Swedish Fire Research Board, Stockholm, Sweden, 92178 AR GH/AB, Sept. 1992.
Hopkinson, J., "Venting a Smoky Fire in a Hotel Stairway," *Fire Safety Journal*, Vol. 10, No. 1, 1986, pp. 11–18.
Hughes, L. D., "Positive-Pressure Ventilation in a Test Setting," *Fire Engineering*, Vol. 142, No. 12, Dec. 1989, pp. 56–59.

Ishino, O., and Tanaka, T., "Efficiency of Smoke Control of Atria by Natural Venting," Twelfth Joint Panel Meeting of the U. S./Japan Government Cooperative Program on Natural Resources (UJNR) on Fire Research and Safety, Oct. 27–Nov. 2, 1992, Tsukuba, Japan, Building Research Inst., Ibaraki, Japan, 1992, pp. 200–207.

Ishino, O., and Tanaka, T., "Efficiency of Smoke Control of Atria by Mechanical Smoke Venting," *Proceedings* of the First International Conference on Fire Science and Engineering, ASIAFLAM '95, March 15–16, 1995, Kowloon, Hong Kong, 1995, pp. 323–334.

Kadoya, M., *et al.*, "Study on Design Methodologies of Smoke Venting Systems. Part 1. Field Measurements of Air Flow Rates and Pressure Differences," *Bulletin of Japanese Association of Fire Science and Engineering*, Vol. 41, No. 2, 1993, pp. 19–28.

Kandola, B., "Effects of Atmospheric Wind on Flows Through Natural Convection Roof Vents," *Fire Technology*, Vol. 26, No. 2, May 1990, pp. 106–120.

Kim, A. K., "Review of Venting of Compartment Fires to Aid Firefighting," National Research Council of Canada, Ottawa, Ontario, Internal Report 627, April 1992.

Klote, J. H., and Cooper, L. Y., "Model of a Simple Fan-Resistance Ventilation System and Its Application to Fire Modeling," NIST 89-4141, Sept. 1989, National Institute for Standards and Technology, Gaithersburg, MD.

Langdon-Thomas, G. J., and Hinkley, P. L., "Fire Venting—Small-Story Industrial Buildings," *Fire Note No. 5*, 1965, Ministry of Technology, Industrial, and Fire Offices Committee, Joint Fire Research Organization, Her Majesty's Stationery Office, London, England.

Lathrop, J. K., "Atrium Fire Proves Difficult to Ventilate," *Fire Journal*, Vol. 73, No. 1, 1979.

Lyons, R. J., "Automatic Heat and Smoke Venting," *Progressive Architecture*, Apr. 1972.

Morgan, H. P., "Roof Venting—Similarities Between Large and Small-Scale Calculations," *Fire Prevention*, No. 198, Apr. 1987, pp. 32–33.

Papoutsis, T., and Reichenbach, S., "Venting Single-Ply Roofs," *Fire Engineering*, Vol. 145, No. 2, Feb. 1992, pp. 29–30, 34, 36–37.

Poole, G. A., "Contribution of Emergency Vent Design to Loss Prevention in Process Plants," *Fire Prevention*, No. 197, Mar. 1987, pp. 28–31.

Schneider, U., Lebeda, C., and Max, U., "Entrauchung von grossen Geschossflächen und unterirdischen Parkdecks. [Smoke Venting of Large Storeys and Underground Car-Parks.]," *VFDB*, Vol. 4, Nov. 1994, pp. 138–149.

Searle, B., "Efficient Smoke Extraction: Saving Lives and Money," *Civil Engineering* (London), July–Aug. 1988, pp. 16–17.

Shea, L. E., "Flammable Vapor Exhaust Systems," *Building Standards*, Vol. 56, No. 1, 1987, pp. 4–8.

Skelton, R. E., "A Remotely Controlled Fire-Venting System in a Shopping Complex," *Fire Journal*, Vol. 72, No. 1, Jan. 1978, p. 76.

Swift, I., "NFPA 68 Guide for Venting of Deflagrations: What's New and How It Affects You," *Proceedings* of the American Institute of Chemical Engineers National Meeting, AIChE, 1988, p. 30.

Talbert, J. H., "Smoke and Heat Venting for Sprinklered Buildings: A Different Perspective," *Fire Journal*, Vol. 81, No. 3, 1987, p. 26.

Teter, W. J., "A Blow to a Killer," *Industrial Fire World*, Vol. 3, No. 4, Aug. 1987, pp. 12–14.

Tewarson, A., "Fire Ventilation," *Combustion and Flame*, Vol. 53, Nos. 1–3, Nov. 1983, pp. 123–134.

Tewarson, A., and Steciak, J., *Fire Ventilation*, Elsevier, New York, 1983.

Than, C. F., "Smoke Venting by Gravity Roof Ventilators Under Windy Conditions," *Journal of Fire Protection Engineers*, Vol. 4, No. 1, Jan./Feb./March 1992, pp. 1–4.

"Venting Plus Sprinklers—The Case Against," *Fire International*, Vol. 10, No. 101, 1986, pp. 31–32, 34.

Waterman, T. E., "Fire Venting of Sprinklered Buildings," *Fire Journal*, Vol. 78, No. 2, Mar. 1984, pp. 30–39, 86.

Waterman, T. E., *et al.*, "Fire Venting of Sprinklered Buildings," IITRI Project J08385 for Fire Venting Research Committee, July 1982, IIT Research Institute, Chicago, IL.

Williams, F. W., *et al.*, "Post-Flashover Fires in Simulated Shipboard Compartments. Phase 3. Venting of Large Shipboard Fires," Naval Research Laboratory, Washington, DC, Hughes Associates, Inc., Columbia, MD, Naval Sea Systems Command, Washington, DC, Naval Sea Systems Command, Washington, DC, NRL MR 6180-93-7338, June 9, 1993.

ADDENDUM

AUTOMATIC HEAT AND SMOKE VENTING IN SPRINKLERED BUILDINGS

This addendum supplements the venting practices in this chapter as written by Dr. Heskestad. It presents in summary the various views on the controversial and unresolved subject of automatic venting in sprinklered buildings. The history of research testing on the subject is traced. Contributors include Ernest E. Miller, Thomas E. Waterman, and Edward J. Ward.* The author has added a new section, "New Sprinkler/Vent Research," for this edition of the handbook.

The Different Views

There has been controversy for more than two decades over automatic venting for sprinklered buildings.

Proponents of automatic venting in sprinklered buildings claim that venting will aid manual fire fighting by delaying loss of visibility, reduce the risk of structural failure by venting gases with dangerously high temperatures, vent the equivalent of an unsprinklered fire if the installed sprinklers are defeated by human failure or mechanical damage, and substitute for manual roof venting by fire service personnel.

Opponents believe that, in some cases, venting may even be detrimental because it draws in fresh makeup air, keeping oxygen levels higher than they would be otherwise, causing more vigorous combustion with an increase in fuel consumption, and potential for loss of sprinkler control. They also believe sprinkler discharge will cool the fire gases, perhaps to the extent that the vent discharge is reduced, hence reducing venting effectiveness. Further, the sprinkler sprays will entrain ambient smoke and air, transporting smoke to the floor level and possibly drawing gas from the vent exhaust wherever sprinklers are located close to vents and further reducing venting effectiveness. Additionally, opponents see little, if any, evidence from research testing of the benefit to fire control obtained from the use of vents in a sprinklered building, and believe that the high cost of installation does not justify their use.

Sprinkler/Vent Research

Research testing on the subject includes:

1. *Large scale tests:* In 1956 Factory Mutual Research Corporation (FMRC) ran a series of large-scale tests from which it concluded that the effects of venting in sprinklered buildings are different from unsprinklered buildings. The research project[1] was conducted in a 120- by 60-ft (36.6- by 18.3-m) test building equipped with 5-ft (1.5-m) draft curtains and vent areas ranging up to 32 sq ft (3 m²) within a curtained area of 2280 sq ft (212 m²). This is a vent ratio of about 1:70—that is, 1 sq ft (0.093 m²) of vent area for every 70 sq ft (6.5 m²) of floor area. A 5-gal-per-min (18.9-L/min) gasoline spray fire was used as the exposure, and protection consisted of 160°F (71°C) automatic sprinklers installed on a 10- by 10-ft (3- by 3-m) spacing.

*Ernest E. Miller was an executive engineer with Industrial Risk Insurers (IRI) at the time he contributed his views. Thomas E. Waterman was employed by the IIT Research Institute and conducted the IITRI test series of 1980–1981. Edward J. Ward was employed by Factory Mutual Research Corporation.

The tests in the series were conducted using various combinations of vents, draft curtains, and sprinklers. Six of these were sprinklered tests. (See Table 7-7F.)

TABLE 7-7F.   FMRC Test Results

Test No.	Curtain ft (m)	Vent sq ft (m²)	Sprinkler Discharge gpm (L/min)	Number of Sprinklers Operated
1	None	None	15 (57)	48
2	None	32 (3.0)	15 (57)	44
3	5 (1.5)	32 (3.0)	15 (57)	24
4	5 (1.5)	16 (1.5)	15 (57)	24
5	5 (1.5)	None	15 (57)	28
6	5 (1.5)	None	25 (95)	15

Comparison of average temperatures indicates that vents contributed significantly to temperature reductions in the unsprinklered tests in the series; however, they were of marginal value in the sprinklered tests. The greatest reduction in sprinkler operation was attributed to increased sprinkler discharge.

2. *FMRC model study:* In the early 1970s, an FMRC study was conducted.[2] The objective was to investigate experimentally the performance of automatic heat and smoke vents in sprinklered fires in one-story buildings, principally in terms of sprinkler water demand, but also in terms of visibility conditions and fuel consumption. The study was performed at FMRC's Norwood, MA, laboratory and involved a 1:12.5 scale model of FMRC's fire test facility at West Gloucester, RI. Automatic, individually fused vents on 50-ft (15.2-m) spacing were employed, often in combination with a draft curtain encompassing a 10,000-sq-ft (929-m²) curtained area.

Of primary interest were venting installations conforming to recommendations of standards-setting groups that were state-of-the-art at the time. For a fire in cellulosic material that opened about fifty 212°F (100°C) sprinklers at a density of 0.27 gpm per sq ft [11 (L/min)/m²], distributed vents alone (without draft curtains) had no effect on the water demand, but delayed loss of visibility from 13.1 to 15.7 min and increased fuel consumption by 31 percent. Vents and draft curtains caused a 35 percent increase in water demand relative to the unvented fire, delayed loss of visibility from 13.1 to 20.2 minutes, and increased fuel consumption by 66 percent.

Favorable effects of venting were observed for a series of large, heptane fires that opened an average of one hundred twelve sprinklers at 0.27 gpm per sq ft [11 (L/min)/m²]. Vents alone reduced the water demand by 8 percent and markedly improved visibility conditions; vents and draft curtains reduced the water demand by 18 percent, but did not improve visibility conditions as much as without draft curtains.

The test results were based on ignition points that were equidistant from the four nearest vents in the vent matrix. Experiments with ignition directly under a vent indicated that some reduction in water demand could be expected for fires that opened fifty sprinklers or less. It was concluded, however, that for randomly started fires, less than about 13 percent of the fires would benefit, respective to water demand, from closeness to a nearby vent.

Critics counter that the 0.08-scale-model experiments, although carefully designed and executed, incorporated numerous uncertainties. Of particular concern was modeling of a full-scale fuel bed (rack storage configuration), by using $^1/_2$-in. (12.7-mm) thick triwall cardboard. Some ten years later, the modeling of room fire development has proceeded to an advanced state in all respects, ex-

cept for the burning fuel; both analytical and experimental models address only the simplest of fuel configurations. Yet the model study assumed that both flame spread and burning rate of the fuel bed were modeled under conditions of depleted oxygen and impacting water droplets.

The FMRC model study was also criticized because of the configuration of the water discharge devices that were used. The study pioneered the use of nondescript miniature nozzles (representing sprinklers) arranged to be operated in small open-head deluge systems of groups of five to eleven nozzles controlled by individual sensors. Critics claim this arrangement magnified the known imprecision in measuring comparative sprinkler performance. It required multiplying the normal spread experienced in full-scale sprinkler operation by a variable factor of 5 to 11. Due to the small deluge systems, which operated as groups, sprinkler skipping occurred in units of 5 or more open nozzles, rather than the skipping of individual sprinklers as would be customary in full-scale fire research.

Lastly, the critics agreed that, although up to three times as much combustible material was consumed in the vented tests as compared to the unvented tests, the implication of detriment was improper. They claimed the explanation for the increased fuel consumption in the vented tests lies in the customary manner of conducting fire tests involving sprinklers. The test fires are allowed to continue far beyond sprinkler control to give the fires every opportunity to escape. During the additional minutes of these tests, the oxygen reduction in the unvented tests limited the fuel consumption, whereas oxygen remained abundant in the vented tests and resulted in greater fuel consumption. Since sprinkler-controlled fires continue to smolder and burn, it was felt that the reported increased fuel consumption was attributable to more complete burning of the rubble and did not represent increased fire damage. Had the fires spread to involve a significantly larger area, additional sprinklers would have operated.

3. *Other FMRC large-scale testing:* More than eighty full-scale rack storage tests were conducted between 1968 and 1975. Only three tests employed perimeter ventilation (not to be confused with roof venting directly over a fire) and only one of these three (No. 72) was identical to two other tests (Nos. 65 and 66), except for the ventilation variable. Results of these tests are given in Table 7-7G.

Comparing Test Nos. 65 and 66 to Test No. 72, researchers discovered that venting caused a dramatic detriment to the performance of the sprinklers. However, it is pointed out that Test No. 72 had a significantly different initial fire growth behavior compared to the other tests, which may have affected the outcome. FMRC's position on the effects of ventilation in these tests is that they are not conclusive.

Venting proponents argue that, although the influence of venting was inconclusive because the perimeter windows were too remote, the ventilated tests maintained visibility for 48:40 and 33:00 (minutes:seconds), as compared to 10:30 and 18:00 for the unventilated tests.

Venting opponents claim the effect of ventilation on fire control was best demonstrated during FMRC's large-scale fire testing

TABLE 7-7G.   Rack Storage Tests

Test No.	Ventilated	Number of Ceiling Sprinklers Operated
65	No	45
66	No	48
72	Yes	92

of rubber tires in 1970. This test was made to evaluate sprinkler protection for automobile tires stored in portable steel racks. Storage was located in a single pile 35 by 50 by 18 ft (11 by 15 by 5.4 m) high. Protection consisted of 286°F (141°C), 1/2-in. (12.7-mm) sprinklers, spaced 50 sq ft (4.6 m²), each with a controlled discharge of 0.6 gal per min per sq ft [24.4 (L/min)/m²]. There was no ventilation at the start of the test.

The fire was started at the base of a rack. The first sprinkler operated 2 min and 15 sec after ignition. By 8 min and 20 sec, forty-three sprinklers had operated and controlled the fire. Only one additional sprinkler operated at 28 min. The fire remained under control until 60 min into the test, when all doors and windows were opened to ventilate the building and sprinklers were left on. The fire then began to spread and grow in intensity. At 117 min, when it was apparent that sprinklers were failing to control the fire, all doors and windows were closed. A total of ninety-five sprinklers operated, and water was discharged for over 5 hr. Only the ninety-fifth sprinkler operated after doors and windows were closed, and that was at 118 min, 1 min after closure.

Critics reappraised the ventilated rubber tire fire test[3] and contended that the loss of initial sprinkler control was not due to remote window ventilation, as reported. Ceiling temperature gradients revealed that during the hour delay before hose extinguishment of the sprinkler-controlled fire was attempted, the fire had burrowed to the other end of the pile where sprinklers were not operating and had erupted with great intensity. The tightly stacked tires, stored on tread, formed horizontal tunnels that offered avenues of fire spread that were sheltered from sprinkler discharge.

4. *IITRI full-scale venting research testing:* In 1977, the IITRI (IIT Research Institute) was commissioned by the intra-industry Fire Venting Research Committee to review past research and fire experience related to vent/sprinkler interactions in large-area single-story structures. Based on IITRI's review and other considerations, the committee ultimately funded some forty-five large-scale segment experiments in 1980–1981.[4] The experiments were conducted in a 75- by 25- by 17-ft high (23- by 7.6- by 5.2-m) portion of IITRI's fire laboratory. Fires were placed in a corner to represent one-quarter of a larger fire in the center of a 150- by 50-ft (46- by 15.2-m) room. To simulate the test area as part of an even larger area, two garage doors on the wall farthest from the fire were partially opened to represent a draft curtain. The area beyond the curtains was also enclosed by vertical air stacks to minimize extraneous wind effects.

To allow variations in vent area, two pairs of automatic roof vents were installed in the laboratory roof. A diagonal sprinkler pattern was chosen to eliminate aisles between sprinklers that were located between the test fires and the vents. Experiments were conducted using propane diffusion burners or stacked pallets as the fuel source. For each fuel source, fire size and sprinkler water supply pressure were varied until a condition of "marginal" sprinkler control was achieved. Then, vent areas were varied to assess effects of venting on sprinkler performance (number of operating sprinklers and temperature levels at various locations). The laboratory size and excessive recirculation of smoke due to the substitution of vertical stacks for added rooms precluded useful measures of smoke obscuration.

Propane-source fires produced results suggesting a slight reduction in water demand for vented fires when either 165 or 286°F (74 or 141°C) sprinklers were employed [vents were operated either with 165°F (74°C) links or by simulated smoke detector or water-flow switch actuation].

All pallet fire experiments employed 165°F (74°C) sprinklers and the same vent arrangement previously described. These experiments demonstrated that, under marginal sprinkler protection, the configuration of the test facility was extremely sensitive to variations in fire intensity from any cause. Once this possibility appeared, a final series of ten "replicate" experiments was conducted, alternating between vented and unvented configurations. These experiments proved to produce widely varying results, apparently independent of the presence of automatic venting. The number of operating sprinklers varied from seven to twenty-two per test, and the pattern of sprinkler operating times appeared unrelated to the time that vents opened. Other measured factors (temperature, O_2) varied in the same way. Although the number of operating sprinklers varied widely in these experiments with marginal sprinkler control, a total of eighty-four sprinklers operated in the five unvented tests, eighty-five in the five vented tests. Earlier tests at higher water supply pressures produced good replication, independent of venting.

On the basis of the IITRI experiments, the test sponsors concluded that automatic roof venting did not detract from current state-of-the-art automatic sprinkler performance. Likewise, no particularly significant benefit to sprinkler performance was noted. Thus, it was suggested that the role of automatic vents, if used in sprinklered properties, will be to:

1. Vent unsprinklered fires, if installed sprinklers are defeated by human failure or mechanical damage, and
2. Substitute for manual roof venting upon arrival of fire service personnel.

Primarily on the basis of the first listed application, the researchers recommended that, where they are used for sprinklered properties, the design and installation of automatic roof vents for unsprinklered properties follow guidance provided by NFPA 204M, *Guide for Smoke and Heat Venting* (hereinafter referred to as NFPA 204M), until further research, testing, or experience suggests more beneficial alternate configurations.

Critics of the IITRI research effort do not agree that the experimental results support the main conclusion of the investigation, which is that automatic roof vents do not impair sprinkler control of fires capable of growth.[5]

The critics also claim that the finding that automatic roof vents do not impair the ability of 165°F (74°C) sprinklers to control fires capable of growth contradicts the studies described above, which have indicated that the increased availability of oxygen associated with open vents may lead to increased fire intensity (of fires capable of growth), water demand, and consumption of combustibles.

The critics concluded that the reported results do not justify the overall conclusions and recommendations of the IITRI researchers. Their critique suggested the investigation suffered from a number of defects: (1) situations claimed to be unvented were not; (2) one of the most important principles of venting was violated (i.e., the provision of adequate openings for inlet air); and (3) inadequate control was exercised over the experimental conditions to the extent that major variations observed in fire behavior could not be attributed to the parameter under study.

Conclusions

The proponents of automatic venting in a sprinklered building cite the IITRI research work as substantiation of the merit of vents in sprinklered buildings. On the basis of this data, they believe that automatic roof venting can provide two major contributions to 165°F (74°C) sprinkler-protected (large one-story) properties:

1. Perform as vents in unsprinklered properties should installed sprinklers be defeated by human failure or mechanical damage.
2. Substitute for manual roof venting upon arrival of the fire services.

The opponents of automatic venting in a sprinklered building believe that the high installation cost of vents is difficult to justify when one considers the limited benefits and possible detriment. They are not cost-effective because it is unlikely that a large loss will be averted solely due to the presence of vents when automatic sprinkler protection is inadequate or impaired.

Venting may or may not be detrimental, depending upon many factors, such as the location of the fire origin in relation to the location of vents. More often than not, vents will increase the water demand.[2] Even when automatic vents result in some improvement, the improvement is not required for sprinklers to gain control. Vents can also create conditions that will inhibit or prevent sprinkler control.[5]

Installing automatic vents in unsprinklered buildings is acceptable. It would certainly be more beneficial to the building owner, however, to install sprinklers rather than vents to obtain active rather than passive fire control.

Installing manual vents in a sprinklered building is also acceptable, if desired by the building owner. These vents can be used during manual overhaul or at locations that would be expected to produce dense smoke. However, a similar effect can be achieved by venting through windows, by fire fighters cutting holes in the roof, or by ventilation equipment or smoke exhausters.

Through NFPA's consensus-making standards process, NFPA members agreed to the following statements for the 1985 edition of NFPA 204M:

Paragraph 6-3: "A series of tests was conducted to increase the understanding of the role of automatic roof vents simultaneously employed with automatic sprinklers [Section 6-5(c)]. The data submitted did not permit consensus to be developed as to whether sprinkler control was impaired or enhanced by the presence of automatic (roof) vents of typical spacing and area."

Paragraph 6-4: "While the use of automatic venting in sprinklered buildings is still under review, the designer is encouraged to use the available tools and data referenced in this document for solving problems peculiar to a particular type of hazard control."

Even though the issue has been argued for years, proponents maintain venting critics have been unable to cite a single fire in which automatic venting was blamed for loss of fire control by sprinklers. Thus, to proponents it is paradoxical that fire research, which is typically motivated only by adverse fire experience, should be the sole basis of suspecting venting to be a detriment. They urge that serious students of venting should obtain and carefully examine this research and compare it with related research, both direct and indirect.

New Sprinkler/Vent Research

During the latter part of 1989, the Fire Research Station of the Building Research Establishment (U.K.) and Colt International (together with Verband der Sachversicherer eV and City of Ghent Fire Brigade) undertook a collaborative test program in a unique test building constructed in Ghent, Belgium, the Multifunctioneel Trainingcentrum.[6] The main objective was to validate a theoretical model for predicting the effect of venting on sprinkler operations, but the direct data are being generally used to assess the efficacy of venting in sprinklered buildings.

The test building in Ghent is designed for fire service training purposes and incorporates a hall of the dimensions 164 ft × 59 ft × 33 ft high (50 m × 18 m × 10 m). For the tests, a 10-ft (3.1-m) deep draft curtain was hung from the ceiling, creating a curtained, flat-ceiling area of 89 ft × 59 ft (27m × 18 m) under which test fires were burned. A total of 20 roof vents, each with an area of 18 sq ft

(1.67 m²), were installed in this area, centered on a grid measuring 15.9 ft × 14.8 ft (4.85 m × 4.5 m). Sprinklers on a spacing of 12 ft × 8 ft (3.7 m × 2.4 m) were also installed in this area [temperature rating of 155°F (68°C)]. The ceiling area on the other side of the draft curtain was provided with 20 roof vents that were always open. A total of 16 inlet vents were installed at ground level in the sides of the building, each having an aerodynamic free area of 33.7 sq ft (3.1 m²). As fire sources were used, n-Hexane floated on water near the floor level. One source had a steady output of 5100 Btu/sec (5.4 MW). The other source provided an exponentially growing heat release rate (roughly equivalent to a "continuous-growth fire" of 50 sec growth time) until the first sprinkler operated, at which time the heat releaser rate was leveled off [8,700 to 12,300 Btu/sec (9.2 to 13.0 MW)].

Note that the vents were very closely spaced (although not all of the vents were used in all the tests) and, furthermore, that the building space on the side of the draft curtain away from the fire was amply ventilated, which may have been factors in the outcome. Other unusual features were that the vents were in effect short chimneys about 5 ft (1.5 m) high and, when venting was used, the vents were open from the start of the test.

The first test series involved the steady fire source. Combinations of 0, 1, and 5 operating sprinklers [at approximately 30 psi (2 bar)] with 0, 10, and 20 open vents were investigated. With no sprinklers operating, the ceiling gas temperatures decreased significantly with the number of open vents, from 220°F (104°C) to 153°F (67°C) at a 20 ft (6 m) radius from the fire axis when conditions changed from no vents to 20 open vents. With 5 sprinklers operating, the drop in gas temperature at the same location was from 203°F (95°C) with no vents to 130°F (54°C) with 20 open vents. Extensive temperature, velocity, and CO_2 concentration data were recorded.

The second test series involved the growing-fire source and automatic operation of the sprinklers. The number of open vents in a test were mostly 0, 10, or 20, but a few tests were conducted using special patterns of 9 and 16 open vents. The water supply in the first tests was very weak. For the test providing data for this review, the water supply was boosted with a fire pump, yielding a pressure of about 88 psi (6 bar) for one open sprinkler, decaying to about 15 psi (1 bar) for 40 open sprinklers (roughly linear decay with number of open sprinklers). The time of the first sprinkler operation ranged from an average of 148 sec for no open vents to an average of 160 sec for 20 open vents. Subsequent sprinkler operations depended on the particular test scenarios, i.e., the original, Scenario A, where the fire size was maintained constant to the end of the test at the value when the first sprinkler operated (leading to operations of all installed sprinklers with no open vents), or the revised, Scenario B, where the fire size was maintained constant for 30 sec and then reduced by 20 percent. In Scenario B, an initial group of 6 sprinklers operated within an interval of 60 sec for both 10 and 20 open vents, compared to an initial group of 9 sprinklers in the same interval with no vents. With 20 open vents, only one additional sprinkler operated (no additional operations for 10 open vents). With no vents, additional sprinkler operations began about 2 min after the first sprinkler operated, accumulating to a final total, nearly 6 min from ignition, of 37 sprinklers.

With respect to the issue of visibility through smoke, the report offers the following observation: "In the experiments with growing fires and the pumped supply, the water pressure was higher and the number of operating sprinklers greater than in the experiments with steady fires. There was then some tendency for the fire gases to be pulled down in both vented and unvented experiments." On other issues, the experiments provided no evidence that the flow through the vents was affected by the wind [wind speeds up to 84 ft/sec (7.8 m/s)], and there were no observable effects of nearby sprinklers on the vent flow (nearest sprinkler about $^3/_4$ vent width from any part of the vent opening). The author often finds fair agreement between his theory and the experimental results, but identifies areas needing further study, which are issues beyond the scope of this review.

On the effect of venting on sprinkler operations, the report concludes that (1) the prior opening of vents will have little effect on the operation of the first sprinkler with fast growing fires; (2) venting substantially reduced the total number of sprinkler operations; and (3) there was no indication that venting could increase the total number of sprinkler operations.

These conclusions of the report are deemed, by this reviewer, to be highly dependent on the special circumstances of the experiments, primarily (1) the use of a fire source that did not respond in heat output to the sprinkler spray per se, and (2) the fact that even unvented fires were supplied with unvitiated air, arriving from the adjacent, amply ventilated part of the building.

Further to the issue of responsiveness of the fire source to sprinkler spray, Gustafsson[7] observed from the sprinkler operation maps in the report that, in vented tests, sprinklers near the fire source often were delayed or did not operate altogether. Gustafsson makes the point very strongly: "It is clearly seen that the effect of ventilation on the operation of sprinklers was strong and detrimental in all cases. It must be appreciated that prevented or substantially delayed operation of any sprinkler close to, or directly above, the fire must be avoided."

Additional research and testing will be needed before definitively establishing the benefits and detriments of automatic venting in a sprinklered building.

BIBLIOGRAPHY

References Cited

1. FMRC, "Heat Vents and Fire Curtains, Effect on Operation of Sprinklers and Visibility," Factory Mutual Research Corporation, Norwood, MA, 1956.
2. Heskestad, G., "Model Study of Automatic Smoke and Heat Vent Performance in Sprinklered Fires," Technical Report FMRC No. 21933 RC74-T-29, 1974, Factory Mutual Research Corporation, Norwood, MA, 1974.
3. Miller, E. E., "Reappraisal of Ventilated Rubber Tire Fire Tests," *Fire Venting Mini-Study No. 8*, Industrial Risk Insurers, Chicago, IL, 1982.
4. Waterman, T. E., "Fire Venting of Sprinklered Buildings," *Fire Journal*, Vol. 78, No. 2, Mar. 1984, pp. 30–39, 86. (Also see "Letters to the Editor," *Fire Journal*, Vol. 78, No. 5, Sept. 1984, p. 6.)
5. Heskestad, G., "Review of 'Fire Venting of Sprinklered Buildings by T. E. Waterman, et al.,' Interoffice Correspondence to E. J. Ward," Factory Mutual Research Corporation, Norwood, MA, 1983. (Also see "Letters to the Editor," *Fire Journal*, Vol. 78, No. 5, Sept. 1985, p. 6.)
6. Hinkley, P. L., *et al.*, "Experiments at the Multifunctioneel Trainingcentrum, Ghent, on the Interaction Between Sprinklers and Smoke Venting," Fire Research Station, Building Research Establishment, Borehamwood, Herts., 1992.
7 Gustafsson, N. E., "Smoke Ventilation and Sprinklers—A Sprinkler Specialist's View," Seminar at the Fire Research Station, Borehamwood, Herts., 1992.

STRUCTURAL FIRE SAFETY: ONE– AND TWO– FAMILY DWELLINGS

Revised by Ronald Brave and Robert Elliott

In 1993, 73.8 percent of U.S. occupied housing units were in single-family dwelllings, including manufactures homes.[1] Another 9.8 percent were in buildings having 2–4 housing units, so one- and two-family dwellings represented 73.8–83.6 percent of all occupied housing units in 1993.

This chapter is confined to a discussion of the structural elements of one- and two-family dwellings as they affect the fire safety of the structure. Manufactured (mobile) homes will also be included in this discussion of one- and two-family dwellings. This chapter does not cover the influence of interior finish or energy conservation measures, the hazards of building facilities (heating and electrical installations), or the role of detection and extinguishing systems as life safety features. These subjects are included in other chapters of the handbook; see Section 3, Chapter 1, "Electrical Systems and Appliances"; Section 3, Chapter 5, "Heating Systems and Appliances"; Section 4, Chapter 18, "Effects of Energy Conservation in Buildings"; Section 5, Chapter 6, "Household Fire Warning Equipment"; Section 6, Chapter 13, "Fast-Response Sprinkler Technology"; and Section 7, Chapter 3, "Interior Finish."

CONVENTIONAL TYPES OF CONSTRUCTION

This part of the chapter addresses the various types of construction used in the design and erection of both conventional and manufactured one- and two-family dwellings. Specific requirements are dictated by local building codes and, in the case of manufactured homes, by federal regulation. In general, most one- and two-family residential buildings erected are of Type V (frame) or Type III (ordinary) construction as defined in NFPA 220, *Standard on Types of Building Construction*. There are, of course, exceptions where some dwellings may be erected of Type II (noncombustible) or Type I (fire resistant), and possibly of Type IV (heavy timber) construction.

Ronald M. Brave, P.E., is a custom homebuilder in Vail, CO. He is a member of SFPE, the NFPA Life Safety Code Subcommittee on Residential Occupancies, the NFPA Technical Committee on Household Fire Warning Equipment Signaling Systems, and the NFPA Technical Committee on Carbon Monoxide and Fuel Gas Detectors.

Robert F. Elliott is an engineer for the National Association of Home Builders' Construction, Codes and Standards Department. He is a member of the NFPA Technical Committee on Automatic Sprinklers and the NFPA Technical Committee on Chimneys, Fireplaces, and Venting Systems for Heat-Producing Appliances.

Type V (Frame) Construction

The principal categories of Type V (frame) construction are: (1) plank and beam, (2) balloon, (3) platform or western, (4) braced, and (5) veneer (a combination of one of the previous four with a facing wythe of brick, concrete masonry units, or stone on the exterior walls). The buildings may be constructed with either basements, cellars, crawl spaces, or slab on grade. Geographical peculiarities or the needs of the owner generally dictate which foundation type is used. Where basements, cellars, or crawl spaces are provided, they play a definite role insofar as residential fire safety is concerned, since in these areas, heating, air-conditioning, and water heater equipment is usually installed. These areas are also used for storage by the building's occupants. Roof framing may be either of the rafter or trussed typed. Where trusses are used, interior bearing partitions may not be required.

Plank and beam framing: In plank and beam framing, a few large members replace the many small wood members used in typical wood framing; i.e., large dimensional beams more widely spaced (4 by 10 in. or 5 by 12 in. on 4- or 6-ft* center) replace the standard floor and/or roof framing of smaller dimensioned members (2 by 8 in. or 2 by 10 in. joists or rafters on 16-in. center). The decking for floors and roofs is planking in minimal thicknesses of 2 in. or more as opposed to 1/2-, 5/8-, or 3/4-in. plywood sheeting. Instead of 2- by 4-in. bearing partitions supporting the floor or roof joist or rafter systems, the beams are supported by posts. Plank and beam framing results in a saving in the number of members and allows for more rapid site assembly.†

Balloon, platform or western, and braced framing: The basic difference among balloon, platform or western, and braced framing is in the manner in which the floor, roof, and bearing partitions are supported and/or erected.

In platform or western framing, stud-bearing partitions are set on soles that are set on top of the joists, with the joists bearing on sills at the foundation level and girts at intermediate levels.

In both balloon and braced framing, the stud-bearing partitions are set on sills at the foundation level. In balloon framing, the studs in exterior-bearing partitions extend for two stories, with the joists supported by 1- by 6-in. ledger strips or ribbon boards at the intermediate

---

*Nominal dimensions; for comparison, 1 ft = 0.305 m.

†Nominal sizes; for comparison, 1 in. = 25.5 mm.

level. In braced framing, the bearing partition extends for one story only with joists at the second level of bearing partitions bearing on 4- by 6-in. girts. In both framing types, the interior bearing partitions are erected in one-story heights. (For SI units: 1 in. = 25.4 mm.)

The use of wood foundation walls has been recently introduced as an alternate construction procedure, and crawl spaces have been utilized as return air plenums for a number of years.

The exterior wall finish of buildings of frame construction, except masonry veneer, generally consists of stucco, wood, or plastic siding, or metal siding having either plastic coating, baked enamel coating, or, in the case of aluminum, an anodized finish.

Masonry veneer exteriors utilize one of the wood framing systems, with the addition of a wythe of masonry secured by metal ties to the wood frame exterior walls. Usually an insulating board of some type is provided between the studs and the masonry. Although the exterior exposed surfaces are masonry, veneer construction is considered a combustible frame construction because the load-bearing members are wood.

Many builders today are adopting steel framing for residential building. As cold-formed steel construction is becoming more prevalent in residential building, the model building codes are addressing the structural and fire safety characteristics of steel framing. Steel framing has many similarities to conventional wood framing construction Types III and V. Steel framing methods are available for stick-built (balloon or platform), panelized, and pre-engineered systems. Steel, like masonry of Type III construction, is noncombustible. However, steel framing can lose its structural capacity, under severe exposure to heat.

Limited testing has been conducted on steel framing wall and floor assemblies. United States Gypsum Company (USG) has completed fire resistance testing of floor and wall assemblies utilizing combinations of metal studs; various building products, such as concrete and plywood; and gypsum wallboard.

Type III (Ordinary) Construction

The significant difference between Type III (ordinary) and Type V (frame) construction lies mostly with the construction of the exterior walls. While in frame construction the load-bearing components of the walls are wood, in ordinary construction the exterior walls are masonry or other noncombustible materials. The interior partitions, floor, and roof framing systems are of wood and, in general, utilize either the platform or braced framing methods previously described.

Manufactured Homes (Mobile Homes)

Four major components or subassemblies—i.e., (1) the chassis, (2) floor system, (3) wall system, and (4) roof system—comprise the structural systems of the manufactured home, which, in essence, is designed specifically to meet the requirements of an efficient assembly-line production.

The chassis is the structural base of the manufactured home, receiving all vertical loads from the walls and floor and transferring them to a substructure of either the wheel assembly (when the home is in transit) or to a foundation (when the home is finally erected at a permanent site). The chassis generally consists of two longitudinal steel beams, braced by steel cross members. Steel outriggers cantilevered from the outsides of the main beams bring the width of the chassis to the approximate overall width of the superstructure. The running gear used when the unit is in transit is also considered part of the chassis.

The floor system consists of its framing members, generally conventional 2- by 6-in. or 2- by 8-in. wood joists, with plywood decking glued and nailed to the joists, glass fiber insulation blankets installed between the joists, and an asphalt-impregnated insulation board sealing the bottom of the floor system. (For SI units: 1 in. = 25.4 mm.) Ductwork and piping are installed longitudinally within the floor system, often requiring the cutting of the joists. The floor finish is generally carpeting or resilient flooring, such as linoleum or tile.

The wall system, utilizing the stressed-skin principle, consists of an interior skin of plywood that is glued and nailed to wood studs, and an exterior skin of aluminum siding. The actual sizing of members is dictated by the structural loads. Insulation batts are installed in the stud spaces.

The roof system consists of either the framed wood roof rafter and ceiling joist system or a wood truss system. Roof decking is generally a rigid insulation board attached to the top of the roof rafters or trusses. Finished roofing is galvanized steel or aluminum decking. The ceiling is either acoustical planks or gypsum wallboard attached directly to the bottom of the ceiling joists or the bottom chords of the trusses. Insulation blankets provide the roof insulation.

Steel tie plates reinforce connections between wall and floor systems; diagonal steel strapping binds the floors, walls, and roof into a complete unit.

CONSTRUCTION CONSIDERATIONS FOR FIRE SAFETY

Manufactured Housing (Mobile Homes)

In 1993, manufactured homes constituted 6.0 percent of U.S. occupied housing units.[1] This share has been steadily growing, as households sought affordable housing in a rapidly increasing real estate market. For example, in 1982, manufactured homes constituted only 4.4 percent of U.S. occupied housing units.[2]

This rapid growth in popularity was viewed with some concern within the fire service and the fire protection community, because manufactured homes were considered to pose a greater life safety risk than stick-built homes. In response to this concern, in 1976 the U.S. Department of Housing and Urban Development (HUD) promulgated the "Mobile Home Construction and Safety Standards," later termed the "Manufactured Home Construction and Safety Standards," which was based on NFPA 501B, Mobile Homes, which has since been withdrawn. These standards have been amended several times to strengthen requirements for interior finish and firestopping, and to identify acceptance criteria limiting the use of materials that generate smoke and gases more toxic than those produced by untreated wood.[3]

The passage of the HUD standards coincided with the advent of a national fire data base sufficient to support analysis of their effectiveness and of the actual risk in manufactured homes. HUD sponsored the first of a series of studies [4] which have continued every one or two years since then.[5] NFPA's Fire Analysis and Research Division has conducted its own statistical analysis of manufactured home fires for the last several years.[6]

The common findings of these studies are that manufactured homes, taken as a group, have more than twice the rate of deaths per hundred fires found in other one- and two-family dwellings. This, partly offset by a lower rate of reported fires per thousand units, gives manufactured homes roughly twice the rate of fire deaths per million housing units as occur in other dwellings. However, the same analysis also shows a dramatic improvement in units built since the introduction of the HUD standards, so much so that it appears the fire risk in post-standard manufactured homes is comparable to that in dwellings other manufactured homes.

The manner in which the HUD standards achieved these results, following the direction set by NFPA 501B, leans heavily on smoke detector requirements and requirements for interior finish, throughout the structure and especially in the vicinity of heating and cooking equipment. The effect of the latter construction requirements is that fires in post-standard manufactured homes are much more likely than fires in pre-standard manufactured homes to have flame damage confined to the room of fire origin.

Since the majority of one- and two-family dwellings and manufactured homes are erected of frame or ordinary construction, the physical nature of these types of construction suggest the addition of certain built-in features that will inhibit rapid fire spread. The large concentrations of combustible materials in the structural framework of these construction types, compounded by voids in concealed spaces connected to other voids, are basic conditions that contribute to the propagation and migration of fire throughout the structure. These conditions are addressed in the various building regulations used in this country. These regulations are clear in their requirements for the installation of draft stops and firestopping.

Single-Family Detached Dwellings

Of all the occupancy use group classifications regulated by building codes, single-family dwelling requirements are the least restrictive insofar as fire safety is concerned. Building code interior finish requirements for one-and two-family dwellings are comparable for other use groups, except for exits and exit access areas. With the exception of fire resistance ratings for exterior walls, which depend upon location of the wall with regard to separation distance from property lines or other buildings on the same lot and upon the separation between garages and habitable space, building codes do not mandate fire resistance ratings for building elements. Fire and party walls of attached single-family dwellings require a minimum 2-hr fire-resistance rating. Building codes regulate the allowable area of unprotected and protected openings in exterior walls and require that penetrations for mechanical, plumbing, and electrical work be firestopped.

Prior to 1985, the minimum property standards (MPS) for one-and two-family dwellings, promulgated by HUD, established specific fire-resistance ratings for bearing walls and partitions, separations between garage and habitable space, and for all floor/ceiling assemblies. These regulations, however, were applicable only to those buildings for which Federal Housing Authority (FHA) or Veterans Administration (VA) loan assistance was provided.

On November 1, 1985, HUD, as part of the administration's efforts to deregulate, abolished the MPS in favor of recognizing the model building codes and local code enforcement.

Each of the three model building codes[7,8,9] and the *CABO One-and Two-Family Dwelling Code*[10] contain firestopping and draftstopping criteria to reduce the spread of fire through concealed horizontal and vertical spaces. These code requirements identify particular locations where firestopping and draftstopping are required to be installed.

Firestopping: The interconnections of vertical and horizontal spaces, such as found at junction points between walls and floors and ceilings, require firestopping, as do soffits over cabinets, drop ceilings, and cove ceilings. Concealed spaces of stud walls and partitions, as well as stairway stringers, require firestopping. Firestopping is also required in ceiling and floor openings, exterior cornices, and behind combustible finish and trim.

Figures 7-2X and 7-2Y in Section 7, Chapter 2, of this handbook illustrate points to be firestopped in balloon and platform framing, respectively. Figure 7-2Z provides expanded views of the points of firestopping.

Draftstopping: Draftstopping is required within all open-web floor systems to reduce the size of the concealed spaces. Additionally, where ceilings are suspended below wood joists or attached to the bottom of open-web floor systems, the space between the floor and ceiling is required to be draftstopped.

One key to structural fire safety is to confine the fire to the room or area of origin, i.e., compartmentalize the structure so as to preclude the extension of the fire to areas adjacent to the room or area of origin. This is not easily achievable in typical one- and two-family dwellings where the structural elements are not required to provide any specific fire rating, and rooms are accessible through non-rated doors or openings with no doors at all.

Unrated gypsum wallboard, the material most commonly used as partition, wall, and ceiling material in the housing industry, affords a definite degree of fire protection when unpenetrated or when its penetrations are properly firestopped. But, since fire resistance ratings are not generally mandated for one- and two- family dwellings, the inspection of the installation of this material is not required, and unprotected penetrations may be made for heating ducts, electrical wiring and appurtenances, and plumbing. To compound the problem, quite often custom residential designs not only challenge a builder's skill to properly install firestopping, but require that, to permit installation of electrical, plumbing, and mechanical work, the firestopping be removed, cut, or drilled. However, with proper inspection and attention to detail, breaches in the compromised firestopping can be mitigated or eliminated altogether. *Building Construction for the Fire Service*[11] addresses wood-frame and ordinary construction in depth, pointing out both their attributes and problems regarding fire safety. The U.S. Army Corps of Engineers (USACE) has a technical manual that further addresses firestopping of penetrations of fire-resistance-rated assemblies.[12]

A study of fires in multifamily residential buildings of frame or ordinary construction reveal a number of built-in contributing factors to the spread of fire that may also be relevant to some one- and two-family dwellings.[13]

1. The use of nonrated soffit materials, both combustible and noncombustible, and the use of soffit or eave vents can allow fires that break through an exterior wall opening to spread into the attic or roof space by the flue action of the attic or roof ventilation system. (See Figure 7-8A.)

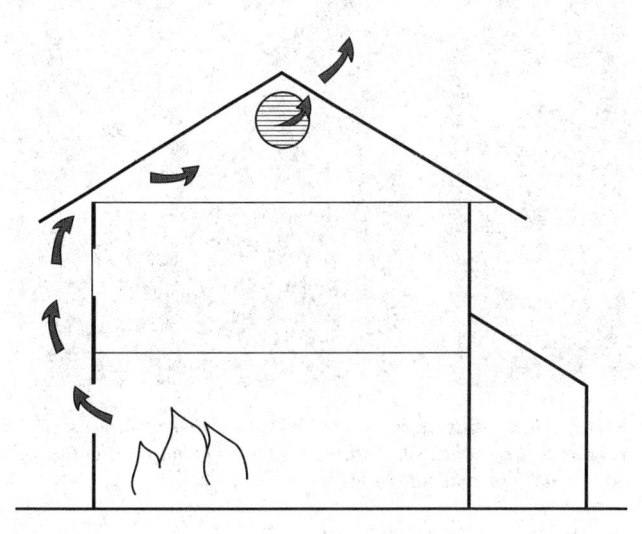

FIG. 7-8A. Attic vents under the eaves have serious potential for outside extension of fire into the attic space.[14]

2. Mansard roofs extending below the top-story windows with the windows recessed. The soffit extending from the face of the window to the edge of the roof is typical of nonrated materials. In addition, the wall construction is not extended in back of the plane of the mansard roof, thus providing an interconnected void. A fire breaking through the window impinges on this soffit, burning through in a relatively short time and involving the attic and roof spaces.

3. Unfirestopped penetrations of the gypsum wallboard partitions by plumbing, mechanical, and electrical work.

4. The widespread use of wood-trussed roof and floor framing precludes the need for the interior bearing partitions normally required where solid sawn joists and rafters are provided. Regulations typically required joists to be doubled under interior bearing partitions running parallel to joists, and bridging to be installed under partitions running perpendicular to joists. This results in a minimum amount of firestopping of joist spaces. The current model building codes compensate for this by requiring draftstopping of the trussed floor systems to reduce the spread of fire.

5. Installing cabinets and other fixtures directly to stud framing, without an intervening protective diaphragm, presents a situation where the only firestopping between a room fire and the concealed stud spaces are the materials that constitute the bottoms and backs of the cabinets or fixtures. (See Figure 7-8B.) This same practice exists in the installation of bathtubs or shower stalls.

BIBLIOGRAPHY

References Cited

1. *Statistical Abstract of the United States* 1995, Washington: U.S. Bureau of the Census, 1996, Table 1230.
2. *Statistical Abstract of the United States* 1986, Washington: U.S. Bureau of the Census, 1987, Table 1317.
3. HUD, " Manufactured Home Construction ans Safety Standards," 1984, U.S. Department of housing and Urban Development (rev. Oct 1984), Washington, DC.

FIG.7-8B.   Soffits (arrow) over kitchen cabinets connect wall voids to ceiling voids, permitting fire extension both vertically and horizontally in concealed spaces.[14]

4. *Fire Performance Evaluation of the Federal Mobile Home Construction and Safety Standard*, Washington: Federal Emergency Management Agency, 1980.
5. John R. Hall, Jr., *Final Report—Manufactured Home Fire Experience through 1992 Fires*, prepared by National Fire Protection Association, Washington: U.S. Department of Housing and Urban Development, 1995.
6. John R. Hall, Jr., *Manufactured Home Fires in the U.S., 1980–1994*, Quincy, Massachusetts: National Fire Protection Association, Fire Analysis and Research Division, Oct. 1996.
7. BOCA, *National Building Code*, Building Officials and Code Administrators International, Country Club Hills, IL, 1996.
8. ICBO, *Uniform Building Code*, International Conference of Building Officials, Inc., Whittier, CA, 1994.
9. SBCCI, *Standard Building Code,* Southern Building Code Congress International, Inc., Birmingham, AL, 1994.
10. CABO, *CABO One- and Two-Family Dwelling Code—1995*, Council of American Building Officials, Falls Church, VA, 1995.
11. Brannigan F. L., *Building Construction for the Fire Service*, 3rd ed. National Fire Protection Association, Quincy, MA, 1992.
12. USACE, "Firestopping," TM-5-812-2, 1985, U.S. Army Corps of Engineers.
13. Vogel, B. M., "A Study of Fire Spread in Multi-Family Residences: The Causes—The Remedies," NBSIR 76-1194, 1977, Center for Fire Research, National Bureau of Standards, Washington, DC.

Reference

HUD, "Minimum Property Standards for One- and Two-Family Dwellings," U.S. Department of Housing and Urban Development, Washington, DC.

NFPA Codes, Standards, and Recommended Practices

Reference to the following NFPA codes, standards, and recommended practices will provide further information on structural fire safety for one- and two-family dwellings discussed in this chapter. (See the latest version of *The NFPA Catalog* for availability of current editions of the following documents.)

NFPA 220, *Standard on Types of Building Construction*

Additional Reading

Anderson, J., and Ranby, A., "Brandskydd av stalkonstruktioner: Projektering Bostadsoch Kontorsbyggnader. [Fire Engineering of Steel Structures: Design Dwelling and Office Buildings.]," Stalbyggnadsinstitutet, Stockholm, Sweden, Publication 132, 1994, 46 pages.

Donegan, H. A., Taylor, I. R., and Meehan, R. T., "Expert System to Assess Fire Safety in Dwellings," *Proceedings* of the Third International Symposium on Fire Safety Science, Elsevier Applied Science, New York, 1991, pp. 485–494.

Estepp, M. H., "Strategies to Implement One- and Two Family Dwellings Sprinkler Initiatives," Prince George's County Fire Dept., MD, International Association of Fire Chiefs (IAFC), Fire-Rescue International Conference Proceedings, Annual Conference, 1993. August 28–September 1, 1993, Dallas, TX, 1993, pp. 171–177.

Gaiser, K. E., "Challenge: Sprinklers in Hise-Rises, Manufactured Homes, One- and Two Family Dwellings...Bull or No Bull," Georegetown City Fire Dept., SC, International Association of Fire Chiefs (IAFC), Fire-Rescue International Conference Proceedings, Annual Conference, 1993. August 28-September 1, 1993, Dallas, TX, 1993, pp. 161–170.

Hare, C. W., "Protecting Residential Construction," *Fire Engineering*, Vol. 140, No. 12, 1987, pp. 38–40.

King, J. A., "Thatched Roofs—Fire Hazards Precautions," *Fire Prevention*, No. 218, Apr. 1989, pp. 29–31.

Mimura, Y., "Fire Safety Standards for Three-Storied Wood-Frame Collective Dwelling Building," Twelfth Joint Panel Meeting of the U. S./Japan Government Cooperative Program on Natural Resources (UJNR) on Fire Research and Safety, October 27-November 2, 1992, Tsukuba, Japan, Building Research Inst., Ibaraki, Japan, 1992, pp. 353–356.

Norman, J., "Fireproof Multiple Dwellings, Apartments, Dormitories, Hotels," *Firehouse*, Vol. 17, No. 1, March 1992, pp. 16, 18.

Pressler, B., "Bread and Butter" Operations: Multiple Dwellings. Part 1. Construction," *Fire Engineering*, Vol. 147, No. 3, March 1994, pp. 53, 56, 58–59.

Rickman, L. M., Luciani, F. S., and Dacquisto, D. J., "Retrofitting Fire Safety Systems in Multi- and Single-Family Dwellings," NAHB National Research Center, Upper Marlboro, MD, National Conference of States on Building Codes and Standards, Council of American Building Officials, Department of Housing and Urban Development, National Association of Home Builders, National Fire Protection Association, National Fire Sprinkler Association and National Institute of Standards and Technology, Sprinklers in Residential and Commercial Buildings, August 19-20, 1990, Charleston, SC, Natl. Conf. States on Bldg. Codes and Standards, 1990, pp. 139–154.

Solomon, R. E., "Fire Problem in One- and Two-Family Dwellings," National Fire Protection Assoc., Quincy, MA, National Conference of States on Building Codes and Standards, Council of American Building Officials, Department of Housing and Urban Development, National Association of Home Builders, National Fire Protection Association, National Fire Sprinkler Association and National Institute of Standards and Technology, Sprinklers in Residential and Commercial Buildings, August 19-20, 1990, Charleston, SC, Natl. Conf. States on Bldg. Codes and Standards, 1990, pp. 25–41.

Tamim, A. S., "Cost-Benefit Analysis of Residential Sprinkler Systems in Single Family Dwellings," Worcester Polytechnic Inst., MA, Thesis, Dec. 1992.

VENTING SYSTEMS FOR COMMERCIAL COOKING EQUIPMENT

Joseph N. Knapp

Fires in commercial cooking establishments are a serious concern, due to the quantity of fuel that can be involved with some appliances. The ventilation system is another component of these sites that can also have a serious effect on a fire. Ventilation systems are composed of: (1) the exhaust system that conveys the effluent of the cooking operations to the outside and (2) the supply air system that provides the makeup air for the exhaust system.

This chapter discusses those aspects of ventilation systems in commercial food establishments that can have an effect on a fire relative to its initiation, spread, containment, and suppression. Definitions and an overview of typical ventilation systems and their components and a breakdown of aspects of these that often cause problems due to poor design or maintenance are provided.

Information relating to this subject can be found in Section 7, Chapter 7, "Venting Practices," and Section 7, Chapter 14, "Air-Conditioning and Ventilating Systems."

TERMINOLOGY

Ventilation

Ventilation is the process of moving air through a space in such a manner as to render the air in the space of such quality as to make it acceptable for the people, products, or processes that occupy the space. The moving of the air may replace the original air with new outside air, or it may clean and recirculate the original air, and it will usually temper the air to some general or specific condition. There may be more than one ventilation process going on within one space, and one ventilation process may affect more than just one space. To whatever degree either of the above is true, the multiple processes of ventilation must be designed to act in concert as one single system if there is to be economy and effective performance of the ventilation system. Common ventilation processes include heating and air conditioning by forced air, exhaust systems of all types, and air distribution and transfer systems.

Natural ventilation: Natural ventilation is that which is accomplished by natural means, such as openings to the outside, draft stacks, and natural pressure differentials due to height differences.

Joseph N. Knapp is staff director of restaurant systems for the McDonald's Corporation, based in Oak Brook, IL. He is a member of NFPA's Technical Committees on Venting Systems for Cooking Appliances and Dry and Wet Chemical Extinguishing Systems.

Natural draft exhaust systems are not addressed in this chapter, as none are known to exist commercially, unless a rare fireplace is so used. One may occasionally find a natural draft makeup/supply system in the field, which would be merely a screened open door or window to the outside. This chapter also does not address these, except to mention here that if the opening is closed, the makeup air is not available, and the exhaust system will not work properly. And, if a strong wind is blowing in through the opening, the draft could also cause the exhaust system not to work properly. Either condition could put a lot of grease on the surfaces of the room, and increase the potential hazard and spread of a fire should one occur.

Mechanical ventilation: Mechanical ventilation is that which is accomplished with fans to create the pressure differentials necessary to produce the desired flows of air. These systems are specifically addressed in this chapter.

Exhaust

Exhaust is the ventilation process that draws noxious air-entrained particulates and vapors from a local point or process, collects them initially in a hood, and contains them in a duct for transport to an outside atmospheric point, or to a processing station that cleans the air before discharging it to the outdoor ambient or returning it to the area of origin. If originating in a closed area, such as a kitchen, the exhaust cannot operate at the flows required without having an equal supply of makeup air available. In the context of this chapter, exhaust primarily applies to the commercial cooking exhaust systems that are designed to capture, contain, and remove from the kitchen the effluents of the cooking operations. Other exhaust systems existing in commercial cooking establishments can be for removal of heat and fumes from restrooms, dishwashers, and refrigeration condensers.

Makeup Air/Replacement Air

Makeup air and replacement air are the more common names given to the air that has to be brought into a space for as long as the exhaust process is continuing, in order to maintain the pressure in the space so the exhaust process can operate as designed. This air may be brought directly into the space via ducts, or indirectly via openings from adjacent areas. The quantity of this air must equal or slightly exceed the exhaust quantity. It must be air that is in addition to any recirculation air requirement for the space, as the air cannot be available for both purposes at once. Separate entering and leaving air requirements within a space are additive, and all must be balanced.

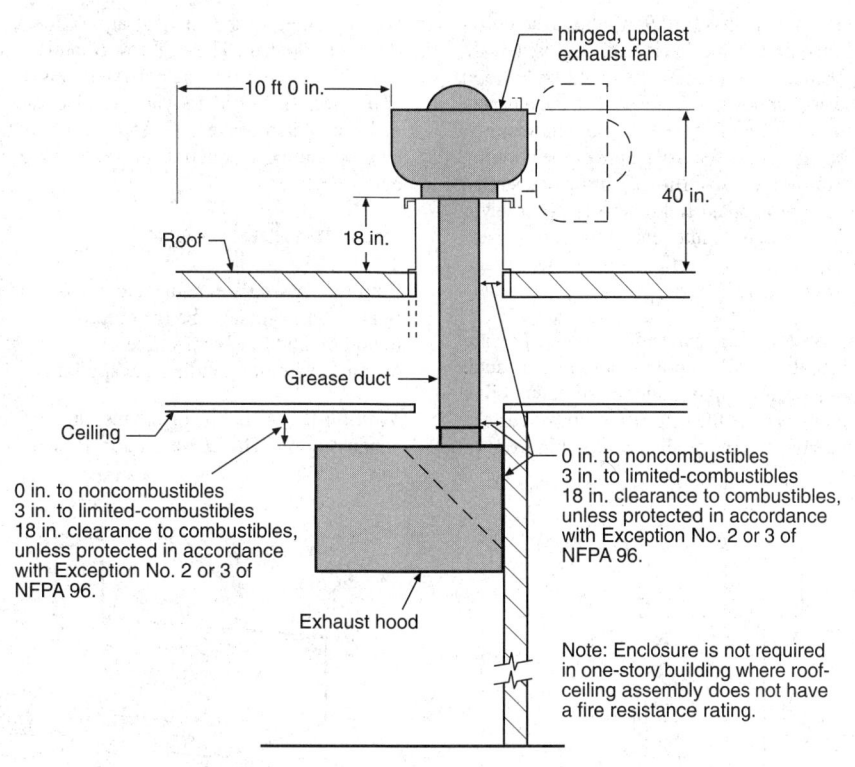

FIG. 7-9A.   Cooking exhaust system.

COMPONENTS

Exhaust

An exhaust system consists of the hood over the cooking appliance(s), the ductwork connecting the hood to the exhaust fan, the exhaust fan itself, and all grease filtration devices and other effluent control devices within the system. The concept of the exhaust system is that it will capture all the cooking effluent within the hood, contain it there until it is removed by the duct system, and transport it through the duct system to the out of doors. At one or more places within the system, the effluent is treated for removal of various parts and percentages of the effluent contaminants (grease, gases, odors). (See Figure 7-9A.) In some newer systems (designated recirculating units) for commercial cooking, the effluent contaminants are considered removed to such a thorough degree that the final air is able to be returned to the same space as the cooking operation instead of having to be discharged out of doors.

Canopy hood: A canopy hood is one whose lower edges overhang the outer edges of the cooking appliance by a minimum of 6 in. (152 mm) on all sides of the cooking appliance that are not closed off between the cooking surface and the lower edge of the hood. If at least the rear side of the hood is closed off to the cooking appliance by a wall or panel, it is then called a wall canopy. If no side of the hood is closed off to the cooking appliance, it is called an island canopy. Very often island canopy hoods are double wide when installed over two rows of cooking appliances that are back to back with each other. Canopy hoods are required to be at a minimum of 6.5 ft (2 m) above the floor to prevent most people hitting their heads on them. Such height above the cooking surface generally requires that canopy hoods operate at fairly high levels of exhaust. (See Figures 7-9B and 7-9C.)

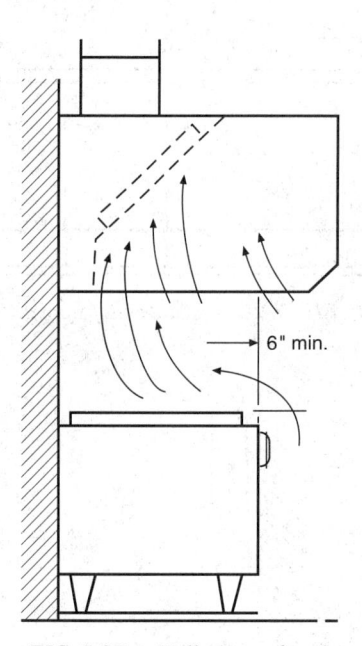

FIG. 7-9B.   Wall canopy hood.

Noncanopy hood: Noncanopy hoods come in many styles. They are noncanopy in that they do not meet the overhang and minimum height above the floor requirements of the canopy hoods. Noncanopy is a poor definition, since it defines what they are not, rather than what they are. They have many different names, such as backshelf, high-sidewall, eyebrow, cap, high-velocity, etc. In general, they are mounted much closer to the cooking surface, and are usually required to have the front lower lip no more than 3 ft (0.9 m) above the cooking surface. This is especially true of the backshelf, high-sidewall, and high-velocity hoods, which are primarily the same style hood with different names. Since noncanopy hoods are so low, they almost always have a setback from the front of the cooking appliance; i.e., the front lower lip of the hood is behind the front edge of the cooking surface. This is done to minimize the possibility of hitting one's head when bending over the cooking appliance. If there is a conveyor chain or other feeding/retrieval device at either end of the cooking appliance, it generally is not covered by any portion of the hood, though local codes may alter this. (See Figure 7-9D.)

Two variations of a canopy hood are cap and eyebrow. They are generally mounted on or just above taller cooking appliances, such as multideck ovens and broilers. The cap hood looks somewhat like a baseball cap, with the main part of the cap (the hood) over the cooking appliance to gather heat, and the smaller bill part in front of

the appliance to gather heat and effluent from the front when the doors are opened. The eyebrow hood is equivalent to the bill part of a cap hood, but is just the portion that is over the front of the cooking appliance. Both hood styles are also used over chain broilers and grills, and dishwashers. Canopy hoods, at times, are also used in every application where noncanopy hoods are used. (See Figures 7-9E and 7-9F.)

Exhaust Ductwork

Single: A single exhaust duct is one that serves one exhaust hood only. It will typically be of the same cross-section from one end to the other, and have one single fire suppression system for the entire exhaust system, including the appliances, hood, and all ductwork.

Manifold: A manifold exhaust duct system serves more than one exhaust hood. The cross-section is at its smallest as it leaves each

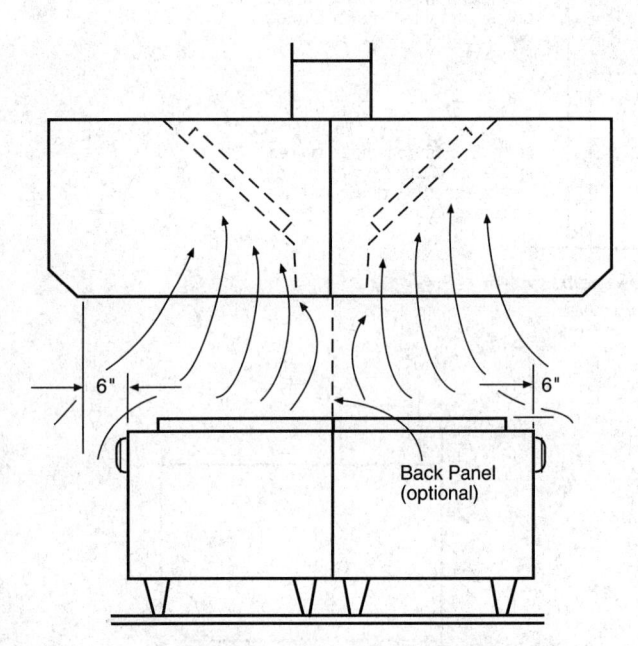

FIG. 7-9C. Island canopy hood.

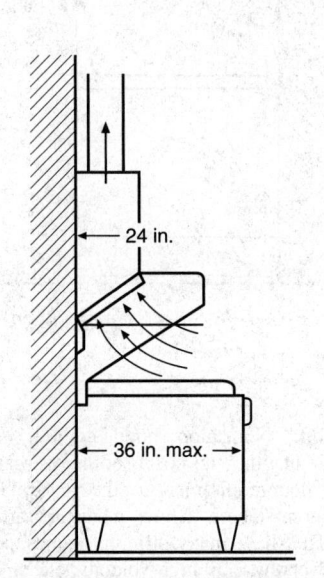

FIG. 7-9E. Cap exhaust hood.

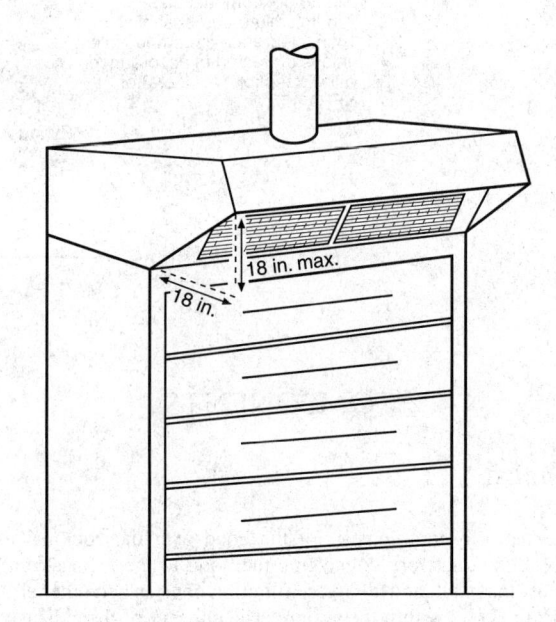

FIG. 7-9D. Noncanopy backshelf hood.

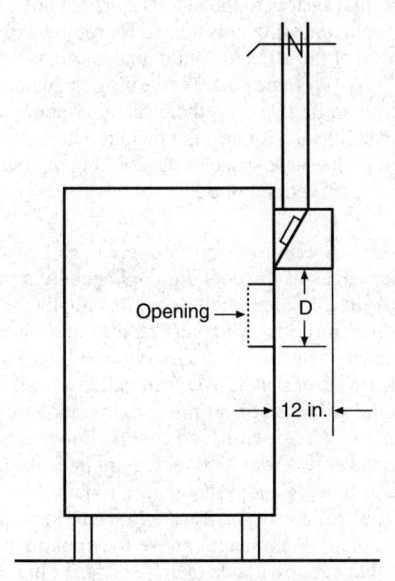

FIG. 7-9F. Eyebrow exhaust hood.

single hood and increases in cross section as it joins with another branch-duct, thereby forming a common duct, and increases further in cross-section as each additional branch duct enters the common duct. These typically have separate fire suppression systems for each exhaust hood and its appliances, and also for each branch duct. All the individual suppression systems will operate simultaneously for the protection of the common duct. However, there may be separate suppression systems for the branch ducts, or for all or portions of the common exhaust duct. In these cases, there are multiple ways to actuate the individual systems and provide for shutdown of the individual appliances. For more information, see NFPA 17, *Standard for Dry Chemical Extinguishing Systems*; NFPA 17A, *Standard for Wet Chemical Extinguishing Systems*; and NFPA 96, *Standard for Ventilation Control and Fire Protection of Commercial Cooking Operations* (hereinafter referred to as NFPA 96). (See Figure 7-9G.)

Exhaust Fan

Upblast: An upblast fan is one that terminates an exhaust duct system. It is located on the roof, sitting on a curb, with the inlet at its base, and the discharge at its top pointing upward. For grease exhaust applications, it is required to be listed for the application. It is also required to be restrained and hinged to tilt back for easy and direct access to the exhaust duct for inspection and cleaning. Typically, it will only operate on single exhaust duct systems or small manifold exhaust duct systems, not operating much over 1.5 in. of water column static pressure (373 Pa). (See Figure 7-9H.)

Utility: A utility fan is one that typically sits on the roof at the termination of the exhaust duct system, but it may also be found inside the building above the kitchen, with discharge ductwork leading it to the outside of the building. When installed on the roof, it sits beside the curb, with the duct coming out of the curb and leading into one side of the fan. The fan discharge is out of an opening that is perpendicular to the inlet and able to rotate the inlet so it can dis-

charge to either side or up or down. The discharge is sometimes ducted to get it more than 10 ft (3.1 m) away from any fresh air inlet, or put in a box to separate much of the grease from the air. It is capable of operating on both large single and manifold duct systems with high static pressure. (See Figure 7-9I.)

In-line: An in-line fan is almost always found inside the building at the same level as the kitchen, but in a back room. It is typically round in shape like a large duct section, with the inlet duct connection at one end, and the outlet duct connection at the opposite end. The vane-axial fan blade is located in the center so the whole fan is "in line" with the ductwork, and thus its name. It must be flanged and bolted to the ductwork at each end, in order that it can be removed for service. In some rare cases, only the fan and motor section are flanged and bolted. In either case, in-line fans are more difficult to service and are typically used only where other types of fans are not practical. (See Figure 7-9J.)

Makeup System

The makeup air system contains all the components (fans, ducts, registers, etc.) that provide for the delivery of the outside makeup air to the immediate or near vicinity of the exhaust hood.

Integral: Integral makeup air is the method of delivering a portion of the outside makeup air directly to a separate chamber of the

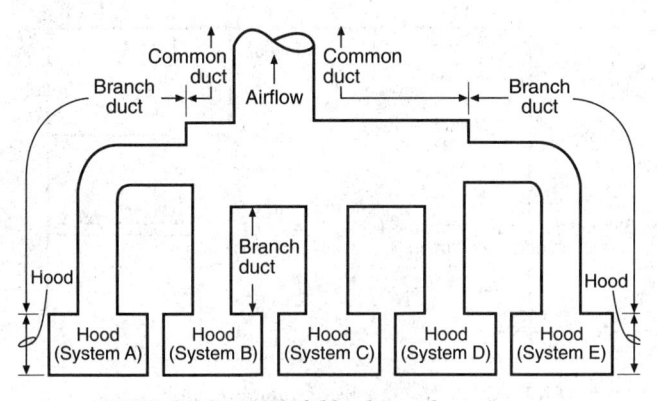
FIG. 7-9G.   Mainfold exhaust duct system.

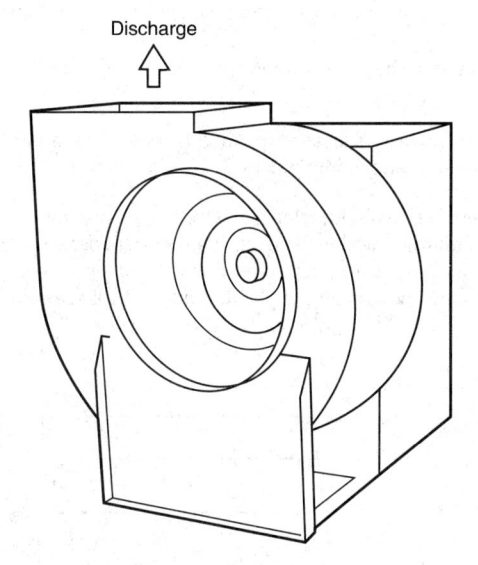
FIG. 7-9H.   Upblast exhaust fan.

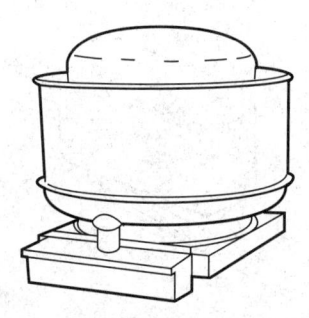
FIG. 7-9I.   Utility exhaust fan.

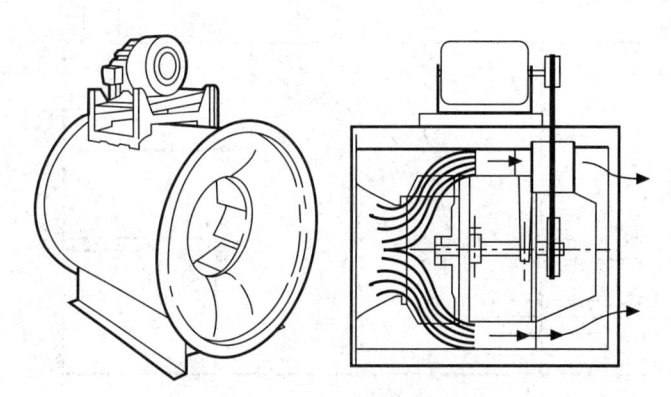
FIG. 7-9J.   In-line exhaust fan.

exhaust hood cabinet. Within the hood cabinet, this makeup air is introduced through registers or slots into the inside of the hood (short circuit), downward out the front lower edge of the hood (air curtain), or horizontally out the front of the hood (face discharge), or in multiple combinations of these. The makeup air is often provided without tempering of the short circuit and air curtain air flows. Hoods without tempered makeup air are often subject to performance variations, as the seasonal changes bring a change to the temperature and density of the makeup air. This method provides only a portion of the makeup air required, with the rest supplied from other sources. (See Figure 7-9K.)

Nonintegral: Nonintegral makeup air is the method of delivering the makeup air to the kitchen space or adjoining area(s) through ducts and registers. Such air is always tempered in all but the mildest of climates. (See Figure 7-9L.)

Controls

Control circuits may be manual or automatic. Automatic control circuits are called interlock control circuits. Interlock control circuits provide for *interdependent* operation of more than one component of a system in order to provide for the proper operation of the entire system. Interlocks provide for the safe conduct of the cooking operation by ensuring that:

1. Cooking cannot operate unless the fire suppression system is armed;
2. Cooking cannot operate unless the exhaust fan and makeup air fans are on; and
3. Cooking is shut down in the event of operation of the fire suppression system.

Only the last of these is required by codes, but all are recommended and often put into practice.

Operating controls/interlocks: Operating systems use many types of controls. Among them are controls that operate cooking appliances and exhaust system components, and safety controls that are built into the cooking appliances and exhaust system compo-

nents to control or shut down the components or system under emergency conditions. Many restaurants still operate with manual controls, but it is far safer to have interlocked control systems. A typical operating interlock control system will not allow cooking to operate unless the exhaust is operating, and will not allow the exhaust to operate unless the makeup air is also operating. A wide variety of controls exist that may be optionally included with these, but most often the control systems are quite simple and use few of the optional controls. These extra controls may be as complicated as controls that regulate the appliance and room temperatures during high- and low-volume periods, the amount of fresh and exhaust air, and the economizer performance of the heating, ventilating, and air-conditioning (HVAC) equipment. With some gas stoves and ovens, and with solid-fuel cooking there are often a number of manual controls also.

Fire controls/interlocks: There are fire detectors to sense a fire and initiate action by the suppression system, and there is a minimum interlock consisting of a switch to shut off the fuel to the cooking appliance in the event of the suppression system operation. There may be additional controls to sound a local fire alarm, or to sound a central alarm in a larger building, or even to notify a local fire department. There is always a manual means of operating the built-in fire suppression system. (See Figures 7-9M and 7-9N.)

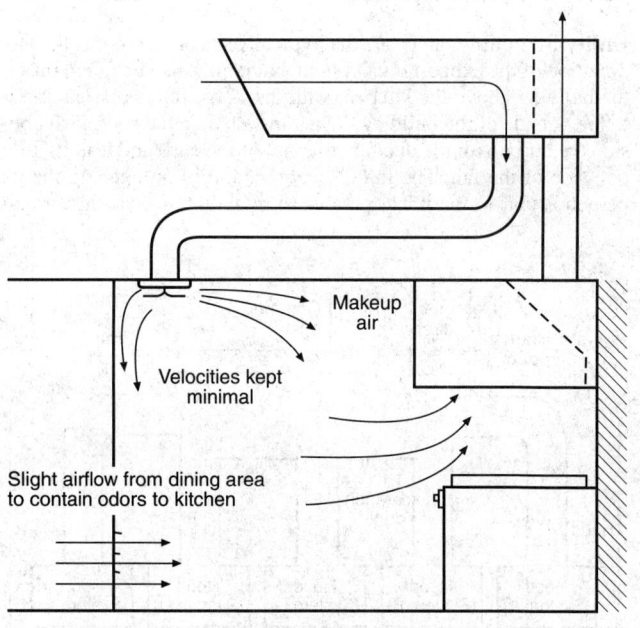

FIG. 7-9L. Nonintegral "standard" makeup air.

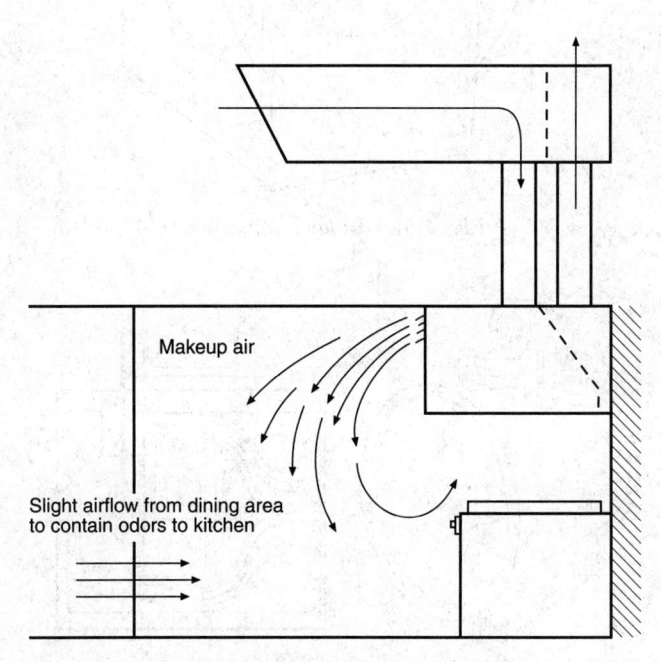

FIG. 7-9K. Integral makeup air hood.

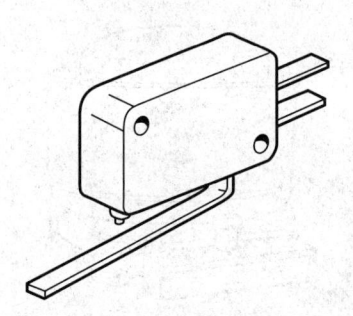

FIG. 7-9M. Fire system interlock switch.

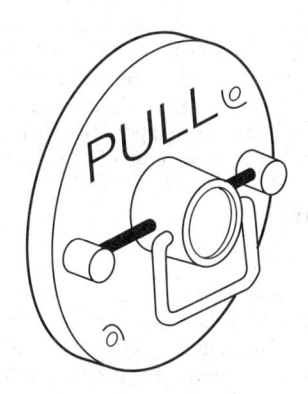

FIG. 7-9N.   Fire system manual pull.

FIRE CONCERNS WITH SYSTEM OPERATION DESIGN

Ventilation Flows

Exhaust: Exhaust flows must be adequate to capture and contain the effluents of the cooking process. If so, they are also adequate to capture and contain the small or short-term fire. But this does not mean they will be adequate to capture and contain the effluents of a serious and longer term grease fire. The expansion of vapors due to the higher temperature fires will ultimately exceed the capacity of the hood if the fire burns long enough, and the fire will escape from the hood. The smaller the appliance on fire, and the larger the hood, the longer it will take for the fire to escape the hood. The larger the appliance on fire, and the smaller the hood, the sooner the fire will escape the hood. Hoods operating with airflows that are lower than designed will not only have some grease effluent leaking into the space, but will also have flames projecting into the space sooner in the event of a fire.

Makeup/replacement air: The makeup/replacement air is one of the more critical airflows related to fire. If this air quantity is inadequate to match the exhaust airflow design, the exhaust will be forced to operate at a lower volume, and the fire will sooner be able to escape the hood. If the makeup/replacement air is introduced at such a location or velocity as to cause the exhaust air and effluents to escape the hood, clearly any fire will also escape the hood. A makeup airflow that is too strong, especially in a short-circuit hood, can also interfere with the proper amount of fire suppressant reaching an appliance fire.

Controls

Operating: If there is no interlock between the cooking appliances, exhaust, and makeup air, or if the interlock fails, it increases the chance of noncontainment of the cooking effluent, and increases the potential hazard in the event of a fire. Failed thermostats and high-limit temperature controls that can no longer shut off the cooking appliance cause many fires in cooking appliances, due to appliance overheating and autoignition of the oil or shortening in the cooking process. The same scenario occurs on manually controlled cooking equipment when pots of oil are left unattended and overheat to the point of autoignition. Where no interlock exists between the exhaust fan and cooking appliance, or if it fails, a fire can start in the hood, especially over gas-operated appliances, due to the flue gases overheating grease buildups in the hood to the point of autoignition.

Fire interlock: The most serious fire interlock failures are the ones that fail to permit the system to operate. These include grease-plugged detector line elbows that do not permit the cable to move, detector links that are too grease coated and insulated to respond in time to the fire, or detector links that are held closed with a wire from some previous maintenance operation. Another serious fire interlock failure is one that will not shut down the fuel that heats the appliance, due to a failed interlock switch, or a jammed mechanical gas valve. Other potential failures include defective switches or open wires in alarm circuits. These situations are the basis for the required 6-month control inspection.

FIRE CONCERNS WITH SYSTEM COMPONENT DESIGN

Exhaust Hoods

The three primary fire concerns with the exhaust hood are: (1) that it is able to extract and drain the maximum amount of grease possible from the effluent to minimize any subsequent fire, (2) that it is able to contain a fire within itself should one occur, and (3) that it is protected from adjacent combustible surfaces so that it will not ignite them by radiant effect. This requires that the hood be of adequate size and operate at adequate airflow and have a reasonably efficient filtration system, that it be of such structure to stay intact during a fire so that the fire cannot escape, and that it be installed in such a way as to provide protection between itself and adjacent combustibles so that it cannot radiate heat sufficient to ignite them.

Filters

Hoods may contain removable baffle filters, or they may have larger filtration baffles that may be removable in cassette form, or filters may be built in. The removable filtration devices must be removed from the hood to be cleaned. The built-in units are cleaned in place with a detergent and water wash system. It is very important for fire safety that all filtration devices be cleaned at least once a day. Further, while dirty filters are being cleaned, removable devices must have replacement units installed at all times when the appliances are still turned on. Operating the exhaust with filters removed will put much more grease into the hood and duct systems. It will also lower the static pressure so much that it will dramatically increase the amount of air being exhausted and unbalance the site. In the case of manifold exhaust systems, removing filters from one hood will increase the exhaust at that hood, but dramatically lower it at all the other hoods connected to that manifold system. This will cause the other hoods to leak grease vapors into the space, which defeats the purpose of the hoods and increases the fire hazard potential. Filters in the hood must be tightfitting on all four sides in order to preclude any grease or fire bypassing them. (See Figure 7-9O.)

Exhaust Ducts

There are two essential requirements for exhaust ducts, relative to fire protection. The first is that the duct must be so solidly welded or assembled (listed ducts only) from one end to the other that a fire within a duct cannot escape through the duct wall to a hidden combustible area. The second is that there must be adequate clearance or protection around the ductwork from one end to the other, so that, should a fire exist in the duct, the radiant heat from the duct wall will not be able to ignite any combustible products that may be close to the duct. This requirement prevents the duct from causing a fire to start in some hidden area of a building that might not be noticed until hours later, causing renewed hazard to unsuspecting tenants.

Proper ductwork structure also includes that all ductwork be designed without dips or traps that can hold grease and be pitched

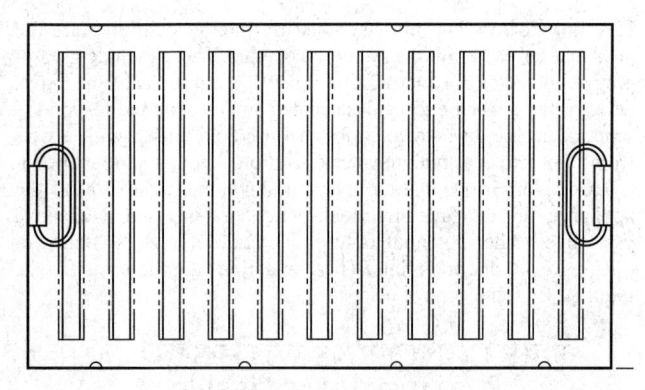

FIG. 7-9O.   Grease filter.

to drain back to hoods or to special grease drains that are acceptable to the local Authority Having Jurisdiction. The exhaust fan must also be designed to drain grease to a closed external container. Discharge from the fan, or the duct that terminates the system, must not direct the flow toward the roof or any combustible surfaces, nor toward any operable windows or doors. These precautions serve two purposes: (1) prevent a fire from igniting any combustible surfaces and (2) keep a fire and any grease discharge from blocking proper ventilation or egress from building openings.

Exhaust Dampers

A fire damper sometimes is placed in a hood just at the duct connection to the hood. Its intent is to prevent fire from getting into the ductwork. This is a controversial aspect of exhaust systems. The issues involved, for and against, are as follows, and each local Authority Having Jurisdiction should consider the special situations existing in each building to determine if fire dampers are the proper choice or not.

Damper advantages: Fire dampers are helpful in commercial kitchen hood/duct systems because:

1. They keep fire out of existing ducts that may be substandard and would not contain the fire.
2. They keep the fire contained to the kitchen level of multistory buildings and, thereby, reduce the potential hazard of fire spreading to other floors and tenants.
3. They provide increased safety in the event the duct suppression system does not function.

Damper disadvantages: Fire dampers are problematic in commercial kitchen hood/duct systems because:

1. A properly designed and installed duct system can exhaust the fire and smoke out of the building in a safe manner if the system is designed to handle such fires, making a damper unnecessary.
2. Closing the damper will back the fire and smoke more quickly into the kitchen space, spreading the fire more quickly, and the smoke will make it much more difficult for the fire service to even see to fight the fire.
3. A fire kept out of the duct system and in the kitchen space can spread quickly to combustible or grease-coated ceiling and wall surfaces, and from there to other hidden areas, increasing the hazard to other tenants and building areas.
4. The exhaust ducts may be so dirty that the damper cannot fully close so that the fire spreads to the ductwork anyway, or the fire gets into the ductwork before the damper closes, so the damper has only provided a false sense of security, and makes it more difficult to access the fire in the duct.

It is clear that much of the concern with dampers has to do with the proper condition of the ductwork beyond the dampers, including its integrity and clearance to combustibles. The local Authority Having Jurisdiction should take all these matters into consideration when deciding to accept fire dampers or not at the hood/duct connection in any particular application. (See Figure 7-9P.)

Exhaust Fans

There are two primary concerns with exhaust fans relative to fire hazard. First, the fan must not trap grease within itself, but drain it to an external covered fireproof container of not more than 1-gal (3.785-L) capacity. This minimizes the chance of the fan contributing fuel to any fire, and places the drained grease in a sealed container of limited capacity that is not likely to catch on fire. Second, the fan must not discharge its effluent toward any combustible surface that might catch fire, or toward any operable door or window or vent that might be a point of egress or fresh air for the occupants. This limits a fire that may come from the discharge of a fan, or the terminal section of ductwork, and prevents fire spread to other combustible surfaces. Further, it allows building reentry and a path of potential egress.

Additional Grease Extraction/Control Devices

Many commercial cooking establishments now have additional grease extraction or control devices downstream of the hood in order to more completely remove the grease and odors from the effluent stream. This is required in many more sites today in order to reduce local environmental concerns and complaints, and, in many cases, just to create a better atmosphere for potential clients. These devices may be installed in ceiling spaces or mezzanines, but most are installed on rooftops immediately before the fan, or they may even contain the fan.

These units consist of various stages of filtration devices that may be metal baffles, metal mesh filters, dry bed or pleated filters, electrostatic precipitators, afterburners, or even water-wash devices, and all are designed to remove the remaining grease from the system. The afterburners also remove odors. The remaining system uses activated carbon or some other chemical catalyst, or even some odor control spray in order to control odors.

Each of these devices (except the afterburner) carries with it the possibility of igniting or providing fuel for a fire. (The after-

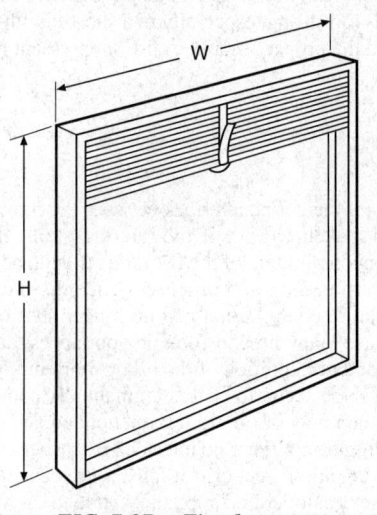

FIG. 7-9P.   Fire damper.

burner has built-in safety controls and cabinetry to preclude a runaway fire in it.) Therefore, each of these units should be provided with a fire detection and suppression system that provides protection in the event of a fire. The cabinetry of the units should be of the same gage as the ductwork to which it is attached and the access panels should match those of the ductwork, or the unit should be listed for equivalency with these requirements and applications in a grease-laden environment. (See NFPA 96 for fire safety regulations for such units.)

Grease Gutters, Cans, and Containers

Grease gutters may exist under the filter rack of a hood, around the lower perimeter of a hood, and on one or more sides of a cooking appliance. They are pitched to drain to a remote grease container. Grease cans or containers collect the grease from the gutters. They are also found as open or closed pans or containers under certain sections of cooking appliances and at the drain of grease fans. The obvious fire hazard with all of these is that they contain extra fuel in the event of a fire. For increased fire safety, the amount of grease that is contained in these at any time must be kept to a minimum. Therefore, the gutters should be pitched as much as practical to minimize the grease they can hold. Containers, cans, and pans should be kept at the minimum volume necessary to get through one major rush period, and not be expected to hold the grease for the entire day in a high-volume operation. Many codes limit the container to 1-gal (3.785-L) capacity. It is also best that the container be capped on top, with only the minimum opening necessary to permit the grease to enter. Should a fire get into such a container, it is most likely to smother itself rather than burn all the fuel. This is the very reason grease gutters are pitched to drain the grease so that it is carried to a remote sealed container rather than left in an open trough.

Appliance Position

It is important that appliances be positioned under the hood in their designated position for three reasons. First, it ensures that the effluent generated by the appliance will be captured and contained by the hood and exhaust system, and not escape into the room where it can extend the fire hazard potential. Second, in the event of a fire in the appliance, it ensures that the flames will be contained in the hood for the longest time possible, limiting the fire hazard potential. Third, it ensures that the fire suppression system nozzle will be able to apply the chemical suppressant to the fire as designed for that appliance, and thus extinguish the fire.

Some appliances are supplied with appliance locks to ensure that they maintain their positions. This is a very good design that is worth encouraging. Locator blocks on the floor or rear wall that position the appliances are also good. But even where these do not exist, it is important that operating personnel realize the importance of keeping the appliances in their designated positions under the hood.

Fire Detectors

Fire detectors, whether mechanical link or electronic sensor, must be applied at a rating and location that permits them to respond *promptly* to any abnormal temperatures in the effluent from the cooking operation. A detector that responds too late may as well not respond at all. The fire may have already spread to areas beyond the range of the suppression system, or the ignited surfaces may be so hot, especially in the case of deep fryers, that even if the suppression system initially puts out the fire, there will be reignition because the foam will break down before the surfaces can cool below the new lower reignition temperature. Whatever the autoignition temperature of the greasy surfaces initially, once the suppression system has discharged, the surface temperatures must cool to at least 50°F

(10°C) below the original autoignition temperature if the surface is to be sure not to reignite.

The best way to choose the proper link rating and location is to monitor the temperatures of the normal effluent airstream under maximum cooking conditions. This is not often possible or practical, so a "best guess" is made and a rating and location selected. Even if this is all that can be done initially before the restaurant opens, temperature monitoring with a remote recorder can always be done later, and then within a very short time the actual maximum normal temperature can be known and the link rating adjusted accordingly. Remember, a "fast" response is necessary, not just a response. The longest pre-burn time for systems listed to the new UL 300, *Standard for Safety Fire Testing of Fire Extinguishing Systems for Protection of Restaurant Areas*, is 2 min. Most of the older systems were listed for only 1 min pre-burn. But links are normally installed behind filters, many feet above the cooking surface. It could easily take much more than 2 min for a fire in a single vat under a canopy hood to raise the temperature of a link to its rating and through its soak time to trip the link. This is especially true with the likelihood that the link is not directly in the line of flame and that air flowing over it is diluted by the cooler air from the rest of the cooking operation. It is not unusual for fire to completely bypass links for 5 min or more. It is definitely better fire safety to install more links than thought necessary, since the path of the flames and hot air will be strongly affected by the airflows created by the exhaust system. These tend to hug the sidewalls of the hood or go straight toward the duct, and leave voids in between. (See Figure 7-9Q.)

Makeup/Replacement Air

Integral: Where makeup air is integral to the hood, it is possible that some grease can also get into this makeup air compartment whenever it might be off, or when it is coming out only a portion of the slot and creating a venturi inlet at the rest of the slot. Grease deposited in hidden areas can spread a hood fire to this point. Therefore, makeup air plenum insulation is required to be nonabsorbent and easily cleanable. Any buildup of grease in this area must prompt the operator to provide makeup air and keep the plenum free of grease.

Nonintegral: Makeup replacement air can disturb the capture of the hood to the extent that grease leaves the hood and enters the room. Therefore, if a fire develops and grows rapidly, it will also leave the hood and track along the grease path provided by the escaping cooking effluent. This can create uncontrollable fire spread.

Operation during fire: There is a tendency in some jurisdictions to have all supply air shut down in the event of fire. This is not always

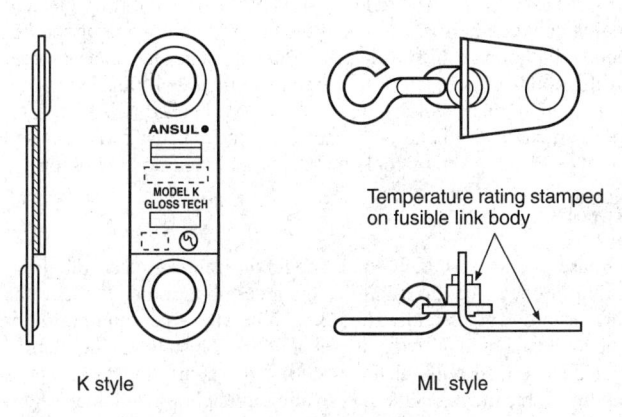

K style   ML style

FIG. 7-9Q.   Detector links.

proper. If the hood is able to capture all grease vapors properly, and if there is no fire damper at the hood/exhaust duct connection, it is better to keep the makeup air operating in the event of an appliance or hood fire, as there is an increased chance the hood will be able to contain the fire and smoke. If the makeup air is shut off, the hood will be short of available air and will not be able to continue exhausting properly.

FIRE CONCERNS WITH SYSTEM MAINTENANCE

Maintenance is essential to ensure that a system performs as designed. Both components and the system as a whole must be kept physically and operationally in the condition for which they were designed. Adjustments, repairs, and replacements must be made as necessary. Maintenance must identify each part of proper system design and performance, and review these on a regular basis to catch and correct the first slip from "normal." This will keep the system working as designed.

Maintenance is not intended to bring back to normal a system after it has broken down. Those commercial cooking and ventilating systems that involve potential fire hazards require that the design keep working, or the fire hazard can seriously increase to the point where almost no safety is present.

Ventilation Flows

The purpose of ventilation systems in commercial cooking applications is to keep greasy effluents captured, contained, and evacuated through the exhaust system, and keep the kitchen environment grease-free. Any decay in the performance or flows of the systems may cause grease vapors to leak from the hood into the kitchen space, contaminating the area with fuel. In the event of fire in a faulty system, the flames may leave the hood more quickly and ignite the grease deposits in the room, seriously increasing the fire hazard. Airflows of the exhaust and makeup air should be checked at least once a year, and more often if it is noticed that the hood is not capturing the effluent properly.

Exhaust Hoods

The primary fire concern with hoods is that they be kept: (1) in the same structural condition in which they were installed and (2) clean so they will not sustain a fire and transport it into the duct. No holes can be cut in the hood, no service lines run through it, and no notching made in end panels to provide additional clearance. The welded shell of the hood is supposed to be able to contain a fire. When holes are put in it, or its perimeter is cut and notched, it can no longer provide this function and the fire hazard increases.

The hood must be kept clean. The filters should be cleaned daily, and the interior surfaces of the hood (both in front of and behind the filters) should be cleaned to bare metal at least once a week. If the hood surfaces and filters are clean, there is little chance of a fire in an appliance spreading to the ductwork. (See NFPA 96 for a recommended frequency of cleaning for the exhaust system, based on type of fuel, type of cooked product, and volume of cooking.)

Grease Filters

Grease filters installed in hoods should typically be cleaned a minimum of once a day to minimize the grease buildup and reduce the fire hazard. The cleaning should be done after operating hours, or the facility should have a spare set of filters to install. The hood must not be operated without filters during the hours the restaurant is open, as this increases the fuel loading in the hood and duct, and it will unbalance the ventilation.

Filters should be in good physical condition, with flat squared frames and no baffles damaged or missing. Damaged filters should be replaced immediately. They cannot filter effectively, and more importantly they cannot stop flame penetration into the hood.

Some hoods have permanent grease filtration devices, and these will have built-in detergent and water-wash systems to clean the grease filtration devices on a daily basis. It is important that these cleaning systems be maintained in good working order to keep the grease filtration devices clean and reduce the fire hazard. It is common to find these cleaning systems without any detergent, and also with the controls or water shut off so they cannot operate. This violates the listing of the unit, thus increasing the fire hazard. Hoods with automatic washdown should be set to self-clean a minimum of once a day, and more frequently if conditions warrant. The washdown system should be inspected for proper operation at least once a month to ensure that all nozzles are open, pressures are adequate, and the controls are cycling properly.

Additional Grease Extraction/Control Devices

Some exhaust systems have additional grease removal devices somewhere in the duct system after the hood. They may be in the false ceiling space, in a mezzanine, or on the roof. These extraction devices also contain various types of filters. These filters must also be cleaned or replaced according to the manufacturer's listed instructions. Failure to clean the filters builds up unnecessary fuel in the system, seriously reducing the airflow, and may cause the hood not to capture the vapors as designed. It is also necessary with these units to make sure that the filters are in good physical shape so the grease cannot bypass them, and make sure that they are in place so they can function properly. Failure to keep these components operating as designed seriously increases the fire hazard. These units should be inspected a minimum of once a month to be sure the filters are still performing properly, and that all panels, seals, and suppression system components are in good shape.

Exhaust Ducts

These ducts should be fully welded, or possibly bolted if a listed duct system. Decay is typically not a problem. However, subsequent remodeling of the building, or other trades working in the vicinity of the ducts, can cause damage or changes to the ducts that reduces their integrity. Any time work is performed in the vicinity of the ducts, they should be inspected again after the work is complete.

Many ducts also have bolted or gasketed connections at the hood or fan. These require more attention, as they are more subject to decay from normal use and vibration. Any grease leaks seen around these points are reason to have the connection checked and repaired, as necessary.

If, for some reason, a facility has an existing duct that is not fully welded, or in other ways does not fully meet code, it is all the more important to establish regularly scheduled inspection of the duct to ensure that grease leaks do not exist and that other damage has not occurred.

Makeup Air Systems

The makeup air system is needed to provide replacement air to balance that taken out by the exhaust system. The makeup air system must operate whenever the exhaust is operating, be matched to the quantity of air taken out by the exhaust system, and not interfere with the performance of the exhaust hood.

The makeup air system usually operates at considerably higher air velocities than the exhaust system, and usually approaches the hood at different angles than those established by the exhaust airflows. Where these angle differences exist, they should be neutralized by distance before the makeup air nears the hood so as to not adversely affect the capture and containment effect of the exhaust airflows.

Clearly, makeup air introduced integrally at the exhaust hood is more likely to have a negative effect on capture and containment than makeup air introduced at the far side of the room from the exhaust hood. Therefore, a more sensitive case of integral makeup air needs more frequent attention to maintain proper performance. Some hoods with integral makeup air have operator-adjustable diffusers that can be set to operator preference for the season, but this will often adversely affect the capture and containment performance of the hood.

If ambient conditions are not acceptable to facility personnel, this should be brought to the attention of the design engineer. Fire safety should not be compromised for the sake of employee comfort.

Service Panels/Gaskets/Hardware for Hoods and Ducts

Inspection panels are often installed in ductwork and in some hoods, in order to inspect and clean the ducts and hoods. When in place, these panels must have integrity equivalent to welded ductwork. This requires that: (1) the panel be ridged and conform to the shape of the opening, (2) the gasket be complete and adequately thick to seal the full perimeter of the opening, and (3) the hardware used to hold the panel closed be tightened securely in order to create a grease-tight seal. If a panel cannot be sealed as described above, it must be repaired immediately. Panels that cannot seal will leak grease, possibly allowing burning grease to escape from the opening to ignite combustibles in the vicinity. This can spread a fire to relatively hidden areas that are difficult to locate. The integrity of these components should be closely inspected each time they are used.

Clearance to Combustibles

All hoods, ducts, fans, and other components of exhaust systems are required by codes to have adequate clearance to combustibles so as not to ignite the combustibles from the radiant heat of the metal shell in the event of an internal fire. Sometimes this clearance is provided by insulation and air gaps, sometimes it is provided by a non-combustible or fire-rated chase, and sometimes it is provided by only a physical distance to the closest combustible surface. Any protective system provided around the exhaust system must be kept intact, in order to maintain this crucial clearance. Chases must have access panels to enable inspection and cleaning. Access panels in the chases must be kept in good shape and secured in place in order to maintain the designed protection. These panels should be inspected and maintained, with attention equivalent to that given the panels in the ductwork itself. In cases where there is only open space between the exhaust system and combustible construction, it is important to maintain that space. It is not uncommon in commercial cooking operations to use false ceiling and attic spaces for storage of combustible products. Where this is done, and the same spaces contain unprotected exhaust ductwork, it is best to put a wire mesh or expanded metal fence around the ductwork to ensure that no combustible materials can come closer than 18 in. (457 mm) to the ductwork. Note: This type of storage is allowed in sprinklered spaces in fully sprinklered buildings.

Appliances

Grease effluent is produced by the products cooked on appliances. Much of the grease effluent is carried into the exhaust system, but one must remember that a large quantity of grease stays with the cooking appliance. Fryers hold shortening in the vats. Grills collect grease on the surface, then it drains (or is scraped) into scrap cans on the side or front of the grill. Ovens and some broilers usually have grease baked on the interior surfaces, but this grease can run out of drain openings to cups. Charbroilers usually burn grease in the fire or on the hot elements, but much of it just settles at the bottom of the bed. Fire safety requires that, where grease collects inside appliances or in containers from appliances, it be frequently removed during the day and not allowed to spill over onto the floor. Inspections and cleanings should occur multiple times each day to minimize the immediate hazard, and to facilitate the cleanup of the grease. The floor, wall, and rear area of appliances should be cleaned at least once a day.

Grease often condenses on the outside of appliances and hoods, or on lower hood or wall parts below any grease gutter. Most of this grease will drain onto the floor, and a path will show on the wall or side of the equipment. It is important for fire safety that such occurrences be corrected with additional gutters and remote covered grease cups, or that they be frequently addressed during each day with manual cleaning. Otherwise, there will be a large buildup of mostly hidden grease that will create a large fuel source in the event of fire.

As many appliances age, they develop internal grease leaks that permit a grease buildup in the immediate vicinity of the gas burner or electrical elements. This can quickly become the source of a fire, and there is no detection or suppression system within this appliance area to suppress the fire when it occurs. Therefore, it is essential to inspect such areas and keep them grease-free. All grease leak sources should be located and repaired immediately. This is especially important for fry vats, as a leaking vat can drain grease onto a burner and cause an immediate and serious fire.

Appliance Changes

Fire suppression systems are installed in hoods to protect the specific appliances that are under the hood at the time of original installation of the system. If these appliances are later repositioned, or changed out for other appliances, the suppression system may no longer be able to provide suppression for the new arrangement. When appliances are repositioned or changed out for other appliances, even if the change only involves a different input rating or a different fuel, the local fire department must review the system and see if it is still adequate, or if a detector, nozzle type, or nozzle location needs to be changed. All restaurant operators should be aware of this concern. Installation education notices, through inspection reports and other correspondence, can provide guidance.

Appliance Seals

Some appliances, notably closed ovens, must be sealed during normal cooking operations to contain the grease and other effluents. If a fire should occur in such an appliance and the seal is not complete, fire can escape from the appliance rather than be contained and smothered. It is important that all such seals be kept in good order to ensure fire safety.

Fire Detectors

Fire detectors should be inspected a minimum of every six months, and mechanical links should be changed a minimum of once a year. Inspection must include that the rating and position of the detector is still appropriate for the application, especially for appliances that may have changed since the last inspection. Detectors covered with an insulating layer of grease or carbon can have a much slower response time than designed. All detectors must have clean surfaces; otherwise, they should be changed out for new detectors.

Fire Dampers

Fire dampers in grease exhaust systems should only be installed in listed hoods at the hood to duct connection, or they may be a component of a listed grease extraction system that is installed downstream of the hood. In either case, they need to follow the same inspection schedule as the suppression system detectors, and they have the same problems if the detector link becomes grease or carbon coated. Exhaust hoods with integral makeup air supplies usually have fire dampers installed in the supply duct to hood connection, or at one or more of the makeup air register openings. Their links have the same inspection and replacement schedule.

Fire dampers also need to be checked for proper and full operation. The damper should be able to close fully and rapidly. If it moves sluggishly or doesn't close fully, its track is too coated with grease buildup, and it must be cleaned. If the fire damper cannot close quickly and completely, it cannot be effective as designed.

Detection Lines/Elbows

Detection line conduits and the pulley elbows in them need to be inspected as required every six months, including doing a trip of the actuator by cutting the terminal link. This will show that the cable is free to move in the conduit system. In grease exhaust applications, it is not unusual for grease to get into the horizontal conduits or into the pulley elbows and harden there to such an extent that the cable cannot be moved. When this occurs, the system cannot operate automatically, and this presents a serious fire hazard. If a cable will not move, or moves only sluggishly, the conduits and elbows should be dismantled to find the problem. If they cannot be readily cleaned, they should be replaced.

Another concern with pulley elbows is when they are located in the hot flue gas from gas-fired appliances. When this occurs and they are not rated for such applications, they will melt in such a way as to seize the cable in the elbow, and, again, the cable will not be able to move when needed. Visual inspection will usually show this problem, but it can also be noted with a terminal detector test.

Gas Valves

Any gas valve can stick open, but it is more likely to occur with mechanical gas valves, as they have openings that permit grease to get inside part of the mechanism which causes it to seize or slow down. These must be operationally checked every six months. If any delay or incomplete closure is noted, the valve must immediately be cleaned or replaced. With mechanical gas valves, it is necessary to ensure that the cable line is not binding in any way and causing the problem. (See Figures 7-9R and 7-9S.)

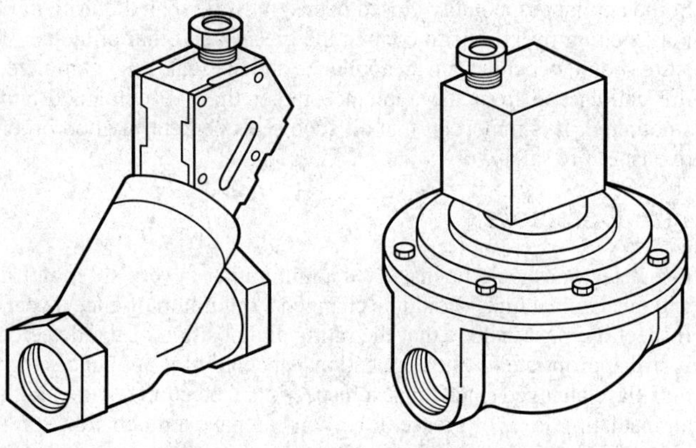

FIG. 7-9R. Mechanical gas valves.

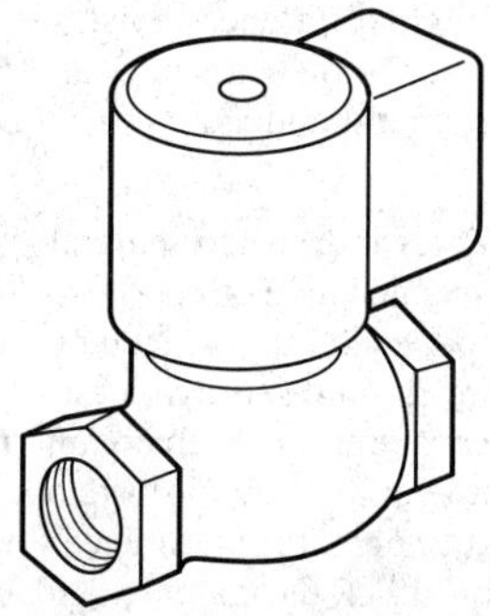

FIG. 7-9S.   Electrical gas valves.

SUMMARY

Ventilation systems for commercial cooking operations provide invaluable services to the industry and permit such operations in many locations where the potential for fire hazard is great due to the fuel, product, and greasy effluent that are inherent in commercial cooking operations. Ventilation systems can be designed to not only provide their prime function of capturing, containing, and removing the greasy effluent, but can also contribute positively to a fire-safe environment. Knowledge exists for safe design and should be conveyed to facility operators, service personnel, and local inspectors. It is important that these people know the value of each component of these systems for operational and fire safety, and how to keep the components and the system in proper working order so that it can continue to safely provide its intended functions. As more education is provided on these issues, the fire safety record will improve, and the industry will better prosper.

BIBLIOGRAPHY

Reference

UL 300, *Standard for Safety Fire Testing of Fire Extinguishing Systems for Protection of Restaurant Areas*, Underwriters Laboratories Inc., Northbrook, IL, 1992.

NFPA Codes, Standards, and Recommended Practices

Reference to the following NFPA codes, standards, and recommended practices will provide further information on venting systems for commercial cooking equipment discussed in this chapter. See the latest version of *The NFPA Catalog* for availability of current editions of the following documents.)

NFPA 17, *Standard for Dry Chemical Extinguishing Systems*
NFPA 17A, *Standard for Wet Chemical Extinguishing Systems*
NFPA 96, *Standard for Ventilation Control and Fire Protection of Commercial Cooking Operations*

Additional Reading

Grosso, J., "Restaurant Grease Duct Fires," *Fire Engineering*, Vol. 148, No. 6, June 1995, pp. 16, 21.

SPECIAL STRUCTURES

Revised by John G. Degenkolb

Special structures are designed for specific purposes that are different from the typical day-to-day uses of most structures. Examples of special structures are towers, tunnels, underground buildings, bomb shelters, highway and railroad bridges, grandstands, tents, membrane structures, signs, mechanical parking garages, marine terminals, piers and structures surrounded by water used for other-than-vessel mooring or cargo handling, some agricultural buildings, and air-right structures. Because these structures represent unusual uses or solutions to design problems, they also present unusual or special fire protection design problems.

Additional relevant information can be found in Section 4, Chapter 5, "Fibers and Textiles"; Section 6, Chapter 25, "Special Systems and Extinguishing Techniques"; Section 7, Chapter 1, "Building and Site Planning for Fire Safety"; Section 8, Chapter 2, "Concepts of Egress Design"; Section 9, Chapter 2, "Occupancies in Special Structures and High-Rise Buildings"; Section 9, Chapter 3, "Assembly Occupancies"; and Section 9, Chapter 34, "Fixed Guideway Transit System."

TOWERS

The major fire problems with towers are the difficulty of extinguishing fires that may occur high in the tower; limited means of egress for occupants of lookout, air traffic control, and observation towers, which prompts the use of noncombustible materials throughout and good fire prevention practices; and vulnerability to exposure fires. When towers are built of unprotected steel or wood, a ground fire can cause ignition or structural failure, or both. Towers are frequently appended or attached to other structures of lower height. In such cases, egress from the tower should be directly outside where possible. If this is not feasible, the building exits should be designed so that occupants of the tower may also use them.

Apart from fire damage, but extremely important from a life safety standpoint, is the possibility of windstorm damage to towers. The Federal Aviation Agency (FAA) has been active in developing building code requirements for the towers it occupies.

The federal government has encouraged its various agencies to comply with local building codes. The FAA has worked with the model codes to develop acceptable requirements for airport or aviation traffic control towers. The height of the tower is subject to the height *vs.* area limitations of the local code for the type of construc-

tion employed. A maximum of 1500 sq ft (140 m^2) area is imposed. One exit stairway is permitted, provided the occupant load per floor does not exceed 15. The stairway and elevator accesses must be separated from each other a distance equal to no less than one-half of the length of the maximum overall diagonal dimension of the area served. The model building codes designate as high-rise buildings only hotels, apartment houses, and office buildings having a floor used for human occupancy more than 75 ft (23 m) above the lowest level of fire department vehicle access. Such high-rise buildings are required to be sprinklered. NFPA *101*®, *Life Safety Code*®, (hereinafter referred to as NFPA *101*), defines a high rise as a building more than 75 ft (23 m) in height as measured from the lowest level of fire department vehicle access to the floor of the highest occupiable story. Chapter 30 of NFPA *101* requires that high-rise buildings be sprinklered. Such towers need not be accessible to the handicapped.

Water-Cooling Towers

Functionally, a water-cooling tower is designed to transfer heat from water to air. This is accomplished by passing the water in small particles through the air, where heat is removed from the water partly by an exchange of latent heat, or the heat required to pass from liquid to a gas due to the evaporation of some of the water particles, and partly by a transfer of sensible heat, or the heat required to change the temperature. Figure 7-10A shows a cross-section of a typical mechanical-induced draft water-cooling tower and some of its significant parts.

Water-cooling towers may be all metal, metal-frame with wood or plastic fillers and wood or plastic draft eliminators, all wood, or ceramic construction. Some towers have a noncombustible exterior covering. Wood, generally redwood, is used in cooling tower construction primarily because of its durability, its lack of corrosion problems, and the relatively low cost of both original and replacement materials.

Water-cooling towers are of atmospheric and mechanical draft types. Mechanical draft towers are of two general types: (1) forced and (2) induced draft. They may be further classified as counterflow and crossflow. (See Figure 7-10B.)

The foreword to NFPA 214, *Standard on Water-Cooling Towers* (hereinafter referred to as NFPA 214), contains this statement: "The fire record of water-cooling towers indicates the failure to recognize the extent or seriousness of the potential fire hazard of these structures both while in operation or when temporarily shut down. Cooling towers of combustible construction, especially those of the

---

John G. Degenkolb is a fire protection engineer and code consultant in Carson City, NV.

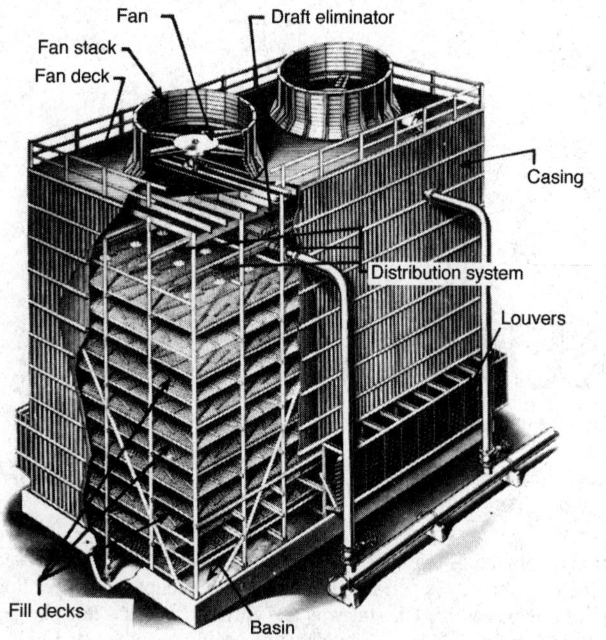

FIG. 7-10A. *A typical mechanical, induced-draft, counterflow water-cooling tower. Water is pumped to the distribution system where it is cascaded over the fill decks and collected in the basin at the base of the tower. The fan draws air in through the louvers and the casing of the tower. Draft eliminators minimize the loss of water through the fan stacks.*

induced draft type, present a potential fire hazard even when in full operation because of the existence of relatively dry areas within the towers."

Among the causes of water-cooling tower fires are sparks from welding and cutting, either on the tower or on adjacent structures; sparks from trash fires, incinerators, chimneys, or industrial stacks; carelessly discarded smoking materials; lightning; and exposure fires. Another problem is the windblown leaves and trash that collect around towers and are easily ignited.

Fires caused by tower equipment may be the result of mechanical failures in gear reduction boxes and bearings, misaligned fan blades, or metal fatigue, which can result in localized heating. Electrical breakdowns in motors, short circuits in wiring, and other electrical faults also have led to cooling tower fires.

The construction of the water-cooling tower is a primary factor in reducing the possible fire hazard. Small towers or those whose main structure does not exceed 2,000 cu ft (57 m^3) in volume are generally completely noncombustible and therefore are seldom a problem. Larger towers, however, generally contain considerable amounts of combustible materials. In addition, towers on roofs may be inaccessible to automotive fire-fighting equipment. Those in yards may require fencing and weed control or protection to prevent exposure damage to or from an adjacent building.

Once the degree of hazard has been determined, any one or a combination of the following protection measures may be considered when dealing with essentially combustible towers:

1. Adequate separation from exposures.
2. Noncombustible exterior coverings.
3. Division of areas by fire partitions between cells.

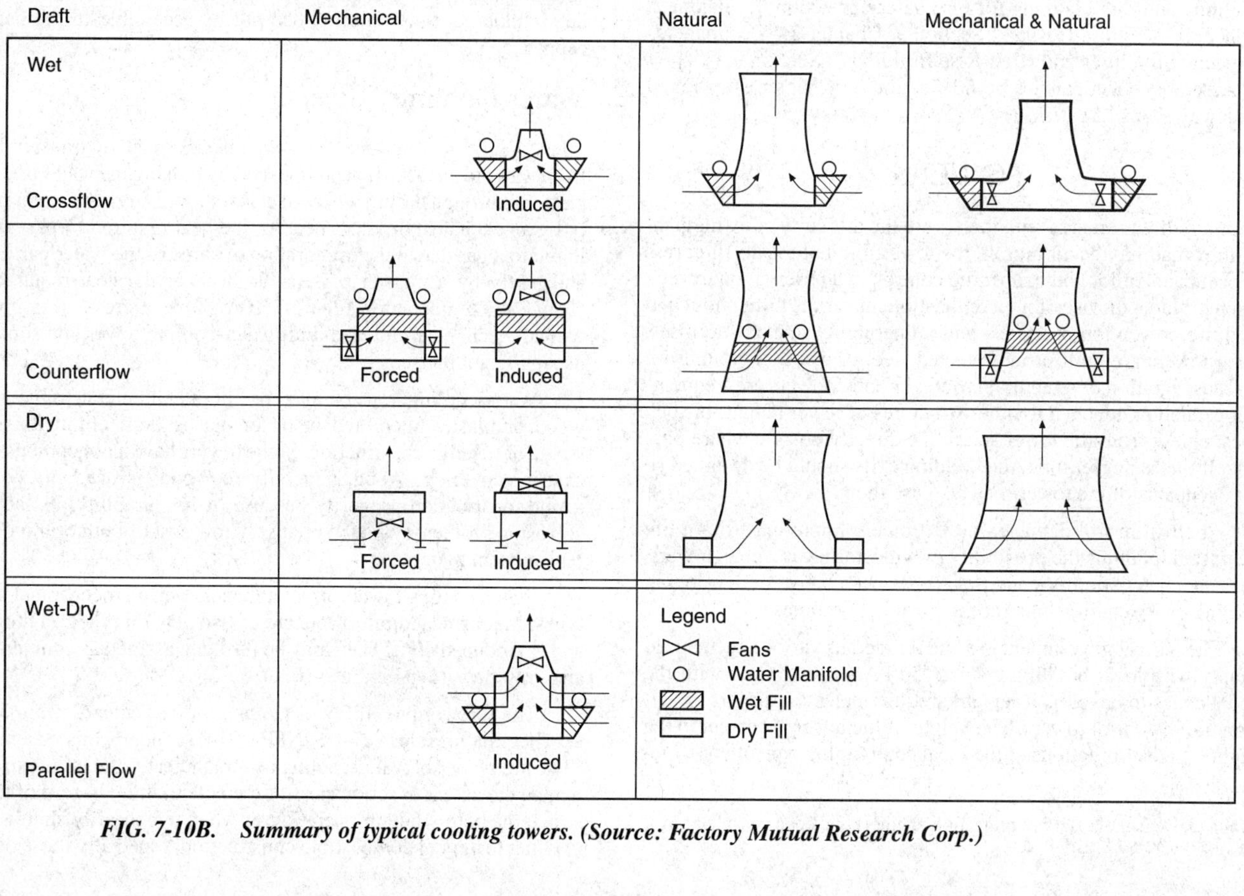

FIG. 7-10B. *Summary of typical cooling towers. (Source: Factory Mutual Research Corp.)*

4. Automatic fire protection systems, such as deluge systems, dry- or wet-pipe automatic sprinkler systems, or a closed-head pre-action system. The wet-pipe system should be used in warm climates only.
5. Automatic water spray systems for the exterior of the tower only.
6. Hydrants and standpipes.
7. Lightning protection.

The key element in fire protection is to provide a combination of prudent, engineered tower design with sufficient consideration given to automatic extinguishment and good preventive maintenance. This is all directed at ensuring that the fan deck, fan, and related motor and drive can survive the effects of a fire intact. This is recommended for two reasons:

1. The entire tower is considerably damaged when the fan deck collapses during a fire, dropping its heavy equipment down through the structure.
2. The tower's downtime is considerably reduced when only the louvers and fill deck material need to be replaced. This is critical since downtime on a water-cooling tower may cause downtime on a key income-producing manufacturing process.

Water-cooling towers, like other outside structures, must be designed and constructed to prevent damage from windstorms. Care also should be taken that the towers are not under, or adjacent to, high-voltage wires or transformers.

Fire prevention also involves a regular maintenance program. Scheduled maintenance of mechanical parts for overheating, excessive wear, inadequate lubrication, and so on is an essential part of this program. Preventing fires and fire damage in water-cooling towers is covered in detail in NFPA 214.

No standard can be promulgated that will guarantee the elimination of fires in water-cooling towers. Technology in this area is under constant development. The designer is cautioned that NFPA 214 is not a design handbook and does not eliminate the need for the engineer or competent engineering judgment. In all cases, the designer is responsible for demonstrating the validity of the fire protection approach.

Television, Microwave, and Radio Transmitting Towers

The electronic equipment in television and radio transmitting towers is frequently at such heights that fire fighting from the ground is impossible. In such situations, an automatic carbon dioxide, halon, or dry chemical extinguishing system, with an interlock to cut off power to the electronic equipment, should be considered. In any case, it is good practice to have a means of shutting off the power to the tower from the ground.

Microwave towers, frequently located in remote areas, sometimes have equipment buildings near their bases for relays, standby generators, and batteries. These buildings should be constructed of noncombustible materials, be protected by automatic detection and extinguishing systems, and be able to alarm to a base station.

Forest Fire and Other Observation Towers

Forest fire and other observation towers should be of noncombustible construction throughout, not only because of the possibility of ignition of the base from a ground fire, but also to avert any fire in the upper structure of the tower. The problem of fire protection in these towers is especially important since they normally are occupied by observation personnel and are remote from fire suppression facilities.

UNDERGROUND BUILDINGS

The inaccessibility of underground structures results in some unique fire problems. Primary among these are the difficulty of venting smoke and gases from fires and the difficulty in fire fighting and evacuating occupants. The inaccessibility of underground structures eliminates any thought of partial fire protection, which might suffice in an aboveground building that depends on the public fire service. Protecting occupants and property in underground structures prompts maximum utilization of good fire protection.

The solution to these fundamental problems has been introduced into building codes and standards. Building code requirements, as well as NFPA 101, include regulations for underground buildings. In 1995 NFPA established a technical committee to develop a standard for subterranean spaces. The underground buildings currently regulated are those having an occupied level with an occupant load of 50 persons. They must be provided with automatic sprinkler protection.

For underground buildings having an occupant load of more than 100 persons and a floor level used for human occupancy more than 30 ft (9.1 m), or more than one level, below the lowest level of exit discharge, smoke venting means must also be provided. To ensure protection against the spread of smoke, standby power is required to serve the smoke exhaust system as well as fire pumps and lighting.

Many of these protective measures are also applicable to windowless buildings, since the only advantage such buildings have over underground buildings is that direct access to the outside can be provided. Windowless buildings provide the same problems of venting, fire fighting, and rescue as underground buildings. These measures are not intended to be all-inclusive or to eliminate the usual fire protection considerations typical of buildings.

Means of egress from underground buildings are provided for in NFPA 101. Those recommendations take into consideration the panic that may result when there is no direct access to the outside and no windows to permit fire department rescue and ventilation.

Automatic sprinklers must be installed in all underground and windowless buildings even if construction is noncombustible, because of combustible contents and concealed spaces. Underground mines and caverns that are being converted for use as buildings without the incorporation of specific building construction features are considered to be of a basic noncombustible construction.

Emergency lighting for safe evacuation is needed, since artificial light is necessary in underground or windowless buildings.

The use of combustible construction and other-than-Class A interior finish should be prohibited. Foam plastics should never be left exposed but should be protected by a thermal barrier acceptable to the Authority Having Jurisdiction, unless significant fire test results show that the thermal barrier is not needed.

Compartmentalization by fire barriers will help limit the extent and severity of a fire and provide areas of refuge for occupants.

Manual fire-fighting equipment, such as standpipe and hose systems and portable fire extinguishers, can provide quick extinguishment of fires when handled by trained personnel who may discover a fire in its incipient stage.

Drainage of water from sprinkler discharge or from hose streams can be a serious problem in underground structures and, therefore, requires early planning. If drainage is required for reasons other than fire protection, it may be possible to use one drainage system for all purposes, provided it is designed to handle the expected maximum flow.

TUNNELS

Tunnels for automotive and tracked vehicles, such as trains and subways, present fire problems similar to those of underground buildings, except that the cargo of the vehicles using the tunnels may present any number of hazards. The transportation of certain cargoes, such as high explosives, through tunnels by automotive vehicles is prohibited in some jurisdictions. Some jurisdictions also prohibit vehicles with LP-Gas on board, including RVs, from using tunnels.

Because gasoline is commonly involved in vehicle fires, handling the hazard of gasoline spill fires is a typical consideration in providing fire protection for tunnels. Another primary consideration is a means for giving a fire alarm in a tunnel. This may be satisfied by fire alarm boxes or telephone stations spaced at intervals along the tunnel. This alarm system can also be tied into the traffic control system so that incoming traffic can be stopped.

Automatic sprinklers provide good tunnel protection. In addition, an adequate water supply with hose connections should be provided on both sides of the tunnel. Supplemental portable fire extinguishers for handling small fires can be installed in wall cabinets for use by attendants and motorists.

NFPA 130, *Standard for Fixed Guideway Transit Systems* (hereinafter referred to as NFPA 130), provides information on fire safety for subways, trains, and so on. NFPA 130 also provides guidelines for the vehicles themselves and for passenger stations.

Fire apparatus equipped to pull disabled vehicles from tunnels, to combat vehicle fires, and to extricate accident victims should be stationed within a reasonable distance of each end of tunnels. Personnel assigned to such locations should have special training and specialized equipment for these situations.

As with underground buildings, ventilation in tunnels is important to dissipate smoke and fire gases. An existing mechanical ventilation system may be satisfactory for emergency use if it can handle the volume required to exhaust smoke and fire gases. Low points in the tunnel should have suitable drainage to handle flammable liquid spills and the water used for extinguishment.

NFPA 502, *Recommended Practice on Fire Protection for Limited Access Highways, Tunnels, Bridges, Elevated Roadways, and Air Right Structures* (hereinafter referred to as NFPA 502), provides more detailed information.

Pedestrian Tunnels

Pedestrian tunnels, normally used to enter and leave buildings, terminals, subway stations, and so on primarily involve safety to life. These tunnels are commonly used to satisfy exit requirements for large industrial and commercial facilities. Detailed requirements for such tunnels are contained in NFPA *101.*

BOMB SHELTERS

Shelters designed to protect people from the effects of bombing, especially nuclear bombs, are underground structures and, therefore, are subject to the same general fire protection factors as other underground structures.

Years ago the federal government encouraged the building codes to include regulations governing fallout shelters. They may still be included in some editions of the model codes.

Areas of refuge designated as bomb shelters in buildings may be considered temporary expedients. Nevertheless, they should be chosen with care, taking into account alternate means of egress, noncombustible construction, sprinklered areas, and proximity to

gas mains, water pipes, and other building services that may be damaged by explosion or by fires resulting from a bombing. Such factors could either help protect or unnecessarily expose people seeking refuge. Storage of emergency supplies should be carefully planned to avoid the problem of exposed combustibles.

BRIDGES

Highway Bridges

Highway bridges are continually exposed to accidents that may cause direct damage from impact or, occasionally, fires that may follow that impact. When flammable cargoes are involved, either on the roadway or on adjacent lands or water, there is a potential for a serious fire. For special fire protection to the structures themselves, fire-resistant or noncombustible construction and the means to transmit emergency alarms on longer structures generally suffice. Established response to accidents should include fire-fighting capability.

Fire alarm boxes or telephones spaced along a bridge provide a means of notifying the fire department or bridge attendants of an emergency. As in tunnels, this system can be tied into a traffic control system to stop traffic from driving onto the bridge and to alert vehicles already on the bridge.

Bridges in critical seismic areas may require special anchoring to ensure continued use in times of severe earthquake activity.

The failure of some interstate highway bridges has focused attention on the safety of bridges, and stricter design requirements are currently being imposed.

Since fire fighting is a necessity on bridges, an adequate water supply from standpipes is good practice, preferably on both sides of the bridge or, on long bridges, at intervals along the length.

The location of apparatus to combat a bridge fire can be a problem if the bridge is located in a rural area. Apparatus should be stationed as near the bridge as is practical. In addition to carrying fire-fighting equipment, this apparatus should be designed to move disabled vehicles and should carry the equipment needed for rescue. NFPA 502 provides more detailed information.

Railroad Bridges

Railroad bridges are a fire problem because so many of them are constructed of wood, which may have been treated with a flammable wood preservative. In addition, many railroad bridges are located in remote areas.

There are two kinds of wood bridges and trestles: (1) open deck and (2) ballast deck. The fire potential is considerably more severe in open-deck design.

Wood railroad bridges may be ignited by such sources as the friction sparks and hot metal particles generated when brakes on heavily loaded freight trains are applied on steep downgrades and by burning waste from overheated journal boxes. Exposure fires from burning grass and brush around supporting piles are another hazard. (See Figure 7-10C.)

Some fire-retardant treatments are being used on timbers for bridges, but wood bridges, even when treated, should not be considered an adequate substitute for fire-resistant or noncombustible bridges. Material treated with a fire retardant may lose its fire retardancy over time, depending upon the environment to which it is exposed.

Fighting railroad bridge fires can present many problems, particularly in areas where bridges are inaccessible to organized fire-fighting facilities. One method of providing minimal fire extin-

FIG. 7-10C. Fire fighters were forced to pull hose lines on hand cars across thousands of feet of track to fight a fire in this wood trestle in a remote area of Long Island, NY. Water around the trestle was too shallow for fire boats. High winds drove the fire rapidly through the trestle and destroyed about 2500 ft (763 m) of the structure. (Source: United Press International)

FIG. 7-10D. More than 5000 people escaped without serious injury when a fire occurred in this wood grandstand in Louisville, KY, in the early 1950s. The evacuation was not orderly, and several occupants received minor bruises in their haste to escape. A cigarette is believed to have fallen through a crack into a concealed space below the wood flooring, where it ignited debris. (Source: Wide World Photos)

guishing capacity is to provide water casks at the approaches to bridges or at intervals along their length for use by train crews. Some railroads keep a magazine of explosives nearby to blow fire breaks in bridges, if necessary. In other instances, pre-emergency planning calls for tearing down sections of platform-type bridges to stop the spread of fire.

PLACES OF OUTDOOR ASSEMBLY

Places of outdoor assembly include amusement parks, athletic fields and stadiums, grandstands, tents, and similar places where crowds gather to watch an event or take part in recreational activities. The risks associated with fire in any place where crowds congregate, reinforces the need for providing adequate and accessible ways of escape and for providing good construction features that reduce the probability of fire spread. NFPA 102, *Standard for Grandstands, Folding and Telescopic Seating, Tents, and Membrane Structures* (hereinafter referred to as NFPA 102), provides reasonable measures for the construction of such occupancy uses. Means of egress requirements have been removed from the standard and are now found in NFPA *101*.

Features that result in construction weaknesses and can accelerate fire spread include temporary wiring; inadequate maintenance and poor housekeeping, particularly when exhibitions involving displays are present; use of considerable combustible materials, including decorations and flammable liquids; and the general lack of proper fire prevention practices.

A good supply of portable fire extinguishers as well as a reliable method of quickly communicating with the closest fire department and other appropriate emergency organizations are necessary for fire protection in places of outdoor assembly.

Tents

Tents became a major concern to fire protection authorities following the disastrous circus fire in Hartford, CT, on July 6, 1944, in which 168 lives were lost. This fire clearly demonstrated the need for flame-retardant materials for tents in places of outdoor assembly. Flame-resistant materials are now generally required for all tents. Decorative materials used in tents are generally required to be flame retardant. The use of open-flame and heating devices is closely regulated or prohibited.

Tents are combustible, even when made of flame-resistant material, and it is good practice to limit their number and size in a location and to keep them separated from each other and from other structures. The area around tents must be kept free of combustible material, including vegetation. Combustible material used inside a tent should be kept to a minimum, and smoking in tents occupied by the public should be prohibited.

Grandstands

Grandstands are typical examples of structures found in places of outdoor and indoor assembly and of structures that constitute a safety problem. These structures must have adequate aisles, the proper distance between rows of seats, strong guard railings, and other exit features to ensure rapid and orderly evacuation to areas of safety. Without these features, exiting problems can develop during an emergency. (See Figure 7-10D.) Requirements applicable to such installations are found in NFPA *101*. Grandstands should also be designed to withstand predetermined deadweight and live loads, as well as anticipated wind loads.

Since many grandstands are constructed entirely or partially of wood, basic fire protection principles of division of areas and separation from exposures apply. Long, combustible grandstands must be divided by fire partitions, or they may be built in smaller sections with adequate spaces between the sections.

Frequently, the space underneath a grandstand is used to store materials and for concession booths, and it becomes littered with refuse. This space should be kept clear of such activities. Refuse under the grandstand was a major factor in the Bradford, England, Soccer Stadium fire on May 11, 1985, which claimed 56 lives. NFPA *101* contains provisions affecting crowd management as a result of the April 1989 Sheffield, England, fire in which 94 persons died.

Amusement Park Structures

Structures supporting amusement rides are generally built of wood and contain considerable electrical wiring, gear boxes for pulleys, and similar friction-producing machinery. They may be poorly maintained, as well. Life safety is the primary consideration in evaluating these structures, so adequate and accessible means of escape in an emergency is vitally important. Fire prevention, noncombustible construction, exposure hazards, maintenance, and housekeeping are other factors to be considered.

There are structures other than the so-called "rides," such as the merry-go-round, roller coaster, and ferris wheel. Some are for amusement purposes, and some are even educational in nature. Others simulate a bobsled run, for example, or a car traveling through a building. Still others operate in semidarkness, such as mystery houses or haunted castles.

Since most of these structures are windowless—or at least the public spaces are without windows—they should be required to have automatic sprinkler systems using the newly developed quick-response sprinklers. Egress provisions almost necessarily are not standard, and some innovations are needed to meet the intent of NFPA *101* and the model building codes. Of primary importance is emergency lighting and good illumination and marking of paths of egress travel. Sprinkler protection is probably the most important requirement in an amusement structure, whether it is a permanent structure or one made up of semi-permanent trucks or trailers.

A fire in a "haunted castle" attraction made up of interconnected commercial trailers claimed eight lives in a Jackson Township, NJ, amusement park on May 11, 1984. Major factors contributing to the loss of life were lack of properly designed fixed detection and suppression systems, ignition of synthetic foam material from which fire spread to other combustible contents and interior finish, and the difficulty occupants had in trying to escape along a convoluted smoke-filled path.

Low-level exit signs and pathway markings are required by some codes for amusement buildings where required exits are not apparent due to theatrical distractions, are disguised, or are not readily available.

MEMBRANE STRUCTURES

NFPA 102 is not applicable solely to places of assembly. The term "membrane" replaces "air-supported" because it better reflects all the various types of structures, not just those supported by air and those in which the occupants are in a pressurized atmosphere. Some structures are of the air-inflated type, where the air pressure is within a series of tubes which then form the structure and keep it upright. In this case, the occupants are not within a pressurized atmosphere. Also included are cable-supported membrane structures, such as some newer sports stadiums, where the roof is supported by cables,

and air pressure may or may not be provided. In other types of structures, the membrane is tensioned, and weather protection is usually provided without enclosing walls. Finally, there are the framework-type structures in which the membrane is stretched over the frame to form enclosing walls and a roof.

By definition, a membrane is a thin, flexible, water-impervious material capable of being supported by an air pressure of 1.5 in. (3.7 Pa) of water column. The membrane may be of very thin metal or glass fiber coated with a fluorocarbon and is considered to be noncombustible or limited combustible. Others are of a glass, nylon, or similar fibrous material coated with vinyl or silicone. These are considered to be combustible structures, even though they are flame-resistant when treated in accordance with NFPA 701, *Standard Methods of Fire Tests for Flame-Resistant Textiles and Films.*

An air-supported structure is a "balloon-like" shelter constructed of flexible, coated fabrics supported and stabilized against wind loads by a small amount of internal pressure, usually 1 to 1.5 in. of water pressure (2.5 to 3.7 Pa), supplied by continuously operated centrifugal fans. (See Figure 7-10E.) There are no beams, columns, girders, or other structural members involved. Variations on the air-supported structure include structures with an air-inflated, double-walled roof and side panels and combination structures that are rib-supported but air-supported in heavy wind.

A plastic-coated synthetic fiber is used in air-supported structures. The synthetic fiber will not absorb flame-resistant solutions, but the plastic coating can contain flame retardants. If the fan serving an air-supported structure fails or if a hole is made in the fabric, by heat from an exposure fire, for example, it takes a relatively long time for the structure to deflate. The normal opening and closing of doors will not affect the structure. In fact, breather holes are provided in the skin for air changes.

PIERS AND WATER-SURROUNDED STRUCTURES

When piers or related structures surrounded by water are used for places of amusement, for restaurants, or for passenger terminals, they fall out of the normal use category—that is, mooring of vessels and handling of cargo—and fall into the special structure category. The question of life safety is of primary concern, since these structures, which may contain large numbers of people, may have limited egress facilities due to their location or arrangement on the water.

For piers falling outside the normal use category and extending more than 150 ft (45.7 m) from the shore, the following recommendations are made. The criteria arise from the recognized maximum

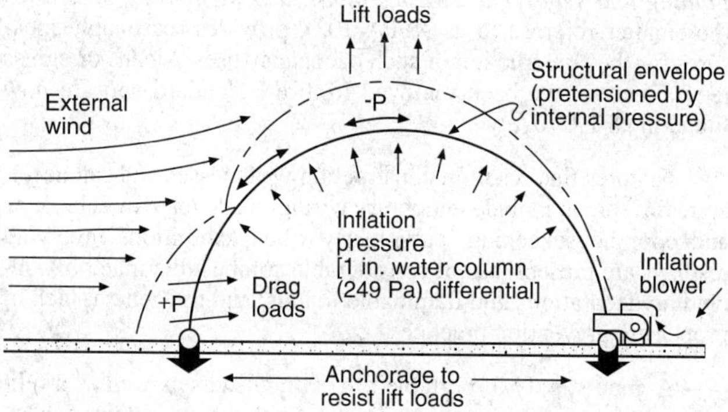

FIG. 7-10E. The aerodynamic and inflation loading of a typical single-walled, air-supported structure.

travel distance requirement found in NFPA *101*. These design approaches are considered equivalent:

1. The pier may be arranged to have two separate ways of travel to shore, as by two well-separated walkways or independent structures.

2. The pier deck may be designed to be of open, fire-resistant construction, resting on noncombustible supports.

3. If combustible construction is used for the pier deck, superstructure, and/or substructure, they should be protected by an automatic sprinkler system.

4. If a pier is completely open and unobstructed and is more than 500 ft (152.5 m) long, it may be considered generally safe if the minimum width is not less than 50 ft (15.2 m) wide. If a pier is open and unobstructed and longer than 500 ft (152.5 m), a 10:1 ratio must be maintained for the relationship of length to width.

Regardless of whether the four items above are applicable, individually or jointly, to a particular pier, the minimum requirements of NFPA *101* apply concurrently for providing means of egress. These requirements cover the number, arrangement, location, and protection of means of egress for the type of occupancy involved, plus related questions of hardware, construction, and interior finish for the means of egress. Further guidance for piers is given in NFPA 307, *Standard for the Construction and Fire Protection of Marine Terminals, Piers, and Wharves.*

When a building or structure, such as a typical lighthouse, or an oil-drilling platform and related housing facilities, is completely surrounded by water, several considerations must be given to life safety from fire. Either sufficient outside ground area, such as an island, is needed for refuge, or a fire-resistant platform must be constructed for this purpose. This must be coupled with a communications system and a means of transportation for evacuating the occupants from the refuge area. Such transportation will be limited to either boat or helicopter, due to the obvious nature and location of these special structures. However, such a system of refuge, communication, and rescue is only as good as the planning and availability of the transportation on a standby basis. Delays imposed by poor planning or by intolerable weather conditions can lead to disaster. Thus, such factors must be considered before an emergency, not after.

Permanently Moored Vessels

Where a sizable vessel is permanently moored and used as a restaurant, hotel, museum, or convention center, it is, for all practical purposes, similar to any permanently constructed building occupied for those purposes.

Because the vessels were not built as "buildings," but under completely different standards, it is almost impossible to try to make them comply literally with the requirements of NFPA *101* or the model building codes. Passageways are not as wide as would be required for exit access corridors. Doors are smaller than those found in conventional buildings. The rise and run of stairs, originally known as "ladders," probably do not conform with stair requirements in buildings. As the tide rises and falls, the land connection changes in slope. So, if the vessel is to be accepted for use as other than a ship, some concessions are usually made. A prime means for accepting nonconforming construction details is to require sprinkler protection for the vessel, to limit the occupant load, to require strict compliance with interior finish requirements, and so on. Generally, the power plant for the vessel is inactivated and shore power is substituted after the operating machinery and fuel have been removed.

AIR-RIGHT STRUCTURES

The economics of purchasing choice urban space, preparing the site, demolishing existing structures, and solving the complex problem of foundation design and caisson construction, all without adversely affecting surrounding structures or exceeding stringently applied cost limitations, has led to the development of a structure that either totally or partially solves many of those problems—the air-right structure. (See Figure. 7-10F.)

In 1992 NFPA first published NFPA 502, which included requirements for air-right structures. The air-right structure is generally a facility built over existing rail lines, highways, or even other buildings. From a fire safety standpoint, its design is not unlike that which would exist if it were isolated on its own site, with one key exception. The air-right structure is more likely to be adversely affected by an exposing event on or in the facility above which it is erected.

Air-right structures also present two fire protection problems. One relates to the persons and property in the air-right structure. The other relates to the persons and property using the roadway that passes under or adjacent to the air-right structure.

While protection of life is the primary consideration, there are other important concerns. The structural members that support the air-right building could be subjected to very high temperatures, particularly in a flammable liquids fire or explosion. Damage to these members could have a serious effect on the building. In addition, openings from the roadway, such as ventilation shafts, drainage systems, and walkways, could permit passage of flammable liquids or vapors to the air-right structure, with subsequent damage from fire or explosion.

An alarm system similar to that used for a tunnel should be considered. Means should be provided whereby emergency alarms may be transmitted by the general public. Coded alarm or outdoor-type telephone boxes should be installed at intervals of not more than 200 ft (60 m) and should be made conspicuous by indicating lights or other suitable markers. Air-right structure roadways with heavy traffic volumes should be equipped with a traffic monitoring system that would automatically transmit alarms to a central control at any time normal traffic flow is interrupted. Such a system is most effective when it is integrated with a closed-circuit television system. Some method of early detection of smoke and/or fire should be provided.

FIG. 7-10F. An air-right hotel and office building shown spanning an expressway and a two-track railroad right-of-way.

A traffic control system should be provided. It may be interlocked with the fire alarm system. The system should be capable of operation from a remote-control source or from either end of the roadway passing under the air-right structure. The traffic control system should be designed for use by authorized personnel only.

Fire apparatus assigned to companies responsible for air-right structure roadways must be equipped to deal with flammable liquid and hazardous materials fires and incidents effectively.

The water supply for the air-right structure roadway should resemble that used for a tunnel. The primary water supply should preferably be a strong municipal water supply system with connections at each end of the air-right structure roadway.

Hose connections should be installed and conspicuously marked so that no point within the air-right structure roadway is more than 150 ft (45.8 m) from a hose connection.

Air-right structure roadways in excess of 200 ft (60 m) in length should be designed with a positive ventilation system. The ventilation equipment should be heat resistant so that it is capable of operating even under fire conditions.

A complete drainage system should be provided for the air-right structure roadways. Sumps with automatic pumps should be provided where necessary.

The Authority Having Jurisdiction over an air-right structure roadway, especially those in excess of 200 ft (60 m) in length, should adopt rules and regulations found in Title 49, "Transportation," *Code of Federal Regulations*, Parts 100–199, applicable to the transportation of hazardous materials.

All structural elements that support buildings over roadways and/or provide separation between the buildings and roadways should have a 4-hr fire-resistance rating.

Buildings above roadways should be designed with the consideration that the roadway below the air-right structure is a potential source of heat, smoke, and toxic gases. The structural elements should be designed in such a way as to shield the air-right building from these potential hazards. The design of the building should neither increase nor create any risk to the patrons of the roadway below.

OTHER SPECIAL STRUCTURES

Signs

Outside signs on buildings are not a serious fire problem if they invove a limited amount of combustibles and a location that does not pose a threat of flame spread. However, certain types of signs, such as those illuminated by neon tubing, may cause fires because of the wiring and transformers connected to them.

An outside sign may suffer considerable damage from wind and may expose other properties to damage if torn loose by wind. If a sign is of such size, construction, and location as to be considered a fire hazard, it should be protected like any other structure; for example, automatic sprinklers may be installed.

A combustible sign attached to the exterior and covering a large percentage of one side of a fire-resistant or noncombustible building may significantly change the fire safety considerations of the building. This is a fire exposure problem not generally contemplated in building design. A large sign can also be hazardous to fire fighters if it falls during fire-fighting operations. For these reasons, some building codes limit the size and construction type of combustible signs that can be installed on buildings.

The model building codes have extensive and detailed requirements applicable to signs. The particular code in effect in the jurisdiction where the sign or signs are to be erected should be consulted.

NFPA *101* is primarily concerned only with signage having to do with means of egress.

Parking Garages

At one time, parking garages that permitted drivers to drive their own vehicles to a parking stall were required to meet much more severe fire-resistant requirements. Despite the presence of gasoline and combustible tires, the potential for rapid fire development in automobiles is low. Where these structures have permanently open exterior walls that can dissipate smoke and hot gases, the likelihood of a serious fire is even lower.

Tests here and abroad have demonstrated these conclusions. One such test was conducted in 1972 by the American Iron and Steel Institute (AISI) in a multistory open parking structure with exposed steel structural members.[1] Three automobiles parked next to each other, each with a fuel tank containing 10 gal (38 L) of gasoline, were used as the test ignition and fuel source. The center car was gutted 48 min after crumpled newspapers in its rear seat were ignited. The contents of its fuel tank were spilled or consumed, but the fire did not spread to the adjacent cars, and the overhead structural steel was essentially unaffected. The temperature of the steel remained far below critical levels throughout the test. A similar test conducted by the British at Borehamwood fire test station yielded similar results.[2] These conclusions do not necessarily apply to aboveground enclosed garages or to underground garages.

Building code requirements permit noncombustible parking structures of unlimited area up to 75 ft (22.9 m) high to have no structural fire protection, as long as they are used for no other purpose and have ample permanent wall openings for the free dissipation of smoke and gases.

BIBLIOGRAPHY

References Cited

1. Gewain, R. G., "Fire Experience and Fire Tests in Automobile Parking Structures," *Fire Journal*, Vol. 67, No. 4, July 1973.
2. Butcher, E. G., *et al.*, "Fire and Car-Park Buildings," *Fire Note No. 10*, Ministry of Technology and Fire Offices' Committee/Joint Fire Research Organization, London, UK, 1968.

NFPA Codes, Standards, and Recommended Practices

Reference to the following NFPA codes, standards, and recommended practices will provide further information on special structures discussed in this chapter. (See the latest version of *The NFPA Catalog* for availability of current editions of the following documents.)

NFPA *101*®, *Life Safety Code*®
NFPA 102, *Standard for Grandstands, Folding and Telescopic Seating, Tents, and Membrane Structures*
NFPA 130, *Standard for Fixed Guideway Transit Systems*
NFPA 214, *Standard on Water-Cooling Towers*
NFPA 220, *Standard on Types of Building Construction*
NFPA 307, *Standard for the Construction and Fire Protection of Marine Terminals, Piers, and Wharves*
NFPA 502, *Recommended Practice on Fire Protection for Limited Access Highways, Tunnels, Bridges, Elevated Roadways, and Air-Right Structures*
NFPA 701, *Standard Methods of Fire Tests for Flame-Resistant Textiles and Films*

Additional Reading

Appleton, B., "Development of Fire Safety on London Underground," Joint Conference of the Tyne and Wear Metropolitan Fire Department and The Institution of Fire Engineers, Northern Branch on Fire Risk Assessment: Opportunity or Problem? March 26, 1993, pp. 1–11.

Ayres, D. J., "Treatment of Shallow Underground Fires," *Proceedings* of the Symposium on Failures in Earthworks, Thomas Telford, London, UK, 1986, pp. 470–472.

Beard, A. N., "Predicting Fire Spread in Tunnels," *Fire Engineers Journal*, Vol. 55, No. 9, Nov. 1995, pp. 13–15.

Beard, A. N., "Prediction of Major Fire Spread in a Tunnel," *Proceedings* of the Second International Conference on Safety in Road and Rail Tunnels, April 3–6, 1995, Granada, Spain, 1995, pp. 337–344.

Beard, A. N., Drysdale, D. D. and Bishop, S. R., "Non-Linear Model of Major Fire Spread in a Tunnel," *Fire Safety Journal*, Vol. 24, No. 4, 1995, pp. 333–357.

Beech, J., "Kent Re-Equips Fire-Fighters in Response to Tunnel Risk," *Fire*, Vol. 83, No. 1024, Oct. 1990, pp. 28, 31.

Bengtson, S., "Brandskydd I underjordsanlaggningar forskningsbehov slutrappot 1990-09-10. [Need for Fire Research in Underground Facilities.]," (Abstract in English), Swedish Fire Protection Assoc., Stockholm, Mar. 20, 1990, 101 pages.

Bengtson, S., "Forskningsbehov brandskydd av overglasade gardar. [Technical Overview of Fire Protection in Malls.],"National Defence Research Establishment, Sundbyberg, Sweden, Sultrapport 1992 01 20, 1992.

Bengtson, S. and Ramachandran, G., "Fire Growth Rates in Underground Facilities," *Proceedings* of the Fourth International Symposium on Fire Safety Science, Intl. Assoc. for Fire Safety Science, Boston, MA, 1994, pp. 1089–1099.

Bettis, R., "Controlling Smoke in Tunnel Fires," *Fire Prevention*, No. 280, June 1995, pp. 19–22.

Beyler, C. and DiNenno, P., "Letters to the Editor. 'Smoke Management for Covered Malls and Atria' by J. A. Milke," *Fire Technology*, Vol. 27, No. 2, 160-169, May 1991, pp. 160–169.

Blackburn, R. B., "County Mall: A Shopping Center Fire Safety Strategy for the 1990's," *Fire Engineers Journal*, Vol. 53, No. 168, Mar. 1993, pp. 18–27.

Bosley, K., "Escaping From Shopping Precincts," *Fire Research News*, No. 14, Autumn 1992, pp. 2–3.

Burton, R. C., *et al.*, "Automatic Fire Detection Underground," *Colliery Guardian*, Vol. 235, No. 5, 1987, pp. 166, 168.

Catchpole, L., "Going Underground," *Fire Prevention*, Vol. 265, December 1993, pp. 24–25.

Catchpole, L., "Integrating Systems at Lakeside Shopping Center," *Fire Prevention*, No. 271, July/Aug. 1994, pp. 20–22.

Charters, D. A. and McIntosh, A. C., "Tunnel Fire Safety Assessment," *Proceedings* of the First International Conference on Fire Science and Engineering, ASIAFLAM '95, March 15–16, 1995, Kowloon, Hong Kong, 1995, pp. 87–97.

Charters, D. A., Gray, W. A. and McIntosh, A. C., "Computer Model to Assess Fire Hazards in Tunnels (FASIT)," *Fire Technology*, Vol. 30, No. 1, First Quarter 1994, pp. 134–154.

Charters, D. and McIntosh, A., "FASIT: A Computer Model for Predicting Fires in Tunnels," *Fire*, Vol. 88, No. 1082, August 1995, pp. 13–14, 16.

Charters, D., "Fire Risk Assessment of Rail Tunnels," First International Conference on Safety in Road and Rail Tunnels, Nov. 23–25, 1992, Basel, Switzerland, 1992, pp. 239–253.

Chengming, H., "Study of Electric Fire Protection in Underground Space Buildings," First Asian Conference on Fire Science and Technology, (ACFST), October 9–13, 1992, Hefei, China, International Academic Publishers, China, 1992, pp. 279–284.

Chow, W. K., "Simulation of Car Park Fires Using Zone Models," *Journal of Fire Protection Engineering*, Vol. 7, No. 3, 1995, pp. 65–74.

Chow, W. K. and Chu, C. K., "Application of HAZARD 1 to Study the Fire Risk of a Shopping Mall," *Proceedings* of Fire Safety Modelling and Building Design, the One-Day Conference to Review the Potential and Limitations of Fire Safety Models for Building Design, March 29, 1994, Salford, UK, 1994, pp. 90–97.

Coates, M. and Loader, K., "FPA Casebook of Fires: Tunnel Under Construction, Los Angeles, California," *Fire Prevention*, No. 239, May 1991.

Corbett, G. P., "Strip Malls: The Taxpayers of Today," *Fire Engineering*, Vol. 149, No. 5, May 1996, pp. 58–62.

Corbin, E., and Sale, J. W., "Fire Protection at the Grand Coulee Hydro Project," *Proceedings* of the International Conference on Hydropower: Waterpower '87, ASCE, New York, 1988, pp. 2091–2099.

Courcier, G. and Huggett, R., "Improving the Safety of Underground Railways With Cleaning Trains," First International Conference on Safety

in Road and Rail Tunnels, November 23–25, 1992, Basel, Switzerland, 1992, pp. 95–110.

Crossland, B., "King's Cross Underground Fire and the Setting Up of the Investigation," *Fire Safety Journal*, Vol. 18, No. 1, 1992, pp. 3–11.

d'Albrand, N. and Bessiere, C., "Fire in a Road Tunnel Comparison of Tunnel Sections," First International Conference on Safety in Road and Rail Tunnels, November 23–25, 1992, Basel, Switzerland, 1992, pp. 439–449.

Denda, D., "What About Parking Garage Fires," *American Fire Journal*, Vol. 45, No. 2, February 1993, pp. 22–25, 27.

Diaz, J. and Jazbec, D. J., "Fire Control System in Highway Tunnels," First International Conference on Safety in Road and Rail Tunnels, November 23–25, 1992, Basel, Switzerland, 1992, pp. 55–61.

Egilsrud, P. E. and Kile, G. W., "Design for Tunnel Safety I-90 Tunnels—Seattle, Washington," First International Conference on Safety in Road and Rail Tunnels, November 23–25, 1992, Basel, Switzerland, 1992, pp. 545–559.

Eisner, H., "Tunnel Segregation," *New Civil Engineer*, Vol. 887, Mar. 15, 1990, p. 26.

Eloranta, E., "Numerical Simulation of Tunnel Fires With the BR12 Zone Model Programme," *Palontorjuntatekniikka*, January 1991, pp. 18–21.

"Fire Protection—A Shopping Center's Best Buy," *Record*, Vol. 67, No. 3, May/June 1990, pp. 16–19.

Foster, R., "Fire Protection in the Sydney Harbor Tunnel," *Fire International*, Vol. 16, No. 132, Dec. 1991/Jan. 1992, pp. 28–31.

"FPA Casebook of Fires: Multi-Storey Car Park, Preston, Lancashire," *Fire Prevention*, No. 240, June 1991, p. 33.

Fransson, C. and Vilselius, H., "Application of a General-Purpose CFD Code to Field Modelling of Tunnel Fires," National Defence Research Establishment, Sundbyberg, Sweden, FOA Report B 20131-2.7, June 1994.

Gardner, J., "Comparing Smoke Vent Systems in Shopping Malls," *Fire Prevention*, No. 236, Jan./Feb. 1991, pp. 23–27.

Gardner, J. P., "Smoke Control in Shopping Malls," *Building Services*, Vol. 13, April 4, 1991, pp. 53–54.

Haack, A., "Fire Protection in Traffic Tunnels: Initial Recognitions From Large-Scale Tests," First International Conference on Safety in Road and Rail Tunnels, November 23–25, 1992, Basel, Switzerland, 1992, pp. 349–362.

Haack, A., "Brandschutz in unterirdischen Verkehrsanlagen EUREKA-Projekt/BMBF-Forschungsprojekt. [FireSafety in Underground Traffic Structures.]," *VFDB*, Vol. 45, No. 1, Feb. 1996, pp. 2–7.

Haerter, A., "Ventilation: Fighting Fires in Road Tunnels," *Tunnels and Tunneling*, Vol. 20, No. 4, 1988, pp. 55–58.

Haksever, A., "Fire Engineering Design of Steel Framed Car-Park Buildings," Helsinki University of Technology, Espoo, Finalnd, CIB W14/91/01 (FL), Report 112, 1990.

Henson, D. A. and Lowndes, J. F. L., "Smoke Control in Subway Tunnels," First International Conference on Safety in Road and Rail Tunnels, November 23–25, 1992, Basel, Switzerland, 1992, pp. 171–181.

Heskestad, G., "To the Editor. 'Smoke Management for Covered Malls and Atria' by J. A. Milke," *Fire Technology*, Vol. 27, No. 2, May 1991, pp. 174–185.

Hettinger, J. C., "Analysis of tunnel Conditions During a Fire Using the Subway Environment Simulation Program," Parsons, Brinckerhoff, Quade and Douglas, Inc., New York, NY, Society of Fire Protection Engineers and Worcester Polytechnic Institute, *Proceedings* of Computer Applications in Fire Protection, June 28–29, 1993, Worcester, MA, 1993, pp. 1–6.

Hoffmann, J., "Kansas: Underground Storage Center Challenges Firefighters. On the Job," *Firehouse*, Vol. 17, No. 6, June 1992, pp. 78–80.

Hoffmann, J., "Trouble Underground," *Fire Prevention*, No. 252, Sept. 1992, pp. 41–43.

Hoj, N. P. and Bennick, K., "Fire and Explosion Safety in Design of Railway Tunnel Under the Great Belt," *Proceedings* of the Second International Conference on Safety in Road and Rail Tunnels, Apr. 3–6, 1995, Granada, Spain, 1995, pp. 551–559.

Hosser, D. and Dobbernack, R., "Konzepte fur die Rauchfreihaltung von uberdachten Innenhofen und Ladenpassagen im Brandfall. [Concepts to Keep Roofed-In Atria and Shopping-Centers Smoke-Free in Cases of Fire.]," *VFDB*, Vol. 41, No. 2, Mar. 1992, pp. 55–59.

Hough, E., "Gatwick Airport: 'A Shopping Mall With Aircraft'?," *Fire*, Vol. 87, No. 1076, Mar. 1995, pp. 23–24.

Iida, T., Kobayashi, S., and Kenmochi, T., "Facilities of the Seikan Tunnel," *Japanese Railway Engineering*, No. 106, June 1988, pp. 12–16.

Johnson, P. F. and Timms, G. R., "Performance Based Design of Shopping Center Fire Safety," *Proceedings* of the First International Conference on Fire Science and Engineering, ASIAFLAM '95, March 15–16, 1995, Kowloon, Hong Kong, 1995, pp. 41–49.

Josephson, R. A. and Shafer, P. R., "Foam Systems for Highway Tunnels," *Industrial Fire Safety*, Vol. 1, No. 1, November/December 1992, pp. 43–44, 46–48, 50.

Kampmann, J., *et al.*, "Risk Analysis of the Railway Tunnel Under the Great Belt," First International Conference on Safety in Road and Rail Tunnels, November 23–25, 1992, Basel, Switzerland, 1992, pp. 273–285.

Kandola, B. S. and Bamford, G. J., "Modelling of Fire Hazards in Tunnels," *Proceedings* of the First International Conference on Fire Science and Engineering, ASIAFLAM '95, March 15–16, 1995, Kowloon, Hong Kong, 1995, pp. 99–111.

Kasahara, I. and Hara, T., "Life Safety Design of an Atrium and an Air Dome With an Applied Zone Model," Taisei Corp., Tokyo, Japan, CIB W14/96/1 (J); Building Research Institute, Mini-Symposium on Fire Safety Design of Buildings and Fire Safety Engineering, June 12, 1995, Tsukuba, Japan, 1995, pp. VIa/1–3.

Keller, E., "Reliability of Road Tunnel Safety Systems," First International Conference on Safety in Road and Rail Tunnels, November 23–25, 1992, Basel, Switzerland, 1992, pp. 303–313.

"Kent Fire Brigade's 'Formidable Range of Unique Problems': Update on Tunnel Fire Safety," *Fire*, Vol. 85, No. 1048, Oct. 1992, p. 6.

Keski-Rahkonen, O., "Tietunneleiden tulipalot. [Fires in Highway Tunnels.]," *Palontorjuntatekniikka*, Vol. 1, 1994, pp. 26–27.

Kodaira, A. and Uehara, S., "Fire Safety Design for a large-Scale Retractable Dome," Takenaka Corp., Chiba, Japan, CIB W14/96/1 (J); Building Research Institute, Mini-Symposium on Fire Safety Design of Buildings and Fire Safety Engineering, June 12, 1995, Tsukuba, Japan, 1995, pp. VIe/1–4.

Lagasco, F., Diamantidis, D. and Righetti, G., "Risk Analysis and Safety Systems Evaluation of a New Alpine Railway Tunnel," First International Conference on Safety in Road and Rail Tunnels, Nov. 23–25, 1992, Basel, Switzerland, 1992, pp. 327–337.

Loeb, D. L., "Fire and the Automotive Repair Garage," *Fire Chief*, Vol. 37, No. 9, Sept. 1993, pp. 38–40, 42.

Lombardo, M. and Pressler, B., "Firefighting in Covered Malls. Part 1," *Fire Engineering*, Vol. 146, No. 11, Nov. 1993, pp. 71–74.

Lombardo, M. and Pressler, B., "Firefighting in Covered Malls. Part 2," *Fire Engineering*, Vol. 147, No. 1, Jan. 1994, pp. 52–56.

Luchian, S. F., "Central Artery/Tunnel Project Memorial Tunnel Fire Test Program," First International Conference on Safety in Road and Rail Tunnels, Nov. 23–25, 1992, Basel, Switzerland, 1992, pp. 363–377.

Luzik, S. J., "Protection of Aluminum Overcast Construction Against Fire," *Fire Technology*, Vol. 24, No. 3, 1988, pp. 227–244.

Mangs, J. and Keski-Rahkonen, O., "Fire Safety for Open Car Park Buildings," VTT-Technical Research Center of Finland, Espoo, TNO Building and Construction Research, *Proceedings* of the Third CIB/W14 Workshop, Fire Engineering Workshop on Modelling, Jan. 25–26, 1993, Delft, The Netherlands, 1993, pp. 206–213.

Marchant, E., "Time is Opportune to Revise Fire Safety Standards for Multi-Storey Car Parks," *Fire*, Vol. 83, No. 1020, June 1990, 21, 24.

Milke, J. A., "Author's Response. 'Smoke Management for Covered Malls and Atria.'," *Fire Technology*, Vol. 27, No. 2, May 1991, pp. 169–174.

Milke, J. A., "Smoke Management for Covered Malls and Atria," *Fire Technology*, Vol. 26, No. 3, Aug. 1990, pp. 223–243.

Milke, J. A., "Providing a Performance-Based Design Method for Smoke Management Systems in Atria and Covered Malls," *SFPE Bulletin*, May/June 1992, pp. 6–8.

Moodie, K. and Jagger, S. F., "Technical Investigation of the Fire at London's King's Cross Underground Station," *Journal of Fire Protection Engineering*, Vol. 3, No. 2, 1991, pp. 49–63.

Moodie, K., "King's Cross Fire: Damage Assessment and Overview of the Technical Investigation," *Fire Safety Journal*, Vol. 18, No. 1, 1992, pp. 13–33.

Morgan, H. P., "Design Principles for Smoke Ventilation in Enclosed Shopping Centers," Fire Research Station, Borehamwood, England, BRE BR 186, 1990.

Nishimori, S. and Haga, H., "Fire Detector for Tunnels (CO$_2$ Resonance Flicker Type)," *Proceedings* of the Second International Conference on Safety in Road and Rail Tunnels, April 3–6, 1995, Granada, Spain, 1995, pp. 405–410.

Obermayer, R., "Handling a Garage Fire," *Firehouse*, Vol. 17, No. 7, July 1992, pp. 69–70, 72–73.

Peduzzi, E.; Notz, R., "Automatic Fire Detection in Tunnels Linear Fire Detection System Fibro Laser," *Proceedings* of the Second International Conference on Safety in Road and Rail Tunnels, April 3–6, 1995, Granada, Spain, 1995, pp. 419–427.

Proulx, G. and Sime, J. D., "To Prevent 'Panic' in an Underground Emergency: Why Not Tell People the Truth?," *Proceedings* of the Third International Symposium on Fire Safety Science, Elsevier Applied Science, New York, 1991, pp. 843–852, 1991.

Rockett, J. A., Howe, J. M. and Hanbury, W. L., "Modeling Fire Safety in Multi-Use, Domed Stadia," *Journal of Fire Protection Engineering*, Vol. 6, No. 1, 1994, pp. 11–22.

Ruohonen, E., "Kaksi ja puoli hehtaaria myyntitilaa samassa palo-osastossa. [Giant Shopping Center in Kuopio—Two and One Half Hectares in a Single Fire Compartment.]," *Palontorjuntatekniikka*, Apr. 1995, pp. 6–10.

Schneider, U., Lebeda, C. and Max, U., "Entrauchung von grossen Geschossflachen und unterirdischen Parkdecks. [Smoke Venting of Large Storeys and Underground Car-Parks.]," *VFDB*, Vol. 4, Nov. 1994, pp. 138–149.

Shentao, R., *et al.*, "Oil Tanker Explosion in a Long Tunnel," First International Conference on Safety in Road and Rail Tunnels, November 23–25, 1992, Basel, Switzerland, 1992, pp. 339–345.

Shirai, H. and Hirai, S., "New Fire Detector for Road Tunnel," *Proceedings* of the Second International Conference on Safety in Road and Rail Tunnels, April 3–6, 1995, Granada, Spain, 1995, pp. 497–503.

Simcox, S., Wilkes, N. S. and Jones, I. P., "Computer Simulation of the Flows of Hot Gases From the Fire at King's Cross Underground Station," *Fire Safety Journal*, Vol. 18, No. 1, 1991, pp. 49–73.

Smith, J. P., "Fighting Strip Mall Fires," *Firehouse*, Vol. 20, No. 10, Oct. 1995, pp. 18, 20, 22.

Stephens, J., "Using Sprinklers in Bus Garages," *Fire Prevention*, No. 253, Oct. 1992, pp. 26–29.

Stutzman, P. E. and Clifton, J. R., "Diagnosis of Causes of Concrete Deterioration in the MLP-7A Parking Garage," National Institute of Standards and Technology, Gaithersburg, MD, NISTIR 5492, June 1994.

Thomas, P. H., "On the Upward Movement of Smoke and Related Shopping Mall Fires," *Fire Safety Journal*, Vol. 12, No. 3, 1987, pp. 191–204.

Thurston, J. W., "Hazards of Multi-Model Railway Tunnel Transport and Assessment of Design Features," First International Conference on Safety in Road and Rail Tunnels, November 23–25, 1992, Basel, Switzerland, 1992, pp. 213–226.

"Tunnel Fire Safety," *Fire*, Vol. 86, No. 1067, May 1994, pp. 16–17.

"Underground Fire Regulations Met by BR Code of Practice," *Fire*, Vol. 83, No. 1024, Oct. 1990, p. 34.

Webb, B., Sr., "Evaluating Equivalencies to code for a Four Level Shopping Mall," Rolf Jensen and Associates, Inc., Fairfax, VA, Video; Federal Fire Forum, Performance-Based Design Issues of Concern to the A/E Community, Nov. 6, 1995, Gaithersburg, MD, 1995.

Webb, W. A., "Using FPETOOL to Evaluate Fire Safety of a Four-Level Shopping Mall," Rolf Jensen and Associates, Inc., Deerfield, IL, ISBN 0-9527398-3-6; Institution of Fire Engineers; University of Sunderland, Fire Research Station, CIB W14; Tyne and Wear Metropolitan Fire Brigade, Fire Safety by Design, Conference Proceedings, Volume 3, Research Papers, July 10–12, 1995, UK, 1995, pp. 33–37.

Webb, W. A., "Using FPETool to Evaluate Fire Safety of a Four-Level Shopping Mall," *Proceedings* of the International Conference on Fire Research and Engineering, September 10–15, 1995, Orlando, FL, SFPE, Boston, MA, 1995, pp. 365–370.

Wharton, R. K., "Studies Relating to Ignition of the Fire at King's Cross Underground Station," *Fire Safety Journal*, Vol. 18, No. 1, 1992, pp. 35–47.

Whitaker, E., "Tunnel Fire Safety: The Role Played by Vehicles That are Travelling Through," *Fire*, Vol. 87, No. 1070, August 1994, pp. 29–30.

Wintworth, G., "Channel Tunnel—Vital Need to Design a Safe System," *Fire Engineers Journal*, Vol. 47, No. 145, 1987, pp. 5–12.

Woods, D. K., "BP's Use and Experience of Passive Fire Protection Offshore," *Proceedings* of the Conference on Offshore Passive Fire Protection, Plastics and Rubber Institute, London, England, 1987, pp. 1–23.

Wu, P. K., Delichatsios, M. M. and Harriman, R. B., "Cooling Tower Fire Hazards," *Proceedings* of the International Conference on Fire Research and Engineering, September 10–15, 1995, Orlando, FL, SFPE, Boston, MA, 1995, pp. 421–426.

Wynn, S. C. W., "Alarm Handling for Road Tunnels," First International Conference on Safety in Road and Rail Tunnels, Nov. 23–25, 1992, Basel, Switzerland, 1992, pp. 37–53.

Yang, K. H. *et al.*, "Fire Safety Analysis of a Submerged Vehicle Tunnel in Taiwan," *Proceedings* of the First International Conference on Fire Science and Engineering, ASIAFLAM '95, March 15–16, 1995, Kowloon, Hong Kong, 1995, pp. 569–575.

Zhao, Y., Yang, G., and Wang, X., "Investigation of the Dynamic Process in the Ventilation Network During Underground Fire," *Archiwum Gornictwa*, Vol. 34, No. 3, 1989, pp. 503–515.

EVALUATING STRUCTURAL DAMAGE

Revised by John C. Stevenson

This chapter covers the factors that should be considered in evaluating the integrity of structures for further use and occupancy after exposure to fire. Special attention is directed to the techniques for evaluating the post-fire strengths and weaknesses of the basic building materials, i.e., steel, concrete, masonry, and wood, in structural assemblies.

Other chapters in this section that address structural fire damage are Chapter 2, "Building Construction"; Chapter 4, "Structural Integrity During Fire"; and Chapter 5, "Confinement of Fire in Buildings."

All structural materials, whether classified as combustible or noncombustible, inherently possess a degree of fire resistance. However, they can be adversely affected when exposed to elevated temperatures during a fire. The degree of damage may vary with the basic materials and building configuration. Where there is no visible cosmetic damage to the building after a fire, such as glass breakage, concrete spalling, or slight steel distortion, there is little chance of serious damage, although this is no guarantee. Current public interest in life safety, legal implications, and the designer's reputation may warrant a careful inspection to verify no further damage occurred due to excessive expansion and instability in unit design.

THE POST-FIRE INSPECTION

Immediately after a fire, several interested groups, such as the fire department, the building inspection department, and insurance interests, may insist on a fact-finding inspection. The fire department will be interested in the cause of the fire, whether accidental or arson, as well as possible code violations. The building official, also interested in code violation, joins with the insurance concern of property damage, as well as residual structural stability and reusability.

A 1988 *Smithsonian Magazine*[1] article referenced an engineering discipline, called "building diagnosticians," a field that is slowly but progressively gaining a reputation as viable analytical experts on inherent structural deficiencies, including fire damage.

It is the responsibility of the building department to inspect the damaged structure for evidence of potential collapse. Inspections may be accomplished on a gross basis by inspecting the major load-bearing elements to determine prime weakness. Distorted or par-

tially collapsed bearing walls, columns that are misaligned by heat (to the point where loads are no longer supported axially), partial collapse of critical supports, and unsymmetrical loadings are important aspects of post-fire inspection. Evidence of removal of lateral supports of flexural members, either by the fire or by subsequent cleanup operations, is also of critical concern. Temporary shoring or lateral supports may be required prior to admitting any inspection parties.

Buildings damaged by fire may be classified as dangerous. The International Conference of Building Officials' (ICBO) *Uniform Code for the Abatement of Dangerous Buildings*, 1994, in Section 302, lists 18 conditions that can be used to declare a building as dangerous. Condition #11 states: "whenever the building or structure, exclusive of the foundation, shows 33 percent or more damage or deterioration of its supporting member or members, or 50 percent damage or deterioration of its nonsupporting members, enclosing or outside wall, or coverings, it is within the jurisdiction of the building official to declare a building as dangerous and suggest repair, evacuation, or demolition. Repaired buildings shall comply with current building code or shall be demolished at the option of the building owner."[2] The Southern Building Code Congress' *Standard Building Code* and the Building Officials and Code Administrators *National Building Code* have similar requirements.

It should be noted that the building official may make slight revisions to the above ruling. (ICBO is not authorized to approve or disapprove a building official's ruling.) Also, costs might well discourage rebuilding. New zoning requirements may also prohibit reconstruction.

If the building is to be renovated, the capacity of the structural assemblies that were exposed to the fire to continue to perform their intended structural function must first be determined. Assessment of damage will involve both evaluation of the structure as a unit, and evaluation of specific structural members and assemblies. The evaluation should be conducted by a qualified structural engineer, as the influence of distortion and residual stresses may have a significant influence on the load-carrying capacity of the structure. The engineer should make every effort to obtain original plans and specifications. These may be available at the building inspection department or possibly from the architect. Cooperation with the fire department is essential to establish the path and progress of fire, including the method of extinguishment. It may be possible to establish maximum temperatures reached during the fire by examining debris for melted material and by inspecting structural members for spalling, cracking, discoloration, deflection, etc.

John C. Stevenson, AIA, is president of John C. Stevenson Architect, Inc., in San Diego, CA.

The influence of temperature on the mechanical properties of structural materials raises the question of the "use ability" of the material after it has cooled down to ambient temperature. The evaluation of the structural behavior of steel, concrete, masonry, and wood is generally described below. A more detailed evaluation procedure and a case study of renovations is given in Chapters 12 and 13 of *A Complete Guide to Fire and Buildings.*[3]

Next, it is important to determine if the fire-resistance rating of specific structural elements or a structural assembly has been reduced to a value less than that required. While a building may remain structurally sound after a fire, i.e., retain its ability to withstand normal loadings, its fire-resistance rating may have been impaired to the extent that occupancy cannot be permitted.

Fire-resistance requirements for different types of building construction are contained in NFPA 220, *Standard on Types of Building Construction.* Construction type is indicated by a Roman numeral. The Arabic numerals following the construction type designation indicate the fire-resistance ratings (in hr) of exterior bearing walls, structural frame or columns and girders supporting loads of more than one floor, and floor construction, in that order. In a multistory structure, for example, fire damage may have been confined to a single story with no reduction in the tensile strength of structural members but with damage to the enclosure or protection of a beam or column. If a Type I (443) structure is required in this example and the fire-resistance rating of a column supporting more than one floor is reduced to less than 4 hr, the structure can no longer be defined as a Type I (443). Under these conditions, occupancy of any portion of the structure would violate building code requirements.

STEEL

The use of several grades of steel in buildings and bridges is accepted in modern building codes and design standards. Although different grades of steel have a wide range of strengths, tensile and yield strengths of all steel grades are similarly affected by the temperatures that may be expected in fires in buildings.

The behavior of beams or other members subject to bending stresses under fire conditions is complex. In a building fire, the steel temperature may vary considerably over a single cross-section, as well as along the length of a structural member. Information relating to the strength properties of steel at elevated temperatures has been derived from the results of tests on small specimens heated so that the entire specimen was at or close to the measured temperature. Conclusions relating to strength characteristics from such small-scale tests may not be applicable to steel structural members. Moreover, plastic action results in the redistribution of stresses in steel members that are loaded close to the design limit. This characteristic permits steel members to sustain loads greater than those calculated to be safe on the base of yield strength alone; therefore, the strength of a given steel member cannot be determined only from isolated temperature data. (See Figure 7-11A.)

In a building fire, parts of the structure may have been exposed to heating followed by abrupt cooling by water from hose streams. This temperature change is usually less severe than what is accepted practice in heat treating of steel during manufacture. If a steel member shows no distortion due to heat, or if it can be straightened, its physical properties are generally unchanged. Connections between members should be checked, however, for cracks around rivet or bolt holes.

Occasionally, steel exposed to a fire will have a somewhat roughened appearance due to excessive scaling and grain coarsening. The coarsening is caused by exposure of steel to temperatures around 1,600°F (870°C) or higher. The steel will usually have a dark

FIG. 7-11A. Unprotected steel damage on 34th floor of office high rise. Fire load was minimal, the loss extensive.

gray color, although other colors may be present if certain chemicals have been involved in the fire. Steel so modified is commonly called "burnt" steel. Members that have become burnt will usually be severely corroded as well, and their suitability for further use is a matter for individual judgment.

Many instances have been reported where straightening of structural steel members distorted due to fire has been both feasible and economical. Following the McChord Air Force Base fire in 1957 near Tacoma, WA, in which two aircraft hangars were seriously damaged, 1,486 structural steel members were either straightened or replaced in the steel frame. Of that number, only 46 members were replaced; the remaining 1,440 members were flame straightened, using welding torches.

The use of cast iron as structural material has all but ceased. Cast-iron buildings, if found, are undoubtedly archaic, historical preservations. Cast iron exposed to temperatures of 800°F (427°C) will deteriorate and fracture if struck with water from a fire hose.

Excluding fire protection features, such as extinguishing systems, low fire loads, compartmentation, etc., damage to structural steel members may be minimized by application of protective coatings that provide a period of fire resistance. Generally, a fiber or cementitious mixture is applied by spray application; however, the adherence and durability of such coatings, even under normal circumstances, has been the subject of some concern. Too often, questionable application resulted in unreliable fire resistance and, subsequently, considerable structural damage. (See Figure 7-11B.)

Sprayed-on protective coverings can be expected to be easily dislodged during fire fighting and overhaul operations. Usually injury of this kind is easily detected, but a complete examination of the structure must be made to determine the extent of damage. Also, before occupancy is permitted, it is imperative to determine that fire damage to a specific section of a building does not impair the integrity of zones of safety, means of egress, and smoke towers, or the operation of fire doors, fire dampers, or other protective devices or systems in areas not damaged by fire.

About 20 years ago several buildings were constructed with hollow exposed steel columns and beams filled with water—apparently an economic endeavor that developed a passing interest. The concept was valid but lacked standard fire tests or actual fire experience. Fire-resistance ratings could revert to unprotected steel in the event of water leakage, improper venting, or damage during an earthquake.

FIG. 7-11B. Vulnerability of steel in a fire. These large unprotected steel girders were damaged beyond repair.

The ASTM E119, *Standard Test Methods for Fire Tests of Building Construction and Materials*, fire-resistance test may not provide factual data of materials and assemblies under actual fire conditions. At best, it is a destructive comparative test with little relation to actual fire.

Heavy damage to roofs, floors, beams, and trusses can occur if fire test listings are misinterpreted. For fire-resistive purposes, many floor/ceiling and roof/ceiling assemblies are considered as a single fire-resistive element. The ceiling by itself does not provide a specific fire-resistance rating. It should be noted that many floor or roof and ceiling tests provide an assembly rating, as well as a beam rating. Untrained persons may assume the beam rating as protection afforded by the ceiling. Unfortunately, since the ceiling membrane does not provide a fire-resistance rating, structural components above the ceiling are subject to damage.

CONCRETE

In a fire lasting 1 to 2 hr, concrete will be generally only moderately damaged, and so routine cutting and patching procedures will usually be adequate before repairs. In intense and lasting fires, such as those which may occur in large, heavily stocked warehouses and department stores, severe damage to concrete may occur. Restoration may entail the removal of severely damaged areas and patching in areas less severely damaged. Experienced engineering judgment is required for evaluating the residual strength of those areas that are somewhere between moderately and severely damaged. An example of one approach to the problem of evaluating the residual strength of apparently marginally damaged concrete structural units can be found in the article "Prestressed Concrete Resists Fire."[4]

A waiting period of perhaps several weeks should elapse after the fire has been extinguished before careful study of structural damage is initiated. This delay will allow any damage to the concrete, such as cracking, layering, calcination, and discoloration [a change from the natural gray color to pink or brown is indicative of heating to temperatures in excess of 450°F (232°C)], to become more discernible. The thickness of fire-damaged concrete in structural members can be determined by chipping with a pick or geologist's hammer. Unsound concrete may be colored and will be more or less soft and friable, while sound concrete will give a distinctive ring when struck with a pick or hammer. Cored concrete samples for compressive strength tests and reinforcing steel for tensile strength determinations will enable a closer evaluation of the residual strength of damaged members. Load tests may be applied, but

should be conducted only under the supervision of a registered structural engineer.

Two other methods for evaluating fire damage are worthy of mention, but each requires an experienced operator. These make use of the Schmidt hammer and the soniscope. The Schmidt hammer has a spring-loaded plunger, which is caused to strike the test surface by the release of a trigger. The rebound of the plunger is a measure of concrete strength. The rebound numbers should not be considered more than a qualitative evaluation of the concrete strength, as there are several variables other than strength that may affect rebound numbers. The soniscope is an electronic device that has been used to gage the soundness of relatively heavy concrete sections, such as highway pavement slabs, bridge components, and dams. A high-frequency pulse is directed by an electronic sender through the section in question and is picked up by an electronic receiver. The speed at which the pulse travels through the section from sender to receiver, in feet or meters per second, is a gage of the integrity of material in the pulse path. Severe exposure of heavy concrete sections to fire may cause "layering," i.e., partial separation of the outer 1 to 2 in. (25.4 to 51 mm) of concrete of a building member from the interior mass. Such "layering" can absorb the full energy of the pulse, so interpretative experience is necessary.

A complete report dealing with a survey of fire damage to a multi-story reinforced-concrete building and the subsequent repair procedure can be found in the paper "Fire Damage to General Mills Building."[5]

A severe fire occurred in 1951 in an unsprinklered paper warehouse. This fire lasted for more than 44 hr, and it is estimated that heat intensities of more than 1,600°F (870°C) existed for 3 hr or longer. Even after severe fire, the structure was still standing and the concrete floors prevented the spread of fire and major water damage. An inspection of the structure after the fire showed that the concrete roof and columns appeared to have suffered little damage, except that the roof showed deflection up to a maximum of approximately $2^1/_2$ in. (64 mm) in a 20-ft (6.1-m) span. When holes were cut through the roof, it was found that the concrete was calcined and had a light brown color in varying depths. This damaged concrete had almost no strength and began to disintegrate after a period of several weeks. As a result, the portions of the roof slab that showed appreciable amounts of calcine were replaced. Many of the columns required removal of the concrete down to the reinforcing steel spiral, and, in some cases, calcining appeared inside the spiral. In these cases, the entire column was replaced. The column caps were replaced completely if more than one-half of the column showed evidence of calcining. Reinforcing rods showed some reduction in tensile strength, but most were salvaged and reused and downgraded to about three-quarters of their original strength.

In 1982 a fire-damaged rapid transit terminal in Chicago was scheduled for demolition. Due to the inflated cost of rebuilding and time lost, a cost-effective study suggested restoration. The $18.4 million building was restored at a cost of $200,000, and a minimum of time.

If a decision is made to repair the damage, the usual procedure is to shore up the structure, if necessary, and to remove the fire-damaged concrete, using a hammer, chisel, or a lightweight mechanical hammer. (Use of heavy hammers is discouraged, since they might aggravate the damage by inadvertently chipping good concrete.) The steel reinforcements are then cleaned and, if necessary, additional reinforcements are added and, finally, the concrete is built up with Gunite™. Durable concrete repairs also have been made with epoxy resins.

The significant difference between conventional reinforced concrete and stressed concrete in fires is the performance of the high-tensile-steel wire or rods used for pre- or post-stressing. Under fire conditions, the stressed steel units are liable to rapid loss of strength at temperatures in excess of 752°F (400°C). A load test is recommended in post-inspection for significant pronounced sag or an indication of excessive temperature in a range where stress loss may have been sufficient to seriously affect the strength.

Tilt-up construction is a popular, simple, and effective type of concrete construction. As the name implies, slabs are precast and lifted into place. Reinforcing steel projecting from adjoining slabs are welded and encased in concrete. Weaknesses can occur and can lead to fires if, for example, reinforcing ties between panels are not welded or joints are grouted with a flexible, combustible mastic instead of concrete (See Figure 7-11C.). Complete collapse and highly damaged panels may preclude any effort of cost-effective repair.

BRICK

During production, clay bricks are exposed to temperatures in excess of 2,000°F (1093°C), hence their strength is retained in actual fires. Reinforcing steel embedded in the center of a clay brick wall would normally be protected by a minimum of 3 to 4 in. (76 to 102 mm) of brick and not be affected. Spalling should be expected, especially if hot bricks are drenched with water from a fire hose. In most cases, temporary repairs can be made with epoxy cement or Gunite. Permanent repairs are more appropriately made by replacement, in kind, of the defaced brick.

WOOD

Wood is one of the oldest and most widely used building materials. Its behavior in fire conditions varies considerably, depending upon the species of wood and the design configuration, i.e., solid sawn lumber, glue laminates, plywood, wood chipboard, etc. The effect of fire on glue laminates may be considered the same as that on solid sawn lumber (heavy timber), assuming that the adhesives used were not affected by heat and that metal plates joining members and their attachment are not damaged. Generally, the phenol-resorcinal and melamine adhesives are not affected by heat.

Plywood and chipboard are also dependent upon proper adhesives, the difference potentially being slow burning *vs.* a hazardous flash fire caused by delamination.

FIG. 7-11C. Fire in tilt-up construction. (Source: Oakland, CA, Fire Department)

Redwood, found on the west coast of the United States, withstands high fire exposures. It is reported that the high resistance is due to the lack of the usual volatile resins and oils found in other woods.

When wood structural members are subjected to fire, the ability to withstand the imposed loads is dependent to a degree upon the remaining undamaged cross-sectional area. The average rate of penetration of char when flame is impinged upon an exposed wood member is approximately $1^1/_2$ in. (38 mm) per hr. Beyond the char area to a distance not more than $1/_4$ in. (6 mm), the structural properties of wood may be affected by its exposure to high temperatures. The degree of strength loss in this small zone adjacent to the char is not exactly known but is presumed to be insignificant.

Fire tests made on two solid sawn wood joists, 4 by 14 in. (102 by 356 mm), nominal size, at the Southwest Research Institute showed that, after 13 min. of fire exposure, 80 percent of the original wood section remained undamaged and available to carry the load.[6] In another test of two 7- by 21-in. (178- by 533-mm) glued laminated beams, after 30 min fire exposure, 75 percent of the original wood section remained and continued to support the design load.[7]

The previous tests substantiate the fact that large-dimension wood members will remain in place under fire conditions and continue to support design loads. It is usually in the larger or heavy timber members that char can be scraped clear and an evaluation made by a qualified engineer or architect to determine the remaining load-supporting capacity of the wood member.

PLASTICS

Plastic construction is limited to small structures, student housing domes, cabins, and radomes. The use of plastics in buildings has steadily increased,[8] as well as the resultant fire load. The Munich Reinsurance Company reported extensive losses due to corrosion from gaseous decomposition of burning plastic. Twelve buildings were listed as having extensive corrosion-damaged structural steel members. In several cases, it was necessary to destroy the buildings. Steel members moderately damaged by corrosion may be sandblasted, but it is suggested that a stress analysis be conducted before reparations are started. Damage to reinforcement concrete is more complex, since corrosion, in time, is apt to penetrate to the reinforcing. Preliminary analysis is necessary before power hammers are used to inspect extent of damage to reinforcing steel.

CONSTRUCTION—HIGH-RISE FIRE-RESISTIVE AND REINFORCED-CONCRETE BUILDINGS

During the application of fireproofing to structural steel members, coating the lower first and second stories may be postponed until final stages of construction. The application of fireproofing to these lower floors may be delayed due to the potential damage in these high-traffic areas. These are the heavy-working areas and contain storage of materials and equipment, much of which is highly combustible. (See Figure 7-11D.) A fire at this stage could result in a tragic loss.

Buildings of reinforced concrete depend upon wood shoring supports to retain the shape of the building until final curing time. As above, at this stage of construction the building is highly vulnerable to fire damage. (See Figure 7-11E.)

An extreme example of this problem was demonstrated by the St. Mary Cathedral in San Francisco, CA, built at a cost of $10 million to replace a former church destroyed by fire. The cupola walls

FIG. 7-11D. *Lower floor used for storage of materials and equipment. The combustible material and density would undoubtedly provide a fire load capable of destroying the building.*

FIG. 7-11E. *Timber used as shoring and support of reinforced-concrete construction, essentially a forest of wood. A fire before the concrete curing process would result in a total loss.*

extended some 150 ft (46 m) above the ground floor. The supporting scaffolding filled the interior with $1^1/_2$ million board feet of lumber at a cost of $100,000. The chief building superintendent, aware of the fire loss potential, ruled that a temporary sprinkler system be installed.

BIBLIOGRAPHY

References Cited

1. Larson, E., "A New Science Can Diagnose Sick Buildings—Before They Collapse," *Smithsonian Magazine*, May 1988.
2. ICBO, *Uniform Code for the Abatement of Dangerous Buildings*, International Conference of Building Officials, Whittier, CA, 1994.
3. Marchant, E. W., ed., *A Complete Guide to Fire and Buildings*, Barnes and Noble, New York, 1973, p. 268.
4. Zollman, L., and Garavaglin, M., "Prestressed Concrete Resists Fire," *Civil Engineering*, 1960, pp. 36-41.
5. Fruchtbaum, J., "Fire Damage to General Mills Building," *Proceedings* of American Concrete Institute, 1941, Detroit, MI.
6. "Comparative Fire Tests on Wood and Steel Joists," Technical Report No. 3, 1961, National Forest Products Association, Washington, DC.
7. "Comparative Fire Tests of Timber and Steel Beams," Technical Report No. 3, 1961, National Forest Products Association, Washington, DC.
8. Schaden Spiegel, "Losses and Loss Prevention," April 1968, Munich Reinsurance Co., Munich, Germany.

NFPA Codes, Standards, and Recommended Practices

Reference to the following NFPA codes, standards, and recommended practices will provide further information on the processes and protective systems evaluating structural damage discussed in this chapter. (See the latest version of *The NFPA Catalog* for availability of the current editions of the following documents.)

NFPA 220, *Standard on Types of Building Construction*
NFPA 251, *Standard Methods of Tests of Fire Endurance of Building Construction and Materials*

Additional Reading

"Assessment of Concrete Structures Damaged by Fire," *Indian Concrete Journal*, Vol. 61, No. 2, 1987, pp. 29-30.
"BLEVE Causes Massive Damage in Gas Depot, Sydney, Australia, April 1, 1990," *Fire Protection*, Vol. 18, No. 3, Sept. 1991, pp. 13–15.
"Coping With Water Damage," *Fire Prevention*, Vol. 227, Mar. 1990, p. 29.
"Importance of Structural Protection," *Fire*, Vol. 88, No. 1082, Aug. 1995, p. 20.
"Salvage and Damage Control," *Fire Prevention*, No. 255, Dec. 1992, p. 18.
ASTM E119, *Standard Test Methods for Fire Tests of Building Construction and Materials,* American Society for Testing and Materials, W. Conshohocken, PA, 1995.
Brode, H. L., "Fire Damage to Urban/Industrial Targets. Volume 1. Executive Summary," February 5, 1988-July 25, 1989, Pacific-Sierra Research Corp., Los Angeles, CA, Defense Nuclear Agency, Alexandria, VA, PSR Report 1936, Aug. 1, 1991.
Brode, H. L., "Predictions of Fire Damage From Nuclear Attack on Urban Targets," Technical Report, Sept. 15, 1986-Feb. 29, 1988, Pacific-Sierra Research Corp., Los Angeles, CA, Defense Nuclear Agency, Alexandria, VA, DNA-TR-88-75, PSR Report 1802, May 1, 1990.
Burgess, I. W., Olawale, A. O. and Plank, R. J., "Failure of Steel Columns in Fire," *Fire Safety Journal*, Vol. 18, No. 2, 1992, pp. 183–201.
Chandrasekaran, V. and Suresh, R., "Analysis of Steel columns Exposed to Fire," Third International Conference and Exhibition on Fire and Materials, Oct. 27-28, 1994, Arlington, VA, 1994, pp. 41–50.
Chandrasekaran, V., Suresh, R. and Grubits, S., "Performance Mechanisms of Fire Walls at High Temperatures," Third International Conference and Exhibition on Fire and Materials, Oct. 27-28, 1994, Arlington, VA, 1994, pp. 51–62.
Clifton, C., "HERA: Fire Resistance of Structural Steel," *BUILD*, February 1991, pp. 28–29.
Cooke, G. M. E., Virdi, K. S. and Jeyarupalingam, "Thermal Bowing of Brick Walls Exposed to Fire On One Side," Fire Research Station, Garston, England, City University, London, England, ISBN 0-9516320-9-4; *Proceedings* of the Seventh International Interflam Conference, Interflam '96, March 26-28, 1996, Cambridge, England, Interscience Communications Ltd., London, England, 1996, pp. 915–919.
Dickey, W. L., "Fire Endurance of Brick Walls and Veneer," *Building Standards*, Vol. 65, No. 2, March/April 1996, pp. 6–10.
Eisner, H., "Evaluating Structural Stability: Specialists Observe, Confer to Maintain Safety," *Firehouse*, Vol. 20, No. 10, Oct. 1995, pp. 67–69.
Grant, C. C., "When the Walls Came Tumbling Down," *NFPA Journal*, Vol. 87, No. 5, Sept./Oct. 1993, pp. 94–101,103.
Hammer, H., "Auswirkung von Schadenverhutungsmassnahmen in der Feuerversicherung. [Consequences of Damage Prevention in Fire Insurance.]," *VFDB*, Vol. 40, No. 3, September 1991, pp. 132–140.
Huttu, I., "Kantavien terasbetonirakenteiden palovauriot. [Fire Damage to Bearing Reinforced Concrete Structures.]," *Palontorjuntatekniikka*, Mar. 1995, pp. 14–16.
Jirsa, J. O., "Proceedings of a Workshop on Evaluation, Repair, and Retrofit of Structures," U.S./Japan Panel on Wind and Seismic Effects, UJNR, Gaithersburg, Maryland, USA, May 12-14, 1990; University of Texas, Austin, NISTIR 4515, April 1991.

Kodur, V. K. R. and Lie, T. T., "Fire Performance of Concrete-Filled Hollow Steel Columns," *Journal of Fire Protection Engineering*, Vol. 7, No. 3, 1995, pp. 89–98.

Kodur, V. K. R. and Lie, T. T., "Structural Fire Protection Through Concrete Filling for Hollow Steel Columns,"*Proceedings* of the International Conference on Fire Research and Engineering, Sept. 10-15, 1995, Orlando, FL, SFPE, Boston, MA, 1995, pp. 527–532.

Kramps, N., "Alternative Approach to Fire Damage," *Fire International*, Vol. 11, No. 104, 1987, pp. 66-67.

Kramps, N., "Damage Control—Limiting Secondary Losses after Fire," *Fire Prevention*, No. 2-5, Dec. 1987, pp. 26-29.

Lawson, R. M., "Behavior of Steel Beam-to-Column Connections in Fire," *Structural Engineer*, Vol. 68, No. 14, July 17, 1990, pp. 263–271.

MacIsaac, K. S., "Warning: Risk of Sudden Roof Collapse," *Fire Chief*, Vol. 39, No. 2, Feb. 1995, pp. 23–25.

Malanga, R., "Fire Endurance of Lightweight Wood Trusses in Building Construction," Technical Note, *Fire Technology*, Vol. 31, No. 1, First Quarter 1995, pp. 44–61.

Malhotra, H. L., "Specifying the Performance of Elements of Construction," FIREX, First International Conference on Fire Safety in Buildings–The Influence of the EEC Construction Products Directive, Conference Papers, June 12-14, 1990, Birmingham, England, 1990, pp. 1–9.

Massa, R. J., "Vulnerability of Buildings to Blast Damage and Blast-Induced Fire Damage," *Fire Engineering*, Vol. 146, No. 12, Dec. 1993, pp. 103–105.

Meyer-Ottens, C., "On the Problem of Spalling of Concrete Structural Components Under Fire-Induced Stress," Foreign Technology Div., Wright-Patterson AFB, OH, FTD-ID(RS)T-0392-90, June 1990.

Phan, L. T., Hendrickson, E. M. and Marshall, R. D., "Measurement of Structural Response Characteristics of Full-Scale Buildings: Analysical [sic] Modeling of the San Bruno Commercial Office Building," National Institute of Standards and Technology, Gaithersburg, MD, NISTIR 4782, March 1992.

Raufaste, N. J., "Research Lab Plays Role in Reducing Fire Damage," *American Consulting Engineer*, Vol. 3, No. 3, Summer 1992, pp. 27–28.

Read, D. L., *et. al.*, "Expedient Repair of Structural Facilities (ERSF). Volume 1. ERSF System Development. Volume 1 of 2," Final Report, Dec. 1987-Oct. 1990, Applied Research Associates, Inc., Tyndall AFB, FL, Air Force Engineering and Services Center, Tyndall AFB, FL, ESL-TR-91-10, Jan. 1992.

Scawthorn, C., "Structural Damage," *Fire Engineering*, Vol. 147, No. 8, August 1994, pp. 68, 71-72,74,76,78,80–82.

Schneider, U. and Nagele, E., "Repairability of Fire Damaged Structures--CIB W14 Report," *Fire Safety Journal*, Vol. 16, No. 4, 1990, pp. 251-336, 1990;   CIB W14/89/04 (FRG), May 1989.

Sharples, J. R., Plank, R. J. and Nethercot, D. A., "Load-Temperature-Deformation Behaviour of Partially Protected Steel Columns in Fires," *Engineering Structures*, Vol. 16, No. 8, November 1994, pp. 637–643.

Skowronski, W., "Load Capacity and Deflection of Fire-Resistant Steel Beams," *Fire Technology*, Vol. 26, No. 4, November 1990, pp. 310–328.

Stone, W. C. and Taylor, A. W., "ISDP: An Integrated Approach to the Seismic Design, Retrofit, and Repair of Reinforced Concrete Structures," *Journal of Structural Engineering*, 1993.

Sullivan, P. J. E. and Sharshar, R., "Performance of Concrete at Elevated Temperatures (As Measured by the Reduction in Compressive Strength)," *Fire Technology*, Vol. 28, No. 3, Aug. 1992, pp. 240–250.

Talamona, D. R., Franssen, J. M. and Recho, N., "Buckling of Eccentrically Loaded Steel Columns Submitted to Fire," *Proceedings* of the International Conference on Fire Research and Engineering, Sept. 10-15, 1995, Orlando, FL, SFPE, Boston, MA, 1995, pp. 533–538.

Talamona, D., *et. al.*, "Factors Influencing the Behavior of Steel Columns Exposed to Fire," *Journal of Fire Protection Engineering*, Vol. 8, No. 1, 1996, pp. 31–43.

Tewarson, A., "Thermal and Nonthermal Damage in Fires," *Proceedings* of the International Conference on Fire Research and Engineering, September 10-15, 1995, Orlando, FL, SFPE, Boston, MA, 1995, pp. 455–460.

Turley, J., "Damage Management," *Fire Surveyor*, Vol. 21, No. 3, June 1992, pp. 20–23.

Uesugi, H., "Load Bearing and Deformation Capacity of Fire Resistance Steel Tubular Columns at Elevated Temperature," *Proceedings* of the Fourth International Symposium on Fire Safety Science, Intl. Assoc. for Fire Safety Science, Boston, MA, 1994, pp. 1149–1158.

Wright, R. N., "Structural Analysis in Context," National Institute of Standards and Technology, Gaithersburg, MD, American Society of Civil Engineers (ASCE), Structural Engineering in Natural Hazards Mitigation, Structures Congress '93 *Proceedings,* Volume 2, April 19-21, 1993, Irvine, CA, Am. Soc. of Civil Engineers, New York, NY, 1993, pp. 1657-1662.

CONSTRUCTION, ALTERATION, AND DEMOLITION OF BUILDINGS

Revised by Roger A. Neal

Buildings and other structures, regardless of construction type or construction method, are more vulnerable to fire when they are under construction, alteration, or demolition than when completed or when the demolition is finished. The reason for this is that when fires occur, they are likely to spread more rapidly due to the absence or impairment of fire suppression and detection systems, lack of compartmentation, and sometimes the presence of heavier concentrations of combustibles than would be expected during normal use. Furthermore, the fact that fires may be more likely to be discovered in their advanced stage means rapid fire development will already have occurred. In new construction, there is a special potential for severe damage because of incomplete structural systems, lack of applied fire-resistant materials, and the exposed condition of the structure. Fires can also cause the destruction of building materials stored on site, lengthy delays in project completion, long-term business interruption, and the possible loss of life.

Analysis by NFPA's Fire Analysis and Research Division found that reported U.S. structure fires in buildings and structures that were vacant or under construction, renovation, or demolition caused an average of $129.6 million and 21 deaths per year during 1989–1993. Most of the 18,300 structure fires a year and associated losses were due to incendiary or suspicious causes.

This chapter discusses the fire hazards associated with the construction, alteration, and demolition of buildings. Special situations, such as asbestos removal and underground structures, are also discussed. Practices for the prevention of fires, pre-fire planning, storage of combustible materials, and installation of fire protection systems are specifically addressed in the chapter.

Additional information on hazards associated with building construction, fire protection systems, and pre-fire planning can be found in Section 1, Chapter 2, "Fundamentals of Fire Safe Building Design"; Section 3, Chapter 14, "Welding and Cutting"; Section 3 Chapter 21, "Storage of Flammable and Combustible Liquids"; Section 3, Chapter 22, "Storage of Gases"; Section 6, Chapter 3, "Water Distribution Systems"; Section 6, Chapter 10, "Automatic Sprinkler Systems"; Section 6, Chapter 16, "Standpipe and Hose Systems"; Section 7, Chapter 1, "Building and Site Planning for Fire Safety"; Section 7, Chapter 2, "Building Construction"; Section 10, Chapter 6, "Pre-Incident Planning for Industrial and Com- mercial Facilities"; and Section 10, Chapter 12, "Fire Loss Prevention and Emergency Organizations."

PLANNING

Construction operations can be made reasonably safe in new construction, rehabilitation, or demolition projects with adequate planning, proper allowances in job estimates, and proper on-site supervision and control. Although new construction, rehabilitation, and demolition projects differ greatly in the scope of work, they have many fire safety items in common. These common items relate to establishment and maintenance of on-site fire protection, provisions for temporary fire protection, and maintenance of means of escape for workers and others in the building. NFPA 241, *Standard for Safeguarding Construction, Alteration, and Demolition Operations*, contains measures that, with pre-fire planning, will help prevent fire or at least minimize damage when fire occurs.

Security guard service on construction sites is recommended during nonworking hours. Patrols should be planned so that no portion of the job site is overlooked. The use of guard clock and time recording systems is recommended to ensure patrols have been made. Guards should be briefed on the fire protection measures available on the job site, on developments during the workday that affect fire hazards and fire protection capabilities, and on emergency procedures. Provision (either a telephone or a fire alarm box) should be made for notifying the fire department in the event of a fire. Provision for security service should be considered as a part of the project planning stage. NFPA 601, *Standard for Security Services in Fire Loss Prevention*, should be consulted for planning security service.

In any project, whether new construction, rehabilitation, or demolition, the fire department should be kept aware of construction progress. The fire department can thereby provide for command posts and pre-fire plans and be ready to deal with unusual problems encountered during the course of the project.

NEW CONSTRUCTION

Site Preparation

Site preparation is the first item necessary in any new construction project. This involves the removal of brush, trees, and debris from the site prior to the start of construction. Although such land clearing provides for control of natural cover fuel fires, it mainly provides site accessibility for construction. Today's methods of site

Roger A. Neal, CFPS, is assistant vice president of M&M Protection Consultants, Seattle, WA, and member of the NFPA Technical Committee on Construction and Demolition, and the NFPA Technical Committee on Loss Prevention Procedures and Practices.

preparation differ from the "scorched earth" methods of several years ago in that the preservation of as much natural vegetation as practical is desired. This should present no serious fire problem as long as temporary buildings and materials storage are properly isolated from any natural cover fuel by fire breaks of at least 30-ft (9.2-m) width.

Of prime importance for site preparation is the establishment of roadways and the installation of a water supply for fire protection. Many jurisdictions now require that permanent all-weather roads and site utilities be provided before construction begins. Construction site access by temporary or permanent all-weather roadways should be provided, as well as water service for fire protection, prior to the start of construction.

While it is usually more economical to provide permanent water service to serve both the finished building needs and the construction fire protection needs early in the project, it is sometimes necessary to provide temporary water service in the form of aboveground water mains, tanker trucks, or ponds.

Temporary Buildings

Temporary buildings are commonly used on construction sites for the storage of tools and equipment; and serve as temporary offices for the construction management team. The use of mobile structures, such as converted mobile homes, trailers, or manufactured buildings, is common at large construction projects. The location of these temporary buildings can create a serious exposure hazard either to the building under construction or to the group of temporary buildings themselves. Often mobile structures are attached or installed adjacent to each other, creating a larger building with greater hazards than one structure alone.

The exposure protection necessary may be overlooked if several temporary or mobile structures are on the same site. Single-story temporary buildings of combustible construction, or with combustible contents, should be located according to Table 7-12A. When the separation distance between temporary structures is less than the minimum separation distance as specified in the table, then the exposed wall length is the sum of the individual exposed wall lengths of the temporary structures. Providing automatic sprinkler protection in the temporary buildings can be used to reduce the separation distance between temporary structures by 75 percent.

TABLE 7-12A. *Separation Distances for Temporary Structures*

Temporary Structure Exposing Wall Length, ft (m)	Minimum Distance, ft (m)
20 (6)	30 (9)
30 (9)	25 (11)
40 (12)	40 (12)
50 (15)	45 (14)
60 (18)	50 (15)
More than 60 (18)	60 (18)

Source: NFPA 241, *Standard for Safeguarding Construction, Alteration, and Demolition Operations.*

The fire hazards are much the same whether a mobile structure or a temporary structure is used. Overloaded temporary wiring, incorrectly installed portable or temporary heating equipment, and inadequate control of open-flame devices and smoking materials are examples of the common hazards. The high level of fuel load in the form of construction materials, construction plans, specifications, and other documents further increases the fire hazards. In addition, the use of space (sometimes concealed) beneath these temporary structures for miscellaneous storage can be hazardous. Such spaces and storage can be subject to vandalism and arson. Vegetation growing under, or left in, these spaces can also create hazards.

The protection of these structures, especially mobile structures, is usually possible by the installation of a residential-type sprinkler system using listed plastic pipe that can be connected to the on-site water supply. The cost of this protection is low for the mobile structures, since there is only a one-time installation cost, and connection costs on the site are minimal.

Loss Prevention During Construction

Normal construction practices require considerable amounts of combustible building materials, forms, and scaffolding, which tend to accumulate in congested areas. Temporary storage of combustible materials in the yard or on the roof should be kept to a minimum and limited to 6 ft (1.8 m) in height and one or two pallet widths. Where space permits, storage should be located at least 25 ft (7.6 m) from the structure being exposed. Those areas of a building where the fire-resistant coatings for structural steel have not been completed should not be used for the storage of combustible items. Foamed plastic materials should only be stored to a maximum height of 5 ft (1.5 m) and only in areas where automatic sprinkler protection is in service.

Water service at the site should be adequate not only for the final building construction needs, but also for fire protection during construction, in view of the amount of combustibles present. In multistory structures, either temporary or permanent standpipes should be put into service as each floor is constructed to provide fire-fighting capability. Temporary standpipe service is usually a dry standpipe with proper valves at each floor. A fire department siamese connection, easily accessible to fire department pumpers, should be provided at the street. If the Siamese connection is located inside the construction site, clear access to the Siamese connection via all-weather roads should be maintained. Large signs indicating the location of the fire department connection and the top level of the standpipe should be installed to assist the local fire department.

Temporary enclosure of buildings is common and should only be done with care. Only flame-resistant tarpaulins or materials of similar fire-retardant characteristics should be used. If temporary enclosure is required, metal or fire-retardant-treated wood forms spaced about every 4 ft (1.2 m) should be used to support the material. The enclosure materials should be securely fastened or guarded to prevent contact with heaters or other sources of ignition. Fabric or plastic film should meet the requirements of NFPA 701, *Standard Methods of Fire Tests for Flame-Resistant Textiles and Films.* (See Figure 7-12A.) Tarpaulins that appear weathered may have lost their fire-retardant qualities and should not be used.

Concrete should be poured as quickly as possible after combustible forms have been constructed, and the forms removed as soon as possible after the concrete has set properly. During the time the forms are in place, storage and construction operations on that floor should be held to a minimum and ignition sources should be eliminated. Noncombustible forms should be used whenever possible. (See Figure 7-12B.) Yard storage of combustible forms should be at least 30 ft (9 m) from the building.

Where structural "fireproofing" is specified, it should be applied as soon as possible to afford fire protection to beams, columns, and other structural elements early in the construction phase.

Rubbish and trash accumulations can present a serious fire hazard on the construction site. Daily cleanup of scrap lumber, cardboard, and other debris removes the fire hazard and provides a safer working environment. Trash containers should be provided on site

FIG. 7-12A.   *Large sheets of plastic are sometimes used to enclose buildings under construction to confine heat. (Source: Francis L. Brannigan)*

FIG. 7-12B.   *Combustible formwork is a constant source of danger to concrete buildings under construction. Here fire consumes wood formwork for cast-in-place floor and wall panels in a parking garage.*

for proper storage of debris and should be located at a safe distance from the building.

Cutting and Welding

Cutting and welding operations must be controlled carefully, and proper safeguards utilized. Careless use of welding and cutting torches and plumbing torches is a leading cause of construction fires than any other ignition source. The use of a permit system that specifies controls for safe welding should be used. Combustible materials within 35 ft (11 m) of any cutting or welding operation need to be protected by either being relocated; covered with noncombustible materials; or, as a last resort, by being wet down prior to and during the operation. Follow the fire protection procedures in NFPA 51B, *Standard for Fire Prevention in Use of Cutting and Welding Processes.*

The cutting and welding permit serves several functions: evaluation of the fire hazards in the immediate area where "hotwork" is

to be performed; specification of requirements for fire suppression; elimination of hazardous activities; and notification of the security service of the location of the hot work. The construction manager should designate an individual, other than the operator of the heat-producing equipment, to complete the permit. (See Figure 7-12C.)

A qualified individual should be assigned the responsibility of being the fire watch during cutting and welding operations. It is important that the fire watch be located in a position to observe where the falling hot materials land. This may be in a location not directly within the cutting and welding operation area. During roofing operations, it may be necessary for a fire watch to also be located under the roof. Particular attention should be paid to areas, such as around flashing, that are being heated by torches. The fire watch should have a portable fire extinguisher or water-charged hose line that may be used to extinguish small fires that start from hot slag or sparks. The fire watch should not have any other duties during this time and should remain as the fire watch for at least one-half hour after the hot work has been completed. Security guard services should pay additional attention to areas where hot work was conducted during the previous shift.

Temporary Heating

Salamanders and other temporary heating devices have been responsible for numerous fires that have resulted in millions of dollars in property losses. When salamanders are in use, they should be visually inspected every one-half hour to ensure that combustibles have not blown over or fallen near the temporary heating device. This may require the use of security guards during nonconstruction hours. During windy periods, it may be necessary to constantly attend the temporary heating devices being used. Any object that is too hot to touch should be moved or the heating device relocated. Liquefied petroleum gas (LPG) cylinders used to fuel temporary heating devices should be securely fastened to prevent falling. Visual inspection should also ensure that the heating appliance is operating properly. Requirements in NFPA 58, *Standard for the Storage and Handling of Liquefied Petroleum Gas*, should be followed for LPG heating device operation.

Flammable and Combustible Liquids

Flammable and combustible liquids present obvious fire protection problems. They should be stored and handled in approved metal containers or in their original shipping containers. Containers should be tightly closed when not in use. Some materials, such as contact cement, are extremely hard to pour, and should be left in their original containers. Care must be taken to ensure that these liquids are not stored in exits or stairways and that they do not block access to firefighting equipment. Quantities of more than 25 gal (95 L) of flammable or combustible liquids should be stored in an approved flammable liquid storage cabinet or other acceptable structure. See Section 4-3 of NFPA 30, *Flammable and Combustible Liquids Code* (hereinafter referred to as NFPA 30), for maximum storage quantities for cabinets. NFPA 30 also provides information for the safe handling, dispensing, and storage of flammable and combustible liquids. Outside storage of Class I liquids should be at least 50 ft (15 m) from buildings under construction. The area around the outside flammable liquid storage area should be maintained free of combustible materials, such as weeds, debris, and construction materials.

Internal combustion engines for pumps, air compressors, etc., should be placed to avoid the discharge of the exhaust near or contacting combustibles. A minimum of a 6-in. (152-mm) clearance to combustibles should be maintained if the exhausts are piped to the exterior. The engine should be shut down and allowed to cool prior to refueling to prevent flash fires. Electric, hydraulic,

CUTTING AND WELDING PERMIT

(Work is not permitted unless this card is filled in and posted in work area.)

DATE _____/_____/_____ TIME _____

BUILDING _____ AM
                                                  PM

DEPT. _____ FLOOR _____

_____

WORK TO BE DONE _____

_____

SPECIAL PRECAUTIONS _____

_____

FIRE WATCH REQUIRED? _____

The location where work is to be done has been examined by me, the necessary precautions taken (see back of permit) and permission is granted for this work.

PERMIT EXPIRES _____/_____/_____ TIME _____ AM
                                                              PM
_____

SIGNED _____
               INDIVIDUAL RESPONSIBLE FOR WORK AUTHORIZATION

TIME STARTED _____ COMPLETED _____

FINAL CHECK

(where fire watch is required)

Work area and all adjacent areas where sparks might have spread were inspected for at least 30 minutes after the work was completed and no fire conditions were noted.

SIGNED _____

Return this permit after work is completed to plant manager for filing and review by insurance representative.

☐ Fire protection system(s) in service (sprinklers, CO_2, foam).

☐ Cutting and welding equipment in good condition.

☐ Floor/ground clean (and wet down when necessary).

☐ Combustibles at least 35 feet from welding area.

☐ Flammable liquids and other hazards removed from area.

☐ All floor and wall openings within 35 feet covered.

☐ Non-combustible covers used to protect nearby combustibles and equipment.

☐ Containers, tanks, ducts, and other enclosures cleaned and purged of flammable vapors, liquids, dusts, and other hazardous materials.

☐ Fire extinguishers or small standpipe fire hose provided.

☐ All hazardous operations discontinued in area.

☐ Fire watch should be present during and at least one-half hour after welding or burning has ceased.

☐ Location of nearest fire alarm box identified.

When possible, do work in a non-combustible area.

An individual should generally be assigned to watch for dangerous sparks in the area and the floor below.

MARSH &
McLENNAN

M&M Protection Consultants

WBP-2-93-2.5

FIG. 7-12C.   Cutting and welding permit.

or pneumatic-driven equipment should be used in underground or basement locations or windowless buildings.

Roofing Operations

Tar and asphalt kettles should be located outdoors, on a noncombustible floor or roof of the building away from combustibles. A layer of gravel at least 12 in. (305 mm) thick can provide protection between a tar kettle and a combustible roof. Uninsulated fuel containers should be kept at least 10 ft (3 m) from the burner flame and should be secured to prevent tipping or falling over. Metal covers (at least 14 gage) should be provided to smother potential fires in asphalt and tar kettles. At least one 20-B-rated fire extinguisher should be located within 30 ft (9.2 m) of horizontal travel distance of every roofing kettle at all times. If the roof is of combustible construction, at least one 2A:20-B:C-rated fire extinguisher should be available on the roof. Should a fire occur, it is important to remember that water thrown into the tar kettle may cause the tar to froth and may overflow the container. Roofing mops should never be left indoors near ignition sources or combustible materials. Roofing mops soaked with tar have been known to be susceptible to spontaneous ignition; however, modern roofing materials are less susceptible to this problem. To reduce this problem, mops should be spun to remove excess

tar and hung or placed so that the mop head is not in contact with combustible materials or other mop heads.

Torch-applied and single-ply roofing systems should be installed using extreme caution. Care must be taken so that the membrane is not overheated or smoking. The flame of the torch should not come in contact with wood nailers, cant strips, or metal flashing. Nor should the torch be used where the flame impingement cannot be fully observed. A base ply should be used to protect wood deck; combustible insulation (such as foam, kraft-faced glass fiber, or wood fiber); small crevices; or plastic fastener plates. Small torches may be used to heat the underside of the membrane away from these areas before securing the membrane. Hot trowels should be used to feather seams at laps and flashings. Torch-applied roofing is considered hotwork, and fire watches should be maintained for at least one hour after the torches have been extinguished. The area under the roofing operation should also be checked frequently by the fire watch.

Permanent Safeguards

In most construction operations, the progress of the work toward completion gradually enhances the safety of the structure. As the need for combustible formwork and scaffolding, materials storage, temporary buildings and trailers, and other inherent hazards diminishes, they should be removed from the site.

Fire fighter access to the structure is important. Installation of permanent stairways and stairway enclosures should be completed as quickly as possible. Fire barriers and other essential features that provide for the confinement of fire should be installed as early as possible. Penetrations of fire-resistant construction made for building services should be sealed with a fire-rated material immediately after the building services are installed. The fire rating of the sealant should be the same as the structural element that has been penetrated. The integrity of all fire cutoffs should be maintained.

In buildings without automatic sprinkler systems, or those in which the installation of water tanks and other fire protection equipment must wait for the completion of the structural elements, standpipe and hose systems may provide the only reliable source of water for fire fighting during the construction of the project. In buildings where automatic sprinkler protection is to be a part of the permanent protection for the building, the sprinkler system should be installed as building construction progresses. Many fires have occurred only a few days prior to the anticipated completion of a sprinkler system. Sprinkler systems should be installed, tested, and operating prior to significant combustibles being introduced into the structure, and immediately following combustible construction. Blank gaskets and blind flanges used to activate sprinkler protection by sections, as construction work is completed, should be provided with lugs that are painted red, or otherwise clearly marked when installed. This will help ensure that the gasket or flange is removed when the system is extended.

Permanent standpipe and hose systems should be installed as construction work progresses and should not wait for the final structural shell to be completed.

Permanent wiring systems should replace temporary wiring as rapidly as possible, and certainly as soon as formwork is removed from concrete construction.

Supervision of Fire Protection

All construction, renovation, and demolition projects should have a qualified person functioning as the fire protection manager and assigned responsibility for fire prevention and protection activities. On smaller projects, this responsibility may be shared with other responsibilities by one individual. This person should also have full authority to enforce fire prevention procedures required for the project. In renovations of occupied buildings, a fire protection manager should be appointed by the building owner. Although responsibility for fire protection rests with the owner, steps to accomplish effective loss prevention are normally carried out by the contractor.

The fire protection manager should develop a written pre-fire plan in cooperation with the responding local fire department and the owner's fire brigade if present. At a minimum, the pre-fire plan should identify emergency access roads, fire-fighting water supplies, nature of the construction project, site-specific hazards, emergency contact lists, and on-site fire-fighting capabilities. Since construction projects are constantly changing, the pre-fire plan should be flexible. The pre-fire plan should be shared with the local fire response agency at least every six months, or more often as changes in construction warrant.

Even with a local fire department response, contractors and security guards should be trained in the use of portable fire extinguishers and standpipe hose lines. On large projects, consideration should be given to establishing a fire brigade using either contractor employees or dedicated fire brigade members. NFPA 600, *Standard on Industrial Fire Brigades,* provides additional information on the equipment and training requirements for fire brigades and employee

fire fighting. The fire brigade may also be involved in other emergency response functions, such as first-aid, confined space entry, or technical rescue (e.g., trench, low angle, or high-angle rescue).

Fire prevention should be a regular topic at contractor safety meetings, which are often called "tail-gate talks." All fires should be investigated by the fire protection manager, and needed changes in construction-related activities or fire prevention requirements communicated to all workers.

Permanent wiring systems should replace temporary wiring as rapidly as possible. Permanent heating plants and temporary non-fired heat exchangers can often be used to supplant the more hazardous salamanders.

DEMOLITION OPERATIONS

If not properly controlled, demolition operations can create more serious problems than construction operations. Demolition operations should include precautions to ensure maintenance of fixed fire protection systems for as long as possible.

Fire Protection Systems

Automatic sprinkler systems should be modified to allow their removal on a floor-by-floor or area-by-area basis. The conversion of a wet system to a dry system fed only by a fire department connection is sometimes possible and provides good protection for the building as well as for fire-fighting personnel.

Standpipe and hose systems should also be removed floor by floor and should be the last fire protection feature removed. This maintains fire-fighting capability to otherwise demolished floors. A wet standpipe can be easily converted to a dry standpipe.

Trash Removal

It is common practice to provide chutes for the removal of trash and debris from the upper floors. If chutes are used, they should be non-combustible and erected outside the building. By installing the chute as straight as possible, the potential for obstructions in the chute will be minimized.

Structural Integrity

Cutting holes through floors to provide an inside chute should be avoided, since this creates a vertical opening through which fire can spread rapidly from floor to floor. Holes in floors also cause potential safety problems for fire fighters. Walls should remain intact as long as possible.

Figure 7-12D shows the results of a demolition operation that violated these principles. The results were disastrous; not only was the building being demolished destroyed by fire, but the fire spread to four other buildings.[1]

Utilities

Prior to the start of demolition, gas supplies should be shut off and capped outside the building. Electrical circuits should be reduced to a minimum, and those circuits that remain energized should be labeled. Discontinued lines should be physically disconnected.

Explosives

Use of explosives requires special precautions. Explosives should be used and stored in accordance with NFPA 495, *Explosive Materials Code.*

FIG. 7-12D. Fire originating on the top floor quickly spread throughout this partially demolished building and involved four other buildings. The sprinkler system had been shut down about a week before the fire—instead of being shut down in stages as demolition progressed. Contrary to good practice, holes had been made in the floors to facilitate the removal of debris.

REHABILITATION PROJECTS

There has been a great deal of rehabilitation of older buildings in recent years. This effort is partly the result of a desire to preserve historic structures and partly a result of rising construction costs. From a fire protection standpoint, rehabilitation projects can produce fire safe buildings if the extensive compartmentation usually present in older buildings is at least partially preserved.

Code-Related Issues

From a building code standpoint, existing buildings being rehabilitated usually must meet current code requirements for means of egress, fire protection, and light and ventilation.

The use of equivalency concepts to provide the level of safety intended by a particular code is usually possible when it is impractical to meet the letter of the code. In buildings with historic significance, it is sometimes undesirable to provide all of the code-required safeguards, as these changes may destroy the historic features. In such cases, the safety of the occupants takes precedence, and the proper safeguards must be installed to ensure life safety. While improvement of exits to total code compliance may not be possible, providing other compensating features in the building to control fire and smoke may result in equivalent safety. These compensating features may involve the installation of automatic sprinkler protection, smoke detection systems, and smoke-control systems. Design equivalency should be evaluated carefully to ensure that a comparable level of life safety to that required by the applicable code is being maintained.

The U.S. Department of Housing and Urban Development (HUD) has produced guidelines for the rehabilitation of buildings. An interesting feature of these guidelines is a volume that provides fire resistance information on older methods of building construction.[2] This information is extremely valuable when attempting to evaluate the fire resistance ratings for existing building structural systems and fire barriers.

Life Safety

In some rehabilitation projects, the renovations must be made while the building is partially occupied. Special precautions must be taken in these cases. The most important of these is the maintenance of proper exits from the building. Not only must exits be provided for building occupants, but at least one exit must also be provided for building trades workers in the construction areas. The accumulation of trash and other debris must be kept to a minimum, and refuse removed daily.

If fire suppression, detection, or alarm systems are being renovated, their downtime must be kept to a minimum. Proper planning can eliminate unnecessary downtime. Isolation of parts of systems is usually possible to ensure operation in unaffected areas of the building. If it becomes necessary to shut down entire systems, a fire watch should be posted near the main control to restore service in case of fire. Some form of reliable communication should be established if the main control is remote from the work area. This precaution is especially important in sprinklered buildings. Removing the entire sprinkler system as part of a rehabilitation project can have disastrous results.

Means of egress should be readily available to building occupants during normal working hours. Therefore, alterations to means of egress should be accomplished during off-hours. In some cases, rerouting of traffic may be a practical alternative. In all cases, exit systems and fire protection systems should be restored at the end of each work day.

In cases where additions to buildings block existing exits, provision must be made for alternate means of egress or for a protected path of travel. The use of combustible materials in temporary exit construction should be avoided.

ASBESTOS REMOVAL

Fires during asbestos removal operations present several additional problems in addition to the issues previously discussed. The potential for exposure of fire fighters and construction workers to hazardous materials outside the containment area impacts access and firefighting operations. Maintenance of the containment area may reduce fire detection and will lengthen the amount of time before manual fire suppression by workers outside the containment area possible. The widespread use of plastic film and tarping may also contribute to the fire load and fire spread.

With the exception of minor asbestos removal, the fire department and any on-site emergency responders should be notified at least 24 hr prior to the start of an asbestos removal operation. The pre-fire plan developed for the project should also address any asbestos removal operations. Plastic film used in the project should be flame resistant. Reflective signs indicating the presence of asbestos should be posted at the entrances, exits, decontamination areas, and waste disposal areas.

UNDERGROUND OPERATIONS

Fires in underground construction projects, especially related to tunnels, are particularly difficult to suppress and can cause significant damage and delays in the completion of construction. A leading cause of these fires is improper precautions taken during cutting and welding operations.

Hotwork Permits

Standard hotwork precautions, including a fire watch and removing combustible materials, should always be followed. Particular attention should be paid when a combustible film membrane, such as high-density polyethylene, is used as a moisture barrier behind the rib structure of a tunnel. Cutting of rib spikes used to hold hoses and cables can transmit sufficient heat to ignite the plastic membrane.

Fire Protection

Water for fire fighting should be provided by means of a standpipe system with fire hose connections a maximum of every 300 lineal ft (92 m). Provisions for additional air supplies for fire-fighting personnel should also be made. Some tunnels use an emergency car that is equipped with wheels the same gage as the muck car tracks. This car can be used to either transport fire fighters or emergency equipment into the tunnel.

Loss Prevention

Mobile mining equipment used underground should be equipped with a fire suppression system. The system should be either automatically or manually activated. See NFPA 123, *Standard for Fire Prevention and Control in Underground Bituminous Coal Mines*, for further information.

Storage of combustible materials should be limited and properly protected, and arranged to prevent ignition or contribution to a fire.

Monitoring of methane is a high priority during underground operations. This is particularly important at the face of any tunnel bore, or where the geology indicates the presence of organic material.

All personnel, including fire fighters who enter a tunnel or underground construction site, must use the sign-in, sign-out board that is usually located adjacent to the underground entrance.

BIBLIOGRAPHY

References Cited

1. Sharry, J. A., "Group Fire, Indianapolis, Indiana," *Fire Journal*, Vol. 68, No. 4, 1974, pp. 13–16.
2. HUD, "Guideline on Fire Ratings of Archaic Materials and Assemblies," Vol. 8, Rehabilitation Guidelines 1980, U.S. Department of Housing and Urban Development, Washington, DC.

NFPA Codes, Standards, and Recommended Practices

Reference to the following NFPA codes, standards, and recommended practices will provide further information on the fire hazards of construction, alteration, and building demolitions discussed in this chapter. (See the latest version of *The NFPA Catalog* for availability of current editions of the following documents.)

NFPA 13, *Standard for the Installation of Sprinkler Systems*
NFPA 13D, *Standard for the Installation of Sprinkler Systems in One- and Two-Family Dwellings and Manufactured Homes*
NFPA 30, *Flammable and Combustible Liquids Code*
NFPA 51, *Standard for the Design and Installation of Oxygen-Fuel Gas Systems for Welding, Cutting, and Allied Processes*
NFPA 51B, *Standard for Fire Prevention in Use of Cutting and Welding Processes*
NFPA 58, *Standard for the Storage and Handling of Liquefied Petroleum Gas*
NFPA 241, *Standard for Safeguarding Construction, Alteration, and Demolition Operations*
NFPA 395, *Standard for the Storage of Flammable and Combustible Liquids at Farms and Isolated Sites*
NFPA 495, *Explosive Materials Code*
NFPA 600, *Standard on Industrial Fire Brigades*
NFPA 601, *Standard for Security Services in Fire Loss Prevention*
NFPA 701, *Standard Methods of Fire Tests for Flame-Resistant Textiles and Films*

Additional Reading

Beason, D., *et al.*, "Fire Exposure to an Office Trailer Complex," UC-707, 1988, Lawerence Livermore National Laboratory, Livermore, CA.
Becker, W., "Role of Intumescents for Construction Products with Improved Fire Safety," *Fire and Materials*, Vol. 15, No. 4, Oct./Dec. 1991, pp. 169–173.
Brannigan, F. L., *Building Construction for the Fire Service*, 3rd ed., National Fire Protection Association, Quincy, MA, 1989.
Brannigan, F. L., "Building Is Your Enemy. Part 3. An Overview of Building Construction Hazards," International Association of Fire Chiefs (IAFC), Fire-Rescue International Conference Proceedings, 1993 Annual Conference, Aug. 28–Sept. 1, 1993, Dallas, TX, 1993, pp. 41–43.
Factory Mutual Research Corporation, Loss Prevention Data Sheet 1-0, *Safeguards During Construction, Alteration, and Demolition*, Factory Mutual Research Corporation, Norwood, MA, Apr. 1992.
Fire Journal, "Sparks from Welder's Torch Ignite Exercise Mats in Athletic Complex," Vol. 88, No. 5, Oct. 1994, p. 30.
Fisher, G. L., "Getting Through Construction," *Fire Prevention*, No. 248, Apr. 1992, pp. 20–22.
Gray, J. A., and Arditi, D., "Fire Prevention and Protection During Construction," *Journal of Applied Fire Science*, Vol. 4, No. 1, 1994–1995, pp. 53–68.
Gustin, B., "Cutting Reinforced Concrete," *Fire Engineering*, Vol. 145, No. 8, Aug. 1992, pp. 32–33, 36–38.
Holemann, H., and Worpenberg, R., "Brandursache Schweissen, Brennschneiden und Loten—Reichweite und Zundpotential gluhender Partikel. [Causes of Fire Welding, Flame Cutting and Soldering—Range and Ignition Potential of Glowing Particles.]," *VFDB*, Vol. 41, No. 2, Mar. 1992, pp. 79–89.
Holemann, H., and Worpenberg, R., "Brandursache Schweiben, Brennschneiden und Loten: Zundmechanismen gluhender Partikel. [Cause of Fire Welding, Flame-Cutting and Soldering: Ignition Mechanism of Glowing Particles.]," *VFDB*, Feb. 1994, pp. 59–64.
Hopkinson, J., and Barber, C., "Managing Construction Sites," *Fire Prevention*, No. 286, Jan./Feb. 1996, pp. 28–30.
"Insurer Loading Spurs Fire Code (Building/Construction Sites)," *New Civil Engineer*, Mar. 7, 1991, p. 10.
Kander, K. A., "Avoiding Costly Construction Fires," *Constructor*, July 1992, pp. 39–41.
Muirhead, J., "Making It Safely through Construction: A Recipe From the Disaster Cookbook," *Fire Surveyor*, Vol. 22, No. 1, Feb. 1993, pp. 5–6.
NFPA, Fire Data Package 19, "Vacant, Abandoned, Under Demolition Buildings," NFPA Fire Analysis—Research Division, National Fire Protection Association, Quincy, MA, 1990.
NFPA Inspection Manual, Chapter 4, "Construction, Alteration, and Demolition Operations," 7th ed., NFPA, Quincy, MA, 1994.
Oba, K., Duhrin, U., and Bjork, F., "Welding Methods With Reduced Fire Hazard for Single-Ply Polymer-Modified Bituminous Roofing Materials," *Fire and Materials*, Vol. 18, No. 6, 1994, pp. 351–358.
Ow, C. S., and Thin, C. C., "Upgrading Fire Protection in Existing Buildings," *Fire International*, Vol. 12, No. 112, Aug.–Sept. 1988, pp. 39–40.
Rule, C. H., "Fire Destroys Philadelphia Building under Construction," *Fire Journal*, Vol. 79, No. 3, May 1985, pp. 42–46, 105–111.
SFPE TR 84-7, "Fire Protection During Construction. . .A Critical Time," Society of Fire Protection Engineers, Boston, MA, 1984.

Scoones, K., "Fires During Construction," *Fire Prevention*, No. 248, April 1992, p. 19.

Scoones, K., "Serious Fires in Unoccupied Buildings and Those Under Refurbishment or Construction in 1994," *Fire Prevention*, No. 286, Jan./Feb. 1996, pp. 18–19.

Steenbeigen, H. E., "In a Fire, Seconds Count," *Fire Command*, Vol. 56, No. 7, July 1989, pp. 15–17.

Uniform Fire Code, Article 87, International Fire Code Institute, Whittier, CA, 1994.

"Welding Technique Averts Fires," *Engineer*, March 8, 1990, p. 37.

MISCELLANEOUS BUILDING SERVICES

John E. Kampmeyer

Building services are those systems that connect a building to utilities and public services to provide a livable and functional environment for people and equipment. To understand how building services may affect fire behavior, and the fire hazards that could result from these systems and their components, a basic knowledge of the individual systems, and their relationship to each other, the structure, and the type of occupancy is needed. In short, not only must one understand how these systems affect the building, but how people will affect the building and these systems.

This chapter presents an overview of building services in the context of fire hazard. These hazards are affected in the three major phases of a building's life, and this chapter is, therefore, organized around a discussion of three stages: (1) design and installation, (2) operation and maintenance, and (3) modification and renovation.

Within the discussion of the three stages, building systems can be classified into five major groups:

1. *Energy production.* These are systems that produce heating, cooling, and electric energy for buildings and machinery. They include boilers, air heaters, chillers, refrigeration, cooling towers, evaporative coolers, domestic water heaters, and electric generators.
2. *Environmental systems.* These systems move air through buildings to provide comfort for occupants and control contaminants in the air. They include central air handling equipment, unitary air-handling equipment, duct systems, building exhaust, process exhaust, and air makeup systems.
3. *Piped services.* These systems connect external mechanical utilities to building systems and interconnect between mechanical energy production and environmental services. They include storm drainage, sanitary drainage, domestic water, fuel gas, fuel oil, medical gases, heated water, chilled water, condenser water, compressed air, glycol, and steam.
4. *Electrical/electronic systems.* These systems connect external electric utilities to building systems and distribute electricity and communications through buildings. They include high-voltage switchgear, electrical distribution, lighting, telephone systems, data transmission systems, building automation, and security systems.

5. *Transportation and conveyance.* These services provide for the movement of people and materials vertically and horizontally in buildings. They include elevators, escalators, conveyors, and people movers, and mail, laundry, and refuse chutes. In this same category are enclosures for mechanical and electrical systems. While these latter enclosures may not seem to be the same as elevators and conveyors, they exhibit many of the same characteristics, in that they cut through fire boundaries in order to "convey" the systems they enclose from one part of the building to another. Included are vertical shafts for ducts, pipe, and conduit, as well as horizontal enclosures, such as those for kitchen exhaust ducts that may pass through fire barriers but are not permitted to have internal obstructions, such as fire dampers.

Fire hazards from these five major groups of systems include situations where systems, or their components, represent potential heat (energy) sources, fuel sources, or heat (energy)-fuel transport systems.

More detailed information about specific systems can be found in the following chapters: Section 4, Chapter 19, "Air-Moving Equipment"; Section 7, Chapter 1, "Building and Site Planning for Fire Safety"; Section 7, Chapter 5, "Confinement of Fire in Buildings"; Section 7, Chapter 6, "Smoke Movement in Buildings"; and Section 7, Chapter 14, "Airconditioning and Ventilating Systems." Related information can be found in: Section 3, Chapter 1, "Electrical Systems and Appliances"; Section 3, Chapter 2, "Control of Electrostatic Ignition Sources"; Section 3, Chapter 3, "Lightning Protection Systems"; Section 3, Chapter 5, "Heating Systems and Appliances"; Section 3, Chapter 6, "Boiler Furnaces"; Section 3, Chapter 7, "Heat Transfer Systems (Nonwater Media)"; Section 3, Chapter 22, "Storage of Gases"; Section 3, Chapter 28, "Refrigeration Systems"; and Section 3, Chapter 31, "Waste Handling and Control." The following chapters also relate to this topic: Section 4, Chapter 8, "Medical Gases"; and Section 4, Chapter 18, "Effects of Energy Conservation in Buildings." Section 8, Chapter 3, "Building Transportation Systems," addresses issues related to elevators, escalators, lifts, and similar systems. Finally, Section 9, Chapter 20, "Materials-Handling Equipment," addresses other conveyance systems. Additional references on these systems, their design, and operation are listed under additional reading at the end of this chapter.

DESIGN AND INSTALLATION

The design and installation of building services in a new structure presents a challenge to architects, engineers, and constructors to in-

John E. Kampmeyer, P.E., is a project sponsor for Maida Engineering, Inc., a Philadelphia-based architectural/engineering firm. He is a member of NFPA's Smoke Management Committee and a member of the Board. He is also a member of N.S.P.E., SFPE, ASHRAE, BOCA, and ASME, and he teaches fire protection courses for the Philadelphia Engineers Club.

tegrate owners' and end-users' occupancy needs and financial limitations with a building's structural system, environmental conditions, and the limits of technology. At the same time, it represents the best opportunity to select elements for inclusion that will provide a safe, functional, and livable environment, since the constraints of an existing structure are not present. The most obvious goal of the designer and installer is to create systems that will meet the users' daily operational needs in a cost-effective and efficient manner. The less obvious and sometimes minimized goal is to meet these needs in a manner that will not present a hazard to the occupants or to the use of the building.

The concept of "performance-based design," which has recently been introduced into fire protection terminology, has long been in evidence in the design of building services. Building heating can be from boilers fired with gas, oil, or electricity, producing steam, hot water, or superheated hot water. Heating can also be produced by gas-, oil- or electric-fired heaters in air-handling equipment.

Cooling for buildings can be produced in central chillers powered by electricity, steam, or natural gas. The heat from these chillers can be transferred to water or directly to the outside air. Cooling can also be achieved through the use of unitary equipment that is mounted in windows, through walls or on roofs. It may also be achieved through the use of water-cooled unitary equipment that transfers the heat to cooling towers, or through the use of split evaporator and condenser units connected together by refrigerant piping.

Air-handling methods also vary widely. It can be delivered at a constant volume, and the temperature of the air raised or lowered to meet the space demands; or it can be delivered at variable volume, and the space demands met by the quantity of air delivered. It can be delivered at high pressures and run through pressure-reducing boxes prior to introduction into the space, or at low pressures and connected directly to the air distribution devices. The ducts through which the air passes can be metal with external insulation or sound-absorbing lining, or they may be constructed of fibrous glass material that provides the multiple functions of containing the air, providing insulation, and absorbing sound.

Electrical services in buildings can be delivered at varying voltages, either at higher voltages and transformed at the points of use, or transformed in central locations and delivered through the building at lower voltages. The main electrical distribution can use wires in conduits, or can be delivered through bus ducts to which panels are connected.

Increasingly, electronic technology is being used in conjunction with building services to monitor and control their operation. This technology is becoming a building service in itself, thereby creating what can be described as the "intelligent buildings revolution."[1] This trend focuses on the use of advanced technologies to create integrated systems designed to enhance productivity by controlling the comfort of building occupants; monitoring and modifying environmental quality (particularly indoor air quality); and providing flexibility in occupant services, e.g., communications, computing, and clerical support. An added benefit to owners is the potential energy cost savings, and return on investment gained through improved system operation. This revolution may not necessarily be limited to commercial buildings. The so-called "smart house" describes the residence of tomorrow as a facility that combines modular construction techniques with state-of-the-art telecommunications and microprocessor technology, to produce a safer, more secure, and energy-efficient home.

Another movement influencing building design and building services stresses is the development of "accessible" design that incorporates building features that provide equal access to everyone, including persons with disabilities. By removing the barriers that limit access to many areas of buildings, the need has been created to pay closer attention to the hazards presented to persons who may have limitations on their perception and/or movement.

With so many choices available to designers and installers, it is easy to see why basic fire safety code requirements are often overlooked. To this end, Authorities Having Jurisdiction should participate in the decision-making process. Soliciting their input—early and often—in the design and construction process may prove invaluable. It is important to note that fire and life safety codes are not specifications, but rather prescribe the minimum acceptable practices for ensuring safety. Economy, efficiency, aesthetics, reliability, and maintainability issues not addressed directly by these regulations are important factors in the selection, operation, and maintenance of building services.

While considering each building service and system and their potential hazards separately, it is important to note that each system functions as a component of the larger building system. Hazards associated with individual systems may take on a larger dimension when considered in the context of the building as an overall system. While these individual building systems may represent a heat or energy source, a fuel source, or a fuel or energy transport mechanism, usually the system under consideration fits more than one of these descriptions, as clarified in the following discussion.

Energy Production

Energy production systems are the primary point of contact between building services and the public utilities being introduced into the building. In some cases, they serve as a point where the utility is metered and transformed from one energy level to another. Such is the case with building electrical substations, district steam pressure reducing stations, district chilled water metering stations, and domestic water meter rooms. In other situations, there is a conversion from one energy source to another. This is illustrated by boilers that use oil, gas, electricity, coal, and even wood or waste products to produce steam and hot water; chillers that convert electricity, steam, or gas into chilled water; and generators that convert oil, gas, or steam into electricity.

These systems are typically located in the basements of buildings, but they can often be found in other locations. For example, the electrical substations in One Meridian Plaza, Philadelphia, PA, site of a 1991 fire that cost a third of a billion dollars, were on the 38th floor, and high-voltage risers traversed the entire height of the building. In many high-rise buildings, boilers and chillers are located at the midpoint of the building to keep pressures in piping systems lower and minimize the need for high-pressure piping and equipment.

Heat (energy) sources: The primary heat source in energy production systems comes from fired equipment that burns fuel to convert heat into other energy forms. The combustion process is contained within the firebox of the unit, and presents minimal hazard when properly controlled. Arrangement of the combustion controls in accordance with NFPA 8501, *Standard for Single Burner Boiler Operation,* is important to minimizing the hazard from this equipment.

A less obvious source of heat is from hot surfaces associated with many energy production systems. Hot-water boilers typically heat water to 180°F in low-temperature systems, to 230°F in high-temperature hot-water systems. Low-pressure steam boilers operate up to 15 psig or 250°F, while medium-pressure, 60 psig boilers produce 300°F steam. District steam systems typically operate at 125 psig or 350°F. Pressures are reduced to low-pressure levels in reducing stations located in the buildings served by the system. While the hot surfaces are generally insulated, portions, such as flanges, valves, and other appurtenances, may not be insulated. Maintaining adequate clearances between these hot surfaces and combustible materials is an important design consideration.

Fuel sources: The most apparent fuel sources in energy production equipment are the fuels that are burned in fired equipment and electrical generators. Liquid and gaseous fuels are typically stored external to the building housing the equipment, and the movement of the fuel into the facility is covered under piped services. However, environmental concerns about leakage from underground fuel storage tanks has lead to an increase in aboveground fuel oil storage facilities. In some cases, the tanks are located inside buildings housing the fired equipment. These storage facilities should be arranged and protected in accordance with NFPA 30, *Flammable and Combustible Liquids Code.*

Refrigerants are of concern because they: (1) operate under pressure and (2) can have both flammable and toxic properties. The most commonly used refrigerants are noncombustible and present little or no fuel hazard. They are also only slightly toxic. Most of the refrigerants that are in the flammable or highly flammable categories, and have higher toxicity levels, are only used in special applications and will not be found in most facilities. Table 7-13A lists the properties of common refrigerants in use today.

With the phaseout of chlorinated fluorocarbons (CFCs), as a result of the Montreal Protocol, new refrigerants are being developed to replace those no longer in production. The development of these new refrigerants is causing a balance between providing good refrigeration characteristics while attempting to minimize the hazards of flammability and toxicity. The hazards of these new refrigerants will have to be considered as they are introduced into common usage.

One refrigerant that has both flammable and toxic properties and is occasionally encountered is ammonia (R-717). This is typically used in cold storage applications and ice-skating facilities. The refrigerant exists as a gas at normal temperatures, so that it must be stored in pressurized containers. The main hazard comes from leaks in piping, seals, and valves. Relief vents from refrigeration equipment should be discharged to locations that do not present a hazard to other building areas, occupants, or fire fighters.

Cooling towers also present a source of fuel. The fill used in cooling towers may be noncombustible (metal or ceramic) or combustible [wood or polyvinyl chloride (PVC)]. During normal operation, the towers are flooded and the fuel is not readily ignitable, although there are dry sections of the tower that are exposed and overheated fan bearings and faulty wiring can provide a source of ignition. Many towers do not operate in the winter months when cooling can be achieved by the use of outside air, and all of the combustible portions of a cooling tower provide a fuel source. For more information, refer to Section 7, Chapter 10, "Special Structures," and NFPA 214, *Standard on Water-Cooling Towers.*

Heat (energy source)—fuel transport.   Stacks from fired equipment transfer combustion products to the outside atmosphere. In multistory buildings, they run up through shafts to the exterior of the structure. In these shafts, they present multiple hazards. They must

TABLE 7-13A. *Basic Hazard Data for Some Common Refrigerants*

Classification of Refrigerants	Formula	Boiling Point °F	Boiling Point °C	Calculated Density Gas (air = 1)	Autoignition Temperature °F	Autoignition Temperature °C	Flammable Limits, Percent by, in Air Lower	Upper	Toxicity
Group 1 (nonflammable, except as noted):									
Carbon dioxide (R-74)	CO_2	−109.0	−78.3	1.52					Slightly toxic
Monochlorodifluoromethane (R-22)	$CHClF_2$	−42.0	41.1	2.98	1,170	632	Very weakly flammable		Slightly toxic
Dichlorodifluoromethane (R-12)	CCl_2F_2	−22.0	−30.0	4.17					Nontoxic for ordinary exposure
Dichlorofluoromethane (R-21)	$CHCl_2F$	48.0	8.9	3.55	1,026	552	Very weakly flammable		Slightly toxic
Dichlorotetrafluoroethane (R-14)	$C_2Cl_2F_4$	38.4	3.5	5.89					Nontoxic
Trichlorofluoromethne (R-11)	CCl_3F	74.8	23.8	4.79					Slightly toxic
Methylene chloride (R-30)	CH_2Cl_2	105.2	40.7	2.93	1,139	615	Very weakly flammable		Slightly toxic
Trichlorotrifluoroethane (R-113)	$C_2Cl_3F_3$	117.6	47.5	6.46	1,256	680	Very weakly flammable		Slightly toxic
1,1,1,2-Tetrafluoroethane (R-134A)	$C_2H_2F_4$	−15.7	−26.5	3.51	1,369	743	Non-flammable		Slightly toxic
2,2-Dichloro-1,1,1-Trifluoroethane (R-123)	$C_2HCl_2F_3$	81.7	27.6	5.3	1,317	714	Non-flammable at STP		Slightly toxic
Group 2 (flammable):									
Ammonia (R-717)	NH_3	−28.0	−33.3	0.59	1,204	651	16	25.0	Toxic
Dichloroethylene (R-1130)	$C_2H_2Cl_2$	99–141	37–61	3.35	856	458	9.7	12.8	Moderately toxic
Ethyl chloride (R-160)	C_2H_3Cl	54.0	12.2	2.22	966	519	3.8	15.4	Moderately toxic
Methyl formate (R-611)	$C_2H_4O_2$	90.0	32.2	2.07	869	465	5.0	23.0	Toxic
Group 3 (highly flammable):									
Butane (R-600)	C_4h_{10}	31.0	−0.5	2.01	900–1,000	482-538	1.55	8.6	Slightly toxic
Ethane (R-170)	C_2H_6	−128.0	−88.8	1.04	959	515	3.0	12.5	Slightly toxic
Propane	C_3H_8	−44.0	−42.2	1.52	920–1,120	493-604	2.2	9.6	Slightly toxic
Ethylene (R-1150)	C_2H_4	−154.8	−103.8	0.98	914	490	2.7	36.0	Slightly toxic

be enclosed in rated construction to prevent the spread of fire and smoke, under fire conditions. The stacks must be adequately insulated to minimize buildup of heat within the shafts. They also must be built with tight construction to minimize the possibility of leakage of toxic and combustible gases into the building.

Environmental Systems

Environmental systems control air movement in buildings to regulate temperature and maintain safe, comfortable, and livable environments. While this involves moving large quantities of air in the building, the amount of air moved by any one system varies considerably. Systems, such as small unitary air conditioners and small exhaust fans, may only move a few hundred cu ft/min of air, and the overall building system incorporates a large number of them. On the other hand, systems may consist of large central air-handling equipment that moves tens of thousands cu ft/min.

Systems may be used to exhaust odors or heat from building areas, or they may be used to supply outside air to replace that removed from the building. The major concern is in recirculating systems that take air from building spaces, clean it to remove particulate or other contaminants, dilute or exchange it with outside air, heat and cool it, and then return the majority of it to the spaces served to provide comfort conditioning.

Several NFPA standards describe construction and installation standards for systems and their components, to minimize the contribution of environmental services to building fire problems. These standards incorporate techniques intended to: (1) control fuels by regulating the materials used in the construction of these systems; (2) control heat sources by prohibiting installations in hazardous locations; and (3) limit heat (energy) source and fuel interactions, by separating and enclosing system elements, and minimizing the transport of products of combustion through the system, among other methods. In some extreme cases, or in lieu of other methods, automatic fire suppression systems are required to control the development, and/or spread, of fires involving these systems.

Heat (energy) sources: Environmental services involve rotating machinery in an air stream so that heat sources from bearings and electrical connections are always present. Systems that heat air also involve a heat or energy source. Generally, the source of heat is from hot water or steam coils that provide heat generated by energy production equipment. However, many types of heating equipment involve the fired heaters located in the air-handling units. These heaters may be in many different forms:

1. Gas-, oil-, coal- or wood-fired furnaces for residential or small commercial applications
2. Fired unit heaters suspended from roof structure for industrial and warehouse applications
3. Fired radiant heaters for heating of large spaces
4. Fired roof-mounted equipment for heating and air-conditioning of commercial and industrial applications
5. Fired roof-mounted equipment for providing large quantities of makeup air to spaces

The type and location of the energy source selected is a function of the local climate; fuel availability; fuel cost; and building environment, use, and occupancy. Most equipment is indirect fired, wherein the energy source is contained within the firebox and heat is transferred to the air through the use of heated surfaces. In some applications, such as direct-fired air makeup units or construction site heaters, the fuel is fired directly into the moving airstream.

The function of these heaters is to raise the air temperature to a modest, comfortable level. Heated, convected air, then, would only represent a potential ignition source under the most extreme circumstances. Exceptionally few substances ignite under these conditions, and this reaction is usually pyrophoric, related more to the oxygen content of the air than its temperature. However, many substances, unlike air, respond vigorously when they come into direct (radiation or conduction) contact with the energy source itself. Metal flue pipes and ductwork near or in direct contact with combustible structural elements, electric baseboards, or gas-fired unit heaters in the vicinity of flammable vapor-air mixtures, or steam radiators in contact with an unprotected human hand, may generate unfortunate results.

The location of the energy source and the type and location of potential fuels, then, must be carefully considered. Fuels present in the immediate environment should always be a prime consideration. Whenever flammable vapor-air mixtures, combustible dusts, or fibers are present, then energy sources capable of igniting them should not be. Designers should specify only indirect heating systems, or direct heating systems designed and approved for hazardous locations, when flammable or explosive gases, vapors, dusts, or fibers are present in the atmosphere. Likewise, combustible building elements should be protected or separated from system components, such as metal ductwork and flue pipes. Many tragic fires have resulted from failure to maintain clearances between building elements and such components.

Fuel sources: Ductwork, duct connectors, duct coverings, linings, vapor barrier facings, tapes, equipment panel, plenums, and filters are among the components of environmental systems, where the use of combustible construction is carefully controlled. However, these systems accumulate combustible materials commonly suspended in air. Dust, dirt, and debris picked up in the systems can coat ductwork, infiltrate linings, and become lodged in filters. They become potential fuel sources that can spread flame and smoke throughout a building, should they be ignited.

Limited combustible materials used to cover or connect ductwork must have a flame-spread rating of less than 25 (when tested in accordance with NFPA 255, *Standard Method of Test of Surface Burning Characteristics of Building Materials*/ASTM E84), with a smoke-developed rating of less than 50. Likewise, combustible materials may not extend through fire-resistance-rated assemblies, or interfere with the operation of fire dampers or fire doors.

Filters are commonly installed in systems to trap combustible matter suspended in air. While filters may not be combustible, their contents usually are. For this reason, filters must be replaced or removed and cleaned regularly. The use of plenums for purposes unrelated to air movement, such as concealment of piping, wiring, and other building services, must not interfere with their use as a plenum, or increase the fire hazard. Therefore, combustible materials must possess low flame-spread and smoke-developed ratings, and most spaces containing materials of this nature must be draft-stopped or protected by automatic sprinklers.

Heat (energy) source—fuel transport: Perhaps the greatest danger associated with environmental systems is that they are usually hidden from view. Located behind walls, above ceilings, and under floors, ducts travel throughout the building to provide and maintain a controlled environment. Large concealed spaces, e.g., plenums and shafts, are also used to permit the movement of air horizontally and vertically. Recognizing the hazards of moving air, people, and fire in common spaces, codes and standards, including NFPA 90A, *Standard for the Installation of Air Conditioning and Ventilating Systems*, have restricted or prohibited the use of means of egress components as return-air plenums.[2] This has forced designers to use other spaces for the movement of this air. While resulting in an improvement in life safety, a more complex fire problem may be created. Fires in concealed spaces can easily grow and travel undetected for long periods of time. To control these potential avenues of fire and smoke spread, vertical shafts must be segregated from the remainder of the building by fire-resistance-rated construction. Large concealed spaces are usually draftstopped to maintain manageable fire areas. However, this cannot be done when the space is used to move air. Therefore, combustibles in unsprinklered plenums and horizontal concealed spaces must be eliminated.

Even when concealed spaces are not used for air movement, the ductwork and air-handling equipment is often concealed. Installation of fire dampers, smoke leakage-rated dampers, or fire doors where ductwork penetrates fire-resistance-rated assemblies can control this potential avenue of fire spread. Additionally, the use of automatic system controls such as smoke detectors in supply, return, or supply and return air plenums, and arranged to stop fans, are required by NFPA 90A. However, shutting down fans alone may not always control the movement of smoke within a building or fire area. Phenomena, such as stack effect, buoyancy, and wind, affect the spread of smoke through supply and return air ducts, air shafts, plenums, and other building openings.[3] In these instances, active smoke control measures should be implemented to control the movement of smoke.

Piped Services

Piped services collect or distribute fluids, usually liquids or compressed gases, but can also include solid/liquid combinations. Common piped distribution systems include water, fuel gases, fuel oil, compressed air, medical gases, and vacuum and sanitary sewer lines. Hazards associated with these services involve the following: (1) the ability of the piping or conduit to conduct heat or electric current, including static electricity; (2) combustibility of the fluid, the conduit, or both; (3) the ability of the fluid to support combustion; or (4) the ability of the system, or the space in which it is installed, to provide an avenue for fire and smoke spread.

Heat (energy) sources: Piped services transporting steam, hot water, or heated gases can generate sufficient energy to dry nearby combustible structural members or building materials, thereby making them susceptible to ignition. Metallic piping systems are usually good conductors of heat and electricity. For this reason, many piped services associated with indirect heating systems are insulated to prevent heat loss. This may make them good conductive ignition sources when exposed to external fires. Perhaps more dangerous, however, is the likelihood of ignition of flammable contents when metallic piping is not properly grounded during flammable liquid transfer operations. Whenever sections of pipe carrying a flammable product are mechanically and electrically isolated, each individual section must be independently grounded, or bonded, to another grounded section.

Fuel sources—contents: Piped services containing flammable liquids and flammable compressed gases are a widely recognized and well-regulated hazard. NFPA 54, *National Fuel Gas Code* (hereinafter referred to as NFPA 54), and NFPA 58, *Standard for the Storage and Handling of Liquefied Petroleum Gases*, regulate storage and distribution of the common fuel gases. Both standards are widely recognized and adopted. NFPA 30 and NFPA 31, *Standard for the Installation of Oil-Burning Equipment*, specify requirements related to piped distribution of flammable and combustible liquids for nearly all applications. NFPA 99, *Standard for Health Care Facilities*, however, prohibits the piped distribution of flammable anesthetics and other flammable medical gases. The design and planning of piped systems containing flammable compressed gases must consider the demands and location of the equipment supplied. In most cases, cooking, heating, medical, or laboratory equipment

is involved. Keeping runs of flammable liquid and compressed gas piping short and direct satisfies the objectives of simplicity and economy, and can contribute to safety, as well. Clearly identifying pipe contents and valves minimizes the likelihood of inadvertent release. Valves critical to isolating or interrupting the flow of gas, in the event of inadvertent release, should be accessible and clearly marked. Charts and diagrams indicating the locations of valves, as well as the systems and areas controlled, should be readily available at all times.

Fuel source—conduit: The widespread use of plastics for distribution and collection piping has focused interest on fire and toxicity hazards of piped services. Interest has been particularly keen in residential occupancies where both the fire death toll is the highest, and use of plastic plumbing fixtures, and drain, waste, and vent (DWV) piping has been accepted the longest. These products include materials fashioned from acrylonitrile-butadiene-styrene (ABS), polyvinyl chloride (PVC), chlorinated polyvinyl chloride (CPVC), and polybutylene (PB), and are widely used throughout the building industry for domestic water, automatic sprinkler, and DWV, among other uses. Services using these materials can be subject to failure when exposed to fire, and can become involved themselves. Many plastic materials, though listed as self-extinguishing, present a fuel hazard when placed under fire conditions; they may greatly add to the smoke development and increased fuel contribution to the fire. Thus, use of these materials is controlled in larger structures and not permitted in plenums handling environmental air. When these services are installed in concealed spaces, automatic sprinkler protection may be advisable, and possibly mandatory if these materials are installed in substantive quantities.

The contribution of these materials to fire hazard has been widely investigated from as early as 1966. Much of the attention has focused on the ability of these materials to propagate fire through fire-resistance-rated assemblies. With the introduction of through-penetration firestop assemblies, much of this concern may also be receding. Zicherman, reviewing the history of the plastic pipe debate, summarizes a discussion of the fire hazard of these materials, as follows:

> Rarely have products used in fire-resistive construction been subjected to as extensive an evaluation as PVC-based piping products. The results of this testing and evaluation, as well as many years of field use, indicate that properly installed PVC-based products do not constitute a fire hazard in structures.[4]

Heat (energy) source—fuel transport: Although fires spreading through piped service conduits themselves may be an infrequent occurrence, structural collapse of these services may permit fire to compromise fire barriers. (See Figure 7-13A.) Like environmental services, piped services travel in concealed spaces, thereby making fire detection and control difficult. Unlike environmental systems, though, some piped service through-penetrations cannot be protected by simple opening protectives, such as fire doors and dampers. Concern over the contribution of pipe and wire penetrations in fire-resistance-rated assemblies to the failure of these assemblies to control horizontal and vertical fire spread has led to new requirements for through-penetration firestopping systems. Most of these systems are either chemical firestops or pipe sleeve protection systems, employing a variety of new materials and technologies to seal the voids around such openings. (See Figures 7-13B and 7-13C.)

The development of through-penetration firestop systems began following the Brown's Ferry Nuclear Power Plant (near Decatur, Alabama) fire in which unprotected cable penetrations allowed the spread of fire and smoke throughout the plant. The first systems developed consisted of intumescent silicone foams that found wide acceptance in the construction industry. Intumescent tapes, foams, putties, and caulks, introduced in the 1980s, used independently or in combination with inorganic insulation materials and retaining collars or pipe sleeves, have become popular solutions due to their flexibility and ease of installation. Today, there are hundreds of systems listed covering horizontal and vertical penetrations.

In addition to the chemical and sleeve protection designs, simple mechanical designs have been developed for application in plastic pipe through-penetration situations.[5] (See Figures 7-13D and 7-13E.) These systems are not widely used in the field, due to their more involved construction and installation requirements.

Regardless of the technique employed, through-penetration firestop systems are finding their way into new and amended code

requirements, referencing ASTM E814, *Standard Test Method for Fire Tests of Through-Penetration Fire Stops.* (See also UL 1479, *Standard for Safety Fire Tests of Through-Penetration Fire Stops.*) These improved methods of protecting openings around pipes are expected to reduce the problem of fire spread and smoke movement between fire areas and in concealed pipe chases and shafts. In addressing the subject of PVC piping and fire safety in detail, it is noted that extensive testing, using ASTM E119, *Standard Test Methods for Fire Tests of Building Construction and Materials,* (NFPA 251, *Standard Methods of Tests of Fire Endurance of Building Construction and Materials*) and ASTM E814 (UL 1479) test methods, suggests that these piping systems and materials present little threat to the integrity of fire-resistance-rated assemblies when proper construction techniques and through-penetration firestop systems are used.[4]

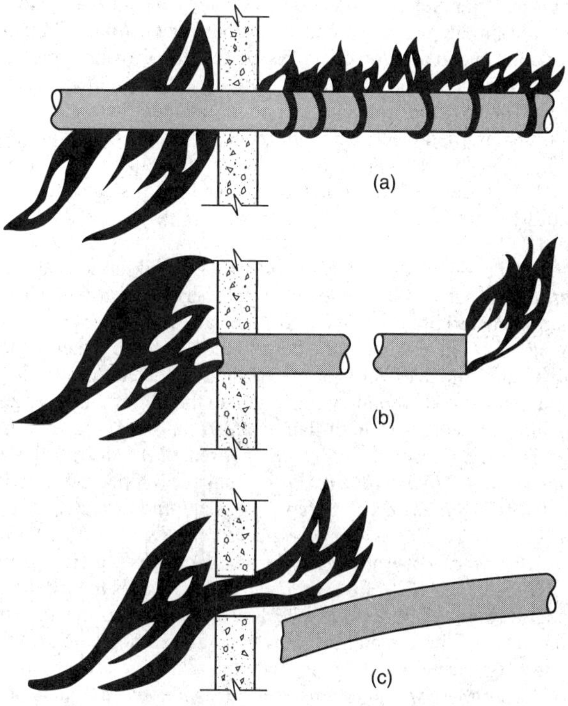

FIG. 7-13A. Fire hazards associated with plastic pipe.

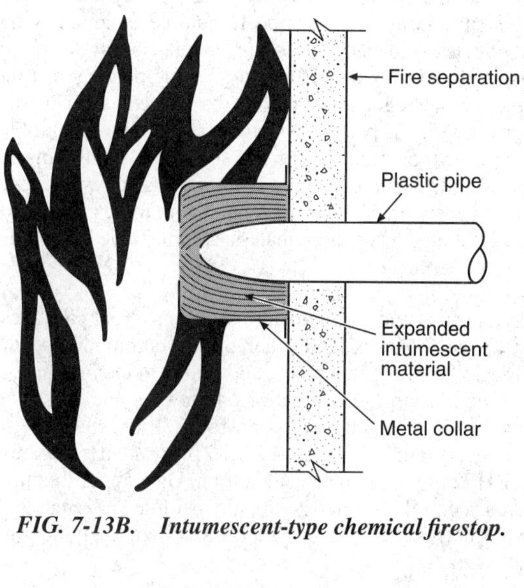

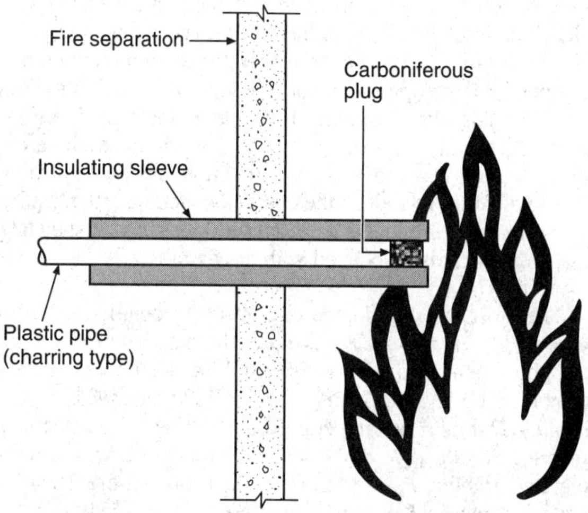

FIG. 7-13C. Protective sleeve for PVC pipe.

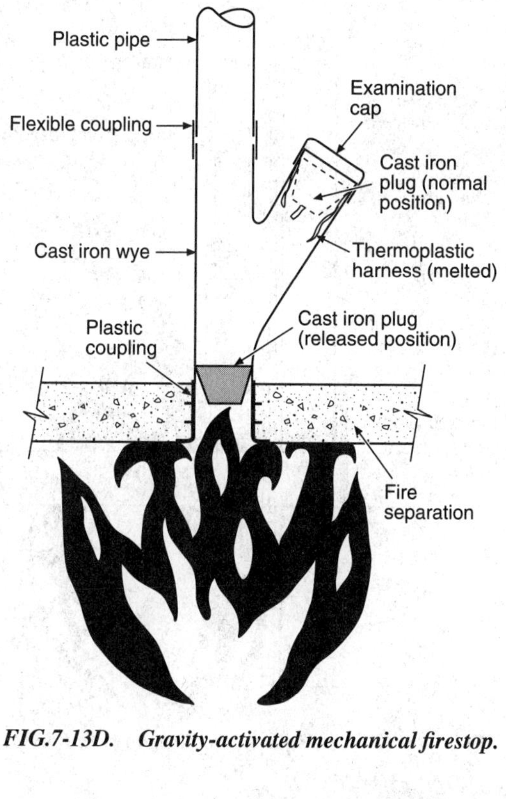

FIG. 7-13B. Intumescent-type chemical firestop.

FIG.7-13D. Gravity-activated mechanical firestop.

Electrical/Electronic Systems

Electrical systems supply electric voltage to power lights, motors, appliances, and other conveniences, or convey signals to controls, or communicate with other people or machines. As such, they are also integral with the other building service systems discussed in this chapter. Electrical conductors and equipment may also be installed to collect and safely dissipate static electricity, radio frequency interference, and other sources of noise.

A major use of electrical systems in buildings is for telecommunications. With a telecommunications revolution well underway, the need to access and control information has led to increased dependence upon electronic technology. Associated with this technology is an immense amount of wire. This has created a challenge in the area of fire protection. It is important to consider the following fire protection concerns associated with these changes: (1) wire and cable insulation, and the equipment itself, represent a potential fuel load; (2) installation of telecommunications equipment commonly accompanies large open areas or a reduction in building compartment; (3) unsightly communications cable and wire is frequently concealed; and (4) the loss of service from a telecommunications system, in many cases, represents a business' most significant interruption exposure. To reduce the risk associated with the last item—business interruption—adequate, appropriate steps must be taken to manage the hazards associated with the first three.

The most widely adopted of all NFPA documents, NFPA 70, *National Electrical Code®*, regulates the hazards of electrical devices, equipment, and wiring. Other NFPA documents in wide usage include NFPA 75, *Standard for the Protection of Electronic Computer/Data Processing Equipment;* NFPA 77, *Recommended Practice on Static Electricity;* and NFPA 110, *Standard for Emergency and Standby Power Systems.* The principal fire hazard associated with electrical systems is their potential as an ignition source. Resistance heating and electrical arcs, incidental to the operation of electrical devices, equipment, and appliances; improper control of

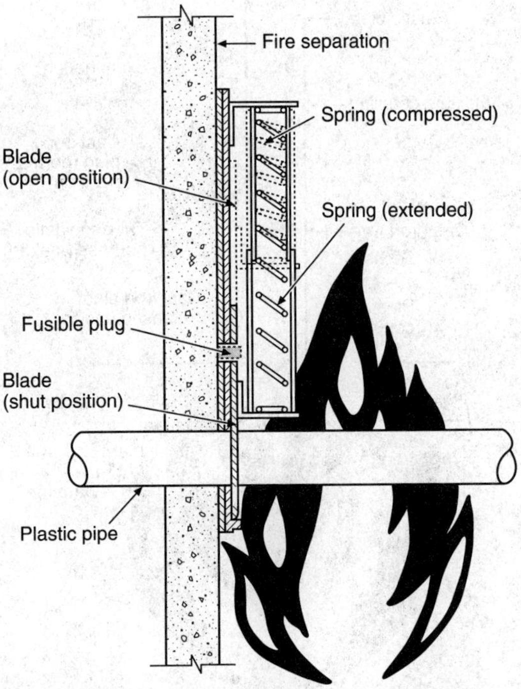

FIG. 7-13E. Spring-activated mechanical firestop.

static charges; and faulty installation or maintenance are significant causes of accidental fires.

Heat (energy) sources: Proper design of electrical installations must consider the atmosphere involved, the loads connected, the anticipated needs of building service equipment and building occupants, and provision of adequate clearances for maintenance and safety. Many electrical and electronic system components produce concentrated or localized heat loads. The following are listed under this heading: (1) switching devices, including fuses and circuit breakers; (2) distribution equipment, such as transformers, panelboards, switchboards, and motor controls; and (3) utilization equipment and devices, such as lighting, resistance and induction heaters, and motors.[6] Care must be taken to maintain clearances from combustibles. Widespread use of aluminum wiring has raised concerns about the hazards associated with this material and common installation and maintenance practices. Greater care must be exercised with aluminum wiring to ensure proper connections. Damage or poor contact can develop due to formation of corrosion—particularly where copper and aluminum conductors or contacts meet—or "creep" due to cold-flow of the conductors under heavy loads. Arcs from unclassified electrical equipment and static sparks are potential ignition sources when flammable liquids and gases are involved. NFPA 70 specifies requirements for electrical installations in hazardous locations.

Fuel sources—electrical systems: Like piped services, the use of plastics has been a boon to the electrical service industry. Plastic insulation materials, conduits, and raceways are coming into widespread use. Concern about the contribution of these materials to fire growth and smoke toxicity has been widely debated. As a result, new and amended code requirements regulate the use of wire, cable, and conduit materials with high-flame spread, or smoke-developed ratings. NFPA 262, *Standard Method of Test for Fire and Smoke Characteristics of Wires and Cables* [originally developed and published as UL 910, *Standard for Safety Test for Flame-Propagation and Smoke-Density Values for Electrical and Optical-Fiber Cables Used in Spaces Transporting Environmental Air*—an adaptation of the NFPA 255 (ASTM E84) Steiner Tunnel Test], is the most widely recognized test method used to evaluate the fire hazard and smoke production of these materials. However, as yet, no standard screening test has been adopted to evaluate the toxicity of combustion byproducts of these insulation materials. Despite this, the State of New York adopted legislation requiring manufacturers who wish to sell electrical conduit and electrical wire insulation; pipe, duct, and thermal insulation; and interior wall, ceiling, and floor finishes to submit toxicity data for inclusion in a state-wide toxicity data bank. Manufacturers submitting data have been required to submit samples of their products to independent testing laboratories for destructive testing, using the University of Pittsburgh toxic potency test method.[7] This pioneering regulatory activity may be expected to spread as concern grows in the private sector over common workplace hazards. Concern over the toxicity issue has stimulated continuing research efforts related to this question. These efforts focus on developing ways to quantify the hazards associated with toxic products of combustion generated when these materials burn. Table 7-13B illustrates some of these scenarios.

In an office telecommunications systems, fuel packages represented by workstations and other plastic commodities, including telephones, facsimile machines, and photocopiers, represent a significant fire hazard. Investigations following the First Interstate Bank fire in Los Angeles, CA, on May 4, 1988, illustrate some of these points. Figures 7-13F and 7-13G illustrate the rate of heat release (RHR) of a computer workstation. Data about this fuel package was central to the National Institute of Standards and Technology (NIST)/Code of Federal Regulations modeling efforts

TABLE 7-13B. *Example of Potential Electrical Fire Scenarios That May Produce Smoke Toxicity Hazards* [8]

Fire Scenario & Electrical/ Electronic Component(s) Involved	Potential Impacts
Electronic component fails and burns	1 to 2 persons in immediate vicinity exposed for several seconds
Switchgear fails and burns	At most, several persons in or near electrical closet /control room exposed for minutes
Larger electrical control(s) or building overloads/heats	Depending on paths for smoke travel, may expose total building population for minutes during escape/evacuation
Wall cavity wiring or receptacle overloads/heats	One or more residents of home may be exposed for minutes
Electrical wiring/tubing in wall cavity involved by room fire	Possible increase in smoke hazard to occupants in adjacent space for minutes during escape
Computer/communications cables in plenum involved by room fire	Possible increase in smoke hazard to perhaps large fraction of building occupants exposed for 5 to 15 min or so
Wiring concentrations in electricity/utility tunnels, etc., involved by fire	Exposure of building occupants for duration of event and evacuation process

for the First Interstate fire, and correlated well with available evidence. Using the detailed investigatory record and engineering calculations, Nelson and other NIST researchers concluded that concentrated groupings of computer workstations, combined with low ceilings, led to flashover conditions.[9] It was suggested that similar circumstances involving fuel and room geometry could lead to similar results. These findings further suggested that traditional assumptions about the light-hazard nature of office environments may need to be reconsidered.

The role of telecommunications wire and cable in significant fires is less conclusive. Certainly, combustible insulation materials can degrade and even burn when subjected to internal heat from overloading or external heat from a fire. Concern about the contribution of combustible insulation materials focuses on two hazards: (1) fuel and (2) toxicity. As a fuel, wire and cable sheathed in combustible insulations, usually plastic and, in particular, PVC, bundled in risers and cable trays, above ceilings, and in air plenums has been a significant concern. In recent years, however, new requirements mandating use of fire-resistant insulation materials have abated much of the concern about flame spread associated with these materials. Likewise, concern about the smoke produced by insulation materials, especially in supply and return air plenums, has been addressed by new requirements for low-smoke-producing wire and cable products. These changes are intended to supplant concern about PVC insulation. PVC, unlike new insulating materials containing polytetrafluoroethylene (PFTE) and other heat-resistant materials, produces hydrogen chloride when it undergoes pyrolysis. Telecommunications, signaling and control, optical-fiber, and coaxial antenna wires and cables carry little or no current or voltage. Consequently, when properly designed, installed, and maintained, these circuits represent little hazard of self-ignition due to overheating from misuse or damage. This combination of power-limited application and fire-resistant/low-smoke-producing insulation promises to maintain the fire hazard of new wire and cable installations within a manageable and generally acceptable level. (See Table 7-13C.)

Heat (energy) source—fuel transport: The mechanism of fuel transport in electrical and electronic systems is much the same as in piped services. Therefore, that discussion should be referred to for this aspect of electrical systems.

Electric service failures: While not in the framework of the discussion of this chapter, the impact of electrical and electronic systems failures can be monumental. Nearly all building services discussed in this chapter depend on electric service for safe, efficient operation. NFPA *101*®, *Life Safety Code*®, requires emergency power for exit signs and means of egress lighting. NFPA 72, *National Fire Alarm Code*, also requires an emergency power supply for the fire alarm system when required by another code or standard. ASME A17.1, *Safety Code for Elevators and Escalators*, requires at least one car in each bank of elevators to be supplied with

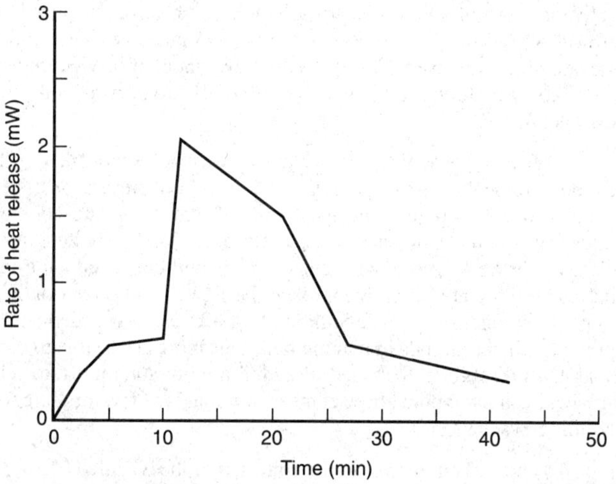

FIG. 7-13F. *Rate of heat release at a typical work station.*

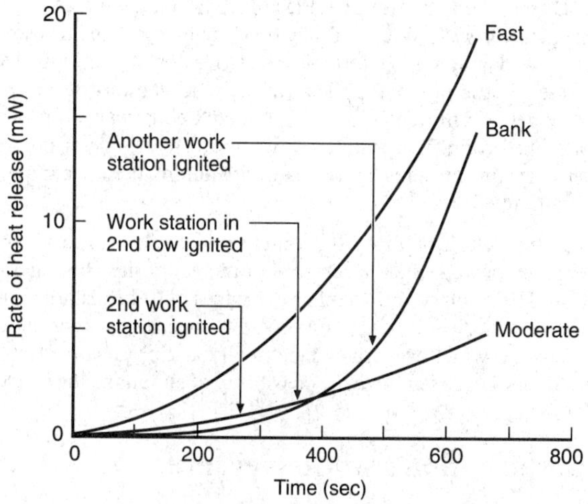

FIG. 7-13G. *Estimated free-burn rate of heat release of groups of work stations on the 12th floor.*

TABLE 7-13C. Cable Markings

Cable Markings	Type
CL2P	Class 2 plenum cable
CL2R	Class 2 riser cable
CL2	Class 2 cable
CL2X	Class 2 cable, limited use
CL3P	Class 3 plenum cable
CL3R	Class 3 riser cable
CL3	Class 3 cable
CL3X	Class 3 cable, limited use
PLTC	Power-limited tray cable
NPLFP	Non-power-limited fire protective signaling circuit plenum cable
NPLFR	Non-power-limited fire protective signaling circuit riser cable
NPLF	Non-power-limited fire protective signaling circuit cable
FPLP	Power-limited fire protective signaling circuit plenum cable
FPLR	Power-limited fire protective signaling circuit riser cable
FPL	Power-limited fire protective signaling circuit cable
MPP	Multipurpose plenum cable
CMP	Communications plenum cable
MPR	Multipurpose riser cable
CMR	Communications riser cable
MP	Multipurpose cable
CM	Communications cable
CMX	Communications cable, limited use
CMUC	Undercarpet communications cable
CATVP	CATV plenum cable
CATVR	CATV riser cable
CATV	CATV cable
CATVX	CATV cable, limited use

emergency power, with a capability available to manually transfer power to other cars in each bank. The arrangement of such systems must minimize the possibility that the emergency power supply will be disabled by a fire, or other emergency, involving the primary power supply. All electrical and electronic components are subject to resistance heating, and can eventually fail. Therefore, consideration should be given to reducing hazards from such failures.

During the One Meridian Plaza Fire (Philadelphia, PA)—that burned for nearly 19 hours, killing three fire fighters, and destroying or heavily damaging 8 floors of a 38-story downtown high-rise— separate electric utility supplies from two substations became disabled when the fire entered an unprotected electrical room on the floor of fire origin. This resulted in disruption to the operation of fire pumps, lighting systems, and elevator operation, thereby hampering fire-fighting efforts.[10]

Article 700 of NFPA 70 permits separate electric services for emergency power when acceptable to the Authority Having Jurisdiction. The primary and secondary services must be installed in accordance with Article 230, with separate service drops, or laterals, and must be widely separated electrically and physically from the normal service to minimize the possibility of simultaneous interruption of supply.

Transportation and Conveyance

Transportation and conveyance systems are those systems provided to facilitate the vertical or horizontal movement of people and materials, bulk or packaged. Shafts and horizontal enclosures for build-

ing service ductwork, piping, and conduit are also included in this category.

Vertical shafts are common in most buildings with stories more than one level above or below grade. These communications perform a variety of functions, including a supply or return air path; a route for piping, wire, or cable; or a means for conveying trash, linen, mail, and other materials from operating areas to service areas. Although these vertical shafts are essential, they present a predictable and persistent fire hazard. Fundamental problems in any fire, e.g., controlling the travel of smoke, heated gases, and flame, are complicated by such communicating spaces. Preventing vertical spaces from contributing to fire spread requires forethought and planning. Fortunately, concern about the contribution of vertical spaces to the transporting of heat and smoke has prompted considerable research and code attention.

Perhaps no other building service has had as significant an impact on the building environment as the elevator. Without a safe, efficient means of moving people vertically in buildings, the modern urban skyline would look much different than it does today. This innovation has also made even buildings of moderate height more accessible to those with temporary or permanent mobility impairments. Like other building services, great debate has centered on the safety of elevators, escalators, and other conveyance systems during fire emergencies. Two dangers are commonly associated with these building services: (1) their association with fire and smoke transport and (2) dangers to persons who may use elevators during fire emergencies. Fire-resistance-rated separation assemblies, automatic fire suppression, smoke management systems, good housekeeping, and management can mitigate the effects related to the first problem. However, the solution to the second problem remains a topic of significant discussion in fire protection and mechanical engineering communities.

Increased sensitivity and attention to the legitimate needs and rights of people with mobility impairments have led to widespread support and acceptance of barrier-free design. A principal component of this design strategy is the installation of special lifts and elevators in multistory buildings. While barriers to ingress are gradually being eliminated through the integration of barrier-free concepts, a major hurdle remains: barrier-free egress. Once mobility-impaired occupants get into a building, the concern arises as to how they get out when fire occurs. ASME A17.1 provides requirements for automatic and manual recall of elevator cars to the designated floor (usually the level of exit discharge or primary level of fire fighter access) during fire emergencies that may threaten elevator occupants. Phase I fire fighter service is, therefore, initiated upon activation of automatic detectors (smoke or heat, as required in ASME A17.1) in elevator lobbies, the elevator machine room, or the elevator hoistway itself. This takes the elevator out of service for use by disabled building occupants, until Phase II service (manual control) is initiated.

NFPA *101*, provides that an elevator must not be considered an element of the means of egress, and the model fire prevention codes and ASME A17.1 require the installation of signs at elevator lobbies directing occupants to use stairs, not elevators during fire emergencies. Ensuring a safe environment for all building occupants requires an integrated design approach. Designers and code officials must work together to ensure all building occupants are adequately protected. These trends in making buildings more accessible to persons with disabilities has lead to additional investigation into this subject in an attempt to arrive at means for making elevators safe for building evacuation.[11]

To date, when interest in the concerns of the disabled has been translated into action, the predominant approach has been to move those persons with mobility impairments to areas of refuge, or "areas of evacuation assistance," to wait for rescuers. In some cases, emergency planning has focused on occupant assistance to persons with disabilities. Meanwhile, persons without mobility impairments are free to descend the stairs and exit out the doors to safety. As a result of the efforts of life safety experts and design professionals dedicated to barrier-free design and advocacy for persons with disabilities, increased attention is once again focused on this concern. The question arises as to whether elevators and lifts will become accepted components of barrier-free egress systems in the future. While there is no definitive answer, it is certain that, as long as tragedies associated with well-intentioned, but poorly reasoned use of elevators during fire emergencies persist, this solution will be resisted by many.

Investigators concluded that the sole fatality in the Los Angeles First Interstate Bank Building fire of 1988 was killed by exposure to fire products when he took an elevator to the floor of fire origin.[12] Preliminary reports from the One Meridian Plaza fire (Philadelphia, PA) indicate that a similar tragedy nearly occurred there. Had another building occupant (a guard in the building lobby) not heard the building engineer's call for help over his portable radio and initiated Phase II fire fighter service, the survivor might have been yet another victim of a poor decision.[13]

Heat (energy) sources: Like other mechanical systems, the main source of heat in elevators and other conveyance systems is the electrical energy used to drive the mechanical equipment. This source is usually located in enclosed machinery rooms, and is present in both the power and motor circuitry and the relay or electronic control equipment. The industry trend is to more electronic control of elevators, resulting in greater heat being generated from the elevator controls. As this occurs, it is increasingly important that ventilation of machine rooms is kept adequate to remove the heat generated by the equipment, thereby reducing the probability of breakdown of electrical circuitry.

Electrical portions of conveyance systems may not always be confined to machine rooms. Escalator motors are located in pits at the base of the moving stairs. Mail-handling systems using motorized containers operate on tracks that travel through enclosed ceilings and shafts to all parts of the buildings served. Electrically operated switching equipment can be located in many areas to direct the containers to their proper destination.

Fuel sources: In addition to plastics associated with electrical equipment, there are other combustible materials associated with conveyance systems. Elevators and escalators involve the use of large quantities of lubricants to keep elevator cars moving smoothly on their guideways, and escalator drive mechanisms operating smoothly. These lubricants are combustible and provide a ready mechanism for collecting dirt, which increases their combustibility. Additionally, they are located in shafts and other concealed portions of buildings.

Another source of fuel is the debris that often gathers in elevator pits. These out-of-sight locations become gathering points for materials that fall out of elevators or are introduced by mechanics. Also, mail-conveying systems that travel through buildings carry significant amounts of combustible materials through concealed areas. Materials that fall out of the systems may remain for many years without being discovered.

Heat (energy) sources—fuel transport: Like any other vertical communication, elevator and dumbwaiter hoistways may be significant avenues of fire and smoke spread, due to stack effect. Similarly, floor openings for escalators may provide an ideal path for fire and smoke spread. The doors used to enclose these floor openings are not tightfitting, due to the need for them to operate under a vari-

ety of conditions. Building pressure differences due to stack effect and air-handling systems can greatly affect the operations of these doors, interfering with their need to be tightfitting fire separation assemblies.

Enclosure of elevators and machine rooms with fire-resistance-rated assemblies; self-closing fire doors at lobbies; ventilation or pressurization of elevator hoistways and machine rooms; installation of automatic sprinklers in elevator machine rooms and shafts; and automatic fire detectors in lobbies, shafts, and machine rooms are all intended to limit fire and smoke spread in these spaces. Escalators can be protected in a number of ways, including complete or partial enclosure by walls, partitions, draftstops or draft curtains, rolling shutters, or moving partitions; automatic sprinklers; and water-spray fixed systems. Code requirements vary regarding which features are required, and designers often have considerable latitude to combine different protection techniques. This ensures that nearly all installations are protected in some way.

Vertical shafts, chases, and chutes, regardless of their purpose, must be segregated from other spaces by fire-resistance-rated enclosures of appropriate fire endurance. Successfully containing a fire and its by-products requires adequate fire-resistance to protect the area exposed and the structural integrity of the assembly itself, in concert with an automatic sprinkler system. Construction of a fire-rated assembly must, therefore, consider the rating required, continuity of the assembly, the structural systems involved, and the function of the spaces on either side of the assembly. NFPA 251 (ASTM E119) provides the standard method of determining the fire endurance of walls, partitions, floor/ceiling assemblies, and roof/ceiling assemblies.

The greatest problem affecting the performance of fire-resistance-rated enclosures may be continuity. Two circumstances repeatedly contribute to failure of fire-resistance-rated enclosures: (1) failure to protect through-penetrations adequately, especially those created after occupancy; and (2) overlooked construction deficiencies. Through-penetration firestop systems and devices are commercially available. When properly selected and installed, they can provide proven protection where ductwork, piping, wire, cable, and other devices or systems must pass through fire-resistance-rated assemblies. These systems or devices may employ heat-passive protection using generic construction materials, such as mortar, grout, and thermal insulation. They may also employ active systems and devices that typically consist of proprietary, heat-actuated chemical and mechanical systems and devices. Mechanical devices in this category include fire dampers and fire doors, as well as guillotine-style and gravity-actuated plug devices for plastic pipe. Chemical systems and devices usually employ intumescent technology to seal irregular openings created around wire, cable, and plastic piping when they collapse or fail during a fire. Construction deficiencies can include inadequate structural fire-resistance supporting rated wall or floor assemblies, non-rated construction materials, substitution of unapproved materials in rated systems and assemblies, and incomplete construction or hidden voids.

The structural system itself must be a prime consideration when designing fire-rated enclosures. A structure supporting a fire-resistance-rated enclosure must remain capable of supporting the enclosure assembly under fire conditions. If the structure involved supports other building elements, these loads must not compromise the integrity or endurance of the fire-rated enclosure. Too frequently, the fire-resistance-rated assemblies are considered separately from the structural system involved. Passive and active fire protection features, like other building services and systems, are often imposed upon the basic architectural and structural design, after the fact. This arrangement can too easily result in construction and maintenance difficulties that will ultimately compromise either passive or active approaches.

The use of the shaft itself is the last, and perhaps most apparent, factor involved in the selection and construction of a fire-resistance-rated enclosure. It must be determined on which side is the hazard and the exposure. Under most circumstances, concern focuses on controlling fires originating outside the shaft that could void the protection and thus spread vertically through these spaces. However, some vertical spaces, particularly service chutes (e.g., linen, laundry, and trash) should be regarded as potential areas of fire origin.

Cranes, lifts, and mechanical and pneumatic conveyors are commonly used in industrial and storage occupancies to move bulk or packaged materials primarily horizontally and relatively short distances vertically. Modern warehouses increasingly turn to labor-saving, computer-controlled lifts and cranes to automate warehousing operations. Protecting horizontal openings in fire barriers, through which such devices pass, is a significant concern. Means must be provided to permit opening protectives to operate, without interference or obstruction from either the stock or the conveyor. Common methods of protecting fire-wall openings include providing mechanical or electrical interlocks on conveyors or installing water-spray fixed systems. Even with these protections, burning objects and smoke can readily pass through these openings, as evidenced in the K-Mart fire (Falls Township, PA). Burning aerosol containers became projectiles that spread the fire throughout the warehouse structure, resulting in a total loss of the building.

OPERATION AND MAINTENANCE

Fires involving heating or electrical distribution equipment, together account for about one-fourth of all structure fires reported to U.S. fire departments.[14] These systems were not designed to start fires or fail. Design flaws, faulty products, and poor installation are certainly responsible for their share of fires. However, improper operation, maintenance, or alteration of mechanical and electrical systems represent significant causes of accidental fires. Proper maintenance and conscientious management of building services can significantly reduce the fire threat in both new and older existing buildings.

Building owners, occupants, insurers, and code officials working together can significantly reduce the fire problem due to operational and maintenance problems. An innovative program run by Mead Loss Control Consultants, Inc., of Dayton, OH, regularly surveys risks covered by the Mead Corporation's captive insurance organization for electrical and mechanical hazards using an infrared scanner and videotape equipment. "Hot spots" detected by the infrared equipment, e.g., worn bearings, faulty transformers, or loose electrical connections, are scheduled for immediate preventive maintenance. Mead's loss control engineers believe this program has contributed significantly to the company's excellent loss history.[15]

Energy Production

Boiler rooms, mechanical rooms, electrical rooms, and generator rooms can become convenient locations for storage of materials associated with the maintenance of this equipment or other maintenance and building operating supplies. Thus, housekeeping plays an important role in keeping adequate separation between combustion and operating equipment and storage.

Regular maintenance of energy production equipment and fuel-handling systems plays an important part in fire prevention. Fuel-handling systems under pressure can develop leaks from joints and seals. If left unattended, they can present major fuel sources in the same vicinity of ignition sources. Burners should be tuned and maintained on a regular basis to prevent firebox explosions. Damaged insulation on hot piping should be repaired to preclude the likelihood of overheating of combustible materials.

Environmental Systems

NFPA *101* requires periodic tests and maintenance of systems serving as part of engineered smoke management systems to ensure proper operation. However, environmental systems that are not part of such an emergency system may become neglected. Systems that move air always move anything suspended in the air. This point is illustrated by the dust that can be seen in a beam of light. This dust and dirt is transported and can accumulate in a building and its systems. This accumulation can be to such a degree that one of the principal functions of an environmental system is to control this dust and dirt. Filters used to remove these substances from the air may present potential problems. Grease, HEPA, and other special-purpose filters are particularly susceptible to fire involvement when exposed to external fires. Consequently, NFPA 96, *Standard for Ventilation Control and Fire Protection of Commercial Cooking Operations*, requires an automatic fire extinguishing system in all such installations. NFPA 91, *Standard for Exhaust Systems for Air Conveying of Materials*, recommends similar protection, compatible with the hazard involved, in systems used for the removal or control of hazardous vapors or substances. Systems moving ordinary environmental air may also be prone to similar hazards. Fires associated with heating systems rank among the most common causes of home fires in the United States. In 1989–1993, heating systems ranked second among causes associated with home fires in the United States.[14] These figures include all fires of all types involving heating equipment. Fires involving furnaces or water heaters fueled by piped fuel gas are a relatively small share of the total, but this still leaves room for a significant absolute number of failures. Nonetheless, fire fighters who must respond to the spate of heating equipment fires at the beginning of each heating season can attest to the results of neglected maintenance.

Periodic maintenance, particularly cleaning and inspecting, will identify and eliminate hazards before they become problems, and extend the life of building environmental system components. Appendix B of NFPA 90A describes the elements of a HVAC system maintenance program, including steps for reducing fire hazard potential. Duct and plenum cleaning, filter replacements, bearing inspections, and tests of automatic and manual controls are among the most important elements of these programs. The appendix notes, "Failure to maintain proper conditions of cleanliness in air duct systems and carelessness in connection with repair operations have been important contributing causes of several fires which have involved air conditioning systems." Storage of combustible materials in plenums and near fresh air intakes can also be an important contributor to fires involving HVAC systems. Misalignment of motor bearings causes overheating and belt wear. The heat produced from such motor failures is sufficient to ignite dirt and debris—including grease and lubricants associated with the motor and bearings—in the system ductwork. Flammable or combustible solvents should never be used to clean system ductwork or filters. Great care should be exercised when open-flame devices are used to perform repairs inside ductwork and plenums.

Piped Services

Three fire scenarios are notable involving piped services: (1) ignition of building structure, insulation, dirt, debris, or other nearby combustibles, while using a torch or other heat-producing device for repair; (2) piping leaks caused by defective or damaged system components; and (3) ignition of flammable piping contents when piping or containers are not properly grounded. In the first scenario, the maintenance activity, not the piped service, is responsible for the ignition. Leaks involving noncombustible, oxidizing gases distributed by pipe may be just as dangerous as the flammable compressed gases when ignition sources are present. Flammable liquids especially can produce prolific amounts of static electricity. Ignitions during transfer operations remain a significant cause of accidental fires involving flammable liquids.

NFPA 51B, *Standard for Fire Prevention in Use of Cutting and Welding Processes*, describes the precautions that should be taken to prevent ignitions while using open-flame devices. These include operation of such equipment only by trained operators under the direct supervision of competent management, once inspections have determined that combustible materials will not be ignited by conduction, sparks, or slag. Combustibles within 35 ft (11 m) must be removed from the work area, which must then be swept clean. Sprinklers—if installed—must be operational, and portable fire extinguishers or small hoselines must be available nearby. The standard also requires adequate protection for workers, including smocks, aprons, gloves, helmets, face shields, and goggles, and proper maintenance of equipment. Following the conclusion of welding or cutting operations, a fire watch must be maintained in the area for at least 30 min to detect and extinguish smoldering fires.

Many safeguards are provided to prevent the inadvertent release of piping contents under conditions of normal use and operation. However, despite the mechanical and physical protection provided, gas and liquid leaks involving flammable and oxidizing compounds routinely occur. To prevent and minimize the hazards presented by such events, piping, valves, and system controls should be clearly identified and readily accessible. When a leak occurs, immediate care should be taken to control or eliminate potential ignition sources and disrupt the flow of pipe contents. Fire-fighting equipment should be readied for service, or summoned while controlling the release. Often, locating and stopping a release at the source is not practical. In such cases, valves or controls remote from the source may have to be shut, possibly impairing other services.

As noted, flammable liquids can produce prolific quantities of static electricity, especially when atmospheric humidity is low. NFPA 77 treats this subject in detail, and NFPA 30 requires the bonding and grounding of flammable and combustible liquid piping systems. NFPA 54 requires grounding of aboveground natural gas piping, as well. However, neither flammable nor combustible liquid piping nor fuel gas piping should be used as a grounding electrode. Bonding during liquid transfer operations will help ensure that static electricity formed by pouring such liquids is safely dissipated without generating a spark capable of igniting escaping vapors.

Electrical/Electronic Systems

Electrical failures resulting in fires may be attributed to one of two causes: (1) arcing or (2) overheating. All electrical equipment and devices are subject to wear and deterioration from prolonged use. Heating and arcing caused by normal operation are responsible for much of this wear and tear. While all devices are designed taking these factors into account, it remains the responsibility of the user, or owner, to determine when these conditions have resulted in an unsafe device, equipment, or installation. Additionally, users and owners must take care not to exacerbate these conditions by placing unreasonable demands on electrical installations or modifying conforming installations in a manner that makes them unsafe. Periodic inspection of electrical equipment, devices, and installations from

damage, defects, and misuse will help prevent these situations from developing into fires.

The importance of maintaining adequate clearances between combustible and heat-producing electrical equipment and devices, noted previously, deserves further attention. Overheating of electrical devices is generally caused by the inability of equipment or devices to adequately dissipate heat generated during normal operation. All electrical and electronic components are subject to resistance heating to some degree. This heat may be caused by insulation, friction, resistance, or induction. Either excess heat is produced faster than it can be safely dissipated, or heat is trapped by thermally insulating materials. Electrical insulation is designed to limit mechanical damage and prevent arcing, but not to restrict the liberation of heat. Coaker and Hirschler have modeled the insulating effect of combustible dusts, e.g., bagasse, coal dust, and sawdust, on electrical transmission cables and concluded that practically attainable, critical thicknesses of these materials (usually on the order of several centimeters) deposited on cables could produce insulation failures resulting in short circuits, if not ignite the dust itself.[16] The possibility of such failures points up the importance of housekeeping and routine inspections of electrical installations.

The following scenarios are not intended to include all the possible conditions likely to produce a fire, but rather to illustrate some common maintenance and operation failures. Loose connections caused by improper maintenance and normal wear and tear on switches or contacts are significant sources of arcing. Turning switches on and off generates a small arc within the switch enclosure. Each time this arc occurs, a small amount of the material from one contact is deposited on the other, reducing the electrical clearance between the conductive paths. Eventually, this situation will result in a short-circuit if not corrected. Contacts in electrical receptacles may similarly wear out from accumulated wear resulting from inserting and removing plugs or routinely unplugging energized equipment, thereby causing a failure. Likewise, damaged conductors, particularly flexible cords, may be subject to resistance heating and eventual arcing from damage, misuse, or neglect. In fact, misuse of flexible cords remains a significant cause of electrical fires. Misaligned bearings in electric motors may result in excessive friction between motor bushings and the motor housing. When the motor seizes up from such damage, often current continues to flow to the equipment until a short-circuit develops, causing the overcurrent protection to trip. However, if sufficient heat is generated before the activation of overcurrent protection, a fire may result. Once again, clearance from combustibles will help avert disaster.

Transportation and Conveyance

Fortunately, fires associated with the normal operation of elevators are rather infrequent. Hazards associated with normal operation of elevators include: (1) housekeeping, (2) repair operations in the elevator hoistway or machine room, and (3) maintenance of motors and equipment. Nearly all elevator hoistways seem to accumulate dirt, trash, and other debris at the bottom. In the case of hydraulic elevators, this may include some accumulation of oil or grease, as well. Obviously, the ignition of such debris can present a serious hazard. All elevators operate on some form of track, or rail, inside the hoistway. Repair or maintenance on this track or rail, using welding or cutting equipment, has been the source of serious fires inside elevator hoistways. One such fire occurred at the U.S. Embassy in Moscow, Russia, in March 1991. Workers using welding and cutting equipment apparently started an accidental fire in the hoistway, thereby heavily damaging the upper floors of the building.[17] Similarly, poorly maintained elevator drive motors may overheat, igniting nearby combustibles. This can include hydraulic fluid in the elevator fluid reservoir.

MODIFICATION AND RENOVATION

If the design and construction of a new building presents a challenge, an even greater one is presented when that building is modified. Buildings are rarely built and occupied by a single occupant, or for a single purpose, through their entire lifespan. Rather, they are constantly undergoing modification to meet the needs of new occupants or the desires of existing ones to avail themselves of the latest technology. However, the codes and standards under which the building was originally built have been superseded with new ones that may not always be met without major changes to the structure. This presents the dilemma to the participants in the renovation process of compromise between updating to current standards and keeping the project within a reasonable budget.

The hazards associated with renovation of existing systems reflect a combination of the problems commonly associated with original design and inadequate, or improper, maintenance. Additions to existing nonconforming systems should, at the very least, not contribute further to these deficiencies. Rather, whenever possible, the opportunity to correct existing weaknesses—whether a result of technological advances, new regulations, design deficiencies, or poor maintenance—should be advanced. In fact, most building and fire safety codes require the correction of past defects or related deficiencies when modifications are undertaken. Where a new, completely objective systems approach is impractical, unwarranted, or otherwise unsuitable, it remains wise to take the opportunity presented by modifications or renovations to detect and correct maintenance deficiencies or operational inadequacies.

In the context of modifications—substantial changes to original installations—electrical system modifications standout as a source of significant and continuing concern. Nonconforming changes are the leading causes of electrical fires in a survey conducted by the U.S. Consumer Product Safety Commission (CPSC), in which 149 fires were examined in closer detail. Table 7-13D illustrates the findings of this study.[18] Electrical distribution system fires are an exception to the general rule that fire risk is not significantly related to age of home.

SUMMARY

Controlling and eliminating hazards associated with building services requires recognition of the hazards associated with these systems, within the context of their contribution to a safe, comfortable, and efficient building environment. As human needs evolve, thereby placing greater demands on the built environment, the state of the art in buildings services is bound to evolve, too. Owners, occupants, architects, engineers, builders, fire protection consultants, and Authorities Having Jurisdiction must work together to keep abreast of these changes to meet the needs of comfort, efficiency, and economy, while, at the same time, control and eliminate hazards in the built environment.

BIBLIOGRAPHY

References Cited

1. Hartman, T., "Intelligent Building Design Demands," *Heating/Piping/Air Conditioning*, Vol. 61, No. 11, Nov. 1989, pp. 67–70.
2. BOCA, *National Building Code 1990*; ICBO *Uniform Building Code*, 1988; SBCCI *Standard Building Code*, 1988.
3. Klote, J. H., "Inspecting and Testing Air Moving Systems for Fire Safety," *Heating/Piping/Air Conditioning*, Vol. 60, No. 4, Apr. 1988, pp. 77–87.
4. Zicherman, J. B., "Is PVC Piping Firesafe," *Fire Journal*, Vol. 84, No. 6, Nov./Dec. 1990, pp. 37–42.
5. Choi, K. K., "Firestops for Plastic Pipe," *Fire Technology*, Vol. 23, No. 4, Nov. 1987, pp. 267–279.
6. Whittington, B. W., "Electrical Installations in Industrial Locations," *Industrial Fire Hazards Handbook*, 3rd ed., Cote, A. E., ed., National Fire Protection Association, Quincy, MA, 1990, p. 1104.
7. Teague, P.E., "The Analysis of the Toxicity Problem Continues," *Fire Journal*, Vol. 83, No. 3, May/June 1989, pp. 62–64.
8. Snell, J. E., "Measuring Hazards of Products of Combustion from Electrical Systems," *Fire Journal*, Vol. 81, No. 5, Sept./Oct. 1987, p. 108.
9. Nelson, H. E., "An Engineering View of the Fire of May 4, 1988, in the First Interstate Bank Building, Los Angeles, California," NISTIR 89-4061, National Institute of Standards and Technology, Gaithersburg, MD, 1989.
10. "Negligence Cited in Lack of Water at Meridian Fire," *Philadelphia Inquirer*, Vol. 323, No. 72, Wed., Mar. 13, 1991, pp. 1A, 14A. (Although the circumstances as of yet are unclear, this or a related emergency power failure also appears to have affected operation of the fire pump.)
11. Klote, J. H., Alvord, D. M., Levin, B. M., and Groner, N. E., *Feasibility and Design Considerations of Emergency Evacuation by Elevators*, NISTIR 4870, National Institute of Standards and Technology, Gaithersburg, MD.
12. Klem, T. J., "Los Angeles High-Rise Bank Fire," *Fire Journal*, Vol. 83, No. 3, May/June 1989, pp. 72–91.
13. "Negligence Cited in Lack of Water at Meridian Fire," *Philadelphia Inquirer*, Vol. 323, No. 74, Wed., Mar. 13, 1991, pp. 1A, 14A.
14. Hall, J. H., *The U.S. Fire Problem Overview Report*, Quincy, MA, NFPA Fire Analysis and Research Division, April 1996.
15. Mead Loss Control, Inc., *Mead Loss Control Engineering Services*, Dayton, OH, The Mead Corp., 1983.
16. Coaker, A. W., and Hirschler, M. M., "Effect of Deposition of Combustible Matter onto Electric Power Cables," *Fire Safety Journal*, 16 (6), pp. 459–467.
17. *Washington Post*, Vol. 114, No. 114, Fri., Mar. 29, 1991, p. A1, A13.
18. Smith, L. E., and McCoskrie, D., "What Causes Wiring Fires in Residences?" *Fire Journal*, Vol. 84, No. 1, Jan./Feb. 1990, pp. 18–24, 69.

TABLE 7-13D. *Contributing Factors Involved in Residential Electrical Distribution Systems Fires by Age of Dwelling*[18]

Contributing Factors	Number	Percent	≤10	11–20	21–30	31–40	>40	Unknown
Improper alterations	55	37	1	2	9	7	29	7
Improper initial installations	30	20	4	4	8	3	7	4
Deterioration due to aging	25	17	2	3	—	5	9	6
Improper use	23	15	1	1	2	5	13	1
Inadequate electrical capacity	22	15	—	—	4	1	16	1
Faulty product	17	11	2	3	2	1	8	1
Unknown	9	6	—	—	—	2	4	3
Total	149	100	10	13	25	24	86	23

NOTE: Frequencies add to more than number of investigations because multiple factors were cited in 31 cases.

Reference

ASTM E814, *Standard Test Method for Fire Tests of Through-Penetrating Fire Stops*, American Society for Testing and Materials, W. Conshohocken, PA (1994).

NFPA Codes, Standards, and Recommended Practices

Reference to the following NFPA codes, standards, and recommended practices will provide further information on the safeguards for hazards of miscellaneous building services discussed in this chapter. (See the latest version of *The NFPA Catalog* for availability of current editions of the following documents.)

NFPA 30, *Flammable and Combustible Liquids Code*
NFPA 31, *Standard for the Installation of Oil-Burning Equipment*
NFPA 51B, *Standard for Fire Prevention in Use of Cutting and Welding Processes*
NFPA 54, *National Fuel Gas Code*
NFPA 58, *Standard for the Storage and Handling of Liquified Petroleum Gases*
NFPA 70, *National Electrical Code®*
NFPA 72, *National Fire Alarm Code*
NFPA 75, *Standard for the Protection of Electronic Computer/Data Processing Equipment*
NFPA 77, *Recommended Practice on Static Electricity*
NFPA 82, *Standard on Incinerators and Waste and Linen Handling Systems and Equipment*
NFPA 90A, *Standard for the Installation of Air Conditioning and Ventilating Systems*
NFPA 90B, *Standard for the Installation of Warm Air Heating and Air Conditioning Systems*
NFPA 91, *Standard for Exhaust Systems for Air Conveying of Materials*
NFPA 96, *Standard for Ventilation Control and Fire Protection of Commercial Cooking Operations*
NFPA 99, *Standard for Health Care Facilities*
NFPA 101®, *Life Safety Code®*
NFPA 110, *Standard for Emergency and Standby Power Systems*
NFPA 211, *Standard for Chimneys, Fireplaces, Vents, and Solid Fuel-Burning Appliances*
NFPA 214, *Standard on Water-Cooling Towers*
NFPA 251, *Standard Methods of Tests of Fire Endurance of Building Construction and Materials*
NFPA 255, *Standard Method of Test of Surface Burning Characteristics of Building Materials*
NFPA 262, *Standard Method of Test for Fire and Smoke Characteristics of Wires and Cables*
NFPA 550, *Guide to the Fire Safety Concepts Tree*
NFPA 780, *Standard for the Installation of Lightning Protection Systems*

Additional Reading

Armstrong, J. H., *Maintaining Building Services: A Guide for Managers*, Mitchell, London, England, 1987.
Belles, D. W., "Reducing Fire Hazards with Fire Retardants," *Fire Journal*, Vol. 83, No. 3, May/June 1989, pp. 52–60, 129.
Birmingham, J. A., "Firestops Protect Property—and Lives," *Fire Journal*, Vol. 80, No. 2, Mar./Apr. 1986, pp. 41–44, 63.
Cooper, P., "Heating Tapes and Cables Can Cause Mobile Home Fires," in Letters, *Fire Journal*, Vol. 84, No. 2, Mar./Apr. 1990, p. 9.
Cooper, L. Y., "Overview of a Theory for Simulating Smoke Movement Through Long Vertical Shafts in Zone-Type Fire Models," National Institute of Standards and Technology, Gaithersburg, MD, NISTIR 5499, September 1994; National Institute of Standards and Technology, Annual Conference on Fire Research: Book of Abstracts, October 17–20, 1994, Gaithersburg, MD, 1994, pp. 93–94.
Cooper, L. Y., "Simulating Smoke Movement Through Long Vertical Shafts in Zone-Type Compartment Fire Models," National Institute of Standards and Technology, Gaithersburg, MD, NISTIR 5526, Nov. 1994.
Cote, A. E., ed., *Industrial Fire Hazards Handbook*, 3rd ed., National Fire Protection Association, Quincy, MA, 1990.
Donoghue, E. A., "Elevators and Sprinklers: The Controversy Continues," in Members' Forum, *Fire Journal*, Vol. 84, No. 2, Mar./Apr. 1990, pp. 15–17.
Donoghue, E. A., "Elevator Code Revisions Protect the Public and Firefighters," in Letters, *Fire Journal*, Vol. 83, No. 6, Nov./Dec. 1989, p. 6.
Doyle, M., "Life Safety and Security Systems Grow in Sophistication," *Building Design and Construction*, Vol. 27, No. 8, Aug. 1986, pp. 108–110.
Earley, M. W., "Minimizing the Hazards of Transformer Fires," *Fire Journal*, Vol. 82, No. 1, Jan./Feb. 1988, pp. 73–74.
Ehrenkrantz, E., *Architectural Systems: Needs, Resources, Design*, McGraw-Hill, New York, 1989.
Gewain, R. G., *et al.*, "Evaluation of Duct Opening Protection in Two-Hour Fire Walls and Partitions," *Fire Technology*, Vol. 27, No. 3, Aug. 1991, pp. 204–218.
Gewain, R. G., *et al.*, "Protection of Duct Openings in Two-Hour Fire Resistant Walls and Partitions," Hughes Associates, Inc., Wheaton, MD, EG&G Rocky Flats, Golden, CO, 1990.
Gibbons, J. A. M., and Stevens, G. C., "Limiting the Corrosion Hazard from Electrical Cables Involved in Fires," *Fire Safety Journal*, 15 (2), 1989, pp. 183–190.
Gray, B. F., *et al.*, "Effect of Combustible Matter onto Electric Power Cables," *Fire Safety Journal*, 16 (6), 1990, pp. 459–467.
Haines, R. W., *Heating, Ventilation and Air Conditioning Systems Design Handbook*, Tab Professional and Research Books, Blue Ridge Summit, PA, 1988.
Harris, C. M., *Building Utilities and Services Handbook: Planning, Design, and Installation*, McGraw-Hill, New York, 1990.
Hartman, T., "Intelligent Building Design Demands," *Heating/Piping/Air Conditioning*, Vol. 61, No. 11, Nov. 1989, pp. 67–70.
Hicks, H. D., "The Fire Protection Engineer and the Authority Having Jurisdiction," *Building Official and Code Administrator*, Vol. XXIII, No. 6, Nov./Dec. 1989, pp. 28–32.
Isner, M. S., "Five Die in High-Rise Office Building Fire," *Fire Journal*, Vol. 84, 4, July/Aug. 1990, pp. 50–59.
Kaufman, S., "Fire Tests and Fire-Resistant Telecommunications Cables," *Fire Journal*, Vol. 80, No. 6, Nov./Dec. 1986, pp. 33–38.
Klote, J. H., "A Computer Model of Smoke Movement by Air Conditioning Systems (SMACS)," *Fire Technology*, Vol. 24, No. 4, Nov. 1988, pp. 299–311.
Klote, J. H. and Forney, G. P., "Zone Fire Modeling With Natural Building Flows and a Zero Order Shaft Model," National Institute of Standards and Technology, Gaithersburg, MD, NISTIR 5251, Sept. 1993.
Ley, M., *Grading Methods of Building Fire Safety*, M.S. Thesis, Worcester Polytechnic Institute, Worcester, MA, 1987.
Mace, R. L., "Smart Planning for Accessibility," *Buildings*, Vol. 83, No. 4, Apr. 1989, p. 72.
Malhotra, H. L., *Fire Safety in Buildings*, Building Research Establishment, Borehamwood, UK, 1987.
Matsushita, T., "Smoke Control Calculation for Pressurized Elevator Shaft," Building Research Institute, Ibaraki-ken, Japan, NISTIR 4449; Eleventh Joint Panel Meeting of the U. S./Japan Government Cooperative Program on Natural Resources (UJNR) on Fire Research and Safety, October 19–24, 1989, Berkeley, CA, 1990, pp. 188–192.
Mercier, G. P. and Jaluria, Y., "Fire-Induced Flow of Smoke and Hot Gases in Open Vertical Shafts," *Proceedings* of the Thermal Science and Engineering Symposium in Honor of Chancellor Chang-Lin Tien, Nov. 1995, Berkeley, CA, 1995, pp. 261–268.
Mercier, G. P., Jaluria, Y. and Wu, G. L., "Fire-Induced Flow of Smoke and Hot Gases in Open Vertical Shafts," Rutgers Univ., New Brunswick, NJ, NISTIR 5499, September 1994; National Institute of Standards and Technology, Annual Conference on Fire Research: Book of Abstracts, October 17–20, 1994, Gaithersburg, MD, 1994, pp. 91–92.
Murray, R. H., and Lathrop, J. K., "Fire Barrier Penetrations Must Be Firestopped," *Fire Journal*, Vol. 81, No. 6, Nov./Dec. 1987, pp. 66–67.
Nireus, K., *et al.*, "Brandventilation genom schakt: experiment. [Fire Smoke Ventilation Through Shafts: Experiments.]," National Defence Research Establishment, Sundyberg, Sweden, FOA Report A 20061-2.6, June 1994, 27 pages.
Olsen, C., "Have Smart Buildings Flunked the Test?" *Building Design and Construction*, Vol. 27, No. 9, Sept. 1986, pp. 72–74.
Pasek, F., "Door Controls Complete Compartmentalization Design," *Building Design and Construction*, Vol. 27, No. 11, Nov. 1986, p. 77.
Plumer, J. A., "We Need Better Lightning Protection," *Fire Journal*, Vol. 81, No. 1, Jan./Feb. 1987, pp. 41–45, 73.

Scurry, P. S. G., "Suggested Topics for Research in Building Services Engineering," *Building Services Engineering Research and Technology*, Vol. 9, No. 2, 1988, pp. 87–88.

Shields, T. J., *Building and Fire*, Wiley, New York, 1987.

Sumathipala, K., "Spatial Separations and Unprotected Openings: Future Directions," *Fire Research News*, No. 70, Fall 1993, pp. 1–3.

Tamura, G. T., "Determination of Critical Air Velocities to Prevent Smoke Backflow at a Stair Door Opening of the Fire Floor—600 RP," *Journal of Applied Fire Science*, Vol. 2, No. 1, 1992/1993, pp. 5–21; American Society of Heating, Refrigerating and Air-Conditioning Engineers, Inc. (ASHRAE), Annual Meeting, June 22–26, 1990, Indianapolis, IN, 1993.

Todd, C. S., "Lack of Maintenance Jeopardizes the Benefits of Active Systems," *Fire Survival*, Vol. 16, 1987, pp. 6–11.

Transue, R. E., "Fire Management in Intelligent Buildings," *Fire Journal*, Vol. 84, No. 2, Mar./Apr. 1990, pp. 74–76.

Watanabe, Y., Matushima, S. and Yamada, T., "Prediction of Smoke Layer in a Fire Compartment Having a Smoke Shaft," *Bulletin of Japanese Association of Fire Science and Engineering*, Vol. 43, No. 1–2, 1995, pp. 21–29.

Zicherman, J. B., "Is PVC Piping Firesafe?" *Fire Journal*, Vol. 84, No. 6, Nov./Dec. 1990, pp. 36–42.

AIR CONDITIONING AND VENTILATING SYSTEMS

Revised by William A. Schmidt

The term "air conditioning" has been defined by the American Society of Heating, Refrigerating, and Air Conditioning Engineers (ASHRAE) as "the process of treating air to simultaneously control its temperature, humidity, cleanliness, and distribution to meet the comfort requirements of the occupants of the conditioned space." Air conditioning and ventilating systems, except for self-contained units, invariably involve the use of ducts for air distribution. The ducts, in turn, present the possibility of spreading fire, fire gases, and smoke throughout the building or area served.

This and other potential hazards of air conditioning and ventilating systems are the subjects of this chapter. Installation and protection of air heating systems are discussed in Section 3, Chapter 5, "Heating Systems and Appliances." Fire confinement, smoke movement, and venting practices are addressed in other chapters in Section 7, Chapter 5, "Confinement of Fire in Buildings"; Chapter 6, "Smoke Movement in Buildings"; and Chapter 7, "Venting Practices."

Details on safeguarding against these hazards are presented in NFPA 90A, *Standard for the Installation of Air Conditioning and Ventilating Systems* (hereinafter referred to as NFPA 90A). The bibliography and additional reading list at the end of this chapter contain additional references from which much useful information may be obtained.

SYSTEM TYPES AND OPERATION

The several types of air conditioning systems include: (1) systems in which air is filtered or washed, cooled and dehumidified in summer, and heated and humidified in winter; (2) systems where air is filtered, cooled, and dehumidified; and (3) systems where air is filtered, heated, and humidified. Figure 7-14A depicts a typical arrangement of the various components of a central air conditioning system. A ventilating system simply supplies or removes air by natural or mechanical means to or from a space. The air may or may not be conditioned.

Referring to the typical system depicted in Figure 7-14A, a fresh air intake duct connects directly to the system's return duct. From this point, the mixture of fresh and recirculated air passes through the air conditioning equipment. The air is subjected to several operations in this equipment, including filtration or cleaning, heating or cooling, and humidification. The conditioned air is then circulated throughout the area served via the duct system. (See Fig-

William A. Schmidt, P.E., is a consultant in private practice in Bowie, MD.

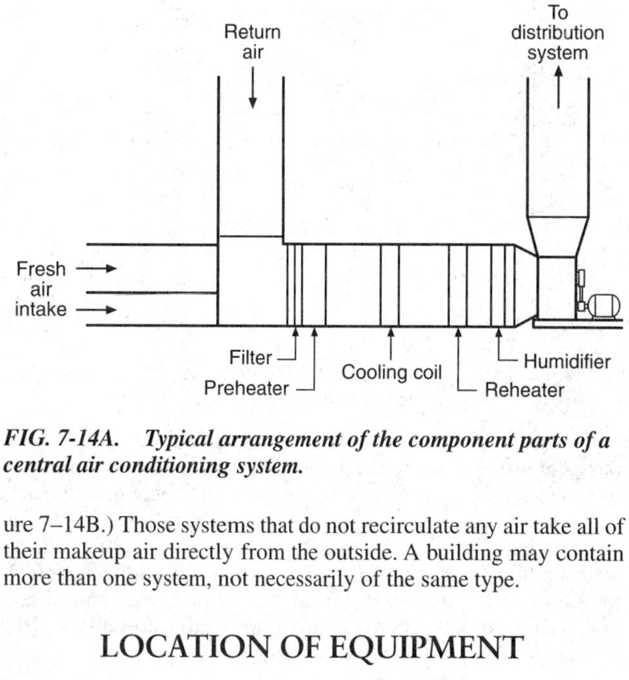

FIG. 7-14A. *Typical arrangement of the component parts of a central air conditioning system.*

ure 7–14B.) Those systems that do not recirculate any air take all of their makeup air directly from the outside. A building may contain more than one system, not necessarily of the same type.

LOCATION OF EQUIPMENT

Fans, heaters, filters, and associated equipment that make up a central system air conditioning unit are preferably located in a room separated from the rest of the building area by walls, floor, and floor-ceiling assembly providing a minimum 1-hr fire-resistance rating. This arrangement prevents a fire involving the equipment from immediately spreading to adjacent areas of the building. Such an arrangement also prevents access to the equipment by unauthorized persons. Ideally, service rooms housing air conditioning equipment are protected by automatic sprinklers. At minimum, smoke detectors should be provided and arranged to initiate an alarm and shut down the air conditioning system. No combustible storage should be allowed in the equipment room.

FRESH AIR INTAKES

The location selected for the fresh air intakes of a system is critical since fire, fire gases, or smoke originating outside the building can easily be drawn in through these intakes and spread throughout the building. Where such a hazard may exist, protection can be provided

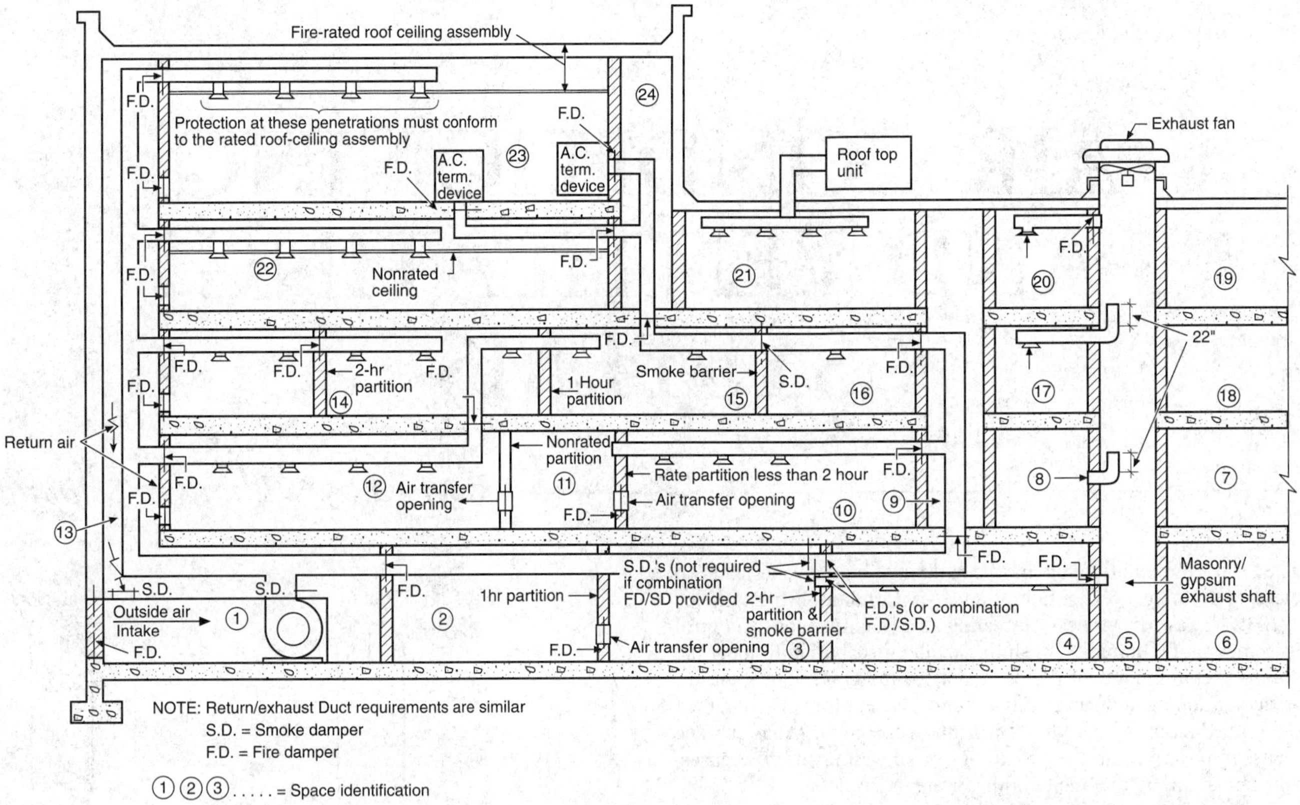

FIG. 7–14B. *Typical building duct installation illustrating required protection of walls, ceilings, floors, and shafts.*

by the installation at the intakes of fire dampers or smoke dampers that are controlled by fire and smoke detectors.

When considering the location of exterior fresh air intakes, thought must be given to the possibility of sparks or other products of combustion from an exposing fire chimney, or incinerator stack, being drawn into the system. Consideration should not be limited to adjacent buildings or combustible storage. The possibility of fire from a different section of the same building or an adjoining building exposing the intakes is a factor that must also be considered. Since smoke usually rises, the lower the intakes are installed, the less the possibility of drawing in smoke.

All air intakes must be provided with screens of corrosion-resistant material not larger than $1/2$-in. (13-mm) mesh to prevent material from entering the system. Proper maintenance includes periodic removal of any accumulated rubbish or other waste from the immediate vicinity of the intakes. Some fresh air intakes are fitted with a filter that is subject to the same hazards mentioned later in this chapter, under "Air Filters and Cleaners." These filters should be protected and maintained accordingly.

AIR COOLING AND HEATING EQUIPMENT

Air cooling equipment presents two basic classes of hazards: (1) those of the electrical equipment and (2) those of the particular refrigerant used. Fire experience is generally good where the cooling equipment is properly installed and maintained. Installation of all electrical equipment should follow both the manufacturer's recommendations and NFPA 70, *National Electrical Code®* (hereinafter referred to as NFPA 70).

One of the problems associated with refrigeration units is the explosion hazard due to overpressurization of the refrigerant. Proper safety controls can mitigate this problem.

As a result of the Montreal Protocol and the ban on some ozone-depleting refrigerants, the industry is actively seeking to develop acceptable substitutes.

At present, there are an array of refrigerants that are being used by manufacturers. Their properties as substitute refrigerants are being thoroughly tested. However, some appear to have flammability, explosive, and toxicity characteristics that have not been adequately addressed. In addition, there are requirements for maintenance personnel to capture and reprocess ozone-depleting refrigerants. Venting to the atmosphere is no longer permissible. Properties of refrigerants can be found in the ASHRAE "Safety Code for Mechanical Refrigeration."[1]

AIR FILTERS AND CLEANERS

The types of air filters and cleaners used in air conditioning and ventilating systems fall into three general categories: (1) fibrous media unit filters, (2) renewable media filters, and (3) electronic air cleaners. The first two are true filters; the last is a static precipitator. Air filters and cleaners pose a potential hazard due to their function of removing entrained dust and other particulate matter from the air stream. This material builds up on the filter media or precipitator collection plates and, if ignited, may burn, producing large volumes of smoke. The smoke and other combustion gases can be circulated throughout the building by the air handling system, thus posing a direct threat to life safety. Not to be overlooked is the possibility that a filter medium may be coated with a combustible adhesive or may itself be combustible.

Underwriters Laboratories Inc. (UL) lists two classes of filter media:[2]

Class 1:   Filters that, when clean, do not contribute fuel when attacked by flame and emit only negligible amounts of smoke.

Class 2:   Filters that, when clean, burn moderately when attacked by flame or emit moderate amounts of smoke, or both.

Both of these classes include renewable (washable and reuseable) media and replaceable (disposable) media.

Class 1 filter media are preferred, particularly for systems serving places of assembly, such as theaters, auditoriums, department stores, etc. Class 1 filter media are also the obvious choice for systems serving occupancies whose contents are easily susceptible to smoke damage. Any filter media, however, if not cleaned or replaced regularly, may become hazardous due to accumulation of combustible dust, etc.

Fibrous Media Unit Filters

Fibrous media unit filters are placed into the air stream and remain there until the pressure drop across the filter reaches some critical point, due to the buildup of entrained material. At this point, the filter is either removed, cleaned, and reinstalled or simply discarded and replaced with a new filter. Fibrous media unit filters are of either the viscous-impingement type or the dry-media type.

The viscous-impingement type of filter is characterized by flat or pleated panels of relatively coarse fiber mats. Porosity of the panel is high. The fibers are coated with an adhesive that traps any material entrained in the air stream. The mats are $1/2$ to 4 in. (13 to 102 mm) thick, 1 and 2 in. (25 and 51 mm) being most common. This type of filter is most effective when air velocity ranges from 400 to 700 ft/min (122 to 213 m/min). Most of these unit filters are ready for replacement when the pressure drop reaches $1/2$ in. of water (124 Pa). The renewable panels must be washed with steam or a hot detergent solution, then recoated with fresh adhesive.

Dry media unit filters are flat or pleated panels of relatively fine fibers, usually $1/2$ to 2 in. (13 to 51 mm) thick. As their name implies, these filters have no adhesive coating. Porosity is not as great as for the viscous-impingement-type filter; hence, particulate removal is greater, depending upon fiber size and porosity. These filters are most effective with an air velocity of 90 to 500 ft/min (27 to 152 m/min). Pressure drop may reach as much as 2 in. of water (500 Pa) before replacement is necessary. Disposable dry fibrous media unit filters, particularly of the deep-pleated design, are fast becoming the standard for large central air-conditioning systems. The dry media classification of air filters also includes membrane and high-efficiency types, which are used to ultraclean air for clean rooms and similar areas.

Renewable Media Filters

Renewable media air filters can more accurately be termed "moving curtain" air filters, since they operate in exactly such a fashion. The medium, whether viscous coated or dry, is supplied on a large roll and extends across the air stream. Either a timer, light-sensitive device, or pressure switch may be used to activate a motor drive which feeds fresh media across the air stream and winds the dirty media onto a takeup spool. Generally, controls are provided that deenergize the drive motor when the media supply is almost exhausted, thus preventing the media from being completely wound onto the takeup spool. A signal indicates that a fresh roll is needed. Operating velocities range from 200 to 500 ft/min (70 to 152 m/min), with the dry media types requiring the lower velocities. Filter thicknesses vary from $1/2$ to $2 1/2$ in. (13 to 63 mm).

Electronic Air Cleaners

Electronic air cleaners utilize the principle of electrostatic precipitation to remove entrained dust and particulate matter. There are several types of electronic air cleaners, but their basic operation is similar. Entrained particles in the air stream pass through intense, nonuniform electrostatic fields and, due to electric polarization, are collected either on a filter or on charged plates. In some units, equipment is provided to pre-ionize the particulate matter.

Special Industrial Filters and Air Cleaners

There are many industrial applications of air cleaning that involve more than the usual filtering and cleaning methods previously cited. Exhaust systems for flammable vapors, dusts, and other materials in suspension require equipment for recovering the suspended material.

Air cleaning to limit contamination and pollution is of growing concern to industry and public health authorities. Whatever the reason for specialized air cleaning, and whatever the type of equipment used, the selection and protection of the equipment are of concern as the fire hazards are similar to more common types of air cleaning equipment.

One current industrial application of air filtering is in clean rooms where high-efficiency particulate air (HEPA) filters are used to filter the atmosphere. These filters must pass an operational efficiency test of 99.97 percent efficiency with 0.3 micron particles. The filters are tested and labeled by UL in accordance with UL 586, *Standard for Safety High-Efficiency, Particulate, Air Filter Units.*[3]

Protection for Air Filters and Cleaners

Fire in the filters or air cleaning equipment can release copious quantities of smoke or fire gases that can be distributed by the air handling equipment throughout the area served. Adequate protection to minimize the possibility of such an occurrence must be designed into and around the air conditioning system.

Some method is necessary to prevent smoke and fire gases from being distributed by the main supply fan to all areas served by the system. To accomplish this, detectors (either heat or smoke, depending on the size of the system) are located in the main supply duct, downstream of the filters or air cleaner, to sense smoke or particles of combustion. The detectors are interlocked so as to immediately shut down the entire air-conditioning system upon activation. The detectors also control the operation of smoke dampers located in the main supply and return ducts, thus isolating the entire air conditioning section of the system.

With specific regard to viscous impingement filters, the adhesive used should have a flash point of not less than 325°F (163°C) via the Pensky-Martens closed tester.[4] Also, combustible adhesives should be stored in a safe location, remote from the equipment room housing the air conditioning equipment.

Fire protection should always include facilities for manual fire fighting, usually portable fire extinguishers. Fans, motors, pumps, control circuits, etc., all point to the need for extinguishers suitable for electrical (Class C) fires; however, dry chemical (Class B) extinguishers would be the most suitable for use on filters, especially the viscous-impingement type, since the chemical powder would readily adhere to the coated media.

The operational voltage and amperage of electronic air cleaners is lethal; therefore, interlocks should be provided on every access door and panel so that the equipment is immediately shut down if any one of them is opened. An alarm should also be incorporated into this safety interlock circuit.

DUCTS

Ducts are to an air conditioning or ventilating system what pipes are to a water system—a means of distribution. Unfortunately, in fire situations, the ducts may transport deadly smoke and products of combustion instead of breathable air. If proper design and installation precautions are not taken, smoke, fire gases, heat, and even flame can spread throughout the area served by the duct system. Exit corridors used as plenums, lack of smoke detection activated control equipment in the system, and lack of required fire and smoke dampers in appropriate walls, ceilings, or partitions can lead to tragic situations.

Duct Construction

Ducts may be fabricated of metal, masonry, or other noncombustible material. The thickness of materials used in metal ducts of various sizes and methods of bracing, reinforcing, and hanging are covered in the duct manuals of the Sheet Metal and Air Conditioning Contractors' National Association, Inc. (SMACNA).[5] SMACNA also publishes manuals on the installation of glass fiber ducts.

UL tests and lists duct materials in accordance with UL 181, *Standard for Safety Factory-Made Air Ducts and Air Connectors.*[6] This standard sets limits on such characteristics as flame spread and flame penetration. Materials are classified as follows:

Class 0: Air ducts and connectors having surface burning characteristics of zero.

Class 1: Air ducts and connectors having a flame spread rating of not over 25, without evidence of continued progressive combustion, and a smoke developed rating of not over 50.

Class 2: Air ducts and connectors having a flame spread rating of not over 50, without evidence of continued progressive combustion, and a smoke developed rating of not over 50 for the inside surface and not over 100 for the outside surface.

In addition to the above, Class 0 and Class 1 materials must pass a 30-min flame penetration test, and Class 2 materials must pass a 15-min flame penetration test.

NFPA 90A gives other mandatory provisions for duct linings and coverings, duct tapes, and bands. It also places limitations on the use of flexible duct connectors that must pass through walls, partitions, floors, or ceilings intended to afford fire resistance or smoke control.

NFPA 90A recently added a provision where gypsumboard may be used for negative pressure or return air ducts, and for shafts subject to specific conditions. Gypsumboard shafts must be constructed in accordance with the Gypsum Association fire-resistance design manual, be constructed airtight, and must be designed to handle structural and air pressure loadings.

Duct Installation

By their very nature, ducts provide a means for transferring heat, fire gases, and flame, thus resulting in the spread of fire from one area to another. In addition, at some point or another, a duct probably will pass through a wall, partition, floor, or ceiling which is designed specifically to provide fire resistance. Literally, a hole has been poked through a fire-rated design. Theoretically, of course, the duct fills the opening. But if the installation has been made without proper regard for firestopping, this is not valid. Also, under severe fire exposure, the duct will eventually collapse, creating an opening in the fire barrier. An effective method of protecting such penetrations is by the installation of fire dampers.

In the gages commonly used, some sheet metal ducts, if properly hung and adequately firestopped, may protect an opening in a building construction assembly for up to 1 hr. Therefore, ducts passing through fire barriers having a rating of up to 1 hr of fire resistance can possibly present no extraordinary hazard. If the wall, partition, ceiling, or floor is required to have a fire-resistance rating of more than 1 hr, a fire damper is required to properly protect the opening. Where it is necessary for a duct to pierce a fire barrier, an automatic closing fire damper or fire door assembly should be installed at the wall opening. NFPA 90A provides comprehensive recommendations and requirements for the location of fire dampers in rated assemblies.

Fire dampers and ceiling dampers are tested and listed for use in air conditioning and ventilating ducts by UL in accordance with UL 555, *Standard for Safety Fire Dampers,*[7] and UL 555C, *Standard for Safety Ceiling Dampers.*[8] These dampers include single-blade, multiblade, and interlocking-blade types, all actuated by fusible links. Most fire dampers are designed for vertical installation, although some can be installed horizontally as well.

SMOKE CONTROL—PASSIVE

Both NFPA *101*®, *Life Safety Code*®, and NFPA 90A recognize two approaches to smoke control: (1) passive and (2) active. The passive approach recognizes the long-standing compartmentation concept, which requires that fans be shut down and fire and smoke dampers in ductwork be closed in fire conditions. The active approach utilizes the building's heating, ventilating, and air conditioning (HVAC) systems to create differential pressures to prevent smoke migration from the fire area and to exhaust the products of combustion to the outside.

Smoke Dampers

Smoke dampers are required in air conditioning or ventilating ducts that pass through required smoke barrier partitions. NFPA 90A requires smoke dampers to operate automatically upon detection of smoke and must function so that smoke movement through the duct is halted. Basically, they simply interrupt airflow through the duct. NFPA 90A permits the use of fire dampers for smoke control purposes. Obviously, a combination fire and smoke damper must meet the requirements for both. Smoke dampers are tested and classified by UL in accordance with UL 555S, *Standard for Safety Leakage Rated Dampers for Use in Smoke Control Systems.*[9]

Smoke Detectors

Smoke detectors must be installed in all air conditioning or ventilating systems over 15,000 cu ft/min (425 m³/min) capacity. Further, smoke detectors for use in ducts must include features in their design specifically for this purpose. Smoke detectors may shut down the air conditioning or ventilating system, sound alarms, operate smoke control dampers, activate fire suppression equipment, or initiate active smoke control functions. Preference is for detectors with adjustable sensitivity, thus minimizing false alarms due to incidental dust or other particulate matter.

These smoke detectors are located in the main supply ducts, downstream of air filters or air cleaners, and in the main return ducts prior to exhausting from the building or joining the fresh air intake ducts. Smoke detectors at these locations are primarily intended to prevent smoke circulation throughout the entire area served by shutting down the air-handling system. Also they prevent circulation of smoke from a fire in the air filters, air cleaners, belts, and motors.

Due to dilution effects of air picked up from various branch return ducts, these smoke detectors cannot be expected to reliably provide "early warning" smoke detection. For this reason, additional smoke detectors may be provided, e.g., in the main branch return duct from each floor or major area division. Such detectors could also be tied into a zoned fire alarm system; however, such an arrangement is not a substitute for a smoke detection system for building protection. For guidance on smoke detector location, see NFPA 72, *National Fire Alarm Code*.

SMOKE CONTROL—ACTIVE

The preceding material covered passive control of smoke and fire gases—control designed to prevent products of combustion from moving out of the area of its source. It is an inescapable fact that attempts to completely confine smoke and fire gases are seldom successful. Recognizing this, smoke movement in high-rise buildings, and its threat to life safety, has been the object of much investigation. A discussion of modes of smoke movement, causes, and effects is outside the scope of this chapter. However, it is worth noting that high-rise buildings do present several unique problems, e.g., lengthy evacuation time, inability of conventional fire department equipment (aerial ladders, snorkels, and monitor nozzles) to reach upper floors, and the reinforcement of existing upward airflow patterns within the building via vertical shafts (stack effect).

In view of the increasing trend toward high-rise buildings, it becomes attractive to utilize the air conditioning or ventilating system for smoke control and removal in case of fire. Tests have been conducted on the feasibility of pressurizing emergency exit stairwells and elevator shafts to prevent smoke migration from the fire floor to other parts of a building during evacuation.[10,11]

The ASHRAE manual *Design of Smoke Management Systems* provides guidelines for designers who wish to provide active smoke control systems for buildings.[12] This information is generally intended to provide systems that exhaust smoke from the immediate fire area, and provide pressurized outside air to adjacent areas, access corridors, and stairwells. It is fully recognized that this approach would apply more to large HVAC units serving individual floors, or large systems with volume control dampers at each floor. For additional guidance, see NFPA 92A, *Recommended Practice for Smoke-Control Systems*, and NFPA 92B, *Guide for Smoke Management Systems in Malls, Atria, and Large Areas*.

The smoke control system must maintain safe exit routes with sufficient exiting time for building occupants to either leave or move to designated safe refuge areas.

Exhaust of smoke from the fire area is of primary concern. To replace this exhausted air, the HVAC system should provide 100 percent outside air to adjacent areas, thereby providing a clean atmosphere in exitways through the area immediately above the fire, and preventing smoke from flowing up through any holes in the floor slab and to the area immediately below the fire to provide safe areas for the fire fighters.

It must be fully understood that the smoke-control systems mentioned previously are intended to provide a smoke-free means of exiting the building. If a fire gets out of hand, there is danger of ducts burning through and falling from the ceiling, or fans being subjected to intense heat so they may cease to operate. It can then be assumed that initial efforts at extinguishing the fire have failed and that occupants remaining in the building would probably be in serious danger. It is not intended that special electrical generation units, heavy-duty fans, or heavy ductwork would be used.

FANS, CONTROLS, ETC.

The fan unit on an air conditioning system should not in itself present undue hazard if properly installed and firmly supported on a rigid foundation. It should also be readily accessible for cleaning, servicing, and lubrication. Fans should be provided with excess vibration switches, wired to sound an alarm and initiate system shutdown when bearing failure is imminent.

Fan motors installed inside ducts or plenums need protective devices to cut off power before temperatures reach a point where smoke may be generated. Most fractional horsepower motors have over-temperature protective devices. Thermal overload relays are recommended for fan motors of one horsepower or larger.

Electric wiring to all components should follow NFPA 70. All electrical equipment should be listed by a qualified electrical testing laboratory. Wiring installed in ducts, plenums, and other spaces used for environmental air should meet the requirements of NFPA 262, *Standard Method of Test for Fire and Smoke Characteristics of Wires and Cables*.

All air conditioning or ventilating systems need manual shut-offs for use in case of fire or other emergency. This shutoff should be well identified and located where it is readily accessible, such as near building exits. In systems of 2,000 to 15,000 cu ft/min (57 to 425 m³/min) capacity, it is good practice to provide automatic shut-off controls triggered by thermostatic devices in the same locations as recommended for smoke detectors. A setting of 136°F (58°C) for the thermostatic device in the return air stream and a setting of 50°F (10°C) above maximum operational temperature for the device in the supply stream are recommended.

Energy Conservation

Energy conservation practices require a careful evaluation of the design and control of HVAC systems. To reduce electrical costs, fans are turned off during various portions of the day and during unoccupied hours. Without airflow, the duct fire and smoke detectors may not detect fire and thereby will not provide the protection that may be assumed as inherent in HVAC systems. The use of variable volume air systems is increasing as these systems have an overall lower energy use than most other systems. The variation in duct airflow in such systems can create a problem with the sensitivity of the duct-mounted detectors and cause false readings and alarms. Fan cycling and variable volume systems can also cause variations in air pressure relationships of the occupied spaces. An area may have a positive pressure relationship to another area when the fans are operating in one mode, and when the fans are cycled a completely different relationship can occur which can affect the performance of detection and control systems, and calls for careful design evaluation.

UNIT AIR CONDITIONERS

The term "unit air conditioner" may include any factory-produced unit serving one room or area. Such units do not involve the hazard of spreading fire from floor to floor or area to area as do ducted systems. However, if improperly designed, installed, or maintained, they may involve the other hazards associated with larger systems. For example, any window-sill-type unit can overheat through failure of its thermal relay. This could lead to ignition of wire insulation or filter media, and the fire could spread to curtains or the wood sash of the window. Defective wiring could lead to the same condition. A problem particularly worth noting involving these units is plugging them into outlets on branch circuits not designed to carry the design load or the circuit is already overloaded. This practice directly violates principles of NFPA 70.

MAINTENANCE

Maintenance and cleaning are of utmost importance to safe operation of any air conditioning or ventilating system. Filters must be changed or cleaned as frequently as necessary. Ducts, particularly on the return side of the system, are cleaned out periodically to prevent hazardous accumulations of combustible dust and lint. Evidence of any defect of wiring or electrical equipment must be checked out immediately and corrected. Repairs on ducts or equipment casings that require welding or cutting should be protected by complete shutdown of the system and thorough cleaning of the area. If possible, the duct section to be repaired should be isolated from the rest of the system.

Proper maintenance includes periodic testing of all fire protection devices, including fire suppression equipment, smoke control and fire dampers, fire detectors, and alarms, and even vibration switches on fans. A regular testing program should be established. If such testing is outside the function of regular maintenance staff, it should be contracted to an outside agency.

BIBLIOGRAPHY

References Cited

1. ASHRAE, "Safety Code for Mechanical Refrigeration," No. 15-94, American Society of Heating, Refrigerating, and Air Conditioning Engineers, Atlanta, GA.
2. UL 900, *Standard for Safety Air Filter Units,* Underwriters Laboratories Inc., Northbrook, IL, 1994.
3. UL 586, *Standard for Safety High-Efficiency, Particulate, Air Filter Units*, Underwriters Laboratories Inc., Northbrook, IL, 1990.
4. ASTM D93, *Standard Test Methods for Flash Point by Pensky-Martens Closed Tester*, American Society for Testing and Materials, Philadelphia, PA, 1994.
5. SMACNA, *Fibrous Glass Duct Construction*, 1992; *HVAC Duct Construction Standards—Metal and Flexible*, 1985; Sheet Metal and Air Conditioning Contractors' National Association, Inc., Chantilly, VA.
6. UL 181, *Standard for Safety Factory-Made Air Ducts and Air Connectors*, Underwriters Laboratories Inc., Northbrook, IL, 1994.
7. UL 555, *Standard for Safety Fire Dampers,* Underwriters Laboratories Inc., Northbrook, IL, 1995.
8. UL 555C, *Standard for Safety Ceiling Dampers*, Underwriters Laboratories Inc., Northbrook, IL, 1992.
9. UL 555S, *Standard for Safety Leakage Rated Dampers for Use in Smoke Control Systems*, Underwriters Laboratories Inc., Northbrook, IL, 1996.
10. Koplon, N. A., *Report of the Henry Grady Fire Tests,* City of Atlanta Building Department, Atlanta, GA, 1973.
11. McGuire, J. H., and Tamura, G. T., "The Pressurized Building Method of Controlling Smoke in High-Rise Buildings," Technical Paper No. 394 (NRCC 13365), 1973, National Research Council of Canada, Division of Building Research, Ottawa, Ontario, Canada.
12. ASHRAE, *Design of Smoke Management Systems,* American Society of Heating, Refrigerating, and Air Conditioning Engineers Inc., Atlanta, GA, 1992.

References

ASHRAE Handbook 1992 Systems and Equipment, American Society of Heating, Refrigerating, and Air Conditioning Engineers Inc., Atlanta, GA.

Gypsum Association Fire Resistance Design Manual, 14th ed., Gypsum Association, Evanston, IL, 1994.

NFPA Codes, Standards, and Recommended Practices

Reference to the following NFPA codes, standards, and recommended practices will provide further information on air conditioning and ventilating systems discussed in this chapter. (See the latest version of *The NFPA Catalog* for availability of current editions of the following documents.)

NFPA 70, *National Electrical Code®*

NFPA 72, *National Fire Alarm Code*

NFPA 80, *Standard for Fire Doors and Fire Windows*

NFPA 90A, *Standard for the Installation of Air Conditioning and Ventilating Systems*

NFPA 92A, *Recommended Practice for Smoke-Control Systems*

NFPA 92B, *Guide for Smoke Management Systems in Malls, Atria, and Large Areas*

NFPA 101®, *Life Safety Code®*

NFPA 252, *Standard Methods of Fire Tests of Door Assemblies*

NFPA 262, *Standard Method of Test for Fire and Smoke Characteristics of Wires and Cables*

Additional Reading

Associate Committee on the National Building Code, *Measures for Fire Safety in High Buildings*, National Research Council of Canada, Ottawa, Ontario, Canada, 1977.

Butcher, E. G., and Parnell, A. C., *Smoke Control in Fire Safety Design*, E & F. N. Spon Ltd., London, England, 1979.

Day, T., "Development of a European Test Method for Fire and Fire/Smoke Dampers," *Fire Prevention*, No. 220, June 1989, pp. 26–28.

"Fire and Smoke Control," *ASHRAE Handbook*, HVAC Systems and Equipment, American Society of Heating, Refrigerating, and Air-Conditioning Engineers Inc., Atlanta, GA, 1992.

Gewain, R. G., "Protection of Duct Openings in Two-Hour Fire-Resistant Walls and Partitions," *ASHRAE Journal*, Vol. 32, No. 5, May 1990, p. 16.

Gewain, R. G., "Fire Protection Criteria for Air Ducts," ASHRAE *Transactions* 1986, Vol. 92, 1986, pp. 550–555.

Hinkley, P. L., "Case for Combining Venting and Sprinkler Systems," *Fire Engineers Journal*, Vol. 47, No. 145, June 1987, pp. 17–22.

Irwin, W. T., "Duct Penetrations of One-Hour Fire Resistance Rated Nonloadbearing Walls/Partitions," *ASHRAE Transactions*, Vol. 92, 1986, pp. 556–565.

Kirson, D., "Automatic Sprinklers in Exhaust Systems," *Heating, Piping, and Air Conditioning*, Vol. 61, No. 1, Jan. 1989, pp. 163–166.

Klote, J. H., "A Computer Model of Smoke Movement by Air Conditioning Systems (SMACS)," *Fire Technology*, Vol. 24, No. 4, 1988, pp. 299–311.

Klote, J. H., "Fire Safety Inspection and Testing of Air Moving Systems," NBSIR 87-3660, Nov. 1987, Center for Fire Research, Gaithersburg, MD.

Klote, J. H., and Cooper, L. Y., "Model of a Simple Fan-Resistance Ventilation System and Its Application to Fire Modeling," NIST 89-4141, Sept. 1989.

Mansergh, R., "Ductwork Integrity Is Vital," Fire, Vol. 83, No. 1027, Jan. 1991, pp. 38, 40.

McCarthy, J. H., "Changing Regulations Over Fire Dampers," *Australian Refrigeration, Air Conditioning, and Heating*, Vol. 42, No. 12, Dec. 1988, pp. 28–31, 32.

Mountain, S. C., "Safety in Air Conditioning," *H&V Engineer*, Vol. 61, No. 687, p. 22.

Newman, J. S., and Tewarson, A., "Flame Propagation in Ducts," *Combustion and Flame*, Vol. 51, 1983, pp. 347–355.

Perzak, F. J., Lazzara, C. P., and Kubala, T. A., "Fire Tests of Rigid Plastic Ventilation Ducts," Investigation Report 9085, 1987, U.S. Bureau of Mines, Pittsburgh, p. 16.

Reddaway, L. N., and Smith, D. J., "Fire Safety—The Future of Air Conditioning," *Australian Refrigeration, Air Conditioning, and Heating*, Vol. 40, No. 12, 1986, pp. 28–30.

Richards, P. E., "Air Ducts: The Uses and Abuses of Flexible Air Ducts," *Australian Refrigeration, Air Conditioning, and Heating*, Vol. 42, No. 9, 1988, p. 4.

Schwartz, J. K., "The Role of Fire Dampers in a Total Fire Protection System," TR 86-2, 1986, Society of Fire Protection Engineers, Boston, MA.

Schwartz, J. K., Jensen, R. H., and Antell, J., "Role of Dampers in a Total Fire Protection System Analysis," ASHRAE *Transactions*, Vol. 92, 1986, pp. 566–576.

Semple, J. B., and Bruce, D. O., "Review of Current Fire Damper Requirements in NFPA Standard 90A and Model Codes," ASHRAE *Transactions*, Vol. 92, 1986, pp. 541–549.

Wild, J., "Fans for Fire Smoke Venting," *Fire Surveyor*, Vol. 18, No. 2, Apr. 1989, pp. 5–9.

Woods-Ballard, W. R., "HT Series of Smoke Venting Fans," *Fire Surveyor*, Vol. 18, No. 3, June 1989, pp. 11–14.

EVACUATION OF OCCUPANTS

Most of the fire prevention and protection strategies addressed in the previous section have been tied directly to the timeline of fire growth and development. Whether the system element is fire prevention; control of materials, products, and environments; suppression; or confining fires, the aim is to slow down or stop the fire and its potentially devastating effects. However, life safety also depends on using the extra time provided by a delayed fire to move occupants out of harm's way. Effective evacuation involves understanding and using the latest knowledge regarding human behavior in fires—how people tend to react and how that affects how they can be trained to react—and building design features that can accommodate or shape occupant behavior toward effective action. This section covers these essential bodies of knowledge.

Chapter 1 covers what we know about human behavior and fire. Forget about panic. When the term is used precisely, such behavior very rarely occurs, even in the most severe fire conditions. Chapter 1 also addresses the means by which people investigate "ambiguous threat cues" and decide that there really is a fire—"recognition, validation, definition, evaluation, commitment, and reassessment."

As Chapter 1 says, "Behavior in fires can be understood as a logical attempt to deal with a complex, rapidly changing situation in which minimal information upon which to act is available." The inescapable conclusion is that people need to be given more and better information. Topping the list is the kind of information people acquire when they practice exit drills effectively.

Chapters 2 and 3 address the design of building features to facilitate evacuation. Chapter 2 provides overall concepts and principles, while Chapter 3 addresses specific building transportation systems, such as elevators and escalators.

Chapter 2 reviews concepts that underlie the provisions of NFPA *101®, Life Safety Code.®* These include the sizes of people and other key factors in flow speed. They also include characteristics of the building that can protect or threaten means of egress, such as interior finish that promotes rapid flame spread. Formulas and models for calculating egress are covered in Chapter 2, as well, which concludes that, "Maintaining the means of egress in safe operating condition at all times is as important to the prevention of loss of life as the proper construction of the building and the elimination of fire hazards."

Also look for these: Chapters 3-12, 3-13, and 3-14 of *The SFPE Handbook of Fire Protection Engineering* provide more detailed treatment of the science of human behavior in fire and the modeling of people movement. NFPA's E.D.I.T.H. program (Exit Drills in The Home) remains the best training program for home exit drills, while other well-designed training materials are available from NFPA to design and implement evacuation plans and drills in schools, health care facilities, and other settings. In the *Fire Protection Handbook*, the early chapters of Section 7 will be of interest, because the building design principles for confining fire and for protecting egress routes need to be considered together. Section 2, Chapter 1, is relevant because the methods of public fire safety education are also essential to teaching people safe ways to respond to fire, particularly prompt, effective evacuation. Section 1, Chapters 2 and 3, include fundamental concepts that help put evacuation in its fire safety systems context. And special evacuation requirements for specific property-use classes, if needed, are addressed in the appropriate chapters of Section 9.

—Ron M. Coté
December 1996

HUMAN BEHAVIOR AND FIRE

Revised by John L. Bryan

How one reacts during a fire is related to the role assumed, previous experience, education, and personality; the perceived threat of the fire situation; the physical characteristics and means of egress available within the structure; and the actions of others who are sharing the experience. Post-event analysis of behavior has described actions as adaptive or nonadaptive, participative or inhibited, and altruistic or individualistic. Detailed interview and questionnaire studies over 30 years have established that instances of nonadaptive, panic-type behavior are rare events that occur under specific conditions. Most behavior in fires is determined by information analysis, resulting in cooperative and altruistic actions.

INTRODUCTION

The characteristics of the behavior of people individually and within groups have been determined primarily by research studies in which individuals were interviewed at the time of the fire by fire department personnel.[1,2] It must be recognized that an individual's behavior in a fire is affected by the variables of the building in which the fire occurs and by the appearance of the fire at the time of detection. For example, the occupants' response will vary if they smell smoke rather than see flames or dark, acrid smoke completely obscuring a corridor. Variables of the fire protection provided for the building also may be critical to the individual's perception of the threat involved. Obviously, the most important individual decisions and behavior in life-threatening situations occur before the arrival of the fire department, in the early stages of the fire. Studies of health care facilities have indicated the importance of this early behavior:

> In the process of investigating these case studies we have come to believe that the period between detection of the fire and the arrival of the fire department is the most crucial life-saving period in terms of the first compartment (the area in direct contact with the room of origin and the fire).[3]

Thus, the behavior of the individuals intimately involved with the initiation of the fire is critical not only for themselves, but often for the other occupants of the building. It should be recognized that the altruistic behavior observed in most fires with the interaction of the occupants and the fire environment in a deliberate, purposeful manner, appears to be the general mode of reaction. The nonadap-

tive flight or panic-type of behavioral reaction is apparently an unusual behavior in fires.

Awareness of the Fire

Obviously, the way in which an individual is alerted to the presence of a fire may determine the degree of threat perceived. With vocal alerting systems in buildings, variations in voice quality, pitch, or volume, as well as the content of the message, tend to provide threat cues.[4]

Proulx and Sime[5] in their study involving evacuation drills in an underground rapid transit station found the use of directive public announcements with an alerting alarm bell was most effective in creating an immediate effective evacuation. Ramachandran[6] in his review of the research on human behavior in fires in the United Kingdom since 1969 summarized the effectiveness of alarm bells as awareness cues:

> The response to fire alarm bells and sounders tends to be less than optimum. There is usually skepticism as to whether the noise indicated a fire alarm and if so, is the alarm merely a system test or drill?[6]

Ramachandran[7] indicated that the development of "informative fire-warning systems," which utilize a graphic display with a computer-generated message and a high-pitched alerting tone have reduced the observed delay times in the initiation of practice evacuations. Cable[8] in his study of the response times of staff personnel to the fire alarm signal in Veterans Administration hospitals found the greatest delay in response time with the coded alarm-bell type systems. Kimura and Sime[9] found in the evacuation of two lecture halls with college students that the lecturer's verbal instructions were the determining factor in choosing to use the fire exit over the normal entrance and exit. Research literature developed from practice evacuations tends to indicate that the use of verbal directive informative messages may be most effective in reducing delay in evacuation initiation.

However, note that if verbal directive messages conflict with other awareness cues, such as the odor or sight of smoke, occupants may question the credibility of the message and disregard the information. One of the few documented cases of this type of situation occurred in the South Tower of the World Trade Center on April 17, 1975. Lathrop[10] reported that the fire occurred in a trash cart in a storage area on the fifth floor, adjacent to an open stairway door that allowed smoke to infiltrate the ninth through twenty-second floors at approximately 9:04 a.m. Occupants of these floors moved into the

John L. Bryan is professor emeritus of the Department of Fire Protection Engineering, College of Engineering, University of Maryland, and currently a consultant in Frederick, MD.

core area of the building, and the building communications center monitoring the core lobby areas verbally directed the people in these areas to remain calm and return to their office areas at 9:10 a.m. Despite this announcement, occupants remained in the core lobby areas and became more concerned about smoke conditions. Thus, with occupants on the affected floors becoming more anxious, an evacuation message was announced at 9:16 a.m.

As Burns[11] reported, simultaneous occupant evacuations occurred in the explosion and fire of February 26, 1993, which severely affected both towers and the Vista Hotel of the World Trade Center. The explosion disrupted the Center's communications center, and the occupants, having experienced within minutes the explosion, loss of power, and smoke infiltration of the floor areas, evacuated without the established verbal directional announcements utilized in previous practice evacuations.

Proulx and Fahy[12] in their questionnaire study of 382 trained fire warden personnel located in both towers of the World Trade Center at the time of the explosion and fire of February 26, 1993, found these personnel were alerted primarily in the following manner:[12]

> Respondents mentioned the following cues, either singly or in combination that something was occurring: hearing or feeling the explosion, loss or flickering of lights or telephones, smoke or dust, sirens and alarms, information from others, and people movement.

Most of the participants in residential occupancy studies were alerted initially to the fire by the odor of smoke. When the two categories "notified by family" and "notified by others" are combined, however, personal notification becomes the most frequently reported means of initial perception of fire, as indicated in Table 8-1A.[1] The category "noise" includes noise from persons moving downstairs and through corridors, plus miscellaneous noise sources, including the breaking of glass and the arrival of fire apparatus.

Table 8-1B presents a comparison of the means of awareness from British study participants[2] and U.S. studies' participants.[1] The number of stimuli was reduced because the British study had fewer categories. The U.S. responses have been adapted to the British categories. There was only one significant difference in the means of awareness between the two groups: 15 percent of the British study population became aware of the fire upon observing flame, contrasted with 8.1 percent of the U.S. participants.

A study of the NFPA-recommended smoke detector noise level of 75 dBA indicates that individuals with hearing impairments, those taking sleeping pills, or those on medication may require a detector noise level exceeding 100 dBA.[13] (See NFPA 72, *National*

TABLE 8-1A. *Means of Awareness of the Fire Incident (U.S. Studies)*

Means of Awareness	Participants	Percent
Smelled smoke	148	26.0
Notified by others	121	21.3
Noise	106	18.6
Notified by family	076	13.4
Saw smoke	052	09.1
Saw fire	046	08.1
Explosion	006	01.1
Felt heat	004	00.7
Saw/heard fire department	004	00.7
Electricity went off	004	00.7
Pet	002	00.3
N = 11	569	100.0

TABLE 8-1B. *Comparison of British and U.S. Study Results Relative to Means of Awareness of Fire Incident*

Means of Awareness	British Percent	U.S. Percent	$P_1 - P_2$	$SE_{P_1 - P_2}$§	CR
Flame	15.0	8.1	6.9	1.64	*4.21**
Smoke	34.0	35.1	1.1	2.27	0.48
Noises	9.0	11.2	2.2	1.41	1.56
Shouts and told	33.0	34.7	2.7	2.25	1.20
Alarm	7.0	7.4	0.4	1.23	0.33
Other	2.0	2.8	0.8	0.70	1.14
	2193	569			

* Critical ratio (CR) significant at or above the 1 percent level of confidence.
§ Standard error.

Fire Alarm Code.) Flashing or activated lights are effective fire signals in occupancies populated primarily by hearing-impaired persons.[14] The 1981 edition of NFPA *101®*, *Life Safety Code®*, permitted the flashing of exit signs along with activation of an audible fire alarm system for the first time.

A study of 24 male subjects designed to determine whether they were awakened by a smoke detector's audible alarm signal and could identify fire cues found that the subjects slept through the alarm signals at a signal-to-noise ratio of 10 dBA and consistently failed to identify the awakening cue or radiant heat and smoke odor cues as fire warnings.[15] Other researchers have indicated the alarm-signal-to-noise ratio is attenuated by physical surroundings.[16] A signal passing through a ceiling or wall may be reduced by 40 dBA, while a signal passing through a door may be reduced by 15 dBA. In addition, the signal could be masked by a typical residential air conditioner noise level of 55 dBA.

The acknowledgment of ambiguous threat cues as signaling an emergency may be inhibited by the presence of other people. Recognition of this phenomenon resulted in an experiment involving college students.[17] While the students were completing a written questionnaire, the experimenter introduced smoke into the room through a small vent in the wall. If the students left the room and reported the smoke, the experiment was terminated. If the students did not report the presence of the smoke within 6 min from the time they first noticed, the experiment was considered complete. Students alone in the room reported the smoke in 75 percent of the cases. When two passive, noncommittal persons joined each student, only 10 percent of the students reported the smoke. When the experimental group consisted of three naive subjects, one individual reported the smoke in only 38 percent of the groups. Of the 24 persons involved in the eight naive-subject groups, only one reported the smoke within the first 4 min of the experiment. In the single-subject situation, 55 percent of the subjects reported the smoke within 2 min and 75 percent in 4 min.

The study reported that reaction to the smoke was apparently delayed by the presence of other persons, with the median being 5 sec for single subjects but 20 sec in both group conditions. These results undoubtedly reflect constraints that people accept regarding their behavior in public places. The performance of naive subjects in the passive-confederate situation was reported as follows:

> The other nine stayed in the waiting room as it filled up with smoke, doggedly working on their questionnaires, and waving the fumes away from their faces. They coughed, rubbed their eyes, and opened the window but did not report the smoke.

It has been suggested that, while trying to interpret the emergency potential of ambiguous threat cues, the individual is influenced by the reactions of others. Should these others remain passive and seem to interpret the situation as a nonemergency, the individual tends to modify his or her own interpretation according to this inhibiting social influence.[17] This behavioral experiment may help explain the reported tendency of people to disregard threat cues or interpret them as nonthreatening when they occur in places where there are many other people, such as restaurants, movie picture theaters, or department stores. These experimental results may help explain the calls received by fire departments minutes or even hours after an incident is first detected. In the report of the Arundel Park Hall fire,[18] several of the sample population indicated that when they entered the hall after observing the fire from outside the building, they warned their friends and suggested they leave but were laughed at, their warnings apparently disregarded.

The processes of social inhibition, diffusion of responsibility, and mimicking were indicated to be primarily responsible for the inhibition of adaptive and assistance behavior in emergencies. The inhibition of behavior in the early stages of a fire, when the cues are relatively ambiguous, may lead to nonadaptive flight behavior because the time available for evacuation has been expended. It is sometimes difficult to get the occupants of a building to evacuate because of social inhibition and diffused responsibility. The tendency to adopt cues for behavior from others is well documented in fires in restaurants, other public assembly occupancies, and hotels.

DECISION PROCESSES OF THE INDIVIDUAL

Seven processes have been identified that an individual may use in trying to structure and evaluate situational threat cues.[19] Six of these—recognition, validation, definition, evaluation, commitment, and reassessment—are presented in Figure 8-1A. The seventh, a lattice involving the failure of successive defensives and a hierarchy of defenses, is not relevant to the decision process in fires.

Recognition

The process of recognition occurs when the individual perceives cues that indicate a threatening fire. The cues may be very ambiguous and not clearly indicative of a severe fire. However, the cues usually are continuous, with an increasing intensity due to the dynamics of flame, heat, and smoke production. An individual is pre-

disposed to recognize threat cues in terms of the most probable occurrences, usually in relation to past experience and in the form of an optimistic wish. The optimistic wish aspect of the response may be a direct result of the individual's concept of his or her personal invulnerability.

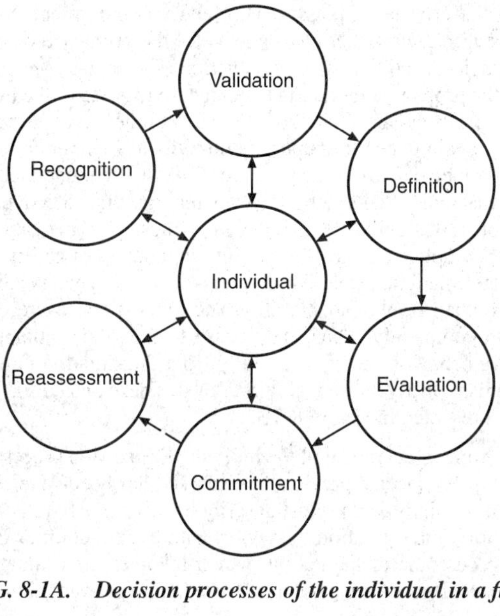

Threat recognition is important for fire protection. The adaptive action involved in the initiation of the fire alarm, the evacuation of building occupants, and the suppression of the fire may be delayed or postponed if individuals do not perceive the cues as indicative of a fire. The ambiguous nature of threat cues indicates that individuals who do not have specialized fire prevention or fire protection education and experience recognize only large amounts of smoke or sudden and threatening flames as indicative of a threatening fire.

Validation

The process of validation consists of attempts by an individual to determine the seriousness of the threat cues, usually by reassurance of the mild nature of the threat and its improbability. When the cues are significantly ambiguous, however, the individual tries to obtain additional information. In other words, the person is aware that something is happening but is not sure exactly what. This process of validation may be conducted by questioning other nearby individuals. Studies of the explosion of a fireworks plant in Houston, TX, found that, of the 139 persons interviewed, 85 individuals—or 61 percent of the population—obtained information on the source and nature of the explosion and smoke from someone they saw or from someone who telephoned and told them.[20] The presence of others during the threat recognition and validation process was found possibly to inhibit or influence the behavioral responses of the individual.

Definition

The definition process essentially consists of an attempt by the individual to relate the information concerning the threat to some of the variables, such as the qualitative nature of the threat, the magnitude of deprivation of the threat, and the time context. The individual's stress and anxiety appear to be most severe before he or she has determined the situation's structure or meaning, although it is apparent that the situation requires interpretation. The individual's role (described at the end of the "Evaluation" section) is one of the critical factors in the situation, relative to the personalization of the threat and the physical environment. The most important physical aspects in the definition process are the generation, intensity, and propagation of the smoke, flames, and thermal exposure.

Evaluation

The process of evaluation may be described as the cognitive and psychological activities required for the individual to respond to the threat. The individual's ability to reduce his or her stress and anxiety levels becomes the essential psychological factor. In the threat situation created by a fire, evaluation is the process involved in the decision to react by fight or flight. With evaluation, an initial decision to make an overt behavioral response is completed. Because of the time context of the generation and propagation of the fire, the mental processes up to and including the process of evaluation may have to be accomplished within several seconds.

Sime[21] emphasized the importance of the individual's perception of the time available for evacuation or to reach a refuge area as being the individual's estimation of the fire threat. He indicates that the "perceived time available" depends upon the information and communication provided to the occupants concerning the fire's location and development. Variables of the physical environment are

FIG. 8-1A. *Decision processes of the individual in a fire.*

an important source of information for individuals involved in formulating adaptation, escape, or defense plans. Additional determinants may be the location of the individual relative to the egress routes, other people, the untenable effects of the fire, and the behavior of other individuals.

During the process of evaluation, an individual may decide to leave the building—flight—or to use a portable fire extinguisher—fight. During this time, he or she is particularly susceptible to the actions and communications of others. Thus, the individual may mimic the behavioral reactions of observed individuals, resulting in mass adaptive or nonadaptive behavior rather than selective, individualized behavior. The following situation at an auto sales and service agency in 1971 demonstrates what may have been an instance of mimicked behavior becoming normative group behavior:[22]

> About 10 p.m., the fire department received an alarm from a street fire alarm box. When fire fighters arrived, the 150 by 200 ft (46 by 61 m), one- and two-story building of wood frame and hollow block construction was well alight and nearly 300 spectators were watching the fire in 10°F (–12°C) weather. An investigation revealed the fire had been burning for about 90 minutes before the fire department was notified.

In studies of nonadaptive group behavior, the concept that this mode of behavior depends directly on the individual's perception of the reward structure of a situation has been developed.[23] People in a building who are confronted with a fire probably would initially perceive a reward structure that would encourage cooperative and adaptive behavior; in such cases, everyone should be able to reach and proceed through the available exits. However, the reward structure perceived by some individuals more remote from the egress routes could result in competitive behavior. Such individuals would perceive that cooperative behavior would make it impossible for them to reach an exit in time to escape the fire. Once the pattern of competitive behavior is initiated, the behavior pattern of the group may become one of intense individual competition for the escape routes.

In the evaluation process, an individual's cultural influences and assumption of a particular role may be very important in formulating defense or escape plans. It is believed that an individual playing a familiar role that is also suitable for the threat situation experiences less anxiety and responds with more adaptive behavior than an individual in an unfamiliar role confronted with an unfamiliar threat.

Jones and Hewitt[24] conducted detailed interviews with 40 occupants of a 27-story office building who had evacuated the building during a fire. It should be noted the fire occurred at 9:00 p.m. when the fire management plan was not in effect due to the building's reduced occupancy. In this situation it appeared that leadership and evacuation group formation were related to the occupants' fire training and their normal roles. The investigators found the relationship of the occupancy roles and the normal or emergent leadership of the occupants were critical factors in successful evacuation with the following variables:

> The social and organizational characteristics of the occupancy, including what a person knows (or believes) of the situation, whether the person is alone or part of a group, the normal roles that people hold within the occupancy, and the organizational structure or framework. One factor that appears to be related to the chosen evacuation strategy of an occupant is the presence of leadership and the form which that leadership takes.

Horiuchi, Murozaki, and Hokugo[25] reported on a questionnaire study of 458 occupants of an 8-story office building involved in a fire incident. The researchers found significant differences between occupants who were familiar with the building and occupants attending training sessions who were unfamiliar with the building, relative to their actions, selection of evacuation routes, and effectiveness in exiting. The regular occupants of the building engaged in fire-fighting actions and alerted or assisted other occupants, while occupants unfamiliar with the building primarily engaged in evacuation behavior.

Commitment

Commitment consists of the mechanisms the individual uses to initiate the behavioral activity required to fulfill the defense plans conceptualized in the evaluation process. This overt response to the threat of fire results in success or failure. If the response fails, the individual immediately becomes involved in the next process of reassessment and commitment. If the action succeeds, the anxiety and stress aspects of the situation are reduced and relieved, although the severity of the general fire situation may have increased.

Reassessment

The processes of reassessment and overcommitment are the most stressful of the individual's processes because previous attempts to adjust to the threat have failed. Thus, more intense effort goes into the behavioral reactions, and the individual tends to become less selective in the choice of response. Encountering successive failures, the individual becomes more frustrated. The possibility of injury and risk increases with a greater activity level and with less probability of success, as was demonstrated in the Arundel Park Hall fire. There, the number of those who selected windows as a means of escape increased as people became involved in their second escape attempts.[18]

In analyzing the behavior of an individual involved in the processes of recognition, validation, definition, evaluation, commitment, and reassessment, it must be remembered that these are dynamic processes. They are constantly being modified in relation to their magnitude, velocity, and intensity. A person's usual psychological and physiological activities probably will be below normal during the recognition process, when he or she is concentrating on perceiving the threat cues. During the process of validation and definition of the threat, adjacent members of the threatened population communicate overtly. The period of hyperactivity appears to occur initially during the process of commitment and to become intense during the process of reassessment and recommitment. Stress increases with each successive stage, since the primary motivation of the behavioral activity is stress reduction. The appearance, the proximity, the propagation, the time, and the toxic gases of the fire threat also tend to predispose the individual to a higher level of behavioral activity, again depending upon the individual's perception of these threat variables. During the process of reassessment and recommitment, the individual's activity level may assume the hyperactive mode of frantic activity, or it may be expressed in the catastrophic state of complete physical immobility with a loss of ability to communicate coherently. These individuals appear to perceive the threat as above their level of adaptability. The stress is too severe, and they give up completely. Thus, they cease to make any attempt at an adaptive behavior and retreat totally from the situation through the mechanism of psychological withdrawal. These behavioral dynamics are presented in Figure 8-1B.

A conceptual model of the individual's decision processes similar to some concepts previously discussed has been developed. Instead of the six processes just described, only three have been used: recognition/interpretation; behavior, with either action or inaction; and the outcome of the action, which involves the evaluation and long-term effect of the behavior.[26] The behavior evaluation is simi-

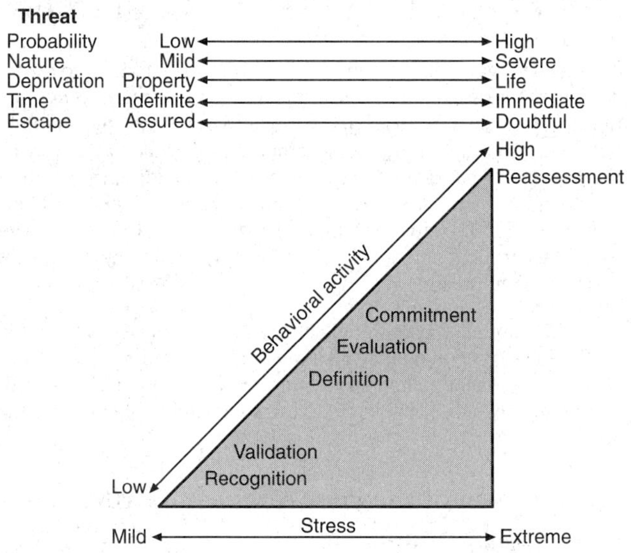

Threat

Probability	Low ← →	High
Nature	Mild ← →	Severe
Deprivation	Property ← →	Life
Time	Indefinite ← →	Immediate
Escape	Assured ← →	Doubtful

FIG. 8-1B.   *Dynamics of behavioral activity of the individual.*

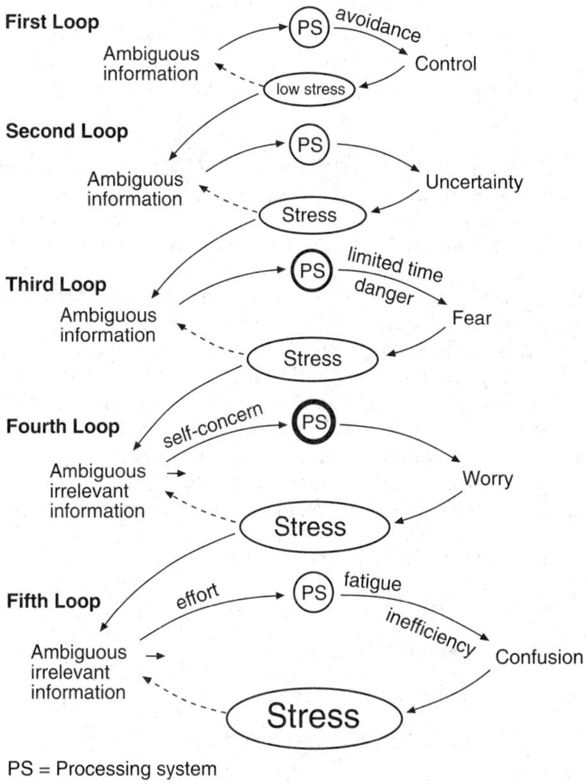

PS = Processing system

FIG. 8-1D.   *Stress model of people in a fire situation.*[28]

lar to the process of reassessing the decision model. Both the recognition/interpretation concept and the behavior concept involve factors critical to the decision processes. Experience and immediate circumstances all have an impact on the recognition/interpretation concept. It has been emphasized that individuals in a fire may not know right away that they are involved in a fire and may not know where, in relation to their location, the fire is developing or where the egress routes are. This conceptual model is presented in Figure 8-1C.

The conceptual model just described has been modified into one involving three phases: detection of cues, definition of the situation, and coping behavior. In addition, tentative determinants of the behavior have been developed that increase the probability of detection and of fire suppression.[27]

Proulx[28] developed a stress model to demonstrate various levels of stress generation within an individual involved in the decision process during a fire incident. Figure 8-1D illustrates this stress model, which should be compared with the behavioral activity dynamics of the individual in a fire incident presented in Figure 8-1B. The left side of Figure 8-1D indicates the information the individual must process; the right side indicates the emotional state that results. Proulx describes the five loops in the model as follows:

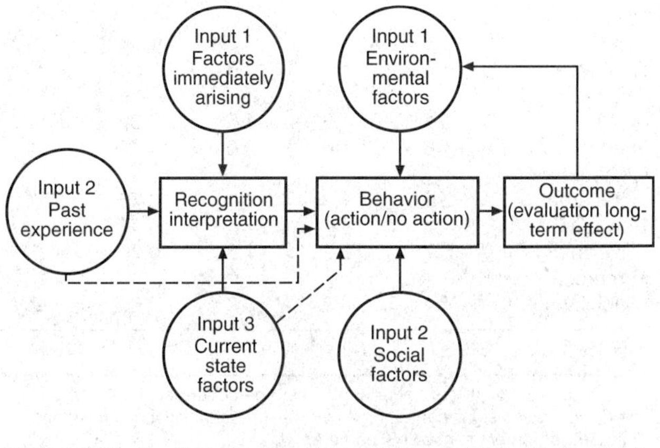

FIG. 8-1C.   *Preliminary heuristic systems model of behavior in fires.*

The first loop of the stress model starts with the perception of ambiguous information. This information is decoded in the processing system (PS in the figure) for interpretation. Given that the available information may not allow for a straightforward assessment of the situation, people will at first minimize or deny the situation. These defensive strategies of avoidance lead to an absence of reaction.

Although individuals may vary considerably in their appraisal of the same event, the repeated perception of ambiguous information will eventually generate a state of uncertainty which will then induce a feeling of stress. Some time can be spent going repeatedly through the second loop of the stress model.

The third loop of the stress model is related to the interpretation of the situation as an emergency. The thicker line around the processing system expresses the pressure of the overload of information with which the person tries to deal at once. The fear felt by the person is a manifestation of a specific appraisal of the environment.

The fourth loop of the model relates to the person's processing of irrelevant information and is represented by the very thick line around the processing system. This irrelevant information creates worry and more stress. The irrelevant information, created by the person, is caused by concern for his or her own performance in coping with the situation. Perceived feelings of arousal and fear, uncertainties regarding how to proceed with the problem, difficulties in interpreting what exactly is going on and self-estimation of the efficiency of already-applied actions all become additional information to process.

The fifth and last loop of the model supposes an investment of more mental effort to master the problem, momentarily reducing the pressure on the processing system but resulting in fatigue and inefficiency manifested in a state of confusion.

Proulx indicates that definitive, valid, and directive information provided to occupants of a building in a fire incident most effectively reduces stress and thus tends to minimize response delays created in the first and second loops of the stress model.

Chubb[29] proposes that the model of the decision processes that fire department officers use in incident command procedure be adopted as the decision process of building occupants in a fire situation. The decision model was developed from the theory of naturalistic decision-making, which evolved from studies of decision-makers in complex, time-critical situations. The critical variables of naturalistic decision-making theory appear to share many environmental and psychological features of fire situations involving building occupants. Chubb identified these critical variables:

Ill-defined goals and ill-structured tasks

Uncertainty, ambiguity, and missing data

Shifting and competing goals

Dynamic and continually changing conditions

Real-time reactions to changed conditions

Time stress

High stakes

Organizational goals and norms

Experienced decision makers

Figure 8-1E illustrates the Recognition-Primed Decision Model (RPD) developed by Klein[30] from studies of fire department officers. Chubb correctly indicated this model's limitation when applied to building occupants: They lack the dynamic abilities—suitable training and previous experience—in building fire situations that fire officers have. Static abilities relative to building occupant's mental and physical capabilities also appear to be more varied and limited than fire officers'. Chubb indicates that successful recognition-primed decision making depends on occupant training and practice of fire safety plans, with the decision support system in the building consisting of egress signs, emergency lighting, and vocal communication systems.

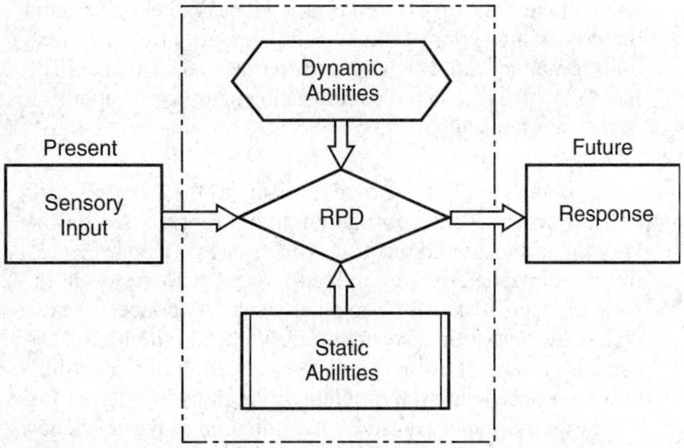

FIG. 8-1E. *Recognition-primed decision model.*

BEHAVIOR ACTIONS OF OCCUPANTS

A study involving 952 fires and 2193 individuals interviewed by fire department personnel at the fire scene in Great Britain found that the most frequent responses to fire involved evacuating the building, fighting or containing the fire, and alerting other individuals or the fire brigade.[2] The identical type of broad behavior categorization was found in a similar study, which involved interviewing 584 participants in 335 fire incidents in the U.S. Interviews were conducted by fire department personnel, who used a structured questionnaire at the scene of the fires.[1]

The examination of initial actions to fire is presented in Table 8-1C. The behavior of the individuals also varied by sex. Males were predominately active in fighting the fire, while females were predominately concerned with alerting and helping others leave the building.

Comparison of the First Actions of British and U.S. Study Populations

It should be noted that there were ten statistically significant differences between the British and U.S. populations. The U.S. study population was more likely to report five categories of first actions: "notified others," "got dressed," "got family," "left area," and "entered the building." A greater percentage of the British population reported as first actions "fought fire," "went to fire area," "closed door to fire area," "pulled fire alarm," and "turned off appliances."

TABLE 8-1C. *Comparison of the First Actions of British and U.S. Study Population*

Actions	British Percent	U.S. Percent	$P_1 - P_2$[††]	$SE_{P_1 - P_2}$[§]	CR[‖]
Notified others	8.1	15.0	6.9	1.38	*5.00†*
Searched for fire	12.2	10.1	2.1	1.51	1.39
Called fire dept.	10.1	9.0	1.1	1.40	0.79
Got dressed	2.2	8.1	5.9	0.85	*6.94†*
Left building	8.0	7.6	0.4	1.27	0.31
Got family	5.4	7.6	2.2	1.11	*1.98**
Fought fire	14.9	10.4	4.5	1.63	*2.76†*
Left area	1.8	4.3	2.5	0.70	*3.57†*
Nothing	2.1	2.7	0.6	0.69	0.87
Had others call fire dept.	2.8	2.2	0.6	0.76	0.79
Got personal property	1.2	2.1	0.9	0.55	1.64
Went to fire area	5.6	2.1	3.5	1.01	*3.47†*
Removed fuel	1.2	1.7	0.5	0.53	0.94
Entered building	0.1	1.6	1.5	0.30	*5.00†*
Tried to exit	1.6	1.6	0.0	0.0	0.0
Closed door to fire area	3.1	1.0	2.1	0.76	*2.76†*
Pulled fire alarm	2.7	0.9	1.8	0.70	*2.57**
Turned off appliances	4.1	0.9	3.2	0.85	*3.20†*
N = 18	2193	580			

*Critical ratios significant at or above the 5 percent level of confidence.
† Critical ratios significant at or above the 1 percent level of confidence.
†† British percent minus U.S. percent.
§ Standard error.
‖ Critical ratio.

The general classification of the three early actions for British and the U.S. study populations alike were categorized as "evacuation," "reentry," "fire fighting," "moved through smoke," and "turned back" behavior. Comparisons of the two populations are presented in Table 8-1D. There was a statistically significant difference in every category except "moved through smoke."

TABLE 8-1D. *Comparison of the Behavior of the British and U.S. Study Populations*

Behavior	British Percent	U.S. Percent	$P_1 - P_2$[†]	$SE_{P_1 - P_2}$[§]	CR[‖]
Evacuation	54.5	80.0	25.5	2.30	*11.09**
Reentry	43.0	27.9	15.1	2.30	*6.57**
Fire fighting	14.7	22.9	8.2	1.74	*4.71**
Moved through smoke	60.0	62.7	2.7	2.29	1.18
Turned back	26.0	18.3	7.7	2.01	*3.83**
	2193	584			

* Critical ratios significant at or above the 1 percent level of confidence.
† British percent minus U.S. percent.
§ Standard error.
‖ Critical ratio.

Comparison of the Behavior of the British and U.S. Study Populations

An interesting aspect of the U.S. study population involves variation in the first, second, and third actions reported by the participants. Table 8-1E presents the three actions for the group, totaling 584 individuals. "Notifying others" accounted for 15 percent of the first actions, but, by the time of the third actions, it accounted for only 5.8 percent. A similar reduction in frequency can be observed for "searching for the fire." This activity decreases from 10.1 percent as the first action to 0.8 percent as the third action. The actions "getting dressed" and "got family" also diminished in frequency with the progression from the first to the third action as time passed during the fire. In contrast, "left building," "fought fire," and "called fire department" increased in frequency from the first to the third actions.

Summary of the First, Second, and Third Actions of the Occupants

Canter, Breaux, and Sime[31] developed a decomposition diagram of the acts of 41 persons in 14 domestic fires. This study, conducted in the United Kingdom, covers home fires, as do the studies by Wood[2] and Bryan[1] discussed previously. Figure 8-1F presents this decomposition diagram and should be compared with Tables 8-1C through 8-1E. The sequence of the first, second, and third actions of the U.S. study population are generally similar to the sequence of actions in the decomposition diagram.

Summary of the First, Second, and Third Actions of the Occupants According to Sex

Differences between the first actions of the participants according to their sex have been examined. Table 8-1F presents the reported initial actions of the U.S. study population relative to the sex of the participants.

TABLE 8-1E. *Summary of the First, Second, and Third Actions of the Occupants*

Actions	1st Action (percent)	2nd Action (percent)	3rd Action (percent)
Notified others	15.0	9.6	5.8
Searched for fire	10.1	2.4	0.8
Called fire dept.	9.0	14.6	12.7
Got dressed	8.1	1.8	0.3
Left building	7.6	20.9	35.9
Got family	7.6	5.9	1.4
Fought fire	4.6	5.7	11.5
Got extinguisher	4.6	5.3	1.6
Left area	4.3	2.8	1.1
Woke up	3.1	0.0	0.0
Nothing	2.7	0.0	0.0
Had others call fire dept.	2.2	4.0	4.1
Got personal property	2.1	3.8	0.8
Went to fire area	2.1	1.0	0.0
Removed fuel	1.7	1.0	1.1
Entered building	1.6	0.8	1.1
Tried to exit	1.6	2.4	0.5
Went to fire alarm	1.6	1.8	1.1
Telephoned others	1.2	0.6	1.1
Tried to extinguish	1.2	1.8	1.9
Closed door to fire area	1.0	0.2	0.3
Pulled fire alarm	0.9	0.6	0.5
Turned off appliances	0.9	0.6	0.3
Checked on pets	0.9	1.4	0.5
Awaited fire dept. arrival	0.0	1.0	3.6
Went to balcony	0.2	0.8	2.7
Removed by fire dept.	0.0	0.0	1.6
Opened doors/windows	0.2	0.4	1.1
Other	3.9	8.0	6.6
N = 29	100.0	100.0	100.0
Range	0–87	0–106	0–131
Percent of participant population	99.3	86.6	62.9

The First Actions of the Occupants Relative to the Sex of the Occupant

Statistical differences between males and females are significant in the categories "searched for fire," "called fire department," "left building," "got family," and "got extinguishers." Male participants more frequently reported investigation and fire-fighting activities. For example, 14.9 percent of the males "searched for fire," as opposed to 6.3 percent of the females, and 6.9 percent of the males "got extinguishers," as opposed to 2.8 percent of the females. Females more frequently reported warning and evacuation activities. For example, 11.4 percent of the females "called fire department" as their initial action, as opposed to 6.1 percent of the male participants. In relation to evacuation behavior, 10.4 percent of the females reported "left building" as the first action, contrasted with 4.2 percent of the male participants. The influence of cultural roles is probably indicated explicitly in the concern for other family members— 11 percent of the females "got family" as the first action, while only 3.4 percent of males engaged in this initial action. It should be noted that the male actions of "searched for" or "fought fire" were matched by the female actions of "alarm initiating" and "evaluation behavior."[1] This behavior also has been observed in health care and educational situations.

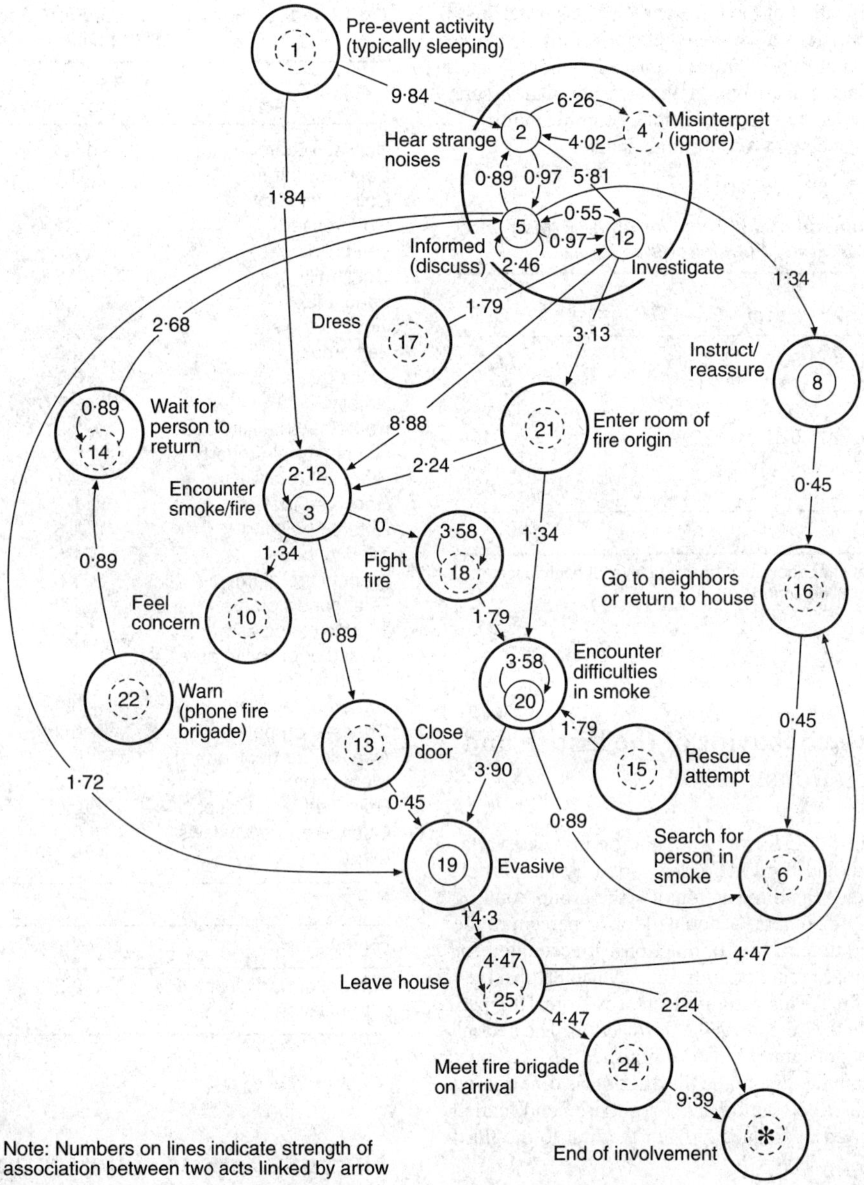

Note: Numbers on lines indicate strength of
association between two acts linked by arrow

FIG. 8-1F.   Decomposition diagram—domestic fires.

Behavior in Hotel Fires

Fire protection of high-rise buildings and their occupants was tested severely by the MGM Grand Hotel fire in Clark County, Nevada, on November 21, 1980,[32] and by the subsequent fire at the Las Vegas Hilton Hotel on February 10, 1981.[33] Both these hotel fires resulted in injuries and fatalities among guests. NFPA conducted an intensive questionnaire study of the guests registered in the MGM Grand Hotel on the evening of November 20–21, 1980.[34]

The MGM Grand Hotel fire was discovered by a hotel employee who entered the unoccupied deli-restaurant located on the casino level of the hotel at approximately 7:10 a.m. November 21, 1980. As instructed, the hotel telephone operator immediately notified the Clark County Fire Department at approximately 7:18 a.m. The telephone operators were forced from their switchboard by the smoke immediately after they had made an announcement over the public address system, at approximately 7:20 a.m., to evacuate the

casino area. The fire quickly reached flashover in the deli, immediately spread from east to west through the main casino area, and extended out the west portico doors on the casino level immediately following the arrival of the first fire department personnel.

An addition to the hotel was being constructed adjacent to the west end of the building, and construction workers helped warn and evacuate guests and assisted in fire fighting. The heat and smoke rapidly extended from the casino area through the seismic joints, elevator shafts, and stairways throughout the 21 residential floors of the hotel. The heat was intense enough on the 26th floor, a top floor, to activate automatic sprinklers in the lobby adjacent to the elevator shafts.

Due to the rapid early evacuation of the telephone staff, guests were not alerted by the hotel public address system or the local fire alarm system. Guests warned early in the fire, and those already awake and dressed, were able to escape before the smoke became untenable on the upper floors. Guests alerted later remained in their

TABLE 8-1F.   *The First Actions of the Occupants Relative to the Sex of the Occupant*

First Action	Male (percent)	Female (percent)	$P_1 - P_2$††	$SE_{P_1 - P_2}$§	CR‖
Notified others	16.3	13.8	2.5	2.98	0.83
Searched for fire	14.9	6.3	8.6	2.51	*3.43 †*
Called fire dept.	6.1	11.4	5.3	2.41	*2.19**
Got dressed	5.8	10.1	4.3	2.30	1.87
Left building	4.2	10.4	6.2	2.22	*2.79 †*
Got family	3.4	11.0	7.6	2.22	*3.42 †*
Fought fire	5.8	3.8	2.0	1.77	1.13
Got extinguishers	6.9	2.8	4.1	1.77	*2.31**
Left area	4.6	4.1	0.5	1.70	0.29
Woke up	3.8	2.5	1.3	1.45	0.90
Nothing	2.7	2.8	0.1	1.38	0.72
Had others call fire dept.	3.4	1.3	2.1	1.23	1.71
Got personal property	1.5	2.5	1.0	1.17	0.85
Went to fire area	1.9	2.2	0.3	1.20	0.25
Removed fuel	1.1	2.2	1.1	1.08	1.02
Entered building	2.3	0.9	1.4	1.02	1.37
Tried to exit	1.5	1.6	0.1	1.05	0.09
Went to fire alarm	1.1	.19	0.8	1.02	0.78
Telephoned others	0.8	1.6	0.8	0.91	0.87
Tried to extinguish	1.9	0.6	1.3	0.91	1.43
Closed door to fire area	0.8	1.3	0.5	0.87	0.57
Pulled fire alarm	1.1	0.6	0.5	0.75	0.66
Turned off appliances	0.8	0.9	0.1	0.79	0.12
Checked on pets	0.8	0.9	0.1	0.79	0.12
Other	6.5	2.5	4.0	1.70	*2.35**
N = 25	262	318			

* Critical ratios significant at or above the 5 percent level of confidence.
† Critical ratios significant at or above the 1 percent level of confidence.
†† Male percentage minus female percentage.
§ Standard error.
‖ Critical ratio.

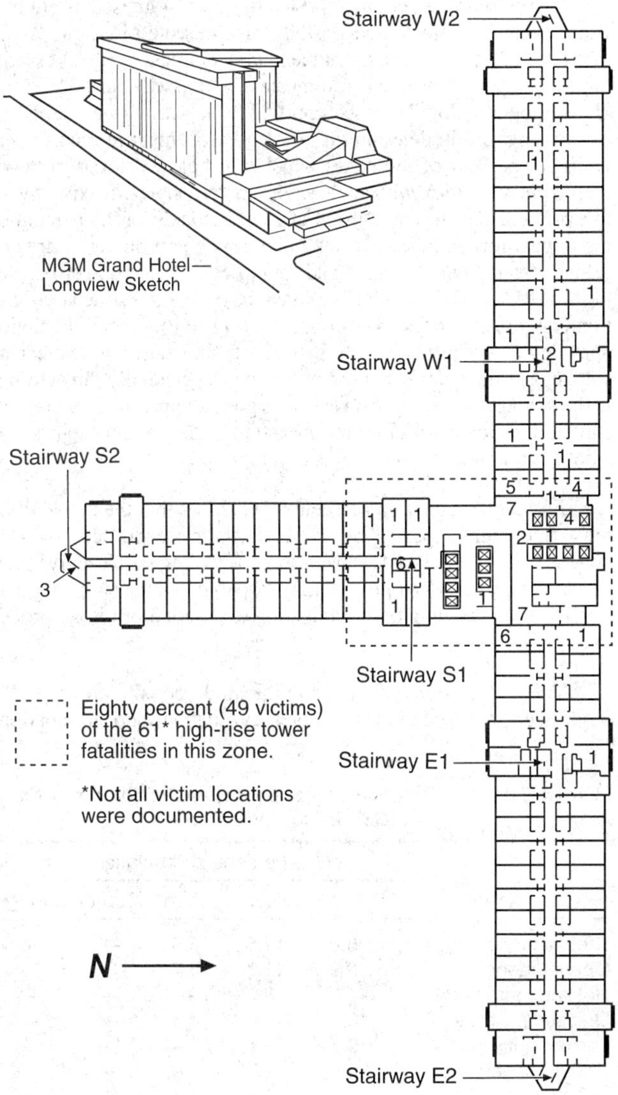

FIG. 8-1G.   Residential floor diagram of the MGM Grand Hotel.

rooms or moved to other rooms, usually with other occupants. The fire itself did not extend above the casino level, except in a rather minor nature, into two guest rooms on the fifth floor. The fire resulted in 85 fatalities and injured 778 guests and 7 hotel employees. Seventy-nine body locations were documented: 18 on the casino level, 25 in guest rooms, 22 in corridors and lobbies, 9 in stairways, and 5 in elevators. The victims were found on the casino level and on the 16th and upper floors, with the majority between the 20th and the 25th floors.

Figure 8-1G is a diagram of the guest floor of the MGM Grand Hotel that was used in the engineering study conducted by NFPA.[32] Of the nine victims found in the stairways, 2 were in Stairway S2 at the extreme south end of the south wing on the 17th floor; 6 were between the 20th and 23rd floor in Stairway S1 at the central end of the south wing; and 1 victim was found at the ground-floor level of Stairway W2 at the extreme west end of the west wing. There are various estimates of the number of guests and fire department personnel who suffered injuries at the MGM Grand Hotel fire. Morris indicated that 619 people were taken to hospitals and another 150 were treated at the Las Vegas Convention Center, where the survivors were transported from the hotel.[35]

The MGM Grand Hotel tragedy was a unique fire, especially from two aspects: it was the second most serious hotel fire in U.S. history, surpassed only by the Winecoff Hotel fire in Atlanta, GA, on December 7, 1946, which killed 119; and it was the first high-rise fire in the U.S. in which helicopters evacuated large numbers of people. About 300 were evacuated in this manner; the fire department rescued approximately 900 people by other means.

Shortly after the MGM Grand Hotel fire, NFPA prepared a 4-page, 28-item questionnaire that included the floor plan of the guest rooms. A total of 1960 questionnaires were mailed, and 554, or approximately 28 percent, of these were returned. Four hundred fifty-five of the responding individuals indicated a willingness to be interviewed.

The age of the questionnaire population ranged from 20 to 84 years, with an average age of 45. The population consisted of 331 males and 222 females; one respondent did not indicate a sexual classification. One hundred and three guests indicated that they were alone at the time they became aware of the fire in the hotel. The presence of other people, especially if they belong to the individual's primary group, appears to be a determinant of the response of many individuals in residential fires.[1]

The initial five actions reported by the 554 guests, as elicited from the NFPA questionnaire study, are presented in Table 8-1G. Notice that the five most frequent first actions were "dressed," "opened door," "notified roommates," "dressed partially," and "looked out window." Guests reporting these actions were predominately engaged in determining the degree of threat to themselves. Only 7.9 percent of the study population began or tried to begin their own evacuation with such actions as "attempted to exit," "went to exit," and "left room." A total of 16 individuals, or 2.9 percent of the population, initiated actions to improve the room as an area of refuge: "wet towels for face" and "put towels around door." The actions of the guests could be classified, in general, as evacuation actions or refuge processes. Actions relating to evacuation behavior appeared to be initiated early if the egress passages were clear of smoke or if the smoke was not perceived as personally threatening. If the smoke was heavy, however, the guests apparently decided to stay in their rooms or other rooms and to initiate actions to prevent smoke migration into the rooms of refuge.

Further examination of Table 8-1G shows that the five actions most frequently reported by guests as their second actions were "opened door," "dressed," "went to exit," "dressed partially," and "secured valuables." Approximately 19 percent of the study population reported dressing before initiating evacuation or refuge procedures.

The third actions of guests in the study population generally progressed to evacuation, attempted evacuation, and notification.

TABLE 8-1G. *First Five Reported Actions of Guests in the MGM Grand Hotel Fire*

Actions	Percent of Population				
	First	Second	Third	Fourth	Fifth
Dressed	16.8	11.6	6.5	—	—
Opened door	15.9	11.7	6.7	3.4	—
Notified roommates	11.6	3.0	—	—	—
Dressed partially	10.1	7.5	4.5	—	—
Looked out window	9.7	5.7	—	—	—
Got out of bed	4.5	—	—	—	—
Left room	4.3	5.4	8.1	2.4	2.0
Attempted to phone	3.4	3.6	—	2.8	—
Went to exit	2.5	10.3	9.5	16.1	6.7
Put towels around door	1.6	2.5	3.0	6.8	7.7
Felt door for heat	1.3	2.3	—	—	—
Wet towels for face	1.3	3.7	6.3	4.6	7.9
Got out of bath	1.1	—	—	—	—
Attempted to exit	1.1	3.0	5.8	4.3	—
Secured valuables	—	6.8	4.3	—	—
Notified other room	—	3.4	2.2	—	—
Returned to room	—	—	3.9	8.4	4.1
Went down stairs	—	—	3.9	5.4	21.3
Left hotel	—	—	3.4	2.6	2.0
Notified occupants	—	—	3.0	—	—
Went to another exit	—	—	—	3.6	4.8
Went to other room	—	—	—	3.6	3.6
Went to other room/ others	—	—	—	3.4	8.7
Looked for exit	—	—	—	2.4	—
Broke window	—	—	—	—	4.3
Offered refuge in room	—	—	—	—	1.8
Went up stairs to roof	—	—	—	—	2.9
Went to balcony	—	—	—	—	1.8
Other	14.8	19.5	28.9	30.2	20.1
Total (percent):	100.0	99.1	96.9	90.6	79.6
No. of guests:	554	549	537	502	441

Approximately 25 percent of this population was involved in evacuation actions, and approximately 10 percent attempted evacuations, as identified by the third actions of "attempted to exit" and "returned to room." The alerting and notification actions are identified as "notified occupants" and "notified other room."

The fourth actions of the guests in the study population indicate a progression to evacuation, attempted evacuation, and self-protection or room refuge actions. The most frequently reported fourth action was "went to exit"; approximately 16 percent of the population indicated that they did this. However, when one combines the guests involved in this action with those who "went down stairs," "went to another exit," "left hotel," and "left room," a total of 151 guests, or approximately 30 percent of the fourth action guest population, were involved in evacuation actions. The process of forming convergence clusters was noted in this hotel fire. This action involved individuals clustering together in rooms they considered areas of refuge with individuals they usually characterized as strangers before the fire. The fourth actions of "went to other room" and "went to other room/others" are explicit indicators of the formation of convergence clusters.[36]

The fifth actions of the guests were primarily for self-protection and included improving the room as an area of refuge and evacuation behavior. The evacuation actions were: "went up stairs to roof," "left hotel," and "left room." Involved in these evacuation actions were 175 individuals, or approximately 40 percent of the study population. Those unable to evacuate, and thus vitally concerned with refuge procedures, reported the fifth actions of "went to other room/others," "wet towels for face," "put towels around door," "broke window," "returned to room," "went to other room," "offered refuge in room," and "went to balcony." Approximately 40 percent of the fifth-action study population was involved in refuge procedures and self-protection actions.

Convergence Clusters

The phenomenon of convergence cluster formation was first noticed in a study of occupant behavior in a high-rise apartment building fire in 1979.[37] The clusters appear to involve occupants of the building who converged into specific rooms they perceived as areas of refuge. In the MGM Grand Hotel fire, guests tended to select rooms on the north side of the east and west wings and rooms on the east side of the south wing. In addition, guests reported that people had converged in the rooms with balconies and doors leading out to the balconies because of the ventilation, reduced smoke, improved visibility, and communication that the balconies offered. The guests who reported this behavior either estimated the number of persons in the room or connecting rooms, or they indicated only that "others" or "other persons" were present.

Table 8-1H lists the rooms identified by guests as areas of refuge for numerous persons other than the original occupants. This table also presents estimates of the length of time that the clusters were maintained in the rooms—usually until the individuals were evacuated, or until the occupants were notified by fire or rescue personnel that evacuation was possible. The numbers in the two right-hand columns indicate the total number of persons in the clusters for the total number of rooms identified on the floor. The smallest number of people identified as a cluster was 3, and the largest was 35.

The greatest number of rooms used by convergence clusters, and the largest population participating in convergence clusters, was located on the 17th floor of the hotel. No convergence clusters were identified by guests on the 6th, 21st, or 26th floors. The clusters appear to serve as an anxiety- and tension-reducing mechanisms for individuals confronted with a threatening situation. The action of "offered refuge in room," previously identified in the discussion of

TABLE 8-1H. *Summary of Rooms, Time Duration, and Number of Guests Reported in Convergence Clusters*

Floor	Room Number(s)	Time (hr)	Persons No.	Persons Percent
7	731	0.6	3	0.7
8	827, 840	1.5–1.75	14*	3.3
9	927	2.5	5	1.2
10	1009A, 1025, 1034, 1060	1–2	53	12.7
11	1129, 1115	1.5–2	30*	7.2
12	1261, 1225, 1233A	2–3	53	12.7
14	1433A, 1461A, 1451, 1416A	1.5–2	8*	1.9
15	1501, 1533A, 1510	2–3	38*	9.1
16	1643, 1625, 1633, 1629, 1627, 1615	2–3.5	35*	8.4
17	1725, 1775, 1731, 1719, 1762, 1756, 1733A	2–2.5	84	20.1
18	1819, 1802, 1850	2–3	20	4.8
19	1929, 1919, 1962A, 1962, 1964, 1925	2–3.5	13*	3.1
20	2027, 2013, 2030	2.5–3.5	25	6.0
22	2213, 2221,2229	2–3	13	3.1
23	2329, 2314, 2342, 2331,2308, 2340	2.5–3.25	20*	4.8
24	2446	3.5	4	0.9
25	2512, 2509A	3.5	*	0
Total: 17	57		418	100.0
Range: 7–25	1–7	0.6–3.5	3–84	0–20.1

* Persons indicated only as "Others."

the fifth actions, is a positive indication of the occurrence of a convergence cluster.

In addition to the detailed human behavior study of the MGM Grand Hotel fire,[38] NFPA conducted a questionnaire study of guests' behavior in the Westchase Hilton Hotel fire in Houston, TX, on March 6, 1982, in which 12 people died.[39]

Figure 8-1H presents the decomposition diagram for 8 multiple occupancy fires covering the actions of 96 persons.[31] These multiple occupancy fires in the United Kingdom involved hotel occupancies. A comparison of Figure 8-1H with Table 8-1G shows the similarities between the five actions of the guests in the MGM Grand Hotel fire and the occupants' behavior in the British study.

The classic types of nonadaptive behavior in a fire ignore adaptive actions that might facilitate the evacuation of others or limit the propagation of smoke, heat, or flame. Nonadaptive behavior ranges from the single act of leaving a room of fire origin without closing the door, thus allowing the fire to spread throughout the structure and endanger the lives of all the occupants, to the more generalized behavior of fleeing from a fire without regard for others and perhaps injuring others in what is often termed "panic."

Nonadaptive behavior may be an omission, such as forgetting to close a door, or it may involve an action that, although well-meaning, results in negative consequences. When the results of behavior are extinguishing the fire and eliminating the threat, the behavior may be said to be adaptive. However, the same behavior may be ineffective because the fire was more severe than was first perceived. In such cases, the time spent trying to extinguish it might have been used more effectively to warn others and to notify the fire department. Thus, some behavior that appears to be nonadaptive really is behavior that would have seemed most adaptive if it had been

successful. The injuries people suffer in relation to a fire may be cues to their nonadaptive or risk behavior.

Panic Behavior

One concept always discussed following a fire in which multiple fatalities occur, such as the Beverly Hills Supper Club fire,[40] is panic behavior. One classic definition of panic is:

A sudden and excessive feeling of alarm or fear, usually affecting a body of persons, originating in some real or supposed danger, vaguely apprehended, and leading to extravagant and injudicious efforts to secure safety.[41]

According to this definition, panic is a flight or fleeing type of behavior that involves extravagant and injudicious effort and is likely not to be limited to a single individual, but to be transmitted to and adopted by a group of people. From simulation experiments, a panic-type of behavior reaction has been defined in the following manner:

A fear-induced flight behavior which is nonrational, nonadaptive, and nonsocial, which serves to reduce the escape possibilities of the group as a whole.[42]

The concept of panic is often used to explain the occurrence of multiple fatalities in fires even when there is no physical, social, or psychological evidence showing that competitive, injudicious flight behavior actually took place. The media and public officials often label various types of fire behavior as panic. The evidence accumulated from interviews with participants in the Beverly Hills Supper Club fire, and questionnaires completed by occupants, provided no evidence of the classic group-type of panic behavior with competitive flight for the exits.[43]

Panic as a concept is primarily a description rather than an explanation of behavior. The concept is used to support the introduction of requirements in fire and building laws or ordinances to provide for the fire safety of occupants. There also has been shown to be a difference between use of the concept to describe other persons' behavior in a fire and the use by someone engaged in the behavior to indicate his or her own state of concern and anxiety.[44] Just because an individual identifies behavior as associated with panic does not necessarily identify the behavior as the classic panic-type response. The outcome of the behavior, as previously discussed, affects its labeling: the behavior of people in a fire is most likely to be misinterpreted when the outcome of the fire has been unfortunate.

The use of the concept of panic must be separated from use of the terms "anxiety" or "fear." The concept of self-destructive or animalistic panic responses to stimuli, such as the presence of smoke, has not been supported by the research on human behavior in fires. As has been pointed out, it is rare to have panic behavior in which the flight is characterized by competition among the participants, with resultant personal injuries.[1,2,4,44,45,46]

In an interview study of 100 participants in single-family dwelling fires, no instances of panic behavior were found. Primarily altruistic, helpful behavior was found instead.[46]

Ramachandran[6] in his review of studies of human behavior in fires in the United Kingdom came to this conclusion about nonadaptive behavior:

In the stress of a fire, people often act inappropriately and rarely panic or behave irrationally. Such behavior, to a large extent, is due to the fact that information initially available to people regarding the possible existence of a fire and its size and location is often ambiguous or inadequate.[6]

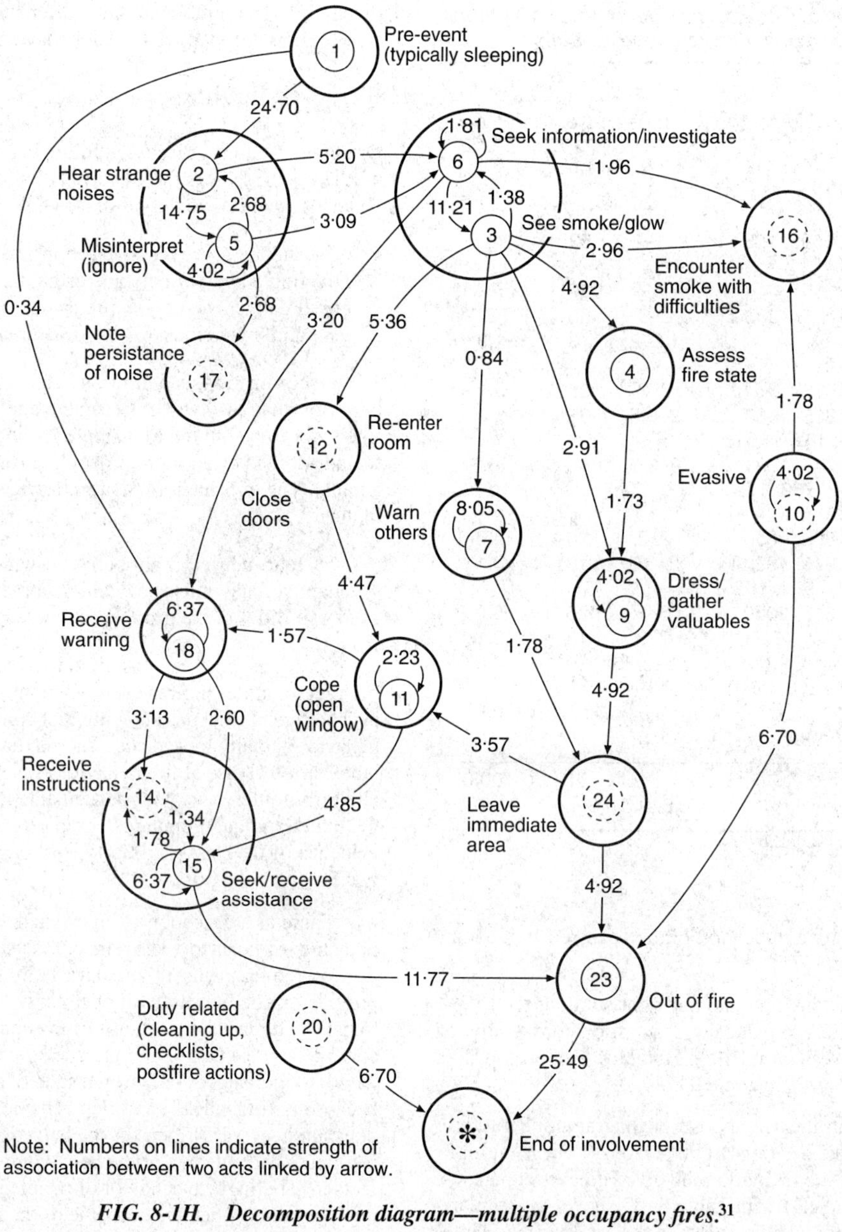

FIG. 8-1H. *Decomposition diagram—multiple occupancy fires.*[31]

Reentry Behavior

The study of the 1956 Arundel Park Hall fire was the first to document the phenomenon of reentry behavior.[18] Some older codes and regulations affecting design of the means of egress appear to have been based on the assumption that pedestrian traffic only moves away from a fire and away from the area or floor of the building involved. However, the Arundel Park Hall study indicated that approximately one-third of the survivors interviewed had reentered the building.

Thus, it has become apparent that doors, stairways, and corridors often will be subjected to two-way movement of occupants and others. The occupant who, after leaving the building safely, turns around and reenters is often completely aware of the fire in the building and of the specific portions of the building involved in fire and smoke propagation. Based on interviews with 61 persons, Table 8-1I presents the number of participants who reentered Arundel

Park Hall during the fire. Note the reasons for the reentry behavior and the fact that those who reentered were predominately male.

TABLE 8-1I. *Behavior of Occupants Who Reentered in the Arundel Park Hall Fire, Relative to Sex and Reentry Reasons*

Sex	Reentered and Left Same Exit	Reentered and Left Different Exit	Stated Reason for Reentrance
M	1		Turn off kitchen stoves
M	1	1	Tell people to leave
M	3		To help
M	1		Assist people
M	2	3	Find wife
M	2	2	Assist fire fighting
M and 1 F		5	No stated reason
N=21 M and 1 F	10	12	

The Arundel Park Hall fire occurred in an assembly occupancy being used for a church-sponsored oyster roast, a family-type affair. Thus, the primary group cultural role of father or husband apparently was a critical variable in the reentry behavior of the population interviewed and may have resulted in the fact that the reentry participants were mostly male. It can reasonably be argued that reentry behavior is not a nonadaptive behavior, since it is often used to assist or rescue persons remaining or believed to be remaining in the building. This type of behavior is often used by parents whose children are missing during a fire. The behavior is often undertaken in a rational, deliberate, and purposeful manner, without the emotional anxiety and self-doubt usually associated with nonadaptive behavior. However, reentry behavior has been considered nonadaptive, since people going back into a burning building often hinder the efficient and effective evacuation of others through the same means of egress.

The reasons elicited from participants in reentry behavior in the Project People study[1] in the U.S. are presented in Table 8-1J. It would seem that 162 people from the total study population of 584, or 27.9 percent, engaged in reentry behavior. The most popular reason given was to "fight fire," followed by "obtain personal property," "check on fire," "notify others," "assist fire department," and "retrieve pets." These six reasons accounted for approximately 73 percent of this reentry behavior.

Table 8-1K compares the reentry behavior of the British and U.S. study populations. Note that all of the reasons were significantly different, with the exception of the item "save personal effects." The reentry reasons of the U.S. population were predominantly "fight fire," "save personal effects," "observe fire," "call fire department," "rescue pets," "notify others," and "assist fire department." The British population's were predominantly "fight fire," "observe fire," "save personal effects," "shut doors," "await fire department," and "fire not severe."

TABLE 8-1J. *Reasons for Reentry of Occupants*

Reasons	Participants	Percent
Fight fire	36	22.2
Obtain personal property	28	17.2
Check on fire	18	11.0
Notify others	13	8.0
Assist fire dept.	12	7.4
Retrieve pets	12	7.4
Call fire dept.	9	5.5
Assist evacuation	4	2.5
Taken to hospital	3	1.8
Turn power back on	2	1.2
Rescue from balcony	1	0.6
Help injured family member	1	0.6
Turn off gas	1	0.6
Open windows	1	0.6
Close door	1	0.6
No apparent danger	1	0.6
Entered non-danger area	1	0.6
Job responsibility	1	0.6
Due to fire	1	0.6
Told to by others	1	0.6
Not reported	16	9.8
N = 21	163	100.0
Range = 1–36	Percent of participant population = 27.9	

TABLE 8-1K. *Comparison of Reasons for Reentry Behavior of British and U.S. Study Populations*

Reasons	British (percent)	U.S. (percent)	$P_1 - P_2$††	$SE_{P_1 - P_2}$§	CR‖
Fight fire	36.0	22.2	13.8	4.02	*3.43†*
Observe fire	19.0	11.0	8.0	3.25	*2.46**
Save personal effects	13.0	17.2	4.2	2.91	1.44
Shut doors	10.0	0.6	9.4	2.38	*3.95†*
Await fire dept.	9.0	0.0	9.0	2.26	*3.98†*
Call fire dept.	2.0	5.5	3.5	1.32	*2.65†*
Rescue pets	2.0	7.4	5.4	1.40	*3.86†*
Fire not severe	5.0	1.2	3.8	1.74	*2.18**
Notify others	0.0	8.0	8.0	0.92	*8.69†*
Assist fire dept.	0.0	7.4	7.4	0.88	*8.41†*
Assist evacuation	0.0	2.5	2.5	0.54	*4.63†*
N = 11	943	163			

* Critical ratios significant at or above 5 percent level of confidence.
† Critical ratios significant at or above the 1 percent level of confidence.
†† British percent minus U.S. percent.
§ Standard error.
‖ Critical ratio.

Occupant Fire-Fighting Behavior

Occupants who engaged in fire-fighting behavior were predominately male, and this behavior now appears to be a culturally determined and expected aspect of the male role. However, it should be noted that, in the Project People[1] study of 335 U.S. fires, approximately 23 percent of the study population of 584 individuals was involved in occupant fire-fighting behavior. Of these, 37.3 percent were female. Of the 134 individuals who participated in fire-fighting behavior, 50 were female and 84 male. They ranged from a 7-year-old girl to an 80-year-old man. Distribution of the participants by sex and age is presented in Table 8-1L. Most of those involved in fire-fighting behavior, or approximately 30 percent of the fire-fighting behavior population, were between 28 and 37 years old.

A higher proportion of males than females reported "got extinguisher" and "fought fire," and this difference is statistically

TABLE 8-1L. *Age and Sex of Occupants Engaged in Fire-Fighting Behavior*

Sex	Participants	Percent
Male	84	62.7
Female	50	37.3
Total	134	100.0
Age		
7–17	8	5.9
18–27	31	23.1
28–37	41	30.6
38–47	27	20.1
48–57	16	11.9
58–67	2	1.5
68–80	3	2.2
Unknown	6	4.7
Total	134	100.0
Percent of participant population = 22.9		

significant. (See Table 8-1M.) Approximately 16 percent of the male population reacted by obtaining extinguishers. Similarly, approximately 26 percent of the male population fought the fire when they became aware of it, as contrasted with approximately 10 percent of the female participants. A higher proportion of women notified the fire department: 33 percent of the females, compared with 26 percent or approximately one quarter of the males, reacted to the fire by notifying the fire department, as indicated in Table 8-1M, but this difference is not statistically significant.

TABLE 8-1M. *Sexual Differences of Occupants Engaged in Fire Fighting and Notifying the Fire Department*

Action	Male (percent)	Female (percent)	$P_1 - P_2$††	$SE_{P_1 - P_2}$§	CR‖
Searched for fire	17.2	9.1	8.1	4.23	1.91
Got extinguisher	15.6	6.0	9.6	3.95	*2.43**
Fought fire	25.6	9.7	15.9	4.83	*3.29†*
Removed fuel	03.4	3.1	0.3	2.17	0.14
Tried to extinguish	05.3	2.8	2.5	2.49	1.00
Went to fire area	3.1	2.8	0.3	2.07	0.14
Total	70.2	33.5	36.7	6.01	*6.11†*
N = 6	184	101			
Called fire dept.	25.6	33.0	7.4	5.83	1.27
Had others call fire dept.	9.2	7.5	1.7	3.27	0.52
Went to fire alarm	3.8	3.8	0.0	0.00	0.00
Pulled fire alarm	1.9	1.6	0.3	1.65	0.18
Total	40.5	45.9	5.4	6.31	0.85
N = 4	106	146			

* Critical ratios significant at or above the 5 percent level of confidence.
† Critical ratios significant at or above the 1 percent level of confidence.
†† Male percent minus female percent.
§ Standard error.
‖ Critical ratio.

Occupant fire-fighting behavior appears most prevalent in occupancies in which the individuals are emotionally and economically involved—that is, in their homes or where such behavior is an assigned role as a result of training.[47] At some time during the fire, 285 individuals engaged in one of the six actions defined as fire-fighting behavior, and 252 individuals participated in one of the four actions defined as notifying the fire department.

In the study of residential fire incidents in Berkeley, CA, 180 persons were involved in extinguishing and fire-fighting behavior. This study surveyed a population different from that of Project People, since the 1411 Berkeley households, with 208 fires, included fires not reported to the fire department; these accounted for approximately 80 percent of the total 208 fires. The majority of the unreported fires were extinguished by the occupants alone or with the help of neighbors.[48] Six percent of these fires self-extinguished, and 52 percent were extinguished by the individual who had started the fire. Thus, it appears that only the fires in the Project People study that were judged uncontrollable by the occupants resulted in notification of the fire department. Similarly, approximately 85 percent of the fires in the National Fire Prevention and Control Administration National Household Survey[49] were not reported to the fire department.

In the Project People study, 107 of the 584 participants did not leave the building voluntarily after becoming aware of the fire. Their reasons for staying in the building are presented in Table 8-1N. Fifty-two of the participants, or approximately 49 percent of the population, who stayed in the building reported that they remained because they wished to engage in fire control or fire-fighting activities. The other most frequent reasons were to notify others of the fire or because the occupant's way out of the building was blocked by smoke.

TABLE 8-1N. *Reasons Elicited from Occupants for Not Leaving the Fire Building*

Reason	Participants	Percent
Fight fire	52	48.7
Notify others	7	6.5
Blocked by smoke	7	6.5
Blocked by fire	5	4.7
Overcome by smoke	5	4.7
Search for fire	3	2.8
Needed help	2	1.9
Secure property	2	1.9
Afraid of fire spread	2	1.9
No fire in area	1	0.9
Help others	1	0.9
Does not know	1	0.9
No response to fire dept.	1	0.9
Home	1	0.9
Return to area	1	0.9
Not reported	16	15.0
N = 15	107	100.0

Range = 1–52 Percent of participant population = 15.6

Occupants' Movement Through Smoke

Often related to fire-fighting behavior, and a definite component of evacuation behavior in many fires,[1,2] is the movement of occupants through smoke. The principal variables influencing an occupant's decision to move through smoke appear to be recognition of the location of the exit and thus of the travel distance, the appearance of the smoke, the smoke density, and the presence or absence of heat.[38,39] To achieve evacuation, occupants have moved through smoke, even for extended distances under conditions of extremely limited visibility at personal risk, and sometimes have been forced to turn back without completing the evacuation.[1,2,38,39]

Jin and Yamada[50] reported on a study involving 31 subjects (14 males and 17 females) traveling a maximum distance of 10.5 m in a corridor while exposed to smoke from smoldering cedar crib chips. The smoke extinction coefficient varied from 0.1 to 1.2 (m-1). Subjects were also exposed to increasing heat from radiant heaters at the end of the corridor, where the mean temperature was 82°C. At five points in the corridor the subjects stopped and were asked mental arithmetic questions. Both walking speed in the corridor and mental arithmetic capability decreased as smoke density and radiant heat exposure increased.

Proulx and Fahy[12] in their questionnaire study of 382 employees in the World Trade Center explosion and fire found that 94 percent of the respondents in Tower 1 and 70 percent of the respondents in Tower 2 moved through smoke. In addition, the study reported that approximately 75 percent of these individuals turned back during their evacuation because of smoke, crowding, locked doors, breathing difficulty, fear, and poor visibility. It was also reported that some occupants continued to move through smoke, even when they perceived the smoke to be worsening and believed that they may have been moving toward the fire.

Table 8-1O compares the distance moved through smoke for the 1316 persons in the British study and the 322 persons in the U.S. study. Sixty percent of the British study population and 62.7 percent of the Project People participants reported that they moved through smoke. It is thus apparent that building occupants will move through smoke in an evacuation process. An important variable may be the smoke density or the visibility distance of the occupants during the evacuation process.

TABLE 8-1O.   Comparison of the Distance Moved Through Smoke for the British and U.S. Populations

Distance Moved (ft)#	British (percent)	U.S. (percent)	$P_1 - P_2$††	$SE_{P_1 - P_2}$§=	CRǁ
0–2	3.0	2.3	0.7	1.02	0.69
3–6	18.0	8.4	9.6	2.23	4.30†
7–12	30.0	17.1	12.9	2.71	4.76†
13–30	19.0	45.5	26.5	2.62	10.11†
31–36	5.0	2.0	3.0	1.25	2.40*
37–45	4.0	4.1	0.1	1.19	0.08
46–60	5.0	11.0	6.0	1.47	4.08†
60+	15.0	9.6	5.4	2.10	2.57*
	1316	322			

* Critical ratios significant at or above the 5 percent level of confidence.
† Critical ratios significant at or above the 1 percent level of confidence.
†† British percent minus U.S. percent.
§ Standard error.
ǁ Critical ratio.
1 ft = 0.3 m.

Table 8-1P presents the visibility distance reported by the British and U.S. occupants as they moved through smoke while evacuating a fire building. They reported their movement through smoke under relatively high smoke density conditions, with visibility under 12 ft (3.7 m) for 64 percent of the British population and for 47.6 percent of the U.S. population.

TABLE 8-1P.   Comparison of the Visibility Distance for the British and U.S. Populations When Moving Through Smoke

Visibility Distance (ft)#	British (percent)	U.S. (percent)	$P_1 - P_2$††	$SE_{P_1 - P_2}$§	CRǁ
0–2	12.0	10.2	1.8	1.99	0.90
3–6	25.0	17.2	7.8	2.65	2.94†
7–12	27.0	20.2	6.8	2.73	2.49*
13–30	11.0	31.7	21.7	2.24	9.69†
31–36	3.0	2.2	0.8	1.03	0.78
37–45	3.0	3.7	0.7	1.08	0.65
46–60	3.0	7.4	4.4	1.21	3.64†
60+	17.0	7.4	9.6	2.24	4.20†
	1316	322			

*Critical ratios significant at or above the 5 percent level of confidence.
† Critical ratios significant at or above the 1 percent level of confidence.
†† British percent minus U.S. percent.
§ Standard error.
ǁ Critical ratio.
1 ft = 0.3 m.

Visibility distance for the British and U.S. populations at the time participants were forced to turn back is presented in Table 8-1Q. Comparison with Table 8-1P reveals that very few participants

turned back when they could see more than 31 ft (9.4 m). The greater percentage of participants turned back at shorter visibility distances. When the visibility distance was below 12 ft (3.7 m), 91 percent of the British study population and 76.4 percent of the U.S. study population turned back. (See Table 8-1Q.)

TABLE 8-1Q.   Comparison of the Visibility Distance for the British and U.S. Populations Relative to the Turn Back Behavior

Visibility Distance (ft)#	British (percent)	U.S. (percent)	$P_1 - P_2$††	$SE_{P_1 - P_2}$§	CRǁ
0–2	29.0	31.8	2.8	5.31	0.53
3–6	37.0	22.3	14.7	5.57	2.64*
7–12	25.0	22.3	2.7	5.02	0.54
13–30	6.0	17.6	11.6	3.07	3.78*
31–36	0.5	1.2	0.7	0.90	0.77
37–45	1.0	0.0	1.0	1.10	0.91
46–60	0.5	4.7	4.2	1.16	3.62*
60+	1.0	0.0	1.0	1.10	0.91
	570	85			

* Critical ratios significant at or above the 5 percent level of confidence.
†† British percent minus U.S. percent.
§ Standard error.
ǁ Critical ratio.
1 ft = 0.3 m.

HANDICAPPED OR IMPAIRED OCCUPANTS

Fire problems involving occupancies designed for permanently or temporarily disabled persons, such as nursing homes and hospitals, appear to be matched on the basis of building design, adequate staff training, and ability to protect the occupants in place until evacuation is possible. An extensive study of human behavior in health care facilities[47] indicated that the nursing staff performed their professional roles toward their patients even when they were at risk.

The few fires studied involving handicapped persons in occupancies other than health care facilities primarily have been in residential occupancies. In two of these cases, handicapped individuals were helped by other occupants to evacuate successfully. One instance involved a wheelchair user[38] and the other a blind person.[51]

Handicapped people may have a variety of limitations that increase their risk in a fire: sensory problems, such as deafness and blindness; mobility problems that may entail the need for a wheelchair; and intellectual problems, such as mental retardation. Many handicapped persons with mobility problems also are concerned about their personal risk in high-rise office and residential buildings where the use of elevators is not allowed in a fire. In such situations, adequate areas of refuge must be provided for handicapped, as well as nonhandicapped, occupants.[52]

In their reports of the explosion and fire in the World Trade Center on February 26, 1993, Isner and Klem[53,54] indicated that when the explosion occurred at approximately 12:18 p.m., normal power was lost and the emergency generators failed about 20 minutes later; all remaining power to the World Trade Center complex was disconnected at approximately 1:32 p.m. Thus, the simultaneous evacuations of both able and disabled occupants from Towers 1 and 2 were conducted in darkness with varying smoke conditions in the stairways. These simultaneous evacuations may have involved the largest number of occupants and the longest evacuation times of any fire-induced evacuations of buildings in the U.S.

Juillet,[55] in one of the first documented studies of this type, reported on the interview study of 27 occupants with disabilities who were evacuated from one of the two towers in the World Trade Center during the explosion and fire of February 26, 1993. The disabilities of the interviewees included fourteen with mobility impairments, three with sight or hearing impairments, three who were pregnant, two with cardiac conditions, and seven with respiratory conditions. Juillet[56] indicated that the total disability population in both Tower's 1 and 2 at the time of the incident was believed to be between 100 and 200 persons, approximately 100 of whom had been previously identified. The average evacuation time of the 27 study participants was 3.34 hours with evacuation times reportedly ranging from 40 minutes to over 9 hours. The predominant means of evacuation was through the stairs with assistance from other evacuees or emergency personnel. The altruistic behavior seen in many fire incidents with large populations[37,38] appeared to have been exhibited in this fire incident in relation to the disabled occupants as reported by Juillet.

However, in the absence of communications by authorities, they gladly accepted assistance—from colleagues and even from complete strangers—in evacuating. These caring groups of people who assisted the disabled protected their "charges" until they were safely evacuated and moved away from the building.[55]

Klote, Alvord, Levin, and Groner[57] examined the design considerations needed to enable elevators in tall buildings to be utilized for the evacuation of disabled occupants. In the World Trade Center explosion and fire, the loss of power in both Towers 1 and 2 (including emergency power) trapped occupants in elevators in both buildings.

Burns[11] indicated Tower 1 had 99 elevator cars, many of them occupied. When one 6×8 ft car was opened, 9 occupants were found unconscious. It was estimated that they had been exposed to the smoke in the shaft for approximately two hours at the ninth floor. Sherwood[58] reported that one 9×12 ft elevator car was stuck for six hours at the 41st floor of Tower 2 with 72 occupants, 62 elementary school children and 10 adults.

A study of a number of evacuation drills in high-rise office buildings in Canada indicated that approximately 3 percent of the occupants will be unable to use the stairs due to permanent or temporary conditions limiting their mobility.[59] The study population included occupants with heart conditions and individuals recovering from surgery, other illnesses, and accidents.

FIRE EXIT DRILLS

Well-marked exits do not ensure life safety during a fire. Exit drills are necessary so that occupants will know how to make an efficient and orderly escape according to NFPA *101*, which contains detailed information on exit drills in individual occupancies. Exit drills are required in schools, board and care facilities, and health care facilities, and are common in industries with high hazards. Employee training and drills are required in assembly, hotel, mercantile, and large business occupancies. Some form of exit drill should be conducted wherever or whenever possible to avoid confusion and ensure the evacuation of all occupants during a fire.[60] Personnel should be assigned to check exits for availability, to search for stragglers, to count occupants once they are outside the fire area, and to control reentry into the building before it is safe.

This reentry behavior for rescue purposes of persons involved in fire incidents should be noted. In a study of 335 fire incidents involving 584 persons, it was found that 163 persons, approximately 28 percent of the study population, reentered the fire buildings following evacuation.[1] Approximately 10.5 percent of those persons who reentered the buildings did so to alert or assist other persons.

Determining when and what area to evacuate is probably the most important decision in a fire emergency. Any area at all affected by heat, flame, or smoke should be evacuated; in case of doubt, the entire building should be evacuated.

The fire loss prevention and control management staff are responsible for planning exit drills. Plans should be discussed with both middle and line management to ensure understanding and cooperation. If there is no fire loss prevention and control manager, the plant, facility, or building manager should assume this responsibility or assign it to a staff member.

All employees should recognize the evacuation signal and know the exit route they are to follow.[61] Upon hearing the signal, they should shut off equipment and report to a predetermined assembly point. Primary and alternate routes should be established, and all employees should be trained to use either route.[60,62]

The problem with audible evacuation signals is the conditioning of the population within the occupancy to ignore the signal due to numerous false alarms. An investigation of an apartment building fire found that the alarm system actuation to initiate evacuation behavior was ignored by many of the building occupants due to the conditioning effect of numerous prior false alarms: 44 percent of the building occupants believed the alarm signal was a false alarm.[63]

When employees are assembled, the manager or supervisor of each area should account for all personnel. Missing employees should be immediately reported to the fire loss prevention and control manager and responding fire department personnel so that search and rescue efforts can be initiated. Only trained fire-fighting search and rescue personnel with adequate protective equipment should be permitted to reenter an evacuated area.

After each exit drill, a meeting of the responsible managers should be held to evaluate the success of the drill and to solve any problems that may have arisen.

One significant improvement to the traditional concept of fire drills in educational occupancies was suggested by a study of fire drills conducted in such occupancies.[64] The concept of smoke drills has been established, whereby the occupants are instructed to move through the simulated smoke areas in a crouched position. Students have transferred the smoke drill concept to fire incidents in residential occupancies, with effective results. Obviously, the utilization of smoke drill training may be effective in a fire incident and should be used where applicable.

The timing of drills depends upon the nature of the operation in the facility. Generally, drills conducted a few minutes before the lunch break have been found to minimize loss of time and production. The frequency of drills should be determined by the degree of hazard present and by the complexity of shutdown or evacuation procedures.

If a facility does not maintain a security organization that is responsible for daily inspection of emergency exits and designated evacuation routes, one employee in each area should be assigned this task. Maintenance of doors, panic hardware, exit lights, and emergency illumination should be given high priority, and repairs must be made without delay.

Research has found that for multi-story office buildings, a trained group of floor wardens is the most effective means of monitoring the evacuation of occupants.[60] Adequate training for floor wardens or other personnel is necessary and must be specifically developed to include the procedures of the emergency evacuation plan for the facility.

The lecture method has been used to convey the essential features of the emergency plan to employees in health care facilities.[62]

However, it was also found that the emergency plan in the facility studied was too general and ambiguous.

The most serious problem with using building monitors is the turnover of personnel due to employee transfers, reassignments, or resignations. Effective evacuation planning and preparation should assign specific responsibilities to staff positions (rather than individuals) within an organization. This ensures continuity of performance despite personnel changes.

A content and time evaluation of fire drill behavior by staff in six nursing homes concluded that a training program of the most modest type can produce changes in both knowledge and behavior of evacuation and fire emergency procedures.[65] A total of 339 nursing home staff participated in the study that matched a group of 37 persons receiving the training with a control group of 49 persons not receiving the training. The high rate of personnel turnover, which appears to be rather typical in nursing home facilities, was noted. Following the presentation of the training program, staff members were evaluated by a written knowledge test, and their behavior was observed during the conduct of a drill of the emergency plan.

The evacuation signal should be familiar to all employees. Vocal alarm systems (VAS)[60,61] reduce the need for employee perception and recognition of a signal, since the system provides vocal communication to the areas designated for evacuation. NFPA *101* first recommended the use of the VAS in assembly occupancies in 1981. An evaluation of VAS systems in nine buildings found that familiarity with the system or initial activation did not significantly affect the egress behavior of populations.[60] In addition, the investigation determined the evacuation drills were valuable because they gave floor or area wardens an opportunity to rehearse their procedures.

The use of an alerting tone in the frequency range of 2,000 to 4,000 Hz for the VAS is recommended prior to the verbal announcement, which should be specific for the audience and the facility.[60,66]

Health Care Facilities Drills

Fire drills in health care institutions are usually conducted as a part of the orientation program for new employees. Later, the drills are supplemented with in-service training for the staff personnel, including the emergency procedures. Fire drills in many facilities are conducted once a month on each shift. The training for the drills typically involves instruction and practice for the staff personnel in the various means of moving nonambulatory patients, procedures for alerting the facility staff, and the method of notifying the fire department. Once or twice a year, many facilities have fire departments provide training in the operation of portable fire extinguishers. Some fire departments actually provide staff personnel with experience in the operation of extinguishers on external fires.

However, most health care facilities prefer to train their personnel. Most adopt the philosophy that it is the staff's responsibility to ensure the safe evacuation of patients initially to an area of refuge and then to the exterior if necessary. The control of the fire is limited to preventing the spread of heat and smoke by closing doors. This protects the occupants and inhibits or restricts the propagation of the smoke and heat throughout the facility. Staff personnel have effectively evacuated numerous patients under fire conditions or have protected the patients in their rooms by closing doors.[67]

Thus, the evacuation process may be considered in four sequential phases: (1) the personnel supply phase, (2) the patient preparation phase, (3) the patient removal phase, and (4) the rest and recovery phase.[68] This approach focuses on the occupants in the fire-threatened area and the patients in or adjacent to the fire area. Removing immediately threatened patients, and closing doors to the room of fire origin and to adjacent patient rooms, would be compatible with this four-phase approach to evacuation.

A detailed report on the fire evacuation organization, training, and drills involved in a 502-bed acute-care teaching hospital with 2,500 employees has been published.[69]

In 1985, NFPA *101* included a new chapter devoted to the fire protection and life safety requirements for board and care facilities. That chapter requires the evaluation and classification of the population of the facility, according to their evacuation capability.

Evaluating Fire Drill Plans

The ultimate evaluation of fire drill and emergency plans has two factors: (1) the performance of the occupants in a fire incident and (2) the effectiveness of the behaviors used in accordance with the fire drills or the fire emergency plan. In a World Trade Center fire incident report,[10] building occupants tended to attempt to verify their beliefs about the threat of fire by physical clues, primarily smoke in the occupants' area. The report also indicated public address messages were not sufficient to alleviate spontaneous evacuations when occupants saw smoke on their floor. Further, occupant evacuation was reported (9th through 22nd floors) due to the perception and concern that a valid fire threat existed. In actuality, the fire did not require such an extensive evacuation.

Successful evacuation of personnel from the two floors above and below a fire in a 28-story high-rise college dormitory has been reported.[70] To allow free evacuation flow of the occupants down the stairways and allow fire department personnel to move up the stairs, stairs have been marked for occupant and fire department movement. The fire department movement stair has a red circle, 6 in. (152 mm) in diameter, on the door, and is also utilized for ventilation. The occupant stair is marked with a 6-in. (152-mm) green circle.

In the safe evacuation of a high-rise hotel fire, with 190 guests, 110 guests were assisted by the fire department. The success of the evacuation was made possible by the hotel employees' fire safety education and practice of evacuation procedures.[71]

Dynamic Evaluation of Exit Design

Exit design may be evaluated on a dynamic basis using calculated evacuation times. Dynamic evaluation allows for the establishment of goals in terms of evacuation times and certain other key criteria.

For example, the U.S. General Services Administration (GSA) has established a goal-oriented system approach to building fire safety that uses specific sets of goals for normal federal office operations.[72] The established goals are:

1. All occupants exposed to the fire environment must be able to evacuate to a safe area within 90 sec of alarm.
2. A portion of this time, not to exceed approximately 15 sec, can be involved in traveling in a direction toward the fire, such as in a dead-end corridor.
3. All occupants must reach an area of refuge within 5 min of downward vertical travel or within 1 min of upward vertical movement.

Further criteria established by the GSA for evaluation uses 3.5 ft/sec (1.07 m/sec) as the rate of horizontal movement. The exit flow rate for level travel is calculated at 60 persons/min/24 in. (610 mm) width, 45 persons/min/24 in. (610 mm) width down stairs, and 40 persons/min/24 in. (610 mm) width up stairs.

Fatigue is judged to become important after 5 min of travel in a downward direction or 1 min travel in an upward direction. Thus, fatigue becomes a human factor consideration affecting design.

Evacuation times[73,74] have been compared with other data[75] for both observed and predicted total evacuation times for tall buildings, as presented in Figure 8-1J. Pauls' data[75] is based on measurements obtained from 29 evacuation drills, primarily in office buildings ranging from 8 to 21 stories high. Pauls observed evacuation times varying from approximately 10 sec/per story for buildings with small populations to approximately 20/sec per story for buildings with large populations. The evacuation equations indicated in Figure 8-1I were developed from these observed evacuation times. The first equation,

$$T = 0.70 + 0.0133p$$

is to be applied to predict evacuation times in buildings with large populations exceeding 800 persons/m of effective stair width. T is the minimum time, in min, to complete an uncontrolled total evacuation by stairs, and p is the actual evacuation population/m of effective stair width, measured immediately above the discharge level of the stair.

It should be recognized that "effective stair width," as used by Pauls, is defined in the following manner:

This empirically based model describes flow as a linear function of a stair's effective width—the width remaining once the edge effects are deducted (6 in. or 150 mm from each wall boundary and 3.5 in. or 90 mm from each handrail centerline). It takes into account the propensity of people to sway laterally—especially when walking slowly in a crowd—and therefore to arrange themselves in a staggered traditional unit-width model based on presumed static dimensions of people's shoulders.

The second equation in Figure 8-1I,

$$T = 2.00 + 0.0117p$$

is to be applied when the population per meter of effective stair width is less than 800 persons.

Pauls[75] also examined the relationship between the speed or velocity of evacuation and the density on the stairs during the uncontrolled total evacuation, as indicated in Figure 8-1J. It should be remembered this movement would be in the vertical, downward direction. The vertical movement appears to be most effective at a density of approximately 0.9 persons/sq m. Pauls indicated that a

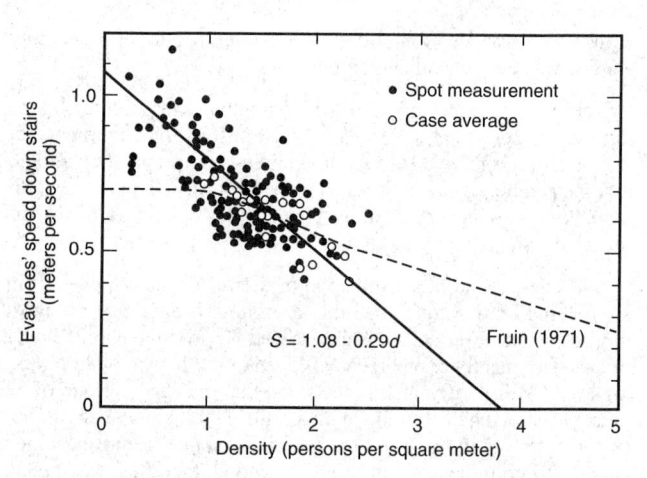

FIG. 8-1J. *Relation between speed and density on stairs in uncontrolled total evacuations.*[75]

speed on the stairs of approximately 2.1 ft/sec approximates traveling 8 to 12 stories/min in office or residential buildings.

HYDRAULIC FLOW CALCULATIONS[76]

The estimation of modeled evacuation time utilizes a series of expressions that relate data acquired from tests and observations to a hydraulic approximation of human flow. While the expressions indicate absolute relationships, there is considerable variability in the data. Figure 8-1J shows a typical relationship between the source data and the derived equation. The equations and relationships presented in the following paragraphs can be used independently or collected to solve a complex egress problem. Such a coordinated collection of equations is demonstrated in the sample problem.

Effective Width, W_e

Persons moving through the exit routes of a building maintain a boundary layer clearance from walls and other stationary obstacles they pass. This clearance is needed to accommodate lateral body sway and assure balance.

Discussion of this crowd movement phenomena is found in the works of Fruin,[77] Pauls,[78] and Habicht and Braaksma.[79] The useful (effective) width of an exit path is the clear width of the path less the width of the boundary layers. Figures 8-1K and 8-1L depict effective width and boundary layer. Table 8-1R is a listing of boundary layer widths. The effective width of any portion of an exit route is the clear width of that portion of an exit route less the sum of the boundary layers.

Clear width is measured:

1. From wall to wall in corridors or hallways.
2. As the width of the treads in stairways.
3. As the actual passage width of a door in its open position.
4. As the space between the seats along the aisles of assembly arrangement.
5. As the space between the most intruding portions of the seats (when unoccupied) in a row of seats in an assembly arrangement.

The intrusion of handrails is considered by comparing the effective width without the handrails, and the effective width using a clear width from the edge of the handrail. The smaller of the two ef-

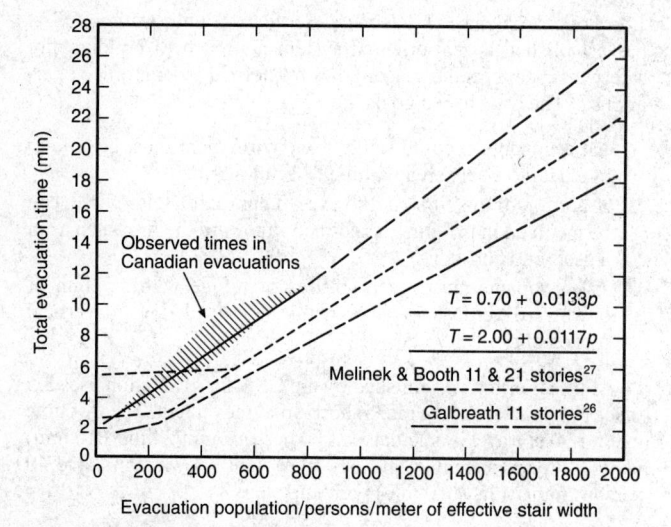

FIG. 8-1I. *Predicted and observed total evacuation times for tall office buildings.*[75]

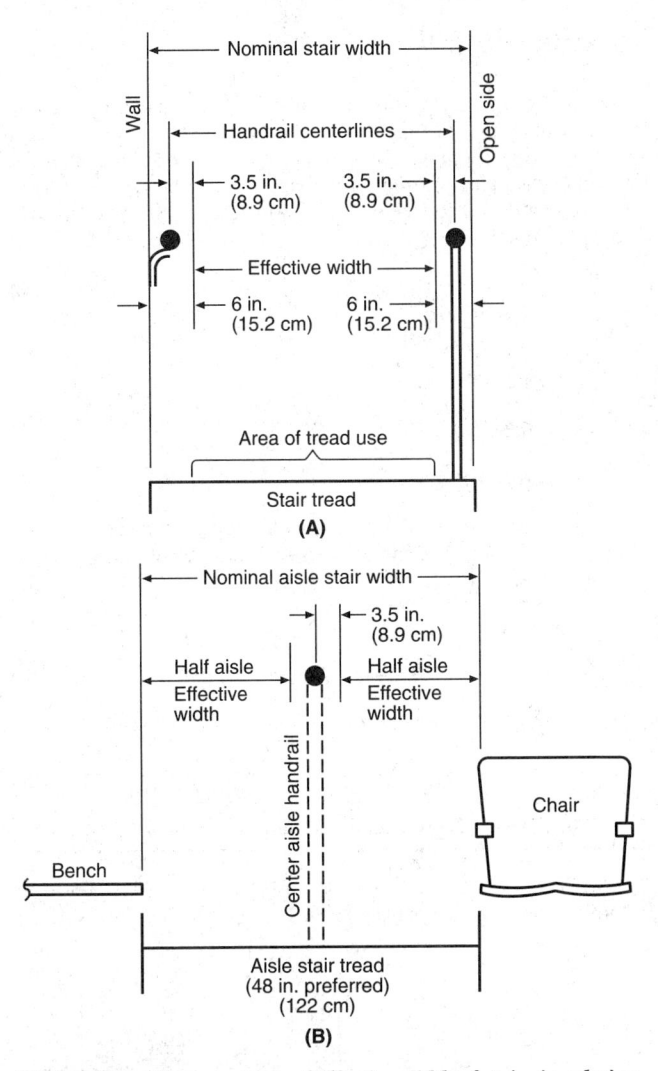

FIG. 8-1K.   *Measurements of effective width of stairs in relation to walls, handrails, and seating.*

fective widths then applies. Using the values in Table 8-1R, only handrails that protrude more than 2.5 in. need be considered. Minor midbody height or lower intrusions such as panic hardware are treated in the same manner as handrails. Where an exit route becomes either wider or narrower, only that portion of the route has the appropriate greater or lesser clear width.

TABLE 8-1R.   *Boundary Layer Widths*

Exit Route Element	Boundary layer	
	in.	cm
Stairways—walls or side of tread	6	15
Railings, handrails*	3.5	9
Theater chairs, stadium benches	0	0
Corridor, ramp walls	8	20
Obstacles	4	10
Wide concourses, passageways	up to 18	46
Door, archways	6	15

* Where handrails are present, use the value if it results in a lesser effective width.

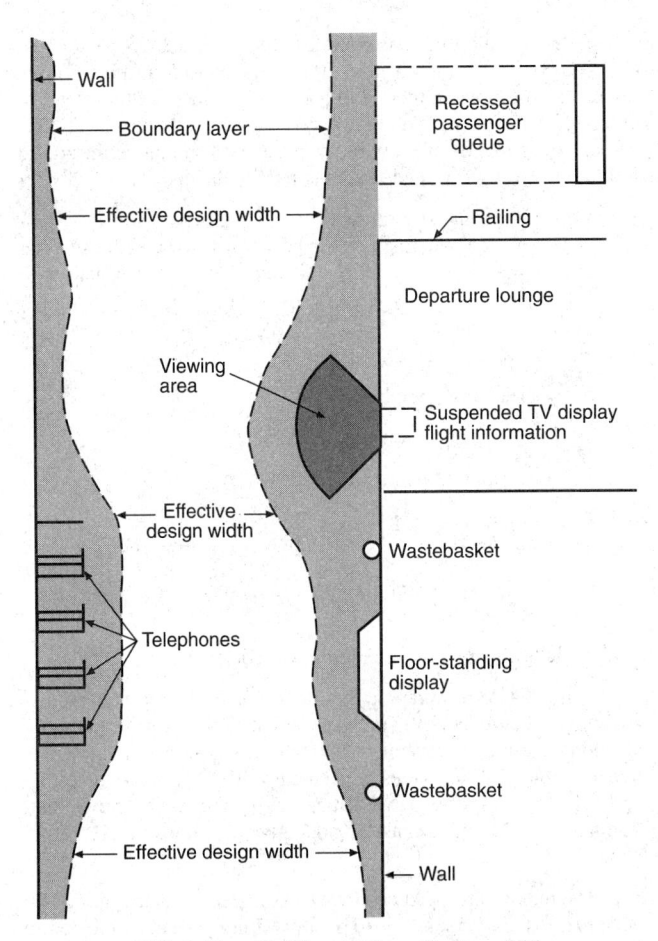

FIG. 8-1L.   *Public corridor effective width.*

Density, *D*

Density is the measurement of the degree of crowding in an evacuation route and is expressed in persons/unit area. The calculations in this chapter are based on density expressed in persons/ft² (or persons/m²).

Unless information on the dispersion of occupants indicates otherwise, the density of the first exit element (aisle, corridor, ramp, etc.) is based on all of the served occupants. This will demonstrate the capacity limits of the route element and produce a value representing the maximum capacity of the element.

Conversely, if the egressing population is widely dispersed, in terms of reaching the exit route element, the calculation is based on an appropriate time step. At each time increment, the density of the exit route is based on those who have entered the route minus those who have passed from it.

The density factors in subsequent portions of the egress system are determined by calculation. The calculation methods involved are contained in the section of this chapter titled "Transitions."

Speed—Movement Velocity of Exiting Individuals, *S*

Observations and experiments have shown that evacuation flow speed of a group is a function of the population density. The relationships presented in this section have been derived from Fruin[77], Pauls,[78] and Predtechenskii and Milinskii.[80]

If the population density is less than about 0.05 persons/ft² (0.54 persons/m²) of exit route (20 ft²/person; 1.85 m²/person), individuals will move at their own pace, independent of the speed of others. If the population density exceeds about 0.35 persons/ft² (3.8 persons/m²), no movement will take place until enough of the crowd has passed from the crowded area to reduce the density.

Between the density limits of 0.05 and 0.35 persons/ft² (0.54 and 3.8 persons/m²) the relationship between speed and density can be considered as a linear function. The equation of this function is

$$S = k - akD \qquad (1)$$

where

S = speed along the line of travel;

D = density in persons per unit area;

k = constant, as shown in Table 8-1S;

k = k_1, and a = 2.86 for speed in ft/min and density in persons/ft²; and

k = k_2, and a = 0.266 for speed in m/s and density in persons/m².

Table 8-1S shows evacuation speed constants.

Figure 8-1M is a graphic representation of the relationship between speed and density. The speeds determined from Equation 2 are along the line of movement; for stairs this is along the line of the treads. Table 8-1T provides convenient multipliers for converting vertical rise of a stairway to a distance along the line of movement. The travel on landings must be added to the values derived from Table 8-1T.

The maximum speed occurs when the density is less than 0.05 persons/ft² (0.54 persons/m²). These maximum speeds are listed in Table 8-1U.

Within the range listed in Tables 8-1S, 8-1T, and 8-1U, the evacuation speed on stairs varies approximately as the square root of the ratio of tread width to tread height. There is not sufficient data to appraise the likelihood that this relationship holds outside this range.

Specific Flow, F_s

Specific flow, F_S, is the flow of evacuating persons past a point in the exit route/unit of time/unit of effective width, W_e, of the route involved. Specific flow is expressed in persons/min/ft of effective width (if the value of $k = k_2$, from Table 8-1S, or persons/s/m of effective width (if the value of $k = k_2$ from Table 8-1S). The equation for specific flow is

$$F_S = SD \qquad (2)$$

where

F_S = specific flow;

D = density;

S = speed of movement; and

F_S is in persons/min/ft² when density is in persons/ft² and speed is in ft/min; F_S is in persons/s/m² when density is in persons/m² and speed is in m/s.

Combining Equations 1 and 2 produces

$$F_S = (1 - aD)kD \qquad (3)$$

where k is as listed in Table 8-1S.

TABLE 8-1S. *Constants for Equation 2, Evacuation Speed*

Exit Route Element		k_1	k_2
Corridor, aisle, ramp, doorway		275	1.40
Stairs			
riser (in.)*	Tread (in.)*		
7.5	10	196	1.00
7.0	11	212	1.08
6.5	12	229	1.16
6.5	13	242	1.23

*1 in. = 25.4 mm.

TABLE 8-1T. *Conversion Factors for Relating Line of Travel Distance to Vertical Travel for Various Stair Configurations*

Stairs Riser (in.)	Tread (in.)	Conversion Factor
7.5	10.0	1.66
7.0	11.0	1.85
6.5	12.0	2.08
6.5	13.0	2.22

TABLE 8-1U. *Maximum (Unimpeded) Exit Flow Speeds*

		Speed—Along Line of Travel	
Exit Route Element		ft/min	m/s
Corridor, aisle, ramp, doorway		235	1.19
Stairs			
Riser (in.)	Tread (in.)		
7.5	10	167	0.85
7.0	11	187	0.95
6.5	12	196	1.00
6.5	13	207	1.05

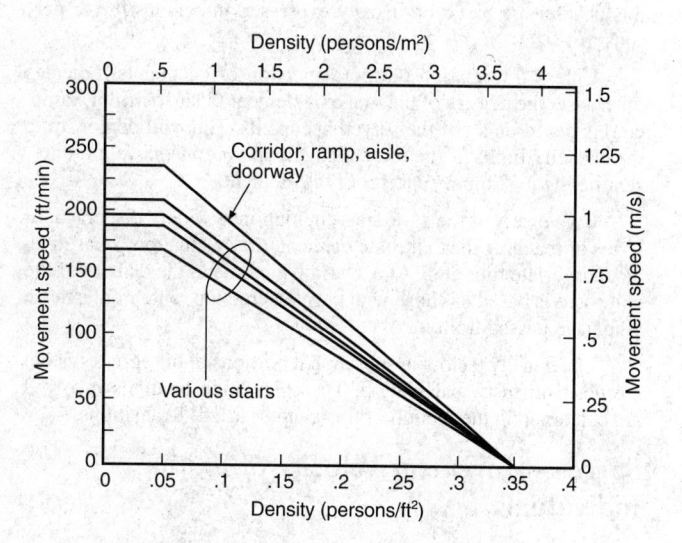

FIG. 8-1M. *Evacuation speed as a function of density.* $S = k - akD$, *where* D = *density is persons/ft² and k is given in Table 8-1S. Note that speed is along line of travel.*

TABLE 8-1V. *Maximum Specific Flow, F_{sm}*

Exit Route Element		Maximum Specific Flow	
		Persons/min/ft of Effective Width	Persons/s/m of Effective Width
Corridor, aisle, ramp, doorway		24.0	1.3
Stairs			
Riser (in.)	Tread (in.)		
7.5	10	17.1	0.94
7.0	11	18.5	1.01
6.5	12	20.0	1.09
6.5	13	21.2	1.16

The relationship of specific flow to density is shown in Figure 8-1N. In each case the maximum specific flow occurs when the density is 0.175 persons/sq ft (1.9 persons/m²) of exit route space. There is a maximum specific flow associated with each type of exit route element; these are listed in Table 8-1V.

Calculated Flow, F_c

The calculated flow, F_c, is the predicted flow rate of persons passing a particular point in an exit route.

The equation for actual flow is

$$F_c = F_s W_e \tag{4}$$

where

F_c = calculated flow,

F_s = specific flow, and

W_e = effective width.

Combining Equations 3 and 4 produces

$$F_c = (1 - aD)kDW_e \tag{5}$$

F_c is in persons/min when $k = k_1$ (from Table 8-1S), D is persons/sq ft, and W_e is ft.

F_c is in persons/sec when $k = k_2$ (from Table 8-1S), D is persons/m², and W_e is m.

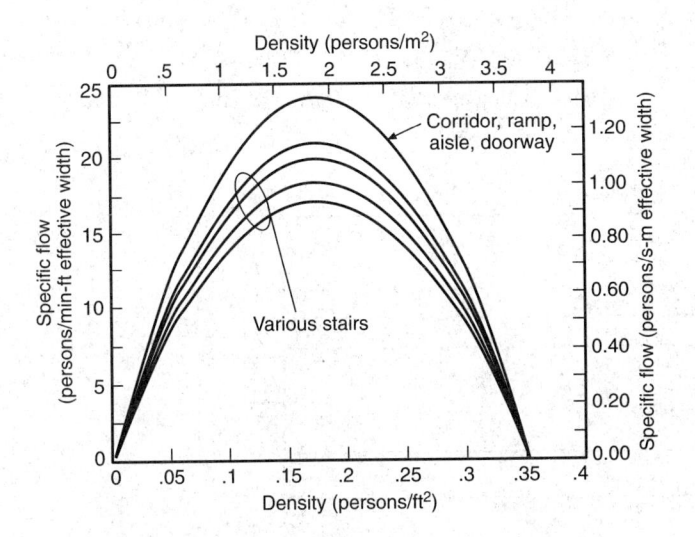

FIG. 8-1N. *Specific flow as a function of density.*

Time for Passage, T_p

Time for passage, T_p, i.e., time for a group of persons to pass a point in an exit route, can be expressed as

$$T_p = P/F_c \tag{6}$$

where T_p is time for passage (T_p is in minutes where F_c is persons/min; T_p is in seconds where F_c is persons/s). P is population in persons.

Combining Equations 5 and 6 yields

$$T_p = P/(1 - aD)kDW_e \tag{7}$$

Transitions

Transitions are any point in the exit system where the character or dimension of a route changes or where routes merge. Typical examples of points of transition include:

1. Any point where an exit route becomes wider or narrower. For example, a corridor may be narrowed for a short distance by an intruding service counter or similar element. The calculated density, D, and specific flow, F_s, differ before reaching, while passing, and after passing the intrusion.
2. The point where a corridor enters a stairway. There are actually two transitions: one occurs as the egress flow passes through the doorway; the other as the flow leaves the doorway and proceeds onto the stairs.
3. The point where two or more exit flows merge. For example, the meeting of the flow from a cross aisle into a main aisle that serves other sources of exiting population. It is also the point of entrance into a stairway serving other floors.

The following rules apply to determining the densities and flow rates following the passage of a transition point:

1. The flow after a transition point is a function, within limits, of the flow(s) entering the transition point.
2. The calculated flow, F_c, following a transition point cannot exceed the maximum specific flow, F_{sm}, for the route element involved multiplied by the effective width, W_e, of that element.
3. Within the limits of rule 2, the specific flow, F_s, of the route departing from a transition point is determined by the following equations:
   (a) For cases involving one flow into and one flow out of a transition point:

$$F_{s(out)} = F_{s(in)} W_{e(in)} / W_{e(out)} \tag{8a}$$

   where
   $F_{s(out)}$ = specific flow departing from transition point
   $F_{s(in)}$ = specific flow arriving at transition point
   $W_{e(in)}$ = effective width prior to transition point
   $W_{e(out)}$ = effective width after passing transition point
   (b) For cases involving two incoming flows and one outflow from a transition point, such as that which occurs with the merger of a flow down a stair and the entering flow at a floor:

$$F_{c(out)} = \{[F_{c(in-1)} W_{e(in-1)}] + [F_{c(in-2)} W_{e(in-2)}]\} / W_{e(out)} \tag{8b}$$

   where the subscripts (in–1) and (in–2) indicate the values for the two incoming flows.
   (c) For cases involving other merger geometries, the following general relationship applies:

$$[F_{c(in-1)} W_{e(in-1)}] + ... + [F_{c(in-n)} W_{e(in-n)}] = [F_{c(out-1)} W_{e(out-1)}] + ... + [F_{c(out-n)} W_{e(out-n)}] \tag{8c}$$

   where the letter n in the subscripts (in-n) and (out-n) is a number equal to the total number of routes entering (in-n) or leaving (out-n) the transition point.

4. Where the calculated specific flow, F_s, for the route(s) leaving a transition point, as derived from the equations in rule 3, exceeds the maximum specific flow, F_{sm}, a queue will form at the incoming side of the transition point. The number of persons in the queue will grow at a rate equal to the calculated flow, F_c, in the arriving route minus the calculated flow leaving the route through the transition point.

5. Where the calculated outgoing specific flow, $F_{s(out)}$, is less than the maximum specific flow, F_{sm}, for that route(s), there is no way to predetermine how the incoming routes will merge. The routes may share access through the transition point equally, or there may be total dominance of one route over the other. For conservative calculations, assume that the route of interest is dominated by the other route(s). If all routes are of concern, it is necessary to conduct a series of calculations to establish the bounds on each route under each condition of dominance.

EXAMPLE:

Consider an office building (Figure 8-1O) with the following features:

1. Nine floors, 300×80 ft.
2. Floor-to-floor height is 12 ft.
3. Two stairways, located at ends of building (no dead ends).
4. Each stair is 44 in. wide (tread width) with handrails protruding 2.5 in.
5. Stair risers are 7 in. high, treads are 11 in. wide.
6. There are two 4×8 ft landings per floor of stairway travel.
7. There is one, 36 in. clear width, door at each stairway entrance and exit.
8. The first floor does not exit through stairways.
9. Each floor has a single 8 ft wide corridor extending the full length of each floor. Corridors terminate at stairway entrance doors.
10. There is a population of 300 persons/floor.

Solution A—First Order Approximation

1. *Assumptions.*

The prime controlling factor will be either the stairways or the door discharging from them. Queuing will occur; therefore, the specific flow, F_s, will be the maximum specific flow, F_{sm}. All occupants start egress at the same time. The population will use all facilities in the optimum balance.

2. *Estimate flow capability of a stairway.*

From Table 8-1R, the effective width, W_e, of each stairway is $44 - 12 = 32$ in. (2.66 ft). Also, the effective width, W_e, of each door is $36 - 12 = 24$ in. (2 ft). The maximum specific flow, F_{sm}, for the stairway (from Table 8-1V is 18.5 persons/min/ft effec-

tive width). Specific flow, F_s, equals maximum specific flow, F_{sm}. Therefore, using Equation 4, the flow from each stairway is limited to $18.5 \times 2.66 = 49.2$ persons/min.

3. *Estimate flow capacity through a door.*

Again from Table 8-1V, the maximum specific flow through any 36-in. door is 24 persons/min/ft effective width. Therefore, using Equation 4, the flow through any door is limited to $24 \times 2 = 48$ persons/min. Since the flow capacity of the doors is less than the flow capacity of the stairway served, the flow is controlled by the stairway exit doors (48 persons/stairway exit door/min).

4. *Estimate the speed of movement for estimated stairway flow.*

From Equation 1 the speed of movement down the stairs is $212 - (2.86 \times 212 \times 0.175) = 105$ ft/min. The travel distance between floors (using the conversion factor from Table 8-1T is $12 \times 1.85 = 22.2$ ft on the stair slope plus 8 ft travel on each of the two landings, for a total floor-to-floor travel distance of $22.2 + (2 \times 8) = 38.2$ ft. The travel time for a person moving with the flow is $38.2/105 = 0.36$ min/floor.

5. *Estimate building evacuation time.*

If all of the occupants in the building start evacuation at the same time, each stairway can discharge 48 persons/min. The population of 2400 persons above the first floor will require approximately 25 min to pass through the exit. An additional 0.36 min travel time is required for the movement from the second floor to the exit. The total minimum evacuation time for the 2400 persons located on floors 2 through 9 is estimated at 25.4 min.

Solution B—More Detailed Analysis

1. Assumptions. The population will use all exit facilities in the optimum balance; all occupants start egress at the same time.

2. Estimate flow density, D, speed, S, specific flow, F_s, effective width, W_e, and initial calculated flow, F_c, typical for each floor.

Divide each floor in half to produce two exit calculation zones, each 150 ft long. Determine the density, D, and speed, S, if all occupants try to move through the corridor at the same time. That is, 150 persons moving through 150 ft of an 8 ft wide corridor.

Density, $D = 150$ persons/1,200 ft^2 corridor area = 0.125 persons/ft^2

From Equation 1, speed, $S = k - akD$.

From Table 8-1S, $k = 275$.

$S = 275 - (2.86 \times 275 \times 0.125) = 177$ ft/min.

From Equation 3, specific flow, $F_s = (1 - aD)kD$.

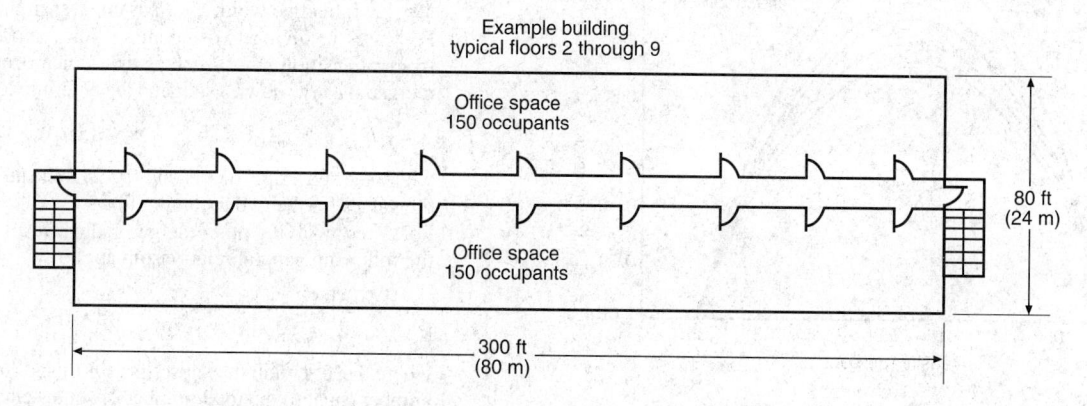

Example building
typical floors 2 through 9

Office space
150 occupants

Office space
150 occupants

80 ft
(24 m)

300 ft
(80 m)

FIG. 8-1O. Floor plan for example.

$F_s = [1 - (2.86 \times 0.125)] \times 275 \times 0.125 = 22$ persons/ft effective width/min.

From Table 8-1V, the specific flow, F_s, is less than the maximum specific flow, F_{sm1}; therefore, F_s is used for the calculation of calculated flow.

From Table 8-1R, the effective width of the corridor is $8 - (2 \times 0.5) = 7$ ft.

From Equation 5, calculated flow, $F_c = (1 - aD)kDW$.

$F_c = [1 - (2.86 \times 0.125)] \times 275 \times 0.125 \times 7 = 154$ persons/min.

Note: At this stage in the calculation, calculated flow, F_c, is termed "initial calculated flow" for the exit route element (i.e., corridors) being evaluated. This is because the calculated flow rate can be sustained only if the discharge (transition point) from the route can also accommodate the indicated flow rate.

3. Estimate impact of stairway entry doors on exit flow.

   Each door has a 36 in. clear width. From Table 8-1R, effective width, W_e, is $30 - 12 = 24$ in. (2 ft).

   From Table 8-1V, the maximum specific flow, F_{sm}, is 24 persons/min/ft effective width.

   From Equation 8, $F_{s(door)} = [F_{s(corridor)} W_{e(corridor)}]/W_{e(door)}$ $F_{s(door)} = (22 \times 7)/2 = 77$ persons/min/ft effective width. Since F_{sm} is less than the calculated F_s, the value of F_{sm} is used. Therefore, the effective value for specific flow is 24.

   From Equation 4 the initial calculated flow, $F_c = F_s W_e = 24 \times 2 = 48$ persons/min through a 36-in. door.

   Since F_c for the corridor is 154 while F_c for the single exit door is 48, queuing is expected. The calculated rate of queue buildup will be $154 - 48 = 106$ persons/min.

4. Estimate impact of stairway on exit flow.

   From Table 8-1R, effective width, W_e, of the stairway is $44 - 12 = 32$ in. (2.66 ft).

   From Table 8-1V, the maximum specific flow, F_{sm}, is 18.5 persons/ft effective width.

   From Equation 8, the specific flow for the stairway, $F_{s(stairway)}$, is $24 \times 2/2.66 = 18.0$ persons/ft effective width. In this case, F_s is less than F_{sm}, and F_s is used.

   The value of 18.0 for F_s applies until the flow down the stairway merges with the flow entering from another floor.

   Using Figure 8-1M or Equation 3 and Table 8-1S, the density of the initial stairway flow is approximately 0.146 persons/ft² of stairway exit route.

   From Equation 1 the speed of movement during the initial stairway travel is $212 \times (2.86 \times 212 \times 0.146) = 123$ ft/min.

   From Solution A, the floor-to-floor travel distance is 38.2 ft. The time required for the flow to travel one floor level is $38.2/123 = 0.31$ min (19 s).

   Using Equation 4, the calculated flow, F_c, is $18.0 \times 2.66 = 48$ persons/min.

   After 0.31 min, $48 \times 0.31 = 15$ persons will be in the stairway from each floor feeding to it. If floors 2 through 9 exit all at once, there will be $15 \times 8 = 120$ persons in the stairway. After this time, the merging of flows between the flow in the stairway and the incoming flows at stairway entrances will control the rate of movement.

5. Estimate impact of merger of stairway flow and stairway entry flow on exit flow.

   From Equation 8, $F_{s(out-stairway)} = \{[F_{s(door)} \times W_{e(door)}] + [F_{s(in-stairway)} \times W_{e(in-stairway)}]\}/W_{e(out-stairway)} = [(24 \times 2) + (18 \times 2.66)]/2.66 = 36$ persons/ft effective width.

From Table 8-1V, F_{sm} for the stairway is 18.5 persons/min/ft effective width. Since F_{sm} is less than the calculated F_s, the value of F_{sm} is used.

6. Track egress flow.

   Assume all persons start to evacuate at time zero. Initial flow speed is 177 ft/min. Assume that congested flow will reach the stairway in 30 s. At 30 s, flow starts through stairway doors. F_c through doors is 48 persons/min for the next 19 s. At 49 s, 120 persons are in each stairway, and 135 are waiting in a queue at each stairway entrance door.

   Note: Progress from this point on depends on which floors take dominance in entering the stairways. Any sequence of entry may occur. To set a boundary, this example estimates the result of a situation where dominance proceeds from the highest to the lowest floor.

   The remaining 135 persons waiting at each stairway entrance on the 9th floor enter through the door at the rate of 48 persons/min. The rate of flow through the stair is regulated by the 48 persons/min rate of flow of the discharge exit doors. The descent rate of the flow is 19 s/floor.

   Thus:

   @ 218 s (3.6 min)—all persons have evacuated the 9th floor
   @ 237 s (4.0 min)—the end of the flow reaches the 8th floor
   @ 401 s (6.7 min)—all persons have evacuated the 8th floor
   @ 420 s (7.0 min)—the end of the flow reaches the 7th floor
   @ 584 s (9.7 min)—all persons have evacuated the 7th floor
   @ 603 s (10.1 min)—the end of the flow reaches the 6th floor
   @ 767 s (12.8 min)—all persons have evacuated the 6th floor
   @ 786 s (13.1 min)—the end of the flow reaches the 5th floor
   @ 950 s (15.8 min)—all persons have evacuated the 5th floor
   @ 969 s (16.2 min)—the end of the flow reaches the 4th floor
   @ 1,133 s (18.9 min)—all persons have evacuated the 4th floor
   @ 1,152 s (19.2 min)—the end of the flow reaches the 3rd floor
   @ 1,316 s (21.9 min)—all persons have evacuated the 3rd floor
   @ 1,335 s (22.3 min)—the end of the flow reaches the 2nd floor
   @ 1,499 s (25.0 min)—all persons have evacuated the 2nd floor
   @ 1,518 s (25.3 min)—all persons have evacuated the building

COMPUTER SIMULATION AND MODELING OF EGRESS DESIGN

Computer simulation and modeling have become important tools in designing adequate means of egress under a variety of occupancy and structural conditions. This section summarizes activity to date in applying simulation and modeling techniques to solving egress design problems.

In one study of the critical variables for fire safety in relation to buildings used to house the elderly, the problem of fire development and evacuation as a time-structured problem was considered.[81] This study analyzed the variables related to the occurrence and spread of fire relative to the population's ability to reach an area of refuge from the effects of the fire. The study established the variables for the fire on a continuum of a "critical time," and the parameters for the survival of the occupants on a continuum of a "reaction time."

The definition adopted for critical time is the elapsed time from the start of the fire to the attainment of intolerable levels. The definition for reaction time is reported as the amount of time used by the occupant to react to the fire and to achieve safety by evacuating the fire area or by obtaining a place of refuge from the effects of the fire. Thus, within the framework of these definitions, both the parameters of the problem of the human behavior factors involved in a fire in a building and the variables of the problem are considered.

This conceptualization of a critical time for fire development within a building and of the reaction time required for occupants to perceive and respond to the fire threat, either by evacuation or movement to an area of refuge, has been studied thoroughly. It has resulted in multiple computer models of this essential egress behavior since Caravaty and Haviland's study in 1967.[81]

A computer model that simulates the movement of people during evacuation through the floor of fire origin has been developed.[82] A Markov-based model focused on the movement of people through a fire floor from the time of alarm until their safe exit or until they became casualties of the fire. Six variables affecting their movement have been identified:

1. Objective location of the threatening stimuli in time and space.
2. Occupants' prior knowledge of effective egress routes.
3. Occupants' perception of the location and severity of the threat.
4. Occupants' perception of available alternatives.
5. Occupants' threat-reducing experiences prior to the current movement decision.
6. The interjection of sudden interpretations to the occupants' goal-directed behavior.

Stahl has indicated that his decade of effort in modeling the movement of occupants has resulted in an improved computer model and an appreciation of the value of simulation in egress design.[83,84] Simulation modeling may be appropriate for problems that would otherwise require costly, time-consuming, and tedious manual effort; that cannot be solved through experimentation because of high costs or unacceptable risks to human participants; and for which past experience, intuition, or available data do not provide the proper insight.

A modeling technique has been developed that would appear to have potential for comparing the evacuation behavior of occupants assisted by trained personnel with evacuation without assistance.[85] This model determines the escape potential from an area of a building based on a deterministic approach with a computer algorithm, and it evaluates the effects of fire growth and smoke movement on occupants as they try to egress. Occupant evacuation is simulated with evaluation of the effects of the egress design, the density of the occupants in the means of egress, the congestion at door or other restrictions, and the effects of the combustion products on occupants.

The basic components of the original Building Firesafety Model have been applied to the specific problem of the evacuation of board and care homes during a fire.[86] This simulation model has added an evaluation of the residents' physical and mental characteristics that affect evacuation. These characteristics are movement variables or impairments and stress behavior, including resistance to assistance, mobility, the need for assistance, and the residents' responsiveness. In addition to residents' characteristics, this model evaluates the staff characteristics that are critical to providing the required assistance for effective evacuation. Simulations with this model have indicated that residents should be located on the ground floor near exits. The quickest egress routes should be identified. With increased staff/resident ratios, rescuing those farthest away from exits first can reduce evacuation time.

The development and the current applications of the building fire simulation model (BFSM) and its five submodels have been reviewed. The applications are the parameter simulation conversion logic, the fire development model, the combustion products model,

the escape routing model, and the estimated escape time escape routing model.[87] The relationship of these submodels is illustrated in Figure 8-1P.

The concepts of critical time and reaction time as originally formulated by Caravaty and Haviland[81] have been adapted into a determination of the available safe egress time (ASET) model.[88] This model is a mathematical procedure for simulating of the conditions that develop between the time of fire ignition and the onset of untenable conditions for human occupancy. Cooper's[88] model thus addresses the fire development and combustion products submodels of the BFSM.

A simulation model has been developed to evaluate the evacuation plans and procedures within a specific building.[88] This model requires input data on the physical dimensions of the building areas, the means of egress, and the specified evacuation routes with the number and location of the occupants. The model provides an estimated average evacuation time and an estimated total evacuation time.

Building network evacuation models similar to the model of O'Leary and Gratz[89] have been developed that evaluate the egress environment within the building and the characteristics of the population relative to density and location.[90] The model predicts evacuation flows and times, and it identifies queueing problems. The model has been validated with evacuation data from high-rise federal office buildings and college dormitories.

An analytical queueing network model (QNM) has also been developed to be used in the analysis of a building design relative to the suitability of the egress system.[91] This model provides estimates of the evacuation times, the average queue lengths along egress routes, potential impairments, and total egress probability.

Fahy[92] has developed a FORTRAN model called EXIT89 to predict evacuation times for high-rise buildings. The model has been validated with a practice "fire drill" evacuation involving 700 occupants from a high-rise office building. This evacuation model

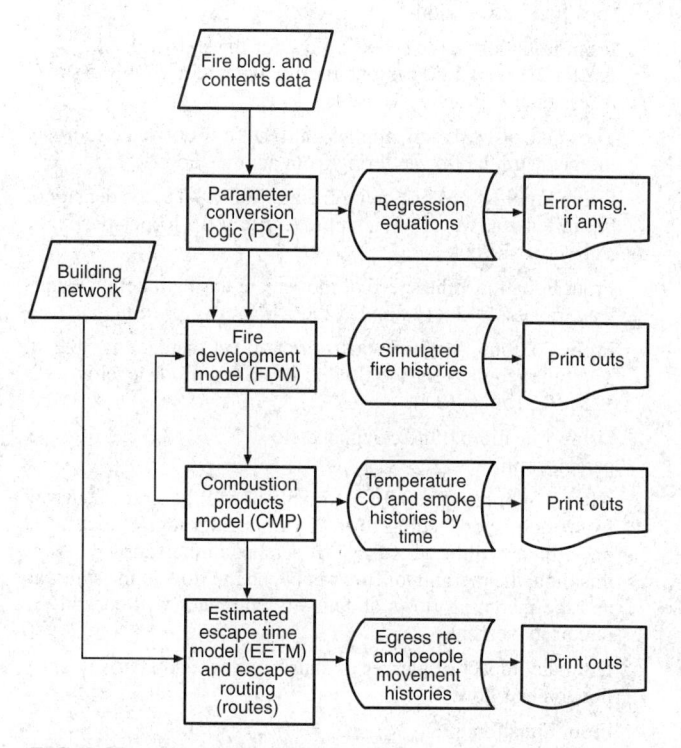

FIG. 8-1P. *The building fire simulation model design with the submodels relationship.*[87]

was designed to work with two of the subprograms of HAZARD I,[93] FAST and TENAB, to estimate the hazard to the occupants from the modeled fire incident. This model calculates the occupants' movement along the shortest evacuation routes or user-defined "familiar" routes, and the queuing of the occupants along these routes. The walking speeds in this model are derived from the occupant densities values developed by Predtechenskii and Milinskii.[80] This promising model is still being developed, and Fahy plans to incorporate aspects of occupants' behavior: assisting others or being assisted, crawling through smoke, reversing direction, delaying response to evacuation alarm signals or other cues, and selecting longer but less crowded evacuation routes.

The improvement and use of computer models for evacuation analysis have been examined and reviewed by Watts.[94] Network, queuing, and simulation models of building evacuations are currently available and are being applied in the design process.

SUMMARY

Behavior in fires can be understood as a logical attempt to deal with a complex, rapidly changing situation in which minimal information upon which to act is available. It is suggested that the goals of codes should be "reoriented to increase the likelihood of informed decisions being made by people in fires."[95] Examination of behavior in the Beverly Hills Supper Club fire led to the recommendation that "fire safety education should consider and be based on people's erroneous conceptions about distance being related to safety, and the time needed to escape from a fire emergency."[96] More than a decade of detailed systematic research on human behavior in fires has resulted in the following consensus[44] on the behavior of most persons:

Despite the highly stressful environment, people generally respond to emergencies in a "rational," often altruistic manner, insofar as is possible within the constraints imposed on their knowledge, perceptions, and actions by the effects of the fire. In short, "instinctive panic" type reactions are not the norm.

The relationship between the physical and social environment in which behavior occurs is complex. The situation is complicated by the individual's perception of ambiguous fire cues, which is primarily influenced by the person's relevant training and previous fire experience, if any. It must be recognized that fire cues are a product of a rapidly changing dynamic process that is constantly altering the decisions of the building occupant. This dilemma has been summarized: "What is an appropriate action at one stage may be quite inappropriate a minute later."

BIBLIOGRAPHY

References Cited

1. Bryan, J. L., *Smoke as a Determinant of Human Behavior in Fire Situations*, Department of Fire Protection Engineering, University of Maryland, College Park, MD, 1977.
2. Wood, P. G., "The Behavior of People in Fires," *Fire Research Note 953*, 1972, Building Research Establishment, Fire Research Station, Boreham Wood, Hertfordshire, England.
3. Lerup, L., *et al.*, "Human Behavior in Institutional Fires and Its Design Complications," NBS-GCR-77-93, 1978, Center for Fire Research, National Bureau of Standards, Washington, DC.
4. Keating, J. P. and Loftus, E. F., "The Logic of Fire Escape," *Psychology Today*, 1981, pp. 14–19.
5. Proulx, G., and Sime, J. D., "To Prevent Panic in an Underground Emergency: Why Not Tell People the Truth?" *Fire Safety Science—Proceedings of The Third International Symposium*, Elsevier Applied Science, New York, 1991, pp. 843–852.
6. Ramachandran, G., "Human Behavior in Fires—A Review of Research in the United Kingdom," *Fire Technology*, Vol. 26, No. 2, May 1990, pp. 149–155.
7. Ramachandran, G., "Informative Fire Warning Systems," *Fire Technology*, Vol. 27, No. 1, Feb. 1991, pp. 66–81.
8. Cable, E. A., "Cry Wolf Syndrome: Radical Changes Solve the False Alarm Problem," Department of Veterans Affairs, Albany, NY, Jan. 1994.
9. Kimura, M., and Sime, J. D., "Exit Choice Behavior During the Evacuation of Two Lecture Theatres," *Fire Safety Science—Proceedings of The Second International Symposium*, Hemisphere Publishing Corp., Washington, DC, 1989, pp. 541–550.
10. Lathrop, J. K., "Two Fires Demonstrate Evacuation Problems in High-Rise Buildings," *Fire Journal*, Vol. 70, No. 1, Jan. 1976, pp. 65–70.
11. Burns, D. J., "The Reality of Reflex Time," *WNYF*, Vol. 54, No. 3, 1993, pp. 26–29.
12. Fahy, R. F., and Proulx, G., "Collective Common Sense: A Study of Human Behavior during the World Trade Center Evacuation," *NFPA Journal*, Vol. 87, No. 2, March/April 1995, p. 61.
13. Berry, C. H., "Will Your Smoke Detector Wake You?" *Fire Journal*, Vol. 72, No. 4, 1978, pp. 105–108.
14. Cohen, H. C., "Fire Safety for the Hearing Impaired," *Fire Journal*, Vol. 76, No. 1, Jan. 1982, pp. 70–72.
15. Kahn, M. J., "Human Awakening and Subsequent Identification of Fire-Related Cues," *Fire Technology*, Vol. 20, No. 1, 1984.
16. Nober, E. H., *et al.*, "Waking Effectiveness of Household Smoke and Fire Detector Devices," *Fire Journal*, Vol. 75, No. 4, July 1981, pp. 86–91, 130.
17. Latane, B., and Darley, J. M., "Group Inhibition of Bystander Intervention in Emergencies," *Journal of Personality and Social Psychology*, Vol. 10, No. 3, 1968, pp. 215–221.
18. Bryan, J. L., *A Study of the Survivors' Reports on the Panic in the Fire at Arundel Park Hall, Brooklyn, Maryland, on January 29, 1956*, Fire Protection Curriculum, University of Maryland, College Park, MD, 1957.
19. Withey, S. B., "Reaction to Uncertain Threat," *Man and Society in Disaster*, Baker, G. W., and Chapman, D. W., eds., Basic Books, New York, 1962.
20. Killian, R. M., *et al.*, "A Study of Response to the Houston, Texas, Fireworks Explosion," Disaster Study No. 2, *Publication 391*, 1956, National Academy of Science, Washington, DC.
21. Sime, J. D., "Perceived Time Available: The Margin of Safety in Fires," *Fire Safety Science—Proceedings of The First International Symposium*, Hemisphere Publishing Corp., Washington, DC, 1986, pp. 561–570.
22. National Fire Protection Association, "Bimonthly Fire Record," *Fire Journal*, Vol. 65, No. 3, May 1971, p. 51.
23. Mintz, A., "Nonadaptive Group Behavior," *Journal of Abnormal and Social Psychology*, Vol. 46, 1951, pp. 150–159.
24. Jones, B. K., and Hewit, J. A. "Leadership and Group Formation in High-Rise Building Evacuations," *Fire Safety Science—Proceedings of The First International Symposium*, Hemisphere Publishing Corp., Washington, DC, 1986, pp. 513–522.
25. Horiuchi, S., Murozaki, Y., and Hokugo, A., "A Case Study of Fire and Evacuation in a Multi-Purpose Office Building," Osaka, Japan, *Fire Safety Science—Proceedings of The First International Symposium*, Hemisphere Publishing Corp., Washington, DC, 1986, pp. 523–532.
26. Breaux, J., *et al.*, *Psychological Aspects of Behavior of People in Fire Situations*, University of Surrey, Guilford, England, 1976.
27. Bickman, L., *et al.*, "A Model of Human Behavior in a Fire Emergency," NBS-GCR-78-120, 1977, National Bureau of Standards, Washington, DC.
28. Proulx, G., "A Stress Model for People Facing a Fire," *Journal of Environmental Psychology*, Vol. 13, 1993, pp. 137–147.
29. Chubb, M., "Human Factors Lessons for Public Fire Educators: Lessons from Major Fires," National Fire Protection Association, Education Section, Phoenix, Nov. 1993.
30. Klein, G. A., and Klinger, D., "Naturalistic Decision Making," CSERIAC Gateway, Crew System Ergonomics Information Analysis Center, Wright-Patterson AFB, 1991, pp. 1–4.
31. Canter, D., Breaux, J., and Sime, J., "Domestic, Multiple Occupancy and Hospital Fires," Canter, D., ed., *Fires and Human Behavior*, John Wiley & Sons, New York, 1980, pp. 117–136.
32. Best, R., and Demers, D. P., "Fire at the MGM Grand," *Fire Journal*, Vol. 76, No. 1, Jan. 1982, pp. 19–37.

33. Demers, D. P., "Investigation Report on the Las Vegas Hilton Hotel Fire," *Fire Journal*, Vol. 76, No. 1, Jan. 1982, pp. 52–57.

34. Bryan, J. L., "Human Behavior in the MGM Grand Hotel Fire," *Fire Journal*, Vol. 76, No. 2, March 1982, pp. 37–41, 44–48.

35. Morris, G. P., "Preplan Was the Key to MGM Rescue as EMS Helped Thousands of Hotel Fire Victims," *Fire Command*, Vol. 48, No. 6, 1981, pp. 20–21.

36. Bryan, J. L., "Convergence Clusters: A Phenomenon of Human Behavior Seen in Selected High-Rise Building Fires," *Fire Journal*, Vol. 74, No. 6, Nov. 1985, pp. 27–30, 86–90.

37. Bryan, J. L., and DiNenno, P. J., "An Examination and Analysis of the Dynamics of the Human Behavior in the Fire Incident at the Georgian Towers on January 9, 1979," NBS-GCR-79-187, 1979, National Bureau of Standards, Washington, DC.

38. Bryan, J. L., *An Examination and Analysis of the Dynamics of the Human Behavior in the MGM Grand Hotel Fire*, revised edition, National Fire Protection Association, Quincy, MA, 1983.

39. Bryan, J. L., *An Examination and Analysis of the Dynamics of the Human Behavior in the Westchase Hilton Hotel Fire*, National Fire Protection Association, Quincy, MA, 1983.

40. Best, R. L., "Tragedy in Kentucky," *Fire Journal*, Vol. 72, No. 1, Jan. 1978, pp. 18–35.

41. English, H. B., and English, A. C., *A Comprehensive Dictionary of Psychological and Psychoanalytical Terms*, Longmans, Green and Company, New York, 1948.

42. Schultz, D. P., "Contract Report NR 170-274," 1968, University of North Carolina, Charlotte.

43. *Investigative Report to the Governor, Beverly Hills Supper Club Fire*, 1977, Kentucky State Police, Frankfort, KY.

44. Sime, J. D., "The Concept of Panic," *Fire and Human Behavior*, Canter, D., ed., John Wiley & Sons, New York, 1980.

45. Quanrantelli, E. L., *Panic Behavior in Fire Situations: Findings and a Model from the English Language Research Literature*, Disaster Research Center, Ohio State University, Columbus, OH, 1979.

46. Keating, J. P., "The Myth of Panic," *Fire Journal*, Vol. 76, No. 3, May 1982, pp. 57–61, 147.

47. Bryan, J. L., et al., "The Determination of Behavior Response Patterns in Fire Situations, Project People II, Final Report—Incident Reports, Aug. 1977 to June 1980," NBS-GCR-80-297, 1980, National Bureau of Standards, Washington, DC.

48. Crossman, E. R., et al., "FIRRST: A Fire Risk and Readiness Study of Berkeley Households," UCBFRG/WP 75-5, 1975, University of California, Berkeley, CA.

49. *Highlights of the National Household Fire Survey*, National Fire Prevention and Control Administration, United States Fire Administration, Washington, DC, 1976.

50. Jin, T., and Yamada, T., "Experimental Study of Human Behavior in Smoke Filled Corridors," *Fire Safety Science—Proceedings of The Second International Symposium*, Hemisphere Publishing Corp., Washington, DC, 1989, pp. 511–519.

51. Bryan, J. L., et al., "An Examination and Analysis of the Dynamics of the Human Behavior in the Fire Incident at the Taylor House on April 11, 1979," NBS-GCR-80-200, 1979, National Bureau of Standards, Washington, DC.

52. Levin, B. N., and Nelson, H. E., "Firesafety and Disabled Persons," *Fire Journal*, Vol. 75, No. 5, Sept. 1981, pp. 35–40.

53. Isner, M. S., and Klem, T. J., "Fire Investigation Report World Trade Center Explosion and Fire, New York, New York, February 26, 1993." National Fire Protection Association, Quincy, MA, 1993.

54. Isner, M. S., and Klem, T. J., "Explosion and Fire Disrupt World Trade Center," *NFPA Journal*, Vol. 87, No. 6, Nov./Dec. 1993, pp. 91–104

55. Juillet, E., "Evacuating People with Disabilities," *Fire Engineering*, Vol. 126, No. 12, Nov. 1993, pp. 100–103.

56. Juillet, E., personal communication, January 18, 1994.

57. Klote, J. H., et al. "Feasibility and Design Considerations of Emergency Evacuation by Elevators," NISTIR 4870, 1992, NIST, Building and Fire Research Laboratory, Gaithersburg, MD.

58. Sherwood, J., "Darkness and Smoke," *WNYF*, Vol. 54, No. 3, pp. 56–60.

59. Pauls, J. L., "Movement of People in Building Evacuations," *Human Response to Tall Buildings*, Conway, D. J., ed., Dowden, Hutchinson and Ross, Stroudsburg, PA, 1977.

60. Keating, J. P., et al., "An Evaluation of the Federal High-Rise Emergency Evacuation Procedures," Department of Psychology, University of Washington, Seattle, 1978.

61. Keating, J. P., and Loftus, E. F., "Vocal Emergency Alarms in Hospitals and Nursing Facilities: Practice and Potential," NBS-GCR-77-102, 1977, Center for Fire Research, Gaithersburg, MD.

62. Herz, E., et al., "The Impact of Fire Emergency Training on Knowledge of Appropriate Behavior in Fires," NBS-GCR-78-137, 1978, Center for Fire Research, Gaithersburg, MD.

63. Scanlon, J., "Human Behavior in a Fatal Apartment Fire—Research Problems and Findings," 1978, Emergency Communications Research Unit, Carleton University, Ottawa.

64. Phillips, A. W., *To Keep Them Safe*, National Smoke, Fire and Burn Institute, Inc., Brookline, MA, 1979.

65. Bickman, L. E., et al., "An Evaluation of Planning and Training for Fire Safety in Health Care Facilities—Phase Two," NBS-GCR-79-179, 1979, Center for Fire Research, Gaithersburg, MD.

66. Shavit, G., "Evacuation: Testing the Effect of Voice-Message Formats," *ASHRAE Journal*, 1978, pp. 38–41.

67. Bryan, J. L., and DiNenno, P. J., "Human Behavior in a Nursing Home Fire," *Fire Journal*, Vol. 73, No. 3, 1980, pp. 82–87, 126–127, 141–143.

68. Archea, J., "The Evacuation of Non-Ambulatory Patients from Hospital and Nursing Home Fires: A Framework for a Model," NBSIR 79-1906, 1979, Center for Fire Research, Gaithersburg, MD.

69. Elliott, S., and Scheidt, J., "Hospital and Fire Department Unite to Design New Code Red Program," *Fire Journal*, Vol. 73, No. 4, 1983, pp. 47–54.

70. Nygren, R. G., "Alarm Signaling and Evacuation in a High-Rise University Resident Hall," *Fire Journal*, Vol. 66, No. 2, 1972, pp. 5–6, 11.

71. Timoney, T., "Howard Johnson's Hotel Fire, Orlando, Florida," *Fire Journal*, Vol. 78, No. 5, 1984, pp. 37–45, 88.

72. Pauls, J. L., "Management and Movement of Building Occupants in Emergencies," *Research Paper No. 788*, 1978, Division of Building Research, National Research Council of Canada, Ottawa.

73. Galbreath, M., "Time of Evacuation by Stairs in High Buildings," *Ontario Fire Marshal*, Vol. 5, No. 2 (First Quarter), 1969.

74. Melinek, S. J., and Booth, S., "An Analysis of Evacuation Times and the Movement of Crowds in Buildings," CP 96175, 1975, Building Research Establishment, Fire Research Station, Borehamwood, England.

75. Pauls, J. L., "Calculating Evacuation Times for Tall Buildings," *Fire Safety Journal*, Dec. 1987.

76. Nelson, H. E., and MacLenna, H. A., "Emergency Movement," *The SFPE Handbook of Fire Protection Engineering*, 2nd ed., National Fire Protection Association, Quincy, MA, 1995, pp. 3-285–3-295.

77. Fruin, J. J., *Pedestrian Planning Design*, Metropolitan Association of Urban Designers and Environmental Planners, Inc., New York, 1971.

78. Pauls, J. L., "Effective-Width Model for Evacuation Flow in Buildings," *Proceedings of the Engineering Applications Workshop*, 1980, Society of Fire Protection Engineers, Boston.

79. Habicht, A. T., and Braaksma, J. P., "Effective Width of Pedestrian Corridors," *Journal of Transportation Engineering*, Vol. 110, No. 1, Jan. 1984.

80. Predtechenskii, V. M., and Milinskii, A. I., *Planning for Foot Traffic in Buildings* (translated from the Russian), Russian publication, Stroizdat Publishers, Moscow, 1969; English translation published for the National Bureau of Standards and the National Science Foundation, Washington, Amerind Publishing Co. Pvt., Ltd., New Delhi, India, 1978.

81. Caravaty, R. D., and Haviland, D. S., "*Life Safety from Fire, A Guide for Housing the Elderly*," Architectural Standards Division, Federal Housing Administration, Washington, DC, 1967.

82. Stahl, F. I., "Simulating Human Behavior in High-Rise Building Fires: Modeling Occupant Movement Through a Fire Floor from Initial Alert to Safety Egress," NBS-GCR-77-92, 1975, Center for Fire Research, National Bureau of Standards, Washington, DC.

83. Stahl, F. I., "B Fires II: A Behavior-Based Computer Simulation of Emergency Egress During Fires," *Fire Technology*, Vol. 18, No. 1., Feb. 1982, pp. 49–65.

84. Stahl, F. I., "B Fires II: A Behavior-Based Computer Simulation of Emergency Egress During Fires," *Fire Technology*, Vol. 18, No. 2, May 1982, pp. 63.

85. Berlin, G. N., "A Simulation Model for Assessing Building Fire Safety," *Fire Technology*, Vol. 18, No. 1, Feb. 1982, pp. 66–75.

86. Berlin, G. N., *et al.*, "Modeling Emergency Evacuation from Group Homes," *Fire Technology*, Vol. 18, No. 1, Feb. 1982, pp. 38–48.

87. Fahy, R., "Building Fire Simulation Model," *Fire Journal*, Vol. 77, No. 4, July 1983, pp. 93–95, 102–105.

88. Cooper, L. Y., "A Mathematical Model for Estimating Available Safety Egress Time in Fires," *Fire and Materials*, Vol. 6, Nos. 3 & 4, Sept./Dec. 1982, pp. 135–144.

89. O'Leary, T. J., and Gratz, J. M., "An Analysis of Fire Evacuation Procedures Using Simulation," *Fire Journal* Vol. 76, No. 3, May 1982, pp. 119–121.

90. Chalmet, L. G., *et al.*, "Network Models for Building Evacuation," *Fire Technology*, Vol. 18, No. 1, Feb. 1982, pp. 90–113.

91. Smith, J. M., "An Analytical Queuing Network Computer Program for the Optimal Egress Problem," *Fire Technology,* Vol. 18, No. 1. Feb. 1982, pp. 18–33.

92. Fahy, R. F., "EXIT89: An Evacuation Model for High-Rise Buidlings," *Fire Safety Science—Proceedings of The Third International Symposium*, Elsevier Applied Science, New York, 1991, pp. 815–823.

93. Bukowski, R. W., *et al.*, *HAZARD I—Volume I: Fire Hazard Assessment Method*, NBSIR 87-3602, July 1987, NIST, Center for Fire Research, Gaithersburg, MD.

94. Watts, J. M., "Computer Models for Evacuation Analysis," *Fire Safety Journal*, Dec. 1987, pp. 237–245.

95. Canter, D., *et al.*, *Human Behavior in Fires*, Department of Psychology, University of Surrey, Guilford, England, 1978.

96. Pauls, J. L., and Jones, B. K., "Research in Human Behavior," *Fire Journal*, Vol. 74, No. 3, May 1980, pp. 35–41.

References

Bryan, J. L., *Human Behavior Factors and the Fire Occurrence in Buildings*, International Fire Protection Engineering Institute, Department of Fire Protection Engineering, University of Maryland, College Park, MD, 1971.

Holton, D., "Boarding Homes—The New Residential Fire Problem?" *Fire Journal*, Vol. 47, No. 2, 1981, pp. 53–56.

Paulsen, R. L., "Human Behavior and Fires: An Introduction," *Fire Technology*, Vol. 20, No. 2, 1984, pp. 15–27.

Swartz, J. A., "Human Behavior in the Beverly Hills Fires," *Fire Journal*, Vol. 73, No. 3, May 1979, pp. 73–74, 108.

NFPA Codes, Standards, and Recommended Practices

Reference to the following NFPA codes, standards, and recommended practices will provide further information on human behavior and fire discussed in this chapter. (See the latest version of *The NFPA Catalog* for availability of current editions of the following documents.

NFPA 72, *National Fire Alarm Code*
NFPA *101®*, *Life Safety Code®*

Additional Reading

Ballast, D. K., *Egress from Buildings in Emergencies: A Bibliography*, Vance Bibliographies, Monticello, IL, 1988.

Beck, K. H., "A Canonical Correlation of Fire Protective Behaviors and Beliefs," *Fire Technology*, Vol. 25, No. 1, Feb. 1989, pp. 41–50.

Bodamer, M., "How People Behave in Fires," *Fire Prevention*, No. 224, Nov. 1989, p. 20, 22–23.

Booker, C. K., Powell, J., and Canter, D., "Understanding Human Behavior During Fire Evacuation," Chapter 6, Council on Tall Buildings and Urban Habitat, *Fire Safety in Tall Buildings. Tall Building Criteria and Loading*, Committee 8A, McGraw-Hill, Inc., Blue Ridge Summit, PA, 1992, pp. 93–104.

Bryan, J. L., "Behavioral Response to Fire and Smoke," *The SFPE Handbook of Fire Protection Engineering*, 2nd ed., National Fire Protection Association, Quincy, MA, 1995.

Canter, D., *Fires and Human Behavior*, 2nd ed., Fulton, London, 1988.

Chalmet, L. G., *et al.*, "Network Models for Building Evacuation," *Management Science*, Vol. 28, No. 1, Jan. 1990, pp. 86–105.

Collins, B. L., Dahir, M. S., and Madrzykowski, D., "Visibility of Exit Signs in Clear and Smoky Conditions," *Fire Technology*, Vol. 24, No. 2, 1993, pp. 154–182.

Cooper, L. Y., and Stroup, D. W., "ASET: A Computer Program for Calculating Available Safe Egress Time," *Fire Safety Journal*, Vol. 9, 1985, p. 29.

Cote, R., ed., *Life Safety Code Handbook*, sixth edition, National Fire Protection Association, Quincy, MA, 1994.

Frantzich, H., "Evacuation Capability of People," Lund Univ., Sweden, TNO Building and Construction Research, CIB/W14 Workshop, *Proceedings of the Third Fire Engineering Workshop on Modelling*, Jan. 25–26, 1993, Delft, The Netherlands, 1993, pp. 224–231.

Fraser-Mitchell, J. N., and Pigott, B. B., "Modelling Human Behavior in the Fire Risk Assessment Model 'CRISP II'," Fire Research Station, Borehamwood, England, University of Ulster and Fire Research Station. CIB W14: Fire Safety Engineering, International Symposium and Workshops Engineering Fire Safety in the Process of Design: Demonstrating Equivalency, Part 3, Symposium: Engineering Fire Safety for People with Mixed Abilities, Sept. 13–16, 1993, Newtownabbey, Northern Ireland, 1993, pp. 1–10.

Fruin, J. J., *Pedestrian Planning and Design*, revised edition, Elevator World, Mobile, AL, 1987.

Golton, C. J., Golton, B. J., and Hinks, A. J., "Human Behavior in Fire: A Background Review for Modelling," Fire Safety Modelling and Building Design, *Proceedings of the One-Day Conference to Review the Potential and Limitations of Fire Safety Models for Building Design*, March 29, 1994, Salford, UK, 1994, pp. 48–62.

Jin, T., and Yamada, T., "Experimental Study on Human Emotional Instability in Smoke Filled Corridor. Part 2," *Journal of Fire Sciences*, Vol. 8, No. 2, March/April 1990, pp. 124–134.

Keating, J. P., "Human Resources During Fire Situations: A Role for Social Engineering," *General Proceedings Research and Design*, American Institute for Architects Foundation, Washington, DC, 1985.

Kendik, E., "Methods of Design of Means of Egress: Towards a Quantitative Comparison of National Code Requirements," *Proceedings of The First International Symposium of Fire Safety Science*, Hemisphere, Washington, DC, 1986.

Kisko, T. M., and Francis, R. L., "EVACNET+, A Computer Program to Determine Optimal Building Evacuation Plans," *Fire Safety Journal*, Vol. 9, No. 2, 1985, pp. 211–220.

Klevan, J. B., "Modeling of Available Egress Time from Assembly Spaces or Estimating the Advance of the Fire Threat," SFPE TR 82-2, May 1982, Society of Fire Protection Engineers, Boston, MA.

Ling, W. C. T., and Williamson, R. B., "Use of Probabilistic Networks for Analysis of Smoke Spread and Egress of People in Buildings," *Proceedings of The First International Symposium for Fire Safety Science*, Hemisphere, New York, 1986, pp. 953–962.

Maclennan, H. A., "Towards an Integrated Egress/Evacuation Model Using an Open Systems Approach," *Proceedings of The First International Symposium on Fire Safety Science*, Hemisphere, 1986, pp. 581–590.

Magawa, M., Kose, S., and Moushita, Y., "Movement of People on Stairs During a Fire Evacuation Drill—Japanese Experience in a High-Rise Office Building," *Proceedings of the First International Symposium on Fire Safety Science*, Hemisphere, Washington, DC, 1986.

Ozel, F., *Way Finding and Route Selection in Fires*, School of Architecture, New Jersey Institute of Technology, Newark, NJ, 1986.

Pauls, J., "Development of Knowledge About Means of Egress," *Fire Technology*, Vol. 20, No. 2, 1984, pp. 28–40.

Pauls, J., Gatfield, A. J., and Juillet, E., "Elevator Use for Egress: The Human-Factors Problems and Prospects," National Research Council of Canada, Ottawa, Ontario, National Task Force on Life Safety and the Handicapped American Society of Mechanical Engineers, Council of American Building Officials and National Fire Protection Association, Elevators and Fire, Feb. 19–20, 1991, Baltimore, MD, 1991, pp. 63–75.

Pauls, J., "Movement of People," *The SFPE Handbook of Fire Protection Engineering*, 2nd ed., National Fire Protection Association, Quincy, MA, 1995.

Paulsen, R. L., "Human Behavior and Fires: An Introduction," *Fire Technology*, Vol. 20, 1984, pp. 15–27.

Poon, L. S., and Beck, V. R., "Numerical Modelling of Human Behavior during Egress in Multi-Storey Office Building Fires Using EvacSim—Some Validation Studies," *Proceedings* of the First International Conference on Fire Science and Engineering, ASIAFLAM '95, March 15–16, 1995, Kowloon, Hong Kong, 1995, pp. 163–174.

Predtechenskii, V. M., and Milinskii, A. I., *Planning for Foot Traffic Flow in Buildings*, Amerind Publishing Co., New Delhi, India, 1978.

Proulx, G., "Human Factors in Fires and Fire Safety Engineering," *SFPE Bulletin*, Winter 1995, pp. 13–15.

Robertson, J. C., "Instilling Proper Public Fire Reaction," *Introduction to Fire Prevention*, Third edition, Macmillan, New York, 1989, pp. 219–248.

Takahashi, K., and Tanaka, T., "An Evacuation Model for the Use in Fire Safety Designing of Buildings," 9th Joint Panel Meeting of the UJNR Panel on Fire Research and Safety, NBSIR 88-3753, National Bureau of Standards, Gaithersburg, MD, April 1988.

Van Bogaert, A. F., "Evacuating Schools on Fire," *Proceedings of The First International Symposium on Fire Safety Science*, Hemisphere, 1986, pp. 551–560.

Weinroth, J., "An Adaptable Microcomputer Model for Evacuation Management," *Fire Technology*, Vol. 25, No. 4, 1989, pp. 291–307.

CONCEPTS OF EGRESS DESIGN

Revised by James K. Lathrop

Means of egress and their design should be based upon an evaluation of a building's total fire protection system and an analysis of the population characteristics and hazards to the occupants of that building. The means of egress design should be treated as an integral part of the total system that provides reasonable safety to life from fire.

This chapter covers the fundamental concepts of good egress design that are the basis for NFPA *101®, Life Safety Code®.* NFPA *101* governs good practices to provide life safety features that can be designed as integral parts of new and existing buildings and structures to provide reasonable safety to occupants in fires. The components of good means of egress are discussed in some detail, with their functions and relationships in the total concept of proper egress design. Computer modeling and simulation to assist the egress design process also are discussed.

FUNDAMENTALS OF DESIGN

The approach to designing means of egress first requires a familiarity with the reaction of people in fire emergencies. These reactions can differ widely, depending upon the physical and mental capabilities and conditions of building occupants. The psychological and physiological factors affecting the use of exits during emergencies are being identified and measured in research studies.

Patterns of movement of people, singly and in crowded conditions, also must be understood. In buildings used as schools or theaters housing highly mobile occupants, for example, there are certain reproducible flow characteristics from persons exiting the buildings. These predictable flow characteristics have fostered computer simulation and modeling to aid the egress design process. However, no number of practical exit facilities can prevent injury or loss of life if the occupant egress flow is inhibited or prevented by the building itself, by personnel, or by fire and smoke conditions.

Human Factors

The design and capacity of passageways, stairways, and other components in the total means of egress are related to the physical dimensions of the human body. The tendency of people to avoid bodily contact with others should be recognized as a major factor in determining the number of persons who will occupy a given space at any given time. Given a choice, people usually automatically establish "territories" to avoid bodily contact with others.

Studies have shown that most adult men measure less than 20.7 in. (520 mm) across at the shoulder, with no allowance for additional thicknesses of clothing.[1] A "body ellipse" concept is used to develop the design of pedestrian systems. The major axis of the body ellipse measures 24 in. (609 mm), whereas the minor axis is considered 18 in. (457 mm). This ellipse equals 2.3 sq ft (0.21 m^2), which is assumed to help determine the maximum practical standing capacity of a space.

The movement of persons results in a swaying action that varies from male to female and, depending upon the type of motion, varies with movement on stairs, on level surfaces, or in dense crowds. Body sway has been observed to range $1^1/_2$ in. (38 mm) left and right during normal free movement. Where movement is reduced to shuffling in dense crowds and to movement on stairs, a total sway range of almost 4 in. (101 mm) has been observed. In theory, this indicates that a total width of 30 in. (762 mm) would be required to accommodate a single file of pedestrians traveling up or down stairs.[2]

Crowding people into spaces where less than 3 sq ft (0.28 m^2) of space per person is available under nonemergency conditions may create a hazard. When the average area occupied per person is reduced to $2^3/_4$ sq ft (0.25 m^2) or less, contact will be unavoidable. Needless to say, under the psychological stresses imposed during a fire, such crowding and contact could contribute to crowd pressures, resulting in injuries. When a queue occurs because of an artificial, temporary situation or because of some permanent design feature, crowd control becomes difficult, and the well-being of individuals is threatened.

Factors Affecting Movement of People

There are several factors that determine how quickly people may pass through the means of egress.

In level walkways an average walking speed of 250 ft/min (76 m/min) is attained under free-flow conditions, with 25 sq ft (2.3 m^2) of space available per person. Speeds below 145 ft/min (44 m/min) show shuffling, which restricts motion. Figure 8-2A, adapted from *Research Report No. 95* of the London Transport Board,[3] shows the rate of speed reduction for space concentrations of less than 7 sq ft (0.65 m^2) per person. Speeds of less than 145 ft/min (44 m/min) result

James K. Lathrop is a principal in the firm Koffel Associates of Ellicot City, MD, and Niantic, CT. He is a member of the Technical Committee on Fundamentals, Residential Occupancies; Board and Care Occupancies; and an alternate on the Technical Committee on Means of Egress.

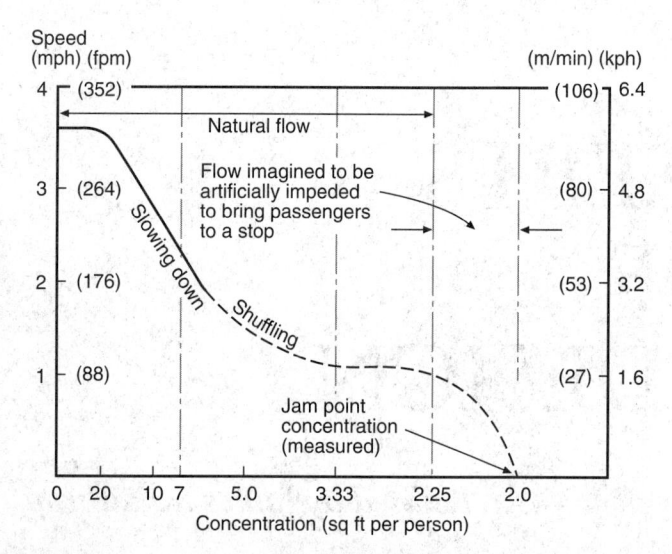

FIG. 8-2A.   Speed in level passageways.[3] (SI units: 1 ft/min = 0.305 m/min; 1 sq ft = 0.093 m².)

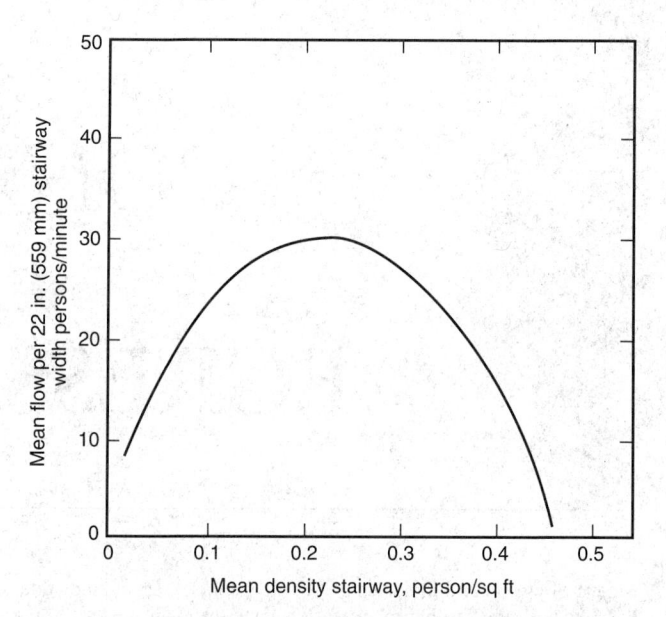

FIG. 8-2B.   Effect of density on flow down exit stairways in evacuations of high-rise office buildings.[2] (SI units: 1 sq ft = 0.093 m².)

in shuffling, and, finally, a jam point is reached with one person every 2 sq ft (0.18 m²). The possibility of a significant nonadaptive behavior exists whenever egress movement is restricted, and the problem becomes urgent under fire exposure conditions, especially when there is more than one person every 3 sq ft (0.28 m²).

Calculations of flow rates using velocity [ft/min (m/min)] and density [persons/sq ft (m³)] will reveal flow [persons/min/ft (m) of width], which increases as the pedestrian area decreases. The flow increases will continue until forward movement becomes restricted to the point that the flow begins to drop. Interestingly, observations of flow rates in one study noted the same flow rate sometimes occurred even though walking speeds of people were significantly different. Investigation revealed that the rate of decrease in speed, accompanied by an increase in density, results in uniform flow rates over a wide range of conditions.

A study of footways indicates that, for passageways over 4 ft (1.2 m) wide, flow rates are directly proportional to width. The London Transport Board *Research Report No. 95*[3] determined the flow rate in level passages to be 27 persons/min/ft (0.30 m) of width. Travel down stairways was determined at 21 persons/min/ft (0.30 m) of width, whereas upward travel was reduced to 19 persons/min/ft (0.30 m) of width. Where the width of a footway is less than 4 ft (1.2 m), the flow rate depends upon the number of possible traffic lanes. Absolute maximum flow rates occur when approximately 3 sq ft (0.28 m) is occupied per person, which is applicable to both level walkways and stairs. In observed and measured evacuations, however, it has been empirically determined that the maximum flow rates down stairs in high-rise buildings occur when from 4 to 5 sq ft of space (0.37 to 0.46 m²) is occupied per person, as shown in Figure 8-2B.[2] When flow in opposite directions takes place in a passageway up to the point where the two flows are of equal magnitude, there is no significant reduction in total flow below that which would be predicted on the basis of unidirectional flow in the same passageway.

Further, flow can be 50 percent greater in short passageways less than 10 ft (3.05 m) long than through a long passageway of the same width. Minor obstructions within a passageway do not appear to have a significant effect on flow. Within a 6 ft (1.82 m) wide passageway, there is no effect on flow rates when a 1-ft (0.45-m) projection is introduced. A 2-ft (0.61-m) projection resulting in a 33 percent reduction in width reduces the flow rate approximately 10

percent. A major obstruction, though, such as that which occurs at a ticket booth or turnstile, may interrupt the movement of people and reduce flow rates.

Corners, bends, and slight grades up to 6 percent are apparently not factors in determining flow rates. A slight reduction in speed does occur, but the flow rate is maintained by an increased concentration of persons.

A center handrail or mullion, which may divide a passageway into narrower sections, can further reduce the capacity of the passageway. The observed capacity of a 6 ft (1.82 m) wide stairway reveals a reduction from 130 to 105 persons/min after installation of a center handrail.

Except for the very young and the very old, age does not appear to be a significant factor in determining travel speed. Studies have shown a significant reduction in walking speeds for persons over 65 years of age. Studies have further revealed that a 40 percent increase is possible in the normal walking speed, which tends to discount this factor as a major influence on flow rates.[1]

For additional information see Section 3, Chapter 13, of *The SFPE Handbook of Fire Protection Engineering.*

Methods of Calculating Exit Width

Two major principles are used to determine the necessary exit width. They are based on anticipated population characteristics identified with a specific occupancy.

The flow method: This method uses the theory of evacuating a building within a specified maximum period of time. Flow rates have traditionally been set at 60 persons per 22 in. (559 mm) width/min through level passageways and doorways. In older editions of NFPA *101* this 22 in. (59 mm) width was referred to as 1 "unit" of exit width. Credit was given only for whole units or ¹/₂ unit, a ¹/₂ unit being 12 in. The flow method may be applied in assembly occupancies, such as theaters, and educational occupancies where people are alert, awake, and assumed to be in good physical condition. Figure 8-2C illustrates the flow time in seconds relative to the effective stair width per person and the units of width.

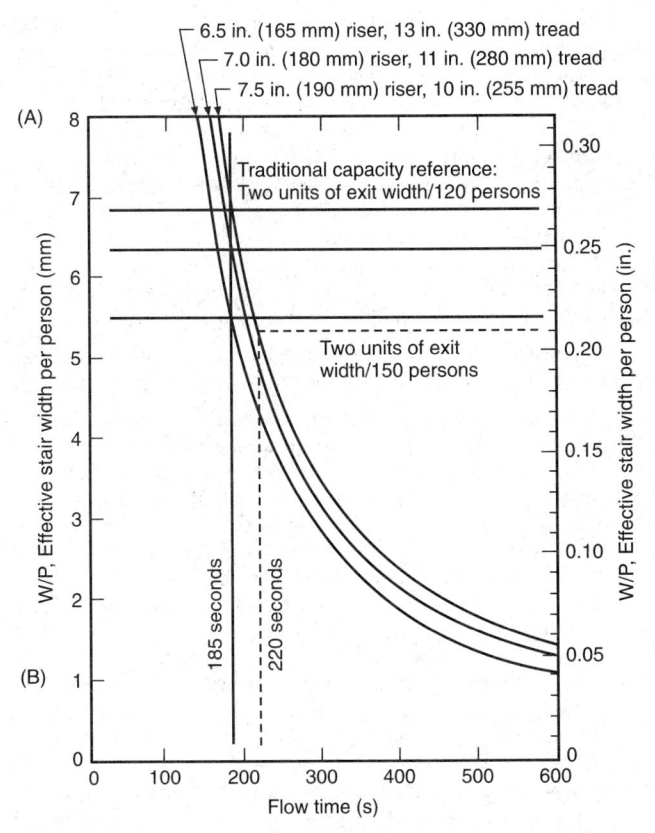

FIG. 8-2C. *Relationship between effective stair width and units of exit width per person and flow time for three stair geometries.*[4]

Pauls'[4] effective stair width concept advocates the consideration of only the portion of the stair used in effective movement by the occupants, as observed in functional and practice evacuations. This width is established with 6-in. (150-mm) clearance from each side wall of the stair. The effective width concept of exit design, for both stairs and level egress components, is included in NFPA 101A, *Alternative Approaches to Life Safety.*

The capacity method: This method is based on the theory that sufficient numbers of stairways should be provided in a building to adequately house all occupants of the building without requiring any movement, or flow, out of the stairways. In theory, assuming a stairwell provides a safe and protected area for all occupants within the protective barrier created by the stairway enclosure, evacuation of the building may then be more leisurely, permitting people to travel at a rate within their physical ability. The capacity method recognizes that evacuation from high-rise buildings is physically very demanding. Further, evacuation of a health care facility is likely to be slow. Thus, design criteria are established to permit holding occupants within exits or areas of refuge.

Application

The capacity and flow methods may both be applied to efficient egress design, depending upon specific circumstances. Where people are expected to be physically or mentally sick, aged, asleep, or incapacitated in any way, evacuation and use of the flow method is unwise. Therefore, the capacity method that provides a place for everyone within an area of refuge is the appropriate method.

There is little time between an alert and the use of an exit in assembly occupancies, and maximum flow rates that cause reductions

in the area used by each person may result in reduced traffic flows. On the other hand, the control of children in an educational setting, coupled with their familiarity with the surroundings, their presumed high physical capabilities, and their experience with a program of drills, should allow rapid evacuation times. The flow method appears to have its application in those occupancies where people are considered to be alert, awake, and of normal physical ability. Pauls has reviewed the historical and current literature relative to the principles of people movement, exit width determination, and the design of the means of egress.[5]

Design of Means of Egress

Designing a means of egress involves more than numbers, flow rates, and densities. Safe exit from a building requires a safe path of escape from the fire environment. The path is arranged for ready use in case of emergency and should be sufficient to permit all occupants to reach a safe place before they are endangered by fire, smoke, or heat. Proper exit design permits everyone to leave the fire-endangered areas in the shortest possible time with efficient exit use. If a fire is discovered in its incipient stage and the occupants are alerted promptly, effective evacuation may take place.

Evacuation travel distances are related to the content fire hazard. The higher the hazard, the shorter the travel distance to an exit.

Depending upon the physical environment of the structure, the characteristics of the occupants, and the fire detection and alarm facilities, fire or smoke may prevent the use of one means of egress. Therefore, at least one alternate means of egress remote from the first is essential. Provision of two separate means of egress is a fundamental safeguard, except where a building or room is small and arranged so that a second exit would not provide an appreciable increase in safety. There are fewer or no advantages to separate means of egress if there is travel through a common space or use of common structural features that result in the loss of the two distinct and physically separate means of egress.

One example of a "common" structure is a multi-story building where scissors stairs are used. These are two stairs enclosed within a common shaft, separated by a partition common to both stairs. Scissors stairs are sometimes used to provide the required exit capacity while minimizing the loss of valuable floor space. Where a set of scissors stairs is the only means of egress when two remote exits are required, however, the fundamental principle of two separate means of egress design may be violated. If the common partition between the stairs fails, it would result in the simultaneous loss of both exits during a fire, leaving no alternate means of egress. With scissors stairs, the validity of the two separate means of egress, therefore, depends upon the design characteristics and construction of the common partition. (See Figure 8-2D.)

In some proposed egress designs, all the exits discharge through a single lobby at street level, even though this procedure results in egress travel through a common space. This design philosophy presumes that the lobby may be considered a safe area for all future egress needs during the life of the building. Where two remote means of egress are required, this type of egress design is unsuitable.

NFPA *101* limits openings in exit enclosures to those necessary for access to the enclosure from normally occupied spaces and for egress from the enclosure. Penetration of enclosures by ducts or other utilities constitutes a point of weakness and may result in contamination of the enclosure during a fire and should not be permitted. Furthermore, it is not good practice to use exit enclosures for any purpose that could interfere with their value as exits. For example, exit stair enclosures should not be used for storage or any other use not associated with egress or areas of refuge for mobility impaired persons.

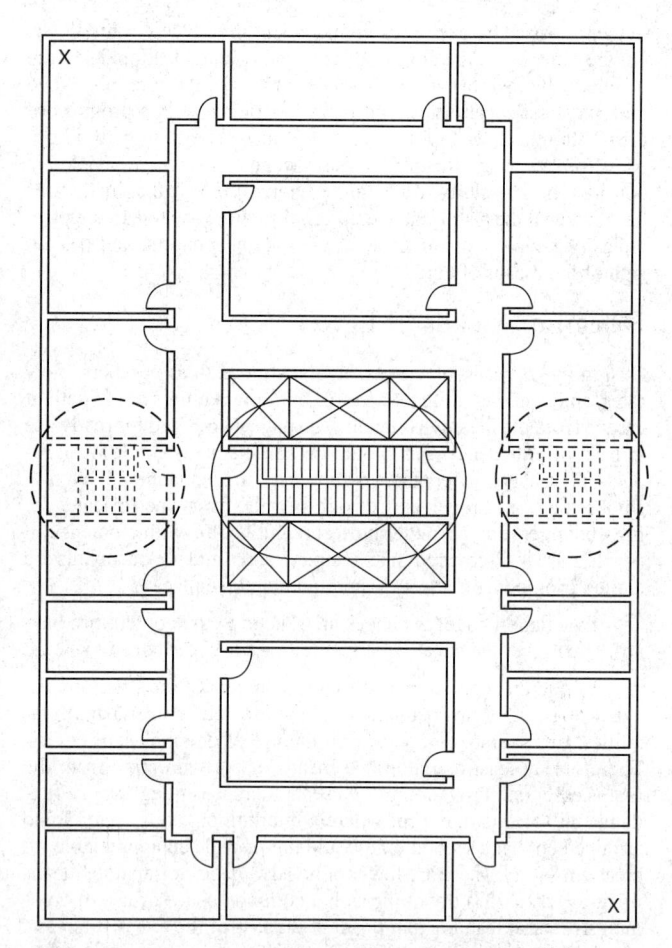

FIG. 8-2D. *Advantages and disadvantages of scissors stairs vs. conventional stairs. This set of scissors stairs provides the same degree of remote exit or entrance doors as the circled stairs shown by dotted lines—travel distance for all occupants is the same, even if the dotted exit stairs were located at opposite corners as denoted by the cross marks. Space is saved; however, the integrity of the separation of the two scissors stairs may remain in question.*

The removal of handicapped persons is an important consideration in the design of an emergency means of egress from a building. A 32-in. (813-mm) doorway is considered the minimum width to accommodate a person in a wheelchair. Since handicapped employees or visitors may be found in all types of buildings, special life safety considerations are indicated. The 1994 edition of NFPA *101* contains several additional provisions to protect mobility impaired individuals.

LIFE SAFETY CODE

NFPA *101*, introduced in 1927 and revised and reissued in successive editions, is developed by several committees under the oversight of the Technical Correlating Committee on Safety to Life, a representative group dedicated to safety of life from fire. NFPA *101* is primarily concerned with the control of conditions that threaten the lives of individuals in building fires. This objective is different from fire protection provisions in building codes, which are concerned with the preservation of property, in addition to the preservation of life.

Adequate means of egress alone are not a guarantee of life safety from fire. They do not protect an individual whose own care-

lessness causes a threat to life, such as setting his or her own clothes on fire. Neither do sufficient means of egress alone provide adequate protection in occupancies such as hospitals, nursing homes, prisons, and mental institutions, where occupants are confined or are physically or mentally unable to escape without effective and immediate assistance. NFPA *101* does recognize such situations and provides life safety measures, including low-flame-spread and reduced-smoke-producing materials for interior finish. In addition, automatic sprinkler and smoke control systems called for by NFPA *101* are designed to restrain the spread of fire and smoke and thus help to defend the occupants within an area of refuge until they are able to use the exits or until the fire has been extinguished.

In general, saving building occupants from a fire requires the following, all of which are identified in NFPA *101*:

1. A sufficient number of properly designed, unobstructed means of egress of adequate capacity and arrangement.
2. The protection of the means of egress against fire, heat, and smoke during the egress time determined by the occupant load, travel distance, and exit capacity.
3. The provision of alternate means of egress for use if one means of egress is blocked by fire, heat, or smoke.
4. The subdivision of areas by proper construction to provide areas of refuge in those occupancies where total evacuation is not a primary consideration.
5. The protection of vertical openings to limit the operation of fire protection equipment to a single floor.
6. The provision of detection or alarm systems to alert occupants and notify the fire department in case of fire.
7. The adequate illumination of the means of egress.
8. The proper marking of the means of egress, and the indication of directions.
9. The protection of equipment or areas of unusual hazard that could produce a fire capable of endangering the egressing occupants.
10. The initiation, organization, and practice of effective drill procedures.
11. The provision of instructional materials and verbal alarm systems in high-density and high-life-hazard occupancies to facilitate adaptive behavior.
12. The use of interior finish materials that prevent a high flame spread or dense smoke production that could endanger egressing occupants.

Figure 8-2E illustrates some of the principles of exit safety.

NFPA *101* recognizes that full reliance cannot be placed upon any single safeguard, since any single protective feature may not function due to mechanical or human failure. For this reason, redundant safeguards, any one of which will result in a reasonable level of life safety, should be provided. NFPA *101* also recommends the special protection of hazardous areas and specifies where automatic sprinkler, detection, and other protective systems are required.

NFPA *101* is used widely as a guide to good practice and as a basis for local laws or regulations. It differs from building codes since it generally provides little distinction among the different classes of building construction. However, where total evacuation of a building is not practical, due either to the occupant characteristics or the building environment, the construction type becomes an important variable and should be considered.

NFPA *101* also recognizes that all habitable buildings contain sufficient quantities of combustible contents to produce lethal quantities of smoke and heat.[6,7] In addition, casualty studies have established that the toxic properties of smoke are the principal hazard to life,[8] and this hazard is recognized in NFPA *101*.

NFPA *101* is intended to be applied to both new and existing buildings and is designed to provide a reasonable level of life safety

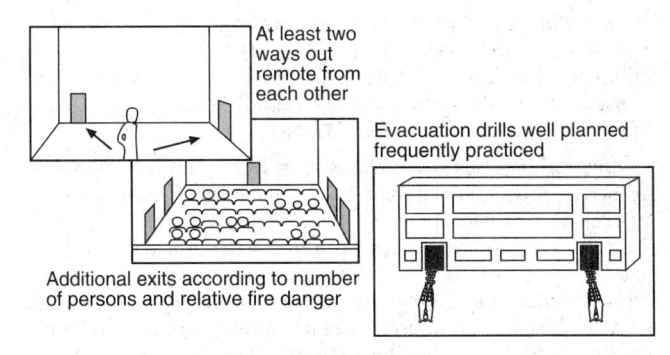

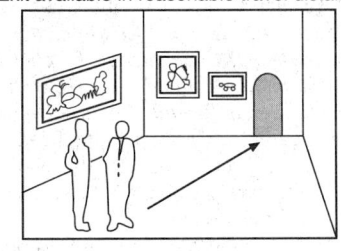

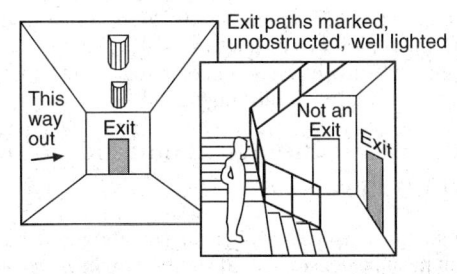

FIG. 8-2E.   Principles of exit safety.

from fire in both types of buildings. The Authority Having Jurisdiction is given considerable latitude in achieving conformance with existing buildings. Each existing building represents a special situation that requires individual attention for the most effective and economical method of achieving a reasonable level of life safety.

The argument that buildings constructed many years ago according to all the legal requirements are sufficiently safe now should not necessarily be accepted. If the economic cost of reasonable life safety is judged to be prohibitive, the occupancy or the structure should be changed or prohibited because there is no justification for subjecting building occupants to an unreasonable level of peril from a fire.

There may be a variety of differing opinions as to what constitutes reasonable life safety from a fire in any given case. It is not possible to guarantee occupants 100 percent life safety from a fire; beyond certain conditions, a building becomes hazardous to the life safety of the occupants in a fire. How should the Authority Having Jurisdiction establish the minimum conditions? NFPA *101* provides guidance for such decisions with the help of studies of major-loss-of-life fires,[9,10] fire development research,[7,11] personnel evacuation,[12,13,14] and human behavior.[15,16]

NFPA *101* examines the various occupancy populations according to their perceived life safety hazard, which includes psychological and sociological variables, in addition to the physiological and environmental factors. These occupancy classifications are assembly, educational, health care, detention/correctional, residential, mercantile, business, industrial, and storage. Additional provisions for special-purpose and high-rise structures are also included.

Separate and distinct means of egress provisions are made for each occupancy classification, with the various occupancy subgroups included. These classifications, based on the perceived hazard to life safety from a fire, often differ from older building code occupancy classifications. For example, mercantile and office occupancies were often grouped together in previous editions of building codes. However, there appears to be an increased hazard to life in mercantile properties, resulting from the displays of combustible merchandise, the greater density of the population, and the transient character of most of the occupants. These factors are not usually found in office and educational buildings, which have a relatively low combustibility content, a lower population density, and normally alert occupants who are in the building daily and presumably have the opportunity to familiarize themselves with the means of egress through functional use and evacuation drills.

INFLUENCES ON EGRESS

Influence of Hazard of Contents

An evaluation of the hazard of the building contents must take into account the relative probability of the ignition of combustibles, the spread of flames and heat, the probable smoke and gases expected to be generated by the fire, and the possibility of a fire-related explosion or other structural failure endangering occupants. The degree of hazard is usually determined by the flammability or toxicity of the contents and by the processes or operations conducted in the building. Most NFPA *101* requirements are based on the exposure created by contents with an ordinary hazard. Special requirements for areas with high-hazard contents usually consist of special protection systems, isolation of the hazard area via fire-rated construction, reduced travel distances, and additional means of egress.

To assist in evaluating the contents hazards, NFPA *101* establishes three classifications of contents: (1) low-, (2) ordinary-, and (3) high-hazard. They are discussed next. These should not be confused with the classifications established by NFPA 10, *Standard for Portable Fire Extinguishers,* or NFPA 13, *Standard for the Installation of Sprinkler Systems,* nor with those established by the model building codes.

Low-hazard contents: These are contents of such low combustibility that no self-propagating fire can occur in them. Consequently, the only probable danger requiring the use of emergency exits will be from smoke or from fire from some external source. These are extremely unusual. The storage of sheet metal without combustible packing is one example.

Ordinary-hazard contents: These are contents that are liable to burn with moderate rapidity and to give off a considerable volume of smoke. This class includes most buildings and is the basis for the general requirements of NFPA *101.*

High-hazard contents: These are contents that are liable to burn with extreme rapidity or from which explosions are to be feared in the event of fire. Examples are occupancies in which flammable liquids or gases are handled, used, or stored; in which combustible dust explosion hazards exist; in which hazardous chemicals or explosives are stored; in which combustible fibers are processed or handled in a manner that produces combustible flyings; and similar situations.

Influence of Building Construction and Design

A building of fire resistance-rated construction is designed to permit a burnout of contents without structural collapse. Fire resistance-rated design does not ensure the life safety of the occupants of such buildings.[9,15] However, the ability of a structural frame to maintain building rigidity under fire exposure is important to the maintenance of the fire-resistance protection of exit enclosures. Where a 2-hr fire-rated exit enclosure is required, a fire-resistance-rated structural frame capable of withstanding stresses imposed by fire for a similar period is also necessary. It is inconsistent to provide a 2-hr exit enclosure in a building with a structural frame rated at less than 1 hr, for example, unless special construction precautions are taken to prevent structural failure of the building from adversely affecting the protective construction of the exit enclosures.

The protection of vertical openings is one of the most significant factors in the design of multi-story buildings, from the standpoint of life safety and exit design. Because of the natural tendency of fire to spread upward in a building, careful attention to details of design and construction are required to minimize this effect. One of the greatest hazards to life safety results from fires that start below the occupants and the means of egress, such as in basements or on the level of exit discharge. Similarly, fires in multi-story buildings may result in smoke spread into enclosed exits before evacuation.[9,10,15] Conversely, escape from fires that occur above the occupants is relatively simple, provided sufficient warning is given and adequate means of egress are available.

The influence on the life safety of the occupants by the materials used in building construction depends primarily upon whether the materials will propagate flame, support combustion, or create dense amounts of smoke when exposed to a fire initially involving the building contents. Some materials used as insulation, for example, could contribute to rapid flame development and dense smoke production spread. Masonry walls enclosing a wood-frame interior provide no increased occupant life safety compared with a total wood-frame structure.

Exit requirements are based on buildings of conventional design. Unusual buildings, such as those without windows or those with unopenable windows, call for special consideration. Windows provide a number of advantages in a fire. Persons at openable windows have access to fresh air, can see fire department rescue operations in progress, are able to communicate verbally and visually with rescue personnel, and thus may be less subject to stress and anxiety. Windows provide a means of escape and accessibility to the building by the fire department for rescue and fire fighting. Automatic sprinklers are considered a primary requirement for life safety in windowless buildings, buildings with unopenable windows, and underground structures.

Influence of Interior Finish, Furnishings, and Decorations

The rapid spread of flame over the surface of walls, ceilings, or floor coverings may prevent occupant use of the means of egress. In general, NFPA *101* limits the flame-spread index classification of interior finish materials on walls and ceilings to a maximum of 200, based on the results of tests conducted in accordance with NFPA 255, *Standard Method of Test of Surface Burning Characteristics of Building Materials*, also known as ASTM E84. Lower ratings are prescribed for the interior finish materials used in exits and in exit accesses. Materials classified as having a lower flame-spread index are also required in certain areas in individual occupancies. A fire-retardant coating may be used on existing interior finish materials to reduce the rate of flame spread. In areas protected with automatic sprinklers, the use of materials with higher flame-spread index classifications sometimes is permitted. Table 8-2A summarizes the interior finish requirements contained in NFPA *101* for the various occupancy classifications.

The flame spread of floor coverings is evaluated by NFPA *101*, through the use of NFPA 253, *Standard Method of Test for Critical Radiant Flux of Floor Covering Systems Using a Radiant Heat Energy Source*, also known as ASTM E648. Two classes of floor coverings are established: (1) Class I finishes, with a minimum critical radiant flux of 0.45 W/cm^2; and (2) Class II finishes, with a minimum critical radiant flux of 0.22 W/cm^2.

Furnishings and decorations—particularly furnishings—play an increasingly important role in loss of life by fire. Decorations can be treated with a flame retardant. Furnishings, on the other hand, are difficult to control and regulate as a fire hazard, since they are not attached to, or part of, the building construction or of the interior finish materials. Furnishings are moved, refurbished, and replaced. However, there are now test procedures for measuring the combustibility of upholstered furniture and its susceptibility to ignition.[17] Two NFPA standards address furniture combustibility: NFPA 260, *Standard Methods of Test and Classification System for Cigarette Ignition Resistance of Components of Upholstered Furniture;* and NFPA 261, *Standard Method of Test Determining Resistance of Mock-Up Upholstered Furniture Material Assemblies to Ignition by Smoldering Cigarettes*. The "Operating Features" chapter of NFPA *101*, 1994 edition, specifies that all new upholstered furniture introduced into health care occupancies must meet the requirements of NFPA 260, with a char length under 1.5 in. (3.8 cm). Some occupancies also have upholstered furniture restrictions. The U.S. Consumer Products Safety Commission (CPSC) also has a standard for evaluating the ignitibility of mattresses.[18] A number of fires have been documented in which severe conditions resulted from fire involvement of only a few furnishing items.[6,7,11]

Influence of Psychological and Physiological Factors on Egress

The psychological and physiological conditions of the occupant population must be considered, in addition to the physical configuration factors of the building, in planning means of egress. Studies indicate people usually behave adaptively and often altruistically in the stress of a fire.[16,19] A heterogeneous collection of persons under the influence of alcohol or drugs, as may be present in a place of public assembly, may pose a greater probability of nonadaptive group behavior, with a competitive flight, panic-type behavior the likely result. Historically, this type of nonadaptive behavior has been documented, although studies indicate that the phenomenon is rare and depends upon unique, predetermined conditions involving both the population and the physical environment of the structure.[16,19,20]

In some cases, evacuation procedures and the creation of areas of refuge within high-rise buildings encourage occupant movement upward within the building. The effectiveness of this concept has not been completely validated in actual fires. Because of the orientation of some people toward total evacuation and escape from the building, it is possible that they may attempt to evacuate a building in the conventional "down and out" approach despite instruction to the contrary.[19]

The evacuation procedures in federal high-rise office buildings, as directed by vocal alarm systems, have continually obtained the selective movement of personnel in both upward and downward directions.[12] In both of two serious high-rise office building fires in São Paulo, Brazil, the occupants moved upward to the roof when their downward movement was inhibited by smoke and heat.[9] In the MGM Grand Hotel fire in Las Vegas, NV, in November 1980, there also was upward movement to areas of refuge in the stairways to the roof and to rooms on upper floors when downward travel was made untenable by smoke and heat.[10,16]

All exits need to be conspicuously marked, since people are likely to be unfamiliar with the various exits from an area under fire conditions and thus to neglect alternate means of egress. It is also

TABLE 8-2A. *Summary of Life Safety Code Requirements for Interior Finish‡*

Occupancy	Exits	Access to Exits	Other Spaces
Assembly—New			
> 300 people	A	A or B	A or B
≤ 300 people	A	A or B	A, B, or C
Assembly—Existing			
> 300 people	A	A or B	A or B
≤ 300 people	A	A or B	A, B, or C
Educational—New	A	A or B	A or B
			C on low partitions
Educational—Existing	A	A or B	A, B, or C
Day-Care Centers—New	A	A	A or B
	I or II	I or II	N.R.
Day-Care Centers—Existing	A or B	A or B	A or B
Group Day-Care Homes—New	A or B	A or B	A, B, or C
Group Day-Care Homes—Existing	A or B	A, B, or C	A, B, or C
Family Day-Care Homes	A or B	A, B, or C	A, B, or C
Health Care—New A.S. Mandatory	A or B	A or B	A or B
		C on lower portion of corridor wall	C in small individual rooms
Health Care—Existing	A or B	A or B†	A or B†
Detention and Correctional—New	A	A	A, B, or C
	I	I	
Detention and Correctional—Existing	A or B	A or B	A, B, or C
	I or II	I or II	
Residential, Hotels and Dormitories—New	A	A or B	A, B, or C
	I or II	I or II	
Residential, Hotels and Dormitories—Existing	A or B	A or B	A, B, or C
	I or II	I or II	
Residential, Apartment Buildings—New	A	A or B	A, B, or C
	I or II	I or II	
Residential, Apartment Buildings—Existing	A or B	A or B	A, B, or C
	I or II	I or II	
Residential, 1- and 2-Family, Lodging or Rooming Houses	A, B, or C	A, B, or C	A, B, or C
Residential, Board and Care—See Chapters 22 and 23			
Mercantile—New	A or B	A or B	A or B
Mercantile—Existing Class A or B	A or B	A or B	Ceiling A or B existing on walls A, B, or C
Mercantile—Existing Class C	A, B, or C	A, B, or C	A, B, or C
Office—New	A or B	A or B	A, B, or C
	I or II	I or II	
Office—Existing	A or B	A or B	A, B, or C
Industrial	A or B	A, B, or C	A, B, or C
Storage	A or B	A, B, or C	A, B, or C

† See occupancy chapters in NFPA *101* for details.
Notes:
Class A Interior Wall and Ceiling Finish—flame spread 0–25, (new) smoke developed 0–450.
Class B Interior Wall and Ceiling Finish—flame spread 26–75, (new) smoke developed 0–450.
Class C Interior Wall and Ceiling Finish—flame spread 76–200, (new) smoke developed 0–450.
Class I Interior Floor Finish—critical radiant flux, minimum 0.45 watts per sq cm.
Class II Interior Floor Finish—critical radiant flux, minimum 0.22 watts per sq cm.
‡Automatic Sprinklers—where a complete standard system of automatic sprinklers is installed, interior wall and ceiling finish with flame spread rating not over Class C may be used in any location where Class B is required and with rating of Class B in any location where Class A is required; similarly, Class II interior floor finish may be used in any location where Class I is required and no critical radiant flux rating is required where Class II is required. This does not apply to new health care facilities.
Exposed portions of structural members complying with the requirements for heavy timber construction are permitted.

important that the means of egress from a building be used as a matter of daily routine, so the occupants will be familiar with their location and operation. NFPA *101* requires that the main exit of assembly occupancies, which also serves as the entrance, be sized to handle at least one-half of the total occupant load of the building.

There are three critical parameters in the effective use of the zoned evacuation of personnel to areas of refuge within a building.[21] They are

1. Proper construction to provide compartmented areas that are protected from the effects of fire and smoke.

2. An effective verbal alarm system giving clear and comprehensive instructions, with provision for originating on-scene instructions from the fire department.[12]
3. Effective evacuation drills to familiarize the occupants with the way the system functions.

It has been advocated that occupants in fire-resistant, compartmented buildings used as hotels, motels, apartments, condominiums, dormitories, hospitals, and other health care facilities should stay in their rooms rather than evacuate, since the rooms are the most adequate area of refuge.[22] This method has not been adopted by NFPA *101* or by the model building codes. However, the concept of areas of refuge has been used by NFPA *101* extensively in occupancies such as health care and detention and correctional facilities for many years and, more recently, to protect occupants with mobility impairments in all occupancies.

Influence of Fire Protection Equipment

It is unsuitable to rely totally on manual or automatic fire extinguishing systems in place of adequate means of egress, since fire extinguishing systems are subject to both human and mechanical failure. In addition, building areas may become untenable for human occupancy before the fire extinguishing systems are effective. Under no condition can manual or automatic fire suppression be accepted as a substitute for the provision and maintainance of proper means of egress.

Where a complete standard system is installed, automatic sprinklers are sufficiently reliable to have a major influence on life safety. In addition to providing an automatic alarm of fire, they quickly discharge water on the fire before smoke has spread dangerously. While automatic sprinklers should never be used in place of adequate means of egress, they are recognized in various ways by NFPA *101*. When totally automatic sprinklers are provided, NFPA *101* permits increased travel distance to exits, the use of interior finish of greater combustibility, and, in health care occupancies, the use of combustible construction in situations where it would otherwise be prohibited. Provisions for areas of refuge are significantly easier to comply with in buildings protected throughout by automatic sprinklers. Sprinklers are particularly valuable in dealing with problems in existing buildings.

Automatic fire detection, or fire alarm, systems are valuable in notifying building occupants of a fire so they may evacuate promptly. Automatic fire detection systems only provide a warning of fire and do nothing themselves to suppress or limit the spread of fire and smoke. An automatic fire detection system is not a substitute for adequate means of egress.

Smoke detection systems can be useful to help mitigate problems in existing buildings. They can be especially useful where earlier egress may help solve problems, such as existing excessive common paths of travel, dead ends, and travel distance.

DEFINITION OF THE TERM "MEANS OF EGRESS"

NFPA *101* and most of the model building codes use the term "means of egress." A means of egress is a continuous path of travel from any point in a building or structure to a public way that is in the open air outside at ground level. Egress consists of three separate and distinct parts: (1) the exit access, (2) the exit, and (3) the exit discharge.

An exit access is defined as that portion of a means of egress that leads to the entrance of an exit.

An exit is that portion of a means of egress that is separated from the area of the building from which escape is to be made by walls, floors, doors, or other means that provide the protected path necessary for the occupants to proceed with reasonable safety to the exterior of the building. An exit may comprise vertical and horizontal means of travel, such as doorways, stairways, ramps, corridors, and passageways.

Exit discharge is that portion of a means of egress between the termination of the exit and a public way.

Figure 8-2F illustrates the relationship among these three parts of an exit in a building.

The Exit

The types of permissible exits are doors leading directly outside or through a protected passageway to the outside at ground level, smokeproof towers, protected interior and outside stairs, exit passageways, enclosed ramps, enclosed escalators and moving walkways in existing buildings. Elevators are not accepted as exits; however, they can be used to provide a way of removing mobility impaired individuals from areas of refuge. See Figures 8-2G and 8-2H for illustrations of some common types of exit arrangements.

The specific placement of exits is a matter of design judgment, given the specifications of travel distance, allowable dead ends, common path of travel, and exit capacity. NFPA *101* states that exits must be remote from each other, thus providing two separate means of egress so located that occupants can travel in either of two opposite directions to reach an exit. This concept is important when it is necessary for occupants to leave a fire or smoke-contaminated area and move toward an exit. If occupants have no choice but to enter the fire area to reach an exit, it is doubtful whether they will be able or willing to do so.

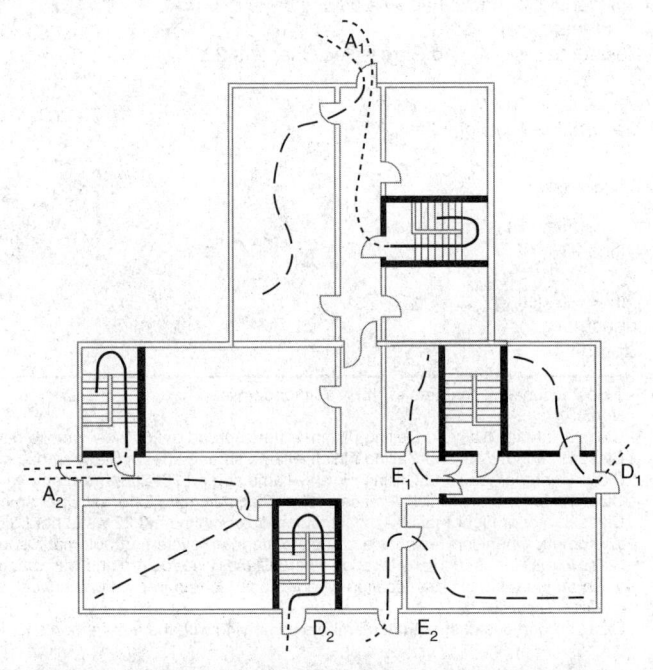

FIG. 8-2F. Examples of exit access, exit, and exit discharge. To the occupant of the building at the discharge level, the doors at A_1, A_2, E_1, and E_2 are exits, and the path denoted by dashes is the exit access. To the person emerging from the exit enclosure, doors A_1 and A_2 and the paths denoted by dotted lines are the exit discharge. Doors D_1 and D_2 are exit discharge doors. Solid lines are within the exit.

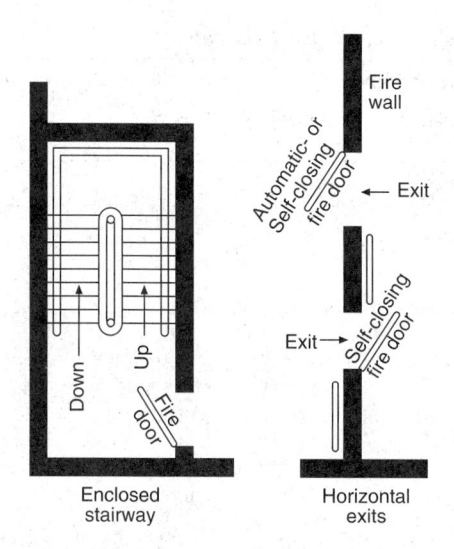

FIG. 8-2G. *Plan views of types of exits. Stair enclosure prevents a fire on any floor from trapping the persons above. A smokeproof tower is better, as it opens to the air at each floor, largely preventing the chance of smoke in the stairway. A horizontal exit provides a quick refuge and lessens the need for a hasty flight down stairs. Fire-rated doors must be arranged to be self-closing or automatic-closing by smoke detection.*

The Exit Access

The exit access is that portion of a means of egress that leads to the entrance of an exit. The exit access may be a corridor, aisle, balcony, gallery, room, porch, or roof. The length of the exit access establishes the travel distance to an exit, an extremely important feature of a means of egress, since an occupant might be exposed to fire or smoke during the time it takes to reach an exit. The average recommended maximum travel distance is 200 ft (61 m), but this distance varies with the occupancy, depending upon the fire hazard and the physical ability and alertness of the occupants. (See Table 8-2B.) The travel distance must be measured from the most remote point in a room or floor area to an exit.

In most cases, the travel distance can be increased up to 50 percent if the building is completely protected with a standard supervised automatic sprinkler system.

A dead end is an extension of a corridor beyond an exit or an access to exits that forms a pocket in which occupants may be trapped. Since there is only one direction of travel to an exit from a dead end, a fire in a dead end between the exit and an occupant prevents the occupant from reaching the exit. Another problem with dead ends is that, while traveling toward an exit in a smoke-filled atmosphere, an occupant may pass by the exit and be trapped in the dead end. In good egress designs, dead-end corridors are not used. However, NFPA *101* permits dead ends in most occupancies, within reasonable limits. Two dead-end corridors are illustrated in Figure 8-2I.

The width of an exit access should be at least sufficient for the number of persons it must accommodate. In some occupancies, the width of the access is governed by the character of activity in the occupancy. One example is a new hospital, where patients may be moved in beds or in gurneys. The corridors in the patient areas of the hospital must be 8 ft (2.4 m) wide to allow for a bed to be wheeled out of a room and turned 90 degrees.

A fundamental principle of exit access is the provision of a free and unobstructed way to the exits. If the access passes through a room that can be locked or through an area containing a fire hazard

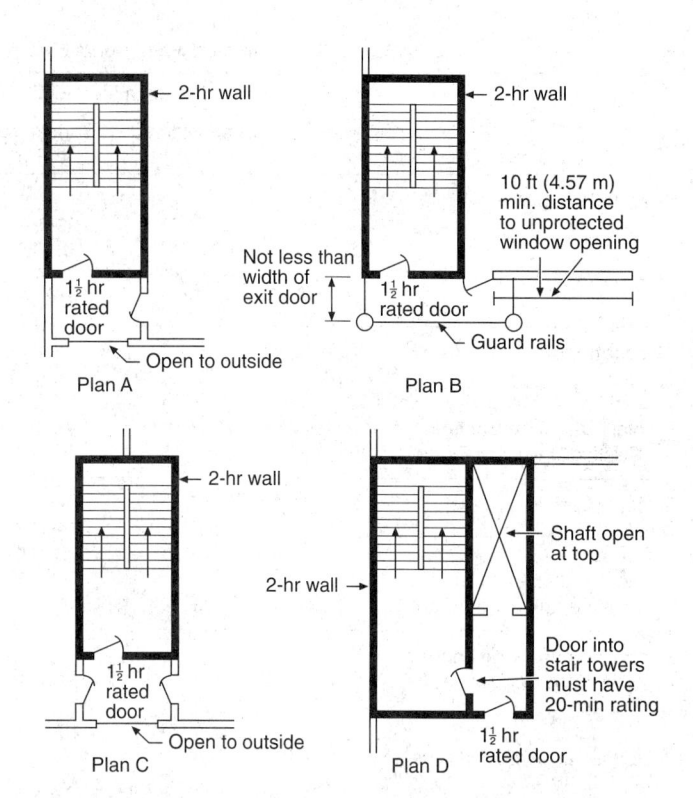

FIG. 8-2H. *Four variations of smokeproof towers. Plan A has a vestibule opening from a corridor. Plan B shows an entrance by way of an outside balcony. Plan C could provide a stair tower entrance common to two areas. In Plan D, smoke and gases entering the vestibule would be exhausted by a natural or induced draft in the open air shaft. In each case, a double entrance to the stair tower with at least one side open or vented is characteristic of this type of construction. Pressurization of the stair tower in the event of fire provides an attractive alternate for tall buildings and is a means of eliminating the entrance vestibule.*

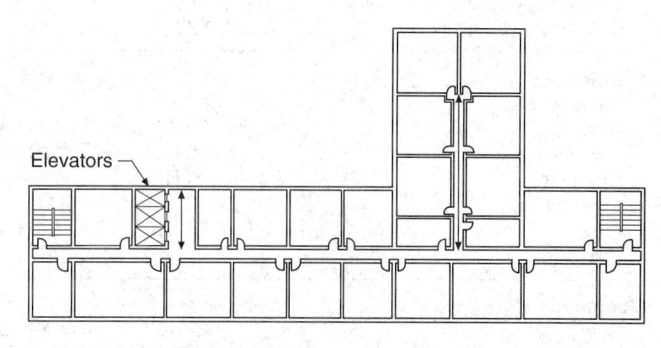

FIG. 8-2I. *Examples of two types of dead-end corridors.*

more severe than is typical of the occupancy, the principles of free and unobstructed exit access are violated.

The floor of an exit access should be level. If this is not possible, small differences in elevation may be overcome by a ramp and large differences by stairs. Where only one or two steps are necessary to overcome differences in level in an exit access, a ramp is preferred, because people may trip in a crowded corridor and fall on the stairs if they do not see the steps or notice that those in front of them have stepped up.

TABLE 8-2B. *Common Path, Dead-End, and Travel Distance Limits (By Occupancy)*

Type of Occupancy	Common Path Limit Unsprinklered ft (m)	Common Path Limit Sprinklered ft (m)	Dead-End Limit Unsprinklered ft (m)	Dead-End Limit Sprinklered ft (m)	Travel Distance Limit Unsprinklered ft (m)	Travel Distance Limit Sprinklered ft (m)
Assembly						
New	20/75[a,b] (6.1/23)	20/75[a,b] 6.1/23)	20[b] (6.1)	20[b] (6.1)	150[c] (45)	200[c] (60)
Existing	20/75[a,b] (6.1/23)	20/75[a,b] (6.1/23)	20[b] (6.1)	20[b] (6.1)	150[c] (45)	200[c] (60)
Educational						
New	75 (23)	75 (23)	20 (6.1)	20 (6.1)	150 (45)	200 (60)
Existing	75 (23)	75 (23)	20 (6.1)	20 (6.1)	150 (45)	200 (60)
New Day-Care Center	N.R.[d,e]	N.R.[d,e]	20 (6.1)	20 (6.1)	150[f] (45)	200[f] (60)
Existing Day-Care Center	N.R.[d,e]	N.R.[d,e]	20 (6.1)	20 (6.1)	150[f] (45)	200[f] (60)
Health Care						
New	N.R.[d]	N.R.[d]	30 (9.1)	30 (9.1)	N.A.[g]	200[f] (60)
Existing	N.R.[d]	N.R.[d]	N.R.[d]	N.R.[d]	150[f] (45)	200[f] (60)
New Ambulatory Care	N.R.[d]	N.R.[d]	30 (9.1)	30 (9.1)	150[f] (45)	200[f] (60)
Existing Ambulatory Care	N.R.[d]	N.R.[d]	50 (15)	50 (15)	150[f] (45)	200[f] (60)
Detention and Correctional						
New - Use Conditions						
II, III, IV	50 (15)	100 (30)	50 (15)	50 (15)	150[f] (45)	200[f] (60)
V	50 (15)	100 (30)	20 (6.1)	20 (6.1)	150[f] (45)	200[f] (60)
Existing - Use Conditions						
II, III, IV, V	50[h] (15)	100[h] (30)	N.R.[d]	N.R.[d]	150[f] (45)	200[f] (60)
Residential						
Hotels and Dormitories						
New	35[i] (10.7)	50[i] (15)	35 (10.7)	50 (15)	175[f,j] (53)	325[f,j] (99)
Existing	35[i] (10.7)	50[i] (15)	50 (15)	50 (15)	175[f,j] (53)	325[f,j] (99)
Apartments						
New	35[i] (10.7)	50[i] (15)	35 (10.7)	50 (15)	175[f,j] (53)	325[f,j] (99)
Existing	35[i] (10.7)	50[i] (15)	50 (15)	50 (15)	175[f,j] (53)	325[f,j] (99)
Board and Care						
Small, New and Existing	N.R.[d]	N.R.[d]	N.R.[d]	N.R.[d]	N.R.[d]	N.R.[d]
Large, New	NA	125k (38.1)	NA	15 (15)	NA	325[f,j] (99)
Large, Existing	110 (48.8)	160 (48.8)	50 (15)	50 (15)	175[f,j] (53)	325[f,j] (99)
Lodging and Rooming Houses	N.R.[d]	N.R.[d]	N.R.[d]	N.R.[d]	N.R.[d]	N.R.[d]
One- and Two-Family Dwellings	N.R.[d]	N.R.[d]	N.R.[d]	N.R.[d]	N.R.[d]	N.R.[d]
Mercantile						
Class A, B, C						
New	75 (23)	100 (30)	20 (6.1)	50 (15)	100 (30)	200 (60)
Existing	75 (23)	100 (30)	50 (15)	50 (15)	150 (45)	200 (60)
Open Air	N.R.[d]	N.R.[d]	0 (0)	0 (0)	N.R.[d]	N.R.[d]
Covered Mall						
New	75 (23)	100 (30)	20 (6.1)	50 (15)	100 (30)	400[l] (120)
Existing	75 (23)	100 (30)	50 (15)	50 (15)	150 (45)	400[l] (120)
Business						
New	75[m] (23)	100[m] (30)	20 (6.1)	50 (15)	200 (60)	300 (91)
Existing	75[m] (23)	100[m] (30)	50 (15)	50 (15)	200 (60)	300 (91)
Industrial						
General	50 (15)	100 (30)	50 (15)	50 (15)	200[n] (60)	250[n] (75)
Special Purpose	50 (15)	100 (30)	50 (15)	50 (15)	300[p] (91)	400[p] (122)
High Hazard	0 (0)	0 (0)	0 (0)	0 (0)	75 (23)	75 (23)
Aircraft Servicing Hangars, Ground Floor	50[p] (15)	50[p] (15)	50[p] (15)	50[p] (15)	Note o	Note o
Aircraft Servicing Hangars, Mezzanine Floor	50[p] (15)	50[p] (15)	50[p] (15)	50[p] (15)	75 (23)	75 (23)

TABLE 8-2B. *(Continued)*

Type of Occupancy	Common Path Limit		Dead-End Limit		Travel Distance Limit	
	Unsprinklered ft (m)	Sprinklered ft (m)	Unsprinklered ft (m)	Sprinklered ft (m)	Unsprinklered ft (m)	Sprinklered ft (m)
Storage						
Low Hazard	N.R.[d]	N.R.[d]	N.R.[d]	N.R.[d]	N.R.[d]	N.R.[d]
Ordinary Hazard	50 (15)	100 (30)	50 (15)	100 (30)	200 (60)	400 (122)
High Hazard	0 (0)	0 (0)	0 (0)	0 (0)	75 (23)	75 (23)
Parking Garages, Open	50 (15)	50 (15)	50 (15)	50 (15)	200 (60)	300 (91)
Parking Garages, Enclosed	50 (15)	50 (15)	50 (15)	50 (15)	150 (45)	200 (60)
Aircraft Storage Hangars, Ground Floor	50[p] (15)	100[p] (30)	50[p] (15)	50[p] (15)	Note o	Note o
Aircraft Servicing Hangars, Mezzanine Floor	50[p] (15)	75[p] (23)	50[p] (15)	50[p] (15)	75 (23)	75 (23)
Underground Spaces in Grain Elevators	50[p](15)	50[p] (15)	N.R.[d,p]	N.R.[d,p]	200 (60)	400 (122)

[a]20 ft (6.1 m) for common path serving > 50 persons; 75 ft (23 m) for common path serving <50 persons.
[b]See Chapters 8 and 9 for special considerations for aisle accessways, aisles, and mezzanines.
[c]See Chapters 8 and 9 for special considerations for smoke-protected assembly seating in arenas and stadia.
[d]No requirement.
[e]See Sections 10-7 and 11-7 for requirement for second exit access based on room capacity or area.
[f]This dimension is for the total travel distance, assuming incremental portions have fully utilized their allowable maximums. For travel distance within the room, and from the room exit access door to the exit, see the appropriate occupancy chapter.
[g]Not applicable.
[h]See Chapter 15 for special considerations for existing common paths.
[i]This dimension is from the room/corridor or suite/corridor exit access door to the exit; thus it applies to corridor common path.
[j]See appropriate occupancy chapter for special travel distance considerations for exterior ways of exit access.
[k]See Section 22-3 for requirement for second exit access based on room area.
[l]See Sections 24-4 and 25-4 for special travel distance considerations in covered malls considered pedestrian ways.
[m]See Chapters 26 and 27 for special common path considerations for single tenant spaces.
[n]See Chapter 28 for industrial occupancy special travel distance considerations.
[o]See Chapters 28 and 29 for special requirements on spacing of doors in aircraft hangars.
[p]See Chapters 28 and 29 for special requirements if high hazard.

The Exit Discharge

Ideally, all exits in a building should discharge directly to the outside or through a fire resistance-rated passageway to the outside of the building. NFPA *101* permits a maximum of 50 percent of the exit stairs to discharge onto the street floor. The obvious disadvantage of this arrangement is that, if a fire occurs on the street-level floor, it is possible for people using the exit stairs discharging to that floor to be discharged into the fire area. If any exits discharge to the street floor, NFPA *101* therefore requires that such exits discharge to a free and unobstructed way to the outside of the building, that the street floor be protected by automatic sprinklers, and that the street floor be separated from any floors below by construction having a 2-hr fire resistance rating.

Discharging an exit to the outside is not necessarily discharging to a safe place. If the exit discharges into a courtyard, an exit passageway must be provided from the courtyard through the building so that the occupants can get away from the building. If the exit discharges into a fenced yard, the occupants must be able to get out of the yard to get away from the building. If the exit discharges into an alley, the alley must be of sufficient width to accommodate the capacity of all the exits discharging into it, and any openings in the building walls bordering the alley should be protected to prevent fire exposure to the occupants proceeding through the alley.

When exit stairs from floors above the street floor continue on to floors below the street floor, occupants evacuating the building may miss the exit discharge door to the street level, continue down the stairway, and enter a floor below the level of exit discharge. Therefore, NFPA *101* requires a physical barrier or other effective means at the street floor landing to prevent evacuees from passing the level of exit discharge.

CAPACITY OF EXITS

The capacity of exits is calculated using a capacity factor provided in NFPA *101*. This capacity factor is given as in./person and varies with the occupancy. (See Table 8-2C.) The total exit capacity for each component of the means of egress, such as doors, stairs, ramps, corridors, and so on, is calculated based on its clear width. For example, one 34-in. clear-width door in an office occupancy would have an exit capacity of 170 persons (34 in. ÷ 0.2 in./person = 170 persons). The reason for these variations in exit capacity factors is to establish a consistent total evacuation time in different occupancies, based on the physical ability, mental alertness, age, and sociological roles of the occupants. In occupancies where people are housed for care, the time taken to reach exits will be greater than in some other occupancies, so the exits must be sufficiently wide to allow nonambulatory occupants to egress and to prevent any waiting to get into the exit.

The capacity of exits was traditionally used to establish a consistency of evacuation time on the basis of the rate of travel through a door of 60 persons/min and down a stairway of 45 persons/min/22 in. (558.8 mm) of exit width, respectively. These figures were established by evacuation counts conducted primarily in federal office buildings.[23] More recent studies of evacuations in high-rise office buildings indicate peak flows of 30 persons/min and mean flows of 24 persons/min/22 in. (558.8 mm) of exit width down stairways.[2,13,14]

TABLE 8-2C. *Summary of NFPA 101®, Life Safety Code®, Provisions for Occupant Load and Exit Capacity*

Occupancy	Occupant Load, sq ft (m²) per person	Level Components (Doors, Corridors, Horizontal Exits, Ramps)		Stairs	
Assembly					
Less concentrated use without fixed seating	15 net (1.4)	0.2		0.3	
Concentrated use without fixed seating	7 Net (.65)	0.2		0.3	
Fixed seating	Actual Number of Seats	0.2		0.3	
Educational					
Classrooms	20 Net (1.9)	0.2		0.3	
Shops and vocational	50 Net (4.6)	0.2		0.3	
Care centers	35 Net (3.3)	0.2		0.3	
Health care		NAS	AS	NAS	AS
Sleeping departments	120 Gross (11.1)	0.5	0.2	0.6	0.3
Treatment departments	240 Gross (22.3)	0.5	0.2	0.6	0.3
Residential	200 Gross (18.6)	0.2		0.3	
Board and care	200 Gross (18.6)	0.2		0.4	
Mercantile					
Street floor and sales basement	30 Gross (3.7)	0.2		0.3	
Multiple street floors—each	40 Gross (3.7)	0.2		0.3	
Other floors	60 Gross (5.6)	0.2		0.3	
Storage—shipping	300 Gross (27.9)	0.2		0.3	
Malls	See Code	0.2		0.3	
Business	100 Gross (9.3)	0.2		0.3	
Industrial	100 Gross (9.3)	0.2		0.3	
Detention and correctional	120 Gross (11.1)	0.2		0.3	

Notes: NAS = nonsprinklered; AS = sprinklered.

Occupant Load

Occupant load, or the number of people to be expected in a building or an area within a building at any time for whom exits must be provided, is determined by the actual anticipated occupant load but not less than that number obtained by dividing the gross area of the building or the net area of a specific portion of the building by the area in sq ft (m²) projected for each person. The amount of floor area projected for each person varies with the occupancy. (See Table 8-2C.) These figures are based on actual counts of people in buildings and on reviews of architectural plans. In some situations, the maximum number of people in a building above the calculated occupant load can be determined at the design stage, in which case this number should be used in the design of the exits. A typical example is an assembly occupancy in which fixed seating is installed. Counting the number of seats provided, and calculating the standing or waiting areas by the occupant load factor, would obviously give a more accurate figure than multiplying a sq ft (m²)/person figure by the net floor area.

Computing Required Egress Width

To compute the minimum required egress widths from the individual floors of a building, it is necessary to

1. Calculate the floor area, either net or gross, whichever is applicable;
2. Determine from NFPA *101* the estimated number of sq ft (m²)/ person, or occupant load factor;
3. Divide the number of sq ft (m²)/person (occupant load factor) into the floor area to determine the minimum number of people for whom exits must be provided for that floor;
4. Measure the clear width of each component in the means of egress;
5. Determine the capacity factor from NFPA *101* for each exit component for the appropriate occupancy;
6. Divide the clear width of each exit component by the capacity factor to determine the exit capacity for each component;
7. Determine the most restrictive component in each egress system;

8. Determine the total capacity for the floor; and
9. Ensure that the total egress capacity equals or exceeds the total occupant load.

In multi-story buildings, the exit capacity for each floor is calculated separately. In other words, the capacity of the stairs need only be wide enough to serve each floor, but it must not be less than the minimum width required by NFPA *101*. It must be noted, also, that the required egress capacity cannot be decreased in the direction of egress travel.

Street-floor exits may require special treatment, depending upon the occupancy. Some occupancies require that street-floor exits be sized to handle not only the occupant load of the street floor but also the occupant load of the exits discharging to the street floor from floors above and below. In addition, in those occupancies where floors above and/or below the street floor are permitted to have unenclosed stairs and escalators connecting them with the street floor, the exits must be sufficient to provide simultaneously for all the occupants of all communicating levels and areas. In other words, all communicating levels in the same fire area are considered a single floor area for the purposes of determining the required exit capacity. This identical, single fire area factor can have a considerable effect on the sizing of the street-floor exits.

Should two or more exits converge into a common exit, the common exit should never be narrower than the sum of the width of the exits converging into it.

Generally, the minimum number of exits is two. In certain limited situations, however, one exit may be permitted in some occupancies if there is a very low occupant load, low fire hazard, and a limited travel distance.

EXIT FACILITIES AND ARRANGEMENTS

The following exit facilities are covered in NFPA *101*.

Doors

Doors should swing in the direction of exit travel, except in small rooms. Horizontal sliding, vertical, or rolling doors are recognized for use as means of egress in some occupancies. In places of assembly and in schools, panic hardware should be installed on all egress doors equipped with latches that serve rooms with an occupant load of 100 or more.

Where doors protect exit facilities, as in stairway enclosures and horizontal exits, they normally must be kept closed to the spread of smoke. If open, they must be closed immediately in case of fire. Although ordinary, fusible-link-operated devices to close doors in case of fire are designed to close in time to stop the spread of fire, they do not operate soon enough to stop the spread of smoke and are not permitted by NFPA *101*. At relatively low temperatures, the smoke accumulation could continue and could reach untenable levels long before the fusible link melts, allowing the door to close.

Sometimes, people keep self-closing doors open with hooks or with wedges under the door. Doors also can be blocked open to provide ventilation, for the convenience of building maintenance personnel, or to avoid the accident hazard of swinging doors. The following measures have been provided in the NFPA *101* to alleviate this undesirable situation:

1. Doors that are normally kept open can be equipped with door closers and automatic hold-open devices that release the door and allow them to close when an automatic sprinkler system, the fire alarm system, an automatic fire detection system, and

smoke or other products of combustion detection devices operate.
2. Doors that are normally closed can be equipped to open electrically or pneumatically when a person approaches the door.
3. Doors that normally are closed can be opened and held open manually by monitors, as in schools.
4. Use of smokeproof towers that protect against smoke, even if the doors are open.

Qualifications and limitations are applicable to each of these measures. One is that, in the event of electrical failure, the door must close and remain closed unless it is opened manually for egress purposes.

Another major maintenance difficulty with exit doors is the exterior door that is locked to prevent unauthorized access or for other reasons. NFPA *101* specifies that, when the building is occupied, all doors must be kept unlocked from the side from which egress is made.

NFPA *101* allows a delayed releasing device on some egress doors, provided this is permitted by the requirements of the occupancy in question. Where the devices are allowed, the following provisions apply:

1. The building must be protected throughout by an approved and supervised automatic fire detection system or automatic sprinkler system.
2. The release devices are installed only in low- or ordinary-hazard areas.
3. The devices must unlock when the fire detection system, alarm, or automatic sprinkler system operates.
4. The devices must unlock upon loss of power.
5. The devices must initiate an irreversible process that will free the latch within 15 sec whenever a force of not more than 15 lb (6.8 kg) is applied to the releasing device, and the door must not relock automatically. Operation of the releasing device must actuate a signal near the door.
6. A sign must be placed adjacent to the door that reads: PUSH UNTIL ALARM SOUNDS. DOOR CAN BE OPENED IN 15 SECONDS!
7. Emergency lighting must be provided at the door.

NFPA *101* also provides "Access Controlled Egress Doors." The code spells out several limitations for these. One of the limitations included is that when an occupant approaches the door, a sensor must unlock it.

Locks on a door that let people exit but not enter are satisfactory, but even this type of lock may not be satisfactory for security purposes. Possible measures to prevent unauthorized use of exit doors include:

1. An automatic alarm that rings when the door is opened.
2. Visual supervision such as wired-glass panels, closed-circuit television, and mirrors, which may be used where appropriate.
3. Automatic photographic devices to provide pictures of users.

So-called exit locks, with a break-glass unit actuated by striking a handle with the hand, are not permitted by NFPA *101* unless installed in conjunction with panic bars. Otherwise, they do not comply with the NFPA *101* provision that reads: "A latch or other fastening device on a door shall be provided with a lever, knob, handle, panic bar, or other simple type of releasing device, having an obvious method of operation and/or all lighting conditions."

Other types of break-glass locks and electrical controls for releasing exits from a central point are not permitted by NFPA *101*.

The exception is an occupancy where controls may be necessary, as in health care, and detention and correctional occupancies.

A single door in a doorway should not be less than 32 in. (813 mm) wide in new buildings and 28 in. (711 mm) in existing buildings. Doors in the required means of egress are limited to a maximum width of 48 in. (1.2 m). To prevent tripping, the floor on both sides of the door should have the same elevation for the full swing of the door.

Panic Hardware

Exit doors in assembly and educational occupancies, such as schools or movie theaters, normally are equipped with panic hardware. Basically, panic hardware devices are designed to facilitate the release of the latching device on the door when a pressure not to exceed 15 lb (6.8 kg) is applied in the direction of exit travel. Such releasing devices are bars or panels extending not less than one-half of the width of the door and placed at a height not less than 30 in. (762 mm) or more than 44 in. (1.1 m) above the floor.

Panic hardware that has been tested and listed for use on fire-protection-rated doors is termed "fire exit hardware." If panic hardware is needed on fire-protection-rated doors, only fire exit hardware is to be used.

Panic hardware is available for use on single and double doors, with variations for rim-mounted hardware and mortise or vertical rod devices.

Horizontal Exits

A horizontal exit is a means of egress from one building to an area of refuge in another building on approximately the same level, or a means of egress through a fire barrier or partition to an area of refuge at approximately the same level in the same building that affords safety from fire and smoke. With a horizontal exit, it is obvious that space must be provided in the area or building of refuge for the people entering the refuge area. NFPA *101* recommends 3 sq ft (0.28 m²) of space per person, with the exception of health care and detention and correctional occupancies, where 6 to 30 sq ft (0.56 to 2.79 m²) of space is recommended. Horizontal exits cannot comprise more than one-half the total required exit capacity, except in health care facilities, where horizontal exits may comprise two-thirds of the total required exit capacity, and in detention and correctional facilities, where horizontal exits can comprise 100 percent of the total exit capacity. Horizontal exits have been applied universally in health care facilities where the evacuation of patients over stairs is slower and more difficult than taking them through a horizontal exit to a safe area of refuge. A horizontal exit arrangement within a single building and between two buildings is illustrated in Figure 8-2J.

A swinging door in a fire wall provides a horizontal exit in one direction only. Two openings, each with a door swinging in the direction of exit travel or sliding doors, are needed to provide horizontal exits from both sides of the wall. Where property protection requires fire doors on both sides of the wall, a normally open, automatic, fusible-link-operated, horizontally sliding fire door may be used on one side, with a swinging fire door on the other.

Stairs

Exit stairs are arranged to minimize the danger of falling, because one person falling on a stairway may result in the complete blockage of an exit. Stairs must be wide enough for two persons to descend side by side, thus maintaining a reasonable rate of evacuation, even though aged or infirm persons may slow the travel on one side. There must be no decrease in the width of the stair along the path of travel, since this may create congestion.

Steep stairs are dangerous. Stair treads must be wide enough to give good footing. NFPA *101* specifies a minimum 11 in. (279 mm) tread and a maximum 7 in. (178 mm) riser for new stairs. Landings should be provided to break up any excessively long individual flight. Continuous railings are now recommended for new stairs. New stairs more than 60 in. (1.5 m) wide should have one or more center rails.

Two classes of stairs are permitted in NFPA *101* for existing buildings, with a single class of stairs for new stairs. There are Class A and Class B stairs for existing buildings. The requirements for each class are given in Table 8-2D.

Stairs can serve as exit access, exit, or exit discharge. When used as an exit, they must be in an enclosure that meets exit enclosure requirements or outside the building and properly protected. Exit access stairs that connect two or more stairs are vertical openings and must be protected as a vertical opening.

Stairways may be inside the building where the NFPA *101* generally specifies protective enclosures. They also may be outside if they comply with the requirements for exterior stairs and are arranged so that persons who fear heights will not be reluctant to use them, are not exposed to fire conditions originating in the building, and, where necessary, are shielded from snow and ice. Exterior stairs should not be confused with fire escape stairs. (See Figure 8-2K.)

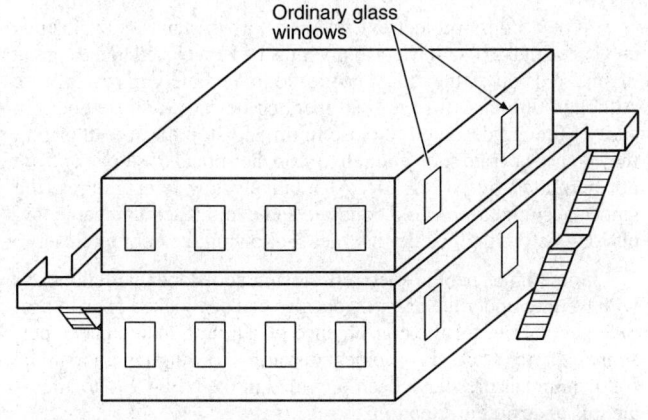

FIG. 8-2K. *An example of using outside stairs to provide direct exits to the outside for all rooms in a multi-story building. There are no interior corridors through which smoke and flame could spread. This method has application in many types of occupancies, such as schools, motels, small professional buildings, and so on. Note that there are two means of egress, remote from each other, from the second-story balcony.*

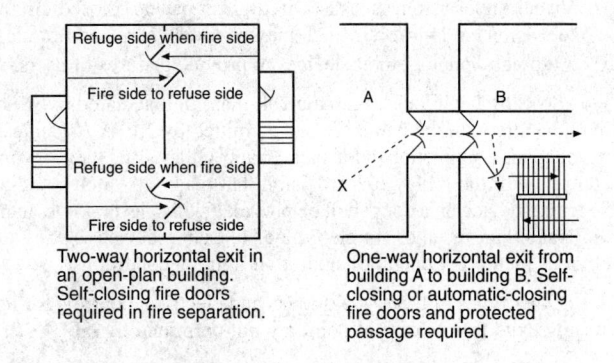

FIG. 8-2J. *Types of horizontal exits.*

TABLE 8-2D. *Requirements for New and Existing Building Stairs*

	New Stairs	Existing Stairs	
		Class A	Class B
Minimum width clear of all obstructions except projections not exceeding 3½ in. (0.89 mm) at and below handrail height on each side	44 in. (1.12 m) 36 in. (0.91 m) where total occupant load of all floors served by stairways is less than 50.	44 in. (1.12 m) 36 in. (0.91 m) where total occupant load of all floors served by stairways is less than 50.	44 in. (1.12 m) 36 in. (0.91 m)
Maximum height of risers	7 in. (178 mm)	7½ in. (191 mm)	8 in. (203 mm)
Minimum height of risers	4 in. (102 mm)		
Minimum tread depth	11 in. (279 mm)	10 in. (244 mm)	9 in. (229 mm)
Minimum headroom	6 ft 8 in. (2.03 m)	6 ft 8 in. (2.03 m)	6 ft 8 in. (2.03 m)
Maximum height between landings	12 ft (3.7 m)	12 ft (3.7 m)	12 ft (3.7 m)
Minimum dimension of landings in direction of travel	Stairways and intermediate landings shall continue with no decrease in width along the direction of exit travel. In new buildings every landing shall have a dimension, measured in direction of travel, equal to the width of the stair. Such dimension need not exceed 4 ft (1.22 m) when the stair has a straight run.		
Doors opening immediately on stairs, without landing at least width of door	No	No	No

Construction details of stair enclosures involve the principles of limiting fire and smoke spread. Doors on openings from each story are essential to prevent the stairway from serving as a flue. In general, stairway enclosures should include not only the stairs, but also the path of travel from the bottom of the stairs to the exit discharge, so that occupants have a protected, enclosed passageway all the way out of the building. The stair enclosure should be of 1 hr construction when connecting three or fewer floors and of 2 hr construction when connecting four or more floors.

Smokeproof Towers

Smokeproof towers provide the highest protected type of stair enclosure recommended by NFPA *101*. Access to the stair tower is only by balconies open to the outside air, vented vestibules, or mechanically pressurized vestibules, so that smoke, heat, and flame will not spread readily into the tower even if the doors are accidentally left open. (See Figure 8-2H.)

Ramps

Ramps, enclosed and otherwise arranged like stairways, are sometimes used instead of stairways where there are large crowds and to provide both access and egress for nonambulatory persons. To be considered safe, exits ramps must have a very gradual slope.

Exit Passageways

A hallway, corridor, passage, tunnel, or underfloor or overhead passageway may be designated an exit passageway, providing it is separated and arranged according to the requirements for exits.

The use of a hallway or corridor as an exit passageway introduces some unique considerations. The use of these spaces for purposes other than exiting may violate fundamental design considerations. In an industrial situation, for example, the use of a gasoline-powered forklift in a corridor designated as an exit passageway would violate the principles of exit design. NFPA *101* specifies that an exit enclosure should not be used for any purpose that could interfere with its value as an exit and is strictly limited by the code. Furthermore, penetration of the enclosure by ducts and other utilities may violate the protective enclosure.

Each opening in an exit enclosure introduces a point of weakness that could allow fire contaminants to spread into the exit and prevent its use. The typical corridor used as an exit with numerous door openings could result in fire contamination of the enclosure if a door fails to close and latch. The door openings in exit enclosures should be limited to those necessary for access to the enclosure from normally occupied spaces. Therefore, doors and other openings to spaces such as boiler rooms, storage spaces, trash rooms, and maintenance closets are not allowed into an exit passageway.

An exit passageway should not be confused with an exit access corridor. Exit access corridors do not have the construction protection requirements of exit passageways, since they provide access to an exit rather than being an extension and component of the exit.

Fire Escape Stairs

Fire escapes should be stairs, not ladders. Fire escapes are, at best, a poor substitute for standard interior or exterior stairs. NFPA *101* only permits existing fire escapes in existing buildings.

The same principles of design apply to fire escapes that apply to interior stairs, though requirements for width, pitch, and other dimensions are generally less strict. NFPA *101* gives the following criteria for fire escape stair design.

Fire escape stairs ideally extend to the street or to ground level. When sidewalks would be obstructed by permanent stairs, swinging stair sections designed to swing down under the weight of a person may be used for the lowest flight of the fire escape stairs. The area below the swinging section must be kept unobstructed so the swinging section can reach the ground. A counterweight of the type that balances on a pivot should be provided for swinging stairs; cables should not be used. Fire escapes that end on balconies above the ground level and provide no way to reach the ground, except by portable ladders or jumping, are unsafe.

Many persons who fear heights are reluctant to use fire escapes. As far as possible, design should provide a sense of security, as well as suitable railings and other details actually needed for safety. Fire escapes must be well-anchored to building walls and kept painted to prevent rust.

Preferred access to fire escapes is through doors leading from the main building area or from corridors, never through rooms that may have locked doors except where every room or apartment has separate access to a fire escape. Although preferred access to fire escapes is by doors, windows may be used, in which case sills should not be too high above the floor. Windows should be of ample size, and, if insect screens are installed, they should be of a type that can be opened or removed quickly and easily. Decorative grilles or security bars should not be installed over windows that provide access to fire escapes.

Fire escapes can create a severe fire exposure to people if flames came out windows beneath them. (See Figure 8-2L.) The best location for fire escapes is on exterior masonry walls without exposing windows, with access to fire escape balconies by exterior fire doors. Where window openings expose fire escapes, fixed wired-glass in metal sashes should be used. Where there is a complete standard automatic sprinkler system in the building, the fire exposure hazard to personnel on fire escapes is minimized.

In northern climates, outside fire escapes may be obstructed by snow and ice.

Escalators, Moving Walkways, and Elevators

Escalators may be recognized as exits in existing buildings, if they have enclosures similar to exit stairs and meet the requirements for stairs as to tread width and riser height. However, they are seldom installed in such a way that they would qualify as exits, and it is common to find escalator installations with unprotected floor openings. Escalators are not recognized as an acceptable component in a means of egress in new construction.

Moving walkways also may be used as means of egress if they conform to the general requirements for ramps, if inclined, and for passageways, if level.

Elevators are not recognized as exits. However, elevators are permitted to be used, under limited conditions, to serve areas of refuge for the mobility impaired.

Ropes and Ladders

Ropes and ladders generally are not recognized in codes as a substitute for standard exits from a building. This is proper since there is no excuse for permitting their use except possibly in existing one- and two-family dwellings where it is economically impractical to add a secondary means of escape. In this case, a suitable rope or chain ladder or a folding metal ladder may be suitable. However, the homeowner should recognize that aged, infirm, very young, and physically handicapped persons cannot use ladders and that, if the

FIG. 8-2L. *Fire escapes are makeshift, often dangerous. Fire may make fire escapes useless as this picture, drawn from a photograph of an actual fire, shows.*

ladder passes near or over a window in a lower floor, flames from the window can prevent the use of the ladder.

Windows

Windows are not exits. They may be used as access to fire escapes in existing buildings if they meet certain criteria concerning the size of window opening and the height of the sill from the floor. Windows may be considered a means of escape from certain residential occupancies.

Windows are required in school rooms subject to student occupancy, unless the building is equipped with a standard automatic sprinkler system, and in bedrooms in one- and two-family dwellings that do not have two separate means of escape. These windows are for rescue and ventilation and must meet the criteria for size of opening, method of operation, and height from the floor.

EXIT LIGHTING AND SIGNS

Exit Lighting

In buildings where artificial lighting is provided for normal use, exit lighting and the illumination of the means of egress are required to ensure that occupants can evacuate the building quickly. The intensity of the illumination of the means of egress should be not less than 1 footcandle (10.77 lu/m^2) measured at the floor. It is desirable that such floor illumination be provided by lights recessed in the wall and located approximately 1 ft (30.5 cm) above the floor because such lights are then unlikely to be obscured by the smoke that might occur during a fire. In auditoriums and other places of public assembly where movies or other projections are shown, NFPA *101* permits a reduction in this illumination for the period of the projection to values of not less than $^1/_5$ footcandle (2.2 lu/m^2).

Emergency Lighting

NFPA *101* requires emergency power for illuminating the means of egress in some occupancies. For example, emergency lighting is required in places of assembly; in certain types of educational buildings; in health care facilities; in detention and correctional facilities; most hotels and apartment buildings; in Class A and B mercantiles; in business buildings based on occupant load and number of stories; in most industrial and stroage buildings; and in underground or windowless structures subject to occupancy by more than 100 persons.

Well-designed emergency lighting using a source of power independent from the normal building service automatically provides the necessary illumination in the event of an interruption of power to normal lighting. The failure of the public utility or other outside electric power supply, the opening of a circuit breaker or fuse, or any manual act, including accidental opening of a switch controlling normal lighting facilities, should result in the automatic operation of the emergency lighting system.

Reliability of the exit illumination is most important. NFPA 70, *National Electrical Code*®, details requirements for the installation of emergency lighting equipment. Battery-operated electric lights and portable lights normally are not used for primary exit illumination, but they may be used as an emergency source under the restrictions imposed by NFPA *101*. Luminescent, fluorescent, or other reflective materials are not a substitute for required illumination, since they are not normally sufficiently intense to justify recognition as exit floor illumination.

Where electric battery-operated emergency lights are used, suitable facilities are needed to keep the batteries properly charged. Automobile-type lead storage batteries are not suitable because of

Fahy, R. F., "EXIT89: An Evacuation Model for High-Rise Buildings," Sixth International Fire Conference, Interflam '93, Fire Safety, Mar. 30–Apr. 1, 1993, Oxford, England, Interscience Communications Ltd., London, England, 1993, pp. 519–528.

Fahy, R. F., "EXIT89: An Evacuation Model for High-Rise Buildings. Model Description and Example Applications," *Proceedings* of the Fourth International Symposium on Fire Safety Science, Intl. Assoc. for Fire Safety Science, Boston, MA, 1994, pp. 657–668.

Fahy, R. F., "EXIT89: An Evacuation Model for High-Rise Buildings—Recent Enhancements and Example Applications," *Proceedings* of the International Conference on Fire Research and Engineering, Sept. 10–15, 1995, Orlando, FL, SFPE, Boston, MA, 1995, pp. 332–337.

Francis, R. L., "Network Models of Building Evacuation: Development of Software Systems," NBS-GCR-85-489, March 1985, National Bureau of Standards, Gaithersburg, MD.

Fruin, J. J., *Pedestrian Planning and Design*, revised edition, Elevator World, Mobile, AL, 1987.

Galea, E. R., Owen, M., and Lawrence, P. J., "Emergency Egress From Large Buildings Under Fire Conditions Simulated Using the EXODUS Evacuation Model," ISBN 0-9516320-9-4; *Proceedings* of the Seventh International Interflam Conference, Interflam '96, March 26–28, 1996, Cambridge, England, Interscience Communications Ltd., London, England, 1996, pp. 711–720.

Hagiwara, I., and Tanaka, T., "International Comparison of Fire Safety Provisions for Means of Escape," Twelfth Joint Panel Meeting of the U. S./Japan Government Cooperative Program on Natural Resources (UJNR) on Fire Research and Safety, Oct. 27–Nov. 2, 1992, Tsukuba, Japan, Building Research Inst., Ibaraki, Japan, Fire Research Institute, Tokyo, Japan, 1992, pp. 224–231; *Proceedings* of the Fourth International Symposium on Fire Safety Science, International Association for Fire Safety Science, 1994, pp. 633–644.

Jin, T., "Evaluation of Fire Exit Sign in Fire Smoke," '93 Asian Fire Seminar, Oct. 7–9, 1993, Tokyo, Japan, 1993, pp. 167–175.

Jin, T., and Yamada, T., "Experimental Study on Effect of Escape Guidance in Fire Smoke by Travelling Flashing of Light Sources," *Proceedings* of the Fourth International Symposium on Fire Safety Science, Intl. Assoc. for Fire Safety Science, Boston, MA, 1994, pp. 705–714.

Jin, T., *et al.*, "Evaluation of the Conspicuousness of Emergency Exit Signs," *Proceedings* of the Third International Symposium on Fire Safety Science, Elsevier Applied Science, New York, 1991, pp. 835–841.

Johnson, P. F., Beck, V. R., and Horasan, M., "Use of Egress Modelling in Performance-Based Fire Engineering Design: A Fire Safety Study at the National Gallery of Victoria," *Proceedings* of the Fourth International Symposium on Fire Safety Science, Intl. Assoc. for Fire Safety Science, Boston, MA, 1994, pp. 669–680.

Jones, B. K., and Hewitt, J. A., "Leadership and Group Formation in High-Rise Building Evacuations," *Proceedings* of the First International Symposium on Fire Safety Science, Hemisphere, 1986, pp. 513–522.

Kendik, E., "Assessment of Escape Routes in Buildings and Design Method for Calculating Pedestrian Movement," TR 85-4, 1985, Society of Fire Protection Engineers, Boston, MA.

Kendik, E., "Methods of Design of Means of Egress: Towards a Quantitative Comparison of National Code Requirements," *Proceedings* of the First International Symposium of Fire Safety Science, Hemisphere, Washington, D.C., 1986.

Kisko, T. M., and Francis, R. L., "EVACNET+, A Computer Program to Determine Optimal Building Evacuation Plans," *Fire Safety Journal*, Vol. 9, No. 2, 1985, pp. 211–220.

Kisko, T. M., and Francis, R. L., "Network Models of Building Evacuation: Development of Software Systems," NBS-GCR-85-489, 1985, National Bureau of Standards, Gaithersburg, MD.

Klevan, J. B., "Modeling of Available Egress Time from Assembly Spaces or Estimating the Advance of the Fire Threat," SFPE TR 82-2, 1982, Society of Fire Protection Engineers, Boston, MA.

Klote, J. H., "Elevators as a Means of Fire Escape," NBSIR 82-2507, May 1982, National Bureau of Standards, Gaithersburg, MD.

Koffel, W. E., "Evaluating Occupant Load for Egress," *NFPA Fire Journal*, Vol. 88, No. 1, Jan./Feb. 1994, pp. 14, 93.

Kostreva, M., "Optimization Models for Fire Egress Analysis," BFRL Video Seminar, Clemson Univ., SC, Video, Aug. 20, 1991.

Kostreva, M. M., "Mathematical Modeling of Human Egress from Fires in Residential Buildings," *Fire Technology*, Vol. 30, No. 3, Third Quarter 1994, NIST-GCR-94-643, June 1994, pp. 338-340.

Ling, W. C. T., and Williamson, R. B., "Use of Probabilistic Networks for Analysis of Smoke Spread and Egress of People in Buildings," *Proceedings* of the First International Symposium for Fire Safety Science, Hemisphere, New York, 1986, pp. 953–962.

MacLennan, H. A., "Towards an Integrated Egress/Evacuation Model Using an Open Systems Approach," *Proceedings* of the First International Symposium on Fire Safety Science, Hemisphere, 1986, pp. 581–590.

Ouellette, M. J., "Exit Signs in Smoke: Design Parameters for Greater Visibility," *Lighting Research and Technology*, Vol. 20, No. 4, 1988, pp. 155–160.

Ouellette, M. J., "Visibility of Exit Signs," Institute for Research in Construction, Canada, Construction Practice, 1994.

Ozel, F., *Way Finding and Route Selection in Fires*, School of Architecture, New Jersey Institute of Technology, Newark, NJ, 1986.

Pauls J., "Development of Knowledge About Means of Egress," *Fire Technology*, Vol. 20, No. 2, 1984, pp. 28–40.

Pauls, J., "Calculating Evacuation Times for Tall Buildings," *Fire Safety Journal*, Vol. 12, No. 3, 1987, pp. 213–236.

Pauls, J., "Egress Time and Safety Performance Related to Requirements in Codes and Standards," Hughes Associates, Inc., Columbia, MD, Building Officials and Code Administration International, Inc. (BOCA) and OBOA, Workshop Landout, June 1990, Ontario, Canada, 1990, pp. 1–10.

Pauls, J., Gatfield, A. J., and Juillet, E., "Elevator Use for Egress: The Human-Factors Problems and Prospects," National Research Council of Canada, Ottawa, Ontario, National Task Force on Life Safety and the Handicapped American Society of Mechanical Engineers, Council of American Building Officials and National Fire Protection Association, Elevators and Fire, Feb. 19–20, 1991, Baltimore, MD, 1991, pp. 63–75.

Pauls, J., "Movement of People," *The SFPE Handbook of Fire Protection Engineering*, 2nd ed., National Fire Protection Association, Quincy, MA, 1988, pp. 3-263–3-285.

Pigott, B. B., "Fire Intelligence: Key to the Design of Economical Means of Escape," Sixth International Fire Conference on Fire Safety, Interflam '93, March 30–April 1, 1993, Oxford, England, Interscience Communications Ltd., London, England, 1993, pp. 181–188.

Poon, L. S., "EvacSim: A Simulation Model of Occupants With Behavioral Attributes in Emergency Evacuation of High-Rise Building Fires," *Proceedings* of the Fourth International Symposium on Fire Safety Science, Intl. Assoc. for Fire Safety Science, Boston, MA, 1994, pp. 681–692.

Ruskouski, C., "The New Emergency Lighting Technology," *Fire Journal*, Vol. 84, No. 1, Jan./Feb. 1990, p. 48.

Semple, J. B., "Vertical Exiting: Are Elevators Another Way Out?" *NFPA Journal*, Vol. 87, No. 3, May/June 1993, pp. 49–52.

Shields, T. J., and Dunlop, K. E., "Emergency Egress Models and the Disabled," Sixth International Fire Conference on Fire Safety, Interflam '93, March 30–April 1, 1993, Oxford, England, Interscience Communications Ltd., London, England, 1993, pp. 143–150.

Shields, T. J., Silcock, G. W., and Dunlop, K. E., "Methodology for the Determination of Code Equivalency With Respect to the Provision of Means of Escape," *Fire Safety Journal*, Vol. 19, No. 4, 1992, pp. 267–278.

Sime, J. D., "Perceived Time Available: the Margin of Safety in Fires," *Proceedings* of the First International Symposium on Fire Safety Science, Hemisphere, 1986, pp. 561–570.

Stevens, R. E., "Scissors Stairs as Exits," *Fire Journal*, Vol. 59, No. 1, Jan. 1965, p. 40.

Stevens, R. E., "What Is an Exit?" *Fire Journal*, Vol. 59, No. 6, Nov. 1965, pp. 44–45.

Stevens, R. E., "Smokeproof Towers," *Fire Journal*, Vol. 60, No. 1, Jan. 1966, pp. 54–55.

Takahashi, K., and Tanaka, T., "An Evacuation Model for the Use in Fire Safety Designing of Buildings," 9th Joint Panel Meeting of the UJNR Panel on Fire Research and Safety, NBSIR 88-3753, Apr. 1988, National Bureau of Standards, Gaithersburg, MD.

Tanaka, T., "Study for Performance Based Design of Means of Escape in Fire," *Proceedings* of the Third International Symposium on Fire Safety Science, Elsevier Applied Science, New York, 1991, pp. 729–738.

Templer, J., *The Staircase: Studies of Hazards, Fills and Safer Design*, MIT Press, Cambridge, MA, 1992.

Van Bogaert, A. F., "Evacuating Schools on Fire," *Proceedings* of the First International Symposium on Fire Safety Science, Hemisphere, 1986, pp. 551–560.

Wade, C. A., "Means of Escape in Multi-Storey Buildings. Study Report," Building Research Association of New Zealand, Judgeford, BRANZ Study Report SR38, July 1991.

Watts, J. M., Jr., "Angle of Exit Remoteness. Technical Note," *Fire Technology*, Vol. 32, No. 1, Jan./Feb. 1996, pp. 76–82.

Webber, G., and Hallman, P., "Photoluminescence for Aiding Escape," *Fire Surveyor*, Vol. 17, No. 6, Dec. 1988, pp. 17–29.

Weinroth, J., "An Adaptable Microcomputer Model for Evacuation Management," *Fire Technology*, Vol. 25, No. 4, 1989, pp. 291–307.

Yoshida, Y., "Evaluating Building Fire Safety Through Egress Prediction: A Standard Application in Japan," *Fire Technology*, Vol. 31, No. 2, May 1995, pp. 158–174.

BUILDING TRANSPORTATION SYSTEMS

Edward A. Donoghue

Because of their impact on fire safety, building transportation systems must be considered during the design stage and should be integrated with the total building fire protection system. Hazards associated with these systems must be included in any prefire planning and fire protection survey.

Building transportation systems, such as elevators and escalators, make high-rise buildings feasible. Today, accessibility requirements mandate elevators in most buildings with two or more stories. While it is unrealistic for multistory buildings to function without elevators, elevator hoistways can contribute to the spread of smoke and/or fire. On the other hand, an elevator is a necessary tool for firefighting operations in high-rise buildings.

The discussion of elevators, dumbwaiters, escalators, and moving walks in this chapter is intended to complement NFPA *101®, Life Safety Code®,* requirements that equipment for elevators and so on be installed in accordance with ASME A17.1, *Safety Code for Elevators and Escalators.*[1] The three model building codes—*National Building Code* (BOCA), *Standard Building Code* (SBCC), and *Uniform Building Code* (ICBO)—also refer to ASME A17.1.[1]

In Canada the *Safety Code for Elevators,* CAN/CSA B4417, applies in lieu of ASME A17.1.[1] The A17 and B44 Elevator Safety Code Committees are currently developing a harmonized code to be administered by a Binational Committee of the American Society of Mechanical Engineers (ASME).

See Section 9, Chapter 20, "Materials-Handling Equipment," for other-than-freight elevators and dumbwaiters. Section 7, Chapter 5, "Confinement of Fire in Buildings," and Section 7, Chapter 6, "Smoke Movement in Buildings," are pertinent to the subject matter of this chapter; and they should be reviewed for a more comprehensive view of fire problems associated with building transportation systems.

ELEVATORS

An elevator can be defined as a hoisting and lowering mechanism equipped with a car or a platform that moves in guide rails and serves two or more landings.[1] The two major elevator types, classified by their driving means, are electric and hydraulic.

Edward A. Donoghue, CPCA, is president of Edward A. Donoghue Associates, Inc., code and safety consultants, Salem, NY. He is a member of the ASME Committee on Elevators and Escalators, NFPA Technical Committee on Safety to Life, and NFPA Technical Committee on Fire Doors and Windows.

Electric Elevators

An electric elevator's power source is an electric motor. The most common electric elevator is the electric traction elevator that employs a grooved traction drive sheave, over which pass suspension ropes that are attached to the car and counterweight. (See Figure 8-3A.) This arrangement is very simple and safe, allowing its use in buildings of any height. An inherent safety feature is traction loss when the car or counterweight is at its travel limit. The car or counterweight cannot be pulled into the overhead structure.

Another type of electric elevator is the winding drum machine, found on older elevators. In these elevators, suspension ropes attach to, and wind onto, a winding drum. Counterweight ropes also may be fastened to, and wind onto, the drum. Winding drum machines typically have no counterweights, but older designs may have a car counterweight and/or a drum counterweight. It is possible for either the elevator car or counterweight to be pulled dangerously into the overhead structure.

Hydraulic Elevators

Hydraulic elevators are powered by a hydraulic plunger or piston. (See Figure 8-3B.) The rise, or travel, of a direct plunger hydraulic elevator is limited to the length of its piston, usually six floors or less.

Another type of hydraulic elevator is the roped hydraulic elevator. This elevator type has its plunger, or piston, connected to the car with wire ropes or indirectly coupled to the car by wire ropes or sheaves.

Classification of Elevators by Use

Elevators are classified according to use as either passenger or freight elevators. A passenger elevator's primary use is to carry persons other than an operator and persons who load and unload. A freight elevator is used for freight; only the operator and persons who unload and load the freight are permitted to ride without obtaining special permission from the Authority Having Jurisdiction. "Service elevator" is another term used to describe elevators. A service elevator is officially classified as a passenger elevator that is designed to carry freight.

HOISTWAY CONSTRUCTION

Elevators travel in a vertical space or opening, called a hoistway or shaft, that extends from the pit floor of a building to the floor, or

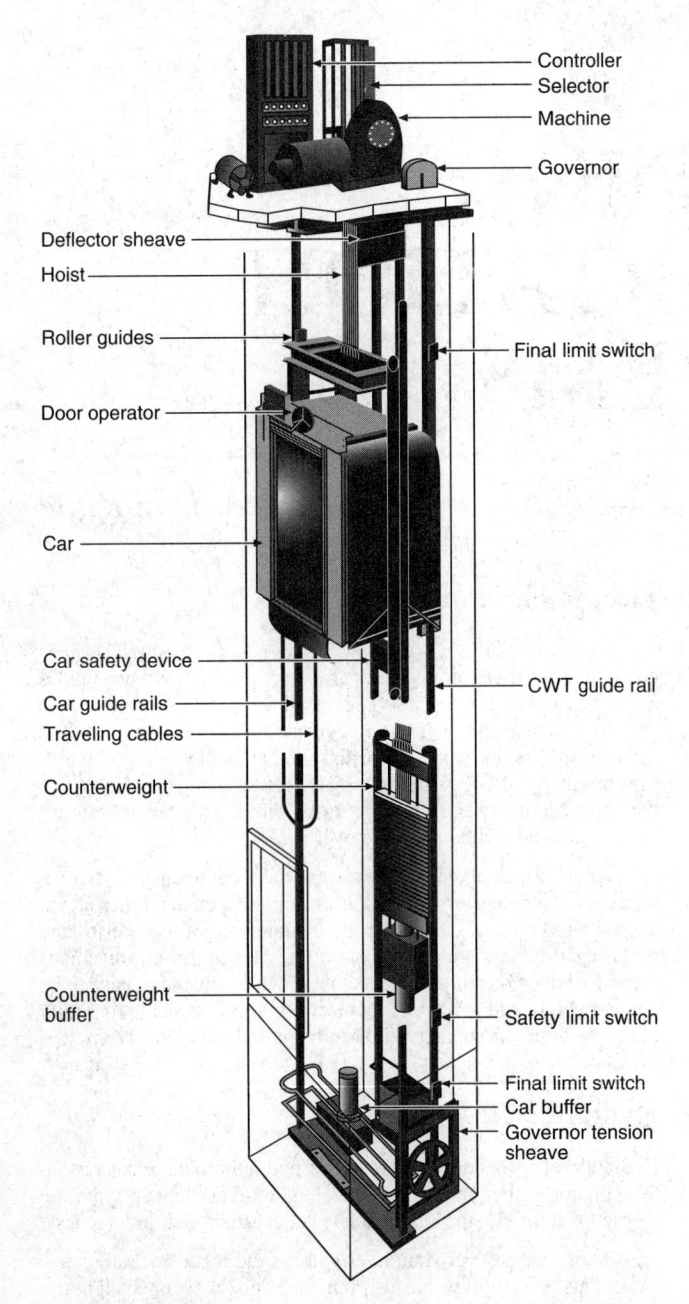

Controller
Selector
Machine
Governor

Deflector sheave
Hoist
Roller guides
Door operator
Car
Car safety device
Car guide rails
Traveling cables
Counterweight

Counterweight buffer

Final limit switch

CWT guide rail

Safety limit switch

Final limit switch
Car buffer
Governor tension sheave

FIG. 8-3A.   Electric traction elevator. (Source: National Elevator Industry, Inc.)

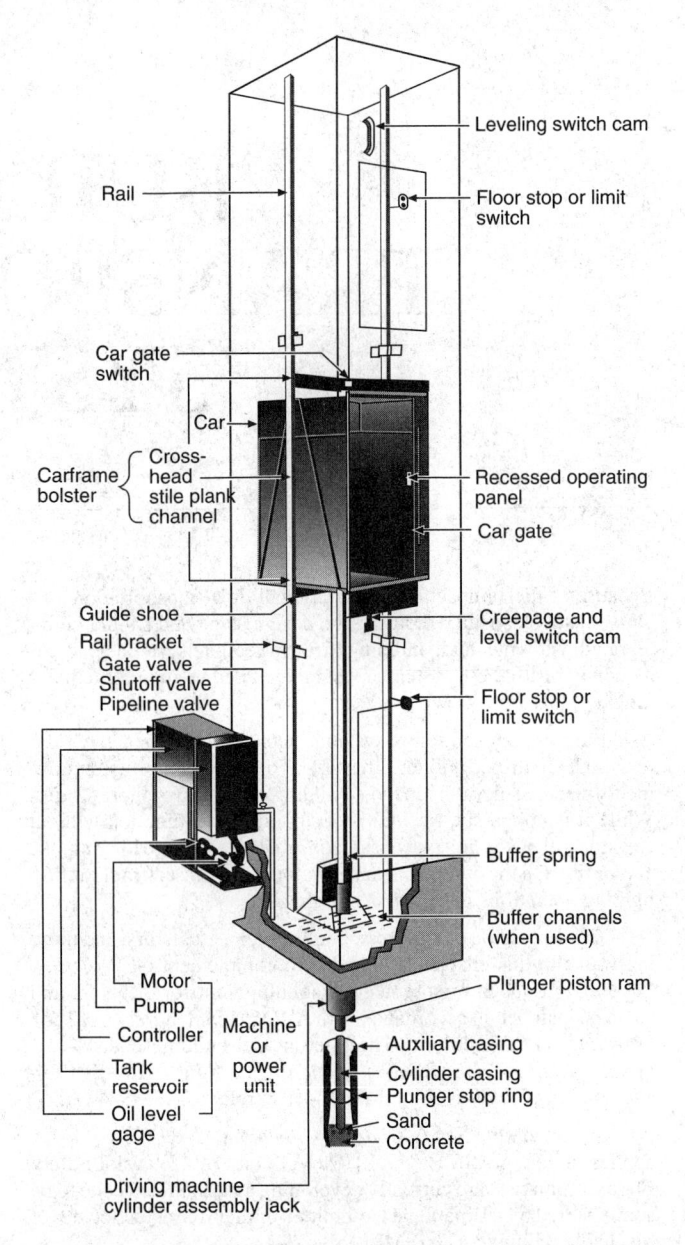

Rail

Car gate switch

Car

Carframe bolster
Cross-head
stile plank
channel

Guide shoe
Rail bracket
Gate valve
Shutoff valve
Pipeline valve

Motor
Pump
Controller
Tank reservoir
Oil level gage

Machine or power unit

Driving machine cylinder assembly jack

Leveling switch cam

Floor stop or limit switch

Recessed operating panel
Car gate

Creepage and level switch cam

Floor stop or limit switch

Buffer spring

Buffer channels (when used)

Plunger piston ram

Auxiliary casing
Cylinder casing
Plunger stop ring
Sand
Concrete

FIG. 8-3B.   Direct plunger hydraulic elevator. (Source: National Elevator Industry, Inc.)

roof, above. ASME A17.1[1] requires that hoistway construction conform to the applicable building code or, in absence of local code, to provisions of one of the three model building codes.

Although elevators made high-rise buildings practical, they also contributed to dangerous and complicated fire safety problems. Hot smoke and gases from fire, if allowed to accumulate at the top of the elevator hoistway, result in mushroom fires on upper floors served by the elevator. This occurs because an enclosed elevator hoistway acts as a chimney or flue due to a smokestack effect. This generally results in the movement of smoke and combustion products from lower to upper building levels.

The stack effect may also adversely affect elevator door closing, even under normal conditions. ASME A17.1[1] specifies the maximum door closing force and kinetic energy for safe automatic door closing. Pressure differentials must be taken into account to safely regulate door closing operations.

All model building codes require venting of elevator hoistways. Most require venting of the hoistway for elevators serving three or more stories. Some local codes permit venting through the machine room, with mechanical or natural venting from the machine room to the outside. Premature loss of elevator service may result in a fire emergency due to elevated machine room temperatures and/or smoke damage to the control system. Other codes prohibit this practice and require that cable slots and other openings between the machine room and hoistway be sleeved from the machine room floor to a point not less than 12 in. (305 mm) below the hoistway vent. Most codes restrict the size of the opening in machine room floors. For fully sprinklered buildings that have no sleeping rooms, some codes allow sprinklers in the hoistways in lieu of hoistway venting.

Some local codes require pressurization of the hoistway to a specified minimum positive pressure above the pressure of the elevator lobby. When a pressurized hoistway is provided, it should contain a vent that opens automatically in case of power failure. A manual means should also be provided to vent smoke that penetrates the hoistway. Due to numerous variances and rapid developments in this area, local code requirements should be checked.

Elevator hoistway enclosure walls are normally fire partitions with a fire-resistant rating of 2 hr. Some situations require no fire-resistant rating. For example, an elevator that serves only open balconies without penetrating separate fire-resistant areas of a building does not require a fire-resistant hoistway. In other cases, the fire-resistant rating may be required on only one or two walls. An example is an observation elevator that permits exterior viewing by passengers, where entrances to the elevator penetrate fire-resistant walls.

The maximum number of elevators permitted in a single hoistway is controlled by building code requirements, which attempt to limit the potential of a fire knocking out all elevator service in a structure. For buildings with three or fewer elevators, codes generally permit all elevators to be in the same hoistway enclosure. Four elevators require at least two separate hoistway enclosures, and, where there are more than four elevators, no more than four can be enclosed in the same hoistway enclosure.

Hoistway Entrances

If the hoistway wall is load bearing, it can support the hoistway entrance operating mechanisms and locking devices. If the walls are not load bearing, other building structures must support the hoistway entrance operating mechanisms and locking devices. Elevator entrance frames are not designed to carry the weight of the wall above, so a lintel must be provided.

Elevator hoistway entrance assemblies customarily have been rated at $1^1/_2$ hr, but they may be rated at 1 hr or $^3/_4$ hr. A labeled entrance ensures that the hoistway entrance door assembly can meet the fire safety requirements specified by building codes and NFPA 80, *Standard for Fire Doors and Fire Windows*. The label is visible from the hoistway. Arrangements must be made with the elevator company so the label can be safely seen from the top of the car.

In masonry construction, passenger elevator doors typically overlap the wall by $^3/_4$ in. (19 mm). When this requirement is met, the frames, if supplied, are not required to be labeled. Masonry-type walls generally consist of brick, concrete block, or reinforced concrete construction. All three materials are fire resistant.

Drywall construction with stone veneers is available but has not been investigated by testing laboratories under fire conditions; therefore, the performance of such construction cannot be accurately predicted. Engineering judgment can be applied when comparing stone-faced drywalls with solid masonry walls. In drywall construction, the application of the facing material could affect the interface between the door and steel frame.

Drywall Enclosure Construction

Drywall construction largely supplanted masonry construction as the most prevalent hoistway construction used today throughout the United States. Early elevator entrance installations in drywall were similar to masonry installations. However, some design changes were made to ensure the integrity of the interface between the entrance assembly and the drywall construction. Brackets were placed within the frame of the entrance assembly to accept "J" struts furnished by the drywall contractor.

The "J" strut is vitally important in ensuring a proper entrance to drywall interface. Drywall frames must be labeled unless the door

panel overlaps the wall by $^3/_4$ in. (75 mm), an atypical condition. There is no difference in the door panels. Figure 8-3C shows a cutaway view of a drywall entrance assembly interface.

In general, the drywall entrance assembly is more critical with respect to fire performance. As a result of the performance of drywall entrance assemblies, the same assemblies can be used in masonry construction.

The classification marking, or label, on passenger elevator door panels covers the design and construction of the door panels only. In addition to the door panels, when the panel does not overlap the wall $^3/_4$ in. (75 mm), the fire-door frames and transom, if provided, must be labeled.

Hardware also must be labeled for a completely listed entrance assembly. Passenger elevator door hardware consists of a header; track hangers; pendant bolts; floor sill with guides, sill support plates, and brackets; retaining angles; and closure assemblies. Each component must be labeled individually. For hardware packaged and shipped as a single unit, ASME A17.1[1] permits one label for all components. ASME A17.1[1] also permits a single label for the entire entrance when it is shipped as a single package and duplicates the tested assembly. Typically freight entrances and dumbwaiter entrances are provided with a single label.

Door Gasketing Materials

Listing agencies have classified gasketing materials for use on elevator fire door and frame assemblies. Individual gasketing materials are investigated for use on particular frame or door types and for specific fire duration periods. The basic standards used to investigate gasketing materials are Underwriters Laboratories Inc.'s *Fire Test of Door Assemblies* (UL 10B)[2] and *Standard for Air Leakage Tests of Door Assemblies* (UL 1784).[3] For additional information,

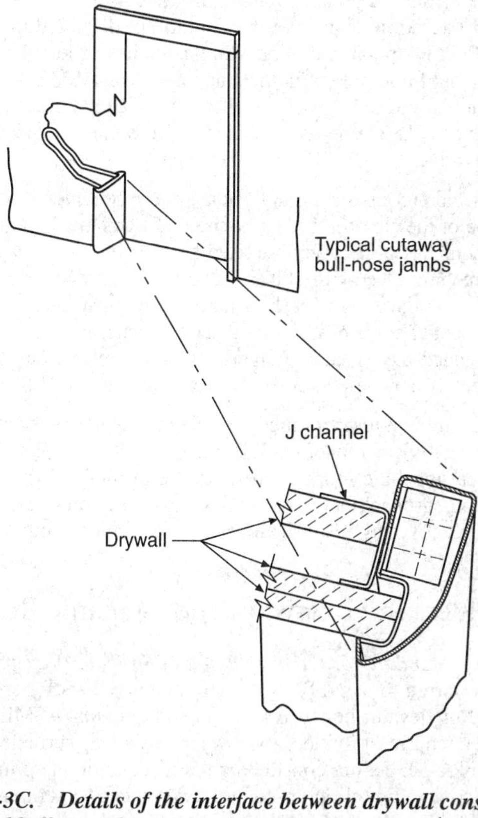

Typical cutaway bull-nose jambs

J channel

Drywall

FIG. 8-3C. *Details of the interface between drywall construction and bull-nose door jamb of an elevator entrance.*[1]

see NFPA 105, *Recommended Practice for the Installation of Smoke-Control Door Assemblies.* However, the tests do not evaluate the potential adverse effect on door operation.

Fire departments consider in-car elevator operation crucial in a high-rise fire. A question arises: Does gasket material affect elevator door operation? The kinetic energy and door closing forces of power-operated doors are regulated for passenger safety by ASME A17.1.[1] When the doors are almost closed, the closing speed must be reduced. Friction from the gasket material also may prevent complete closure, immobilizing the elevator and exposing the elevator hoistway to smoke, hot gases, and fire. Another question is: What effect will this seal have on doors designed to close automatically if the car leaves the landing zone for any reason.[4] ASME A17.1[1] requirements address the safety needs of passengers, including fire fighters, when gasketing is provided on elevator entrances.

Although enclosing elevator lobbies with smoke and draft control doors is perhaps not as aesthetic as the architect and building owner would like, it is the most practical substitute for gasketing elevator entrances, as a means of controlling the movement of smoke into the elevator hoistway. An enclosed lobby also could serve as an area of rescue assistance for the disabled in a fire and as a safe staging area for fire fighting.

Machine Room Construction

When fire-resistant hoistways are required for electric elevators, the machine room must be constructed of equally fire-resistant material. When a non-fire-resistant hoistway is constructed for an electric or hydraulic elevator and when the building code permits, the machine room enclosure must be of noncombustible material extending to a height of not less than 6 ft (1.83 m). Machine rooms are to be kept closed and locked, and entry should be restricted to authorized personnel familiar with the precautions that are necessary around moving elevator machinery. Machine room floors should be kept free of oil and grease. Material not necessary for the maintenance or operation of the elevator should not be stored in the machine room. Flammable liquids with flashpoints of less than 110°F (43°C) should not be kept in the machine room. ASME A17.1[1] requires an extinguisher for use on Class A, B, and C fires in all machine spaces; the extinguisher should be conveniently located near the access door.

Machine rooms should not contain equipment inessential to the operation of the elevator, because a fire that starts in a basement machine room has a clear unobstructed path to the hoistway. ASME A17.1[1] requires electric machinery to be in a room that contains only elevator equipment. Previously a machine room located above the hoistway could contain other equipment essential to building operation, provided the elevator equipment was separated from the other equipment by a metal grill or solid partition at least 6 ft (1.83 m) high.

The ambient temperature in the elevator machine room is critical, especially for modern solid state controls. ASME A17.1[1] requires that machine room temperature be maintained in the range the equipment manufacturer specifies. Loss of elevator service in a fire emergency may result if the elevator machine room HVAC is shut down.

Sprinklers in Hoistways and Machine Rooms

NFPA 13, *Standard for the Installation of Sprinkler Systems* (hereinafter referred to as NFPA 13), outlines provisions for sprinklers installed in elevator hoistways and machine rooms. ASME A17.1[1] does not require sprinklers, but, when installed, sprinklers must meet NFPA 13. Before considering the installation of sprinklers in spaces containing elevator control equipment, certain potential problems should be analyzed. One should consider whether any wa-

ter discharge on the control equipment may adversely affect the elevator's safe operation and note the effect of water on safety circuits. Water may short safety circuits and allow an elevator to run with an open car or hoistway door. In addition, wet brakes may be unable to stop and hold an elevator.

ASME A17.1[1] requires that, when a sprinkler in the machine room or hoistway is discharged, the main line power source to the elevator be disconnected. NFPA 72, *National Fire Alarm Code*, contains details on how to implement this requirement. In a fire, the sequence of events would probably be as follows:

1. A smoke detector in the machine room, hoistway, or elevator lobby recalls the elevator.
2. The main line power automatically disconnects.
3. The sprinkler discharges.

Even if the smoke detector is not activated or if all elevators have not been returned to the lobby, stopping the car and containing the passengers is considered safer than allowing operation to continue with the sprinkler discharging. Fire fighters should be aware that, if fire is in the elevator machine room, the elevators with machinery in that room will not be available for their use.[5]

Other ASME A17.1[1] requirements pertain to the location of sprinkler piping in elevator machine rooms and hoistways. Sprinkler activation by a smoke detector is prohibited. Previous editions of NFPA 13 required that sprinklers be installed at the top and bottom of elevator shafts. This provision always resulted in controversy due to the potential for water spraying the elevator equipment. NFPA 13 now permits the elimination of sprinklers at the bottom of the shaft when the elevator shaft is enclosed with noncombustible materials and combustible hydraulic fluids are not present in the shaft. (See NFPA 13 Exception to 4-5.5.1.) Clearly this provision applies to hydraulic elevators. A potential problem is the interpretation of this provision relative to an electric traction elevator installation with oil buffers. NFPA 13 requires that sprinklers installed at the bottom of the shaft must be positioned so that they do not interfere with the platform guard. (See NFPA 13 Appendix A-4-13.5.1.) In addition, sprinklers may be omitted from the top of non-combustible elevator hoistways of passenger elevators complying with ASME A17.1[1] (See NFPA 13 Exception to 4-13.5.3.)

Elevator Car Construction

ASME A17.1[1] requires that all material exposed to the car interior and the hoistway be metal or glass, or have a flame spread rating of 0 to 75 and smoke development rating of 0 to 450. Materials must be tested in their end use configuration. Padded protective linings for temporary use in passenger cars during freight handling also must meet these criteria.

The smoke development rating for materials in cars is equal to NFPA *101*, requirements for an exit passageway in a fire-resistant building. Smoke development characteristics of materials may be more crucial than flame spread, as passengers in a confined enclosure box may not be able to escape rapidly.

Napped, tufted, woven, looped, and similar materials must be subjected to a vertical burn test as specified by ASME A17.1.[1] This test is based on applicable portions of FAA Regulation 25.853, in addition to the flame spread and smoke development test (ASTM E84) and NFPA 255, *Standard Method of Test of Surface Burning Characteristics of Building Materials.* Such material, located on elevator car walls, is extremely susceptible to accidental ignition. Tufted material, such as carpeting, may accumulate dust, which could increase the possibility of ignition and flame spread.

Floor covering, underlayment, and adhesive are required to have a critical radiant flux of not less than 0.45 W/cm², as measured by the ASTM E648, *Test for Critical Radiant Flux of Floor Covering Systems*,[6] or NFPA 253, *Standard Method of Test for Critical Radiant Flux of Floor Covering Systems Using a Radiant Heat Energy Source*. The level of critical radiant flux acceptable for floor systems in elevator cars is based on the assumption that occupants of an elevator may not be able to escape rapidly. The flux level equals the requirements recommended for health care facilities.

Handrails, operating devices, ventilating devices, signal fixtures, audio and visual communications devices, and their housings are not required to meet any of these criteria because they contribute little to fire load in relation to overall elevator car assembly.

Accessibility Considerations

Two national standards address accessibility requirements for elevators: (1) ADAAG, *Americans with Disabilities Act Accessibility Guidelines*;[7] and (2) CABO/ANSI A117.1, *American National Standard for Accessibility*.[8] Both standards require all passenger elevators to be made accessible to the disabled. Planned and practiced procedures are needed to evacuate the disabled in a fire. Additional information on accessibility requirements for elevators can be found in *ADA and Vertical Transportation*.[16]

Emergency Lighting, Alarm, and Communications

ASME A17.1[1] requires emergency lighting on all elevators. The power source must originate on the elevator to ensure that passengers are not thrust into total darkness if building power fails. The building emergency power source is not a recognized alternative to this requirement, because this source generally would not respond to a branch circuit failure and usually is activated only upon loss of the building power supply. Moreover, it would rely on the elevator's traveling cables to supply lighting power, which would be useless if the lighting problem were in the traveling cables.

All elevators must have an audible signaling device, an alarm, and a means of two-way conversation to an accessible point outside the hoistway. (See Figure 8-3D for a diagrammatic description of these requirements.) The two-way communication system does not require a means of activation from within the car. Most communication systems, excluding telephones, are activated in the main lobby or building emergency control center. ADAAG[7] and CABO/ANSI A117.1[8] specify that the emergency signaling device shall not be limited to voice communications.

In addition to these requirements, the three model building codes require all elevators in high-rise buildings to have a communication system from the elevator lobby, car, and machine rooms to the building's central control station.

Electrical Requirements

ASME A17.1[1] requires electrical wiring for elevators, escalators, moving walks, and dumbwaiters to conform to NFPA 70, *National Electrical Code®*. Electrical equipment must be certified and labeled as complying with CSA B44.1/ASME A17.5.[1,15]

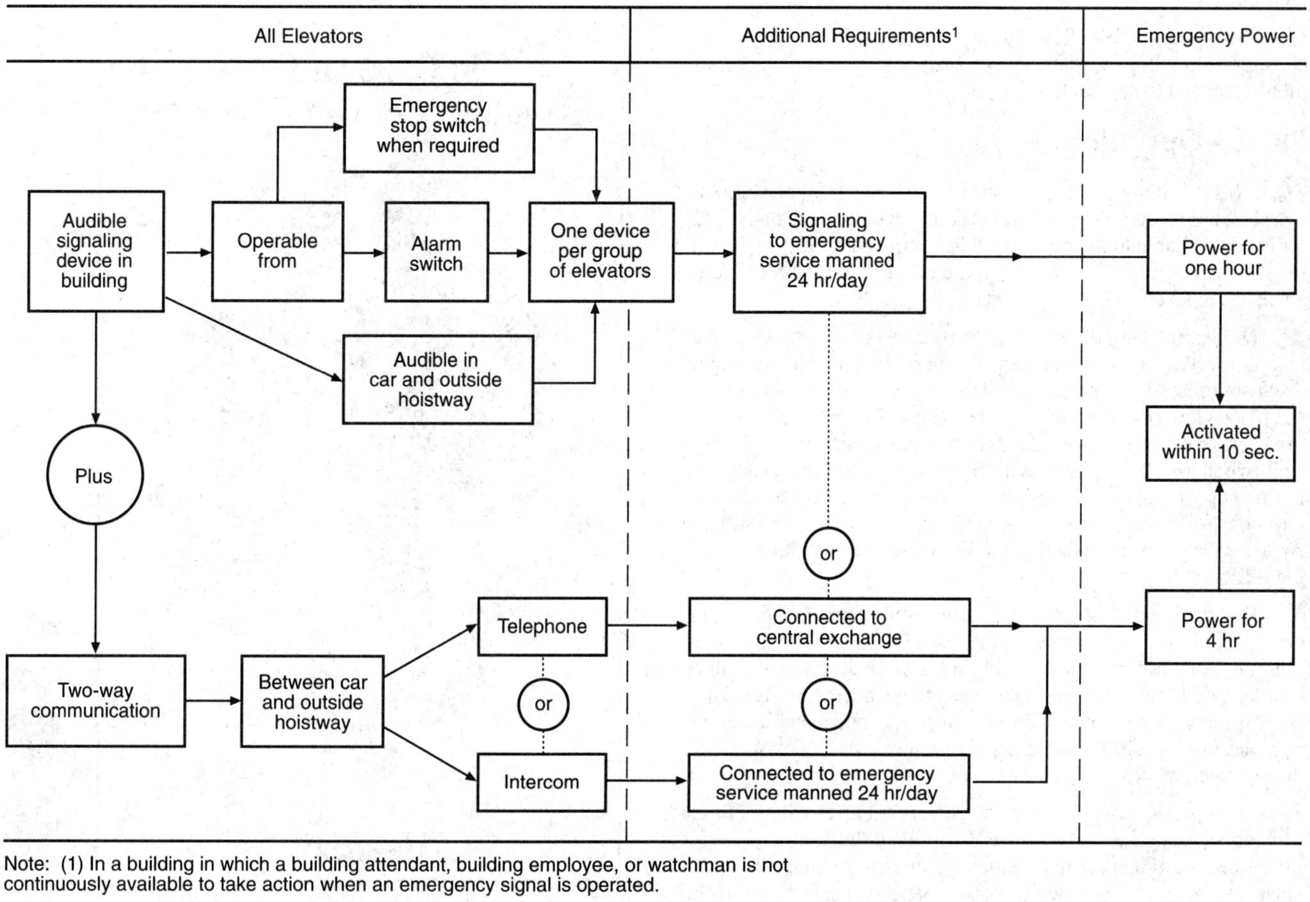

Note: (1) In a building in which a building attendant, building employee, or watchman is not continuously available to take action when an emergency signal is operated.

FIG. 8-3D. Emergency elevator signaling device requirements.[1]

ELEVATORS IN FIRE EMERGENCIES

High-rise buildings present new and different problems to fire suppression forces, including elevator use during emergencies. Elevators are unsafe in a fire for several reasons:

1. Persons may push a corridor button and wait for an elevator that may never respond, losing valuable escape time.
2. Elevators do not prioritize car and corridor calls, and one of the calls may be at the fire floor.
3. Elevators cannot start until the car and hoistway doors close, and panic could lead to elevator overcrowding and door blockage, which would thus prevent closing.
4. Power can fail at any time during a fire, thus leading to entrapment.

The elevator can be delivered to the fire floor when:

1. An elevator passenger presses the car button for the fire floor.
2. One or both of the corridor call buttons is pushed on the fire floor.
3. Heat melts or deforms the corridor push button or its wiring at the fire floor.
4. The elevator functions normally at the fire floor, as in high or low call reversal.

ASME A17.1[1] recognizes all these unsafe conditions and mandates elevator recall, more commonly referred to as *Phase I, Emergency Recall Operations.* ASME A17-1[1] requires fire fighters' service on all new elevators. NFPA 72, *National Fire Alarm Code,* contains requirements for the smoke detector system that provides the signal to the elevator control system for initiation of Phase I operation.

The provision that allows fire-fighting personnel to operate the elevator from within the car, or emergency in-car operation, is commonly referred to as *Phase II.*

Phase I Operations

Phase I is the automatic or manual return, or recall, of elevators to a designated level or recall floor. Phase I also ensures that an elevator is not available to the general public during a fire. If the fire is in the designated level elevator lobby, the elevator will be recalled to an alternate level.

Designated level is defined as the main floor or some other level that best serves the needs of emergency personnel for fire-fighting or rescue purposes.[1] Emergency personnel can use a three-position on/off, bypass key-operated switch in the designated-level lobby to control each group of elevators. Another two-position on/off, key-operated switch may be provided in the building's central control station. In the on position, the switch instructs all controlled elevators to return nonstop to the recall floor. All registered calls will be canceled. An ascending elevator will stop and reverse direction, without opening its doors.

Hoistway doors on the recall floor should remain open. Upon arrival, fire fighters have little time to learn whether the elevators are present. Open hoistway doors will reveal those elevators that have been recalled, as well as those that need to be located to ensure that no passengers are trapped. To help emergency personnel locate errant elevators, position indicators are required to remain operative during emergencies.

The open hoistway door at the designated level will admit air to the hoistway. If the fire is on a floor below the neutral pressure plane, the additional air will reduce the amount of smoke that can enter the hoistway from the fire floor, a beneficial side effect. If the fire is above the neutral pressure plane, flow of air from the hoistway to the floor will increase, which is also beneficial.

Visual and audible signals alert passengers in an automatically operated elevator to an emergency and can help minimize any apprehension that the passengers have as the elevator returns to the recall floor. In an attendant-operated elevator, this signal alerts the attendant to the emergency so that he or she can return the car immediately to the recall floor.

Elevator recall systems combine a key-operated switch with an automatic smoke detection system, using detectors at each floor lobby. Elevator recall by the smoke detector is the same as placing the key switch in the on position. The only difference is that the smoke detector at the designated level will recall the elevator to an alternate level. If the automatic recall sent the elevator to the alternate level, the key switch will override the designated level smoke detector and recall the elevator to the designated level. The key switch will not recall an elevator to the alternate level.

To advise the public that elevators may not be available during a fire, the pictograph in Figure 8-3E should be posted above all elevator hall call buttons. Some jurisdictions may require different wording and pictographs, but the message is similar. A proposal under consideration is a back-lighted sign visible only when illuminated that advises when Phase I recall has been activated.

Phase II Operations

Phase II benefits fire fighters who must have the elevators for their own use. It is standard operating procedure in high-rise structures for fire fighters to use elevators not only to carry equipment for fire-fighting or evacuation purposes, but also to deliver fire personnel to nonfire floors.

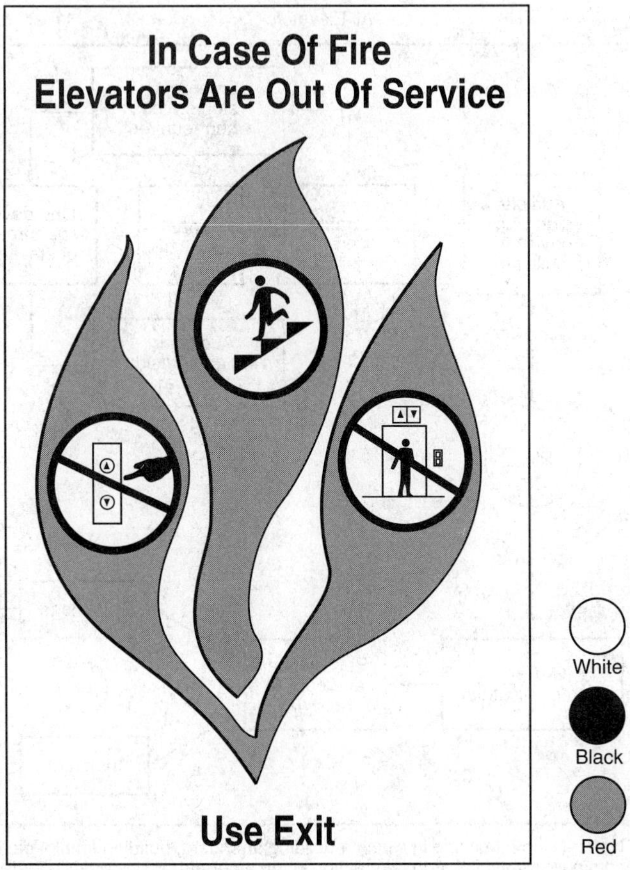

FIG. 8-3E. *An example of a pictographic warning message for elevator use.*[1]

The current ASME A17.1[1] calls for Phase II operation on all new elevators. The rule considers the need for fire fighters to evacuate disabled persons during an emergency. This code has also standardized operation procedures, regardless of the elevator manufacturer.

The switch keys required for Phase I and Phase II operation must be the same to ensure a speedy response to an emergency. The use of lock boxes to hold Phase I and Phase II keys is preferred over the use of a jurisdiction-wide uniform key, because the lock box method provides greater assurance that the keys are restricted. The uniform key method has disrupted normal elevator service, because it provides no means of controlling the distribution of elevator keys.

Operating procedures must be incorporated with, or adjacent to, the Phase I and Phase II key-operated switches to provide emergency personnel quick access to instructions during an emergency. Fire fighters should also be aware that when the visual signal required by ASME A17.1[1] is flashing, it signals an impending problem and continued use of the elevator is not advised. (See Figure 8-3F.)

ASME A17.1[1] does not require standby power for elevators. This requirement is found normally in the building code. However, the safety code does require that elevator operation be the same when standby power is supplied as it is when normal power is supplied. ASME A17.1[1] does require that a manual selection switch be provided for fire fighters to select the car(s) to which standby power is to be directed.

Evacuation from a Stalled Elevator

Trained elevator personnel should perform any evacuation of passengers from a stalled elevator car. Their expertise ensures that evacuation is carried out in a manner that is safe for both the passengers and the rescue party. However, waiting for trained elevator personnel may be impossible in a life-threatening emergency. Under such conditions, passenger evacuation must be performed by carefully selected and trained personnel. Planned and practiced procedures are needed to ensure the safety of rescued passengers and members of the rescue party. Any rescue party should be cautioned never to try to move an elevator from the machine room. Only trained elevator personnel should attempt this.

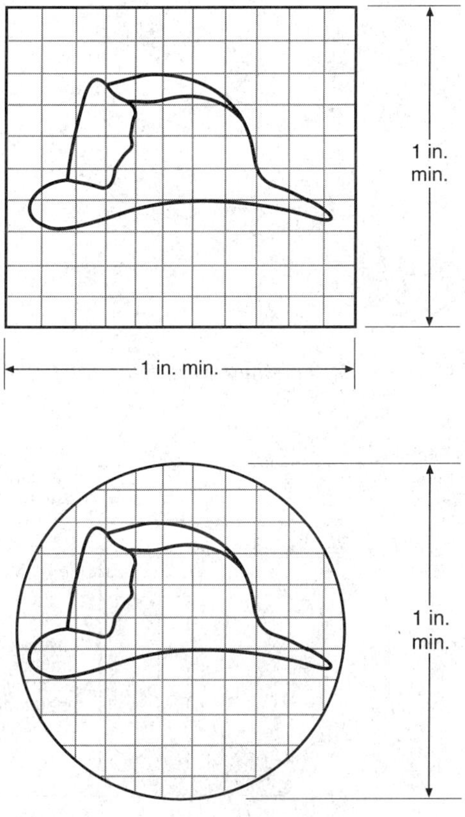

ASME A17.4, *Guide for Emergency Evacuation of Passengers from Stalled Elevator Cars,*[9] details procedures to follow in performing a safe evacuation.

DUMBWAITERS

A *dumbwaiter* is defined as a hoisting and lowering mechanism, used exclusively for carrying materials, with a limited size car that moves in guides in a substantially vertical direction.[1] Dumbwaiters must conform to requirements of ASME A17.1.[1] A dumbwaiter is actually a small elevator used for a variety of purposes, such as mail distribution in office buildings; distribution of medication, food, and supplies in hospitals; and moving books from floor to floor in a library. A dumbwaiter can be loaded and unloaded manually or automatically at the floor or at a counter. Dumbwaiter car size is restricted to discourage riding. When dumbwaiters penetrate fire-resistant floors, their hoistway construction is critical. Requirements for a dumbwaiter hoistway are the same as those for an elevator. The section on elevator hoistway construction earlier in this chapter details fire-resistant hoistway enclosures and entrances.

ESCALATORS

The escalator is another type of vertical people mover. An escalator may be defined as a power-driven, inclined, continuous stairway used for raising or lowering passengers.[1] In the early days, escalators were found principally in department stores and in bus and airline terminals. Today, they are common in office buildings, hospitals, banks, sports arenas, and other locations where many people must be moved quickly from one floor to another.

NFPA *101* requires that escalator floor openings be enclosed or protected like other vertical openings. In completely sprinklered buildings, escalators need not be enclosed if they are protected by alternate means.

The sprinkler draft curtain method is detailed in NFPA 13 and consists of surrounding the escalator opening with an 18-in. (457-mm) deep draft stop located on the underside of the floor to which the escalator ascends. A row of automatic sprinklers located outside the draft stop surrounds the escalator well. Figure 8-3G shows a typical installation.

Two methods recognized in previous editions of NFPA *101* and found in existing buildings are described next.

Sprinkler Vent Method

An automatic fire or smoke detection system, an automatic exhaust system, and an automatic water curtain are combined in this method. When a detector senses a fire, the automatic damper opens, thereby venting smoke and gases. Replacement air is drawn through the outdoor air intake above the escalator opening. To control smoke, an exhaust rate of about 60 air changes per hour is required. The sprinkler curtain provides a thermal barrier. Figure 8-3H shows a typical installation.

FIG. 8-3F. Visual signal.

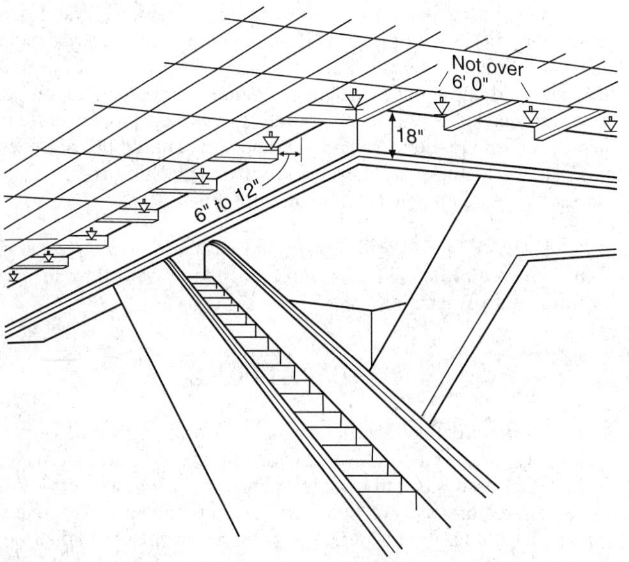

FIG. 8-3G.   Sprinklers around an escalator.

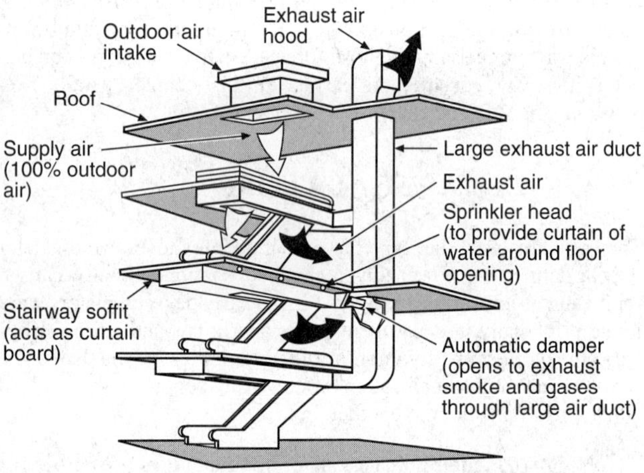

FIG. 8-3H.   Sprinkler vent method installation.

Spray Nozzle Method

This method combines an automatic fire or smoke detection system and a system of high-velocity water spray nozzles. There must be enough spray nozzles with discharge angles such that the escalator opening between the top of the wellway housing and the treadway will be completely filled with a dense water spray when the system operates.

Escalator Enclosures

Two methods of enclosing the escalator opening are used in both new and existing buildings:

1. *Rolling shutter method.* In this method, escalators above the street floor may be protected by automatic self-closing rolling shutters that completely enclose the top of the escalator wellway. To avoid injury, the escalator must stop automatically before the shutter begins to close. The shutter should close slowly

and have a pressure-sensitive leading edge that will stop the shutter upon contact. Figure 8-3I shows a typical installation.

Rolling steel shutters are not recommended at the tops of escalators or between basements and street floors, even though the stairways are not required exits under the provisions of NFPA *101*. In an emergency, persons seeking egress from basements served by escalators could be trapped by fully closed rolling shutters at street level, even though other means of egress are available to them. Also, quite a different psychological reaction is exhibited by those confronted with a closed shutter above their heads, as they try to climb to safety from a basement, than is seen in those faced with a closed shutter at an exit on the same level as the floor they are trying to leave. In the latter case, the operation of the shutter is clearly visible, and some other means of egress could be readily found and used if the requirements of NFPA *101* are followed.

2. *Partial enclosure method.* Two fire-rated enclosures, one for the down escalator and one for the up escalator, can prevent the spread of smoke and fire. Automatic fire doors release when fire is detected. The enclosure should have a fire-resistant rating equal to that of the floor/ceiling assemblies, and there should be no unprotected escalator penetration of more than two floors. Figure 8-3J shows a typical installation.

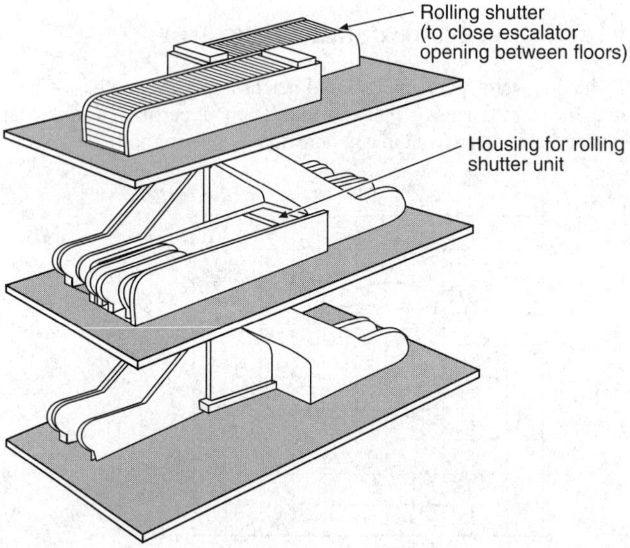

FIG. 8-3I.   Rolling shutter escalator installation.

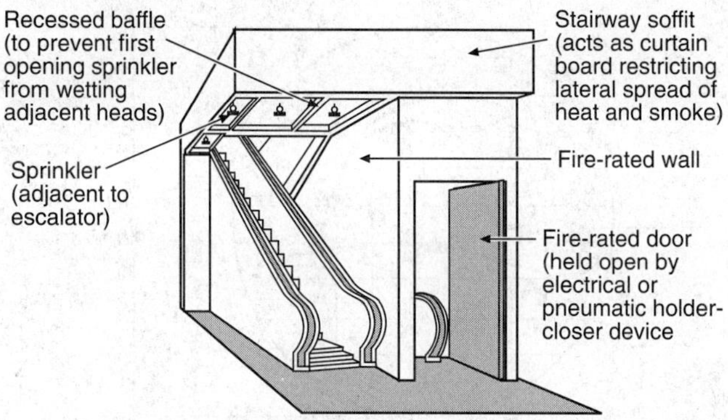

FIG. 8-3J.   Partial enclosure escalator installation.

In new buildings, escalators cannot be used as a means of egress. Modern escalator design requires the removal of steps for normal maintenance, thus rendering the escalator unusable. A stopped escalator also presents a tripping hazard, due to the nonuniform riser height of the first and last few steps. In some existing buildings, escalators are part of the means of egress. These escalators should be arranged to operate only in the direction of egress, or they should be stopped during a fire. Where the evacuation plan calls for escalator use, efficient crowd control is required to avoid serious crowding and potential panic. NFPA *101* requires the openings for escalators used as required exits to be enclosed in the same manner as exit stairs.

Another important factor in escalator fire prevention is keeping truss and machine areas free from oil, grease, and dust. A periodic cleaning schedule should be implemented.

MOVING WALKS

A *moving walk* is defined as a moving device on which passengers stand or walk. Its passenger-carrying surface remains parallel to its direction of motion and is uninterrupted. Moving walks may be installed at an incline between two floors. The same fire prevention features described for escalators apply to moving walks. Some moving walks use a combustible belt, essentially the same as a large conveyor belt. This belt has the potential to spread a fire by itself. All moving walks can carry burning items along their lengths.

EXISTING ELEVATORS AND ESCALATORS

At minimum, existing elevators and escalators should be required to conform to the requirements of ASME A17.3, *Safety Code for Existing Elevators and Escalators.*[10] NFPA and BOCA codes require existing equipment to conform to this code. Existing installations that do not meet this code must be upgraded. When an existing installation is upgraded, at a minimum it must comply with ASME A17.1,[1] Part XII; and ASME A17.3.[10] Existing installations required to meet more stringent requirements must continue to meet those requirements.

TESTS AND INSPECTIONS

Elevators, dumbwaiters, escalators, moving walks, and other passenger conveyance equipment that are required to conform to the requirements of ASME A17.1[1] must be tested and inspected periodically. ASME A17.2.1, *Inspectors' Manual for Electric Elevators;*[11] ASME A17.2.2, *Inspectors' Manual for Hydraulic Elevators;*[12] or ASME A17.2.3, *Inspectors' Manual for Escalators and Moving Walks,*[13] should be used as guides for these inspections. The quality of inspections depends upon the inspector's competence. ASME QEI-1, *Qualifications of Elevator Inspectors,*[14] establishes uniform criteria for this purpose and serves as the standard for detailing the expertise necessary to perform inspections.

BIBLIOGRAPHY

References Cited

1. ASME A17.1, *Safety Code for Elevators and Escalators*, ASME, Fairfield, NJ, 1996.
2. UL 10B, *Fire Test of Door Assemblies*, UL, Northbrook, IL, 1988.
3. UL 1784, *Standard for Air Leakage Tests of Door Assemblies*, UL, Northbrook, IL, 1990.
4. Donoghue, E. A., "Smoke and the Elevator Hoistway," *Building Standards*, May/June 1983; *Elevator World*, May 1983.
5. Donoghue, E. A., "Elevators and Water—A Volatile Combination," *Elevator World*, Aug. 1989.
6. ASTM E648, *Test for Critical Radiant Flux of Floor Covering Systems*, ASTM, New York, 1994.
7. ADAAG, *Americans with Disabilities Accessibilities Guidelines.*
8. CABO/ANSI A117.1, *American National Standard for Accessibility*, CABO, Falls Church, VA, 1992.
9. ASME A17.4, *Guide for Emergency Evacuation of Passengers from Stalled Elevator Cars*, ASME, Fairfield, NJ, 1991.
10. ASME A17.3, *Safety Code for Existing Elevators and Escalators*, ASME, Fairfield, NJ, 1996.
11. ASME A17.2.1, *Inspectors' Manual for Electric Elevators*, ASME, Fairfield, NJ, 1993.
12. ASME A17.2.2, *Inspectors' Manual for Hydraulic Elevators*, ASME, Fairfield, NJ, 1993.
13. ASME A17.2.3, *Inspectors' Manual for Escalators and Moving Walks*, ASME, Fairfield, NJ, 1994.
14. ASME/QEI-1-1990, *Qualifications of Elevator Inspectors*, ASME, Fairfield, NJ, 1993.
15. CAN/CSA-B44.1-M91 ASME-A17.5, *Elevator and Escalator Electrical Equipment*, ASME, Fairfield, NJ, 1996.
16. Donoghue, E. A., "ADA and Vertical Transportation," *Elevator World*, Elevator World, Inc., Mobile, AL.
17. CAN/CSA B44, *Safety Code for Elevators*, CSA, Rexdale (Toronto), ON, Canada, 1994.

NFPA Codes, Standards, and Recommended Practices

Reference to the following NFPA codes, standards, and recommended practices will provide further information on building transportation systems discussed in this chapter. (See the latest version of *The NFPA Catalog* for availability of current editions of these documents.)

NFPA 13, *Standard for the Installation of Sprinkler Systems*
NFPA 70, *National Electrical Code®*
NFPA 72, *National Fire Alarm Code*
NFPA 80, *Standard for Fire Doors and Windows*
NFPA *101®*, *Life Safety Code®*
NFPA 105, *Standard Method of Test forSmoke-Control Door Assemblies*
NFPA 253, *Recommended Practice for the Critical Radiant Flux of Floor Covering Systems Using Radiant Heat Energy Source*
NFPA 255, *Standard Method of Surface Burning Characteristics of Building Materials*

Additional Reading

ASTM E84, *Standard Test Method for Surface Burning Characteristics of Building Materials*, ASTM, Philadelphia, PA, 1994.
Aikman, A. J. M., "Elevator Operation During Fire Emergencies in High Buildings," National Research Council of Canada, Ottawa, Ontario, American Society of Mechanical Engineers, Council of American Building Officials and National Fire Protection Association, Elevators and Fire, Feb. 19–20, 1991, Baltimore, MD, 1991, pp. 16–21.
Barker, F. H., Jr., "Multi-Purpose Elevators for Persons With Disabilities and Firefighters in High-Rise Office Buildings in the U.S. (With Concepts for Accessibility and Other Buildings and Locations)," *Proceedings* of the Second Symposium of the American Society of Mechanical Engineers (ASME) on Elevators, Fire and Accessibility, Apr. 19–21, 1995, Am. Soc. of Mechanical Engineers, New York, NY, 1995, pp. 207–219.
Bathurst, D., and Blazek, B., "Elevators," VHS Video Tape, General Services Administration, Washington, DC, National Institute of Standards and Technology, Federal Fire Forum on Technology Update, Nov. 1, 1990, Gaithersburg, MD, 1990.
Black, B. D., "Limited-Use/Limited Application Elevators: A New Option for Vertical Accessibility," *Proceedings* of the Second Symposium of the American Society of Mechanical Engineers (ASME) on Elevators, Fire and Accessibility, Apr. 19–21, 1995, Am. Soc. of Mechanical Engineers, New York, NY, 1995, pp. 233–236.
Building Material Directory, UL, Northbrook, IL.
Caldwell, J., "Standby Power and Elevator Safety in Fire Emergencies," *Proceedings* of the Second Symposium of the American Society of Mechanical Engineers (ASME) on Elevators, Fire and Accessibility,

Apr. 19–21, 1995, Am. Soc. of Mechanical Engineers, New York, NY, 1995, pp. 100–103.

Chapman, E. F., "Elevator Design for the 21st Century," *Proceedings* of the Second Symposium of the American Society of Mechanical Engineers (ASME) on Elevators, Fire and Accessibility, Apr. 19–21, 1995, Am. Soc. of Mechanical Engineers, New York, NY, 1995, pp. 157–162.

Chapman, E. F., "Elevator Design for the 21st Century: Design Criteria for Elevators When Used as the Primary Means of Evacuation During Fire Emergencies," *Journal of Applied Science*, Vol. 1, No. 4, 1990–1991, pp. 339–347; Fire Safety Directors Association of Greater New York's, Annual Fall Seminar, Elevators and Fire, Nov. 14, 1991, 1990.

Chapman, E. F., "Utilization of Elevators During Fire Emergencies," Nassau County Fire Academy, New York, American Society of Mechanical Engineers; Council of American Building Officials and National Fire Protection Association, Elevators and Fire, Feb. 19–20, 1991, Baltimore, MD, 1991, pp. 8–15.

Cockram, I., "Analysis of Fire Protection and Detection for Escalators," *Fire*, Vol. 83, No. 1024, Oct. 1990, pp. 24, 27, 32, 34.

Corbett, G. P., "Restriction Devices on Elevator Car Doors," *Fire Engineering*, Vol. 143, No. 5, May 1990, pp. 10, 12.

DeCicco, P. R., "Elevators for Evacuation of Occupants and Firefighter Access," Editorial Note, *Journal of Applied Fire Science*, Vol. 2, No. 1, 1992/1993, pp. 3–4.

Degenkolb, J., "Elevator Usage During a Building Fire," American Society of Mechanical Engineers, Council of American Building Officials and National Fire Protection Association, Elevators and Fire, Feb. 19–20, 1991, Baltimore, MD, 1991, pp. 76–79.

Donoghue, E. A., "Member's Forum—Elevators and Sprinklers: The Controversy Continues," *Fire Journal*, Vol. 84, No. 2, Mar./Apr. 1990, pp. 15–16.

Donoghue, E. A., "Elevators and Fire," *Journal of Applied Fire Science*, Amityville, NY, 1991.

Donoghue, E. A., *Handbook A17.1 Safety Code for Elevators and Escalators*, ASME, Fairfield, NJ, 1993.

Dunn, V., "Elevator Use at High-Rise Fires," *Firehouse*, Vol. 19, No. 11, Dec. 1994, pp. 20–21.

"Elevators, Fire and Accessibility Symposium," *Papers*, ASME, New York, 1994.

Elevator World Educational Package and Reference Library, Elevator World, Inc., Mobile, AL.

Elevator World, Published monthly by Elevator World, Inc., Mobile, AL.

Faup, J. J., "Alarm Systems Ensure Safe Elevator Recall," *NFPA Journal*, Vol. 87, No. 3, May/June 1993, pp. 28, 104.

Favro, P. C., "In Consideration of Elevators as Part of a Building Evacuation Scheme," *Proceedings* of the Second Symposium of the American Society of Mechanical Engineers (ASME) on Elevators, Fire and Accessibility, Apr. 19–21, 1995, Am. Soc. of Mechanical Engineers, New York, NY, 1995, pp. 181–185.

Fox, C. C., "Handicapped Use of Elevators," American Society of Mechanical Engineers, Council of American Building Officials and National Fire Protection Association, Elevators and Fire, Feb. 19–20, 1991, Baltimore, MD, 1991, pp. 80–82.

Fraser, B., "Applying Advanced Fire Detection and Controls Technology to Enhance Elevator Safety During Fire Emergencies," *Proceedings* of the Second Symposium of the American Society of Mechanical Engineers (ASME) on Elevators, Fire and Accessibility, Apr. 19–21, 1995, Am. Soc. of Mechanical Engineers, New York, NY, 1995, pp. 111–117.

Gatfield, A. J., "Elevators for Emergencies: There Are Many Codes—But Where Is the Right One?" *Proceedings* of the Second Symposium of the American Society of Mechanical Engineers (ASME) on Elevators, Fire and Accessibility, Apr. 19–21, 1995, Am. Soc. of Mechanical Engineers, New York, NY, 1995, pp. 121–130.

Gatfield, A. J., "Elevators and Fire: Designing for Safety," American Society of Mechanical Engineers, Council of American Building Officials and National Fire Protection Association, Elevators and Fire, Feb. 19–20, 1991, Baltimore, MD, 1991, pp. 95–107.

Gittleman, E. L., "Elevators and Sprinklers: Have We Made the Solution Too Complex?" *Fire Journal*, Vol. 84, No. 5, Sept./Oct. 1990, pp. 13–15.

Groner, N. E., "Selecting Strategies for Elevator Evacuations," *Proceedings* of the Second Symposium of the American Society of Mechanical Engineers (ASME) on Elevators, Fire and Accessibility, Apr. 19–21, 1995, Am. Soc. of Mechanical Engineers, New York, NY, 1995, pp. 186–189.

Groner, N. E., and Levin, B. M., "Human Factors Considerations in the Potential for Using Elevators in Building Emergency Evacuation Plans," George Mason Univ., Fairfax, VA, National Institute of Standards and Technology, Gaithersburg, MD, NIST-GCR-92-615, Aug. 1992, 61 pages.

Groner, N. E., and Levin, B. M., "Will Building Occupants Use Elevators for Evacuation? Factors Affecting Compliance With the Emergency Plan," *Proceedings* of the Second Symposium of the American Society of Mechanical Engineers (ASME) on Elevators, Fire and Accessibility, Apr. 19–21, 1995, Am. Soc. of Mechanical Engineers, New York, NY, 1995, pp. 194–196.

Hodgens, J., "Perspective on Elevator Safety During Fires," *Journal of Applied Science*, Vol. 1, No. 4, 1990–1991, pp. 361–373; Fire Safety Directors Association of Greater New York's, Annual Fall Seminar, Elevators and Fire, Nov. 14, 1991, 1990.

Holland, G. A., "Sprinklers in Elevator Hoistways and Machine Rooms," Dover Elevator, Memphis, TN, American Society of Mechanical Engineers, Council of American Building Officials and National Fire Protection Association, Elevators and Fire, Feb. 19–20, 1991, Baltimore, MD, 1991, pp. 50–52.

Holt, L., "High-Rise Building Emergencies: Education, Evacuation by Elevator, Firefighter's Elevators, External Platform," *Proceedings* of the Second Symposium of the American Society of Mechanical Engineers (ASME) on Elevators, Fire and Accessibility, Apr. 19–21, 1995, Am. Soc. of Mechanical Engineers, New York, NY, 1995, pp. 197–204.

Howkins, R. E., "Use and Design of Elevators for People With Disabilities," *Proceedings* of the Second Symposium of the American Society of Mechanical Engineers (ASME) on Elevators, Fire and Accessibility, April 19–21, 1995, Am. Soc. of Mechanical Engineers, New York, NY, 1995, pp. 240–244.

Isman, K. E., "Elevator Dilemma—Solved at Last?" *Sprinkler Quarterly*, No. 88, Fall 1994, p. 10.

Klote, J. H., "Design of Smoke Control Systems for Elevator Fire Evacuation Including Wind Effects," *Proceedings* of the Second Symposium of the American Society of Mechanical Engineers (ASME) on Elevators, Fire and Accessibility, Apr. 19–21, 1995, Am. Soc. of Mechanical Engineers, New York, NY, 1995, pp. 59–77.

Klote, J. H., "Elevators for Fire Evacuation," VHS Video Tape, National Institute of Standards and Technology, Gaithersburg, MD, Video; Federal Fire Forum, Meeting Today's Challenges on Federal Fire Safety for the Disabled, May 15, 1995, Gaithersburg, MD, 1995.

Klote, J. H., "Feasibility of Fire Evacuation by Elevators," VHS Video Tape, National Institute of Standards and Technology, Gaithersburg, MD, Video; Federal Fire Forum, Current Issues in Fire Protection, June 15, 1992, Gaithersburg, MD, 1992.

Klote, J. H., "Method for Calculation of Elevator Evacuation Time," *Journal of Fire Protection Engineering*, Vol. 5, No. 3, 1993, pp. 83–96.

Klote, J. H., "Use of Elevators for Egress," VHS Video Tape, National Institute of Standards and Technology, Gaithersburg, MD, Video; Federal Fire Forum, Providing Fire Safety for Equipment Use in Buildings, [This Forum Will Also be a Part of the BFRL Building Technology Symposia Series.] June 6, 1994, Gaithersburg, MD, 1994.

Klote, J. H., and Alvord, D. M., "Routine for Analysis of the People Movement Time for Elevator Evacuation," National Institute of Standards and Technology, Gaithersburg, MD, General Services Administration, Washington, DC, NISTIR 4730, Feb. 1992,

Klote, J. H., and Fowell, A. J., "Fire Protection Challenges of the Americans Disabilities Act: Elevator Evacuation and Refuge Areas," National Institute of Standards and Technology, Gaithersburg, MD, University of Ulster and Fire Research Station, CIB W14: Fire Safety Engineering, International Symposium and Workshops Engineering Fire Safety in the Process of Design: Demonstrating Equivalency, Part 1, Symposium: Engineering Fire Safety for People With Mixed Abilities, Sept. 13–16, 1993, Newtownabbey, Northern Ireland, 1993, pp. 79–95.

Klote, J. H., and Tamura, G. T., "Design of Elevator Smoke Control Systems for Fire Evacuation," *ASHRAE Transactions*, Vol. 97, No. 2, 1991, pp. 634–642.

Klote, J. H., and Tamura, G. T., "Smoke Control Systems for Elevator Fire Evacuation," National Institute of Standards and Technology, Gaithersburg, MD, National Research Council of Canada, Ottawa, Ontario, American Society of Mechanical Engineers, Council of American Building Officials and National Fire Protection Association, Elevators and Fire, Feb. 19–20, 1991, Baltimore, MD, 1991, pp. 83–94.

Klote, J. H., *et al.*, "Feasibility and Design Considerations of Emergency Evacuation by Elevators," National Institute of Standards and Technology, Gaithersburg, MD, General Services Administration, Washington, DC, NISTIR 4870, Feb. 1992.

Klote, J. H., *et al.*, "Fire Evacuation by Elevators," *Elevator World*, Vol. 41, No. 6, June 1993, pp. 66–70, 72–75.

Klote, J. H., Levin, B. M., and Groner, N. E., "Emergency Elevator Evacuation Systems," *Proceedings* of the Second Symposium of the American Society of Mechanical Engineers (ASME) on Elevators, Fire and Accessibility, April 19–21, 1995, Am. Soc. of Mechanical Engineers, New York, NY, 1995, pp. 131–150.

Klote, J. H., Levin, B. M., and Groner, N. E., "Feasibility of Fire Evacuation by Elevators at FAA Control Towers," National Institute of Standards and Technology, Gaithersburg, MD, George Mason Univ., Fairfax, VA, Department of Transportation, Washington, DC, NISTIR 5445, May 1994.

Koffel, W. E., "Using Elevators as a Means of Egress," *NFPA Journal*, Vol. 90, No. 1, Jan./Feb. 1996, p. 30.

Koffel, W. E., "Sprinklers in Elevator Hoistways and Machine Rooms," Koffel Associates, Inc., Ellicott City, MD, American Society of Mechanical Engineers, Council of American Building Officials and National Fire Protection Association, Elevators and Fire, Feb. 19–20, 1991, Baltimore, MD, 1991, pp. 53–55.

Kose, S., and Hokugo, A., "Elevators and People With Disabilities in Japan: From Accessibility to Egressibility?" *Proceedings* of the Second Symposium of the American Society of Mechanical Engineers (ASME) on Elevators, Fire and Accessibility, Apr. 19–21, 1995, Am. Soc. of Mechanical Engineers, New York, NY, 1995, pp. 151–156.

Levin, B. M., and Groner, N. E., "Human Factors Considerations for the Potential Use of Elevators for Fire Evacuation of FAA Air Traffic Control Towers," George Mason Univ., Fairfax, VA, National Institute of Standards and Technology, Gaithersburg, MD, NIST-GCR-94-656, Aug. 1994, 23 pages.

Levin, B. M., and Groner, N. E., "Some Control and Communication Considerations in Designing an Emergency Elevator Evacuation System," *Proceedings* of the Second Symposium of the American Society of Mechanical Engineers (ASME) on Elevators, Fire and Accessibility, Apr. 19–21, 1995, Am. Soc. of Mechanical Engineers, New York, NY, 1995, pp. 190–193.

Levy, S. J., "Protecting Elevator Systems From Fire Incidents," General Services Admin., Washington, DC, February 20, 1991, 16 pages.

Levy, S. J., "General Services Administration Policy Sprinklers in Elevator Hoistways and Machine Rooms," General Services Admin., Washington, DC, American Society of Mechanical Engineers, Council of American Building Officials and National Fire Protection Association, Elevators and Fire, Feb. 19–20, 1991, Baltimore, MD, 1991, pp. 59–60.

Madison, R. W., "Elevator Operation in a High Temperature Environment," Robert W. Madison and Associates, American Society of Mechanical Engineers, Council of American Building Officials and National Fire Protection Association, Elevators and Fire, Feb. 19–20, 1991, Baltimore, MD, 1991, pp. 32–36.

Marchitto, N., "High Temperature Operation of Elevators," Otis Elevator Co., American Society of Mechanical Engineers, Council of American Building Officials and National Fire Protection Association, Elevators and Fire, Feb. 19–20, 1991, Baltimore, MD, 1991, pp. 25–31.

Marchitto, N., "Operation of Elevators During High Ambient Machine Room Temperature Conditions," *Proceedings* of the Second Symposium of the American Society of Mechanical Engineers (ASME) on Elevators, Fire and Accessibility, Apr. 19–21, 1995, Am. Soc. of Mechanical Engineers, New York, NY, 1995, pp. 93–95.

Matsushita, T., "Smoke Control Calculation for Pressurized Elevator Shaft," Building Research Institute, Ibaraki-ken, Japan, NISTIR 4449; Eleventh Joint Panel Meeting of the U. S./Japan Government Cooperative Program on Natural Resources (UJNR) on Fire Research and Safety, Oct. 19–24, 1989, Berkeley, CA, 1990, pp. 188–192.

McDonald, B., and Rearick, J., "Renovation and Maintenance of Elevators for Use by People With Disabilities," *Proceedings* of the Second Symposium of the American Society of Mechanical Engineers (ASME) on Elevators, Fire and Accessibility, Apr. 19–21, 1995, Am. Soc. of Mechanical Engineers, New York, NY, 1995, pp. 220–223.

Milke, J. A., "Fire ResistanceEvaluation of Building Assemblies Associated With Elevators," *Proceedings* of the Second Symposium of the American Society of Mechanical Engineers (ASME) on Elevators, Fire and

Accessibility, Apr. 19–21, 1995, Am. Soc. of Mechanical Engineers, New York, NY, 1995, pp. 78–84.

Morehart, J., "Selection of Heat Detectors for Elevator Power Disconnection," VHS Video Tape, National Institutes of Health, Bethesda, MD, Video; Federal Fire Forum, Current Issues in Fire Protection, June 15, 1992, Gaithersburg, MD, 1992.

NISTIR 4993, Klote, J. H., *et al.*, "Workshop on Elevator Use During Fires," National Institute of Standards and Technology.

National Building Code, BOCA, Homewood, IL, 1993.

Nelson, H. E., "Areas of Refuge and Elevators," *Proceedings* of the Second Symposium of the American Society of Mechanical Engineers (ASME) on Elevators, Fire and Accessibility, Apr. 19–21, 1995, Am. Soc. of Mechanical Engineers, New York, NY, 1995, pp. 31–58.

Nichols, T., "Detecting Escalator Hazards," *Fire*, Vol. 88, No. 1082, Aug. 1995, p. 18.

Pauls, J., Gatfield, A. J., and Juillet, E., "Elevator Use for Egress: The Human Factors Problems and Prospects," National Research Council of Canada, Ottawa, Ontario, National Task Force on Life Safety and the Handicapped American Society of Mechanical Engineers, Council of American Building Officials and National Fire Protection Association, Elevators and Fire, Feb. 19–20, 1991, Baltimore, MD, 1991, pp. 63–75.

Pedestrian Planning and Design, Elevator World, Inc., Mobile, AL.

Peelle, H. E., Jr., "Accessibility and the Freight Elevator Under A17.1 Rules and ADA Regulations," *Proceedings* of the Second Symposium of the American Society of Mechanical Engineers (ASME) on Elevators, Fire and Accessibility, Apr. 19–21, 1995, Am. Soc. of Mechanical Engineers, New York, NY, 1995, pp. 245–248.

Robibero, V. P., "High Temperature Operation of Elevators," Schindler Elevator Corp., Morristown, NJ, American Society of Mechanical Engineers, Council of American Building Officials and National Fire Protection Association, Elevators and Fire, Feb. 19–20, 1991, Baltimore, MD, 1991, pp. 37–43.

Semple, J. B., "Elevator System Integrity Under Fire Conditions," *Proceedings* of the Second Symposium of the American Society of Mechanical Engineers (ASME) on Elevators, Fire and Accessibility, Apr. 19–21, 1995, Am. Soc. of Mechanical Engineers, New York, NY, 1995, pp. 96–99.

Semple, J. B., "Vertical Exiting: Are Elevators Another Way Out?" *NFPA Journal*, Vol. 87, No. 3, May/June 1993, pp. 49–52.

Standard Building Code, SBCC, Birmingham, AL, 1994.

State Elevator Code Authority Directory, Edward A. Donoghue Associates, Inc., Salem, NY.

Strakosch, G. R., "Fire and Elevators: The Building Environment," *Proceedings* of the Second Symposium of the American Society of Mechanical Engineers (ASME) on Elevators, Fire and Accessibility, Apr. 19–21, 1995, Am. Soc. of Mechanical Engineers, New York, NY, 1995, pp. 23–27.

Sumka, E. H., "Alternate Floor Revall Provisions for Elevators Should Be Maintained," American Society of Mechanical Engineers, Council of American Building Officials and National Fire Protection Association, Elevators and Fire, Feb. 19–20, 1991, Baltimore, MD, 1991, pp. 5–7.

Sumka, E., "Presently, Elevators Are Not Safe in Fire Emergencies," *Elevator World*, Vol. 36, No. 11, 1988, p. 5.

Swerrie, D. A., "Dilemma: Providing Vertical Transportation for the Mobility Impaired Using Normal Operation, or Even Firefighters' Service, Following an Earthquake," *Proceedings* of the Second Symposium of the American Society of Mechanical Engineers (ASME) on Elevators, Fire and Accessibility, Apr. 19–21, 1995, Am. Soc. of Mechanical Engineers, New York, NY, 1995, pp. 104–110.

Tamura, G. T., and Klote, J. H., "Experimental Fire Tower Studies of Elevator Pressurization Systems for Smoke Control," *Elevator World*, Vol. 37, No. 6, 1989, pp. 80–89.

USG Cavity Shaft Wall Systems, United Gypsum Company, Chicago, IL.

Uniform Building Code, ICBO, Whittier, CA, 1994.

Vener, A. S., "Elevators: From Liabilities to Assets to Building Fires," *Journal of Applied Fire Science*, Vol. 2, No. 2, 1992–1993, pp. 157–171.

Vertical Transportation Standards, NEL, Fort Lee, NJ, 1992.

Webb, W. A., "Elevator Fire Safety," Rolf Jensen and Associates, Inc., Deerfield, IL, American Society of Mechanical Engineers, Council of American Building Officials and National Fire Protection Association, Elevators and Fire, Feb. 19–20, 1991, Baltimore, MD, 1991, pp. 56–58.

SYSTEMS APPROACHES TO PROPERTY CLASSES

Sections 2 to 8 presented specific fire safety elements—fire prevention and fire protection—arrayed along the timeline of fire development and building and occupant response. For the most part, these elements were presented in a general form that made them applicable to all occupancies and kinds of properties. But as Section 1 stated, fire safety requires a systems approach, which means that the special characteristics of the specific occupancy or property must be addressed in the approach. Each property within a class of occupancy has its own needs, and each class of occupancy has some needs that are different from the general needs of all occupancies.

Section 9 combines all the chapters that address special requirements of particular occupancy classes or particular areas, processes, or activities of properties within an occupancy class. Chapter 1 introduces this systems approach from the perspective of life safety assessment in traditional buildings. Chapter 2 adds the complexity introduced by placing a usual occupancy within an unusual surrounding, such as locating it underground or on the upper floors of a high-rise building.

Chapters 3 to 8 cover a variety of occupancies including assembly, mercantile, business, educational, detention and correctional, and health care. Each of these chapters defines the occupancy's fire problem. For each occupancy the building's arrangement for functional needs, the characteristics of the occupants, and their activities dictate the needed system of life safety protection.

Chapters 9 to 14 cover each of the residential occupancy classes: board and care, hotels, apartment buildings, lodging or rooming houses, one- and two-family dwellings, and manufactured housing and recreational vehicles. Again, each of these chapters defines the occupancy's fire problem. Although all residential occupancies share some common fire protection problems, each has its own characteristics related to building size and number of persons at risk, occupant familiarity with the building and its egress arrangement, presence of staff and its ability to mitigate a fire's effect, and the recognition and enforceability of fire regulations. The differences justify the treatment of these occupancies in separate chapters.

Chapters 15 to 20 cover storage occupancies by addressing general and specific storage uses. Covered are library and museum collections, warehouse operations, general indoor storage, outdoor storage practices, and materials-handling equipment.

Chapters 21 to 31 cover numerous industrial occupancies, mainly those used for processes for which specialized NFPA standards are available. Included are chapters on industrial occupancies, motion picture studios, food processing facilities, solvent extraction facilities, wastewater treatment plants, laboratories using chemicals, telecommunications equipment facilities, electronic equipment, electric generating plants, nuclear facilities, and mining methods and equipment.

In the last group of chapters, Chapters 32 to 37, fire protection and life safety are addressed for transportation systems. Covered are motor vehicles, alternative fuels for vehicles, fixed guideway transit systems, rail transportation systems, aviation, and marine.

Also look for these: In this section most chapters reference related material in all the previous sections. The *Life Safety Code Handbook* and the *Industrial Fire Hazards Handbook* are essential references for additional detail on fire protection for specific occupancies and industrial processes.

—Ron M. Coté
December 1996

ASSESSING LIFE SAFETY IN BUILDINGS

Revised by John M. Watts, Jr.

Assessment of life safety is the process of estimating the quality of security against fire and its effects. There is no well-defined method of assessing life safety from fire in buildings. Life safety is a concept, and no formula can identify or guarantee that a building is safe from fire.

First of all, assessment requires an understanding of the fundamentals of the life safety concept. Then an evaluation can be made of the parameters that create hazards to life and those that tend to offset or control a portion of the life hazard. A wide variety of tools has been devised to relate these parameters to assess life safety in buildings.

Every general type of fire safety practice and fire protection strategy discussed in this handbook may be applicable to life safety in buildings. Specific additional information that can be helpful in a life safety assessment can be found in Section 1, Chapter 1, "America's Fire Problem and Fire Protection"; Section 1, Chapter 2, "Fundamentals of Fire Safe Building Design"; Section 1, Chapter 3, "Systems Concepts for Building Fire Safety"; Section 4, Chapter 2, "Combustion Products and Their Effects on Life Safety"; Section 8, Chapter 1, "Human Behavior and Fire"; Section 8, Chapter 2, "Concepts of Egress Design"; and Section 10, "Organizing for Fire Protection."

LIFE SAFETY CONCEPTS

Concern for life safety implies avoiding exposure to harmful levels of products of combustion. This goal is usually achieved by controlling the fire process or by separating endangered individuals from the harmful effects of fire by time, distance, or shielding. Thus, life safety is accomplished if endangered individuals can move away from potential danger or if harmful effects of fire can be delayed indefinitely from reaching people in the building.

Details of the fire's development, along with characteristics of exposed occupants, determine the magnitude of the hazard. In addition, specific safety measures may be employed to reduce the hazard. Understanding the interrelationships of these components is the first step in assessing life safety from fire in buildings.

John M. Watts, Jr., Ph.D., is director of the Fire Safety Institute, a not-for-profit information, research, and educational corporation, located in Middlebury, VT. He also serves as editor of NFPA's quarterly technical journal, *Fire Technology*.

The Importance of Time

As a fire develops over time, smoke and heat build up to create an environment that is hazardous to life. The rate at which the environment will deteriorate is difficult to predict, since a great many variables are involved. Figure 9-1A is a generalization of the way life hazard increases over time. At ignition (lower left corner), the environment is normal. Most fires develop slowly, so the initial hazard is small. Eventually, fire intensity will increase more rapidly, building up the level of dangerous products of combustion.

At some level of accumulation of the products of combustion, the fire will be discovered by automatic fire detectors or by personal detection, e.g., smell or sight. The level of hazard at which discovery occurs corresponds, in Figure 9-1A, to a specific time in the course of the fire, the discovery time. The onset of lethal or incapacitating conditions (critical time) typically occurs later, either a short time later or after a period sufficient to permit meaningful actions.

Fire Hazard

The critical time will depend on the interaction of three sets of conditions: (1) elevated environmental temperatures, (2) toxic conditions resulting from pyrolysis or combustion products, and (3) pre-existing or current psycho-physiological attributes of the occupants. Elevated temperatures can cause burns or heat stress ranging from trivial to lethal. Specific chemical pyrolysis or combustion products, decreased oxygen concentration, or particulate smoke can cause

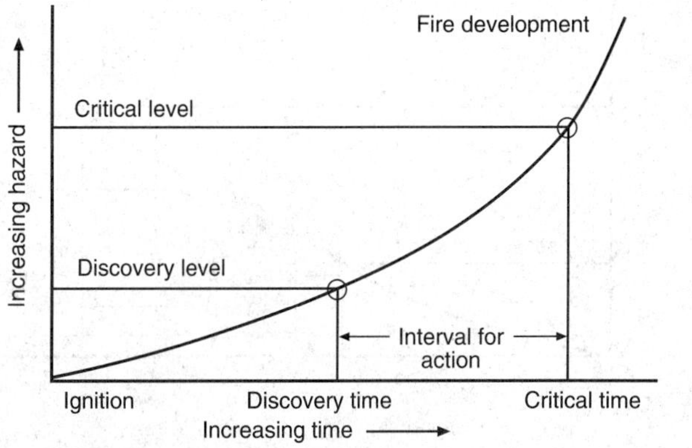

FIG. 9-1A. *Representative rate of deterioration of the environment as a fire progresses.*

toxic effects. The clinical consequences of these physical conditions also will reflect the psycho-physiological conditions of the occupants. For example, pre-existing heart disease can be a contributory factor in fire deaths. Assessment of the variable interactions of these three sets of conditions is, obviously, extremely complicated.

Interval for Action

The interval between discovery and criticality is the time available to undertake action to prevent occupants from being exposed to the critical hazard level. This action may take various forms, such as activation of automatic equipment, evacuation of occupants, or both. As Figure 9-1A shows, lowering the level where discovery occurs will increase the interval of time available for action. On the other hand occupants with greater susceptibility and, hence, a lower personal critical level, will have less time to react.

Fire growth and hazard are not the same in every fire. If conditions are insufficient to sustain a fast-spreading fire, the rate of deterioration of the environment is reduced, resulting in a decreased slope of the fire development curve. For example, curve B in Figure 9-1B represents a slower-developing fire than curve A, with a resultant increase in the interval of time available for action. This could be effected by a decrease in the combustibility of interior finish materials.

Alternatively, curve C might represent the effect of an automatic extinguishing system used alone or in combination with a smoke control system. Activated after fire detection, the system maintains the atmosphere below the critical level for an indefinite period of time.

CHARACTERISTICS OF OCCUPANTS

The most difficult component of life safety to evaluate is the population at risk (i.e., the occupants of the building). The difficulty is due to wide variations among building occupants. It is necessary to assess their susceptibility to fire and fire products and their ability to undertake and follow procedures necessary for their safety (i.e., anticipated response to a fire emergency). Indications of these abilities may be found in physical and mental characteristics of the occupants, as individuals and as a group. Three characteristics of the occupants of a building influence their ability to survive in a fire: (1) reaction, (2) response, and (3) rate of travel.

Reaction

Reaction refers to the ability of a person or persons to become aware of a fire emergency. A major factor is whether occupants are awake and alert (for example, in offices) or asleep (for example, in residential occupancies). In addition, limited awareness may be found among those persons who have taken drugs. The influence of alcohol and narcotics contributes to a significant number of fire deaths.[1,2] Most residential and institutional occupancies contain individuals whose degrees of awareness vary at different times of day. Assessment of life safety in these buildings is a function of how well alerting devices or procedures correspond to the reaction characteristics of the occupants.

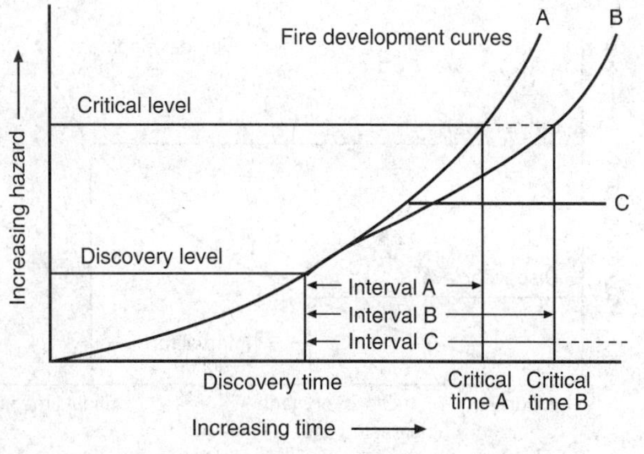

Response

Response is the ability of occupants to receive, comprehend, and follow survival instructions provided during an emergency or in previous training. A major consideration is impaired comprehension due to age or other factors.

Prior knowledge can be a vital component of self-preservation, as can instinct. Training and drills may increase the level of occupants' self-preservation knowledge. This can be extended to include societal self-preservation, whereby certain individuals are trained, or respond spontaneously, to assist others. A less direct aspect of knowledge involves occupants' familiarity with the premises. Regular occupants of a particular building are more likely to know of the location of exits than transient visitors are.

Discipline is characteristic of occupants as a group more than as individuals. In general, people regularly exposed to disciplinary control and training are apt to respond to authority in a fire emergency and less likely to engage in endangering behavior. Control is evident in occupancies such as schools and certain industries.

Rate of Travel

Rate of travel is usually a direct function of mobility, and for most people, mobility depends on age: the very young and very old are less mobile. However, many other classes of persons are immobile or have limited mobility. As access to public facilities increases, disabled persons are likely to be present in more and more locations. Thus, it is necessary to consider the limitations of people with impaired mobility in assessing the life risk. In occupancies such as hospitals, many people are temporarily incapacitated; they can be relocated only with assistance. In other occupancies such as psychiatric institutions or prisons, individuals of average agility may be restrained or incarcerated. Thus, lack of mobility is not necessarily tied to age or even physical capability.

Population density, or the number of persons in a given area, is an important factor in the magnitude of risk and the safe relocation of occupants. The greater the number of people in a given area, the greater the potential loss of life. Studies have shown the relationship of occupant density to speed of movement.[3,4] Phenomena such as formation of arches of people, which obstructs passage through a doorway, are dependent upon occupant density.

EXAMPLE: Chapter 5, "Evacuation Capability Determination for Board and Care Occupancies," of NFPA 101A, *Guide on Alternative Approaches to Life Safety* (hereinafter referred to as NFPA 101A), addresses some characteristics of occupants. The chapter describes a method for determining evacuation capability from a specific type of occupancy. This method calculates an evacuation capability score using information on residents, staff, and vertical egress travel required. Rating of residents assesses several characteristics of occupants that are particularly important to board and care occupancies. Table 9-1A summarizes these characteristics and illustrates that for some types of building use, more detailed information on occupants is necessary.

FIG. 9-1B. Difference in deterioration rates between a faster-developing and a slower-developing fire.

TABLE 9-1A. *Risk Factors for Rating Evacuation Capability of Residents in a Board and Care Occupancy (as addressed in NFPA 101A)*

Risk Factor	Description
Risk of resistance	During an emergency evacuation, the resident might resist leaving the facility
Impaired mobility	Physical ability of the resident to leave the facility
Impaired consciousness	A resident could experience partial or total loss of consciousness in a fire emergency
Response to instructions	Residents' ability to receive, comprehend, and follow simple instructions during a staff-directed evacuation
Waking response to alarm	Fire alarm might fail to awaken the resident

FIG. 9-1C. Simplified portion of the fire safety concepts tree.

Surrogate Descriptors

Occupant age is often used as surrogate descriptor for characteristics that determine life safety risk. Relatively easy to identify, age is often directly associated with characteristics such as mobility, awareness, and knowledge. Statistics showing that the very young and the very old suffer higher fatality rates from fire indicate variation of life risk with age.

NFPA *101®*, *Life Safety Code®*, and model building codes, categorize building use or occupancy classifications to simplify the application of regulations governing construction, fire protection, and other life safety requirements. In application, occupancy class, once determined, becomes the basis for most code requirements. In effect, we use occupancy classification as a surrogate for life safety risk. This generalization is inadequate for more performance-based assessment of life safety in buildings.

LIFE SAFETY STRATEGIES

Ideally, building design considers the risk factors associated with occupants and fire, and it includes safety features to mitigate the risks. The discussion in this chapter is intended as a synthesis of the various safety measures that may be applied to reduce the danger of fire to occupants. Major categories of safety strategies have been identified by NFPA 550, *Guide to the Fire Safety Concepts Tree*, as (1) prevent fire, (2) manage fire, and (3) manage exposed (occupants). A simplified portion of the fire safety concepts tree showing the relationships of these strategies is presented as Figure 9-1C.

Fire Prevention

No harm can come from fire if no fire takes place. Fire prevention, therefore, has the potential to eliminate the need for other fire safety measures. However, no fire prevention strategy is ever totally effective, so it is not prudent to rely on fire prevention alone.

Fire prevention considerations relate to heat energy sources, fuels, and the mechanisms which bring them together. For example, major potential ignition energy sources, such as electrical power, can be controlled largely through regulation, such as by following the requirements of NFPA 70, *National Electrical Code®*. In addition, some fuels, such as flammable liquids and interior finish, also may be addressed by codes and standards. However, not all energy sources nor all fuels can be regulated. This is also true of people, the primary causal factor of ignition.

Assessment of the probability of fire occurrence involves analysis of ignition potential, especially in relation to the activities of building occupants, and of factors that will decrease the likelihood of ignition.

Fire Management

Since it is impossible to avoid all ignitions, managing fires that occur also is important. The strategies of fire management may be seen as reducing the slope of the fire development curves portrayed in Figures 9-1A and 9-1B. These approaches strive to control the rate of smoke and heat production by altering fuel and/or the environment, controlling the combustion process by manual or automatic suppression, and controlling the products of combustion through venting and/or containment.

Objectives of the manage fire strategy are to reduce risks associated with fire growth and to reduce fire and smoke spread. Together, these reductions diminish the impact of a fire on building occupants.

Occupant Management

This life safety strategy is the most complex, dealing with the risk factors of both fire and people. Occupant management involves undertaking emergency action appropriate to the expected fire development and to characteristics of the occupants.

To initiate occupant management action, there first must be detection and alerting activities. Automatic equipment may perform these functions. Actions for managing exposed persons involve evacuation, refuge, or rescue. Evacuation is the most common approach when occupants are alert and mobile. In other cases, areas of refuge from fire and smoke within a building are employed, along with movement assisted by emergency personnel.

LIFE SAFETY ASSESSMENT TOOLS

The components of a generalized approach to assessment of life safety from fire in buildings are shown in Figure 9-1D. The first step in this approach is analysis of the fire hazard and of the safety measures corresponding to fire prevention and fire management. This analysis should yield a concept of the expected fire in the building. The second step is consideration of this expected fire in conjunction with occupant risk factors and management of exposed occupants. The life safety assessment is then evaluated; if found lacking, step

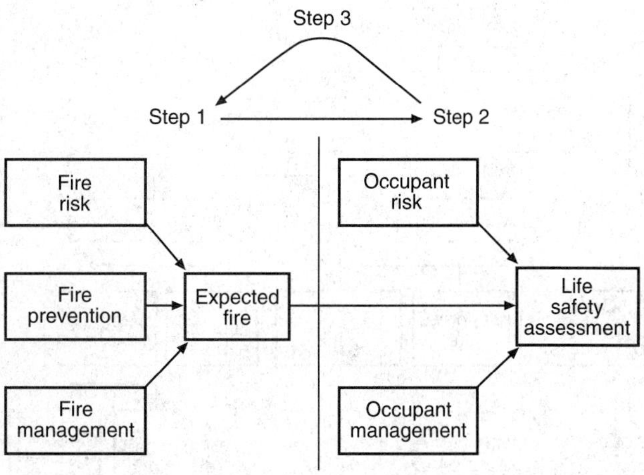

FIG. 9-1D.   General approach to assessing life safety in buildings.

three is a return to the beginning to consider enhanced fire safety strategies. Thus, improving life safety in buildings is an iterative process of matching risk and safety factors.

Specific techniques can be used in an assessment of life safety from fire in buildings including codes, life safety evaluation, exit analysis, fire safety evaluation systems (FSES), fire scenarios, computer simulation, and a performance-based approach.

Codes

NFPA *101*® in particular and building codes in general are the predominant overall guides to life safety from fire in buildings in the United States. Assessments based on codes focus on code compliance as surrogates for a direct assessment of life safety from fire. This approach has important limitations:

1. The stated purpose of NFPA *101* and most building codes is to establish minimum requirements for public welfare. Other tools are required when a higher degree of life safety is desired for altruistic or liability reasons.
2. While NFPA *101* and the building codes contain a provision on equivalency concepts to foster a performance-based approach to fire protection, the codes do not usually specify practical means of establishing equivalent safety to life.
3. In situations where several buildings are under one ownership or authority, it may be of interest to identify the "worst" life safety risks. Some form of numeric rating schedule or other risk-ranking technique is necessary in these cases.
4. NFPA *101* identifies circumstances under which a life safety evaluation is required.

Life Safety Evaluation

NFPA *101* defines a life safety evaluation as a written review dealing with the adequacy of life safety features relative to fire and other perils. A qualified person or persons acceptable to the Authority Having Jurisdiction should conduct the review. The evaluation will include documentation confirming that products of combustion produced by all probable scenarios will not endanger occupants using the means of egress in the facility. Moreover, it should establish that means of egress, together with facility management, will be adequate to cope with scenarios where an egress route is blocked for some reason.

In addition to making realistic assumptions about the capabilities of persons in the facility—for example, an assembled crowd in-

cluding many disabled persons and/or persons unfamiliar with the building—the life safety evaluation should include a factor of safety of at least 2.0 in all calculations relating hazard development time and required egress time. This takes into account the possibility that some of the egress routes may not be usable in an emergency.

Topics to be considered in preparing a life safety evaluation include, but are not limited to, human behavior, exit system, building construction, contents and finishes, fire detection, emergency notification, smoke management, fire suppression systems, and emergency preparedness management.

Exit Analysis

Exit analysis is an important component of a life safety evaluation, as well as a tool for other forms of life safety assessment. The analysis uses anthropometric and biokinetic data on occupants, together with hydraulic flow analogies or other models, to calculate the time required to evacuate a building or space. The result can be compared to a calculated time for deterioration of the environment. Some basic engineering approaches to exit analysis have been developed and are presented elsewhere.[4,6,7,8] It is critical that results of exit analysis calculations, whether manual or computer, be examined with respect to the vagaries of human behavior that accompany an emergency. These concepts are discussed by Bryan in Section 8, Chapter 1, of this handbook and elsewhere.[9,10]

Emergency egress systems should be assessed in terms of adequacy and reliability. Adequacy refers to structural components that determine the capacity to evacuate part or all of a building within a safe egress time span. Reliability considers how efficiently the egress capacity will be utilized. Factors of reliability include alerting messages and instructions, signage and emergency lighting, and protection of egress routes from fire, smoke, and toxic gases. Figure 9-1E, a logic diagram, enumerates some features of egress adequacy and reliability.

FSES

Most of NFPA 101A deals with FSES. The FSES is a fire risk ranking approach to determining equivalencies to NFPA *101* for certain occupancies. The Center for Fire Research, National Bureau of Standards (now the Building and Fire Research Laboratory, National Institute of Standards and Technology) developed the technique in cooperation with the U.S. Department of Health and Human Services in the late 1970s.[11,12] The FSES originally was created to provide a uniform method of evaluating health care facilities and to identify measures that would provide a level of fire safety equivalent to NFPA *101*.

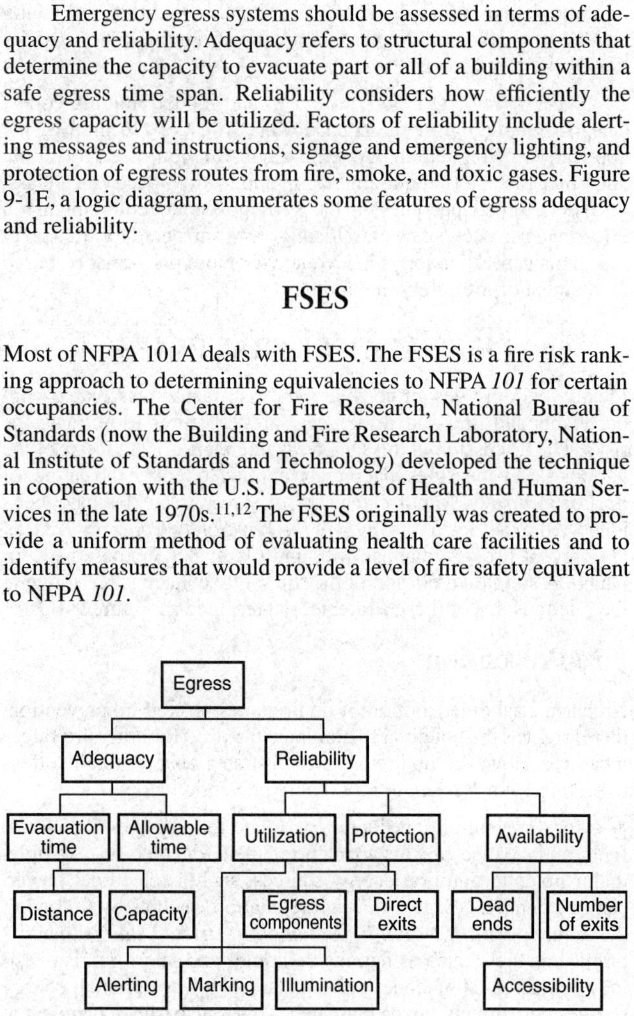

FIG. 9-1E.   Logic diagram of egress features.

Occupant risk: Unlike NFPA *101*, the FSES begins with a determination of relative risk derived from occupant characteristics. Variations of selected characteristics are assigned relative weights, as Figure 9-1F shows, for health care occupancies. Values in this figure were determined from the experienced judgment of a group of fire professionals. Occupancy risk is then calculated as the product of the risk factor values appropriate for the particular occupancy.

Safety parameters: The calculated risk must be offset by safety features. Thirteen possible safety parameters are considered. These parameters and their respective ranges of values for health care occupancies are based on opinions of the same expert panel.

Safety strategies: An important concept of the FSES is redundancy through the simultaneous use of alternative safety strategies. Fire safety strategies considered in the FSES are containment, extinguishment, and people movement. Figure 9-1G indicates the expert panel's opinion of which safety parameters apply to each strategy. Values for a particular occupancy are selected from Figure 9-1F, entered in the appropriate places on Figure 9-1G, and then added for each column. The resulting sums are compared to predetermined minimum values and to the occupancy risk previously calculated using Figure 9-1F.

A method of minimizing the cost of retrofitting for compliance with the FSES for health care occupancies has been adapted for personal computers.[11] This approach is of practical interest to persons with fiscal responsibilities for health care facilities. The linear programming algorithm should also be studied for applicability to other occupancies.

Fire Scenarios

In the theater, a scenario outlines the action of a play. Used in forecasting and political analysis, scenarios are sequences of events that lead to some specified circumstances. Scenarios are particularly suited to examining related conditions (such as building occupancy) and events (such as ignition). A fire scenario is a generalized, detailed description of an actual or hypothetical, but credible, fire incident. Such a scenario identifies a chain of events leading to deaths and other fire losses.

Each fire scenario includes all details relevant to a fire's development and the subsequent behavior of people and protection mechanisms. A properly developed fire scenario describes all essential elements of a fire incident. Scenario development is one of the most useful devices for anticipating a largely uncertain future. Fire protection engineering evaluation of a building or facility involves creating and analyzing fire scenarios. In the simplest form, a fire protection engineer hypothesizes plausible fire scenarios using professional experience and scientific principles of fire dynamics. A more sophisticated form uses computer models to predict fire scenario events.

Fire scenarios are particularly useful in assessing life safety in buildings. By considering various fire scenarios, it often is possible to identify and evaluate the potential for life loss. Plausible fire scenarios are an essential component of some other techniques of life safety assessment, such as the performance-based approach.

Computer Simulation

Computer models for the assessment of life safety in buildings are developing rapidly, concurrent with technical knowledge and hardware

Risk Parameters	Risk Factor Values				
1. Patient Mobility (M)	Mobility Status	Mobile	Limited Mobility	Not Mobile	Not Movable
	Risk Factor	1.0	1.6	3.2	4.5
2. Patient Density (D)	No. of Patients	1-5	6-10	11–30	> 30
	Risk Factor	1.0	1.2	1.5	2.0
3. Zone Location (L)	Floor	1st	2nd or 3rd	4th to 6th	7th and Above	Basements
	Risk Factor	1.1	1.2	1.4	1.6	1.6
4. Ratio of Patients to Attendants (T)	Patients / Attendant	1-2 / 1	3-5 / 1	6-10 / 1	>10 / 1	One or More† / None
	Risk Factor	1.0	1.1	1.2	1.5	4.0
5. Patient Average Age (A)	Age	Under 65 Years and Over 1 Year		65 Years and Over 1 Year and Younger		
	Risk Factor	1.0		1.2		

†A risk factor of 4.0 is charged to any zone that houses patients without any staff in immediate attendance.

(For use with NFPA 101A-1995/NFPA *101*-1994)

FIG. 9-1F. Occupancy risk parameter factors.

Safety Parameters	Containment Safety (S_1)	Extinguishment Safety (S_2)	People Movement Safety (S_3)	General Safety (S_4)
1. Construction			■	
2. Interior Finish (Corr. and Exit)		■		
3. Interior Finish (Rooms)		■	■	
4. Corridor Partitions/Walls		■	■	
5. Doors to Corridor		■		
6. Zone Dimensions	■			
7. Vertical Openings	■	■		
8. Hazardous Areas			■	
9. Smoke Control	■	■		
10. Emergency Movement Routes	■	■		
11. Manual Fire Alarm	■		■	
12. Smoke Detection and Alarm	■			
13. Automatic Sprinklers			÷ 2 =	
Total Value	$S_1 =$	$S_2 =$	$S_3 =$	$S_4 =$

(For use with NFPA 101A-1995/NFPA 101-1994)

FIG. 9-1G. Individual safety evaluations for fire safety evaluation of health care occupancies.

capability. These advances occur internationally in acknowledgment of their potential effect on reducing loss of life from fire.

Simulation is the process of representing item by item and step by step the system's essential features and then predicting what is likely to happen by running the model case by case. A great advantage of computer simulation is that by using computer-generated random numbers, the model can represent with desired precision, processes for which satisfactory analytical approximations do not exist. A system that is not well enough understood to permit mathematical relations between variables to be formalized may often be modeled as a simulation. In studying behavioral systems, the need for simulation becomes acute. When objective data are very difficult to obtain, most predictions must be derived from vague, tentative, and intuitive conjectures. Such ill-defined structures and laborious analyses make computers indispensable tools.

The flexibility of simulation modeling and the complexity of the evacuation problem have resulted in numerous applications. Kostreva et al.[12] identify computer simulation models of building evacuation as having one or more of these components:

1. A network description of the building.
2. A set of heuristics for determining evacuee decisions.
3. A quantity that the evacuees are interested in minimizing.
4. Input from a separate simulation that tracks the fire's progress and smoke propagation.
5. An algorithm.

These models differ in complexity, input requirements, underlying assumptions, specific application areas, and validation. Computer simulations of emergency egress range from simple travel time models to detailed representations of human decision making.

Watts describes a survey of early computer models for evacuation analysis.[13] It includes a discussion and example of EVAC-NET+,[14,15] one of the first evacuation simulation models for a personal computer made publicly available. EVACNET+ is used internationally to evaluate building evacuation plans[16] and to evaluate other egress models.[17]

At the Fourth International Symposium on Fire Safety Science, a significant number of new and developing computer simulation models of building evacuation were presented. These include EVACSIM,[18] SIMULEX,[19] EXODUS,[20] WAYOUT,[21] EXIT89,[22] and EvacSim.[23] Particularly notable is the use of real-time animation as in SIMULEX and the continuing development of EXIT89, which will be publicly available. EXIT89 is being proposed as the evacuation module to extend the use of the software package Hazard I[24] to larger, more complex buildings.

Performance-Based Approach

Most building codes maintain only a tenuous relationship between fire safety requirements and fire safety objectives. For example, the number of exits has an intuitively positive correlation with life safety, but no explicit relationship and no functional association for determining cost-benefit.

Concepts of performance criteria are beginning to focus on fire safety objectives.[25,26] This implies an examination of how all materials, products, components, and assemblies in a building work together as a system to meet fire safety objectives. For example, in determining adequacy of egress routes, it is desirable to consider the overall effect of number, size, length, and reliability of use, as well as the interdependencies among fire growth, fire resistance, and fire suppression.

The performance-based approach is perceived as establishing fire safety objectives and leaving means for achieving objectives to the design professional. Carrying out this significant departure from traditional code practice requires the capability to determine whether fire safety objectives are being met. This requires measurement. Three measurement components need to be considered to properly implement a performance-based approach: (1) an acceptable level of performance, (2) a unit of measure, and (3) appropriate measurement tool(s).

An acceptable level of performance is a performance criterion framed or formulated so that it identifies a measurable level of success in achieving fire safety objectives. It is likely that conservative safety factors need to be included in these criteria until our knowledge enables us to measure fire safety more accurately.

Unit of measure can be an elusive concept. In some situations, time may be an appropriate measure, for example, when the time required to escape is less than the time available to escape. Ultimately we will measure fire safety success in probability terms because fire and its effects are stochastic in nature. However, for the present, the measurement tools available limit our choice of a unit of measure.

Measurement tools are the analytical schedules, equations, formulae, and models, with which we calculate a level of fire safety. In the area of fire resistance of structural members, such tools are sophisticated and reliable. In the area of life safety, tools are still developing. Interrelationships among the areas of fire resistance, life safety, and suppression are complex and often ill defined.

Appendix C of NFPA *101* outlines a performance-based design option for life safety in buildings. Five specific components of this approach are: (1) a performance objective, (2) survivability criteria, (3) applicable fire scenarios, (4) fire models and calculation methods, and (5) assumptions regarding occupants. The objective of additional work on the development of these components is to produce a complete performance-based design option for the next edition of NFPA *101*.

BIBLIOGRAPHY

References Cited

1. Halpin, B. M., *et al.*, "A Fire Fatality Study," *Fire Journal*, Vol. 69, No. 3, 1975.
2. Karter, M. J., Jr., and Miller, A. L., "Patterns of Fire Casualties in Home Fires by Age and Sex, 1983–1987," February 1990, Fire Analysis and Research Division, National Fire Protection Association, Quincy, MA.
3. Peschl, I.A.S.Z., "Passage Capacity of Door Openings in Panic Situations," *Bouw*, Vol. 2, No. 9, January 1971.
4. Pauls, J., "Movement of People," *The SFPE Handbook of Fire Protection Engineering*, 2nd ed., DiNenno, P. J., *et al.*, eds., National Fire Protection Association, Quincy, MA, 1995, pp. 3-263–3-285.
5. Kendik, E., "Designing Escape Routes in Buildings," *Fire Technology*, Vol. 22, No. 4, Nov. 1986, pp. 272–294.
6. Nelson, H. E., and MacLennan, H. A., "Emergency Movement," *The SFPE Handbook of Fire Protection Engineering*, 2nd ed., DiNenno, P. J., *et al.*, eds., National Fire Protection Association, Quincy, MA, 1995, pp. 3-286–3-295.
7. Pauls, J. L., and Fruin, J. J., *People Movement in Buildings and Public Places: Design, Management*, and Safety for Individuals and Crowds, Butterworth, Stoneham, MA, 1991.
8. Bryan, J. L., "Behavioral Response to Fire and Smoke," *The SFPE Handbook of Fire Protection Engineering*, 2nd ed., DiNenno, P. J., *et al.*, eds., National Fire Protection Association, Quincy, MA, 1995, pp. 3-241–3-262.
9. Benjamin, I. A., "A Fire Safety Evaluation System for Health Care Facilities," *Fire Journal*, Vol. 73, No. 2, March 1979.
10. Nelson, H. F., and Shibe, A. J., "A System for Fire Safety Evaluation of Health Care Facilities," NBSIR 78-1555, Nov. 1978, National Bureau of Standards, Washington, DC.
11. Weber, S. F., and Lippiatt, B. C., "ALARM 1.0: Decision Support Software for Cost-Effective Compliance with Fire Safety Codes," NISTIR 5554, Dec. 1994, National Institute of Standards and Technology, Gaithersburg, MD.
12. Kostreva, M. M., Wiecek, M. M., and Getachew, T., "Optimization Models in Fire Egress Analysis for Residential Buildings," *Fire Safety Science—Proceedings of the Third International Symposium*, Elsevier, London, 1992, pp. 308–319.
13. Watts, J. M., Jr., "Computer Models for Evacuation Analysis," *Fire Safety Journal*, Vol. 12, 1987, pp. 237–245.
14. Kisko, T. M., Francis, R. L., and Noble, C. R., *EVACNET+ User's Guide*, Department of Industrial and Systems Engineering, University of Florida, Gainsville, 1984.
15. Fahy, R. F., "Software Review: EVACNET+, A Building Evacuation Computer Program," *Fire Technology*, Vol. 22, No. 1, 1986, pp. 75–76.
16. Johnson, P. F., Beck, V. R., and Horasan, M., "Use of Egress Modelling in Performance-Based Fire Engineering Design—A Fire Safety Study at the National Gallery of Victoria," *Fire Safety Science—Proceedings of the Fourth International Symposium*, IAFSS, Gaithersburg, MD, 1994, pp. 669–680.
17. Donegan, H.A., Pollock, A. J., and Taylor, I. R., "Egress Complexity in a Building," *Fire Safety Science—Proceedings of the Fourth International Symposium*, IAFSS, Gaithersburg, MD, 1994, pp. 601–612.
18. Drager, K. H., Løvås, G. G., and Wiklund, J., "EVACSIM: A Comprehensive Evacuation Simulation Tool," *Proceedings* of the 1992 International Emergency Management and Engineering Conference, The Society for Computer Simulation, 1992, pp. 101–108.
19. Thompson, P. A., and Marchant, E. W., "SIMULEX: Developing New Computer Modelling Techniques for Evacuation," *Fire Safety Science—Proceedings of the Fourth International Symposium*, IAFSS, Gaithersburg, MD, 1994, pp. 613–624.
20. Galea, E. R., Perez Galparsoro, J. M., and Pearce, J., "A Brief Description of the EXODUS Evacuation Model," 18th International Conference on Fire Safety, San Francisco, 1993.
21. Shestopal, V. O., and Grubits, S. J., "Evacuation Model for Merging Traffic Flows in Multi-Room, Multi-Storey Buildings," *Fire Safety Science—Proceedings of the Fourth International Symposium*, IAFSS, Gaithersburg, MD, 1994, pp. 625–632.

22. Fahy, R. F., "EXIT89—An Evacuation Model for High-Rise Buildings—Model Description and Example Applications," *Fire Safety Science—Proceedings of the Fourth International Symposium*, IAFSS, Gaithersburg, MD, 1994, pp. 657–668.

23. Poon, L. S., "EvacSim: A Simulation Model of Occupants with Behavioral Attributes in Emergency Evacuation of High-Rise Building Fires," *Fire Safety Science—Proceedings of the Fourth International Symposium*, IAFSS, Gaithersburg, MD, 1994, pp. 681–692.

24. Peacock, R. D., *et al.*, "Software User's Guide for the HAZARD I Fire Hazard Assessment Method," *NIST Handbook 146*, Vol. 1, National Institute of Standards and Technology, Gaithersburg, MD, July 1991.

25. Richardson, J. K., "Moving Toward Performance-Based Codes," *NFPA Journal*, Vol. 88, No. 3, 1994, pp. 70–78.

26. Meacham, B. J., "International Development and Use of Performance-Based Building Codes and Fire Safety Design Methods," *SFPE Bulletin*, March/April 1995, pp. 7–16.

NFPA Codes, Standards, and Recommended Practices

Reference to the following NFPA codes, standards, and recommended practices will provide further information on manufactured housing and recreational vehicles discussed in this chapter. (See the latest version of *The NFPA Catalog* for availability of current editions of the following documents.)

NFPA 70, *National Electrical Code®*

NFPA *101®*, *Life Safety Code®*

NFPA 101A, *Guide on Alternative Approaches to Life Safety*

NFPA 550, *Guide to the Fire Safety Concepts Tree*

Additional Reading

Clarke, F. B., and Birky, M. M., "Fire Safety in Dwellings and Public Buildings," *Bulletin of the New York Academy of Medicine*, Vol. 57, No. 10, Dec. 1981, pp. 1047–1060.

Committee on Fire Toxicology, *Fire and Smoke: Understanding the Hazards*, National Academy Press, Washington, DC, 1986.

Coté, R., ed., *Life Safety Code Handbook*, 6th ed., National Fire Protection Association, Quincy, MA, 1994.

Lerup, L., *et al.*, *Learning from Fire: A Fire Protection Primer for Architects*, University of California, Berkeley, CA, 1977.

Lyons, J., *Fire*, Scientific American, New York, 1985.

Nelson, H. E., "Concepts for Life Safety Analysis," *Fire Safety Journal*, Vol. 12, No. 3, 1987, pp. 167–178.

Rasbash, D. J., "The Definition and Evaluation of Fire Safety," *Fire Prevention Science and Technology*, No. 16, March 1977.

Shields, T. J., and Silcock, G. W. H., *Buildings and Fire*, Longman, Esses, UK, 1987.

Takahashi, J., Tanaka, T., and Kose, S., "An Evacuation Model for Use in Fire Safety Design of Buildings," *Fire Safety Science—Proceedings* of the Second International Symposium, Hemisphere, NY, 1989, pp. 551–560.

OCCUPANCIES IN SPECIAL STRUCTURES AND HIGH-RISE BUILDINGS

Revised by Carol Caldwell

This chapter is intended to give a brief overview of the relevant issues and features of fire protection for: (1) special structures and high-rise buildings as addressed in NFPA *101*®, *Life Safety Code* ® (hereinafter referred to as NFPA *101*), Chapter 32 of 1997 edition, and (2) traditional occupancies that are located in special or unusual structures.

Special structures, as defined in NFPA *101* include towers, vehicles, vessels, underground structures, windowless structures, water-surrounded structures, open structures, and high-rise buildings. Piers are included in the definition of a special structure but are subject to the specific requirements of NFPA 307, *Standard for the Construction and Fire Protection of Marine Terminals, Piers, and Wharves* (hereinafter referred to as NFPA 307).

Conventional occupancies are often located in special structures that are not normally designed for that occupancy classification. As a result, the terms "unusual occupancy" and "special structure" could both apply to the same situation. Common examples of unusual occupancies in special structures would be:

1. A restaurant located on top of a tower (assembly occupancy).
2. A specialty retail shop located in a fixed railroad car, boat, or airplane (mercantile occupancy).
3. A residential occupancy located in a lighthouse tower.

The structure is the key to Chapter 30 of NFPA *101*. It defines and places specific requirements upon the listed special structures. The code also recognizes that these structures are sometimes used for unusual purposes and, as a result, may require special considerations. It must be remembered that an "unusual use" of a "special structure" still must meet the requirements that would otherwise apply to that occupancy as regulated by NFPA *101*, Chapters 8 through 31, in addition to the "special structure" requirements.

In addition, Chapter 32 of NFPA *101* covers high-rise buildings. Although these structures are not unusual, they are special due to their height and resulting requirements for fire protection.

STRUCTURE CHARACTERISTICS

Occasionally, an owner chooses to locate a place of business in a structure that is not traditionally used for that line of business. The uniqueness of the structure could be used to attract customers and provide a competitive edge. While perhaps giving an owner a competitive edge, the structure can also create difficult life safety problems not found in more traditional structures.

NFPA *101* details provisions for each individual occupancy class that provide the minimum code requirements for those occupancies when they are located in special structures. It is important to note that placing an occupancy in a special structure does not justify reducing the minimum code standards for the given occupancy classification. Designers and code officials must ensure that the structure meets the traditional building code occupancy requirements, as well as the requirements found in NFPA *101*.

Acceptable Design

It is noted that the development of an occupancy in a special structure is limited only by the owner's and designer's imaginations. As a result, it is impossible for a model code such as NFPA *101* to anticipate precisely all of the conditions and hazards that such a design could create. It is incumbent upon building owners, designers, and code officials to include those additional (beyond the code) life safety features that may be deemed necessary due to the uniqueness of the special structure and to the atypical life safety problems that it may create. This can only be done using sound design and engineering judgment, common sense, and cooperation between the designers and code officials.

When it is not possible to follow prescriptive building code and NFPA *101* requirements, alternative designs are necessary. In order to develop an appropriate alternative design, basic information must be agreed upon. The designer, building owner, and Authority Having Jurisdiction need to jointly discuss design philosophy and specific performance requirements. The addition of an automatic sprinkler system can be very beneficial. Issues like the following should be considered:

1. Building geometry and use.
2. Location of adjacent properties.
3. Fuel load and distribution.
4. Number and location of occupants.
5. Proximity and likely response of the fire service.
6. Available water supply.
7. Strategy for achieving performance requirements.[1]

Finally, the proposed alternative design should be reviewed. If the Authority Having Jurisdiction (AHJ) is uncomfortable reviewing and accepting it, there are several options to consider. One is to have another competent fire protection engineer review the design. Perhaps the review cost can be included as part of the building fees.

Carol Caldwell is president of Caldwell Consulting Ltd., a fire protection engineering and research firm in Christchurch, New Zealand.

Or the AHJ could work jointly with another competent fire engineer on the review. The design could be sent elsewhere within the same state for competent review. However it is done, the design, particularly one involving an extensive, complicated submittal, should be thoroughly reviewed prior to acceptance.

Ignition Sources

A special structure may have ignition sources to be considered, in addition to those normal for the occupancy. Once the basic occupancy classification—that is, business, assembly, educational, and so on—has been determined, many of the ignition sources will be typical for that occupancy class. One also must consider the unique features of the structure and determine what additional ignition sources may have to be considered. For example, an assembly occupancy located on top of a tower may be more susceptible to lightning as an ignition source than a traditional structure housing a place of assembly. In this case, the uniqueness of the structure requires extra protection that normally would not be indicated. Utility systems, such as electrical, heating, and cooling, may require special modification to adapt and serve the special structure. This may result in atypical ignition sources.

Fuel Load

Like ignition sources, fuel loads are primarily dependent upon the occupancy classification, as well as the construction classification that is being used in the special structure. A study of the appropriate chapter for the occupancy classification under consideration provides a basic insight into fuel loads that should be anticipated. The fuel load then determines the hazards of contents classification per NFPA *101*. When considering the fuel load of any structure, one must take into consideration that it is normally the contents, or fuel, in the building that create the fire problem. Typically, the building is a "victim" of the fire, not necessarily the cause of the fire. An exception to this rule is combustible construction with concealed combustible spaces or combustible interior finish, which constitute a fuel load.

The contents should be the primary concern when evaluating the fuel. The following items should be taken into consideration in the evaluation:

1. The amount of fuel, measured in lb/sq ft (kg/m^2).
2. Btu output per lb of fuel per sq ft (mJ/m^2).
3. Rate of heat release.
4. Configuration of fuel.
5. Burning characteristics of the fuel.
6. Amount of fuel per individual fuel station.
7. Configuration of fuel stations.
8. Burning characteristics of fuel stations.

In many cases, the actual fuel loading in a special structure exceeds that of traditional occupancies. In addition, the configuration and combustibility of the fuel load may be atypical and hazardous in itself. One example would be the use for aesthetic purposes of a combustible cellular foam that is not protected with adequate thermal barriers. Excessive untreated wood surfaces and mill work on walls and ceilings are other examples.

Once the fuel and its arrangements have been evaluated, the relationship of the fuel load to the unique features of the special structure must be assessed. It must be ensured that the fuel type and loading is at least consistent with the traditional occupancies, or extra protection may be required. The fuel type and loading must not diminish, or have an impact upon, the following:

1. Fundamental requirements in Chapter 2 of NFPA *101*.
2. Egress requirements in Chapter 5 of NFPA *101*.

3. Features of fire protection in Chapter 6 of NFPA *101*.
4. Minimum construction requirements of the occupancy chapters of NFPA *101*.

Construction

Special structures can involve any type of construction classification and often include a combination of two or more construction types. Although a facility may be classified as an occupancy in a special structure, this does not minimize the need to conform to the minimum construction classification requirements as specified for that occupancy classification. In addition, many model building codes impose the minimum construction classification fire ratings, height, and area limitations.

Because of the problems associated with the structure and its occupancy classifications, increasing the fire-resistant characteristics of the construction classification may be desirable to compensate for unique hazards that may be created. An example would be increasing the integrity of the egress systems through fire-resistant enclosures or passageways, or requiring additional fire separations and smoke divisions to allow additional time for safe egress.

Many times, the conceptual aesthetic design of an occupancy located within a special structure will overlook even the most basic elements of good fire protection and prevention. Key components commonly overlooked are:

1. Concealed combustible spaces.
2. Protection of vertical openings.
3. Remoteness of exits.
4. Enclosure of exits.
5. Illumination and marking of exits.
6. Controlled interior finishes.
7. Automatic suppression systems.
8. Smoke management systems.
9. Fire alarm and detection systems.
10. Enclosure of hazardous areas.

SPECIFIC SPECIAL STRUCTURES

Towers

Chapter 32 of NFPA *101* defines a tower as an independent structure or part of a building that is used for signaling or some similar limited purpose. This definition does not include towers used for routine industrial, business, or assembly purposes. When a tower is used for these or other traditional occupancy purposes, it must comply completely with the appropriate occupancy classification and code provisions in NFPA *101* and other applicable codes.

When a tower is being used in accordance with its definition, Chapter 32 of NFPA *101* applies. In this case, the means of egress from a tower is usually a single exit and, in some cases, may be only a ladder. If only one means of egress exists, caution must be taken to ensure that occupants cannot be trapped. The occupancy of the tower is limited in number. A tower should not be used for sleeping or living purposes. Fire-resistant or noncombustible construction should be used, with either Class A or B interior finish.

The amount of combustible materials in, under, or around a tower must be carefully controlled. Combustibles should be totally eliminated except for necessary furniture and operating supplies. Careful judgment should be used in determining what combustible materials to allow in the immediate vicinity of the tower. Similar care should be used in evaluating a tower located near high-hazard occupancies. The tower should be separated by a reasonable distance, especially when it has only a single means of egress.

Towers surrounded by water or offshore structures require special consideration and are strictly controlled by the U.S. Coast Guard. Coast Guard regulations and those of similar regulatory bodies in foreign countries normally take precedence over NFPA *101* requirements.

In small towers where there will be no more than three occupants at any one time, the exit may be a ladder. Ladders should be arranged according to requirements of ANSI A14.3, *Safety Requirements for Fixed Ladders,* and must provide continuous access to grade at all times. In towers with a single exit ladder, vertical openings need not be protected as such a requirement does not significantly increase protection for occupants.

Towers that are occupied on the top floor only often do not need enclosed stairs since the lack of operations at lower levels minimizes fire exposure. Alternatively, fire escape stairs may be used on open structures.

It should be noted that NFPA *101* requirements for towers anticipate a traditional tower used for signaling or some similar limited purpose. Towers that house assembly, educational, business, or some similar occupancy must comply both with traditional occupancy requirements and with special tower requirements.

Windowless Buildings and Underground Structures

The hazards of windowless buildings and underground structures are very similar. (See Figure 9-2A.) Therefore, the same concerns found in underground structures also would be related to the windowless building. The significant difference from the standpoint of NFPA *101* is that underground structures are not required to have smoke venting facilities. Obviously, the feasibility of ventilating the structure should be looked at and related to the classification, nature, and character of the occupancy. In the event that the occupancy is of such a character that a significant occupant load could occur therein, then smoke control and smoke venting capabilities will have a serious impact on the evaluation of the level of life safety.

Typically, a good exit system can be provided in a windowless building. Additionally, this type of structure lends itself to the use of exterior access panels for fire department access and emergency use. It is noted that access panels that do not meet the requirements of the means of egress cannot be considered an exit.

Features of a windowless building that make it a special structure follow:

1. Due to the lack of windows, there are fewer points of entry into the building for manual fire suppression forces.

2. Ventilation during emergencies and removal of products of combustion, such as heat and smoke, depend on mechanical systems and door openings. If there are no mechanical systems designed for ventilation and smoke removal, then emergency operations may be hindered or additional damage/downtime may occur, as compared to buildings with windows.

3. Natural illumination is limited or nonexistent. This issue has an impact upon egress and emergency operations within the building.

4. Windowless buildings may make the fire department's initial size-up more difficult due to the lack of visibility from the exterior. Many times, size-up will not be possible until an initial interior observation can be made.

A common problem with respect to windowless buildings is the determination as to whether the building does, or does not, have windows. It is unusual to find a building that does not have at least a few windows. It is necessary to review the definition of windowless structures and to understand its intent in order to determine whether or not the structure meets the definition of "windowless."

The definition from NFPA *101* is divided into two parts that address two considerations. The first part is, "A building lacking openings to the outside for rescue." This portion of the definition looks at the ability to egress the building, which usually is not a problem unless the designer intentionally deleted exits along the exterior of the building. Exits are often deleted for access control and security purposes. While security and access control are obviously important, they cannot be allowed to interfere with a minimum level of life safety in a building.

The second portion of the definition addresses "Outside openings for ventilation." The problem thus becomes, "How many windows or openings must be provided to not consider a building a windowless structure?" This determination must be based on the judgment of the Authority Having Jurisdiction, but the definition does give some insight into the intent of NFPA *101*. NFPA *101* provides detailed definitions for determining whether a structure is windowless. These definitions address buildings with specific provisions for ventilation and rescue. A review of the definitions should give reasonable guidance to assess the classification.

Judgment must be exercised in determining whether a building is truly windowless, taking into consideration the nature and character of the occupancy and the actual occupancy classification. For example, a windowless building in a storage occupancy would not be as critical as a windowless building in a place of assembly.

FIG. 9-2A.   A typical windowless structure.

Underground structures present additional problems not experienced in most other structures. The two primary issues are accessibility or lack of it, and below-grade fires.

As with windowless buildings, it can be difficult to identify and classify underground structures. Many times, the difficulty occurs when a building or structure is constructed on a site with varying grades. This may result in some levels of the building being below grade on one or more sides and at, or above, grade on other sides. One example is a four-story structure built into the side of a steep hill. Although this arrangement creates a building in which more than 50 percent of the total wall surface is below ground, the building is not considered an underground structure because the levels of exit discharge and adequate accessibility are above grade.

NFPA *101* gives specific guidance in determining when a building is considered an underground structure. The lowest floor used as a level of exit discharge may be considered the demarcation point for above and below grade. This may be true even if much of the wall surface of that floor is below grade. In this specific case, the building levels in question might be considered equivalent to a windowless building *vs.* an underground structure.

From a practical standpoint, a windowless building has fire protection problems similar to an underground structure, but the windowless building can be constructed more easily to meet natural ventilation and rescue accessibility requirements. An underground structure must meet the requirements of a windowless building, and, in addition, some form of smoke management system must be provided. If a windowless building cannot meet the ventilation and rescue requirements, it should be considered equivalent to an underground structure, and the same code requirements should be applied.

Water-Surrounded Structures

This category includes structures fully surrounded by water, including lighthouses and offshore platforms. Occupant loads could vary from a limited number of people in some cases to 200 to 300 persons on offshore oil platforms. A large water-surrounded structure could include areas that contain assembly, industrial, or residential occupancies. Egress from the entire structure is limited because of its location.

Other than the fact that the structure is surrounded by water, traditional occupancy requirements and fire protection features remain the same with respect to the actual occupancy classifications being evaluated. The occupancy should be addressed as if it were not surrounded by water, and the particular circumstances created by the special structure should then be applied. The special structure features should be evaluated as they relate to the occupancy at hand.

Often, the structure is subject to severe environmental conditions, including wind, wave action, and heavy rain. For those reasons, NFPA *101* excludes water-surrounded structures from its provisions and recommends compliance with U.S. Coast Guard regulations.

Piers

NFPA 307 includes a definition of a pier. NFPA 307 deals mostly with property conservation features of piers and wharves. Further, it references the applicable occupancy chapter in NFPA *101* for occupancies built on piers. This includes Chapter 32, "Special Structures and High-Rise Buildings," which addresses piers. This section of NFPA *101* specifically addresses egress and the special circumstances for piers, including the fact that the occupancy still must meet the traditional occupancy chapters of NFPA *101*.

Open Structures

Open structures are most often found in industrial operations, where the structure would come under the guidelines of an industrial occupancy. (See Figure 9-2B.) It is often difficult to determine whether the occupancy should be classified as general, special purpose, or high-hazard industrial. The hazard classification usually is determined by the classification of the hazards of the contents that can be present. Open-air structures are commonly classified under special-purpose industrial as defined in Chapter 28 of NFPA *101*.

Chapter 28 of NFPA *101* makes several provisions for special-purpose industrial occupancies. The special-purpose industrial occupancy is to be characterized by a relatively low density of employee population, where most of an open-air structure is occupied by machinery or equipment. Typically, the open structure facilitates access to the equipment by use of platforms, gratings, stairs, and ladders.

It is sometimes difficult to determine when a structure is open and when it is enclosed. An open structure may include a roof to provide some protection from the elements. Additionally, open structures are provided with partial walls intended to shield the operations from environmental conditions or to segregate the operations. The Authority Having Jurisdiction must determine whether the open-air structure truly meets the definition of an open structure or whether the facility is a building. As a guide, the authority could determine whether the structure would react as an enclosed building in the event of fire.

Within a building, walls acting in conjunction with a roof enclose the combustion process and products of combustion, thereby allowing fire to spread both horizontally and vertically to other areas. The advantage of an open structure is that it allows the products of combustion to vent to the atmosphere, instead of spreading to unaffected areas of the structure. This depends on wind and climatic conditions, in addition to the design of the structure.

If the roof and walls could direct or channel products of combustion to other unaffected portions of the structure, thereby preventing their free venting to the atmosphere, then that portion of the open structure could be considered a building. It is not uncommon to have buildings constructed within the framework of an open structure, resulting in a typical mixed classification.

FIG. 9-2B. A typical open industrial structure. This particular building is a solvent extraction plant.

High-Rise Buildings

Definitions: NFPA *101* defines a high-rise building as "A building more than 75 ft (22.5 m) in height." This definition is consistent with many model building codes, but it should be noted that many different definitions exist in local jurisdictions that use varying height and measurement criteria. These height changes can range from 40 ft (12 m) to as high as 150 ft (45 m). When designing, inspecting, or reviewing a high-rise building it is important to review local codes to determine height and measurement definitions.

Along with the varying heights used to determine the classification of a high-rise, local codes also vary in the method used to measure the height of a building. NFPA *101* states, "Building height shall be measured from the lowest level of fire department vehicle access to the floor of the highest occupiable story."

Much emphasis has been placed on defining the characteristics of a high-rise building. The traditional approach has evolved around fire department accessibility with aerial apparatus, with many code definitions predicated on the length of the responding aerial ladder that may be available. Such an approach has several inherent weaknesses that may not serve either the designer or the servicing fire department. Most fire officers recognize that the actual reach of an aerial apparatus is more important than the length of the ladder. The reach depends upon several factors including the placement of the apparatus and the site/building configuration. Taking these into consideration, many two-story buildings could be considered "highrise" because they exceed the ability of the aerial device to reach occupied floors and portions of the building. Therefore, the figure of

75 ft, or 40 ft, or any other measurement is a compromise that provides a defined number for the designer and code official.

Fire loss history: Only recently have fire statistics been able to indicate a clear picture on high-rise fire incidents. An article in the *NFPA Journal*, March/April 1994, entitled "U.S. High-Rise Fires: The Big Picture,"[2] documented for the first time an overall perspective of the problem. The information that follows is from that article.

High-rise buildings account for a very small number of the total buildings within each occupancy class. However, they account for a large share of the people and property exposed. Table 9-2A provides information on four classes of high-rise occupancies and the incidence of fires, civilian deaths and injuries, and property damage. A substantial portion of property losses in high-rise buildings comes from other than offices, hotels and motels, apartment buildings, and health care facilities. Storage, manufacturing, and industrial facilities can be considered high-rise buildings, on the basis of total structure height, even though industrial high-rise buildings may have a limited number of floors.

Figures suggest that high-rise buildings have a lower risk of fire/ft^2 of floor area. Sprinklers have been shown to be beneficial in preventing deaths from fire and in significantly reducing property losses. Of the four occupancy types in Table 9-2A, sprinklers cut average losses by at least half. Typically, use of several fire protection systems and features is widespread in high-rise buildings (see Table 9-2B), so fatalities are more likely due to special risk factors. Some of these risk factors are drugs, alcohol, and mental and physical disabilities.

TABLE 9-2A.   *High-Rise Building Fire Totals and Percentage of Fires in Buildings of All Heights*
(1987–1991 Structure Fires Reported to U.S. Fire Departments)

Year	Fires		Civilian Deaths		Civilian Injuries		Direct Property Damage (in Millions)	
Office Buildings								
1987	910	(10.5%)	4	(100.0%)	8	(7.4%)	$ 7.0	(6.5%)
1988	1,040	(12.7%)	0	(0.0%)	18	(14.1%)	$ 33.1	(12.2%)
1989	920	(12.5%)	0	(*)	23	(28.1%)	$ 20.8	(18.2%)
1990	940	(13.5%)	4	(50.0%)	8	(11.9%)	$ 15.6	(10.1%)
1991	820	(11.7%)	0	(*)	12	(11.7%)	$ 11.3	(8.5%)
Hotels and Motels								
1987	1,500	(19.8%)	5	(10.9%)	38	(10.5%)	$ 6.1	(9.2%)
1988	1,840	(24.0%)	8	(24.9%)	55	(15.5%)	$ 19.9	(24.4%)
1989	1,550	(21.8%)	5	(22.2%)	60	(20.5%)	$ 4.5	(7.1%)
1990	1,610	(24.2%)	7	(15.5%)	120	(25.5%)	$ 6.0	(9.3%)
1991	1,300	(21.2%)	0	(0.0%)	91	(28.1%)	$ 6.1	(8.8%)
Apartment Buildings								
1987	8,860	(7.7%)	45	(5.1%)	521	(8.6%)	$ 23.4	(3.0%)
1988	10,270	(8.8%)	83	(8.2%)	639	(9.4%)	$ 48.1	(5.4%)
1989	10,990	(9.8%)	97	(9.8%)	611	(9.5%)	$ 30.1	(3.4%)
1990	9,410	(8.8%)	77	(9.4%)	460	(7.0%)	$ 22.4	(2.5%)
1991	9,860	(9.2%)	23	(3.1%)	588	(8.3%)	$128.7	(11.4%)
Facilities That Care for the Sick								
1987	1,710	(32.5%)	0	(0.0%)	72	(43.2%)	$ 1.4	(26.4%)
1988	1,500	(34.8%)	2	(33.3%)	67	(39.5%)	$ 1.1	(9.6%)
1989	1,370	(34.5%)	6	(50.0%)	105	(56.9%)	$ 2.6	(23.0%)
1990	1,300	(37.4%)	3	(50.0%)	37	(25.7%)	$ 4.2	(37.1%)
1991	1,140	(35.7%)	2	(100.0%)	53	(47.4%)	$ 4.0	(31.7%)

*The four fire deaths in office buildings in these years all occurred in buildings of unknown height, so the estimates are based on the share of fires.
   Note: Fires are estimated to the nearest ten and direct property damage to the nearest hundred million dollars. Fires in buildings of unknown height have been proportionally allocated. The percentages shown in parentheses are the high-rise share of all fires in buildings of that type and of known height.

TABLE 9-2B. *High-Rise Buildings vs. All Buildings of Known Height by Presence of Fire Protection Features and Equipment* (Percent of 1987–91 Structure Fires Reported to U.S. Fire Departments)

	Percent of Fires Reporting Presence of		
	Automatic Fire Suppression Equipment	Automatic Fire Detection Equipment	Fire-Resistive Construction
Office Buildings			
High-rise	48.9%	71.7%	53.1%
All known height	21.4%	41.3%	19.4%
Hotels and Motels			
High-rise	63.3%	84.1%	54.1%
All known height	30.0%	71.1%	22.5%
Apartment Buildings			
High-rise	23.7%	74.1%	55.9%
All known height	4.9%	62.9%	11.1%
Facilities that Care for the Sick			
High-rise	62.1%	88.3%	59.5%
All known height	61.1%	85.8%	51.3%

Note: Fires in buildings of unknown height are not included. Fires in which the performance of automatic suppression or detection equipment or the type of construction was unknown have been proportionally allocated. Fire protection equipment is not coded as to type, extent of coverage, operational status, or compliance with applicable codes and standards.

Unique features: It is important to review the major issues that make high-rise buildings different from other buildings. Some of the major fire protection factors in high-rise buildings are as follows:

1. Fire department accessibility, a major issue, takes several forms:
   (a) The limitations of present-day fire apparatus in reaching upper floors from the exterior of the building.
   (b) The height of the fire and the number of personnel required to deliver adequate types and amounts of equipment to the fire area. Due to the possible height of the fire area, equipment transport and force deployment efforts could exact an exhaustive toll on fire-fighting forces before they can even mount a fire attack. This has a significant impact when available fire fighters and equipment are limited. This condition can also occur in a large low-rise building with the same distance problems as a high-rise, although the elevational difference in a high-rise exaggerates the problem.
   (c) Inherent delays in deploying equipment and fire fighters can indirectly affect fire growth, thereby resulting in a fire of greater magnitude by the time manual forces are in place to attack.
   (d) The height and location of a building further restrict the fire department's ability to approach the area of origin from more advantageous directions. As a result, the fire department may only be able to approach a fire area from below. In many cases where the area of origin has no automatic suppression and compartmentalization, the area of origin may not be approachable at all, resulting in a defensive approach that depends upon fuel consumption for control and extinguishment. This approach can easily result in a large or unacceptable loss.

2. Egress and people movement systems within a high-rise building also make the building unique. Due to building height, traditional features of life safety, such as means of egress, are limited. It is possible that stairs may slow evacuation and create queuing times. Much information is available on egress from high-rise buildings.[3]

To mitigate such evacuation issues, other concepts, such as passive and active people movement systems, have evolved. Concepts such as areas of refuge, defend in place, pressurized exits, enclosed/pressurized elevators, increased fire ratings, communication systems, and life support systems all apply when mitigating the egress conditions a high-rise building presents.

3. The high-rise building often has natural forces affecting fire and smoke movement that are not normally significant in lower buildings. Stack effect and the impact of winds can be very significant, and very different, in high-rise buildings.

The stack effect results from the temperature differential between two areas, which creates a pressure differential that results in natural air movements within a building. In a high-rise building, this effect is increased due to the height of the building. Many high-rise buildings have a significant stack effect, capable of moving large volumes of uncontrolled heat and smoke through the building.

No manual fire-fighting techniques are known to counter stack effect or to mitigate its effect during a fire. Stack effect cannot be eliminated because of the temperature differential and building height. As a result, potential stack effect will exist and may vary with climatic conditions. The only way to mitigate the potential of stack effect is to design and construct the building to minimize the effect. A thorough review of available resources on the subject will enhance design and construction. Basic concepts include:
   (a) Airtight compartmentalization from floor to floor and wall to wall. This involves sealing penetrations and exterior building tightness.
   (b) Limiting continuous shaft heights by constructing air-lock vestibules on stairs, elevators, and other vertical shafts.
   (c) Use of vestibules and gasketing on exterior doors at both the bottom and top of a building.
   (d) Eliminating naturally ventilated shafts and floors that could contribute to stack effect during a fire. One of the most commonly encountered situations is elevator shafts with normally open vents. These designs should be active so that the venting will terminate during a fire, thus reducing the impact of stack effect.
   (e) Zone compartmentalization and control of mechanical systems, which could contribute to, or be affected by, stack effect.

Winds can profoundly effect fire and smoke movement within a building. Wind patterns around high-rise buildings have been the subject of study due to their effect upon structural issues, but very little has been studied with regard to fire protection. In high-rise buildings, winds can vary radically in intensity and direction at different floor levels or sections. It is known that wind velocity and direction can be very different at the street level than they are in the upper sections of the structure, and there is no known method of predicting or preliminarily measuring their effects. However some information can be found in reference 4.

The primary mitigating design approach is to make the exterior walls of a building as tight as possible. This approach is highly dependent upon the design criteria, materials chosen, and integrity of workmanship. The result should be that wind direction and velocity have a reduced impact upon air movement within the structure, thereby allowing conventional smoke management systems to overcome the effects of wind to some degree. It is noted that the degree of the smoke management system's ability to overcome the wind effect is based upon many variables, including system design, system power, tightness of the building, wind velocity, and wind direction.

It is reasonable to assume that, if wind is not taken into consideration in the design and construction of the high-rise building and the exterior of the building is not reasonably tight, any fire may be significantly affected by wind velocity and direction. Wind direction can complicate the accessibility and preferred direction of fire attack and can increase fire and smoke impingement upon areas of refuge, possibly exceeding their design criteria.

4. Due to their nature and design, high-rise buildings significantly increase the occupant, equipment, and material load in a given building. Stacking floor upon floor dramatically increases the number of occupants and potential fuel load that could be exposed to a given event when compared to lower-height buildings. This higher density of occupants and fuel load results from stacking many floors on the same building footprint. Fire has a natural tendency to move upward, where additional occupants and fuel are stacked.

5. One factor to be considered is the combination of occupancies often located in a single high-rise building. These occupancies range from residential to business, offices, stores, restaurants, and places of assembly; often several levels of underground parking are located below the high-rise structure. Many designs include within-the-structure atriums, building interconnections, and public transportation systems such as trains and subway service. Mixing these occupancies and structures complicates the requirements for fire protection, necessitating a sophisticated, and sometimes complex, approach to fire safety.

6. The internal utility services for high-rise buildings typically are arranged to provide for the height of the building. Many high-rise buildings have equipment or mechanical levels servicing every ten, or so, floors. These arrangements can result in multiple levels of fire pumps, additional pressure-reducing devices on sprinkler and standpipe systems, and the host of complex problems that go along with such elaborate arrangements.

Even when HVAC systems are zoned and compartmented, they can share common exhaust shafts and fresh-air intake shafts. These shafts and duct runs often penetrate multiple fire zones and require special attention regarding design and protection. Many high-rise buildings integrate smoke management systems into their conventional HVAC systems. Although this is acceptable, and in many cases preferred, it requires special design considerations, including safe and adequate controls, acceptance testing, and ongoing maintenance.

Items 1 through 6 are primary issues that make high-rise buildings different from conventional, lower buildings. As a result, a high-rise building requires different, and additional, design considerations and features when compared to other types of structures. These considerations should not be based only upon the reach of an aerial ladder, but upon the unique features of the particular building and its specific fire protection features.

NFPA *101* includes specific requirements for high-rise buildings, including automatic sprinkler protection, detection, alarm, communications, emergency power systems, and command stations. NFPA *101* also addresses other special features of high-rise buildings, such as atriums and mixed occupancies.

SUMMARY

Special structures and high-rise buildings are unique with regard to their use, arrangement, fire hazards, and fire protection features. One must always differentiate between a special structure as defined and a special structure being used as a non-traditional occupancy classification.

When evaluating a special structure, many traditional concepts of fire safety, fire protection, and sound engineering judgment apply. In addition, the arrangement of the structure and occupancy may require new approaches to remedy life safety and fire protection problems.

BIBLIOGRAPHY

References Cited

1. Buchannan, A. H., *Fire Engineering Design Guide*, 1st ed., Centre for Advanced Engineering, Christchurch, England, 1994, pp. 9–11.
2. Hall, J. R., Jr., "U.S. High-Rise Fires: The Big Picture," *NFPA Journal*, March/April 1994, pp. 47–53.
3. Zicherman, J., *Fire Safety in Tall Buildings*, McGraw-Hill, USA, 1992.
4. Tamura, G. T., *Smoke Movement and Control in High-Rise Buildings*, National Fire Protection Association, Quincy, MA, 1994.

NFPA Codes, Standards, and Recommended Practices

Reference to the following NFPA codes, standards, and recommended practices will provide further information on occupancies in unusual structures. (See the latest version of *The NFPA Catalog* for availability of the current editions of the following documents.)

NFPA 92A, *Recommended Practice for Smoke-Control Systems*
NFPA *101*®, *Life Safety Code*®
NFPA 307, *Standard for the Construction and Fire Protection of Marine Terminals, Piers, and Wharves*

Additional Reading

Appleton, B., "Development of Fire Safety on London Underground," Joint Conference of the Tyne and Wear Metropolitan Fire Department and The Institution of Fire Engineers, Northern Branch on Fire Risk Assessment: Opportunity or Problem? Mar. 26, 1993, pp. 1–11.

Barker, F. H., Jr., "Multi-Purpose Elevators for Persons with Disabilities and Firefighters in High-Rise Office Buildings in the U.S. (with Concepts for Accessibility and Other Buildings and Locations)," *Proceedings* of the Second Symposium on Elevators, Fire and Accessibility, April 19–21, 1995, Am. Soc. of Mechanical Engineers, New York, 1995, pp. 207–219.

Beard, A. N., "Predicting Fire Spread in Tunnels," *Fire Engineers Journal*, Vol. 55, No. 9, Nov. 1995, pp. 13–15.

Beard, A. N., "Prediction of Major Fire Spread in a Tunnel," *Proceedings* of the Second International Conference on Safety in Road and Rail Tunnels, April 3–6, 1995, Granada, Spain, 1995, pp. 337–344.

Beard, A. N., Drysdale, D. D., and Bishop, S. R., "Non-Linear Model of Major Fire Spread in a Tunnel," *Fire Safety Journal*, Vol. 24, No. 4, 1995, pp. 333–357.

Bengtson, S., and Ramachandran, G., "Fire Growth Rates in Underground Facilities," *Proceedings* of the Fourth International Symposium on Fire Safety Science, Intl. Assoc. for Fire Safety Science, Boston, MA, 1994, pp. 1089–1099.

Bettis, R., "Controlling Smoke in Tunnel Fires," *Fire Prevention*, No. 280, June 1995, pp. 19–22.

Beyler, C., and DiNenno, P., "Letters to the Editor. 'Smoke Management for Covered Malls and Atria' by J. A. Milke," *Fire Technology*, Vol. 27, No. 2, May 1991, pp. 160–169.

Blackburn, R. B., "County Mall: A Shopping Center Fire Safety Strategy for the 1990's," *Fire Engineers Journal*, Vol. 53, No. 168, Mar. 1993, pp. 18–27.

Bosley, K., "Escaping From Shopping Precincts," *Fire Research News*, No. 14, Autumn 1992, pp. 2–3.

Brannigan, F. L., "Can High-Rise Columns Fall?" *Fire Engineering*, Vol. 144, No. 11, Nov. 1991, pp. 63–68.

Brubaker, A. F., and Schwanzl, R. J., "Conducting a High-Rise Mock Disaster Drill," *Fire Engineering*, Vol. 143, No. 4, Apr. 1990, pp. 27–28, 30–34.

"Building System Failures Mark Fatal Philadelphia High-Rise Blaze," *Building Official and Code Administration*, Vol. 25, No. 2, Mar./Apr. 1991, pp. 8–9.

Campbell, J. A., and Varley, R. B., "Smoke Control in High-Rise Buildings," *Consulting-Specifying Engineer*, Vol. 17, No. 1, Jan. 1995, pp. 53–54, 56.

Cassacio, E. K., "High-Rise Fire Protection: A Tall Order," *Record*, Vol. 68 No. 4, July/Aug. 1991, pp. 3–11.

Catchpole, L., "Going Underground," *Fire Prevention*, Vol. 265, December 1993, pp. 24–25.

Catchpole, L., "Integrating Systems at Lakeside Shopping Center," *Fire Prevention*, No. 271, July/Aug. 1994, pp. 20–22.

Chapman, E. F., "Guidelines for Strategic Decision Making at High-Rise Fires," *Fire Engineering*, Vol. 148, No. 9, 67–70, Sept. 1995.

"Safety Requirement for Fixed Ladders," ANSI A14.3, 1992, American National Standards Institute, New York.

ASSEMBLY OCCUPANCIES

Revised by John A. Sharry

Assembly occupancies generally can be defined as structures in which groups of people gather for purposes such as deliberation, worship, or entertainment. Examples of assembly occupancies are large meeting rooms, restaurants, nightclubs, dance halls, bars, cocktail lounges, auditoriums with fixed or loose chair seating, libraries, concert halls, theaters, cinemas, multipurpose rooms, exhibition halls, convention centers, sports arenas, field houses, and passenger terminals for air or surface transportation.

Because this type of occupancy is concerned with the safety and hazards of large numbers of people gathered in one place, an established minimum number of occupants must be reached before one of the above examples constitutes an assembly occupancy. Although this minimum can vary, most building codes and NFPA *101*®, *Life Safety Code*® (hereinafter referred to as NFPA *101*), set it at 50. Occupancies with similar uses but occupant loads below 50 are classified as other occupancy categories, often mercantile.

OCCUPANCY CHARACTERISTICS

Inasmuch as an assembly occupancy involves the safety of a large group of people, the density of the occupant population presents the major safety problem. No other structural occupancy experiences occupant loads of such density. It is not uncommon for occupant densities in assembly occupancies to approach one person per 5 sq ft (0.46 m²).

This density produces problems in the physical movement and behavior of the occupants, the capacity of the exits, the maintenance of adequate aisles of proper width, and the method of alerting occupants in case of emergency. Assembly occupancies are usually inhabited by persons who do not use the building frequently and, therefore, are not familiar with exit locations, exit paths, or other safeguards that may be present. In addition, many assembly occupancies, such as theaters, concert halls, nightclubs, lounges, and some restaurants, are used in conditions of near-total darkness.

All these factors—high occupant density, unfamiliarity with the building, and darkness—are common to the many and varied forms of assembly occupancies. Provisions must be made to deal with all of these factors to avoid panic when an emergency occurs.

John A. Sharry is the fire chief at Lawrence Livermore National Laboratory in Livermore, CA.

SPECIAL CONSIDERATIONS

Restaurants and Nightclubs

Both restaurants and nightclubs share common problems, the most prevalent of which is aisles that provide inadequate access to exits. There is a tendency in these assemblies to add tables in an effort to increase profits. While the resultant crowding is simply uncomfortable under normal conditions, it can be disastrous in an emergency.

Proper aisles must be maintained to facilitate exiting. Aisles should be wide enough to accommodate the movement of people, taking into consideration the position of chairs during normal use and during emergencies. Proper aisle widths may actually improve profits by making patrons feel more comfortable and making table service easier and more efficient. Nightclubs, on the other hand, generally do not value ease and efficiency in table service because service is less needed in nightclubs than in restaurants. To improve the profit of their admission fees, nightclubs tend to squeeze as many people into the building as possible, especially when dinner service is not offered. Certain techniques, such as using small, round tables rather than larger, square tables, help achieve maximum patron capacity.

In both restaurants and nightclubs, the decor can be elaborate and thematic. Unless it is properly designed, decor can obstruct or obscure exits, exit markings, emergency lighting, automatic suppression systems, and other safety features.

Methods of alerting patrons to an emergency and methods of notifying the fire department are common problems in these occupancies. Restaurants and nightclubs may be reluctant to notify patrons of an emergency before the last possible minute because it may destroy the "ambiance" of the facility, may result in lost revenue from patrons who have not settled their bills, and may give the establishment a bad reputation.

Reluctance can result in delayed notification and shorten the available evacuation time in case of a real emergency. Restaurants and nightclubs may also be reluctant to notify the fire department for the same reasons, which will result in delayed fire department response.

Fixed Seating

Added to the common problems of adequacy of exits, exit marking, emergency illumination, and notification of patrons in an emergency are special problems related to seating. Occupancies that use fixed seating present the usual, as well as some unique, problems.

Fixed seating arrangements put a large number of people into a confined environment where there is a limited ability to move. People must move within a row before they can reach an aisle, and the aisle then limits their movements in reaching an exit. The design of the seating in these occupancies is critical to ensuring the rapid egress of all occupants in a fire.

Before 1988, fixed seating was classified into two types: (1) normal and (2) continental. Normal seating involves a limited number of seats, usually a maximum of 14 or 15, between aisles; limited length of dead-end rows; and use of mid-aisles and cross-aisles to ensure rapid exiting. (See Figure 9-3A.) The spacing of the seats to provide adequate width between rows is important for ease of movement. In general, 12 in. (305 mm) of clear space for movement between rows is considered the minimum. (See Figure 9-3B.) The aisles must be wide enough to accommodate the influx of patrons as they leave their rows and enter the aisles. The proper aisle width must continue toward the exit. If cross aisles are used, they must be wide enough to accommodate the flow of people from tributary aisles and to permit movement to the exits. Finally, the exits must be sized properly to accommodate the flow of people from the cross aisles and their tributary aisles. Dead-end aisles should be limited in length or avoided altogether.

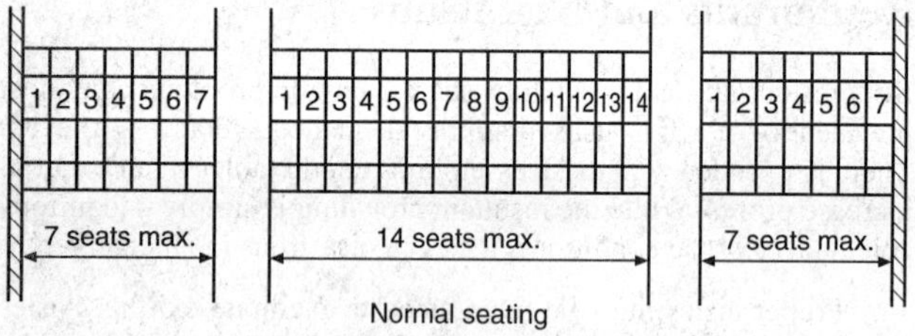

Continental seating is characterized by long rows of chairs discharging into side aisles which, in turn, discharge directly to the exits. Mid- and cross-aisles are not used. (See Figure 9-3C.) Although most codes limit rows to 100 seats each, they are usually functionally limited to fewer seats by the available sight lines. Because the flow of people is less complicated than it is in normal seating, and because row spacing is greater, the rows themselves function as mini-aisles. This results in efficient people movement and superior evacuation times compared with those for normal seating. Continental seating also has the advantage of greater patron comfort and more seats in the prime viewing areas, due to the increased row spacing and longer row length.

In 1988, NFPA *101* was the first model code to adopt new requirements for seating in new construction. These new requirements recognize the advantages of continental seating and apply those concepts to all seating. Rather than dividing seating into two categories, in 1988 NFPA *101* established one set of requirements for all seating. These new requirements can be summarized as follows:

1. All seating arrangements are permitted to extend to 100 seats per row. The old limit on the number of seats between aisles and the limit on the number of seats being served by an aisle on one side only were eliminated.
2. The spacing between rows, now called aisle accessway width, was changed by increasing the aisle spacing above the minimum 12 in. (305 mm). The aisle accessway width served by aisles on both sides is increased by 0.3 in. (8 mm) for each seat over 14 across. The aisle accessway width served by an aisle on one side only is increased by 0.6 in. (15 mm) for each seat over 7 across. The maximum required aisle accessway width is 22 in. (560 mm).
3. Aisle widths are determined by a formula that takes into account the fire protection of the building, especially its ability to protect the occupants against smoke; the geometry of the aisle, that is, either steps or ramps; the riser height of the steps in a stepped

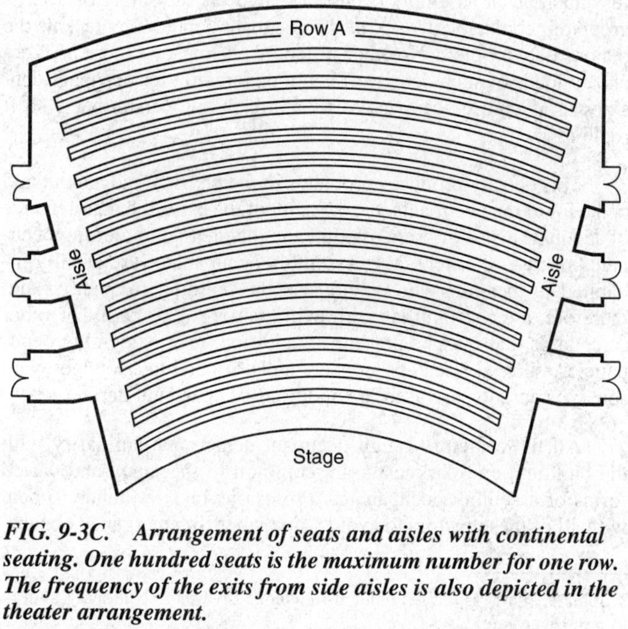

FIG. 9-3A. *Arrangement of seats and aisles in an assembly occupancy with normal seating.*

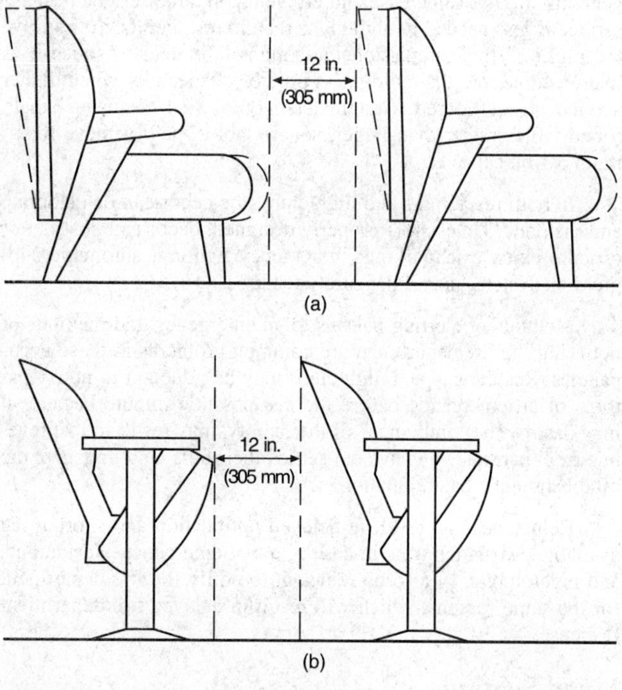

FIG. 9-3B. *Correct measurement of minimum spacing between rows of seats (now called aisle accessway width) without self-rising seats and with self-rising seats.*

FIG. 9-3C. *Arrangement of seats and aisles with continental seating. One hundred seats is the maximum number for one row. The frequency of the exits from side aisles is also depicted in the theater arrangement.*

aisle; the availability of handrails; the slope of the ramp in ramped aisles; and an assumed flow pattern of the occupants.
4. The requirements for cross-aisles and mid-aisles were eliminated.

These changes have resulted in better seating arrangements for life safety, audience site lines, and viewer comfort.

Where loose chairs are used in lieu of fixed seating, care must be taken to keep the chairs from being moved into a haphazard arrangement that could impair exit access. Generally, loose chairs must be arranged similarly to fixed seating to maintain proper aisle accessway and aisle widths. To prevent chair movement during use, most codes require that loose chairs be fastened together in groups too large to permit easy repositioning, which might block egress. Most chair manufacturers produce chairs with interlocking features to fulfill this requirement. These features usually require no tools to join or separate the chairs, which can also be stacked for storage.

Festival Seating

A concept called "festival seating" is popular for certain entertainment events. Festival seating is an audience or spectator accommodation in which no seating, other than the floor or ground surface, is provided for the audience gathered to observe a performance. This type of seating has the potential to result in overcrowding and high audience density that may compromise occupant safety. Also notable is the potential for injury prior to the actual event as patrons rush into the venue seeking the best locations. These types of seating arrangements should be used with caution. NFPA *101* requires the use of a Life Safety Evaluation for these and other unique assemblies. Crowd managers are also required for occupancies with a large number of occupants (over 1,000).

Multipurpose Rooms

Ballrooms, meeting rooms, cafeterias, and gymnasiums are usually multipurpose rooms and may be used for dining, dancing, sporting events, exhibits, meetings, or receptions. In general, local codes contain requirements for each use. Overall, the key to occupant safety is adequate exiting, together with automatic suppression systems if exhibit hall fuel loading is expected.

Where large areas are subdivided by movable walls or partitions, each subdivision must have proper exit accesses and exits, a factor that facility planners frequently overlook. Care must be taken to ensure that proper exiting is designed into the arrangement.

Exhibit Halls

Exhibit halls are large assembly areas that are really multipurpose in nature. Special diligence regarding control of display and chair arrangement must be exercised to ensure unimpeded egress and to maintain the proper travel distances to exits. It is recommended that trade shows or exhibitions be required to submit layouts for displays and seating in advance so that they may be approved before the use date.

Exhibit halls might have excessive quantities of combustible materials unless provisions are made for remotely storing the packing materials needed to transport the displays and for the surplus quantities of literature. The importance of this cannot be overemphasized, since failure to remove this heavy fuel load to proper storage areas can result in large fires that spread rapidly. The McCormick Place fire of January 1967, in Chicago, IL, is an example of the fire potential of this occupancy.

Recent trends in exhibit booth design are toward larger and multilevel booths. Special attention must be paid to the travel distance to the aisle within an exhibit booth or display; it is recommended that this distance be limited to 50 ft (15 m). Multilevel exhibit booths present problems in both egress and fire protection.

This type of booth, as well as large exhibit booths with ceilings, can shield the booth contents from the building sprinkler system. In these cases, consideration should be given to providing sprinkler protection for the booth itself. In all cases, booths should be constructed of noncombustible, limited combustible, or fire-retardant-treated materials. NFPA *101* contains special requirements for exhibit halls.

Because of the fuel load and the large number of occupants, moderate and larger size exhibit halls should be provided with an automatic fire suppression system.

Sports Arenas

Sports arenas and field houses usually have fixed seating for large numbers of spectators, as well as a large area. It is not uncommon for such occupancies to hold more than 10,000 people. Here, occupant density presents the major problem to life safety from fire. Proper aisles and travel distances are difficult to maintain, unless care is taken in the planning stage. On the positive side, these occupancies usually include very large open areas, permitting smoke to be diluted in the early stages of a fire, and smoke control measures can keep these facilities reasonably safe for egress. If they are used for exhibitions, the special considerations for exhibit halls must also be followed.

Passenger Terminals

Passenger terminals require special consideration for fire safety because only general rules can be written for their regulation. Dense populations are localized, especially at departure and arrival gates. NFPA *101* recommends special occupant load factors for airport terminals. However, most terminals have large, open areas with significant fuel loads; this can lead to large losses if a fire is not extinguished in its incipient stage. As in most assembly occupancies, the occupants of passenger terminals are generally unfamiliar with the building, its exit routes, and its protection features. Proper marking of exits and exit routes and maintenance of proper travel distances are particularly important.

Special Amusement Buildings

Special amusement buildings are structures, either temporary, permanent, or mobile, that contain a conveying system or walking space to permit the viewing of a display for the purposes of amusement. These structures are of particular concern because the egress path is not readily apparent due to low lighting levels, visual or audio distractions, and intentionally confounded paths of travel. In addition, the amusement display usually involves a large amount of combustibles. The Haunted Castle fire at the Great Adventure Amusement Park in Jackson Township, NJ, on May 11, 1984, is an example of all of these concerns. Eight teenagers, ranging in age from 15 to 19 years, died in that fire.

Special amusement buildings require unique protection provisions. Consideration should be given to automatic sprinkler protection, special controls that will increase lighting levels and turn off confusing or confounding sounds and visual displays when smoke is detected, the strategic placement of exit signs, and control of interior finish materials. NFPA *101* contains special provisions for special amusement buildings that involve all the above considerations.

OCCUPANCY HAZARDS

Cooking and Open Flames

Cooking is associated with the use of many assembly occupancies. The hazard of cooking in the traditional kitchen can be handled by automatic extinguishing systems and proper vapor removal, as outlined in NFPA 96, *Standard for Ventilation Control and Fire Protection of Commercial Cooking Operations.* The widespread use of "display cooking" presents new problems in vapor removal and in protecting the cooking surfaces and humans in case of fire. However, display cooking on properly designed and fixed equipment can be made reasonably safe with fixed systems, proper aisles, and a trained staff.

The use of open-flame devices in tableside and flambé cooking should be carefully reviewed before it is permitted. Specially designed equipment and a properly trained staff are necessary to ensure safety.

Candles and other open flames require the same degree of awareness as tableside cooking. Several commercial devices are available that make open-flame lighting safe, even in the event of a tipover. Coordination between facility operators and the fire department can help maintain an acceptable level of safety.

Theatrical Stages

Many of the most serious theater fires have begun in the stage area. The 1903 Iroquois Theater fire in Chicago, IL, is probably the most vivid example of a stage fire and its tragic consequences.

Traditional proscenium stages have scenery and lighting "flown" above the stage and scenery on the back and sides of the stage; shops along the back and sides of the stage; and storage, props, trap doors, and stage lifts under the stage floor. The potential for a fire is great with this combination of fuel and potential ignition sources. Since the Iroquois Theater fire, many safeguards have been required to control these hazards, including:

1. Automatic sprinklers at the ceiling level, below the gridiron, in usable spaces under the stage, and in all auxiliary spaces surrounding the stage.
2. Ventilation over the stage, operable manually or automatically by fusible links.
3. An automatically closing fire-resistant curtain to cut off the stage from the audience chamber.

All of these features are necessary with a full theatrical stage.

Modern stages are generally modified versions of the classic theatrical stage. More scenery is now moved horizontally, eliminating the need for the high scene loft or fly space. Modern stages also tend to be variations of the traditional proscenium stage, thrusting the stage farther out into the audience or having it open to the audience on three sides, as in an arena theater—often called "theater in the round."

Whatever the arrangement of scenery on the stage, scenery-handling equipment and allied shops need to be protected to prevent a large stage fire from threatening the audience, either directly or indirectly.

Projection Booths

Projection booths refer to any booth housing equipment for the transmission of light onto a screen, curtain, or stage. The hazard is not with the projection of motion pictures or slides made on flammable film, as some assume, but with the mechanism used to project light. Electric arc, xenon, and other light sources generate hazardous gases, dust, or radiation, and can fail with explosive force under certain circumstances. Where these light sources are used, a booth around the projection equipment is required to protect the audience. Proper ventilation also is required. Incandescent light sources do not produce the same hazard and, therefore, need not have the same

enclosure. The enclosure of projection rooms in modern motion picture theaters is for sound control, not fire protection purposes.

Modern motion pictures and slides use safety film, which does not introduce the hazards of nitrocellulose found in early movies. Therefore, modern projection rooms are not designed for, and cannot safely project, nonsafety film.

Storage

In assembly occupancies, as well as many others, fire and life safety problems often arise from improper storage practices. Storage of tables and chairs in exit ways, storage of scenery so it blocks exits, and storage of supplies in aisles and exit paths are all too common in assembly occupancies. In large measure, this problem is related to improper building design. Little consideration is given to storage needs in the design process. It is essential that space be allocated for proper storage of materials, supplies, props, tables, chairs, and other items associated with assembly occupancies to avoid the creation of unnecessary hazards.

LIFE SAFETY CONCEPTS

Construction

Assembly occupancies have a high occupant density under variable conditions which, to one degree or another, makes alerting and evacuating the occupants more difficult than in other occupancies. Because of this, building construction is an important element in the life safety system.

In general, the building should be constructed to allow the structure to remain intact under fire conditions and not contribute significantly to the fire development for at least the time necessary for evacuation. Obviously, for a one-story building, this is relatively easy to accomplish when occupant loads are moderate in size (under 1,000). But even in a one-story building with proper exits, a large occupant load (over 1,000) means higher degrees of structural protection are necessary to ensure everyone's safety.

As building heights increase, the level of construction and protection must increase accordingly to permit large assembly occupancies at higher stories of a building. NFPA *101* specifies, for both new and existing buildings, sizes of assembly occupancies that can be located in buildings based on construction, height of the assembly occupancy above the level of exit discharge, and fire protection provided.

Alarm Systems

Prompt notification of occupants that there is an emergency cannot be overemphasized. However, the sounding of a fire alarm without accompanying verbal instructions might introduce the threat of panic. To overcome this possibility, an appropriate alarm system—perhaps one using prerecorded or electronically synthesized voice messages—can be installed.

Panic Hardware or Fire Exit Hardware

Because of the threat of a large group of occupants rushing to an exit, leaning together on the door, and making the use of traditional knob-activated latching devices difficult or impossible, assembly occupancies usually require panic hardware or fire exit hardware. Panic hardware and fire exit hardware consist of a door-latching assembly incorporating a wide bar-like or push pad-like device that releases the latch upon application of force in the direction of egress travel. Some security-conscious owners in the past have objected to panic hardware or fire exit hardware because of the potential for slipping the latch from the outside for purposes of gaining unauthorized entry. However, modern panic hardware designs provide for both life, safety and security.

There is no excuse for chained or padlocked doors, given today's availability of the right kind of exit hardware.

Interior Finish

Interior finish is a key element to provide for fire control and life safety. Materials used in an assembly occupancy must limit the rate of flame spread and smoke production. Generally, materials are limited to flame spreads of 75 or less (Class A or B) for all but the smallest assembly uses.

Drapes, hangings, tapestries, and other decorative materials must be fire retardant if they constitute a significant part of the decor or if they would provide a continuous path for fire travel. Inadvertent ignition of such materials has been responsible for panic in densely occupied spaces, such as the Cocoanut Grove nightclub fire in Boston, MA, in 1942. Proper treatment to make materials fire retardant is necessary to avoid disaster.

Employee Training

Employees of assembly occupancies are critically important to the life safety of patrons. If the staff is trained properly, they can prevent serious injury or death to many patrons. The Beverly Hills Supper Club fire in 1977 at Southgate, KY, showed the importance of staff assistance in fire survival. Although the building lacked proper exits, staff members saved hundreds of lives by leading patrons to means of escape unknown to the patrons.

It is recommended that management of every assembly occupancy devise an emergency plan in conjunction with the local fire department, deal with the many contingencies associated with fires, and properly train staff members in safe emergency procedures.

Assembly occupancies can be safely occupied when the following conditions are avoided:

1. Overcrowding.
2. Blocked or impaired exits or means of exit access.
3. Chained or locked exits.
4. Storage of combustibles in improper locations.
5. Improper use or control of open flames.
6. Disregard for fire characteristics of materials and decorations.

BIBLIOGRAPHY

NFPA Codes, Standards, and Recommended Practices

Reference to the following NFPA codes, standards, and recommended practices will provide further information on assembly occupancies discussed in this chapter. (See the latest version of *The NFPA Catalog* for availability of current editions of the following documents.)

NFPA 96, *Standard for the Ventilation Control and Fire Protection of Commercial Cooking Operations*

NFPA 101®, *Life Safety Code*®

NFPA 101A, *Guide on Alternative Approaches to Life Safety*

Additional Reading

"Blaze at Theatre Royal Devastates Drury Lane," *Fire Prevention*, No. 262, September 1993, pp. 46–47.

Bukowski, R. W., and Spetzler, R. C., "Analysis of the Happyland Social Club Fire With HAZARD I," *Fire and Arson Investigator*, Vol. 42, No. 2, March 1992, pp. 36–47; *Journal of Fire Protection Engineering*, Vol. 4, No. 4, 1992, pp. 117–131.

Choppen, B., "Voice of Alarm in Public Areas," *Fire Prevention*, No. 266, Jan./Feb. 1994, pp. 28, 30.

Chow, W. K., "Use of the ARGOS Fire Risk Analysis Model for Studying Chinese Restaurant Fires," *Fire and Materials*, Vol. 19, No. 4, July/Aug. 1995, pp. 171–178.

Chow, W. K., "Zone Model Simulation of Fires in Chinese Restaurants in Hong Kong," *Journal of Fire Sciences*, Vol. 13, No. 3, May/June 1995, pp. 235–253.

Chubb, M., "Indianapolis Athletic Club Fire, Indianapolis, Indiana, (February 5, 1992)," USFA Fire Investigation Technical Report Series, TriData Corp., Arlington, VA, Federal Emergency Management Agency, Washington, DC, Report 063, 1992.

Crossland, J. W., "Fire Precautions in Public Houses and Restaurants," *Fire Surveyor*, Vol. 18, No. 6, Dec. 1989, pp. 9–16.

Edwards, J., "Theatre Fires—The Importance of Smoke Ventilation," *Fire Prevention*, No. 198, April 1987, pp. 37–38.

Ferris, M., "Brooklyn Theatre Fire," *Firehouse*, Vol. 16, No. 2, Dec. 1991, pp. 86–88.

Fesman, G. and Jacobs, B. A., "Regulatory Flammability Testing for Public Assembly Furniture," Akzo Chemicals Inc., Dobbs Ferry, New York, Fire Retardant Chemicals Association, International Conference on Fire Safety—Fire Retardant Technology and Marketing, March 25–28, 1990, New Orleans, LA, Fire Retardant Chemicals Assoc., Lancaster, PA, 1990, pp. 213–221.

"Fires in Places of Public Assembly," *Fire International*, Vol. 10, No. 100, 1986, pp. 25-27, 30-31, 34-36.

Grosso, J., "Restaurant Grease Duct Fires," *Fire Engineering*, Vol. 148, No. 6, June 1995, pp. 16, 21.

Harvey, C. S., "Polo Club Condominium Fire, Denver, Colorado," *Journal of Applied Fire Science*, Vol. 2, No. 1, 1992/1993, pp. 85–95.

Hodge, J., "Managing Stadiums More Effectively," *Fire Prevention*, No. 272, Sept. 1994, pp. 31–33.

Hough, E., "Gatwick Airport: 'A Shopping Mall With Aircraft'?" *Fire*, Vol. 87, No. 1076, Mar. 1995, pp. 23–24.

Isner, M. S., "3 Die, 8 Injured in Athletic Club Fire," *NFPA Journal*, Vol. 86, No. 6, Nov./Dec. 1992, pp. 31–36, 38.

Isner, M. S., "Private Club Fire, Indianapolis, Indiana, February 5, 1992," Fire Investigation Report, National Fire Protection Association, Quincy, MA, 1992.

Isner, M. S., "Stadium Fires Demonstrate Unique Protection Problems," *NFPA Journal*, Vol. 88, No. 4, July/Aug. 1994, pp. 49–54.

Isner, M. S., "Two Stadium Fires, Atlanta, Georgia, July 20, 1993 and Irving, Texas, October 13, 1993," Fire Investigation Report, National Fire Protection Association, Quincy, MA, 1993.

Kasahara, I., and Hara, T., "Life Safety Design of an Atrium and an Air Dome With an Applied Zone Model," Taisei Corp., Tokyo, Japan, CIB W14/96/1 (J); Building Research Institute, Mini-Symposium on Fire Safety Design of Buildings and Fire Safety Engineering, June 12, 1995, Tsukuba, Japan, 1995, pp. VIa/1–3.

Klem, T. J., "56 Die in English Stadium Fire," *Fire Journal*, Vol. 80, No. 3, May/June 1986, p. 128.

Kodaira, A. and Uehara, S., "Fire Safety Design for a Large-Scale Retractable Dome," Takenaka Corp., Chiba, Japan, CIB W14/96/1 (J); Building Research Institute, Mini-Symposium on Fire Safety Design of Buildings and Fire Safety Engineering, June 12, 1995, Tsukuba, Japan, 1995, pp. VIe/1–4.

McGrail, D. M., "Denver's Polo Club Condo Fire: Atrium Turns High-Rise Chimney," *Fire Engineering*, Vol. 145, No. 3, Mar. 1992, pp. 66–71, 74.

Meland, O. and Lintorp, S., "Fire Safety and Escape Strategies—Rock Cavern Stadium," Sixth International Fire Conference on Fire Safety, Interflam '93, March 30–April 1, 1993, Oxford, England, Interscience Communications Ltd., London, England, 1993, pp. 171–180.

"New York—Bronx Social Club Arson Claims 87 Lives. On the Job," *Firehouse*, Vol. 15, No. 6, June 1990, pp. 46–48.

O'Rourke, J. J., "Woodstock '94: Fire Planning for Large Public Events. Part 1: Fire Safety Preparations," *Fire Engineering*, Vol. 148, No. 1, January 1995, pp. 74–78.

Orr, R., "Evacuating Hospital Theatres," *Fire Prevention*, No. 280, June 1995, pp. 28–29.

Pauls, J., "Movement of People," *The SFPE Handbook of Fire Protection Engineering*, National Fire Protection Association, Quincy, MA, 1988, pp. 246-268.

Rockett, J. A., "Modeling Fire Safety in a Stadium Environment," Fire Modeling and Analysis, Washington, DC, Society of Fire Protection Engineers and Worcester Polytechnic Institute, *Proceedings* on Computer Applications in Fire Protection, June 28–29, 1993, Worcester, MA, 1993, pp. 7–13.

Rockett, J. A., Howe, J. M., and Hanbury, W. L., "Modeling Fire Safety in Multi-Use, Domed Stadia," *Journal of Fire Protection Engineering*, Vol. 6, No. 1, 1994, pp. 11–22.

"Social Clubs—New York's Happy Land Fire Drew Attention to Code Enforcement Efforts in Social Clubs," *Building Official and Code Administrator*, Vol. 24, No. 3, May/June 1990, pp. 28–33.

Stevens, J. B., "Indiana: Flashovers During Indianapolis Athletic Club Fire Kill 3 On the Job," *Firehouse*, Vol. 17, No. 4, Apr. 1992, pp. 30–32, 34.

Taylor, K. T., "What's Cooking in Restaurants. . . Besides the Food," *Fire Journal*, Vol. 83, No. 5, Sept./Oct. 1989, p. 72.

Yorke, T., *et al.*, "New Goals for Stadiums," *Building*, Vol. 255, No. 7640, Mar. 23, 1990, pp. 47–52.

MERCANTILE OCCUPANCIES

Ed Schultz

This chapter addresses the development of mercantile occupancies from small stores operated by families to large shopping malls and multi-use projects being developed today. It also describes changes in building codes in response to the increased complexities of mega-projects.

Because in many instances a mercantile use may be an ancillary or accessory use in a building with a different occupancy classification, this chapter covers only those facilities whose primary occupancy classification is mercantile.

MERCANTILE OCCUPANCY CLASSIFICATIONS

Mercantile occupancies are facilities in which a wide variety of goods and services are displayed and sold. Examples of such goods and services include clothing and footwear; jewelry; beauty, health, and fitness products; home furnishings; luggage, books, cards, and stationery; music, electronic equipment, and photography supplies; sport specialties; toys, hobbies, and pets; drugs, varieties, and tobacco; food and food specialties; and hardware.

Many items displayed in mercantile occupancies are combustible and spread over the available floor where customers circulate to view and purchase the items. This places many occupants in an unfamiliar environment that is conducive to rapid fire development after a fire begins. This condition raises various levels of concern because the size of the mercantile occupancy and the number of occupants vary greatly. Therefore, it is important to categorize mercantile stores by size.

Small Mercantile Occupancies

In the past, smaller shops were constructed principally in cities and towns, along business thoroughfares, next to each other, often with basements for stock, and sometimes with an office or dwelling unit immediately above them. Many such stores still exist today, especially in older business sections of cities; they are sometimes called "ma and pa" stores because their proprietors often live in the dwelling units directly above them. However, this type of store is slowly fading out of existence.

Because of more restrictive zoning regulations, larger stores are now constructed without other occupancies above them, unless

the mercantile occupancy serves an ancillary or accessory use such as a retail shop in a hotel lobby. In addition, basement space, even for stock space, is being phased out because stock space on the same level as the sales area is more efficient and functional.

Originally, these stores were constructed primarily of wood- or masonry-bearing walls with wood joists, floors, and roofs. Subsequent establishment of fire districts in urban areas, plus the adoption of more stringent building code requirements, led to construction with noncombustible materials, such as masonry walls, load bearing or nonload bearing; steel joists; concrete floors; and metal roof decks.

Until recently, these small stores were seldom sprinklered. NFPA *101*®, *Life Safety Code*®, currently requires sprinkler protection based on height and area and, in certain instances, whether the stores are located below the level of exit discharge.

Medium Mercantile Occupancies

Medium size mercantile occupancy stores, such as supermarkets and discount drug stores, did not become numerous until the mid to late 1940s. In many areas, they replaced the neighborhood grocer and the corner druggist, who could not compete with high volume chains.

Medium mercantile occupancy stores generally range in size from 10,000 to 40,000 sq ft (2,800 to 3,700 m²), are usually one story high with mezzanine management offices, and are usually required to be of noncombustible unprotected construction. Until the mid-1970s sprinklers were not required in these stores.

These stores are generally free-standing buildings, unless they are part of a shopping center complex, and they may have relatively large contiguous parking areas surrounding them.

Medium size mercantile occupancy stores usually have at least two principal areas: (1) the stock space or "back of the house" employee area and (2) the sales space or customer area. Approximately one-quarter to one-third of their area is apportioned to stock and two-thirds to three-quarters to sales. These stores typically include non-mercantile accessory uses, such as management offices, that occupy a small area in relation to sales and stock space. From a retailer's viewpoint, the distinction between retail stock space and storage space is important. Employees enter the retail stock areas frequently during the day to bring merchandise onto the sales floor. Storage areas, on the other hand, are less frequently accessed because goods are set aside in these areas for extended time periods. Today, if one

---

Ed Schultz is president of Code Consultants, Incorporated, St. Louis, MO.

corporation operates many stores, its storage areas are generally in a separate location, essentially a distribution warehouse.

Large Mercantile Occupancies

The multiple category merchandise store, such as a department store, is as common as other types of mercantile occupancies. Generally from one to four stories high, its area normally exceeds 40,000 sq ft.

Earlier department stores were constructed exclusively in business and retail districts in cities. Stores were typically either of exterior masonry-bearing wall construction with protected wood joists and floors, or of reinforced concrete. Many had no sprinkler protection. Today, most department stores are of protected and unprotected noncombustible construction and fully sprinklered.

As their name implies, department stores are divided into departments or sections for different kinds of merchandise, with or without partition walls. Departments are laid out in a particular way to influence customer movement patterns vertically and horizontally through the store according to a preconceived merchandising plan. Vertical movement in multistory stores is most effectively accomplished using escalators and stairs in open wells in highly visible and accessible areas.

At one time basements were quite commonly used as either sales or stock areas. This practice generally is no longer followed because basement space is not as desirable as the same area constructed at or above ground level.

Unlike the vast majority of small and medium mercantile occupancies where stock space is located in the rear of the store, stock space in large mercantile occupancies is often distributed throughout the store in many small stockrooms. These stockrooms often represent 10 to 15 percent of the total floor area, with the remaining area used for retail sales.

Another example of a large mercantile store is the discount store, which grew out of the well-known "five-and-dime" variety store. It is usually one story high and laid out like a retail food supermarket, with shopping carts, self-service, and checkout counters. However, the same sectioning of merchandise occurs as in a department store.

Accessory Occupancies

To round out the full range of goods and services a customer needs and expects, mercantile occupancies sometimes include accessory or ancillary occupancies, in addition to administrative office areas. Examples are restaurants, beauty shops, travel agencies, and optometry services.

OCCUPANT LOAD

Occupant load varies with the seasons, generally being higher before school begins; around Thanksgiving, Christmas, and Easter; and during occasional clearance sales. NFPA *101* takes these primary merchandising seasons into account in establishing the occupant load factor for calculating egress requirements. Many model codes, including NFPA *101,* require that occupant load for mercantile occupancies be calculated on the basis of a factor of one occupant/30 sq ft (2.8 m^2) of street floor or basement sales areas, and a factor of one occupant/60 sq ft (5.6 m^2) of other upper sales floors. The occupant load for stock or storage space that is not open to the public and other accessory occupancies is calculated using different factors, depending upon the particular use.

FIRE LOAD/CONTENTS

As in most other occupancies, contents represent the major potential hazard. Sprinklers have proven effective as a means of combating fires in mercantile occupancies. In terms of life safety, NFPA *101* addresses the issue of cost-benefit in installing sprinklers. Its requirements for sprinklers, based on height and area limits, let many small shops be unsprinklered (unless they are accessory uses to an occupancy requiring sprinklers), while requiring medium and large mercantile occupancies to be protected with automatic sprinklers.

COVERED MALL SHOPPING CENTERS

A covered mall is commonly defined as a covered or roofed interior area used as a pedestrian way, which connects buildings or portions of a building housing single and/or multiple tenants. Covered mall shopping centers became very popular in a relatively short time period, and today most large shopping centers either planned or under construction are of this type. In addition, many covered malls are now being built in urban areas as multi-use projects in conjunction with parking decks, office buildings, hotels, recreational facilities, and public or private transit systems.

The covered mall may be considered as either (1) a way of exit access from the connected buildings or (2) a pedestrian way permitting an increase in the distance of travel from each of the tenant stores to a mall exit.

The first instance treats the entire complex as one building or one compartmented retail facility; the covered mall is simply an extension of the sales space. In terms of exit access, the covered mall is an aisle common to the various tenant stores that extends (continues) exit access aisles of tenant stores to mall exits. This treatment, in terms of NFPA *101,* leads to the strict application of all design factors as though the shopping center were a large single mercantile occupancy or department store.

The second instance employs a systems approach. The mall is considered to offer a higher degree of safety than a typical aisle serving as an exit access if certain minimum conditions and systems are provided. The covered mall should be of at least sufficient clear width to accommodate egress requirements for all mercantile tenants, but in no case less than 20 ft (6 m) wide at its narrowest dimension. The minimum width dimension of 20 ft (6 m) is particularly significant in terms of protecting occupants from direct exposure to a fire within a tenant store while those occupants utilize the exit access paths within the covered mall. It should not be construed as a requirement to provide minimum separation between storefronts.

Design of Covered Mall Shopping Centers

The covered mall shopping center is generally designed as one building connecting several sizeable department stores and composed of numerous smaller specialty stores (mall tenant shops), all interconnected by a covered, climate-controlled, public pedestrian way. The complex may also include ancillary occupancies, such as movie theaters, bowling lanes, ice arenas, project management offices, and other customer service areas accessed from the mall.

The mall may be designed as a single- or multilevel structure, depending upon such factors as topography, size of the land parcel, and number and size of department stores and retail shops.

When a mall has two levels, the parking areas surrounding it usually are shaped to provide grade access to each level of the department stores and the small shops. This permits each level to serve as an entrance/exit, thus providing stores on all levels increased and equalized customer traffic as customers enter and leave the building. The two-level shopping center differs from the two-story shopping center in this important respect. (See Figure 9-4A.) When department stores are part of a two-level covered mall shopping center, they are designed and merchandised in a similar manner.

Building codes have recognized the construction of covered mall buildings up to a maximum of three stories high with floor openings within the structures complying with the code mall provisions and not being considered atriums. This is based on special code provisions that address covered mall buildings.

Exiting concept: The covered mall proper is designed primarily to meet requirements for occupant flow and egress. Storefronts along the public pedestrian way (the covered mall) must be designed to encourage unrestricted customer flow and traffic control from the mall to individual shops and to any connected department stores. This system of customer traffic flow and traffic control reflects the merchandising concept of covered mall shopping centers and the exiting concept of the codes. The exit concept is that the mall is an inherently safe space with controlled uses, adequate width, smoke venting, and sprinkler protection. This then allows egress capacity design to be based on the assumption that 100 percent of the occupants in the covered mall buildings will exit via the mall proper. This allows the back exit system for covered mall shops to be sized for the most restrictive of:

1. The code-stated minimum dimension.
2. The exit width required for the largest single tenant occupant flow exiting into the back exit system.
3. The exit width required for occupants exiting from the mall, if the exit passageway also serves as a mall exit.

This is the case even if the mall uses the same exit passageway that is used for a second exit by several mall tenants shops. Again this concept is based on the fact that occupants tend to exit the way they entered a store, via the covered mall itself. Therefore, the back exit system is truly a secondary exit system, and the mall is the primary egress path for covered mall building occupants.

Open storefronts between tenant mall shops are critical because swinging doors, whether glass or solid, not only make exiting less easy but also promote what the retail industry calls "threshold resistance"—that is, customers' reluctance to push, pull, or slide doors to enter. The storefront closures ideally suited to flow requirements and maintaining security are those that can be fully recessed, such as rolling overhead grilles, side-coiling grilles, and horizontal sliding doors. Because the ambient temperatures in the mall and the

tenant space is the same, the storefront closure needs only satisfy a security requirement when the store is unoccupied.

Occupant allowance requirements: Covered mall shopping centers are constructed before tenants are known. When construction begins, the ultimate ratio of sales space to storage space for tenant stores and what changes may occur in the life of the building are also unknown. Hence, occupant load must be projected to determine the required egress width well before negotiations with prospective tenants.

Egress width from the covered mall need only satisfy the calculated occupant load based on all connected tenant occupancies. It is not necessary to provide egress width from the covered mall commensurate with the minimum mall width. Doing so would necessarily result in wall-to-wall doors.

The occupant load factor used to determine egress width for the mall proper varies with the aggregate area of the connected stores. This sliding scale of occupant load factors reflects the relationship between an increase in a shopping center's size and an increase in its occupant load, which is not linear but parabolic, as Figure 9-4B shows. Once determined, the occupant load factor is applied to the gross leasable area (GLA) to calculate the covered mall building total occupant load for egress design. Again, this factor is not applied to gross building area.

The covered mall proper—a systems approach: In most cases, two- or three-level covered malls are designed for visual communication among levels through openings in the floor of the upper level(s). Wherever practicable, visual communication between levels is important in maximizing exposure of storefronts and merchandise to customer traffic. The floor areas between openings in the upper level(s) serve as bridges that promote cross-mall shopping convenience for the customer. (See Figures 9-4C and 9-4D.)

When construction starts, not everything about the shopping center is known, even to the developer. Mechanical, electrical, and fire protection systems are designed for main distribution only in the tenant store areas, with branch distribution added to the construction documents as lease agreements define tenant space.

The covered mall proper may contain other amenities for sales promotion, customer service, and convenience, such as seating areas, art objects, planting areas, pools, specialty retail kiosks, directories, openings in upper levels of multilevel structures, and areas

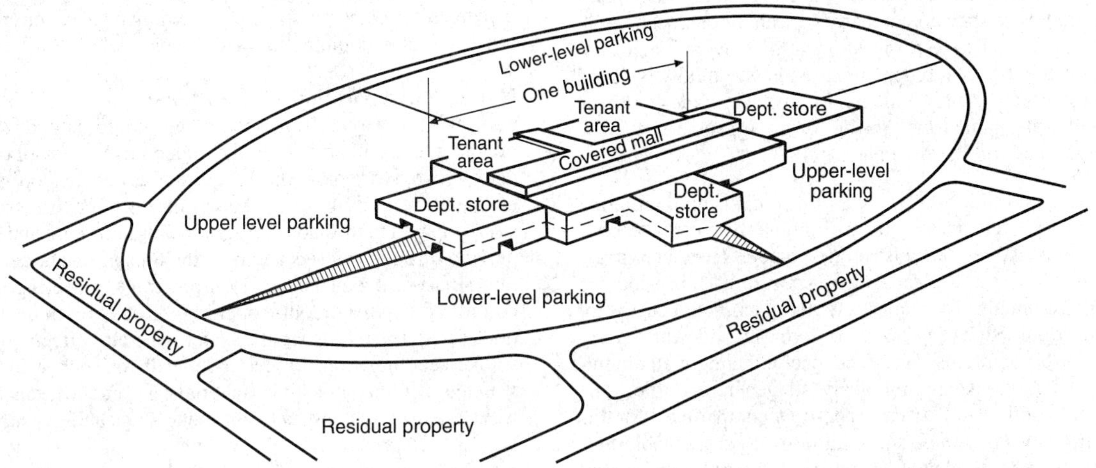

FIG. 9-4A. Many shopping centers today are purposely designed (and merchandised) to provide an equal distribution of customer traffic on all levels.

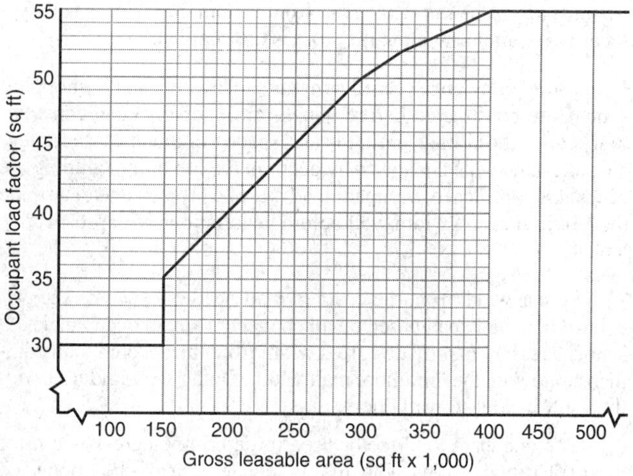

FIG. 9-4B. *Relationship of an increase in the size of a covered mall to an increase in the occupant load.*

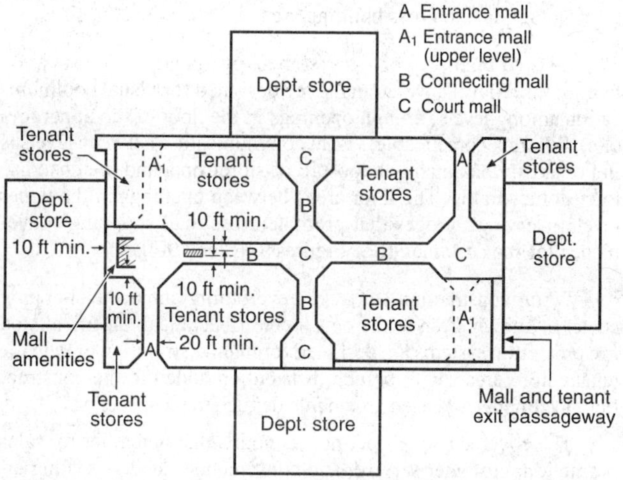

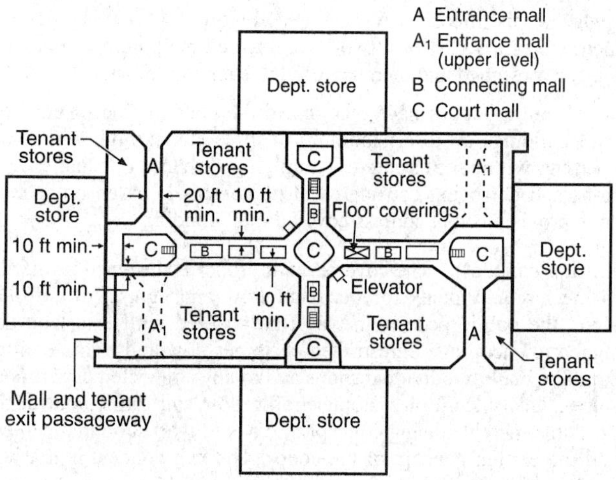

FIG. 9-4D. *Typical two-level shopping center (upper level).*

FIG. 9-4C. *Typical two-level shopping center (lower level).*

designed for promotional activities. Care must be exercised in locating these amenities so they do not interfere with, or encroach upon, required exit access. However, the width of the covered mall connecting courts at entrances to department stores generally is found well in excess of the 20-ft (6-m) minimum to provide for these amenities and at the same time provide for an aggregate of 20 ft (6 m) of exit access width. (See Figure 9-4C.)

Stores located on either side of entrance malls and not along the direct path to department stores (commonly called "anchor stores" or "magnet stores" because of their greater customer attraction) do not enjoy as much customer exposure as those located along connecting malls. To compensate and maximize convenient cross-mall customer shopping, stores on opposite sides of the entrance mall are brought closer together by eliminating mall amenities that by themselves are not likely to generate traffic. The minimum mall width of 20 ft (6 m) is entirely compatible with this merchandising concept and does not impose an unreasonable burden on developers or merchants. This minimum width requirement has been applied successfully in shopping centers throughout the U.S.

The covered mall should have an unobstructed exit access of not less than 10 ft (3 m) clear width on each side of the mall floor area, parallel and adjacent to the mall storefronts. The exit access should lead to an exit having a minimum of 66 in. (168 cm) of width. The purpose of requiring 10 ft (3 m) of clear exit access is to provide a higher degree of safety in the covered mall commensurate with its use as a continuation of exit access from the tenant stores. Safe, continuous, and unobstructed exit access from each tenant store is provided, while the owners' merchandising and operational requirements for mall amenities are recognized.

The minimum requirement of 10 ft (3 m) of clear exit access parallel and adjacent to the mall storefronts relates directly to the minimum mall width of 20 ft (6 m). None of the mall amenities mentioned earlier should encroach on this minimum dimension of 10 ft (3 m). In effect, this requirement precludes the installation of mall amenities in areas of the covered mall that are the minimum 20 ft (6 m) width. In life safety terms, this requirement has proven reasonable and workable and has the added advantage of not being unduly restrictive or burdensome to developers.

Sprinkler protection: The covered mall and all its connected buildings must be provided with an electrically supervised automatic sprinkler system throughout. This has proven the most effective way to protect life and property from fire. Sprinklers allow the replacement of many passive long-standing code requirements by active systems in combination with specified design requirements.

Horizontal separation of mall tenant stores: Walls dividing mall building stores from each other should extend, to the extent practicable, from the floor to the underside of the roof deck or floor deck above. No separation is required between a store and the covered mall. The purpose of this requirement is to confine fire and smoke spread to its place of origin by inhibiting fire and smoke from passing through ceilings and over the tops of tenant demising walls into adjacent occupancies. However, the engineering of a smoke-control system may require openings or deletion of the tenant partitions above the ceilings so the system operates effectively to prevent smoke accumulation. Model codes call for such walls to be constructed of noncombustible materials; in practice, tenant demising walls are usually constructed of metal studs and gypsum board.

In effect, tenant demising walls represent a degree of compartmentation to the extent that no separation is provided at the storefront. A requirement for separation at the storefront would defeat the merchandising purpose of the covered mall: to optimize unrestricted customer flow between the covered mall and the tenant store. This form of compartmentation, in concert with an electrically supervised automatic sprinkler system, eliminates the need for any fire separation wall requirements at the mall storefronts. In shopping centers, additional fire separation is unnecessary and, in fact, burdensome from an operational standpoint. In the leasing and subsequent re-leasing of shopping centers, maximum flexibility to rearrange space is important to continued success.

Smoke management: The covered mall should have an effective smoke-control design. Due to the open storefront configuration, under fire conditions any buildup of smoke in a tenant store could find its way quickly into the covered mall. If a tenant shop has no smoke management capabilities, then effective smoke management in the covered mall proper is essential to ensure the mall's continuous use as a tenable primary way of exit access for the time necessary for occupants to exit.

Sectionalized heating, ventilation, and air-conditioning (HVAC) system service for multiple tenants, as opposed to a separate system for each tenant, is increasingly used to maximize efficiency and control and to minimize energy consumption. An added benefit of such systems is their effective use for smoke-control in the event of a fire in any given tenant space. Thus, with the help of the sprinkler system, combustion products can be confined to their area of origin and subsequently discharged to the exterior.

General requirements: In addition to the special code provisions, certain exit conditions should be present in all covered malls:

1. Every floor and every store of a covered mall except those less than 50 ft (15 m) deep should have at least two exits remotely located from each other.
2. Every covered mall should have unobstructed exit access parallel and adjacent to the connected tenant spaces. The exit access should extend to the mall exits.
3. Exits from the covered mall should be arranged so that travel distance from any mall storefront entrance to an exit is no more than 200 ft (61 m).

GENERAL LIFE SAFETY FACTORS APPLYING TO MERCANTILE OCCUPANCIES

A number of factors related to life safety are common to all types of mercantile facilities.

Prompt discovery is one of the most important factors associated with minimizing loss due to fire. This is especially important considering how many reported losses in mercantile occupancies occur during nonbusiness hours, from late night to very early morning. Many fires occur in stock areas, as opposed to sales areas. Poor housekeeping and smoking are two important causes in stock areas. For example, workers may remove merchandise from cartons for display on the sales floor and not properly dispose of the cartons and packaging, thus increasing the hazard.

Properly engineered sprinklers can be expected to respond to a fire to control its development, if not extinguish it entirely.

The need to renovate existing facilities as a result of new trends in fashion or store design is ever present in mercantile facilities as is the need to reconstruct existing facilities when re-leasing space, especially in shopping centers. Extra care should be taken to mitigate the potential problem of fire during these times, since reconstruction and renovation are very likely to increase the hazard.

Certain accessory uses, such as food preparation areas in restaurants, often found in mercantile facilities and more particularly in department stores and shopping centers, warrant special attention. They are very often where fire originates—much more so than in public eating space. Such areas should be segregated with appropriate construction, if not sprinklered, and appropriate fire suppression devices should be installed in food preparation equipment.

CONCLUSION

An important purpose of this chapter on the nature of mercantile occupancy is to form a basis for considering or reconsidering code requirements that let the mercantile trade function optimally and yet provide the necessary degree of life safety for the facilities' occupants.

All requirement related items discussed in this chapter have been practically applied in many stores and covered mall shopping centers throughout the U.S. They have answered the need for life safety without imposing burdensome or unreasonable restrictions on either the developers or the merchandising needs of the covered mall shopping centers.

In addition to NFPA *101*, model building codes address the code issues presented in this chapter in efforts to update and improve the codification of requirements for covered mall shopping centers. This effort's success will depend upon how successfully the life safety concern is addressed in harmony with the retail merchandising purpose of the covered mall shopping center.

Overreaction in the formulation of code requirements is possible if the fire record is not reviewed and used as a guide to the need for additional requirements. This is especially important when operational needs of the building type being considered are not fully understood. Therefore, code revisions should be tested to see if they are unreasonably restrictive or unfounded in terms of any real threat to life safety, compared to practicability in application and cost-benefit.

BIBLIOGRAPHY

NFPA Codes, Standards, and Recommended Practices

Reference to the following NFPA codes, standards, and recommended practices will provide further information on mercantile occupancies discussed in this chapter. (See the latest version of *The NFPA Catalog* for availability of current editions of the following documents.)

NFPA *101*®, *Life Safety Code*®

Additional Reading

Bengtson, S., "Forskningsbehov brandskydd av overglasade gardar. [Technical Overview of Fire Protection in Malls.]," National Defence Research Establishment, Sundbyberg, Sweden, Sultrapport 1992 01 20, 1992.

Beyler, C., and DiNenno, P., "Letters to the Editor. 'Smoke Management for Covered Malls and Atria' by J. A. Milke," *Fire Technology*, Vol. 27, No. 2, 160–169, May 1991, pp. 160–169.

Blackburn, R. B., "County Mall: A Shopping Center Fire Safety Strategy for the 1990's," *Fire Engineers Journal*, Vol. 53, No. 168, Mar. 1993, pp. 18–27.

Bosley, K., "Escaping From Shopping Precincts," *Fire Research News*, No. 14, Autumn 1992, pp. 2–3.

Catchpole, L., "Integrating Systems at Lakeside Shopping Center," *Fire Prevention*, No. 271, July/Aug. 1994, pp. 20–22.

Chow, W. K., and Chu, C. K., "Application of HAZARD 1 to Study the Fire Risk of a Shopping Mall," *Proceedings of Fire Safety Modelling and Building Design*, the One-Day Conference to Review the Potential and Limitations of Fire Safety Models for Building Design, Mar. 29, 1994, Salford, UK, 1994, pp. 90–97.

Corbett, G. P., "Strip Malls: The Taxpayers of Today," *Fire Engineering*, Vol. 149, No. 5, May 1996, pp. 58–62.

"Fire Protection—A Shopping Center's Best Buy," *Record*, Vol. 67, No. 3, May/June 1990, pp. 16–19.

Gardner, J., "Comparing Smoke Vent Systems in Shopping Malls," *Fire Prevention*, No. 236, Jan./Feb. 1991, pp. 23–27.

Gardner, J. P., "Smoke Control in Shopping Malls," *Building Services*, Vol. 13, Apr. 4, 1991, pp. 53–54.

Gardner, J. P., "Unsprinklered Shopping Centers: Smoke Ventilation," *Fire Surveyor,* Vol. 17, No. 6, Dec. 1988, pp. 44–47.

Heskestad, G., "To the Editor. 'Smoke Management for Covered Malls and Atria' by J. A. Milke," *Fire Technology*, Vol. 27, No. 2, May 1991, pp. 174–185.

Hosser, D., and Dobbernack, R., "Konzepte fur die Rauchfreihaltung von uberdachten Innenhofen und Ladenpassagen im Brandfall. [Concepts to Keep Roofed-In Atria and Shopping-Centers Smoke-Free in Cases of Fire.]," *VFDB*, Vol. 41, No. 2, Mar. 1992, pp. 55–59.

Hough, E., "Gatwick Airport: 'A Shopping Mall With Aircraft'?" *Fire*, Vol. 87, No. 1076, Mar. 1995, pp. 23–24.

Johnson, P. F., and Timms, G. R., "Performance Based Design of Shopping Center Fire Safety," *Proceedings* of the First International Conference on Fire Science and Engineering, ASIAFLAM '95, Mar. 15–16, 1995, Kowloon, Hong Kong, 1995, pp. 41–49.

Law, M., "A Note on Smoke Plumes from Fires in Multi-Level Shopping Malls," *Fire Safety Journal*, Vol. 10, No. 3, 1986, pp. 197–202.

Lombardo, M., and Pressler, B., "Firefighting in Covered Malls. Part 1," *Fire Engineering*, Vol. 146, No. 11, Nov. 1993, pp. 71–74.

Lombardo, M., and Pressler, B., "Firefighting in Covered Malls. Part 2," *Fire Engineering*, Vol. 147, No. 1, Jan. 1994, pp. 52–56.

Milke, J. A., "Author's Response. 'Smoke Management for Covered Malls and Atria.'," *Fire Technology*, Vol. 27, No. 2, May 1991, pp. 169–174.

Milke, J. A., "Smoke Management for Covered Malls and Atria," *Fire Technology*, Vol. 26, No. 3, Aug. 1990, pp. 223–243.

Milke, J. A., "Providing a Performance-Based Design Method for Smoke Management Systems in Atria and Covered Malls," *SFPE Bulletin*, May/June 1992, pp. 6–8.

Morgan, H. P., "Design Principles for Smoke Ventilation in Enclosed Shopping Centers," Fire Research Station, Borehamwood, England, BRE BR 186, 1990.

Morgan, H. P., and Marshall, N. R., "Depth of Void-Edge Screens in Shopping Malls," *Fire Engineers Journal*, Vol. 48, No. 152, 1989, pp. 7–9.

Platt, S., "Life Safety in Covered Shopping Complexes, Part 2: Means of Escape and Alarms," *Fire Surveyor*, Vol. 17, No. 1, Feb. 1988, pp. 4–8.

Platt, S., "Life Safety in Covered Shopping Complexes, Part 3: Exits," *Fire Surveyor*, Vol. 17, No. 3, June 1988, pp. 24-29.

Platt, S., "Life Safety in Covered Shopping Complexes, Part 4: Public Areas," *Fire Surveyor*, Vol. 17, No. 5, Oct. 1988, pp. 20–25.

Ruohonen, E., "Kaksi ja puoli hehtaaria myyntitilaa samassa palo-osastossa. [Giant Shopping Center in Kuopio—Two and One Half Hectares in a Single Fire Compartment.]," *Palontorjuntatekniikka*, Apr. 1995, pp. 6–10.

Smith, J. P., "Fighting Strip Mall Fires," *Firehouse*, Vol. 20, No. 10, Oct. 1995, pp. 18, 20, 22.

Thomas, P. H., "On the Upward Movement of Smoke and Related Shopping Mall Problems," *Fire Safety Journal*, Vol. 12, No. 3, 1987, pp. 191–203.

Webb, B., Sr., "Evaluating Equivalencies to code for a Four Level Shopping Mall," Rolf Jensen and Associates, Inc., Fairfax, VA, Video; Federal Fire Forum, Performance-Based Design Issues of Concern to the A/E Community, November 6, 1995, Gaithersburg, MD, 1995.

Webb, W. A., "Using FPETOOL to Evaluate Fire Safety of a Four-Level Shopping Mall," Rolf Jensen and Associates, Inc., Deerfield, IL, ISBN 0-9527398-3-6; Institution of Fire Engineers; University of Sunderland, Fire Research Station, CIB W14; Tyne and Wear Metropolitan Fire Brigade, Fire Safety by Design, Conference Proceedings, Volume 3, Research Papers, July 10–12, 1995, UK, 1995, pp. 33–37.

Webb, W. A., "Using FPETool to Evaluate Fire Safety of a Four-Level Shopping Mall," *Proceedings* of the International Conference on Fire Research and Engineering, September 10–15, 1995, Orlando, FL, SFPE, Boston, MA, 1995, pp. 365–370.

BUSINESS OCCUPANCIES

Donald G. Bathurst

As the service side of the economy continues to grow, more and more of us have become part of the many industries that deal in information. Recordkeeping, data processing, and information management are central to businesses that are physically located in office buildings, and office buildings are the prototype for the class of properties defined as business occupancies in NFPA *101®*, *Life Safety Code®*.

Most, or nearly 5,000 per year in 1990-1994, of the fires each year in business occupancies occur specifically in general business offices. Large numbers of business occupancy fire problems also occur in college classroom and administrative buildings, medical offices, and bank buildings. Other examples of this occupancy class include city or town halls, courthouses, laboratories not involving hazardous chemicals, and ambulatory outpatient clinics. In each of these properties, information storage dominates the building's distinctive fuel load.

This chapter aims to describe the potential effects of a fire in a business occupancy on people in the building. The business occupancy has long been considered a low-hazard occupancy and, therefore, not worthy of serious protection consideration. However, this is not the case, as this chapter will show. It examines building construction, occupancy characteristics, potential fire scenarios, means of egress, effects of sprinklers, effects of detection systems, and special cases of existing buildings.

Rather than recite requirements available for reference in codes and standards, this chapter approaches the subject of protecting life in a business occupancy from a more philosophical standpoint. It challenges the reader to look beyond the codes and to observe, measure, and understand the impact that building and occupancy conditions actually have on persons in them.

To be aware of the concept of risk is to begin to understand the fire problem. The business occupancy is certainly of a lower hazard class than many industrial or hazardous materials handling operations. But when more people occupy a single building than reside in a small town, as is often the case in the modern high-rise, the magnitude of the challenge faced in protecting the building, its contents, and, most importantly, its occupants, is realized.

This chapter examines how the contents and the building itself affect fire growth and severity, how growth and severity affect notification and egress time of occupants, and how the conditions of oc-

cupants affect their abilities to escape. A format and issues for analysis are presented.

Additional information on determining necessary levels of protection can be found in Section 1, Chapter 3, "Systems Concepts for Building Fire Safety," and in Section 11, "Information and Analysis for Fire Protection," as well as in the referenced and suggested reading material. More information on fire safe building design can be found in Section 1, Chapter 2, "Fundamentals of Fire Safe Building Design." Information on interior finish can be found in Section 7, Chapter 3, "Interior Finish."

BUILDING CONSTRUCTION

Construction Type

Business occupancies can be found in a variety of building types. Although a two- or three-story noncombustible structure or a fire-resistant high-rise may be most familiar, many business occupancies are housed in ordinary, frame, and even heavy timber buildings.

Each construction type affects the business occupancy differently because fires behave differently in each. This discussion and its examples are confined to noncombustible and fire-resistive construction because it is most prevalent. The reader must understand that, in other construction types, the availability of fuel from the structure itself increases fire severity and, therefore, decreases potential structural integrity.

In noncombustible construction, construction materials do not burn, thereby providing a measure of increased occupant safety. In reviewing the life safety needs of a business occupancy in a building of noncombustible construction, the designer or reviewing official can be primarily concerned with fuel load from the contents, furnishings, and interior finish, and with the means of egress. Although the potential fire's effect on the structure also must be assessed, this is usually simpler to do than it is in the case of combustible construction materials.

Fire-resistive structures are similar to noncombustible structures except that the material or combinations of materials used inherently resist failure due to fire or to the passage of fire. These added protections help ensure sufficient time for persons to egress and/or receive other forms of assistance.

Donald G. Bathurst is deputy administrator of the U.S. Fire Administration, Federal Emergency Management Agency, Washington, DC.

Interior Finish

Interior finish comprises the various aesthetic treatments to the structure that make it more comfortable to work in and more pleasing to the eye. Interior finish includes the materials used to cover walls, ceilings, and floors, such as paint, wallpaper, vinyl wallcovering, paneling, acoustic panels, hardwood floors, vinyl tile, terrazzo, and carpet. Each behaves differently in a fire, and various combinations can make for an interesting analysis.

The main fire performance measure of an interior finish is flame spread. This is the ability of the material to propagate a flame front across its surface. Wall and ceiling materials are ranked by relative ratings based on a value of 100 given for a specially prepared sample of red oak. Floor finishes are rated by the amount of heat energy imparted to the surface from an external source that is necessary to sustain combustion; this is called critical radiant flux. The better the flame-spread characteristics—that is, the lower the rating for wall and ceiling material or the higher the critical radiant flux—the more occupant protection can be assumed.

It is important to note that flame-spread ratings are based on tests of materials under controlled conditions and with specific backing materials. A wallcovering mounted on gypsum wallboard would be expected to perform differently than a wallcovering installed over brick or block.

A most important but often overlooked point is that interior finish materials are also fuels. When materials are relatively thin—two coats of paint or one layer of vinyl wallcovering, for example—the additional fuel may be insignificant. In general, wall covering materials with a thickness of $1/28$ in. (.91 mm) or less are typically excluded from any minimum criteria. But when several layers of wallcovering, carpet, floor coverings, or other materials are present, the hazard and risk change dramatically. It should not be assumed that a low flame-spread rating alone results in a significant risk reduction. It is important to understand the impact this additional fuel has on the entire environment.

Furnishings and Contents

Furnishings and contents make up the bulk of the potential fuel load in a business occupancy. Furnishings, for the purpose of this discussion, include desks, chairs, sofas, files, office supplies, and paper.

Over the past several years, there have been many changes in the furnishings put into buildings. At one time, desks and chairs were routinely wood. Then metal became popular. Now, any combination of wood, metal, thermoplastics, and foamed plastics can be found. In addition, the increased use of computers has also added to the fuel load.

Business occupancies are considered lower hazard than industrial or other occupancies handling hazardous materials. However, this perception sometimes confuses familiarity with safety. The quantity (in lb/sq ft) of fuel load in business occupancies can be a source of significant hazard, even if hazardous materials are involved. In addition, some relatively benign materials may have burning characteristics—such as peak intensity of heat when fully involved in fire—comparable to some hazardous materials.

The designer and code official must be aware of the characteristics of the materials in a building, such as construction materials, interior finish materials, and furnishing materials. Each provides a degree of protection and a degree of hazard; their combined effect influences risk. Without understanding the potential fire in a business occupancy, one cannot begin to understand its potential effect on people.

OCCUPANCY CHARACTERISTICS

Business occupants are generally characterized as awake, alert, and familiar with their surroundings. While this is true of most offices, designers and code officials must assess the actual characteristics of a particular business occupancy, such as a doctor's office or an outpatient clinic, in which a minority of the occupants are familiar with the surroundings. In business occupancies, generalities do not always apply.

Support services often are found within business occupancies. Each provides a challenge in the analysis of potential fire impact because these service areas present their own special hazards. Examples include cafeterias, auditoriums, print shops, storage areas, parking facilities, small retail outlets, and banks or credit unions. People usually gather in these areas during certain times, often in large concentrations relative to the general occupant load for the entire building. Many service areas increase the potential for fire ignition because of their operation; one example is a cafeteria kitchen.

Another growing service area in business occupancies is the child care center. With more families having both parents, or the only parent, working, employer-provided child care is becoming more common. This poses special concerns: the occupants' ages, their lack of familiarity with their surroundings, and the fact that they sleep in the space from time to time. Special protection must be considered for child care centers, including smoke detection, sprinkler protection, fire-resistive enclosures, location at grade level with direct access to the outside, and sufficient adult supervision per child.

Another growing segment of the population in business occupancies is the disabled. Persons with disabilities are contributing members of the work force, and their numbers in the workplace are increasing with increased requirements for access for them. Also increasing is awareness of the large number of people with ability limitations of some kind to some degree that could affect their response to fire. Disabilities cover a wide range, including sight, hearing, and mobility impairments. Also many persons have hidden disabilities, such as heart disease. A disability also can be temporary or transient, such as pregnancy or a broken leg. Disabilities can increase the time needed to alert individuals or the time necessary to safely egress an area or the building. Designers and code officials must consider the presence of persons with known or hidden disabilities when evaluating a life safety system for a business occupancy.

While persons in a business occupancy are usually awake and alert, their responsiveness differs considerably. Actual occupancy conditions must be reviewed in detail. Only when one understands who is being protected can one decide how to protect them from potential fire.

POTENTIAL FIRE SCENARIOS

Fire safety design for a particular property must begin with a consideration of characteristics of most likely fires and the most serious fire the occupancy might encounter. This may be done by defining potential fire scenarios to use as a basis for analysis.

Fuels for a fire are readily available. They include furnishings, paper, books, and other office supplies. But the significant contribution to fire severity that interior finish materials can provide must not be forgotten. The wide use of plastics also must be recognized as a factor in the potential fuel load in a business occupancy. This includes furnishings and a variety of office equipment. Even the structure itself can provide fuel for fire, especially if it is of ordinary or wood-frame construction.

Transient fuels in a building represent a significant and often overlooked fuel load. These fuels, which include furniture and carpet deliveries, mail operations, and trash disposal, often expose the egress system, so special care should be taken to identify them.

Leading accidental sources of heat for ignition include electrical distribution systems, portable and area heating equipment, other appliances, torch and other maintenance operations, matches and lighters, and smoking materials. The design professional and code official must know where potential ignition sources are and where they expose potential fuels.

One of the two leading causes of fire in business occupancies is not accidental, however; it is arson. Arson prevention is partly a matter of understanding the business and those who might wish to harm it, such as competitors, disgruntled employees, angry customers, or economically strapped owners. But arson more often occurs as a result of less direct motives, such as emotional disturbance, so it is important to emphasize security controls and arrangements, such as housekeeping, that will reduce the range of attractive and easy targets for potential arsonists. Also, the importance of arson underscores the limits of fire prevention as a strategy, so post-ignition fire protection provisions are needed, too.

The value at risk from a fire in a business occupancy can easily be underestimated. With the increase in regular use of computers, facsimile machines, copy machines, and printers, and similar equipment, and the purchase of telephone systems for offices and buildings, the value of an average workstation has grown appreciably over the past 15 years. The value of loss of business and other interruption costs, including the occupancy of temporary space and reduced employee productivity during space restoration, also must be considered.

The designer and code official must look carefully at the business occupancy for available fuels and potential ignition sources. The impact of fire on the occupancy and occupants can then be predicted. Computer models and other engineering tools can translate the occupancy's physical and occupant characteristics into a timeline of fire development and its effects on people and property.

MEANS OF EGRESS

In analyzing the potential fire scenario and choosing strategies to deal with the results of the analysis, it is useful to think in terms of major groups of fire protection provisions. Means of egress will be the first discussed here. Before the means of egress needs can be assessed, however, an objective must first be established: every occupant should be able to reach a safe location from anywhere in the building without incurring serious harm from fire along the way.

For a long time, this objective was fulfilled by evacuating building occupants when fire was discovered. As buildings get larger and taller, this is not always practical or possible. Some high-rise building populations are the size of a small town; total evacuation may take hours. Nor does it make sense to move people from the 40th floor of a building in response to a small trash fire on the first floor, especially if the fire is being controlled or extinguished by an automatic sprinkler system. Persons with severe mobility impairments may not have this option available without assistance.

Where total evacuation can be accomplished, it is desired. However, the designer and code official must be aware of conditions that warrant a staged or delayed evacuation, or relocation within the building. In any event, the objective of keeping people in a tenable environment must control all actions.

One method of establishing an acceptable level of protection for persons from fire is the application of NFPA *101*. It provides guidance for occupancy loads, arrangement of exits, capacity of exits, and exit travel distances, as well as guidance for the design of doors, stairs, and interior finish materials. NFPA *101* recognizes differences in existing buildings and new construction, and it provides

different requirements for each. In fact, all codes and standards have equivalency clauses so as not to exclude sound engineering approaches that accomplish the same or similar results. The code official must be aware of the application of these equivalency clauses.

Total Evacuation

If a building is small, total evacuation is preferred. In this case, persons should be able to leave the building before conditions in any part of it become untenable. Fire growth, temperature rise, and the production of toxic combustion products must be estimated. Then the time needed to evacuate the building must be estimated. The difference between the evacuation time and the time to untenable conditions is the time allowed for fire detection, notification of occupants, and occupant reaction.

The alarm system, either manual or automatic, must be chosen based on the time available to notify occupants. In most business occupancies, a manual alarm system is sufficient, but the designer and code official must be aware of conditions that warrant the use of an automatic detection and alarm system. Where automatic sprinkler protection is provided, the designer should take advantage of its presence and use it as an alarm-initiating device.

Alternatives to Total Evacuation

In very large or very tall buildings, total evacuation is neither practical nor possible, but the need to protect the occupants is similar to that of smaller buildings. It is important to ensure that persons can get to an area of relative safety before conditions become untenable.

Relocation: Fire conditions and necessary egress time must be estimated. The area of relative safety must then be chosen. This may be a trial-and-error design process because estimated fire conditions and the structure's ability to resist fire affect where the safer areas will most likely be. Furthermore, persons must be able to reach those areas in a reasonable time. Typically, one or two floor levels above or below the fire floor are considered sufficient in a high-rise building. The designer and code official are cautioned not to take these suggestions, or others in codes and standards, blindly. A measure of the actual expected conditions must be taken before a plan for protection can be devised.

Other egress or occupant protection strategies are similar to relocation and include defend in place, staged evacuation, and use of horizontal exits.

Defend in place: Defend-in-place strategies are generally used in health care occupancies where supervisory staff and persons cannot be readily moved. While not preferred for business occupancies, its existence must be noted and understood.

Staged evacuation: Staged evacuation can take several forms. One is staged total evacuation where the fire floor occupants are notified, and, after a determined time period, other occupants are notified. This allows those on the fire floor the first opportunity to escape. Persons in areas farther from the fire start to evacuate a bit later.

Another method of staged evacuation is the use of "hardened" areas where persons may seek temporary refuge until they are able to egress. These staging areas pose several questions:

1. How should the area be hardened? Fire resistance is not enough. Ventilation and pressurization are critical to maintaining a tenable environment when the area is exposed to post-flashover conditions.
2. How many people will use the area? These areas have been suggested for use by the disabled. With the population of disabled

growing in business occupancies, a minimum number of such occupants must be able to be safely protected in such areas. NFPA *101* provides guidance on this topic.

3. How will people use these areas? Practice is the only answer. If provided, these areas must be addressed in the fire evacuation plan for the building. This would suggest that these staging areas have only limited impact on persons who are not familiar with the building.

4. How will persons eventually leave the building? These areas must not be considered as a place to wait out a complete fire. Persons generally must leave the fire floor. These areas, if provided, should be adjacent to stairs or elevators so that fire department personnel or building fire wardens can assist persons in these areas to other areas of relative safety.

The use of this type of staging area takes much planning, practice, maintenance, and coordination. Unfortunately, it is of little benefit to persons unfamiliar with the building and its features. Therefore, a preferred method would be to establish a way to protect the occupants with little or no special instructions.

In potential impact and effectiveness of staging areas (also known as areas of refuge) for disabled occupants, complete automatic sprinkler protection (properly designed, installed, and maintained) afforded occupants sufficient safety without actually needing to egress the floor. If the sprinkler system uses quick response sprinklers, there will likely be no perception of risk on the part of building occupants.

Horizontal exit: Another method of staged evacuation is through horizontal exits. Horizontal exits can be provided by subdividing a building with fire-resistive walls or partitions. Persons passing through doors in these walls move to an area of relative safety. In some cases, the horizontal exit alone may provide enough egress protection. In others, the horizontal exit may be used to provide enough protection to stage occupants, thereby allowing more time for total evacuation or assisted evacuation, as in the case of persons with disabilities. Special instructions or drills are not necessary in the use of a horizontal exit.

In either relocation or staged evacuation, fire control is important. To ensure the success of the egress or occupant protection strategy, fire growth and combustion products must be held to a minimum. This can be done in several ways. Fuel can be limited, room geometry can be controlled, and ventilation of fire areas can be limited. A proven effective method of controlling fire growth and smoke production is through the use of automatic sprinklers. Therefore, where occupant relocation or staging is an intended strategy, the building should be protected with automatic sprinklers, although individual conditions dictate the actual protection necessary.

EXIT LIGHTING

The means of egress should be continuously illuminated by a reliable lighting system. This usually means spacing lighting fixtures and providing primary and emergency power to ensure a lighting level of at least 1 footcandle (10 lux), measured at the floor level.

Emergency power can be provided in many ways: batteries, inverters, generators, separate building substations with automatic transfer switches, and power distribution system network grids. The concept behind emergency power for egress lighting is to ensure that the exit path is illuminated if a fire in the building disrupts the primary power source. Because a complete power failure at the utility would rarely occur at the same time a fire in the building would, it is sufficient to have only one of the recommended emergency power sources. However, other design arrangements may be necessary, depending on specific local conditions and utility reliability.

AUTOMATIC SPRINKLERS

As discussed, automatic sprinkler protection can increase the level of fire safety in a business occupancy. From the life safety standpoint, sprinklers limit fire growth and the production of smoke, thereby increasing the time available for persons to egress. If designed to do so, automatic sprinklers can even keep conditions in close proximity to the fire in a tenable range.

Another reason sprinklers should be considered is property and mission protection. Fire department suppression capabilities are limited. Even the best department can only be expected to have a suppression capability of about 5,000 sq ft (464.5 m²). If fire fighters must rescue occupants, their capability decreases. As property values and mission criticality increases, the typical business cannot afford this potential exposure.

The potential exposure, of a business occupancy to fire can be considerable when size, value, and mission occupant load are factored in.

Sprinkler protection automatically detects a fire and starts suppression actions, providing additional time for occupants to exit. The sprinkler system can even notify occupants of the fire. Sprinklers also limit fire and water damage, thereby decreasing business interruption. With the expanded use of quick-response sprinklers, these occupant protection factors and property protection features can be expected to increase.

Retrofitting sprinkler systems is becoming easier with the use of special materials such as lightweight steel pipe, copper tube, and plastic pipe. In addition, sprinklers can be safely installed in buildings with sprayed-on asbestos containing fireproofing through a process of local encapsulation and control. See the references in the bibliography on this subject for more information.

EXISTING BUILDINGS

Business occupancies in existing buildings pose several concerns from a life safety standpoint. Many do not meet the requirements of codes, standards, or design practices for new construction. The question should not be, "Do they comply?" Rather, we should ask, "Is the occupant protection objective fulfilled?" In this case, occupant protection should be no different than that previously discussed. The objective is to get people to an area of relative safety before conditions become untenable.

The difference is that an existing building constrains the protection strategy. Several resources can be used to help find solutions to egress problems in existing buildings.

NFPA *101* has a separate chapter for existing business occupancies. Requirements in that chapter are similar to those for new buildings, but they reflect the necessity of accepting those things that cannot be changed, provided a reasonable degree of safety is present.

Another good reference is NFPA 101A, *Guide on Alternative Approaches to Life Safety*. Chapter 7 of NFPA 101A is a fire safety evaluation system (FSES) for business occupancies. This numeric method provides a way of ranking building conditions. The result indicates whether the conditions, as measured by the FSES, provide a level of safety equivalent to NFPA *101*.

Other analytical methods exist that may be beneficial in evaluating the risk to life in an existing business occupancy. These include hand calculations and computer-based predictive fire models. (See Section 11, "Information and Analysis for Fire Protection.") In any event, the process is the same—determine the fire exposure, the time available for safe egress, and the time needed to egress.

A special category of existing buildings must be mentioned: historic or architecturally significant buildings. Society puts additional constraint on these structures, as to the way alterations must be done. Here, historic preservation and life safety objectives must be carefully outlined and communicated. These objectives are not necessarily at cross purposes. Innovative design techniques and engineered solutions, rather than code compliance, are usually called for. A reference on the process and design tips is included in the list of recommended reading following this chapter.

SUMMARY

Business occupancies comprise a large segment of the communities protected, and they can be found in a variety of building types. Construction type, occupancy characteristics, potential fire scenarios, means of egress, the effect of sprinklers, and existing conditions all affect the ability to protect occupants from the effects of unwanted fire. An objective has been set for protecting these occupants by keeping them in a tenable environment or by allowing them to escape to an area of relative safety before untenable conditions are reached. This area may be outside the building, on another floor, or through a horizontal exit.

Several references can be used to determine the potential fire exposure and the necessary protection strategies. NFPA 101, the FSES in NFPA 101A, and the various calculation methods have been discussed.

The designer and code official must look deeper than codes and standards alone to understand actual conditions and measure exposures. They will then be able to recommend ways of fulfilling the objective of protecting occupants.

BIBLIOGRAPHY

References

Madrzykowski, D., "The Hazard of Fire Gas Exposure from Sprinklered Rooms to Corridors and Adjoining Areas," National Institute of Standards and Technology, Gaithersburg, MD, to be published as an NISTIR.
Matthey, R. G., et al., "Procedures for Sprinkler Anchor Installation on Surfaces with Fireproofing Materials," NBSIR 87-3676, Feb. 1988, National Bureau of Standards, Gaithersburg, MD.
Nelson, H. E., "Fireform: A Computerized Collection of Convenient Fire Safety Calculations," NBSIR 86-3308, Apr. 1986, National Bureau of Standards, Gaithersburg, MD.
Nelson, H. E., "FPETOOL: Fire Protection Engineering Tools for Hazard Estimation," NISTIR 4380, Oct. 1990, National Institute of Standards and Technology, Gaithersburg, MD.
"Report from the 1987 Workshop on Analytical Methods for Designing Buildings for Fire Safety," National Research Council.
"Retrofitting Firesafety Features to Historic Buildings," Advisory Council on Historic Preservation, Washington, DC, Aug. 1989.

Stroup, D., and Madrzykowski, D., "Conditions in Quarters and Adjoining Areas Exposed to Post-Flashover Room Fires," National Institute of Standards and Technology, Gaithersburg, MD, 1991.

NFPA Codes, Standards, and Recommended Practices

Reference to the following NFPA codes, standards, and recommended practices will provide further information on business occupancies discussed in this chapter. (See the latest version of _The NFPA Catalog_ for availability of current editions of the following documents.)

NFPA _101®_, _Life Safety Code®_

NFPA 101A, _Guide on Alternative Approaches to Life Safety_

Additional Reading

Clarkson, G., Taylor, J. H., and Carey, P. W., "Study and Review of Fire Safety in High Rise Office Buildings," _Fire Engineers Journal_, Vol. 50, No. 156, Mar. 1990, pp. 13–15.
Hadjisophocleous, G. V. and Yung, D., "Parametric Study of the NRCC Fire Risk-Cost Assessment Model for Apartment and Office Buildings," _Proceedings_ of the Fourth International Symposium on Fire Safety Science, Intl. Assoc. for Fire Safety Science, Boston, MA, 1994, pp. 829–840.
"High-Rise Office Building Fires," _Fire and Arson Investigator_, Vol. 44, No. 1, Sept. 1993, pp. 15–19.
"High-Rise Office Building Fires," National Fire Protection Association, Quincy, MA, 1992, 16 pages.
Hughes Associates, Incorporated, "Fire Safety Evaluation System (FSES) for Business Occupancies Software (ver 1.0 for Windows) Users' Manual," Hughes Associates, Inc., Baltimore, MD, National Institute of Standards and Technology, Gaithersburg, MD, General Services Administration, Washington, DC, NIST-GCR-96-692, Mar. 1996.
Jones, A., "Phased Evacuation From Office Buildings," Fifth International Fire Conference on Fire Safety, Interflam '90, September 3–6, 1990, Canterbury, England, Interscience Communications Ltd., London, England, 1990, pp. 331–335.
Madrzykowski, D., "Fire Tests of Modular Office Furniture," VHS Video Tape, National Institute of Standards and Technology, Gaithersburg, MD, Video; Federal Fire Forum, Current Issues in Fire Protection, June 15, 1992, Gaithersburg, MD, 1992.
Stroup, D. W., "Fire Hazards of Open Office Cubicles," VHS Video Tape, General Services Administration, Washington, DC, Video; Federal Fire Forum, Building Fire Safety, Nov. 1, 1994, Gaithersburg, MD, 1994.
Taylor, K. T., "Office Building Fires. . . A Case for Automatic Fire Protection," _Fire Journal_, Vol. 84, No. 1, Jan./Feb. 1990, pp. 52–54.
Walton, W. D., "Suppression of Wood Crib Fires with Sprinkler Sprays: Test Results," NBSIR 88-3696, Jan. 1988, National Bureau of Standards, Gaithersburg, MD.
Walton, W. D., and Budnick, E. K., "Quick-Response Sprinklers in Office Configurations: Fire Test Results," NBSIR 88-3695, Jan. 1988, National Bureau of Standards, Gaithersburg, MD.

EDUCATIONAL OCCUPANCIES

Catherine Stashak

Educational occupancies can be generally defined as structures where six or more persons gather for purposes of instruction. Examples of educational occupancies include schools, academies, nursery schools, day-care centers, kindergartens, and other facilities whose purpose is primarily educational even though their students may be of preschool age. Colleges and universities, which will be discussed later, are excluded from educational occupancies. Recent actions by model code groups have defined educational occupancies as places where students are present for instruction for more than 12 hr per week or more than 4 hr per day through grade 12. This definition excludes facilities in which instruction is incidental to some other occupancy. In such cases, the requirements of the predominant occupancy apply. Examples of excluded facilities include:

1. Rooms located within places of worship used for purposes of religious education while services are being held in the building.
2. Rooms located within places of worship used as nurseries while services are being held in the building.
3. Rooms used for temporary child care during short-term recreational activities of the child's relative or guardian.
4. Rooms used for extracurricular purposes such as piano instruction, dance instruction, and martial arts classes.

OCCUPANCY CHARACTERISTICS

Nature of Occupants

People in educational occupancies vary in their ability to deal with an emergency condition, depending upon their age and mental and physical condition, as well as the facilities' physical characteristics. Generally, regulations concerning educational occupancies are based on abilities of children in third through eighth grade, with special provisions made for younger children. High-school occupants gain extra safety because fire safety design is based on younger childrens' capabilities. Children younger than third grade require special consideration because of their limited ability or inability to traverse stairs, or even evacuate, in an emergency. Because there is a potential for younger children to be overrun on stairs by other students, most codes require that kindergarten and first-grade students be housed on the first floor or level of exit discharge to facilitate egress. Second-grade students are usually limited to one level above

the first floor or level of exit discharge in recognition of their limited movement ability.

Design Considerations

Schools of only one story, with each room having a direct exit to the outside, represent the most conservative design for life safety. However, economics often dictate other designs, which also can be made safe by adhering to certain design principles noted elsewhere in this chapter.

The need to accommodate physically and mentally disabled students within the general school population is a recent development affecting school design. Access and egress must be considered for these students by providing facilities such as elevators, horizontal exits, and areas of refuge in multiple story buildings. Provisions must be made for using elevators or other special facilities to evacuate these students. It is important that the local fire department be aware of any special arrangements for evacuating physically disabled students, as it may change their approach to rescue efforts.

Flexible and Open Plan Design

Flexible and open plan design concepts sometimes used in schools differ from conventional designs in that the walls are demountable and easily moved without major reconstruction. The ability to easily rearrange walls to suit changing educational needs is valuable, but proper consideration to fire safety must be included in planning.

Open plan buildings delineate spaces and corridors by using movable fixtures and low-height partitions [usually maximum height of 5 ft (1.5 m)]. While this design provides great flexibility, it diminishes the building's fire safety by omitting features that could confine fire and smoke to a single, smaller compartment long enough to permit evacuation. The compensating factor is occupants' ability to observe the entire area over the low-height partitions and presumably detect any fire in its incipient stage. Because occupants are awake and alert, their natural faculties serve as fire and smoke detectors. Early detection should make prompt evacuation and fire extinction possible.

College and Universities

Requirements of educational occupancies are not applicable to colleges, universities, or levels of education beyond grade 12. Students at these levels are generally presumed to be capable of adult behavior because of their age.

Catherine Stashak is a fire fighter inspector with the Des Plaines Fire Department, Des Plaines, IL. She is chair of the NFPA Technical Committee on Educational and Day-care Occupancies.

Because college and university buildings are actually combinations of other occupancies, codes now consider the actual use of the space and apply the respective occupancy requirements. For example, codes treat classrooms for fewer than 50 occupancies as business occupancies, classrooms with an occupant load of 50 or more as assembly occupancies, instructional laboratories as business occupancies, and noninstructional laboratories as industrial occupancies. (Depending on the quantities of chemicals found in noninstructional laboratories, NFPA 45, *Standard on Fire Protection for Laboratories Using Chemicals,* may apply.)

Day-Care Facilities

Day-care facilities have traditionally been associated with child care and thus have been historically considered as educational occupancies. Recent changes in society have indicated a need to expand the definition of day-care facilities to include elderly people. Because many physical limitations that apply to children also apply to older adults, this approach seems logical from fire protection and life safety standpoints. Design criteria for these facilities should focus less on age group classification and more on clients' ability to evacuate and staff members' ability to evacuate clients. This approach is used in the Residential Board and Care chapters of NFPA *101®, Life Safety Code®.* For example, a common method of evacuating infants is to place several infants in one crib and roll the crib out of the facility. Additionally, consideration needs to be given as to whether an adult confined to a wheelchair can evacuate without assistance or whether that adult needs to be wheeled out by a staff member. Toddlers may be able to walk but when asked to climb up or down stairs, they may drop to their knees to traverse stairs. Thus, client mobility is an important consideration in developing life safety requirements. NFPA *101* describes new location/construction type limitations. Table 9-6A shows these requirements.

TABLE 9-6A. *Day-Care Facilities' Location and Construction Type Limitations*

Location of Day Care	Sprinklered Building	Construction Type Permitted
1 story below LED	Yes	Any type other than III(200) and V(000)
Story of exit discharge	No	Any type
1 story above LED	Yes	Any type
	No	I(443), I(332), II(222)
2 or 3 stories above LED	Yes	Any type other than III(200), IV(2HH), and V(000)
> 3 stories above LED but not high rise	Yes	I(443), I(332), II(222), or II(111)
High rise	Yes	I(443), I(332), or II(222)

Mandatory staff-to-client ratios have been eliminated because, from the standpoint of the local authority, ratios are very difficult to enforce. Most state licensing requirements include staff-to-client ratios; any additional requirements are considered a duplicate effort.

It should be recognized that some day-care facilities operate 24 hr a day to care for dependents of people who work at night. These facilities present the greatest hazard because their occupants may be asleep when fire starts. Day-care staff members are required to be awake and alert at all times; however, it's always possible that some staff may be napping when an emergency occurs. Therefore, day-care facilities that operate at night require special provisions.

OCCUPANCY HAZARDS

Mixed Occupancy

Often, more than one type of occupancy exists within an educational facility. In a mixed-occupancy situation, care must be taken to ensure that the fire and life safety provisions for all occupancies are met. It is important to provide sufficient protection to prevent the educational occupancy from being exposed to hazards of other occupancies in the structure.

Hazardous Areas and Arrangements

Areas such as vocational shops, laboratories, home economics rooms, kitchens, storage areas, and stages must be separated from educational occupancies. Usually, fire-rated walls and doors provide the necessary degree of protection unless the hazard is severe, in which case automatic suppression systems may also be required.

Clothing, personal effects, and other combustible materials should not be stored in corridors. If automatic detection or suppression systems protect the corridors, or if metal lockers are used, clothing and personal effects can be stored in corridors. Metal lockers should not reduce required egress width.

Child-prepared artwork and teaching materials attached directly to walls should not cover more than 20 percent of the wall area. It is advantageous not only to limit the quantity of child-prepared materials displayed but also to avoid placing such materials near a room's exit access doors. Because the artwork's combustibility cannot be effectively controlled, the quantity, in terms of the percentage of the overall wall area covered, is regulated to avoid creating a continuous combustible surface that will spread flame across the room.[1] Hanging banners in corridors should be avoided unless they meet requirements for flame resistance as given in NFPA 701, *Standard Methods of Fire Tests for Flame-Resistant Textiles and Films* (small- and large-scale tests). Because they frequently obstruct exit signs, consideration should be given to problems associated with hanging banners before allowing their use.

LIFE SAFETY CONCEPTS

Means of Egress

There are no substitutes for properly designed exit access paths, exits, and exit discharges. Travel distance requirements and dead-end restrictions are especially important in educational occupancies because of the nature of the occupants. Designs involving circuitous exit access paths should be avoided. Proper exit enclosures are important to prevent smoke and heat contamination.

Precautions must be taken to prevent corridors from becoming untenable. The use of fire-rated corridor construction, generally with a rating of 1 hr, is common practice. If the corridor ceiling is an assembly that also has a 1-hr fire resistance rating (when tested as a wall), the corridor walls are normally permitted to end at the corridor ceiling. NFPA *101* does not require lavatories to be separated from corridors, provided the lavatories are separated from all other spaces by walls having a fire resistance rating of not less than 1 hr. This is shown in Figure 9-6A.

Proper door controls are also necessary to protect the corridor. Although many codes require self-closing doors in corridors, operational necessities make them impractical in many schools. Therefore, staff training is necessary to ensure doors are properly closed at appropriate times.

NFPA *101* requires the installation of specially sized windows for fire department use for rescue and ventilation. (See Figure 9-6B.)

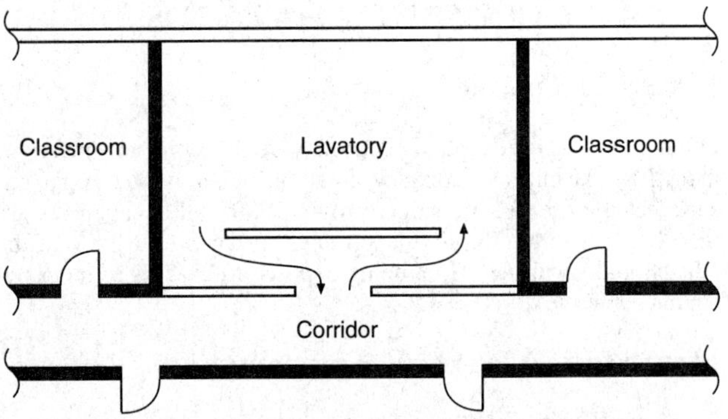

FIG. 9-6A. *Lavatory permitted to be open to corridor.*

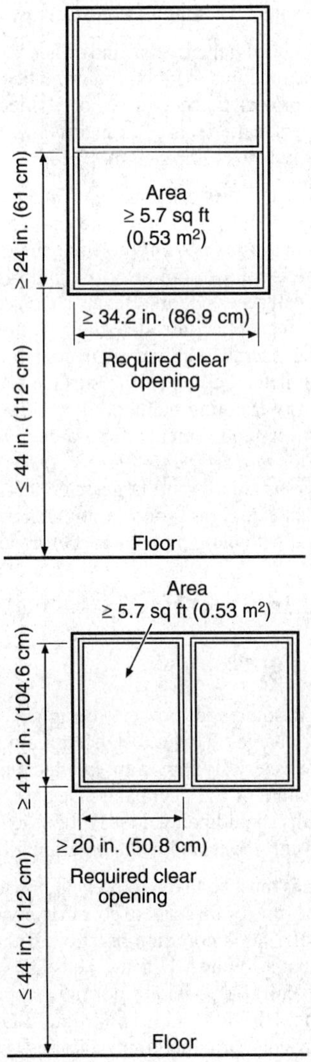

FIG. 9-6B. *Minimum dimensional criteria for rescue and ventilation.*

One of these windows is required in any room larger than 250 sq ft that is subject to occupancy by students. The minimum dimension requirement allows students to escape through the window and fire fighters with air packs on their backs to enter the room for rescue or fire suppression purposes.

Exit Hardware

Because of the potential for a large number of occupants to use the same exit doors and to prevent crushing against unopened doors, panic hardware or fire exit hardware is required in schools, generally when the occupant load exceeds 100 persons. In no instance are chains or padlocks permitted on exit doors when the building is occupied. If such security devices are used when the building is empty, provisions for their removal must be made to the satisfaction of local authorities.

Alarm Systems

In general, schools minimally require a manual fire alarm system. Because of school vandalism, fire alarm systems sometimes have been turned off or otherwise disabled to prevent malicious false alarms from interrupting the learning process. Recent changes in some codes permit the deletion of manual pull stations if two-way communication is maintained in all rooms and there is a continuously staffed central location where an alarm can be sounded.

In school structures that contain single classrooms, fire alarm systems are not usually required. As with flexible and open plan design areas, the entire area can be observed and therefore a fire would be easily detected in its incipient stage. A minimum separation distance of 50 ft between the single classroom building and any other structure should be maintained.

In day-care occupancies where clients sleep or are incapable of self-preservation, early notification of fire by use of detection is imperative. Emergency forces notification is also critical because first arriving fire department personnel can assist in evacuation procedures.

Staff Training and Fire Exit Drills

Educational occupancies should develop emergency plans in conjunction with the local fire department. The local fire department needs to be made aware of the school's plan regarding evacuation of disabled students or clients incapable of self-preservation. Areas of refuge (as required by NFPA *101* or the Americans with Disabilities Act Accessibility Guidelines or any state accessibility laws) need to be identifiable from the exterior so fire department personnel can access these areas.

Frequent fire exit drills are critical in achieving proper response by occupants during an emergency. Fire exit drills should include assignment of staff or more mature students to hold doors open in the line of march or to close doors where necessary to prevent spread of fire or smoke; to search the lavatories or other rooms; to account for all occupants if possible; and to achieve brisk, quiet, and orderly evacuation of the building or relocation to areas of refuge. Fire exit drills should be held not less than once per month when the building is open for occupancy. In climates where winter weather is severe, most drills should be run before bad weather starts. This ensures that staff and clients understand the fire exit drill procedure.

Drills should be unannounced and held at different hours of the day, evening, or night; during changing of classes; when the school is at assembly; and during recess and gym periods. This is to avoid distinction between drills and actual fire situations. As drills simulate an actual fire condition, the fire alarm signaling device should be used so occupants recognize the signal and respond appropriately. Repetitive training in fire drills will benefit the client when he or she is within another occupancy type. When a fire alarm is activated within the other occupancy, the client should recognize the signal and respond appropriately.

Occupants should not be allowed to obtain clothing after the alarm sounds, due to the confusion that would result in forming lines and the danger of tripping over dragging apparel.

BIBLIOGRAPHY

References Cited

1. Coté, R., ed., *Life Safety Code Handbook*, 6th ed., National Fire Protection Association, Quincy, MA, p. 771.

NFPA Codes, Standards, and Recommended Practices

Reference to the following NFPA codes, standards, and recommended practices will provide further information on educational occupancies discussed in this chapter. (See the latest version of *The NFPA Catalog* for availability of current editions of the following documents.)

NFPA 45, *Standard on Fire Protection for Laboratories Using Chemicals*
NFPA *101®*, *Life Safety Code®*
NFPA 701, *Standard Methods of Fire Tests for Flame-Resistant Textiles and Films*

Additional Reading

Adams, R., "Be Prepared for School Hazards," *Firehouse*, Vol. 20, No. 10, Oct. 1995, p. 16.

Collier, P., "Burn-Up at Avondale College," *Build*, Oct. 1990, p. 11.

Cowan, D., and Kuenster, J., "To Sleep with the Angels: The Story of Fire," Chicago: Ivan R., Dec., 1996.

Endthoff, G., "U. S. School Fire Problem," *Fire Prevention*, No. 242, Sept. 1991, pp. 21–23.

"Fire at Tonbridge School Chapel on 17 September 1988," *Fire Prevention*, No. 278, Apr. 1995, p. 17.

Horasan, M., and Bruck, D., "Investigation of a Behavioral Response Model for Fire Emergency Situations in Secondary Schools," *Proceedings* of the Fourth International Symposium on Fire Safety Science, July 13–17, 1994, Intl. Assoc. for Fire Safety Science, Boston, MA, 1994, pp. 715–726.

Horasan, M., and Johnson, P., "Comparison of Evacuation Models for Occupant Populations With Different Levels of Independence and Mobility—Case Studies on Primary Schools and a Special Care Center for Disabled Persons," *Proceedings* of the First International Conference on Fire Science and Engineering, ASIAFLAM '95, Mar. 15–16, 1995, Kowloon, Hong Kong, 1995, pp. 175–186.

Leech, D., "Risk Surveyor Examines How Automatic Sprinklers Can Limit School Fires," *Fire*, Vol. 85, No. 1050, Dec. 1992, pp. 15–16.

Powers, D., "Installing Sprinklers in Existing School Buildings," *Fire Prevention*, No. 277, Mar. 1995, pp. 16–18.

"Sprinklers: The Solution to School Fires?" *Fire Prevention*, No. 255, Dec. 1992, p. 17.

Scoones, K., "Serious Fires in Educational Establishments During 1992," *Fire Prevention*, No. 263, Oct. 1993, pp. 19–21.

Scoones, K., "Serious Fires in Educational Establishments During 1993," *Fire Prevention*, No. 283, Oct. 1995, pp. 35–37.

Scoones, K., "Serious Fires in Educational Establishments During 1994," *Fire Prevention*, No. 290, June 1996, pp. 31–33.

Taylor, K. T., "Burning Down the School. . .The Lessons We Don't Learn Can Hurt Us," *Fire Journal*, Vol. 84, No. 3, May/June 1990, pp. 60–64, 66, 68–69.

"University Fire Protection," *Record*, Vol. 69, No. 6, Nov./Dec. 1992, pp. 3–8.

Williams, B. E., "St. John's School of Alberta: A Fire Sprinkler Case Study," *Fire Fighting in Canada*, Vol. 38, No. 10, Dec. 1994, pp. 4–5.

Williams, D. J., "Fire Safety Education in Schools in the United Kingdom," Mid Glamorgan Fire Service, UK, CIB W14/95/1 (J); Tokyo Fire Department, Firesafety Frontier '94, International Fire Conference and Exhibition in Tokyo on Creating a Safe Tomorrow, Oct. 18–22, 1107–112, Tokyo, Japan, 1994, pp. 299–303.

DETENTION AND CORRECTIONAL FACILITIES

Wayne G. Carson

Detention and correctional occupancies include buildings and facilities in which persons are restrained by locks they do not control. Such occupancies include adult correctional institutions, adult local detention facilities, adult community residential centers, juvenile detention facilities, juvenile training schools, juvenile community residential centers, adult and juvenile work camps, and adult and juvenile substance abuse centers. Some mental hospitals and nursing homes may have similar security provisions, but this chapter addresses only facilities where controlled movements are not based entirely on health care considerations. Security is a major operational consideration in these facilities and must be included in the fire protection design process.

Large correctional facilities may contain other occupancies, including industrial occupancies, such as vocational shops; business occupancies, such as classrooms and offices; assembly occupancies, such as dining halls, auditoriums, and gymnasiums; and storage occupancies, such as warehouses. The principal concern of Chapters 14 and 15 of NFPA *101*®, *Life Safety Code*®, is the residential portion of correctional facilities. Except for the locking of doors, other uses within the detentional and correctional occupancy must comply with the appropriate occupancy chapter of NFPA *101*. If locking of doors in these occupancies is required, Chapters 14 and 15 of NFPA *101* should be reviewed for guidance. These chapters do permit locking of doors in other occupancies if staff is available with keys to unlock the egress doors. The other occupancies are discussed elsewhere in this section.

Additional information relevant to fire safety in correctional institutions can be found in Section 1, Chapter 3, "Systems Concepts for Building Fire Safety"; Section 7, Chapter 3, "Interior Finish"; Section 7, Chapter 6, "Smoke Movement in Buildings"; and the portable fire extinguisher chapters in Section 6.

OCCUPANCY CHARACTERISTICS

The major difference between detention and correctional occupancies and other residential occupancies is that the occupants are not capable of significant self-preservation actions until someone on the staff unlocks the doors. However, the occupants are capable of such actions when the doors are unlocked.

Wayne G. Carson, P.E., is principal of Carson Associates, Inc., Fire Protection Engineers & Code Consultants, Warrenton, VA. He is a member of NFPA's Safety to Life Committee.

The degree of locking can have a significant impact on the risk to the occupants. For example, a facility with only two locks that can be remotely released to allow the occupants out of the building into a controlled exercise yard generally would present less risk to the occupants than one that required 15 locks to be unlocked manually with 10 different keys.

The protection provided generally depends on the degree of locking. Five types of restraint or locking systems are generally found in detention and correctional occupancies. These different locking conditions, referred to as "Use Conditions" by NFPA *101*, are illustrated in Figure 9-7A and defined as follows.

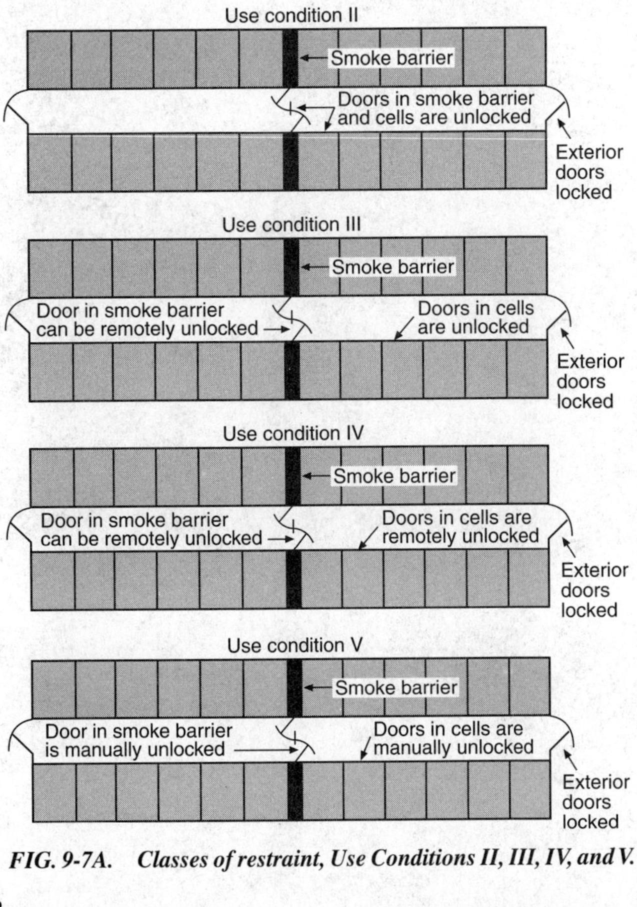

FIG. 9-7A. *Classes of restraint, Use Conditions II, III, IV, and V.*

Use Condition I—Free Egress

Free movement is allowed from sleeping areas and other spaces where access or occupancy is permitted to the exterior through means of egress meeting the requirements of NFPA *101*. For example, a work release center, where the doors are not locked, would be considered Use Condition I. Use Condition I is not considered a detention and correctional facility and is addressed in other residential chapters of the code, such as dormitories in Chapters 16 and 17.

Use Condition II—Zoned Egress

Free movement is allowed from sleeping areas and any other occupied smoke compartments to other smoke compartments, but the exterior doors are locked.

The individual cell door may be locked from the inside by the occupant and can be unlocked by the occupant. Also, the individual cell door may be locked by the occupant when leaving the cell. This is sometimes done to provide security for the occupants' property while occupants are out of their cells during work detail, for example.

Use Condition III—Zoned Impeded Egress

Free movement is allowed within individual smoke compartments—for example, within a residential unit composed of individual sleeping rooms and within group activity space—with egress from that smoke compartment to another smoke compartment provided by the remote-control unlocking of doors. The exterior doors may be manually locked. Again, the individual cell doors may be locked by the occupant as noted in the "Use Condition II" section.

Use Condition IV—Impeded Egress

Free movement is restricted from an occupied space. Remote-control release is provided to permit movement from all sleeping rooms, activity spaces, and other occupied areas within the smoke compartment to other smoke compartment(s). Exterior doors may be manually locked.

Use Condition V—Contained

Free movement is restricted from an occupied space. Staff-controlled manual release at each door is provided to permit movement from all sleeping rooms, activity spaces, and other occupied areas within the smoke compartment to other smoke compartments.

When evaluating the degree of locking for determination of Use Condition, any operable locks, even those not normally used, need to be considered lockable.

THE FIRE PROBLEM

Each year NFPA statistically analyzes the fire problem in correctional facilities.[1] The expanded version of this analysis, which appears in Table 9-7A, identifies the significant common characteristics of correctional facility fires and groups them into several categories.

Most fires in detention and correctional facilities are incendiary or suspicious in nature. Motives may include:

1. Increased chances of escape.
2. Malicious damage as a protest against conditions.
3. Show of force during a riot.
4. Suicide attempt.
5. Divert attention from other activities.

In nearly all intentionally set fires for which the ignition source could be identified, the source was a match, smoking materials, or a cigarette lighter. These are the ignition sources most readily avail-

TABLE 9-7A. *Patterns of Fires in Prisons and Jails, 1989–1993**

Category	Fires per Year	
Major Cause		
Incendiary or suspicious causes	1,700	67.8%
Smoking materials (i.e., lighted tobacco products)	200	6.3
Appliances, tools, or air-conditioning	100	5.8
Dryers	100	3.6
Cooking equipment	100	4.6
Electrical distribution system	100	3.4
Open flame, embers, or torches	100	4.6
Matches	100	2.5
Other equipment	100	3.3
Heating equipment	0	1.7
Natural causes	0	1.3
Other heat source	0	0.6
Exposure (to other hostile fire)	0	0.4
Child playing	0	0.2
Total	2,600	100.0
Leading Areas of Fire Origin		
Cell or other bedroom	1,600	64.1
Laundry room	100	5.4
Kitchen	100	4.8
Hallway or corridor	100	3.9
Leading Areas First Ignited		
Mattress or bedding	1,000	40.3
Trash	300	10.1
Paper	200	6.1
Clothing on or not on a person	200	6.0
Wire or cable insulation	100	4.3
Linen other than bedding	100	3.4

* Source: 1989—93 NFIRS, NFPA survey.

Fires are expressed to the nearest hundred. Percentages are calculated on the actual estimates, so two figures with the same rounded-off estimates may have different percentages.

The 12 major cause categories are based on a hierarchy developed by the U.S. Fire Administration. Sums may not equal totals because of rounding errors.

able to inmates. The typical correctional facility fire is incendiary in origin; starts in a cell; and involves bedding, trash, or papers.

THE SYSTEMS APPROACH TO FIRE SAFETY IN CORRECTIONAL FACILITIES

To deal with the fire safety problem in correctional facilities, two concerns need to be addressed continually: security for the public and safety from fire for the inmates. The systems approach to fire safety helps to achieve these two objectives through a systematic analysis of each problem area and application of available technologies. Fire safety should be an integrated subsystem of building design, construction, and operation. This method of designing fire safety is known as the systems approach or systems concept of fire-safe building design. (See Section 1, Chapter 3, "Systems Concepts for Building Fire Safety.") A system organizes interacting components in such a way that, working together, they perform a predetermined function or reach a specified objective. The use of the systems approach first requires that fire safety goals be clearly identified. These goals should describe the amount of protection a building should provide its occupants, contents, and operations. These goals should be quantified wherever possible. Once these fire safety goals are established, the methods used to achieve them should work collectively to achieve and maintain the prescribed level of fire safety.

The systems approach can be illustrated by the simplified fire safety system for correctional facilities. The box at the top of Figure 9-7B shows the four objectives of the simplified fire safety system:

Life safety: The protection of life is paramount in any fire. Occupants can be protected by providing a safe area of refuge from the fire within the building or by providing a safe and readily available path of travel to the outside.

Property protection: The protection of property is also of concern, particularly with today's higher replacement costs. Some areas may be considered more important than others—for example, a control room vs. a storage room—and therefore require more protection.

Limited downtime: This involves providing protection to limit fire damage so an area can be returned to use quickly. For example, the loss of a storage area may not seriously affect the prison's operation, as trailers or other temporary storage units can be obtained quickly. However, loss of the kitchen could seriously affect a prison's operation.

Security: The primary function of a correctional facility is to maintain a secure perimeter. Fire safety must be compatible with this function.

These objectives are achieved through a series of goals: (1) ignition control, (2) fuel control, (3) occupant protection, (4) detection and suppression, and (5) planning and training. Success in meeting the overall objectives will depend on the degree of success in meeting these five goals, and the relationships between goal success and objective success may be mathematically complex. In practical terms, however, no one goal is likely to be sufficient by itself to meet the objectives, so it is safest to assume that all five goals must be addressed.

Some methods to attain each goal are listed beneath the goals shown in Figure 9-7B. For the goal of ignition control, for example, the methods available to reduce ignitions include: (1) controlling smoking materials, (2) controlling electrical ignition sources, and (3) using alternate grievance procedures to reduce intentionally set fires. When certain methods for achieving the goal are not practical or not particularly effective, other parts of the system may have to be enhanced to compensate. For example, when ignition sources cannot be effectively reduced, fuel control becomes more important, or when the quantity of combustible materials cannot be effectively controlled, sprinklers may offset the hazard.

Because different goals do not have an equal impact on the ultimate level of fire protection, the concept of equivalent protection, also known as tradeoffs, may be necessary. For example, under certain conditions, NFPA *101* permits the use of higher flame spread materials as interior finish if automatic sprinklers are provided. The systems approach provides for analysis of the interacting components of the fire safety system so that alternate means of protection can be evaluated. This evaluation may include initial costs, operational and maintenance costs, and impact on operations.

Ignition Control

Ignition control means greater and more dependable protection of heat sources and separation from fuel sources. In a property where most fires are deliberately set, this primarily means eliminating unnecessary heat sources. Unfortunately, total ignition control is nearly impossible to maintain and may therefore be the least effective fire defense measure. Sources of ignition are nearly always available because cigarettes and matches are, in most cases, available to inmates, and electrical appliances are sometimes allowed in cells. However, an effort must be made to control unwanted or unnecessary ignition sources. Some jails and prisons are prohibiting smoking to reduce ignition sources. If smoking is permitted, the types of lighters or matches can be specified, and the overall use of electrical power must be carefully supervised. The number and type of heat-producing appliances, such as toasters, hot plates, and space heaters, must be controlled by the facility administrator to prevent overloading circuits. If electrical devices are allowed in the cell, adequate electrical outlets should be provided to discourage the use of extension cords.

Fuel Control

Fuel control means controlling the type, arrangement, and burning characteristics of potential fuels. These fuels may include the build-

Life safety	Security
Property protection	Limited downtime

Ignition control	Fuel control	Occupant protection	Detection and suppression activities	Planning and training operations
1. Control smoking materials 2. Control electrical ignition sources 3. Use alternate grievance procedures	1. Control quantity of fuel 2. Control types of fuel 3. Control fuel arrangement	1. Provide reliable evacuation to secure area 2. Provide features for "defend in place" occupancy	1. Provide early warning detection 2. Provide reliable alarm system 3. Provide reliable suppression 4. Provide manual suppression	1. Provide staff education and training 2. Provide inmate education 3. Plan emergency procedures 4. Conduct drills

FIG. 9-7B. Simplified fire safety system for correctional facilities.

ing structural system, interior finish materials, and combustible contents, such as furnishings and inmates' personal property. Assuming that heat sources will be available, the likelihood of fire and the rate of fire growth can be controlled by limiting the type, quantity, and arrangement of the fuels available. Limiting the speed at which fire develops and spreads reduces its impact on life safety. Slow-burning fires give people and suppression equipment time to react.

However, fast fires—those with a rapid rate of smoke development, flame spread, and production of toxic gases—are possible and pose a particular risk of fire deaths in correctional facilities. Ordinary combustible materials, such as low-density fiberboard ceilings, thin plywood paneling, and some plastics used in furnishings and interior finish, can contribute to fast fire development.

Achieving the goal of fuel control means keeping fuel quantity at a minimum, controlling the type of fuel, eliminating those fuels that are easy to ignite or result in fast fires, and separating fuels from readily available ignition sources. One part of fuel control is to eliminate highly combustible furnishings. Since mattresses and bedding are the fuel sources most readily available to inmates, they are most frequently involved in fires. Certain types of polyurethane foam mattresses have been a significant fuel source in many prison fires because they typically result in fast fire development, high heat release, and heavy smoke production.

Mattresses can be made to resist ignition from cigarettes, which is a low-heat ignition source. The federal "Flammability Standard for Mattresses"[2] requires that mattresses resist ignition by cigarettes placed at 18 specified locations. This test does not include sheets or blankets on the mattress and does not indicate the resistance to ignition from open flames or the intensity of the resulting fire. A series of tests of mattresses by the U.S. Department of Agriculture[3] concluded that "the data show that mattresses containing polyurethane foam easily comply with the federal standard[2] where cigarettes are the igniting source; however, such mattresses present a significant hazard where the igniting source is an open flame. Mattresses containing 100 percent polyurethane foam usually burn vigorously until little char remains. Burning in such cases is accompanied by the release of copious amounts of black smoke."

In 1977, the National Bureau of Standards (now the National Institute of Standards and Technology) conducted fire tests of mattresses and issued a report[4] which categorized mattresses into four groups in order of safety.

Group A: Mattresses that did not exceed any of the tenability criteria for the duration of the 30-min test. This group included two treated, cotton-batting mattresses.

Group B: Mattresses that only exceeded the smoke obscuration criterion. Two neoprene mattresses were in this category.

Group C: Mattresses that exceeded all tenability criteria, but did not cause full room involvement. This group included three polyurethane foam-core mattresses and one of mixed fiber construction. The best-performing polyurethane mattress was of the type associated with a multiple-life-loss prison fire.

Group D: Mattresses that exceeded all criteria. Included were one styrene-butadiene latex foam core and one polyurethane foam-core mattress. The latex mattress was of the type associated with a multiple-life-loss fire in a health care institution.

These tests indicate that the best mattresses for use in correctional facilities are those with padding material of cotton treated with boric acid. Mattresses of cotton treated with boric acid pass the federal standard[2] and perform well when exposed to open-flame ignition sources. Such mattresses will burn but at a much slower rate than urethane foam mattresses.[5] Some promising new materials are

entering the market for mattress padding, including treated neoprene. Since there is no national standard for open-flame ignition testing of mattresses, however, each jurisdiction must evaluate new mattress materials carefully and thoroughly.

Inmate possessions and furnishings in the cell must be controlled in order to limit the size of fire that can be expected. Books, clothing, and other combustible personal property allowed in cells should be stored in closable metal lockers or fire-resistant containers. Combustible decorations should be prohibited unless they have been treated with flame retardants. Draperies, curtains, or other decorative and acoustical materials should be noncombustible or rendered and maintained flame resistant according to NFPA 701, *Standard Methods of Fire Tests for Flame-Resistant Textiles and Films.* There is no maximum limit of combustibles permitted in NFPA *101.* The code does provide a cautionary statement for management to control the combustibles.

Interior finish materials contribute fuel to a fire, they may accelerate room flashover, and they can result in rapid fire spread. Materials that have high flame spread rates or produce hazardous concentrations of smoke or noxious gases should not be used. This includes plastic wall finish material, fiberboard, thin paneling, and untreated plywood. (See Table 9-7B.) Interior finish alternatives with good fire characteristics include paneling treated to provide Class A interior finish rating with a flame spread of 25 or less, painted steel, gypsum board, or masonry block.

TABLE 9-7B.   *Flame Spread of Some Building Materials**

Material	Flame Spread
Ceilings	
Glass-fiber sound-absorbing blankets	15 to 30
Mineral-fiber sound-absorbing panels	10 to 25
Shredded wood fiberboard (treated)	20 to 25
Sprayed cellulose fibers (treated)	20
Walls	
Aluminum (with baked enamel finish on one side)	5 to 10
Asbestos cement board	0
Brick or concrete block	0
Cork	175
Gypsum board (with paper surface on both sides)	10 to 25
Northern pine (treated)	20
Southern pine (untreated)	130 to 190
Plywood paneling (untreated)	75 to 275
Plywood paneling (treated)	10 to 25
Red oak (untreated)	100
Red oak (treated)	35 to 50
Floors	
Carpeting	10 to 600
Concrete	0
Linoleum	190 to 300
Vinyl asbestos tile	10 to 50

*For a comprehensive list of flame spread and smoke developed ratings for proprietary materials, see the current edition of *Building Materials Directory* published by Underwriters Laboratories Inc.

Occupant Protection

Occupant protection means providing life safety to building occupants in case of fire. This may be accomplished either by evacuating them to a secure area or by defending them in place. This goal is the most controversial in the fire safety system for correctional facilities because "evacuation" is often interpreted by administrators and staff to mean "escape." While this assumption is not valid, means of egress does directly affect security.

With reference to the evacuation of inmates, most detention and correctional facilities can be grouped into two general types, based on their physical arrangement. In the first group are larger institutions that have an outdoor secure area such as an outdoor exercise yard or courtyard. It is possible to evacuate inmates rapidly from the fire area to this outside area and still maintain security. In the second category are the smaller city or county facilities where there are no separate, secure outdoor areas. In these facilities, which have no enclosed courtyards or other areas to evacuate inmates and still maintain security, it becomes necessary to defend the inmates in place from fire.

A defend-in-place occupancy is a workable and accepted means of providing fire safety for building occupants and is commonly used to protect persons in hospitals and nursing homes. The technology exists to defend inmates in place so that only a few cells need to be evacuated—possibly only the cell of fire origin. A defend-in-place occupancy, as well as an occupancy using evacuation to a secure area of refuge, needs a reliable means of egress. Although means for defending the occupants in place are provided, it is always considered necessary to provide reliable exits. According to NFPA *101*, this means that two distinct paths of travel must be provided from each cell block or area. The exits should be remote from each other so that a single fire cannot readily block both. In addition, travel distances must not exceed those specified by NFPA *101*. The travel distance can be increased if the building is sprinklered. (The maximum distance is 150 ft without sprinklers and 200 ft with sprinklers in detention and correctional occupancies.) The exit access needs to be free and unobstructed, and it should not be circuitous. An exit access that requires travel through a room or an area containing a fire hazard higher than usual for the occupancy violates the principles of safe egress. Dead-end corridors are not good practice because a fire in a dead end between an exit and an occupant can prevent the occupant from reaching the exit. Exitways need to be illuminated continuously, and emergency lighting must be provided. Exits and the paths to them should be clearly marked. However, NFPA *101* permits the elimination of exit signs in inmate housing areas where no visitors are permitted. The rationale for this is that the inmates cannot leave until someone unlocks the doors and that the inmates are generally familiar with the door locations.

Compartmentation is an important aspect of occupant protection, whether inmates are evacuated or defended in place. Compartmentation limits the number of persons exposed to a single fire. The walls, ceilings, and floors of fire compartment boundaries must be fire-resistant. Openings must be protected with self-closing, fire-rated doors. The degree of fire protection these elements provide depends on the degree of hazard the occupancy presents, or its fuel loading, and the type of building construction—for example, fire resistive *vs.* combustible—and the function served, such as a load-bearing or nonload-bearing wall.

Controlling smoke means controlling the type and quantity of materials used in the building construction and building contents, and providing a means of restricting smoke movement within the building. Dilution, which is reducing smoke concentration by introducing massive quantities of uncontaminated air into a building, is not always satisfactory when used alone. Smoke may be generated in such quantities that adequate dilution may not be obtained. Prudent use and placement of automatically opening vents can provide some relief from smoke and combustion products in small, one-story buildings. However, smoke confinement can best be achieved by providing a physical barrier, such as a wall with self-closing doors and dampers that restricts smoke movement, and by using a pressure differential across the physical barrier to prevent smoke from entering the nonfire area. The building's heating, ventilating,

and air-conditioning (HVAC) system can be designed to provide this pressurization. Physical barriers alone usually are not effective in controlling smoke movement because they have many penetrations, and the fire, the building HVAC, and/or the weather conditions create uncontrolled pressures. NFPA 92A, *Recommended Practice for Smoke-Control Systems*, provides information on the design and installation of smoke-control systems.

Since the primary objective of detention and correction facilities is security, reliable locking systems are essential. However, problems with locking systems have played a major role in some of the deadliest correctional fire tragedies. Therefore, whatever their type, locking systems must function reliably in emergencies, whether a facility is designed to evacuate its inmates or to defend them in place. Individual key locks have proven to be the most unreliable for several reasons, including the time required to unlock each door, the loss of keys or the breaking of a key in a lock during an emergency, heat and smoke prohibiting entry into the cell block, and confusion due to the number of keys required to release the inmates.

It is important that the number of keys needed to evacuate the occupants to the outside of an area of refuge be limited to reduce the confusion. Also, all keys needed for evacuation should be marked by sight and touch for quick identification in an emergency.

Detection and Suppression Activities

Detection and suppression activities deal directly with the fire—detecting its presence, sounding an alarm to alert occupants, and inhibiting fire growth by active fire suppression. To achieve this goal, the following must be provided: an early warning fire detection system, a reliable alarm system, and a reliable fire suppression system.

There are three separate phases of detection and suppression activities, each of which can be performed automatically or manually. (See Table 9-7C.) Automatic and manual systems can also be combined. For example, detection can be provided by smoke detectors installed throughout the facility with manual fire fighting by fire brigade personnel using a standpipe hose. However, any system requiring manual intervention requires that someone make a decision, requires time for human action, and requires trained personnel with the required equipment to do the job. The human element is generally the most unreliable part of a fire safety system.

TABLE 9-7C. *Stages of Detection and Suppression*

Stage	Manual	Automatic
Detection	Inmate or guard detects fire	Smoke detectors or sprinklers detect fire
Alarm	Guard verbally transmits information to other guards	Fire alarm system activates audible and visible alarms
Suppression	Guards use extinguishers to suppress fire	Sprinklers begin extinguishing immediately

The effectiveness of fire detection and suppression systems depends on the rate of the fire growth and the time from ignition until detection and, eventually, suppression activities begin. If the fuel available is capable of producing a fast fire, as it is in padded cells, the extinguishing agent, usually water, must be applied to the fire very quickly after ignition. This is very difficult to do with manual fire suppression even if early warning fire detection, such as automatic smoke detectors, is available. Automatic sprinklers provide reliable fire detection and suppression, and they need no human intervention. With an automatic sprinkler system, the extinguishment phase begins almost simultaneously with detection and alarm, a response difficult to achieve with manual suppression.

Sprinklers have been, and are being, installed quite successfully in detention and correctional facilities throughout the U.S. and Canada. Sprinklers control fire development, thereby reducing the production of heat and smoke. They also give fire service personnel a better opportunity to begin rescue and extinguishment operations. Automatic sprinklers can reduce property loss and limit downtime. In addition, they offer several economic advantages:

1. Insurance premiums may be reduced with the installation of sprinkler protection.
2. Many building codes offer tradeoffs when sprinklers are used, thereby reducing building costs.
3. Sprinklers can be installed at a discount in correctional facilities by inmates under the supervision of a qualified fire protection engineer.
4. Sprinklers may also result in fewer persons being evacuated, reduce the burden for manual fire fighting, and, in effect, increase security.

NFPA *101* requires smoke detectors in the sleeping areas of detention and correctional occupancies. Smoke detectors may cause needless or nonfire emergency alarms. Therefore, in housing areas, they may alarm at a constantly attended location, such as a control room, and are not required to sound a general evacuation alarm nor call the fire department. A manual fire alarm system is also required. NFPA *101* does permit the manual stations to be locked, provided the staff has the keys. In addition, the alarm systems should be connected directly to the local fire department. Manual suppression ability should include portable fire extinguishers of the proper type and number, readily accessible fire hose stations for use by correctional officers, and a means to stretch fire hoses through double security gates, or sallyports, so the gates can be closed and security maintained. All automatic detection and suppression systems and manual equipment need to be inspected and tested regularly.

Planning and Training Operations

Planning and training means conducting training activities among inmates and staff and planning emergency operating procedures. To achieve this goal, the following methods are used: staff education and training, inmate education, planning emergency procedures, and conducting drills. Planning and training require little capital investment in equipment, but they can make a big difference in reducing the impact of a potentially disastrous fire, especially in a system where fire defenses are marginal. In addition, planning and training are important to maintain a high level of security during a fire.

Because fire safety should be everyone's concern, both staff and inmates should be involved in a fire safety program. Each shift should practice the fire emergency plan at least quarterly, and new employees should be briefed routinely on the fire emergency plan. Inmates should be instructed in emergency procedures. Safety information may also be included in an inmate information booklet.

The fire safety program should be designed to meet the needs of the facility. The size and age of the buildings and the security classification of the facility, as well as its proximity to municipal fire departments should be considered. All training programs should include a short description of the background and evolution of the problem; how to recognize, prevent, and reduce fire hazards; information on the fire protection technology available for application at the facility; hands-on training for staff and inmate fire brigades; emergency operating procedures; and identification of potential problems that affect fire safety.

Fire brigades should be equipped and trained to deal effectively with a fire emergency. Brigade responsibility and functions will vary with the size of the facility and the size and location of the nearest fire department, but they should include calling the fire department, safeguarding lives, providing manual suppression to control the fire until the fire department arrives, and protecting equipment.

Each building in a correctional facility should have a written fire emergency plan detailing staff action during a fire emergency. The plan provides a guide for evaluating the particular problems and coordinating the response of the fire brigade, fire department, staff, and inmates. Emergency plans should be simple yet comprehensive, specific, flexible, and workable. Preparing the plan can be accomplished in five steps:

1. Define the potential fire protection problems in the particular correctional facility.
2. Set objectives for what can be accomplished with the plan in an emergency.
3. Determine the facility's capability for controlling an emergency.
4. Define the roles of the responding agencies, especially the fire department and fire brigade.
5. Put the information in written form.

The local fire department should be involved in formulating the fire emergency plan and briefed on building conditions, contents, and fire-fighting equipment within the complex. Fire departments with correctional facilities within their jurisdictions should conduct surveys of correctional facilities and have their own pre-fire plans for each building.

BIBLIOGRAPHY

References Cited

1. Hall, J. R., Jr., *The U.S. Fire Problem Overview Report Through 1988*, NFPA Fire Analysis and Research Division, Quincy, MA, 1989.
2. "Flammability Standard for Mattresses," FF4-72, June 1973, U.S. Department of Commerce, Washington, DC.
3. "Resistance of Mattresses Containing Boric Acid Treated Cotton Batting to Open Flame Ignition," *Journal of Consumer Product Flammability*, Vol. 4, June 1977, pp. 169–188.
4. "Combustion of Mattresses Exposed to Flaming Ignition Sources/Part I—Full-Scale Tests and Hazard Analysis," NBSIR-77-1290, 1977, National Bureau of Standards, Washington, DC.
5. "Firesafety Bulletin on Penal Institutions," National Fire Protection Association, Quincy, MA, 1977.

NFPA Codes, Standards, and Recommended Practices

Reference to the following NFPA codes, standards, and recommended practices will provide further information on detention and correctional facilities discussed in this chapter. (See the latest version of *The NFPA Catalog* for availability of current editions of the following documents.)

NFPA 13, *Standard for Installation of Sprinkler Systems*
NFPA 72, *National Fire Alarm Code*
NFPA 90A, *Standard for the Installation of Air Conditioning and Ventilating Systems*
NFPA 92A, *Recommended Practice for Smoke-Control Systems*
NFPA *101*®, *Life Safety Code*®
NFPA 251, *Standard Methods of Tests of Fire Endurance of Building Construction and Materials*
NFPA 255, *Standard Method of Test of Surface Burning Characteristics of Building Materials*
NFPA 701, *Standard Methods of Fire Tests for Flame-Resistant Textiles and Films*

Additional Reading

Coffey, J. L., Sr., "Correctional Facility Fire Departments," *Firehouse*, Vol. 17, No. 7, July 1992, pp. 78, 81–83.

"Deadliest Prison Fire," *NFPA Journal*, Vol. 89, No. 5, Sept./Oct. 1995, pp. 84–85.

"Inter-Service Liaison Helps at Leyland Prison Incident," *Fire*, Vol. 86, No. 1062, Dec. 1993, pp. 37, 40.

Morrison, L., "Evaluation of Life Safety in Correctional Institutions," Professional Loss Control, Canada, Society of Fire Protection Engineers and Worcester Polytechnic Institute, *Proceedings* of Computer Applications in Fire Protection, June 28-29, 1993, Worcester, MA, 1993, pp. 15–19.

National Fire Protection Association, "Prisons. Case Histories," National Fire Protection Assoc., Quincy, MA, 1990.

National Fire Protection Association, "Special Data Information Package: Prisons and Jails," National Fire Protection Association, Quincy, MA, 1993.

Newsome, A., "Prison Design Moves Downtown," *Consulting-Specifying Engineer*, Vol. 18, No. 4, October 1995, pp. 58–60, 62, 64.

Occupational Safety and Health Administration (OSHA), *Code of Federal Regulations,* Part 29, Section 1910.156, Fire Brigades.

Rice, M.E., and Harris, G. T., "Firesetters Admitted to a Maximum Security Psychiatric Institution," *Journal of Interpersonal Violence*, Vol. 6, No. 4, Dec. 1991, pp. 461–475.

Ruikar, V. G., "Customized Tests for Detention and Correctional Facilities," *ASTM Standardization News*, Vol. 24, No. 5, May 1996, pp. 16–19.

Teague, P. E., "Protecting Our Prisoners: The Challenge of Keeping Prisons Safe From Fire," *NFPA Journal*, Vol. 88, No. 5, Sept./Oct. 1994, pp. 49–55.

Tucker, R. W., "Early Detection in Jails—While Minimizing Unwanted Alarms," *SFPE Bulletin*, September/October 1990, pp. 6–8.

vanderDoorn, S. G. J., "Requirements for Staff in Institutions to be Able to Evacuate Persons in Case of Fire," Sixth International Fire Conference on Fire Safety, Interflam '93, March 30–April 1, 1993, Oxford, England, Interscience Communications Ltd., London, England, 1993, pp. 809–826.

Vogt, B. M., "Evacuation of Institutionalized and Specialized Populations," Oak Ridge National Lab., Oak Ridge, TN, Department of Energy, Washington, DC, ORNL/SUB-7685/1-T23, Feb. 1990.

Yancey, C. W. C., "Test Methods for Detention and Correctional Facility Locks," National Institute of Standards and Technology, Gaithersburg, MD, National Institute of Justice, Washington, DC, NISTIR 4975, Nov. 1992.

HEALTH CARE OCCUPANCIES

Revised by Thomas W. Jaeger

Health care facilities are used for the treatment or care of persons suffering from physical or mental illness, disease, or infirmity, and for the care of infants, convalescents, or aged persons. These facilities provide sleeping accommodations for occupants who may be incapable of self-preservation because of physical or mental disability or age. Some buildings that house health care occupants have security measures that limit freedom of movement.

In recent years, facilities—sometimes called ambulatory health care facilities—have been developed to provide medical treatment on an outpatient basis. Although patients might be placed under general anesthesia or other treatment that would render them incapable of self-preservation, they are not housed overnight. Occupants exhibit some characteristics of people in business occupancies and some characteristics typical of people in health care facilities.

In the last several years, tremendous growth has occurred in an occupancy related to health care: residential board and care facilities. These facilities are commonly called assisted living facilities, personal care facilities, and so on. Residential board and care facilities provide personal care to occupants in a residential setting. These residents' ability to respond to a fire threat differs greatly from patients in health care facilities. The safeguards appropriate for health care facilities should only be applied when a facility meets the definition in NFPA *101*®, *Life Safety Code* ®, of a health care facility.

This chapter describes the characteristics and fire hazards of health care facilities. It specifically discusses the characteristics of the health care occupant and the fire safety features that all health care facilities should provide.

Additional information relating to health care facilities can be found in Section 7, Chapter 3, "Interior Finish"; Section 7, Chapter 5, "Confinement of Fire in Buildings"; Section 9, Chapter 9, "Board and Care Facilities"; and Section 9, Chapter 28, "Protection of Electronic Equipment."

OCCUPANCY CHARACTERISTICS

Occupants of health care facilities are generally presumed to be incapable of self-preservation. A significant percentage of occupants in hospitals and nursing homes are incapable of self-evacuation or are ambulatory but incapable of perceiving a fire threat and choosing a rational response.

Thomas W. Jaeger, P.E., is president of Gage-Babcock & Associates, Inc., fire protection consultants, Fairfax, VA.

There are three types of care in most modern hospitals: (1) ambulatory, (2) general, and (3) intensive care. Given proper directions, unless smoke or heat is intense, ambulatory patients can make their own way to safety. General care patients may be transported on stretchers or in wheelchairs with some difficulty; horizontal and even some vertical movement is generally possible, although independent evacuation is not. Patients in intensive care are likely to be connected to various life support devices, making movement for even short distances very difficult and evacuation almost impossible without further endangering these patients' lives.

Occupants of nursing homes vary from geriatric residents who are capable of evacuation with limited assistance to residents who are comatose and require close supervision with evacuation being very difficult. Today's nursing home resident is less ambulatory and requires more medical care than a resident of even 10 years ago. The significant increase in the number of assisted living facilities has made the typical intermediate care nursing home resident of yesterday an assisted living resident today.

Because some occupants are incapable of movement or slow to evacuate, a health care facility resembles a ship at sea: it is better to keep the fire from the patient than to remove the patient from the fire. Thus, occupants must be defended in place. As a result, health care facility design and operation must incorporate methods by which a fire can be detected early, contained, and fought rapidly and successfully. Accomplishing this requires careful planning of the health care facility and its day-to-day operation.

NFPA *101*, widely used to establish minimum requirements for life safety from fire within health care facilities, sets forth criteria based upon the following general principles:

1. Fire-resistive construction.
2. Subdivision of spaces, known as compartmentation.
3. Protection of vertical openings.
4. Provision of adequate means of egress.
5. Provision of exit marking, exit illumination, and emergency power.
6. Limits on the use of interior finish materials.
7. Fire alerting facilities.
8. Control of smoke movement.
9. Protection of hazardous areas.
10. Adequate protection of building service equipment.
11. Control of fuel loads.
12. Operational features.

Total building fire protection for life safety is more necessary in health care facilities than in other occupancies because of the nature of the occupants. At the same time, exits are slightly less important. The first principle of designing a fire safe health care facility must be that safety must not depend wholly upon any single safeguard.

IGNITION SOURCES

In 1989–1993, smoking materials, both lighted tobacco products and implements used to light them, and incendiary and suspicious acts accounted for 59–64 percent of deaths and 46–66 percent of injuries in facilities that care for the sick, and for 71 percent of deaths and 51 percent of injuries in facilities that care for older adults. These causes totally dominate the ignition side of the life safety problem in these properties, although other causes, notably failures in electrical distribution systems, appliances, and other equipment, are also important in property damage. (See Tables 9-8A through 9-8D.)

More and more health care facilities are establishing smoke-free environment policies. In addition, the Joint Commission on Ac-

TABLE 9-8A. *Causes of Fires and Direct Property Damage Care-of-the Sick Facility Fires, 1989–1993 Annual Average Unknown-Cause Fires Allocated Proportionally**

Cause	Fires		Property Damage (in millions)	
Other equipment	600	(16.9%)	$2.4	(21.2%)
Separate motors or generators	100	(2.6%)	$0.1	(1.2%)
Telephones, computers, X-ray machines, or other electronic equipment	100	(2.3%)	$1.1	(9.7%)
Smoking materials	500	(15.8%)	$1.0	(8.4%)
Cooking equipment	500	(14.8%)	$0.9	(8.0%)
Stoves	100	(4.3%)	$0.1	(0.7%)
Portable cooking or warming devices	100	(4.2%)	$0.7	(6.1%)
Ovens or microwave ovens	100	(2.0%)	$0.0	(0.2%)
Appliances, tools, or air-conditioning	500	(13.7%)	$0.8	(7.4%)
Dryers	200	(5.4%)	$0.2	(1.8%)
Incendiary or suspicious causes	400	(12.5%)	$2.9	(25.3%)
Electrical distribution	400	(11.1%)	$1.4	(12.8%)
Light fixtures, ballasts, or signs	100	(3.2%)	$0.0	(0.2%)
Fixed wiring	100	(1.6%)	$0.2	(2.2%)
Open flame, embers, or torches	300	(8.0%)	$0.7	(5.8%)
Torches	100	(3.9%)	$0.4	(3.2%)
Heating equipment	100	(4.2%)	$0.7	(6.0%)
Natural causes	0	(1.2%)	$0.1	(1.2%)
Other heat source	0	(1.0%)	$0.3	(2.4%)
Exposure (to other hostile fire)	0	(0.6%)	$0.2	(1.3%)
Child playing	0	(0.2%)	$0.0	(0.1%)
Total*	3,400	(100.0%)	$11.3	(100.0%)

Source: NFIRS, NFPA Survey.
Note: Fires are expressed to the nearest hundred, property damage to the nearest hundred thousand dollars. Property damage figures have not been adjusted for inflation. Percentages are calculated on actual estimates, so two figures with the same rounded-off estimates may have different percentages.

*The 12 major cause categories are based on a hierarchy developed by the U.S. Fire Administration. Sums may not equal totals because of rounding errors.

TABLE 9-8B. *Causes of Civilian Deaths Care-of-the Sick Facility Fires, 1989–1993 Annual Average Unknown-Cause Fires Allocated Proportionally**

Cause	Civilian Deaths		Civilian Injuries	
Incendiary or suspicious causes	2	(31.8%)	33	(24.2%)
Smoking materials	1	(27.3%)	30	(22.2%)
Other equipment	1	(18.2%)	20	(14.4%)
Electrical distribution	0	(9.1%)	9	(6.9%)
Other heat source	0	(9.1%)	4	(2.7%)
Open flame, embers, or torches	0	(4.5%)	26	(19.3%)
Appliances, tools, or air conditioning	0	(0.0%)	8	(5.9%)
Cooking equipment	0	(0.0%)	5	(3.5%)
Heating equipment	0	(0.0%)	1	(0.6%)
Child playing	0	(0.0%)	0	(0.2%)
Natural causes	0	(0.0%)	0	(0.0%)
Exposure (to other hostile fire)	0	(0.0%)	0	(0.0%)
Total*	5	(100.0%)	136	(100.0%)

Source: NFIRS, NFPA Survey.
Note: Civilian deaths and injuries are expressed to the nearest one. Percentages are based on the actual estamates, which include fractions of casualty per year, so two figures with the same rounded-off estimates may have different percentages.
* The 12 major cause categories are based on a hierarchy developed by the U.S. Fire Administration. Sums may not equal totals because of rounding errors.

creditation of Healthcare Organizations (JCAHO) requires facilities to establish a no-smoking policy. As the result of prohibiting or controlling smoking in health care facilities, a new phenomenon has developed: visitors giving smoking materials to patients and residents without the health care staff's knowledge.

Most of fatalities in health care fires are so close to the point of origin that their locations are coded as "intimate with ignition." (See Table 9-8E.) Most fatalities that occur in the room where a fire begins turn out to have been intimate with ignition.

Most intimate-with-ignition fire deaths begin on the victims' clothing or in the mattress, bedding, or upholstered furniture on which they were lying when the fire began. Fires such as these typically begin near the upper torso, since the materials ignited are near the area where the victims dropped a lighted tobacco product or lighting implement. Only a couple of cases in the NFPA's files involve points of origin well away from the victim's upper torso. For example, one victim died in a fire that began when a lamp fell on the bottom of her bed and ignited the sheet.

FIRE LOADS

Studies show relatively low fuel loads within most spaces in health care facilities. Fire duration varies from approximately 20 min for patient areas to several hours, depending upon the space involved.

A study by the National Bureau of Standards [now the National Institute of Standards and Technology (NIST)] conducted in 1942 involving three hospitals revealed an average fuel load of 5.7 lb per sq ft (27.8 kg/m²).[1] Fuel loads for typical spaces are indicated in Table 9-8F. A more recent study of fuel loads in a U.S. Navy hospital confirms the relatively low fuel loads for general hospital areas but indicates that higher fuel loads may be anticipated in some spaces such as medical libraries, X-ray file rooms, linen storage, and general storage rooms. Further, the increased use of disposables and modern medical record storage practices are likely to result in above average fuel loads. These areas may contain fuel loads suffi-

TABLE 9-8C.   Causes of Fires and Direct Property Damage Care-of-Aged Facility Fires, 1989–1993 Annual Average Unknown-Cause Fires Allocated Proportionally*

Cause	Fires		Property Damage (in Millions)	
Appliances, tools, air conditioning	1,000	(30.4%)	$1.5	(22.1%)
Dryers	700	(20.0%)	$0.7	(10.8%)
Room air conditioners	100	(1.8%)	$0.1	(0.8%)
Cooking equipment	600	(16.4%)	$0.5	(7.0%)
Stoves	300	(8.3%)	$0.1	(1.3%)
Portable cooking or warming devices	100	(3.1%)	$0.1	(1.3%)
Ovens	100	(2.6%)	$0.3	(4.1%)
Smoking materials	500	(13.7%)	$1.7	(24.8%)
Other equipment	300	(9.4%)	$0.3	(4.5%)
Electrical distribution	300	(9.4%)	$0.8	(12.4%)
Lighting fixtures, ballasts, or signs	100	(2.4%)	$0.1	(1.8%)
Fixed wiring	100	(1.6%)	$0.2	(3.1%)
Heating equipment	300	(8.5%)	$0.8	(11.2%)
Fixed-area heaters	100	(4.0%)	$0.3	(4.9%)
Central heating units	100	(2.2%)	$0.4	(5.6%)
Incendiary or suspicious causes	200	(5.7%)	$0.5	(7.4%)
Open flame, embers, or torches	100	(3.0%)	$0.3	(5.0%)
Natural causes	100	(2.0%)	$0.2	(3.1%)
Spontaneous ignitions	100	(1.6%)	$0.2	(2.8%)
Other heat source	0	(0.8%)	$0.1	(1.8%)
Exposure (to other hostile fire)	0	(0.4%)	$0.1	(0.7%)
Child playing	0	(0.1%)	$0.0	(0.1%)
Total*	3,400	(100.0%)	$3.8	(100.0%)

Source: NFIRS, NFPA Survey.

Note: Fires are expressed to the nearest hundred, property damage to the nearest hundred thousand dollars. Property damage figures have not been adjusted for inflation. Percentages are calculated on the actual estimates, so two figures with the same rounded-off estimates may have different percentages.

* The 12 major cause categories are based on a hierarchy developed by the U.S. Fire Administration. Sums may not equal totals because of rounding errors.

TABLE 9-8D.   Causes of Civilian Deaths and Injuries Care-of-Aged Facility Fires, 1989–1993 Annual Average Unknown-Cause Fires Allocated Proportionally*

Cause	Civilian Deaths		Civilian Injuries	
Smoking materials	4	(40.5%)	84	(32.1%)
Open flame, embers, or torches	2	(19.0%)	15	(5.7%)
Lighters	1	(10.8%)	4	(1.6%)
Matches	1	(8.3%)	9	(3.5%)
Electrical distribution	1	(14.3%)	29	(11.0%)
Cords or plugs	1	(9.5%)	13	(4.9%)
Incendiary or suspicious causes	1	(11.9%)	36	(13.6%)
Cooking equipment	1	(7.1%)	19	(7.3%)
Other equipment	0	(4.8%)	11	(4.4%)
Appliances, tools, or air conditioning	0	(2.4%)	40	(15.1%)
Heating equipment	0	(0.0%)	20	(7.6%)
Natural causes	0	(0.0%)	5	(1.8%)
Other heat source	0	(0.0%)	3	(1.0%)
Child playing	0	(0.0%)	1	(0.5%)
Exposure (to other hostile fire)	0	(0.0%)	0	(0.0%)
Total*	9	(100.0%)	263	(100%)

Source: NFIRS, NFPA Survey.

Note: Civilian deaths and injuries are expressed to the nearest one. Percentages are based on the actual estimates, which include fractions of a casualty per year, so two figures with the same rounded-off estimates may have different percentages.

* The 12 major cause categories are based on a hierarchy developed by the U.S. Fire Administration. Sums may not equal totals because of rounding errors.

TABLE 9-8E.   Locations of Victims of Fatal Health Care Facility Fires

Property Class	Intimate with Ignition	Same Room as Fire but Not Intimate	Not Same Room as Fire
Facilities that care for the sick	100%	0%	0%
Hospitals	100	0	0
Other or unknown type	100	0	0
Facilities that care for older adults	71	20	9
With nursing staff	83	17	0
Other or unknown type	45	27	27
All facilities combined	81	13	6

Source: 1989–1993 NFIRS and NFPA survey.

cient for fires from 1 to 3 hr duration. However, such spaces represent a small percentage of total floor area.[5]

Doctor office areas are intermediate fuel load spaces having combustibles in sufficient quantity to produce fires of 30 to 90 min duration.

A major portion of the floor area of health care facilities is used for patient sleeping or treatment rooms. Fuel loads for such spaces are low. For example, fuel loads in nursing homes have been estimated at 2.5 to 3 lb per sq ft (9.72 to 14.64 kg/m²).[2] NIST surveys indicate that fuel loads for hospital patient rooms approximate 4 lb per sq ft (19.52 kg/m²) and that the combustible load in patient sleeping areas ranges from approximately 3 to 4.5 lb per sq ft (14.64 to 22.96 kg/m²).[5] A study of a Navy hospital indicates fuel loads for sleeping rooms average 1 lb per sq ft (4.88 kg/m²) or less.[13] Assuming standard time/temperature conditions, fire duration for patient areas and a majority of other occupied spaces would be less than 30 min.

Disposables

The use of combustible disposable equipment and supplies is on the increase in health care facilities. Disposables include bedding, gowns, gloves, drapes, collection bags, tubing, dishes, glasses, syringes, needles, and many diagnostic and therapeutic instruments. All such items require packaging, which adds to the combustible load both before and after use.

Data Processing and Medical Records

Data processing centers are becoming increasingly important within health care facilities. Used to satisfy general business needs as well as to store patient records, the centers contain quantities of combustibles. These combustibles could expose high-value equipment, result in loss of vital records, and produce a fire that threatens occupants outside the data processing areas. Special protection should be considered for data processing centers.

TABLE 9-8F. *Fuel Loads in Typical Health Care Facilities*

Location	Average Combustible Contents (psf)		
	Movable Property	Exposed Woodwork and Floors*	Total
Rooms (single)	0.5	3.2	3.7
Corridors	0.0	2.6	2.6
Waiting rooms	1.7	1.5	3.2
Janitors' closets and supplies	3.1	3.4	6.5
Doctors' offices	5.7	2.9	8.6
Nurses' offices and rooms	3.1	1.9	5.0
Nurses' infirmary	0.8	2.2	3.0
Diet kitchens and dining rooms	1.2	2.4	3.6
Laundries	4.4	0.6	5.0
Laundries and clothes storage	12.5	0.6	13.1
Dormitories	0.8	2.0	2.8
Pharmacy, dispensary, and stores	5.8	1.9	7.7
Lockers, toilets, and barber shops	0.2	1.2	1.4
Approximate average for entire usable floor area of three hospital buildings surveyed			5.7**

*Combustible floor finish, where present, was 1/4 in. linoleum; it was assumed to be the equivalent of 1 lb of combustible material, such as wood, per sq ft of floor area (4.88 kg/m²). Doors, windows, trim, mouldings, baseboards, etc., are included.

**This approximate average weight was computed from Table 16 on page 25 of the Bureau Report BMS 921. The value is somewhat high because the highest weight in each bracket of combustible contents was used in figuring it, i.e., in the bracket "0 to 4.9 lb/ft² (23.9 kg/m²)," the value 4.9 lb (2.22 kg) was applied to the indicated area, etc.

Health care facilities generate and store medical files in considerable quantity. Files stored in closed steel cabinets do not represent any significant increase in hazard over that typical of most areas. However, files stored on open shelving create a serious fire hazard. Large quantities of open-file storage should be treated as a severe hazard and should be separated by fire-rated construction and protected by automatic sprinklers.

FIRE SEVERITY

Determining relative fire hazard involves considerations beyond total fuel loads. The arrangement of combustibles, their chemical makeup, and their physical state are all factors to be evaluated, in addition to room geometry, ventilation rates, fire compartment size, and fire protection facilities. Generally, the faster a fire develops, the greater its threat.

Although the total fuel load in patient rooms remains low, the nature of the combustibles is changing. The introduction of foam plastic decubitus pads, upholstered furniture, and polyurethane foam mattresses may not affect the length of time a fire will burn, but such items have affected the rate of fire growth. These types of products can cause fire to grow to a large size quickly.

Fires that reach flashover produce acutely lethal atmospheres generating thousands of cubic feet of smoke per minute [1,000 cu ft per min (28 m³/min)]. Such fires threaten fire-rated barriers and produce enough energy to drive smoke to remote areas. A fire that grows to full room size in a health care facility represents an unacceptable level of risk; there is a high likelihood that such a fire will result in injuries or fatalities if its origin is in the patient sleeping area. Therefore, every effort should be made to recognize and elim-

inate or adequately protect fuel arrangements that might produce such fires.

Fire tests and actual fire experience have shown that certain common arrangements and types of fuels in patient rooms create especially hazardous situations because they are able to produce large fires in short time periods. In January 1976, for example, multiple-death fires in both the Wincrest and Cermak House nursing homes involved wooden wardrobes within patient sleeping rooms.[3,4] Tests show that combustible wardrobe fires can result in acutely hazardous environments in as little as 120 sec.[5]

A 1989 fire in a Norfolk nursing home resulting in 12 deaths originated in a foam plastic decubitus pad on a mattress in a patient sleeping room. The fire grew to flashover in less than 5 min.[6] A 1985 fire in a Michigan hospice initially involved an upholstered chair. The fire caused eight deaths.[7] Analysis of this fire established that flashover could have occurred in the room of origin in under 4 min. See Figure 9-8A for documentation of the analysis for the hospice fire.

Any combination of finishes, combustible building materials, or contents and furnishings that could result in full room involvement or flashover in 5 min or less represents a severe fire hazard in a health care facility. Such spaces should always be protected by automatic sprinklers and separated by fire-rated construction.

Computer fire models developed at NIST have been used to establish the rate of heat release required to produce flashover in a typical patient sleeping room. Under the "worst" circumstances—where the room door is partially open—a 1-MW fire can cause flashover. When the room door is fully open, approximately a 2-MW fire is necessary to cause flashover. The Appendix of the 1997 edition of NFPA *101* suggests that upholstered furniture, mattresses, and wardrobes be constructed to produce a maximum rate of heat release of 250 kW. It also has been determined that when a fire grows beyond 250 kW in size, hazard thresholds are exceeded, and patients must be removed from sleeping rooms.

It is a well-established fact that furnishings are frequently major contributors to fire growth. Recent developments make it possible to determine whether furnishings in a given environment can produce sufficient energy to cause full room involvement.

1 – From NFPA study of the hospice fire, Southfield, Michigan, December 1985 and CFR/NBS computer fire model analysis.
2 – From NFPA 72, Appendix B.
A – Room flashover.

FIG. 9-8A. Comparison of the hospice fire to NFPA 72, Appendix B, growth curves.

A simplified equation for estimating the rate of heat release required for flashover to occur in a room[8] is given by the expression:

$$Q \geq 750\,A\,\sqrt{h}$$

where Q is the rate of heat release (kW), A is the ventilation opening area in sq ft (m[2]), and h is its height in ft (m).

Once the heat release rate required to produce flashover is known for a typical room geometry, such information can be compared to actual heat release rates for typical furnishings, as determined by tests conducted in a calorimeter such as the NIST furniture calorimeter[9] or by Underwriters Laboratories Inc. (UL) Subject 1056, "Outline of Investigation for Upholstered Furniture." Figures 9-8B and 9-8C are idealized curves developed by NIST that illustrate the rate of heat release for typical furnishings as a function of time. Such information can be used to establish the probability of flashover; it also allows estimation of the time required to reach critical fire size.

Interior Finish

A successful fire protection strategy requires fires to remain small. Any large fire in a confined space creates a potentially lethal atmosphere. A fire's initial growth may be affected significantly by the interior finish of walls, ceilings, and floors. Interior finishes, therefore, deserve special attention. Combustible wall and ceiling finishes can act as a fuse, causing a fire to spread to objects remote from the fire origin. Alternatively, wall and ceiling finishes provide a large continuous surface over which fire may spread. Thus, an interior finish may, by releasing energy, cause a fire that would otherwise have remained small to become large.

The relative hazard of an interior finish is usually determined by a test conducted in accordance with NFPA 255, *Standard Method of Test of Surface Burning Characteristics of Building Materials,* commonly called the "tunnel test."

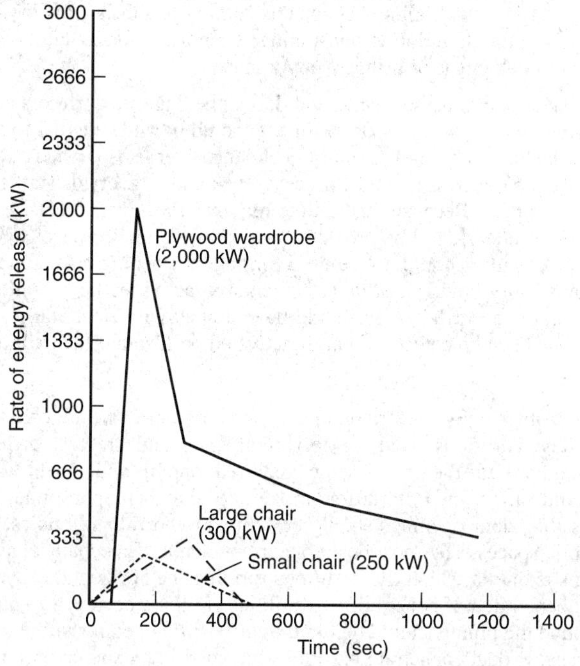

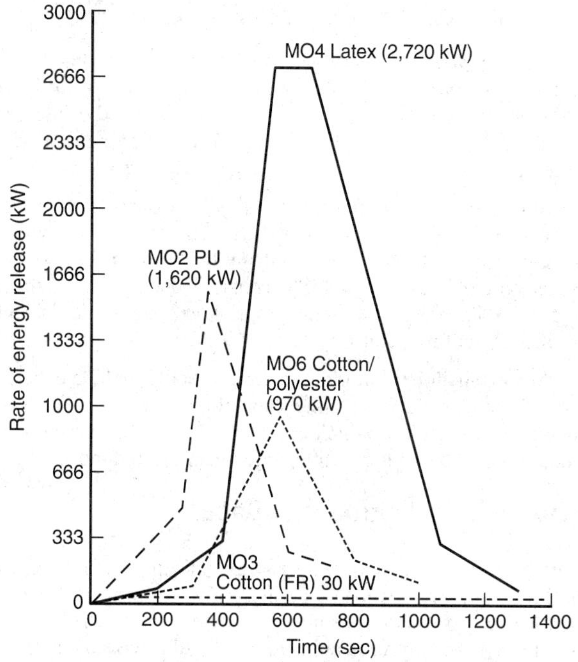

FIG. 9-8C. *Heat release rates for beds.*

FIG. 9-8B. *Heat release rates for miscellaneous furniture.*

Interior finish on walls and ceilings within the means of egress and any room should be limited to Class A materials, which have a flame-spread rating of 25 or less. Class B materials, or those with a flame-spread rating of 25 to 75, are considered tolerable, but they should be limited to small rooms. Full-scale experiments show that automatic sprinklers can limit fire growth in rooms with Class C wall and Class D ceiling finishes.[2,5] Accordingly, in buildings with automatic sprinkler protection, it has been judged acceptable practice to allow the use of materials having higher flame spread classifications than would otherwise be permitted. For example, Class B materials are sometimes used where Class A is normally specified, and Class C materials where Class B is normally required.

Further, it has been shown that the performance of interior finishes also is related to location.[10] Finishes on the upper portions of walls and ceilings contribute more significantly to flame propagation than finishes on the floor or on the lower half of a wall. Therefore, where finishes are limited to the lower half of a wall and are less than 4 ft (1.2 m) above floor level, materials having a higher flame-spread rating than would otherwise be allowed might be used without significantly affecting fire growth. For example, where Class A materials are judged necessary, a Class B material might be allowed on the lower portion of a wall.

In the past, floor finish materials were excluded from requirements for interior finishes, based on favorable experience and the assumption that limited exposure exists at the floor level during an actual fire. However, a fire on January 9, 1970, in the Harmer House Convalescent Home in Marietta, OH, significantly changed this attitude.[11] Thirty-two persons died in this fire. Carpeting with foam-rubber backing was judged to have played a significant role.

Interior floor finishes in the corridors of nonsprinklered health care facilities should be limited to Class I materials or those with a minimum critical radiant flux of 0.45 W/cm[2]. Where automatic sprinklers are provided, interior floor finishes need not be regulated beyond the federal flammability standard.

BUILDING CONSTRUCTION

Because occupants of health care facilities must be defended in place, construction is an important factor, especially in multi-story buildings. Buildings preferably should be constructed of noncombustible materials that resist the effects of fire and maintain structural integrity.

Buildings of two or more stories should be constructed of noncombustible materials, with major structural members having at least 2-hr fire resistance. Materials that either burn or support combustion, although less desirable, are considered acceptable if special precautions are taken. An automatic sprinkler system is an essential part of the total fire defense system for combusitble buildings.

Any evaluation of building materials should include consideration of their smoke-generating capabilities. In addition, plastic construction materials, which are becoming more common, are sometimes capable of generating large quantities of smoke.

Subdivision of Building Spaces

Separation of patient sleeping rooms: Because it may not be possible to remove occupants during a fire, sleeping rooms other than the room of fire origin sometimes must serve as temporary areas of refuge. Therefore, sleeping rooms should be isolated from all other building spaces by fire-rated construction. Partitions should be continuous from the floor slab to the floor or roof above through any concealed spaces, such as those above suspended ceilings. If the building is protected by sprinklers, NFPA *101* allows these walls to be nonrated, provided the walls resist the passage of smoke, and they are permitted to terminate at the ceiling, provided the ceiling resists the passage of smoke. (See Figure 9-8D for typical floor plan for a health care facility.)

There has been much discussion in the past about the operational considerations *vs.* the fire safety considerations of equipping patient room doors with door closures. All three model building codes and NFPA *101* do not require door closing devices on patient room doors. The current opinion is that a health care facility's functional needs prevail, and other fire safety and operational features a health care facility provides are adequate alternatives to door closures.

Any penetration of 1-hr partitions by building service equipment should be protected in order to maintain the 1-hr fire-rated separation. All spaces around piping and ducts should be sealed tightly with a noncombustible material having adequate fire resis-

tance and capable of retarding the transfer of smoke. Transfer grills should not be used within such doors or partitions.

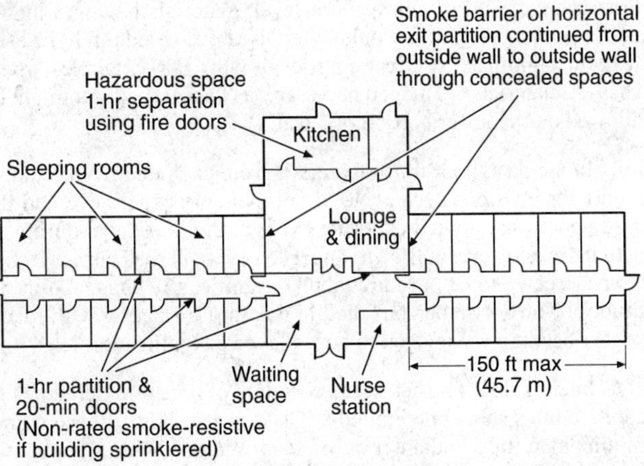

Smoke barriers: Every floor used by inpatients should be subdivided into at least two compartments by 1-hr partitions capable of retarding smoke. A horizontal exit, when constructed to satisfy the additional criteria imposed upon construction of smokestop barriers, is a desirable alternative to smokestop partition. Smoke barriers and smoke compartments play a very important fire safety role in health care facilities. Subdividing each floor minimizes the number of occupants exposed to a single fire. More importantly the barriers allow for horizontal evacuation of occupants to an area of refuge on the same floor. In a protect in place occupancy where evacuation is difficult, having the capability to evacuate horizontally is extremely important.

Protection of Vertical Openings

Fire and fire-produced contaminants tend to spread vertically within a building. Special effort is required to prevent fire on one level from threatening the occupants above; this is especially important in health care facilities.

All shafts should be provided with fire-rated enclosures. Vertical openings not connecting more than three floors should have a minimum of 1-hr fire resistance rating. Two-hour fire resistance ratings should be provided for vertical openings connecting more than three floors. Openings to shaft wells are limited to those necessary, and these openings must be protected.

When designing partitions to enclose vertical shafts, consideration should be given to the varying durability of materials. In spaces where partitions may be subject to mechanical injury, materials used to provide floor-to-floor separation should be able to resist damage in order to maintain the required fire resistance.

Exit Design

Exits in health care facilities should be limited to doors leading directly outside of the building, interior stairs and smokeproof towers, ramps, horizontal exits, outside stairs, and exit passageways.

Vertical evacuation of occupants within a health care facility is, at best, difficult and time-consuming. Therefore, horizontal movement of patients is of primary importance.

Horizontal passageways and doors opening into corridors and rooms used for sleeping or treatment should be wide enough to allow the horizontal movement of occupants, even those in beds. Relocation of patients is a slow process, even under favorable staff-to-patient ratios. Because of the time required to move patients, exit access routes should be protected against fire effects. Spaces open to the corridor should not be used for patient sleeping or treatment rooms, nor should hazardous contents or activities be permitted within them. Such spaces should be equipped with electrically supervised smoke detectors that, if activated, will sound the building fire alarm.

Horizontal exits: Horizontal exits are common in health care facilities. Partitions used as horizontal exits and smoke barriers should provide the fire resistance required for exits and, in addition, should satisfy the criteria for smoke barriers when appropriate. If possible, door openings should be limited to corridors, lobbies, or public spaces. The most desirable arrangement of mechanical systems is one in which the partitions forming the horizontal exit are not penetrated. If penetration by utilities or piping occurs, the space around the piping should be filled tightly with noncombustible materials. If ducts penetrate partitions intended to be smoke barriers, fire dampers that close if smoke detectors within the duct activate should be provided.

FIG. 9-8D. Typical floor plan for a health care facility.

Because a horizontal exit implies that occupants will be transferred from one side of a partition to the other (horizontal evacuation), adequate space must be available to house occupants after movement. At least 30 net sq ft (2.79 m^2) per patient should be available on each side of the horizontal exit, allowing for the total number of patients in adjoining compartments.

Interior stairs: Exit stairs should be designed to satisfy the criteria for interior stairs. Stairs should be enclosed with fire-resistive materials, with stair openings limited to those necessary for access and discharge purposes. Stairs must be properly protected from the effects of fire.

Exit Features

Life safety from fire in health care facilities relies upon a "defend in place" principle. Horizontal exits or smoke barriers are required to subdivide each story of a health care facility to provide an area of refuge on each floor.

The original concept for the design of exits for health care facilities was derived from studies conducted in the early part of this century. A practice evolved of using exit stair enclosures as areas of refuge. Exit capacity for health care facilities was set conservatively to offset slow travel rates and to create space within exit enclosures for "storing" patients on litters and in wheelchairs.

Exit design concepts have changed over the years. For example, NFPA *101* approaches life safety from fire in health care facilities using a "defend in place" principle. Means for horizontal evacuation are provided for each story, using either a smoke barrier or a horizontal exit. A barrier subdivides each story to provide an area of refuge for patients without traversing stairs. Furthermore, it is common—indeed, it is required in the 1994 edition of NFPA *101*—to install automatic sprinklers throughout all new health care facilities. Automatic sprinklers limit fire size and complement the defend-in-place strategy currently in use.

Exit capacity in nonsprinklered health care facilities is set at 0.6 in. (15.24 mm) per person for travel over stairs, while the capacity through doors and level passageways is calculated at 0.5 in. (12.7 mm) per person. Where automatic sprinklers are provided throughout a building, exit capacity is increased to 0.3 in. (7.62 mm) per person for travel over stairs and 0.2 in. (5.08 mm) per person for travel over level passageways.

Capacity is calculated using a "flow rate" principle. Flow rates assumed are for able-bodied persons because it is presumed that evacuation over stairs will involve only staff, visitors, and ambulatory patients. Nonambulatory occupants will be expected to remain in the building under the defend-in-place concept, with those patients on the floor of fire origin being moved horizontally to an area of refuge.

Limits on travel distance reflect anticipated slow movement. Travel distance normally should not exceed the following:

1. One hundred feet (30 m) between an exit and any room door intended for use as an exit access.
2. One hundred and fifty feet (46 m) between an exit and any point in a room.
3. Fifty feet (15 m) between any point in a sleeping room or suite of sleeping rooms and the exit access door of that room.

Increases in corridor travel distance are permitted in sprinklered buildings.

Because smoke barriers and areas of refuge play such an important role in fire safety, travel distances to smoke barriers is also restricted in NFPA *101*. Travel distance from any point on a floor to reach a door in a smoke barrier is limited to 200 ft.

In addition, facilities should be arranged to limit travel in the direction of the fire to less than 30 ft (9 m). Elevators are not usually counted as required exits because they possess numerous shortcomings that may prevent their use during a fire. In the cases of critically ill patients, patients in body casts or balkan frames, and others who would be difficult to move, however, elevators provide the only practical evacuation method from upper stories of the facility. If separate banks of elevators are located in separate smoke compartments, and if the staff are well trained, it may be possible to devise a plan in which using elevators during fires is safe.

Exit Marking and Exit Illumination

Readily visible signs should mark all exits. Where access to exits is not immediately visible, access routes also should be marked.

The entire means of egress must be continuously illuminated whenever the building is occupied. In some cases, normal street lighting is adequate for illumination of exit discharge. However, consideration should be given to the conditions that would result from a power failure.

Emergency power also is required to illuminate the means of egress and exit marking. In hospitals, this power should be supplied by the life safety branch of the hospitals' essential electrical system. Luminescent, fluorescent, or reflective material should not be substituted for required lighting.

Emergency power supplies should maintain illumination automatically in the event of a power failure, without any appreciable interruption during the changeover from normal to emergency power. Where a generator is provided, the delay should not be more than 10 sec, per NFPA 99, *Standard for Health Care Facilities* (hereinafter referred to as NFPA 99). Where emergency power is supplied by a central system with an engine-driven generator, the design should minimize the possibility of any single emergency simultaneously interrupting both normal and emergency power supplies. The switch(es) that transfers power from normal to emergency circuits is one place where normal and emergency circuits are required to merge. If this switch(es) is exposed to fire, it could simultaneously interrupt power to both normal and emergency circuits. The transfer switch and other electrical distribution panels and switch gear should be separated from the generator, as well as from the remainder of the building.

Fire Alarms

Every building should be equipped with an electrically supervised, manually operated, fire alarm system. When actuated, the system should sound alarms that can be heard above ambient noise levels throughout the facility. Usable alarm indicating devices are permitted in critical areas in lieu of audible alarm devices. The fire alarm should also transmit automatically to the fire department. Any fire detection or fire suppression system that activates should automatically activate the building alarm system.

Alarm systems, including detection devices, should be provided with an emergency power supply and designed according to NFPA 72, *National Fire Alarm Code.*

Any alarm that is activated should provide automatically, without delay, a general alarm. Presignal systems are not considered suitable for health care occupancies. Zoned systems with a coded signal have certain desirable characteristics and should be considered for use.

FIRE SUPPRESSION EQUIPMENT

A proven effective and most practical and reliable approach to life safety in health care facilities is to use automatic fire suppression—in particular, automatic sprinkler systems. Although persons in the area of origin may still be seriously threatened, those in adjoining spaces—and, in a number of cases, in the same room—are protected.[2,5,6,12] NFPA *101* now requires all new health care facilities to be protected by automatic sprinklers.

The ability of automatic sprinklers to provide a survivable environment for building occupants has been debated at length. Full-scale tests have shown that standard sprinklers can extinguish many fires while maintaining a survivable atmosphere outside the room of fire origin.[2,5] In addition, sprinklers have been shown to be effective in limiting carbon monoxide to nonlethal levels outside the room. Tests have also shown that privacy curtains may interfere with sprinkler discharge.[13] To obtain full sprinkler effectiveness, the influence of any building design feature or furnishing that would impair sprinkler discharge should be evaluated carefully.

The quick-response or fast-response sprinkler has been shown in most instances to maintain a survivable atmosphere within the room of fire origin. Recent full-scale tests at NIST have documented this performance.[13]

Where fire defenses are to be based upon automatic sprinkler systems, fast-response sprinklers should be considered for the entire building. As a minimum, fast-response sprinklers should be used throughout compartments having sleeping rooms. This is now required for new health care facilities by NFPA *101.*

Automatic sprinklers should adhere to NFPA 13, *Standard for the Installation of Sprinkler Systems.* Sprinklers' operation should sound the building fire alarm automatically. The sprinkler system and components should be supervised electrically to ensure reliable operation; this should include gate valve tamper switches with a local alarm at a constantly attended location when the valve is closed. If pressure tanks are the primary source of water, air pressure, water level, and temperature should be supervised. If fire pumps are provided, electrical supervision should monitor the fire pump in accordance with NFPA 20, *Standard for the Installation of Centrifugal Fire Pumps.*

Portable fire extinguishers should be placed in all buildings, as they may provide an opportunity to control a fire during its early stages. In all cases, however, the fire department should be notified before or at the same time that occupants are attempting to fight the fire. Delayed alarms have allowed fires to grow to large scale and, in turn, threaten occupants before fire department notification and arrival.

Smoke Control

Although operable windows have been required in every patient sleeping room for many years, NFPA *101* no longer requires them. Outside windows are required but need not be operable. Windows in atrium walls are considered outside windows in health care facilities.

Special forced-air systems or, in some cases, the adaptation of conventional building air-handling systems can permit venting of products of combustion early in the fire. Such systems also may make it possible to create a pressure differential across physical barriers, such as floors or partitions, and prevent smoke transfer. The effectiveness of fire partitions often improves significantly by the use of such systems.

Where adaptation of the building air-handling system is contemplated for smoke removal, their design should be as NFPA 92A, *Recommended Practice for Smoke Control Systems,* suggests. Consideration should be given to alternate power supplies and electrical supervision of critical system components.

PROTECTION OF HAZARDOUS AREAS

Areas with contents more hazardous than those normally found in health care facilities should be arranged to minimize occupants' exposure if a fire occurs. Hazardous areas, such as boiler and heater rooms, laundries, storage rooms, and repair shops, should be separated by 1-hr fire-rated construction, with openings protected by fire doors. Alternatively, the area should be protected by automatic sprinklers and, if the hazard is judged severe, 1-hr separation should also be provided. Typical spaces requiring both separation and sprinklers are soiled linen rooms; paint shops; and rooms or areas, including repair shops and trash collection rooms, used to store combustibles, supplies, and equipment in hazardous quantities.

Laboratories that contain quantities of flammable, combustible, or hazardous materials should be protected and separated from other building spaces. Pertinent information is found in NFPA 30, *Flammable and Combustible Liquids Code,* which is applicable to repair shops as well as laboratories, and in Chapter 10 of NFPA 99.

Cooking equipment should be arranged and protected in accordance with NFPA 96, *Standard for Ventilation Control and Fire Protection of Commercial Cooking Operations.*

Nitrous oxide and oxygen are usually piped through hospital spaces from a central distribution point. Although these gases do not burn, they do significantly accelerate combustion. The special precautions necessary are detailed in NFPA 50, *Standard for Bulk Oxygen Systems at Consumer Sites,* and in NFPA 99.

Building Service Equipment

Building service equipment should be installed and maintained in accordance with appropriate NFPA standards. Special consideration should be given to the design and installation of heating and air conditioning systems. Because rubbish chutes can be an important factor in fire, due regard should be given the design of rubbish and linen chutes, including pneumatic systems.

Portable heating devices are unsafe in patient-occupied portions of health care facilities. All heating devices should be designed and installed to prevent ignition of combustible materials. Approved suspended unit heaters may be used, except in means of egress and patient sleeping areas, if they are high enough to be out of the reach of persons using the area.

Combustion and ventilation air for boilers, incinerators, or heater rooms should be taken directly from, and discharged directly to, the outside of the buildings.

OPERATING FEATURES

Careless smokers cause many fires in health care facilities. Adoption and enforcement of suitable smoking regulations is essential.[7] Smoking should be prohibited in any room, ward, or compartment where flammable liquids, combustible gases, or oxygen is used or stored. Such areas should be posted with suitable signs. Smoking by patients who are under sedation or who are not considered responsible should be prohibited. Smoking should be permitted in a sleeping room only when authorized by medical staff and then only under direct staff supervision. Metal containers with self-closing covers should be available for disposal of smoking materials in all areas where smoking is allowed.

Most hospitals and many nursing homes have established no smoking policies. JCAHO requires hospitals to be smoke free. Nursing homes are becoming smoke free on a voluntary basis. These are very positive steps in improving the level of fire safety in health care facilities. Those who oppose no-smoking policies argue that they only drive smoking underground, creating a greater fire safety hazard. Because smoke-free health care facilities are relatively new, there is no clear evidence that smoking has gone underground.

However, evidence shows that visitors sometimes give patients smoking materials without the staff's knowledge. Health care facilities need to address this very serious new issue. In 1989, 12 residents died in a Norfolk, VA, nursing home fire attributed to a visitor giving a resident smoking materials.

Window draperies and curtains should be made of noncombustible material or material that has been rendered and maintained flame retardant. These window hangings should be capable of passing both the large- and small-scale tests required by NFPA 701, *Standard Methods of Fire Tests for Flame-Resistant Textiles and Films*. Furnishings, decorations, and other objects should not obstruct exits or exit access routes. Combustible decorations should be prohibited.

Exits and mechanical devices provided to control or limit fire should be maintained to ensure reliable operation. Inspections and tests should be performed, as required, to verify satisfactory performance.

In facilities with locked exits, adequate staff should always be present to release occupants and direct them away from the fire area to a place of safety during an emergency.

Care should be exercised during construction and repair operations to ensure that such activities do not reduce life safety.

Adequate preventive maintenance for mechanical systems, including tests and periodic inspections, are necessary to ensure their reliability.

EMERGENCY PLANNING

In no occupancy is staff training and emergency planning more important than a protect-in-place occupancy housing a significant number of occupants incapable of self-preservation. Every health care facility must have a fire and evacuation plan, including a disaster plan with which all personnel must be familiar. NFPA 99 provides guidance for both internal and external disasters that affect health care facilities. In addition, personnel should be trained to use fire extinguishers and hose cabinet lines if provided. They must also know how to sound an alarm, move or evacuate patients, and contain the fire.

Each facility should have a safety officer whose primary responsibility is to recognize hazards, act as liaison with the fire service, and arrange for training personnel. While training health care personnel is straightforward and can be accomplished on the job, orienting members of the fire service to health care facility problems is more difficult.

Copies of the fire and evacuation plan should be available to all personnel. The plan should contain specific instructions for key—that is, supervisory—personnel if there is a fire. A copy of the plan also should be posted for reference. All employees should be periodically trained to ensure readiness.

Emergency drills should include transmission of a fire alarm signal and simulation of emergency conditions insofar as possible without jeopardizing occupants. Drills should be conducted on each shift at least quarterly, with at least 12 drills held every year. The drills should be varied to test the alertness of all shifts and, if possible, should be unannounced. Use of the building alarm during drills also verifies its normal operation.

The fire and evacuation plan should include the following fundamentals:

1. Training personnel to use the alarm and alarm equipment.

2. Transmission of alarm to the fire department.
3. Details of the fire location.
4. Evacuation practices for all areas.
5. Preparation of building spaces for evacuation.
6. Fire extinguishment.

During drills, emphasis should be on immediate notification of the fire department; many fires have spread because of delayed alarms.

BIBLIOGRAPHY

References Cited

1. "Building Materials and Structures," Report BMS 92, 1942, National Bureau of Standards, Washington, DC.
2. "Full-Scale Fire Tests in a Nursing Home Patient Room," Report 7463, 1975, U.S. Department of Health, Education and Welfare, HEW Contract HSA 105-74-116. Prepared by American Health Care Association.
3. Best, R., "The Wincrest Nursing Home Fire," *Fire Journal,* Vol. 70, No. 5, Sept. 1976.
4. Best, R., "The Cermak House Fire," *Fire Journal,* Vol. 70, No. 5, Sept. 1976.
5. O'Neill, J. G., *et al.*, "Full-Scale Fire Tests with Automatic Sprinklers in a Patient Room, Phase II," NBSIR 80-2097, 1980, National Bureau of Standards, Gaithersburg, MD.
6. Hall, J. R., Jr, "The Elderly, the Sick, and Health Care Facilities," *Fire Journal,* Vol. 84, No. 4, July/Aug. 1990.
7. NFPA Fire Analysis Division, "Fatal Fire Risks and Hazards in Health Care Facilities," *Fire Journal,* Vol. 81, No. 5, Sept./Oct. 1987.
8. Babrauskas, V., "Estimating Room Flashover Potential," *Fire Technology,* May 1980.
9. Babrauskas, V., *et al.*, "Upholstered Furniture Heat Release Rates Measured with a Furniture Calorimeter," NBSIR 82-2604, 1982, National Bureau of Standards, Washington, DC.
10. Christian, W. J., and Waterman, T. E., "Flame Spread in Corridors: Effects of Location and Area of Wall Finish," *Fire Journal,* Vol. 65, No. 4, July 1971.
11. Sears, A. B., Jr., "Nursing Home Fire, Marietta, OH," *Fire Journal,* Vol. 64, No. 3, May 1970.
12. Boettcher, E. N., M.D. "Hospital Fire Defense: People and Sprinklers," *Fire Journal,* Vol. 61, No. 4, July 1967, pp. 93-96.
13. Notarianni, K. A., "Five Small Flaming Fire Tests in a Simulated Hospital Patient Room Protected by Automatic Fire Sprinklers," Report of Test FR-3982, Oct. 31, 1990, National Institute of Standards and Technology, Gaithersburg, MD.

Reference

UL Subject 1056, "Outline of Investigation for Upholstered Furniture," Underwriters Laboratories Inc., Northbrook, IL.

NFPA Codes, Standards, and Recommended Practices

Reference to the following NFPA codes, standards, and recommended practices will provide further information on health care facilities discussed in this chapter. (See the latest version of *The NFPA Catalog* for availability of current editions of the following documents.)

NFPA 13, *Standard for the Installation of Sprinkler Systems*
NFPA 20, *Standard for Installation of Centrifugal Fire Pumps*
NFPA 30, *Flammable and Combustible Liquids Code*
NFPA 50, *Standard for Bulk Oxygen Systems at Consumer Sites*
NFPA 72, *National Fire Alarm Code*
NFPA 90A, *Standard for the Installation of Air Conditioning and Ventilating Systems*
NFPA 92A, *Recommended Practice for Smoke-Control Systems*
NFPA 96, *Standard for Ventilation Control and Fire Protection of Commercial Cooking Operations*
NFPA 99, *Standard for Health Care Facilities*
NFPA 101®, *Life Safety Code®*

NFPA 255, *Standard Method of Test of Surface Burning Characteristics of Building Materials*

NFPA 701, *Standard Methods of Fire Tests for Flame-Resistant Textiles and Films*

Additional Reading

Abrahams, J., "More Changes Afoot for Fire Safety in Hospitals and Care Premises," *Fire*, Vol. 84, No. 1043, May 1992, pp. 23–24.

Ainley, J., and Charters, D., "Fire Precautions in New Hospitals," NHS Estates, UK, ISBN 0-9527398-3-6; Institution of Fire Engineers, University of Sunderland, Fire Research Station, CIB W14, Tyne and Wear Metropolitan Fire Brigade, Fire Safety by Design, Conference Proceedings, Volume 3, Research Papers, Jul. 10–12, 1995, UK, 1995, pp. 221–233.

"Architect's View of Fire Protection for Old and New Hospital Buildings," *Fire*, Vol. 84, No. 1043, May 1992, p. 27.

Bartlett, G., "Achieving Hospital Safety Through Fire Engineering," *Fire Prevention*, No. 276, Jan./Feb. 1995, pp. 29–32.

Bartlett, G., "Hospital Fire Officers Must Cultivate a Co-Ordinating Role," *Fire*, Vol. 87, No. 1077, Apr. 1995, pp. 15–16.

Bentley, R., "Hospital Fire Safety Hampered by Plethora of Ageing Buildings," *Fire*, Vol. 82, No. 1018, Apr. 1990, pp. 41–42.

Bentley, R., "Human Aspects in Fire Safety—The Hospital Fire Officier's View," *Fire Prevention*, No. 246, Jan./Feb. 1992, pp. 20–22.

Cable, E. A., "Analysis of Delay in Staff Response to Fire Alarm Signals in Health Care Occupancies," Worcester Polytechnic Inst., MA, Thesis, Jan. 1993.

Cass, P., "Hospital Managements Need Guidance From Specialist Fire Safety Advisers," *Fire*, Vol. 84, No. 1043, May 1992, p. 34.

Chandler, S. E., "Trends in Hospital Fires," *Fire Prevention*, No. 246, Jan./Feb. 1992, pp. 14–15.

Charters, D. A., "Fire Risk Assessment in the Development of Hospital Standards," *Proceedings* of the First International Conference on Fire Science and Engineering, ASIAFLAM '95 Mar. 15–16, 1995, Kowloon, Hong Kong, 1995, pp. 481–489.

Charters, D. A., "Quantified Assessment of Hospital Fire Risks," NHS Estates, UK, ISBN 0-9516320-9-4; *Proceedings* of the Seventh International Interflam Conference, Interflam '96, Mar. 26–28, 1996, Cambridge, England, Interscience Communications Ltd., London, England, 1996, pp. 641–651.

Charters, D. A., and Ainley, J., "Fire Risk Assessment in the Development of Hospital Standards," NHS Estates, UK, ISBN 0-9527398-3-6; Institution of Fire Engineers, University of Sunderland, Fire Research Station, CIB W14, Tyne and Wear Metropolitan Fire Brigade, Fire Safety by Design, Conference Proceedings, Volume 3, Research Papers, Jul. 10–12, 1995, UK, 1995, pp. 201–210.

Clarke, B., "Will the Lifting of Hospitals' Crown Immunity Mean Better Fire Safety?" *Fire*, Vol. 82, No. 1018, Apr. 1990, pp. 27, 34.

Cox, A., "Red Labels Mark Bad Fire Safety for Hospitals in West Birmingham Area," *Fire*, Vol. 85, No. 1054, Apr. 1993, p. 27.

Cox, A., "Story of a Hospital Fire," *Fire Surveyor*, Vol. 21, No. 6, Dec. 1992, pp. 22–23.

Creak, J., "Hospital Laundry Gets a Safe Evacuation System," *Fire*, Vol. 87, No. 1077, Apr. 1995, p. 22.

Crowley, M. A., "Meeting Hospital Fire Codes: Mission Impossible?" *Consulting-Specifying Engineer*, Vol. 17, No. 2, Feb. 1995, pp. 57–58, 60.

"Fire Doors Prevent Blaze in Hospital Becoming 'Serious'," *Fire*, Vol. 85, No. 1051, Jan. 1993, p. 48.

Fire Protection for Hospitals, P8404, Factory Mutual Engineering and Research, Norwood, MA.

"Fires in Hospitals and Health Care Premises During 1992," *Fire Prevention*, No. 276, Jan./Feb. 1995, pp. 27–28.

Hagglund, B. and Wickstrom, U., "Smoke Control in Hospitals—A Numerical Study," *Fire Safety Journal*, Vol. 16, No. 1, 1990, pp. 53–63.

Hall, J. R., "A Brief Look at Fire Experience in Health Care Facilities," *Fire Journal*, Vol. 80, No. 6, Nov. 1986, p. 51.

Hall, J. R., Jr., "Elderly, the Sick, and Health Care Facilities," *Fire Journal*, Vol. 84, No. 4, Jul./Aug. 1990, pp. 32–35, 37–42.

Hospital Hazardous Materials, ECRI, Plymouth Meeting, PA, 1987.

Isner, M. S., "Hospital Fire, Sprinkler Success, Weymouth, Massachusetts, Jan. 24, 1993. Summary," Fire Investigation Report, National Fire Protection Association, Quincy, MA, 1993.

Isner, M. S., "Sprinklers Prevent Tragedy in Two Health Care Facility Fires," *NFPA Journal*, Vol. 87, No. 5, Sept./Oct. 1993, pp. 49–52, 54–55.

James, R., "Important Role of Exercises in Hospitals' Fire Safety Plans," *Fire*, Vol. 86, No. 1066, Apr. 1994, p. 19.

Klein, B. R., ed., *Health Care Facilities Handbook*, 2nd ed, National Fire Protection Association, Quincy, MA, 1987.

Klein, B. R., "Health Care Firesafety: A Prognosis for the '90s," *NFPA Journal*, Vol. 84, No. 3, May/June 1992, pp. 89–92, 94.

Koffel, W. E., "Health Care Facilities: Do Sprinklers and Detectors Save Lives?" *Fire Protection*, Vol. 17, No. 1, Mar. 1990, pp. 6–8, 10–12.

Klote, J. H., "Smoke Control at Veterans Administration Hospitals," NBSIR 85-3297, Jan. 1986, National Bureau of Standards, Gaithersburg, MD.

Lipschultz, F., and Klein, B., "Managing Disasters in Health Care," *Fire Prevention*, Vol. 230, June 1990, pp. 31–33.

Marchant, E. W., "Hospitals 88," *Fire Technology*, Vol. 24, No. 4, 1988, pp. 369–373.

Marchant, E. W., "Hospital Safety," *Fire Prevention*, No. 213, Oct. 1988, pp. 27–30.

Mawn, P., "Major Fire Safety Upgrade for Cardiff Hospital," *Fire*, Vol. 87, No. 1077, Apr. 1995, pp. 20, 24.

Meland, O., and Sharet, E., "Smoke Control in Hospitals," *Fire Technology*, Vol. 22, No. 1, 1986, pp. 33–44.

Murphy, J. J., Jr., "Health-Care Facility Preplans," *Fire Engineering*, Vol. 145, No. 6, June 1992, pp. 75, 78, 80, 82.

Notarianni, K. A., "Five Small Flaming Fire Tests in a Simulated Hospital Patient Room Protected by Automatic Fire Sprinklers," National Institute of Standards and Technology, Gaithersburg, MD, National Institutes of Health, Bethesda, MD, FR 3982, Oct. 31, 1990.

Notarianni, K. A., "Measurement of Room Conditions and Response of Sprinklers and Smoke Detectors During a Simulated Two-Bed Hospital Patient Room Fire," National Institute of Standards and Technology, Gaithersburg, MD, National Institutes of Health, Bethesda, MD, NISTIR 5240, Jul. 1993.

Orr, R., "Evacuating Hospital Theatres," *Fire Prevention*, No. 280, June 1995, pp. 28–29.

Orr, R., "Hospital Evacuation: A Unique Fire Safety Challenge," *Fire Prevention*, No. 266, Jan./Feb. 1994, pp. 24–25.

Palmer, K., "Fire Protection on Health Care Premises," *Fire Prevention*, No. 209, May 1988, pp. 27–31.

Patterson, M., "Alarming Voices in the Hospital," *Fire Prevention*, No. 259, May 1993, p. 35.

Platt, S., "Loss of Crown Immunity—The Effects on Health Care Premises," *Fire Prevention*, No. 246, Jan./Feb. 1992, pp. 16–19.

"Rx for Hospital Loss Control," Factory Mutual Engineering, Norwood, MA, P9412, Jan. 1995; Factory Mutual Engineering, Business Under Fire! A Manager's Guide to Fire Prevention Programs, 1995, pp. 1–10.

Sime, J., Breaux, J., and Canter, D., "Human Behavior Patterns in Domestic and Hospital Fires," Fire Research Station, Borehamwood, England, BRE Occasional Paper 59, May 1994.

Stollard, P., "Use of Mathematical Fire Risk and Fire Safety Audits in Health Care Facilities," Fifth International Fire Conference on Fire Safety, Interflam '90, Sept. 3–6, 1990, Canterbury, England, Interscience Communications Ltd., London, England, 1990, pp. 323–329.

Stollard, P., Abrahams, J., and Ainley, J., "Fire Risk Assessments of Hospitals," *Fire Safety Engineering*, Vol. 2, No. 1, Feb. 1995, pp. 8–10.

Sugunendran, S., "Hospital Fire Safety: New Developments," *Fire International*, Vol. 136, Aug./Sept. 1992, pp. 27, 32.

Sugunendran, S., "New Developments in Hospital Fire Safety," *Fire Prevention*, No. 204, Nov. 1987, pp. 33–34.

Todd, C., "Fire Saftey in Health Care Premises," *Fire Surveyor*, Vol. 18, No. 1, Feb. 1989, pp. 33–40.

Weiss, P., "Safety, Health, and Fire Prevention Guide for Hospital Safety Managers. Technical Guide," Army Environmental Hygiene Agency, Aberdeen Proving Ground, MD, Technical Guide 152, Mar. 1, 1993.

BOARD AND CARE FACILITIES

Revised by James K. Lathrop

Although it has been with us for more than a century, the residential board and care home has become an American phenomenon of the 1970s, 1980s, and 1990s.

Reasons for the growth of this occupancy are numerous. The population of state hospitals was reduced sharply in the ten-year period from 1969 to 1979. Social, judicial, economic, and therapeutic pressures to remove patients from the restrictive environments of psychiatric hospitals, nursing homes, state schools, and other institutional settings helped create a need for board and care facilities. This occupancy provides the deinstitutionalized person with a more normal setting to live in, and furnishes the support services the resident needs to cope with life's daily requirements.

Today's older adults—those over age 65—often live alone, even after they cannot care for all their own needs. They might need to be in a setting where someone can look after their nutrition, personal hygiene, and so on.

Alcohol and drug rehabilitation programs often use the halfway house concept, in which the individual is somewhat sheltered from the pressures of returning too quickly to the general population for independent living.

Shelters for battered persons, unwed mothers, and homeless children also became more common in the 1980s.

The fire experience in what NFPA statistical analysts call board and care homes was striking during 1979–1983 and remains a concern today. (See Table 9-9A.) During the first two weeks of April 1979, fires in three board and care homes resulted in 45 fatalities. Many of the multiple-death fires reported in Table 9-9A killed over half of the residents, and it is not unusual for these largest fires to totally destroy facilities. At the same time, these fires have been considerably rarer in the past few years. Fires involving 10 or more fatalities in particular are declining, and this is a hopeful development.

This chapter examines what makes this occupancy so susceptible to fire and loss of life and what has to be done to improve the fire record.

DEFINITION

A residential board and care facility is defined in NFPA *101®*, *Life Safety Code®* (hereinafter referred to as NFPA *101*), as a residence that provides shelter and overnight accommodations, often provides

TABLE 9-9A. *Board and Care Home Fires Causing Five or More Deaths (1971–1995)**

Date	Occupancy	Civilians Killed
1/73	Rest Home, Pleasantville, NJ	10
4/79	Wayside Inn Boarding House, Farmington, MO	25
4/79	Halfway House, Washington, DC	10
4/79	Foster Home, Connellsvile, PA	10
11/79	Coats Boarding Home, Pioneer, OH	14
7/80	Brinkley Inn Boarding Home, Bradley Beach, NJ	24
11/80	Adult Foster Care Home, Detroit, MI	5
1/81	Beachview Rest Home, Keansurg, NJ	31
10/82	Domicilary Care Home, Pittsburgh, PA	5
2/83	Group Home, Eau Claire, WI	6
3/83	Adult Foster Care Home, Gladstone, MI	5
4/83	Community Home, Worcester, MA	7
8/83	Boarding House, Lawrenceville, GA	8
12/83	Group Home for Mentally Retarded, Cincinnati, OH	6
11/84	Board and Care Facility, Lexington, KY	5
3/91	Board and Care Facility, Colorado Springs, CO	10
6/92	Adult Foster Care Facility, Detroit, MI	10
10/94	Boarding Home, Mobile, AL	6
12/94	Board and Care Facility, Fort Lauderdale, FL	6

*The classification as a board and care home is based on the needs of the occupants, specifically the presence of tenants who were over age 65 or had chronic mental illness. The classification is not based on the level of care the facility offered or the level of care it claimed to offer.

Specific facility names are given for only the four largest and best-known incidents.

meals, and, as distinct from an ordinary boarding house, provides residents with some form of personal care. In fire incident data bases, a board and care facility is defined by its occupants' needs. It is important to keep these two definitions in mind, because there is no guarantee that the occupants' needs match the care the facility provides. In particular, the care needs of occupants that qualify a property as a board and care facility for fire reporting purposes will often be greater than the care delivered by a facility legitimately constituted as a board and care facility under NFPA *101*. Therefore, Table 9-9A may not properly describe the fire experience of facilities that NFPA *101* would have classified as board and care homes or that claimed to be board and care homes.

James Lathrop is a principal in the firm Koffel Associates Inc., of Ellicott City, MD, and Niantic, CT.

The NFPA *101* definition could easily describe a hospital or a nursing home. The primary difference is in the definition of personal care. The care given in a hospital or a nursing home is certainly personal, but its primary purpose is to meet medical needs of patients. In the board and care home, the care provided is not medical in nature, nor does it require the services of doctors and nurses. It is generally a kind of assistance that can be rendered by laypersons who help residents cope with the rigors of daily living. This care might include looking after residents' personal hygiene, cleaning their living quarters, helping those who are physically handicapped to move around, preparing meals, looking after residents' financial matters, seeing that appointments are kept, and numerous other tasks.

As this discussion implies, NFPA *101* seeks to identify distinct populations in terms of needed levels of care and then define distinct categories of facilities that are constituted to provide the level of care required by some group of potential tenants. A separate problem must be solved; making sure that tenants are not put in facilities that cannot or will not serve all their needs. This problem, implicit in most fires in Table 9-9A, is one for enforcement authorities and probably the most important key to fire safety for all populations needing care.

Although there are thought to be more than 300,000 board and care homes in the U.S., few such facilities call themselves by that name. Some facilities that fit into the definition of this occupancy are:

1. Group homes for mentally retarded persons.
2. Group homes for released psychiatric patients.
3. Rest homes for the aged.
4. Foster care homes.
5. County poor houses.
6. Orphanages.
7. Shelters for battered persons.
8. Shelters for unwed mothers.
9. Halfway houses for rehabilitated alcoholics.
10. Halfway houses for rehabilitated drug abusers.
11. Halfway houses for prison parolees.
12. Shelters for runaways.
13. Rescue missions with overnight accommodations.

Whatever their names, a common thread ties these facilities together: they provide residents with more than just food and shelter. The residents usually are furnished some form of assistance, support, or care that is not of a medical nature.

OCCUPANCY CHARACTERISTICS

Characteristics of Buildings

A board and care facility falls into the ordinary hazard category by the traditional methods of evaluation under NFPA *101*. There are rarely any massive concentrations of combustible materials and usually no use of flammable liquids or gases. No processes pose "special hazards." Ignition sources are typical of those found in residential occupancies, and fuel loads are generally light. The most significant fire threat in this occupancy is the enormous potential for loss of life.

Structural characteristics of buildings that house residential board and care facilities vary enormously. They range from single-family dwellings, in which a family takes in a few boarders, to highrise, reinforced-concrete apartment buildings for the elderly operated by religious organizations, to commercial rest homes built for the purpose. Construction is practically any type.

Characteristics of Residents

A description of the characteristics of the "typical" resident of a board and care facility is just as elusive as a description of a "typical" construction type. Residents of board and care homes do not fit a stereotype—they may be children or elderly, physically impaired or physically sound, mentally handicapped or brilliant. They may be capable of self-preservation or totally dependent upon others to provide for their safety. No one description fits the residents of all board and care homes.

Variations that exist in the buildings and the range of residents' capabilities in these facilities pose the most significant challenge to fire protection professionals.

FIRE PROBLEMS

Problems of board and care facilities generally fall into two categories: (1) structural and (2) behavioral.

Structural Problems

It is probable that the vast majority of the board and care facilities are single-family dwellings or single-family dwellings to which space has been added to accommodate a number of boarders. Consequently, it is probable that the vast majority of facilities are in structures built to standards intended for housing a "normal" family with normal abilities to detect, respond to, and escape from a fire. What follows are some of the specific fire safety characteristics for those kinds of facilities.

Building construction: Most facilities are believed to be housed in one-, two-, and three-story wood-frame structures. In older buildings, balloon construction and unprotected wood structural members are common structural hazards. Under the best circumstances, wood-frame construction can offer some measure of protection, but fire endurance is limited and the combustible structure can eventually contribute fuel to a fire.

Exit facilities: Generally, there is only a single route that occupants can use from any given room to get to safety outdoors. This escape route frequently consists of a single unenclosed stairway that might take escaping residents through the area of the house that is on fire. Often, the only alternative escape route is through a window, which in the case of a three-story building, may be more than 20 ft (6 m) above the ground. Typical of single-family dwellings, exit doors often are equipped with locking devices that impede egress, particularly for those unfamiliar with the locks or incapable of operating them.

Compartmentation: Separation of rooms from corridors and rooms from each other is either nonexistent or of construction that provides little or no resistance to the passage of fire and smoke. Consequently, a fire in any room can quickly fill the entire building with smoke and toxic combustion products, limiting visibility and making the environment intolerable for human survival.

Vertical openings: As with many multi-story dwellings, the stairway, which also serves as the primary means of escape, is not enclosed. The stairs are usually open to the corridor on the upper floors and the living spaces on the first floor. The spread of fire and smoke from one floor to the next can occur very rapidly.

Interior finishes: Frequently, walls and ceilings are finished with paneling and acoustical ceiling tiles, without any thought to the combustibility of these materials. The facility owner, trying to improve the environment in which the residents live by providing attractive decor, may be placing them in jeopardy without knowing it.

Highly combustible wall paneling and ceiling tiles can contribute to rapid horizontal and vertical fire spread and can be a major factor contributing to flashover. They also contribute to the speed with which exit pathways become unusable.

Electrical wiring and appliances: Wiring is generally typical of that found in single-family dwellings. In older buildings, circuits may be inadequate in number and capacity, insulation may be old and deteriorating, fuses may be the wrong sizes for the circuits they protect, and wiring modifications, if any, may have been made by unqualified people. Use of extension cords, portable heaters, and other residential-type electrical appliances may be common and contribute to the risk of fire.

Furnishings: The fire hazard posed by furniture and furnishings varies widely. Practically all such products can contribute fuel to a fire, with some, of course, being worse than others. Generally no regulations govern the kinds of furnishings that a board and care facility can use. Furniture that burns fiercely or gives off inordinately large quantities of thick smoke or toxic combustion products would be a major hazard. Often, residents are permitted to bring an unlimited quantity of personal furnishings into the facility.

Behavioral Problems

Much concern over life safety in board and care facilities arises from the fact that they house a number of people who, in many cases, have diminished physical and/or mental capacity. Although this is not true of all facilities, or all residents of a given facility, the very fact that impaired people are residents of a facility that assists them in coping with daily life suggests that they may have problems fending for themselves in an emergency.

The impact that behavioral problems create generally falls into two categories: (1) evacuation capability and (2) safety consciousness.

Evacuation capability: Because populations housed in board and care facilities differ widely, a broad statement describing evacuation capability in all such facilities cannot be made.

In the case of homes for the elderly, some residents may be slow and unsteady, and unable to manage stairs, to see well, or to remember procedures to be followed in a fire. Some may be so feeble that they cannot be evacuated without a great deal of assistance.

Residents of some facilities may be physically handicapped to such an extent that self-evacuation is not possible.

Evacuation capability can be affected by the facility's staff. In a facility housing deaf residents, for example, the presence of one staff member who can awaken and alert the residents to the emergency might be sufficient to normalize the evacuation capability of the group. In facilities housing more debilitated handicapped persons, a large staff might still be incapable of effecting a prompt evacuation.

Safety consciousness: One factor that makes this occupancy more at risk to fires is that residents often are unaware of or unconcerned about the dangers of their actions. In many small facilities, even the management personnel may have little awareness of good life safety and fire prevention practices and procedures. Careless smoking habits and improper use of portable heating units, electrical extension cords, and appliances are fairly common hazards in many facilities as they are in many ordinary homes. In addition, some residents of board and care facilities have been known to set fires intentionally.

FIRE PROTECTION STRATEGIES

The systems approach to fire safety in correctional facilities can be modified to apply to board and care facilities as well. (See Section 9, Chapter 7, "Detention and Correctional Facilities.") With respect to life safety, the fire problems in board and care facilities can be addressed by ignition control, fuel control, occupant protection, detection and suppression activities, and planning and training operations.

Ignition Control

Many fatal fires in board and care facilities are started by smoking materials. Each facility should have a smoking policy that addresses such issues as where smoking is permitted and the proper disposal of smoking materials. Clearly, the policy should attempt to prevent ignitions that occupants cause by falling asleep while smoking or improperly disposing of smoking materials. Proper ash receptacles should be provided in areas where smoking is permitted. In those facilities that permit smoking, designated smoking areas that are usually supervised or in which an occupant is rarely alone can work well.

Other ignition sources, such as heat-producing appliances, portable heating devices, and electrical equipment, should be controlled and maintained properly. The use of such equipment should be limited to those who are capable of operating the appliance properly or to those who are being trained under supervision. Adequate electrical outlets should be provided to eliminate the use of extension cords.

Fuel Control

In many board and care facilities, residents are permitted to bring their own furnishings into their sleeping rooms. Most of these furnishings have not been subjected to any type of test to indicate their ease of ignition or heat-release rates. Where furnishings can be controlled, consideration should be given to items that resist ignition by both cigarettes and small flames. Furnishings also should be evaluated to determine their heat-release rates. Chairs commonly used in board and care facilities can result in heat-release rates capable of producing flashover in a typical sleeping room without ignition of any other item.

If furnishings cannot be controlled, other protection features, such as early detection, suppression, and compartmentation, should be evaluated to ensure that an acceptable level of protection is provided.

Other fuel sources, such as interior finish materials, also should be evaluated with respect to their potential to spread fire.

Occupant Protection

Setting minimum criteria for the protection of life from fire in this broad occupancy type is not simple. As indicated, this is not a homogeneous occupancy. What is minimum protection in one facility may be overprotection or inadequate protection in another. In few, if any, other property classes is it so difficult to ensure over time that tenants in residence require all the care and fire protection the facility provides, but no more. However, it is always appropriate to require a certain level of protection for the occupants, particularly for those who are not intimate with the fire from its onset.

Perhaps residents' evacuation capability is the most significant factor in determining the fire protection features needed to achieve an acceptable level of life safety in a given board and care home. A facility that houses a group with a fairly normal ability to respond to a fire may need relatively little protection beyond that of other residential occupancies. In buildings housing people who can respond to an emergency, but who do so slowly, a protection system that provides a little extra time to react and evacuate may be adequate. Groups that cannot be expected to evacuate to safety unaided may need a defend-in-place protection system.

One might suspect that, over a period of time, all facilities may house residents who are incapable of evacuation, so that protection for the worst situation should be provided. This approach is valid in the instance of homes for the aged, where deteriorating physical and mental conditions are predictable. However, many types of board and care facilities, because of their size and the nature of the services they provide, never deal with people incapable of self-preservation. Such facilities should not be required to provide the same degree of built-in fire protection as those that house less capable populations. Therefore, it is reasonable to establish fire protection criteria based upon the evacuation capability of the facilities' residents. More or less built-in protection would be required, depending upon the residents' ability to fend for themselves.

For a population with relatively normal evacuation capability, the most basic system of protection consists of:

1. Controls to prevent ignition.
2. Prompt detection.
3. Means to alert the residents to the existence of a fire.
4. Possibly a means of automatically suppressing the fire.
5. At least two routes by which to escape from the endangered area.
6. Prohibition of the use of highly combustible interior finish materials. (This limits the rate of fire spread and minimizes the risk of flashover.)
7. Some form of enclosure of vertical openings. (This limits the vertical spread of fire and smoke.)

For a population that is able to evacuate, but does so slowly or with difficulty, a reasonable system of protection would include all the items listed above, plus:

1. Tighter controls to prevent ignition.
2. Early detection and alarm.
3. A greater need for automatically suppressing the fire.
4. Exits that are easier to use than those that might be adequate for a more capable population.
5. Tighter controls on the combustibility of interior finishes.
6. Inclusion of a degree of fire endurance to maintain building integrity during a longer evacuation period.
7. Inclusion of a degree of compartmentation to confine fire and smoke, thus keeping exit pathways usable during the longer evacuation period.
8. Greater resistance to the vertical spread of fire and smoke.

For a population that is not capable of evacuation, a defend-in-place protection concept would include all the features listed above, plus:

1. Sufficient structural fire endurance to allow the building and the internal compartmentation to outlast the fire.
2. A means of automatically suppressing the fire before it seriously threatens the structure and the compartmentation.
3. Controls on fuels, such as furnishings.

These protective concepts are the same as those used in various other occupancies where people sleep overnight. NFPA *101* criteria is based on three different levels of evacuation capabilities: prompt, slow, and impractical. The criteria are also based on three types of facilities: small (≤ 16 residents), large (> 16 residents), and units within an apartment building. For a detailed set of criteria for life safety in this occupancy, see Chapters 22 and 23 of NFPA *101*. The *Life Safety Code Handbook* provides extensive diagrams and tables to illustrate the requirements of NFPA *101*.

Detection and Suppression Activities

The means by which ignition controls and fuel controls are implemented, as well as the occupants' evacuation capability, are critical factors in determining the appropriate level of detection and suppression necessary. In sleeping areas, residential sprinklers should be considered a means of increasing life safety. In accordance with NFPA 101, either an NFPA 13, *Standard for the Installation of Sprinkler Systems,* NFPA 13D, *Standard for the Installation of Sprinkler Systems in One- and Two-Family Dwellings and Manufactured Homes,* or NFPA 13R, *Standard for the Installation of Sprinkler Systems in Residential Occupancies up to and Including Four Stories in Height* system may be used in board and care facilities, depending on the size of the facility and the evacuation capability of the occupants. With respect to NFPA 13D, it should be recognized that the water supply needs to last only 10 minutes and the domestic water demand is not necessarily included. With respect to NFPA 13D and NFPA 13R, it should be recognized that sprinklers may be omitted from certain areas such as small closets, bathrooms, attics, and other areas in which the data indicate that fatal fires are less likely to originate. Due to this the *Life Safety Code* does add criteria when using NFPA 13R or NFPA 13D in certain facilities. A system installed in accordance with NFPA 13 results in sprinklers installed throughout the building. The latest edition of NFPA 13 now mandates the use of either quick response or residential sprinklers for these types of facilities.

In addition to the desired level of protection, insurance carriers and applicable codes should be consulted before selecting detection and suppression systems. The preventive maintenance requirements of each type of system should also be considered during the selection and design process.

Planning and Training Operations

The facility management should ensure that an emergency plan is developed that covers various emergencies or disasters, including fires. The plan should be realistic and capable of being implemented with the minimum resources that may be available, typically during the night. The plan should be coordinated with the local fire department to ensure that their expectations and the expectations of the facility are consistent. Periodically during the year, the plan should be re-evaluated with an exercise that tests its various aspects.

Depending on their capabilities, residents should participate in the training and drills. As discussed above, residents need to be familiar and comfortable with various egress paths available to them. Local fire departments, fire training schools, and fire marshal offices may have personnel and audio-visual materials that can assist in staff and resident training. This is one area that board and care facilities have an advantage over other residential occupancies. There can be significantly better control over occupants in these facilities, and in many cases these residents perform above average during fire drills.

BIBLIOGRAPHY

NFPA Codes, Standards, and Recommended Practices

Reference to the following NFPA codes, standards, and recommended practices will provide further information on board and care facilities discussed in this chapter. (See the latest version of *The NFPA Catalog* for availability of current editions of the following documents.)

NFPA 13, *Standard for the Installation of Sprinkler Systems*
NFPA 13D, *Standard for the Installation of Sprinkler Systems in One- and Two-Family Dwellings and Manufactured Homes*
NFPA 13R, *Standard for the Installation of Sprinkler Systems in Residential Occupancies up to and Including Four Stories in Height*

NFPA *101®*, *Life Safety Code®*
NFPA 101A, *Guide on Alternative Approaches to Life Safety*

Additional Reading

Alvord, D. M., "The Escape and Rescue Model—A Simulation Model for the Emergency Evacuation of Board and Care Homes," NBS-GCR-83-453, Dec. 1983, National Bureau of Standards, Gaithersburg, MD.

Ayotte, T. E., "Fire Safety in Special Care Facilities," *Fire Engineering*, Vol. 144, No. 6, June 1991, pp. 83–84, 86.

Berlin, G. N., *et al.*, "Modeling Emergency Evacuation from Group Homes," *Fire Technology*, Vol. 18, No. 1, Feb. 1982, pp. 38–48.

Blye, P., and Yess, J. R., *Fire Safety in Boarding Homes: A Self-Study Guide for Owners and Operators*, National Fire Protection Association, Quincy, MA, 1985.

"Board and Care Homes: A Study of Federal and State Actions to Safeguard the Health and Safety of Board and Care Home Residents," U.S. Department of Health and Human Services, Washington, DC, 1982.

Bonnie Walker and Associates, "Conducting a Fire Drill in a Group Home," Publication of the National Fire Safety Certification System, Bonnie Walker and Associates, Bowie, MD, 1990.

Cote, R., and Timoney, T., "Boarding House Fire Causes Fifteen Deaths," *Fire Journal*, Vol. 79, No. 1, Jan. 1985, pp. 44–48, 50–54.

Cote, R., ed., *Life Safety Code Handbook*, 6th edition, National Fire Protection Association, Quincy, MA, 1994.

Deal, S., "Evaluating Small Board and Care Homes: Sprinklered *vs.* Nonsprinklered Fire Protection," National Institute of Standards and Technology, Gaithersburg, MD, Fire Administration, Emmitsburg, MD, NISTIR 5302, Nov. 1993.

Groner, N. E., "A Matter of Time—A Comprehensive Guide to Fire Emergency Planning for Board and Care Homes," NBS-GCR-82-408, Nov. 1982, National Bureau of Standards, Gaithersburg, MD.

Groner, N. E., *Fire Safety in Board and Care Homes*, Project SHARE, Rockville, MD, 1986.

Groner, N. E., "Guide to Board and Care Fire Safety Requirements in the 1991 Edition of the Life Safety Code," George Mason Univ., Fairfax, VA, National Institute of Standards and Technology, Gaithersburg, MD, NIST-GCR-93-629, July 1993.

Hall, J. R., Jr., "U.S. Fires in 'Board and Care' Homes Matrix Display of Selected Fatal Fires. Special Analysis," National Fire Protection Association, Quincy, MA, National Institute of Standards and Technology, Gaithersburg, MD, NIST-GCR-93-627, April 1993.

Hallberg, G., "Defining Care and Residential Buildings with Respect to the Evacuation Capability of Occupants," *Proceedings* of the Third International Symposium on Fire Safety Science, Elsevier Applied Science, 1991, pp. 825–834.

Hallberg, G., "Residents' Evacuation Capacity in Buildings Where Care and Housing Are Combined," Royal Institute of Technology, Sweden, University of Ulster and Fire Research Station, CIB W14: Fire Safety Engineering, International Symposium and Workshops Engineering Fire Safety in the Process of Design: Demonstrating Equivalency,

Part 1, Symposium: Engineering Fire Safety for People with Mixed Abilities, Sept. 13–16, 1993, Newtownabbey, Northern Ireland, 1993, pp. 117–131.

Henricsson, L., "Measures for the Safety of the Lives of the Elderly and the Physically Handicapped," Stockholm Fire Dept., Sweden, CIB W14/95/1 (J); Tokyo Fire Department, Firesafety Frontier '94, International Fire Conference and Exhibition in Tokyo, Creating a Safe Tomorrow, Oct. 18–22, 1107–112, Tokyo, Japan, 1994, pp. 281–286.

Isner, M. S., "Board and Care Facility Fire, Colorado Springs, Colorado, March 4, 1991. Summary Fire Investigation Report," National Fire Protection Assoc., Quincy, MA, Summary Fire Investigation, 1991.

Isner, M. S., "Ten (10) Die in Detroit Board and Care Facility," *NFPA Journal*, Vol. 87, No. 1, Jan./Feb. 1993, pp. 29–35.

Isner, M. S., and Smith, R., "Fire in Boarding Home: A Success Story," *Fire Journal*, Vol. 80, No. 2, March 1986, p. 75.

Kakegawa, S., *et al.*, "Evaluation of Fire Safety Measures in Care Facilities for the Elderly by Simulating Evacuation Behavior," *Proceedings* of the Fourth International Symposium on Fire Safety Science, Intl. Assoc. for Fire Safety Science, Boston, MA, 1994, pp. 645–656.

Klem, T. J., "Board and Care Fire Claims Four," *Fire Command*, Vol. 57, No. 2, Dec. 1990, pp. 13–15.

Levin, B. M., ed., "Fire and Life Safety for the Handicapped," NBS-STP-585, July 1980, National Bureau of Standards, Gaithersburg, MD.

Levin, B. M., and Nelson, H. E., "Firesafety and Disabled Persons," *Fire Journal*, Vol. 75, No. 5, Sept. 1981, pp. 35–40.

Levin, B. M., Groner, N. E., and Paulsen, R., "Affordable Fire Safety in Board and Care Homes. A Regulatory Challenge. Final Report," George Mason Univ., Fairfax, VA, National Institute of Standards and Technology, Gaithersburg, MD, National Institute of Disability and Rehabilitation Research, Washington, DC, Administration on Aging, Washington, DC, Social Security Administration, Baltimore, MD, Administration on Developmental Disabilities, Washington, DC, Health Care Financing Administration, Baltimore, MD, National Institute of Mental Health, Rockville, MD, NIST-GCR-93-632, July 1993.

Mabon, G., "Protection People in Care-Type Premises," *Fire*, Vol. 88, No. 1080, June 1995, pp. 35, 36.

Pearson, R. G., and Joost, M. G., "Egress Behavior Response Times of Handicapped and Elderly Subject to Simulated Residential Fire Situation," NBS-GCR-83-429, 1983, National Bureau of Standards, Gaithersburg, MD.

Waller, M. B., and Vreeland, R. G., "Report of a Conference on Fire Emergency Plans in Group Homes for the Developmentally Disabled," NBS-GCR-81-315, March 1981, National Bureau of Standards, Gaithersburg, MD.

Yates, J., and Kirby, R., "Four-Fatality Fire in Residential Board and Care Facility, Bessemer, Alabama, September 19, 1990," USFA Fire Investigation Technical Report Series, TriData Corp., Arlington, VA, Federal Emergency Management Agency, Washington, DC, Report 043, 1991.

HOTELS

James R. Bell

Hotels are buildings or groups of buildings under the same management having more than 16 sleeping accommodations primarily used by transients who are lodged, with or without meals, for a period of less than 30 days. These facilities may be designated as hotels, inns, clubs, motels, guest quarters, suite hotels, or by other names. So-called apartment hotels also are classified as hotels, because they are subject to transient occupancies like those of hotels.

OCCUPANCY CHARACTERISTICS

This occupancy may include guest rooms that combine living, sleeping, sanitary, and storage facilities within compartments, or guest suites. Guest suites consist of accommodations with two or more contiguous rooms comprising a compartment, with or without doors between such rooms, providing living, sleeping, sanitary, and storage facilities. Additionally, some apartments or extended stay hotels may have rooms or suites which include kitchen or cooking facilities for guest use.

These facilities are unique in that they almost always combine several different occupancies under one roof. In addition to guest rooms or guest suites, which are residential occupancies, hotels usually provide space for assembly occupancies such as ballrooms, meeting rooms, exhibition halls, and restaurants; or mercantile occupancies such as shopping areas, gift shops, and other retail areas; and offices and commercial establishments, or business occupancies.

Larger hotels in major cities may have several floors available for sale as condominium apartments. Many large hotels also contain industrial-type laundries and dry cleaning facilities, as well as kitchens, which use large broilers, ovens, and deep-fat fryers.

Larger-scale hotel building configurations often have atria two or three stories high and sometimes up to sixty stories high, which are often the focal point of building design. Atrium areas themselves may include several occupancies or mixed functions associated with the hotel operations.

Some structures also may include parking facilities which may be open, enclosed, above- or below-ground, and often directly beneath or adjacent to the hotel itself. These arrangements may require special types of fire protection, and the building codes may require fire separations.

Since hotel guests are transients, special consideration must be given to the potential threat to their life safety from fire. For example, occupants of the residential portion of a hotel sleep in unfamiliar surroundings and could possibly become disoriented when trying to evacuate under heavy smoke conditions. Likewise, persons in various ballrooms, lounges, and restaurants could become disoriented due to low-level lighting, crowd size, and unfamiliarity with evacuation routes.

Furthermore, it is not unusual to find the residential portion of a hotel full at the same time the banquet halls, ballrooms, lounges, or restaurants are operating at capacity. Without proper design consideration, this situation could tax the egress facilities of a hotel and present a logistical problem for fire fighters trying to locate and extinguish a fire, while simultaneously attempting rescue and evacuation. Added to this are inherent problems in evacuating high-rise buildings with sealed windows, inaccessibility to fire department ladders, and the potential smoke problems in corridors and stairwells.

CONSTRUCTION FEATURES

Current model building codes address these problems through the application of construction requirements, protection of means of egress, fire suppression, local and area fire detection and notification, and other features designed to limit the effects of fire within a building. In recent years, installation of sprinkler protection throughout all portions of hotel buildings, regardless of their height, has become a general requirement in most building codes.

In addition, some model building codes, and many local ordinances, require that the sprinklers in hotel guest rooms, suites, or other portions of the building be fast response or residential-type sprinklers that stop a fire while still small.[1] Thus, the possibility of a large fire and the production of large volumes of blinding smoke, non-survivable temperatures, and fatal toxic gas levels is held to a minimum.

Of particular importance for the protection of hotels four stories or fewer in height is the recent development of NFPA 13R, *Standard for the Installation of Sprinkler Systems in Residential Occupancies up to and Including Four Stories in Height* (hereinafter referred to as NFPA 13R), which provides design and installation requirements of sprinkler systems for small, low-rise residential occupancies.

Corridor compartmentation, smokeproof stairways, smoke barriers, and enclosed elevator lobbies are some of the features used

James R. Bell is a senior fire protection specialist for the Marriott International, based in Bethesda, MD.

to prevent the propagation of smoke within the means of egress. Automatic shutdown of air circulation or the activation of smoke evacuation equipment and stairwell pressurization by smoke detectors or sprinkler waterflow are also used to help ensure safe use of the means of egress by guests.

In addition to building code requirements, the potential fire and life safety problems that facilities of this type present are addressed in the various occupancy sections of NFPA *101®*, *Life Safety Code®* (hereinafter referred to as NFPA *101*). For example, ballrooms, restaurants, lounges, and exhibition halls are dealt with in the chapter of NFPA *101* that addresses assembly occupancies, whereas the portion of the hotel providing sleeping rooms is dealt with in the residential chapters of NFPA *101*. NFPA *101* does not generally require occupancy separation; however, the model building codes may require separations between different occupancy classifications. This fire barrier essentially separates the occupancies into different buildings. This occupancy separation applies unless one use is considered incidental or the uses are intermingled. In the latter case, the more restrictive life safety requirements of the occupancies involved apply.

LIFE SAFETY FEATURES

NFPA *101* provides guidelines for minimum standards regarding various life safety features in the building(s).

Means of Egress

This section deals with the type, capacity, number, arrangement, illumination, marking, and measurement of travel distances to exits. These provisions greatly enhance the chance for a safe means of egress from the building in emergencies. Because the risk to life safety is viewed as different in unsprinklered and sprinklered hotels, factors such as common paths of travel, dead-end corridors, travel distance from guest room doors to the nearest exit, and travel distances within a room or suite may be increased if sprinklers are provided.

Protection

Enclosures of vertical openings, such as stairways, elevator shafts, and atria, are required by NFPA *101* to impede the spread of smoke and heat within the building. Hazardous areas such as employee locker rooms, gift and retail shops, laundries, storage rooms, trash rooms, and similar areas require separation by 1-hr rated separation walls and/or sprinkler protection, depending on the severity of the hazard. This section also deals with protection from other potential hazards, such as boilers, transformers, and other service equipment subject to explosion.

Interior Wall, Ceiling, and Floor Finish

Due to the flammability and toxicity of modern fabrics and interior finish materials on walls and ceilings, flame spread and smoke production in these materials are regulated in exit enclosures, corridors, lobbies, and all other spaces. Requirements for floor finish in corridors and exits are also addressed.

Detection and Alarm Systems

Because hotels are occupied by persons who are normally expected to be asleep as part of the standard conditions of occupancy and who are unfamiliar with the building, provisions outlined in NFPA *101* require early warning or notification of occupants and annunciation of fire alarm signals. This section outlines requirements for the initiation of audible and visible alarms from manual fire alarm stations, automatic sprinkler systems, and any required automatic detection systems. In high-rise buildings, notification of occupants

is required to be through a voice/alarm system operated from a central fire command room. Single-station smoke detector(s) installed in each guest room or in each sleeping room and each living area within a guest room or guest suite are required to alert the occupants of that room or suite, but are not usually connected to the building fire alarm system.

Extinguishment Requirements

NFPA *101* requires that approved automatic sprinkler protection be installed in all new hotels, except non-high-rise buildings that have guest room doors that open directly to the outside at ground level or have exterior exit accesses arranged in accordance with NFPA *101*. The sprinkler systems in buildings up to, and including, four stories may be installed according to NFPA 13R.

NFPA *101* also requires that all existing high-rise hotels be sprinklered, unless the exceptions for exterior exit access, arranged in accordance with Chapter 5 of NFPA *101* are met.

Corridor Walls and Doors

Another provision of NFPA *101* requires rated corridor walls and self-closing doors in guest room areas. (See Figures 9-10A and 9-10B.) Not only are door-closing mechanisms effective life safety features of NFPA *101*, but they are also inexpensive to install and maintain. Several major multiple-death hotel fires have illustrated the importance of self-closing doors.[2,3]

In most cases, fire and smoke can be contained in the room of origin as long as the door remains closed. This is especially true in sprinklered rooms. Fast-acting sprinklers are not only effective in suppressing fire and minimizing smoke in guest rooms, but they will, in conjunction with door closers, contain smoke and heat in the room of origin, leaving the egress corridors tenable.

Staff Training

Even though these built-in life safety features are paramount in the maintenance of a fire safe hotel, a well-developed fire emergency plan and a well trained and drilled emergency organization are of equal importance. The emergency organization consists of on-duty employees who are trained to conduct fire prevention inspections

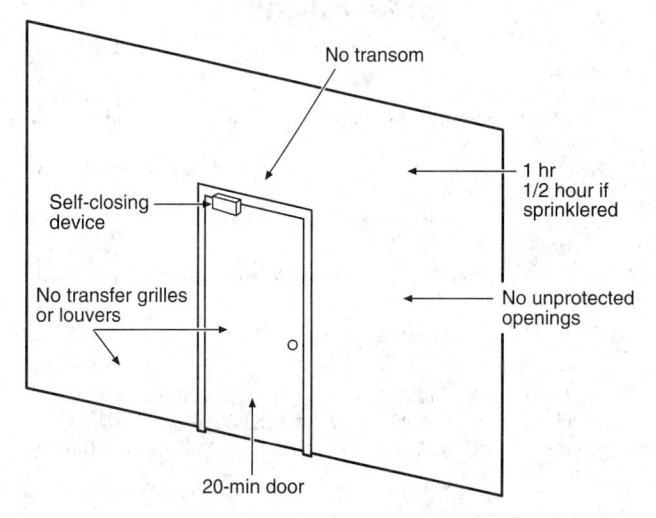

FIG. 9-10A. Provisions for the protection of guest rooms (corridors) in new hotels. The need for the self-closing, 20-min door cannot be overemphasized. There are no exceptions or options to the self-closing door.

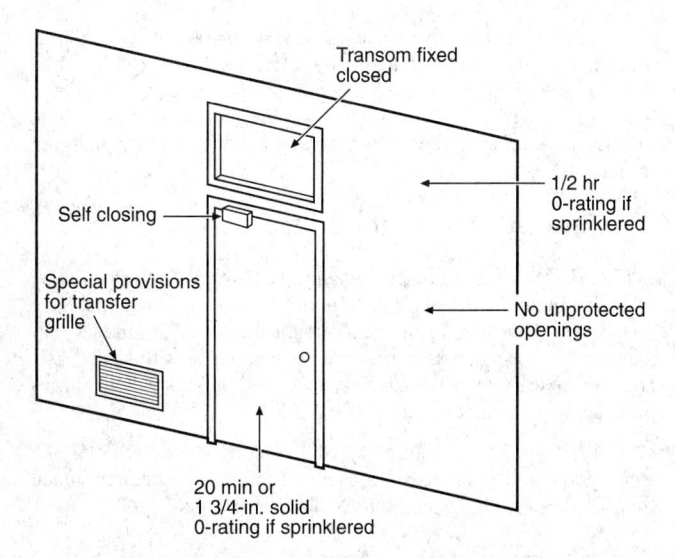

Transom fixed
closed

Self closing

1/2 hr
0-rating if
sprinklered

Special provisions
for transfer
grille

No unprotected
openings

20 min or
1 3/4-in. solid
0-rating if sprinklered

FIG. 9-10B. *Provisions for protection of guest rooms (corridors) in existing hotels. The need for the self-closing door cannot be overemphasized. There are no exceptions or options to the self-closing device. Although no rating is required when sprinkler protection is provided, the wall must be substantial and must resist the passage of smoke.*

and tests of life safety systems on a regular basis. In addition, they should hold regularly scheduled fire and disaster drills. In the event of an emergency, their responsibilities may include notifying the fire department, incipient-stage fire suppression, evacuation, and sprinkler/fire pump monitoring, as well as salvage and overhaul after extinguishment. Organization members should also be trained in first aid and certified to perform emergency medical treatment, such as cardiopulmonary resuscitation (CPR).[4]

An effective emergency organization, along with automatic sprinklers, compartmentation, effective egress design, and detection and alarm systems, will provide hotel guests with a degree of protection that will greatly increase the probability that the hotel in which they sleep will be safe from serious loss in fire.

BIBLIOGRAPHY

References Cited

1. *The Uniform Building Code*, International Conference of Building Officials, Whittier, CA, 1994.
2. Klem, T. J., and Best, R. L., "Four Hotel Fires: Lessons Learned," *Fire Journal*, Vol. 76, No. 6, Nov. 1982, p. 72.
3. Klem, T. J., "Investigation Report on the Dupont Plaza Hotel Fire," *Fire Journal*, Vol. 8, No. 3, May/June 1987.
4. *Hotel Fire Emergency Organization*, Marriott International, Inc., Washington, DC.

References

The BOCA Basic/National Building Code, Building Officials and Code Administrators International, Inc., Country Club Hills, IL, 1993.
Standard Building Code, Southern Building Code Congress, Inc., Birmingham, AL, 1994.

NFPA Codes, Standards, and Recommended Practices

Reference to the following NFPA codes, standards, and recommended practices will provide further information on hotels discussed in this chapter. (See the latest version of *The NFPA Catalog* for availability of current editions of the following documents.)

NFPA 13, *Standard for the Installation of Sprinkler Systems*

NFPA 13R, *Standard for the Installation of Sprinkler Systems in Residential Occupancies up to and Including Four Stories in Height*

NFPA 25, *Standard for the Inspection, Testing, and Maintenance of Water-Based Fire Protection Systems*

NFPA 72, *National Fire Alarm Code*

NFPA *101*®, *Life Safety Code*®

Additional Reading

Bell, *et al.*, "Investigation Report on the Westchase Hilton Hotel Fire," National Fire Protection Association, Quincy, MA, 1982.

Best, R., and Demers, D., "Investigation Report on the MGM Grand Hotel Fire," National Fire Protection Association, Quincy, MA, 1982.

Bill, R. G., "A Life Safety Team: Smoke Detectors and Sprinklers in Hotels," *Fire Journal*, Vol. 83, No. 3, May/June 1990, p. 28.

Bill, R. G., and Kung, H. C., "Evaluation of an Extended Coverage Sidewall Sprinkler and Smoke Detector in a Hotel Occupancy," *Journal of Fire Protection Engineering*, Vol. 1, No. 3, 1989.

Bryan, J. J., *An Examination and Analysis of the Dynamics of the Human Behavior in the MGM Grand Hotel Fire*, revised edition, National Fire Protection Association, Quincy, MA, 1983.

Coakley, D., *et al.*, *The Day the MGM Grand Hotel Burned*, Lyle Stuart, 1982.

Cote, R. M., *et al.*, "Investigation Report of the Ramada Inn Central Fire," National Fire Protection Association, 1984.

Daly, T. G., "Hotel Firesafety—An Update," *Fire Journal*, Vol. 79, No. 1, Jan. 1985, pp. 56–58.

"Hotel and Motel Fire Safety Act of 1990 (Public Law 101-391)," Federal Emergency Management Agency, Washington, DC, L-201, Dec. 1992.

Hotel Fires: Behind the Headlines, National Fire Protection Association, Quincy, MA, 1982.

Klem, T. J., "Ninety-Seven Die in Arson Fire at Dupont Plaza Hotel," *Fire Journal*, Vol. 81, No. 3, May/June 1987.

"Operation Life Safety," The Operation San Francisco Smoke/Sprinkler Test, International Association of Fire Chiefs, Washington, DC, 1984.

Rule, C. H., "Increased Fire Safety in Hotel/Motel Guest Rooms," *Fire Engineering*, Vol. 138, No. 12, 1985, pp. 40–43.

Rutes, W. A., and Penner, R. H., *Hotel Planning and Design*, Whitney Library of Design, Watson-Guptill Publications, 1985.

Schaenman, P. S., and Daly, T. G., "The Facts about Fire Protection in the Lodging Industry," *Fire Journal*, Vol. 83, No. 1, Jan./Feb. 1989, pp. 23–26.

Thomas, J., "Dupont Plaza. . . MGM Grand: Could They Happen Again," *Fire Journal*, Vol. 82, No. 3, May/June 1988.

Zeiss, W. E., *Active Fire Protection Concepts for Hotels*, Cerberus A. G., Switzerland, 1985.

APARTMENT BUILDINGS

Kenneth Bush

Apartment buildings are identified as those structures containing three or more living units with independent cooking and bathroom facilities, whether designated as apartment houses, tenements, condominiums, or garden apartments. Apartments differ from multi-unit residential occupancies that are not considered homes, such as hotels and boarding homes, by the provision of individual cooking facilities, the number of sleeping rooms, and the less transient nature of the occupants.

The adequate protection of residential buildings and their occupants from the effects of fire has historically been a major concern of fire protection professionals. The construction of multiple dwelling units makes this job more complex. Such construction not only adds to building size and occupant numbers, but also adds major complexity due to building design, construction, and fire protection features.

OCCUPANCY CHARACTERISTICS

In 1993, 24,776,000 (26.2 percent) of all housing units occupied year round were in structures with two or more housing units—that is, in duplexes or apartments. (Duplexes cannot be isolated in these statistics.) The breakdown by size was 9.8 percent of housing units in structures with 2 to 4 units, 5.0 percent in structures with 5 to 9 units, 4.4 percent in structures with 10 to 19 units, 3.3 percent in structures with 20 to 49 units, and 3.6 percent in structures with 50 or more units. Of the 24,776,000 units, 14 percent were owner occupied.[1]

Residential properties in the United States continue to dominate fire losses in this country, and apartment buildings share in that dominance. During 1989–1993, 75 percent (roughly 479,000) of all reported structure fires occurred in residential properties. Of these, 109,000 fires per year were reported in apartment buildings, accounting for 17 percent of all structure fires and causing $924 million per year in property loss.

In 1989–1993, 21,500 civilians per year were injured in residential fires. This meant 87 percent of civilian injuries in structure fires occurred in residential properties. Of these injuries, 6,900 per year occurred in apartments, or 28% of civilian injuries in structure fires. In 1989–1993, 3,950 civilians per year died in residential fires, 95% of deaths in structure fires. The number of deaths from fires in apartment buildings averaged 810 per year, 19% of civilian deaths in structure fires.

Comparisons of fire risk between apartments and one- and two-family dwellings, including manufactured homes, are difficult because statistics on duplexes are grouped with apartments for housing units and with single-family properties for fire experience. Since, as noted, apartments accounted for 17–28 percent of housing units in 1991 and accounted for 21 percent of house fire deaths, it appears the risk of fatal fire is comparable in apartments and dwellings.

OCCUPANCY HAZARDS

The occupancy hazards of apartment buildings include all risk factors that may arise in particular segments of the population. Preschoolers and older adults, if present, have a higher risk of dying in fires because of the mental or physical limitations associated with their age. Older children and younger adults may be at risk if they are physically or mentally handicapped, or as a result of drug or alcohol abuse. And fatal fires are more common at night, when people are asleep.

The pattern of causes for fatal apartment fires is different in many ways from the pattern for one- and two-family dwellings. Because apartment buildings are more likely to use central systems and professional supervision and maintenance for their equipment, most equipment-related causes rank lower for apartment buildings than they do for one- and two-family dwellings. For example, using 1989–1993 fire deaths, heating equipment ranked fifth, with 4.9 percent of civilian fire deaths in apartments, compared to second and 16.4 percent in one- and two-family dwellings. Electrical distribution equipment ranked sixth, with 4.6 percent of civilian fire deaths in apartments, compared to fourth and 10.8 percent in one- and two-family dwellings. By contrast, smoking, incendiarism, and child fire play ranked first, second, and third, respectively, as causes of apartment fire deaths, resulting in 69.5 percent of fire deaths, compared to first, third, and fifth, and 46.2 percent for one- and two-family dwellings. Other NFPA analyses show that one-third of fatal apartment fires begin in bedrooms, compared to one-fourth of fatal fires in single-family dwellings, duplexes, or manufactured homes. This probably means apartment victims are closer to the fire on average.

Smoke detectors have been a great success in apartment buildings, as well as in private homes. Use of detectors continues to grow, but more slowly since 1982. In 1995, 93 percent of all homes had at least one smoke detector, and there was evidence that their use in apartments was higher still, probably because state and local laws requiring detectors often have stronger requirements for apartments. However, NFPA studies indicate that when fire occurs, roughly one-third of apartment detectors are not operational, the same share as for one- and two-family dwellings. These studies also indicate that

---

Kenneth Bush is a senior fire protection engineer with the Office of the Maryland State Fire Marshal. He is a member of three NFPA Safety to Life Technical Committees.

smoke detectors cut the risk of dying in a home fire by roughly 40 percent. However, the estimated reduction is only 16 percent in apartments, townhouses, and condominiums based on fires that occurred between 1985 and 1994. Although available data does not lead to definite conclusions, it appears that the difference in effectiveness levels may be due to the fewer number or longer distance in apartment building escape routes, more success in early detection to prevent fires from growing large enough to report, or the delayed reaction of apartment dwellers to alarms outside of their unit.

Once a potentially serious fire starts, apartment buildings can pose special problems to fire department suppression forces.

The variety of these types of residences, as well as the architectural desire to make buildings aesthetically pleasing yet functional, creates an interesting combination of fire protection demands. Building designs, space considerations, and parking arrangements can create major access problems, particularly for fire-fighting apparatus. (See Figure 9-11A.) Long hose lays and blocked roadway accesses can create problems for fire-fighting operations in addition to existing problems, such as weather, personnel deficiencies, and hampered water supplies. (See Figure 9-11B.) Architectural designs may use false partitioning and veneered construction, giving an unrealistic impression of the type and magnitude of the actual building construction. (See Figure 9-11C.) Although NFPA 101®, *Life Safety Code*® (hereinafter referred to as NFPA *101*), makes no special requirements for the construction of these buildings, many such requirements are found and enforced through the model building codes, as well as through many local codes and ordinances currently in effect.

An additional problem arises in the location and combination of apartments with other occupancy types, particularly those above and behind mercantile and business occupancies. (See Figure 9-11D.)

FIG. 9-11C.   *Architectural designs often use false partitioning and veneered construction, creating an unrealistic impression of the building.*

FIG. 9-11A.   *Many apartment building designs and parking arrangements create major access problems and hinder fire-fighting operations. (Photo courtesy of Stephen C. Leahy.)*

FIG. 9-11B.   *The location and lack of access routes for many new apartment buildings create major fire apparatus placement and fire-fighting operation problems.*

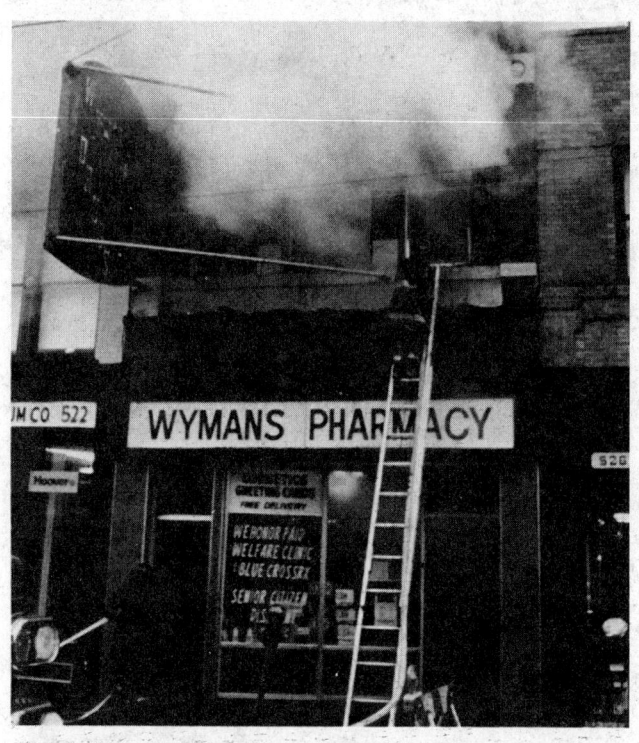

FIG. 9-11D.   *Additional fire safety problems arise in the location and combination of apartments with other occupancy types, particularly above and behind stores and businesses. (Photo courtesy of Stephen C. Leahy.)*

Where multiple occupancy classes occur in the same building and are so intermingled that separate safeguards are impractical, means of egress facilities, construction, protection, and other safeguards must comply with the most restrictive life safety requirements for the occupancies involved.

LIFE SAFETY CONCEPTS

The residential fire death problem is very complex, and several strategies must be presented to help solve it. The installation and proper maintenance of smoke detectors and quick-response residential sprinklers in sleeping areas provide an effective, low-cost, and widely available method of early warning so that evacuation, containment, and suppression can begin during the incipient stages of a fire.

The general public must be taught to protect themselves and others occupying the residential unit from fire through the safe use and installation of alternative heating devices, the proper storage and handling of flammable liquids, and the identification and correction of home fire hazards. In addition, stringent flammability standards for upholstered furniture, such as those previously issued by the U.S. Consumer Product Safety Commission (CPSC) for children's sleepwear and mattresses, will provide long-term fire safety benefits. Specially designed systems that are effective in terms of life safety should be installed to provide a high level of protection to occupants, particularly where large numbers of persons are present, such as in a high-rise apartment building, or where the occupants' ability to escape is hampered, such as in apartments for the elderly.

Although such fire safety precautions are not always practical, particularly within an individual dwelling unit, a more concentrated effort must be made to confine the fire and its effects to a well-defined, limited area of the structure. This task is accomplished through the adoption of accepted fire safety codes, such as NFPA *101*. Since many current laws and regulations prevent fire inspection authorities from entering individual living units, adequate fire protection features must be designed and constructed during the initial phases of the project. The regulation of interior furnishings provides only a partial solution to the fire safety problem, although such consumer items are generally in use for many years before they are replaced. Futhermore, the individual habits and attitudes of building occupants can never be regulated or otherwise altered.

Because apartment buildings, and residential occupancies in general, are so diverse in size, configuration, and resident population, many regulatory codes pay particular attention to construction designs and fire protection features. NFPA *101* covers the presence of hazards such as cooking and heating equipment, as well as the occupant's degree of familiarity with his/her living space. NFPA *101* requires most new apartment buildings to be protected by sprinklers. However, sprinkler protection is not required where each apartment has a door directly to the outside at ground level, such as in townhouse-style apartments, or is served by an interior stair totally enclosed by fire-rated construction, or where an outside stair serves a maximum of two units located on the same level. (See Figure 9-11E.) NFPA *101* has a unique method of regulating the construction and design of existing apartment buildings based upon the type and extent of automatic fire detection and/or fire suppression equipment provided. Recognizing equivalency provisions found elsewhere in NFPA *101*, the residential subcommittee established four equivalent schemes to provide a high degree of flexibility. These designs equally provide a minimum level of life safety for apartments. Some systems have additional protective capability, permitting higher building heights and larger building areas.

This overall approach provides one of the first system design attempts to be codified. Whereas a total system would have many design approaches, this more limited system offers only four differ-

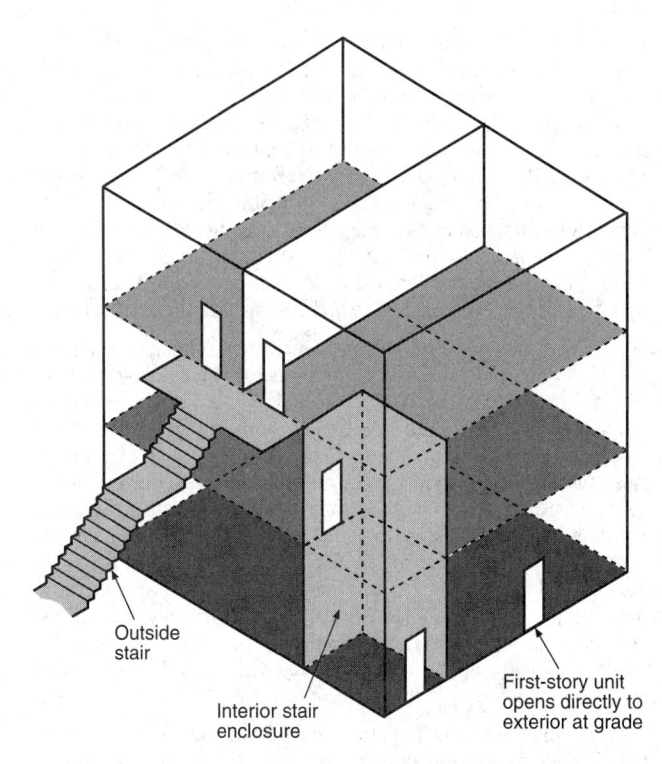

FIG. 9-11E.   Examples of required exiting arrangements for new apartment buildings without sprinkler protection.

ent approaches. Yet an evaluator can identify an appropriate method from the four approaches based on the building size, height, and arrangement. This allows the person evaluating the existing building to formulate a safety approach that best fits the building, rather than fitting the building to a single codified design criterion.

Under special definitions found in Chapter 4 of NFPA *101*, the contents of living units in residential occupancies fall into the ordinary-hazard classification. Characteristics of these hazards include moderately rapid burning and considerable smoke generation. (See Figure 9-11F.) Under normal conditions, no unduly dangerous exposure should develop during the period necessary to escape from

FIG. 9-11F.   Apartment buildings are classified as ordinary hazard occupancies, which include moderately rapid burning of building contents and considerable smoke generation. (Photo courtesy of Stephen C. Leahy.)

the fire area. Generally, this classification represents conditions found in most buildings. Even with the introduction of plastics, foam materials, and other combustible materials in furnishings, the amount and configuration of such materials present a lower hazard than in other occupancies where they may be sold, stored, or displayed. NFPA 13, *Standard for the Installation of Sprinkler Systems*, classifies the contents of apartment buildings as "light" for the design of extinguishing systems. This classification difference is based on the threat to life safety assumed in the ordinary classification *vs.* that assumed where the extinguishing capability of the automatic sprinkler system results in the light-hazard classification.

In designating the number and types of exits required for apartment buildings, NFPA *101* determines the occupant load on the basis of 1 person/200 sq ft (18.6 m²) of gross floor area, or the maximum probable population of any room or section under consideration, whichever is greater. Many buildings, such as dormitories, may produce a greater occupant load. However, even though some portions of these buildings are densely populated, the whole building may not necessarily exceed 1 person/200 sq ft (18.6 m²) of gross area, owing to the space taken for toilet facilities, hallways, closets, and living rooms. These requirements do not preclude the need for providing adequate exit capacities from concentrated use and sleeping areas based on the maximum probable population, rather than on the design figure. The capacities of required exits are as specified in NFPA *101*.

Although seemingly contrary to the basic code requirements of two distinct, remote means of egress from all locations, several exceptions to this rule are found within the requirements for apartment occupancies. While the traditional interior exit corridor arrangement commonly found in larger apartment buildings and condominiums requires two remote exits, several other options in apartment building configurations permit other arrangements. The common townhouse arrangement, which provides an exit directly to the street or yard at ground level or by way of an outside or enclosed stairway serving that apartment only, permits the front door to be the only required exit. Therefore, a rear door, which is often provided, would not be required and could be a sliding door to a porch or patio.

Another exception is a basic design approach for garden apartments where the apartment entrances are enclosed around a single protected stairway. This single stairway usually is open to the exterior or is glass enclosed. It must be separated from the building by construction of at least 1-hr fire-resistance rating. The doors should be treated similarly. Note that this exception for a single exit is limited to buildings having four or fewer stories with not more than four living units per story and protected throughout by an approved, supervised automatic sprinkler system, and to existing buildings of three stories or less. Interestingly, the basic design of a typical garden-style apartment has influenced other chapters of NFPA *101*. Previous editions of NFPA *101* required that all exit stairways continuing beyond the floor of discharge be interrupted at that level by partitions, doors, or other effective means to make clear the direction of egress to the street. However, an exception to this requirement allows these stairways to continue one-half story beyond the level of discharge without such physical interruption. This exit arrangement is typically found in garden-style apartments that have apartments or service areas on the ground floor or terrace level. Based on this exception, stairways serving these levels need not be interrupted at the floor of exit discharge for fire safety reasons.

In addition to the number and types of required exits, the arrangement and travel distance to these exits are regulated by the general requirements of Chapter 5 of NFPA *101*, with due consideration to the levels of fire protection features within the building. As in other occupancies, all exits must terminate directly at a public way or at an approved exit discharge, unless certain additional protection features, such as automatic sprinkler protection and vesti-

bule separation and/or enclosures, are provided in accordance with Section 5-7 of NFPA *101* to permit 50 percent of the required exit capacity to discharge through areas on the level of discharge. A recent modification to NFPA *101* also recognizes the use of areas of refuge, particularly for spaces accessible to persons with severe mobility impairments. Because of the high level of satisfactory performance and response to fire conditions by supervised automatic sprinkler systems, any story of an apartment building equipped with this level of protection may be considered an area of refuge even if an occupant does not have access to any of the apartment units. Because locked apartment unit doors create inaccessibility to spaces other than the corridor, the corridor maintains its tenability due to the effectiveness of the required fire protection systems and serves as the area of refuge.

Appropriate illumination must be provided for public spaces, hallways, stairways, and other means of egress. Emergency lighting is required for apartment buildings that have more than 12 living units or are more than three stories high, except in buildings where every living unit has a direct exit to the outside at grade, such as in townhouses. Approved exit signs are required in all apartment buildings with more than one required exit. However, discretion must be used in placing such signs where exits are obvious, or all occupants are familiar with the building.

Because a higher degree of transient occupancy in some apartments is a definite possibility, special exiting considerations are made within the individual living units, in addition to those found in the public access spaces. All individual living units with two or more rooms must comply with the minimum provisions for operable exterior openings to be used as a secondary means of escape, or with the specialized separation or protection requirements of Chapter 21 of NFPA *101*. Not only must these openings be of a minimum size, but they also must be accessible and easy to operate. In addition, every dwelling unit greater than 2,000 sq ft (180 m²) or having a travel distance to the primary means of escape greater than 75 ft (22.5 m) must have two primary means of escape remote from each other, or it must be protected by approved supervised automatic sprinkler protection. Special multilevel units, such as penthouses, are permitted to have regular, curved, or spiral stairways that connect to one level above or below the entrance to the apartment.

Additional requirements prohibit any door in an apartment building, including individual unit doors, from being locked against egress while the building is occupied. This means that a door may be locked so that it can be opened from within the building but not from the outside. Ordinarily, dead bolts, double-cylinder locks, and chain locks would not meet these provisions. Multiple-death fires occur if a key can not be found to unlock the door or the chain is not removed. The use of delayed egress locks, however, is permitted in apartment buildings protected throughout by automatic sprinkler systems or fire detection systems. The limitation of one such locking device in any one egress path and the 15- or 30-sec delayed release provides the security needed to infrequently used doors, but, at the same time, ensures that the door remains available for emergency use.

In apartment buildings, the protection of vertical openings, such as stairways, elevator shafts, and atria, follows the requirements found in NFPA *101*. (See Figure 9-11G.) Smokeproof towers are required in existing high-rise apartment buildings except those that are totally sprinklered. This limits the possibility of smoke contamination of exits and ensures a safe, protected means of egress. The protection from hazards, as well as the classification of interior finishes, is also based upon the levels of fire detection and/or protection within the building.

Perhaps the best form of fire protection in a residential occupancy is early warning. All living units in an apartment building

FIG. 9-11G.   The vertical spread of fire and other products of combustion is of primary concern to most fire safety codes.

must have approved automatic smoke detection. These detectors are usually located in hallways that have access to the sleeping areas although new unsprinklered apartment buildings must also be equipped with single-station smoke detectors in every sleeping room. In multiple-level units, a detector also should be located at the tops of stairways. Detection systems must be installed in accordance with basic engineering practices. Special consideration should be given to room configuration, air movement, and stagnant air pockets that would reduce detector efficiency, and to cooking areas, fireplaces, or other normally smoke-producing appliances that cause an excess of nuisance alarms. All detectors required for installation in multiple-dwelling units must be powered by house current. In addition, where a single apartment needs more than one detector, the detectors must be electrically interconnected. Note that this type of detection system is required in addition to the requirements for any other detection or suppression system.

Manual fire alarm systems are required in apartment buildings according to the building's size and height and the occupants' evacuation capabilities. As building designs become higher and more complex, additional special features are required, such as annunciator panels, visible alarm signals, voice communications systems, and automatic transmission of building fire alarms. Voice communications systems often are used as part of a multi-faceted system that allows the fire department to contact building occupants during a fire.

As mentioned before, NFPA *101* requires almost all new apartment buildings to be protected by automatic sprinklers. In existing apartment buildings, the extent of automatic sprinkler protection is again governed by the building's size and height, as well as by the four protection options discussed in Chapter 19 of NFPA *101*. To the extent permitted by the applicable sprinkler system installation standard, the use of residential or quick-response sprinklers is required for new installations of sprinkler systems in dwelling units, apartments, and guest rooms. The increased technology, cost effectiveness, and appearance of these systems have provided numerous additional incentives for mandating this protection option.

NFPA *101* requires portable fire extinguishers to be located in hazardous areas only, not throughout the entire building. Many building owners place these extinguishers in individual living units, under the direct responsibility of the occupant.

The design, installation, and maintenance of all such fire detection and protection systems are based on referenced fire protection standards, as well as on sound engineering judgment. For residential occupancies of four or fewer stories, see NFPA 13R, *Standard*

for the Installation of Sprinkler Systems in Residential Occupancies up to and Including Four Stories in Height.

When fires cannot be confined to the area of origin, the means of egress must be sufficiently protected to allow all building occupants enough time for safe evacuation. This basic concern for providing safety for the occupant in a room during a fire led to a minimum corridor wall construction to keep fire in a corridor from entering a room or to keep fire in a room from entering the corridor. Doors protecting openings in these enclosures must provide a level of protection commensurate with the expected fuel load in the room and with the fire resistance of the corridor wall construction. Contrary to the general rule for door assembly considerations, the requirements for doors in residential occupancies apply to the leaf only, with a $1^3/_4$-in. (45-mm) solid, bonded, wood-core door considered equivalent to a door with a 20-min fire protection rating. The door rating requirement establishes a minimum quality of construction and does not require a complete listed fire door assembly. However, the requirement that doors between apartments and corridors be self-closing reduces the risk of fire fatalities. Fire spread can occur mainly due to the door being left open as the occupant flees the fire. People may die after opening the door to a fully involved room, or after causing a room to become fully involved by opening a door and introducing oxygen to the fire. Under both circumstances, a door equipped with spring-loaded hinges or a door closer would prevent smoke or fire from spreading down the corridor and exposing other occupants.

During a fire, the evacuation of all residents from a large apartment building at best a time-consuming process. This problem is compounded by the introduction of larger and taller structures, complex and intricate building designs, and a larger population of residents who are incapable of rapid evacuation, especially by stairways. For these reasons, the most expedient movement of a sizable number of residents may be horizontally. The introduction of required areas of refuge serves three purposes fundamental to the protection of building occupants by (1) limiting the spread of fire and fire-produced contaminants, (2) limiting the number of occupants exposed to a single fire, and (3) providing for horizontal evacuation by creating safe and tenable space on the same level.

Accessible means of egress and associated areas of refuge are required in apartment buildings based upon the evacuation capabilities of the building's occupants and on the construction and design characteristics of the building and levels of fire protection features. Since the majority of apartment residents are assumed to be capable of vertical evacuation using stairways, the location of stairways provides both the required area of refuge as well as the required means of egress. Where travel time to stairways increases due to excessive horizontal distances and/or where a majority of residents have reduced mobility, specialized areas are required on the same floor level.

Typically, all building service equipment and utilities in apartment buildings are required to be designed and installed in accordance with sound engineering practices and all applicable NFPA standards. The use of a public corridor as part of the supply, return, or exhaust air system is not permitted. In fact, in certain apartment building designs, these corridors must be pressurized against smoke movement. The standard requirements for the reliability of the means of egress and fire retardancy for interior furnishings and decorations, and the proper maintenance and testing of fire protection equipment also apply equally to apartment occupancies.

Occupants of each living unit must be given emergency instructions on a yearly basis, indicating the location of alarms, exiting paths, and actions to be taken in response to a fire in the living unit and in response to the sounding of an alarm. However, other operating features, such as specialized fire safety training and fire exit drills, have been deleted due to the nontransient nature of most building occupants.

The statistical data previously cited provide clear evidence that the fire problem in apartment buildings and in residential occupancies as a whole is complex, significant, and not easily solved. The increasing demand for adequate, yet low-cost, functional and aesthetically pleasing housing has led to an increasing demand for adequate, safe, and economical fire protection. The problem intensifies with the demand for higher and larger buildings with a higher density of occupants and with the demand for more personal services for building occupants. Although the solution to these fire problems undoubtedly requires much future study, the obvious short-term solution of effective code formulation, adoption, and enforcement is the immediate responsibility of every fire safety professional.

BIBLIOGRAPHY

Reference Cited

1. U.S. Bureau of the Census, *Statistical Abstract of the United States: 1995*, 115th edition, Washington, DC, 1996, Table 1230.

References

Hall, J. R., Jr., *The U.S. Experience with Smoke Detectors*, NFPA Fire Analysis and Research Division, Quincy, MA, August, 1996.
Hall, J. R., Jr., *The U.S. Problem Overview Report*, NFPA Research Division, Quincy, MA, April 1996.
Karter, M. J., Jr., "Fire Loss in the United States in 1993," *Fire Journal*, Vol. 88, No. 5, Sept./Oct. 1994, pp. 57–65.

NFPA Codes, Standards, and Recommended Practices

Reference to the following NFPA codes, standards, and recommended practices will provide further information on apartment buildings discussed in this chapter. (See the latest version of *The NFPA Catalog* for availability of current editions of the following documents.)

NFPA 13, *Standard for the Installation of Sprinkler Systems*
NFPA 13R, *Standard for the Installation of Sprinkler Systems in Residential Occupancies up to and Including Four Stories in Height*
NFPA *101®*, *Life Safety Code®*

Additional Reading

(Note: All major causes of fires in apartment buildings and use of smoke detectors are addressed in annual statistical reports from NFPA's Fire Analysis & Research Division.)
Alvard, D. M., "The Fire Emergency Evacuation Simulation for Multifamily Buildings," NBS-GCR-84-483, Dec. 1984, National Bureau of Standards, Gaithersburg, MD.
"Apartment Building Suffers Unusual Fire Spread," *Fire Fighting in Canada*, Vol. 35, Sept. 1991, pp. 6–7, 43.
"Balanced Design Approach to Firesafety for Low-Rise Multifamily Construction. Fire Protection Planning Report No. 18 Of a Series," Concrete and Masonry Industry Firesafety Committee, Skokie, IL, Fire Protection Planning 18, 1990.
Beck, V. R., and Yung, D., "Cost-Effective Risk-Assessment Model for Evaluating Fire Safety and Protection in Canadian Apartment Buildings," *Journal of Fire Protection Engineering*, Vol. 2, No. 3, 1990, pp. 65–74.
Blye, P., and Yess, J. P., "Firesafety in Elderly Housing," *Fire Journal*, Vol. 81, No. 6, Nov. 1987, p. 28.
Fire in the United States, fourth edition, Federal Emergency Management Agency, Washington, DC, 1982, pp. 25–27.
"Fire Safety in Multi-Family Housing," U.S. Congressional Hearing, Subcommittee on Science, Research and Technology, July 18, 1988.
Gautier, F., "Simulation of a Three-Story Wooden Apartment Building with CFAST 1.5.0," Centre Scientifique et Technique du Batiment, Cedex, France, Oct. 1991/May 1992.

Hadjisophocleous, G. V., and Yung, D., "Fire Risk and Protection Cost Assessment Model for Highrise Apartment Buildings," National Research Council of Canada, Ottawa, Ontario, ASTM STP 1150; American Society for Testing and Materials, Fire Hazard and Fire Risk Assessment, Sponsored by ASTM Committee E-5 on Fire Standards, ASTM STP 1150, December 3, 1990, San Antonio, TX, ASTM, Philadelphia, PA, 1990, pp. 224–233.
Hadjisophocleous, G. V., and Yung, D., "Parametric Study of the NRCC Fire Risk-Cost Assessment Model for Apartment and Office Buildings," *Proceedings* of the Fourth International Symposium on Fire Safety Science, Intl. Assoc. for Fire Safety Science, Boston, MA, 1994, pp. 829–840.
Karter, M. J., Jr., "Fire Loss in the United States in 1988," *Fire Journal*, Vol. 83, No. 5, Sept. 1989.
Katajamaki, J., "Puukerostaloilla poliittinen tuki. [Wooden Apartment Buildings Win Political Support.]," *Palontorjuntatekniikka*, Jan. 1995, pp. 12–13.
Kirby, R. E., "Nine-Fatality Apartment House Fire, Ludington, Michigan (February 28, 1993)," USFA Fire Investigation Technical Report Series, TriData Corp., Arlington, VA, Federal Emergency Management Agency, Washington, DC, Report 072, 1993.
Luo, M., and Beck, V., "Study of Fire Behavior in a Full Scale Apartment: Prediction and Experiment," *Proceedings* of the First International Conference on Fire Science and Engineering, ASIAFLAM '95, March 15–16, 1995, Kowloon, Hong Kong, 1995, pp. 525–530.
Norman, J., "'Fireproof' Multiple Dwellings, Apartments, Dormitories, Hotels," *Firehouse*, Vol. 17, No. 1, March 1992, pp. 16,18.
Norman, J., "Garden Apartment and Townhouse Fires," *Firehouse*, Vol. 21, No. 3, March 1996, pp. 18, 20, 92.
Proulx, G., "Evacuation Time and Movement in Apartment Buildings," *Fire Safety Journal*, Vol. 24, No. 3, 1995, pp. 229–246.
Proulx, G., and McQueen, C., "Evacuation Timing in Apartment Buildings. Internal Report," National Research Council of Canada, Ottawa, Ontario, Internal Report 660, June 1994.
Reid, J. K., "Apartment Owner Finds Sprinkler Retrofit Affordable," *Sprinkler Age*, Vol. 11, No. 5, May 1992, pp. 10, 12.
"Report of the Real Scale Burning Test of a Wooden 3-Stories Apartment House. [Wooden 3-Storey Apartment Building Shake and Burn Test Report.]," Building Research Inst., Tsukuba, Japan, The Foundation, Japan Building Center, Tokyo, Japan, Jan. 1992.
Rickman, L. M., Luciani, F. S., and Dacquisto, D. J., "Retrofitting Fire Safety Systems in Multi- and Single-Family Dwellings," Sprinklers in Residential and Commercial Buildings, August 19–20, 1990, Charleston, SC, Natl. Conf. States on Bldg. Codes and Standards, 1990, pp. 139–154.
Salamone, R., "Retrofitting High-Rise Dorms with Alarm and Detection Systems," *Fire Journal*, Vol. 84, No. 1, Jan./Feb. 1990, pp. 37–39.
Santavuori, A., "Sprinklerilaitteistot puukerrostaloissa. [Sprinkler Systems in Wooden Apartment Buildings.]," *Palontorjuntatekniikka*, Vol. 4, 1994, pp. 13–15.
Schaenman, P., "Apartment Complex Fire, 66 Units Destroyed, Seattle, Washington, September 21, 1991," USFA Fire Investigation Technical Report Series, TriData Corp., Arlington, VA, Federal Emergency Management Agency, Washington, DC, Report 059, 1992.
Sultan, M. A., and Halliwell, R. E., "Optimum Location for Fire Alarms in Apartment Buildings," *Fire Technology*, Vol. 26, No. 4, Nov. 1990, pp. 342–356.
Takeda, H., and Yung, D., "Simplified Fire Growth Models for Risk-Cost Assessment in Apartment Buildings," *Journal of Fire Protection Engineering*, Vol. 4, No. 2, 1992, pp. 53–66.
Tammisaari, M., "Huoneistopalon sammuttaminen pienella vesimaaralla. [Extinguishing Apartment Fires With Small Quantities of Water.]," *Palontorjuntatekniikka*, Vol. 4, 1991, pp. 24–27.
Tammisaari, M., "Huoneistopalon sammuttaminen pienella vesimaaralla. Osa 2: Suihkuputkien vertailukokeet. [Extinguishing Apartment Fires With Small Quantities of Water. Part 2: Nozzle Comparisons]," *Palontorjuntatekniikka*, Vol. 1, 1994, pp. 10–15.
Tammisaari, M., "Huoneistopalon sammuttaminen pienella vesimaaralla. Osa 3: Llmanvaihto-olosuhteiden vaikutus sammutukseen ja koulutuskayttoon soveltuva vesisammutuksen simulointiohjelma. [Extinguishing Apartment Fires With Small Amount of Water. Part 3: The Effect of Air Conditioning on Extinguishing and Simulation Training for Water Extinguishing.]," *Palontorjuntatekniikka*, Feb. 1995, pp. 6–11.

Tuomisaari, M., "Huoneistopalon sammuttaminen pienella vesimaaralla. Osa 1. Sammutusveden tehokkaimpaan kayttoon vaikuttavia tekijoita. [Putting Out an Apartment Fire Using a Minimum Amount of Water. Part 1. Factors Affecting Efficient Use of Water in Firefighting.]," *Palontorjuntatekniikka*, Feb. 1993, pp. 30–34.

Yang, D., Hadjisophocleous, G. V., and Takeda, H., "Comparative Risk assessments of 3-Storey Wood-Frame and Masonry Construction Apartment Buildings," Sixth International Fire Conference on Fire Safety, Interflam '93, March 30–April 1, 1993, Oxford, England, Interscience Communications Ltd., London, England, 1993, pp. 499–508.

Yung, D., and Hadjisophocleous, G. V., "Use of the NRCC Risk-Cost Assessment Model to Apply for Code Changes for 3-Storey Apartment Buildings in Australia," National Research Council of Canada, Ottawa, Ontario, Society of Fire Protection Engineers and Worcester Polytechnic Institute, *Proceedings* of Computer Applications in Fire Protection, June 28–29, 1993, Worcester, MA, 1993, pp. 57–62.

LODGING OR ROOMING HOUSES

Revised by Alfred J. Longhitano and Mario A. Antonetti

Lodging or rooming houses include buildings in which separate sleeping rooms are rented and sleeping accommodations are provided for 16 or fewer persons on either a transient or a permanent basis with or without meals, but without separate cooking facilities for individual occupants. One- and two-family dwellings in which a room or rooms are rented to no more than three persons are not considered lodging or rooming houses. Facilities with sleeping accommodations for more than 16 persons are classified as hotels, while those with individual cooking facilities are classified as apartment buildings.

Occupants of a lodging or rooming house must be capable of self-preservation. The presence of four or more occupants who are incapable of self-preservation reclassifies the facility as a health care, or possibly a board and care, facility.

Based on these restrictions, the classification "lodging or rooming house" is effectively limited to facilities providing sleeping accommodations without separate cooking facilities for 4 to 16 persons who are capable of self-preservation.

OCCUPANCY CHARACTERISTICS

In the 1970s and 1980s two entirely new markets spurred a growth in the number and nature of lodging and rooming houses, which were once considered budget-conscious alternatives to expensive hotels. First, landmark court decisions required that psychiatric patients capable of living in less restrictive environments than psychiatric hospitals be released to live in the general population. Many deinstitutionalized people preferred not to assume all the responsibilities of independent living, meal preparation, and the like, and chose to live in rooming houses. The second phenomenon affecting the growth of rooming houses is the increasing popularity of the bed-and-breakfast and country inns as alternative travel accommodations.

Very few new buildings are being classified as lodging or rooming houses. Most are existing facilities in converted dwellings, motels, or small apartment houses. Mixed occupancies generally are not encountered, except in cases where mercantiles are located on the lower floor or floors and sleeping rooms are located above. Code does not permit such an arrangement unless the dwelling occupancy and exits are separated from the mercantile occupancy by

construction with a fire-resistance rating of at least 1-hr or the mercantile occupancy is protected by automatic sprinklers.

Lodging or rooming houses can be in buildings of any construction type permitted by local building codes. NFPA *101*®, *Life Safety Code*® (hereinafter referred to as NFPA *101*), does not identify any minimum construction requirements. Because most lodging or rooming houses are located in converted dwellings or share space with mercantile operations, construction is generally limited to Type III or V, as defined by NFPA 220, *Standard on Types of Building Construction*.

The hazard of the contents is considered "ordinary," a trait consistent with most residential occupancies.

Occupancy Conversion or Remodeling

It is common for fire hazards to be inadvertently created during conversion or remodeling. Existing wall and ceiling finishes are removed, exposing combustible structural components. Concealed combustible spaces may be created by installing drop ceilings and new wall finishes. These concealed combustible spaces permit rapid fire development and spread and limit accessibility for fire fighting. The wall and ceiling finishes in new facilities are limited to Class A, B, or C, but in existing buildings, finish materials typically exceed this limitation and significantly affect fire spread and smoke generation.

Lack of Firestopping

As a result of a lack of building codes or their enforcement at the time of construction, many older buildings lack firestopping in walls and ceilings. This can provide pathways for the rapid, undetected spread of fire and smoke.

Unprotected Vertical Openings

Unprotected vertical openings are not limited to stairs. They include shafts, ducts, heat grates, or simply old floor openings that were never covered. Each represents a path for fire and smoke movement, which may compromise an exit that is considered protected.

Building Services

In lodging or rooming houses, building services generally are limited to taking care of the electrical systems, heating, and possibly air conditioning. Primary consideration should be given to the proper maintenance of heating equipment and to the prevention of electrical circuit overloading. Some tenants may be permanent so they are

Alfred J. Longhitano, P.E., is a principal with Gage-Babcock & Associates, Inc., Mount Kisco, NY. Mario A. Antonetti, P.E., is supervisory engineer and secretary of Gage-Babcock & Associates, Inc., Mount Kisco, NY.

likely to accumulate more electrical appliances than transient tenants. Consideration should be given to providing sufficient electrical receptacles in tenant spaces to prevent tenants from overloading receptacles or misusing extension cords.

Storage Areas for Personal Belongings

Storage areas normally are located in basements, under stairs, and in rooms off egress corridors. Because they frequently store a high concentration of combustible contents, storage areas present a serious hazard.

Supplemental Heating Devices

Inadequate heating can lead to the use of supplemental heating devices, such as portable electric, gas, and kerosene space heaters. Each represents a specific hazard and has its own accompanying set of requirements. For example, one specific code requirement prohibits installing a stove or heater where its malfunction would block escape in case of fire.

LIFE SAFETY CONSIDERATIONS

Due to the limited number of sleeping accommodations and the lack of separate cooking facilities for individual occupants, lodging or rooming houses are viewed, from a life safety standpoint, as falling somewhere above one- and two-family dwellings and below dormitories, hotels, and apartment buildings. NFPA *101* requirements for this occupancy center on four key elements essential to life safety.

Means of Escape

For life safety purposes, safe egress from the building is always of paramount concern. In lodging or rooming houses, the egress system consists of a "means of escape," which, by definition, is a way out of a building or structure that does not conform to the strict definition of means of egress but does provide a safe way out. This term, separate and distinct from means of egress, refers to a lower level of egress for this specific occupancy in relation to all other occupancy classes and a higher level of egress than required in one- and two-family dwellings.

Sleeping rooms and living areas above and below the level of exit discharge are required to have a primary and secondary means of escape, one of which must be either an enclosed interior stairway, an exterior stairway, a horizontal exit, or an existing fire escape. The primary means of escape must provide a safe path of travel from the building, one that does not expose occupants to an unprotected vertical opening, except in completely sprinklered buildings of three or less stories. (See Figures 9-12A, 9-12B, and 9-12C.)

The specified means of escape need not meet the requirements of Chapter 5 of NFPA *101*, with the exception of width, stair riser, and tread. Details about the second means of escape are provided in Section 21-2 of NFPA *101*, and include either a door or stairway that provides unobstructed travel to the outside of the building at street or ground level, or an outside operable window. Note that a second means of escape is not required, provided the room has a door leading directly outside to grade or an approved exterior stair.

Sleeping rooms located on the level of exit discharge are required to have two unspecified means of escape, one of which may be an operable window that meets the requirements of NFPA *101*, Section 21-2. Exceptions to the requirement for two means of escape apply to rooms with doors leading outside to grade and buildings protected by automatic sprinklers.

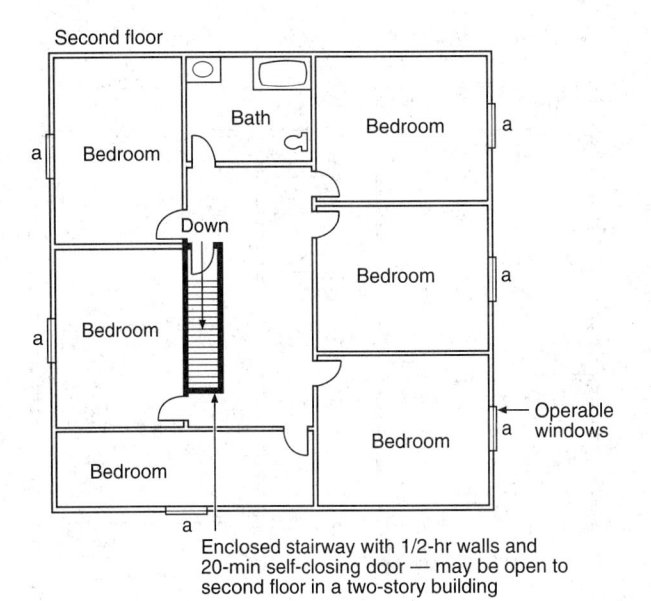

FIG. 9-12A. *Means of escape from a lodging or rooming house. This example shows an enclosed interior stairway that discharges directly outside. Access to the interior stairway is by an interior corridor that is not exposed to an unprotected vertical opening. The second means of escape could be through windows. The bottom of the stairway can be enclosed several ways; this is only one example.*

No required means of escape may be routed through an area that is not under the control of the occupant who is leaving the building. Nor may it be routed through a door that may be locked.

Detection and Alarm Systems

Detection and alarm systems in residential occupancies play a critical role in the early warning and evacuation of occupants. A manual fire alarm system installed in accordance with NFPA 72, *National Fire Alarm Code*, is required in all cases, except buildings that have approved smoke detection or sprinkler systems.

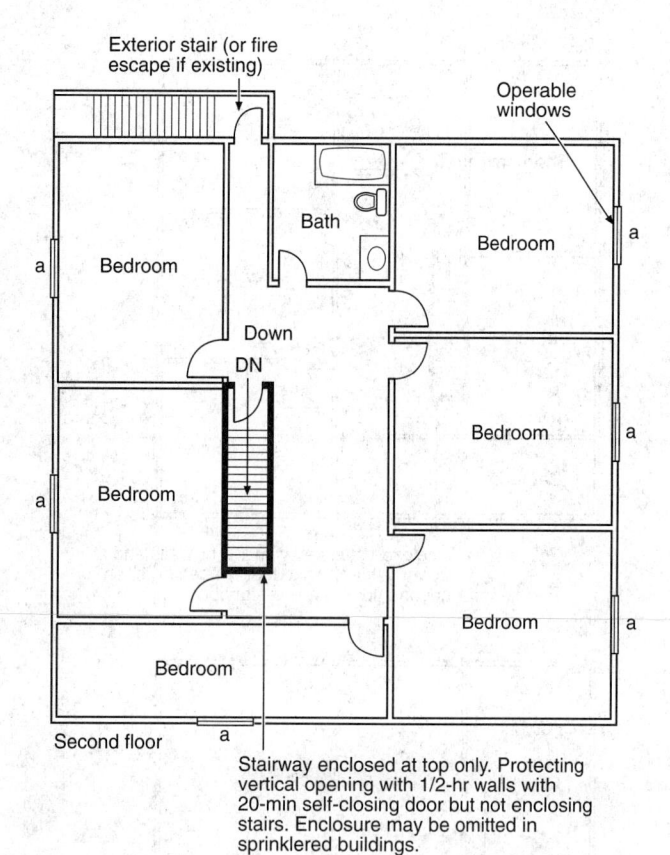

FIG. 9-12B.   Means of escape from a lodging or rooming house. This example shows an exterior stairway. Tenants reach exterior stairway through an interior corridor that is not exposed to an unprotected vertical opening. The second means of escape is through windows.

Smoke detectors installed in accordance with NFPA 72, Chapter 2, and powered by the house electrical service are required on each floor, including basements but excluding crawl spaces and unfinished attics. When activated, the detectors should initiate an alarm audible in all areas. This requirement is generally satisfied by interconnecting detectors with the manual fire alarm system.

In addition to smoke detectors in the corridors on each level, a single-station smoke detector powered by the building electrical system is required in each sleeping room to alert the occupant to a fire or smoke in the room.

It is critical that the fire alarm system will be audible in all sleeping rooms.

Separation of Sleeping Rooms

Rooming houses are required to be compartmented to restrict the spread of smoke. Specifically, sleeping rooms must be separated from corridors by smoke-resistant walls and self-closing doors. This feature helps to prevent smoke from compromising the egress corridors in case of a fire in a room; if the corridor does become smoke-logged before all occupants can escape, it allows an occupant to re-

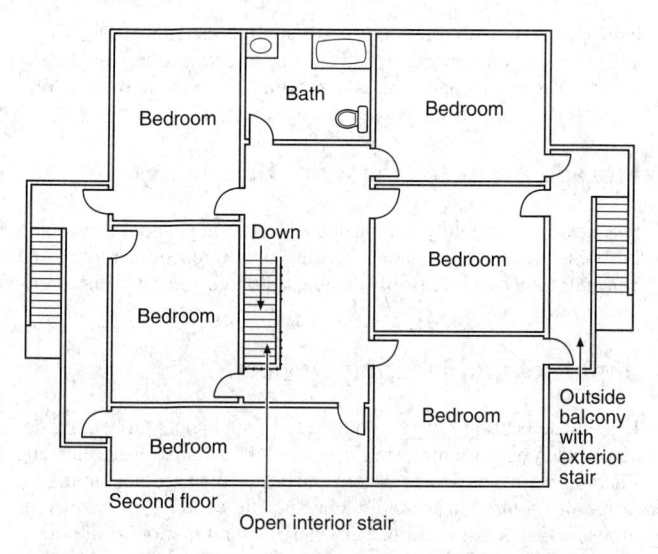

FIG. 9-12C.   Means of escape from a lodging or rooming house. This example shows an outside stairway with exterior exit access. Access is not through the interior corridor, which is exposed by the open stairway. The second means of escape is through the open interior stairway.

main within the room for a limited period of time until the occupant can be rescued or alternate escape strategies can be developed.

Automatic Sprinkler Protection

NFPA *101* requires most new lodging and rooming houses to be protected by automatic sprinklers. Exceptions to other requirements can be made if sprinklers have been installed.

BIBLIOGRAPHY

NFPA Codes, Standards, and Recommended Practices

Reference to the following NFPA codes, standards, and recommended practices will provide further information on lodging or rooming houses discussed in this chapter. (See the latest version of *The NFPA Catalog* for availability of current editions of the following documents.)

NFPA 72, *National Fire Alarm Code*

NFPA *101*®, *Life Safety Code*®

NFPA 220, *Standard on Types of Building Construction*

Additional Reading

Birth, K., "Fire Safety for Lodging Houses Review After Berlin Deaths," *Fire International*, No. 123, June/July 1990, pp. 39, 41.

Crossland, J. W., "Fire Precautions in Public Homes and Restaurants," *Fire Surveyor*, Vol. 18, No. 6, Dec. 1989, pp. 9–16.

Haley, G., "Lodging or Rooming Houses," *Conducting Fire Inspections*, second edition, National Fire Protection Association, Quincy, MA, 1989.

ONE- AND TWO-FAMILY DWELLINGS

Harry L. Bradley

One- and two-family dwellings are defined as residential structures in which not more than two households reside. These dwelling units may be attached to or detached from one another, and typical styles include detached single-family dwellings, townhouses, and duplexes. There are many different designs, such as the one-story ranch, two-story colonial, split level, and three-level townhouse. Wood frame and masonry are the most common construction types.

In 1993, 67.9 percent of all year-round occupied housing units were single-family detached or attached units, and another 6.0 percent were manufactured homes or trailers.[1] Two-family houses were not tabulated separately, but housing units in buildings with two to four housing units accounted for another 9.8 percent of the total. Therefore, one- and two-family housing accounts for somewhere between three-fourths and five-sixths of all housing in the United States.

Each year, the majority of all U.S. fire deaths occur in one- and two-family dwellings. In 1995, for example, 66 percent or 3,035 deaths,[2] occurred in such dwellings. They represented 83 percent of the 3,640 home fire deaths, a share comparable to their share of all housing units. Thus, it is important to consider every element of fire safety for its potential value in this setting.

NFPA *101*®, *Life Safety Code*® (hereinafter referred to as NFPA *101*), includes requirements for one- and two-family dwellings. In particular, NFPA *101* addresses the need for two means of escape from every room for occupant evacuation; the selection of interior finish materials to prevent rapid early fire growth and spread and to protect occupant evacuation routes; and smoke detectors to provide detection and alarm. Additional information on smoke detectors and residential sprinkler systems can be found in Section 5, Chapter 6, "Household Fire Warning Equipment," and Section 6, Chapter 13, "Fast Response Sprinkler Technology."

The philosophy of life safety and fire protection for one- and two-family dwellings is based on evacuation of the building in the event of a fire. The defend-in-place concept in other occupancy types, which relies on fixed fire protection systems and fire-resistant construction, does not lend itself to one- and two-family dwellings. To allow residents time to evacuate, they must be alerted to a fire that has been detected at its incipient stage, they must know what action to take, and they must have adequate means of escape.

Proper installation and use of residential-type smoke detectors can accomplish early detection. NFPA 72, *National Fire Alarm Code*, recommends smoke detectors be installed on each level of the dwelling.

After the occupants have been alerted, they must promptly evacuate. A written fire plan should be made for each dwelling, showing two means of escape from every room, especially the bedrooms, and the entire family should practice this plan. Programs such as NFPA's "EDITH" (Exit Drills in the Home) explain how to write and practice a home fire plan.

Adequate means of escape from each room should be provided in accordance with the requirements of NFPA *101*. Two means of escape are required from each room. One must be a door or stairway, and the second way must be an operable window. (See Figures 9-13A and 9-13B.)

Residential-type automatic sprinkler systems provide a high degree of protection for one- and two-family dwellings. These relatively new types of automatic systems are not widely installed at present, but the technology is gaining in acceptance and use.

FIRE AND LIFE SAFETY HAZARDS

Many other fire protection strategies are under discussion, but they have not yet been reduced to codes or regulations. These address the contribution of various types of contents, furnishings, and construction elements to fire growth; smoke and gas generation; and confinement after a fire begins.

Most fire fatalities in one- and two-family dwellings are a result of toxic gases, not burns, and most victims die in a room that is not the room or area of fire origin. In other words, they die as a result of a fire that grew large enough to spread beyond the room in which it began. In most fatal fires in dwellings, duplexes, and manufactured homes, the physics and chemistry of fire development can be quite complex. Prevention or action early in the fire is needed to control the development of these lethal conditions.

Prevention of Fire Ignition

Fatal fires in one- and two-family dwellings are so numerous that nearly every imaginable cause may be worthy of attention. The leading cause of fire deaths in one- and two-family dwellings is smoking materials, or lighted tobacco products, which caused 630 civilian deaths a year from 1989 to 1993, or one-fifth of all fire deaths in dwellings and duplexes.[3] (All statistics are based on a hierarchical

Harry L. Bradley, P.E., is fire protection engineer, State Fire Marshal's Office, Baltimore, MD.

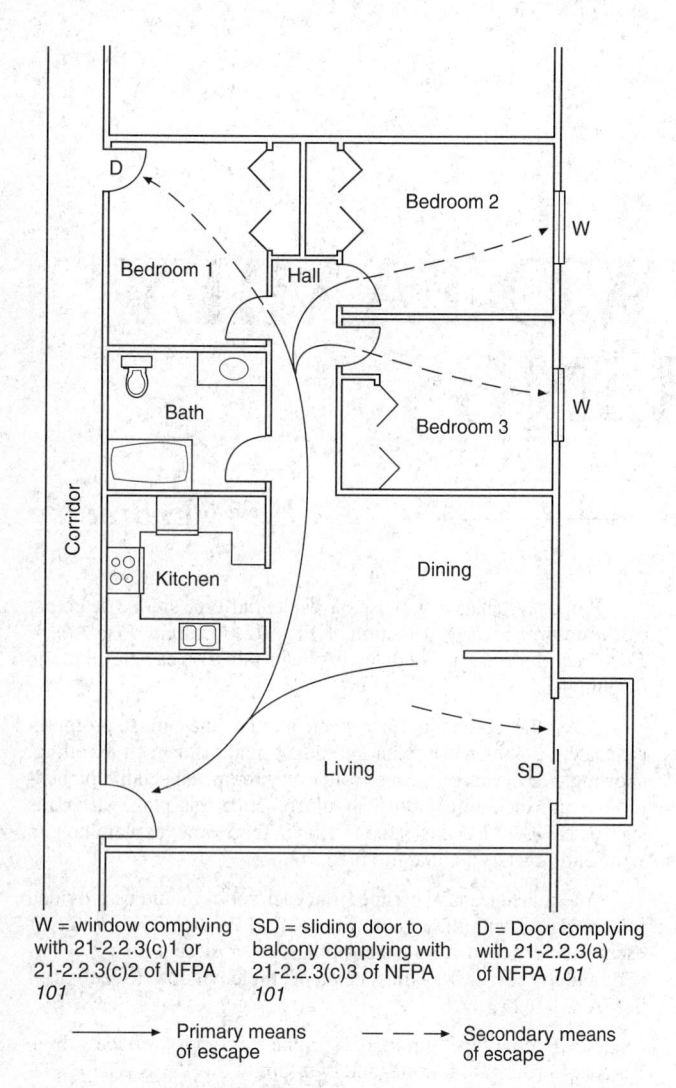

W = window complying with 21-2.2.3(c)1 or 21-2.2.3(c)2 of NFPA *101*

SD = sliding door to balcony complying with 21-2.2.3(c)3 of NFPA *101*

D = Door complying with 21-2.2.3(a) of NFPA *101*

⟶ Primary means of escape

⤍ Secondary means of escape

FIG. 9-13A. *Means of escape from a three-bedroom single-family home. The bedroom in the upper left has no window, but it has a door directly to the outside. The alternate means of escape from the other bedrooms is through windows and must comply with the requirements of NFPA 101.*

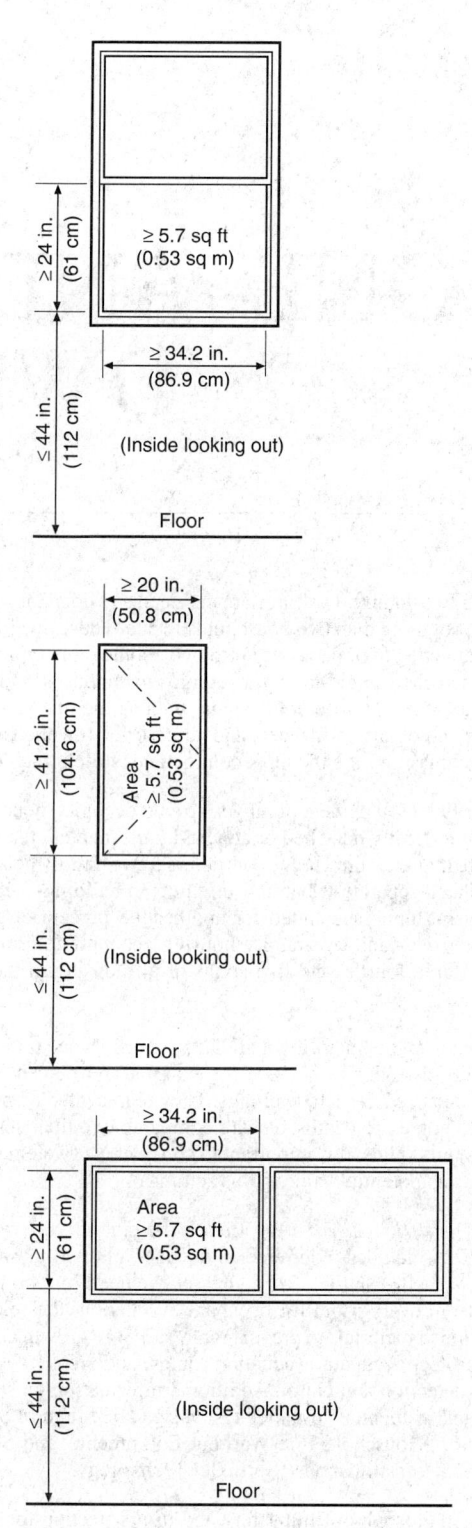

FIG. 9-13B. *Minimum size and dimensions of outside windows used as a second means of escape.*

sorting procedure developed by the U.S. Fire Administration and include a proportional allocation of fires of unknown causes. Higher estimates are obtained if the causes are considered individually outside the hierarchy.) Fatal smoking-material fires usually involve the ignition of upholstered furniture, mattresses, or bedding, which also rank highest as combustible items first ignited in fires. Fires begun by people smoking in bed are less common than those that start in living rooms, family rooms, and dens, and it is not unusual for a fire to smolder for hours before open burning begins. This is why many fatal fires occur at night when everyone is asleep.

The second-ranking cause of fire deaths in one- and two-family dwellings is heating equipment, which caused 500 deaths a year. Heating equipment was also the leading cause of fires: 79,500 fires a year. Portable or area heating devices and associated chimneys and connectors, more than water heaters and furnaces and their associated ducts and pipes, caused most of these fires and fire deaths. Each portable or area unit has its own rules for safe installation, maintenance, and use. Solid-fuel appliances, in particular, are covered by NFPA 211, *Standard for Chimneys, Fireplaces, Vents, and Solid Fuel-Burn-*

ing Appliances. Adequate clearance to combustibles, whether fixed, such as walls and structural members, or moveable, such as furniture, bedding, and clothing, is a common problem for all types of portable and area heating equipment. Safe fueling practices are especially important for liquid-fueled units, as is adequate venting for gas-fueled units and regular inspection and cleaning for solid-fueled units.

Arson and suspected arson ranked third in fire deaths in one- and two-family dwellings, at 490 per year, and first in direct property damage, averaging $629 million per year from 1989 to 1993. Incendiary and suspicious fires have been declining for the past decade; however, juvenile firesetters account for more than half of all arson arrests overall and are the largest single contributors to the problem.

Electrical distribution system components ranked fourth as a cause of fire deaths, resulting in 330 fatalities per year. Many people mistakenly believe that electrical system fires represent a larger share of fires, but the widespread and long-term use of NFPA 70, *National Electrical Code®*, has reduced them to a lower-order problem that usually occurs because of failure to observe the Code. Electrical system fires occur for several reasons including damaged components, such as frayed cords, broken switches, or plugs; bypassed protection systems, such as oversized fuses or pennies used in place of fuses; overloaded cords, circuits, or outlets; and cords running through traffic areas or pinned down by furniture. Knowledgeable, experienced people should always perform repairs and modifications in strict accordance with applicable codes.

The fifth leading cause of fire deaths in one- and two-family dwellings in the United States was children playing with fire, usually with matches or lighters. This type of fire resulted in 290 deaths a year in 1989–1993. Children playing with fire also is the leading cause of fire deaths for children under age six, and has been one of the few causes of fire deaths to increase in recent years. Children must be monitored closely and taught at an early age that matches and lighters are tools for adults, not toys. In an effort to address the issue of children playing with lighters, the Consumer Product Safety Commission adopted regulations requiring butane lighters manufactured in the United States to be child resistant by July 1, 1994. Matches and lighters must be kept out of the reach and sight of young children.

Cooking equipment ranked sixth as a cause of fire deaths, resulting in 260 per year in 1989–1993, but first in fire injuries at 3,150 per year. Such equipment also accounts for a majority of unreported fires, which injure tens of thousands. Unattended cooking is a principal cause. Others include combustibles, such as rags, pot holders, or bags, placed too close to heat; failure to clean stove tops and ovens; and overhanging curtains, drapes, or clothing.

Of the appliances and heat sources not already mentioned, clothes dryers, hot embers and ashes, and lightning stand out. Each caused at least 1 percent of fires in one- and two-family dwellings in 1989–1993. Care is needed in installing and using or handling all potential heat sources.

BIBLIOGRAPHY

References Cited

1. *Statistical Abstract of the United States 1995*, 115th edition, U.S. Department of Commerce, Washington, DC, 1996.

2. Karter, M. J., Jr., "Fire Loss in the United States During 1995," *NFPA Journal*, Vol. 90, No. 5, Sept./Oct. 1996, pp. 57–65.

3. Hall, J. R., Jr., *The U.S. Fire Problem Overview Report Through 1994*, Fire Analysis and Research Division, National Fire Protection Association, Quincy, MA, April 1996.

NFPA Codes, Standards, and Recommended Practices

Reference to the following NFPA codes, standards, and recommended practices will provide further information on one- and two-family

dwellings discussed in this chapter. (See the latest version of *The NFPA Catalog* for availability of current editions of the following documents.)

NFPA 13D, *Standard for the Installation of Sprinkler Systems in One- and Two-Family Dwellings and Mobile Homes*
NFPA 70, *National Electrical Code®*
NFPA 72, *National Fire Alarm Code*
NFPA *101®*, *Life Safety Code®*
NFPA 211, *Standard for Chimneys, Fireplaces, Vents, and Solid Fuel-Burning Appliances*

Additional Reading

"A Study of Fatal Residential Fires," National Fire Protection Association, Quincy, MA.

"A Study of One- and Two-Family Dwelling Fires," NFPA No. FR75-1, National Fire Protection Association, Quincy, MA.

CABO One and Two Family Dwelling Code, 1989 edition, The Council of American Building Officials, Falls Church, VA, 1989.

Donegan, H. A., Taylor, I. R., and Meehan, R. T., "Expert System to Assess Fire Safety in Dwellings," *Proceedings* of the Third International Symposium on Fire Safety Science, Elsevier Applied Science, New York, 1991, pp. 485–494.

Estepp, M. H., "Strategies to Implement One- and Two Family Dwellings Sprinkler Initiatives," Prince George's County Fire Dept., MD, International Association of Fire Chiefs (IAFC), Fire-Rescue International Conference Proceedings, Annual Conference, 1993. Aug. 28–Sept. 1, 1993, Dallas, TX, 1993, pp. 171–177.

Fire In The United States, 1983–1990, eighth edition, National Fire Data Center, United States Fire Administration, Emmittsburg, MD.

Gaiser, K. E., "Challenge: Sprinklers in Hise-Rises, Manufactured Homes, One- and Two Family Dwellings. . . Bull or No Bull," Georegetown City Fire Dept., SC, International Association of Fire Chiefs (IAFC), Fire-Rescue International Conference Proceedings, Annual Conference, 1993. Aug. 28–Sept. 1, 1993, Dallas, TX, 1993, pp. 161–170.

Hall, J. R., Jr., "Update on the Auxiliary Heating Fire Problem," *Fire Journal*, Vol. 80, No. 2, March 1986, p. 52.

"Home Safe Home," P7838, Factory Mutual Research Corp., Norwood, MA.

Kennedy, T., "Energy-Efficient Windows in Multiple Dwellings," *Fire Engineering*, Vol. 144, No. 1, Jan. 1991, pp. 14, 16, 18.

Levin, B. M., "EXITT—A Simulation Model of Occupant Decisions and Actions in Residential Fires: Users Guide and Program Description," NBSIR 87-3591, July 1987, Center for Fire Research, Gaithersburg, MD.

McNeil, M., "Fatal Dwelling Fire, Wellington County, Ontario, 8 August 1991. Fire Investigation Report," Office of the Fire Marshall, Canada, Fire Investigation Report, June 28, 1993, 21 pages.

Mischutin, V., "Fire in the Home—A Viewpoint from the USA," *Fire International*, Vol. 10, No. 100, 1986, pp. 51–53.

Rickman, L. M., Luciani, F. S., and Dacquisto, D. J., "Retrofitting Fire Safety Systems in Multi- and Single-Family Dwellings," NAHB National Research Center, Upper Marlboro, MD, National Conference of States on Building Codes and Standards, Council of American Building Officials, Department of Housing and Urban Development, National Association of Home Builders, National Fire Protection Association, National Fire Sprinkler Association and National Institute of Standards and Technology, Sprinklers in Residential and Commercial Buildings, August 19–20, 1990, Charleston, SC, Natl. Conf. States on Bldg. Codes and Standards, 1990, pp. 139–154.

Scoones, K., "Serious Fires in Dwellings During 1993," *Fire Prevention*, No. 281, 28–31, July/Aug. 1995, pp. 28–31.

Shields, T. J., Silcock, G. W., and Bell, Y., "Fire Safety Evaluation of Dwellings," *Fire Safety Journal*, Vol. 10, No. 1, 1986, pp. 29–36.

Solomon, R. E., "Fire Problem in One- and Two-Family Dwellings," National Fire Protection Assoc., Quincy, MA, National Conference of States on Building Codes and Standards, Council of American Building Officials, Department of Housing and Urban Development, National Association of Home Builders, National Fire Protection Association, National Fire Sprinkler Association and National Institute of Standards and Technology, Sprinklers in Residential and Commercial Buildings, August 19–20, 1990, Charleston, SC, Natl. Conf. States on Bldg. Codes and Standards, 1990, pp. 25–41.

Tamim, A. S., "Cost-Benefit Analysis of Residential Sprinkler Systems in Single Family Dwellings," Worcester Polytechnic Inst., MA, Thesis, December 1992.

Todd, C., "Detecting Fires in Dwellings: The Proposed New Code," *Fire Safety Engineering*, Vol. 1, No. 3, June 1994, pp. 25–30.

U. S. House of Representatives, "U. S. Congress To Require the Secretary of Housing and Urban Development to Promulgate Regulations Requiring Smoke Detection Devices in Residential Dwelling Units Financed or Assisted by any Department of Housing and Urban Development Program," 102d Congress, 1st Session, H. R. 2099, House of Representatives, Washington, DC, Congress, Washington, DC, H. R. 2099, April 25, 1991.

U. S. Senate and U. S. Congress, "To Require the Secretary of Housing and Urban Development to Promulgate Regulations Requiring Smoke Detection Devices in Residential Dwelling Units Financed or Assisted by any Department of Housing and Urban Development Program," 102d Congress, 1st Session, S. 977, Senate, Washington, DC, Congress, Washington, DC, S. 977, April 25, 1991.

Willey, A. E., "Reducing Home Fire Deaths in the 1990s," *Fire Journal*, Vol. 81, No. 5, Sept./Oct. 1987, p. 121.

MANUFACTURED HOUSING AND RECREATIONAL VEHICLES

Travis Pitts

This chapter covers manufactured housing and recreational vehicles, including their associated regulations, utility systems, fire loss history, and fire safety requirements.

MANUFACTURED (MOBILE) HOMES

Amendments made in 1983 to the National Manufactured Housing Construction and Safety Standards Act of 1974 changed the definition of manufactured homes from "mobile" to "manufactured." Manufactured housing is not to be confused with modular or factory-built housing.

Manufactured homes are factory-assembled structures that can be transported in one or more sections. Sections are typically 8 ft (2.5 m) or more wide and 32 to 80 ft (10 to 24 m) long, each built on a permanent chassis. When assembled on site for occupancy as a dwelling, manufactured homes are connected to required utilities and placed on stabilizing devices, which may be piers and footings rather than permanent foundations. The increasing cost of conventional housing and the improved quality of manufactured homes make these homes the choice of residence for more than one-quarter million families each year. Forty percent of new manufactured homes are purchased by persons over the age of fifty.[1]

Some manufactured homes have been employed for unintended uses, such as multi-family occupancies, dormitories, and multi-story structures. Basic manufactured homes are also used in some states as temporary banks or offices, as field offices at construction sites, and for many other purposes. Manufactured homes are not designed for such uses, and special attention must be paid to fire safety when they are so utilized.

NFPA 501A, *Standard for Fire Safety Criteria for Manufactured Home Installations, Sites, and Communities*, should be consulted for fire safety at the installation site.

NFPA 501B, *Standard for Mobile Homes* (hereinafter referred to as NFPA 501B) (last published in 1977 and subsequently withdrawn) had been the guiding construction standard for manufactured homes since 1968. It addressed the unique construction characteristics, or the body and frame design; plumbing systems; heating, cooling, and fuel burning systems; and electrical systems of transportable homes. Almost all states adopted the NFPA 501B standard for manufactured homes, and concerned federal agencies,

such as the Federal Housing Administration and the Veterans Administration, used it in connection with home purchases under loan programs they administered.

However, both the manufactured housing industry and the federal government were concerned about the lack of enforcement of NFPA 501B by states, counties, and cities. To ensure a safe, low-cost form of housing of uniform quality throughout the U.S., Congress passed the National Manufactured Housing (formerly Mobile Home) Construction and Safety Standards Act of 1974 (Title VI of the Housing and Community Development Act of 1974). This act established the U.S. Department of Housing and Urban Development (HUD) as the sole source of standards governing the construction of these homes in the U.S. It gave HUD the authority to establish procedural and enforcement regulations necessary to ensure that makers of manufactured homes uniformly met the construction standards.

HUD fully implemented the act in mid-1976, substantially adopting NFPA 501B as the construction standard and naming the National Conference of States on Building Codes and Standards (NCSBCS) as the contract agent for ensuring uniform enforcement. Approximately 4.5 million manufactured homes have been built under the act, adding to the estimated existing manufactured housing stock of 4.3 million built before mid-1976.[2,3,4] The infusion of safer new manufactured homes into the U.S. housing stock has dramatically affected life safety and fire incidents in manufactured homes built after 1976.

During 1985–1994, manufactured homes built before 1976 averaged 2.5 civilian fire deaths per 100 fires, compared to 1.0 civilian fire deaths per 100 fires for manufactured homes built after 1976.[5] The 59 percent lower fire death rate per 100 fires in post-1976 units was also strengthened by a significantly lower rate of reported fires per thousand units, although precise calculations are not possible because the census reporting on year of manufacture breaks at 1974–1975 not at 1976. Manufactured homes built after 1976 also averaged 3.8 civilian fire injuries per 100 fires, 26 percent lower than the 5.1 civilian fire injuries per 100 fires averaged by manufactured homes built before 1976. Despite these large differences and the steady shift toward post-1976 units in the manufactured home inventory, the total number of fire deaths and injuries in manufactured homes have shown fairly modest declines, considerably less than the declines for other types of dwellings, which have had nothing comparable to the Federal standards driving their progress. It is not clear why this is the case.

The dramatic difference in the injury and death rates is directly attributable to HUD's uniformly applying and enforcing the

---

Travis Pitts is deputy director of the Department of Housing and Community Development, Sacramento, CA.

updated version of NFPA 501B (i.e., HUD Manufactured Home Construction and Safety Standards, CFR 24, Part 3280) as the core document for the portion of the HUD Standards that applies to fire safety. Both the mandatory smoke detector provision and the restrictions on flame spread ratings, throughout the unit and especially around heating and cooking equipment, are major contributors to the effects of the HUD Standards.[5] Distressingly, as late as 1994, one third of post-1976 units having fires were reported as having no detectors present when fire occurred and others had no working detectors.[5]

Some other pertinent features of the HUD standard that positively affect fire safety are discussed in the following paragraphs.

The HUD standard requires that the interior finish be regulated for flame spread characteristics and, in certain areas of the home, for combustibility. Firestopping is required in concealed wall and partition spaces. At least two exterior doors are required, and they must be arranged to provide a means of unobstructed travel to the outside. The minimum width of doors is specified, as is the ability to operate locking mechanisms without a key from inside to facilitate egress. Each room designed expressly for sleeping, unless it has a door to the outside, must have at least one outside window large enough to permit emergency egress. The window must be openable without the use of special tools.

Utilities

Where LP-Gas is used for heating or cooking, the LP-Gas containers are limited in size, number, and location. They must be mounted on the home's A-frame or in a vented compartment that is vapor tight to the inside of the home and accessible only from the outside. Requirements are included to ensure that containers are securely mounted so they will not jar loose, slip, or rotate when the home is in transit. Special provisions are included to protect shutoff valves on containers while the home is in transit or storage. LP-Gas regulators must be connected directly to the container shutoff valves or mounted on adjacent support brackets connected by a listed high-pressure connector to the valve. LP-Gas safety relief devices are required, and the discharge from these devices is regulated as to the proximity to any opening in the manufactured home. Allowable pressures of gas piping systems are also regulated, as are piping materials, routing, sizing, anchoring, and the provision of shutoff valves. Appliance connectors cannot run through walls, floors, ceilings, or partitions, and the manufacturer must perform specific gas piping leakage tests before delivering the home.

Where oil is used as the heating fuel, there are specific requirements for oil tanks, fill and vent pipes, liquid level gages, and shutoff valves. Materials for oil piping systems are specified as to size, type of joints, couplings, grading, hangers, and leakage tests.

A unique feature of the HUD standard concerns the installation of heat-producing appliances. Especially noteworthy is the requirement that all fuel burning appliances except ranges, ovens, illuminating appliances, clothes dryers, solid fuel burning fireplaces, and solid fuel burning fireplace stoves be installed so the combustion system is completely separate from the home's interior. Combustion air inlets and flue gas outlets must be listed or certified components of the appliance. The required separation may be obtained by installing direct vent systems (sealed combustion systems) or by installing the appliance within an enclosure so its appliance combustion and venting systems are separated from the manufactured home's interior.

No forced air appliance, nor its air return system, can allow the creation of a negative pressure that will affect either its combustion air supply or that of any other appliance, nor can they act to mix products of combustion with circulating air. Other requirements for heat-producing appliances include special provisions relating to solid fuel-burning fireplaces or fireplace stoves. The size and air

tightness of circulating air duct materials are specified, and the design and combustibility of duct materials are regulated. Special provisions are included for registers and grills, and ducts in expandable or multiple transportable section connections.

The HUD standard also specifies the minimum construction requirement, thickness, and gage of the metal ventilating hood to be installed over a kitchen range. In addition, the HUD standard limits the flame spread of the wall and cabinet finishes in the immediate vicinity of the kitchen range. This combination of requirements effectively negates or minimizes the direct exposure of the wall and cabinet surfaces nearest the kitchen range in the event of a cooking surface fire. Fire experience statistics show that these provisions produce smaller cooking and heating fires in post-1976 manufactured homes and that a much smaller share of such fires spread beyond their room of origin than they do in pre-1976 units.[5]

The HUD standard includes specific requirements for electrical installations. It concentrates on those provisions that differ from provisions applicable to all buildings under NFPA 70, *National Electrical Code®* (hereinafter referred to as NFPA 70). Both documents are coordinated. Some special features relate to the compactness of manufactured homes, to the fact that they are factory-assembled, to the lighter construction techniques commonly used, to the under chassis wiring requirements, to the wiring of expandable units with multiple transportable sections, to the need for outdoor outlets, to the need to bond sometimes extensive amounts of noncurrent-carrying metal parts, and to the grounding of services and appliances.

HUD issued the first update to its standard in 1985. Due to the standard's success in the area of fire safety, HUD "fine tuned" the standard in lieu of making wholesale changes. HUD lowered the maximum height above the floor of the egress window latch from 60 to 54 in. (1.5 to 1.4 m). More importantly, HUD issued a stricter limit on ceiling interior finish flame spread—to a maximum rating of 75—and expanded its criteria on kitchen cabinet protection. Other minor changes included further clarification of firestopping construction techniques in concealed wall, ceiling, and floor construction; recognition of $^5/_{16}$-in. (8-mm) thick gypsum board as firestopping material; and clarification of smoke detector placement instructions.[6]

RECREATIONAL VEHICLES

Recreational vehicles are primarily designed as temporary living quarters for recreational, camping, travel, or emergency use. They either have their own motor power or are mounted on or drawn by another vehicle. Basic vehicles are travel trailers, camping trailers, truck campers, van conversions, and motor homes. Together they account for about 5,000 fires a year reported to U.S. fire departments. There are several variations of these recreational vehicles, and new equipment is being developed constantly.

In general, federal motor vehicle safety standards cover automotive features of recreational vehicles. However, federal standards do not currently apply to plumbing, heating, and electrical systems in recreational vehicles. Nor do they apply, except in multi-purpose passenger vehicles such as van conversions and motor homes, to construction materials for interior walls, partitions, ceilings, exit facilities, and fire protection. NFPA 501C, *Standard on Recreational Vehicles*, (hereinafter referred to as NFPA 501C) or ANSI A119.2, concentrates on plumbing, heating, electrical, fire safety, and life safety considerations. Its requirements for installing fuel-burning appliances provide for complete separation of the combustion system from the interior. This applies to all types of fuel-burning appliances used in recreational vehicles except ranges, ovens, and illuminating appliances. NFPA 501C likewise gives attention to ventilation and combustion air systems, clearances between heat producing appliances and adjacent surfaces, circulating air systems ducts, and their sizing, duct supports, registers, and grills.

NFPA 501C also covers electrical systems, as does Article 551 of NFPA 70. The bulk of the provisions relate to electrical equipment and materials required in a recreational vehicle for connection to a wiring system nominally rated 115 V (two wires with ground) or a wiring system nominally rated $^{115}/_{230}$ V (three wires with ground). Other special requirements cover low-voltage systems other than circuit supply lines subject to federal or state regulations, combination electrical systems for connection to a battery or to a direct current supply that may also be connected to a 115-V source, and generator installations.

Fire and life safety requirements in NFPA 501C cover interior walls, partitions, ceilings, and exit facilities. They stipulate that each recreational vehicle equipped with fuel-burning appliances or an internal combustion engine must have a listed portable fire extinguisher with a specified minimum rating.

To facilitate proper use of recreational vehicles, another standard, NFPA 501D, *Standards for Recreational Vehicle Parks and Campgrounds*, or ANSI A119.4, has been developed. This standard's intent is to provide minimum construction requirements for parks designed primarily for use by owners of recreational vehicles. It devotes attention to park design and construction, recreational vehicle stand construction, environmental health and sanitation, fuel gas systems, storage of flammable and combustible liquids, electrical systems, and fire safety.

The 1977 edition of NFPA 501B is being reissued as the 1997 edition of NFPA 501, *Manufactured Dwelling Code.*

BIBLIOGRAPHY

References Cited

1. Letter from the American Association of Retired Persons to the California Department of Housing and Community Development, March 31, 1995.

2. "Shipment Figures," National Conference of States on Building Codes and Standards, 1995.

3. Budnick, E. K., and Klein, D. P., "Mobile Home Fire Studies, Summary and Recommendations," NBSIR 79-1720, 1979, National Bureau of Standards, Gaithersburg, MD.

4. Klem, T. J., "Interim Report: The Mobile Home Fire Problem in the United States," 1978, National Fire Prevention and Control Administration (now U.S. Fire Administration), Washington, DC.

5. Hall, J. R., Jr., "Manufactured Home Fires in the U.S., 1980–1994," NFPA Fire Analysis and Research Division, Quincy, MA, Oct. 1996.

6. Memorandum: "HUD Standards on Formaldehyde, Firesafety, and Ventilation," Manufactured Housing Institute, Arlington, VA, 1984.

NFPA Codes, Standards, and Recommended Practices

Reference to the following NFPA codes, standards, and recommended practices will provide further information on manufactured housing and recreational vehicles discussed in this chapter. (See the latest version of *The NFPA Catalog* for availability of current editions of the following documents.)

NFPA 70, *National Electrical Code®*

NFPA 501, *Manufactured Dwelling Code*

NFPA 501A, *Standard for Fire Safety for Manufactured Home Installations, Sites, and Communities*

NFPA 501B, *Standard for Mobile Homes*

NFPA 501C, *Standard on Recreational Vehicles*

NFPA 501D, *Standard for Recreational Vehicle Parks and Campgrounds*

Additional Reading

Bill, R. G., Jr., and Kung, H. C., "Limited-Water-Supply Sprinklers for Manufactured (Mobile) Homes," *Fire Technology*, Vol. 29, No. 3, Third Quarter 1993, pp. 203–225.

Bill, R. G., Jr., *et al.*, "Development of a Limited-Water-Supply Sprinkler for Mobile, Manufactured Homes," Twelfth Joint Panel Meeting of the U. S./Japan Government Cooperative Program on Natural Resources (UJNR) on Fire Research and Safety, October 27–November 2, 1992, Tsukuba, Japan, Building Research Inst., Ibaraki, Japan, 1992, pp. 121–130.

Bowie, G. R., "Sylmar Mobile Home Park Fires," Fire Engineering, Vol. 147, No. 8, August 1994, pp. 52–54, 56, 58–59, 62.

Carpenter, D. J., Jr., "Nine-Fatality Mobile Home Fire, Maxton, North Carolina, November 18, 1989," USFA Fire Investigation Technical Report Series, TriData Corp., Arlington, VA, Federal Emergency Management Agency, Washington, DC, Report 037, 1990, 20 pages.

Franck, H., "Mobile Home Fires: Live Burn Exercise," *Fire and Arson Investigator*, Vol. 46, No. 3, March 1996, pp. 20–25.

Hall, J. R., Jr., "Fire Experience in Manufactured—Don't Call Them Mobile—Homes," *NFPA Journal*, Vol. 84, No. 3, May/June 1992, pp. 62–68.

Hall, J. R., Jr., "Mobile Home Fires in the U. S., 1980–1987," National Fire Protection Assoc., Quincy, MA, May 1990, 34 pages.

McPhee, R. A., "Questions About Limited-Water-Supply Sprinklers for Manufactured (Mobile) Homes," *Fire Technology*, Vol. 30, No. 1, First Quarter 1994, pp. 179–182.

Nowak, J., "Mobile Homes: Preventing Fires. Simple Precautions Can Save Lives," *Workbench*, Vol. 51, No. 5, October/November 1995, pp. 46, 48.

Obermayer, R., "Mobile Home Fires," *Firehouse*, Vol. 19, No. 3, March 1994, pp. 54, 56, 58, 60, 62.

STORAGE OCCUPANCIES

Bruce W. Hisley

This chapter discusses fire hazards and fire protection factors that apply to storage occupancies, including the assessment of life safety risk within such occupancies. It also discusses occupancy characteristics and hazards associated with storage occupancies as related to contents and their arrangement, as well as operational hazards and fire prevention practices, life safety considerations, building construction, fire protection systems, and uses requiring special considerations.

For additional information, see Section 3, Chapter 26, "Storage and Handling of Grain Mill Products"; Section 5, Chapter 1, "Fire Alarm Systems"; Section 5, Chapter 7, "Fire Alarm Systems as a Management Tool"; Section 6, Chapter 5, "Water Supply Requirements for Fire Protection"; Section 6, Chapter 8, "Theory of Automatic Sprinkler Performance"; Section 6, Chapter 9, "Automatic Sprinklers"; Section 6, Chapter 11, "Water Supplies for Sprinkler Systems"; Section 6, Chapter 17, "Care and Maintenance of Water-Based Extinguishing Systems"; Section 6, Chapter 23, "Fire Extinguisher Use and Maintenance"; Section 9, Chapter 1, "Assessing Life Safety in Buildings"; and Section 9, Chapter 17, "Warehouse Operations."

OCCUPANCY CHARACTERISTICS

Storage occupancies are buildings or structures used to store or shelter goods, merchandise, products, vehicles, or animals. Storage facilities can be separate and distinct facilities or part of a multiple-use occupancy. "In no other area does one occupancy present such a wide range of hazards, particularly as facilities are increasing in size and contain an increasing variety of products."[1] According to Factory Mutual Research Corporation (FMRC), examples of storage occupancies are warehouses, freight terminals, parking garages, aircraft hangars, grain elevators, barns, and stables.

Storage occupancies are unique in that they can be classified as low, ordinary, or high hazard, or as a combination of these where mixed commodities are stored together and not effectively separated. Factors affecting the hazard classification are burning characteristics of the material stored, combustibility of packaging, method of storage and packaging, and quantity stored. Modern developments in material handling have brought rapid change to storage oc-

cupancies, including high-rack storage areas as tall as 50 to 100 ft (15 to 30 m).

Computer-controlled stacker cranes are now being used to move and place commodities in rack storage areas. Super regional distribution warehouses now being developed cover several acres under one roof and may contain two- or three-level mezzanines. Storage occupancies usually have a small number of people in relation to total floor area. Work patterns usually require employees to move through the structure using industrial trucks. Totally computer-controlled warehouses have even fewer occupants, which can complicate the means of egress design.

Mini-storage complexes now being developed consist of several small rental spaces ranging from 40 to 400 sq ft (8 to 80 m²) in one building. These complexes are unique. There is very little control over the type and amount of hazardous materials that may be stored in rental areas because they are accessible to renters on a 24-hr basis. Rental spaces usually are not separated from each other by fire resistance-rated walls. A building may consist of 50 or more rental spaces.

Storage occupancies can consist of raw materials, finished products, or goods in an intermediate production stage. These materials can be found in bulk storage, solid piling, palletized storage, and rack storage arrangements. Bins and narrow shelves are also used, usually in stockrooms, to hold only small amounts of materials for immediate use. The main difference among these storage arrangements that affects fire behavior and control is the nature of horizontal and vertical air spaces or flues, that storage configurations create. Parking garages, grain elevators, and aircraft hangars have other unique problems caused by the nature of operations performed in them. This chapter addresses these problems later.

HAZARDS ASSOCIATED WITH OCCUPANCY

Contents

It is not unusual to find large, densely stored quantities of inherently combustible materials, such as textiles, stored in combustible packing materials, such as wood, plastic, and cardboard. The materials in storage also may contain highly flammable materials and gases. Plastics are now increasingly used as part of the product as well as the packaging. In determining the hazard classification of contents, consideration should be given to the product, product container, and

Bruce W. Hisley is a program chair for Fire Protection Technical Programs at the National Fire Academy, U.S. Fire Administration, Emmitsburg, MD. He was division chief, Fire Prevention Bureau of the Anne Arundel County Fire Department, MD, before joining the Academy staff.

packaging materials. Fire behavior depends on ease of ignition, fire spread rate, and heat release rate of the stored commodity.

Commodities are often complex items whose fuel content, arrangement, shape, and form affect their performance in a fire. A packaged commodity must be considered as a whole—that is, the way it burns. For example, bottled beer would be considered a low hazard, but when stored in cardboard or wood containers, it becomes an ordinary hazard. Washing machines have very few combustible components. Packaged in cardboard boxes and surrounded with plastic packing, however, they can burn rapidly and produce large amounts of smoke. NFPA 231, *Standard for General Storage* (hereinafter referred to as NFPA 231), divides stored commodities into general burning hazard classifications, based upon the fire behavior of typical items.

In areas where mixed items are stored, hazard classification should be based upon the most severe hazard category. This mixed condition is common in hardware or automotive parts warehouses. High-hazard materials or materials that require special consideration can sometimes be separated to keep them from influencing the overall hazard classification. Examples of these materials are rubber tires, plastic products, combustible fibers, paper and paper products, hanging garments, carpeting, pesticides, flammable liquids and gases, and reactive chemicals. The requirements of NFPA 231D, *Standard for Storage of Rubber Tires*; NFPA 231E, *Recommended Practice for the Storage of Baled Cotton*; and NFPA 231F, *Standard for the Storage of Roll Paper*, should be considered.

Aerosol cans containing flammable products stored at high pressure present a very high hazard risk. When ignited, they can explode, producing fireballs, and can rocket through a given area, starting multiple fires. Bulk storage of these containers should meet the requirements of NFPA 30B, *Code for the Manufacture and Storage of Aerosol Products.*

Arrangement of Contents

The arrangement of stored materials greatly impacts fire spread. Fire behavior depends on storage height, aisle widths, and whether storage is in bulk, solid piles, or palletized piles. Fire development depends on the combustible surfaces of stored goods and the horizontal and vertical flue spaces between the surfaces of materials stored. Where combustible fibers are stored, a fast-traveling surface fire is possible if fiber bales have multiple exposed fibers.

Fire usually spreads in a fan shape, as fire at the base preheats the material above, which then ignites and burns. As storage height increases, fire growth can be expected to intensify as the fire moves rapidly upward. Aisles at least 8 ft (2.5 m) wide are needed to help restrict the spread of fire across an aisle. Horizontal air spaces formed by the pallets in palletized storage can cause fast spreading fires. Rapid fire spread can also be expected in rack storage areas due to the rack height and narrow aisle width usually present when automatic materials-handling equipment is used.

OPERATIONAL HAZARDS AND FIRE PREVENTION PRACTICES

Storage occupancies can consist of large floor spaces on several levels occupied by only a few personnel. Because of this, fires often go undetected for an extended time period, particularly in spaces not protected by automatic fire detection or suppression systems.

Common hazards and fire causes generally associated with storage occupancies are incendiarism, careless handling of hot embers or ashes or open fires, sparks from cutting or welding, exposure to fire from another building, deficiencies or piling stored too close or usage error in electrical systems or heating equipment, lightning,

careless disposal of smoking materials, poor housekeeping practices, packing and unpacking of goods, storage of idle pallets, and ignition from industrial trucks or other mobile equipment. Specially designed materials-handling systems may also present unique hazards.

Incendiarism

Storage occupancies are vulnerable to incendiary acts because they often present an available fuel supply and a large potential loss. Often an arsonist can enter and start a fire without being detected. Even where security personnel are present, arsonists may ignite a fire in a location beyond their sight, which may grow to major proportions before detection.

Welding and Cutting Operations

Precautions should be taken at all times during welding and cutting operations. Such operations should not begin until all combustible materials have been removed from the affected area or covered with a fire-retardant cover. Portable fire extinguishers or small hoselines should be readied before the operation begins. A fire watch should be present at all times during operation and for at least 30 min after its completion. Consideration should be given to using mechanical fastening devices and saws when repairing or replacing steel racks to avoid cutting or welding. All cutting and welding should be conducted in accordance with NFPA 51B, *Standard for Fire Prevention in Use of Cutting and Welding Processes.*

Heating Equipment and Electrical Appliances

Material should be stored so that adequate clearance is maintained from heating air ducts, unit heaters, duct furnaces, flue pipes, radiant space heaters, lighting fixtures, electrical appliances, and other heat-producing equipment. Special care is needed when using temporary or portable heating appliances. Relevant NFPA codes and standards must be strictly observed for all types of heating and electrical distribution system equipment.

Smoking

Smoking should always be controlled. This requires management's sincere effort to enforce observance of permissible and prohibited smoking areas for employees and visitors. Smoking should be confined to specific areas set aside from storage, kept clean, and provided with adequate disposal containers.

Housekeeping Practices

The importance of good housekeeping cannot be overstressed. Packing or unpacking goods usually requires or results in the presence of quantities of loose combustibles, such as foamed polystryene beads, cocoons of foamed plastic, shredded paper, excelsior, and straw. Other combustibles, such as baled fibers and detached labels or tags, can act as kindling to ignite stored materials.

Packing and Unpacking Operations

Areas where packing and unpacking, sorting, repairing, refinishing, painting, and general maintenance take place often invite accumulation of combustible materials. Such areas are comparable to an industrial occupancy. It is good practice either to locate this type of operation away from the storage area or to provide a fire resistance rated separation.

Storage of Idle Pallets

Piles of wooden pallets introduce a severe fire hazard incidental to many storage occupancies. A small ignition source can easily ignite aging pallets. Even in sprinkler-protected areas, the pallets' undersides provide dry areas where a fire can grow and spread. The process of fire jumping to other pallets can continue until fire spreads through the top of the stack. With the strong updraft of flame, very little water from an ordinary hazard sprinkler system can reach the seat of the fire. Pallets should be stored in accordance with NFPA 231. Plastic pallets and shipping containers often present similar hazards.

Industrial (Forklift) Trucks

Misuse of industrial forklift equipment can lead to large-loss fire. Hazards associated with the use of industrial trucks are found in refueling operations, maintenance, and storage (when not in use). The type of industrial truck used should be approved for use within the building for the hazardous material stored. NFPA 505, *Fire Safety Standard for Powered Industrial Trucks Including Type Designations, Areas of Use, Maintenance, and Operation*, designates the type of truck that can be used in a hazardous area. All refueling operations should be conducted outside the building. Fuel for trucks must be properly stored and handled. Areas used for maintenance and battery recharging for electrical trucks should be separate from storage areas.

LIFE SAFETY CONSIDERATIONS

In storage occupancies, potential loss of life depends on the fire hazard of the stored material and special characteristics of the storage building. Fire in high-piled or rack storage can be expected to spread rapidly with some types of materials. In single-story buildings with high ceilings, smoke and heat have a place to accumulate before the means of egress becomes obscured or impassable. In buildings with low ceilings, the opposite occurs. In multiple-level buildings with open grating floor type mezzanines, smoke and heat from fires below can rapidly block a means of egress.

Particular attention to the means of egress and exit locations is required. In older multiple-floor warehouses, unprotected vertical openings for freight elevators and other materials-handling systems are common. These allow rapid smoke and heat spread to upper floors.

Factors that affect life safety in storage occupancies are the stored commodity, which consists of the product, the packing, and the container; the commodity's fire behavior, which is based on ease of ignition, fire spread rate, and heat release rate; the storage arrangement, which includes storage height and aisle width; the maximum single fire area; smoke and heat venting; exterior fire-fighting access; and the form of the stored material.

NFPA 101®, *Life Safety Code*® (hereinafter referred to as NFPA 101), addresses life safety from fire and similar emergencies. It covers features necessary to minimize danger to life from fire, smoke, fumes, and panic. NFPA 101 also deals with egress and other considerations essential to life safety. Its minimum requirements should be followed in storage occupancies.

Storage occupancies can be classified as low, ordinary, or high hazard contents, or as a combination where mixed commodities are stored together. Noticeable differences exist among the life safety requirements for different contents hazard classifications in NFPA 101. Where these different classifications exist in the same structure and cannot be effectively fire separated, requirements for the most hazardous classification apply. Sound judgment must be used when determining the contents hazard classification. NFPA 101 uses the ordinary hazard contents classification as the basis for general requirements; most storage occupancies fall into this classification. Only a very few fall into the low or high hazard contents classification.

Special concern for the storage arrangement's stability is required when storage is high-piled. High-piled storage can collapse during the early stages of fire development, due either to inherent instabilities or to excessive height and weight, which crushes the lower units. Pile collapse may block egress routes from the structure.

Egress Design

Means of egress in storage occupancies are arranged like those in other occupancies. The chief difference is that fewer exits are typically provided in storage occupancies. The exits are provided based on travel distance considerations and on the minimum number of exits required from a given area. Unlike other occupancies, no occupant load factor is considered. Rather, occupant load is based on not less than the maximum probable number of occupants present at any time. Only a small number of people usually occupy a storage occupancy at any time. Work patterns usually require employees to move throughout the structure, which complicates the egress design for life safety. Every storage occupancy classified as ordinary or high hazard contents must have access to at least two separate means of egress as remote from each other as practicable. A single means of egress is permitted from ordinary hazard contents storage, provided that occupants can reach the exit within 50 ft (15 m) of common path travel or 100 ft (30 m) of common path travel when an approved supervised automatic sprinkler system protects the entire building. These two means of egress should be available from all parts and levels of the occupancy and must be within the travel distances specified by NFPA 101. Exits must be arranged so that occupants can reach them using different paths of travel from different directions. However, a common travel path is allowed in ordinary hazard contents areas. Low hazard contents areas require only a single means of egress, with no limitation on travel distance to reach an exit. In areas that contain high hazard contents, occupants must be able to travel at least two separate routes remote from each other to reach an exit. These areas need special consideration to make sure that occupants are not limited to a single egress path.

Exit doors must remain unlocked from the egress side of the door when the building is occupied. In low and ordinary hazard contents areas, special listed locking devices are allowed if an approved automatic fire detection or supervised sprinkler system protects the entire building. Access controlled egress doors are permitted, provided they meet specific requirements in NFPA 101. Egress doors from high hazard contents areas must swing in the direction of egress travel. The continual movements of stock within a storage occupancy demands special attention to prevent required means of egress aisles and exit doors from being blocked.

Exit Identification

Signs must identify the means of egress and exit locations and be readily visible from all directions of exit access. In high-stocked storage areas exit identification can be a problem, so other measures, such as painting exit travel paths on the floor, may be needed. Adequate illumination for the means of egress is necessary, especially in windowless buildings without openings for natural light. Buildings that are occupied at night or lack openings for natural light should be equipped with emergency lighting. This provides illumination for required egress stairs, aisles, corridors, ramps, and passageways.

Vertical Openings

Vertical openings are permitted in low or ordinary hazard contents storage occupancies, provided that they do not extend more than three floors, that an automatic sprinkler system protects the entire building, and that the openings do not serve as required exits. Exits from all floor levels must be protected in accordance with NFPA 101.

Fire Alarm Systems

Due to the large open floor areas and few employees who may be working in many different parts of a storage building, a fire alarm system should be considered to alert all occupants and allow for timely exiting and initiation of emergency action. If the storage occupancy has a trained fire brigade or an emergency pre-fire plan, there should be a means of notifying people in all building areas so that the emergency plan can be initiated. Because of the low fire risk, occupancies with low hazard contents do not require an alarm system. NFPA *101* does require a fire alarm system in ordinary and high hazard contents storage occupancies that are not protected by an automatic sprinkler system and whose total floor area exceeds 100,000 sq ft (9290 m²). In buildings equipped with a fire alarm system, provisions must be made so that the alarm system, when activated, provides audible and visible signals to a constantly attended location for purposes of notifying occupants and initiating emergency action. Required fire alarm systems need to be installed, tested, and maintained in accordance with NFPA 70, *National Electrical Code®*; and NFPA 72, *National Fire Alarm Code* (hereinafter referred to as NFPA 72).

Fire alarm systems are of utmost importance in warning occupants of fire, especially in storage areas that contain only a few employees in a large floor area and in multiple-story buildings. Alarm systems should emit distinctive sounds. Devices should be distributed so that they are audible in every room or building area. Notification signals for evacuation need to be audible and visible in accordance with NFPA 72 and CABO/ANSI A117.1, *American National Standard for Accessible and Usable Buildings and Facilities*. It is important that the alarm pull stations be clear of storage and well marked for easy identification.

BUILDING CONSTRUCTION AFFECTING HAZARDOUS CONDITIONS

Storage occupancies can be located in any type of construction. In addition, storage is found in mixed-use structures, because all types of occupancies need storage space. Most construction types are suitable for storage. However, combustible construction adds to the fire load and can create combustible concealed space that contributes to fire spread.

The combustibility, flame spread rate, smoke production, heat release potential, and structural stability of materials used in building construction should all be considered. One must consider the type of covering used on steel deck roofs, plastic panels used for insulation within wall panels and roofing, and the flame spread rating of the vapor barrier for exposed interior insulation. Certain expanded plastic insulation core panel walls should be used only in buildings with automatic sprinkler protection and be covered with fire-resistant material. The lightweight steel and wood truss construction now commonly used in newer buildings may be susceptible to early structural collapse. In some buildings with rack systems, the structural framework of the racks themselves supports the building's exterior walls and roof.

Buildings up to 100 ft (30 m) high used for rack storage are not uncommon. Fully automated rack systems may have narrow aisle widths of 4 ft (1.2 m) between racks. Providing complete fire extinguishment within the racks may be difficult even in buildings protected by automatic sprinkler systems.

Fire involving upper portions of high-rack storage can present unusual problems and severe risks to fire fighters. Access to the upper portion for complete extinguishment and overhaul is very limited. Using ground ladders inside buildings is extremely difficult and dangerous and is limited to 35 ft (10.5 m). Heat produced by the fire may cause structural failure of the rack system or collapse of stored material into aisle space. Falling material blocks the aisle, making access to the fire area even more difficult.

Buildings or structures with limited exterior access for fire fighting and rescue operations pose a serious life safety hazard and can cause large areas to become lethal fire environments.

Installing heat and smoke vents in storage occupancies should be considered. Some storage occupancies are extremely large with very few windows or doors that can be opened to remove smoke and heat. In buildings without automatic sprinkler protection, venting can reduce the horizontal smoke and heat spread along the ceiling, which aids in evacuation. In buildings protected by automatic sprinkler systems, venting aids in smoke removal, which, in turn, aids fire-fighting operations and rescue. Smoke and heat venting facilities, as well as automatic sprinkler protection, are essential for life safety in windowless buildings and underground structures.

Special mechanical ventilation systems should be considered for large-area storage occupancies to remove smoke and combustion products. Ventilating large area structures can be particularly difficult when automatic sprinkler systems have controlled a fire and cooled the atmosphere. Deaths can occur if fire fighters became lost in zero-visibility conditions created by "cold smoke" and run out of air. Combustion products may easily overcome persons without respiratory protection trapped in the building. Unnecessary smoke damage may result if effective ventilation is not provided.

FIRE PROTECTION SYSTEMS

Although automatic sprinklers are considered the main line of fire defense in storage occupancies, they are by no means the only method of protection. Standpipe hose systems, high-expansion foam systems, portable fire extinguishers, manual fire-fighting operations, and alarm systems also have a role in protecting storage occupancies.

Goods stored to great heights, mechanical stacking equipment, and narrow aisles are highly conducive to vertical fire spread. Even with automatic sprinkler protection, these types of storage arrangements pose a challenge for control.

Automatic Sprinklers

Automatic sprinklers are particularly effective for life safety because they warn of the existence of fire and, at the same time, apply water to the burning area. Properly installed and maintained sprinklers provide a highly effective safeguard against life and property loss.

The model building codes or locally adopted codes usually require automatic sprinkler systems in a storage occupancy classified as high hazard. Some codes require sprinkler protection when materials are stored in racks or piles more than 12 ft (3.6 m) high. The model building codes usually allow storage buildings to be constructed with unlimited floor area when they are protected by sprinkler systems and are not used for high hazard storage.

Disastrous fires can occur in sprinklered warehouses if combustible materials are stored to heights over 50 ft (15 m). The systems' design must be compatible with the commodity stored and with the storage heights. A major factor in large fire loss in storage occupancies can be an overtaxed automatic sprinkler system, which was not designed for the type of material stored or for the type of storage arrangement. It is important to keep detailed records of what type of hazard is in the storage occupancy. Records should indicate the type of material stored, pile arrangement, storage methods, and storage height. Automatic sprinkler systems should be designed and installed in accordance with NFPA 13, *Standard for the Installation* of Sprinkler Systems; NFPA 231, Standard for General Storage; and NFPA 231C, Standard for Rack Storage of Materials. Automatic sprinkler systems shall be inspected, tested, and maintained in accordance with NFPA 25, Standard for the Inspection, Testing and Maintenance of Water-Based Fire Protection Systems.

Standpipe Hose Systems

Standpipe hose systems should be installed in storage occupancies that contain rack or high-piled storage. Connections for $1^1/_2$-in. (38-mm) hoselines should be provided for initial fire-fighting and mop-up operations. Only fire department personnel or members of a properly trained and equipped industrial fire brigade should use the standpipe hose connections. Connections should be so located that all portions of the area can be reached. Providing hose connections for high-rack systems requires special consideration. In addition to normal floor-level connections, additional connections may be needed at different heights to enable hoselines to reach all areas.

Fire Extinguishers

All employees should be familiar with the fire extinguishers' locations and use. Employee training should stress the importance of sounding an alarm and notifying the fire department as soon as a fire is discovered.

Extinguishers should be kept in readily visible locations where storage does not prevent access at any time. All employees should be familiar with the operation of extinguishers provided. (See NFPA 10, Standard for Portable Fire Extinguishers.)

USES REQUIRING SPECIAL CONSIDERATIONS

Parking Garages

Parking garages require special consideration because their operation and use affects life safety requirements. There are two types of parking garages: (1) those in which vehicles are parked exclusively by attendants or mechanical means and (2) those in which customers park the vehicles. NFPA 101 addresses the egress requirements of garages in which the customer parks the vehicle.

Occupancy characteristics: Parking garages can be on several levels above or below grade and enclosed or open. A garage is considered open when at least 25 percent of the enclosing walls are open to the atmosphere on two sides of each level.

Parking garages are often found in conjunction with other uses, such as hotels, apartments, and office buildings. In garages used for both parking and repairs, the repair area should be separated from the parking area and treated as an industrial occupancy.

Hazards associated with garages: Motor vehicles generally contain fast-burning fuels, including gasoline, diesel fuel, or LP-Gas. Major parts of newer model vehicles are made of plastic. These vehicles can create severe fires that produce large amounts of smoke. Gasoline is dispensed in some garages. NFPA 30A, Automotive and Marine Service Station Code; and NFPA 88A, Standard for Parking Structures, list requirements for this type of operation. In addition, NFPA 101 has special requirements for garages where gasoline is dispensed. Garages used for vehicle storage are classified as ordinary hazard contents areas.

Only a few persons at any one time usually occupy parking garages. However, structures that are located near large assembly occupancies, such as civic centers or theaters, can be occupied by a large number of persons at one time after a show or event ends. During this time, several hundred people move throughout the structure, either going to their vehicles or waiting in their vehicles to exit the parking garage.

Life safety considerations: To ensure life safety conditions in a garage, every floor level must have access to at least two separate exits, arranged so that the travel path is in two different directions. Vehicle ramps can serve as exit access under special conditions. Travel distance allowed to reach an exit depends upon whether the building is open or enclosed and whether automatic sprinklers protect the garage.

Special consideration must be given to the gasoline pumps' location to avoid trapping occupants if there is a fire or explosion at the pumps. Travel in any direction must lead to an outside exit at the same level or to properly arranged egress stairs. Where parking areas are located below gasoline dispensing pumps, each floor must have direct access to the outside. This arrangement eliminates the possibility of gasoline vapor accumulation in enclosed exit stairs.

The means of egress to exits should be provided with illumination and emergency lighting when the garage is used at night or has no natural lighting from windows or other openings during the day. Most people using a parking garage are unfamiliar with the location of exits. Exits and access to the exits should be marked and identified.

Enclosed parking garages must have a fire alarm system when the total aggregate floor area exceeds 100,000 sq ft (9290 m²) and the building is not protected by automatic sprinkler systems. In general, automatic sprinkler protection is required in enclosed garages located below grade and over 50 ft (15 m) high with a combustible roof or floor assembly, and in garages inside or below buildings used for other occupancies.

Grain and Other Bulk Storage Elevators

These structures move, store, and process grain, such as wheat, corn, oats, soy beans, and sunflower seeds. (See Figure 9-15A.) Handling the commodity generates some dust due to kernel breakage and so on, making the grain stream susceptible to fire and explosion. Commodities are stored in bulk form in concrete and steel silos, wooden bins, and steel tanks. Dust explosions are the primary hazard in the grain industry. NFPA 61, Standard for the Prevention of Fires and Dust Explosions in Agricultural and Food Product Facilities, lists fire protection and safety requirements for the installation and operation of grain elevators.

Structures of this type are unique and particularly dangerous, posing a serious life safety hazard for all occupants. The hazards of both fire and explosion need special consideration. Grain elevators have accounted for some of the largest life-loss incidents in storage occupancies. In most instances, they store and handle materials that create highly combustible dust. In many dust explosions, the primary or only means of exiting from the upper parts of an elevator may be damaged or destroyed, trapping workers on top of the structure.

NFPA 101 addresses life safety considerations based on the occurrence of fires, not on the possibility of a grain dust explosion. Basic life safety requirements require that at least two means of egress be available and located as remotely as possible from each other. These means of egress should be available from all working areas and from the headhouse located on top of the storage structure. All working areas should have access to at least one stairway leading to an exit discharge point at ground level. This stairway should be constructed within a dust and fire resistive rated enclosure.

In addition to the stairway, an alternate means of escape is required. This secondary means should provide a continuous path to the ground from the top of the elevator structure, but it need not

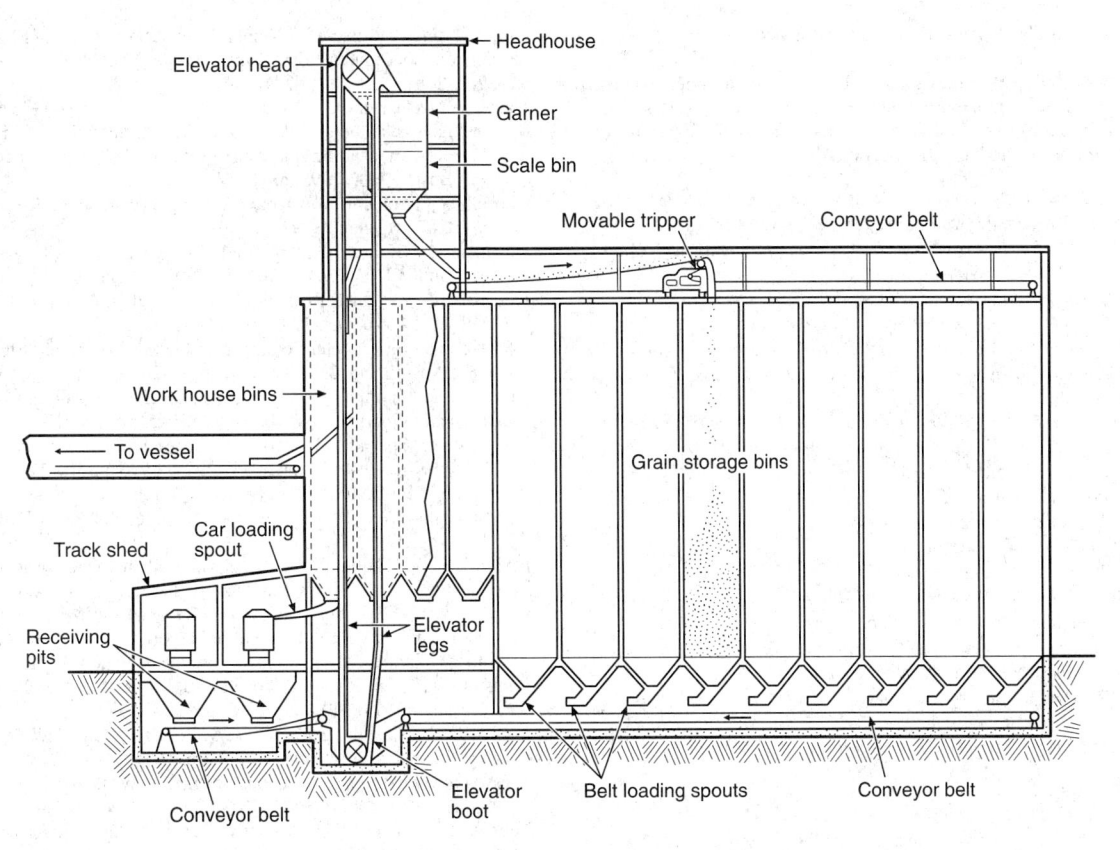

FIG. 9-15A. *Cutaway view of a grain elevator.*

meet all provisions for a means of egress. This secondary escape can consist of exterior stairs or basket-type ladders accessible from all working levels of the headhouse. This escape means provides passage to the ground or to the top of adjoining structures.

An exterior stair or basket-type ladder should be provided from the top of storage structures such as silos, conveyors, galleries, and gantries. This secondary escape should be located at the end of the structure opposite the headhouse.

Most grain elevators have underground passages beneath silos for conveyors. These spaces should have at least two means of egress so located that no dead-end condition is present. One means of egress can be a ladder or other escape method that leads to a window, hatch, or panel that can be opened readily to allow escape.

Aircraft Hangars

An aircraft hangar is a large structure built to provide weather protection and shop space for aircraft during maintenance and storage. Because it is impractical to remove all fuel from a craft before moving it into a hangar, large quantities of flammable liquids are always possible inside a hangar. Fire during maintenance and servicing can create great exposure to personnel in the hangar and endanger aircraft that may be worth millions of dollars. Many maintenance procedures use highly flammable solvents and sometimes unstable toxic chemicals. Special precautions for ventilation and control of ignition sources are needed.

Aircraft hangars provide a particular challenge when designing adequate exits for life safety. The large open area needed for aircraft, especially jumbo jets, requires extended travel distances to reach exits. The presence of flammable and combustible liquids compounds this problem. Exit locations must be located at intervals not exceeding 150 ft (46 m) along all exterior walls of the hangar. Each part of the hangar or servicing area must have access to at least two exit locations. If approved interior horizontal exits (2-hr fire barriers) are provided, then exit locations along the wall must be provided at intervals not exceeding 100 ft (30 m). Since a large hangar door cannot be left open in poor weather, small access doors can be installed in large hangar doors. To use additional space, mezzanine floors can be constructed in the hangar. Egress from these levels should be arranged so that an exit can be reached within a maximum travel distance of 75 ft (23 m). These exits must lead directly to an approved enclosed stairway that discharges directly to the outside, to another cutoff fire-separated area, or to outside stairs.

Because hangars encompass such large areas, they can contain several hazard classifications at one time. In areas classified as high-hazard contents or where a high-hazard operation is conducted, exits must be available and located where no dead-end conditions are present. For life safety purposes, a fire alarm system is required in hangars not protected by automatic sprinkler systems and whose total floor area exceeds 100,000 sq ft (9290 m²). For additional requirements related to constructing and protecting aircraft hangers from fire, refer to NFPA 409, *Standard on Aircraft Hangars.*

BIBLIOGRAPHY

References Cited

1. *Record*, Factory Mutual Research Corporation, Norwood, MA, July/August 1982.

NFPA, Codes, Standards, and Recommended Practices

Reference to the following codes, standards, and recommended practices will provide further information on storage occupancies discussed in this chapter. (See the latest version of *The NFPA Catalog* for availability of current editions of the following documents.)

NFPA 10, *Standard for Portable Fire Extinguishers*

NFPA 13, *Standard for the Installation of Sprinkler Systems*

NFPA 25, *Standard for the Inspection, Testing, and Maintenance of Water-Based Fire Protection Systems*

NFPA 30, *Flammable and Combustible Liquids Code*

NFPA 30A, *Automotive and Marine Service Station Codes*

NFPA 30B, *Code for the Manufacture and Storage of Aerosol Products*

NFPA 51B, *Standard for Fire Prevention in Use of Cutting and Welding Processes*

NFPA 61, *Standard for the Prevention of Fires and Dust Explosions in Agricultural and Food Products Facilities*

NFPA 70, *National Electrical Code®*

NFPA 72, *National Fire Alarm Code*

NFPA 88A, *Standard for Parking Structures*

NFPA 101,® *Life Safety Code®*

NFPA 231, *Standard for General Storage*

NFPA 231C, *Standard for Rack Storage of Materials*

NFPA 231D, *Standard for Storage of Rubber Tires*

NFPA 231E, *Recommended Practice for the Storage of Baled Cotton*

NFPA 231F, *Standard for the Storage of Roll Papers*

NFPA 409, *Standard on Aircraft Hangars*

NFPA 505, *Fire Safety Standard for Powered Industrial Trucks Including Type Designations, Areas of Use, Maintenance, and Operation*

Additional Reading

ANSI A117.1, *American National Standard for Buildings and Facilities Providing Accessiblity and Useability for Physically Handicapped People*, American National Standards Institute, New York, 1992.

Atkinson, G. T., and Jagger, S. F., "Assessment of Hazards from Warehouse Fires Involving Toxic Materials," Sixth International Fire Conference on Fire Safety, Interflam '93, March 30–April 1, 1993, Oxford, England, Interscience Communications Ltd., London, England, 1993, pp. 343–354.

Bacon, M. M., "Warehouse Fire Challenges Firefighters. On the Job: Rhode Island," *Firehouse*, Vol. 19, No. 2, Feb. 1994, pp. 46–48.

Before the Fire—Fire Prevention Strategies for Storage Occupancies, NFPA Ad Hoc Task Force, National Fire Protection Association, Quincy, MA, 1987.

Beitel, J. J., "National Quick Response Sprinkler Research Project: Group 1 Retrofit Test Series. [Group 1 Warehouse Retrofit Test Series.]," Southwest Research Inst., San Antonio, TX, National Fire Protection Research Foundation, Quincy, MA, Group 1, June 1990.

Cote, A. E., ed., *Industrial Fire Hazards Handbook*, 3rd ed., National Fire Protection Association, Quincy, MA, 1990.

"Data Sheet 8-9: A New Approach to Storage Protection," *Record*, Vol. 70, No. 5, Sept./Oct. 1993, pp. 3–70.

Denda, D., "What About Parking Garage Fires," *American Fire Journal*, Vol. 45, No. 2, Feb. 1993, pp. 22–25, 27.

Dickinson, C., "Cold Storage Warehouse Turns Up the Heat," *Fire Engineering*, Vol. 143, No. 7, July 1990, pp. 32–34, 38.

European Insurance Commission, Fire Commission, Subcommitee 42P, *Notes on the Recommendation Concerning Fire Protection in Warehouse Stocking Dangerous Materials*, Zurich, Switzerland, 1988.

Factory Mutual Engineering Corp, *Loss Prevention Data 8-0*, March 1977.

Fields, T., "Warehouses: A Case Study for Control," *Fire Prevention*, No. 215, Dec. 1988, pp. 24–29.

Gibson, R., "Warehouse Protection," *Fire Surveyor*, Vol. 21, No. 1, Feb. 1992, pp. 5–8.

Goring, G., "Fire Protection in Low Temperature Warehouses," *Fire International*, No. 139, May 1993, pp. 47–48.

Grant, G., and Drysdale, D., "Application of the Cone Calorimeter to the Problem of Fire Growth in Warehouses," Sixth International Fire Conference on Fire Safety, Interflam '93, March 30–April 1, 1993, Oxford, England, Interscience Communications Ltd., London, England, 1993, pp. 539–548.

Grant, G., and Drysdale, D., "Numerical Modelling of Early Flame Spread in Warehouse Fires," *Fire Safety Journal*, Vol. 24, No. 3, 1995, pp. 247–278.

Hamer, A., and Beyler, C., "Modeling the Activation of Dry Pipe Sprinklers in High Rack Storage," *Proceedings* of the International Conference on Fire Research and Engineering, Sept. 10–15, 1995, Orlando, FL, SFPE, Boston, MA, 1995, pp. 65–70.

Hansell, G., "Problems of Effective Venting in Warehouses," *Fire Prevention*, No. 262, Sept. 1993, pp. 40–42.

Hoffmann, J., "Kansas: Underground Storage Center Challenges Firefighters. On the Job," *Firehouse*, Vol. 17, No. 6, June 1992, pp. 78–80.

Hoffmann, J. W., "One Jurisdiction's New Warehouse Codes," *Firehouse*, Vol. 18, No. 10, Oct. 1993, pp. 88–90.

"IRI Promotes Safer Metal Container for Flammable Liquids Storage," *Sentinel*, Vol. 50, No. 3, Third Quarter 1993, p. 9.

Isman, K. E., "Sprinkler Protection for General Storage Occupancies," *Sprinkler Quarterly*, No. 93, Winter 1995, pp. 6, 8, 10.

Isman, K. E., "Sprinkler Protection for General Storage Occupancies. Part 2," *Sprinkler Quarterly*, No. 94, Spring 1996, pp. 10, 12.

Isner, M. S., "$100 Million Fire Destroys Warehouses. Summary Investigation Report," *NFPA Journal*, Vol. 85, No. 6, November/December 1991, pp. 37–41.

Isner, M. S., "49 Million Loss in Sherwin-Williams Warehouse Fire," *Fire Journal*, Vol. 82, No. 2, March/April 1988.

Isner, M. S., "Cold Storage Warehouse Fire, Madison, Wisconsin, May 3, 1991. Summary Fire Investigation Report," National Fire Protection Assoc., Quincy, MA, Summary Fire Investigation, 1991.

Johnston, D., "High-Rise Fire Protection at Monsanto's Warehouse," *Fire Prevention*, Vol. 229, May 1990, pp. 28–30.

Koffel, W. E., "Designing Warehouse Sprinkler Systems," *NFPA Journal*, Vol. 87, No. 4, July/August 1993, pp. 16, 90.

"Lesson of Gwent Blaze: Warehouse's Fire Safety 'Inadequate'," *Fire*, Vol. 84, No. 1036, Oct. 1991, p. 13.

Loeb, D. L., "Fire and the Automotive Repair Garage," *Fire Chief*, Vol. 37, No. 9, Sept. 1993, pp. 38–40, 42.

Mawhinney, J. R., "Development of Regulations in the 1990 National Fire Code of Canada on Storage of Dangerous Goods," *Fire Technology*, Vol. 26, No. 3, Aug. 1990, pp. 266–280.

Miles, S. D., *et al.*, "Modelling the Environmental Consequences of Fires in Warehouses," *Proceedings* of the Fourth International Symposium on Fire Safety Science, Intl. Assoc. for Fire Safety Science, Boston, MA, 1994, pp. 1221–1232.

Moore, J., "Understanding Warehouse Fire Problems," *Firehouse*, Vol. 17, No. 9, Sept. 1992, pp. 70, 72–73.

Moulton, G. A., ed., *NFPA Inspection Manual*, sixth edition, National Fire Protection Association, Quincy, MA, 1989.

Moulton, G. A., ed., NFPA Inspection Manual, 7th ed., National Fire Protection Association, Quincy, MA, 1994.

Mulhaupt, R., "Fire Protection for Flammable and Combustible Liquids in Warehouses," First International Conference on Fire Suppression Research, May 5–8, 1992, Stockholm, Sweden, 1992, pp. 393–399.

Murrell, J. V., and Field, P., "Sprinkler Protection of Post Pallet Storage in High Racks," *Fire Surveyor*, Vol. 19, No. 1, Feb. 1990, pp. 16–20.

Nugent, D. P., "Directory of Fire Tests Involving Storage of Flammable and Combustible Liquids in Small Containers," Schirmer Engineering Corp., Deerfield, IL, Directory, Feb. 1994.

Nugent, D. P., "Fire Tests Involving Storage of Flammable and Combustible Liquids in Small Containers," *Journal of Fire Protection Engineering*, Vol. 6, No. 1, 1994, pp. 1–10.

Persson, H., "Sprinkler Protection of Warehouses: A New Method for Classification of Commodities," Sixth International Fire Conference on Fire Safety, Interflam '93, March 30–April 1, 1993, Oxford, England, Interscience Communications Ltd., London, England, 1993, pp. 489–497.

Scoones, K., "Serious Fires in Warehouses and Storage Premises During 1991," *Fire Prevention*, No. 252, Sept. 1992, pp. 22–24.

Scoones, K., "Serious Fires in Warehouses and Storage Premises During 1992," *Fire Prevention*, No. 262, Sept.1993, pp. 29–31.

"Spotting Fires on Storage Sites," *Fire Prevention*, No. 262, Sept. 1993, p. 38.

Stephens, J., "Using Sprinklers in Bus Garages," *Fire Prevention*, No. 253, Oct. 1992, pp. 26–29.

Thomas, P. H., "Some Comments on Fire Spread in High Racked Storage of Goods," Lund Univ., Sweden, 93295RA/BK/AB, 1993.

Troup, J. M. A., "Protection of Warehouse Retail Occupancies with Extra Large Orifice (ELO) Sprinklers," *Journal of Fire Protection Engineering*, Vol. 8, No. 1, 1996, pp. 1–12.

"Uncontrollable Blaze Devastates Tooley Street Warehouses," *Fire Prevention*, No. 256, Jan./Feb. 1993, pp. 27–28.

United National Environment Programme, "Storage of Hazardous Materials—A Technical Guide for Safety Warehousing of Hazardous Materials," 3rd draft, prepared by UNEP/IPO Secretariat, Paris, France, Jan. 1989.

Wackerlig, H., "Risk Assessment of Chemical Warehouses," *Fire Protection*, Vol. 17, No. 4, Dec. 1990, pp. 4–6.

"Which Sprinkler Will Best Protect Your Warehouse?" *Record*, Vol. 68, No. 2, March/April 1991, pp. 3–10.

Williamson, R. B., and Schroeder, R. A., "Fire Safety Assessment of the Scrap Tire Storage Methods," *Proceedings* of the International Conference on Fire Research and Engineering, September 10–15, 1995, Orlando, FL, SFPE, Boston, MA, 1995, pp. 431–436.

Williamson, R. B., and Schroeder, R. A., "Fire Safety Assessment of the Scrap Tire Storage Methods," California Univ., Berkeley, ISBN 0-9516320-9-4; *Proceedings* of the Seventh International Interflam Conference, Interflam '96, March 26–28, 1996, Cambridge, England, Interscience Communications Ltd., London, England, 1996, pp. 89–101.

Woolhead, F., "The Warehouse Fire Threat to Profits and Prospects," *Fire Surveyor*, Vol. 17, No. 5, Oct. 1988, pp. 13–15.

Yung, D., and Mehaffey, J. R., "Fire Resistance Requirements for Rubber-Tire Warehouses," *Fire Technology*, Vol. 27, No. 2, May 1991, pp. 100–112.

LIBRARY AND MUSEUM COLLECTIONS

Revised by Stephen E. Bush and Danny L. McDaniel

This chapter discusses the incidence of fire in libraries and museums, analyzes the differences between fires in which large and minimal losses occur, and describes principles and practices for managing these cultural resources to reduce the risk of fire occurring in them and to minimize losses when they do occur.

At first glance libraries and museums seem quite different because of the nature of their collections, the concentration of combustibles, and the degree of public access. Nevertheless, most museums contain libraries, many libraries feature exhibits of art or artifacts, and both contain archival records essential to documenting and understanding the collections. Their fire protection problems are similar. Library and museum collections include large quantities of combustible, highly valuable, often irreplaceable materials. Another striking similarity is that while library and museum buildings are public assembly occupancies, their extraordinarily high fuel loads are more typical of warehouse occupancies.

From a fire protection viewpoint, libraries and museums are storage facilities. They provide a controlled environment to preserve cultural and technological heritage and, at the same time, provide public access to these resources. The fire risk in these occupancies should not be underrated. (See Table 9-16A.) Public areas, such as library reading rooms and museum exhibit galleries, generally are no smoking areas with low fuel loads and high housekeeping standards, and they are monitored constantly by staff or security officers. Nevertheless, library reading rooms and museum exhibit galleries can suffer costly fires.

Nonpublic areas, such as bookstacks, collection storage, work areas, exhibit preparation areas, and conservation laboratories present additional special hazards. Fuel loads are much higher, and housekeeping and smoking are more difficult to control.

A review of museum and library fire experience in NFPA 910, *Recommended Practice for the Protection of Libraries and Library Collections* (hereinafter referred to as NFPA 910), and NFPA 911, *Recommended Practice for the Protection of Museums and Museum Collections* (hereinafter referred to as NFPA 911)[1,2] clearly demonstrates the risk to library and museum collections. Recent fires wor-

Stephen E. Bush, CSP, ASSE, was chief of safety services for the U.S. Library of Congress, Washington, DC. Now semi-retired, he is a consultant and chairman of the Library Task Group of the NFPA Cultural Resources Committee. Danny L. McDaniel, CPP, CSP, ASSE, ASIS, ICOM, AAM, is director of security, safety, and transportation for the Colonial Williamsburg Foundation, Williamsburg, VA. He is chairman of the NFPA Cultural Resources Committee.

TABLE 9-16A. *U.S. Library and Museum* Fires and Associated Losses by Year, 1980–1993*

Year	Fires	Civilian Deaths	Civilian Injuries	Direct Property Damage**
1980	484	0	0	$ 3,050,100
1981	386	0	0	1,311,300
1982	381	0	0	14,407,600
1983	266	0	0	4,384,700
1984	355	0	10	4,926,800
1985	384	0	0	9,728,300
1986	343	0	7	44,903,300
1987	278	0	2	15,761,400
1988	235	0	0	845,300
1989	253	0	3	9,549,200
1990	235	0	2	1,076,600
1991	259	0	1	459,100
1992	238	0	4	2,634,700
1993	202	0	2	4,701,100
Annual average 1980–1993	307	0	2	8,410,000

* Refers to the whole building, not library or museum room, and includes art galleries, aquariums, and planetariums.
** Direct property damage is rounded to the nearest hundred.
Source: 1980–1993 NFIRS and NFPA survey.

thy of special attention include those at the Los Angeles Central Public Library (two in 1986 and one in 1988), the Louisiana State Museum in New Orleans in 1988, the Library of the Russian Academy of Sciences in St. Petersburg, Russia, in 1988 and 1995, and the Norwich Central Library in Norwich, England, in 1994.

CAUSES OF LIBRARY AND MUSEUM FIRES

Arson dominates as the most significant cause of fires in libraries, both in the number of fires and in the value of property damaged. (See Table 9-16B.) In 1980–1993, arson accounted for 71 percent of the property loss in library fires,[3] including the disastrous Los Angeles Central Public Library fire in 1986. Arson also was the second most frequent cause of museum fires during the period 1980 through 1993. (See Table 9-16C.) The second most significant cause

TABLE 9-16B. *Major Causes of U.S. Library Fires, 1980–1993 Annual Average*

Major Cause	Fires	Civilian Deaths	Civilian Injuries	Direct Property Damage*
Incendiary or suspicious causes	82	0	0**	$4,415,900
Electrical distribution system	40	0	0**	278,900
Other equipment	15	0	0	6,200
Smoking material (e.g., cigarette)	13	0	0**	70,400
Heating equipment	11	0	0**	156,400
Open flame	11	0	0	13,700
Appliance, tool, or air conditioning	8	0	0**	34,400
Cooking equipment	7	0	0	11,200
Child playing	5	0	0	500
Natural causes	3	0	0	8,100
Exposure (to other hostile fire)	3	0	0	86,800
Other heat source	2	0	0	0*
Unknown	13	0	0	1,122,900
Total+	213	0	0**	6,205,300

*Direct property damage is rounded to the nearest hundred.
**Not zero but rounds to zero.
+Sums may not equal totals due to rounding error.
Source: 1980–1993 NFIRS and NFPA survey.

TABLE 9-16C. *Major Causes of U.S. Museum* Fires, 1980–1993 Annual Average*

Major Cause	Fires	Civilian Deaths	Civilian Injuries	Direct Property Damage**
Electrical distribution system	20	0	0+	$629,100
Incendiary or suspicious causes	18	0	0	622,400
Open flame	8	0	0+	22,100
Other equipment	8	0	0+	162,300
Heating equipment	7	0	0+	507,700
Smoking material (e.g., cigarette)	7	0	0	15,800
Cooking equipment	5	0	0	18,400
Appliance, tool, or air conditioning	4	0	1	2,300
Natural causes	4	0	0	24,100
Exposure (to other hostile fire)	4	0	0	58,500
Other heat source	1	0	0	2,700
Child playing	1	0	0	0+
Unknown	8	0	0	139,300
Total++	95	0	1	2,204,700

**Direct property damage is rounded to the nearest hundred.
+Not zero but rounds to zero.
++Sums may not equal totals due to rounding error.
Source: 1980–1993 NFIRS and NFPA survey.

of fires in libraries, and the leading cause in museums, has been failures in electrical distribution system equipment. Together, these two causes accounted for 52 percent of the fires and 71 percent of the property damage in cultural properties from 1980 through 1993.

FIRE PREVENTION PRINCIPLES

Fire prevention is the cultural institution's first line of defense against fire. It consists of controlling and separating ignition sources and the supply of fuel.

Ignition Control

Ignition control involves identifying all heat sources that could cause ignition in combustibles and isolating these sources from fuel.

Arson: The single most common source of ignition for fires in libraries and a somewhat lesser source in museums is arson. Between 1980 and 1993 arson accounted for 71 percent of the average annual property damage losses attributed to fires in libraries and for 28 percent of the fire losses in museums. (See Tables 9-16B and 9-16C.)

Arson is a difficult crime to prevent because the arsonist is an unpredictable variable in the fire prevention program. A determined arsonist can circumvent even the most carefully defined fire prevention program. However, most arsonists are not so determined and have no special knowledge of fire or safety systems. While preventing all deliberately set fires may be impossible, a fire prevention program that incorporates a strong security program can limit risk.

Fundamental precautions against arson are a strong access control program backed by detailed opening and closing procedures and strong measures to deter and detect unauthorized entry. At a minimum a library or museum security program should include:

1. A detailed, written, access control policy with two objectives: (1) to control (define) legitimate access and (2) to prevent illicit or unnecessary access. These objectives can be met by establishing levels of access to spaces and degrees of access to protected areas or objects. Legitimate access to grounds or buildings does not mean access to non-public areas. Access to storage does not necessarily mean access to protected objects. Likewise, access does not mean uncontrolled access. Visitor escorts are a legitimate means of access control.

   An effective access control policy includes these provisions:

   (a) Identification of those who may obtain access (both routine and occasional) to nonpublic spaces and under what circumstances, for example, employees, scholars, researchers, visitors, service vendors, emergency response personnel, and others.

   (b) Procedures to obtain legitimate access to non-public areas. The policy should identify which staff can grant access, how much access they may grant, and on what basis. The policy should identify who must be escorted in non-public areas.

   (c) Responsibility for keeping access lists up to date and for making sure they are used routinely.

   (d) Specification of at least the following minimum parcel control procedures: (1) limit the size of parcels visitors may bring into the building, (2) search parcels larger than 11×15 in. (27.9×38.1 cm) taken into research or non-public areas by non-employees, and (3) property passes for employee-owned parcels leaving the building.

2. The best access and key control program is of little value if the facility is improperly opened or closed. Following written opening and closing procedures is the safest way to make the transition from one condition to the other. Standardizing how the facility is opened and closed establishes a baseline condition of the facility during each period. In the risk assessment process, these baseline conditions dictate the additional physical security tools needed to achieve acceptable security levels.

Opening and closing procedures should:

(a) Identify who may open or close. As the basis of the facility's access control program, designating, in writing, who has authority to open a building or controlled area establishes both the responsibility and authority to control access.

(b) Establish locking, unlocking, entry sequences, and paths. When opening or closing a building, a clearly defined entry and exit procedure should assure continuity and complete coverage. Opening and closing procedures should be designed so the person opening the building starts at one selected point and the person closing the building finishes at the same point. This procedure assures that all critical areas are consistently checked.

(c) Establish procedures to inspect the building for stay-behinds. Large buildings usually have many places where someone intent on remaining in the building can avoid detection. The first step in dealing with stay-behinds is to make it as difficult as possible for unauthorized persons to get into areas where they can hide. Nonpublic areas such as utility closets, mechanical areas, office spaces, and the like, should be locked while visitors are in the building. In some cases isolated spaces should have local daytime intrusion detectors to alert staff if an unauthorized person enters one of them. At the end of the day, it is important to search the building in a systematic way, usually starting at the top and working down and out. If possible, lock or secure building areas as they are inspected to prevent an intruder from moving back into a space after inspection. Opening and closing procedures should highlight areas that are particularly vulnerable to stay-behinds for special consideration. Where installed, the intrusion detection system may be zoned so that it can be activated in sections to help keep cleared areas secure.

(d) Identify other specific features of concern in the building.

3. Windows and doors should be secured with good quality locking devices. Panic hardware on emergency exit doors allows doors to be opened from the inside and remain locked from the outside. Doors into nonpublic areas should be kept locked to prevent uncontrolled access by the general public.

4. Windows and doors should be equipped to provide an alarm upon any unauthorized opening. In some communities intrusion alarms may be connected directly to the police or fire department. Alarm monitoring services also are available from commercial central stations.

5. Closed circuit television can be extremely useful for monitoring little used spaces, exhibit areas, print rooms, and other collection study facilities, as well as emergency exits. In addition, the presence of a camera often is a psychological deterrent to illegal behavior.

6. Exterior lighting is an important element in the security program. Illuminating concealed building areas, potential entry points, and large windows let police, security patrols, and others in the neighborhood observe activity around the building. Where public utilities do not provide it, lighting should be added at all concealed approaches to the building.

7. Libraries should either eliminate book drops used for after-hours book return, isolate them from the building by fire-resistant construction, or equip them with an automatic fire suppression system. Book drops are favorite targets for arsonists and vandals.[4]

8. If feasible, security patrols should inspect the building along well planned and monitored routes at irregular intervals.

9. Staff should regularly and closely inspect ash trays, waste receptacles, soiled linen hampers, electric heating appliances, and other places that breed fire. Although not specifically an arson prevention measure, this routine can uncover hazards that could involve incendiarism, such as a filled wastebasket hiding a time-delay igniter, or other unusual conditions.

Electrical system: The electrical distribution system—wiring, panel and junction boxes, lighting fixtures, receptacles, switches, and lamps—is the second leading cause of fires in libraries and the leading cause in museums. Electrical system fires accounted for nearly one-third of the property damage losses in museums during the period 1980–1993.

Wiring installed in accordance with the requirements of NFPA 70, *National Electrical Code®*, is equipped with overcurrent protection that opens the circuit before current flow reaches a level that dangerously overheats the conductor or its insulation. Nevertheless, overcurrent devices may fail, and under some conditions overloaded or partially grounded wiring may generate enough heat to ignite combustibles without blowing fuses or tripping circuit breakers.[5] For example, circuits subjected to flooding, when a water line breaks or a steam line leaks, for example, must be thoroughly dried. Otherwise, water can partially ground the circuit, generating excessive heat when current is applied.[6] The most common example, however, involves improperly altering the circuit, by installing an improperly sized fuse or circuit breaker.[7] A comprehensive preventive maintenance program for the building's electrical distribution system is an essential element in the fire prevention program.

The proliferation of information processing equipment (computer terminals, word processors, PCs, modems, printers, and telefacsimile and photocopying machines) adds significant electrical loads that can exceed the existing building wiring system's capacity. More often, however, the problem is either a shortage of electrical outlets to accommodate the equipment or a shortage of outlets where they are needed. The temptation is to make the existing arrangement work by adding multiple outlet adapters to expand the capacity of an existing outlet or to use extension cords to bring power from a less conveniently placed outlet. Neither solution is acceptable. Multiple outlet adapters encourage overloading of the electrical circuit and should never be permitted. Extension cords are not an acceptable substitute for fixed wiring: (1) they are subject to mechanical damage, and (2) they can be overloaded easily. Extension cords usually are made with smaller gage wire than the fixed wiring in the building. Electrical resistance increases significantly as wire size decreases and length increases. Where permitted, extension cords must be for temporary, short-term use only. The conductor must be rated to carry the current load expected over the distance it will be used. In addition, where extension cords have multiple outlet assemblies, they should have a circuit breaker at the output end to provide overcurrent protection for the extension cord. Extension cords must be protected from mechanical damage and never concealed under carpets.[8] (See Section 3, Chapter 1, "Electrical Systems and Appliances.")

Surface temperatures of incandescent lamps vary widely. Spotlights, for example, easily can reach the ignition temperature of paper or textiles. Incandescent bulbs near, or in contact with, combustible materials may ignite them even if their surface temperature is below the ignition temperature of the combustible material. Steady application of a heat source below a combustible material's ignition temperature can generate enough heat to ignite the material, given enough time.

The cooler surface temperatures of fluorescent lamps do not present the ignition hazards of incandescent bulbs. Nevertheless, ballasts in fluorescent light fixtures generate heat and are potential ignition sources, particularly when installed in a way that traps heat. For example, ballasts mounted flush against combustible, low density, cellulose fiberboard or other materials of similar combustibility present a potentially severe fire hazard. Fluorescent light ballasts also may be a source of ignition when they fail. At a minimum, a failing ballast may generate significant quantities of acrid smoke.

Heating equipment: Heating equipment malfunction or misuse is an important potential cause of fires. Boilers and furnaces of central heating systems should be isolated in a fire-resistant enclosure, either separate from or attached to the library or museum. Unit space heaters and portable heaters should not be used in combustible collection storage areas.

Because cellulosic materials, such as wood, paper, or textiles, store heat, they may ignite after prolonged contact with hot objects that are well below the materials' usual ignition temperature. Such materials should not be in contact with steam piping, heat ducts, or other low grade heat sources.

Electrical appliances: If not prohibited entirely, space heaters, hot plates, microwave ovens, coffee makers, and other small appliances should be rigidly regulated and closely monitored. Where permitted, each appliance should satisfy these conditions:

1. Listed by an independent testing laboratory.
2. Placed on a noncombustible base.
3. Located so they are adequately separated from combustible materials.
4. Located where air circulation prevents heat buildup.

Using extension cords to connect heating devices to electric outlets also should be prohibited. Those responsible for closing the building or area should have a listing of the location of each heating appliance and should inspect them as part of the building closing procedure. The inspection should ensure that the appliance works properly, that combustible materials are not touching or close to the appliance, and that appliances not necessary for building protection are turned off. If the facility has security officers, they, too, should have a listing of each heating appliance and should check each appliance during their rounds. Where feasible, heating appliances should be unplugged when not in use.

Smoking: Increasingly, museums and libraries are becoming smoke-free facilities. Where permitted, smoking should be limited to those places, such as administrative offices, staff rest areas, and designated visitor lounges, where fire safe practices can be monitored and controlled. As more facilities ban smoking, however, it is important to recognize that a total ban may result in surreptitious smoking in unsupervised locations, thus actually increasing the risk of fire from careless smoking practices. Where smoking is banned inside the building, adequate facilities should be provided outside the building to accommodate those who wish to smoke.

Open flame devices: Open flames used in building repairs, laboratory operations, or in the interpretation of historic environments are severe ignition hazards, as are hot metal sparks or slag from welding and cutting. Use of open flame devices should require the written approval of a responsible management official. Before signing such an authorization, the official should evaluate the risk to life safety and collection materials, and should prescribe protective measures as conditions of the authorization. Management conditions for authorization should include at least:

1. Relocating exhibit or collection materials and other combustibles a safe distance from the ignition source.
2. Requiring a fire watch of one or more trained persons with appropriate fire extinguishing equipment. The fire watch should be on site continuously, with no other duties, while the hazardous activity is performed and should remain in the area long enough to ensure the discovery of any hidden fire after interruption or termination of the activity. A fire watch should monitor the area

for at least 30 min after welding and cutting torches have been used.

Fuel Control

Fuel control must include all combustible contents, including the collection and its ancillary records, as well as furniture, interior finish, packaging materials, and flammable liquids. Control measures include arrangement of storage to limit fire spread and separation of fuel from ignition sources. Much of this can be accomplished through good housekeeping practices. (See Section 3, Chapter 33, "Housekeeping Practices.")

Flammable liquid storage and handling: Where possible, flammable and combustible liquids should be stored in fire-resistant structures isolated from the library or museum building. Where this is not feasible, small quantities of flammable or combustible liquids may be stored in approved, self-closing, flammable liquid storage cabinets. All flammable or combustible liquids permitted inside a library or museum should be contained in safety cans. When in use, only that quantity of flammable or combustible liquid required during an 8-hr shift should be permitted outside the flammable liquid storage cabinet. It, too, must be in safety cans until actually used. Special attention is required to ensure adequate ventilation, and ignition sources must be eliminated or carefully controlled. (See Section 3, Chapter 21, "Storage of Flammable and Combustible Liquids.")

Spontaneous heating: Oil and solvent soaked cloths and rags, such as those used with oil-based paint, mineral spirits, and some cleaning products, can heat spontaneously. They should be placed in a metal can with a tightly fitting metal top immediately after use and completely removed from the building no later than the end of each day.

Fuel geometry and overcrowding: A useful way to consider the combustible materials contained in a library or museum is to see them as arranged in fixed fuel packages: exhibit cases, bookstacks, desks, tables, chairs, and so on. These fuel packages also may contain quantities of movable fuel, such as loose paper, which can vary the total fuel load and radically change the ignition characteristics of individual fuel packages. Although these fuel packages usually are not contiguous, and indeed may be housed in a noncombustible building shell, they may become contiguous because of combustible interior finishes. Increasing the distance between the fuel packages and using interior finish materials, such as carpet, wall coverings, and draperies, with a low flame spread rating, minimizes the risk of fire spreading from one package to another.

Interior finish: Interior finish materials with high flame spread rates, such as fiberboard and some carpets, especially when vertically oriented, negate the effectiveness of distance as a tool to isolate fuel packages. For example, flame can spread in low density cellulosic fiberboard and acoustic ceiling tile, as shown in Figure 9-16A, as fast as people can run. Flame spread index ratings for low density cellulosic fiberboard and acoustic ceiling tile products have been reported to range from 225 to 350.[11,12]

Although building test fires in 1959 demonstrated this life safety problem, these materials are still found in libraries and museums across North America.[9] Table 9-16D lists 19 fire incidents from 1958 through 1990 involving ceiling tile; 16 resulted in 352 fatalities (3 of these were fire fighters). The list does not include all fires involving ceiling tile during this period, only those fires NFPA actually investigated.

Where such low density combustible finish materials still exist, they should be replaced. During removal it is especially important

TABLE 9-16D. *Occupancy and Fatality List of Fire Incidents Involving Ceiling Tile*

City	State/ Country	Date	Occupancy/ Use Group	Civilian Fatalities	Fire Fighter Fatalities	Total Fatalities
Chicago	IL	12/01/58	Educational	95	0	95
Hartford	CT	12/08/61	Health care	16	0	16
Montgomery	AL	02/07/67	Assembly	25	0	25
Salt Lake City	UT	09/15/71	Health care	6	0	6
Texas Township	PA	10/19/71	Health care	15	0	15
Montreal	Canada	09/01/72	Assembly	37	0	37
Kearney	NO	11/27/72	Health care	4	0	4
New Orleans	LA	11/29/72	Business	4	0	4
Pleasantville	NJ	01/29/73	Board and care	10	0	10
Manassas	VA	10/27/73	Educational	1	0	1
Indianapolis	IN	01/01/74	Apartment	0	0	0
Rio de Janeiro	Brazil	01/15/74	Business	0	0	0
Bradley Beach	NJ	07/26/80	Board and care	24	0	24
Williamsburg	VA	01/20/83	Dormitories	0	0	0
Gladstone	MI	03/12/83	Board and care	5	0	5
Dundalk	MD	10/22/84	Mercantile	0	3	3
Jacksonville	IL	05/27/86	Board and care	4	0	4
Johnson City	TN	12/24/89	Apartment	16	0	16
Bronx	NY	03/25/90	Assembly	87	0	87
Total incidents found	19					

FIG. 9-16A. **Typical low density cellulosic acoustical tile.**

to also remove any adhesive used to attach the tile to ceilings and/or walls. The adhesive has been identified as a key factor in the rapid flame spread over ceiling tile in several fatal fires.[39] If replacement is not feasible, these materials should be encapsulated behind fire-rated gypsum board. Treatment with intumescent paint may not provide the intended protection unless this application is approved by the listing laboratory and maintained as the manufacturer prescribes.[10] Carpet may present a similar hazard when used in exhibits or applied to wall surfaces unless it is specifically tested to establish an acceptable rate of flame spread when vertically installed. (See Section 7, Chapter 3, "Interior Finish.")

Inadequate space: Museums and libraries collect; therefore, growth of the collection is inevitable. As collections grow, staffing levels tend to grow also. Adding space is expensive and usually the last resort. These conditions cause overcrowding of both collections and staff, reducing the space between fuel packages. Overcrowding increases the risk that a fire will grow by spreading to adjacent fuel packages rather than burning itself out. Periods of overcrowding, therefore, require compensating emphasis on other elements of the fire prevention program, especially control of ignition sources.

FIRE PROTECTION

Experience shows that early detection followed by quick action to stop fire growth can prevent large fire losses. A common element in most large loss fires is the absence of automatic suppression systems to control or extinguish fire in its early stages. Some museum curators, librarians, and archivists believe water to be a greater hazard to collections than fire. They emphasize fire prevention almost exclusively, limiting fire protection to installation of automatic detection systems. Museum and library fires continue to occur despite this heavy emphasis on fire prevention. (See Table 9-16A.) Table 9-16E shows that 24 percent of all reported library fires occur between 9 p.m. and 9 a.m.; an additional 23 percent start between 5 p.m. and 9 p.m.—times when libraries have little or no staff on duty.[13] These off-hour periods also account for about half of the property loss.

Total dependence upon fire prevention and manual extinguishment places a library or museum at serious risk of catastrophic loss when prevention fails. The 1986 fire in the Los Angeles Central Library illustrates the point clearly. In this fire, smoke detectors warned building occupants, who evacuated safely. Nevertheless, despite timely notification to the Los Angeles Fire Department, lack of an automatic suppression system allowed the fire to spread. Bringing it under control took more than 7 hr. The fire destroyed more than 400,000 objects, and 700,000 wet books were placed in cold storage to await restoration.

Because fire prevention efforts can lessen, but not eliminate, the possibility of fire in a museum or library, knowledge of the options available to limit fire spread is essential. Local fire and building codes provide minimum requirements for the survival of the building and for the life safety of its occupants. Nevertheless, sole reliance on code provisions will not adequately preserve building contents, especially highly valuable collections. Fire protection features to protect the building contents are needed over and above minimum requirements that local codes impose.

The Systems Approach to Fire Safety

A systems approach to fire safety is a useful management tool for selecting the appropriate level of fire protection to limit fire loss to an acceptable level. The systems approach uses decision tree analy-

TABLE 9-16E. *Major Causes of U.S. Library Fires and Direct Property Damage by Time of Day, 1980–1993 Average*

Time of Day	Fires	Percent of Fires	Direct Property Damage	Percent of Damage*
Midnight to 1 AM	3	1.4%	$ 6,700	0.1%
1 AM to 2 AM	3	1.5	35,300	0.6
2 AM to 3 AM	5	2.4	891,500	14.4
3 AM to 4 AM	2	0.8	1,125,700	18.1
4 AM to 5 AM	2	0.7	150,700	2.4
5 AM to 6 AM	2	1.0	6,600	0.1
6 AM to 7 AM	4	1.7	11,300	0.2
7 AM to 8 AM	4	1.8	27,800	0.4
8 AM to 9 AM	10	4.5	24,200	0.4
9 AM to 10 AM	10	4.5	11,200	0.2
10 AM to 11 AM	12	5.8	2,704,700	43.6
11 AM to Noon	12	5.8	24,900	0.4
Noon to 1 PM	12	5.7	3,800	0.1
1 PM to 2 PM	15	7.2	77,800	1.3
2 PM to 3 PM	12	5.6	194,700	3.1
3 PM to 4 PM	16	7.3	12,400	0.2
4 PM to 5 PM	23	10.7	35,900	0.6
5 PM to 6 PM	16	7.7	80,600	1.3
6 PM to 7 PM	13	6.3	401,800	6.5
7 PM to 8 PM	9	4.4	38,500	0.6
8 PM to 9 PM	10	4.7	29,000	0.5
9 PM to 10 PM	7	3.3	219,000	3.5
10 PM to 11 PM	5	2.4	4,900	0.1
11 PM to Midnight	6	2.6	86,300	1.4
Total**	213	100	6,205,300	100
Selected Time Periods				
9 AM to 5 PM	112	52.6	3,065,400	49.5
5 PM to 9 PM	48	23.1	549,900	8.9
9 PM to 9 AM	53	24.1	2,590,000	41.7

*Direct property damage is rounded to the nearest hundred.
**Sums may not equal totals due to rounding error.
Source: 1980–1993 NFIRS and NFPA survey.

sis, failure modes, effects analysis, and other systems techniques to identify hazards, assess probability, establish maximum acceptable fire loss, and identify fire defense alternatives.

A critical element in the systems approach is to develop fire safety objectives for the facility based on life safety considerations, the value and vulnerability of the collections, and obligations for continuity of service. The fire safety objectives adopted by the Library of Congress in Washington, DC, may provide a useful reference point.[14] The General Firesafety Objectives and Critical Space Objectives, shown graphically in Figure 9-16B, are similar to those the U.S. General Services Administration (GSA) originally adopted for federal office buildings nationwide.[15] This graph shows the fire safety objectives for Library of Congress buildings and represents the desired minimum probability of success in limiting a fire's growth from its start to full building involvement. Adoption of acceptable limits of fire loss should, of course, reconcile the cost of the fire defense system to achieve those limits with the resources available to the institution.[42] (See Section 1, Chapter 3, "Systems Concepts for Building Fire Safety.")

Designing the Fire Safety System

The specific design selected for the fire protection system in a typical library or museum depends upon the value of the collection and the value and characteristics of the building and its fuel load. At a minimum, the system should include: fire detection, alarm notification, emergency communication, manual extinguishment, automatic fire suppression, protection for special hazards and other passive construction elements for safe occupant egress, limiting fire spread, and isolating hazardous operations and building systems.

Detection and alarm: A fire detection and alarm system should be installed throughout a library or museum building. The system should include smoke detectors in all exhibit galleries, reading rooms, and collection storage areas; and smoke, heat, or flame detectors, as appropriate, in other parts of the building.

A detection and alarm system should alert building occupants of an unwanted fire before it becomes a threat. A properly designed and installed system will alert occupants in time to evacuate in an orderly manner. The system also will facilitate prompt notification of the fire service. Ideally the detection system also provides an opportunity for trained building staff to extinguish the fire before automatic fire suppression equipment activates or the fire service arrives. (Section 5 of this handbook describes types of fire detection and alarm systems.) Care and good judgment are imperative, however. Fire in concealed spaces can burn for a considerable period before activating detection equipment. Those responding to investigate an alarm must be alert to signs that fire is well established. If there is any doubt, evacuate. Attempting to manually suppress a well established concealed space fire can lead to tragedy. When the fire breaks out of the concealed space, it can spread with blinding speed.

An automatic sprinkler system also can provide the alarm when the first sprinkler opens and water begins to flow in the system (e.g., when the fire grows to the size of a person). However, this detection occurs at a later stage of fire development when fire and

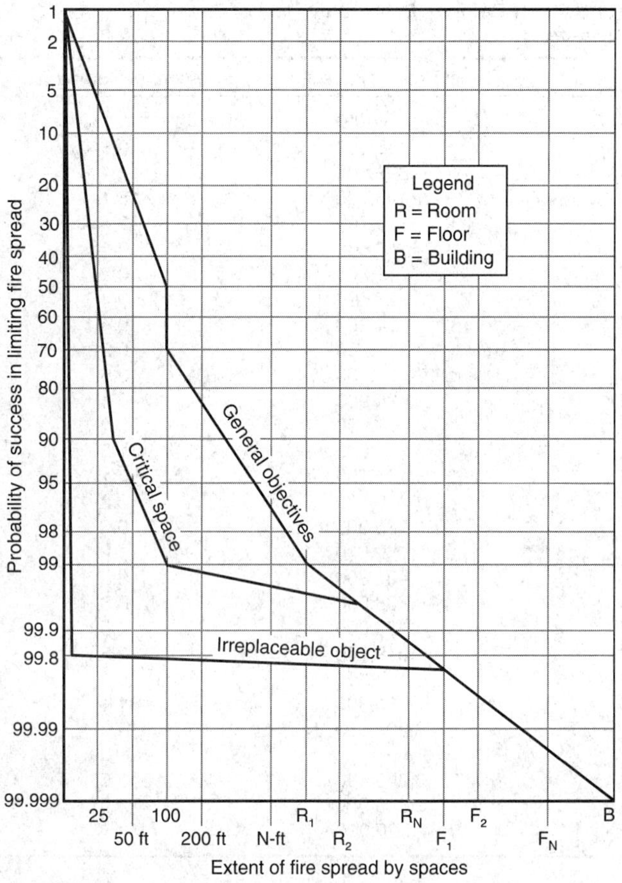

FIG. 9-16B. Graphic representation of fire safety objectives.

After installation and acceptance tests have been completed, a regular test and maintenance program should be established. This is very important because testing is the only way to determine that no part of the detection and alarm system has failed.[16] Furthermore, because the sensitivity of smoke detectors can change as dust and dirt accumulate, periodic inspection and cleaning is needed to keep the detector working within its intended sensitivity range. (See Section 5, Chapter 2, "Automatic Fire Detectors.")

Communication: A public address system should be considered, especially in very large facilities, to provide emergency voice instructions and information to facilitate evacuation. Library and museum buildings often have complicated egress routes. Unfamiliar with building layout, the general public may experience difficulty finding appropriate means of egress in spite of properly marked egress routes.

Manual extinguishment: Portable extinguishers in the hands of trained users constitute an institution's first line of fire attack. To prevent the use of the wrong class of extinguisher on a particular fire, some institutions provide only multipurpose portable extinguishers. These should be appropriately located, and regularly inspected and maintained, so they are fully charged and operational when needed. (See Section 6, Chapter 23, "Fire Extinguisher Use and Maintenance," and Section 6, Chapter 24, "The Role of Extinguishers in Fire Protection.")

Occupant use standpipe hose lines require specific hands-on training and drills to avoid personal injury and unnecessary property damage. Hose lines must be inspected at least annually and should be service tested when doubtful conditions are identified. Unfortunately, because many building occupants are not trained to properly use a fire hose and because of a general tendency not to inspect and test hoses as required, municipal fire departments often arrive at a fire to find the occupant hose used ineffectively, sometimes dangerously, or not at all. Many fire departments, taught by experience not to trust the integrity of occupant use hoses, have established as departmental policy a procedural requirement to replace occupant use fire hoses with their own hoses as soon as they arrive at the scene of a fire. Regarding this delay as unwarranted, many fire departments have recently recommended that occupant use fire hose be removed from existing occupancies and not provided in new occupancies.

Obviously, the fire department also provides manual extinguishment. In buildings without automatic sprinkler systems, it follows occupant fire suppression efforts as the second and final line of fire suppression defense to control loss from fire. It is generally recognized that if the fire area has reached 2,500 to 3,000 sq ft (232 to 279 m²) at the time of water application, the fire may be confined but not readily extinguished. Figure 9-16C suggests that the probability of fire department success actually begins to drop rapidly when fire areas exceed 750 to 1,500 sq ft (70 to 140 m²).[17]

The most significant concern with manual extinguishment is that once fire enters the free burning phase, fire damage increases exponentially until the fire is under control.[18] Fire protection systems that depend entirely upon manual extinguishment risk disaster if human response is delayed or unavailable.

Automatic sprinklers: As stated earlier, lack of an automatic suppression system is a major factor contributing to large loss fires in museums and libraries. Sprinkler systems have been shown to reliably and effectively minimize the risk of large loss fires in cultural properties.[18] A Factory Mutual Research Corporation study found that the average dollar loss in fires studied with adequate sprinkler protection in the five-year period 1984 through 1988 was $126,000 (indexed to 1988 dollars). By contrast, the average loss during the same period in fires studied without adequate sprinkler protection

smoke may hamper orderly evacuation. Quick response sprinklers (QRS) may be an exception because they respond to heat at an earlier stage of fire development than other standard sprinklers—in some configurations only seconds later than smoke detectors. For this reason some libraries, including the National Library of Canada, the National Library of Scotland, the New British Library, the U.S. Library of Congress, and the U.S. National Archives are installing QRS to protect their collection storage areas. (See Section 6, Chapter 13, "Fast-Response Sprinkler Technology.")

In addition to sounding an alarm throughout the building, the detection and alarm system should be connected to a central station, the local fire department, or some other acceptable monitoring facility. The value of a detection system is as a communications tool. The detection system itself takes no action to control or extinguish fire. A local alarm may be adequate for evacuating building occupants, but a local alarm alone does not summon the fire department or other emergency assistance. This is particularly true when the building is unoccupied. For this reason connection of the detection system to a central station or equivalent is recommended.

A qualified fire protection specialist with experience in library and museum protection should design the fire detection and alarm systems to ensure proper building coverage and to minimize the risk of unwanted alarms. The designer must have knowledge of, and experience in, fire behavior in museum and library buildings to properly select fire detectors for the threat and the ambient environment. The designer must also be well versed in the requirements of applicable codes and standards.

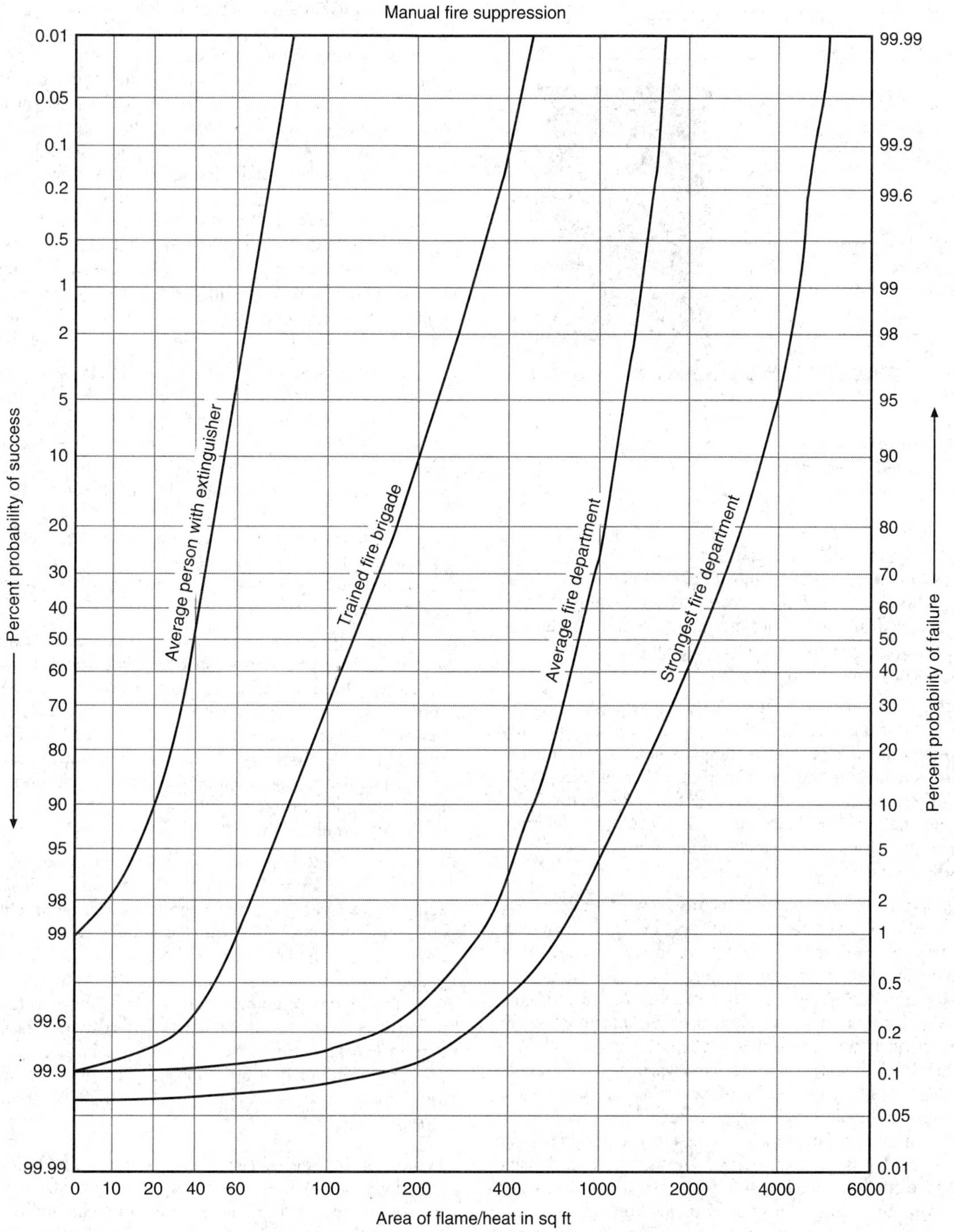

Manual fire suppression

FIG. 9-16C.   The probability of success at different levels of manual extinguishment.[17]

was $732,600, or 5.8 times more. For this study, fires where adequate sprinkler protection existed were those reported to FMRC in which (1) sprinklers were in service, (2) there were no sprinkler system or water supply deficiencies, and (3) sprinkler density was appropriate for the risk.[19] (See Figure 9-16D.)

For many years, library and museum professionals opposed automatic water sprinkler protection, fearing water damage even more than fire damage. Many library and museum planning references reflect this attitude.[20,21,22] Through the years libraries and museums

have experienced water damage from various sources: roof leaks, storm damage, plumbing problems, rivers overflowing, and so on. By comparison, water damage from sprinklers is rare.

Water damage is more likely however, when fire department hose lines are used to fight fires. Automatic sprinklers actually minimize the potential for water damage. An automatic sprinkler releases 20 to 25 gpm (76 to 95 L/min) in a spray pattern, in contrast to the 150 to 250 gpm (568 to 946 L/min) of pressurized water from each fire fighter's hose stream. Most fires in buildings with automatic

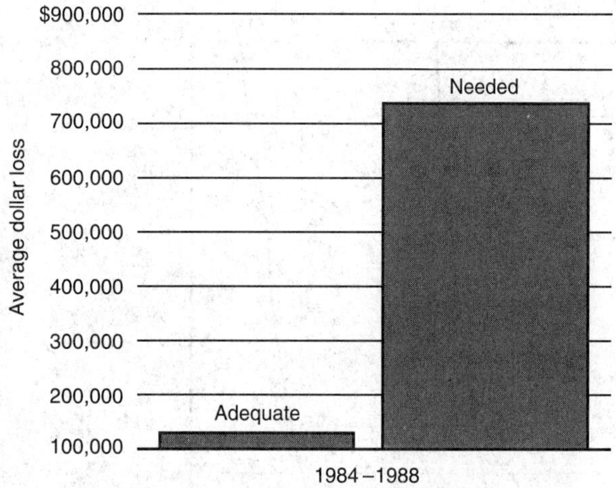

FIG. 9-16D.   Sprinklers adequate vs. sprinklers needed.
(Courtesy Factory Mutual Record)[19]

sprinklers are controlled with fewer than five sprinklers opening, which collectively release less water per minute than one hose stream. Hoses can sweep materials from shelves and displays into a disorganized mass on a flooded floor. Prospects of salvaging wet materials that remained in place under the relatively gentle action of sprinklers are much better than for materials knocked to the floor, soaked, and possibly trampled upon.

Another consideration is that sprinklers discharge water directly in the area of the fire. Sprinklers respond to heat, and only those in the direct fire area become sufficiently hot to open. Fire department hose streams, on the other hand, must be directed to the general area of the fire until smoke clears sufficiently for fire fighters to find and soak the concentration of fire. Therefore, whereas sprinklers tend to restrict damage to the room of origin, fire department hose streams can spread water damage much more widely.

Finally, sprinklers operate early in fire growth, reducing the amount of water needed to extinguish the fire. Fire department hose streams are used much later in the growth of the fire, usually after free burning begins. Therefore, extinguishing fire with a fire department hose stream requires significantly more water. The combination of more water flowing, more area to cover, and more water needed for extinguishment means that fire department hose streams pour huge quantities of water on a fire. This water has to go somewhere: it floods out of the fire area through every available drain and crack and drains to the building's lowest point. Therefore, fires on upper floors of a museum or library can result in extensive water damage to materials many floors below. The use of high density compact storage modules on lower floors and in basements has magnified this problem in recent years. Compact movable storage modules significantly increase floor load requirements, and, especially in retrofits, only the ground floor or basement of a building has sufficient floor loading capacity. This concentrates collections in a building's lowest areas, so fighting a fire on an upper floor, or on the roof, can result in significant water damage.

Alternatives to the standard wet pipe sprinkler system, such as on/off sprinklers and automatic recycling systems, minimize water damage after a fire has been extinguished. NFPA 910, Section 7-2.1, provides more information.

Another alternative, the pre-action sprinkler system, may be used for library and museum applications where fear of water damage to collections from damaged sprinklers or broken piping is a significant concern. The pre-action sprinkler system piping is dry until a signal from an automatic fire detection system opens the wa-

ter supply valve. Failure of the fire detection system, however, eliminates the automatic operation of a pre-action sprinkler system and requires manual operation of the water supply valve. The fire detection system, then, is a single point of failure in a pre-action system, reducing its reliability below that of a wet pipe sprinkler system. Furthermore, pre-action systems require higher levels of regular maintenance to preclude additional potential failure modes. This further reduces their reliability relative to wet pipe systems. Higher installation and maintenance costs, and less reliability, are the trade offs for any increased confidence the staff may have that a pre-action system is less likely to accidentally release water on a collection than a wet pipe system. For reasons of reliability, maintainability, and cost, many libraries and museums have elected wet pipe sprinkler systems, and some have converted their pre-action systems to wet pipe.

Sprinkler systems retrofitted into library or museum collection areas can pose a problem when the system is initially charged with water. Final acceptance testing of a sprinkler system requires a 2-hr 200 psi (1034 kPa) hydrostatic test on all piping, fixtures, and sprinklers. Most newly installed sprinkler systems leak, at least a little, when the system is first charged. This is unacceptable where collections are still in place. One solution, used by the Office of the Architect of the U.S. Capitol in the installation of wet pipe sprinkler systems in Library of Congress bookstacks, requires pneumatic testing of each system before proceeding with normally required hydrostatic testing.

Where sprinkler systems are used in special or high value collection areas, those items especially vulnerable to water damage should be stored in cabinets, manuscript boxes, or other appropriate containers. This practice serves several purposes: (1) containerized materials are more difficult to ignite, (2) the containers not only shield the collections from water spray in the event of sprinkler discharge but also from intrusion of water from other sources (i.e., roof leaks, plumbing problems, etc.), and (3) containers protect collections from smoke damage.[8]

Since about 1970, a growing number of new and existing libraries and museums have installed automatic sprinkler systems to achieve fire safety objectives and to keep fire risk within manageable, and often insurable, limits. The reduction in insurance premiums can pay off the sprinkler system's installation cost in a few years.[23] Even if the presence of automatic sprinklers does not reduce insurance premiums, their presence may make the facility a more attractive and marketable risk. This can be a significant advantage in getting competitive bids on a facility's property insurance program. (See Section 6, Chapter 8, "Theory of Automatic Sprinkler Performance," and Section 6, Chapter 10, "Automatic Sprinkler Systems.")

Protection for special risks: Areas containing highly valuable or especially vulnerable collections or equipment, or areas essential for continuity of operations, may require a higher level of fire protection to minimize damage or to keep interruption of service within acceptable limits than the combination of a smoke detection system with automatic sprinklers can provide.[23] NFPA 550, *Guide to the Fire Safety Concepts Tree* (hereinafter referred to as NFPA 550), is an excellent tool to help managers assess the effectiveness of fire protection strategies, such as installing special fire protection systems or other techniques to achieve higher protection levels within the limits of affordability.[24]

In devising protection strategies for special risks, particular attention should be given to passive fire protection, for example, using special fire and water resistant vaults, cabinets, manuscript boxes, or other containers to protect highly valuable collection items. Passive features may be less expensive to install and maintain and may, over time, be more reliable than more complicated alternatives. Nevertheless, relatively low levels of heat can destroy film and other heat sensitive materials. Therefore, passive features are not a substitute for timely detection and suppression; rather, a combination of active and passive fire protection elements increases the likelihood that sensitive, highly valuable collections can survive a fire intact. (See Section 3, Chapter 25, "Storage and Handling of Records.")

Parts of the museum or library collection may be so valuable, rare, or vulnerable that their protection becomes one of the institution's highest priorities: the only acceptable level of loss is zero. When this is true, the institution may opt for a special protection system in addition to automatic sprinklers to provide nearly instantaneous response. Such special systems may include sophisticated air sampling smoke detection, flame detectors, linear heat detectors, or other rapid detection devices that automatically activate special fire suppression systems, using a gaseous agent or a water mist system. Everything else being equal, a fire in a space protected with a special protection system will be extinguished in the first seconds of combustion, thus limiting damage to the lowest possible level.

Unless it at least matches the automatic sprinkler system's reliability and maintainability, the special system installed for this purpose should be used in combination with, and not as a substitute for, the sprinkler system. The installation cost of a special system to augment a standard automatic sprinkler system should be justified based on the extraordinary value and/or irreplaceability of the collections or because it is required for continuity of operations that cannot accept any down time, such as in the control room for the facility fire protection/security systems and building utilities. The justification should demonstrate a need for a very high probability of success in confining fire growth to an area significantly smaller than would be possible using standard automatic sprinklers alone—for example, less than 100 sq ft (9.3 m^2). A sprinkler system is a vital backup to the special protection system, however, if the special system fails to extinguish the fire.

Halon 1301 often was used to provide this extra level of protection. Cultural institutions found it particularly attractive because it can be used in normally occupied areas, rapidly extinguishes the flaming portion of fires, leaves no residue, and is not known to harm objects chemically or physically.

During the 1980s scientists linked chlorofluorocarbons (CFCs) and halon fire suppressants to the depletion of stratospheric ozone.[25] The signatories of the 1987 Montreal Protocol agreed to phase out halon suppression systems over a 10 to 20 year period. However, increasing public concern about the impact of ozone depletion on the environment accelerated the schedule. The signatories of the Montreal Protocol, including the United States, ceased commercial production of halon for fire suppression at the end of 1993. Halon manufactured and stored before the end of 1993 is still available, and existing Halon 1301 is being recycled as systems are dismantled. While the Montreal Protocol recognized, but did not specify, some critical uses of Halon 1301 for which there are no viable alternatives, the Halons Technical Options Committee Report states that "social benefit and human safety considerations are considered to be the only justifications to offset the environmental risk associated with halon use."[26] A matrix approach has been developed that managers of cultural heritage institutions can use to identify essential uses of Halon 1301 in their facilities.[27]

Research efforts to develop an environmentally acceptable substitute for Halon 1301 began as soon as it became apparent that its use would be curtailed. Alternatives for Halon 1301 suppression systems in cultural properties fall into three groups: (1) water (sprinklers and water mist), (2) inerting gases, and (3) synthetic Halon-like replacement gases. Each has merits.

Water mist: Water closely fits all requirements of the ideal fire suppression agent. It is safe for the environment and for people, it has excellent extinguishment properties, and it has the advantages of being both cheap and abundant. However, all items in the room or compartment where it is released are exposed to the mist because, like gaseous suppression agents, it functions as a total flooding agent. A recent development in fire protection, water mist technology promises to provide all advantages of water with less subsidiary damage and much faster response. Already used in Europe, water mist technology experience appears very positive.

The water mist system combines high pressure and finely machined discharge heads to create a water fog or mist. While any fire detection device can activate the system, use of smoke detectors (cross-zoned or matrix to minimize false signals) facilitates extinguishment very early in a fire's growth cycle. The combination of a very fast response and small water droplets reduces the amount of water needed to suppress the fire and limits water damage. A possible drawback for a cultural institution is that all the heads in a protected area operate at once, totally flooding the room, compartment, or zone with a very fine fog and exposing all contents whether involved in fire or not. Another concern is that because the smoke detection system controls the system, it is very important for the detection system not to send unwanted or nuisance alarms. While small water droplets are an attractive way to limit water damage, they may not be as effective as a sprinkler system in controlling a fast developing fire (such as an arson fire where an accelerant is present). Small droplets may not penetrate a high energy fire plume sufficiently to suppress it. Finally, water mist systems may be less effective on a deep seated fire in normal combustibles such as wood or paper. Here, again, the small size of the water droplets limits their ability to penetrate, although it will cool the outside.

The University of Maryland Department of Fire Protection Engineering at the Maryland Fire and Rescue Institute is conducting a developmental program to test these and other end-user issues. Reporting on its evaluation of tests in library stack scenarios, it has identified the possibility of limiting the activation of the water mist system to the aisle in which the fire originates.[28] Two branch lines were installed at different elevations in the flue between back-to-back ranges instead of around the room's perimeter at the ceiling. After a 30-sec pre-burn time, the water mist system controlled the fire in 32 sec with a total flow of 0.4 gal (1.5 L) of water. Test results in these library bookstack scenarios favor local application of the mist over total flooding and suggest the use of linear heat detectors or an air sampling smoke detection system to zone the compartment on an aisle-by-aisle basis rather than totally flooding the entire compartment.

While water damage to items not involved in fire appears minimal, a consortium of cultural resource institutions will be using a water mist system that has been installed in a room in the Library of Congress for the purpose of empirical testing (without fire) to evaluate the extent of water damage to collection items placed in the room for exposure scenarios to the mist. (See Section 6, Chapter 15, "Water Mist Suppression Systems.")

Gaseous agents: Two classes of gaseous fire suppression agents are viable alternatives for Halon 1301. Inerting agents, such as carbon dioxide, argon, and Inergen™, lower the oxygen content in the protected area to below the level that supports combustion. Synthetic gases, such as FM 200™ and FE 13, resemble Halon 1301 and interrupt the chain reaction that supports combustion.

Among the inerting agents, carbon dioxide is best known and was the standard gaseous fire suppression agent before halon. The concentration of carbon dioxide needed to suppress a fire, however, is three times the concentration that renders a person unconscious. Carbon dioxide total flooding systems are not generally recommended for normally occupied areas. Argon has similar drawbacks. Inergen, a newly developed fire suppression agent, is mostly nitrogen. It lowers a room's oxygen content to about 12.5 percent. While oxygen concentrations below 16 percent can affect the average person, Inergen contains a small quantity of carbon dioxide that acts to increase the respiration rate, so a person can remain in a room for a short period after Inergen has been released. This does, however, presuppose that no one in the room has breathing difficulty to begin with, emphysema, for example.

NFPA 2001, *Standard on Clean Agent Fire Extinguishing Systems*, limits the application of clean agents in normally occupied spaces to those with design concentrations equal to or less than the "no observed adverse effects level". Carbon dioxide is not approved for use in a normally occupied space. Inergen may be.

Synthetic gases act like Halon 1301 and suppress fire by breaking up the chemical chain reaction that must occur to support oxidation. Several gases are available commercially, FM-200 and FE 13 being two more popular alternatives. Both have an ozone depletion potential of 0. Synthetic gases now available, however, are more likely than Halon 1301 to break down into acidic and toxic byproducts under fire conditions.

Replacements for existing Halon 1301 systems: Compared to Halon 1301, the inerting and synthetic alternatives require 50 to 100 percent more agent to achieve an adequate fire-fighting concentration. Replacement agents, too, may demand high pressure piping and larger discharge heads than a comparable Halon 1301 system to discharge the agent in the same time. Therefore, none of the available alternative gases is a straight drop-in replacement for Halon 1301.

The key element in making a rational decision about when and how to replace an existing Halon 1301 system is to conduct a detailed risk analysis that evaluates the type and level of fire threat to the building and the collection and identifies ways to reduce the threat. In some cases the best answer may be providing more compartmentation to reduce the area at risk, or perhaps changing the way the building operates, how operations are conducted, or the type and location of potentially hazardous operations. As previously mentioned, NFPA 550 is an excellent tool for systematically evaluating fire safety alternatives.

After identifying potential threats and examining alternatives, consider how best to provide protection. These deliberations require the assistance of someone experienced in fire protection. An independent fire protection engineer is a good choice. A thorough and objective evaluation of threats and alternatives requires input from an unbiased professional who understands the problems inherent in protecting cultural properties.

Other special hazards: While the same fire safety principles are applicable to the requirements of museum and library collections, several activities common to museums and some libraries require special attention. These include:

1. Storing art and other valuable objects.
2. Packing and unpacking traveling exhibits.
3. Storing shipping crates and packing materials.
4. Cleaning and restoration.
5. Constructing and renovating displays.

Areas where these activities occur should be isolated and sprinklered. Good practice limits storage of combustible materials used for exhibit space rearrangement to those required for the current exhibit(s) under preparation or being dismantled, with an off-site warehouse for storing all other materials.

CONSTRUCTION CONSIDERATIONS

Intrinsic design features of library and museum buildings should include fire safety aspects of site planning, fire-resistant construction, interior finish smoke control, water supply for fire protection, and special requirements for underground structures. Topics deserving special attention in this chapter are compartmentation, means of egress, lighting, and gallery flexibility.

Compartmentation

A well-protected building is subdivided into compartments where walls provide barriers to the spread of fire. Compartments should be as small as possible, and walls should be designed to resist the estimated fire duration, based on maximum fuel loads in the compartments. Where small fire compartments would restrict flexible use of space, the installation of automatic sprinkler systems permits larger compartment areas and compensates for heavy fuel loads in collection storage areas and bookstacks.

Smoke damages collections and harms occupants. Fire barriers should confine smoke to the compartment of fire origin. Fire doors protecting openings in these walls must be self-closing and should remain closed at all times. Where it may be necessary for efficient operations to hold open these doors, this may be done only where automatic devices, such as electromagnetic hold-open devices installed in compliance with NFPA *101*®, *Life Safety Code*® (hereinafter referred to as NFPA *101*), automatically release the door upon a signal from a smoke detector or from the fire detection system. In the past, some building codes allowed fusible link assemblies to be used to hold doors open with automatic door closers. They are not acceptable for smoke control or life safety because they depend upon heat at the door to release them. By the time the fusible link reaches its release temperature, the fire has already made the door opening unusable, and smoke, which precedes the fire, has already spread beyond the space.

High hazard contents areas where large quantities of combustible material are used or stored should be separated by fire walls from exhibit galleries, reading rooms, other public areas, bookstacks, and collection storage areas. These high hazard contents areas include carpenter shops; display or exhibit preparation shops; research, restoration, and conservation laboratories; paint rooms; packing rooms; and restaurant kitchens, among other places. When possible, hazardous materials such as flammable liquids, including some paints, should be housed in a structure separate from the museum or library building. Fire walls protecting high hazard areas should be designed to withstand fires involving the maximum fuel loads they will enclose, rather than the minimum building codes require.

Means of Egress

Library and museum security arrangements commonly require all patrons and employees to enter and leave the building at one or two closely monitored points. Unfortunately, this often means that other doors required by NFPA *101* for exit access or egress are locked. This problem most often arises in museums and libraries with valuable collections and in libraries with heavily used reference collections, such as college and university libraries. Use of electromagnetic and electromechanical door locking systems is gaining acceptance as a way to resolve this conflict between life safety and security.[41] Such systems should include features that provide a level of life safety equivalent to that NFPA *101* requires. NFPA *101* allows delayed release locks, with strict limitations, in assembly occupancies, provided automatic sprinklers or fire detectors protect the entire building. Also see NFPA 910 and NFPA 911.

Attendance in exhibition galleries and reading rooms varies with the exhibit's popularity and the season. For example, a blockbuster museum exhibit draws thousands of visitors each day. During periods of high attendance, museums and libraries must control occupant loading of the building to ensure egress capacity is not exceeded.[43] Particularly at peak times, a complete emergency plan and a well trained staff are critical. Staff must take the lead in helping visitors evacuate safely, especially those with disabilities that might make emergency evacuation difficult. As mentioned previously in this chapter, the use of voice communications should be considered for large facilities or those with high attendance or complex egress paths. Voice communications facilitate coordination and reduce occupants' fears and anxieties during evacuations.

Lighting

Art museums typically use as much natural light as possible. They often have features such as glass roofs, light diffusing glass ceilings, and window walls, which all increase the museum's vulnerability to fire exposure. Conversely, to protect collections from ultraviolet radiation and achieve complete light control, some museums have minimal wall and roof openings. Complex wiring systems carrying unusually heavy electrical loads compensate for the absence of natural light in these buildings. In addition to general area lighting, museums use independent light sources for case and exhibit lighting. Therefore, museums need numerous electrical outlets in floors, walls, and ceilings to provide flexibility for frequent exhibit changes and to avoid dangerous use of extension cords.

Gallery Flexibility

Movable walls, temporary furring, and lightweight partitions facilitate rearranging exhibit spaces and changing exhibits. Such features, however, should not be allowed to impede occupants' movement in an emergency, nor should materials used increase fire risk by adding significant quantities of highly combustible material to the building's fuel load.

FIRE EMERGENCY MANAGEMENT

Managing a library or museum fire involves decisions and actions necessary to mitigate the consequences before, during, and immediately following an incident. Effective management includes continuous monitoring to ensure that fire protection systems and features are always operational and never degraded or defeated by space changes or other building modifications. This also includes providing emergency electric power to support emergency lighting, communications, security systems, and fire protection systems. (See Section 3, Chapter 4, "Emergency and Standby Power Supply.")

At a minimum, a viable fire emergency plan should provide for preplanning with the fire department and pre-emergency training of all building staff in evacuation and salvage operations. (See Section 10, Chapter 6, "Pre-Incident Planning for Industrial and Commercial Facilities.") The decision to establish fire brigades should depend upon whether the building(s) are close enough to a municipal or county fire department for timely response and whether employees are expected to use portable fire extinguishers and standpipe hose lines. In deciding whether to require employees to use fire-fighting equipment, management must be prepared to comply with state and federal regulations for protective clothing, training, and maintenance of fire-fighting equipment.[29] (See Section 10, Chapter 12, "Fire Loss Prevention and Emergency Organizations," and Section 10, Chapter 18, "Fire Service Protective Clothing and Protective Equipment.") In addition to safety considerations, emphasis should be placed on minimizing water damage and protecting or removing endangered, highly valuable collections. Pre-fire planning also must provide security for valuable collection materials after their removal from the building.

SALVAGE

Using freeze/vacuum drying techniques, library preservation specialists have been successful in salvaging water-soaked library, archival, and other collection materials. However, water is only one hazard to which collections are vulnerable during a fire. Others include smoke, acids contained in combustion products, and chemical agents used in fire fighting. These hazards perhaps endanger museum collections more than water.

Museum Salvage

With much higher values per item and fewer items than libraries, museums must concentrate salvage efforts upon removing endangered collections from the fire area to a safe location as quickly and carefully as possible. Conservation-oriented fire-fighting strategies and techniques enhance salvage efforts and minimize collection losses.

The Winterthur Museum in Wilmington, DE, has developed fire-fighting techniques that other museums and libraries might adapt.[30,40] Fire-fighting efforts are divided into three concurrent phases that the fire marshal and staff direct: (1) collection protection and salvage operations, led by the housekeeping supervisor and staff; (2) security for collections removed from the fire area, which is the responsibility of the security supervisor and staff; and (3) overall coordination, provided by the building superintendent or a designee.

Library Salvage

The success and cost of salvaging water-soaked library collections depends upon advance planning and how wet the materials are. Freeze/vacuum drying very effectively removes water and smoke odors, while using ethylene oxide, a sterilizing agent, during the process arrests the mold growth. Freeze/vacuum drying cannot, however, restore physical damage inflicted by hose streams, trampling, or mishandling.

Before the fire: Planning for salvage efforts before a fire must include:

1. Establishing salvage priorities. This facilitates decision making under emergency conditions by identifying, in advance, the items or collections to save first.
2. Identifying preservation specialists in salvage operations who can provide needed guidance.
3. Identifying sources of salvage materials, equipment, and facilities that are available on short notice.
4. Organizing and training employees who will be needed in the salvage effort during and after the fire.
5. Orienting and training the fire service to understand where vulnerable or especially valuable collections or objects are stored. The fire department should prepare a pre-fire plan that takes these factors into account and provides ongoing salvage and damage minimization during fire suppression operations.

During the fire:

1. Removing endangered highly valued collections, if this can be done without placing staff at risk. Where this is not possible, the fire service may perform this function with guidance.
2. Using tarpaulins or plastic sheeting to protect from water materials that remain in or near the fire area but are not directly involved in the fire. Here, too, staff must be alert to hazards and must not place themselves at risk to save objects. A well-developed fire department pre-fire plan will include provisions to accomplish these things.

After the fire:

1. Wrapping and freezing damaged articles.
2. Obtaining guidance from a preservation specialist experienced in salvage.
3. Determining appropriate restoration procedures.

Much less salvage work should be needed in facilities protected by automatic sprinklers. Water absorption is minimized when materials are left on shelves undisturbed by the gentler action of automatic sprinklers, as opposed to hose streams' more forceful action. An automatic sprinkler system also greatly reduces the quantity of materials requiring salvage by limiting the size of the fire area to that covered by the operation of one or just a few sprinklers.[31]

FIRE SAFETY IMPLICATIONS OF NEW DEVELOPMENTS

Automation of Library Catalogs and Museum Accession Records

Automation is leading to replacement of traditional card catalogs and paper museum accession records. As this occurs, continuity of public service increasingly depends upon the computer and necessitates a high level of fire protection including automatic suppression to ensure continuity of operation. Duplicates of magnetic media containing a library's computer-based catalog or a museum's accession records should be maintained in a safe, remote, off-site location.

Bookstacks

Multitier bookstacks: Bookstacks, modeled after those developed for the U.S. Library of Congress in 1893, have been used for several decades in many libraries. (See Figure 9-16E.) Typically, unprotected iron or steel columns and beams at intervals of 7 ft (2.14 m) support decks or tiers. Vertical openings between decks permit heating of bookstack areas by convection or air currents. (See "Deck slit" in Figure 9-16F.) Under fire conditions, these same openings function as flues to accelerate vertical fire spread to the tiers above.

Smoke barriers should be installed in these vertical openings, or deck slits, to facilitate smoke detection on the tier or deck of fire origin. This was not done when the smoke detection system was installed in the multitier bookstack of the Los Angeles Central Public Library because a ducted ventilation system had not been installed for bookstacks. When an arsonist set a fire on the fifth tier, the first detectors to annunciate were on the sixth tier, one tier above the tier of fire origin. A delay in finding the fire resulted, and occupants lost the opportunity to stop the fire with portable extinguishers while it was small, with catastrophic results.[32] It took nearly 350 fire fighters and over 70 pieces of fire apparatus 7 hr to extinguish the ensuing inferno.[33]

Modern versions of these systems are still sold but without vertical openings in the modules. (See Figure 9-16G.) Unprotected steel support structures are vulnerable to collapse in a fire, however. Cast iron and steel structural members lose strength at high temperatures encountered in fires and may collapse. For this reason, the fire service may not be able to enter a multitier bookstack for manual fire suppression. This also applies to any variation on this bookstack design that uses exposed steel structural elements. Because it may be impossible for the fire service to provide manual fire suppression in bookstacks of this or similar design, it is essential that the bookstacks be equipped with automatic fire suppression systems to defend against fire.

In new library construction, multitier bookstacks replace freestanding bookstack ranges installed on structural fire-resistant floors. The difference in fuel load between multitier stacks and freestanding stacks is substantial. All levels in a multitier system are usually considered as one fire compartment, due to the absence of vertical fire barriers between tiers. Fire loads depend on the number of tiers in the bookstack. With shelf loading at 70 percent of capacity, the expected fire load in each tier would be 55 to 65 lb/sq ft (268 to 317 kg/m^2). Thus, for a multitier bookstack with 10 tiers in a single fire compartment, fire duration is estimated at 70 to 85 hr. Conversely, a fire in a free-standing bookstack that is seven shelves high and set on a fire-resistant floor, would be limited to a single tier and a fire duration of 7 to 8 hr.

Compact storage or track files: Bookstack shelving units, or ranges, mounted on carriages that roll on tracks are common in new library and museum construction and in renovations. In this space-saving storage system, one moving aisle serves several ranges. The track systems are modular, with two or more mobile ranges between two fixed-end ranges in each module. (See Figure 9-16H.) This storage method can create fire loads exceeding 120 lb/sq ft (586 kg/m^2) and fire durations exceeding 15 hr. This is more than enough to challenge the most fire-resistant barriers and construction prescribed by building and fire codes.

The NFPA Technical Committee on Protection of Cultural Resources recommends automatic sprinkler protection for compact storage installations. Without such protection, a fire in a facility with this type of storage system endangers not only the collection, but may cause severe structural failures and perhaps building collapse as well. An independent analysis[34] of fire tests conducted to evaluate sprinklers' effectiveness of sprinklers on a compact storage system[35] concluded that sprinklers installed over a compact storage module, will most probably prevent a fire that originates in the module from damaging building structural elements outside the module. Furthermore, the tests showed (1) there is a good probability that sprinklers will prevent fire spread across 4-ft (1.22-m) aisles between compact storage modules, and (2) sprinklers may effectively prevent fire spread between back-to-back shelves of adjacent modules when a sheet metal panel with a 1-in. (25-mm) air space between adjacent panels backs each shelf. Thus, automatic sprinklers can be expected to limit fire development to that portion of the module of fire origin between the fixed-end range and the open aisle. Nevertheless, total loss of contents within that portion of the storage module is possible. (See Figure 9-16H.)

Facility designers must carefully consider these points in designing compact storage systems:

1. More efficient use of space results in greater density or concentration of values. That is, it increases the number of items subject to fire damage per unit volume of storage space. This risk should be considered in determining the maximum number of ranges to include in each module.
2. Existing automatic fire detection and fire suppression systems may have to be modified, and gaseous agents may be ineffective. In an existing facility with sprinkler protection where compact storage systems are retrofitted, particular attention must be given to water density, ceiling clearances, and structural features of the space. Tests conducted for the National Library of Canada clearly showed that for compact storage systems installed where "explicit compliance with current installation standards" is not possible, it may be necessary to go well beyond normal sprinkler design criteria to achieve adequate protection.[36]
3. Compact storage modules may conceal the origin of smoke, thereby compounding the difficulty in locating and extinguishing a fire.

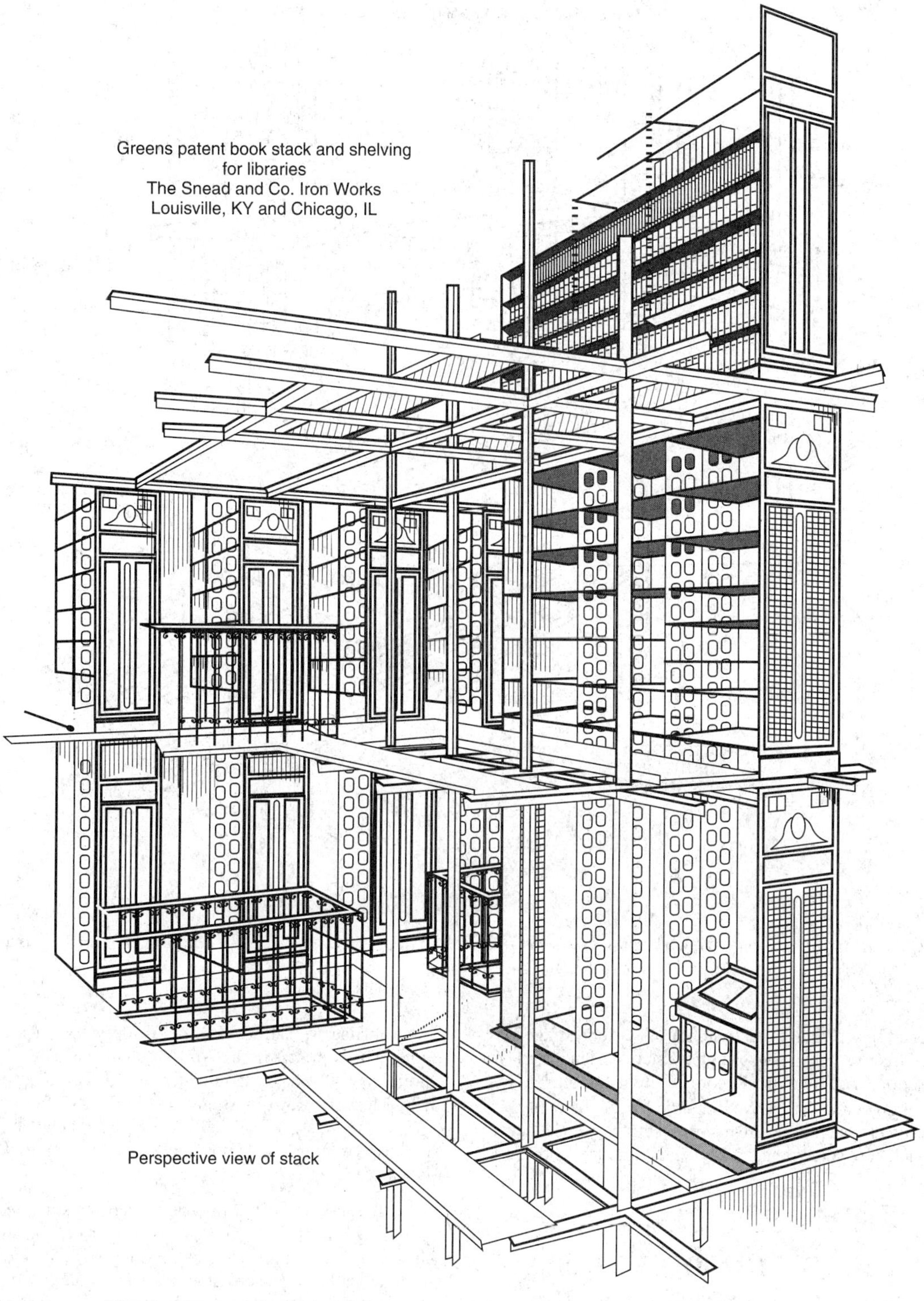

Greens patent book stack and shelving
for libraries
The Snead and Co. Iron Works
Louisville, KY and Chicago, IL

Perspective view of stack

FIG. 9-16E.   Multitier bookstack as installed in the Library of Congress, circa 1893.

4. Compact storage modules prevent penetration of water from hose streams for fire extinguishment and delay detection by trapping smoke and heat. Consider designing the system so that ranges can be parked with a 5-in. (12.7-cm) gap between them during inactive storage periods. Underwriters Laboratories tested this arrangement with satisfactory results for the U.S. National Archives and Records Administration using quick response sprinklers. Paper record loss was limited to 10 cubic ft (0.28 m^3) and computer record loss to 3 cubic ft (0.08 m^3).[37]

5. Water mist fire suppression test results in bookstack scenarios show promise for special application to the burrowing fire problem in compact storage. It may be possible to design water mist

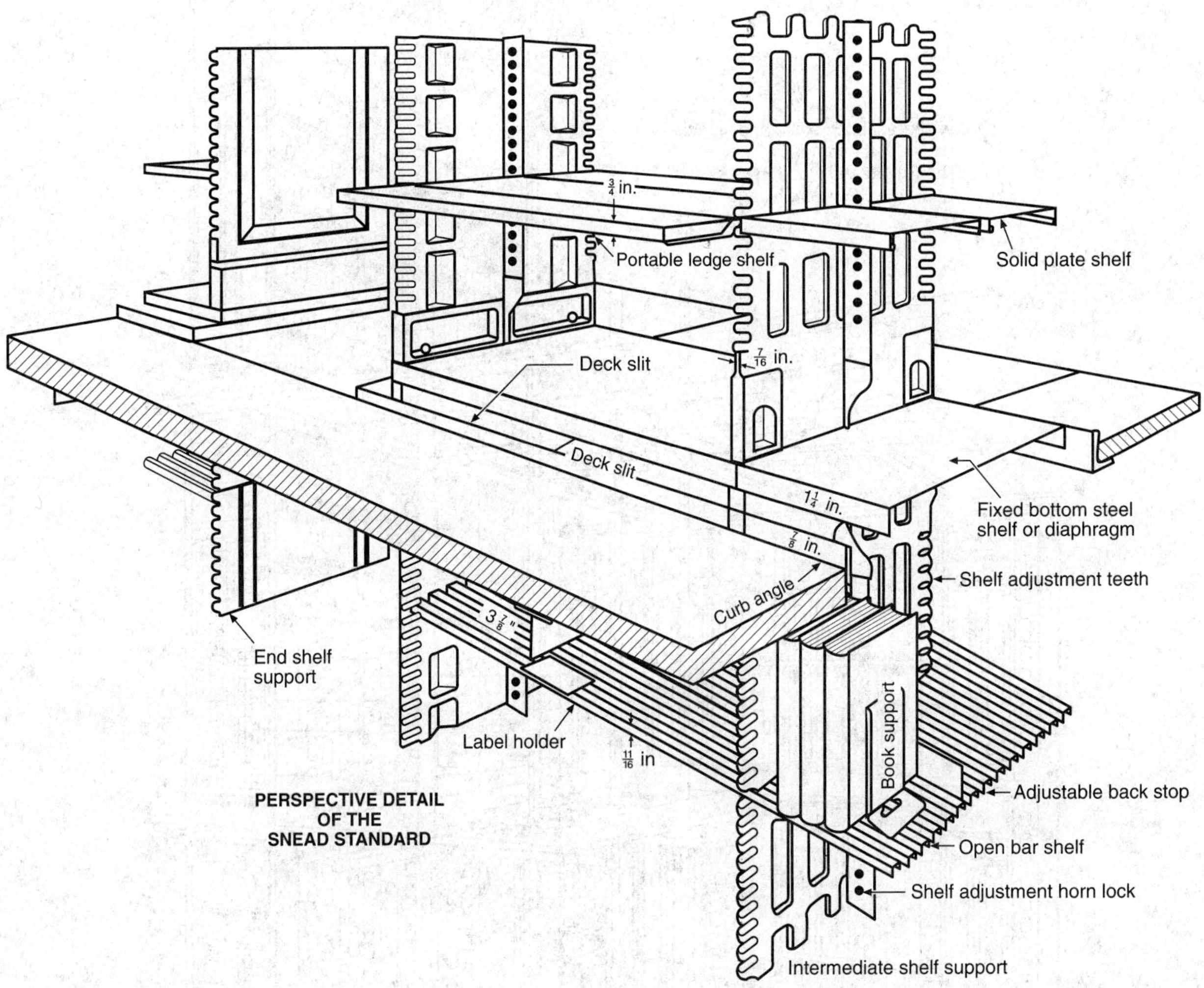

THE SNEAD AND COMPANY IRON WORKS, INC.

FIG. 9-16F. Multitier bookstack—view showing section through deck flooring, deck slit, fixed bottom shelf, and adjustable shelves.

systems for compact storage that limit fire and water damage loss even more effectively than the sprinkler system design tested for the U.S. National Archives and Records Administration.[28,37]

6. Bookstack ranges may be moved electrically or manually. In compact storage systems that are moved electrically, the electric motor in each bookstack range is a potential ignition source.

FIRE PROTECTION DECISIONS

The elapsed time between a fire's discovery and the beginning of effective manual suppression can easily be too long to prevent unacceptable loss, especially when the building is unoccupied. Faced with this reality, directors of museums and libraries and their governing bodies are deciding to incorporate automatic fire suppression systems as the primary defense against fire loss.

In a 1989 survey of museums of various sizes, including 30 of the largest in the U.S., 77 percent reported that they use automatic sprinkler protection.[18] The list of museums that made this decision includes the 15 Smithsonian museums: National Air and Space Museum, National Museum of American History, National Museum of Natural History, National Portrait Gallery, the Sackler Gallery, National Museum of African Art, National Museum of the American Indian, and so on. It also includes the J. Paul Getty Museum, the High Museum of Art in Atlanta, the Phillips Gallery in Washington, DC, and the Dewitt Wallace Decorative Arts Gallery and Abby Aldrich Rockefeller Folk Art Center in Colonial Williamsburg, VA, to name a few.

Similarly, the list of libraries, large and small, equipped to defend their collections against fire loss by installing automatic sprinkler systems continues to grow. Examples include city libraries in Atlanta; Dallas; Houston; Los Angeles; Portland, ME; and Portland, OR, as well as many academic libraries, the State Library of Michigan, and national libraries, such as the Saudi Arabian National Library, the U.S. Library of Congress, the National Library of Canada, and the National Library of Scotland. The new British Library at St. Pancras, with four floors below grade level, 180 mi (300 km) of shelving, 864,000 sq ft (80,000 m^2) of floor area, and plans for later expansion to 2,160,000 sq ft (200,000 m^2) will be added to the list when it opens.[38]

NFPA 910, NFPA 911, and other chapters of this handbook provide guidance for procuring, installing, testing, and maintaining fire protection systems.

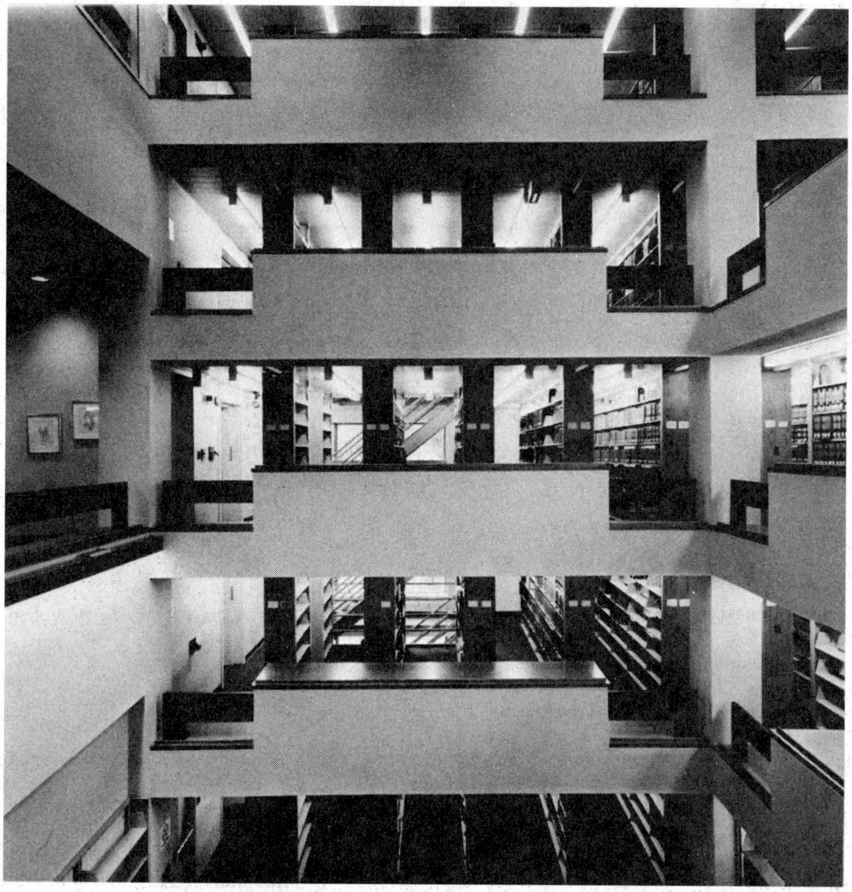

FIG. 9-16G. A modern multitier bookstack installation.

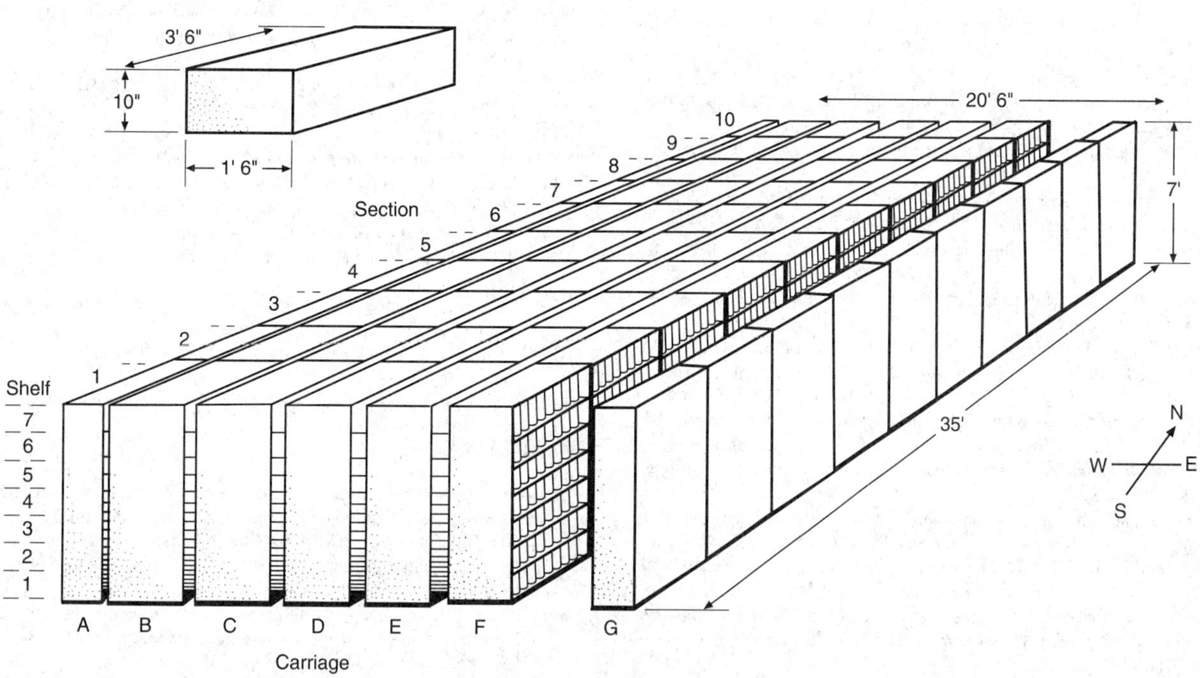
TYPICAL INDIVIDUAL HALF SHELF IN INDIVIDUAL SECTION
3' 6"
10"
1' 6"
20' 6"
7'
35'
Section
1
2
3
4
5
6
7
8
9
10
Shelf
7
6
5
4
3
2
1
A
B
C
D
E
F
G
Carriage
N
W
E
S

FIG. 9-16H. Compact mobile shelving.

BIBLIOGRAPHY

References Cited

1. Morris, J., *Managing the Library Fire Risk,* second edition, University of California, Berkeley, CA, 1979.
2. Johnson, E. M., ed., *Protecting the Library and Its Resources*, American Library Association, Chicago, IL, 1963.
3. Morris, J., *The Library Disaster Preparedness Handbook*, American Library Association, Chicago and London, 1986.
4. Morris, J., "Is Your Library Safe From Fire?" *American School and University*, April 1980, p. 61.
5. McElroy, F. E., ed., *Accident Prevention Manual for Industrial Operations*, seventh edition, National Safety Council, Chicago, IL, 1974.
6. Factory Mutual Research Corp., *Handbook of Industrial Loss Prevention*, second edition, McGraw-Hill, Inc., NY, 1967.
7. McKinnon, G. P., ed., *Industrial Fire Hazards Handbook*, first edition, National Fire Protection Association, Quincy, MA, 1979.
8. Johnson, E. V., and Horgan, J. C., *Museum Collection Storage*, United Nations Educational Scientific and Cultural Organization, Paris, France, 1979.
9. *Operational School Burning: Official Report on a Series of School Fire Tests Conducted April 16 to June 30, 1959, by the Los Angeles Fire Department*, National Fire Protection Association, Quincy, MA, 1959.
10. Juillerat, E., Jr., "The Hartford Hospital Fire," *NFPA Quarterly*, Vol. 55, No. 3, January 1962, pp. 295–303.
11. *Fire Protection Handbook*, twelfth edition, Table 8-143, National Fire Protection Association, 1962, pp. 8–145.
12. Lathrop, J. K., "Building Under Construction," *Fire Journal*, Vol. 68, No. 5, Sept. 1974, pp. 37, 39.
13. Harvey, B., "Fire Hazards in Libraries: Part I—Clearing the Smoke," *Library Security Newsletter*, Vol. 1, No. 1, Jan. 1975.
14. Firepro, Inc., *Fire Defense Alternatives: Library of Congress*, Washington, DC, 1977, unpublished report.
15. *Building Firesafety Criteria*, PBS P5920.9, Appendix D, 1979, General Services Administration, Washington, DC.
16. Zimmerman, C. E., ed., *Fire Alarm Signaling Handbook*, NFPA SPP 82, 1987, National Fire Protection Association, Quincy, MA.
17. Fitzgerald, R. W., and Wilson, R., *The Systems Technique for Evaluating Building Firesafety*, Worcester Polytechnic Institute, Worcester, MA, 1976.
18. Wilson, J. A., "Fire Fighters," *Museum News*, Nov./Dec. 1989, pp. 68–72.
19. Kirsch, A., "Preserving Today's Treasures for Tomorrow," *Factory Mutual Record*, Vol. 67, No. 1, Jan./Feb. 1990, pp. 4–8.
20. Myller, R., *The Design of the Small Public Library*, Bowker, NY, 1966.
21. Thompson, G., *The Museum Environment*, Butterworths, London, England, 1978.
22. Brawne, M., *Libraries: Architecture and Equipment*, Praeger, New York, 1970.
23. Thomas, S., ed., "Retrofitting with Sprinklers for Added Protection," *Factory Mutual Record*, Vol. 59, No. 3, May/June 1982, pp. 13–19.
24. Grant, C. C., P.E., "Life Beyond Halon," *Fire Journal*, Vol. 84, No. 3, May/June 1990.
25. Grant, C. C., P.E., "Halons and the Ozone Layer," *Fire Journal*, Vol. 83, No. 5, Sept./Oct. 1989.
26. "Final Report of the Halons Technical Options Committee," United Nations Environment Programme, *Montréal Protocol Assessment Technology Review*, August 11, 1989. Available from ICF Incorporated, 9300 Lee Highway, Room 1171, Fairfax, VA, USA 22031-1207.
27. DiNenno, P. J., and Taylor, G. M., *Best and Essential Halon Use, A Methodology*, National Fire Protection Research Foundation, Quincy, MA, Dec. 1989.
28. Milke, J., Gerschefski, C. E., and Veltre, J. V., *Evaluation of Water Mist Suppression System Designs—Library Stack Test Scenarios,* Report Prepared for Reliable Sprinkler Company, April 1995, Department of Fire Protection Engineering, University of Maryland at College Park.
29. U.S. Department of Labor, Occupational Safety and Health Administration (OSHA), *Occupational Safety and Health Standards* (29 CFR 1910), Subparts E & L.
30. Fennelly, L. J., ed., *Museum, Archive and Library Security*, Butterworths, London, England, 1983.
31. Fuhlrott, R., and Dewey, M., *Library Interior Layout and Design*, Saur, Munich, Germany, 1982.
32. Morris, J., "Fire Protection for the Library," *The Construction Specifier*, Oct. 1989.
33. Isner, M. S., "Fire in Los Angeles Central Library Causes $22 Million Loss," *Fire Journal*, Vol. 81, No. 5, March/April 1987.
34. Cutler, H. R., *Engineering Analysis of Compact Storage Fire Tests*, Library of Congress, Washington, DC, 1979.
35. Chicarello, P. J., et al., *Fire Tests in Mobile Storage Systems for Archival Storage*, General Services Administration, Washington, DC, 1978.
36. Lougheed, G. D., Mawhinney, J. R., and O'Neill, J., "Full-Scale Fire Tests and Development of Design Criteria for Sprinkler Protection for Mobile Shelving Units," *Fire Technology*, Vol. 30, No. 1, First Quarter 1994.
37. Underwriters Laboratories Inc., "Report on Archives II Mobile High Density Shelving Fire Protection System." A copy of this report, dated November 30, 1989, is available from the National Archives and Records Administration (NARA), Washington, DC 20408.
38. Cartwright, N. K., "Fire Protection at the New British Library," *Fire Prevention*, No. 203, Oct. 1987.
39. Brannigan, F. L., "Fire Growth," *Firehouse,* July 1995, pp. 80-85.
40. Cash, J., "Sophisticated System Protects Historic Museum," *NFPA Journal*, Vol. 86, No. 5, Sept./Oct. 1992, pp. 63–67.
41. Bordenaro, M., "Library is Textbook of Systems Integration," *Building Design and Construction*, Vol. 33, No. 11, Nov. 1992, pp. 42, 45.
42. Marchant, E. W., "Fire Engineering Strategies," *Fire Science and Technology*, Vol. 11, No. 1–2, 1991, pp. 13–19.
43. Shields, T. J., Dunlop, K. E., and Silcock, G. W. H., "Management Strategy to Establish Life Safety Equivalency for Historic Buildings," *Fire Science and Technology*, Vol. 11, No. 1–2, 1991, pp. 21–26.

NFPA Codes, Standards, and Recommended Practices

Reference to the following NFPA codes, standards, and recommended practices will provide further information on library and museum collections discussed in this chapter. (See the latest version of *The NFPA Catalog* for availability of current editions of the following documents.)

NFPA 1, *Fire Prevention Code*
NFPA 10, *Standard for Portable Fire Extinguishers*
NFPA 11, *Standard for Low-Expansion Foam*
NFPA 12, *Standard on Carbon Dioxide Extinguishing Systems*
NFPA 12A, *Standard on Halon 1301 Fire Extinguishing Systems*
NFPA 13, *Standard for the Installation of Sprinkler Systems*
NFPA 14, *Standard for the Installation of Standpipe and Hose Systems*
NFPA 40, *Standard for the Storage and Handling of Cellulose Nitrate Motion Picture Film*
NFPA 45, *Standard on Fire Protection for Laboratories Using Chemicals*
NFPA 70, *National Electrical Code®*
NFPA 101®, *Life Safety Code®*
NFPA 232, *Standard for the Protection of Records*
NFPA 232A, *Guide for Fire Protection for Archives and Record Centers*
NFPA 550, *Guide to the Fire Safety Concepts Tree*
NFPA 600, *Standard on Industrial Fire Brigades*
NFPA 910, *Recommended Practice for the Protection of Libraries and Library Collections*
NFPA 911, *Recommended Practice for the Protection of Museums and Museum Collections*
NFPA 914, *Recommended Practice for Fire Protection in Historic Structures*
NFPA 2001, *Standard on Clean Agent Fire Extinguishing Systems*

Additional Reading

Artim, N., "Fire Protection Strategy for Library Bookstacks," National Fire Protection Association and International Association of Fire Chiefs, in-FIRE (international network for Fire Information and Reference Exchange.), Annual Conference, *Proceedings*, April 28–30, 1993, Norwood, MA, 1993, pp. 67–78.
Beilicke, G., "Basic Thoughts on Structural Fire Protection of Protected Historic Buildings," *Fire Science and Technology*, Vol. 11, No. 1–2, 1991, pp. 57–59.
Bochnak, P. M., *Fire Loss Control: A Management Guide,* second edition, rev. and expanded, M. Dekker, New York, 1991.

Braidwood, J., "Historic Blaze: Houses of Parliament Burned to the Ground," *Fire Prevention*, No. 255, Dec. 1992, pp. 28–29.

Bryan, J. L., *Fire Suppression and Detection Systems,* third edition, Maxwell Macmillan International, New York and London, 1993.

Burgess, D., "The Library Has Blown Up!" *Library Journal*, Oct. 1, 1989.

Bush, S. E., "How a Library Can Overcome Its Fear of Frying," inFIRE, the international network for Fire Information and Reference Exchange and Society of Fire Protection Engineers and International Association of Fire Chiefs, Annual Conference, *Proceedings*, May 11–13, 1994, Fairfax, VA, 1994, pp. 41–51.

Cassidy, K. A., *Fire Safety and Loss Prevention*, Butterworth-Heinemann, Boston, 1992.

Cockcroft, D., "Possible Terrorist Attacks Must Now Be Considered in Museum Fire Safety," *Fire*, Vol. 86, No. 1056, June 1993, p. 43.

Coon, J. W., *Fire Protection: Design Criteria, Options, Selection*, R. S. Means Co., Kingston, MA, 1991.

DeCandido, G. A., "Special Report: Fire at the USSR Academy of Sciences Library," *Library Journal*, June 15, 1988.

Donnally, H., "Saving Library Collections," *Fire Prevention*, No. 254, Nov. 1992, pp. 26–27.

GSA, "Facility Standards for Public Buildings Service," PBS P 3430.1A, 1989, General Services Administration, Washington, DC.

Geissler, V., "Preservation of Historical Monuments and Fire Safety," *Fire Science and Technology*, Vol. 11, No. 1–2, 1991, pp. 5–8.

Gibbon, D., "Novel Approach to Library Fire Safety," *Fire Prevention*, No. 267, March 1994, pp. 28–30.

"Heritage Under Fire. A Guide to the Protection of Historic Buildings," Fire Protection Assoc., London, England, 1990.

Hillbruner, M., "Historical Buildings Demand Special Retrofit Measures," *Consulting/Specifying Engineer*, Vol. 9, No. 3, March 1991, pp. 32–34.

"Historic Buildings Under Fire Again," *Fire Prevention*, No. 257, March 1993, pp. 38–39.

James, D. W. B., *Fire Prevention Handbook,* Butterworths, London and Boston, 1986.

Kidd, S., "Preparing for Disasters in Museums and Galleries," *Fire Prevention*, No. 254, Nov. 1992, pp. 22–25.

Kristensen, O. B., "Historic Building Fires Spur Danish Action," *Fire International*, No. 141, Nov. 1993, pp. 14, 24.

Lee, S., "National Library of Scotland: Protecting a Nation's Heritage," *Fire Prevention*, No. 249, May 1992, pp. 22–26.

Lougheed, G., and Mawhinney, J., "Automatic Sprinkler Protection for Compact Mobile Shelving," *Fire Research News*, No. 67, Winter 1993, pp. 1–3.

Malhotra, H. L., and Papaioannou, K., "Framework for a CIB Guide on Fire Safety for Historic Buildings," *Fire Science and Technology*, Vol. 11, No. 1–2, 1991, pp. 69–72.

Marryatt, H. W. *FIRE, A Century of Automatic Sprinkler Protection in Australia and New Zealand 1886–1986,* rev ed. Australian Fire Protection Association, North Melbourne, Australia, 1988.

Milke, J. A., and Gerschefski, C. E., "Overview of Water Mist Research for Library Applications," *Proceedings* of the International Conference on Fire Research and Engineering, Sept. 10–15, 1995, Orlando, FL, SFPE, Boston, MA, 1995, pp. 133–138.

Morris, J., "Protecting Libraries and Museums From Fire," *Fire Science and Technology*, Vol. 11, No. 1–2, 1991, pp. 35–43.

Morris, J., "Restoring Water-logged Books," *Cleaning and Restoration*, Nov. 1989.

Morris, J., "Mopping Up after Library Fires," *Fire Prevention,* No. 220, June 1989.

Papaioannou, K. K., "Fire Safety in Historic Buildings and Sites," *Fire Science and Technology*, Vol. 11, No. 1–2, 1991, pp. 1–4.

Papaioannou, K. K., "Fire Safety in Historic Buildings," Aristotle Univ., Thessaloniki, Greece, CIB 92; Conseil International du Batiment, Congres Mondial du Batiment, World Building Congress CIB 92, *Proceedings [Compte-rendu]*, Tome 1, Session 2A and 2B, Decision-Making and Rehabilitation, May 18–22, 1992, Montreal, Canada, 1992, pp. 217–218.

Robertson, J. C., *Introduction to Fire Prevention,* fourth edition, Prentice Hall, Englewood Cliffs, NJ, 1995.

Scoones, K., "Serious Fires in Libraries and Museums, 1986–1991," *Fire Prevention*, No. 254, Nov. 1992, pp. 20–21.

Sharma, T. P., *et al.*, "Insight to Museums From Fire Hazards Assessment and Fire Protection Point of View," *Fire Science and Technology*, Vol. 11, No. 1–2, 1991, pp. 61–68.

Shields, T. J., Dunlop, K. E., and Silcock, G. W. H., "Emergency Evacuation of a Mixed Ability Population From a Museum Building," University of Ulster, UK, TNO Building and Construction Research, CIB/W14 Workshop, *Proceedings* of the Third Fire Engineering Workshop on Modelling, Jan. 25–26, 1993, Delft, The Netherlands, 1993, pp. 214–223.

Tillotson, R. G., *Museum Security*, International Council of Museums, Paris, France, 1977, p. 46.

Todd, C., "Protecting Historic Buildings," *Fire Prevention*, No. 246, Jan./Feb. 1992, pp. 28–32.

Tweedie, A. T., "Freeze-Dried Books," *Library Journal*, Jan. 15, 1974, p. 82.

Waters, P., *Procedures for Salvage of Water-Damaged Library Materials*, Library of Congress, Washington, DC, 1975.

Wilson, R., "The L-Curve: Evaluating Fire Protection Tradeoffs," *Specifying Engineer*, May 1977, pp. 110–113.

WAREHOUSE OPERATIONS

Jeffrey Moore and Larry Davis

In 1989–1993, there were 23,900 reported U.S. structure fires per year in storage facilities other than dwelling garages, or one every 22 min. The highly combustible loading in a warehouse and the storage of highly valued commodities result in fires that are severe and costly, and the fires cited above collectively caused nearly half a billion dollars in direct loss per year. Recent warehouse fire experience has shown the need for great care in designing warehouse fire protection systems and in maintaining the conditions upon which the design was based.

This chapter covers general fire protection concerns related to storage of combustible commodities in warehouses. It also covers areas of concern when developing a pre-incident plan for dealing with a warehouse fire and the strategic and tactical considerations faced on the fire ground during a warehouse fire.

See Section 9, Chapter 18, "General Indoor Storage," for an overview of the basic principles of fire protection and storage practices in the context of NFPA storage standards. Additional information on storage can be found in the following chapters in Section 4: Chapter 6, "Flammable and Combustible Liquids"; Chapter 7, "Gases"; Chapter 12, "Explosives and Blasting Agents"; and Chapter 15, "Dusts."

WAREHOUSE FIRE PROTECTION PROBLEMS

Warehouses represent a unique fire challenge to both fixed fire suppression systems and manual fire-fighting forces called upon to deal with a fire. Modern warehouses and storage occupancies are especially subject to rapidly developing fires of great intensity because complex configurations of storage are conducive to rapid fire spread, presenting numerous obstacles to manual fire suppression efforts. The only proven method of controlling a warehouse fire is with properly designed and maintained automatic sprinkler systems. If sprinkler protection is not provided, the likelihood of controlling a significant fire in a warehouse is minimal, at best.

It is essential to recognize the significant difference between *control* of a fire and *extinguishment* of a fire in a warehouse. A fire is said to be under control when enough sprinklers have operated to stop the spread of the fire by wetting fuel ahead of the fire. It also means that temperatures within the building have been reduced to a

point where the structure is no longer in danger of collapsing. Typical warehouse sprinkler designs anticipate 2,000 to 5,000 sq ft (186 to 465 m^2) of sprinkler operation before the fire is controlled. At this time, although the fire may be under control, it is far from extinguished. Whereas only a few minutes may be required for the sprinklers to control the fire, several hours or even days may be required to actually extinguish the fire. It is important to realize that a severe warehouse fire is an emergency of major proportions and requires a planning effort and emergency response of equal magnitude to both control and eventually extinguish it. Factors that must be considered when developing a pre-incident plan to deal with a warehouse fire will be discussed later.

Any warehouse presents the problem of two competing interests with differing requirements. Much that is done to improve the efficiency of the warehouse in terms of material handling needs and storage capacity is detrimental to the building's fire protection interests. Likewise, much that is done to improve the fire protection is likely to adversely affect the capacity of the building and the efficiency of material-handling equipment. The key is to strike a balance between these two interests.

Automatic Sprinklers

Even a sprinklered warehouse or storage area can experience a catastrophic fire. Sprinklers that are properly designed for the occupancy can be expected to bring a fire under control and maintain control until the fire brigade, the public fire department, or both can accomplish final extinguishment. Improperly designed automatic sprinklers will not control the fire. In fact, in such cases, the fire may be far beyond control by the time the fire department arrives. An inadequately designed sprinkler system is less likely to hold a fire in check, and manual fire suppression efforts cannot be expected to compensate for an inadequately designed sprinkler system.

No one should assume that a building equipped with automatic sprinklers is "safe" from a major fire. High piling in a warehouse or materials with a high rate of heat release may present a fire challenge far beyond the hazard the sprinklers were originally designed to control. This can easily occur as the type, amount, and arrangement of storage in a warehouse changes over time.

For warehouse sprinkler system design and protection, see "Automatic Sprinkler Protection," "Supplemental Means of Fire Protection for Warehouses," and "Special Storage Facilities" in Section 9, Chapter 18, "General Indoor Storage."

Jeffrey Moore, P.E., and Larry Davis, CFPS, are senior loss prevention training consultants for Industrial Risk Insurers, Hartford, CT.

Unrecognized Fire Protection Deficiencies

Building owners may not recognize protection deficiencies in their warehouse. The business situations at hand may distract them. For example, if excessive storage heights compromise sprinkler protection in a warehouse, the owner may be busy worrying about the large amount of inventory on hand. If business is very good and large amounts of inventory are moving through the warehouse, the owner may be completely concentrating on the good business climate, instead of protecting the warehouse.

Building owners may assume that if a problem exists in their warehouse, the public fire department will point it out. The reality is that, while the fire department may generally know the type and arrangement of storage in a particular warehouse, they seldom have the personnel or time to track all changes that occur. In some instances, storage in a warehouse may change significantly from month to month, or even from week to week. Few fire departments have the resources required to fully analyze the protection in a warehouse, to determine the capabilities of the sprinkler system, or to recognize how changes in storage affect the sprinklers' ability to control a fire. As a result, critical protection deficiencies that can spell disaster during a fire may go unrecognized.

Most fire departments and building code officials are only responsible for enforcing the local fire code and building code. If the minimum requirements of these codes are met, local code officials are not obligated, nor do they generally have the time and resources, to provide guidance beyond the minimum protection level the codes require. Although the minimum requirements of the local codes are all that may be necessary to meet the statutory requirements to obtain an occupancy certificate, this in no way ensures the warehouse is protected from a fire. Other protection measures may be needed to fully protect the warehouse.

Insurance Does Not Solve the Problem

If the warehouse is insured by a major insurance company, many people may assume it must be properly protected. Insurance does not guarantee that a building is properly protected. Insurance companies have several means of protecting themselves financially when they insure a poor risk. These include: (1) charging very high insurance rates for the property, (2) using very high deductibles (which can exceed several million dollars), (3) purchasing reinsurance on the property, and (4) providing only limited amounts of insurance for the property. So a major property insurance company may insure a building because its financial risk is limited.

Storage Configuration Problems

Sprinkler designs are based on open aisles that act as fire breaks and permit sprinkler discharge to fall between what is burning and what is not. Storage present in the aisles (even one pallet load) forms a bridge of fuel that enables fire to extend across aisle space more quickly.

Storage height must not be increased beyond the design of the sprinkler system. Not only does it increase the fuel load, but it can also block discharge from the sprinkler heads. Even one additional layer of storage may be enough to surpass a sprinkler system's capability to control the fire.

Commodities

Classification: For a complete discussion of commodity classifications, storage arrangement, and application of NFPA storage standards, see Section 9, Chapter 18, "General Indoor Storage."

High-hazard commodities: Extensive use of plastic materials in both finished products and the product packaging has significantly increased the hazard of commodities stored in most warehouses. Any increase in the hazard of the storage requires a corresponding increase in the level of protection. No proven rules determine how much of a higher hazard material may be introduced before sprinkler protection is compromised. For some materials, such as flammable liquids or aerosols, less than one pallet load is more than enough to burn a warehouse to the ground. For other materials, such as expanded foam plastics used as packing materials inside cartons, the increase in hazard may not be readily apparent. If there is any question as to the hazard level of the storage, fire protection should be designed or reinforced to provide for the most hazardous commodity.

WAREHOUSE FIRE-FIGHTING OPERATIONS AND PRE-INCIDENT PLANNING

The following information is intended primarily for fire brigade officers and public fire officials who must develop and implement a pre-incident plan for a warehouse. It will also assist fire officials in command positions who are responsible for tactical and strategic decisions made on the fire ground.

Because automatic sprinkler protection is the most effective, and, in many instances, the only method to successfully control a warehouse fire, all fire-fighting operations should be geared toward support of the sprinkler system. This support not only increases the likelihood of fire control but also maximizes fire fighter safety because it is unnecessary for fire fighters to be exposed to the hazards of an interior fire attack.

After effective support for the sprinklers has been established, the fire emergency can be considered stabilized. This provides sufficient time to carefully organize a safe and successful fire suppression operation. In addition, sprinklers buy time that helps maximize safety and operational effectiveness—hurried fire-fighting operations are not only disorganized but are also unsafe and ineffective.

The overall success of a warehouse fire-fighting operation is determined long before the first smoke is visible—in fact, it is determined before actual ignition occurs. The operation's success is determined as the warehouse pre-incident plan develops.

The pre-incident plan must be a cooperative effort among the building owner, plant emergency teams, and the public fire department that will respond to the fire. Other agencies may also provide valuable input to the pre-fire plan. Such agencies include plant security forces, plant engineering staff, public utility companies, and the insurance carrier.

Developing a Pre-Incident Plan

The pre-incident planning process begins with a warehouse tour. Arrangements should be made with the building owner for someone with a working knowledge of the facility and its operations available to accompany the fire department representatives on the tour. Building owners should cooperate fully with the public fire department when they develop a pre-incident plan. Architectural drawings and diagrams of the facility showing the building layout, construction, storage configurations, and the design of the fire protection systems and equipment should be provided for examination prior to the tour.

The tour's purpose is to gather information that the fire department will use to predict the type of fire anticipated and the possible complications encountered during fire suppression operations. Information gathered should include at least:

- Evaluation of building size and construction characteristics to determine a fire's probable effect. For example, a roof supported by lightweight steel bar joists or trusses can be expected to collapse early during a serious fire.

- Location of firewalls, fire barriers, and fire doors that separate the storage area from the surrounding operations. They should be checked for proper operation because they must be relied upon during an emergency to check the spread of the fire.

- Identification of the commodities stored within the warehouse. Knowledge of their burning characteristics is essential. For example, burning aerosol products, flammable liquids, or similar materials may be beyond the capabilities of the sprinkler system.

- Determination of the method of packaging materials in the warehouse. For example, cardboard cartons can be expected to collapse quickly into the aisle, and plastic straps attaching materials to pallets can be expected to melt quickly.

- Identification of storage method, e.g., racks, solid piles on the floor, or palletized.

- Evaluation of normal and maximum storage levels. Stock levels must be assessed as to relative stability from season to season. Pre-fire plans should be based on the worst case. For example, the worst case fire in a department store warehouse occurs in September or October when it is filled to capacity with holiday sale merchandise.

- Determination of the type and design of the automatic sprinkler systems. Obtaining complete information on the capability of the sprinkler systems is very important so it can be compared to the hazards the storage in the warehouse pose. If sprinkler design is inadequate for the storage type and arrangement, fire control is unlikely.

- Determination of minimum flow and pressure required to supply sprinkler systems. To evaluate the adequacy of the available water supplies, water supply demands for the sprinklers must be known. If the water supply is inadequate, sprinklers will not perform as designed. If current water supply test results are not available, tests should be conducted to obtain current information.

- Identification of the type and number of water supplies available for fire protection and information on how to start fire pumps so that fire fighters may start them manually, if necessary.

- Location of public hydrants and fire department sprinkler connections. Fire apparatus should not steal water from the sprinkler systems by connecting to yard hydrants. The location of suitable public hydrants and static water sources to supply fire apparatus must be located and evaluated.

- Arrangement of interior standpipes and hose connections. Are they fed from overhead sprinklers? Can they remain in service with the sprinklers shut off? All hose connections must have the same hose threads as the public fire department.

- Assessment of type, capacity, and method of operation of any automatic or manual heat and smoke venting facilities. How can exhaust fans be activated? How is the power supply to the fans arranged? Will power be lost when electric power for the warehouse shuts down during an emergency?

- Determination of primary and alternate means of access to the warehouse. What security features make access more difficult? Are security guards or guard dogs normally in the warehouse during off hours?

- Determination of type of training provided for the plant emergency organization. Also assess how many members will be available during a fire, and what procedures they follow in a fire.

- Assessment of types of manual fire suppression equipment available to the plant emergency organization, and whether the public fire department can utilize this equipment.

- Assessment of general storage practices within the building. Are piles neat and orderly? Are they sloppy and in disarray, indicating poor housekeeping practices conducive to rapid fire spread?

- Investigation of aisle storage. Even so-called "temporary" storage blocks access to the aisles in an emergency and can cause rapid bridging of a fire across aisles.

- Determination of the material-handling equipment available during overhaul operations to move stock out of the building. Will employees be available to operate the equipment or must fire fighters be trained to use the equipment?

After this basic information has been gathered, development of the actual pre-incident plan begins. The plan, of course, should be developed in conjunction with the building owner so that everyone knows his or her specific responsibilities during an emergency incident. Employees and the plant emergency organization should be instructed in the procedures for reporting a fire, making sure that employees are evacuated, checking to ensure that sprinkler control valves are open, and closing fire doors to prevent a fire's spread.

Strategic Fire-Fighting Considerations

The strategy employed in fighting structure fires is to confine the fire to the smallest possible area after arrival of the first fire department unit. In most cases, this means confining the fire to the room or compartment of origin. Even in large commercial buildings, fire fighters can usually depend on at least some degree of compartmentation to at least slow a fire's spread and intensity. When compartmentation works as intended, fires are "simple" room-and-contents fires.

Unfortunately, modern warehouses often lack such compartmentation. Warehouses with several hundred thousand sq ft (100,000 sq ft equals 9290 m^2) of floor area under a single roof with few, if any, true fire walls or fire barriers are not uncommon. Conveyor openings, drive-through doors, and holes for utilities to pass through often compromise existing fire walls and fire barriers. Fire-stopping or automatic closing fire doors may or may not properly protect these openings.

Sizing up a warehouse fire can, and most likely will, be very difficult. A serious fire may be burning in a large warehouse, yet very little smoke may be visible from outside. Smoke visible from the building's exterior may not indicate the location of the fire within. Because most warehouses have few if any windows, there may be no easy method of determining the fire's size, location, or current status. This makes having an up-to-date pre-incident plan even more important.

Sprinkler System Support

If the sprinkler system cannot control the fire, it is probably uncontrollable. Sprinklers begin fighting the fire before the fire department arrives, discharge water directly into the fire area, and are unaffected by poor visibility, heat, and toxic gases. The best course of action is to support the sprinkler system's operation by pumping into the fire department sprinkler connection at full capacity. The higher the pressure available to the sprinkler system, the more water discharged from the operating sprinklers, and the better the chance of controlling the fire. After support for the sprinklers has been established, the fire ground commander has time to carefully size up the situation to determine the best use of available resources. (See Section 9, Chapter 18, "General Indoor Storage," for sprinkler system requirements.)

Support of sprinklers means that the fire department must pump into the fire department sprinkler connections to ensure that proper pressures are maintained within the system. Steps must also be taken to ensure that water for manual fire suppression operations is not taken from the sprinklers or that apparatus connecting to yard hydrant systems do not rob water from sprinklers. In cases where water supplies were marginal, fire departments have taken water from the system, causing pressure within the sprinkler system to drop to critically low levels. Fire fighters should understand that, while a flow of 1,000 gpm (63 L/s) may be a high rate of flow for many fire types, the sprinkler system in a warehouse may require from 1,500 to over 3,000 gpm (95 to 190 L/s) to operate effectively.

Arriving fire companies may face a number of different situations that hinge mostly on the effectiveness of the automatic sprinkler system. Such situations include the following:

No sprinklers provided: If the warehouse has no automatic sprinklers and the fire is beyond the incipient stage, controlling it will be very difficult. Saving anything in the compartment of origin may be unlikely, and efforts should be directed toward preventing the fire's spread beyond the compartment currently involved. This can be accomplished by closing fire doors, using hose lines to prevent the fire's spread through unprotected openings in fire walls and fire barriers, and using master streams through doorways to contain the fire. Entry into the fire area can be extremely dangerous due to the speed with which the fire spreads and the probability of early structural collapse. A good water supply should be secured as soon as possible to supply the large-caliber hose streams required to control the fire and protect exposures.

Sprinklers provided but not operating: If sprinklers are provided but not yet operating when fire companies arrive, one of two situations has occurred:

1. It is possible that the fire has not yet grown large enough to fuse any sprinkler heads. In this case, hose lines can be advanced into the area to control the fire.

2. Sprinklers may not be operating because the sprinkler system is impaired. One or more valves may be closed for maintenance or repair. Also someone may have intentionally closed the sprinkler valves in an incendiary incident. Personnel should always be assigned to check that sprinkler control valves and water supplies are in service. At least one engine company should be assigned to support the sprinkler systems by pumping into the sprinkler connection, which may be arranged to bypass the closed valve.

Sprinklers operating and controlling the fire: If sprinklers are operating on arrival and appear to be controlling the fire, a first arriving engine company should be assigned to pump into the sprinkler connection. Increasing the pressure available to the sprinklers results in quicker, more effective, fire control. After the sprinklers have controlled the fire, there will be time to plan operations needed for final extinguishment, overhaul, and salvage.

Sprinklers operating but not controlling the fire: If the sprinklers are not controlling the fire on arrival, pressure in the sprinkler system must be increased immediately by connecting to and pumping into the sprinkler connection. There will be very little time to establish support for the sprinklers before the sprinkler systems are overwhelmed. If only one sprinkler connection is available, it should be used to maximum advantage by connecting multiple lines to it. Additional water can be supplied to the sprinkler systems by pumping into wall hydrants or private yard hydrants, if available.

Hazards Posed by Stock

Personnel at a warehouse fire must be familiar with specific hazards of stock in the warehouse. Materials, such as flammable liquids, flammable aerosols, and toxic chemicals, pose obvious hazards. Hazards of other materials are often overlooked. For example, baled materials, such as scrap paper or paper pulp, can be deadly. These materials absorb water from fire suppression operations and can collapse on fire fighters operating inside the building. Another hazard of materials as they absorb water is that they may expand when wet. If materials, such as paper pulp, are packed into a building against masonry walls, it is possible that, as the material expands, it may push walls outward, causing local or total collapse of the structure. When developing pre-incident plans, such possibilities must always be considered.

Materials stored in cardboard cartons also may collapse as the cartons become wet and weak. In fact, in some cases a pile collapse is considered advantageous from a fire control standpoint: it exposes fire in the pile's interior to discharge from the sprinkler system. Fire fighters should understand that fire protection engineers may want a pile in a warehouse to collapse to facilitate fire control.

An example of the hazards of collapsing stock occurred in Idaho where three members of an industrial fire department and two other employees died during a fire in a sprinklered paper warehouse. The noncombustible storage building was 46 ft × 72 ft (14 × 23 m) and fully protected by automatic sprinklers. Part of a paper mill, it stored baled scrap paper for recycling to a height of about 15 ft (4.6 m). An employee discovered a fire started by welding and grinding operations, and the plant fire department was notified. Arriving fire fighters found the building heavily charged with smoke and sprinklers operating at the ceiling. They entered the building with hose lines to find the seat of the fire. One crew of fire fighters was only 10 to 12 ft (3 to 3.7 m) into the building when bales of paper collapsed on them and trapped two fire fighters. Other fire fighters and employees outside the building rushed in to assist the trapped fire fighters. A second collapse trapped seven would-be rescuers. By the time the incident ended, two employees and three fire fighters were killed by the collapsing bales. All victims died of suffocation. Seven other fire brigade members were injured in the incident. The NFPA investigative report on the incident said, "This fire clearly illustrates the hazard to fire fighters of the collapse of storage materials which expand with the absorption of water. The hazard is particularly acute in sprinklered warehouses and must be considered in the pre-fire plans of fire departments that protect plants or storage facilities involving such commodities."[1]

Personnel must also be aware of the potential of stock falling from racks. In many cases, racks may be over 30 ft (9.1 m) high. Some warehouses have storage heights exceeding 100 ft (30.5 m). Burning stock falling from such heights could easily be fatal to anyone unfortunate enough to be working in the aisles. Regardless of the type of stock, the storage arrangement, or the design of the fire protection systems, fire fighters should always be alert to the possibility of collapsing stock and protect themselves accordingly.

Visibility

Visibility inside a warehouse during a fire incident is usually limited. Most warehouses are built with limited access and few windows. Natural lighting may not exist, resulting in poor visibility, especially if electric power to the building is cut as often happens when a fire occurs. When smoke fills the building, visibility is reduced even further.

Like a hose line discharged into an enclosed area, sprinklers operating in the fire area disrupt the thermal layers normally present in a fire. Contrary to what some believe, sprinklers do not "drive" smoke to the floor. As they operate, steam forms that tends to cool the atmosphere in the fire area. As every fire fighter knows, cool, damp smoke is difficult to vent. However, the automatic sprinklers should never be shut down in an attempt to improve visibility.

Ventilation

Ventilation is important in a warehouse fire, as in any other structure fire. Automatic smoke and heat vents or mechanical smoke exhaust systems may aid fire fighters. If no heat and smoke venting facilities are provided, ventilation may be accomplished by more traditional methods, such as opening the roof. Cross-ventilation may be accomplished by opening overhead doors or breaking eave-line windows, if they are present.

It is important that fire-fighting personnel not be committed to the roof of a steel-frame building unless the sprinklers have positively controlled the fire. If the fire is not under control, the lightweight steel structural members could quickly fail, causing the roof to collapse. Personnel should also never be committed to the roof of a wood-truss-supported structure. Wood-truss buildings can collapse quickly and without warning during a fire. Fire fighters have died in incidents when both wood-truss and steel-frame structures collapsed.

Charged hose lines should be available before venting because ventilation, although necessary to improve visibility, may increase fire intensity. As mentioned in the visibility discussion, properly operating sprinklers cause steam to form in the fire area, and smoke saturated with steam becomes difficult to move. Cross-ventilation may be the best method of clearing the building. Positive pressure ventilation may also be used.

Fire-Fighter Orientation

Warehouses are big places. Aisles and storage pile arrangements can become confusing. If fire-fighting personnel are not careful, they can easily become disoriented and lost in the building, especially when visibility is limited. Personnel operating inside a warehouse should always maintain contact with a hose line or a life line that they can follow outside to safety.

Logistics

The logistics of a warehouse fire-fighting operation must not be underestimated. Even under ideal circumstances, successfully fighting a fire requires large numbers of personnel and supplies. Physical demands on fire fighters due to the building's sheer size requires regular rotation of personnel out of the fire area for rest and rehabilitation. Pre-incident plans should contain provisions for assembling a large pool of trained personnel to assist in fire-fighting operations.

Other required supplies include air cylinders. Most self-contained breathing apparatus (SCBA) have only a 30-min rating and probably last only half that long during strenuous fire-fighting operations. Fire fighters who must walk 300 ft (91.5 m) into the building to the actual fire area may only be able to spend 5 to 7 min fighting the fire before they must replenish their air supply.

Material-handling equipment for moving stock during overhaul and salvage operations must also be available, as well as qualified personnel who can operate it under emergency conditions. This may require that the operators be trained to use self-contained breathing apparatus so they can assist with overhaul operations.

Fire Control Priorities

After ensuring life safety, the primary problem confronting the fire department on arrival at a warehouse fire is controlling the fire. Establishing control simply means preventing the fire from spreading any farther. Assuming that sprinklers are operating, water damage

should not be a concern until the fire has been extinguished. Anything that is going to get wet was probably wet before the fire department arrived.

If building owners or tenants complain about water flowing into the building, fire fighters may feel pressure to shut off the sprinklers to prevent further water damage. They must ignore such pressure.

Overhaul

As effectively as sprinklers control a warehouse fire, final extinguishment must be accomplished manually. This means that personnel must break open smoldering pallets, extinguish residual fire inside loads, and then remove them from the warehouse. If possible, personnel should accomplish this under the protection of operating sprinklers. At the very least, multiple-charged hose lines must be available in the area.

Overhauling a warehouse fire can present several challenges, especially if rack storage is involved. For example, how can fire fighters and hose lines reach the 30-ft (9.1-m) or the 80-ft (24.4-m) level in a high-rack storage area for final extinguishment and overhaul? The pre-fire plan must address such problems and should be practiced during on-site drills.

Rekindling

Fire may rekindle during overhaul operations when cartons of stock are broken open. If the sprinklers are shut down, someone with a radio should be stationed at the sprinkler control valve(s) to restore the system quickly to service if rekindling occurs.

In general, basic rules for shutting down the sprinkler system are:

- Do not shut the sprinklers off until fire extinguishment is positively determined. If there is any doubt, the sprinklers should remain in operation.

- Charged hose lines should be available in the fire area(s) before sprinklers are shut down.

- Station someone with a radio at the sprinkler control valve(s) to reopen the valve(s) if rekindling occurs.

- Restore the sprinkler system to full operation as soon as possible.

Limitations of Manual Fire Suppression

The best method of protecting a warehouse is to install a properly designed automatic sprinkler system and maintain it in proper working order. If sprinklers are not provided or they are not in service due to poor maintenance, the chances of saving a warehouse involved in a fire are very slim. Personnel should not be committed to dangerous positions in an attempt to control an uncontrollable fire. The best equipped fire department can never compensate for the lack of an effective sprinkler system.

SUMMARY

A severe warehouse fire is a very serious occurrence. Only properly designed automatic sprinkler systems can control these fires. After control has been established, the difficult task of overhauling and extinguishing the fire falls to the manual fire suppression forces of the plant emergency organization and the municipal fire department. The effectiveness of their actions largely depends on the thoroughness of pre-incident planning efforts conducted in cooperation with the warehouse owner.

BIBLIOGRAPHY

Reference Cited

1. Best, R. L., "Storage Collapse Kills Five," *Fire Command*, Nov. 1980.

NFPA Codes, Standards, and Recommended Practices

Reference to the following NFPA codes, standards, and recommended practices will provide further information on warehouse operations discussed in this chapter. (See the latest version of *The NFPA Catalog* for availability of current editions of the following documents.)

NFPA 10, *Standard for Portable Fire Extinguishers*

NFPA 11A, *Standard for Medium- and High-Expansion Foam Systems*

NFPA 13, *Standard for the Installation of Sprinkler Systems*

NFPA 13E, *Guide for Fire Department Operations in Properties Protected by Sprinkler and Standpipe Systems*

NFPA 14, *Standard for the Installation of Standpipe and Hose Systems*

NFPA 24, *Standard for the Installation of Private Fire Service Mains and Their Appurtenances*

NFPA 30, *Flammable and Combustible Liquids Code*

NFPA 30B, *Code for the Manufacture and Storage of Aerosol Products*

NFPA 51B, *Standard for Fire Prevention in Use of Cutting and Welding Processes*

NFPA 72, *National Fire Alarm Code*

NFPA 102, *Standard for Grandstands, Folding and Telescopic Seating, Tents, and Membrane Structures*

NFPA 204M, *Guide for Smoke and Heat Venting*

NFPA 231, *Standard for General Storage*

NFPA 231C, *Standard for Rack Storage of Materials*

NFPA 231D, *Standard for Storage of Rubber Tires*

NFPA 231F, *Standard for Storage of Roll Paper*

NFPA 307, *Standard for the Construction and Fire Protection of Marine Terminals, Piers, and Wharves*

NFPA 505, *Fire Safety Standard for Powered Industrial Trucks, Including Type Designations, Areas of Use, Conversions, Maintenance, and Operation*

NFPA 600, *Standard on Industrial Fire Brigades*

NFPA 601, *Standard for Security Services in Fire Loss Prevention*

NFPA 1420, *Recommended Practice for Pre-Incident Planning for Warehouse Occupancies*

Additional Reading

"Designing to Limit Loss in High-Rise Rack Warehouses," *Kemper Group Report*, Vol. 10, No. 3, Sept. 1981, pp. 2–9.

"Fire-Fighters Pour Cement on Blazing Chinese Warehouses," *Fire International*, No. 141, Nov. 1993, p. 9.

"Lesson of Gwent Blaze: Warehouse's Fire Safety 'Inadequate'," *Fire*, Vol. 84, No. 1036, Oct. 1991, p. 13.

"Uncontrollable Blaze Devastates Tooley Street Warehouses," *Fire Prevention*, No. 256, Jan./Feb. 1993, pp. 27–28.

"Warehouse Storage: Large-Scale Solutions for Large-Scale Problems," P8324, Factory Mutual Engineering Corp., Norwood, MA.

"Which Sprinkler Will Best Protect Your Warehouse?," *Record*, Vol. 68, No. 2, March/April 1991, pp. 3–10.

Ad Hoc Committee for Emergency Operations in High Rack Storage, *Emergency Operations in High Rack Storage*, 1982. Available from International Fire Service Training Association, Stillwater, OK.

Air Structures Institute, "Air Structures Design and Standards Manual," ASI-77, 1977 Air Structures Institute, St. Paul, MN.

Atkinson, G. T., and Jagger, S. F., "Assessment of Hazards from Warehouse Fires Involving Toxic Materials," Sixth International Fire Conference on Fire Safety, Interflam '93, March 30–April 1, 1993, Oxford, England, Interscience Communications Ltd., London, England, 1993, pp. 343–354.

Bacon, M. M., "Warehouse Fire Challenges Firefighters. On the Job: Rhode Island," *Firehouse*, Vol. 19, No. 2, Feb. 1994, pp. 46–48.

Barritt, J. S., "Fire Fighting Tactics for Racked Storage," *Fire Journal*, Vol. 67, No. 6, Nov. 1973.

Barritt, J. S., "Fire Protection and Today's Warehouses," *The Sentinel*, First Quarter 1988.

Before the Fire—Fire Prevention Strategies for Storage Occupancies, NFPA Ad Hoc Task Force, 1987.

Beitel, J. J., "National Quick Response Sprinkler Research Project: Group 1 Retrofit Test Series. [Group 1 Warehouse Retrofit Test Series.]," Southwest Research Inst., San Antonio, TX, National Fire Protection Research Foundation, Quincy, MA, Group 1, June 1990.

Best, R., "$100 Million Fire in K-Mart Distribution Center," *Fire Journal*, Vol. 77, No. 2, March 1983, pp. 36–42.

Best, R., "Firefighter Killed in Warehouse Fire," *Fire Command*, June 1980.

Brunacini, Alan V., *Fire Command*, National Fire Protection Association, Quincy, MA, 1985.

Corbett, G. P., "$100-Million Warehouse Fire," *Fire Engineering*, Vol. 138, No. 6, 1985, p. 36.

Davis, L., and Moore, J., *Industrial Fire Control Concepts*, International Society of Fire Service Instructors, Ashland, MA, 1988.

Deacon, F. C., "Designing Fire Protection to Limit Monetary Loss," SFPE TR 80-2, 1980, Society of Fire Protection Engineers, Boston, MA.

Dickinson, C., "Cold Storage Warehouse Turns Up the Heat," *Fire Engineering*, Vol. 143, No. 7, July 1990, pp. 32–34,38.

Field, P. and Murrell, J., "The Fire Hazard and Protection of Bin Storage," *Fire Surveyor*, Vol. 17, No. 6, Dec. 1988, pp. 5–15.

Fields, T., "Warehouses: A Case Study for Control," *Fire Prevention*, No. 215, Dec. 1988, pp. 24–29.

Gibson, R., "Warehouse Protection," *Fire Surveyor*, Vol. 21, No. 1, Feb.1992, pp. 5–8.

Goring, G., "Fire Protection in Low Temperature Warehouses," *Fire International*, No. 139, May 1993, pp. 47–48.

Grant, G., and Drysdale, D., "Application of the Cone Calorimeter to the Problem of Fire Growth in Warehouses," Sixth International Fire Conference on Fire Safety, Interflam '93, March 30–April 1, 1993, Oxford, England, Interscience Communications Ltd., London, England, 1993, pp. 539–548.

Grant, G., and Drysdale, D., "Numerical Modelling of Early Flame Spread in Warehouse Fires," *Fire Safety Journal*, Vol. 24, No. 3, 1995, pp. 247–278.

Hansell, G., "Problems of Effective Venting in Warehouses," *Fire Prevention*, No. 262, Sept. 1993, pp. 40–42.

Hoffmann, J. W., "One Jurisdiction's New Warehouse Codes," *Firehouse*, Vol. 18, No. 10, Oct. 1993, pp. 88–90.

Industrial Risk Insurers, "Commodity Classification Guide," IM.10.0.1, June 1992, Hartford, CT.

Industrial Risk Insurers, "Maximum Reliability for High Rack Storage Protection," IM.10.1.2.1, June 1994, Hartford, CT.

Industrial Risk Insurers, "Pre-Incident Planning for Fire Emergencies Involving Storage Racks at Heights of 25 ft (7.6 m) or Greater," IM.10.1.2.2, Sept. 1994; Hartford, CT.

Industrial Risk Insurers, *Overview*, IRI, Hartford, CT, 1986.

Isner, M. S., "$100 Million Fire Destroys Warehouses. Summary Investigation Report," *NFPA Journal*, Vol. 85, No. 6, Nov./Dec. 1991, pp. 37–41.

Isner, M. S., "$30 Million Fire Destroys Jet Fuel Storage Tanks," *NFPA Journal*, Vol. 86, No. 1, Jan./Feb. 1992, pp. 60–64,66.

Isner, M. S., "Cold Storage Warehouse Fire, Madison, Wisconsin, May 3, 1991. Summary Fire Investigation Report," National Fire Protection Assoc., Quincy, MA, Summary Fire Investigation, 1991.

Jenaway, W. E., *Pre-Emergency Planning*, International Society of Fire Service Instructors, Ashland, MA, 1986.

Jihu, Q., Kejun, X., and Changhao, H., "Retrospective Discussion on the Shenzhen Warehouse Explosion," *Proceedings* of the First International Conference on Fire Science and Engineering, ASIAFLAM '95, March 15–16, 1995, Kowloon, Hong Kong, 1995, pp. 69–75.

Johnston, D., "High-Rise Fire Protection at Monsanto's Warehouse," *Fire Prevention*, Vol. 229, May 1990, pp. 28–30.

Khan, M., "Evaluation of Fire Behavior of Packaging Materials," FMRC RC87-TP-7, presented at Defense Fire Protecton Association Symposium, 1987.

Klem, T. J., "Explosion in Cold Storage Kills Fire Fighter," *Fire Journal*, Vol. 79, No. 2, March 1985, p. 43.

Koffel, W. E., "Designing Warehouse Sprinkler Systems," *NFPA Journal*, Vol. 87, No. 4, July/Aug. 1993, pp. 16,90.

Kunkelmann, J., "Brandausbreitung bei Verschiedenen Stoffen, die in Lagermabiger Anordnung Gestapelt Sind. Teil 8: Simulation der Wechselwirkungen eines Tropfenschwarmes mit einer Heissgasstromung. [Flame Spread of Various Materials in Warehouses. Part 8. Simulation of the Interaction of Droplets and Hog Gas Streams.]," Karlsruhe Univ., West Germany, Report 80, Dec. 1991.

Kunkelmann, J., "Brandausbreitung bei Verschiedenen Stoffen, die in Lagermabiger Anordnung Gestapelt Sind. Teil 9: Simulation der Wechselwirkungen eines Sprinklers. [Flame Spread of Various Materials in Warehouses. Part 9. Simulation of Sprinkler Water Spray.]," Karlsruhe Univ., West Germany, Report 83, Dec. 1992.

Kunkelmann, J., "Brandausbreitung bei Verschiedenen Stoffen, die in Lagermassiger Anordnung Gestapelt Sind. Teil 10. Weiterfuhrende Literaturubersicht uber die Brandausbreitung Sowie uber die Wechselwirkungen des Tropfenschwarmes eines Sprinklers mit einer Heissgasstromung. [Flame Spread on Various Materials in Warehouses. Part 10. Additional Literature Survey about Flame Spread and the Interaction of Stream of Droplets from a Sprinkler and Hot Gases.]," Karlsruhe Univ., West Germany, Report 84, FA. Nr. 143, ISSN 0170-0060, June 1993.

Kunkelmann, J., "Brandausbreitung bei Verschiedenen Stoffen, die in Lagermassiger Anordnung Gestapelt Sind. Teil 11. Grossbrandversuche 6 - Brandausbreitung in Palettenlagern und Vergleich mit Gitterbox— und Blocklagerung. [Flame Spread of Various Materials in Warehouses. Part 11. Large Scale Fire Experiments 6—Flame Spread in Pallet Storge and Comparison with Large Containers and Block Storage.]," Karlsruhe Univ., West Germany, Report 88, FA. NR. 147, ISSN 0170-0060, Sept. 1994.

Lee, J. L., "Extinguishment of Rack Storage Fires of Corrugated Cartons Using Water," *Proceedings* of the First International Symposium on Fire Safety Science, Hemisphere, Washington DC, 1986, pp. 1177–1186.

Lee, J. L.,"The Effect of Paper Construction and Polyethylene Coating on the Flammability and Extinguishability of Roll Paper Products," FMRC JI ONOE9.RA, July 1986.

Linville, J. L., *ed.*, "Industrial Storage Practices," *Industrial Fire Hazards Handbook*, Third Edition, National Fire Protection Association, Quincy, MA., 1990.

Miles, S. D., *et. al.*, "Modelling the Environmental Consequences of Fires in Warehouses," *Proceedings* of the Fourth International Symposium on Fire Safety Science, Intl. Assoc. for Fire Safety Science, Boston, MA, 1994, pp. 1221–1232.

Moore, J., "Firefighter Safety During Warehouse Fire Fighting Operations," *The Voice*, April 1989.

Moore, J., "Understanding Warehouse Fire Problems," *Firehouse*, Vol. 17, No. 9, Sept. 1992, pp. 70,72–73.

Mulhaupt, R., "Fire Protection for Flammable and Combustible Liquids in Warehouses," First International Conference on Fire Suppression Research, May 5-8, 1992, Stockholm, Sweden, 1992, pp. 393–399.

Pedrotti, D. R., "Firefighters Down!," *Firehouse*, Jan. 1986.

Schatz, H., "Loscheinsatz bei gelagerten Stoffen. Teil 10. Literaturauswertung–Tropfenverteilungen–Loschversuche. [Fire Suppression of Materials in Warehouses. Part 10. Literature Search–Droplet Distribution–Extinction Experiments.]," Karlsruhe Univ., West Germany, Report 85, FA. NR. 144, ISSN 0170-0060, Aug. 1993.

Schatz, H., "Loscheinsatz bei gelagerten Stoffen. Teil 11. Literaturauswertung–Sprinklereinsatz bei Palettenlager. [Fire Suppression of Materials in Warehouses. Part 11. Literature Search–Sprinkler Activation for Materials on Pallets.]," Karlsruhe Univ., West Germany, Report 87, FA. NA 148, ISSN 0170-0060, July 1994.

Schatz, H., "Loscheinsatz bei Gelagerten Stoffen. Teil 8.Literaturauswertung Sprinkler. [Extinction in Warehouses. Part 8. Literature Search on Sprinklers.]," Karlsruhe Univ., West Germany, Report 78, Sept. 1991.

Schatz, H., "Loscheinsatz bei Gelagerten Stoffen. Teil 9. Messung und Simulation der Wasser–Beaufschlagung–Flussigkeitsverteilungern–Bestimmung von Tropfengroben. [Flame Spread of Materials in Storage. Part 9. Measurement and Simulation of Sprinklers - Liquids - Determination by Droplets.]," Karlsruhe Univ., West Germany, Report 82, Dec. 1992.

Scoones, K., "Serious Fires in Warehouses and Storage Premises During 1991," *Fire Prevention*, No. 252, Sept. 1992, pp. 22–24.

Scoones, K., "Serious Fires in Warehouses and Storage Premises During 1992," *Fire Prevention*, No. 262, Sept. 1993, pp. 29–31.

Tomes, W. J., and Golinveaux, J., "Fire Test Performance of Extra Large Orifice Sprinklers for Rack Storage of Group A Plastics in Warehouse-Type Retail Occupancies," *Proceedings* of the International Conference on Fire Research and Engineering, Sept. 10–15, 1995, Orlando, FL, SFPE, Boston, MA, 1995, pp. 415–420.

Troup, J. M. A., "Protection of Warehouse Retail Occupancies with Extra Large Orifice (ELO) Sprinklers," *Journal of Fire Protection Engineering*, Vol. 8, No. 1, 1996, pp. 1–12.

Troup, J. M. A., Tomes, W. J., and Golinveaux, J., "Full-Scale Fire Test Performance of Extra Large Orifice Sprinklers for Protection of Rack Storage of Group A Plastic Commodities in Warehouse-Type Retail Occupancies," Factory Mutual Research Corp., Norwood, MA, Tomes, Van Rickley and Associates, San Diego, CA, Central Sprinkler Co., Lansdale, PA, ISBN 0-9516320-9-4; *Proceedings* of the Seventh International Interflam Conference, Interflam '96, March 26–28, 1996, Cambridge, England, Interscience Communications Ltd., London, England, 1996, pp. 405–413.

Wackerlig, H., "Risk Assessment of Chemical Warehouses," *Fire Protection*, Vol. 17, No. 4, Dec. 1990, pp. 4–6.

Yung, D., and Mehaffey, J. R., "Fire Resistance Requirements for Rubber-Tire Warehouses," *Fire Technology*, Vol. 27, No. 2, May 1991, pp. 100–112.

GENERAL INDOOR STORAGE

Revised by Milosh Puchovsky

This chapter addresses fire protection for storage and warehousing operations. Due to the nature, quantity, and arrangement of inventory in these occupancies, fires can develop quite rapidly into an uncontrollable state. Developing and implementing a comprehensive fire protection plan are essential.

Modern storage practices aim to maximize available space by piling products high and leaving minimal aisle space between storage piles or racks. As a result, enormous amounts of material are often located in a single fire area or building. In addition, the contents of many warehouses and retail outlets contain or substantially consist of plastics. Plastics have the potential to burn rapidly, release heat, and generate large amounts of combustion products.

This chapter reviews basic fire protection principles primarily in the context of NFPA storage standards. It begins with a discussion of sprinkler systems and associated design factors unique to storage operations. Then it presents an overview of other fire protection features that a comprehensive fire protection program needs. The remaining sections highlight materials and storage occupancies that pose unusual fire protection challenges.

This chapter covers fire protection for a broad range of materials and storage arrangements but not the protection of materials that present an explosion-type hazard. For information on these subjects, refer to the chapters in this handbook that address aerosols, blasting agents, chemicals, exotic metals, fireworks, flammable and combustible liquids, and gases.

NFPA STORAGE STANDARDS

NFPA has developed several standards that present effective means for preparing a fire protection plan for various types of storage occupancies. These standards include NFPA 231, *Standard for General Storage* (hereinafter referred to as NFPA 231); NFPA 231C, *Standard for Rack Storage of Materials* (hereinafter referred to as NFPA 231C); NFPA 231D, *Standard for Storage of Rubber Tires* (hereinafter referred to as NFPA 231D); NFPA 231E, *Recommended Practice for the Storage of Baled Cotton* (hereinafter referred to as NFPA 231E); and NFPA 231F, *Standard for the Storage of Roll Paper* (hereinafter referred to as NFPA 231F). In addition, NFPA 13, *Standard for the Installation of Sprinkler Systems* (hereinafter referred to as NFPA 13), addresses fire protection for miscellaneous storage.

Milosh Puchovsky, P.E., is senior fire protection engineer in the NFPA general engineering division.

Miscellaneous storage typically does not exceed 12 ft (3.7 m) in height and is incidental to a building's main operation. Refer to the individual storage standards for a description of the types of miscellaneous storage that NFPA 13 covers.

NFPA storage standards are based on full-scale fire tests and past loss experience. The standards do not address storage arrangements where test data are not available or where conclusions could not be drawn from the extrapolation of available data, judgment, or experience. The reader is advised to check with the appropriate storage standard for specific information. For storage arrangements beyond the scope of a specific NFPA document, engineering judgment and other sources should be utilized. However, NFPA storage documents and this chapter outline basic fire protection concepts for storage operations and can provide guidance where specific standards are not available.

AUTOMATIC SPRINKLER PROTECTION

Overall, automatic sprinkler protection, supplemented by manual fire-fighting operations and sound storage and housekeeping practices, is a proven effective and practical means of fire protection. Properly designed, installed, and maintained systems will perform their intended tasks. However, a fire protection plan incorporating an inadequate, impaired, or partial sprinkler system cannot be expected to produce good results.

Recent advances in sprinkler technology provide three basic approaches for sprinkler protection of storage operations. These include the use of (1) spray sprinklers, (2) large drop sprinklers, and (3) early suppression fast response sprinklers (ESFRS). The spray sprinkler category typically includes standard orifice, large orifice, and extra large orifice (ELO) sprinklers. Either standard response or fast response sprinklers are permitted. Large drop sprinklers produce much larger water droplets that better penetrate a fire plume and reach the seat of the fire. The ESFR sprinkler, which incorporates a fast response operating element and a large orifice, discharges considerable amounts of water on a fire while it is still relatively small. Currently the ESFR sprinkler is the only sprinkler intended to extinguish a fire. The spray sprinkler and large drop sprinkler are intended to control a fire until the fire-fighting response team arrives. Refer to the chapters in Section 6 on sprinklers for more detailed discussions regarding various types of sprinklers available.

Because automatic sprinkler protection has proven to most effectively control and, in certain cases, extinguish fires in storage occupancies, the NFPA storage standards are comprised mainly of sprinkler system criteria.

A number of factors need consideration when preparing a fire protection plan, but particularly when selecting and designing a sprinkler system for a storage operation. The following sections summarize these factors, of which commodity classification and storage arrangement are most critical.

Commodity Classification

The first step in preparing or evaluating a fire protection plan is to categorize the materials being stored according to their burning characteristics. This can prove a difficult task because a wide variety of products can be found in storage commodities. NFPA 231 and NFPA 231C provide an approach that categorizes materials into one of seven major categories. These include Classes I through IV and Group A, B, and C plastics.

The commodity designation describes the nature of stored goods, including their packaging materials and storage aids such as pallets. Classifications are based on heat of combustion, rate of heat release, and rate of flame spread, as supported by full-scale fire test results. The standards describe and give examples of the types of products that constitute a specific classification. Because the examples are not intended as a comprehensive list, engineering judgment must be applied to properly classify a specific material that the standards do not identify.

Class I: These are essentially noncombustible products arranged on combustible pallets, in ordinary corrugated cartons with or without single-thickness dividers, or in ordinary paper wrappings with or without pallets. Such products may have negligible amounts of plastic trim, such as knobs or handles.

Class II: These are Class I products placed in slatted wood crates, solid wood boxes, multiple-thickness paperboard cartons, or equivalent combustible packaging material with or without pallets. In rack storage, only wooden pallets are contemplated.

Class III: These include wood, paper, natural fiber cloth, or Group C plastics with or without pallets. In rack storage, only wooden pallets are contemplated. The products may contain a limited amount of Group A or B plastics. (See the following section, "Plastics.") A metal bicycle with plastic handles, pedals, tires, and seat, or a wooden dresser with plastic drawer glides, handles, and trim are examples of a commodity with a limited amount of plastic.

Class IV: These commodities are Class I, II, or III products containing an appreciable amount of Group A plastics in corrugated cartons and Class I, II, and III products in ordinary corrugated cartons, with Group A plastic packaging with or without pallets. In rack storage, only wooden pallets are contemplated. This class also includes Group B plastics and free-flowing Group A plastics. An example of packing material is a foamed (expanded) plastic cocoon that holds a typewriter in an ordinary corrugated carton.

Plastics: Plastics present unique fire protection challenges because their combustion can produce about twice as much heat per unit of weight as cellulosic materials, i.e., wood, paper. In addition, plastics burn at a much faster rate. As a result, controlling or extinguishing a fire involving plastics is likely to be much more challenging.

Plastic materials are categorized into one of three groups: A, B, or C. Group A plastics present the most severe hazard, while Group C plastics present the least severe. Group B plastics behave similar to Class IV commodities, and Group C plastics behave similar to Class III commodities. Therefore Class IV and III commodities include Group B and C plastics, respectively.

The following list identifies a number of plastics according to their potential fire severity. The list contains unmodified plastic materials. The use of flame retarding modifiers or a different physical form of the material, e.g., shreds, can impact classification. Although fire-retardant modifiers and other additives inhibit ignition, once ignited, heat release is usually the same as for unmodified plastic. Full-scale testing is the best means to determine a modified material's classification.

Group A

ABS (acrylonitrile-butadiene-styrene copolymer)

Acrylic (polymethyl methacrylate)

Acetal (polyformaldehyde)

Butyl rubber

EPDM (ethylene-propylene rubber)

FRP (fiberglass-reinforced polyester)

Natural rubber (if expanded)

Nitrile rubber (acrylonitrile-butadiene rubber)

PET (thermoplastic polyester)

Polybutadiene

Polycarbonate

Polyester elastomer

Polyethylene

Polypropylene

Polystyrene

Polyurethane

PVC (polyvinyl chloride—highly plasticized, e.g., coated fabric, unsupported film)

SAN (styrene acrylonitrile)

SBR (styrene—butadiene rubber)

Group B

Cellulosics (cellulose acetate, cellulose acetate butyrate, ethyl cellulose)

Chloroprene rubber

Fluoroplastics (ECTFE—ethylene-chlorotrifluoroethylene copolymer, ETFE—ethylene-tetrafluoroethylene copolymer, FEP—fluorinated ethylene-propylene copolymer)

Natural rubber (not expanded)

Nylon (nylon 6, nylon 6/6)

Silicone rubber

Group C

Fluoroplastics (PCTFE—polychlorotrifluoroethylene, PTFE—polytetrafluoroethylene)

Melamine (melamine formaldehyde)

Phenolic

PVC (polyvinyl chloride—rigid or lightly plasticized, e.g., pipe, pipe fittings)

PVDC (polyvinylidene chloride)

PVF (polyvinyl fluoride)

PVDF (polyvinylidene fluoride)

Urea (urea formaldehyde)

In determining the classification of a commodity that consists of a variety of different materials, the amount of the most hazardous material and its effect on a fire must be considered. If the most hazardous material dominates the behavior of the burning commodity, then the product needs to be classified the same as the most hazardous material. For instance, if the Group A plastic packing material surrounding a Class IV commodity in a corrugated carton has the greatest influence on the fire, the whole commodity needs to be classified as a Group A plastic.

Currently, engineering judgment and experience must be employed to determine the classification of a commodity consisting of various materials. At the time of this writing, however, the Technical Committee on General Storage is working to establish a more formal approach to making this determination.

Idle pallets: Pallets used as storage aids are considered part of the commodity classification. Pallets are approximately 4 in. (102 mm) high and usually composed of wood or plastic, although some are metal or cardboard. Figure 9-18A shows a conventional pallet. The pallets' main purposes are to aid storage practices by allowing products to be moved easily by fork lift trucks and to provide stability. The commodity classification does not include idle pallets because of their unique combustion characteristics.

Stockpiling idle pallets indoors or outdoors presents a very severe fire hazard. Wooden pallets tend to dry out so their edges become frayed and splintered. In this condition, a relatively small ignition source can ignite them. Their dried-out condition, high rate of heat release, and overall arrangement result in a quickly developing fire. The undersides of the slats or planks, shielded from the discharge of automatic sprinklers, make controlling or extinguishing a fire difficult.

The fire hazard idle pallets present can be mitigated by strictly limiting storage arrangements and employing an appropriate sprinkler system, or by moving idle pallets outside and away from the building. Table 9-18A lists protection criteria found to be effective for pallets stored indoors. However, the provisions of Table 9-18A need not be applied if the idle pallets are:

1. No higher than 6 ft (1.8 m).
2. Stored in piles not exceeding four stacks.
3. In piles separated by a distance of 8 ft (2.4 m) from one another or by 25 ft (7.6 m) of commodity.

The storage of idle pallets in racks further increases the inherent fire risk. Therefore, it is strongly recommended that idle combustible pallets not be stored in racks.

Plastic pallets other than nonexpanded polyethylene solid deck pallets present an even greater fire hazard. Additional precautions must be taken. Refer to NFPA 231 for details.

When stored outdoors, all types of pallets should be located away from a building at the distances indicated in Table 9-18B.

Storage Arrangement

The manner of storing a commodity has a major impact on how a potential fire burns. Therefore, a commodity's storage arrangement

FIG. 9-18A. Conventional wooden pallet.

TABLE 9-18A. Protection for Indoor Storage of Wooden Idle Pallets or Nonexpanded Polyethylene Solid Deck Idle Pallets

Height of Pallet Storage, ft (m)	Sprinkler Density Requirements, gpm/sq ft [(L/s)/m²]	Area of Sprinkler Demand, sq ft (m²)	
		Sprinkler Temperature Rating	
		286°F (141°C)	165°F (74°C)
Up to 6 (1.8)	.20 (.14)	2000 (186)	3000 (279)
6 to 8 (1.8 to 2.4)	.30 (.20)	2500 (232)	4000 (557)
8 to 12 (2.4 to 3.7)	.60 (.41)	3500 (325)	6000 (557)
12 to 20 (3.7 to 6.1)	.60 (.41)	4500 (418)	—

TABLE 9-18B. Recommended Clearances Between Outside Idle Pallet Storage and Building

Wall Construction		Minimum Distance [ft (m)] of Wall from Storage of		
Wall Type	Openings	Under 50 Pallets	50 to 200 Pallets	Over 200 Pallets
Masonry	None	0	0	0
	Wired glass with outside sprinklers, 1-hr doors	0	10 (3.0)	20 (6.1)
	Wired or plain glass with outside sprinklers, ³/₄-hr doors	10 (3.0)	20 (6.1)	30 (9.1)
Wood or metal with outside sprinklers		10 (3.0)	20 (6.1)	30 (9.1)
Wood, metal, or other		20 (6.1)	30 (9.1)	50 (15.2)

must be considered when preparing the fire protection plan. Materials are most commonly arranged as bulk storage, solid pile, palletized pile, or rack storage. Bins and narrow shelves also are also used, more commonly in smaller stockrooms holding moderate quantities of product. Most warehouses combine these various storage arrangements.

The most critical factor among the various storage arrangements that affect fire behavior and the difficulty of controlling a fire is the presence of flue spaces. Storage configurations create these horizontal and vertical air spaces. Air passes through flue spaces, oxidizing the fire. Although flue spaces are not desirable, they should not be obstructed when present. Obstructed flue spaces prevent heat and combustion products from reaching a fire detector and also prevent water or other fire protection mediums from reaching the fire.

Bulk storage: Bulk storage consists of piles of loose, free-flowing materials, including powder, granules, pellets, or flakes, and agricultural items, such as peanuts. The materials are typically stored in silos, bins, tanks, or in large piles on the floor. Flue spaces are typically not present.

Fires in large piles tend to burrow down into the pile, making them difficult to extinguish. This type of fire requires prolonged soaking to reach the seat of the fire. Bulk storage arrangements are also subject to spontaneous ignition. Fires starting inside the piles are difficult to locate, unless heat sensors immersed in the pile continually monitor internal heating.

Material-handling equipment, such as belt conveyors, air-fluidizing ducts, and bucket conveyors ("legs") are often employed to transfer the material to and from its storage location. This moving process agitates and disturbs material. Airborne combustible material present an explosion hazard. This concern is significant, especially in grain storage facilities. In addition, conveyor belts and other components can be combustible and can burn together with the commodity in inaccessible places high above the floor or in tunnels. Automatic sprinklers are frequently needed in the housings around conveying equipment.

Solid piling: Solid piling consists of cartons, boxes, bales, bags, etc., in direct contact with each other. Air spaces, or flues, exist only where contact is imperfect, or where a pile is close to, but not touching, another pile. Because pallets are not typically used, stacking is done by hand or by lift trucks using side clamps or prongs, which are pried between packages or bales without damaging the product. (See Figure 9-18B.)

Relative to palletized pile and rack storage arrangements, solid piling gives fire the least opportunity to develop and spread and gives a sprinkler system the greatest opportunity to be effective, because flue spaces are minimal. Still, high piling presents a significant fire hazard, especially where the pile's surface consists of a material with rapid flame spread properties. Also, the higher the pile, the more difficult it is to break up and separate during a fire.

Palletized storage: Palletized storage consists of unit loads placed on pallets that are then stacked on top of one another. A pallet load usually takes the form of a cube, with dimensions of about 4 to 5 ft (1.2 to 1.5 m) in height, and consists of a single package or multiple packages. The top surface of the pallet load must adequately support other pallet loads so that the commodity is not crushed or so the pile does not become unstable. Due to these considerations, the maximum height of palletized storage usually does not exceed 30 ft (9.1 m).

Pallets contain horizontal spaces that allow the prongs of a fork lift truck to be inserted into the pallet to lift and move the pallet load. These horizontal spaces make this storage arrangement functional, but they increase the fire hazard. Air spaces usually continue in one direction along the entire width of a pile. Like vertical flue spaces, horizontal spaces let air easily pass to the fire. In addition their configuration lets a fire burn in the space while shielding it from the water from an overhead sprinkler system.

Rack storage: A storage rack is a structural framework in which a commodity is placed, usually as a pallet load. The design of rack storage systems maximizes vertical storage capability. The ceiling height or the vertical reach of material handling equipment limits storage heights. Some rack storage arrangements are over 100 ft (30 m) high.

The most common racking configurations are single-row and double-row racks. Single-row racks include racks up to 6 ft (1.8 m) wide, separated from other storage by at least 3.5-ft (1 m) aisles. Double-row racks consist of two single-row racks placed back-to-back, with a combined width up to 12 ft (3.7 m) and aisles at least 3.5 ft (1 m) wide on each side. (See Figures 9-18C and 9-18D.) Multiple-row racks, which utilize a flow-through or drive-in configuration, are also becoming more widely used. They consist of racks greater than 12 ft (3.7 m) wide, or single- or double-row racks separated by aisles less than 3.5 ft (1 m) wide whose overall width exceeds 12 ft. (3.7 m). (See Figures 9-18E and 9-18F.)

Other types of storage racks are also utilized. These include cantilever racks in which arms that extend horizontally from columns support the load. The load can rest on the arms or on shelves the arms support. (See Figure 9-18G.) Movable racks are connected to fixed rails or guides. They can be moved back and forth only on a horizontal, two-dimensional plane. A moving aisle is created as abutting racks are loaded or unloaded, then moved across the aisle to abut other racks. (See Figure 9-18H.) Portable racks are also utilized. They are not fixed in place and can be arranged in a number of configurations. (See Figure 9-18I.) A combination of rack and palletized storage can also be utilized. (See Figure 9-18J.)

Depending on the height and type of the rack structure, various types of material handling equipment can be utilized. These include manual and automated devices. (See Figures 9-18K through 9-18M.)

Rack structure inherently creates flue spaces within the storage arrangement. Racks also help to smother the fire and provide stability during a fire so that the burning commodity cannot collapse upon

FIG. 9-18B. *A type of solid piling in which spaces left between cartons accommodate lift-truck prongs. Wide aisles give the lift truck plenty of room to maneuver. Dark pipes at the ceiling are sprinkler lines.*

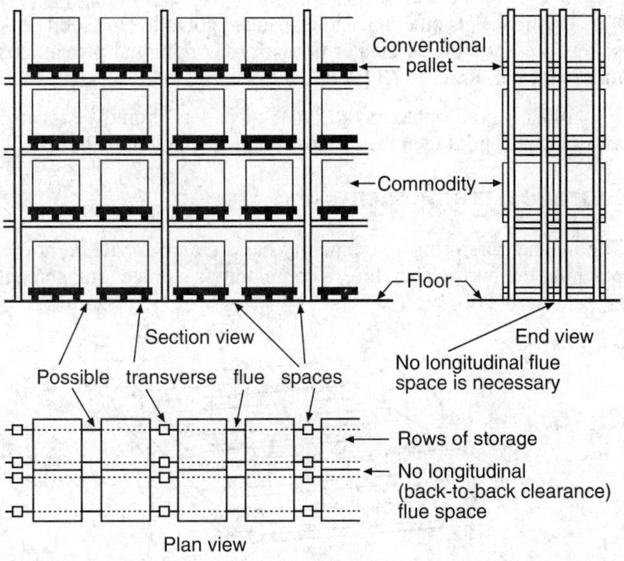

FIG. 9-18C. *Typical double-row rack with back-to-back loads.*

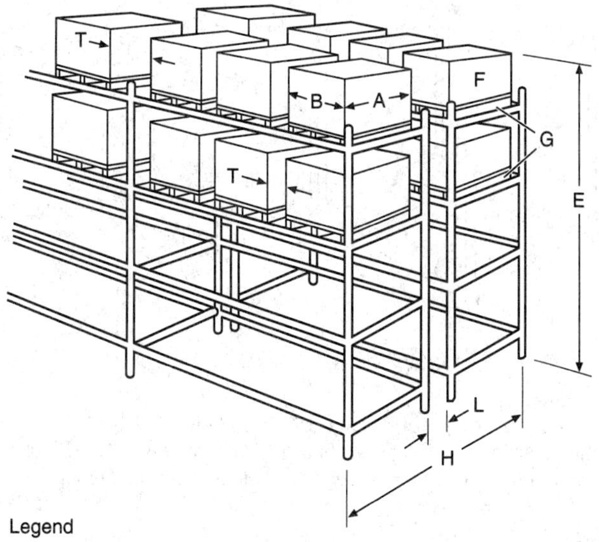

Legend
A = Load depth   G = Pallet
B = Load width   H = Rack depth
E = Storage height   L = Longitudinal flue space
F = Commodity   T = Transverse flue space

FIG. 9-18D. Double-row rack holding pallet loads.

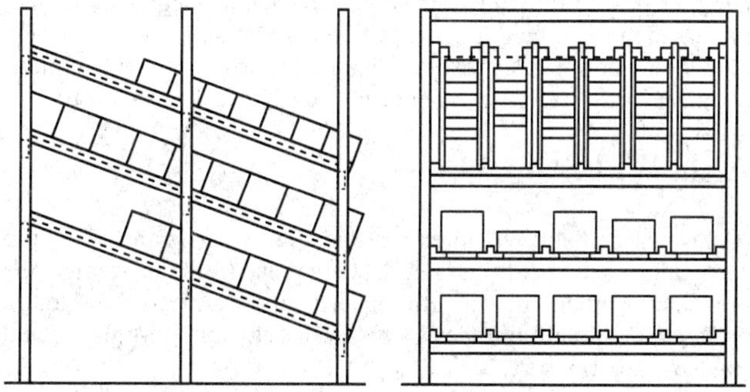

End View   Aisle View

FIG. 9-18E.   Multiple-row flow-through rack.

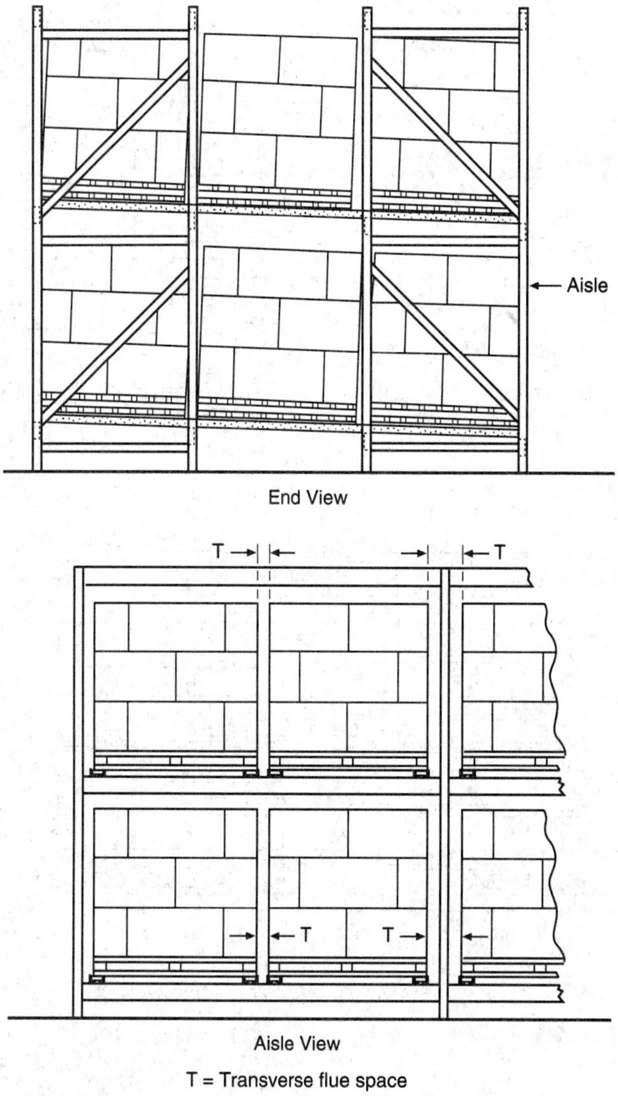

End View

Aisle View

T = Transverse flue space

FIG. 9-18F.   A multiple-row, drive-in storage rack, two or more pallets deep. Fork-lift trucks drive into the rack to deposit and withdraw loads in the depth of the rack.

itself. As a result, fires in rack storage configurations can be the most difficult to control and extinguish.

Fires in racks typically burn up through flue spaces while consuming products on the racks. However, flue spaces should be kept unobstructed because they let heat from the fire reach and activate sprinklers. Flue spaces also provide a means for sprinkler discharge to penetrate the rack structure, pre-wet the uninvolved commodity, and reach the fire.

Even if content damage is not severe in high-rack storage fires, heat from the fire makes the rack structure vulnerable to distortion and warping. This adversely affects material-handling equipment because it must be accurately aligned to function properly. Such damage could impair the entire warehouse operation.

Fortunately, storage racks are usually permanent structures that can support in-rack sprinklers. In-rack sprinklers significantly contain fire to a small, localized area by reducing the horizontal and vertical spread of fire. Without in-rack sprinklers, fires in high-rack configurations pose an impossible situation for a conventional

sprinkler system installed at the ceiling. (Refer to the discussion on in-rack sprinklers later in this chapter.)

Storage Height

After commodity classification and storage arrangement, probably no other factor more profoundly influences a fire and its controllability than storage height does. As expected, the higher the storage arrangement, the more difficult controlling the fire is. In-rack storage fire tests, fire intensity varied proportionally with storage height. For example, in a rack storage configuration without in-rack sprinklers, a 10 percent increase in storage height from 20 to 22 ft (6.1 to 6.7 m) required a 35 percent increase in the overhead sprinkler discharge density to effectively control the fire.

NFPA 231 and 231C provide sprinkler densities for the 20-ft (6.1 m) storage of certain commodities as indicated in Figure 9-18N. For other than 20 ft (6.1 m) storage heights, sprinkler densities need adjustment according to Figures 9-18O or 9-18P. These figures illustrate the effect of storage height on sprinkler density for

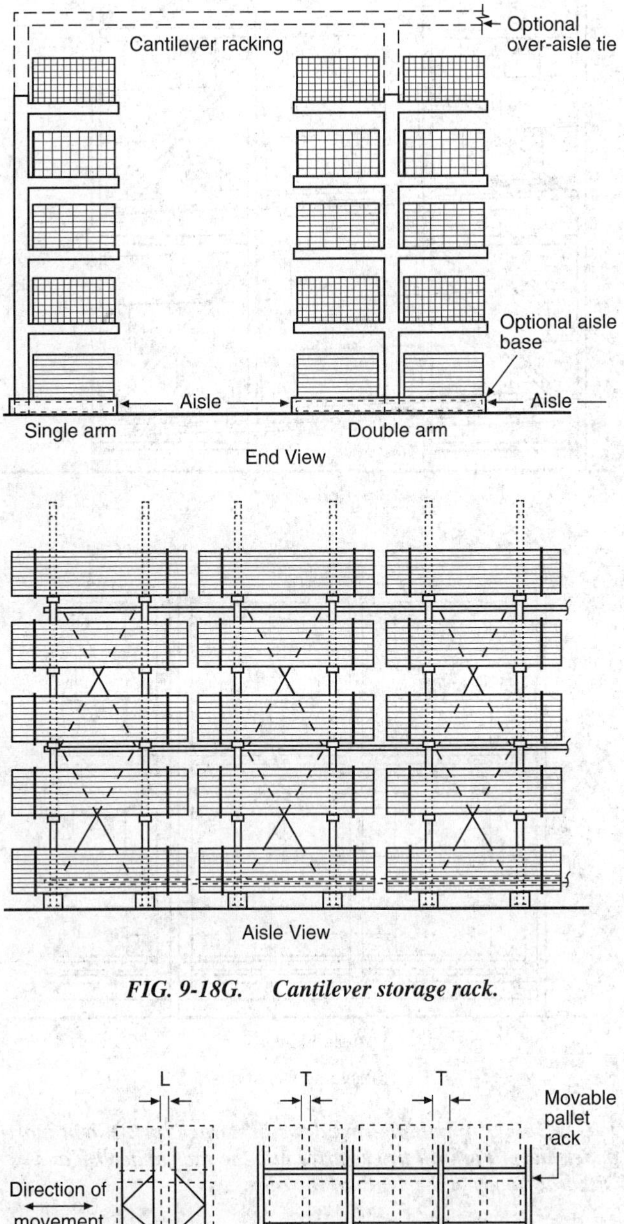

FIG. 9-18G.   Cantilever storage rack.

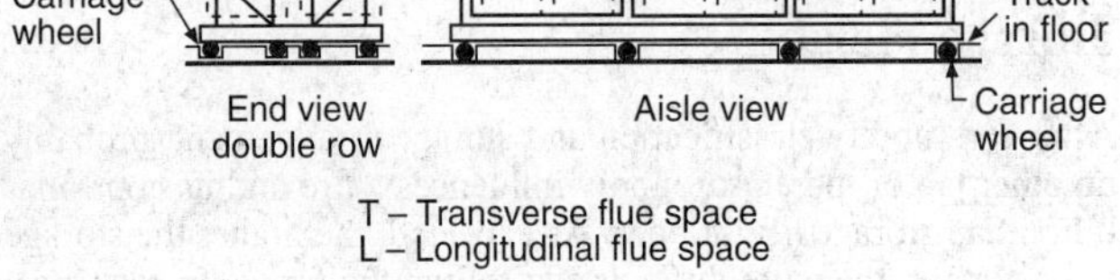

T – Transverse flue space
L – Longitudinal flue space

FIG. 9-18H.   Movable racks.

spray sprinklers. However, these relationships do not apply to most Group A plastics arrangements and where in-rack, large drop or ESFR sprinklers are used.

Although increasing storage height increases fire intensity, storage heights up to 12 ft (3.7 m) in piled storage and up to 15 ft (4.6 m) in rack storage need not be adjusted. For instance, palletized

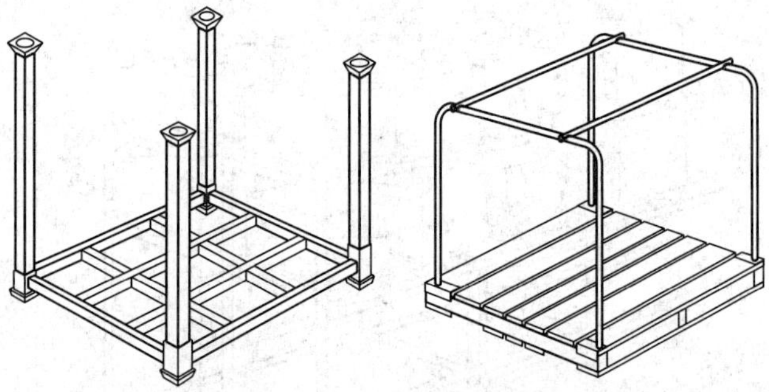

FIG. 9-18I.   Portable racks.

pile storage of Class I through IV commodities 12 ft (3.7 m) high require no more protection than 5-ft (1.5-m) high storage of the same commodity. In this situation, the density for the higher storage height is used, anticipating the eventual use of higher storage heights.

When designing a sprinkler system, using the maximum possible storage height in the building is advisable, even if the maximum height is not currently used. Although the lower storage height requires a less demanding sprinkler system, it limits storage height available in the building because the sprinklers cannot capably protect a higher storage arrangement. For example, in a 30-ft (9-m) high building, a ceiling sprinkler system designed to protect storage heights up to 15 ft (4.5 m) cannot be expected to adequately protect increased storage height of 25 ft (7.6 m). Although lower storage height may permit a less expensive initial sprinkler system, upgrading an existing sprinkler system can become much more costly.

Aisle Width

Aisles spaces between stored materials let water from the ceiling sprinklers reach the fire and mitigate its spread from one storage pile or rack to another. The wider the aisle, the more advantageous to fire protection. In addition, aisles provide access for fire-fighting and salvage operations.

For rack storage up to 25 ft (7.6 m) high, aisle widths must be considered when determining sprinkler system requirements. NFPA 231C provides sprinkler criteria for single- and double-row racks with 4-ft (1.2 m) and 8-ft (2.4 m) aisles as indicated in Figure 9-18Q. Interpolation is required for aisle widths between 4 ft (1.2 m) and 8 ft (2.4 m).

Racks incorporating 4-ft (1.2 m) aisles require greater sprinkler protection than those using 8-ft (2.4 m) aisles. Aisles wider than 8 ft (2.4 m) provide no more benefit than an 8-ft (2.4 m) wide aisle and therefore do not further reduce the sprinkler criteria. If aisles are narrower than 3.5 ft (1 m), then the rack arrangement is considered multiple row. Multiple-row racks demand a more powerful sprinkler system than double-row racks: the narrower aisles increase the likelihood of a fire spreading from rack to rack.

Aisles between palletized storage piles should not be more than 50 ft (15.2 m) wide. This lets streams from small hose lines penetrate 25 ft (7.6 m) into the center of the pile. Although NFPA 231 does not specially consider wide or frequent aisles, it assumes the presence of aisles. The aisles should be at least 8 ft (2.4 m) wide. This requirement is not unreasonable, considering that aisles are necessary for material-handling operations and 8 ft (2.4 m) are typically needed to maneuver a lift truck. When judging the adequacy of existing sprinkler protection, aisle spacing and frequency need consideration.

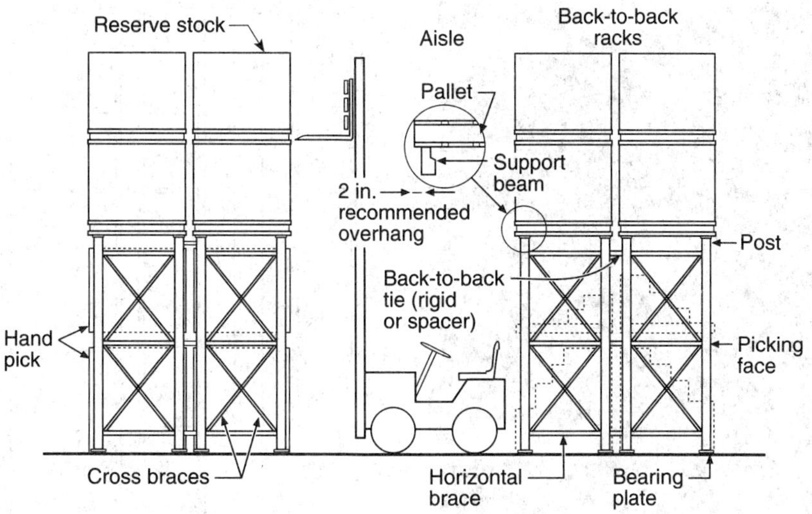

FIG. 9-18J. *Common arrangement of double-row storage racks with palletized storage atop. (Source: Unarco Materials Storage)*

In-Rack Sprinklers

Rack structures inherently create obstructions to rising heat and combustion products from a fire, and to water discharge from a ceiling sprinkler system. As a result, a fire in racks can become critical, overpowering most ceiling sprinkler systems. This phenomenon becomes more pronounced when storage heights increase and when both horizontal and structural flue spaces become obstructed. Sprinklers installed in racks alleviate this concern.

In-rack sprinklers work to contain a fire in a small, localized area. Once activated, they discharge water on or near the fire's source, without the delay associated with conventional ceiling sprinklers. In-rack sprinklers also pre-wet the area directly near the fire, thus reducing horizontal and vertical fire spread. In-rack sprinklers also minimize the potential water damage to stock because water is only applied to the localized area of the fire, usually eliminating the activation of the ceiling sprinkler system. Using in-rack sprinkler systems also reduces the demand on the ceiling sprinkler system and uses water supply more efficiently.

In-rack sprinklers are installed in the longitudinal flue spaces of the rack structure. As rack height increases, additional in-rack sprinklers may also be needed in the transverse flue spaces within 18 in. (203 mm) of the aisle. This type of in-rack sprinklers, referred to as "face sprinklers," protects against fire spread over the rack's vertical face and helps prevent a fire from jumping across the aisle to an adjacent rack. In some cases solid horizontal barriers covering the rack's entire length and width are also used with in-rack sprinklers. These barriers help activate of the in-rack sprinklers and reduce the number of in-rack sprinklers needed. Figure 9-18R illustrates these concepts.

Sprinklers installed in racks can be standard or fast response type with standard or large orifices. Ordinary-temperature rated sprinklers should be used. However, intermediate- and high-rated temperature sprinklers are required near heat sources. Water shields are required to guard against cooling or cold soldering of unactivated sprinklers by the water spray from in-rack sprinklers located higher in the rack. (See Figure 9-18S.)

The need for in-rack sprinklers depends on a number of factors, including the condition and type of ceiling sprinkler system, commodity type, and storage height. Generally when spray sprinklers are used, in-rack sprinklers are required for rack storage over 25 ft (7.6 m), regardless of the commodity. For storage up to 25 ft (7.6 m), the need for in-rack sprinklers varies. When ESFR sprinklers are used, in-rack sprinklers are not required. (See the discussion of sprinkler response time later in this chapter.)

NFPA 231C provides criteria to determine if in-rack sprinklers are necessary, their number, and arrangement. Various combinations of in-rack sprinkler installations have been found effective, so several design options are available.

Temperature Rating of Ceiling Sprinklers

In storage occupancies of low height, typically 12 ft (3.7 m) and under, ordinary-temperature rated [135 to 170°F (57 to 76°C)] sprinklers are generally used. Sprinklers with higher temperature ratings, i.e., intermediate [175 to 225°F (79 to 107°C)] and high [250 to 300°F (121 to 149°C)] ratings, are used in areas where ambient temperatures exceed 100°F (37.8°C). This reduces the likelihood of accidental sprinkler activation. Refer to the chapters in Section 6 on automatic sprinklers for discussions of temperature ratings of sprinklers.

In high-piled storage arrangements up to 30-ft (9.1 m) and in-rack storage up to 25-ft (7.6 m), high-temperature sprinklers can reduce sprinkler operating area by 40 percent, as compared with ordinary- or intermediate-temperature rated sprinklers. Due to the high heat release rates associated with the commodities stored, ordinary- and intermediate-rated sprinklers not within the vicinity of the fire will operate. However, their spray will not reach the fire, so that their effectiveness is limited. The operation of sprinklers that are not close to the fire robs water from sprinklers near the fire source. This increases the system water demand and area of operation. Using high-temperature rated sprinklers for these storage heights reduces the number of sprinklers operating unnecessarily beyond the actual fire area.

Somewhat surprisingly, the situation is reversed for rack storage over 25-ft (7.6 m). In-rack sprinklers operated before ceiling sprinklers, indicating that in racks over 25-ft (7.6 m) high, in-rack sprinklers convert what normally would be a rapidly developing fire to a more slowly developing fire with a reduced heat release rate. This result differed from tests involving storage 20-ft (6.1 m) high:

FIG. 9-18K. *Unusual storage racks with three pallets between uprights, hence three transverse flues. (Source: Clark Industrial Truck Division)*

Legend

A = Load depth	G = Pallet
B = Load width	L = Longitudinal flue space
E = Storage height	T = Transverse flue space
F = Commodity	

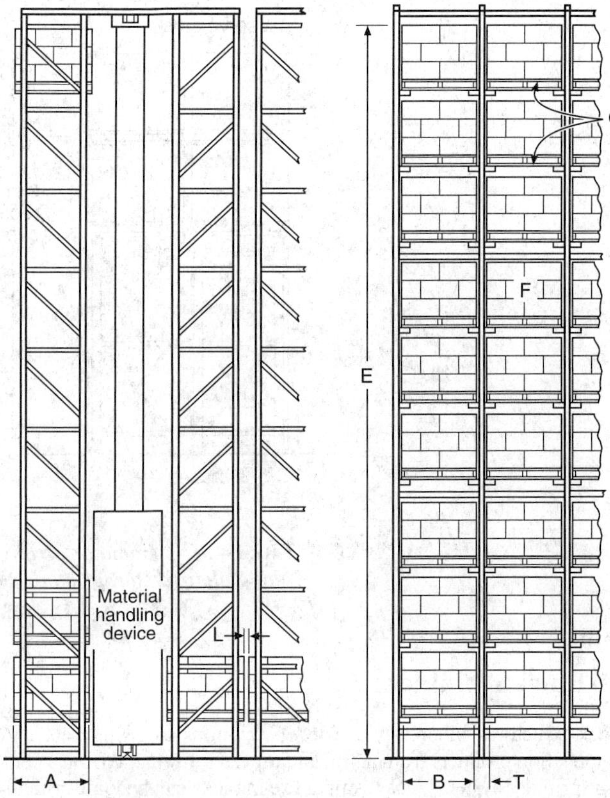

FIG. 9-18M. *Automatic storage-type rack.*

FIG. 9-18L. *53-ft (16-m) high storage racks with narrow aisle. Stacker machine rides on floor rail, guided from the top. (Source: Unarco Materials Storage)*

ceiling sprinklers operated before the in-rack sprinklers. Therefore, in rack storage over 25 ft (7.6 m) high, ordinary-temperature ceiling sprinklers are preferred because they require a distinctly lower density than high-temperature ceiling sprinklers. For example, for Class IV commodities, a density of 0.35 gpm/sq ft [14.3 (L/min)/m²] is required for ordinary-temperature sprinklers *vs.* 0.45 gpm/sq ft [18.3 (L/min)/m²] for the high-temperature type. (See Table 9-18C.)

Clearance below Ceiling Sprinklers

The minimum clearance of 18 in. (0.46 m) required below ceiling sprinkler deflectors allows the proper spray pattern to develop. When ESFR or large drop sprinklers are used, the minimum clearance required is 36 in (1 m). However, there is a limit to how much clearance is desirable over high-piled or rack storage, especially for commodities that produce a high-heat release rate. As clearance increases, the ceiling sprinklers' effectiveness changes. The greater the clearance, the more likely that water droplet size will decrease. This allows droplets to remain suspended in the air or be carried away laterally by the updraft of the fire plume. Thus sprinkler discharge does not reach the fire.

There are two conflicting variables: (1) under a given roof height, higher storage creates more severe fire potential, while (2) lower storage creates greater difficulty for sprinklers if the clearance exceeds $4\frac{1}{2}$ ft (1.37 m). These two variables may not neutralize each other, so the designer needs to explore the worst condition where storage heights vary, including the effect of a sloping roof. NFPA storage standards adjust sprinkler design criteria according to

TABLE 9-18C. *Storage over 25 ft (7.6 m) High*

Commodity Class	Encapsulated	Double-Row Racks without Solid Shelves, Aisles Wider Than 4 ft (1.2 m)						Multiple-Row Racks					
		Ceiling Sprinklers Operating Area		Ceiling Sprinkler Density				Ceiling Sprinklers Operating Area		Ceiling Sprinkler Density			
				gpm/sq ft		(L/min)/m²				gpm/sq ft		(L/min)/m²	
		sq ft	m²	165°F	286°F	74°C	141°C	sq ft	m²	165°F	286°F	74°C	141°C
I	No	2000	186	0.25	0.35	10.2	14.3	2000	186	0.25	0.35	10.2	14.3
	Yes	2000	186	0.25	0.35	10.2	14.3	2000	186	0.31	0.44	12.6	17.9
I, II, III*	No	2000	186	0.30	0.40	12.2	16.3	2000	186	0.30	0.40	12.2	16.3
	Yes	2000	186	0.30	0.40	12.2	16.3	2000	186	0.37	0.50	15.0	20.4
I, II, III, IV**	No	2000	186	0.35	0.45	14.3	18.3	2000	186	0.35	0.45	14.3	18.3
	Yes	2000	186	0.35	0.45	14.3	18.3	2000	186	0.44	0.56	17.9	22.8

*Good in-rack sprinkler arrangement.
**Not as good in-rack sprinkler arrangement.

FIG. 9-18N. Sprinkler system design curves, 20-ft (6.1-m) high storage—ordinary-temperature rated sprinklers.

clearance. In general, test data and protection criteria are not available for clearances over about 10 ft (3 m).

Dry-Pipe *vs.* Wet-Pipe Sprinkler Systems

In a wet-pipe sprinkler system, the piping attached to the automatic sprinklers contains water under pressure at all times. In a dry-pipe system, the piping contains air or nitrogen under pressure. When a sprinkler operates, air pressure is reduced, a dry-pipe valve opens, water enters the piping, and after a short delay flows out the open sprinklers. Dry-pipe systems should only be used in areas subject to freezing temperatures. Refer to the chapters in Section 6 of this handbook for discussions of various types of sprinkler systems.

Where dry-pipe systems are used, their time delay lets a fire grow and spread heat that activates more sprinklers beyond the immediate fire area. To compensate, the required design area of sprinkler operation usually increases by 30 percent for dry-pipe systems.

Density should be selected so that, after the 30 percent increase, the design area does not exceed the maximum allowable design area of sprinkler operation.

Pile Stability

Unstable piles often are undesirable. Piles collapse into the aisles, providing a bridge for fire spread, impeding fire-fighting operations, and adversely impacting structural building elements such as columns and walls. However, tests show that, in some cases, sprinklers are more effective after piles collapse, contents spill, or stacks lean across flue spaces. Evidently in these situations, a collapse chokes the upward flow of fire in flue spaces and exposes otherwise concealed fire to water from sprinklers.

Under fire conditions, anticipating pile stability is difficult, but some guidelines are available. Compartmented cartons containing stiff cardboard dividers were found stable under fire test conditions, while noncompartmented cartons tended to be unstable. Examples of stable storage include storage on pallets within height limits suitable for the commodity to withstand compressive forces, and items held in place by materials that do not readily deform under fire conditions. Leaning stacks, crushed bottom cartons, or reliance on combustible bands for stability are predictors of pile instability under fire conditions. Racks, in particular, should be adequately cross braced, anchored to the floors, and tightly connected because rack collapses tend to cause chain reactions.

Encapsulation

Encapsulation is a method of packaging. A plastic sheet completely encloses the sides and top of a pallet load containing a combustible commodity or a group of combustible commodities. Combustible commodities individually wrapped in plastic sheeting and stored exposed in a pallet load are also considered encapsulated. If holes or voids in the plastic or waterproof cover on top of the carton exceed more than half the cover's area, the packaging is not considered encapsulated. Removing the plastic covering the top converts an encapsulated pallet load to an unencapsulated load.

The plastic covering prevents water from reaching a fire that penetrated a pallet load from below, a likely occurrence in rack

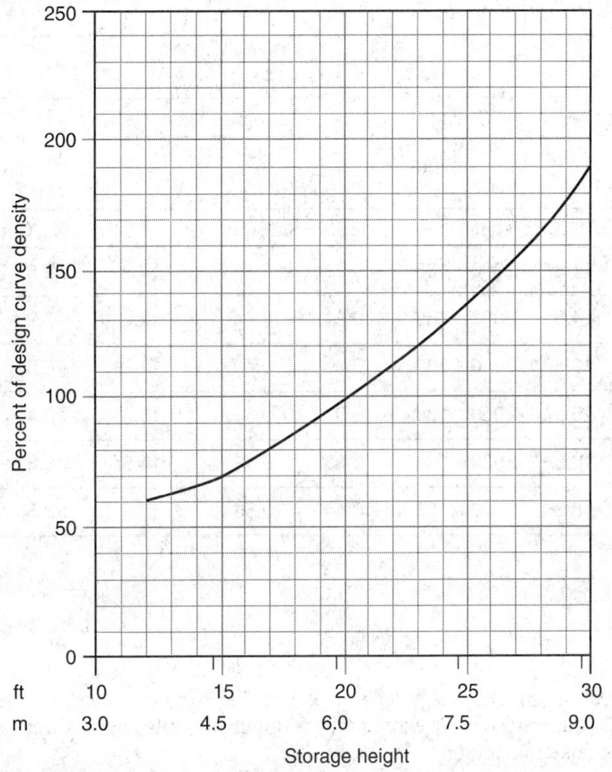

FIG. 9-18O. *Ceiling sprinkler density vs. storage height for piled storage.*

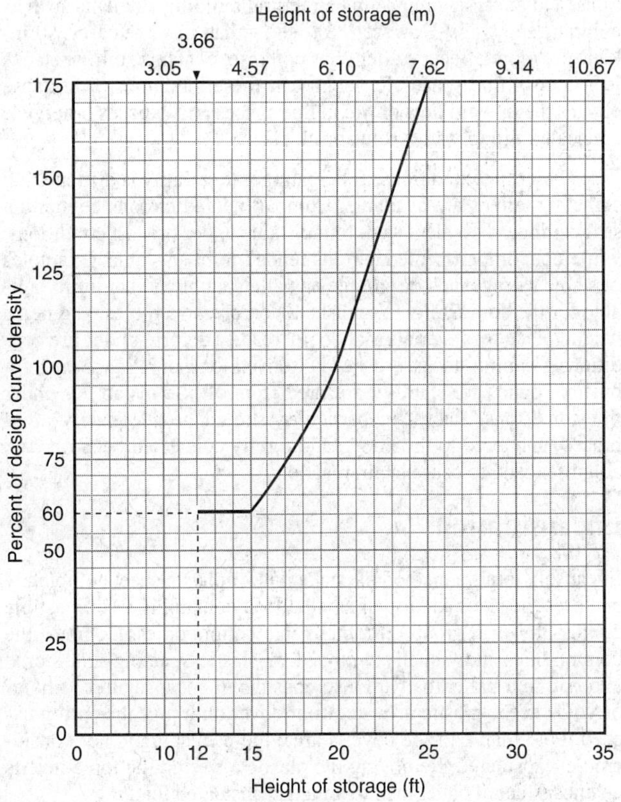

FIG. 9-18P. *Ceiling sprinkler density vs. storage height for rack storage.*

storage. In addition, encapsulation prevents pre-wetting of materials that are not burning. Tests show that, with encapsulation, a higher ceiling sprinkler density is needed to control the fire for rack storage at all heights, even with in-rack sprinklers.

Full-scale tests of piled storage show no appreciable difference in the number of sprinklers that operate for either unencapsulated or encapsulated products piled up to 15 ft (4.6 m) high. Test data are not available for encapsulated products stored higher than 15 ft (4.6 m). No difference in the ceiling sprinkler density is required for encapsulated or unencapsulated commodities in nonrack storage.

Smoke Venting

Smoke removal is important to manual fire-fighting and overhaul operations. Ideally, ventilation activities should not begin until automatic sprinkler operation reduces temperatures in the building to a safe level. However, this guideline becomes difficult to follow: many fire fighters are uncomfortable with waiting for the building to cool. Hence, they will cut ventilating holes in the roof and enter the building to direct hose streams at the fire.

The benefit of automatic roof vents in sprinklered storage occupancies is a topic much debated among fire protection professionals. Automatic roof vents open and let smoke and hot gases escape from the building, especially near the fire's source. They are quite effective in unsprinklered buildings. NFPA 204M, *Guide for Smoke and Heat Venting*, provides guidance on this subject.

When roof vents operate before sprinkler activation, they let heat escape. This could delay the sprinkler system operations and result in failure to control the fire. On the other hand, the operation of automatic sprinklers prior to the roof vents opening can interrupt the vents' activation and the stack effect that lets hot smoke escape. Also, the initial effect of sprinkler operation is often to reduce the smoke layer's height, obscuring vision within the building.

Based on fire tests that did not use smoke venting, NFPA storage standards do not require smoke removal equipment. However, the standards indicate that venting through eave-line windows, doors, monitors, gravity vents, or mechanical exhaust systems is essential for smoke removal after the fire is controlled.

High-Expansion Foam

While automatic high-expansion foam extinguishing systems can independently suppress a fire, some are reluctant to use them as the sole means of automatic fire control in certain cases. They are usually more expensive and more complicated than sprinkler systems; do not protect the roof structure until foam reaches it; involve the entire contents of a protected area regardless of the fire's size; and leave foam residue on unburned materials. However, they can be desirable when used with automatic sprinklers for certain high-challenge storage occupancies.

A high-expansion foam system uses a series of foam-makers at ceiling level. When activated by the fire detection system throughout the area, the foam system sprays water mixed with special foam concentrate in each foam-maker and strikes a screen while a fan blows through it. The system produces many uniformly sized bubbles, which cascade down into and gradually fill the warehouse area. The foam's expansion ratio is as high as 1,000:1. It flashes to steam on contact with burning material and engulfs other material, keeping it from burning. Building doors should automatically close when the system activates. NFPA 11A, *Standard for Medium- and High-Expansion Foam Systems*, contains requirements for installing these systems.

When high-expansion foam systems combine with ceiling sprinkler systems, the ceiling sprinkler discharge density may be re-

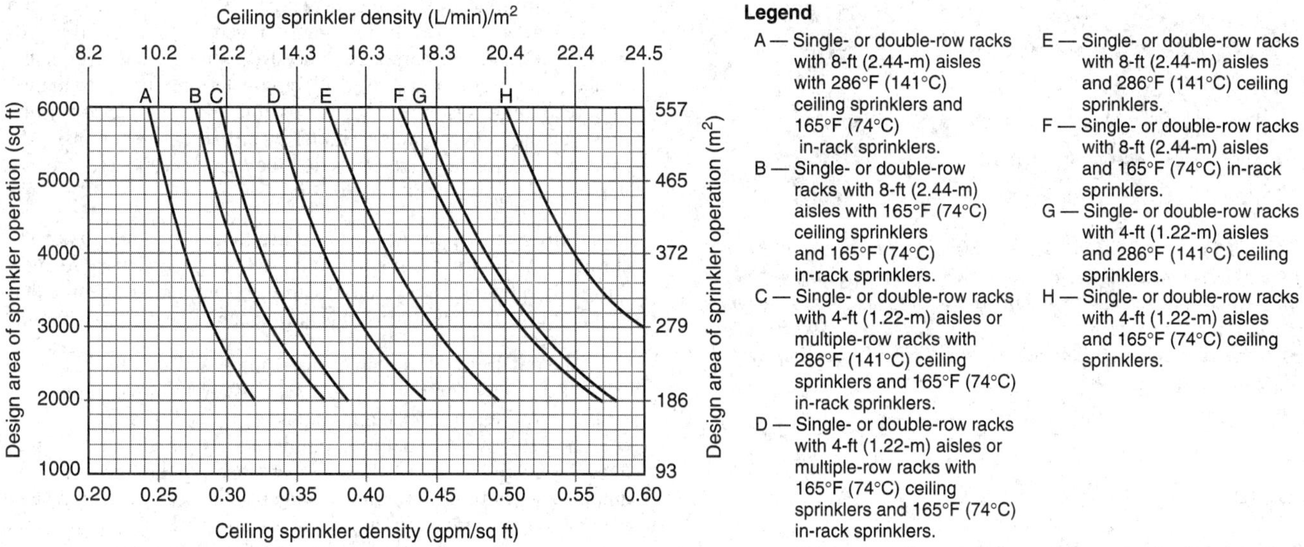
Ceiling sprinkler density (L/min)/m²
8.2 10.2 12.2 14.3 16.3 18.3 20.4 22.4 24.5
A B C D E F G H
Design area of sprinkler operation (sq ft)
6000 5000 4000 3000 2000 1000
557 465 372 279 186 93
Design area of sprinkler operation (m²)
0.20 0.25 0.30 0.35 0.40 0.45 0.50 0.55 0.60
Ceiling sprinkler density (gpm/sq ft)
Legend
A — Single- or double-row racks with 8-ft (2.44-m) aisles with 286°F (141°C) ceiling sprinklers and 165°F (74°C) in-rack sprinklers.
B — Single- or double-row racks with 8-ft (2.44-m) aisles with 165°F (74°C) ceiling sprinklers and 165°F (74°C) in-rack sprinklers.
C — Single- or double-row racks with 4-ft (1.22-m) aisles or multiple-row racks with 286°F (141°C) ceiling sprinklers and 165°F (74°C) in-rack sprinklers.
D — Single- or double-row racks with 4-ft (1.22-m) aisles or multiple-row racks with 165°F (74°C) ceiling sprinklers and 165°F (74°C) in-rack sprinklers.
E — Single- or double-row racks with 8-ft (2.44-m) aisles and 286°F (141°C) ceiling sprinklers.
F — Single- or double-row racks with 8-ft (2.44-m) aisles and 165°F (74°C) in-rack sprinklers.
G — Single- or double-row racks with 4-ft (1.22-m) aisles and 286°F (141°C) ceiling sprinklers.
H — Single- or double-row racks with 4-ft (1.22-m) aisles and 165°F (74°C) ceiling sprinklers.

FIG. 9-18Q. Sprinkler system design curves for 20-ft (6.1-m) high rack storage of Class IV unencapsulated commodities using conventional pallets.

duced to one-half the density otherwise required. However, the density cannot be less than:

1. 0.15 gpm/sq ft [6.1 (L/min)/m^2] for Class I through IV solid-piled or palletized commodities, including idle pallets or plastics.
2. 0.24 gpm/sq ft [9.8 (L/min)/m^2] for rubber tire storage.
3. 0.25 gpm/sq ft [10.2 (L/min)/m^2] for roll paper storage.

In-rack sprinklers are not required for rack storage when high-expansion foam is installed. In addition, ceiling density can be reduced to 0.2 gpm/sq ft [8.1 (L/min)/m^2] for Class I, II, and III commodities, and to 0.25 gpm/sq ft [10.2 (L/min)/m^2] for Class IV commodities.

Sprinkler Orifice Size

With the exception of rolled paper storage, which requires $^{17}/_{32}$-in. (13.5 mm) orifice sprinklers, the minimum orifice size for sprinklers used in storage occupancies is $^1/_2$ in. (12.7 mm). However, larger orifice sizes are permitted in certain instances and in many cases are more desirable. Refer to the NFPA storage standards previously cited.

Through hydraulic theory, any required sprinkler discharge density can be achieved with any size automatic sprinkler by coordinating sprinkler spacing and available pressure. Sprinklers with larger orifices require less pressure to produce the required discharge densities. This is especially beneficial where the available water supply cannot meet the demand of a system using $^1/_2$-in. (12.7 mm) orifice sprinklers. Conversely, if the same pressure applied to a $^1/_2$-in. (12.7 mm) orifice sprinkler is also applied to a $^{17}/_{32}$-in. (13.5 mm) orifice sprinkler, the sprinkler will discharge more water with a more penetrating water spray. Refer to Section 6, Chapter 6, "Hydraulics," for additional information.

Large drop sprinklers are also permitted. Their design is not based upon the area/density method discussed above. They require a certain operating pressure depending on the height, arrangement, and type of commodity stored. In some cases in-rack sprinklers are also required. (See Table 9-18D.) Overall, large drop sprinklers generate much larger water droplets than standard spray sprinklers. This larger drop and the momentum associated with its discharge make it much more likely for water spray to penetrate fire plumes and reach the seat of the fire.

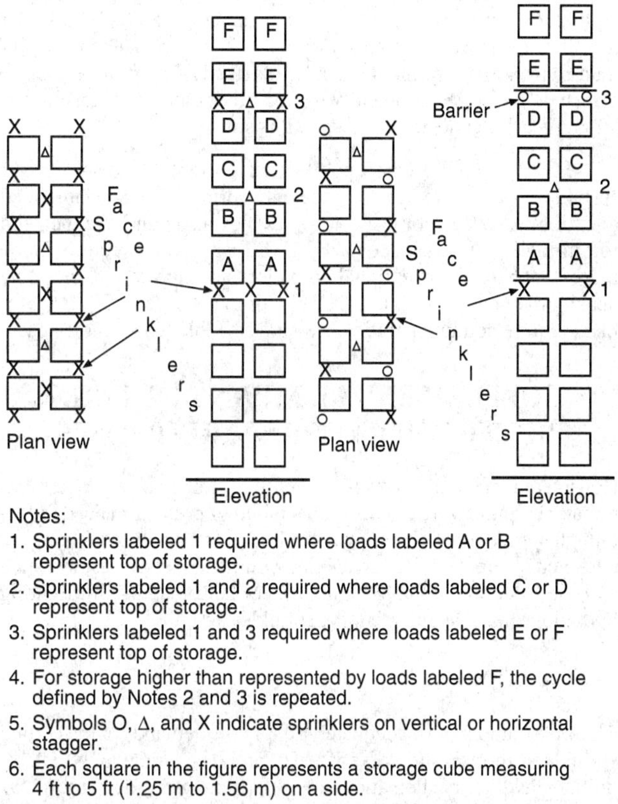
F F
E E
3
D D
C C
2
B B
A A
1
Face sprinklers
Plan view
Elevation
Barrier
Plan view
Elevation
Notes:
1. Sprinklers labeled 1 required where loads labeled A or B represent top of storage.
2. Sprinklers labeled 1 and 2 required where loads labeled C or D represent top of storage.
3. Sprinklers labeled 1 and 3 required where loads labeled E or F represent top of storage.
4. For storage higher than represented by loads labeled F, the cycle defined by Notes 2 and 3 is repeated.
5. Symbols O, Δ, and X indicate sprinklers on vertical or horizontal stagger.
6. Each square in the figure represents a storage cube measuring 4 ft to 5 ft (1.25 m to 1.56 m) on a side.

FIG. 9-18R. In-rack sprinkler arrangement for Class I, II, or III commodities, with storage heights over 25 ft (7.6 m). Note how horizontal barriers within the racks (right) reduces the number of in-rack sprinklers required (left). The symbols O, Δ, or X indicate sprinklers installed in a vertical or horizontal alternating or "staggered" arrangement on the sprinkler piping.

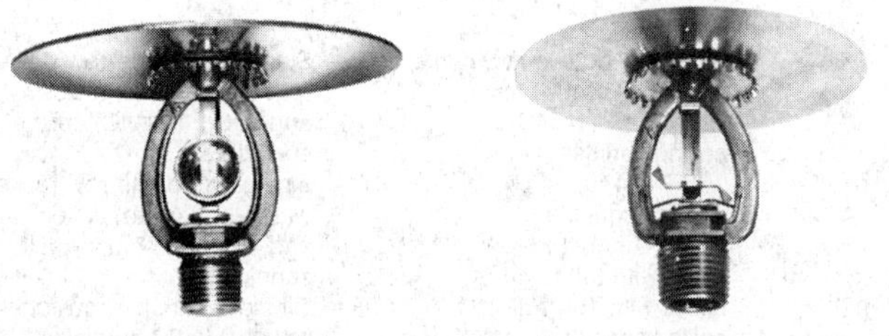

FIG. 9-18S.   Typical in-rack sprinklers: upright standard response (left) and upright fast response (right). Pendant sprinklers can be field-equipped with water shields. The fast-response type is appropriate for protection under open-girded catwalks between racks, shelves, or bins. (Source: Grinnell Fire Protection Systems Co., Inc.)

Sprinkler Response Time

Standard response or quick response sprinklers in ceiling or in-rack systems can protect storage occupancies. (Refer to the chapters on sprinklers in Section 6 for a discussion of various types of fast response sprinklers, which include quick response and ESFR sprinklers.) However, no test data indicate that quick response sprinklers are more effective than standard response sprinklers for storage occupancies. Therefore using quick response sprinklers does not decrease the sprinkler discharge density required or area of operation.

The development of ESFR sprinklers was a major advance for sprinkler technology for storage occupancies. The ESFR sprinkler has a fast response operating element that lets it operate much faster than conventional sprinklers. In addition, the ESFR sprinkler's large orifice delivers greater quantity of water to the fire. As a result, water can reach the fire quickly while it is still small.

ESFR sprinklers can suppress certain fires. As a result, ESFR sprinkler systems are currently the only sprinkler systems intended to extinguish a fire, not just control it. They also eliminate the need for in-rack sprinklers. However, ESFR sprinklers require much more water than conventional sprinklers and do not use the area/density method for design. Like large drop sprinklers, ESFR sprinklers require certain operating pressure as Table 9-18E indicates.

SUPPLEMENTAL MEANS OF FIRE PROTECTION FOR WAREHOUSES

Considered the main line of fire defense in storage occupancies, automatic sprinklers are not the only method of protection needed. Although sprinklers often extinguish a fire, most sprinkler systems are designed to control a fire for a period of time. Other provisions, such as hose systems, portable extinguishers, and manual fire-fighting operations, must supplement the sprinkler system.

Water Supply

The water supply must adequately supply the demands of the ceiling sprinklers, in-rack sprinklers if used, and hose streams. NFPA 13 outlines hydraulic calculations for determining the demand of the sprinkler system, including in-rack sprinklers. Sprinkler demand is expressed as a flow rate, such as gpm (L/min), at a corresponding minimum residual pressure. Typically 500 gpm (1894 L/min) is added to the sprinkler demand for large and small hose streams. Hose streams are intended to operate concurrently with the sprinkler system, so the residual pressure demand for the sprinklers is typically used.

The sprinkler system piping design, i.e., pipe size, fittings, and material, affects the residual pressure. In general, at least 30- to 50-psi (207- to 345-kPa) pressure is needed at the ceiling level distribution point or top of the riser. When greater pressure is available, more economic system design is possible. However, exposing the most remote conventional sprinkler to pressures over 60 psi (414 kPa) is undesirable. High pressures may create very fine, but not very effective, spray patterns.

The water supply needs to be available long enough for the sprinklers to control the fire, manual fire fighting, and mop-up operations. Two hours are typically required. However, if a central station monitors the fire alarms, this period can be reduced to 90 min.

The reliability of available water supplies also needs consideration, especially for high-value locations. A multiple supply arrangement, i.e., a fire pump connected to a suction tank in conjunction with an adequate public supply, might be advantageous. Such redundancy is based on engineering judgment. For example, a $100-million warehouse may be connected to a single city water main. Although the supply may be adequate, it's possible that supply could be out of service for some time or deteriorate.

The layout of yard mains, tanks, pump houses, power supply to pumps, hydrants, and valves warrants qualified fire protection engineering attention in the planning stage. To correct water supply deficiency, alternate methods can be considered. The factors discussed earlier offer such alternatives. Adding a booster fire pump may be an economical alternative for a sprinkler system requiring high pressure. On the other hand, it may be more economical to either:

1. Upgrade the piping layout only in that system.
2. Replace ceiling sprinklers for that system with ones of a different temperature rating.
3. Install in-rack sprinklers in that area.
4. Lower pile heights in that area (not a good long-range solution).
5. Combine methods.

Large and Small Hose

Warehouses of moderate area close to city hydrants present no special fire suppression problems. But when a warehouse's smaller dimension exceeds about 200 ft (61 m), hose lays from public hydrants to the building's far side can be difficult. Private hydrants then become important around the perimeter, with access doors to the warehouse near them. Private hydrant spacing of 250 ft (76 m) is common in such situations. NFPA 24, *Standard for the Installation of Private Fire Service Mains and Their Appurtenances,* provides guidance on the number and location of hydrants. Indoor small hose is recommended in high-piled or rack storage buildings for occupants to use on incipient fires. Their relatively good reach and long-lasting supply make them more effective than portable fire extinguishers in certain situations. However, they do not eliminate the need for portable fire extinguishers.

Portable Fire Extinguishers

Portable fire extinguishers should be located throughout the premises. They can be effective during a fire's incipient stages if the fire is detected early and the operator is properly trained. NFPA 10, *Standard for Portable Fire Extinguishers*, provides guidance in selecting types and sizes of fire extinguishers and locating them properly. Up to one half of the portable fire extinguishers restorage areas where fixed small hose lines are available and reach all parts of the storage area.

TABLE 9-18D. *Large-Drop Sprinkler Data*
Pressure and Number of Design Sprinklers Required for Various Hazards for Large-Drop Sprinklers

| Hazard | Type of System | Minimum Operating Pressure,* psi (bar) | | | Hose Stream Demand, gal/min (dm³/min) | Water Supply Duration, hr |
| | | 25 (1.7) | 50 (3.4) | 75 (5.2) | | |
		Number Design Sprinklers				
Palletized† storage						
Class I, II, and III commodities up to 25 ft (7.6 m) with maximum 10-ft (3.0-m) clearance to ceiling	Wet	15	§	§	500 (1900)	2
	Dry	25	§	§		
Class IV commodities up to 20 ft (6.1 m) with maximum 10-ft (3.0-m) clearance to ceiling	Wet	20	15	§	500 (1900)	2
	Dry	Does not apply	Does not apply	Does not apply		
Unexpanded plastics up to 20 ft (6.1 m) with maximum 10-ft (3.0-m) clearance to ceiling	Wet	25	15	§	500 (1900)	2
	Dry	Does not apply	Does not apply	Does not apply		
Expanded plastics commodities up to 18 ft (5.5 m) with maximum 8-ft (2.4-m) clearance to ceiling	Wet	Does not apply	15	§	500 (1900)	2
	Dry	Does not apply	Does not apply	Does not apply		
Idle wood pallets up to 20 ft (6.1 m) with maximum 10-ft (3.0-m) clearance to ceiling	Wet	15	§	§	500 (1900)	1½
	Dry	25	§	§		
Solid piled† storage						
Class I, II, and III commodities up to 20 ft (6.1 m) with maximum 10-ft (3.0-m) clearance to ceiling	Wet	15	§	§	500 (1900)	1½
	Dry	25	§	§		
Class IV commodities and unexpanded plastics up to 20 ft (6.1 m) with maximum 10-ft (3.0-m) clearance to ceiling	Wet	Does not apply	15	§	500 (1900)	1½
	Dry	Does not apply	Does not apply	Does not apply		
Double-row rack storage‡ with minimum 5.5-ft (1.7-m) aisle width and multiple-row rack storage with minimum 8.0-ft (2.5-m) aisle width						
Class I and II commodities up to 25 ft (7.6 m) with maximum 5-ft (1.5-m) clearance to ceiling	Wet	20	§	§	500 (1900)	1½
	Dry	30	§	§		
Class I and II commodities up to 30 ft (9.2 m) with maximum 5-ft (1.5-m) clearance to ceiling	Wet	20 plus one level of in-rack sprinklers⁵	§	§	500 (1900)	1½
	Dry	30 plus one level of in-rack sprinklers⁵	§	§		
Class I, II, and III commodities up to 20 ft (6.1 m) with maximum 10-ft (3.0-m) clearance to ceiling	Wet	15	§	§	500 (1900)	1½
	Dry	25	§	§		
Class I, II, and III commodities up to 25 ft (7.6 m) with maximum 10-ft (3.0-m) clearance to ceiling	Wet	15 plus one level of in-rack sprinklers⁵	§	§	500 (1900)	1½
	Dry	25 plus one level of in-rack sprinklers⁵	§	§		
Class IV commodities up to 20 ft (6.1 m) with maximum 10-ft (3.0-m) clearance to ceiling	Wet	Does not apply	20	15	500 (1900)	2
	Dry	Does not apply	Does not apply	Does not apply		

TABLE 9-18D. *(Continued)*

Hazard	Type of System	Minimum Operating Pressure,* psi (bar)			Hose Stream Demand, gal/min (dm³/min)	Water Supply Duration, hr
		25 (1.7)	50 (3.4)	75 (5.2)		
		Number Design Sprinklers				
Class IV commodities up to 25 ft (7.6 m) with maximum 10-ft clearance to ceiling	Wet	Does not apply	20 plus one level of in-rack sprinklers[5]	15 plus one level of in-rack sprinklers[5]	500 (1900)	2
	Dry	Does not apply	Does not apply	Does not apply		
Unexpanded plastics up to 20 ft (6.1 m) with maximum 10-ft (3.0-m) clearance to ceiling	Wet	Does not apply	30	20	500 (1900)	2
	Dry	Does not apply	Does not apply	Does not apply		
Unexpanded plastics up to 25 ft (7.6 m) with maximum 10-ft (3.0-m) clearance to ceiling	Wet	Does not apply	30 plus one level of in-rack sprinklers§	20 plus one level of in-rack sprinklers§	500 (1900)	2
	Dry	Does not apply	Does not apply	Does not apply		
Class IV commodities and unexpanded plastics up to 20 ft (6.1 m) with maximum 5-ft (1.5-m) clearance to ceiling	Wet	Does not apply	15	§	500 (1900)	2
	Dry	Does not apply	Does not apply	Does not apply		
Class IV commodities and unexpanded plastics up to 25 ft (7.6 m) with maximum 5-ft (1.5-m) clearance to ceiling	Wet	Does not apply	15 plus one level of in-rack sprinklers[4]	§	500 (1900)	2
	Dry	Does not apply	Does not apply	Does not apply		
On-end storage of roll paper†						
Heavyweight paper in closed array, banded in open array, or banded or unbanded in a standard array, up to 26 ft (7.9 m) with maximum 34-ft (10.4-m) clearance to ceiling	Wet	Does not apply	15	§	#	#
	Dry	Does not apply	Does not apply	Does not apply		
Any grade of paper, except lightweight paper with stacks in closed array, or banded or unbanded in a standard array, up to 20 ft (6.1 m) with maximum 10-ft (3.0-m) clearance to ceiling	Wet	Does not apply	15	§	#	#
	Dry	Does not apply	25	§		
Medium weight paper completely wrapped (sides and ends) in one or more layers of heavyweight paper, or lightweight paper in two or more layers of heavyweight paper, with closed array, banded in open array, or unbanded in a standard array, up to 26 ft (7.9 m) with maximum 34-ft (10.4-m) clearance to ceiling	Wet	Does not apply	15	§	#	#
	Dry	Does not apply	Does not apply	Does not apply	Does not apply	Does not apply
Record Storage						
Paper records and/or computer tapes in multitier steel shelving up to 5 ft (1.5 m) in width and with aisles 30 in. (76 cm) or wider, without catwalks in the aisles, up to 15 ft (4.6 m) with maximum 5-ft (1.5-m) clearance to ceiling	Wet	15	§	§	500 (1900)	1½
	Dry	25	§	§		

TABLE 9-18D. *(Continued)*

| Hazard | Type of System | Minimum Operating Pressure,* psi (bar) | | | Hose Stream Demand, gal/min (dm³/min) | Water Supply Duration, hr |
| | | 25 (1.7) | 50 (3.4) | 75 (5.2) | | |
		Number Design Sprinklers				
Same as above, but with catwalks of expanded metal or metal grid with minimum 50 percent open area in the aisles	Wet	Does not apply	15	§	500 (1900)	1¹/₂
	Dry	Does not apply	15	§		

For SI units: 1 ft = 0.3048 m; 1 psi = 0.0689 bar.

*Open wood joist construction. Fully firestop each joist channel to its full depth at intervals not exceeding 20 ft (6.2 m). In unfirestopped open wood joist construction, or if firestops are installed at intervals exceeding 20 ft (6.1 m), increase the minimum operating pressures of Table 9-18D by 40 percent.

†See NFPA 231, *Standard for General Storage.*

‡With rack storage, use conventional wood pallets only; no slave pallets.

§The high pressure may be used, but the required number of design sprinklers may not be reduced from that required for the lower pressure.

¶Install in-rack sprinklers in accordance with NFPA 231C, *Standard for Rack Storage of Materials.*

#Hose stream demands and water supply durations may vary for roll paper storage depending on local conditions. See NFPA 231F, *Standard for the Storage of Roll Paper.*

TABLE 9-18E. *ESFR Sprinkler Data*

Type of Storage	Commodity	Maximum Ht. of Storage in ft (m)	Maximum Ht. of Building in ft (m)*	Nominal K Factor	Sprinkler Design Pressure in psi (bar)	Commodity Limitation
Palletized and solid pile storage and single-, double-, and multiple-row rack storage (no open top containers or solid shelves)	Cartoned, unexpanded plastic; cartoned, expanded plastic; uncartoned, unexpanded plastic; and Class I, II, III, or IV commodities encapsulated or unencapsulated	25 (7.6)	30 (9.1)	13.5–14.5	50 (3.4)	
	Cartoned, unexpanded plastic; and Class I, II, III, or IV commodities, encapsulated or unencapsulated	35 (10.7) 20 (6.1)	40 (12.2) 25 (7.6)	13.5–14.5 11.0–11.5	75 (5.2) 50 (3.4)	†
Roll paper on end, open/ standard or closed array, banded or unbanded	Heavy weight or medium weight	20 (6.1)	30 (9.1)	13.5–14.5	50 (3.4)	
Aerosol storage	See NFPA 30B‡					

*Maximum building height is to be measured to the underside of the roof deck or ceiling.

†Only ESFR sprinklers specifically listed for 40-ft (12.2-m) high buildings should be used in buildings higher than 30 ft (9.1 m) up to 40 ft (12.2 m).

‡Code for the Manufacture and Storage of Aerosol Products.

Fire Fighting

Even with adequate fire control by sprinklers, manually fighting storage fires presents significant challenges. Spray from operating sprinklers should reach, surround, and generally control the fire, making it safer for occupants to evacuate. However, as burning continues, fire fighters must cope with the volume of smoke and steam being generated. As sprinkler discharge generates steam, visibility tends to decrease. (See the discussion on smoke venting earlier in the chapter.)

Other manual fire-fighting issues include accessing fires located in the upper reaches of high racks, gaining access through aisles blocked by collapsed storage, and removing burning items, such as fiber bales or rubber tires, that are not easily extinguished. A well-planned fire protection system should simplify manual fire-fighting tasks. Automatic alarms to the fire department from operating sprinkler systems, as well as adequate hydrants, 24-hr emergency access, audible/visual indication of which sprinkler system is operating, and clear identification of sprinkler valves, hydrants, interior small-hose stations, and siamese pumped connections, all contribute to successful fire-fighting operations. Pre-fire inspections and planning are also strongly recommended. NFPA 600, *Standard on Industrial Fire Brigades,* and NFPA 1420, *Recommended Practice for Pre-Incident Planning for Warehouse Occupancies,* provide additional guidance.

Plant Emergency Organization

A well-organized plan is essential for successfully handling fire emergencies. Even if the plan for small warehouse operation designates an individual to notify the public fire department and use a fire extinguisher, decisions must be made to identify the individual and determine how to perform tasks. Larger properties frequently warrant in-plant assignments for communication, using extinguishers and small hoses, checking the condition of valves, operating fire pumps, advising responding fire fighters, salvage operations, and more. These can be a private fire brigade's duties. NFPA 600 provides guidance on the organization, duties, and training of such brigades.

Fire Prevention

Some commodities, such as baled fibers, easily ignite, sometimes merely by friction sparks. Goods in cartons need a substantial or prolonged ignition source, unless inadequate housekeeping permits scrap materials to accumulate where they can easily ignite and kindle stored items.

More common causes of fires in storage are smoking, electrical causes, industrial trucks (gasoline, LP-Gas, diesel, or electrical power), and sparks from flame cutting and welding. It's preferable to bolt steel racks together rather than weld them; this avoids flame cutting during possible future disassembly. Cutting and welding sparks tend to start fires beyond the range of convenient portable fire-fighting equipment.

Arson fires pose special problems and are typically combated by superior automatic sprinkler protection and appropriate security measures. However, disgruntled employees who start fires may know how to override fire safety systems. Familiarity with business operations, practices, and schedules let them cause the most damage possible with the least effort. In addition to knowing how to start a fire, they may also have access to control valves, which can shut down the sprinkler systems, as well as other means to inhibit fire control.

SPECIAL COMMODITIES

Many commodities have been studied separately because of the fire properties associated with their physical form, method of storage, and difficulty of fire control. In addition to the brief descriptions of the fire characteristics that follow, Table 9-18F compares sprinkler system density requirements for 15-ft (4.6-m) high storage of various materials.

Rubber Tires

Fires in rubber tires occur infrequently, but, once ignited, they are very hot, smoky, and difficult to control and extinguish. High sprinkler densities are needed to control the fire and protect the building. (See NFPA 231D.)

Tires are stored in piled form, in compact portable racks referred to as "pallets," and in racks. They may be arranged on tread, on side, or in a laced configuration. (See Figures 9-18T, 9-18U, and 9-18V.) Because tires are not packaged, they contribute their own circular air space to the storage, forming considerable horizontal or vertical flues. Interiors of tire carcasses can flame vigorously, mostly beyond the reach of water from sprinklers. Final extinguishment involves laboriously applying water to individual tires, usually after removing them from the building.

Combined with sprinklers, high-expansion foam is very effective on rubber tires and causes minimal water damage or contamination. To fully penetrate the inner reaches of the tires, an additional hour of foam soaking is advised after the sprinklers are shut off to maintain the foam level.

Roll Paper

Roll paper is stored on its side or, more commonly, on end. One might assume that the latter arrangement permits easier paths for sprinkler water to penetrate the vertical air spaces and slow the fire, but experience with roll paper shows otherwise.

In storage on end, peeling or exfoliation during a fire is a major concern. This phenomenon can be abated somewhat by metal banding, tightly applying steel baling wire by hand, covering ends as well as sides with fire-retardant-treated tight paper wrappers, or closely spacing the columns of rolls. In effective close spacing, stacks or columns are less than 4 in. (101 mm) apart.

NFPA 231F prescribes sprinkler density area requirements according to the rolls' arrangement, whether they are banded or unbanded, the piles' height, and the paper's weight, i.e., heavy, medium, light, or tissue. Because rolled paper tends to absorb water from a discharging sprinkler and swells in size, it should be stored with sufficient clearance from walls.

Carpet Storage

Rugs or carpeting in rolls are generally stored in racks that accommodate 12- to 15-ft (3.7- to 4.6-m) long rolls. Therefore, racks are at least that deep and sometimes placed back to back, forming a total width of 24 to 30 ft (7.3 to 9.1 m) between aisles. Of course, small carpet pieces and rolls may be placed in cartons and stored in regular solid or palletized piles, or conventional racks.

Long carpet rolls would deflect if their length were not supported, so solid or slatted shelves are used in wide racks. An exception is made for individually encased carpets in strong paperboard tubes that have sufficient bending strength to be supported in open-style racks without shelving. Note that in-rack sprinklers are needed under each solid or slatted shelf in racks, unless transverse flue spaces are at least 6 in. (152 mm) wide. (See the discussion on in-rack sprinklers earlier in this chapter.)

With carpet racks having as many as ten tiers, the number of sprinklers and the amount of piping for in-rack sprinklers can be prohibitive. Therefore, protection focuses on providing a minimum 2-in. (51-mm) wide transverse flue at all vertical supports, which are normally 10 to 12 ft (3 to 3.7 m) apart, and providing a level of in-rack sprinklers, including face sprinklers for the wider racks, about every third tier for storage over 12 ft (3.7 m) high. Tier height varies from 24 to 40 in. (0.61 to 1 m), so the sprinkler is installed in a tight area. (See Figure 9-18W.) Some engineers recommend vertical transverse barriers at every third set of vertical supports instead of relying upon 2-in. (51-mm) wide transverse flues.

For back-to-back racks, a 12-in. (0.3-m) space should be provided between vertical members to create a longitudinal flue. If rolls tend to project into this space, toe plates or other mechanisms fastened to the rack's rear face physically stop the rolls. Fire does not tend to spread rapidly in normal carpet rolls along the rack's length unless a large quantity of carpet has attached expanded (foam) rubber backing. Nevertheless, because the racks often extend for several hundred feet (100 ft = 30.5 m), providing clear spaces 8 ft (2.4 m) wide and about 100 ft (30.5 m) apart along the rack's length restrict fire spread. This assists in-rack sprinklers or acts as a fire-stop where no in-rack sprinklers are used, usually for storage under 12 ft (3.7 m) high.

TABLE 9-18F. *Protection for Various Specific Materials Stored 15 ft (4.6 m) High**

	Sprinkler System Density		Temperature Rating of Head in °F‡	Additional gpm for Hose Streams§	Duration of Flow (hr)	Area Adjustment for Dry Pipe	Reference Standard	Remarks
	gpm/sq ft¶	Area ft†						
Rubber tires								
Palletized & fixed Rack storage	0.48	3550	165	750	3	30%	NFPA 231D	
with pallets	0.48	2100	286	750	3	30%	NFPA 231D	
Fixed rack storage without pallets								
or shelves	0.60	5000	286	750	3	30%	NFPA 231D	Water supply to meet both requirements
	0.90	3000	286	750	3	30%		
Baled combustible fibers	0.15	6000	212	500	4	None	FM 8-7	
Hanging garments	0.60	3000	165	250	1	Avoid DP	FM 8-18	Max. 3 ft (1.5 m) below sprinklers
Roll paper								
Rolls on end, close spaced	0.30	2000	286	500	2	30%	NFPA 231F	Max. 5 ft (1.5 m) below sprinklers
Banded rolls, 4 in. (100 mm) or more apart; heavy wt. paper	0.30	2500	286	500	2	30%	NFPA 231F	" "
Banded rolls, 4 in. (100 mm) or more apart; med. wt. paper	0.45	2500	286	500	2	30%	NFPA 231F	" "
" "	0.60	2300	286	500	4–6	About 75%	FM 8-21	
Baled waste paper	0.20	3200	160	500–1000	4–6	35%	FM 8-22	Water supply
Baled waste paper	0.20	2700	212–286	500–1000	4–6	40%	FM 8-22	from curves
Carpets								
Cubicle racks	0.25	3000	286	500	2	None	FM 8-30	
Baled wool	0.15	2500	212	500	4	None	FM 8-7	
Baled cotton	0.25	3000	165	500	2	30%	NFPA 231E	Water supply
Rack storage	0.33	3000	165	500	2	30%	NFPA 231E	from curves
Wood pallets	0.60	4500	286	500	2	None	NFPA 231	

*Full requirements are in the referenced NFPA standards. These data are abstracted from the standards for comparison purposes.
†1 sq ft = 0.093 m^2
‡$^5/_9$(°F − 32) = °C
§1 gpm = 3.78 L/min
¶1 gpm/sq ft = 40.746 (L/min)/m^2

Carpet Padding

Padding or backing, usually made of expanded plastic, expanded rubber, or jute, generally is stored in carpet warehouses. These materials typically burn much faster than carpeting and should be segregated, and specially arranged and protected. Ceiling sprinkler density that is adequate for the warehouse as a whole is probably inadequate for padding unless the storage is kept low, less than 5 ft (1.5 m) for expanded rubber or plastic.

Baled Fibers

The main hazard of baled cotton and other fibers of vegetable origin comes from the multitude of exposed minute fibers on the surfaces of the bales. Fire flashes quickly over these vertical surfaces, loose particles on the floor, and lint on overhead piping and structure. Fire on one side of an automatic-closing fire door can progress through the open doorway on loose floor scraps before the door can close automatically. Housekeeping, therefore, is especially important.

Fire also tends to penetrate between and into bales, requiring their removal from the building for extinguishment. Considerable smoke emission complicates fire fighting. Certain fibers, such as jute, swell when wet, so they should be piled with regard to stability in a fire and with proper clearance from walls.

NFPA 231E limits the height of tiered or rack storage to 15 ft (4.6 m) and pile sizes to 700 bales, with a 12-ft (3.7-m) main aisle and 4-ft (1.2-m) cross aisles, in maximum fire divisions of 10,000 bales. The standard calls for large operating areas for sprinklers. (Factory Mutual Research Corporation's *Loss Prevention Data Sheet 8-7* covers baled combustible fibers generally, including wool, which burns much slower than most fibers.)

Ordinary dry chemical (BC-type) fire extinguishers very effectively knock down surface fires on bales. However, they should be backed up by water spray from small hose or garden hose, or by water-type extinguishers with spray nozzle attachments to extinguish smaller fires that may have penetrated the bales. Preferably, "wet water," a chemical additive that increases water's penetrating and spreading ability, should be used.

FIG. 9-18T. On-tread storage of tires in open, portable racks. (Source: Goodyear)

FIG. 9-18U. On-side storage of rubber tires in palletized racks. (Source: Goodyear)

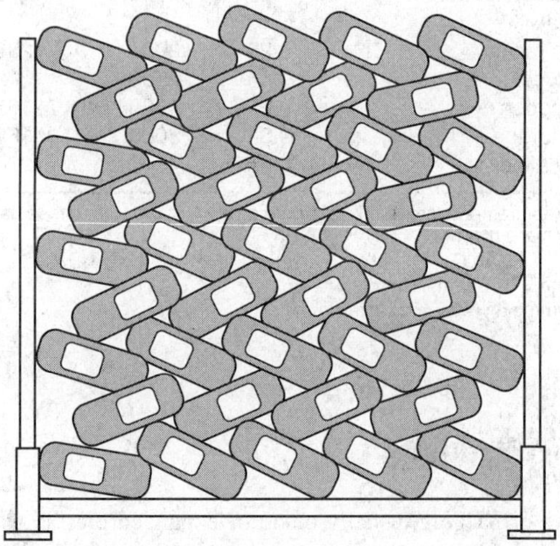

Baled Wastes

Paper: Like baled fibers, baled wastepaper is stored in solid piles, in which fire tends to burrow. Fire can flash over the surface of finely shredded paper bales, as it does over baled fiber. Wastepaper becomes mushy and difficult to handle when wet; in fact, the bales' integrity slowly disappears as the hose stream application progresses, making removal of burning bales difficult. Smoke can also be troublesome. Baled paper sometimes is stored at great heights because of its low unit value, but fire protection standards utilizing conventional sprinklers do not apply to storage over 30 ft (9.1 m) high, except in racks.

Pesticides

Poisonous to insects and small animals, pesticides harm and even kill humans when ingested or breathed as combustion products. They are stored as solutions in flammable or combustible liquids, as powders or granules in combustible packages, or as compressed gases for fumigation. Special attention must be given to the personnel hazard, in addition to the commodity's combustibility. Pesticide fires can endanger persons fighting the fire or standing nearby, and poisonous runoff from fire fighting can pollute water or soil in the area.

Automatic sprinkler protection is advisable, with provision for safe accumulation and disposal of water runoff. Fire-fighting strategy should include plans to use protective clothing and respiratory equipment, spray nozzles rather than straight streams to reduce wa-

FIG. 9-18V. Laced tire configuration.

ter runoff, and specialized medical backup services; and to avoid container breakup and pesticide dispersal.

NFPA 43D, *Code for the Storage of Pesticides*, gives general requirements for both inside and outside storage of pesticides.

SPECIAL STORAGE FACILITIES

Cold Storage

Cold storage warehouses are used primarily for extended storage of food products at temperatures that prevent or retard spoilage. Prod-

FIG. 9-18W. Racks for storing carpeting on slatted shelves. The tiers' closeness to each other, plus their depth, complicates problems of providing in-rack sprinkler protection. (Source: Clark Industrial Truck Division.)

ucts such as pharmaceuticals, antibiotics, and unstable chemicals may also require refrigerated storage.

Depending upon the products or processes, warehouse temperatures range from 35°F (1.7°C) for initial freezing up to 65°F (18.3°C). Despite low temperatures, cold storage warehouses are not immune to fire hazards. In fact, low temperatures present unusual fire prevention and control problems that may be aggravated for warehouses located away from municipal water distribution systems. Combustible materials in such warehouses include cork or expanded plastic insulation; wood dunnage, pallets, boxes, and baskets; fiberboard containers; paper and cloth wrapping; plastic containers and boxes; and grease-impregnated materials. These present the potential for large monetary losses.

Construction: Refrigerated warehouses may be constructed of combustible or noncombustible material, or a combination of both. The current trend is to build larger and higher buildings in which automated storage racks are structural units supporting the walls and roofs of the buildings. Such construction is acceptable if the rack structure adequately supports wind and snow loads. Although the structural materials themselves are noncombustible, such buildings must be adequately sprinklered to protect both the structure and the stored material. When large amounts of cork and/or foam insulation ignite, they can burn beyond control, generating temperatures high enough to distort steel and cause structural collapse.

Polyurethane foam, either in sheets or foamed-in-place, and polystyrene foam are two common insulation materials. When used in walls or ceilings, these materials should be protected by an approved thermal barrier or by a $^{1}/_{2}$-in. (12.7-mm) coat of cement plaster on metal lath attached to the building's framing. For polystyrene, the barrier also may be either $^{1}/_{2}$-in. (12.7-mm) Type X gypsum wallboard or $^{3}/_{4}$-in. (19-mm) fire-retardant plywood supported by studs or furring attached to the framing. To meet sanitary standards, these thermal barriers must be coated with a United States Department of Agriculture (USDA) approved washable finish. If used in the floor, the insulation should be covered with concrete and the joint between floor and wall provided with a cove or curb. Automatic sprinklers should be provided below suspended ceilings, and the space above the ceiling should be protected against fire spread and sprinklered if combustible.

Refrigerant gases: Refrigerant gases present two basic hazards: toxicity and flammability. Sprinklers are very effective because the spray dilutes, disperses, and cools escaping gases.

Storage arrangements: As in other warehouses, storage arrangements in refrigerated warehouses can be bulk, solid piling, palletized, or rack type, with bulk storage minimally used. One exception to this arrangement is the hook storage of meat and poultry products awaiting further processing. This storage method presents minimal fire hazard because of the low temperature and absence of combustible packaging.

Housekeeping practices: Housekeeping in refrigerated warehouses is usually excellent because of the possibilities of contamination.

Sprinkler protection: An adequate sprinkler system should protect a cold storage warehouse. The same sprinkler system design used for nonrefrigerated buildings is appropriate for this occupancy, except that it must be designed not to freeze. Room-flooding carbon dioxide or high-expansion foam systems are not acceptable alternatives to automatic sprinklers. In chill rooms and coolers, sprinkler systems of the electrically operated preaction type are preferable to dry pipe systems. In freezers or holding rooms, systems should combine deluge and dry pipe features; Figure 9-18X shows such a system. In this system, a $^{1}/_{8}$-in. (3.2-mm) orifice in a bypass maintains water pressure under the dry pipe valve. A deluge valve holds back the main water supply. A separate fire detection system electrically activates the valve.

Where high racks are used for storage, it is necessary to use in-rack sprinklers designed to reach spaces that overhead sprinklers cannot reach. Provisions must be made for thawing and draining after system activation.

All sprinklers in refrigerated warehouses should be pressurized and equipped as described in NFPA 13. They should be designed so that they can be easily inspected and disassembled to remove ice plugs.

The combination dry pipe deluge system shown in Figure 9-18X helps prevent ice formation due to accidental tripping of dry valves. When this combination is used:

1. The dry system should be manifolded to the deluge valve.
2. The protected area should not exceed 40,000 sq ft (3,716 m²).
3. The distance between valves should be as short as possible.
4. The chamber between deluge and dry pipe valves should be pressurized from the supply side of the deluge valve.
5. An approved ball drip valve should be installed on the dry pipe valve.

Properly located heat detecting devices in the protected area should automatically operate the deluge valve. An electrically activated deluge valve is preferred.

Figure 9-18Y shows an acceptable alternative to the deluge valve/dry pipe valve combination. Approved equipment should be used throughout, and all electrical equipment should be compatible. The piping to the heat-activated devices should be supervised. The control piping should be arranged so both loss of sprinkler piping air pressure and operation of a heat-activated device occur before the deluge valve release mechanism operates.

Piers and Wharves

Warehouses often form the superstructure of piers and wharves. They may be called "sheds" and, on piers, project great distances over the water. They share the same general characteristics as other warehouses, but present more fire protection and suppression problems. Among these are reduced accessibility for land-based fire department vehicles; the frequently combustible nature of their substructures, which can be exposed to floating burning debris or flammable liquids; the danger of water supply mains freezing; and the hazard of ships colliding into piers. More favorably, commodities tend to be piled low because of the temporary, transfer nature of the storage, i.e., ship-to-shore or shore-to-ship.

NFPA 307, *Standard for the Construction and Fire Protection of Marine Terminals, Piers, and Wharves*, requires a water supply for both hose streams and sprinkler systems to be available for at least 4 hr—about double the usual requirement. To protect a combustible substructure's underside, sprinklers should be either standard pendant sprinklers in the upright position, or old-style sprinklers in the upright position, to wet the undersurface of the substructure.

Limiting the area between transverse fire walls in a pier is desirable. It makes these walls continuous with substructure fire walls, which, in turn, should extend to low water. The walls should be not more than 450 ft (137 m) apart. Below a combustible substructure, transverse firestops should be not more than 150 ft (45 m) apart.

Storage Garages

Motor vehicles, including automobiles and trucks in "dead storage," are housed in storage garages or outdoors. Fires in individual vehicles can be severe because of their generally fast-burning upholstery and gasoline or diesel-oil content. However, the overall building fire loading is moderate, evidently because the considerable amount of metal in the vehicles absorbs heat, and the average weight of combustibles per sq ft (m²) is not high.

According to NFPA 88A, *Standard for Parking Structures*, storage garages and enclosed parking structures (not open-air parking structures) need automatic sprinkler protection if they are in basements, in buildings more than 50 ft (15 m) high having combustible roof/floor assemblies, or inside or below buildings used for other purposes, typically high-rise apartments, hotels, or offices. In the last case, an alternative to sprinklers is a supervised smoke detection system combined with a mechanical ventilating system capable of exhausting smoke. Dispensing flammable liquid fuels from underground tanks sometimes is a feature inside such garages. The pumps should be at street level within 50 ft (15 m) of a vehicle exit or entrance. (See NFPA 30, *Flammable and Combustible Liquids Code*.) Many municipal building codes require automatic sprinkler protection in noncombustible and fire-resistive public parking garages that exceed certain areas and/or heights, so applicable codes should be consulted.

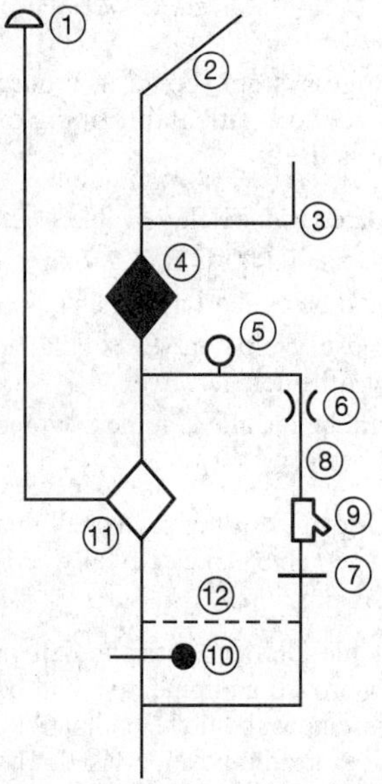

Isolated Storage Buildings

Some high-value warehouses are needed for only a few years, such as those for construction projects in remote areas. At times, these warehouses can contain quite a bit of electrical and electronic equipment, wire, cables, fixtures, pipe fittings, etc. Experimental or mining sites may also have large warehouses. These remote areas often lack public water mains or fire departments, making adequate private protection more costly. Several factors must be weighed, such as whether the building is critical to the project and the time needed to replace vital contents. Standards and building codes do not apply rigidly in these situations. Suitable decisions could be to separate unprotected warehousing into small units, to provide fully adequate private protection, or to compromise and use protective systems with a limited water supply. Increased vigilance over guard service, housekeeping, maintenance, and smoking control should accompany any compromised physical protection.

Underground Storage

Large underground warehouses can be a boon to energy conservation because rock caverns have a nearly constant temperature. In the United States, some caverns have stored compressed gas, and some old mines have stored records.[1] Caves developed for warehousing include automatic sprinkler protection.

In general, the problems of underground structures include exits, venting smoke and heat, and access for fire fighting. Automatic sprinkler protection installed under constructed floors and ceilings is vital to protect combustible storage and associated personnel.

Air-Supported Structures

For warehousing in temporary or remote locations, such as at construction sites, or as a low-cost adjunct to more permanent conventional buildings, large air-supported structures are sometimes used, based on convenience and the economics of the risk involved. (See Figures 9-18Z and 9-18AA.) These structures are plastic-coated fabric envelopes resembling balloons and kept rigid by internal low positive air pressure. When used for warehousing the structure may have loading dock arrangements, such as air locks with electric roll-up doors. Structures less than 150 ft (46 m) in width or diameter should comply with NFPA 102, *Standard for Grandstands, Folding and Telescopic Seating, Tents, and Membrane Structures*, which deals with wind resistance, strength, load distribution, and pressurizing air-supported structures. A trend toward using a cable harness net system to encapsulate the envelope in all directions increases the structure's stability during wind load.[2]

Because these structures do not support overhead piping or wiring, the usual need for automatic fire control systems cannot be met. Therefore, they are unsuitable for long-term combustible storage of highly valued or highly hazardous materials. When reasons

1. Heat-actuated device
2. To sprinklers
3. To air supply
4. Dry pipe valve
5. Pressure gage
6. ⅛in. (3 mm) brass restricting orifice
7. Valve (normally open)
8. ½in. (12.7 mm) brass pipe
9. Strainer
10. Main water supply valve
11. Deluge valve
12. Alternate connection to main water supply

FIG. 9-18X. Piping to chamber between deluge valve and dry pipe valve.

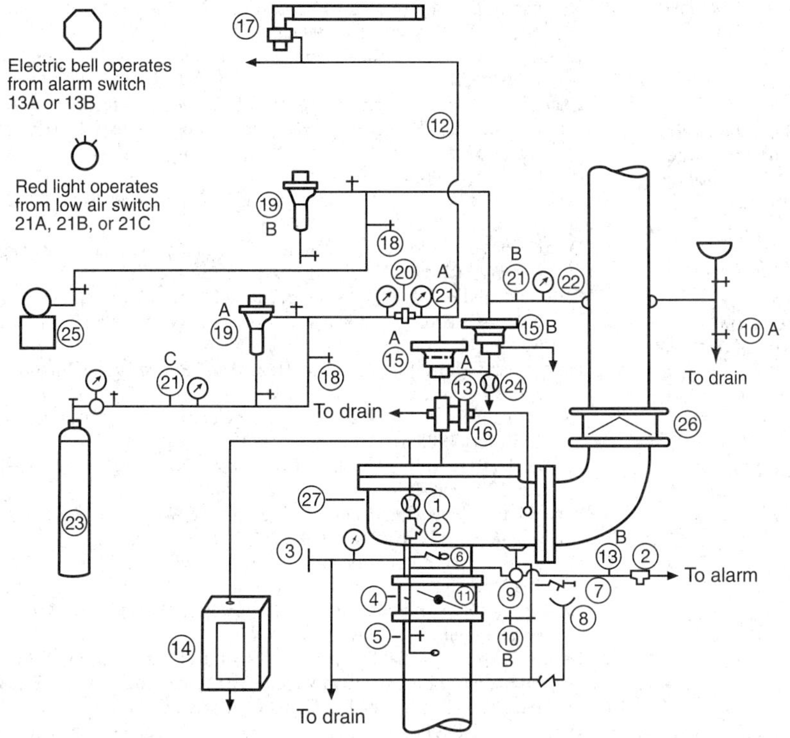

Electric bell operates from alarm switch 13A or 13B

Red light operates from low air switch 21A, 21B, or 21C

1. 0.125 in. (3.2 mm) restrictor orifice
2. Strainer
3. Flow test valve*
4. Priming pipe
5. Priming water valve*
6. Check valve
7. Drip check
8. Drip cup
9. Alarm test valve
10. System drain valve*
11. Main water supply valve
12. $\frac{1}{2}$ in. (12.7 mm) thermostatic release piping
13. Alarm switch
14. Emergency trip valve
15. Diaphram bypass valve
16. Pilot-operated relief valve
17. Model "C" thermostatic release
18. Air bypass valve*
19. Air pressure maintenance device
20. $\frac{1}{32}$ in. (0.8 mm) restriction plug
21. Low air pressure switch
22. Circle seal check
23. Nitrogen cylinder & regulator
24. 0.046 in. (1.2 mm) restrictor orifice
25. Air compressor
26. Sprinkler system check valve
27. Deluge valve

* Valves are normally closed

FIG. 9-18Y.   Pneumatic dual-release automatic water control valve.

to use these structures are compelling, a number of separate smaller units are preferable to one large structure.

CONSTRUCTION

Any important building, warehouse or otherwise, merits durable construction that resists wind and snow loads. Many construction types meet these criteria but are combustible or vulnerable to fire temperatures. Automatic sprinklers are virtually always needed to protect storage contents and compensate for these hazards.

Buildings used for both manufacturing and warehousing should have a good barrier wall, preferably a true fire wall, between these areas because the greater hazard in the former exposes the highly valued material in the latter to a fire's damaging effects. Fire partitions should separate incidental adjoining areas, such as boiler, machinery, or service rooms.

FIG. 9-18AA.   Interior of a 100- × 300-ft (30- × 91-m) air-supported warehouse, showing rack storage. (Source: Air-Tech Industries, Inc.)

Steel columns within storage racks over 15 ft (4.6 m) high that have no in-rack sprinklers require the additional means of fire protection described in NFPA 231C.

BIBLIOGRAPHY

References Cited

1. Holdcraft, R. L., "Fire Protection Criteria for Caves," Fire Journal, Vol. 79, No. 3, May/June 1985, pp. 35–37, 121.

FIG. 9-18Z.   A typical air-supported structure suitable for warehousing purposes. (Source: Air-Tech Industries, Inc.)

2. Air Structures Institute, "Air Structures Design and Standards Manual," ASI-77, 1977, Air Structures Institute, St. Paul, MN.

NFPA Codes, Standards, and Recommended Practices

Reference to the following NFPA codes, standards, and recommended practices will provide further information on general indoor storage discussed in this chapter. (See the latest version of *The NFPA Catalog* for availability of current editions of the following documents.)

NFPA 10, *Standard for Portable Fire Extinguishers*
NFPA 11A, *Standard for Medium- and High-Expansion Foam Systems*
NFPA 13, *Standard for the Installation of Sprinkler Systems*
NFPA 14, *Standard for the Installation of Standpipe and Hose Systems*
NFPA 24, *Standard for the Installation of Private Fire Service Mains and Their Appurtenances*
NFPA 25, *Standard for Inspection, Testing, and Maintenance of Water-based Fire Protection Systems*
NFPA 30, *Flammable and Combustible Liquids Code*
NFPA 30B, *Code for the Manufacture and Storage of Aerosol Products*
NFPA 43D, *Code for the Storage of Pesticides*
NFPA 51B, *Standard for Fire Prevention in Use of Cutting and Welding Processes*
NFPA 72, *National Fire Alarm Code*
NFPA 88A, *Standard for Parking Structures*
NFPA 102, *Standard for Grandstands, Folding and Telescopic Seating, Tents, and Membrane Structures*
NFPA 204M, *Guide for Smoke and Heat Venting*
NFPA 231, *Standard for General Storage*
NFPA 231C, *Standard for Rack Storage of Materials*
NFPA 231D, *Standard for Storage of Rubber Tires*
NFPA 231E, *Recommended Practice for the Storage of Baled Cotton*
NFPA 231F, *Standard for the Storage of Roll Paper*
NFPA 307, *Standard for the Construction and Fire Protection of Marine Terminals, Piers, and Wharves*
NFPA 505, *Fire Safety Standard for Powered Industrial Trucks Including Type Designations, Areas of Use, Maintenance, and Operation*
NFPA 600, *Standard on Industrial Fire Brigades*
NFPA 601, *Standard on Guard Service in Fire Loss Prevention*

Additional Reading

Alvares, N. J., *et al.*, "Analysis of a Run Away High Rack Storage Fire," Extended Abstracts of the Society of Fire Protection Engineers Engineering Seminars on Large Fires: Causes and Consequences, November 16–18, 1992, Dallas, TX, Society of Fire Protection Engineers, Boston, MA, 1992, pp. 29–33; *Proceedings* of the Fourth International Symposium on Fire Safety Science, International Association for Fire Safety Science, Boston, MA, 1994, pp. 1267–1278.
Andrews, C. L., "Rebuilding a Warehouse Complex," *Fire Prevention*, No. 204, Nov. 1987, pp. 31–32.
Atkinson, G. T., and Jagger, S. F., "Assessment of Hazards from Warehouse Fires Involving Toxic Materials," Sixth International Fire Conference on Fire Safety, Interflam '93, March 30–April 1, 1993, Oxford, England, Interscience Communications Ltd., London, England, 1993, pp. 343–354.
Bacon, M. M., "Warehouse Fire Challenges Firefighters. On the Job: Rhode Island," *Firehouse*, Vol. 19, No. 2, Feb. 1994, pp. 46–48.
Barry, T. F., "Application of Fire and Explosion Risk Assessment to an LPG Bulk Storage Facility," Professional Loss Control, Inc., Oak Ridge, TN, ASTM STP 1150; American Society for Testing and Materials, Fire Hazard and Fire Risk Assessment, Sponsored by ASTM Committee E-5 on Fire Standards, ASTM STP 1150, Dec. 3, 1990, San Antonio, TX, ASTM, Philadelphia, PA, 1990, pp. 183–206.
Before the Fire—Fire Prevention Strategies for Storage Occupancies, NFPA Ad Hoc Task Force, 1987.
Beitel, J. J., "National Quick Response Sprinkler Research Project: Group 1 Retrofit Test Series. [Group 1 Warehouse Retrofit Test Series.]," Southwest Research Inst., San Antonio, TX, National Fire Protection Research Foundation, Quincy, MA, Group 1, June 1990.
Beyler, C. L., and Hunt, S. P., "Fire Development in and Hazard Analysis of Concrete Hazardous/Radioactive Waste Storage Facilities," *Proceedings* of the Society of Fire Protection Engineers Seminars on Perfor-

mance-Based Fire Safety Engineering, Nov. 15–17, 1993, Phoenix, AZ, SFPE, Boston, MA, 1993, pp. 45–55.
Brannigan, F. L., "High Rack Storage," *Fire Engineering*, Vol. 139, No. 5, 1986, p. 32.
Chan, T. S., *et al.*, "Experimental Study of Actual Delivered Density for Rack-Storage Fires," *Proceedings* of the Fourth International Symposium on Fire Safety Science, Intl. Assoc. for Fire Safety Science, Boston, MA, 1994, pp. 913–924.
"Data Sheet 8-9: A New Approach to Storage Protection," *Record*, Vol. 70, No. 5, Sept./Oct. 1993, pp. 3–70.
Dickinson, C., "Cold Storage Warehouse Turns Up the Heat," *Fire Engineering*, Vol. 143, No. 7, July 1990, pp. 32–34, 38.
El-Iskandarani, B. M. K., "Emergency Relief Venting Analysis for Flammable Liquid Storage Tanks," Worcester Polytechnic Inst., MA, Thesis, May 1994.
Field, P., and Murrell, J., "The Fire Hazard and Protection of Bin Storage," *Fire Surveyor*, Vol. 17, No. 6, Dec. 1988, pp. 5–15.
Fields, T., "Warehouses: A Case Study for Control," *Fire Prevention*, No. 215, Dec. 1988, pp. 24–29.
"Fire and Safety Arrangements in a Warehouse Complex," *Fire Prevention*, No. 199, May 1987, pp. 21–22, 24–27.
"Fire-Fighters Pour Cement on Blazing Chinese Warehouses," *Fire International*, No. 141, Nov. 1993, p. 9.
Gibson, R., "Warehouse Protection," *Fire Surveyor*, Vol. 21, No. 1, Feb. 1992, pp. 5–8.
Goring, G., "Fire Protection in Low Temperature Warehouses," *Fire International*, No. 139, May 1993, pp. 47–48.
Grant, G., and Drysdale, D., "Application of the Cone Calorimeter to the Problem of Fire Growth in Warehouses," Sixth International Fire Conference on Fire Safety, Interflam '93, March 30–April 1, 1993, Oxford, England, Interscience Communications Ltd., London, England, 1993, pp. 539–548.
Grant, G., and Drysdale, D., "Numerical Modelling of Early Flame Spread in Warehouse Fires," *Fire Safety Journal*, Vol. 24, No. 3, 1995, pp. 247–278.
Hall, J. R., Jr., "Fire Risk Analysis Model for Assessing Options for Flammable and Combustible Liquid Products in Storage and Retail Occupancies," *Fire Technology*, Vol. 31, No. 4, 4th Quarter, Nov. 1995, pp. 290–308.
Hall, J. R., Jr., "Fire Risk Analysis Model for Assessing Options for Flammable and Combustible Liquid Products in Storage and Retail Occupancies," National Fire Protection Association, Quincy, MA, ISBN 0-9516320-9-4; *Proceedings* of the Seventh International Interflam Conference, Interflam '96, March 26–28, 1996, Cambridge, England, Interscience Communications Ltd., London, England, 1996, pp. 591–600.
Hamer, A., and Beyler, C., "Modeling the Activation of Dry Pipe Sprinklers in High Rack Storage," *Proceedings* of the International Conference on Fire Research and Engineering, September 10–15, 1995, Orlando, FL, SFPE, Boston, MA, 1995, pp. 65–70.
Hansell, G., "Problems of Effective Venting in Warehouses," *Fire Prevention*, No. 262, September 1993, pp. 40–42.
Hoffmann, J., "Kansas: Underground Storage Center Challenges Firefighters. On the Job," *Firehouse*, Vol. 17, No. 6, June 1992, pp. 78–80.
Hoffmann, J. W., "One Jurisdiction's New Warehouse Codes," *Firehouse*, Vol. 18, No. 10, October 1993, pp. 88–90.
Hunt, S. P., "Concrete Hazardous/Radioactive Waste Storage Facilities: Fire Development and Hazard Analysis," Society of Fire Protection Engineers, *Selected Readings in Performance-Based Fire Safety Engineering*, 1995, Boston, MA, Society of Fire Protection Engineers, Boston, MA, 1995, pp. 117–128.
Ingason, H., "Fire Experiments in a Two Dimensional Rack Storage," BRANDFORSK Project 701-917, Swedish National Testing and Research Institute, Boras, Sweden, SP REPORT 1993:56, 1993.
Ingason, H., "Modelling of Two Dimensional Rack Storage Fires," BRANDFORSK Project 701-917, Swedish National Testing and Research Institute, Boras, Sweden, SP REPORT 1993:57, 1993.
Ingason, H., "Two Dimensional Rack Storage Fires," *Proceedings* of the Fourth International Symposium on Fire Safety Science, Intl. Assoc. for Fire Safety Science, Boston, MA, 1994, pp. 1209–1220.
"International Foam-Water Sprinkler Research Project. Task 4. Palletized Storage Fire Tests 1 Through 13," Technical Report, Underwriters Laboratories, Inc., Northbrook, IL, National Fire Protection Research Foundation, Quincy, MA, Feb. 1992.

"International Foam-Water Sprinkler Research Project. Task 5. Rack Storage Fire Tests," Underwriters Laboratories, Inc., Northbrook, IL, National Fire Protection Research Foundation, Quincy, MA, Nov. 1992.

"IRI Promotes Safer Metal Container for Flammable Liquids Storage," *Sentinel*, Vol. 50, No. 3, Third Quarter 1993, p. 9.

Isman, K. E., "Sprinkler Protection for General Storage Occupancies," *Sprinkler Quarterly*, No. 93, Winter 1995, pp. 6, 8, 10.

Isman, K. E., "Sprinkler Protection for General Storage Occupancies. Part 2," *Sprinkler Quarterly*, No. 94, Spring 1996, pp. 10, 12.

Isner, M. S., "$100 Million Fire Destroys Warehouses. Summary Investigation Report," *NFPA Journal*, Vol. 85, No. 6, Nov./Dec. 1991, pp. 37–41.

Isner, M. S., "$30 Million Fire Destroys Jet Fuel Storage Tanks," *NFPA Journal*, Vol. 86, No. 1, Jan./Feb. 1992, pp. 60–64, 66.

Isner, M. S., "Cold Storage Warehouse Fire, Madison, Wisconsin, May 3, 1991. Summary Fire Investigation Report," National Fire Protection Assoc., Quincy, MA, Summary Fire Investigation, 1991.

Jihu, Q., Kejun, X., and Changhao, H., "Retrospective Discussion on the Shenzhen Warehouse Explosion," *Proceedings* of the First International Conference on Fire Science and Engineering, ASIAFLAM '95, March 15–16, 1995, Kowloon, Hong Kong, 1995, pp. 69–75.

Johnston, D., "High-Rise Fire Protection at Monsanto's Warehouse," *Fire Prevention*, Vol. 229, May 1990, pp. 28–30.

Khan, M., "Evaluation of Fire Behavior of Packaging Materials," FMRC RC87-TP-7, presented at Defense Fire Protection Association Symposium, 1987.

Koffel, W. E., "Designing Warehouse Sprinkler Systems," *NFPA Journal*, Vol. 87, No. 4, July/Aug. 1993, pp. 16, 90.

Kukkola, T., "Palavan nesteen varastoalueet. [Storage Areas for Flammable Liquids.]," *Palontorjuntatekniikka*, April 1995, pp. 26–27.

Kukkola, T., "Palavien nesteiden sailiorakenteet. [Storage Structures for Inflammable Liquids.]," *Palontorjuntatekniikka*, Feb. 1993, pp. 23–24.

Kunkelmann, J., "Brandausbreitung bei verschiedenen Stoffen, die in lagermabiger Anordnung gestapelt sind. Teil 8: Simulation der Wechselwirkungen eines Tropfenschwarmes mit einer Heissgasstromung. [Flame Spread of Various Materials in Warehouses. Part 8. Simulation of the Interaction of Droplets and Hog Gas Streams.]," Karlsruhe Univ., West Germany, Report 80, Dec. 1991.

Kunkelmann, J., "Brandausbreitung bei verschiedenen Stoffen, die in lagermabiger Anordnung gestapelt sind. Teil 9: Simulation der Wechselwirkungen eines Sprinklers. [Flame Spread of Various Materials in Warehouses. Part 9. Simulation of Sprinkler Water Spray.]," Karlsruhe Univ., West Germany, Report 83, Dec. 1992.

Kunkelmann, J., "Brandausbreitung bei verschiedenen Stoffen, die in lagermassiger Anordnung gestapelt sind. Teil 10. Weiterfuhrende Literaturrubersicht uber die Brandausbreitung sowie uber die Wechselwirkungen des Tropfenschwarmes eines Sprinklers mit einer Heissgasstromung. [Flame Spread on Various Materials in Warehouses. Part 10. Additional Literature Survey About Flame Spread and the Interaction of Stream of Droplets From a Sprinkler and Hot Gases.]," Karlsruhe Univ., West Germany, Report 84, FA. Nr. 143, ISSN 0170-0060, June 1993.

Kunkelmann, J., "Brandausbreitung bei verschiedenen Stoffen, die in lagermassiger Anordnung gestapelt sind. Teil 11. Grossbrandversuche 6—Brandausbreitung in Palettenlagern und Vergleich mit Gitterbox- und Blocklagerung. [Flame Spread of Various Materials in Warehouses. Part 11. Large Scale Fire Experiments 6—Flame Spread in Pallet Storge and Comparison With Large Containers and Block Storage.]," Karlsruhe Univ., West Germany, Report 88, FA. NR. 147, ISSN 0170-0060, Sept. 1994.

Kunkelmann, J., "Brandausbreitung in Palettenlagern und Vergleich mit Gitterboxund Blocklagerung. [Fire Spread in Pallet Storage and Comparison with Grillbox and Compact Storage.]," *VFDB*, Vol. 45, No. 1, Feb. 1996, pp. 24–31.

Lee, J. L., "The Effect of Different Storage Configurations on Required Delivered Density," FMRC J.I. OMOER.RA, April 1986.

Lee, J. L., "The Effect of Paper Construction and Polyethylene Coating on the Flammability and Extinguishability of Roll Paper Products," FMRC JI ONOE9.RA, July 1986.

Lee, J. L., "Extinguishment of Rack Storage Fires of Corrugated Cartons Using Water," *Proceedings* of the First International Symposium on Fire Safety Science, Hemisphere, Washington DC, 1986, pp. 1177–1186.

"Lesson of Gwent Blaze: Warehouse's Fire Safety 'Inadequate'," *Fire*, Vol. 84, No. 1036, Oct. 1991, p. 13.

Linville, J. L., ed., "Industrial Storage Practices," *Industrial Fire Hazards Handbook*, third ed., National Fire Protection Association, Quincy, MA., 1990.

Little, R., *et al.*, "Flammability and Storage Studies of Selected Fire-Resistant Fluids," *Fire Safety Journal*, Vol. 17, No. 1, 1991, pp. 57–76.

Marsh, J. W., Hugger, E. J., and Zimmerman, L. A., "Storage Tank Design Faces Environmental Issues," *Consulting-Specifying Engineer*, Vol. 13, No. 5, April 1993, pp. 60–62, 64.

Maslansky, S. P., and Cullen, J. M., "Proper Storage the Key to Acetylene Tank Safety," *Fire Engineering*, Vol. 149, No. 6, June 1996, pp. 99–101.

Mawhinney, J. R., "Development of Regulations in the 1990 National Fire Code of Canada on Storage of Dangerous Goods," *Fire Technology*, Vol. 26, No. 3, Aug. 1990, pp. 266–280.

Miles, S. D., *et al.*, "Modelling the Environmental Consequences of Fires in Warehouses," *Proceedings* of the Fourth International Symposium on Fire Safety Science, Intl. Assoc. for Fire Safety Science, Boston, MA, 1994, pp. 1221–1232.

Mock, L. L., "Covered Floater in Jacksonville: Steuart Petroleum Bulk Storage Tank Fire," *Fire Engineering*, Vol. 146, No. 11, Nov. 1993, pp. 28–32, 34, 38, 40, 42, 44–46, 48–50.

Moore, J., "Understanding Warehouse Fire Problems," *Firehouse*, Vol. 17, No. 9, Sept. 1992, pp. 70, 72–73.

Mulhaupt, R., "Fire Protection for Flammable and Combustible Liquids in Warehouses," First International Conference on Fire Suppression Research, May 5–8, 1992, Stockholm, Sweden, 1992, pp. 393–399.

Murrell, J. V., and Field, P., "Sprinkler Protection of Post Pallet Storage in High Racks," *Fire Surveyor*, Vol. 19, No. 1, Feb. 1990, pp. 16–20.

Nugent, D. P., "Directory of Fire Tests Involving Storage of Flammable and Combustible Liquids in Small Containers," Schirmer Engineering Corp., Deerfield, IL, Directory, Feb. 1994, 75 pages.

Nugent, D. P., "Fire Tests Involving Storage of Flammable and Combustible Liquids in Small Containers," *Journal of Fire Protection Engineering*, Vol. 6, No. 1, 1994, pp. 1–10.

Nugent, D. P., "Full Scale Sprinklered Fire Tests Involving Storage of Flammable and Combustible Liquids in Small Containers," *Proceedings* of the International Conference on Fire Research and Engineering, Sept. 10–15, 1995, Orlando, FL, SFPE, Boston, MA, 1995, pp. 427–430.

Persson, H., "Sprinkler Protection of Warehouses: A New Method for Classification of Commodities," Sixth International Fire Conference on Fire Safety, Interflam '93, March 30–April 1, 1993, Oxford, England, Interscience Communications Ltd., London, England, 1993, pp. 489–497.

Persson, H., Nahlinder, J., and Tuovinen, H., "Brandrisker vid lagring och hantering av brandfarlig vara. [Fire Risks in Various Oil Tank Storage Areas.]," (Abstract in English), Swedish National Testing Inst., Boras, Sweden, SP Report 1990:18, 1990.

"Roll Paper Storage," *Loss Prevention Data Sheet 8-21*, Factory Mutual Engineering Corp., Norwood, MA.

Schatz, H., "Brand- und Loschversuche mit Wassernebel an Pappkartons und Kasten aus Polypropylen im Hochregal. [Fire- and Extinguishing Tests with Water Fog on Cardboard Boxes and Boxes of Polyethylene in High Rack Storage.]," *VFDB*, Vol. 45, No. 1, Feb. 1996, pp. 31–36.

Schatz, H., "Loscheinsatz bei gelagerten Stoffen. Teil 10. Literaturauswertung—Tropfenverteilungen-Loschversuche. [Fire Suppression of Materials in Warehouses. Part 10. Literature Search—Droplet Distribution -Extinction Experiments.]," Karlsruhe Univ., West Germany, Report 85, FA. NR. 144, ISSN 0170-0060, Aug. 1993.

Schatz, H., "Loscheinsatz bei gelagerten Stoffen. Teil 11. Literaturauswertung—Sprinklereinsatz bei Palettenlager. [Fire Suppression of Materials in Warehouses. Part 11. Literature Search—Sprinkler Activation for Materials on Pallets.]," Karlsruhe Univ., West Germany, Report 87, FA. NA 148, ISSN 0170-0060, July 1994.

Schatz, H., "Loscheinsatz bei gelagerten Stoffen. Teil 8.Literaturauswertung Sprinkler. [Extinction in Warehouses. Part 8. Literature Search on Sprinklers.]," Karlsruhe Univ., West Germany, Report 78, Sept. 1991.

Schatz, H., "Loscheinsatz bei gelagerten Stoffen. Teil 9. Messung und Simulation der Wasser—beaufschlagung—Flussigkeitsverteilungern—Bestimmung von Tropfengroben. [Flame Spread of Materials in Storage. Part 9. Measurement and Simulation of Sprinklers —Liquids—Determination by Droplets.]," Karlsruhe Univ., West Germany, Report 82, Dec. 1992.

Scoones, K., "Serious Fires in Warehouses and Storage Premises During 1991," *Fire Prevention*, No. 252, Sept. 1992, pp. 22–24.

Scoones, K., "Serious Fires in Warehouses and Storage Premises During 1992," *Fire Prevention*, No. 262, Sept. 1993, pp. 29–31.

Seeger, P. G., "Rack Storage Fire Suppression by Sprinklers. Work at the Research Station for Fire Protection Technology," First International Conference on Fire Suppression Research, May 5–8, 1992, Stockholm, Sweden, 1992, pp. 83–101.

Shebeko, Y. N., *et al.*, "Fire and Explosion Risk Assessment for LPG Storages," *Fire Science and Technology*, Vol. 15, No. 1/2, 1995, pp. 37–45.

"Solid, Palletized, and Rack Storage of Plastics," *Loss Prevention Data Sheet 8-9*, Factory Mutual Engineering Corp., Norwood, MA.

"Spotting Fires on Storage Sites," *Fire Prevention*, No. 262, Sept. 1993, p. 38.

Staggs, K. J., *et al.*, "Development of Flammable Liquid Storage Wooden Cabinets for Chemical Laboratories," Lawrence Livermore Lab., CA, Department of Energy, Washington, DC, UCRL-ID-115605, Nov. 1993.

Stahl, K. H., "VdS-Brandversuche mit Kunststoff-Lagerbehaltern. [(German Property Insurers' Association) Informs About Fire Tests of Plastic Storage Tanks.]," *VFDB*, Feb.1994, pp. 15–23.

"Stored Plastics—The New High Challenge Risk," P7422, Factory Mutual Engineering Corp., Norwood, MA.

Thomas, P. H., "Some Comments on Fire Spread in High Racked Storage of Goods," Lund Univ., Sweden, 93295RA/BK/AB, 1993.

Tomes, W. J., and Golinveaux, J., "Fire Test Performance of Extra Large Orifice Sprinklers for Rack Storage of Group A Plastics in Warehouse-Type Retail Occupancies," *Proceedings* of the International Conference on Fire Research and Engineering, Sept. 10–15, 1995, Orlando, FL, SFPE, Boston, MA, 1995, pp. 415–420.

Troup, J. M. A., "Protection of Warehouse Retail Occupancies With Extra Large Orifice (ELO) Sprinklers," *Journal of Fire Protection Engineering*, Vol. 8, No. 1, 1996, pp. 1–12.

Troup, J. M. A., Tomes, W. J., and Golinveaux, J., "Full-Scale Fire Test Performance of Extra Large Orifice Sprinklers for Protection of Rack Storage of Group A Plastic Commodities in Warehouse-Type Retail Occupancies," Factory Mutual Research Corp., Norwood, MA, Tomes, Van Rickley and Associates, San Diego, CA, Central Sprinkler Co., Lansdale, PA, ISBN 0-9516320-9-4; *Proceedings* of the Seventh International Interflam Conference, Interflam '96, March 26–28, 1996, Cambridge, England, Interscience Communications Ltd., London, England, 1996, pp. 405–413.

Tuovinen, H., and Ingason, H., "CFD Simulation of Fires in Two Dimensional Rack Storage," Swedish National Testing and Research Institute, Boras, Sweden, ISBN 0-9516320-9-4; *Proceedings* of the Seventh International Interflam Conference, Interflam '96, March 26–28, 1996, Cambridge, England, Interscience Communications Ltd., London, England, 1996, pp. 209–215.

"Uncontrollable Blaze Devastates Tooley Street Warehouses," *Fire Prevention*, No. 256, Jan./Feb. 1993, pp. 27–28.

"Underground Storage Tanks—EPA Technical Rules in Effect But Insurance Rules Delayed Until 1991," *Building official and Code Administrator*, Vol. 24, No. 4, July/Aug. 1990, pp. 13–17.

VanRensselaer, S., "Fire Protection at Petrochemical Bulk Storage Facilities: Lessons From Australia," *Industrial Fire Safety*, Vol. 2, No. 1, Jan./Feb. 1993, pp. 32–37.

Wackerlig, H., "Risk Assessment of Chemical Warehouses," *Fire Protection*, Vol. 17, No. 4, Dec. 1990, pp. 4–6.

"Warehouse Storage: Large-Scale Solutions for Large-Scale Problems," P8324, Factory Mutual Engineering Corp., Norwood, MA.

"Which Sprinkler Will Best Protect Your Warehouse?" *Record*, Vol. 68, No. 2, March/April 1991, pp. 3–10.

Williamson, R. B., and Schroeder, R. A., "Fire Safety Assessment of the Scrap Tire Storage Methods," *Proceedings* of the International Conference on Fire Research and Engineering, Sept. 10–15, 1995, Orlando, FL, SFPE, Boston, MA, 1995, pp. 431–436.

Williamson, R. B., and Schroeder, R. A., "Fire Safety Assessment of the Scrap Tire Storage Methods," California Univ., Berkeley, ISBN 0-9516320-9-4; *Proceedings* of the Seventh International Interflam Conference, Interflam '96, March 26–28, 1996, Cambridge, England, Interscience Communications Ltd., London, England, 1996, pp. 89–101.

Wilson, M., "Protecting Against Storage Tank Bund Fires," *Fire Surveyor*, Vol. 22, No. 4, Aug. 1993, pp. 8–11.

Woolhead, F. E., "Meeting the Challenge of Fires in Warehouses," *Fire International*, Vol. 10, No. 101, 1986, pp. 23–25.

Yu, H. Z., and Stavrianidis, P., "Transient Ceiling Flows of Growing Rack Storage Fires," *Proceedings* of the Third International Symposium on Fire Safety Science, Elsevier Applied Science, New York, 1991, pp. 281–290.

Yu, H. Z., Lee, J. L., and Kung, H. C., "Suppression of Rack-Storage Fires by Water," *Proceedings* of the Fourth International Symposium on Fire Safety Science, Intl. Assoc. for Fire Safety Science, Boston, MA, 1994, pp. 901–912.

Yung, D., and Mehaffey, J. R., "Fire Resistance Requirements for Rubber-Tire Warehouses," *Fire Technology*, Vol. 27, No. 2, May 1991, pp. 100–112.

OUTDOOR STORAGE PRACTICES

Revised by Milosh Puchovsky

Truly engineered fire protection is extremely difficult to achieve for outdoor storage, as it is for indoor unsprinklered storage. However, depending upon the hazard, monetary value, and storage area, a basic fire protection plan can be developed. Thus, for small amounts of incidental and transitory outdoor storage, appropriate protection might consist merely of good security fencing and outdoor lighting. On the other hand, very large, high-value storage yards warrant elaborate safeguards.

This chapter outlines good general practices, developed from fire experience, for sizable storage areas, and references certain types of storage that have received special study. (See NFPA 231, *Standard for General Storage.*)

Additional information on storage practices can be found in Section 3, Chapter 14, "Welding and Cutting"; Section 3, Chapter 24, "Storage and Handling of Solid Fuels"; Section 6, Chapter 5, "Water Supply Requirements for Fire Protection"; Section 9, Chapter 17, "Warehouse Operations"; and Section 9, Chapter 18, "General Indoor Storage."

CHOOSING THE SITE

Size

The storage area should be large enough to hold the quantity of stored material, with allowance for spacing between piles and possible future expansion. If the area is inadequate, additional sites should be found. Congestion can be a major factor in fire spread.

Terrain

The ground should be leveled, if possible. Sloping terrain is undesirable because it makes piles unstable and complicates fire-fighting operations. Refuse areas, sawdust-filled land, swampy ground, or areas prone to underground fires are considered unsuitable.

Exposure

Adequate clearances to adjacent properties should be maintained to minimize mutual fire exposure. Zoning regulations often specify clear distances to a property line. If no such regulations exist, the na-

ture of future development on adjacent property should be considered, for it may affect the storage area's size.

Other exposure considerations include protection from grass, brush, or forest fires; and protection from potential ignition sources, such as sparks generated by rolling railroad equipment, incinerator stacks, electrical transformers on poles, or lighted cigarettes discarded from nearby cars.

Fire Protection

Because fixed automatic systems are usually not an option, the manual protection available must influence the arrangement of outdoor storage, including pile size, and the clearances required from adjacent piles, buildings, and property boundary lines. Local fire protection features, such as public water supply and its reliability, the available fire flow, the fire department's location and travel time, and the means available for notifying the fire department, should be evaluated.

Floods and Windstorms

Areas subject to flooding or windstorms are to be avoided. Windstorms present a particular problem in some coastal areas. Fires occurring during high winds pose an exposure problem against which protection is almost impossible. Winds carry burning materials and brands over considerable distances, even though the fire may be confined to a relatively small area.

PREPARING THE SITE

Clearing the Site

Vegetation should be cleared away so that it cannot dry out and become fuel for ignition or fire spread. The site should be leveled as flat as possible and then paved. If the entire yard is not surfaced, care should be taken to ensure that fire apparatus can easily maneuver throughout the yard. This requires properly developed roadways capable of supporting heavy trucks. Proper site drainage is also essential.

It is good practice to paint lines within the yard to show aisles and roadways, as well as the location of yard hydrants, extinguishers, water buckets, and hose houses.

Milosh Puchovsky, P.E., is senior fire protection engineer in the NFPA general engineering division.

Layout of the Site

A site plan is advised. It should detail where specific materials are stored, the proper separation distances between piles, and access points to all site areas, including fences and their gates.

The density of stored materials tremendously influences the manner in which the material burns. Lumber stacked in solid, orderly piles to a height of 20 ft (6 m) does not present the same problem as lumber stored in "stickered" piles. In the latter case, fire develops more rapidly because of air spaces in the piles. Pulpwood is sometimes stored in piles 200 ft (61 m) in diameter and 100 ft (30 m) high; these piles contain air spaces and pockets, made by the logs' rough and varied shapes, which accelerate burning.

Certain materials burn quickly and produce a great deal of heat. Baled cotton, hay, lumber, packing materials, pallets, plywood, pulpwood, and rubber are examples of such materials, and considerable clearance should be provided between and around these piles. Aisle widths equal to the height of the pile are desirable for these types of commodities.

Both the yard itself and the piles within the yard must be easily accessible to fire fighters. Access to storage close to railroad spurs is often difficult because of boxcars and other railroad equipment left on the tracks. Driveways at least 15 ft (4.5 m) wide let fire equipment reach all portions of the storage area. Aisles of 10 ft (3 m) or more reduce the likelihood of fire spreading from pile to pile. In unusually large storage yards or moderate-size yards with valuable commodities, main aisles or firebreaks subdivide the storage much like fire walls do indoors. Overall, ideal aisle width depends upon a number of factors, including the type of material, how it is stored, storage height, anticipated wind conditions, fire-fighting capabilities, and fire-fighting equipment.

In some cases, packaging and palletizing affect storage methods, particularly of irregularly shaped or small materials. Combustible pallets may present an added hazard. Some materials are packed in heavy crates. Usually, they do not increase hazard except after severe weathering or where handling has caused breakage or splintering. Tarpaulins and other weather-resistant materials used to cover stock should be of fire-retardant-treated material.

A stable pile under normal conditions may present a severe hazard in a fire. It is wise to anticipate how fire and water affect the stability of piles when planning their sizes and configurations. Collapsing piles may cause severe fire spread and hamper fire fighting, particularly if there is danger from flying brands.

Areas used to service and maintain equipment should be designated and properly marked or cordoned off. Flammable liquid fuel, if used, should be stored in an underground tank, if possible. If not, fuel storage tanks should be located so spills do not flow under or around storage, and conversely, so that a fire in a storage area will not ignite a tank of flammable or combustible liquid.

Installation of Fire Protection

Fire protection features and equipment should be installed during site preparation. Depending upon the site's size and location, installing private water mains and hydrants throughout the yard may be necessary.

In planning for hydrants and yard mains, the water supply, whether from public mains or private sources, should be able to furnish the needed fire flow in gallons or liters per minute.

Using accumulated fire experience, the Insurance Services Office (ISO) developed formulas to calculate fire flow for large unsprinklered buildings, based on the area, combustibility of stored goods, and a construction factor that, of course, does not apply to outdoor storage. It is important to realize that fire flows needed for hose stream protection of indoor unsprinklered storage, as well as outdoor storage, are far greater than those needed for a well-designed automatic sprinkler system and hose streams. Larger flows do not necessarily minimize the fire damage as well as lower flows utilized in sprinklered buildings, but they represent reasonable protection.

Table 9-19A contains calculated fire flows for hose stream protection of outdoor storage. They assume early detection, adequate hydrant availability, and access.

The four occupancy classes are defined as:

1. Limited combustible: Low combustibility with limited concentrations of combustible material.
2. Combustible: Moderate combustibility.
3. Free burning: Examples include cotton bales, furniture stock, wood, and wood products.
4. Rapid burning: Commodities which either:
   (a) Burn with great intensity.
   (b) Ignite spontaneously and are difficult to extinguish.
   (c) Emit flammable vapors or dusts at ordinary temperatures.

Properly marked portable fire extinguishers, preferably the nonfreezing water type, should be placed so that the distance to the nearest unit is 75 ft (23 m) or less. For active and normally occupied sites, fully equipped hose houses should be provided and personnel trained to use them in a fire incident.

A means for notifying the fire department is essential. A system that directly signals the fire department is best; minimally, a telephone should be accessible.

Security Measures

Because they can attract youths, outdoor storage yards may have problems with trespassing, vandalism, and theft. Measures should be taken, such as installing fencing and lighting, to deal with these problems. This also protects against possible arson, sometimes attempted by children and vagrants. A fence, however, must not prevent ready access to the yard by the fire department. The fire

TABLE 9-19A. *Fire Flow for Hose Stream Protection of Outdoor Storage**

Area		Limited Combustible		Combustible		Free Burning		Rapid Burning	
sq ft	m²	gpm	L/min	gpm	L/min	gpm	L/min	gpm	L/min
10,000	929	1,500	5,678	1,750	6,624	2,000	7,571	2,250	8,517
20,000	1,848	2,000	7,571	2,500	9,464	3,000	11,356	4,000	15,142
50,000	4,645	3,500	13,429	4,000	15,142	4,500	17,034	6,000	22,712
100,000	9,290	5,000	18,927	5,500	20,820	6,500	24,606	8,000	30,283
200,000	18,580	6,750	25,551	8,000	30,282	9,250	35,015	10,000	37,854

*Adapted by author from ISO formulas. Assumes early detection, adequate hydrants, and access.

department should be consulted about incorporating remote gates in the fencing to provide additional access under serious fire conditions.

Security guards can very effectively protect outdoor storage yards. NFPA 601, *Standard for Security Services in Fire Loss Prevention*, covers the selection, duties, instruction, and training of such personnel.

UTILIZING THE SITE

Management

When a yard is selected and laid out, the maximum amount of material to be stored in it is also established. This quantity should not be exceeded because excess material could reduce the fire safety available. Frequent periodic inspections should monitor the quantity of the material and other fire prevention and protection features, including disposal of waste materials.

In general, management's attitude toward the yard's fire protection provides an example that employees are likely to follow.

Fire Protection and Prevention

All fire protection equipment, including hydrants, fire pumps, fire extinguishers, and any suppression, detection, or alarm systems on the storage site, should be properly maintained and tested. All material-handling equipment at the site should be equipped with a portable fire extinguisher suitable for any fire, i.e., a Class ABC extinguisher.

The locations of all hydrants, hose houses, portable extinguishers, alarm boxes, and other fire protection equipment should be properly marked, preferably with elevated signs. Arrows and signs painted on the pavement are the minimum markings required.

If material to be torch cut or welded cannot be moved from the storage yard to a remote location, work should be done only after taking adequate precautions. (See NFPA 51B, *Standard for Fire Prevention in Use of Cutting and Welding Processes*.)

Depending upon the nature of the material stored, smoking should be prohibited or restricted to specific areas. In either case, adequate signs should be displayed and regulations strictly enforced.

A private fire brigade comprised of yard employees provides initial fire-fighting capabilities. If there are enough employees to organize a fire brigade, the availability and ability of a public fire department normally governs the level of organization, training, and equipment provided to the brigade. The larger the storage yard and the more valuable its contents, the greater the need for a private fire brigade. NFPA 600, *Standard on Industrial Fire Brigades*, gives helpful information on this subject.

Perhaps the most important duty of a brigade or security guard is to immediately notify the fire department of fire. The fire department should be called even if brigades can handle the emergency without outside help. It is important that notifying the public fire department not be delayed while brigades or security guards attempt to control fires, as this may result in higher losses than necessary.

Maintenance and Housekeeping

Repairs to fencing, lighting, and the yard surface should be made promptly, and material handling equipment should be kept in good condition. Suitable safeguards should be provided to minimize the hazard of sparks from equipment, such as refuse burners, boiler stacks, vehicle exhausts, and locomotives.

Good housekeeping also includes controlling weeds and vegetation. Weeds, grass, and other vegetation should be sprayed with an acceptable herbicide or ground sterilizer, or grubbed out. Dead weeds should be removed after destruction, but weed burners should not be used in removal.

In many instances, scrap lumber, broken pallets, broken containers, bales, or pieces of material stored on the site are discarded into the clear space around the yard. Daily checks should be made for such discarded materials; where found, they should be properly removed. Care should be taken that wind-blown debris and other combustible materials do not accumulate under material piles.

If a pile of material falls into an aisle or clear space, the stock should be repiled immediately.

OUTDOOR STORAGE OF SPECIFIC MATERIALS

A discussion of safe storage practices for a few common materials follows. As a source for good practices, see the NFPA publications referenced and industry or insurance company standards. Many times, inside storage guidelines for a commodity are a good starting point for devising outdoor storage guidelines.

Wood and Wood Products

Most wood and wood products are stored outdoors. Because wood is moved from the forests only at certain times of the year, large quantities of logs are stored at sawmills, paper mills, and pulp mills. Large circular piles with logs dumped on the top are called "stacked piles." Logs stacked parallel, like matches in a box, are called "ranked piles." Stacked piles are usually much larger, with the piles' height often approaching 100 ft (30.5 m). Ranked piles are usually 10 to 15 ft (3 to 5 m) high. Because maintaining separation and aisles is much easier for ranked piles, fire protection is also much easier for ranked piles. (See Figure 9-19A.)

Controlling log yard fires requires large quantities of water, perhaps many thousands of gallons or liters per minute. This is especially true if many portable turrets and monitor nozzles, each discharging about 1,000 gpm (3785 L/min) operate simultaneously. A looped underground fire main system with hydrants spaced throughout the yard is good practice. Monitor nozzles on towers are advantageous in log yards, especially those with high stacked piles. Burrowing fires in large stacked piles are difficult to extinguish even after surface burning is under control. NFPA 46, *Recommended Safe Practice for Storage of Forest Products* (hereinafter referred to as NFPA 46), provides additional guidelines for storing logs.

Chip storage has replaced log storage at many pulp and paper mills. Two completely different types of fires occur in chip piles:

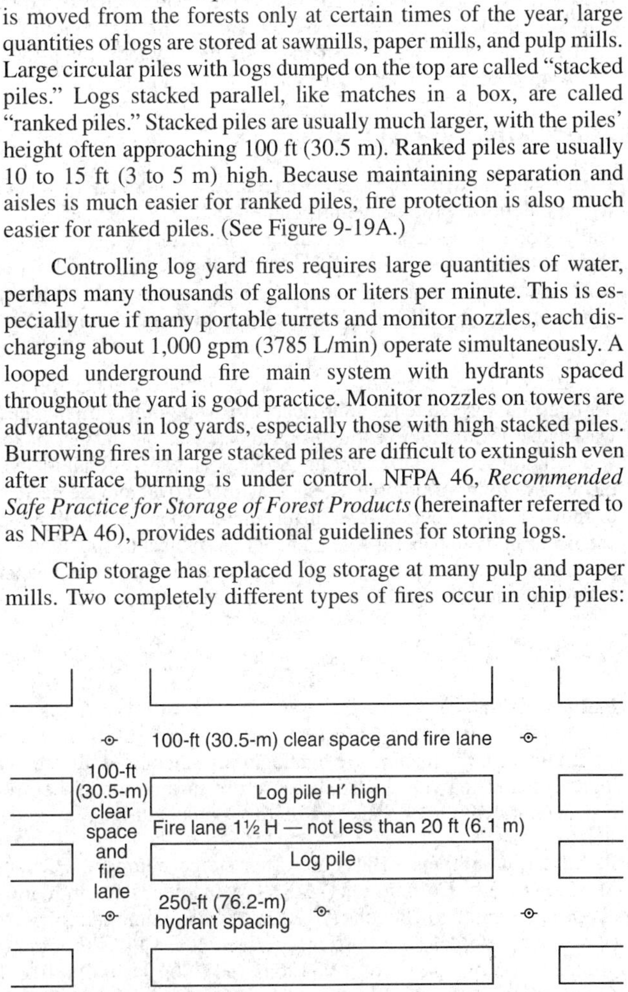

FIG. 9-19A.   *A good layout for ranked piles of logs in a storage yard.*

surface fires and internal fires. NFPA 46 provides guidelines for fire protection of chips. The amount of water needed to control a chip pile fire varies substantially, depending upon the pile's size. Weather conditions, operating methods, geographic location, wood chip type, and the degree to which wetting may be employed all influence fire conditions. Experience indicates that exposure to long periods of hot, dry weather with no regular surface wetting creates conditions under which fast-spreading surface fires can occur. Many hose streams would be required.

Likewise, frequency of pile turnover and operating methods affect prospects for serious internal fires. Piles built using methods that allow a concentration of fines, and piles stored for long time periods with no turnover, are subject to internal heating which, if undetected, can create intense internal fires.

Sawdust storage is not common now but was once prevalent at sawmills. Fires in sawdust can smolder unless the piles are broken open and the fire is completely extinguished.

Lumber storage can occur in urban areas, even though local ordinances generally prohibit storing any appreciable quantity in zoned areas, because lumberyards originally located in unzoned areas may come to border heavily developed areas. Exposure of stored wood to adjoining areas is a major consideration. Adequate aisles and good housekeeping, along with proper procedures for controlling ignition sources, such as salamanders, smoking, etc., reduces the fire hazard considerably. Large quantities of water may be needed to bring a lumberyard fire under control after it has had time to develop. Flying brands can readily spread fire throughout the yard and into adjoining areas, particularly under windy conditions. NFPA 46 provides guidelines for safely storing lumber.

Paper and Paper Products

Some paper and paper products are stored outdoors; most are stored indoors to protect them from the weather. Roll paper is normally stored in large rolls weighing 2 tons (1814 kg) or more. Tightly wound paper is very difficult to ignite; however, roll paper stored outdoors is more vulnerable. Once ignited, fire in roll paper stored outside is difficult to extinguish with manual fire-fighting methods.

Baled wastepaper does not produce as intense a fire as roll paper does, but instead tends to produce burrowing fires. Fire in baled wastepaper is difficult to extinguish and can generate heavy smoke, which inhibits manual fire fighting. Broken bales are difficult to handle after wetting. A motorized vehicle with a shovel may be needed to move debris to complete extinguishment. Spontaneous ignition can occur in wastepaper if foreign materials subject to spontaneous heating become trapped in bales. Fire protection for wastepaper storage is generally the same as for other outdoor storage.

Rubber

Rubber tires: Tire storage presents a severe hazard. Tires can burn rapidly, emitting intense heat and large quantities of dense smoke that hamper manual fire fighting. Large quantities of water at high pressures may be needed, because tire casings shield the fire. Narrow piles with good aisles are best for outdoor storage.

Baled crude rubber: Bales are often stored outdoors. Factory Mutual Research Corporation (FMRC) recommends that individual piles be limited to 100 tons (90 720 kg) with a minimum of 30 ft (9 m), but preferably 50 ft (15 m) between piles.[1] Piles should be grouped to a maximum of 1,000 tons (907 000 kg) with 100-ft (30-m) aisles separating them. It is important that burlap used to wrap the bales not be contaminated with oil; oil in a tightly packed pile may result in spontaneous heating.

Baled Combustible Fibers

Although storing baled combustible fibers outdoors is not good practice, much of this material, such as baled cotton in transit from the gin to a warehouse, is temporarily stored outdoors. FMRC recommends that piles be limited to 500 bales per pile with a minimum of 30 ft (9 m), and preferably 50 ft (15 m), of clear space between individual piles, from an exterior fence, or from buildings.[2] Piles of uncleaned baled flax straw should be limited to 300 tons (272 000 kg) and a height of 20 ft (6 m), and should be located 200 ft (60 m) from buildings and 100 ft (30 m) from each other or potential ignition sources. Storing bales of combustible fibers off the ground on pallets provides ventilation and prevents excessive damage from ground moisture. Covering the tops and sides of all piles with a securely fastened tarpaulin of fire-retardant material protects them from weather and mildew.

Coal

Bituminous coal can heat spontaneously. Anthracite coal is not subject to spontaneous heating. Freshly mined coal, however, is more susceptible because it absorbs oxygen more rapidly. Moisture and coal dust also make coal more susceptible to spontaneous ignition.

BIBLIOGRAPHY

References Cited

1. Factory Mutual Research Corporation, "Storage of Baled Crude Rubber," Loss Prevention Data Sheet 8-1, Factory Mutual Research Corp., Norwood, MA.
2. Factory Mutual Research Corporation, "Baled Fiber Storage," Loss Prevention Data Sheet 8-7, Factory Mutual Research Corp., Norwood, MA.

NFPA Codes, Standards, and Recommended Practices

Reference to the following NFPA codes, standards, and recommended practices will provide further information on outdoor storage practices discussed in this chapter. (See the latest version of *The NFPA Catalog* for availability of current editions of the following documents.)

NFPA 10, *Standard for Portable Fire Extinguishers*
NFPA 24, *Standard for the Installation of Private Fire Service Mains and Their Appurtenances*
NFPA 46, *Recommended Safe Practice for Storage of Forest Products*
NFPA 51B, *Standard for Fire Prevention in Use of Cutting and Welding Processes*
NFPA 80A, *Recommended Practices for Protection of Buildings from Exterior Fire Exposures*
NFPA 231, *Standard for General Storage*
NFPA 505, *Fire Safety Standard for Powered Industrial Trucks, Including Type Designations, Areas of Use, Maintenance, and Operation*
NFPA 600, *Standard on Industrial Fire Brigades*
NFPA 601, *Standard for Security Services in Fire Loss Prevention*

Additional Reading

Chubb, M., "Aboveground Storage Tank Debate," *Fire Engineering*, Vol. 148, No. 6, June 1995, pp. 73–74, 76, 78–81.
"Coal and Coal Storage," Loss Prevention Data Sheet 8-10, Factory Mutual Engineering Corp., Norwood, MA.
Devonshire, J., "Protecting Against Storage Tank Fires," *Fire International*, No. 144, Aug./Sept. 1994, pp. 23–24.
El-Iskandarani, B. M. K., "Emergency Relief Venting Analysis for Flammable Liquid Storage Tanks," Worcester Polytechnic Inst., MA, Thesis, May 1994.
"General Storage Safeguards," Loss Prevention Data Sheet 8-0, Factory Mutual Engineering Corp., Norwood, MA.

Howard, H. A., "Tires Burning by the Acre," *Fire Engineering*, Vol. 141, No. 6, 1988, pp. 22–34.

Marsh, J. W., Hugger, E. J., and Zimmerman, L. A., "Storage Tank Design Faces Environmental Issues," *Consulting-Specifying Engineer*, Vol. 13, No. 5, April 1993, pp. 60–62, 64.

Mock, L. L., "Covered Floater in Jacksonville: Steuart Petroleum Bulk Storage Tank Fire," *Fire Engineering*, Vol. 146, No. 11, Nov. 1993, pp. 28–32, 34, 38, 40, 42, 44–46, 48–50.

"Outdoor Storage of Wood Chips," Loss Prevention Data Sheet 8-27, Factory Mutual Engineering Corp., Norwood, MA.

Persson, H., Nahlinder, J., and Tuovinen, H., "Brandrisker vid lagring och hantering av brandfarlig vara. [Fire Risks in Various Oil Tank Storage Areas.]," (Abstract in English), Swedish National Testing Inst., Boras, Sweden, SP Report 1990:18, 1990.

"Roll Paper Storage," Loss Prevention Data Sheet 8-21, Factory Mutual Engineering Corp., Norwood, MA.

"Sawmills and Lumber Yards," Loss Prevention Data Sheet 7-25, Factory Mutual Engineering Corp., Norwood, MA.

Stahl, K. H., "VdS-Brandversuche mit Kunststoff-Lagerbehaltern. [(German Property Insurers' Association) Informs About Fire Tests of Plastic Storage Tanks.]," *VFDB*, Feb. 1994, pp. 15–23.

Williamson, R. B., and Schroeder, R. A., "Fire Safety Assessment of the Scrap Tire Storage Methods," *Proceedings* of the International Conference on Fire Research and Engineering, Sept. 10–15, 1995, Orlando, FL, SFPE, Boston, MA, 1995, pp. 431–436.

Williamson, R. B., and Schroeder, R. A., "Fire Safety Assessment of the Scrap Tire Storage Methods," California Univ., Berkeley, ISBN 0-9516320-9-4; *Proceedings* of the Seventh International Interflam Conference, Interflam '96, March 26–28, 1996, Cambridge, England, Interscience Communications Ltd., London, England, 1996, pp. 89–101.

Wilson, M., "Protecting Against Storage Tank Bund Fires," *Fire Surveyor*, Vol. 22, No. 4, Aug. 1993, pp. 8–11.

MATERIALS-HANDLING EQUIPMENT

Revised by Richard E. Munson

Materials-handling equipment, particularly industrial trucks and automatic guided vehicles (AGVs), mechanical and pneumatic stock conveying systems, and cranes, provides essential services to industry and commerce. This chapter describes the basic types of equipment and systems for materials handling that are encountered and discusses the fire and explosion hazards inherent in them. It also discusses protection methods.

Readers interested in and/or concerned with materials-handling equipment should also consult Section 3, Chapter 1, "Electrical Systems and Appliances"; Section 3, Chapter 2, "Control of Electrostatic Ignition Sources"; Section 3, Chapter 26, "Storage and Handling of Grain Mill Products"; Section 4, Chapter 14, "Explosion Prevention and Protection"; Section 4, Chapter 15, "Dusts"; Section 4, Chapter 16, "Metals"; Section 4, Chapter 19, "Air-Moving Equipment"; and Section 6, Chapter 12, "Water Spray Protection."

INDUSTRIAL TRUCKS

Industrial trucks, which include lift trucks and AGVs, are among the most common types of materials-handling equipment. Lift trucks are available in many special designs to suit the many uses and types of load to be handled, with the fork and squeeze clamp types being most common. They may be propelled electrically or by diesel, gasoline, or LP-Gas engines. AGVs are usually propelled electrically. Unless these vehicles are of a type listed by a testing laboratory and properly maintained and used, they may introduce serious fire dangers. All vehicles should be selected, outfitted, maintained, and operated in accordance with the hazards of the locations.

TYPE DESIGNATIONS AND AREAS OF USE

NFPA 505, *Fire Safety Standard for Powered Industrial Trucks, Including Type Designations, Areas of Use, Maintenance, and Operation* (hereinafter referred to as NFPA 505), lists 15 different type designations of industrial trucks or tractors. These designations are divided by power types: diesel, electric, gasoline, and LP-Gas. In addition, dual-fuel trucks are available. The designation letter reflects the power type:

1. Type D units are diesel-powered units having minimal acceptable safeguards against inherent fire hazards.
2. Type DS units are diesel-powered units that meet all requirements for the Type D units and have additional safeguards in the exhaust, fuel, and electrical systems.
3. Type DY units are diesel-powered units that have all the safeguards of Type DS units and, in addition, have no electrical equipment, including ignition. They are also equipped with temperature limitation features.
4. Type E units are electric-powered units having minimum acceptable safeguards against inherent fire and electrical shock hazards.
5. Type ES units are electric-powered units that meet all requirements for Type E units and additional safeguards for the electrical system to prevent emission of hazardous sparks and to limit surface temperatures.
6. Type EE units are electric-powered units that meet all requirements for Type E and ES units, and have electric motors and all other electrical equipment completely enclosed.
7. Type EX units are electric-powered units that differ from the Type E, ES, or EE units in that the electrical fittings and equipment are so designed, constructed, and assembled that the units may be used in atmospheres containing specifically named flammable vapors, dusts, and, under certain conditions, fibers. Type EX units are specifically tested and classified for use in the Class I, Group D, or in Class II, Group G, hazardous locations defined in NFPA 70, *National Electrical Code®* (hereinafter referred to as NFPA 70).
8. Type G units are gasoline-powered units having minimum acceptable safeguards against inherent fire hazards.
9. Type GS units are gasoline-powered units that meet all requirements for the Type G units and have additional safeguards to the exhaust, fuel, and electrical systems.
10. Type LP units are liquefied-petroleum-gas-powered units having minimum acceptable safeguards against inherent fire hazards.
11. Type LPS units are liquefied-petroleum-gas-powered units that meet requirements for Type LP units and are provided with additional safeguards to the exhaust, fuel, and electrical systems.
12. Type G/LP units operate on either gasoline or liquefied petroleum gas and have minimum acceptable safeguards against inherent fire hazards.
13. Type GS/LPS units operate on either gasoline or liquefied petroleum gas, meet all requirements for the Type G/LP units, and have additional safeguards for the exhaust, fuel, and electrical systems.

Richard E. Munson is a principal consultant in the Safety, Health & Environmental Excellence Center of E. I. DuPont de Nemours Company and former chairman of NFPA's Industrial Trucks Technical Committee.

14. Type G/CN units operate on either gasoline or compressed natural gas having minimum acceptable safeguards against inherent fire hazards.
15. Type GS/CNS units operate on either gasoline or compressed natural gas and, in addition to meeting all the requirements for Type G/CN units, are provided with additional safeguards to the exhaust, fuel, and electrical systems.

Various types of industrial trucks are limited to the locations specified in Table 9-20A. Fires can result if less than minimum type trucks are operated in hazardous locations.

A potential fire source for gasoline-, diesel-, and LP-Gas-powered trucks is fuel leaks that are ignited by the hot engine, hot muffler, ignition system, other electrical equipment, or other sparks. This danger is somewhat lower for diesel trucks because of the higher flash point of diesel fuel. However, a special hazard in LP-Gas trucks is that the vapors are difficult to disperse and tend to gravitate toward lower spots or pits. Care must be exercised with LP-Gas trucks to avoid high temperatures near ovens, furnaces, and similar heat sources. Special safeguards in the design of some types of gasoline, diesel, and LP-Gas trucks help to reduce these fire hazards, but the hazards cannot be completely avoided and areas of use should be strictly controlled. (See Table 9-20A.)

To facilitate identification of truck types and their areas of use, a uniform system of marking has been developed and is described in NFPA 505. (See Figures 9-20A and 9-20B). Using these truck and building markers allows easy and quick recognition of vehicles approved for the location.

Fire Hazards and Prevention

Industrial trucks can contribute to property loss through careless operation or operation by inadequately or improperly trained operators. Collision with sprinkler piping, fire doors, and other fire protection equipment; too-high tiering of rack storage; and careless handling of loads, such as containers of flammable liquids, can contribute directly to fires and make fire fighting more difficult. A comprehensive instruction course for truck operators can reduce the risk of such accidents. Adequate, clear passageways for truck travel and clear warnings of overhead and exposed piping also reduces the number of incidents. A good source for safe operating

rules is ANSI B56.1, *American National Safety Standard for Low Lift and High Lift Trucks.*

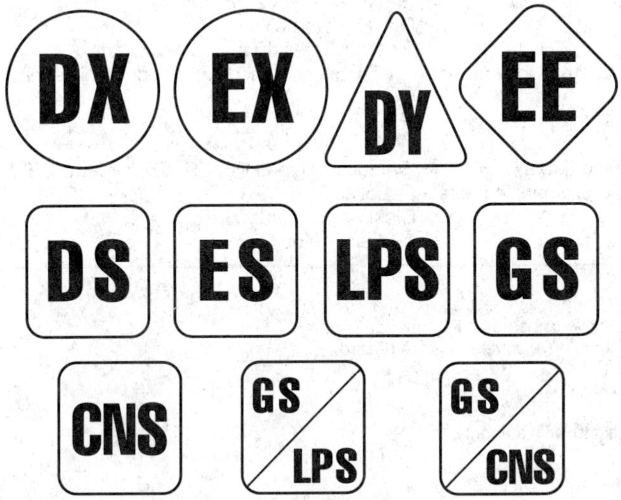

Fires involving industrial trucks can be caused by equipment failure due to inadequate or improper maintenance. Regularly scheduled maintenance should be based on engine-hour or motor-hour experience, to reduce the risk of these fires. Because many trucks operate on a nearly continuous basis, maintenance programs must be carefully followed. Especially important is detecting faulty or damaged fuel connections on gasoline, diesel, and LP-Gas trucks and removing accumulated grease, lint, and dirt. Portable extinguishers suitable for the hazards presented by the truck and the environment are recommended.

LP-Gas fuel containers require special care to prevent fires. They must not be overfilled. Use a soap solution to check for leaks, never a match or an open flame. Means should be provided in the fuel system to minimize the escaping fuel when exchanging removable LP-Gas containers, for example, closing the valve on the LP-Gas container and using an approved automatic quick-closing coupling in the fuel line. Removable LP-Gas containers must be securely mounted to keep them from jarring loose, slipping, or rotating. The pressure relief device inlet should always be in contact with the vapor space at the top of the container. An indexing pin is provided to assure that the LP-Gas container is positioned correctly. It is good practice to examine all LP-Gas containers for defects or damage before refilling.

Fires involving battery-powered trucks can occur due to electrical short circuits, hot resistors, arcing and fused contacts, lint accumulation, or exploding batteries.

Two types of batteries commonly used are lead and nickel-iron. They contain corrosive chemical solutions, either acid or alkali, and therefore present a chemical hazard. On charge, they give off hydrogen and oxygen which, when combined in certain concentrations, can explode. Battery-charging installations should be located in areas specifically set aside for that purpose. The facilities should include means for flushing and neutralizing spilled electrolyte; barriers for protecting charging apparatus from damage by trucks; adequate ventilation for dispersal of fumes from gassing batteries; and adequate fire protection. A carboy filler or siphon is needed for dispensing acid from carboys. Avoid open flames, sparks, or electric arcs in battery-charging areas when charging batteries. Also, keep the vent caps in place to avoid electrolyte spray, and ensure that vent caps are functioning. The battery or compartment cover(s) should remain open during charging to dissipate heat and gas.

Recharging and Refueling

Refueling and battery-charging operations should be performed by trained and designated personnel, and only in specified, well-ventilated areas away from manufacturing and service areas; outdoor refueling is recommended where practicable. Smoking should be prohibited in these areas. Scales for weighing LP-Gas containers should be calibrated periodically to ensure accurate filling.

Trucks using liquid fuels, such as gasoline and diesel fuel, should only be refueled from approved dispensing pumps in safe locations away from heat and ignition sources. Care must be taken not to spill fuel or overfill the vehicle fuel tank, as many fires involving trucks using liquid fuels result from spillage during refueling.

Maintenance and Repair

Trucks should never be repaired in Class I, II, and III hazardous locations. Repairs to the fuel and ignition systems of industrial trucks that involve fire hazards should be conducted in locations designated for such repairs. Repairs to the electrical system of battery-powered

FIG. 9-20A. Markers used to identify types of industrial truck.

TABLE 9-20A. *Recommended Types of Trucks for Various Occupancies*

Location	Typical Occupancies	Types of Trucks†‡ Approved and Listed
Indoor or outdoor locations containing materials of ordinary fire hazard	Grocery warehouse Cloth storage Paper manufacturing and working Textile processes except opening, blending, bale storage, and other Class III locations Bakery Leather tanning Foundries and forge shops Sheet-metal working Machine-tool occupancies	Electrical—Type E Gasoline—Type G Diesel—Type D LP-Gas—Type LP Dual Fuel—Type G-LP
Class I, Division 1.* Locations where explosive concentrations of flammable gases or vapors may exist under normal operating conditions or where accidental release of hazardous concentrations of such materials may occur simultaneously with electrical equipment failure	Few areas in this division in which trucks would be used	Electrical—Type EX# Gasoline, diesel, and LP-Gas—not recommended for this service
Class I, Division 2.* Locations where flammable liquids or gases are handled in closed systems or containers from which they can escape only by accident or locations where positive mechanical ventilation normally prevents hazardous concentrations	Paint mixing, spraying, or dipping Storage of flammable gases in cylinders Storage of flammable liquids in drums or cans Solvent recovery Chemical processes using flammable liquids Paper and cloth coating using flammable solvents in closed equipment Rubber-cement mixing	Electrical—Types EE, EX** Diesel—Type DY** Gasoline, diesel Types D & DS, and LP-Gas—not recommended for this service
Class II, Division 1.* Locations where explosive mixtures of combustible dusts may be present in the air under normal operating conditions or where mechanical failure of equipment might produce such mixtures simultaneously with arcs or sparks from electrical equipment or where electrically conductive dusts may be present	Grain processing Starch processing Starch molding (candy plants) Wood-flour processing	Electrical—Type EX¶ Diesel—Type DY Gasoline, diesel Types D & DS, and LP-Gas—not recommended for this service
Class II, Division 2.* Locations where explosive mixtures of combustible dusts are not normally present or likely to be thrown into suspension through the normal operation of equipment but where deposits of such dust may interfere with the dissipation of heat from electrical equipment or where such deposits may be ignited by arcs or sparks from electrical equipment	Storage and handling of grain, starch, or wood flour in bags or other closed containers Grinding of plastic molding compounds in tight systems Feed mills with tightly enclosed equipment	Electrical—Types EE, EX, ES‖ Gasoline—Type GS‖ Diesel—Type DY, DS‖ LP-Gas—Type LPS‖
Class III, Division 1.* Locations where easily ignitible fibers or materials producing combustible flyings are handled, manufactured, or used	Opening, blending, or carding of cotton or cotton mixtures Cotton gins Sawing, shaping, or sanding areas in wood-working plants Preliminary processes in cordage plants	Electrical—Types EE, EX Diesel—Type DY Gasoline, diesel Types D & DS, and LP-Gas—not recommended for this service
Class III, Division 2.* Locations where easily ignitible fibers are stored or handled (except in process of manufacture)	Storage of textile and cordage fibers Storage of excelsior, Kapok, or Spanish moss	Electrical—Types EE, ES, EX, preferred; E‖ Gasoline Type GS Diesel—Type DS, DY LP-Gas—Type LPS

NOTES:
*Hazardous location as classified in the *National Electrical Code.*
†Type G (gasoline), Type D (diesel), and Type LP (LP-Gas) trucks are considered to have comparable fire hazard.
‡Type GS (gasoline), Type DS (diesel), and Type LPS (LP-Gas) trucks are considered to have comparable fire hazard.
¶Acceptable for Groups G and F, but subject to special investigation.
‖Acceptable but subject to special investigation.
#Class I, Division 1, Group D, only; no truck should be used in Groups A, B, and C.
**Class I, Division 2, Group D, only; may be used in Groups A, B, and C areas subject to local authority.
(Source: Factory Mutual Research Corp.)

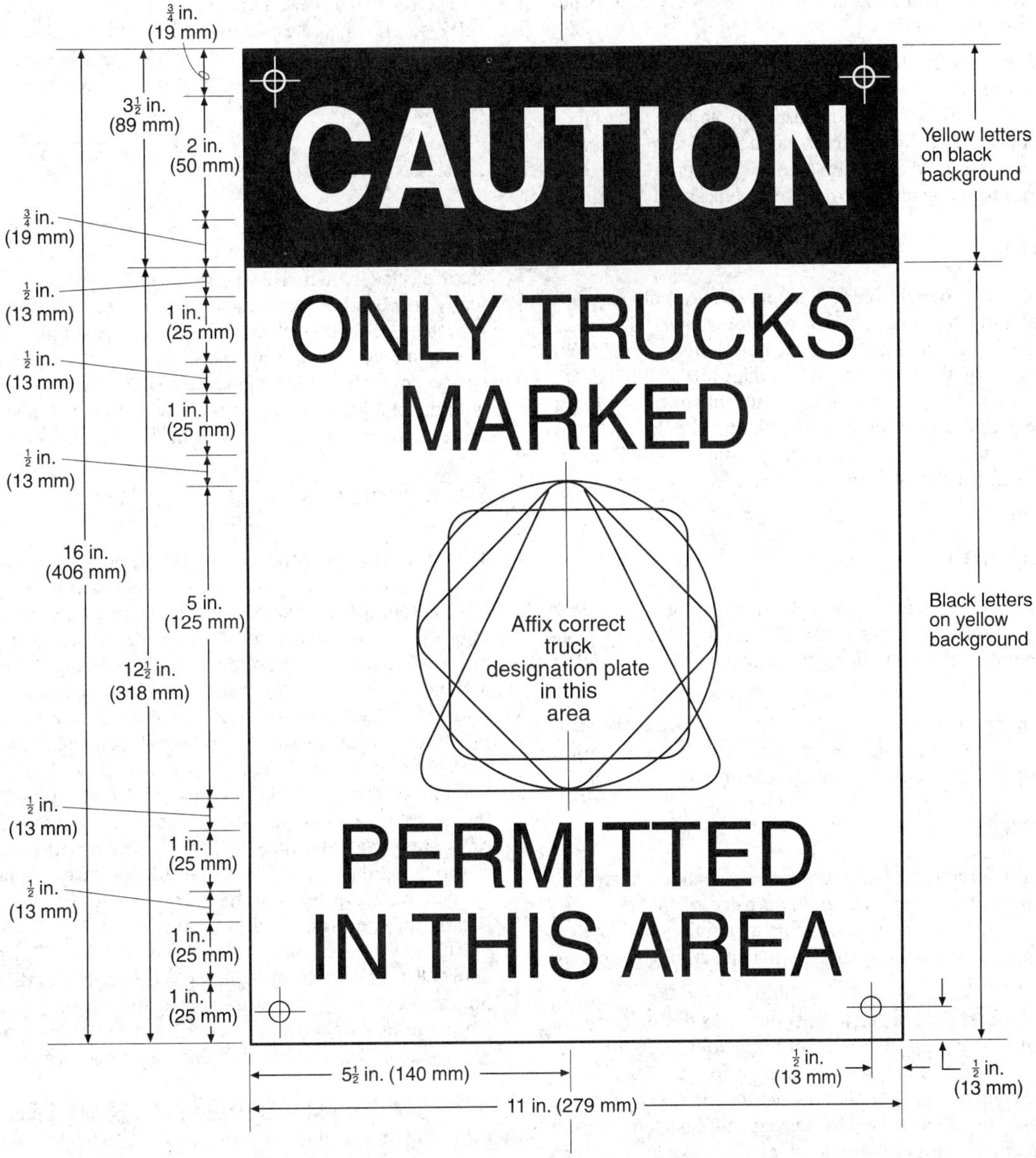

FIG. 9-20B. *Building signs for posting at entrances to hazardous areas.*

industrial trucks should be performed only after the battery has been disconnected. All parts of any industrial truck that need to be replaced should be replaced only with parts providing the same degree of fire safety as those used in the original design. Water mufflers should be filled daily or as frequently as needed to maintain the water supply at 75 percent of filled capacity or above. Do not operate vehicles with clogged mufflers, screens, or other parts. Immediately remove from service any vehicle that emits hazardous sparks or flames from its exhaust system, and do not return it to service until the cause of sparks and flames has been eliminated. When the temperature of any part of any truck exceeds normal operating temperature, a hazardous condition is created, and the vehicle must be removed from service until the cause of overheating has been eliminated.

It is good practice to keep industrial trucks clean and free of lint, excess oil, and grease. The conditions under which the truck operates should govern the frequency of cleaning. Noncombustible agents should be used for cleaning trucks; low-flash-point [below 100°F (38°C)] solvents must not be used. Precautions regarding tox-

icity, ventilation, and fire hazard should be appropriate for the agent or solvent used. When the cooling system requires antifreeze, a glycol-based material should be used.

MECHANICAL CONVEYORS AND ELEVATORS

Commonly used equipment in materials handling are mechanical conveyors and elevators that efficiently transport bulk and packaged materials. There are many classes and designs of mechanical conveyors. The primary selection considerations are the distance and inclination over which material is to be conveyed; the types of atmosphere and location through which material is to be conveyed; and the lump size, density, fluidity, abrasiveness, toxicity, corrosiveness, and other properties, of the materials to be conveyed. Dust materials, or dust atmospheres, or both, require equipment that can be enclosed adequately. The lump size screw and *en masse* conveyors can handle high temperatures that often rule out belt conveyors;

long distances usually require belt conveyors. Bucket elevators best handle substantial vertical lifts.

Important fire protection considerations around conveyors and elevators include temperature, dust, and the material being conveyed. For belt conveyors both the belt's combustibility and the material being conveyed must be considered; both can contribute to fire and its spread. When designing and installing conveyors, protecting openings and controlling static charges must be given careful attention.

Temperature

Screw, vibrating, and certain pan conveyors are normally the best choices for handling very hot materials or for use in very hot atmospheres. While the upper limit for most belt conveyors is 200°F (93°C), they are normally restricted to less than 150°F (65°C).[1] The highest rate of flame propagation occurs with rubber belts but that the ignition characteristics of neoprene, rubber, and polyvinyl chloride differ little. Thus, most belt conveyors are not suitable for moving of hot or molten materials and are not used in high-temperature atmospheres.

Fire Protection

Automatic sprinkler or water spray protection is generally needed for important belt conveyors, particularly if they are enclosed. Controls should be provided to shut down the conveyor when sprinklers operate.

Hose stream protection to reach all portions of the conveyor is also needed. This protection can be provided by $1^1/_2$-in. (38-mm) hose connections to sprinkler lines or from hydrants.

Dust Control

Factors in fires involving mechanical conveyors include dusty material, a dust atmosphere, and dust usually created by the materials-handling process. Dust is produced in most materials-handling systems, especially at inlet and discharge points and at long chutes for free-falling materials.

Dusty material should be fed to belt conveyors through a choke feed to prevent dust clouds. Where dusty conditions prevail, adequate aspiration should be provided to remove dust to collectors in safe locations. If dust clouds cannot be avoided, it is better to use spiral or enclosed conveyors where the escape of dust can be more readily prevented. When combustible or explosive dusts are handled, conveyor piping should be strong enough to withstand the maximum pressure produced in explosions of the dust involved, and adequate static grounding is essential. Avoid sharp changes in direction wherever possible, and provide vent pipes to outdoors at any necessary changes in direction and at the ends of lines. (This should not be construed as prohibiting the use of explosion-relief vents.)

Screw conveyors should be fully enclosed in tight, noncombustible housings with free-lifting covers at the discharge end and over each shaft coupling. En masse or drag-type conveyors should be of substantial metal construction designed to prevent the escape of dust. Covers on cleanout, inspection, and other openings should be securely fastened. Conveyors should be designed and constructed to withstand anticipated explosion pressures, considering the pressure release that explosion-relief vents afford. A choke or seal of proper design or a suppression system should be installed in a conveyor (other than pneumatic) to prevent an explosion from propagating from one building to another or from one portion of a building to another separated by a fire wall.

Dust collectors should be located outdoors or in detached rooms with adequate explosion vents for the collectors and the rooms. Where it is necessary to use bag-type collectors, they should be enclosed in a metal housing. A single dust collector should not serve several separate processes. Use individual collectors for progressive stages in any one process. Water spray collectors may be located within buildings, but they are not recommended for certain types of dust.

Static Protection

All machinery and conveyor parts should be thoroughly bonded and grounded to minimize static discharge. The chances of generating static electricity increase when heated or dry materials are pneumatically conveyed or a conveyor belt is operated in a heated or dry atmosphere.[2] Static electricity can be controlled by using belts made of conductive material, by applying conductive dressing to the belt surface, or by installing a grounded static collector nearly in contact with the belt just beyond the point where the belt leaves the pulley.[2] Pulleys, guards, and other metal bodies should also be grounded.

Protection of Conveyor Openings

Water spray method: Protecting wall and floor openings through which conveyors pass presents difficulties in installing ordinary forms of closures due to the objects carried through the opening. Where fire doors or shutters are impractical for conveyor openings, a method of protection incorporating the pressure effect and the cooling action of water spray from directed spray nozzles is available. (See Figure 9-20C.) With proper nozzle design and water pressure to provide suitable water velocity and droplet size, the pressure effect from the nozzles will overcome draft due to the temperature difference between one side of the wall and the other and the height of the opening above floor level, unless adverse air currents are prevalent. Since the spray's cooling effect is directly proportional to the time of exposure of hot gases in the draft to the spray, the effectiveness of the heat absorption may be increased by adding an enclosure to the opening. (See Figure 9-20D.) Figure 9-20E shows the cooling effect of four $1/_2$-in. (12.7-mm) nozzles on an 8×8-ft (2.4×2.4-m) opening with various draft velocities and nozzle pressures in tests conducted by Factory Mutual Research Corporation. The nozzles discharged water at 28 gpm (106 L/min) each, with an effective angle of about 65 degrees.

The Factory Mutual Research Corporation recommendations for installation include:

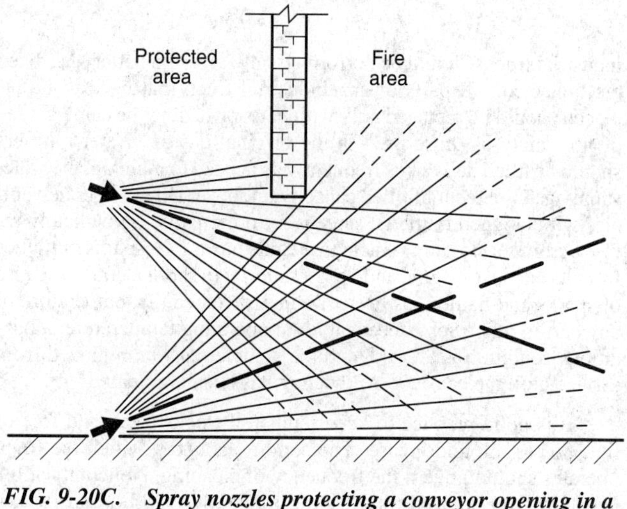

FIG. 9-20C. *Spray nozzles protecting a conveyor opening in a fire wall.*

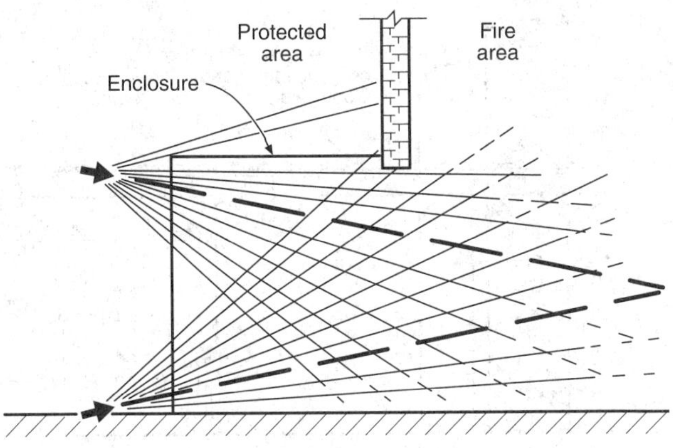

Protected area — **Enclosure** — **Fire area**

FIG. 9-20D.   *Greater heat absorption is possible by enclosing the opening for a conveyor.*

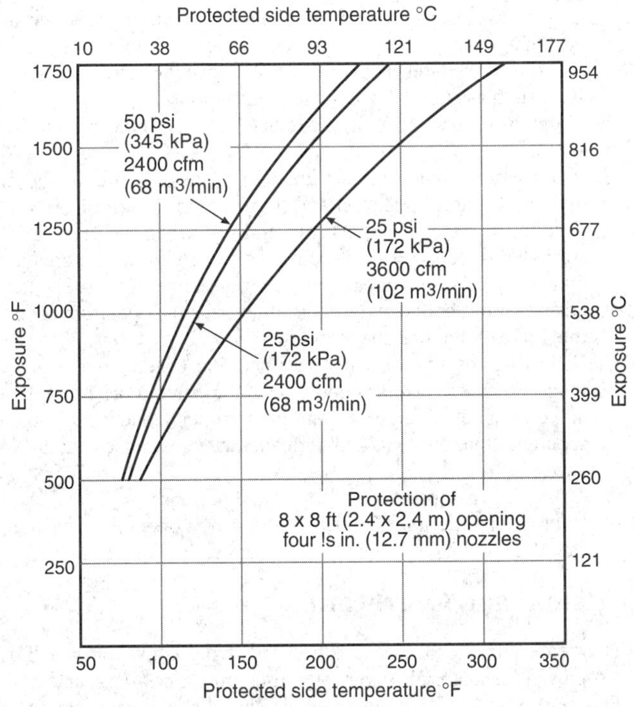

Protected side temperature °C

50 psi (345 kPa) 2400 cfm (68 m³/min)

25 psi (172 kPa) 3600 cfm (102 m³/min)

25 psi (172 kPa) 2400 cfm (68 m³/min)

Protection of 8 x 8 ft (2.4 x 2.4 m) opening four !s in. (12.7 mm) nozzles

Exposure °F — Exposure °C

Protected side temperature °F

FIG. 9-20E.   *Exposure temperatures vs. protected side temperatures at various draft velocities and nozzle pressures.*

1. Where fire may be expected to originate on either side of the opening, install nozzles on both sides. An automatic valve actuated by a heat detector controls the nozzles. Four nozzles per side are recommended to completely cover the opening. Water discharge rates between 2 and 4 gpm/sq ft [81 and 163 (L/min)/m²] or more, depending upon the opening's height and unfavorable draft effects, are desirable. Nozzles are located at an angle not more than 30° between the center line of nozzle discharge and a line perpendicular to the plane of the opening. To prevent the nozzle counterdraft from forcing air from the fire area into other areas, all communicating openings to the fire area are protected in a standard manner.

2. This method may also protect conveyor openings through floors, provided an enclosure is constructed around the convey-

or from the floor up to, or slightly beyond, the spray nozzles and draft curtains, extending 20 to 30 in. (0.5 to 0.8 m) below and around the floor opening. (See Figure 9-20F.)

3. The effectiveness of the protection system, of course, depends upon rapid detection and appropriate interlocks between the detection system and the machinery.

Fire doors: That conveyorized openings cannot be protected by fire doors is a misconception. Where possible, of course, conveyor penetration of a fire wall is avoided by rerouting or, as is sometimes feasible with a one-story building, by running the conveyor through the roof, over the fire wall, and down an inverted "V" housing arranged to vent fire readily to the atmosphere. (See Figure 9-20G.) Any cutout of a labeled fire door done in the field to allow closure around a conveyor track or other components voids its label. This practice should be avoided, if possible. Where notching is distinctly advantageous and inspection finds the notched door in compliance

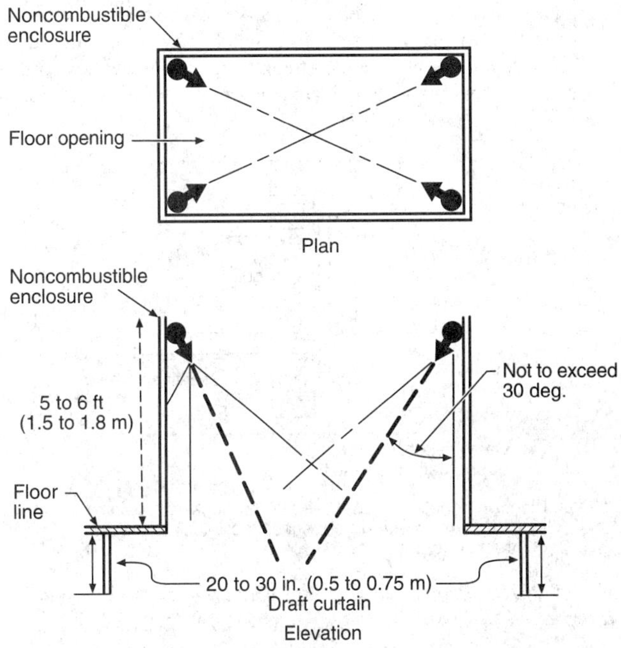

Noncombustible enclosure — Floor opening — Plan

Noncombustible enclosure — 5 to 6 ft (1.5 to 1.8 m) — Floor line — Not to exceed 30 deg. — 20 to 30 in. (0.5 to 0.75 m) Draft curtain — Elevation

FIG. 9-20F.   *Spray nozzle protection for floor opening.*

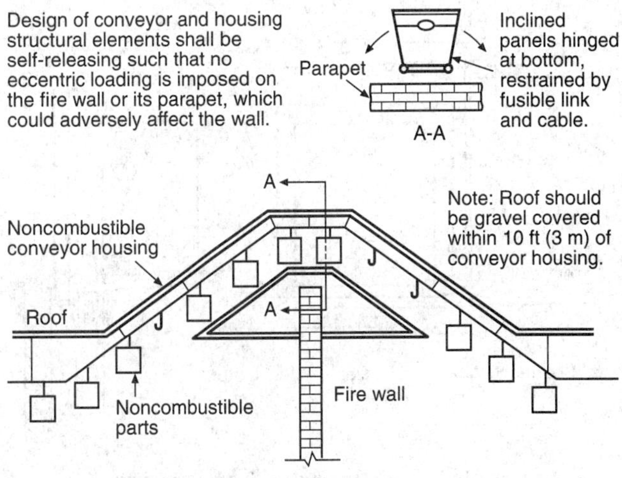

Design of conveyor and housing structural elements shall be self-releasing such that no eccentric loading is imposed on the fire wall or its parapet, which could adversely affect the wall.

Parapet — Inclined panels hinged at bottom, restrained by fusible link and cable. — A-A

Noncombustible conveyor housing — Roof — Noncombustible parts — Fire wall

Note: Roof should be gravel covered within 10 ft (3 m) of conveyor housing.

FIG. 9-20G.   *A conveyor carried over a fire wall.*

with the laboratory standards in all other aspects, a certificate may be furnished by the testing laboratory affixing the label.

Figures 9-20H through 9-20J illustrate various conveyor designs and/or programming devices that minimize or eliminate the threat of obstruction to complete fire door closure by the conveyor or conveyed stock.

The illustrations can only show basic concepts. Proper performance depends equally on conservative design, good workmanship in installation, operating inspection, and maintenance. Guidelines to observe are:

1. Select a design that acts as simply and directly as possible. Emphasis is on "fail safe" operation.
2. Program the sequence of operating steps and interlocks so that obstruction (conveyor or conveyorized material) to the door closure is positively and permanently—until manually reset—removed from the door's path before the door is released to close.
3. Design structural and mechanical components, linkages, clearances, etc., conservatively. Counterweights, springs, and other operating forces that are uninterruptable by initial fire stages must have enough reserve strength to handle overload introduced by reasonably anticipated minor changes in configuration

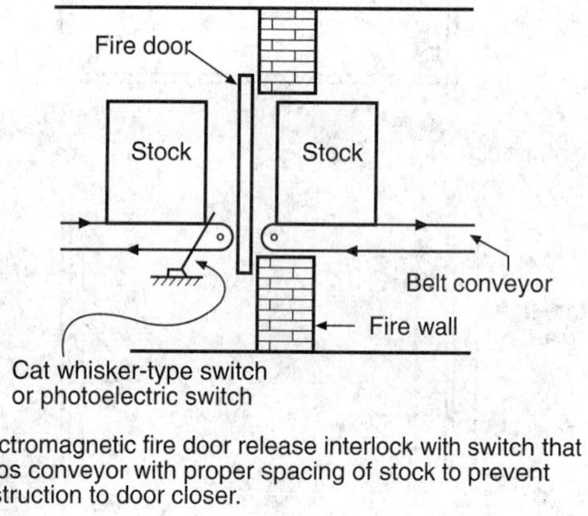

FIG. 9-20H.   *Protecting openings when a belt conveyor can be interrupted.*

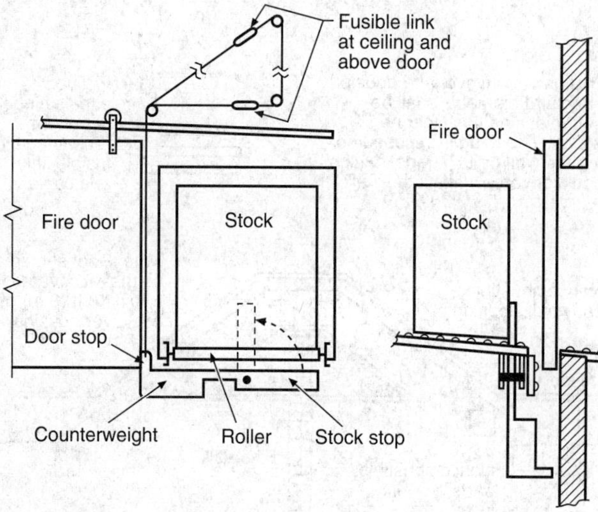

FIG. 9-20I.   *Method of stopping stock on a gravity roller conveyor.*

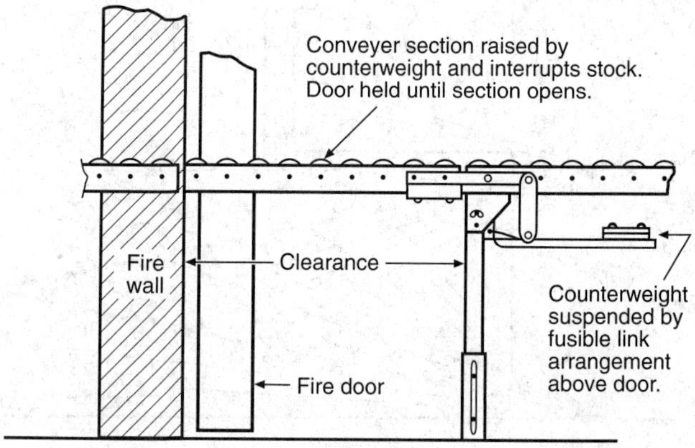

FIG. 9-20J.   *Counter-weighted, hinged section of a roller conveyor.*

and weight of conveyorized material, normal wear, and friction. Major changes necessitate complete re-engineering to ascertain design adequacy.
4. Incorporate self-releasing features in conveyor components' design, such as trolley track, chain, and supports.
5. Maintain $3/_8$-in. (9.5-mm) clearance between the door and the sill.
6. If advisable, provide another fire door on the opposite side of the opening to increase the reliability of the protection of the wall opening in the event of a fire. Similarly, when the property is sprinklered, consider reinforcing the protection of the opening by a water curtain from automatic sprinklers. Care must be taken that sprinkler discharge does not impede the operation of fusible links on the fire door assembly.
7. Following installation, conduct operating tests that reflect the range of varied adverse conditions anticipated to ascertain that all components operate smoothly, in proper sequence, within a specified time interval, and with adequate clearances and tolerances.
8. Close all fire doors during inoperative periods.
9. Routine closure that simulates emergency operation can indicate the continued adequacy of the protection of the opening.

Friction and Overheating

The heat of friction can cause fires involving conveyors, such as those used for materials like raw cotton, grains, powders, and coal. This results from accumulated grease and dirt and overheating of defective or unlubricated parts, especially rollers. This common fire hazard is easily reduced and controlled by patrolling belts, by frequent equipment inspection, and by the removal of dirt and grease buildups.[3] Early replacement of old and worn parts is also important.

Bucket Elevators

These elevators are used in bulk processing plants to convey loads vertically and are susceptible to the same fire hazards as mechanical conveyors. The same precautions need to be taken for temperature, dust control, protection of openings, and elimination of friction and overheating. Elevators should be enclosed in substantial dust-tight casings of noncombustible construction, extend without reduction in size through the roof, and be fitted with light, weatherproof covers designed to lift readily and relieve explosion pressure.

Elevators should be installed close to an outside wall of the building and provided with short direct vents through the wall at 20-ft (6-m) intervals on tall elevator legs. Doors should be installed to provide access to head and boot pulleys. All doors in elevator leg casings should be dust-tight. Ample clearance around elevator boots should be provided for cleaning and oiling. Safeguard elevators against overheating or choking by automatic releases actuated by overloading or a reduction in operating speed.

PNEUMATIC CONVEYORS

Pneumatic conveying is a common method of transferring powders and small solid particles from place to place in a building or from one building to another. This process presents an explosion hazard due to the ease and rapidity by which explosive proportions of dust can be introduced into the air.[2]

Description of Systems

Pneumatic conveying systems consist of an enclosed piping system that normally transports material by an air stream with a velocity sufficient to keep the conveyed material in motion. Noncombustible gases may be used in place of, or mixed with, air. These systems are of two principal types—pressure or suction—or a combination of the two types.

Pressure systems: Pressure systems transport material using air at greater than atmospheric pressure. These systems basically consist of a blower that draws air through a filter, an airlock feeder that introduces materials into the system, piping or ducts, and a suitable air-material separator.

Suction systems: Suction systems transport material using air at less than atmospheric pressure. These systems basically consist of a material and air intake, piping or ducts, a suitable air-material separator, and a suction fan or blower.

Figure 9-20K is a schematic drawing of a typical combination pressure and suction system.

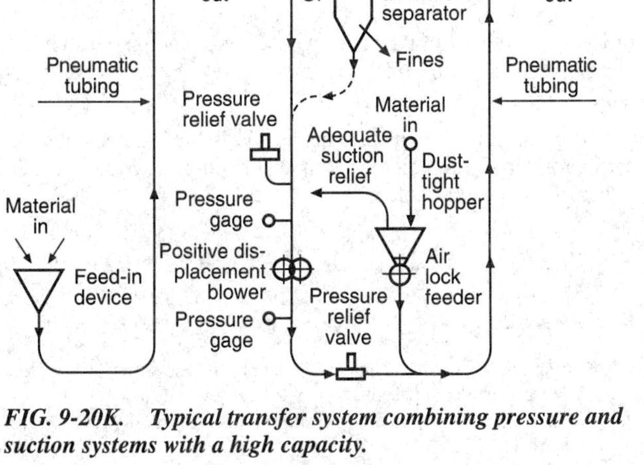

FIG. 9-20K. *Typical transfer system combining pressure and suction systems with a high capacity.*

Conveyor Ducts

Conveyor ducts are fabricated of nonferrous, minimum-sparking metal or of nonmagnetic, minimum-sparking stainless steel. They are electrically bonded and grounded. Plastic or other nonconductive liners are not desirable.

Inert Atmospheres

Inert gas is recommended in conveying systems wherever the concentration of powder or dust may come within the explosive range. The inert gas, such as nitrogen, argon, or helium with a limiting oxygen concentration based on the explosive characteristics of the particles in the system, is essential for preventing fires and explosions. Basically, it contains insufficient oxygen to permit combustion and no carbon monoxide, and it has a dew point such that no free moisture can condense or accumulate at any point in the system. The type of material being conveyed and the character of the inert gas dictate further limitations. For example, the inert gas for magnesium dust systems should not contain carbon dioxide. Oxygen limits of 3 to 5 percent should be maintained in aluminum powder systems using a controlled type of flue gas. Other limits are applicable where other inert gases and dusts are present.[4]

A continuous monitor is needed to sound an alarm if the oxygen content of the inert gas is not within the established safe range.

Light metal conveying: Various mechanical means of particle size degration produce light metal and light metal alloy powders. These processes, as well as certain finishing and transporting operations, tend to expose a continuously increasing area of new metal surface. Most metals immediately experience a surface reaction with available atmospheric oxygen, which forms a protective oxide coating that then serves as an impervious layer to inhibit further oxidation. This reaction is exothermic, producing sensible heat. If a fine or thin lightweight particle having a large area of new surface is suddenly exposed to the atmosphere, enough heat will be generated to raise its temperature to the ignition point.

A completely inert gas will not promote operational safety and is not used for transport of light metal powder in a pneumatic or fluidized transfer device. This practice would be very unsafe because somewhere in the process of manufacture, packaging, or ultimate use, the powder is eventually exposed to the atmosphere where the unreacted surfaces react suddenly with available oxygen to produce sufficient heat to cause either a fire or an explosion. To provide maximum safety, a means for controlling oxidation of newly exposed surfaces is provided—as soon as they are exposed—by regulating the oxygen content of the inert gas. An inert gas containing a limited and controlled amount of oxygen is effective for this purpose. This mixture controls the oxidation rate, thus providing an environment that materially reduces the fire and explosion hazard.

Air Conveying

If the conveying gas is air, as it often is in atomizing, the dust-air ratio throughout the conveying system is held below the minimum explosive concentration of the metal dust, as determined by data from the U.S. Bureau of Mines.[4] Although the metal-dust-air suspension may be held below the explosive concentration in the conveying system, the suspension necessarily passes through the explosive range in the collector at the end of the conveying system unless the dust is collected in a liquid, such as in a spray tower. Such wet collection is not always possible or desirable. Any liquid used must be nonflammable, nonreactive with metal dust, or reactive at a controlled minimum rate under favorably controlled conditions. The liquid remaining in or on the product must be compatible with subsequent processing requirements.

In an air conveying system, any dry collector must be considered an explosion hazard containing a dust-air mixture in the explosive range. It is, therefore, located in a safe place and provided with requisite barricades or other design means of protecting personnel. Construction is of nonferrous, nonsparking metal or of nonmagnetic, nonsparking stainless steel. The entire system, particularly the collector, is thoroughly and completely bonded and grounded. When checked with an ohmmeter, the ground system should show less than 5 ohms resistance to ground.

Where the conveying duct is exposed to weather or moisture, it should be moisture-tight, since any moisture entering the system can react with the dust, generating heat and serving as a potential ignition source.

A minimum conveying velocity is employed throughout the conveying system to prevent the accumulation of dust at any point and to pick up any powder that might drop out during an unscheduled system stoppage.

If the conveying gas is inducted into the system in a relatively warm environment and the ductwork and collectors are relatively cold, the gas temperature may drop below the dew point, causing condensation. To avoid this, the ducts and collectors are insulated or provided with a heating means.

Relief Vents for Conveyor Ducts

Vents of sufficient area connected to ducts or openings protected with antiflashback swing valves and extending to the outside of the building can provide explosion relief. Care should be taken to limit the inertia of swing valves to the minimum required. Rupture diaphragms can be used in place of swing valves. Wherever damage may result from a ruptured duct, the duct is designed for a minimum internal working pressure of 100 psi (690 kPa) in case the relief vent fails to offer sufficient pressure relief. Where the duct is so located that no damage will result if it bursts, it may be of very light construction to intentionally fail as an auxiliary vent for the system.

Fan Construction and Arrangement

Fan blades and housings that are used to move air or inert gas in conveying ducts are constructed of conductive, nonsparking metal, such as bronze, nonsparking stainless steel, or aluminum. In no case should the design be such that it draws the dust through the fan before entering the final collector. Personnel should not be permitted within 50 ft (15 m) of the fan during operation. This means that the fan and associated equipment are shut down for oiling, inspection, or preventive maintenance. If the area must be approached during operation for a pressure test or other technical reasons, it must only be done under the direct supervision of competent technical personnel and with the knowledge and approval of the operating management. Ultimately, all fans in dust collector systems accumulate enough dust to be a potential hazard; for this reason, locating them outside of all manufacturing buildings is preferable.

It is good practice to equip fan bearings with suitable instruments for indicating the temperature or to monitor vibration. Such instruments are wired with an alarm device to warn of over-temperature or imbalance.

Sight Glasses

Avoid sight glasses in pneumatic systems whenever possible. If installed, they should be of noncombustible material that is not readily subject to physical damage. The ducting is supported above and below each sight glass so that the sight glass does not carry any of the system weight and is not, in itself, subject to resulting stresses or strains. The system's electrical bonding must be continuous around all sight glasses. The strength of the sight glass, its mounting mechanism, and its inside diameter are equal to the adjoining tube system. In pressure-type systems, connections between sight glass and tubing are butted squarely and fastened together with rigid, airtight couplings, connectors, or comparable devices. In suction systems, connections between the sight glass and tubing are butted squarely and sealed with approved sleeves extending a minimum of 3 in. (76 mm) above and below the sight glass and tubing connections. The sleeves are of a material with elastic properties that provide an airtight seal.

Air-Material Separators

Air-material separators should preferably be located outside the building and provided with lightning protection. Material discharge outlets are provided with positive choke devices. The separator should be electrically conductive and bonded. Exhaust air should be always discharged to the outside, except where provision is made to recirculate transport air directly back into the pneumatic conveying system. The air-material separators should be constructed of noncombustible materials and the cloth filters of low-hazard materials. Where using combustible filter media is necessary, flame-retardant treatment is desirable. In addition, the cloth filters should be housed in grounded metal enclosures, with provision for cleaning the filters. The separators' construction should eliminate ledges or other areas where dust may accumulate.

CRANES

Cranes are an essential and necessary part of many industrial or manufacturing operations where heavy materials must be lifted or moved. Many cranes, such as overhead traveling, gantry, tower, and bridge-type cranes, move along rails. Overhead traveling can be used effectively indoors or out, whereas gantry, tower, and bridge cranes are used mainly outdoors. Mounting other cranes on wheels makes them mobile. Cranes can be powered by electricity or by internal combustion engine drive. Normal safeguards and maintenance must be observed with these power systems.

Outdoor cranes are susceptible to damage by high winds, unless a positive means of anchorage is provided, in addition to the operating or service brakes. Large cranes that move along rails are generally provided with automatic or manual rail clamps. Other means of anchorage are crane traps, wedges, and cables. Anchorage and support are particularly necessary and advisable when a crane is not in use and unattended.

Crane operators should be thoroughly trained in the equipment's proper operation, including emergency action that may be necessary in a fire or other hazardous incident. The crane operator's cab should be of noncombustible construction and provided with an emergency means of egress. It should be kept free of oily waste, rubbish, and other combustibles. All electrical equipment should be in conformance with NFPA 70, and securely mounted and maintained.

Portable fire extinguishers may be desirable in some operators' cabs. For large equipment, automatic fixed-piping extinguishing systems may be advisable.

BIBLIOGRAPHY

References Cited

1. Buffington, M. A., "Mechanical Conveyors and Elevators," *Chemical Engineering*, Oct. 13, 1969, pp. 33–49.2.
2. *Handbook of Industrial Loss Prevention*, second edition, McGraw-Hill, New York, 1976.

3. Mitchell, D. W., *et al.*, "Fire Hazard of Conveyor Belts," RI 7053, 1967, U.S. Bureau of Mines, Washington, DC.
4. Jacobson, M., Cooper, A. R., and Nagy, J., "Explosibility of Metal Powders," RI 6516, 1964, U.S. Bureau of Mines, Pittsburgh, PA.

Reference

Hartmann, I., Nagy, J., and Brown, H. R., "Inflammability and Explosibility of Metal Powders," RI 3722, 1943, U.S. Bureau of Mines, Pittsburgh, PA.

NFPA Codes, Standards, and Recommended Practices

Reference to the following NFPA codes, standards, and recommended practices will provide further information on materials-handling equipment discussed in this chapter. (See the latest version of *The NFPA Catalog* for availability of current editions of the following documents.)

NFPA 15, *Standard for Water Spray Fixed Systems for Fire Protection*
NFPA 70, *National Electrical Code*®
NFPA 80, *Standard for Fire Doors and Fire Windows*
NFPA 505, *Fire Safety Standard for Powered Industrial Trucks Including Type Designations, Areas of Use, Maintenance, and Operation*
NFPA 651, *Standard for the Manufacture of Aluminum Powder*

Additional Reading

Anderson, A. E., "Review of Fire Resistant Belting Requirements," *Proceedings* of the International Conference on Fire Resistant Hose, Cables and Belting, Plastics and Rubber Institute, 1986, pp. 1–12.
ANSI B56.1, American National Safety Standard for Low Lift and High Lite Trucks, American National Standards Institute, New York, 1993.
"Conveyor Belt Flammability: Small-Scale Test Produces Large-Scale Benefits," *Record*, Vol. 71, No. 1, Spring 1994, pp. 10–12.
Egan, M. R., "Smoke, Carbon Monoxide, and Hydrogen Chloride Production From the Pyrolysis of Conveyor Belting and Brattice Cloth," Bureau of Mines, Pittsburgh, PA, IC 9304, 1992, 18 pages.
"Fire Protection for Belt Conveyors," FMRC Loss Prevention Data Sheet 7-11, Factory Mutual Engineering Corporation, Norwood, MA.
Grannes, S. G., Ackerson, M. A., and Green, G. R., "Preventing Automatic Fire Suppression System Failures on Underground Mining Belt Conveyors," Bureau of Mines, Minneapolis, MN, IC 9264, 1990, 22 pages.
International Conference on Fire Resistant Hose, Cables and Belting, Plastics and Rubber Institute, London, 1986.
Janssens, L. J., "Solid Woven Belts Convey Their Message Underground," *Colliery Guardian*, Vol. 240, No. 2, 1992, pp. 56–58.
Kallioniemi, P., "Hihnakuljettimien paloturvallisuus. [Fire Safety of Conveyor Belts.]," *Palontorjuntatekniikka*, Vol. 4, 1994, pp. 6–9.

Litton, C. D., Lazzara, C. P., and Perzak, F. J., "Fire Detection for Conveyor Belt Entries," Bureau of Mines, Pittsburgh, PA, RI 9380, 1991, 27 pages.
May, J. P., "User's Guide for INSLOPE3: A Computer Code to Analyze the Effect of Haulage Truck Operation on Dump Point Stability," Bureau of Mines, Pittsburgh, PA, RI 9291, 1991, 25 pages.
Melissi, S., and McDonough, D., "Lifting Debris By Crane. Heavy Construction Equipment Aids In Search for Victims," *Firehouse*, Vol. 20, No. 9, September 1995, pp. 78, 80.
Mintz, K. J., " Long-Term Reproducibility of Small-Scale Tests Used for Measuring the Flammability of conveyor Belting," *Fire and Materials*, Vol. 15, No. 1, 1991, pp. 43–46.
Mintz, K. J., "Evaluation of Laboratory Gallery Fire Tests of Conveyor Belting," *Fire and Materials*, Vol. 19, No. 1, January/February 1995, pp. 19–27.
Nakagawa, Y., "Studies on Fire-Resistance Evaluation of Rubber Conveyor Belts With Fabric Skeletons," National Institute for Resources and Environment, Tsukuba-Shi, Japan, NIRE No. 2, Thesis, 1992, 120 pages.
Nakagawa, Y., and Komai, T., "Correlation Between the Thermogravimetric Analysis and Some Other Flammability Tests on Covers of Rubber Conveyor Belts," *Journal of Fire Sciences*, Vol. 8, No. 6, November/December 1990, pp. 455–476.
Nakagawa, Y., *et al.*, "Studies on Fire-Resistance Evaluation of Rubber Conveyor Belts With Fabric Skeletons. [With Revisions]," National Institute for Resources and Environment, Tsukuba-Shi, Japan, NIRE No. 2, Thesis, 1992, 120 pages.
Nakagawa, Y., Komai, T., and Kohno, M., "Laboratory-Scale Gallery Fire Test on Rubber Conveyor Belts With Fabric Skeletons," *Fire and Materials*, Vol. 15, No. 1, 1991, pp. 17–26.
Perzak, F. J., *et al.*, "Hazards of Conveyor Belt Fires. Report of Investigations," Pittsburgh Research Center, Pittsburgh, PA, RI 9570, May 1995, 33 pages.
Potts, J. D., and Jankowski, R. A., "Dust Considerations When Using Belt Entry Air to Ventilate Work Areas," Bureau of Mines, Pittsburgh, PA, RI 9426, 1992, 16 pages.
Schenkel, W. K., "Flame-Resistant Conveyor Belts Reinforced with Kevlar for Underground Coal Mining," *Proceedings* from the International Conference on Fire Resistant Hose, Cables and Belting, Plastics and Rubber Institute, London, 1986, pp. 1–17.
Szymanski, J., and Mintz, K. J., "Small-Scale Flame Tests on Fire-Retardant Conveyor Belting," *Journal of Fire Science*, Vol. 4, No. 4, July/Aug. 1986, pp. 231–236.

INDUSTRIAL OCCUPANCIES

Carol A. Caldwell

Industrial occupancies encompass a broad spectrum of purposes. Some employ many people to work in a single building. Other facilities house large items of equipment but have very few employees. Industrial occupancies are generally divided into two categories: general-purpose industrial, which are low- and moderate-hazard facilities, and high-hazard facilities. Examples of industrial occupancies include factories, laboratories, power plants, refineries, and laundries.

This chapter discusses the life loss history of industrial occupancies; general risk assessment; occupancy and occupant characteristics; fire protection features available; NFPA *101*®, *Life Safety Code*® (hereinafter referred to as NFPA *101*) criteria; and pre-fire planning.

For additional information relevant to life safety in industrial occupancies, see Section 3, Chapter 2, "Control of Electrostatic Ignition Sources"; Section 7, Chapter 3, "Interior Finish"; Section 7, Chapter 5, "Confinement of Fire in Buildings"; and Section 9, Chapter 2, "Occupancies in Special Structures and High-Rise Buildings." Explosion suppression devices are covered in Section 4, Chapter 14, "Explosion Prevention and Protection." Section 11, Chapter 5, "Deterministic Computer Fire Models," provides information on computer fire models.

LOSS HISTORY

According to NFPA's Fire Analysis and Research Division, industrial and manufacturing facilities averaged 18,500 structure fires per year in 1989-1993, with associated annual average losses of 37 civilian deaths, 738 civilian injuries, and $847.1 million in direct property damage. Fires, deaths, and injuries all declined from 1980-1984, by nearly one-half for fires and by nearly one-third for injuries. Nevertheless, these loss rates remained needlessly and unacceptably high, and industrial and manufacturing facilities still accounted for nearly three of every ten dollars lost to non-residential structure fires and more than one of every five civilian injuries occurring in nonresidential structure fires.

The past decade, 1987–1996, saw four U.S. industrial facility incidents with property damage severe enough to rank among the 20 most costly fires and explosions of all time in the U.S. after adjusting for inflation. These were a 1989 polyolefin plant fire in Texas, a 1995 textile mill fire in Massachusetts, a 1988 petroleum refinery

fire in Louisiana, and a 1987 chemical plant fire in Texas. Each incident cost more than $200 million in direct property damage. The previous decade, 1977–1986, had produced no industrial fires of such historic size.

A North Carolina chicken processing plant fire provided a recent example of the continued potential for life loss in industrial plant incidents, as 25 people died, making that fire one of the deadliest U.S. fires anywhere in the year it occurred.[1] However, the deadliest recent industrial fire in the world by far occurred elsewhere in the world, in a toy factory in Thailand in 1992.[2] That fire killed 188 workers, far surpassing the death toll of 145 in the Triangle Shirtwaist Company fire of 1911, which is still the deadliest manufacturing plant incident not involving an explosion in U.S. history.

Another recent fire, this one in 1992 in China, occurred in a doll factory and took the lives of 81 occupants.[3] These and other, less deadly incidents around the world underline the need for a higher standard of global industrial fire safety. Setting that standard starts with recognizing the many ways in which serious fires can occur in industrial and manufacturing facilities. Table 9-21A shows the cause patterns for fires and associated property damage in industrial and manufacturing properties; Table 9-21B shows the cause patterns for associated civilian deaths and injuries. The wide variety of potentially hazardous industrial and manufacturing process is clearly reflected in these figures, but so are more familiar hazards associated with heating, electrical service, arson, and smoking.

Even when they occur, industrial fires need not be deadly and need not be costly. Many factors typically contributed to turn an industrial fire into a major loss of life or property. One such factor is the lack of functioning fire suppression systems in a structure. The absence of appropriate fire detection systems is another factor.

Yet another factor is improperly designed egress systems. Unfortunately, locked exits still contribute to fatalities. At the North Carolina fire just cited, locked and blocked exits significantly contributed to the 25 deaths. An important cause of the loss of life in the Thailand fire was inadequate exit arrangements. It is only one example of how major industrial fires tend to involve a failure to follow well-established fire safety requirements. NFPA *101* provides a method for designing appropriate egress systems.

HAZARD ASSESSMENT AND EVALUATION

To provide adequate life safety for an industrial occupancy, the relative degree of fire and life safety risk must be evaluated. Since the

---

Carol A. Caldwell, P.E., is president of Caldwell Consulting Ltd, a fire protection engineering and research firm in Christchurch, New Zealand.

TABLE 9-21A.   *Causes of Fires and Direct Property Damage, Industry and Manufacturing Fires, 1989–1993 Annual Average, Unknown-Cause Fires Allocated Proportionally*

Cause†	Fires*		Property Damage* (in Millions)	
Other equipment	6,400	(34.7%)	$435.9	(51.5%)
Furnaces, ovens, or kilns	900	(4.8%)	15.4	(1.8%)
Working or shaping machines	700	(3.9%)	9.2	(1.1%)
Unclassified processing equipment	600	(3.0%)	306.0	(36.1%)
Separate motors or generators	400	(1.9%)	6.7	(0.8%)
Unclassified special equipment	300	(1.8%)	12.2	(1.4%)
Heat-treating equipment	300	(1.7%)	5.1	(0.6%)
Casting, molding, or forging equipment	200	(1.2%)	5.5	(0.6%)
Waste recovery equipment	200	(1.1%)	2.6	(0.3%)
Open flame, embers, or torches	2,500	(13.4%)	47.5	(5.6%)
Torches	1,400	(7.6%)	32.0	(3.8%)
Bonfires, campfires, or other open fires	300	(1.4%)	2.6	(0.3%)
Rekindles or reignitions	200	(1.1%)	2.1	(0.2%)
Electrical distribution	2,000	(10.7%)	68.6	(8.1%)
Fixed wiring	600	(3.2%)	19.5	(2.3%)
Overcurrent protection devices	300	(1.4%)	10.2	(1.2%)
Lighting fixtures, ballasts, or signs	200	(1.2%)	6.4	(0.8%)
Transformers	200	(1.0%)	5.7	(0.7%)
Incendiary or suspicious causes	1,500	(7.9%)	99.8	(11.8%)
Natural causes	1,400	(7.8%)	45.8	(5.4%)
Spontaneous ignitions	1,000	(5.3%)	32.7	(3.9%)
Lightning	300	(1.4%)	8.1	(1.0%)
Heating equipment	1,400	(7.7%)	71.9	(8.5%)
Fixed-area heaters	400	(1.9%)	5.6	(0.7%)
Central heating units	200	(1.3%)	3.3	(0.4%)
Appliances, tools, or air conditioning	1,000	(5.3%)	12.8	(1.5%)
Dryers	400	(2.0%)	2.7	(0.3%)
Other heat source	600	(3.4%)	16.1	(1.9%)
Exposure (to other hostile fire)	600	(3.2%)	13.3	(1.6%)
Cooking equipment	600	(3.1%)	7.6	(0.9%)
Smoking materials	500	(2.5%)	26.7	(3.2%)
Child playing	100	(0.5%)	0.9	(0.1%)
Total	18,500	(100.0%)	847.1	(100.0%)

Source: NFIRS, NFPA survey.

†The major cause categories are based on a hierarchy developed by the U.S. Fire Administration.

*Fires are expressed to the nearest hundred, property damage to the nearest hundred thousand dollars. Property damage figures have not been adjusted for inflation. Percentages are calculated on the actual estimates, so two figures with the same rounded-off estimates may have different percentages. Sums may not equal totals because of rounding errors.

industrial classification is so broad, the hazards involved vary greatly. The first concern is the amount of time available for evacuation. Both the contents of the building and the interior construction affect this, as well as the materials of the building itself if they contribute to the fire during occupant evacuation. Examples of items to consider are interior finish, rate of fire spread, burn rate, toxic fume evolvement, potential ignition sources, fuel load, and smoke generated. After employees evacuate, the problem becomes one of property protection.

Many resources exist to aid in hazard assessment. Examples include talking to plant operators, reviewing building blueprints, and so on. The NFPA codes contain valuable information on various hazards. Insurance companies and records of previous fires also provide useful information. Section 9, Chapter 1, "Assessing Life Safety in Buildings," describes methods for assessing life safety from fire in buildings, including exit analysis, available safe egress time (ASET), and scenarios.

Computer models can assist in assessing an occupancy's hazard. After determining the appropriate potential fire, a computer model can help predict the impact of fire on a facility. This is valuable in determining the possible time available for occupants to exit the area or facility before conditions become untenable. The ASET,

FPETOOL, and CFAST models do this. The assessment can be done in conjunction with egress analysis. Required safe egress time is a model to analyze egress. Combined models can evaluate whether there is sufficient time for the occupants to exit before conditions become untenable. This particular method of assessing hazard and evaluating life safety is common practice in code-enforcement systems that accept performance-based codes.

Because explosions are probably the most destructive industrial accidents in terms of life and property loss, they require special attention. Flammable dust, vapor, mist, or gas presents an explosion hazard. A carefully designed system is needed to ensure life safety from explosive forces. Two primary approaches to the problem exist: prevent the explosion—obviously the best choice—and control the damage from the explosion. Specialized equipment designed to detect, suppress, and control damage from explosions is available. NFPA 69, *Standard on Explosion Prevention Systems*, provides additional information.

BUILDING CONSTRUCTION

NFPA *101* does not include construction requirements for industrial occupancies. The varying types of industrial occupancies use many

TABLE 9-21B. *Causes of Civilian Deaths and Injuries, Industry and Manufacturing Fires, 1989–1993 Annual Average, Unknown-Cause Fires Allocated Proportionally*

Causes†	Civilian Deaths*		Civilian Injuries	
Other equipment	14	(37.1%)	311	(42.1%)
Unclassified processing equipment	9	(25.8%)	41	(5.5%)
Natural causes	11	(30.8%)	86	(11.7%)
Spontaneous ignition	10	(26.2%)	51	(7.0%)
Static discharges	2	(4.6%)	30	(4.1%)
Open flame, embers, or torches	3	(8.2%)	84	(11.4%)
Torches	2	(6.1%)	56	(7.7%)
Appliances, tools, or air conditioning	2	(6.3%)	39	(5.3%)
Hand tools	2	(5.3%)	12	(1.6%)
Incendiary or suspicious causes	2	(6.3%)	27	(3.6%)
Electrical distribution	2	(5.0%)	61	(8.3%)
Overcurrent protection devices	1	(2.7%)	17	(2.3%)
Heating equipment	1	(2.5%)	42	(5.7%)
Other heat source	1	(2.5%)	23	(3.2%)
Smoking materials	0	(1.3%)	21	(2.9%)
Cooking equipment	0	(0.0%)	32	(4.3%)
Exposure (to other hostile fire)	0	(0.0%)	9	(1.2%)
Child playing	0	(0.0%)	3	(0.4%)
Total	37	(100.0%)	738	(100.0%)

*Civilian deaths and injuries are expressed to the nearest one. Percentages are calculated on the actual estimates, which include fractions of a casualty per year, so two figures with the same rounded-off estimates may have different percentages.

†The 12 major cause categories are based on a hierarchy developed by the U.S. Fire Administration. Sums may not equal totals because of rounding errors.

Source: NFIRS, NFPA survey.

different types of construction. Smaller facilities might be constructed of nonrated wood, corresponding to Type V in NFPA 220, *Standard on Types of Building Construction*, while larger buildings might be of fire-resistive construction, Type I. Additionally, buildings can range from totally enclosed high-rises and windowless buildings to open structures providing equipment protection from weather. The type of construction is generally based on local building codes for the type of use, rather than on life safety concerns.

OCCUPANT CHARACTERISTICS

In general, occupants of an industrial facility are able-bodied adults. Statistically less likely to be fire victims, they are ambulatory and capable of responding quickly to a fire. These occupants should be familiar with facility hazards and emergency exit locations. The likelihood also exists that employees in an industrial setting have received some emergency training. As density of the occupant load varies significantly, it must be considered in exit system design. Facilities employing many physically and mentally disadvantaged people require supplemental pre-fire planning to ensure safe evacuation.

FIRE PROTECTION FEATURES

A primary life safety feature of an industrial occupancy should be an automatic sprinkler system. Originally a property protection feature, it has proved effective in preventing loss of life. Automatic sprinklers mitigate the effects of fire, thereby allowing employees sufficient time to evacuate the building. In addition, the sprinkler system water flow alarm can alert building occupants of the fire earlier and thereby give them more time to react. Several sources have documented the effectiveness of sprinkler systems. The Department of Energy of the U.S. government reported on the effectiveness and reliability of the sprinkler systems protecting DOE facilities during the years 1952–1980.[4] Several conclusions resulted from the study:

- Sprinkler systems are over 98 percent effective in controlling or extinguishing fires.

- Considering only building fires, sprinklers reduce the loss by at least a factor of 5.

- About one-third of all fires may be completely extinguished with only one sprinkler head operating.

Additionally, a book from the Australia Fire Protection Association on the performance of automatic sprinklers in Australia and New Zealand, 1886–1968,[5] includes a detailed analysis of fire involving different occupancies. The section on industrial occupancies is broken down into different categories. Sprinklers operated effectively over 98 percent of the time. Life loss in these facilities was minimal, nil in the majority of incidents.

Both these studies were limited to special groups of facilities in which design, usage, or maintenance may tend to be better than the average U.S. practice. NFPA analysis of industrial and manufacturing facility fires reported to U.S. fire departments indicates sprinklers reduce the average loss per fire by one-half to two-thirds.

To be effective, a system must be operable. This was a factor in fatalities at a cogeneration facility in 1993.[3] The sprinkler system was not functioning. Programs must be in place to provide for inspection of control valves to ensure they are open, and for maintenance of associated trim for alarm piping, dry-pipe valves, and preaction and deluge systems. For preaction and deluge systems, the associated detection systems must also be operational.

NFPA *101*

Classification of Occupancy

To design the appropriate exit system for an industrial occupancy, the life safety industrial occupancy hazard classification per NFPA *101* must be established. The three basic classifications include general industrial, special purpose industrial, and high hazard.

General-purpose industrial occupancy: General-purpose industrial occupancies involve low- and ordinary-hazard manufacturing operations occurring in any type of building, be it single-story, multistory, or multiple tenant. These occupancies normally have a higher density of employees than other classifications. Examples are a laundry or an auto assembly plant having low- or moderate-hazard operations.

Special-purpose industrial occupancy: Special-purpose industrial occupancies also involve low- and ordinary-hazard manufacturing operations. They are characterized by low employee density, with most of the floor area occupied by equipment. An example is a structure built to protect large processing equipment from weather, where people associated with the process work in a separate control room, thereby reducing employee exposure to fire. These occupancies have slightly different egress requirements that are generally more lenient, such as an increase in the travel distance allowed. Also, egress is designed for the actual number of employees in the structure, rather than the calculated occupant load. A telephone switch building or a turbine generator building at a power plant meets the definition of low or moderate hazard and very reduced number of occupants.

High-hazard industrial occupancy: A high-hazard industrial occupancy differs from low- and ordinary-hazard industrial occupancies by the materials used and the potential results and byproducts of their use. High-hazard industrial occupancies include structures in which there are processes involving highly combustible, highly flammable, or explosive materials, or structures in which materials are likely to burn with extreme rapidity or to produce poisonous fumes or gases. Also included are industrial facilities in which flammable liquids are routinely handled, used, or stored in large quantities or those in which explosive dusts from grain, wood, flour, plastic, aluminum, magnesium, or other explosive-dust-generating materials are produced. Plants such as a grain handling facility or a gasoline refinery are examples of high-hazard industrial occupancies.

Structures with adequately protected, incidental high-hazard uses do not have to be classified as high-hazard facilities. These structures may include specific fire prevention measures such as a small flammable liquid storage room designed to meet NFPA *101* and NFPA 30, *Flammable and Combustible Liquids Code*. The restricted use of flammable liquids does not require a high-hazard classification. An example is a metalworking plant, normally classified as low-hazard, that has a flammable solvent dip tank coater. While adequate means of egress away from the coater are required, the classification of the entire facility does not change to high-hazard.

The classification of the life safety hazard of an industrial occupancy must be carefully considered. The goal is to evaluate the hazard of the contents, process, and other factors affecting fire development. The time available for occupants to evacuate can be very much affected by the growth of the fire. Many resources are available to assist in the determination. Review of pertinent NFPA literature, including the codes and this handbook, is helpful. Officials in

other jurisdictions with similar occupancies and plant operators can provide additional valuable information.

Means of Egress Design Requirements

Designing the egress system for a building requires consideration of many factors. Some involve distance requirements, such as travel distance, common path of travel, and dead ends. These distances vary based on the life safety hazard classification and the absence or presence of automatic sprinklers. Egress distance limitations are shown in Table 9-21C. To use some of the greater distances involves special requirements, such as application to one-story buildings only, with automatic sprinklers mandatory and smoke and heat venting.

High-hazard occupancies are very restrictive on means of egress distances allowed, with no dead ends or common paths of travel permitted. Generally, two exits are required from every story or section. However, NFPA *101* provides exceptions to this requirement. By careful design, the listed requirements can be met. The addition of exit tunnels, overhead passageways, or horizontal exits can aid in meeting required distances.

Emergency Lighting

The potential hazards of an industrial occupancy make adequate lighting of the means of egress critical. Under normal conditions, illumination is required for the means of egress. If the building is occupied only during daylight hours, and skylights and windows are arranged to provide the required level of illumination, then electrical lighting is not required.

During a power failure, emergency lighting is required for the means of egress. No appreciable delay is permitted when changing from one energy source to another. Emergency lighting is designed to operate for $1^{1}/_{2}$ hr to allow occupants to evacuate and to perform search and rescue as necessary. Emergency lighting is not designed to provide illumination for emergency repairs or equipment shutdown. Again, if adequate exterior light is available from skylights and windows and the facility is occupied only during daylight hours, then supplemental emergency lighting is not required. In special-purpose occupancies in areas not subject to routine personnel use, lighting the means of egress is not required.

Fire Alarm Systems

Due to the size and complexity of most industrial facilities, a fire alarm system is a valuable tool for life safety. It can alert designated personnel, such as an industrial fire brigade, to a fire and allow for prompt response to evacuate occupants, fight the fire, shut down equipment, and other actions, as required. NFPA *101* requires a fire alarm system for most industrial occupancies. Exceptions are based on occupant load, where total capacity is less than 100 persons and fewer than 25 persons are located above or below the level of exit discharge. In low- and ordinary-hazard industrial occupancies, the

TABLE 9-21C. *Egress Distance Limitations*

Industrial Occupancy	Travel Distance			Dead End	Common Path of Travel	
	Unsprinklered ft (m)	Sprinklered ft (m)	Special ft (m)	ft (m)	Unsprinklered ft (m)	Sprinklered ft (m)
General purpose	200 (60)	250 (75)	400 (122)*	50 (15)	50 (15)	100 (30)
Special purpose	300 (91)	400 (122)	—	50 (15)	50 (15)	100 (30)
High hazard	75 (23)	75 (23)	—	0	0	0

*Special requirements in NFPA *101* for this travel distance in general-purpose industrial occupancies

fire alarm must sound an audible alarm at a continuously attended location to initiate emergency action. At high-hazard industrial occupancies, in addition to the above, the fire alarm must initiate an occupant evacuation signal. In these occupancies, early notification can be critical to life safety.

PRE-FIRE PLANNING

Different preplanning is appropriate for property protection. Training in the use of fire extinguishers for incipient fires must be provided if employees are expected to use them. Designated employees need to be trained to check that fire protection equipment is operating as designed—for example, that sprinkler valves are open. The responding fire department should be made familiar with the facility, and an employee should be selected as their contact. Hazardous materials should be identified and their location noted on a building plan. Full participation in the drills is necessary to completely test the plan.

The pre-fire plan should be evaluated routinely through simulated fire exercises. Such exercises are critical to identify deficiencies in the plan. If possible, request the participation of the responding fire department.

If there are actual fire brigades at the site, they must be adequately trained. Training should include brigade organization and responsibility, the basics of fire behavior, special on-site fire and health hazards, and hands-on training with the fire equipment they are expected to use. Available equipment might include fire extinguishers, standpipes, hoses, and hydrants. Appropriate use of this equipment on incipient fires can contribute to a major reduction in the risk of fire loss and loss of life in industrial occupancies. Restricting the spread of fire also reduces the threat to life.

SUMMARY

Industrial occupancies must be adequately evaluated to determine fire hazard. Fire hazard affects the potential for life loss and property damage. Loss history can be used to see whether current designs and requirements are effective. NFPA *101* requirements for industrial facilities are briefly discussed. It is important that building design and fire protection systems allow for safe occupant egress in the event of fire. The effectiveness of sprinklers has been documented from two sources in particular.

BIBLIOGRAPHY

References Cited

1. Klem, T. J., "25 Die in Food Plant Fire," *Fire Journal,* Vol. 86, No. 1, Jan./Feb. 1992, pp. 29–35.
2. Grant, C. C., and Klem, T. J., "Toy Factory Fire in Thailand Kills 188 Workers," *NFPA Journal,* Vol. 88, No. 1, Jan./Feb. 1994, pp. 42–49.
3. Lathrop, J. K., "Life Safety Code Key to Industrial Fire Safety," *NFPA Journal,* Vol. 88, No. 4, July/Aug. 1994, pp. 36–46.
4. "Automatic Sprinkler System Performance and Reliability in United States Department of Energy Facilities 1952–1980," DOE/EP-0052, UC-41, June 1982, Washington, DC.
5. Marryatt, H. W., "General Occupancy Analysis," *Fire—Automatic Sprinkler Performance in Australia and New Zealand 1886–1968,* McCarron Bird Pty, Melbourne, Australia, 1971, pp. 103–118.

References

Fire Protection Guide to Hazardous Materials, ninth ed., National Fire Protection Association, Quincy, MA, 1986.
Hall, J. R., Jr., "Selections from the U.S. Fire Problem Overview Report Through 1988—Leading Causes and Other Patterns and Trends, Industrial and Manufacturing Facilities," National Fire Protection Association, Quincy, MA, 1989.

NFPA Codes, Standards, and Recommended Practices

References to the following NFPA codes, standards, and recommended practices will provide further information on the industrial occupancies discussed in this chapter. (See the latest version of *The NFPA Catalog* for availability of current editions of the following documents.)

NFPA 30, *Flammable and Combustible Liquids Code*
NFPA 69, *Standard on Explosion Prevention Systems*
NFPA *101®, Life Safety Code®*
NFPA 220, *Standard on Types of Building Construction*
NFPA 600, *Standard on Industrial Fire Brigades*

Additional Reading

Adolphs, W., "Fire Protection Covering the Risks Occurring in Industrial Plants," *Proceedings* of the Conference on New Technology to Reduce Fire Losses and Costs, Elsevier, New York, 1986, pp. 329–331.
Alfert, F., and Fuhre, K., "Venting of Dust Explosions in Filters and Integrated Systems. Final Report for a Research Program on Sizing Dust Explosion Vents for the Process Industry," CHR Michelsen Inst., Fantoft, Norway, CMI-92-A25021, CMI Project 25810, Apr. 1992.
Altamura, S. J., "Emergency Lighting: A Key Element in Industrial Plant Safety," *Industrial Fire World,* Vol. 5, No. 3, June/July 1990, pp. 12–14.
Arata, M. J., Jr., "Codes and Their Effect on Industry," *Industrial Fire Safety,* Vol. 2, No. 2, Mar./Apr. 1993, pp. 12–13.
Bailey, B., "Foam: A Fire Fighting Primer for the Industrial Sector," *Fire Fighting in Canada,* Vol. 38, No. 2, Mar. 1994, pp. 8–9.
Barry, I., "Vermiculite Cements and Passive Fire Protection in the Hydrocarbon and Chemical Industries," *Fire Surveyor,* Vol. 20, No. 5, Oct. 1991, pp. 4–7.
Bouvier, K. J., "Overview of an Industrial Response Team's Capabilities," *Fire Chief,* Vol. 36, No. 1, Jan. 1992, pp. 42–43.
Buchanan, A. H., and Cosgrove, B. W., "Fire Design of Industrial Buildings," University of Canterbury, Christchurch, New Zealand, ISBN 0-9516320-9-4; *Proceedings* of the Seventh International Interflam Conference, Interflam '96, March 26–28, 1996, Cambridge, England, Interscience Communications Ltd., London, England, 1996, pp. 951–957.
Capaul, T., "Fire Risk Management for an Industrial Plant," Worcester Polytechnic Inst., MA, Thesis, May 1991.
Chow, W. K., and Cheung, K. C., "Industrial Fire Simulation with Zone Models and Review on the Current Regulations," Hong Kong Polytechnic Univ., Hong Kong, ISBN 0-9516320-9-4; *Proceedings* of the Seventh International Interflam Conference, Interflam '96, March 26–28, 1996, Cambridge, England, Interscience Communications Ltd., London, England, 1996, pp. 601–620.
Cote, A. E., ed., *Industrial Fire Hazards Handbook,* third ed., National Fire Protection Association, Quincy, MA, 1990.
Connolly, J. M., "Renewed Focus on Industrial Fire Departments," *NFPA Journal,* Vol. 86, No. 2, Mar./Apr. 1992, pp. 46–48.
Coutts, D. A., "Estimating the Fire Risk for an Industrial Facility," Westinghouse Savannah River Co., Aiken, SC, NISTIR 5499, Sept. 1994; National Institute of Standards and Technology, Annual Conference on Fire Research: Book of Abstracts, Oct. 17–20, 1994, Gaithersburg, MD, 1994, pp. 41–42.
Davis, L., and Moore, J., *Industrial Fire Control Concepts,* International Society of Fire Service Instructors, Ashland, MA, 1988.
Fender, B. G., "Fighting Fires in High Rise Industrial Buildings," *Fire International,* Vol. 12, No. 109, Feb./Mar. 1988, pp. 25–27.
Fowell, A. G., "Industrial Fire Suppression," *NBSIR 86–3360,* April 1986, NBS Workshop on Fire Protection Technology.
Frydenlund, M. M., "Protecting Industrial Plants Against Lightning's Dangers," *Consulting/Specifying Engineer,* Vol. 8, No. 3, Sept. 1990, pp. 60–62, 64, 66, 68.
Gheorghe, A. V., "Risk Assessment of Large Industrial Complexes in Eastern Europe: A Comparative Prospective," *PSAM-II Proceedings,* An International Conference Devoted to the Advancement of System-Based Methods for the Design and Operation of Technological Systems and Processes, Vol. 3, Sessions 73–108, Mar. 20–25, 1994, San Diego, CA, 1994, pp. 074/13–18.

Gustin, B., "Preplanning Special Industrial Hazards: Metro-Dade Takes on the Deep Fryer," *Fire Engineering*, Vol. 147, No. 11, Nov. 1994., pp. 39–40, 42, 44–45

Haase, R., "Structure Fires in Industrial Parks," *Fire Engineering*, Vol. 145, No. 7, July 1992, pp. 14–17, 22.

Herigodt, G. J., and deVries, D., "Industrial Fire Protection Requires a Head Start," *Consulting-Specifying Engineer*, Vol. 15, No. 5, May 1994, pp. 32–34, 36, 38, 40.

Hultsten, B., and Ellnebo, M., "Automatiserat Industribrandforsvar: Projekt 1.46. Slutrapport. [Automatic Industrial Fire Protection: Project 1.46. Summary Report.]," Communicator Sensornik AB, Sweden, SR-92010, Sept. 9, 1992.

Jenaway, W. F., "Safety Dance: Are There Dueling Priorities Between Fire Departments and Industrial Plant Safety Officials?" *Building Official and Code Administrator*, Vol. 28, No. 5, Sept./Oct. 1994, pp. 28–31.

Koffel, W. E., "Applying NFPA 101 to Industrial Occupancies," *NFPA Journal*, Vol. 88, No. 4, July/Aug. 1994, pp. 16, 110.

Koffel, W. E., "Establishing Industrial Fire Safety Programs," *NFPA Journal*, Vol. 87, No. 2, Mar./Apr. 1993, pp. 12, 86.

Lathrop, J. K., "Life Safety Code Key to Industrial Fire Safety," *NFPA Journal*, Vol. 88, No. 4, July/Aug. 1994, pp. 36–46.

Long, M. H., and Croce, P. A., "Process Industry: Protection Based on Risk Quantification," Little, (Arthur D.) Inc., Cambridge, MA, Worcester Polytechnic Institute, Conference on Firesafety Design in the 21st Century, Part 1, Conference Report, Part 2, Conference Papers, May 8–10, 1991, Worcester, MA, 1991, pp. 236–245.

Loss Prevention Council, "Code of Practice for the Construction of Buildings. Insurers' Rules for the Fire Protection of Industrial and Commercial Buildings," Loss Prevention Council, London, England, ISBN-0-90216700-6, 1992.

Mangs, J., "Prosessiteollisuuden polyrajahdyksien paineenalentamisaukkojen mitoitus II. [Dimensioning Pressure Outlets to Avert Dust Explosions in the Processing Industry.]," *Palontorjuntatekniika*, Mar. 1992, pp. 30–31.

McDonald, S., "Reducing Risk and Exposure in the Process Industries," *Fire Prevention*, No. 279, May 1995, pp. 33–34.

McQuaid, J., "Industrial Fire Problems: An Overview," *Proceedings* of the Third International Symposium on Fire Safety Science, Elsevier Applied Science, New York, 1991, pp. 61–81.

Miller, C., "Early Suppression—Fast Response Sprinklers: A New Level of Industrial Fire Protection," *National Engineer*, Vol. 93, No. 9, Sept. 1989, pp. 18–19.

"NFPA Investigates Thailand Industrial Company Fire," *Building Official and Code Administrator*, Vol. 47, No. 6, Nov./Dec. 1993, p. 12.

Nolan, P. F., and Bradley, C. W. J., "A Simple Technique for the Optimization of Lay-Out and Location for Chemical Plant Safety," *Plant/Operation Progress*, Vol. 6, No. 1, Jan. 1987, pp. 57–61.

"Protecting Hazardous Locations from Fire Disasters," *Consulting Specifying Engineer*, Vol. 2, No. 3, 1987, pp. 104–108.

Rozman, T., "Industrial Fire Protection in Yugoslavia and Eastern Europe," *Industrial Fire World*, Vol. 5, No. 3, June/July 1990, pp. 16–17.

Rubin, D. L., and Avsec, R. P., "Virginia: Large Industrial Plant Fire Strikes the Southeast. On the Job," *Firehouse*, Vol. 18, No. 8, Aug. 1993, pp. 82–84.

Sax, N. I., and Lewis, R. J., *Dangerous Properties of Industrial Materials*, Van Nostrand Reinhold, New York, 1989.

Schram, P. J., and Earley, M. W., *Electrical Installations in Hazardous Locations*, National Fire Protection Association, Quincy, MA, 1989.

Scoones, K., "Serious Fires in Manufacturing Industries in 1993," *Fire Prevention*, No. 285, Dec. 1995, pp. 31–33.

Stronach, I., "Industrial Fire Brigades: Alive and Well?" *Firehouse*, Vol. 19, No. 1, Jan. 1994, pp. 44, 46.

Sudasna, C., "Major Industrial Fire Kader Industrial (Thailand) Co. Ltd., Buddha Monthon, Nakhon Pathom Province, Thailand on Monday, May 10, 1993," Royal Thai Police Dept., Bangkok, Thailand, CIB W14/95/1 (J); Tokyo Fire Department, Firesafety Frontier '94, International Fire Conference and Exhibition in Tokyo, Creating a Safe Tomorrow, Oct. 18–22, 1994, Tokyo, Japan, 1994, pp. 81–86.

Tamanini, F., "Explosion Suppression for Industrial Applications," Factory Mutual Research Corp., Norwood, MA, NISTIR 5766, November 1995; National Institute of Standards and Technology, Solid Propellant Gas Generators: *Proceedings* of the 1995 Workshop, June 28–29, 1995, Gaithersburg, MD, 1995, pp. 208–217.

Tewarson, A., and Khan, M. M., "New Standard Test Method for Fire Propagation Behavior of Electrical Cables in Industrial and Commercial Occupancies," Fifth International Fire Conference on Fire Safety, Interflam '90, Sept. 3–6, 1990, Canterbury, England, Interscience Communications Ltd., London, England, 1990, pp. 101–113.

Vincent, B. G., Stavrianidis, P., and Kung, H. C., "Use of Quick Response Sprinklers for Industrial Fire Protection Applications," *Journal of Fire Protection Engineering*, Vol. 2, No. 4, 1990, pp. 99–110.

Wang, H., "Fire Science and Industrial Firesafety in China," *Proceedings* of the USA/China Bilateral Conference on Fire Safety Engineering, Aug. 14–15, 1995, Sacramento, CA, 1995, pp. 1–10.

White, D. A., Beyler, C. L., and DiNenno, P. J., "Performance-Based Engineering of Industrial Facilities," Hughes Associates, Inc., Columbia, MD, Society of Fire Protection Engineers and Worcester Polytechnic Institute, *Proceedings* of Computer Applications in Fire Protection, June 28–29, 1993, Worcester, MA, 1993, pp. 51–55.

Woodward, C. D., "The Industrial Fire Problem," *Fire Safety Journal*, Vol. 15, No. 5, 1989, pp. 348–366.

Woodward, D., "The Industrial Fire Problem—Part I," *Fire Prevention*, No. 218, Apr. 1989, pp. 24–28.

Woodward, D., "The Industrial Fire Problem—Part II," *Fire Prevention*, No. 219, May 1989, pp. 28–33.

MOTION PICTURE AND TELEVISION STUDIOS AND SOUNDSTAGES

James R. Quiter and Raymond A. Grill

NFPA is developing a new standard, NFPA 140, *Standard on Motion Picture and Television Industry Facilities.* The standard defines motion picture and television studios as buildings, portions of buildings or groups of buildings designed and constructed for use by the entertainment industry to produce motion pictures, television shows, or commercials, or to broadcast television programs utilizing a soundstage. This covers all pre- and post-production facilities including construction shops, warehouses, soundstages, control rooms, editing facilities, and other exterior filming locations on the studio property. Actual production occurs in a soundstage; on location, which includes any location not specifically designed for production; or on a studio lot, which can consist of exterior sets.

The motion picture and television industry is dynamic in nature and needs flexibility. As a result, the fire protection needs of these types of facilities must be evaluated based on the hazards present, which change from production to production, and the industry's operational needs. Many occupancies and hazards found at a studio are similar to those in more typical industrial facilities.

This chapter, therefore, focuses primarily on fire protection issues relating to soundstages. It does not deal with locations or exterior production areas that might be found on a studio lot. For instance, construction shops for building sets contain the same hazards as many other shops or factories and should be treated as such. Warehouses on studio property can also be treated like other warehouses, while bearing in mind their contents' dynamic nature. Administrative areas, control rooms, and commissaries can also be treated like other occupancies. However, soundstages, which are specifically designed for filming, have special operational needs, and an understanding of their specific fire protection hazards and features is required.

OCCUPANCY CHARACTERISTICS

A soundstage is a building or a portion of a building, usually insulated from outside noise and natural light, used to produce motion pictures, television shows, or commercials. Soundstages like the one shown in Figure 9-22A are specifically designed and constructed to create a controlled environment that can be isolated from exterior noise and light.

James R. Quiter, P.E., is senior vice president at Rolf Jensen & Associates, Inc., Walnut Creek, CA. Raymond A. Grill, P.E., is senior vice president at Rolf Jensen & Associates, Inc., Fairfax, VA. He is a member of the NFPA Technical Committee on the Motion Picture and Television Industry.

Soundstages are often stand-alone buildings, but can also be a room within a larger mixed-used building. Mixed-use is often found in television production and broadcasting. Multiple soundstages within the same building are also possible.

According to a summary prepared by Wylie Carter Architects, there are about 300 soundstages over 5,000 sq ft (465 m^2) in Southern California alone; their average area is about 12,000 sq ft (1115 m^2). The largest is about 42,000 sq ft (3900 m^2). Clear height of these facilities ranges from 14 to 65 ft (5 to 20 m), with an average height of 32 ft (10 m).

The motion picture and television production industry commonly converts existing buildings to soundstages. This process may or may not include a change of occupancy for the existing facility. Vacant warehouses have been the primary type of existing facility converted into soundstages because of their openness. Its contents make a soundstage unique.

Live broadcast facilities, such as network news, incorporate production activities within other activities, such as newsrooms and offices. Some productions also include live audience environments within the soundstage. Introducing a live audience, depending on its size, can impact occupancy classification and other life safety and fire protection issues. These issues are discussed later in this chapter.

Sets within soundstages can be very elaborate. The duration of their use varies significantly from production to production. Some sets for television series may be used for years; others are specifically constructed for an episode. A set may be developed to simulate an entire building or series of buildings within the soundstage. Sets are typically constructed of wood and generally treated as temporary structures.

HISTORICAL FIRE EXPERIENCE

In 1993, the National Fire Protection Association analyzed fire losses in motion picture and television studios based on annual averages between 1987 and 1991.[1] The information gathered was based on data from NFPA's annual stratified random sample survey and the United States Fire Administration's National Fire Incident Reporting System (NFIRS).

Between 1987 and 1991, an estimated 34 reported structure fires annually in motion picture studios resulted in an average total fire loss of $290,700 in direct property damage per year. During the same period, an average of 50 fires in radio or television studios annually resulted in a yearly average of $1,225,800 in direct property damage. These fires were not limited to soundstages. They included

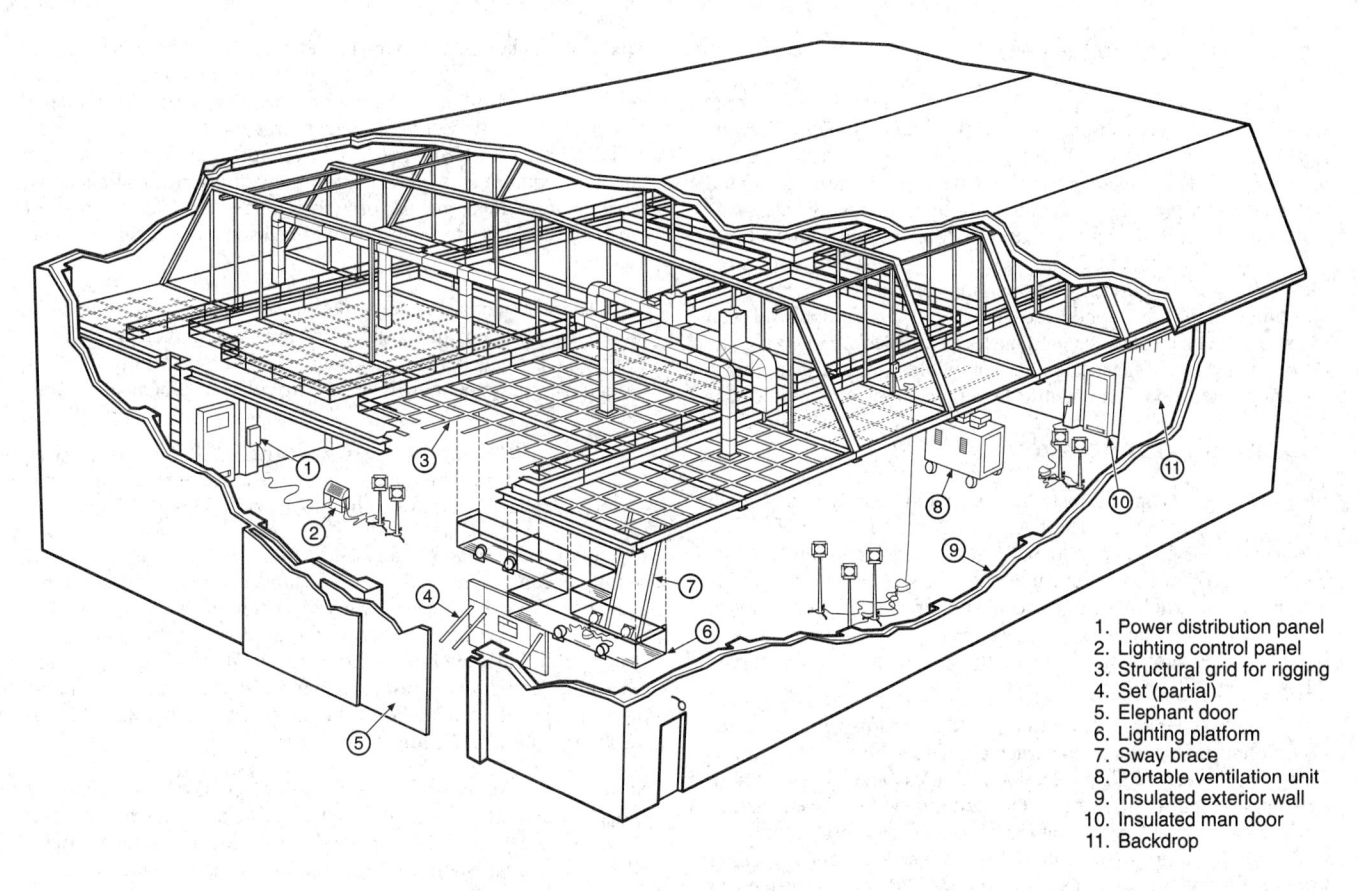

1. Power distribution panel
2. Lighting control panel
3. Structural grid for rigging
4. Set (partial)
5. Elephant door
6. Lighting platform
7. Sway brace
8. Portable ventilation unit
9. Insulated exterior wall
10. Insulated man door
11. Backdrop

FIG. 9-22A.   A typical soundstage.

allied services and facilities, such as motion picture film processing, editing, wardrobe, warehousing, etc. No deaths were reported and injuries (one per year) were reported only for radio or television studio fires. The type of material first ignited was predominantly wood or paper in all types of studio fires. The second most predominant ignition materials were plastics and fabrics.

Motion Picture Studios

More than one-third of motion picture studio fires and three-quarters of the associated dollar loss involved electrical distribution equipment. Lighting fixtures, lamp holders, ballasts, signs, and fixed wiring accounted for the largest share of electrical distribution equipment fires. Electrical equipment was also a leading factor among forms of heat of ignition. Nearly 37 percent of fires in motion picture studios were attributed to heat from arcing or overloaded electrical equipment.

Power transfer equipment or fuel was involved in the second largest share of fires but accounted for the largest share of direct property damage.

Radio and Television Studios

Mechanical or electrical failure or malfunction was responsible for the largest share of fires as well as direct property damage for radio or television studios. Short circuits or ground faults caused roughly half of the fires attributed to mechanical or electrical failure or malfunction. Fires involving mechanical or electrical failure or malfunction also resulted in the largest share of direct property damage. Electrical distribution equipment accounted for 29 percent of fires.

Heat from arcing or overloaded electrical equipment was the leading form of heat of ignition in radio and television studio fires. This heat source contributed to half the fires and 55 percent of direct property damage.

OCCUPANCY HAZARDS

Due to the unique nature of motion picture and television production, the contents of a soundstage and its combustible load vary from time to time. Soundstages themselves may also include combustible materials as part of their construction. Supplemental and temporary equipment, as well as set construction, contribute to the potential hazard within the soundstage. Introduction of occupants (performers and audiences) also needs to be addressed from a life safety exposure and egress perspective.

Combustion Loading

Production sets consist of simple platforms, backdrops, or very elaborate structures. Sets are typically constructed of wood framing. They may consist of a simple platform to provide for appropriate camera angles or may include walls and ceilings to simulate a room or building. The model building codes generally exempt motion picture and television sets from permit requirements. Sets are typically considered temporary because changes occur regularly.

Set construction might obstruct sprinkler protection and other fire protection features. Ceilings of sets may be movable or built with materials that are not self-venting and, therefore, obstruct heat flow and sprinkler operation.

Set Construction/Hotwork

The use of arc welders, oxyacetylene torches, and grinders is common during set construction. NFPA 51B, *Standard for Fire Prevention in Use of Cutting and Welding Processes* (hereinafter referred to as NFPA 51B), provides comprehensive criteria for managing the potential hazards of these operations. Requirements include management responsibilities as well as operators' responsibilities. Fires caused by hotwork can best be prevented by separating combustibles from ignition sources or by shielding combustibles. Regular inspections should be conducted to verify that compressed gas cylinders are adequately secured, hoses and cables are in good condition, combustible materials and flammable liquids are appropriately separated, and adequate fire-fighting equipment is provided.

Electrical Considerations

Electrical equipment and installations should comply with Article 530 and applicable sections of Article 520 of NFPA 70, *National Electrical Code®*. Article 530 and Article 520 address theaters, audience areas of motion picture and television studios, and similar locations. These articles include specific requirements for equipment, switchboards, lighting, and grounding. They address portable wiring, which is extensively used in motion picture and television production.

Mobile generators are routinely used in motion picture production. Soundstages are not constructed to accommodate the temporary power needs of all production activities. Therefore, portable generators often supplement normal building power. When routing auxiliary power cables, care should be taken not to obstruct or interfere with fire ratings of walls or windows. For cables passing through fire-rated assemblies, penetrations should be protected in accordance with NFPA 221, *Standard for Fire Walls and Fire Barrier Walls*. Temporary electrical equipment might also obstruct egress and fire-fighting operations. When laying cables and locating portable generators, egress routes should be carefully avoided.

The location of portable power generating equipment should include consideration of:

- Potential exposure to the soundstage and adjacent structures
- Potential obstruction of egress or fire-fighting access
- Service and fueling needs

Portable power generators should be regularly inspected for leaks and adequate portable fire extinguishers should be available for use on this equipment.

Portable Heating, Venting, and Air Conditioning

Auxiliary heating, venting, and air conditioning equipment often supplements the built-in capacity of a soundstage. These units should be listed and located so they do not obstruct egress or access. Flexible ducts used with these systems should be noncombustible and also arranged not to block egress and access.

Special Effects

Special effects might significantly impact the safety of soundstage occupants. Special effects include pyrotechnic materials, gas, or other open flames, smoke, or fuel-powered vehicles. Clearly, special effects' impact is greater when there is an audience.

Pyrotechnics: NFPA 1126, *Standard for the Use of Pyrotechnics before a Proximate Audience*, applies to the indoor display of pyrotechnic special effects in television and movie production only if the production is before a proximate audience. Special effects are partially defined as visible or audible effects for entertainment purposes, frequently an illusion. NFPA 1126 defines proximate audience as "an audience closer to pyrotechnic devices than allowed by NFPA 1123, *Code for Fireworks Display*." This NFPA standard contains requirements on storage, permits, operator qualifications, labeling of pre-loads and use of pyrotechnics. It should be consulted when pyrotechnic special effects will be used indoors before an audience.

Another standard is being developed for the use of pyrotechnics in the motion picture and television industry where 49 or fewer people are present. This standard will be similar to NFPA 1126, although its requirements will be less strict because of smaller audiences.

NFPA *101®*, *Life Safety Code®* (hereinafter referred to as NFPA *101*) addresses pyrotechnic devices in its chapters on assembly occupancies by requiring compliance with NFPA 1126.

Open flame: NFPA *101* also addresses open flame devices in assembly occupancies. It allows open flame on stages and platforms as a necessary part of a performance, when precautions satisfying the Authority Having Jurisdiction are taken to prevent ignition of combustible material or injury to occupants. It permits gas lights under similar conditions. Prior to the performance, use of open flame or gas lights before an audience should be worked out with the Authority Having Jurisdiction.

Smoke: Smoke used as a special effect should not obscure exit signs, particularly if an audience is present. If no audience is present, and dense smoke is integral to the production, those affected by the smoke should be trained and drilled regarding egress routes.

Decorations

Combustible drapes, drops, hangings, cycloramas, cut greenery, and other decorative materials utilized on sets should be maintained in a flame-retardant condition. Foam plastic materials used for decorative purposes, scenery, or props should be limited to materials with a maximum heat release rate of 100 kW when tested in accordance with UL 1975, *Standard for Fire Tests for Foam Plastics Used for Decorative Purposes*.

Housekeeping/Smoking

Combustible waste and other materials should be stored in a manner that limits their potential of coming in contact with ignition sources. Approved self-closing metal containers should be provided for containing these materials. Housekeeping hazards include accumulated sawdust or wood scraps; accumulated litter; combustible storage below platforms; storage of props, equipment, and material that might block exits; fire protection equipment; and electrical equipment.

Smoking not required for the production should be limited to designated safe areas.

LIFE SAFETY/EGRESS

A soundstage may have two separate and distinct uses, depending upon the presence of an audience. If no audience is anticipated, the soundstage can be considered an industrial occupancy. Means of egress should meet the requirements of Chapter 28 of NFPA *101*. Soundstages exceeding 1,500 sq ft (139 m²) should have an aisle along the perimeter, clear to 7 ft (2.1 m) height above the floor. Motion picture soundstages often maintain a 4-ft (1.2-m) width aisle, with the floor painted yellow to clearly indicate that it is to be kept open. If this type of aisle cannot be maintained, special arrangements should be made.

If an audience of 50 or more people might be present, the requirements of Chapter 8 or 9 of NFPA *101* apply. These chapters address new and existing assembly occupancies. Other portions of this handbook address egress from assembly spaces, and the reader should refer to them for guidelines and requirements on number of exits, stair and handrail criteria, aisle and seating arrangements, emergency lighting, and the need for sprinkler and alarm systems.

If an audience is present, maintaining clear and obvious egress paths is important, especially if special effects or other visual illusions are used.

FIRE PROTECTION

Soundstages should be provided with automatic sprinkler protection to control fires. The design of automatic sprinkler systems needs to consider the special circumstances that platforms and solid-ceiling sets create. In recognition of the temporary nature of platforms and sets it might be necessary to permit omission of sprinklers in these areas when alternative protective means are provided. Such means might include limiting the the size of platforms, and solid ceilings, limiting use below platforms and using heat or smoke detectors. When any of these alternatives are used, the design of the building's automatic sprinkler system should be increased to Extra Hazard, Group 2. This design criteria is in NFPA 13, *Standard for the Installation of Sprinkler Systems.* When a fire is shielded from sprinklers, it might develop beyond its incipient stage before sprinkler activation. The Extra Hazard, Group 2, design criteria are necessary to mitigate the effects of a fire of this type.

Soundstages have unique needs with respect to audible and visible alarms. Audible and visible notification appliances within soundstages might be deactivated during production provided:

1. Alarm system activation causes notification appliances to activate within a constantly attended location.
2. The attendants of the location have a means of communicating with the building fire command center, if one is provided, and with the occupants of the soundstage.
3. Deactivation of notification appliances on the soundstage activates an appropriately labeled visible signal at an approved location that remains illuminated while notification appliances on the soundstage are reactivated.

This approach is most likely to be utilized in a soundstage that is part of another building having a fire alarm system.

Portable fire extinguishers should be provided throughout soundstages in accordance with the requirements of NFPA 10, *Standard for Portable Fire Extinguishers.*

OPERATING FEATURES

Several other NFPA standards may apply to soundstage operations. These include NFPA 1, *Fire Prevention Code*, regarding maintenance of systems, waste, and refuge accumulation, and elimination or mitigation of hazards. NFPA 30, *Flammable and Combustible Liquids Code*, and NFPA 58, *Standard for the Storage and Handling of Liquefied Petroleum Gases*, address flammable and combustible liquids and liquefied petroleum gases, respectively. Welding is addressed by NFPA 51, *Standard for the Design and Installation of Oxygen-Fuel Gas Systems for Welding, Cutting, and Allied Processes*, and NFPA 51B.

Of primary importance in the operation of a motion picture or television production facility is the need to continually monitor and maintain the fire protection provided. Occupancies, uses, and hazards change daily. The operator should assign a person to maintain proper fire protection and life safety who can apply good fire protection practice and common sense to day-to-day operations. This person needs to interface effectively with Authorities Having Jurisdiction and must have enough internal authority to correct or mitigate hazards as they arise.

BIBLIOGRAPHY

Reference Cited

1. "Filming in California," California State Fire Marshal, Nov. 1993. Memorandum from Alison Miller of NFPA to Chuck Smeby dated August 12 1993. Subject: Final report on fire in motion picture studios and radio or television studios.

NFPA Codes, Standards, and Recommended Practices

Reference to the following NFPA codes, standards, and recommended practices will provide further information on the motion picture and television studios and soundstages discussed in this chapter. (See the latest version of *The NFPA Catalog* for availability of the current edition of the following documents.)

NFPA 1, *Fire Prevention Code*

NFPA 10, *Standard for Portable Fire Extinguishers*

NFPA 13, *Standard for the Installation of Sprinkler Systems*

NFPA 30, *Flammable and Combustible Liquids Code*

NFPA 51, *Standard for the Design and Installation of Oxygen-Fuel Gas Systems for Welding, Cutting, and Allied Processes*

NFPA 51B, *Standard for Fire Prevention in Use of Cutting and Welding Processes*

NFPA 58, *Standard for the Storage and Handling of Liquefied Petroleum Gases*

NFPA 70, *National Electrical Code®*

NFPA *101®*, *Life Safety Code®*

NFPA 140 (Proposed), *Standard on Motion Picture and Television Industry Facilities*

NFPA 221, *Standard for Fire Walls and Fire Barrier Walls*

NFPA 1123, *Code for Fireworks Display*

NFPA 1126, *Standard for the Use of Pyrotechnics before a Proximate Audience*

Additional Reading

Bro, S., "Filmland Fire Protection: The Burbank Studios Fire Department," *American Fire Journal*, Vol. 42, No. 3, March 1990, pp. 24–25, 40–41.

FOOD PROCESSING FACILITIES

Jane I. Lataille

The food processing industry is especially sensitive to damage from fire, smoke, heat, or water due to their contaminating aspects. Food contamination is the outstanding loss potential common to all food processing and handling facilities. Fires that cause relatively minor damage in other industries can cause extremely high amounts of damage to food items, since health authorities often condemn items exposed to smoke, heat, or water, even when no damage is visible.

This chapter discusses causes of fire loss in the food processing industry, primarily those associated with combustible dusts, flammable and combustible liquids, conveyors, refrigeration systems, fuel-fired equipment, and general storage. After a discussion of the raw materials that undergo food processing, it discusses the fire hazards of process equipment, gases, liquids, combustible dusts, conveyors, and storage. Finally, it presents safeguards, including firesafe construction, sprinkler systems, and fixed extinguishing systems.

For hazards associated with electrical control systems, see Section 3, Chapter 1, "Electrical Systems and Appliances." Section 3, Chapter 21, "Storage of Flammable and Combustible Liquids," and Section 3, Chapter 22, "Storage of Gases" provide in-depth discussions of the storage practices and hazards of flammable and combustible liquids and gases. For a discussion of hazards of aerosol food production, see Section 3, Chapter 20, "Aerosol Charging Operations."

RAW MATERIALS

Fruits, vegetables, grains, beef, poultry, fish, and milk are common raw materials in the food processing industry. Other materials are powders, such as flour, salt, sugar, cornstarch, and cocoa; liquids, such as water, oil, alcohol, molasses, and vinegar; and miscellaneous items, such as butter, chocolate, flavorings, colorings, chemicals, and spices. In addition to product ingredients, food processsing uses oils and fats for cooking; various gases for carbonation, fumigation, and ripening; and fluids for heat transfer associated with heating and cooling systems.

Flour, salt, sugar, and other powdered or granulated raw materials are often pneumatically conveyed, e.g., from railroad cars or into storage silos, bins, or hoppers. These materials are highly susceptible to damage by water and other liquids and must be stored in well-sealed containers. In multiple-story plants, they are usually stored on upper floors or in penthouses.

Cooking oils, alcohol, vinegar, and liquids used at room temperature are stored in tanks of various shapes and sizes. Molasses and other liquids requiring heating are stored in steam-jacketed tanks. Most liquids storage is found on lower floors, sometimes even in basements. Wherever stored, adequate curbing and drainage must be provided to reduce existing loss potentials, and the chance of damage to other materials, equipment, storage, and the building itself. Generally, basement storage is not advisable, due to the chance of flooding from surface water runoff, water main breaks, spills from upper floors, reduced fire department accessibility, and the difficulty of venting heat and smoke.

PROCESSES

Dairy Production

The dairy industry depends heavily upon refrigeration. Dairy products must be kept within specified temperature ranges, or they may be declared unfit for use. The consequences of lost refrigeration capability are therefore important considerations in planning and protecting refrigeration systems.

In pasturization, milk is heated to kill bacteria and other microorganisms. It then must be cooled within a specified amount of time and kept within acceptable temperature ranges. Milk is also homogenized, or mixed, to give it uniform consistency. Other dairy products are cream, skim milk, butter, cream cheese, dips, ice cream, buttermilk, yogurt, and cheese. These products also must be kept within specified temperature ranges for acceptable quality.

Manufacture of Convenience Foods

Another facet of the food industry is manufacture of convenience foods. Examples are instant soup, powdered milk, instant coffee, egg and meat substitutes, powdered drinks, instant potatoes, and instant rice. The advantages of these products include economy, ease of preparation, and long shelf life.

The processes most often used in producing convenience foods are flash drying and flash freezing. During these processes, food products are converted into their desired forms by rapid exposure to extreme heat or extreme cold. Some products also contain various chemicals and additives for preservation, texture or flavor improvement, coloring, and moisture control.

Jane I. Lataille, P.E., ARM, is a research consultant with Industrial Risk Insurers, Hartford, CT.

Preserving

Since most crops can only be grown during limited times, it is necessary to preserve them to achieve year-round availability. Meat, poultry, and fish also are preserved. Canning, freezing, and pickling are common means of food preservation. In both canning and freezing, the food is usually cooked in steam-jacketed vessels. Salt and vinegar are the primary materials used for pickling.

Baking

Continuous fuel-fired ovens are representative of food baking process equipment. The fuel, number of burners, number of zones, temperature, and amount of ventilation will vary for this equipment. Occasionally, electric ovens are used. Baked products include bread, pastry, pies, cakes, cookies, and crackers. In pie and pastry making, fruit must be prepared and cooked before baking. Some ovens are arranged to spray vegetable oil into pans before baking or onto the product during baking.

Candy Making

Most candy making involves cooking or heating, e.g., melting chocolate to pour into molds and cooking flavored sugar mixtures to make hard candy. Process equipment is likely to include kettles, ovens, and cooling tunnels. Many candies are covered with coatings, such as chocolate, icings, fruit mixtures, and glazes. The coatings are either melted and mixed or carried in a solvent.

Meat Processing

In processing, meat is cleaned and cut into desired sections. Meat preservation consists mainly of drying, salting (soaking in brine), and smoking. Smoking is performed in smoke chambers. Other processing is usually done by plants that use meat in their products, such as those making soups, stews, hash, canned meat, meat rolls, or even pet foods.

Distribution Warehousing

Some food facilities simply store large quantities of many different food products for distribution to grocery stores. As with nonfood warehousing facilities, some grocery warehouses are equipped with automatic retrieval systems (automated warehousing). Automatic retrieval systems usually allow the storage of products at a greater density than conventional racks. This increases fuel loading, lessens access to stock, and increases the amount of food exposed to contamination from a single loss. Many other locations may depend on grocery warehouses for food products; consequently, loss prevention is very important for these warehouses.

Miscellaneous Processes

Some food processes, such as frying, do not fall into the previous categories. Frying processes include making potato chips, corn chips, french fries, onion rings, cheese curls, pork rinds, and other snacks. Fryers containing hot fat or oil are used to make these products. Other miscellaneous processes include making nondairy beverages, such as carbonated beverages, fruit drinks and juices, beer, and liquors.

FIRE HAZARDS

Process Equipment

Refrigeration equipment within the food processing industry presents hazards that are inherent to compressed gases, potential vapor release, moving mechanical parts, and electrical control systems. Ammonia refrigerant is volatile and toxic. (See Figure 9-23A.) Freon is less volatile but still considered toxic. Both can contaminate food and otherwise damage equipment and stock. Refer to Section 3, Chapter 28, "Refrigeration Systems," for more information. Vessels for cooking or melting are usually heated with electricity or steam. Their fire hazards include those of electrical systems and overheating. Boilers that produce steam for these vessels present the same fire and explosion hazards as those associated with fuel-fired equipment.

Baking ovens also present fuel firing hazards. In addition, ovens have interior moving parts and fixed ductwork that become coated with oils and can be involved in a fire. Electrically heated ovens sometimes contain concealed oil-filled transformers that are a concern. Another feature of ovens that requires attention is combustible insulation. The primary hazards introduced by cooling tunnels are combustible construction and/or insulation. (See Figure 9-23B.) Fryers and associated ductwork involve heating of combustible liquid. The degree of hazard increases as the temperature approaches the liquid's flash point.

Another characteristic of food processing equipment that contributes to its loss potential is a high level of automation. Automated processing equipment is often custom made and can take a long time to replace. It is so efficient that production lost while the equipment is out of service usually cannot be made up. In fact, if the

FIG. 9-23A.   Result of an anhydrous ammonia explosion in Houston, TX, on December 11, 1983. (Source: Houston Fire Department)

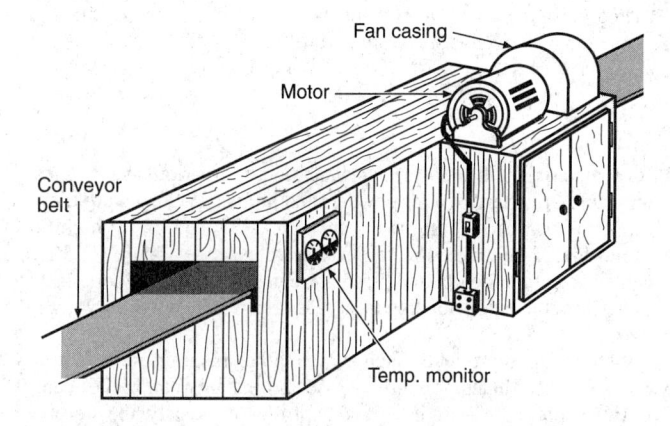

FIG. 9-23B.   Cooling tunnels are sometimes made of wood.

equipment is out of service, lost business sometimes cannot be made up. For this reason, highly automated processes should be reviewed for duplication, alternate production methods, and outside assistance. If they exist, duplicate equipment lines should be separated so they are not exposed to the same potential loss incident.

Gases and Liquids

Some gases used in the food industry are hazardous. Cocoa and other powdered products are treated with gaseous fumigants which, depending upon their composition, could be flammable. Ethylene oxide is a flammable gas used to ripen fruit that has been shipped "green" to distribution centers. Whenever gases are used, their physical characteristics should be checked.

Combustible and flammable liquids are commonly found in food processing plants. Combustible liquids used include lubricating oils, cooking oils, certain alcoholic beverages, and solvent-based extracts. Alcohol, high-proof alcoholic beverages, and solvents are the most commonly used flammable liquids. In the candy industry, ether is used to dissolve the coating that makes jelly beans shiny. In many types of plants, alcohol is used to dissolve dried, crushed materials, or liquids, to make flavorings and colorings. The production of vanilla extract is one example. Waxes, which are liquids when heated, also are used in food processing.

Combustible Dusts

Dust hazards are not unusual in the food processing industry. Under the right conditions, any organic dust can explode. Among the many potentially explosive dusts are those from sugar, flour, grains, starches, and cocoa. Because the explosibility of a dust depends partly upon its fineness, finer dusts, such as confectioners' sugar and cornstarch, present the most severe explosion potential.

Conveyors

Conveyors, which can be used anywhere in a process from raw material transfer to processing, packaging, and storage, pose many hazards. Their drives can jam, misalign, or overheat, and cause friction, which can lead to fires. The conveyors themselves can spread fire from one area to another; so may openings in the walls they pass through. Many belt conveyors have combustible belts. Rubber or plastic belts produce large quantities of smoke when they burn, resulting in a high potential for contamination. Chain conveyors can carry plastic pans. Wide conveyors shield the areas below from ceiling sprinkler protection. Finally, conveyors are part of highly automated systems that cannot function without them. A typical conveyor-type oven is shown in Figure 9-23D.

Storage

Raw materials and food products are stored in various ways. Bins and tanks are often used for powders and liquids. Packaged products are commonly stored on pallets or in racks 80 ft (24.4 m) high or more, including automatic retrieval systems. (See Figure 9-23C.) Since wood, cardboard, and plastics are used to package food products, the combustible loading in storage areas can become very high.

Some food products, such as vegetable oil spray, are stored in aerosol cans. Under pressure, aerosol propellants can rocket cans during a fire, spreading it beyond control. Necessarily, the storage of this type of product must be carefully planned.

FIG. 9-23C.   Erecting storage racks in a grocery warehouse. (Source: Giant Food)

SAFEGUARDS

Construction Considerations

Since food products are highly susceptibile to smoke contamination, noncombustible construction of buildings is preferred. Fire cutoffs should be provided to separate manufacturing and warehousing areas, separate duplicate production lines, and isolate fuel-fired equipment, refrigeration systems, and storage of more hazardous materials such as flammable liquids, paper, cardboard, plastics, pallets, and aerosols. Fire cutoffs need to be preserved where conveyors pass through them. To do this, conveyor lines are specially designed to break open at, or hinge away from, the point where a fire door or shutter closes. The use of water spray systems is sometimes proposed as an alternative to cutoffs. This is less desirable because water spray by itself cannot prevent the radiation of heat through an opening. Water spray systems also depend upon a water supply and therefore are subject to impairments.

Heating and cooling requirements in the food processing industry call for the widespread use of insulating materials in buildings and process equipment. Noncombustible insulation should be chosen whenever possible. When combustible insulation is used, it should have a low flame spread, as well as low fuel-contributed and low smoke-developed ratings established by a testing laboratory. Sprinklers and thermal barriers, such as gypsum or dry wall should protect combustible insulation. Process equipment should be noncombustible.

Sprinkler Systems

Because combustible materials are commonly encountered in the food processing industry, sprinkler protection is needed in all buildings. Sprinklers also should be provided at intermediate levels of

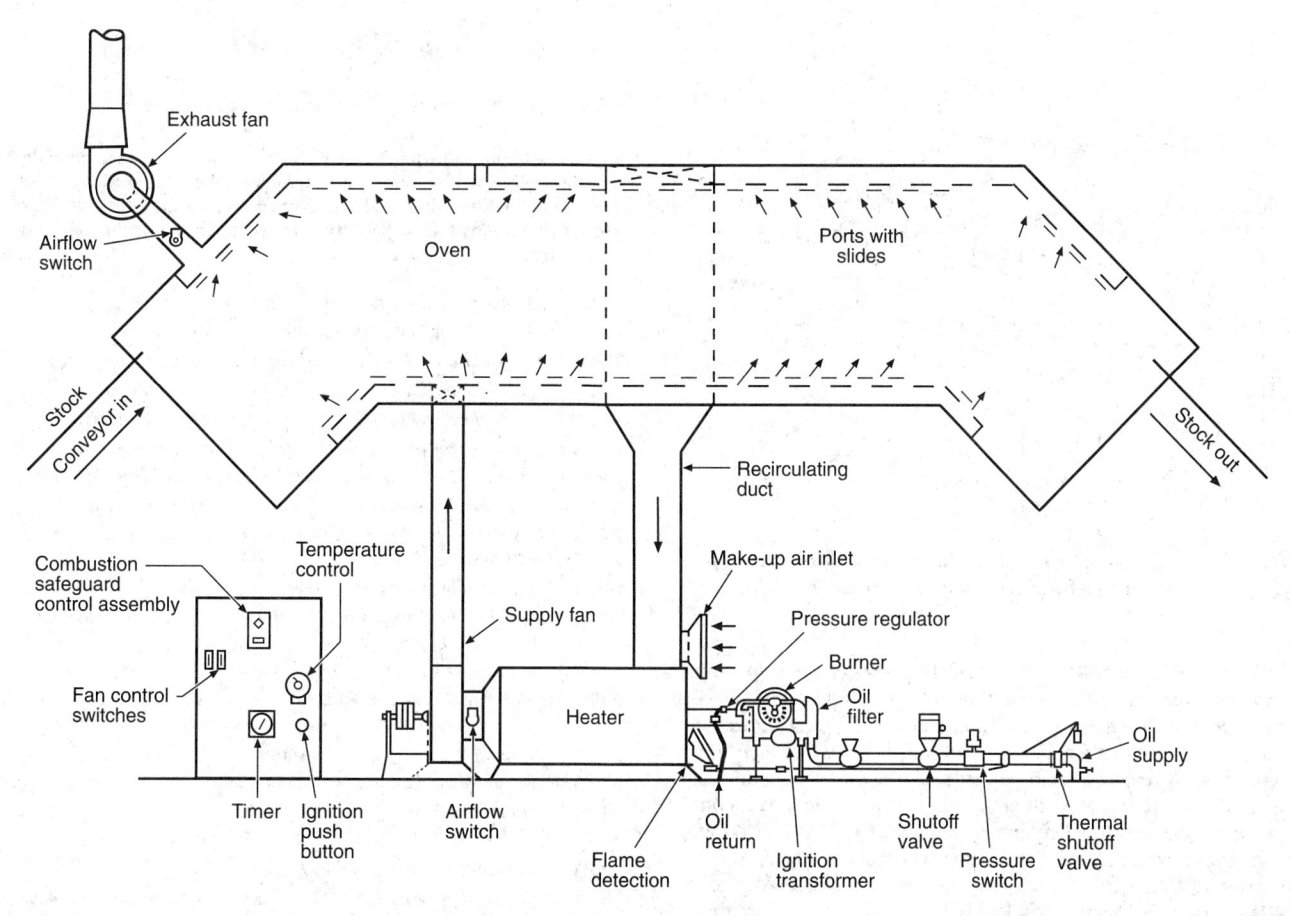

FIG. 9-23D. *Conveyor-type oven with the controls labeled. (Source: Industrial Risk Insurers)*

high storage racks; inside process enclosures handling or built of combustible materials; under obstructions over 4 ft (1.22 m) wide, such as conveyor belts; and inside coolers, freezers, and heating enclosures. Hose connections supplied from sprinkler systems should be provided in storage areas and wherever large amounts of combustibles may be present. Water supplies should be sufficient to supply the sprinkler demand plus hose streams for the amount of time prescribed by the hazard for the highest demand area.

Despite low temperatures, fires can occur in freezers. Providing sprinkler protection in them, however, requires special design. A combination dry-pipe valve and deluge valve replaces the usual wet-pipe sprinkler system. (See Figure 9-23E.) This system minimizes the chance of water damage by requiring two means of detection before the system operates.

Hose connections are more complicated to install in freezers. They must be kept drained and be equipped to both release the air in the sprinkler piping and open the deluge valve. Subfreezing temperatures make these complex systems difficult to maintain.

Fixed Extinguishing Systems

A fixed extinguishing system most effectively protects some hazards found in the food processing industry. Deep-fat fryers and their associated ductwork can be protected with carbon dioxide, or dry or wet chemical extinguishing systems. (See Figure 9-23F.) Large outside tanks of flammable liquids should be protected with water spray or foam systems. Explosion suppression systems are desirable for large or critical equipment handling combustible dusts or for such equipment with insufficient explosion relief. Fixed extinguish-

ing systems can similarly protect other hazards involving flammable liquids, heated combustible liquids, and combustible dusts.

Electrical Equipment

Electrical equipment in food processing plants should be suitable for the location in which it is installed. [See NFPA 70, *National Electrical Code®*; NFPA 497A, *Recommended Practice for the*

FIG. 9-23E. *A combined dry-pipe and deluge system. (Source: Giant Food)*

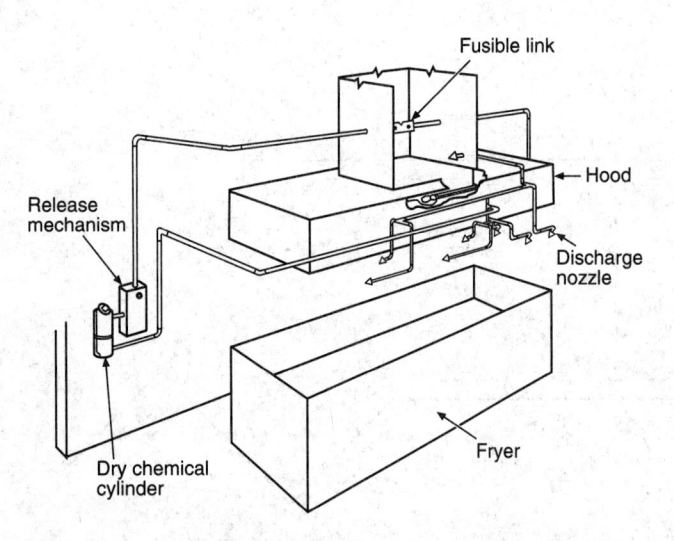

FIG. 9-23F. *Layout of a wet chemical extinguishing system for a fat fryer. (Reprinted with permission of Wormald U.S., Inc.)*

Classification of Class I Hazardous (Classified) Locations for Electrical Installations in Chemical Process Areas; and NFPA 497B, *Recommended Practice for the Classification of Class II Hazardous (Classified) Locations for Electrical Installations in Chemical Process Areas.*] A review should be made of possible vapors or dusts that could be present in each area. Where liquids such as alcohol may be present, electrical equipment suitable for Class I, Group D, locations is recommended. Where ether is used, the electrical classification is Class I, Group C. Ethylene oxide requires the use of Class I, Group B, equipment. Class II, Group G, locations are those where nonconductive dusts are likely to be present.

Other Protection Features

Areas handling flammable liquids should be well ventilated to prevent significant vapor buildup. Enclosures that may contain dusts or vapors should be provided with explosion relief designed to minimize damage. Magnets to extract ferrous debris are used wherever powdered materials enter a storage or distribution system. Portions of systems handling powdered materials from which dust can escape should be provided with permanent dust collection systems. For equipment that handles either dusts or vapors, proper bonding and grounding practices must be observed.

High-piled storage areas are a fire-fighting challenge even with properly designed sprinkler protection. For these areas, there should be good outside hydrant coverage. Inside hose stations and heat and smoke vents, either manual or automatic, should be used to augment automatic sprinkler protection.

Interlocks should be provided for automated processes to run them safely. Fuel-fired equipment should be provided with standard combustion safeguards and air and fuel train controls. Ovens and other heated enclosures should have high temperature cutouts. Enclosures that could contain hazardous vapors need combustible-gas detection systems. Boilers producing steam require excess pressure cutouts, and those producing hot water require high temperature cutouts. When process safety requires a flowing liquid, such as cooling water, the liquid flow and level should be monitored. Conveyor belts should be monitored for low speed or excessive friction to prevent malfunction from causing a fire. Continuous-feed ovens or dryers requiring ventilation should be interlocked to shut down upon loss of ventilation. Potentially unsafe conditions in any process should be monitored and interlocked to avoid hazardous conditions.

BIBLIOGRAPHY

NFPA Codes, Standards, and Recommended Practices

Reference to the following NFPA codes, standards, and recommended practices will provide further information on the safeguards for food processing facilities discussed in this chapter. (See the latest version of *The NFPA Catalog* for availability of current editions of the following documents.)

NFPA 10, *Standard for Portable Fire Extinguishers*

NFPA 11, *Standard for Low-Expansion Foam*

NFPA 11A, *Standard for Medium- and High-Expansion Foam Systems*

NFPA 12, *Standard on Carbon Dioxide Extinguishing Systems*

NFPA 12A, *Standard on Halon 1301 Fire Extinguishing Systems*

NFPA 13, *Standard for the Installation of Sprinkler Systems*

NFPA 14, *Standard for the Installation of Standpipe and Hose Systems*

NFPA 15, *Standard for Water Spray Fixed Systems for Fire Protection*

NFPA 16A, *Standard for the Installation of Closed-Head Foam-Water Sprinkler Systems*

NFPA 30, *Flammable and Combustible Liquids Code*

NFPA 34, *Standard for Dipping and Coating Processes Using Flammable or Combustible Liquids*

NFPA 49, *Hazardous Chemicals Data*

NFPA 61, *Standard for the Prevention of Fire and Dust Explosions in Agricultural and Food Products Facilities*

NFPA 68, *Guide for Venting of Deflagrations*

NFPA 69, *Standard on Explosion Prevention Systems*

NFPA 70, *National Electrical Code®*

NFPA 77, *Recommended Practice on Static Electricity*

NFPA 86, *Standard for Ovens and Furnaces*

NFPA 96, *Standard for Ventilation Control and Fire Protection of Commercial Cooking Operations*

NFPA 204M, *Guide for Smoke and Heat Venting*

NFPA 231, *Standard for General Storage*

NFPA 231C, *Standard for Rack Storage of Materials*

NFPA 325M, *Guide to Fire Hazard Properties of Flammable Liquids, Gases, and Volatile Solids*

NFPA 496, *Standard for Purged and Pressurized Enclosures for Electrical Equipment*

NFPA 497A, *Recommended Practice for Classification of Class I Hazardous (Classified) Locations for Electrical Installations in Chemical Processing Areas*

NFPA 497B, *Recommended Practice for the Classification of Class II Hazardous (Classified) Locations for Electrical Installations in Chemical Process Areas*

NFPA 497M, *Manual for Classification of Gases, Vapors, and Dusts for Electrical Equipment in Hazardous (Classified) Locations*

NFPA 650, *Standard for Pneumatic Conveying Systems for Handling Combustible Materials*

NFPA 8501, *Standard for Single Burner Boiler Operation*

NFPA 8502, *Standard for the Prevention of Furnace Explosions/Implosions in Multiple Burner Boiler-Furnaces*

Additional Reading

Catchpole, L., "Fires in the Food Industry," *Fire Prevention*, No. 285, Dec. 1995, pp. 21–25.

Cote, A. E., ed., *Industrial Fire Hazards Handbook*, third edition, National Fire Protection Association, Quincy, MA, 1990.

"Fire Risks of Insulated Panels in the Food Industry," *Fire Prevention*, No. 279, May 1995, pp. 25–27.

Food Processing and Packaging Equipment, nineteenth edition, Business Trend Analysts, Dix Hill, NY.

Strouppe, D. E., "Ammonia Contamination: An Avoidable Loss," *Locomotive*, Vol. 66, No. 5, Spring 1989.

SOLVENT EXTRACTION

Revised by C. Louis Kingsbaker

Oilseeds that contain approximately 20 percent oil by weight, such as soybeans, are usually extracted directly. Seeds containing more than 20 percent oil usually are prepared by prepressing to reduce their oil content to 10 to 20 percent. Facilities employing this step are called prepress solvent extraction plants. Some major oilseeds processed this way are cottonseed, rapeseed, flaxseed, corngerm, sunflower, safflower, peanuts, and copra.

Additional information relevant to the solvent extraction process can be found in Section 3, Chapter 26, "Storage and Handling of Grain Mill Products"; Section 3, Chapter 31, "Waste Handling and Control"; Section 4, Chapter 6, "Flammable and Combustible Liquids"; and Section 4, Chapter 15, "Dusts."

THE SOLVENT EXTRACTION PROCESS

Solvents Used

Commercial hexane is the solvent exclusively used in oilseed extraction because it is relatively easy to desolventize from the extracted meal and oil products, making them nontoxic; it is essentially immiscible with water; and it can be produced from petroleum at a relatively low cost. Commercial hexane consists of a mixture of these petroleum solvents: normal hexane with a weight ranging from about 53 percent to 85 percent (60 percent by weight is a normal average); methylcyclopentane, about 13 percent to 15 percent; methylpentanes, about 22 percent to 25 percent; and dimethylbutanes, about 1 percent. The major disadvantage of commercial hexane, i.e., the danger of fire and/or explosion, is overcome by careful design, construction, and operation of the solvent extraction plant.

Other flammable solvents, such as mixtures of aromatic hydrocarbons, are used for extracting lignite, Douglas fir bark, and other special raw materials whose end products are not used by humans for internal consumption.

Description of the Process

Unit operations for extracting all oilseeds are similar, and usually the same equipment can handle the flakes or cakes from these materials. A flow diagram for a typical extraction process is shown in Figure 9-24A.

C. Louis Kingsbaker is president of C. L. Kingsbaker, Inc., consultants to the oilseed industry.

For safe storage, oilseeds are brought into the processing plant, checked for moisture content, and dried in a flue gas dryer to about 10 percent or 11 percent moisture content by weight. To eliminate fire hazards in both the dryer and in the preparation step of the extraction process, seeds are cleaned to remove trash, dirt, and foreign material before they are dried.

The solvent extraction process is continuous. Its rate is usually controlled by measuring at the weigh scale and adjusting the cracking mill feed rolls. The cracking mill splits the hulls and breaks the seeds. Screens and aspiration remove hulls. The seeds are flaked between rollers after passing through a rotary steam tube conditioner. Here they are heated to temperatures between 140 and 165°F (60 and 74°C) and the moisture content is adjusted to between 10 percent and $10^1/_2$ percent.

Due to the demand for high protein soybean meal for poultry and other animal feeds, most U.S. extraction plants remove hulls from cracked soybeans using sifters and aspirators in the separation step. To achieve good separation of hulls from bean meats, typically the bulk moisture is reduced to about 10 percent by weight. The beans are then stored for several days in "tempering" silos to allow the residual moisture to equilibrate. A newer method is to process the beans directly from storage at their normal moisture content by heating them quickly in a fluidized bed dryer. Then the beans are cracked and dehulled while still hot. The meats are then conditioned in a second fluidized bed dryer prior to flaking.

Within the past 10 years, many processors have installed equipment to produce pellets or collets from soybean flakes and other oilseeds. This machine is called an expander and consists of a tubular barrel, a feeding device that takes material into the barrel, a cut flight screw inside the barrel, bolts that protrude from the barrel into the spaces between the discontinuous flights to add shear to the material inside, a die plate at the outlet of the barrel through which the extruded material leaves the machine, and a large motor and drive that powers the unit. Live steam injected into the barrel heats and cooks the oilseed and raises its moisture content. The purpose of using an expander is to rupture additional oil cells and increase the pellet's density when compared to the flake. However, the most important benefit is to greatly improve the drainage of the solvent and miscella (mixture of extracted oil and hexane) streams added to the extractor. With better drainage in the extractor, material leaving the extractor has a lower solvent holdup, which reduces the desolventizing duty in the desolventizer-toaster and saves considerable steam energy. Some plants only expand about 20 percent to 30 percent of the incoming flakes, while others expand 100 percent of the

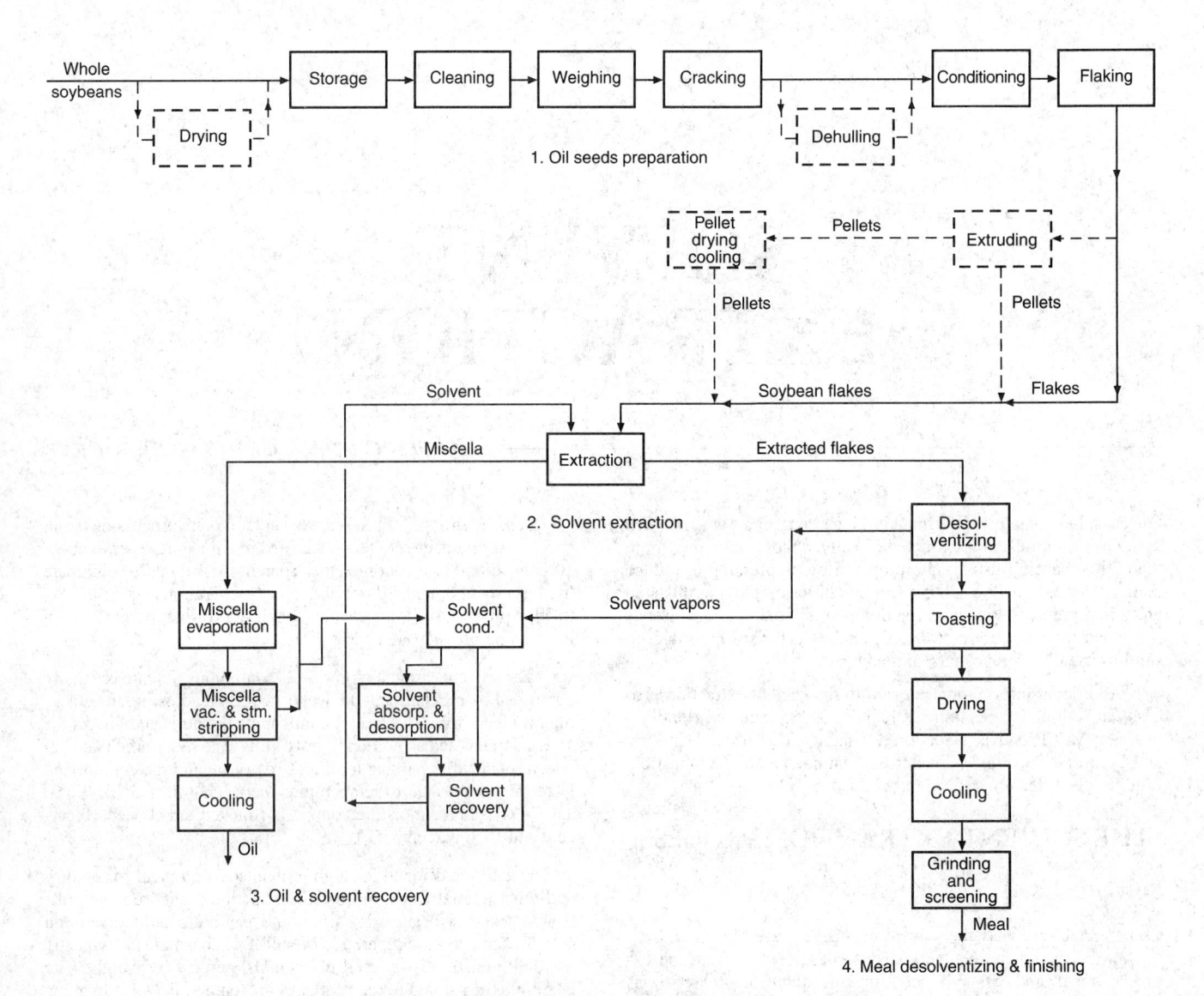

FIG. 9-24A.   Flow diagram of the soybean solvent extraction process.

stream. As long as no more than about 40 percent of the flakes are extruded, the pellets can be fed directly back into the remaining flake stream. But if more than 40 percent of the flakes are pelletized, a dryer-cooler must be provided to reduce the pellets' moisture content and cool them to about 165°F (74°C). Higher oil content oil-seeds, such as cottonseed, can be extruded and then extracted without first prepressing them in a screw press.

Hot flakes are delivered to the extractor, which is usually of the continuous percolation type. The liquid solvent enters above the flake bed and drains or percolates through the bed to a compartment below the flakes to be recycled by a pump to another stage of the extractor. Most percolation extractors have six or more extraction stages.

The many types of percolation extractors include vertical basket, rotary, stationary bottom, sliding cell, perforated belt, and rectangular loop extractors. Figure 9-24B shows a rectangular loop extractor. Fresh, recovered hexane solvent, heated to about 140°F (60°C), contacts flakes near the extractor's discharge end to remove the remaining oil from the flakes. This solvent, which now contains some oil, is pumped from stage to stage in a direction counter to the

flakes' flow and becomes richer in oil with each successive stage. Oil extracted from the flakes and mixed with hexane, the miscella, usually has a concentration of 25 percent to 30 percent by weight when it exits the extractor. The amount of hexane solvent added to the extractor is one part solvent to one part flake, by weight. This is the solvent ratio and ranges from 0.85 to 1.1. The extractor normally operates at 140°F (60°C) and at either atmospheric pressure or at a slight vacuum of $\frac{1}{2}$-in. water gage (gage pressure of 124 Pa). A variable-speed drive, or variable-speed motor, controls the extractor speed for changes of plant rate and also bed height in the extractor.

After the flakes move beyond the solvent addition point in the extractor, time is provided for hexane to drain. Extracted and drained flakes are conveyed to equipment to desolventize, toast, dry, and cool the material. Two processing systems are used to accomplish this. The first, which is now being phased out in older plants, requires three individual equipment items: (1) the desolventizer-toaster (D-T unit), (2) a rotary steam tube dryer (D), and (3) a rotary cooler (C) using ambient air to cool the meal. The second system, the counter current D-T-D-C, is normally installed in extraction plants. This single piece of equipment achieves the four processing steps. (See Figure 9-24C.) It consists of eight trays: two predesolventizing trays (1 and 2); two hollow staybolt trays, often called countercurrent trays (3 and 4); one spare tray (5); two drying trays (6 and 7); and one cooling tray (8). Additional trays can be provided for each processing step as needed, depending upon plant capacity, or the type of oilseed processed.

The processing principles are the same for each type of seed. Open or sparge steam in the desolventizer strips solvent from the flakes. Toasting is accomplished by indirect contact with high-pressure steam in trays 1, 2, 3, and 4 and the open steam. The D-T-D-C is a vertical, cylindrical unit consisting of multiple trays or decks. Flakes drop from tray to tray. Open steam in the top five trays removes solvent. Steam condenses in the flakes, raising the flakes' moisture content. Toasting is also done in these trays, which are also heated by indirect steam.

Usually, the desolventizing step uses about 10 percent excess sparge steam. The D-T-D-C dome's normal operating temperature is about 158 to 165°F (70 to 74°C). The outlet temperature from tray 5 must be at least 221°F (105°C). Lower flake outlet temperature indicates that not all solvent has been removed, which could present a potential safety hazard. Operating pressure at the top of the D-T-D-C should be lower than the pressure in the extractor. This prevents steam vapor from flowing back to the extractor and

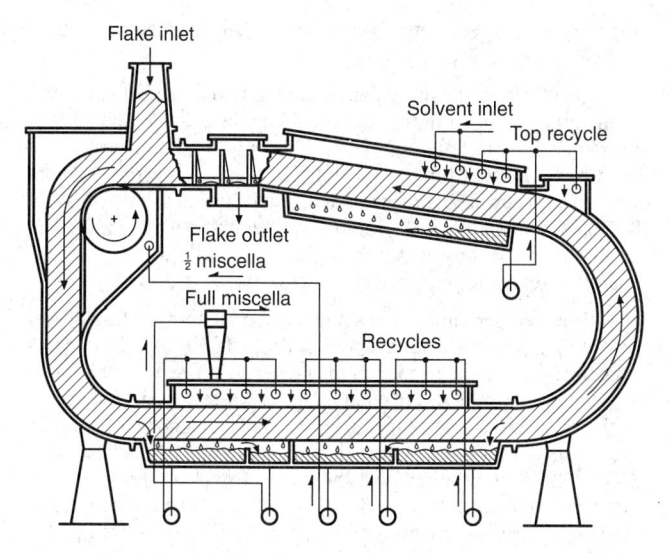

FIG. 9-24B. Rectangular loop extractor.

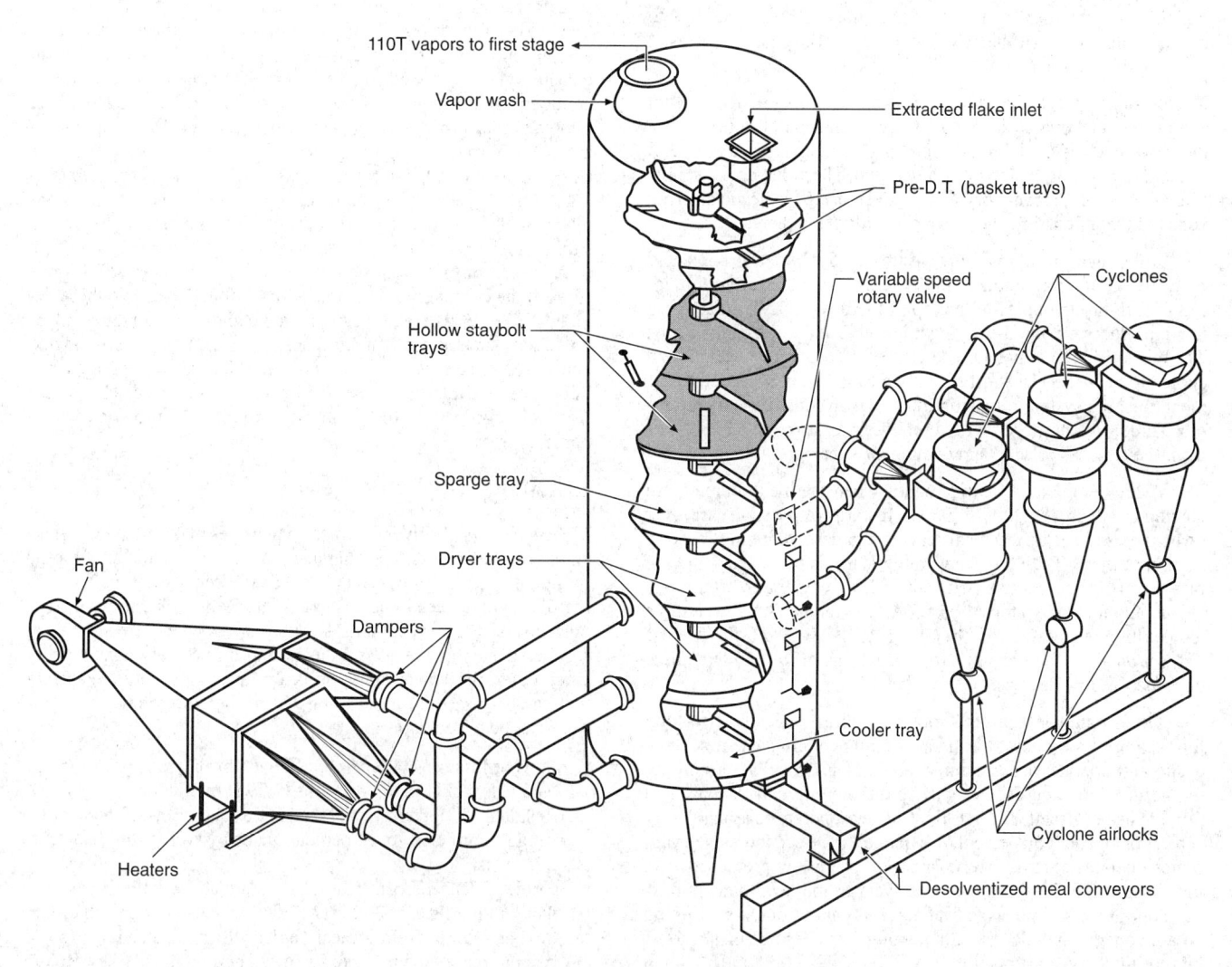

FIG. 9-24C. Counterflow D-T-D-C.

stops solvent vapor from flowing downward through the outlet of the toasting section. Solvent and excess sparge steam vapors leaving the top of the D-T-D-C are scrubbed with liquid hexane to remove any fine particulates that might be entrained in these vapors, and are condensed, recovered, and reused in the extraction step.

The specially designed rotary valve located at the outlet of tray 5 has a variable speed drive. Its speed controls the operating levels in trays 1 through 5. It also seals tray 5 from tray 6 to prevent drying air from blowing into tray 5 from tray 6. Tray 6 has a much higher operating pressure than tray 5. The rotary valve is a safety device that prevents explosive mixtures of hexane and air from forming in the unit's top five trays.

The desolventized material drops into trays 6 and 7, where it contacts hot air. As the water-moist meal proceeds downwards, heated air is injected through holes in the top of these two trays. The hot air evaporates excess moisture in the meal, and as water evaporates from desolventized meal, its temperature drops. A fan sends atmospheric air through a steam coil air heater and into the trays for drying.

The outlets from trays 6 and 7 each have a gate and sail arrangement. The gate is normally closed until the level in each tray rises high enough to contact a sail at the top of its tray. The sail, located inside the tray, has a shaft that extends through the D-T-D-C shell; the end of the shaft connects externally to the gate at the tray's outlet. When the meal contacts the sail, the sail opens the gate to permit material to exit the tray. This automatically maintains meal levels in the trays.

A steam purge line admits a small quantity of open steam against the rotary valve at the outlet of tray 5. Its purpose is to purge the valve pockets of the hot drying air picked up from tray 6 as it turns in its cycle at the top of drying tray 6. The valve pockets, now filled with steam, prevent dryer air from entering the desolventizing zones of the unit and forming explosive mixtures there.

Dried meal from tray 7 drops into tray 8 where it is contacted by a stream of cool, ambient air, in similar countercurrent flow as tray 7 and through holes in the top of the tray deck. The same fan that introduces air into the drying trays sends air to the cooling tray. The meal, dried to a moisture content of 12 percent by weight and cooled to between 115 and 120°F (46 and 49°C), is conveyed to the meal finishing system for sizing by the desolventized meal conveyors. Three cyclones, each with a rotary air lock, remove meal fines from the D-T-D-C drying and cooling air streams.

A low temperature alarm sounds when the meal temperature leaving tray 5 drops to 205°F (96°C). It warns of possible carryover of hexane solvent in the meal leaving the desolventizing-toasting section of the D-T-D-C. If the warning alarm is not heeded and the meal temperature continues to fall, a low temperature shutdown switch automatically stops the inlet feed to the D-T-D-C when this temperature reaches 195°F (90°C). NFPA 36, *Standard for Solvent Extraction Plants* (hereinafter referred to as NFPA 36), requires both these safety devices.

Oil is removed from the solvent in three stages by two long tube rising film evaporators and a stripping column. The first stage evaporator operates at a vacuum of about 15 in. (51 kPa) of mercury on the tube side and uses vapors from the desolventizer-toaster to provide heat. Miscella from the first stage evaporator has an oil concentration of about 50 percent to 60 percent. The second-stage evaporator operates at atmospheric pressure, using low-pressure steam in the shell side. Final removal of solvent from the 92 percent miscella out of the second stage evaporator is carried out in a stripping column operating under vacuum ranging from 22 to 27 in. (74 to 91 kPa) of mercury at a temperature of 210 to 240°F (99 to 115°C) and using open or sparge steam at the bottom of the column. The vac-

uum dries or removes water from the oil, and the sparge steam strips the final traces of hexane solvent.

All solvent is condensed and returned to the solvent separator, along with solvent from the flake-desolventizing step. The solvent separator's 15-min holding interval allows water condensed in the process to separate from the solvent. The water flows from the separator to a wastewater evaporator, where it is heated to at least 200°F (93°C) with open steam to remove any entrained solvent, before leaving the plant for a wastewater trap and finally to a sewer. The separated solvent returns to the extractor for reuse.

A low temperature alarm sounds when the temperature of water leaving the waste water evaporator drops 185°F (85°C). It warns that carryover of hexane solvent may enter public sewer systems. If the warning alarm is not heeded, and the waste water temperature continues to fall, a low temperature shutdown switch automatically stops the extractor drive motor when the temperature reaches 175°F (80°C). NFPA 36 requires both these safety devices. There was no such alarm and shutdown device at a soybean plant in Louisville, KY, when a massive quantity of hexane entered the sewers in 1981 and blew up six miles (10 km) of the sewer system.[1]

The extractor and all other tanks and condensers in the process are vented to a separate header and into a vent recovery system. This system, which uses an edible mineral oil to absorb vent gases, consists of two columns filled with ceramic packing. One column absorbs hexane solvent vapors with a flow of cool mineral oil counter to the flow of the vapors. The other column strips the absorbed hexane solvent vapors from the mineral oil with open steam. A condensing system condenses these vapors. The mineral oil recirculates continuously for reuse. A blower on top of the mineral oil absorber removes unabsorbed water and air vapors that discharge into the atmosphere through a flame arrestor. The blower also provides a slight vacuum for all equipment in the extraction system. Maintaining the process at a low vacuum prevents solvent vapors from escaping from the system and entering the extraction building. To prevent similar incidents, extraction plants are now equipped with high pressure alarms and high pressure shutdown switches. High pressures in the extractor and the top of the D-T-D-C activate these devices. NFPA 36 also requires extraction plants to install relief valves on vessels such as the extractor, solvent work tank, miscella tank, and solvent-water separator. This provides emergency venting to relieve excessive internal pressures in case of a fire. The discharge line from the relief valves to the atmosphere must be provided with a flame arrestor.

Efficiency of the Process

Most designers of solvent extraction facilities make efficiency guarantees for the process. The guarantees are important not only for the profitability of the process but also for safe plant operation. Exceeding these guarantees indicates potential safety problems, and steps must be taken to correct them. The guarantees given here are typical for the soybean processing industry and do not vary considerably when other oilseeds are processed. In a process having a capacity that can be measured in U.S. tons/24-hr day, the following conditions can be expected:

1. Solvent loss of 0.095 percent by weight of beans processed or 0.35 U.S. gal/U.S. ton of beans (2.29 L/metric ton).
2. Residual oil content of flakes from extractor of 0.5 percent by weight, corrected to 12 percent water by weight on a solvent-free basis.
3. Crude oil 0.15 percent of total moisture and volatiles by weight that will not flash at 300°F (150°C) in a closed-cup flash tester.
4. Soybean flakes from a meal cooler will not exceed a 0.15 pH rise in urease enzyme activity, measured by the Caskey-Knapp method.[2]

HAZARDS OF SOLVENT EXTRACTION

The principal hazards specific to solvent extraction operations are the highly flammable solvents used and combustible dusts associated with storing and handling oilseeds. Ignition sources include arcing electrical equipment, static electricity, and open flame. Ignition can occur during regular operations, maintenance operations, or an emergency. Attempts to operate a facility beyond its design capacity can introduce hazards of abnormal operation.

Normal Operations

Solvent loss: To prevent fires, it is necessary to keep solvent liquid or vapor from leaving the extraction plant and controlled areas. Hexane solvent vapor is about three times heavier than air and, if not dispersed, could possibly enter areas where an ignition source is present. To be ignitable, the vapor concentration in air must be within its flammable limits—1.2 percent to 6.9 percent by volume.

Solvent loss in an extraction plant is a key indicator to possible hazards during normal operation. Solvent loss of 0.095 percent by weight of the material processed is considered within the bounds of efficient operation. If the loss exceeds 0.14 percent by weight, reasons for the increase must be found and corrective action taken. Table 9-24A shows the five sources of solvent loss from the extraction process and the percentage of loss attributed to each.

TABLE 9-24A. *Sources of Solvent Loss from the Solvent Extraction Process*

Source	Loss lb/ton*	Loss (% by weight)
Flakes from D-T unit	1.0	0.04
Leaks	0.3	0.04
Vent gas from vent system	0.2	0.04
Oil from oil stripper	0.3	0.02
Effluent from process waste water	0.1	0.01
Total	1.9	0.15

*Pounds of solvent per ton of material processed.

It is of primary importance to prevent flammable solvents from leaving the extraction building and entering the preparation building. This can happen if solvent remains in the flakes after they have been removed from the desolventizer-toaster and sent to the preparation building to be ground, sized, and finished. Because the equipment in this building is not explosionproof, ignition could occur with disastrous results.

Whenever flammable solvents are transferred, immediate precautions must be initiated to protect the building until the material is removed.

Startup, shutdown, and purging: The startup or shutdown of an extraction plant introduces a transient hazard. The equipment, especially the extractor, is full of air before startup. As hexane solvent is added, the concentration of solvent vapor in the atmosphere passes through the flammable range of 1.2 percent to 6.9 percent by volume. Two methods eliminate the possibility of ignition by static electricity when using hot liquid to purge the extractor of air. One method is to design the plant so hot vapors from the second stage evaporator flow into and rapidly heat the extractor to 104°F (40°C), thus replacing the air. The advantage of this procedure is to shorten extractor heat-up time, usually about 4 hr, to about $^1/_2$ hr. Stage pumps do not start until the extractor is hot. The second method is

to replace extractor air with carbon dioxide or nitrogen gas before adding hot liquid hexane to the extractor.

Vapor-proof slide gates located upstream and downstream of the extractor are closed during warm-up to prevent hexane solvent vapors from flowing back to the preparation building.

A normal or controlled shutdown is accomplished by emptying the extraction plant of all oilseed flakes and hexane liquid and purging the empty extractor or desolventizer-toaster unit with air until the unit cools. Steam purging is used if normal maintenance is required inside the extractor.

If welding is necessary in the extraction area, personnel must take these precautions:

1. Empty all solids and liquids.
2. Store hexane from the plant in solvent tanks isolated from the plant.
3. Steam the purge extractor, desolventizer-toaster unit, and all equipment containing hexane or miscella.
4. Fill all vessels with water.
5. Use water to seal flake connections from the extractor and desolventizer-toaster unit to the preparation building.
6. Use a portable gas analyzer to locate possible flammable mixtures.

Inert gas is satisfactory for purging an empty extractor before normal startup procedures. Inert gas can also be used for purging during normal shutdown, but steam purging is a surer method than inert gas purging for removing solvent vapor.

Abnormal Operations

Emergency breakdowns: A serious hazard in the operation of a solvent extraction plant is an emergency breakdown in the extractor when it is full of solvent-laden flakes and cannot be emptied. Quite often the problem can neither be determined nor corrected without someone entering the unit.

The extractor should be cooled to about 104°F (40°C). The required time to do this depends upon the extractor size and its specific method of operation. During this time, a plan for correcting the problem should be devised and approved by a supervisor. The extractor must then be purged with air, using the extractor purge blower provided with the unit. The purge blower air flow capacity should be 1 extractor volume air change every 3 min, based upon an empty extractor. The purge blower must be of sparkproof construction. Inert gas purging of an extractor that has failed under load does not ensure a nonflammable condition will exist. If the extractor is not equipped with a purge blower, the unit should be steam purged for 24 to 48 hr or longer to be sure all hexane solvent inside the extractor has boiled off.

While air purging, air entering the purge blower must be checked with a meter to detect the explosive limits of hexane. Air purging continues until the air reaches 0.25 percent or less of the lower explosive limit (LEL) of hexane in air by volume. The extractor can be entered, but the purge blower keeps operating during the repair process. Persons entering the extractor must wear a safety harness and an air mask, and be supervised at all times. The LEL of the purged air leaving the extractor must be constantly monitored to ensure it is below 0.25 LEL of hexane. External and internal ignition sources must be eliminated because a flammable mixture of hexane and air might exist.

If a D-T-D-C fails under load, the material in the heating trays may overheat and cause a fire. It is important to immediately turn off the steam to these trays when such a failure occurs. Normally, sparge or open steam can be added to the unit's trays for several hours to desolventize the flakes and to steam smother smoldering material inside the trays. Doors on each tray are opened individually and their contents raked until the unit is empty and the problem corrected.

Failures: The supply of emergency cooling water should be sufficient to operate the condensers long enough for a safe shutdown. If cooling water fails, the process must be stopped, the steam shut off, and the vapor slide gates closed. Similar steps should be taken in the event of steam or electrical failures.

System overloads: If a solvent extraction plant capacity increases above its design rating, this creates a hazard, because normally the solvent condensing capacity is insufficient to recover the solvent. This will result in higher vent pressures and the inherent danger that solvent vapors may be forced into the atmosphere or carried out with flakes to form flammable mixtures.

FIRE PROTECTION

Fire protection for solvent extraction plants involves building construction and services, plant layout, fire protection equipment and systems, and safe operating procedures. Detailed guidance can be found in NFPA 36 and in other related standards, guides, and recommended practices listed at the end of this chapter.

Layout and Construction

Structural elements of solvent extraction plants should be laid out so that separation distances adequately isolate or dissipate flammable concentrations of solvent vapor from ignition sources. A typical layout is shown in Figure 9-24D.

The extraction and preparation buildings should be of fire-resistive or noncombustible construction and equipped with some form of explosion relief. It is important that electrical wiring and equipment be suitable for use in the presence of flammable vapors. All vessels, pipes, tubes, and hoses containing or carrying flammable solvent should be electrically bonded together and grounded to prevent the buildup of a static charge. Portable flammable gas detectors should be available for monitoring the atmosphere.

For a process involving combustible dust, electrical wiring and equipment should be suitable, a dust collecting system should be provided, and protection against static electricity discharges should be installed.

Programmable Logic Control (PLC) systems along with computers are now being installed to monitor and control the operations of extraction plants. Locating these systems in the extraction building control room is convenient, but since they are not of Class I, Groups C and D, Division 1, construction, special control room design must be provided. In such a design, the extraction plant control room must be designed in accordance with NFPA 496, *Standard for Purged and Pressurized Enclosures for Electrical Equipment,* Chapter 3, "Purged Control Rooms in Class I Locations." Type X purging must be used; this design reduces the classification within the control room from Division 1 to nonhazardous.

Fire Protection Equipment

Water spray, deluge, or foam-water systems—singly or in combination—are suitable for use in extractor buildings. Automatic sprinkler systems are appropriate for use in preparation buildings. Portable fire extinguishers of the proper types and sizes may adequately protect the solvent unloading and storage area, if it is isolated from exposure hazards.

Portable fire extinguishers should be strategically located. The facility may also have standpipe and hose systems equipped with combination nozzles. A system of yard hydrants is an essential part

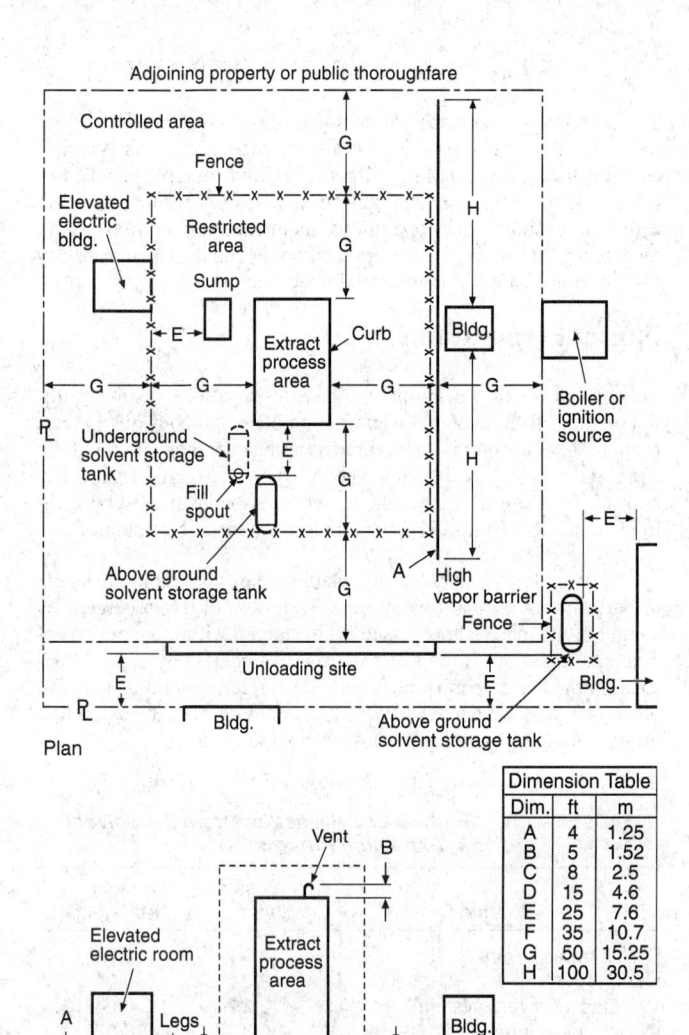

Dimension Table		
Dim.	ft	m
A	4	1.25
B	5	1.52
C	8	2.5
D	15	4.6
E	25	7.6
F	35	10.7
G	50	15.25
H	100	30.5

FIG. 9-24D. *Typical distance diagram.*

of protection. Finally, the available water supply must adequately meet the needs of fixed protection and yard hydrant systems.

Policies and Procedures

It is imperative that solvent extraction equipment be operated in the manner the manufacturer prescribes and that no attempt be made to bypass or defeat safety devices and systems. Company policy should limit the output demanded of the extraction plant and not overtax its design capacity. Policy should also provide for shutdown in the event of excessive solvent loss and for regular maintenance and cleaning operations. If solvent loss increases from 0.095 to 0.135 weight percent per ton, 0.35 to 0.50 U.S. gal/ton (1.5 to 2.1 L/metric ton) of material processed, an abnormal hazard exists and preventive action must be taken.

Fire prevention is a key factor in any fire protection program. It includes conspicuously posting safety rules and emergency procedures throughout the plant, safety education for new employees, appointment of a plant emergency organization, and a compulsory monthly safety program. A safety system audit program must be implemented to regularly check the workability of all safety devices and controls.

If a fire occurs, appropriate action is to activate fire extinguishing systems (if they are not automatically controlled), activate the plant interlock system to shut down all solids-conveying equipment in the plant, turn off the main steam valve to the extraction plant, leave the area, and notify the fire department.

BIBLIOGRAPHY

References Cited

1. LeBlanc, P., and Redding, D., "1981 Large-loss Fires in the United States," *Fire Journal,* Vol. 76, No. 5, p. 40.
2. "Urease Activity, ACOS Method #Ba9-58," *Official Methods and Recommended Practices of the American Oil Chemists' Society,* 1993.

NFPA Codes, Standards, and Recommended Practices

Reference to the following NFPA codes, standards, and recommended practices will provide further information on the safeguards for solvent extraction discussed in this chapter. (See the latest version of *The NFPA Catalog* for availability of current editions of the following documents.)

NFPA 10, *Portable Fire Extinguishers*

NFPA 13, *Standard for the Installation of Sprinkler Systems*

NFPA 14, *Standard for the Installation of Standpipe and Hose Systems*

NFPA 15, *Standard for Water Spray Fixed Systems for Fire Protection*

NFPA 16, *Standard for the Installation of Deluge Foam-Water Sprinkler and Foam-Water Spray Systems*

NFPA 24, *Standard for the Installation of Private Fire Service Mains and Their Appurtenances*

NFPA 30, *Flammable and Combustible Liquids Code*

NFPA 36, *Standard for Solvent Extraction Plants*

NFPA 61, *Standard for the Prevention of Fires and Dust Explosions in Agricultural and Food Products Facilities*

NFPA 68, *Guide for Venting of Deflagrations*

NFPA 70, *National Electrical Code®*

NFPA 91, *Standard for Exhaust Systems for Air Conveying of Materials*

NFPA 385, *Standard for Tank Vehicles for Flammable and Combustible Liquids*

NFPA 496, *Standard for Purged and Pressurized Enclosures for Electrical Equipment*

Additional Reading

Michaud, P., Granier, A., and Crooker, R. M., "FORANE 141b: State of the Art After One Year of Industrial Use in a Solvent Area," *Proceedings of the 1992 International CFC and Halon Alternatives Conference on Stratospheric Ozone Protection for the 90's,* Sept. 29–Oct. 1, 1992, Washington, DC, 1992, pp. 445–449.

Pohl, K. D., *et al.,* "Systematische Untersuchungen zur Ermittlung der Zundfahigkeit fein vernebelter Wasser-Losemittel-Gemische. [Systematic Analysis to Determine the Ignition Point of Atomized Water-Solvent-Mixtures.]," *VFDB,* Feb. 1994, pp. 23–28.

Pohl, K. D., *et al.,* "Systematische Untersuchungen zur Ermittlung der Zundfahigkeit fein vernebelter Losemittel-Wasser-Gemische (II). [Systematic Investigations to Determine the Inflammability of Finely Atomized Solvent-Water-Mixtures (II)]," *VFDB,* Vol. 4, Nov. 1994, pp. 161–163.

Sawyer, W., "Relationship of Flash Points of Solvents, Resin Solutions, and Paints," *Journal of Coatings Technology,* Vol. 49, No. 627, pp. 52–55.

Welker, J. R., Martinseen, W. E., and Johnson, D. W., "Effectiveness of Fire Control Agents for Hexane Fires," *Fire Technology,* Vol. 22, No. 3, 1986, pp. 329–340.

PROTECTION OF WASTEWATER TREATMENT PLANTS

James F. Wheeler

This chapter discusses wastewater treatment processes, process equipment, and their fire and explosion hazards. Traditionally, wastewater treatment process equipment is classified according to the various unit operations being performed. The basic unit processes fall into three categories: (1) collection and pumping, (2) liquid treatment systems, and (3) solid treatment systems. Specific unit processes to consider are pumping station wet wells that handle raw sewage or unstabilized solids, preliminary screening and grit removal processes, primary sedimentation, anaerobic digestion, scum collecting processes, and fuel and chemical handling and storage areas. A considerable variety of makes and models of equipment provide the necessary processing. Consequently, the fire and explosion hazards associated with wastewater treatment processes are much too broad a subject to examine in a single chapter. Therefore, this chapter discusses some basic principles of risk evaluation, mitigation, and fire protection that can be applied generally to individual unit processes and equipment.

The collection and transportation of liquid wastes, known as wastewater, from a community creates a disposal problem that demands a solution to protect the public's health and comfort. The series of closed conduits (pipes) conveying liquid wastes in a municipality is commonly called a sewer system. Among the first historical traces of wastewater collection is the sewer arch at Nippur, India, which was probably constructed about 3950 BC. Although other examples of ancient sewers exist, the development of the modern wastewater collection system did not begin in the United States until 1805, with the construction of the first large sewer in New York City.[1]

With the collection of wastewater comes the requirements to provide treatment. All methods aim to produce an effluent that can be discharged without nuisance and that can protect human health and the environment. Modern cities and present-day metropolitan centers could not exist as they do without protecting the health and conveniences that comprehensive collection systems and sophisticated treatment facilities afford. Today, in the U.S., over 17,000 municipal wastewater treatment facilities serve nearly 200 million people and treat approximately 35 trillion gallons (130 trillion liters) of wastewater per day. (See Figure 9-25A.) Thousands of industries and businesses also provide their own wastewater treatment services. In addition, millions of people, most living in rural areas, are still served by individual on-site treatment units, such as septic

tanks and small aerobic package plants.[2] As development expands into these rural areas, and as on-site septic systems begin to fail and pollute ground water supplies, new collector sewers must be constructed and larger, more complex treatment plants will be required.

Wastewater treatment covers any artificial processes that remove or alter wastewater's objectionable constituents to make it less offensive or dangerous. Early treatment plants included anaerobic lagoons or cesspools, which provided barely primary treatment, or provided treatment simply by dilution. Modern wastewater treatment began in the United States in 1886 with the introduction of the intermittent sand filter. The wastewater treatment industry grew rapidly during the early 1900s with the development of aerobic processes, such as the trickling filter and activated sludge treatment.[1]

FIG. 9-25A. *Aerial view of a modern wastewater treatment plant.*

James F. Wheeler, P.E., is an environmental engineer for the U.S. Environmental Protection Agency, Washington, DC.

In addition to conventional pollutants, such as biochemical oxygen demand (BOD) and suspended solids (SS), today's wastewater industry must be prepared to treat hazardous and toxic industrial waste or explosive and flammable materials that may be illegally or accidentally discharged to the collection system. The treatment of these wastes involves highly specialized knowledge of the design, construction, maintenance, and operation of sophisticated equipment and processes. Wastewater treatment is a viable and growing industry that is continually changing as demands for better removal of toxic and hazardous materials increase and more sophisticated and complex chemical and biological processes evolve.

OPERATIONS AND EQUIPMENT

Treatment of wastewater can be broken into three major components: (1) collection and pumping systems, (2) liquid treatment systems, and (3) solids treatment systems. Each component utilizes different processes and equipment. Each also has specific hazards associated with its particular function.

Collection and Pumping Systems

Collection systems gather wastewater from houses, factories, or other sources and transport it to a central facility for processing. Where the terrain is flat, the collection system may consist solely of gravity piping; however, in most areas the collection system includes pumping. Among the many different kinds of collection systems, the sanitary sewer and the combined sewer are most common. The sanitary sewer carries wastewater from individual homes and excludes storm water, surface water, and ground water. Sanitary sewers also collect wastewater from industrial and commercial businesses in the service area. Combined sewers carry both domestic wastewater and storm water, as well as industrial wastewater. Although today only separate sanitary sewers are being constructed, many older cities still have thousands of miles of large combined sewers.

Sanitary sewers are generally smaller than combined sewers. Sanitary sewers with diameters of 8 to 48 in. (0.2 to 1.2 m) are common. Combined sewers generally are very large, often 20 ft (6 m) in diameter or larger. These collection systems also include other appurtenances, such as manholes, valve boxes, catch basins, curb inlets, and a variety of other structures. Combined sewers may also have regulators or diversion structures that allow overflow directly into rivers or streams during major storms.

Pumping wastewater has long been a problem because of fibrous and stringy material in the wastewater. Rotary and reciprocating pumps are generally not suited for wastewater pumping, unless the wastewater has been treated to remove solids. Even many centrifugal pumps used to pump wastewater experience problems with handling solids. Today, solids-handling centrifugal pumps with single-suction volutes pump most wastewater.[3] These pumps have open impellers, thus creating large passageways, as shown in Figure 9-25B. These centrifugal pumps can also be equipped with bladeless impellers for pumping unscreened wastewater and sludge. Pneumatic ejectors have also been used successfully for handling small volumes of unscreened wastewater.

Most pumping stations have two parts, a dry well and a wet well. The wet well receives and temporarily stores wastewater. Wet wells often contain screens that remove large solids that could damage the pumps. Wet wells may also contain mechanical equipment, fans, lights, etc. However, most electrical equipment, including motors, pumps, and switches, are placed in the dry well. Most pumping stations have both above and below ground structures, although some smaller stations are placed entirely underground, as shown in

Figures 9-25C and 9-25D. Figure 9-25E shows a suction lift pump that can be used in either station configuration.

Most stations have a number of pumping units with various capacities designed to handle changes in flow rates. The pumps are

FIG. 9-25B. *Open non-clog impeller for pumping raw sewage.*

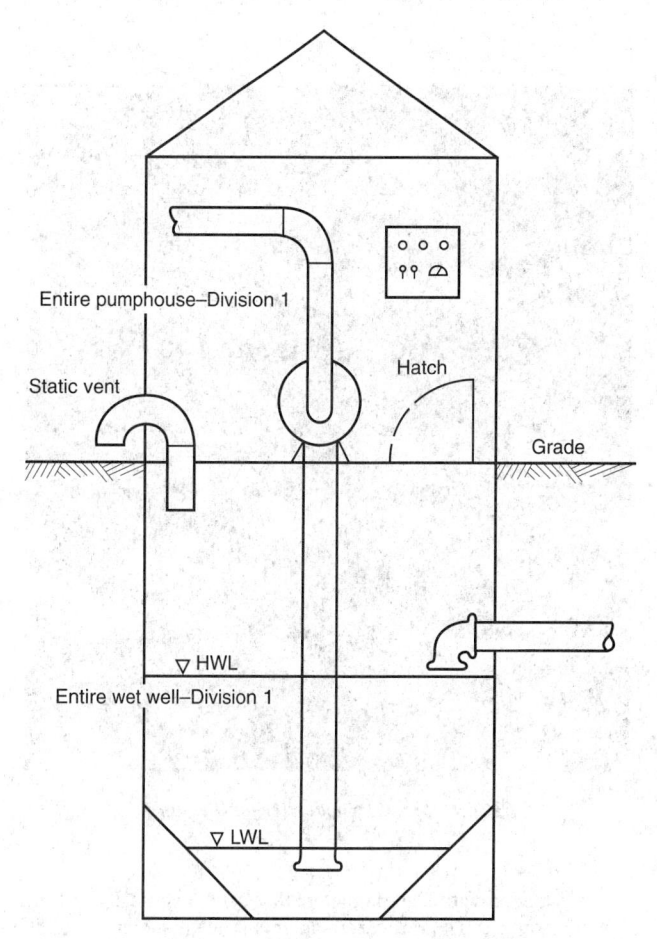

FIG. 9-25C. *Typical above/below ground pumping station.*

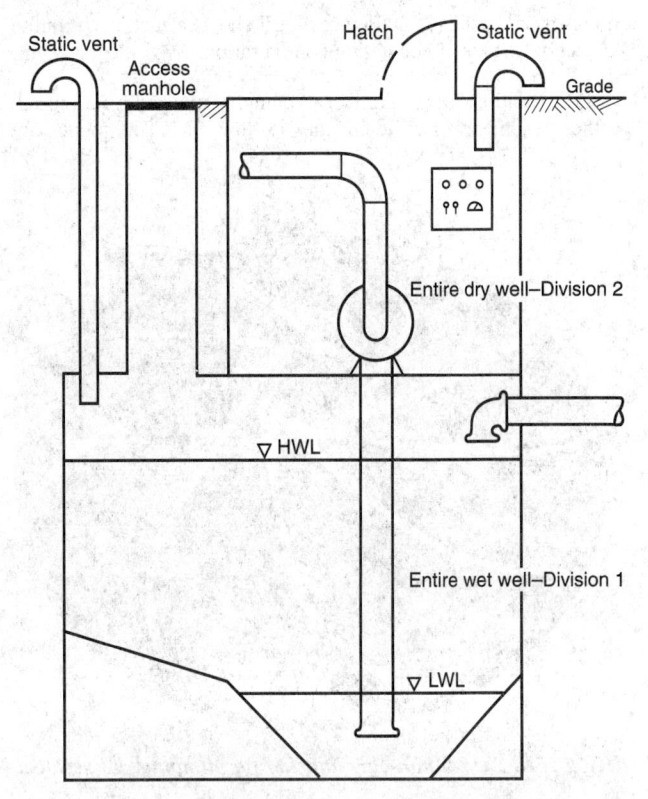

FIG. 9-25D. *Typical below ground pumping station.*

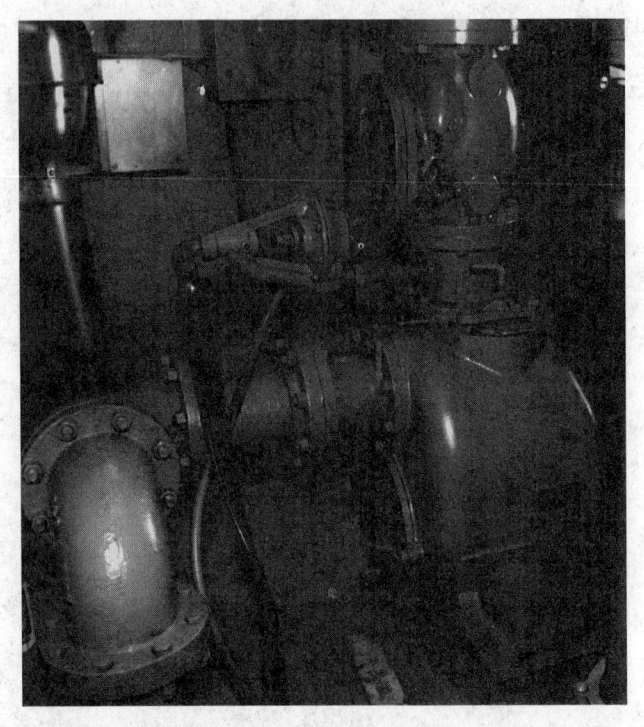

FIG. 9-25F. *Typical horizontal mounted pump.*

FIG. 9-25E. *Typical suction lift pump.*

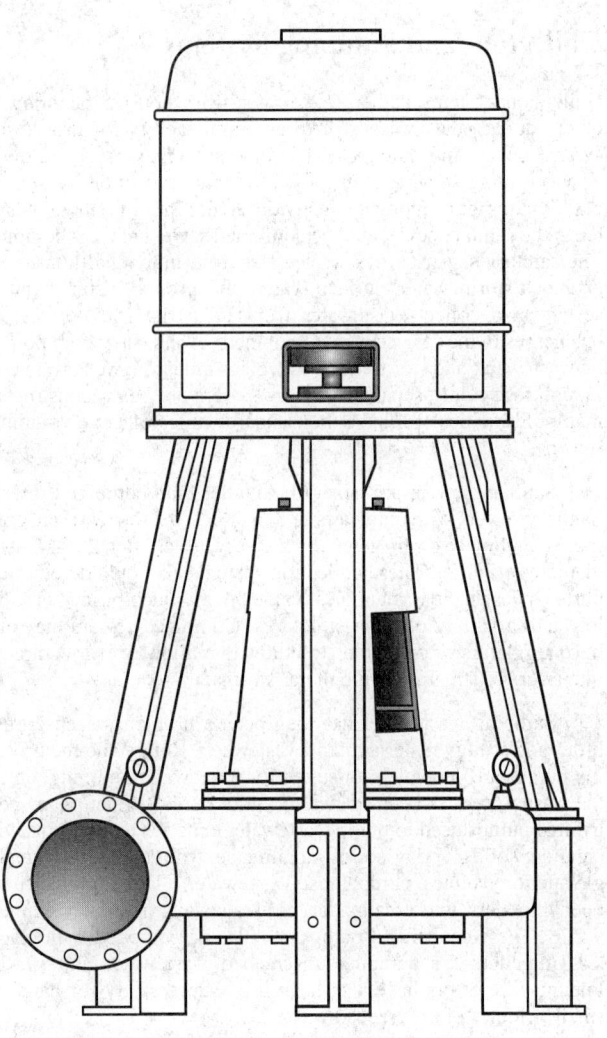

FIG. 9-25G. *Typical vertical mounted pump.*

generally driven with electric motors that can be vertically or horizontally mounted. The vertical arrangement shown in Figures 9-25G and 9-25H has the advantage of permitting the motor to be placed above the pump, to safeguard it against flood damage. Today, many pump stations use submersible pumps. These pumps are located in the wet well, as shown in Figure 9-25H, and can be equipped with quick disconnects so the pumping units can be removed from the wet

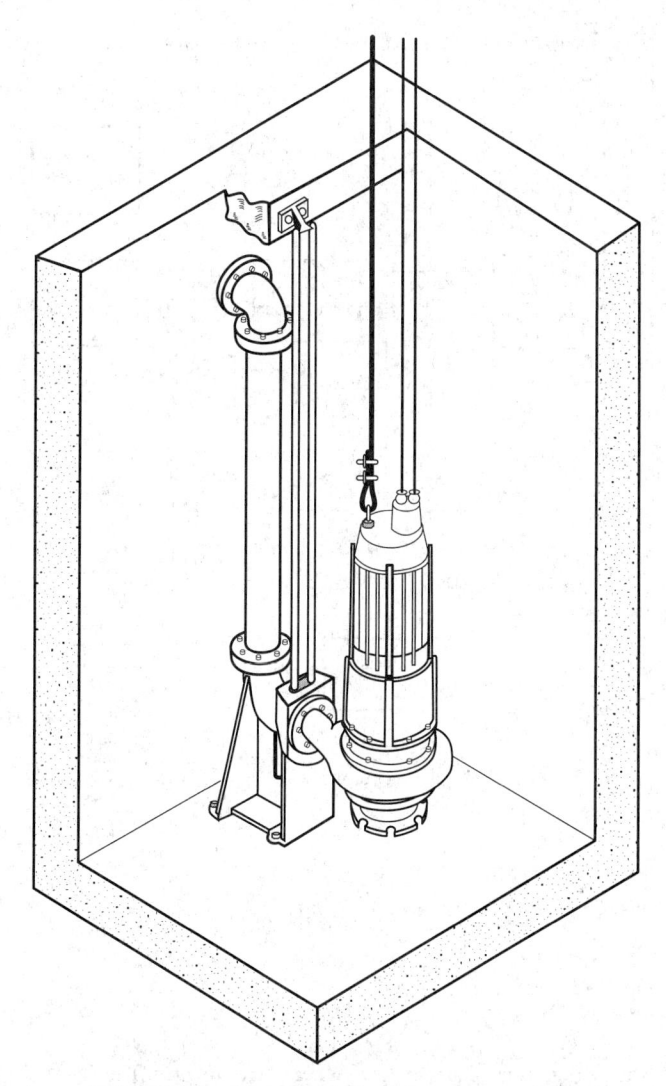

FIG. 9-25H.   *Typical submersible pumping station.*

well for servicing. A flat efficiency curve is desirable for sewage pumps, because the head against which the pumps must work is variable. Because of special nonclog impellers used in sewage pumps, efficiency is generally not as high as those used for pumping potable water. Because dependability is more important than efficiency with most wastewater pumping applications, slow-speed operation is desirable because it increases equipment life.

Where interrupting the electric power source is possible, an auxiliary power source is often required. In larger stations, electric generators or gasoline or natural gas engines are provided. The engines connect directly to the pump shaft by means of a right-angled gear drive unit so that either the engine or electric motor can be used by operating a clutch. Some gas engines are duel-fuel compatible and can use either natural gas or methane gas produced at the treatment plant by anaerobic digestion of sewage sludge.

Liquid Treatment Systems

Liquid treatment processes remove objectionable constituents from the wastewater so that treated water can be returned to the environment without being offensive or dangerous to public health. The principal components of liquid treatment are preliminary treatment, primary treatment, secondary treatment, tertiary treatment, and disinfection. A typical schematic flow and process diagram for wastewater treatment is shown in Figure 9-25I.[4]

Preliminary treatment conditions wastewater as it enters the treatment plant. Preliminary treatment removes materials that might harm or adversely affect the operation of the treatment plant or process equipment. Such material may include lumber, cardboard, rags, stones, sand, plastic, and grit. Methods and equipment used to remove these materials include comminutors, mechanically or manually cleaned bar racks, and bar screens. Particularly in combined sewers, grit and sand washed into the collection system from the street can damage mechanical equipment and adversely affect downstream units, such as sedimentation tanks and digesters. Grit is generally removed by gravity or in aerated grit chambers.

Primary treatment includes first-stage sedimentation, which removes settleable suspended solids, and floating material, such as grease and scum, from the wastewater. Figure 9-25J shows a typical primary clarifier. Well-operated primary treatment facilities may remove as much as 60 percent of the influent-suspended solids and 30 percent of the influent biochemical oxygen demand. However, primary treatment does not remove colloidal or dissolved solids. Some treatment plants add chemicals, such as lime, alum, or ferric chloride, to increase colloidal solids removal.

Secondary treatment intends to reduce the concentrations of remaining suspended solids and dissolved and colloidal organic matter in wastewater. Such material is not removed to any significant degree in primary treatment. A wastewater treatment plant having secondary treatment following primary treatment commonly can remove 85 percent of the influent-suspended solids and biochemical oxygen demand of the raw wastewater. Secondary treatment processes may be either biological or physical-chemical.

Most municipal secondary treatment processes are biological. These processes may be classified as fixed film or suspended growth. Fixed film processes include trickling filters, bio-towers, and rotating biological contactors (RBC). In these processes a thin biological slime (film) grows on a fixed surface. Suspended growth processes include aerobic ponds, oxidation ditches, and a variety of air- and oxygen-activated sludge systems. In these processes the biological organisms are suspended in liquid. In each process, a mixed population of microorganisms establishes in the presence of oxygen. These microorganisms metabolize dissolved organic matter in wastewater and form a biological mass. The effluent from fixed film or suspended growth processes contains suspended biological solids. A portion of these solids called waste sludge is removed from the treated wastewater in a secondary sedimentation tank for further processing. A portion of these solids called return sludge may be recycled to the head of the suspended growth process for additional treatment, as shown in Figure 9-25I.

In a small number of cases, physical-chemical treatment is used in place of, or combined with, biological unit processes. Physical-chemical processes can include chemical addition, coagulation, precipitation, and filtration to remove suspended matter, and activated carbon absorption to remove soluble organics. In addition to lime, alum, and ferric chloride, a variety of anionic and cationic polymers also enhance settling in the physical-chemical process.

Tertiary treatment is used, as required, to reduce the concentration of inorganic and organic constituents below concentrations achievable through secondary treatment. Tertiary treatment also includes removing nitrogen and phosphorous by additional unit process operations. Tertiary treatment processes may be physical, chemical, biological, or a combination. A wide variety of chemicals are used in tertiary treatment processes. Most common are lime, alum, ferric chloride, polymers, methanol, chlorine, carbon, ozone, and oxygen.

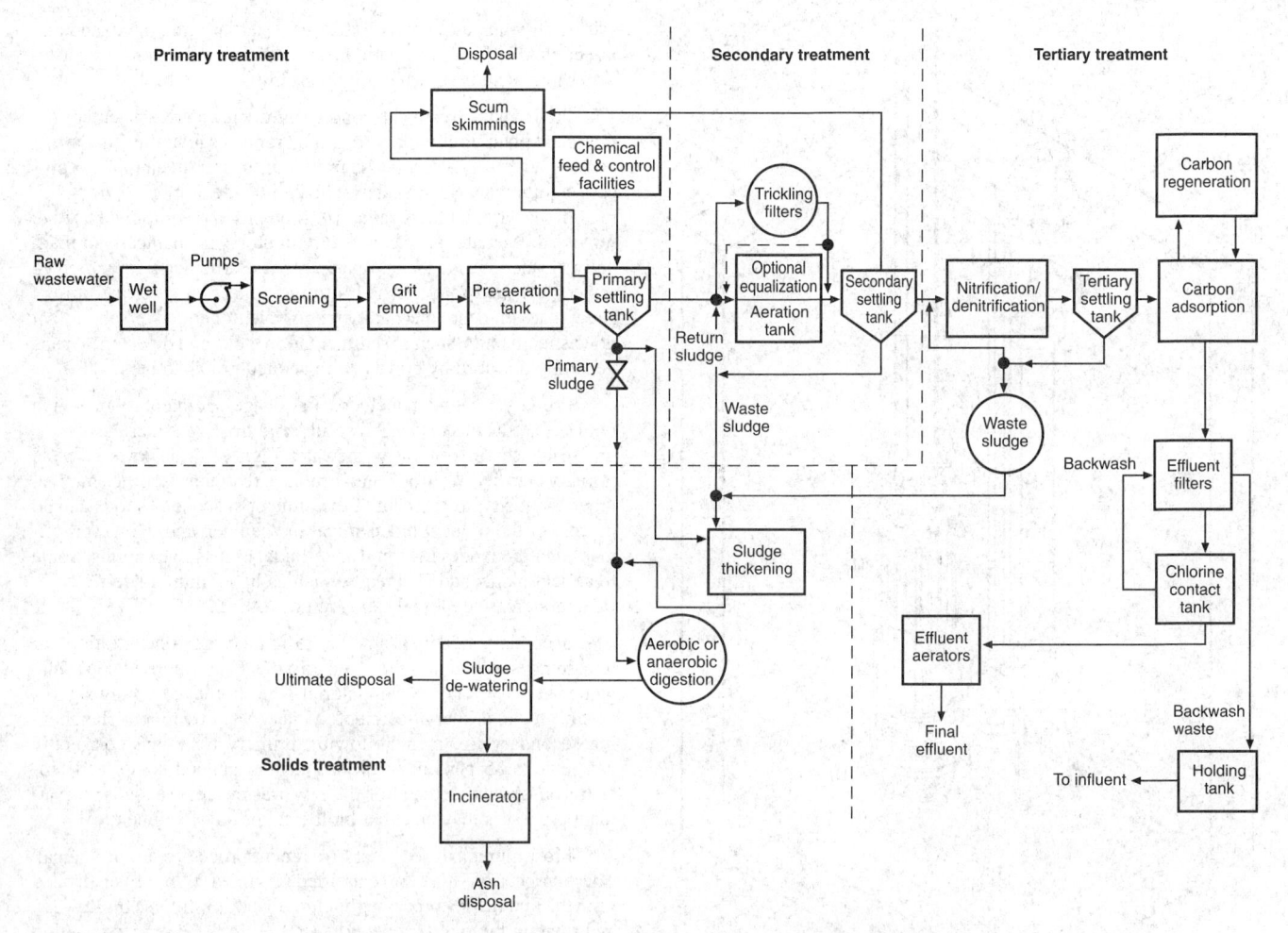

FIG. 9-25I.   *Typical schematic flow and process diagram.*[4]

Disinfection is required to destroy pathogenic bacteria, viruses, and amoebic cysts commonly found in wastewater. Disinfection processes may be chemical (chlorination or ozonation) or physical (ultraviolet irradiation). Chemical disinfection using chlorine, and infrequently ozone, is the most widely used means of wastewater disinfection. Ozone and ultraviolet irradiation are produced on site. However, chlorine is generally supplied as a gas or liquid. Smaller plants use chlorine gas in 150-lb to 1-ton (23- to 907-kg) cylinders, while larger plants may use chlorine in bulk from a tank truck or railroad truck car. Today many treatment plants also require dechlorination to protect aquatic plants and animals from the toxic effects of any chlorine remaining in the effluent. The most common chemicals used in dechlorination are sulfur dioxide or activated carbon.

Solids Treatment Systems

Solids treatment processes further treat the sludge or biosolids removed from wastewater by the liquid treatment processes. Sludges withdrawn from the treatment processes are still largely water, as much as 97 percent. The principal components of solids treatment are digestion, thickening, dewatering, and disposal. Figure 9-25I shows a flow schematic and process diagram for a typical solids treatment system.[4]

Sludge handling and disposal are often the most difficult, important, and costly parts of wastewater treatment. Sludge may be stabilized either under anaerobic or aerobic conditions. Anaerobic

sludge digestion occurs in the absence of free oxygen. The digested solid end product of anaerobic digestion is relatively nonputrescible and inoffensive. The off-gas produced in anaerobic sludge digestion contains about 65 percent methane and may be collected and used as fuel.

Both anaerobic and aerobic digestion result in a reduction in the total volume and weight of excess organic matter. It is often desirable, before final disposal, to reduce further the volume and weight of sludge and to change it from a liquid that is more than 95

FIG. 9-25J.   *Typical primary clarifier.*

percent water to a semisolid form. This dewatering may be accomplished by using drying beds, vacuum filters, centrifuges, filter presses, or mechanical gravity units. The dewatering operation often is enhanced by chemically conditioning the sludge with lime or polymers before dewatering. Conditioning may include a thickening step, which could use gravity or air flotation. Thermal conditioning may also be used to prepare sludge for dewatering.

In smaller plants much of the dewatered and stabilized sludge is trucked to sanitary landfills. Municipal sludge, however, contains essential plant nutrients and useful trace elements. Some treatment plants take advantage of these nutrients and apply the sludge to the land. The sludge can be composted and used as a soil conditioner, or the sludge can be applied as a liquid fertilizer. For larger plants, or where land is not available, sludge is often incinerated. The two most common types of incineration systems are the multiple-hearth furnace and the fluid-bed incinerator. Exhaust gas from the incinerator must be treated carefully to avoid air pollution. The incinerator ash can be toxic and may require special handling and disposal methods.

PROCESS HAZARDS

Many potential hazards may be found in wastewater treatment plants and in the operations they perform. These include explosion and fire hazards, health hazards, and environmental hazards. This discussion, however, deals only with explosion and fire hazards.

Three principal sources of hazards are associated with the wastewater treatment industry. These include materials and chemicals that enter the collection system, materials and chemicals used in the treatment processes, and materials and chemicals produced as byproducts of the treatment processes. Hazard sources commonly associated with the wastewater industry include fuel gases, sewer and sludge gases, specialty gases, liquids, solids, dusts, and construction materials.

Fuel Gases

Fuel gases include natural gas, manufactured gas, sludge gas, liquefied petroleum gas-air mixtures, liquefied petroleum gas in the vapor phase, and mixtures of these gases. Some of these gases have specific gravities lower than that of air, so when released they rapidly rise and diffuse above a leakage point. Flammable mixtures are produced when these gases are mixed with air within certain limits, and, since the oxygen content of each gas (when not mixed with air) is generally low, they may also be considered as suffocating gases. In addition, the wastewater treatment plant uses a variety of gaseous, solid, and liquid chemicals that by themselves or mixed with oxygen or other chemicals may be potential sources of fire and/or explosion.

Sewer and Sludge Gases

Sewer and sludge gases are flammable gases resulting from fermentation or decomposition of organic matter. Explosive conditions (especially concerning the anaerobic digestion process) may result when these gases mix with air within certain limits.

Specialty Gases

Specialty gases commonly found in the wastewater treatment plant are used in a variety of operations including: (1) laboratory analysis and instrumentation calibration, e.g., hydrogen, methane, etc.; (2) wastewater treatment plant unit processes, e.g., chlorine, ozone, etc.; and (3) welding operations, e.g., acetylene, oxygen, etc. These gases may provide flammable and/or explosive conditions when either acting alone or mixed with other chemicals.

Liquids

Liquid waste chemical products discharged into the collection system and transported to the wastewater treatment plants may be potential sources or contributing causes of fire and/or explosive conditions. Hydrocarbon liquids, such as gasoline, kerosene, oils, and other solvents, either discharged to drains or used for various applications at wastewater treatment plants, may also provide fire and/or explosion hazards at certain locations.

Combustible Solids and Dusts

Solids, such as granulated or powdered activated carbon, and other combustible dust byproducts produced by various wastewater treatment processes, may be combustible and may cause potential fire and/or explosive conditions.

Construction Materials

Some construction materials used in wastewater treatment plants, such as plastic, fiberglass-reinforced plastics (FRP), paints and coatings, insulating materials, and furnishings, may either be combustible or limited combustible under certain conditions. Some of these materials present a considerable fuel load, if ignited.

Ignition Sources

In addition to a combustible or flammable fuel source, there must also be an ignition source to cause a fire or explosion. Certain fundamental conditions limit the potential ignition of flammable gases, liquids, and solids (including dusts) found at a wastewater treatment plant. Gases and generated vapors must mix with sufficient air or oxygen to form a flammable mixture and be provided with heat of sufficient intensity for ignition. Ignition may result from one or more of these causes: (1) open flames and hot surfaces; (2) electrical arcs and sparks; or (3) chemical reactions.

Open flames and hot surfaces: These potential means of ignition may be encountered during operations and repairs or with malfunctioning equipment and appliances. Ignition sources can include welding equipment, boilers, incinerators, kerosene-type lanterns, internal combustion engines, and smoking by humans.

Electrical arcs and sparks: Sustained arcing faults can extensively damage electrical switchgear and motor control centers. This can provide sufficient heat to ignite flammable gases or vapors present or generated as a result of the arc, e.g., pyrolysis of insulating material.

Sparks generated by defective/worn electrical and mechanical equipment, activities performed by personnel, and static electricity may be ignition sources for gases or flammable vapors.

Chemical reaction: Fire and explosion may result from the chemical reaction of substances that are (1) introduced by the wastewater treatment plant influent, (2) used for laboratory analyses, (3) required in various unit processes, and (4) produced as byproducts. Chemical reactions can cause hazardous conditions ranging in severity from the generation of flames, e.g., spontaneous combustion, to explosions.

Many collection systems collect not only wastewater but various hazardous materials and chemicals. Materials and chemicals entering the collection system are the most numerous type of in-plant hazards and the least controllable, and, therefore, probably present the most risk of fire and explosion.

Some industries discharge combustible or explosive products, such as alcohol, industrial solvents, hexane, and other liquids and gases. Sewers also collect natural gas or gasoline from leaking pipe-lines or underground storage tanks. Combined sewers are particularly dangerous because they collect hazardous liquids and gases from highway spills, and fire departments often wash flammable hydrocarbons from traffic accidents into the combined sewer. In addition, illegal discharge of explosive or flammable material into the sewer system is always possible.

Pumping stations that handle raw wastewater are subject to the same hazards as the collection system. When confined in a wet well, many hydrocarbons float on the surface of the wastewater. Gases that are lighter than air diffuse rapidly through the structure. Other liquid treatment processes, such as screening equipment and grit chambers that are housed in buildings or in below-grade pits, are often subject to the same fire and explosion hazards as the pumping station wet well. Unit processes, such as primary sedimentation, are generally open and allow lighter gases to escape but may collect and concentrate floating flammable liquids. Secondary and tertiary sedimentation tanks and aeration tanks not preceded by primary sedimentation may also collect and concentrate floating flammable liquids.

Liquid treatment processes are not only subject to the hazardous materials and chemicals in the incoming wastewater. They are also subject to a variety of chemicals used as additives in treating wastewater. Some of these materials are flammable or explosive, some are nonflammable but accelerate combustion, and some are not hazardous by themselves but react with other materials or chemicals. Gases commonly used in wastewater treatment and their physical properties are shown in Table 9-25A. Their uses are described below.[4]

Natural gas and gasoline are used as vehicle fuels, to heat buildings and unit processes, to provide hot water for personnel and cleaning operations, and for engine-driven pumps or generators. Methanol is used as a carbon source in denitrification processes. Oxygen is used in certain types of biological processes, such as oxygen-activated sludge. Chlorine and ozone are used for disinfection, and sulfur dioxide is used in dechlorination processes. Lime, alum, ferric chloride, and polymers are used in sedimentation processes; and granulated and powdered carbon are used in some activated sludge processes or in carbon columns to remove organic compounds.

Solids treatment processes not only use gases and chemicals, but some processes also produce explosive or flammable materials or gases. Although several anaerobic processes produce sludge gas, the principal gas-producing unit process is anaerobic digestion. When organic matter decomposes in the absence of oxygen, sludge gas is produced. This sludge gas is approximately 65 percent methane. Large plants may use this gas as a fuel source. Sludge decomposing in sewers, wet wells, pits, or other tanks can produce a form of sludge gas sometimes called sewer gas. Although never used as a fuel source, sewer gas can contain up to 5 percent methane and can reach explosive limits if confined in improperly ventilated spaces.[3]

As with liquid treatment processes, natural gas is used to heat solids treatment unit processes, such as anaerobic digesters and other heat treatment processes. Natural gas is also used as a fuel in sludge incinerators. Ammonia is used as a nutrient in some activated sludge processes and is a byproduct of air-stripping processes and some composting and sludge dewatering processes. Chlorine and ozone are used in sludge stabilization processes. Hydrogen sulfide, found in sewer gas and sludge gas, is a byproduct of some sludge

TABLE 9-25A. *Gases Commonly Found in Wastewater Treatment Plants*

Name (chemical formula)	Explosive Limits (% vol.)		Density* Heavier/Lighter Than Air	Sources
	LEL	UEL		
Ammonia¶ (NH_3)	16	25	L	Storage tank leaks
Chlorine† (Cl_2)	Nonflammable		H	Disinfection processes Storage tanks
Gasoline¶ ($C_3H_{12} - C_9H_{20}$)	1.3	7.1	H	Storage tanks Tank truck spills
Hydrogen chloride (HCl)	Nonflammable		H	Storage tank leaks Ceramic diffuser cleaning
Hydrogen sulfide¶, # (H_2S)	4.0	44	H	Sewer gas Sludge gas
Natural gas¶	3.8–6.5	13–17	L	Gas piping leaks
Nitrogen (N_2)	Nonflammable		L	Storage tanks Oxygen generation processes Denitrification processes
Oxygen† (O_2)	Nonflammable		H	Generation of oxygen on-site Activated sludge processes Storage tanks Sludge processes
Ozone† (O_3)	Nonflammable		H	Disinfection processes On-site generation processes
Sewer gas‡	5.3	19.3	H	Sewer systems
Sludge gas§	5	15	L	Sludge digestion processes
Sulfur dioxide (SO_2)	Nonflammable		H	Dechlorination processes Storage tanks

NOTES:
*The table lists the physical properties at standard temperature and pressure. In actual field conditions, these gases might disperse and might be present throughout the structure.
†These gases accelerate combustion.
‡Contains approximately 70 percent carbon dioxide, 5 percent methane, and 25 percent other gases (source: United States Environmental Protection Agency).
§Contains approximately 65 percent methane, 30 percent carbon dioxide, and 5 percent other gases (source: United States Environmental Protection Agency).
¶Source: NFPA 325, *Guide to Fire Hazard Properties of Flammable Liquids, Gases, and Volatile Solids.*
#Rarely reaches explosive concentration in wastewater treatment plants.

stabilization and dewatering processes. Nitrogen, used as a medium to transport very dry organic material in certain sludge dehydration processes, is a byproduct of on-site oxygen generation processes. Some sludge stabilization processes and reaeration processes use oxygen.

Of course, many other chemicals, fuels, and materials used in or around the wastewater treatment plant might present a fire or explosion hazard, if not handled properly. These chemicals and gases include solvents, oxidizing materials, hydrogen, acetylene, kerosene, oils, and various specialty gases used in laboratory analyses and instrumentation calibration. The handling of these gases and chemicals is beyond this chapter's scope, but it is covered in other NFPA documents.

FIRE AND EXPLOSION MITIGATION

General Practice

Control procedures used to minimize potential fire and explosion incidents at wastewater treatment plants include (1) hazard evaluation, (2) process and equipment controls, (3) ventilation, (4) construction materials, and (5) education. These control procedures also include proper electrical classification of hazardous locations and proper selection, installation, and operation of electrical equipment, motors, and devices that are suitable for these locations. Effective control procedures cannot be implemented and enforced without the complete cooperation of the public, private, and government sectors, and plant management and personnel, in comprehensive safety programs.

Hazard Evaluation

Hazard evaluation should be initiated early in the design process and hazard prevention and protection recommendations integrated into plant process specification or design considerations. The evaluation should be reviewed and updated periodically as conditions at the plant change. All wastewater processes should be given some kind of hazard evaluation before plants are built and whenever significant changes are made. Some evaluations can be quite informal, while others are highly structured with complete documentation and attempts to quantify hazard inherent in the operation.

The first and most important consideration is the identification of potential hazards. The potential hazard associated with raw wastewater entering the plant, materials used in the treatment processes, and materials produced by the treatment processes must be completely understood and evaluated. Effects of temperature and pressure outside of normal ranges must be studied. The possibility of creating a hazardous byproduct and gradually increasing its concentration in a closed system must be considered. In other words, there must be complete hazard identification for all aspects of the treatment. With this information, potential severity of the process hazards can be determined. Then the means to control the anticipated hazards can be identified. Hazard analysis begins with complete, comprehensive hazard recognition.[5]

Risk Evaluation

Risk evaluation takes into account the likelihood of the hazard resulting in a fire or explosion. Numerical quantification of the risk requires estimating failure rates of equipment and people. A judgment must be made about the necessary degree of maintenance of all important equipment and instruments, as well as the degree of training and judgment required of operators.

Risk control demands that wastewater treatment processes and equipment be well designed and well maintained to minimize potential for leaking and spilling. Redundant instrumentation, emergency controls, and fail-safe design are some risk-control tools. Venting and pressure-relief systems are essential for preserving process equipment. Damage-limiting construction should be used where processes are housed in a structure. Equipment and piping should be well supported to prevent vibrations that could cause leaks.

Control of Fire and Explosions

The three basic ways to control explosions and fire are: (1) remove the fuel or explosive material, (2) remove oxygen required for combustion, or (3) remove the ignition source. In most wastewater treatment plants, controlling fuel or oxygen sources is difficult; therefore, controlling the ignition source is more practical. Each part of the system—collection, liquid treatment, solids treatment, or fuel and chemical handling—has different potential sources of ignition to be considered. In general, the potential for ignition within the collection system is restricted to sites where personnel entry is possible. Therefore, manholes and pumping stations are likely to present the greatest concern. The introduction of ignition sources at these sites must be limited to those occasions following life safety procedural mandates and always with adequate ventilation. In the event that a foreign combustible material enters the sewer system, removal by vacuum or coverage with foam may be necessary.

Potential for ignition within the liquid stream of a treatment plant is restricted to flammable materials entering the early stages of treatment and to flammable gases generated as byproducts of wastewater decomposition in nonflowing or septic portions of the liquid stream. Introduction of ignition sources into these areas should be limited, and adequate ventilation should always be provided.

Conditions in the solids treatment stream are often similar to liquid treatment processes. Potential for ignition within the solids treatment process is restricted to floating flammable materials skimmed from tanks and to flammable gases generated as byproducts of anaerobic decomposition of sewage solids. Introduction of ignition sources into these areas should be limited, and adequate ventilation should always be provided.

Fuel and chemical handling and storage areas, including laboratories, require special consideration. In general, potential for ignition is restricted to reactive or flammable chemicals, liquid fuels, flammable natural fuel gases, and fuels generated as byproducts for use in various unit processes. Introduction of ignition sources into these areas should be limited, and adequate ventilation should always be provided. Chemicals and fuels should always be stored in accordance with safety codes and the manufacturer's instructions. Reactive chemicals should always be stored separately.

Ventilation

Mitigation of explosions and fire hazards is best achieved by copious flushing with fresh air, i.e., ventilation. Ventilation rates for enclosed spaces containing wastewater exposed to the atmosphere should be based upon the calculated vaporization rate of the most volatile liquid anticipated to be present in the wastewater. In considering ventilation requirements, the designer should base calculations upon the surface area of channels, tanks, or other vessels containing wastewater exposed to the atmosphere; the temperature of the wastewater; the ambient air temperature; and the vaporization rate of the volatile liquid. Allowances should be made for vaporization rate of the liquid from the free water surface in connecting sewers within a reasonable distance from the structure, turbulence that may accelerate vaporization of the volatile liquid, inefficiency of the ventilation system in purging the enclosures, and any other factors that the designer could reasonably expect to affect the release rate of the flammable vapor to the structure. Table 9-25B presents typical ventilation rates for various areas in wastewater treatment plants. Ventilation rates should conform to Table 9-25B or be based on actual calculations, whichever is greater.

Construction Materials

Construction materials can be divided into three basic categories: (1) combustible, (2) noncombustible, and (3) limited combustible. NFPA 220, *Standard Types of Building Construction*, defines these terms. Construction materials for wastewater treatment plants should be selected based upon the fire hazard and fire risk evaluation. Proper application of construction materials will not prevent explosion or fire hazards. Proper application may, however, reduce or eliminate effects of fire or explosion by maintaining structural integrity, controlling flamespread and smoke generation, preventing the release of toxic products of combustion, and maintaining the facility's serviceability and operation.

Construction materials meeting the definition of noncombustible should be used in all building or unit processes considered *critical* to the integrity of the treatment plant (e.g., raw wastewater pumping station), and, if out of service for even a few hours, could permanently or unacceptably damage the environment or endanger public health by allowing the release of raw wastewater or sludge. All buildings should be constructed in accordance with all national and local fire codes.

Construction materials meeting the definition of noncombustible or limited combustible should be used in building or unit processes that are considered *important* to the integrity of the treatment plant (e.g., biological treatment), and, if out of service for short periods of time, would not permanently or unacceptably damage the environment or endanger public health, but would become critical if out of service for several days.

Any appropriate construction materials can be used in other areas of the treatment plant, where being out of service for a week or more would not permanently or unacceptably damage the environment or endanger public health.

For separate sanitary sewers serving residential areas, where there is limited possibility that significant quantities of flammable or hazardous materials will enter the system, all appropriate material may be considered. For sanitary sewers that may handle flammable or hazardous materials and for all combined sewers, only materials meeting the definition of noncombustible or limited combustible should be used.

Structures entered by personnel who handle raw or partially treated wastewater should be constructed only of materials meeting the definition of noncombustible or limited combustible. When combustible materials and coatings are used, consideration should be given to flame spread rates, smoke generation, and release of toxins during fire events.

TABLE 9-25B. *Minimum Ventilation Rates*

| Description | Ventilation Rate, Air Changes per Hour, or Velocity as Noted | | |
	Class I, Div. 1	Class I, Div. 2	Unclassified
1. Wet wells, screen rooms, and other enclosed spaces with wastewater exposed to the room atmosphere	< 12 air changes/hr	≥ 12 air changes/hr	
2. Below-grade spaces (such as dry wells, equipment rooms, tunnels, or galleries),			
(a) With equipment using or processing flammable gas	< 12 air changes/hr < 74 ft/min (22.2 m/min) velocity in tunnels or galleries	≥ 12 air changes/hr ≥ 74 ft/min (22.2 m/min) velocity in tunnels or galleries	
(b) With gas piping		< 6 air changes/hr < 37 ft/min (11 m/min) velocity in tunnels or galleries	≥ 6 air changes/hr ≥ 37 ft/min (11 m/min) velocity in tunnels or galleries
(c) Without gas piping		< 6 air changes/hr for dry wells NR for tunnels and galleries	≥ 6 air changes/hr for dry well
3. Above-grade spaces (such as equipment rooms and galleries)			
(a) With equipment using or processing flammable gas	< 12 air changes/hr < 74 ft/min (22.2 m/min) velocity for galleries	≥ 12 air changes/hr ≥ 74 ft/min (22.2 m/min) velocity in galleries	
(b) With gas piping		< 6 air changes/hr < 37 ft/min (11 m/min) velocity in galleries	≥ 6 air changes/hr, ≥ 37 ft/min (11 m/min) velocity in galleries
(c) Without gas piping		NR for galleries	

NR = No requirement.

Noncombustible materials should be used for air-supply and exhaust systems. However, when combustible or limited combustible materials are used to control corrosion, approved smoke dampers and fire dampers should be installed. Separate smoke ventilation systems are preferred; however, smoke venting can be integrated into normal ventilation systems using automatic or manually positioned dampers and motor speed control.

Cellular or foamed plastic material should not be used as interior finish. Roof coverings should be Class A. Construction materials and interior finishes should provide a maximum degree of fire resistance, with a minimum flame spread rate and smoke generation for a particular application.

Plastic or fiberglass-reinforced plastic products are often used as construction materials in unit processes, such as rotating biological contactors (RBCs), bio-towers, trickling filters, inclined plate (tube) settlers, ventilation ducts, and other equipment that may be subject to corrosion. Under normal operating conditions, these plastic or fiberglass-reinforced plastic materials may be submerged; however, they may become exposed during maintenance and repair. Extreme care should be taken with open flame, such as repair cutting torches, during any maintenance or repair operation, because exposed plastic or fiberglass-reinforced plastic materials present a considerable fuel load, if ignited.

Education

Another important and often overlooked mitigation measure is education. In-house training programs, such as plant emergency organizations (PEOs), housekeeping, and maintenance, should be established so that all employees understand and can identify, prevent, and handle hazardous sources or situations relating to potential fire, explosion, and toxicity. Close liaison should be implemented between the local fire department (including other authorized emergency personnel) and wastewater treatment plant safety personnel, so that mutually approved emergency procedures (including familiarity of the plant) can be established.

FIRE PROTECTION MEASURES

Wastewater treatment plants receive flammable and combustible liquids and gases in three ways: (1) through the collection system, (2) through introduction into the treatment processes, or (3) as byproducts of the processes. In spite of considerable care taken in design and operation, fires and explosions do occur, and fire protection and suppression equipment are necessary. Generally, fire protection and suppression measures used in the wastewater treatment industry include: (1) automatic sprinkler systems, (2) chemical suppression systems, (3) standpipes and hydrants, (4) portable fire extinguishers, and (5) special fire protection measures.

Automatic Sprinkler Systems

Automatic sprinkler systems should be installed where appropriate in buildings or structures located at a wastewater treatment plant. The fire hazard evaluation should include detailed hazard classifications, with system density and application rates sufficient to suppress any anticipated fire. In certain areas of the wastewater treatment plant, such as chemical storage areas, underground tunnels or structures, or areas where electrical hazard is the principal concern, the use of other appropriate fire protection measures should be carefully considered and evaluated.

Chemical Suppression Systems

Foam, halon, and dry chemical systems should be installed or used where appropriate in buildings or structures located at a wastewater treatment plant. Foam systems are most effective for wet wells, grit and screening processes, primary clarifiers, and pits and tanks where floating flammable liquid may collect on the surface. Halon systems are generally used where computer equipment is located. Dry chemical extinguishing systems should be considered in chemical storage areas, underground tunnels or structures, or where electrical hazard is the principal concern and water damage would seriously impair the treatment plant's integrity.

Standpipes and Hydrants

Standpipes, hose streams, and hydrants should be provided where appropriate in buildings or structures located at a wastewater treatment plant. Water supplies should be capable of delivering the total demand of sprinklers, hose streams, and foam systems. In areas with no public water supply or an inadequate public water supply, treatment plant effluent can be used for fire protection if all contaminents have been eliminated. In treatment plants with both a public water supply and a plant water system, the plant water system can be used as the principal water source for the fire protection system or as a backup to the public water supply system.

When the plant water system is the principal source for fire protection, the system must be capable of providing an adequate quantity and pressure and have sufficient standby capacity to meet all fire waterflow requirements. If the plant water system is used as a backup to the public water supply, the system should provide easy access and connection for pumper equipment. Where dual-hydrant systems are used, cross connections (where allowed) between the public water supply and the plant water system must be installed in accordance with local requirements. When fire pumps are the sole source of supply used in the plant water system, multiple pumps with sufficient capacity to meet fire flow requirements with the largest pump out of service should be provided. Pumps should be automatic starting with manual shutdown and should not be subject to common failure of either electrical or mechanical equipment.

Portable Fire Extinguishers

Portable fire extinguishers should be provided in buildings where appropriate. In some areas of the treatment plant, such as basements, underground pipe galleries connecting buildings, and other areas not continuously occupied, existing codes and standards may be inappropriate. In these cases, the Authority Having Jurisdiction should consider local optional requirements for portable fire extinguishers, depending on frequency of occupancy, intended use, equipment contained in the space, and hazard potential for fire and/or explosion.

Special Fire Protection Measures

Some wastewater treatment plants and processes have unique problems that require special fire protection considerations. For example, fire protection measures during construction at both new and existing wastewater facilities should consider life safety, property protection, and potential construction delays, as well as plant or unit process startup. Lightning protection should be provided for those structures with a risk index (R) of 4 or greater, according to NFPA 780, *Standard for the Installation of Lightning Protection Systems*, Appendix I. In addition, arrangements should be made to permit the municipal fire department, police department, or other authorized personnel to rapidly enter in case of fire or other emergency. The plant emergency organization (PEO), if present, should be properly instructed and trained in the use of all fire protection equipment located at the treatment plant.

Early detection and reporting of fires and explosions is also important in protecting the integrity of the buildings and unit processes.

Protective measures that detect flame, heat, or smoke should be selected and installed where needed. The location of proper types of detection equipment, e.g., smoke, heat, flame, etc., is a function of the hazard severity or electrical classification. Where appropriate, central station, local protective auxiliary, remote station, or proprietary sprinkler waterflow alarms should be provided.

BIBLIOGRAPHY

References Cited

1. Babbitt, H. E., *Sewerage and Sewage Treatment*, seventh edition, John Wiley & Sons, Inc., New York, 1956, pp. 1–5.
2. "1988 Needs Survey Report to Congress," EPA 430/9-89-001, Feb. 1989, U.S. Environmental Protection Agency, Office of Municipal Pollution Control, Washington, DC.
3. Steel, E. W., *Water Supply and Sewerage*, fourth edition, McGraw-Hill, New York, 1960, pp. 396–402.
4. NFPA 820, *Standard for Fire Protection in Wastewater Treatment and Collection Facilities*, National Fire Protection Association, Quincy, MA.
5. Miller, R. L., "Chemical Processes," *Industrial Fire Hazards Handbook*, 2nd ed., National Fire Protection Association, Quincy, MA, 1984.

NFPA Codes, Standards, and Recommended Practices

Reference to the following NFPA codes, standards, and recommended practices will provide further information on the processes and safeguards for wastewater treatment operations. (See the latest version of *The NFPA Catalog* for availability of current editions of the following documents.)

NFPA 1, *Fire Prevention Code*
NFPA 10, *Standard for Portable Fire Extinguishers*
NFPA 11, *Standard for Low-Expansion Foam*
NFPA 11A, *Standard for Medium- and High-Expansion Foam Systems*
NFPA 11C, *Standard for Mobile Foam Apparatus*
NFPA 12, *Standard on Carbon Dioxide Extinguishing Systems*
NFPA 12A, *Standard on Halon 1301 Fire Extinguishing Systems*
NFPA 14, *Standard for the Installation of Standpipe and Hose Systems*
NFPA 15, *Standard for Water Spray Fixed Systems for Fire Protection*
NFPA 17, *Standard for Dry Chemical Extinguishing Systems*
NFPA 20, *Standard for the Installation of Centrifugal Fire Pumps*
NFPA 22, *Standard for Water Tanks for Private Fire Protection*
NFPA 24, *Standard for the Installation of Private Fire Service Mains and Their Appurtenances*
NFPA 30, *Flammable and Combustible Liquids Code*
NFPA 31, *Standard for the Installation of Oil Burning Equipment*
NFPA 37, *Installation and Use of Stationary Combustion Engines and Gas Turbines*
NFPA 45, *Standard on Fire Protection for Laboratories Using Chemicals*
NFPA 49, *Hazardous Chemicals Data*
NFPA 50, *Standard for Bulk Oxygen Systems at Consumer Sites*
NFPA 51, *Standard for the Design and Installation of Oxygen-Fuel Gas Systems for Welding, Cutting, and Allied Processes*
NFPA 51B, *Standard for Fire Prevention in Use of Cutting and Welding Processes*
NFPA 53, *Guide on Fire Hazards in Oxygen-Enriched Atmospheres*
NFPA 54, *National Fuel Gas Code*
NFPA 55, *Standard for the Storage, Use, and Handling of Compressed and Liquefied Gases in Portable Cylinders*
NFPA 58, *Standard for the Storage and Handling of Liquefied Petroleum Gases*
NFPA 59A, *Standard for the Production, Storage, and Handling of Liquefied Natural Gas (LNG)*
NFPA 61, *Standard for the Prevention of Fires and Dust Explosions in Agricultural and Food Products Facilities*
NFPA 68, *Guide for Venting of Deflagrations*
NFPA 69, *Standard on Explosion Prevention Systems*
NFPA 70, *National Electrical Code®*
NFPA 70B, *Recommended Practice for Electrical Equipment Maintenance*
NFPA 70E, *Standard for Electrical Safety Requirements for Employee Workplaces*

NFPA 72, *National Fire Alarm Code*
NFPA 77, *Recommended Practice on Static Electricity*
NFPA 80, *Standard for Fire Doors and Windows*
NFPA 82, *Standard on Incinerators and Waste and Linen Handling Systems and Equipment*
NFPA 90A, *Standard for the Installation of Air Conditioning and Ventilating Systems*
NFPA 91, *Standard for Exhaust Systems for Air Conveying of Materials*
NFPA 101®, *Life Safety Code®*
NFPA 204M, *Guide for Smoke and Heat Venting*
NFPA 220, *Standard on Types of Building Construction*
NFPA 231, *Standard for General Storage*
NFPA 241, *Standard for Safeguarding Construction, Alteration, and Demolition Operations*
NFPA 251, *Standard Methods of Tests of Fire Endurance of Building Construction and Materials*
NFPA 253, *Standard Method of Test for Critical Radiant Flux of Floor Covering Systems Using a Radiant Heat Energy Source*
NFPA 255, *Standard Method of Test of Surface Burning Characteristics of Building Materials*
NFPA 256, *Standard Methods of Fire Tests of Roof Coverings*
NFPA 259, *Standard Test Method for Potential Heat of Building Materials*
NFPA 327, *Standard Procedures for Cleaning or Safeguarding Small Tanks and Containers Without Entry*
NFPA 328, *Standard Methods for the Control of Flammable and Combustible Liquids and Gases in Manholes, Sewers, and Similar Underground Structures*
NFPA 329, *Recommended Practice for Handling Underground Releases of Flammable and Combustible Liquids*
NFPA 395, *Standard for the Storage of Flammable and Combustible Liquids at Farms and Isolated Sites*
NFPA 430, *Code for the Storage of Liquid and Solid Oxidizers*
NFPA 491M, *Manual of Hazardous Chemical Reactions*
NFPA 496, *Standard for Purged and Pressurized Enclosures for Electrical Equipment*
NFPA 600, *Standard on Industrial Fire Brigades*
NFPA 601, *Standard for Security Services in Fire Loss Prevention*
NFPA 780, *Standard for the Installation of Lightning Protection Systems*
NFPA 1231, *Standard on Water Supplies for Suburban and Rural Fire Fighting*

Additional Reading

40 CFR761.30, July 1989, *Code of Federal Regulations*, Office of the Federal Register, National Archives and Records Administration, U.S. Government Printing Office, Washington, DC.

ASTM E136, *Standard Test Method for Behavior of Materials in a Vertical Tube Furnace at 1,382°F (750°C)*, American Society for Testing and Materials, Philadelphia, PA.

ASTM E814, *Standard Method of Fire Tests of Through-Penetration Fire Stops*, American Society for Testing and Materials, Philadelphia, PA.

Chlorine Manual, fifth edition, Chlorine Institute, Washington, DC, 1986.

Noll, G. G., "Handling Haz Mats: Subsurface Spills and Releases Into a Sewer System," *Fire Engineering*, Vol. 148, No. 4, Apr. 1995, pp. 54–55, 58, 60.

OSHA 1910.156, July 1989, *Code of Federal Regulations*, Office of the Federal Register, National Archives and Records Administration, U.S. Government Printing Office, Washington, DC.

Weres, G., and O'Donnell, H., "Treatment Process for Firefighter School Wastewater Utilizing Electrochemically Generated Hydroxyl Free Radical," Final Technical Report, May 1994–December 1994, Sonoma Research Co., Vineburg, CA, Armstron Laboratory, Brooks AFB, TX, AL/EQ-TR-1995-0002, Feb. 10, 1995.

FIRE PROTECTION OF LABORATORIES USING CHEMICALS

Revised by Marian H. Long

This chapter addresses risk management of laboratories using chemicals. Risk can be defined as the potential for the occurrence of an undesirable consequence arising from some activity. Risks can be described in terms of their expected frequency of occurrence (probability) and their consequent impact (severity). A hazard is an event or physical condition that has the potential for causing fatalities, injuries, property damage, business interruption, or other types of harm or loss.

Risk management involves the systematic identification, evaluation, and control of potential losses that may arise from uncertain future events. The practice of risk management permits decision makers to anticipate such losses and to evaluate their potential impacts, so they can be managed effectively. This requires recognizing possible risks, evaluating their frequency and the magnitude of their consequences, and determining appropriate measures for preventing or reducing these risks from a cost/benefit viewpoint.

Critical to risk management of laboratories using chemicals are design and construction, fire protection, explosion hazard protection, ventilation, safe storage and handling of chemicals, and judicious use of compressed and liquefied gases. These topics will be addressed throughout this chapter. In addition, the chapter includes a brief discussion of occupational safety and health risk, such as biohazards and radiation.

DESIGN AND CONSTRUCTION

The discussion of design and construction presented lays the foundation for a more detailed assessment. Each laboratory is unique in its operation, risks, and associated needs and should be evaluated as such. NFPA 45, *Standard on Fire Protection for Laboratories Using Chemicals* (hereinafter referred to as NFPA 45), contains detailed information on classifying laboratory units and work areas based on flammable and combustible liquid content, construction, and the presence of explosion hazards. Other good resources for risk management guidance are NFPA 99, *Standard for Health Care Facilities*, Chapter 10; and various federal, state, and municipal regulatory agencies.

Design

Laboratory layout and furnishing should be based on critical design criteria. They include traffic flow, which is important to means of

Marian H. Long, P.E., CSP, is a senior consultant at Arthur D. Little, Inc., Cambridge, MA. She is a former long-term member of the NFPA Technical Committee on Chemical Laboratories, and is presently a member of the NFPA Technical Committee on Disaster Management.

egress (discussed below), and minimizing the transport of hazardous materials. Safe storage of hazardous materials is also key to protecting against fires and other types of accidents. An ergonomics evaluation can address worker performance and safety, thus minimizing the risk of accidents. An additional consideration is the possible need for decontamination, as such features as finish resistance can minimize the consequences of a fire or spill event.

Construction

Laboratory construction depends on a variety of considerations—the type and quantity of potential hazards involved, the area or location with respect to adjacent occupancies, and the type of fire protection to be provided. Well-located laboratories are separated from other laboratories and nonlaboratory space by fire-resistive construction. Interior finish used in laboratory construction and the access to exits from them must comply with the requirements of NFPA 101®, *Life Safety Code®* (hereinafter referred to as NFPA *101*).

Egress

Life safety considerations for laboratory buildings, laboratory units, and laboratory work areas must comply with the requirements of NFPA *101*, unless otherwise specified in NFPA 45.

Two exits, or means of egress, from a laboratory work area are required for all of the following:

- Laboratories with a work area larger than 1,000 sq ft (93 m²) with moderate quantities of flammable and combustible liquids;

- Laboratories containing large quantities of flammable and combustible liquids within a laboratory unit in excess of 500 sq ft (46 m²);

- Laboratories containing an exhaust hood located adjacent to the primary exit;

- Laboratories containing compressed gas cylinders that are larger than a lecture bottle size, that contain a gas that is flammable or has a Health Hazard Rating of 3 or 4, and that could prevent safe egress in the event of accidental release of the cylinder contents;

- Laboratories that use cryogenic containers that contain a flammable gas or have a Health Hazard Rating of 3 or 4 and that could prevent safe egress in the event of accidental release of the container contents;

• Laboratories that contain a laboratory work area with an explosion hazard so located that an incident would block escape from or access to the laboratory work area.

Required exit access doors from high-hazard or moderate-hazard laboratories must swing in the direction of egress. Furniture and laboratory equipment should be arranged so that the means of access to an exit may be reached easily from any point in the laboratory work area without undue obstruction.

Electrical Equipment

All electrical installations, appliances, and equipment must comply with NFPA 70, *National Electrical Code®* (hereinafter referred to as NFPA 70). Circuits should be designed for all electrical equipment to be installed, and sufficient overcurrent protection devices should be provided. Emergency power for operating exhaust fans, exit lighting, exit signs, evacuation alarms, and other emergency or fire and safety equipment is essential and also required by the Occupational Safety and Health Administration (OSHA) in its laboratory regulations.

All electrical appliances and equipment should display a listing with, or approval by, a recognized testing laboratory for its intended use, unless no test standards have been established for a device. All electrical appliances and equipment should have an operator's manual, together with a circuit diagram, available for qualified maintenance personnel. It is also important that all portable electrical equipment intended for laboratory use be properly grounded by recognized methods.

Installing a sufficient number of electrical outlets at convenient locations eliminates the use of extension cords. Several outlets should be provided with emergency power to prevent loss or interruption of electrical power to critical laboratory equipment or experimentation.

FIRE PROTECTION

All laboratories must have fire protection. Minimally, they should have portable fire extinguishers appropriate to the hazard, an adequate fire alarm system, and an emergency evacuation plan. In addition, automatic sprinkler systems are often required as a function of the hazards present and the combination of the laboratory unit. Heat detection systems can also be very effective in the early detection of fire.

Automatic Fire Extinguishing/Protection Systems

It is common practice to protect laboratories, including associated storage rooms and enclosures, with automatic sprinkler systems designed and installed according to NFPA 13, *Standard for the Installation of Sprinkler Systems*. In some instances, special-hazard laboratories may require automatic detection systems and special extinguishing agents, such as foam, carbon dioxide, and so on. When such systems are used, they must be installed, inspected, and maintained in accordance with appropriate NFPA standards. Periodic inspection of the systems ensures that they will activate in an emergency.

The activation of a fire detection system or the discharge of a fire extinguishing system must sound an alarm to ensure prompt response. The alarm system should be arranged to sound alarms both inside the laboratory building and at a constantly attended location, such as a telephone switchboard, communications center, or security office, and the fire department. The alarm system should also be arranged to activate other equipment, such as electromagnetic door closers that affect the fire protection of the facility. NFPA 72, *Na-tional Fire Alarm Code,* gives requirements on the various alarm systems and detection equipment that can be used.

Standpipe and Hose Systems

Laboratory buildings with two or more stories above or below ground level should be provided with $2^1/_2$-in. (64-mm) hose connections with hose and combination straight stream and fog nozzle. NFPA 14, *Standard for the Installation of Standpipe and Hose Systems*, is the standard to follow.

Portable Fire Extinguishers

Portable fire extinguishers, suitable to a laboratory's particular hazards, should be installed and located so that they are readily available to laboratory personnel. Class A fire extinguishers suitable for fires in ordinary combustible materials, such as wood, paper, and rags, should be provided and supplemented by enough extinguishers suitable for use on Class B (flammable liquid) and Class C (electrical) fires. Special Class D fire extinguishers are needed where the use of metals or metal hydrides indicates a need for them. Multipurpose extinguishers for use on Class A, B, and C type fires are frequently chosen because of their versatility and ease of use. A reference source is NFPA 10, *Standard for Portable Fire Extinguishers*.

Explosion Hazard Protection

The hazards of explosion, rupture of apparatus and systems from overpressure, implosion due to vacuum, sprays, or emission of toxic or corrosive materials, or flash ignition of escaping vapors require substantial physical protection for exposed personnel. Analysis of the characteristics of the type of hazards and their potential effects is critical to selecting effective and economical mitigating measures. They can include limiting the amount of hazardous chemicals, explosion venting, explosion suppression, explosion resistant construction, and/or isolation of the hazard.

Ventilation

Duct systems for laboratory heating and ventilating, including warm air heating systems, general ventilating systems, air cooling systems, and laboratory exhaust and hood systems deserve special attention. NFPA 45; NFPA 90A, *Standard for the Installation of Air Conditioning and Ventilating Systems*; and NFPA 91, *Standard for Exhaust Systems for Air Conveying of Materials*, are good standards to follow.

Air handling systems for laboratories deserve special attention. The fresh air supply system must not draw in flammable or toxic materials or combustion byproducts emitted from the laboratory building itself or other sources, such as vehicle exhaust. Exhaust discharges for laboratories should be located away from all air intakes, including those from other buildings, and high enough above roof level to provide atmospheric dilution. High-velocity discharge also assists in exhaust dilution.

Air handling systems should be balanced to provide a negative pressure with respect to the corridors and surrounding nonlaboratory spaces. An exception is those laboratories housing areas such as clean rooms that preclude a negative pressure in relation to the surrounding space. A slightly positive pressure is required in these areas, and precautions must be taken to prevent atmospheric contaminants in the laboratory space from escaping to surrounding spaces.

Air exhausted from laboratory areas in which highly infectious or radioactive materials are used or handled must pass through high-efficiency filters before being discharged into the atmosphere.

Laboratory hoods are the most commonly used means of removing gases, dust, mist, vapors, and fumes from laboratory operations and of preventing toxic exposures and flammable concentrations. To be effective, a laboratory hood and its associated components must confine contaminants within the hood, dilute and remove them through the ductwork, and disperse them so that they do not return to the building through the air supply system. Hood face velocities must be sufficient to ensure the capture of contaminants, especially in the presence of cross drafts. Recommended face velocities are published by the American Conference of Governmental Industrial Hygienists as well as the Chemical Rubber Company (CRC) in its *Handbook of Laboratory Safety.* Laboratory hoods for hazardous operations, such as perchloric acid, radioactive materials, explosion hazards, and so on, require special design and must be constructed of suitable materials that meet rigid criteria. Fume hood controls should be so arranged that shutting off the ventilation in one hood does not reduce the exhaust capacity or create an imbalance for any other hood. All shutoff valves for services should be located outside the hood enclosure in a readily accessible location.

STORING AND HANDLING CHEMICALS

Flammable and Combustible Liquids

Flammable and combustible liquids present a potentially serious fire and explosion hazard in a laboratory. The severity of the hazard from a flammable liquid process or the storage of these liquids depends on conditions such as the liquid's quantity, volatility, and flammable range; whether it is in a closed or open system exposed to air; the method of containment to control leakage or overflow; the location in relation to other equipment and outside ignition sources; the building construction; and the adequacy of fire protection.

Storage and use: Flammable and combustible liquids should be stored in and used from containers of a size and material appropriate to the potential hazard. Refer to NFPA 45 for specific information on container size and type. Established laboratory practices are the limiting factors in determining the working quantity of flammable and combustible liquids present. All nonworking quantities should be stored in well-constructed storage cabinets or storage rooms. [See NFPA 30, *Flammable and Combustible Liquids Code* (hereinafter referred to as NFPA 30), for construction details.] Laboratory personnel must be thoroughly familiar with the properties and hazards of flammable and combustible liquids used.

Handling and transfer: Only trained personnel should be involved in receiving, transporting, unpacking, and dispensing flammable and combustible liquids. These activities should be performed in locations and in a manner that minimize risks. Transfer from bulk stock containers to smaller containers must be performed in a separate area outside the building, a laboratory hood, an area with adequate ventilation to prevent the accumulation of flammable vapor, or a separate inside storage area that meets the requirements of NFPA 30.

Disposal: Applicable governmental regulations must be observed in disposing of flammable and combustible liquids. Disposal of these hazardous materials must be accomplished by a competent disposal specialist who knows the basic character and hazards of the waste.

Hazardous Chemicals

Storage and use: Hazardous chemicals must not be brought into a laboratory space unless the design, construction, and fire protection of the receiving, using, and storage facilities are commensurate with the quantities and hazards of the chemicals involved. All laboratory personnel must be aware of the hazards of all chemicals in use. Special storage facilities are needed for materials having unique physical or hazardous properties, such as temperature sensi-

tivity, water reactivity, or explosivity. A reference source is NFPA 49, *Hazardous Chemicals Data.*

Disposal: Applicable governmental regulations must be observed in the disposing of waste chemicals. A competent disposal specialist who knows the character and hazards of the waste should be assigned to this activity.

COMPRESSED OR LIQUEFIED GASES

The use of compressed or liquefied gases in a laboratory presents special problems. Precautions must be based on the particular properties of the materials to be used.

Storage, Handling, and Use

Containers designed, constructed, and tested in accordance with U.S. Department of Transportation (DOT) specifications and regulations must be used for storing compressed or liquefied gases. Compressed gas container storage should be outside and protected from the effects of weather and sun. An alternate location is a well-ventilated louvered room equipped with an automatic sprinkler system. (See NFPA 55, *Standard for the Storage, Use, and Handling of Compressed and Liquefied Gases in Portable Cylinders;* and Compressed Gas Association Pamphlet P-1, *Safe Handling of Compressed Gases in Containers.*) Compressed gas cylinders must be securely positioned and stored in racks, away from heat or other ignition sources. Flammable gases should be separated from other oxidizing gases by construction that has a fire resistance rating of at least 1 hr. Electrical equipment in flammable gas storage areas must comply with NFPA 70 provisions for Class I, Division 2, hazardous locations.

Handling and transporting gas cylinders must be consistent with established safety procedures. Compressed gas cylinders must never be used without pressure regulators, and adapters must not be permitted between the cylinder and the regulator.

Piping systems: When laboratory equipment is operated routinely with flammable gases supplied from compressed gas cylinders, the cylinders should be located outside the building and connected to laboratory equipment by a permanently installed piping system. Pressure-reducing valves should be connected to each cylinder and adjusted to limit pressure in the piping system to the minimum required gas pressure. Pressure regulators must be compatible with the gases for which they are used. Supply and discharge terminals of piping systems should be permanently marked at both ends with the name of the gas to be piped through them. Never use piping systems for gases other than those for which they were designed and identified. Do not attempt to transfer compressed or liquefied gases from one gas container to another.

Additional requirements are covered in NFPA 58, *Standard for the Storage and Handling of Liquefied Petroleum Gases.*

General Safety Rules

The following guidance is provided for handling compressed or liquefied gas in a laboratory:

- Know and label cylinder contents.
- Know properties of contents.
- Handle cylinders carefully.
- Store cylinders in a well-ventilated area away from heat.
- Secure cylinders during use, transit, or storage.
- Never tamper with valves, safety plugs, or packing nuts.

- Do not strike an electric arc on cylinders.
- Do not use cylinders without a regulator.
- Close cylinder valves when not in use.
- Never attempt to refill a cylinder.

BIOLOGICAL HAZARDS

Laboratory work done with animals, infectious diseases, or toxic materials presents a variety of safety problems both to laboratory workers and others whose duties may be required in the laboratory. Fire, explosions, and other accidents present two types of problems: the release of biological hazards, and the hazards encountered by fire-fighting personnel entering the space. Laboratory fire brigades and local fire-fighting personnel must be educated about the hazards they may encounter in laboratories engaged in biological work.

Where biological hazards are used, stored, or handled, most laboratories label equipment, storage containers, and rooms with the universal biohazard sign and the word "biohazard." The label may also contain information on the biological hazard in use and the criteria for entry (e.g., personal protective equipment, vaccinations, etc.).

Additional signs may indicate one of four biosafety levels, with level 1 representing the lowest risk and level 4 representing the highest risk. These biosafety levels describe the level of containment, including standard and special practices, safety equipment, and facility design criteria, that is appropriate for operations performed and biological hazards present within the laboratory.

Sufficient protective equipment, breathing apparatus, clothing, gloves, and so on, must be readily available for use by emergency response personnel. Laboratories engaged in this type of work must be identified appropriately, and clear and concise safety instructions must be posted.

RADIATION HAZARDS

Laboratories using radioactive materials present hazards to both laboratory workers and others who must enter the facility. While rigid safety and health programs have been established for laboratory workers, it is wise to institute a similar program for other personnel, such as fire fighters. Good communication between laboratory radiation safety and health officers and public fire-fighting personnel can help keep fire fighters informed of all laboratories actively engaged in radioactive work. Such rapport can be the basis for establishing an educational program planned to cope with disasters and emergencies. See NFPA 801, *Standards for Facilities Handling Radioactive Materials*, for further discussion of the hazards and safeguards associated with nuclear reactors, radiation machines, and radioactive materials.

PERSONNEL PROTECTION

Personal protective equipment is essential to an effective safety program in any laboratory. Where engineering and administrative controls are not feasible, are being installed, or during emergencies, personal protective equipment shields or isolates individuals from chemical, physical, and biological hazards. Protective clothing and equipment reduce the probability and severity of injury if an inadvertent release occurs.

Laboratories must perform hazard assessments to identify operations where protective clothing and equipment are needed to prevent employee exposure. For instance, protective clothing and equipment must be provided when there is a recognized risk of fire or explosion involving materials used in laboratory. The nature of the equipment necessarily depends upon the particular hazard. Once

appropriate protective clothing and equipment have been selected, employees must be trained on their use, care, and limitations, and must demonstrate their knowledge before using the clothing and equipment.

To minimize inadvertent contact with various chemical, physical, and biological hazards, or contaminants that may be present in laboratories, following good personal hygiene practices is essential:

- Wash hands, face, and other exposed body parts, or shower, after working with hazardous materials, especially before eating, drinking, applying cosmetics, etc.
- Remove exposed work clothing and ensure its proper disposal or decontamination after working with hazardous materials.
- Do not eat, drink, smoke, or apply cosmetics in laboratory areas.
- Do not store foods, beverages, cosmetics, cigarettes, chewing gum, etc., in areas where hazardous materials are present.
- Avoid contacting potentially contaminated substances.
- Use and maintain appropriate personal protective equipment.
- Know the location of emergency decontamination equipment, such as emergency showers and eyewashes.

Where applicable, training laboratory personnel to use respiration equipment, including the incorporation of a comprehensive respirator fit testing program, is an essential component of a comprehensive personal protective equipment program.

Respirators are essential for protecting laboratory personnel working directly with or required to enter areas during an emergency where infectious diseases, radioactive materials, harmful dusts, mists, fogs, fumes, sprays, or vapor are present. Respirators, or self-contained breathing apparatus (SCBA) suitable for the specific potential problem are necessary. Training laboratory personnel, to use respirator equipment is essential.

Emergency shower installations should be provided for minimizing injury to personnel from clothing fires and other emergencies, such as contact with caustics or cryogenic fluids. Emergency showers supply large quantities of water needed for diluting, warming, cooling, or flushing chemicals, or extinguishing clothing fires. They should be located outside the laboratory space in conspicuous areas, preferably in usual traffic patterns but not more than 25 ft (8 m) from any laboratory entrance. Eye wash fountains are sometimes installed adjacent to emergency showers. Avoid locating showers near electrical apparatus and power outlets. The location of showers should be plainly marked on the floor. Access to showers and eyewashes must be unobstructed, so that the equipment is readily available for immediate emergency use. Showers and eye wash fountains should be tested and flushed, preferably quarterly, to ensure their proper operation.

HAZARD IDENTIFICATION

Signs should be posted at entrances to laboratories, storage areas, and associated facilities to warn personnel, especially emergency personnel, of unusual or severe hazards inside that may or may not be directly related to an emergency situation. These include signs relating to particularly unstable chemicals, radioactive chemicals, pathogenic or infectious materials, water-reactive chemicals, and explosives. NFPA 704, *Standard System for the Identification of the Hazards of Materials for Emergency Response*, is a good system to use. Periodic review is necessary to ensure that the particular warning label being displayed properly indicates the nature of the material being used within the identified laboratory area. Severe hazards should be discussed with fire-fighting personnel to make certain they know what to anticipate in the event of an emergency.

Contents of all individual containers within a specific laboratory area must be identified. Unlabeled containers present a serious potential hazard, not only to laboratory workers but also to other personnel required to enter laboratory areas, especially during emergencies.

BIBLIOGRAPHY

NFPA Codes, Standards, and Recommended Practices

Reference to the following NFPA codes, standards, and recommended practices will provide further information on protection for laboratories discussed in this chapter. (See the latest version of *The NFPA Catalog* for availability of current editions of the following documents.)

NFPA 10, *Standard for Portable Fire Extinguishers*

NFPA 11, *Standard for Low-Expansion Foam*

NFPA 11A, *Standard for Medium- and High-Expansion Foam Systems*

NFPA 12, *Standard for Carbon Dioxide Extinguishing Systems*

NFPA 12A, *Standard for Halon 1301 Fire Extinguishing Systems*

NFPA 13, *Standard for the Installation of Sprinkler Systems*

NFPA 14, *Standard for the Installation of Standpipe and Hose Systems*

NFPA 15, *Standard for Water Spray Fixed Systems for Fire Protection*

NFPA 30, *Flammable and Combustible Liquids Code*

NFPA 45, *Standard on Fire Protection for Laboratories Using Chemicals*

NFPA 49, *Hazardous Chemicals Data*

NFPA 50, *Standard for Bulk Oxygen Systems at Consumer Sites*

NFPA 50A, *Standard for Gaseous Hydrogen Systems at Consumer Sites*

NFPA 50B, *Standard for Liquefied Hydrogen Systems at Consumer Sites*

NFPA 53, *Guide on Fire Hazards in Oxygen-Enriched Atmospheres*

NFPA 55, *Standard for Storage, Use, and Handling of Compressed and Liquefied Gases in Portable Cylinders*

NFPA 58, *Standard for the Storage and Handling of Liquefied Petroleum Gases at Utility Gas Plants*

NFPA 68, *Guide for Venting of Deflagrations*

NFPA 70, *National Electrical Code®*

NFPA 72, *National Fire Alarm Code*

NFPA 77, *Recommended Practice on Static Electricity*

NFPA 80, *Standard for Fire Doors and Fire Windows*

NFPA 86, *Standard for of Ovens and Furnaces*

NFPA 90A, *Standard for Air Conditioning and Ventilating Systems*

NFPA 91, *Standard for Exhaust Systems for Air Conveying of Materials*

NFPA 99, *Standard for Health Care Facilities*

NFPA 101®, *Life Safety Code®*

NFPA 220, *Standard on Types of Building Construction*

NFPA 251, *Standard Methods of Tests of Fire Endurance Building Construction and Materials*

NFPA 259, *Standard Test Method for Potential Heat of Building Materials*

NFPA 325, *Guide to Fire Hazard Properties of Flammable Liquids, Gases, and Volatile Solids*

NFPA 491M, *Manual of Hazardous Chemical Reactions*

NFPA 496, *Standard for Purged and Pressurized Enclosures for Electrical Equipment*

NFPA 704, *Standard Systems for the Identification of the Fire Hazards of Materials for Emergency Response*

NFPA 801, *Standard for Facilities Handling Radioactive Materials*

Additional References

American Conference of Governmental Industrial Hygienists, *1995–96 Threshold Limit Values (TLVs™) for Chemical Substances and Physical Agents and Biological Exposure Indices (BEIs™)*. American Conference of Governmental Hygienists, Cincinnati, OH, 1994–95 (revised annually).

ASHRAE Publications, American Society of Heating Refrigeration and Air Conditioning, Inc., 1791 Tullie Circle, N.E., Atlanta, GA 31329. *ASHRAE Handbook*, Fundamentals Volume, Chapter 14, "Airflow Around Buildings," 1993. ASHRAE 110, *Method of Testing Performance of Laboratory Fume Hoods*, 1995.

ANSI Publications, American National Standards Institute, 1430 Broadway, New York, NY 10018. ANSI/AHIA Z9.5, *Laboratory Ventilation*, 1992.

CGA Publications. Compressed Gas Association, 1725 Jefferson Davis Highway, Suite 1004, Arlington, VA 22202-4102. CGA Pamphlet P-I, *Safe Handling of Compressed Gases in Containers*, eighth edition, 1991.

Committee on Industrial Ventilation, *A Manual of Recommended Practice*, American Conference of Governmental Industrial Hygienists, Cincinnati, OH.

Additional Reading

ACS Laboratory Safety Video Series, American Chemical Society, Washington, DC: "Things That Burn—Flammable and Combustible Chemicals," "Compressed Gases: Safe Handling Procedures," "Compressed Gases: Compressed Hazards," "Introduction to Chemical Laboratory Safety."

Benedetti, R. P., *Flammable and Combustible Liquids Code Handbook*, fourth edition, National Fire Protection Association, Quincy, MA, 1990.

Bretherick, L., *Hazards in the Chemical Laboratory*, fourth edition, Royal Institute of Chemistry, London, 1986.

DeRoos, J. L., "The Safe Use of Refrigerator and Freezer Appliances for Storage of Flammable Materials," *Fire Journal*, Vol. 67, No. 2, Mar. 1973, pp. 63–64.

DiBerardinis, L. J., *et al.*, *Guidelines for Laboratory Design*, second edition, Wiley, New York, 1993.

Furr, A. K., ed., *CRC Handbook of Laboratory Safety*, fourth edition, CRC Press, Boca Raton, FL, 1995.

Gerlach, R., "Things That Burn—Flammable and Combustible Chemicals," ACS Laboratory Safety Service, American Chemical Society, Washington, DC, 1995.

Le, N. B., Santay, A. J., and Jabrenski, J. S., "Laboratory Safety Design Criteria," *Plant/Operation Progress*, Vol. 7, No. 2, Apr. 1988, pp. 87–94.

Matheson Gas Data Book, fourth edition, Matheson Co., East Rutherford.

Miller, B. M., ed., *Laboratory Safety; Principles and Practices*, American Society for Microbiology, Washington, DC, 1986.

Morehart, J., "QR-AS Sprinkler Test in Chemical Laboratories," VHS Video Tape, National Institutes of Health, Bethesda, MD, Federal Fire Forum, Current Technical Fire Issues, Nov. 1, 1993, Gaithersburg, MD, 1993.

National Research Council, "Prudent Practices in the Laboratory—Handling and Disposal of Chemicals," National Academy Press, Washington, DC, 1995.

Nelson, H. E., *Fire Safety Evaluation System for NASA Office/Laboratory Buildings*, NBSIR 86-3404, Nov. 1986, National Bureau of Standards, Gaithersburg, MD.

NIOSH, "Pocket Guide to Chemical Hazards," U.S. Department of Health and Human Services, Public Health Service, Centers for Disease Control and Prevention, National Institute for Occupational Safety and Health, Superintendent of Documents, U.S. Government Printing Office, Washington, DC 20402, June 1994.

Orvos, D. R., "Biohazard," *Fire Engineering*, Vol. 141, No. 4, 1988, p. 49.

OSHA Laboratory Standard, Title 29, Code of Federal Regulations, Part 1910.

Standard on Laboratory Fume Hoods (SEFA 1-1992), Scientific Equipment and Furniture Association, Washington, DC.

Standard Specification for Laboratory Fume Hoods, Environmental Protection Agency, Washington, DC 20460, ATTN: Chief, Facilities Engineering and Seal Property Branch (PM-215).

Stricoff, R. S., and Walters, D. B., *Handbook of Laboratory Health and Safety*, second edition, Wiley, New York, 1995.

Walton, W. D., *Quick Response Sprinklers in Chemical Laboratories: Fire Test Results*, NISTIR 89-4200, Nov. 1989, National Institute of Standards and Technology, Gaithersburg, MD.

FIRE PROTECTION FOR TELECOMMUNICATIONS CENTRAL OFFICES

Lawrence A. McKenna, Jr.

This chapter discusses the basic layout and function of the U.S. telecommunications network. Further, it describes that network's various components and provides guidelines for providing fire protection for those components. It also provides recommendations for fire department response to telecommunications industry fires and for code compliance issues. It discusses both the public telecommunications network and private telecommunications equipment.

Since the initial version of this chapter in the 17th edition of the *Fire Protection Handbook*, dramatic changes have occurred in the North American telecommunications network, driven by both technological advances and political (regulatory) changes that have nothing at all to do with fire safety. The impact of these changes on society and on fire risk is not yet clear. This chapter describes many of these changes, their potential impact on fire risk, and some mitigating techniques.

Since Alexander Graham Bell made the first phone call on March 10, 1876, telephones have played an increasingly important role in people's lives. Once an oddity, then a convenience, the telephone is now a necessity of modern life. Telecommunications, which touch almost every facet of daily life, include: (1) automatic bank teller machines, (2) point of sale transaction processing terminals, (3) electronic funds transfer functions, (4) facsimile machines, (5) teleconferencing and video-conferencing, (6) video services, (7) cable TV, and (8) standard telephone service.

Continuing evolution in the uses of telecommunications services has driven technologies that provide service to ever-increasing levels of sophistication. Four important developments in the industry have resulted:

1. Long-distance transmission of signals between points in the network has been converted from analog signaling, which transmits a continually variable electrical signal, to digital signaling, which transmits on-and-off pulses of electricity or light.
2. Transmission routes have been extensively converted to fiber optics. This technology has found its way into the local loop, cable TV, and other wide bandwidth transmission facilities in many communities. Extension of fiber optic technology to the home is the next logical step in this development, although advances in signal compression technology may make the tremendous cost involved in such an undertaking unnecessary.

3. The development of advanced radio-based telecommunications systems is on the telecommunications horizon. This evolution would allow personal, pocket-sized telephones capable of sending and receiving calls anywhere in the world by satellite. In this technology's complete and "perfect" implementation, as viewed by some proponents, land-based telecommunications facilities may not be needed. It is unlikely that we will see such developments in our lifetime.
4. The last and most recent development that will radically change the telecommunications industry is unrestrained competition. The Telecommunications Reform Act of 1996 promises to remove all artificial barriers between long-distance carriers and local carriers, traditional telecommunications service providers and new entrants, such as cable television companies and others.

THE U.S. TELECOMMUNICATIONS NETWORK

Currently, approximately 1,600 service companies provide telecommunications services in the United States. These companies operate networks ranging in size from small rural companies serving up to 2,000 customers, to large companies operating worldwide networks and serving millions. Buildings called central offices (COs) house primary control functions for these networks. Approximately 21,000 central offices and some 30,800 other structures are required to operate the U.S. telecommunications network. More than 300 of these are over 100,000 sq ft (9300 m^2) in floor area.[2]

Between 1984 and the signing of the 1996 Telecommunications Reform Act passed by Congress, the U.S. telecommunications network had two segments: (1) the portion local exchange carriers (LECs), or local phone companies, provided; and (2) the portion interexchange carriers (IECs), or long-distance phone companies, provided. The concept of the local access and transport area (LATA), established at the court-ordered divestiture of the Bell System in 1984, distinguished these two types of phone companies. The enactment of the 1996 Telecommunications Reform Act blurred the distinction between them. Their disappearance over the next 5 to 10 years is further ensured by the pending entry of other industries such as cable TV, into the telecommunications service provider market.

Figure 9-27A is a simplified schematic of a local network, provided by an LEC. In terms taken from sprinkler system design, the local network is a "tree" network. Each customer telephone connects to one, and only one, central office or exchange. Failure in any

---

Lawrence A. McKenna, Jr., is senior engineer at Hughes Associates, Inc., Baltimore, MD.

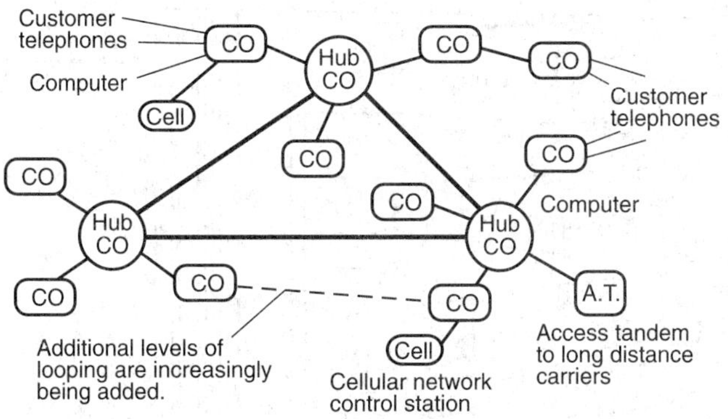

FIG. 9-27A. *Typical configuration of a local telecommunications network.*

CO or the cables linking COs isolates all customers or services on one side of the failure from those on the other.

Although lacking redundancy, this design is deliberate, made in compliance with the overall network design criteria directed by the Federal Communications Commission (FCC), as well as various state regulatory bodies. The criteria intended to provide universal service at a low price, not uninterrupted service. However, much effort is currently being expended to provide redundancies in certain high-traffic routes, such as links between hub COs. Customers generally have the option of paying more for additional redundancies for personal use.

Paths between the end user, or customer, and the CO are generally composed of twisted-pair copper wires bundled into cables. In some cases, radio, coaxial copper, or fiber optic cable may be provided for certain large customers. Paths connecting various COs to one another may be composed of copper or fiber optic cables, or of microwave radio, depending on local conditions and terrain.

Figure 9-27B is a simplified diagram of a typical IEC network configuration. Some important differences exist between this diagram and Figure 9-27A. In sprinkler parlance, many IEC networks are designed as "looped and gridded" systems. Multiple paths between each switching center provide for redundant paths between points in the network in the event of failure. The number and diversity of these paths are two key factors that predict reliability of a carrier's service. Note that this discussion does not directly apply to "resellers" of either long-distance or local telecommunications service. These companies purchase large blocks of time on the networks of telecommunications companies that own physical networks.

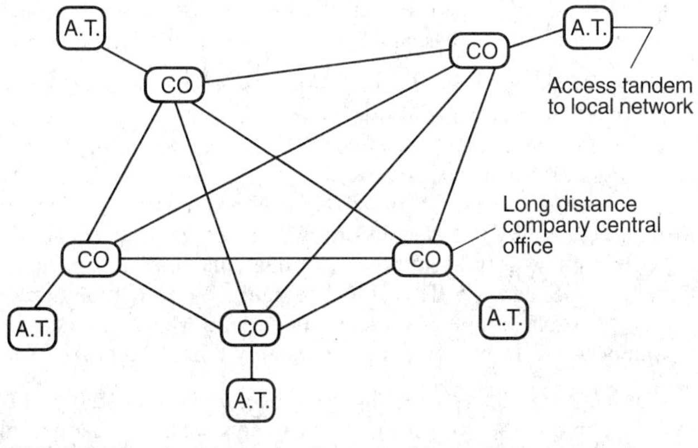

FIG. 9-27B. *Architecture of a typical long-distance telephone network.*

Paths connecting the COs in Figure 9-27B are composed of various transmission technologies, depending on traffic loads, topography, prevailing weather, and redundancy requirements. They may include one or more of the following: (1) copper cable, (2) fiber optic cable, (3) microwave radio, or (4) satellite radio. Along each path, as well as those in the LEC networks, a number of repeater or regeneration stations amplify and retransmit signals. These stations compensate for signals' deterioration over distance.

FIRE RECORD

There have been two studies of the fire record of telecommunications facilities since 1985. The first study, based on national projections using National Fire Incident Reporting Systems (NFIRS) and NFPA survey data, estimated that 109 fires occur annually in approximately 20,000 U.S. telephone exchanges.[1] Most are too small to activate detection and suppression systems.

A more detailed study on the fire history of the telecommunications industry was done for the FCC in 1992 and 1993.[2] This study surveyed local exchange and interexchange carriers in the United States, major telecommunications equipment manufacturers, and many other sources, representing more than 93 percent of the telephone lines in the nation. The study covered the fire record for the years 1988 to 1992.

During the study period, a total of 189 fire events occurred in telecommunications facilities. On average, there were 38 incidents per year; only 8 affected service. (For comparison, NFIRS projections for 1990–1994 show 58 reported structure fires per year in telephone exchanges or central offices.) Effects on service ranged from several minutes to one instance of several weeks of service outage. The study included an exhaustive root cause analysis and made recommendations to prevent recurrence. Other findings from this landmark study are discussed later in this chapter.

The fire record that the telecommunications industry has achieved is due, in large part, to exhaustive fire prevention efforts by the industry, described in greater detail later in this chapter. The studies do not show one significant fact: even "large" telecommunications fires may involve small quantities of materials burned, area of burn damage, and so on. Even the Hinsdale, IL, fire of May 1988, which burned for over 3 hr, was confined to cable trays in a 30 × 40 ft (9 × 12 m) area.[3] Though small by most measures, these fires' impact on the community can be quite large.

THE CENTRAL OFFICE

The nodes in each of these networks are located in buildings called central offices (COs), or telephone exchanges. Designed and sized to meet customers' service needs, COs range in size from small, single-story buildings to large multistory buildings in large cities. Despite size differences, the equipment layout and function are similar. Figure 9-27C shows typical components of a central office. In remote areas, all these components may be in a single small room. In large cities, each component may use one or more floors of a building.

Most major metropolitan central office buildings are constructed of reinforced concrete with rather massive structural elements. Older (pre-1980) suburban and rural central offices tend to be of fire-resistive construction. In addition to loss prevention incentives for this type of construction, the buildings need to be strong enough to support the large masses of heavy copper cable the network requires. Although changes in equipment technology, described below, somewhat diminish cabling needs, the massive structural elements remain. Central offices and other telecommunications facilities constructed since the advent of competition in the telecommunications industry have generally been constructed in

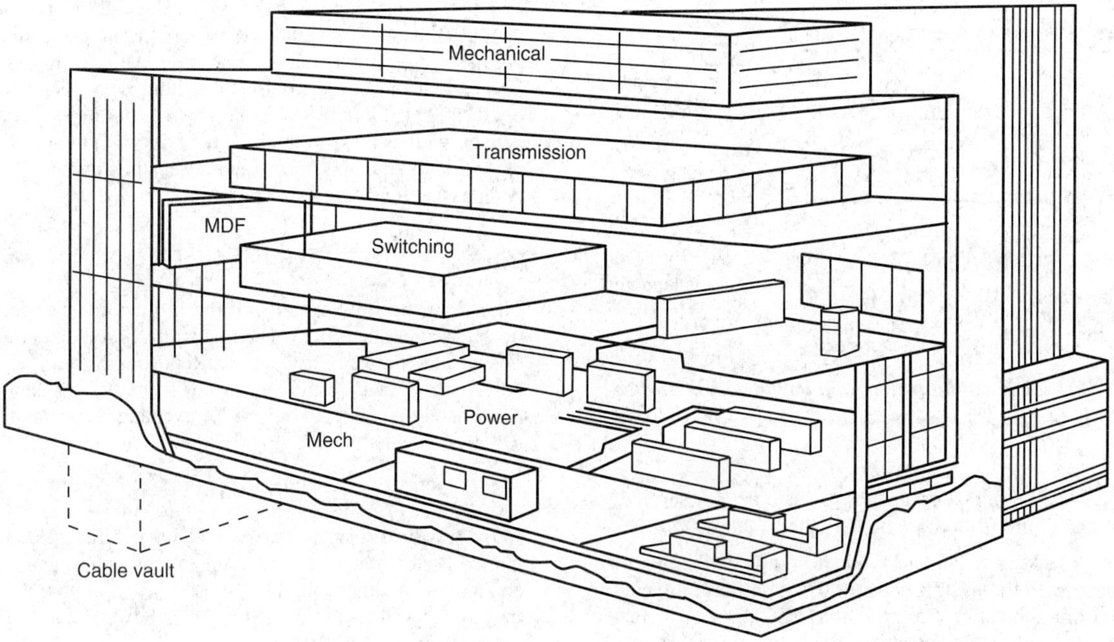

FIG. 9-27C.   Typical central office components.

accordance with applicable building codes, not in excess of those codes, as had been prior industry practice.

Ceilings in COs are generally much higher than those in commercial buildings. Typical slab-to-slab heights in central offices are 14 to 16 ft (4.3 to 4.9 m). Few ceilings are dropped or suspended in the equipment areas so vertical space remains unobstructed. Clear space was necessary to accommodate 11-ft (3.5-m) equipment frames commonly used when the buildings were constructed, together with required cable racks, air-conditioning ducts, and other service equipment located above them.

Cables enter and leave most COs through the cable vault, where they split into smaller, more fire-resistant cables for distribution in the building. Cables then rise through the building to the main distributing frame (MDF) where they break into individual pairs of wire for each circuit and cross connect to cables leading to the switching systems. After switching, signals pass to the transmission systems, which multiplex and boost the signals. After this is accomplished, the signal travels back out of the building through the cable vault on other cables.

In addition to standby generators, battery systems, and associated power control and distribution equipment, a variety of standard mechanical and building service systems support the telecommunications equipment.

TELECOMMUNICATIONS EQUIPMENT

Description

Telecommunications equipment is fairly standardized in both function and appearance. It is arranged in parallel rows, or "line-ups," 19 to 26 in. wide (48 to 66 cm). The latest generations of digital electronic equipment generally range from 6 to 7 ft (1.83 to 2.14 m) in height.

Earlier generations of equipment were mounted in open frames without enclosures. This allowed easy access for maintenance and permitted heat dissipation by natural convection. Modern equipment is often enclosed in metal cabinets and looks much like ordinary computers. This arrangement, combined with miniaturization of equipment and attendant higher energy densities, requires forced ventilation (internal fans and extensive room air conditioning systems) to dissipate heat generated during operation.

Providing for heat dissipation is important because the equipment is sensitive to temperature and humidity changes in the operating environment. Although absolute operating limits are rather wide, rapid changes of either a few degrees of temperature or a few percentage points of humidity can cause serious problems. Consequently, extensive environmental controls are used to ensure that ambient conditions remain within required limits.

Differences from Standard Computers

On first examination, modern digital telecommunications equipment resembles modern computers or data processing equipment. Although there are similarities, a number of important differences exist. Some were deliberately created to reduce the incidence, or effect, of potential loss-causing events. Other differences, necessitated by various technical requirements, have the opposite effect. Some of these differences present challenges that most traditional loss prevention measures cannot adequately mitigate.

One of the most significant differences is the type of information processing performed. Telecommunications systems are on-line information exchange systems. The telecommunications system does not store or process customer data; it merely transfers data from point A to point B. In a disruption, all information in transit is lost, and the ability to transmit information ceases. Data cannot be reloaded as long as the disruption continues, and missed opportunities cannot be made up. This ability to transmit information in real time makes telecommunications so important. Many transactions routinely conducted over the network simply cannot wait until later.

In contrast, data processing systems generally process information stored in the system's memory subsystems. In a disruption, current memory is lost, along with any calculation results that have not been placed in permanent memory. However, all stored information remains on disks or tapes, depending on the storage media. Calculations can usually be reproduced after correcting the cause of disruption.

Another important difference between telecommunications equipment and data processing equipment is their power supplies. Data processing equipment is supplied with ac power to the cabinet. The ac to dc conversion takes place within the equipment, which generally operates at 5 to 10 V dc. On the other hand, dc power supplies telecommunications equipment. Conversion from commercial ac power to dc power takes place outside the equipment, often in a different building compartment. The power plant supplies the equipment with 12, 24, 48, or 130 V dc, depending on system requirements. Power supply segregation is discussed later in this chapter.

Another important difference lies in the redundancies usually provided for telecommunications equipment. In most networks, for instance, equipment operating in a duplex configuration performs all switching and call routing functions. This means that two separate processors simultaneously handle each call. In the unlikely event that one processor fails, the other continues to process the call, without interrupting the customer. In fact, the customer is unaware that the event occurred.

The final difference, which significantly impacts loss prevention, is the control programs' size. Programs that control various networks are among the largest and most complex. Programs involving millions of lines of program code are not uncommon. Their size naturally requires much more time to reload and initialize in the event of a disruption than common data processing programs. This is one reason why it can take many hours to restart an electronic

switching system in the event of a power loss. The thermal shock created by suddenly restoring power to the entire system is another.

Telecommunications equipment provided for use on a customer's premises, known as Private Branch Exchange (PBX) equipment, differs in many respects from equipment used in COs. Major differences that impact a loss prevention program are discussed later in this chapter.

FIRE PROTECTION FOR THE PUBLIC NETWORK

A systematic approach is essential in determining proper fire protection measures applicable to telecommunications equipment facilities. (See Section 1, Chapter 3, "Systems Concepts for Building Fire Safety.") Such an approach helps clearly define the challenges at hand and provides a framework for identifying all elements that impact fire safety. One example of such an approach is the fire safety concepts tree, found in NFPA 550, *Guide to the Fire Safety Concepts Tree*, and illustrated in Figure 9-27D.

Figure 9-27D shows an example fire safety concepts tree for telecommunications central offices. Modeled after NFPA 550, this tree has been customized for this application. Boxed terms in Figure 9-27D correspond to headings in the remainder of this section.

In concept, the fire safety objective can be achieved by providing one path through the decision tree. In practice, however, multiple paths provide redundancies that increase confidence in accomplishing the objective. Most U.S. telecommunications providers follow a program that employs many, if not all, paths through this tree. A risk management decision to use multiple paths through

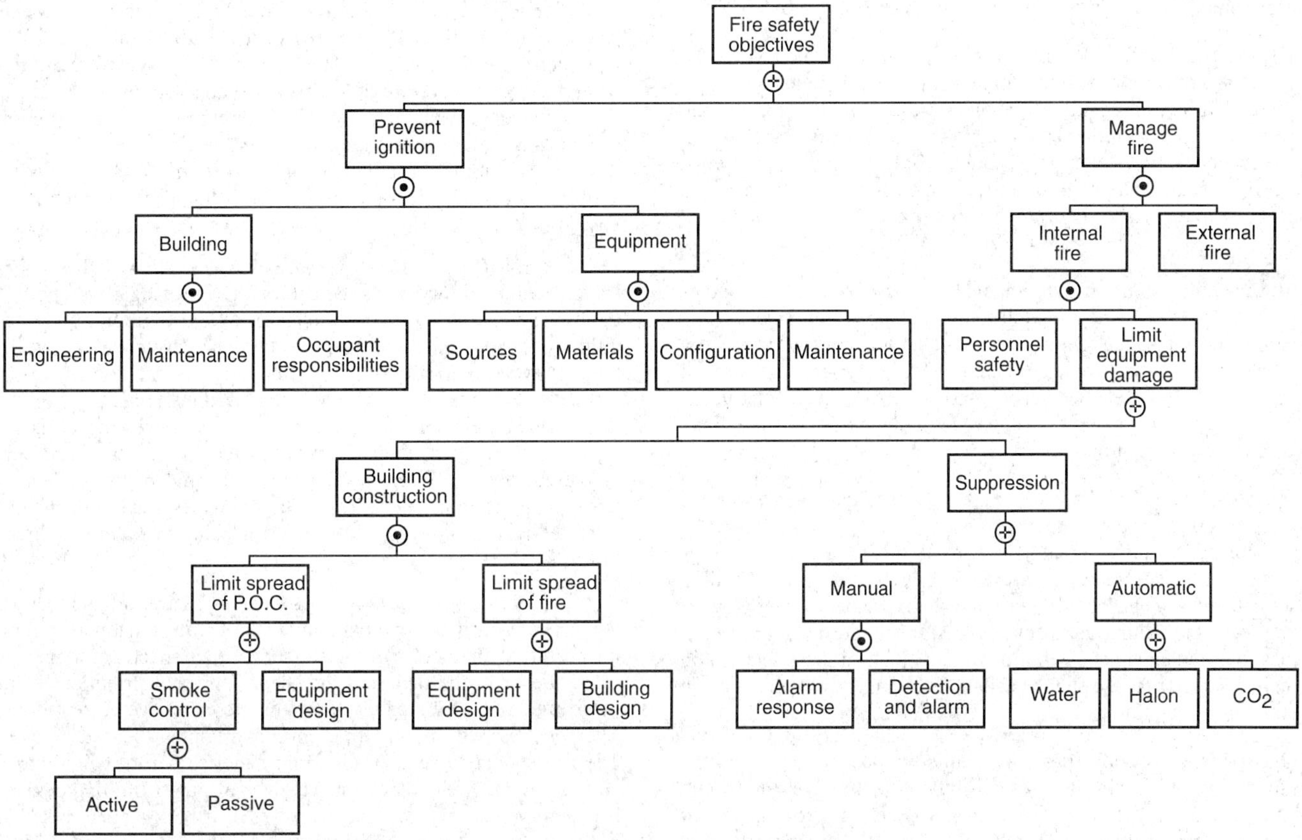

FIG. 9-27D.   *Fire safety concepts tree for telecommunications central office.*

the tree effectively changes all the gates in the tree to AND gates. The tree in Figure 9-27D includes both AND and OR gates for the purpose of illustrating concepts.

Before wholly focusing on fire protection, it is important to note that fire is just one of many potential causes of telecommunications service disruption. Some techniques employed to mitigate other types of risk negatively impact fire risk. This effect can work in the opposite direction as well.

Fire Safety Objectives

At the top of the decision tree are the fire safety objectives to be accomplished through the fire risk management program. In the case of a CO, all telecommunications providers want to minimize service disruptions from any cause. Customer demands, as well as various regulatory agencies governing telecommunications service providers, influence actual quantification of this goal. In the United States, the FCC and various state regulatory agencies perform this function.

Desired fire safety objectives can be accomplished by preventing ignition or by managing the fire once ignition has occurred. Considering that ignition prevention is the most desirable scenario in an industry that provides vital services, that branch of the tree will be discussed first.

Prevent Ignition

In the context of a CO, ignition prevention can be separated into two areas of concern: (1) those internal to the telecommunications equipment and (2) those external to the equipment or in other areas of the building. Modern telecommunications equipment is susceptible to service-interrupting damage from both the direct insult of a fire, as well as from indirect insults, such as smoke. Therefore, preventing both is necessary to prevent equipment damage.

Within the building: Ignition prevention associated with the building, i.e., not in the telecommunications equipment, should be handled fairly traditionally, although it warrants more rigorous than normal implementation. The design of major energy transmission and mechanical systems, i.e., heating systems, electrical systems, emergency and standby generators, elevators, etc., must be in accordance with applicable codes and should exceed code to the extent practical.

Engineering: The building should be engineered and designed as a single-use occupancy for telecommunications equipment. This eliminates the people, ignition sources, and fuel packages typically found in other occupancies. Where this is not possible, substantial fire-rated and smoke-rated barriers should separate telecommunications equipment from other occupancies and the risks that they may pose to service continuity.

Maintenance: Building service equipment, electrical systems, fire protection systems, and other systems in the building should be maintained in a fashion that reduces or eliminates the chance of ignition. The widespread use of digital control systems for building services equipment provides the opportunity for decreasing maintenance-related failures as a cause of fires. It is important that companies that implement such control systems include these features in their systems, because such installations often lead to a decrease in maintenance staff.

Occupant responsibilities: All building occupants should be thoroughly trained in their responsibilities of fire prevention and emergency preparedness. This includes proper procedures for housekeeping, storage of combustibles, and emergency actions. Other chapters of this handbook cover these issues in greater detail.

Smoking should be prohibited in all equipment and storage areas and throughout the building if possible.

Within the equipment: Ignition prevention within the telecommunications equipment is especially important since any *established* fire in the equipment is likely to result in service disruption. A fire is generally considered an established fire when it becomes self-sustaining, i.e., when flame height reaches 12 in. (30 cm). To prevent ignition within the equipment, four issues are addressed: (1) sources of ignition energy, (2) selection of materials that would constitute the fuel package for a fire, (3) the configuration of these materials in the equipment, and (4) equipment maintenance.

Sources of ignition energy: All unnecessary sources of ignition energy should be removed from the building. Any necessary sources of ignition energy should be separated from the telecommunications equipment as much as possible. A common source of ignition energy is power equipment such as the rectifiers, transformers, and converters that compose a power supply system. Physical separation reduces the risk of fire damage to the more sensitive circuitry in the telecommunications equipment, should a fire occur in the power supply.

Potentially dangerous energy sources in a central office are battery systems found throughout most telecommunications facilities. There are no current limiting devices (i.e., fuses) in the line between the battery string and the power distribution board. (See Figure 9-27E.) The nature of direct current (dc) electrical power further compounds the risk. An arcing fault or a short circuit in a dc power cable will likely be "seen" by a fuse as simply "work," and the fuse will not open. Two recent fire incidents illustrate both dangers.

In May 1988, a fire in the Hinsdale, IL, central office was initiated by a short circuit between a plastic insulated dc power cable and a metal jacketed (BX) cable.[3] The short energized the metal jacket, which overheated due to resistance heating and ignited adjacent plastic insulated cables. The fuse for the power cable did not open under this short circuit. One contributing factor was that telecommunications cables are often sized for current drop (i.e., friction loss) instead of ampacity and fusing is selected based on the ampacity rating given in NFPA 70, *National Electrical Code®*. It is critically important that fusing for all telecommunications circuits be based on actual electrical loads, not on the cable's ampacity rating.

Shortly after this fire, there was a flurry of activity in the industry to develop prototype "arc detectors." The investigation report of the fire and an FCC report on fire safety in telecommunications facilities each recommended such detectors be developed and used.[2] At least two organizations succeeded in this effort. These arc detectors sense faults, such as the one that caused the Hinsdale fire, and can automatically shut down the circuit, thus preventing or extinguishing the fire. The industry has not widely accepted, nor generally used, these devices despite their potential benefits. Several technical issues must be addressed before widespread implementation, but as of this writing, little effort is being made in this direction.

In March 1994, a fire in a Los Angeles central office involved the dc power distribution board and battery strings in a power plant on the thirteenth floor.[4,5,6] A "dead" short circuit in temporary power cables between the batteries and the dc power distribution board overheated one of the battery posts due to excessive current draw. The post heated to the melting point and ignited the battery case. The fire spread to several adjoining batteries before being extinguished by the fire department. The lack of power limiting devices between the batteries and the dc power distribution board resulted in the excessive current flow that melted the battery post. Smoke from this fire damaged 8 to 10 other power plants in the room, which

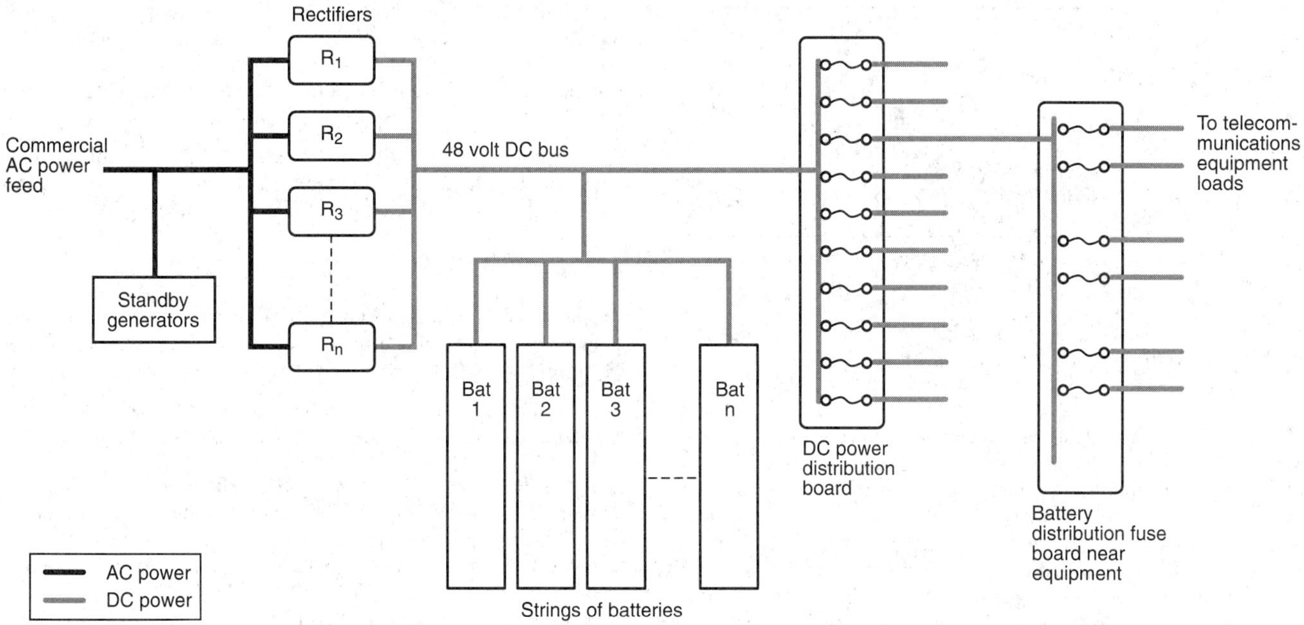

FIG. 9-27E. *Typical telecommunications power supply system design.*

eventually had to be replaced to retain the telecommunications system's reliability.

Materials selection: The selection and configuration of materials used in the manufacture of telecommunications equipment has traditionally been controlled in the United States by the Network Equipment-Building System (NEBS) Generic Equipment Requirements: Physical Protection.[7] This document references many national and international fire test standards, and, in doing so, applies their requirements to telecommunications equipment. The NEBS standard is considerably more stringent than any other fire tests required of electronic equipment by listing laboratories such as Underwriters Laboratories. The advent of competition in the telecommunications industry led many companies to minimize the importance of the following requirements. The most notable example can be found in cellular telephone base station equipment, which does not meet these fire resistance requirements.

Telecommunications equipment requires three types of tests: (1) flammability tests of components, wiring, and subassemblies; (2) equipment flame spread tests; and (3) equipment extinguishability tests.

Individual electronic components must meet requirements of the International Electrotechnical Commission (IEC) Standard 695-2-2.[8] This test exposes the component to application of a needlelike flame for a time period determined by the component's volume. If the component does not ignite or self-extinguish within 30 sec after flame removal, *and* if any plastic drippings do not ignite a piece of tissue paper placed 2 in. (5 cm) below the sample location, it passes the test. The test determines whether energizable components tested in a nonenergized state can resist flame spread.

Structural elements, such as circuit boards, must evidence an oxygen index of 28 percent or greater, as determined by ASTM D2863-77, *Standard Method for Measuring the Minimum Oxygen Concentration to Support Candle-Like Combustion of Plastics (Oxygen Index).*[9] They must also achieve a 94 V-1 or better rating under UL 94, *Tests for Flammability of Plastic Materials for Parts in Devices and Appliances.*[10]

Other nonmetallic parts and components must also meet flammability requirements established by numerous other standard test methods, some of which are listed in the bibliography at the end of this chapter.

Materials configuration: After individual components satisfy these requirements, the completed assembly is subjected to two large-scale tests:

1. The equipment assembly fire test is used to classify fire propagation characteristics of the equipment assembly and to assure that an assembly fire does not spread beyond the structural elements of the frame or cabinet containing the assembly. In the equipment assembly fire test, a fully loaded equipment assembly is subjected to a severe fire insult from a methane line burner, developed to simulate the burning rate of a fully involved circuit board. The line burner is inserted at the bottom of the lowest shelf of circuit cards, in the center of the shelf (the worst case location). The fire is then permitted to burn until it goes out by itself. Temperature, radiant heat flux, and heat release rate are measured throughout the test. If flames spread beyond the frame being tested, the equipment fails the test. If the peak rate of heat release exceeds 150 kW, if the average heat release rate exceeds 100 kW, or if the radiant flux exceeds 15 kW/m², the assembly fails the test.

2. In the equipment assembly extinguishability test, the same fire insult is applied to an equipment frame at least 36 in. (91 cm) high. After a 15 min pre-burn, the resulting fire must be capable of being extinguished by a 10B:C rated fire extinguisher.

Together, these test requirements assess the ignitability and subsequent flame spread within the equipment. Equipment passing these tests is minimally susceptible to ignition, fire growth, and subsequent service disruption. Figures 9-27F and 9-27G present evidence of this.

Figure 9-27F shows heat release rates from two typical telecommunications equipment configurations. By itself, this chart does little to illustrate the relative fire hazard of telecommunications equipment. Looking at Figure 9-27G is more useful for comparison

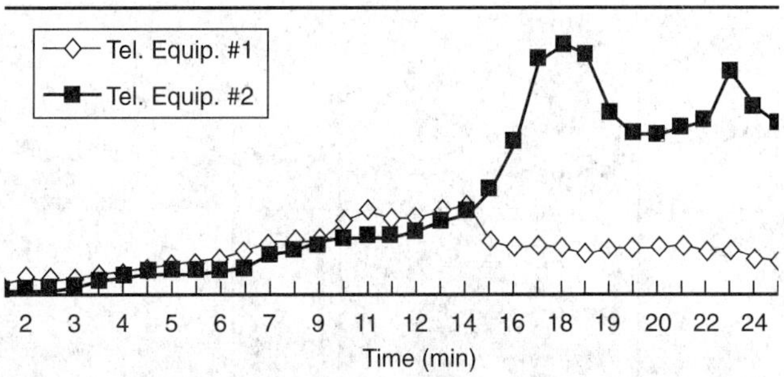

FIG. 9-27F. *Heat release rate histories for selected fuel packages: two telephone equipment configurations.*

purposes. It plots the same typical telecommunications equipment assemblies against some more common fuel configurations. These include a typical upholstered chair, a typical residential sofa, and an exhibit package typical of those used at conventions, such as the NFPA annual meeting. Major points illustrated by these comparisons are that the telecommunications equipment (1) has *much* lower heat release rates, (2) takes longer to reach its peak burning rate than other ordinary fuels, and (3) clearly represents a much smaller risk than the ordinary fuels that we all live with daily.

Equipment maintenance: After the equipment is designed, manufactured, and installed to meet these exacting requirements, it must be maintained in a fashion that eliminates the chance of ignition. Electronic equipment is fairly simple to maintain from a fire prevention perspective. The major maintenance fire hazards are related to the large quantities of cables associated with the equipment.

The majority of cables in a telecommunications facility are low energy signal cables. A number of higher energy power cables also run between the power equipment and the telecommunications equipment. Not inherently hazardous, high-energy power cables present a somewhat greater fire risk, if damaged, than low-energy signal cables.

Continuous upgrades and changes to CO equipment result in constant addition, and frequent abandonment, of cables. Unused cables do not present an ignition source when disconnected at both ends. However, they do contribute fuel load for a potential fire, and for this reason, removing unused cables may be desirable.

Removal creates certain hazards that must be evaluated against its benefits. Removal activities like cable mining are inherently hazardous from both fire protection and service continuity perspectives. Damage caused during cable mining operations was identified as a possible cause of the Hinsdale, IL, central office fire.[3]

Removing cables from the lower levels of large cable bundles is extremely difficult. Trained technicians must carefully perform these operations to decrease potential service interruptions.

Manage the Fire

Despite the best efforts to prevent ignition in the equipment and elsewhere in the building, fire may still occur. Thus, provisions must be made to manage a fire's impact and byproducts.

As the fire safety concepts tree in Figure 9-27D shows, fires outside the building must be managed as well as those within.

External fires: External fire exposure should be mitigated in a number of traditional ways. The CO site should be carefully selected so that adjacent properties or operations do not constitute a risk to network equipment. At a minimum, building construction should meet the requirements of NFPA 220, *Standard on Types of Building Construction*, for Type II construction. Where necessary, exposure protection systems should be provided.

The lack of windows in many CO buildings is a prominent example of an exposure protection system. It protects against exposure fires, as well as sabotage or terrorist activities. The lack of windows does, however, make fire fighter access and ventilation more difficult in the event of a fire.

Internal fires: The safety of building occupants must come first when managing fire. Strict adherence to NFPA *101*®, *Life Safety Code®*, helps ensure this. After personnel safety is ensured, equipment safety, or, more correctly, limiting damage to the telecommunications equipment, can be considered.

Limit equipment damage: Equipment damage limitations should combine safeguards built into the building and the telecommunications equipment, as well as actual fire suppression.

Telecommunications equipment can suffer damage that affects service from both direct and indirect fire exposure, e.g., from the products of combustion (poc). Both must be managed to achieve the

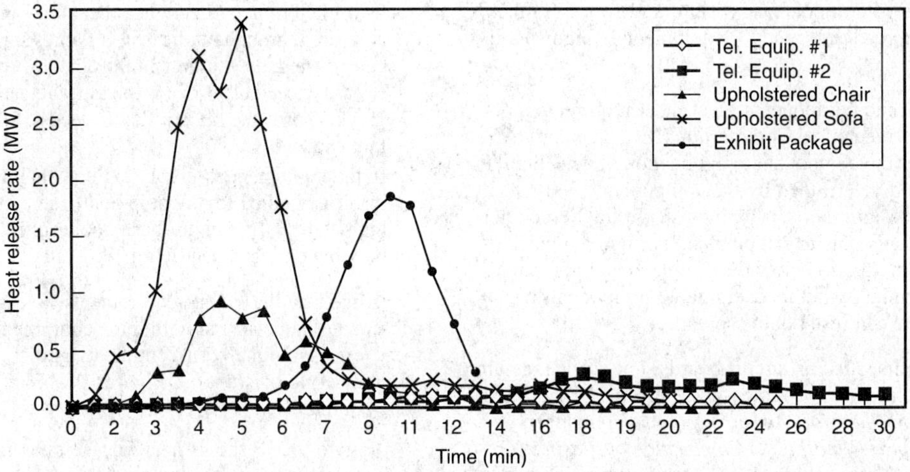

FIG. 9-27G. *Heat release histories for selected fuel packages: telecommunications equipment compared to ordinary combustible materials.*

fire safety objective. Both building design/construction and equipment design/construction can be profitably used to limit damage.

Limit smoke (poc) spread: Two broad approaches are useful in limiting smoke damage. The first is to select materials that minimize smoke generation and to limit materials that generate smoke that is particularly damaging to the equipment. The second approach is to control the smoke that is produced. When selecting materials, the use of polyvinyl chloride (PVC) as insulation is of particular interest.

PVC is widely used as insulation on cables and wires in COs. This material has excellent electrical and fire resistance properties. When PVC burns, however, it releases gaseous hydrogen chloride (HCl), which can combine with atmospheric water vapor to form hydrochloric acid. This, in turn, can be deposited on, and destructively react with, metallic parts of the equipment. This phenomenon can cause serious damage in fires involving electronic equipment.

A potential solution to this acidic smoke concern involves using nonhalogenated insulation material, which recently was made available. This material produces no HCl when burning and passes all applicable flammability tests. Widespread retrofitting of existing COs with this cable would pose a large risk of service outages and is not generally recommended. However, cable with nonhalogenated insulation should be used wherever possible in new equipment installations or in existing equipment modifications.

When considering the control of smoke produced by a fire, both passive and active measures apply. Passive smoke control measures are generally adaptations of measures used to prevent fire spread, i.e., fire and smoke barriers. Compartmentation, augmented by fire/smoke dampers in duct penetrations, smoke seals on fire doors, and good firestopping of other penetrations, is widely considered a successful control measure. Active smoke control measures combine fans and ducts that pressurize some building areas and exhaust others. (See Section 7, Chapter 6, "Smoke Movement in Buildings.") The 1993 FCC study included the recommendation that telecommunications companies provide smoke control for central offices.[2] Several companies are currently considering various means of implementing this recommendation.

Limit fire spread: Building construction and equipment design should limit fire spread. Fire spread within the equipment should be controlled using the equipment design and material selection criteria discussed above. Equipment in the network that does not meet these criteria should be isolated from all other equipment with fire-rated barriers. Additional protection may be appropriate for such nonconforming equipment if it presents risks to other equipment or services.

An example of the appropriate use of fire barriers can be found in cable vaults. Cables exposed to weather or earth are usually insulated with polyethylene plastic, which provides protection from these elements. Because polyethylene has rather poor fire-retardant properties, its use inside the CO increases fire risk. To mitigate this risk, outside cables enter the building in a room with a substantial fire resistance rating. Inside this vault, the polyethylene outside plant cables terminate and are spliced to cables with acceptable fire resistance properties. This fire retardant cable is then used throughout the building.

Vaults also serve as collection areas for vapors or gases that may enter the building along the path of the buried cable, creating a health or explosion hazard. Appropriate detection and exhaust systems should be provided where such incidents could occur.

Fire detection: Reliable and sufficiently early detection is particularly challenging in modern telecommunications equipment environments. High air flow rates, high ceilings, horizontal air currents, and numerous overhead obstructions combine, severely challenging traditional smoke detection strategies. Air-sampling type detection systems have demonstrated capabilities to meet these challenges and should be considered for use in telecommunications facilities.[11,12]

Additional early warning may be obtained through the use of chemical species monitoring of selected spaces. Current COs use large quantities of PVC insulated cables. PVC releases HCl gas in detectable quantities when heated. This release begins at temperatures well below PVC's ignition temperature and before combustion begins. An installed HCl detection system can detect an overheated cable before smoke detection systems can activate.[11] This type of system may be desirable in certain instances.

Another strategy some companies employ is multilevel installation of point-type detectors programmed to give a pre-alarm when less smoke is present than standard smoke detectors can sense. This strategy works reasonably well in low-to-moderate air flow environments but not as effectively as air sampling detection systems in high air-flow environments.

Fire suppression: Two options are available to suppress a fire in a CO: (1) manual action by trained personnel or the public fire service and (2) automatic suppression systems. When considering a fire originating within the telecommunications equipment, note that an established fire results in failure to meet the fire safety objective. This is the primary reason for the extraordinary efforts made to prevent ignition.

Manual fire suppression: Detection systems only signal fire's presence, alerting personnel who must respond. Employees expected to perform fire suppression and related duties must be fully trained, and training should be demonstrated regularly through exercises and drills.

One of the most important emergency duties is shutting off power. Fires involving any telecommunications equipment also involve energized electrical equipment and/or circuits. The most common extinguishing action for electrical fires is to remove power from the circuit by unplugging the appliance or throwing the switch or circuit breaker. This is much more difficult in telecommunications facilities.

The power system in telecommunications facilities is designed with multiple redundancies because power interruption disrupts service quickly for an extended period. Typically, generators and large battery systems back up the commercial power supply. These power systems are designed much like the power supply for an electric fire pump, so that the system runs until total failure. The complexities of these power systems make it difficult to turn the power off to fight a fire. The importance of an easy method to shut off the power was clearly demonstrated in the investigation report on the Hinsdale, IL, CO fire.[3] The FCC report also recommended that companies provide a means of disconnecting power during emergencies.[2]

Although current technology does not facilitate reliable and easy power disconnects for telecommunications equipment, procedures should be developed to guide emergency responders through the power down process. These procedures should clearly identify steps to be taken to remove power to various areas within the building, and they should be updated frequently.

Automatic fire suppression: Current model building codes permit the omission of automatic suppression systems in telecommunications equipment areas in COs under a restrictive set of conditions. This exception is based on the very low probability of ignition, the low fire growth rates, and the excellent fire record of the equipment. The fire resistance tests described above measure these important attributes. Telecommunications equipment that does not pass these tests does not qualify for the exception granted by the model building codes.

Selecting a suppression system for areas containing telecommunications equipment is difficult and must be carefully weighed against a clear understanding of loss prevention objectives. The mere presence or functioning of an automatic suppression system does not ensure that fire safety objectives are met. Such a system should not be substituted for the fire safety measures already described.

Three options are available for automatic suppression systems in electronic equipment areas: (1) gaseous halon alternatives, (2) CO_2, and (3) automatic sprinklers. Several researchers have devoted considerable effort to adapt water mist suppression systems for this application, but the effort still has far to go to overcome technical obstacles.

Any automatic suppression system selected for use in a telecommunications facility should meet several criteria: (1) rapid initiation of suppression activity, (2) capability for sustained application of agent, (3) ability to control damage to acceptable limits, and (4) permitting safe fire fighter access to the area.

No current suppression system option meets all these criteria. It is important that the selection and use of any suppression system also include a full analysis of potential damage the system could cause if activated during a fire or inadvertently.

CO_2 systems can be designed to activate early enough to limit fire damage, given an appropriate detection system. However, several disadvantages generally preclude their widespread use. It is very difficult to seal such large areas to prevent CO_2 leakage and to reliably maintain the seal. Dozens of large cable penetrations through the floor and ceiling are common; all must be tightly sealed. Given the high ceilings commonly used and the location of the major fuel load—the cables, 12 ft (3.7 m) or more above the floor in racks—it becomes apparent that enormous quantities of agent would be necessary. Personnel safety hazards in a large, complex floor area would be substantial.

Gaseous suppression agents pose other technical difficulties. When released, compressed gases are very cold and will cause the room's temperature to suddenly drop. Telecommunications equipment is very sensitive to temperature changes; much of it is limited to a maximum change of 15°F (9°C) per hour. Changes exceeding this can substantially damage the affected equipment.

Concurrent with a rapid temperature drop is a rapid increase in relative humidity. Experimental evidence shows that the dew point can be reached quickly. The condensation that then forms on many equipment components may create a host of electrical problems and subsequent damage.

Automatic sprinkler systems offer the advantages of sustained suppression capability and better fire fighter access than gaseous suppression systems. Disadvantages include relatively slow activation time and potentially significant collateral damage to telecommunications equipment, during both fire induced and inadvertent system operations. Inadvertent operation causes extensive damage to equipment and results in an extended period (days or weeks) of service outage.

Measurements taken during equipment assembly flame spread tests show that telecommunications equipment meeting the standards described have a very low heat release rate and a slow fire growth rate. (See Figure 9-27F.) Tests at a major fire-testing laboratory confirmed very slow fire growth. In those tests, a 165°F (74°C) fast-response sprinkler located directly over the equipment frame where ignition occurred did not activate until 55 min after ignition.[13]

Sprinkler systems cannot respond to a fire before it grows well beyond the established burning phase. Therefore they cannot respond to and extinguish a fire before it significantly damages equipment. Hence, automatic sprinklers in telecommunications equipment areas do not meet service continuity fire safety objectives.

Collateral damage due to water release in equipment areas include numerous electrical short circuits, chemical corrosion of circuit paths from entrained contaminants, and serious electrolytic corrosion. Although this damage pales in comparison to damage from a serious fire, it can be significantly greater than damage from a small fire. This must be considered carefully when selecting a suppression system for an occupancy with an extremely low fire incidence rate.

In practical terms, electrolytic corrosion in telecommunications equipment is nearly instantaneous when water is present from any source, including suppression systems, condensation, and leaks. The rate of electrolytic corrosion varies exponentially with the voltage applied. Hence, telecommunications equipment operating at 48 V has a corrosion rate many times that of data processing equipment operating at 5 V.

Removing power from equipment before applying the agent can negate these effects. If the equipment is de-energized before system activation, aqueous systems present minimal risk of collateral equipment damage. No automatic means of de-energizing telecommunications equipment provides assurances against inadvertent tripping necessary for modern telecommunications networks.

Private Telecommunications Equipment

Equipment provided by public telecommunications companies terminates at a network interface on the customer's property. The customer is responsible for providing all on-site telephone equipment and wiring. Customers accomplish this in a variety of ways. Commonly, they purchase and maintain their own telecommunications system, lease the equipment from either a phone company or a third party vendor, or contract with the local telephone company for a central office based service such as Centrex. Numerous other possibilities exist.

Large commercial or business occupancies typically have large telecommunications systems; many are more complex than those provided for most towns. These systems can provide a wide array of voice, data, and video information to their users. Usually, they are controlled by private branch exchange (PBX) equipment located in the building or complex.

At first glance, modern PBX equipment appears to be very similar in design to modern central office equipment. PBX equipment performs many functions that CO equipment does but only for the building or complex it serves.

From a fire protection viewpoint, three significant differences exist between CO and PBX equipment:

1. The most important difference is that PBX equipment does not have to conform to the very rigorous flammability standards that CO equipment meets. Listed by independent agencies, PBX equipment does meet a variety of other tests, including flammability tests, required to achieve that listing. The implications of this are discussed below.
2. Replacement PBX equipment is more readily available than replacement CO equipment. This allows more flexibility in risk management decision making in some instances.
3. PBX systems are usually easier to turn off than CO equipment. PBX power supplies often have redundancies similar to those of the CO, but the systems are smaller and less complex. In some cases, a switch or circuit breaker can be used to power down the equipment in the event of an emergency.

Protection Fundamentals

Most protection measures indicated for central offices are appropriate for PBX equipment as well. In general, the PBX should be protected in the same fashion as an important computer room. Several elements are worthy of special note:

1. The equipment should be located in a room separated from the rest of the building by at least a 1-hr rated fire barrier.
2. Access to the room should be limited to authorized persons.
3. Good fire prevention and housekeeping measures should be in place at all times. Storage of combustible materials in the room should be strictly prohibited.
4. The room should be equipped with a smoke detection system that alarms at a constantly attended location.
5. Power shut-down procedures should be developed and made available for responding fire fighters to use.

Automatic suppression: Exceptions found in the model building codes that allow a central office to omit automatic sprinklers in equipment areas do not apply to PBX equipment or to telephone closets. Those exemptions are based in part on ignition prevention efforts described earlier in this chapter, and these efforts are not fully utilized in PBX equipment.

If code requires a building to be equipped with an automatic suppression system, the system should be provided in the PBX room. An appropriate system should be selected with a clear understanding of risks involved and advantages that each candidate system offers. The system may be omitted if the local Authority Having Jurisdiction grants a variance or exemption. The system's owners may decide to install additional fire protection features to enhance the reliability of their system, since model code requirements do not address business interruption concerns. Such additional features might include air-sampling smoke detection systems or gaseous agent suppression systems.

Fire Department Concerns

Despite all prevention measures many telecommunications companies take, fires still occur. Although these fires have historically been small in relation to fires in other industries, fire fighters face some clear differences when responding to such a fire.

1. Fires in telecommunications equipment tend to produce very large quantities of dense black corrosive smoke. Visibility becomes nearly impossible, even with hand-held lamps. When combined with unfamiliar building configurations, the prospect for disorientation and the dangers that follow are clear. Appropriate precautions must be planned in advance. Fire departments should consider requesting that the owner install manual smoke exhaust systems, which differ from smoke control systems, to assist in fire control efforts. The portable smoke extractors or smoke fans that fire departments carry are totally inadequate to handle the large quantities of smoke the fires generate. Experience has shown that even 10 to 12 smoke fans are inadequate. Smoke exhaust systems have been proven highly effective in at least one actual telecommunications fire incident.[4,5,6]
2. A fire involving cables will probably involve power cables. It is critical that responding fire fighters be familiar with the power disconnect procedures established by the owner. If the owner does not provide such a plan, fire departments should demand one. Multiple telecommunications industry studies have called for the provision of these plans and procedures. The State of Illinois, site of the Hinsdale fire in 1988, requires a power shut-down plan for every telecommunications facility. This is not something new, and there are established procedures for developing these plans.
3. Two physical hazards are present in telecommunications facilities that are not present in the same degree and quantity in other occupancies: lead-acid batteries and electrical power. Both are generally well engineered and maintained; neither presents extraordinary dangers in comparison to other dangers fire fighters commonly encounter. Responding personnel must be aware of these hazards' presence and must be familiar with appropriate safety precautions before entering a smoke-filled telecommunications facility.

These issues are not insurmountable. Most telecommunications companies are happy to provide tours to familiarize fire departments with their facilities. It is important that these be done, if for no other reason than fire fighter safety.

SUMMARY

Precautions discussed in this chapter provide a sound, proven basis for fire risk management for telecommunications equipment and services. Like any loss prevention program, the fire risk management program for telecommunications networks and equipment must be ongoing. There will be a great many changes both in the technology employed in the equipment and in the uses of telecommunications services in the near future.

Many changes envisioned today will serve to increase redundancy in networks, thereby decreasing overall risk. Other changes may increase the net level of risk and present unique new challenges for risk managers. Attendant risks must be evaluated systematically, so that effective risk reduction measures that truly reduce overall risk levels can be implemented.

BIBLIOGRAPHY

References Cited

1. Taylor, K. T., "Temporarily Disconnected," *Fire Journal*, Vol. 83, No. 3, May/June 1989.
2. *Network Reliability: A Report to the Nation, Fire Prevention in Telecommunications Facilities*, Federal Communications Commission Network Reliability Council, Washington, DC, June 1993.
3. Forensic Technologies International Corporation, "Hinsdale Central Office Fire Final Report," Joint Report of the Illinois Office of State Fire Marshall and Illinois Commerce Commission, March 1989.
4. McKenna, L. A., "Investigation and Reconstruction of the Power Plant Fire at the Pacific Bell Central Office at 420 South Grand Avenue, Los Angeles, California, on March 15, 1994," AT&T Environment and Safety Engineering Center, Basking Ridge, NJ, Nov. 1, 1994.
5. Isner, M. S., "E911 Fails in Telephone Exchange Fire," *NFPA Fire Journal*, Vol. 89, No. 5, Sept./Oct. 1995.
6. Isner, M. S., *Fire Investigation Report: Telephone Exchange Fire, Los Angeles, California, March 15, 1994*, National Fire Protection Association, Quincy, MA, 1994.
7. "Network Equipment-Building System (NEBS) Generic Equipment Requirements: Physical Protection," GR-63-CORE, Issue 1, October, 1995, Bellcore, Morristown, NJ.
8. "Fire Hazard Testing, Part 2: Test Methods, Needle Flame Test," IEC Standard 695-2-2, International Electrotechnical Commission, Geneva, Switzerland.
9. ASTM D2863-77, *Standard Method for Measuring the Minimum Oxygen Concentration to Support Candle-like Combustion of Plastics (Oxygen Index)*, American Society for Testing and Materials, Philadelphia, PA.
10. UL 94, *Tests for Flammability of Plastic Materials for Parts in Devices and Appliances*, Underwriters Laboratories Inc., Northbrook, IL.
11. Forssell, E. W., and Budnick, E. K., "*In Situ* Tests of Smoke Detection Systems for Telecommunications Central Office Facilities," American Telephone and Telegraph Company, Basking Ridge, NJ, October 1990.
12. Johnson, P. F., "Fire Detection in Computer Facilities," *Fire Technology*, Vol. 22, No. 1, Feb. 1986, pp. 14–32.

13. Edgar, J. A., *et al.*, *Investigation of Automatic Sprinklers to Protect Digital Multiplex System Telephone Central Office Equipment*, Bell Northern Research, Ottawa, Canada, December 1987.

NFPA Codes, Standards, and Recommended Practices

Reference to the following NFPA codes, standards, and recommended practices will provide further information on protection of telecommunications equipment facilities as discussed in this chapter. (See the latest version of *The NFPA Catalog* for availability of current editions of the following documents.)

NFPA 75, *Standard for the Protection of Electronic Computer/Data Processing Equipment*

NFPA 220, *Standard on Types of Building Construction*

NFPA 262, *Standard Method of Test for Fire and Smoke Characteristics of Wires and Cables*

NFPA 550, *Guide to the Fire Safety Concepts Tree*

Additional Reading

ANSI T1.304-1989, "Telephone Central Office Equipment—Ambient Temperature and Humidity Requirements," July 1989, American National Standards Institute, New York.

Anschutz, K., "Fire in a Telephone Office," *Firehouse*, Vol. 15, No. 7, July 1990, pp. 92, 94.

Bengtson, S., and Lundin, K., "Heat Release Rates in Telecommunication Tunnels," *Proceedings* of the First International Conference on Fire Science and Engineering, ASIAFLAM '95, March 15–16, 1995, Kowloon, Hong Kong, 1995, pp. 127–138.

Botting, R. E., "Telephone Exchange Fire Study Looks at MDF Hazards," *Fire Engineers Journal*, Vol. 51, No. 163, December 1991, pp. 19–22.

Budnick, E. K., "In Situ Tests of Smoke Detection Systems for Telecommunications Central Office Facilities," International Symposium on Fire Protection for the Telecommunications Industry, May 14–15, 1992, New Orleans, LA, 1992, pp. 1–27.

Duyck, T. O., "Recovery From Fire in a Central Office," International Symposium on Fire Protection for the Telecommunications Industry, May 14–15, 1992, New Orleans, LA, 1992, pp. 1–8.

H. H. Angus and Associates Limited, "Special Need for a Smoke Exhaust System to Minimize Secondary Damage to Electronic Telephone Switching Equipment, During a Fire Within a Switch Unit," International Symposium on Fire Protection for the Telecommunications Industry, May 14–15, 1992, New Orleans, LA, 1992, pp. 1–24.

Hill, A., "Fire Suppression in Telephone Switching Centers Utilizing Automatic Sprinklers," International Symposium on Fire Protection for the Telecommunications Industry, May 14–15, 1992, New Orleans, LA, 1992, pp. 1–32.

Hills, A. T., Simpson, T., and Smith, D. P., "Water Mist Fire Protection Systems for Telecommunication Switch Gear and Other Electronic Facilities," Fire and Safety International Research, Slough, England, NISTIR 5207, June 1993; National Institute of Standards and Technology, *Proceedings* of the Water Mist Fire Suppression Workshop, March 1–2, 1993, Gaithersburg, MD, 1993, pp. 123–142.

IEC Standard 332, *Tests on Electric Cables Under Fire Conditions*, International Electrotechnical Commission, Geneva, Switzerland.

IEEE Standard 383, *Type Test of Class IE Electric Cables, Field Splices, and Connections for Nuclear Power Generating Stations*, Institute of Electrical and Electronic Engineers, New York.

Isner, M. S., "E911 Fails in Telephone Exchange Fire," *NFPA Journal*, Vol. 89, No. 5, September/October 1995, pp. 109–112, 115–117, 119–120.

Isner, M. S., "Telephone Exchange Fire, Los Angeles, California, March 15, 1994. Fire Investigation Report," National Fire Protection Association, Quincy, MA, Fire Investigation Report, 1994, 36 pages.

Isner, M. S., "Telephone Switching Station Fire Creates Widespread Emergency," *Fire Journal*, Vol. 84, No. 3, May/June 1990, pp. 72–73, 75–77, 79, 81, 83, 95.

Johnson, P. F., "High Sensitivity Smoke Detection for the Telecommunication Facilities," International Symposium on Fire Protection for the Telecommunications Industry, May 14–15, 1992, New Orleans, LA, 1992, pp. 1–15.

Johnson, P. F., "Risk Assessment in Telephone Exchanges," International Symposium on Fire Protection for the Telecommunications Industry, May 14–15, 1992, New Orleans, LA, 1992, pp. 1–19.

Karydas, D. M., "Probabilistic Methodology for the Fire Smoke Hazard Analysis of Electronic Equipment in Telephone Central Offices," International Symposium on Fire Protection for the Telecommunications Industry, May 14–15, 1992, New Orleans, LA, 1992, pp. 1–8.

Kidde-Fenwal, "Water Mist Fire Protection in Telecommunications Facilities," International Symposium on Fire Protection for the Telecommunications Industry, May 14–15, 1992, New Orleans, LA, 1992, pp. 1–12.

Kushler, B. D., Simpson, J., and Budnick, E. K., "Approach to Fire Risk Assessment for Telecommunications Facilities," International Symposium on Fire Protection for the Telecommunications Industry, May 14–15, 1992, New Orleans, LA, 1992, pp. 1–3.

Mawhinney, J. R., "Water-Mist Fire Suppression Systems for the Telecommunication and Utility Industries," *Fire Research News*, No. 74, Fall 1994, pp. 1–3.

Nishikata, K., "Description of Telephone Cable Fire in a Cable Tunnel and the Response of the Fire Force," Tokyo Fire Dept., Japan, CIB W14/95/1 (J); Tokyo Fire Department, Firesafety Frontier '94, International Fire Conference and Exhibition in Tokyo, Creating a Safe Tomorrow, October 18–22, 1994, Tokyo, Japan, 1994, pp. 75–80.

Notz, R., "Fire Protection Concept for EDP and Telecommunication Systems," International Symposium on Fire Protection for the Telecommunications Industry, May 14–15, 1992, New Orleans, LA, 1992, pp. 1–15.

O'Connor, D. J., and Nugent, D. P., "Achieving Zero Downtime in Telecommunication Facilities," *Consulting-Specifying Engineer*, Vol. 19, No. 5, May 1996, pp. 32–34, 36–39.

"Protecting Telephone Central Offices," *Record*, Vol. 70, No. 4, July/August 1993, pp. 13–14.

Ramachandra, S., *et al.*, "Improved Processability in Reduced Emission Wire and Cable Materials for the Telecommunication Industry," Union Carbide Chemicals and Plastics Company, Inc., Somerset, NJ, U. S. Army Communications Electronics Command (CECOM), Thirty-ninth International Wire and Cable Symposium, November 13–15, 1990, Reno, NV, 1990, pp. 655–660.

Shaw, J. S., "Flammability Standards for Central Office Equipment," International Symposium on Fire Protection for the Telecommunications Industry, May 14–15, 1992, New Orleans, LA, 1992, pp. 1–6.

Simpson, J., Kushler, B., and Schubert, R., "Fire Calorimetry for Telecommunications Equipment," Second International Conference on Fire and Materials, September 23–24, 1993, Arlington, VA, 1993, p. 263.

Simpson, T., and Smith, D. P., "Fully Integrated Water Mist Fire Suppression System for Telecommunications and Other Electronics Cabinets," Fire and Safety International, UK, SP Report 1994:03; Swedish National Testing and Research Institute, *Proceedings* of the International Conference on Water Mist Fire Suppression Systems, November 4–5, 1993, Boras, Sweden, 1993, pp. 153–166.

Simpson, T., and Smith, D. P., "Suppressing Fires in Telecommunications Switchgear," *Fire Prevention*, No. 267, March 1994, pp. 17–20.

"Telecommunications Symposium. Conference Report," *Fire Technology*, Vol. 28, No. 3, August 1992, pp. 280–282.

"Telephone Central Offices," Loss Prevention Data Sheet No. 5-14, Jan. 1985, Factory Mutual Engineering Corp., Norwood, MA.

UL 510, *Insulating Tape*, Underwriters Laboratories Inc., Northbrook, IL.

UL 910, *Test Method for Fire and Smoke Characteristics of Electrical and Optical Fiber Cables Used in Air Handling Spaces*, Underwriters Laboratories Inc., Northbrook, IL.

UL 1441, *Coated Electrical Sleeving*, Underwriters Laboratories Inc., Northbrook, IL.

UL 1459, *Telephone Equipment*, Underwriters Laboratories Inc., Northbrook, IL.

UL 1581, *Reference Standard for Electrical Wires, Cable, and Flexible Cords*, Underwriters Laboratories Inc., Northbrook, IL.

UL 1666, *Test for Flame Propagation Height of Electrical and Optical Fiber Cables Installed Vertically in Shafts*, Underwriters Laboratories Inc., Northbrook, IL.

UL 1950, *Information Technology Equipment Including Business Equipment*, Underwriters Laboratories Inc., Northbrook, IL.

Yokel, F. Y., "Earthquake Resistant Construction of Electrical Transmission and Telecommunication Facilities Serving the Federal Government," National Institute of Standards and Technology, Gaithersburg, MD, Federal Emergency Management Agency, Washington, DC, NISTIR 89-4213, FEMA-202, Earthquake Hazard Reduction 56, February 1990, 47 pages.

PROTECTION OF ELECTRONIC EQUIPMENT

Robert J. Pearce

Now more than ever, our society depends upon computers and electronic equipment. The personal computer's increased speed and storage capacity and its reduced size and cost have caused a metamorphosis of electronic computing. No longer restricted to government, business, and industry, computers have entered everyone's personal life. Some type of computer or electronic data processing machine is now found in everyone's work station. The computer's strategic importance in business depends on its uninterrupted use. Should a vital computing operation be damaged, businesses can be paralyzed. Computers now control many processes such as semiconductor fabrication, petrochemical, basic steel, and pulp and paper. We have reached the point where humans can no longer replace computer-controlled robotics. Many businesses depend on special sophisticated electronic equipment such as communication systems or test equipment. Electronic or laboratory equipment, such as production prototypes, simulators, computer-aided design, and measuring devices, may be vital to business development. This equipment is very difficult to replace, and its destruction could have a very adverse impact.

Because of businesses' dependence on computers and other electronic equipment, the importance of fire protection cannot be overemphasized. This includes security measures because computers can be attractive targets for arson and sabotage. The introduction of remote work stations further hinders the security of the mainframe computer room. Because of the immense cost of large computer systems as well as space considerations, providing remote hot sites or even libraries of all records and data may seem economically impossible. Yet, only this can ensure continued operations in the event of a catastrophe. Custom-building a system for its applications compounds the problems of restoration because no compatible replacement will be available. Even so, many companies enlist partners to share the cost of hot remote sites. When catastrophic incidents occur, such as the bombing of the World Trade Center, the Northridge earthquake, or Hurricane Andrew, limited time is available at these hot sites. Owners often do not contemplate failure at more than one partners' premises. Official information relevant to protecting computers and electronic equipment from fire may be found in Section 3, Chapter 25, "Storage and Handling of Records"; Section 5, Chapter 1, "Fire Alarm Systems"; Section 5, Chapter 2, "Automatic Fire Detectors"; and Section 6, Chapter 25, "Special Systems and Extinguishing Techniques."

PHYSICAL FACILITIES, LOCATIONS, AND CONSTRUCTIONS

Because mainframe computer systems and electronic data equipment are expensive, sensitive to damage, and often critical to operations, they should be installed in specifically designed areas. These areas provide security, fire protection, and a controlled environment. Such installations may be located in computer rooms adequately separated from other parts of the building or even in separate structures. Local area networks (LANs) and wide area networks (WANs) commonly use host mainframes 3,000 miles (4,828 km) away. Offices and support areas of light combustible loading and ordinary hazard occupancies, such as light manufacturing, should be separated from mainframe computer rooms with walls having a 2-hr rating. Listed fire doors should be incorporated in the walls, shutters, and dampeners in the air ducts. The computer room should not be located above, below, or adjacent to areas housing hazardous processes, unless reliable protective separation is provided. These rooms should be of noncombustible fire-resistive construction with noncombustible interior finish. Most fires damaging computer rooms do not originate in the specialized equipment itself. These fires originate in and from the ordinary equipment, exposing the computer room. Rooms for related necessities, such as programmers, maintenance, paper supplies, recorded data, super printers, should be separate from the computer equipment room. Only the computer and electronic equipment and necessary auxiliary equipment should be permitted the computer room.

With the popularity of PCs and their tolerance for various controlled environments, mainframe computer rooms can be smaller, better protected, and more secure.

Small supervisors' or operators' offices strictly related to electronic equipment may be located in the computer room. Combustible materials should be kept at minimal levels. Service transformers, motor generator sets, switchgear, and battery backup should not be permitted. Supplies for paper and repair services shops should be strictly limited to the minimum needed for efficient operation and preferably located outside the computer room. Paper should be disposed of in waste disposal units designed to extinquish a fire, and trash should be removed each shift.

Records generated by the electronic computer system should be protected according to their importance and difficulty to reproduce. In most cases, information stored in computer systems or stored as input or output data is more valuable than computer hardware. Protecting this information is essential. Keeping backup copies of extremely

---

Robert J. Pearce is assistant area vice president—area loss prevention at Industrial Risk Insurers, San Francisco, CA. He is a member of the NFPA Technical Committee on Cleanrooms.

critical information in a safe and/or remote location is important. The information itself may be on magnetic tapes, disks, paper, or a variety of other media that are susceptible to damage from fire, heat, combustion products, or water. Any backup system should be regularly tested. Compression techniques allow a great deal of media to be stored on very little disk space. It is important that these files can be recovered in a usable fashion.

Mainframe computers are generally installed on raised floors that have electrical wiring and serve as air handling plenums. The floors, including structural support, should be concrete, steel, aluminum, or some other noncombustible material. Where openings to another area under the floor are needed, fire wall separations must be maintained under the floor. Openings on raised floors for cable penetration should be kept minimal to prevent debris and other combustibles from collecting beneath the floor. Carpeting should be avoided on raised floor systems. Carpeting and the adhesive attaching it to the floor could increase the damage to computer equipment involved in a fire. Periodic inspections and maintenance of underfloor space are required to eliminate the accumulation of combustibles. One major problem identified with an underfloor area is that unused wire and cable equipment accumulates as equipment is relocated and updated. This material adds excessive potential fuel for a fire in the form of wire insulation and connectors. A cable management program should be applied to all underfloor areas.

EQUIPMENT AND CONSTRUCTION FEATURES

Power communications and signal cables as well as plugs and connectors should be listed and properly sized and applied in accordance with NFPA 70, *National Electrical Code®*. These should be suitable for installation under raised floors. Floor openings should be smooth to prevent damaged cables. Power cables should run in conduit. No splices or connections should be permitted in the power or wiring under the floor, except within listed junction boxes or approved receptacles or connectors. When used, the number of junction boxes should be minimal.

Cables should be arranged to avoid serious multiple failures because of fault or overheating in any one conduit. Very large computers and some other electronic data processing equipment produce a great deal of heat and require a large amount of air to pass through the equipment for cooling. Air conditioning system design should be based on the requirements of NFPA 90A, *Standard for the Installation of Air Conditioning and Ventilation Systems*. Each computer room should have its own air conditioning system. Air intake should be located to minimize its exposure to flame, smoke, and undesirable fumes. All filters should be Class 1, noncombustible. Duct work and its installation should be noncombustible. The system should be capable of going to 100 percent makeup and exhaust automatically upon detection of smoke. The air handling system should be capable of purging the room of combustion products. Shutting down minimizes recirculation contamination. As technology advances, computers and other electronic systems become more dense and more powerful. As a result, they generate more heat. In plans for upgrading a computer system, cooling requirements may need modification for proper operation. The modern supercomputer may use coolant media other than air, such as water and nitrogen.

Most large computer systems rely on an uninterruptable power source (UPS) for protection against power loss and voltage fluctuation. The UPS is actually a large rechargeable battery system that is normally "on charge" while it provides power to the computer. The UPS has the capacity to keep the system running in the event of a power loss until regular emergency power supplies are restored or systematic shutdown can be accomplished. Electrical power supplies and other electrical equipment, such as transformers, motor generators, distribution controls, and circuit breakers, should have a disconnect switch and be maintained in a separate cutoff room. Larger battery banks should be kept in a separate, well-ventilated detached room. Diesel light or natural or liquid petroleum gas generators for emergency power supplies should be in separate machinery rooms or outside the building.

FIRE PROTECTION

Computers and other electronic equipment are particularly susceptible to fire damage and the accompanying heat, steam, and combustion products. Elevated temperatures may even damage equipment and data. Above 120°F (49°C), permanent damage can begin. At 125°F (52°C), magnetic tapes, floppy disks, and similar materials start to lose information. A sustained ambient temperature of 150°F (66°C) damages hard disks. Equipment components begin to fail at 174°F (79°C), with major component failures occuring between 300 and 500°F (149 and 200°C). Microfilm damage begins at 225°F (107°C) in the presence of high humidity. Even paper products can be damaged at 350°F. Finally, between 650 and 750°F (310 and 399°C), polystyrene cases and reels degrade, producing flammable styrene monomer.

Portable Fire Extinguishers

Computer rooms should be equipped with clean agent, Class BC extinguishers in accordance with NFPA 10, *Standard for Portable Fire Extinguishers*. Dry chemical extinguishers should not be used. If paper or ordinary combustibles are present, a sufficient number of Class A pressurized water extinguishers should also be provided. Occupants need to be instructed on the proper use of extinguishers.

AUTOMATIC SPRINKLER SYSTEM

A fixed automatic sprinkler system should be installed in accordance with NFPA 13, *Standard for the Installation of Sprinkler Systems*. The sprinkler protection should be installed on a hydraulic design basis arranged as a wet pipe or preaction system. With the increased delay time, a double interlock preaction system should not be used. Water distribution density should be evaluated for each installation. Few computer room installations meet the definition of light hazard occupancy, based on the amount of combustibles found in the room. Computer facilities with super printers that often handle a large amount of paper should be considered Group 2, ordinary hazard. Special media storage arrangements could affect sprinkler location and discharge densities.

SPECIAL EXTINGUISHING SYSTEMS

When the value and risk of interruption of computer equipment has been deemed critical, total Halon 1301 systems have been installed in these rooms. Serious environmental concern involving Halon 1301 and its effect on the ozone layer demands consideration of alternate means of protection. Replacement agents now in the market can capably handle fires below raised floors. FM 200® and Inergen® have both been installed in extremely important computer rooms. There is a trend to eliminate the further use of halon. International halon proposals:

- Require that all noncritical halon systems be decommissioned at their first discharge—or routine maintenance—not later than July 1, 1997.

- Prohibit installation of new halon systems after January 1, 1996, except for extremely critical uses.

- Restrict the use of substitutes and replacements that have a global environmental impact. The government is purchasing much of the excess recycled halon for essential military uses. Although halon recycling and banking is still an option, elimination of its controlled use appears to be just a matter of time.

SMOKE DETECTORS

Smoke detection should be installed at the ceiling level and under the raised floor system. Protection devices installed under floor areas should be listed specifically for this application. Air flow characteristics of underfloor plenums make many types of detectors unsuitable. The use of high-efficiency sampling type detectors has proved effective in many cases. Alarms and detection systems should be designed and installed in accordance with NFPA 72, *National Fire Alarm Code* (hereinafter referred to as NFPA 72).

EMERGENCY PLANS AND PROCEDURES

An emergency plan's primary purpose is to prepare key personnel to initiate actions during and after any emergency. Supervisory personnel should develop and regularly review plans and procedures. The plan should incorporate contingency factors for emergencies such as power failures, severe weather conditions, tornadoes, floods, earthquakes, and so on.

The plan should include actions to safeguard systems from damage and provisions to ensure that backup copies of program software and data are available in case damage occurs. There should be a contingency plan to relocate operations or to provide alternate means to complete critical functions if a computer center is temporarily out of service or permanently damaged.

Alternate equipment should be available, either within the organization or from an outside organization, if the operation is to continue during the restoration of damaged equipment. A prearranged plan should consider and test the compatibility of backup equipment.

The plan should include local emergency personnel such as security and/or local fire response organizations and outside fire departments. Local fire departments should be invited to visit the computer and electronic equipment rooms to familiarize them with the emergency plan. It is important that they be aware of any fixed automatic fire extinguishing systems and know how they operate and what procedures to follow, for example, where to enter the building if responding to a fire alarm.

Restoration companies can successfully restore much of the hardware involved in incidents. An effective preplan includes preliminary visits by a restoration company. This will allow them to preplan for the chemicals, manpower, and equipment needed for fast and effective restoration.

The second part of the plan is an emergency personnel evacuation procedure. The primary purpose of this plan is life safety for all persons in the computer room or electronic equipment area. The procedure should be posted and updated periodically—every six months—so that all employees and others entering the area know what to do when the alarm signal sounds. The alarm signal must be audible throughout the entire area, and when it is initiated, all occupants should immediately assemble at a predesignated location, for example, a safe, interior part of an adjacent building. Employees should remain in the assembly area until all are accounted for, and they are advised otherwise. Several persons should be assigned to check their areas to make certain all persons have exited. Laboratories, washrooms, storage rooms, and so on should be checked. The plan should include provisions for those who may need physical assistance. Visible notification appliances should be provided for people who cannot hear the alarm, in accordance with NFPA 72.

Emergency lighting also should be provided for the entire interior of the building that houses the computers or electronic equipment and offices. Auxiliary battery lighting packs are permitted.

BIBLIOGRAPHY

NFPA Codes, Standards, and Recommended Practices

Reference to the following NFPA codes, standards, and recommended practices will provide further information on protecting the electronic equipment discussed in this chapter. (See latest version of *The NFPA Catalog* for availability of the current edition of the following documents.)

NFPA 10, *Standard for Portable Fire Extinguishers*
NFPA 13, *Standard for the Installation of Sprinkler Systems*
NFPA 70, *National Electrical Code®*
NFPA 72, *National Fire Alarm Code*
NFPA 90A, *Standard for the Installation of Air Conditioning and Ventilating Systems*

Additional Reading

Ahola, H., "High Sensitivity Smoke Detectors for the Fire Protection of Computer Rooms: An Alternative to Halons," Third International Symposium on Fire Protection of Buildings, May 10–12, 1990, Eger, Hungary, 1990, pp. 153–162.

Ahola, H., "Performance of Smoke Detectors in Computer Rooms," VTT-Technical Research Center of Finland, Espoo, VTT Research Reports 681, Project PAL6007, May 1990.

Allianz Risk Service, "Fire and Extinguishing Tests on Computer Equipment," *Fire Systems*, Vol. 5, No. 2, 1991, pp. 61–77.

Anschutz, K., "Fire in a Telephone Office," *Firehouse*, Vol. 15, No. 7, July 1990, pp. 92, 94.

Bengtson, S., and Lundin, K., "Heat Release Rates in Telecommunication Tunnels," *Proceedings* of the First International Conference on Fire Science and Engineering, ASIAFLAM '95, March 15–16, 1995, Kowloon, Hong Kong, 1995, pp. 127–138.

Botting, R. E., "Telephone Exchange Fire Study Looks at MDF Hazards," *Fire Engineers Journal*, Vol. 51, No. 163, Dec. 1991, pp. 19–22.

Budnick, E. K., "In Situ Tests of Smoke Detection Systems for Telecommunications Central Office Facilities," International Symposium on Fire Protection for the Telecommunications Industry, May 14–15, 1992, New Orleans, LA, 1992, pp. 1–27.

Cider, L., "Brandskadad elektronik: sanering och tillforlitlighet. BEST-projektet. [Reconditioning Surface Mount Electronics After Smoke Contamination.]," IVF-Insitutet for Verkstadsteknisk Forskning, Stockholm, Sweden, IVF-Internrapport 92/20, 1992.

Cider, L., "Cleaning and Reliability of Smoke-Contaminated Electronics," *Fire Technology*, Vol. 29, No. 3, Third Quarter 1993, pp. 226–245.

Congram, G. E., "Fire Retardancy: New Horizons in the Computer Industry," Recent Advances in Flame Retardancy of Polymeric Materials: Materials, Applications, Industry Developments, Markets, Vo. 2, May 14–16, 1991, Stamford, CT, Business Communications Co., Inc., Norwalk, CT, 1991, pp. 254–261.

Congram, G. E., "Fire Retardant Chemicals in Computer Parts: A New Concern for the Future," Technical and Marketing Issues Impacting the Fire Safety of Electrical, Electronic and Composite Applications, Oct. 20–23, 1991, Coronado, CA, Fire Retardant Chemicals Assoc., Lancaster, PA, 1991, pp. 1–5.

Cooper, R. J., Johnson, P. F., and Timms, G., "Optimizing Sampling Arrangements for High Sensitivity Smoke Detectors in Computer Facilities," *Fire Safety Journal*, Vol. 19, No. 2–3, 1992, pp. 141–157.

Cote, A. E., "Computer and Fire Protection Engineering—25 Years of Progress," *Fire Technology*, Vol. 26, No. 3, August 1990, pp. 195–201.

Duyck, T. O., "Recovery From Fire in a Central Office," International Symposium on Fire Protection for the Telecommunications Industry, May 14–15, 1992, New Orleans, LA, 1992, pp. 1–8.

Faulkner, G., "Fire Detection in Computer Suites is Enhanced by Australian Invention," *Fire*, Vol. 85, No. 1044, June 1992, p. 42.

"Fire Surveillance and Security for Electronic Equipment," *Cerberus Alarm*, No. 112, Sept. 1991, pp. 1–5.

Foehrkolb, J., "Fire Protection in Computer Room Subfloors," *Fire Systems*, Vol. 5, No. 2, 1991, pp. 78–83.

H. H. Angus and Associates Limited, "Special Need for a Smoke Exhaust System to Minimize Secondary Damage to Electronic Telephone Switching Equipment, During a Fire within a Switch Unit," International Symposium on Fire Protection for the Telecommunications Industry, May 14–15, 1992, New Orleans, LA, 1992, pp. 1–24.

Hadley, P., "How Fast Action and Forward Planning Will Reduce Fire Damage to Computers," *Fire*, Vol. 85, No. 1044, June 1992, pp. 22, 24.

Hill, A., "Fire Suppression in Telephone Switching Centers Utilizing Automatic Sprinklers," International Symposium on Fire Protection for the Telecommunications Industry, May 14–15, 1992, New Orleans, LA, 1992, pp. 1–32.

Hills, A. T., Simpson, T., and Smith, D. P., "Water Mist Fire Protection Systems for Telecommunication Switch Gear and Other Electronic Facilities," Fire and Safety International Research, Slough, England, NISTIR 5207, June 1993; National Institute of Standards and Technology, *Proceedings* of the Water Mist Fire Suppression Workshop, March 1–2, 1993, Gaithersburg, MD, 1993, pp. 123–142.

Isner, M. S., "E911 Fails in Telephone Exchange Fire," *NFPA Journal*, Vol. 89, No. 5, Sept./Oct. 1995, pp. 109–112, 115–117, 119–120.

Isner, M. S., "Telephone Exchange Fire, Los Angeles, California, March 15, 1994. Fire Investigation Report," National Fire Protection Association, Quincy, MA, Fire Investigation Report, 1994.

Isner, M. S., "Telephone Switching Station Fire Creates Widespread Emergency," *Fire Journal*, Vol. 84, No. 3, May/June 1990, pp. 72–73, 75–77, 79, 81, 83, 95.

Johnson, P. F., "High Sensitivity Smoke Detection for the Telecommunication Facilities," International Symposium on Fire Protection for the Telecommunications Industry, May 14–15, 1992, New Orleans, LA, 1992, pp. 1–15.

Johnson, P. F., "Risk Assessment in Telephone Exchanges," International Symposium on Fire Protection for the Telecommunications Industry, May 14–15, 1992, New Orleans, LA, 1992, pp. 1–19.

Karydas, D. M., "Probabilistic Methodology for the Fire and Smoke Hazard Analysis of Electronic Equipment," Sixth International Fire Conference on Fire Safety, Interflam '93, March 30–April 1, 1993, Oxford, England, Interscience Communications Ltd., London, 1993, pp. 509–517.

Karydas, D. M., "Probabilistic Methodology for the Fire Smoke Hazard Analysis of Electronic Equipment in Telephone Central Offices," International Symposium on Fire Protection for the Telecommunications Industry, May 14–15, 1992, New Orleans, LA, 1992, pp. 1–8.

Kidde-Fenwal, "Water Mist Fire Protection in Telecommunications Facilities," International Symposium on Fire Protection for the Telecommunications Industry, May 14–15, 1992, New Orleans, LA, 1992, pp. 1–12.

Kroushl, P., "Flame Retardant PVC and Polyethylene-Based Polymer Systems for Electronic Wire and Cable Applications," Recent Advances in Flame Retardancy of Polymeric Materials: Materials, Applications, Industry Developments, Markets, Vol. 4, May 18–20, 1993, Stamford, CT, Business Communications Co., Inc., Norwalk, CT, 1993, pp. 207–206.

Kushler, B. D., Simpson, J., and Budnick, E. K., "Approach to Fire Risk Assessment for Telecommunications Facilities," International Symposium on Fire Protection for the Telecommunications Industry, May 14–15, 1992, New Orleans, LA, 1992, pp. 1–3.

Mangs, J., and Keski-Rahkonen, O., "Full Scale Fire Experiments on Electronic Cabinets," VTT-Technical Research Center of Finland, Espoo, VTT Publications 186, June 1994.

Mawhinney, J. R., "Water-Mist Fire Suppression Systems for the Telecommunication and Utility Industries," *Fire Research News*, No. 74, Fall 1994, pp. 1–3.

McChesney, C. E., "Liquid Crystal Polyesters and Polyphenylene Sulfide Polymers for the Electrical/Electronics Market," Special Conference on Recent Advances in Flame Retardancy of Polymeric Materials—Materials, Applications, Industry Developments, Markets, May 15–17, 1990, Stamford, CT, 1990, pp. 1–8.

Meacham, B. J., "Factors Affecting the Early Detection of Fire in Electronic Equipment and Cable Installations," International Symposium on Fire Protection for the Telecommunications Industry, May 14–15, 1992, New Orleans, LA, 1992, pp. 1–38.

Nishikata, K., "Description of Telephone Cable Fire in a Cable Tunnel and the Response of the Fire Force," Tokyo Fire Dept., Japan, CIB W14/95/1 (J); Tokyo Fire Department, Firesafety Frontier '94, International Fire Conference and Exhibition in Tokyo, Creating a Safe Tomorrow, October 18–22, 1994, Tokyo, Japan, 1994, pp. 75–80.

Notz, R., "Fire Protection Concept for EDP and Telecommunication Systems," International Symposium on Fire Protection for the Telecommunications Industry, May 14–15, 1992, New Orleans, LA, 1992, pp. 1–15.

O'Connor, D. J., and Nugent, D. P., "Achieving Zero Downtime in Telecommunication Facilities," *Consulting-Specifying Engineer*, Vol. 19, No. 5, May 1996, pp. 32–34, 36–39.

Peck, L., "Selective Automatic Fire Extinguishers for Computers (SAFE COMP)," Air Force Joint Technology Applications Office, Wright-Patterson AFB, OH, ALD/JTT-TT-91074; XF-ALD-JTT, Aug. 2, 1991.

Peterson, D., and Schwalbe, K., "Preventing Fire Disasters in Computer Rooms," *Consulting/Specifying Engineer*, Vol. 8, No. 3, Sept. 1990, pp. 96–98, 100, 102.

"Post-Halon Computer Protection," *Record*, Vol. 70, No. 5, Nov./Dec., pp. 3–9.

Price, B. J., "Computer Room Fire Protection," *Library Hi Tech*, Vol. 29, No. 1, April/June 1990, pp. 43–56.

"Protecting High-Value Computer Systems," *Fire Prevention*, No. 228, April 1990, pp. 17–20.

"Protecting Telephone Central Offices," *Record*, Vol. 70, No. 4, July/Aug. 1993, pp. 13–14.

"Protection for Computers Without Using Halons," *Fire*, Vol. 87, No. 1072, Oct. 1994, pp. 35–36.

Ramachandra, S., *et al.*, "Improved Processability in Reduced Emission Wire and Cable Materials for the Telecommunication Industry," Union Carbide Chemicals and Plastics Company, Inc., Somerset, NJ, U. S. Army Communications Electronics Command (CECOM), Thirty-ninth International Wire and Cable Symposium, Nov. 13–15, 1990, Reno, NV, 1990, pp. 655–660.

Reagor, B. T., "Impact of Smoke Corrosivity on Electronic Equipment," Fire Safety Developments and Testing: Toxicity—Heat Release—Product Development—Combustion Corrosivity, October 21–24, 1990, Ponte Vedra Beach, FL, Fire Retardant Chemicals Assoc., Lancaster, PA, 1990, pp. 205–213.

Shaw, J. S., "Flammability Standards for Central Office Equipment," International Symposium on Fire Protection for the Telecommunications Industry, May 14–15, 1992, New Orleans, LA, 1992, pp. 1–6.

Simmons, F., "Computer Room Protection—The Debate Continues," *Fire Prevention*, No. 232, Sept. 1990, pp. 19–21.

Simpson, J., Kushler, B., and Schubert, R., "Fire Calorimetry for Telecommunications Equipment," Second International Conference on Fire and Materials, Sept. 23–24, 1993, Arlington, VA, 1993, p. 263.

Simpson, T., and Smith, D. P., "Fully Integrated Water Mist Fire Suppression System for Telecommunications and Other Electronics Cabinets," Fire and Safety International, UK, SP Report 1994:03; Swedish National Testing and Research Institute, *Proceedings* of the International Conference on Water Mist Fire Suppression Systems, Nov. 4–5, 1993, Boras, Sweden, 1993, pp. 153–166.

Simpson, T., and Smith, D. P., "Suppressing Fires in Telecommunications Switchgear," *Fire Prevention*, No. 267, March 1994, pp. 17–20.

"Telecommunications Symposium. Conference Report," *Fire Technology*, Vol. 28, No. 3, Aug. 1992, pp. 280–282.

Todd, C., "Safeguarding Computer Installations From Fire," *Fire Prevention*, No. 256, Jan./Feb. 1993, pp. 20–24.

U.S. Department of the Air Force, "Engineering Technical Letter (ETL) 93-5: Fire Protection Engineering Criteria —Electronic Equipment Installations," Department of the Air Force, Tyndall AFB, FL, ETL 93-5, Dec. 22, 1993.

Wadehra, I. L., "Flame-Retarded Materials for Computer and Business Machine Applications," Special Conference on Recent Advances in Flame Retardancy of Polymeric Materials—Materials, Applications, Industry Developments, Markets, May 15–17, 1990, Stamford, CT, 1990, pp. 1–5.

Yokel, F. Y., "Earthquake Resistant Construction of Electrical Transmission and Telecommunication Facilities Serving the Federal Government," National Institute of Standards and Technology, Gaithersburg, MD, Federal Emergency Management Agency, Washington, DC, NISTIR 89-4213, FEMA-202, Earthquake Hazard Reduction 56, Feb. 1990.

ELECTRIC GENERATING PLANTS

Leonard R. Hathaway

Today's society increasingly relies on the socioeconomic benefits of electric energy for residential, commercial, and industrial applications. Over the years in our everyday lives, our dependence on electrically powered labor-saving devices has escalated. In 1940 average annual U.S. consumption was about 1,100 kWh per person, and in the early 1990s consumption rose to 11,000 kWh per person. In the next 50 years, consumption is expected to rise to 25,000 kWh per person.[1] Industry turns toward electrification in new ways to increase efficiency and control costs in a highly competitive global marketplace. Electrification of developing countries is an important goal in improving the standard of living and attracting investment in these countries.

The availability and reliability of the electrical supply impacts our lives in subtle ways. These include safety, security, and the creation of employment opportunities. Certainly, today's information-hungry, service-oriented, and computer-oriented society could not function without an adequate and reliable electrical supply.

The process of producing and delivering electrical energy to customers includes three distinctive elements: generation, transmission, and distribution. This chapter addresses fire protection for generation of electricity from a variety of fuels. One section discusses high voltage direct current converter stations. Section 9, Chapter 30, "Nuclear Facilities," discusses some unique fire protection challenges of nuclear fueled generation. Note that many hazards and fire protection concepts discussed in this chapter may apply to all means of electric generation, including nuclear.

Over 10,000 electric generating units operate in the United States.[2] Over the years the generation industry has scaled up: in the 1920s units had a capacity of about 20MW. Plants designed in the last two decades have capacities over 1,000 MW. Capacity expanded most rapidly in the 1950s and 1960s as technological advancements enabled production of larger and more efficient turbines, generators, boilers, transformers, etc., with accompanying lower capital, operating, and maintenance costs per megawatt. An electric generating plant is capital intensive, with the cost spread over its 40 to 50 year life.

The increase in energy prices in the early 1970s changed the economic structure of the utility industry. Driven by inflation, regulation, and environmental concern, the cost of building electric generating units began to escalate rapidly. Rising costs undermined traditional economies of scale, requiring optimization of plant size and configuration to control costs. Therefore, many recently constructed electric generating plants are smaller than those constructed from the 1960s to the 1980s.

The U.S. electric utility industry is changing profoundly as it takes the first steps down the path of deregulation. Changes began with the passage of the Public Utility Regulatory Policies Act of 1978 (PURPA), which created a new class of preferred power producers. The act intended to encourage alternative sources of electric generation. As a result, a host of newly formed entrepreneurial enterprises erected electric generating plants to compete with traditional utilities that were required to purchase their electrical output. These new non-utility generators became known as independent power producers.

In 1994 independent power producers generated 10 percent (60,000 MW) of U.S. power.[3] In the future, independent power producers will construct a considerable portion of the new capacity, traditional utilities will construct the balance.

Continuing increase in competition among electric producers will elevate reliability issues in the eyes of the stakeholders in future years. Avoiding forced outages will become more important in a deregulated environment. Good fire protection practice will contribute to the reliability of an electric generating plant, operator safety, and protection of physical assets.

OWNERSHIP AND FUELS FOR ELECTRIC GENERATING PLANTS

In the U.S., electric energy is produced by investor-owned utilities (71 percent), publicly owned utilities (8 percent), federally owned utilities (7 percent), cooperatives (4 percent), and nonutility generators (10 percent).[4]

In its most basic form, a fossil fuel electric generating plant converts fuel to produce steam to spin a generator to produce electrical energy. A hydroelectric plant uses kinetic energy to spin a generator for similar purposes.

In 1993 the means used to produce electric energy in the U.S. were coal (56.9 percent), oil (3.5 percent), natural gas (9.0 percent), nuclear (21.2 percent), hydro (9.2 percent), and alternative fuels (0.3 percent).[4] This chapter addresses the fire-related hazards for each fuel, as well as hazards unique to electric generating plants. All references to monetary damage in this chapter are in 1994 U.S. dollars.

Leonard R. Hathaway is senior vice president of M&M Protection Consultants, Atlanta, GA.

MANAGEMENT POLICY AND DIRECTION

Best practice guidance is available in NFPA 850, *Recommended Practice for Fire Protection for Electric Generating Plants and High Voltage Direct Current Converter Stations,* and NFPA 851, *Recommended Practice for Fire Protection for Hydroelectric Generating Plants* (hereinafter referred to as NFPA 850 and 851, respectively). These documents provide information on the design, construction, operation, and protection of various types of electric generating plants.

A key element in a well-protected plant occurs when management establishes a policy and follows through on a program to protect lives, conserve property, and ensure continuity of operations. For an operating electric generating plant this includes a written plant fire prevention program (i.e., control of combustibles, housekeeping, hotwork controls, etc.); a fire emergency plan; testing, inspection, and maintenance programs; and related policies and procedures.

When an electric generating plant is being designed or undergoing changes, a Fire Risk Evaluation should be initiated to assure that all fire hazards are identified and appropriate strategies are established to cope with these hazards. The Fire Risk Evaluation should address the fire prevention and fire protection concepts offered in NFPA 850 and NFPA 851.

Management's decision on the establishment of a plant fire brigade should be made after considering a number of variables, including requirements specified by the U.S. Occupational Safety and Health Administration (OSHA). Some electric generating plants are located in areas remote from immediately available professional or volunteer fire departments. For these plants the establishment of a fire brigade may be necessary. The trend in the generating industry today is to reduce the size of the plant operating staff. Obviously, this directly impacts the pool of qualified personnel who can perform fire brigade duties. Some smaller plants operate unmanned or with as few as three people on a shift.

In all cases a workable fire emergency plan needs to be developed. Fire fighters (plant fire brigade or local fire department) who are expected to respond to an electric generating plant should be trained in the unique hazards that they could face under fire conditions.

GENERAL PLANT DESIGN

The "power block" typically houses the steam generator (boiler-furnace), turbine-generator, and auxiliary equipment. Electric generating plants are usually large open structures, perhaps several stories high. (See Figure 9-29A.) Building construction is typically of heavy structural steel frame with insulated metal panel walls in cold climates. In warm climates the structure may be open, including the turbine-generator operating floor.

Some hydroelectric and pumped storage plants may be located the equivalent of several stories below grade in reinforced concrete and unprotected steel structures. Generators may be housed in a single undivided building or structure.

Life Safety

Life safety design is a critical component of the overall design of a new electric generating plant. In general, locally adopted codes govern life safety design for new construction; however, at a minimum the design should meet the requirements of NFPA *101®, Life Safety Code®.* NFPA 850 provides guidance for applying NFPA *101* requirements to specific areas of electric generating plants.

Other standards impacting the design of new domestic electric generating plants are OSHA and the Americans with Disabilities Act (ADA). All design criteria should be approved by the Authority Having Jurisdiction, including identification of ADA "accessible" areas. These areas typically include administration and office areas not associated with plant operations.

Electric generating plants are unique structures due to the equipment arrangement required for their operation and maintenance. These equipment requirements often present challenges in designing for life safety and egress.

Large enclosed boiler buildings can be up to 250 ft (76 m) high and typically consist of multiple grating levels providing access to equipment for maintenance, inspection, testing, and repairs. This arrangement requires a continuous vertical opening up the boiler and, in the event of fire, presents a potential smoke hazard to any personnel performing these operations.

In emergency situations, performance of emergency shutdown procedures often delays egress of control room personnel. Therefore, control facilities should have a means of egress separate from other plant areas.

At many existing plants the control room's location has the disadvantage of exposure to smoke, heat, and fire if it is too close to the turbine-generator. Adequate fire barriers and egress facilities are required. Fire can trap control room personnel, with lethal consequences, or block them so that escape is achieved only with difficulty. (See Figure 9-29B.)

NFPA 850 recommends special hazards protection in turbine-generator buildings and boiler buildings in lieu of a more restrictive

FIG. 9-29A. Aerial view of typical fossil fuel electric generating unit.

FIG. 9-29B. Partial view of a modern control room. (Source: Duke Power Company)

philosophy of complete automatic sprinkler protection. In general, this philosophy is accepted as an equivalent to complete sprinkler protection, allowing extended travel distance, common path of travel, and dead-end limits in accordance with NFPA *101*. The Authority Having Jurisdiction should review and approve this interpretation before designing for life safety and egress.

Water Supplies

An adequate and reliable water supply should be provided. For many electric generating plants this is typically two independent supplies, each sized to meet the simultaneous demands of the hydraulically most demanding sprinkler system and 500 gal/min (1,900 L/min) for hose streams for 2 hr. Older plants have combination service water-fire protection systems, while newer plants have dedicated fire protection water supplies and underground fire water delivery systems.

Site conditions dictate the supply source. Rivers, lakes, cooling tower basins, cooling lakes and ponds, suction tanks, and municipal water supplies have all been used successfully to meet fire protection needs.

Fuel Hazards

Fuels utilized at an electric generating plant have unique hazards that the Fire Risk Evaluation should address.

Coal: Today's coal-fired plants typically consume bituminous coal, sometimes called Eastern coal, and subbituminous coal, sometimes called Western coal. Lignite is another type of coal available in certain geographic areas. For environmental reasons many electric generating plants burn lower sulfur coals. Coal's principal hazards are combustibility, spontaneous ignition, and propensity to create fugitive dust. All should be carefully considered at the outset of any new project or coal conversion project.

Coal is transported by a variety of means and stored or "stacked out" in open coal yards. (See Figure 9-29C.) For economic reasons the trend is to limit storage to 30–45 days' supply, whereas in the past storage of over 90 days' supply was common. Coal stored in open piles should be compacted to prevent spontaneous heating. Coal pile fires are rarely the cause of a major fire exposure to the electric generating plant. Additional information on coal can be found in Section 3, Chapter 24, "Storage and Handling of Solid Fuels."

A coal handling system moves the coal from the storage pile to the boiler-furnace where it is consumed to produce steam. The typical coal conveyor belt is combustible and can ignite from friction between the conveyor belt and the roller, belt slippage, or static electricity. Uncontrolled hotwork has also caused fires in coal handling areas. Housekeeping and dust control are vitally important. An automatic sprinkler system is needed to protect areas above and below the conveyor belt. Fire retardant conveyor belts are recommended, but they burn under certain circumstances and therefore cannot substitute for automatic sprinkler protection. Electrical equipment in coal handling areas should meet NFPA 70, *National Electrical Code*®, criteria in Article 500 for Class II, Division 1 or 2, Group F occupancies. Crusher buildings, transfer buildings, and tripper areas (galleries) are all considered parts of the coal handling system that require automatic sprinkler protection.

The largest coal handling fire in the U.S., identified in a 30-year study of large-loss fires, occurred in a crusher building because of a serious dust condition. An unknown ignition source caused an explosion that engulfed the unoccupied crusher house. The ensuing fire burned several thousand feet of conveyor belt, causing widespread damage. The result was several injuries, $27 million in property damage, and a 4-month outage. There were no fixed automatic fire suppression systems.[5]

In-plant coal storage may be in silos (600 to 1,000 tons) or bunkers (up to 5,000 tons). During plant outages the coal should be emptied to preclude spontaneous ignition. Coal fires can lead to explosions and injuries and are difficult to extinguish. These fires can last for several days or even weeks. NFPA 850 discusses various strategies to deal with these fires.

A pulverized coal system has significant explosion potential. Applicable references are NFPA 8503, *Standard for Pulverized Fuel Systems*; and Electric Power Research Institute Project 1883-1, February 1987, *Prevention, Detection, and Control of Coal Pulverizer Fires and Explosions.*

Natural gas: Beginning in the 1980s and continuing through the 1990s, natural gas has become the fuel of choice for much of the electric generating capacity that has or will come on-line. This is particularly true for combustion turbines operating in either simple-cycle or combined-cycle configurations. The gas supply piping system should be designed, installed, and maintained in accordance with NFPA 54, *National Fuel Gas Code*, and NFPA 58, *Standard for the Storage and Handling of Liquefied Petroleum Gases.*

Oil: Very little oil-fired capacity is being added to the U.S. generation mix for economic reasons. A number of areas in the U.S. continue to depend on oil-fired capacity—particularly on the West and East coasts. Also note that many coal-fired plants use oil for ignition. NFPA 30, *Flammable and Combustible Liquids Code,* and NFPA 31, *Standard for the Installation of Oil-Burning Equipment,* should be followed.

Fuel oil is often heated before it is pumped to the boiler burner tip. Heating equipment may be installed inside the power block. Pipe failure and human error can result in fires involving pumping and heating equipment that expose vital equipment. NFPA 850 calls for automatic fire suppression systems to protect the hazard and a minimum 2-hr fire barrier to isolate it.

Alternative fuels: NFPA 850 defines alternative fuels as refuse-derived fuel, municipal solid waste, and biomass. The latter includes wood chips, wood waste, rice hulls, and sugar cane stalks, among others. Each fuel presents a unique hazard and needs to be addressed as part of the Fire Risk Evaluation. For additional information refer to Section 3, Chapter 24, "Storage and Handling of Solid Fuels."

FIG. 9-29C.   Coal conveyor.

Boiler-Furnaces

The steam generator or boiler-furnace presents a major exposure due to the potential for a fire box explosion. At the burner, an uneven or low fuel flow can result in loss of flame in a portion of the boiler. Continued introduction of fuel can result in accumulated raw fuel that another burner or hot surface can ignite. These rare events can be very destructive. Between 1977 and 1993, eight explosions alone resulted in an average of $41 million in property damage, with forced outages at a rate of about one year.[5] Undoubtedly, there have been numerous "puffs" or explosions with relatively minor damage.

The NFPA Technical Committee on Boiler Combustion System Hazards develops guidance on preventing explosions and implosions through properly controlling boiler operations. See NFPA 8501, *Standard for Single-Burner Boiler Operations,* and NFPA 8502, *Standard for the Prevention of Furnace Explosions/Implosions in Multiple-Burner Boilers.*

Burner front fires are a significant fire hazard associated with boiler-furnaces. These occur when oil inadvertently released from the fuel system or ignition system ignites on the hot surface of the boiler's exterior face. The oil pumping system feeds the fire, which may extend several stories on the boiler's entire vertical face. NFPA 850 recommends fire suppression systems to protect the fuel and ignition piping, and igniter and burner areas within a 20-ft (6.1-m) horizontal distance from the boiler face on all firing levels.

Flue Gas Treatment

Environmental requirements have become more stringent in an effort to control the release of sulfur dioxide, nitrogen oxide, and particulate matter—particularly from coal-fired plants. To deal with this challenge, the flue gas stream is erected with equipment that meets environmental requirements. This equipment includes flue gas desulfurization systems (scrubbers), electrostatic precipitators, fabric filters (bag houses), and chimneys to release the off-gases. All have unique fire hazards that are addressed in NFPA 850.

Of particular concern is the flue gas desulfurization system, which may contain large quantities of plastics and other combustibles, particularly duct linings for corrosion control purposes. Some chimneys erected with combustible materials have presented challenging fire situations.

Steam Turbine-Generators

A study of 20 significant turbine-generator fires (See Figure 9-29D.) that occurred worldwide over 10 years at both fossil and nuclear power plants found four fatalities, several injuries, and an average of $42 million in property damage with an average forced outage of

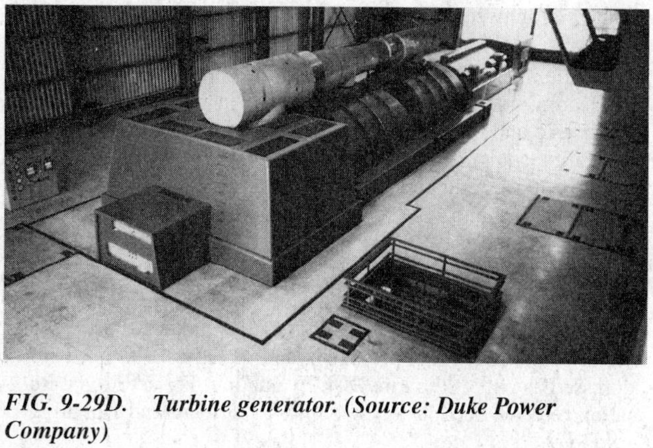

FIG. 9-29D. Turbine generator. (Source: Duke Power Company)

200 days. Two fires led to the permanent decommissioning of plants. Extensive vibration, overspeed, instrumentation failure, human error, blade failure, and filtration failure caused these fires.[6]

The heart of the fire challenge is the pressurized lubrication oil system that uses a Class III B combustible liquid with a flashpoint ranging from 400 to 500°F (204 to 260°C) and an autoignition temperature of 700°F (371° C). Lubrication oil is released if the lubrication oil piping system is breached. Ignition may occur due to a hot surface, which can cause a spray fire, three-dimensional fire, pool fire, or a combination of all three, producing large quantities of smoke and heat. Turbine-generators can take from 20 to 40 min to coast down. Turbine shaft-driven pumps shut down at 90 to 95 percent of rated speed. However, motor-driven backup pumps are designed to continue to supply lubricating oil while the shaft rotates and therefore they could feed oil to a fire.

Strategies to deal with a turbine-generator fire are outlined in NFPA 850 and Electric Power Research Institute Project 1843-2, July 1985, *Turbine-Generator Fire Protection by Sprinkler System.* Automatic sprinklers and heat detectors should be provided over all bearings. All lubrication oil piping, including instrumentation lines above the turbine operating floor and its lagging or skirt area, should be protected by automatic sprinklers. Protecting areas where oil can spread beneath the turbine-generator with automatic sprinklers is vitally important. The extent of coverage is critical and should include areas beneath the mezzanine and in condenser pit areas.

Fire suppression is also recommended for the oil storage reservoir and filtration equipment. Ideally, this equipment should be installed in a room with a minimum fire barrier rating of 2 hr.

Turbine-generator fires can also expose the turbine hall roof to excessive heat exposure and possible structural failure. There is no requirement for automatic sprinklers throughout the roof area of a large turbine hall, but roof level automatic sprinklers may be critical for newer, more compact plants without solid turbine operating floors. Smoke and heat venting is important in preventing structural roof failures.

Another potential hazard is the hydraulic control system if it is common with the lubrication oil discussed above. These systems operate at high pressures (up to 1600 psi) and, therefore, can result in spray oil fires. Newer units, particularly those built after the early 1970s, utilize a separate electro-hydraulic control system that has nearly eliminated this hazard by using a less flammable synthetic fluid such as a phosphate ester.

Concentric or guarded piping for lubrication oil systems is recognized as good design practice. It does not, however, replace the automatic sprinkler systems discussed earlier. Fire experience shows that smaller lines that cannot be practically placed in concentric piping have been known to fail. In one case, an instrumentation line failure caused $16 million in damage. In another case, oil escaped from a concentric pipe, resulting in a $15 million fire, while in yet another case, human error caused oil to be released from a filter system, resulting in a $45 million fire.[5]

The vast majority of generators are hydrogen cooled. In North America hydrogen is received in bulk and may be located on tube trailers or in high-pressure cylinders. Some foreign plants produce hydrogen on site. Hydrogen can contribute to turbine-generator fires and cause explosion damage, expected to be minor in comparison to the more severe fire damage associated with turbine lubrication oil. Hydrogen explosions inside the generator are considered unlikely. Rugged generator casings have minimized the amount of damage.

Hydrogen systems should conform to NFPA 50A, *Standard for Gaseous Hydrogen Systems at Consumer Sites,* and NFPA 50B, *Standard for Liquefied Hydrogen Systems at Consumer Sites.*

Hydrogen seal oil systems should also be protected by automatic sprinklers.

Historically, one reason for not providing fire protection for turbine-generators was the belief that water released from the fire protection system would cause thermal shock to turbine components and auxiliary equipment. The previously mentioned Electric Power Research Institute (EPRI) report studied this subject.[7] The study focused on utility experience with respect to the application of water around hot turbine parts, including both intended and inadvertent fire protection system activation. Of the 24 events studied, the application of water caused no significant damage. No cases of thermal shock have been reported in the decade since the Electric Power Research Institute report's release. A dissertation also examined this issue: it corroborated the EPRI report and found that fire risk far exceeds any damage that a fire protection system causes.[8]

Combustion Turbines

Traditionally, combustion turbines or gas turbines were used in peaking applications during high demand periods, usually in the summer. The bulk of the peaking capacity was installed during the decade following the 1965 Northeast blackout. These machines are typically in the 10 to 40 MW range.

The advent of independent power producers, coupled with advances in gas turbine technology and the availability of inexpensive natural gas, has led to a worldwide population of 5,500 units.[9] Further, the size of today's machines has been scaled up to as much as 160 MW. In combined-cycle installations, high efficiencies ranging from 54 percent to 58 percent can be obtained.[10]

Combustion turbines are typically purchased as a package, and the seller, not the buyer, usually determines the type of fire suppression system. The overwhelming majority have been protected with carbon dioxide or halon systems. Fear of thermal shock is the chief reason for avoiding water-based systems. Currently, there is interest in protecting these machines with the emerging fine water spray technology.

Gas turbines are frequently installed in combined-cycle facilities utilizing heat recovery steam generators (HRSG). Steam produced by a HRSG drives a steam turbine with the same fire hazards as described earlier. The NFPA Technical Committee on Boiler Combustion System Hazards has produced NFPA 8506, *Standard on Heat Recovery Steam Generator Systems.*

Hydroelectric Plants

Hydroelectric plants are addressed in NFPA 851. Hydroelectric plants lack many fire hazards addressed above with respect to combustible fuels and large pressurized lubrication oil systems that operate under high pressure. Nevertheless, hydroelectric plants do present unique fire protection challenges. Among them are life safety (particularly when facilities are several hundred feet below grade), oil-filled cable, grouped electrical cable, minimal staffing or unattended operations (even in plants as large as 1,300 MW), transformer exposure, hydraulic control and lubrication oil systems for penstock or wicket gates, and generator windings.[11]

One of the largest hydroelectric plant fires occurred in 1981. A fault in an oil-filled cable led to a hot smoky fire, complicated by difficult access into the cable tunnel. An inadequate water supply hampered fire fighters. Property damage was $29 million, and the 700-MW unit was out of service for one year.[5]

Electrical Cable

Electric generating plants use insulated, jacketed cable for power, instrumentation, and control. Cables are typically routed in open cable trays throughout the power block. Plants built in the 1960s through the 1990s used significant quantities of cable. Future electric generating plants may use less cable due to computer advances and changes in technology, including the use of fiber optic cable.

Many electric generating plants are traditionally arranged to concentrate cable in areas immediately below the control room. The arrangement may typically include a control room at the top level with a cable spreading room below and a burner management room below that, with motor control centers and relay rooms on lower elevations. All these areas should be erected within walls and floor/ceiling assemblies of 2-hr fire resistance and with suitably protected openings.

Grouped electrical cables constructed of organic materials present fire hazards of various magnitude. A few very severe fires involved PVC jacketed cable, including the 1975 fire in the cable spreading room at the Browns Ferry Nuclear Power Plant. Electricians used an open candle flame to check a cable penetration seal made of polyurethane. The flame accidentally ignited the sealant material in the wall between the cable spreading room and reactor building. Fire then spread to the grouped electrical cable system. No serious injuries resulted, but the fire caused $227 million in property damage, a $1^1/_2$ year outage for the two 1,065 MW units, and worldwide reappraisal of fire protection for nuclear power plants.[5]

At a foreign fossil-fueled plant, an electrical fault ignited the cables, consuming 5 tons of PVC in 10 hr and causing $33 million in property damage. The plant was out of service for nearly one year.[5]

NFPA 850 recommends installation of cable that passes the flame propagation test of IEEE 383. The insulation and/or jacketing of many newer cables are of materials that are difficult to ignite. However, under certain circumstances even this cable burns. A Fire Risk Evaluation should be performed to determine the need for automatic fire sprinkler systems in areas with significant amounts of grouped electrical cable.

Transformers

Electric generating plants utilize both indoor and outdoor transformers. Usually indoor transformers are relatively small, air insulated, and used for individual plant service. Larger, oil-cooled transformers used for main generator voltage step up and for station service are located outdoors. These transformers are usually grouped in an area close to the generator to minimize the length of the isolated-phase bus duct from the generator leads to the main power transformer. The amount of cooling oil can range from 20,000 to 25,000 gal (7.6 to 9.5 km^3). Internal high energy faults can be violent, expending tremendous power, rupturing the transformer casing, releasing oil (and possibly missiles), and subsequently igniting.[12]

For best practice, installation should include lightning protection, a drainage system, fire barriers, and automatic water spray systems. (See Figure 9-29E.) The latter should be designed according to NFPA 15, *Standard for Automatic Water Spray Fixed Systems for Fire Protection.*

NFPA 850 provides guidance on transformer, spatial separation, and fire walls. For example, transformers with an oil capacity of 500 to 5,000 gal (1.9 to 19 m^3) should be separated from adjacent transformers and from buildings by a minimum distance of 25 ft (7.6 m) or by a fire wall. NFPA also provides criteria for the fire resistance and height of fire walls.

The purpose of an automatic water spray system is exposure protection. The benefit of such systems under catastrophic conditions has been debated. Some contend that the violent nature of a transformer failure renders automatic water spray systems useless. Although this may have occurred, in many cases the automatic water

FIG. 9-29E. *Water spray system protecting a power transformer.*

spray systems did survive explosions and were credited with controlling fire, limiting damage, and minimizing plant downtime.

High-Voltage Direct Current Converter Stations

A high-voltage direct current converter station and transmission line effectively control electrical power systems that may not be synchronized, such as those between the United States and Canada. High voltage direct current transmission links also transmit large blocks of power more efficiently over long distances. At one end of the line, the converter station converts alternating current, and at the other end, the converter station functions as an inverter, changing power back to alternating current for consumption by the local grid.

A 30-year study of the approximately 75 converter stations in the world identified about 30 minor fires and 3 major fires between 1989 and 1993.[5] The major fires caused property damage of $23 million, $25 million, and $30 million, respectively. First utilized in these facilities, mercury arc technology has now been replaced by thyristor technology. All three major fires have been in thyristor valves. The valves are stacked in an array of energized electrical components with combustible dielectric fluids and plastics. The stacks can be suspended from the ceiling or mounted on the floor.

This technology presents a unique fire problem. Under normal shutdown conditions, isolating and properly grounding the equipment may take up to an hour. Under emergency conditions this can be done more rapidly, but a delay can be expected in manual fire fighting operations. High temperatures at the ceiling or thermal shock from fire water may cause the valve stack supports to fail. Major damage can then be expected from smoke and contamination of sensitive electronic equipment.

Air sampling smoke detection systems should be considered as a means to detect smoke in a fire's incipient stage. The use of automatic sprinklers is expected to be limited. There is a need for early warning fire detection and a pre-planned fire fighting strategy.

Summary

This chapter addresses conventional means of producing electrical energy from fossil fuels, biomas, and hydro sources. Other technologies, such as geothermal, solar, wind, and fuel cells, also exist. A variety of economic and technological reasons limit their contribution to the total electric production mix. It is highly likely that technological advances will bring forth new and unknown electric production technologies.

As the demand for electricity increases, there will be a need to build new electric generating plants, to renovate older plants, and to retire and replace plants that have outlived their usefulness. The trend is to deregulate the electric utility industry in the U.S. and to privatize in many foreign countries. All these issues increase the importance of fire protection for electric generation plants globally.

BIBLIOGRAPHY

References Cited

1. "Electricity Demand Poised for Next Stage of Growth," *Power Engineering*, February 1994.
2. *EIA Inventory of Power Plants in US*, 1993.
3. Smok, R. W., "Repeal PURPA's Mandatory Purchase Requirement," *Power Engineering*, May 1995, p. 11.
4. *Electric Power Annual*, National Energy Information Administration, 1993.
5. Hathaway, L. R., *A 30-Year Study of Large Losses in the Gas and Electric Utility Industry*, fifth edition, M&MPC Protection Consultants, March 1995.
6. Hathaway, L.R., "Don't Be Caught Off-Guard When It Comes to Fire Protection," *Power*, July 1994.
7. "Turbine-Generator Fire Protection by Sprinkler System," Report No. NP-4144, July 1985, Electric Power Research Institute, Palo Alto, CA.
8. Drewry, D., "A Quantitative Analysis of the Thermal Effect of Water Spray Protection When Applied on Hot Turbine Components," Ph.D. Worcester Polytechnical Institute, January 1986.
9. Natole, R., "Gas Turbine Components—Repair or Replace," *Global Gas Turbine News*, May/June 1995.
10. Kuehn, S. E., "Advancing Gas Turbine Technology: Evaluation and Revolution," *Power Engineering*, May 1995, p. 25.
11. Hathaway, L. R., "Recommended Practice for Hydroelectric Plant Fire Protection," M&M Protection Consultants: *Power Engineering*, February 1988.
12. Biggins, J. B., "Protect Transformers from Fire Hazards," M&M Protection Consultants, *Electrical World*, August 1993.

NFPA Codes, Standards, and Recommended Practices

References to the following NFPA codes, standards, and recommended practices will provide further information on electric generating plants discussed in this chapter. (See the latest version of *The NFPA Catalog* for availability of current editions of the following documents.)

NFPA 15, *Standard for Water Spray Fixed Systems for Fire Protection*
NFPA 30, *Flammable and Combustible Liquids Code*
NFPA 31, *Standard for the Installation of Oil-Burning Equipment*
NFPA 50A, *Standard for Gaseous Hydrogen Systems at Consumer Sites*
NFPA 50B, *Standard for Liquefied Hydrogen Systems at Consumer Sites*
NFPA 54, *National Fuel Gas Code*
NFPA 58, *Standard for the Storage and Handling of Liquefied Petroleum Gases*
NFPA 70, *National Electrical Code®*
NFPA 101®, *Life Safety Code®*
NFPA 850, *Recommended Practice for Fire Protection for Electric Generating Plants and High Voltage Direct Current Converter Stations*
NFPA 851, *Recommended Practice for Fire Protection for Hydroelectric Generating Plants*
NFPA 8501, *Standard for Single-Burner Operation*
NFPA 8502, *Standard for the Prevention of Furnace Explosions/Implosions in Multiple Burner Boilers*
NFPA 8503, *Standard for Pulverized Fuel Systems*
NFPA 8506, *Standard on Heat Recovery Steam Generator Systems*

Additional Reading

Adcock, R. C., *et al.*, "Linking Fire Protection Choices with Clean Air Act Compliance," *Power Engineering*, Mar. 1992.
"Belt Conveyors," Loss Prevention Data 7-11, Factory Mutual Engineering Corp., Norwood, MA.
Biggins, J. B., "Fire Protection Considerations When Switching Fuels," *Power Engineering*, Aug. 1994.

Chingo, S. J., "Report on the NFPA HVDC Task Group," International High Voltage Direct Current Conference, Apr. 1995, Burnaby, British Columbia, Canada.

Clayton, T. C., "Air Preheater Fire Protection for the Rotary Regenerative Air Preheater," WATTec '79 Energy Conference, Knoxville, TN.

Clayton, T. C., "Minimizing Coal Pulverizer Fires and Explosions Symposium on Coal Pulverizers," Nov. 1985, Electric Power Research Institute, Denver, CO.

Clayton, T. C., and Hall, D. T., "Turbine-Generator Fire Protection Overview," 12th Annual WATTec Energy Conference, Feb. 12–15, 1985, Knoxville, TN.

"Fire Protection for Steam Turbines and Electric Generators," Loss Prevention Data 7-101, Factory Mutual Engineering Corp., Norwood, MA.

Grant, C. C., "Newark Cogeneration Facility Fire," NFPA Special Report, Quincy, MA.

Hathaway, L. R., "Nuclear Power Plant Turbine Hall Fires," Fire Protection and Prevention for Nuclear Facilities Conference and Exhibition, Dec. 5–7, 1994, Barcelona, Spain.

Hathaway, L. R., "Turbine-Generator Fire Prevention Programs Cut Costs," M&M Protection Consultants: *Power Engineering*, Feb. 1989.

"PURPAS Power," *Wall Street Journal*, May 19, 1995.

Prasso, C. R., "Electrical Systems and Sprinklers Do Mix," *NFPA Journal*, Vol. 87, No. 6, Nov./Dec. 1993, pp. 31, 120.

"Transformers," Loss Prevention Data 5-4/14-8, Factory Mutual Engineering Corp., Norwood, MA.

Werley, D. R., "Power Transformer Fire Protection," WATTec 80 Energy Conference, Feb. 20, 1980, Knoxville, TN.

Edison Electric Institute

References to various good practice reports developed by the EEI Fire Protection Committee will provide further information on electric generating plants discussed in this chapter.

"Air Preheaters and Electrostatic Precipitators, 1980."

"Combustion Turbine Fire Prevention and Protection, 1993."

"Electric Utility Facilities, 1981."

"Fire Brigades and Training for Electric Utilities, 1985."

"Fire Doors and Fire Dampers, 1986."

"Fire Prevention and Protection in Computers and Simulators, 1983."

"Fire Protection and Prevention in Baghouse Dust Collectors, 1983."

"Fire Protection and Prevention in Coal Handling Systems, 1981."

"Fire Protection Systems for Conveyors, 1989."

"Fossil Generating Station Administrative Controls for Fire Prevention and Fire Protection, 1989."

"Steam Generator (Boiler) Area Fire Prevention and Protection, 1983."

"Suggested Guidelines for Completing a Fire Hazards Analysis, 1981."

"Use, Handling, and Storage of Liquid Fuels in Electric Utility Generating Plants, 1980."

"Water Distribution Systems for Fire Protection, 1986."

NUCLEAR FACILITIES

Revised by Wayne D. Holmes

Since the 1950s the use of nuclear materials has dramatically increased worldwide. In addition to nuclear fuels used for the production of electric power, nuclear materials are widely used for medical purposes, industrial controls, household uses (e.g., ionization smoke detectors), basic research, and weapons operations. Global use of nuclear materials continues to expand.

The use and storage of nuclear materials present special fire protection concerns. While the general fire hazards in facilities that use or store nuclear materials are the same as those of similar facilities that do not include nuclear materials, fires in nuclear facilities may have more significant consequences. Employees and emergency-response personnel must be protected against direct exposure to nuclear radiation and contamination. The environment and general public should not be unnecessarily exposed to radiation hazards. Accidental nuclear criticality (self-sustaining nuclear fission reactions) must be avoided. In addition, fires and explosions may compromise circuits, mechanical systems, and equipment necessary to safely contain and confine nuclear materials and radiation.

This chapter discusses some special practices and hazards associated with facilities handling and using nuclear materials. Among the special occupancies this chapter addresses are nuclear reactors, radiation machines, and other facilities handling nuclear materials. This chapter also discusses hazards associated with radiation exposure and special fire protection procedures for nuclear facilities. Other chapters of this handbook discuss general fire hazards, construction practices, and fire control practices, common to all facilities and not unique to nuclear facilities.

NUCLEAR REACTORS

Nuclear power is well established as a global source of nuclear energy. While nuclear power production is not currently increasing in the United States, worldwide nuclear power production continues to expand. Nuclear power, a significant part of many countries' energy programs, is expected to continue to increase as long as nuclear power continues to be a safe, cost-efficient, and reliable power source. Because fire safety is not a static concept, improvements are always possible and necessary in facilities where high-consequences accidents are also possible.

A nuclear reactor is a device or assembly for initiating and maintaining a controlled nuclear chain reaction in a fissionable fuel

(uranium or plutonium). Nuclear reactors are used to produce energy, to study the fission process, or to produce radioactive materials within the reactor or in a material exposed by the reactor's radiation or radioactive particles. Basically, nuclear reactors may be divided into two categories: (1) large size nuclear power reactors, up to 3,500 MW (megawatts thermal) or 1,100 MW(e) (megawatts electrical), and (2) research reactors that operate at power levels from a few watts to many megawatts. As of 1994, there were 109 operating power reactors in the United States and 319 operating power reactors outside the United States. The number of power reactors continues to increase globally. Dozens of research reactors are now operating, but, in general, their number is not increasing.

Nuclear reactors that include a containment vessel, generating equipment, and heat-removal equipment can be as large as the largest fossil-fueled electrical generating plants. Most research reactors, however, are so small that they may occupy only one small corner of a room in a typical laboratory building at a research facility or college campus. Therefore, the magnitude of hazard each kind of reactor presents varies considerably.

Various national and international groups have addressed the need for fire protection requirements for nuclear power reactors and research reactors. These groups include the National Fire Protection Association (NFPA), the American Nuclear Society, the American National Standards Institute (ANSI), the United States Nuclear Regulatory Commission (NRC), the MAERP Reinsurance Association, the American Nuclear Insurers, Nuclear Mutual Limited, and the International Atomic Energy Agency. (See the bibliography at the end of this chapter.)

Primary objectives for achieving and maintaining adequate nuclear reactor safety include:

1. Provide means to safely shut down the reactor and maintain it in safe shutdown condition during and after accident conditions.
2. Provide means to remove residual heat from the reactor core after reactor shutdown, including during and after accident conditions.
3. Provide means to reduce the potential for radioactive releases within acceptable limits.

To meet these safety objectives, different reactor designs provide redundant safety systems so that initiating events, such as fires, do not prevent safety systems from performing their required functions. Advanced reactor designs include a high degree of safety system redundancy. Earlier designs of reactors include less redundancy. As redundancy decreases, the need to protect each redundant system

---

Wayne D. Holmes, P.E., is vice president of HSB Professional Loss Control and secretary of the NFPA Technical Committee on Atomic Energy.

from the effects of fire and explosion increases. Improved passive fire protection, increased physical separation, and greater use of fire detection and suppression systems often provide this protection.

A high degree of fire protection is necessary to assure that fires will not significantly degrade the level of nuclear safety. The "defense-in-depth" concept can achieve this. It includes three principle objectives:

1. Preventing fires from occurring;
2. Detecting and quickly extinguishing those fires that do start, thus limiting fire damage; and
3. Preventing the spread of those fires that have not been extinguished, thus minimizing their effect on essential plant functions.

The first objective requires that plant design and operation be such that the probability of a fire is minimized. The second objective concerns early detection and extinguishing of fires by a combination of automatic and/or manual fire-fighting techniques and relies on active fire protection measures. The third objective places particular emphasis on passive fire barriers and physical separation. This is the last line of defense if the first two objectives are not met. For the defense-in-depth concept to be effective, fire safety design and procedures must include all three objectives and implement them simultaneously.

Heat Removal

All nuclear reactors, even very low-power training or research reactors, produce heat while operating. This heat either must be dissipated or used, depending upon the amount produced and the reactor's intended purpose.

Reactor Control

Reactor control systems and safety systems are of utmost importance. The control system design is fitted to the technical characteristics of the reactor and is capable of producing power changes at acceptable rates. The control system design also makes it possible to produce and maintain the desired power level within the reactor so that excessive temperatures are avoided. The safety system, an addition to the control system, is adapted to the characteristics of the reactor in the instrument and control system. It automatically operates in response to signals from the instruments to prevent operational variables from exceeding safe limits. On appropriate signals, the safety system warns of incipient performance changes and, if necessary, shuts down the reactor.

A reactor becomes "critical" when the total rate of production of neutrons, under control conditions, is such that self-sustaining reactions occur. Control methods must be rapid, sensitive, and protected from fire. Since the control system is vital to the reactor's adequate functioning and safe operation, protection of the control room, cableways, emergency power supply, and electrically or hydraulically operated equipment is of prime importance. Protection for these areas should be fully consistent with that used in computer rooms containing vital records.

Construction Problems

Fire records indicate that one of the more vulnerable periods for fire damage in the lifetime of a large reactor system, such as found in nuclear power plants, is the construction stage. (See Figure 9-30A.) Because the construction of such a plant usually requires many years, construction hazards acquire a nearly permanent status and should be considered permanent.

Power reactors present unique fire protection problems during construction, since they require a containment vessel as a final pro-

FIG. 9-30A.   *Nuclear facilities are most vulnerable to fire damage during construction, a period of several years, and thus require a good construction fire safety program.*

tection system. Construction techniques require the containment structure to be built early in the project. This means that much of the subsequent reactor construction takes place inside a large vessel that has limited exits for evacuation and limited fire-fighting access. In addition, the vessel confines smoke and other products of combustion, which greatly increases the difficulties of evacuation and manual fire fighting. A good construction fire safety program should ensure that all penetrations of the containment vessel suitable for evacuation remain open and usable while other construction takes place within the vessel.

Due to the limited access for fire fighting and the lack of normal venting possibilities for smoke and gases, it is imperative to severely limit all combustible materials needed for construction. Metal formwork, scaffolding, platforms, stairways, etc., are preferable to wood. Wood use should be limited to those species that are appropriately treated to reduce inherent combustibility and flame-spread ratings.

Installing utilities and equipment in the containment vessel requires special care to maintain a low level of combustibles. Since reactor equipment must meet high quality assurance levels, reactor equipment that has been subjected to fire and smoke damage is much more likely to require replacement than similarly exposed equipment in normal industrial installations. Special efforts are necessary to reduce the usual accumulation of packing cases, cartons, insulation, etc., to an acceptable level. This may take the form of conducting all uncrating operations outside the containment vessel and providing special handling devices to transport unpackaged items into the vessel.

Research and Production Reactors

In general, these may present particular problems due to their location in existing buildings (which are sometimes of combustible construction), and the use of much combustible loading in the form of paraffin shielding, boronated polyethylene, wall finishes, and instrumentation. Combustibles should be eliminated wherever possible. For example, water can be used in place of paraffin. Where combustibles cannot be eliminated, automatic sprinkler protection can be used to control potential fire exposure. Supervision of construction/operating personnel is especially important because a number of different operating groups may be involved with a single reactor.

A general lack of containment for research and production reactors further increases the need for effective fire protection design and procedures. Large power reactors often include passive containment systems and structures, typically of massive concrete and steel construction, to contain nuclear materials; research and production reactors may lack such containment structures. In these cases, confinement systems control and limit the undesirable release of nuclear contaminants. Ventilation systems become a key component in preventing nuclear releases. Special precautions must be taken to assure that fires will not directly damage ventilation and confinement systems necessary to prevent the release of nuclear contaminants. Further, dynamic pressures created by fires and explosions must be considered and protected against to prevent air flow reversals that might significantly impact the spread of nuclear contamination.

RADIATION MACHINES

Radiation machines include mechanical and electrical devices that produce or use subatomic particles or electromagnetic radiation, or both. X-ray machines are used in radiography, therapeutic treatment, and studies of the behavior of radiation and its effect upon materials. Particle accelerators, while radiation sources, are primarily devices for imparting extremely high energies to subatomic particles that enter and alter atomic nuclei, thereby providing the means for developing basic information concerning the structure and behavior of matter. Gamma ray sources in the form of radioactive isotopes are employed in radiography equipment. This equipment may produce intense radiation. The use of isotopes as a source of radiation is not discussed in this chapter.

X-Ray Machines

Except for the radiation hazard while in use, X-ray machines' hazards are mainly hazards of high potential and high-energy electrical equipment. When shut down, there is seldom any appreciable residual radiation to interfere with fire-fighting or salvage operations. Although no flammable gases or hazardous amounts of flammable liquids are ordinarily needed to operate X-ray machines, they are frequently encountered in studies of the effects of radiation on a wide variety of substances.

Particle Accelerators

Particle accelerators include Van de Graaff generators, linear accelerators, cyclotrons, synchrotrons, betatrons, or bevatrons. The machines are used, as their names imply, to accelerate various charged atomic particles to tremendous speeds and, consequently, to high-energy levels. Particle accelerators furnish scientists with atomic particles, in the form of a beam, which may be utilized for fundamental studies of atomic structure. In addition, accelerators furnish high-energy radiation, which may be utilized for radiography, therapy, or chemical processing.

Particle accelerators emit appreciable radiation only while operating. Attempts to extinguish a fire in the immediate vicinity of the machine should be delayed until the machine's power supply can be disconnected.

Certain "target" materials become radioactive when bombarded by atomic particles, and for this reason monitoring equipment should be used during fire-fighting operations to estimate the radiation hazard. The usual hazard presented by particle accelerators is largely that of electrical equipment. There are, however, some important exceptions. Some installations use such hazardous materials as liquid hydrogen or other flammable materials in considerable quantities. The large amounts of paraffin used in the past for neutron shielding have been replaced, in many cases, by boronated polyethylene. In fires, polyethylene materials have a very high heat release rate, which can further increase the fire hazard. In some cases, fire retardant chemicals added to polyethylene decrease its ignition potential. Once ignited, however, even "fire retardant" polyethylene may burn intensely when exposed by other burning materials. Limited, small-scale fire testing of combustible moderator and shielding materials does not ensure that these materials will not burn in the configurations in which they are used. Another factor is the possible presence of combustible oils used for insulating and cooling.

Industrial applications for particle accelerators include chemical activation, acceleration of polymerization in plastics production, and the sterilization and preservation of packaged drugs and sutures. The general fire protection and prevention measures for these machines should include the use of noncombustible or limited combustible (Type I or Type II) construction housing, noncombustible or slow-burning wiring and interior finishing, and the elimination of as much other combustible material as possible. (See NFPA 220, *Standard on Types of Building Construction.*) Automatic sprinkler protection should be provided in areas containing hazardous amounts of combustible material or equipment. Special fire protection should be provided for high-voltage electrical equipment.

FACILITIES HANDLING RADIOACTIVE MATERIALS

The type of equipment used to process radioactive materials depends not only upon the work to be performed, but also upon the degree of hazard associated with the material and the process it is to undergo. Materials with low levels of radioactivity and with little or no inherent fire or explosion hazards require less protective equipment than others. To protect personnel, the amount and kind of shielding required will depend upon the types of radiation emitted, as well as the activity involved. In addition, the chemical and physical nature of the radioactive materials will dictate the degree of containment necessary, as well as the construction materials necessary in the containment system. All equipment used for handling and processing radioactive materials should be designed to minimize fire and explosion potentials, as well as to protect personnel against harmful radiation exposure and prevent damage to property by contamination. Many types of equipment and systems handle radioactive materials; most can be classified as either benches, hoods, glove boxes, or hot cells.

Benches

Benches are used generally for handling relatively small amounts of alpha- or beta-emitting materials requiring little or no shielding when handled with gloved hands or tongs. No special ventilation for the bench is provided in most instances, and its use is thereby restricted to materials that will not easily become airborne.

Benches should be of noncombustible construction with a nonporous continuous working surface that can be decontaminated easily. Usually, one or two layers of blotting paper on the bench top to absorb small spills will not materially increase the fire hazard.

Hoods

Hoods, sometimes referred to as "fume hoods," are similar to benches except for an enclosure and exhaust system added to remove vapors. The nature of the operations conducted within the hood may require a filter system to prevent the spread of radioactive materials. Filters with a low degree of combustibility are desirable. Fire protection and ventilation requirements for hoods are similar to those described for glove boxes.

Glove Boxes

The term "glove box" refers to a system designed to contain materials, generally alpha radiation emitters, which present little or no external radiation hazard but which can present a serious problem if they become airborne. Glove boxes may be large and used in a wide variety of operations involving flammable liquids and gases, combustible solids, and toxic materials. The sides are fitted with long rubber-like gloves that permit manual operations to be conducted without personal contact with the hazardous materials. (See Figure 9-30B.)

Confinement of radioactive materials that may be contained in hoods and glove boxes greatly depends upon the enclosure materials and the ventilation systems. Therefore, fire hazards, fire protection, construction materials, and ventilation must receive special attention to prevent a breach of the confinement system.

Burning of combustible materials that may be present in the hood or glove box could produce dynamic pressures that might over-pressurize the enclosure or cause direct fire damage to the enclosure. Where combustible material must be used inside the hood or glove box, special extinguishing agents, such as dry powders, clean fire suppression agents, or water sprinklers, may be necessary to prevent internal fires from breaching the containment enclosure. This could lead to release of nuclear materials. Additionally, extreme care must be exercised to assure that introducing the fire suppression agent will not overpressurize the enclosure nor increase the potential for accidental nuclear criticality where fissile materials

FIG. 9-30B.   *A typical glove box showing gloves extending from the ports in the viewing window. Note the portable fire extinguisher adapted for discharge into the box through a fixed piping arrangement.*

may be present. The design of ventilation systems—whose purpose is to maintain a negative pressure in the hood or glove box or to maintain a specific airflow pattern—should anticipate and control pressure and airflow changes due to fires or fire suppression systems.

Hot Cells

A hot cell is a heavily shielded enclosure where gamma-emitting radioactive materials can be handled by persons using remote manipulators while viewing the operation through shielded windows or periscopes. Hot cells are preferably constructed of noncombustible materials and contain the minimum amount of combustibles consistent with operational requirements.

The high-gamma radiation hazard associated with hot cells may be further complicated by the need to handle combustible materials in the hot cells, the use of combustible moderators and shielding materials, combustible oils in shielded viewing windows, and the lack of internal accessibility within the cell. Fire protection and ventilation concerns for hot cells are similar to those described for hoods and glove boxes.

In addition to all the fire and explosion hazards of glove boxes, hot cells also present increased damage potential due to the nature of the high-gamma ray producing materials used. Where gamma radiation levels are very high, the possible failure of containers as a result of radiation damage must also be considered.

RADIATION EXPOSURE

Radiation Injury

How radiation actually damages living cells is not exactly known. Unfortunately, the human body has no defense against radiation; nor can any of the five senses detect nuclear radiation. Thus, it is possible for an individual to be severely exposed to radiation without knowing it. This danger requires some form of instrumentation for detecting radioactivity.

The amount of injury to a person from radiation varies with the type of radiation, how much of the body has been exposed, and whether it is a one-time exposure or an accumulation of exposures to small amounts of radiation. Injuries from excessive exposures may not become apparent for days, weeks, months, or years.

Similar to many potentially hazardous materials, any exposure to radiation, either external or internal, has some element of risk, which may be too low to be measurable. Certain exposures are believed to be of reasonable risk balanced against the benefits from any activity using radiation exposure, such as medical and dental uses of radiation. Since the human race has evolved on a planet which has always been radioactive from naturally occurring radioisotopes in the air, soil, and water, it is probable that a low "background" radiation is tolerable. The controversy on radiation exposure concerns the additional amounts that humans may tolerate. Standards have been set with the objective of lowering risks to the best practicable levels.

OCCUPATIONAL EXPOSURE FROM RADIATION IN AIR AND WATER

Radiological Authorities Having Jurisdiction have set extremely low limits on concentrations of radioisotopes in air and water, based on quantities that may be inhaled or ingested. Complete treatment of the subject would require consideration of maximum permissible concentrations in both air and water, but the problem of airborne concentrations is of particular concern in fire situations, as fire fighters and other emergency personnel may be confronted with such material in the fumes, dust, smoke, and gases liberated by a fire involving radioactive materials.

A hard-and-fast rule should be mandated that self-contained breathing apparatus (SCBA) is to be used whenever exposure to airborne radiation is a possibility. Maximum permissible concentrations (MPC) of various radioisotopes in air and water are commonly stated as limits intended to apply to persons who are continuously exposed to the concentrations named. Consequently, the concentrations are far below the exposures that could be tolerated for infrequent exposure, as would most often be the case with fire fighters or emergency workers.

Apparatus is available for a sampling an atmosphere. The number of radioactive emissions in the air can be counted with an appropriate counting instrument. From these counts, the extent to which the particular atmosphere is contaminated can be determined. (See Figure 9-30C.)

Any space normally occupied by persons not primarily engaged in radiation work should not be subjected to unreasonable radiation levels; also a person outside the installation should not be subjected to excessive radiation level exposure through contact with

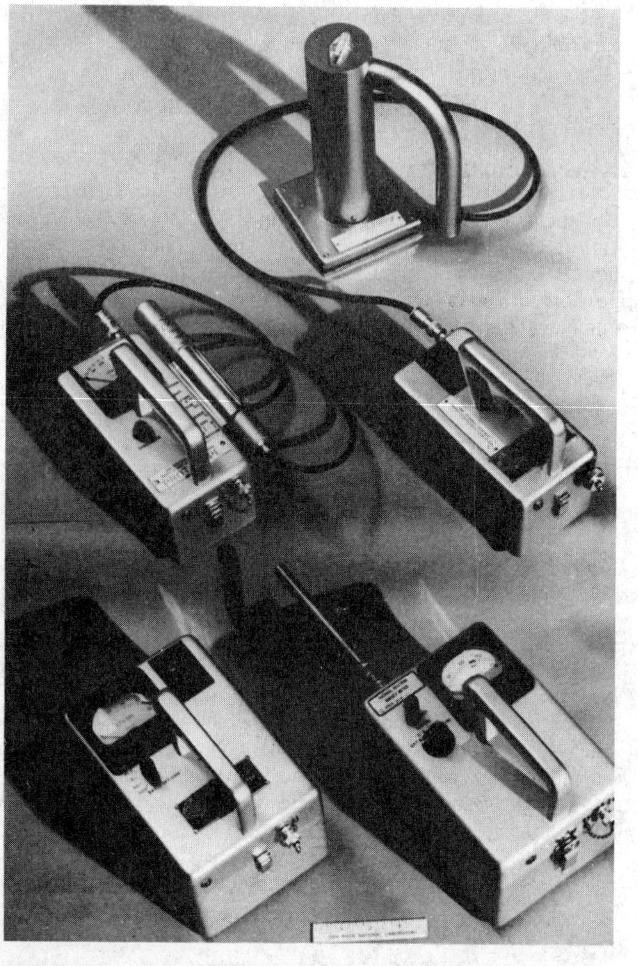

FIG. 9-30C.   *Portable instruments developed at Oak Ridge National Laboratory for measuring radioactive emissions. The meters are for fast neutrons (bottom row at left), thermal neutrons (bottom row at right), beta-gamma (top row at left), and alpha particles (top row at right), with an alpha scintillation detector at the top of the picture.*

radioactive waste or by other means. Air should be monitored continuously for the presence of radiation from fixed sources or from airborne radioactive matter. In emergencies, alarms should sound, and radiation levels be recorded by available commercial instruments.

Fire Department Radiation Exposures

Emergency exposures are usually allowed to exceed those tolerable to persons who work continuously with radioactive materials. In an emergency, such as a necessary rescue operation, raising the exposure, within limits, for single doses is considered acceptable. The National Council on Radiation Protection and Measurement has recommended that in a life-saving action, such as search for and removal of injured persons or for entry to prevent conditions that would injure or kill numerous persons, the planned dose to the whole body should not exceed 100 rems. During less stressful circumstances, where it is still desirable to enter a hazardous area to protect facilities, eliminate further escape of effluents, or control fires, the recommended planned dose to the whole body should not exceed 25 rems. These rules may be applied to the fire fighter for a single emergency; further exposure is not recommended. Internal radiation exposure may be guarded against by adequate respiratory equipment.

External exposure at the time of a single fire emergency can be judged by the use of commercial radiation survey meters, which measure radiation in roentgens or by counted disintegration rates, or by the close observation of the dosimeter indicators individuals carry. Pocket-sized dose rate alarms, which can be carried on the person, are also available. These give an audible signal dependent upon radiation intensity. Film badges do not provide immediate information.

A rescue procedure that would combine external and internal radiation exposure is usually not attempted. Self-contained breathing apparatus should be used when instruments indicate the presence of airborne radiation.

FIRE PROTECTION

As noted previously, substances and operations involving radioactive substances or devices presenting radiation hazards have the same fire and explosion features as those of similar materials and operations without radiation hazards. The presence of radiation or of radioactive substances affects the loss caused by fire, explosion, and accident in the following ways:

1. Possible interference with manual fire fighting due to the presence of harmful radiation or possible criticality of radiation levels.
2. Possible increased delay in salvage and in normal resumption of operations due to the necessity of decontaminating of buildings, equipment, or materials.

Property Contamination

Entire buildings, land, and important equipment can be rendered unusable for long time periods because of severe radioactive contamination from the accidental escape of radioactive substances.

Radiological contamination may not stay in one area—it can sift through openings or ventilating systems in the form of dust or vapor and spread throughout a structure. Careless movement of persons through a contaminated area could also spread contamination to an uncontaminated area.

Once a surface has become contaminated, a decision must be made as to how to remove the particular contaminating material, if this is possible. Vacuum cleaning can sometimes remove radioactive dust from building surfaces. If vacuum cleaning is used, however, absolute filters must be used on the exhaust. Some surfaces can be hosed with water. Cleaning with soap and detergents is often a hand operation. The amount of exposure that persons doing the cleaning may tolerate must be continuously checked. Sand or vacuum blasting can be used on some surfaces, and paint may cover alpha contamination.

Plant Fire Protection Organization

In properties where atomic energy is a factor, an in-plant fire protection force is recommended. In nuclear reactors, and many other such plants, 24-hr/day routines must be maintained for handling fires and emergencies.

Plan for Handling Fires

In plants involving a nuclear reactor, radiation machines, and in other facilities handling radioactive materials, the problems affecting decisions on how best to deal with a fire or other emergency are not those types of problems that can be solved by simply calling the public fire department. As many decisions as possible must be made with respect to the types of fire or emergency to be expected—and these decisions must be made well in advance. The particular fire-fighting and personnel safety measures to be taken may involve shutting down or isolating parts of the plant or individual equipment items. The areas where special procedures are necessary must be identified, and the procedures for these special areas thoroughly understood by all plant/facility personnel.

Fire emergency arrangements should include provisions for prompt notification of the plant emergency organization and public fire department, usually through a public fire alarm signal box. However, the plant fire protection department must preplan fire-fighting operations with the local fire department so that the local department will be properly coordinated with the plant's own emergency plans. Emergency planning should include measures to prevent the spread of contamination and to promptly decontaminate the area in case of accidental release of radioactive substances.

Fire fighters and other emergency personnel operating in areas where radiation exposure is a danger must be fully trained and provided with suitable protective clothing. Respiratory protective equipment is a must, and competent radiological advisors, equipped with instruments for measuring area and local exposure, are necessary to guide emergency personnel. Dosimeters or other instruments for recording each individual's accumulated radiation exposure are helpful.

A nuclear reactor site must have a generous water supply to facilitate fire control and decontamination operations. Facilities must also be prearranged for safe disposal or storage of water that may be contaminated. The use of noncombustible materials for reactor buildings and equipment will help to avoid complications of fire hazards. For example, all finish materials used for decorative, acoustical, or insulation purposes should be both noncombustible and easy to decontaminate.

The hazard of a reactor structure exposing other buildings to radiation is prevented by appropriate distance separation or fire barriers. To prevent exposure to the reactor, it is always appropriate to separate shops and service spaces from the reactor equipment and structure itself. Wiring ducts in floors introduce an opportunity for fire or contaminated liquid or gas to spread from one space to another. Good duct seals separate one space from another. Subassembly or other operations in the preparation of fuel elements for reactors is carried on in work areas separated from the reactor so that fire cannot reach the reactor space.

Equipment for Fighting Fires

Automatic sprinkler systems or specially designed piped water spray systems are the first choice for fire protection in any location where fires may occur in nuclear reactor plants, properties housing radiation machines, and facilities handling radioactive materials. Sprinklers can operate with full effectiveness under radiation or contamination conditions that would make approach by fire fighters impossible.

In spaces where water used in fire fighting is subject to possible contamination, the collection and disposal of this water must be provided for in the local facilities; this means the facilities should have waterproofed floors and controlled floor drainage. Substantial capacity of such drainage systems would be required if hose streams and manual fire fighting were necessary. By contrast, sprinklers or a specially designed spray system would require relatively modest amounts of water for fire fighting.

If a fire occurs in a containment vessel during construction, the difficulties of access and visibility warrant the provision of temporary fixed automatic extinguishing systems when combustibles cannot be effectively controlled. Temporary interior hose stations and an ample supply of portable extinguishing equipment should be within easy reach in all portions of the vessel. Because of the smoke confinement potential, only very fast manual response may be effective; hence the available manual fire-fighting equipment should be in excess of normal construction practice to ensure the earliest response.

Incompatible Materials

Careful design analysis is required to reduce the fire protection problems inherent in the use of materials that are incompatible in fire situations. As an example, the contemplated use of liquid metal as a reactor coolant/moderator requires special extinguishing systems not compatible with water; in fact, the possibility of a water-liquid metal reaction may justify the exclusion of water systems from the area. Such a decision, however, imposes severe limitations on the presence of flammable oils, plastics, foam insulations, and other materials that generally require copious quantities of water for fire extinguishment. Where such mixed hazards exist, careful consideration must be given to the potential for a failure in one system to cause a failure in the incompatible system. In such cases, either protection systems must be provided that can ensure the extinguishment of fire in either system before it can rupture the other system, or a single protection system (such as inerting) must be developed that is adequate for either hazard. The difficulties inherent in such problems warrant the most thorough hazards analysis at the earliest design stages.

Water is a preferred fire suppression agent in many cases due to its high effectiveness in suppressing fires, its low cost, and its general availability. Water from sprinkler systems, water spray systems, and hose streams can be effectively used in nuclear facilities to reduce the nuclear safety impact of fires. However, the positive fire control contribution of water as a suppression agent in nuclear facilities must be balanced against the detrimental effects of water, if used indiscriminately in nuclear facilities. Because of the very positive fire suppression capabilities of water and the potentially highly detrimental effects of uncontrolled fires in nuclear facilities, the use of water should not be dismissed just because its use may have some undesirable effects.

The use of water as a fire suppression agent introduces the potential for spread of water-born contamination. Means of confining potentially contaminated fire-fighting water must be provided. In addition to providing for hold-up of potentially contaminated water, radwaste processing systems should have sufficient capacity to process the amounts of water that may be discharged for fire fighting. Automatic sprinkler systems provide the advantage of discharging only limited amounts of water and only in locations where fires are actually burning, thus limiting the total amount of water that may become contaminated. Further, automatic sprinklers have a demonstrated high-effectiveness in suppressing fires and a very low frequency of inadvertent water discharge.

Where combustible and fissile materials may be present, it may be desirable to eliminate water suppression systems due to the concern for the neutron moderating and reflecting effects of the water and the potential for accidental nuclear critically (i.e., accidental, self-sustaining, nuclear fission reactions). Some nuclear criticality specialists are reluctant to allow the use of water to control fires where fissile materials are present, even when the degree of fire hazard dictates that providing automatic sprinklers would be prudent. Such reluctance is certainly understandable, but not necessarily justifiable in all cases. In many, if not most, situations where both combustible and fissile materials are present, water spray or sprinkler protection can and should be provided. Nuclear criticality safety analyses can predict and include the neutron moderating effects of sprinklers. A safety system that simultaneously maintains fire safety and criticality safety at acceptable levels can include automatic sprinkler protection.

BIBLIOGRAPHY

NFPA Codes, Standards, and Recommended Practices

Reference to the following NFPA codes, standards, and recommended practices will provide further information on the safeguards for nuclear facilities discussed in this chapter. (See the latest version of *The NFPA Catalog* for availability of current editions of the following documents.)

NFPA 10, *Standard for Portable Fire Extinguishers*

NFPA 72, *National Fire Alarm Code*

NFPA 220, *Standard on Types of Building Construction*

NFPA 255, *Standard Method of Test of Surface Burning Characteristics of Building Materials*

NFPA 259, *Standard Test Method for Potential Heat of Building Materials*

NFPA 480, *Standard for the Storage, Handling, and Processing of Magnesium*

NFPA 481, *Standard for the Production, Processing, Handling, and Storage of Titanium*

NFPA 482, *Standard for the Production, Processing, Handling, and Storage of Zirconium*

NFPA 801, *Standard for Facilities Handling Radioactive Materials*

NFPA 802, *Recommended Fire Protection Practice for Nuclear Research and Production Reactors*

NFPA 803, *Standard for Fire Protection for Light Water Nuclear Power Plants*

NFPA 804, *Standard for Fire Protection for Advanced Light Water Reactors Electric Generating Plants*

Additional Reading

10CFR50.48 and Appendix R, *Fire Protection for Light Water Reactors*, Code of Federal Regulations, U.S. Nuclear Regulatory Commission, Washington, DC.

50-SG-D2, *Safety Guide on Fire Protection in Nuclear Power Plants*, International Atomic Energy Agency, Vienna, Austria, 1992.

Apostolakis, G., "Some Probabilistic Aspects of Fire Risk Analysis for Nuclear Power Plants," *Proceedings* of the First International Symposium for Fire Safety Science, Hemisphere, 1986, pp. 1039–1046.

Brandsjo, K., "Being There: A Firsthand Account of the Nuclear Accident at Chernobyl," *Fire Journal*, Vol. 82, No. 5, 1988, p. 67.

Chavez, J., "An Experimental Investigation of Internally Ignited Fires in Nuclear Power Plant Control Cabinets, Part I," NUREG/CR-4527/1 of 2, April 1987, Sandia National Labs, Albuquerque, NM.

Clarke, M. C., and Kememy, L. G., "Proposal for the Handling of Out of Control Supercritical Nuclear Reactors," *Chemical Engineering in Australia*, Vol. 12, No. 1, 1987, pp. 11–13.

Cooper, L. Y., and Steckler, K. D., "Methodology for Developing and Implementing Alternative Temperature-Time Curves for Testing the Fire Resistance of Barriers for Nuclear Power Plant Applications," National Institute of Standards and Technology, Gaithersburg, MD, Nuclear Regulatory Commission, Washington, DC, NISTIR 5842, May 1996, 116 pages.

Green, B. J., "Protection in a Nuclear Research Establishment," *Fire International*, Vol. 11, No. 105, 1987, pp. 24–26, 29.

Holmes, W. D., "Fire Protection by Sprinklers Where Fissile Materials Are Present," *Proceedings* of the International Symposium on Fire Protection and Fire Fighting in Nuclear Installations, International Atomic Energy Agency, Vienna, Austria, 1989.

Holmes, W. D., "Fire Risk Separation for Nuclear Facilities," *Proceedings* of the International Symposium on Fire Protection and Fire Fighting in Nuclear Installations, International Atomic Energy Agency, Vienna, Austria, 1989.

Hosser, D., and Sprey, W., "A Probabilistic Method for Optimization of Fire Safety in Nuclear Power Plants," *Proceedings* of the First International Symposium for Fire Safety Science, Hemisphere, Washington DC, 1986, pp. 1047–1056.

Inspection of Fire Protection Measures and Fire Fighting Capability at Nuclear Power Plants, International Atomic Energy Agency, Vienna, Austria, 1994.

Evaluation of Fire Hazard Analyses for Nuclear Power Plants, International Atomic Energy Agency, Vienna, Austria, 1996.

Assessment of the Overall Fire Safety Arrangements at Nuclear Power Plants, International Atomic Energy Agency, Vienna, Austria.

Kench, R. L., "Fire Safety in Nuclear Power Stations," *Fire Prevention*, No. 206, 1988, pp. 18–22.

Kench, R. L., "Fire Safety in Atomic Power Plants," *Fire Prevention*, No. 207, 1988, pp. 32–37.

Lalloo, R., "Fire Service Involvement with Nuclear Reactors," *Fire Engineers Journal*, Vol. 49, No. 154, Sept. 1989, pp. 17–21.

Lambright, J. A., *et al.*, "Analysis of the LaSalle Unit 2 Nuclear Power Plant: Risk Methods Integration and Evaluation Program (RMIEP). Internal Fire Analysis," Sandia National Labs., Albuquerque, NM, Nuclear Regulatory Commission, Washington, DC, SAND92-0537-VOL-9, March 1993.

Lambright, J. A., *et al.*, *Fire Risk Scoping Study: Investigation of Nuclear Power Plant Fire Risk, Including Previously Unaddressed Issues*, NUREG/CR-5088, 1989, U.S. Nuclear Regulatory Commission, Washington, DC.

Lehto, M., "Fire Data Sources and Associated Ignition Probabilities of Nuclear Power Plants in the United States," IVO International Ltd., Vantaa, Finland, VTT-Technical Research Center of Finland and Forum for International Cooperation in Fire Research, Nordic Fire Safety Engineering Symposium, Development and Verification of Tools for Performance Codes, Aug. 30–Sept. 1, 1993, Espoo, Finland, 1993, pp. 1–4.

Majumdar, K. C., Alesso, H. P., and Altenbach, T. J., "On Fire Risk/Methodology for the Next Generation of Reactors and Nuclear Facilities," Lawrence Livermore National Laboratory, CA, Department of Energy, Washington, DC, UCRL-JC-110420, Oct. 15, 1992; American Nuclear Society Meeting, Jan. 27–29, 1993, Clearwater, FL, 1992.

Malet, J. C., *et al.*, "Fire-Fighting Procedure and Estimation of Fire Consequences in Nuclear Plants," CEA/IPSN/DRS/SESRU, Cedex, France, *Proceedings* of the Third International Symposium on Fire Safety Science, Elsevier Applied Science, New York, 1991, pp. 1019–1028.

Matsuoka, T., Miyazaki, K., and Kondo, M., "Development of the Fire Risk Analysis Methodology for Nuclear Power Plants," *PSAM-II Proceedings*, An International Conference Devoted to the Advancement of System-Based Methods for the Design and Operation of Technological Systems and Processes, Vol. 3, Sessions 73–108, March 20–25, 1994, San Diego, CA, 1994, pp. 102/1–7.

Mawby, F. D., and Haighton, A. P., "Approach Adopted in England and Wales in Defining the Level of Fire Safety at Nuclear Power Stations," Nuclear Electric plc, Manchester, England, Fire Research Station and the Institute of Fire Safety in association with the Fire Service Research and Training Trust, *Proceedings* of the Technical Conference on Fire Safety by Design: A Framework for the Future, Featuring the Combined 1993 Fire Research Lecture and IFS Inaugural Lecture, Nov. 10, 1993, Borehamwood, England, 1993, pp. 1–14.

Mawby, F. D., and Haighton, A. P., "Approach Adopted in England and Wales in Defining the Level of Fire Safety at Nuclear Power Stations," *Fire Safety Journal*, Vol. 23, No. 2, 1994, pp. 185–192.

Mowrer, D. S., "Summary of Fire Protection Audits in Twenty Nuclear Power Plants," *Nuclear Plant Journal*, Vol. 7, No. 5, Sep./Oct. 1989, pp. 82–84.

Notley, D. P., "Fire Protection in Nuclear Power Plants: Understanding Competing Requirements for Safety," *Proceedings* of the International Symposium on Fire Protection and Fire Fighting in Nuclear Installations, International Atomic Energy Agency, Vienna, Austria, 1989.

Nowlen, S. P., "Nuclear Power Plants: A Unique Challenge to Fire Safety," *Fire Safety Journal*, Vol. 19, No. 1, 1992, pp. 3–18.

Nuclear Safety (bimonthly), U.S. Government Printing Office, Washington, DC.

Payne, A. C., Jr., Brown, T. D., and Miller, L. A., "Integrated Level III Risk Assessment for the LaSalle Unit 2 Nuclear Power Plant," Sandia National Labs., Albuquerque, NM, Nuclear Regulatory Commission, Washington, DC, SAND91-1668C, 1991.

Payne, A. C., Jr., *et al.*, "Analysis of the LaSalle Unit 2 Nuclear Power Plant: Risk Methods Integration and Evaluation Program (RMIEP). Integrated Quantification and Uncertainty Analysis," Sandia National Labs., Albuquerque, NM, Nuclear Regulatory Commission, Washington, DC, SAND92-0537C, July 1992.

Payne, A. C., Jr., *et al.*, "Analysis of the LaSalle Unit 2 Nuclear Power Plant: Risk Methods Integration and Evaluation Program (RMIEP). Initiating Events and Accident Sequence Delineation," Sandia National Labs., Albuquerque, NM, Nuclear Regulatory Commission, Washington, DC, SAND92-0537-VOL-4, Oct. 1992.

Preston, T., "Fire Safety is Major Concern at UK's Only Fast Breeder Nuclear Reactor," *Fire*, Vol. 85, No. 1044, June 1992, p. 18.

Rasmuson, D. M., *et al.*, "Development of a Database of Common Cause Failure Events at U.S. Nuclear Power Plants," *PSAM-II Proceedings*, An International Conference Devoted to the Advancement of System-Based Methods for the Design and Operation of Technological Systems and Processes, Vol. 1, Sessions 1–36, March 20–25, 1994, San Diego, CA, 1994, pp. 004/9–14.

Ravindra, M. K., and Baron, H., "Analysis of the LaSalle Unit 2 Nuclear Power Plant: Risk Methods Integration and Evaluation Program (RMIEP). External Event Scoping Quantification," Sandia National Labs., Albuquerque, NM, Nuclear Regulatory Commission, Washington, DC, SAND92-0537, July 1992.

Regan, J. D., "Design for Fire Protection," *Nuclear Engineering*, Vol. 28, No. 1, 1987, pp. 5–9.

Restrepo, L. F., "Fire Accident Analysis Modeling in Support of Non-Reactor Nuclear Facilities at Sandia National Laboratories," Sandia National Labs., Albuquerque, NM, Department of Energy, Washington, DC, SAND-93-0463C, 1993.

Sawyer, E. A., "Failure Analysis of Inadvertent Operations of Fire Protection Systems Including Design Recommendations," Master of Science Thesis, Worcester Polytechnic Institute, Worcester, MA, 1986.

Scotford, G. B., "For Protection for the Nuclear Industry," *Nuclear Engineer*, Vol. 27, No. 4, 1986, pp. 106–111.

Scotford, G. B., "Particular Problems for the Fire Service in the Nuclear Industry," *Fire Engineers Journal*, Vol. 46, No. 142, 1986, pp. 8–9.

Sedge, L. A., and Lakey, J. R. A., "Fire Training, Drills and Exercises in the Nuclear Industry," *Nuclear Engineering*, Vol. 28, No. 1, 1987, pp. 16–20.

Sherman, D. L., "Fire Hazards in Nuclear Power Plants," *Proceedings* of the First International Fire Information Conference, Cosponsored by International Fire Information Conference (IFIC) and International Network for Fire Information and Reference Exchange (inFIRE), April 27–May 1, 1992, Moreton-in-Marsh, England, Society of Fire Protection Engineers, Boston, MA, 1993, pp. 93–99.

Siu, N., and Apstolakis, G., "Methodology for Analyzing the Detection and Suppression of Fires in Nuclear Power Plants," *Nuclear Science and Engineering*, Vol. 94, No. 3, Nov. 1986, pp. 213–226.

Standards of the U.S. Nuclear Regulatory Commission, Code of Federal Regulations, Part 20, Chapter 1, Title 10, U.S. Government Printing Office, Washington, DC.

Sullivan, K., "Fire Protection of Safe Shutdown Capability at Commercial Nuclear Power Plants," Brookhaven National Laboratory, Upton, NY, Nuclear Regulatory Agency, Washington, DC, BNL-NURG-49137, 1993.

Vecchio, F. J., and Sato, J. A., "Drop, Fire and Thermal Testing of a Concrete Nuclear Fuel Container," *ACI Structural Journal*, Vol. 85, No. 4, July/Aug. 1988, pp. 374–383.

MINING METHODS AND EQUIPMENT

William H. Pomroy

Mining is an industrial activity characterized by inherent and significant fire and explosion risks. At the start of the twentieth century, mine fires and explosions resulted in loss of life and property damage on a scale unmatched in other industrial sectors and possibly unmatched in the history of industrialized society. However, marked progress has been achieved in controlling these hazards, as evidenced by the precipitous decline in mine fires and explosions reported in recent decades. This chapter's purpose is to describe mining methods and equipment, fire and explosion hazards, and safeguards involved in mining coal, metal and nonmetal ores, sand and gravel, and stone.

The following chapters contain additional information relevant to mining processes and hazards: Section 1, Chapter 6, "Explosions"; Section 3, Chapter 13, "Fluid Power Systems"; Section 3, Chapter 21, "Storage of Flammable and Combustible Liquids"; Section 4, Chapter 12, "Explosives and Blasting Agents"; Section 4, Chapter 15, "Dusts"; and Section 5, Chapter 8, "Gas and Vapor Detection Sytems and Monitors."

Mining is the process of gaining access to, and extracting minerals from, the earth. Mining operations range in scale from tiny exploration prospects worked on a part-time basis by an individual miner with simple tools, to huge industrial complexes employing thousands of workers and modern, high-technology systems and equipment. This chapter focuses on mining methods and equipment, fire and explosion hazards, and safeguards involved in larger scale, industrial mining operations.

Mining can be grouped into two broad categories: surface mining and underground mining. Surface mining requires removing all overlying dirt and rock strata, or overburden, before excavating the desired minerals. Underground mining is practiced where the overburden thickness precludes surface mining. It permits selective extraction of desired minerals, with minimal disturbance to the overlying strata.

The economics of large-scale production and mechanization strongly favor surface mining. Over 95 percent of all mine production is accomplished by surface methods.[1]

SURFACE MINING METHODS

Numerous surface mining methods have evolved in response to varying surface and subsurface geological conditions and to continual improvement in mining equipment technology.

Methods of surface coal mining, or strip mining, include area strip mining and various forms of contour mining.[2,3] Area strip mining is practiced where terrain is relatively flat and coal seams are moderately thick [3 to 30 ft (0.9 to 9 m)]. Most surface coal mining in the midwestern states and the less mountainous areas in the eastern and western coal fields employs the area strip mining method. It begins with removal of overburden from above the coal seam in a roughly linear strip usually 100 to 170 ft (30 to 52 m) wide and several miles long. Overburden depths of 60 to 120 ft (18 to 37 m) are common. Coal uncovered is loaded into trucks and hauled to storage or processing facilities. When an entire strip has been mined, the process is repeated, with the overburden above the adjoining strip of coal cast onto spoil piles in the cut previously mined. (See Figure 9-31A.)

Specialized mining systems resembling metal and nonmetal open pit methods (see below) have been developed for multiple, steeply pitched, and/or thick seams. Some of the largest and most productive surface coal mines in the world are located in Wyoming and Montana where these methods are used to mine coal seams 200 ft thick or more.

Contour strip mining is most common where flat-lying coal seams occur in rolling or mountainous terrain, typical in many parts of the Appalachian coal field. In contour mining, an initial cut is established along a hillside at the point where the coal seam outcrops, or is exposed. Successive cuts then are made into the hillside until an economic stripping limit is reached. Mining proceeds laterally along the hillside, following the outcrop. Contour mining follows the same overburden removal/coal loading cycle as area strip mining, but generally employs much smaller equipment.

Most surface metal and nonmetal mining is by the open pit or open cast method.[4] The overburden is first stripped to expose the top of the ore body; stripping and mining then proceed concurrently. The pit expands in depth and area until an economic pit limit is reached. (See Figure 9-31B.) Pit geometries vary, owing to ore bodies' irregular shape and orientation, but dimensions of $\frac{1}{2}$ mi (0.8 km) to several miles across and 100 to over 1,000 ft (30 to 305 m) deep are common.

Specialized mining systems have been developed for certain metal and nonmetal ore deposits. Examples include dragline stripping and hydraulic removal of phosphates, placer mining, and dredging of alluvial beach sands, gold, and tin. Sand and gravel generally are mined from shallow pits and small surface excavations, or with draglines and dredges from natural or artificial ponds.

Crushed stone and dimension stone are mined from rock quarries. Stone for crushed aggregate generally is mined by conven-

William H. Pomroy is a former group supervisor of the Mine Fire Safety Group, Twin Cities Research Center, Bureau of Mines, U.S. Department of the Interior.

FIG. 9-31A.   *Area strip mining.*

FIG. 9-31B.   *Open pit mine.*

tional open pit methods. Dimension stone is removed in large blocks using wire saws, channeling machines, or similar means.

Other than maintenance facilities, the surface mining process involves few fixed structures. Rather, surface mining is performed outdoors by specialized mobile and portable mining equipment. Major equipment types and their primary functions include blasthole drills to drill holes for explosives (used in both overburden and coal or ore); stripping machines, such as draglines, stripping shovels, and bucketwheel excavators, for overburden removal; equipment, such as shovels, wheeled loaders, and hydraulic excavators,

for loading overburden, coal, or ore at the working face; and haulage trucks, trains, or conveyors to transport materials from the working face to storage or processing facilities.[5]

The mining process requires support from a wide variety of miscellaneous equipment. This equipment includes scrapers, explosives haulers, dozers, graders, water trucks, pumps, power substations, generators, crushers, cranes, backhoes, and maintenance and utility vehicles.

UNDERGROUND MINING METHODS

Access to underground mine workings from the surface is by shaft, slope, or adit. A shaft is a vertical or nearly vertical opening, usually equipped with hoisting facilities for transporting personnel, supplies, and mined products. A slope—also called an incline or decline—is an inclined opening that may be designed to accommodate a belt conveyor for materials haulage, a hoist for personnel and supplies, or diesel trackless haulage. An adit mine, or drift mine, is entered through a horizontal opening in a hillside. An individual mine may employ any or all of these access means, depending upon the depth and orientation of its coal seam or ore body and the requirements of its production and materials-handling systems.

Where a single, uniform, relatively flat-lying coal seam or ore body is worked, the entire mine is generally confined to a single working level. Where multiple, irregularly shaped, and/or steeply inclined coal seams or ore bodies are worked, mining may be spread over many levels. Large mines may comprise hundreds of miles of drifts or entries, which are horizontal openings in metal or coal mines respectively, on many levels and interconnected with shafts, slopes, raises (vertical openings between levels), and ramps (inclined openings between levels).

The predominant constraint influencing the underground mining method selected is the amount of support required to prevent failure of the roof and/or walls. Factors related to this constraint are the physical character, size, shape, orientation, and depth of the coal seam or ore body, and the characteristics of the surrounding and overlying strata. In addition, planned production capacity and the presence of water, methane, high-rock temperatures, radon, or other environmental factors may be important considerations in selecting a mining method.

Mine designs and operating practices vary widely, since nearly infinite variability exists among these factors from deposit to deposit and even within a single mine. However, the broad spectrum of underground mining methods can be subdivided into three generic classifications:

1. Methods that produce openings that are naturally supporting or that require minimal artificial support.
2. Methods that provide for substantial artificial roof support.
3. Caving methods, where roof failure is integral to the mining process.

With few exceptions, coal mining systems differ from noncoal mining systems, and this chapter treats them separately. Similarities that do exist are limited to softer, bedded nonmetallics, such as potash, salt, and trona, which are often amenable to coal mining systems and equipment.

Underground Coal Mining

The two basic coal mining methods are (1) room-and-pillar, which produces naturally supportive openings, and (2) longwall, which requires substantial artificial support.[2] As its name implies, room-and-pillar mining involves extracting coal in a regular pattern of rooms separated by pillars of in-place coal to support the roof. (See Figure 9-31C.)

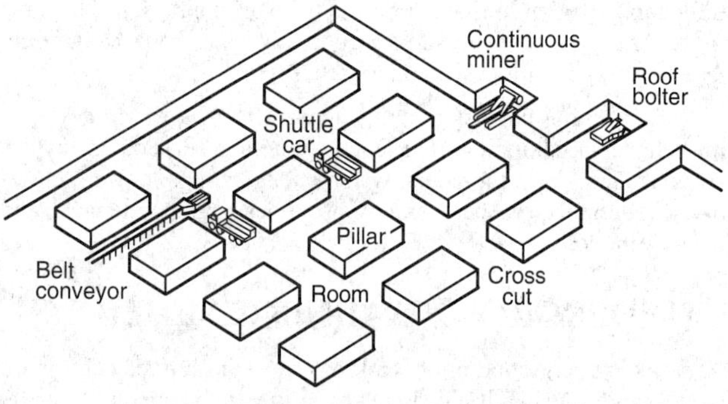

FIG. 9-31C.   Room-and-pillar coal mining.

Two systems are used for room-and-pillar coal mining: (1) conventional and (2) continuous. Of the two methods, continuous mining is much more common in the United States. In conventional mining, coal is extracted in a sequence of operations, and executing each step requires specialized equipment. A cutting machine resembling a large chain saw cuts a slot into the coal face across the entire width of the room along the floor, center, or roof line. Next, a drill bores holes for explosives in the coal face. A loading machine gathers coal broken by the explosives. The machine either feeds a self-propelled shuttle car or may be loaded and hauled by a self-propelled diesel or electric scoop. In either case, coal is discharged onto a conveyor belt or into rail cars for transport to the surface or to hoisting facilities.

In continuous mining, a single machine called a "continuous miner" mechanically cuts coal from the face using a rotating drum fitted with dozens of cutting picks. The continuous miner also has a means for gathering broken coal and loading it into a shuttle car or self-propelled scoop. After each drill-blast-load cycle in conventional mining, or each machine advance in continuous mining, the mining equipment is withdrawn, and a machine called a "roof bolter" secures the roof. Holes are drilled into the roof, and long bolts are inserted and anchored to strengthen the roof span and prevent roof failure.

Longwall mining involves extracting coal in a nearly continuous operation using a specialized, integrated mining and roof support system. (See Figure 9-31D.) Using standard room-and-pillar techniques, blocks of unmined coal called longwall panels are prepared. These blocks are typically 300 to 1,000 ft (90 to 305 m) by 3,000 to 6,000 ft (915 to 1,830 m); however, "super longwalls" with lengths of 20,000 feet (6,100 m) or more are possible. Each longwall panel is mined in linear slices at full seam height by a mechanical cutting machine or plow drawn back and forth across the short face. Cut coal is forced onto a chain conveyor that extends the face's entire length. Large, self advancing hydraulic jack units called "longwall shields" support the roof immediately adjacent to the face. As the roof support line advances with the coal face, the roof behind it is permitted to cave. Longwall mining is more productive and generally considered safer than conventional or continuous mining. It accounts for about 35 percent of U.S. underground coal mine production.

Most coal mining equipment is electrically powered. Electrically powered mobile equipment, such as undercutting machines, drills, loading machines, continuous miners, shuttle cars, scoops, and roof bolters, generally receive power through a trailing cable. Belt conveyors, pumps, and trolley haulage systems are generally wired directly to a permanent or semipermanent power center. Battery powered scoops, personnel transports, and utility vehicles are also common. Some mobile haulage equipment, notably scoops, load-haul-dump (LHD) vehicles (low-profile front-end loaders, also called "scoop trams"), and rail locomotives, can also be diesel-pow-

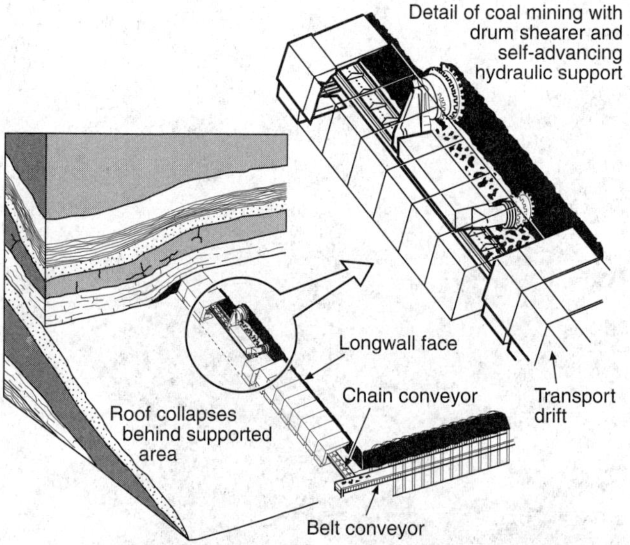

FIG. 9-31D.   Longwall coal mining. (Source: Atlas Copco)

ered. Although significant and growing, diesel powered equipment is not as extensively used as electrically powered equipment in underground coal mining.

Numerous ancillary operations must be coordinated with the mining process. The main haulage network, whether rail or belt, must function smoothly to prevent materials-handling bottlenecks from interfering with production. The roof in active areas is constantly monitored for failures or signs of impending failure. In heavily traveled areas, such as main haulageways and important ventilation routes, heavy timbers or steel arches or beams are often installed to help support the roof. Water supply, power distribution, and ventilation systems require periodic attention; equipment and facilities inspection, maintenance, and repair are continuous. Dust control also receives high priority, both to protect miners' health and to prevent explosive conditions.

Underground Metal and Nonmetal Mining

Underground metal and nonmetal mining methods are more diverse than underground coal mining methods because the physical character, size, shape, and orientation of these ore bodies vary more greatly than most coal seams.[6] The following paragraphs describe six standard mining methods. In actual practice, great variability, combination, and adaptation of standard methods are required to suit local conditions.[7] Room-and-pillar and sublevel stoping methods are used in extraction areas that require minimal artificial roof support. Shrinkage stoping and cut-and-fill are used in extraction areas that require substantial roof and wall support. The two most common caving methods are (1) sublevel caving and (2) block caving.

The room-and-pillar method is used for relatively flat-lying deposits of moderate thickness with a competent roof. With the exception of the softer, bedded nonmetallics, metal and nonmetal mine openings are generally larger than those in typical room-and-pillar coal operations, permitting the use of larger, more productive mining equipment. (See Figure 9-31E.)

Sublevel stoping, used in vertical or nearly vertical ore bodies of highly competent rock, requires considerable pre-mine development to prepare the ore body for production mining. Sublevel drifts along the longitudinal centerline of the ore body are driven in a vertical column between the mine's main levels. (See Figure 9-31F.) On the main haulage levels, drawpoints perpendicular to the sub-

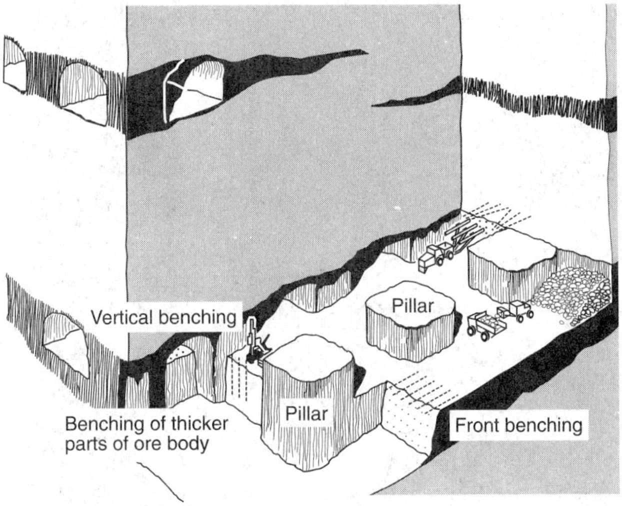

*FIG. 9-31E.   Room-and-pillar metal and nonmetal mining.
(Source: Atlas Copco)*

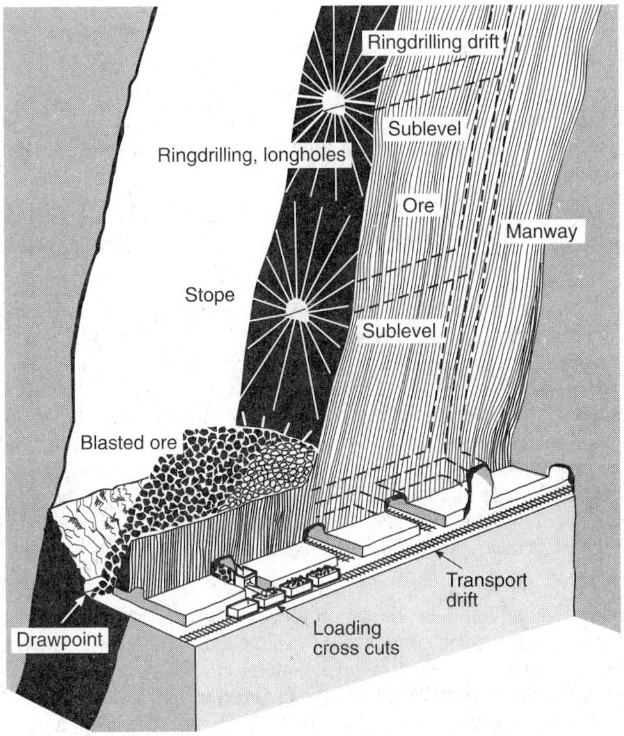

FIG. 9-31F.   Sublevel stoping. (Source: Atlas Copco)

level drifts are driven into the ore body. A complete radial pattern of blastholes, or parallel rows of vertical blastholes, are then drilled at regular intervals along the entire length of each sublevel drift. Each ring or row of holes is loaded individually with explosives and blasted. Every blast vertically slices ore, which falls to the bottom of the open cavity, or stope, where it is loaded out through the draw-points.

If the ore or surrounding rock is too weak or the rock pressures too great, open stopes are not feasible. Such ore bodies require arti-ficial support of the stopes during mining. Shrinkage stoping and

cut-and-fill stoping are most commonly employed to extract such ore bodies. Shrinkage stoping requires very little preproduction de-velopment. Drawpoints are driven into the ore body at regular inter-vals along the main levels, and raises are driven into the ore between levels. (See Figure 9-31G.) The ore is extracted in horizontal slices, starting at the bottom of the stope and advancing upward by drilling and blasting the roof. Just enough broken ore is removed from the stope to provide a work platform for miners. Ore remaining in the stope supports the walls and prevents adjoining rock from breaking loose and contaminating and diluting the ore. The process of mining horizontal slices repeats until the upper limit of the stope is reached. The stope is then emptied of ore.

Preproduction development for cut-and-fill mining is similar to that for shrinkage stoping, except it requires only one or two drawpoints. (See Figure 9-31H.) Mining is similar as well, with the ore body mined in horizontal slices from bottom to top. However, the stope is emptied of broken ore after each slice, or cut, and the stope is filled with waste material before mining resumes. The fill provides a work platform for miners and supports the stope walls. Ore is removed from the stope through timber or steel ore chutes embedded in the fill.

A high degree of mechanization, high production capacity, and low production costs characterize caving systems. In sublevel cav-ing, preproduction development is similar to sublevel stoping, with sublevel drifts driven into the ore at 30- to 50-ft (9- to 15-m) inter-vals. (See Figure 9-31I.) Drawpoints on the main levels are not re-quired, however, because the ore is retrieved through the sublevel drifts. Fan-shaped blasthole patterns are drilled upward at regular intervals along the entire length of each sublevel drift. Drill fans are blasted individually, with broken ore caving into the sublevel drifts. LHDs load the ore and transport it to an ore pass system.

The principle of block-cave mining is the natural fracturing and caving of ore that has been undercut over a large area. The ore extraction phase requires minimal drilling and blasting, because the ore fractures itself as gravity acts on the undercut rock masses.

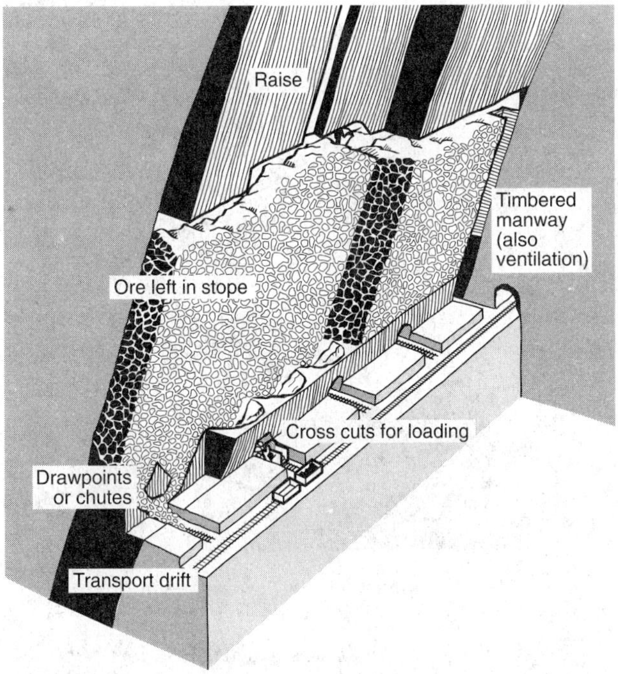

FIG. 9-31G.   Shrinkage stoping. (Source: Atlas Copco)

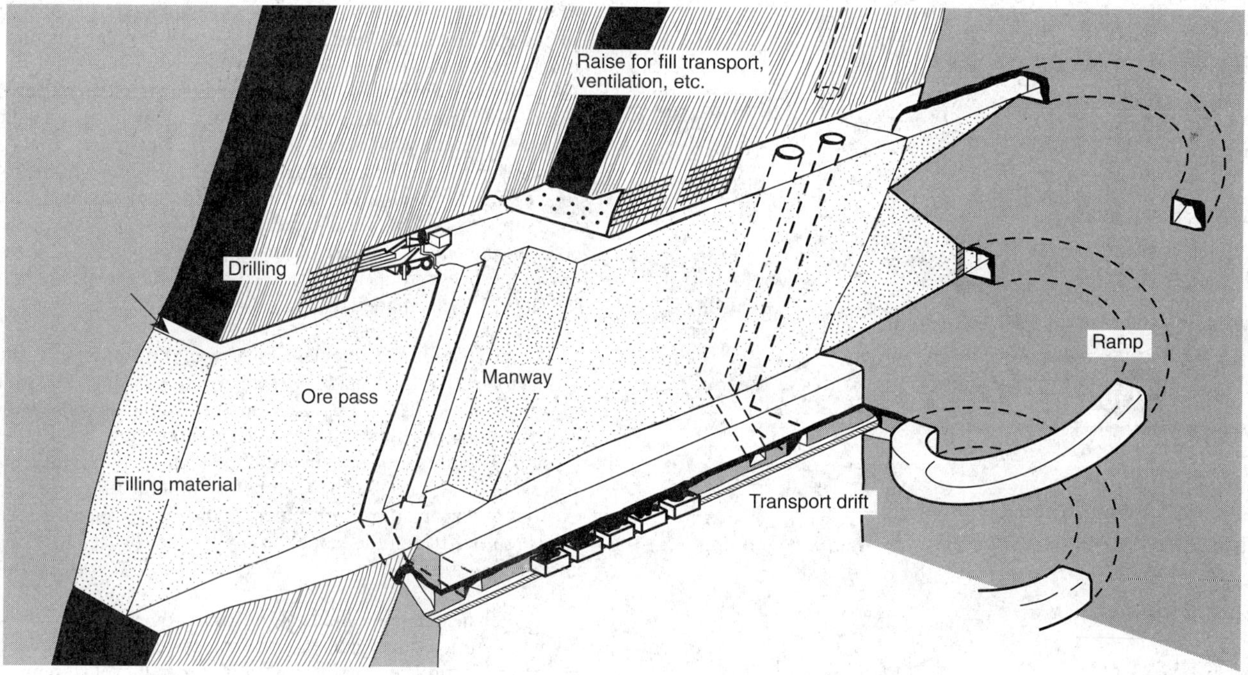

FIG. 9-31H. Cut-and-fill mining. (Source: Atlas Copco)

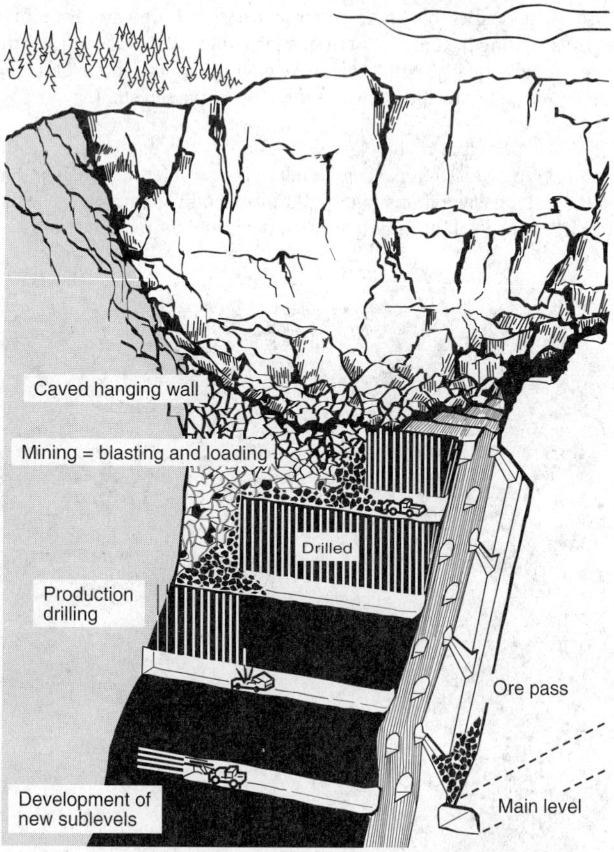

FIG. 9-31I. Sublevel caving. (Source: Atlas Copco)

Preproduction development is undertaken in four steps. (See Figure 9-31J.) First, a network of haulageways is developed beneath the ore horizon. Next, finger raises are driven up to a set of grizzly drifts. From the grizzly drifts, a second set of finger raises is driven, gradually widening to form funnels at the undercut level. Finally, the entire ore block is undercut.

The weight of overlying ore and overburden creates enormous stresses on the ore at the bottom of the block, causing it to fracture and cave into the funnels.

A wide variety of equipment is used in underground metal and nonmetal mining. Where space permits, drilling is performed with mobile "drill jumbos" fitted with two or more rock drills. Diesel-powered LHDs are used to load and haul broken ore from the draw-point or working face. Where space is limited, small, portable, man-ually operated rock drills perform drilling. Broken ore is loaded out by small pneumatic loading machines or by a scraper blade sus-pended on a cable between a hoist and a pulley, which gathers ore as it is dragged over a pile. Main haulage usually is accomplished with belt conveyors, rail cars, or haulage trucks.

As with underground coal mining, a vast array of ancillary oper-ations supports the ore extraction process. Primary among these are the inspection, maintenance, and repair of mine equipment, which is performed at most mines in large, underground shops. Mines relying on diesel equipment also must have facilities in which to store and/or transfer diesel fuel. In addition, power distribution, ventilation, hoist-ing, roof control, and dewatering are important ancillary operations.

With the dual goals of increasing productivity and removing miners from hazardous workplaces, emerging mining technologies in the coal and metal and nonmetal sectors employ modern automa-tion and robotics technology. In the coal sector, significant research is aimed at both fully automated and tele-operated remotely-con-trolled extraction and haulage systems. The metal and nonmetal sector has developed and deployed fully automated haulage systems based on both rail and trackless equipment.

FIRE AND EXPLOSION HAZARDS

Fires and explosions pose a constant threat to miners' safety and mines' productive capacities. Mine fires and explosions historically

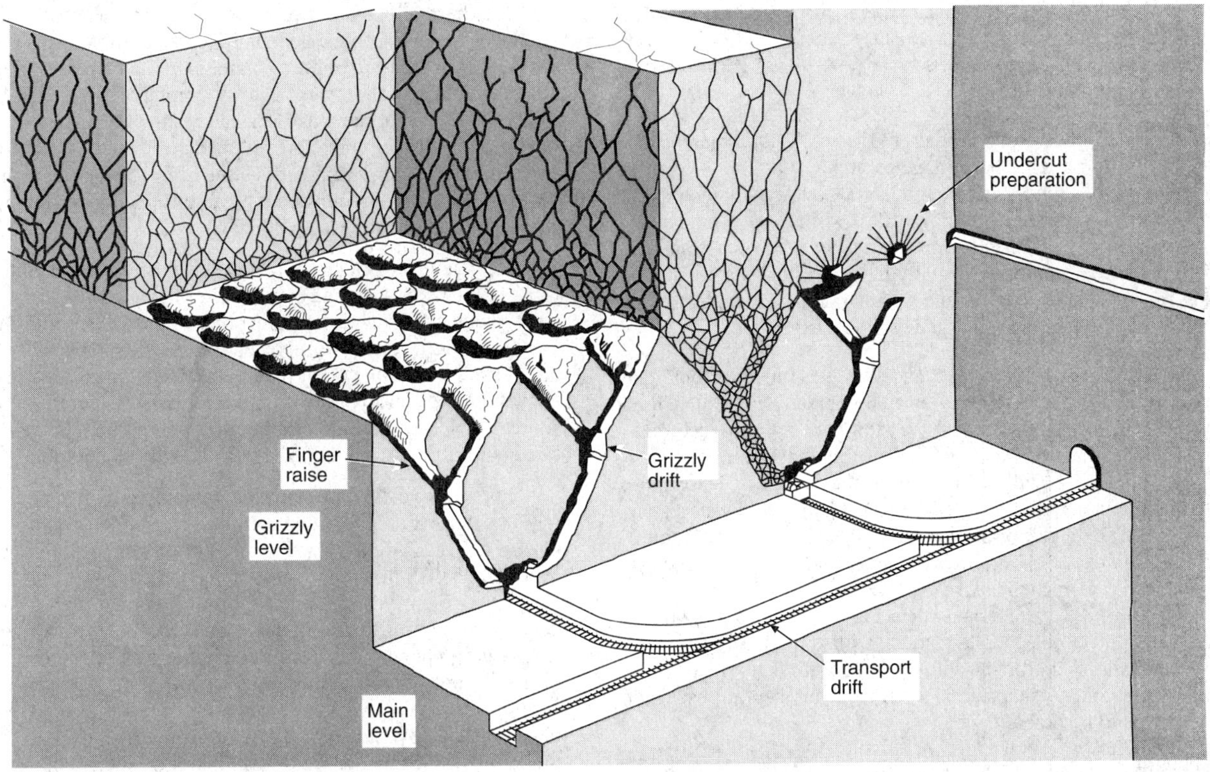

FIG. 9-31J. Block caving. (Source: Atlas Copco)

rank among the most devastating industrial disasters: over 10,000 miners died in coal mine explosions alone from 1880 to 1940.[8] More recent advances in mining and safety technology have reduced losses to a tiny fraction of their former levels. However, fires and explosions continue to occur, challenging the best efforts of mine safety, production, and engineering personnel to minimize losses.[9] The following sections discuss major fire and explosion hazards in mining.

Mine Fire Hazards

During the 1960s and 1970s, about 50 fires per year were reported to federal mine safety authorities, nearly all in underground mines. Nonreportable fires, defined as those lasting less than 30 min and not resulting in injury, were estimated at another 200 per year.[10–14] In the mid-1970s, fire frequency steadily declined, probably due to stricter mine safety regulations and the adoption of improved mining and safety technology, such as roof bolts for roof support, ac-powered face equipment in coal mines, and so on. Reported fires in underground mines now average less than 20 per year; however, the downward trend has leveled off.[15,16] Another 10 to 15 fires per year are reported in surface mines.

From 1968 to 1994, mine fires resulted in over 180 fatalities. Although nearly one-half of the fires occurred in surface mines, over 80 percent of the fatalities occurred in underground mines, where smoke and toxic fire gases present the primary life safety hazards. Even more significant than the toxic effects of smoke may be its effect in limiting visibility, causing disorientation, confusion, and difficult travel. Table 9-31A lists underground fires by equipment type involved for both coal and noncoal mines. Table 9-31B shows fires in surface mines.

More fires originate on mobile equipment than at any other source. In underground coal mines, most mobile equipment is elec-

TABLE 9-31A. Equipment Involved, Injuries, and Fatalities in Underground Mine Fires

	Coal*	Noncoal†
Belt conveyor/conveyor drive	33	7
Mobile equipment	57	36
Stationary electrical	14	12
Welding/cutting	23	8
Explosives	0	6
Other/none/unknown	37	12
Total fires	164	81
Injuries	43	51
Fatalities	30	0

*1978–1992 (Source: U.S. Bureau of Mines)[15]
†1980–1994 (Source: MSHA)[16]

TABLE 9-31B. Equipment Involved, Injuries, and Fatalities in Surface Mine Fires*

	Coal	Noncoal
Belt conveyor/conveyor drive	17	3
Mobile equipment	65	11
Welding/flame cutting	4	5
Structures/other	64	14
Total	150	33
Injuries	15	46
Fatalities	5	2

*1980–1994 (Source: MSHA)[16]

trically powered, and most fires result from an electrical fault. During the 1960s and 1970s, about 40 percent of underground coal mine electrical fires involved face equipment powered through a trailing cable, and faults in the cable caused about half of these fires. In recent years, conversion from dc to ac power has significantly reduced face equipment fires.

In noncoal underground mines, most mobile equipment is diesel powered, and a large percentage of fires result from broken or leaking hydraulic fluid, fuel, or lubricating oil lines, which spray combustible liquid mist onto a hot surface, such as an engine manifold. In recent years, diesel equipment usage has grown rapidly in underground coal mining, especially in the western states. In coal mines, the presence of coal, coal dust, and methane compounds fire hazards normally associated with diesel operations.[17]

As noted earlier, prototype and first-generation commercial mining systems incorporating automated, remotely controlled, teleoperated, and robotic equipment are slowly being introduced into underground mines.[18–20] Presently quite limited in number, such systems are likely to be widely used in the future. Because these systems contain both fuel sources (diesel fuel, hydraulics, hose, wiring, tires, etc.) and ignition sources (diesel engines, electrical equipment, conveyors, brakes, etc.), risks exist for onboard fires. Because these systems are unattended, risks are compounded by the absence of an operator who could quickly detect and suppress a fire while it is still small. In a mine operated largely with automated equipment, the underground work force would likely be quite small. In the event of a large fire, quickly marshaling a sufficient number of fire brigade members could be problematic.

In surface mines, mobile equipment is also a frequent source of fires. (See Figure 9-31K.) Like fires on underground diesel equipment, fires on diesel powered surface mining equipment typically involve leaking high-pressure hydraulic lines that spray hot, highly combustible liquid mist onto an ignition source, such as a hot exhaust manifold or turbocharger. Fire can grow so quickly on this equipment that the operator's cab can become engulfed in flame before the driver is even aware of a fire. On large power shovels, blasthole drills, and draglines, faults in electrical equipment most frequently cause fires.

Welding and cutting operations, a leading cause of mine fires, have been increasing in frequency in recent years. Welding and cutting operations are the primary cause of shaft fires in both coal and noncoal underground mines. The potential seriousness of shaft fires cannot be overemphasized, because a fire in a shaft may contaminate the mine's intake air supply and/or block a critical escape route from the mine.

FIG. 9-31K. Haulage truck fire.

Exothermic oxidation reactions occur in both coal and metal sulfide ores.[21,22] When the heat these reactions generate is not dissipated, the temperature of the rock mass or pile increases. If temperatures become high enough, rapid combustion of coal, sulfide minerals, and other combustibles can result.[23] Although spontaneous ignition fires occur relatively infrequently, they are generally quite disruptive to mine operations and difficult to extinguish. Spontaneous ignition fires often occur in remote, abandoned mine sectors where fire-fighting access is difficult.

Heat generated by friction between a conveyor belt and a drive roller or idler was a leading cause of fires in underground coal mines until about 1970, when requirements for sequence and slippage switches became effective. In recent years, with conveyor belt haulage replacing rail haulage systems in older mines, fires involving conveyors and conveyor drives have increased. Other frictional heat sources include overheated rubber tires and parking brakes on mobile equipment that are engaged during vehicle operation.

For many years, timber most commonly provided both temporary and permanent roof support for underground openings. Its extensive use resulted in a large number of timber fires. Before 1967, over half of all underground metal and nonmetal mine fires involved timber. In recent years, the use of roof bolts has significantly reduced the need for timber underground. But older mines still contain large amounts of timber, and timber still supplements other means of roof support in most mines. Only about 10 percent of underground fires involve wood. However, timber fires tend to constitute a greater overall hazard than this figure indicates, because timber fires spread more rapidly, burn longer, and are more difficult to extinguish than fires involving most other extraneous combustible material.

The ultimate objective of underground coal mine fire protection efforts is preventing ignition of the coal. The amount of combustible material present generally limits equipment fires. Once coal ignites, however, the entire mine represents the available fuel supply. Over one-third of fires involving burning coal last longer than 24 hr, with an average duration of 6 days. Failure to extinguish a developing fire results in a 50-percent chance that more than 8 hr will be required to bring it under control and a 5-percent chance that part or all of the mine will be sealed.[24,25]

The storage, handling, and use of flammable and combustible liquids pose fire hazards for all sectors of the mining industry. More fires in surface mines involve flammable and combustible liquids than any other combustible material. Most occur in mobile equipment. Hydraulic fluid at elevated temperature and pressure is a primary hazard.

In underground mines, the storage of flammable and combustible liquids is an especially important concern. These materials ignite more easily and propagate more rapidly than ordinary combustibles involved in fires. Both flammable and combustible liquids are stored underground in most noncoal mines. In some mines, the main storage facility for diesel fuel, lubricating oil and grease, and hydraulic fluid is underground, with storage facility capacities reaching 30,000 gal (11.4 m³). The potential seriousness of a fire in an underground flammable and combustible liquid storage area prompted extreme care in storage areas' design, plus the implementation and strict enforcement of safe operating procedures. As a result, few fires have been reported in such storage areas.

Mine Explosion Hazards

Over 1,400 ignitions and explosions have occurred in U.S. mines since 1980.[16] Underground coal mines accounted for over 90 percent of the total. These ignitions and explosions killed over 100 miners and injured over 200.

Practically all underground coal mine explosions start with methane ignition. Methane is a colorless, odorless gas usually present in coal seams and sometimes in the rock strata above and below coal seams.[26] It is also found in some noncoal mines. During mining, methane escapes into the mine and is transported by the mine's ventilation air flows. In a properly designed and operated ventilation system, the air flows dilute the methane, render it harmless, and carry it away. However, if ventilation is inadequate, or if a large amount of methane releases suddenly (in a methane outburst, for example), dangerous and explosive methane-air mixtures can form. Even when the average methane concentration is below dangerous levels, explosive mixtures may form in pockets or layers.

The high number of ignitions and explosions can be attributed to the small amount of methane necessary to produce an explosive mixture in air and to the low energy required to ignite a methane-air mixture. The lower explosive limit (LEL) for methane in air is 5 percent. As little as 0.3 millijoule of electrical energy can ignite methane under proper conditions. This is equivalent to $1/50$ of the static electricity the average person accumulates walking across a carpeted floor on a dry day.[8]

If an explosion propagates more than a few hundred feet, coal dust, swept off the floor, roof, and side walls of the mine entry, becomes the most important fuel source. The LEL for a suspension of high-volatile bituminous coal dust in air is 0.05 oz/cu ft (0.28 g/m³). Methane's presence in the atmosphere increases the hazard by producing a linear reduction in the LEL for coal dust. In the absence of methane, electrical or frictional sparks, or explosives can ignite coal dust. Much more energy is required for ignition if methane is not present. Methane ignitions generally trigger coal dust explosions. The flame provides the necessary ignition source, and the shock wave causes settled coal dust to become airborne.[27]

The destructive potential of ignitions and explosions results from the high-temperature flame fronts and the static and dynamic pressures produced. Flame temperatures of 3,600°F (1,982°C) and static pressures in excess of 50 psig (345 kPa) are possible. Dynamic, or "wind," pressure is directional and depends upon the air velocity the explosion produces, which, in turn, depends upon the static pressure.

Table 9-31C shows ignition sources of recent mine explosions and methane ignitions. Frictional sparking between rock surfaces and cutting bits, continuous miners, roof bolters, cutting machines (undercutters), face drills, and longwall equipment causes over 85 percent of ignitions and explosions in underground coal mines. Other possible sources of frictional sparking are drill steel striking machine frames, sandstone hitting sandstone or other hard rock during a roof failure, and mobile vehicle brake linings.

Welding and cutting operations are the second leading cause of mine explosions and ignitions. In underground coal mines, the trend since 1970 has been increasing numbers of explosions and ignitions caused by welding and cutting operations.

About 2.5 percent of underground coal mine ignitions and explosions result from electric arcs. This figure is surprisingly low, considering the extensive use of electrical equipment in underground coal mine face areas and the low electrical energies required to ignite methane.

One cause of explosions in noncoal mines is sulfide dust ignition during a blast.[28] As noted earlier, ore minerals in many metal mines are metallic sulfides and are combustible under certain conditions. In some cases, the blast may create a combustible dust that the flame produced from the explosive or blasting agent can ignite.

In surface mines, the predominant ignition and explosion sources are welding and flame cutting, and kilns, furnaces, and dryers. Surface welding and cutting explosions usually involve hot

TABLE 9-31C. *Mine Ignition and Explosion Sources,* Injuries, and Fatalities*

	Underground		Surface	
	Coal	Noncoal	Coal	Noncoal
Electrical	33	1	8	0
Explosives	26	0	3	0
Welding/cutting	96	4	18	8
Friction				
Continuous miner	814	0	1	0
Roof bolter, drill	47	1	0	0
Longwall	211	0	0	0
Cutting machine	19	0	0	0
Auger	0	0	0	8
Kiln/dryer /furnace	0	0	40	15
Other	30	1	18	33
Total ignition and explosions	1276	7	88	64
Injuries	147	5	43	37
Fatalities	90	2	5	9

*1980–1994 (Source: MSHA)[16]

work on vessels that had contained flammable or combustible gasses or liquids but were not properly inerted. Some explosions also result from welding or cutting pressurized vessels, including rims of inflated tires.

SAFEGUARDS

Fire and explosion safety in the mining industry is based on the general principles of fire and explosion prevention, early fire detection and warning, fire and explosion suppression, limiting fire propagation, and providing for miners' safety during a fire. The following sections discuss implementation of these general principles.

Most states and the federal government have promulgated mandatory regulations addressing fire protection for both surface and underground mines.[29] In addition, many NFPA codes, standards, recommended practices, and guides apply to mining operations; Authorities Having Jurisdiction have adopted some of them. The following discussion of general fire and explosion safety principles and practices is for informational purposes only, not as a comprehensive list of mandatory safety requirements.

Fire Prevention

Fire prevention practices fall into three categories: (1) limiting fuel sources, (2) limiting ignition sources, and (3) limiting fuel and ignition source contact.

Limiting fuel sources: Where fires involving certain combustible materials are of sufficient frequency and/or severity, and where suitable and less hazardous substitutes for those materials are available, using less hazardous materials is preferred, or, in some cases, mandated by law. Examples of less hazardous materials include fire resistive hydraulic fluids, conveyor belting, hydraulic hoses, and ventilation tubing; all are required in underground mines under certain conditions.[30,31,32] In some cases, highly toxic products of combustion resulting from the burning of certain materials may prompt substitution with less hazardous materials. For example, many countries have banned polyurethane foam, once widely used in underground mines for ventilation seals. Even more basic is good housekeeping to prevent unsafe accumulations of trash, oily rags, coal dust, and other combustible materials.

Limiting ignition sources: Fire sources nonessential to the mining process can be banned altogether. Examples are prohibitions on open fires and on smoking in underground coal mines and in certain areas of underground noncoal and surface mines. Means to prevent unwanted heat buildup, such as slippage and sequence switches on conveyors and thermal cutouts on electric motors, are also common.

Limiting fuel and ignition source contact: Many combustible materials and potential ignition sources are essential to mining operations. Fire safety in these instances depends upon preventing contact between the ignition source and the fuel source. For example, precautions are taken during welding and cutting operations. When welding cannot be performed in fire safe enclosures, areas are wet down, and, if practical, nearby combustibles are covered with fire resistive materials or relocated. Fire extinguishers must be readily available and a fire watch posted for as long as necessary to guard against smoldering fires. On mobile equipment, the hydraulic fluid, fuel, and lubricant lines can be rerouted away from hot surfaces, electrical equipment, and other possible ignition sources. Spray shields also can be installed to deflect sprays of combustible liquid from a broken fluid line away from a potential ignition source. Areas with a high loading of combustible materials, both underground and surface, such as timber storage areas, explosives magazines, flammable and combustible liquid storage areas, and shops, can be designed to minimize possible ignition sources. Potential sources of ignition, such as power substations, battery chargers, and electric motors, can be kept clear of extraneous combustible material.

Explosion Prevention

Explosion prevention practices fall into the same three categories as fire prevention practices: (1) limiting fuel sources, (2) limiting ignition sources, and (3) limiting fuel and ignition source contact.

Limiting fuel sources: Coal dust and methane are the primary fuels involved in underground coal mine explosions. Methane also may be present in noncoal mines.[33] Although every attempt is made to minimize the generation of coal dust in mining processes, the tiny amount of dust sufficient to propagate a coal dust explosion is virtually unavoidable. Indeed, a layer of dust on the floor just 0.005 in. (0.012 mm) thick, if suspended in air, propagates an explosion; less is necessary if methane is present. Bituminous coal dust properly mixed with an inert material, such as pulverized limestone, dolomite, or gypsum (rock dust), does not propagate an explosion.[8] Hence, regulations require that the incombustible content of dust in coal mines be frequently checked and followed by rock dusting if necessary to insure incombustibility.

Methane is treated in two ways. Most commonly, it is diluted by ventilation air and exhausted from the mine. However, "degassification" in advance of mining may be practiced where levels of methane are unusually high.[34,35]

Limiting ignition sources: Every effort is made to eliminate extraneous ignition sources from the mining environment. Smoking is prohibited in coal mines, and electrical equipment operating where methane may be present must be "permissible" or "intrinsically" safe. Permissible electrical equipment has been tested by the federal government and designed, constructed, and installed so that its operation will not cause a mine fire or explosion. Use of explosion-proof enclosures, specially designed plugs and receptacles, automatic circuit interrupting devices, and voltage limitations are typical. Intrinsically safe equipment has been tested by the federal government and determined to be incapable, under normal or abnormal conditions, of releasing sufficient electrical or thermal energy

to ignite an explosive gas mixture. Some success has been achieved in reducing frictional sparking at the face by using slower bit rotation, novel bit designs and bit materials, and specially engineered water sprays.[36–40] In coal mines, explosives must be permissible and used in strict conformance with special blasting regulations. Where sulfide dust explosions are possible in metal mines, special anti-incendive blasting agents are available.[28] In both coal and noncoal mines where methane is present, diesel powered equipment must also be permissible.

Limiting fuel and ignition source contact: Because of geologic nonuniformity, varying mining rates, atmospheric pressure, and other factors, methane is not uniformly liberated into mine ventilation air from the workings. In addition, methane and ventilation air may not mix completely and may form explosive pockets or layers, even though the average methane concentration is below the LEL. Thus, diluting and draining methane cannot always be relied upon to produce nonexplosive atmospheres. Great effort has been devoted to developing safer procedures for ventilating active coal mine working areas, both longwall and room-and-pillar, to minimize the occurrence of localized pockets of explosive methane/air mixtures.[41–43] Methane checks are made frequently with handheld and machine-mounted methanometers. If the methane levels reach or exceed a predetermined point, all electrical equipment is shut down.

Early Fire Detection and Warning

The time elapsed between fire's onset and its detection is critical, because fires tend to grow in size and intensity with time. Early fire detection and warning permit the initiation of an emergency plan in the mine (evacuation, fire fighting, etc.) while the fire is still small or, ideally, while it is still in the incipient stage. Over 70 percent of fires detected within 15 min only lightly damage a mine.[11] Advanced fire detection and warning systems using sensitive heat, flame, smoke, and gas analyzers provide the most rapid and reliable indication of fire.[44–48]

Thermal fire detection systems are commonly installed over conveyor belts, belt drives and takeups, and other unattended equipment. In localized, high-hazard areas, such as flammable and combustible liquids storage areas, refueling areas, and shops, faster-acting fire detection devices may be appropriate. Optical flame detectors, which sense either the ultraviolet or infrared radiation a fire emits, have been used successfully.

Smoke and/or gas detection is the most cost-effective approach to providing large area, or even whole mine, fire detection coverage.[49-63] (See Figure 9-31L.) Instruments sampling the mine ventilation air streams can detect increases in particulates, or smoke, of less than one micrometer (1 μm) in size in concentrations as low as 1 mg/m³ and of distinguishing diesel smoke from fire smoke.[56] Instrumentation also is available that can detect very small increases in combustion gases. For example, instruments intended for permanent underground installation can reliably detect carbon monoxide (CO) at levels as low as 2 parts/million (2 ppm).

After fire detection and notification of responsible mine officials, each miner must be warned. Telephones and messengers are sometimes used, but miners are often too remote from telephones and too widely scattered for these means to provide a timely warning. In coal mines, the most common means of fire warning are electric power shutdown and/or notification by telephone and messengers. In noncoal mines, electric power shutdown is generally ineffective because so little equipment is electrically powered. Stench warning is the most common method of emergency communication in noncoal underground mines.[64]

Wireless radio frequency communication systems have been used successfully in both coal and noncoal mines.[56] Ultrahigh frequency systems utilize a "leaky feeder" transmission antenna that is installed wherever underground signal reception is desired. Miners use beltheld "walkie talkie" type or vehicle mounted transceivers for two-way voice communication. Ultralow frequency systems utilize antennae deployed on the surface that transmit signals through the earth to receivers built into miners' cap lamp battery cases or to dashboard mounted receivers installed on mobile equipment. These systems enable one-way transmission of messages, which appear on LCD displays.[65-67]

FIG. 9-31L.   Fire detection instruments in an underground mine.

Fire Suppression

The types of fire suppression equipment most commonly used in underground coal mines are portable hand extinguishers, generally multipurpose dry chemical; water hoselines and sprinkler systems; rock dust, applied manually or from a rock dusting machine; and foam generators. Fire suppression equipment in noncoal underground mines includes portable hand extinguishers and water hoselines. Fire suppression systems, manual or automatic, are becoming more common for mobile equipment, combustible liquids storage areas, conveyor belt drives, and electrical installations.[56,64,68-70] (See Figure 9-31M.) Automatic fire suppression is especially important for unattended, automated, or remote control equipment where personnel are not present to detect a fire, activate a fire suppression system, or initiate fire-fighting operations.

NFPA 123, *Standard for Fire Prevention and Control in Underground Bituminous Coal Mines,* contains fire protection requirements for underground coal mines. NFPA 122, *Standard for Fire Prevention and Control in Underground Metal and Nonmetal Mines,* covers requirements for underground metal and nonmetal mines.

Where large areas of coal or timber are burning in an underground mine and fire fighting is complicated by extensive roof falls, ventilation uncertainties, and explosive gas accumulations, the only practical alternatives are inerting with nitrogen, carbon dioxide, or the combustion products of an inert gas generator; flooding with water; and/or sealing all or parts of the mine.[71] All underground miners are trained to use basic fire-fighting equipment. However, only highly trained and specially equipped fire-fighting teams can fight advanced fires. To assist these teams in fighting advanced fires,

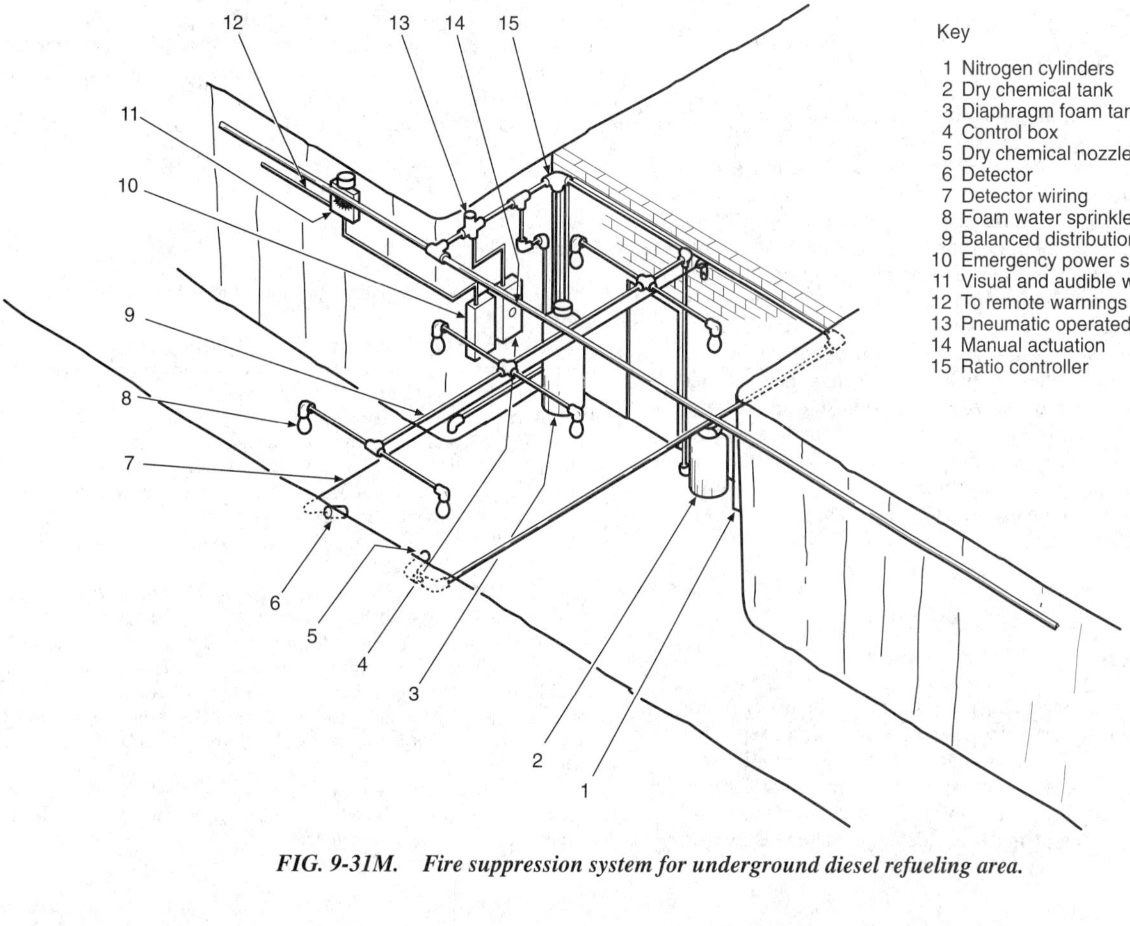

Key

1 Nitrogen cylinders
2 Dry chemical tank
3 Diaphragm foam tank
4 Control box
5 Dry chemical nozzle
6 Detector
7 Detector wiring
8 Foam water sprinkler head
9 Balanced distribution piping
10 Emergency power supply
11 Visual and audible warnings
12 To remote warnings
13 Pneumatic operated main control valve
14 Manual actuation
15 Ratio controller

FIG. 9-31M.   Fire suppression system for underground diesel refueling area.

research has investigated the technological feasibility of tele-operated remote-controlled fire-fighting vehicles capable of approaching fires at close range to apply fire suppressing agents.

Mobile equipment is the primary fire hazard in surface mines. Consequently, most surface mining fire suppression equipment is specifically designed for maximum effectiveness on typical mobile equipment fires. Mobile equipment is commonly provided with one or more portable hand extinguishers. Manually or automatically actuated suppression systems may be installed for added protection.[72-77] NFPA 121, *Standard on Fire Protection for Self-Propelled and Mobile Surface Mining Equipment*, contains fire protection requirements for such mining equipment.

Explosion Suppression

On a limited basis, some European coal mines use explosion suppression technology in the form of passive or triggered barriers. Passive barriers consist of rows of large tubs containing water or rock dust that are suspended from the roof of a mine entry. In an explosion, the pressure front that precedes the flame creates wind forces that dump the tubs' contents. The dispersed suppressants quench the flame as it passes through the entry protected by the barrier system. Triggered barriers utilize an electrically or pneumatically operated activation device that the heat, flame, or pressure of the explosion triggers to release suppressing agents stored in pressurized containers.[78-80]

Ignition suppression devices have also been developed for quenching methane ignition at its origin point, usually on a continuous mining machine or longwall shearer. These systems typically consist of a rapid response optical detector. Upon sighting a fire ball of sufficient size and intensity, the detector triggers the release of a suppressing agent. The systems are designed to totally flood the area surrounding the cutting drum (roof to floor, full room width), the likely ignition source.

Limiting Fire Propagation

Means for confining or limiting the propagation rate of an underground mine fire can help ensure a safer mine evacuation and lessen fire-fighting hazards. In underground coal mines, oil and grease must be stored in closed, fire-resistant containers, and storage areas must be of fire resistant construction. Transformer stations, battery charging stations, air compressors, substations, shops, and other installations must be housed in fire resistant areas or fireproof structures. Unattended electrical equipment must be mounted on a noncombustible surface and separated from the coal or protected by a fire-suppression system. Materials for building bulkheads and seals, including wood, brattice cloth, saws, nails, hammers, plaster or cement, and rock dust, must be readily available to each working section. In underground noncoal mines, oil, grease, and diesel fuel must be stored in tightly sealed containers in fire resistive areas at safe distances from explosives magazines, electrical installations, and shaft stations. Ventilation control stoppings and doors, and fire doors are required in certain areas to prevent the spread of fire, smoke, and toxic gas.[81-83]

Providing for Miner Safety During a Fire Emergency

The primary concern during an underground fire is the safety of underground personnel. Because smoke and toxic fire gases represent the greatest threat, the mine ventilation system and its relationship to various working areas, potential fires, primary and secondary escape routes, refuge chambers, and so on, are critically important. Detailed emergency planning must consider these factors. However,

the complex behavior of mine fires, the flow of contaminated air through the ventilation system, and the interactions between a fire and a ventilation system that has been altered by fire-induced flow changes are normally far too complex to understand and control without specialized analytical methods. Using computer models that simulate mine ventilation systems can be very helpful in understanding these complex processes, developing and optimizing emergency plans, and designating escape routes. Particularly valuable are advanced models that calculate ventilation changes due to a mine fire (heat-induced buoyant forces causing throttling, reversals, etc.) and can track contaminants' distribution and concentration.[56,84-86] To ensure smooth implementation of the emergency plan, miners are comprehensively trained and annually retrained in emergency procedures. Frequently performed fire drills, complete with the activation of the mine warning system, reinforce training and identify the plan's weaknesses.

Although every attempt is made to avoid such situations, evacuating miners must sometimes travel through areas contaminated by smoke and toxic fire gases. These combustion products present two hazards: (1) toxic atmospheres and (2) impaired visibility. Appropriate respiratory protection is required to safeguard miners from toxic combustion products. In coal mines, self-contained breathing apparatus, supplying a minimum of 1 hr of oxygen, must be provided.[87-89] In noncoal mines, 1-hr-rated self-rescue devices that convert carbon monoxide to carbon dioxide must be provided.

Toxic atmospheres are the principal cause of death in mine fires. However, recent research demonstrates that impaired visibility due to smoke accumulation in escapeways can become quite significant long before the concentration of combustion products becomes acutely toxic. As a result, miners lost in or disoriented by the smoke may succumb to the toxic atmosphere when they deplete their respiratory protection, even though they had adequate warning and time to escape. For this reason, smoke-free escapeways and frequent fire drills and evacuation training are especially important to familiarize miners with escape routes and evacuation procedures. To further combat the visibility problem smoke buildup causes, some mines have installed rope lifelines in escapeways. They enable miners to "feel" their way to safety through smoke-filled entries.

BIBLIOGRAPHY

References Cited

1. *Mineral Commodity Summaries*, U.S. Bureau of Mines, Washington, DC, 1993.

2. Cassidy, S. M., ed., *Elements of Practical Coal Mining*, Society of Mining Engineers of the American Institute of Mining, Metallurgical, and Petroleum Engineers, Inc., Port City Press, Baltimore, MD, 1983.

3. U.S. Bureau of Mines, "Economic Engineering Analysis of U.S. Surface Coal Mines and Effective Land Reclamation," BuMines Contract Final Report S0241049, 1975, Skelly and Loy, Engineers-Consultants.

4. Kennedy, B. A., ed., *Surface Mining*, 2nd edition, Society for Mining, Metallurgy, and Exploration, Littleton, CO, 1990.

5. Martin, J. W., *et al.*, *Surface Mining Equipment*, Martin Consultants Inc., Golden, CO, 1982.

6. Hartman, H. L., ed, *SME Mining Engineering Handbook*, 2nd edition, Society for Mining, Metallurgy, and Exploration, Littleton, CO, 1992.

7. Hustrilid, W. A., *Underground Mining Methods Handbook*, Society of Mining Engineers of the American Institute of Mining, Metallurgical, and Petroleum Engineers, Inc., Port City Press, Baltimore, MD, 1982.

8. Nagy, J., "The Explosion Hazard in Mining," U.S. Mine Safety and Health Administration Informational Report 1119, 1981, Washington, DC.

9. Richmond, J. K., *et al.*, "Historical Summary of Coal Mine Explosions in the United States 1959–81," Information Circular 8909, 1983, U.S. Bureau of Mines, Washington, DC.

10. McDonald, L. B., and Baker, R. M., "An Annotated Bibliography of Coal Mine Fire Reports," Contract Final Report J 1275008, 1979, U.S. Bureau of Mines, Washington, DC.

11. McDonald L. B., and Pomroy, W. H., "A Statistical Analysis of Coal Mine Fire Incidents in the United States from 1950 to 1977," Information Circular 8830, 1980, U.S. Bureau of Mines, Washington, DC.

12. Stephan, C. R., "Summary of Underground Coal Mine Fires, Jan. 1, 1978 to July 31, 1986," Bruceton Safety Technology Center Report No. 06-357-86, 1986, U.S. Mine Safety and Health Administration, Washington, DC.

13. Butani, S. J., and Pomroy, W. H., "A Statistical Analysis of Metal and Nonmetal Mine Fire Incidents in the United States from 1950 to 1984," Information Circular 9132, 1987, U.S. Bureau of Mines, Washington, DC.

14. Baker, R. M., Nagy, J., and McDonald, L. B., "An Annotated Bibliography of Metal and Nonmetal Mine Fire Reports," Contract Final Report J0295035, 1980, U.S. Bureau of Mines, Washington, DC.

15. Pomroy, W. H., and Carigiet, A. M., "Analysis of Underground Coal Mine Fire Incidents in the United States from 1978 through 1992," Information Circular 9426, 1995, U.S. Bureau of Mines, Washington, DC.

16. MSHA, *Mine Fires and Explosions Incident Tabulation*, Bruceton Safety and Health Technology Center, Pittsburgh, PA, 1995.

17. Bickel, K. L., "Analysis of Fires on Diesel-Powered Mine Equipment," *Proceedings* of the Bureau of Mines Technology Transfer Seminar, Diesels in Underground Mines, Information Circular 9141, 1987, U.S. Bureau of Mines, pp. 9–17.

18. Udd, J. E., and Pathak, J., "Mining Automation in Canadian Hardrock Mines—A Progress Report," *Proceedings* of the International Symposium on Mine Mechanization and Automation, Golden, CO, June 10–13, 1991.

19. Schnackenberg, G. H., and Sammarco, J. J., "Overview of the U.S. Bureau of Mines Computer-Assisted Mining Research Program," *Proceedings* of the International Symposium on Mine Mechanization and Automation, Golden, CO, June 10–13, 1991.

20. Eriksson, G., and Kitok, A., "Automatic Loading and Dumping Using Vehicle Guidance in a Swedish Mine," *Proceedings* of the International Symposium on Mine Mechanization and Automation, Golden, CO, June 10–13, 1991.

21. Kuchta, J. M., Rowe, V. R., and Burgess, D. S., "Spontaneous Combustion Susceptibility of U.S. Coals," Report of Investigations 8474, 1980, U.S. Bureau of Mines, Washington, DC.

22. Smith, A. C., and Thompson, C. N., "Development and Application of a Method for Predicting the Spontaneous Combustion Potential of Bituminous Coals," *Proceedings* of the 24th International Conference of Safety in Mines Research Institutes, Makeevka State Research Institute for Safety in the Coal Industry, Makeevka, Russia, 1991.

23. Ninteman, D. J., "Spontaneous Oxidation and Combustion of Sulfide Ores in Underground Mines," Information Circular 8775, 1978, U.S. Bureau of Mines, Washington, DC.

24. Mitchell, D. W., and Burns, F. A., "Interpreting the State of a Mine Fire," 1979, U.S. Mine Safety and Health Administration Informational Report 103, Washington, DC.

25. Mitchell, D. W., *Mine Fires*, Maclean Hunter Publishing Co., Chicago, IL, 1990.

26. Trevits, M. A., Finfinger, G. L., and Lascola, J. C., "Evaluation of U.S. Coal Mine Emissions," *Proceedings* of the 5th U.S. Mine Ventilation Symposium, Y. J. Wang, ed., Society for Mining and Exploration of the American Institute of Mining, Metallurgical, and Petroleum Engineers, Littleton, CO, 1991.

27. Cybulski, W., *Coal Dust Explosions and Their Suppression* (translated from the Polish), National Technical Information Service, Springfield, VA, 1975.

28. Weiss, E. S., *et al.*, "Low Incendive Explosives for Sulfide Ore Blasting," *Journal of Explosives Engineers*, Vol. 5, Jan./Feb. 1992.

29. Office of the Federal Register, *Code of Federal Regulations, Title 30, Mineral Industries* (Chapter 1), Mine Safety and Health Administration, U.S. Government Printing Office, National Archives and Records Administration, Washington, DC, 1987.

30. "Coal Mine Fire and Explosion Prevention," Information Circular 8768, 1978, U.S. Bureau of Mines, Washington, DC.

31. Sapko, M. J., "Fire Resistance Test Method for Conveyor Belts," Report of Investigations 8521, 1981, U.S. Bureau of Mines, Washington, DC.

32. Schenkel, W. C., "Flame Resistant Conveyor Belts Reinforced with Kevlar for Underground Coal Mining," *Proceedings* of the International Conference on Fire Resistant Hose, Cables and Belting, Plastics and Rubber Institute, London, 1986, pp. 1–17.

33. Thimons, E. D., Vinson, R. P., and Kissell, F. N., "Forecasting Methane Hazards in Metal and Nonmetal Mines," Report of Investigations 8392, 1979, U.S. Bureau of Mines, Washington, DC.

34. Hartman, H. L., Mutmansky, J. M., and Wang, Y. J., *Mine Ventilation and Air Conditioning*, John Wiley & Sons, New York, 1982.

35. Duel, M., and Kim, A. C., "Methane Control Research: Summary of Results, 1964–1980," U. S. Bureau of Mines Bulletin 687, 1988, U.S. Bureau of Mines, Washington, DC.

36. Agbde, R. O., *et al.*, "Frictional Ignition Suppression by the Use of Cutter Drum Mounted Sprays," Bituminous Coal Research Inc. Contract Final Report J0395040, 1982, U.S. Bureau of Mines, Washington, DC.

37. Hanson, B. D., "Cutting Parameters Affecting the Ignition Potential of Conical Bits," BuMines RI 8820, 1983, U.S. Bureau of Mines, Washington, DC.

38. Roepke, W. W., and Hanson, B. D., "Bit Ignition Potential with Worn Carbide Tips," Technical Progress Report 121, 1983, U.S. Bureau of Mines, Washington, DC.

39. Courtney, W. G., "Frictional Ignition with Coal Mining Bits," Information Circular 9251, 1990, U.S. Bureau of Mines, Washington, DC.

40. Watson, R. W., "Prevention of Frictional Ignition on Longwall Sections," *Journal of Mines, Metals and Fuels*, Vol. 37, No. 6-7, June/July, 1989, pp. 261–265.

41. Cecala, A. B., Rajan, S. R., and Jankowski, R. A., "Cut Longwall Ignitions by Lowering Methane Concentrations Around Longwall Shearers," *Coal Age*, Vol. 91, No. 8, 1986, pp. 66–67.

42. Cecala, A. B., Organiscak, l. A., and Jankowski R. A., "Improving Dust and Methane Control" *Coal Mining*, Vol. 24, No. 10, 1987, pp. 51–52.

43. Divers, E. F., and Volkwein, I. C., "Guidelines for Choosing Face Ventilation Systems," *Coal Mining*, Vol. 24, No. 10, 1987, pp. 46–59.

44. Griffin, R. E., "In-Mine Evaluation of Underground Fire and Smoke Detectors," Information Circular 8808, 1979, U.S. Bureau of Mines, Washington, DC.

45. Kacmar, R. M., "Reliability of Computerized Mine-Monitoring Systems," Information Circular 8882, 1982, U.S. Bureau of Mines, Washington, DC.

46. Johnson, C. A., and Forshey, D. R., "In-Mine Tests of Mine Shaft Fire and Smoke Protection Systems," Information Circular 8788, 1978, U.S. Bureau of Mines, Washington, DC.

47. Welsh, l. H., "Computerized, Remote Monitoring Systems for Underground Coal Mines—Fires and Explosive Atmospheres," Information Circular 8875, 1982, U.S. Bureau of Mines, Washington, DC.

48. Dobroski, H., Jr., and Conti, R. S., "Distributed Temperature Sensing for Underground Belt Lines," *Mine Health and Safety* (Chapter 3), Society for Mining and Exploration of the American Institute of Mining, Metallurgical, and Petroleum Engineers, Littleton, CO, 1992.

49. Hertzberg, M., and Litton, C. D., "Multipoint Detection of Products of Combustion with Tube Bundles," BuMines RI 8171, 1976, U.S. Bureau of Mines, Washington, DC.

50. Litton, C. D., "Guidelines for Siting Product-of-Combustion Fire Sensors in Underground Mines," Information Circular 8919, 1983, U.S. Bureau of Mines, Washington, DC.

51. Litton, C. D., "Design Criteria for Rapid Response Pneumatic Monitoring Systems," Information Circular 8912, 1983, U.S. Bureau of Mines, Washington, DC.

52. Stevens, R. B., "Mine Shaft Fire and Smoke Protection Systems," FMC Corp.—BuMines Contract Final Report H0242016, 1975, U.S. Bureau of Mines, Washington, DC.

53. Thomas, E. C., "A Pneumatic Sampling Fire Detection System in an Underground Haulageway," *IEEE Transcript on Industry Applications*, Vol. IA-l9, No. 3, May/June 1983.

54. U.S. Bureau of Mines, "Underground Metal and Nonmetal Mine Fire Protection," Information Circular 8865, 1981, U.S. Bureau of Mines, Washington, DC.

55. "Metal Mine Fire Protection Research," Information Circular 8752, 1977, U.S. Bureau of Mines, Washington, DC.

56. "Recent Developments in Metal and Nonmetal Mine Fire Protection," Information Circular 9206, 1988, U.S. Bureau of Mines, Washington, DC.

57. Conti, R. S., and Litton, C. D., "Response of Underground Fire Sensors: An Evaluation," Report of Investigations 9412, 1992, U.S. Bureau of Mines, Washington, DC.

58. Litton, C. D., Lazzara, C. P., and Perzak, F. J., "Fire Detection for Conveyor Belts," Report of Investigations 9380, 1991, U.S. Bureau of Mines, Washington, DC.

59. Morrow, G. S., and Litton, C. D., "In-Mine Evaluation of Smoke Detectors," Information Circular 9311, 1992, U.S. Bureau of Mines, Washington, DC.

60. Pomroy, W. H., Griffin, R. E, and Ackerson, M. A., "Rapid Response Pneumatic Fire Detection for Multilevel Metal Mines: System Design and In-Mine Testing," Report of Investigations 9305, 1990, U.S. Bureau of Mines, Washington, DC.

61. Burton, R. C., *et al.,* "Automatic Fire Detection Underground," *Colliery Guardian*, Vol. 235, No. 5, 1987, pp. 166, 168.

62. Litton, C. D., "Fire Detectors in Underground Mines," Information Circular 8786, 1979, U.S. Bureau of Mines, Washington, DC.

63. Pomroy, W. H., "Spontaneous Combustion Fire Detection for Deep Metal Mines," Report 9144, 1987, U.S. Bureau of Mines, Minneapolis, MN.

64. Pomroy, W. H., and Muldoon, T. L., "A New Stench Gas Fire Warning System," *Proceedings* of the 1983 MAPAO Annual General Meeting and Technical Sessions, Mines Accident Prevention Association of Ontario, North Bay, Ontario, Canada, 1983.

65. Hjelmstad, K. E., and Pomroy, W. H., "Novel Fire Warning System for Underground Mines," *Mining Engineering*, Vol. 42, No. 1, Jan. 1990, pp. 107–112.

66. Pomroy, W. H., and Hjelmstad, K. L., "Underground Mine Fire Warning Alarm System," *Mining Technology International*, June 1991.

67. Hjelmstad, K. E., and Pomroy, W. H., "Ultralow Frequency Electromagnetic Fire Warning Alarm System for Underground Mines," Report of Investigations 9377, 1991, U.S. Bureau of Mines, Washington, DC.

68. Pomroy, W. H., and Johnson, G. A., "Improved Fire Protection for Underground Fuel Storage and Fuel Transfer Areas," Information Circular 9032, 1985, U.S. Bureau of Mines, Washington, DC.

69. Grannes, S .G., Ackerson, M. A., and Green, G. R., "Preventing Automatic Fire Suppression Systems Failures on Underground Mining Belt Conveyors," Information Circular 9264, 1990, U.S. Bureau of Mines, Washington, DC.

70. Johnson, G. A., "Automatic Fire Protection for Mobile Underground Mining Equipment," Information Circular 8954, 1983, U.S. Bureau of Mines, Washington, DC.

71. Ramaswatny, A., and Katiyar, P. S., "Experiences with Liquid Nitrogen in Combating Coal Fires Underground," *Journal of Mines, Metals, and Fuels*, Vol. 36, No. 9, Sept. 1988, pp. 415–424.

72. Johnson, G. A., and Forshey, D. R., "Automatic Fire Protection Systems for Large Haulage Vehicles," Information Circular 8683, 1975, U.S. Bureau of Mines, Washington, DC.

73. Pomroy, W. H., Goodwin, N., and Lynch, F., "Economic Analysis of Surface Mining Mobile Equipment Fire Protection Systems," Report of Investigations 8698, 1982, U.S. Bureau of Mines, Washington, DC.

74. Pomroy, W. H., and Bickel, K. L., "Automatic Fire Protection Systems for Surface Mining Equipment," Information Circular 8832, 1980, U.S. Bureau of Mines, Washington, DC.

75. "Dragline Fire Protection," *U.S. Bureau of Mines Technology News*, 106, U.S. Bureau of Mines, Washington, DC, 1981.

76. Pomroy, W. H., "Fire Protection for Mobile Mining Equipment," *Holmes Safety Association Bulletin*, U.S. Mine Safety and Health Administration, Arlington, VA. (Part I, "An Introduction and Background to Mobile Equipment Fires," Oct. 1993; Part II, "Fire Suppression Systems for Rubber Tired and Tracked Vehicles," Nov. 1993; Part III, "Fire Suppression Systems for Large, Enclosed Surface Mining Machinery," Dec. 1993).

77. Pomroy, W. H., "Automatic Fire Suppression for Mobile Mining Equipment," *Stone Review*, National Stone Association, Washington, DC, Feb. 1994.

78. Hertzberg, M., *et al.*, "Inhibition and Extinction of Coal Dust and Methane Explosions," Report of Investigations 8708, 1982, U.S. Bureau of Mines, Washington, DC.

79. Liebman, I., and Richmond, J. K., "Suppression of Coal Dust Explosions by Water Barriers in a Conveyor Belt Entry," Report of Investigations 8538, 1981, U.S. Bureau of Mines, Washington, DC.

80. Liebman, I., *et al.*, "Extinguishing Agents for Mine Face Gas Explosions,'" Report of Investigations 8294, 1978, U.S. Bureau of Mines, Washington, DC.

81. "Improved Fire Doors for Noncoal Underground Mines," *U.S. Bureau of Mines Technology News*, 188, U.S. Bureau of Mines, Washington, DC, 1983.

82. Bickel, K. L., and Pomroy, W. H., "Fire Doors for Noncoal Mines," Information Circular 9165, 1987, U.S. Bureau of Mines Washington, DC.

83. Ng, D., and Lazzara, C. P., "Performance of Concrete Block and Steel Panel Stoppings in a Simulated Mine Fire," *Fire Technology*, Vol. 26, No. 1, Feb. 1990, pp. 51–76.

84. Edwards, J. C., and Greuer, R. E., "Real-Time Calculation of Product-of-Combustion Spread in a Multilevel Mine," Information Circular 8901, 1982, U.S. Bureau of Mines, Washington, DC.

85. Chang, X., Laage, L. W., and Greuer, R. E., "A User's Manual for MFIRE: A Computer Simulation Program for Mine Ventilation and Fire Modeling," Information Circular 9245, 1990, U.S. Bureau of Mines, Washington, DC.

86. Greuer, R. E., "Modeling the Movement of Smoke and the Effect of Ventilation Systems in Mine Shaft Fires," *Fire Safety Journal*, Vol. 9, No. 1, 1985, pp. 81–88.

87. Watson, R. W., *et al.*, "Evaluation of the Safety of the CSE SR-100 Self-Contained Self-Rescuer," Report of Investigations 9333, 1991, U.S. Bureau of Mines, Washington, DC.

88. Kovac, J. G., and Kravitz, J. H., "Person Wearable Self-Contained Self-Rescuer (PW-SCSR)," *Proceedings* of Specialized Meeting of International Section of ISSA for the Prevention of Occupational Risks in the Mining Industry, *ISSA-AISS Bulletin*, No. 7, 1990.

89. Kovac, J. G., Vaught, C., and Brnich, M. J., Jr., "Probability of Making a Successful Escape While Wearing a Self-Contained Self-Rescuer—A Computer Simulation," Report of Investigations 9310, 1990, U.S. Bureau of Mines, Washington, DC.

NFPA Codes, Standards, and Recommended Practices

Reference to the following NFPA codes, standards, and recommended practices will provide further information on mining methods and equipment discussed in this chapter. (See the latest version of *The NFPA Catalog* for availability of current editions of the following documents.)

NFPA 10, *Standard for Portable Fire Extinguishers*

NFPA 11A, *Standard for Medium- and High-Expansion Foam Systems*

NFPA 12A, *Standard on Halon 1301 Fire Extinguishing Systems*

NFPA 13, *Standard for the Installation of Sprinkler Systems*

NFPA 17, *Standard for Dry Chemical Extinguishing Systems*

NFPA 30, *Flammable and Combustible Liquids Code*

NFPA 51, *Standard for the Design and Installation of Oxygen-Fuel Gas Systems for Welding, Cutting, and Allied Processes*

NFPA 51B, *Standard for Fire Prevention in Use of Cutting and Welding Processes*

NFPA 70, *National Electrical Code*®

NFPA 120, *Standard for Coal Preparation Plants*

NFPA 121, *Standard on Fire Protection for Self-Propelled and Mobile Surface Mining Equipment*

NFPA 122, *Standard for Fire Prevention and Control in Underground Metal and Nonmetal Mines*

NFPA 123, *Standard for Fire Prevention and Control in Underground Bituminous Coal Mines*

Additional Reading

Ayres, D. J., "Treatment of Shallow Underground Fires," *Proceedings* of the Symposium on Failures in Earthworks, Thomas Telford, London, 1986, pp. 470–472.

Banerje, S. C., and Acharya, A. K., "High Expansion Foams—Its Use in Mine Fires," *Journal of Mines*, Metals and Fuels, Vol. 34, No. 10, 1986, pp. 448–451.

Banerjee, S. C., *et al.*, "Approach to Assessing the Status of Sealed Off Fires by Examination of Fire Indices," *Mining Science and Technology*, Vol. 10, No. 1, Jan. 1990, pp. 37–51.

Bagley, S. T., Baumgard, K. J., and Gratz, L. D., "Polynuclear Aromatic Hydrocarbons and Biological Activity Associated With Diesel Particulate

Matter Collected in Underground Coal Mines," Michigan Technological Univ., Houghton, MI, Bureau of Mines, Minneapolis, MN, IC 9324; U. S. Department of the Interior, Bureau of Mines, Diesels in Underground Mines: Measurement and Control of Particulate Emissions, Bureau of Mines Information and Technology Transfer Seminar, September 29–30, 1992, Minneapolis, MN, 1992, pp. 40–48.

Chang, X., Laage, L. W., and Greuer, R. E., "User's Manual for MFIRE: A Program Computer Simulation Program for Mine Ventilation and Fire Modeling," Xian Mining Institute, The Peoples Republic of China, Bureau of Mines, Minneapolis, MN, Mining Technological Univ., Houghton, MI, IC 9245, 1990, 175 pages.

Chaudhari, P. D., "Reducing the Risk of Accidents in the Mining Industry by Use of Automatic Fire Protection Systems on Mining Equipment," Journal of Mines, Metals and Fuels, Vol. 63, No. 9, Sept. 1988, pp. 442–447.

Chekan, G. J., and Listak, J. M., "Design Practices for Multiple-Seam Longwall Mines," Bureau of Mines, Pittsburgh, PA, RI 9360, 1993, 39 pages.

Conti, R. S., and Litton, C. D., "Comparison of Mine Fire Sensors. Report of Investigations," Bureau of Mines, Pittsburgh, PA, RI 9572, May 1995, 14 pages.

Conti, R. S., Zlochower, I. A., and Sapko, M. J., "Rapid Sampling of Products During Coal Mine Explosions," Combustion Science and Technology, Vol. 75, No. 4–6, 1991, pp. 195–209.

Dalverny, L. E., and Chaiken, R. F., "Mine Fire Diagnostics and Implementation of Water Injection With Fume Exhaustion at Renton, PA," Bureau of Mines, Pittsburg, PA, RI 9363, 1991, 46 pages.

De Rosa, M. I., and Litton, C. D., "Determining the Relative Toxicity and Smoke Obscuration of Combustion Products of Mine Combustibles," Report of Investigation 9274, 1989, U.S. Bureau of Mines, Pittsburgh, PA.

DeRosa, M. I., and Litton, C. D., "Embedded Hydrogen Chloride and Smoke Particle Characteristics During Combustion of Polyvinyl Chloride and Chlorinated Mine Materials," Bureau of Mines, Pittsburgh, PA, RI 9368, 1991, 17 pages.

De Rosa, M. I., and Litton, C. D., "Hydrogen Cyanide and Smoke Particle Characteristics During Combustion of Polyurethane Foams and Other Nitrogen-Containing Materials: Development of a Test Parameter," Report of Investigations 9367, 1991, U.S. Bureau of Mines, Washington, DC.

De Rosa, M. I., and Litton, C. D., "Primary Gas Toxicities and Smoke Particle Characteristics During Combustion of Mine Brattices," Report of Investigations 9262, 1989, U.S. Bureau of Mines, Pittsburgh, PA.

Edwards, J. C., and Morrow, G. S., "Evaluation of Smoke Detectors for Mining Use. Report of Investigations," Pittsburgh Research Center, Pittsburgh, PA, RI 9586, September 1995, 17 pages.

Egan, M. R., "Smoke, Carbon Monoxide, and Hydrogen Chloride Production from the Pyrolysis of Conveyor Belting and Brattice Cloth," Information Circular 9304, 1992, U.S. Bureau of Mines, Washington, DC.

Egan, M. R., "Summary of Combustion Products From Mine Materials: Their Relevance to Mine Fire Detection," Bureau of Mines, Pittsburgh, PA, IC 9272, 1990, 16 pages.

Egan, M. R., "Summary of Combustion Products from Mine Materials: Their Relevance to Mine Fire Detection," Information Circular 9272, 1990, U.S. Bureau of Mines, Washington, DC.

Engel, W., and Kihl, H., "Flame-Resistant and Non-Noxious HFD Hydraulic Fluids for Mining," Glueckauf and Translation, Vol. 123, No. 22, Nov. 19, 1987, pp. 625–628.

Fowkes, R. S., and Aiken, E. G., "Human Factors in Mining Search System," Bureau of Mines, Pittsburgh, PA, IC 9243, 1990, 24 pages.

Gall, P., "Safety Research for the Coal Mining Industry Develops Advanced Systems," Australian Mining, Vol. 81, No. 6, June 1989.

Gerrit, V. R., and Kissel, F. N., "Important Factors for Escaping a Mine Fire," Coal Magazine, March 1990, pp. 75–77.

Goodman, G. V. R., and Kissell, F. N., "Evaluating Those Factors That Influence Escape From Coal Mine Fires," Transactions AIME/SME, Vol. 286, July 1990, pp. 1801–1805.

Grannes, S. G., Ackerson, M. A., and Green, G. R., "Preventing Automatic Fire Suppression System Failures on Underground Mining Belt Conveyors," Bureau of Mines, Minneapolis, MN, IC 9264, 1990, 22 pages.

Hanson, L. C., Hodges, J. E., and Redmond, K. M., "Determining Effectiveness of Past Mine Fire Surface Seal Abatement Methods. Final Report," L. C. Hanson Co., Helena, MT, Department of Interior, Denver, CO, Final Report, December 1990, 230 pages.

Herbert, M., "Beating Fire and Explosion Risks in Coal Mines," Fire Prevention, No. 231, July/August 1990, pp. 25–28, 30–31.

Hertzberg, M., "A Critique of the Dust Explosibility Index," U.S. Bureau of Mines, Pittsburgh, PA, 1987.

Hjelmstad, K. E., and Pomroy, W. H., "Novel Fire Warning System for Underground Mines," Mining Engineering, Vol. 42, No. 1, January 1990, pp. 107–112

Hjelmstad, K. E., and Pomroy, W. H., "Ultra Low Frequency Electromagnetic Fire Alarm System for Underground Mines," Bureau of Mines, Minneapolis, MN, RI 9377, 1991, 15 pages.

Huntley, D. W., et al., "Underground Coal Mine Fire, Wilberg Mine," I.D. No. 42-00080, Emery Mining Corporation, Orangeville, Emery County, Utah, 1987, U.S. Department of Labor, Mine Safety and Health Administration.

Jankowski, R. A., Jayaraman, N. I., and Potts, J. D., "Update on Ventilation for Longwall Mine Dust Control," Bureau of Mines, Pittsburgh, PA, IC 9366, 1993, 15 pages.

Kim, A. G., "Laboratory Determination of Signature Criteria for Locating and Monitoring Abandoned Mine Fires," Bureau of Mines, Philadelphia, PA, RI 9348, 1991, 23 pages.

Kim, A. G., and Chaiken, R. F., "Fires in Abandoned Coal Mines and Waste Banks," Bureau of Mines, Pittsburgh, PA, IC 9352, 1993, 63 pages.

Kuchta, J. M., et al., "Diagnostics of Sealed Coal Mine Fires," Report of Investigations 8625, 1982, U.S. Bureau of Mines, Washington, DC.

Lee, C. F. T., "Mining Dredge Fire in Malaysia," Fire International, No. 145, Winter 1994, p. 39.

Luczik, S. J., "Development of a Fire-Resistant Cardboard Compressible Block for Use in Stopping Construction in Underground Mines," Fire Technology, Vol. 29, No. 2, May 1993, pp. 131–153.

Mitchell, D. W., "Mine Fires—Prevention, Detection, Fighting," 1990, 216 pages.

Morris, R., and Badenhorst, C. P., "Analysis of Underground Fires in South African Coal Mines from 1970–1985," Journal of the Mine Ventilation Society of South Africa, Vol. 40, No. 10, 1987, pp. 134–139.

Morrow, G. S., and Litton, C. D., "In-Mine Evaluation of Smoke Detectors," Bureau of Mines, Pittsburgh, PA, IC 9311, 1992, 17 pages.

Nelson, H. E., "A Qualitative Analysis of the Inherent Fire Safety/Fire Risk in a Coal Mine," NBSIR 86-3502, Dec. 1986, National Bureau of Standards, Gaithersburg, MD.

Ng, D., and Lazzara, C. P., "Performance of Concrete Block and Steel Panel Stoppings in a Simulated Mine Fire," Fire Technology, Vol. 26, No. 1, February 1990., pp. 51–74

Organiscak, J. A., Page, S. J. and Jankowski, R. A., "Sources and Characteristics of Quartz Dust in Coal Mines," Bureau of Mines, Pittsburgh, PA, IC 9271, 1990, 25 pages.

Peters, R. H. and Randolph, R. F., "Miners' Views About Why People Go Under Unsupported Roof and How To Stop Them," Bureau of Mines, Pittsburgh, PA, IC 9300, 1991, 63 pages.

Pomroy, W. H., and Ackerson, M. A., "Remotely Controlled Mine Fire Fighting Concepts," Proceedings of the International Symposium on Mine Mechanization and Automation, Golden, CO, June 10–13, 1991.

Pomroy, W. H. and Laage, L., "Method of Locating Underground Mine Fires," Department of the Interior, Washington, DC, PAT-APPL-7-657 627, February 21, 1991, 13 pages.

Pomroy, W. H., Griffin, R. E., and Ackerson, M. A., "Rapid Response Pneumatic Fire Detection for Multilevel Metal Mines: System Design and In-Mine Testing," Bureau of Mines, Philadelphia, PA, RI 9305, 1990, 20 pages.

Sammarco, J. J., "Field Evaluation of the Modular Azimuth and Positioning System (MAPS) for a Continuous Mining Machine," Bureau of Mines, Pittsburgh, PA, IC 9354, 1993, 17 pages.

Schibe, I. A., "The Development of a Fire Evaluation System for Underground Coal Mines," NBSIR 86-3425, August 1986, National Bureau of Standards, Gaithersburg, MD.

Sharma, D., "Indian Mine Tragedy: Poor Safety Record," Fire International, No. 143, May 1994, p. 27.

Sinha, A. K., Rajwar, D. P., and Ghosh, A. K., "Underground Fire Hazard Detection through Carbon-Monoxide Monitors," Journal of Mines, Metals and Fuels, Vol. 35, No. 8, August 1987, pp. 365–367.

Smith, A. C., *et al.*, "A Method to Evaluate the Performance of Coal Fire Extinguishants," Report of Investigations 9392, 1991, U.S. Bureau of Mines, Washington, DC.

Stewart, B. A., "Development of a Fire Protection Consensus Standard in the Mining Industry," *Proceedings* of the Earthmoving Industry Conference, Society of Automotive Engineers, 1987, p. 5.

Timko, R. J., and Derick, R. L., "Applying Atmospheric Status Equations to Data Collected From a Sealed Mine, Postfire Atmosphere," Bureau of Mines, Pittsburg, PA, RI 9362, 1991, 27 pages.

Timko, R. J., Derick, R. L., and Thimons, E. D., "Analysis of a Fire in a Colorado Coal Mine—A Case Study," *Proceedings* of the 3rd Mine Ventilation Symposium, Society of Mining Engineers of AIME, Littleton, CO, 1987, pp. 444–452.

Unger, R. L., Glowacki, A. F., and Stein, R. R., "Evacuation Simulation for Underground Mining," Bureau of Mines, Pittsburgh, PA, Institute of Electrical and Electronics Engineers, Inc. (IEEE), International Emergency Management and Engineering Conference, 10th Anniversary, Research and Applications, 1993, pp. 75–80.

Vaught, C., Wiehagen, W. J., and Brnich, M. J., Jr., "Self-Contained Self-Rescuer Donning Proficiency at Eight Eastern Underground Coal Mines," Bureau of Mines, Pittsburgh, PA, RI 9383, 1991, 21 pages.

von Velsen-Zerweck, R., "Ways of Improving Fire and Explosion Protection in Underground Coal Mining," *Glueckauf and Translation*, Vol. 124, No. 3, Feb. 4, 1988, p. 76.

Zhao, Y., Yang, C., and Wang, X., "Investigation of the Dynamic Process in the Ventilation Network During Underground Fire," *Archiwum Gornictwa*, Vol. 34, No. 3, 1989, pp. 503–515.

MOTOR VEHICLES

Revised by Neill Darmstadter

Many factors influence motor vehicle fire safety. Among these are vehicle design and construction; materials from which a vehicle is constructed or which are carried by a vehicle, particularly upholstery, plastic, wood, fuel, and hazardous materials cargoes; vehicle maintenance, including safeguards against fire during vehicle repair; vehicle operation, including avoidance of collisions and of other accidents, such as those involving smoking materials; and garaging or storage of vehicles.

The degree of fire hazard depends upon the vehicle type, such as passenger car, motorcycle, motor home, or tank truck; the use of the vehicle for pleasure driving, commercial service, or off-road; the climate and environment in which it is used; the age, condition, and maintenance of the vehicle; the material and construction standards to which it is built, particularly the fuel and electrical systems; and the type of fuel used—gasoline, diesel fuel, or propane—and the way it is contained in the vehicle.

Research into causes of motor vehicle fires based on individual crash investigations and collections of crash and other loss data indicate the magnitude of the motor vehicle fire problem and some specific causal factors in motor vehicle fires.

Motor vehicle fires represent about 20 percent of all reported fires, and only 2 percent of vehicle fires result from crashes. Yet serious injury from vehicle fires occurs overwhelmingly from crash-related fires—fires that occur at a rate of approximately 1 per 1,000 crashes. Nearly two-thirds of fire deaths in motor vehicle fires come in postcrash incidents.

In 1995 motor vehicle fires were an estimated 20 percent of all reported fires, a decrease of 4.0 percent from the previous year. The estimated 386,000 motor vehicle fires caused property damage estimated at $1.013 billion, with an average loss of $2,600.[1]

THE NATURE OF VEHICLE FIRES

As with other fires, motor vehicle fires require a flammable substance, an ignition source, and oxygen. Virtually all motor vehicles carry flammable fuel—gasoline, gasohol, diesel fuel, or another hydrocarbon compound. Vehicle upholstery, insulating and sound-deadening materials, electrical wiring insulation, and plastic body and trim also can fuel a motor vehicle fire. Materials carried as cargo and fare in or on motor vehicles, particularly trucks, also may be flammable.

Ignition sources include electrical short circuits or other electrical malfunctions that cause excessive heating of conductors or components; sparks from an engine ignition system; hot exhaust system components; engine backfire; overheating of tires, brakes, and wheel bearings; friction-generated sparks from a collision or from metal components scraping against the pavement; and careless use of cigarettes and other smoking materials. Generally, oxygen in the ambient atmosphere is sufficient for a motor vehicle fire.

Table 9-32A summarizes 1989–1993 passenger vehicle fires by area of origin of fires. Table 9-32B summarizes 1989–1993 freight vehicle fires by area of origin. More detailed but dated analyses have been developed by the office of motor carriers (OMC) of the U.S. Department of Transportation (DOT).

Death and injury in motor vehicle fires result from direct exposure to heat and from inhalation of toxic combustion products. In major vehicle crashes involving fire, determining with certainty whether death resulted from the crash or from the fire may not be possible. In some cases, injury results from or is exacerbated by the inability of a vehicle's occupant to get out of the burning vehicle—doors or seats are jammed by a crash, or the occupant is injured or stunned.

Hazardous materials in motor vehicles present a particular fire problem. The National Transportation Safety Board (NTSB), the OMC, and NFPA all carried out a number of detailed accident investigations into such fires. These investigations led to changes in regulations for hazardous materials containers and in other aspects of the transportation of such cargoes. Despite such precautions, hazardous materials cargo fires and explosions occur, primarily when vehicles carrying such cargoes are involved in major collisions.

The NTSB has issued investigative reports on accidents in which hazardous materials exploded or burned following a crash. (These and other NTSB reports are available from the National Technical Information Service, Springfield, VA 22161.) As a result of these and other highway crash investigations, the NTSB made numerous recommendations to the OMC, the Research and Special Programs Administration (RSPA) of DOT, and to other agencies concerning modification and enforcement of safety and hazardous materials regulations.

Neill Darmstadter is senior safety engineer, American Trucking Associations, Alexandria, VA, and a member of the NFPA Technical Committee on Motor Vehicle and Highway Fire Protection.

TABLE 9-32A. *Passenger Road Transport Vehicles, by Area of Origin, 1989–1993 Annual Averages*

Area of Origin	Fires		Civilian Deaths		Civilian Injuries		Direct Property Loss (in Millions)	
Engine area, running gear, wheel area of transportation equipment	216,650	(67.4%)	164	(39.4%)	1,207	(56.7%)	$335.4	(55.9%)
Passenger area	61,480	(19.1%)	92	(22.2%)	392	(18.4%)	$158.4	(26.4%)
Exterior exposed surface of transportation equipment	8,230	(2.6%)	9	(2.1%)	48	(2.3%)	$14.4	(2.4%)
Unclassified vehicle areas	6,950	(2.2%)	25	(6.0%)	53	(2.5%)	$15.8	(2.6%)
Fuel tank or fuel line area of transportation equipment	5,790	(1.8%)	8	(1.9%)	72	(3.4%)	$10.1	(1.7%)
Trunk or load carrying area of transportation equipment	4,960	(1.5%)	81	(19.4%)	190	(8.9%)	$13.7	(2.3%)
Operating or control area of transportaton equipment	4,290	(1.3%)	3	(0.8%)	25	(1.2%)	$12.0	(2.0%)
On or near a highway, public way or street	3,240	(1.0%)	11	(2.6%)	35	(1.6%)	$7.5	(1.2%)
Other known areas	10,000	(3.1%)	23	(5.6%)	108	(5.1%)	$32.8	(5.5%)
Total	321,570	(100.0%)	416	(100.0%)	2,130	(100.0%)	$600.0	(100.0%)

Source: Laurence J. Stewart, *U.S. Vehicle Fire Trends and Patterns*, NFPA Fire Analysis and Research Division, Quincy, MA, August 1996.
*Sums may not equal totals due to rounding error.
Fires are estimated to the nearest ten, civilian deaths and injuries to the nearest one, and direct property loss to the nearest hundred thousand dollars.
Unknowns have been allocated.

TABLE 9-32B. *Freight Road Transport Vehicles, by Area of Origin, 1989–1993 Annual Averages*

Area of Origin	Fires		Civilian Deaths		Civilian Injuries		Direct Property Loss (in Millions)	
Engine area, running gear, wheel area of transportation equipment	25,590	(59.3%)	47	(38.8%)	186	(41.9%)	$80.1	(548.5%)
Passenger area	6,370	(14.8%)	18	(15.3%)	54	(12.2%)	$26.1	(15.8%)
Trunk or load carrying area of transportation equipment	4,380	(10.2%)	6	(5.0%)	78	(17.6%)	$20.9	(12.7%)
Exterior exposed surface of transportation equipment	1,510	(3.5%)	5	(4.0%)	18	(4.1%)	$6.7	(4.0%)
Unclassified vehicle areas	1,300	(3.0%)	5	(4.2%)	14	(3.2%)	$5.2	(3.1%)
Operating or control area of transportaton equipment	1,090	(2.5%)	4	(3.1%)	12	(2.6%)	$7.1	(4.3%)
Fuel tank or fuel line area of transportation equipment	870	(2.0%)	27	(22.7%)	54	(12.2%)	$8.0	(4.9%)
On or near a highway, public way or street	440	(1.0%)	5	(4.0%)	6	(1.3%)	$2.6	(1.6%)
Other known areas	1,630	(3.4%)	4	(53.0%)	22	(5.1%)	$8.5	(5.1%)
Total	43,160	(100.0%)	120	(100.0%)	445	(100.0%)	$165.2	(100.0%)

Source: Laurence J. Stewart, *U.S. Vehicle Fire Trends and Patterns*, NFPA Fire Analysis and Research Division, Quincy, MA, August 1996.
*Sums may not equal totals due to rounding error.
Fires are estimated to the nearest ten, civilian deaths and injuries to the nearest one, and direct property loss to the nearest hundred thousand dollars.
Unknowns have been allocated.

FEDERAL REGULATIONS AFFECTING MOTOR VEHICLE FIRE SAFETY

Federal Motor Vehicle Safety Standards

The National Traffic and Motor Vehicle Safety Act of 1966 authorized the National Highway Traffic Safety Administration (NHTSA) of DOT to set minimum safety standards applicable to new motor vehicles, trailers, and motor vehicle equipment manufactured after the effective date of the standard, or any subsequent amendment. To date, only two fire-related standards have been issued for new vehicles, and none has been issued for vehicles in use. The NHTSA also has the authority to investigate motor vehicle defects, some of which may present fire hazards, and to order manufacturers to notify owners and repair defects.

In 1993, 224 recalls affected 10,939,458 vehicles. Of these, 33 recalls affecting 4,338,513 vehicles were for potential fire hazards. The actual number of fires that prompted the recall campaigns is not available. Table 9-32C lists the general nature of the hazard, the number of recall campaigns, and the number of vehicles affected.

TABLE 9-32C. *Motor Vehicle Recalls, 1993*

Vehicle System	No. of Compaigns	No. of Vehicles
Fuel	14	2,336,629
Automatic transmission fluid leakage	2	1,706,630
Electrical	12	261,883
All other	5	33,371
Total	33	4,338,513

Federal Motor Vehicle Safety Standard (FMVSS) 301 sets requirements for the fuel system integrity of passenger cars, trucks, buses, and multipurpose vehicles having a gross vehicle weight rating (GVWR) of 10,000 lb (4,536 kg) or less. The standard also applies to school buses of more than 10,000 lb (4,536 kg) GVWR using fuel with a boiling point of more than 32°F (0°C).

FMVSS 301 is intended specifically to reduce the fire hazard resulting from fuel leakage in crashes. It was first applied to passenger cars manufactured on and after January 1, 1968. Since that time, the standard has been made increasingly stringent.

Under present provisions, which became effective on September 1, 1977, all vehicles with a GVWR of 10,000 lb (4,536 kg) or less must pass prescribed front, side, and rear crash barrier tests and a rollover test. Front and rear tests are conducted at impact speeds of 30 mph (48 km/hr), while lateral impact speed is 20 mph (32 km/hr). In the crash barrier tests, fuel leakage cannot exceed 1 oz (28 g) by weight from impact until the vehicle stops moving; 5 oz (142 g) by weight in the 5 min following cessation of motion; or 1 oz (28 g) by weight per min for 25 min.

In the rollover test, the vehicle is turned about its longitudinal axis in 90° increments and is held for 5 min in each position. Rotation occurs every 1 to 3 min. Fuel leakage cannot exceed 5 oz (142 g) by weight in the first 5 min of each segment or 1 oz (28 g) per min for the remainder of the segment.

The school bus test procedure for buses over 10,000 lb (4,536 kg) GVWR uses a special, moving, contoured barrier that strikes the test vehicle at the front, rear, and sides. No rollover test is prescribed for these buses. Maximum permissible fuel leakage is the same as indicated above for front, side, and rear barrier crash tests.

The other NHTSA fire safety standard, *Federal Motor Vehicle Safety Standard 302*, sets flammability limits for materials used in interiors of vehicles in the driver and passenger space. Its requirements are directed at reducing the hazards of interior fires caused by smoking and matches. Vehicle appointments specifically covered by the standard include seat cushions, seat backs, seat belts, headlining, arm rests, trim panels, compartment shelves, head restraints, floor covering, engine compartment covering, padding, and the surface of crash protection components.

The standard specifies a horizontal flame test for all materials and incorporates detailed requirements for the test procedure. Material under test must not burn or transmit a flame across the surface at a rate exceeding 4 in. per min (1.7 mm/s).

Federal Motor Carrier Safety Regulations

OMC, a part of the Federal Highway Administration (FHWA) of DOT, administers the Federal Motor Carrier Safety Regulations (FMCSRs). These regulations apply to vehicles in use, as contrasted with the Federal Motor Vehicle Safety Standards promulgated by the National Highway Traffic Safety Administration (NHTSA) that apply to new vehicles.

The FMCSRs are set forth in Title 49 of the Code of Federal Regulations, Parts 382–399, and apply to commercial vehicles in interstate or foreign commerce. Because most states have adopted these regulations by reference, they also apply, *de facto*, to most commercial vehicles in intrastate commerce.

Provisions of these regulations that address fire safety issues for commercial vehicles include:

- A requirement for fire extinguishers and for the driver to check the extinguisher at the start of each trip.
- A prohibition against attaching a flame-producing warning device to a vehicle and a prohibition against using flame-producing devices on vehicles transporting certain explosives and tank vehicles transporting flammable liquids or gases (loaded or empty).
- Fueling precautions including a prohibition against smoking, a requirement to maintain contact between the tank and the nozzle, and restrictions on fueling buses when passengers are aboard.
- Standard for installing and protecting the electrical system, including the battery, and standards for constructing and testing fuel tanks.
- Special driving and parking rules applicable to drivers transporting hazardous materials including requirements to periodically check for overheated tires, a requirement that someone be present at the nozzle when the vehicle is being fueled, and a prohibition against smoking in or within 25 ft (7.6 m) of a vehicle transporting specified classes of hazardous materials.

Hazardous Materials Regulations

Federal hazardous materials regulations, other than those governing driving and parking, are published in Title 49, Code of Federal Regulations, Parts 170–189. By law, hazardous materials regulations apply to both intrastate and interstate transportation. Many provisions of these regulations have been promulgated specifically to meet the needs of emergency responders by providing essential information pertaining to the materials in transport and to warn the public of those vehicles transporting hazardous materials in excess of specified quantities. For further information on the federal hazardous materials regulations, refer to Section 8, Chapter 10, "The Role of the Fire Protection Community in Management of Hazardous Substances"; and Section 9, Chapter 10, "Managing the Response to Hazardous Materials Incidents."

Provisions of the hazardous materials regulations address:

- Preparation of shipping documents to facilitate the recognition and identification of hazardous materials including a 24-hr emergency telephone that can be contacted for additional information, if needed.
- Job-specific training requirements for employees involved in transporting hazardous materials at any level.
- Emergency instructions to be carried in vehicles and kept in motor carrier terminals where hazardous materials are handled.
- Marking and labeling of hazardous materials packaging to identify its contents and to indicate that the packaging complies with applicable specifications.
- Specifications for bulk packaging such as cargo tanks and portable tanks, and for nonbulk packaging such as cylinders, boxes, drums, and bags used for hazardous materials.
- Requirements for shippers that specify the types of packaging to be used for hazardous materials shipments based on the commodity's specific hazard(s).

State Activity

Most states have adopted by reference the provisions of the Federal Motor Carrier Safety Regulations and the DOT hazardous materials regulations. A number of other states have regulations similar to the federal requirements. As these trends continue, they will foster increasing uniformity of legal requirements.

The National Transportation Safety Board

The National Transportation Safety Board, an independent federal agency, oversees safety for all types of transportation. Probably best known for its investigations of aircraft accidents, the agency is also concerned with fire-related highway accidents. The NTSB is responsible for making recommendations for improved safety, based on its investigations. The majority of its recommendations deal with suggested changes in the regulations and practices of federal and state government agencies, but they may also be directed at others, such as vehicle manufacturers and users.

The Role of Congress

In Congress, the responsibility for legislation and oversight of transportation rests with the Senate Committee on Commerce and the House Committee on Transportation and Infrastructure. These committees take a keen interest in the activities of the federal government in the field of motor vehicle safety. They regularly hold legislative and oversight hearings on NHTSA activities and occasionally on OMC and NTSB activities. Reports of these hearings are available from the U.S. Superintendent of Documents or from the committees.

DESIGN AND CONSTRUCTION SAFEGUARDS

Many items in the interior of vehicles are flammable, such as upholstery, paneling, trim, carpeting, or floor mats. If ignition occurs from any source and prompt control is not possible, the vehicle can be destroyed.

In large truck bodies and in freight-carrying trailers, wood is used as flooring. It withstands torsional stresses, impacts, and abrasions encountered in this type of service, and replacement of worn or damaged sections is relatively economical. Wood provides safety benefits because blocking and bracing can be nailed in place to prevent shifting of cargo. It offers a high coefficient of friction for safe forklift operation and a slip-resistant footing for personnel who handle cargo.

Plywood is used for interior lining material in van-bodied trucks and trailers because it effectively protects other structural components from damage that may occur during loading, unloading, and transportation of freight. As with wood flooring, replacing damaged sections is relatively inexpensive.

Plywood flooring is used in buses for essentially the same reasons. Metal flooring would create "tin canning" under torsional stress and a resulting noise level unacceptable to passengers.

These uses of wood do not create any significant fire hazards. In the forms used, the wood is relatively difficult to ignite and burns slowly. Specific provisions of the Federal Motor Carrier Safety Regulations require that flooring be in good condition. The use of flooring permeated with oil or gasoline is prohibited.

In major accidents involving fire, wood can be a contributing factor, but usually only in accidents so severe that their outcomes would have been little affected if wood had not been present.

When the cargo must be covered during transportation on flatbed, open-top, and other types of nonenclosed vehicles, the use of tarpaulins is the only practicable means of providing needed protection. Such tarpaulins normally are not given flame-retardant treatment, but they pose little risk of fire.

Vehicle Fuel Tanks and Systems

The location, construction, and security of fuel tanks are important design features for fire safety in motor vehicles. Liquid fuel tanks for general passenger car use are generally of thin-gage steel of various shapes and dimensions, dependent upon other body and chassis characteristics. The vast majority of U.S.-built cars, trucks, and buses have liquid fuel tanks located at the rear of the vehicle, frequently

in such a position that they are not entirely enclosed in the body. The greatest risk of loss of life and property therefore occurs in fires following rear-end collisions.

The Office of Motor Carriers has developed fuel tank regulations applicable to commercial vehicles operating in interstate or foreign commerce. These regulations closely parallel Underwriters Laboratories (UL) standards referred to elsewhere. In 1973, requirements that previously had been limited to gasoline tanks were extended to include tanks for diesel fuel. This action also required protection for diesel fuel crossover lines located below the bottom of a tank or sump. The DOT regulations include requirements for tank construction and installation and for adequate fillpipe closure. Tanks must pass drop, rupture, vent, and spillage tests. Required safety vent systems must limit pressure rise to 50 psig (345 kPa) in a specified fire test.

Tanks must be marked by a manufacturer's certificate to indicate that they meet DOT-FHWA specifications for sidemounted or nonsidemounted tanks, as applicable.

Efforts have been made to gain DOT approval for plastic fuel tanks for truck service. Despite certain advantages their manufacturers claim, doubts about the safety of such tanks in commercial vehicle operations have not been satisfactorily resolved.

Vehicles using flammable compressed gas as a motor fuel must be equipped with fuel systems constructed and installed in accordance with applicable provisions of NFPA 58, *Standard for the Storage and Handling of Liquefied Petroleum Gases*, or NFPA 52, *Standard for Compressed Natural Gas (CNG) Vehicular Fuel Systems*.

Fuel systems must be maintained so they are free of leaks. Fuel tanks must be capped, and connections must be checked regularly. Metal fuel lines are subject to vibration-induced fatigue, and flexible lines may deteriorate with age. Chafing, cutting, or proximity to hot surfaces also can damage lines. In a gasoline system, carburetor flooding due to dirt in the float valve may cause gasoline to overflow onto a hot surface and result in a fire.

These problems are somewhat less critical on diesel-powered vehicles because diesel fuel has a higher flash point. In recent years, however, diesel engines have been designed to operate at higher temperatures to control exhaust emissions. This, in turn, led to higher fuel temperature in the return flow line, thus making the fuel more prone to ignition if the warmed, leaking fuel contacts a hot surface.

NFPA AND RELATED STANDARDS ON MOTOR VEHICLES

Three NFPA standards contain requirements for transportation by truck of specific hazardous materials. NFPA 495, *Explosive Materials Code*, details fire and explosion prevention related to the transportation of explosive materials, including driver qualifications, vehicle design, placarding, fire-fighting equipment, and vehicle operation. This code is widely used as the basis for state regulations.

NFPA 385, *Standard for Tank Vehicles for Flammable and Combustible Liquids,* and NFPA 407, *Standard for Aircraft Fuel Servicing,* both contain requirements for design and construction of cargo tanks of tank vehicles, for auxiliary equipment, and for operation of tank vehicles. NFPA 407 covers only special requirements for aircraft fuel servicing tank vehicles and aircraft fuel servicing hydrant vehicles.

NFPA 58 contains requirements for transporting liquefied petroleum gases, including requirements for transporting portable containers, for cargo tanks, and for parking and garaging vehicles used to carry LP-Gas. NFPA 58 also includes requirements for installing LP-Gas systems in vehicles fueled by LP-Gas. It has been adopted for regulatory purposes by 47 states and by the DOT for vehicles in interstate commerce.

NFPA 52 covers design and installation of compressed natural gas (CNG) engine fuel systems on vehicles of all types and their associated fueling systems. NFPA 512, *Standard for Truck Fire Protection*, covers property-carrying motor vehicles.

NFPA 58 covers parking and garaging of LP-Gas cargo vehicles. Two other NFPA standards include provisions relating to parking trucks: NFPA 513, *Standard for Motor Freight Terminals*, and NFPA 498, *Standard for Safe Havens and Interchange Lots for Vehicles Transporting Explosives*. Primarily concerned with fire protection of freight in a terminal, NFPA 513 also contains recommendations for parking vehicles at terminals. Parking vehicles hauling explosives is the subject of NFPA 498.

The dependence of the fire service upon motor vehicles is well recognized, and it is particularly important that fire apparatus be safely constructed. NFPA 1901, *Standard for Automotive Fire Apparatus*, and NFPA 414, *Standard for Aircraft Rescue and Fire Fighting Vehicles*, detail practices that should be followed on these types of apparatus.

Manufactured homes are sometimes classified as "vehicles," but modern practice tends to remove them from this category. For 11 years, NFPA issued NFPA 501B, *Standard on Mobile Homes*, which covered body and frame design and construction, as well as plumbing, heating, and electrical systems for these vehicles. That standard was withdrawn in 1979, and authority for regulating manufactured homes now rests with the federal government through its *Manufactured Home Construction and Safety Standards*, Part 280 of the Code of Federal Regulations, Title 24. Most states also have manufactured home regulations, some with requirements more stringent than federal regulations.

Recreational vehicles can be divided into four categories: travel trailers, motor homes, camp trailers, and campers. NFPA 501C, *Standard on Recreational Vehicles*, covers details of plumbing, heating, and electrical systems for travel trailers, motor homes, truck campers, and camping trailers. It includes life safety and exiting requirements for these vehicles when they are used as temporary living quarters.

This standard currently calls for a 5B:C-rated portable fire extinguisher in any such unit containing fuel-burning equipment. The standard is expected to be amended to require a 10B:C portable fire extinguisher in each powered vehicle.

According to the Recreational Vehicle Institute of America (RVIA), twelve states have adopted NFPA 501C. Recreational vehicles sold in those states must be equipped with portable fire extinguishers. This is not necessarily the case for vehicles sold in other states.

Requirements for plumbing are developed by the ANSI A119 Committee, of which the Recreational Vehicle Industry Association is Secretariat. Those requirements are published and distributed under one cover as ANSI A119.2, *Recreational Vehicles*.

Underwriters Laboratories Inc.

Certain over-the-road motor vehicle components and signaling appliances have special fire hazard significance. Underwriters Laboratories Inc. tests and lists them in UL's *Accident, Automotive, and Burglary Protection Equipment* directory. Specific items listed include:

1. Electrical equipment such as automobile fuses and switches.
2. Fill and vent fittings.
3. Fuel equipment such as backfire deflectors, fuel feed systems, electric gasoline gages, automotive-type LP-Gas accessories, automotive fuel tanks, and tubing.
4. Automotive heaters of the combustion type.
5. LP-Gas automotive vehicles—that is, farm and road tractors incorporating fuel systems designed for use of LP-Gas as engine fuel.
6. Mufflers for automobiles.
7. Signals and signaling appliances, or highway emergency signals.

UL also has its own standard, UL 395, *Standard on Automotive Fuel Tanks*, for liquid fuel tanks mounted outside the frame or in other exposed locations on gasoline- or diesel-powered trucks, tractors, or trailers. It has issued UL 307(a), *Standard on Liquid Fuel-Burning Heating Appliances for Mobile Homes and Recreational Vehicles*, and UL 307(b), *Standard for Gas Burning Heating Appliances for Mobile Homes and Recreational Vehicles*.

Society of Automotive Engineers, Inc.

The Society of Automotive Engineers, Inc. (SAE) is concerned with any vehicle that moves under its own power. SAE's standardization program covers passenger cars, trucks and buses, farm and earthmoving machinery, marine propulsion units, aircraft, and space vehicles, as well as the materials and components that go into them. Content of the SAE standards is limited to environmental and operating problems with these vehicles. The *SAE Handbook* contains all the society's surface vehicle documents.

Other Cooperating Agencies Interested in Highway Safety

American Petroleum Institute (API): Active in the field of truck transportation of petroleum products, API publishes a number of publications of interest and value from the fire safety viewpoint.

American Trucking Association, Inc.: The association reproduces the federal hazardous materials regulations in their entirety as "Hazmat Transport Regs." The association's safety department sets trucking industry standards for selecting, training, and supervising employees and develops a variety of materials for the use of trucking companies dealing with general safety and safe transportation of hazardous materials.

Center for Auto Safety: The center is a consumer advocate organization dedicated to improving the safety and value of automobiles, other motor vehicles, highways, and mobile homes. The center regularly issues reports, books, comments on federal rule making, and other materials.

Chemical Manufacturers Association (CMA): The association has an active technical service on transportation and packaging. CMA's *Chemical Safety Data Sheets*, with listings available on request, give details on bulk transportation of chemicals. CMA also operates the Chemical Transportation Emergency Center Chemtrec information on hazardous materials involved in emergencies through its WATS line, 1-800-424-9300.

Compressed Gas Association, Inc. (CGA): CGA provides technical services on products of interest and concern to its membership.

Consumers Union of U.S., Inc. (CU): A nonprofit consumer information organization with a large technical staff, CU regularly tests automobiles—and occasionally trucks and other motor vehicles—and publishes its findings monthly in *Consumer Reports* magazine.

Cooperative Hazardous Materials Enforcement and Development (COHMED): COHMED develops and promotes training programs for personnel responding to hazardous materials emergencies.

Hazardous Materials Advisory Council: The council is concerned with all aspects of hazardous materials transportation affecting shippers, all modes of transportation, and manufacturers of packaging. It also conducts training in the safe transportation of hazardous materials.

Insurance Institute for Highway Safety: The institute is an independent, nonprofit, scientific, and educational organization dedicated to reducing losses resulting from crashes on the nation's highways. The American Insurances Service Group, the American Mutual Insurance Alliance, the National Association of Independent Insurers, and several individual insurance companies support the institute.

Manufactured Housing Institute: The institute represents mobile home manufacturers east of the Rocky Mountains.

National LP-Gas Association: This association develops LP-Gas standards and provides safety programs for the general public.

National Safety Council (NSC): A nonprofit organization dedicated to accident prevention, the council has an extensive program on highway safety geared principally to the motoring public. NSC produces numerous publications on transportation and highway safety.

National Tank Truck Carriers, Inc. (NTTC): A Conference of the American Trucking Associations, Inc., NTTC helps members, who are operators of tank truck fleets, solve problems associated with safe handling of flammable liquids, gases, and chemicals.

Physicians for Automotive Safety: This organization of medical doctors who are concerned with preventing death and injury in motor vehicle accidents has taken a particular interest in the safety of children in motor vehicles.

Professional Drivers Council: The council is a membership group of truck and bus drivers with an interest in their occupational safety and health.

Recreational Vehicle Institute, Inc.: The institute represents the recreational vehicle industry. Manufacturing members produce travel trailers, truck campers, camping trailers, and motor homes.

The Truck Body and Equipment Association, Inc. (TBEA): This is a trade association of the manufacturers of truck bodies, suppliers of component parts, and distributors. It has a Fire Apparatus Manufacturers Division, formerly a separate organization known as Fire Apparatus Manufacturers Association. The TBEA represents its members by dealing with DOT on safety legislation.

Truck Trailer Manufacturers Association: A trade association of manufacturers of truck trailers, it cooperates, through its various committees and representatives, with NFPA, other allied industry associations, and governmental agencies in matters affecting tank transport design and fabrication.

Western Manufactured Housing Institute: The institute represents manufacturers of mobile homes and recreational vehicles whose equipment is made and/or distributed on the west coast.

PREVENTING MOTOR VEHICLE FIRES

Fire prevention in motor vehicles requires the attention of everyone involved. Vehicle designers must be aware of fire hazards in their products. Sources of heat and ignition must be kept away from flammable materials as much as possible. Consideration should be given to the vulnerability of fuel system components, in particular—to their basic integrity, their location away from the perimeter of the vehicle and from crash-collapsible and intrusive parts, and their location away from potentially hot exhaust and electrical components.

Examples of location hazards were found in 1971–1976 Ford Pintos and 1975–1976 Mercury Bobcats. These cars' fuel filler necks and tanks could separate in a rear-end collision, resulting in fire. Manufacturers recalled 1.4 million of these vehicles for retrofitting of longer fuel filler necks and for installation of protective shields on the front of fuel tanks. Improperly routed fuel lines to the carburetor have been the subject of other recalls because of potential rupture and underhood fires. Other recalls have involved routing fuel close to hot surfaces, which could result in deterioration and rupture; fuel lines subject to physical damage; and potential leakage of oil onto hot surfaces.

Because electrical system failures can cause fires resulting in property damage, more attention must be paid to routing of wiring, aging of insulation and components, and failure modes of components. (That is, do they short circuit or open circuit when they fail?)

To minimize potential fuel leakage in any crash, extraneous plumbing and openings should be eliminated in fuel and hazardous liquid and gaseous cargo containers. Necessary plumbing should be designed so it is protected from potential crash damage.

In vehicle construction, care must be taken to ensure that fuel and electrical lines are correctly routed and connections secure. In addition, location of sharp or pointed components near fuel or hazardous cargo tanks should be avoided.

A major hazard during the vehicle repair process is the use of torches, usually in the repair of exhaust systems and crash damage. Some items used in vehicle repair are particularly flammable, such as paint, solvents, adhesives, and oily rags. Operating an engine with the carburetor air cleaner removed may result in a fire if the engine backfires through the carburetor.

Care must be taken when maintaining and rebuilding damaged vehicles to ensure that their fuel and electrical systems are free of hazards not present in the original vehicle.

Vehicle Electrical Systems

Automotive lighting and accessory circuits are usually 6 or 12 V, but vehicle ignition systems have high voltage and low amperage. Automobile electrical system fire safety is largely a matter of proper installation, fusing, and maintenance. Important check points include the location and protection afforded the battery; battery cable integrity and protection of exposed battery terminals; adequate flame-retardant insulation on all wiring; proper fusing of normal lighting and accessory circuits; proper support, location, and security of all wiring, with adequate protection by rubber insulating bushings where wiring passes through metal; and good ignition system maintenance.

Electrical fires may be fed by oily deposits in and around the engine or by combustible materials, such as interior fabric linings and upholstery. In collision or upset, electrical short circuits may occur and may ignite fuel vapors. It is desirable, particularly on buses and trucks, to provide an approved type of manual battery/generator disconnect switch, preferably one incorporating short circuit supervision. However, use of such devices on vehicles is rare.

Miscellaneous Vehicle Hazards

Proper exhaust system installation is important because the system's hot surfaces may ignite nearby combustible components. Hot carbon particles and hot gases discharged from the exhaust can ignite flammable liquids, grease, or similar exposed materials, such as insulation on electrical wiring.

Catalytic converters in exhaust systems can cause wildland fires, if automobiles are driven through dry grass or vegetation.

Overuse of vehicle brakes can result in sufficient overheating to cause severe smoldering and sometimes fire. Improper adjustment that prevents full release of a brake can lead to similar conditions.

The problem of overheating brakes increases with the vehicle's size and gross weight. On long downgrades, a driver must rely on the braking effect of the engine for primary control of the vehicle, limiting brake use to slowing for sharp turns or unanticipated, localized, hazardous conditions and to stop. The engine's maximum braking effect is achieved only in the transmission lower gears. Particularly on trucks and buses, it is essential for the driver to get into a low gear at the top of the grade before the vehicle reaches a speed at which downshifting is no longer possible. In recent years, engine braking devices have been used widely on diesel-powered trucks, especially in mountainous regions. Other types of electrically and hydraulically operated retarders are used in special situations.

Another serious common problem arising from overuse of brakes is sufficient heat buildup to cause brake fade. When this happens, brakes become ineffective and a major crash is possible, whether or not fire ensues.

If one tire of a dual-tire installation runs flat or is seriously underinflated, flexing may cause sufficient internal heat to build up to cause ignition. This possibility also exists if tires are seriously overloaded and may be exacerbated by sustained high-speed operation and high ambient temperature. Occasionally, fire results if dual tires are spaced so closely together that they rub or chafe against each other when the vehicle is in motion.

Provisions of the Federal Motor Carrier Safety Regulations for all heating systems used on vehicles in interstate commerce have been upgraded. This led to a significant reduction in the number of fires combustion heaters caused on vehicles. Reducing such fires has also been aided by technological advances in heaters. The few fires that do occur generally result from unforeseen heater malfunctions, or from a shift in freight loaded too close to a heater.

Alcohol-based antifreeze is no longer widely used in motor vehicles. Therefore, vehicle fires involving its vapors are seldom a problem.

Fire Hazards of Special Vehicles

Certain types of special vehicles, such as farm tractors, motorcycles, all-terrain vehicles, and others, have special problems resulting from the close spacing of such components as the fuel tank and other parts of the fuel system, the engine, the ignition system and other electrical components, and the exhaust, coupled with minimal protective bodywork that increases the susceptibility of these fire-sensitive components to damage.

The need for care to avoid fuel spillage and sparks from static electricity is especially important when refueling in the field from portable containers.

The advent of vapor recovery nozzles on gasoline pumps has introduced a potential fire hazard in fueling motorcycles and other vehicles with small fuel tanks. The depth to which the nozzle must be inserted in order to activate vapor recovery control and permit fuel flow is so great that effective refueling cannot be accomplished in the normal manner. It has become necessary to use a device to bridge the fill-pipe opening. Otherwise, the operator must pull back the bellows with one hand and activate the nozzle trigger with the other to obtain fuel.

Use of Fire Extinguishers

Commercial vehicles operating in interstate commerce must be equipped with fire extinguishers to comply with Federal Motor Carrier Safety Regulations. These regulations provide two minimum levels of protection: a 5-B:C extinguisher is required if the power unit is not transporting hazardous materials in sufficient quantity to require marking or placarding the vehicle, and a 10-B:C extinguisher is required on a power unit used for transporting hazardous materials.

Many states have similar requirements for commercial vehicles, through adoption of applicable provisions of federal regulations or their own regulations. Some state laws require additional or larger extinguishers on tank vehicles transporting flammable liquids or gases, combustible liquids, or Class A and/or Class B explosives.

There are no legal requirements for carrying fire extinguishers in private passenger cars, vans, or light trucks, although many people do keep them in vans and motor homes. As a minimum, the vehicle owner should consider buying a portable extinguisher that can deal with Class B (flammable liquids) and Class C (electrical) fires. Also on the market are extinguishers using multipurpose agents for use on Class A fires (ordinary combustibles), as well as Class B and C fires. Anyone equipping a vehicle with a fire extinguisher must understand that the best chance for its successful use lies in prompt action in the fire's early stages and proper use to avoid wasting the agent and exhausting the extinguisher's contents prematurely. NFPA 10, *Standard for Portable Fire Extinguishers*, gives additional information on ratings and use of fire extinguishers.

NFPA 512 references pertinent federal requirements to guide truck owners who otherwise may not be subject to regulatory requirements.

Other Fire Hazards

Carrying extra gasoline in a can is extremely dangerous because of the possible accumulation of fumes. The hazard increases if glass or plastic containers are used, because they're more likely to break and leak.

Smoking by a driver or passengers presents the potential for fire if sparks, hot ashes, or incompletely extinguished matches get into upholstery or other interior materials. Water or an agent suitable for Class A fires can extinguish such fires; the burn area then must be watched in case it continues to smolder and eventually rekindles.

While the vehicle is operating, the driver should be alert for any indication of fire or a fire hazard, such as the smell of something hot, of unusual fumes, or of leaking fuel or lubricant. The driver should check for conditions that can create sufficient heat buildup to cause a fire, such as exhaust leaks and dragging brakes.

In an accident, the engines and all electrical accessories of all involved vehicles should be turned off to minimize the danger of fire. At night, headlights can be left on to protect the scene, but persons in the vicinity should be alert for indication of overheated wiring or fire. Smoking and all use of open flames or lights must cease, particularly if leaking fuel is detected. If a travel trailer or motor home is involved, the probable presence of liquefied petroleum gas should be kept in mind, smoking should be prohibited, and open flames and lights should be kept well away from the scene.

When a vehicle fire occurs, the following basic steps should be taken, regardless of the type of vehicle involved:

1. Shut off the engine and all electrical systems that can be reached.
2. Get everyone out of the burning vehicle and well away from it, preferably uphill and upwind. The potential for fire is one

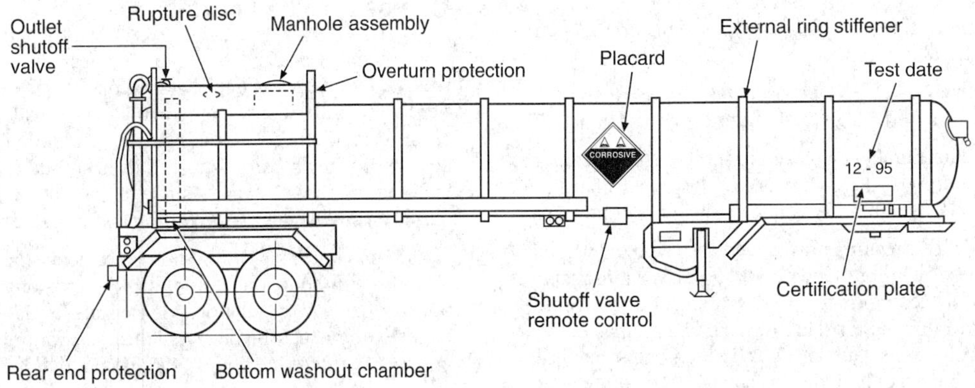

FIG. 9-32A. *Corrosive materials cargo tanks must be equipped with an internal shutoff valve if unloaded from the bottom.*

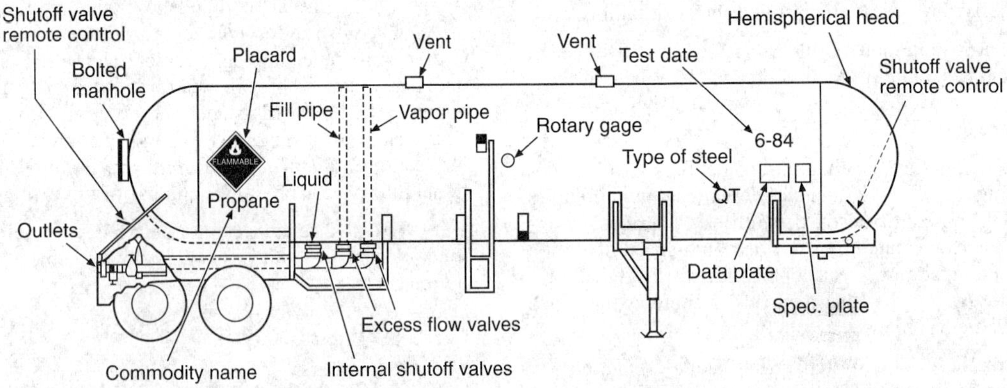

FIG. 9-32B. *Compressed gas cargo tanks must be equipped with internal shutoff valves at liquid and vapor discharge openings.*

justification for quickly moving injured persons from the scene of an accident.

3. Summon the fire department by the best available means.
4. Attempt to fight the fire only if it can be done without endangering anyone's personal safety.

Basic principles to be observed in attempting to fight a vehicle fire with a portable fire extinguisher are:

1. First, see item 4 above. The fire should be fought from the windward side so that wind, if any, will help carry the fire extinguishing agent toward the fire and help blow smoke and flame away from the person attempting to extinguish it.
2. The fire extinguisher should be aimed at the base of the flames.
3. If leaking flammable liquid is involved, work toward the source of the leak in putting out the flames. Ultimate control may depend on the ability to shut off leaking fuel at its source.
4. If flammable gas is burning, allow it to continue to burn until the source of any leak is controlled. Otherwise, extinguishing the fire may permit leaking gas to accumulate, with later fire or explosion likely. If flames impinge on a flammable gas container, the container should be kept cool to avoid a flame-induced weakening and possible rupture or a BLEVE (boiling liquid expanding vapor explosion). If fire-fighting hose streams are not available, RUN. Leave IMMEDIATELY.
5. If no extinguisher is available, throw sand or dirt on a fire to control flames.

6. If stopping to help at the scene of a vehicle fire, park uphill, preferably upwind, and at a safe distance from the burning vehicle.
7. After a fire has been extinguished, do not operate the vehicle until all burned parts are cool and the cause of the fire has been corrected.
8. During an electrical fire, turn off the ignition switch and lights to try to cut off the flow of current and permit faster extinguishment. If tools are available and it can be done safely, disconnect the battery.
9. In fighting an underhood fire, use great care in opening the hood to avoid injury in case of a sudden flareup.
10. Inspect and recharge all extinguishers without delay.

The following special considerations apply to fires and potential hazards involving commercial vehicles:

1. Any hazardous situation should be evaluated to avoid endangering lives and wasting extinguishing agent.
2. If a vehicular combination unit is involved in a fire, the power unit should be unhooked and moved a safe distance away from the vehicle if it can be done safely.
3. Drivers must be alert for the hard pulling characteristics that may indicate dragging brakes and must readjust brakes before a fire develops.
4. Drivers should check tire inflation and temperature every 2 hr or 100 mi (161 km), whichever comes first, and at any other stop. OMC regulations require this practice for a vehicle placarded for transportation of hazardous materials.

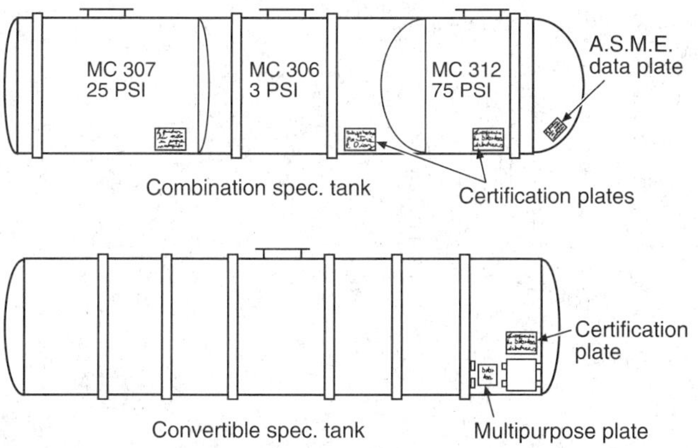

FIG. 9-32C. *Profiles of combination (multiple compartment) and convertible (single compartment) cargo tanks.*

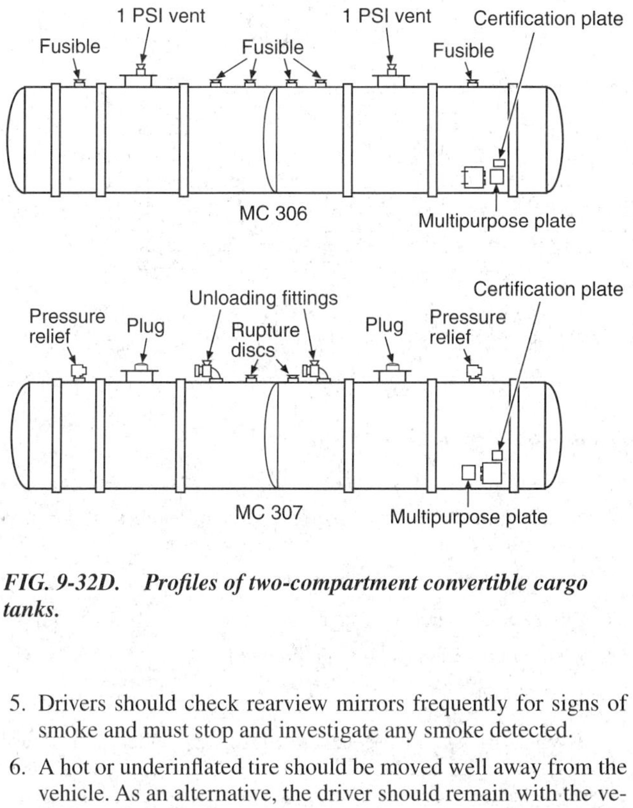

FIG. 9-32D. *Profiles of two-compartment convertible cargo tanks.*

5. Drivers should check rearview mirrors frequently for signs of smoke and must stop and investigate any smoke detected.

6. A hot or underinflated tire should be moved well away from the vehicle. As an alternative, the driver should remain with the vehicle until the tire is cool to the touch and then make repairs. If a vehicle is left with a hot tire, the tire may burst into flames, destroying the vehicle and its load. With proper tire-changing equipment and a pair of heavy leather gloves, a driver can safely remove a hot tire that has not burst into flames.

7. Because a tire fire results from internal heat buildup, portable extinguishers normally cannot do more than control open flames. The cooling effect of large quantities of water is required for extinguishment.

8. If fire is suspected inside a closed van body—usually indicated by smoke seeping out around the doors—the doors should be left closed until fire department assistance is at hand. Should the doors be opened, oxygen reaching the fire may cause a sudden flareup. As long as the doors are closed, the fire can only smolder from lack of oxygen.

In the interest of preventing cargo fires, motor carriers generally prohibit smoking while vehicles are being loaded or unloaded, so sparks or hot ashes do not lodge in the cargo and burst into flame later.

Federal Motor Carrier Safety Regulations prohibit smoking during refueling; while loading, unloading, or transporting explosives, flammable liquids, gases, or solids, or oxidizing materials requiring marking or placarding of the vehicle; and while driving an empty tank vehicle that last contained a flammable liquid or gas and that was required to be marked or placarded.

CARGO TANKS (TANK TRUCKS)

Cargo tanks and portable tanks are large containers for hauling liquids and gases in bulk. Cargo tanks are sometimes called "tank motor vehicles" or "tank trucks." Any person offering a hazardous material in a container for transportation must determine that the container used is authorized for the commodity. Specifications for cargo tanks, portable tanks, and all other hazardous materials packaging other than rail tank cars are in 49 CFR, Part 178. Authorized packaging prescribed in 49 CFR, Part 173, the shipper regulations for the particular material, as listed in columns 5(a) and (b) of 49 CFR 172.101. If a carrier supplies a container, such as a cargo tank, the shipper still must determine that the tank supplied is a proper container for the material being offered. The shipper must either examine the manufacturer's specification marking on the tank or obtain papers from the carrier certifying that the tank is proper for the material, per 49 CFR, Part 173.22.

Cargo tanks are the most common transport vehicles used to move combustible, flammable, and corrosive liquids, as well as compressed gases, both flammable and nonflammable. Hazardous materials in powder or granular form are transported in hopper-type dump-bodied vehicles. When these hazardous materials are transported in bulk containers, such as cargo tanks, and an accident occurs, the disaster potential is substantial.

To minimize risks involved in transporting hazardous materials and to reduce hazards should an accident occur, the federal government specifies requirements for the manufacture, operation, maintenance, inspection, testing, and repair of cargo tanks.

DOT's hazardous materials regulations apply to interstate and intrastate commerce and are enforced by both federal and state authorities. DOT is phasing in new cargo tank regulations designed to reduce spillage and to ensure that tanks are properly constructed, maintained in safe condition, and properly repaired. The new cargo tank regulations are designated in 49 CFR, Part 180. In addition to upgrading the specifications of certain cargo tanks, the regulations provide new specification numbers listed in Table 9-32D.

Leakage is likely to occur in both contact and noncontact accidents, leading to high cost in terms of personal injury and property damage. The major causes of unintentional release of hazardous materials are poor maintenance and deviation from tank specifications.

TABLE 9-32D. *Specification Numbers for Hazardous Materials*

Current Specification Number	New Specification Number
MC-306	DOT-406
MC-307	DOT-407
MC-312	DOT-412
MC-331	No change
MC-338	No change

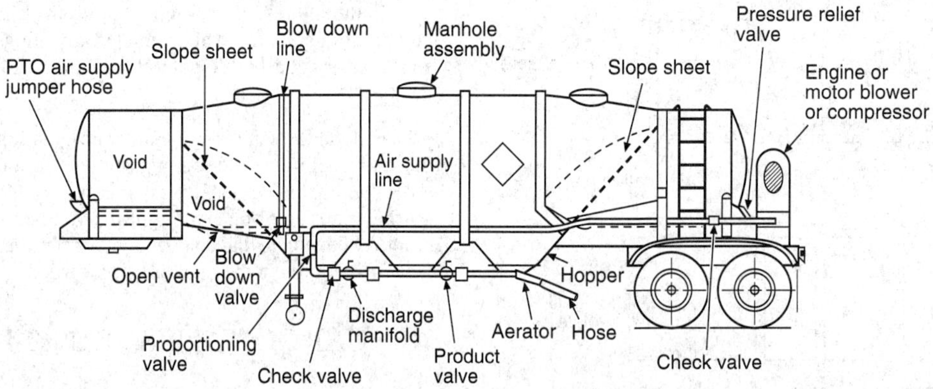

FIG. 9-32E. *Profile of a dry bulk cargo tank. Examples of cargo carried are ammonium nitrate, an oxidizer, or corrosive solids. Tanks of similar appearance are also widely used to transport cement and other non-hazardous dry bulk materials.*

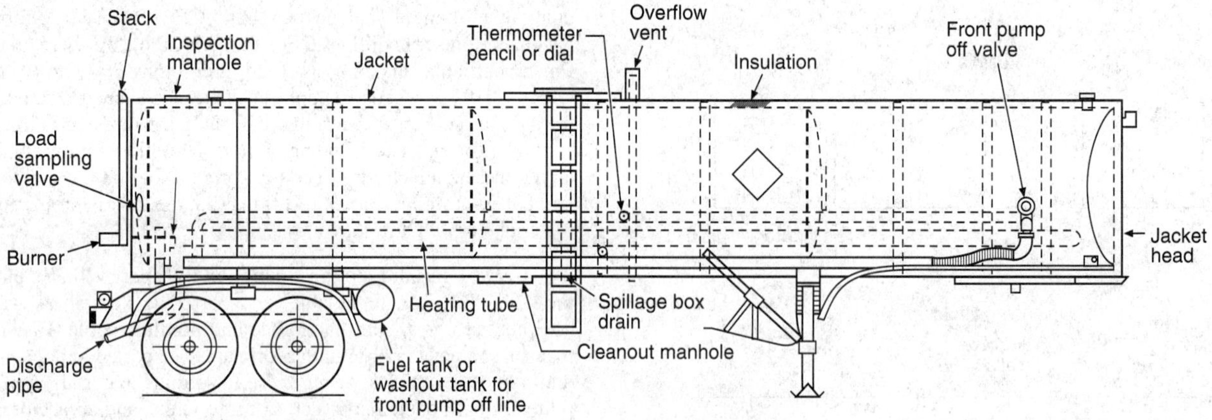

FIG. 9-32F. *A nonspecification asphalt cargo tank.*

Potential hazards to fire fighters dealing with cargo tanks transporting hazardous materials are as follows:

1. Multipurpose tanks holding improper commodities for the specific container.
2. Multicompartment containers hauling materials in more than one hazard class (corrosive/flammable/poison/oxidizer).
3. Defective or missing safety controls.
4. Release of large quantities of hazardous materials.
5. Increase of potential fires—extra equipment required to fight fires and control spills.
6. Potential for a BLEVE.
7. Incorrect or missing placarding and identification markings on exterior of cargo tank.
8. Missing, incomplete, or incorrect shipping papers.

Persons receiving a report of an actual or potential highway transportation hazardous materials emergency should obtain as much of the following information as possible:

1. Location of the emergency.
2. Type of commercial vehicle(s) involved, such as cargo tank, flatbed, van, and so on.
3. Class(es) of hazardous material(s) as indicated by placards, if visible, including the four-digit identification number, if visible.
4. Proper shipping name(s), hazardous class(es), and identification number(s) of hazardous materials, if shipping papers are available.

5. Nature of the emergency. Is it a fire? Is there leakage? Are persons affected or injured by hazardous materials?
6. Name and address of trucking company.
7. Other vehicles and/or property involved.
8. Weather conditions and wind direction.
9. Type of area.

Upon arriving at the scene, emergency responders should take time to assess the situation and make every effort to ascertain what type(s) of hazardous materials and packaging is involved. The scene should be approached from upwind. Responders must avoid contact with spilled or leaking materials that may penetrate protective or other clothing, resulting in death or injury.

Figures 9-32A through 9-32F display tank construction, denoting areas where leakage is most likely. All these diagrams should be considered as guidelines rather than constants. Figures 9-32G and 9-32H show common tank shapes and internal/external inspection items.

Many tank vehicles contain hazardous materials of one or more types. These units, which include farm "nurse" tanks, tank trucks, tank trailers, and railroad tank cars, must be treated with equal respect. Each emergency must be sized up quickly and individually, taking into consideration the safety of every person at the scene.

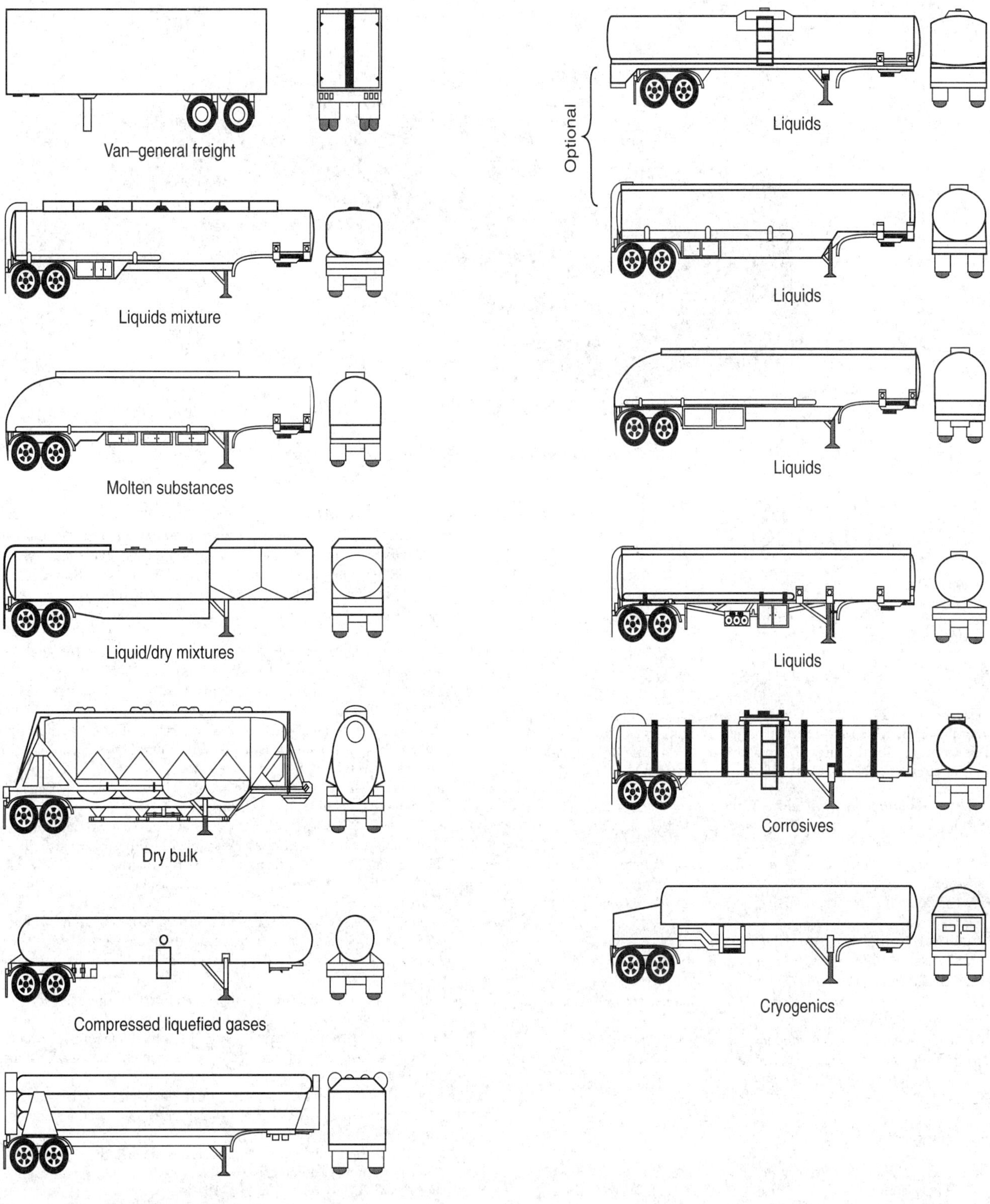

Van–general freight

Liquids mixture

Molten substances

Liquid/dry mixtures

Dry bulk

Compressed liquefied gases

Compressed gases tube trailer

Optional

Liquids

Liquids

Liquids

Liquids

Corrosives

Cryogenics

FIG. 9-32G.   *This chart depicts only the most general shapes of road trailers. Emergency response personnel must be aware that many variations of road trailers, not illustrated here, are used for shipping chemical products. Suggested guides are for the most hazardous products that these trailer types may transport. Recommended guides should be considered as a last resort if the product cannot be identified by any other means.*

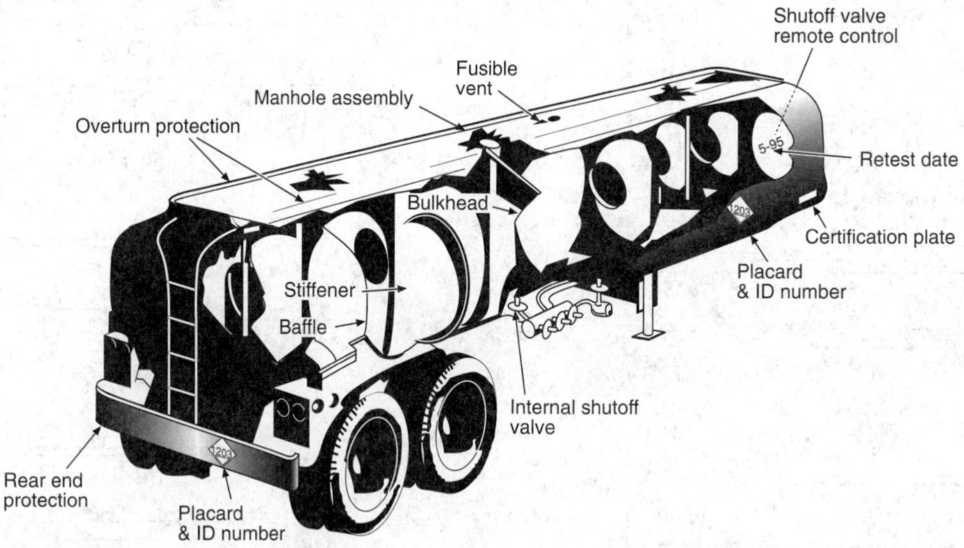

FIG. 9-32H.   Cargo tank inspection guide (MC 306).

BIBLIOGRAPHY

Reference Cited

1. Karter, M. J., Jr., "Fire Loss in the United States in 1995," *NFPA Journal*, September/October 1996.

NFPA Codes, Standards, and Recommended Practices

Reference to the following NFPA codes, standards, and recommended practices will provide further information on motor vehicles discussed in this chapter. (See the latest version of *The NFPA Catalog* for availability of current editions of the following documents.)

NFPA 10, *Standard for Portable Fire Extinguishers*
NFPA 52, *Standard for Compressed Natural Gas (CNG) Vehicular Fuel Systems*
NFPA 58, *Standard for the Storage and Handling of Liquefied Petroleum Gases*
NFPA 385, *Standard for Tank Vehicles for Flammable and Combustible Liquids*
NFPA 407, *Standard for Aircraft Fuel Servicing*
NFPA 414, *Standard for Aircraft Rescue and Fire Fighting Vehicles*
NFPA 495, *Explosive Materials Code*
NFPA 498, *Standard for Safe Havens and Interchange Lots for Vehicles Transporting Explosives*
NFPA 501C, *Standard on Recreational Vehicles*
NFPA 512, *Standard for Truck Fire Protection*
NFPA 513, *Standard for Motor Freight Terminals*
NFPA 1901, *Standard for Automotive Fire Apparatus*

Additional Reading

An Analysis of Fires in Passenger Cars, Light Trucks, and Vans, National Highway Traffic Safety Administration, DOT H5 808 208. Available through National Technical Information Service, Springfield, VA, 22161.

Anderson, B. G., "Car Engine Fires: Tactics for Quick Knockdowns," *Fire Engineering*, Vol. 147, No. 2, February 1994, pp. 37, 40, 42–44.

Bennett, J., "How to Stop Car Fires," *Car and Travel*, July/August 1995, pp. 10–11.

Betroni, J., "Vehicle Fires: Meeting the Challenge of the Modern Vehicle," *Fire and Arson Investigator*, Vol. 46, No. 2, December 1995, pp. 27–30.

Bomgardner, Paul M., *Handling Hazardous Materials*, Department of Safety, American Trucking Association, Inc., Alexandria, VA, 1994.

Bomgardner, Paul M., *Hazardous Materials Shipper's Guide,* Department of Safety, American Trucking Association, Inc., Alexandria, VA, 1994.

Clark, A., "Redesigning a Modern Car to Test Updated Fire Service Expertise," *Fire Engineers Journal*, Vol. 51, No. 161, June 1991, p. 19.

"Complete Guide to Car Fires. Photos," *Fire Prevention*, Vol. 229, May 1990, pp. 16–17.

Cox, G., and Kumar, S., "Numerical Model of Fire in Road Tunnels," *Tunnels and Tunnelling*, Vol. 19, No. 3, 1987, pp. 55–57, 59–60.

Donahue, M. L., and Campbell, C. A., "Investigation of Motor Vehicle Fires. Part 1," *Firehouse*, Vol. 15, No. 8, August 1990, pp. 58–60.

Donahue, M. L., and Campbell, C. A., "Investigation of Motor Vehicle Fires. Part 2. Recommended Investigative Procedures," *Firehouse*, Vol. 15, No. 11, November 1990, pp. 42, 44–46.

Drivers Guide to Hazardous Materials, Safety Department, American Trucking Association, Inc., Alexandria, VA.

Ekman, L., "Full-Scale Fire Testing of Automobiles," Volvo Car Corp., Gothenburg, Sweden, Plastics and Rubber Institute and British Plastics Federation, Flame Retardants 1990, Fourth International Conference, January 17–18, 1990, London, England, Elsevier Applied Science, New York, 1990, pp. 126–133.

Emergency Response Guide, Research & Special Programs Administration, U.S. Department of Transportation, Washington, DC.

Facts for Drivers, Safety Department, American Trucking Association, Inc., Alexandria, VA.

Fanning, D., "Car Fire Research Points to Danger of Wiring Defects and Plastic Tanks," *Fire*, Vol. 85, No. 1049, November 1992, pp. 9, 12.

Federal Emergency Management Agency, "Motor Vehicle Fires. . .What You Need to Know," Federal Emergency Management Agency, Washington, DC, L-202, February 1993, 6 pages.

Fire Protection Association, "Investigation of Vehicle Fires. Fire Safety Data Arson Dossier AR5," Fire Protection Assoc., London, England, AR5, 1991, 12 pages.

Gallagher, G. A., and McCone, S. W., "Lessons in Hazardous Materials Transportation Based on Case Histories," *Plant/Operation Progress*, Vol. 5, No. 3, July 1986, pp. 186–191.

Gowecke, K., "Brandverhalten von Pkw nach Verkehrsunf allen. [Fire Behavior of Cars After Traffic Accidents.]," *VFDB*, No. 3, February 1994, pp. 111–124.

Gustin, B., "New Fire Tactics for New-Car Fires," *Fire Engineering*, Vol. 149, No. 4, April 1996, pp. 43–46, 48, 50, 52, 54.

Hazardous Materials Transportation: Citations from the NTIS Bibliographic Database, National Technical Information Service, Springfield, VA, 1988.

Hazmat Transport Regs, a reprint of federal hazardous materials transportation regulations, 49 *CFR*, Parts 106, 107, 110, 130, 170–180, Department of Safety, American Trucking Associations, Inc., Alexandria, VA, reprinted annually.

Higgins, M., "Vehicle Fires: A Practical Approach," *Fire and Arson Investigator*, Vol. 43, No. 3, March 1993, pp. 57–60.

Hoffman, J., "Investigating Vehicle Fires," *Firehouse*, Vol. 17, No. 1, March 1992, pp. 22, 24–25.

Holmes, S., "Car Airbags: Safety of Emergency Personnel Needs Consideration," *Fire*, Vol. 87, No. 1070, August 1994, pp. 41–42.

"Improve Designs for Fire Safety Car Manufacturers Are Urged," *Fire*, Vol. 85, No. 1046, August 1992, p. 4.

Kletz, T. A., "Transportation of Hazardous Substances: The U.K. Scene," *Plant/Operation Progress*, Vol. 5, No. 3, July 1986, pp. 160–164.

Kuhner, R. E., "Propane Incident," *Fire Engineering*, Vol. 141, No. 11, 1988, pp. 26–29.

Magnus, S., "Aspects of Vehicle Fire Investigation," *Fire Engineers Journal*, Vol. 55, No. 176, March 1995, pp. 34–37.

"Malicious Car Fires Shown as Biggest Rise in Statistics," *Fire*, Vol. 86, No. 1056, June 1993, p. 37.

Moore, G., "Growing Problem of Road Vehicle Fires," *Fire Prevention*, No. 222, Sept. 1989, pp. 29–31.

Navesey, F., "Are Natural Gas Vehicles Fire Safe?" *Fire Prevention*, No. 280, June 1995, pp. 24–27.

Noll, G., "The Gasoline Tank Truck: Part I," *Fire Engineering*, Vol. 140, No. 6, 1987, p. 39.

Noll, G., "The Gasoline Tank Truck: Part II," *Fire Engineering*, Vol. 140, No. 7, 1987, p. 37.

Norman, J., "Car Fire Is a Car Fire. . . Is a Car Fire?" *Firehouse*, Vol. 20, No. 5, May 1995, pp. 16, 18.

Sandel, P., "Vehicle Fires: More Accurate Ways of Reporting Causes Are Needed," *Fire*, Vol. 84, No. 1036, October 1991, pp. 22, 28, 30.

Schappert, R. J., III, "Putting a Stop to Vehicle Fires, Safely," *Fire Chief*, Vol. 38, No. 7, July 1994, pp. 26, 28–29.

Shipp, M., and Spearpoint, M., "Measurements of the Severity of Fires Involving Private Motor Vehicles," *Fire and Materials*, Vol. 19, No. 3, 1995, pp. 143–151.

Suefert, F. J., "Automobile Engine Fires," *Fire and Arson Investigator*, Vol. 45, No. 4, June 1995, pp. 23–26.

Suefert, F. J., "Automobile Engine Fires. Part 2," *Fire and Arson Investigator*, Vol. 46, No. 1, September 1995, pp. 23–26.

Truck Drivers Handbook, Safety Department, American Trucking Associations, Inc., Alexandria, VA, 1994.

U. S. General Accounting Office, "Motor Vehicle Safety: Information on Accidental Fires in Manufacturing Air Bag Propellant. Report to the Chairman, Committee on Energy and Commerce, House of Representatives," General Accounting Office, Washington, DC, GAO/RCED-90-230, September 1990, 23 pages.

U.S. Department of Transportation, "Flammability of Interior Materials. Motor Vehicle Safety Standard No. 302," Department of Transportation, Washington, DC, *Code of Federal Regulations*, Vol. 49, No. 400-999, October 1, 1994, pp. 606–608.

Whitaker, E., "Vehicle Fires: Looking Behind the Statistics," *Fire Prevention*, No. 222, Sept. 1989, pp. 33–38.

Wilson, T. W., "Vehicle Air Bags: Just the Facts," *Fire Engineering*, Vol. 144, No. 1, January 1991, pp. 53–54, 57.

The following reports are sponsored by the National Highway Traffic Safety Administration.

An Assessment of Automotive Fuel System Fire Hazards, DOT HS-800 624, Dec. 1971, Dynamic Science, Phoenix, AZ.

Escape Worthiness of Vehicles and Occupant Survival, DOT HS-800 428, Dec. 1970, University of Oklahoma Research Institute, Norman, OK.

Flammability Characteristics of Vehicle Interior Materials, DOT HS-800 205, May 1969, Engineering Mechanics Division, IIT Research Institute, Chicago, IL.

Heavy Truck Fuel System Safety Study, DOT HS 807 484, September 1989, University of Maryland, College Park, MD. Available from National Technical Information Service, Springfield, VA 22161.

Prevention of Electrical Systems Ignition of Automotive Crash Fire, DOT HS-800 392, March 1970, Dynamic Science, Phoenix, AZ.

ALTERNATIVE FUELS FOR VEHICLES

Ron Buys, Cora Urgena, and Scott Stookey

The use of alternative fuels for vehicles is becoming more predominant. Forces driving their use are increasing air pollution and the United States' dependence on foreign oil. It is evident that new federal regulations such as the Energy Policy Act (EPACT) of 1992 and the Clean Air Act Amendments (CAAA) of 1990 were specifically designed to increase the driving public's use of alternative fueled vehicles to benefit the environment and to increase the use of natural resources available in North America. This chapter discusses several areas including properties and hazards associated with different alternative fuels, vehicle fueling stations, vehicle component designs and operation, emergency response to alternative fuel incidents, and federal regulations that may impact alternative fuels users.

A variety of alternative fuels are currently used in the United States. However, the review of alternative fuels in this chapter is limited to compressed natural gas (CNG), liquefied natural gas (LNG), and liquefied petroleum gas (LP-Gas), specifically propane. The scope is limited to these three gaseous fuels because the hazards of alternative fuels like flammable liquids—alcohols and reformulated gasoline—are well known and generally accepted. On the other hand, gaseous fuels have recently gained popularity and more information is needed on their use as motor vehicle fuel.

CAAA defines alternative fuels as methanol, ethanol, and other alcohols, reformulated gasoline, reformulated diesel (for trucks only), natural gas, propane, hydrogen, or electricity. EPACT includes all fuels listed under CAAA except for reformulated gasoline and diesel. EPACT also defines other alternative fuels: fuels derived from biomass, liquid fuels derived from coal, and alcohol blended with other fuels containing at least 85 percent alcohol by volume.[1]

Alternative Fuel Users

Many entities such as private, state, federal, and local government fleets and the general public use alternative fuels. Fleets using alternative fuels may be small or large but are commonly located predominantly in urban areas.

Ron Buys, P.E., is a hazardous materials engineer and a graduate of Texas A&M University with a B.S. in civil engineering and an M.S. in environmental engineering. Cora Urgena is a hazardous materials engineering associate and a graduate of the University of Houston with a B.S. in chemical engineering. Scott Stookey is a hazardous materials engineering associate and graduate of the Oklahoma State University School of Fire Protection and Safety Engineering Technology. They are members of the Austin, Texas, Fire Department Hazardous Materials Response Team.

The Clean Fleets Program established by CAAA states that by 1998, serious, severe, and extreme ozone nonattainment areas must purchase alternative fuel vehicles. This applies to fleets with 10 or more vehicles capable of being centrally refueled and located in a area with a population of 250,000 or more. The law excludes emergency, farm, construction, rental, demonstration, and privately garaged vehicles.[1]

EPACT mandates the purchase of alternative fuel vehicles with the intent of reducing the United States' dependence on oil imports. Federal and state fleets, as well as private-sector companies that produce alternative fuels, will be required to purchase alternative fuel vehicles. This can be a phased schedule for fleets of 50 or more vehicles, with at least 20 that can be centrally refueled, and located in an area with a population of 250,000 or greater. The law can extend to include municipal and private fleet owners if the U.S. Environmental Protection Agency (EPA) determines that EPACT's goals cannot be met.

Clean Air Act Amendments of 1990

Alternative fuel use has increased since the EPA established maximum concentration levels called National Ambient Air Quality Standards (NAAQs) for certain pollutants in the ambient air. The standards targeted three pollutants: (1) automotive exhaust pollutants such as ozone, (2) nitrous oxides (NO_X), and (3) carbon monoxide (CO). Two other pollutants exhausted from diesel fuel in trucks and buses include (1) sulfur dioxide (SO_2) and (2) particulate matter (PM). The phaseout of leaded gasoline is controlling lead, the sixth criteria pollutant. Both mobile and stationary sources generate these pollutants: aboveground storage tanks, industrial processes that introduce pollutants into the atmosphere, and diesel-fueled generator sets.

A community or area can be designated as a nonattainment area when concentrations of these pollutants exceed NAAQ limits. Ozone and CO nonattainment are most predominant. Individual states are responsible for attaining each EPA NAAQ. States accomplish this by utilizing State Implementation Plans, which outline specific actions to reduce emissions. One method of reducing mobile sources of criteria pollutants is using alternative fuels instead of more traditional fuels such as gasoline or diesel.

The use of alternative fuels significantly reduces ozone production. Alternative fuels can reduce ozone generation by reducing both the amount of volatile organic compounds (VOCs) in tailpipe emissions and the photochemical reactivity of VOC emissions. Ozone, or smog, is not a direct product of the combustion process. Atmospheric photochemical reactions between VOCs and the oxides of nitrogen (NO_X) created ozone. NO_X is primarily produced when nitrogen and oxygen in combustion air combine at high temperatures. VOCs result from incomplete combustion of hydrocarbon fuels used in motor vehicles.[2]

Simple compounds such as methane and propane predominantly comprise LNG, CNG, and LP-Gas fuels. The amount of VOCs these fuels generate is lower than the amount gasoline generates. Furthermore, ozone formation increases when constituents of tailpipe emissions have high photochemical reactivity.[2] Fuels composed of small molecules such as methane generate less ozone because they are unreactive in the atmosphere. Propane is more reactive than methane, but significantly less reactive than large molecules of hydrocarbon fuels, such as the constituents of gasoline.

Currently, Denver, Colorado, is the only nonattainment area for CO. Approximately 80 percent of CO is from mobile sources, and the majority of that is from emissions from gasoline-powered vehicles. Alternative fuels composed of methane and propane emit much less CO than gasoline because they have lower carbon to hydrogen ratios. In addition, alternative fuel vehicles do not need to run fuel rich during cold weather like gasoline-fueled vehicles do.[1] This lowers CO emissions.

PM emissions can be reduced by using alternative fuels because the primary mobile source of these emissions is diesel-powered vehicles.[2] Diesel fuels' high sulfur content and aromatic hydrocarbons contribute to high PM emissions. In gaseous alternative fuels, sulfur is not present in appreciable amounts and aromatic hydrocarbons are not present at all.

As mentioned earlier, nonattainment of ozone and CO are the most widespread. Studies show that using alternative fuels such as CNG, LNG, and LP-Gas can significantly reduce these pollutants.

Risk Management Programs

The phase-in schedules established by EPACT and CAAA will rapidly increase the purchase and use of alternative-fueled vehicles in the United States. The number of fueling stations will also increase as convenience and accessibility become major issues. As the number of fueling stations grow, owners and facility operators must become more aware of federal, state, and local requirements that could impact the stations' construction or operation.

One regulation the EPA proposes could greatly impact some alternative fuel stations: the Prevention of Accidental Releases of CAAA of 1990, Title III, Section 112 R. This federal legislation aims to prevent accidental chemical releases at storage locations and to minimize their consequences.

Section 112 R of Title III of CAAA does not exempt fueling facilities. Its proposed requirements apply to existing and new stationary sources in excess of threshold quantities. The List of Regulated Flammable Substances includes methane and propane with threshold storage quantities of 10,000 lb. LNG is composed of 95 to 99 percent methane, while LP-Gas is predominantly propane. Assuming commercial automotive grades of LNG and LP-Gas, facilities using stationary containers with capacities greater than 2,780 gal of LNG or 2,380 gal of LP-Gas may be required to prepare a risk management plan when the EPA rules are final. The proposed rule should not impact most CNG fueling stations because storage is normally below the threshold storage quantity. Natural gas from a distribution line is usually compressed at the facility, so it is usually unnecessary to store large quantities.

The October 20, 1993 edition of the *Federal Register*[3] outlines proposed requirements for the development of a Risk Management Program to meet the CAAA Chemical Accidental Release Prevention requirements. The proposed Risk Management Program and its implementation consist of three components: (1) a hazard assessment, (2) a program for preventing accidental releases, and (3) an emergency response program. A Risk Management Plan must be submitted to the implementing agency, the State Emergency Planning Committee, the Local Emergency Planning Committee, and the proposed Chemical Safety and Hazard Investigation Board.

Hazard Assessment

Hazard assessment is a determination of the worst-case release scenario, identification of other more likely significant releases of propane or LNG, an analysis of the off-site consequences, and a history of accidental releases of propane or LNG.

Under proposed CAAA rules, worst-case releases are determined by assuming that in an instantaneous release, all mitigation systems fail. Other more likely accidental releases can include failure of transfer hoses, excess flow valve, emergency shutoff controls, piping, and other significant equipment failures. The proposed rule also states that consequences must be determined from the rate and quantity of the substance lost in the air, the event's duration, the distance to exposures such as single- and multiple-family dwellings, the populations within these distances, and the potential for environmental damage. The owner or operator must maintain a 5-year history that includes such information as the date and time of a release and its quantity and duration.

Prevention Program

The prevention program's intent is to evaluate potential hazards present at a storage and dispensing station and to find the best way to control them. The program consists of a management system, process hazard analysis, process safety information, standard operating procedures, training, maintenance, pre-startup reviews, management of change, safety audits, and accident investigations.

The proposed rule states that the process hazard analysis must be conducted using quantitative or qualitative methodologies such as what-if, fault tree analysis, hazard and operability study (HAZOP), or other accepted practices. The process hazard analysis should address the hazards of the process; identify previous incidents that impacted offsite exposures, engineering, and administrative controls; predict the consequences of control or human failures, and stationary source sitting; and qualitatively evaluate possible safety and health effects if controls fail. It is anticipated that industry associations will conduct generic studies for LNG and LP-Gas fueling stations. These studies can be made available to independent operators to use in preparing the required studies and plans.

Emergency Response Program

Of particular importance to emergency response personnel is the proposed rule that will require the owner or operator of an LNG or LP-Gas fueling station to establish and implement an emergency response program for responding to and mitigating accidental releases. The emergency response must include information on evacuation routes, employee instructions, descriptions of response and mitigation technologies at the fueling station, and procedures for informing the public and emergency response agencies of a release.

The proposed rule states that procedures for testing, maintenance, and use of response equipment must be established. In addition, the owner or operator will be responsible for training employees in emergency response procedures and documenting all training activities. The proposed rule states that drills or exercises will be conducted and documented, but does not specify the frequency of these drills.

The proposed emergency response program will require facilities storing LNG or LP-Gas to document first aid and emergency medical treatment procedures necessary to treat persons exposed. The emergency response plan must coordinate with other local emergency response plans developed by the Local Emergency Planning Committee (LEPC) or other local emergency response agencies.

The three elements of the risk management program must be summarized in a Risk Management Plan and submitted to the aforementioned four entities. The report must be reviewed every 5 years.

Analysis of the Chemical and Physical Properties and Hazards of Alternative Fuels

When a vehicle is being fueled or repaired, release from the vehicle to the fueling or repair facility is a reasonable scenario that requires fire fighters' consideration.

The properties of each fuel require consideration in determining its potential hazard. Whether the hazard becomes real depends on various time-dependent conditions, the fuel system design, and the quality of the fuel system's manufacture and installation. This is due to the properties and hazards of the materials, whether release occurs inside or outside of a building, and the release scenario. This chapter discusses the relationships between identified properties and hazards of CNG, LNG, and LP-Gas.

Mixing and dispersion of gases depends on discharge velocities, vapor density, diffusion characteristics of the gas, and conditions and location of release. CNG can be expected to disperse readily, mix, and quickly form combustible mixtures. Under unconfined conditions, methane gas mixtures dilute rapidly to concentrations below the lower flammable limit. In contrast, the ability to form combustible mixtures rapidly generates a hazardous situation if fuel leaks in confined areas without significant ventilation.

LNG is natural gas liquefied by cryogenic refrigeration and stored at a temperature of –260°F. The vapor cloud above a pool of released LNG is fairly difficult to ignite. However, the very low temperature of LNG causes normally buoyant methane to become heavier than air, creating large volumes of flammable methane-air mixtures that can spread significantly beyond the area of the initial spill. The possibility of ignition increases, and the potentially large volume of combustible gas mixture constitutes a major safety risk in substantial LNG spills.[4] When LNG vaporizes, the gas vapor cloud begins to warm and comes into thermal equilibrium with the atmosphere.

One relative safety comparison of the three alternative fuels discussed here found that safety ranking of LP-Gas appears to be similar to gasoline's.[2] This can be attributed to heavier than air LP-gas vapors that do not disperse as quickly as other alternative fuels'.

Automotive grade natural gas contains 95 to 99 percent methane and a mixture of several other gases.[5] Methane is the simplest hydrocarbon, with one carbon molecule bonded to four hydrogen molecules. Other hydrocarbons present include ethane, propane, and butanes, in addition to trace amounts of metal oxides, nitrogen, carbon dioxide, water, helium, and hydrogen sulfide. LNG and CNG are primarily composed of natural gas; however, prior to liquefaction of LNG, constituents such as carbon dioxide, water hydrogen sulfide, the odorant, and trace metal oxides are removed because they solidify at LNG storage temperatures.[5]

LP-Gas can be either butane, propane, or a mixture of both. However, LP-Gas used as an alternative fuel is composed of commercial propane. Commercial propane is not pure. It contains varying percentages of butane, ethane, ethylene, propylene, isobutane, and butylene. Propane is a straight-chain saturated hydrocarbon composed of three carbon atoms and eight hydrogen atoms. Propane is a gas at room temperature but is liquefied when stored under a pressure equal to its vapor pressure.

Fire Hazards

The need to understand the fire hazards associated with the three fuels covered in this chapter grows as the number of alternative fueled vehicles increases in the community. Emergency response personnel need to recognize that these fuels present significantly different fire hazards than traditional fuels such as gasoline and diesel.

Table 9-33A summarizes properties of the three alternative fuels. CNG and LNG have similar fire hazards because some natural gas properties do not change between the two different physical states. Flammability limits and autoignition temperature of natural gas are the same in the two physical storage states of CNG and LNG. Its flammable range is narrow, and its autoignition temperature is high. LP-Gas produces a flammable mixture in air at lower concentrations than natural gas but has a narrower flammable range.

Natural gas, which has a vapor density of less than one, tends to rise at ambient conditions. This is one important property of natural gas to consider because vapor density varies under different temperatures. CNG released into the atmosphere at normal temperatures and pressures rapidly rises and disperses into the air. CNG-fueled vehicles provide a high degree of safety if a release occurs outdoors because the compressed methane diffuses rapidly in unconfined locations. If a container relief valve operates or the fuel line piping breaks, the duration of the release is limited because the gas discharges at high pressures.

When CNG vents from a relief device or pipe opening, it forms a gas jet that "blows" into the atmosphere in the direction the hole faces, continually entraining and mixing with air. If the CNG encounters an ignition source, a flame jet of considerable length may form. Flame jets pose a thermal radiation hazard to the vehicle and

TABLE 9-33A. Selected Properties of Alternative Fuels

	CNG	LNG	LP-Gas (Propane)
Molecular weight	16.04	16.04	44.1
Boiling point @ atmospheric pressure	–258.68°F	–258.68°F	–43.73°F
Storage pressure	2,500–3,600 psig	10–50 psig	127 psig @ 70°F
Storage state	Compressed gas	Cryogenic fluid	Liquefied compressed gas
Vapor density (air = 1)	0.65 @ 68°F	0.65 @ 68°F	1.50 @ 68°F
Flammable range in air	5.3–15%	5.3–15%	2.15–9.60%
Minimum energy for ignition in air	0.29 mJ	0.29 mJ	0.25 mJ
Autoignition temperature	1,003°F	1,003°F	920–1,120°F
Net heat of combustion	23,000–34,000 Btu/gal	75,000 Btu/gal	91,502 Btu/gal

persons nearby, and are particularly hazardous if they impinge on the exterior of adjacent containers. This can lead to the operation of additional pressure-relief devices, increasing the fire's thermal flux. Compressed methane gas releases are of relatively short duration. The ensuing fire usually involves only combustibles on the vehicle because natural gas is rapidly consumed during the jet fire.

A CNG release in an enclosed space can present a higher risk for a deflagration because the contained gas may not readily diffuse. Because methane has a vapor density of 0.6, it rises rapidly. Light fixtures and air handling equipment normally installed at a building's ceiling or roof can serve as potential ignition sources.

Studies have evaluated the deflagration hazard resulting from a leak in a CNG vehicle fuel system or relief device operation inside parking garages and transit tunnels. A gas utility corporation commissioned a study that examined the consequences of a fuel pipe leak or the operation of a container pressure relief device inside a partially open public parking garage. The study found that a leak in the fuel piping was the most credible event, while the operation of the container relief valve the worst-case event. Recognizing that the three model codes and NFPA 88A, *Standard for Parking Structures*, require a method for mechanically ventilating carbon monoxide exhaust, and that all parking garages have fairly open spaces, vapor dispersion models were created to evaluate selected leak scenarios. Analyses of the most probable event resulted in no flammable concentrations beyond the leak point's immediate vicinity. The leak did not result in accumulated flammable gas in the garage. The operation of the container relief valve did result in the brief formation of a gas puff that was within the flammable limits of methane. However, the vapor puff was effectively mitigated by the rapid diffusion of the gas in open space and its removal from the garage by the mechanical exhaust system. The study showed that a CNG vehicle poses no extraordinary risk; that is, the CNG vehicle's risk is equal to or less than the risk a gasoline fueled vehicle poses.[6]

One study compared the flammable vapor clouds formed following a fuel line failure of a hypothetical CNG-powered and gasoline-powered van inside ventilated and unventilated tunnels with various types of ceiling construction. Depending on tunnel ventilation, the hypothetical CNG vehicle would be expected to produce both a smaller and shorter-lived flammable gas region than the equivalent gasoline van. Effective ventilation could be mechanical ventilation by fans or ventilation induced by a vehicle's motion.[7] The study recommended that transportation authorities develop policies limiting the size of vehicles allowed in tunnels.

LNG is stored at an approximate temperature of –260°F. Studies show that at temperatures less than –180°F, LNG is more dense than air at 60°F.[5] Because LNG is a flammable cryogen, the integrity of the container and piping is of primary concern; its failure can result in a cryogen liquid spill. However, loss of liquid from a container requires that the outer steel shell, insulation, and inner vessel all fail. All three components are less likely to fail than are single-wall, atmospheric storage tanks found on conventional gasoline- or diesel-fueled vehicles. Nonetheless, the probability of a spill is still a consideration.

The liquid from an LNG spill begins to vaporize rapidly and forms a white cloud as cold vapors condense in air. Initially, the vapor is negatively buoyant and sinks to the ground until it warms to approximately –180°F. At this temperature the vapor becomes positively buoyant and begins to rise.

The vaporization rate and distance the gas cloud travels and disperses depends on four factors:

1. *Wind speed.* Wind speed is influential because as velocity increases, the LNG evaporation rate tends to increase. Increased wind velocity helps to induce mixing of methane vapors with air, causing the vapor cloud to disperse into smaller clouds, or puffs. This reduces the flammable atmosphere's area.

2. *Surface area of the spill.* The larger the spill area, the greater its vaporization rate. This can result in larger vapor clouds, increasing the boundary of the flammable atmosphere.

3. *Surface type.* The surface where the LNG spills influences its vaporization rate. Vaporization rates are high on wet surfaces, such as muddy soil, because water acts as a heat sink. Asphalt promotes vaporization because its dark color acts as a heat radiator. If LNG spills onto a good heat-insulating surface, boiling may eventually slow down, but the remaining liquid pool continues to evaporate rapidly. Evaporation maintains the remaining liquid near its boiling point as it absorbs heat from its surroundings.

4. *Time of the incident.* At night, lack of atmospheric heating by the sun and ground reduces the evaporation rate of LNG. While this influences the vaporization rate, it also influences the distance from the source spill in which the vapor puff remains in its flammable range. At night, the distance is greater because the time needed to heat the vapor until it is lighter than air increases.

Fire fighters should remember that LNG vapor formed during the phase change is colorless and odorless at normal temperature and pressures. They cannot rely on visual or olfactory indicators to establish safe boundaries. Only gas detection equipment designed to detect methane can establish safe exclusion and operating zones when dealing with LNG releases.

LP-Gas is stored as a liquefied gas at its vapor pressure and at ambient temperature. At ambient conditions, propane's vapor density is $1^1/_2$ times that of air. This means that propane released to the atmosphere sinks to the ground and forms a vapor cloud that can follow surface contours over a large area. Because of the vapor density and expansion ratio of LP-Gas, this class of materials can challenge fire fighters. Incidents involving LP-Gas containers may result in jet fires, vapor cloud ignition, or BLEVEs. Because the alternative-fueled vehicles on the road are predominantly LP-Gas fueled, more fire fighters can expect to respond to incidents involving this alternative fuel.

Two types of LP-Gas releases can occur. The first involves loss of LP-Gas in the vapor phase. A vapor phase release can occur if a component installed above the liquid line is damaged or the container is punctured or otherwise damaged above the liquid line. During a vapor release, LP-Gas vents at a high velocity from the opening, possibly creating some liquid droplets during the process. The velocity of the release reduces with time as boiling inside the container cools the liquid mass. The tank's surface forms a layer of frost due to the refrigeration effect of the liquid being cooled. The gas may continue venting for a considerable time until no liquid is left in the container.

The second type of LP-Gas release is a liquid phase release. It occurs if an LP-Gas component installed below the liquid line is damaged or if the container is punctured or damaged below the liquid line. When the release occurs below the liquid level, the liquefied gas may jet from the hole. The velocity of the liquid discharge depends on the amount of liquid in the container and the gas vapor pressure. During such a release, large amounts of the liquid flash into gas or vapor. Depending on the temperature and pressure, a propane release in its liquid phase may discharge a large mass of vapor mixed with small droplets suspended as an aerosol. Sometimes liquid propane pools on the ground surface and typically boils off into gas.

During a vapor phase or liquid phase release, volatile gases and vapors, if not ignited immediately, form a plume that moves downwind of the container. If the plume contacts an ignition source, and the vapor plume is within the flammable range, the flame may flash back toward the source of the release. Unprotected people within the cloud may be severely injured. If the velocity of the flame front is great enough, air, products of combustion, and unburned fuel can compress to a point where a vapor cloud explosion could occur. In this case, the deflagration's pressure can also impact persons and property.

Boiling Liquid Evaporation Vapor Explosions

LP-Gas and LNG containers are susceptible to boiling liquid expanding vapor explosions (BLEVEs). BLEVEs occur when liquids are stored in a container under pressure at a temperature above their boiling points. BLEVE is defined as a major container failure, into two or more pieces, at the moment in time when the contained liquid is well above its normal boiling point at atmospheric pressure.[8]

LNG and LP-Gas both possess high expansion ratios. For instance, 1 cu ft of LNG expands to 618 cu ft of natural gas at ambient conditions, and 1 cu ft of liquid propane expands to approximately 265 cu ft of propane gas at normal temperature and pressure. Expansion of a liquefied gas increases the pressure of tank walls if pressure relief valves fail to open, or if relief valves are incapable of adequately relieving internal tank pressure. A puncture in the tank wall or heating of the container surface due to an exposure or pool fire could cause expansion. As the tank walls' strength decreases and liquefied gas within the tank continues to expand, the tank fails, breaking into two or more pieces. The flammable gas ignites simultaneously with tank failure.

The insulation of code constructed LNG containers helps protect the inner tank from exposure to high temperatures resulting from flame impingement. Storage vessels used at LNG fueling sites are similar to those used for storing cryogenic liquids. LNG containers' outer metal jacket protects an insulating material maintained under a vacuum. This assembly surrounds an inner steel vessel that holds the cryogenic natural gas liquid. This outer jacket and insulation, combined with the liquid's extreme cold, are expected to provide substantial protection from radiant heat exposure due to a pool fire. Fire tests conclude that the pressure relief valves for an uninsulated cryogenic vessel activate within one hour. If the tank insulation and vacuum are intact, the relief valve does not actuate for several days.[9]

Additional information on BLEVEs involving LP-Gases can be found in Section 4, Chapter 7, "Gases."

Health Hazards

Health risks of the three alternative fuels discussed in this chapter generally result from two different exposure routes: (1) skin contact and (2) inhalation. Ingestion is usually not a likely route of exposure.

Potential for skin contact with CNG is very low. Almost immediate dispersion results when a container releases CNG stored at pressures in excess of 2,400 psi. LNG, a cryogenic liquid stored at –260°F, can cause burns or frostbite upon contact. Hypothermia can also occur from prolonged exposure to LNG vapor.

Asphyxiation occurs when other gases displace oxygen in the air and the oxygen concentration decreases to less than 10 percent.[5] LNG vapor is not toxic; however, there is potential for lung damage or asphyxiation. Lung damage could occur from prolonged breathing of cold vapors. Neither CNG or LNG has an established time-weighted average because they are simple asphyxiates; LP-Gas has a time-weighted average of 1,000 PPM. All the gaseous alternative fuels may cause asphyxiation if released in a confined space.

Odorization

Natural gas and propane are both odorless gases. Odorization of a gas is an important safety characteristic. It indicates, by odor, a gas leak or a release event. Natural gas is odorized within the pipeline prior to general distribution.

CNG taken from a distribution line and compressed contains the odorants typically added to natural gas. Natural gas is normally odorized at a level allowing detection at 20 to 25 percent of the lower flammable limit of methane. One major difference between CNG and LNG is that LNG has no odor. Consequently, LNG leaks cannot be detected by smell. San Diego Gas and Electric Company developed an odorization technique for LNG in the 1970s. The technique utilized the same odorants used in natural gas but blended them with a carrier gas such as liquid propane. A Gas Research Institute study revealed that odorants such as ethyl mercaptan, i-propyl mercaptan, and t-butyl mercaptan could be used with carrier gases such as liquid propane and i-butane to odorize LNG.[10] Although the study indicates viable options to odorization of LNG, neither NFPA 59A, *Standard for the Production, Storage, and Handling of Liquefied Natural Gas (LNG)*, NFPA 54, *National Fuel Gas Code*, nor NFPA 57, *Standard for Liquified Natural Gas (LNG) for Vehicle Fuel Systems*, requires it.

LP-Gas is odorized before delivery to a bulk plant.[11] Like natural gas, it is detectable by odor at a concentration in air at less than $1/_5$ of the lower flammability limit. Typical odorants used include ethyl mercaptan and thiophane.

It is not possible to warn everyone of a gas leak using odorants. Often odorants fade in transmission lines and containers due to oxidization by oxygen, rust, or iron oxides. Furthermore, olfactory senses diminish with increasing age, allergies, and during smoking and eating.

Applicable NFPA Standards

NFPA standards have been developed for all alternative fuels described in this chapter. NFPA 52, *Standard for Compressed Natural Gas (CNG) Vehicular Fuel Systems*, contains requirements for storing and dispensing CNG as well as vehicle requirements. NFPA 58, *Standard for the Storage and Handling of Liquefied Petroleum Gas*, contains requirements for LP-Gas.

NFPA 59 has requirements for storing and constructing LNG containers. However, the standard was originally written for site-erected LNG containers similar to those found at peak-shaving plants. Conversely, LNG stations normally utilize shop-constructed containers. Many station designers and operators wanted a separate standard that specifically applies to vehicle fuel systems and LNG dispensing. The National Fire Protection Association responded by developing NFPA 57, *Standard for Liquefied Natural Gas (LNG) Vehicle Fuel Systems*.

VEHICLE FUELING STATIONS

Compressed Natural Gas Vehicle Fueling Stations

CNG fueling stations can be generally categorized according to the time needed to fuel. The method used depends on the filling time the user desires. Available equipment combinations result in facilities being categorized as either "fast-fill dispensing" or "slow-fill dispensing."

Refueling a standard passenger vehicle at slow-fill CNG stations can commonly take up to 8 hr. These stations connect to the natural gas service line that typically supplies fuel for building heat. CNG slow-filling systems utilize a small compressor; consequently, high gas flow and large gas supply piping are not necessities. The small compressor usually combines with the dispensing equipment to form a compact module mounted next to the vehicle's parking space. Typically, no storage containers are associated with these units, which contributes to longer filling times. Mostly utilized for overnight fueling, these units are typically used by fleets of less than 30 vehicles that do not operate continuously. Compact unit design and smaller supply requirement makes slow-fill fueling possible at private residences, depending upon local regulations. (See Figure 9-33A.)

Fast-fill CNG stations normally fuel a conventional passenger vehicle in less than 5 min. Sometimes fixed storage vessels are used as supply sources and must be transfilled from a high pressure, transport tube trailer that is brought to the station site. More commonly, a connection to the nearest domestic natural gas distribution system supplies the natural gas. A pressure cascade system compresses and dispenses low pressure natural gas (60 to 100 psi) at the higher pressures (2,500 to 4,500 psi) the CNG vehicle requires. These systems can be found at locations typically serving more than 30 vehicles.

CNG siting considerations: CNG fueling is performed at public service stations, small private fleet operations, metropolitan mass transit facilities, school districts, cab companies, government agencies, and residences (if local regulations allow). CNG fueling typically is performed outdoors. CNG compression, storage, and dispensing systems should take into consideration setback distances from property lines, on-site adjacent buildings, overhead power lines, nonclassified overhead electrical lighting, and ignition sources. Protective canopies and open structures installed to protect equipment should be of noncombustible construction with a roof design that ensures adequate ventilation and prohibits pocketing or collection of CNG due to the buoyancy of natural gas. Typical accidental releases of CNG will be completed in a matter of seconds due to very high compression and storage pressures. The quick release and buoyancy of CNG minimize the potential threat from open-air flammable vapor cloud ignitions. Ignition of leaking high-pressure gas is possible, and the resulting jet fire can extend from 10 to 30 ft

FIG. 9-33A.   *Vehicle refueling appliances (VRAs) are listed as slow-fill compressors. This installation used VRAs to fuel CNG-powered vans overnight. Average fill time for an automobile or van is 4 to 8 hr, depending on the size and number of cylinders. (Source: Fuelmaker, Inc.)*

in length, depending upon the leak's size and CNG pressure. This makes the orientation of storage cylinders and relief valves an important site consideration.

With the addition of protective measures and depending upon local regulations, CNG fueling can be accomplished indoors. This is a potential consideration when weather- or freeze-protection is needed for the compression, storage and dispensing systems. Indoor fueling with CNG can only be accomplished inside structures dedicated to this purpose or within rooms or additions provided with at least 2-hr fire resistive walls as separation from other occupancies. When dispensing CNG indoors, overpressure protection, such as explosion venting, is necessary. Mechanical ventilation that is continuous or activated by a gas detection system must be installed. To allow for delays in the mechanical ventilation startup, the gas detection system should activate ventilation at a level not exceeding $^1/_5$ the lower flammable limit. An audible alarm should also sound at the 20 percent lower flammability level. Activation of the gas detection system must result in automatic shutdown of compression and dispensing equipment by placement of adequate solenoid switches and automatic shutoff valves. Although not required by NFPA 52, some model building and fire codes may require the installation of an automatic sprinkler system for indoor dispensing and storage of CNG. Local approvals for indoor dispensing may vary, but public structures, such as parking garages, are commonly unacceptable for fueling of any type of conventional or alternative fuel.

When considering the siting of CNG fueling stations, the general public is probably most familiar with the hazard potential of natural gas. Their expectation or acceptance of hazard is commonly based on the low pressure natural gas used for home cooking and heating. Mentioning high-pressure natural gas used in CNG fueling is likely to conjure up consequences attributed to large diameter, high-pressure natural gas transmission pipelines. Major CNG fueling facilities that utilize large capacity compressors sometimes require the extension of a larger and higher pressure natural gas pipeline. Those who live near large CNG fueling stations may question their safety and acceptability because they fear CNG fueling components and the increased size of natural gas mains in the public right-of-way.

CNG compressor components: The very large range of CNG compressors' capacity makes fill time vary. Small, slow-fill units with a compressor capacity of approximately 5 to 10 SCF/min are small enough to fit in the front seat of a car. NFPA 52 defines these compressors as residential refueling appliances or vehicle refueling appliances. (See Figure 9-33A.) They generally require only a normal domestic natural gas supply line. Medium sized, fast-fill compressors typically found at public fueling stations are capable of supplying approximately 1,000 SCF/min. A normal domestic supply line can also supply these medium-sized compressors. Like the slow-fill units, these fast-fill compressor sets are sometimes packaged with other fueling components and are available as portable, premanufactured units that fit in the space required for a passenger vehicle. Large, industrial compressors can compress approximately 5,000 SCF/min at about 3,600 psi and can fill the space needed to park a bus. Depending on the number of large compressors needed, supply from a 6-in. diameter natural gas pipeline is not uncommon.

Compressors require pressure relief devices set for each stage pressure. A four-stage compressor might have relief valves set for 250, 1,250, 1,400, and 4,150 psi depending on the compression chamber's size and associated piping into which they are installed. CNG compressors are also required to have automatic shutdown capability in the event of high discharge pressure or low suction pressure. Compressor controls must be manually reset if an automatic shutoff or power failure occurs. Internal combustion engines, modified to operate using natural gas, usually drive compressor sets. NFPA 52 requires the design, construction, and maintenance of these compressor drivers to be in accordance with NFPA 37, *Standard for the Installation and Use of Stationary Combustion Engines and Gas Turbines.*

CNG storage components: Fast-fill dispensing transfers CNG from a bank of pressure vessels to the vehicle in a fairly short time, usually less than 5 min. These pressure vessel banks, or cascades, are usually multiple compressed gas cylinders, spherical vessels, or horizontal tubes. Normally three pressure banks are involved in a cascade system. (See Figure 9-33B). To "top off" a vehicle container, dispensing begins from a storage vessel having lower pressure until pressure equalizes, then automatic controls switch to a vessel of higher storage pressure. This automatic switching between cascade banks continues sequentially until vehicle fueling is complete. Subsequent vehicles can be filled until cascade pressure drops too low. Then the system automatically switches to the compressor. This, however, increases fill time, so the capacity of storage containers at a site usually reflects the number of vehicles to be served.

Bulk CNG storage is infrequently required when obtaining an onsite natural gas supply to compress is not possible or when the economics of installation does not justify the cost of a compressor. Bulk storage is usually accomplished by transporting gas compressed at another site, then transfilling from the transport tube trailer into pressure vessels. Bulk vessels do not draw down at the same rate when fueling a vehicle but their automated systems operate like the cascade system. Bulk compressed gas is usually dispensed from at least three vessels. (See Figure 9-33B.)

CNG cascade and bulk storage vessels must be adequately supported on no more than two longitudinal points and should be adequately protected from vehicle impact. A check valve should be installed so that back pressure cannot bleed back to the supply, and adequate pressure relief devices must be installed in each vessel. The pressure relief devices should be vertically vented and protected against weather. Pressure gages installed on each vessel or pressure bank determine the fullness of a storage vessel. Sometimes outlets for storage vessels are fitted with excess flow valves. These devices restrict flow from the vessel if the tubing or piping leaks. A check type valve closes when flow exceeds that rated for the piping system. Usually carbon or stainless steel tubing and piping connect

the cascade or bulk vessels, and approved flexible metallic hose is installed for expansion, contraction, jarring, vibration, and settling where necessary.

CNG dispenser components: Dispensing equipment can look like conventional gasoline dispensing equipment or it can simply be a hose and nozzle rated for high-pressure service and mounted on a post. (See Figure 9-33C.) If located at a public service station, the dispenser usually measures "equivalent gallons" of fuel; otherwise the unit of measure varies depending on the owner's needs or local weights and measures standards. Underground piping usually connects CNG dispensers to the storage and compression systems. Just as at a conventional service station, multiple CNG dispensers are not typically installed on one supply line. Instead each has individual, dedicated piping from the storage or supply.

CNG dispensing systems require electronic or pneumatic controls that sense vehicle container fill pressure and automatically stop the flow. Settings for these controls are based on outside temperature. In addition, an emergency shutdown device commonly installed at the storage and compressor activates if there is a problem at the dispenser. Even so, a "quarter turn" manual shutoff valve is located at the dispenser just upstream from the dispensing hose and its required breakaway device. The hose breakaway device, located outside the dispenser end, closes if a vehicle moves away from the dispenser without breaking the vehicle fueling connection. CNG breakaway devices release and seal at approximately 44 lb of force from any direction.

The CNG fueling hose must be rated for the system's service pressures. Like flammable liquids dispensing hose, the CNG hose must be nonconductive. A significant feature of the CNG fueling hose is the inclusion of a second hose, usually designed as integral to the supply hose. When fueling is complete, this second hose collects natural gas from the vehicle fueling connection and supply hose. Some dispensing units vent the hose automatically; others require a manual valve to accomplish this venting. In either case, the vehicle connection does not release until the connection and hose is

FIG. 9-33B. Pressure vessels are a common method of storing compressed natural gas (CNG). NFPA 52 requires pressure vessels to be constructed in accordance with the ASME Boiler and Pressure Vessel Code.

FIG. 9-33C. CNG dispensers have an appearance similar to conventional fuel dispensers. Fuel dispensers should be provided with protection from vehicle impact, using islands, pipe bollards or other appropriate designs.

depressurized. After CNG fueling is complete, vented gas returns to the dispenser and ultimately to storage or an approved vent stack, depending upon local requirements. Typical CNG fueling hose is easy to handle and is approximately $\frac{1}{2}$ to $\frac{3}{4}$ in. in diameter. CNG dispensing hose cannot be more than 25 ft long and must be supported above ground, usually by a retractable cord like those used on conventional fuel dispensers.

The CNG dispensing nozzle that connects the fueling hose to the vehicle is similar to a typical quick-connect coupling used on compressed air hoses, except it includes breakaway protection. The vehicle fueling receptacle, fuel control valve, and dispenser nozzle are designed in accordance with ANSI/American Gas Association NGV 1, *Basic Requirements for Compressed Natural Gas Vehicle Fueling Connection Devices.* The NGV 1 standard includes requirements for the vehicle fueling receptacle, the dispenser nozzles, and the fuel control valve installed at the dispenser. Depending on the component's style, devices are rated for service pressure between 2,400 to 3,600 psig.

The NGV 1 standard defines three types of nozzles. Each must be designed so that the fueling system is depressurized before the nozzle can be removed.

- Type 1 nozzles utilize an integral valve operating mechanism that vents the gas between the vehicle's fill line check valve and the nozzle inlet valve, and then disconnects the nozzle from the vehicle. (See Figure 9-33D.)

- Type 2 nozzles utilize an external valve that must be vented before disconnection.

- Type 3 nozzles utilize controls that automatically depressurize the fill line upon dispenser shutdown. They are normally found on vehicle refueling appliances or residential refueling appliances.

Liquefied Natural Gas Vehicle Fueling Stations

LNG fueling stations can be characterized according to the method utilized for obtaining liquid transfer pressure. Some use gravity available from a vertical bulk storage vessel to fill a smaller transfer vessel at 75 to 100 psi. A portion of the LNG passes through a vaporizer and back into the smaller transfer vessel to warm the liquid. The 618:1 expansion ratio of the LNG in the smaller vessel increases the pressure to as much as 230 psi to force liquid into the LNG vehicle. Transfer rates of 40 to 50 gal/min are not uncommon.

The other common type of LNG fueling station utilizes a centrifugal cryogenic pump supplied from a vertical or horizontal bulk LNG storage vessel. Centrifugal cryogenic pumps typically provide LNG fueling rates of 30 to 50 gal/min, depending on the LNG vehicle tank pressure required. The net positive suction head that the

centrifugal pump requires is important. Some installations experience cavitation and poor performance when storage design does not provide enough suction head. Fueling stations that use underground LNG bulk storage vessels are available but not as common. Underground installations use a slow action, positive displacement pump that lifts LNG about 25 ft, provides 30 to 50 gal of liquid per min, and delivers at pressures as high as 200 psi.[12] LNG fueling facilities utilizing cryogenic pumps typically operate on electricity.

LNG fueling facilities can also be designed to supply CNG vehicles. LNG fueling stations normally create some vapor due to external heat absorption of the stored LNG, priming of pumps, and temperature equalization of equipment. In addition to normal vaporization, LNG is sent to a vaporizer to create gas specifically for fueling CNG vehicles. Under some conditions, LNG to CNG fuel supply or liquefied-compressed natural gas (LCNG) may be cheaper than CNG supplied from domestic natural gas lines. Facilities that utilize vapor expansion for LNG fueling pressure provide a compressor and cascade system to fuel CNG vehicles. Systems with high-pressure LNG pumps combine LNG pump capacity and gas expansion, through the vaporizer, to fill the CNG cascade storage. A method of odorizing the CNG is also required.

LNG siting considerations: LNG vehicle fueling is typically associated with large fleet operations in heavily populated urban areas. These include such facilities as mass transit, cab companies, and government agencies. Rural fleets that utilize large vehicles, such as mining operations or transport trucking, have also converted to LNG use. Because a significant number of fleet vehicles is necessary for LNG fueling to be economical, self-serve general public vehicle fueling with LNG is not expected to become commonplace.[9] In addition, indoor fueling with LNG is not typically performed. However, it has been proposed that this activity be limited to the use of less than 125 gal containers when necessary gas detection, ventilation, fire protective, and building construction features are provided.

Lacking a specific standard for LNG fueling, some Authorities Having Jurisdiction have applied the siting requirements of NFPA 59A. NFPA 59A requires that the containment area for LNG containers be sited in accordance with the equation:

$$d_h = 0.8(A)^{1/2}$$

where

d_h = the horizontal distance to the property line that may be built upon from the nearest edge of the impounded liquid (ft), and

A = surface area of impounded liquid (sq ft).

This data, together with a specified allowable level of thermal radiation at the property line, serve as the basis for LNG siting requirements in NFPA 59A.

Siting LNG fueling facilities generally requires at least 50 to 75 ft clearance between the fueling station and adjoining property lines. A setback is prudent due to the potential of spilled LNG. Normally buoyant, natural gas vaporizing from a supercooled liquid state remains initially heavier than air. While cool, the natural gas forms a vapor cloud that follows the ground surface until it heats sufficiently to regain its lighter-than-air characteristics. In addition to separation, site consideration should include containment of the fueling facility. Containment helps limit the surface area of an LNG spill, reduces the spill vaporization rate, and thereby reduces the distance a potential vapor cloud travels before becoming buoyant. Separation and containment are important considerations when locating an LNG fueling facility near other land uses that are densely populated or provide potential ignition sources.

FIG. 9-33D. *CNG nozzles are similar to quick-connect couplings used on compressed air hoses. The Type 1 nozzle shown has an integral valve which vents the gas between the vehicle's fill line check valve and nozzle inlet before it can be disconnected from the vehicle.*

LNG storage and use is not something that very many communities have experienced. Members of the general public who are presently familiar with LNG facilities usually live in areas that have bulk LNG supply for domestic natural gas distribution or power plant peak shaving needs. Cryogenic gas storage and liquid transfer have a long history, but bulk storage and dispensing of LNG vehicle fueling is a relatively new use of an adapted technology. Compared with the other types of alternative fuels, LNG use in private vehicles does not seem to be as well demonstrated. This could result in hesitation to accept the additional hazards associated with LNG storage and vehicle fueling in densely populated urban areas.

LNG storage components: LNG is normally transported from a liquefaction plant to the fueling facility using cryogenic bulk transport trailers. Transfilling may occur at night because LNG fueled fleets need to be filled before business hours. The number of vehicles to be filled and the capacity of on-site storage dictate the frequency of LNG transport and transfilling. Bulk LNG at typical fueling sites is stored in containers having a total capacity of approximately 15,000 gal. This quantity allows unloading of one transport that usually carries 11,000 to 13,000 gal. (See Figure 9-33E.) Depending on delivery schedules and demand, more than one container might be located at a fueling site. Containers can be horizontal or vertical, depending on owner preference, aesthetics, and available space.

Transfer of LNG from the supply truck trailer to bulk storage is accomplished by increasinging the trailer's normal storage pressure of 10 psi to approximately 40 to 50 psi. If a vertical bulk storage vessel is used, transfer pressure may be as high as 250 psi. This is accomplished by boiling off some LNG in an auxiliary line and

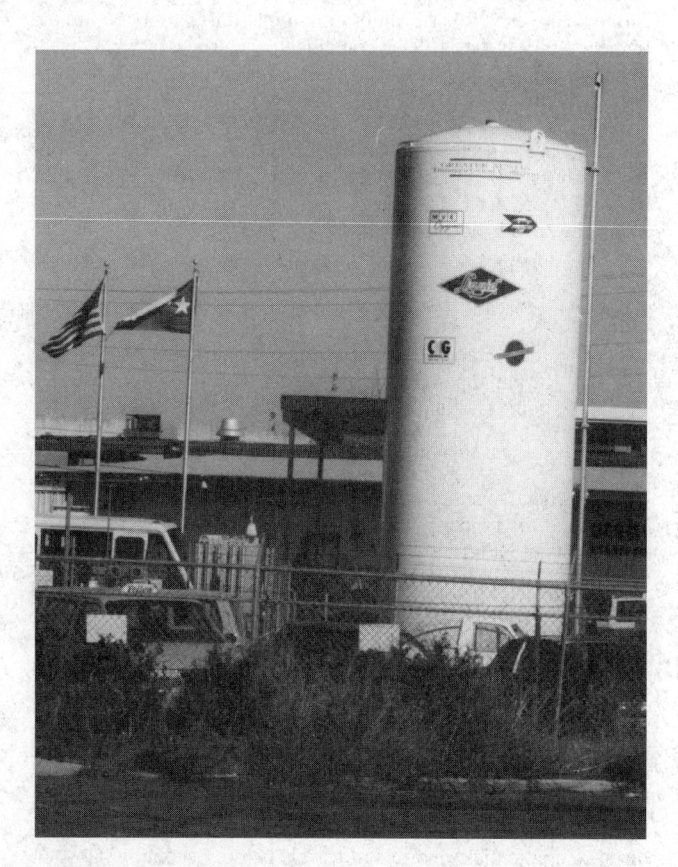

FIG. 9-33E. A 13,500-gal LNG container used for vehicle refueling. Siting of containers should be based on the requirements of NFPA 59A.

allowing it back into the trailer. A small amount of LNG may leak out of the hose connection to the bulk vessel because of the temperature difference between the inside and outside of the fueling connection. LNG bulk storage containers have liquid level gages and a high liquid level alarm or indicator that warns the operator to stop filling the bulk vessel. When transfer is complete, a pump on the supply truck transfers any LNG remaining in the hose to the storage vessel. During transfer, a pressure differential of at least 15 psi is maintained between the transport trailer and the bulk storage vessel. In the event of a backfill condition, a backflow check valve is required on the transfer fill piping of an LNG bulk storage vessel.

Secondary containment for the vessel should be provided to ensure that accidental LNG spills do not flow towards adjacent property, buildings, drains, sewers, or waterways. Secondary containment around LNG facilities can also be designed to limit the consequences of flammable vapor clouds that can form when the liquid spill warms. Containers placed in below grade pits act to insulate the cryogenic spill and slow the warming rate. The sides of abovegrade secondary containment can also be constructed of low-density cement-based concrete or resin-based materials. The use of insulated materials considerably reduces vapor dispersion distances. The containment system's design and the type of insulating material used can reduce the vaporization rate as much as 30 percent.[13] Equipment within the containment should be limited to LNG related apparatus. CNG equipment such as compressors, vaporizers, and CNG storage cylinders should not be located within the LNG secondary containment, due to potential fire exposure and damage from high pressure.

Pressure relief valves must be designed with a capacity to not only vent vaporized natural gas adequately if fire exposure occurs but also must include capacity for the flow of liquid in the event an overfill of the bulk storage occurs. Pressure relief valves are typically located in the storage vessel, the insulation space of the vessel, and any insulated piping that can be isolated between two block valves. Typically a system of air-operated valves controls the flow of liquid for vehicle fueling. Air supply lines for the flow control valves commonly consist of low melt-point plastic tubing. If exposed to fire, these air supply lines will melt and cause the valves to shut in a fail-safe manner.

LNG dispenser components: Dispensers available for LNG fueling can look like conventional fuel dispensers. Typically they do not. Instead they are an assembly of a cryogenic coupling nozzle, insulated cryogenic hose, shutoff valves, and panel or pole mounted gages. Sometimes the fueling hose attaches to a rotating arm that supports the hose as it is attached to the vehicle. If not, a retractable stainless steel cable supports the cryogenic hose. LNG fueling hoses are insulated and can be 3 to 4 in. in diameter. They are fitted with a large cryogenic coupling, about 1 in. in diameter. Compared to conventional fuel hoses and nozzles, this assembly can be cumbersome.

A shutoff valve is attached on the open end of the cryogenic hose, just before the LNG coupling that connects to the vehicle. This coupling includes an interlock device that does not allow release during transfer. LNG couplings operate like sexless fittings commonly used on fire hose. A manual emergency shutoff valve is placed in the transfer system, near the connection to supply piping at the storage vessel. This manual shutoff valve must be readily accessible by the operator. Bleed or vent connections incorporated into the coupling let dispenser hoses be drained, depressurized, and sometimes purged with nitrogen. These connections usually drain to a vaporizer, and natural gas either vents or is transferred to a CNG fueling process. LNG dispensing units may have electronic level sensors that automatically activate the dispensing shutoff valve. More often, a level gage provided on the vehicle requires the operator to measure the vehicle tank level and then manually stop dispensing. Even with these controls, providing drainage should be considered so that accidental LNG spills drain into the storage vessel secondary containment.

Liquefied Petroleum Gas Fueling Stations

Types of LP-Gas fueling stations: LP-Gas used for vehicle fueling is typically propane. Propane has a long history of use in industrial applications and predominately rural areas as a home-and-business heating fuel. While most domestic propane heating appliances use vapor as fuel, vehicle fueling transfers propane in liquid form. The equipment needed has been utilized for many years at bulk plant facilities for transfer to bobtail propane delivery trucks and for container filling. This equipment has also been extensively used for forklift truck fueling at manufacturing and storage occupancies.

Historically, propane vehicle fueling in the United States has been predominantly at bulk plant locations or with the other traditional propane supply services. Dedicated propane vehicle fueling is, however, becoming more common in some parts of the United States at fleet facilities and public retail service stations. Some consider LP-Gas vehicle fueling the least complicated of the three alternative fuels discussed in this chapter. The mechanics of supplying liquefied gases for fueling vehicles does not vary significantly from the processes used for supplying LP-Gas for heating. LP-Gas at vehicle fueling locations is typically moved to the fueling site by highway transport truck and then transfilled into the bulk fueling storage container.

There are three common ways to transfer LP-Gas in the liquid phase: (1) by use of an LP-Gas pump, (2) by pressure differential or gravity, and (3) by use of a compressor. Using LP-Gas pumps to transfer liquid is the most common method. This is because of the flexibility afforded by not having a required tank elevation or hazards associated with vessel pressurization. LP-Gas pumping has been used extensively to fill LP-Gas bobtail delivery trucks from bulk storage vessels. The same pump technology has been adapted for use in fueling vehicles with propane. Positive displacement pumps, such as sliding vane, gear pumps, or centrifugal pumps are common with the pump discharge pressure limited to a maximum of 350 psig. Electricity usually powers the LP-Gas pumps used at vehicle fueling stations.

LP-Gas vehicle fueling facilities can utilize horizontal, vertical, or sometimes underground pressure vessels. Vehicle fueling at LP-Gas bulk storage operations are typically in horizontal pressure vessels. LP-Gas vehicle fueling at locations with space limitations sometimes uses vertical pressure vessels. Underground pressure vessels dedicated for LP-Gas vehicle fueling are not as common because additional cost, maintenance responsibilities associated with corrosion protection, and sometimes prohibitive soil conditions.

LP-Gas siting considerations: Vehicle fueling with LP-Gas is limited to outdoors. The liquid-vapor expansion ratio of propane is approximately 260:1,[14] so an accidental liquid discharge releases a greater quantity of vapor than a leak from the LP-Gas vessel's vapor space. In addition, vapor release from emergency relief valves due to accidental container overfilling makes LP-Gas vehicle fueling hard to justify as an enclosed, indoor activity. Even outdoors, propane tends to form a vapor cloud that, under the right conditions, can travel downwind at flammable concentrations until it encounters an ignition source. This is because LP-Gas has a vapor density greater than air. Less common, but possible, is an LP-Gas container BLEVE caused by an emergency relief device failure or rapid container exposure to intense radiant heat.

Setback distances from adjoining property lines and adjacent buildings should be evaluated not only for building exposure from leaks at the LP-Gas fueling station but also for building fire exposure on the fueling station. Considering mutual exposure is sometimes complicated. Many published setback requirements intend to prevent smaller, more common LP-Gas vapor releases from reaching an ignition source. These distances also provide space to buy time for implementing emergency activities such as applying water to a propane pressure vessel to protect it from an adjacent fire's radiating heat. Typically these distances offer protection for releases, such as those from emergency relief valves, that result from container overfilling. A large, continuous liquid leak from a damaged or malfunctioning container connection, transfer piping, or pump casing can produce more serious vapor cloud, BLEVE, and shrapnel hazards. The hazards associated with these situations may possibly exceed the predominant setback distances specified for LP-Gas applications.

Historically, the location of LP-Gas applications has been been predominantly in rural or suburban areas, or in commercial or industrial areas. In these locations, hazards are commensurate with other hazardous operations and exposure to the public has been limited accordingly.[15] The addition of LP-Gas vehicle fueling stations to heavily populated urban areas offers LP-Gas suppliers a new market, sometimes near congested residential areas that historically supported conventional public fueling stations. Additional setbacks and protective equipment should be considered in these situations. These help limit urban public exposure to a risk level enjoyed before the alternative fuels market expanded. (Figure 9-33F shows a CNG and LP-Gas fueling facility in an urban area.)

LP-Gas storage components: LP-Gas vehicle fueling has historically been accomplished at bulk storage facilities where large volumes are typically stored in horizontal pressure vessels. With LP-Gas vehicle fueling increasing at existing industrial and commercial fleet sites, limited available land area has made the use of vertical pressure vessels more common. The size of pressure vessel used for storage varies because some installations are not dedicated solely to LP-Gas vehicle fueling. Typically, installations dedicated for public vehicle fueling use storage pressure vessels having 1,000 to 2,000 gal water capacity.

These storage vessels must meet standards addressed in the ASME *Boiler and Pressure Vessel Code or the API-ASME Code for Unfired Pressure Vessels for Petroleum Liquids and Gases.* LP-Gas pressure vessels must meet design or service pressures that depend upon the vapor pressure of the LP-Gas stored. Propane storage vessels used for fueling require a minimum design pressure of 250 psig. Structural support for LP-Gas storage vessels containing propane can vary from concrete saddles on horizontal vessels, to structural metal supports protected with a fire-resistive coating on vertical tanks. LP-Gas fueling vessels should be permanently fastened to paved surfaces or concrete pads, adequately protected from vehicle impact, and protected from tampering by chain-link fencing. Secondary containment is not provided at LP-Gas vehicle fueling facilities because when accidentally spilled or released, nonrefrigerated liquid propane immediately vaporizes.

Bobtail tanker trucks with 2,000 to 4,000 gal capacity most commonly supply LP-Gas vehicle fueling stations. The operator transfers liquid using a pump mounted on the transport vehicle. Transfer hose connected to the pump discharge is attached to a threaded fill pipe connection located a short distance away from the storage vessel. This fill pipe allows the operator to top fill the fueling station storage container without having to climb on top. For bulk LP-Gas storage, transport hose connections at the fill pipe include breakaway protection. At bulk plants, bulkhead devices provide this protection with cable activated or pneumatic emergency shutoff

FIG. 9-33F.   In densely populated and congested urban areas, the location of aboveground hazardous materials containers should consider the impact of a release or fire on surrounding fire exposures and the impact of such an event to general public and emergency responders. The station shown stores and dispenses gasoline, CNG, and LP-Gas.

valves. On smaller vessels dedicated for LP-Gas vehicle fueling, these are not typical. Instead, their design usually incorporates a breakaway valve in the vessel fill piping. In addition, a required backflow check valve at the inlet to the storage vessel prevents pressurized contents from flowing from the fill connection if the piping is damaged.

Filling the storage vessel proceeds until the operator notes that the liquid level has reached the maximum allowed. Several options exist for monitoring when the storage vessel is full. Using a fixed maximum liquid level gage is a popular method. A No. 54 drill hole bleed valve in this device connects to an internal dip tube. When vessel filling is to begin, the valve is opened and normally vents invisible vapor. The operator notes refrigerated air moisture through the bleed valve when vaporizing liquid reaches the predetermined interior dip tube height. The operator then stops the liquid transfer. Variable liquid level gages, such as a slip tube or rotary tube, operate similarly but indicate liquid height on a scale that the operator reads to determine when the vessel is properly filled.

Emergency shutoff valves are required at LP-Gas vehicle fueling facilities. These consist of emergency devices associated with dispensing from storage into the vehicle. Manual shutoff valves, excess-flow check valves, backflow check valves, and quick closing internal valves can be used at liquid bypass, vapor recycle, liquid withdrawal, and equalizing connections. Storage vessel discharge connections located at the bottom of the vessel can be provided with shear protection in the event the discharge piping is damaged. To protect the vessel against significant fire exposure in the event of a leak, a fusible link shutoff valve is commonly installed at the vessel liquid opening, usually in conjunction with an excess flow check valve.

The excess flow check valve for discharge from the vessel is usually installed internal to the vessel; some are manufactured to be installed in-line. A spring normally holds excess flow check valves open. The valves close automatically when the device's rated liquid flow is exceeded. Excess flow check valves operate only if a break in the vessel discharge line is sufficiently large for flow to exceed the rated valve setting. Pump bypass liquid lines and vapor equalization lines, commonly connected to LP-Gas storage, also have excess flow check valves.

LP-Gas storage vessels used for vehicle fueling must have one or more pressure relief devices installed in their vapor space. ASME vessels greater than 1,200 gal water capacity must have spring-loaded pressure relief valves that directly terminate into the vessel vapor space. These self-actuated relief valves help to limit the accidental release of LP-Gas because they normally close when excess pressure is relieved. Other devices, such as frangible or rupture disks, let the entire contents of the pressure vessel discharge and, therefore, are not used in LP-Gas pressure vessels. In addition to spring-loaded relief valves, fusible plugs not exceeding $1/_4$ in. diameter may be installed in pressure vessels smaller than 1,200 gal water capacity.

LP-Gas dispenser components: LP-Gas dispensing devices typically include a meter, vapor separator, hydrostatic relief valves, check valve, solenoid valve, pressure differential valve, and manual shutoff valves. As with the other types of alternative fuels, dispensers can be the simple assemblies mounted on a post that are common at fleet operations or the manufactured dispensing units for public fueling locations. Manufactured dispensing units can look like those used for conventional gasoline or diesel fueling. Excess flow check valves in a shear plate are generally installed below the dispensing unit and function like shear valves under conventional flammable liquid fuel dispensers.

LP-Gas dispensers can fill at a maximum rate of approximately 18 gal/min.[16] Pursuant to NFPA 30A, *Automotive and Marine Service Station Code*, they should be located at least 20 ft from gasoline or diesel dispensers. LP-Gas fuel hoses must be fitted with a listed emergency breakaway device to prevent propane releases in the event a driver accidentally pulls away while the fueling hose is still connected. The LP-Gas vehicle fueling hose, commonly known as "wet" hose, is required to be continuously marked "LP-GAS," "Propane," "350 PSI WORKING PRESSURE." Fueling hose typically connects to the vehicle via a U.S. standard $1^3/_4$-in., acme-threaded connection. These connectors are designed with a mechanism that prohibits liquid flow unless connected to a vehicle. A "quarter turn" valve located before the threaded connection controls flow. Flow control valves are available with a lever action similar to conventional gasoline dispensing nozzles. The main difference is that LP-Gas lever valves do not presently shut off automatically when the liquid level is full. Instead, an outage gage or outage vent valve is located on the vehicle, near the hose connection.

The vehicle outage gage operates like bulk vessel liquid level gages. When the vehicle tank is approximately 80 percent full, a small amount of liquid propane is released through the outage vent. The operator recognizes that the vehicle container is full and manually disengages the dispensing flow valve. This venting to atmosphere at each refueling has been recognized as a potential problem. Automatic shutoff devices, such as a float valve inside the vehicle tank, are available, and computer controlled overfill systems are being developed. When perfected, these systems may be required on all LP-Gas vehicles or dispensers. The 1995 edition of NFPA 58 requires automatic means of overfill protection on containers manufactured after January 1, 1984.

VEHICLE COMPONENTS, DESIGN, AND OPERATION

The appearance and shape of alternative fueled vehicles are the same as conventional fueled vehicles. How the fuel is carried and consumed differs in the vehicles. While vehicle storage containers and delivery systems differ for each type of fuel, the concept is the same for all hydrocarbon alternative fuels. CNG, LNG, and LP-Gas vehicles store and consume a mixture of flammable gas. Understanding fuel storage and consumption is important from the perspectives of the vehicle's operator, the individual fueling the vehicle and the emergency responder.

There are two types of alternative fueled vehicles: (1) those converted from another fuel, called bi-fueled conversions, and (2) those designed to be exclusively powered by CNG, LNG, or LP-Gas. Bi-fuel converted vehicles may operate on natural gas or propane and a conventional fuel, such as gasoline or diesel, because the liquid fuel system remains intact. Only CNG, LNG, or LP-Gas fuels dedicated alternative fueled vehicles. They can cost more than conventionally fueled vehicles because of the vehicle fuel tank expense.

Compressed Natural Gas Vehicles

CNG vehicles include conventional passenger vehicles, light- and medium-duty trucks and heavy-chassis vehicles such as transit buses. CNG vehicles store natural gas compressed at pressures of 2,500 to 4,000 psig in compressed gas cylinders equipped with a pressure relief device, a filling connection, and a shutoff valve. The vehicle fuel container connects to either a manifold or single fuel line equipped with a manual shutoff valve. Natural gas is dispensed to the engine via a pressure regulator that reduces the gas to the service pressure the engine uses. After its pressure is reduced, the gas enters a gas-air mixer to produce a flammable mixture that is injected into the internal combustion engine.

NFPA 52 applies to the design and installation of CNG fuel systems on all types of vehicles and their fuel dispensing systems. The standard's scope includes the CNG element of bi-fueled vehicles and vehicles originally manufactured to operate using CNG.

Fuel is delivered from a CNG dispenser through its nozzle and introduced into the vehicle's container through the vehicle fueling receptacle. The attachment is like a typical quick-connect coupling used on compressed air hoses. Vehicle fueling receptacles are designed in accordance with ANSI/American Gas Association NGV 1.

CNG containers are designed to store compressed methane gas at high pressures. The gas is stored in a compressed state to increase the quantity that can be stored onboard the vehicle, thus increasing the vehicle's driving range. American Natural Standards Institute/American Gas Association NGV2, *Basic Requirements for Compressed Natural Gas Vehicle Fuel Containers,* addresses the design and construction of CNG fuel containers. NGV2 containers have service pressures between 2,400 and 4,350 psi at 70°F, capacities not greater than 35.4 cu ft, and a service life of 15 yr. NGV2 containers are either constructed of metal (NGV2-1), a metal liner reinforced with hoop-wrapped (NGV2-2) or full-wrapped (NGV2-3) resin-impregnated continuous filament, or a full-wrapped continuous filament with a nonmetallic liner (NGV2-4). All construction materials must be capable of operating throughout a temperature range of –40 to 180°F. Containers installed in vehicles require identification using a label that reads "CNG ONLY." This label is required to prevent the installation of cylinders not designed for CNG service.

CNG vehicle fuel containers are subject to nondestructive pressure and leak tests. To achieve certification, metal containers and liners must successfully pass destructive tests that evaluate the material's yield strength, tensile strength, elongation, and its performance under mechanical impact. Other requirements that must be met include drop, fire exposure, gunfire, gas permeation, pressure cycling, and burst tests. These tests ensure the container maintains its integrity over its service life.

Vehicle containers can be located inside, below, or above the driver or passenger compartment, provided that all connections to the container are either external or sealed and vented from the area where the driver or passengers sit. Containers are installed in a location that minimizes the probability of damage resulting from a collision. Neither containers nor container components can be located ahead of the vehicle's front axle or behind its rear bumper.

The mounting mechanism must secure the CNG vehicle fuel container to the vehicle's frame, body, or cargo box to prevent damage resulting from road hazards, container slippage, and loosening or rotating of the container. The method used for mounting the container must withstand a static force of eight times the weight of a fully pressurized cylinder applied in six directions of loading. When subjected to this load test, the container is allowed a maximum displacement of $1/2$ in. in any direction of loading. Unless shielded, containers require a minimum 8-in. clearance from vehicle exhaust systems. If metallic clamps or clamping bands are used, they cannot be in direct contact with the container. This necessary provision protects the container from exterior corrosion resulting from moisture accumulation or mechanical vibration. To satisfy the requirement, a resilient, water resistant material is commonly installed between the clamp and the container supports.

Vehicle fuel containers are required to be fitted with one or more pressure-relief devices. Pressure-relief devices protect the container from catastrophic failure due to overpressure resulting from fire exposure.

Pressure-relief devices for CNG cylinders are constructed in accordance with Compressed Gas Association Pamphlet S-1.1, *Pressure Relief Device Standards, Part 1—Cylinders for Compressed Gases.* Pressure-relief devices can be rupture disks, fusible plugs, a combination rupture disk/fusible plug, or a pressure relief valve. A rupture disk is the component of a pressure relief device designed to burst at a predetermined pressure and allow to gas discharge. Rupture discs are preformed metal designed to fail at a particular pressure or temperature. When a rupture disc operates, it vents the entire contents of the container. Rupture disks do not close. Fusible plug devices are also nonreclosing pressure relief devices designed to function when a plug of fusible material melts at a suitable temperature. Fusible plugs are designed to operate at a nominal yield temperature of either 165°F or 212°F, depending on the material selected.

Pressure relief valves are safety valves designed to relieve excessive pressure. They reclose and reseal to prevent the further flow of gas from the container after the reseating pressure has been achieved. Pressure relief devices are installed so the relief device is in direct communication with the fuel. Additionally, a shutoff valve should not be installed between the pressure relief device and the cylinder, nor after the pressure relief devices.

When containers are installed inside vehicles, such as in the cargo trunk of a passenger car, there are requirements concerning how the pressure relief device is installed and terminated to the vehicle's exterior:

1. Installation in the same vehicle compartment of the container.
2. Termination to the exterior using piping or tubing of at least the same diameter as the pressure relief device.

3. Termination of the vent opening so that road debris will not block it.

4. Enclosure of the container neck and all CNG fittings within a gastight compartment.

Valves located at the container and engine isolate the piping system from the container and the engine. A vehicle manual shutoff valve is also located along the frame rail. These valves are provided for the safety of the vehicle user, persons performing repairs, and emergency responders.

A valve is required for the vehicle container. The container valve can be operated manually or with a normally closed remotely activated shutoff valve. When the container has a remotely activated valve, the valve must be equipped with a means that allows the cylinder to be depressurized. Regardless of the valve type used, Authorities Having Jurisdiction should ensure that personnel performing maintenance activities on CNG vehicles clearly understand the manufacturer's procedures for depressurizing a charged cylinder of CNG. In Texas in March of 1993, the Austin Fire Department responded to a fire involving a CNG vehicle. The fire resulted from a mechanic improperly depressurizing the container by not electrically bonding the depressurizing tool to the vehicle frame. The high-pressure discharge of natural gas created a static accumulation and spark that ignited the gas jet, forming a flame jet approximately 15 ft long, as witnessed by arriving fire fighters. Although the fire did not result in any fire fighter or civilian injuries, the vehicle of fire origin was destroyed and an exposed vehicle suffered damage from the thermal flux of the flame jet.[17]

A second valve is required between the engine and the fuel cylinder that permits fire fighters and repair personnel to isolate the containers from the remainder of the fuel system. This manual shutoff valve must be installed in an accessible location and can have no more than a 90° rotation from open to closed position. Labels or stencils stating "MANUAL SHUTOFF VALVE" identify the valve location. The manual shutoff valve requires protection from vibration and road hazards. (See Figure 9-33G.)

Finally, a third valve is required in the fuel system that automatically prevents the flow of CNG when the engine is not running but the ignition switch is in the ON position. This ensures that methane does not flow during repairs while the vehicle's electrical system is energized.

Piping is used in a CNG fueling system because it offers a reliable, safe, and affordable method of moving high-pressure gas from the container to the engine. Normally, rigid or flexible stainless steel fuel lines connect the fuel containers, engine, and vehicle fueling receptacle. The use of certain materials for CNG piping service are prohibited: plastic piping and fittings for high-pressure service and galvanized piping and fittings. Even so, other considerations, such as temperature and vibration, must be taken into account.

FIG. 9-33G. *NFPA 52 requires the location of the manual shutoff valve between the fuel containers and engine to be identified by a legible label or stencil.*

Piping installed on vehicles is subjected to a fairly rigorous service life. As a result, the piping installed between the container and the secondary shutoff valve located along the vehicle's frame must capably withstand a hydrostatic pressure of at least four times the rated service pressure without structural failure. NFPA 52 does not clearly state the required temperature and pressure rating of piping from the manual shutoff valve to the engine; however, the same section *does* reference ASME/ANSI B31.3, *American National Code for Chemical Plant and Petroleum Refinery Piping.* Given the temperature extremes that may be encountered, ANSI B31.3 requires piping to meet the standard's requirements for severe cyclic service. This generally results in only stainless steel tubing being used for CNG vehicle fuel lines.

Pipe connections can be welded, mechanically bent, or mechanically connected. When piping or tubing is threaded, the pipe sealing material must be impervious to natural gas and be applied to all male pipe threads before assembly. When mechanical joints or connections are used, joints must be located in an accessible location.

Pipe routing must consider the effects of vibration, corrosion, damage from road objects and collisions, and normal service life wear and tear. Piping routed through engine compartment fire walls or other vehicle panels requires protection from abrasive wear through the use of grommets or other acceptable means.

When the engine requires fuel, the gas leaves the storage containers and flows to a high-pressure regulator. The gas pressure from the fuel container reduces to an intermediate pressure. As the gas expands from a fuel container pressure of 2,500 to 4,000 psig to the intermediate pressure of 120 to 300 psig, its temperature drops by as much as 100°F. If the gas contains too much moisture, it can freeze, and damage the fuel system and engine. Also, cold gas can harden pipe seals in the fuel system and result in leaks. For these reasons, most CNG vehicles have pressure regulators heated by engine coolant.

From the high-pressure regulator, the methane gas flows toward the engine through another line to the low-pressure regulator. Located in the engine compartment, the low-pressure regulator reduces gas pressure to approximately 100 psi. Engine coolant does not normally heat this regulator because the refrigeration effect is negligible. From the low-pressure regulator the gas flows to the engine, entering the fuel injectors and the cylinder's combustion chamber.

Liquefied Natural Gas Vehicles

LNG vehicles can be used for such applications as midsize and heavy-duty truck chassis vehicles, like refuse trucks and transit vehicles. However, LNG is also a viable fuel for railroad locomotives and marine vessels. LNG vehicle fuel systems have several components, such as a pressure regulator and gas-air mixer, that are like those on CNG vehicles. Because engines cannot consume cryogenic fluids, the liquid must be converted to gas, which requires a heat exchanger.

NFPA 57 applies to the design and installation of LNG fuel systems on vehicles and their fuel dispensing systems. The standard scope includes the CNG element of bi-fueled vehicles and vehicles originally manufactured to operate using LNG.

Vehicles fueled by LNG have components designed to operate at temperatures and pressures encountered in cryogenic service. These components are on the system's liquid side. Major components on the cryogenic side of the fuel system include the vehicle's LNG storage tank and fuel lines. When LNG vaporizes into a gas, its components are designed for compressed gas service. Major components on the vapor phase side of the system include the heat exchanger and pressure regulator.

LNG vehicle containers commonly use a vacuum-jacketed Dewar type design, based on the Dewar flask. This double-walled vessel has reflective surfaces on the container's exterior where vacuum is maintained between the two walls. This helps prevent the LNG loss when the container and liquid become warm. An important consideration in LNG vehicle containers' design is hold time. Hold time is the length of time a vehicle can remain unused until pressure in its tank builds to a point that it must be vented. Hold time depends on the liquid/vapor ratio in the tank, as well as the tank's construction. The liquid/vapor ratio depends on the amount of tank ullage. Space is always required to allow for vaporization and liquid expansion that occurs as temperature and pressure increase. The ullage is typically 5 to 10 percent of the vehicle's tank volume. Ullage is necessary in the event the tank reaches a vent pressure. When this occurs, methane is discharged, not LNG.[18]

LNG vehicle fuel containers are constructed with an inner tank and outer tank. The inner tank is normally constructed of stainless steel. Because it serves as a cryogenic pressure vessel, the inner tank's design meets Section VIII, Division 1 of the ASME *Boiler and Pressure Vessel Code* or U.S. DOT regulations 49 CFR 178, e.g., DOT 4L tank. It normally contains a baffle and method of indicating the level of liquid in the container. The outer tank is constructed of either stainless or carbon steel. The inner tank connects to the outer tank with supports designed to minimize heat conduction but maintain the container's structural integrity. To minimize radiation heat transfer, insulation consisting of multiple layers of aluminum/polyester film separated by nylon netting is placed in the container's interstice. A vacuum drawn in the interstice minimizes the effect of convective heat transfer. Common LNG vehicle containers are cylindrically shaped with elliptical heads. The structural integrity of an LNG vehicle fuel container during normal operation and in accident conditions is significantly considered and analyzed in the tank design. The structural integrity of the LNG container must be evaluated against vibration, acceleration, and shock loads to determine crashworthiness.

LNG vehicle fuel containers are required to have identification features for individuals performing repairs and emergency responders. With the exception of openings for pressure relief or liquid gaging, all openings are designated as to whether they open into the vapor or liquid space. The total volume of the container is identified, and signs indicate that the container can only be used for LNG service. Valves to control liquid and vapor flow must be readily accessible and operable without the use of tools.

Location of the LNG storage container on a vehicle is an important consideration. Containers should not be located ahead of the vehicle's front axle, beyond its rear bumper, or on its roof. Containers may be installed in the interior of vehicles provided the compartment is not in direct communication with the driver, passenger, or any space containing radio transmitters or spark producing equipment. This can be accomplished by providing gastight enclosures over the equipment and venting the enclosure outside the vehicle. When LNG is not odorized, a continuous monitoring methane gas detection system is required in the compartment holding the container. The detection system should be designed to sound an alarm when the atmosphere in the compartment exceeds 20 percent of the lower flammable limit for methane. The gas detection system is not required to stop the flow of fuel or turn off the engine's ignition.

LNG storage containers require pressure relief devices. Most vehicle containers are equipped with pressure relief valves and rupture discs. Pressure relief valves relieve the pressure that develops during the LNG hold time in the container. If the pressure relief device fails, pressure increased until the rupture disc fails. The burst pressure of the disc is higher than the relief valve setting, but lower than the container's design pressure. Unlike conventional relief valves, rupture discs are not designed to close. If a rupture disc operates, LNG vapor is released until the tank is empty. If the vehicle is involved in an accident in which the container is overturned and the rupture disc operates, LNG discharges from the pressure relief opening.

LNG must be converted from liquid to gas for proper combustion in reciprocating engines. A heat exchanger performs the phase change. Commonly called a vaporizer, the heat exchanger raises the temperature of LNG vapors so the phase change from liquid methane to methane gas is efficient. A commonly used heat transfer medium is engine coolant recirculated between the heat exchanger and the engine's radiator. At maximum fuel flow rate, the heat exchanger must have the capacity to completely vaporize the LNG. In addition, it must heat the gas to the appropriate temperature before the gas enters the fuel regulator. Identifying the maximum allowable working pressure for heat exchangers is a necessity to protect the device against possible overpressurization resulting from the installation of an improperly sized relief device.

From an emergency response perspective, the heat exchanger is the component in the fuel system dividing the cryogenic side and the vapor phase, or noncryogenic side. A pressure regulator ahead of the heat exchanger controls the pressure of gas consumed at the mixer or carburetor.

Piping transfers LNG from the vehicle container's liquid withdrawal connection, through the heat exchanger and pressure regulator, to the carburetor or fuel injector. As LNG travels from the fuel container through the fuel line to the heat exchanger, it begins to warm, causing condensation on exterior surfaces. Most fuel lines are insulated to retard heat loss. Piping designed for LNG vehicles is required to withstand pressure of at least twice the container's maximum allowable working pressure. Typical construction materials include stainless steel, brass, and copper that are fitted using threaded, welded, or brazed joints. Where valves are arranged such that LNG could be trapped between them, a pressure relief device must be installed in the pipe section to relieve any excess pressure the warm LNG causes.

Liquefied Petroleum Gas Vehicles

LP-Gas vehicles include industrial trucks, farm equipment, and vehicles ranging in size from automobiles to heavy-duty trucks. LP-Gas has the longest history of all the alternative fuels. Propane used to heat rural homes has been readily available on the farm for use as a vehicle fuel. According to industry sources, approximately 350,000 vehicles in North America are fueled using LP-Gas.

LP-Gas vehicles normally use propane, a flammable, liquefied petroleum gas stored at pressures of approximately 120 to 200 psig, depending on its specific gravity and storage temperature. The vehicle container connects to either a manifold or single fuel line that dispenses propane to the engine via expansion of the liquefied gas. The gas enters a propane carburetor and produces a flammable mixture that is injected into the engine's combustion chamber.

NFPA 58 applies to the design, installation, and maintenance of LP-Gas fuel systems on all vehicles and fuel dispensing systems. NFPA 58 covers the installation of fuel systems supplying engines used to propel vehicles such as passenger cars, taxicabs, multipurpose passenger vehicles, buses, recreational vehicles, trucks (including tractors, tractor semi-trailer units, and truck trains), and farm appliances.

NFPA 58 regulates the design, fabrication, and construction of LP-Gas containers designed for installation on vehicles. LP-Gas vehicle fuel containers must meet either the requirements of ASME

Boiler and Pressure Vessel Code, Section VIII, Division I, "Rules for Construction of Unfired Pressure Vessels," or U.S. DOT requirements. Normally, pressure vessels used for LP-Gas service have a minimum design pressure of 250 psig, based on NFPA 58 and ASME requirements for gases with a vapor pressure of 215 psig at 100°F. However, when the pressure vessel is intended for installation on a motor vehicle, such as an automobile or bus, the minimum design pressure for the container increases to 312.5 psig. The higher design pressure ensures the pressure relief device has an increased operating setting because the container may be subjected to operating temperatures not normally found at stationary LP-Gas installations. In addition, premature operation of a relief valve on a school bus filled with children introduces a variety of concerns to vehicle operators and emergency responders. Establishing a higher starting pressure for opening a relief valve reduces the probability of premature LP-Gas release.

Particular attention must be paid to the location of vehicle LP-Gas containers. Containers must be located to minimize the possibility of damage to them and their fittings. Suitable locations may include the trunk of a passenger car, directly attached to the frame of a vehicle, or in the cargo box of a pickup truck. Containers should be installed in a location to minimize damage resulting from a collision. Containers and their components cannot be located ahead of the vehicle's front axle, behind its rear bumper, or on its roof.

A container installed inside a vehicle should be installed so that the driver or passengers are not exposed to any release of LP-Gas, nor should it be within any space containing radio transmitters or other spark producing equipment. NFPA addresses three options to satisfy this concern:

1. Locate the container and its appurtenances outside the passenger compartment.
2. Completely enclose the container appurtenances in a gastight enclosure or housing.
3. Store the container and its appurtenances in the trunk of an automobile, and seal the compartment to form a separation from the passenger carrying space.

The size of container on vehicles is also limited. Vehicles designed to transport passengers, such as buses or automobiles, are limited to an aggregate of not more than 200-gal water capacity, or approximately 160-gal of LP-Gas. Vehicles not designed to carry passengers, such as farm machinery or mail trucks may have an aggregate 300-gal water capacity of volume, or approximately 240-gal of LP-Gas. In both instances more than one tank can be used and manifolded, but the total volume cannot exceed the aggregate quantity.

A method of preventing the overfilling of vehicle LP-Gas containers is required. Overfilled containers can be hazardous when liquid LP-Gas expands to the point where the containers become liquid-full, causing the pressure relief valve to open and discharge LP-Gas. To minimize the hazard, containers manufactured after January 1, 1984 and installed in vehicles must be equipped with a device that prevents overfilling. Listed stop fill valves designed to satisfy this requirement are available. Other means, such as a solenoid valve, may be used.[15]

Vehicle LP-Gas fuel containers require a method of relieving excess pressure resulting from an external heat source. Spring-loaded pressure relief devices accomplish this. The relief valve is recessed or flush with the surface of the pressure vessel. For ASME containers, the relief valve is normally designed to meet the requirements of Underwriter Laboratories Standard 132, *Safety Relief Valves for Anhydrous Ammonia and LP-Gas,* Section UG-125 of the ASME *Boiler Pressure Vessel Code,* or other standard acceptable to the authority having jurisdiction. DOT containers require spring-loaded pressure relief valves in accordance with CGA Pamphlet S-1.1. NFPA 58 prohibits the use of relief devices that utilize fusible plugs because the vehicle container can be subjected to considerable heat while garaged or in use.

LP-Gas pressure relief devices must be installed in direct communication with the vapor space of the vehicle fuel container. The relief device must also be located at the container's uppermost point. When this is not possible, the device should be internally piped to the uppermost practical point in the container's vapor space. Because propane has an approximate expansion ratio of 1:260, any liquid discharge from a relief device results in the development of a gas vapor cloud. The requirement for locating the relief valve at either the uppermost point of the container or the vapor space is based on experience with relief valves located at or directly above the 80 percent liquid level, in which the relief valve sometimes discharged liquid LP-Gas. When the container is installed inside the vehicle, its discharge outlet must terminate outside the vehicle. It must also be directed so it does not directly impinge on the vehicle fuel containers, the exhaust system, or any other part of the vehicle; it cannot be directed into the passenger compartment.

A manual shutoff valve is required for LP-Gas vehicle fuel containers. This valve has an additional protective feature known as the internal excess-flow check valve. It closes and stops the flow of gas from the container in the event the shutoff valve is sheared or broken off the container. The excess-flow check valve does not operate if the shutoff valve is only partially damaged. Also, the valve does not stop the flow of LP-Gas if a fuel pipe leaks downstream of the valve. Manual shutoff valves are normally installed at the container's liquid and vapor openings. These valves must also be readily accessible so they can be shut off using tools or other means.

The construction materials for piping, hose, and fittings used to connect the fuel container to the vaporizer and carburation equipment are the same as for fixed installations. Piping must be corrosion-resistant or protected against corrosive action, free of leaks, and securely mounted on the vehicle. Any fuel piping that passes through a vehicle's floor must be installed so the pipe is beneath or adjacent to the container and connections are outside the vehicle. Hydrostatic relief valves are required between any section of pipe or hose in which LP-Gas could be isolated between shutoff valves.

Carburetion equipment for LP-Gas vehicles is designed to complete the phase change of propane from liquefied phase to gas phase. Equipment is required to be listed for vehicle service and should be acceptable to the Authority Having Jurisdiction. Carburetion equipment is normally rated with a design pressure of 250 psig though operating pressures are normally limited to the 125 to 250 psig. Like CNG and LNG fuel systems, LP-Gas fuel systems require an automatic shutoff valve provided as close as practical to the inlet of the pressure regulator. The valve prevents the flow of fuel to the carburetor when the engine is not running but the ignition switch is in the ON position. Pressure relief requirements for carburetion equipment are the same as for containers installed on vehicles; relief devices must be spring-loaded pressure relief valves; fusible plugs are not acceptable.

EMERGENCY RESPONSE TO ALTERNATIVE FUEL INCIDENTS

Highway operation of vehicles that use CNG, LNG, and LP-Gas is increasingly common. Fire fighters must be prepared to respond to hazards involved with the release of these alternative fuels. Emergency response to a vehicle or fueling station releasing alternative fuel requires an understanding of the fuel's hazards release mechanisms, methods available to safely stop the release, and, in the event an ignition source is introduced, how to safely and effectively initiate a successful fire attack. Three plausible scenarios for emergencies involving the release of alternative fuels are:

1. The vehicle is being fueled.
2. The vehicle is involved in an accident.
3. The vehicle is under repair.

Vehicle Accident Data

Limited historical incident data is currently available to accurately validate the types of risks fire fighters face when responding to accidents or fires in vehicles powered by natural gas or propane. The relatively small number of vehicles operating in the United States accounts for the lack of data for LNG and CNG.

In 1992, the American Gas Association performed a survey of accident data comparing the relative safety of CNG and LNG vehicles.[19] The report analyzed the accident history for approximately 8,300 natural gas vehicles, which cumulatively traveled 278.3 million miles. The data shows that the injury rate per vehicle mile traveled was 37 percent lower than the rate for gasoline fleet vehicles and 34 percent lower than the rate for the entire population of registered gasoline vehicles. The same report shows no deaths recorded for natural gas vehicles in the 278.3 million miles driven. The rate for gasoline fleet vehicles in 1990 was 1.28 deaths per 100 million miles traveled, and the national average for all vehicles registered in the U.S. in 1990 was 2.2 deaths per 100 million vehicle miles traveled. In addition, CNG and LNG vehicles showed a fire incidence rate of 2.9 incidents per 100 million vehicle miles traveled. Of these 2.9 fires, only 1 per 100 million vehicle miles traveled was directly attributable to a failing of CNG fuel system. Data comparing CNG and LNG vehicles and gasoline-powered vehicles is unavailable.

Identifying Alternative-Fuel Vehicles

Components of alternative fuel vehicles are normally protected and hidden from view to protect these components in an accident. Fire fighters approaching alternative fueled vehicles cannot usually distinguish these vehicles from those fueled with gasoline or diesel. Fortunately, measures have been taken to identify these vehicles to responding fire fighters.

All three NFPA standards require that vehicles be identified with a weather-resistant, diamond-shaped label on the exterior vertical surface on the lower right rear of the vehicle. (See Figure 9-33 H.) Most labels are placed on the trunk lid or rear cowl; they

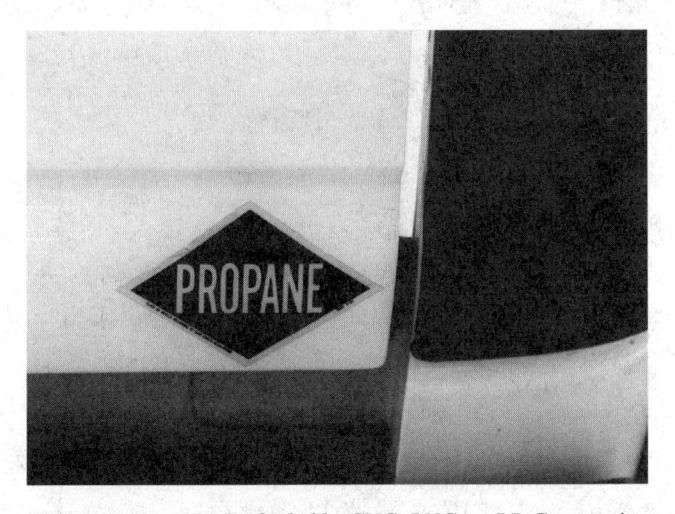

FIG. 9-33H.   Vehicles fueled by CNG, LNG, or LP-Gas require a diamond shaped label that identifies the fuel used by the vehicle. The diamond is affixed to the right rear side of the vehicle. The diamond shown is for a propane fueled vehicle.

cannot be affixed to the bumper because they may be covered by bumper stickers or damaged in minor vehicle accidents. For CNG vehicles, NFPA 52 requires white or silver letters "CNG" on a blue background. NFPA 58 and NFPA 57 require white or silver letters "propane" and "LNG," respectively, on a blue background. The retroreflective labels are visible at night or in fog.

Response to Vehicle Fires and Accidents

Many variables outside fire fighters' control impact how an emergency response will be handled. It is important to understand that alternative fuels have somewhat predictable characteristics when released to the atmosphere, and all the fuels are flammable gases at normal temperatures and pressures. These facts, along with an understanding of the basics of fireground safety, will result in the safe conclusion of an incident.

Apparatus should be placed upwind of a vehicle involved in an accident or burning. If the vehicle is LNG fueled, the apparatus should be located away from the path of flowing LNG. However, this is not as critical as wind direction because flammable LNG or LP-Gas vapors can travel significant distances downwind. LNG vapors will eventually become positively buoyant. Conversely, LP-Gas vapors always remain heavier than air due to the molecular weight of the gas. Fire fighters should also recognize that LNG is normally unodorized, but LP-Gas is odorized.

Fire fighters should accurately assess the scene before initiating a fire attack. Maintaining an adequate distance, fire fighters should walk completely around the vehicle, observing it and paying special attention to the rear so they can identify the fuel used. Fire fighters should look for the blue or black retroreflective diamond that identifies which fuel is used.

For CNG and LP-Gas vehicles, water from attack streams can be used to control fire and enhance vapor dispersion. However, for LNG releases, applying water from fire streams greatly increases the vaporization rate for a pool of liquid methane. Fire fighters should attack LNG fires with potassium bicarbonate, ammonium phosphate, or sodium bicarbonate dry chemical fire extinguishers until they confirm that the fire does not involve fuel.

Regardless of the fuel, the primary concern at vehicle accidents is to ensure the safety of fire fighters and accident victims. The strategy is to control or stop the release of fuel and eliminate ignition sources.

The fuel systems for CNG, LNG, and LP-Gas can be stabilized by closing the primary shutoff valve at the storage containers. CNG vehicles normally have a second fuel isolation valve installed along the frame rail. NFPA 52 requires a method of identifying this valve's location. All alternative fuels containers are required to vent outside the vehicle. The fire fighter assigned the task of closing the primary valve should also locate and inspect the tank vent line to ensure it is not blocked.

Before fire fighters attempt to isolate the fuel supply on LNG vehicles, they should check the vehicle's on-board gas detection system to determine if fuel is leaking. Because an accident could damage this system, fire fighters should also survey the vehicle and accident scene using gas detection equipment. After checking the vehicle and scene for leaks, fire fighters can proceed to close the liquid product valve. LNG containers have two valves, one for liquid product and the other for the tank vent. The vent line should not be closed because it lets the LNG vehicle container vent gas if tank pressure increases, as it does in a fire. If labels are missing or obscured, fire fighters can always locate the liquid valve because it will be covered with ice and frost formed from the moisture in atmosphere. The vent valve will not.

Response to Vehicle Fueling Stations

Response considerations for fueling stations are the same as those for vehicles. The difference between the fueling station and vehicle is the amount of fuel stored at the site and the types of safety features.

Fire fighters who respond to a vehicle fueling station should prepare a written plan documenting how to manage and control the response to the facility in the event of a fire or fuel release. The pre-incident plan should consider points of access, required and available water supply, capabilities of fire department resources and personnel, buildings that could be exposed, safe evacuation of occupants, the location of fuel delivery valves and the method of closing them, and safety features or controls provided. The evaluation should also consider probable and worst-case event scenarios. For example, the overfilling of a LP-Gas container or the operation of a relief valve on a pressure vessel storing compressed methane are not uncommon and will probably occur over the life of a facility. Conversely, worst-case scenarios can include a catastrophic container failure or a BLEVE. While catastrophic incidents are rare, such an event can occur. Its consequences to the public and fire fighters must be considered before any response to the fueling station. Where available, fire fighters should consider using a vapor dispersion model such as CHARM (Chemical Hazard Airborne Release Model) to determine the anticipated size and boundary of the exclusion zones. One model available in the public domain is AR-CHIE, or Automated Resource for Chemical Hazard Incident Evaluation. This software and its companion document, *Handbook of Chemical Hazard Analysis Procedure*, was jointly produced by the Federal Emergency Management Agency, DOT, and EPA. The menu-driven software is capable of predicting downwind plume distances of CNG, LNG, and LP-Gas; thermal flux energy from jet fires, BLEVEs and vapor cloud explosions; and overpressure values in the event of a deflagration or detonation.

All vehicle fueling stations require emergency shutoff devices. These can be simple electrical kill switches or valves designed with integral fusible-links that close if subjected to flame impingement or thermal flux. These devices' location and method of operation should be known and understood. LNG facilities normally require a gas detection system because of the lack of odorant. The gas detection system features and its sequence of operation should be included in the pre-incident plan.

Facility management should be cognizant of the hazards of the alternative fuel stored and should develop its own plan in the event of a release or fire. Further, the fire department and facility management should work jointly to prepare and communicate the pre-incident plan, so realistic action plans can be formulated and the consequences of probable and worst-case events are understood.

BIBLIOGRAPHY

1. J. E. Sinor Consultants, Inc., *Comparative Analysis of Transportation Fuels*. Clean Fuels Consulting, Inc., Canada 1994.
2. Klausmier, R., "Assessment of Environment, Health, and Safety Issues Related to the Use of Alternative Transportation Fuels," Radian Corporation, Austin, TX, 1989.
3. *Risk Management Programs For Chemical Accidental Release Prevention* (proposed rule), *Federal Register*, Vol. 58, No. 201, Wednesday, October 20, 1993. pp. 54190–54219.
4. Krupka, M. C., Peaslee, A. T., and Laquer, H. T., *Gaseous Fuel Safety Assessment for Light Duty Automotive Vehicles*, Los Alamos National Laboratory, Los Alamos, NM 87545, Nov. 1983, p. 12.
5. *Introduction to LNG Vehicle Safety*, Gas Research Institute, Chicago, IL, September 1992.
6. Grant, T. J., Shaaban, S., and Zalak, V., *Hazard Assessment of Natural Gas Vehicles in Public Parking Garages*, Ebasco Services Incorporated, July 1991, pp. 53–55.
7. Zalosh, R., Pilette, Y., and Weining, W., *Hazard Analysis of Alternative Fueled Vehicles in Tunnels—Part 1: CNG Fueled Vehicles*, Center for Firesafety Studies, Worcester Polytechnic, Worcester, MA, April 1994, pp. 8-1–8-2.
8. Walls, W., "Just What Is a BLEVE?" *Fire Journal*, November, 1978, p. 46.
9. Beale, J., "Natural Gas Liquefaction Primer," *Natural Gas Fuels*, November 1993, p. 33.
10. Greene, T., Williams, T., and Blazek, C., *An Investigation of the Use of Odorants in Liquefied Natural Gas Used as a Vehicle Fuel*, Gas Research Institute, Chicago, IL.
11. Lemoff, T., *Liquefied Petroleum Gas Handbook*, National Fire Protection Association, Quincy, MA, 1994, pp. 428–429.
12. Lewis, J., "LNG Explained," *Natural Gas Fuels*, May 1995, p. 59.
13. Moorhouse, J., and Roberts, P., "Cryogenic Spill Protection and Mitigation," *Cryogenics*, December 1988, pp. 838–846.
14. Lemoff, T., "Gases," *Fire Protection Handbook*, 17th ed., National Fire Protection Association, Quincy, MA, July 1991, pp. 3–78.
15. Lemoff, T., *Liquefied Petroleum Gases Handbook*, National Fire Protection Association, Quincy, MA, June 1992, p. 169.
16. *Butane-Propane News*, March 1995, p. 24.
17. Tremblay, K. J., "Fire Destroys Alternative Fuel-Vehicle During Repair Work," *NFPA Fire Journal*, July/August, 1994, p. 29.
18. Powars, C., *et al.*, *A White Paper: Preliminary Assessment of LNG Vehicle Technology, Economics and Safety Issues*, Gas Research Institute, Chicago, IL, January 10, 1992, pp. 2–16.
19. *Natural Gas Vehicle Safety Survey—An Update*, Issue Brief 1992-3, March 20, 1992, American Gas Association, Arlington, VA.

NFPA Codes, Standards, and Recommended Practices

Reference to the following NFPA codes, standards, and recommended practices will provide further information on alternative fuels for vehicles discussed in this chapter. (See the latest version of *The NFPA Catalog* for availability of the current edition of the following documents.)

NFPA 30A, *Automotive and Marine Service Station Code*

NFPA 37, *Standard for the Installation and Use of Stationary Combustion Engines and Gas Turbines*

NFPA 51, *Standard for the Design and Installation of Oxygen-Fuel Gas Systems for Welding, Cutting, and Allied Processes*

NFPA 52, *Standard for Compressed Natural Gas (CNG) Vehicular Fuel Systems*

NFPA 54, *National Fuel Gas Code*

NFPA 57, *Standard for Liquefied Natural Gas (LNG) Vehicle Fuel Systems*

NFPA 58, *Standard for the Storage and Handling of Liquefied Petroleum Gases*

NFPA 59A, *Standard for the Production, Storage, and Handling of Liquefied Natural Gas (LNG)*

NFPA 88A, *Standard for Parking Structures*

Additional Reading

Kerwin, R., "Addressing the Fire Hazards of Alternative Fuels for Public Transit Buses. Part 1," *Industrial Fire Safety*, Vol. 2, No. 5, September/October 1993, pp. 39–45.

Kerwin, R., "Addressing the Fire Hazards of Alternative Fuels for Public Transit Buses. Part 2," *Industrial Fire Safety*, Vol. 3, No. 1, January/February 1994, pp. 19–28.

Navesey, F., "Are Natural Gas Vehicles Fire Safe?" *Fire Prevention*, No. 280, June 1995, pp. 24–27.

Shanbhag, S., *et al.*, "Comparative Study of Organic Byproducts From the Thermal Degradation of Alternative Automobile Fuels," University of Dayton Research Institute, OH, University of California, Berkeley, Toxic Toxic Combustion Byproducts, 4th International Congress, ABSTRACTS ONLY, June 5–7, 1995, Salt Lake City, UT, 1995, p. 112.

Skuggevig, W., "Electric Shock Considerations for Electric Vehicle Charging Systems," *On the Mark*, Vol. 1, No. 2, Summer 1995, pp. 13–15.

Stookey, S. A., "Handling Incidents Involving Natural Gas-Fueled Vehicles," *NFPA Journal*, Vol. 88, No. 4, July/August 1994, pp. 84–89.

FIXED GUIDEWAY TRANSIT SYSTEMS

Revised by Lawrence M. Engleman

The environment of a rapid transit system lends itself to some peculiar and difficult fire safety problems. Large numbers of people who move through these subterranean enclosures and elevated structures compound the life safety dangers. Fires within subways have attracted national attention over the years due to the danger facing passengers who must evacuate trains through smoke-filled trackways and other emergency exitways.

Among the causes of fire in a mass transit system are mechanical failure of undercar components, high-energy electrical short circuits, accumulation of combustible debris along the trackways and within vent shafts, accumulation of road film and oils on the transit vehicle underside, combustible materials in car components, combustible construction of trainways including crossties, and hazards associated with the human factor, such as carelessness, arson, and pranks.

This chapter covers areas in a rapid transit system requiring special attention, including subways, underground stations, and surface-level and elevated transit systems.

SYSTEM CHARACTERISTICS

Although subways and underground stations pose the most serious threat to transit users from fire, certain basic considerations must be recognized throughout the transit system.

Traction Power

Traction power, or the electrical energy needed to run trains, may be carried throughout the system by means of a third rail or overhead wiring and is typically a nominal 600 to 750 V dc. For modern transit systems with traction power provided by a third rail, a coverboard of electrical insulating material installed throughout the system prevents people from inadvertently coming in contact with the energized rail. The coverboard should be strong enough to hold a 250-lb (113-kg) weight applied at any point without deflection and should be supported so that a fire department ladder can rest against it safely. Warning signs should be affixed, as required, to the top of the coverboard in any area accessible to the public.

Disconnect switches should be located strategically throughout the system so traction power can be cut as quickly as possible

under emergency conditions. Otherwise, alternate prompt and reliable means for disconnecting power should be provided.

Access/Egress to Stations

When determining required egress capacity, future occupancy needs must be considered. After a station has been constructed, it becomes very difficult, if not impossible, to place additional stairs or escalators where they best serve the public. Complying with required egress width may present major difficulty to designers. NFPA *101®*, *Life Safety Code®*, as modified by NFPA 130, *Standard for Fixed Guideway Transit Systems,* is a good source of guidance.

To establish required egress widths for any structure, it is first necessary to determine occupancy classification. For example, the State of California, other jurisdictions, and NFPA *101* determined that places where people await transportation should be classified as assembly occupancies. Effective June 1983, however, NFPA 130 established requirements for life and fire safety in transit stations, trainways, vehicles, and outdoor vehicle maintenance areas. The dynamic exiting concept in NFPA 130 determines occupancy based on estimated transit patronage waiting in stations and alighting from trains. NFPA 130 considers only those transit stations accommodating passengers and employees of the fixed guideway system and incidental occupancies. Where large numbers of people other than transit patrons are involved, such as where transit stations are incorporated into shopping malls, sports facilities, or other occupancies, NFPA *101* establishes egress requirements.

Because rapid transit systems are designed to move people rapidly, most new systems have installed escalators as the primary means to enter and exit stations. Escalators that are designed to be fully reversible, depending upon commute conditions, are acceptable as emergency exits. However, escalators should not account for more than half of the egress capacity from any one level.

When escalators are designed and accepted as required egress components, their usability in evacuating a station under emergency conditions must be seriously studied. For example, provisions must be made to stop all platform-bound escalators so that they can be used as stairs. Consideration also must be given to methods of stopping escalators so as to prevent injury to patrons. One possibility is to notify patrons, using a recorded public address announcement, which the fire alarm system may activate, stating that the escalator will stop within a specified period. This allows patrons to grasp the handrail and obtain a firm footing or may even alert patrons not to board escalators until they stop.

Lawrence M. Engleman is technical director, system safety and security, Parsons Transportation Group, Washington, DC.

In many instances, escalators penetrate two or more floor levels to access the intended platform. This can pose special design problems because many designers prefer to maintain an open environment throughout a station complex. Vertical penetration of more than one floor without enclosures has been permitted in some rapid transit systems by increasing other fire protection components.

Exit Barriers

Rapid transit stations contain two basic areas to separate those individuals who have paid their fare from those who have not. These are commonly referred to as "paid areas" and "free areas." Patrons in the free area can move directly to the outside without encountering barriers. Patrons within the paid area must go through fare barriers to access the free area. During emergency evacuation, either all fare barriers should open automatically, allowing patrons to evacuate the station rapidly, or specially designed access gates should be installed for the same purpose. With increased emphasis being directed to handicapped patrons, access gates become more useful in basic station design.

Procedures and methods that allow operators to quickly control or limit the number of patrons allowed into a station in the event of a delay or a malfunction are most important. A public address system that notifies and advises patrons during an emergency or a malfunction also is considered essential.

Grilles or steel doors completely close off many stations after normal revenue hours. Installing small doors in the grilles can provide easy access for fire department personnel, as well as serve as an acceptable means of egress for any employees who may be working in the station.

Emergency Procedures

Due to the variety of hazards involved in a rapid transit system, local fire authorities must establish a direct line of communication with the system's top management. Good inspection techniques must be developed, and coordinated emergency response procedures planned. Access points to all parts of the system must be pinpointed, and engine and truck company response patterns tailored to meet special access problems. Information regarding the system, such as floor plans, electrification, vehicle design, structures, fire protection systems, and communications, is needed to plan emergency procedures properly. Specially developed map books, known as "Fire Maps," for fire department use may prove valuable. A wise course for a fire department is to assign one command officer the responsibility of coordinating all emergency activities required to protect against fires on rapid transit trains, on the right-of-way, and adjacent to the right-of-way.

Communication also must extend in the other direction. Fires and other emergencies in adjacent properties may require modified transit operation. In downtown areas, fire suppression activities in nearby buildings can cause water drainage into subways, or flammable liquid spills can endanger below grade stations. Elevated structures may block fire department access to adjacent buildings.

The Transit Vehicle

The construction and use of the transit vehicle has become a fire and life safety problem. With a reasonable degree of attention, other elements of the transit system—guideway, stations, wayside equipment, and maintenance facilities—can be designed with an acceptable level of fire protection. Construction materials and methods can provide a noncombustible, if not fire-resistive, profile. These elements are static, and many solutions for achieving a degree of fire resistance are already in use in related structures, such as buildings and bridges, and in support subsystems, such as elevators, escalators, and electrical and mechanical systems.

Because of economic and comfort considerations, the modern vehicle has made extensive use of manufactured materials. The vehicle includes many attractive and serviceable plastics, such as foams, fibers, and other synthetic materials. These are often quite combustible and may emit toxic combustion products while burning. These problems are not unique to the fixed guideway transit vehicle. Combustible plastics are used in passenger compartments of other modes of transportation, including aircraft, buses, railroad trains, and the family car. However, the transit vehicles can accommodate several thousand people—far more than even a jumbo jet—and it therefore presents the potential for great loss of life.

Much attention must be focused on the materials and finishes of every item that goes into each transit vehicle, including insulation for propulsion motors and other electrical wiring. The subway transit vehicle is the principal source of toxic smoke generated by, and fuel contributed to, a fire.

Therefore, the transit vehicle is a primary concern in a fire and life safety program for a transit system. Unfortunately, many existing fire test methods used to measure performance of materials in fire test conditions may be inadequate to predict actual fire performance. A guideline developed by the Federal Transit Administration (formerly Urban Mass Transit Administration) deals with the vehicle fire problem and recommends specific test methods and tests be used. It has been incorporated into NFPA 130 as Table 4-2-4. In addition, some transit systems have been able to use a so-called "Spartan" vehicle that has fewer combustible or toxic materials.

Considering all this, a complete fire risk assessment for new or renovated transit vehicles should be made and include fire propagation resistance, smoke emission, ease of ignition, and heat and smoke release rate. Vehicles' total combustible loading should be limited to as low a value as possible. For example, NFPA 130 advises that the combustible loading of the vehicle above the floor should not exceed 45,000 Btu/sq ft (511 mJ/m^2). Values as low as 30,000 Btu/sq ft (341 mJ/m^2) can be achieved. (See Appendix B-2.4.3. of NFPA 130.)

Newer individual test methods and appropriate use of test results are part of an overall fire safety effort. One test detailed in NFPA 253, *Standard Method of Test for Critical Radiant Flux of Floor Covering Systems Using a Radiant Heat Energy Source*, provides a realistic test to evaluate flame spread on floor covering systems. Existing tests, such as those in NFPA 255, *Standard Method of Test of Surface Burning Characteristics of Building Materials;* NFPA 251, *Standard Method of Tests of Fire Endurance of Building Construction and Materials;* NFPA 258, *Standard Research Test Method for Determining Smoke Generation of Solid Materials;* ASTM E-162; ASTM E-84; and others also can be used if results are interpreted in relation to the transit environment. More appropriate tests are being developed to better predict the performance of materials when subjected to elevated temperatures, especially with respect to the release of toxic products.

The difficulties of removing a person from the fire environment, particularly in subways, demands extreme care in evaluating materials for toxic hazard and their use in composite assemblies. To date, the use of test results related to smoke development and generalized statements on avoidance of "recognized" materials generating toxic products of combustion, have been the way to address this problem. The development and use of toxicity tests of materials during fire exposure remains an unanswered need for transit safety.

SUBWAYS

Fires within the confines of a subway system, especially those in trainways, are among the most difficult to extinguish because fire fighters have limited room in which to operate. Access points for fire personnel, emergency exit locations for passengers, availability of water supply, and ventilation capabilities are among the fundamentals that must be considered in providing adequate safeguards for passengers and facilities for fire fighters coping with emergencies.

A direct liaison between rapid transit authorities and fire officials is essential to keep emergency forces apprised of current conditions in underground installations. Floor plans of stations and subways must be reviewed continually. Fire protection included in the original system design may have changed, thus requiring fire-fighting plans to be updated.

Ventilation

Ventilation of underground subways is of primary importance. The removal of smoke under fire conditions must be seriously studied. Several computer programs are available to model various smoke removal scenarios. Exiting schemes that allow passengers to exit rapidly into fresh air must also be developed. Some rapid transit systems have installed fully reversible fans within vent shafts. They move large quantities of air by intake or exhaust and thus purge smoke from the subway in a predetermined direction so passengers can evacuate to the nearest station. Ventilation shafts and fans have been placed at each end of an underground station and, in some instances, between stations. When vertical shafts are included in the ventilation criteria, the possibility of flammable liquid spills entering the shaft at street level must be considered. Shafts should terminate in roadways only when absolutely no other alternative exists. Adequate sumps equipped with flammable vapor detectors must be developed to retain such spills. Electrical equipment installed within shafts and fan rooms should meet local authorities' requirements for expected fire exposure.

Access/Egress

Walkways: Primary consideration must be given to evacuation of large numbers of passengers to safety if a train is stopped between stations. "Safety" could be a station platform or direct access to the open air through vertical exitways. Many older, existing systems evacuate passengers by having them climb from the train to the trackway, and then walk between the rails on crossties until they reach a vertical exit or station platform. In modern transit systems passengers evacuate the train to a walkway constructed alongside the trackway opposite the third-rail power source. (See Figure 9-34A.) This lets passengers step directly from the train to the walkway, keeping them off the trackway and allowing specially equipped vehicles to respond to train problems.

For trackways separated by concrete walls and fire doors, the walkway system also provides easy access to a train on the opposite trackway and an acceptable horizontal egress path that remains relatively free of smoke and heat (See Figure 9-34B.). Obviously, adequate emergency lighting must be provided to allow safe evacuation.

Tunnel or bore: Many rapid transit systems currently operate underground on trackways completely separated from each other by concrete walls and fire doors. (See Figure 9-34B.) As previously mentioned, this concept provides a refuge for passengers, as long as adequate train control and planning is practiced. Separation walls should have no less than a 2-hr fire-resistance rating, and openings in them should be protected by $1^1/_2$-hr fire doors. Cross-passageways between tunnels can be utilized in lieu of emergency exit stairways to the surface. When used, cross-passageways should not be

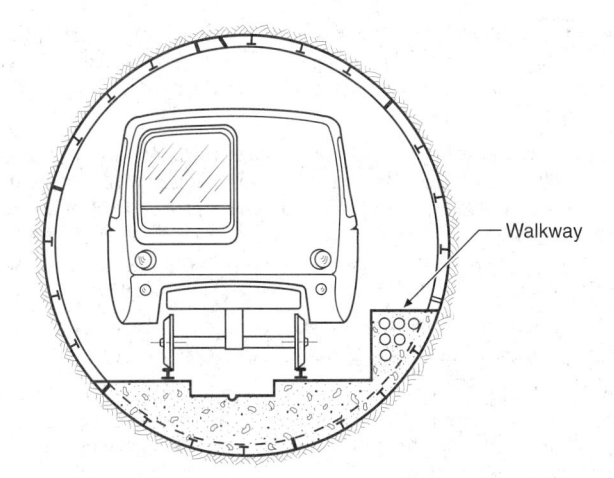

FIG. 9-34A. *Sectional view of a single-tube subway tunnel, showing the location of the emergency passenger walkway to the floor of the car.*

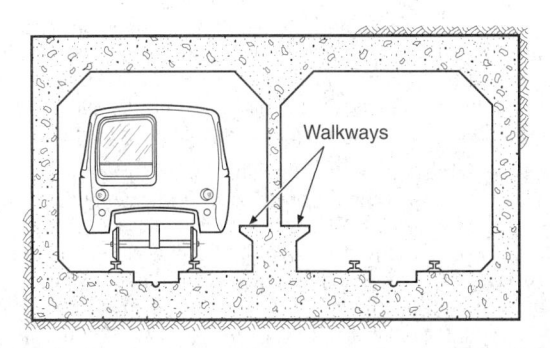

FIG. 9-34B. *Sectional view of a subway tunnel having trackways separated from each other by a concrete wall. Openings through the wall, protected by fire doors, can provide an acceptable horizontal exit path from the trackway on which an accident has occurred to the other trackway, which remains relatively free of smoke and heat.*

more than 800 ft (244 m) apart. Hardware normally used in the installation of fire doors and frames has been found inadequate for the pressures generated by a train moving through the tunnel and for the damp environment found in tunnels. Special study and consideration must be given to the type of door, frame, hinges, locking device, and method of securing the frame to the opening. Sliding fire doors with counterweights require special attention.

Emergency egress: A number of factors must be considered to determine egress requirements from an underground subway system:

1. Train length.
2. Depth of trackway below grade.
3. Accessibility to exit at grade.
4. Accessibility to exit at trackway.
5. Time and distance needed for people to move to surfaces.
6. Available ventilation and controls.
7. Access to trackway for fire service.

Due to modern transit systems' ability to control smoke using sophisticated ventilation systems, and since unlimited numbers of vertical shafts are impractical, NFPA 130 allows emergency exit shafts to be placed so that maximum travel distance does not exceed 1,250 ft (381 m) in tunnels. The intent and basics of NFPA 101, as modified by NFPA 130, should be followed as closely as possible in designing exit stairs. Hatches can be used at the surface, if provisions prevent obstructions that would keep them from opening in an emergency.

Many transit systems have created emergency exits by constructing conventional stairways in ventilation shafts. Unless the stairway is completely separated from the vent shaft by 2-hr fire-resistant construction and adequate control of air currents within the bore is maintained, this is extremely risky because smoke and heat may contaminate the vertical exitway.

Under no circumstances should required exitways contain straight, vertical ladders.

Internally illuminated exit signs with two power supply sources should be installed at the trackway level to direct passengers, and emergency lighting should be provided within the exit enclosure.

Communication

Probably one of the most serious problems encountered during underground emergencies involves communications. Emergency telephones normally are identified throughout the subway system by a distinctive light: in most systems blue lights designate emergency phone locations. Communication from the trackway directly to a central control room is accomplished merely by picking up the phone. Some systems use the Centrex principle for calls.

Blue light stations identifying the location of emergency telephones also can serve multiple purposes. For example, on the Bay Area Rapid Transit (BART) System headquartered in Oakland, CA, stations include a third-rail or power-disconnect button, a maintenance jack, a 110-V outlet, and a 20-lb (9-kg) multipurpose dry chemical portable extinguisher. Stations are spotted throughout the underground system at 1,000-ft (30-m) intervals or in line of sight, whichever is shorter.

Communication through a subway system is sometimes very difficult, particularly radio communication. Many newer subway systems have installed equipment that lets fire department radios be used in subsurface stations and tunnels. Where this capability does not exist, fire authorities should make every effort to test their communications systems under actual conditions throughout the subway system. Consideration should be given to using hard-wired communication systems that may already exist. Portable radios or emergency phones with a direct line to the transit system's central headquarters or to train radios are both useful. Every system should have a radio network providing two-way communication connecting personnel on trains and other vehicles anywhere in the system.

During emergencies, radio frequencies and telephone lines quickly become overwhelmed. Portable hardware telephones, dedicated to fire department use, should be available for point-to-point communications.

Fire Protection in Underground Trainways

A principal fire protection device in fixed guideway transit systems is the standpipe hose system. NFPA 14, *Standard for the Installation of Standpipe and Hose Systems*, provides good guidance for installations, and Class I or Class III service should be specified.

Standpipes: Standpipes should be located in stations and throughout the underground trainway system. Standpipes in tunnels should be at least 4 in. (101.6 mm) in diameter or sized by hydraulic calculations, and may be of the dry type. Generally speaking, local fire authorities determine locations of standpipe outlets, with consideration given to fire department access at tunnel level, available

vertical access shafts for emergency forces, available street access to the fire department, and Siamese connections at grade level.

Some rapid transit systems require that dry standpipe outlets be no more than 200 ft (61 m) apart throughout the underground. Others have determined that maximum separation distances of 300, 400, and 500 ft (91, 122, and 152 m) are acceptable.

Some rapid transit systems place bins or boxes containing $2^1/_2$-in. (64-mm) hose near dry standpipe outlets and leave inspection and testing of such hose to rapid transit system personnel. Fire authorities operating under this principle should seriously consider assigning personnel to witness hose inspections and required waterflow testing. Other transit systems and fire departments have agreed to only place hose and appliances in lockers at the base of access shafts. In areas where local fire authorities wish to use only their own hose, provisions must be made to transport hose and other equipment to a fire site or to store necessary equipment in tunnels and stations.

Special apparatus: In lieu of supplying the fire hose required at standpipe stations, some systems have taken innovative measures to ensure that hose is available when needed. BART, for example, with approval of fire authorities, removed all fire hose from its subway system, except the $1^1/_2$-in. (38-mm) hose located within stations. In turn, BART agreed to furnish five fully-equipped, specially-designed fire engines capable of traveling by street or rail to previously designated rendezvous points. The responding fire department then boards the apparatus and goes to the fire or emergency location within the underground. Each piece of apparatus includes, as part of its inventory, 1,200 ft (366 m) of $2^1/_2$-in. (64-mm) hose and 400 ft (122 m) of $1^1/_2$-in. (38-mm) hose, as well as nozzles, a portable emergency generator, a power saw, a cutting torch, extension cords, flood lights, demand breathing apparatus, life lines, miscellaneous tools, a high-expansion foam unit and nozzle, a 300-gal (1136-L) water tank, and a small pressure pump.

A more practical solution is manual pushcarts designed to run on rails and stored in stations. These carts range from small commercially available maintenance carts to specially designed folding multilevel carts constructed using light weight alloys.

Fire department operational plans: Fire departments that respond to emergencies in underground systems must be extremely familiar with the layout, access, exit, communications, ventilation, and other special factors that must be part of an operational plan for both the fire department and the transit system. Such plans must be reviewed and practiced regularly, and all fire department members and system employees must be trained to respond appropriately to underground incidents.

Every fire fighter entering an underground system to fight a fire must have self-contained breathing apparatus (SCBA) to meet the requirements of NFPA 1500, *Standard on Fire Department Occupational Safety and Health Program*. One challenge facing the fire department is how to maintain an adequate supply of replacement air cylinders for members working below grade level. This may require long-duration breathing apparatus and trucks to refill or deliver quantities of SCBA air cylinders.

Fire extinguishers: Fire extinguishers should be placed on board each vehicle. In addition, fire extinguishers should be placed near fixed equipment throughout the underground system where ignition sources and/or combustibles exist. Locations should include electrical equipment, sumps, ventilation equipment, and so on. If the blue light or the emergency station is suitably located, one or the other would be a preferred place for extinguishers. The size and type of exposure expected dictates the extinguisher's size and type.

Drainage: Adequate drainage of water accumulated during a fire requires careful consideration. Fire authorities should be made aware of the underground system's drainage capabilities and recognize that large amounts of water could seriously impair other required rescue operations. When fighting structural fires at street level near station entrances or street ventilation grills, fire officers should alert rapid transit officials that water intrusion into their system may be expected and take steps to prevent water intrusion using tarps.

Stations

Underground rapid transit stations should be constructed of minimum approved noncombustible materials—that is, Type I, Type II, or combinations thereof—as outlined in NFPA 220, *Standard on Types of Building Construction*. Many modern rapid transit systems use reinforced concrete construction for the station's outer shell and protected steel beams, girders, and columns within the shell. This results in a structure with a substantial fire-resistive capability.

A rare fire of major proportion within a subway station was the King's Cross Station fire, which occurred in London, England, on November 18, 1987. Thirty patrons and one fire fighter were killed in that fire. Station fire hazards include poor housekeeping and electrical fires in escalator machine rooms. The primary fire hazards in underground stations are concessions and trains. Small concessions that sell newspapers and magazines normally have a low-hazard classification, while major concessions, such as restaurants or theaters, have a high hazard classification for smoke and heat development.

Newspaper and magazine concessions should be limited in size and must not block required egress routes. Automatic sprinkler protection should be installed to protect the entire concession area. Possible future expansion of the area should be considered in designing the fire protection system.

Station public areas should be separated from nontransit areas with a fire barrier having at least a 3-hr fire rating. Major concessions or openings from public areas of the station into existing commercial structures should be protected by fire door assemblies having a 3-hr fire protection rating. Even if the rapid transit system owns the commercial activity and the station, good separation between the two areas is needed. Fire barriers can be established by installing fire doors or automatic fire extinguishing systems to prevent fire and smoke spread from the commercial occupancy into the station or from the station into the commercial occupancy. Before making any opening into a rapid transit system, existing conditions on both sides of the proposed opening must be reviewed carefully so an effective fire protection plan can be developed.

Protection for Underground Stations

An underground rapid transit station may be compared to a multiple-basement structure, except that a rapid transit station generally has a low fire-loading factor when compared to a commercial basement. Strict application of building or fire code requirements may cause exceptionally high and not necessarily justified construction costs. If minimal combustible storage or concession areas are involved, large, open public areas may be left with little or no risk of fire involvement. If such is the case, limiting automatic fire protection systems to those areas containing combustibles and designing the systems for possible expansion may have some merit, although the limitations of partial protection must be clearly understood.

Automatic sprinkler systems: All areas in transit stations that are used for concessions, storage, and fare collection, including other similar areas with combustible loadings, as well as the steel truss area of all escalators in a single-entry station should be protected by automatic sprinklers equipped with flow alarms and supervision.

Trainways are excluded. NFPA 13, *Standard for the Installation of Sprinkler Systems*, should be used for guidance. Other approved fire extinguishing systems, such as clean agent extinquishng systems in train control rooms, could also be used in lieu of the automatic sprinkler system, but only with the approval of the Authority Having Jurisdiction.

Deluge systems: Train fires may start in under-car equipment, and train operators are instructed to take trains with fires onboard to stations for passenger evacuation. When possible, many newer systems have opted for under-car deluge systems. These systems consist of fixed piping connected to a water supply, with open sprinklers located between rails. They are activated manually from the station attendant's booth or at the ends of the platform.

Over-car deluge systems are designed to create a water curtain between a burning train and the platform. They can be activated automatically by heat detectors in the trainway or manually as described above. Examples of both types of deluge installations are found in the Baltimore Metro System.

Emergency communications: Subway stations typically do not have traditional fire alarm systems with pull stations and bells. Instead, the public address system alerts and directs passengers in the event of an emergency. Passengers can report emergencies through emergency voice alarm reporting devices located on the platform and throughout the station. Emergency communication systems should conform to NFPA 72B, *National Fire Alarm Code*.

Fire detection systems: All nonpublic areas commonly classified as support or ancillary areas, if not protected by an automatic fire suppression system, should be fully protected by a fire detection system conforming to NFPA 72. Listed smoke detectors are preferable, although rate-of-rise heat detectors may be used if authorized by the fire authority.

It is good practice to provide supervision for all fire detection and automatic fire suppression systems. Supervisory signals should be received at the system's command center and annunciated by zone at the station agent's booth or kiosk.

Wet standpipes: Where separate standpipe and automatic sprinkler systems are provided, wet standpipes usually are supplied by a piping system that is separate and distinct from that used for automatic sprinkler systems. Fire department Siamese connections, properly labeled and installed at grade level, let the fire department supply additional water to either system.

Fire hose cabinets may contain $1^1/_2$-in. (38-mm) and $2^1/_2$-in. (64-mm) control valves, as recommended in NFPA 14, *Standard for the Installation of Standpipe and Hose Systems*. A maximum of 100 ft (30 m) of attached $1^1/_2$-in. (38-mm) hose with nozzle and/or $2^1/_2$-in. (64-mm) hose and a spanner wrench may be stored in cabinets for use by the fire department at the discretion of the local Authority Having Jurisdiction. A suitable portable fire extinguisher may also be placed in the cabinet. To reduce pilferage of fire equipment, cabinets should be locked and a breakaway glass panel installed to allow ready access under emergency conditions. Some rapid transit systems have installed intrusion alarms, which send a signal to the station agent if a fire hose cabinet has been opened.

SURFACE OR ABOVEGROUND GUIDEWAYS AND STATIONS

Generally, combatting fires in transit systems at or above grade is no more difficult than fighting any other type of fire at grade. Primary consideration always must be given to the traction power. Even when assured that the power has been shut off, fire department personnel should operate under the assumption that the third rail or overhead wire is still energized.

For fires in trains on surface track or aerial guideways design must address certain basic considerations, such as means of escape from elevated trainways. As in all other rapid transit configurations, communication is paramount in any emergency operation. The ability to direct emergency forces to the scene of an incident and to place the system in a safe condition for access is necessary for patrons and emergency personnel.

Access to a trainway also may require special provisions. Locating trainways in a joint corridor with a railroad or freeway may limit access for long distances of the trainway. Trainways extending from population centers through undeveloped areas sometimes extend where surface roads and water mains do not exist. Trainways in congested areas, trainways elevated to unusual heights, and trainways in other sections not accessible to emergency vehicles or paramedic units, could create other problems. Trainway construction also may cause problems of fire exposure and access to adjacent structures. For example, an elevated trainway above existing city streets and adjacent to existing buildings may interfere with fire department operations at those buildings.

Similar problems may exist at stations. However, permanent exits provide both patron evacuation and emergency personnel access. Recent systems have incorporated stations in the lower elevations of high-rise buildings or within mall complexes that have divergent occupancies. In these cases, detailed plans for coordination and compatible design are necessary.

Stations that can be considered to be buildings would normally be reviewed through normal code procedures. If a transit system passes through more than one jurisdiction, particularly if different basic codes apply, development of a common set of requirements is desirable to achieve uniformity throughout the system.

Most existing building codes do not usually address the subject of trainways. The exception is NFPA *101*, which does deal with such facilities. Items that should be considered during a plan review of a guideway include:

1. Designated mileage markers to provide a response location for fire personnel.
2. Locations where overhead trainways may intersect with known street locations.
3. Access gates allowing patrons to exit the system under fire department direction.
4. Available wayside communications allowing the fire department to communicate directly with the system's command center.
5. Locations of substations, switching stations, gap breaker stations, or other major electrical installations of concern to local fire authorities.

BIBLIOGRAPHY

NFPA Codes, Standards, and Recommended Practices

Reference to the following NFPA codes, standards, and recommended practices will provide further information on fixed guideway transit systems discussed in this chapter. (See the latest version of *The NFPA Catalog* for availability of current editions of the following documents.)

NFPA 10, *Standard for Portable Fire Extinguishers*
NFPA 13, *Standard for the Installation of Sprinkler Systems*
NFPA 14, *Standard for the Installation of Standpipe and Hose Systems*
NFPA 30, *Flammable and Combustible Liquids Code*
NFPA 72, *National Fire Alarm Code*
NFPA *101®, Life Safety Code®*

NFPA 130, *Standard for Fixed Guideway Transit Systems*
NFPA 220, *Standard on Types of Building Construction*
NFPA 251, *Standard Method of Tests of Fire Endurance of Building Construction and Materials*
NFPA 253, *Standard Method of Test for Critical Radiant Flux of Floor Covering Systems Using a Radiant Heat Energy Source*
NFPA 255, *Standard Method of Test of Surface Burning Characteristics of Building Materials*
NFPA 258, *Standard Research Test Method for Determining Smoke Generation of Solid Materials*
NFPA 259, *Standard Test Method for Potential Heat of Building Materials*
NFPA 1500, *Standard on Fire Department Occupational Safety and Health Program*

ASTM Publications

Available from American Society of Testing Materials, 100 Barr Harbor Dr., West Conshohocken, PA 19428-2959.

ASTM C542-78, *Vertical Fire Test Procedures* (FAA-FAR 853)
ASTM E83-81a, *Surface Burning Characteristics of Building Materials*
ASTM E119-82, *Fire Tests of Building Construction*
ASTM E136-81, *Standard Method of Test for Noncombustibility of Elementary Materials*

Additional Reading

Best, R. L., and Engleman, L. M., "Fact Sheet—King's Cross Station Fire," *Fire Command*, January 1988.
Cheong, S. Y., "Fire Protection and Fire Safety Features in Singapore's Mass Rapid Transit System," *Fire Engineers Journal*, Vol. 52, No. 164, March 1992, pp. 25–28.
Cleary, T. G., and Fowell, A. J., "Report on Material Tests Associated With the Los Angeles Subway Fire," National Institute of Standards and Technology, Gaithersburg, MD, December 1990, 33 pages.
Cogan, M., and Pressler, B., "Union Square Subway Crash: The Rescues," *Fire Engineering*, Vol. 145, No. 5, May 1992, pp. 29–30, 32, 34, 36–38.
Cox, G., Chitty R., and Kumar, S., "Fire Modeling and the King's Cross Fire Investigation," *Fire Safety Journal*, 1989, pp. 103–106.
Dodman, N., and Ransley, P., "Safety Case Development for a New Underground Mass Transit Railway," *Proceedings* of the Second International Conference on Safety in Road and Rail Tunnels, April 3–6, 1995, Granada, Spain, 1995, pp. 585–592.
Fennell, D., *Investigation into the King's Cross Underground Fire*, Her Majesty's Stationary Office, London, November 1988.
"Fire Down Below," *Railway Gazette International*, Vol. 144, No. 1, 1988, pp. 27–28.
Fire Map Books, Washington Metropolitan Area Transit Authority, Washington, DC.
"Fire Safety in Mass Transit Railway," *Fire International*, Special Issue 1993, pp. 26–27.
"Fire Safety in Underground Railway Systems," *Fire Surveyor*, Vol. 18, No. 1, Feb. 1989, pp. 42–43.
Friedman, D. M., Williams, T. A., and Farkhondehpay, D., "Safety Assessment for Indoor Refueling of Compressed Natural Gas Mass Transit Buses," *Industrial Fire Safety*, Vol. 2, No. 1, January/February 1993, pp. 56–60.
Fruin, J. J., *Pedestrian Planning and Design*, revised edition, Elevator World, Mobile, AL, 1987.
Gaffney, P., and Kynaston, R., "Smoke Control System of the Hong Kong Mass Transit Railway," First International Conference on Safety in Road and Rail Tunnels, November 23–25, 1992, Basel, Switzerland, 1992, pp. 79–93.
Hastbacka, M., and Levesque, J., "Compounds That Meet Smoke Density Guidelines for Aircraft and Mass Transit Applications," Recent Advances in Flame Retardancy of Polymeric Materials: Materials, Applications, Industry Developments, Markets, Volume 3, May 19–21, 1992, Stamford, CT, Business Communications Co., Inc., Norwalk, CT, 1992, pp. 227–233.
Heffels, P., "Improving Fire Protection in Tunnels for Commuter Rail Traffic," *Tunnelling Underground Space Technology*, Vol. 2, No. 3, 1987, pp. 311–314.

Henson, D. A., and Lowndes, J. F. L., "Smoke Control in Subway Tunnels," First International Conference on Safety in Road and Rail Tunnels, November 23–25, 1992, Basel, Switzerland, 1992, pp. 171–181.

Kennedy, W. D., Bayne, M. L., and Sanchez, J. G., "Safety Issues in the Design of Inflatable Barriers for the Control of Smoke in the Washington Subway," First International Conference on Safety in Road and Rail Tunnels, November 23–25, 1992, Basel, Switzerland, 1992, pp. 479–489.

"King's Cross Disaster was Foreseeable in Law," *Fire Prevention*, No. 212, Sept. 1988, pp. 18–21.

Malanga, R., "Fire Protection Systems Development to Protect Subway System Token Booth Clerks From Incendiary Attack," *Journal of Fire Protection Engineering*, Vol. 3, No. 4, October/December 1991, pp. 109–122.

National Materials Advisory Board, "Fires in Mass Transit Vehicles: Guidelines for the Evaluation of Toxic Hazards," National Materials Advisory Board, Washington, DC, Department of Transportation, Washington, DC, NMAB-462, Project DC-06-0559-01, June 15, 1991, 80 pages.

Parker, A. J., "Fire Testing of Mass Transit Assemblies," *Proceedings* of the Fourth International Conference and Exhibition on Fire and Materials, November 15–16, 1995, Crystal City, VA, 1995, pp. 235–244.

Trabanco, A. M., "Safety in the Subways," *Fire Engineering*, Vol. 143, No. 10, October 1990, pp. 66–68.

Transportation Research Board, "Emerging Concepts and Products for Transit Systems. Annual Progress Report 1," Transportation Research Board, Washington, DC, Annual Progress Report 1, October 1995, 59 pages.

RAIL TRANSPORTATION SYSTEMS

Revised by Roger K. Fitch and Charles J. Wright

Railroad trains transport billions of dollars of merchandise annually. With approximately 155,000 Class I route miles of mainline track operated in freight service in the United States and Canada, the rail network is a vital link in our transportation system. Anything that incapacitates even a portion of the rail system affects much more than transportation.

This chapter introduces the reader to physical aspects of the rail transportation industry. It identifies different types of rail rolling equipment used by North American railroads, both freight and passenger, and discusses fire hazards associated with them. It also identifies fire problems involving rights of way and specialized railroad facilities, and presents safety precautions, both general and situation-specific, that all persons should follow while on railroad properties or near rail activities. The chapter generally discusses the rail transportation of hazardous materials. It also discusses in detail the transportation of hazardous materials in rail tank cars and in intermodal tank containers, and describes emergency response capabilities of the rail industry.

The business of rail transportation touches upon a wide variety of topics. Most chapters in this handbook provide useful assistance in dealing with at least some aspect of the industry. However, the reader is specifically referred to these chapters for discussions of related topics: Section 9, Chapter 4, "Mercantile Occupancies"; Section 9, Chapter 34, "Fixed Guideway Transit Systems"; Section 10, Chapter 8, "The Role of the Fire Protection Community in Management of Hazardous Substances"; and Section 10, Chapter 9, "Managing the Response to Hazardous Material Incidents."

Two important rail transportation terms are used in this chapter: "rolling equipment," or "rolling stock," is an all-inclusive term that includes locomotives, freight cars of all types, cabooses, and on-rail track maintenance (maintenance-of-way) equipment. "Right-of-way" refers to the (usually) continuous strip of land upon which tracks are laid and trains operate. Right of way also, technically, refers to the contiguous property areas upon which shops, offices, switching yards, and other railroad facilities and activities are located or take place. The rail transportation company usually owns and maintains the right of way.

Roger K. Fitch, formerly manager of fire protection for the Union Pacific Railroad Company, is now a fire protection consultant based in Omaha, NE. Charles J. Wright is manager of training in the Chemical Transportation Safety group at Union Pacific Railroad, Omaha, NE.

RAILROAD FIRE LOSSES

Railroads since their inception have been associated with fires. Steam engines powered the earliest locomotives; burning wood made the steam. Engine exhaust sparks commonly ignited nearby dry vegetation. A published account of the first passenger train trip in the United States describes how these sparks ignited passengers' clothing and parasols. In today's rail operations, fires still cost the industry about $20 million annually, according to statistics furnished by U.S. and Canadian railroads and compiled annually by the NFPA Rail Transportation Systems section. In 1989–1993, rail vehicle fires reported to U.S. fire departments averaged 810 per year and $15.9 million per year in direct loss.

The rail transportation industry has fire risks primarily in three separate areas:

1. The company's fixed physical plant, which ranges from high-rise offices to locomotive heavy repair shops to diesel fuel storage and dispensing facilities.
2. Fires on trains, involving locomotives, rolling stock, passengers, or cargo.
3. Fires caused by trains or right-of-way maintenance activities that cause damage to railroad property or to adjacent, privately owned properties.

Railroads are unique in that they typically operate over a continuous right-of-way that they both own and maintain.

RAILROAD OPERATIONS

Today, on all but the smallest of railroads, trains are dispatched from remote dispatching centers. On several of the largest railroads, one central dispatching center controls all trains on the entire system. Crews on the trains follow explicit instructions at all times and are in continual radio and electronic contact with a dispatcher. Some locomotives have full-time computer contact both with the dispatcher and with their company's central computer system. One railroad tracks its trains' locations by satellite.

Advances in electronics and their increased use permit other railroad operational efficiencies as well. One or more locomotives (the "locomotive consist") at the front of the train still pull most trains, but it is now practical in certain circumstances to place locomotives at intervals throughout the train, providing important operational and control efficiencies. However, with such advances come potential problems: a train is no longer the predictable entity it used to be. The emergency responder faces more uncertainties, as well as genuine benefits: advanced computer systems have made a tremendous amount of information on a train's cargo, its proper handling, and emergency response protocols instantly available even at remote locations. Thus, it is now more imperative than ever for non-railroaders involved in rail transportation activities or responses to communicate with, and to coordinate their activities with, proper railroad officials.

COMMUNICATING WITH A RAILROAD

Communication with railroads today is simple, effective, and very necessary. On most railroads, a call to any office can be quickly connected to any other, including dispatching centers, maintenance crews, railroad emergency response teams, and, often, even the train itself. Similarly, the crew on a train can immediately notify dispatchers of an emergency on or off railroad property, and advise in detail as events unfold. Dispatchers are specially trained to make appropriate emergency contacts both inside and outside of their company. See the sections "Hazardous Materials in Rail Transportation" and "Emergency Response Teams" later in this chapter for information on railroad emergency response capabilities.

SAFETY ON RAILROAD PROPERTY

All who visit rail transportation properties under emergency or routine conditions need to exercise particular caution. The hazards are many and unfamiliar. The potential consequences are severe. Always assume the tracks are "live"—expect a train to appear from either direction at any time. Rails are slippery—always step over, not on, a rail, after first making sure the way is clear.

More and more often, a person at a distant dispatching center electrically operates track switches, the devices that physically move the rails and route a train from one track to another. Unlike the person who stood near the switch and manually operated it, a remote operator cannot see the switch and is unaware of activities in its vicinity. Keep feet, hands, and equipment away from the moveable switch components.

Do not block tracks or lay items across them, without positive assurance from the railroad that all trains in the area have been stopped. Even then, minimize exposure to trains and clear the tracks as soon as practical. In many incidents, hose lines have been laid across a railroad track, only to be cut by a passing train.

Until the railroad has given positive assurance that all trains in the area have been stopped, do not work around tracks or place equipment on or near tracks, until flaggers with red flags (in daylight) or red fusees (in darkness) have been deployed a minimum of $1\frac{1}{2}$ mi (2.4 km) down the track in each direction. Where trains may be moving rapidly, an additional flagger should be placed a mile far-

ther away in each direction. Fast moving trains, freight or passenger, take at least that long to stop.

Rail emergencies on bridges and in tunnels with multiple tracks pose special hazards to the responder, as there usually is no work space except on tracks. Until positive assurance is received that all train activity in the area has stopped, there is no safe workplace for emergency activities and equipment and no safe haven for workers should a train appear on an adjacent track.

The railroad should be contacted at first opportunity whenever emergency activities occur on its property. This not only helps to maximize responder safety; it also activates the railroad's considerable emergency response capacity.

Maintaining regular contact with each rail transportation entity in their area ensures that emergency responders are fully aware of the transportation company's emergency response resources that are available to them.

MOTIVE POWER

Diesel Electric Locomotives

What is generally referred to as a diesel locomotive is actually a diesel electric locomotive. (See Figure 9-35A.) The power is developed by the diesel engine, which drives a main traction generator (or alternator on the latest ac locomotives). The output of this generator or alternator is then transmitted to the traction motors, which are mounted under the locomotive in massive frames called trucks and geared to the axles and wheels. The traction motors actually drive the locomotive. The numerous support systems of a locomotive include a battery that provides power to start the diesel engine; an auxiliary generator or alternator that charges the battery and supplies low-voltage current for control and lighting circuits; and an auxiliary alternator, mounted integrally with the main traction generator or alternator that furnishes field current for the main alternator and power for the radiator cooling fan motors and for a blower motor, which cleans the inertial carbody filters.

Main line locomotives have a normal service life of more than 30 years. Thousands of units still in service have operated far longer. Locomotives in main line service on a major railroad will average between 150,000 and 250,000 miles (241,403 and 402,338 km) per year.

The diesel electric locomotive is an enclosed, self-contained piece of equipment. Fire prevention in the engine room is almost exclusively limited to housekeeping and maintenance, which includes stopping and repairing fuel oil, lubricant oil, or exhaust leaks and ensuring that debris and products of leakage do not accumulate. Improved carbody filtration, through the development of self-cleaning

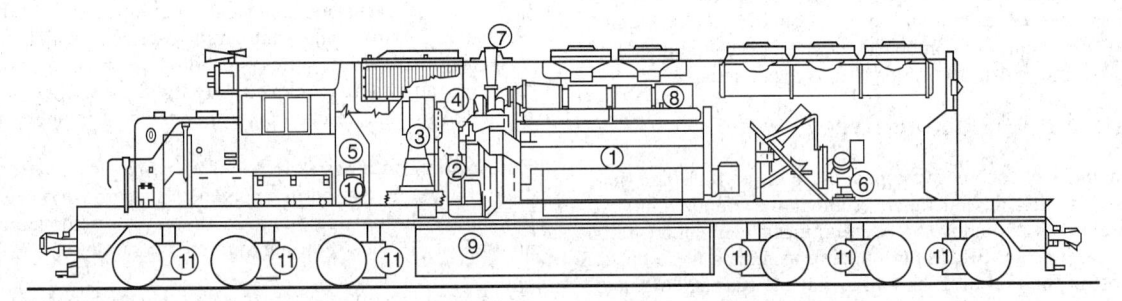

FIG. 9-35A.   A diesel electric locomotive showing the location of principal components of the power system: 1. engine, 2. generator-alternator, 3. traction motor-generator blower, 4. auxiliary generator, 5. electrical control cabinet, 6. air compressor, 7. engine exhaust stack, 8. exhaust manifold, 9. fuel tank, 10. electrical cabinet air filter, 11. traction motor, 12. radiator cooling fans (turbocharged locomotives only), and 13. dynamic brake grid cooling fans (turbocharged locomotives only).

inertial-type carbody filters and carbody pressurization has also helped to improve engine room safety in modern diesel electric locomotives.

Other recent improvements have resulted in virtually "fireproof" electrical cabinets on newer locomotives. Among these improvements are the use of fire-retardant wire and cable insulation and the practice of ventilating electrical cabinets with filtered air. This prevents flammable gases from collecting and keeps dirt out of the electrical cabinet. Other improvements include better wiring techniques, increased use of circuit breakers and fuses, and the pressurization of electrical cabinets. Design improvements, particularly in the now extensive use of electronics throughout the locomotive, have also reduced fire hazards.

Fire protection for diesel locomotives includes the following:

Hot engine protection: To protect against excessive temperatures in the engine cooling system and hot engine oil, devices are installed to prevent excessive heat buildup in the engine and supporting systems.

Engine crankcase protection: This device is designed to detect a dangerous buildup of pressure in the engine and to shut down the engine when necessary.

High-voltage ground protection: If a high-voltage fault develops on the locomotive and goes to ground, the ground relay detects the fault and disconnects the electrical load to the locomotive.

Emergency fuel cutoff switches: In an emergency, the fuel supply to the engine can be stopped by pressing any one of three emergency fuel cutoff buttons. Two buttons are located on the underframe near the fuel filler, one on either side of the locomotive, and a third is found on the engine control panel in the cab.

Main battery knife switch: This large, single-throw knife switch is located in the engine cab at the lower portion of the fuse panel. It connects the battery to the locomotive low-voltage system.

Personnel protection from high voltage: Modern locomotives have their high voltage electric componentry isolated in a labeled panel. This panel has interlocks and other design features to ensure that opening the panel door automatically de-energizes the electric components. Older units may not have this feature, however, so extreme caution is advised when working around locomotive electrical systems.

Fire protection equipment: Each diesel electric locomotive is equipped with portable fire extinguisher protection, generally ordinary dry chemical. The quantity and type depends upon the size of the locomotive and its use in yard, local, or main line service.

Preventive maintenance and inspections: Diesel electric locomotives are inspected, tested, and serviced on a regular schedule, based on life expectancy of component parts, empirical failure data, and Federal Railroad Administration (FRA) regulations.

Alternative Fuel Locomotives

The vast majority of locomotives operates on diesel fuel. The railroad industry is, after the federal government, the country's largest user of diesel fuel. Locomotive fuel is a significant element of a railroad's operating costs. Efforts are ongoing to reduce these costs, both the cost of the fuel itself as well as its storage, handling, and dispensing costs.

One railroad is now extending the range of some of its long-distance trains by placing a tank car of diesel fuel in its locomotive consist. This practice also permits the train to take on large quantities of fuel where the price is lowest. The larger quantity of locomotive fuel potentially involved in a train collision or derailment should be recognized by the fire community.

A few railroads are experimenting with liquefied natural gas (LNG) as a locomotive fuel. A handful of experimental units are now or soon will be in use on selected assignments. Some LNG locomotives designed only for use in switching yards use a relatively low-powered, spark-ignition engine that permits LNG to be the only fuel on board.

Other experimental LNG locomotives are high-horsepower units intended for long-distance hauling. They use engines with traditional diesel-type, compression-ignition technology. These units require that a percentage of diesel fuel be mixed with the LNG primary fuel to achieve combustion, so both diesel and LNG fuels are found on board. At present there is no uniformity of application: some units use high pressure LNG technology; others use low-pressure designs.

Refueling arrangements for LNG units are also in the experimental stage. Installations and practices vary widely, both within a single railroad and between different companies. Some locations have fixed LNG fueling facilities operated by railroad personnel. At other locations fueling may be by a vendor's crew and tank truck.

The LNG-powered locomotive consists for long-distance service in use or contemplated now have this basic similarity: all have two high-horsepower locomotives with an LNG fuel tank car, or tender, between them. Equipment is well protected from all known hazards, including break-apart. Considered part of the consist—that is, multiple locomotives in a train—the LNG-containing tender is not required to display the customary hazardous materials placard.

All experimental locomotives and locomotive consists, along with LNG fueling operations, can be expected to be well marked.

Electric Locomotives

Electric locomotives today are used almost exclusively on passenger trains in major metropolitan areas.

Figure 9-35B illustrates the general arrangement of an electric locomotive. This particular unit can deliver 5100 hp (3.8 MW) at the rail and operates on 25,000 or 50,000 V alternating current from an overhead wire.

Emergency stop buttons: These buttons, one on each side of the locomotive under the operator's cab, can be reached from the ground. Pushing either button trips the vacuum circuit breaker on the roofs of all locomotives in the consist and removes the power.

Manual pantograph grounding switch: Usually located in the right rear of the locomotive, this switch grounds the pantograph before personnel perform maintenance or climb to the roof through the rear hatch. Grounding the pantograph (catenary), which is described in special instructions, varies depending upon locomotive type and the overhead system employed. Fire fighters should obtain this information from railroad officials and include it in prefire plans.

Emergency shutdown switch: This switch, located on the master control housing, is used in emergencies to remove power from the locomotive during single-unit operation or to remove power from all units of the consist during multiple operation.

Electric locomotives, by their very nature, operate with voltages considerably higher than diesel locomotives. Modern electric locomotives have vacuum breakers that trip in cycles, thereby considerably reducing short circuits that could cause fires on older motive power.

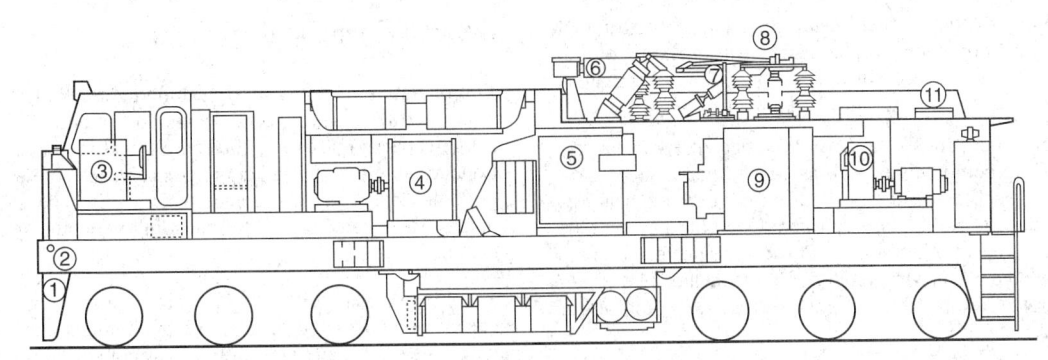

FIG. 9-35B.   An electric locomotive showing the location of some essential pieces of apparatus: (1) proximity switch, (2) emergency stop button, (3) operator's console, (4) equipment blower, (5) main rectifier compartment, (6) vacuum circuit breaker, (7) pantograph grounding hook, (8) pantograph (in folded position) (9) transformer, (10) air compressor, and (11) roof hatch. Because the electric locomotive derives its power from an electrical circuit, it is necessary to deenergize the circuit to extinguish any electrical fire safely.

Preventive maintenance: Preventive maintenance is mandatory to preclude downtime caused by fire or equipment failure. Electrical circuits are tested for grounds or other defects when equipment is at major shops or at intervals of no more than three months. All tests and scheduled maintenance are performed in compliance with U.S. Department of Transportation (DOT) rules, as a minimum.

Fire protection equipment: Electric locomotives are equipped with portable fire extinguishers. Gaseous extinguishers are normally located in each cab and in the equipment compartment. The number of onboard extinguishers varies with equipment type and company policy. To minimize damage to delicate equipment found on electric locomotives, every effort should be made to use a gaseous agent to extinguish a fire.

Safety Precautions

Certain precautions must be observed when boarding an electric locomotive in an emergency:

1. Never climb on top of any railroad equipment under an overhead catenary system until obtaining positive assurance that the wires have been de-energized and that locomotive grounding switches are engaged. Failure to follow this procedure can result in severe burns and even electrocution.
2. It is imperative that all persons on or near electric locomotives be aware of hazards associated with the high voltages involved. Electricity supplied by the overhead catenary system is typically 11,000 V ac; from the third rail, 600 V dc. Voltages within locomotive electrical cabinets are 600 V dc and 480 V ac. If the catenary wire falls on the train body, it energizes the entire train with high voltage. Do not assume that safety features on the power system or locomotive have functioned and de-energized the circuits. Make certain the power is off, or use suitable precautions.
3. Unless human life is threatened, no fire-fighting activity should be undertaken on an electric locomotive until the catenary power has been removed and the overhead wire grounded by a qualified electric traction (ET) representative from the railroad. If action is required before an ET representative arrives, consult the train crew's conductor or engineer. These individuals are trained in the proper procedures to lower and secure the pantograph and to ground the electrical equipment.
4. Never enter any high-voltage compartment when the locomotive is energized.
5. Never touch motors, switches, protective barriers, or other electrical apparatus without knowing their exact purpose and function.

6. To protect against serious or fatal personal injury, be sure the pantograph is actually latched down before opening the roof hatch.

FREIGHT CARS AND EQUIPMENT

Major causes of fire in freight equipment, including motive equipment, are collision/derailment, suspected arson, electrical, brake shoe sparks or overheated brakes, welding and cutting, hot journals, and heating appliances.

Design advances and preventive maintenance are keys to reducing or eliminating fires caused by brake shoe sparks, exhaust sparks, hot journals or "hot boxes," electrical failures, and hot particles emitted from internal combustion engines. Recent design advances that should lessen the frequency of fires include increased use of turbochargers; renewed attention to the design and maintenance of eductor tubes, long a source of hot carbon particles on turbocharged locomotives; spark retention exhaust manifolds; nonsparking brake shoes; roller bearings; heat detectors, sensors, and indication lamps that supervise electrical circuits and components; and increased use and efficiency of trackside devices to detect overheated journals and dragging equipment.

Cotton bales now are shrink-wrapped, which eliminates the need for metal bale bands, and virtually eliminates cotton bale fires. In the past, bale fires were a common occurrence, caused by friction heat, which resulted when the bale bands rubbed against and ignited loose strands of cotton.

Fires resulting from collisions and derailments are a major concern of the railroad industry. To reduce these incidents, operating rules and safety programs are undergoing major revisions. A program to inventory all grade crossings was completed recently, and the inventory is being used by individual states to establish priorities for crossing projects. This systemized approach to reducing crossing accidents, combined with the Federal Aid Highway Act of 1976 and the "Operation Lifesaver" educational program, is keeping the public alert to the dangers associated with railroad crossings.

Refrigerator Cars

Three types of cars commonly transport perishable and semiperishable commodities that require refrigeration, heater service, or protection from the extremes of heat or cold.

1. The standard ice car (RS) is virtually obsolete as a refrigerator car. Few railroads still furnish ice for shipment in such cars. However, some cars still may be used to keep commodities from

freezing. Portable liquid fuel heaters that burn a mixture of methanol and isopropanol can be installed in the end ice bunkers. Alcohol foam is recommended as an extinguishing agent for these fuels because water may be ineffective.

2. The insulated bunkerless car (RB) is widely used to protect semiperishable commodities such as canned goods, beer, grocery products, and drugs and medicines from extreme heat and cold. These cars provide no refrigeration, but they can be equipped with portable liquid fuel heaters. The heaters are usually suspended by chains from hooks, four per heater, in the ceiling of doorway areas. Heat shields protect the ceiling above each heater.

3. The mechanically refrigerated car (RP) is the most modern and versatile of the freight cars used for perishables. (See Figure 9-35C.) It provides automatically controlled temperatures from 0 to 70°F (–18 to 21°C). Foamed-in-place polyurethane insulation helps to ensure consistent and dependable car temperatures. A compressor-evaporator is powered by a self-contained diesel-engine generator set located in a separate compartment within the car. The compressor-evaporator also can be powered directly from the alternator output of a diesel electric unit, or from any suitable 220 V standby power supply. The fuel tank for the diesel engine is suspended below the car and normally carries 500 to 550 gal (1.9 to 2.1 m³) of No. 1 or No. 2 diesel fuel oil. Fire fighting can be complicated by burning insulation and by the fact that fire fighters have not learned the location of the diesel engine stop control, which is clearly identified. Fire fighters should wear self-contained breathing apparatus when they enter a burning insulated or refrigerator car.

Box Cars

The design of all box cars is essentially the same; only the capacity, length, and details of the cars vary. Additions, such as special racks, dunnage devices, and nailable steel floors, are made to prevent vibration damage to the contents of the car without adding combustibility.

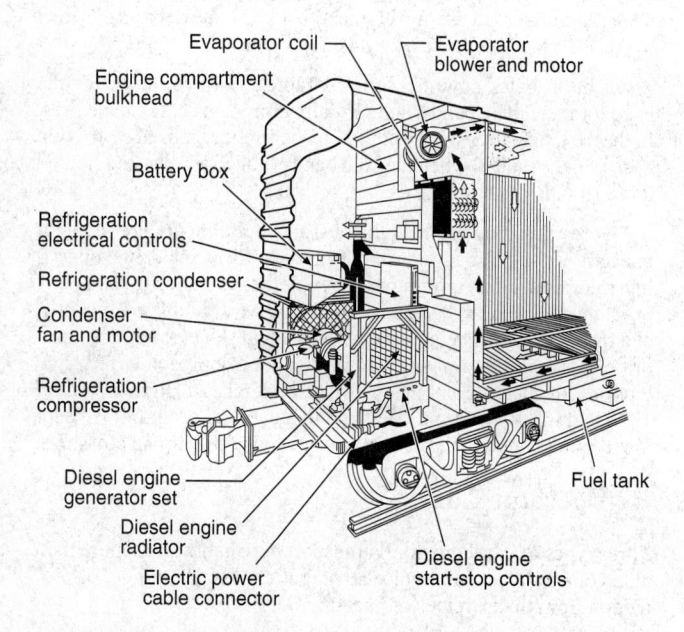

FIG. 9-35C. *The engine compartment of a typical mechanically refrigerated car. Note the start-stop control for the diesel engine at the lower left, mounted on the side of the car.*

Special Purpose Rail Cars

Equipment used to meet special shipping needs includes articulated cars, multilevel and enclosed cars for automobiles, "high cube" (large-capacity) box cars, 100-ton (90 metric ton) capacity covered hoppers, side-loading "all door" box cars, and aerated cars to handle bulk materials that require pneumatic loading and unloading mechanisms to "inhale" cargo at the loading point and "exhale" it at the destination.

"Intermodal service," in which specially designed flatcars carry loaded cargo containers, intermodal tank containers, or highway-type truck trailers, is an increasingly important part of the rail transportation business. Loaded and sealed, containers and trailers are delivered to the railroad at special intermodal terminals, where they are placed, usually by crane, onto the flatcar. Intermodal cars usually are part of a regular train, but intermodal-only trains are increasingly common. A single intermodal flatcar carrying two trailers, or as many as four cargo containers, usually has a wider variety of cargo than a single boxcar.

Cabooses

Once every train had a locomotive at one end and a caboose at the other. The caboose was the traveling office of the conductor, the person in charge of the train, and a rear brakeman. Today, cabooses are a rarity; end-of-train electronic devices have taken over their "train-completion" function. Now only two or three persons riding in the lead locomotive customarily staff long-haul trains. Cabooses today are found only on an occasional branch-line or short-line train.

Cabooses' chief fire hazards are their combustible furnishings and heating systems. Most cabooses are heated with oil or coal. Fires can start because stoves and stove pipes do not have adequate clearances from combustible materials. Fire protection varies from water extinguishers to portable dry chemical extinguishers, although some carriers have discontinued this protection because of continuing problems with extinguisher theft.

HAZARDOUS MATERIALS IN RAIL TRANSPORTATION

Hazardous material shipments represent less than 5 percent of the total carloads of freight moving in rail transportation. They are strictly regulated by the U.S. DOT through the Research and Special Programs Administration (RSPA). The FRA enforces these regulations.

Hazardous Material Traffic

In 1995, U.S. and Canadian railroads transported 1,785,547 hazardous material shipments. (See Figure 9-35D.)

During 1995, twenty-five commodities accounted for 72 percent of all hazardous materials shipped by rail. (See Table 9-35A for a list of these materials by hazardous material response code, name, and volume.) The highest ranked shipment by rail is mixed loads in nonbulk packaging transported in trailers and containers. These shipments vary from a single package of a single hazardous material to an entire load of a variety of hazardous materials.

Although hazardous materials are transported in various types of rail cars, i.e., covered hoppers, gondolas, box cars, and trailers and containers on flat cars, tank cars transport 64 percent of hazardous material shipments.

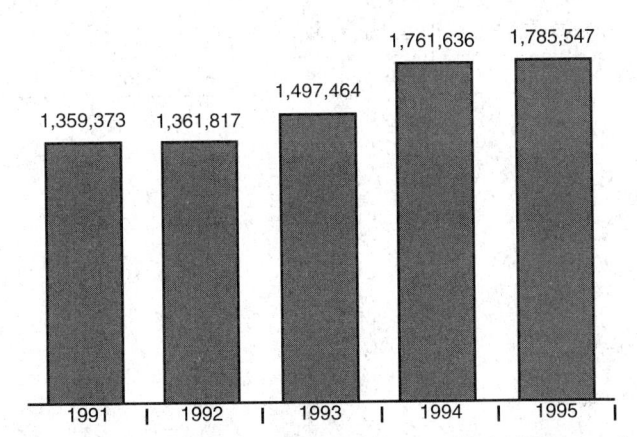

FIG. 9-35D. *Hazardous material shipments, United States and Canada.*

Hazardous Material Releases

In 1995, the AAR and the Bureau of Explosives (BOE) reported 1347 releases of hazardous materials in rail transportation in the United States. These statistics were reported as either accident releases or nonaccident releases. (See Figure 9-35E.)

An accident release is the release of a hazardous material in an accident. Accidents include collisions, derailments, fires, explosions, acts of nature, or other events involving rail equipment and resulting in more than $6300 in damages (1995). An accident does not always involve hazardous materials.

A nonaccident release is any other release of hazardous materials during rail transportation, e.g., a release from unsecured fittings or venting from pressure relief devices.

People tend to recall the few, well publicized derailments resulting in loss of life (Waverly, TN, 1978), environmental damage (Dunsmire, CA, 1991), or mass public evacuation (Superior, WI, 1992). While these releases are certainly newsworthy, they do not represent the typical rail hazardous material release.

A detailed examination of FRA and AAR statistics reveals that the "typical" rail hazardous material release involves a:

• Nonpressure tank car

• Class 8 (corrosive material) or Class 3 (flammable liquid)

• Nonaccident release

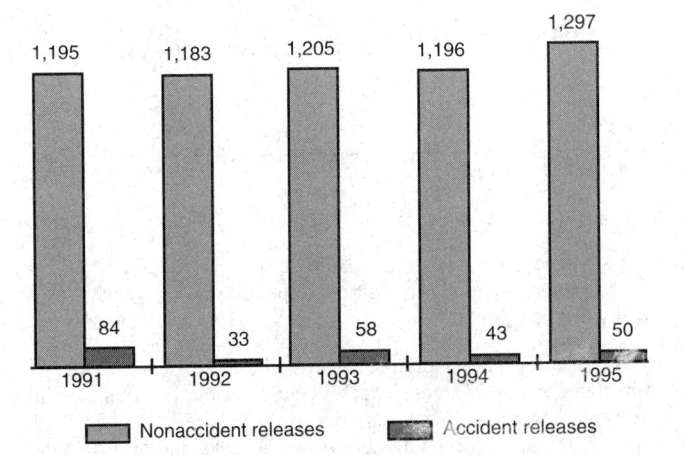

FIG. 9-35E. *Hazardous material releases by rail, United States. (Source: AAR/Bureau of Explosives)*

• Release of a small amount [generally less than 100 gal (378 L)] of hazardous material

• Fitting that is readily accessible and easily repairable

Handling Hazardous Material Releases

With the increased danger associated with emergencies involving hazardous materials, shippers, carriers, and local, state, and federal agencies, i.e., fire, police, health, and environmental agencies, respond. The immediate purpose of the response is to minimize harm that would otherwise occur: death, injury, property damage, environmental damage, and system disruption. To accomplish this task, responders must identify hazardous materials involved in the emergency.

Sources for detecting the presence of hazardous materials include placards and markings and shipping papers. Placards are affixed to both sides and both ends of the rail car, trailer, or container. They alert response personnel to a hazardous material's presence and indicate the hazard class. The DOT identification number, also found on both ends and both sides of the rail car, trailer, or container, may cross-reference the specific material involved, along with emergency response information. Another marking, the name of the contents, may be stenciled on the sides of the rail car, trailer, or container. Due to the orientation of the cars, placards and markings may not be visible in some emergencies. Because of the nature of the emergency, it may not be prudent to approach the car closely enough to read the placards and markings. Aided vision (binoculars, etc.) may be helpful in some cases.

Rail cars, trailers, or containers transporting hazardous materials must be documented on shipping papers. The typical shipping paper for a train movement is a train consist. This document lists the location of each rail car, trailer, or container in the train. It indicates the contents of each car, including the hazardous material shipments. Shipping papers may differ slightly among railroads; therefore, response personnel should be familiar with the railroads that move through their jurisdiction.

The rail shipping paper typically includes the following entries: proper shipping name, hazard class, identification number, packing group, quantity, and hazardous material response codes [previously the 49 series Standard Transportation Commodity Code (STCC)]. A 24-hr emergency response telephone number and detailed emergency response information are included with the shipping papers.

Shipping papers are available from the conductor on the train crew on site or from the train dispatcher. Although traditionally found on the caboose, the conductor will be found in the lead locomotive of trains without cabooses. However, in an emergency involving hazardous material, the conductor and the rest of the train crew are likely to move to a safe location one half mile upwind, uphill, or upstream of the problem.

Upon notification of an accident or other occurrence involving rail equipment, response personnel should take the following actions:

1. Contact the responsible railroad. The number should be identified in pre-emergency planning activities. Obtain a copy of the shipping papers and emergency response information—whether hard copy or from the railroad verbally.
2. Obtain as much information about the incident as is available: identify hazardous materials involved by name, DOT identification number, or placard applied; determine the types of containers involved; determine the topography, bodies of water, population density, etc.; obtain specific emergency location and best avenues of approach; and ascertain whether any injuries have occurred; determine when the emergency occurred.

TABLE 9-35A. *Top 25 Hazardous Materials in Rail Transportation, 1995*

Rank	Haz Mat Code	Class	Commodity Name	Shipments	Total
1	4950150	—	Mixed loads	278,447	511,072
	4950130	—	Mixed loads	188,377	
	4950110	—	Mixed loads	3,792	
2	4905752	2.1	Liquefied petroleum gas	111,693	157,424
	4905721	2.1	Petroleum gases, liquefied	5,715	
	4905782	2.1	Propylene	5,676	
	4905791	2.1	Liquefied petroleum gas	5,144	
	4905784	2.1	Propylene	4,714	
	4905723	2.1	Petroleum gases, liquefied	3,748	
	4905766	2.1	Petroleum gases, liquefied	3,271	
	4905789	2.1	Butane	2,985	
	4905741	2.1	Petroleum gases, liquefied	2,786	
	4905724	2.1	Petroleum gases, liquefied	1,970	
	4905781	2.1	Propane	1,606	
	4905748	2.1	Isobutylene	1,593	
	4905730	2.1	Petroleum gases, liquefied	1,585	
	4905767	2.1	LPG (butylene/isobutane)	1,484	
	4905768	2.1	LPG (butane/butylene)	1,193	
	4905753	2.1	Isopentane	1,182	
	4905739	2.1	Liquefied petroleum gas	1,076	
3	4935240	8	Sodium hydroxide *(caustic soda)*		88,818
4	4945770	9	Molten sulfur		69,355
5	4930040	8	Sulfuric acid		68,708
6	4941196	9	Elevated temperature material	52,588	60,660
	4966311	9	Elevated temperature material	3,748	
	4941139	9	Elevated temperature material	3,131	
	4941273	9	Elevated temperature material	1,193	
7	4904210	2.2	Anhydrous ammonia		59,239
8	4904523	2.3	Chlorine		48,351
9	4909230	3	Methyl alcohol		30,404
10	4905778	2.1	Vinyl chloride	13,849	27,208
	4905792	2.1	Vinyl chloride	13,179	
11	4930247	8	Phosphoric acid		25,773
12	4912210	3	Fuel oil	22,255	25,167
	4912620	3	Fuel oil	2,912	
13	4909152	3	Denatured alcohol	18,153	20,261
	4909151	3	Denatured alcohol	2,108	
14	4907265	3	Styrene monomer, inhibited		18,362
15	4930228	8	Hydrochloric acid		17,497
16	4904509	2.2	Carbon dioxide, refrigerated liquid		16,833
17	4915164	CL	Fuel oil, diesel	11,324	14,150
	4914851	CL	Fuel oil, diesel	1,356	
	4914847	CL	Fuel oil, diesel	1,470	
18	4918311	5.1	Ammonium nitrate fertilizer		11,373
19	4918723	5.1	Sodium chlorate		10,696
20	4908175	3	Gasoline		10,744
21	4921598	6.1	Phenol		9,175
22	4905704	2.1	Butadiene		9,016
23	4908224	3	Methyl tert butyl ether		8,960
24	4910165	3	Crude oil, petroleum		7,878
25	4966110	9	Adipic acid		6,969
Total top 25 commodities					1,293,667

*Source: AAR/Bureau of Explosives Annual Report of Hazardous Materials Transported by Rail—1995

3. Upon arrival at the scene, establish contact with the senior railroad official or representative before attempting to deal with the emergency.
4. Initiate command and control activities, including protecting yourself and others, initiating an incident command system, controlling access to the site, and making appropriate notifications.

Other rail-oriented assistance available includes:

AAR Bureau of Explosives: BOE personnel will respond to major rail emergencies involving hazardous materials on request, typically from a railroad.

Carrier emergency response teams: Some carriers have developed emergency response capabilities to handle emergencies involving hazardous materials. Such teams are equipped with specialized tools and equipment to repair leaks and even transfer contents of damaged tank cars.

Many rail carriers prepare guides and recommendations for handling rail emergencies involving hazardous materials and make them available to local emergency response agencies along their right-of-way. Railroads also present various training programs to local response agencies and assist at various fire school programs.

Tank Car and Intermodal Tank Container Fire Hazards

Fire hazards of tank cars and intermodal tank containers are minimal unless a car is damaged in an accident. If a fire does occur, it is essential to learn the contents of the car or the container before attempting to fight the fire. A tank car or a container exposed to radiant energy from a fire, particularly direct flame impingement, should be approached cautiously. If a commodity absorbs too much heat, vapor pressure builds up and causes the safety valve to operate or the tank to rupture. If any of the commodity is released, personnel should be prepared to deal with corrosive, toxic, or otherwise harmful effects. Direct contact with vapors discharged from tank cars or containers under emergency conditions should be avoided, unless appropriate protective clothing and self-contained breathing apparatus are worn.

RAIL TANK CARS

More than 200,000 tank cars transport 64 percent of the hazardous materials moving in rail transportation. Nonrailroad companies own most of these tank cars.

A tank car is a type of bulk packaging in which the tank is an integral part of a rail transport vehicle. Tank cars transport a diverse range of commodities—both hazardous and nonhazardous—including a variety of compressed and liquefied gases, liquids, and solids. The number and types of materials transported in a tank car depends on the size, construction, fittings, linings, and other physical features of the tank.

When faced with a tank car problem, communicating an accurate, detailed description of its contents, its condition, and other circumstances to tank car specialists not on the scene may be the most beneficial contribution an emergency responder can make. To help provide that information, the following paragraphs describe tank car construction and features, tank car markings, types of tank cars, and tank car fittings.

Tank Car Construction and Features

Typically, tank cars are enclosed longitudinal cyclinders (with or without compartments) with rounded ends, called heads. Tank cars may or may not have underframes. If not, the tank must bear the stresses of train movement. Tank cars typically have railroad running gear and safety appliances. Figure 9-35F shows the shape and components of typical tank cars.

Tank cars may have such features as insulation, thermal protection, linings, and claddings. Manufacturers add these features to a basic tank, depending on the physical properties of commodities the tank is to carry. Safety features are applied to tank cars to reduce the potential for a release. Older tank cars have been retrofitted with these newer safety features.

Tank cars are built to mechanical standards common to all rail freight cars. In addition, they must meet the requirements of both the *Hazardous Materials Regulations of the Department of Transportation* (49 CFR, Part 179) and the AAR *Manual of Standards and Recommended Practices, Section C—Part III, Specifications for Tank Cars.* They also meet or exceed the ASME standards for pressure vessels. Builders must obtain design approval from the AAR Tank Car Committee before building a tank car. Transport

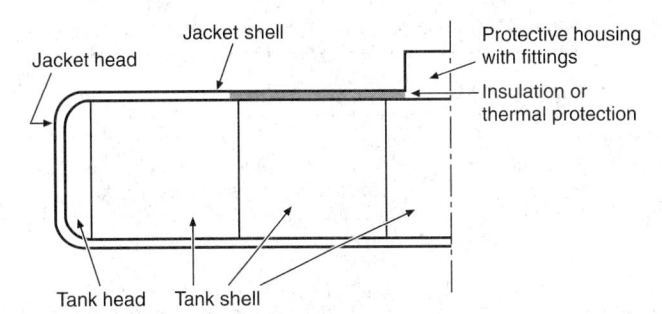

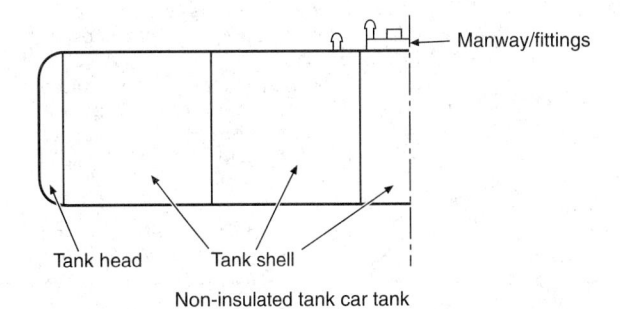

FIG. 9-35F. Tank car tank design and nomenclature.

Canada's *Regulations for the Transportation of Dangerous Commodities by Rail* contains Canadian tank car requirements.

Repairs of tank car tanks requiring welding, riveting, removal of damage, alterations, or conversions must occur only at AAR-certified facilities, using procedures and materials the AAR Tank Car Committee has approved.

Tank car tanks: Each tank car tank consists of a cylindrical shell closed at the ends by heads. The shell is constructed of two to seven rings formed from metal plates. Heads are made from plates pressed into a flanged-and-dished or an ellipsoidal shape; 2 to 1 ellipsoidal heads are the strongest. The rings and heads are fusion-welded together. In some older nonpressure tank cars, the tank parts are riveted together.

The completed tanks are heated to 1100°F (593°C) for 1 hr to relieve metal stresses caused by welding. Representative samples are X-rayed to identify possible flaws in the metal or the welds. Finally, they are hydrostatically tested.

Carbon steel is used in over 90 percent of tank car tanks, with aluminum making up most of the remainder. Stainless steel, referred to in the regulations as alloy steel, is used in a smaller number of cars. Nickel or nickel alloy is used for some tanks in acid or food services. Regulations specify the plate thickness of materials used to construct tank car tanks. (See Table 9-35B.)

The capacities of tank car tanks vary from 4,000 gal (15 m³) to approximately 45,000 gal (170 m³) in jumbo tank cars. Since 1970, however, the maximum capacity of new DOT specification tank cars transporting regulated hazardous materials is limited to 34,500 gal (130 m³), or not more than 263,000 lb (119,295 kg) gross weight on rail.

Tank car tanks may be divided into as many as six compartments, each of which is actually a separate tank. Compartments may be of different capacities and may transport different commodities. (See Figure 9-35G.)

TABLE 9-35B. *Minimum Tank and Jacket Plate Thickness*

Minimum Plate Thickness After Forming	Common Use of Plate Thickness
Steel	
11 gage (approximately $1/8$ in.), also aluminum	Jacket of insulated tank cars, or jacket for thermally protected cars
$7/16$ in.	Tank for nonpressure tank cars, outer tank for nonpressure tank within a tank, or shell portion of outer tank for cryogenic liquid tank cars
$1/2$ in.	Head puncture resistance (head shield), or head portion of outer tank for cryogenic liquid tank cars
$9/16$ in.	Tank for steel pressure tank cars with tank test pressures of 200 psi and below
$11/16$ in.	Tank for steel pressure tank cars with tank test pressure of 300 psi and greater
$3/4$ in.	Tank for steel pressure tank cars in chlorine service
Aluminum	
$1/2$ in.	Tank for nonpressure aluminum tank cars
$5/8$ in.	Tank for aluminum pressure tank cars

Notes:
1. If high-tensile-strength steels are used, plate thickness for pressure tank car tanks may be reduced but in no case should that thickness be less than $1/2$ in.
2. Plate thickness for nonpressure steel tank cars with expansion domes is a function of where the plate is used in the tank and the diameter of the tank. Thickness ranges from $1/4$ to $1/2$ in. For tank cars built after 1969, minimum plate thickness is $7/16$ in., except for tank cars with a diameter of 112 to 122 in. where the thickness is $1/2$ in.
3. For SI units: 1 in. = 2.54 cm; 1 psi = 6.9 kPa.

Car structure: The completed tank is mounted on a car structure. The car structure consists of body bolsters attached to a framework (either continuous or stub sill). The framework rests on trucks at each end. The structure of the tank car includes the following components:

1. *Underframe:*
   a. Either a continuous underframe in which a one-piece assembly bridges the trucks of a tank car and assumes the draft and buff forces associated with train movement; or

   b. Stub sill (underframeless), in which a short, longitudinal structural member welded to each end of the tank accommodates the coupler and draft gear assemblies and the attachment of the tank to its trucks.

2. *Bolster.* In which a structural crossmember is mounted transversely to cradle the tank. Tank cars have two body bolsters, one at each end of the tank.

3. *Truck assembly.* Which consists of wheels, axles, sideframes, springs, truck bolster, and center bowl and pin. It supports one end of the car and allows it to be moved on the rail.

Features: To increase their usefulness for a wide range of materials, tank cars may be equipped with various features, such as insulation, thermal protection, jackets, linings and claddings, and heater coils.

1. *Insulation.*   Insulation is used on some tank cars to moderate the effects of temperature on the cars' contents. Fiberglass or polyurethane foams are the most common insulating materials. However, cork insulates tank cars carrying hydrocyanic acid, and perlite insulates cars transporting refrigerated liquid gases.

2. *Thermal protection.*   Thermal protection, not to be confused with insulation, is used on some tank cars to protect the tank from flame impingement. It is designed to keep tank metal temperatures below 800°F (427°C) for 100 min (pool fire impingement) or 30 min (torch fire impingement). Thermal protection is used primarily on tank cars transporting liquefied flammable gases, and on some nonpressure tank cars, such as the DOT 111J100W4 tank car for ethylene oxide. Two types of thermal protection are used:
   a. Jacketed thermal protection, in which mineral wool or various manmade ceramic fiber blankets are held in place by a metal jacket.
   b. Sprayed on thermal protection, in which rough-textured coating sprayed onto the tanks or the jacket in some jacketed cars protects by expanding on exposure to fire.

3. *Jacket.*   Both insulation and jacketed thermal protection are held in place by a jacket, which also keeps out the elements. The jacket is constructed of a minimum of 11-gage (approximately $1/8$-in.) steel and serves as a heat-radiation shield. Wood blocks hold the jacket away from the tank.

   Jacketed cars can be recognized by a metal protective cover, or flashing, over the body bolster or tank bands; by the flattened appearance of the ends of the tank car; by flat sections on the

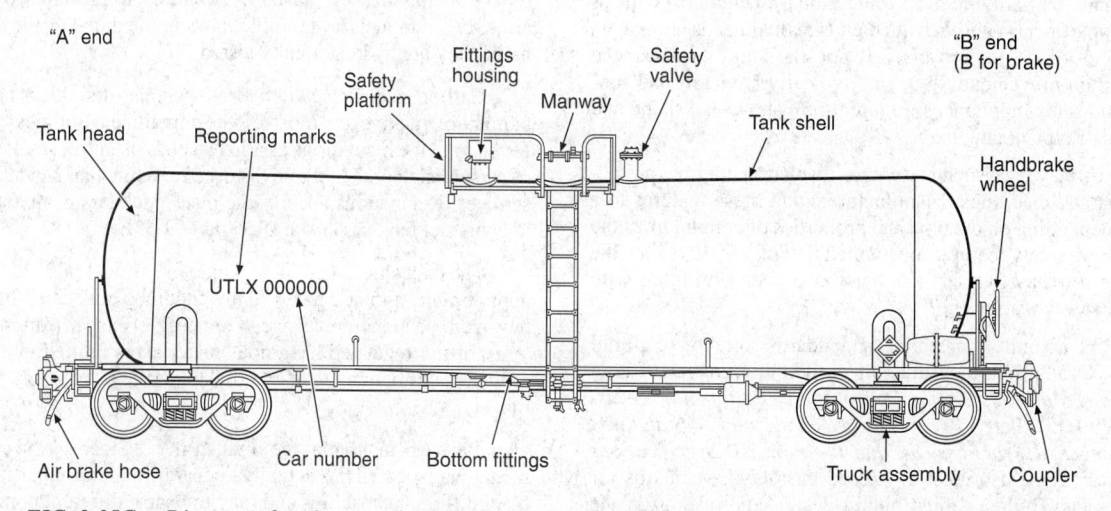

FIG. 9-35G. Diagram of tank car tank with some basic features identified. (Source: Union Tank Car Company)

sides of the car; or by the rough appearance of visible welds, even lap welds, which are generally thinner than the tank weld.

4. *Linings and claddings.*   Some tank cars are lined to protect the tank from the corrosive or reactive effects of the contents or to maintain the contents' purity.
   a. Lining materials are applied after the tank is constructed. Rubber is most commonly used in hazardous materials service, but glass, lead, nickel, polyurethane, polyvinyl chloride, and stainless steel are also used.
   b. Cladding materials are applied to the tank base metal before the plate is formed. Nickel and stainless steel are used as cladding materials.

5. *Heater coils.*   Some tank cars are equipped with heater coils, at most 16, either inside or outside the tank. Steam, water, or hot oil from an external source runs through these coils to heat thick or solidified materials, such as asphalts, fused solids, heavy fuel oils, phenol, metallic sodium, or petroleum waxes to make them flow more easily when unloading. Outlets and inlets for heater coils have caps, which are in place when the car is loaded so that any damage to the coils does not release contents. Caps are off when the car is emptied.

Safety Programs

Various safety programs have been instituted for tank cars to improve the safety of new and existing cars. The next few sections provide examples of these programs.

Head puncture resistance (head shields):   The head shield is applied to the lower portion of both ends or may cover the entire head to protect it against punctures. Some head shields are built into the jacket, while others are visible either as "half head" or trapezoidal-shaped plates of steel mounted on both ends of the tank car.

Top and bottom shelf couplers: A top and bottom shelf coupler is a type of coupler with a vertical restraint mechanism, top and bottom, to reduce potential coupler disengagement in an accident.

Lifting lugs: The lifting lugs are reinforced lifting points or lifting provisions for uprighting tank cars involved in derailments.

Bottom discontinuity, or skid, protection: Skid protection is a safety device applied to the bottom of tank cars to protect bottom outlets, washouts, and sumps in a derailment.

Tank Car Markings

Various markings found on tank containers include initials, or reporting marks, and a tank number; a specification marking; a DOT exemption marking; and tank and valve test dates. (See Figure 9-35H.)

Initials, or reporting marks, and tank number: Tank cars, like any other freight car, are identified by the car's initials, which are called reporting marks, and tank number. The initials and number are found on both sides and both ends of a tank car. As one faces the side of the car, the initials and number are found on the left.

The initials indicate ownership of the car. An "X" as the last initial represents private ownership; that is, the railroad does not own the car. The number following the initials identifies the specific car.

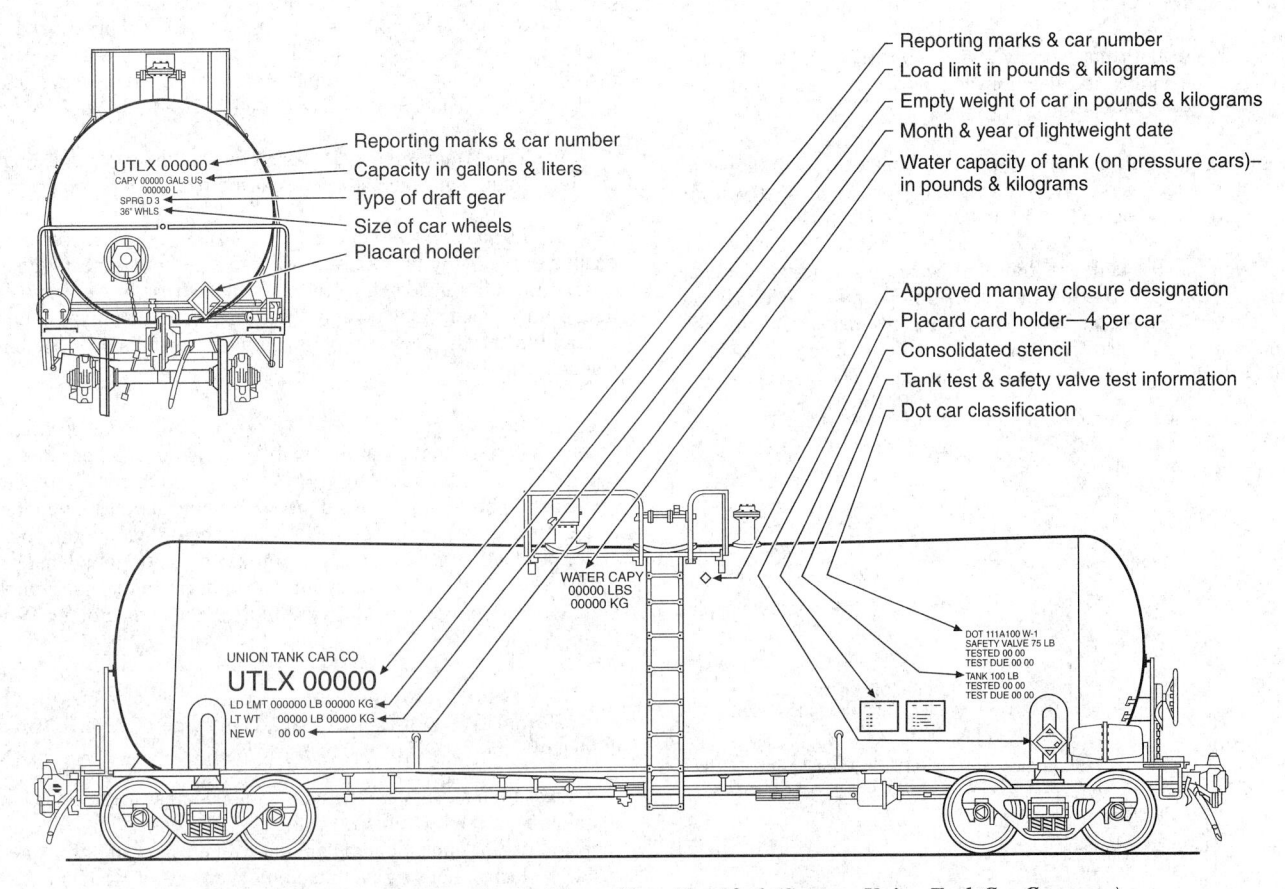

FIG. 9-35H.   Diagram of tank car with markings identified. (Source: Union Tank Car Company)

Using the car's initials and number, shippers, carriers, and emergency response personnel can identify a tank car's contents with shipping papers or computer data. They can then determine whether its contents are hazardous.

Specification marking: A specification marking identifies the standards to which a tank car is built. The specification marking is stamped into the head of the tank and stenciled on the car's sides. As one faces the side of the car, the specification number is to the right, opposite its initials and number.

The specification marking consists of a three-letter prefix, followed by a series of numerals and/or letters. Specification markings may be as simple as DOT-103 or as detailed as DOT-111A60ALW1. A specification marking seldom identifies the car's contents, but does indicate how it was constructed. Specification information can be obtained from the railroad's computer using the car's initials and number. (See Figure 9-35I.)

DOT exemption marking: When exemptions from the regulations for packages and containers, and the preparation and offering of hazardous materials for shipment are made, the outside of each package must be plainly and durably marked "DOT-E," followed by the number assigned, such as "DOT-E8623." On portable tanks, cargo tanks, and tank cars, the marking must be in 2-in. (50.8-mm) letters.

Tank and valve test dates: Tank and safety valves, if installed, have varying retest intervals, depending on the car's specification and/or age. Retest and test due dates must be marked or stenciled on the tank or on the dataplate. Interior coils are also tested.

Tank Car Types

Tank cars are classed according to their construction, features, and fittings. The tank's specification determines the product it may transport. (See Table 9-35C.)

Nonpressure tank cars: Nonpressure tank cars, also known as general-service or low-pressure tank cars, transport a wide variety of hazardous and nonhazardous materials at low pressures.

Nonpressure tank cars transport hazardous materials, such as flammable and combustible liquids, flammable solids, oxidizers, organic peroxides, poison B materials, corrosive materials, and molten solids. They also transport nonhazardous materials, such as tallow, fruit and vegetable juices, tomato paste, and caramel.

Tank test pressures for nonpressure tank cars range from 35 to 100 psi (251 to 689 kPa). Capacities range from 4000 to 45,000 gal (15 to 170 m³).

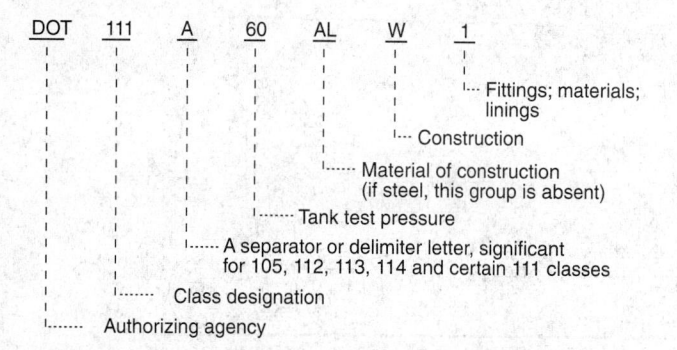

FIG. 9-35I. *Explanation of specification marking.*

TABLE 9-35C. *Tank Car Classes*

Nonpressure tank car classes

DOT-103	DOT-111	AAR-201	AAR-206
DOT-104	DOT-115	AAR-203	AAR-211

Pressure tank car classes

DOT-105	DOT-112	DOT-114	DOT-120
DOT-109			

Cryogenic liquid tank car classes

DOT-113	AAR-204

Miscellaneous tank car classes

DOT-107	High pressure tube car
AAR-207	Pneumatically unloaded hopper car

Nonpressure tank car tanks are cylindrical with rounded heads. Newer nonpressure tank cars have at least one manway to allow access to the car's interior. Some older nonpressure tank cars have at least one expansion dome with a manway. (See Figures 9-35J and 9-35K.) Fittings for loading and unloading, pressure and/or vacuum relief, gauging, and other purposes are visible at the top and/or bottom of the car.

Nonpressure "tank-within-a-tank" tank cars. One nonpressure tank car is built as a tank within a tank. It consists of an inner tank of steel, alloy steel, or aluminum covered with thick insulation and enclosed within an outer tank. This tank car transports temperature-sensitive materials, such as food products.

Tank Train®. The Tank Train® is a series of nonpressure tank cars interconnected with flexible hoses that allow the cars to be loaded and unloaded from one end. A spring-loaded butterfly valve arrangement on each car is controlled pneumatically from the loading/unloading point. After loading, hoses are purged of liquid, and valves close automatically, isolating each car.

The Tank Train® transports hazardous and nonhazardous materials, including gasoline, fuel, oil, caustic soda, and pesticides.

Multicompartmented nonpressure tank cars. Multicompartmented nonpressure tank cars have up to six compartments. Each compartment is considered a separate tank because of its construction. Compartments may be of different capacities and may contain different materials. Each compartment has its own loading and unloading fittings.

Pressure tank cars: Pressure tank cars typically transport hazardous materials, including flammable, nonflammable, or poisonous gases at higher pressures. However, pressure tank cars can transport other commodities, depending upon the characteristics of the product or the process for loading and unloading the tank. Other products transported in pressure tank cars are ethylene oxide, pyrophoric liquids, sodium metal, motor fuel anti-knock compounds, bromine, anhydrous hydrofluoric acid, and acrolein. (See Figure 9-35L.)

Tank test pressures for these tank cars range from 100 to 600 psi (689 to 4137 kPa). Pressure tank cars range in capacity from 4000 to 45,000 gal (15 to 170 m³).

Pressure tank cars are cylindrical, noncompartmented steel or aluminum tanks with rounded heads. They are top-loading, with fittings for loading and unloading, pressure relief, and gauging located inside protective housing mounted on a manway, of which there is typically one.

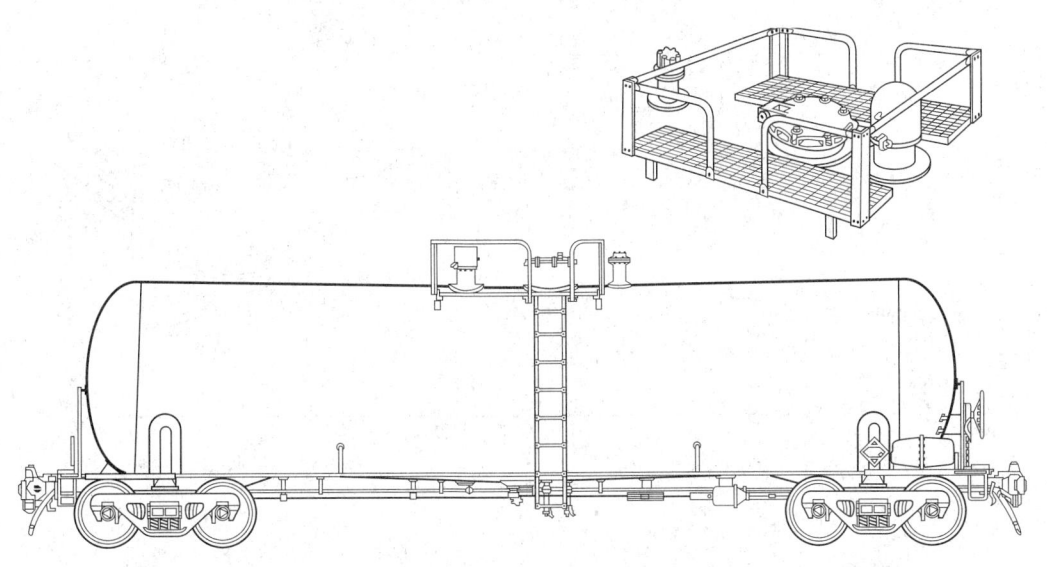

FIG. 9-35J.   Diagram of a typical nonpressure tank car without an expansion dome.

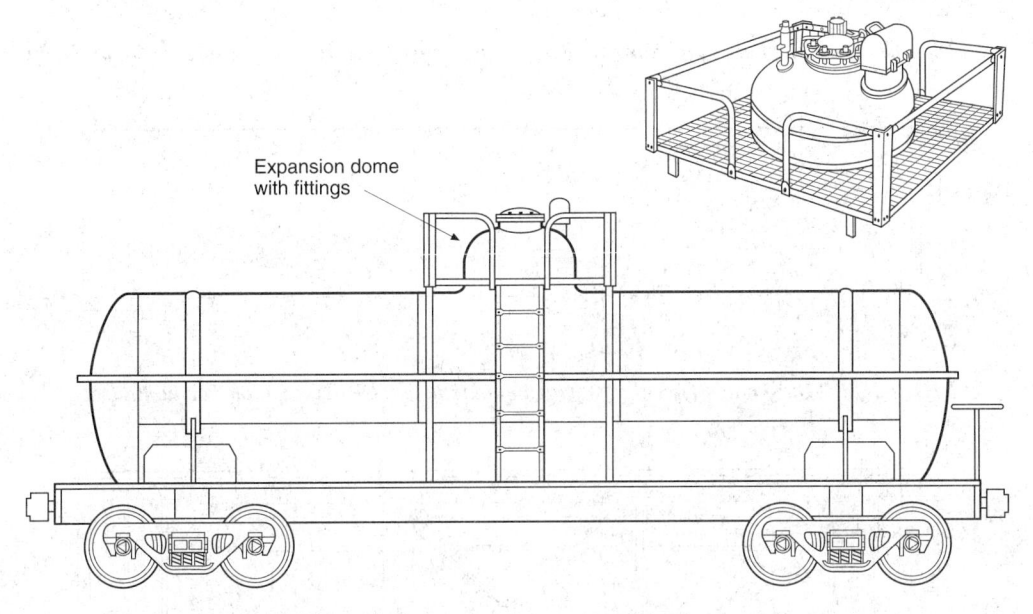

Expansion dome
with fittings

FIG. 9-35K.   Diagram of a typical older style nonpressure tank car with an expansion dome. (Source: Union Pacific Railroad)

Pressure tank cars may be insulated and/or thermally protected. The top two-thirds of pressure tank cars without insulation and without jacketed thermal protection will be painted white.

Specialized tank cars: Several other types of rail equipment listed under the tank car regulations do not fit the usual image of a tank car. These include cryogenic tank cars, high-pressure tube cars, and pneumatically unloaded covered hopper cars.

Cryogenic tank cars.   Cryogenic tank cars carry low-pressure—usually 25 psi (172 kPa) or lower—cold liquids refrigerated to −155°F (−104°C) and below. Typical contents of these tanks are argon, ethylene, hydrogen, nitrogen, and oxygen. (See Figure 9-35M.)

Cryogenic tank car tanks are of the tank-within-a-tank style, with an alloy (stainless or nickel) steel inner tank inside an outer

tank. The space between the inner tank and the outer tank is filled with insulation and under a vacuum. The combined vacuum and insulation protect the contents for only 30 days, making shipments in these cars time sensitive.

Fittings for loading and unloading, pressure relief, and venting are located in ground-level cabinets at diagonal corners of the car or in the center of one end of the car.

High-pressure tube cars.   The high-pressure tube car is a 40-ft (12-m) box-type, open-frame car. Inside the frame is a visible cluster of 30 seamless steel, uninsulated cylinders, arranged horizontally and permanently attached to the car. Tank test pressures of these tube cars range up to 5,000 psi (345,000 kPa). These tube cars transport helium, hydrogen, or oxygen in gaseous form. (See Figure 9-35N.)

A steel case encloses each end of the tube car. Loading and unloading fittings and safety devices are located in a walk-in cabinet

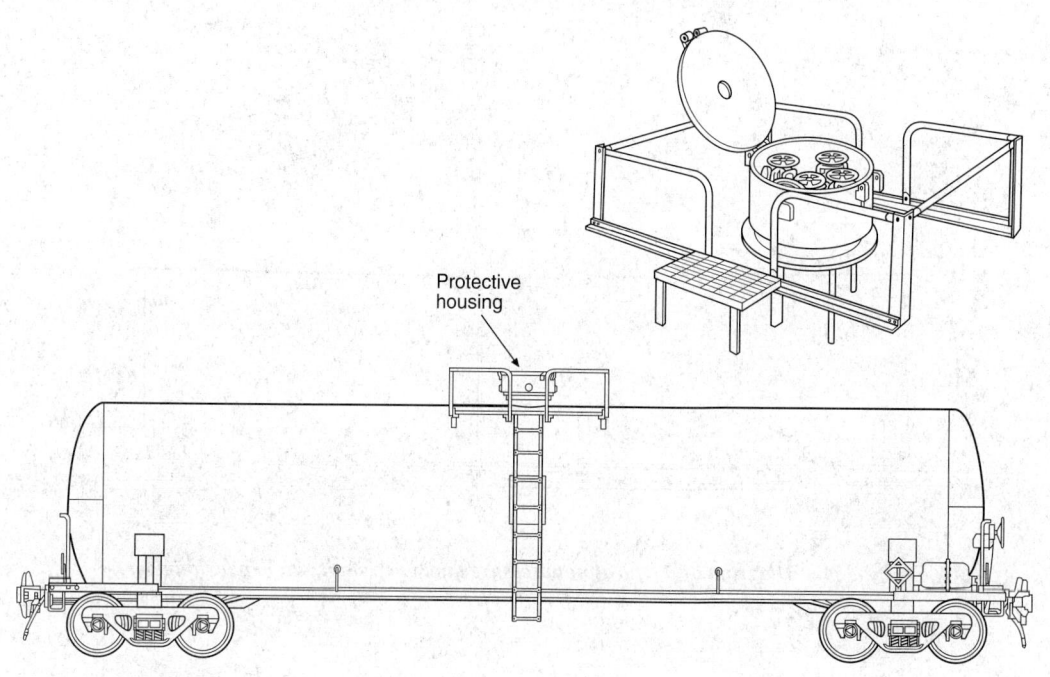

Protective
housing

FIG. 9-35L. *Diagram of a typical pressure tank car. Note the protective housing (inset). (Source: Union Pacific Railroad)*

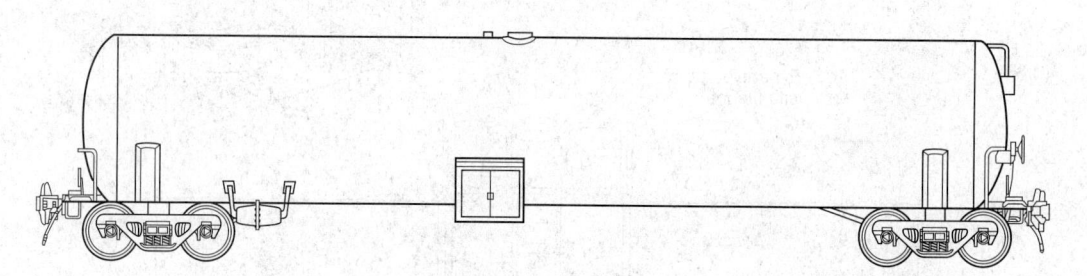

FIG. 9-35M. *Diagram of typical cryogenic liquid tank car. (Source: Union Pacific Railroad)*

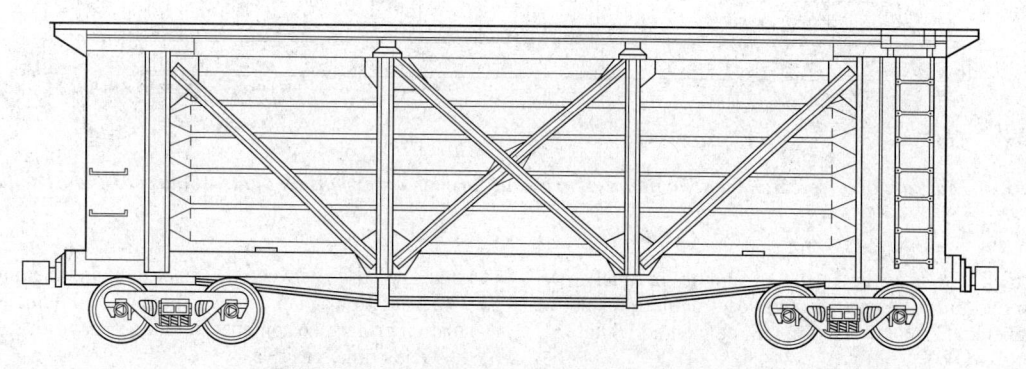

FIG. 9-35N. *Diagram of a high-pressure tube car.*

at one end of the car. Safety relief devices, either safety relief valves or safety vents, are found on each cylinder. For cars transporting flammable gases, safety relief devices are equipped with ignition devices to burn off any released material.

Pneumatically unloaded covered hopper cars. The pneumatically unloaded covered hopper car is built to tank car specifications. It is an covered hopper car unloaded through pressure differential, or pneumatics, by applying air pressure. Even though the pressure is used only during unloading, tank test pressures for

the car range from 20 to 80 psi (138 to 551 kPa). Dry caustic soda is one commodity transported in this type of car. (See Figure 9-35O.)

Tank Car Fittings

Numerous fittings used on tank cars allow for loading and unloading and provide for safe transportation of their contents. All fittings are not found on all tank cars.

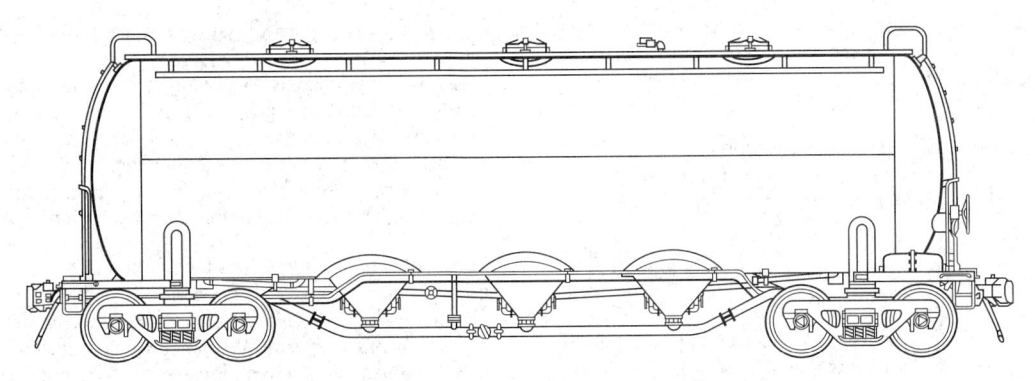

FIG. 9-35O. *Diagram of pneumatically unloaded covered hopper car.*

This section addresses tank car fittings for nonpressure and pressure tank cars from a very general standpoint.

Fittings for loading and unloading tank cars: Various fittings are used to load and unload tank cars, including the following.

Manway. This large opening on top of the tank car allows access to the tank's interior. It is closed either with a permanently mounted cover plate with fittings attached or with a hinged-and-bolted cover opened for loading the car.

Liquid or vapor valves. These valves control the passage of liquid or vapor into or out of the tank. (See Figure 9-35P.)

Liquid valves, usually two, are used to load or unload liquids from tank cars. Unloading occurs as a result of pressure generated by the contents or when the tank is pressurized with air, nitrogen, or other gas. An eduction pipe reaches from the liquid valve to the tank's bottom.

One or two vapor valves are used to remove vapor from the tank or to pressurize the tank. On pressure tank cars, they are called vapor valves; on nonpressure tank cars, they are called air inlets or air valves.

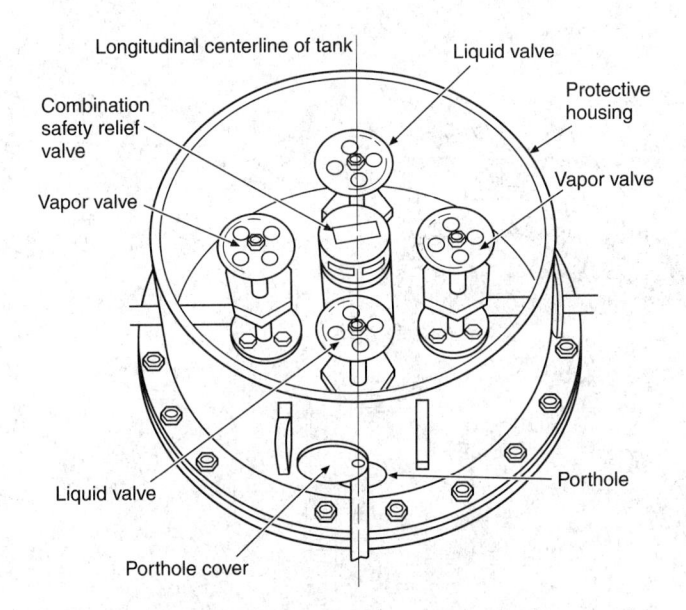

FIG. 9-35P. *Details of the fitting arrangement in the protective housing on a liquid chlorine tank car.*

(Note: Excess flow valves found under various fittings shut when flow exceeds a designed amount.)

Bottom outlet valves. These valves are used to unload the tank from the bottom. Examples are a plug-type valve operated from the top of the tank, called a "top-operated bottom unloading valve"; a ball-type valve either external to or inside the tank and operated from ground level; or a wafersphere, or butterfly valve, mounted outside the tank.

Safety relief devices: These devices reduce the buildup of excess internal pressure that commodity vaporization and expansion cause.

Safety relief valve. This relief device has an operating part held closed by one spring or multiple springs. It is designed to open and reclose at predetermined pressures. The device can be mounted with the spring inside or outside the tank.

Safety vent. This device uses a frangible disk, called a rupture disk, that seals the vent opening under normal conditions. The disk is designed to rupture at predetermined pressures and, once ruptured, does not reclose. Rupture disks are made of steel, plastic, rubber, or a combination of metal, plastic, or rubber. Lead is no longer authorized. Safety vents are used on certain nonpressure tank cars.

Combination device. This device uses a rupture disk and a breaking pin combined with a spring operated safety relief valve.

Fittings for gauging the innage and outage in tank cars: Gauging devices are used to determine the amount of lading in a tank. These devices allow one to measure the amount of liquid in the tank, such as liquefied or cryogenic gases (innage) or the amount of vapor space left in the tank (outage). Gauging devices usually measure outage. Therefore, two types of gauging devices:

1. *Open-type gauging devices.* This device ejects liquid once the liquid level reaches the bottom of a fixed- or adjustable-length tube. It reads outage.
2. *Closed-type gauging devices.* This device uses a float, coupled with a magnet on a measuring rod or a dial indicator to show the liquid level of the commodity.

Another method of determining the amount of lading in a tank car is to insert a measuring pole while using the appropriate innage/outage tables. Outage is also determined by gauge bars or "T" bars on most nonpressure cars.

Other fittings: Various other fittings may be found on tank cars.

Sample line. This device is used to obtain a sample of the commodity without opening the tank.

Thermometer well. This closed tube, filled with a permanent-type antifreeze, extends into the tank. By inserting a thermometer into the tube, the temperature of the lading inside may be determined.

Bottom washout. This permanently closed opening in the bottom of the tank car is used only when a tank is cleaned. There is no way to control the lading flow once the closure is removed.

Sump. This closed projection in the bottom of the tank allows the liquid eduction pipe to extend slightly below the bottom of the tank to unload the tank more completely.

Vacuum relief valve. This valve is designed to prevent excessive internal vacuum buildup [greater than 0.75 psi (5.5 kPa)] negative pressure in some nonpressure tanks. The vacuum relief valve used during unloading opens to admit air when a vacuum occurs and then recloses after normal conditions have been restored.

INTERMODAL TANK CONTAINERS

In the mid-1960s, intermodal tank containers or, simply, tank containers, were introduced in Europe. Currently, they are used worldwide, and projections indicate that the total number of tank containers may reach 90,000 by the year 2000.

Since the early 1980s, the use of tank containers in North America has increased greatly. Factors accounting for this increase are improved safety, portability, lower transportation costs, and benefits of a multimodal transport system.

Advantages over alternate methods of transport, such as drums, tank trucks, and trailers, include reduction in the amount of handling required, faster transfer at ports and terminals, convenient storage options at the destination, and ready use of conventional container-handling equipment already in place.

Built according to many domestic and international standards, tank containers are used to transport diverse commodities, including an increasing number of hazardous materials.

Problems involving tank containers are best handled by giving an accurate, detailed description of the contents and condition of the tank container, and the circumstances to those with specialized knowledge and skills necessary to help handle the problem.

Tank Container Construction and Features

Tank containers consist of a single, metal tank mounted inside a sturdy, metal supporting frame. This unique frame structure, built to international standards, makes tank containers multimodal (intermodal). Thus, they are used interchangeably in two or more modes of transport, such as rail, highway, or water. (See Figure 9-35Q.) Tank containers may be equipped with a variety of features.

Tank container tank: The tank container tank is generally built as a cylinder enclosed at the ends by heads. Other tank shapes—rectangular tanks—and configurations—tube modules—exist but are rare. Tanks with multiple compartments, each built as a separate tank, are also rare. A tank's capacity does not ordinarily exceed 6,340 gal (about 24 m³).

Ninety percent or more of the tanks are built of stainless steel; the rest are constructed of mild steel. Stainless steel is used because it resists cold temperatures. Aluminum and magnesium alloy tanks are available, but they cannot be used in water transport mode.

Minimum head and shell thickness is measured in terms of "equivalent thickness in mild steel" after forming. For steel tanks, the minimum thickness for nonregulated commodities is $1/4$ in. (6.35 mm). For regulated materials, the minimum thickness is $3/8$ in. (9.5 mm). For stainless steel tanks, the minimum thickness for nonregulated materials is slightly less than $1/8$ in. (3.2 mm). For regulated commodities, the minimum thickness is just under $3/16$ in. (4.8 mm).

Most tanks are built according to the pressure-vessel standards of ASME, and the welds are X-rayed. Welds on carbon steel tanks are post-weld heat-treated.

Supporting frame: The supporting frame of a tank container protects the tank and provides for stacking, lifting, and securing the container. It also supports the walkways and ladders. Like other types of freight containers, supporting frames are built to the requirements of the International Standards Organization (ISO).

The most common size of supporting frame is 20 ft (6 m) long, 8 or $8^1/_2$ ft (2.4 or 2.55 m) wide, and from 8 to $9^1/_2$ ft (2.4 to 2.85 m) high. Except for the Sea-Land Company's half-height tank contain-

FIG. 9-35Q. Diagrams of tank containers in various modes of transportation. (Source: Union Pacific Railroad)

ers, which are 35 ft (10.5 m) long, very few tanks longer than 20 ft (6 m) are used in the United States.

There are two basic supporting frame types: the box type, which encloses the tank in a cage-like framework; and the "beam type," which uses frame structures only at the ends of the tank.

Corner Casting: Supporting frames for all intermodal containers, including tank containers, are built with standard corner fittings, called corner castings. They are used to secure the tank and to lift it with standard container-handling equipment. Corner castings must conform to ISO standards. Cast-iron corner castings are prohibited. In an accident where proper lifting devices are not available, the corner castings may be used to lift or move the tank only after consulting the tank's owner or manufacturer.

Features: To increase their use for a wide range of cargoes, tank containers may be equipped with various features, such as linings that protect the tank from the lading, refrigeration units, electrical or steam heating, and insulation to maintain the desired temperature of the contents.

Foamed plastic, both polyurethane and polystyrene, is the predominant insulating material. However, mineral wool or fiberglass are also used. Insulation, generally 3 to 4 in. (76.2 to 101.6 mm) thick, is always covered with a weathertight jacket. Jackets are made of metal, at least $1/32$ in. (1 mm) thick, or of an equivalent thickness of plastic reinforced with either glass or fiber.

Tank Container Markings

Various markings may be found on tank containers, including the tank initials and number, the specification marking, the DOT exemption marking, the AAR 600 marking, the country and size/type markings, the dataplate, and the tank and valve test dates. (See Figure 9-35R.)

Tank initials and number: Tank containers must be registered with the International Container Bureau in France. They must be marked with initials, or reporting marks, and with a tank number. The initials indicate the tank's ownership, and the tank number identifies the specific tank.

The initials and tank number are generally on the right side of the tank as one faces it from either side, and on both ends. They may be displayed either on the tank itself or on the frame.

With this information, shippers, carriers, and emergency response personnel can identify the tank's contents, using shipping papers or computer data. They can then determine if the contents are hazardous.

Specification marking: The design, construction, and use of tank containers must conform to strict standards and regulations of many agencies, both in the United States and abroad. Markings on the tank container indicate the standards to which the tank was built.

In the United States, tank containers must meet the design, construction, and safety standards set forth by the DOT. Tanks meeting these standards have the specification markings or an exemption number displayed on both sides (generally the tank's initials and number).

Examples of specification markings are: IM 101, IM 102, and Spec. 51.

DOT exemption marking: When DOT grants exemptions from the regulations governing packages and containers, and the preparation and offering of hazardous materials for shipment, the outside of each package must be plainly and durably marked in 2-in.

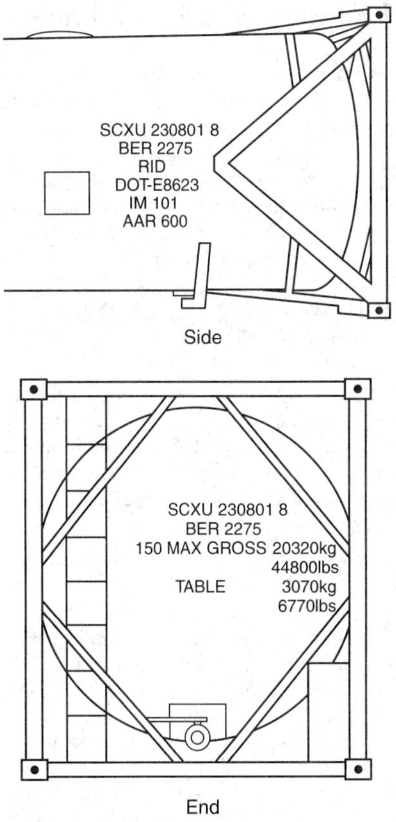

FIG. 9-35R. *Typical markings on tank containers.*

(50.8-mm) letters and numbers—that is, "DOT-E" followed by the number assigned, such as "DOT-E8623."

AAR-600 marking: For interchange purposes in rail transportation, tank containers should conform to the requirements of Section 600, "Specification for Acceptability of Tank Containers" of the AAR *Manual of Standards and Recommended Practices, Section C—Part III, Specification for Tank Cars.* Tanks meeting these requirements display the AAR 600 markings in 2-in. (50.8-mm) letters on both sides, near the tank's initials and number. The AAR-600 marking identifies tanks that can be used for regulated materials, whereas the AAR-600NR marking identifies tanks that cannot be used for regulated materials.

Country and size type markings: In addition to the initials and number and the specification marking, the tank also displays a country code and a size/type code. The two or three letter country code indicates the tank's country of registry.

The four-digit size/type code follows the country code. Its first two numbers jointly indicate the container's length and height. The second pair of numbers is the type code, which indicates the tank's pressure range.

Dataplate: For further technical, approval, and operational data, dataplates are permanently attached to the tank containers' tanks or frames.

Tank and valve test dates: Tanks and safety valves, if installed, must have a retest interval not greater than 5 years. Retest and test due dates must be marked or stenciled on the tank or on the dataplate.

General Classes of Tank Containers

Tank containers are classed according to the specification of the tank and fittings. The classification of a tank container determines which products it may carry.

Nonpressure tank containers: Also called intermodal portable tanks or IM portable tanks, nonpressure tank containers comprise over 90 percent of the total. (See Figure 9-35S.) They usually transport liquid and solid materials at maximum allowable working pressures (MAWP) of up to 100 psi (690 kPa). Tanks are tested to at least 1.5 times the MAWP. In the United States, two groups of non-pressure tank containers are common:

1. *IM 101 portable tanks.*   These tanks are built to withstand MAWPs of from 25.4 to 100 psi (175 to 690 kPa). They may transport both nonhazardous and hazardous materials, including toxics, corrosives, and flammables, with flashpoints below 32°F (0°C). Internationally, an IM 101 portable tank is called an IMO type 1 tank container.

2. *IM 102 portable tanks.*   These tanks are designed to handle lower MAWP—that is, from 14.5 to 25.4 psi (100 to 175 kPa). They transport materials such as liquor, alcohols, some corrosives, pesticides, insecticides, resins, industrial solvents, and flammables with flashpoints between 32 and 140°F (0 and 60°C). More often they transport various nonregulated materials, such as food-grade commodities. Internationally, an IM 102 tank is called an IMO type 2 tank container.

Pressure tank containers: Also known as DOT Spec. 51 portable tanks, these containers are less common in transport. Designed to handle internal pressures ranging from 100 to 500 psi (690 to 3,450 kPa), and they generally transport gases liquefied under pressure, such as LP-Gas and anhydrous ammonia. They may also carry liquids, such as motor fuel anti-knock compounds or aluminum alkyls. Internationally, they are called IMO type 5 tank containers. (See Figure 9-35T.)

Specialized tank containers: There are several other variations of tank-type containers. Among them are:

1. *Cryogenic tank containers.*   These containers carry refrigerated liquid gases, such as argon, oxygen, and helium, and are built

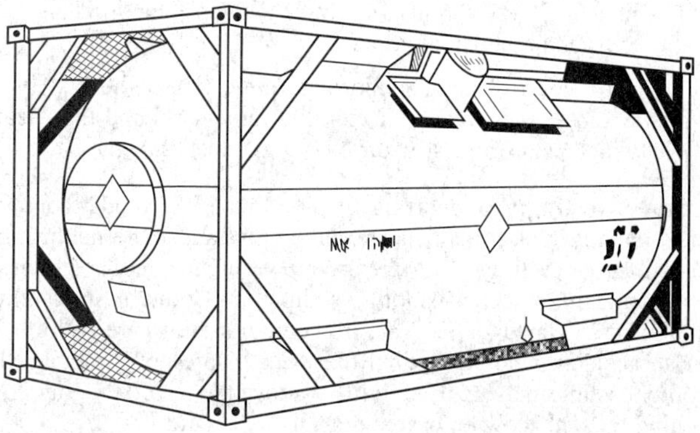

FIG. 9-35T.   Diagram of a typical pressure tank container. (Source: Union Pacific Railroad)

to DOT Spec. 51 standards. Internationally, they are called IMO type 7 tank containers.

2. *Tube modules.*   These containers transport gases in high-pressure 3T cylinders tested to 3,000 or 5,000 psi (20,700 or 34,500 kPa), permanently mounted within an ISO frame. (See Figure 9-35U.)

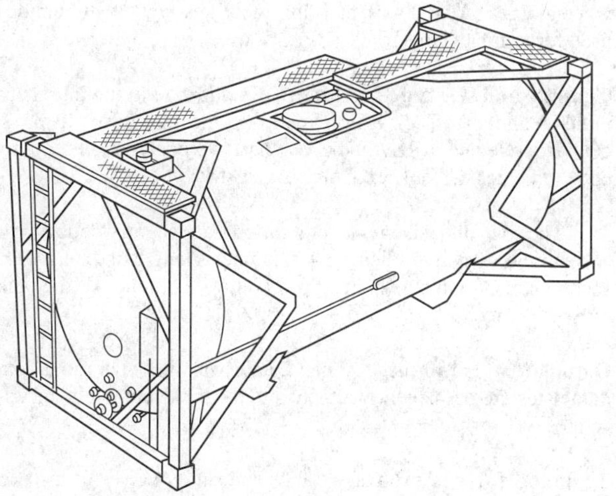

Tank Container Fittings

Nonpressure tank container fittings: Various fittings make non-pressure tank containers functional and safe.

Spillbox.   On nonpressure tanks, a spillbox encloses the top fittings. It protects the tank's shell from any spillage. Any spilled material, as well as rainwater in the spillbox, drains through an open pipe.

Manhole, cover, and dipstick.   The manhole, located on top center of the tank, is secured by six or eight wing nuts on a hinged and bolted lid fitted with a replaceable gasket.

The dipstick may be inside the manhole. Used in conjunction with a calibration chart, known as a strapping chart, it measures the amount of commodity in the tank. For liquids other than compressed gases, a minimum outage of 2 percent of total tank capacity must be provided in rail transportation.

Top-loading valves.   Top-loading valves attached to a removable eduction pipe—dip leg, dip tube, or siphon tube—run into the tank.

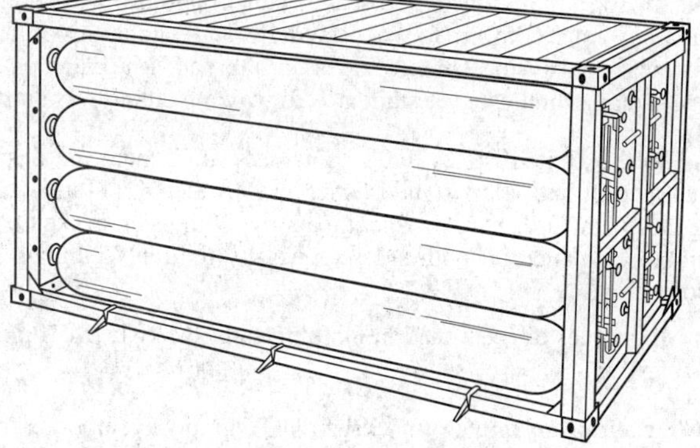

FIG. 9-35S.   Illustration of nonpressure tank container. (Source: Union Pacific Railroad)

FIG. 9-35U.   Diagram of tube module. (Source: Union Pacific Railroad)

Bottom-outlet valves. When tank containers are used for hazardous materials, two externally operated, bottom-outlet valves are required. Connected in series with a replaceable gasket between them, they are located at one end of the tank, the so-called discharge end. A liquidtight closure on the external valve is also required. It may be a blind flange, which is required on international shipments; a screw cap; or a cam-lock cap attached to the external valve. A positive lock is required on the external bottom valve to keep it closed.

In an emergency, the internal valve, or foot valve, can be shut off from a remote location. As one faces the discharge end of the tank, the emergency shutoff is on the right side near the far end.

Air-line connection. An air-line connection, located on the tank's top, can be used for pressure unloading, vapor return, and blanketing contents with an inert gas.

On occasion, blind flanges may replace the top-loading valve and air-line connection. In this situation, the shipper or supplier of the tank must provide appropriate fittings for loading and unloading.

Pressure/vacuum relief valves. Generally found in pairs on non-pressure tank containers, a combination pressure/vacuum relief device serves dual purposes of protecting the tank from both overpressure and from a vacuum of more than 0.75 psi (5 kPa) negative pressure. In many cases, this relief valve has a rupture, or bursting, disk between the safety relief valve spring and the commodity. It protects the spring from the commodity.

Thermometer. Some tanks have a built-in thermometer to measure the temperature of the lading.

Pressure tank container fittings: Like nonpressure tank containers, pressure tank containers are equipped with various fittings to make them functional and safe. A wide variety of fitting arrangements exists for pressure tank containers. Pressure intermodal tanks may have fittings on top, at the end, or on the bottom. Generally, fittings are recessed or enclosed with a cover to protect them from mechanical damage.

Loading/unloading valves. Liquid and vapor valves are used to fill and empty the tank. The liquid valve reaches into the lading by means of an eduction pipe, which may be fitted with an excess-flow check valve. Vapor valves, which also may have an excess-flow check valve, are used to remove vapor from the tank or to pressurize the tank for unloading.

Safety relief devices. Safety relief valves, mounted on top, protect the tank from overpressure under abnormal conditions. In many cases, a rupture disk between the safety relief valve spring and the commodity protects the spring from the contents.

Gauging devices. Gauging devices may be installed to measure how much liquid is in the tank.

Other fittings. A sample line for sampling the lading and a thermometer well for measuring lading temperature may also be installed.

PASSENGER TRANSPORTATION

Safety Precautions for Passenger Equipment

1. Passenger trains, including commuter trains, typically use 480 V ac power generated from a locomotive or a power car for onboard electrical systems, including heating, air conditioning, lighting, and food service. Generally, cables located beneath and between rail cars on both sides transmit this power. In an emergency situation, these cables may be damaged or loosened. Never attempt to disconnect any cables without assistance from the train crew. Never open any empty electrical receptacle until assured that the 480 V ac system has been de-energized. Contacts in the receptacle can be energized without the presence of cables and can produce an arc that can result in serious injury or death. Typically, 480 V ac systems are stepped down by transformers located under the car to provide 220/110 V ac power for onboard electrical equipment. Some electrical components use capacitors, which store residual electrical energy for a time after the system has been shut down.

2. Typical passenger trains employ a 72 V dc emergency battery backup system for operation of lighting, electric doors, and public address systems when the 480 V ac power is lost. These batteries, typically nickel-cadmium and generally located under the car, can present a hazardous materials danger if damaged in an accident. In addition, train cars may be connected by 72 V dc jumper cables for public address and other systems. These cables must be approached with the same caution as 480 V ac cables.

3. Typical passenger trains also employ a pressurized air system, up to 140 psi (965 kPa), for operation of brakes, doors, and lavatories. Compressors in locomotives or power cars provide air pressure, and most passenger cars are equipped with at least one pressurized air reservoir. Some passenger cars ride on air-pressurized bellows instead of springs for improved passenger comfort. Emergency response personnel should be aware of the air pressure hazard and remain clear of any components where they hear the sound of leaking air, until assured that the system has been depressurized.

Passenger Equipment

Modern cars are constructed of steel and aluminum. Materials used for the interiors of today's passenger coaches, diners, and sleeping cars are, at the least, tested to the performance criteria outlined in Table 4-2.4 in NFPA 130, *Standard for Fixed Guideway Transit Systems*. Interiors designed to this criteria should show limited fire involvement when exposed to fire. Major causes of fires on passenger trains include careless smoking and electrical faults. However, extensive fire involvement can and has occurred when cars have been exposed to an external fire source such as a flammable liquids fire. In diners, hazards are identical to those inherent in stationary restaurants, namely, grease and exhaust system fires. These, plus air conditioning and heating systems, must be properly safeguarded, and regular preventive maintenance is essential.

The newest Amtrak sleeping cars are equipped with smoke detection systems which, when activated, shut down the HVAC system, sound an alarm within the car, and provide evacuation instructions to the car's passengers.

Modern rail passenger equipment is designed to withstand tremendous physical forces that would occur in a collision. This construction defeats typical rescue or forcible entry tools. Rescue and fire-fighting operations should be conducted through a car's doors and windows. The windows in passenger equipment generally are made of polycarbonate material that is far stronger and more flexible than safety glass—it cannot be smashed with an axe. Attempts to break windows result in tool rebound, which can easily cause serious injury to the rescuer. Windows can be removed from inside or outside cars following the procedures given on labels or placards generally found adjacent to the windows.

Emergency response personnel must always assume that other trains are operating nearby. Modern passenger trains typically operate at 79 mph (127 kph), with 125 mph (201 kph) being the top speed in Amtrak's Northeast Corridor between Washington, DC, and New York City. The emergency stopping distance of a train traveling at 79 mph (127 kph) is approximately $1^1/_2$ mi (2.4 km) after brakes are applied, $2^1/_2$ mi (4 km) at 125 mph (201 kph).

Emergency responders must protect the scene by providing flagging protection, as detailed earlier in this chapter, on the track in both directions from the incident at a distance of $1^1/_2$ mi (2.4 km). In the high speed corridor, this distance should be increased to $2^1/_2$ mi (4 km). Flagging protection should be maintained until positive assurance has been received that the system operator or dispatcher knows that emergency response personnel are on the right-of-way and that other rail traffic has been notified. Even with these assurances, it is recommended that flagging protection be maintained for the event's duration.

Because of the weight and size of railroad passenger cars, stabilization in a derailment cannot be obtained using the highway-grade cribbing materials, portable rachet pullers, or air bags that emergency response personnel typically carry. It is strongly recommended that damaged or derailed cars be stabilized by using railroad ties (readily available at most derailment sites) or other suitably heavy duty materials.

Most cars in a passenger train are those owned by the train's operator, usually Amtrak. Occasionally, however, privately owned passenger cars are attached to the train. While most commercial passenger cars use electrical power for onboard lighting, food service, door operation, and heating/air conditioning systems, privately owned passenger cars may employ LP gas or diesel fuel for heating, cooking, or a generator system. For this reason, responders must be prepared for such materials when dealing with a privately owned passenger car.

Rail transportation companies typically provide training for emergency response personnel on the proper procedures for dealing with a rail incident. Emergency responders with rail operations in their area should make certain that they take full advantage of the rail operators' training resources.

MISCELLANEOUS RAILROAD FACILITIES

Many railroad facilities are similar to industrial facilities, with familiar hazards. However, other facilities are peculiar to the railroad industry, such as diesel locomotive running repair and heavy repair shops, freight car repair shops, paint shops, and locomotive fueling facilities. Hazards associated with a diesel locomotive repair shop include cutting, welding, poor housekeeping, and fuel leakage.

Fueling facilities need particular protection, not only because of the economic value of the systems and of the stored fuel, but also because of the value of locomotives and the cost of interruption of service. Fuel pump houses should be of noncombustible construction and, preferably, equipped with automatic fire protection and with light fixtures and devices suitable for hazardous locations. Because leakage is a prime hazard, careful housekeeping and preventive maintenance are essential. Portable and wheeled dry chemical fire extinguishers should be located throughout the fueling area. It is most desirable to have a high-volume water supply nearby.

Other facilities are critical to railroad operation and, if involved in fire, could stop the movement of trains in a given territory. These facilities include computer centers, microwave communications facilities, centralized train control equipment, and the many support elements in classification yards. Private detection and protection systems should protect these facilities. Typical fire suppression systems range from sprinklers to carbon dioxide and other clean-agent designs. Nearby adequate water and hydrant systems are essential.

Bridges and Trestles

Fire prevention is the key to continued operation of these structures. Regular maintenance should be done to provide and maintain a 25- to 50-ft (7.5- to 15-m) clearance between combustible structures and nearby vegetation. Susceptibility to fire damage can be reduced by treating the structure with a fire retardant and by providing water barrels, fire detection systems, standby pipe and hose systems, and guard service.

At some particularly valuable and critical bridges, fireboats and water buckets that can be attached to helicopters have been provided, and noncombustible bridge segments have been constructed at regular intervals to limit damage should fire occur. Most railroads have been replacing combustible bridges and other track structures with noncombustible construction at every opportunity, thus reducing their exposure to fire-related losses.

Exposure Risks for Rail Equipment

Exposure fires that destroy loaded and unloaded freight equipment reflect the need for close analysis of terminal fire hazards. These fires can start both on and off railroad property. When classification or switching yards are designed, adequate fire protection, both private and public, should be a prime consideration.

Lack of nearby fire mains and hydrants can contribute to fire loss in large yards. In a major incident, fire-fighting activities should be confined to containment and protection of exposed property. A prefire plan should map fire mains and hydrants both in and near the yard, because protection in the yard may be rendered useless or inaccessible during a major event.

Railroad sidings present a similar exposure hazard, particularly at grain elevators, lumber yards, flammable liquid or LP-Gas unloading facilities, and similar industrial properties.

Self-Propelled Fire Fighting and Derrick Equipment

Most carriers are well prepared to clear derailments and rerail freight equipment with their own forces or by standing arrangements with specialty contractors. They use lifting equipment with capacities ranging from 150 to 250 tons (136 to 227 metric tons). These may be mounted on a work train, that also may carry water cars and fire pumps powered by the train air line or powered independently by an internal combustion engine. It is increasingly common to find wreck-clearing equipment with rubber tires or crawler tracks transported to the scene by highway. Carrier equipment usually includes fire hoses, nozzles, foam extinguishing compound, play pipes, and a quantity of extinguishers.

Work trains are especially valuable in incidents that occur in remote and inaccessible areas. Fire-fighting needs for such emergencies can be handled by driving a fire department pumper onto a flat car and taking it to the emergency by train.

Some railroads have self-propelled fire-fighting cars that can be used independently of the work train. The availability of such fire-fighting equipment is often overlooked in pre-emergency planning.

RIGHTS OF WAY

A difficult problem is the right-of-way fire, or fire on railroad land adjacent to the roadbed. These fires can spread off railroad property. Such fires have occurred since the first days of railroading, but their frequency and severity have been greatly reduced by improved railroad equipment, controlled burning, chemical control of vegetation, and installation of fire breaks. Cooperation with the U.S.D.A. Forest

Service has led to development of spark arresters, spark arrester exhaust manifolds, and sparkless brake shoes. Planned vegetation, planting the right-of-way with fire resistive vegetation, is a practice that, if properly maintained, can further reduce these fires.

NFPA Rail Transportation System Section

Rail Transportation Systems, a membership section of the National Fire Protection Association, was organized in 1963 as the Railroad section. It is dedicated to promoting interest in, and improving the methods of, fire prevention and protection in the railroad industry and to encouraging proper safeguards and recommended practices against fire through the interchange of ideas and experience.

Rail Transportation Systems' annual meeting is held each May in connection with the NFPA Annual Meeting. This event provides an excellent forum for the exchange of ideas between the fire service and railroad fire safety specialists.

ACKNOWLEDGMENTS

Arthur Candenquist, manager, Emergency Preparedness and Regulatory Compliance, and James P. Gourley, P.E., senior fire protection engineer, Amtrak, for their assistance in updating sections on electrical locomotives and passenger equipment.

Union Pacific Railroad Company for materials on tank cars and tank containers, and for a variety of factual updates and general information.

BIBLIOGRAPHY

References

"Annual Report of Hazardous m Materials Transported by Rail," Bureau of Explosives, Association of American Railroads, Washington, DC, 1996.
GATX Tank and Freight Car Manual, 6th edition, General American Transportation Company, Chicago, IL, 1994.
General Guide to Tank Cars, Union Pacific Railroad, Omaha, NE, 1996.
General Guide to Tank Container Operation, Sea Containers Group, London, England, 1983.
General Guide to Tank Containers, Union Pacific Railroad, Omaha, NE, 1996.
"Hazardous Material Regulations of the Department of Transportation," Bureau of Explosive Tariff No. BOE-6000, Bureau of Explosives, Association of American Railroads, Washington, DC, revised annually and updated quarterly.
Inspection, Repair, and Test Requirements for Tank Containers, Sea Container Group, London, UK.
Manual of Standards and Recommended Practices, Section C—Part III, Specification for Tank Cars, Operation and Maintenance Department, Association of American Railroads, Washington, DC, 1987.
Maty, A. D., "Guide to the Identification of Tank Cars and Fittings Utilized for the Transportation of Hazardous Materials," 1985, unpublished.

"Report of Railroad Tank Car Leaks of Hazardous Materials by Commodity by Source of Leak for the Year 1995," Bureau of Explosives, Association of American Railroads, Washington, DC, 1996.
Tank Car Book, Union Tank Car Company, Chicago, IL, 1996.
Wright, C. J., and Student, P. J., "Understanding Railroad Tank Cars," *Fire Command*, Vol. 52, Nos. 11 and 12, Nov. and Dec. 1985.
Wright, C. J., and Hand, W. T., "Intermodal Tank Containers," *Fire Command*, Vol. 55, Nos. 6 and 7, June and July 1988.

NFPA Codes, Standards, and Recommended Practices

Reference to the following NFPA codes, standards, and recommended practices will provide further information on rail transportation systems discussed in this chapter. (See the latest version of *The NFPA Catalog* for availability of current editions of the following documents.)

NFPA 130, *Standard for Fixed Guideway Transit Systems*
NFPA 471, *Recommended Practice for Responding to Hazardous Materials Incidents*

Additional Reading

Davis, J. S., "All Aboard: Preplanning for Rail Incidents," *Fire Command*, Vol. 56, No. 5, May 1989, pp. 32–34.
Dee, A. W., "Fire Protection on Trains," *Proceedings* of the Institution of Mechanical Engineers, Part D: Transport Engineering, Vol. 201, No. 3, 1987, pp. 217–227.
Gaffney, P., and Kynaston, R., "Smoke Control System of the Hong Kong Mass Transit Railway," First International Conference on Safety in Road and Rail Tunnels, November 23–25, 1992, Basel, Switzerland, 1992, pp. 79–93.
Goransson, U., and Lundqvist, A., "Fires in Buses and Trains, Fire Test Methods," Swedish National Testing and Research Institute, Boras, Sweden, SP REPORT 1990:45, CIB W14/92/11 (S), 1990, 63 pages.
Iida, T., Kobayashi, S., and Kenmochi, T., "Facilities of the Seikan Tunnel," *Japanese Railway Engineering*, No. 106, June 1988, pp. 12–16.
Jameel, H. M., "Rail Rapid Transit Safety Standards: A Feasibility Study," Report No. UMTA-CA-MT-82-106, 1982, California Public Utilities Commission, San Francisco, CA.
"Passenger Fire Safety in Transportation Vehicles," A. D. Little Report No. C-78203, May 1975, U.S. Department of Transportation, Washington, DC.
Peacock, R. D., and Braun, E., "Fire Tests of Amtrak Passenger Rail Vehicle Interiors," NBS Tech Note 1193, May 1984, National Bureau of Standards, Gaithersburg, MD.
"Recommended Fire Safety Practices for Rail Transit Materials Selection," Urban Mass Transit Administration, U.S. Department of Transportation, *Federal Register*, Vol. 47, No. 228, Nov. 26, 1982, p. 53559.
Ryczkowski, J. J., "Railroads for the First Responder: LNG Locomotive," *American Fire Journal*, Vol. 46, No. 11, November 1994, pp. 12–15.
Smith, D. A., "Some Aspects of Fire Safety Design on Railways," *Fire and Materials*, Vol. 8, No. 1, Mar. 1984, pp. 6–9.

AVIATION

Revised by John F. Rooney

Among the many factors that contribute to aviation fire safety are requirements that the aircraft be airworthy and fire safe. Hazards of certain maintenance practices must be recognized and proper control measures applied so operating personnel, aircrews, cabin attendants, and ground mechanical and servicing personnel are well trained in normal and emergency conditions.

Airborne and ground navigational aids, weather, air traffic control, proper design of runways with adequate "overrun" areas free of hazardous obstructions, and proper maintenance of this equipment and facilities are of extreme importance; however, they are beyond the scope of this chapter. This chapter discusses only the problem of fire and life safety in aircraft in the operational and ground environment modes.

AIRCRAFT FIRE SAFETY

Fire safety in aircraft starts on the drawing board, and aeronautical engineers bear the brunt of responsibility for fire prevention and control. Aircraft require large quantities of fuel, lubricating oils, hydraulic fluids, and Class A combustibles, which are in proximity to potential ignition sources such as power plants, auxiliary power units, electrical systems, and heaters. Many aircraft carry oxygen systems (liquefied and gaseous), and some use oxidizers for auxiliary power units or are equipped with small rocket units for additional takeoff thrust. Consequently, aircraft fire safety requires skillful blending of reasonable safeguards that are lightweight and do not interfere unduly with the use and mission of the particular aircraft.

Although most nations that have a major aircraft production industry also have extensive regulations governing the use and arrangement of the foregoing items, it is extremely important that the designer be thoroughly versed in the letter and intent of these regulations.

Installed fire detection and extinguishing equipment are normally required for those aircraft areas that possess inherent fire hazards and ignition potentials. Other fire prevention techniques used are separations of systems containing flammable fluids from potential ignition sources, compartmentation of fire hazard areas from critical structural components and flight control systems, and judicious use of other materials that are fire and heat resistive or of very low flammability. Fire safety must receive high priority because aircraft fires in flight present tragic life hazard potentials. Fires following impact accidents—which can result in the loss of many lives—are also an aircraft design concern.

In commercial, light general aviation, and military reciprocating engine aircraft, the lightweight structure needed to allow the aircraft to perform its mission also permits impact energy to be transmitted to the aircraft occupants. Thus, more deaths occur due to impact trauma than fire.

In modern turbine aircraft that operate in a more demanding flight envelope, heavier gage metals, advanced composite materials, and computer-assisted design techniques strengthen airframe structures. These aircraft structures absorb a greater amount of impact energy, thereby transmitting less of it to its occupants. Subsequently, in accidents involving this type of aircraft, the post-accident fire is a more significant cause of serious injuries or death to the occupants.

As a result of aircraft accident investigations, much research has been conducted, and is still in progress, on methods of minimizing aircraft fatalities. Improved seat belt design, heat-resistant evacuation slides, elimination of injury-causing projections, and floor mounted emergency lighting systems are some more recent developments that have improved occupants' chances of survival when an aircraft accident occurs.

In July of 1993 the FAA proposed to revise its crash-worthiness and flammability standards for commuter aircraft (60 passenger seats or fewer). The proposal included upgrading seat/restraint systems, static load factors affecting fuselage stability during emergency landings, and an increase in flammability standards of passenger seats to a level commensurate with those of larger commercial aircraft. The effective date of the revised standards has not yet been announced.

Many methods are being studied to reduce the post-impact fire hazard. They include, but are not limited to, the following:

1. Segregating flammable-fluid containers and systems from ignition sources.

2. Improving fuel containment methods.

3. Reducing the rate of evaporation and the speed of flame spread over the surface of the spilled fuel.

4. Improving the materials used for interior decor, insulation, sound attenuation, and cushioning to reduce ease of ignition, flame spread, and smoke and toxic gas generation.

John F. Rooney is a retired battalion chief of the Los Angeles City Fire Department and past chairman of the NFPA Technical Committee on Aircraft Rescue and Fire Fighting.

5. Compartmentation with lightweight fire walls in the occupied portions of the cabin, blind spaces, lavatories, and trash containers.

6. Onboard cabin fire suppression systems using a variety of extinguishing agents.

7. Early warning fire/smoke detection systems in occupied portions of the fuselage.

8. Inerting fuel tank vapor space to reduce the possibility of explosion from static electrical discharge and lightning strikes. This can be effective in flight but does little to contain the post-impact fire, because the inert atmosphere would be lost almost immediately on impact, with the disruption of the tank structure.

AIRCRAFT POWER PLANTS

Civil aircraft in the United States are subject to extensive federal regulation through the U.S. Department of Transportation (DOT), Federal Aviation Administration (FAA). The Code of Federal Regulations (CFR), Title 14, "Aeronautics and Space," contains the Federal Air Regulations. Other nations have similar requirements. The following list summarizes the regulations that apply to aircraft power plants:

1. All reciprocating engines, auxiliary power units, fuel-burning heaters, or other combustion devices intended for operation in flight must be separated from other areas of the aircraft by fire walls, shrouds, or equivalent means so that no hazardous quantities of air, fluid, or flame can pass from these compartments to other portions of the aircraft. These regulations also apply to the combustion, turbine, and tailpipe sections of turbine engines. The fire walls and shrouds must be made of a material that withstands heat at least as well as steel. When applied to power plants, the material must perform under the most severe conditions of fire and duration likely to occur in such zones. All openings in fire walls and shrouds must be sealed with close-fitting fire-resistive grommets, bushings, or fittings. In reciprocating engine nacelles, a fire seal or bulkhead is used to isolate the engine power section and exhaust system from the engine accessory section. In addition, a main fire wall segregates the complete engine assembly (power section, exhaust system, and accessory section) from the remainder of the nacelle section and, as applicable, the wheel well. In gas turbines, the fire seal or bulkhead separates the combustion, turbine, and tailpipe section from the compressor and accessory section. An additional main fire wall isolates the engine assembly from the support pylon or remainder of the aircraft, as applicable.

2. The cowling and nacelle skin are designed to prevent fire from circumventing the fire seals and main fire walls.

3. Tanks containing a flammable fluid cannot be located in a fire zone except when it can be proved that construction, connecting lines, controls, and shutoff means provide protection equivalent to segregation. A specified air gap is required between flammable fluid tanks and a fire wall, and materials that can absorb flammable fluids are prohibited from the area of the tank or any other system containing flammable fluids.

4. Emergency fuel shutoffs are required for each engine, auxiliary power unit, or combustion heater. The emergency shutoffs must be fire resistive or located so that a fire in any fire zone does not affect their operation. Operation of the emergency shutoff must not affect other emergency functions, nor the operation of any other engine.

5. Flammable liquid lines in fire zones must be fire resistive and, where required, flexible. This also applies to drain and vent lines for flammable fluids or vapor.

6. Reciprocating and turbine engine air inlets must be arranged to prevent backfire flames from entering the fire zone and designed so discharge from vents and drains cannot enter the air induction system.

7. Exhaust systems must discharge in a safe manner, not expose any portion of a flammable fluid system, and be provided with heat shields wherever they may impinge on other portions of the aircraft.

8. Drains must be provided for all fire zones, with the discharge arranged so that drained fluids will not be reingested into any portion of the aircraft.

AVIATION FUELS

The fire hazard properties of aviation fuels are identified according to susceptibility or ease of ignition, flash points, flammability limits, distillation range (initial and end boiling point), and electrostatic susceptibility. (Note: Octane rating has no relation to the degree of fire hazard of a fuel.) Table 9-36A summarizes characteristics of more common aviation fuels. The fire hazard properties of aviation fuels vary. NFPA 407, *Standard for Aircraft Fuel Servicing,* requires that the same fire safety precautions apply to all types.

Two basic differences must be considered when evaluating fuel in an operating aircraft and fuel stored in a tank at a refinery or bulk plant:

1. Because an aircraft travels at different altitudes and through varying ambient temperatures, conditions of flammability in the tank vapor space can change rapidly. Thus, a fuel type such as Jet A—which is normally too lean in a tank vapor space at sea level with a fuel temperature of 70°F (21°C)—can move into the flammable range as the aircraft gains altitude. Other factors contribute to this change, such as aircraft skin heating from air friction, outgassing of dissolved oxygen, and sloshing of the fuel from air turbulence. Figure 9-36A illustrates changes in flammability limits of aircraft fuels in tank vapor space due to altitude and temperature changes.

2. After an impact when major structural damage occurs to aircraft fuel tanks, the fuel may be released as a mist due to forward momentum, splashing, and wind shearing. Regardless of the type of fuel involved, this mist can ignite easily from disrupted electrical circuits, hot engine surfaces, or ignition sources on the ground. The resulting fireball then acts as the ignition source for other combustibles in the area, including pools of high flash point Jet A fuel. In some aircraft accidents where deceleration forces are low, liquid fuel flowing from ruptured fuel tanks or broken fuel lines can be vaporized and ignited by hot engine surfaces, hot brakes, heavy electrical arcs, etc.

Relative Safety of Jet Fuels

Jet A (kerosene grade) turbine fuels apparently offer a safety advantage over other types of aviation fuels, especially during fueling operations and aircraft fuel systems maintenance. Fires are less likely in impact-survivable accidents when the aircraft involved had Jet A fuel in the tanks. However, once ignition occurs, all fuels exhibit similar behavior, and control measures must be instituted quickly to prevent injury and fatalities.

In some portions of the world, Jet A (kerosene grade) fuel is not readily available, and Jet B (JP-4) is used. Jet B is easier and more economical to produce. Some nations also have reduced the minimum flash point of Jet A to increase yield of aviation fuel per barrel; the minimum flash point has been in the 80°F (27°C) range. The actual flash point has been above 100°F (38°C). However, if a worldwide fuel shortage develops again, wider use of reduced flash point kerosene (RFK) and Jet B can be expected.

TABLE 9-36A. *Summary Data on the Fire Hazard Properties of Aviation Fuels*

Characteristics	Gasoline AVGAS	Kerosene Grades JET A JET A-1 JP-5, JP-6, JP-8	Blends of Gasoline and Kerosene JET B and JP-4
Freeze point*	−76˚F	−40˚F to −58˚F	−60˚F
Vapor pressure† (Reid-ASTM D323-58)	5.5 to 7.0 psi	0.1 psi	2.0 to 3.0 psi
Flash point* (by closed-cup method at sea level)	−50˚F	+95˚F to +145˚F	−10˚F to +30˚F
Flash point* (by air saturation method)	−75˚F to −85˚F	None	−60˚F
Flammability limits Lower limit Upper limit Temp. range for flam. mixtures*	1.4% 7.6% −50˚F to +30˚F	0.74% 5.32% +95˚F to +165˚F	1.16% 7.63% −10˚F to +100˚F
Autoignition temperature*	+825˚F to +960˚F	+440˚F to +475˚F	+470˚F to +480˚F
Boiling points* Initial End	110˚F 325˚F	325˚F 450˚F	135˚F 485˚F
Pool rate of flame spread‡§	700–800 fpm	100 fpm (or less)	700–800 fpm

Note: Figures vary for some of these values in different data sources. Those shown here are average figures based on the latest available information.
* $5/9$ (˚F−32) = ˚C.
† 1 psi = 6.9 kPa.
‡ 1 fpm = 0.3 m/min.
§ In mist foam, rate of flame spread in all fuels is very rapid.

FIG. 9-36A. *Flammable ranges of aviation gasoline and Jet A (kerosene) and Jet B (JP-4) turbine fuels, showing variations with altitude. According to laboratory studies, some sea-level flammable limits shown on this chart for Jet A and Jet B turbine fuel do not agree exactly with Table 9-36A. Test data used to develop this chart were taken from a single source; Table 9-36A combines and averages data from all available sources.*

Electrostatic Susceptibility

The degree to which a static charge may be acquired by aviation fuels depends upon many factors: the amount and type of residual impurities, dissolved water, the linear velocity through piping systems, and the types of filters and water separators used. Jet A and

Jet B fuels are better static generators than AVGAS. While all fuels generate static charges, the electrical conductivity of the fuel directly relates to the speed of charge relaxation (dissipation).

Antistatic additives have been developed. However, rather than preventing the formation of static charges, they actually increase fuel conductivity, thus considerably shortening relaxation time. Antistatic additives increase static charge, since, by nature, they are an impurity in the fuel. The charge generation increases as fuel passes through filter/water separators and other equipment. An unbonded object in the vicinity of the charge generation point before the charge has an opportunity to relax can act as a collector and cause high-energy static discharge.

Although possibly misnamed, antistatic additives have greatly increased the safety of aircraft fueling operations in all but extenuating circumstances (e.g., the unbonded collector in close proximity to the static generation point).

AIRCRAFT FUEL SYSTEMS

Most aircraft make extensive use of the internal wing volume to store fuel. In larger aircraft, the wing structure is sealed and forms the fuel tank. This is commonly called integral tank, or wet wing, construction. Older and light aircraft may incorporate a flexible bladder to contain the fuel within the wing structure, using the wing structure only for support. Separate metal tanks or fiber-reinforced nonmetallic tanks are not widely used, and they are mainly located in light aircraft and older transports.

With integral tank construction, and only to a slightly lesser extent with thin-wall bladder tanks, disruption of the wing structure by ground impact or other damage results in the release of fuel and potential for ignition. This may occur even though occupied portions of the aircraft may be only slightly damaged. Thus, fire becomes a threat to the occupants. Figure 9-36B illustrates typical fuel-occupied spaces in a large turbine-powered aircraft.

One solution proposed is crash-resistant fuel tank construction with dry break fittings and automatic fuel shutoffs. While the technology to develop this tank has been highly advanced, it has not been widely adopted due to weight limitations and reduction in volume available for fuel storage in long-range, military, and commercial aircraft. However, certain helicopters operated by the U.S. Army are equipped with such crashworthy fuel systems.

Open-celled foam blocks, cut to fit and placed meticulously in the tank, are used on some military aircraft. While primarily intended for explosion protection after projectile penetration (incendiary bullets, etc.) of the vapor space, they also were tested for improvement of the post-crash fire situation. Tests showed some resulting improvement, but the problems of removal and replacement for tank interior maintenance and loss of available fuel volume preclude their use in large, long-range, civil and military aircraft.

Some aircraft equipped with these foam blocks have had internal fuel tank fires. Charred foam, discovered when the tank was opened for inspection and repair, led to the theory that foam increases the static charging tendency and/or acted as the unbonded collector. However, the presence of the remaining foam in the tank vapor space prevented further propagation of the fire/explosion.

Many aircraft utilize the wing center section (where it passes through the fuselage) for additional fuel storage. In some aircraft, such as the Boeing 727 and DC-9, this is an integral tank with a bladder liner. Newer aircraft designs utilize a double-wall tank of metal with a honeycomb core. While such tanks are well protected by the center wing section box structure (the heaviest on the aircraft), they directly expose the fuselage interior. To extend normal operating range, some aircraft utilize fuselage fuel tankage of the double-wall type outside of the wing center section box, and thus are deprived of heavy structural protection. As a reasonable compromise, all fuel tanks (at least those within the fuselage) should be crash-resistant with dry break fittings and automatic shutoff valves.

Certain models of existing aircraft contain fuel tanks well forward or aft in the fuselage. Fuel storage in the horizontal stabilizer and vertical fin is being used on certain aircraft to extend range.

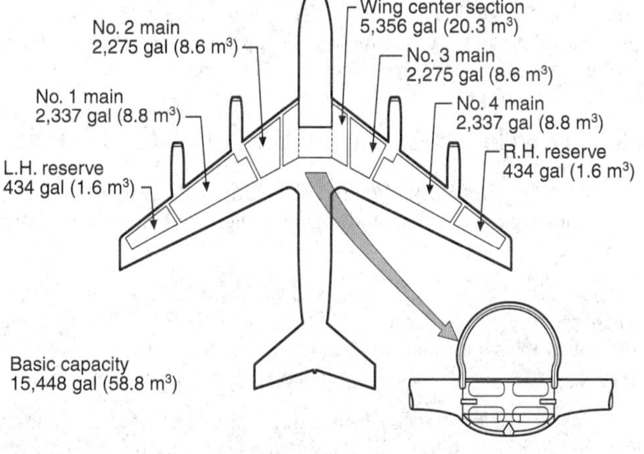

FIG. 9-36B. *Arrangement of fuel tanks in a typical turbine-powered air carrier aircraft. The reserve tanks and Nos. 1, 2, 3, and 4 are "integral" tanks in this particular airplane, and the wing center section tank is of the "bladder" type.*

OTHER DESIGN CONSIDERATIONS

The Federal Air Regulations mention several design principles that affect basic aircraft fire safety and crashworthiness (excluding cabin furnishings and evacuation systems). Developed as a result of accident investigations, they include:

1. When the same structure supports fuel tanks and landing gear, shear pins must be incorporated in the landing gear support structure, allowing the gear to be wiped off without applying structural loads to the fuel tanks.
2. The metal aircraft structure must have electrical continuity to prevent accumulation of static electrical charges, particularly in the fuel tank areas. This is extremely important when designing an all-metal aircraft, since the aircraft structure acts as a Faraday Cage, shielding all contents from lightning strikes.
3. Static discharge devices and lightning divertors must be located and installed correctly.
4. Fuel lines supplying rear fuselage-mounted engines must be designed to ensure proper fire resistance and have the flexibility to resist rupture in a crash situation.
5. Fuel lines and main electrical leads must be segregated.
6. Main electrical power cables in the fuselage must be shrouded in fire resistive flexible conduit.
7. The hydraulic system must be designed properly and use fire resistive fluids. Such fluids presently in use have autoignition temperatures in excess of 1,000°F (538°C) and are only slightly flammable. (Older types of mineral oil-based hydraulic fluids are still widely used, particularly in the military, and have been responsible for many serious fires.)

AIRCRAFT CABIN FIRES

Cabin fires in flight are infrequent and usually discovered in their incipient stage and rapidly extinguished. The potential life loss, however, is severe if fire progresses beyond the point of control by use of handheld fire extinguishers. The fire reaction of aircraft cabin interior materials, once ignited, can be crucial from a life safety viewpoint and can result in severe structural damage to the aircraft.

Prior to the mid-1940s, few, if any, regulations concerned the flammability of cabin furnishings. Old aircraft at air shows and in museums show wicker chairs, gauze fabric curtains, and other combustibles. Following World War II, a qualification test was required that involved igniting small horizontally mounted samples by a short exposure to a Bunsen burner flame. If a sample did not continue to burn after removal of the igniting flame, or burned less than a specified distance in the specified time, it was classified as self-extinguishing or slow burning. Although this test procedure was developed for laboratory comparisons, it created a widespread impression in the aviation industry that cabin furnishings were fire resistive.

After several disastrous fires in flight and on the ground, research to improve the fire-resistance criteria began. Various public and private organizations conducted tests that proved the inadequacies of the existing fire-resistance criteria and, further, disclosed the problem of heavy smoke and toxic gas generation.

In the late 1960s and early 1970s, the fire-resistance criteria were upgraded to require a material sample to be self-extinguishing when tested in the vertical position, with exceptions such as floor carpeting. Although aircraft fires indicate reduction in flame spread of materials qualified by the new vertical burn standards, more improvement is still required and the problem of smoke/toxic combustion products remains.

In the late 1970s and early 1980s, researchers developed a method of "fire blocking" aircraft seat cushions, which are usually made of polyurethane foam. Inserted between the cushion and decorative upholstery, this fire blocking layer is intended to delay or prevent seat cushion involvement in a cabin fire. The FAA now requires all commercial aircraft seat cushions to be fire blocked.

In addition to fire blocking, research is continuing toward the development of an improved, fire-resistant aircraft cabin window material to delay or prevent post-impact fire entry into the fuselage via the windows.

Many aircraft carry supplemental oxygen for the potential failure of cabin pressurization and in portable cylinders for first-aid purposes. In a crash, localized fire intensification can be expected if the oxygen cylinder fuse plug operates due to fire exposure and releases oxygen in the cabin atmosphere.

Possible causes of aircraft cabin fires include:

1. Ignition of the cabin interior from a post-impact fuel fire or a fire resulting from a fueling operation malfunction. In such a fire, although the interior materials have a certain fire resistance, heat from the exposing fire overwhelms the materials' resistance.
2. Electrical failure within concealed spaces, often behind the cabin decorative liner. Such fire can remain undetected for a considerable time period and grow beyond control of handheld fire extinguishers before it is discovered.
3. Improper use of flammable liquids during cabin cleaning and refurbishing.

Aircraft cabin fires caused by careless use of smoking materials which have been a significant hazard, are expected to lessen considerably due to the trend toward "no smoking" flights. Aircraft lavatories aboard U.S. carriers are now required to be equipped with smoke detectors specifically designed for those areas. In addition to its functional purpose, the requirement deters persons who might wish to smoke in a lavatory. Tampering with or rendering aircraft lavatory smoke detectors inoperative is a federal offense.

AIR TRANSPORT OF DANGEROUS GOODS

Hazardous cargoes (hazardous materials, nuclear materials, and restricted articles) occasionally present a problem to aircraft in flight, on the ground, during ground handling, and in a post-crash situation. This is an international problem, since many shipments cross national boundaries. Many agencies regulate hazardous cargoes, including the U.S. Department of Transportation, the International Air Transport Association, the International Civil Aviation Organization (an agency of the United Nations), and others.

The regulations cover type of cargo that may be shipped, quantities allowed, packaging method, and aircraft type (passenger or cargo only) in which this material may be carried. Flight crews must be informed of the presence of hazardous cargoes on the manifest and, in some cases, can refuse to carry the material.

If the rules are followed rigidly, there is little hazard to safety of flight or to ground personnel, whether in normal handling or a post-crash situation. Problems arise when the shipper, either in ignorance or in willful contempt of the regulations, misrepresents the contents of the shipment to a carrier or does not package the material adequately to ensure containment during transit.

After several serious incidents, the regulations were revised and further restrictions adopted. Better training is now required for each carrier's receiving agents, standard labels have been adopted, and penalties are assessed against shippers who violate the law. Detailed information regarding the air transport of dangerous goods and a listing of the United Nations warning label system may be found in Appendix B of NFPA 402, *Guide for Aircraft Rescue and Fire-fighting Operations.*

AIRCRAFT FIRE DETECTION AND EXTINGUISHING SYSTEMS

Fire detection and extinguishing systems are required in certain zones of the aircraft, notably the power plant areas. These systems are defined in design regulations of the nation of aircraft origin.

These systems are similar to those used in ground installations, but there are significant differences. In most aircraft installations, the flight crew manually activates the extinguishing system. The detection system must be very sensitive, very stable (i.e., free of false alarms), and able to withstand extremes of the environment in which it must function—temperature, vibration, high airflows, and structural flexing. The system must also reset automatically to notify the flight crew when the fire has been extinguished, or when the temperature in the protected zone drops below the set point. System response must be within seconds of the beginning of abnormal conditions.

Extinguishing agent hardware must be lightweight, rugged, and, to overcome the high airflows of flight, capable of higher rates of agent discharge than would be expected in a similar ground installation.

The most common extinguishing agents used in aircraft are bromotrifluoromethane (Halon 1301) and bromochlorodifluoromethane (Halon 1211). Older aircraft still use carbon dioxide, methyl bromide, and chlorobromomethane. All are effective on flammable liquid and electrical fires.

Cargo Compartment Fire Detection and Extinguishing Systems

Cargo compartments in passenger aircraft often are equipped with fire detection systems. Air starvation is one of the more common techniques used in fire extinguishment in cargo spaces. Upon actuation of fire detection systems, all ventilation to the compartment is sealed off and the fire is allowed to self-extinguish due to lack of oxygen. Fire extinguishing systems usually utilizing Halon 1301 or Halon 1211 are also installed for cargo compartments.

Recent regulations have been adopted to improve the fire resistance and fire penetration-resisting capabilities of cargo compartment liners. An upper limit has been placed on the volume of a cargo compartment; above this limit, a fire suppression system must be installed regardless of compartment construction.

AIRCRAFT HAND FIRE EXTINGUISHERS

The FAA requires portable hand fire extinguishers in all air transport category aircraft. In some smaller general aviation aircraft, they are recommended but not required.

NFPA 408, *Standard for Aircraft Hand Portable Fire Extinguishers*, requires portable hand fire extinguishers on all aircraft and describes the type, quantity, minimum capacity, installation, location, and spacing of the extinguishers.

While the FAA recognizes carbon dioxide, water, dry chemical, Halon 1211, Halon 1301, and halon combinations, at least two extinguishers installed on large transport aircraft must be of the Halon 1211 or combination Halon 1211/1301 type. Due to environmental problems associated with halogenated agents, their use is discouraged except in specific circumstances.

NFPA 408 no longer recognizes carbon dioxide and dry chemical as suitable for use on aircraft. Carbon dioxide is not recognized as suitable for the following reasons:

1. Ineffectiveness on Class A materials.
2. Chilling effect of solid carbon dioxide (snow) may damage delicate electronic components.
3. Weight of container and agent *vs.* similar agent capacities of the halon type.

Dry chemical has long been banned from aircraft in NFPA 408 for the following reasons:

1. Obscuration of the operator's vision, particularly in confined spaces.
2. Possibility of an insulating chemical layer forming on delicate electrical contacts, which could affect continued safety in flight.
3. Dry chemical can irritate the eyes and mucous membranes even though it is not considered toxic.
4. If not immediately cleaned up, certain types of dry chemical can be highly corrosive to aircraft metals.

Some concern has been expressed over the toxicity of both the halon agents and the products of decomposition in the event of a fire. Figure 9-36C shows the effect of discharging Halon 1211 in both a ventilated and unventilated cockpit of a Boeing B-737. These data were generally confirmed by tests the FAA Technical Center conducted in small aircraft.

Products of halon decompostion are indeed toxic; however, compared to the toxic products of combustion of cabin and flight deck furnishings, the toxicity problem is ncgligible. (See NFPA 12A, *Standard on Halon 1301 Fire Extinguishing Systems.*)

The addition of a flexible discharge device, as recommended by NFPA 408, greatly enhances the capability of the extinguisher to reach hard-to-get areas behind the instrument panels, under seats, etc. The device enables the operator to hold the unit upright during discharge, gaining full use of the agent charge. Figures 9-36D and 9-36E illustrate the advantage of using an extinguisher equipped with a flexible hose and nozzle over one with only a short nozzle.

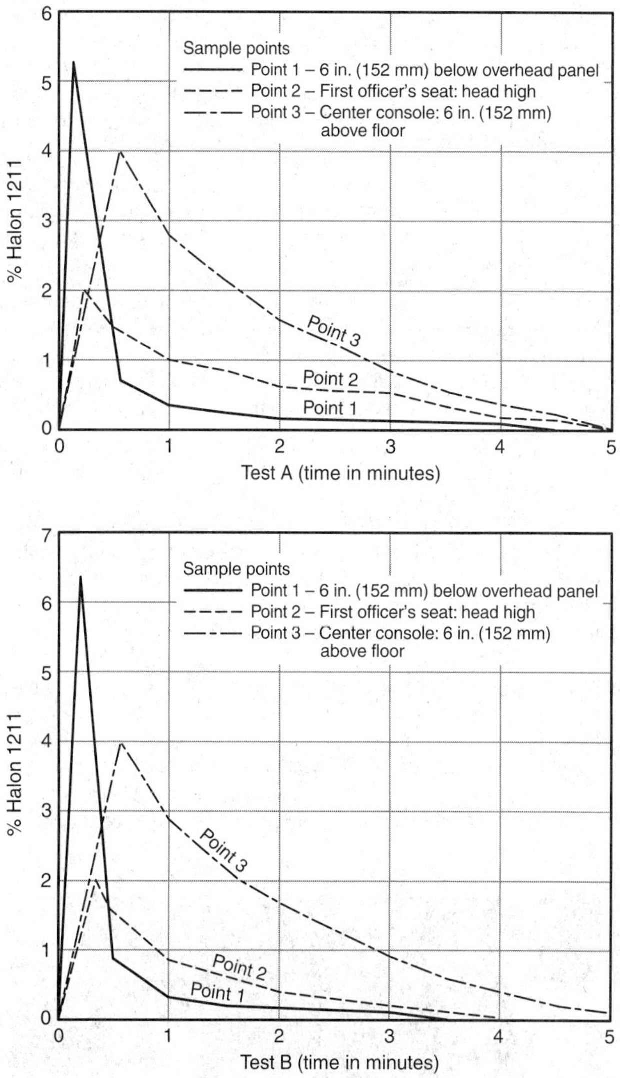

MEANS OF EGRESS FROM AIRCRAFT

In aircraft, emergency exit facilities are particularly important because postcrash fire exposure can be severe and safe evacuation time for the occupants is limited.

U.S. Federal Air Regulations set the number of exits required in transport category aircraft, depending upon passenger capacity. The manufacturer or operator must demonstrate that evacuating a full load of passengers can be accomplished in 90 sec or less, with half the available exits inoperative. These tests require that the test subjects represent an average passenger load, with certain percentages of able-bodied males and females, elderly, children, and handicapped. Tests are accomplished without normal cabin lighting; emergency lighting, however, is functional. The exits to be blocked are not preannounced to crew or occupants.

While the data obtained are of value, the tests have some shortcomings:

1. Smoke, which could obscure emergency lights or exit signs, is not used.
2. The test subjects know this is an emergency exit certification test and will seek any usable exit. (In real life, most passengers may pay little attention to preflight announcements and, consequently, tend to attempt leaving the aircraft by the same door through which they entered.)

FIG. 9-36C. *Halon concentration in a B-737 cockpit (240 cu ft or 7 m³) during and after discharge of a 3-1b (1.4-kg) Halon 1211 extinguisher. Altitude sea level, temperature 68°F (20°C), 11.5-sec discharge time, cockpit doors and windows closed. Test A, no forced ventilation. Test B, calculated ventilation rate 1.9 air changes per min, the minimum for this aircraft. (Source: United Airlines)*

3. Again, many test subjects have taken part in previous demonstrations and, while there is a required minimum time period since last participation, retain a certain knowledge of "what to do in the test."

Escape Systems and Devices

It is acceptable and permissible to utilize the normal entrance and passenger service doors as part of the emergency exit system. If the door is power assisted in its normal mode of operation and operates relatively slowly, it is equipped with a high-speed power assist for the emergency mode.

Modern passenger aircraft are equipped with inflatable evacuation slides at all emergency exits that have a normal door sill height that exceeds 5 ft (1.5 m). The distance from door sill to ground level, however, can drastically change in the event of a collapse of the main gear or nose gear. (A deployed slide under these conditions would not provide an effective angle for safe descent.) FAA regulations require the slide to deploy and inflate automatically when an exit door opens in emergency mode. In many instances evacuation slides can also serve as life rafts. Figure 9-36F shows one type of slide deployed.

The openings designated as emergency exits must be operable from the outside and designed to resist jamming due to fuselage distortion. The slides must deploy properly, regardless of wind direction or other influences.

The fire resistance of the slide material used to be very low, and even slight exposure to flame caused deflation of the slide. However, slide material has now been developed that incorporates a reflective coating, reducing the possibility of damage and deflation caused by radiant heat. At present, not all aircraft have been upgraded with this new material.

Obscuration of exit signs and emergency exit lighting by smoke can seriously impede evacuation of the aircraft. As of 1985, all air transport aircraft are required to have exit signs installed at a lower level and the exit access must be visible from a minimum distance of 35 ft (11 m), assuming a smoke level above the seat back height. In addition, illuminated exit path lights installed in aisle seat armrests or in the floor must be provided.

SPECIAL MILITARY AIRCRAFT HAZARDS

Among the special hazards of military aircraft are ejection seats and canopy ejectors; armaments, consisting of machine guns and automatic rapid-fire cannons, bombs, rockets, nuclear weapons, pyrotechnics, and rocket propellants; and special high-energy fuels.

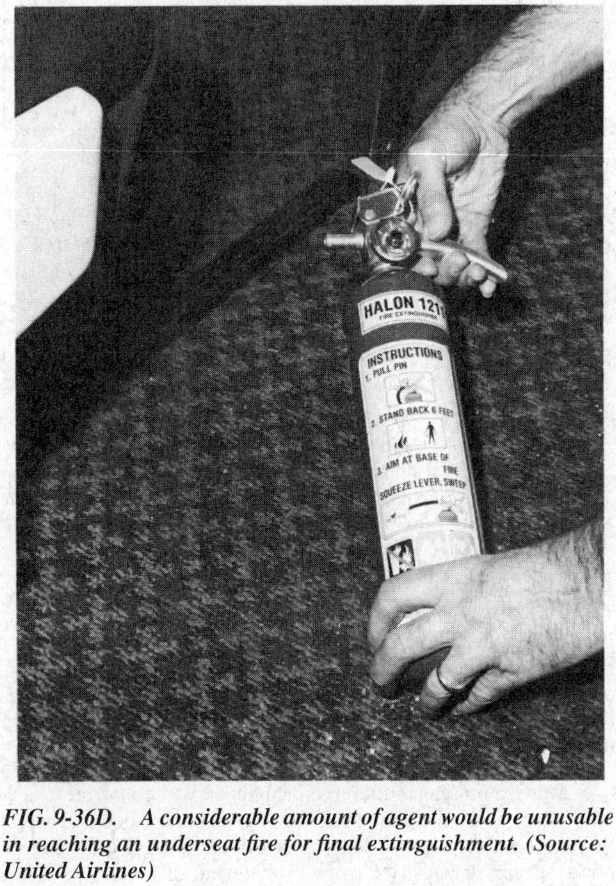

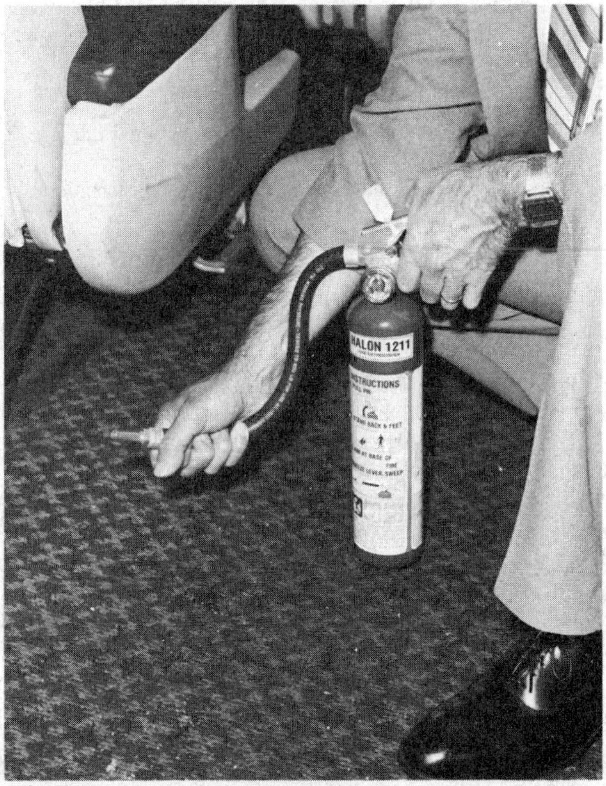

FIG. 9-36E.   *In the same situation, full use of the contents is obvious. (Source: United Airlines)*

FIG. 9-36D.   *A considerable amount of agent would be unusable in reaching an underseat fire for final extinguishment. (Source: United Airlines)*

FIG. 9-36F.   *One type of inflatable, overwing ramp/slide escape device used on wide-body aircraft. (Source: United Airlines)*

Information concerning fire and rescue for U.S. military aircraft is available from military aviation installations and is given freely to organized fire departments. Information is only limited on subjects that may be sensitive to national military security. Excerpts appear in NFPA 402, *Guide for Aircraft Rescue and Fire Fighting Operations.*

AIRCRAFT RESCUE AND FIRE FIGHTING

Two distinct types of ground accidents are part of the flight environment of the aircraft (excluding fires from fueling accidents, malfunction of ground servicing equipment, or fires originating during maintenance procedures).

The first type of ground accident is the high-speed and/or high-angle impact with the ground or other object that results in such structural breakup of the aircraft that survival of the occupants is highly unlikely (i.e., fatalities occur due to impact trauma).

The second accident type involves relatively low-speed and/or shallow ground impact angles. In this type of accident, occupant survival rate can be very high, especially if there is no postcrash fire, or if the fire is controlled quickly in the area immediately surrounding the occupied portion of the aircraft. Figure 9-36G illustrates an impact-survivable accident.

Fire can be anticipated in almost all abnormal aircraft landings as well as incidental occurrences such as overheated brakes or undetected fuel leaks. It is therefore essential that airports maintain a well-equipped and trained rescue and fire-fighting department during all hours that flight operations are in progress. At smaller airports the department is often made up of persons having other airport duties such as security, fueling operations, maintenance, or baggage handling. However staffed, personnel need to participate in frequent and meaningful training sessions, be equipped with proper protective gear, and be provided with the means to respond immediately to the scene of any emergency on or adjacent to the airport.

The term "rescue" used in the context of aircraft rescue and fire-fighting means the control of life-threatening fire in the critical area during the time necessary for all physically able aircraft occupants to self-evacuate. The continuance of fire control is necessary in order to permit the removal of those who are severely injured or pinned in the wreckage. Assisted rescue of individuals from small aircraft is often routine; however, in aircraft having a large number of occupants, such assistance is usually not possible in the time required.

Survivable accidents tend to occur within a certain area related to the runway centerline and threshold lights. This "critical rescue and fire-fighting access area" (CRFFAA) is clearly defined and illustrated in NFPA 402. (See Figure 9-36H.)

To be effective, the demonstrated response time of the first responding fire-fighting vehicle to reach any point on the operational runway must be two minutes or less, and to any point remaining in the CRFFAA, three minutes or less. These conditions must be met when flight operations are in progress; however, NFPA 402 recognizes that mitigating circumstances can make it impossible to meet these times.

Preplanning for declared emergency landings on airports is extremely important. Pre-positioning of personnel and vehicles greatly reduces response time and increases the chance of survival of the aircraft occupants should the landing become a major emergency. Figure 9-36I illustrates one system of pre-planning for this type of emergency.

Specialized vehicles have been developed that have the ability to transport rescue and fire-fighting personnel, equipment, and extinguishing agents to the scene of aircraft accidents within the required response time regardless of terrain surface or condition. Detailed information regarding such vehicles are contained in NFPA 414, *Standard for Aircraft Rescue and Fire Fighting Vehicles.*

The number and size of vehicles and the amounts of extinguishing agents required at a given airport depends on the types of aircraft used and the frequency of air carrier operations. The minimum level of protection necessary was determined in part by the use of NFPA

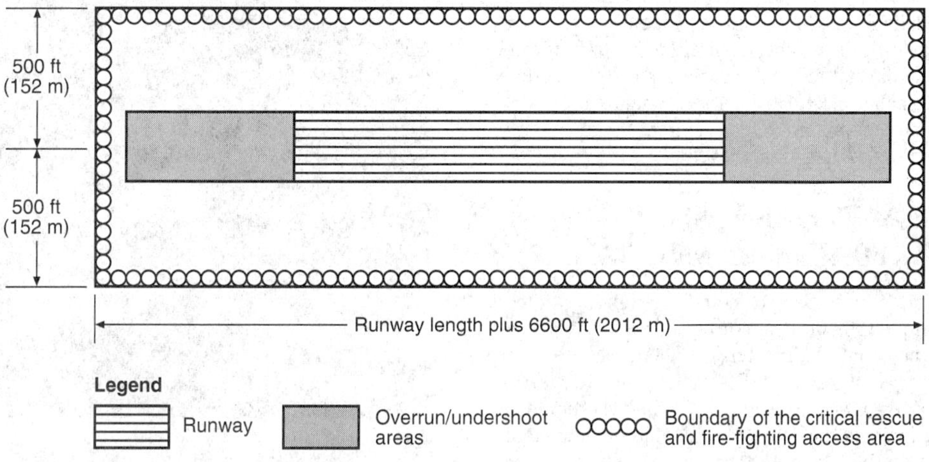

FIG. 9-36G. An impact-survivable accident followed by fire. (Photograph from James J. Brenneman collection)

500 ft (152 m)

500 ft (152 m)

Runway length plus 6600 ft (2012 m)

Legend

Runway

Overrun/undershoot areas

Boundary of the critical rescue and fire-fighting access area

FIG. 9-36H. Critical rescue and fire-fighting access area (CRFFAA).

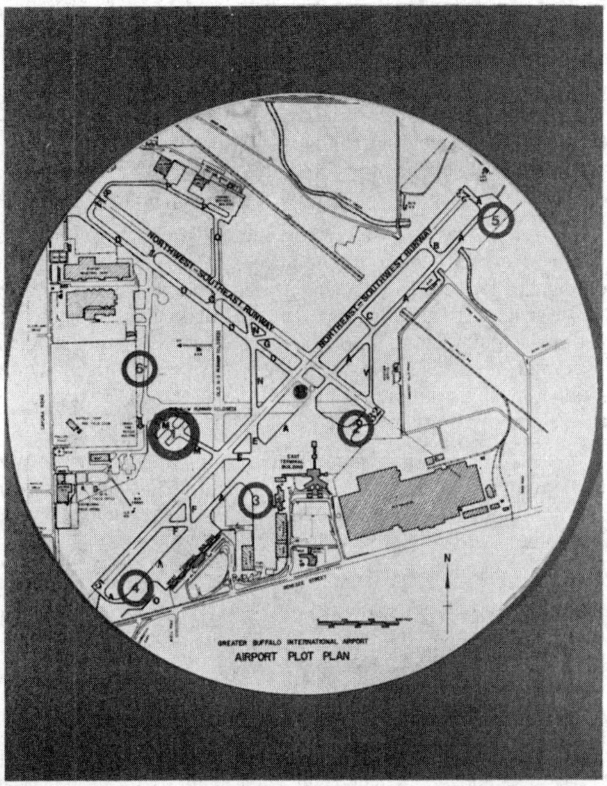

FIG. 9-36I. *Typical airport plot plan indicating pre-planned standby positions for rescue and fire-fighting (RFF) vehicle.*

standards and private consultant studies. Part 139 of Title 14 of the Code of Federal Regulations (CFR) contains these minimum levels. In addition, the FAA Advisory Circular 150/5210-6B gives recommended levels of airport protection that are considerably higher than the minimum levels in Part 139.

Other information on the subject of aircraft rescue and fire-fighting protection is published in NFPA 403, *Standard for Aircraft Rescue and Fire Fighting Services at Airports*; and the International Civil Aviation Organization (ICAO) in Annex 14 (Aerodromes) to the Convention on Civil Aviation.

There are interesting similarities of requirements among NFPA, ICAO, and the FAA Advisory Circular 150/5210-6B, as compared to the minimums of CFR Part 139. One major difference among the three is that ICAO and the FAA allow a remission factor so an airport may drop to a lower index if the number of movements (takeoffs and landings) of the largest aircraft to use the airport is below a certain daily or yearly operational level. NFPA does not recognize the remission factor, requiring protection for the largest aircraft that regularly uses the airport, regardless of operational levels.

RESCUE AND FIRE-FIGHTING PERSONNEL, TRAINING, AND SAFETY

After requirements for an airport's rescue and fire-fighting protection have been determined, the next major consideration is the selection of qualified personnel. Airport fire fighters should, at a minimum, meet standards set forth in NFPA 1003, *Standard for Airport Fire Fighter Professional Qualifications*. Candidates should be in top physical condition, and the department should maintain a required physical conditioning program to ensure the best possible performance of its personnel. Physical and mental stresses encoun-

tered when dealing with a major aircraft accident can be very demanding, and efficient performance is crucial.

In addition to physical conditioning, frequently scheduled and meaningful training sessions covering all phases of aircraft rescue and fire fighting are extremely important. Such sessions are necessary to maintain a high degree of competence and readiness. One of the most effective "hands-on" type of training is the "live fire." (See Figure 9-36J.) This realistic training scenario lets fire fighters experience an actual aircraft fuel fire. Training of this type produces confidence in fire fighters that with proper knowledge and equipment a fire of this nature can be successfully controlled in the time required to save lives.

Recently "live fire" training using hydrocarbon fuels and other flammable liquids has been suspended at many airports due to local environmental constraints. As a result, airport rescue and fire-fighting crews are being denied one of the most effective training exercises. Currently the FAA, the U.S. military, and private industry are all in the process of developing "live fire" training facility standards intended to duplicate realistic fuel fire atmospheres and, at the same time, meet all necessary environmental requirements.

Special aircraft informational charts developed by airframe manufacturers are available to airport fire fighters as well as mutual aid fire departments adjacent to an airport. These charts indicate locations and operation of emergency exits, areas that normally contain flammable and hazardous materials, oxygen storage, and other pertinent information. Combined with pre-incident planning inspections of aircraft that normally use the airport, periodic review of the charts keep fire fighters current on pertinent information on various types and models of aircraft. Most airline operators encourage familiarization with their aircraft and schedule tours when flight schedules permit. A variety of aircraft information charts are contained in Appendix A of NFPA 402M.

Anywhere in the vicinity of operating aircraft extreme danger exists, especially with regard to rotating propellers and rotors and jet engines. Severe accidents can occur when persons do not remain alert and do not maintain the necessary safe distance. Figure 9-36K depicts typical jet engine intake and exhaust danger areas.

AIRPORT AND HELIPORT DESIGN AND SAFETY

Many factors influence the safe operation of aircraft at airports: runway length and construction, navigational aids, clearances adjacent

FIG. 9-36J. *Large scale "live fire" training session. (Photograph from James J. Brenneman collection)*

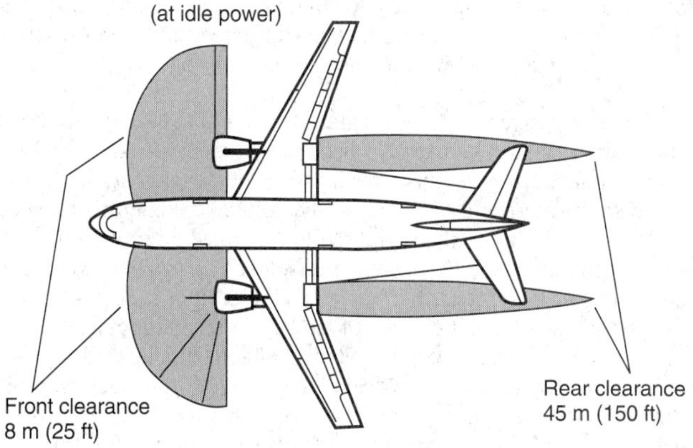

FIG. 9-36K. *Engine intake and exhaust danger areas of a typical modern jet aircraft. Crosswinds would considerably affect contours shown. (Drawing based on Boeing Commercial Airplane Co. data)*

(at idle power)

Front clearance
8 m (25 ft)

Rear clearance
45 m (150 ft)

to runways, off-pavement terrain conditions, zoning regulations, and fire protection.

Serious accidents can occur off the ends of runways due to insufficient clearance from various objects. The problem may be compounded where zoning regulations permit construction in these areas. Dikes, structures supporting navigational aids, and blast fences may be struck by aircraft operating in conditions of poor visibility.

Atmospheric conditions can also contribute to serious aircraft accidents. Fog, rain, sleet, and snow all adversely affect runway surfaces, aircraft performance, and visibility from the flight deck. Abnormal wind conditions, particularly the phenomenon called windshear, also can be factors. Normally, unusual atmospheric conditions are handled routinely. When they are involved in combination with too many other factors, accidents can occur.

During airport construction and repair, especially on movement areas, presence of construction and resurfacing equipment can be a factor in accidents. Another principal area where accidents can occur is in the overrun portions of runways when aircraft are unable to brake effectively, running off the runway, striking airport boundary fences, or bogging down while traversing unimproved ground surfaces. Unimproved ground surfaces between runways can result in aircraft being inaccessible to fire and rescue equipment, especially equipment not specifically designed for off-highway use.

Heliports present special problems because they are frequently located in congested areas of cities, on roofs of buildings, near hospitals, and on piers adjacent to water. NFPA 418, *Standard for Heliports,* covers landing deck construction, drainage, egress, and fire protection requirements.

PRE-PLANNING FOR AIRCRAFT ACCIDENT EMERGENCIES

All commercial airports serving passenger aircraft are required to develop and maintain an emergency plan that can be immediately activated should an aircraft accident occur on or near the airport. The plan needs to be flexible and encompass all aspects of the most difficult of anticipated emergencies. All local emergency services, civic organizations, and resources need to be included in the plan. To be effective, a full-scale exercise of the plan should be conducted at least once a year to test and update its effectiveness. Refer to

NFPA 424, *Guide for Airport/Community Emergency Planning,* for further guidance.

Airports located adjacent to large bodies of water must give special consideration to the possibility that an aircraft accident could occur in the water. In these instances, special water craft equipment dependent on the specific need should be provided. In addition, mutual aid agreements should be established with all local water rescue resources such as the Coast Guard, Navy, and fire boats.

SPECIAL AIRPORT FACILITIES

Hangars

An aircraft hangar is simply a building built to provide weather protection and shop space during aircraft maintenance and storage. Hangars pose unusual fire protection problems due to their occupancy's special nature. Since removing all fuel from an aircraft before moving it into a hangar is often impractical and economically unsound, potential exists for having large quantities of flammable liquids, mainly aviation fuel, inside the hangar. Quite often, aircraft contained in the hangar, particularly large aircraft, are several times more valuable than the hangar, necessitating some measure of protection for the aircraft as well as the hangar structure.

Fixed hangar fire protection systems of the past provided protection for the hangar structure and consisted mainly of water-type sprinkler systems using the extra-hazard pipe schedule, or hydraulically calculated systems, primarily of the deluge type. In aircraft hangars, these water systems are no longer recognized as viable protection by NFPA 409, *Standard on Aircraft Hangars.* With increased size and fuel loading of the aircraft, using foam-water deluge sprinkler systems is now common practice. Current designs incorporate foam-producing and oscillating monitor nozzles, in conjunction with previously mentioned systems, to provide fire control/extinguishment for the shadow area beneath the wings of large aircraft. (Figure 9-36L illustrates this shadow area.) Adequate

FIG. 9-36L. *A typical wide-body aircraft maintenance and overhaul facility and structure. Foam-water deluge is the primary protection, with fixed-pipe foam-water deluge protection installed beneath the work platforms for the shadow area. Swing joints and flexible connections are provided to allow limited maintenance structure movement to position the aircraft. Structure above the wing and flexible connectors are for fuel tank ventilation. (Source: United Airlines)*

protection of the hangar structure can be achieved as well as providing increased protection for the aircraft, primarily due to more rapid fire control and/or extinguishment.

In older hangars being modernized, a different technique has more recently been used with the approval of the authority having jurisdiction. This technique utilizes overhead closed-head type sprinkler systems hydraulically designed to deliver a specified minimum water density over a specified area. It's capable of delivering aqueous film-forming foam (AFFF) solution through standard (nonaspirating) sprinklers by utilizing wide-range proportioning devices.

Primary fire extinguishing capability affording some degree of aircraft protection is provided by oscillating monitors that completely blanket the floor of the entire hangar with foam within 20 to 30 sec of actuation.

Whether the foam delivery system is used only for shadow area protection or total floor coverage, foam monitor location, oscillation speed, foam pattern, and barrel centerline height above the floor are critical. Monitors should be located so that all areas to be covered can be easily reached. Experience indicates that, depending upon the arc of coverage, no more than 15 sec should be necessary for a complete cycle. The foam pattern should be dispersed rather than a straight stream. This minimizes turbulence at the point of stream impact with objects or the floor. Barrel centerline height should be a minimum of 12 ft (4 m) above the floor to extend range and to minimize obstruction by aircraft maintenance equipment, etc. Figure 9-36M illustrates an installation meeting these criteria.

Due to hangar size or multiple aircraft positions, monitor nozzles can be mounted at the bottom of the roof truss space to provide coverage for these void spaces. Arrangements should be made to exercise and test the oscillating mechanism from the hangar floor or other easily accessible location. (See Figure 9-36N.)

The principal means of fire detection for actuation of overhead sprinkler systems in hangars has been rate-of-rise type devices; however, as hangar roof heights have increased to as much as 150 ft (46 m) from the floor to the roof deck, the ability of this type of device to perform within the time required to minimize aircraft damage is limited. In climates with moderate to severe winter conditions, it is becoming increasingly difficult to maintain stability of such systems, due to the opening and closing of hangar doors and the installation of high-recovery-rate heating systems, and still maintain the sensitivity necessary for rapid fire detection. Optical type detectors of ultraviolet/infrared type or dual function can be used for high-speed detection of relatively small fires and actuation of monitor nozzles. Rapid temperature changes do not affect those devices.

Care must be taken in selecting the type of detector and control devices for optical detection. They must have demonstrated operational integrity, yet not be overly sensitive and subject to false or unwanted operation. Considerable progress has been made recently in developing optical detection systems that automatically ignore or are insensitive to routine aircraft maintenance procedures involving welding, high-intensity lights, boroscopes, radioisotopes, X-ray inspections, etc. Whenever an optical system is installed, its sensitivity to fire should be tested. Further, a recording device should be attached to provide a hard-copy record of what detectors "see" during routine aircraft maintenance operations. During the initial period after installation (usually 60 to 90 days), the detection systems should be in the "alarm only" mode and a careful record made of the recording device printout. Aircraft maintenance records often can determine what operation affected the optical detectors.

Floor drainage is another overlooked portion of the hangar fire protection system. The drainage system removes excess fire protection water and also removes large quantities of liquid fuel from the fire scene. To prove their efficiency, these systems should be flushed thoroughly with high volumes of water at least annually. Oil/water separators should be meticulously maintained to prevent pollution of lakes and streams and to reduce hazards in the airport drainage system to which the hangar system may be connected.

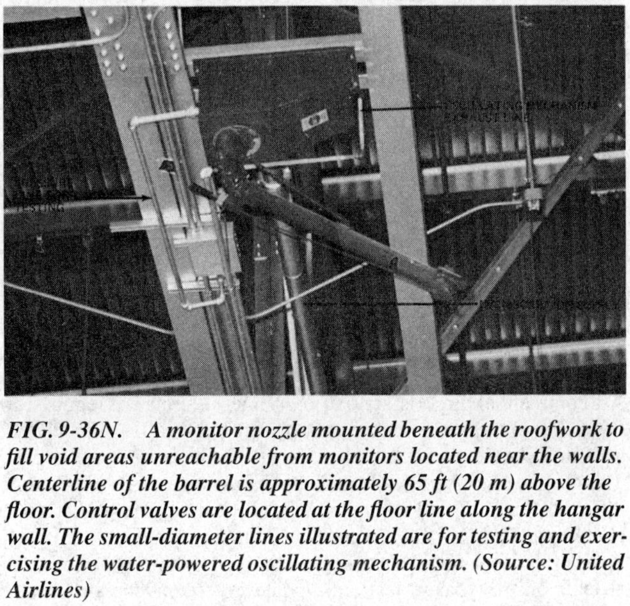

FIG. 9-36M.   A foam monitor nozzle. Centerline of this particular nozzle is approximately 12 ft (4 m) above floor level, to clear obstructions. (Source: United Airlines)

FIG. 9-36N.   A monitor nozzle mounted beneath the roofwork to fill void areas unreachable from monitors located near the walls. Centerline of the barrel is approximately 65 ft (20 m) above the floor. Control valves are located at the floor line along the hangar wall. The small-diameter lines illustrated are for testing and exercising the water-powered oscillating mechanism. (Source: United Airlines)

In recent years environmental concerns have necessitated more restrictions on the disposal of AFFF solution subsequent to discharge. Therefore, new system designs should consider the provision of bypasses and retentional areas for initial and periodic system testing.

Aircraft hangars are divided into Types I, II, or III, which roughly correspond to the types of aircraft they house and their anticipated fuel loadings. NFPA 409 defines these types and outlines requirements for construction, protection, drainage, fire cutoffs, and other provisions for each type of hangar.

Airport Terminal Buildings

Airport terminal buildings include any fully enclosed extensions that function as passenger concourses (sometimes called fingers or piers) and satellite buildings that serve passenger-handling functions. The satellite building may be connected to the main terminal building via tunnels beneath the aircraft operating ramp or via people-mover transit systems operating on a fixed guideway, above or below ground. Terminals also contain airline-related businesses, such as offices, restaurants, and gift shops.

Special attention is given to airport terminal buildings in NFPA 415, *Standard on Airport Terminal Buildings, Fueling Ramp Drainage, and Loading* (hereinafter referred to as NFPA 415). These buildings have a severe fire exposure potential from immediately adjacent aircraft fueling and servicing operations and the high-occupancy (i.e., people) load in terminals.

Walls and glass areas of terminal buildings and other airport structures that may be exposed to jet blast or explosion require design for wind loads in excess of that which building codes normally require.

Aircraft Loading Bridges or Walkways

This equipment is frequently installed at the terminal area for passenger convenience and to protect passengers from the weather while they move from the terminal loading gate area to the aircraft. Loading equipment also increases safety by reducing ramp congestion and segregating passengers from the aircraft servicing equipment and personnel.

In the event of fire caused by a fueling malfunction or other reason, a loading bridge or walkway can serve as a means of egress from the aircraft to a safe refuge, e.g., the terminal building. Most people naturally tend to exit an area (aircraft or building) the same way they entered. Because the emergency situation requiring evacuation may occur before the preflight safety and evacuation briefing, the loading bridge may be the only means of egress familiar to passengers. For this reason, NFPA 415 requires 5 min of safe passage from the aircraft to the terminal. Construction features, special fire protection, or a combination of both may accomplish this. Safe egress must be provided during severe fire exposure to the bridge.

Doors, if any, at the terminal end of the bridge should be equipped with panic hardware and swing in the direction of travel from the aircraft to the terminal. The loading bridge should not be considered part of the terminal egress system, unless a vestibule and stairs conforming to NFPA *101*®, *Life Safety Code*®, are provided at the terminal end, even though auxiliary stairs may be located near the aircraft end. Auxiliary stairs are strictly for the use of aircraft servicing personnel or flight crew.

Training of loading bridge operators in positioning the equipment and obtaining the best possible "seal" with the aircraft in *all* weather conditions is very important. Unless the bridge and closure curtain are properly positioned and utilized, the integrity of the loading bridge as a means of egress from the aircraft is seriously compromised. The operator must be aware that the closure curtain is a *safety* item, not merely weather protection.

Aircraft Fueling Ramp Drainage

The large amounts of flammable liquids (in the form of fuel) handled on aircraft servicing ramps demand that special attention be paid to the drainage system provided to prevent exposing the terminal or aircraft unnecessarily in the event of accidental release. NFPA 415 specifies the amount of slope and direction regarding ramp geometry, location of drainage system inlets, water seal traps, and oil/water separators.

AIRCRAFT MAINTENANCE AND SERVICING

To keep an aircraft airworthy, every part must be inspected, repaired, or overhauled periodically. The frequency of these procedures varies. Operations may be performed all at once in a single maintenance visit, or as part of a progressive overhaul system where a certain portion of a total overhaul is conducted periodically. In addition, periodic inspections are required between major overhauls to repair minor discrepancies, inspect critical structures, and test various components. Nonroutine maintenance can occur at any time to replace or repair a defective or prematurely failed part.

Many maintenance procedures involve the use of highly flammable solvents, sometimes unstable or toxic chemicals, and personnel entry into integral fuel tanks. When situations such as these occur, special procedures must be developed to ensure the safety of the operation. It is sometimes possible to pressurize equipment housings with shop air, using appropriate power/pressure interlock to prevent penetration of flammable vapors. Localized ventilation and curtailment of operations utilizing flammable or combustible liquids are also helpful. In any event, each potentially hazardous procedure should be developed or reviewed by a person who is thoroughly familiar with the basics of fire safety and aircraft maintenance requirements and procedures. NFPA 410, *Standard on Aircraft Maintenance*, should be used to develop local procedures for the more common hazardous aircraft maintenance operations, such as welding, spray painting, fuel tank ventilation, etc.

[NFPA 70, *National Electrical Code*®, defines certain areas in hangars and around aircraft as Class I, Group D, Division I or II locations. In this age of sophisticated testing using electronics, eddy currents, X-rays, etc., it is often impossible to design or obtain equipment listed or approved for these locations. Welding (usually inert gas-shielded arc) is frequently performed on installed engines or aircraft structures.]

Testing fuel tanks for system leaks is commonly performed in hangars using high flash point Jet A fuel or a special, usually combustible, test fluid. Toward this end, many hangars have a fixed system to board and unload fuel or test fluid. It has also been a practice to park a standard fuel servicing tank truck outside the hangar and run long hoses to the aircraft connection points. This practice is to be discouraged, because the hose is the weak point in any system. Although hoses are necessary for flexible connections, they should be kept as short as possible.

Painting the interior and exterior of aircraft is also accomplished in hangars. Although a few hangars have been constructed specifically for aircraft exterior painting, it is more common to control ignition sources and rely on the sheer volume of the hangar and natural ventilation to dissipate flammable vapors. Latex paints, which contain little if any solvent, are being used for aircraft interiors. Interior panel partitions and equipment can be painted in place rather than removed to a standard paint spray booth. Whenever using flammable paints inside an aircraft is necessary, special precautions for ventilation and control of ignition sources must be followed.

Aircraft Fueling

The potential exists for spills and fires each time an aircraft is fueled. Preventing such incidents requires use of NFPA 407, *Standard for Aircraft Fuel Servicing*, a document accepted worldwide. The standard outlines specific precautions to control the extent of this hazard and gives recommendations on the design of aircraft fueling, hose, aircraft fuel servicing tank vehicles, airport fixed fueling systems utilizing fuel hydrants, hydrant vehicles (to connect the fuel hydrant to the aircraft), as well as safety procedures to follow in the event of an accidental spill. A common type of spill is at the aircraft tank vent point, when an aircraft internal shutoff valve fails or closes slowly, allowing an overfill condition to develop.

BIBLIOGRAPHY

NFPA Codes, Standards, and Recommended Practices

Reference to the following NFPA codes, standards, and recommended practices will provide further information on the safeguards for aviation discussed in this chapter. (See the latest version of *The NFPA Catalog* for availability of current editions of the following documents.)

NFPA 12A, *Standard on Halon 1301 Fire Extinguishing Systems*
NFPA 70, *National Electrical Code®*
NFPA 101®, *Life Safety Code®*
NFPA 402, *Guide for Aircraft Rescue and Fire Fighting Operations*
NFPA 403, *Standard for Aircraft Rescue and Fire Fighting Services at Airports*
NFPA 407, *Standard for Aircraft Fuel Servicing*
NFPA 408, *Standard for Aircraft Hand Portable Fire Extinguishers*
NFPA 409, *Standard on Aircraft Hangars*
NFPA 410, *Standard on Aircraft Maintenance*
NFPA 412, *Standard for Evacuating Aircraft Rescue and Fire Fighting Foam Equipment*
NFPA 414, *Standard for Aircraft Rescue and Fire Fighting Vehicles*
NFPA 415, *Standard on Airport Terminal Buildings, Fueling Ramp Drainage, and Loading*
NFPA 418, *Standard for Heliports*
NFPA 422, *Guide for Aircraft Accident Response*
NFPA 424, *Guide for Airport/Community Emergency Planning*
NFPA 1003, *Standard for Airport Fire Fighter Professional Qualifications*

Additional Reading*

Published works, including those of many foreign governments and scientists, are available from the National Technical Information Service, Springfield, VA 22161, under the headings "Transportation Safety" and "Air Transportation." In addition, NFPA has, from time to time, published reports of tests, aircraft accidents, etc., which have significant lessons for, or impact upon, the aviation community. The serious researcher, fire investigator, student, or aviation fire safety enthusiast should contact these organizations for an index of materials available.

Federal Aviation Administration Advisory Circulars

The Federal Aviation Agency of the U.S. DOT publishes circulars with the following designations and titles:

Aircraft Fire and Rescue Communications, 150/5210-7B, April 30, 1984.

---

*To attempt to list all relevant published work in the limited space available is impossible. Research and testing are being conducted continually throughout the world in aviation fire safety, primarily funded by various governmental agencies. This research, testing, etc., encompasses all aspects of aviation fire safety, emergency evacuation, new concepts and devices in detection and extinguishment, extinguishing agents, etc.

Aircraft Fire and Rescue Facilities and Extinguishing Agents, 150/5210-6B, Jan. 25, 1973.

Airport Fire and Rescue Vehicle Specification Guide, 150/5220-14, March 15, 1979.

Analysis of Test Criteria for Specifying Foam Firefighting Agents for Aircraft Rescue and Firefighting, August 1994.

Babrauskas, V., "Role of Aircraft Panel Materials in Cabin Fires and Their Properties," DOT/FAA/CT-84/30, June 1985, Department of Transportation.

Bennett, M., "Halon Replacement for Aviation Systems," *Proceedings* of the 1992 International CFC and Halon Alternatives Conference, Stratospheric Ozone Protection for the 90's. Sept. 29–Oct. 1, 1992, Washington, DC, 1992, pp. 667–670.

Berg, H. D., "Future Needs in the Development of Materials for Aircraft Interiors and Equipment," Deutsche Aerospace Airbus GmbH, West Germany, DOT/FAA/CT-93/3; International Conference for the Promotion of Advanced Fire Resistant Aircraft Interior Materials, Feb. 9–11, 1993, Atlantic City, NJ, 1993, pp. 333–337.

Blake, D. R., "Development and Growth of Inaccessible Aircraft Fires Under Infligh [sic] Airflow Conditions. Final Report," Federal Aviation Admin., Atlantic City Airport, NJ, DOT/FAA/CT-91/12, Feb. 1991.

Boeing Commercial Airplanes, "Aircraft Material Fire Test Handbook," Final Report, Oct. 1987–Sept. 1990, Boeing Commercial Airplanes, Seattle, WA, Federal Aviation Admin., Atlantic City International Airport, NJ, DOT/FAA/CT-89/15, ACD-240, Sept. 1990.

Bottomley, D. M., Muir, H. C., and Lower, M. C., "Aircraft Evacuations: The Effect of a Cabin Water Spray System Upon Evacuation Rates and Behavior," Civil Aviation Authority, London, England, CAA Paper 93008, March 1993.

Civil Aviation Authority, "Improving Passenger Survivability in Aircraft Fires: a Review," Civil Aviation Authority, London, England, CAP 586, April 1991.

DeBaggis, A., "Polyimide Foam for Advanced Fire Resistant Aircraft Applications," Illbruck Inc., Minneapolis, MN, DOT/FAA/CT-93/3; International Conference for the Promotion of Advanced Fire Resistant Aircraft Interior Materials, Feb. 9–11, 1993, Atlantic City, NJ, 1993, pp. 37–48.

Denny, S. A., and Mowrer, F. W., "Development of a Hypermedia Knowledge Base of Aircraft Fire Safety Regulations," Maryland Univ., College Park, DOT/FAA/CT-93/3; International Conference for the Promotion of Advanced Fire Resistant Aircraft Interior Materials, Feb. 9–11, 1993, Atlantic City, NJ, 1993, pp. 321–331.

Diehl, R. G., "Applications of Continuous Fiber Reinforced Thermoplastics in Aircraft Interiors," Fokker Aircraft B.V., Amsterdam, The Netherlands, DOT/FAA/CT-93/3; International Conference for the Promotion of Advanced Fire Resistant Aircraft Interior Materials, Feb. 9–11, 1993, Atlantic City, NJ, 1993, pp. 93–104.

Fire and Rescue Service for Certificated Airports, 150/5210-12, March 2, 1972.

Fire Department Responsibility in Protecting Evidence at the Scene of an Aircraft Accident, 520/5200-12A, April 8, 1985.

Fire Fighting Exemptions under the 1976 Amendment to the Federal Aviation Act, 150/5280-3, Feb. 4, 1977.

Fire Hazards of Mixed Fuels on the Flight Deck, Naval Research Labratory, Washington, DC April 28, 1993.

Fire Prevention During Aircraft Fueling Operations, 150/5230-3, April 8, 1969.

Grosshandler, W. L., *et al.*, "Assessing Halon Alternatives for Aircraft Engine Nacelle Fire Suppression," *Journal of Heat Transfer*, Vol. 117, May 1995, pp. 489–494.

Heliport Design Guide, 150/5390-1B, August 22, 1977.

Klem, T. J., and Conroy, M. T., "NFPA Alert Bulletin: The Crash of United Flight 232," *Fire Command*, Vol. 56, No. 10, Oct. 1989, pp. 22–24.

Lodge, J., "Manchester Jet Fire—Remarkable Evidence," *Fire International*, Vol. 13, No. 116, April/May 1989, pp. 34–35.

McCaffrey, B. J., and Rinkinin, W. J., "Fire Environment in Counterflow Ventilation (The In-Flight Cabin Aircraft Fire Problem)," NBSIR 88-3806, June 1988, Center for Fire Research, Gaithersberg, MD.

McCaffrey, B. J., *et al.*, "Model Study of the Aircraft Cabin Environment Resulting From In-Flight Fires. Final Report," Maryland Univ., College Park, National Institute of Standards and Technology, Gaithersburg, MD, Federal Aviation Administration, Atlantic City International Airport, NJ, DOT/FAA/CT-90/22, Nov. 1992.

Moussa, N. A., "Flammability of Aircraft Fuels," BlazeTech Corp., Winchester, MA, 901949; SAE, Aerospace Technology Conference and Exposition, SAE Technical Paper Series, Oct. 1–4, 1990, Long Beach, CA, 1990, pp. 1–10.

Murray, T. M., "Airplane Accidents and Fires," *Proceedings* of Improved Fire- and Smoke-Resistant Materials for Commercial Aircraft Interiors, Nov. 8–10, 1994, Washington, DC, National Academy Press, Washington, DC, 1994, pp. 7–23.

National Transportation Safety Board, "Aircraft Accident Report. Fuel Farm Fire at Stapleton International Airport, Denver, Colorado, November 25, 1980," National Transportation Safety Board, Washington, DC, NTSB/AAR-91/07; Notation 5576, Oct. 1, 1991.

Quaglia, C., and Grubits, S. J., "Fire-Safety Engineering Design of Airport Terminals," *Proceedings* of the First International Conference on Fire Science and Engineering, ASIAFLAM '95, March 15–16, 1995, Kowloon, Hong Kong, 1995, pp. 561–567.

Response to Aircraft Emergencies, 150/5210-11, April 15, 1969.

"Safety from Fire in Aircraft," *Aircraft Engineering*, Vol. 59, No. 3, 1987, pp. 19-20, 22.

Scheffey, J. L., *et al.*, "Comparative Analysis of Film Forming Fluoroprotein Foam (FFFP) and Aqueous Film Forming Foam (AFFF) for Aircraft Rescue and Fire Fighting Services," Hughes Associates, Inc., Wheaton, MD, Naval Research Lab., Washington, DC, Naval Sea Systems Command, Washington, DC, Report 2108-A01-90, June 1990.

Water Supply Systems for Aircraft Fire and Rescue Protection, 150/5220-4A, Dec. 11, 1985.

Webb, S., West Sussex Fire Brigade, *Faro DC-10 Crash*, 21 December, 1992.

Williams, C., and Laughlin, J., eds., *Aircraft Fire Protection and Rescue Procedures*, International Fire Service Training Association, Oklahoma State University, Stillwater, OK, 1978.

Wright, J., Hampton, L., and Do, D., "Driver's Enhanced Vision System," FAA Technical Center, Atlantic City, NJ, January 1995.

MARINE

Revised by Marjorie Murtagh and Robert Loeser

According to United States Coast Guard (USCG) statistics from 1983 to 1993, fires and explosions caused almost 10 percent of fatalities and 22 percent of injuries in marine vessel casualties. During that same 10-year period, fires and explosions were responsible for 7,505,500 gal (28,500 m³) [approximately 17 percent] of maritime oil spills in the United Sates. The International Maritime Organization's analysis of serious casualties to oil and chemical carriers during the 15-year period from 1977 to 1991 concluded that 25 percent were the result of fires and explosions, as were the loss of 801 lives, which represented 77 percent of the total fatalities.

This chapter discusses marine fire hazards, as well as the design, construction, and shipboard fire prevention systems normally installed to deal with them, on small pleasure craft, small commercial boats, and commercial vessels. It first covers pleasure and small commercial craft; large commercial vessels are the subject of the balance of the chapter.

Additional information about extinguishing systems suitable for use on vessels is found in Section 6, Chapter 10, "Automatic Sprinkler Systems"; Section 6, Chapter 12, "Water Spray Protection"; Section 6, Chapter 20, "Carbon Dioxide and Application Systems"; Section 6, Chapter 18, "Halogenated Agents and Systems"; Section 6, Chapter 22, "Foam Extinguishing Agents and Systems"; and Section 6, Chapter 23, "Fire Extinguishers Use and Maintenance."

PLEASURE AND SMALL COMMERCIAL BOAT HULLS

The latest boating safety statistics from the USCG (1994) indicate there are more than 20 million pleasure and small commercial boats in the United States. Statistics for the 5-year period from 1990 to 1994 show that annually there were approximately 6,450 accidents, 840 fatalities, and 3,800 injuries involving such boats. The report indicates that approximately 320 fire and explosion accidents caused, on average, about 10 fatalities and 168 injuries per year. These accidents also resulted in a 5-year total property damage of almost $40 million.

The introduction of the 1960 edition of NFPA 302, *Fire Protection Standard for Pleasure and Commercial Motor Craft*, con-

tained this statement, "There are few other uses of petroleum fuels by the public in which the fire and explosion hazards parallel those possible in motor craft." In spite of technical developments since that time, the statement remains true today.

Boats and vessels of all sizes are constructed to create a form that is closed on the bottom and all sides. This basic hull form is necessary for the boat to float. In the case of trains, aircraft, and all land vehicles (other than amphibious types), openings can be provided where needed to permit both liquid fuel leakage and heavier-than-air vapors to drain by gravity to the open atmosphere. Gasoline is commonly used for both propulsion engines and for onboard generators, and liquefied petroleum gas (LPG) is commonly used for cooking. Any leakage of these fuels and any gas or vapors released as a result of using those fuels is retained inside the hull. The two means of discharging gas and/or vapor "overboard" are: (1) the slow process of natural ventilation, which requires airflow over the boat (from boat-motion or wind-speed), and (2) mechanical ventilation, which is used before starting any engine. Exhaust blowers tend to be effective only in the general area of the exhaust blower pickup. By comparison, automobiles and trucks are normally completely open on the bottom, providing effective ventilation in all areas of the fuel system. Seepage of liquid fuel and vapors can freely drain to the open atmosphere along the road from any point in the system. Aircraft, likewise, have drains in their engine nacelles and under their wings, allowing vapors to be dispelled to the atmosphere in flight or on the ground.

Since bottom drains are not possible in a boat, safety on boats depends on a carefully maintained, leak-free fuel system and on ventilation capable of discharging heavier-than-air fuel vapors overboard. When vapors are discharged overboard from lower levels, fresh air is automatically introduced into the hull's upper levels. The fresh air tends to dilute vapors in the upper level. An adequate removal and dilution process keeps the vapor and/or gas mixture below the lower explosive limit.

Consider that 1 oz (0.3 L) of gasoline fuel can create about 11 cu ft (0.3 m³) of explosive vapor. Gasoline vapor is approximately four times as heavy as air and tends to sink to the bottom of the bilge. Natural or mechanical ventilation cannot render a boat safe in any reasonable time, if any liquid leakage occurs. An operating marine exhaust blower with a pickup duct placed 1 ft (0.3 m) above an open container of gasoline does not materially reduce the time required for the fuel to evaporate. Because an exhaust blower, or any ventilation, does not materially accelerate the fuel's vaporization rate, power ventilation should not be used when liquid fuel is

Marjorie Murtagh is chief of the National Transportation Safety Board's Marine Division. She is based in Washington, DC. Robert Loeser is head of Underwriters Laboratories' marine safety section. He is based in Palm Harbor, FL.

known to be present. Ventilation, in addition to removing some flammable vapors in the duct intake areas, introduces fresh air that can bring an overly rich mixture into the explosive range. Whenever liquid fuel is present, no electrical circuit should be turned on or off. The battery should not be connected or disconnected. An arc can occur when making or breaking a circuit.

Boat Hull Arrangement

The hull of a pleasure boat or small commercial vessel must be arranged so that all compartments are accessible as practicable for inspection and maintenance. Access must always be provided to sea cocks for emergency use. There must be two means of escape from all accommodation spaces in case fire in the engine space or the galley blocks one means of egress. Escape hatches must be unobstructed, readily accessible, and adequately sized and shaped for their designed purposes. Congestion of an engine compartment is unsafe; it may make sea cocks inaccessible and discourages adequate inspection and maintenance of engines and equipment. Adding internal combustion auxiliary machinery increases ventilation requirements.

Engine spaces (for propulsion engines, generators, and other auxiliary engines) must be separated from accommodation spaces by fore, aft, and transverse bulkheads and decks. The separation must serve as an effective fire barrier and minimize the escape of fire extinguishing media that may be discharged into the space. Unsealed bulkheads with a $^1/_4$-in. (6.4-mm) annular space around items penetrating the bulkhead are considered adequate for ignition-protection purposes and for confining fire extinguishing agents. However, these same bulkheads do not prevent carbon monoxide that may be present in the engine space from passing through the bulkhead to accommodation spaces.

Materials

Boat hulls, bulkheads, and superstructures are constructed of a wide variety of materials, including aluminum, steel, reinforced fiberglass laminates (FRP), and wood. In addition to solid fiberglass laminates there are fiberglass laminates with balsa wood or plastic honeycomb core materials and wooden structures that have been covered with a fiberglass laminate on both sides, making a sandwich construction. The combustibility, flame spread rate, smoke production, and heat release characteristics of these various flammable materials have not been investigated and classified for use on boasts. Thinner outer laminations of cored construction burn through relatively quickly, exposing core materials. Choice of material is determined by boat size, type, intended service, and cost. Horizontal deck surfaces, such as the deck above the engines, are very quickly involved in fire spread because of the low height and small volume of the compartments above which the decks are located. To minimize fire spread, fire-retardant resins and fire-retardant coatings suitable for marine service may be used. Because fire-retardant resins may reduce the structural strength of the laminate in which they are used, changes may be needed to compensate for strength differences. Upholstery and carpeting used in cockpits and accommodation spaces might have high-flame spread ratings. The flame spread ratings of all materials and fire-retardant coatings used should be carefully considered. As in engine spaces, ceiling heights in accommodation spaces are low compared to those in buildings, and the height reduction effectively accelerates horizontal flame spread.

Ventilation

Ventilation is defined as the positive changing of air, by natural or mechanical means, within a compartment. Boat ventilation is primarily designed to remove heavier-than-air flammable vapors from the bottom of the compartment. When air is removed from the compartment, fresh air is simultaneously introduced at the upper level.

The fresh air dilutes the flammable vapors in the upper levels where the intake ducts terminate.

Natural ventilation's effectiveness depends on outside air velocity and direction. Without wind or boat motion, there is virtually no natural ventilation other than that created by thermal air currents. The most important factors affecting air movement inside a hull are (1) the shape of the boat's superstructure (cabin or, in small open boats, the windshield), (2) the location of the boat cabin or windshield, and (3) the type and location of the ventilation fittings. With a bow wind, a boat cabin's shape makes it function like a ventilating clam shell, reducing pressure under the cabin. In runabouts, pressure is reduced behind the windshield. Reduced pressure results in an airflow to that area. For maximum ventilation efficiency, the engine space airflow that ventilating fittings create should be used to reinforce the airflow the cabin creates. Neither natural nor power ventilation effectively removes the hazard of vapors created by the presence of liquid gasoline because flammable vapors can be generated faster than they can be removed. Airflow created by the ventilating system only slightly affects the time necessary to evaporate liquid gasoline. Therefore, in the presence of liquid gasoline leakage, introducing fresh air by operating blowers may create, rather than eliminate, an explosive mixture. The boat's blowers should not be used.

Ventilation in an engine space plays an important role in controlling heat levels in that space. Ventilation for temperature control depends on natural airflow into the engine space when the boat is underway and on the efficiency of external ventilating fittings and ducts. These fittings and ducts must be large enough to supply the air required for the engines. The compartment pressure depression must be measured with all engines operating at full power. Exhaust blowers are not used to control temperature or to supply air to the compartment. The engines operate with block temperatures of about 140 to 180°F (60 to 80°C). Because of anticipated high engine room temperatures, electrical conductors are derated in those spaces.

Of all cooking fuels, LPG creates the greatest ventilation problem because the gas is approximately 1.6 the density of air; it tends to sink and must be removed with positive ventilation.

Ventilation and Carbon Monoxide

Carbon monoxide (CO) is a flammable, gaseous poison having the same vapor density as air. Large quantities of CO are produced as a component of gasoline engine exhausts and, to a lesser degree, from diesel exhausts. Cooking appliances produce smaller quantities. This gas must be kept out of accommodation spaces to the greatest degree possible. Control of CO depends on awareness of the following basic factors:

1. The separation of the engine exhaust system discharge fittings from ventilation system fittings, scuppers, drains, and any other openings into the hull is vital to prevent the gas from reentering the boat. The direction and velocity of airflow over the boat must be considered. It affects the flow of air into and from all the ventilation intakes and exhausts.

2. It is not possible to prevent CO gas from reentering the boat under some operating conditions. One common condition, for instance, would be operating a boat at slow speed with a following wind. The operator must recognize and avoid these conditions.

3. Even engine room bulkheads that meet the ignition protection requirements of USCG regulation 33 CFR 183.410 do not effectively prevent CO gas from moving from the engine space into accommodation spaces.

4. A CO detector can help warn the boater of the presence of CO. CO is a flammable gas, but its greatest danger is that it combines

with blood hemoglobin in the same manner as oxygen. CO is the leading cause of gas poisoning deaths in the United States.

Lightning Protection

Lightning hits many boats every year. USCG statistics consistently show that people killed by lightning on boats are usually in small open motorboats, not sailboats. Even if a sailboat does not have a complete lightning system, its metallic mast inherently serves as the major down conductor. It tends to conduct the lightning past the occupants.

The probability of a lightning strike on a boat varies with geographic location and season. Lightning protection can be provided on a boat that has a conductive mast with a conductivity equal to a No. 4 AWG conductor and enough height that an imaginary line drawn at a 45 degree angle from its tip does not intersect any part of the boat. A pointed lightning rod should be installed at the tip. The mast's base must be connected vertically and as directly as possible to an external grounding plate or grounding strip, located below the lightning mast. An external rectangular (1 sq ft [0.09 m²] or larger) grounding plate or an external horizontal grounding strip of the same approximate area may be used, running fore and aft from below the lightning mast to a position aft below the engine(s). Large metallic masses, such as metallic guard rails, engines, and tanks near the lightning ground conductor, should be connected to the grounding strip to prevent side flashes. If the mast being used is stepped on deck, the connection between the mast and the external grounding plate or strip should be made with a No. 4 AWG conductor or two parallel No. 6 conductors routed on both sides of an opening. The normal horizontal electrical system bonding conductor in a boat may be connected to a lightning conductor system, but should not be relied on as a lightning conductor. To be effective, lightning conductors must provide an essentially vertical path to the ground plate or strip.

Figures 9-37A through 9-37D show normal boat lightning protection systems. Spiral wrapped radio transmitter antennas will not function as lightning masts, but steel antennas may help, provided the antennas' coils are bypassed with a lightning arrester. If an external grounding strip is used, as illustrated in Figure 9-37B, the sharp edges of the grounding strip should not be rounded. In both sailing auxiliaries and cruising power boats, the external grounding strip most effectively connects items in the boat to the ground without long horizontal conductors inside the boat. The mast and bow rails can be connected to the forward end, and the engine and back stay (sailboats) to the aft end of the strip. In larger boats it is difficult to avoid installing a number of horizontal lightning conductors to connect large metal masses to the system. Instead, an equalization bus may be installed inside the boat parallel to the external lightning grounding strip. The internal equalization bus and external grounding strip are interconnected at the forward and aft ends and must be connected to the lightning ground system. They must never be connected so that they become part of the main down conductor. If a seacock or through-hull fitting carries the full lightning current, any water in the seacock turns to steam instantaneously, breaking the fittings and possibly sinking the boat.

Occasionally, a sailboat's aluminum mast hits an overhead power line, and the question arises: Should masts or sections of masts be made of a nonconductive material to avoid such accidents? The answer: A boat's lighting system should be maintained.

Engines

Liquid-cooled marine gasoline and diesel engines may be cooled directly by seawater or with fresh water using a raw water cooled heat exchanger. Liquid-cooled engines use water-jacketed blocks, heads, and exhaust manifolds. A self-priming, engine-driven pump circulates the cooling water. In a raw water-cooled engine, water circulates through the engine and then discharges into the exhaust pipe to create a raw water-cooled exhaust. The water discharges into the exhaust pipe through a water-jacketed riser at the engine. The riser's design prevents water from backing up into the engine through the exhaust manifold. Fresh water-cooled engines using a heat exchanger require two separate water pumps. One pump recirculates the fresh cooling water through the engine, and the second circulates water through the heat exchanger and provides raw water for the water-cooled exhaust. Fresh water-cooled engines can operate efficiently at approximately 180°F (82°C), while raw water-cooled engines, using seawater, must operate less efficiently at 140°F (60°C). The lower temperature prevents excessive salt precipitation in the block, manifold, and head of the engine. Raw water-cooled engines and engines using a heat exchanger can overheat if the pump fails or the water intake becomes blocked. Means to indicate

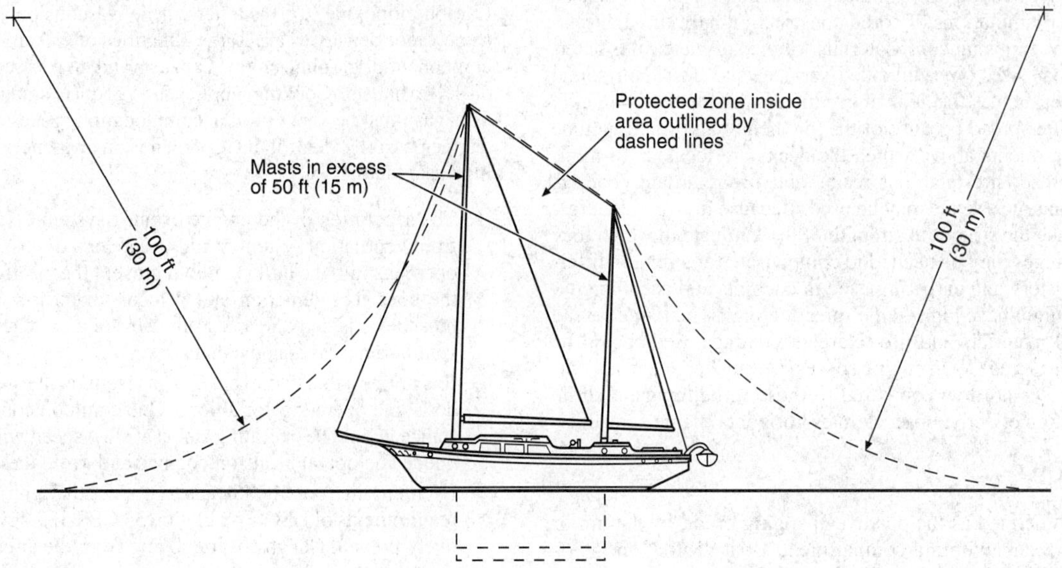

FIG. 9-37A. *Proper lightning protection for a boat with masts more than 50 ft (15 m) above the water. Protection is based on a lightning striking distance of 100 ft (30 m).*

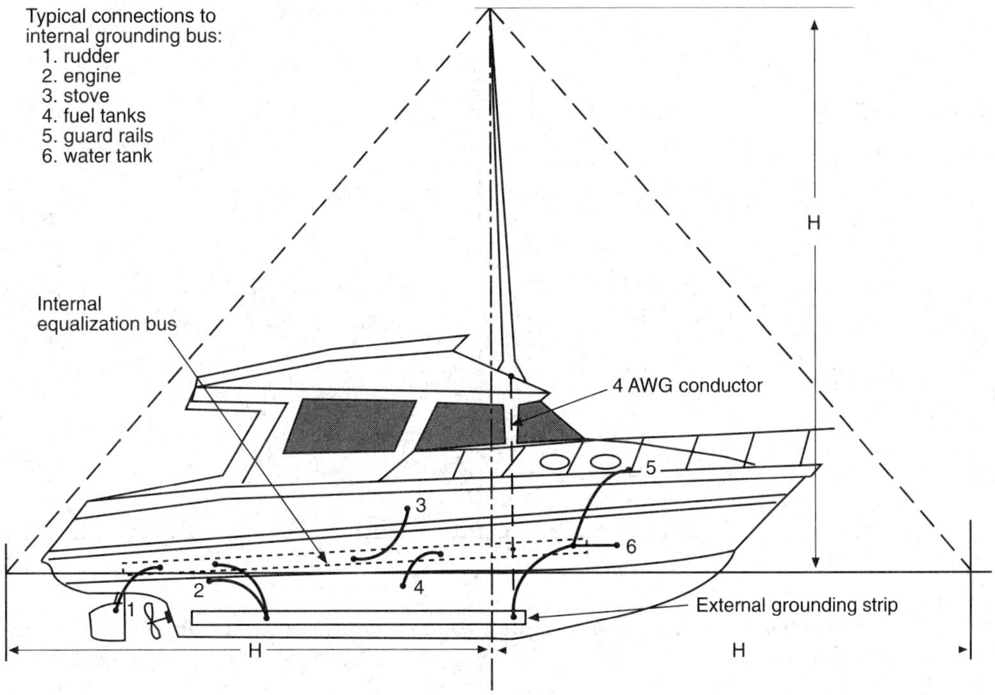

Typical connections to
internal grounding bus:
1. rudder
2. engine
3. stove
4. fuel tanks
5. guard rails
6. water tank

Internal
equalization bus

4 AWG conductor

H

External grounding strip

H   H

FIG. 9-37B.   *Diagram of a boat with mast not more than 50 ft (15 m) connected to external ground strip, with optional equalization bus.*

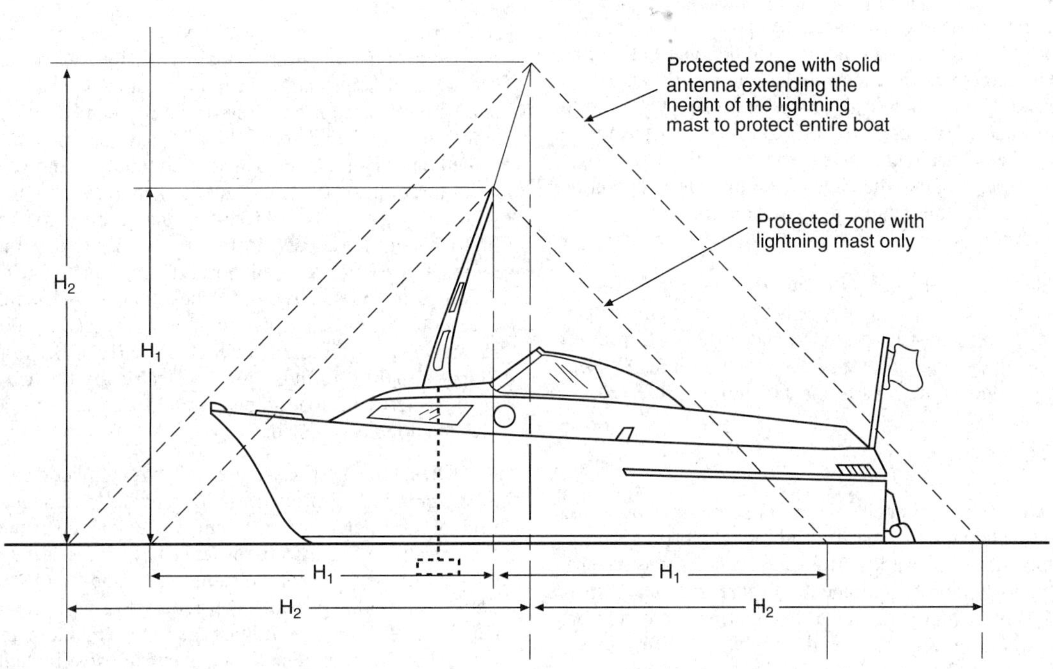

Protected zone with solid
antenna extending the
height of the lightning
mast to protect entire boat

Protected zone with
lightning mast only

H_2

H_1

H_1 H_1

H_2 H_2

FIG. 9-37C.   *Proper lightning protection for a small sailboat with a mast not more than 50 ft (15 m) above the water.*

loss of engine cooling water must be provided at all helm positions for propulsion engines. Nonpropulsion engines must automatically shut down with the loss of lubricating oil pressure, excessive engine temperature, and excess heat in the exhaust system.

Air-cooled engines are not recommended. However, if they are used in an enclosed space, a separate air duct must be provided for effective cooling. Because these ducts are large, they must be located to minimize the intake of excessive seawater, which could af-

fect boat stability. A device must be provided to warn of excessive temperatures.

Technical advancement in automotive and marine engines has caused rapid replacement of carburation with fuel injection. The more efficient fuel injection systems eliminate boil-over and float-leakage problems of carburetors but are still subject to rap or lock. However, injection systems with return fuel lines present a major hazard not encountered in carburetor engines. Fuel injection gasoline

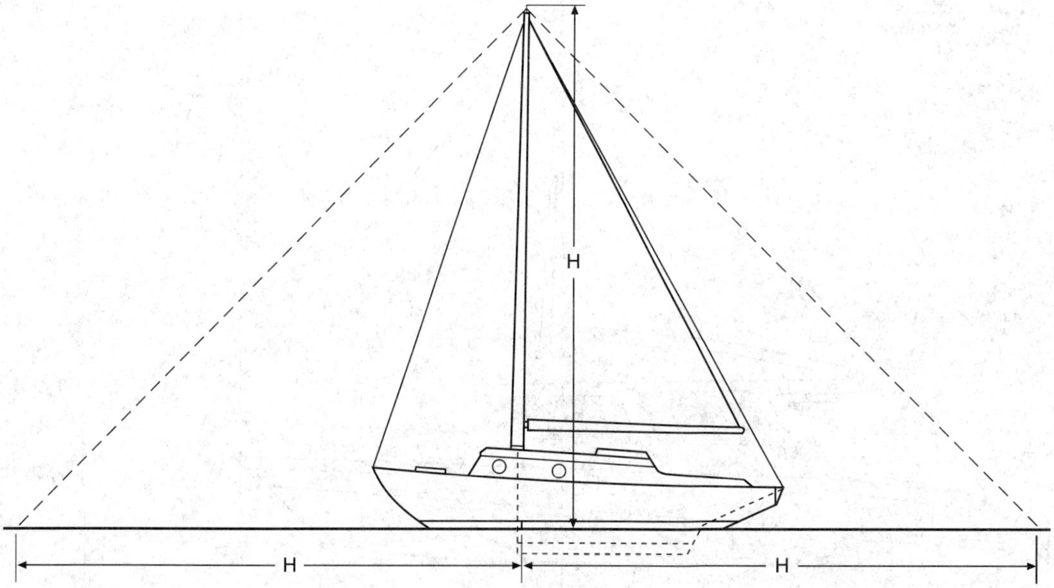

FIG. 9-37D. *Proper lightning protection for a boat with a mast not more than 50 ft (15 m) above the water.*

engines and diesel engines may use a return fuel line that is warm and under pressure. In comparison, fuel in the distribution line from the fuel tank to the engine operates at reduced pressure, minimizing the hazard. When the engine is not running and the distribution line is at a reduced pressure, a pinhole may not leak any fuel at all. When the engine is running, a pinhole leak in the return line will tend to spray fuel, creating an immediate hazard. Consider that the return line can totally fail and not affect engine operation. The operator may be unaware of fuel spilling into the bilge unless there is an operating vapor detector. The fuel line requirements remain the same for distribution and return lines.

Backfire flame arresters are still required to meet the same requirements, but they no longer require a USCG approval number. Diaphragm fuel pumps must be designed and installed to prevent the release of fuel to the bilge in case of diaphragm failure. All electrical equipment in the engine space must be ignition protected.

Ignition Protection

All ac and dc electrical equipment in a gasoline engine space and any space with fuel line fittings must be ignition protected. Ignition-protected equipment has the same performance objective as hazardous location (explosionproof) equipment on shore and on commercial vessels. However, the requirements for ignition protection are less stringent and do not require hazardous location wiring. Accident statistics indicate that ignition protection has effectively reduced explosions in boats. Unfortunately, investigations of fire and explosion accidents have revealed that ignition-protected equipment is frequently replaced with non-ignition-protected equipment, eliminating the protection originally provided.

To eliminate the need for ignition-protected equipment in accommodation spaces, a bulkhead must separate the engine space from accommodation spaces. The bulkhead must be watertight for 12 in. (30.5 cm) or one-third the bulkhead height, whichever is less. Above that level the annular space around anything passing through the bulkhead, (wires, cables, ducts, etc.) must not exceed $^1/_4$ in (6.3 mm).

Fuel Systems

In aircraft and land vehicles, gravity drains liquid leakage thereby preventing accumulation; this is not possible in a boat. Because both liquid fuel leakage and fuel vapors accumulate in a boat, fuel system components' design must resist vibration and shock failure, corrosion, and chemical deterioration. The system must be designed, installed, and maintained so as to be liquidtight and vaportight to the hull interior at all times. All fuel system parts should be capable of withstanding exposure to free-burning gasoline for 2.5 min without fuel leakage. Hose used instead of copper tubing must be USCG Type A1 or A2 fire-resistant hose. [Exception: fire-resistant fuel distribution lines are not required if less than 5 oz (150 mL) of fuel will leak when the distribution line is cut at the lowest point.] When fire resistance is not required, a USCG Type B1 or B2 hose can be used. Metallic fuel fill fittings must be electrically bonded to the engine negative terminal to dissipate static electrical charges. Fill hoses must be double-clamped.

Diesel fuel systems and gasoline fuel injection systems require the installation of efficient fuel filters and water separators to prevent dirt and water from reaching the fuel injector pump. Diesel and some gasoline fuel injector systems have pressurized return lines to the tank that must meet the same requirements as the fuel distribution line to the engine. Heated return fuel will gradually raise the temperature of the fuel in the tank. At normal temperatures, diesel fuel does not vaporize and create an explosive fuel air mixture, but diesel fuel has a greater tendency to accumulate in the bilges and create a fire hazard. Pressurized gasoline lines create an immediate hazard at any leakage level.

Appliances

Open flame devices must be designed specifically for marine use and be installed to minimize personal and physical hazards. Open flame devices must be designed for attended use. The manufacturer must furnish printed instructions for their proper installation, operation, and maintenance. Portable devices and any device that cannot be secured in position when in use or stored are not permitted. A durable and permanently legible instruction sign describing safe operation and maintenance must be provided. The sign should be installed on or adjacent to the appliance where it can be quickly read. Air consumption and exhaust product venting from galley stoves and other appliances must be considered in the design and layout of compartments where they are installed. Lack of permanently open vents to supply air to the appliances can result in oxygen depletion in the compartment. Exposed materials around galley stoves and other heat-producing appliances must have a flame spread index (rating) of less than 75. Fabrics must not be used within 39 in. (1 m) of a galley stove.

Pilot lights and other automatic igniters are prohibited unless they are used in sealed combustion chamber type devices with outside air supply and exhaust discharge. Gas-fueled appliances must be equipped with flame failure devices to prevent gas flow if the flame is extinguished. Unattended devices, such as water heaters, refrigerators and cabin heaters, cannot have automatic ignition unless they are of the sealed combustion chamber type, with the internal flame completely separated from accommodation spaces. Open flame water heaters and refrigerators are not permitted because they could be left on during fueling and because undetected leaks could occur in the fuel system.

Gasoline is not permitted for use as a galley fuel.

Coal, charcoal, and wood burning stoves must be installed on a hollow tile base or mounted on a noncombustible base with a clearance of 5 in. (13 cm) between the stove bottom and any deck. The deck must be sheathed with a noncombustible material. The sides and back of a stove must have a clearance of not less than 9 in. (23 cm) and must have a flame spread index of not more than 75.

Liquefied Petroleum Gas and Compressed Natural Gas Systems

The properties of both liquefied petroleum gas (LPG) and compressed natural gas (CNG) must be understood to use them safely. Both are frequently used in the galley for cooking and are popular because they do not require priming like pressure alcohol stoves. At ambient temperature, LPG can be liquefied at well under 250 psig (1724 kPa), the rated pressure of an LPG cylinder. CNG does not liquefy at ambient temperatures and is stored in cylinders under a pressure of about 2,000 psig (13,790 kPa). LPG is a two-phase (liquid/vapor) fuel with a higher calorific value than CNG. LPG, when released as a gas, is approximately 1.6 times the density of air, and tends to sink to a boat's bilges. CNG is released as a lighter-than-air gas and tends to collect at the top of a compartment if not vented. When confined to a space, the gases will fill that space in a short time through diffusion. Diffusion and convection cause the gases to pass through and over bulkheads in all directions. An odorant added to both gases facilitates detection. A pressure regulator lowers the stored pressure of both gases to a system working pressure. LPG, normally stored at a pressure of 150 psig (1034.35 kPa) or less, is delivered to the appliance at approximately 14 in. (36 cm) water column, or about .735 psig (5.0 kPa gage). CNG is reduced to about 6 in. (15 cm) water column, or about 0.22 psig (1.5 kPa gage) at the appliance.

NFPA 302 permits LPG and CNG containers and regulators to be located on an open deck or on a cabin top in a housing, open at the top and bottom so escaping gas does not reenter the hull or accumulate in a cockpit or any enclosed space. The housing must protect the valves and the regulator assembly from mechanical damage. If not located on an upper deck or cabin top, the LPG containers and regulator assemblies must be located in a vaportight locker. The locker must be located above the waterline, be vaportight to the hull interior, be provided with a latched gasketed cover, and be vented overboard. It must be located so that when the cover is open, escap-

ing vapor does not flow to the bilges, machinery spaces, accommodation spaces, or other enclosed spaces. Normal locations are along the side of the cabin in a molded-in locker or in an open cockpit in a vaportight locker. LPG lockers are vented with a 1/2 in. (13 mm) line led overboard with traps. The locker vent discharge must be separated from any hull opening and from any exhaust discharge fittings. Because CNG is lighter than air, CNG lockers must be vented with a 1/2 in. (13 mm) vent to the open atmosphere at a level above the cylinders and regulator assembly. CNG tanks, like LPG tanks, may be installed on an open deck, in a vaportight locker with an overboard vent or in a compartment that has no openings to an engine space above the level of the regulator. Engine room bulkheads that meet the ignition-protection requirements are not vaportight and therefore might communicate with other spaces.

Low-pressure distribution lines from the system regulator assembly to an appliance may be either copper tubing or hose. Copper tubing can be used with LPG without a problem, but copper tubing used with CNG must be internally tinned. LPG hose with permanently attached fittings must meet the requirements of UL Standard 21. CNG hose must meet the requirements of NFPA 52, *Standard for Compressed Natural Gas (CNG) Vehicular Fuel Systems*. Distribution lines must be installed in continuous lengths from the regulator assembly to the appliance. The only exception is the installation of a flexible hose connection between copper tubing or copper piping and a gimbaled appliance.

Standards prohibit the use of pilot lights or other automatic ignition forms on gas-consuming appliances, but do permit a flame control for an oven in use. The flame control must be off when the oven control is off. All gas-fueled appliances must have a flame failure control to prevent gas flow if the flame is extinguished. Stoves with gas containers having less than 8 oz (0.24 L) of fuel are exempt from this requirement.

LPG or CNG systems installed in accordance with NFPA 302 can be simply, quickly, and positively checked for leakage in about 10 min. The method, which requires a pressure gage, should be used frequently. The standard leakage check should be done after the system has been serviced, after the containers have been changed, and after any storm or grounding that might affect the system. NFPA 302 recommends that the check be done at the start of any cruise or voyage during which the system will or may be used. The standard specifies the following procedure:

With the appliance valves closed and all other valves open, note the pressure on the gage. Close the container valve. The pressure should remain constant for at least 10 min. If pressure drops, locate leakage by application of soapy water solution at all connections. Repeat test for each container in multicontainer systems. NEVER USE FLAME TO CHECK FOR LEAKS. NEVER USE SOAP CONTAINING AMMONIA.

Electrical Systems under 50 Volts

A boat's low voltage dc system is intended to be a two-wire system with insulated conductors running parallel to each other to and from the power source, to each item of electrical equipment. The two-wire system minimizes the creation of magnetic fields that can affect the compass and other magnetically sensitive equipment, and also minimizes the chances for stray dc current flow between metallic underwater components. Negative ground has been adopted as standard for dc systems, and the engine negative terminal is established as the single point for the dc system ground. When stray dc current flows between any two metallic components in contact with an electrolyte such as seawater, the more positive (anodic) metal suffers accelerated corrosion. The more negative (cathodic) metal tends to be protected. Making the engine's lower unit or propeller and shaft purposely more negative or cathodic protects them and any other metallic objects connected to the same ground from accelerated corrosion due to stray currents. Although a true two-wire system would best serve the boating industry, most marine engines use converted automotive single-wire components. The automotive industry uses a common chassis ground return system so that all engine-mounted components have a negative conductor grounded to the frame. Because of the engine, the two-wire dc system on a boat is something of a compromise: mostly a two-wire system, it remains a single-wire system at and near the engine. The single-wire system has been extended off the engine by permitting an accessory negative bus for accessories near the engine. Accessories connected to a panelboard may have their negative conductors connected to the accessory bus near the engine, instead of the panelboard negative terminal strip. Whenever the system becomes a single-wire system and the feed and return conductors are not run in a pair, a magnetic loop forms. Standards do permit a completely ungrounded system, provided the engine's electrical components are all ungrounded.

Even though the engine's system is single wire, metal hulls and boat bonding systems must never be used as ground return paths. The boat bonding system—used to connect the non-current-carrying metal housing of dc equipment to the engine negative terminal—serves three functions.

1. Because the non-current-carrying parts of dc electrical equipment are connected to the engine negative terminal, stray current leaking from this equipment returns to the engine negative terminal (and battery negative) on the bonding conductor inside the hull. Electrical flow inside the hull minimizes current flow outside the hull between metallic appendages that would result in stray current corrosion of the equipment. A stray current leak of less than an ampere through a 1-in. (2.5-cm) through-hull fitting can destroy the fitting and sink the boat in three days.
2. A boat bonding system may be interconnected with and become part of the lightning protection system. When used in the lightning protection system to interconnect large metallic objects, the conductor size must be at least a No. 6 AWG conductor.
3. The bonding system conductor connects underwater metallic hardware to a galvanic anode (zinc) for galvanic protection.

If a boat has two or more propulsion engines and more than one battery bank, its engine blocks must be interconnected with two conductors capable of carrying the cranking current of the largest engine. The conductors prevent heavy damaging stray current corrosion due to current flow between propellers and shafts or outdrives when cranking one engine from the battery of another. Positive interconnection of all engines also prevents a current path through metallic fuel lines and control cables.

All dc conductors in the boat, except cranking motor conductors, require overcurrent protection at the power source for any circuit, generally within 7 in. (17.8 cm) of the source. USCG regulations and NFPA 302 permit a wire that is directly attached to the battery to be 72 in. (1.8 m) long without overcurrent protection. Using this exception is not recommended. NFPA 302 specifically requires the underground battery charger leads to be protected at the battery point of connection to the dc system. Under the regulations and NFPA 302, it is also possible to extend an unprotected sheathed wire's length from 7 in. (17.8 cm) to 40 in. (101.6 cm). The exception is used between the engine cranking motor solenoid hot terminal and the engine circuit breaker to position the circuit breaker high on the engine.

Boats have a special problem because many dc motors used on boats have no internal thermal protectors and by design cannot have the equivalent impedance protection. Lacking internal protection against overload, branch-circuit breakers or fuses must protect these motor-operated devices (pumps, blowers, etc.) against a "locked-rotor" condition. A locked-rotor condition exists if the pump becomes jammed and cannot turn. The motor manufacturer cannot easily specify the branch-overcurrent protection because it depends on the total wire length in that branch circuit.

Electrical Systems over 50 Volts

Three different sources may power the 120/240 V ac electrical system on a boat: (1) shore power, (2) an onboard ac generator, or (3) an onboard dc to ac inverter. Using shore power invites a special problem because the ac grounded neutral conductor is grounded to the earth on shore, but must never be grounded on the boat. When powered from shore, ac leakage currents return to the earth ground on shore. In contrast, both the onboard ac generators and the onboard inverters are grounded on the boat so that all leakage current remains on the boat. When two or all three sources of ac power are available, a switching arrangement must be provided to prevent any two sources of power from feeding the same circuit at the same time. This switch must disconnect one ac power source before connecting the circuits to another power source.

As on shore, under the *National Electrical Code*®, a green grounding-wire system keeps the non-current-carrying housings of ac appliances and ac equipment at ground potential to prevent electrical shock. When ac current is from an ac generator or an inverter on the boat, the green grounding-wire connects to the generator or inverter ground on the boat. With shore power, the green ac grounding-wire to the boat engine's negative terminal connects the underwater hardware of the boat (propellers, shaft, outdrivers, etc.) to other boats or any grounded metal on shore. Accordingly, aluminum- and steel-hulled boats and boats with aluminum outboard engines or aluminum stern drives suffer accelerated corrosion when connected to more noble metals (cathodic materials) on other boats or the dock. The interconnection with other boats or the dock can result in severe continuous galvanic corrosion of the anodic metals. This inter-boat corrosion could be stopped by not connecting the green grounding-wire to the boat engines, but any ac leakage on the boat can charge the surrounding water, electrocuting any swimmer near the boat. The solution requires the installation of a marine "galvanic isolator" or the installation of an isolation transformer in accordance with NFPA 302. The galvanic isolator is a device installed in series with the green grounding-wire on the boat. The device blocks the flow of dc galvanic currents while permitting the flow of ac current in the grounding conductor. The galvanic isolator must be sized to carry the system's full current because if it fails, the green grounding-wire stops working.

Fire Extinguishers

Testing laboratories normally evaluate the fire extinguishment potentials of portable fire extinguishers. The USCG also classifies portable fire extinguishers based on the Underwriters Laboratories Inc. classification of fires but uses a different method to indicate extinguishment potential. Table 9-37A lists these designations.

PLEASURE AND SMALL COMMERCIAL BOAT OPERATIONS

General Maintenance

Operation and maintenance of pleasure and small commercial boats consist of general good housekeeping. Safety requires periodic maintenance of all equipment. When conducting checks, unprotected electric lights should not be taken into areas of possible vapor accumulation from gasoline or liquids used for cleaning. The following, and other inspections and tests, are needed:

TABLE 9-37A. *Number and Distribution of Fire Extinguishers*

Boat Type	No. of Existing	Minimum ANSI/UL Rating*†	Minimum USCG Rating‡	Location
Open boats under 16 ft (4.8 m) with fiberglass or metal hulls and a light load of flammable Class A materials	1	5 B:C	B-1	Steering position
Open boats under 16 ft (4.8 m)	1	1A: 10B:C	B-1	Steering position
Boats 16 ft (4.8 m) up to, but not including, 26 ft (8 m)	2	1A: 10B:C	B-1	Steering position and galley, when onboard, or cockpit
Open boats 16 ft (4.8 m) up to, but not including, 26 ft (8 m)	2	1A: 10B:C	B-1	Steering position and galley or cockpit
Boats 26 ft (8 m) up to, but not including, 40 ft (12 m)	3	1A: 10B:C	B-1	Outside engine compartment, steering position, and near galley or passenger cockpit
Boats 40 ft (12 m) up to, but not including, 65 ft (20 m)	4	1A: 10B:C	B-1	Outside engine compartment, steering position, crew quarters, and galley, when onboard, or cockpit

*If a discharge port is installed, a USCG Type B-1 portable fire extinguisher might not be adequate.
†Extinguishers intended for machinery space protection are not required to have a Class A rating.
‡Boats under 26 ft (8 m) in length without enclosed accommodation spaces or enclosed galleys shall be permitted to be equipped with a bucket with attached lanyard in lieu of Class A-rated portable fire extinguishers.

1. *Rags.* Clean rags should be kept in covered metal containers. Dirty waste and rags should be kept in separate covered metal containers. These should be disposed of each time the boat is docked.
2. *Fuel system.* The entire fuel system, from the fuel tank to the engine, must be visually examined periodically. Determine if the fuel tank shows any evidence of corrosion or leakage. Once a year have the entire system pressure checked at 3 psi (20.7 kPa). The fuel hoses should be checked to determine that the connections (fittings and clamps, etc.) are tight and that hoses show no signs of deteriorating or cracking. The fuel pump should be checked to determine that it is a marine type, with a hose back to the intake manifold or carburetor in case of diaphragm failure. Fuel filters must be clean and not leaking. Replace used gaskets with appropriate new ones whenever a fuel filter is cleaned.
3. *Detectors.* All detectors must be checked to determine if they are working.
4. *Exhaust blowers.* Ventilating ducts should be checked to determine that they are installed without excessive bends and are neither kinked nor torn. Overcurrent protection should include locked-rotor protection.
5. *Ignition protection.* Check to be sure that all electrical equipment in the engine space, or any space with fuel line fittings, is ignition protected. Replacement with automotive parts eliminates such protection.
6. *Painting.* When repainting interior areas, fire-retardant paints should be used.

Proper maintenance and sensible fuel system operation, used together with ignition-protected electrical equipment, are probably a boat owner's most important fire prevention duties.

Fueling

Utmost care must be exercised during fueling operations. Some general guidelines to be observed during fueling operations are:

1. Fueling should not be undertaken at night, except under well-lighted conditions.
2. Smoking must be forbidden onboard the boat and in the area of the fuel pumps and hose.
3. Before opening fuel tanks:

   a. Shut down all engines, motors, and fans.
   b. Extinguish all open flames.
   c. Close all ports, windows, doors, and hatches.
   d. Determine the quantity of fuel to be taken aboard before fueling operations start.
   e. Point out to the fuel attendant the correct deck fuel fitting for each tank.
4. The fuel delivery nozzle must be in contact with the fuel fill deck fitting before fuel delivery begins. This contact should be maintained continuously until flow has stopped. Static discharge is a serious hazard if this rule is not followed. The fuel-fill fitting must be of a material that conducts electricity and must be electrically connected to the engine's negative terminal to ground the static charge.
5. Fuel tanks should never be completely filled because fuel expands. A minimum of 5 percent tank space must be allowed for expansion.
6. After fuel flow stops:

   a. Tightly secure the fill fitting cap.
   b. Completely wipe up spillage.
   c. Ventilate all spaces and check for gasoline vapors before starting any engine or operating any appliance.

Storage between Voyages

Before storing a boat, even for a short time period, several precautionary actions should be taken. The entire vessel should be thoroughly inspected. All combustible trash and rags should be removed, as well as any paint or other flammable liquids. Bilges should be inspected and pumped dry. All underwater hardware, including propellers, outdrives, rudders, through-hull fittings, and the hull should be checked and necessary repairs made. All seacocks should be checked and operated to make sure they can operate in an emergency. Liquid level of batteries and specific gravity of battery cells should be checked. If possible, lockers and bilge hatches should be secured in a partially open position to ventilate the spaces.

COMMERCIAL VESSELS

Almost any accident involving a vessel presents either a direct or secondary threat of fire. Collision or stranding can rupture tanks or containers of hazardous materials and provide enough frictional heat to ignite the contents. The same hazards can result from the heavy pounding a ship receives in bad weather. Even water leaking into a cargo hold can cause accelerated oxidation and eventual combustion of some organic substances. In discussing marine fire hazards, both cargo transfer and vessel operation must be considered in addition to traditional fire problems.

The steadily increasing world population and demands for foreign merchandise have resulted in an ever-increasing volume of commercial water-borne tonnage. The majority of vessels carrying passengers and freight on international voyages are foreign flag ships. Foreign and domestic commercial vessels on an international voyage must comply with the international requirements contained in the Safety of Life at Sea (SOLAS) Convention. Several international conventions relate directly or indirectly to fire protection and prevention aboard commercial vessels; these are concerned with construction, operation, staffing, training, and inspection. The United States is signatory to many of these conventions, including SOLAS, whose key points are stressed in this section. In addition, several international codes relate to the carriage of hazardous materials. Although they are not conventions, the codes carry the weight of a convention due to their acceptance by a large number of nations.

Domestic commercial vessels and vessels of foreign countries not signatory to the SOLAS Convention that operate from U.S. ports are regulated by the U.S. Coast Guard for compliance with domestic regulations. The Code of Federal Regulations, Titles 33, 46, and 49, specify the design, construction, maintenance, and operational requirements for these ships. The majority of U.S. flag vessels that must comply with the fire protection requirements in these regulations are those operating between U.S. ports for either loading or discharging freight or passengers, e.g., those operating on the inland waterways.

MARITIME REGULATIONS AND STANDARDS

The fire protection requirements for ships are not identical to those required for similar installations ashore; however, the marine community has had a longstanding relationship with the NFPA in the development of both international and domestic shipboard fire safety standards. In order to understand the philosophy employed in their development, a brief history is helpful.

History

On Sunday June 15, 1904 the US passenger vessel General Slocum burned in New York's East River killing 1,020 of the mostly women and children passengers aboard for a picnic excursion. In 1919 the NFPA formed a marine committee to prepare regulations to be used in shipbuilding. Other committees such as the sprinkler committee, the committee on internal combustion engines and the electrical committee were identified as having land practices that could be modified for marine use. The marine committee developed a standard for shipboard fire safety and in 1922 at the NFPA annual meeting in Atlantic City, the Association adopted the first NFPA regulations for the construction and maintenance of ships. The new regulations were published as an NFPA pamphlet and covered subjects including separation of combustible materials from boilers, materials of construction, and confined space tank entry procedures.

The first international conference to discuss shipboard safety was held in 1914 as a direct result of the sinking of the *Titanic*. That tragedy heightened public interest, and it became apparent that standards had to be developed internationally to ensure safety at sea. Subdivision, adequate lifeboat capacity, and the use of radio equipment for communications were key areas of interest in this first SOLAS Convention; no mention was made of fire protection.

The 1914 SOLAS Convention, however, never entered into force due to World War I.

A second conference, held in 1929, addressed fire protection for the first time. The Convention required the fitting of fire-resisting bulkheads above the weather deck to restrict fire to its zone of origin by dividing the ship into main vertical zones, in general not to exceed 131 ft (40 m) in length. However, the Convention did not specify by definition or test what a "fire-resisting bulkhead" must be.

By 1929, the NFPA's requirements were more stringent than the new requirements of the 1929 SOLAS. By the end of 1934, the marine member section had subcommittees on vessel construction, tankers and handling of oil, internal combustion motors, building and lay-up, sprinkler systems, alarm and detection, extinguishing apparatus, marine electric wiring, hazardous cargo, inspection and maintenance, and power boat safety. Later that year, the country was shocked as the US passenger vessel Morro Castle burned off the New Jersey coast while enroute from Havana to New York. The vessel's open stairs and combustible construction including wood bulkheads and finishes were cited as a chief cause of the rapid fire spread which killed 104 people.

Numerous fire tests conducted aboard a test ship, the *S.S. Nantasket,* ultimately singled out a construction technique that the Marine Section of NFPA had recommended. Class A-1 bulkheads were steel, asbestos lined, and resisted fire spread for 1 hr; Class B bulkheads were "incombustible" materials (now called "noncombustible" materials) and resisted fire spread for 30 min. At the 1936 Association meeting the marine committee reported good progress building on the 1929 SOLAS requirements requiring main vertical zones every 131 ft (40 m), a length which corresponded to every second main watertight subdivision bulkhead. Under the direction of Dr. Ingberg of the National Bureau of Standards the Marine Section developed time temperature performance requirements from testing aboard the ship *S.S. Nantasket*. In 1937 the NFPA put all of the work of the marine committees at the disposal of the US Congress and the Bureau of Marine Inspection and Navigation in the wake of the Morro Castle tragedy. By 1938, ships built to NFPA requirements for steel, fire resistive decks and bulkheads were referred to as fireproof. By 1941, the Federal Government had codified into law the fire safety requirements developed by the NFPA. Test results were so favorable that the U.S. Senate ratified the 1929 SOLAS Convention and amended the U.S. Code to require that U.S. vessels be similarly constructed.

A third international conference in 1948 upgraded the 1929 SOLAS Convention. The United States proposed the fire protection techniques in the U.S. Code, but the international community did not agree that this should be the only construction method. Materials were not readily available worldwide, and several nations felt that active fire protection systems were equivalent to passive fire protection systems.

The 1948 SOLAS Convention allowed three alternate methods of shipboard fire protection. Method I was the "incombustible" U.S. construction technique. Method II was the United Kingdom proposal to require sprinklers with no limitation on combustibles. Method III was the proposal by France to use a limited amount of fire-resistive bulkheads with the addition of fire detection systems. Cargo ships and some passenger ships built in accordance with the 1948 SOLAS and some later Conventions still have these three construction methods. In 1948, the United Nations maritime conference also agreed to form a special UN agency, the International Maritime Consultative Organization (IMCO, now IMO), to develop internationally acceptable standards to improve safety at sea and prevent pollution.

IMCO held its first SOLAS conference in 1960 to address issues arising from the 1956 collision near Nantucket between the ocean liners *Stockholm* and *Andrea Doria.* However, in the early 1960s, a series of tragic passenger ship fires with multiple loss of life prompted the U.S. to call for a special diplomatic conference to address fire protection features of existing ships. In 1966 the international community developed requirements for the retrofit of some structural fire protection features, such as main vertical zone bulkheads and upgraded means of escape, on existing passenger ships. The ratification process for international adoption at that time was extremely lengthy, and the amendments were not internationally enforced. The United States unilaterally adopted the requirements for ships operating from U.S. ports for the safety of its own citizens. These requirements put the famous *Queen Mary* out of service as a passenger liner.

The international community continued its work to improve the fire safety standard for new ships and, in 1974, agreed to accept only "incombustible" construction as the single construction method for all new passenger ships. In addition, new passenger ships had to be fitted with either fire detection systems or sprinkler systems. This latest 1974 SOLAS Convention was ultimately ratified internationally and took effect on May 25, 1980. Still in effect, it has been amended several times.

In the 1980s, investigations into another series of passenger fires prompted the National Transportation Safety Board (NTSB) to recommend that IMO require that all passenger vessels, regardless of construction type, have sprinkler systems installed. IMO was developing these requirements for all new passenger vessels when, on April 12, 1990, the Bahamian passenger ship, *Scandinavian Star,* caught fire off the coast of Sweden; 150 lives were lost. The ship had been built to noncombustible standards, and the results of an international accident investigation supported the need for sprinklers in addition to noncombustible construction. The international community responded quickly and, in a highly unusual move, in April 1992 agreed to make sweeping retroactive requirements on all passenger ships. Known as the 1992 Amendments to the SOLAS Convention, they require that the retrofit of sprinklers and smoke detectors be completed on passenger ships no later than October 1, 1997.

International Maritime Regulations and Standards

The International Maritime Organization (IMO) a specialized agency of the United Nations headquartered in London, addresses matters relating to maritime activities. IMO is the depository of all treaties dealing with commercial vessel safety and has developed mandatory requirements as well as recommended standards relating to that same subject. The conventions and their amendments must be ratified by member countries (Administrations) signatory to IMO before they may take effect. They are generally referred to by the dates they are developed, even though they may come into force several years later (e.g., SOLAS '74 took effect on May 25, 1980). The conventions, amendments, and codes are stand-alone documents that remain in effect even when new versions are developed. For example, SOLAS '48 requirements apply to ships whose keels were laid on or after November 19, 1952. Additional requirements may only apply if later conventions retroactively applied them. Retroactive requirements must specify which ships they apply to, either by referring to specific dates and conventions or by applying to *all* ships. Thus, to determine the requirements that apply to any ship, you need copies of each of convention and amendments and must know the date the keel was laid.

Under the terms of the conventions, all signatory countries accept a certificate attesting to a ship's safety issued by one signatory country. Usually, one of the classification societies performs the survey on behalf of the Administration associated with issuing the certificate, i.e., the Flag State. The certificate intends to ensure a minimum level of safety and to permit ships to operate freely from one country to another without risking discrimination due to their safety features. However, requirements usually are not very specific and are open to interpretation. As a result, some countries exert their rights as Port States to review safety features before permitting ships to operate from their ports. Scandinavian countries and the United States have exerted their rights more freely in the 1990s, due to the large numbers of passengers at risk on foreign ships operating from their ports.

A list of conventions and codes follows, with a brief description of how they relate to fire protection and prevention.

Safety of Life at Sea (SOLAS): The most important of all international conventions dealing with maritime safety is the International Convention for the Safety of Life at Sea, better known as SOLAS, which covers a wide range of measures designed to improve the safety of shipping. More than 100 countries, controlling more than 95 percent of the world's shipping fleet, have adopted the latest convention, SOLAS '74. Beginning with SOLAS '48, various SOLAS Conventions have addressed, among other things, fire protection aboard commercial vessels. Although the conventions cover all types of vessels, they address passenger vessels in greatest detail, including construction of bulkheads separating various compartments. The inclusion of the first and second set of amendments to SOLAS '74 increased emphasis on cargo ships and tank vessels. For example, one fire protection feature of SOLAS '74 is the requirement for an inert gas system for new oil tankers over 100,000 deadweight tons (DWT) (95,256 metric tons) and for combination oil/bulk solid carriers over 50,000 DWT (47,628 metric tons).

Several changes to SOLAS '74 have been developed since it was finalized. These have been expressed in the 1978 Protocol Relating to the International Convention for the Safety of Life at Sea–1974, and in Amendments in 1981, 1983, 1988, 1989, 1991, and 1992. The majority of changes to Chapter II-2, which deals with fire protection, were made in 1981, 1991, and 1992. There were also significant changes to Chapter III, Life-Saving Appliances and Arrangements, in the 1983 Amendments and the 1991 Amendments, which specify requirements for fire drills and crew training in fire safety.

The 1978 Protocol: The 1978 Protocol grew out of the 1978 Tanker Safety and Pollution Prevention (TSPP) Conference. This conference directly resulted from U.S. initiatives in response to a tragic series of tanker casualties that occurred in 1976 and 1977. Since the SOLAS '74 treaty had not yet been ratified, it could not be amended. Therefore, the changes were issued as a Protocol. (The Protocol has since been added to Chapter II-2 of the Convention.) The 1978 Protocol Relating to the International Convention for the Safety of Life at Sea–1974, came into force May 1, 1981. It mandates steering gear improvements, collision avoidance aids, dual radars, inert gas systems (IGS) for new and existing tank ships down to 20,000 DWT (19,000 metric tons), deck foam systems for new tank ships down to 20,000 DWT (19,000 metric tons), and closed ullage systems for all tank ships fitted with IGS. Finally, it requires IGS for tank ships capable of crude oil washing.

The 1981 amendments: The first set of amendments to SOLAS '74, developed between 1974 and 1981 by various IMO subcommittees, came into force September 1, 1984. Changes to SOLAS contained in the SOLAS '78 Protocol are embodied and partly updated in this first set of amendments. Changes to the fire protection chapter, Chapter II-2, included provisions for halogenated hydrocarbon extinguishing systems and a new Regulation 62 on inert gas systems. The latter regulation was adopted by the eleventh IMO Assembly as a recommendation in 1979. The changes were so extensive that Chapter II-2 was essentially reorganized and rewritten.

The 1983 amendments: These amendments became effective July 1, 1986. The most important changes concerned lifesaving appliances and the incorporation by reference of the International Bulk Chemical Code and the International Gas Carrier Code into SOLAS. These two codes became mandatory in the 1983 Amendments.

The 1991 amendments: These amendments entered into force on January 1, 1994, and specified the international requirements for atria on passenger ships. The requirements are similar to those for land-based designs in NFPA *101®*, *Life Safety Code®*, and include sprinklers, smoke detection, a smoke extraction system, and a separate enclosed vertical means of escape (stairwell).

The 1992 amendments: Two sets of fire safety amendments were adopted in 1992. The first, in April 1992, are aimed at improving the fire safety of existing passenger ships; the second, in December 1992, are specific to new ships. Both sets of amendments entered into force on October 1, 1994.

The new passenger ship requirements:

- Improve fire mains.
- Protect fire pumps.
- Provide safety release mechanisms for CO_2 systems.
- Prohibit the installation of new halon systems.
- Require more sophisticated fire detection and alarm systems.
- Provide greater fire protection for main and secondary fire boundaries (bulkheads and decks).
- Prohibit dead-end corridors.
- Upgrade, protect, and increase the means of escape.
- Include a provision for floor proximity lighting and markings.
- Improve remote and automatic release of fire doors.
- Improve the fire safety of windows facing escape routes, embarkation stations, and lifeboats.
- Increase the fire protection for galley exhaust ducts.
- Restrict the use of combustible furniture in escape routes.
- Require the fitting of smoke detection and sprinkler systems.
- Protect spaces in which vehicles are carried.
- Provide for appropriate control panels in continuously attended control stations.

The existing ship requirements:

- Upgrade basic fire-fighting equipment.
- Substantially upgrade fire safety with remote release of fire doors, the retrofit of floor proximity lighting, a general emergency alarm system and a public address system, a smoke detection system, an automatic sprinkler system, and an alarm system.
- Ultimately remove all combustible construction materials from the structure of the ship.

International Convention on Standards of Training, Certification, and Watchkeeping for Seafarers, 1978 (STCW 1978 or STCW Convention): The STCW Convention sets qualifications for masters, officers, and watchkeeping personnel on seagoing merchant ships. In July 1995, the Conference of Parties to the STCW Convention adopted a new set of amendments. These amendments

came into force on February 1, 1997, and require candidates for licenses and other certificates to establish competence through both subject-area examinations and demonstration of skills. The amendments also allow for review of qualifications by Port States, mutual oversight by Flag State/Port State, and assign responsibility to all state parties to ensure seafarers meet objective standards of competence.

Dangerous Goods Code

As a result of the Safety of Life at Sea Conference of 1960, IMO began to develop a code covering the safe transportation by sea of dangerous goods or hazardous materials. Many dangerous goods have as their primary or secondary hazard a potential for fire. If fire is not a hazard, dangerous goods, by their presence, complicate fire-fighting and rescue efforts. Today, numerous maritime nations recognize the IMO Dangerous Goods Code, which has become the seafarers' "bible." The IMO Dangerous Goods Code is compatible with the method for classifying hazardous materials that the United States used for years before the code's development. As a result, the code is recognized as a substitute for the safe carriage of hazardous materials aboard foreign and U.S. vessels while in U.S. waters. Appropriate IMO committees constantly update the IMO Dangerous Goods Code.

Bulk Chemical Code

As a result of increased demand for chemicals among industrialized trading nations, the method of shipping chemicals has changed from packages to larger quantities. Standard petroleum tankers were modified and shifted to this trade. Several nations, among them the United States, became convinced that these modified tankers did not have the safety features necessary to protect the crew, the port, and the environment against chemicals with properties other than flammability. IMO was petitioned to begin developing a code covering the design, construction, and operation of chemical carriers. On October 12, 1971, IMO adopted a code for bulk chemical carriers that includes special fire protection systems using suppressants other than water that do not react with the chemical cargo, yet are effective when applied. On September 26, 1977, the U.S. Coast Guard adopted the IMO Bulk Chemical Code as part of its regulations. In addition to upgrading various systems aboard these vessels, the code requires increased levels of stability and cargo containment to protect the cargo and the ship when they are involved in an accident.

The International Bulk Chemical Code, which is applicable to new ships, became effective internationally on July 1, 1986. This code, reflecting improvements and previously agreed to amendments to the IMO Bulk Chemical Code, became mandatory under the second set of amendments to SOLAS '74.

Liquefied Gas Code

While the IMO Bulk Chemical Code was being developed, it became obvious that the code should specifically address the carriage of gases. Gases are transported in other than their natural state, using low temperature, high pressure, or a combination of both. Methane, for example, cannot be liquefied by pressure alone. To transport it economically as a liquid, it must be refrigerated to −162°F (−108°C), and very specialized containment systems are needed to maintain that temperature for the long sea voyages involved. Specialized fire-fighting systems and techniques are also required to put out any fire that may develop aboard one of these bulk gas vessels. IMO now has two bulk gas codes, one for existing ships and one for newly built liquefied bulk gas ships. The United States adopted the IMO code for new ships on October 4, 1976. The U.S. Coast Guard requires existing ships to upgrade where the IMO code also requires it and maintains the existing U.S. standards when they are more strict than existing IMO code.

As noted above, the International Gas Code, which applies to new ships, came into force internationally July 1, 1986. This code contains improvements and previously agreed to amendments to the *Liquefied Gas Code*. It became mandatory under the second set of amendments to SOLAS '74.

Domestic Regulations

U.S. Coast Guard regulations covering commercial vessels are contained in the following parts of the Code of Federal Regulations:

1. 46CFR Parts 24-28, *Uninspected Vessels.*
2. 46 CFR Parts 30-40, *Tank Vessels.*
3. 46 CFR Parts 70-80, *Passenger Vessels.*
4. 46 CFR Parts 90-105, *Cargo and Miscellaneous Vessels.*
5. 46 CFR Parts 107-109, *Mobile Offshore Drilling Units.*
6. 46 CFR Parts 147-148, *Dangerous Cargoes.*
7. 46 CFR Parts 150-154, *Certain Bulk Dangerous Cargoes.*
8. 46 CFR Parts 159-164, *Equipment, Construction and Materials: Specifications and Approval.*
9. 46 CFR Parts 166-169, *Nautical Schools.*
10. 46 CFR Parts 175-185, *Small Passenger Vessels (under 100 gross tons).*
11. 46 CFR Parts 188-196, *Oceanographic Research Vessels.*
12. 49 CFR Parts 171-180, *Hazardous Materials Regulations.*

These regulations delineate the fire protection standards that a ship owner/operator must meet from concept through the design, construction, and operational stages of their vessel. They include specific areas of structural fire protection, detection, and suppression systems for each ship type. These requirements are sometimes quite limited: for example, fire protection requirements for Uninspected Vessels, including commercial fishing industry vessels, are limited to handheld portable fire extinguishers. In some cases, requirements are limited because casualty statistics or cost benefit have not justified higher standards, or because the concept is new and the regulatory process could not keep pace with shipowners' needs.

For example, in recent years, some land-based concepts have been transferred to the marine environment. Modern cruise ships now incorporate atria in their design, an idea borrowed from hotels. A major city in need of more prison space built a floating detention facility. These are novel ideas for ships, and the marine regulatory community has had to respond rapidly to allow for their safe development. Fortunately, land-based standards exist, and, although not directly transferable, they have formed the basis for the latest marine guidance for designing and building these ships. NFPA *101* was used as the basis for the structural fire protection and fire protection system requirements aboard floating detention facilities and provided the basis for new requirements for ships with atria. In order to keep pace with technology and provide for more flexibility in design, a general code for fire safety for merchant ships, based upon land standards, has been proposed. NFPA is currently developing the code in cooperation with the U.S. Coast Guard and the maritime industry.

United States Coast Guard

The U.S. Coast Guard (USCG) is the federal agency responsible for enforcing regulations covering the safety of U.S. commercial vessels, both ocean-going and inland. The Coast Guard also has authority over foreign vessels operating into and out of U.S. ports, under the authority of the Ports and Waterways Safety Act as modified by the Tanker Safety Act. The Coast Guard's inspection program for foreign tank vessels is directed toward preventing fires and explo-

sions such as the one that occurred aboard the *S.S. Sansinena* in Los Angeles Harbor on December 17, 1976. Inspections cover the integrity of various cargo containment and fire-fighting systems aboard a tanker. Included in the cargo containment system are the tanks, piping, and venting system that contain the cargo and its vapors. The Coast Guard also inspects foreign vessels to ensure compliance with the various international treaties and standards described earlier. As enforcement agency for SOLAS '74, the Coast Guard has the authority and responsibility to ensure that U.S. ships and foreign ships visiting U.S. ports comply with the treaty and its amendments.

In most ports, the USCG Captain of the Port (COTP) now heads the Ports Operations Division of the Marine Safety Office (MSO). The COTP is the U.S. Coast Guard official responsible in a port for ensuring that commercial vessels operate, load, and discharge cargo safely and that they are maintained in accordance with the regulations. The COTP or MSO also is responsible for ensuring that waterfront facilities are maintained and operated in a safe manner, especially regarding fire prevention. Lacking detailed regulations in this area, the COTP or MSO frequently develops local regulations covering waterfront facilities. The COTP or MSO is responsible for protecting waterfront facilities and required to have a detailed contingency plan to be implemented in the event of a fire involving a vessel or waterfront facility. This official is also responsible for responding to pollution incidents involving navigable waters and should have a contingency plan covering this type of incident response. Since pollution can involve petroleum products and chemicals, contingency plans for both are usually incorporated into one plan. These plans customarily include the local fire department because the Coast Guard's resources are devoted primarily to fire prevention, limiting its capability to fight fires.

The U.S. Coast Guard Officer in Charge of Marine Inspection (OCMI) is responsible for ensuring that U.S. vessels are built and maintained according to U.S. regulations, most of which are related to the treaties described previously. In nearly every U.S. port, functions of both the COTP and OCMI are now combined in one office, that of Commanding Officer of the Marine Safety Office. The Port Operations Division handles the COTP functions, and the Inspection Division handles most OCMI functions. Additionally, the OCMI may conduct limited inspections of foreign flagged vessels to ensure compliance with applicable international requirements.

GENERAL SHIP CONSTRUCTION

Basically, a ship may be considered a huge box girder consisting of the hull plating and the main deck. These parts are, in turn, strengthened by such members as the keel, frames, beams, keelsons, stringers, girders, pillars, lower decks, and transverse bulkheads. To appreciate a ship in its entirety, one must understand the functions of each of its parts.

The keel is primarily the backbone of the ship. It consists of a rigid fabrication of plates and structural shapes that run fore and aft along the ship's centerline. The stem connects to the forward end, and at the after end is the stern frame, which supports both the rudder and the propeller.

The frames are the ribs of the ship. Their lower ends are attached at intervals along the keel, and their upper ends are attached through brackets to the beams that support the deck. Keelsons and stringers running fore and aft provide internal bracing. The frames determine the ship's form, and they support and stiffen the shell plating.

The shell plating, necessary for watertightness, also is one of the principal strength members of the ship. Running continuously from the stem to the stern frame and from the keel to the weather deck, the shell plating forms three sides of the box girder. The plating, aided by the frames, must be able to withstand exterior water pressure, stress from buffeting waves, and rubbing and bumping against docks.

The main deck of the ship forms the fourth side of the girder and, for this reason, must be of strong construction. The plating connects to beams that extend from side to side across the ship. The deck is strengthened by doubling plates in regions weakened by openings, such as hatchways and companionways, and under all deck machinery, chocks, and bits. Girders and pillars support the deck from below.

The ship's bottom, sides, and main deck would not be strong enough to withstand the stresses of an ocean voyage without some internal stiffening. In a cargo vessel, the lower decks and the main transverse bulkheads provide this. For tank vessels, stiffening comes from transverse and longitudinal bulkheads that separate the vessel into numerous cargo tanks extending from the keel to the main deck. In newer tank vessels, including all chemical and gas carriers, double bottoms and sides provide added protection from cargo release, should a crack occur in the shell plating.

In addition to supporting the shell and decks, the main transverse bulkheads are watertight. Subdividing the vessel into watertight compartments confines water in case of damage. All doors through these bulkheads must be fitted with gaskets to make them watertight. Doors must also be kept clear at all times so they can be closed instantly. The bulkheads also are fire-resistant and can limit fire spread. This degree of compartmentation can complicate fire-fighting efforts, unless fire fighters know a ship's layout. In an emergency, the general arrangement plan for the vessel should be examined to locate access and egress routes for fire fighting. Figure 9-37E shows some typical general arrangement plans. Both domestic and international regulations also require a fire-fighting plan (fire control plan), which shows the type and location of all fire protection, to be maintained aboard the vessel, readily available for land-based fire fighters. For a more detailed discussion, see the section "Fire-Fighting Plans" later in this chapter.

The first bulkhead aft of the stem is known as the collision bulkhead, because its purpose is to limit flooding that might occur after a collision. No doors or other openings are permitted below the main deck in this bulkhead.

In a cargo vessel, except for Great Lakes bulk carriers, the engine and boiler rooms usually are located amidship. For a tank vessel, the engine room is located in the after section, aft of the cargo tank. For fuel efficiency, modern cargo and tankships are usually fitted with diesel engines rather than steam-powered boilers and turbines. Special foundations are necessary to support heavy engines and boilers. To provide sufficient headroom for propelling machinery, it usually is necessary to omit one or more of the decks in this region. To maintain the vessel's strength in the absence of these decks, several extra-heavy web frames and transverse beams are fitted.

The propeller shaft extends through the after holds from the engine to the stern gland. Because the propeller shaft must be accessible at all times for inspection and lubrication, it is enclosed in a narrow tunnel known as the shaft alley. The entrance to the shaft alley from the engine room is closed by a watertight door, and the sides are of watertight construction so that a fracture of the tail shaft or similar accident results in only the tunnel being flooded.

The necessity for good drainage requires special attention in ship design and construction. Free water on the decks, in a hold, or in the bilges is detrimental to the vessel's stability. Therefore, the drainage system must be as efficient as possible. The decks should be cambered to permit drainage to the scuppers that lead the water either overboard or to the bilges. Sufficient scuppers and suctions must be provided so drainage will be effective under any condition of list or trim of the ship. Solid bulwarks, where fitted around a deck, should be pierced by large freeing ports to allow any water that is shipped to escape quickly. For a discussion on the effects of applying fire-fighting water, see the section "Stability."

Ventilation

The type of ventilation used in a cargo vessel depends upon the nature of the space and the service of the ship. It can be natural, mechanical, or a combination of the two, and may be extended to air heating, cooling, cleaning, humidifying, and dehumidifying. Air movement in natural ventilation systems is created by the difference in density between inside and outside air and depends upon relative air temperatures. Natural systems use ventilators dependent on wind direction and velocity for induction of air currents. Natural ventilation generally is limited to a few ships, lockers, and storerooms, depending upon their location, and to some dry cargo holds. It is also sometimes used for engine and boiler rooms. In dry cargo holds, ventilation generally is accomplished with vents in the forward end for exhaust, and cowls at the after end for supply. Large ducts are required because of the low velocity necessary for airflow. In a mechanical system, various types of fans driven by electric motors move air. These systems provide positive circulation of air at desired temperature and volume, function regardless of outside atmospheric conditions, and are easily controlled to meet variations in requirements.

The ventilation for living accommodations on both cargo and tank vessels is by means of forced air through a duct system similar

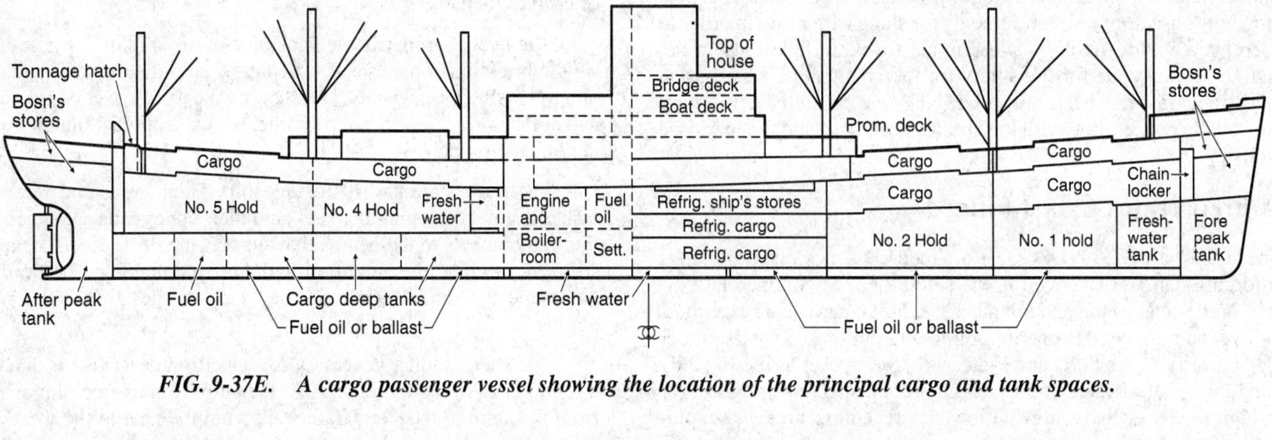

FIG. 9-37E.   A cargo passenger vessel showing the location of the principal cargo and tank spaces.

to that used on shore. All vents interconnect through these ducts. A damper provided in each duct isolates one compartment from another to prevent the spread of fire. Dampers can operate manually or automatically using a fusible link. In some modern passenger ships, dampers may be remotely operated from a central control station to prevent the spread of smoke.

Cargo tanks on tank vessels are not ventilated, but they use a vent system to allow the tank to relieve excess pressure or to allow air ingress if a vacuum develops. The section "Tank Vessels" explains this venting arrangement.

Stability

The interaction of two opposing forces controls a ship's stability, the ability of a floating body to remain at rest in a stable position. The center of gravity (point G in Figures 9-37F, 9-37G, and 9-37H) is the point where the weight of the ship and any other weight aboard

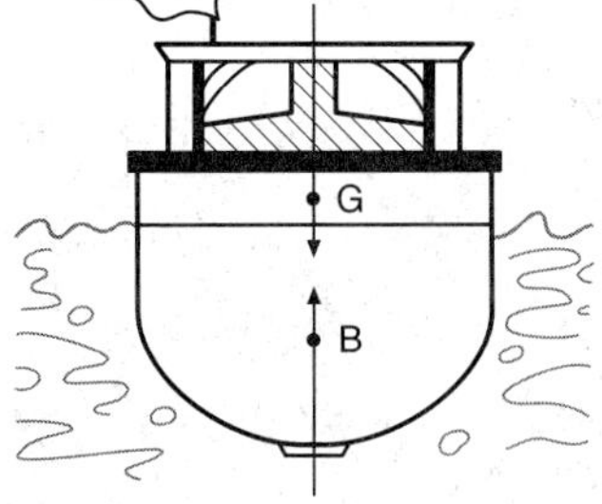

FIG. 9-37F. Ship in state of vertical equilibrium.

FIG. 9-37G. Ship inclined by an external force.

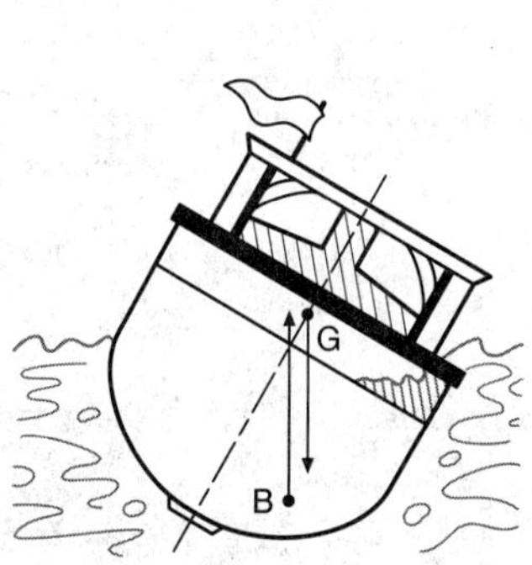

FIG. 9-37H. The effect of adding water (weight) to the upper portion of a ship.

is concentrated. It remains in a constant position until the weight distribution aboard the vessel changes, and it acts with a downward force. The center of buoyancy is defined as the center of gravity of the body of water displaced by a floating vessel. It is the center of the immersed part of the ship, and it acts with an upward force, perpendicular to the water's surface.

When a ship is in a state of vertical equilibrium, as shown in Figure 9-37F, the downward force of gravity (G) and the upward force of buoyancy (B) lie in the same vertical line. When the ship is inclined by some external force, the center of gravity (G) remains in its original position. The center of buoyancy (B), since it is the center of the immersed part of the ship, shifts toward the lower side, as shown in Figure 9-37G. In that situation, the central force of gravity (G) downward, combined with the outer force of buoyancy (B) upward from the low side, act together to return the vessel to an upright position.

Adding water to the upper portion of the hull, as shown in Figure 9-37H, produces an entirely different situation. Adding weight at a high point within the hull raises the center of gravity. The fact that water runs to, and stays at, the lowest point inclines the vessel and holds it in an inclined position. The center of buoyancy shifts to the lower side, as it did before. If the addition of weight (water) stops while the downward force of gravity (G) is inboard—that is, toward the centerline—of the upward force of buoyancy (B), the vessel settles in a new stabilized position, inclined toward the side of the additional weight. If sufficient water is added to bring the downward force of (G) outboard—that is toward the ship's side-shell—of the upward force of (B), as shown in Figure 9-37H, the vessel continues to roll over and eventually capsizes.

From this explanation of stability, it is easy to see how adding water, especially high in the vessel, can dramatically affect the vessel's stability. Whenever a fire is fought on board a vessel, this problem must be recognized. A means of removing water from a vessel during fire fighting must be established, since there is a limit to how much water can be pumped aboard before the vessel rolls over on its side, as did the *S.S. Normandie* in New York Harbor in 1942. To prevent this, a person who understands the principle of stability (i.e., a naval architect) should be consulted while a shipboard fire is being fought. Otherwise, the incident could cause loss of life, as well as loss of a vessel.

Passenger Ships

The primary goals of passenger ship fire protection requirements are to limit a fire to the compartment of origin, and to provide for safe egress routes and refuge areas. This is typically achieved via a combination of passive and active measures. Spaces on a ship are classified according to occupancy. Minimum barrier ratings are stipulated between different occupancies, and certain occupancies are required to be protected by active systems (e.g., an engine room may be required to be protected with a total flooding system or a passenger accommodation area may be required to be protected with an automatic sprinkler or water mist system). Egress routes are protected by limiting interior finish materials, stipulating minimum fire endurance ratings, and limiting the types of spaces that can directly access these routes. Egress routes typically culminate at a space where passengers can be expected to be sheltered from the effects of fire while the crew attempts suppression measures, or life saving appliances such as lifeboats or liferafts are prepared for passenger evacuation.

Tank Vessels

Older tank vessels include pump rooms, one forward and one aft, between the last cargo tanks and the engine room. The cargo and ballast pumps are located within the pump room. All cargo and ballast water passes through the pump rooms into and out of the cargo tanks. A pump room can accumulate flammable vapor from any cargo leakage into the bilges from the various piping and valves within the room. For this reason, the room should be well ventilated before anyone enters it, and someone topside should be aware of persons entering a pump room in case they are exposed to an acute concentration of vapors. Tank vessels of newer design, especially the chemical and gas carriers, use deep-well pumps in each cargo tank, thereby eliminating the need for pump rooms.

Each cargo tank on a tank vessel is fitted with a venting system to regulate pressure within the tank. On crude oil carriers and petroleum product carriers, each tank connects to a common venting system. On a chemical or gas carrier, vents are independent of each other. This arrangement is required to ensure that vapors from different chemicals do not come in contact with each other and to prevent incompatible reactions or contamination.

The venting system contains flame screens or flame arrestors. These devices prevent any flame that develops outside the cargo tank from entering it and causing an explosion, as happened on the *S.S. Sansinena*. In that case, it was theorized that flame entered the cargo tank through a wasted section of the vent piping below the flame screen. A missing flame screen was also cited in the accident investigation of explosions and fires aboard the U.S. tankship *OMI Yukon* in October 1986.

It is extremely important that the various piping systems aboard a tank vessel, including the vent system, be maintained in good condition, and that flame screens be placed over any openings into a cargo tank. During cargo loading or when ballasting a tank that previously contained cargo, copious amounts of flammable vapors are generated. These vapors are vented to the atmosphere to ensure that the cargo tanks are not overpressurized. In the vicinity of a tanker conducting these operations, the air contains a flammable mixture, especially if low winds are present. Should the flammable mixture ignite, the flame will be prevented from entering the tank if flame screens or arrestors are in place and all the piping systems that penetrate the cargo tank are intact.

For some toxic chemicals and for all gases, a vapor recovery system is generally employed to return the vapors to shore during loading and offloading. With this arrangement, no venting to the atmosphere takes place unless an emergency occurs. The cargo transfer operations and vapor return are balanced to prevent over- or under-pressurizing of the cargo tank. This procedure eliminates the potential flammable atmosphere around the vessel during cargo transfer operations. However, the various piping systems must be intact to prevent release of any flammable vapors or liquids.

As a result of SOLAS '74 and the 1978 TSPP Conference, inert gas systems are now installed aboard tankers. These systems are intended to ensure that the atmosphere within the cargo tanks does not contain a flammable mixture. Without an inert gas system, flame screens or arrestors must protect all openings into a cargo tank, and all piping must be in good condition. Otherwise, as discussed previously, an ignition source can be introduced into the cargo tanks. Experience shows that precluding entry of all ignition sources is difficult. Consequently, the trend is toward inerting cargo tanks whenever they contain hydrocarbon vapors. Design, construction, and operation of the inert gas system must render the atmosphere of cargo tanks nonflammable and keep it that way at all times, except when the tanks are required to be gas-free. The system must be capable of:

1. Inerting empty cargo tanks by reducing the oxygen content of the atmosphere in each tank to a level that cannot support combustion.

2. Maintaining the atmosphere in any part of any cargo tank with an oxygen content not exceeding 8 percent by volume and at a positive pressure at all times in port and at sea, except when it is necessary for such a tank to be gas-free.

3. Eliminating the need for air to enter a tank during normal operations, except when it is necessary for such a tank to be gas free.

4. Purging empty cargo tanks of hydrocarbon gas so that subsequent gas release operations will at no time create a flammable atmosphere within the tank.

Figure 9-37I is a simplified illustration of tanker operations using inert gas systems. The rule is never to mix both air and hydrocarbons during operation, but to insert inert gas between the air and hydrocarbon phases of the operation. Specific practices for these systems are described in 46 CFR 32.53; Regulation 62 of Chapter II-2 of SOLAS '74, as amended, which describes international re-

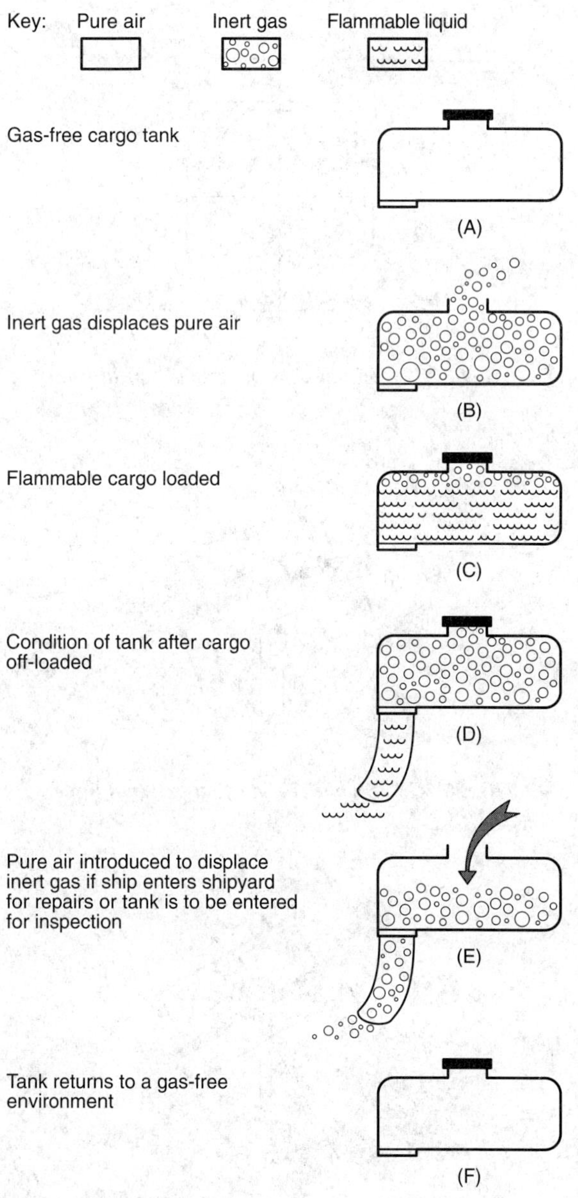

Key: Pure air   Inert gas   Flammable liquid

Gas-free cargo tank

(A)

Inert gas displaces pure air

(B)

Flammable cargo loaded

(C)

Condition of tank after cargo off-loaded

(D)

Pure air introduced to displace inert gas if ship enters shipyard for repairs or tank is to be entered for inspection

(E)

Tank returns to a gas-free environment

(F)

FIG 9-37I.   A tanker inert gas system.

quirements that are very similar to U.S. criteria; and in the IMO Guidelines for Inert Gas Systems. Use of inert gas systems enhances tanker safety significantly by greatly reducing the risk of fire and explosion within cargo tanks.

Barges

Almost any commodity can be shipped by water. The inland waterways industry has developed various types and sizes of barges for the efficient handling of products, ranging from coal in open hopper barges to chemicals and "thermos bottle" barges, and from dredged rock in dump scows to railroad cars on car floats. Barging is the only practical mode of marine transportation to move outside machinery, tanks, kilns, and some space vehicles for long distances.

The most versatile, inexpensive, and common type of barge in the United States is the hopper barge. With minor modifications, it can be adapted to transport literally any solid commodity in bulk or package. Basically a single skinned, open-top box, its inner shell forms a long hopper or cargo hold. The bottom, sides, and ends of the hold are free of appendages. It is generally of welded plate construction, with a double bottom for greater safety. Hopper barges are braced to resist the heaviest external blow, as well as to absorb the impact of buckets used in loading and unloading. The covered dry cargo barge is used for bulk-loading commodities that need weather protection. Generally, the only difference between these vessels and open hopper barges is that covered dry cargo barges are equipped with watertight covers over the entire cargo hold.

Three basic types of tank barges are used to transport liquid commodities: (1) single-skinned tank barges, (2) double-skinned tank barges, and (3) cylindrical tank barges.

Single-skinned tank barges have bow and stern compartments separated from the midship by transverse collision bulkheads. The vessel's entire midship shell constitutes the cargo tank. Hydrodynamic and stability considerations require that this huge tank be divided by bulkheads. The usual compartmentation consists of a centerline bulkhead and three transverse bulkheads forming four separate cargo compartments on either side. The entire framing structure is inside the cargo tanks.

Double-skinned tank barges have, as the term implies, an inner and an outer shell. The inner shell forms cargo tanks that are free of appendages and are therefore easier to clean and to line with a protective coating. Poisons and other hazardous liquids require the protection of void compartments between the inner and outer shells.

Barges with independent cylindrical tanks transport liquids under pressure, or, when pressure is in use, discharge the cargo. Because highly efficient linings and insulation can be incorporated in them, cylindrical tank barges can be used to carry cargoes at or near atmospheric pressures. Because cylindrical cargo tanks are generally mounted in the barge hopper, they are free to expand or contract independently of the entire structure. For this reason, they are preferred for high-temperature cargoes, such as liquid sulphur at 280°F (138°C), or refrigerated cargoes, such as anhydrous ammonia at −28°F (−33.3°C).

Barges carrying hazardous materials must have a 2- × 3-ft (0.6- × 0.9-m) warning sign facing outboard without obstruction, on which is printed "WARNING—DANGEROUS CARGO. NO VISITORS, NO SMOKING, NO OPEN LIGHTS." The sign also must specify names and locations of all cargoes aboard the vessel. In addition to the warning sign, barges carrying hazardous materials must have an information placard giving details about each hazardous cargo aboard. These placards are required to be displayed in a waterproof container as near to the warning sign as practical.

Information placards for all hazardous materials in an entire tow must be maintained in the towboat wheelhouse where they are readily accessible to the operator. In case of any problem, the towboat operator on watch should be able to furnish specific details via radio concerning any hazardous cargo in tow.

Tank Cleaning and Gas-Freeing of Cargo Tanks

One of the most hazardous processes in the operation of tank ships and tank barges is that of tank cleaning and gas-freeing. This procedure is performed in varying degrees for a change of cargo, on or off charter survey, for periodic U.S. Coast Guard inspection, and for repairs. Although tank ships are usually cleaned and gas-freed at sea and barges are cleaned in port, the procedure for both is very similar.

The biggest problem in tank cleaning and gas-freeing is the necessity of reducing the flammable vapors in the atmosphere from a point above the upper explosive limit (UEL) to a point below the lower explosive limit (LEL). In other words, the atmosphere of the space must be brought down through the full flammable range of the previously stored material. A spark or other ignition source within the space while the atmosphere is within the explosive range can cause an explosion.

The most common and most feared source of ignition during cleaning and gas-freeing operations is a spark caused by static electricity discharge. During washing operations, water mist forming in a tank's atmosphere takes an electrical charge from the cleaning water jet that penetrates the mist. Should any object not properly bonded to the tank be lowered into the tank space, a static discharge similar to a small lightning bolt could travel from the mist cloud to the ungrounded object.

Another severe fire and explosion hazard exists while a vessel is being repaired. Even though the tanks may have been thoroughly cleaned and gas-freed at sea, there is no guarantee that they will remain in the same condition throughout the repair process. The U.S. Coast Guard and Department of Labor regulations require that each space be tested and certified as "safe for hot work" by an NFPA-certificated marine chemist before any hot work can begin in port. A cargo tank that has been certified completely safe by a marine chemist will not necessarily remain in that condition. Sun beating down on a tank with a misty interior can cause sufficient flammable vapors to be released inside the tank, changing the atmosphere from safe to well within the explosive range. Therefore, the marine chemist must not only evaluate the atmospheric condition at the time of the inspection, but must also evaluate any weather conditions and interior vessel conditions that might cause flammable vapors to regenerate. If there is any doubt, the marine chemist should recommend that the repairer take additional precautionary measures and should specify circumstances under which a marine chemist should be recalled for further testing. Timely and regular use of the services of an NFPA-certificated marine chemist has helped to virtually eliminate some of the most severe fire and explosion hazards in the marine industry.

However, the requirements for a marine chemist only apply in port. At all other times, the ship's officer may serve as the certifying official. Coast Guard regulations specifically mandate that NFPA Publication 306, *Standards for the Control of Gas Hazards on Vessels,* be used as a guide in conducting required inspections and issuing certificates when making alterations, repairs, or other operations involving welding, burning, or similar fire-producing actions. When this guidance is not fully understood, explosions and fires can still occur, such as those on board the U.S. tankship *OMI Charger* in October 1993 near Galveston, TX. In that case, questionable gas-freeing procedures may have been used and were not followed up with proper procedures to ensure the tanks being welded were gas-free.

SHIPBOARD FIRE EXTINGUISHING EQUIPMENT

Because each ship is different, it follows that fire protection systems are rarely identical from one ship to another. Under Coast Guard regulations governing U.S. merchant vessels, several alternative fire protection systems may be either permitted or required, depending upon their designed purpose. Foreign vessels are likely to have different systems or arrangements. Similar systems manufactured and installed by competing companies can vary widely. Most fire extinguishing and fire detection equipment for use aboard U.S. vessels is approved by the Coast Guard, as are most structural fire protection materials used on U.S. vessels. Testing laboratories should list and lable for marine use other acceptable fire extinguishing equipment, such as portable fire extinguishers.

Firemain System

The firemain system is the line that provides fire-fighting and flushing water throughout a ship. It is the backbone of all fire-fighting systems aboard ships. Only the capacity of the fire pumps supplying the system limits the quantity of water available for fire fighting. Most vessels are required to have two fire pumps, with all suction inlets, sources of power, and so on, for each pump separately located so that one fire will not put both pumps out of operation. An alternative, which the Coast Guard must approve, is to install both pumps in the same space and to protect the space with a carbon dioxide extinguishing system. This arrangement is permitted only in unusual circumstances, where separation of pumps would not increase safety, a course usually acceptable for only small vessels. The required pumps' size depends upon the vessel's size and service, as well as the arrangement of the pumps and piping aboard the vessel.

Hydrants on the firemain system are located so that two hose streams may be directed into all portions of the vessel accessible to passengers and crew during navigation. Because one stream must flow from a single length of hose, it is essential that the hose be long enough to direct water into all portions of the space, not just long enough to get the nozzle to the door.

It is also essential that fire hoses be connected to the firemain outlets at all times. Critical time can be lost during fire-fighting operations if this is not done.

International Shore Connection

One particular item of equipment SOLAS requires for all vessels deserves special mention: the international shore connection. Vessels must be fitted with a special bolted plate that can be attached to the firemain. The flange must have a flat face on one side, and on the other side must have a permanently attached coupling that fits the ship's hydrants and hose. The connection must be kept aboard ship together with a gasket suitable for 150-psi (1,034-kPa) service, together with four $5/_8$-in. (16-mm) bolts, 2 in. (50 mm) long, and eight washers. This fitting eliminates the possible mismatch of hose threads aboard vessels and in port facilities. Provision of such connections is not mandatory for shore installations but can prove invaluable to organizations having marine fire-fighting responsibilities. A diagram of an international shore fire hose connection is shown in Figure 9-37J.

Carbon Dioxide Systems Aboard Vessels

Carbon dioxide as an extinguishing agent aboard ship has many desirable properties. It does not, of itself, damage cargo or machinery, and it leaves no residue to be cleaned after a fire. Even on a ship without power, a carbon dioxide system can deliver the agent to all parts of a space because it is pressurized. Basically, there are two carbon dioxide systems for protecting compartments in a ship: fixed CO_2 systems aboard commercial vessels are classified either as (1) cargo systems or (2) total flooding systems.

Cargo systems: Fires in Class A materials (ordinary combustibles) carried in cargo holds usually start with smoldering and production of large quantities of smoke. Rapid burning occurs only when sufficient heat develops to reach the temperature at which solid combustibles give off enough gases to support continued rapid combustion. Until that time, the combustion rate is relatively slow. After a smoldering fire in a ship's hold is discovered, time to flaming combustion might be at least 20 min, depending upon oxygen availability and other circumstances. This allows time to prepare for fire-fighting operations. The cargo hold is sealed, and the CO_2 normally is released several bottles at a time, so that the oxygen level falls below the rate that supports combustion.

Total Flooding Systems: Fires in machinery and similar spaces are generally Class B (flammable liquids). In this type of fire, heat builds up rapidly. Since a ship's safety depends to a great extent upon the equipment in the machinery space, introducing the extinguishing gas quickly is important. Rapid release of an extinguisher also prevents heat from causing failure of bulkheads, which makes maintaining a sufficient concentration of carbon dioxide impossible and prevents structural members from reaching high temperatures. Quick release of the extinguishing agent also prevents heat updraft from the fire from carrying away the carbon dioxide and limits the extent of equipment damage. Discharge of 85 percent of the required quantity of carbon dioxide in these systems should be completed within 2 min. Slow release, as in the cargo system, might result in no extinguishment. Systems designed to protect roll-on/roll-off cargo spaces require discharging two-thirds of the required quantity within 10 min. On all domestic vessels and on foreign vessel installations on or after October 1, 1994, two separate and deliberate controls are required to avoid unintentional release of the gas. One releases the minimum required amount of carbon dioxide, and the other operates the stop valve or direction valve.

Shipboard carbon dioxide extinguishing systems also carry inherent dangers. In 1993, two people were killed in a shipyard during a test of a CO_2 system when the system unexpectedly discharged [NFPA Journal, March/April 1996, p83.] In 1996, six people died when a CO_2 system was discharged during an engine room fire [Lloyds List, 10/12/96.] Personnel protection is typically provided through three means: (1) pre-discharge delays, (2) pre-discharge alarms, and (3) two distinct operations necessary to release the CO_2. These provisions may not be necessary for small, normally unoccupied spaces. The pre-discharge delay and alarms should be set such that there is sufficient time for all personnel to evacuate from all portions of the space. Of the two valves, one should control the release of CO_2 from the storage, and one should control the discharge of CO_2 into the protected space.

Also see Section 6, Chapter 20, "Carbon Dioxide and Application Systems."

Halon Systems

Bromotrifluoromethane, or Halon 1301, is one of several halogenated hydrocarbon extinguishing agents introduced aboard ships in the

International Shore Connection

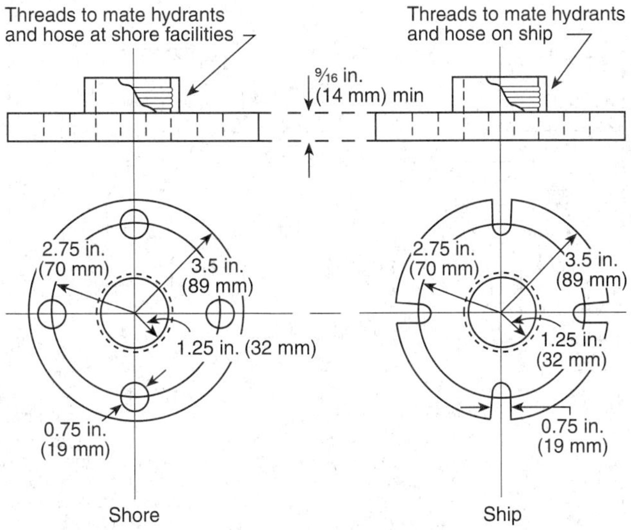

FIG. 9-37J. *International shore fire hose connection.*

1980s. In general, halogenated agents have high extinguishing efficiency per unit of weight. This makes them particularly suitable for installations that are weight-critical, such as hydrofoils.

Title 46 of the Code of Federal Regulations does not contain any provisions for the design or requirements of halon systems aboard commercial vessels. Instead, approval of these systems has been based upon a provision that grants the commandant of the U.S. Coast Guard authority to allow equivalent systems.

No new halon has been manufactured in the United States since January 1, 1994. Recycled halon as well as halon "banked" prior to that date may be used to recharge an existing system. Existing systems are permitted to remain in service, provided they are maintained properly.

Currently, both Halon 1301 and Halon 1211 portable extinguishers carry Coast Guard approval for marine use.

Internationally, IMO developed detailed requirements compatible with U.S. requirements for halon systems for machinery spaces, as well as for roll-on/roll-off vehicle spaces. A halon piping diagram for the engine room of a foreign flag, 80,000 DWT tanker appears in Figure 9-37K. The diagram appeared in the May 1984 volume of *Chevron's Safety Bulletin.*

The first set of amendments to the SOLAS '74 Convention permits installation of, and sets out international requirements for, halon systems on commercial vessels. However, recent events have reversed the trend toward the installation of halon systems in lieu of CO_2 systems. The Second Meeting of the Parties to the Montreal Protocol on Substances that Deplete the Ozone Layer, held in June

1990 in London, adopted amendments to that protocol. The target completion date set for phasing out halons is the year 2000. IMO has agreed to a requirement prohibiting new installations of halon on vessels engaged in international trade which took effect October 1, 1994. Additionally, the Subcommittee on Fire Protection of IMO has been charged with drafting a plan for phasing out existing halon systems aboard commercial vessels engaged in international trade.

Mechanical Foam Systems Aboard Ship

Mechanical foam is produced by introducing foam concentrate in proper proportions into a flowing stream of water and then aspirating it with air. Aboard ship, foam concentrate is normally introduced through proportioning equipment located near the foam concentrate storage container at a central location on the vessel. The foam solution is pumped through fixed piping to foam nozzles, monitors, and so on, in the area to be protected. Air mixes with the foam solution at the discharge nozzle.

Deck Foam Systems

Deck foam systems are required aboard tank vessels constructed after January 1, 1962. This system is intended to protect any deck area, using a predetermined flow rate from foam stations—that is, monitors or hose stations—aft of the area to be protected. Piping and foam stations are arranged so that ruptured sections of piping in a fire's path can be isolated. With this arrangement, it should be possible to fight the fire effectively wherever it occurs by working forward from the after house, assuming machinery, foam pumping, and proportioning equipment are located aft, and are accessible under

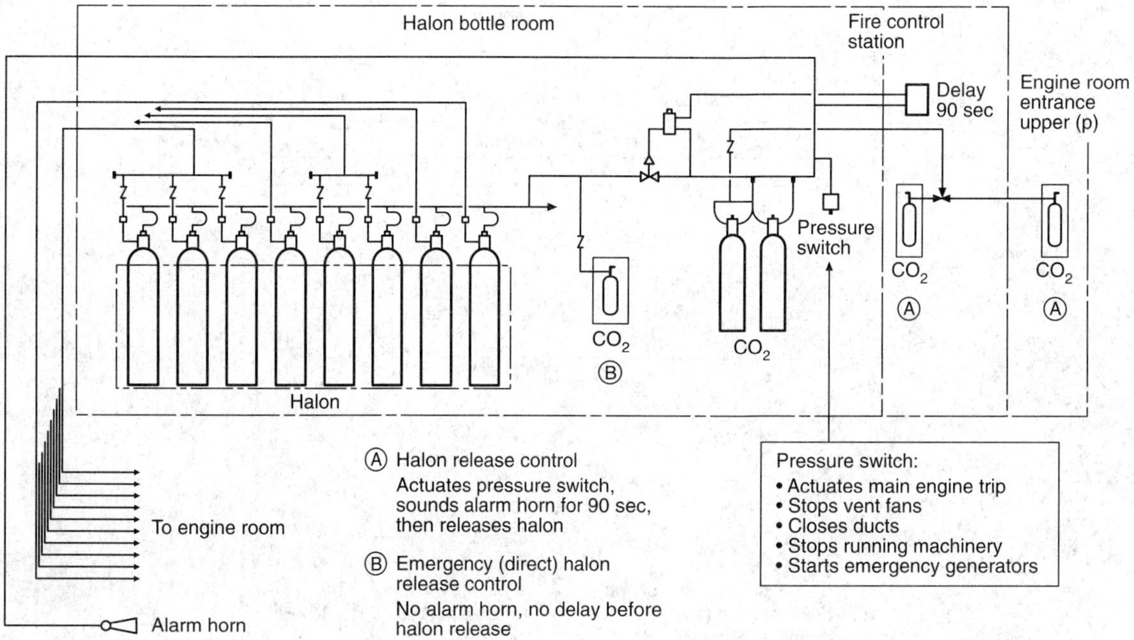

FIG. 9-37K. *Halon piping diagram for engine room (System No. 1).*

emergency conditions. The importance of availability under emergency conditions was highlighted in the accident investigation of the explosions and fires aboard the tankship *Surf City* when the crew could not reach valves to activate the system.

Concentrate normally supplied for mechanical foam is suitable for use on most, but not all, flammable liquids carried aboard tankers. For example, it is impossible for foam made from ordinary concentrate to form a blanket on alcohols, esters, ketones, or ethers, commonly called water-soluble or polar solvents. An alcohol type foam concentrate is available for shipboard fires involving water-soluble flammable liquids. An alternate method of combating alcohol type fires is using water spray to dilute the flammable liquid and cool the surrounding areas. A foam concentrate effective on both polar and nonpolar solvents is also available.

Chemical Carriers

Chemical carriers may be required to have more fire-fighting systems than ordinary tank vessels. The systems required for a chemical carrier will be based on the chemical involved; 46 CFR 153 describes the type of fire protection required. A foam system will be specified as polar or nonpolar solvent, depending on whether or not the chemical is water-soluble. A water spray system or a dry chemical system may also be required. Each system must be designed to cover each part of the cargo containment system on the weather deck with the exception of the vent riser. The cargo containment system includes the cargo tank, piping system, venting system, and gaging system.

Gas Carriers

Gas carriers are required to have more fire-fighting systems than those required for ordinary tank vessels. An exterior water spray system must be installed to protect all exposed cargo tank surfaces, each boundary of an accommodation space or navigational area that faces the cargo area, and the cargo loading and discharge area. The spray system is activated in the event of a fire to cool down and protect these areas. It is not intended to extinguish a liquid flammable gas fire. On the other hand, a dry chemical system must be installed to protect the cargo area and cargo piping, and to extinguish a liquefied flammable gas fire that might result from a broken cargo handling line or a fire in the vent system. Foam systems are not effective against a liquefied gas fire. If installed, they only augment, not replace, the systems just described.

Sprinkler Systems

Traditionally, automatic sprinkler systems were not used on U.S. flagged vessels; although manual deluge type sprinklers have been used for some time on certain vehicle decks of ferry vessels. However, the use of automatic sprinkler systems in this country has increased over the last few years. Sprinkler system design on vessels presents some challenges which are not found in systems installed in buildings. For example, a fire pump which takes suction from the sea may not be considered sufficiently reliable because of concerns with failure mode of the power source and the possibility of pulling mud into the suction of vessels which may operate in shallow water. Typically, a water tank containing a pressurized charge of water equal to one minute of demand over the hydraulically most demanding design area is provided.

International requirements for automatic sprinkler, fire detection, and fire alarm systems aboard both new and existing passenger ships only recently came into effect. The 1992 Amendments to the 1974 SOLAS Convention now mandate their use and incrementally require their retrofit on all passenger ships. The Convention allows the use of both conventional and equivalent water mist systems.

Also see Section 6, Chapter 10, "Automatic Sprinkler Systems"; and Section 6, Chapter 12, "Water Spray Protection."

Fire-Fighting Plan

NFPA 1405, *Guide for Land-Based Fire Fighters Who Respond to Marine Vessel Fires*, provides guidance for shoreside responders to shipboard fires. Despite the information the standard provides, many shoreside fire fighters are ill prepared for the complexity of fighting a shipboard fire, as found in the accident investigation of the tankship *Seal Island,* which occurred in the U.S. Virgin Islands on October 8, 1994. Ships' decks and main bulkheads are steel, retain the heat of combustion for a long period of time, and can make locating the fire more difficult; spaces and access are more restricted than in buildings, thus making fire fighting and retreating confusing; and the accumulation of water on decks can worsen the situation. Most important, the command structure for fighting a shipboard fire is not always understood. At sea only the crew is available to respond to fire emergencies, while in port shoreside fire fighters are likely to be called by the Coast Guard.

One readily available piece of information critical for fire fighting is located aboard the ship. All U.S. vessels of 1,000 gross tons and over, and all vessels subject to SOLAS, must permanently display a fire-fighting plan or fire control plan of the ship. Although the plan's primary purpose is to guide shipboard personnel, it should contain the following information to assist port fire-fighting groups:

1. Fire control stations for each deck.
2. Sections enclosed by fire-resistive divisions.
3. Fire alarms, detection systems, and sprinkler systems.
4. Fire extinguishing appliances.
5. Access to different compartments, decks, and so on.
6. Ventilating systems, including master fan controls, positions of dampers, location of remote controls, and identification numbers of ventilating fans serving each section.

Instead of being posted on a plan, the information listed can be described in a fire-fighting booklet. The booklet is required to be aboard ship at all times.

Whether posted as a fire-fighting plan or presented in booklet form, information should be available on all watertight compartments, openings therein, and means of closure, controls, and arrangements for correcting any list due to flooding.

SOLAS '74 requires that a duplicate set of fire control plans be permanently stowed in a prominently marked, weathertight enclosure outside the deckhouse to assist shoreside fire-fighting personnel. Figure 9-37L is an illustration of a fire-fighting plan for the aft end of a super tanker.

The international community has also agreed upon standard symbols for fire control plans. (See Figure 9-37M.) Most symbols are to be shown in red on the general arrangement plan to highlight their location for shipboard and shoreside fire fighters.

NFPA MARINE STANDARDS

In 1994, the U.S. Coast Guard, and the marine community embarked on an ambitious project to shift development and maintenance of fire protection regulations applicable to U.S. flagged commercial ships to consensus standards making organizations. The cornerstone of this effort is NFPA 301 (proposed), *Code for Safety to Life from Fire on Merchant Vessels,* a "life safety code" for ships. Additionally, chapters on marine application have been added to NFPA 13, *Installation of Sprinkler Systems* and NFPA 750, *Water Mist Fire Protection Systems.* Additionally, chapters on marine application are currently proposed or planned for NFPA 10, *Portable Extinguishers;* NFPA 11, *Low-Expansion Foam;* NFPA 12, *Carbon Dioxide Systems,* and NFPA 2001, *Clean Agent Systems.*

Other NFPA standards relating to marine vessels are contained in NFPA 306, *Standard for the Control of Gas Hazards on Vessels;* and NFPA 312, *Standard for Fire Protection of Vessels During Construction, Repair, and Lay-up.* NFPA 306 describes the conditions necessary for safety before repairing any vessel carrying or having carried fuel as cargo, combustible or flammable liquids, flammable compressed gases, and bulk chemicals. NFPA 306 is also the basis for confined space safety in the maritime industry. It also contains a section on flammable cryogenic liquid carriers. NFPA 312 covers measures for preventing and controlling fires on vessels in the building or repair yard, or on vessels that are laid up. Provisions for each circumstance appear separately in the standard.

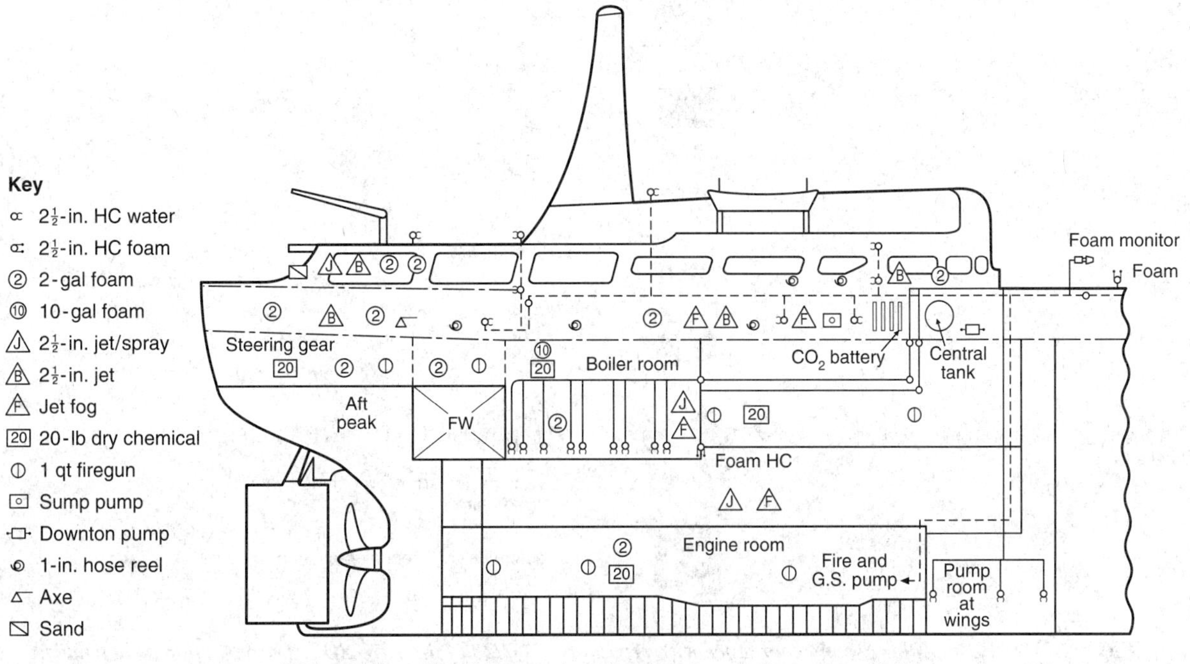

Key

∝ 2½-in. HC water
∝ 2½-in. HC foam
② 2-gal foam
⑩ 10-gal foam
⚠ 2½-in. jet/spray
Ⓑ 2½-in. jet
Ⓐ Jet fog
[20] 20-lb dry chemical
① 1 qt firegun
⊡ Sump pump
-⊡- Downton pump
◔ 1-in. hose reel
△ Axe
▨ Sand

FIG. 9-37L. Fire-fighting plans for the aft end of a super tanker.

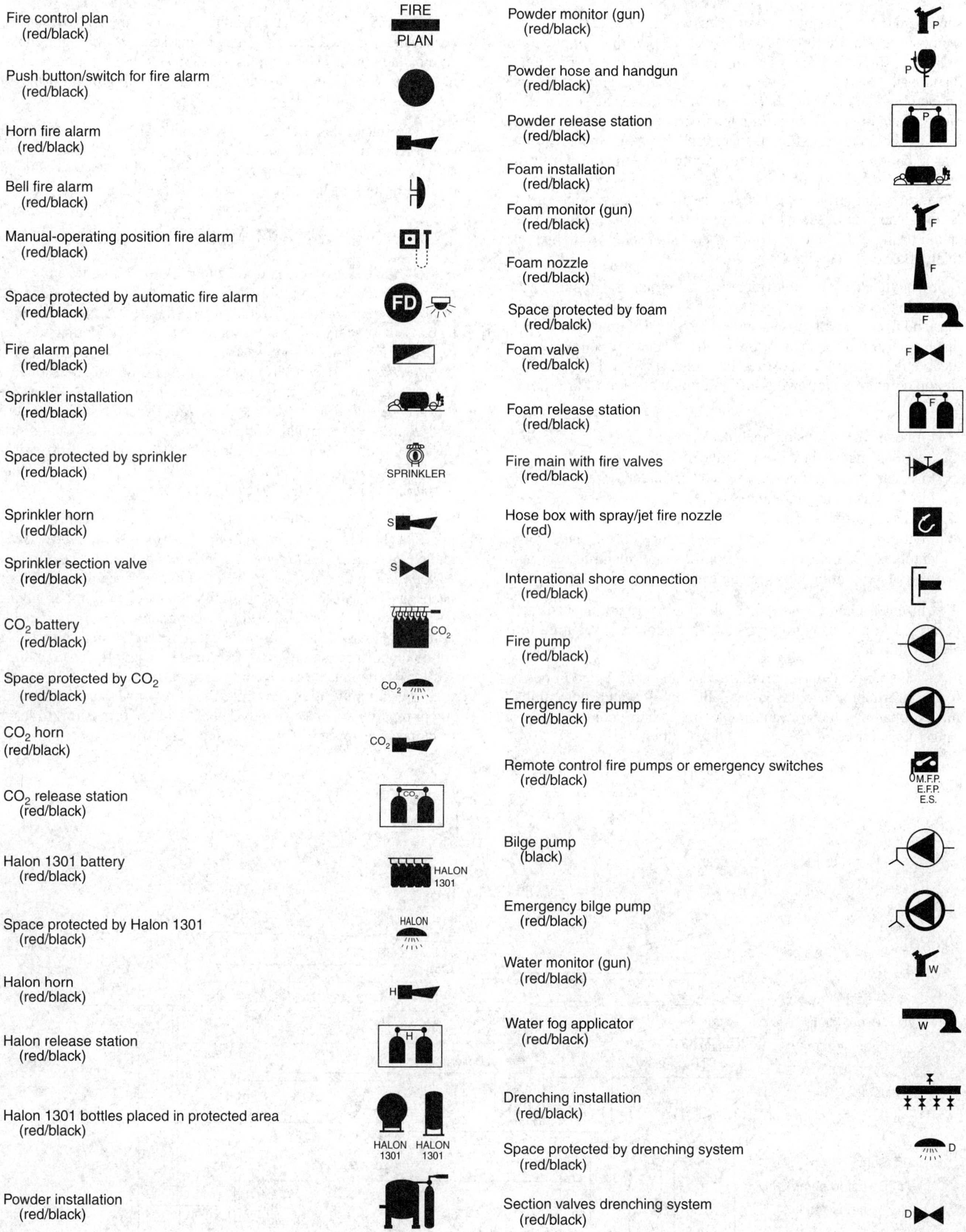
Fire control plan
(red/black)
FIRE
PLAN
Push button/switch for fire alarm
(red/black)
Horn fire alarm
(red/black)
Bell fire alarm
(red/black)
Manual-operating position fire alarm
(red/black)
Space protected by automatic fire alarm
(red/black)
FD
Fire alarm panel
(red/black)
Sprinkler installation
(red/black)
Space protected by sprinkler
(red/black)
SPRINKLER
Sprinkler horn
(red/black)
S
Sprinkler section valve
(red/black)
S
CO_2 battery
(red/black)
CO_2
Space protected by CO_2
(red/black)
CO_2
CO_2 horn
(red/black)
CO_2
CO_2 release station
(red/black)
CO_2
Halon 1301 battery
(red/black)
HALON
1301
Space protected by Halon 1301
(red/black)
HALON
Halon horn
(red/black)
H
Halon release station
(red/black)
H
Halon 1301 bottles placed in protected area
(red/black)
HALON
1301
HALON
1301
Powder installation
(red/black)
Powder monitor (gun)
(red/black)
P
Powder hose and handgun
(red/black)
P
Powder release station
(red/black)
P
Foam installation
(red/black)
Foam monitor (gun)
(red/black)
F
Foam nozzle
(red/black)
F
Space protected by foam
(red/balck)
F
Foam valve
(red/balck)
F
Foam release station
(red/black)
F
Fire main with fire valves
(red/black)
Hose box with spray/jet fire nozzle
(red)
International shore connection
(red/black)
Fire pump
(red/black)
Emergency fire pump
(red/black)
Remote control fire pumps or emergency switches
(red/black)
M.F.P.
E.F.P.
E.S.
Bilge pump
(black)
Emergency bilge pump
(red/black)
Water monitor (gun)
(red/black)
W
Water fog applicator
(red/black)
W
Drenching installation
(red/black)
Space protected by drenching system
(red/black)
D
Section valves drenching system
(red/black)
D

FIG. 9-37M. Symbols for use in accordance with regulation 11-2/20 of the 1974 SOLAS Convention as amended.

Portable fire extinguishers (red/black)	Class A fire door (red/black)
Wheeled fire extinguishers (red/black)	Class A fire door self-closing (red/black)
Fire station (red)	Class A fire door self-closing (red/black)
Locker with fireman's outfit (red)	Class B fire door (yellow/black)
Locker with additional breathing apparatus (blue/black)	Class B fire door (yellow/black)
Locker with additional protective clothing (red/black)	Class B fire door self-closing (yellow/black)
Primary means of escape (green/black)	Class B fire door self-closing (yellow/black)
Secondary means of escape (green/black)	Closing appliance for exterior ventilation inlet or outlet (red/black)
Class A division (red/black)	
Class B division (yellow/black)	Remote ventilation shutoff (black)
Fire damper in vent duct (red/black)	
Remote control skylights (black)	Main vertical zone (red/black)
Remote control fuel/lubricating oil valves (red/black)	
Control station (red)	Smoke detector (red/black)
Portable foam applicator (red/black)	Heat detector (red/black)
Inert gas installation (red)	Gas detector (red/black)
High-expansion foam supply truck (red/black)	Flame detector (red/black)
CO_2/Nitrogen bulk installation (red/black)	Emergency telephone station (red/black)
Emergency generator (red/black)	
Emergency switchboard (red/black)	Fire ax (red/black)
Class A fire door (red/black)	

FIG. 9-37M. Continued.

OTHER ORGANIZATIONS CONTRIBUTING TO MARINE FIRE SAFETY

The American Bureau of Shipping

The American Bureau of Shipping (ABS) is the American classification society whose technical committees develop rules for the dimensions and use of materials in the construction and conversion of marine vessels. After a vessel is completed according to the rules and under the constant supervision of an ABS field inspector, the bureau classes and lists it as having met all design and structural safety requirements.

The American Petroleum Institute

In the area of marine fire fighting and fire protection, the American Petroleum Institute (API) is active in developing domestic and international standards, codes, and regulations. Under the policy direction of the API Central Committee on Transportation by Water, most actions are directed by the API Committee on Tank Vessels. *Ad hoc* groups are sometimes formed for special studies of a nonrepetitive nature.

The Marine Chemist Association, Inc.

This is a professional organization of chemists certificated for marine work by NFPA in accordance with provisions of NFPA 306. Established in May 1938 as the Marine Chemists' Subsection of the NFPA Marine Section, the Marine Chemists organized its present association in 1948 after the Marine Section disbanded.

The National Cargo Bureau, Inc.

The Bureau is a nationwide, nonprofit membership organization established in 1952 as a private, nongovernmental agency that formulates recommendations to various governments for regulating the safe stowage of dangerous goods and other cargoes and for regulating related cargo-handling gear.

The National Transportation Safety Board

The National Transportation Safety Board (NTSB) is an independent federal agency dedicated to promoting aviation, marine, railroad, highway, pipeline, and hazardous materials safety. Established in 1967, the agency is mandated by Congress through the Independent Safety Board Act of 1974 to investigate transportation accidents, including those involving fire, explosion, and pollution; determine the probable cause of accidents; issue safety recommendations; study transportation safety issues; and evaluate the safety effectiveness of government agencies involved in transportation. The Safety Board makes public its actions and decisions through accident reports, safety studies, special investigation reports, safety recommendations, and statistical reviews.

USCG Advisory Committee on Hazardous Materials

This committee of the National Academy of Sciences, National Research Council, is charged with advising the U.S. Coast Guard on scientific and technical matters relating to safe maritime transportation of hazardous materials.

BIBLIOGRAPHY

NFPA Codes, Standards, and Recommended Practices

Reference to the following NFPA codes, standards, and recommended practices will provide further information on marine fire hazards discussed in this chapter. (See the latest version of *The NFPA Catalog* for availability of the current edition of the following documents.)

NFPA 13, *Standard for the Installation of Sprinkler Systems*
NFPA 30A, *Automotive and Marine Service Station Code*
NFPA 52, *Standard for Compressed Natural Gas (CNG) Vehicular Fuel Systems*
NFPA 101®, *Life Safety Code®*
NFPA 302, *Fire Protection Standard for Pleasure and Commercial Motor Craft*
NFPA 303, *Fire Protection Standard for Marinas and Boatyards*
NFPA 306, *Standard for the Control of Gas Hazards on Vessels*
NFPA 312, *Standard for Fire Protection of Vessels During Construction, Repair, and Lay-up*
NFPA 750, *Standard on Water Mist Fire Protection Systems*
NFPA 1405, *Guide for Land-Based Fire Fighters Who Respond to Marine Vessel Fires*

Additional Reading

Ahmed, A. K., and Horsley, M. E., "Mathematical Modelling of Fires in Ships Cabins With Forced Ventilation," Sixth International Fire Conference on Fire Safety, Interflam '93, March 30–April 1, 1993, Oxford, England, Interscience Communications Ltd., London, England, 1993, pp. 385–399.

Allison, D. M., Marchand, A. J., and Morchat, R. M., "Fire Performance of Composite Materials in Ships and Offshore Structures," *Marine Structures*, Vol. 4, No. 2, 1991, pp. 129–140.

American Society for Testing and Materials, "Standard Test Method for Flammability of Marine Surface Finishes," American Society for Testing and Materials, Philadelphia, PA, ASTM E 1317-90, 1990, 18 pages.

An, M., Sousa, A. C. M., and Venart, J. E. S., "Numerical Study of a Water-fog Fire Suppression System in a Shipboard Machinery Room," *Proceedings* of the First International Conference on Fire Science and Engineering, ASIAFLAM '95, March 15–16, 1995, Kowloon, Hong Kong, 1995, pp. 243–253.

Arvidson, M., "Efficiency of Different Water Mist Systems in a Ship Cabin," Swedish National Testing and Research Inst., Boras, Sweden, SP Report 1994:03; *Proceedings* of the International Conference on Water Mist Fire Suppression Systems, November 4–5, 1993, Boras, Sweden, 1993, pp. 45–62.

Back, G. G., Darwin, R. L., and Leonard, J. T., "Full Scale Tests of Water Mist Fire Suppression Systems for Navy Shipboard Machinery Spaces," Hughes Associates, Inc., Baltimore, MD, Naval sea Systems Command, Washington, DC, Naval Research Laboratory, Washington, DC, ISBN 0-9516320-9-4; *Proceedings* of the Seventh International Interflam Conference, Interflam '96, March 26–28, 1996, Cambridge, England, Interscience Communications Ltd., London, England, 1996, pp. 435–444.

Back, G. G., *et al.*, "Full Scale Tests of Water Mist Fire Suppression Systems for Navy Shipboard Machinery Spaces. Part 1. Unobstructed Spaces. September 1993–December 1994," Hughes Associates, Inc., Baltimore, MD, Naval Sea Systems Command, Washington, DC, Naval Research Laboratory, Washington, DC, Naval Sea Systems Command, Washington, DC, NRL/MR/6180-96-7830, March 8, 1996, 24 pages.

Back, G. G., *et al.*, "Full Scale Tests of Water Mist Fire Suppression Systems for Navy Shipboard Machinery Spaces. Part 2. Obstructed Spaces. September 1993–December 1994," Hughes Associates, Inc., Baltimore, MD, Naval Sea Systems Command, Washington, DC, Naval Research Laboratory, Washington, DC, Naval Sea Systems Command, Washington, DC, NRL/MR/6180-96-7831, March 8, 1996, 29 pages.

Bailey, J. L., and Tatem, P. A., "Validation of Fire/Smoke Spread Model (CFAST) Using Ex-USS SHADWELL Internal Ship Conflagration Control (ISCC) Fire Tests," Naval Research Laboratory, Washington, DC, Office of Naval Research, Arlington, VA, NRL/MR/6180-95-7781, September 28, 1995, 51 pages.

Bluhme, D. A., "Transport Sector—Shipbuilding," Dantest, Copenhagen, Denmark, European Commission DG 3, EGOLF and CFPA Europe, Fire and Furnishing in Buildings and Transport, November 6–8, 1990, Luxembourg, Belgium, 1990, pp. 1–10.

Brown, J. R., et al., "Fire-Retardant Performance of Some Surface Coatings for Naval Ship Interior Applications," Fire and Materials, Vol. 19, No. 3, 1995, pp. 109–118.

Budnick, E. K., and Leonard, J. T., "Feasibility of Fuel Spray Fire Extinguishment With Existing Shipboard Manual Firefighting Capabilities," First International Conference on Fire Suppression Research, May 5–8, 1992, Stockholm, Sweden, 1992, pp. 301–317.

Budnick, E. K., et al., "Preliminary Evaluation of High Expansion Foam Systems for Shipboard Applications. March 1991–October 1991," Hughes Associates, Inc., Baltimore, MD, Naval Research Laboratory, Washington, DC, Naval Sea Systems Command, Washington, DC, NRL/MR/6180-95-7785, November 27, 1995, 66 pages.

Caloggero, J. M., "Shocking Truth About Boat Repairs," NFPA Journal, Vol. 89, No. 5, September/October 1995, p. 40.

Chiguo, S., et al., "Analysis of Heavy Fires Occurring in Chinese and Foreign Ships in Chinese Territorial Waters," First Asian Conference on Fire Science and Technology (ACFST), October 9–13, 1992, Hefei, China, International Academic Publishers, China, 1992, pp. 245–250.

Colonna, G. R., "Marine Chemist Program: NFPA and Industry at Work to Promote Maritime Safety," Third International Conference on Management and Engineering of Fire Safety and Loss Prevention: Onshore and Offshore, February 18–20, 1991, Aberdeen, Scotland, Elsevier Applied Science, New York, 1991, pp. 121–131.

Coxon, P., "Materials in the Commercial Marine Industry: The Coast Guard Approval Process," Second International Conference on Fire and Materials, September 23–24, 1993, Arlington, VA, 1993, pp. 131–141.

Cromer, P. B., "Shipboard Firefighting Operations and Ship Stability," Industrial Fire Safety, Vol. 3, No. 1, January/February 1994, pp. 40–42.

Crook, J., "Insurance Underwriting: The Key to Marine Safety?" Fire Prevention, No. 260, June 1993, pp. 21–23.

Croquette, J., "Safety on Board Ships: Fire Resistance and Composites," Composite Polymers, Vol. 5, No. 4, 1992, pp. 282–291.

DiNenno, P. J., et al., "Shipboard Tests of Halon 1301 Test Gas Simulants. February 1989–September 1989," Naval Research Lab., Washington, DC, NRL Memorandum Report 6708, August 22, 1990, 113 pages.

Dolph, B. L., "Modeling Fires on Ships," Coast Guard Research and Development Center, Groton, CT, Society of Fire Protection Engineers and WPI Center for Firesafety Studies, Proceedings of Computer Applications in Fire Protection Engineering, Technical Symposium, Final Program, June 20–21, 1996, Worcester, MA, 1996, pp. 1–4.

Dolph, B. L., "Ship Fire Safety Engineering Methodology: A Performance Based Fire Safety Analysis," Third International Conference and Exhibition on Fire and Materials, October 27–28, 1994, Arlington, VA, 1994, pp. 199–206.

Dow, R. S., and Bird, J., "Use of Composites in Marine Environments," Ironmaking and Steelmaking, Vol. 21, No. 6, 1994, pp. 431–444.

Drache, T., and Hahne, J., "Untersuchungen zur Ansprechwahrscheinlichkeit von Branderkennungssensoren unter realen Entdeckungsbedingungen im Schiffsbetrieb. [Investigations of the Probability of Response of Fire Detector Sensors Under Actual Detection Conditions in Ships.]," Proceedings of the Tenth International Conference on Automatic Fire Detection "AUBE '95," [Internationale Konferenz uber Automatischen Brandentdeckung.] April 4–6, 1995, Duisburg, Germany, 1995, pp. 472–481.

Durkin, A. F., et al., "Post-Flashover Fires in Shipboard Compartments Aboard ex-USS SHADWELL. Phase 4. Impact of Navy Fire Insulation," Naval Research Laboratory, Washington, DC, Hughes Associates, Inc., Columbia, MD, Naval Sea Systems Command, Washington, DC, Naval Sea Systems Command, Washington, DC, NRL MR 6183-93-7335, June 9, 1993, 148 pages.

Egglestone, G. T., and Turley, D. M., "Flammability of GRP for Use in Ship Superstructures," Fire and Materials, Vol. 18, No. 4, July/August 1994, pp. 255–260.

Fisher, K. J., "Is Fire a Barrier to Shipboard Composites?" Advanced Composites, May/June 1993, pp. 20–26.

Garcia, C. S., "Training for Shipboard Fire Fighting," Firehouse, Vol. 15, No. 10, October 1990, pp. 52–54.

Gibson, A. G., and Hume, J., "Fire Performance of Composite Panels for Large Marine Structures," Plastics, Rubber and Composites Processing and Applications, Vol. 23, No. 3, 1994, pp. 175–183.

Gibson, G., "Fire Performance of Marine Composites Examined," Fire and Flammability Bulletin, July 1995, pp. 7–9.

Gracik, T. D., "Intumescent Coatings for Fire Protection of Shipboard Structure," Naval Surface Warfare Center, Annapolis, MD, Fire Retardant Chemicals Association, Industry Speaks Out on Flame Retardancy: Coatings; Polymers and Compounding; Test Method Development; New Products, 1992 Fall Conference, October 27–30, 1992, Charleston, SC, Technomic Publishing Co., Lancaster, PA, 1992, pp. 219–227.

Grant, C. C., "Munitions Ship Explosion Levels Halifax," NFPA Journal, Vol. 84, No. 3, May/June 1992, pp. 96–98, 100, 102, 105–107.

Gustin, B., "Fire in the Hold: Shipboard Drill," Fire Engineering, Vol. 145, No. 4, April 1992, pp. 41–42, 44–48.

Hansen, B., and Campbell, D., "Responders Use CO2 on Difficult Shipboard Fire," NFPA Journal, Vol. 86, No. 6, November/December 1992, pp. 48–52, 54–55.

Hovde, P. J., and Hansen, A. S., "New Fire Test Methods and Classification Criteria for Interior Surface Products for Ships and Offshore Installations," Third International Conference and Exhibition on Fire and Materials, October 27–28, 1994, Arlington, VA, 1994, pp. 139–149.

Jacobsen, S. E., "Approval of Water Mist Systems on Ships. Consideration of Equivalency to Sprinkler and Water Spray Systems," Det Norske Veritas Classification AS, Norway, SP Report 1994:03; Proceedings of the International Conference on Water Mist Fire Suppression Systems, November 4–5, 1993, Boras Sweden, 1993, pp. 171–175.

Jansson, R., Bengtson, S., and Bergstrom, G., "Smoke Distribution on Large Passenger Ships and the Effect of the Ventilation System," Swedish National Defence Research Establishment, Sundbyberg, Sweden, Brandskyddslaget, Stockholm, Sweden, Flakt Marine, Gothenburg, Sweden, FOA Report A20049.2.4, November 1991, 74 pages.

Kennett, S. R., et al., "Fire Risk Assessment Methodology for Naval Ships," Fire Safety '91 Conference, Volume 4, November 4–7, 1991, St. Petersburg, FL, Am. Technologies Intl., Pinellas Park, FL, 1991, pp. 384–396.

Leonard, J. T., DiNenno, P. J., and Beyler, C. L., "Methods for Evaluating Flammability Characteristics of Shipboard Materials," Naval Research Lab., Washington, DC, Hughes Associates, Inc., Wheaton, MD, NRL/MR/6184-94-7451, February 28, 1994, 153 pages.

Leonard, J. T., et al., "Post-Flashover Fires in Simulated Shipboard Compartments. Phase 1. Small Scale Studies," Naval Research Lab., Washington, DC, Naval Sea Systems Command, Washington, DC, Hughes Associates, Inc., Wheaton, MD, Naval Sea Systems Command, Washington, DC, NRL Memorandum 6886, September 3, 1991, 169 pages.

Leonard, J. T., et al., "Post-Flashover Fires in Simulated Shipboard Compartments. Phase 2. Cooling of Fire Compartment Boundaries," Naval Research Lab., Washington, DC, Naval Sea Systems Command, Washington, DC, Hughes Associates, Inc., Wheaton, MD, NRL Memorandum 6896, September 19, 1991, 44 pages.

Leonard, J. T., et al., "Project HULVUL: Propellant Fires in a Shipboard Compartment," Naval Research Laboratory, Washington, DC, Naval Sea Systems Command, Washington, DC, Naval Weapons Center, China Lake, CA, Hughes Associates, Inc., Columbia, MD, NRL Report 9363, November 29, 1991, 58 pages.

"Marine Accident Report—Explosions and Fires Aboard the U.S. Tankship OMI Yukon in the Pacific Ocean about 1,000 Miles West of Honolulu, Hawaii, October 28, 1986" (NTSB/MAR-87/06).

"Marine Accident Report—Explosion and Fire on Board the U.S. Tankship OMI Charger at Galveston, Texas, October 9, 1993" (NTSB/MAR-94/04).

"Marine Accident Report—Fire on Board the U.S. Fish Processing Vessel All Alaskan Near Unimak Island, Alaska, Bering Sea, July 24, 1994." (NTSB/MAR-95/02).

"Marine Accident Report—Engine-room Fire on Board the Liberian Tankship Seal Island While Moored at the Amerada Hess Oil Terminal in St. Croix, U.S. Virgin Islands, October 8, 1994," (NTSB/MAR-95/04).

Mawhinney, J. R., "Design of Water Mist Fire Suppression Systems for Shipboard Enclosures," National Research Council of Canada, Ottawa, Ontario, SP Report 1994:03; Proceedings of the International Conference on Water Mist Fire Suppression Systems, November 4–5, 1993, Boras, Sweden, 1993, pp. 16–44.

McKenna, P., "Fire Safety Factor on Passenger and Cargo Ships," Fire International, No. 120, December/January 1989/1990, pp. 26–27,80.

Morchat, R. M., and Hiltz, J. A., "Fire-Safe Composites for Marine Applications," Defence Research Establishment Atlantic, Nova Scotia, Canada, Society for the Advancement of Material and Process

Engineering, *Proceedings* of Advanced Materials: Meeting the Economic Challenge, October 20–22, 1992, Toronto, Canada, 1992, pp. T153–164.

Nash, R. A., "International Passenger Ship Fire Protection Design," Worcester Polytechnic Inst., MA, Thesis, May 1991, 182 pages.

National Transportation Safety Board, "Marine Accident Report. Brief Format Issue Number 8. Reports Issued June 14, 1991," National Transportation Safety Board, Washington, DC, NTSB/MAB-91/01, June 1991, 112 pages.

Naval Air Station, "Navy and Marine Corps Annual Fire Report. Fiscal Year 1990. Incidents; Injuries; Deaths; and Property Loss," Naval Air Station, Norfolk, VA, Annual Fire Report, 1990, 133 pages.

Ohnstad, S. J., "Marine Regulatory Applications of the Computer Fire Model HAZARD I," Worcester Polytechnic Inst., MA, Thesis, March 1990, 180 pages.

Olsson, S., and Ryderman, A., "Brander i fartygsmaskinrum kan slackas med vatten- eller skum-sprinkler. [Fires in Ship Engines Extinguished by Sprinklers Studied.]," (Abstract in English), Swedish National Testing and Research Institute, Sweden, SP Report 1990:31, 1990.

Peatross, M. J., Beyler, C. L., and Back, G. G., "Validation of Full Room Involvement Time Correlation Applicable to Steel Ship Compartments. Final Report. October 1992–November 1993," Hughes Associates, Inc., Columbia, MD, Worcester Polytechnic Inst., MA, Department of Transportation, Washington, DC, Coast Guard, Groton, CT, USCG-D-16-94, HAI-1117-004-93, November 1993, 431 pages.

Pepi, J. S., "Fire Test Status Report on the Evaluation of the AquaMist Fixed Water Mist Deluge System in Ventilated Marine Machinery Spaces," *Proceedings* of the Technical Symposium on Halon Alternatives, June 27–28, 1994, Knoxville, TN, 1994, pp. 85–102.

Porth, D., "Regional Shipboard Firefighting Partnership Protects the Columbia," *American Fire Journal*, Vol. 48, No. 1, January 1996, pp. 12–14.

Rodionov, A. A., "Investigation of Mathematical Models of Ship Structure Optimization," *Sudostroyeniye* (Shipbuilding Journal of the Ministry of Industry of the Russian Federation), No. 8–9, August/September 1992, pp. 11–18.

Schaenman, P., "Major Ship Fire Extinguished by CO2, Seattle, Washington, September 16, 1991," USFA Fire Investigation Technical Report Series, TriData Corp., Arlington, VA, Federal Emergency Management Agency, Washington, DC, Report 058, 1992, 96 pages.

Sevat, J. L., "Flammability and Toxicity of Composite Materials for Marine Vehicles," *Naval Engineers Journal*, September 1990, pp. 45–54.

Sheinson, R. S., *et al.*, "Conducting Shipboard Total Flooding Fire Testing With Halon 1301 Replacements," Naval Research Laboratory, Washington, DC, GEO-CENTERS, Inc., Fort Washington, MD, NAVSEA 03G2, Crystal City, VA, University of South Alabama, Mobile, MPR Associates, Inc., Alexandria, VA, Hughes Associates, Inc., Columbia, MD, Naval Sea Systems Command, Washington, DC, NISTIR 5499, September 1994; National Institute of Standards and Technology, Annual Conference on Fire Research: Book of Abstracts, October 17–20, 1994, Gaithersburg, MD, 1994, pp. 61–62.

Sheinson, R. S., *et al.*, "Halon Replacement Considerations for Shipboard Machinery Spaces," *Proceedings* of the Twentieth International Conference on Fire Safety, Volume 20, January 9–13, 1995, Millbrae, CA, Product Safety Corp., Sunnyvale, CA, 1995, pp. 112–119.

Sheinson, R. S., *et al.*, "Shipboard Total Flooding Fire Testing With Halon 1301 Replacements: Test Results," Naval Research Laboratory, Washington, DC, GEO-CENTERS, Inc., Fort Washington, MD, Worcester Polytechnic Inst., MA, Hughes Associates, Inc., Columbia, MD, Naval Sea Systems Command, Washington, DC, NISTIR 5499, September 1994; National Institute of Standards and Technology, Annual Conference on Fire Research: Book of Abstracts. October 17–20, 1994, Gaithersburg, MD, 1994, pp. 63–64.

"Ship Fire That Started in Gibraltar Fought in Suffolk," *Fire*, Vol. 87, No. 1071, September 1994, p. 43.

Sorathia, U., Dapp, T., and Kerr, J., "Flammability Characteristics of Composites for Shipboard and Submarine Internal Applications," David Taylor Research Center, Annapolis, MD, Society for the Advancement of Material and Process Engineering, *Proceedings* of How Concept Becomes Reality, Volume 36, Part 2, April 15–18, 1991, San Diego, CA, 1991, pp. 1868–1878.

Sprague, C. M., Richards, R. C., and Blanchard, M. A., "Methodology for Evaluation of Ship Fire Safety," *Naval Engineers Journal*, Vol. 104, No. 3, May 1992, pp. 104–113.

Steward, F. R., Morrison, L., and Mehaffey, J., "Full Scale Fire Tests for Ship Accommodation Quarters," *Fire Technology*, Vol. 28, No. 1, February 1992, pp. 31–47.

Stronach, I., "Unusual Fire Boat: Busy Port of Montreal Finds Economical Way to Meet Fire Protection Needs," *Firehouse*, Vol. 20, No. 11, November 1995, pp. 114–116, 118.

Stuart, D., "Case Study: Shipboard Explosion and Fire," *Fire and Arson Investigator*, Vol. 44, No. 4, June 1994, 17 pages.

Sundstrom, B., "Brandhardiga produkter pa fartyg—nulaget och forslag till forskningsinsatser. [Fire Safety Requirements and Test Methods for Ships.]," Swedish National Testing and Research Inst., Boras, Sweden, SP Report 1993:41, 1993, 40 pages.

Sundstrom, B., "IMO: Uses Modern Fire Testing Technology for Evaluation of Materials for High Speed Passenger Ships," Third International Conference and Exhibition on Fire and Materials, October 27–28, 1994, Arlington, VA, 1994, pp. 151–157.

Sundstron, B., "High Speed Passenger Ships—Fire Calorimetry Is Used to Identify High Performance Products," *Proceedings* of the First International Conference on Fire Science and Engineering, ASIAFLAM '95, March 15–16, 1995, Kowloon, Hong Kong, 1995, pp. 473–479.

Thorell, C., "New Thinking on Passenger Ship Safety After Two Big Fires," *Fire International*, Vol. 137, December 1992, pp. 24–25.

Turner, A. R. F., "Water Mist in Marine Applications," Marioff Hi-Fog Oy, Vantaa, Finland, NISTIR 5207, June 1993; *Proceedings* of the Water Mist Fire Suppression Workshop, March 1–2, 1993, Gaithersburg, MD, 1993, pp. 105–122.

"Two Marine Fires Within Weeks Tackled by Dyfed Fire Departments," *Fire*, Vol. 85, No. 1055, May 1993, pp. 19–20.

U.S. General Accounting Office, "Coast Guard: Additional Actions Needed to Improve Cruise Ship Safety. United States General Accounting Office Report to the Congressional Requesters," General Accounting Office, Washington, DC, GAO/RCED-93-103, B-248714, March 1993, 46 pages.

Underwriters Laboratories Inc., "Carbon Monoxide Gas Detectors for Marine Use. Standard for Safety," Underwriters Laboratories, Inc., Northbrook, IL, UL 1524, 1st Edition, February 20, 1992, 17 pages.

Underwriters Laboratories, Inc., "Tube Fittings for Flammable and Combustible Fluids, Refrigeration Service, and Marine Use. Standard for Safety," Underwriters Laboratories, Inc., Northbrook, IL, UL 109, 5th Edition, April 28, 1993, 13 pages.

White, D. A., "Computer Modeling of Smoke Transport Onboard Ships," Worcester Polytechnic Inst., MA, Thesis, June 1990, 116 pages.

White, D. A., *et al.*, "Modeling the Impact of Post-Flashover Shipboard Fires on Adjacent Spaces," Hughes Associates, Inc., Baltimore, MD, Naval Research Laboratory, Washington, DC, ISBN 0-9516320-9-4; *Proceedings* of the Seventh International Interflam Conference, Interflam '96, March 26–28, 1996, Cambridge, England, Interscience Communications Ltd., London, England, 1996, pp. 225–233.

Wikman, J., "Ship Fire Safety Engineering. Enclosures," Lund Univ., Sweden, LUTVDG/TVBB-1012-SE, ISSN 1102-8246, 1995, 156 pages.

Wilhelmi, G. F., "Composite Materials for Ship Machinery Applications," David Taylor Research Center, Bethesda, MD, NRL Publication 161-ONT-263, May 1990; Chief of Naval Research and The American Defense Preparedness Association, First Navy TransFair, August 29–31, 1989, Kansas City, MO, Office of Naval Technology, Washington, DC, 1990, 12 pages.

Williams, F. W., *et al.*, "Post-Flashover Fires in Simulated Shipboard Compartments. Phase 3. Venting of Large Shipboard Fires," Naval Research Laboratory, Washington, DC, Hughes Associates, Inc., Columbia, MD, Naval Sea Systems Command, Washington, DC, Naval Sea Systems Command, Washington, DC, NRL MR 6180-93-7338, June 9, 1993, 57 pages.

Williams, F. W., *et al.*, "Shipboard Smoke Control Tests Using Forced Counterflow Air Supply," Naval Research Laboratory, Washington, DC, Naval Sea Systems Command, Washington, DC, NRL/MR/6180-94-7616, September 8, 1994, 72 pages.

Yoshida, K., "Evaluation of Sprinkler Systems for Ships Based on Heat Release Rate," Research Institute of Marine Engineering, Higashimurayama, Tokyo, CIB W14/93/2 (J); First *Proceedings* of the Japan Symposium on Heat Release and Fire Hazard, Session 3, Scope for Next-Generation Fire Safety Testing Technology. May 10–11, 1993, Tsukuba, Japan, 1993, pp. III/57–64.

ORGANIZING FOR FIRE PROTECTION

Today's "fire fighter" is part of a broad-based code enforcement and emergency-response team and system. The challenges faced by these men and women are of unprecedented technical complexity. More than $10 billion is spent each year to operate paid local fire departments. The equivalent cost is several times that figure if one includes the costs of paid or donated time of all career and volunteer fire fighters, whether in public or private emergency organizations, at the national, state, or local level. As the range of emergencies covered by the "fire" service grows, as the sophistication of equipment and methods to do the job increases, and as our understanding of what works and what doesn't work evolves, it is ever more important to refine our set of standards, practices, and core knowledge to keep pace. This section provides that foundation for the modern fire protection organization.

Section 10 presents a broad overview of the issues and considerations involved in the delivery of fire protection services. Several new chapters have been added and others have been extensively revised to provide valuable guidance for fire service managers in both the public and private sectors. The effective organization and delivery of fire protection services is a critical issue for government entities and corporate organizations around the world. Regardless of whether you provide fire protection services in the public or private sector, it is necessary to provide an adequate fire protection program that includes fire prevention activities, pre-incident organization and planning, fire station location planning, and the delivery of emergency services. This section provides valuable guidance from recognized experts to help fire service managers understand the complexities they may face and to design programs that best utilize their fire protection resources.

Section 10 also addresses the organization of fire service delivery systems at the community level, including administration and operations, evaluation and planning, information systems, and legal aspects. It provides guidance relating to fire fighter occupational safety and health, fire fighter protective clothing, and fire department apparatus and equipment. Several chapters review operational issues such as hazardous materials response, wildland fire management, rescue operations, and the delivery of emergency medical services.

Chapters on fire loss prevention for emergency organizations, pre-incident planning for industrial and commercial facilities, along with a chapter on fire prevention and code enforcement, provide an excellent overview on how fire protection organizations can deliver valuable community services aimed at minimizing or preventing loss from fire or other emergencies.

—Gary O. Tokle
December 1996

FIRE DEPARTMENT ADMINISTRATION AND OPERATIONS

Revised by Robin Paulsgrove

This chapter provides an overview of the elements involved in the organization, administration, management, and operations of a fire department, from the earliest fire protection services of ancient Rome to today's diverse safety organization.

There are currently more than 30,000 fire departments in the United States that are organized in various ways to meet the specific needs of the communities they serve. Although public fire protection is normally a function of local government, state, provincial, or federal properties may also have organized fire departments for protection. Likewise, large industries will often provide organized private fire departments at their industrial complexes. The organization and objectives of public fire departments vary according to resources available, and range from simple to complex.

Almost all fire departments were administered by clearly defined organizational structures long before systems techniques were applied to industry and business. A system of task allocation to engine and ladder crews was developed whereby each person on the apparatus performed certain functions in sequence so the team operated as a coordinated unit, without duplication of effort.

Although most public fire departments are structured around the traditional mission of fire suppression, more organizations are beginning to emphasize fire prevention and public education as the most effective way to protect life and property. Many fire departments also have multiple roles in their community, providing emergency medical response, hazardous materials mitigation, fire code inspections, plans reviews, and technical rescues, in addition to the more familiar fire suppression services.

Most fire suppression and rescue activities are organized around a system of decentralized fire stations so that personnel and equipment can respond quickly and effectively to emergency incidents. This organization may be staffed by career, part-time, or volunteer personnel and may reflect a variety of characteristics derived from local tradition, needs, and structure.

Effective fire department management requires specific decisions related to specific communities, as well as general practices relating to objectives, structure, budgeting, purchasing, planning, intergovernmental relations, and analysis.

Robin Paulsgrove is chief of the Austin, TX, Fire Department, and a member of the board of directors of the National Fire Protection Research Foundation.

EARLY FIRE SUPPRESSION AND FIRE REGULATIONS

Public fire departments evolved from the level of community effort to professional organizations in five basic steps:

1. The establishment of nightwatch services.
2. The drafting of fire prevention regulations and the appointment of fire protection officers.
3. The organization of societies to salvage building contents from loss by fire.
4. The formation of voluntary fire-fighting companies.
5. The appointment of fire-fighting officers and personnel.

Possibly the first organized fire protection occurred when Augustus became ruler of Rome in 24 B.C. A "vigil," or watch service, was created, and regulations for checking and preventing fires were issued. Night-patrolling and night-watch forces were the principal services, and some of the vigils had duties more like those of police or soldiers than fire-fighters. It is clear from the history of that period, however, that fires were a major problem and that the vigils were provided with fire-fighting tools and equipment (buckets, axes, etc.)

One of the earliest recorded fire protection regulations dates back to 872 A.D. in Oxford, England, when a curfew was adopted requiring hearth fires to be extinguished at a fixed hour. Later, William the Conqueror established a general curfew law in England, enforced to prevent both fires and revolt.

Fire Brigades in Great Britain

Until 1830, Great Britain did not provide statutory authority to units of local government for night or other watch services with fire apparatus. *The History of the British Fire Service*[1] mentions only two cases where this subject was introduced in England before the establishment of civil police forces. One was during the English Civil War in 1643 when a company of fifty women was organized to patrol the town of Nottingham at night.

The second involved the organization of private fire-fighting forces. Fire insurance brigades were formed in England primarily as a result of the Great Fire of London in 1666. These brigades were formed by the insurance companies without statutory authority or obligations; and the insurance company offices, not the government authorities, decided where the brigades would be located. In London, the insurance office fire brigades were consolidated into the

London Fire Engine Establishment in 1833, which was taken over by the Metropolitan Fire Brigade in 1865.

It was not until Edinburgh's 1824 Fire Brigade Establishment that public fire services began to develop modern standards of operation. A surveyor named James Braidwood was appointed chief of the brigade. He selected eighty part-time aides between the ages of 17 and 25, and required regular drills and night training. Braidwood wrote the first comprehensive handbook on fire department operation in 1830. His handbook included some 396 standards and explained for the first time the kind of service a good fire department should perform.

Fire Protection in Early America

Following a disastrous Boston, Massachusetts, fire in 1631, the first fire ordinance in the new world was adopted. The ordinance prohibited thatched roofs and wood chimneys, and was enforced by the colonial city's board of selectmen. In 1648, New Amsterdam appointed five municipal "fire wardens" with fire prevention responsibilities; this event is often considered to be the origin of the first public fire department in North America.

In Boston, after a 1679 conflagration destroyed 155 principal buildings and a number of ships, laws were adopted requiring stone or brick walls for buildings and slate or "tyle" roofs for houses. That fire also led to the establishment of the first paid municipal fire department in North America, if not the world. Boston imported a fire engine from England and employed twelve fire fighters and a fire chief. From the start, Massachusetts used paid municipal fire fighters on an on-call basis, as contrasted with the unpaid volunteer fire companies that were later organized in other colonies.

Colonial communities required each householder to keep two fire buckets and, when the church bells rang an alarm, to report to the scene of a fire to form lines for passing water from wells or springs. As late as 1810, Boston citizens were subject to a $10 fine for failure to respond to alarms with their buckets. When hand-pump fire engines were obtained, teams were organized to operate them. The laws in a number of states still impose penalties on citizens who refuse to assist in fighting fires upon orders from fire officers.

Fire wardens were appointed in Boston in 1711. With the members of their staffs, fire wardens responded to fires and supervised citizen bucket-brigades. By 1715, Boston had six fire companies with engines.

Mutual fire societies: The first of a number of mutual fire societies was formed in Boston in 1718 when some of the more affluent citizens organized to assist each other in salvaging goods from fires in their homes or businesses. Each person's equipment comprised a bag in which to collect valuables, a screwdriver, and a bed key to help disassemble and remove beds from burning buildings. About a century later, these mutual fire societies became inactive when fire insurance became available to the more prosperous citizens. The mutual fire societies were forerunners of the salvage corps.

Salvage corps: After the principal cities of the United States organized paid fire departments equipped with steam engines around 1850, insurance interests formed salvage corps to reduce water damage at fires. Gradually, improved fire-fighting procedures and the increased expense of operation led to the disbandment of salvage corps. Today, public fire departments handle salvage operations at fires.

The growth of paid departments: Lack of discipline of volunteer fire fighters, coupled with their resistance to the introduction of steam-pump engines, led to the organization of paid fire departments. Following serious disorders at fires, a paid fire department

with horse-drawn steam pumpers was placed in service in Cincinnati, Ohio, on April 1, 1853. In 1855 two steamers were delivered to New York City, but the volunteer fire fighters would not use them. Ten years later the "Metropolitan Fire Department," using steamers, replaced New York's volunteer force.

Duty Systems and Training

Until World War I, paid members of fire departments worked a continuous duty system, with limited off-duty time. By World War II, most paid fire departments had adopted some form of the two-platoon system, which allowed additional days off and reduced the average number of hours worked per week, although in most cities, fire fighters work more hours per week than workers in other municipal employment or private industry. Although many fire fighters continue to work 24-hr duty shifts, the 1975 Fair Labor Standards Act has had a significant impact on the working hours of fire fighters, allowing up to a 53-hr work week.

In 1889, the Boston Fire Department established the first drill school where basic training and uniform company drills were performed. Today most departments provide some degree of training for personnel.

In 1914, New York City established a Fire College for advanced officer training. During that same year, the first state fire school was organized in North Carolina. In 1925, Illinois and Iowa started state fire schools for the training of volunteer fire fighters, and by 1950 most states were providing systematic fire service training. In 1937, Oklahoma A&M College (now Oklahoma State University) initiated a two-year (later four-year) college program in fire protection. The initial objective of the Oklahoma program was to provide trained individuals for the fire service. However, the emphasis was later shifted to the industrial and insurance aspect of fire protection. Currently, approximately 200 to 250 institutions in the United States offer two-year programs in fire-protection education. One four-year program (offered by the University of Maryland) offers a full curriculum in fire-protection engineering. Several colleges offer four-year programs of study leading to degrees in fire technology or fire service management, and one (Worcester Polytechnic Institute in Massachusetts) offers a master's degree in fire-protection engineering.

FIRE SERVICE ORGANIZATIONS

Most public fire service organizations in the states and provinces of North America are established by local agencies of government or special independent local agencies that provide fire protection. For the overwhelming percentage of people in North America, local government is responsible for providing adequate fire protection and the framework within which the protection operates.

Types of Organizations

One of the most common types of public fire protection is the *public fire department,* a department of municipal government, with the head of the department responsible to the chief administrative officer of the municipality. Most large municipalities, as well as numerous small communities, operate with this type of organization.

Less common is a *fire bureau,* which is usually a division of a department of public safety. In this type of organization, the public safety department head manages several important functions, including police and fire service.

The *county fire department* has gained considerable acceptance in many areas. With this type of organization, numerous suburban municipalities can enjoy the benefits of a large, professionally administered public fire department with staff and service facilities that, ordinarily, few small communities could afford individually. Frequently, this department begins with a county fire prevention office and a fire communications system. The smaller, often volunteer departments initially remain autonomous for fire suppression purposes. Gradually, more functions, including suppression, are assumed by the county organization.

Another type of public fire service organization is the *fire district,* which is organized under provisions of state or provincial law. It is, in effect, a separate, independent unit of government, having its own governing body composed of commissioners or trustees, and is commonly supported by a tax levied throughout the district. Usually it is organized following a favorable vote of the property owners in the proposed district. The fire district may include portions of one or more townships or other governmental subdivisions.

A fifth common type of fire protection authority is the *fire protection district,* which in some states is a legally established, tax-supported unit that contracts for fire protection from a nearby fire department or even from a voluntary fire association. This type of organization provides the equivalent of municipal fire protection for rural or suburban areas that might find it difficult to maintain their own experienced and effective fire-fighting forces. The fire protection district often provides a source of extra income and special rural fire apparatus for the small municipal fire departments that contract to supply fire protection.

Another type of public fire service organization is the *volunteer fire company* or *association* that raises its own funds by public activities and subscriptions, frequently with contributions of tax dollars or equipment from interested units of government. Many voluntary fire associations maintain excellent equipment and stations, and also serve as centers for various community activities. Often, volunteer organizations prefer to retain their independence from government, especially when purchasing equipment, although in some instances the activities of independent fire organizations are coordinated through special associations and governmental advisory boards.

EXPANDED ROLE OF THE FIRE SERVICE

In recent years, the role of the fire service in many communities has expanded far beyond fire suppression. The name "fire department" doesn't begin to cover the many services that progressive organizations are providing to their communities. Public safety is the "business" of the modern fire department. With this expansion, fire prevention and public education have appropriately begun receiving an increased emphasis as the proactive elements of a fire service delivery system. Citizens are dependent on the fire department to ensure their protection against the dangers of fire, panic, explosion, and any emergency event that may occur in the community. Departments have responded to these demands with a growing range of new services, such as hazardous materials, rescue, and emergency medical services.

The leadership challenge for the fire service is to define what an effective, competitive fire department will look like in the future and establish a steady course in pursuit of that vision. Fire fighters cannot define the business they want to be in based on the missions that have the most appeal or immediate gratification. Neither can the fire chief impose his or her will on the department. The measure of success must be one that satisfies changing community needs in a dynamic environment.

Competition, not only from private companies but from other local departments that rely on taxpayers for funding, is compelling the fire service to move out of its comfort zone and modify its services to match its community's changing needs. The fire department of the future will provide a broad menu of safety services, in many cases including an expanding emergency medical role. The successful fire department will focus on prevention as the most effective way to accomplish the mission of protecting life and property, but also maintain safe and effective emergency response capability.

Fire Prevention

Fire prevention along with code enforcement is actually one of the two major areas of responsibility for the fire service. At one time, governments focused more on prevention, but as they developed the technical ability to deliver additional suppression services, their focus shifted. As fire departments have professionalized, they have historically developed their fire suppression capacity, rather than their fire prevention abilities. Most of a fire department's resources are dedicated to responding to emergencies after the fact; only a small percentage of resources is dedicated to the prevention of those emergencies. Since emergencies by nature cannot be eliminated or predicted, community protection depends on reliable and effective emergency response. Positioning the fire service for the future depends on an affordable balance.

The local fire department should take a leadership position in initiating the adoption of a complete nationally recognized fire and building code. Most fire departments adopt one of the model codes that are developed by regional and national standard development organizations. In the future, cities may have the option of adopting a single national code, increasing consistency. Model codes may be amended to adjust for local concerns and needs.

In the past, most fire prevention activities were limited to a small nucleus of full-time specialists who might be civilian or uniformed personnel. The size of the department and the community served determines whether it is necessary to maintain full-time fire prevention personnel. The prevention responsibilities of the fire department are greater than can be performed strictly by specialists.

Fire suppression personnel have been increasingly active in inspections and code enforcement. With proper training and support, suppression personnel are effective in performing code-enforcement inspections. It is important that all fire department personnel recognize that fire safety education and prevention are a major part of the fire fighter's responsibilities.

Pre-incident planning: Pre-incident planning should involve all fire suppression personnel on a continuing basis. Pre-incident plans should complement standard operation procedures by familiarizing personnel with complex occupancies and special risks. Pre-incident plan information should be documented in a standard format to help the incident commander manage any situation that could occur at the fire or other emergency location. For an example, see NFPA 1420, *Recommended Practice for Pre-Incident Planning for Warehouse Occupancies.* Plans should be sufficiently flexible to allow for varying conditions and should use the framework of standard operating procedures. Plans that are excessively rigid or complex may handicap an operation more than benefit it.

The following steps are part of the pre-incident process:

1. *Information gathering.*   Pertinent information, such as building construction features, occupancy, exposures, utility disconnects, fire hydrant locations, and water main sizes, that might significantly affect fire-fighting operations must be collected at the selected site.
2. *Information analysis.*   The information gathered must be analyzed for what is pertinent and vital to fire suppression operations, so that a plan can be documented in a format that can be used on the fireground. A pre-incident plan usually includes a site plan or map; floor plans and diagrams identifying pertinent features; hazards and fire control equipment; and additional text

outlining special problems, specific tactics, hazardous contents, and information on parties responsible for specific areas.

3. *Information dissemination.*   Pre-incident plans should be assembled and distributed in a standard format for easy use on the emergency scene. Plans may be maintained on paper and carried in fire apparatus, microfilmed and used with viewers in command vehicles, or stored in computer systems accessed by mobile data terminals or fax machines.

4. *Class review and drill.* Any company that might be involved at a location for which a pre-incident plan has been developed should review the plan on a regular schedule. If possible, periodic drills and familiarization tours with all the companies involved should be scheduled on the property.

Pre-incident plans are desirable for all target hazards, for special risks, and for large complexes. Standard operating procedures are usually sufficient for single-family dwellings and smaller occupancies. Pre-incident planning is a necessary adjunct to tactical operations and should aid in efficient operations, reduce fire losses, and help provide an optimum level of emergency services.

Fire Investigation

Many fire departments and community task forces have placed an increased emphasis on fire investigation. It is critical that their focus not be restricted to arson investigation. Fire investigations, not only to identify criminal activity, but to make fire cause determination, can identify factors useful in lessening the number and severity of fires that may occur in the future. Information gained through investigations is a valuable tool in developing an effective fire prevention program, including needed code revisions, public education programs, and planning for future fire protection needs. In addition, a thorough fire investigation of all incendiary or suspicious fires is a powerful deterrent to the crime of arson.

Community-Based Safety Education

The tragedies to which fire departments respond can be prevented most effectively when they are addressed at the community level. While fire stations have always had a community profile, in many cities they have a great deal of unrealized potential. The neighborhood fire station can be the delivery vehicle for fire and emergency prevention efforts throughout the community. The concept is simple. Fire fighters at fire stations can more effectively deliver a comprehensive program of community fire prevention than can a few centrally located individuals. With the effective use of demographic and fire experience data, a department can customize prevention education and inspections to the needs of individual response territories. The task of the central fire prevention and public education staff can then be focused on technical inspections, along with coordination of prevention initiatives by community fire stations.

While community fire education has been a part of the fire service for many years, it has only recently emerged as a major component in fire protection. National conferences that champion the public education effort have been introduced in Oklahoma and Texas. The National Fire Protection Association's (NFPA) *Learn Not to Burn*® program is one component that has been successful in reaching and educating school children on fire safety issues. "Fire safety houses," portable model homes transported on trailers, have proved to be valuable teaching tools in many communities.

Public Information and Community Partnerships

Fire departments supported by public funds need public support for their budget requests. The public's interest and cooperation is required to make fire prevention programs fully effective and to enable the fire department to operate successfully within its particular political environment. Without community support, it can be difficult for the department to obtain the needed funds to operate at the required level. Accordingly, public information and community service programs are an important fire department management activity.

In a good public information program, it is important to develop and to maintain procedures that keep the public informed of departmental activities and programs. Relations with the news media should be cordial, and the fire department should cooperate fully with representatives of the media. This is not always easy, because hundreds of suburban fire departments are located in communities that depend largely upon a metropolitan press, and items that might be of interest to local citizens may not be newsworthy for the entire region. It may be effective to explore options, such as local weekly papers or local-area editions of metropolitan papers, that do have local correspondents.

The amount of publicity fire departments get seems to vary greatly among geographic areas. In some areas, citizens are traditionally interested and involved in local governments and in their fire departments, while in other areas, fire departments are not considered to be newsworthy. In the latter case, greater efforts must be made to get fire-related information to the public. In general, communities whose citizens regularly participate in town affairs seem to give fire departments better support. Communities with substantial numbers of low-income residents subject to fire dangers often are most concerned with the adequacy of fire department services.

It is important that news personnel be given information promptly. Most fire departments specify that only the officer in charge at a fire or emergency or a designated public information officer should give out information. This is to avoid conflicting and inaccurate statements that may be misleading or may compromise fire investigations.

Some large fire departments have a designated public information officer who is assigned to give the press all possible cooperation and information. In smaller fire departments, this function may be one of the duties performed by the fire chief. The public information position is among the designated functions in an effective incident command system.

Staff public information officers often report directly to the fire chief because of the need for regular communication between the two. The public information officer may act as a communications or media advisor to the chief, while the chief should keep the public information officer apprised of all potentially newsworthy department decisions and activities.

Emergency Medical Services Delivery

Emergency medical service (EMS) has become an important function of many fire departments. While many fire departments have operated ambulance service for decades, the major growth in this area has occurred since 1970, stimulated by higher standards for patient care, training, and equipment. Increased interest from the public, the medical community, and local government in emergency medical service delivery has prompted many fire departments to increase their participation in EMS, and the quality of EMS delivery consequently has improved greatly.

This growing service demand has changed the business definition of fire departments throughout the country. Nationally, many fire department managers and organized labor groups are linking the competitiveness of their organizations with emergency medical service delivery.

The addition of EMS delivery to a fire department is a natural extension of responsibilities and often can increase organizational productivity without compromising fire protection. Fire service personnel are already oriented toward delivering emergency service and assisting citizens. Fire stations are decentralized to provide rapid response to all areas, while fire communications systems are designed to receive emergency requests from the public and dispatch assistance without delay, both of which are required to provide effective EMS delivery.

Increasingly, the emergency medical services delivery system may be solely the responsibility of the fire department, from first responder providing basic life support to advanced life support with paramedic-level care, including transport.

Nationwide, fire department emergency medical service calls outnumber fire alarms by nearly five to one. Since many departments do not provide this service, ratios in particular communities can be much higher. The workload can adversely effect volunteer fire companies, causing many seasoned volunteers to drop out. Some volunteer rescue squads solve this problem by attracting members who prefer this work.

Even though responding to EMS calls places an increased workload on fire-fighting companies, it has proved very efficient in many jurisdictions. Providing basic EMS equipment on engine and ladder companies with emergency medical technician (EMT)-certified firefighters allows these units to respond to most life-threatening situations. First responders and EMTs are generally backed up by paramedic personnel responding on ambulances or rescue vehicles. In some fire departments, all personnel are trained and certified at the EMT level, and dual-role paramedics are assigned to engine and ladder companies.

Hazardous Materials

Significant and growing safety risks are associated with the storage, use, transportation, and disposal of hazardous materials and hazardous waste. Changing federal regulations, along with increased sensitivity to environmental protection, have created a new service demand in many communities. In many cases, the fire department is the lead agency in responding to hazardous materials incidents.

Many fire departments find that they neither have the trained personnel nor the funds to support a full Hazardous Materials Emergency Response Team. Regional teams have been organized on a county level, and in some cases multiple counties have cooperated in organizing a Hazardous Materials Emergency Response Team.

In addition, fire departments should have direct involvement as either a lead or participating agency in the process of gathering and organizing information, identifying risks and planning for hazardous materials emergencies, as well as the regulation of the storage, use, transportation, and disposal of hazardous materials and hazardous wastes.

Management Improvement Systems

There are a number of systems that have been developed to improve management decisionmaking through teamwork and enhanced working relationships. Traditionally, these systems have been developed for use by the private sector. However, their application has been found to be effective in improving the management of the public sector organizations. Several systems have been developed, including organizational development, quality circles, and quality and service improvement programs.

Forward-looking organizations are taking a page from the private sector and introducing process improvements, such as total quality management (TQM), to enhance services. Quality management stresses a total organizational approach that makes the quality of service, as perceived by the customer, the number one driving

force for the operation of the organization. The purpose of these systems is to improve the quality of service delivered, improve productivity, increase cost-effectiveness, and develop clearly stated goals and measurable objectives. This is accomplished by increasing the participation of the managers and workers in the organization's decision making.

Every fire department should have a mission statement regarding the scope of services it provides to its jurisdiction. The most progressive fire service organizations begin with a vision. Their vision states where they are headed and where they believe the future lies for their organization. Many fire departments have also developed a set of values. Values say something very individual about an organization. For instance, the mission of the fire department—the menu of services provided to the community—may be very similar from one part of the country to another. Its values, on the other hand, may reveal what things are most important to the department and to the community. They may tell the story of "how" an individual department will achieve its mission. In addition, goals should be established for the department along with measurable objectives for each level of the organization.

Management Information Systems

Increased emphasis is placed on the use of management information systems to provide data that can be analyzed in order to improve decision making. It has been especially effective in the areas of the fire protection system, the emergency medical services delivery system, personnel, and fiscal management.

There are some systems that collect data regionally and nationally. NFPA analyzes and prepares statistical reports from a national system of data collection. Many standard reports (such as *U.S. Fire Fighter Deaths*) and customized analysis are available from the NFPA Fire Analysis and Research Division. This information is available for use by individual departments.

Occupational Safety and Health

Fire fighting has been recognized for many years as one of the most dangerous professions. Between 1978 and 1988, more than 125 fire fighters per year lost their lives and more than 100,000 were injured yearly. In response to this unenviable record, NFPA 1500, *Standard on Fire Department Occupational Safety and Health Program*, was established to guide fire departments in the development of fire service health and safety programs. The standard is dedicated to making the fire service less dangerous by reducing the risk of accident, injury, and death during line-of-duty activities and to make fire fighters healthier and more physically fit.

NFPA 1500 addresses the elements of an occupational safety and health program that have been accepted as the minimum level that today's fire departments should provide. NFPA 1500 was developed by the fire service, for the fire service, and supports the requirements of the federal Occupational Safety and Health Administration (OSHA) fire brigade standard developed for private sector and public fire suppression organizations.

NFPA 1500 addresses health and safety as an overall program to reduce the risks to members of the fire department. The standard applies to all members of the department and specifically identifies management as a provider of as safe a work environment as possible, given the hazardous nature of the work. Fire service management, therefore, has an important role in leading and directing the establishment of rules, regulations, and procedures geared toward the operation of an effective health and safety program.

Entering into a comprehensive safety program can be a difficult process, since a fire department may have to address a variety of issues with varying degrees of difficulty and budgetary impact.

Implementing a health and safety program may be most effectively accomplished on a phased basis tailored to the specific needs and capabilities of a particular fire department. (See NFPA 1500, Section 1–3, "Implementation," for specifics of a "phase-in schedule for compliance.") The development of this comprehensive implementation plan will require each fire department to assess its own strengths and weaknesses, and to determine what would be required to meet the intent of the standard. Major elements of this implementation plan should include vehicles and equipment, training and education requirements, protective clothing and protective equipment, incident command procedures, facility safety, health and fitness standards, and employee assistance programs.

FIRE DEPARTMENT STRUCTURE

Fire departments, like other organizations, are composed of people working together in a coordinated effort to achieve a common set of objectives. For a department to function effectively it must have an organizational plan that shows the relationship between the operating divisions and the total organization. An organizational plan does not preclude the necessity for active leadership; it merely provides the means by which the organization can be managed effectively. Organizational charts with typical structures of small, medium-sized, and large fire departments are shown in Figures 10-1A, 10-1B, and 10-1C, respectively.

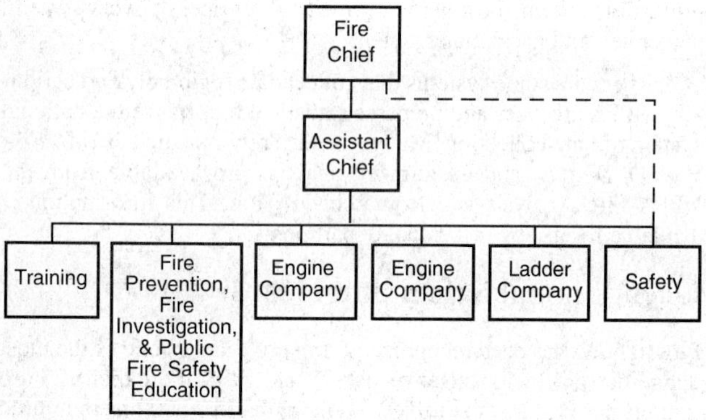

FIG. 10-1A.   *Organizational chart for a small fire department.*

Principles of Organization

One of the most basic organizational principles is that work should be divided among the individuals and operating units according to a plan. The plan should be based on the individual functions that must be performed, such as fire prevention, training, and communications.

The next principle is that, as a department increases in size and complexity, the need for coordination also increases. Small departments allow frequent personal contact among individuals, reducing the need for extensive formal coordination. However, as departments increase in size and complexity, they require more extensive coordination of the operating units in order to achieve their objectives.

The most successful organizations operate as a team. Department leaders are organized as a "system" in which all divisions and sections are equally important to achieve the desired objective: service to the community. A risk of growth and an unfortunate byproduct of an organization large enough to fund specialized staff assignments can be the development of offices or whole divisions that operate without coordination. Successful organizations must also avoid the message that, when they are large enough to fund specialized staff assignments, the mission becomes "their job" exclusively. In departments that operate as a "system," the funding of staff positions to coordinate training, inspection, safety, or public education programs cannot be viewed as removing or distancing these important responsibilities from the field officers and fire fighters.

Line Functions

Line functions in fire departments normally refer to those activities directly involved with fire suppression operations. Fire suppression officers are primarily considered to be line officers. This does not mean, however, that they do not have other functions. As these officers are promoted within the department, their line responsibilities may be divided equally with staff responsibilities. At the highest officer levels within the department, line responsibilities diminish while staff responsibilities increase.

Staff Functions

Staff functions are those activities that do not involve dealing with day-to-day emergency incidents, such as fire suppression, emergen-

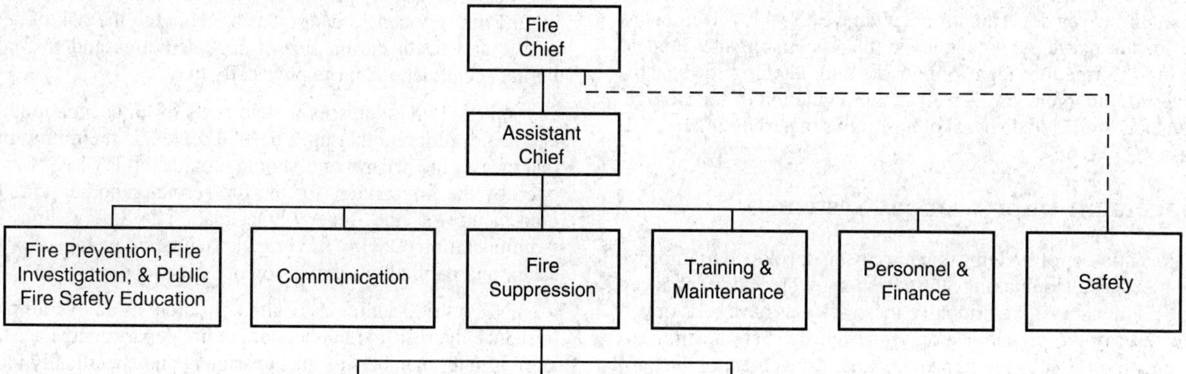

FIG. 10-1B.   *Organizational chart for a medium fire department.*

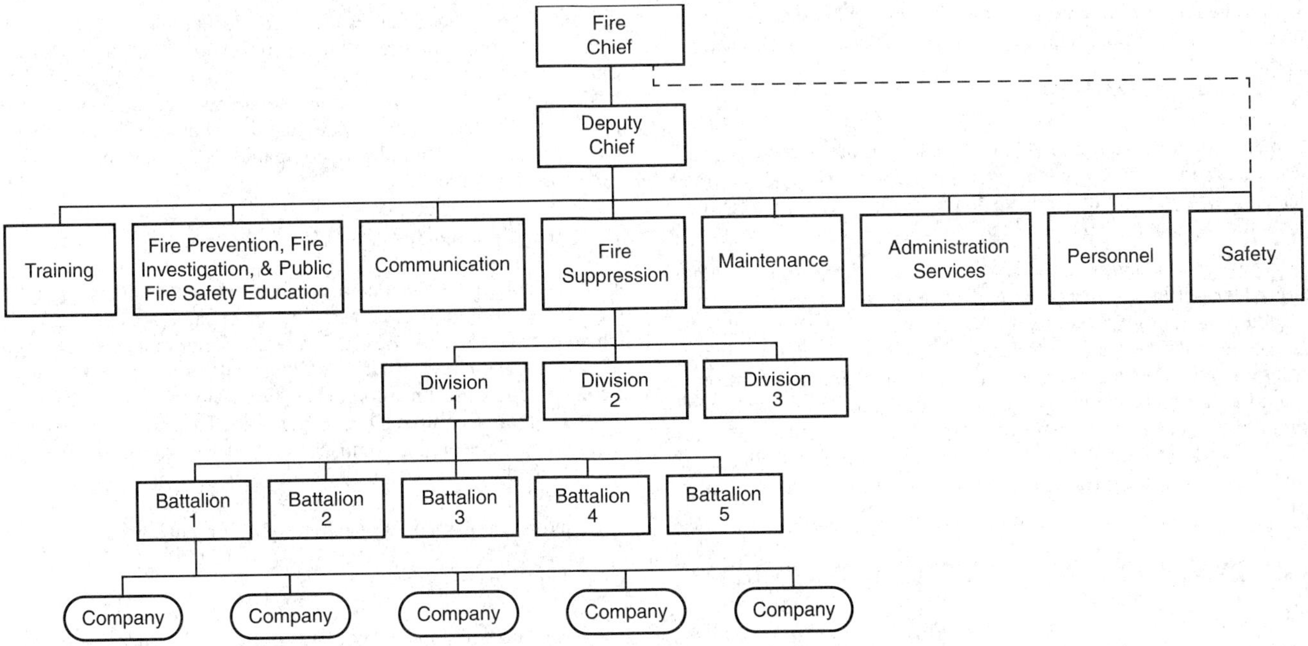

FIG. 10-1C. Organizational chart for a large fire department.

cy medical service, and various hazardous conditions or hazardous materials incidents. Staff functions may include the following activities.

Fire prevention: The inspection of construction and of existing properties for compliance with codes and ordinances; the operation of a public education program; and the investigation of fires to determine the reasons that fires start and spread.

Training: Training of all personnel in their job skills; administering continuing education programs in special subject areas; administering the fire department safety program; and organizing and administering pre-incident planning.

Maintenance: The maintenance of apparatus, equipment, and physical facilities; recommendation of replacement programs for apparatus and equipment; and the development of specifications for the purchase of apparatus and equipment.

Communications: Providing and maintaining adequate facilities for the receipt of alarms from the public and communication with fire companies both in quarters and by radio in the field. Communications functions also include developing and maintaining dispatch policy and response requirements.

Research and planning: Creating the new knowledge that the fire department needs to provide better service; forecasting long-range department goals; and performing the types of analyses necessary for other department functions to assess program effectiveness.

Public information: Maintaining contact with the public and the news media to tell the fire department story; in larger departments, may also be responsible for internal communications efforts.

Financial management: Preparing a budget; monitoring expenses against the budget planning for capital expenditures; and supervising the purchase and inventory of needed materials.

Personnel management: Supervising the recruitment, selection, and promotion of personnel; administering the retirement system and the benefits program; and supervising the administration of discipline within the department.

Fire protection engineering: Reviewing plans for new construction, major renovation, or installation of fire protection systems; developing proposals for code changes; and assisting in the technical aspects of code enforcement and fire investigations. Fire protection engineering is also used in the research and planning functions.

In large fire departments, a staff officer is normally assigned to supervise each of these functions. Such officers are not normally involved in line functions. The number of personnel assigned to a staff officer will vary with the size of the department and the importance ascribed to the function by the chief and the community. In small departments, line officers may also be assigned staff functions, or a single officer may supervise more than one staff function. In the smallest of departments, the fire chief or assistant may directly assume many of the functions.

Organizational Plans

The manner in which fire departments are organized depends on the size of the department and the scope of its operations. Organizational plans are designed to show the relationship of each operating division to the total organization. A good organizational plan that reflects the current status of the department is essentially a blueprint of the organization.

A list of responsibilities or a job description for each position should accompany the organizational plan. In small departments, a single individual may have responsibility for more than one function. For example, a single officer may be responsible for both training and maintenance. This should be detailed in the job description.

The organizational chart should show how the various functions that may demand time and support from other personnel or groups will be coordinated within the fire department. Both the personnel within the ranks of the fire department and the public need to see a clear coordinated effort of providing fire protection to the community.

Rules and Regulations

As with any organization, rules and regulations are needed to govern operations. This is especially true in the fire service, due to the hazardous nature of much of the activity and the need for a clear understanding of expected performance.

Every fire department should have a set of rules and regulations that outline performance expectations for its members, the standard operating procedures for the department, and disciplinary action that may be taken for failure to follow the regulations. These rules and regulations can be, and often are, supplemented by orders from the fire chief who may add to or clarify the rules or change them for a special event or specific purpose. Both the rules and regulations and subsequent orders from the chief should be written and distributed in such a manner as to ensure all persons are made aware of them.

ORGANIZATION FOR FIRE SUPPRESSION

The dominant factor in the structure of most fire departments is the mission of fire suppression. The complexity of the organization is directly related to the size of the department. Many small fire departments operate under the direct supervision of a local fire chief. In larger departments, companies are organized into battalions or districts under the supervision of intermediate-level command officers. (See Figure 10-1D.)

The company is the basic tactical unit of the fire department. A company is a complement of personnel operating one or more pieces of apparatus under the supervision of a company officer. Several companies may operate from a single fire station. A number of

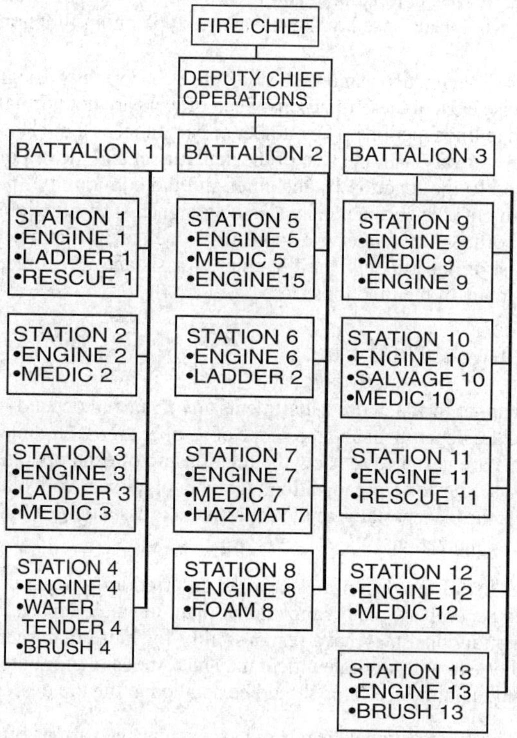

FIG. 10-1D. Organizational chart for the operations division of a medium-sized fire department.

different types of companies are used, depending on local needs. Engine and ladder companies are the most numerous, although a variety of others often are provided to perform combined or specialized functions. These include rescue squads and companies that operate special tactical or support-function apparatus. Some fire departments organize a number of companies operating from the same station as task forces.

Engine Company

The most common type of company in a fire department is normally the engine company. The basic unit of apparatus is the pumper, which carries hose, nozzles, an on-board water tank, and a pump. The engine company's basic role in tactical operations is to deliver water through hose lines to control fires. Most engine companies carry additional equipment to provide medical services. In most cases, at least one engine company is based at each fire station to respond quickly and to begin fire control operations at a fire scene. The engine company is considered the basic unit of a fire department and is supplemented by other types of companies.

Ladder Company

The basic ladder company apparatus is an aerial ladder or elevating platform device, which provides access above ground level or to rooftops, or directs elevated master streams on fires. Ladder trucks also carry a complement of ground ladders and a selection of hand and power tools. Ladder companies perform a supporting role in fireground operations, including search and rescue, forcible entry, ventilation, salvage, and overhaul. They also use ladders to gain access to fires and rescue persons trapped above ground level.

Ladder companies are provided in relation to the degree of urban development and the need for aerial apparatus. In a densely developed city, one ladder company may be provided for every two or three engine companies. In rural and sparsely populated suburban areas, the functions normally performed by ladder companies often are assigned to engine companies carrying additional equipment.

Rescue Company

Many fire departments use separate rescue companies for both firefighting and non-fire-related incidents. Rescue companies specialize in technical rescue, such as extricating victims from vehicles involved in accidents, removing injured persons from perilous locations, and assisting victims of industrial accidents.

In fire-fighting operations, rescue companies are usually assigned primarily to search and rescue and to deliver medical treatment. Additional duties often involve activities similar to those of a ladder company, particularly forcible entry, ventilation, and the use of power tools.

Special Apparatus

In addition to the apparatus normally assigned to engine and ladder companies, fire departments often employ specialized vehicles, including off-road vehicles for brush fires, water tenders, hose wagons, foam pumpers, hazardous materials units, lighting trucks, breathing-air-supply trucks, and command vehicles. These may be organized as individual companies, or they may be assigned to regular companies.

Some fire departments operate special companies with additional personnel. These companies are often referred to as "manpower squads" and operate vehicles designed to carry a minimum of equipment beyond the protective clothing and breathing apparatus used by the crew.

Fire apparatus may be purchased with a variety of options and configurations to suit the needs of a particular community or fire department. These options include aerial devices, water towers, and foam systems on engine-company apparatus; high-volume pumps on ladder trucks; remote-control nozzles; and large electrical generators and air-supply systems.

There is an increasing tendency to purchase apparatus with multiple capabilities and to equip companies to perform the functions of two or more types of companies.

The National Fire Protection Association publishes a series of standards for the design and construction of fire apparatus. Refer to Section 10, Chapter 10, "Fire Department Apparatus and Equipment," for a detailed discussion of fire apparatus.

FIREGROUND OPERATIONS

Incident Management

During fire suppression operations, a fire department uses its resources to combat a fire. The success of a fire-fighting operation depends upon a fire department's ability to use the available resources to protect lives and property effectively and efficiently. The incident commander is responsible for managing the available personnel and equipment to achieve maximum results, depending upon the situation, the associated conditions, and the resources available. The same principles apply to other types of incidents that the department may be expected to handle, including medical, rescue, and hazardous materials situations.

An incident commander is responsible for directing and controlling operations at every incident. From the arrival of the first unit at the fire scene, one identified person should be in command with the responsibility and authority to direct all phases of the operation. The officer in charge of the first-arriving company assumes the role of incident commander until relieved by an officer of higher rank. In complex situations, command may be transferred one or more times, as higher ranking chief officers arrive and assume command.

Many incidents are routinely handled by a single company, with the company officer functioning as incident commander. The officer in charge of the first-arriving company should automatically assume command of an incident, following an established standard operating procedure.

The first-arriving officer may remain in command of the incident, particularly if only one or two companies are used. When the incident requires a larger commitment of resources, a chief officer should respond and assume command. A battalion or district chief will usually take over incidents involving three or more companies.

Higher-level command officers usually will respond to and assume command of multiple-alarm incidents. The incident organization is subdivided according to a standard incident management system to maintain a functional span of control at each level. NFPA 1561, *Standard on Fire Department Incident Management System*, defines the basic requirements for an incident management system. The incident command system (ICS) taught by the National Fire Academy is a model for managing emergency incidents.

Command Responsibilities

The incident commander has the overall responsibility to manage operations at the scene of an incident. Intermediate levels of command report to the incident commander and are responsible for certain sectors, or geographical portions, of the operation or for supervising particular functions. These officers coordinate the operations of a group of companies.

A major incident may require a fairly complex command staff with a number of officers reporting to the incident commander. Several officers may be assigned to directly support the incident commander, at a command post, providing information or further subdividing the span of control into manageable units.

Standard operating procedures are an important aspect of incident management. Every fire department should have a set of procedures that outlines the basic operating principles to be employed in any situation, from the simplest to the most complex. The procedures should be flexible enough to allow fire fighters to react to different situations, and they should be incremental to permit adjustment to the scale of the incident. Standard operating procedures provide a menu of basic functions that can be employed as needed and establish consistent approaches to fire control.

The incident commander must use strategy and tactics to manage a fire suppression or similar incident. The incident commander is responsible for strategic decision making and for translating strategic goals into tactical objectives and task assignments. (See Figure 10-1E and Table 10-1A.)

Strategy: Strategy involves the development of a basic plan to deal with a situation most effectively. The plan must identify major goals and prioritize objectives. Strategic decisions are based upon an evaluation of the situation, the risk potential, and the available resources.

* May also be known as a Division or Group.

FIG. 10-1E. *A common type of fireground organization. The incident commander develops a strategic plan and assigns tactical objectives to the sector officers. The sector officers assign companies to perform specific tasks in order to accomplish the tactical objectives.*

TABLE 10-1A. *Example of Strategic Decision-Making*

Strategy
Offensive—Protect occupants and control fire in the area involved on arrival.

Tactics
Rapid search and rescue.
Attack with interior hose lines.

Tasks
1. Search and rescue in immediate fire area.
2. Extend interior hose lines to attack fire.
3. Provide ventilation.
4. Back up attack hose lines.
5. Search and rescue floor above fire.
6. Protect property.

Strategic decisions identify the priorities for committing resources to various tactical positions and activities, based on standard approaches and prevailing conditions. The generally accepted priorities for strategic decisions are rescue, fire control, and property conservation. The strategic plan identifies where and when the forces will attempt to control the fire and how their activities will be combined and prioritized.

The strategic options available to the incident commander involve some very important, basic decisions. The most fundamental is the choice between offensive and defensive modes of operation, based on the available resources and the risk to personnel. For offensive operations, companies extend hose lines into the interior of an involved fire area and extinguish the fire where they find it. In defensive operations, heavy streams are applied from the exterior to confine or control a fire, conceding the loss of the involved area. Offensive and defensive modes of operation must not be mixed in the same place at the same time, so the incident commander must identify what can and cannot feasibly be saved without undue risk to personnel.

Tactics: Tactics are the methods the incident commander uses to implement the strategic plan. The tactical objectives define specific functions that are assigned to groups of companies operating under sector officers. The achievement of these objectives contributes to the strategic goals and must be compatible with the overall strategic plan.

Tactical activities are based on three distinct phases or priorities. The first is to provide for the safety of the public by extending search-and-rescue efforts to all areas where potential victims may be located. The second priority is to control the fire, and the third is to conserve property. Fireground tactics usually involve a mixture of tasks performed under the supervision and coordination of sector officers.

Tasks: The translation of tactical objectives to tasks results in assignments to individual companies. A company generally undertakes one or two specific tasks at any particular time, which must be coordinated and combined to achieve tactical objectives.

Tactical Functions

Several tactical functions may be employed at each fire and may be carried out simultaneously during multicompany operations. Every company must be trained to carry out all basic functions and to contribute to any or all tactical objectives.

Search and rescue: Rescue is the first and most important consideration at any fire and, until it has been completed, may preclude any fire control efforts. In most fire incidents, the simplest and fastest tactic for rescue is to control or extinguish the fire. The incident commander must initiate fire control activities to protect the rescue operation and to keep the fire away from potential victims. Rescue operations may require only one or two companies, or they may require resources beyond the capabilities of the entire first-alarm assignment. All involved or threatened occupancies should be searched thoroughly for occupants by companies specifically assigned to this task. Every company must be prepared to perform search and rescue as a first priority. Rescue operations may be compounded by the time of the incident, the occupancy, and the height and construction of the structure. Rescue is the only acceptable reason for exposing fire fighters to otherwise unnecessary risks.

Exposure protection: The second fire suppression priority is to control the fire. This begins with confinement of the fire to the property initially involved. The most basic goal of fire departments, with respect to property, is to protect the community from large-loss fires

extending beyond the property involved on arrival. Failure to protect exposed structures adequately may allow a fire to extend beyond the building of fire origin. The problem of exposure protection may be compounded by closely spaced buildings, combustible construction, the type of occupancy, the lack of fire department access to the fire, and the lack of fire department resources. Exposure protection is a vital and necessary tactical consideration and should be the major objective in defensive fire control situations. When the magnitude of a fire requires all of the available fire control forces to be committed to exposure protection, the strategy is considered totally defensive.

Confinement: The confinement of a fire to its area of origin is often a complex problem. Fire control is achieved when the fire is successfully confined to a manageable area. All avenues of possible fire travel must be secured. The concept of surrounding the fire is necessary for successful confinement. Additional factors that influence the success or failure of confinement operations are the type of fuel involved, the location of the fire, building construction features, the presence of built-in fire suppression systems, and the availability of fire department resources. Confinement of a fire may involve offensive or defensive strategies, or a combination of the two options.

Extinguishment: Offensive fire control strategies are aimed at controlling and extinguishing the fire where it is encountered by attack forces. Successful offensive operations are regulated by the type of fuel involved, the location of the fire and degree of involvement, and the ability of fire control forces to apply sufficient extinguishing agents directly on the fire. In some instances, the use of special extinguishing agents may be required.

In defensive operations, final extinguishment may be achieved only when the fire burns down to a size that can be extinguished by the fire department. Defensive tactics depend upon the department's ability to apply a large volume of water or other agents on the fire to confine and eventually extinguish it.

Ventilation: Ventilation operations are the planned and systematic removal of heat, smoke, and fire gases from the structure. In many cases, it may be necessary to initiate ventilation with rescue in order to protect occupants from combustion products and heat and to provide visibility and tenability during rescue operations. Ventilation is also necessary during confinement and extinguishment to limit fire spread, to aid in locating the fire, and to provide safer working conditions for fire suppression personnel, as well as to reduce overall damage to the structure and its contents.

Property conservation: Salvage operations are conducted by fire suppression personnel to conserve property by minimizing damage to the structure and contents due to heat, smoke, and water. This involves, in part, covering contents and removing excess water. Salvage is an integral part of tactical operations and should commence as soon as possible to prevent additional damage to the structure and its contents.

Overhaul: Overhaul operations are required to extinguish the fire completely, make the structure safe, and aid in determining the fire ignition sequence. Overhaul may involve only a few personnel for a short period of time, or it may involve large numbers of personnel over an extended period. Extensive overhaul may require special pieces of equipment in addition to those normally provided by the fire department. It is important that extensive overhaul not be started before a thorough investigation has been conducted to determine the cause of the fire. Once the investigation is complete, overhaul should continue to ensure that the premises are left as safe as possible and that all fires have been extinguished.

Fire-Fighting Safety

Fire fighting has historically been considered to be one of the most dangerous occupations in North America. Safety should be a primary concern for all personnel, from the incident commander to the individual fire fighter.

All fire fighters must be provided with complete turnout clothing and protective equipment, including self-contained breathing apparatus, which comply with the appropriate NFPA standards. Regulations requiring the use of the proper protective equipment in every situation must be implemented and enforced. All fire departments should implement a complete safety program as outlined in NFPA 1500, including a physical fitness program, health monitoring and medical supervision, the provision and use of protective equipment, ongoing training, and the management of situations that present an unusual hazard to personnel. A comprehensive breathing apparatus program for all fires is an essential element of a successful fire department safety program.

Safety must be a primary concern of the incident commander at all times. Safety officers should be assigned specifically to monitor conditions and advise the incident commander and sector officers of hazardous conditions and safety concerns. Sector officers must be particularly concerned with the safety of personnel under their direction. Every tactical option must be evaluated in terms of its risk to fire fighters.

HAZARD REVIEW

Fire hazard review is an essential step in determining the level of resources appropriate for a particular community. A detailed knowledge of the potential fire problem in a fire department's service area is necessary in planning resource deployment and responses and in locating facilities. This hazard review should be conducted systematically throughout the service area to determine an appropriate level of fire suppression resources that will provide adequate public protection. The community's fire suppression capability should be directly related to its hazards.

To ascertain fire potential, it is essential to gain an in-depth understanding of the dynamics and behavior of fire, building construction, human factors, and the properties and hazards of materials.

Assessment of fire hazards, together with a technical knowledge of fire, provides an understanding of the potential that can take some of the uncertainty out of fire suppression activities and fireground decision making.

This information is also critical when a fire officer is called upon to make rapid, often irrevocable decisions affecting the commitment of personnel and apparatus at the fire scene. The fire officer must appreciate the way fire can be expected to behave in any given situation.

This analytical approach can be expanded and used to plan resources and response levels for almost any hazard and for preparing action plans. It also can be used to develop training exercises to prepare officers and officer candidates to think in terms of a systematic assessment for fireground decision making.

The following factors should be considered in determining fire potential for a particular location.

Life safety factors: How many occupants are present? Are they likely to be active, physically capable, and able to help themselves, or are they likely to be infirm, aged, bedridden, handicapped, or otherwise unable to take any action towards rescuing themselves? Are the occupants likely to be asleep and, therefore, more prone to being trapped? Does the building provide early-warning, fire suppression systems, and adequate exits, or are the occupants highly dependent upon the fire department to be found and rescued?

Contents: Does the building have a high fire load? Are the contents highly flammable, producing rapid fire spread and high heat-release rates? Are they liable to explode, produce dense smoke, or give off highly toxic products of combustion? Are the contents extremely valuable or highly susceptible to heat, smoke, or water damage?

Construction: Is the construction fire-resistive and likely to maintain structural integrity, or is it unprotected and likely to fail early in the fire? Are the interior finishes liable to cause rapid flame spread and extension of fire? Is the building compartmented with fire walls and doors, or are there openings in floors, walls, ducts, and shafts that will allow a fire to spread? Is the building old, or has it been remodeled with possible resultant weaknesses? Are there very large structures or groups of buildings that present a situation in which an established fire is virtually impossible to control? Are there unsprinklered high-rise buildings that present an excessive demand for personnel and equipment to support aboveground fire suppression operations?

Built-in protection: Have automatic sprinklers been installed to control the growth and spread of fire? Are there fire doors, compartments, or other fire protection arrangements that would help to confine a fire?

Time: Are occupants alert, and would they discover a fire soon after its inception? Are occupants asleep, or is no one present to call the fire department? Is there a smoke- or fire-detection system? Is the system connected to the fire department or to a central station system? What is the response time from the nearest fire station? Are there railroad crossings, heavy traffic patterns, or other restrictions on the speed of fire department response? Are there any other factors that would delay discovery, alarm, or response?

Suppression resources: Are water or other extinguishing agents available? Is special apparatus required? Will entry or access be a problem? Will there be rescue problems? Will there be a need for long-duration breathing apparatus? Are there any other factors affecting the fire department response or action? Are adequate fire suppression resources available, including mutual aid and reserve personnel and equipment? Are additional resources that are beyond the capability of the fire department needed?

In addition to structural fires, fire departments must be prepared to face a wide variety of incidents, including fires in outside storage, brush and wildlands, transportation systems, and industrial installations. Most fire departments also are responsible for hazardous materials incidents and for rescuing persons trapped by floods, vehicle accidents, building collapses, cave-ins, and numerous other situations. Many fire departments also have assumed a major role in the delivery of emergency medical services, which often involves more emergency activity than fire suppression.

The fire department has become an all-purpose emergency service and requires a diversified array of equipment and training to fulfill this diversified mission. Many of the standard procedures and much of the equipment used for structural fire fighting are transferable to these additional roles, but additional specialized equipment and methods often are needed to deal with the identifiable risks and hazards in a community. Preparedness is an important element of fire department planning.

Transportation Incidents

All public fire departments need to be as well prepared for transportation incidents as they are for structure fires. These incidents may range from automobile fires to aircraft accidents or train derailments, and they may occur anywhere at any time. Planning is essential for large-scale incidents, such as those involving mass-transit systems, airports, and railways.

Fires and accidents that occur on bridges, on freeways, on railroads, in tunnels, on mass-transit systems, and at other locations must be a part of every fire department's planning process since these incidents combine difficult access, limited water supplies, and special requirements for rescue and fire suppression. These situations may require specialized apparatus, equipment, training, and procedures so personnel can deal effectively with the variety of problems that can occur. Incidents ranging from a gasoline tanker burning on a bridge to a subway train collision and fire in a tunnel must be considered, and effective plans must be in place to deal with them. Some of the most disastrous fires have occurred where unanticipated events and circumstances exceeded the capabilities of the emergency response organizations.

Large quantities of flammable liquids and other hazardous materials are constantly in transit by road and rail. Any incident involving trucks or freight trains should be considered a potential hazardous materials incident. Personnel must be familiar with labeling and placarding systems for hazardous materials, with the use of cargo manifests and waybills to identify cargoes, and with resources such as CHEMTREC, a 24-hr emergency service provided by the Chemical Manufacturers Association.

Air: Aircraft crashes that occur on or off airfields often present a major fire and rescue problem. For effective rescue, fire department action must be rapid and effective. Many public fire departments also provide aircraft rescue and fire fighting (ARFF) protection at airports. There are several good sources of information on dealing with aircraft accidents. (Consult the bibliography at the end of this chapter.)

Marine: Water-borne vessels range from small pleasure boats to barges, cargo carriers, warships, cruise ships, and supertankers. Vessel categories present differing problems that require a thorough knowledge of ship construction, design, stability, and the special techniques of shipboard fire fighting.

In the case of cargo ships, the nature of the cargo may vary from inert materials to general combustibles and highly flammable and explosive materials. As with any other transportation fire, restraint is required until positive identification is made of the materials involved. Materials can be identified from the ship's manifest, which is usually available from the ship's master.

Fire departments with port facilities should be trained in fighting shipboard and waterfront fires. Marine fires may require the use of fireboats .and often cannot be handled in the same manner as shore fires.

Some basic factors to consider when fighting marine fires are the stability of the vessel, which may be affected by the application of water; ventilation; water pollution and the possible obstruction of waterways or port facilities should the vessel capsize or sink; the ability to deal with fire away from wharfside; maximum use of a ship's built-in extinguishing systems; and fire-spread through vessel bulkheads. The fire department also has to interact with port authorities, the U.S. Coast Guard, and the master of the vessel concerned. The often unclear divisions of authority and responsibility should be clarified before an incident occurs.

Pipelines: Underground pipelines carry highly flammable products at high pressures through many populated areas. The hazard involved in a rupture of one of these lines may be extremely critical. Fire departments should be aware of the locations of underground pipelines and the products they carry, and they should know how to shut down the pipelines in an emergency.

Mutual Aid and Major Emergencies

The possibility of fires and disasters that overwhelm the local emergency forces must always be considered. For this reason, most fire departments traditionally have rendered mutual assistance to other departments in times of need. Mutual-aid plans establish procedures for requesting and dispatching help between fire departments so that each party will know what is expected. Mutual-aid plans may include immediate joint response of several fire departments to high-risk properties; joint response to alarms adjacent to the boundaries between fire department areas (automatic aid); coverage of vacant territories by outside departments when the resources of the local department are engaged; provision of additional units to assist at major fires that may be too large for the local department to handle; and provision of specialized types of fire-fighting equipment not available locally in adequate quantity for the particular incident. Mutual-aid plans also should include provisions for incident management, standard operating procedures, interdepartmental communications, common terminology, maps, adaptors, and other considerations that directly affect the ability of departments to operate effectively together.

Command responsibility, jurisdictional questions, insurance coverage, and legal constraints should be covered in written agreements, supported by enabling legislation, to establish mutual-aid systems properly for the participating departments.

Some jurisdictions have extended the mutual-aid concept to multi-jurisdictional agreements in which fire department resources are pooled or merged into an integrated system with standardized training procedures and communications. These networks may include shared facilities, joint purchases of specialized apparatus and equipment, and a coordinated approach to long-range planning.

True mutual aid is a relationship in which each member is prepared to assist the parties of the agreement. Most larger fire departments are willing to assist smaller jurisdictions when major incidents occur that obviously exceed local capabilities. In many places, communities or individual properties known to be deficient in fire-fighting resources contract in advance to pay another jurisdiction for certain fire-fighting assistance. These programs are known as outside aid programs. In some instances, the contract covers basic first-alarm response; in other cases, additional assistance is provided for fighting major fires. In the cases of both mutual aid and outside aid, definite agreements should be made in advance according to the legal requirements governing fire department operations outside normal jurisdictional areas.

It should be recognized that some existing mutual-aid agreements contain deficiencies. Each local department may have different operating methods and types of equipment, so maximum coordination may be hampered. The parties to the plan may choose to render assistance only to the extent that they feel they can do so without seriously reducing local protection, although the better regional plans provide for maintaining coverage for all districts dispatching apparatus. Coverage weaknesses should be identified and corrected in advance.

Experience with natural disasters and large-scale incidents has focused attention upon the importance of plans and organizational procedures for systematically mobilizing fire forces. Some states and provinces have established large-area disaster plans involving all emergency services, coordinated under standard incident management systems. For successful disaster operations, it is imperative that such plans integrate with the normal organizational and command procedures used by fire departments. These large-scale mutual-aid networks form a natural basis for smaller-scale, more routine mutual-assistance plans.

COMMUNICATIONS

An effective communications system is a key factor in fire department operations. The communications system is responsible for receiving notification of emergencies from the public; alerting and dispatching personnel and equipment; coordinating the activities of the units engaged in emergency incidents; and providing nonemergency communications for coordinating fire department units.

Receiving Alarms

Most alarms are received from the public over the public telephone system or the municipal alarm systems. In many areas, combined emergency service answering centers receive telephone calls for police, fire, and emergency medical services through a 911 emergency number, while the fire departments in other jurisdictions receive calls at a separate communications office. Enhanced 911 telephone systems are able to identify automatically the telephone number and location from which the call originates. Small departments with low activity levels may receive alarms through a variety of systems that provide 24-hr coverage.

As a result of high false alarm rates, many cities have removed municipal alarm systems or converted from telegraph to systems providing voice contact between the dispatcher and the caller. The majority of fire alarms in most areas are received from private and public telephones.

For more information, see NFPA 1221, *Standard for the Installation, Maintenance, and Use of Public Fire Service Communication Systems.* Additional information is provided in Section 10, Chapter 17, "Fire Department Emergency Communications Systems."

The communications center should provide recording equipment for all telephone lines and radio channels to provide immediate playback capability to verify information and to provide a complete record of all activity. In addition to telephones and municipal alarm systems, many fire departments receive automatic fire alarms from systems installed in buildings connected to the fire department directly or through a private alarm company.

Dispatch Procedures

The second phase of fire department communications is alerting and dispatching personnel and equipment to an incident. The complexity of this phase varies with the size and population of the area served and the number of units under the control of the alarm center.

The dispatcher must identify the units that are to respond to an incident based on the geographic location and the type of situation indicated. The selection criteria for response units must be determined in advance, based on distance from, or response time to, the reported location. The type and number of units due to respond to each type of incident should also be determined in advance, based on risk criteria and unit capabilities.

The dispatcher must know the status of each unit in the system and be able to contact immediately all units that are available to respond to an incident. If regularly assigned units are engaged at another incident or are otherwise unavailable, the system should identify the next units to substitute. This information is also needed if the dispatched units request additional assistance or sound a multiple alarm. Many fire departments and regional systems have installed computer-aided dispatch systems (CADS) that combine the complex set of information required to manage these functions with the speed necessary for emergency dispatch. Smaller systems often rely on printed "running cards," or policies to provide the dispatcher with response information for each zone.

The units assigned to respond to an incident may be alerted by radio, microwave, telephone, telegraph, or other wired systems. In some areas, individual pagers, outside sirens, or horns are used to alert volunteer or call personnel. Most departments use a voice message from the dispatcher to the responding unit, which may be carried over the radio or wired circuits to the fire station. There should be at least two separate means of communication between the alarm center and each fire station for backup in case of equipment failure. In addition to voice messages, some fire departments use printers or telegraph systems to back up the vocal message. Units out of quarters are normally alerted by radio.

Radio Communications

Every fire department vehicle should be equipped with two-way radios. Units responding to or engaged at incidents should be in radio contact with other units and with the alarm center. For larger departments, this may require several radio channels to provide sufficient communications capacity. The alarm center must be able to contact responding units to provide additional information or directions while en route to an incident, while units at the scene must be able to request or return additional resources. The incident commander should be able to contact the alarm center at all times, providing progress reports, advising on the need for assistance, or releasing units from the scene. A fireground channel should provide communications between the incident commander and the units operating at the scene.

Each company officer and sector officer should have a portable two-way radio on a fireground channel.

Units released from an incident by the incident commander should advise the alarm center when they are available to respond to other incidents.

The radio system is essential to maintain radio communications with units that are engaged in activities out of quarters. Fire departments have increased their dependency on radio communications as they have become more mobile and active, with units spending less time in quarters. The radio system must reach units in every geographical area of a jurisdiction and provide sufficient capacity and channels to handle the volume of communications a major emergency generates. Regional mutual-aid channels are essential to coordinate activities involving units from multiple jurisdictions.

Communications Center Staffing

Fire communications centers require trained operators who are familiar with fire department operations and equipment. In very small communities, the fire department telephone number may be arranged to ring in a number of locations where appropriate action can be taken to dispatch fire apparatus. Dispatching may also be handled by the police department, a town office, or other locations that provide 24-hr coverage. Increasingly, small fire departments are being served by regional fire communication centers, which can be properly equipped and staffed.

Larger communities and regional communication centers require trained, capable personnel to be on duty at all times. The number of personnel on duty depends upon the workload. There must be enough operators on duty to handle the volume of communications required by busy activity periods and working incidents. Major emergencies and high-activity situations may necessitate calling in off-duty personnel or additional trained dispatchers.

System maintenance and repair personnel should be on-duty or on-call at all times. All equipment must be maintained properly and tested for maximum reliability, and back-up equipment or systems should be provided in the event that any key component fails. Every fire department should have a back-up facility to which basic communication capability can be transferred if a critical situation makes the primary center unusable.

Information Retrieval and Storage

The communications center should have immediate access to essential information that may be needed in dispatching assistance to fires and emergencies. The data files must include a complete geofile index of all streets, intersections, and related numbering systems in the area. The geofile should also include all target hazards such as schools, hospitals, and major buildings. The system must allow the dispatcher to determine the appropriate response zone and map location for any reported emergency so that the proper units can be dispatched. Maintaining and updating these files is an important ongoing function, whether the information is kept in hard-copy form on cards or index systems or in a computer system. The same approach is necessary to keep pre-incident plan information and maps up to date in the communications center, as well as on responding fire apparatus.

Additional information that should be included in the communications center data files includes telephone numbers for responsible parties for major buildings and businesses, utility companies, and other agencies to contact during fires, as well as a variety of other individuals who may have to be notified of, or asked to respond to, certain incidents.

Electronic data processing equipment speeds information retrieval and makes it possible to store and retrieve considerably more data than are available quickly by more conventional means. Computer-aided dispatch systems automate many or all of the dispatch- and information-management functions, including the provision of specific data on individual occupancies. Information available on a particular location may include data on building arrangements, construction, hazards, code-inspection access, water supplies, and even prior fire experience. The system may provide for direct transmission of this information to terminals in responding vehicles or for storage in portable computers or microfiche systems carried in command vehicles.

Station/Personnel Alerting Systems

Some fire departments maintain a 24-hr watch at a console or control room in each fire station. The person on watch is responsible for receiving dispatch messages and controlling alerting devices within the station. Controls for lights, electrically operated doors, traffic-control devices, and similar equipment are usually located at the watch desk. Instead of maintaining a watch in each station, other fire departments control some or all of these devices from the communication center. Radio signals or hard-wired circuits may be used to control equipment from a remote location.

Where stations are staffed entirely by volunteer or on-call personnel, the communications center may alert personnel by activating sirens or horns or by tone activation of radio pagers carried by personnel or kept in their homes.

MAINTENANCE OF FIRE DEPARTMENT EQUIPMENT

The maintenance of fire department apparatus and equipment at peak operating efficiency is a primary fire department responsibility. The safety of the public and of fire department personnel, as well as fire-fighting efficiency, depend considerably upon the effectiveness of the maintenance program. When properly cared for, fire apparatus can give years of reliable service. Each fire department needs a clearly established policy concerning maintenance responsibilities and routines to avoid possible equipment failure during operation.

Every fire department should assign an officer the responsibility for apparatus and equipment. This may be a part-time job in small fire departments. This officer should be responsible for the condition, status, and department needs for both apparatus and equipment and should help prepare specifications in accordance with accepted standards for the procurement of new items.

Complete records should be maintained for every piece of apparatus and equipment, showing all repairs and service performed. Service tests of pumping engines should be conducted annually, and test records should be maintained and compared with original equipment performance. Service tests and structural examinations also should be conducted regularly on ground ladders, aerial ladders, elevating platforms, and other equipment. Tests should be conducted in accordance with NFPA 1931, *Standard on Design of and Design Verification Tests for Fire Department Ground Ladders,* and NFPA 1904, *Standard for Aerial Ladder and Elevating Platform Fire Apparatus.*

A preventive maintenance program should be set up for fire apparatus and equipment, with all minor repairs made promptly. Since fire apparatus is subject to frequent starts and relatively high-speed operation with cold engines, this approach becomes critical to maintaining dependable apparatus.

Much routine apparatus maintenance is the responsibility of the assigned apparatus operators under the direction of company officers. This maintenance should be carried out on a daily and weekly basis following a set checklist. Among the things for which apparatus operators are responsible are keeping the fuel tank full, checking the engine oil level, checking the batteries, maintaining tire pressure at the specified level, maintaining the water level in the radiator, keeping the apparatus water tanks full, and checking all lights. All items needing repair should be reported immediately and the work completed without avoidable delay. Frequently used parts should be in stock or readily available, and the maintenance or service manuals delivered with each piece of apparatus should be followed.

Apparatus taken out of service for repairs or maintenance should be replaced by reliable, properly equipped reserve apparatus. Every fire department should have at least one reserve pumping engine in good, usable condition. Some fire departments have a reserve fleet ranging from 25 to 33 percent of their first-line equipment, including reserve ladder trucks, rescue trucks, and ambulances. This reserve apparatus also may be placed in service by off-shift personnel for major fires or during periods of peak activity.

Minor equipment and power tools also need routine maintenance and servicing. All equipment should be inspected periodically and repaired or replaced as needed.

Every fire department should have a repair and servicing facility to care for its apparatus and equipment. This shop should have the necessary work benches, ample power outlets, and jacks and hoists to handle heavy equipment. The shop should meet all fire-protection standards for a repair garage, and areas used for painting should be properly protected. A shop facility should include an equipment repair section and a section or area where fire hose and couplings can be repaired properly. In some cases, pump-testing equipment is provided at the shop; in others, these facilities are located at the fire department training center.

Many fire departments employ mechanics to maintain and service apparatus and to respond to fires to oversee the mechanical equipment operation and to perform emergency repairs. Many departments also have maintenance personnel on 24-hr call. Some small fire departments without full-time maintenance personnel have one or more fire fighters who fill in as part-time mechanics in addition to their fire fighter duties.

In some communities, apparatus repair is assigned to a municipal service garage. This arrangement may be satisfactory if a regular fire apparatus repair and service section is provided and staffed by persons qualified for the specialized nature of the work and available on a 24-hr basis. Often, general mechanics do not have the necessary skills to carry out a proper fire department maintenance program, and fire apparatus may receive a low priority in a municipal repair garage. In some cases, it is advisable to send apparatus to the manufacturer for major repairs or to be rebuilt.

TRAINING

The level of performance demonstrated by a fire department is usually a good indication of the type, quantity, and quality of the training provided. In addition to departmental training programs, most states provide some type of fire service training that, depending upon the state, ranges from basic fire fighter training to fire department management programs.

Large city and county fire departments usually have their own training facilities and full-time training staffs. In these larger departments, training takes several forms, starting with continuing in-service training conducted daily by company officers. This may be supplemented by scheduled programs on various subjects at a fire training center and by specialized training for drivers, apparatus operators, and other personnel. Additional officer training may be conducted monthly or annually at a fire college or a staff-and-command school. New fire fighters normally undergo an intensive recruit training class before they are assigned to regular duty. Recruit training normally requires two to three months of full-time participation.

Training Objectives

The goal of any fire department training program should be to teach each person in the department to operate at acceptable performance levels for his or her rank and assignment. Ideally, training courses should have their own instructional objectives, a list of enabling objectives showing how the instructional objectives can be reached, and stated methods that explain how anticipated or desired behavioral changes can be measured.

State Fire Training Programs

Many state fire training programs operate under the umbrella of an educational agency within the state—usually vocational or occupational education—or under the auspices of a state university or college. In other cases, the director of fire training operates under the state fire marshal's office. In Delaware, for example, the state fire training program operates under a commission appointed by, and reporting directly to, the governor.

Some state fire training directors have no full-time instructors, while others have a small cadre of full-time senior instructors or supervisors who, in addition to their teaching functions, may be responsible for assigning and supervising part-time field instructors, developing instructional outlines, developing and producing audiovisual training aids and props, and monitoring the records required for a state program. State programs also may train recruit fire fighters for small departments. Additionally, state programs provide training for specialists within the service, conduct seminars on material of current interest and importance, and serve as a valuable informational reference.

National Fire Academy

The National Fire Academy (NFA) is a center for fire-protection-related education, information, and expertise aimed at improving the professional development of fire service personnel and allied professionals. The academy offers a wide variety of courses, including incident management, fire technology, and fire prevention. These courses are delivered through an on-campus program and an extensive outreach program.

The NFA's resident-student program consists of courses conducted at the National Emergency Training Center (NETC) in Emmitsburg, MD. Students, mainly from the fire service, attend these courses at minimal cost through a stipend program that covers the costs of tuition, travel, and accommodations in campus dormitories. Most of the individual courses last for two weeks and are linked so students can either advance through various levels in a particular subject or develop management expertise in a variety of subjects.

The outreach program provides short-term courses, generally two days long, to students all over the United States. These courses are coordinated through state and local training agencies and are concentrated on weekends to allow maximum participation from both career and volunteer fire service personnel. These courses are designed to be transferred to state and local delivery systems for wider distribution through an increasing network of instructors.

The National Fire Academy's full-time professional faculty and staff is supplemented by a large network of adjunct faculty members who teach courses in their particular areas of expertise. Both resident and adjunct faculty include individuals from the fire service and other backgrounds. The main campus provides classroom space, laboratories, educational resources, and living quarters for several hundred students.

The Open Learning Fire Service Program is an additional aspect of the National Fire Academy educational system. Through a network of accredited colleges and universities, this program coordinates independent-study programs for fire service personnel. Students may earn degrees through these educational institutions by combining a series of upper-division fire-related courses with credits in related subjects. The open learning concept provides a high degree of flexibility for individual students taking correspondence courses and seminars, and it provides them with opportunities to transfer credits earned at other educational institutions.

Fire Service Organization Training Activities

Several fire service organizations sponsor conferences and seminars as a part of an annual meeting or as separate courses on materials of current importance. Examples of these organizations include the National Fire Protection Association (NFPA), the International Society of Fire Service Instructors (ISFSI), the International Association of Fire Chiefs (IAFC), and local and state fire service organizations.

Utility Company and Private Agency Training Programs

At the local and regional levels, training programs are available from major utility companies; state agencies, such as those concerned with forestry, public safety, and civil defense; and private agencies that interact in various ways with the fire service.

College Training Programs

There are hundreds of college fire science programs in the United States, most offered by community and junior colleges on either a full- or part-time basis. In most cases, an associate degree is awarded upon successful completion of the program, and some institutions award certificates for partial completion. A few four-year degree programs that emphasize fire service management are also offered. Master's programs are offered by the University of Maryland and Worcester Polytechnic Institute in Massachusetts.

College fire science programs usually are designed to supplement, though not replace, the training provided by fire departments and state training agencies. The curriculum design for many such programs emphasizes the arts and sciences, with special courses in fire protection subjects that are beyond the scope of basic fire training programs. A majority of programs emphasize supervisory and management skills along with fire science.

Certification Standards

In an effort to lend uniformity to fire service training, the National Professional Qualifications Board for the Fire Service has developed a set of standards based on performance objectives. The standards, developed through NFPA, are as follows:

NFPA 1001, *Standard for Fire Fighter Professional Qualifications;*

NFPA 1002, *Standard for Fire Department Vehicle Driver/Operator Professional Qualifications;*

NFPA 1003, *Standard for Airport Fire Fighter Professional Qualifications;*

NFPA 1021, *Standard for Fire Officer Professional Qualifications;*

NFPA 1031, *Standard for Professional Qualifications for Fire Inspector;*

NFPA 1033, *Standard for Professional Qualifications for Fire Investigator;*

NFPA 1035, *Standard for Professional Qualifications for Public Fire and Life Safety Educator;*

NFPA 1041, *Standard for Fire Service Instructor Professional Qualifications;*

NFPA 1051, *Standard for Wildland Fire Fighter Professional Qualifications; and*

NFPA 1061, *Standard for Professional Qualifications for Public Safety Telecommunicator.*

In addition, the following NFPA standards are being used by many fire service training agencies:

NFPA 472, *Standard for Professional Competence of Responders to Hazardous Materials Incidents; and*

NFPA 600, *Standard on Industrial Fire Brigades.*

These standards are applied by several states and municipalities. Some states offer formal certification. The certification program may be mandatory or voluntary, depending upon the authority having jurisdiction over the process, and may provide for certification in a number of areas and levels. Formal certification provides a standard of preparation and performance for various positions and may be applied to hiring, promoting, retaining, and transferring personnel.

Training Records and Reports

Recent court decisions regarding job-related examinations have placed additional importance on training records and personnel evaluations. Training records and reports are particularly important at the entry or recruit level. Training records should be clear and concise and should document a fire fighter's advancement in the department. The content of training records and reports is particularly important when personnel are evaluated for pay increases and promotions. In larger departments, training records may be integrated into a larger management information system along with other personnel and management data.

TYPES OF FIRE DEPARTMENTS

In 1995, there were an estimated 260,850 career fire-fighting personnel and 838,000 on-call or volunteer fire-fighting personnel serving in the United States. The principal distinction between career and on-call or volunteer personnel is that career personnel are assigned regular periods of duty and are compensated on a regular basis. On-call or volunteer personnel are not normally required to be available except for meetings, training sessions, and emergency responses. They may or may not receive compensation for their services.

While most cities of 50,000 or more people employ career personnel, some use auxiliary personnel or volunteers to supplement their regular forces. Smaller cities or towns tend to use fewer career and more on-call or volunteer personnel. Some communities maintaining their own fire departments may have a career fire chief, officers, and apparatus operators, but rely upon on-call or volunteer personnel to provide the staffing balance necessary for efficient fire-fighting operations. Other communities may use career personnel only during normal daytime working hours and rely upon on-call or volunteer personnel during the night.

A number of important factors can influence the type of personnel used within a fire department. These factors are: (1) the financial resources of the community, (2) the availability of on-call or volunteer personnel, (3) the frequency of fire incidents, (4) the range of services expected from the department, and (5) the type of department preferred by the community.

Community's Financial Resources

Frequently, the financial resources of many smaller communities will dictate that the department be composed entirely of volunteer personnel. Since salaries normally consume a large percentage of a fire department budget, available finances may be sufficient only to purchase and maintain apparatus and equipment. Other communities with adequate financial resources may or may not elect to implement a full career service.

On-Call or Volunteer Personnel Availability

In some cases where a community desires fire protection, there may not be a sufficient number of persons willing or able to serve in a volunteer capacity. This requires that the community finance the operation of a fire department staffed either fully or partially by career personnel. A situation of this type is probably most evident in the rapidly growing suburban communities surrounding major metropolitan areas. Originally, these outlying areas were composed of small communities that provided a climate suitable for volunteer departments. As times changed, however, these communities experienced rapid growth rates in population, housing, and ancillary services. Demands for fire protection services grew accordingly. The original corps of volunteers found themselves unable to meet the increased demand or to recruit new members since many residents in these areas commute out of the community on normal workdays and are unavailable for fire department participation. Therefore, many of these once-volunteer fire departments have added the services of career personnel.

Effect of Incident Frequency

The frequency with which a fire-fighting department responds to incidents may determine the type of personnel chosen to staff the department. Large, congested areas, especially those that are heavily commercial or industrial, are likely to produce an increased need for fire department service compared to more sparsely settled, largely residential areas. A department that responds to a large number of incidents each year will tend to inhibit a volunteer operation unless the department has a large membership and the workload can be apportioned to reduce the commitment required of each individual. A determination on the number of incidents that require the use of career personnel must be made locally based on the ability of the department to perform at a level acceptable to the community supporting the service. Statistics including average number of volunteers per call (weekdays should be a separate statistic) and "turn out" times should be monitored for any adverse trends.

Type of Department Personnel Preferred by the Community

Communities often have preferences for the type of staff members they wish to have serve in their fire departments. In some instances, the volunteer fire department serves as a focal point of community activity, and the community is satisfied with the level of service provided by volunteer staff.

There are many fire departments—career, combination, and volunteer—that provide an acceptable level of service to their respective communities. The success of their operations is not dependent upon whether the personnel are paid or unpaid, but upon their individual and collective ability to perform and to accomplish department objectives. There are no simple guidelines that set forth the requirements as to the type of personnel to be used for fire department operations. The decision is one that must be made at the local level following a careful analysis of all pertinent factors.

MANAGEMENT AND BUDGETING

The operation of a fire department is normally a function of local government—in the case of a fire district, possibly the only function—which supports the service and is responsible for the level of service rendered. As with any governmental or business operation, this involves three major areas of responsibility: (1) fiscal management, (2) personnel management, and (3) productivity. Other areas of responsibility include planning, research, and recordkeeping.

Fiscal Management and Budgeting

In general, fiscal management practices follow those used by the government agency supporting the department. These practices involve budgeting, cost accounting, personnel costs including payroll, and purchasing or procurement costs. The degree to which these factors are a direct responsibility of fire department management varies, depending upon the practices of local government.

Fire department budgets are generally prepared and submitted by the fire chief to the elected body that administers the city or district operations. The type of budget used and the manner in which projections are submitted will vary depending upon the jurisdiction. The personnel costs are generally the most significant costs for most municipal fire departments, accounting for approximately 90 percent of the total expenditures of a fully paid fire department. Personnel costs of combination fire departments may total approximately 40 to 60 percent of their overall budgets.

It is critical for fire department managers to understand thoroughly their jurisdiction's budgeting system. Inadequately prepared budgets can lead to serious monetary shortfalls at the end of the fiscal year. In order to ensure smooth operations, all costs must be estimated realistically and expenditures monitored on a regular basis. An effort should be made to develop a long-range plan that will project capital replacement costs for items such as staff vehicles, fire apparatus, fire stations, and other major pieces of equipment.

Fire apparatus costs normally run from 1 to 2 percent of payroll costs. Some fire departments include an apparatus replacement allowance in their operating budgets, but this item is regularly cut, with the result that apparatus replacement may be included in a capital expenditures budget. While this reduces the fire department's annual budget, it ultimately results in higher taxes due to interest costs. However, such decisions are generally made above the level of the fire department administration. New fire stations usually are included in a capital improvement budget separate from the fire department budget.

In large fire departments, there may be separate budget accounts for staff divisions such as fire prevention, maintenance, and training. Expenditures are charged against specific items in the line budget, and the remaining balance is shown after each expense deduction. Usually, the department head or staff division supervisors have the authority to make emergency transfers of funds between line categories. Transfers between major categories can be made only when authorized by the municipal management, finance officer, or other governing body.

Fire department administrators are required to submit their budget estimates by a specified time for the coming fiscal year. Usually, the budgets are submitted to a finance officer or finance committee, and department heads are then asked to justify specific items. Although the salary total may be governed by contract with the employees, estimates must be included covering all ranks and overtime costs. Quite often, salary increases are not determined before the budget is submitted, but municipal administrators commonly make a percentage allowance for increases based on reasonable assumptions or percentages they project to be accepted in contract negotiations. After a departmental budget has been approved by the city administration, it must be approved by the city or town council. In some municipalities, it must be approved by town meeting. With some municipal charters, the council can reduce, but not increase, the budget. This is to guard against political pressure on the administration. Once approved, the budget takes effect at the beginning of the fiscal year. If not approved in time, it is customary to permit expenditures at the same rate as those made the previous year.

Personnel Management

Fire departments use personnel with specialized skills who are organized into various operational and staff units. Fire department management is involved to some degree in the recruitment, selection, and promotion of personnel to fill various positions in the organization. Largely these processes are governed by local and/or state law; by personnel agencies, including civil service authorities; and by the direct decisions of the governmental agency operating the fire department. The assignment of available personnel to positions provided in the budgeted organizational structure and the supervision of personnel performance are normally the direct responsibility of the fire department management, although certain assignments may be governed by work contract agreements.

Recruiting: Fire departments are becoming more involved in recruiting efforts to fill vacancies in their ranks. A common arrangement is to conduct recruiting efforts jointly with the local government personnel agency. Because of their makeup, most fire districts and volunteer departments recruit their own members exclusively. Many large fire service organizations have recruitment sections.

Fire department management has three recruitment responsibilities. The first is to develop appropriate recruitment standards.

The second is to provide the basic training necessary for the new recruits so they can perform their assigned duties properly. The third is to certify, after providing the basic training, that the new members are ready for appointment as permanent fire fighters or, where individuals prove unable to perform satisfactorily, to recommend that their services be terminated before permanent appointment.

Selection of personnel must meet local, state, and federal standards. U.S. courts have ruled that there must be no discrimination in hiring practices. Many departments administer aggressive recruitment programs designed to increase the representation of women and minorities within their organization. Many communities have identified diversity in the public sector as a value and have established a goal that personnel reflect the diversity of the community they serve. Some rulings prohibit residence requirements for recruits, although fire department rules of employment may stipulate that, because of the emergency nature of the work, employees must reside within a reasonable distance of the community. One court decision has ruled out examinations that require a knowledge of fire department practices and equipment before appointment and in-service probationary training. Many states have adopted, or are in the process of adopting, minimum fire fighter qualifications standards. Selection practices are a sensitive issue, and knowledgeable counsel should be sought to maintain a sound legal foundation.

In most jurisdictions, applications for employment as a fire fighter are obtained from municipal personnel offices or a civil service agency. In at least two states, recruitment is handled by a state civil service commission. Age requirements for entry-level appointments vary and have been impacted by recent federal regulations.

Qualifications: It is imperative that all fire service personnel be fully qualified and capable of efficiently performing the wide range of services necessary to protect life and property. Many states have enacted legislation establishing commissions on fire fighter standards that require that all personnel employed by fire departments satisfactorily complete required basic training before they are given permanent employment.

In 1971, the Joint Council of National Fire Service Organizations (JCNFSO) created the National Professional Qualifications Board for the Fire Service (NPQB) to facilitate the development of nationally applicable performance standards for various levels of the uniformed fire service. On December 14, 1972, the Board established four technical committees to develop those standards using the NFPA standards-making system. The initial committees addressed fire fighter, fire officer, fire service instructor, and fire inspector and investigator.

The original concept of the professional qualification standards, as directed by the JCNFSO and NPQB, was to develop an interrelated set of performance standards specifically for the fire service. The various levels of achievement in the standards were to build upon each other. In the late 1980s, the standards were changed to recognize that the documents should stand on their own merits in terms of job-performance requirements for a given field. Accordingly, the strict career ladder concept was abandoned, except for the progression from fire fighter to fire officer. The later revisions facilitated the use of the documents by other than the uniformed fire services.

In 1990, the NFPA assumed the total responsibility for the appointment of professional qualifications committees and the development of the professional qualifications standards. The NFPA Standards Council appointed the Correlating Committee for Professional Qualifications Standards in 1990 and assumed the responsibility for coordinating the requirements of all of the professional qualifications documents. The chair of each technical committee sits on this committee.

Currently, technical committees are working on the following projects:

NFPA 1000, *Standard for Fire Service Professional Qualifications Accreditation and Certification Systems;*

NFPA 1001, *Standard for Fire Fighter Professional Qualifications;*

NFPA 1002, *Standard for Fire Department Vehicle Driver/Operator Professional Qualifications;*

NFPA 1003, *Standard for Airport Fire Fighter Professional Qualifications;*

NFPA 1021, *Standard for Fire Officer Professional Qualifications;*

NFPA 1031, *Standard for Professional Qualifications for Fire Inspector;*

NFPA 1033, *Standard for Professional Qualifications for Fire Investigator;*

NFPA 1035, *Standard for Professional Qualifications for Public Fire and Life Safety Educator;*

NFPA 1041, *Standard for Fire Service Instructor Professional Qualifications;*

NFPA 1051, *Standard for Wildland Fire Fighter Professional Qualifications; and*

Each of these standards defines the levels of progression within the specified job. The job-performance requirements defined for each level of progression are considered the minimum requirements for individuals at each level.

The intent of the committees is to develop clear and concise job-performance requirements that can be used to determine whether individuals possess the skills and knowledge necessary to perform the related duties and tasks of the job. These job-performance requirements can be used by any fire department in any city, town, or private fire service organization.

Under the new leadership of the National Board on Fire Service Professional Qualifications, the NPQB has assumed the role of third-party accreditor of fire service certification programs and developer of a national registry of those certified. Certification in the fire service is based primarily upon the NFPA professional qualifications standards. The Fire Service Accreditation Congress at Oklahoma State University has also been established as an accreditation agency for the fire service.

The establishment of standards and testing procedures does not in itself ensure that all personnel will achieve the required level of competency; NFPA 1001 is a professional qualifications standard, not a training standard. Training programs are necessary to prepare members of the fire service to acquire the skills and knowledge necessary to meet the job-performance requirements set forth for each grade. However, training should not be random. It should be organized to prepare trainees to meet the specified levels of performance demonstrated by the performance testing described in NFPA 1001.

Courts have ruled that height requirements are discriminatory, both racially and sexually, but candidates may be required to demonstrate their ability to perform the required duties. Recent legislation prohibits discrimination against individuals with a disability. Fire departments must evaluate their hiring process, defining the "essential functions" for fire department positions. NFPA 1001 is commonly used as a valid basis for "essential functions" of a fire fighter.

Fire fighting requires physical strength, and fire fighters depend on each other in fire suppression and rescue operations. When conducted in compliance with the law, departments may evaluate an applicant's physical capacity to perform the essential functions of their positions.

In some jurisdictions, the services of the fire department training division may be used to test recruits and to conduct promotional examinations. In such cases, recognized standards, such as NFPA 1001, should be followed carefully so that the results will not be subject to charges of discrimination and so that qualifications essential to the work will be valid and adequately tested.

Promotion practices: In the vast majority of career fire departments, various officer ranks are filled by personnel serving in the next lower rank or ranks, although more fire departments are recognizing the potential benefits of allowing lateral entry transfers and promotions of well-qualified personnel from other areas and departments. Promotion procedures are designed to take into account technical qualifications for the particular rank and fire department experience. It is essential that examination procedures in the civil service be fully competitive and nondiscriminatory.

In general, promotion procedures are administered by personnel departments or by state or local civil service authorities with the assistance of persons who are experienced and knowledgeable about the particular job classification. Such advisors include fire chiefs, fire fighter organizations, and technical consultants. Their assistance may include guidance as to the relative weight to be given to experience and to the result of written examinations covering the technical qualifications of the position. In many systems, however, the testing process is rigidly defined by civil service law or contract.

In some systems, performance ratings may be included. Some supervisors tend to be much more demanding than others when rating performance, and their subordinates often have less favorable performance grades than other employees, who may actually be less qualified. Because of the lack of uniformity, subjective performance ratings are now used less frequently in the promotion process.

Some promotion processes include an oral review. While this provides an opportunity to evaluate important attributes that are difficult to quantify and test, they can also be very subjective. Few oral interviews are scientifically designed or professionally administered. Therefore, they may reflect the interviewer's bias or are subject to a challenge of bias. As a result, more emphasis is given to written examinations that test technical knowledge exclusively.

Where permitted by law, more fire departments are now relying upon assessment centers as a means of selecting candidates for promotion. Assessment centers have been used for quite some time by industry. The assessment center requires a candidate to demonstrate certain abilities through the use of problem-solving exercises, role playing, and other simulated exercises. Each candidate is observed by a trained assessment team and scored according to performance. Some feel that the assessment center is the most realistic means of determining a candidate's suitability for a particular position.

NFPA 1021, *Standard for Fire Officer Professional Qualifications,* specifies the levels of performance for fire officers. NFPA 1021, written in performance terms, requires an individual to demonstrate competence by knowledge and performance. More fire departments are including educational requirements, such as community college fire science certificates, for all officers or bachelor's degrees for chief officers. It is not uncommon to find fire departments that seek individuals with graduate-level degrees in public administration or management for the position of fire chief. Many jurisdictions specify completion of the National Fire Academy's Executive Fire Officer program in their job qualifications for chief executive officer. The fire department administration, as an arm of the municipal administration, has an important role to play in the promotion process. First, it must advise the personnel agency of the qualifications required in any job or rank to be filled, where such qualifications have not been previously listed. Second, when a list of successful candidates is received from the personnel agency, the administration should advise the promoting authority of the promotions to be made. Usual practice is to fill vacancies from the top of the promotion list, except where the head of the fire department specifies in writing valid reasons for rejecting an individual. Such reasons might be a record of serious disciplinary problems, including disobedience of written orders; frequent bad judgment when performing assigned duties; a record of conflicts with other employees; or other major personality problems. These problems should be relatively current and not something that occurred years ago that the individual has corrected. While such problems might not be serious enough to warrant dismissal from the current job, they do indicate that the particular candidate, though technically qualified, would be less preferable than another candidate on the list. Personnel records should be available to substantiate any such reasons for rejection.

Personnel records: A complete personnel record must be maintained for each member of the fire department. Such records cover all the pertinent facts of each member's fire service career, from probationary appointment through retirement, and include the original application for employment, all assignments, transfers, promotions, commendations, and records of disciplinary action. Records must be carefully maintained. The record should also include information on any special skills an individual possesses that may be useful to the department, as well as the individual's educational background, including any courses that might be of value to the department.

In addition to the general record of an individual's service, a training file should be maintained, as covered in NFPA 1401, *Recommended Practice for Fire Service Training Reports and Records.* This will show training periods and subjects in which the individual has received instruction, such as apparatus operation, emergency medical service, and fire inspection. A medical history must be kept for each member, showing absences due to sickness and service-connected injuries. A new addition to the individual file is a record of exposure to hazardous materials and infectious diseases. This provides an historical record of the fire fighter's exposure to toxic materials or to infectious communicable diseases during his or her employment, as well as any appropriated inoculations he or she has received.

Productivity

Effectiveness in the fire service is the most difficult ingredient for management to measure. The basic objective of the fire service is the protection of life and property. Modern fire service practice involves two major activities: (1) controlling hazards to minimize fire losses and to prevent fires and (2) dealing with actual fires and emergencies to minimize suffering and losses. It is difficult to assess the number of fires and the amount of suffering that fire department activities have prevented. However, experience demonstrates that a lack of effective fire prevention and control measures invites disaster. Likewise, the fact that most fires are suppressed with minimum losses and injuries does not indicate conclusively that an adequate level of fire department service has been provided. Major fires and emergencies often arise from a combination of circumstances that are beyond the immediate control of fire department management, but which must be dealt with effectively to protect the public.

It is imperative that fire department management maintain reasonable standards based upon local and national fire loss experience. Fire department management is responsible for maintaining highly trained and efficient operational units to perform assigned tasks in both the prevention and suppression of fires.

In addition to their strictly fire-oriented activities, many modern fire departments also provide emergency medical services. These services range from responding as "first responders" to assist ambulance crews to actually providing engine and ladder companies and paramedic ambulance service. Fire departments involved in this aspect of public safety have seen significant increases in the use of personnel and equipment and in public support in the community.

The hazardous materials response team is another emerging area for fire department participation.

Planning

Planning for the future needs of a fire department is the most important job of fire department managers. Without adequate planning, an administrator will find he or she is handling one crisis after another and can never seem to get ahead. Long-range planning has often been neglected by fire department managers, but with budget constraints today, it is absolutely essential. The U.S. Fire Administration (USFA) has developed a master planning process, available to local communities, that outlines the steps a community may take to determine long-range goals for its fire department and explains how these goals can be achieved. The NFA Technical Support Program will assist local governments in developing a master planning process.

All departments, small or large, need to develop long-range plans. These plans need to be flexible and continually updated to reflect changes in the community, as well as developments within the fire service.

Research

In the fire service, the term "research" is commonly heard, yet few fire departments are staffed or financed to support any significant research activity. Limitation in research is due largely to the fact that most fire departments are relatively small organizations that do not have sufficient personnel to meet their ongoing fire protection obligations. True research into efficient equipment design is generally beyond the ability of most fire departments. The same is true even for the majority of fire equipment suppliers because this is a relatively small-volume, competitive business with little profit margin for research and development. In recent years, a major part of the engineering effort of fire apparatus builders has been devoted to meeting the increasing vehicle safety standards of the U.S. Department of Transportation (DOT). Some improvements have been made in various features of fire apparatus design, but not on a uniform or planned research basis.

One area in which fire department research can readily show returns is in fire record analysis, using programs such as the National Fire Incident Reporting System (NFIRS), which is administered by the United States Fire Administration in cooperation with state fire marshal offices. Properly applied fire prevention efforts can be directed against the hazards shown to be most dangerous and significant at any given period, and results of these programs can be analyzed effectively.

Research has been used in a number of instances to help determine the optimum locations for fire stations. However, a principal difficulty is that all areas of a community must be covered within reasonable response times. Considerable judgment and experience are required when considering such variables as population densities, valuations at risk, fire frequency and severity, and the number of fire companies required in a given area to apply the required water flow and maintain coverage during fires. These requirements may be at variance with any optimum location for a given fire station. A number of research studies have placed considerable emphasis upon the arrival time of the nearest fire company, rather than the total fire protection requirements of the area. Arrival times may be less significant than time required for actually getting to work at a fire, particularly in large high-rise structures or shopping centers.

Fortunately, the fire service is not without competent research resources. NFPA conducts research designed to help fire departments directly. The International Association of Fire Fighters (IAFF) Research Department is available on request to assist its local affiliates with problems. And the USFA conducts research relating to fire service needs.

Management Records and Reports

A records system should be provided to supply the fire chief and other administrative officers with data indicating the effectiveness of the department in preventing and fighting fires to facilitate management of the department. It is essential to maintain complete records of all fires and inspections. The records system should provide data on fire department activities that should be made available by city officials to the public. The fire chief should specify the records to be kept and methods of gathering data. A records retention and disposal system should be employed. All records should be examined in light of their usefulness.

STAFFING PRACTICES

The principal resource of a fire department is its highly trained personnel. The vast majority of personnel are assigned to the fire-fighting division, and possibly 2 to 3 percent of the personnel are assigned full-time to the fire prevention bureau. Thus, for most effective resource allocation, maximum use must be made of the fire-fighting personnel through careful time-utilization schedules, assigning appropriate allocations of work periods to apparatus and equipment maintenance, fire service training, and scheduled inspections. These programs require close coordination between the demands of the fire fighting, training, and fire prevention programs.

Staffing levels for fire departments vary considerably and are influenced by such things as the population protected, population density, fire fighters' work hours per week, response distances, and fire fighter safety. Generally, fire departments use a three- or four-person platoon system that will accommodate a 56- to 42-hr work week, respectively. A four-platoon system requires about 25 percent more personnel than a three-platoon system. Staffing levels for cities of 250,000 or more population range from 0.5 to 2.7 fire fighters per thousand population, with a median of 1.0 to 1.5 thousand. Cities in the Northeast generally have a higher staffing ratio than cities in the North Central and South, which in turn tend to have higher staffing ratios than cities in the West. Communities must assess their needs to determine the level of staffing that meets their requirements. However, it has been demonstrated that when staffing falls below four fire fighters per company, fireground effectiveness may be compromised. Tests conducted with the Dallas, TX, Fire Department indicated that staffing below a crew size of four can overtax the operating force and lead to higher losses.[1]

Fire departments operating emergency medical service transports need additional personnel to maintain basic fire company strength. In some smaller communities, the staffing ratio per population protected may be relatively high because of the need for sufficient on-duty personnel for effective initial attack and rescue operations. This is especially true in "bedroom communities," where call personnel are not readily available during the workday.

Likewise, fire departments in many core cities protect more lives and property than population figures reflect. A city of 80,000 may be the business center for an area of 200,000 persons and house a high percentage of the low-income groups. The number of high-rise and large-area structures to be protected and the frequency of alarms for fires and emergencies should be considered in determining on-duty fire department staffing.

Some very large fire departments may operate with a lower relative strength per 1,000 population than those in cities of a more average size because, with high population densities, these departments have sufficient companies to provide needed coverage while handling working fires. For example, a large city fire department may operate one engine company per 15,000 to 20,000 population and still have a large number of well-distributed fire companies, whereas a city of 30,000 could not be properly protected with only two engine companies.

Mutual aid plays an important role in providing additional resources. Almost all jurisdictions rely to some extent on mutual aid from surrounding areas to provide fire-fighting resources on a routine or major emergency basis. Some departments use automatic mutual aid on initial response. Even large cities are making increased use of both regularly assigned and automatic mutual aid. Often this is practical because companies from neighboring fire departments may be much nearer to a fire than some of the local fire companies.

It is frequently impossible for small cities to fully staff all of the fire companies they need to handle working fires throughout the community. In many cases, the population density and the values protected per square mile are relatively low. In such communities, some engine companies may respond with only three persons on duty, and ladder trucks with only two. Such low levels of staffing should be backed up promptly to ensure adequate personnel by off-shift or call personnel or by multiple-alarm response. Combination fire departments that use a mix of career and either paid on-call or volunteer fire fighters are found throughout the United States. In some cases, additional apparatus may be assigned to respond, offsetting deficient company strength. In communities with large geographic areas and relatively low concentrations of value, this may be an acceptable arrangement. In general, however, each engine company should have a minimum of four fire fighters on duty, including an officer. This parallels NFPA 1500 and OSHA requirements to have at least four fire fighters on the scene before starting interior structural fire fighting. It would seem inappropriate to dispatch an engine company to a fire if the crew could not start fire fighting and rescue operations because of safety concerns.

Staffing Career Fire Departments

The current approach of fire departments is to determine the essential positions in the fire fighting force that must be covered 24 hr a day throughout a year, and the total number of duty tours involved. With 24-hr shifts, there is one duty tour per position per day. With 10- to 14-hr or similar tours of duty, there are 2 per day, or 730 per year.

Assume that a community has determined that a minimum of 58 officers and fire fighters of various rank should be on duty at all times. With a standard 42-hr average work week, including day and night shifts, each shift or group is on duty 182.5 times during a 365-day year. With 58 persons on duty, this requires 42,340 individual tours of duty per year. A review of fire department records indicates that with vacations, sick and injured leave, and other absences, the individual members of the fire fighting force average not 182.5 but 146.5 tours of duty per year. The required 42,340 tours of duty, divided by 146.5 tours worked per person, shows that rather than 212 officers and fire fighters, 289 members, or 72 per platoon, are actually needed to cover the normal anticipated absences. Even when vacations are carefully scheduled so that only one person from each shift of each company is absent at any given time, there may be times off-duty personnel have to be used on an overtime basis due to sickness or injury. This is less expensive than carrying more relief

fire fighters on the roster than are required to maintain normal minimum coverage. Overtime will be required when members are called back for major fires and emergencies, so budget allowances should be made for such eventualities.

This example is based upon a minimum effective staffing of eight engine and four ladder companies grouped in two fire districts and commanded by an on-duty chief with aides. The minimum staffing for all companies is four persons on duty including the company officer. Five fire fighters are maintained with two ladder companies in districts of high life-hazard and higher-than-average fire duty. To distribute the needed relief personnel, the platoon roster for each company carries one additional fire fighter above the minimum that must be maintained, and one relief officer is assigned to each district headquarters.

In a number of fire departments, the minimum levels are not absolute. When there is an unusual amount of absence, stations housing more than one company may be allowed to operate one person below the normal minimum rather than pay overtime, unless the company minimum strength is specified by contract or city ordinance. Other jurisdictions with a defined minimum staffing level may be forced to take units out of service if the funding level cannot support the impact on personnel overtime.

Generally, a nominal 42-hr work week requires not four but five persons per position. Thus, a twenty-person company is needed to maintain an average of four on duty. Because this does not allow two members to be away from their shifts at one time, overtime or personnel swaps between companies may occasionally be needed.

With a fire department operating on the three-platoon system with 24-hr duty tours, each platoon covers 122 tours in 366 days (a leap year). It requires 21,208 tours to have 58 officers and fire fighters on duty. The records in this example indicate that the average member works 98 tours per year. Thus, 21,208 divided by 98 requires 216 fire fighters, or 72 for each of the three platoons. With a 53-hr maximum work week permitted under federal labor law (FLSA), however, each member working more than 53 hr in each 28-day work period would exceed the allowable maximum by 3 hr and would be eligible for extra compensation, except possibly when the work cycle is broken by absences.

Where minimum company strength of five persons is desired for a 42-hr work week, most fire companies have 24 persons assigned to the four duty shifts and 28 assigned for approximately every fourth company. In some fire departments where more vacations are scheduled in the summer months, a four-person minimum may be maintained in the summer and a five-person minimum maintained in the winter. However, most fire departments prefer to maintain minimum staffing year round rather than attempt to adjust for seasonal trends. If additional personnel are needed during severe winter storms, they are provided on a temporary overtime basis and are paid from the overtime account.

It was customary in the past to allow 10 percent for absences from assigned shifts due to vacations and sickness. Now the figure commonly is 20 percent or more. Vacations have been increased, and employee benefit programs or work contracts often provide for various other compensated absences from duty.

Minimum staffing: During recent years, an increasing number of fire departments have established minimum staffing levels for each fire company or each duty shift. It is a policy in many fire departments not to operate engine or ladder companies with fewer than four fire fighters, including an officer, on duty. In some cases, because of the workload and the population and values protected per company, the minimum is five persons on duty per company. Where a company member is sick or injured while another member is on vacation and no on-duty fire fighter is available to cover the absence, it is the department's policy to employ an off-duty member of the company on an overtime basis to maintain the essential minimum strength.

Decisions by labor boards and at least one court have found that minimum staffing agreements or ordinances are reasonable requirements for the protection of the public and personnel. A number of small fire departments that do not attempt to maintain minimum on-duty company strength have established a minimum for the duty shift while employing off-duty personnel to maintain the predetermined minimum effective strength. Such a plan should account for apparatus that must be operated from the several fire stations before off-shift or call fire fighters can arrive to assist.

Managers often find it more economical to use personnel on overtime to cover duty absences that exceed the average allowed in the organization staffing tables rather than to maintain additional personnel on each fire company duty shift to cover abnormal amounts of absence. In the past, however, there was no allowance in the staffing tables of many fire departments to cover scheduled absences, and fire companies were allowed to run shorthanded, thus seriously compromising their operating efficiency and the safety of fire fighters. In a number of cases, departments also failed to allow for the usual amount of sickness and injury. Calculations for the number of personnel needed should also include members on terminal leave and new recruits assigned to basic training. Otherwise, overtime costs may exceed the cost of having the needed number of members per shift.

Work schedules: The work weeks of career fire fighters average 40 to 56 hr. Most fire departments working an average of more than 50 hr pr week use a 24-hr tour of duty. Most fire departments working 48 hr per week or less have day and night shifts, and the most popular is a 10-hr day shift and 14-hr night shift. Where the law states that anyone working more than 40 hr per week will be paid overtime, many fire departments work a 42-hr four-platoon schedule by paying 2 hr of overtime. Often, this is considerably less expensive than hiring additional personnel, and better teamwork is maintained by keeping crews together on a regular four-shift basis.

Occasionally, municipal administrators who are not knowledgeable about fire department operations have suggested that fire fighters be assigned to an 8-hr-day/40-hr week work schedule similar to police schedules. This has not proved to be practical or desirable. On-duty police staffing properly varies with the time of day and day of the week as required by needs for traffic control, patrolling, details, and so on. Fire fighting is a team effort, and the team includes platoon chief officers, company officers, apparatus operators, and fire fighters working together on a regular basis. Serious fires can occur at any hour of the day or night and on any day of the week. Thus, the constant uniform staffing provided by the three- and four-platoon systems is essential. These are readily scheduled in seven 24-hr or fourteen 10- and 14-hr tours of duty per week on either the 42-hr or 56-hr average work week. This may include a 40-hr pay week or a 54-hr or 56-hr pay week with 2 hr of overtime. When the 168-hr calendar week is divided by 8-hr tours, this requires 21 work shifts that cannot be scheduled on a uniform basis with even platoons. In most instances, where the 8-hr schedule has been proposed, it has been rejected. In the entire United States, there are only a few fire departments using an 8-hr work shift.

Given a choice between 24-hr shifts and increasing staffing by 40 percent or reducing staffing or resources by 40 percent to accommodate 8-hr work shifts, most municipal fire departments have not seriously considered 8-hr work shifts. As the fire service becomes more adept at computerized analysis of emergency incidents and develops a good base of 5 to 10 yr of response data, it is possible

that a staffing arrangement similar to police staffing may be used in the future.

The requirement for overtime has resulted in improved mutual-aid arrangements because on-duty companies from nearby departments can respond much faster than calling back off-shift local personnel, thus keeping overtime costs down. However, many small fire departments rely heavily upon off-duty response on an overtime basis. Usually a minimum of 2- and 4-hr overtime pay is guaranteed for each response. Where alarms for structural fires are infrequent, it may be much more economical for a small municipality to contract for overtime response than to provide full on-duty fire company staffing around the clock. One major drawback may be increased response times that will adversely affect fire losses.

Staffing Combination and Volunteer Fire Departments

Part of the population of both Canada and the United States lives in communities that cannot support a career or even a combination career/on-call fire department. Therefore, volunteer fire departments are essential. In 1995, 17 percent of all the fire departments in the United States served communities with populations of 10,000 or more; 30 percent protected communities with populations from 2,500 to 10,000; 53 percent served communities with populations of less than 2,500, or served the fire protection needs of rural areas.

Three-fourths (74 percent) of 1995 fire departments consisted entirely of volunteers. Even if a town of 2,500 inhabitants had as many as three career fire fighters per 1,000 population, which is considerably above the big-city average, there would be only two members on each duty shift, so a full paid fire department would not be feasible unless a substantial tax base existed to support it.

In most small communities, volunteer fire departments are operated on a "neighbor helping neighbor" basis in which fire department members—except possibly fire station custodians and sometimes a few career fire fighters on duty—receive no compensation other than possibly some reimbursement for personal expenses and uniforms. To replace this free community labor with minimum staffing by fully salaried personnel may involve an added tax burden on the local population. Various states have recognized the contribution made by their volunteer fire fighters by enacting protective legislation and providing statewide training programs, facilities, and retirement systems.

When they receive an alarm, members of many volunteer fire departments report to assigned fire stations from which they respond with apparatus while others allow members to respond directly to the incident. To provide a minimum effective working crew, many such fire departments require that the first piece of apparatus not respond with fewer than three members. Volunteer fire companies should respond to an alarm with a minimum of four members.

Fire department administrators should periodically review response records to determine that enough active fire company members are available to respond in a timely fashion at all times. Where necessary, administrators should recruit and train additional personnel to provide the required minimum response. All essential staff positions in a well-organized all-volunteer fire department are covered by assigned volunteer officers. In many jurisdictions, however, career fire prevention, training, and communication functions are provided by the county or by other units of the government.

Where a community can afford on-duty career fire fighters, response to alarms is faster, and efficiency is increased. The on-duty fire fighters take the apparatus to the fire, and the volunteers notified by radio go directly to the fire. The apparatus operator normally is in charge of the apparatus and of the fire station, but volunteer officers may direct the fire fighting. One difficulty with this arrangement is that there are few, if any, opportunities for advancement for career personnel.

With the increasing legal and technical responsibilities of the fire chief as the principal fire protection officer of a community, the chief should be appointed on the basis of qualifications and experience. Communities should consider adopting NFPA standards for fire officers and fire fighters as a minimum requirement.

An arrangement that works successfully is to have the first responding pumper staffed by a career officer, apparatus operator, and, where staffing permits, an additional fire fighter. This force is supplemented by additional personnel assigned to respond on call. The second-due engine may be staffed by volunteers or paid call personnel, or it may have a career apparatus operator to take the apparatus to the fire, where it is joined by the volunteer or paid call fire fighters. The ladder truck has a career apparatus operator assigned, but is staffed at the fire by volunteer or paid call fire fighters. This arrangement permits a reasonably effective initial fire attack, quickly backed by volunteer or paid call members. In all cases, there should be just one fire department in any jurisdiction, operating under a clearly defined and unified chain of command.

In combination fire departments, career fire fighters may complain about being commanded by volunteer officers whom they feel lack the needed experience and qualifications. All fire officers, whether elected or appointed, should meet the appropriate NFPA qualifications for their rank. Countless volunteer officers meet the technical standards for their duties. When a career apparatus operator is assigned to operate volunteer fire company apparatus, he or she should work under the orders of the volunteer officer of that company on the fireground. It is important that administrators make clear the respective roles and duties of all members, career and volunteer.

Call Fire Fighters

Many fire departments in small communities employ fire fighters who have no regular duty shift in the stations, but are paid by the hour or by the incident for response to alarms and drills. In some cases, such members are loosely termed "volunteers," but, under federal labor and local fire department rules, they are considered paid employees of the fire department and, as such, are subject to the requirements of federal wage and hour regulations. The paid on-call members may also be employed by other municipal agencies, in which case the time they spend on fire department duty may affect their overtime status. Call fire fighters are expected to meet the same standards of performance as career members of the same rank, but they may not be assigned as apparatus operators where there are sufficient career operators on shift duty.

Various methods are used to determine compensation for paid call fire fighters. In many departments, they receive the same hourly wage for the rank they hold as do employees who work regular duty shifts. Time is based upon attendance at fires and training sessions, with a minimum hourly rate specified for response to alarms. Often, upon reaching mandatory retirement age, they also receive a prorated pension, based on their years of service and hours of duty as call fire fighters.

Another method of compensation is a fixed annual salary based on rank, from which deductions are made for excessive numbers of unexcused failures to respond with the assigned companies. Many fire chiefs arrange to excuse members known to be at their regular employment, except in the case of multiple-alarm fires. Chiefs should have the authority to dismiss members who frequently fail to respond to fires and assigned training sessions.

Still another method of compensation is to make an annual appropriation for call fire service, based upon past experience, designed to approximate the hourly wage rate for fire fighters. This is divided on a regular basis among the call members as determined by individual attendance at fires and training sessions, so that members responding most faithfully receive the largest compensation. A fire department should pay for members' insurance, workers' compensation, and all protective equipment.

It is important that accurate individual service records be kept in the personnel files of all fire fighters, both career and volunteer. Response and attendance records are also essential for all members. Members who are habitually late in arriving should be replaced. It is recommended that volunteer members be furnished with night turnout suits so that they will lose no time in getting dressed to respond.

In a number of cases, senior fire fighters who cannot respond regularly to first alarms are assigned to operate reserve apparatus or to pilot mutual-aid fire companies when serious fires occur. Other senior call members may be assigned to cover the alarm desk when all of the paid apparatus operators are out of quarters.

Keeping accurate response records for all fire fighters may point to staffing deficiencies that require fire chiefs to hire additional career personnel. Municipal officials may believe that, because they have the names of several hundred volunteers on the roster, the fire department has ample strength. This can result in serious delays and in the extension of fires that should have been readily controlled. Where such situations persist, additional career personnel may be required.

PROCUREMENT OF EQUIPMENT AND SUPPLIES

In most municipalities, fire department equipment and supplies are procured through purchasing departments. Items that are common to all departments may be requisitioned from the purchasing agency and charged to the appropriate fire department account. Where items are of a specialized nature, such as fire apparatus or fire-fighting tools and equipment, purchasing specifications must be prepared by the fire department, approved by the purchasing department, and advertised for bids. In preparing specifications for such items as fire apparatus and fire hose, current NFPA standards should be followed.

The fire chief, with advice from the apparatus equipment supervisor, should determine whether proposals submitted by bidders meet specifications. In larger jurisdictions, the law requires that a contract be awarded to the lowest responsible bidder. However, bidders may take exception to various details in the specifications or offer substitutes. This requires that a fire chief determine whether such proposals meet the intentions of the specifications. If they do not, the bids should be rejected. But if the proposals conform to the specifications and are within the appropriation, the contract should be awarded. Upon delivery, new equipment should be tested in accordance with the provisions of appropriate NFPA standards.

When emergency purchases must be made, it is customary to require bids from several suppliers. If the amount involved is small and funds are available in the appropriate budget item, the fire chief can authorize the expenditure. If funds are not available in the fire department budget, authorization and funds must be obtained from the municipal management or finance officer.

INTERGOVERNMENTAL RELATIONS

Fire departments are but one agency of local government, and much of their success depends upon their cooperative working relationships with other local, state, and federal agencies. Some of the more important intergovernmental contacts are discussed below.

Building Department

The proper construction and arrangement of buildings is essential to a sound fire protection program. The building department of a community is a key component in ensuring quality control in building construction and compliance with the fire protection features of local building codes.

State laws and local ordinances or agreements between fire and building departments increasingly require that the head of the fire department give written approval of specified fire protection features before building and occupancy permits can be issued. Close cooperation also is needed between these departments to control serious fire hazards that commonly are present while buildings are under construction, before the required fire-resistance or protection features have been installed. In small communities, the fire chief generally must handle this assignment, but in many fire departments, it is one of the responsibilities delegated to the fire prevention bureau.

Law Enforcement

Cooperation between the fire department and law enforcement officials is essential. Regular law enforcement response to fire alarms is necessary to control traffic and crowds. It is also important for fire and enforcement agencies to develop coordinated plans in the event of an incident that requires the evacuation or closure of areas. Law enforcement and fire officials must also develop working relationships to deal with fire investigations. In many areas, a combination of fire and law enforcement fire investigation teams have been developed. Many other fire departments determine the cause, origin, and accomplish the complete investigation with fire investigators that have received the appropriate law enforcement training to obtain "police powers." In these cases, both agencies work together and bring their special expertise to a coordinated effort in cases where arson is suspected.

Water Department

Adequate water supplies, including hydrant service, are essential for fire fighting and are the responsibility of the water department. A knowledgeable fire officer should be assigned as liaison with the water authority. All too often, water authorities have little knowledge of the water-flow requirements of the fire department for various areas and types of property. Fire hydrants may be set up improperly because water crews do not understand the proper location and setting of hydrants required for efficient fire fighting. In some communities, the fire department is responsible, by ordinance, for determining the location and setting of hydrants.

It is important that all hydrants be serviced regularly and after use, especially in cold weather. The fire department should report all hydrants it has used in a particular incident to the water department promptly. Alert fire departments maintain a list of hydrants in each fire company inspection district with flow data on each hydrant. Hydrants should be properly marked for flows and painted for nighttime visibility.

Public Education

In the realm of fire safety education, the local public and private school systems, the parks and recreation department, and the public library system can augment and support the local fire department's educational activities. If public fire safety educational activities are to be successful, the local fire department must be able to serve as a focal point for other community resources and be able to build partnerships across a wide variety of public and private organizations. NFPA's *Learn Not to Burn* program is an excellent example of a fire safety education program with proven results.

Personnel Department

Members of career fire departments are public employees, and, as such, their recruitment and promotion may involve cooperation with the personnel agency, which, in some jurisdictions, is responsible for conducting entrance and promotional examinations. A fire department officer should be assigned as liaison with the personnel office.

Finance Department

Fire departments need to work closely with finance officials when developing and administering budgets. Budgeting is generally very complex, and it is important that fire officials prepare budget documents correctly. After budgets have been prepared and adopted, it is necessary to ensure that proper records of expenditures are maintained and that spending is kept within the adopted budget. Although some fire departments may have a finance officer, liaison with the finance department may be handled by designated staff officers.

Purchasing

All purchases exceeding stipulated amounts must be made according to specifications, usually with competitive bidding by the purchasing department. Close liaison is necessary to ensure that specifications are drawn properly and meet fire department needs and that, when bids are opened, any proposals that deviate from specifications are rejected.

Data Processing

Increasingly, fire departments are using electronic data processing for keeping fire records and payroll records for statistical analysis. Each fire department should have persons knowledgeable in data processing. It may be desirable to appoint one officer to coordinate this activity, although fire departments commonly have an administrative committee that includes representatives of plans and research, where provided; administration; fire prevention; and fire suppression. This same committee may also be involved in long-range planning for the department.

Planning

Fire departments, particularly those in rapidly growing areas, should maintain a close working relationship with local and regional planning groups. These agencies can provide valuable information on growth patterns that will affect the resources of the department. The plans, studies, and reports prepared by the planning agencies can be used to determine whether more personnel, equipment, or station facilities are needed.

BIBLIOGRAPHY

Reference Cited

1. Hendrix, Merrell, C., "Manpower Analysis 1969—Dallas, Texas, Fire Department", International Association of Fire Chiefs, Bulletin No. FSTB-401, Washington, D.C., 1969

NFPA Codes, Standards, and Recommended Practices

Reference to the following NFPA codes, standards, and recommended practices will provide further information about fire department administration and operations. (See the latest version of *The NFPA Catalog* for availability of current editions of the following documents.)

NFPA 1, *Fire Prevention Code*

NFPA 13E, *Guide for Fire Department Operations in Properties Protected by Sprinkler and Standpipe Systems*

NFPA 291, *Recommended Practice for Fire Flow Testing and Marking of Hydrants*

NFPA 295, *Standard for Wildfire Control*

NFPA 298, *Standard on Fire Fighting Foam Chemicals for Class A Fuels in Rural, Suburban, and Vegetated Areas*

NFPA 299, *Standard for Protection of Life and Property from Wildfire*

NFPA 402M, *Guide for Aircraft Rescue and Fire Fighting Operations*

NFPA 403, *Standard for Aircraft Rescue and Fire Fighting Services at Airports*

NFPA 471, *Recommended Practice for Responding to Hazardous Materials Incidents*

NFPA 472, *Standard for Professional Competence of Responders to Hazardous Materials Incidents*

NFPA 704, *Standard System for the Identification of the Fire Hazards of Materials*

NFPA 901, *Standard Classifications for Incident Reporting and Fire Protection Data*

NFPA 902M, *Fire Reporting Field Incident Manual*

NFPA 903, *Fire Reporting Property Survey Guide*

NFPA 921, *Guide for Fire and Explosion Investigations*

NFPA 1000, *Standard for Fire Service Professional Qualifications Accreditation and Certification Systems*

NFPA 1001, *Standard for Fire Fighter Professional Qualifications*

NFPA 1002, *Standard for Fire Department Vehicle Driver/Operator Professional Qualifications*

NFPA 1003, *Standard for Airport Fire Fighter Professional Qualifications*

NFPA 1004, *Standard on Fire Fighter Medical Technicians Professional Qualifications*

NFPA 1021, *Standard for Fire Officer Professional Qualifications*

NFPA 1031, *Standard for Professional Qualifications for Fire Inspector*

NFPA 1033, *Standard for Professional Qualifications for Fire Investigator*

NFPA 1035, *Standard for Professional Qualifications for Public Fire and Life Safety Educator*

NFPA 1041, *Standard for Fire Service Instructor Professional Qualifications*

NFPA 1051, *Standard for Wildland Fire Fighter Professional Qualifications*

NFPA 1061, *Standard for Professional Qualifications for Public Safety Telecommunicator*

NFPA 1201, *Standard for Developing Fire Protection Services for the Public*

NFPA 1202, *Recommendations for Organization of a Fire Department*

NFPA 1221, *Standard for the Installation, Maintenance, and Use of Public Fire Service Communication Systems*

NFPA 1231, *Standard on Water Supplies for Suburban and Rural Fire Fighting*

NFPA 1401, *Recommended Practice for Fire Service Training Reports and Records*

NFPA 1403, *Standard on Live Fire Training Evolutions in Structures*

NFPA 1404, *Standard for a Fire Department Self-Contained Breathing Apparatus Program*

NFPA 1405, *Guide for Land-Based Fire Fighters Who Respond to Marine Vessel Fires*

NFPA 1410, *Standard on Training for Initial Fire Attack*

NFPA 1420, *Recommended Practice for Pre-Incident Planning for Warehouse Operations*

NFPA 1451, *Standard for a Fire Service Vehicle Operations Training Program*

NFPA 1500, *Standard on Fire Department Occupational Safety and Health Program*

NFPA 1521, *Standard for Fire Department Safety Officer*

NFPA 1561, *Standard on Fire Department Incident Management System*

NFPA 1932, *Standard on Use, Maintenance, and Service Testing of Fire Department Ground Ladders*

NFPA 1962, *Standard for the Care, Use, and Service Testing of Fire Hose, Including Couplings and Nozzles*

NFPA 1964, *Standard for Spray Nozzles (Shutoff and Tip)*

NFPA 1971, *Standard on Protective Clothing for Structural Fire Fighting*

NFPA 1972, *Standard on Helmets for Structural Fire Fighting*

NFPA 1973, *Standard on Gloves for Structural Fire Fighting*

NFPA 1974, *Standard on Protective Footwear for Structural Fire Fighting*

NFPA 1981, *Standard on Open-Circuit Self-Contained Breathing Apparatus for Fire Fighters*

NFPA 1982, *Standard on Personal Alert Safety Systems (PASS) for Fire Fighters*

NFPA 1983, *Standard on Fire Service Life Safety Rope and System Components*

NFPA 1991, *Standard on Vapor-Protective Suits for Hazardous Chemical Emergencies*

NFPA 1992, *Standard on Liquid Splash-Protective Suits for Hazardous Chemical Emergencies*

NFPA 1993, *Standard on Support Function Protective Clothing for Hazardous Chemical Operations*

Additional Reading

Beresford, W., "Evolution of the Fire Department," *American Fire Journal*, Vol. 46, No. 11, Nov. 1994, pp. 16–17.

Brannigan, F. L., *Building Construction for the Fire Service*, 3rd ed., National Fire Protection Association, Quincy, MA, 1992.

Callahan, T., *Fire Service and the Law*, 2nd ed., National Fire Protection Association, Quincy, MA, 1988.

Carter, H. R., "Combination Fire Department: A 21st Century Solution," *Firehouse*, Vol. 18, No. 3, March 1993, pp. 60–62, 64.

Carter, H. R., "Tailoring the Fire Department to the Community: An Introduction," *Firehouse*, Vol. 21, No. 5, May 1996, pp. 16, 18, 20.

Carter, H. R., Rausch, E., *Management in the Fire Service*, 2nd ed., National Fire Protection Association, Quincy, MA, 1989.

Davis, P. O., and Daston, C. O., "Physiological Aspects of Fire Fighting," *Fire Technology*, Vol. 23, No. 4, 1987, pp. 280–291.

Federal Emergency Management Agency, "Guide to Funding Alternatives for Fire and Emergency Medical Services Departments," Federal Emergency Management Agency, Washington, DC, FA-141, Dec. 1993.

Fire Protection Guide on Hazardous Materials, 11th ed., National Fire Protection Association, Quincy, MA, 1994.

Granito, J. A., "Expanding Fire Department Public Service Programs," *NFPA Journal*, Vol. 89, No. 1, Jan./Feb. 1995, pp. 70–74.

Grant, N. K., "Achieving Effective Use of Technology in Fire Departments," *Journal of Applied Fire Science*, Vol. 4, No. 2, 1994/1995, pp. 95–103.

Karter, M. J., Jr., "NFPA Surveys U.S. Fire Departments," *NFPA Journal*, Vol. 87, No. 4, July/Aug. 1993, pp. 59–62.

Lavelle, L. G., "Role of Fire Departments in the Application of Technology," International Symposium on Fire Protection for the Telecommunications Industry, May 14–15, 1992, New Orleans, LA, 1992, pp. 1–7.

Murry, L. S., "Training for the Year 2000," *Fire Command*, Vol. 56, No. 3, March 1989, pp. 16–18.

Nielsen, R. D., "Fire Departments and Communities: Partners in Prevention," *Fire Engineering*, Vol. 148, No. 6, June 1995, pp. 101–102.

Paulsgrove, R. F., "Recruiting and Retaining Fire Department FPEs," *NFPA Journal*, Vol. 87, No. 1, Jan./Feb. 1993, pp. 58–63.

Richman, H., *Engine Company Fireground Operations*, 2nd ed., National Fire Protection Association, Quincy, MA, 1986.

Robertson, J. C., "Preparing Fire Service Personnel for Fire Prevention Duties," *Introduction to Fire Prevention*, 3rd ed., Macmillan, New York, 1989, pp. 99–111.

Walsh, M. B., and Wojcik, K., "What Can the Public Fire Service Learn From Private Departments?" *Fire Chief*, Vol. 34, No. 1, Jan. 1990, pp. 37–40, 42, 44.

Wessel, H. E., "TQM, Benchmarking and the Fire Department," *American Fire Journal*, Vol. 45, No. 4, May 1993, pp. 18–20.

The following additional readings are published by the International Fire Service Training Association, Oklahoma State University:

Essentials of Fire Fighting

Forcible Entry

Ladders
Public Fire Education
Hose
Salvage and Overhaul Practices
Fire Stream Practices
Fire Ventilation Practices
Fire Service Rescue
Fire Service First Responder—Medical
Fire Prevention and Inspection Practices
Fire Service Practices for Volunteer Fire Departments
Fire Service Orientation and Indoctrination
Fire Service Training Programs
Photography for the Fire Service

Water Supplies for Fire Protection
Firefighter Safety
Aircraft Fire Protection and Rescue Procedures
Ground Cover Fire Fighting Practices
The Fire Department Officer
Fire Department Facilities, Planning, and Procedures
Fire Service Instructor Training
Leadership in the Fire Service
IFSTA Source Material Reference Guide

EVALUATION AND PLANNING OF PUBLIC FIRE PROTECTION

Revised by John Granito

The purposes of this chapter are to demonstrate that adequate fire protection and related public safety services are essential components of any community and that both the level and types of services provided should be conscious and carefully considered choices of the community. Further, objective evaluation and planning can ensure that a community's desire for a stipulated level of protection is being met in as cost-effective a manner as possible. Several other chapters provide information useful in the self-assessment of a public fire protection system. In addition, certain NFPA publications, including a number of codes, standards, and recommended practices, are applicable. Many of these publications are listed in the bibliography at the end of this chapter.

Public fire protection needs to be carefully planned and requires that certain logical steps be taken to achieve a comprehensive, acceptable, and workable plan. There are two important aspects to any good plan: (1) the plan itself, which must be feasible and directed toward clear goals; and (2) the process by which the plan is developed, which must ensure that all major goals are considered and every constituency to be affected by the plan is reasonably involved in the planning process.

Without adequate involvement of the necessary constituencies, implementation of the plan may likely fail because of a lack of cooperation and commitment. For a satisfactory plan to evolve, the planners must decide the end results they wish to achieve (goals), determine the status of the fire protection system in relation to those goals (evaluation), and calculate how much and what kinds of progress actually can take place over a certain period of time (objectives, tactics, time frame). Each of these three steps—setting goals, evaluation, working out the details—requires the collection and analysis of relevant information. Broad goals are achieved through planned strategies and precise objectives are achieved through implemented tactics. Each must be relevant to the other.

Careful consideration of these important factors, which will vary considerably from one community to another and which must take into account assets, liabilities, challenges, and available resources lends to what often is referred to as a "strategic plan."

Because some degree of public fire protection is almost always in place, it is common for the entire process to begin with an evaluation of the fire protection that is already available. The information obtained from the evaluation, when analyzed in terms of broad, generally recognized public fire protection goals, identifies needs and provides responsible community officials with the approximate parameters of the plan to be developed.

It is important, then, to take into consideration those fire protection and other public safety services that may be provided by a fire/rescue department and that will meet specific local needs. One example of this is hospital transportation by a fire department ambulance.

Typical fire department services include fire prevention programs—such as code administration and enforcement through plans review, inspection services, and public education—as well as fire suppression, rescue, emergency medical, hazardous materials response, and disaster response. Various other public services, such as health screening programs, are conducted more and more often by progressive departments. This approach to broad-based community safety service takes advantage of departmental resources, such as neighborhood fire stations and on-duty personnel.

As already noted, the plan cannot be developed without the involvement of a wide variety of community groups. Even if the various constituencies seem willing to allow fire protection officials to develop and write the comprehensive plan without their consultation, the fire officials should be cautious; the citizens eventually must be willing to accommodate and pay for the implementation. Since a comprehensive plan envisions a larger group or system of integrated parts, a number of organizations and agencies outside the fire department will need to play important roles in implementation of the plan. Local elected officials typically play key roles in plan approval and implementation.

Modern practices make it necessary for fire departments to work cooperatively with a variety of agencies and organizations. These range, for example, from the local building code enforcement staff to the U.S. Coast Guard for hazardous materials spills near navigable waterways. Good planning requires consultation with all operational components on a continuing basis.

One basic aspect of a comprehensive public fire protection plan is the concept that it is infinitely better for a community to prevent fires altogether, or to mitigate them automatically through fire safety education and built-in fire protection features, than to depend solely on the fire suppression capabilities of the community's fire department. The goal of reducing the incidence and effects of fire, then, involves all aspects of fire prevention. Historically, much more energy and many more resources have been devoted to evaluating, planning, and implementing fire suppression/fire-fighting capabilities than fire

---

John Granito, Ed.D., is professor emeritus and vice president for Public Service and External Affairs, State University of New York, Binghamton, NY.

prevention capabilities. Simply stated, the United States as a whole has focused more on "fire engines and fire fighters" than on public awareness and built-in mitigation features. That focus is still very necessary and exceedingly important, but effort must be placed also on measures that do not depend solely on fire suppression personnel to reduce the toll of fire. Planning groups have difficulty, however, in evaluating the degree of effectiveness of such multifaceted fire protection programs.

A reasonably effective method exists for conducting the evaluation, involving two kinds of analysis. The first kind of analysis requires the community to maintain consistent and carefully recorded information each year concerning the number of fires that occur and the cost of those fires in lives and dollars including wages and tax revenues lost. When those human and physical costs are added to the expense of maintaining fire prevention and suppression systems, plus the cost of fire insurance, a standardized total cost of fire to the community can be compared with that same cost in earlier years, or over an average of three or more years. Necessary adjustments for inflation, community growth or decline, and other important variables can be made by local officials, thus permitting a community to compare its present to its prior fire performance.

The second kind of analysis most useful in determining cost-effectiveness involves communities identifying other communities that are similar in ways important to fire protection (such as services desired, size, construction, hazards, and geography) and comparing their own total performance to the total performance of the similar communities. This process is termed "benchmarking" and has become a popular methodology. However, the number of site-specific variables may be so high and the span of differences so great that comparisons prove confusing. Benchmarking requires cautious application.

In the first analysis, the community uses an internal data source (itself) as a yardstick, and in the second analysis it uses an external yardstick (other communities). As more data are collected concerning the total cost of fires in various communities operating with various public fire protection plans, any given community will be able to benefit relatively quickly not only from its own experience, but also from the experiences of others.

Important to this planning process is the ability and willingness of the community to finance and otherwise support the total level of fire protection required by the plan's goals. Examples include legislation dealing with the retrofitting of sprinkler systems, or innovative thinking that produces a core group of senior citizen volunteers who conduct fire safety programs for their peers. In some states built-in fire protection requirements are set at the state level, and local government may not increase the level of fire safety provided by building features (automatic fire sprinklers, non-combustible roofing, etc.).

While fire protection officials must always be concerned with reducing the total cost of fire (fire loss, plus costs of insurance, prevention, and suppression), citizens living in tight economic times will ultimately reserve the right to make decisions, or "trade-offs," concerning the level of protection they wish their tax dollars to purchase.

To assist citizens in making decisions concerning budgets, fire protection officials must accurately describe the effect on total cost if additional or fewer resources are applied to particular prevention or suppression efforts. This kind of technical knowledge and analytical ability of officials provides a crucial element to comprehensive planning and evaluation. It is one of the most important responsibilities of the public fire protection officer.

One of the most serious issues faced by public officials and residents is that of adequate fire department staffing. In communities served by volunteer fire fighters, there may be very serious volunteer recruitment and retention problems, or problems in providing timely response during working hours when many of the volunteers may not be in the community they protect. In communities served by career, full-time fire fighters, restricted budgets often limit the number of fire-fighting vehicles ready to respond, or the crew size for each vehicle and, thus, the total number of trained personnel responding to the incident in time to mount an effective offense against the fire. Fire protection officers and municipal officials need to communicate objectively the level of service being provided and how changes in fire department resources will affect that level of service. That is, the type and level of risk should be expressed in terms understandable to the fire protection nonexpert and neither over- nor understated. Of course, the fire department must bear the responsibility of formulating and operating a cost-effective organization, no matter what level of service is desired.

EVALUATION

In addition to assessing the capabilities of the fire department and related aspects of fire protection, fire officials evaluating the capability of the existing fire protection system must take a number of factors into account. Examples include known combustibles, the life hazard that exists, fire frequency, climatic conditions, demographic and geographic factors, and a basic consideration of the specific role of a public fire department in providing fire protection to the community. Failure to consider each of these factors adequately can lead to a large-loss fire. A fire department's suppression capabilities can never be expected to compensate totally for the deficiency or lack of built-in fire protection systems.

The evaluation of local fire suppression capability, while no more important than the evaluation of local prevention and public safety education efforts, is a necessary effort for two primary reasons. First, most communities have little understanding of either the level of local protection available, or the level necessary for reasonable community well-being. Second, the time availability of volunteers is quite limited, and the cost of full-time career personnel typically is significant. The actual number of available crew members—while exceedingly important—may not be realized by non-fire officials and citizens. This lack of information concerning actual resources can lead to a false sense of security.

Two concepts are useful in local suppression considerations. First is the "capability" of the fire department to respond within a short time with sufficient trained personnel and equipment to rescue any trapped occupants and confine the fire to the room or building of origin. How many crew members and vehicles of various types are necessary is dependent upon the type of call and the conditions of the local community which either aid or hinder fire fighting. These variables, which certainly number more than twenty, range from water supply and sprinkler ordinances to weather conditions and the age of buildings.

The second concept is that of "capacity," which is the ability of the fire department to respond adequately to multiple-alarm incidents and/or simultaneous calls of any type. If alarm patterns are examined, the volume of multiple alarms and simultaneous response demands over a period of time can be approximated. Larger municipalities typically average more demand for capacity and, thus, typically have larger departments; but, obviously, remaining capacity is diminished as suppression units are deployed, even in the largest departments.

In evaluating both local response capability and capacity, officials need to consider the following: In most areas it is relatively easy to increase capacity (through the use of mutual aid and group or entire shift callbacks), but it is much more difficult to improve capability, which requires *immediate* response of nearby forces. A common rule of thumb is that a community using on-duty crews at fire stations should be able to have an initial attack team comprising an entire first-alarm response on the scene within approximately 8 min of receipt of the alarm. This equates to about 6 min of running time.

Another rule of thumb is that, for a totally volunteer, non-staffed station, the average time from when a call is received to the moment a crewed vehicle leaves the station is about 6 min. In those situations, then, the "call to on-scene time" for a 6-min run is about 12 min for one or more vehicles.

Several techniques are used by progressive fire officials to help mitigate these handicaps to achieving adequate response capability and capacity. These include using *automatic*, immediate mutual-aid response from nearby departments, rapid callback systems, volunteer personnel who "bunk in" at stations, part-time on-duty crew members, special shift arrangements, elaborate regional mutual-aid agreements, and other methods. "Doing more with less" is a popular concept, but there are limits that must be realized.

As part of a local evaluation, fire officers and other community officials can determine the status of their response forces by establishing what capabilities and capacities are necessary for adequate protection of their area and then determining what is available.

Following a local hazard analysis, for example, the following types of "what should be" questions concerning *what is needed for adequate response capability* may be posed:

- Maximum time for call processing at dispatch center?
- Minimum number of persons on first-out pumper?
- Minimum number of persons on first-out aerial?
- Minimum number of persons on first-out heavy rescue?
- Minimum number of persons on first-out emergency medical service (EMS) vehicle?
- Weekday get-out time for first-due unit?
- Weekday get-out time for second-due unit?
- Nighttime get-out time for first-due unit?
- Nighttime get-out time for second-due unit?
- Maximum clear weather running time (pumper) in district?
- Maximum clear weather running time (aerial) in district?
- Maximum clear weather running time (EMS) in district?
- Number of qualified fire fighters plus incident commander to be at scene within 10 min of dispatch?
- Number of qualified fire fighters plus incident commander to be at scene within 15 min?
- Number of minutes for mutual aid to arrive at scene?

Figure 10-2A illustrates one method for listing the actual response capability of a department for a simple, single-family-dwelling fire. Figure 10-2A can be expanded to include first-alarm responses to other types of occupancies. Response times can be verified and listed for the components of the first-alarm assignment, and other questions can be posed to identify local response capacity to handle multiple and/or simultaneous alarms.

An important stipulation put into effect in 1995, and continuing beyond that year, is that a minimum of four trained personnel be at the scene of a structure fire before an interior attack can be launched, except under the most extenuating circumstances. Both the U.S. Occupational Safety and Health Administration (OSHA) and NFPA have issued documents dealing with this, the latter in the form of a temporary interim amendment (TIA) to NFPA 1500, *Standard on Fire Department Occupational Safety and Health Program.* Also in the United States, and in certain other countries, various

# of Resources dispatched	Confirmed working single-family-dwelling fire
Chief(s)	
Aides or incident command technicians	
Company officers	
Fire fighters, including "flying squads"	
Single-purpose emergency medical technicians (EMTs)	
Standard pumpers	
Quints	
Aerials	
Ladder tenders	
Ambulances	
Heavy (technical/urban) rescues	
Light/medium rescues	
Mini/midi attack pumpers	
Special operations (air, lighting, etc.) units	
"Flying squad" vehicles	
Mobile command vehicles	
Other (types)	

FIG. 10-2A. Typical, actual, first-alarm fire attack assignment.

governmental agencies have issued regulations and advisories concerning the necessity for certain emergency operations to have present trained incident commanders and standby safety/rescue crews, as well as to have a defined incident command system operating. Local officials are well advised to have current information concerning all applicable federal (national), state, and provincial requirements, as well as applicable NFPA codes, standards, and recommended practices, as issued from time to time.

Rural Fire Protection

One principal difference between operation of rural and urban fire departments is that rural departments must deal with water supply issues with a broader variety of solutions than most urban departments. Rural fire department operations and apparatus emphasize not only fire-fighting requirements, but also the provision of water for fire fighting. Rural fire apparatus must have large water tanks to permit effective initial attack on fires while supplementary water supplies are being brought into action. Supplementary water supplies include drafting sources on or adjacent to rural properties, and mobile water tanker vehicles for transporting water from more distant sources. Also in use are portable folding canvas tanks into which tank trucks quickly discharge their water supply through special dump valves. Rural fire departments often use apparatus and hose to relay water from sources several thousand feet (1000 ft equals 304.8 m) from the emergency. Initial response of pumpers, tankers, and auxiliary apparatus should be adequate for a quick attack on the burning property. With adequate highways and well-designed apparatus, it is often possible to bring substantial firefighting forces to an emergency in rural areas in sufficient time for a properly planned and executed initial fire attack operation to be effective, even though many of the personnel may arrive in private automobiles. Many rural properties are now located in areas that enjoy some level of fire protection. Some properties, of course, may have to depend entirely upon their own private fire protection and whatever help they may obtain from forestry agencies or distant fire departments.

Newer approaches to fire insurance rating give "protected" status to property without a municipal water supply system if a fire station is within five travel miles and a stipulated gal per minute flow can be maintained by the local department.

Minimum protection for a rural area would include a pumper with a large water tank and a water tank vehicle responding on an initial alarm. Properly designed tanks should be able to transport water from a source 1 mi (1.6 km) from the scene so a minimum of 250 gpm (946.25 L/min) can be pumped at the fire scene by the pumper. Since a larger flow often is required to provide adequate fire protection services, additional tankers must be used, or drafting sources within reasonable distance of the fire scene must be identified. Programs that encourage the construction of year-round rural drafting sites are to be encouraged. Rural apparatus should carry $3^1/_2$-in. (89-mm) or larger supply hose to provide adequate water supply at the fire scene. It is always advisable to lay large-diameter fire hose from the water supply source to as near the fire scene as possible in order to avoid extensive friction loss. At the emergency, large-diameter hose is sometimes connected into smaller handlines, or used more often to supply another pumper from which handlines are extended. Other pieces of equipment, such as rescue and aerial ladder vehicles, should be provided as needed to carry out the mission. Elevated master streams are not needed extensively in rural operations, and sometimes ladder truck equipment for rescue, forcible entry, ventilation, and salvage operations is carried on pumpers and equipment vehicles.

To be even minimally effective in controlling a fire, the initial responding apparatus should reach the emergency scene before very rapid fire spread. As is the case in urban fire fighting, this is termed "initial attack" and is aimed at stopping the fire as close to the point of origin as possible. So-called "sustained attack" attempts to reduce the loss to the exposed adjoining or nearby property. Because of longer response times, rural departments may find themselves in the sustained attack, or "defensive," mode upon arrival. Unless sufficient water can be made available within a short time frame, the British thermal units (Btu) generated by the burning material cannot be absorbed so the temperature is not reduced sufficiently to extinguish the fire. The keys to successful rural protection planning usually reduce to response times and water availability.

Of special concern are the sometimes extensive supplies of fertilizer and pesticides located on rural properties. Response crews must be alert also to the possibility of above- and below-ground storage tanks for fuel and other products. The safe operations guidelines for hazardous materials response are applicable in rural as well as urban areas.

In any locale, to be even minimally effective in controlling a fire, the initial responding apparatus should reach the emergency scene in time to prevent "flashover" (a very rapid spreading of the fire due to the heating of room contents and other combustibles). The time from ignition to flashover is controlled by many variables (e.g., the type of combustibles, the ventilation available, the source of ignition, etc.). Recorded times to flashover vary widely, but fire departments should operate on the assumption that prevention of flashover requires a response in less than 10 min. Where fire departments provide emergency medical service, the widely recommended 4-min response for non-breathing or trauma victims is very important. That response time also is most helpful in fire situations. Given the unknown component of fire discovery and reporting its existence to the responsible fire suppression agency, recommending a "generic" response time is difficult.

However, it is possible to give such counsel if one can make certain assumptions that may be controlled by the community. These assumptions relate primarily to the presence of appropriately placed smoke detectors and other types of fire detection devices and a means to relay that alarm to the fire suppression agency, but they also include the existence of enforced building codes that provide some degree of resistance to fire spread from the room of origin.

When there is no public or private fire protection agency, any fire that is beyond the control of portable extinguishing equipment (extinguishers, garden hose, etc.) may be expected to burn until all fuel in a given unsprinklered fire area has been consumed. Minimum levels of fire protection, while serving to confine fires to isolated properties, leave much to be desired by the property owner who suffers the loss and the fire department whose morale is often affected by its inability to control the average fire successfully. Because of distances, lack of water, and a general resource deficiency, small rural communities are especially vulnerable to this kind of heavy fire loss.

Urban Fire Protection

In urban areas, inadequate fire department response to initial alarms can be a major factor in fire losses due to high population and structural densities. The number of simultaneous fire-fighting operations that may need to be conducted at the incident also dictates the total amount of personnel and equipment needed to provide effective fire-fighting operations. In all but the smallest structural fire, several operations must be carried on simultaneously and the fire attack must be made from several points. This cannot be accomplished by the crew of a single fire apparatus. Multiple apparatus must be positioned properly, and adequate waterflow made available to cope with the amount of fuel (fire load) involved or exposed.

In simplest terms, structural fire suppression in an urban setting involves the accomplishment of at least the following tasks, many of which must occur almost simultaneously to ensure effective and safe operations (the proper sequence will vary, depending on circumstances, as will additional tasks):

- Command of the incident (to ensure both effectiveness and safety of the fire fighters);

- Application of water in appropriate quantities (dependent upon the fire environment and other factors);

- Provision of appropriate source of water supply for above;

- Ventilation of smoke and other hazardous products of combustion from the fire area to the outside;

- Search for and rescue of fire victims;

- Forcible entry;

- Control of utilities; and

- Salvage and other property conservation operations.

At large-structure fires, additional fire-fighting personnel are needed to cover the various points of fire attack. In some cases various functions can be handled more efficiently by specially trained crews such as rescue companies and hazardous material teams operating from specially equipped apparatus.

The number of personnel and equipment necessary to accomplish the above will vary with a number of factors [i.e., the expectations placed on the mobile suppression group, the material burning, the construction of the building, the type of built-in protection provided, separation between buildings, availability of water supply, the number and physical and emotional condition of persons in the fire building, the type of equipment available to fire fighters, the level of proficiency of the fire suppression crew (including the commanders), etc.]. Hence, it is difficult to determine a minimum number of fire fighters or equipment required without careful, objective planning, and without considering the important variables. Obviously, personnel needs will differ significantly between a small detached structure fire and a high-rise fire.

Some general considerations may, however, be listed:

* The more arduous the expectations placed on the mobile fire suppression crew, the greater the required resources (e.g., the community that expects its fire department to contain fires to the room of origin should expect to provide more fire suppression resources than the community that expects its fire department only to prevent the spread of fire from one building to another; given the same level of protection demands, the community that leaves all fire protection to the mobile fire suppression force will require a more extensive suppression force than the community that requires a high level of built-in protection).

* The more extensive the concentrated fire potential, the greater the required fire suppression resources (e.g., given the same expectations of its mobile fire suppression force, a community having high-rise buildings, a high population density, and extensive industrial risks will normally require greater fire suppression resources than a largely residential community).

* The broader the services provided by a fire protection agency, the greater the need for resources (e.g., a fire agency providing emergency medical services will, given the same level of expectations for its mobile suppression forces, require more resources than an agency providing only fire protection services, assuming a significantly increased *total* workload demand, including a significant increase in simultaneous calls).

* The greater the geographic area protected, the greater the resource requirement of the mobile fire suppression forces (i.e., given the same service level expectations, a community providing service to 20 sq mi will require more resources than a community providing protection to only 10 sq mi; this is caused by the need for timely arrival at the scene of fire, medical, or environmentally threatening incidents). (For SI units: 1 sq mi = 2.6 km.) Computerized geographic information systems and accompanying computerized response maps have powerful ability to demonstrate visually the response effectiveness of various station locations.

In most smaller and medium-size communities, all initial-response (first-alarm) apparatus will not arrive at the fire scene simultaneously. In many departments with on-duty personnel, apparatus has to respond from more than one station, and some apparatus have longer travel times to the fire scene. In volunteer departments, personnel must travel varying distances to get to the fire station or the fire scene, and thus all apparatus cannot go into operation at the same time. Those fire fighters and vehicles that cannot arrive at the fire scene within the first critical time period have limited impact on the initial attack, regardless of the department's response assignment ("running card"). Communities may have a false sense of security in this regard, until actual response times are tested and working initial-attack personnel counted.

The critical numbers for policy makers to use in planning related to staffing of shifts and apparatus crews—sometimes termed "minimum manning"—are those that describe how many trained personnel can arrive within a stipulated initial attack time frame. Ideally, all or most will arrive as composite crews, and not as individuals.

The minimum fire force recommended for any community is necessarily dependent upon the expectations of the community and the members of the mobile fire force. Objective planning and evaluation are necessary to determine the resources required for effective and safe fire suppression operations that will meet community requirements.

Some fire agencies are addressing the suppression challenge with different arrangements of fire crews within the communities than were heretofore generally accepted (e.g., use of multipurpose apparatus task force assignments, and differential staffing.). It is important to note that, while these different approaches may augment flexibility, they do not appear to reduce the number of suppression personnel required to carry out fire-fighting operations safely.

It does seem reasonable to say that not less than two fire suppression vehicles and a command officer should respond to any structure fire and that the number of personnel responding should be sufficient to carry out the tasks indicated above, and whatever else is typically necessary in local operations, in a timely fashion. Normative data are available from some cities. While that number is dependent on numerous factors, it is relevant to note that, in the broad spectrum of environments protected by 41 of the fire departments making up a portion of the Metropolitan Chiefs section of the International Association of Fire Chiefs, no department in 1995 dispatched fewer than thirteen fire fighters (including a command officer) to a reported fire in a single-family detached dwelling. The average number dispatched was 18.6.[5]

Fewer than eleven fire fighters would be most hard pressed to accomplish safe, effective, initial interior fire attack in a timely manner at a detached single-family dwelling, not including an incident commander.

Where necessitated by fire frequency or response distances, additional pumpers, ladder trucks, and tankers may be needed. Reserve apparatus is desirable not only to permit the repair of first-line equipment without reducing available fireground forces, but to provide additional fire-fighting units during major emergencies. Specialized vehicles for hazardous materials response, heavy-duty and specialized rescue—including boats—lighting equipment, breathing apparatus bottles, emergency medical response, etc., also must be planned for. If these cannot be made available locally, then mutual aid arrangements are needed. Different types of vehicles may be necessary for the various levels of emergency medical response: first-responder or basic life support, advanced life support, and hospital transport.

Commercial, industrial, and mercantile areas generally require additional apparatus, or more, in response to the initial alarm. If properties with considerable life hazard are involved (schools, hospitals, nursing homes, etc.) additional resources should be considered for initial alarms. Especially large numbers of personnel are needed for search and rescue operations in these properties, with several fire fighters needed to "sweep and search" each floor. (See Table 10-2A.)

The required fire-fighting units should arrive on scene close enough in time after the initial alarm to operate as an effective fire-fighting unit following planned tactical procedures.

TABLE 10-2A. *Typical Initial Attack Response Capability Assuming Interior Attack and Operations Response Capability*

High-hazard occupancies (schools, hospitals, nursing homes, explosives plants, refineries, high-rise buildings, and other high life hazard or large fire potential occupancies)

At least 4 pumpers, 2 ladder trucks (or combination apparatus with equivalent capabilities), 2 chief officers, and other specialized apparatus as may be needed to cope with the combustible involved; not fewer than 24 fire fighters and 2 chief officers.

Medium-hazard occupancies (apartments, offices, mercantile and industrial occupancies not normally requiring extensive rescue or fire-fighting forces)

At least 3 pumpers, 1 ladder truck (or combination apparatus with equivalent capabilities), 1 chief officer, and other specialized apparatus as may be needed or available; not fewer than 16 fire fighters and 1 chief officer.

Low-hazard occupancies (one-, two- or three-family dwellings and scattered small businesses and industrial occupancies)

At least 2 pumpers, 1 ladder truck (or combination apparatus with equivalent capabilities), 1 chief officer, and other specialized apparatus as may be needed or available; not fewer than 12 fire fighters and 1 chief officer.

Rural operations (scattered dwellings, small businesses, and farm buildings)

At least 1 pumper with a large water tank [500 gal (1.9 m³) or more], one mobile water supply apparatus [1,000 gal (3.78 m³) or larger], and such other specialized apparatus as may be necessary to perform effective initial fire-fighting operations; at least 12 fire fighters and 1 chief officer.

Additional alarms

At least the equivalent of that required for rural operations for second alarms; equipment as may be needed according to the type of emergency and capabilities of the fire department. This may involve the immediate use of mutual-aid companies until local forces can be supplemented with additional off-duty personnel. In some communities, single units are "special called" when needed, without always resorting to a multiple alarm. Additional units also may be needed to fill at least some empty fire stations.

In evaluating the adequacy of fire protection in any given area, major consideration must be given to the ability of the fire department to handle efficiently any reasonably anticipated work load. This requires an evaluation of the possibility of simultaneous working fires and other emergencies; weather factors that may contribute to the spread of fire, the delay in response, or the possibility of slow operations at the scene; and other demographic or geographic conditions that might affect the frequency, severity, and spread of fire occurrence and the response time of initial fire-fighting units. Where fire frequency is such that any fire company may expect two or three working fires per day, or where structures to be protected require a heavy initial response, closer geographic spacing of or increased personnel assigned to individual fire companies may be necessary. The number of other fire-fighting or related operations such as grass, brush, rubbish, and automobile fires and emergency rescue operations may also require greater-than-normal staffing of equipment and closer spacing of fire companies. Major structural fires may result when the normal first-alarm coverage in a district is depleted through coverage of these other emergencies, making remaining fire-fighting forces inadequate. Staffing fire apparatus at a level below minimum requirements can result in less effective and less safe fire-fighting performance. This factor also has an adverse effect on the number of required fire companies for various alarms, since additional fire companies must be dispatched to the scene of an emergency to provide adequate total staffing.

The desirable practice of assigning emergency medical responsibility to the fire department must be calculated into the staffing formula. It is difficult also to obtain effective teamwork and coordination with understrength crews. Some fire departments have attempted to solve this problem by supplementing their crews with part-time or volunteer fire fighters, or by providing off-duty fire fighters with tone-activated radio receivers and paying them for overtime when they respond to a fire. The on-duty personnel make the initial fire attack and holding action while off-duty personnel provide the additional assistance needed for continuing fire-fighting operations. Efficiency is lost, and increased fire losses can be expected with this arrangement. Such protection should not be relied upon to replace adequately the required staffing and equipment needed immediately at the scene for initial attack and rescue.

Personnel requirements are not merely a matter of numerical strength, but are also based on the establishment of a well-trained and coordinated team necessary to utilize complicated and specialized equipment under the stress of emergency conditions. Attempting to operate more fire companies than can be effectively staffed, even if some response distances must be somewhat increased, is less desirable than fewer but appropriately staffed companies. The effectiveness of pumper companies must be measured by their ability to get required hose streams into service quickly and efficiently. NFPA 1410, *Standard on Training for Initial Fire Attack*, should be used as a guide in measuring this ability. Seriously understaffed fire companies generally are limited to the use of small hose streams until additional help arrives. This action may be totally ineffective in containing even a small fire and in conducting effective rescue operations.

Consideration must be given also to maintaining an adequate concentration of additional forces to handle multiple alarms at the same fire, while still providing minimum fire protection coverage for the other areas under fire department protection. If available personnel prove adequate for routine fires but inadequate for major emergencies, arrangements should be made to supplement the fire protection coverage by calling back the off-shift personnel and by promptly calling nearby fire departments for mutual aid. Off-shift personnel may operate reserve apparatus or relieve or supplement personnel on the fireground. Fire companies not dispatched or utilized on the fire scene should be repositioned throughout the remaining area of the jurisdiction to ensure minimum response times to other alarms.

Reserve apparatus should be properly maintained and equipped, and when placed in service should be staffed to a degree commensurate with standard fire apparatus requirements. Since it may take up to 30 min or more to place reserve units in service with personnel recalled in an emergency, these reserve units should not be completely relied upon to immediately provide an adequate level of fire protection services.

Concern must be shown, under legislation and local policy, for the health and safety of fire fighters and others, for environmental protection, and for the rights of those being served. Officials must demonstrate reasonable and prudent action with establishment of incident command systems, for example, and the appointment of qualified safety officers.

In cases where several fire departments occupy adjacent or contiguous territories, arrangements (often termed "line response") should be made for joint response along common boundaries to high-risk hazards and for assistance in covering vacant fire stations at times of major fires. Mutual aid or mutual response should not be relied upon for routine emergencies, since there could be times when local commitments will preclude the anticipated assistance.

Mutual-aid agreements do not reduce the responsibility of each jurisdiction to maintain adequate facilities to handle normal fire protection needs. It also must be assumed that teamwork and tactical efficiency at a fire will be somewhat less than that expected of equal units from the same department under a united command. Often, however, specialized units (such as hazardous materials response teams) are organized to protect larger areas encompassing several fire departments.

In the past it was a common practice to relate the number of pumping engines and their pumping capacity, and other apparatus personnel requirements, to the population to be protected. With the industrialization of many areas and the construction of commercial shopping centers, hospitals, schools, and nursing homes in residential areas, it is possible that concentrations of life hazard and property value in areas of small or large populations may require substantial fire-fighting forces. Those with the responsibility for providing public fire protection must be prepared to cope with fire potential in any location in the jurisdiction. Fire department response requirements are now based on the water flow in gpm (L/min) that may have to be applied. A rule of thumb is to provide one company for each 250 gpm (946.25 L/min) that may be needed in an interior attack, plus personnel for rescue and other operations that need to be performed simultaneously with the advancing of hose lines.

Some may argue that it is not the public's responsibility to provide adequate fire protection to high-hazard risks that should have built-in fire protection systems. However, failure to attempt to provide fire protection for large taxable values on which the economy of a community may be based would place the community's fiscal viability at risk. Burned-out businesses may not rebuild, and then local people will lose employment. Also, fire spread to other properties is possible.

As already described, time is another critical factor in the evaluation of public fire protection. It is generally considered that the first-arriving piece of apparatus should be at the emergency scene within 5 min of the sounding of the alarm, since additional minutes are needed to size up the situation, deploy hose lines, initiate search and rescue, etc. In dense urban settings the desired response time is often shorter, and 4 min for the first-responding pumper is a rule-of-thumb maximum time for 90 percent of an urban area. An old adage says, "The first 5 min of most fires is the determining factor as to whether that fire will remain a small fire or become a large fire." While this may not always be true, delays in sounding an alarm obviously must be minimized or eliminated, as well as delays in responding and initiating rescue and attack. Time, however, cannot become the all-important factor at the expense of safety.

There are increasingly numerous instances where highly specialized apparatus and equipment must be available to municipalities. One category of specialization concerns apparatus designed to handle hazardous materials, including spills of petroleum products and other chemicals that require special extinguishing agents such as foams or dry powders, and special equipment to apply these agents. These dangerous substances may be present because of airports, marinas, manufacturing or storage facilities, or transportation routes in the district. Another category of specialization includes apparatus and equipment needed because of particular structures or facilities such as research laboratories, hospitals, high-rise buildings, oil and gas wells, and seaports. In some communities, fire departments are expected to conduct specialized functions such as extricating at automobile wrecks and performing water and mountain rescue, as well as providing emergency medical services. These services also require special equipment and possibly special vehicles. The necessity for many departments to deliver at least first-responder emergency medical service, and for the delivery of a wide variety of technical rescue and disaster response services, cannot be overlooked.

As with standard fire-fighting equipment and apparatus, specialized tools cannot be used effectively and safely unless personnel are highly trained in their use under a wide variety of circumstances. Whether personnel are volunteer or career, in rural or urban areas, no plan can be implemented and no reasonable level of protection afforded to the community unless well-designed and well-managed training programs are carried out.

Progressive departments of all types and sizes also concentrate on providing a broad range of community-oriented services.

Fire Prevention

The term *fire prevention* as used here generally includes inspections, education, and equipment meant to reduce the occurrence of fire and to mitigate the effects of that fire prior to the arrival of the mobile suppression force. As an example, the installation of a sprinkler system is not designed to prevent fire but to extinguish it in its very early stages; the "Stop, Drop, and Roll" message of the NFPA's *Learn Not to Burn®* program is clearly a mitigation rather than prevention effort. Other education efforts are directed at stopping the occurrence of fire.

As noted previously, fire prevention activities are somewhat difficult to evaluate. In a real sense, if prevention activities are effective, fires and fire-related tragedies occur with less frequency. There is a reduction or absence of fire activity, and these results are statistically evident although they do not appear in dramatic news clips and photographs. Some departments do report not only dollar amounts of fire loss, but also the value of structures that were threatened by fire and thus "saved." Without careful and systematic long-term record-keeping concerning the incidence of fires, fire losses, and related tragedies, the effect of prevention programs cannot be documented. Inability of fire officials to demonstrate the value of committing some additional community resources to the broad range of possible prevention activities may well result in a withdrawal of resources from prevention programs and a subsequent increase in the need for a much larger suppression budget. Rational decisions and sound recommendations concerning evaluation and planning cannot be made unless fire officials learn what changes there can be to total fire cost by reallocating resources applied to the total fire defense system.

Both evaluation and planning require recognition of the component and integrated parts of a fire prevention system. Until recent years prevention was seen as a narrow band of activities, often greatly limited or nonexistent in most smaller communities. In urban areas it was limited frequently to the periodic inspection of certain types of buildings. More modern approaches to fire prevention recognize that a comprehensive program includes all organized activities, other than suppression, that reduce the incidence of fire and fire-related losses. Ideally, these activities would be carried out in communities of every size, whether rural or urban, with appropriate adjustments made for community size, type, location, and fire history. Community- and neighborhood-based focus programs, often emanating from the local station, appear to have excellent effects.

Prevention activities may be categorized in several ways, but it is usually helpful to group them as follows:

1. Activities that relate to construction, such as building codes, the approval of building and facility plans, and occupancy certification and recertification for new occupants. Also included may be a sign-off for the presence of smoke detectors when new or old properties are sold.
2. Activities that relate to the enforcement of codes and regulations, such as inspections of certain occupancies, the licensure

of certain hazardous facilities, the design of new regulations and codes, and legislation to adopt existing model codes.

3. Activities that relate to the reduction of arson, such as fire investigation, the collection of information, public education, and data related to setting fires. Included may be arson investigation and related court proceedings, and programs such as counseling for juvenile fire setters.

4. Activities that relate to the collection of data helpful in improving fire protection, such as standardized fire reporting, case histories, and fire research.

5. Activities that relate to public education and training, including fire prevention safeguards, evacuation and personal safety steps, plant protection training for industrial and other work groups, hazardous materials and devices safeguards, and encouragement to install early warning and other built-in signaling and extinguishing devices. Very popular are programs for school children, such as NFPA's *Learn Not to Burn* curriculum and self-help classes such as water safety, urban survival, and similar "Stay Alive 'Til We Arrive" projects.

An analysis of the community's fire history, conducted during the evaluation phase of the fire protection plan, will usually indicate to fire experts and citizen groups which categories need strengthening. Comparing the number of fires and fire-related incidents, plus fire loss (property, life, injury) statistics over several years as more prevention activities are phased in, provides an assessment of program effectiveness. Calculating the total cost of fire to a community (fire loss plus prevention costs plus suppression costs plus fire insurance costs) will enable the fire department and the community to estimate the efficiency or cost-effectiveness of a proposed prevention program.

PUBLIC PROTECTION CLASSIFICATIONS

Fire Department Service-Level Analysis

The public, fire and other government administrators, organized labor, fire protection organizations, and the fire service in general have for many years sought to find a generally accepted method for the evaluation of services provided to a community by its fire protection agency. The approaches have been as varied as the fire service agencies being evaluated and the parties doing the evaluation. This has resulted in the application of inconsistent methodology and criteria.

For many years material developed by NFPA has been used in the analysis of services provided by fire protection organizations. However, the application of NFPA materials has been inconsistent, and in 1995 no single organization existed that enabled a comprehensive study or audit of a department.

Since 1990, NFPA has continued its long history of involvement in the field of fire department evaluation through programs to provide criteria, process, and organization for the analysis of service levels provided by fire departments. With the Board of Directors' concurrence, a "Select Committee" was appointed to advise in the "Fire Department Analysis" project, with representation from the fire service, municipal management, organized labor, and the insurance industry. Appropriate recommendations to the NFPA Standards Council have been made. NFPA 1500 was modified and reissued in 1992. NFPA 1201, *Standard for Developing Fire Protection Services for the Public*, was modified and reissued in 1994. A new NFPA Technical Committee covering fire department organization and deployment, including emergency medical services, was created in 1995, and is developing a new NFPA standard, *Organizations, Operation, Deployment, and Evaluation of Fire and Emergency Medical Services* (NFPA 1200).

Initial work of the Select Committee, conducted in 1990 and 1991, determined that the best program for the analysis of services would:

- Consider the total scope of services provided by the fire service agency;
- Set forth uniform definitions;
- Be primarily performance-based and user-friendly;
- Be founded on consensus and other standards to the greatest possible extent;
- Include user input;
- Be designed for specific community application;
- Be capable of self-application; and
- Include a validation process.

The Voluntary Fire Service Accreditation Program, sponsored by the IAPC, uses self-assessment and peer group review processes to assist fire departments in evaluating important aspects of their organization and its operations.

A group of experienced fire service personnel, meeting in October 1996 at the Wingspread IV Conference, listed the following emerging issues of national importance to the fire service: customer service; managed care; competition and marketing; service delivery standards; fire fighter wellness; and labor-management relations.

The following were listed as earlier and ongoing issues of national importance to the fire service: leadership in changed environments; expanded prevention and public education efforts; professionalized training and education; increased use of fire and life safety detection, alarm, and extinguishment systems; the formation of strategic partnerships; the collection of relevant data; and protection of the environment.

ISO/CFRS Fire Suppression Rating Schedule (FSRS)

While not all states in the United States use the existing Insurance Service Office (ISO) grading schedule, and while it is not used in other nations, it is applied to many departments in most states approximately every 10 yr. The purpose is to aid in the calculation of fire insurance rates and is not for property loss prevention or life safety purposes.

As will be seen below, the older form of the grading schedule contains more categories and more items than the schedule in current use. However, the older form is still applied in one or more states and is still used by some community officials as a reference tool in self-evaluations.

It should be noted that, in Section I of the ISO grading schedule, the result and classification apply to properties with a needed fire waterflow of 3,500 gpm (134,248 L/min) or less. Private and public properties with larger needed flows are individually evaluated in Section II.

With improved ratings, fire insurance companies that subscribe to the ISO rating system may lower commercial fire insurance premiums and may lower residential rates in certain instances.

The "Grading Schedule for Municipal Fire Protection," developed originally by the National Board of Fire Underwriters (NBFU) and continued by its successor, the American Insurance Association, and then by the ISO, has provided a guideline for municipalities to classify their fire defenses and physical conditions. The gradings obtained under the schedule were used in establishing base rates for fire insurance purposes. The schedule has been subject to change with the state of the art, and sweeping changes were made in the 1980 edition with the development of a revised FSRS. Some expansion of the current schedule, to include related building code applications, may be forthcoming as of 1995. Under certain circumstances, credit may now be given for property within five road miles of a fire station, even though a municipal water supply system is lacking. Additional modifications may be forthcoming.

The current ISO grading schedule reviews and correlates those features of public fire protection that have a significant effect on minimizing fire damage. Credit is given for existing fire protection, instead of debit for what is not in place.

The FSRS[1] produces ten different Public Protection Classifications, with Class 1 receiving the most rate recognition and Class 10 receiving no recognition. The Fire Suppression Rating Schedule simply defines different levels of public fire suppression capabilities that are credited in the individual property fire insurance rate relativities.

Starting in 1975, on a state-by-state basis following Insurance Department approval, ISO implemented the Commercial Fire Rating Schedule (CFRS), which was a major revision to the method used to develop individual property rate relativities. The CFRS reviews and correlates the construction, occupancy, exposures, and private and public fire protection (represented by the Public Protection Classification number). This correlation allows development of an equitable rate relativity applicable to the individual property. To this rate relativity, statistical experience adjustments are applied either by ISO or by affiliated companies to produce the applicable fire insurance rate. The FSRS represents a revision in the method used to derive the Public Protection Classification number used in the CFRS. The Public Protection Classification number is also used as a rate relativity variable for most class-related properties, in addition to construction and occupancy variables.

The *Grading Schedule for Municipal Fire Protection*,[2] although a much-improved system from previous editions, was not developed as an integral part of the individual property rating system. The previous schedules were somewhat independent primarily due to their historical development by the NBFU, which was not an insurance rating organization. The previous schedules were used more to quantify underwriting information, but did define different levels of public fire protection that could be used for a specific rating.

The FSRS is designed to assist in an objective review of those features of available public fire protection that have a significant influence on minimizing damage once a fire has occurred. This revision ties logically to the review of contributive and causative hazards that can be performed with the CFRS.

Comparison of Details

The following material outlines the specific changes that are incorporated into the FSRS, as compared with the 1974 *Grading Schedule for Municipal Fire Protection*.

In 1995, building, electrical, and fire prevention laws and climatic conditions are not included in the FSRS (Table 10-2B) for the following reasons:

1. Most communities have adopted one of the available model codes; thus differences were not really being measured.
2. Evaluation of codes enforcement and climatic conditions is subjective when trying to review on a citywide basis.
3. Results of the law and its enforcement manifest themselves in the actual conditions found in individual properties when surveyed for application of the CFRS. (See Table 10-2C.)
4. The 1974 schedule and its predecessor schedules contained many standards or benchmarks of evaluation that were detailed specifications. If a condition in a community did not meet the specifications, deficiency points were assigned. Specifications

TABLE 10-2B. *Comparison of Schedules: Major Items*

1974 Schedule	Relative Weight (Percent)	1980 FSR Schedule	Relative Weight (Percent)
Water supply	39	Water supply	40
Fire department	39	Fire department	50
Fire alarm	9	Fire alarm	10
Building laws	1.7		
Electrical laws	0.8		100
Fire prevention laws	10.5		
Climate conditions	(variable)		
	100		

TABLE 10-2C. *Comparison of Schedules: Water Supply*

1974 Schedule	Relative Weight (Percent)	1980 FSR Schedule	Relative Weight (Percent)
Adequacy of supply works	6	Supply works	
Reliability of source of supply	6	Fire flow delivery Distribution of hydrants	35
Reliability of pumping capacity	3	Hydrants—size, type, and installation	2
Reliability of power supply	4	Hydrants—inspection and condition	3
Condition, arrangement, operation, and reliability of system components	4		40
Adequacy of mains	16		
Reliability of mains	2		
Installation of mains	2		
Arrangement of distribution system	2		
Additional factors and conditions relating to supply and distribution	4		
Distribution of hydrants	5		
Hydrants—size, type, and installation	2		
Hydrants—inspection and condition	2		
Miscellaneous factors and conditions	6		
	64		
	(Max = 39)		

decrease a schedule's flexibility in use to evaluate changing conditions and technology properly. Because the FSRS has eliminated essentially all of the specifications that were in the 1974 schedule, it is referred to as a performance schedule. The only criteria in the FSRS that could be considered specifications are the descriptions of the minimum conditions to be reviewed under the schedule. For example, with the 1974 schedule, it was essentially impossible for a community to receive other than a Class 9 rating if it did not have a water supply with conventional underground water mains. However, the FSRS is based upon delivery performance only, without regard to the delivery method used, and is adaptable to crediting the delivery of water by other means. Specific reference is made to water carried and delivered to the fire location by fire department apparatus.

5. As previously mentioned, the FSRS has been designed to interface directly with individual property insurance rate development. As a result, numerous items that were evaluated both in the individual property review and in the previous public fire protection review have been removed from the FSRS.

The review and enforcement of building, electrical, and fire prevention codes have been deleted from the FSRS, because the presence of hazards due to poor codes or lack of enforcement are recognized in an individual property survey and rate review.

Section I of the FSRS can be used to develop a public protection classification that reflects the city's ability to handle fires in small to moderate-size buildings; these are defined as those buildings having a needed fire flow of 3,500 gpm (13,248 L/min) or less. Section II of the FSRS can be used to develop public protection classifications for large individual properties having needed fire flows greater than 3,500 gpm (13,248 L/min).

The two sections can help communities design their fire protection needs around normally expected fires. The Public Protection Classification in Section I is applicable to average buildings, and the influence of the fire protection demands for larger buildings has been removed from that analysis. Section II can be applied individually to each large building to develop an individual public protection classification that reflects the protection available for that specific property.

The 1974 schedule and its predecessors did not provide significant recognition of the decreased need for public fire suppression when large properties are protected by automatic sprinkler systems. The FSRS does recognize this important fact by excluding all properties with standard automatic sprinkler systems from the development of needed fire flow figures.

The water supply items dealing with reliability in the 1974 schedule are not included in the FSRS because its author concluded from recent history that the dependability and replacement capability of water supply equipment is such that measurement of these features has little bearing on the water supply performance. The remaining items of the 1974 schedule that are not included in the FSRS are omitted because they either require subjective review or are not considered significant enough to measure the performance of the water system for fire insurance rating purposes. The emphasis of the FSRS review regarding a water system is the consideration of the actual water supply that is available for fire suppression at representative locations throughout the city.

As in the water supply feature, items that require subjective review and consider minor features are not included in the FSRS. (See Table 10-2D.) The FSRS places emphasis on the review of the first-alarm response by the fire department because of the importance of initial attack to minimize potential losses. Additionally, the entire fire department item has been given increased weight in the overall review in recognition of the value of fire department operations in early stages of fire suppression and the fact that some operations are possible with less-than-optimum water supplies.

As in the water supply and fire department sections, features that are subjective or are minor items dealing with the fire alarm system are not included in the FSRS. (See Table 10-2E.) The emphasis

TABLE 10-2D. Comparison of Schedules: Fire Department

1974 Schedule		1980 FSR Schedule	
	Relative Weight (Percent)		Relative Weight (Percent)
Pumpers	4.8	Engine companies	11
Ladder trucks	3.4	Ladder service companies	6
Distribution of companies and types of apparatus	4	Distribution of companies	4
Pumper capacity	4.4	Pumper capacity	5
Design, maintenance, and condition of apparatus	3	Department manning	15
		Training	9
Number of officers	2		50
Department manning	8		
Engine and ladder company unit manning	6.4		
Master and special stream devices	1		
Equipment for pumpers and ladders	2		
Hose	2.8		
Condition of hose	1.6		
Training	6		
Response to alarms	2		
Fire operations	8		
Special protection	6		
Miscellaneous factors and conditions	6		
	71.4		
	(Max = 39)		

has been placed on the performance of handling and dispatching fire alarms rather than the method of notification.

Rate-Level Effect

A testing program was conducted by ISO to determine any potential rate-level effect resulting from the introduction of the FSRS. The test results indicate that the introduction of the FSRS will have a negligible rate-level impact.

A sample of 500 communities was selected so distribution by protection class and by population group within the protection classes corresponded with the average premium distribution across the United States. Such a selection process was used so the test results would properly reflect any premium change caused by the introduction of the FSRS.

The communities actually tested were randomly selected from across the country. The FSRS was applied to recent ISO file data. To calculate the expected rate-level impact, the redistribution of protection classes was correlated to the current nationwide average protection-class rate differentials. The test results indicate a 95 percent probability that the rate-level effect will be minus 0.08 percent, plus or minus 1.11 percent.

Not considered in the test was the fact that ISO normally resurveys each community only once every 10 yr unless requested to do so sooner by the community. Therefore, the actual rate level impact would be somewhat lower than the calculated impact.[1]

TABLE 10-2E. *Comparison of Schedules: Fire Alarm System*

1974 Schedule	Relative Weight (Percent)	1980 FSR Schedule	Relative Weight (Percent)
Communication center	0.8	Receipt of fire alarm operators	2
Communication center, equipment and current supply	2.8	Alarm dispatch	3
Boxes	1.2	Circuit facilities	5
Alarm circuits and alarm facilities, including current supply at fire stations	2		10
Material, construction, condition, and protection of circuit	1		
Radio	0.8		
Fire department telephone service	1.7		
Fire alarm operators	0.8		
Conditions adversely affecting use	1.3		
Credit for boxes installed in residential districts	−0.4		
	12.0 (Max = 9)		

Monitoring

The FSRS involves many significant changes in the approach to establishing Public Protection Classifications. As in any comprehensive new rating program, the possibility of unforeseen rating situations cannot be overlooked. Consequently, the results are monitored very carefully to ensure that the program is working as intended. Any required changes to the schedule that are identified through the monitoring program are implemented following thorough evaluation.

Implementation

The revised schedule is applied to communities during the normal resurvey program or when significant changes occur in a community that would require a resurvey. Section II of the FSRS pertaining to individual nonsprinklered properties will not be applied unless Section I has been applied to the community; Section II also will not be applied in cities that are statistically rated.

When application of the schedule results in an improvement in the Public Protection Classification, the new classification is implemented as soon as insurers and producers are notified through ISO's manual page-distribution procedures. Should a community retrogress, the community officials will be notified of the change and the possible effect on insurance rates. If within 90 days the community decides to engage in a program that would improve the classifica-

tion, the implementation of the poorer classification will be delayed indefinitely as long as an improvement program begins within 1 yr and continues to completion.

Transition Rule

The usual impact of a change in the Public Protection Classification, for those properties that have classifications such as a rating variable, is approximately 5 percent for each change in class. Even though it is highly unlikely that there would be any significant rate impact for individual property by a change in the Public Protection Classification, the following transition rule is used for any unusual situations that may arise.

When used solely as a result of the application of the FSRS, and when the applicable Public Protection Classification number increases by one or more, the following must apply to any resulting individual property rate increases:

1. The rate increase must be limited to 25 percent for the first publication.
2. If the rate increase is between 25 and 50 percent, the balance of the increase must be included when the rate is next published for the risk, but not sooner than 1 yr after the effective date in Item 1 above.
3. If the rate increase exceeds 50 percent, Items 1 and 2 above must apply, except that an increase over 50 percent must not be published sooner than 2 yr from the effective date in Item 1 above.

PLANNING

Whenever a community—rural, suburban, or urban—considers its fire defenses, it must scrutinize the past and present and make predictions or forecasts for the future. Reviewing the past is called *data analysis* and depends upon good record-keeping. *Evaluation*, which is looking at the present, requires the ability to examine a situation objectively. The process of forecasting future conditions and preparing for them requires that a planning process be followed. This *planning* process results in a plan and its implementation, so that future challenges to the community are met. As the plan is implemented, the process must include the establishment of a *feedback loop*, providing a continuing assessment of how well the plan is contributing to successful completion of goals and objectives, and feeding revised data back into the plan so continuing redesign occurs.

Fire protection organizations, and especially fire departments, need to develop several kinds of plans related to fire prevention and fire suppression. These plans should be quite specific, directed at clearly defined goals, and operational over a relatively brief time period (usually from 1 to 5 yr). Typically, such plans are internal to the department and do not involve broad-based planning groups from the outside. Examples of these types of plans, which are most often technical in nature, are apparatus replacement plans, training program plans, revised initial-response plans, plans for a special hazardous-materials attack unit, and plans for adapting fireground procedures to incorporate the use of larger-diameter hose. However, once department planning begins to consider aspects of fire protection that will have an impact on external groups, those groups will need to be consulted and incorporated into the planning process. Fire department planning, for example, must consider land-use planning (zoning), water department planning, building code enforcement, etc.

Examples of fire department planning that require the early involvement of other groups are station relocations or closings, building inspection programs, public education programs, and changes in the scheduling of work platoons. These plans, while involving some other groups, are still fairly narrow in scope and usually can be formulated over a relatively short time.

A third type of planning, called comprehensive or master planning, addresses the total community fire protection problem, incorporating both prevention and suppression, and obviously involves many community agencies and organizations, perhaps even county, state, and federal agencies. Comprehensive planning is a necessity for communities and is aimed at integrating all community efforts at prevention and suppression, and improving efficiency and cost-effectiveness of those activities. Improved total community fire protection is the goal of this planning. Its degree of success must be measured by figures relating to the total cost of fire to the community, and not just in gains for one subsystem. Comprehensive plans often consist of a number of subplans from various agencies that are developed at the same time as part of a larger, total process, and that fit together to make a comprehensive and integrated plan.

Comprehensive plans have clearly stated goals with agreed-upon ways of measuring their attainment. These overall goals are reached through overall strategies acceptable to all involved agencies and to the citizens who must pay for the fire protection system. Each goal is composed of some number of subgoals or objectives, and for each objective there is a tactic designed to reach that objective. All objectives lead to the accomplishment of the overall strategies. When the objectives and tactics are laid out on a timeline, the overall time required to implement the comprehensive plan is then known, and the timing for attaining each objective is apparent.

Fire protection has been largely a local responsibility, and for good reasons it seems destined to remain so.* Each community has a set of conditions unique to itself; it cannot be assumed a system of fire protection that works well for one community will work equally well for other communities. To be adequate, the fire protection system must respond to local conditions, and especially to changing conditions. Planning is the key: without local-level planning, the fire protection system is apt to be ill-suited to local needs and unadaptable to the changing needs of the community.

Excellent fire protection (for example, in the form of automatic extinguishing systems such as residential sprinklers) is technically available and certainly can be provided with the resources of most communities. Even with considerable public support, however, this protection may require several years to attain. In the meantime, in every fire jurisdiction (whether a municipality, county, or region), fire protection goals must be set and plans made to achieve those goals. The issues below discuss some of the concepts to be defined in setting these standards.

Adequate level of fire protection: The question of "adequacy" is addressed not only in day-to-day needs, but in major contingencies that can be anticipated for future needs as well. A definition of "optimal" protection is needed—in contrast to "minimal" protection, which fails to meet contingencies and future needs, and "maximal" protection, which is usually more expensive than a community can afford.

Comprehensive planning must include contingencies drawn from an analysis of community hazards. This process of hazard identification and analysis is crucial to fire department planning.

Reasonable community costs: Fire, both as threat and reality, has its costs, including deaths, injuries, property losses, hospital bills, and lost tax revenues, plus the costs of maintaining fire departments, paying fire insurance premiums, and providing built-in fire protection. Each community must decide upon an appropriate level of investment in fire protection. Some costs that are beyond the pub-

lic's willingness to bear may be transferred to the private sector (as when buildings over a certain size or height or with a certain occupancy are required to have automatic extinguishing systems).

Acceptable risk: A certain level of fire loss must be accepted as tolerable simply because of limited resources of a community. Conditions that endanger the safety of citizens and fire fighters beyond the acceptable risk must be identified as targets for mitigation.

Consideration of these matters helps to determine what functions and emphasis should be assigned to the fire department, other municipal departments, and the private sector both now and in the future. It helps to define new policies, laws, or regulations that may be needed. Most importantly, consideration of these matters makes it clear that fire safety is a responsibility shared by the public and private sectors. Because the fire department cannot prevent all fire losses, formal obligations to have built-in fire protection fall on owners of certain kinds of buildings. For the same reason, private citizens have an obligation to exercise prudence with regard to fire in their daily lives. But prudence also requires education in fire safety, and the obligation to provide that education appropriately falls in the public sector, chiefly the fire department. The public sector (again, chiefly the fire and building code enforcement departments) also has an obligation to see that requirements for built-in protection in the private sector are being met.

A fire department, then, has more than one responsibility—and the aforementioned responsibilities are not exhaustive.

The following are significant functions for which fire protection agencies typically have primary or significant roles.

Fire suppression: Fire fighters need proper training and adequate equipment to save lives, extinguish fires quickly, and to ensure their own safety.

Specialized emergency and disaster services: These include hazardous-materials incidents, floods, earthquakes, multiple-vehicle accidents, cave-ins, collapsed buildings, volcanic eruptions, searches for lost persons, attempted suicides, a variety of specialized rescue services, etc.

Emergency medical services: Capabilities needed during fires and other emergencies include first aid, resuscitation, and possibly advanced life support (paramedical services). (*Paramedical services* means emergency treatment beyond ordinary first aid, performed by fire service personnel under supervision—through radio communication, for example—of a physician.)

Fire prevention: This includes approval of building plans and actual construction; inspection of buildings, their contents, and their fire protection equipment; and investigation of fire cause and spread to guide future fire prevention priorities and determine when the crime of arson has been committed.

Fire safety and dangerous-situation education: Fire departments have an obligation to bring fire safety and dangerous-situation education not only into schools and private homes but also into occupancies such as restaurants, hotels, hospitals, and nursing homes that have a greater-than-average fire potential or life safety hazard. These programs may include such topics as swimming pool safety and babysitter training.

Deteriorated building hazards: In coordination with other municipal departments, fire departments can work to abate serious hazards to health and safety caused by deteriorated structures or abandoned buildings.

---

*Some of the information that follows in this chapter has been extracted in whole and in part from *America Burning*[3] and *America Burning Revisited*.[4]

Regional coordination: Major emergencies can exceed the capabilities of a single fire department, and neighboring fire jurisdictions should have detailed plans for coping with such emergencies. But effectiveness may also be improved through sharing of day-to-day operations—such as, for example, an areawide communication and dispatch network.

Data development: Knowledge of fire department performance and how practices should change to improve performance depends upon adequate record keeping. Microcomputers and minicomputers now play an important role in fire protection and emergency management planning.

Community relations: Fire departments are representative of the local community that supports them. The impression they make on citizens affects how citizens view their government. Volunteer departments dependent upon private donations must, of course, also be concerned with community relations. Moreover, since fire stations are strategically located throughout the community, they can serve as referral or dispensing agencies for a wide range of municipal services.

As communities set out to improve their fire protection, they must not consider the fire department alone. The police have a role in reporting fires and in handling traffic and crowds during fires. The cooperation of the building department is needed to enforce the fire safety provisions of building codes. The work of the water department in maintaining the water system is vital to fire suppression. In fire safety education, the public schools, the department of recreation, and the public library can augment the work of the fire department. Future community development and planning will influence the location of new fire stations and how they will be equipped.

The nine functions above are just the obvious examples of interdependence. Although it may seem trivial, the manner in which house numbers are assigned and posted, for example, can affect the ability of a fire department to respond quickly and effectively to emergencies.

Master Planning

Fire protection is only one of many community services. Not only must it compete for dollars with other municipal needs such as the education system and the police department, but in planning for future growth the fire protection system must account for the changes in progress elsewhere in the community. For example, if a deteriorated area is to be torn down and replaced with high-rise apartment buildings, the fire protection needs of that area will change. Changes in zoning maps will also change the fire protection needs in different parts of the community.

To cope with future growth, local administrators are turning increasingly to the concept of master (comprehensive) planning of municipal functions. Such plans include an examination of existing programs, projection of future needs of the community, and a determination of methods to fill those needs. They seek the most cost-effective allocations of resources to help ensure that the needs will be met.

A major section of a community's general plan of land use should be a master plan for fire protection, which should be written chiefly by fire department managers. This plan should, first of all, be consistent with and reinforce the goals of a city's overall general plan and its time frame. For example, managers should plan the deployment of personnel and equipment according to the kind of growth and the specific areas of growth that the community foresees. It is critical that the comprehensive plan determine and set goals and objectives for the fire protection system and the fire department in terms that are understandable to the citizens being protected.

Having established goals, the department officers should use the plan to establish some form of management systematic and quality assurance within the fire department. Management is most effective when each person is aware of how tasks fit into the overall goals and is committed to getting specific jobs done in a specified time.

Because fire departments exist in a real world where a variety of purposes must be served with a limited amount of money, it is important that every dollar be invested for maximum return on investment. The fire protection master plan should not only seek to provide the maximum cost-benefit ratio for fire protection expenditures, but it should also establish a framework for measuring the effectiveness of these expenditures. Lastly, the plan should clarify the fire protection responsibility for other groups, both governmental and private, in the community.

Often, as the result of a comprehensive planning effort, a community will formulate a "developmental plan." This document provides a schedule for implementing those changes in the fire protection system deemed appropriate by the community.

Key questions for developmental plans include:

1. What are the fire protection, rescue, emergency medical, disaster response, and safety education needs?
2. What organizational structure and what resources are currently available to meet those needs?
3. Is there a disparity between what is available and what is needed?
4. What will the community profile be like in 5 to 10 yr?
5. What will be needed then to provide adequate protection?
6. What is and what will be the financial resource base?
7. What options are and will be available to enhance protection, or to keep it at an adequate level?
8. How can changes be phased in to gain community goals and cost-effectiveness?

Devising a Fire Protection Plan

Key questions to be asked by those planning for fire protection are:

1. Why is planning necessary for us at this time?
2. What do we need to start the process, and are the necessary groups committed to the process?
3. What are the necessary steps in the planning process?
4. How will the plan be implemented?
5. Are all aspects of the plan legally possible and enforceable?
6. How will the plan be evaluated? Will it be a part of the integrated emergency management plan of the community?
7. How will feedback be gathered and the plan modified and updated?

In *Introductory Summary: Fire Prevention and Control Master Planning,* the U.S. Fire Administration[5] points out the following, in providing its overview of the planning process:

Master planning is a participative process which should result in the establishment of a fire prevention and control system which is goal-oriented, long-term, comprehensive, provides known cost/loss performance, and adapts continually to the changing needs of your community.

Master planning should consider all community elements . . . related to fire prevention and control system elements.

Master planning involves the participation of all parties interested in the development of a defined cost/loss relationship. . . .

Master planning allows you . . . to systematically analyze fire prevention and control through common-sense procedures. . . . Master planning has three phases: preplanning, planning, and implementation. The preplanning phase gets necessary commitments, committees, estimates and schedules, and go-ahead approvals. The planning phase gathers and analyzes data, sets goals and objectives, determines an acceptable level of fire protection service, identifies alternatives, and constructs the plan. The implementation phase never ends, because the plan is ongoing and always being revised and updated.

The following can serve as guidelines to fire department administrators for developing and presenting a master fire protection plan as part of the comprehensive master plan outlined earlier.

Phase I

1. Identify the fire protection problems of the jurisdiction.
2. Identify the best combination of public resources and built-in protection required to manage the fire problem, within acceptable limits:
   a. Specify current capabilities of and future needs for public resources, and
   b. Specify current capabilities and future requirements for built-in protection.
3. Develop alternative methods that will result in trade-offs between benefits and risks.
4. Establish a system of goals, programs, and cost estimates to implement the plan:
   a. Develop department goals and programs, including maximum possible participation of fire department personnel of all ranks;
   b. Provide goals and objectives for all divisions, supportive of the overall goals of the department; and
   c. Strive to develop management development programs that increase acceptance of authority and responsibility by all fire officers as they strive to accomplish established objectives and programs.

Phase II

1. Develop a definition of the roles of other government agencies in the fire protection process.
2. Present the proposed municipal fire protection system to the city administration for review.
3. Present the proposed system for adoption as the fire protection element of the jurisdiction's general plan. The standard process for development of a general plan provides the fire department administrator an opportunity to inform the community leaders of the fire protection goals and system, and to obtain their support.

Phase III

In considering the fire protection element of the general plan, the governing body of the jurisdiction will have to pay special attention to:

1. Short- and long-range goals.
2. Long-range staffing and capital improvement plans.
3. Code revisions required to provide fire loss management.

Phase IV

The fire loss management system must be reviewed and updated as budget allocations, capital improvement plans, and code revisions occur. Continuing review of results should concentrate on these areas:

1. Did fires remain within estimated limits?
2. Should limits be changed?
3. Did losses prove to be acceptable?
4. Could resources be decreased, or should they be increased?

BIBLIOGRAPHY

References Cited

1. National Commission on Fire Prevention and Control, *America Burning: The Report of the National Commission on Fire Prevention and Control*, U.S. Government Printing Office, Washington, DC, 1974.
2. Insurance Services Office, *Grading Schedule for Municipal Fire Protection*, Insurance Service Office, New York, 1974.
3. NCFPC, *America Burning*, U.S. Fire Administration, Washington, DC, 1974.
4. United States Fire Academy, *America Burning Revisited*, U.S. Fire Administration, Washington, DC, 1987.
5. National Fire Safety and Research Office, *Introductory Summary: Fire Prevention and Control Master Planning*, U.S. Department of Commerce, National Fire Safety and Research Office, Washington, DC.

NFPA Codes, Standards, and Recommended Practices

Reference to the following NFPA codes, standards, and recommended practices will provide further information on public fire protection issues discussed in this chapter. (See the latest version of *The NFPA Catalog* for availability of current editions of the following documents.)

NFPA 13E, *Guide for Fire Department Operations in Properties Protected by Sprinkler and Standpipe Systems*
NFPA 295, *Standard for Wildfire Control*
NFPA 297, *Guide on Principles and Practices for Communications Systems*
NFPA 402, *Guide for Aircraft Rescue and Fire Fighting Operations*
NFPA 403, *Standard for Aircraft Rescue and Fire Fighting Services at Airports*
NFPA 414, *Standard for Aircraft Rescue and Fire Fighting Vehicles*
NFPA 471, *Recommended Practice for Responding to Hazardous Materials Incidents*
NFPA 472, *Standard for Professional Competence of Responders to Hazardous Materials Incidents*
NFPA 901, *Standard Classifications for Incident Reporting and Fire Protection Data*
NFPA 902, *Fire Reporting Field Incident Guide*
NFPA 903, *Fire Reporting Property Survey Guide*
NFPA 904, *Incident Follow-up Report Guide*
NFPA 1001, *Standard for Fire Fighter Professional Qualifications*
NFPA 1002, *Standard for Fire Department Vehicle Driver/Operator Professional Qualifications*
NFPA 1003, *Standard for Airport Fire Fighter Professional Qualifications*
NFPA 1021, *Standard for Fire Officer Professional Qualifications*
NFPA 1031, *Standard for Professional Qualifications for Fire Inspector*
NFPA 1033, *Standard for Professional Qualifications for Fire Investigator*
NFPA 1035, *Standard for Professional Qualifications for Public Fire and Life Safety Educator*
NFPA 1041, *Standard for Fire Service Instructor Professional Qualifications*
NFPA 1201, *Standard for Developing Fire Protection Services for the Public*
NFPA 1221, *Standard for the Installation, Maintenance, and Use of Public Fire Service Communication Systems*
NFPA 1401, *Recommended Practice for Fire Service Training Reports and Records*
NFPA 1402, *Guide to Building Fire Service Training Centers*
NFPA 1403, *Standard on Live Fire Training Evolutions*
NFPA 1410, *Standard on Training for Initial Fire Attack*

NFPA 1452, *Guide for Training Fire Service Personnel to Make Dwelling Fire Safety Surveys*

NFPA 1500, *Standard on Fire Department Occupational Safety and Health Program*

NFPA 1521, *Standard for Fire Department Safety Officer*

NFPA 1901, *Standard for Automotive Fire Apparatus*

NFPA 1911, *Standard for Service Tests of Pumps on Fire Department Apparatus*

NFPA 1914, *Standard for Testing Fire Department Aerial Devices*

NFPA 1932, *Standard on Use, Maintenance and Service Testing of Fire Department Ground Ladders*

NFPA 1961, *Standard for Fire Hose*

NFPA 1962, *Standard for the Care, Use, and Service Testing of Fire Hose, Including Couplings and Nozzles*

NFPA 1964, *Standard for Spray Nozzles (Shutoff and Tip)*

NFPA 1971, *Standard on Protective Ensemble for Structural Fire Fighting*

NFPA 1975, *Standard on Station/Work Uniforms for Fire Fighters*

NFPA 1981, *Standard on Open-Circuit Self-Contained Breathing Apparatus for Fire Fighters*

NFPA 1982, *Standard on Personal Alert Safety Systems (PASS) for Fire Fighters*

NFPA 1983, *Standard on Fire Service Life Safety Rope and System Components*

NFPA 1991, *Standard on Vapor-Protective Suits for Hazardous Chemical Emergencies*

NFPA 1992, *Standard on Liquid Splash-Protective Suits for Hazardous Chemical Emergencies*

Additional Reading

*Allcott, B. H., *Manning Level Impact on Initial Fire Attack,* National Fire Academy, Emmitsburg, MD, Jan. 1991. (Executive Fire Officer Program, Applied Research Project.)

*Allcott, B.H., Arwood, R. A., DiLuzio, F., Frentress, K., Latipow, J. K., Peters, J. P., Rose, L. R., Snyder, L. L., *Fire Service Mid-Management Positions: Do We Need Them?* National Fire Academy, Emmitsburg, MD, May 30, 1988. (Executive Fire Development.)

*Amestoy, L., *Determining Shift Relief Factor from Agency Data—Staffing Analysis for Organizational Development,* National Fire Academy, Emmitsburg, MD, Apr. 1990. (Executive Fire Officer Program, Applied Research Project.)

Backoff, R. W., "Measuring Firefighter Effectiveness: A Preliminary Report," School of Public Administration, Ohio State University, Columbus, OH.

Baird, D., "Minimum Manning Requirements," *Fire Fighting in Canada,* Vol. 36, No. 2, Mar. 1992, p. 14.

*Barkman, D. R., Bendilli, J. A., Cooper, S. M., McGee, J. A., Olney, J. J., Williams, D. R., *A Review of Engine Company Staffing,* National Fire Academy, Emmitsburg, MD, Jan. 14, 1991. (Fire Executive Development.)

*Bates, G., *Training Academy Staffing in Metropolitan Fire Departments,* National Fire Academy, Oct. 1993. (Executive Fire Officer Program. Applied Research Project.)

*Becker, B., *Fighting Fires in the 90s, Will You Be Effective?* National Fire Academy, Emmitsburg, MD, Aug. 1990. (Executive Fire Officer Program, Applied Research Project.)

Bennett, J., "Minimal Staffing: How to Do More With Less," *Fire Engineering,* Vol. 142, No. 3, Mar. 1989, pp. 46-48.

Best, A., "Multi-Agency Approach to Children's Fire Education," *Fire,* Vol. 86, No. 1065, March 1994, p. 11.

*Blankenship, R., *Utilizing College Students for the Fire Service,* National Fire Academy, Emmitsburg, MD, Oct. 1993. (Executive Fire Officer Program, Applied Research Project.)

*Blehm, R. E., Breeding, W., Hopkins, E., Meyer, S., Pidala, J. A., and Powell, D., *Fire Departments in Transition: Is Urban Migration Affecting Volunteer Fire Departments?* National Fire Academy, Emmitsburg, MD, May 21, 1990. (Fire Executive Department.)

Bond, A., "Understanding Emergencies," *Fire Prevention,* No. 249, May 1992, pp. 40–42.

Brannigan, F. L., "Disaster Planning," *Industrial Fire World,* Vol. 4, No. 6, Dec. 1989/Jan. 1990, p. 22.

Brunacini, A. V., "Shrinking Resources vs. Staffing Realities," *NFPA Journal,* Vol. 86, No. 3, May 1992, pp. 28, 132.

Budney, J. L., and Duncombe, W. D., "An Economic Evaluation of Paid, Volunteer, and Mixed Staffing Options for Public Services," *Public Administration Review,* Vol. 52, No. 5, Sept.-Oct. 1992, pp. 474-481.

Burns, R. B., "Planning for Community Fire Protection," *Managing Fire Services,* ICMA, Washington, DC, 1986.

Carter, H. R., and Rausch, E., *Management in the Fire Service,* National Fire Protection Association, Quincy, MA, 1989.

Clark, B., "Is There Safety in Numbers?" *Fire Engineering,* Vol. 147, No. 2, Feb. 1994, pp. 24, 27.

*Clark, D. H., *Manpower and Scheduling: Report and Recommendations,* National Fire Academy, Emmitsburg, MD, Feb. 1991. (Executive Fire Officer Program.)

*Clark, H. D., Schaerer, E.J., and Vaughn, J., *Shift Work in the Fire Service,* National Fire Academy, Emmitsburg, MD, Jul. 1990. (Executive Fire Officer Program, Applied Research Project.)

*Clark, P. E., *An Inquiry into Utilization in the Fire Service,* National Fire Academy, Emmitsburg, MD, Jan. 22, 1991. (Executive Fire Officer Program, Applied Research Project.)

*Cooper, S. M., *Developing Engine Company Staffing Recommendations,* National Fire Academy, Emmitsburg, MD, Oct. 1991. (Executive Fire Officer Program, Applied Research Project.)

*Corso, J. B., Ellyson, R. V., Griffith, D. L., Murray, S. P., and Trigg, G. E., *Fire Service Manning Levels and the Impact on Civilian Fire Deaths,* National Fire Academy, Emmitsburg, MD, Dec. 4, 1989. (Fire Executive Development.)

*Curtis, J., *Three Men and a House—Effective Use of Manpower,* National Fire Academy, Emmitsburg, MD, Oct. 1992. (Executive Fire Officer Program, Applied Research Project.)

"Directory of National Community Volunteer Fire Prevention Program. Community-Based Fire Prevention Education Initiatives, 1984–1992.," National Criminal Justice Assoc., Washington, DC, Federal Emergency Management Agency, Emmitsburg, MD, FA-92, April 1993.

Discussion paper on on-site staffing within a comprehensive fire safety effectiveness model. The Office [Ontario, Canada] Sept. 15, 1992.

*Dunkel, R. E., *Apparatus Manning Levels for Paid On-Call Fire Departments,* National Fire Academy, Emmitsburg, MD, Aug. 1994. (Executive Fire Officer Program, Applied Research Project.)

*Esparza, H. R., *A Study to Determine the Impact of Fire Company Deactivations in the Fort Worth Fire Department,* National Fire Academy, Emmitsburg, MD, Feb. 25, 1994. (Executive Fire Officer Program. Applied Research Project.)

Factory Mutual Engineering Corp., "Guide to Planning Your Emergency Organization," Factory Mutual Research Corp., Norwood, MA, P8116, 1994.

Federal Emergency Management Agency, "Public Fire Education Today: Fire Service Programs Across America," Federal Emergency Management Agency, Washington, DC, FA-98, Sept. 1990.

"Fire Departments and Communities: Partners in Prevention," *Fire Engineering,* Vol. 148, No. 6, June 1995, pp. 101–110.

Fire Engineering Books and Videos, *The Fire Chief's Handbook,* Saddle Brook, NJ, 1995.

Gamble, C., "Pre-Planning Rural Water Supplies," *Fire Chief,* Vol. 39, No. 3, March 1995, pp. 68, 70, 72.

*Glenn, G. A., *Suppression Company Staffing Levels on Their Impact on Operations,* National Fire Academy, Emmitsburg, MD, Mar. 1990. (Executive Fire Officer Program, Applied Research Project.)

Granito, J. A., ed., *Literature Review of Fire Suppression Crew Size and Related Topics,* Project Reference Copy, National Fire Protection Association, Office of Fire Service Relations, Quincy, MA, Feb. 1991.

Granito, J. A., and Dionne, J. M., "Evaluating Community Fire Protection," in *Managing Fire Services,* International City Management Association, Washington, DC, 1986.

Grant, N., and Hoover, D., *Fire Service Administration,* National Fire Protection Association, Quincy, MA, 1993.

*Guice, D., *Staffing and Apparatus Study,* National Fire Academy, Emmitsburg, MD, Jan. 1991. (Executive Fire Officer Program, Applied Research Project.)

Harlow, D., "Staffing Alternatives," *Fire Command,* Vol. 53, No. 2, Feb. 1986, pp. 42-44.

---

*Written by students at the National Fire Academy, Emmitsburg, MD.

*Hendges, C. M., *Staffing of Engine and Ladder Companies—Proposals to NFPA 1500 vs. the Real World*, National Fire Academy, Sept. 1992. (Executive Fire Officer Program, Applied Research Project.)

Herzog, C. T., "Department Solves Staffing Problems with Pagers," *Fire Chief*, Vol. 35, No. 3, Mar. 1991, pp. 58, 109-113.

Hickey, H. E., *Fire Suppression Rating Schedule Handbook*, Professional Loss Control Educational Foundation, and Society of Fire Protection Engineers, Boston, MA, 1993.

*Highely, S., *Part-time Firefighters, Are They an Effective Solution for Staffing Problems?* National Fire Academy, Emmitsburg, MD, May 1994. (Executive Fire Officer Program, Applied Research Project.)

*Hoyle, K. E., *Cost-Effective Staffing: Utilizing College Students in Fire Departments*, National Fire Academy, Emmitsburg, MD, Oct. 1991. (Executive Fire Officer Program, Applied Research Project.)

*Hughes, G. M., Miller, L. C., Stinnette, E., Wenzel, J. L., and Williamson, D. R., *An Alternative Solution to Augmenting Staffing of the First-Arriving Engine Company*, National Fire Academy, Emmitsburg, MD, Feb. 3, 1992.

International City Management Association, *Managing Fire Services*, ICMA, Washington, DC, 1986.

International City Management Association, *Municipal Yearbook*, ICMA, Washington, DC, 1994.

Jacobs, D. T., *Physical Fitness for Public Safety Personnel*, National Fire Protection Association, Quincy, MA, 1990.

Jennings, C., "High-Rise Office Building Evacuation Planning: Human Factors versus "Cutting Edge" Technologies," *Journal of Applied Fire Science*, Vol. 4, No. 4, 1994/1995, pp. 289–302

Johnson, M. A., "Burn Aware: A Public Education That Works," *American Fire Journal*, Vol. 43, No. 7, July 1991, pp. 54, 61.

*Jones, K. L., *Hiring Strategies for Staffing up to an ALS System*, National Fire Academy, Emmitsburg, MD, Sept. 1989. (Executive Fire Officer Program, Applied Research Project.)

Jones, R. E., III, "Company and Platoon Response; The Efficient Use of Manpower," *Fire Command*, Vol. 53, No. 3, Mar. 1986.

Kalman, B. J, "Applying Technology to Prefire Planning," *Fire Engineering*, Vol. 146, No. 1, Jan. 1993, pp. 45–46,48–49.

*Kittelson, J., *Assessing Effective and Efficient Fire Suppression Needs*, National Fire Academy, Emmitsburg, MD, Jan. 2, 1992. (Executive Fire Officer Program, Applied Research Project.)

*Lalond, P. G., *Planning for Firefighter Staffing in a Combination Fire Department*, National Fire Academy, Emmitsburg, MD, Dec. 1992. (Executive Fire Officer Program, Applied Research Project.)

Lambert, C. E., McDaniel, M. F., and Santos, S. L., "Effectively Communicating Risk Assessments to the Public," *PSAM-II Proceeding*, An International Conference Devoted to the Advancement of System-Based Methods for the Design and Operation of Technological Systems and Processes, Vol. 3, Sessions 73-108, March 20-25, 1994, San Diego, CA, 1994, pp. 096/5-8.

Larson, R. D., "From Ashes to Education," *Firehouse*, Vol. 18, No. 1, Jan. 1993, pp. 44, 46.

*Lawrence, C., *Optimal Stafffing for Fire Attack*, National Fire Academy, Emmitsburg, MD, Sept. 1991. (Executive Fire Officer Program, Applied Research Project.)

*Leiper, R. D., *Cost-Effective Manning for Fire Attack*, National Fire Academy, Emmitsburg, MD, Oct. 1991. (Executive Officer Program, Applied Research Project.)

Lemire, Chief B. T., "Minimum Staffing: Double-Edged Sword," *Fire Chief*, Vol. 36, No. 11, Nov. 1992, p. 28.

*Lipe, H. L., *Integrating Fire and EMS Operations: A Systems Approach Maximizing Response Capabilities with a Reduction in Staffing Levels*, National Fire Academy, Emmitsburg, MD, Nov. 1991. (Executive Fire Officer Program, Applied Research Project.)

Maxwell, Capt. C., "The True Cost of Personnel," *Fire Chief*, Vol. 38, No. 5, May 1994, pp. 85–89.

Meyer, S., "Volunteer Departments Can Benefit From Master Planning," *Fire Chief*, Vol. 38, No. 10, Oct. 1994, pp. 47–50.

Millsap, S., "Planning for Disasters in Your Community," *Fire Engineering*, Vol. 147, No. 12, Dec. 1994, pp. 28–29, 32–33.

Mitchell, A., "Do We Have to Do Public Fire Safety Education?," *Fire Chief*, Vol. 40, No. 5, May 1996, pp. 54, 56, 58.

*Morrison, R. C., *Manning Levels for Engine and Ladder Companies in Small Fire Departments*, National Fire Academy, Emmitsburg, MD, 1990. (Executive Officer Program, Applied Research Project.)

"National Association of Counties, Multi-Jurisdictional Fire Protection Planning," prepared for the Federal Emergency Management Agency, U.S. Fire Administration, Emmitsburg, MD.

Pilie, G. M. ,and Croxton, G. T, "Effectively Communicating Risk to the Public and to Regulators: Can It Be Accomplished?," *PSAM-II Proceedings*, An International Conference Devoted to the Advancement of System-Based Methods for the Design and Operation of Technological Systems and Processes, Volume 3, Sessions 73-108, March 20–25, 1994, San Diego, CA, 1994, pp. 096/1-4.

Porth, D., "Recipe for a Successful Fire and Life Safety Education Program," *American Fire Journal*, Vol. 45, No. 9, Sept. 1993, pp. 23–26.

*Prater, J. W., *Alternative Staffing Strategies for the Operation of the Fire Prevention Bureau*, National Fire Academy, Emmitsburg, MD, Aug. 1990. (Executive Fire Officer Program, Applied Research Project.)

"Prefire Planning Initiatives in 1992," *Record*, Vol. 69, No. 2, March/April 1992, pp. 6–9.

"Prefire Planning Rewards and Challenges," *Record*, Vol. 70, No. 2, March/April 1993, pp. 9–13.

Research Triangle Institute, *et al.*, "Evaluating the Organization of Service Delivery," Research Triangle Institute, Center for Population and Urban-Rural Studies, Durham, NC, no date.

Research Triangle Institute, *et al.*, *Municipal Fire Service Workbook*, Government Printing Office, Washington, DC, no date.

Rhodes, P., "New Directions in Fire and Life Safety Education," *American Fire Journal*, Vol. 43, No. 10, Oct. 1991, pp. 14–15,42.

Richman, H. C., *Engine Company Fireground Operations*, 2nd ed., National Fire Protection Association, Quincy, MA, 1991.

Richman, H. C., *Truck Company Fireground Operations*, 2nd ed., National Fire Protection Association, Quincy, MA, 1991.

Rielage, R. R., "Alternative Staffing for Combination Departments," *Voice*, Vol. 22, No. 8, Aug.–Sept. 1993, pp. 32–33.

Rousey, G. C., "Some Inexpensive Ways to Bring Fire Safety Education to Your Community," *Fire Chief*, Vol. 37, No. 3, March 1992, pp. 58–59.

Sanders, R. E., "Long-Term Answer Is Public Education," *NFPA Journal*, Vol. 87, No. 6, Nov./Dec. 1993, pp. 14, 26.

*Sands, L. S., *Did the Carrot Work? A Financial Incentive for Working Weekend Hours Is Evaluated*, National Fire Academy, Emmitsburg, MD, Sept. 1990. (Executive Fire Officer Program. Applied Research Project.)

Schaenman, P., *et. al.*, "Proving Public Fire Education Works," TriData, Arlington, VA, 1990.

*Shanklin, B. Y., *Cost-Effective Fire Department Staffing—Minimum Staffing vs. Constant Staffing*, National Fire Academy, Emmitsburg, MD, June 1993. (Executive Fire Officer Program, Applied Research Project.)

Shearer, R. W., "Creating Defensible Staffing Strategies," *Fire Chief*, Vol. 33, No. 10, Oct. 1989, pp. 42–44.

Shields, J., "Fire Safety Education and Training in Transition," *Fire Engineers Journal*, Vol. 50, No. 159, Dec. 1990, pp. 12–15.

Spillman, D., "Community Analysis for Emergency Management Planning," *Speaking of Fire*, Vol. 3, No. 2, Summer 1996, pp. 16–17.

Stittleburg, P. C., "Staffing, NFPA 1500, and the Courtroom," *Fire Chief*, Vol. 36, No. 8, Aug. 1992, p. 28.

Teele, B. W., *NFPA 1500 Handbook*, National Fire Protection Association, Quincy, MA, 1993.

Tokle, G.,ed., *Hazardous Materials Response Handbook*, 2nd ed., National Fire Protection Association, Quincy, MA, 1992.

*Tucker, T. A., *As Nationally Recognized Associations Debate Minimums of 3-, 4-, and 7-Man Minimums for Interior Structural Firefighting, What's Going on in Florida Fire Departments Serving from 8,000–20,000 People?* National Fire Academy, Emmitsburg, MD, June 1993. (Executive Fire Officer Program, Applied Research Project.)

*Williams, H. M., *The Inspector/Firefighter Paradigm*, National Fire Academy, Emmitsburg, MD, June 1993. (Executive Fire Officer Program, Applied Research Project.)

*Written by students at the National Fire Academy, Emmitsburg, MD.

FIRE DEPARTMENT INFORMATION SYSTEMS

Revised by Clifford S. Harvey

Smaller fire departments face many of the same decisions and need much of the same data as large fire departments. Larger fire departments have used computers for many years to capture, store, and quickly recall or summarize data. With the availability of low-cost computer power, many of the applications discussed in this chapter can be automated in smaller departments by using desktop computers. If computer capability is not available, a good manual record keeping system, with periodic analysis of the data, can often support the information needs of the department.

Today, effective management is more critical than ever to the fire service. Economic conditions have reduced local government tax revenues, leading to diminished resources; yet the fire service, like all government units, is being asked to maintain, and even increase, service levels.

The contradiction of "doing more with less" can be resolved only with improved management practices that result in higher productivity. There are many paths to this goal; one such path is information technology.

While one might argue that the computer is not a panacea, information technology has become a pervasive management tool in many fire departments. Specifically, the computer is being used to help:

1. Strategically position capital resources, such as fire stations.
2. Dispatch and manage fire-fighting resources.
3. Manage personnel resources.
4. Control fire department performance.
5. Plan and control budgets.
6. Manage fleet maintenance.
7. Combat arson.
8. Improve organizational communications.
9. Manage energy consumption.
10. Manage fire prevention inspections and permits.
11. Manage emergency medical operations.
12. Manage financial resources.
13. Manage pre-fire planning operations.

Few fire departments have automated all of these applications. However, almost all fire departments protecting urban and suburban areas use a computer in one or more of these areas.

Clifford S. Harvey was an assistant chief (retired) with the Boulder, CO, Fire Department, and is now president of Fire Mark Limited, located in Florence, OR.

For related information, see Section 11, Chapter 2, "Fire Data Collection and Data bases"; and Section 11, Chapter 3, "Use of Fire Incident Data and Statistics."

MANAGEMENT ISSUES

The successful function of a fire department's information resource depends upon how that function is organized and managed. There is, of course, no "single best way" to manage the information resource, since every situation is unique.

There is growing recognition that the information resources of an organization are crucial to that organization's overall effectiveness. Consequently, the information resource is a valuable asset that must be well managed. Many fire departments now have division-chief-level personnel responsible for this resource. The term "information resource," as used in this chapter, includes computers, software, data, and personnel.

A balanced information resource fulfills three needs: (1) strategic planning, (2) management control, and (3) support of daily operations.

Strategic planning is the process by which management allocates long-range resources, such as fire stations, fire-fighting equipment, and fire fighters. Some fire departments have developed sophisticated models for this purpose, while others have contracted for the models and services of outside resource groups. Most fire departments, however, depend upon the statistical information from fire incident and emergency medical reporting systems.

Management control is necessary in several areas: (1) finance, (2) personnel, (3) fire prevention, (4) fire investigation, (5) fire suppression, and (6) energy management. Control information is often provided as a by-product of systems designed to support day-to-day operations. For example, personnel control information may be secondary to the real reason the personnel system was developed—to increase the efficiency of collecting the personnel record data.

Most fire service information systems are designed to support day-to-day operations. The most common examples of this type of system are computer-aided dispatch (CAD) and "enhanced" 911 (E911). Other systems support financial management, fire investigations, equipment inventory and maintenance, inspections, permits, organizational communications, and personnel management.

While systems may be classified as suggested above, they are rarely developed explicitly in that fashion. Rather, the packaging of information is a function of priorities, systems planning, organizational needs, and other situational factors. There are, however, certain recurring patterns to the way information is packaged into application systems.

COMPUTER-AIDED DISPATCH

Fundamentally, computer-aided dispatch (CAD) is oriented toward operations. It is here that resource status is maintained, alarms are received, fire stations are alerted, and situations are managed. Its purpose is to support current activity rather than provide managerial control or strategic planning information. However, the functions of CAD systems often go beyond operational support. In an integrated information systems environment, the CAD system may produce nonoperational information (e.g., incident reporting data) and may use nonreal time inputs (e.g., occupancy data).

CAD systems support several functions of the typical dispatching operation:

1. *Alarm receipt.* Telecommunications management, fire alarm box decoding, central station communications, etc.
2. *Communications and signaling management.* Alerting (unit/station, two-way exchange of data, audit switching, frequency selection, etc.).
3. *Status maintenance.* Resources, equipment, personnel, system supervisor, etc.
4. *Response assignment.* Dynamic assignment based on current availability and relative location of fire-fighting resources to the incident.

Although CAD is usually credited with reducing response time, the payoff is not necessarily in this area. Some departments even report an increase in response time after implementing a CAD system. The real benefits of CAD occur in the management of fire suppression activity. Thus, rapid recall of information, such as apparatus status and occupancy data, is at least as important as response time. Occupancy data coupled with response information can be invaluable to the fire prevention officer, when he/she is planning future inspection programs.

Choosing a correct approach to implementing a CAD system is a multifaceted problem. The first facet of the problem is related to policy. In some cases, dispatching is a shared police-fire responsibility. In others, dispatch serves multiple jurisdictions. The organizational situation can have an important influence on other aspects of the approach as well as the ultimate usefulness of the system.

A major starting point in CAD development is a decision on the scope of the system. Is the system to be developed only for CAD, or will it include other applications? Most departments elect to develop a stand-alone CAD system with the necessary interfaces to other applications, e.g., the fire incident reporting system. Others may elect to more closely integrate their CAD system with other computerized record-keeping systems in the department. It is important, however, that, regardless of how many applications may be bundled together, the CAD operation is the top priority and is designed to continue to operate even if some of the other applications are temporarily unavailable.

Closely related to the issue of scope is the degree of modularity built into the system. A highly modular system permits the system to be expanded to include enhanced CAD functions as well as other applications. It also allows new software to be tested in an off-line mode and operators to be trained without interfering with operations. Modularity is important in all applications, but it is especially critical in CAD systems because the systems are subject to almost constant change.

CAD systems are often developed by an outside contractor, although effective systems have been developed by internal teams. If

an outside contractor is used, the fire department must be extensively involved in developing the system requirements and specifications to ensure that the final system meets their needs. Likewise, when the system is installed, fire department personnel must be involved so they understand all aspects of the system as designed and are properly trained to maintain the system. Good written documentation is extremely important.

Communications is another aspect in selecting an approach. Standard telephone service has typically been used for communications between the dispatch center and the fire stations. Increasingly, other ways are being found to transmit data. The 800 MHz radio frequencies have been used successfully, as have microwave links dedicated data lines. Digital pagers, computer terminals in fire apparatus with both paper printouts and liquid crystal display (LCD) readouts, fax machines, and cellular telephones are also becoming increasingly popular for dispatching apparatus and communicating with company officers and incident commanders.

ENHANCED 911 SYSTEMS

Enhanced 911, or E911, is an emergency phone system that displays the caller's telephone number, address, response district, and other important information automatically on the dispatcher's computer screen, along with dispatch information for fire, police, Emergency Medical Service (EMS), or other emergency responders. This system is invaluable, especially in those cases where the caller is unable to give information fully or clearly because of language barriers, confusion or excitement caused by the incident, age, or even lack of knowledge of the area. Such a system can have a dramatic effect on reducing response times. Because it identifies the actual telephone location from where the call is placed, it also helps reduce false alarms.

This system is being widely installed throughout the United States. Even more sophisticated systems can provide information concerning previous calls to the same address, hazardous materials stored on-site, day care or other life-safety-intensive operations at the address, and other information that is valuable to the responding departments.

PERSONNEL APPLICATIONS

Personnel systems are among the most complex of all fire department applications. They may be part of a community-wide personnel system or may be specific to the fire department. Consisting of a number of subapplications, these systems are important in large departments to support the multiple functions of personnel administration, such as basic personnel records management, time and attendance reporting, payroll (sometimes part of financial management), duty schedules, physical fitness and medical records, employee evaluation, training records, position control, badge history and control, and disability/injury analysis. Without computer support for the personnel administration functions, large fire departments would find it difficult, if not impossible, to maintain all the necessary information, and little or no analysis of the collected data would be available.

The operational benefits are obvious. In a 1,000-person fire department, a large volume of information is needed to carry out standard personnel functions. Fire fighters must be scheduled and personnel balanced daily. Detailed training records are required for each department member. Time and attendance records are necessary for payroll. Employee records must be evaluated and considered for promotion. These functions require an efficient way of storing and retrieving information.

The information that is useful for day-to-day operations also has residual value for strategic planning and managerial control. For example, the roster data may be aggregated in such a way that is useful for position control or personnel resource planning. Time and attendance data may be reviewed by management in an effort to control excessive sick leave or overtime. Information needed for strategic planning and managerial control cannot be defined prior to when it is needed. Therefore, it is useful to have the facility for answering *ad hoc* inquiries using the operational data base.

At the highest level of design, the approach taken will depend upon the relationship of fire department personnel functions to city (or county) personnel functions. For example, the payroll may be prepared by the city from prescribed time and attendance records. In cases like this, it may be difficult for the fire department to develop some of the components of a personnel system.

Another aspect of the design approach is the degree to which the components of the personnel system are integrated. In most cases, the personnel system has been built over several years, with one component implemented at a time. There may not have been an overall design that provides for integration of files and input. A comprehensively planned system provides for integration, even if the components are implemented separately.

FLEET MANAGEMENT

After the personnel costs, capital and operational expenditures for vehicles and equipment represent the greatest use of fire department funds. Fleet management systems provide the information needed for budgeting and controlling those expenditures as well as for ensuring that the fleet is maintained in a high state of readiness. The need for information about the fire department fleet exists at all levels: strategic, managerial, and operational.

To accomplish these purposes, fleet management systems may involve several components: fleet inventory, maintenance records, vehicle replacement, fuel consumption, internal billing for vehicle use (e.g., in the case of citywide carpools), tax reporting, and performance analysis. Some or all of these components may be present in any given system.

At the strategic level, information about the existing fleet is needed to plan and budget for vehicle acquisition. To make sound replacement decisions, it is important to have accurate information on maintenance frequency, operating costs, age, etc., for each major piece of equipment. Indeed, those decisions are often so complicated and have such political overtones that decision support software (operations research models) is needed to help analyze the data.

At the managerial control level, fleet managers need computer support to control preventive maintenance intervals, work in progress, gasoline mileage, and staffing of the maintenance facility. Their staff also need ready access to information, such as vehicle performance, previous repair records, parts inventory and interchangeable parts data, and standard repair times. Likewise, administrative services managers need financial data on expenses for budgetary control.

Personnel at the operational level also need information from a fleet management system. Most important, of course, is the ability to have information on vehicle availability. While this requirement is not always provided by the fleet management system per se, dispatchers need to know what equipment is ready to respond to a given type of emergency.

Few fire departments develop their own fleet management systems. They either use a commercially available system or a citywide fleet management system. Commercial systems may be difficult to interface directly with other department or city systems. Also, added capabilities, such as vehicle replacement models, may not easily integrate with these systems. Yet, this approach can be imple-

mented easily and is probably cost-effective, especially for fire departments that do not have highly integrated systems.

The citywide approach also suffers from some of the same deficiencies. Since citywide systems are typically designed for standard street vehicles, they may ignore certain features that would be useful to fire departments. For example, the city system may not have a provision for capturing and maintaining special data on apparatus components, or may not provide for maintenance records of special equipment unique to the fire department. If these deficiencies are to be eliminated, the fire department must take an active part in defining information requirements. When this is done, the citywide approach can be quite worthwhile.

INCIDENT REPORTING SYSTEMS

Incident reporting systems are used to maintain statistics on fire, medical emergencies, and other emergencies to which a fire department responds. This application is one of the most common uses of information technology.

The methods of capturing data are rapidly changing. Batch data entry, where a form or forms are completed after each incident and the data is input in batches at a computer by someone other than the person completing the form, is giving way to applications where the fire officer enters the data at a computer terminal in the fire station.

Departments with a CAD system usually collect at least some of the incident data as a by-product of a CAD operation. These data can then be displayed on a networked computer or terminal at the fire station where data collected at the incident scene can be added by the company officer or officer in charge. Allowing the officer in charge to enter the data directly has the advantage of ensuring completeness and correctness, because the data can be edited as they are entered and mistakes corrected immediately.

Casualty Reporting

Casualty reporting systems are generally subsets of incident reporting systems. As such, they are linked to the incident record so that cause and effect studies can be done. Typically, records on fire fighter casualties and civilian casualties are maintained separately, as the information desired or level of detail is different for the two.

Data on civilians injured at fires are useful for supporting injury reduction programs. These programs can be educational in nature to correct inappropriate behavior or may be focused at either getting unsafe products off the market or redesigned to make them safer. Injury data also are important to code and standard developers interested in controlling the environment at times of fires so that injuries are reduced.

Fire fighters experience a high number of injuries while doing their jobs. Tracking those injuries is important and may involve more than one part of the fire department's information system. Data about the cause and circumstances of the injury can assist in making decisions for additional training, equipment redesign, or operational changes. Medical data about the injury are often important for worker compensation claims. Data on potential time lost are needed to schedule other fire fighters to fill the shift and could affect the overtime budget.

The Basic Casualty Report, Form 902G, in NFPA 902, *Fire Reporting Field Incident Guide,* provides a method of uniformly gathering casualty data for both civilians and fire fighters. This report provides for the classification of much of the collected data so they can be easily summarized and analyzed to support injury reduction programs. This form is not designed to be a medical form for reporting to workers' compensation, although some of the basic data might be applicable to both needs.

EMERGENCY MEDICAL SERVICE

Some level of emergency medical service (EMS) is now provided by many fire departments. Fire equipment and fire fighters are generally dispatched to perform or support EMS activities. The fire department may or may not be involved in the actual patient transport. Ambulances, whether operated by the fire department, another city agency, or the private sector, are often dispatched by the fire department.

In many jurisdictions, the cost of emergency medical service is billed to the patient. Since this represents a substantial revenue source for the city or county, the billing process is the focus of most EMS systems. Other components of an EMS data system include: (1) patient care issues and (2) resource needs and allocation.

The statistics generated from EMS systems are used in both long-range plans and operational planning and control. Response times, locations of incidents, and units responding are required to plan and budget for resources and determine if the resources are being deployed properly.

In some fire departments, separate systems have been designed to handle both the EMS billing system and the EMS incident records. In other departments, data from the billing system are transferred to the fire incident reporting system from which operational statistics are generated. Having all incident data on the same data base is helpful in studying fire department emergency activity.

A Basic EMS Report, Form 902H, is found in NFPA 902. This "generic" report can be used by fire departments to keep track of their EMS activities and can be amended to reflect their specific EMS reporting needs.

FIRE PREVENTION APPLICATIONS

Fire prevention management information systems are inventory and tracking systems for storing: (1) fire-protection-related information on a city's buildings and (2) information on the fire department's contacts with those buildings. The heart of a fire prevention system is a flexible, address-based file of city properties. The application is useful in planning and managing fire prevention activities, controlling hazardous contents, and pre-fire planning for fire suppression.

Fire prevention records in many communities still consist of long, often handwritten forms and narratives filed in individual folders, one for each property. This makes it virtually impossible to develop any information on patterns of hazards or violations across the city or for particular property classes. Even within a file on a single property, it can be difficult to ascertain whether the fire department has been providing the kind of frequency of contacts called for by the department's policies. This difficulty exists because such files are often only dumping grounds for any and all papers bearing the property's address. Minor, unofficial correspondence and news clips may share space with basic inspection reports and detailed, lengthy construction plans. Files on properties not covered by the fire prevention code (such as single-family dwellings), on fire education contacts, and on contacts by other agencies (such as the building and housing departments) are generally nonexistent or wholly separate and incompatible with the fire inspection files. Fire incident records are rarely tied to files on the violation and hazard histories that may have caused the fires. In fact, it would be difficult to identify any aspect of a fire department's record-keeping that is more in need of refinement and automation than fire prevention.

The first requirement for an adequate fire prevention system is a comprehensive address-based inventory of all properties in the

city. This is a very difficult file to create. Past studies have shown that, despite conscious efforts, fires are reported in properties that should have been subject to fire inspections but were not shown in the inspection records. These were not new businesses or businesses operating illegally out of homes—they were legitimate businesses of some years' standing that simply had been overlooked.

It may be impossible to guarantee file completeness, but this can be improved. Address files should be linked to the fire incident records, preferably through a link to the city's CAD system, so that emergency runs to corresponding unlisted addresses will lead to the creation of an inspection file.

Sources for both the initial building inventory and periodic updates are the city's tax records, the housing department, the building department, any inspection agency, and city-owned utilities, such as the water, sewer, and light departments. On the other hand, some lists (e.g., the telephone company "Yellow Pages," cross-indexes, and mail carrier routes) tend to be unwieldy or poor sources for finding missing addresses.

The fire inspection procedure itself can be a source of effective property list updates. Inspections conducted on a "block by block, always go around the block" basis allow inspectors to see properties they might miss if they were to organize their routes with strict attention to those properties identified only by the computer.

Once the file is created, it can be used as a basis for organizing and scheduling inspections. The system can be set to print inspection reports for properties scheduled for inspection in any given time period (weekly, biweekly, or monthly), thereby allowing inspectors to plan their work. Likewise, the need for reinspections to check that hazards are abated or other violations corrected can be easily tracked. Inspector workloads can be balanced either by the number of inspections or by the total time spent on inspections if statistics are kept on the typical duration of inspections by type and size of property.

Accuracy is important when automating data on buildings in a community. Although the occupants of a building change from time to time, the building itself usually doesn't change much. Some jurisdictions use the Basic Structure Report, Form 903SR, to gather basic information about structures. This form is an excellent way to capture data about buildings that need to be inspected periodically. Furthermore, the information gathered is extremely valuable for pre-fire planning efforts at that structure.

For all buildings, especially those that contain more than one occupancy, the Basic Occupancy Report, Form 903TR, provides a way to gather information (e.g., who is using the building) for both inspection and training functions.

Both of these reports (i.e., Forms 903SR and 903TR) are included in NFPA 903, *Fire Reporting Property Survey Guide*, along with an explanation of each entry. The best asset of such an information-gathering system is that the data is classified in a uniform manner, based on NFPA 901, *Standard Classifications for Incident Reporting and Fire Protection Data*. With the data classified, it is easy to analyze the data, assess the severity of hazards being reported, and retrieve records for properties that have similar characteristics. This allows a supervisor to prioritize a fire department's inspection functions, based on the severity of the potential fire problem, especially when that department may not have sufficient resources to inspect every building as often as might be desired.

Fire prevention systems can also be used in the control of hazardous materials. In addition to supporting license and permit controls on hazardous materials and processes, a fire prevention file can be expanded to include information on the type, quantity, and location of hazardous materials and the type, frequency, and location of hazardous activities. The prime difficulty in making such a system work is that the situation is so changeable. Types, quantities, and locations of materials may fluctuate widely, according to the demands of the business. Annual or even semiannual updating of a central file may be insufficient to give a picture of the city's true hazardous materials profile.

Ownership information on properties can be among the most difficult to obtain. Current ownership probably is not needed until legal proceedings have to be instituted. It is easier, and more reliable, to check legal ownership in those relatively few cases as they occur than to try to develop and maintain a file on legal ownership for all properties. It is important, however, to know who is responsible for the property so that contact can be made in the event of an emergency or to schedule inspections.

ARSON INFORMATION SYSTEMS

The principal advances in the area of arson information systems have been carried out through cooperative neighborhood or community block organizations working with the fire service in their communities. A number of computer-based data systems have been developed to profile the properties in the neighborhood and assist in identifying arson-prone buildings. The community group or public officials are then able to implement intervention strategies appropriate for the situation.

Much of the early work sponsored by the Federal Emergency Management Agency (FEMA) has led to a developed and tested microcomputer-based arson information management system (AIMS). The system provides both identification of arson-prone buildings and a case-management system for fire investigators.

There are many municipal and private data bases or record systems that are typically used to support an arson information management system. These include data from the fire department incident reporting system and code-enforcement program, the building department, the assessor's office, the police department crime reporting system, the prosecutor's office, the tax collector's office, the registry of deeds, mortgage companies, and the insurance industry.

Many jurisdictions are choosing to computerize the incident follow-up report, Form 904I, found in NFPA 904, *Incident Follow-up Report Guide*, in order to enhance the information available on significant fires within these jurisdictions. Analysis of these data may reveal patterns showing types of fires that could point to deliberately set fires and other criminal activity.

Data collection and verification from all available sources are expensive and time consuming. However, the ability for automatic integration of existing computer data bases is becoming more successful, due to improvements in software that can match addresses and other property identifiers.

OFFICE AUTOMATION

Office automation is a term used in connection with a group of applications that support clerical and managerial activities. These applications include word processing, spread sheets for financial and numeric data, data base development for various record-keeping applications, electronic mail, "tickler" systems, teleconferencing and video conferencing, and personal calendar management.

The purpose of these applications is to make the executive and the clerical worker more productive. While one might say the same about traditional information systems, the focus here is different. Traditional information system applications are process oriented; office automation, on the other hand, is more people oriented. The objective is to make office personnel more productive by allowing them to use computers as a natural extension of their minds and bodies.

Executives and clerical workers need office automation to cope with the complexities of the modern office. Any fire agency must contend with an explosion of information in the form of reports, forms, memoranda, and computer files. The efficient production and effective management of this information resource strains even the most dedicated staff. The computer, through automation applications, extends human intellectual capabilities.

To provide the needed integration while retaining the flexibility for stand-alone units, some organizations are turning to shared data systems. These are full-fledged computer systems specifically designed (both in terms of hardware and software) to handle office automation, as well as more advanced systems that support standard data processing chores. As user needs grow, more systems can be added and networked together. The individual office computer is connected to the central computer network. This allows the user to have a stand-alone workstation on a desktop that can perform all the office automation functions and still have access to the data bases retained on the central computer.

There are several variations on these themes. Depending upon how data are captured and used throughout the network, various functions can be performed either by the desktop computers, network servers, or the central computer. For example, a desktop computer might be used for word processing as a stand-alone task but be tied to a network system to transmit electronic mail, link to other computers, and make inquiries into data bases on a central computer. The network provides the mechanism to integrate all these activities.

Internal Data Bases

With the advent of desktop and laptop computers, commercial data base packages are becoming the accepted way of gathering information for a fire department. These generic data base packages are known as relational data bases, as opposed to flat data bases. Where flat data bases are acceptable for mailing lists and other mundane data-gathering chores, relational data bases not only accomplish those tasks, but go much further in their usefulness. Relational means information can be compared in relation to other information; for example, "what if " questions are easily and rapidly answered when data are gathered using a relational data base. If a fire chief is asked to cut the budget and the only way to do so is to close a station for some period of time or send fewer engines, he or she can quickly determine what time period represents the best time to accomplish that reduction in service. The computer will quickly be able to analyze incident data to show the best time to close down a station, based on number of alarms, severity of fires, or other criteria the fire chief may want to use so the closure will have the smallest impact on the services provided by the department.

Relational data bases, even though created for use on desktop computers, are extremely powerful, highly useful programs. The ease of learning and using of these programs has improved dramatically over the past several years, such that most departments, no matter how large or small, would find them very valuable tools for all types of information management.

External Data Bases

Much of the technical and legal information needed by fire departments exists on commercial data bases. These data bases are created for general use, and users are charged for access on the basis of a flat monthly rate and/or the time actually used. In a very real sense, commercial data bases are a public utility.

The source for the information stored in these data bases is primarily current technical literature and legal materials. In some cases, abstracts of documents and books are stored, while in others, the complete text is stored.

Data bases are provided by numerous firms. Literally hundreds of data bases covering various fields are available today through scores of companies, such as *Reader's Digest*, *The Source*, and *The New York Times*.

The data bases of greatest utility to the fire service are probably those dealing with hazardous materials and legal information. Chemical and hazardous materials data bases are now available that allow a user to search for a hazardous material by chemical structure, trade name, generic name, CAS registry number, chemical or physical properties, and textual description. These data bases can be powerful tools in handling chemical spills, fires involving hazardous materials, and some emergency medical situations.

Legal information systems are generally of two types: (1) full-text storage and (2) search and abstracts (headnotes) only. Both types cover case law, statutory law, and some administrative rulings. Mead Data Central's LEXIS* is representative of the full text approach. LEXIS users enter key words, and documents containing the proper sequence, count, and/or combination of those words are displayed. A service of West Publishing Company (WESTLAW)† allows the same type of search, but the object of the search is the West headnotes, rather than the full text.

The Internet is rapidly becoming a source of information on almost any subject. Some of this information can have relevance in supporting programs or in just establishing contacts with other persons dealing with the same problems. One note of caution that needs to be raised with the use of data from the Internet is that its source and validity may need to be verified, as there is generally no professional oversight as to what is posted other than that posted by technical organizations. NFPA maintains a home page on the worldwide web at http://www.nfpa.org.

With the rapid introduction of two-way cable television systems and videotext services, there will likely be more data bases and quick access methods available to fire departments.

Financial Applications

Financial applications include those systems that support budgeting, budget control, general ledger, accounts payable, and billing. Payroll is sometimes included, although more often it is a part of the personnel system. Billing may also be included in specific systems (e.g., emergency medical for patients, and brush clearance for owners needing removal of illegal brush). The implementation of financial systems in fire departments has been slow, partly because the support of staff functions, such as finance, has been secondary to the support of line activities. In addition, municipality-wide financial systems often preempt those in the individual departments.

Perhaps the most critical need for financial information in the fire department occurs during the yearly budget cycle. The arduous and sometimes politically risky process of allocating resources requires considerable analysis. Decisions must be supported by the kinds of facts best stored and manipulated by a computerized financial system.

Budget control is also a high priority for fire agencies. Management needs to know if the department is staying within budget. Even if central systems are available for this purpose, they may not provide data in sufficient detail or in a form that is useful to the fire department.

The remaining needs addressed by financial systems are primarily operation oriented: general ledger, billing, accounts payable, and payroll. The last three requirements more often than not are a part of associated applications (e.g., payroll in the personnel system, billing in the emergency medical system, and accounts payable in the procurement system). Manual or automated "bridges" are then provided to general ledger and budgeting systems.

The most common design approach to financial systems is a strongly centralized one. This is not surprising in view of the fiduciary responsibility of the chief financial officer in a city or county.

Energy Management

Energy management systems provide information for control of energy costs, including information about the consumption of electric power, natural gas, and heating oil. In some cases, the use of motor vehicle fuel may be monitored, although this function is more properly a function of fleet management systems.

The typical approach to the design of energy management systems is to capture energy billing data, update master files, and make various reports available to administrators. This can be done using one of the relational data bases referred to earlier, or a spread sheet. More sophisticated approaches include real-time monitoring systems and even real-time energy control. Computer-based evaluation models are also used to assess the impact of conservation measures. With rising energy costs, the cost of implementing a computerized energy management system can be recovered quickly through increased energy resource efficiency.

SYSTEMS INTEGRATION

Many organizations do not have a systematic approach to information systems development. The approach is more often piecemeal without an overall plan for systems integration. By analogy, an airplane designed this way might have three wings, an engine with insufficient power for takeoff, and a fuel tank where there should be cargo or passengers. In short, while each subsystem might fulfill its intended function, the overall system is a failure.

The answer to this problem is information resource planning. Like other organizations, the fire department needs an overall plan for integrating the various component subsystems. This does not mean that all components have to be implemented at one time to have a workable system. But it does require a strategic plan for eventual integration and an action plan indicating how to get there.

Normally, there are two levels of integration to be considered. The first is integration of fire department systems with citywide systems. Considerable benefits can arise from, say, having a central data base of occupancy data that can be used by the assessor, tax department, and fire department. However, information resource managers in the fire department should be generally wary of data bases over which they have no control. Still, the potential benefits are well worth the necessary effort in working toward a well-developed citywide plan. At minimum, the fire department should think through potential interfaces with city systems and provide the "hooks" for future connections.

The second level of integration is within the fire department itself. Here, there is significant opportunity for controlling the design of systems and data bases and their interfaces. The remainder of this chapter discusses some techniques for planning such an integrated systems environment.

An organization uses information systems to support strategic planning, managerial control, and operational activities. These ac-

---

*For more information on LEXIS, contact Mead Data Central, 200 Park Avenue, New York, NY 10166.

† For more information on WESTLAW, contact West Publishing Company, 50 W. Kellogg Blvd., St. Paul, MN 55165.

tivities may be classified into a number of business functions (e.g., scheduling personnel, dispatching, and budgeting). In turn, each function requires a certain subset of the organizational data base. Obviously, different functions may require overlapping data subsets.

This type of analysis suggests two ways to integrate systems. The first approach is related to the way the business functions are "packaged" into application systems. For example, a typical entry-level application in a fire department is computer-aided dispatch (CAD). In most cases, CAD is implemented on a stand-alone minicomputer. However, the boundary of the system could be expanded to include other functions, such as inspections and incident reporting. The rationale for this would be based on a broader definition of the dispatch function. In the case of inspections, it could be assumed that the fire fighters need information about hazardous materials, occupancy, guard dogs on premises, etc. In the case of incidence data, CAD collects much of the data needed to report fire incident statistics. Thus, while the focus is still on CAD, the systems boundary would be enlarged to include "subordinate" or supporting functions. The supporting functions would be developed as subsystems with their own data bases (or files) and "keys" provided to tie them together.

The other approach is based on a "data model" of the organization. The focus here is on the data and their interrelationships. Whereas the data traditionally have been owned by the application, now the application has become almost insignificant, replaced by "transaction sets" that update and query the corporate data base.

In reality, one does not define a single physical data base but rather groups the data bases according to some coherent scheme. (Relational data bases are appropriate.) Such a scheme should ensure that like entities are grouped together. Although other arrangements are possible, this line of thinking suggests the following groupings:

1. *Personnel data base:* Includes data on scheduling of personnel, training, payroll (if not part of the financial system), accidents, assignments, etc.
2. *Financial data base:* Includes data on budgets, flow of funds, etc. This data base is likely to be citywide in scope.
3. *Incidents data base:* Includes historical data on closed incidents, both fire and emergency medical.
4. *Fire prevention data base:* Includes a building inventory and data on occupancy, ownership, hazardous materials, etc. Arson data may also be included in this data base.
5. *Operational data base:* Includes data about incidents in progress. These data are maintained for relatively short durations. At the close of an incident, or periodically, they may be moved to the incidents data base. This prevents overloading the real-time CAD system with non-real-time data.
6. *Supplies and equipment data base:* Includes supplies inventory and equipment inventory and may include fleet maintenance. This is typically a citywide data base.

The data base planning approach has the potential of providing a higher level of integration and a more stable systems environment than the applications planning approach. It is not without problems, however. Users may not want to relinquish control over their data to a central data administration function. Data modeling requires specialized skills often not available to fire departments. Perhaps most importantly, it can lead to "analysis paralysis," the reluctance to implement anything until the last detail has been planned. While this is true of any planning approach, the tightly coupled nature of data base planning increases the danger of overplanning. Yet, the concept of data models is too powerful to be ignored.

There are numerous methodologies for defining both application and data base boundaries, developing data models, and planning the integrated systems environment. Three methodologies are mentioned here, but there are numerous others.

1. The Business Systems Planning (BSP) methodology was developed by IBM and has been used in a variety of organizations. It is extensive in scope and involves users at all levels to define the business functions and data classes. It is especially useful in providing an overview of the systems environment, but is usually not updated on a regular basis because of the resources required.
2. The comprehensive planning methodology uses an approach where 1-year and 5-year plans are updated annually. The planning process should originate with the central information systems department even if the fire department has its own data processing capability. The fire department can then develop its plan to coordinate with the central information system plan. This will allow for better future integration with citywide systems.
3. Structured methodology is an approach used on a project level.[1] Strictly speaking, this is a project development rather than a strategic planning methodology. The hierarchical nature of the approach permits adaptation for planning purposes, although the level of planning is too detailed for pure long-range planning; however, it can be a useful adjunct to such planning.

These three approaches are just some of the many methodologies that are available to the systems planner. There are many others, some of which are fully automated. Whatever the approach taken, the information systems manager is advised to select one that addresses the problems of systems and/or data base integration. The piecemeal approach is more costly and frustrating in the long run than a carefully planned strategy.

BIBLIOGRAPHY

Reference Cited

1. Game, C., and Sarson, T., *Structured Systems Analysis: Tools and Techniques*, 2nd ed., Prentice-Hall, Englewood Cliffs, NJ, 1979.

NFPA Codes, Standards, and Recommended Practices

Reference to the following NFPA codes, standards, and recommended practices will provide further information on analysis of fire department information systems discussed in this chapter. (See the latest version of *The NFPA Catalog* for availability of current editions of the following documents.)

NFPA 901, *Standard Classifications for Incident Reporting and Fire Protection Data*
NFPA 902, *Fire Reporting Field Incident Guide*
NFPA 903, *Fire Reporting Property Survey Guide*
NFPA 904, *Incident Follow-up Report Guide*

Additional Reading

Barry, T. F., "Fire and Explosion Quantitative Risk Assessment: A Decision Support Methodology," *Proceedings* of the International Conference on Fire Research and Engineering, Sept. 10–15, 1995, Orlando, FL, SFPE, Boston, MA, 1995, pp. 253–257.
Centaur Associates, Inc., "Fire Suppression Crew Size: Literature Review," prepared for the Federal Emergency Management Agency, U.S. Fire Administration, Washington, DC, 1982.
Chapman, E. F., "Guidelines for Strategic Decision Making at High-Rise Fires," *Fire Engineering*, Vol. 148, No. 9, Sept. 1995, pp. 67–70.
Coleman, R. J., and Granito, J. A., eds., *Managing Fire Services*, 2nd ed., International City Management Association, Washington, DC, 1988.
Doerschler, G. K., "Mapping the Future," *Fire Command*, Vol. 56, No. 8, Aug. 1989, pp. 18–22.

Donahue, M. L., "CHEMTREC and CHEMNET Offer Timely Hazmat Data, Response Services," *Fire Chief*, Vol. 34, No. 10, Oct. 1990, pp. 40–43.

Fire Protection Association, "Investigation of Vehicle Fires. Fire Safety Data Arson Dossier AR5," Fire Protection Assoc., London, England, AR5, 1991.

Goliash, B. I., "Ye Cads!" *Fire Command*, Vol. 56, No. 10, Oct. 1989, pp. 16–21.

Griffith, S. J., and Munday, J. W., "Legal Implications of Real Fire Data Collection and Computer Modeling," Metropolitan Police Forensic Science Laboratory, London, England, ISBN 0-9516320-9-4; *Proceedings* of the Seventh International Interflam Conference, Interflam '96, March 26-28, 1996, Cambridge, England, Interscience Communications Ltd., London, England, 1996, pp. 1027–1031.

Hall, J. R., Jr., "Practical Rules for Selecting Fire Science Tools Appropriate to the Decision to be Made," National Fire Protection Association, Quincy, MA, Tyne and Wear Metropolitan Fire Department and The Institution of Fire Engineers, Northern Branch Joint Conference on Fire Risk Assessment: Opportunity or Problem?, March 26, 1993, 1993, pp. 1–9; Sixth International Fire Conference on Fire Safety, Interflam '93, March 30–April 1, 1993, Oxford England, Interscience Communications Ltd., London, England, 1993, pp. 75–82.

Hall, J. R., Jr., *et al.*, *Fire Code Inspections and Fire Prevention: What Methods Lead to Success,* National Fire Protection Association, Quincy, MA, 1979.

Hollis, J. B., "Engineering of a Mobile Data System for the London Fire Brigade," *Electronic Technology*, Vol. 22, No. 1, 1988, pp. 13–15.

Holmes, S., "Extending Information Data Bases for Effective Incident Control," *Fire*, Vol. 86, No. 1067, May 1994, pp. 41–42.

Jackson, D. P., ed., *Fire Department Information Technologies in Nine Western Communities*, Center for Information Resource Management, California State University, Los Angeles, CA, 1982.

Larson, D. A., and Small, R. D., "Analysis of Large Urban Fire Departments—Part I: Theory. Part II: Parametric Analysis and Model City Simulations," Pacific-Sierra Research Corp., Los Angeles, CA, 1982.

Lehner, P. E., and Brannigan, V. M., "Computational Domain Models and Regulatory Decision Making," *Proceedings* of the International Conference on Fire Research and Engineering, Sept. 10–15, 1995, Orlando, FL, SFPE, Boston, MA, 1995, pp. 247–252.

Lewis, R. J., "Fire Station Location Studies Using a Microcomputer," TR 86–1, Society of Fire Protection Engineers, Boston, MA, 1986.

Lippiatt, B. C., and Weber, S. F., "HIST 1.0: Decision Support Software for Rating Buildings by Historic Significance," National Institute of Standards and Technology, Gaithersburg, MD, General Services Administration, Washington, DC, NISTIR 5683, Oct. 1995.

McGarry, F. A., "Toward a Safer Built Environment: New York State's Building Materials and Finishes Fire Gas Toxicity Data File," Special Conference on Recent Advances in Flame Retardancy of Polymeric Materials—Materials, Applications, Industry Developments, Markets, May 15–17, 1990, Stamford, CT, 1990, pp. 1–4.

Meldrum, G., "Data on the Move," *Fire Engineers Journal*, Vol. 53, No. 171, Dec. 1993, p. 32.

Moriatry, M., "Managing Information for the Fire Department," *Fire Engineering*, Vol. 148, No. 4, April 1995, pp. 48–53.

Mowrer, F. W., "Development of the Fire Data Management System," Maryland Univ., College Park, National Institute of Standards and Technology, Gaithersburg, MD, NIST-GCR-94-639, June 1994.

Narayanan, P., "New Zealand Fire Risk Data," Building Research Association of New Zealand, Judgeford, Study Report 64, Dec. 1995.

Pettitt, G. N., Harvey, B., and Worthington, D. R. E., "Use of Quantitative Risk Assessment as a Decision Making Tool," *PSAM-II Proceedings*, An International Conference Devoted to the Advancement of System-Based Methods for the Design and Operation of Technological Systems and Processes, Vol. 1, Sessions 1–36, March 20–25, 1994, San Diego, CA, 1994, pp. 007/1–6.

Portier, R. W., "Fire Data Management System, FDMS 2.0, Technical Documentation," National Institute of Standards and Technology, Gaithersburg, MD, NIST TN 1407, Feb. 1994.

Portier, R. W., "Fire Data Management System, FDMS 2.0," National Institute of Standards and Technology, Gaithersburg, MD, NISTIR 5499, Sept. 1994; National Institute of Standards and Technology, Annual Conference on Fire Research: Book of Abstracts, Oct. 17–20, 1994, Gaithersburg, MD, 1994, pp. 39–40.

Robertson, J. C., "Records, Reports, and Measurement of Results," *Introduction to Fire Prevention*, 3rd ed., Macmillan, New York, 1989, pp. 204–218.

Rowe, W. D., and Beierschmitt, K. J., "Managing Uncertainty in Risk Analysis Decisions," *PSAM-II Proceedings*, An International Conference Devoted to the Advancement of System-Based Methods for the Design and Operation of Technological Systems and Processes, Vol. 1, Sessions 1–36, March 20–25, 1994, San Diego, CA, 1994, pp. 015/7–13.

Seamon, R. P., "A County Opts for ICS," *Fire Command*, Vol. 56, No. 10, Oct. 1989, pp. 32–33.

Shelton, W. G., "DOD Fire Incident Reporting System," Fort Belvoir Fire Dept., VA, International Association of Fire Chiefs (IAFC), Fire-Rescue International Conference *Proceedings*, 1993 Annual Conference, August 28–September 1, 1993, Dallas, TX, Swing, 1993, pp. 379–395.

Simard, A. J., and Eenigenburg, J. E., "METAFIRE: A System to Support High-Level Fire Management Decisions," *Fire Management Notes*, Vol. 51, No. 1, 1990, pp. 10–17.

Steffens, J. T., "RPD Model of Fireground Decision Making," *Firehouse*, Vol. 19, No. 4, April 1994, pp. 75–77,85.

Stringfield, W. H., "Mastering Title III Paper Work," *Fire Command*, Vol. 56, No. 1, Jan. 1989, pp. 24–27.

vanBowen, J., Jr., "Use of In-House Data for Fire Station Location," *Journal of Applied Fire Science*, Vol. 1, No. 1, 1990–1991, pp. 39–44.

Walker, W. E., *et al.*, *Fire Department Deployment Analysis*, Elsevier, New York, 1979.

Watts, J. M., Budnick, E. K., and Kushler, B. D., "Using Decision Tables to Quantify Fire Risk Parameters," *Proceedings* of the International Conference on Fire Research and Engineering, September 10–15, 1995, Orlando, FL, SFPE, Boston, MA, 1995, pp. 241–246.

Watts, J. M., Jr., "Fire Safety Decision Making. Editorial," *Fire Technology*, Vol. 32, No. 2, May 1995, pp. 97–98.

FIRE SERVICE LEGAL ISSUES

Maureen Brodoff

THE LIABILITY OF FIRE SERVICE ORGANIZATIONS FOR NEGLIGENT FIRE FIGHTING

In 1955 a large quantity of gasoline was spilled onto a city street in Lawrence, KS, during the removal of gasoline storage tanks from a gas station. The local fire department was notified and quickly arrived at the scene. In order to determine the extent of the problem, the fire chief who was supervising the scene instructed a fire fighter to touch a cigarette lighter to the ground. Not surprisingly, a conflagration ensued that destroyed several automobiles. In the lawsuit that followed, the court refused to hold the town liable for the foolhardy tactic of its fire chief.[*]

This case and many others from the period reflect the traditional view that local governments were not liable for their failure to provide effective fire protection. Indeed, even extreme carelessness in fighting fires would not give rise to liability. Today, however, in the field of fire fighting as with most modern-day endeavors, the historical limitations on legal liability are eroding, and theories of liability are expanding. Fire service legal liability must now, of necessity, be a concern to the fire service.

This chapter will describe the general legal principles that are used in analyzing the legal liability of fire service organizations for negligence in conducting fire-fighting activities.[†] It should be remembered that the law in this area is not uniform but is governed largely by state and local laws and, therefore, varies from jurisdiction to jurisdiction. The liabilities of individual fire service organizations, therefore, can only be determined by reference to the specific law in its jurisdiction.[‡]

WHAT IS NEGLIGENCE IN THE FIRE-FIGHTING CONTEXT?

As with any other endeavor, particularly one as fraught with danger and uncertainty as fire fighting, things can go wrong. Fires sometimes cause deaths and injury in spite of the best efforts of the fire service. Property damage may result, not only from the effects of fire, but frequently from the activity of fire fighting itself. A tactical decision made in the midst of impending disaster may, in hindsight, turn out to have been terribly wrong. Bad outcomes alone, however, do not make the fire service liable.

The principal theory of liability used in lawsuits for personal injury and property damage is what is known in the law as negligence.[§] The law of negligence does not hold a person liable for *any* damage that results from his or her actions, only damage that results from some act of carelessness in circumstances where the actor had some duty to act with reasonable care. This principle can be understood by way of an illustration drawn from an actual case involving allegations of negligent fire fighting.

---

[*]*Perkins v. City of Lawrence*, 281 P.2d 1077 (Kan. 1955).

[†]There are many potential types of legal liabilities encountered in the modern fire service. This chapter treats only one—liability for negligent fire-fighting activities—and is intended only as a general introduction to the subject. It does not deal with other types of liability a fire department could owe to members of the public, such as for negligent inspection or automobile negligence. It also does not deal with employment law issues, such as workers' compensation, wrongful termination of employment, and the expanding field of anti-discrimination law which has had a widening impact on the fire service in recent years. For the interested reader, there are several works designed for the layperson on these and other legal issues relating to the fire service. See, for example, Callahan, T., *Fire Service and the Law*, second edition (National Fire Protection Association, 1988); Schneid, T. D., *Fire Law* (Van Nostrand Reinhold, 1995); and Grant, N., and Hoover, D., *Fire Service Administration* (National Fire Protection Association, 1993).

[‡]For convenience, this chapter will generally use the terms "fire service organization" or "municipality" in referring to the entity that may bear liability for negligent fire fighting. The reader, however, should be aware that the actual party that is named in a lawsuit alleging negligent fire fighting will vary depending on how the fire service is organized in a particular locale. Most frequently, fire departments are branches of municipal government and, when a lawsuit is brought, it is the city or town that is named in the suit and which is responsible to pay any judgment. In other cases, it may be an independent fire district or a county that is the responsible party.

[§]Sometimes other theories of liability are used in suits against the fire service. For example, lawsuits have been brought under a federal law permitting lawsuits for injuries resulting from the deprivation of some civil right. (See 42 U.S.C. § 1983.) These lawsuits require more than allegations of negligence. Typically, they allege some discriminatory action, such as the withholding of adequate fire protection from a minority neighborhood. Lawsuits also sometimes allege deliberate misconduct, as opposed to mere negligence. A full discussion of these other theories is beyond the scope of this chapter.

Maureen Brodoff is associate general counsel for the National Fire Protection Association.

In 1978 in the city of Lowell, MA, a fire occurred in and destroyed five brick buildings.¶ The fire started on the sixth floor of an unoccupied building. This building had a working sprinkler system and, indeed, the system worked properly in the initial stages of the fire. The fire fighters who responded to the fire, however, chose to use the available water source to operate hoses. This reduced water pressure in the sprinkler system, in effect turning it off. There was evidence that good fire-fighting practice would have been to rely on the building's sprinkler system to fight the fire rather than to have diverted the water from the system to fight the fire with hoses. There was also evidence that the sprinkler system, if allowed to operate, would have put out the fire or contained it until it could have been put out by manual means. Instead, because of the choice of the fire fighters to effectively shut off the sprinkler system, the fire eventually engulfed and destroyed five buildings.

In this case, one can see all of the essential elements of a fire-fighting negligence case. First, under the law, all persons generally have a duty, once they undertake to act, to do so with reasonable care. As explained later in this chapter in the discussion of the "public duty" rule, there is some controversy whether fire fighters owe such a duty of care. In this case, however, it was conceded that once the fire fighters undertook to fight the fire, they had a duty to fight the fire with reasonable care.

Second, the fire fighters breached their duty to act with reasonable care. In lay terms, this simply means that they acted carelessly in fulfilling their duty to fight the fire. Reasonable care, in the context of fire fighting, means that level of care which the reasonably prudent fire fighter would use in similar circumstances. Since the evidence in the case showed that proper fire-fighting practice would have been to leave the sprinklers on, turning them off was viewed in the eyes of the law as negligent.

Third, the fire fighters' breach of their duty to reasonably fight the fire caused the destruction of the buildings. From the evidence, if the sprinklers had been allowed to function, the fire would have been contained. In other words, the fire fighters were the legal cause of the destruction of the buildings, because the destruction of the buildings would not have occurred had the sprinkler system been left on, and the consequences of shutting off the system were reasonably foreseeable.

Finally, the fire fighters' negligence resulted in damages. In this case, the damages roughly equaled the value of the destroyed buildings and their contents.

This case of the turned-off sprinkler is a good example of what any case of negligence will have to prove in order to be successful; i.e., the existence of a duty of care, the breach of that duty, causation, and damages. It is important to remember, however, that this is but one example of what can be alleged as negligent fire fighting. The types of negligence that can be alleged in the fire-fighting context is infinite. Areas of potential liability include fire suppression activities, tactics and strategies, emergency response system failures, operation of fire service vehicles,# hydrant and water supply maintenance, and maintenance of fire-fighting equipment.

¶The fact pattern used for this illustration is drawn from the Massachusetts case of *Harry Stoller & Co. v. Lowell*, 412 Mass. 139 (1992).

#The laws involving the operation of fire service vehicles present something of a special case, since many states have laws aimed specifically at limiting liability for the operation of emergency and fire service vehicles. These statutes vary from state to state. Indiana, for example, provides immunity for the operation of fire service vehicles only when the operator of the vehicle is an employee of the fire service organization and only in the case of authorized emergency vehicles. [See *Indiana Code* § 9-4-1(d).] Other states have additional requirements, such as that emergency sirens or lights be activated.

Actual cases that have been brought illustrate the variety of claims that creative lawyers can allege. In an Indiana case, for example, it was alleged that a fire service organization was negligent in failing to maintain a sufficient number of fire fighters for the equipment intended to be used.* In the same case, it was also alleged that there was negligence in the service's failure to supervise and train its fire fighters in controlling and extinguishing fires under the conditions encountered in a particular fire. In an Alabama case, negligence was claimed in the failure of a fire department to respond to a house fire because the apparatus operator had gone home sick.† In a Maryland case, negligence was alleged in the failure to properly control and extinguish a brush fire that eventually reignited, causing a second fire in which a warehouse was destroyed.‡ And in a Massachusetts case, it was alleged that fire fighters were negligent in fighting a fire burning at the rear of a house by spraying water on the front of the house where there was no fire.§

In one particularly dramatic case in Alaska, liability was alleged and found for negligent failure to rescue a person stranded in an upper floor of a burning building during a fire. The rescue failed because the ladder used in the attempt was too short to reach the victim's window. While the court said this fact alone did not constitute negligence, the fire fighters failed to use other common-sense methods of rescue that were available as an alternative. In particular, the court was deeply disturbed that some spectators who had obtained an extension ladder of sufficient length to reach the victim, and who had raised the ladder and started to extend it, were ordered by a fire official to get away from the building and, when they refused to obey, were driven off by fire hoses.¶

These illustrations would seem to indicate that liability exists around every corner. Although allegations of negligence are easily made, however, not all negligence claims result in a finding that the fire service was liable. There are two broad reasons tending against findings of fire service liabililty for negligent fire fighting.

The first reason is that allegations of negligent fire fighting are generally more difficult to prove than in the typical negligence case. In a typical negligence case, a party is accused of creating a dangerous situation that resulted in injury. In the typical fire fighter negligence case, the dangerous situation, i.e., the fire, already exists when fire fighters enter the picture. A plaintiff, therefore, is usually in a position of having to prove that the fire fighters either made worse or failed to mitigate a harm that they did not cause. This is a difficult task, especially since the unpredictability and destructive power of fire in general often makes it difficult to say with any assurance that some other course of action not taken by the fire fighters would have yielded a better result. Thus, as the Alaska case described above vividly shows, liability is most often found in the extreme case where the conduct of the fire fighters is viewed as foolhardy or outrageous.

The second reason requires some explaining, but it has even greater impact on fire service liability. As discussed earlier, a case of negligence is built by proving that an individual or group by their careless actions violated a duty to act with reasonable care and, thereby, caused damage. If fire fighting were strictly a private enterprise, carried out by and for the benefit of private parties, such proof

---

*See *City of Hammond v. Cataldi*, 449 N.2d 1184 (Ind. App. 3d Dist. 1983).

†See *Williams v. City of Tuscumbia*, 426 So.2d 824 (Ala. 1983).

‡See *Utica Mut. v. Gaithersburg-Washington Grove*, 455 A.2d 987 (Md. App. 1983).

§See *Cryan v. Ware*, 413 Mass. 452, 469 (1992).

¶See *City of Fairbanks v. Schaible*, 375 P.2d 201, 206 (Alaska 1962).

would be all that was required to entitle the injured party to hold the fire service organization liable to pay for all damages.

Fire fighting, however, is not a private enterprise. It is generally a governmental function carried out by cities, towns, and other governmental units for the benefit of the public. Because of this, the fire service is the beneficiary of an elaborate body of law that has been developed to shield the government from liability, even when it has acted negligently. This law of "governmental immunity" greatly complicates the question of whether and when a fire service organization can be held liable for damages caused by negligent fire fighting.

THE FIRE SERVICE AND THE DOCTRINE OF GOVERNMENTAL IMMUNITY

Until the last twenty or so years, the doctrine of governmental immunity fully protected governments from lawsuits aimed at governmental functions. Under this doctrine, the government as "sovereign" could do no wrong and could not be sued. In the case of fire-fighting activities, it meant that no matter how negligent a fire department might be or how much damage to life or property that negligence might cause, the municipality whose fire fighting had caused the damage could not be sued.

The doctrine of governmental immunity left no remedy to individuals who had suffered grievous injuries, as a result of negligent fire fighting or other governmental negligence. As might be expected, the injustice that the doctrine often seemed to impose led to much criticism of the doctrine and calls for reform.

Beginning in the 1970s, the federal government and the states began, either through court decisions or, more often, through the passage of legislation, to severely limit the absolute governmental immunity that governmental entities had enjoyed. The most common type of legislation, now in existence in some form in most states as well as the federal government, is commonly known as a "tort claims act." Each state act is different but, generally speaking, the acts provide that government entities are liable for injury caused by their negligence in the same manner and to the same extent as a private entity.[#] There are important qualifications, however, that provide the fire service with significant protection against liability.

LEGAL PROTECTIONS FOR THE FIRE SERVICE TODAY

Limits on the Amount of Damages

Probably the most important protections for the fire service provided by the various state tort claims acts, are the limitations they impose on the amount of damages. Thus, while government entities can now be held liable for their negligence just like a private individual, the amount they can be required to pay has been limited. In Massachusetts, for example, the amount of liability that a municipality can be required to pay if found negligent is limited to $100,000 per claimant. Frequently, there is an overall cap so that, no matter the number of claimants, the total damages awardable for fire department negligence in any one incident cannot exceed a given amount. Vermont, for example, has a limit of $250,000 per claimant with a maximum aggregate liability of $1,000,000 to all claimants arising out of any given occurence; Maryland has a limit of $200,000 per claimant with an overall cap of $500,000 per occurence.[*] A few states provide lower maximum compensation for

property damage than for personal injury.[†] Finally, punitive damages, i.e., damages designed to punish the wrongdoer rather than compensate the injured party, are generally forbidden.[‡]

These various limits on liability are highly significant protections for fire service organizations, since they cap damages at an amount that may frequently represent only a small fraction of the damages actually awarded by a jury. Particularly in the area of fire suppression, where mistakes can result in millions of dollars of personal injury and property damage, the tort claims acts provide a great deal of protection against potentially huge damage awards.

Exceptions Specifically Aimed at Fire-Fighting and Related Activities

Several state tort claims acts have exceptions that specifically retain governmental immunity for fire-fighting and related activities. North Dakota, for example, retains governmental immunity for failure to provide adequate fire prevention personnel or equipment, except if gross negligence can be proved.[§] Illinois retains governmental immunity for any injury caused by the failure to suppress or contain a fire or while fighting a fire.[¶] Kansas and Texas retain immunity for the failure to provide, or the inadequate provision of, fire protection.[#]

There are other types of specific exceptions that relate to the fire service, as well. Alaska specifically retains governmental immunity for the performance of duties "in connection with an enhanced 911 emergency system," and for the performance of duties upon the request of or by agreement with the state "to meet emergency public safety requirements."[**] Some states retain immunity for claims relating to the provision of or failure to provide emergency services.[†] Also, many states specifically retain immunity for the failure to make an inspection, or the making of an inadequate or negligent inspection.[‡]

The "Discretionary Functions" Exception

In addition to the exceptions in some states expressly relating to fire-fighting and related activities, there exists in most state tort claims acts another type of exception that provides significant protection for the fire service. This is known as the "discretionary function" exception, and it requires some explanation.

Although, in passing tort claims acts, the lawmakers of the various states wished to make it possible for citizens to obtain compensation for injuries caused by governmental entities, they were reluctant to abolish immunity for all governmental activities. They were concerned that lawsuits might be used to second guess every governmental policy decision and that the constant fear of lawsuits could severely hamper the ability of municipal officials to govern and to freely exercise the discretion of their office. In order to address these concerns, lawmakers created an exception in the tort claims acts that preserved governmental immunity for "discretionary functions."

[†]See, for example, Or. Rev. Stat., § 30.270; N.M. Stat. Ann., § 41-4-19.

[‡]See, e.g., Mass. Gen. Laws ch. 258, § 2; Minn. Stat. Ann., § 466.04.

[§]See N.D. Cent. Code, § 32-12.1-03(3).

[¶]See Ill. Rev. Stat. ch. 85, §§ 5-102, 5-103.

[#]See Kan. Stat. Ann., § 75-6104(n); Tex. Civ. Prac. & Rem. Code Ann., 101.055(3).

[**]See Alaska Stat., §§ 09.65.070(d)(6), and (d)(5).

[†]See, for example, Iowa Code, § 613A.4; Tex. Civ. Prac. & Rem. Code Ann., § 101.055(2).

[‡]See, for example, Kan. Stat. Ann., § 75-6104(k).

---

[*]See Vt. Stat. Ann. tit. 12, § 5601(b); Md. Cts. & Jud. Proc., § 5-403(a).

The typical discretionary functions exception merely states that the tort claims act, and the governmental immunity which it abolishes, simply does not apply to any claim based upon a public employee's performance or failure to perform a "discretionary function."§ What is a "discretionary function," however, and what does it mean in the context of fire fighting? Many legal battles have been fought over the meaning of this exception.

The main problem has been in determining the breadth of the discretionary function exception which, read literally, could be quite broad indeed. The word "discretion" implies, in its essence, the exercise of judgment, and, therefore, a literal interpretation of the discretionary function exception might lead to the conclusion that all conduct involving the exercise of judgment is immune from negligence liability. In addition, since virtually all fire-fighting activities involve the exercise of judgment, even if only concerning minor details, one might conclude that all fire-fighting activities are immune from suit. This, however, is not the case.

Courts have generally rejected a too literal reading of the term "discretionary function." They have felt that granting immunity to all acts that involve some exercise of judgment would, in effect, immunize all governmental activity and, therefore, defeat the whole purpose of the tort claims acts, which was, after all, intended to abolish complete immunity. Courts, therefore, have looked to the purpose of the discretionary function exception. It was only intended to protect conduct involving the kinds of broad public policy and planning judgments that governmental actors need to be able to perform without the constant threat of being sued. In keeping with this purpose, most courts have tended to find immunity, not for all discretionary conduct, but only for conduct that involves "policy making or planning."¶

In the context of fire fighting, what does this mean? The legal decisions are far from clear and vary widely in how they treat particular fire service activities. Nevertheless, certain themes emerge.

First, there are aspects of fire service decision-making that have an obvious planning or policy basis. These are administrative policy decisions involving the overall structure and makeup of the fire department, the training and equipping of fire fighters, and the allocation of limited fire-fighting resources within the community. They include decisions about the number and location of fire stations, the amount and type of equipment to purchase, the size of the fire-fighting staff, the type and extent of fire-fighter training, the number and location of hydrants, or the adequacy of the water supply.

When lawsuits blame these types of administrative policy decisions for bringing about injury or death, they are frequently dismissed based on the discretionary function exception. For example, a claim that fire fighters were unable to suppress a fire because the nearest fire station was too far away to make timely fire fighting possible, would generally fail because decisions about where to locate a fire station are, even if patently unreasonable, protected from liability under the discretionary function exception.

Of course, some courts are more stringent in applying the exception than others. Many courts, for example, will view any broad administrative-level policy decision as categorically immune from liability, without any inquiry into the thought processes of the decision-makers. Other courts, however, will require that the fire service organization, in order to claim immunity, present evidence showing that the decision-makers actually went through a weighing of policy choices. In a case alleging that fire fighters were not supplied with a particular type of rescue equipment that would have prevented an injury, such a court would, for example, require that the fire service organization show that its failure to provide such equipment was the result of a conscious policy-making decision involving the balancing of competing interests, rather than the result of a simple failure to consider the question whether the equipment was needed.#

It is clear, therefore, that administrative decisions are generally covered and immune from suit, at least where the decisions involve a conscious weighing of policy considerations. What about operational decisions made in the course of fighting a particular fire? Some have argued, for example, that the initial decision to fight a fire is discretionary, but that all of the subsequent actions in actually fighting the fire are not. Under this view, almost all actual fire suppression activities on the fireground would fall outside the "discretionary function exception" and would therefore be subject to potential liability.

This point of view, however, has mostly been rejected. While many, indeed even most, decisions taken in the course of fire fighting do not involve policy or planning considerations, courts have found some decisions made on the fireground to involve broad public policy choices protected by governmental immunity.

To determine whether a particular operational decision constitutes a discretionary function, courts generally look not at the particular action, but the reasons for it. To illustrate, consider the case, described earlier in this chapter, of the fire fighters who turned off an operating sprinkler system in an unoccupied building because they wanted to fight the fire with hoses. Their decision to turn off the sprinkler system was clearly negligent, in that good fire-fighting practice would, in the circumstances, have dictated using the sprinkler system to extinguish the fire. The decision to shut off the sprinkler system also did not involve any policy or planning considerations, since the fire fighters simply made a careless decision about how to best fight the fire.

In slightly different circumstances, however, the decision to shut off a sprinkler system could easily involve a significant policy choice. Suppose, for example, that the fire fighters had decided to shut off the sprinkler system in the unoccupied buildings, not merely to deploy hoses, but to divert the water supplying the sprinkler to fight a fire in a neighboring building that was occupied. The decision to sacrifice the unoccupied buildings in order to devote limited fire-fighting resources to saving lives next door clearly involves policy choices about the relative value of property and human life. Thus, even if the decision to shut off the sprinkler system was ultimately shown to be negligent in that the occupied building could have been saved without diverting water from the sprinkler system, the decision was grounded in an important policy choice, and many courts would rule that it was immune from liability under the discretionary function exception.*

In summary, although the court decisions in this area vary widely from jurisdiction to jurisdiction, it can be said that the discretionary function exception provides immunity to the fire service from liability for most broad administrative-level decision making and for many operational-level decisions as well. Though far from comprehensive, and less than certain in any individual case, the discretionary function exception still provides the fire service with substantial protection from liability.

---

§See, for example, Mass. Gen. Laws ch. 258, § 10(b).

¶See, for example, *King v. Seattle*, 84 Wash.2d 239, 525 P.2d 228, 233 (1974); *Keopf v. County of York*, 198 Neb. 67, 251 N.W.2d 866, 870 (1977); *Johnson v. State*, 69 Cal.2d 782, 788, 447 P.2d 352 (1968); and *Harry Stoller & Co. v. Lowell*, 412 Mass. 139 (1992).

#As examples of the differing approaches used by courts, see and compare *City of Hammond v. Cataldi*, 449 N.E.2d 1184 (Ind. App. 3 Dist. 193); and *Waldorf v. Shuta*, 896 F.2d 723, 728–730 (3d Cir. 1990).

*See, for example, *Dahlheimer v. City of Dayton*, 441 N.E.2d 534 , 538–539 (Minn. 1989).

A FINAL COMPLICATION: THE PUBLIC DUTY RULE AS AN ADDED PROTECTION FOR THE FIRE SERVICE

So far, the liability for negligent fire fighting, while complicated, can be summarized quite simply. Initially, the doctrine of governmental immunity completely protected fire-fighting operations from liability, even if a fire was fought negligently. Today, however, fire service organizations can, in some circumstances, be successfully sued for negligence. The extent to which they can be sued is limited in most jurisdictions by the tort claims acts and the court decisions that interpret them. In general, those acts will, through the discretionary functions exception and other exceptions aimed at protecting fire and emergency services, immunize fire service organizations entirely for some activities. Where immunity is abolished, however, the acts will still provide a cap on damages that can be won against a fire service organization.

If the law preferred simplicity, even of a relative kind, the issue might end here. There is, however, one more important twist: "the public duty rule."

Even though state legislatures have abolished governmental immunity and permitted local governments to be sued in many situations, courts have continued to be sympathetic to the problems of cities and towns and their difficulty in paying even the limited awards permitted by the tort claims acts. Particularly in the area of fire and other public protective services, some courts have expressed the fear that imposing even limited liability for negligence could pose a crushing burden on municipalities, particulary in busy urban areas. As one judge pointed out, in extreme circumstances, such as the Los Angeles riots where there were hundreds of fires and severe obstacles to effective fire fighting, the potential costs of liability could be catastrophic.[†] This fear has led some courts to develop a judge-made rule to shield the fire service and other public protective services, such as police and inspection. The "public duty" rule offers immunity from negligence liability that goes far beyond the limited immunity retained in the tort claims acts.

Under the public duty rule, fire fighting and other public protective services are viewed as an obligation that governments owe, not to any particular individual, but to the general public as a whole. Based on this view, the public duty rule holds, somewhat paradoxically, that, because fire fighting is for the benefit of all, it is, in effect, for the benefit of no one in particular. Under the rule, therefore, no individual can seek damages for injuries caused by negligent fire fighting because, in essence, the fire service owes no duty to any particular individual to act reasonably.

What is the effect of the public duty rule? Its greatest impact is at the operational level of fire fighting. The immunity offered by most state tort claims acts under the discretionary function exception is quite limited, and most decisions made on the fireground can give rise to potential liability, unless they can be said to involve public policy choices. In states that follow the public duty rule, however, such decisions become largely immune, and any fire-fighting activity, even if negligent and even if devoid of public policy implications, cannot give rise to liability.

It would seem that fire service organizations in states following the public duty rule have nothing to fear from lawsuits claiming negligent fire fighting. This is largely true. However, in the law every rule has its limitations and exceptions, and so does the public duty rule.

First, courts that follow the public duty rule limit its protection to the suppression of fires not caused by the fire department itself.

Thus, if a fire department negligently caused a fire during a fire training exercise or some other non-fire-suppression-related activity, the public duty rule would offer no protection. Also, a few courts would not apply the public duty rule to protect fire fighters in cases where they aggravated an existing fire as opposed to merely failing to suppress it.[‡] These courts, for example, would provide complete immunity where fire fighters negligently failed to put out an existing fire, because they negligently aimed hoses on the wrong part of the building. They would, however, permit liability where fire fighters took some action that actively made matters worse, as, for example, turning off an operating sprinkler system that, left alone, would have contained the fire.

Further, there is an exception to the public duty rule, known as the "special relationship" exception. It holds that, even though fire fighting and other protective services are viewed in general as an obligation owed only to the public at large, a fire or other protective service organization can, by words or actions in a particular case, create a special duty to particular private parties.

How does a fire service organization create for itself this "special relationship," where liability may occur? The special relationship exception has been complicated and often inconsistent in application, and, in practice, courts have been very reluctant to apply it to hold the fire service liable.[§] It seems clear that a fire department does not create a special relationship to an individual property owner merely by responding to the owner's call for assistance and fighting the fire on the property.[¶] In order to expose a fire department to liability under this exception, fire fighters would probably have to either offer a special service or protection to an individual that was not available to the general public, or induce an individual, through specific assurances of assistance or safety, to put him- or herself in danger.

Apart from potential liability opened up by the limitations and exceptions to the public duty rule, there is yet another reason why fire service organizations in states that observe the public duty rule should not rest with complete ease. The public duty rule has been widely criticized as fundamentally inconsistent with the tort claims acts, because the rule creates immunity where legislatures have sought to remove it. The trend, therefore, has been to abolish the rule, and many states have done so.[#]

[†]See *Cryan v. Ware*, 413 Mass. 452, 455 (1992).

[‡]See *Cryan v. Ware, supra*, 413 Mass. 452 (1992).

[§]Contrast, for example, the markedly different treatment the court gave police and fire fighters, respectively, in *Cryan v. Ware, supra*, 413 Mass. 452 (1992); and *Irwin v. Ware*, 392 Mass. 745 (1993).

[¶]See, for example, *Commerce & Indus. Ins. Co. v. City of Toledo*, 543 N.E.2d 1188 (Ohio 1989). But *see Ziegler v. City of Millbrook*, 514 So.2d 1275 (Alabama 1987).

[#]See *Adams v. State*, 555 P.2d 235 (Alaska 1976); *Leake v. Cain*, 720 P.2d 152, 158–159 (Colo. 1986); *Adam v. State*, 380 N.W.2d 716, 724 (Iowa 1986); *Maple v. Omaha*, 222 Neb. 293, 301 (1986); *Schear v. County Comm'rs*, 101 N.M. 671 (1984); *Coffey v. Milwaukee*, 74 Wis.2d 526 (1976); *DeWald v. State*, 719 P.2d 643, 653 (Wyo. 1986); *Ryan v. State*, 134 Ariz. 308, 310 (1982); *Commercial Carrier Corp. v. Indian River County*, 371 So.2d 1010, 1015–1016 (Fla. 1979); and *Brennen v. Eugene*, 285 Or. 401, 407 (1979).

Many states, however, still recognize the public duty rule.* And at least one state, Massachusetts, has, following court abolition of the rule, revived it by incorporating it into the state tort claims act.† It is safe to say, therefore, that the public duty rule, in many states, still offers fire service organizations a large measure of immunity.

CONCLUSION

The degree to which the fire service may be exposed to liability for negligent fire fighting depends, to a large degree, on the law of the state in which the fire service organization is located. Although all states have abolished to one degree or another the total governmental immunity from negligence lawsuits, the amount of protection that remains varies among jurisdictions. In all states, however, it is important to remember that, even in the absence of complete immunity, there are still significant protections offered the fire service. Although the principles defy easy summarization, the following, in general, can be said.

1. In virtually all states, fire service organizations will be protected from suits criticizing broad administrative policy decisions of fire department and municipal administrators. Thus, suits claiming injuries resulting from bad policy decisions about the placement or number of fire stations, the size of fire department staffs, and the adequacy of funding for fire department activities will generally be prohibited under the "discretionary functions" exception or similar legal principles contained in the torts claims acts of most states.

2. In the majority of states, fire service organizations will generally be liable for injuries for negligent fire fighting when the claims are based on bad operational decisions made on the fireground about how to fight a particular fire. Even operational decisions, however, will sometimes be protected under the "discretionary functions" exception, if those decisions involve public policy choices.

3. In the minority of states that still follow the "public duty" rule, fire service organizations will be protected from negligence suits, even those involving operational decisions, unless the department or its employees have by words or deed created a "special relationship" with an injured party thus creating a duty beyond that owed to the public as a whole.

The trend of recent years has been to hold fire service organizations liable when fire fighters have acted negligently. This trend however is far from complete. As the above discussion shows, limited immunity still exists both in the discretionary function exception and in the persistence of the public duty rule, which, despite much criticism, continues to show vitality in many states. In addition, the caps on damages that exist in all states will continue to serve to soften the impact of negligence cases on the fire service. The explosion of liability that has occurred in all areas of the law in recent years has had only mixed results when it comes to the fire service.

*See, for example, *Shore v. Stonington*, 187 Conn. 147 (1982); *Randall v. Fairmont City Police Dep't*, 186 W. Va. 336 (1991). One recent case has identified twenty-one states and one district that follow the traditional public duty rule. They include: California, Connecticut, District of Columbia, Hawaii, Illinois, Indiana, Kansas, Maryland, Michigan, Minnesota, Missouri, Nevada, New Hampshire, New York, North Carolina, Ohio, Pennsylvania, Rhode Island, South Carolina, Utah, Washington, and West Virginia. See *Jean v. Commonwealth*, 414 Mass. 496, 518 n. 1 (O'Connor, J., concurring).

†See Massachusetts General Laws, ch. 258, § 10, as amended by 1993 Stat. Ch. 495, § 57.

FIRE DEPARTMENT OCCUPATIONAL SAFETY AND HEALTH

William Peterson

NFPA 1500, *Standard on Fire Department Occupational Safety and Health Program* (hereinafter referred to as NFPA 1500), was developed by the National Fire Protection Association's Technical Committee on Fire Service Occupational Safety and Health, beginning in late 1984. It was adopted by the membership at NFPA's 1987 Annual Meeting in Cincinnati, OH, and issued by the NFPA Standards Council on July 17, 1987, with an effective date of August 7, 1987. The second edition of NFPA 1500 was adopted by the membership at the 1992 Annual Meeting in New Orleans, LA. Developed within NFPA's consensus standards-making process, NFPA 1500 has given a whole new emphasis and direction to the fire service over the past 10 yrs and will continue to have a favorable impact in the years to come. With the 1992 edition, the Technical Committee expanded on chapters dealing with incident management, infectious disease control, risk management, and emergency scene operations. This new emphasis on fire fighter health and safety has significantly reduced the number of deaths and the severity of injuries experienced in the fire service since its development.

Fire fighting continues to be one of the most hazardous occupations in the United States. Over the past 19 years, an average of well over 100 fire fighter deaths and 100,000 injuries per year have placed fire fighting among those jobs rated highest for occupational death and injury. The losses experienced each year affect career and volunteer fire fighters alike. NFPA 1500 was developed to make fire fighting less dangerous by reducing deaths, injuries, and accidents. The standard also emphasizes the well-being of the fire fighter, especially his or her health and physical fitness.

In order to understand the thrust and direction of NFPA 1500, it is necessary to understand the current situation as it relates to fire fighter death and injury statistics. Each year, the National Fire Protection Association gathers the most complete data available on fire fighter deaths and injuries. By understanding the factors that lead to fire fighter deaths and injuries, it becomes possible to manage the risk of providing fire protection services in a way that reduces the fire fighter's exposure to harm to reasonable levels and provides an acceptable level of occupational safety and health.

FIRE FIGHTER DEATHS

In 1995, 95 fire fighters died while on duty. This figure marked the third lowest number of fire fighter fatalities since NFPA began collecting statistics in 1977. This figure continues a trend over three of

William Peterson is chief of the Plano Fire Department in Plano, TX.

the past four years for fire fighter fatalities to fall below a rate of 100 per year. The exception in 1994 was largely a result of one 14 fatality wildland fire incident. Table 10-5A shows the distribution of fire fighter deaths for the years 1977 through 1995.

Table 10-5B shows the distribution of the 95 deaths by type of duty. As in previous years, nonfireground activities accounted for almost half of the deaths in 1995.

TABLE 10-5A. *Line-of-Duty Fire Fighter Deaths, 1977-1994*

Year	Number of Deaths
1977	157
1978	171
1979	124
1980	138
1981	135
1982	126
1983	113
1984	119
1985	127
1986	120
1987	131
1988	136
1989	117
1990	107
1991	108
1992	75
1993	77
1994	104
1995	95

TABLE 10-5B. *1995 Fire Fighter Deaths By Type of Duty*

Cause	Number of Deaths	Percentage
Fireground	43	45.3
Responding/returning from alarm	29	30.5
Nonfire incident	8	8.4
Training	11	11.6
Other on-duty	4	4.2

Table 10-5C shows the distribution of deaths by cause of fatal injury. The term "cause" refers to the action, lack of action, or circumstances that directly resulted in the fatal injury.

Stress was a major cause of death, resulting in half of the fatalities, as in most previous years. A significant improvement in this category could be anticipated if fire fighters were in better physical condition and better able to handle the strenuous physical activities associated with emergency operations.

The distribution of deaths by nature of fatal injury or illness is shown in Table 10-5D. The most frequently reported fatal injury was heart attack (50.5%). Of the 48 heart-attack victims, at least 20 were known to have had prior heart problems, usually heart attacks or bypass surgery; and medical documentation showed that 8 others had severe arteriosclerotic heart disease. Other major categories of fatal injuries are internal trauma, asphyxiation, and burns.

The fire fighters who died while on duty in 1995 ranged in age from 15 to 72 years. Table 10-5E shows the distribution of fire fighter deaths by age and cause of death. Most years reflect a high proportion of deaths by heart attack for older fire fighters and a low proportion for fire fighters under the age of 35.

Table 10-5F shows the death rates by age categories using estimates of the number of fire fighters in each age group from NFPA's 1993 profile of fire departments and the fatality data from 1991 to 1995. As the data show, the death rate is lowest for fire fighters aged 30 to 39, slightly below the average for fire fighters aged 40 to 49, and significantly higher than average for fire fighters age 50 and over. It is of particular interest to note that only 14 percent of all fire fighters are over age 50, yet that particular age group continues to account for more than a third of the deaths from the 5-yr period from 1991 to 1995.

Table 10-5G shows the distribution by fixed property use of the 43 fireground deaths in 1995. It is interesting to note that, while fire fighters respond to non-residential fires far less frequently than to residential fires, the death rate in nonresidential properties is much higher. This fact is probably due to the size, complexity, familiarity, and special hazards that exist in these types of occupancies, and to the associated problems encountered during fire-suppression activities.

In 1995 six additional fire fighters were killed in wildland fires, and one died in a controlled burn. In addition to these seven fireground deaths, eight fire fighters died while returning from wildland fires, and one died while enroute to a controlled burn. This is a sharp contrast to 1994, when wildland fires and controlled burns resulted in 36 fire fighter deaths. Fourteen of those 36 deaths occurred at the Storm King Mountain Fire in Colorado. This was the deadliest wildland fire in more than 40 years.

Table 10-5H shows the 1990–1994 average fireground death rates for various occupancy types. These rates were calculated using

TABLE 10-5C. *1995 Fire Fighter Deaths By Cause of Fatal Injury*

Cause	Number of Deaths	Percentage
Caught or trapped	19	20.0
Stress	49	51.6
Struck by/contact with object	25	26.3
Fell or jumped	2	2.1

TABLE 10-5D. *Fire Fighter Deaths 1994—By Nature of Fatal Injury or Illness*

Cause	Number of Deaths	Percentage
Heart attack	48	50.5
Internal trauma	19	20.0
Asphyxiation	14	14.7
Burns	5	5.3
Crushing	3	3.4
Drowning	3	3.2
Electrocution	1	1.1
Seizure	1	1.1
Stroke	1	1.1

TABLE 10-5E. *1995 Fire Fighter Deaths by Age and Cause of Death*

Age Group	Heart Attacks	Other Trauma	Total Deaths
20 and under	1	4	5
21–25	0	6	6
26–30	1	4	5
31–35	1	11	12
36–40	3	6	10
41–45	5	5	10
46–50	7	7	14
51–55	9	3	12
56–60	9	1	10
Over 60	10	0	10
Not Reported	1	0	1

TABLE 10-5F. *Average Death Rates per 10,000 Fire Fighters, 1991–1995*

Age Group	Deaths per 10,000 Fire Fighters
Under 20	0.56
20–29	0.67
30–39	0.55
40–49	0.78
50–59	1.75
60 and over	3.43
Overall Average: 0.86 deaths per 10,000 fire fighters	

Note: These figures combine career and volunteer fire fighters. The two groups may have very different age distributions, which are not reflected.

TABLE 10-5G. *1995 Fireground Deaths by Fixed Property Use**

Fixed Property Use	Number of Deaths	Percentage
Residential	17	39.5
Wildland	7	16.3
Manufacturing	6	14.0
Storage	4	9.3
Stores and offices	3	7.0
Street/Highway	2	4.7
Vacant	1	2.3
Museum	1	2.3
Outside recycling center	1	2.3
Not Reported	1	2.3

*There were 43 fireground deaths in 1995.

TABLE 10-5H. *Fire Fighter Death Rates on the Fireground, 1989—1993*

Occupancy Type	Deaths per 100,000 Fires
Vacant, Special*	15.0
Public assembly	15.7
Storage	13.6
Manufacturing	11.1
Stores and offices	10.9
Residential	4.3
Educational	2.2
Institutional	0.0

*Includes idle buildings, buildings under construction, etc.

the updated fire fighter fatality data and the estimates of fires from NFPA's annual fire experience surveys.

In the 19-year period from 1977 to 1995, at least 2,280 fire fighters died while on duty in the United States. Of these, 215 were not municipal fire fighters. Of the remaining 2,065, 1,024 (58%) were volunteers and 861 (42%) were career fire fighters. From 1977–1979 to 1993–1995, using three-yr averages, the career fire fighter fatality total fell 43 percent while the volunteer total fell 62 percent. The 1992–1995 fire fighter fatality totals were the four lowest in the 19 years studied. The 1992–1995 municipal career fire fighter death tolls were four of the five lowest in the 19 years, and the 1992–1995 municipal volunteer fire fighter death tolls were four of the seven lowest in the 19 years. A comparison of deaths per 100,000 fire fighters in 1993–1995 vs. 1983–1985 show the career fire fighter death rate down 51 percent and the volunteer fire fighter death rate down only 10 percent. The latter figure indicates little real progress. Table 10-5I compares career fire fighter fatalities to volunteer fire fighter fatalities over the past 19 yrs.

Over the past 19-year period that NFPA has been analyzing fire fighter deaths, remarkable progress has been made in reducing the number of such deaths annually. In the late 1970s there were an average af 151 deaths a year. During the 1980s that annual average death toll dropped to 126. And in the first six years of this decade, the average annual deaths have dropped even lower, to 94 per year. A combination of improved protective clothing and equipment,

TABLE 10-5I. *Comparison of Career and Volunteer Fatalities, 1977—1995*

Year	Career	Volunteer
1977	70	82
1978	64	99
1979	57	57
1980	61	69
1981	57	65
1982	49	67
1983	54	51
1984	43	59
1985	54	66
1986	51	55
1987	48	68
1988	43	81
1989	42	65
1990	25	62
1991	36	66
1992	24	44
1993	21	53
1994	33	38
1995	29	57

safer fire apparatus, better training, and better incident management can be credited with much of this improvement. But certain stubborn aspects of the problem remain.

A thorough review of fire fighter fatalities shows that three major factors are responsible for the great majority of deaths: (1) heart attacks, (2) motor vehicle accidents, and (3) training fatalities. If substantial improvement is to be accomplished in the next 10 yrs, each of these factors must be addressed by U.S. fire services.

Heart attack continues to be the leading cause of U.S. fire fighter deaths in the period from 1986 through 1995. Of the 1,070 on-duty fire fighter fatalities over the 10-yr period, at least 498 are heart related. The numbers of heart-related fatalities have fluctuated from year to year but seem to be trending downward.

Information on physical condition prior to fatal injury was available for 278 of the 498 heart attack victims. This information was obtained from medical documentation that accompanied fatality reports. Of these 278 fire fighters, 150 (54.0 percent) had had prior heart-related conditions, such as previous heart attacks or coronary bypass surgery. Another 81 of the victims (29.1 percent) had severe arteriosclerotic heart disease.

One third of the victims were career fire fighters, although it is estimated that only 25 percent of the nation's fire fighters are career. Therefore, career fire fighters suffer proportionately more heart attack deaths than volunteer fire fighters. Career fire fighters, in fact, suffer proportionally more deaths of all types than volunteers, possibly because of the higher number of hours and calls per year per person for career fire fighters. Victims over age 60 were mostly volunteer, possibly due to the tendency of volunteer fire fighters to remain active well beyond the retirement ages of career fire fighters. Heart attack deaths among career fire fighters have generally been decreasing since 1986; among volunteers, since 1988.

Steps to reduce the risk of heart attacks among fire fighters, including more detailed medical evaluations and further study in the areas of physical fitness and dietary requirements for fire fighters, must be taken.

Above all, attention must be focused on the significant problem of fire service personnel who have heart problems, yet are allowed to remain active in fire fighting.

Approximately half of the fire fighter deaths experienced each year result from heart attacks and most of the victims for whom medical documentation is available had known or detectable heart problems. Proper screening of applicants to fire service, fitness requirements that are maintained throughout their career, and annual health testing are essential tools in ensuring that fire fighters are ready for the stress of duty. NFPA 1582, *Standard on Medical Requirements for Fire Fighters*, should be used to screen both new and existing fire fighters.

The second factor contributing to a substantial number of fire fighter deaths is motor vehicle accidents. From 1986 through 1995, there were 220 deaths in motor vehicle accidents, accounting for 20.6 percent of the total 1,070 fire fighter deaths that occurred over the period.

The number of deaths each year has been generally decreasing since 1988, but motor vehicle accidents still tend to account for 20 to 25 percent of the deaths each year.

The proportion of career fire fighter deaths resulting from motor vehicle accidents is usually less than 15 percent of the total deaths in a year and is often less than 10 percent. In contrast, among the 185 volunteer members of fire departments, the number of deaths in motor vehicle accidents ranged from 13 to 25 per year and generally accounted for a quarter to a third of the total number of deaths of volunteer fire fighters. Not all deaths in motor vehicle accidents occur while responding to or returning from alarms, although that is when the majority of accidents occurred. Deaths have occurred on the fireground, at the scene of nonfire emergencies, during training activities, and during other on-duty activities.

Of the deaths that occurred while responding to or returning from alarms, 43 percent involved occupants of personal vehicles. All of these victims were volunteer fire fighters and almost all were killed while responding to alarms. Tankers/tenders were the next type of vehicle most frequently involved, followed by engines or pumpers.

Twenty-four victims were struck by nonfire department vehicles while operating at the scene, clearly showing the hazards posed by highway traffic at emergency scenes.

Most of the collisions that resulted in fatalities could be attributed to operator error, frequently driving too fast for road conditions. Failure to use seatbelts was also frequently reported. Four of the drivers were reported to be intoxicated at the time of their fatal accidents.

Motor vehicle accidents continue to claim a substantial share of all fire fighter deaths each year. In many cases, it was failure to heed the rules of the road, not adverse conditions, that led to the fatalities.

Deaths resulting from falls from apparatus back steps have decreased over the last several years with the requirement of enclosed cabs, directives requiring fire fighters to ride seated and seat-belted, and prohibiting the use of back steps, or any exterior riding position.

Many fire service motor vehicle accidents are preventable, either through driver training or through procedural changes in how responses to incidents are handled. Fire departments must stress driver safety and establish an enforced, written policy that will ensure a rapid, adequately staffed response to incidents without jeopardizing the safety of responders or others traveling the roads.

The third factor contributing to fire fighter deaths is training fatalities. Training is, of course, a vital part of fire department operations, but it too often results in deaths and injuries. In the 10-year period from 1986 through 1995, 91 fire fighters died while engaged in training-related activities. Of the 91 fire fighters, 48 were career fire fighters and 43 were volunteers.

Training-related deaths cover a broad range of activities. The category where the most deaths were reported involved apparatus and equipment drills. This category includes ladder climbing, pump and drafting operations, hose evolutions, high-rise drills, rappelling, smoke jumper training, and emergency medical training.

Live fire training claimed 10 lives. Three of these deaths occurred in one incident when the victims were caught by rapid fire progress and died of smoke inhalation. In other incidents, two fire fighters were asphyxiated during live fire training—one became disoriented when attempting to exit the building after a flashover occurred and the other depleted his self-contained breathing apparatus (SCBA) air supply while attempting to exit.

Three fire fighters suffered fatal heart attacks during or just after live fire training. One fire fighter was struck and killed by a collapsing chimney, and one fire fighter was struck by a motor vehicle while directing traffic during training.

More than half of the training deaths in the 10-year period were due to heart attacks. In findings similar to those reported for fire fighter deaths in 1993 and 1994, among other years, almost all of the fire fighters who died of heart attacks during training-related activities, for whom medical documentation was available, had had prior heart attacks, bypass surgery, severe arteriosclerotic heart disease, diabetes, or hypertension.

In training-related fatalities, deaths are more frequently the result of heart attack as age increases. The rates for fire fighters under age 40 are below the average for all fire fighters. The rates then climb as fire fighters get older.

Training is an area where needless deaths occur. Motor vehicle accidents while traveling to and from training sessions represent an area where ordinary precautions and attention to driving rules and road conditions should have an impact. Training exercises themselves, since they are conducted in controlled settings and in compliance with NFPA 1403, *Standard on Live Fire Training Evolutions*, should be designed to not endanger the participants. This requires that recommended safety procedures be followed. That, in combination with competent instruction, should result in the level of safety necessary to protect lives. Training safety is emphasized in Chapter 3 of NFPA 1500. This chapter outlines the training requirements qualified personnel must follow meeting the qualifications of NFPA 1041, *Standard for Fire Service Instructor Professional Qualifications*. This chapter further outlines training for emergency operations, specifics for hazardous material training (as required by OSHA/EPA), and the frequency of training (commensurate with duties.)

In two consecutive years—1992 and 1993, and again in 1995—the number of deaths were substantially below 100. This is most likely due to effects of improvement in fire fighter protective equipment, apparatus design, and fire fighter fitness and training, as well as an increased awareness throughout the fire service that the vast majority of fire fighter deaths do not have to occur, in spite of the very real, inherent dangers of the profession.

There remain, however, two very important areas where improvements must be made. The first, where the largest number of deaths occurs each year, is the heart attack problem. Too many fire fighters are continuing to actively engage in fire-fighting activities after they have experienced health problems that should preclude their participation. Basic life-style issues must be addressed or more fire fighters will be killed at the dinner table than on the fireground.

Second, factors in the motor vehicle accidents were not reported for every incident, but among those that were cited were driver inattention, speeding, and failure of the drivers to observe the rules of the road. These are all issues that can be addressed by a written, enforced policy and better training.

The U.S. fire service has made remarkable progress in the past few years in dramatically reducing the number of fire fighter deaths, demonstrating the effectiveness of recent advances in fire fighter safety and health. These efforts need to continue to further reduce the annual death toll.

FIRE FIGHTER INJURIES

Based on survey data reported to NFPA, 94,500 injuries were experienced by fire fighters on duty during 1995. This decrease in injuries, which began in 1994 is due, in part, to a change in the way NFPA categorizes incidents in which fire fighters are exposed to infectious diseases, such as hepatitis, meningitis, and HIV. Table 10-5J shows the number of fire fighter injuries between 1986 and 1995 and reveals that reported injuries remained within approximately 5 percent of 100,000 each year during this period. Until 1994, there had been no sustained trend in the number of estimated injuries.

Looking at fire fighter injuries relative to emergency incidents to which the fire service responded, 25.8 fireground injuries per 1,000 fires were experienced in 1995. This figure represents almost no change from 1994 and was in the middle of rates in the 10-yr period from 1986 to 1995. Table 10-5K lists the injury rate for this period for both fireground incidents and at nonfire emergencies. Injuries at the fireground decreased 9.6 percent from 55,990 in 1986

TABLE 10-5J. *Number of Fire Fighter Injuries, 1986—1995*

Year	Number of Injuries
1986	96,450
1987	102,600
1988	102,900
1989	100,700
1990	100,300
1991	103,300
1992	97,700
1993	101,500
1994	95,400
1995	94,500

TABLE 10-5K. *Fire Fighter Injuries per 1,000 Incidents*

Year	Injuries	Injuries per 1,000 Fires	Injuries	Injuries per 1,000 Incidents Non-fire Emergencies
1986	55,990	24.6	12,545	1.30
1987	57,755	24.8	13,940	1.41
1988	61,790	25.4	12,325	1.13
1989	58,250	27.5	12,580	1.11
1990	57,100	28.3	14,200	1.28
1991	55,830	27.3	15,065	1.20
1992	52,290	26.6	18,140	1.43
1993	52,885	27.1	16,675	1.25
1994	52,875	25.7	11,810	0.84
1995	50,640	25.8	13,500	0.94

to 50,640 in 1995. The rate of injuries per 1,000 fires increased slightly over the same period because the number of fires decreased 14.5 percent, more than the decrease in fireground injuries.

Fire fighter injuries by the type of duty engaged in at the time of the injury for 1995 are presented in Table 10-5L. Types of duty considered by the NFPA study include responding to or returning from an emergency, fireground activities, nonfire emergencies, training, and other on-duty activities.

Table 10-5M shows 1995 fire fighter injuries by the nature of injury and the type of duty. The four major types of injuries that occur during fireground operations are: (1) strains and sprains; (2) wounds, cuts, bleeding, and bruises; (3) burns; and (4) smoke or gas inhalation. The conscientious use of SCBA and other articles of fire fighter protective clothing, along with comprehensive training and a fireground incident management system, would have a significant impact on the current injury situation.

If any effort to control injury-causing situations is to be successful, it is important that the causes of the injuries are understood. Table 10-5N takes a closer look at fireground injury causes. As de-

TABLE 10-5L. *1995 Fire Fighter Injuries by Type of Duty*

Type of Duty	Number of Injuries
Responding/returning	5,230
Fireground	50,640
Nonfire emergency	13,500
Training	7,275
Other on-duty	17,855

fined here, "cause" refers to the initial circumstance leading to the injury.

Table 10-5O presents the average number of fire fighter injuries and the injuries requiring hospitalization per department by the population of the protected community during 1995. The average number of injuries per department range from a low of 0.37 for communities that serve a population of less than 2,500, to a high of 502.73 for departments that protect a population of 500,000 to 999,000. Generally, the larger the community, the greater the number of injuries.

The average number of injuries requiring hospitalization ranged from 0.03 for communities with populations of less than 2,500, to a high of 21.60 for communities with populations of 500,000 to 999,999. For communities with populations of less than 100,000, an average of fewer than 1 injury requiring hospitalization was experienced for each population category. Generally, the data seem to indicate that larger communities have more fire fighters who therefore suffer more injuries requiring hospitalization. Injuries are a function of many factors, all of which may be present in communities of any size. In fact, the high percentage of injuries requiring hospitalization in communities with populations under 25,000 shows that when these departments sustain injuries they are more likely to be serious.

Table 10-5P shows the average number of fireground injuries per department by the population of the community protected in 1995. These figures suggest that the number of fires to which a fire department responds is directly related to the population protected, and that the number of fireground injuries incurred by a department is directly related to its exposure to fire, i.e., to the number of fires attended by the department.

In order to allow a more direct comparison between fire departments of various sizes, examining the number of injuries per 100 fires attended will reveal whether community size has an effect on the number of fire fighter injuries in a particular community. The number of fireground injuries per 100 fires, shown in column 4 of Table 10-5P, reveals that the overall rate varies much less than the average number of injuries. This indicates that the wide range noted in average fireground injuries by population decreases dramatically when the actual fire rate is considered.

Table 10-5Q displays the average number of fires and fireground injuries in 1995 per department by population of community protected and by region of the country. The results for each region indicate that the number of fires to which a fire department responds tends to be related to the population protected and that the number of fireground injuries incurred by a fire department tends to be related to the number of fires it attends. The Northeast reports a higher number of fireground injuries for all sizes of communities served, except for the two largest population groups for which no sufficient data were available. Table 10-5Q also shows the number of fireground injuries per 100 fires by population served and by region of the country. These results are consistent with the average fireground injury findings, showing the Northeast as highest for all populations served, except for the two largest population groups for which no sufficient data were available.

As the data reported by NFPA show, fire-fighting operations pose a significant safety hazard and risk to the fire fighter. Because of the nature of the work and the types of hazards encountered on the fireground, it is highly unlikely that all fire fighter injuries can be eliminated. Good risk-management practices and the use of protective clothing and safety equipment can have a positive impact on the current experience. Reducing injuries will ultimately result in reduced suffering, less lost time, and lower medical costs.

TABLE 10-5M. *1995 Fire Fighter Injuries by Nature of Injury and Type of Duty*

Nature of Injury	Responding to or Returning from an Incident		Fireground		Nonfire Emergency		Training		Other On-Duty		Total
	Number	Percent	Number	Percent	Number	Percent	Number	Percent	Number	Percent	Number
Burns (fire or chemical)	25	0.5	4,890	9.7	100	0.7	565	7.8	395	2.2	5,975
Smoke or gas inhalation	35	0.7	4,625	9.1	295	2.2	40	0.6	90	0.5	5,085
Other respiratory distress	35	0.7	1,075	2.1	325	2.4	170	2.3	275	1.5	1,880
Eye irritation	210	4.0	3,140	6.2	450	3.3	250	3.4	660	3.7	4,710
Wounds, cuts, bleeding, bruises	1,340	25.6	10,310	20.3	2,310	17.1	1,365	18.8	4,455	25.0	19,780
Dislocation, fracture	180	3.4	1,370	2.7	255	1.9	335	4.6	590	3.3	2,730
Heart attack or stroke	40	0.8	345	0.7	35	0.3	75	0.4	235	1.3	685
Strain, sprain, muscular pain	2,785	53.2	19,280	38.1	7,250	53.7	3,320	53.9	8,265	46.3	41,500
Thermal stress (frostbite, heat exhaustion)	140	2.7	2,935	5.8	135	1.0	245	3.5	105	0.6	3,570
Other	440	8.4	2,670	5.3	2,345	17.4	350	4.7	2,785	15.6	8,585
	5,595		50,640		13,500		6,545		17,855		94,500

NOTE: If a fire fighter sustained multiple injuries for the same incident, only the nature of the single most serious injury was tabulated.
Source: NFPA's *Survey of Fire Departments for U.S. Fire Experience* (1995).

Data presented thus far in this chapter have been obtained for the U.S. fire fighter death and injury reports published by the National Fire Protection Association each year.

NFPA 1500 OVERVIEW

NFPA 1500 was written for the purpose of making the profession less dangerous by reducing the risk of accidents, injuries, and fatalities in the course of conducting emergency operations. NFPA 1500 is also geared toward making fire fighters healthier and more physically able to undertake their duties during their careers and for the remainder of their lives.

The yearly surveys and analyses of fire fighter deaths and injuries clearly show that fire fighters are suffering the consequences of placing their jobs and the welfare of others ahead of their own health and safety. The results of this situation clearly indicate that cancer and heart disease are a major problem for both active and retired fire fighters.

NFPA 1500 was drafted by a cross section of the fire service. The committee consisted of fire chiefs, company officers, line fire

TABLE 10-5N. *1995 Fireground Injury Causes*

Cause of Injury	Percentage
Fell, slipped, jumped	23.9
Overexertion	21.8
Exposure to fire products	14.7
Stepped on/contact with object	12.9
Struck by object	10.4
Exposure to chemicals/radiation	5.1
Extreme weather	3.2
Caught/trapped	1.5
Other	6.5

fighters, and union officials representing both career and volunteer fire departments. Other members of the committee included apparatus and protective equipment manufacturers, independent experts, and representatives of the United States Fire Administration (USFA). Like all NFPA committees, the NFPA 1500 group represented a balanced mixture of interests with diverse views of, and perspectives on, fire fighter safety. However, each member of the committee was committed to improving the health and safety of members of the fire service.

The document was drafted over a period of three-and-one-half years with the input of local fire service members adding to the expertise of the committee members. During the open public comment period, approximately 1,200 public comments were received on the proposed standard. Committee review of these public comments resulted in further refinement and improvements for the final draft. With strong support from both the International Association of Fire Chiefs (IAFC) and the International Association of Fire Fighters (IAFF), the standard was accepted by a near-unanimous vote of the NFPA membership at the 1987 NFPA Annual Meeting.

As an NFPA standard, NFPA 1500 is not a mandatory requirement for any fire service organization until it has been adopted by an authority having appropriate jurisdiction. NFPA develops and publishes standards that express the current thinking of volunteer members on each of more than 200 technical committees. In the case of NFPA 1500, the standard describes the elements of an occupational safety and health program that has been accepted as the minimum level of care that should be provided in the fire service today. All fire departments are being encouraged to comply voluntarily with the standard. As a specific document, NFPA 1500 is also available for legal adoption and enforcement by official entities, such as state and local governments or state occupational safety and health administrations.

Adoption and enforcement procedures vary widely across the country, so it is difficult to predict when and where this standard will

TABLE 10-5O. *Average Number of Fire Fighter Injuries and Injuries Requiring Hospitalization per Department by Population of Community Protected, 1995 for all Types of Duty*

Population of Community Protected	Average Number of Fire Fighter Injuries	Average Number of Injuries Requiring Hospitalization	Number of Injuries Requiring Hospitalization per 100 Injuries
500,000 to 999,000	502.73	21.60	4.30
250,000 to 499,000	152.06	1.22	0.80
100,000 to 249,000	51.57	2.33	4.52
50,000 to 99,000	21.11	0.78	3.69
25,000 to 49,000	9.52	0.31	3.26
10,000 to 24,999	4.04	0.19	4.70
5,000 to 9,999	1.73	0.14	8.09
2,500 to 4,999	0.90	0.11	12.22
Under 2,500	0.37	0.03	8.11

Source: NFPA's *Survey of Fire Departments for U.S. Fire Experience* (1995).

TABLE 10-5P. *Average Number of Fires, Fireground Injuries, and Injury Rates per Department by Population of Community Protected, 1995*

Population of Community Protected	Average Number of Fires	Average Number of Fireground Injuries	Number of Fireground Injuries per 100 Fires	Number of Fireground Injuries per 100 Fire Fighters
500,000 to 999,000	4,834.4	224.0	4.6	17.0
250,000 to 499,000	2,246.2	66.9	3.0	13.3
100,000 to 249,000	977.3	29.2	3.0	11.9
50,000 to 99,000	388.0	9.4	2.4	7.7
25,000 to 49,000	194.4	4.5	2.3	5.3
10,000 to 24,999	101.6	2.0	2.0	4.8
5,000 to 9,999	55.2	1.0	1.8	3.0
2,500 to 4,999	31.1	0.7	1.9	2.2
Under 2,500	13.6	0.2	1.5	1.0

Source: NFPA's *Survey of Fire Departments for U.S. Fire Experience* (1995).

become a legally required document. A survey of more than 450 fire departments conducted by the International Association of Fire Chiefs in 1990 indicated that 55 percent of the departments had adopted a plan to implement NFPA 1500. Some states have already begun to adopt health and safety regulations incorporating parts of NFPA 1500 for the fire service. For an example; there are over 80 communities in South Carolina that have adopted NFPA 1582, *Standard on Medical Requirements for Fire Fighters.* These communities, which participate in a comprehensive workmen's compensation plan, had not required any type of medical exam for their fire fighters. Consequently, their injury/illness rates were taxing the pool. They were required to adopt and use NFPA 1582 or be dropped from the insurance plan. In addition, the newly adopted Occupational Safety and Health Administration/Environmental Protection Agency (OSHA/EPA) training requirements for hazardous materials response contain some of the same requirements of NFPA 1500. One of the basic objectives of NFPA 1500 was to develop a program that would satisfy all of the requirements of OSHA *Fire Brigade Standard* (1910.156 subpart L) and to take those requirements one step further in several important areas. Unlike the OSHA standard, however, NFPA 1500 was written for the fire service, by the fire service.

Establishing a fire service occupational safety and health program can be difficult for a fire department because of the wide variety of issues that need to be addressed. In addition, it may be difficult to deal with program-element costs and with the time and resources required to implement a compliance plan. NFPA 1500 includes a section on implementation that allows for a phased approach which can be tailored to meet the needs and capabilities of a particular fire department. This approach calls for a comprehensive implementation plan and schedule that should be approved by an Authority Having Jurisdiction. This approval should be based on reasonable compliance under anticipated individual circumstances. The development of such a plan will require each fire department to assess where it has achieved compliance, and in what areas of the standard it needs to address to determine what would be required to comply with NFPA 1500. Furthermore, this compliance plan must include a comprehensive risk analysis to determine what types/levels of service are going to be provided. This obviously impacts both on-going fiscal year budgets, and long-range capital planning budgets. Resource allocation is not strictly monetary sources, but must include personnel, and capital equipment purchase as part of the implementation plan. The plan that is then developed must balance the problems that have been identified and the resources available to correct them.

NFPA 1500 addresses fire fighter health and safety as an overall program to create a work environment that will reduce the risk to members of the fire department. The standard considers a department member to be anyone involved in the organization at any level, whether career or volunteer, line or staff. The role of management is to provide as safe and healthy a work environment as possible, considering the scope and work of the organization. Management also is required to establish the necessary rules, regulations, and standard operating procedures that will allow the program to work effectively.

TABLE 10-5Q. *Average Number of Fires and Fireground Injuries per Department and Injuries per 100 Fires by Population of Community Protected and Region, 1995*

Column 1: Average Reported Number of Fires; Column 2: Average Reported Number of Fireground Injuries; Column 3: Number of Fireground Injuries per 100 Fires

Population of Community Protected	Northeast			North-Central			South			West		
	Col. 1	Col. 2	Col. 3	Col. 1	Col. 2	Col. 3	Col. 1	Col. 2	Col. 3	Col. 1	Col. 2	Col. 3
500,000 to 999,000	—	—	—	—	—	—	5,015.9	236.6	4.7	3,571.5	103.8	2.9
250,000 to 499,000	—	—	—	2,477.4	98.6	4.0	2,717.0	74.7	2.7	1,766.8	31.6	1.8
100,000 to 249,999	1,541.1	122.9	8.0	815.0	18.4	2.3	1,118.3	21.5	1.9	699.7	12.2	1.7
50,000 to 99,999	478.5	25.3	5.3	297.6	7.7	2.6	475.4	7.4	1.6	367.1	6.6	1.8
25,000 to 49,999	229.7	7.9	3.4	176.4	3.5	2.0	204.8	5.0	2.4	188.5	2.9	1.5
10,000 to 24,999	92.1	2.9	3.1	92.1	1.7	1.8	127.5	2.1	1.6	101.7	1.9	1.9
5,000 to 9,999	49.8	1.4	2.8	47.9	0.8	1.7	67.8	1.0	1.5	64.6	1.0	1.5
2,500 to 4,999	29.5	1.0	3.4	26.1	0.4	1.5	40.3	0.7	1.7	31.8	0.6	1.9
Under 2,500	11.5	0.4	3.5	12.5	0.1	0.8	20.2	0.3	1.5	11.4	0.2	1.8
Overall regional rate	—	—	4.3	—	—	1.9	—	—	2.2	—	—	1.9

*Insufficient data.
Source: NFPA's *Survey of Fire Departments for U.S. Fire Experience* (1995).

Every member of the organization has the right to be protected by an effective health and safety program, and the responsibility to comply with the established rules and regulations that structure the program. Organization members also have the right to participate or be represented in the research, development, implementation, and enforcement of the program, since the program is designed for their benefit.

NFPA 1500 requires the fire department to identify the services that will be provided, a basic organizational structure, the types of functions members will be expected to perform, and the type, amount, and frequency of training that will be provided to members of the department.

The fire department is required to adopt a written risk management plan which addresses all departmental policies and procedures. The risk management plan must include an identification of anticipated problems, the likelihood of occurrences and severity of consequences for each problem, solutions for elimination or mitigation for each problem, and an on-going evaluation of the effectiveness of the risk control techniques.

The department is required to identify a safety officer. A companion document, NFPA 1521, *Standard for Fire Department Safety Officer,* describes the role of this person in structuring and implementing the program. The position/function of the health and safety program manager should not be confused with the position/function of an incident scene safety officer. The National Fire Academy has developed two 2-day field courses that outline the specific roles and responsibilities of these positions, the training needed to fill these positions, and the relationship of these positions within a departments' occupational safety and health program. These courses reference applicable codes, standards, rules, and regulations.

An occupational safety and health committee must be established in the fire department. This committee should be made up of a cross-section of the department and be responsible for reviewing matters that deal with occupational safety and health issues within the department.

A data-collection system must be established to track accidents, injuries, illnesses, and deaths that are or might be job-related. The data-collection system also should contain records of occupational exposure to toxic substances or contagious diseases. Records

dealing with personnel training and certification and with the inspection, maintenance, and repair of emergency vehicles and equipment should be maintained.

The standard includes basic training requirements intended to ensure that members are trained adequately to provide for their own safety and that they do not create a hazard to others in the performance of their duties. Members who have not been trained fully must be supervised by qualified, experienced personnel, and their emergency scene duties must be limited to those that fall in the scope of their training and capabilities. If members of the department engage in structural fire fighting, training must be held at least monthly. Any training held in the department must be based on written standard operating procedures. Any training that involves live fire fighting exercises must be conducted in accordance with NFPA 1403, *Standard on Live Fire Training Evolutions.* Specialized training also must be provided if the department responds to special hazards, such as hazardous materials responses or belowgrade/confined-space rescue, high-angle rescue, or any other type of special operation.

A major requirement in NFPA 1500 relates to the safe operation of fire apparatus. Riding on tailsteps or in other exposed positions is expressly prohibited. All members must be seated with their seat belts fastened before the fire fighting vehicle begins to move or at any time the vehicle is in motion. As far as NFPA 1500 is concerned, the only safe way to ride fire apparatus is in a seat, in an enclosed area, with a seat belt fastened. An enclosed area is defined as having three sides and a top. Canopy areas that are open to the rear are acceptable on existing apparatus. However, requirements of NFPA 1901, *Standard for Automotive Fire Apparatus,* mandate four-door fully-enclosed cabs on all new fire apparatus.

NFPA 1500 requires the fire department to specify the number of personnel who may ride on a piece of apparatus, and requires the vendor to provide that number of seats, with seat belts, in an enclosed area. The fire department, the driver operator, and the officer, then assume responsibility for making sure that everyone riding in the vehicle is properly secured in a seat, with a seat belt, before the apparatus moves.

NFPA 1500 also requires fire apparatus and equipment to be inspected periodically and after each use in order to identify unsafe conditions. Any deficiencies found during these inspections must be corrected before the vehicle or piece of equipment is placed back in service.

Hearing protection is required for anyone exposed to vehicle or equipment noise in excess of 90 dB, whether they are drivers, passengers, or operators. This may be accomplished by hearing-protection devices, or by insulation, or by a combination of the two.

The fire department is required to provide full protective clothing and self-contained breathing apparatus (SCBA) to all members who will be exposed to the hazards of structural fire fighting. The items provided must comply with appropriate NFPA standards, and the department must ensure that all protective clothing is used properly by all members exposed to those hazards. When the activity is other than structural fire fighting, the department is obligated to provide appropriate protection for the hazards of each particular activity. NFPA 1500 is explicit in requiring full-body protection, including protection for the face and eyes. Full-body protection includes coat and pants, not three-quarter-length boots. The use of existing items of protective clothing is restricted only to those that met the respective NFPA standards in effect at the time the item of protective clothing or equipment was bought.

For interior fire fighting, self-contained breathing apparatus must be used, and operations must be conducted in teams of two or more. During interior operations, one member must be assigned to remain outside the area in which respiratory protection is required to be accountable for the interior crews. In imminently hazardous situations, such as hazardous materials incidents or confined-space or below-grade rescues, a safety officer must be assigned, a back-up rescue team must be ready to assist, and qualified emergency medical support personnel must be standing by with medical equipment and transport capability.

NFPA 1500 requires that operations at the scene of a fire or other emergency be conducted under an incident management system. This incident management system must be based on written standard operating procedures that identify roles and responsibilities relating to the safety of emergency operations. This incident management system requires that the incident commander is responsible for the overall safety of the operation and of all members at the incident scene. NFPA 1561, *Standard on Fire Department Incident Management Systems*, further identifies roles and responsibilities of all the functions within the management system. It states that personnel must be trained before being assigned to those positions. The system also identifies the need for proper span of control (1-5), that the system is both flexible and modular and grows based on the complexity of the incident, not necessarily the size of the incident. The standard outlines the need for inter-agenda coordination and cooperation, the use of an accountability system that tracks personnel at the scene both by location and function, and a system for the rehabilitation of personnel or scene. This rehabilitation process shall include medical evaluation and monitoring, fluid replenishment, and shelter in adverse weather conditions. Incident management is a primary tool that if utilized has a positive impact on fire fighter health and safety at all emergency incidents. Fire fighters and officers alike need to be trained in it's use and then the use of the system is a natural when operating at an incident scene.

Once again the National Emergency Training Center in Emmitsburg, MD, has recognized the need for proper incident scene management, and has included those requirements for not only fires and EMS personnel, but law enforcement, public works, emergency management personnel, and other agencies that must work together during emergencies. This is outlined in the jurisdiction's Comprehensive Emergency Management Plan (CEMP), which carried out by the Local Emergency Planning Committee (LEPC) of which the fire department is a primary participant. These requirements for the private sector, and their role in incident management are further outlined in NFPA 1600, *Recommended Practice for Disaster Management.*

NFPA 1500 emergency scene operations to those that can be accomplished safely with available personnel. The standard requires at least four fire fighters on the scene prior to initiating interior fire-fighting operations at a working structural fire unless an imminent threat to life or serious injury exists for structure occupants. The standard includes a strong recommendation for a minimum of four fire fighters on each engine and ladder company in order to provide adequate personnel to perform standard functions safely and efficiently.

NFPA 1500 provides guidance to limit a significant risk to fire fighters only to those situations where there is a potential to save endangered lives of occupants of a structure. In situations where only property is at risk, actions should be taken to reduce or avoid situations that negatively impact the safety of fire fighters.

In situations where there is no possibility to save lives or property, the standard requires that fire fighters be exposed to no risk to their safety.

Fire stations and other fire department facilities also are regulated by NFPA 1500. All fire department facilities are required to meet all legally applicable health, safety, building, and fire codes. Sleeping/living areas in fire stations must be equipped with smoke detectors, and automatic sprinkler protection is strongly recommended. The use and installation of carbon monoxide detectors in living and sleeping areas are also strongly encouraged. The health and safety of the facility can be enhanced when separate disinfection and cleaning areas are added for protective clothing, equipment, and storage of that equipment. This is also requirement in NFPA 1581, *Standard on Fire Department Infection Control Program*. Fire fighters should be made aware of the hazards associated with contaminated clothing and equipment, and be provided the proper training in it's cleaning and disinfection. Washing machines for that use should be of the commercial type, and be installed in areas separate from living and sleeping quarters. Members should be trained not to bring their contaminated protective clothing and equipment home and not to wash it in their laundering facilities at home. The U.S. Fire Administration has published a document titled *Developing and Implementing an Infectious Disease Control Program*; which may be obtained from the USFA free of charge. Mandatory inspections are required at least annually for code compliance and at least monthly to identify health and safety hazards. The standard also requires a 1-hr separation between vehicle storage areas and sleeping areas, and stations must be designed and equipped with devices to ventilate exhaust emissions from apparatus to prevent fire fighters from being exposed to harmful fumes.

NFPA 1500 addresses both the health and physical fitness of fire fighters, requiring an annual medical evaluation and examination and a structured physical fitness program. The established standards for fitness must be validated by a job-task analysis and should help members detect and correct problems at an early stage. The fire department is required to establish a confidential health data base for all members that can track exposures and research occupational illnesses in the future, while providing for doctor-patient confidentiality.

A subject of increasing interest in the fire service is the area of exposure to contagious disease, particularly when the fire department is involved with the delivery of emergency medical services. NFPA 1500 treats this as an occupational danger and requires a program to provide state-of-the-art protection for all exposed members. NFPA 1581 outlines a comprehensive program that includes the reporting and data tracking required as part of the Ryan White law. The technical committee included in the appendix sample forms for standard operating procedures, exposure tracking processes, and a sample program. The critical component in this document is the need for an infectious disease control person who serves as the point of contact or liaison with the hospital when a member may have been exposed. The documentation is also required by OSHA and must be kept for 30 yrs after the member has retired or resigned from the department.

NFPA 1500 addresses the need for access to employee assistance programs and recommends an approach that will support the overall wellness of the members of the fire department. This program should deal with substance abuse; stress, including critical incident stress debriefing; smoking cessation; and other personal and family problems. If they are to be accepted and successful, employee assistance programs should be provided by qualified professionals and should protect the confidentiality of those seeking assistance.

NFPA 1500 includes an appendix that contains supporting information and explanations for many of the specific provisions of the standard. A complete set of referenced standards also is included to help the user find complete information on many of the subjects covered by the standard.

IMPLEMENTING NFPA 1500

NFPA 1500 was written by the Technical Committee on Fire Service Occupational Safety and Health in response to the growing base of knowledge and information on how fire fighters were being injured and killed in the line of duty. NFPA 1500 currently serves as a consensus standard for a fire service occupational safety and health program. As such, any fire service organization can use NFPA 1500 to evaluate the completeness of an existing program or to design and implement a new program where one does not already exist. An occupational safety and health program developed and implemented in compliance with NFPA 1500 is instrumental in securing the highest possible levels of health and safety, given the hazardous nature of fire fighting.

One of the more common questions asked since the development of NFPA 1500 is, "How do we develop an NFPA 1500 compliance plan?" While it may be true that individual elements of an NFPA 1500 compliance plan may vary from fire department to fire department, the process each department goes through during the plan's development will be fairly standard. Itemized here is a suggested methodology for creating a written document that will identify the necessary elements in a fire department's occupational safety and health program. The methodology may, and should, be modified as needed on an individual basis. A sample compliance plan is included in both the NFPA 1500 handbook and in the Appendix to the document.

Step 1: Obtain a Copy of the Standard

The first step in the process is to obtain a copy of NFPA 1500. The document is available from the National Fire Protection Association, Batterymarch Park, Quincy, MA 02269. NFPA 1500 also is contained in Volume 8 of the *National Fire Codes*, 1995 edition.

After obtaining a copy of the standard, take the time to read it carefully and get a good feeling for the general requirements listed in the ten chapters. The document also contains two appendices, which may be used to determine the intent and meaning of specific requirements in each of the standard's chapters. The information contained in the appendices is not a requirement of the standard. Appendix material is informative in nature only and may be used by the reader as he or she sees fit.

Step 2: Establish a Project Team

A project team should be established to do all of the work required to develop and implement the program. It is important that the project team represent all areas and ranks within the department for the process to be successful. One of the primary responsibilities of the project team members will be to communicate with all members of the organization about the needs and requirements of fire fighter health and safety issues. For the plan to be effective, all members of the organization should feel that they have the opportunity to become involved in developing the plan. Team members selected or appointed to the team should be good communicators. Good written and verbal communications skills will be an extremely important aspect of the process.

Step 3: Compare NFPA 1500 with Established Practices

The next step is a line-by-line review of the specific requirements of the standard against the department's established current practices and procedures. At this time, the comparison need not be very involved. A checklist of the individual section numbers of the standard with an indication of "yes" or "no" with regard to compliance will provide an initial assessment as to where the organization stands with regard to the standard. At this step, many organizations find that they comply with the standard in more ways than they had previously assumed.

Step 4: Identify Sections Where Compliance Is Met

After completing the initial assessment review, it is necessary to do a more detailed review of the requirements of each section. This step is intended to identify and document whether compliance is either met or exceeded. Each section of the standard, including the specific requirement, should be listed on an individual sheet of paper. Adequate documentation in the form of a written policy, procedure, or practice, should be itemized on the same sheet or attached and made a part of the work records. By the time each section of the standard has been reviewed, adequate material will have been gathered to document compliance with the standard's minimum requirements.

Step 5: Identify Sections Where Compliance Is Needed

The remaining sections of the standard where compliance is still needed occupy the time and efforts of the committee for the next few steps. Each section of the standard in this category should be identified on a separate sheet of paper, as before. The current level of effort in each section should be listed in the event of partial compliance.

Step 6: Identify Available Alternatives or Specific Solutions

A goal should be identified for each section to indicate what is needed in order to comply with each specific requirement identified in Step 5. For example, this goal might indicate that a training program is needed or, perhaps, that only a standard operating procedure is required for compliance. In some instances, a number of alternatives might make compliance with a specific requirement of the standard possible. Each of these alternatives should be identified at this point in the process. If only one alternative exists, it should be identified at this time.

Step 7: Estimate Time and Cost for Each Alternative

Each alternative identified in the previous step should now be examined to determine the time and cost necessary for implementation. Some of the alternatives will probably cost little or nothing. These alternatives will typically be those related to the development of a policy or procedure, and the time required to conduct the training needed to implement the solutions. Some alternatives, particularly those that identify apparatus or equipment that would be required for compliance, may be very costly. Care should be taken to be as accurate as possible in estimating time factors for implementation of the various alternatives identified. Compliance with all of the requirements of NFPA 1500 will probably take far more time than might be expected initially. Even if funding is available for noncompliant requirements, the typical fire department is looking at a timeframe of approximately 5 yrs to comply fully with the entire standard.

Step 8: Select the Best Alternatives

For those items for which there were a number of possible alternatives, the best alternative should be selected for inclusion in the plan. The choice will depend on the specific situation that exists in each organization. The method chosen by one department might not be appropriate for another.

Step 9: Prepare a Draft Plan

All of the work accomplished up to this point will now be used to write a draft compliance plan. The written plan should include the existing information, practices, and procedures identified in Step 4 that comply with NFPA 1500 and those elements identified in Steps 5 through 8. The draft plan can be organized any way the department may desire. A number of organizations have found it helpful to use the format of NFPA 1500 as a guide for the development of their own compliance plans. At this point, the plan should identify how the organization intends to comply with the requirements of NFPA 1500, and present a timetable for phasing in those elements that will lead to compliance in the future.

Step 10: Legal/Risk-Management Review

While not actually required by NFPA 1500, it is a good idea at this time to ask the department's legal counsel to review the draft document. It is also a good idea to have the authority's risk manager review it. In the past 5 yrs, many municipalities and other governmental entities have secured the services of professional risk managers as a means of controlling liability, and this person can be a valuable resource for the department. Feedback and suggestions made by either the legal counsel or the risk manager should be reviewed by the project team. Where appropriate, the draft plan should be changed or modified as a result of this legal or risk-management review.

Step 11: Submit the Draft Plan for Adoption

After the draft plan has been finalized, it should be submitted to the Authority Having Jurisdiction for review and discussion. The Authority Having Jurisdiction in this case might be the fire chief, the mayor or city manager, or the city council, town board, or district trustees, among others. Adequate backup information should be submitted to identify the need for the program, the goal of achieving specific levels of fire fighter health and safety, and the anticipated benefits that may be expected to accrue to the individual employees, the fire department, and the community at large. The organization should be prepared to answer questions or issues raised in this step.

Educating all parties responsible for the adoption of the plan is crucial to the success of the program.

Step 12: Adopt the Plan

After the concept of the plan has been accepted and approved, the plan should be adopted. This may be accomplished by informal means, such as a cover letter signed by the fire chief, or it may include formal adoption through a statute, law, or ordinance. No matter how the plan is adopted, there should be a commitment to follow through with the programs, practices, and procedures identified in the document.

Step 13: Organize Implementation Teams

After the plan has been adopted, various implementation teams should be established to carry out the specifics of the plan. Each team should be responsible and accountable for implementing specific sections of the plan.

Step 14: Identify Implementation Strategy

For the plan to have a maximum impact on fire fighter health and safety, an overall implementation strategy must be developed. It is likely that simultaneous implementation of a number of elements will be impossible. A common-sense approach should be used in implementing the plan. Each specific element of the plan should be integrated in such a way that full compliance with the standard is ultimately accomplished. The implementation strategy for a particular department may be different from that of any other organization and should take into consideration specific factors of which the individual department is aware.

Step 15: Implement the Plan

After the implementation strategy has been established, the plan should be implemented. The implementation should follow the step-by-step sequence identified.

Step 16: Monitor Progress

After implementation, progress should be regularly and periodically assessed. Adjustments to the phasing plan should be made, as necessary, to follow the overall implementation strategy. Any problems should be addressed by the project team. Standard project management practices should be employed to maintain steady progress toward completing implementation.

Step 17: Review and Update the Plan Regularly

The plan should be reviewed periodically to ensure that the information and assumptions that were used to develop it are still valid. New information, fire suppression practices, changes and improvements in technology, and a variety of other factors can have an impact on the validity of any identified health and safety plan. This review will help ensure that the intended objective is still valid when accomplished. Specific attention should be given to keeping abreast of the other fire service standards referenced by NFPA 1500 so that the original plan may be modified appropriately during the implementation process.

The implementation of a NFPA 1500 health and safety program is an excellent example of a win-win situation for everyone. The individual fire fighters will benefit from improved levels of health and safety. The fire department will benefit from healthier employees who are better prepared and equipped to carry out their duties. The community will benefit from a better level of service at an efficient cost. While implementation of such a program is neither quick nor necessarily easy, it is one of the most responsible actions that anyone in the fire service today can undertake.

BIBLIOGRAPHY

NFPA Codes, Standards, and Recommended Practices

Reference to the following NFPA codes, standards, and recommended practices will provide further information to assist in the development of a program to reduce fire fighter injuries and deaths. (See the latest version of *The NFPA Catalog* **for availability of the following documents.)**

NFPA 472, *Standard for Professional Competence of Responders to Hazardous Materials Incidents*

NFPA 473, *Standard for Competencies for EMS Personnel Responding to Hazardous Materials Incidents*

NFPA 1001, *Standard for Fire Fighter Professional Qualifications*

NFPA 1002, *Standard for Fire Department Vehicle Driver/Operator Professional Qualifications*

NFPA 1003, *Standard for Airport Fire Fighter Professional Qualifications*

NFPA 1021, *Standard for Fire Officer Professional Qualifications*

NFPA 1041, *Standard for Fire Service Instructor Professional Qualifications*

NFPA 1201, *Standard for Developing Fire Protection Services for the Public*

NFPA 1403, *Standard on Live Fire Training Evolutions*

NFPA 1404, *Standard for a Fire Department Self-Contained Breathing Apparatus Program*

NFPA 1500, *Standard on Fire Department Occupational Safety and Health Program*

NFPA 1521, *Standard for Fire Department Safety Officer*

NFPA 1561, *Standard on Fire Department Incident Management System*

NFPA 1581, *Standard on Fire Department Infection Control Program*

NFPA 1582, *Standard on Medical Requirements for Fire Fighters*

NFPA 1600, *Recommended Practice for Disaster Management*

NFPA 1901, *Standard for Automotive Fire Apparatus*

NFPA 1911, *Standard for Service Tests of Pumps on Fire Department Apparatus*

NFPA 1914, *Standard for Testing Fire Department Aerial Devices*

NFPA 1931, *Standard on Design of and Design Verification Tests for Fire Department Ground Ladders*

NFPA 1932, *Standard on Use, Maintenance, and Service Testing of Fire Department Ground Ladders*

NFPA 1961, *Standard for Fire Hose*

NFPA 1962, *Standard for the Care, Use, and Service Testing of Fire Hose Including Couplings and Nozzles*

NFPA 1964, *Standard for Spray Nozzles (Shutoff and Tip)*

NFPA 1971, *Standard on Protective Ensemble for Structural Fire Fighting*

NFPA 1975, *Standard on Station/Work Uniforms for Fire Fighters*

NFPA 1976, *Standard on Protective Clothing for Proximity Fire Fighting*

NFPA 1981, *Standard on Open-Circuit Self-Contained Breathing Apparatus for Fire Fighters*

NFPA 1982, *Standard on Personal Alert Safety Systems (PASS) for Fire Fighters*

NFPA 1983, *Standard on Fire Service Life Safety Rope and System Components*

NFPA 1991, *Standard on Vapor-Protective Suits for Hazardous Chemical Emergencies*

NFPA 1992, *Standard on Liquid Splash-Protective Suits for Hazardous Chemical Emergencies*

NFPA 1993, *Standard on Support Function Protective Clothing for Hazardous Chemical Operations*

NFPA 1999, *Standard on Protective Clothing for Emergency Medical Operations*

Additional Reading

Aronson, K. J., "Causes of Death Among Firefighters," *Fire Chief*, Vol. 38, No. 2, February 1994, pp. 42–43.

Crume, R. V., "Occupational and Environmental Health Impacts Related to the Use of Replacement Firefighting Agents at U.S. Air Force Training Sites: A Progress Report," *Proceedings* of the 1991 Halon Alternatives Technical Working Conference 1991, April 30-May 1, 1991, Albuquerque, NM, 1991, pp. 153–154.

Dunkel, R., II and Sage, J. E., "Color: An Occupational Safety Issue. Training and Development Research Report," Ohio State Univ., Columbus, Research Report, March 1992, 25 pages.

Dunn, V., "Firefighter Death and Injury: 50 Causes and Contributing Factors," *Fire Engineering*, Vol. 143, No. 8, August 1990, pp. 34–36, 39, 41, 43, 44, 46, 48, 50, 52.

Fornell, D. P., "Understanding Turnout Technology: A Matter of Health and Safety," *Fire Engineering*, Vol. 145, No. 6, June 1992, pp. 105–106, 108–111, 113.

Gabler, W., "Breakneck Developments in Breathing Apparatus," *Fire International*, No. 151, May 1996, pp. 15, 17–18.

Gerner, G.D. and Schaper, F. C., "Why Firefighters Continue to Get Injured and Killed," *Speaking of Fire*, Vol. 3, No. 2, Summer 1996, pp. 6, 20.

Green, C., "Does the Dual Approach Answer the Future Needs in Breathing Apparatus?" *Fire*, Vol. 86, No. 1057, July 1993, p. 12.

Laing, R. M., Burridge, J. D. and Wilson, C. A., "Assessment of the Potential for Reducing Occupational Injury Through Protective Clothing," University of Otago, Dunedin, New Zealand, ASTM STP 1133; American Society for Testing and Materials (ASTM), Performance of Protective Clothing: Challenges for Developing Protective Clothing for the 1990s, Fourth (4th) Volume, ASTM STP 1133, June 18–20, 1991, Montreal, Canada, ASTM, Philadelphia, PA, 1991, pp. 924–932.

Laughlin, J., *Last Alarm—True Stories of Fire Fighters,* National Fire Protection Association, Quincy, MA, 1986.

Norris, D. A., "Where and How of Firefighter Injuries: A Long-Term Study," *Fire Chief*, Vol. 37, No. 2, February 1993, pp. 30–32.

O'Dwyer, D., "Hereford and Worcester Fire Brigade's Response to a Health and Safety Improvement Notice," *Fire Engineers Journal*, Vol. 55, No. 178, September 1995, pp. 25–28.

Occupational Safety and Health Administration, "Occupational Exposure to Bloodborne Pathogens: Precautions for Emergency Responders," Occupational Safety and Health Admin., Washington, DC, OSHA 3130, 1992.

"Physical Stress, Firefighting Kit, Breathing Apparatus and the Firefighting Environment," *Fire Engineers Journal*, Vol. 56, No. 181, March 1996, p. 32.

Putt, W. C., "Health and Safety at Wildland Fires," *Fire Engineering*, Vol. 145, No. 6, June 1992, pp. 60–62, 64, 66, 68, 70, 72–73.

Rosenstock, L., *et al.*, "Respiratory Mortality Among Firefighters," *British Journal of Industrial Medicine*, Vol. 47, 1990, pp. 462–465.

Rubin, D. L. and Foley, S. N., "Evolution of NFPA 1500," *Firehouse*, Vol. 18, No. 5, May 1993, p. 46.

Scott, G., "Physiological Effects of Wearing Breathing Apparatus," *Fire Research News*, Vol. 17, Summer 1994, pp. 3–4.

Sugimoto, R., *et al.*, "Comparative Experiment of Self-Contained Breathing Apparatus at Fire Fighting," Japanese Association of Fire Science and Engineering, Fire Research Annual Conference, May 17–18, 1990, pp. 143–146.

Teele, B. W., ed., *NFPA 1500 Handbook,* National Fire Protection Association, Quincy, MA, 1992.

Titus, J. J., "Respiratory Protection Devices: Missing Ingredient in the 'Fire Tetrahedron'," *Industrial Fire Safety*, Vol. 2, No. 4, July/August 1993, pp. 25, 30–32.

Walsh, M. B. and Wojcik, K., "Coping With NFPA 1500," *Fire Chief*, Vol. 34, No. 2, February 1990, pp. 43–47.

PRE-INCIDENT PLANNING FOR INDUSTRIAL AND COMMERCIAL FACILITIES

Revised by Gary S. Keith

A pre-incident plan is one of the most valuable tools available for aiding the fire department and the on-site fire brigade, if available, in effectively controlling a fire or other emergency incident. Planning for fires in industrial and commercial facilities increases the confidence and ability of fire service personnel to deal with most emergency situations. More important, it increases the potential for saving lives and property. This chapter will discuss the pre-incident planning process, who needs to be involved in this process, and the type of information necessary to gather for industrial and commercial facilities.

PRE-INCIDENT PLANNING: WHAT IT IS AND IS NOT

Pre-incident planning can be defined as a written document resulting from the gathering of general and detailed information/data to be used by public emergency response agencies and private industry for determining the response to reasonably anticipated emergency incidents at a specific facility.

In simple terms, pre-incident planning is ensuring that responding emergency personnel know as much as they can about a facility's construction, occupancy, and fire protection systems *before* an incident occurs. With this knowledge, the fire department can compare a potential incident at the facility with its available resources and plan the department's response accordingly. Pre-incident planning is not restricted to building components. It includes other factors and conditions that may be relevant to an emergency at a particular site.

Pre-incident planning is not easy: It takes considerable effort by the fire department to get a pre-incident planning program up and running. With current limits in budgets and other resources, fire departments are faced with much competition for staff time and money.

The reality, however, is that fire departments can't afford *not* to pre-incident plan. They may be responding to incidents today with fewer people than they were years ago. If not, they still need to meet public expectations, which may grow with the state of the art. Advance knowledge of the facility could mean the difference in managing the incident properly.

Gary S. Keith is director of regional operations at the National Fire Protection Association. He is the former chair of the NFPA Technical Committee on Pre-Incident Planning.

Proper training is also critical to pre-incident planning success. The individuals doing the pre-incident surveys and preparing the resulting documentation must have the proper level of training. Conducting the survey requires much more than walking through the building. It requires asking the right questions about construction features, storage arrangements, water supply adequacy, and sprinkler system components and design. This information must then be converted into a usable format for both fire department training and an actual incident.

The burden for successful pre-incident planning does not rest with the fire service alone: Pre-incident planning requires three-way open communication among facility management, the fire service, and the property insurance industry. Other agencies might also be able to provide valuable input during the development of the pre-incident plan. Such agencies include police, security, public utilities, environmental agencies, and contractors.

Facility management must take a coordinating role in pre-incident planning and be willing to host a joint meeting of all three entities. At this meeting, management must be willing to share information that its property insurance carrier has most likely already gathered about the hazards within the facility and the level of protection that is available. Most highly protected risk (HPR) insurance carriers are willing to provide this information, if they have the consent of facility management. The fire department must then take this information and incorporate it into its response plan for the facility.

When pre-incident planning is successful, everyone benefits. The fire department manages its response effectively, which results in a safer response and minimized property loss. A minimized property loss keeps the facility in business, which maintains the tax base for the fire department's budget. And, the loss cost is kept low for the property insurance carrier.

Conducting pre-incident planning is not the same as inspecting for code violations: When pre-incident planning, the fire department representative takes the facility "as is" and develops a response strategy around existing conditions. Conducting pre-incident planning is not an attempt to improve fire prevention or identify code violations at the facility. Its sole purpose is to prepare for an incident at the facility under existing conditions. Pre-incident planning assumes an incident will occur. It makes no special effort to prevent a fire or eliminate a hazard, but rather it is preparation for an incident, regardless of the likelihood.

Assume for a moment that a facility in the community is known to have several automatic sprinkler system code deficiencies. Also assume that the facility's management has agreed to correct the situation within 90 days. What happens if a fire occurs the day after tomorrow? Just because the code violations have been documented doesn't mean a pre-incident plan isn't needed. Code inspections and pre-incident planning are two distinct functions with two different purposes.

The fire department that is responsible for both code inspections and pre-incident planning needs to handle these types of situations with tact and diplomacy. When conducting code inspections, it is easy to stand behind the delegated authority of the position. But the fire department will never get the cooperation it needs for effective pre-incident planning if an authoritative approach is taken. The difference between each function must be understood by the fire department and communicated effectively to facility management.

Pre-incident planning is continuous: Many major fire incidents can be traced back to a change—often seemingly minor—that took place at a facility. Perhaps the packaging of the products stored in the warehouse was changed from cardboard to plastic. Or maybe some renovations took place that left concealed spaces unprotected by automatic sprinklers. Maybe the on-site suction tank for the fire pump is going to be drained and repainted over the next several weeks and it is the only adequate water supply for the facility.

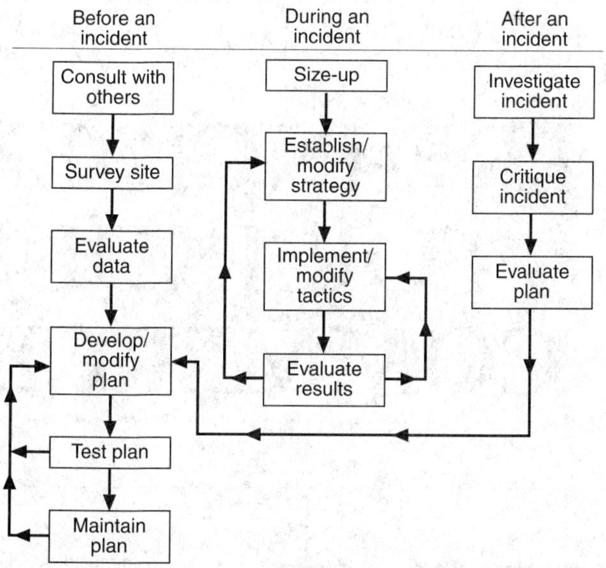

If a pre-incident plan is prepared once and never updated, these changes could go unnoticed until it is too late. Keeping up with changes in a facility requires a high level of commitment, communication, and cooperation among the pre-incident partners.

Determining the best frequency for updating pre-incident plans depends on the facility. A minimum of once per year may be fine at some locations, but it could be woefully inadequate at other locations.

Pre-incident planning requires a willingness by facility management to bring the fire department in and tell them what they need to know (e.g., "The sprinkler system in the warehouse is not currently adequate to protect what we're storing there"). It requires the fire department to know what questions to ask and to react positively when they get the truthful answers to those questions. And it requires the property insurer to act as an information link between the two.

THE PRE-INCIDENT PLANNING PROCESS

From the fire department perspective, once a facility has been selected, the owner/operator should be contacted. The needs and benefits of pre-incident planning might have to be explained in detail. The fire department should explain the nature of the information required, as the facility might have to arrange to have certain specialized employees available when the visit is made.

The pre-incident planning process can be shown as a flowchart. (See Figure 10-6A.) Obviously, much time and effort will be needed with the collection of data. This will involve consulting with the facility's staff and other individuals as necessary. The site will need to be visited to get an overview of the layout, construction, occupancy, and fire protection features.

Most major industrial and commercial facilities have developed an emergency operations plan in which the actions to be taken in the event of an emergency have been specified. A review of the emergency operations plan by the fire department can provide valuable insight into the actions to be coordinated by both facility personnel and the fire department. Incorporating the emergency operations plan into the fire department pre-incident plan is beneficial in case the facility

FIG. 10-6A. Pre-incident planning process flowchart.

personnel are assigned certain responsibilities, but circumstances during the emergency prevent them from performing these duties.

Once the data are collected, decisions will need to be made about what data to keep, and how they should be stored, retrieved, and presented. The method of presentation is best left up to the discretion of each fire department, but it should be based on the fire department's incident management system.

Fire departments may keep certain parts of the plan in the dispatch center for reference. Others may use computers in apparatus to store and retrieve information. Other departments may keep hard copies of the pre-incident plan in the individual fire stations or the apparatus itself. A sample pre-incident planning data form is shown in Figure 10-6B.

Testing and practicing the pre-incident plan will provide an opportunity to fine-tune the data and to revise and update the plan.

During the incident, the pre-incident plan should become the foundation for operations. It will provide important data that will assist the incident commander in developing appropriate strategy and tactics for managing the incident. Changes in conditions might dictate a revised strategy, including adjustments to responses and tactics. Throughout the incident, consulting the pre-incident plan will keep the incident commander informed about factors that might affect the success of any given strategic or tactical adjustment.

Each incident provides an opportunity for improvement. Assumptions, predictions, and accuracy of the pre-incident plan should be evaluated in light of the actual results during the incident. Modifications to the pre-incident plan should be made as necessary, and the plan should be retested, especially if the modifications are extensive.

PRE-INCIDENT PLANNING DATA COMPONENTS

Building Construction

Construction features, such as structural framing systems, building materials, and the interior and exterior finishes, can be key factors in the rate and extent of fire spread within a building. The fire conditions in a building of fire resistive construction will be markedly different from those in a building constructed entirely of combustible materials. Fire development, intensity, and spread can be far better controlled in structures of fire resistive construction. The size of the building, both vertical and horizontal, can also have a drastic effect on the decision process that takes place during an emergency. Awareness of these features by the fire department is important in estimating the potential fire problem and ultimately confining the fire to a limited area.

The pre-incident plan should include information regarding construction materials for exterior walls, such as metal panel, masonry, or wood frame. This information will be helpful in determining the potential for exposure protection, the potential loss of structural integrity, and the ability to access the building interior.

Information on roof construction is also needed. This includes the roof support, such as wood joist, steel joist, or steel beam; the roof deck, such as steel, metal panel, wood plank, or concrete; and the roof covering, such as built-up tar and gravel, or insulated membrane system. The type of roof construction can have an impact on ventilation operations, can create exposure problems, and, most important, can increase the dangers to fire fighters. The decision to place fire fighters on a building roof during a fire should be based partly on advance knowledge of the particular roof construction. Roofs unprotected by automatic sprinklers, even those of fire resistive construction, can collapse very early during a fire incident.

False ceilings and common attics can lead to fire spread within a building and impair the ability of fire fighters to control the fire. These construction features can create safety hazards by permitting fires to burn undetected in areas that are not readily visible. Awareness of these structural features allows fire fighters to prevent or reduce fire spread in them.

Like roofs, similar information should be gathered about the building's floor construction, since the same collapse potential may exist. In addition, information should be obtained regarding the floor's capabilities for drainage, especially where flammable and combustible liquids are used or stored.

The location of a building's entrances and exits, interior hallways, stairs, or other travel paths can be extremely valuable information in developing strategy or directing rescue, fire control, or other tactical operations. Familiarity with building access is often key to directing fire attack efforts that will confine the fire to the involved area, rather than driving it into uninvolved portions of the structure.

Vertical openings can exist in the form of open stairways, conveyor openings, utility shafts, elevator shafts, etc., and can affect the travel of fire, heat, and smoke.

Barriers to horizontal smoke and heat movement can exist in the form of fire walls, smoke barriers, etc. Attention should be given to the protection provided for horizontal openings. The types of fire doors provided and their method of closure should be noted.

The pre-incident plan should also note how venting will be accomplished and the location of any manual or automatic vents.

Occupancy

The type of occupancy is important in establishing incident priorities and tactical operations. An awareness of a facility's operations and contents increases the fire department's ability to fight a fire in that facility effectively and safely.

Life safety considerations should be the first priority for pre-incident planning. Many factors can influence the number and location of people in industrial and commercial facilities: hours of operation, day of the week, time of year, and the physical layout of the

facility all have an impact on the type of life safety concerns the fire fighters may be faced with during an incident.

The operations within a facility can be classified into several general categories, such as office, retail, manufacturing, warehouse, etc. However, each occupancy has its own unique characteristics which should be considered during pre-incident planning. Office buildings may have extensive computer rooms or laboratory areas, retail facilities may have large storage areas, and manufacturing occupancies have many different types of specialized equipment, hazardous materials, and hazardous processes.

Material safety data sheets (MSDS) should be reviewed for information on hazardous materials. Locations of MSDS files and related data, as well as the facility personnel responsible for this information, should be included in the pre-incident plan.

Warehouses have a wide range of occupancy considerations: commodity classifications, storage height and configuration, controlled environments, and limited access areas. Special attention should be paid to warehouses storing large quantities of hazardous or toxic materials, flammable or combustible liquids, aerosols, roll paper, rubber tires, and plastics. All of these types of commodities have proved to be substantial fire protection challenges over the years.

The pre-incident plan should detail salvage needs, as well. Often the largest part of a loss within an industrial or commercial occupancy is due to smoke damage outside of the immediate fire area.

With all occupancies, a critical component of pre-incident planning is keeping up with changes. If the occupancy of an area changes, the adequacy of the existing sprinkler system design and available water supply must be reviewed. This is especially important in warehouse occupancies. Changes in product commodity classification and storage height and configuration can be subtle and occur over a period of time. If the sprinkler system design or the water supply are inadequate for the type and arrangement of storage, fire control in the area of ignition is unlikely.

Protection

Knowledge of the fire protection systems in a building, knowing how these systems operate, and what actions are necessary to supplement these systems are all essential to pre-incident planning.

Automatic sprinklers play a vital role in the protection of industrial and commercial facilities. Pre-incident planning must determine not only whether or not the building is sprinklered, but it must also determine whether or not the sprinkler system (and its available water supply) is adequate for the occupancy of the building. This determination may be beyond the capabilities of the fire department personnel assigned to do pre-incident planning. Resources outside of the fire department must be utilized, if necessary, to make this determination. This is where a team approach to pre-incident planning with the facility's property insurance carrier can be crucial. The facility itself may also have personnel either locally or at the corporate level that have this information. A fire protection engineer might need to be consulted if the information cannot be determined elsewhere. If the sprinkler system is not adequately designed to meet the protection requirements of the occupancy in the building, the building may be little better protected than an unsprinklered building.

Pre-incident planning should include obtaining the minimum sprinkler system flow and pressure requirements for each major area of the facility. If the sprinkler system is a hydraulically designed system, it should have a placard posted on the riser indicating the specific flow and pressure design characteristics of the system. A key point: While the placard indicates what the system was designed for, this design may not be what is needed to protect the current occupancy.

Pre-Incident Data Gathering Form

Date: 7/24/93

Location: 20511 Hicks Rd.

   Sport Fashions

(Name of company and address)

Number of employees: 43

Phone number: 708-634-3035

Owner name/operator name: Fashion Apparel USA/Fred Jones Mfg.

Access to site: Streets —   1000-ft private drive with many large potholes to main west of Hicks Rd.

Lock box: Outside of office door on left side

Annunciator panel — location:

   Outside of office door on right side

Fire department connections — location:

   1. Front of Guard House at Hicks Rd. entrance; 2. W. corner fenced yard, accessible from Ridge Rd.

Door locations/forcible entry notes:

   1. Office-use lock box key; 2. Dock office, center, E. side, 24-hr occy. ex. weekends; 3. Pump house—lock box key at main office.

Storage configuration hazards:

   Clothing warehouses; stock in cardboard boxes on back-to-back double row racks 23 ft high; 8-ft-wide aisles, ordinary

   wood pallets; no solid shelves

Construction: Width (side facing fronting street):   345 ft

No. stories: Depth: 200 ft

   Height: 1 story = 27 ft

Wall construction: Insulated metal

Roof construction: Support structure — Steel joists on unprotected steel beams and columns

   Roof covering — Built-up layers w/tar and gravel

Location of vertical and horizontal cutoffs (space separation—fire walls):

Type of fire doors: 12-in. HCB w/parapet at offices

Utilities:	Gas	Unit heaters	Cutoff points	100 ft w/guard house
	Electrical		Cutoff points	1. 200 ft
	Oil pumps	None	Cutoff points	2. Ridge Rd. side
	Propane tank	None	Cutoff points	

Critical exposures:	Side 1:	Open	Side 2:	Automatic sprinkler whse 90 ft
	Side 3:	Open	Side 4:	Open

Protection (on site): Public water: 1650   gpm

Residual pressure: 51   psi   Normal static at site: 65   psi   Test date: 6/14/93

Sprinkler design: 1380   gpm 57   psi

FIG. 10-6B.   Pre-incident planning form.

In-rack sprinklers provided: [x] Yes   [ ] No

    (#) Booster pump ___1500___ gpm ___40___ psi   Driver type _____Elec._____

    (#) Fire pump __2000__ gpm __75__ psi   Driver type _____Diesel_____

Hose stream supply: On site: __Pond_____   Public: __16-in. main—Hicks Rd.__

    Hose length to reach: __Hydrant—200 ft; Pond—600 ft_____

Water demand: for AS: ___1610_____ for hose streams: ___500 gpm_____

Venting: Manual hatches: [ ] Yes   [x] No

    Automatic: _____   Powered: _____

    Power switch locations _____

Special conditions:

    Employee head count location:_____Guard house — Main Hicks Rd. entrance_____

    Contact person:_____Days — Mgr. Fred Jones; Nights — Shift Foreman_____

    Special units needed: _2nd ladder truck on 1st alarm from Ft. Mudge FD. Shift commander must request dispatch to TX._

Other: _____

Sketch details:

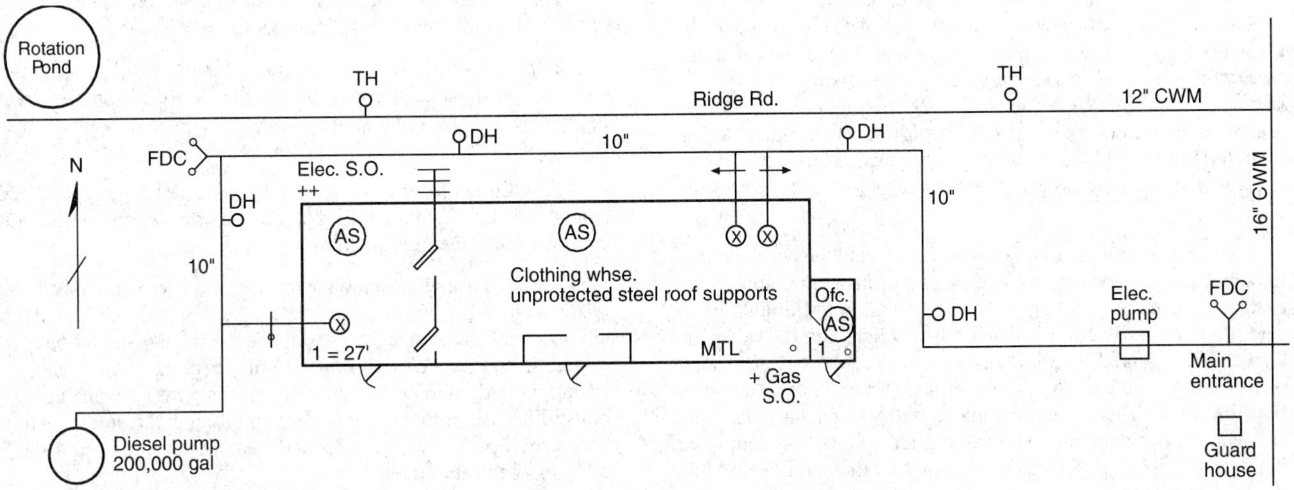

FIG. 10-6B.   Continued

An important part of pre-incident planning for protection features is to anticipate potential sources of sprinkler system failure. These can be summarized into one of three major categories.

1. *Design deficiency.* The water supply feeding the sprinkler system or the sprinkler system itself is not properly designed for the occupancy. The original system design might not have been adequate, or a change to a more hazardous commodity or storage array might have resulted in an inadequate system. Even seemingly minor changes in some occupancies, such as warehouses, can compromise existing sprinkler protection.

2. *Impairment to the sprinkler system before a fire.* This generally occurs when a sprinkler system is actually shut off during new construction or building renovations, or where an obstruction, such as a rock, works its way into sprinkler piping and blocks the flow of water.

3. *Impairment to the sprinkler system during a fire.* The system is impaired when any sprinkler control valve is shut prematurely during a fire. This obviously turns off the water to the sprinklers. Well-meaning facility employees or fire fighters may shut the valve in order to reduce smoke or to control water damage; however, this action only prevents sprinklers from gaining control of a fire in its critical development stage. Even if the valve is turned on again later, the fire might have grown beyond the point where sprinklers can control it.

The location of the fire department (Siamese) connections should be determined. The connections should be identified as to whether they feed the entire site, individual buildings, individual systems, or standpipes. Threads should be physically checked for compatibility with the fire department thread.

The most common source of water for the sprinkler system is a public water supply. In some cases, if the volume of the public supply is adequate, but the available pressure is too low, a booster pump is connected to the public supply to boost the system's pressure to the necessary level. Current water supply results should be available or tests should be conducted to obtain current information.

A static water source would usually require a fire pump to provide the necessary water volume and pressure to the system. Tanks, wells, reservoirs, and rivers are examples of these suction sources.

Where either booster pumps or fire pumps are provided, it is important to note their size, pressure settings, starting arrangement, and power and fuel supply. Fire fighters responding to an incident should check to make sure that required fire pumps have started, particularly when the fire pumps are the only adequate source of water.

Regardless of the source, a sprinkler system's water supply should be capable of meeting not only the sprinkler demand, but also the demand for hose streams. The water source chosen should be carefully selected so that the primary water supply for the sprinkler system will not be reduced. The location and capacity of hydrants, both public and private, should be determined. Where hydrant systems and water mains are not available, alternative resources must be determined to deliver sufficient water to the fire scene.

The location of sprinkler control valves, and the building areas they protect, must be determined during pre-incident planning. Facility management should post signs that will help fire fighters locate these valves upon arrival. Knowing where control valves are located is essential, because checking to make sure they are open is one of the first actions the fire department should take when it arrives on scene. This action, combined with establishing the water supply, is key to ensuring that the designed sprinkler system is doing the job it was intended to do, i.e., protect the building structure and its occupants, including fire fighters entering the building to complete final extinguishment.

Some facilities may have other types of protection systems, such as dry chemical, foam, carbon dioxide, or other gaseous extinguishing systems. The presence of these systems should be noted, as well as the areas they protect. In most cases, these systems will be considered supplementary to automatic sprinkler systems.

The pre-incident plan should note if available standpipe systems are of the wet or dry type, and the size of the hose discharge outlet. Small hose stations may be provided and supplied from the sprinkler system. The pre-incident plan should note if these stations will be impaired if the sprinkler system is shut off in a particular area.

The type of alarm system should be determined as to the type of detection provided (smoke, heat, waterflow) and the method of transmittal to the fire department. The presence of an alarm system should not eliminate the need for a follow-up telephone call from the facility confirming the nature and exact location of the alarm.

Procedures should be established for facility management to notify the fire department when sprinkler systems, fire pumps, special extinguishing systems, or alarm systems are taken out of service, regardless of the duration. If alarms are out of service, a temporary notification procedure should be instituted. When the fire department is notified of an impairment, the pre-incident plan should be referred to and contingency actions considered.

Site Considerations

Familiarity with local streets and roads in the area can provide the best access route to the location. Pre-planning can also provide alternative ways to reach the area should normal routes become unusable due to bridges with weight restrictions, narrow streets, road construction, flooding, drifting snow, railroad grade crossings, etc. Alternative approaches for responding units also allow for the placement of additional apparatus. It is important to determine which units will respond and from what direction they will come. This can help in the safety of apparatus approaching the same intersection from different directions. Assignments for fire companies based on probable order of arrival, including any neighboring companies that may respond on an automatic mutual aid, can increase the effectiveness of the fireground operation.

Data regarding security service at the site should be obtained. The number of security personnel on duty, where they are normally located, areas of "restricted" access, and the presence of guard dogs are among the items that should be reviewed. The location of any lock boxes and their contents should also be noted. An emergency call list of facility staff for off-hours should be obtained.

Exposures to the site can be both structural and nonstructural elements, such as neighboring buildings, fuel storage tanks, standing or tidal waters, vegetation, wooded areas, yard storage, and airport flight paths.

All utilities for the site should be reviewed. The type of fuel, storage location, and quantities should be noted, along with the locations of the emergency shutoffs.

The location of the entrance for electric power should be noted. Also, the location of the nearest disconnecting means within the facility as well as the nearest one outside the facility should be determined. If an electric-driven fire or booster pump is provided, it is important to determine the reliability of the power supply, noting if power lines run through the facility and if an independent feed has been provided so that the pump will remain running even if the facility has been shut down.

The location of emergency generators should be noted, as well as what equipment is powered when normal power is lost. Special consideration is needed if the emergency generator also supplies power to an electric-driven fire or booster pump.

Heating, ventilation, and air-conditioning (HVAC) systems can contribute to the spread of smoke throughout a facility. The ability to engage or disengage these systems as needed and the location of automatic and manual controls for this equipment should be documented.

Facility services, such as domestic water, compressed air, and steam, should also be reviewed.

Runoff of fire-fighting water contaminated with hazardous waste must be considered. Data should be gathered on the location of water drainage collection points, potable water supply locations, and any other potential exposures. The anticipated exposure should be measured against the expected fire scenario, based on the facility's other construction, occupancy, and protection features.

Outside Assistance

During many incidents, fire service personnel are assisted by other agencies or other industrial facilities capable of providing the support functions necessary to control the fire. Interagency and outside assistance can be vital to the outcome of emergency incidents.

If automatic response agreements result in a mutual-aid fire department arriving first on the scene, this department should be involved in the pre-incident planning process and should receive copies of the finished documentation.

Other assistance may include traffic control by police; the use of specialized resources, such as fire-fighting foam stored at another industrial facility; utility service control for electric, gas, and water systems; and other items, such as sand, barricades, and heavy construction equipment.

Because the amount and type of interagency assistance will vary in different areas, pre-incident plans should identify what is available and how to effectively request a timely response when it is needed.

BIBLIOGRAPHY

NFPA Codes, Standards, and Recommended Practices

Reference to the following NFPA codes, standards, and recommended practices will provide further information on pre-incident planning for industrial and commercial facilities discussed in this chapter. (See the latest version of *The NFPA Catalog* for availability of current editions of the following documents.)

NFPA 13, *Standard for the Installation of Sprinkler Systems*
NFPA 13E, *Guide for Fire Department Operations in Properties Protected by Sprinkler and Standpipe Systems*
NFPA 231, *Standard for General Storage*
NFPA 231C, *Standard for Rack Storage of Materials*
NFPA 231D, *Standard for Storage of Rubber Tires*
NFPA 231E, *Recommended Practice for the Storage of Baled Cotton*
NFPA 231F, *Standard for the Storage of Roll Paper*

NFPA 241, *Standard for Safeguarding Construction, Alteration, and Demolition Operations*
NFPA 600, *Standard on Industrial Fire Brigades*
NFPA 1021, *Standard for Fire Officer Professional Qualifications*
NFPA 1201, *Standard for Developing Fire Protection Services for the Public*
NFPA 1231, *Standard on Water Supplies for Suburban and Rural Fire Fighting*
NFPA 1420, *Recommended Practice for Pre-Incident Planning for Warehouse Occupancies*

Additional Reading

Beck, B., "Sprinklered Industrial Buildings: Preplanning and Operations," *Fire Chief*, Vol. 37, No. 8, Aug. 1993, Argus Press, Chicago, IL.

Brannigan, F. L., *Building Construction for the Fire Service*, 3rd ed., National Fire Protection Association, Quincy, MA, 1992.

Carlson, G., "Planning the Use of Private Fire Extinguishing Systems at Target Hazards," *Fire Engineering*, Vol. 147, No. 7, July 1994, Pennwell Publishing, Saddle Brook, NJ.

Davis, L., and Moore, J., *Industrial Fire Control Concepts*, International Society of Fire Service Instructors, Ashland, MA, 1988.

Davis, L., and Moore, J., "Warehouse Pre-Fire Planning and Fire-Fighting Operations," *Industrial Fire Hazards Handbook*, Cote, A. E., ed., 3rd ed., National Fire Protection Association, Quincy, MA, 1990.

Factory Mutual Research Corp., *Warehouse Occupancy: The Effects of Change*, Norwood, MA, 1989.

Factory Mutual Research Corp., *A Pocket Guide to Automatic Sprinklers*, Norwood, MA, 1990.

Factory Mutual Research Corp., *Fighting Fire in Sprinklered Buildings*, Norwood, MA, 1991.

Factory Mutual Research Corp., *Pre-Fire Planning: The Rewards Are Mutual*, Norwood, MA, 1992.

Garner, D., and Keith, G., "Automatic Sprinkler Systems: What the Fire Department Needs to Know," *Fire Engineering*, Vol. 144, No. 4, Apr. 1991, Pennwell Publishing, Saddle Brook, NJ.

Garner, D., and Keith, G., "Sprinkler System Water Supply Analysis," *Fire Engineering*, Vol. 144, No. 12, Dec. 1991, Pennwell Publishing, Saddle Brook, NJ.

Garner, D., and Keith, G., "What Fire-Fighters Should Know About Automatic Sprinkler Choices," *Fire Engineering*, Vol. 147, No. 6, June 1994, Pennwell Publishing, Saddle Brook, NJ.

Gustin, B., "Pre-Planning Special Industrial Hazards," *Fire Engineering*, Vol. 147, No. 11, Nov. 1994, Pennwell Publishing, Saddle Brook, NJ.

Jenaway, W. E., *Pre-Emergency Planning*, 2nd ed., International Society of Fire Service Instructors, Ashland, MA, 1992.

Kalman, B. J., "Applying Technology to Prefire Planning," *Fire Engineering*, Vol. 146, No. 1, Jan. 1993, pp. 4546,48–49.

Keith, G. S., "Understanding the Intricacies of Pre-Incident Planning," *The Times*, Iss. 2, July 1994, National Fire Protection Association, Quincy, MA.

Ludwig, G. G., "Pre-Emergency Planning Through Computers," *Firehouse*, Vol. 18, No. 6, June 1993, pp. 80–81.

WILDLAND FIRE MANAGEMENT

Dan W. Bailey and William C. Teie

Each year, on average, hundreds of thousands of fires burn millions of acres (2.2 ha) of protected forest, brush, and grass-covered lands in the U.S. and Canada. Protection services cost well over a billion dollars annually, with losses approaching $4 billion. These costs do not reflect the services of thousands of volunteer fire fighters in both countries, nor the expenses of the many city fire departments that fight fires on wildlands—lands that are essentially undeveloped—within or near their jurisdictions.

Wildfires include any unwanted fire burning on wildlands. Most are extinguished while smaller than 1 acre (0.4 ha) by a few fire fighters working with hand tools or water-handling equipment. Under extremely adverse conditions, however, a fire can spread to well over 100,000 acres (40,000 ha) and require thousands of fire fighters and hundreds of mechanized units for several weeks to over a month.

The wildfire problem is highly variable depending on location. This is because:

1. Fire ignition is dependent on natural phenomena, such as lightning or volcanic activity, on human activity, on fuel bed characteristics, on the weather, and on the effectiveness of prevention efforts.
2. Fire behavior—that is, how a fire spreads and how intensely it burns—is dependent on local conditions, such as weather, fuels, and topography.
3. The effectiveness of fire suppression is closely related to accessibility, difficulty of control, and the capability and performance of the local protection agency.
4. Public safety is involved when fires threaten recreation areas, mountain homes, or urban areas or when smoke obscures visibility along highways or near airports.

The foregoing considerations vary so much by location that each geographical region must deal with a unique set of circumstances. In the Mediterranean climate of southern California, for example, a long, hot summer dries the chaparral, thus preparing it for explosive burning during the gale-force "Santa Ana" winds. Such conditions pose an enormous threat to human life and property. In the humid subtropical climate of the southeastern United States, early spring and late fall fire seasons threaten extensive forests of pine and hardwood trees. Throughout the boreal forests of central and western Canada, lightning often ignites hundreds of fires during long periods of hot, dry weather, causing extensive damage. Wildfires burning in the tundra of Alaska and northern Canada produce vast ecological effects and, therefore, require special suppression techniques. During droughts, even the marine climate of the Pacific coastal forest, which extends from northern California to Alaska, does not protect these areas from devastating timber fires.

CAUSES OF FIRES

The leading cause of wildfires in the 1.5 billion acres (0.6 billion ha) of protected wildlands of the United States is incendiarism, which accounts for one-quarter to one-third of the fires and burned-over area. Debris burning accounts for an additional one-fourth of the fires, while lightning accounts for another one-eighth to one-seventh of the fires, but a larger share of the burned area. Other prominent causes of fires include equipment use, careless smoking, children playing with matches, campfires, and railroad use.

Nearly all of these fires are controlled at a size of 100 acres (40 ha) or less. However, roughly 2 percent of fires result in two-thirds of the total burned-over area. All together, hundreds of thousands of wildfires occur each year in the U.S., burning millions of acres of forest, brush, and grass-covered lands.

The relative importance of the fire causes mentioned above varies considerably among the different geographical regions of the country. In terms of the number of fires and burned areas, lightning is the leading cause in the northwestern part of the country, as well as in Arizona and New Mexico. For the entire eastern half of the United States, incendiarism and debris burning are the chief causes. In California, equipment use and incendiarism rank highest, while debris burning and lightning are the main problems in the central Rockies. Alaska's wildfires can be blamed chiefly on debris burning and lightning.

In Canada, thousands of wildfires occur each year, burning millions of acres of wildlands. About one-third of these fires are caused by lightning, although fires started by lightning account for most of the burned-over area.

The Role of Lightning

On a global scale, lightning is common: there are eight million cloud-to-ground discharges per day around the world. While the

Dan W. Bailey is the staff officer for Air, Fire, and Aviation Management for the USDA Forest Service's Northern Region, Lolo National Forest, based in Missoula, MT. He is immediate past chair of the NFPA Wildland Fire Management Section, chairs the NFPA 1051 *Committee on Professional Qualifications for Wildfire Suppression Personnel,* and serves as one of the sixteen national Type I incident commanders, responsible for managing America's most complex wildland fires. William C. Teie is a member of NFPA's Technical Committee on Forest and Rural Fire Protection.

energy per lightning bolt varies greatly, 250 kilowatt-hours is a good estimate of the average energy discharged in each stroke. Almost 75 percent of a lightning bolt's energy is converted to heat during discharge. This is more than sufficient to ignite most fuels.

There are two types of discharge: cold and hot stroke. Cold strokes exhibit high voltage of short duration, generally with mechanical or explosive effects. Hot strokes discharge a lesser current for a longer duration, thus starting more fires. Approximately 21 percent of all lightning moves from cloud to ground. Of this amount, 20 percent is hot stroke, according to studies in the northern Rockies. Therefore, in the northern Rockies, one stroke in 25 has the electrical characteristics to start a fire, depending on where or what it strikes and on the local weather.

AGENCIES INVOLVED

In the United States, wildland fire protection is handled by federal, state, county, and in some cases, city agencies, as well as by private corporations and career/volunteer fire departments. The U.S. Forest Service protects about 200 million acres (80 million ha) of national forest and other lands. Approximately 587 million acres (197 million ha) of other federal lands, mostly in the public domain, are protected by the Bureau of Land Management (BLM), the National Park Service (NPS), the U.S. Fish & Wildlife Service, and the Bureau of Indian Affairs. States, local governments, corporations, and career/volunteer fire departments protect about 840 million acres (340 million ha) of the country's essentially undeveloped lands. About 158 million acres (64 million ha), or roughly $9^1/_2$ percent, of the wildlands in the United States are not protected.

In Canada, wildland fire protection is handled by ten provinces, two territories, and Parks Canada, as well as by private corporations and career/volunteer fire departments.

COMPONENTS OF WILDLAND FIRE PROTECTION

Fire prevention, which involves all activities concerned with minimizing the incidence of destructive fires, is a team effort by fire protection agencies and the community. It is accomplished through educating the public and increasing their awareness, regulating the public use of the wildlands, engineering developments, and enforcing fire regulations and laws. Regulating the public use of wildlands limits activities during periods of high fire danger. Such regulation includes area closures, campfire prohibitions, restrictions on burning, equipment operation shutdowns, and the use of spark arrestors on motor-driven equipment. Engineering activities include the reduction of flammable vegetation along roadways and in areas of heavy use and the modification of vegetation to favor less flammable species. Law enforcement activities are aimed mainly at the prevention of incendiarism, the leading cause of wildland fires.

Fire Prevention

Fires caused by people account for the majority of wildfires nationally. Unwanted fires threaten valuable resources, and, for this reason, fire prevention is an ongoing job. Fire prevention is one of the most important parts of a fire protection agency's responsibilities. It is the first line of defense against fire. A fire that can be prevented will save lives, resources, and money.

Fire Prevention Week is the full week—Sunday through Saturday—that includes the date of October 9, the anniversary of the Great Chicago Fire of 1871 which killed 250 people, left 100,000 homeless, and destroyed 17,430 buildings. Also on this day in Peshtigo, WI, more than 1,500 people died when a wildland fire swept through the town. During this period, national and local programs emphasize the fire prevention message. Fire prevention messages are repeated in books, on the radio and television, in public service announcements, on signs and posters, in news articles and brochures, in videos and slide-talks, and by personal contacts.

One of the most effective educational programs in the United States since 1950 has been Smokey Bear. This program is credited with being a major factor in the prevention of thousands of fires, thereby saving millions of dollars in damages and fire suppression costs. Smokey's message has remained the same, but partnerships have been developed to reach new groups. Such partnerships include the American Cowboy, pro sports figures, and amateur athletes, who have joined with Smokey to get the fire prevention message to the public. Specific prevention programs have been developed by different agencies and areas of the country. The Keep Green Program and the Fire-Safe Program are examples of programs developed to meet regional needs.

Wildland/urban interface: The growth of the wildland/urban interface fire problem has focused prevention programs in a new direction. Nationally, the Wildfire Strikes Home campaign has been successful in creating an awareness of the issue. This program has been sponsored by the U.S. Forest Service, the U.S. Department of the Interior, the National Association of State Foresters, the National Fire Protection Association, and the U.S. Fire Administration. Regional and local programs have been developed to meet the different wildland/urban fire needs of each area. More and more homes are being built on scenic sites and slopes in forests and other wildlands. The result is that vast areas of North America now contain high-value properties intermingled with highly flammable native vegetation. Several recent incidents have involved the loss of thousands of homes to fires in these new residential areas. Burned homes are not the only cause for serious concern: according to officials, major losses of life are also a concern.

The current trend of building homes in areas threatened by wildland fires has introduced new factors that seriously complicate the duties of fire fighters. Both wildland and structural fire fighters have developed time-proven methods for controlling the types of fires expected in their jurisdictions. The efficient, successful methods for protecting a forest's natural resources from fire were not developed with the idea that the forest would include numerous homes. And the methods used to extinguish structure fires are not effective against the power of a wildfire, especially as the fire spreads through areas in which water supplies for fire fighting are limited.

Economic constraints have slowed effective cross-training and equipping of all fire fighters to attack these two vastly different types of fire. The areas in which these two types of fires interface can be classified into three general types:

1. *The mixed interface* contains structures scattered throughout rural areas. Usually, there are isolated homes surrounded by areas of undeveloped land. When a fire starts, the individual homes are hard to protect because of the large area that may be burning. While relatively few homes may be at risk, the risk to the individual homes is great.

2. *An occluded interface* is characterized by isolated areas of wildlands, either small or large, within an urban area. An example would be a city park surrounded by homes, where the goal is to preserve some contact with a natural setting. Many homes and other buildings may be at risk, but these relatively small wildland areas are generally less susceptible to uncontrolled raging.

3. *A classic interface* is where homes, especially those crowded onto smaller lots in new subdivisions, press against wildland vegetation along a broad front. These vast wildland areas can propagate a massive flame-front during a wildland fire, causing

numerous homes to be at risk from a single fire. Because the built-up quality of the subdivision may give a false sense of security, this is where the greatest loss of life is possible.

Fire Detection

Discovering, locating, and reporting fires often is accomplished in wildland areas by lookouts in fire towers and by aircraft patrols. Additionally, thousands of fires are reported annually by travelers, campers, woodcutters, hunters, and others. In the western United States and parts of Canada, lightning strikes are routinely detected electronically and their locations plotted by computer. Computer programs have been developed to assess the likelihood of ignition at various locations, as well as the characteristics of the strikes themselves. Automated fixed-point infrared detection systems have been developed but they have yet to be used extensively in North America, although they are used in some European countries.

While fire detection from satellites is technically feasible, the complications of discriminating between legitimate heat sources and incipient fires has not yet allowed the practical use of this technology.

Fire Suppression

Fire suppression includes all work to extinguish or confine a fire, beginning with its discovery. Suppressing or fighting a fire is usually difficult work that is inherently dangerous. However, knowing and applying safety principles and fire-fighting tactics serves to increase the safety and effectiveness of operations.

Standard fire-fighting orders: A series of standard fire-fighting orders provides the basis for safety for those involved in wildland fire suppression. Every fire fighter must learn these orders, understand, and follow each when it applies:

F   Fight all fires aggressively, but first provide for safety.
I   Initiate all action based on current and expected fire behavior.
R   Recognize current weather conditions and obtain forecasts.
E   Ensure that instructions are given and understood.

O   Obtain current information on fire status.
R   Remain in communication with crew members, your supervisor, and adjoining forces.
D   Determine safety zones and escape routes.
E   Establish lookouts in potentially hazardous situations.
R   Retain control at all times.
S   Stay alert, keep calm, think clearly, act decisively.

Fire situation hazards:

- Fire not scouted or sized-up.
- Being in areas that fire fighters have not seen before in daylight.
- Safety zones and escape routes not identified.
- Being unfamiliar with weather and local factors influencing fire behavior.
- Being uninformed about strategy, tactics, and hazards.
- Instructions and assignments not clear.
- No communications link with crew members/supervisor.
- Constructing fireline downhill without safe anchor point.
- Building fireline downhill with fire below.
- Attempting frontal assault on fire.
- Unburned fuel between the fire fighter and the fire.
- Not being able to see main fire; not in contact with anyone who can.

- Being on a hillside where rolling material can ignite fuel below.
- Weather getting hotter and drier.
- Wind increasing and and/or changing direction.
- Getting frequent spot fires across the fire line.
- Terrain and fuels making escape to safety zones difficult.
- Taking a nap near the fireline.
- Personal protective equipment not being available or not being used properly.
- Operating unfamiliar equipment.

Four common denominators: Four major common denominators of fire behavior have been noted in fires where a fire fighter has been killed or from which fire fighters narrowly escaped. Such situations often occur:

1. On relatively small or deceptively quiet sectors of large fires.
2. In relatively light fuels, such as grass, herbs, and light brush.
3. When there is an unexpected shift in wind direction or windspeed.
4. When the fire responds to topographic conditions and runs uphill.

These factors should not be considered all-inclusive. For example, a sudden change of wind may change the direction of fire spread, regardless of topography.

Each set of circumstances has the potential for creating a tragedy or near-miss fire. Often, human behavior is the determining factor. Fire fighters who remain calm when the wind direction changes and move back into a burned area should survive. Those who try to outrun a fire under similar conditions may die. The difference between a tragic fire and a near-miss fire may be due to luck, skill, and/or advance planning. In all cases, it is important to be alert and aware of conditions that may signal a sudden change in fire behavior.

In a few words, be alert, watch out for light fuels, wind shifts, steep slopes, and "chimneys."

Those who remain alert and on the lookout for possible trouble have the best chance of survival.

Suppression of Small Fires

Wildland fire management strategies are aimed at prompt, safe, aggressive suppression action of all wildfires. Small fires pose the same kind of suppression problems and require the same kind of practices as larger fires. Effective suppression of small fires involves the following steps: first attack, line location, line construction, burning out, mop-up, patrol, and declaring the fire out.

Suppression strategies range from prompt control at the smallest acreage possible to containment using a combination of fireline and natural or constructed features, to merely ensuring that the fire remains confined to a defined geographical area.

Initial attack: The principles of initial attack include:

- *Sizing up a fire.* Go around the fire as quickly and safely as possible, or inspect it from a vantage point. But do not go around the head of the fire if it is moving rapidly, as entrapment is likely. Size up a fire from a vantage point or from the flanks of the fire.

- *Selecting a point of attack, and making an attack.* The universal rules are to take prompt action on an attack point; to stay with the fire, and take the most effective action possible with the available forces and equipment; to inform the dispatcher of the situation by radio; and to continue to work day or night, if night work can be done safety.

- *Mopping up.*   After primary line construction work is completed and a fire is called "controlled," many things remain to be done to make the fireline "safe" and put the fire out. This work is called mop-up. The objective of mop-up is to put out all embers or sparks to prevent them from crossing the fireline. A certain amount of mop-up work is done while building the control line. Mop-up becomes an independent part of fire fighting as soon as the spread of the fire has been stopped and all lines have been completed. Ordinarily, mop-up is composed of two actions: putting the fire out and disposing of fuel, either by burning it to eliminate it or by removing it so it cannot burn.

- *Patrolling.*   Patrol is that portion of the mop-up job that consists of moving back and forth over the control line and the edges of burn areas to check for and put out any fire that may burn or blow across the line, and, at the same time, to check for and put out spot fires outside the line.

- *Declaring the fire out.*   Before abandoning a fire, and as a follow-up, the incident commander will take steps to ensure that the fire has been extinguished and that any fireline that has been constructed is adequate should a flare-up occur within the fireline.

PRINCIPLES OF COMBUSTION

Fire is a chemical reaction in which energy is produced. When forest material burns, the oxygen in the air combines chemically with wood, pitch, and/or other burnable elements in the fuel. In the forest, several stages in the fire process are normally encountered: first, the igniting spark; then a period of smoldering; and finally, the more rapid combustion of fuels. This process may continue to involve leaping flames, dense smoke, intense heat, loud noises, and occasional explosions. Throughout its life, the action of a fire is governed by certain natural laws or principles of combustion, as they are described below. An understanding of these principles is basic in judging the effect of various environmental factors on fire behavior.

The Fire Triangle[*]

Three things—heat, oxygen, and fuel—are required in proper combination for ignition and combustion to occur. If any one of the three is absent, or if they are not in proper balance, ignition or combustion will not occur. Variations in balance among heat, oxygen, and fuel govern the violence of a fire, and determine whether the fire will smolder and spread slowly or flame brightly and travel rapidly.

Heat: It is difficult to specify the amount of heat required to ignite forest fuels because of the variation in the nature of the fuels. Vegetative material is high in carbon content and ignites at relatively low temperatures, provided the moisture content is low and the substance is freely exposed to the air. During the forest fire season, a large part of the vegetative matter in a forest—duff, dead limbs, pine needles, tree branches, rotted logs, and so on—is dry enough to be ignited easily. The temperature requirement for ignition of forest fuels varies from approximately 500 to 750°F (260 to 399°C). Many

common ignition sources can provide the required heat, including a burning match, a glowing cigarette, and a lightning bolt.

Ignition often depends upon the length of time the fuel is exposed to heat. Dry pine needles may be ignited in a few seconds by the heat from a flaming match. Moist pine needles also may be ignited by a match if subjected to the heat for several minutes. When wood fuels are exposed to heat for a long time, the normal ignition temperature may be lowered. Studies have shown that wood exposed to a temperature of 400°F (204°C) for approximately 30 min may ignite.

The ease of ignition of forest fuels exposed to heat for a considerable time has an important influence on fire behavior. In a coniferous forest, a hot fire burning in a tangle of downed logs and dry limbs may generate enough heat to make the green tree crowns above the fire easily ignitable. All that is needed to cause the tree crowns to ignite is a flying ember and a gust of wind. Similarly, a fire smoldering in a pile of leaves may generate enough heat to lower the normal ignition temperature of the entire pile. When the leaves are stirred up, the whole pile breaks into flame. The stirring of the pile lets in enough oxygen for the preheated leaves to be ignited easily.

Oxygen: In the forest, there is usually enough oxygen to permit ignition and combustion. However, some of the forest fuels may be arranged so that oxygen is not available in sufficient amount to support a fire. In deep, tightly compacted duff, only the top particles can get enough air to permit a fire to burn. In this situation, a fire will burn from the top down as each layer is exposed to the air. By contrast, in a very loose layer of pine needles, the entire mass is fairly well exposed to air, and consequently, combustion will take place rapidly.

When wind blows on a fire, it usually speeds up the combustion process. Wind forces oxygen around fuel particles where the flow of air normally may be restricted. In addition, wind has a physical reaction on the flames, often bending them into positions that create more favorable situations for the spread of the fire.

Fuel: Under forest conditions, fuel is a major variable in the fire triangle. The fire fighter must become acquainted with the general nature of forest fuels to understand their burning characteristics.

The ease of ignition and rate of combustion of forest fuels depends mainly upon the type of fuel—is it logging slash or dense duff in a green forest? On fuel continuity—is the fuel distributed more-or-less evenly over the area, or is it only present in patches? On moisture content—does it feel damp when touched, or does it crackle and appear very brittle? On fuel temperature—is the fuel exposed to the heat of the sun, or does it lie in cool shade?

Combustion process: When fuels are heated to their ignition temperature, they produce gases which will ignite if combined with oxygen. When a log burns, the flames may appear to be coming directly from the fuel. Actually, the flames come from the ignited fuel gases emerging from the heated log.

In forest fires, the two common stages of combustion are smoldering and flaming. In the smoldering stage, heat is liberated, and a rise in fuel temperature occurs, while flames absent or appear only intermittently. The absence of flames is caused by insufficient oxygen or by moisture in the fuels, which slows the oxidation process.

When a fire is in the flaming stage, all the elements of the fire triangle—that is, heat, oxygen, and fuel—are in proper combination for rapid oxidation to occur.

The flames of very hot fires may be observed intermittently in the smoke at a considerable distance above the fire. This occurs

---

*Although it is recognized that there is a fourth component involved in the combustion process—a chemical chain reaction—for the purposes of this presentation only the components that make up the commonly referred to "fire triangle"—heat, fuel, and oxygen—will be addressed. The process of breaking the chemical chain reaction is not normally addressed in relation to wildland fire suppression methods. For a discussion relating to the fire tetrahedron and the chemical chain reaction process, see Section 1, Chapter 5, "Chemistry and Physics of Fire."

when oxygen is consumed so rapidly near the base of the fire that combustion of the gases is incomplete. As these superheated gases rise, they reach a fresh oxygen supply and break into flames, giving the appearance that the smoke itself is flaming.

Reignition of apparently dead fires is an important factor in forest fuels. Although neither flames nor smoke may be observed, a fire can break out again if sufficient heat remains. Reignition occurs easily where fuels have been subjected to heat for a considerable length of time. To prevent reignition, an experienced fire fighter feels out fuels by hand before leaving the area. If heat continues to be present, the danger of reignition exists, and the control job remains incomplete.

Breaking the Fire Triangle

In fire control operations, the objective is to prevent combustion by breaking the fire triangle. If a sufficient volume of water is applied, the fuel temperature can be lowered below the ignition point. Smothering fires with dirt deprives them of oxygen. Building a line in which all flammable material is removed from the path of the fire prevents further spread by robbing the fire of fuel.

In forest fire fighting, the actual method of suppression is often dictated by the equipment and fire fighters available. The most effective methods for each stage of a suppression operation can be determined by keen observation of the combustion process.

Reducing heat: Water being used to reduce fuel temperatures should be applied directly on the fuels being consumed. It is a mistake to apply water on the flames rather than the fuels. If the liberation of these gases is to be stopped, the fuels themselves must be cooled by the water. While some of the water played on the flames may reach the fuels, a large part will either miss its mark or be lost through vaporization. If water is being used for cooling purposes, the ultimate target is always the fuel.

In a very hot fire accompanied by leaping flames, the nozzle user may not be able to get close enough to the fuels to apply water directly on the trouble spot. In such cases, a fog nozzle should be used to knock down the flames. The water particles help cool the heated gases being liberated by the fuels. Once the flames have been knocked down, the water should be used to thoroughly cool the heated fuels that are generating flammable gases.

Reducing oxygen: Fires burning in forest fuels are difficult to smother completely. Soil thrown on burning forest materials may retard combustion by shielding portions of the fuel surface from the air, but the porous nature of most soils makes it difficult to completely shut off the supply of oxygen in this manner. Throwing dirt on a fire is nevertheless an important means of reducing the rate of combustion, after which the control operation may be continued by attacking one of the other legs of the fire triangle.

In very fine fuels such as dry grass, the oxygen supply may be reduced easily. Fuels of this nature do not retain heat for long periods, and, therefore, combustion normally may be checked by temporarily shielding the fuel surface from the air. Gunnysacks, fire swatters, and shovelfuls of dirt correctly applied are effective implements in smothering grass fires. When larger accumulations of flammable materials, such as dead stems and leaves, lie beneath a stand of grass, the danger of re-ignition exists. The understory material may hold enough heat to cause the fire to reignite once an adequate supply of oxygen becomes available. This danger can be determined by carefully observing and "feeling out" the fuel.

Removing fuel: In wildland fire suppression, the removal of fuel from the path of the fire is a common method of attacking the fire triangle. The fire continues to burn until its fuel is consumed, but it is prevented from spreading any further. A slowly advancing fire burning in sparse ground fuels may be checked simply by constructing a fire line down to mineral soil (soil with little humus or organic matter). A hot or fast-running fire may require several fuel removal operations. Snags of trees or tall brush that could cause spot fires may have to be felled. Thickets of reproduction, or dry brush, may need to be weeded out. Low-hanging limbs may have to be removed to prevent the build-up of a crown fire. Concentrations of limbs and logs may have to be broken up. The objective of these operations is to remove or reduce the flammable substances which could either allow the fire to build in intensity or to continue to spread.

Heat Transfer

The rate and amount of heat transferred influences the rate of spread and the intensity of a fire. Combustion cannot be sustained unless heat continues to be transferred. Heat transfer occurs by three means: radiation, convection, and conduction.

Radiation: In forest fires, radiation is the most significant means of transferring heat from burning fuels to exposed fuels. The location and width of firelines must be governed, in part, by the rate of radiant heat transfer. Potentially hot fires and spots where dangerous flare-ups might occur can often be prevented by breaking up the fuels to reduce the radiant heat transfer between them.

The degree of slope influences the amount of radiant heat transmitted to fuels upslope and downslope of the fire. Most forest fires will burn in the direction of the wind, partly due to the effects of radiant heat. Wind influences radiant heat transfer in two ways: (1) it increases the rate of combustion, thus creating a hotter fire; and (2) it bends the flames, thus decreasing the distance between the heat source and the fuels lying in the path of the fire. The dual effect of the hotter fire and the bent-over flames dries and heats the fuels lying ahead to such an extent that rate or spread may be greatly accelerated.

Often a fire will burn into the wind. This occurs because the wind increases combustion to such a degree that a large amount of radiant heat may be transmitted to the fuels lying on the upwind side. On the upslope side of a fire, the fuels receive more radiant heat. This is one reason fires often spread more rapidly up steep slopes than on level ground.

Convection: In a forest fire, the fuels lying in the path of convection currents are heated and thus transformed into a more ignitable condition. The hot air mass rising from a surface fire transfers a large amount of heat to the tree crowns, thus bringing them nearer to the ignition temperature.

When fires break out in tree crowns and in other aerial fuels, heat transfer by convection is usually increased. Sparks and burning embers from burning tree crowns and snags also will drop and start new fires in fuels on the ground. As these ground fires gain in intensity, more hot air masses rise through the aerial fuels, and a type of chain reaction may be created. Fire fighters must always be alert to the danger of hot surface fires burning underneath a forest canopy. These surface fires must be cooled promptly, and fuel masses must be broken up to reduce convective heat transfer.

Heat transfer through convection also is increased by wind which moves hot masses of air ahead of the fire. Wind both increases the rate of combustion and accelerates the transport of hot air masses. When driven by wind, the hot convection currents may move closer to the ground fuels and thus create more flammable conditions in the understory vegetation of the forest.

Fuels located above the fire on steep slopes also receive heat by the movement of convection currents up the slope and are thus subjected to an accelerated rate of heating and drying. This is one more reason why fires often burn very rapidly up the sides of steep slopes.

Conduction: Wood is a poor conductor of heat, and the conduction method of heat transfer is relatively minor in evaluating forest fire behavior.

TOPOGRAPHY

Topography provides a useful and easily recognized indicator of fire behavior. Fires often have distinctive behavior characteristics according to aspect, elevation, position on slope, steepness of slope, and shape of the surrounding countryside. These topographic features usually are easy to identify in the field and thus are important factors in the evaluation of fire behavior.

Differences in topography may cause local variations in climate and day-to-day weather conditions. These variations, in turn, influence the character of forest growth and the flammability of fuels. In rough terrain, such as the northern Rocky Mountains, topography has a great influence on fire behavior.

Aspect

Aspect, sometimes referred to as exposure, describes the direction a slope faces. Fire conditions vary greatly according to aspect because different aspects receive varying amounts of sunshine and wind.

In general, southern and southwestern slopes provide favorable conditions for the ignition and spread of fires. Because these slopes receive direct sunshine, the air and fuel temperatures are somewhat higher. This causes snow to melt earlier on southern slopes. For these reasons, vegetation on south-facing slopes is not only sparser, but also drier and more flammable than vegetation on north-facing slopes.

There also are variations in temperature and relative humidity on northern and southern slopes. On south-facing slopes, the average July–August temperature is higher, and the relative humidity is lower. The time of day is also critical. South- and west-facing slopes are most dangerous in the afternoon and early evening. When the combined effects of aspect and elevation are considered, some striking indicators of fire behavior become evident.

Elevation

In the Rocky Mountains, there is a vertical difference of more than 10,000 ft (3,048 m) between the lowest valleys and the highest mountains. At Lewiston, ID, the elevation is only 757 ft (231 m) above sea level. Along the Continental Divide, there are many peaks over 10,000 ft (3,048 m), and in northwestern Wyoming, some peaks are over 13,000 ft (3,962 m) above sea level. Between these extremes of elevation are a variety of weather and fuel conditions that create distinctive fire control problems.

Position of Fire on Slope

Fire behavior at various positions on a slope may be influenced not only by aspect and elevation, but also by the magnitude of the fuel body and by topographic barriers. When a fire starts at the bottom of a slope, an entire mountainside of fuels may lie in its path. Once a fire starting at the base of a slope gains headway, the availability of a continuous fuel body makes a large burn possible.

Steepness of Slope

Other conditions being equal, fires burn more rapidly on steep slopes. In general, as the steepness of the slope increases, the rate of fire spread increases. As explained previously, combustion is accelerated on steep slopes primarily due to increased heat transfer through radiation and convection. A fire will double in rate of spread on a 30 percent slope. On a 55 percent slope it will double again.

Shape of Country

In rugged, mountainous areas, the shape of the country is of great importance to the fire fighter who must evaluate fire behavior. Narrow canyons, side drainages, sharp ridges, and massive, irregular slopes all have a bearing on the direction of travel, rate of spread, and general behavior of fires.

Experience has shown the following topographic features to be of special importance:

1. *Narrow canyons.* Wind direction normally will follow the direction of the canyon. Wind eddies and strong upslope air movement may be expected at sharp bends in a canyon. Radiant heat transfer from one slope to another is great, and, as a result, fires may spot across the canyon easily. Near the bottom of the canyon, there is little difference between conditions on various aspects.
2. *Wide canyons.* Prevailing wind direction will not be altered to any great extent by the direction of the canyon. Cross-canyon spotting of fires is not common except in high winds. Strong differences will occur between general fire conditions on northern and southern aspects.
3. *Box canyons.* Fires starting near the base of box canyons will react similarly to a fire in a stove. Air will be drawn in from the canyon bottom, thus creating very strong up-slope drafts. These same conditions may occur at the heads of narrow canyons and at the heads of high mountain valleys.
4. *Ridges.* Fires burning along lateral ridges may change direction when they reach a point where the ridge drops off into a canyon. This change of direction is caused by the flow of air in the canyon. In some cases, a whirling motion by the fire may result from a strong flow of air around the point of a ridge.

FUELS

Keen observation of variations in forest fuels is essential in reliably estimating fire behavior. Fuel is the material of primary concern to fire control agencies. A good fire fighter must be able to evaluate flammability and difficulty of control in the various fuel situations encountered in a forest area.

Flammability Analysis

In a forest, great differences exist in the character of flammable materials. Deep duff, newly fallen dead leaves, clumps of grass, litters of dry twigs and branches, downed logs, low shrubs, green tree branches, hanging moss, snags, and many other types of material are present. Each of these materials has distinctive burning characteristics. The flammability of a particular fuel body is governed by the burning characteristics of individual materials and by the combined effects of the various types of materials present.

Before flammability can be analyzed, the physical characteristics of combustible forest materials must be classified. Such a classification permits the identification of the fuel factors that influence flammability. After the fuel has been classified properly, topographic and weather factors must be considered before the rate of spread and the general behavior of fires in that fuel can be determined.

Because forest fuels are so varied and complex, it is necessary to develop a systematic approach to flammability analysis. First, the fuel body is subdivided into two broad classes, ground fuels and aerial fuels. Then each of these classes is evaluated according to the arrangement, compactness, continuity, volume, and moisture content of the principal materials involved. Aspect also affects the flammability of fuels. Southwest–facing slopes heat up to a greater extent. With practice, this procedure can be performed quickly and easily.

Ground Fuels

Ground fuels include all flammable materials lying on or immediately above the ground or in the ground. The principal materials are duff, tree roots, dead leaves, grass, fine dead wood, downed logs, stumps, large limbs, low brush, and young stands of timbers.

Duff: Duff is composed of partially decayed vegetative matter found on the forest floor. Duff seldom has a major influence on the spread rate of fire as it is typically moist and tightly compressed so that little of its surface is freely exposed to air and its rate of combustion is slow. In forest fires, most of the duff is consumed down to mineral soil. Occasionally, duff contributes to the rate of spread by furnishing a path for the fire to creep along between patches of more flammable material.

The smoldering characteristics of duff fires make them somewhat difficult to control. Firelines must be dug down to mineral soil in duff areas unless the lower layers of duff are too wet or too tightly compressed to burn. Duff fires inside a fireline are best mopped up by turning over the duff with a shovel, loosening the material for better exposure to air, and allowing it to burn out. Water may be used effectively, but great care must be taken to stir the fuel to ensure that all particles are thoroughly soaked. Extinguishing duff fires by smothering them with dirt is very difficult to accomplish.

Roots: Roots are not an important factor in rate of fire spread, as the greatly restricted air supply prevents rapid combustion. However, fires can creep slowly in roots—in fact, some fires have escaped control because a root provided an avenue for the fire to cross the control line. The most flammable roots are the large laterals stemming from dead snags. Root fires are controlled simply by cutting the root where it crosses the fireline. Mop-up of persistent fires may require complete excavation of roots.

Dead leaves: As leaves decay on the ground, they gradually become part of the duff layer. Before this decay takes place, however, leaves are a highly flammable material and should be considered separately in evaluating ground fuels. In northern Rocky Mountain forests, the cover of dead leaves on the ground is composed primarily of needles dropped from coniferous trees. Ponderosa pine needles are the most flammable because their large size and shape lead to a loose arrangement allowing free circulation of air. Smaller needles, such as those from Douglas fir, are generally less flammable as they are more tightly compacted on the ground.

Needles that are still attached to dead branches are especially flammable because they are exposed freely to air and are not typically in direct contact with the more moist material on the ground. Needles remaining on fallen limbs form highly combustible kindling for larger material. For this reason, logging slash containing dry needles is dangerous fuel.

Many techniques are necessary to control fires for which leaves or needles are the primary fuel. The advance of a fast-running fire may be checked temporarily by smothering it with dirt or by cooling it with water. Only a light cover of dirt thrown with the sweeping motion of a shovel is necessary to slow down a needle fire.

When water is used to check the advance, a spray or fog-type stream is most effective, with the water being aimed at the base and directly in front of the flames. A penetrating straight stream of water is most effective for mopping up a needle fire. Careful stirring of all flammable material is required. A fireline is essential around all fires burning in a continuous cover of needles.

Grass: Grass, weeds, and other small plants are important ground fuels that influence rate of fire spread. The key factor in these fuels is the degree of curing. Succulent green grass acts as a fire barrier. During the course of a normal fire season, however, grass gradually becomes drier and more flammable until the stems and leaves die due to lack of moisture. At this time, the major part of the grass cover becomes highly flammable. Cured grass, if present in large and uniform volume, provides the most flammable ground fuel in the region.

Grass and other small plants occur on the floor of almost all forests. Fire fighters need to observe the volume and continuity of the grass cover. In dense forests where little sunlight reaches the ground, very little grass is found. In more open forests, such as in mature stands of ponderosa pine, there may be a large amount. If there is a more-or-less continuous cover of dry grass on the forest floor, the spread rate of a fire will be governed largely by that cover, rather than by the heavier fuels normally associated with a forest.

The ease with which fires in dry grass can be controlled depends mainly on the volume of grass. When the volume is low, the fire is often cool enough to permit fire fighters to work near the edge of the flame front. When the volume of grass is great, however, the resulting hot fire may prevent work close to the fire front.

In grass fires, smothering with dirt or swatters and cooling with a spray or fog stream of water are effective control methods. Both methods require follow-up to ensure that no hangover fires remain. Wherever forest material, such as a mat of dead pine needles, twigs, or small limbs, is intermixed with the stand of grass, a fireline dug down to mineral soil is usually required.

Fires in dry grass often have high rates of spread. For this reason, special safety measures need to be observed.

Fine dead wood: Fine dead wood consists of twigs, small limbs, bark, and rotting material. Normally, the fine dead wood classification is confined to material with a diameter of less than 2 in. (50.8 mm). These fine dead ground fuels are among the most important of all materials influencing the rate of fire spread and general fire behavior in forest areas. Fine dead wood ignites easily and often provides the main avenue for carrying fire from one area to another. It is the kindling material for larger, heavier fuels.

In areas where a great volume of fine dead wood exists, a fire can rapidly develop tremendous heat. The greatest volume of fine dead wood is usually found in areas containing logging slash. Under dry conditions, fires in such areas burn violently, and the strong convection currents created by the intense heat pick up burning embers and throw them out ahead, causing spot fires beyond the main fire front.

In some areas, the occurrence of fine dead wood may be spotty. Troublesome situations may be avoided by breaking up the worst bunches of fine dead wood before a fire reaches them.

Granulated dry rotten wood, while not an especially important factor in the rate of fire spread, is a highly ignitable fuel. Flying embers from the main fire often cause spot fires in rotten wood lying on the ground or in hollow places on old logs or stumps.

Fire fighters searching for spot fires should seek out and carefully check these areas.

In most forest fires, the control job must be aimed largely at checking the spread of fires in fine dead wood and in the associated material—that is, duff and leaves—lying underneath. Unless a large volume of dry grass is present, the fine dead wood is the most important fuel and should receive the first and most concentrated attention. If the fire is very hot, the first action may be to smother it with dirt or cool it with water. Final control requires a fireline dug down to mineral soil in nearly all cases.

Downed logs, stumps, and large limbs: Heavy fuels, such as downed logs, stumps, and large limbs, require long periods of hot, dry weather before they become highly flammable. When such material reaches a dry state, however, very hot fires may develop. The most dangerous heavy fuels are those containing stringers of dry wood, shaggy bark, or many large checks and cracks. Smooth-surfaced material is less flammable as it dries out more slowly, has little surface exposed to air, and contains less attached kindling fuel. Extremely hot fires may develop in piles of downed logs and large limbs or in crisscrossed windfalls as the various fuel components radiate heat to each other. Individual limbs and logs will not burn very hot unless the fire is supported by large accumulations of fine dead wood.

The control of an advancing fire seldom rests on the fire fighters' ability to suppress the fire in heavy fuels. Usually, fires will not advance from log to log unless finer fuels are present or the area is covered with logs that radiate heat to each other.

Once a fire becomes well established in heavy fuels, complete suppression becomes difficult. A control line around the area is essential. Smothering with dirt or cooling with water is effective, but both these methods require very large volumes of suppression agent. In some cases, it is necessary to allow the fire to burn out until the heat has subsided sufficiently to permit final mop-up operations to take place. When a fire is allowed to burn out in logs containing stringers of bark, a careful lookout must be kept for spot fires caused by embers carried in the strong convection currents. If at all possible, hot fires in such areas should be cooled promptly by any available means.

Low brush and reproduction: Low brush, tree seedlings, and small saplings are classified as ground fuels because they are so closely intermixed with the flammable material on the forest floor. This understory vegetation may either accelerate or slow down the spread rate of a fire. During the early part of a fire season the shade normally provided by understory vegetation prevents other ground fuels from drying out rapidly. As the season progresses, however, continued high air temperature and low relative humidity dry out both the fuel lying on the ground and the understory vegetation. When this happens, most of the low vegetation, particularly small coniferous trees, become fire carriers.

The understory vegetation in a forest often provides a link between ground fuels and aerial fuels. The crowns of small trees may catch fire and, in turn, spread the fire to aerial fuels in the forest canopy. Either thickets of reproduction or dead brush may provide the first means for a surface fire to flare up and spread into the crowns of the overstory trees. In anticipation of this possibility, alert fire fighters break up reproduction thickets and dead brush along the fireline or, wherever possible, locate control lines to prevent fire from spreading into these danger areas. It is not uncommon for fire to creep through ground fuels under low brush or reproduction during periods when there is a low burning index, such as during the night. Heat from the creeping fires dries out the leaves and stems of the low brush and reproduction. Then, on the following day, as the burning index increases, a reburn occurs. Often on the reburn, the dried, low vegetation burns hotly and carries the fire to the aerial fuels, causing a crown fire.

Aerial Fuels

Aerial fuels include all green and dead materials located in the upper forest canopy. The main aerial fuel components are tree branches, crowns, snags, moss, and high brush.

Tree branches and crowns: The live needles of coniferous trees are a highly flammable fuel. Their arrangements on the tree branches allows free circulation of air. In addition, the upper branches of trees are more freely exposed to wind and sun than most ground fuels. These factors, plus the volatile oils and resins in coniferous needles, make tree branches and crowns important components in aerial fuels.

Tree branches and crowns are fuels that can flash quickly with changes in relative humidity. Crown fires seldom occur when relative humidity is high. However, coniferous needles dry out quickly when exposed to hot, dry air. The dryness of needles is influenced by the transpiration process in a tree. When the ground is moist, trees pump a large amount of moisture into the air through the leaves. As the ground becomes drier, the transpiration process slows, and, as a result, leaves and branches become drier and more flammable.

Dead branches on trees are an important aerial fuel. Concentrations of dead branches, such as those found in insect- or disease-killed stands, may cause fire to spread from tree to tree. Concentrations of dead branches on the lower trunks of trees may provide an additional avenue for fires to spread from ground fuels into tree crowns. The most flammable dead branches are those still containing needles.

Fires in the upper crowns of trees are extremely difficult to control. The main control method must be aimed at suppressing the fire in ground fuels and preventing the fire from entering the tree crowns. Removal of limbs on the lower trunks of trees is one method of preventing crown fires. Limbs should be removed wherever a concentration of ground fuels makes a crown fire likely.

Snags: Snags, or tree stumps, are one of the most important aerial fuels that influence fire behavior. Although green trees greatly outnumber snags in most forests, more fires start in snags because they are drier and are arranged for easier ignition.

Snags vary widely in character and, consequently, in their effect on fire behavior. Smooth, solid snags that contain very little bark and few checks or cracks are not highly flammable. On the other hand, broken-topped, shaggy-barked, or partially decayed snags ignite easily and have a major influence on the spread of fires. Burning embers blown from shaggy-barked snags are prolific starters of spot fires. In areas containing large numbers of snags, fire may spread from one trunk to another because of the intense radiation of heat.

Fires in burning snags must be controlled promptly. Wherever possible, the main control effort on the fire must be designed to prevent the blaze from getting into snags. When fires become established in snags, the control method usually requires that the snags be felled. Snag-falling is especially important if shaggy bark is carried from the burning trunk by wind or strong convection currents.

Tree moss: Moss hanging on trees is the lightest and flashiest of all aerial fuels. Moss is important principally because it provides a means of spreading fires from ground fuels to other aerial fuels or from one aerial component to another. Like other light fuels, moss reacts quickly to changes in relative humidity. During dry weather, crown fires may develop easily in heavily moss-covered stands. Methods of controlling moss fires are aimed primarily at breaking up ground fuels to prevent fire from entering the tree crowns through hanging moss. In addition, lower limbs containing tree moss should be removed at all danger spots.

High brush: Crowns of high brush are classified as aerial fuels because they are separated distinctly by distance from ground fuels. In the northern Rocky Mountain region, heavy stands of brush may develop in old burns, and they often form the principal vegetative cover in such areas. Crown fires in brush fuels ordinarily do not occur unless heavy ground fuels are present to develop the required heat. In some brush stands, however, a high proportion of dead stems may create a sufficient volume of fine dead aerial fuels to permit very hot and fast-spreading crown fires. Key factors in evaluating the behavior of fires in high brush are volume, arrangement, the general condition of ground fuels, and the presence of fine dead aerial fuels.

Fuel Conditions

Fuel continuity: Fuel continuity describes the distribution of fuels in a given area. Fuel continuity is an important factor in fire behavior because the distribution of fuels influences the potential area where a fire may spread, as well as the rate of spread. If a dangerous fuel is uniformly distributed over an entire area, a high potential exists for a complete burn to occur at a rapid rate of spread. If the fuel body is broken up by patches of bare ground or much less flammable material, both the potential area of the burn and the rate of fire spread are reduced.

A wide range of fuel continuity conditions is found in most forest areas. For the sake of simplicity in making fire behavior estimates, two broad fuel continuity classes are recognized:

1. Uniform fuels, which include all fuels distributed continuously over the area being evaluated. Areas containing a network of stringers, or blocks, which connect with each other to provide a continuous path for the spread of fire, are included in this classification.
2. Patchy fuels, which include all fuels distributed unevenly over the area being evaluated. Definite breaks should be present, such as patches of rocky outcropping or plots where the dominant vegetation is of much lower flammability than the main fuel body.

Volume: As the amount of flammable materials in a given area increases, the amount of heat a fire produces also increases. The hottest fires, as well as those most difficult to control, occur in areas containing the greatest quantity of fuel.

In evaluating fuel volume, the quantity of both small- and large-sized fuel components should be observed. A great volume of small materials, such as fine dead wood, indicates ample kindling for the ignition of other fuels. A great volume of either small- or large-sized material indicates a high potential for a hot fire. Where fires develop in areas containing a great volume of large-sized material, there is intense radiation of heat to fuels lying in the path of the fire.

Fuel compactness: Fuel compactness—that is, the number of individual fuel particles per unit of volume—varies greatly in all kinds of fuel, but it is significant principally in duff and dead leaves lying on the ground. Fires burn rapidly in loosely compacted fuels because more of the individual fuel particles are freely exposed to air.

Live fuel moisture: Two conditions of fuel moisture have major influence on the rating of fuel types. One concerns the greenness, or curing stage, of vegetation. The other relates to the shade and protection furnished by green timber.

In grass fuels, moisture content is a critical factor in determining flammability. Fires spread only at a low rate, or not at all, in grasses that are rank green, but when the same grasses become cured and dry, fires will race through them at an extremely rapid rate. The degree of curing of grass is difficult to evaluate and requires keen observation of the grass stand. For purposes of considering fire behavior, the moisture content of grass can be judged according to the state of curing, as follows:

Green—The condition can be recognized by the green color and the cool, moist feel when crushed in the hand.

Curing—As hot, dry weather prevails, grasses ordinarily progress through a period of gradual curing. This stage is detected by close observation of the individual grass clumps or stems. For cheatgrass, the curing stage is typified by a lavender tinge commonly referred to as "the purple stage." For most other grasses, the curing stage begins when the tips of grass blades become tan or brown, or when individual grass blades take on a cured appearance.

Cured—In the cured stage, grasses are dried completely to a tan or brownish hue, and the stems feel dry or crackly when crushed or rubbed in the hand.

Soils influence curing primarily because of soil moisture relationships. In the deep, moist soils bordering creeks, curing is much slower than in the thin soils on slopes.

Some grasses seldom reach a dangerous cured condition. The shade and protection afforded by timber stands influence fuel type ratings due to the favorable fuel moisture conditions that are created. In a dense forest, ground fuels are protected from the sun and wind. Temperatures and wind velocities are lower so that moisture does not evaporate as rapidly from the dead fuels on the ground. Lower ratings are assigned to fuels situated beneath dense timber canopies.

The moisture content of fuels is influenced also by aspect, altitude, time of year, time of day, and other factors. For the purpose of fuel type classification, these factors are disregarded, and fuel types are classified on the basis of the physical characteristics of the fuels themselves.

Dead fuel moisture: Dead fuel moisture is changed by the moisture content of the air. Time lag is the time it takes for the moisture content of the dead fuels and the surrounding air to equalize. Time lag is expressed as a rate (usually in hours). (See Table 10-7A).

SPECIAL FIRE BEHAVIOR FACTORS

Spot Fires

The development of spot fires depends not only upon topographic and weather factors, but also upon the character of fuels in the main

TABLE 10-7A. *Dead Fuel Moisture: Time Lag Relationship to Fuel Size*

Time Lag	Diameter of Fuel	Examples
1 hr	Less than $1/4$ in.	Annual grass
10 hr	$1/4$ to 1 in.	Coastal sage, juniper, and chaparral
100 hr	1 to 3 in.	Logging slash
1,000 hr	3 to 8 in.	Logs and mature standing timber

For SI units: 1 in. = 25.4 mm.

fire and fuels beyond the main fire. In the main fire, rotten, shaggy-barked snags, such as broken-topped hemlock snags, and large quantities of ground fuels, such as heavy logging slash, are the fuels most likely to cause spot fires. Spot fires frequently are started by crown fires. Widespread crown fires, with their intense heat and strong convection currents, can throw burning embers far out ahead of the main fire.

Fuels that are ignited most readily by embers thrown out ahead of the main fire, listed in the order of susceptibility, include:

1. Rotten wood on the ground, on logs, and in snags.
2. Moss and lichens in tree tops.
3. Slash, particularly when compacted in tight piles.
4. Duff.
5. Cured grass.

Spot fires become more frequent and severe with lower fuel moisture and increased wind. On an average dry, late summer afternoon, the smallest sparks quickly ignite rotten wood or tree moss. In compacted slash, in duff, and in cured grass, larger burning embers are required to start spot fires unless sufficient wind exists to fan smoldering material into flames.

Crown Fires

Many forests contain a combination of aerial and ground fuels that favor the development of crown fires. Fuel conditions that influence the probability and character of crown fires include:

1. Volume and arrangement of the timber canopy.
2. Volume and arrangement of fine dry aerial fuels, such as moss and dead twigs.
3. Position of aerial fuels above ground fuels.
4. Character of ground fuels.

In open forests where single trees or small clumps of trees are well isolated from each other, crown fires are usually confined to single trees or to small groups of trees. Crowning, in such cases, increases rate of spread by intensifying heat and causing spot fires, but it will not result in racing crown fires like those that sometimes occur in dense forests.

When the timber canopy is sufficiently dense and continuous to carry broadcast crown fires, the severity of the crown fire is influenced by the quantity, character, and arrangement of dry fuels in the air and on the ground.

An important consideration in evaluating fuel conditions is the likelihood of reburns in tree crowns over previously burned ground fuels. A combination of fuel, weather, and topographic conditions may confine a fire to ground fuels for a considerable period of time. While the fire is burning on the ground, the aerial fuels may be scorched and singed, and may become critically dry. If conditions change, a crown fire may develop and sweep back over the area. This type of crown fire may develop rapidly and burn more explosively because the aerial fuels have been dried by surface fire.

RESEARCH

Both Canada and the United States maintain fire research programs whose ultimate goals are to reduce the loss of life, property, and forest resources from wildfires and to use prescribed fire better to achieve forest and range management objectives at reduced costs.

The Canadian Forestry Services (CFS) of Canada's Department of the Environment is the principal research organization in Canada. Its forest fire research program is centered at the Petawawa National Forestry Institute in Chalk River, Ontario. Fire research work is also carried out at regional forest research centers in Victoria, British Columbia; Edmonton, Alberta; and Sault Ste. Marie, Ontario. These research programs address six major areas:

1. *Fire behavior:* Fuel moisture physics, fire spread physics, prediction of fire danger by forest type, fire/weather interactions, fire danger rating system, and spatial weather models.

2. *Fire ecology:* Post-fire forest regeneration mechanisms, cyclic forest development from fire to fire, predictors of post-fire forest development, and age-class distribution in fire-cycled forests.

3. *Fire suppression:* Performance testing of fire management equipment, air tankers, fire retardants and water additives, aerial ignition devices, backfiring methods, and new suppression methods.

4. *Prescribed fire:* Tree damage and mortality, use of fire for slash removal, seedbed preparation and vegetation control, design of prescriptions for proper burning conditions, and operational techniques.

5. *Economics:* Estimation of values at risk, effect of fire on timber supply, relation between fire management costs and burned area, allowable cut effect, and ultimate impact on the forest economy.

6. *Fire management systems:* Remote sensing applications, computerized systems for integrating weather, fuel type, and terrain into fire spread and growth models, prediction of lightning and human-caused fires, air patrol routing, resource deployment, and fire management strategies.

Current U.S. fire research is being conducted at six forest and range experiment stations. Much of the work is done at forest fire laboratories located in Macon, Georgia; Missoula, Montana; and Riverside, California. Major research projects include:

1. *Fire behavior:* Fundamental studies in fuel chemistry, combustion and ignition processes, firespread mechanisms, time-temperature heat flux interrelationships, and effects of fuel moisture, wind, and slope. Development of systems to aid fire and land managers in dealing with the growth of large fires, fuel consumption and energy release, probability of ignition by lightning, and fire spread in nonuniform fuels.

2. *Fire suppression:* Development of real-time and planning guidelines for individual suppression activities related to primary fuel and fire variables; integration of production rate knowledge into real-time and planning guidelines for fire suppression strategies and tactics; and development of design criteria for chemical formulations and for aerial and ground delivery systems for primary strategies and tactics.

3. *Fire effects, use, and ecology:* Research to determine how, when, and where prescribed fire may be used to improve tree growth, provide better wildlife habitat, reduce fire hazard, and accomplish other forestry goals. Includes studies on soil to ascertain biological responses, impacts on soil, stand structure, and so on.

4. *Fire management planning:* Studies to relate wildland fires and social benefits desired from wildlands; to determine the effectiveness of fire prevention activities; to determine the influence of weather and climate on fire occurrence, control, and effects; and to determine the productivity and effectiveness of air tanker systems.

5. *Other:* Additional research projects involving smoke management for prescribed fires, reducing residues from forestry activities, evaluating the economics of fire protection systems, and aiding managers with systems for applying knowledge gained through research projects.

SUMMARY

The toll on human life, the financial impact, and the increased value of natural resources make it undesirable to lose thousands of acres of valuable timber, rangeland, and watersheds. Continuing efforts to prepare fire managers and fire fighters to deal with both the beneficial aspects of fire through the use of prescribed fire and the control and extinguishment of unwanted fires is of the utmost importance. In addition, the public must understand its critical role in preventing unwanted fires, and communities must recognize the importance of designing developments that will minimize the threat of fire to real property.

BIBLIOGRAPHY

Additional Reading

"Surviving the Wildland or Intermix Fire. Part 1," American Fire Journal, Vol. 48, No. 5, May 1996, pp. 6–8, 11

"Wildfire: New Challenges, New Innovations," Fire Command, Vol. 56, No. 7, July 1989, pp. 23–28.

"Wildland Fires: 1993 Forest Fire Season Was Well Below Average," Fire Fighting in Canada, Vol. 38, No. 6, Aug. 1994, pp. 52–53.

Abt, R., Kelly, D., and Kuypers, M., "The Florida Palm Coast Fire: An Analysis of Fire Incidence and Residence Characteristics," Fire Technology, Vol. 23, No. 3, 1987, pp. 230–252.

Alkema, J. K., "Red Flags for Wildland Safety," American Fire Journal, Vol. 47, No. 10, Oct. 1995, pp. 6–7.

Anderson, L., "Chemicals Used in Wildland Fire Suppression: A Risk Assessment," Foam Applications for Wildland and Urban Fire Management, Vol. 7, No. 1, May 1995, pp. 30–32.

Aubry, Y., "Passive Detection of Forest Fires," Revue Generale de Thermique, Vol. 27, No. 317, May 1988, pp. 313–318.

Bailey, D. W., "Structural Wildfire Urban Interface. The Wildland/Urban Interface: Implications in the 90's," Lolo National Forest, Missoula, MT, International Association of Fire Chiefs (IAFC), Fire-Rescue International Conference Proceedings, 1993 Annual Conference, Aug. 2–Sept. 1, 1993, Dallas, TX, 1993, pp. 88–89.

Bailey, D. W., "Wildland Fire Fighter Standard Approved," NFPA Journal, Vol. 89, No. 5, Sept./Oct. 1995, p. 39.

Bailey, D. W., "Wildland Fire Management" (Monthly Column), Fire Command, 1988-1990; NFPA Fire Journal 1991, 95.

Bailey, D. W., "Wildland Urban Interface: A National Problem with Local Solutions," Presentation at Interwest Fire Council Meeting, Kamouraska, Quebec, Canada, 1988.

Berry, D., "Seasonal Wildland Crews Give Salt Lake County a Hand," American Fire Journal, Vol. 47, No. 5, May 1995, pp. 6–7.

Bishop, J., "Whys and Wherefores of Wildland Fire Behavior," Fire Chief, Vol. 40, No. 6, June 1996, pp. 51–55.

Blankenship, P. L., "Class A Foam for 1993 Wildland Fires," American Fire Journal, Vol. 45, No. 6, June 1993, pp. 17–18.

Brabander, O. P., et al., "Investigation of the Conditions of Transition of a Surface Forest Fire to a Crown Fire," Combustion, Explosion and Shock Waves, Vol. 24, No. 4, Jan. 1989, pp. 435–440.

Broom, J. S., "New Mobilization Plan Helps Oregon Fight Wildland Fires," Fire Chief, Vol. 36, No. 10, Oct. 1992, pp. 48, 50.

Burgan, R. E., and Hartford, R. A., "Computer Mapping of Fire Danger and Fire Locations in the Continental United States," Journal of Forestry, Vol. 86, No. 1, 1988, pp. 25–80.

Cappellini, V., Mattii, L., and Mecocci, A., "Intelligent System for Automatic Fire Detection in Forests," Proceedings of the Third International Conference on Image Processing and Its Applications, IEE Conference Publication No. 307, Stevenage, England, 1989.

Davis, L., "Class A FoamsNew Technology Brings Class A Foams From Wildland to Structural Firefighting," Firehouse, Vol. 16, No. 4, April 1991, pp. 49–50, 75

DeGrosky, M. T., "Montana Approach to Rating Fire Risk in Wildland Developments," Fire Management Notes, Vol. 53–54, No. 4, 1992/1993, pp. 17–19, 26.

Doolittle, L. and Donoghue, L. R., "Status of Wildland Fire Prevention Evaluation in the United States. Forest Service Research Paper," North

Central Forest Experiment Station, St. Paul, MN, FSRP-NC-298, 1991.

Eisner, H., "Wildland Fires," Firehouse, Vol. 17, No. 11, Nov. 1992, pp. 44–45.

Federal Emergency Management Agency, "Wildland/Urban Interface Fire Protection. Training Kit," Federal Emergency Management Agency, Washington, DC, Training Kit, 1991.

Fire Information Handbook, British Columbia Ministry of Forests and Lands, Canada, 1987.

Gardner, P., and Cortner, H. J., "When the Government Steps In," Fire Journal, Vol. 82, No. 3, May/June 1988, p. 32.

Garza, J., MacDonnell, J. and Bradford, G., "Going Live With Wildland Fire Training," Fire Chief, Vol. 39, No. 7, July 1995, pp. 53–55, 57.

Guth, D., "Organization Meets Wildfire Challenge," Fire Command, Vol. 56, No. 1, June 1989, pp. 19–21.

Guth, R., and Cohen, S. B., Red Skies of 88: The 1988 Forest Fire Season in the Northern Rockies, the Northern Great Plains and the Greater Yellowstone Area, Pictoral Histories, Missoula, MT, 1989.

Haines, J., "Wildfire: A Global Issue," Fire Command, Vol. 57, No. 1, Jan. 1990, pp. 28–30.

Hamilton, M. P., Salajar, L. A., and Palmer, K. E., "Geographic Information Systems: Providing Information for Wildland Fire Planning," Fire Technology, Vol. 25, No. 1, Feb. 1989, pp. 5–23.

Klug, M., "Wildland Safety For Engine Companies," Fire Engineering, Vol. 146, No. 4, April 1993, pp. 78–80, 82–84.

Kluver, M., "Observations From the Southern California Wildland Fires," Building Standards, Vol. 63, No. 1, Jan./Feb. 1994, pp. 12–17.

Lankester, C., "Economic Perspective of the Wildland Fire Problem," Wildfire–International Wildland Fire Conference Proceedings, Meeting Global Wildland Fire Challenges: The People, The Land, The Resources, July 23-26, 1989, Boston, MA, U. S. Forest Service, Washington, DC, 1990, pp. 14–17.

Larson, R. D., "Wildland '94 Features Learning and Burning," American Fire Journal, Vol. 46, No. 12, Dec. 1994, pp. 12–15.

Leschak, P. M., "Preplanning Wildland Interface: A Minnesota Approach," Fire Engineering, Vol. 147, No. 12, Dec. 1994, pp. 58–61.

LeVan, S., et. al., "Assessing the Fire Hazard of Structures in the Wildland-Urban Interface," Forest Products Lab., Madison, W, Aug. 1990.

Maranghides, A., "Wildland Urban Interface Fire Model," Worcester Polytechnic Inst., MA, Thesis, Dec. 1993.

Martin, R. E., et. al., "Wildland Fire Research Laboratory Studies of the Oakland/Berkeley Hills "Tunnel" Fire of Oct. 20, 1991," Extended Abstracts of the SFPE Engineering Seminars on Large Fires: Causes and Consequences, Nov. 16–18, 1992, Dallas, TX, Society of Fire Protection Engineers, Boston, MA, 1992, pp. 1–5.

McFadden, J., Clayton, B., and Day, D., "Wildfires Incident Command System," Fire International, Vol. 11, No. 104, 1987, pp. 33-35.

McIlvenna, M., "Using Structural Apparatus at Wildland Fires. Part 1," Fire Engineering, Vol. 147, No. 2, Feb. 1994, pp. 22–23.

McIlvenna, M., "Using Structural Apparatus at Wildland Fires. Part 2," Fire Engineering, Vol. 147, No. 6, June 1994, pp. 16, 18.

McKenzie, D. W., "Compressed Air Foam Systems for Use in Wildland Fire Applications," Forest Service, San Dimas, CA, Report 9251 1203-SDTDC, Sept. 1992.

McKenzie, D. W., "Proportioners for Use in Wildland Fire Applications," Forest Service, San Dimas, CA, Report 9251 1204-SDTDC, Sept. 1992.

Mercier, J. C., "Overview of the Global Wildland Fire Problem," Wildfire--International Wildland Fire Conference Proceedings, Meeting Global Wildland Fire Challenges: The People, The Land, The Resources, July 23–26, 1989, Boston, MA, U. S. Forest Service, Washington, DC, 1990, pp. 8–9.

National Wildfire Suppression Technology, "FOAMBIB: A Bibliography of Wildland Fire Foam Evaluation and Use. Supplement Number 2," National Wildfire Suppression Technology, Washington, DC, Supplement Number 2, Sept. 1994.

National Wildfire Suppression Technology, "FOAMBIB: A Bibliography of Wildland Fire Foam Evaluation and Use. Supplement Number 1," National Wildfire Suppression Technology, Washington, DC, Supplement Number 1, Sept. 1993.

National Wildfire Suppression Technology, "FOAMBIB: A Bibliography of Wildland Fire Foam Evaluation and Use," National Wildfire Suppression Technology, Washington, DC, Dec. 1992.

Perry, D. G., "Fire Behavior, Tactics and Command," *Wildland Firefighting,* 1987.

Perry, D. G., "Wildland Fire Season '95," *American Fire Journal,* Vol. 47, No. 5, May 1995, pp. 11–13.

Perry, D. G., "Wildland Siege in Santa Barbara County," American Fire Journal, Vol. 42, No. 8, Aug. 1990, pp. 12–15, 46–47.

Proceedings of the Symposium on Wildland Fire 2000, USDA Forest Service Gen. Tech. Rep. PSW-101, 1987, Pacific Southwest Forest and Range Experiment Station, Forest Service, U.S. Dept. of Agriculture, Berkley, CA.

Putt, W. C., "Health and Safety at Wildland Fires," Fire Engineering, Vol. 145, No. 6, June 1992, pp. 60–62, 64, 66, 68, 70, 72–73.

Pyne, S. J., *Fire in America—A Cultural History of Wildland and Rural Fire,* Princeton University Press, Princeton, NJ, 1989.

Pyne, S. J., *World Fire, the Culture of Fire on Earth,* Henry Holt and Company, New York, 1995.

Qianxi, W., Fan, W. and Qingan, W., "Mathematical Model of Severe Wildland Fires," First Asian Conference on Fire Science and Technology, (ACFST), October 9-13, 1992, Hefei, China, International Academic Publishers, China, 1992, pp. 130–135.

Queen, P. L., "Fighting Fire in the Urban/Wildland Interface. Initial Attack Resources," American Fire Journal, Vol. 45, No. 1, January 1993, pp. 12–21.

Queen, P. L., "Fighting Fires in the Urban/Wildland Interface. Handline Construction," American Fire Journal, Vol. 44, No. 12, Dec. 1992, pp. 12–15.

Queen, P. L., "Fighting Fire in the Urban/Wildland Interface. Commanding the Initial Resources," American Fire Journal, Vol. 45, No. 3, March 1993, pp. 19–23.

Queen, P. L., "Fighting Fire With Fire in the Wildland/Urban Interface," American Fire Journal, Vol. 44, No. 11, Nov. 1992, pp. 26–29.

Queen, P. L., "Surviving the Wildland or Intermix Fire. Part 2," American Fire Journal, Vol. 48, No. 6, June 1996, pp. 38–41.

Queen, P. L., "Surviving the Wildland or Intermix Fire. Part 3," American Fire Journal, Vol. 48, No. 7, July 1996, pp. 6–11, 27.

Queen, P. L., "Weather and the Wildland/Urban Interface Fire: The 10 Essential Elements," American Fire Journal, Vol. 44, No. 10, Oct. 1992, pp. 24–27.

Queen, P., "Fighting Fires in the Urban/Wildland Interface. Part 1. Survival," American Fire Journal, Vol. 44, No. 7, July 1992, pp. 12–17.

Queen, P., "Fighting Fires in the Urban/Wildland Interface. Part 2. Structure Protection," American Fire Journal, Vol. 44, No. 8, Aug. 1992, pp. 34–39.

Queen, P., "Wildland Firefighting Tips: Protective Clothing," American Fire Journal, Vol. 44, No. 7, July 1992, p. 6.

Ramsay, G. C., McArther, N.A., and Dowling, V.P., "Preliminary Results from an Examination of House Survival in the 16 February 1983 Bushfires in Australia," *Fire and Materials,* Vol. 11, No. 1, March 1987, p. 49.

Robertson, F. D., "Global Challenge for Wildland Fire Management," Wildfire–International Wildland Fire Conference Proceedings, Meeting Global Wildland Fire Challenges: The People, The Land, The Resources, July 23–26, 1989, Boston, MA, U. S. Forest Service, Washington, DC, 1990, pp. 22–25.

Rochna, R. R., "Hi-Ex Foam Has It Covered in the Wildland," American Fire Journal, Vol. 46, No. 2, Feb. 1995, pp. 12–15.

Rossomando, C., "Wildlands Fire Management: Federal Policies and Their Implications for Local Fire Departments," USFA Fire Investigation Technical Report Series, TriData Corp., Arlington, VA, Federal Emergency Management Agency, Washington, DC, Report 045, 1991.

Salisbury, H. E., *The Great Black Dragon Fire: A Chinese Inferno,* Little, Brown, and Company, Boston, MA, 1989.

Schmoldt, D. L., "Knowledge Management: An Application to Wildfire Prevention Planning," *Fire Technology,* Vol. 25, No. 2, May 1989, pp. 150–164.

Shaw, R., "Influence of Wildland Fire on the Recovery of Endangered Plant Species Study Project. Final Report. Sept. 30, 1994–Sept. 20, 1995," Colorado State Univ., Fort Collins, Army Medical Research and Material Command, Fort Detrick, MD, Final Report, Oct. 1995.

Sicking, L. P., "Wildland Fire Hose Guide," Technology and Development Center, San Dimas, CA, IE01P47, Aug. 1995.

Simard, A. J., Blank, R. W., and Hobria, S. L., "Measuring and Interpreting Flame Height in Wildland Fires," *Fire Technology,* Vol. 25, No. 2, May 1989, pp. 114–133.

Swinford, R. M., and Tokle, G. O., project managers, *Building Interagency Cooperation,* National Fire Protection Association, Quincy, MA, 1988.

Teie, W. C., "Structural Wildfire Urban Interface. The Wildland/Urban Interface Fire Problem: A California Action Plan," Consultant, Rescue, CA International Association of Fire Chiefs (IAFC), Fire-Rescue International Conference Proceedings, 1993 Annual Conference, Aug. 28-Sept. 1, 1993, Dallas, TX, 1993, pp. 95–98.

Teie, W. C., *Firefighter's Handbook on Wildland Firefighting...Strategy, Tactics, and Safety,* Deer Valley Press, Rescue, CA, 1994.

Tran, H. C., Cohen, J. D. and Chase, R. A., "Modeling Ignition of Structures in Wildland/Urban Interface Fires," First International Conference on Fire and Materials, Sept. 24–25, 1992, Arlington, VA, 1992, pp. 253–262.

U.S. Department of Agriculture, "Findings From the Wildland Firefighters Human Factors Workshop. Improving Wildland Firefighter Performance Under Stressful, Risky Conditions: Toward Better Decisions on the Fireline and More Resilient Organizations. June 12-16, 1995. Missoula, Montana," Department of Agriculture, Washington, D, 5100-F&AM, 9551-2855-MTDC, Nov. 1995.

Ventimiglia, M., "Paradise (Almost) Lost: Pebble Beach Fire," *Fire Engineering,* Vol. 141, No. 5, 1988, p. 18–30.

Wang, Q. X., et. al., "Expert and Graphical Systems for Surface Wildland Fires," First Asian Conference on Fire Science and Technology, (ACFST), October 9-13, 1992, Hefei, China, International Academic Publishers, China, 1992, pp. 166–171.

Webster, J., *The Complete Australian Brushfire Book,* Impact Printing Ptg Ltd, Melbourne, Australia, 1986.

Wilson, J. F., "1989: A Minimum Test," *Fire Command,* Vol. 57, No. 1, Jan. 1990, pp. 16–19.

Winston, R. M., "South Canyon Fire: Final Report on the Disaster That Killed 14 Federal Wildland Firefighters," Firehouse, Vol. 20, No. 7, July 1995, pp. 64–66, 68.

Winston, R. M., "Structural Wildland Interzone," Firehouse, Vol. 19, No. 1, Jan. 1994, p. 50.

Winston, R. M., "Structural Wildland Interzone." A Commentary, and Seeking Your Comments Too," Firehouse, Vol. 20, No. 4, April 1995, p. 96.

Winston, R. M., "Structural Wildland Interzone." Oakland, Three Years Later," Firehouse, Vol. 19, No. 10, Oct. 1994, pp. 24-25.

Winston, R. M., "Wanted Now: The Structural Wildland/Urban Interface Fire Specialist," Firehouse, Vol. 18, No. 11, Nov. 1993, pp. 35, 38.

PUBLIC FIRE PROTECTION AND HAZMAT MANAGEMENT

Michael S. Hildebrand and Gregory G. Noll

Historically, public fire departments have played a major role in flammable liquids and gases fire prevention and code enforcement. Flammable hazardous materials (hazmats) code enforcement has fit in well with fire department prevention programs, because it is seen as being "part of the fire problem." In the early 1980s, however, many public safety agencies began finding the line between the fire problem and the hazardous materials problem difficult to distinguish. Public interest in the environment and major catastrophes involving hazardous materials have resulted in promulgation of a wide range of laws, regulations, and standards that mandate fire department involvement.

Today, many laws created at the federal level assign overlapping responsibility for public safety to federal, state, and local agencies. Organizing an effective hazardous materials prevention and enforcement program within a fire department can be a difficult task. Consider the fact that, within the last twenty years, the U.S. Congress has passed six major pieces of legislation concerning hazardous materials and public safety. These new laws have produced seven different major federal agency regulations that contain no fewer than six different legal definitions for a hazardous material. Add the fact that there are also fifty individual state codes and over seventy different voluntary consensus standards, and it is easy to see why the code enforcement issue can be confusing.

This chapter has been written for the fire department prevention and enforcement specialist who wants to learn the fundamentals of hazardous materials laws, regulations, and standards. It begins by defining what a hazardous material is, and provides an overview of the major elements of a hazardous materials prevention and enforcement program. The chapter provides an in-depth overview of the key elements of each of the major federal laws and regulations, as well as a general summary of the primary hazardous materials codes and standards.

DEFINITION OF A HAZARDOUS MATERIAL

The first step in understanding hazardous materials from a prevention and enforcement perspective is to define what a hazardous ma-

Michael S. Hildebrand, CSP, and Gregory G. Noll, CSP, are consultants specializing in hazardous materials emergency response issues. They have served on numerous NFPA, ANSI, and API hazardous materials committees and are the authors of the textbook *Hazardous Materials: Managing the Incident*.

terial actually is. Many of the commonly used hazardous materials terms are sometimes used interchangeably, but they have distinctively different meanings. Various state and federal regulations govern the manufacture, transportation, storage, use, and cleanup of chemicals in the United States and Canada, and use numerous terms, definitions, and lists to convey information.

Hazardous materials: Any substance or material in any form or quantity that poses an unreasonable risk to safety and health and property when transported in commerce. (U.S. Department of Transportation, 49 CFR 171.)

Hazardous substances: Any substance designated under the Clean Water Act and the Comprehensive Environmental Response, Compensation, and Liability Act (CERCLA) as posing a threat to waterways and the environment when released. [U.S. Environmental Protection Agency (EPA), 40 CFR 302.] It should be noted that the term "hazardous substances," as used within Occupational and Safety Health Administration (OSHA) 1910.120, refers to any substance defined by EPA within Section 101 of CERCLA; any biological agent or other disease-causing agent as defined by EPA within Section 101 of CERCLA; any substance listed by Department of Transportation (DOT) as a hazardous material; and any hazardous waste, as defined by EPA in 40 CFR 261.3 or by DOT in 49 CFR 171.8.

Extremely hazardous substances (EHS): Chemicals determined by the EPA to be extremely hazardous to a community during an emergency spill or release as a result of their toxicities and physical/chemical properties. (U.S. EPA, 40 CFR 355.)

Hazardous chemicals: Any chemical that would be a risk to employees if exposed in the workplace. (U.S. OSHA, 29 CFR 1910.) Examples of a "physical hazard" are combustible liquids, compressed gases, explosives, flammables, organic peroxides, oxidizers, pyrophorics, unstables (reactives), or water-reactives. "Health hazard" means a mixture of chemicals or a pathogen for which there is statistically significant evidence, based on at least one study conducted in accordance with established scientific principles, that acute or chronic health effects may occur in exposed employees. The term "health hazard" includes chemicals that are carcinogens, toxic or highly toxic agents, reproductive toxins, irritants, corrosives, sensitizers, hepatoxins, nephrotoxins, neurotoxins, agents that act on the hematopoietic system, and agents that damage the lungs, skin, eyes, or mucous membranes.

Hazardous wastes: Discarded materials regulated by the EPA because of public health and safety concerns. Regulatory authority is granted under the Resource Conservation and Recovery Act (RCRA). (U.S. EPA, 40 CFR 260 - 281.)

Marine pollutant: Materials that have an adverse impact on the marine environment.

Dangerous goods: An internationally used term under the International Maritime Dangerous Good Code (IMDG). In Canadian transportation, the United States term "hazardous materials" has basically the same meaning as "dangerous goods."

HAZARDOUS MATERIALS PREVENTION AND ENFORCEMENT PROGRAMS

A comprehensive and integrated pubic safety hazardous materials program should consist of four major program elements: (1) prevention, (2) preparedness, (3) response, and (4) recovery. The prevention program element usually consists of four subelements: (1) construction and design standards; (2) an inspection and enforcement program; (3) a public education program; and (4) handling, notification, and reporting requirements.

Process, Container Design, and Construction Standards

Almost all hazardous materials facilities, containers, and processes are designed and constructed to some standard. This "standard of care" may be based upon voluntary consensus standards, such as those developed by the National Fire Protection Association (NFPA) and the American Society for Testing and Materials (ASTM), or on government regulations. Many major petrochemical companies, hazardous materials companies, and industry trade associations have also developed their own respective engineering standards and guidelines.

All containers used for the transportation of hazmats are designed and constructed to both specification and performance regulations established by the U.S. DOT. These regulations are referenced in Title 49 of the Code of Federal Regulations (CFR). In certain situations, hazmats may be shipped in non-DOT-specification containers that have received a DOT exemption.

Inspection and Enforcement

Fixed facilities, transportation vehicles, and transportation containers are subject to some form of hazardous materials inspection. Fixed facilities will commonly be inspected by state and federal OSHA and EPA inspectors, in addition to state fire marshals and local fire departments. It should be recognized that many of these inspections will focus upon fire safety and life safety issues, and may not adequately address either environmental or process safety issues.

Transportation vehicle inspection is generally based upon criteria established within Title 49 CFR. The enforcing agency is usually the state police; however, some local fire departments and state fire marshal's offices are also involved in inspecting hazardous materials cargo tank trucks. This will vary according to the individual state, the hazardous materials being transported, and the mode of transportation.

Among the U.S. DOT agencies with primary hazardous materials regulatory responsibilities are the following.

Office of Hazardous Materials Transportation (OHMT) of the Research and Special Programs Administration (RSPA): Responsible for all hazardous materials transportation regulations except bulk shipment by ship or barge. This includes designating and classifying hazmats, container safety standards, label and placarding requirements, and handling, stowing, and other in-transit requirements.

Office of Motor Carrier Safety (OMCS) of the Federal Highway Administration (FHA):   Responsible for inspection and enforcement activities relating to hazardous materials highway transportation and depot trans-shipment points.

Federal Railroad Administration (FRA):   Responsible for enforcement of regulations relating to hazmats carried by rail or held in depots and freight yards.

Federal Aviation Administration (FAA):   Responsible for the enforcement of regulations relating to hazardous materials shipments on domestic and foreign carriers operating at U.S. airports and in cargo handling areas.

U.S. Coast Guard (USCG):   Responsible for the inspection and enforcement of regulations relating to hazmats in port areas and on domestic and foreign ships and barges operating in the navigable waters of the United States.

Public Education

Hazardous materials are a concern not only for industry but also for the community. The average homeowner contributes to this problem by improperly disposing of substances such as used motor oil, paints, solvents, batteries, and other chemicals used in and around the home. As a result, many communities have initiated full-time household chemical waste awareness, education, and disposal programs. In other instances, communities have established used motor oil collection stations and chemical clean-up days in an effort to reclaim and recycle these materials.

Handling, Notification, and Reporting Requirements

These guidelines actually act as a bridge between planning and prevention functions. There are many federal, state, and local regulations that require those who manufacture, store, or transport hazmats and hazardous wastes to comply with certain handling, notification, and reporting rules. Key federal regulations include the facility reporting requirements of Superfund Amendments and Reauthorization Act (SARA), Title III, and the release notification requirements of CERCLA (Superfund). There are also many state regulations that are similar in scope and that often exceed the federal requirements.

FEDERAL HAZARDOUS MATERIALS LAWS

Laws are primarily created through an act of Congress or by individual state legislatures. Laws typically provide broad goals and objectives, mandatory dates for compliance, and established penalties for noncompliance. Federal and state laws enacted by legislative bodies usually delegate the details for implementation to a specific federal or state agency. For example, the U.S. Occupational Safety and Health Act enacted by Congress delegates rule making and enforcement authority on worker health and safety issues to the Occupational Safety and Health Administration.

History shows that hazardous materials laws have been enacted by Congress in response to specific catastrophes, such as Love Canal, New York, and Bhopal, India. Laws now regulate everything from finished products to hazardous waste.

Because of their lengthy official titles, many simply use abbreviations or acronyms when referring to these laws. The following summaries outline some of the more important laws impacting hazardous materials emergency planning and response.

The Comprehensive Environmental Response, Compensation, and Liability Act (CERCLA)

Known as "Superfund," this law addresses hazardous substance releases into the environment and clean-up of inactive hazardous waste disposal sites. It also requires those individuals responsible for the release of the hazardous materials (commonly referred to as the responsible party) above a specified "reportable quantity" to notify the National Response Center.

In 1980, Congress passed the Comprehensive Environmental Response, Compensation, and Liability Act (CERCLA) and established a fund of $1.6 billion called the "Superfund" (PL 96-510) to be administered by the EPA. Under this legislation, the EPA is authorized to inventory all uncontrolled hazardous waste sites in the nation. If the substances at these sites pose an immediate danger to public health and/or the environment, and those responsible for the contamination cannot be identified or cannot pay for the clean-up, the EPA can use the Superfund to clean up chemical spills or toxic wastes.

The Resource Conservation and Recovery Act (RCRA)

In 1976, Congress passed major legislation establishing a uniform national policy for hazardous and solid waste disposal. This act is called the Resource Conservation and Recovery Act (RCRA). Congress intended that states assume responsibility for implementing RCRA, with oversight from the federal government. The rationale was that states are more familiar with the regulated community and are in a better position to administer the programs and respond to specific state and local needs most effectively. The state program must be fully equivalent to the federal program. However, states may impose requirements that are "more stringent" or "broader in scope" than the federal requirements.

There are four major programs established under RCRA: (1) solid waste, (2) underground storage tanks, (3) medical waste, and (4) hazardous waste.

Solid waste: Subtitle D of the Act encourages states to develop and implement solid waste management plans. These plans, among other purposes, are intended to promote recycling of solid wastes and require closing or upgrading of all environmentally unsound dumps.

Underground storage tanks: Subtitle I of the Act regulates petroleum products and hazardous substances (as defined under Superfund) stored in underground tanks. The objective of this section is to prevent leakage to ground water from tanks and to clean up past releases. There are also performance standards for new tanks and regulations for leak detection, prevention, closure, financial responsibility, and corrective action at all underground tank sites.

Medical waste: Subtitle J addresses the problem of medical waste mismanagement. Included are requirements pertaining to medical waste generation, treatment, destruction, and disposal.

Hazardous waste: Subtitle C establishes a program to manage hazardous wastes from "cradle-to-grave." The objective of the program is to ensure that hazardous waste is handled in a manner that protects human health and the environment. These regulations cover the generation, transportation, and treatment, storage, or disposal of hazardous wastes. Any facility generating more than a minimum amount of hazardous waste must follow the RCRA requirements for storage, transportation, and disposal.

Regulations of both the DOT and the EPA govern the transportation of hazardous wastes. A hazardous waste manifest system, among other purposes, ensures that the material is properly identified, states the place of origin and destination for treatment, storage, or disposal; classifies the waste; provides the quantity and flashpoint of the substance; and gives special handling instructions. Each shipper handling a hazardous waste cargo must sign the manifest certifying acceptance.

Comprehensive guidelines have been established by EPA for tracking the movement, treatment, storage, and disposal of these waste products. Among the provisions of the guidelines are the following:

1. Identification and listing of materials classified as hazardous waste;
2. A system of record keeping and labeling;
3. Procedures for providing correct information regarding hazardous waste contents to and by persons transporting, storing, or disposing of this waste;
4. A manifest and permit system regulating the transport of wastes to authorized sites;
5. Requirements that transporters of hazardous wastes properly label hazardous waste shipments and carry them only to treatment, storage or disposal sites licensed under RCRA;
6. Requirements that operators of Toxic Storage Disposal (TSD) facilities maintain records and comply with the manifest system. Only EPA-approved methods of treatment, storage, and disposal can be used as indicated in the permit issued by EPA;
7. A mandate for producers and site operators to develop contingency plans in the event of a hazardous waste emergency. These plans must be the result of consultations with local fire, police, and hospital officials. These organizations must be included as part of the local response team. Contingency plans provide detailed information, such as the industrial site plan, the on-scene coordinator's phone number, and types of hazards present; and
8. Encouragement for states to develop and manage their own hazardous waste programs consistent with RCRA guidelines. In the absence of a state plan, RCRA authorizes EPA to impose a federal plan for the state.

RCRA also requires hazardous waste generators to do the following:

1. Consult the list of hazardous and toxic organic and inorganic compounds published by the EPA to determine if a chemical waste is hazardous;
2. Subject all unlisted wastes to chemical analysis to determine if they possess any of the hazardous waste characteristics established in the regulations;
3. Declare a waste hazardous (based on knowledge of the material or processes used in its production);
4. Obtain an EPA identification number by filing with the EPA's regional administrator. Fire officials and local emergency-planning committees can obtain a list of hazardous waste generators in their area by contacting their EPA regional office. This list is updated on a regular basis and will provide the name and address of the facility, its EPA identification number, and its hazardous waste generator classification. Hazardous waste generators are classified in one of three categories, as follows:
   a. *Large-quantity generator (LQG).* Generates more than 2,200 (1,000 kg) of hazardous waste or more than 2.2 lb (1 kg) of acutely hazardous waste in a month. Once the first 2,200 lb (1,000 kg) has been accumulated, the waste must

be shipped within 90 days. There is no limit to the amount that can be accumulated.

b. *Small-quantity generator (SQG)*. Generates less than 2,200 lb (1,000 kg) of hazardous waste in a month and/or less than 2.2 lb (1 kg) of acutely hazardous waste as listed by the EPA.

c. *Conditionally exempt small-quantity generator (CESQG)*. Generates less than 220 lb (100 kg) of hazardous waste or less than 2.2 lb (1 kg) of acutely hazardous waste in a month.

5. Apply for a facility permit if waste is accumulated on the generator's site for more than 90 days. The waste is considered to be stored at the site when it remains on the generator's property for more than 90 days. If this condition continues, the generator must secure a facility permit for storage of hazardous wastes, under RCRA Section 3005;

6. Treat, store, or dispose of hazardous wastes on site subject to requirements under RCRA Sections 3004 and 3005; and

7. Transport hazardous wastes to other locations for use, storage, treatment, or disposal in properly labeled approved containers, and prepare a hazardous waste manifest.

On November 8, 1984, Congress passed the Hazardous and Solid Waste Amendments Act of 1984, which modified waste-management practices and strengthened RCRA, by:

1. Establishing tighter requirements for hazardous waste land disposal;

2. Regulating underground storage tanks for new or waste products, under RCRA;

3. Bringing small-quantity generators, i.e., those generating between 220 and 2,200 lb (100 and 1000 kg) of hazardous waste per month under RCRA's authority;

4. Giving the authority to order corrective action at RCRA-regulated facilities to the EPA;

5. Setting forth "minimum technological standards," such as double liners, two-leachate collection systems, and groundwater monitoring, for new land-disposal facilities, as well as requiring interim-status facilities to either be retrofitted or stop receiving, storing, or treating hazardous waste; and

6. Establishing new requirements for delisting hazardous wastes.

The Clean Air Act (CAA)

This act establishes requirements for airborne emissions and the protection of the environment. The Clean Air Act Amendments of 1990 addressed emergency response and planning issues at certain facilities with processes using highly hazardous chemicals. This included the establishment of a national Chemical Safety and Hazard Investigation Board; EPA's promulgation of 40 CFR, Part 68, "Risk Management Programs for Chemical Accidental Release Prevention"; and OSHA's promulgation of 29 CFR, 1910.119, "Process Safety Management of Highly Hazardous Chemicals, Explosives, and Blasting Agents." In addition, certain facilities are required to make information available to the general public regarding the manner in which chemical risks are handled within a facility.

Superfund Amendments and Reauthorization Act of 1986 (SARA)

SARA has had the greatest impact upon hazardous materials emergency planning and response operations. As the name implies, SARA amended and reauthorized the Comprehensive Environmental Response, Compensation, and Liability Act of 1980 (CERCLA or Superfund). While many of the amendments pertained to hazardous waste site cleanup, SARA's requirements also established a national baseline with regard to hazardous materials planning, preparedness, training, and response.

Title I of this Act required OSHA to develop health and safety standards covering numerous worker groups who handle or respond to chemical emergencies and led to the development of OSHA 1910.120, "Hazardous Waste Operations and Emergency Response (HAZWOPER)."

Most familiar to the emergency response community is SARA, Title III. Also known as the Emergency Planning and Community Right-to-Know Act (EPCRA), SARA, Title III, led to the establishment of the State Emergency Response Commissions (SERC) and the Local Emergency Planning Committees (LEPC).

The Right-to-Know Act of 1986 includes four major sections affecting public safety. These include: (1) Sections 301 through 303: emergency planning; (2) Section 304: emergency release notification; (3) Sections 311, 312: community right-to-know reporting requirements; and (4) Section 313: toxic chemical release inventory.

Emergency planning (Sections 301 through 303): Sections 301 through 303 of the Act required the governor of each state to appoint a state emergency response commission (SERC) that, in turn, was required to designate local emergency planning districts and appoint a local emergency planning committee (LEPC) for each district. Membership in the LEPC is required to include elected state and local officials; members of fire, police, civil defense, and public health organizations; environmental, hospital, and transportation officials; and representatives of industry, business, the community, and the media.

The primary responsibility of the LEPC is to develop a comprehensive emergency response plan and submit it to the SERC. After being reviewed by the SERC, the plan is returned to the LEPC with approval or recommendations for revision. This emergency response plan is required to be revised and updated annually. The plan must include the following:

1. The identification of specific sites and transportation corridors in which extremely hazardous materials are stored, used, or transported.

2. Emergency response procedures.

3. Designation of a community coordinator and facility coordinator to implement the emergency plan.

4. Emergency notification procedures.

5. Procedures for determining the occurrence of a chemical release and the probable area and population that will be affected.

6. Description of community and industrial emergency response equipment and resources, and the identity of personnel responsible for them.

7. Evacuation plans.

8. Description and schedule of a chemical emergency response training program.

9. Method and schedule for conducting emergency response plan exercises.

Emergency release notification (Section 304): Section 304 requires facility owners or operators to give immediate notification to the community emergency coordinator of the LEPC, and to the SERC, of any release of a listed hazardous substance or its vapor that will extend beyond the facility's property and possibly endanger the general public. In most communities, the initial notification point will be to the emergency 911 telephone number or the local equivalent.

Under Section 304, the term "facility" means any building, structure, installation, equipment, pipe or pipeline (including any pipe into a sewer or publicly owned treatment works), well, pit, pond, lagoon, impoundment, ditch, storage container, motor vehicle, rolling stock, or aircraft, or any site or area where a hazardous substance has been deposited, stored, disposed of, or placed, or otherwise come to be located; this, however, does not include any consumer product use or any waterborne vessel.

Releases that result only in exposure to persons solely within the site on which the facility is located are not required to be reported to the LEPC or SERC, but may require the owner or operator to notify the National Response Center, under Section 103(a) of CERCLA. Hazardous substances that are subjected to these reporting requirements can be found on a list of approximately 360 extremely hazardous substances that was published by the EPA in the Federal Register (40 CFR 355) or on a list of approximately 725 substances subject to emergency notification requirements, under CERCLA, Section 103(a), 40 CFR 302.4. There are additions and deletions to these lists on a periodic basis, and some of the substances can be found on both lists.

Initial emergency notification of a release by the owner or operator can be made by telephone, radio, or in person. To the extent it is known at the time of the notice, and as long as it does not result in a delay in responding to the emergency, the notification must include the following information:

1. The involved substance or substances names or identity;
2. An indication of whether the substance or substances are on the extremely hazardous list;
3. An estimate of the quantity of the substance or substances released into the environment;
4. The time and duration of the release;
5. The medium or media (air, land, water), into which the release occurred;
6. Any known or anticipated acute or chronic health risks associated with the emergency, and, if necessary, advice regarding medical attention for exposed individuals;
7. Proper precautions to take as a result of the release, such as evacuation; and
8. The name and telephone numbers of the person or persons to be contacted for further information.

As soon as possible after a release that requires notice, the owner or operator must provide a written followup emergency notice. As more information becomes available, the owner or operator must update the information originally required and include additional information with respect to the following:

1. Actions taken in responding to and containing the release;
2. Any known or anticipated acute or chronic health risks associated with the release; and
3. Where appropriate, advice regarding medical attention necessary for exposed individuals.

Notification of releases of hazardous substances during transportation, or during storage relating to the transportation of such substances, must be reported by dialing emergency 911 or, in the absence of a 911 emergency telephone number, by calling the operator.

Community right-to-know reporting requirements (Sections 311 and 312): Section 311 requires that the owner or operator of a facility prepare or have available a material safety data sheet (MSDS) for each hazardous chemical covered under the Occupational Safety and Health Act of 1970 (and any regulations promulgated under that act) and submit a MSDS to each of the following:

1. The local emergency planning committee;
2. The state emergency-response commission; and
3. The fire department with jurisdiction over the facility.

If the hazardous chemical is a mixture, the owner or operator of the facility can fulfill the requirements of Section 311 by submitting a MSDS for, or identifying on a list, each element or compound in the mixture that is a hazardous chemical. The owner or operator may also submit a MSDS for, or identify on a list, the mixture itself.

In lieu of an MSDS, the owner or operator may submit a list of the hazardous chemicals grouped in categories of health and physical hazards according to OSHA; or in terms of groups of hazardous chemicals that present similar hazards in an emergency, as designated by the EPA administrator. The list of hazardous chemicals must include: (1) the chemical name or common name of each substance, as provided on the MSDS; and (2) any hazardous component of each substance, as provided on the MSDS.

An MSDS or revised list must be provided when new hazardous chemicals become present at a facility in quantities at or above established threshold levels. Upon request by the local emergency-planning committee, an owner or operator who submits a list of chemicals must provide the appropriate MSDS for any chemical on that list. If requested by any member of the general public, the LEPC must make available to that person a MSDS for any hazardous chemicals required to be submitted to them. This information may be given at a location designated by the LEPC during normal working hours.

Within 3 months following the discovery by an owner or operator of significant new information concerning an aspect of a hazardous chemical for which an MSDS sheet was previously submitted to the LEPC, a revised MSDS must be provided to the LEPC.

Section 312 stipulates that owners or operators of a facility who are required to prepare or have available a MSDS for each hazardous chemical covered under the Occupational Safety and Health Act of 1970 (and any regulations promulgated under that act), prepare and submit an emergency and hazardous chemical inventory form to the local emergency planning committee, the state emergency response commission, and the fire department with jurisdiction over the facility.

The inventory form must include all hazardous chemicals above specified thresholds that were present at the facility at any time during the previous calendar year. The EPA has established threshold quantities for hazardous chemicals covered by Section 312. If the quantities of hazardous chemicals are below that set forth by EPA, a report is not required. Current threshold quantities are as follows:

1. For extremely hazardous substances: 500 lb (227kg) or the threshold planning quantity, whichever is lower.
2. For all other hazardous chemicals that meet or exceed 10,000 lb (4,450 kg) for which OSHA requires an MSDS. It should be noted that there are over 500,000 chemicals and mixtures for which OSHA requires an MSDS be maintained.

Facilities may file either a Tier I or Tier II inventory form. Tier I reporting requires owners or operators to submit the following information for each of the hazard categories set forth under OSHA regulations:

1. Fire, sudden release of pressure, reactivity, acute health hazard, chronic health hazard;
2. An estimate (in ranges) of the maximum amount of hazardous chemicals in each category present at the facility at any time during the preceding calendar year;
3. An estimate (in ranges) of the average daily amount of hazardous chemicals in each category present at the facility at any time during the preceding calendar year; and
4. The general location of hazardous chemicals in each category.

The EPA administrator may modify the OSHA categories of health and physical hazards by requiring information to be reported in terms of groups of hazardous chemicals that present similar hazards in an emergency, or require reporting on individual hazardous chemicals of special concern to emergency response personnel. Tier I inventory forms are required to be submitted on or before March 1 of each year and contain data with respect to the preceding calendar year.

Tier II inventory reporting forms are required if requested by a state emergency response commission, an LEPC, or a fire department with jurisdiction over the facility. If an owner or operator of a facility files a Tier II inventory form, a Tier I form is not required, as long as the same deadlines are followed. The public may also request Tier II information from the LEPC or SERC. Tier II information must include the following:

1. The chemical name or common name of each chemical, as provided on the MSDS;
2. An estimate (in ranges) of the maximum amount of hazardous chemicals present at the facility at any time during the preceding calendar year;
3. An estimate (in ranges) of the average daily amount of hazardous chemicals present at the facility at any time during the preceding calendar year;
4. A brief description of the manner of storage of the hazardous chemical;
5. The location at the facility of the hazardous chemical; and
6. An indication of whether the owner elects to withhold location information of a specific hazardous chemical from disclosure to the public.

Upon request by a fire department with jurisdiction over any facility that files an inventory form under Section 312, the owner or operator must allow the fire department to conduct an on-site inspection of the facility and must provide to the fire department specific location information on hazardous chemicals at the facility.

Toxic chemical release inventory (Section 313):   Section 313 requires owners or operators of certain facilities to file an annual Toxic Chemical Release Inventory Form (Form R) with the EPA and to those officials designated by the governor of the state, on or before July 1 of each year. A release is defined as any spilling, leaking, pumping, pouring, emitting, emptying, discharging, injecting, escaping, leaching, dumping, or disposing into the environment (including the abandonment or discarding of barrels, containers, and other closed receptacles) of any toxic chemical. Section 313 applies to facilities that have 10 or more full-time employees, fall under the Standard Industrial Classification (SIC) Codes 20 through 39, and that manufactured, processed, or otherwise used a listed toxic chemical in excess of specified threshold quantities during the previous calendar year. The threshold quantity is 25,000 lb (11,350 kg) per year for toxic chemical manufacturing or processing. The threshold quantity with respect to use at a facility is 10,000 lb (4,450 kg) of toxic chemical per year. The initial List of Toxic Chemicals subject to the provisions of Section 313 was published in August 1986 for the Senate Committee on Environment and Public Works. Through rule making, the EPA has modified this list to date by adding nine chemicals and delisting six. Chemicals may be added to the list if there is sufficient evidence to establish any one of the following:

1. The chemical is known to cause, or can reasonably be anticipated to cause, significant adverse acute human health effects at concentration levels that are reasonably likely to exist beyond facility site boundaries as a result of continuous, or frequently recurring, releases;

2. The chemical is known to cause, or can reasonably be anticipated to cause, cancer or teratogenic effects, or serious or irreversible reproductive dysfunction, neurological disorders, heritable genetic mutations, or other chronic health effects in humans; or
3. The chemical is known to cause, or can reasonably be anticipated to cause, because of its toxicity, its toxicity and persistence in the environment, or its toxicity to bio-accumulate in the environment, a significant adverse effect on the environment of sufficient seriousness, in the opinion of the EPA, to warrant reporting under Section 313.

The EPA has established and maintains a national toxic-chemical inventory database (TRI database) on the information submitted. The computer database is accessible by the public and government officials via computer telecommunications, and is intended to help answer questions about toxic chemical releases in their community. The EPA is using the data to target problem pollution areas and as a screening tool for the development of regulations, guidelines, and standards. Researchers from the government or universities conducting environmental analysis can also access the data.

Toxic Substances Control Act (TSCA)

The Toxic Substances Control Act (TSCA—pronounced Tosca), passed by Congress in 1976, establishes regulations with respect to the inventory and testing of all chemical substances manufactured or processed in the United States. This act once again makes EPA the central implementation arm of the federal government. TSCA authorizes EPA to do the following:

1. Develop a uniform listing of all chemical substances;
2. Establish a testing procedure for chemicals already in use and any one of approximately 1,000 new chemicals developed each year;
3. Determine if these chemicals present an unreasonable risk to the public's health or the environment;
4. Prohibit or limit the manufacture, processing, use, application, and concentration of such chemicals; and
5. Recall or seize by civil action hazardous substances that are determined to be imminently harmful to the public's health or the environment.

Federal Water Pollution Control Act (FWPCA)

In 1970, the Federal Water Pollution Control Act (PL 92-500) was amended in Section 311 to include Oil and Hazardous Spills. Through this legislation, both the EPA and the U.S. Coast Guard are mandated to regulate spills of oil and/or other hazardous substances that threaten coastal waters and inland waterways.

Spill Prevention, Control, and Countermeasure Plan (SPCC)

This regulation was issued by the EPA in 40 CFR 112, and applies to facilities engaged in drilling, producing, gathering, storing, processing, refining, transferring, distributing, or consuming oil and oil products, which due to location could reasonably be expected to discharge oil in quantities that may be harmful into or upon navigable waterways or adjoining shorelines, or upon the waters on the contiguous zone.

Owners or operators of these facilities must prepare a Spill Prevention, Control, and Countermeasure (SPCC) plan that includes appropriate containment and/or drainage control structures or equipment to prevent discharged oil from reaching navigable water before cleanup occurs.

Facilities that are excluded from being required to have a SPCC plan are facilities that do not have an aboveground storage capacity of at least 1,320 gal (4,996 L) of oil. No single container at these facilities can hold more than 660 gal (2,498 L) of oil.

Oil Pollution Act of 1990 (OPA)

Commonly referred to as OPA, this Act amended the Federal Water Pollution Control Act. Its scope covers both facilities and carriers of oil and related liquid products, including deepwater marine terminals, marine vessels, pipelines, and railcars. Requirements include the development of emergency response plans, regular training and exercise sessions, and verification of spill resources and contractor capabilities. The Act also requires the establishment of Area Committees (AC) and the development of Area Contingency Plans (ACP) to address oil and hazardous substance spill response in coastal zone areas.

HAZARDOUS MATERIALS REGULATIONS

Regulations, sometimes called "rules," are created by federal or state agencies as a method of providing guidelines for complying with a law that was enacted through legislative action. A regulation permits individual governmental agencies to enforce the law through inspections, which may be conducted by federal and state officials.

Laws delegate certain details of implementation and enforcement to federal or state agencies who are then responsible for writing the actual regulations that enforce the legislative intent of the law. Regulations will either: (1) define the broad performance required to meet the letter of the law (i.e., performance-oriented standards) or (2) provide very specific and detailed guidance on satisfying the regulation (i.e., specification standards).

Federal Regulations

The following summary includes several of the more significant federal regulations that affect hazardous materials emergency planning and response.

Hazardous waste operations and emergency response (29 CFR 1910.120):   Also known as HAZWOPER, this federal regulation was issued under the authority of SARA, Title I. The regulation was written and is enforced by OSHA in those 23 states and 2 territories with their own OSHA-approved occupational safety and health plans. In the remaining 27 "non-OSHA" states, public sector personnel will be covered by a similar regulation enacted by the EPA (40 CFR Part 311).

The regulation establishes important requirements for both industry and public safety organizations who respond to hazardous materials or hazardous waste emergencies. This includes fire fighters, law enforcement, emergency medical personnel (EMS), hazardous materials responders, and industrial emergency response team (ERT) members. Requirements cover the following areas:

1. Hazardous materials emergency response plan;
2. Emergency response procedures, including the establishment of an incident management system, the use of a buddy system with backup personnel, and the establishment of a safety officer;
3. Specific training requirements covering instructors and both initial and refresher training;
4. Medical surveillance programs; and
5. Post-emergency termination procedures.

Of particular interest to hazardous materials managers and responders are the specific levels of competency and associated training requirements identified within OSHA 1910.120(q)(6).

First responder at the awareness level.   These are individuals who are likely to witness or discover a hazardous substance release and who have been trained to initiate an emergency response notification process. The primary focus of their hazardous materials responsibilities is to secure the incident site, recognize and identify the materials involved, and make the appropriate notifications. These individuals would take no further action to control or mitigate the release. First responder–awareness personnel must have sufficient training or experience to objectively demonstrate the following competencies:

1. An understanding of what hazardous materials are, and the risks associated with them in an incident.
2. An understanding of the potential outcomes associated with a hazardous materials emergency.
3. The ability to recognize the presence of hazardous materials in an emergency and, if possible, identify the materials involved.
4. An understanding of the role of the first responder–awareness individual within the local emergency operations plan. This would include site safety, security and control, and the use of the *North Amercian Emergency Response Guidebook.*
5. The ability to realize the need for additional resources and to make the appropriate notifications to the communication center.

The most common examples of first responder–awareness personnel include law enforcement and plant security personnel, as well as some public works employees. There is no minimum hourly training requirement for this level; the employee would have to have sufficient training to objectively demonstrate the required competencies.

First responder at the operations level.   Most fire department suppression personnel fall into this category. These are individuals who respond to releases or potential releases of hazardous substances as part of the initial response for the purpose of protecting nearby persons, property, or the environment from the effects of the release. They are trained to respond in a defensive fashion without actually trying to stop the release. Their primary function is to contain the release from a safe distance, keep it from spreading, and protect exposures.

First responder–operations personnel must have sufficient training or experience to objectively demonstrate the following competencies:

1. Knowledge of basic hazard and risk assessment techniques.
2. Knowledge of how to select and use proper personal protective clothing and equipment available to the operations-level responder.
3. An understanding of basic hazardous materials terms.
4. Know how to perform basic control, containment, and/or confinement operations within the capabilities of the resources and personal protective equipment available.
5. Know how to implement basic decontamination measures.
6. Possess an understanding of the relevant standard operating procedures and termination procedures.

First responders at the operations level must have received at least 8 hr of training or have had sufficient experience to objectively demonstrate competency in the previously mentioned areas, as well as the established skill and knowledge levels for the first responder–awareness level.

Hazardous materials technicians.   These are individuals who respond to releases or potential releases for the purposes of stopping the release. Unlike the operations level, they generally assume a more aggressive role, in that they are often able to approach the point of a release in order to plug, patch, or otherwise stop the release of a hazardous substance.

Hazardous materials technicians are required to have received at least 24 hr of training equal to the first responder–operations level and have competency in the established skill and knowledge levels outlined below:

1. Capable of implementing the local employers' emergency operations plan.
2. Ability to classify, identify, and verify known and unknown materials by using field survey instruments and equipment (direct reading instruments).
3. Able to function within an assigned role in the incident management system.
4. Able to select and use the proper specialized chemical personal protective clothing and equipment provided to the hazardous materials technician.
5. Understand hazard and risk assessment techniques.
6. Able to perform advanced control, containment, and/or confinement operations within the capabilities of the resources and equipment available to the hazardous materials technician.
7. Understand and implement decontamination procedures.
8. Understand basic chemical and toxicological terminology and behavior.
9. Understand termination procedures.

Many communities and facilities have personnel trained as emergency medical technicians (EMT), yet do not have the primary responsibility for providing basic or advanced life support medical care. Similarly, hazardous materials technicians may not necessarily be part of a hazardous materials response team. However, if they are part of a designated team as defined by OSHA, they must also meet the medical surveillance requirements within OSHA 29 CFR 1910.120.

Hazardous materials specialists. These are individuals who respond with, and provide support to, hazardous materials technicians. While their duties parallel those of the technician, they require a more detailed or specific knowledge of the various substances they may be called upon to contain. This individual would also act as the site liaison with federal, state, local, and other governmental authorities, with regard to site activities.

Similar to the technician level, hazardous materials specialists must have received at least 24 hr of training equal to the technician level and have competency in the following established skill and knowledge levels:

1. Capable of implementing the local emergency operations plan.
2. Ability to classify, identify, and verify known and unknown materials by using advanced field survey instruments and equipment (direct reading instruments).
3. Knowledge of the state emergency response plan.
4. Able to select and use the proper specialized chemical personal protective clothing and equipment provided to the hazardous materials specialist.
5. Understand in-depth hazard and risk assessment techniques.
6. Able to perform specialized control, containment, and/or confinement operations within the capabilities of the resources and equipment available to the hazardous materials specialist.
7. Able to determine and implement decontamination procedures.
8. Ability to develop a site safety and control plan.
9. Understand basic chemical, radiological, and toxicological terminology and behavior.

Where the hazardous materials technician possesses an intermediate level of expertise and is often viewed as a "utility person" within the hazardous materials response community, the hazardous materials specialist possesses an advanced level of expertise. Within the fire service, the specialist will often assume the role of the safety officer or hazardous materials branch. Further, the specialist must meet the medical surveillance requirements outlined within OSHA 129 CFR 1910.120. The reader should note that a hazardous materials specialist, as defined by OSHA, is different from "specialist employee," as defined by NFPA 472, *Standard for Professional Competence of Responders to Hazardous Material Incidents.* The differences in these terms is often the source of confusion among industrial emergency responders.

On-scene incident commander. Incident commanders, who will assume control of the incident scene, must receive at least 24 hr of training equal to the first responder operations level. In addition, the employer must certify that the incident commander has competency in the following areas:

1. Know and be able to implement the employer's incident management system.
2. Know how to implement the employer's emergency operations plan.
3. Understand the hazards and risks associated with working in chemical protective clothing.
4. Know of the state emergency response plan and of the federal regional response team.
5. Know and understand the importance of decontamination procedures.

Skilled support personnel. These are personnel who are skilled in the operation of certain equipment, such as cranes and hoisting equipment, and who are needed temporarily to perform immediate emergency support work that cannot reasonably be performed in a timely fashion by emergency response personnel. It is assumed that these individuals will be exposed to the hazards of the emergency response scene.

Although these individuals are not subject to the HAZWOPER training requirements, they must be given an initial briefing at the site prior to their participation in any emergency response effort. This briefing must include elements such as instructions in using personal protective clothing and equipment, the chemical hazards involved, and the tasks to be performed. All other health and safety precautions provided to emergency responders and on-scene workers must be used to ensure the health and safety of these support personnel.

Specialist employees. These are employees who, in the course of their regular job duties, work with and are trained in the hazards of specific hazardous substances, and who will be called upon to provide technical advice or assistance to the incident commander at a hazardous materials incident. This would include industry responders, chemists, and related professional or operations employees. These individuals must receive training or demonstrate competency in the area of their specialization annually. The term "specialist employees" should not be confused with hazardous materials specialists described above, since they are two different titles and have different levels of competency.

On August 22, 1994, OSHA issued technical amendments to the existing Appendix B and added a nonmandatory Appendix E to 29 CFR 1910.120, *Hazardous Waste Operations and Emergency Response.* The revisions to Appendix B mainly involved updating various references; however, the addition of Appendix E provides suggested guidelines for a more effective training curriculum and program.

Hazard communication standard (29 CFR 1910.1200): The Hazard Communication Standard (HCS) was originally issued by OSHA in November 1983. OSHA's most recent revision of the HCS was issued on March 11, 1994, and covers manufacturers, importers, distributors, and users of hazardous chemicals. The Hazard Communication Standard is commonly called the "right-to-know" law or "HAZCOM."

The Hazard Communication Standard is different from other OSHA health and safety standards, because it covers *all hazardous chemicals*. The rule incorporates a "downstream flow of information," which means that producers of chemicals have the primary responsibility for generating and disseminating information, whereas users of chemicals must obtain this information and transmit it to their own employees. The basic responsibilities are outlined in Table 10-8A.

TABLE 10-8A. *Responsibilities Outlined by Hazardous Communications Standard*

Chemical manufacturers/ importers	Determine the hazards of each product.
Chemical manufacturers/ importers/ distributors	Communicate the hazard information and associated protective measures downstream to customers through labels and MSDSs.
Employers	Identify and list all hazardous chemicals in their workplace.
	Obtain MSDSs and labels for each hazardous chemical.
	Develop and implement a written hazard communication program, including labels, MSDSs, and employee training.
	Communicate hazard information to their employees through labels, MSDSs, and formal training programs.

Hazard determination. Chemical manufacturers and importers are required to review the scientific information on the chemicals they produce or import, and to report the information they find to their employees and to employers who distribute or use their products. Downstream employers can rely on the evaluations performed by the chemical manufacturers or importers to establish the hazards of the chemicals they use.

The chemical manufacturers, importers, and any employers who choose to evaluate hazards are responsible for the quality of the hazard determinations they perform. Each chemical must be evaluated for its potential to cause adverse health effects and its potential to pose physical hazards, such as flammability. In addition, chemicals that have been evaluated and found to be a suspected or confirmed carcinogen must be reported as such.

Written hazard communication program. Employers that fall under the requirements of HAZCOM must develop, implement, and maintain a written hazard communication program that includes provisions for container labeling, collection, and availability of MSDSs and employee training. It also must contain a list of the hazardous chemicals in the workplace, the means the employer will use to inform employees of the hazards of nonroutine tasks, and the hazards associated with chemicals in unlabeled pipes. If the workplace has multiple employers on-site (e.g., a construction site), HAZCOM requires these employers to ensure that information regarding hazards and protective measures be made available to the other employers on-site, where appropriate.

Labels and other forms of warning. Chemical manufacturers, importers, and distributors must be sure that containers of hazardous chemicals leaving their facilities are labeled, tagged, or marked with the identity, appropriate hazard warning, and the name and address of the manufacturer or other responsible party.

In the workplace, each container must be labeled, tagged, or marked with the identity of the hazardous chemicals contained within, and must show hazard warnings appropriate for employee protection. The hazard warning can be any type of message, words, pictures, or symbols that convey the hazards of the chemical(s) in the container. Labels must be legible, in English (plus other languages, if desired), and prominently displayed.

Material safety data sheets. Chemical manufacturers and importers must develop a material safety data sheet (MSDS) for each hazardous chemical they produce or import, and must provide the MSDS automatically at the time of the initial shipment of a hazardous chemical to a downstream distributor or user. Distributors must also ensure that downstream employers are similarly provided an MSDS.

Each MSDS must be in English and include information regarding the specific identity of the hazardous chemical(s) involved and the common names. In addition, information must be provided on the physical and chemical characteristics of the hazardous chemical, known acute and chronic health effects and related health information, exposure limits, whether the chemical is considered to be a carcinogen, precautionary measures, emergency and first-aid procedures, and the identification of the organization responsible for preparing the MSDS.

Copies of the MSDS for hazardous chemicals in a given work site are to be readily accessible to employees in that area. Although OSHA has developed a model MSDS format (OSHA form 174), there is currently no specific MSDS form required by the regulation. OSHA's model MSDS is divided into nine separate sections: (1) chemical identity, (2) hazardous ingredients, (3) physical and chemical characteristics, (4) fire and explosion hazard data, (5) reactivity details, (6) health-hazard information, (7) precautions for safe handling, (8) use of the material, and (9) any control measures to be taken.

The American National Standards Institute (ANSI) has developed a standard on the preparation of MSDSs. ANSI Z400.1, *Hazardous Industrial Chemicals, Material Safety Data Sheet–Preparation,* is used by many chemical manufacturers as a standardized format for MSDS preparation.

Employee MSDS training must include methods and observations that may be used to detect the presence or release of a hazardous chemical in the work area (e.g., monitoring conducted by the employer, continuous monitoring devices, visual appearance, or odor of hazardous chemicals when released). Physical and health hazards of the chemicals used and measures that the employees can take to protect themselves from these hazards must be covered, as well as specific procedures the employer has implemented to protect employees from exposure to hazard.

Trade secrets. A "trade secret" is something that gives an employer an opportunity to obtain an advantage over competitors who do not know about the trade secret or who do not use it. For example, a trade secret may be a confidential device, pattern, information, or chemical makeup. Chemical industry trade secrets are generally formulas, process data, or a "specific chemical identity." The last is the type of trade secret information referred to in the Hazard Communication Standard. The term includes the chemical name, the chemical abstract services (CAS) registry number, or any other specific information that reveals the precise designation. It does not include common names.

The HCS standard strikes a balance between the need to protect exposed employees and the employer's need to maintain confidentially of a bona fide trade secret. This is achieved by providing limited disclosure to health professionals who are furnishing medical or other occupational services to exposed employees, or employees and their designated representatives, under specified conditions of need and confidentiality.

Medical emergency.   The chemical manufacturer, importer, or employer must immediately disclose the specific chemical identity of a hazardous chemical to a treating physician or nurse when the information is needed for proper emergency or first-aid treatment. As soon as circumstances permit, the chemical manufacturer, importer, or employer may obtain a written statement of need and a confidentiality agreement.

Hazardous materials transportation regulations (49 CFR 100 through 199):   This series of regulations is issued and enforced by DOT. The regulations govern container design, chemical compatibility, packaging and labeling requirements, shipping papers, transportation routes and restrictions, etc. The regulations are comprehensive and strictly govern how all hazardous materials are transported by highway, railroad, pipeline, aircraft, and by water. There are some local fire departments that routinely perform inspections of hazardous materials cargo tank trucks using DOT regulations.

Retention of DOT markings, placards, and labels (29 CFR 1910.1201):   In 1994, OSHA issued requirements that those facilities that receive containers of hazardous materials, which are required to be marked, labeled, or placarded in accordance DOT regulations, retain those markings, labels, or placards on the container until the container is sufficiently cleaned of residue and purged of vapors to remove any potential hazards.

National contingency plan (NCP) (40 CFR 300, Subchapters A through J):   The National Contingency Plan (NCP) outlines the policies and procedures of the federal agency members of the National Oil and Hazardous Materials Response Team [also known as the National Response Team (NRT)]. The regulation provides guidance for emergency responses, remedial actions, enforcement, and funding mechanisms for federal government response to hazardous materials incidents. The NRT is chaired by EPA, while the vice-chairperson represents the U.S. Coast Guard (USCG).

Each of the ten federal regions also has a Regional Response Team (RRT) that mirrors the makeup of the NRT. RRTs may also include representatives from state and local government and Indian tribal governments.

When the NRT or RRT is activated for a federal response to an oil spill or hazardous materials incident, a federal on-scene coordinator (OSC) will be designated to coordinate the overall response. The OSC will represent either EPA or the USCG, based upon the location of the incident. If the release or threatened release occurs in coastal areas or near major navigable waterways, the USCG will usually assume primary OSC responsibility. If the situation occurs inland and away from navigable or major waterways, the EPA will serve as the OSC. Local emergency responders should contact EPA and USCG personnel within their region to determine which agency has primary responsibility and will act as the federal OSC for their respective area.

Process safety management of highly hazardous chemicals (29 CFR, Part 1910.119):   Issued by OSHA in 1992, this regulation applies mainly to hazardous materials manufacturing facilities, such as refineries and chemical plants. Other affected sectors include natural gas liquids, farm product warehousing, electric, gas, and sanitary services.

The requirements apply to companies that deal with any of more than 130 specific toxic and reactive chemicals in EPA-listed quantities. It also applies to flammable liquids and gases in quantities greater than 10,000 lb (4,540 kg).

The regulation includes the following major requirements:

1. *Process safety information* must include information on the hazards of the highly hazardous chemicals used or produced by the process, information on the technology of the process, and information on the equipment in the process.
2. *Process hazards analysis* must include an initial process analysis (hazard evaluation) appropriate for the complexity of the process and must identify, evaluate, and control the hazards involved.
3. *Written operating procedures* must be developed and implemented that provide clear instructions for safely conducting activities involved in each process. Procedures must address initial startup and shutdown, operating limits, and safety and health considerations.
4. *Employee participation* must be solicited on the conduct and development of process management.
5. *Training programs* must include initial employee training on health and safety hazards, refresher training provided at least every three years on current operating procedures, and a training documentation program. The regulation also applies to contractors performing maintenance, repair, renovation, or specialty work on a process.
6. *Pre-startup safety review* must be provided for new facilities and for modified facilities when the modification is significant enough to require a change in process safety.
7. *Mechanical integrity* must be provided for pressure vessels and storage tanks, piping systems, relief and vent systems and devices, emergency shutdown systems, controls and alarms, and pumps.
8. *Management of change* must be provided for written procedures to manage changes to process chemicals, technology, equipment, etc.
9. *Incident investigation* must include a thorough investigation of incidents to identify the chain of events and causes, so that corrective measures can be developed and implemented.
10. *Emergency planning and response procedures* must be provided for handling spills and releases at the facility.
11. *Compliance audit* must be provided that includes mandatory process safety management reviews every 3 years to verify that procedures are adequate and being followed.

Risk management programs for chemical accidental release prevention (40 CFR, Part 68):   Promulgated under amendments to the Clean Air Act, this was a Final Rule at the time of publication. The regulation is similar in scope to OSHA's *Process Safety Management of Highly Hazardous Chemicals* described above, with the primary focus being community safety as compared to employee safety. The regulation requires that an employer establish and implement an emergency action plan for the entire plant in accordance with the provisions of 29 CFR 1910.38(a) (emergency planning).

The Final Rule includes the following major requirements:

1. *Emergency response plan* must include evacuation routes or protective actions for employees, procedures for responding to a hazardous materials release, descriptions of mitigation technologies, and procedures for informing the public and emergency response agencies about the release.
2. *Written procedures for emergency response equipment* must address inspection, testing, and maintenance. The maintenance program must be documented.

3. *First aid procedures and emergency medical treatment guidelines* are required to treat victims exposed to hazardous materials.

4. *Training programs* must be designed to train employees on emergency response procedures.

5. *Drill and exercise programs* must be designed to evaluate and improve the emergency response program.

6. *Coordination requirements* must be arranged with local emergency response agencies. Upon request of the local emergency planning committee, the facility owner must submit information necessary for the LEPC to develop a community emergency response plan.

State Regulations

Each of the 50 states and the 2 U.S. territories maintains an enforcement agency that has responsibility for hazardous materials. The three key players in each state usually consist of the: (1) state fire marshal, (2) state Occupational Safety and Health Administration, and (3) the State Department of the Environment (sometimes known as Natural Resources or Environmental Quality). While there are many variations, the state fire marshal is typically responsible for the regulation of flammable liquids and gases, due to the close relationship between the flammability hazard and the fire prevention code; the state environmental agency would be responsible for the development and enforcement of environmental safety regulations.

While known by various titles, most states have a government equivalent of the federal OSHA. Twenty-three states and two territories have adopted the federal OSHA regulations as state (territory) law. This method of adoption has increased the level of enforcement of hazardous materials regulations such as the HAZWOPER. State governments also maintain an environmental enforcement agency that usually enforces the federal RCRA, CERCLA, and CAA laws at the local level. State involvement in hazardous waste regulatory enforcement has significantly increased the number of hazardous materials incidents reported.

VOLUNTARY CONSENSUS STANDARDS

Voluntary consensus standards are normally developed through professional organizations or trade associations as a method of improving the individual quality of a product or system. Within the emergency response community, NFPA is recognized for its role in developing consensus standards and recommended practices that impact fire safety and hazardous materials operations. In the United States, standards are developed primarily through a democratic process, whereby a committee of subject specialists representing varied interests writes the first draft of the standard. The document is then submitted to either a larger body of specialists or the general public, who then may amend, vote on, and approve the standard for publication. This procedure is known as the consensus standards process.

When a consensus standard is completed, it may be voluntarily adopted by government agencies, individual corporations, or organizations. Many hazardous materials consensus standards are also adopted as a reference in a regulation. In effect, when a federal, state, or municipal government adopts a consensus standard by reference, the document becomes a regulation.

Standards developed through the voluntary consensus process play an important role in increasing both workplace and public safety. Historically, a voluntary standard improves over time, as each revision reflects recent field experience and adds more detailed requirements. As users of the standard adopt it as a way of doing business, the level of safety gradually improves over time.

Consensus standards are also updated more regularly than governmental regulations. For example, since OSHA 29 CFR 1910.120 was released in March 1989, the first edition of NFPA 472, has been revised once and a third revision will be completed in 1997.

In many respects, a voluntary consensus standard provides a way for individual organizations and corporations to self-regulate their business or profession. All of the national fire codes in the United States are developed through the voluntary consensus standards process; the two most active associations are: (1) NFPA and (2) Western Fire Chiefs Association. Standards developed by these two associations address many hazardous materials issues, including hazardous materials storage and handling, personal protective clothing and equipment, and hazardous materials professional competencies.

NFPA Standards

NFPA has over 70 individual hazardous-materials-related voluntary consensus standards. These standards are developed through the committee process, according to NFPA rules and procedures.

NFPA's hazardous materials standards are widely used by both industry and public safety organizations as recommended practices for inspection, safe handling, and installation, etc. They do not have the force of law unless they are adopted by a government agency. For example, state governments that have approved state OSHA plans under Section 18b of the Occupational Safety and Health Act of 1970 must adopt standards to enforce requirements that are at least as effective as federal requirements. There are currently 23 states and 2 territories that adopt federal OSHA regulations as state law. The current OSHA regulations on flammable and combustible liquids adopts NFPA 30, *Flammable and Combustible Liquids Code,* by reference. Consequently, while NFPA does not specifically write regulations, NFPA standards often result as law through a similar adoption process.

Unfortunately, the pace of revising and updating state regulations lags behind the development and approval of revised NFPA standards. Many state regulations that have adopted NFPA standards, such as NFPA 30 and NFPA 58, *Standard for the Storage and Handling of Liquefied Petroleum Gases,* have adopted editions that are 5 to 10 years behind current editions. Most consensus standards reflect current and accepted practices, as opposed to "cutting edge" technology, practices, and procedures. This means that an organization that adopts a 1990 edition of a standard, as a local or state regulation, is actually adopting a standard that may not be based on the most recent technology.

The following list summarizes current NFPA hazardous materials-related standards. They have been categorized by the authors for easier reference by the standards users and are not necessarily grouped by committee responsibility.

Flammable Liquids Related Standards

NFPA 30, *Flammable and Combustible Liquids Code*

NFPA 33, *Standard for Spray Application Using Flammable or Combustible Materials*

NFPA 34, *Standard for Dipping and Coating Processes Using Flammable or Combustible Liquids*

NFPA 321, *Standard on Basic Classification of Flammable and Combustible Liquids (inactive)*

NFPA 325, *Guide to Fire Hazard Properties of Flammable Liquids, Gases, and Volatile Solids*

NFPA 327, *Standard Procedures for Cleaning or Safeguarding Small Tanks and Containers Without Entry*

NFPA 328, *Recommended Practice for the Control of Flammable and Combustible Liquids and Gases in Manholes, Sewers, and Similar Underground Structures*

NFPA 329, *Recommended Practice for Handling Underground Releases of Flammable and Combustible Liquids*

NFPA 385, *Standard for Tank Vehicles for Flammable and Combustible Liquids*

NFPA 386, *Standard for Portable Shipping Tanks for Flammable and Combustible Liquids*

Flammable Gases Related Standards

NFPA 50A, *Standard for Gaseous Hydrogen Systems at Consumer Sites*

NFPA 50B, *Standard for Liquified Hydrogen Systems at Consumer Sites*

NFPA 51A, *Standard for Acetylene Cylinder Charging Plants*

NFPA 54, *National Fuel Gas Code*

NFPA 58, *Standard for the Storage and Handling of Liquified Petroleum Gases*

NFPA 59, *Standard for the Storage and Handling of Liquified Petroleum Gases at Utility Gas Plants*

NFPA 59A, *Standard for the Production, Storage, and Handling of Liquified Natural Gas (LNG)*

NFPA 306 *Standard for the Control of Gas Hazards on Vessels*

Dusts and Powders Related Standards

NFPA 91, *Standard for Exhaust Systems for Air Conveying of Materials*

NFPA 650, *Standard for Pneumatic Conveying Systems for Handling Combustible Materials*

NFPA 651, *Standard for the Manufacture of Aluminum Powder*

NFPA 654, *Standard for the Prevention of Fire and Dust Explosions in the Chemical, Dye, Pharmaceutical, and Plastics Industries*

Explosives and Reactive Chemicals Standards

NFPA 35, *Standard for the Manufacture of Organic Coatings*

NFPA 43A, *Storage of Liquid and Solid Oxidizers (inactive)*

NFPA 43B, *Code for the Storage of Organic Peroxide Formulations*

NFPA 43C, *Storage of Gaseous Oxidizing Materials (inactive)*

NFPA 50, *Standard for Bulk Oxygen Systems at Consumer Sites*

NFPA 51, *Standard for the Design and Installation of Oxygen-Fuel Gas Systems for Welding, Cutting, and Allied Processes*

NFPA 53, *Guide on Fire Hazards in Oxygen-Enriched Atmospheres*

NFPA 68, *Guide for Venting of Deflagrations*

NFPA 69, *Standard on Explosion Prevention Systems*

NFPA 490, *Code for the Storage of Ammonium Nitrate*

NFPA 491M, *Manual of Hazardous Chemical Reactions*

NFPA 495, *Explosive Materials Code*

NFPA 498, *Standard for Safe Havens and Interchange Lots for Vehicles Transporting Explosives*

NFPA 655, *Standard for Prevention of Sulfur Fires and Explosions*

NFPA 1124, *Code for the Manufacture, Transportation, and Storage of Fireworks*

Nuclear Standards

NFPA 801, *Standard for Facilities Handling Radioactive Materials*

NFPA 802, *Recommended Fire Protection Practice for Nuclear Research and Production Reactors*

NFPA 803, *Standard for Fire Protection for Light Water Nuclear Power Plants*

Facilities and Related Manufacturing Standards

NFPA 30B, *Code for the Manufacture and Storage of Aerosol Products*

NFPA 32, *Standard for Drycleaning Plants*

NFPA 36, *Standard for Solvent Extraction Plants*

NFPA 45, *Standard on Fire Protection for Laboratories Using Chemicals*

NFPA 86, *Standard for Ovens and Furnaces*

NFPA 86C, *Standard for Industrial Furnaces Using a Special Processing Atmosphere*

NFPA 86D, *Standard for Industrial Furnaces Using Vacuum as an Atmosphere*

NFPA 231, *Standard for General Storage*

NFPA 231C, *Standard for Rack Storage of Materials*

NFPA 820, *Standard for Fire Protection in Wastewater Treatment and Collection Facilities*

Electrical Standards

NFPA 70, *National Electrical Code®*

NFPA 77, *Recommended Practice on Static Electricity*

NFPA 79, *Electrical Standard for Industrial Machinery*

NFPA 110, *Standard for Emergency and Standby Power Systems*

NFPA 496, *Standard for Purged and Pressurized Enclosures for Electrical Equipment*

NFPA 497A, *Recommended Practice for Classification of Class I Hazardous (Classified) Locations for Electrical Installations in Chemical Process Areas*

NFPA 497M, *Manual for Classification of Gases, Vapors, and Dusts for Electrical Equipment in Hazardous (Classified) Locations*

NFPA 780, *Standard for the Installation of Lightning Protection Systems*

Emergency Response Standards

NFPA 471, *Recommended Practice for Responding to Hazardous Materials Incidents*

NFPA 472, *Standard for Professional Competence of Responders to Hazardous Materials Incidents*

NFPA 473, *Standard for Competencies for EMS Personnel Responding to Hazardous Materials Incidents*

NFPA 704, *Standard System for the Identification of the Hazards of Materials for Emergency Response*

NFPA 1991, *Standard on Vapor-Protective Suits for Hazardous Chemical Emergencies*

NFPA 1992, *Standard on Liquid Splash-Protective Suits for Hazardous Chemical Emergencies*

NFPA 1993, *Standard on Support Function Protective Clothing for Hazardous Chemical Operations*

Relevant Standards

NFPA 43D, *Code for the Storage of Pesticides*

NFPA 49, *Hazardous Chemicals Data*

NFPA 72, *National Fire Alarm Code*

NFPA *101®, Life Safety Code®*

Hazardous Materials Emergency Response Standards

Within the hazardous materials community there are three major consensus standards that have been widely accepted as the baseline standard for hazardous materials response teams. These are:

1. NFPA 471, *Recommended Practice for Responding to Hazardous Materials Incidents;*

2. NFPA 472, *Standard for Professional Competence of Responders to Hazardous Materials Incidents;* and
3. NFPA 473, *Standard for Competencies for EMS Personnel Responding to Hazardous Materials Incidents.*

As the titles imply, these standards primarily address the emergency response and field operational aspects of hazardous materials. Nevertheless, they provide an excellent basis for developing hazardous materials inspection and enforcement personnel.

NFPA 471, *Recommended Practice for Responding to Hazardous Materials Incidents.*   The document covers planning procedures, policies, and application of procedures for incident levels, personal protective clothing and equipment, decontamination, safety, and communications. The purpose of NFPA 471 is to outline the minimum requirements that should be considered when dealing with responses to hazardous materials incidents, and to specify operating guidelines.

NFPA 472, *Standard for Professional Competence of Responders to Hazardous Materials Incidents.*   The purpose of NFPA 472 is to specify minimum competencies for those who will respond to hazardous materials incidents. The overall objective is to reduce the number of accidents, injuries, and illnesses during response to hazardous materials incidents, and to prevent exposure to hazmats to reduce the possibility of fatalities, illnesses, and disabilities affecting emergency responders.

It is important to recognize that NFPA 472 covers all hazardous materials emergency responders from both the public and private sector.

NFPA 472 provides competencies for the following levels of hazardous materials responders. These levels parallel those listed within OSHA 1910.120, with the exception that the hazardous materials specialist has been deleted and the specialist employee has been expanded and clarified upon:

1. *First responder at the awareness level.*   These are individuals who, in the course of their normal duties, may be the first on scene of an emergency involving hazmats. They are expected to recognize hazardous materials presence, protect themselves, call for trained personnel, and secure the area.
2. *First responder at the operations level.*   These are individuals who respond to releases or potential releases of hazmats as part of the initial response to the incident for the purpose of protecting nearby persons, the environment, or property from the effects of the release. They must be trained to respond in a defensive fashion to control the release from a safe distance and keep it from spreading.
3. *Hazardous materials technician.*   These are individuals who respond to releases or potential releases of hazmats for the purpose of controlling the release. Hazardous materials technicians are expected to use specialized chemical protective clothing and specialized control equipment.
4. *Incident commander.*   The person who is responsible for directing and coordinating all aspects of a hazardous materials incident.
5. *Off-site specialist employee.*   These are individuals who, in the course of their regular job duties, work with or are trained in the hazards of specific materials and/or containers. In response to incidents involving chemicals, they may be called upon to provide technical advice or assistance to the incident commander relative to their area of specialization. There are three levels of off-site specialist employee:
   a. Level C are those persons who may respond to incidents involving chemicals and/or containers within their organization's area of specialization. They may be called upon to gather and record information, provide technical advice, and/or arrange for technical assistance consistent with the organization's emergency response plan and standard operating procedures. The individual is not expected to enter the hot/warm zone at an incident.
   b. Level B are those persons who, in the course of their regular job duties, work with or are trained in the hazards of specific chemicals or containers within their organization's area of specialization. Because of their education, training, or work experience, they may be called upon to respond to incidents involving chemicals. The Level B employee may be used to gather and record information, provide technical advice, and provide technical assistance (including working within the hot/warm zone) at the incident consistent with the organization's emergency response plan and standard operating procedures, and the local emergency operations plan.
   c. Level A are those persons who are specifically trained to handle incidents involving chemicals and/or containers for chemicals used in their organization's area of specialization. Consistent with the organization's emergency response plan and standard operating procedures, the Level A employee must be able to analyze an incident involving chemicals within the organization's area of specialization, plan a response to that incident, implement the planned response within the capabilities of the resources available, and evaluate the progress of the planned response.

At the time of publication, the NFPA 472 Technical Committee was working on the development of the third edition of the standard. Consideration is being given to the development of additional competencies for the hazardous materials branch officer, the hazardous materials branch safety officer, and various areas of specialization (e.g., railroad, cargo tank trucks, etc.).

NFPA 473, *Standard for Competencies for EMS Personnel Responding to Hazardous Materials Incidents.*   The purpose of NFPA 473 is to specify minimum requirements of competence and to enhance the safety and protection of response personnel and all components of the emergency medical services system. The overall objective is to reduce the number of EMS personnel accidents, exposures, injuries, and illnesses resulting from hazardous materials (HM) incidents. There are two levels of EMS/HM responders:

1. *EMS/HM Level I.*   Persons who, in the course of their normal duties, may be called on to perform patient care activities in the cold zone at a hazardous materials incident. EMS/HM Level I responders provide care to only those individuals who no longer pose a significant risk of secondary contamination. Level I includes different competency requirements for basic (BLS) and advanced life support (ALS) personnel.
2. *EMS/HM Level II.*   Persons who, in the course of their normal duties, may be called on to perform patient care activities in the warm zone at a hazardous materials incident. EMS/HM Level II responders may provide care to those individuals who still pose a significant risk of secondary contamination. In addition, personnel at this level must be able to coordinate EMS activities at a hazardous materials incident and provide medical support for hazardous materials response personnel. Level II includes different competency requirements for basic (BLS) and advanced life support (ALS) personnel.

Uniform Fire Code, Article 80

The *Uniform Fire Code* (UFC) is a model fire prevention code developed by the International Fire Code Institute (IFCI) and is published by the International Conference of Building Officials (ICBO). The *Uniform Fire Code* is intended to maintain a safe environment in and around a building that is constructed to the standards found in the *Uniform Building Code*.

Article 80 of the *Uniform Fire Code* is intended to protect the general public, employees, emergency responders, and property from an "unauthorized" release of hazardous materials due to leak, spill, or fire within a facility. It does not regulate hazardous materials in transportation and is not designed to protect wildlife or the environment.

Article 80 is only one of 48 articles in the *Uniform Fire Code*. In order to use Article 80 correctly, it must be used in conjunction with other articles and appendices including Article 4, "Permits"; Article 9, "Definitions"; Article 10, "Fire Protection"; Article 51, "Semiconducter Fabrication Using Hazardous Materials"; Article 79, "Flammable and Combustible Liquids"; Article 81, "High-Piled Combustible Storage"; as well as Appendix II-E and Appendix VI-A.

Other Standards Organizations

There are many other important standards-writing bodies that develop hazardous materials safety and emergency-response-related standards. The more significant organizations include the American Society of Mechanical Engineers (ASME), the American Society for Testing and Materials (ASTM), and the American Petroleum Institute (API). Each of these organizations approves or creates standards, ranging from hazardous materials container design to personal protective clothing and equipment. Standards produced by these groups usually follow the standards development guidelines of the American National Standards Institute (ANSI). ANSI is an approval body and is recognized by the International Organization for Standardization (ISO).

BIBLIOGRAPHY

Additional Reading

Barkenbus, B. D., et. al., "Environmental Protection for Hazardous Materials Incidents. Volume 1," Hazardous Materials Incident Management System, Final Report, Oct. 1986–June 1990, Air Force Engineering and Services Center, Tyndall AFB, FL, ESL-TR-89-15 VOL 1, Nov. 1990.

Barkenbus, B. D., et. al., "Environmental Protection for Hazardous Materials Incidents. Volume 2," Appendices, Final Report, Oct. 1986–June 1990, Air Force Engineering and Services Center, Tyndall AFB, FL, ESL-TR-89-15 Vol 2, Nov. 1990.

Eriksen-Rattan, Fluer, L., Grijalva, R., and McLaughlin, A., *Users Guide to Article 80*, Western Fire Chiefs Association, Ontario, California, 1991.

Federal Emergency Management Agency, *et al.*, *Liability Issues in Emergency Management*, National Emergency Training Center, Emmitsburg, MD, 1992.

Fire, F. L., Grant, N. K., and Hoover, D. H., *SARA Title III—Intent and Implementation of Hazardous Materials Regulations*, Fire Engineering Books and Videos, Tulsa, OK, 1990.

National Fire Protection Association, *Hazardous Materials Response Handbook,* 2nd ed., Tokle, G., ed., National Fire Protection Association, Quincy, MA, 1992.

National Response Team, "Hazmat Emergency Planning Guide," NRT-1, 1987, National Response Team, Washington, DC.

Noll, G., Hildebrand, G., Michael, S., and Yvorra, J. G., *Hazardous Materials: Managing the Incident*, 2nd ed., Oklahoma State University, Stillwater, OK, 1995.

Stringfield, W. H., *A Fire Departments Guide to Implementing SARA, Title III, and the OSHA Hazardous Materials Standards*, International Society of Fire Service Instructors, Ashland, MA, 1987.

U.S. Environmental Protection Agency, *Hazmat Team Planning Guidance*, U.S. Environmental Protection Agency, Washington, DC, 1990.

U.S. Environmental Protection Agency, *et al.*, *Handbook of Chemical Hazard Analysis Procedures*, U.S. Environmental Protection Agency, FEMA, DOT, Washington, DC, 1989.

U.S. Environmental Protection Agency, *et al.*, *Technical Guidance for Hazards Analysis—Emergency Planning for Extremely Hazardous Substances*, U.S. Environmental Protection Agency, FEMA, DOT, Washington, DC, 1987.

U.S. EPA Title III, *List of Lists*, "Consolidated List of Chemicals Subject to Reporting Under the Emergency Planning and Community Right-to-Know Act," EPA 560/4-90-011, Jan. 1990, U.S. Environmental Protection Agency, Washington, DC.

U.S. Occupational Safety and Health Administration, *HAZWOPER Interpretative Quips*, OSHA Office of Health Compliance Assistance, Washington, DC, March 1993.

MANAGING THE RESPONSE TO HAZARDOUS MATERIAL INCIDENTS

Charles J. Wright

Hazardous materials may be an important part of our high standard of living, but when a release occurs, it can harm people, including response personnel, property, and the environment. It can disrupt critical systems, damage reputations, and create a residual fear of hazardous materials. The effective management of hazardous materials and other chemicals in the community requires: (1) prevention, (2) preparedness, (3) response, and (4) recovery activities. While this chapter focuses on the response activity, the underlying concepts apply to the other hazardous material management activities.

The purpose of the response activity is "to change the sequence of events constituting the emergency before it has run its course naturally and to minimize harm that would otherwise occur."[1] To accomplish this purpose, four duties are undertaken. (See Figure 10-9A.)

1. Analyze the problem.
2. Plan a response.
3. Implement the planned response.
4. Evaluate progress and adjust accordingly.

Each duty involves a series of tasks and steps that must be considered and resolved by decisions and actions. These duties, when supported by the response community, are the framework for an ap-

propriate, survival-oriented response to hazardous material incidents. Reasoned decisions based on this approach will minimize the harm resulting from a hazardous material incident while reducing the risk to responders.

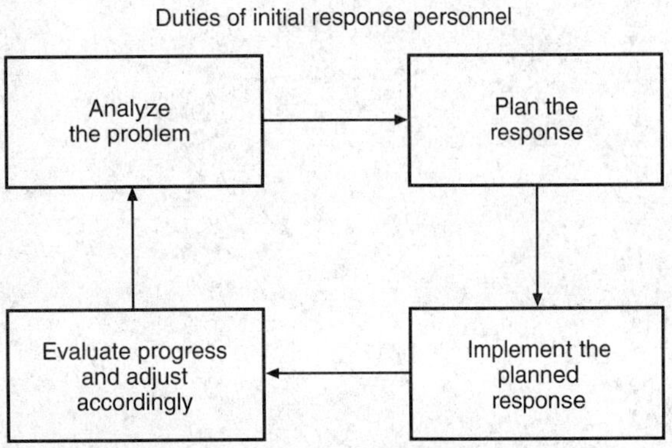

DEFINITIONS

The following terms are used in this chapter.

Bulk packaging is any containment system, including transport vehicles and freight containers, having a capacity meeting one of the following criteria: liquid—internal volume of more than 119 gal (450 L); solid—capacity of more than 882 lb (400 kg); or compressed gas—water capacity of more than 1001 lb (454 kg).

Containment systems (or containers) are a combination of a container and its closures used to isolate the contents from the surrounding environment. A closure is a device for closing the openings in a container.

Elevated temperature material (DOT) means a material that, when offered for transportation or transported in a bulk packaging, is: (1) in a liquid phase and at a temperature at or above 212°F (100°C), (2) in a liquid phase with a flash point at or above 100°F (37.8°C) that is intentionally heated and offered for transportation or transported at or above its flash point, or (3) in a solid phase and at a temperature at or above 464°F (240°C).

Emergency is the term used for a sudden, generally unexpected disruption in a normal sequence of events requiring immediate action.

Hazardous material, for the purpose of this chapter, is a substance that upon release has the potential of causing harm to people, property, or the environment. This definition includes hazardous materials, hazardous substances, hazardous wastes, elevated temperature materials, marine pollutants, and dangerous goods in Canada.

Hazardous material (DOT) means a substance or material, including a hazardous substance, that has been determined by the Secretary of Transportation to be capable of posing an unreasonable risk to health, safety, and property when transported in commerce, and that has been so designated. Table 10-9A lists the hazard classes and divisions associated with hazardous materials in transportation, as defined by the U.S. Department of Transportation (DOT).

Hazardous substance (DOT) means a material, including its mixtures and solutions, in a quantity, in one package, which equals or exceeds the reportable quantity (RQ) listed in DOT's Hazardous Materials Regulations (49 CFR, Parts 171–172).

Duties of initial response personnel

FIG. 10-9A. *A flow diagram of the response duties associated with emergencies involving hazardous materials.*

Charles J. Wright is manager of hazardous materials training in the Chemical Transportation Safety group at the Union Pacific Railroad Company.

TABLE 10-9A. *DOT Hazard Classes*

Class	Division with Explanation	Examples of Materials (by Hazard Class or Division)	General Hazard Properties (Not All-Inclusive)
Class 1	Division 1.1—explosive with mass explosion hazard	Dynamite, TNT, black powder	Explosive; exposure to heat, shock, or contamination could result in thermal and mechanical hazards.
	Division 1.2—explosive with projection hazard	Projectiles with bursting charges	
	Division 1.3—explosive with fire, minor blast, or minor projection hazard	Propellant explosives, rocket motors, special fireworks	
	Division 1.4—explosive device with monir explosion hazard	Common fireworks, small arms ammunition	
	Division 1.5—very insensitive explosive	Ammonium nitrate-fuel oil mixtures	
	Division 1.6—extremely insensitive explosive		
Class 2	Division 2.1—flammable gas	Propane, butadiene (inhibited) acetylene, methyl chloride	Under pressure; container may rupture violently (fire and nonfire); may be a flammable, poisonous, a corrosive, an dasphyxiant and/or thermally unstable.
	Division 2.2—nonflammable, nonpoisonous gases	Carbon dioxide, anhydrous ammonia	
	Division 2.3—poisonous (toxic) gas	Arsine, phosgene, , chlorine, methyl bromide	
	Division 2.4—corrosive gas (Canada)	Anhydrous Ammonia (Canada only)	
Class 3	Flammable Liquids	Acetone, amyl acetate, gasoline,	Flammable; container may rupture violently from heat/fire; may be corrosive, toxic, and/or thermally unstable.
	*Division 3.1—flash point <0°F (17.7°C)	methyl alcohol, toluene	
	*Division 3.2—flash point ≥0°F (17.7°C) to <73°F (22.7°C)		
	*Division 3.3—flash point ≥73° (22.7°C) to <141°F (60.5°C)		
	Combustible liquids	Fuel oils	
Class 4	Division 4.1—flammable solids	Nitrocellulose, matches	Flammable, some spontaneously; may be water reactive, toxic, and/or corrosive; may be extremely difficult to extinguish.
	Division 4.2—spontaneously combustible material	Phosphorus , pyrophoric liquids and solids	
	Division 4.3—Dangerous when wet material	Calcium carbide, potassium, sodium, magnesium	
Class 5	Division 5.1—oxidizer	Ammonium nitrate fertilizer	Supplies oxygen to support combustion; sensitive to heat, shock, friction, and/or contamination.
	Division 5.2—organic peroxides	Dibenzoyl peroxide	
Class 6	Division 6.1—poisonous (toxic) material	Aniline, arsenic, tear gas, carbon tetrachloride	Toxic by inhalation, ingestion, and skin and eye contact; may be flammable.
	Divison 6.2—infectious substances	Anthrax, botulism, rabies, tetanus	
Class 7	Radioactive material	Cobalt , uranium hexafluoride	May cause burns and biologic effects; energy and matter.
Class 8	Corrosive Material	Hydrochloric acid, sulfuric acid, sodium hydroxide, nitric acid	Disintegration of contacted tissues; may be fuming, water-reactive.
Class 9	Miscellaneous hazardous materials	Dry ice, molten sulfur, adipic acid, PCBs	
ORM- D		Consumer commodities	

For SI unit = °C = (°F − 32) × $5/_9$.

Hazardous waste (DOT) means any material that is subject to the Hazardous Waste Manifest Requirements of the U.S. Environmental Protection Agency (EPA) specified in 40 CFR, Part 262.

Hazardous material incident is an emergency involving the release or potential release of hazardous materials with or without fire.

Marine pollutant (DOT) means a material that is harmful to aquatic life (listed in 49 CFR, Part 172.101, Appendix B).

Nonbulk packaging is any containment system having a capacity meeting one of the following criteria: liquid—internal volume of 119 gal (450 L) or less; solid—capacity of 882 lb (400 kg) or less; or compressed gas—water capacity of 101 lb (454 kg) or less.

Outcomes are the direct and indirect results or consequences associated with an emergency. Direct outcomes include death, injury, property damage, and environmental damage. Indirect outcomes include system disruption, damaged reputations, and residual fear of hazardous materials.

Response community is composed of those individuals, agencies, or companies who have resources and can become involved in handling a hazardous material incident. These may include fire-service personnel, law-enforcement personnel, rescue and emergency medical personnel, private-sector response personnel (manufacturers and carriers), and state and federal government response personnel.

DEPARTMENT OF TRANSPORTATION HAZARD CLASSES

The following list presents the general definitions of the Department of Transportation (DOT) hazard classes and divisions. These definitions should not be used to determine compliance with the DOT hazardous materials regulations, as they do not indicate the exceptions and specificity found in the regulations in all cases.[2]

Class 1 *(Explosives and blasting agents)*

Explosive means any substance or article, including a device, that is designed to function by explosion (i.e., an extremely rapid release of gas and heat) or that, by chemical reaction within itself, is able to function in a similar manner even if not designed to function by explosion. Explosives in Class 1 are divided into six divisions. Each division will have a letter designation (compatibility group letter.)

Division 1.1: Consists of explosives that have a mass explosion hazard. A mass explosion is one that affects almost the entire load instantaneously.

Division 1.2: Consists of explosives that have a projection hazard but not a mass explosion hazard.

Division 1.3: Consists of explosives that have a fire hazard and either a minor blast hazard or a minor projection hazard or both, but not a mass explosion hazard.

Division 1.4: Consists of explosive devices that present a minor explosion hazard. No device may contain more than 0.9 oz. (25 g) of a detonating material.

Division 1.5: Consists of very insensitive explosives. These substances have a mass explosion hazard but are so insensitive that there is very little probability of initiation or of transition from burning to detonation under normal conditions of transport.

Division 1.6: Consists of extremely insensitive articles that do not have a mass explosion hazard. These articles contain only extremely insensitive detonating substances and demonstrate a negligible probability of accidental initiation or propagation.

Class 2 *(Gases)*

Division 2.1: *Flammable gas* means any material that is a gas at 68°F (20°C) or less and 14.7 psi (101.3 kPa) of pressure, or a material that has a boiling point of 68°F (20°C) or less at 14.7 psi (101.3 kPa) that:

1. Is ignitable at 14.7 psi (101.3 kPa) when in a mixture of 13 percent or less by volume with air, or
2. Has a flammable range of at least 12 percent in air at 14.7 psi (101.3 kPa) regardless of the lower limit.

Division 2.2: Nonflammable, nonpoisonous compressed gas, including compressed gas, liquified gas, pressurized cryogenic gas, and compressed gas in solution means any material (or mixture) which exerts in the packaging an absolute pressure of 41 psi (280 kPa) at 68°F (20°C).

A *cryogenic liquid* means a refrigerated liquefied gas having a boiling point colder than –130°F (–90°C) at 14.7 psi (101.3 kPa) absolute.

Division 2.3: *Poisonous gas* means a material that is a gas at 68°F (20°C) or less and a pressure of 14.7 psi or 1 atm (101.3 kPa), that has a boiling point of 68°F (20°C) or less at 14.7 psi (101.3 kPa), and that:

1. Is known to be so toxic to humans as to pose a hazard to health during transportation, or
2. In the absence of adequate data on human toxicity, is presumed to be toxic to humans because, when tested on laboratory animals, it has an LC_{50} value not more than 5,000 ppm.

Division 2.4: *Corrosive gas*—Canada only—means a gas that has corrosive qualities.

Class 3 *(Flammable liquids)*

Flammable liquids are any liquids having a flashpoint of not more than 140°F (60°C).

In Canada, there are three divisions of flammable liquids:

Division 3.1: Means a flammable liquid with a flashpoint below 0°F (18°C).

Division 3.2: Means a flammable liquid with a flashpoint equal to or greater than 0°F (–18°C), but less than 73°F (23°C).

Division 3.3: Means a flammable liquid with a flash point equal to or greater than 73°F (23°C), but less than 141°F (61°C).

Combustible Liquids

Combustible liquid means any liquid that has a flashpoint above 140°F (60°C) and below 200°F (93°C).

Class 4 *(Flammable solids and reactive liquids and solids)*

Division 4.1: *Flammable solid* means any of the following three types of materials:

1. *Wetted explosives*—explosives wetted with sufficient water, alcohol, or plasticizers to suppress explosive properties.
2. *Self-reactive materials*—materials that are liable to undergo, at normal or elevated temperatures, a strongly exothermal decomposition caused by excessively high transport temperatures or by contamination.
3. *Readily combustible solids*—solids that may cause a fire through friction and any metal powders that can be ignited.

Division 4.2: *Spontaneously combustible solid* means either of the following materials:

1. *Pyrophoric material*—a liquid or solid that, even in small quantities and without an external ignition source, can ignite within 5 min after coming in contact with air; or

2. *Self-heating material*—a material that, when in contact with air and without an energy supply, is liable to self-heat.

Division 4.3: *Dangerous-when-wet material* means a material that, by contact with water, is liable to become spontaneously flammable or to give off flammable or toxic gas at a rate greater than 1 L/kg of the material, per hr.

Class 5 *(Oxidizers and organic peroxides)*

Division 5.1: *Oxidizer* means a material that may, generally by yielding oxygen, cause or enhance the combustion of other materials.

Division 5.2: *Organic peroxide* means any organic compound containing oxygen (O) in the bivalent -O-O- structure and that may be considered a derivative of hydrogen peroxide, where one or more of the hydrogen atoms have been replaced by organic radicals.

Class 6 *(Poisonous materials and infectious substances)*

Division 6.1: *Poisonous material* means a material, other than a gas, that is either known to be so toxic to humans as to afford a hazard to health during transportation, or in the absence of adequate data on human toxicity, is presumed to be toxic to humans.

Division 6.2: *Infectious substance* means a viable microorganism, or its toxin, that causes or may cause disease in humans or animals. Infectious substance and etiologic agent are synonymous.

Class 7 *(Radioactive materials)*

Radioactive material means any material having a specific activity greater than 0.002 microcuries per gram (μCi/g).

Class 8 *(Corrosive material)*

Corrosive material means either a liquid or solid that causes visible destruction or irreversible alterations in human skin tissue at the site of contact, or a liquid that has a severe corrosion rate on steel or aluminum.

Class 9 *(Miscellaneous hazardous materials)*

Miscellaneous hazardous material means a material that presents a hazard during transport, but that is not included in another hazard class, including the following:

1. Any material that has an anesthetic, noxious, or other similar property that could cause extreme annoyance or discomfort to a flight-crew member so as to prevent the correct performance of assigned duties.
2. Any material that is not included in any other hazard class (a hazardous substance or a hazardous waste).

ORM-D Materials

An *ORM-D material* means a material, such as a consumer commodity, that presents a limited hazard during transportation, due to its form, quantity, and packaging.

Other Definitions

Hazard zones associated with gases (Class 2) listed as poison-inhalation hazards are:

- *Hazard zone A*–LC_{50} less than or equal to 200 ppm.
- *Hazard zone B*–LC_{50} greater than 200 ppm and less than or equal to 1,000 ppm.

- *Hazard zone C*–LC_{50} greater than 1,000 ppm and less than or equal to 3,000 ppm.
- *Hazard zone D*–LC_{50} greater than 3,000 ppm and less than or equal to 5,000 ppm.

Hazard zones associated liquid listed as poison-inhalation hazards in Classes 3, 4, 5, 6, and 8 are:

- *Hazard zone A*–LC_{50} less than or equal to 200 ppm.
- *Hazard zone B*–LC_{50} greater than 200 ppm and less than or equal to 1,000 ppm.

Forbidden means prohibited from being offered or accepted for transportation. Prohibition does not apply if these materials are diluted, stabilized, or incorporated in devices.

ANALYZING THE HAZARDOUS MATERIAL PROBLEM

Background

Responding to emergencies involving hazardous materials is not easy. Just the shear numbers of these materials present an intimidating list for those who respond to the emergencies they create. Consider these examples:

1. A review of the Chemical Abstract Service's (CAS) list indicates that out of 1.5 million chemicals or chemical formulations found outside of the laboratory, 63,000 are considered hazardous.
2. The U.S. Department of Transportation (DOT) regulates the transportation of some 4,000 hazardous materials.
3. The Occupational Safety and Health Administration (OSHA) of the U.S. Department of Labor regulates about 600 hazardous substances on the basis of occupational exposure.

A national U.S. Environmental Protection Agency (EPA) study[3] and subsequent regional studies indicate that the most commonly released chemical is vehicle fuel. Two other findings are of interest:

- 49.5 percent of the rest of the releases involved only 10 chemicals in addition to vehicle fuels. (Table 10-9B lists these chemicals.) These additional 10 chemicals are typically produced in the largest volume.

- 74.8 percent of the releases occurred in facilities (production, storage, and use) while the remaining 25.2 percent occurred in transportation.

Regional differences from this national study may change the makeup of the list locally.

Overview

The process of analyzing the hazardous material problem (sizeup) provides a way of determining the specific hazards and the potential magnitude of the problem in terms of outcomes. An understanding of the magnitude of the problem is the basis for subsequent decisions on the handling of an incident. Failure to consider all the hazards increases the responder's risk.

The analysis process begins when a responder receives notification of a problem and continues throughout the incident, typically at the scene threatened by any hazardous materials involved. Pre-emergency planning, following the same tasks and steps, lessens the time expended to achieve a safe and cost-effective response at a hazardous material incident, and also increases the available response options for that response.

An analysis of the hazardous material problem involves completing the following tasks. (See Figure 10-9B.)

TABLE 10-9B. *List of Most Commonly Released Hazardous Materials with the Percentage of Releases and the Percentage of Deaths and Injuries Associated with the Material[3]*

Chemical Name	Percentage of Releases	Percentage of Deaths and Injuries
PCB (polychlorinated biphenyls)	23.0	2.8
Sulfuric acid	6.5	4.7
Anhydrous ammonia	3.7	6.8
Chlorine	3.5	9.6
Hydrochloric acid	3.1	5.6
Sodium hydroxide	2.6	1.9
Methyl alcohol	1.7	0.4
Nitric acid	1.7	1.5
Toluene	1.4	2.4
Methyl chloride	1.4	0.0

FIG. 10-9B. *A flow diagram of the tasks associated with analyzing a hazardous material problem.*

1. Detect presence of hazard material.
2. Initiate command and control activities.
3. Survey hazardous material incident.
4. Collect and interpret hazard and response information.
5. Assess the extent of damage to the containment system.
6. Predict likely behavior of containment system or contents.
7. Estimate potential outcomes within engulfed area.

Detecting the Presence of Hazardous Materials

The first task in analyzing, and ultimately understanding and solving, hazardous material problems is recognizing those situations where hazardous materials are present. This task begins with the re-

ceipt of the initial notification of the emergency and continues throughout the handling of that emergency. Any emergency should be approached from a direction that will provide protection if hazardous materials are present. As with all tasks in the analysis process, detection should be performed from a safe location—upwind, uphill, and upstream, if possible.

The following steps should be taken when responding to any emergency.

1. *Review the initial information received for indications of hazardous materials:* Clues to the presence of hazardous materials may be provided by the caller or dispatcher.
2. *Review the occupancy, location, or local planning documents for indications of hazardous materials:* Planning documents, such as local emergency response plans, industry site-specific plans, and fire-service pre-emergency response plans, may indicate the presence of hazardous materials. During pre-emergency planning activities, typical occupancies and locations where hazardous materials are manufactured, transported, stored, used, or disposed of in the community, are identified.
3. *Look for containment system characteristics that indicate hazardous materials:* The hemispherical ends on the container or the protective housing around the fittings on top of a pressure tank car are indications of hazardous materials. An awareness of the various types of hazardous material containment systems is useful.
4. *Look for facility and transportation markings and color that indicate hazardous materials:* The use of color is not consistent nationwide; however, color markings may be consistent in local areas. Various markings indicate the presence of hazardous materials. Some examples are as follows.

DOT identification numbers are four-digit numbers assigned to a specific hazardous material or group of hazardous materials. They are used to cross-reference the name of a material in order to access hazard and response information for that material. In transportation, DOT identification numbers must be displayed on the following:

- Nonbulk packages of hazardous materials (except limited quantities) printed adjacent to the required labels on the package; and
- Bulk packages of hazardous materials, such as cargo tanks, portable tanks, tank cars, and other bulk shipments. On bulk packages, DOT identification numbers can be displayed on an orange panel, in the center of the appropriate placard, or in the center of a placard-sized, square-on-point white display. (See Figure 10-9C.)

On shipping papers, DOT identification numbers are preceded by the prefix "UN" (United Nations) for domestic and international shipments or "NA" (North American) for shipments only within North America.

NFPA 704 markings are diamond-shaped symbols used within facilities to alert people to the type and degree of hazard. (See NFPA 704, *Standard System for the Identification of the Hazards of Materials for Emergency Response.*) They may be found on nonbulk packages; however, they are not used on transport vehicles. The diamond-shaped symbol is divided into four color-coded quadrants. (See Figure 10-9D.) The blue in the left quadrant refers to the health hazard; the red in the top quadrant indicates the flammability hazard; the yellow in the right quadrant refers to the reactivity hazard; and the bottom quadrant carries special information, such as "OX" for oxidizers or "w" for water-reactive materials. Each quadrant contains a number from 0 to 4 that represents the relative degree of hazard (0 is low; 4 is high). (See Figure 10-9E.)

Military markings are used to indicate hazardous materials for military shipments.

FIG. 10-9C.   Methods of display of Department of Transportation (DOT) identification numbers. The rectangular box has an orange background. The diamonds have white backgrounds.

FIG. 10-9D.   Sample NFPA 704 marking.

Special hazard communication markings are used in facility situations to indicate hazardous materials.

Warning labels indicate the presence of hazardous materials using words such as warning, danger, caution, and poison.

Pipeline markers mark the location of pipelines, especially where the pipeline crosses a street.

Container markings, such as the stenciled name of the contents, also may indicate the presence of hazardous materials.

5. *Look for U.S. and Canadian placards and labels:* Placards and labels are diamond-shaped symbols that are required by DOT to communicate the presence of hazardous materials in transportation. Placards, $10^3/_4$-sq in. (273 mm), are required on some bulk packages, such as transport vehicles (e.g., cargo tanks, portable tanks, tank cars, and hopper cars). Labels, 4 sq in. (102 mm) or smaller for cylinders, are found on some non-bulk packages. Placards and labels convey information by their color, symbol, hazard class and division number, and either the hazard class wording or four-digit identification number. DOT Chart provides examples of U.S. and Canadian placards and labels. (See Figure 10-9F.)

Transport vehicles and freight containers could contain up to 1,001 lb (454 kg) of certain hazardous materials without a placard being applied. Limited-quantity, division 6.2 (infectious substance/etiologic agents), and ORM-D shipments do not require placards. Residue placards are no longer authorized in rail transportation except for shipments to and from Canada. To determine load/empty status, use the shipping papers. Square white backgrounds are used behind Division 1.1, Division 1.2, Division 2.3, Division 6.1 Packing Group I Zone A, and Class 7 placards to indicate special handling. The dangerous placard indicates a mixed load of several classes of hazardous materials.

6. *Obtain and review facility documents and shipping papers:* Shipping papers accompany each shipment during transportation, and provide detailed information about the contents of the shipment. These shipping papers are used in the various modes of transportation (highway, rail, water, and air) are listed in Table 10-9C by mode of transportation, title of shipping paper, location of shipping papers, and the person responsible. Shipping papers may indicate the presence of hazardous materials through a variety of required entries (all of which are not always present). The most common entries include the proper shipping name, hazard class or division, identification number, packing group, and the total quantity (by weight, volume, or as otherwise appropriate). Additional entries may include technical and chemical group names, the letters RQ (reportable quantity notation) POISON—INHALATION HAZARD, LIMITED QUANTITY notation, DANGEROUS WHEN WET notation, and shipper contact information. Shipping papers also include telephone number and emergency-response information for the specific commodity.

Facility documents include the material safety data sheet (MSDS). The MSDS is a document that provides information about the composition, physical and chemical properties, health and safety hazards, emergency response, and waste disposal of materials required by OSHA in 29 CFR 1910.1200.

7. *Identify clues (other than occupancy/locations, container shape, markings/color, and placards/labels) that use sight, sound, and odor to indicate hazardous materials:* Sensual indications include chemical odors, dead animals and fish, vapor clouds, flames or smoke, irritation of the skin or eyes, the hiss of escaping gas, or the sound of an explosion.

As soon as the presence of hazardous materials is detected, move to the next task in the analysis process.

Initiating Command and Control Activities

Although not part of the analysis process, command and control activities should be initiated upon detecting the presence of hazardous materials and simultaneously with the initial survey of the situation. These command and control actions are taken: (1) to protect the responder and others during completion of analysis and planning duties, (2) to provide for a command structure at the incident, and (3) to provide for notification of resources required to handle the situation. Specific procedures for initiating command and control activities are

NFPA 704 Marking System

SCALE	HEALTH HAZARD	FLAMMABILITY HAZARD	REACTIVITY HAZARD
4	Materials which on a very short exposure could cause death or major residual injury. Too dangerous to be approached without specialized personal protective equipment.	Materials which will rapidly or completely vaporize at atmospheric pressure and normal ambient temperature or which is readily dispersed in air and which will burn readily.	Materials which in themselves are readily capable of detonation or of explosive decomposition or reaction at normal temperatures and pressures.
3	Materials which on short exposure could cause serious temporary or residual injury. Requires protection from all contact.	Liquids and solids that can be ignited under almost all ambient temperature conditions.	Materials which in themselves are capable of detonation or explosive reaction but require strong initiating source or which must be heated under confinement before initiation or which react explosively with water.
2	Materials which on intense or continued exposure could cause temporary but not chronic incapacitation or possible residual injury. Requires use of personal protective equipment with independent air supply.	Materials that must be moderately heated or exposed to relatively high ambient temperatures before ignition can occur.	Materials which in themselves are normally unstable and readily undergo violent chemical change but do not detonate. Also materials which may react violently with water or which may form potentially explosive mixtures with water.
1	Materials which on exposure would cause irritation but only minor residual injury. Requires use of approved air purifying respirator.	Materials that must be preheated before ignition can occur.	Materials which in themselves are normally stable, but which can become unstable at elevated temperatures and pressures or which may react with water with some release of energy but not violently.
0	Materials which on exposure under fire conditions would offer no hazard beyond that of ordinary combustible materials.	Materials that will not burn.	Materials which in themselves are normally stable, even under fire exposure conditions and which are not reactive with water.

(Source: 1985 edition of NFPA 704, *Fire Protection Guide on Hazardous Materials*.)

FIG. 10-9E. *A brief explanation of the significance of the numbers found on the NFPA 704 marking system.*

TABLE 10-9C. *A List of the Shipping Papers by Mode of Transportation, Title of Shipping Paper, Location, and Responsible Person*

Mode of Transportation	Title of Shipping Paper	Location of Shipping Papers	Responsible Person
Highway	Bill of lading or freight bill	Cab of vehicle	Driver
Rail	Waybill and/or consist	With member of train crew (conductor or engineer)	Conductor
Water	Dangerous cargo manifest	Wheelhouse or pipe-like container or barge	Captain or master
Air	Air bill with shipper's certification for restricted articles	Cockpit (may also be found attached to the outside of packages)	Pilot

found in the local emergency response plan or organization's standard operating procedures.

To initiate command and control activities, take the following steps.

1. *Implement an incident command system.* This step provides organization of the incident scene and may begin prior to arrival on the scene.
2. *Locate personnel at a safe distance from the problem.* With consideration of the terrain and weather conditions, response personnel should stop at a safe location from the incident—uphill, upwind, and upstream. Suggested minimum "safe" distances should be specified in the local emergency response plan or the organization's standard operating procedures. At this point, the analysis process is not complete, and sufficient information is not available to determine safe distances.

3. *Initiate the notification process.* Specific procedures and contacts for notification of required resources should be developed prior to an emergency and listed in the local planning documents.
4. *Control access to the site.* Prevent unauthorized persons from entering the area to reduce the negative outcomes in the incident.

The command and control actions taken depend on both the responder's level of training and the resources available. They should be attempted without risk to the responder.

Surveying the Hazardous Materials Incident

After detecting the presence of hazardous materials in an emergency, and while initiating command and control activities, the next

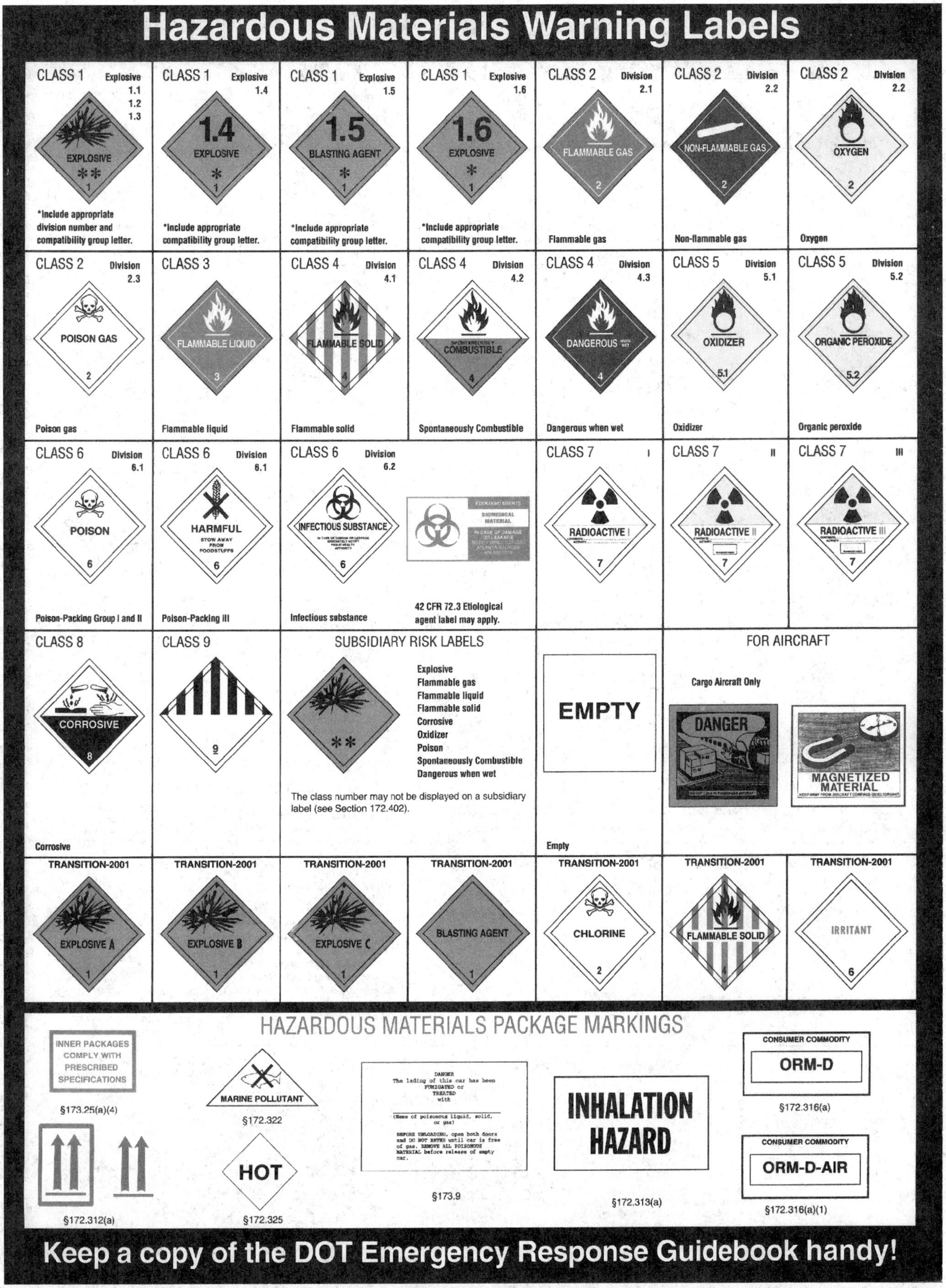

Hazardous Materials Warning Labels

HAZARDOUS MATERIALS PACKAGE MARKINGS

Keep a copy of the DOT Emergency Response Guidebook handy!

FIG. 10-9F.   DOT Chart 10 depicting hazardous materials marking, labeling, and placarding.

Hazardous Materials Warning Placards

CLASS 1

EXPLOSIVES

EXPLOSIVES
*Enter Division Number 1.1, 1.2, or 1.3 and compatibility group letter, when required. Placard any quantity.

CLASS 1

1.4 EXPLOSIVES

EXPLOSIVES 1.4
*Enter compatibility group letter, when required. Placard 454 kg (1,001 lbs) or more.

CLASS 1

1.5 BLASTING AGENTS

EXPLOSIVES 1.5
*Enter compatibility group letter, when required. Placard 454 kg (1,001 lbs) or more.

CLASS 1

1.6 EXPLOSIVES

EXPLOSIVES 1.6
*Enter compatibility group letter, when required. Placard 454 kg (1,001 lbs) or more.

CLASS 2

OXYGEN

OXYGEN
Placard 454 kg (1,001 lbs) or more, gross weight of either compressed gas or refrigerated liquid.

CLASS 2

FLAMMABLE GAS

FLAMMABLE GAS
Placard 454 kg (1,001 lbs) or more.

CLASS 2

NON-FLAMMABLE GAS

NON-FLAMMABLE GAS
Placard 454 kg (1,001 lbs) or more gross weight.

CLASS 2

POISON GAS

POISON GAS
Placard any quantity of Division 2.3 material.

CLASS 3

FLAMMABLE

FLAMMABLE
Placard 454 kg (1,001 lbs) or more.

CLASS 3

GASOLINE

GASOLINE
May be used in the place of FLAMMABLE on a placard displayed on a cargo tank or a portable tank being used to transport gasoline by highway.

CLASS 3

COMBUSTIBLE

COMBUSTIBLE
Placard a combustible liquid when transported in bulk. See §172.504(f)(2)for use of FLAMMABLE placard in place of COMBUSTIBLE placard.

CLASS 3

FUEL OIL

FUEL OIL
May be used in place of COMBUSTIBLE on a placard displayed on a cargo tank or portable tank being used to transport by highway fuel oil not classed as a flammable liquid.

CLASS 4

FLAMMABLE SOLID

FLAMMABLE SOLID
Placard 454 kg (1,001 lbs) or more.

CLASS 4

SPONTANEOUSLY COMBUSTIBLE

SPONTANEOUSLY COMBUSTIBLE
Placard 454 kg (1,001 lbs) or more.

CLASS 4

DANGEROUS WHEN WET

DANGEROUS WHEN WET
Placard any quantity of Division 4.3 material.

CLASS 5

OXIDIZER 5.1

OXIDIZER
Placard 454 kg (1,001 lbs) or more.

CLASS 5

ORGANIC PEROXIDE 5.2

ORGANIC PEROXIDE
Placard 454 kg (1,001 lbs) or more.

CLASS 6

HARMFUL
STOW AWAY FROM FOODSTUFFS

KEEP AWAY FROM FOOD
Placard 454 kg (1,001 lbs) or more.

CLASS 6

POISON

POISON
Placard any quantity of 6.1, PGI, inhalation hazard only. Placard 454 kg (1,001 lbs) or more of PGI or II, other than PGI inhalation hazard.

CLASS 7

RADIOACTIVE 7

RADIOACTIVE
Placard any quantity of packages bearing the RADIOACTIVE III label. Certain low specific activity radioactive materials in "exclusive use" will not bear the label, but RADIOACTIVE placard is required.

CLASS 8

CORROSIVE 8

CORROSIVE
Placard 454 kg (1,001 lbs) or more.

CLASS 9

9

MISCELLANEOUS
Not required for domestic transportation. Placard 454 kg (1,001 lbs) or more gross weight of a material which presents a hazard during transport, but is not included in any other hazard class.

DANGEROUS

DANGEROUS
Placard 454 kg (1,001 lbs) gross weight of two or more categories of hazardous materials listed in Table 2. A freight container, unit load device, motor vehicle, or rail car which contain non-bulk packagings with two or more categories of hazardous materials that require placards specified in Table 2 may be placarded with a DANGEROUS placard instead of the separate placarding specified for each of the materials in Table 2. However, when 2,268 kg (5,000 lbs) or more of one category of material is loaded at one facility, the placard specified in Table 2 must be applied.

SUBSIDIARY RISK PLACARD

CORROSIVE

Class numbers do not appear on subsidiary risk placard.

Required background for placards on rail shipments of certain explosives and poisons. Also required for highway route-controlled quantities of radioactive materials (see §§172.507 and 172.510).

UN or NA Identification Numbers

PLACARDS OR ORANGE PANELS →

1090

Appropriate Placard must be used.

and

MUST BE DISPLAYED ON TANK CARS, CARGO TANKS, PORTABLE TANKS AND OTHER BULK PACKAGINGS

1090

FLAMMABLE 3   1017 2   1993 3

Response begins with identification!

FIG. 10-9F.   Continued.

task is to survey the hazardous material incident. Completion of this task provides an inventory of the containment systems and materials involved, materials released, and surrounding conditions. This incident survey should be conducted from a safe distance, using aided vision, without exposure to the released materials.

The following steps should be completed during the survey of a hazardous material incident.

1. *Identify each containment system by type, identifier, and size.* Containment systems fit into one of three types: (1) nonbulk, (2) bulk, or (3) facility containment systems. (See Table 10-9D.) The containment system identifier (e.g., facility or carrier name and number) is used to differentiate one containment system from another and to allow tracking of that container throughout the incident. The quantity within, or capacity of, the containment system can be obtained from markings on the container or entries on the shipping papers or facility documents. The quantity information will help indicate the magnitude of the problem.

2. *Identify the name, DOT identification number, or placard applied to each hazardous material containment system.* This information provides a means of accessing various sources of hazard and response information.

   For facilities: Sources for identifying the material include pre-emergency planning document, markings and color, contact with the facility manager, and review of appropriate MSDS.

   In transportation: The identity of the hazardous material can be determined from the DOT identification number, commodity stencil, type of placard or label applied, shipping paper entries, and manufacturer, shipper, or consignee contacts using the 24-hr emergency telephone number on the shipping papers. Pipeline markers provide the name of the commodity or at least the name and telephone number of the pipeline company.

3. *Identify leaking containment systems.* Clues indicating leakage include material on the outside of the containment system, taste or smell, presence of vapor clouds, or the operation of a safety relief valve. If possible during the survey, note the form

of the released material (solid, liquid, or gas) and the location of the release.

4. *Identify the surrounding conditions.* Surrounding conditions should be noted when surveying hazardous material incidents. These conditions include topography, land use (including utilities and fiber optic cables), accessibility, weather conditions, bodies of water (including recharging ponds), public exposure potential, and the nature and extent of injuries. If a facility is involved, look for information about floor drains, ventilation ducts, air returns, etc., as appropriate.

For ease of collection, recording, and interpretation of information obtained during the survey of an incident, a hazardous material incident survey form is used. (See Figure 10-9G.)

Finally, the information collected in the incident survey should be reviewed to verify its accuracy. For example, if the shipping paper identifies the material as a gas and a solid is being released, some of the information obtained may not be correct. An understanding of the characteristics of various containment systems will help to verify information on the contents.

Collecting and Interpreting Hazard and Response Information

Once a hazardous material is identified, information about the material's hazards, behavior characteristics, and suggested response options is collected. This information, which may be collected simultaneously with determining the extent of containment system damage, is used to predict the behavior of that material. The information to be collected is divided into six basic groups: (1) material identification information, (2) physical properties, (3) chemical properties, (4) physical hazards, (5) health hazards, and (6) response information. The task of obtaining, recording, and interpreting hazardous material information can be lengthy and rigorous. Various forms are being used to record hazard and response information. (See Figure 10-9H.)

A number of resources are available for this task, including the following.

Printed materials: Printed materials include various response guides and MSDS.

Various *response guides* are available for use in obtaining hazard, behavior, and response information. A listing of the sources of the most common response guides is found in Table 10-9E. As one will note from the X's in Table 10-9F, no one printed material source provides all the information. Therefore, more than one resource may have to be consulted to obtain the needed information. An understanding of the type of information available from each resource is important. These references augment personal knowledge and experience.

Material safety data sheets (MSDS) are available at facilities from the manager or other facility personnel and may be found in transportation. CHEMTREC has more than 250,000 MSDS on file if one is not available locally. MSDS typically provide the following types of information:

1. Physical and chemical characteristics;
2. Physical hazards of the material;
3. Health hazards of the material;
4. Signs and symptoms of exposure;
5. Routes of entry;
6. Permissible exposure limits;
7. Responsible party contact;
8. Precautions for safe handling (including hygiene practices, protective measures, and procedures for clean-up of spills/leaks);

TABLE 10-9D. *A Listing of the Various Types of Nonbulk, Bulk, and Facility Containment Systems[5]*

Transportation		Facility Containment Systems
Nonbulk	Bulk	
Bags	Bulk bags	Buildings
Bottles	Bulk boxes	Machinery
Boxes	Cargo tanks	Open piles (outdoors and indoors)
Carboys	Covered hopper cars	Piping
Cylinders	Freight containers	Reactors (chemical and nuclear)
Drums	Gondolas	Storage bins, cabinets, or shelves
Jerricans	Pneumatic hopper trailers	
Multicell	Portable tanks and bins	
Tanks and storage vessels packages	Protective overpacks for radioactive materials	
Wooden barrels	Tank cars	
	Ton containers	
	Van trailers	

Hazardous Material Incident Survey Form

Location _____

Date ___ - ___ - ___   Time: ___ : ___ ___   Weather Conditions _____

Terrain _____

Closest Populated Buildings _____

Closest Bodies of Water _____

Closest Other Buildings _____

No. Deaths _____   No. Injuries _____   Remedial Actions Taken _____

Containment System			Material Identification			Release		
Container Type	Container ID No.	Qty	DOT ID No.	Placard	Name of Material	Yes	Form	Location
						Y	☐ Solid ☐ Liquid ☐ Gas	
						Y	☐ Solid ☐ Liquid ☐ Gas	
						Y	☐ Solid ☐ Liquid ☐ Gas	
						Y	☐ Solid ☐ Liquid ☐ Gas	
						Y	☐ Solid ☐ Liquid ☐ Gas	
						Y	☐ Solid ☐ Liquid ☐ Gas	

Sketch of scene   N ↑

FIG. 10-9G.   Hazardous material incident survey form. This worksheet is used for recording information collected during the task of surveying a hazardous material incident.[6]

Hazardous Material Data Sheet	Containment System ID.

Material Name _____ DOT ID № _____ STCC № _____
Synonyms _____
Hazard Class _____
NFPA 704 Marking: Health _____ Flammability _____ Reactivity _____ Other _____

Physical Properties

Form	Color	Odor	Chemical Formula	Molecular Wt.
☐ Solid ☐ Liquid ☐ Gas				

Chemical Properties

Actual Temp.	Boiling Point	Melting Point	Vapor Pressure	Expansion Ratio	Specific Gravity	Vapor Density	Soluble?	Degree of Solubility
							☐ Yes	

Physical Hazards

		Actual Temperature	Flash Point	Ignition Temperature
Flammable (heat/fire)	☐ Yes			
Cryogenic (cold)	☐ Yes			
Oxidizer (supports combustion)	☐ Yes	**Actual Concentration**	**Flammable Range**	**Toxic Products of Combustion**
Explosive	☐ Yes			
Reactive	☐ Yes			
To What? _____				

Health Hazards

		Actual Concentration	Non-life Threatening Exposure Limits		
Acute Health Hazards:			TLV-TWA(PEL)	TLV-C	TLV-STEL
Poisonous	☐ Yes				
Corrosive	☐ Yes				
To What? _____		Odor Threshold	Life Threatening Exposure Limits		
Asphyxiation	☐ Yes		IDLH	LC_{50}	LD_{50}
Etiologic	☐ Yes				
Radiation	☐ Yes				
Type: alpha beta gamma					

Route of Entry	☒ = YES	Toxicity Rating	Notes
Inhalation	☐	1 2 3 4 5 6	
Dermal	☐	1 2 3 4 5 6	
Ingestion	☐	1 2 3 4 5 6	

Chronic Health Hazards:
Carcinogen ☐ Yes
Mutagen ☐ Yes
Teratogen ☐ Yes
Aquatic Hazard ☐ Yes

Response Information

Evacuation Distances _____

First Aid _____

_____

Personal Protective Equipment _____

Decontamination _____

Extinguishing Agents _____

Neutralizing Agents _____

FIG. 10-9H.   Hazardous material data sheet. This worksheet is used for recording information obtained during the task of collecting and interpreting hazard and response information.[6]

TABLE 10-9E. *A List of Common Reference Guides for Collecting Hazardous Response Information and Their Sources*[6]

Printed Information	Source
North American Emergency Response Guidebook	U.S. Department of Transportation Research and Special Programs Administration Washington, DC 20590
Fire Protection Guide on Hazardous Materials, National Fire Protection Association (NFPA 325, NFPA 49, and NFPA 491M)	National Fire Protection Association 1 Batterymarch Park Quincy, MA 02269
Emergency Handling of Hazardous Materials in Surface Transportation (EH) *Emergency Actions Guides* (EAG)	Association of American Railroads 50 "F" Street, NW Washington, DC 20001
U.S. Coast Guard Chris Manual Series (CHRIS) *Hazardous Chemical Data* CG-446-2	Superintendent of Documents U.S. Government Printing Office Washington, DC 20402
Farm Chemical Handbook (FCH)	Meister Publishing Company 37841 Euclid Avenue Willoughby, OH 44094
NIOSH/OSHA References (NIOSH) *Pocket Guide to Chemical Hazards*	Superintendent of Documents U.S. Government Printing Office Washington, DC 20402
Threshold Limit Values for Chemical Substances and Physical Agents and Biological Exposure Indices	American Conference of Governmental Industrial Hygienists Kemper Woods Center 1330 Kemper Meadow Drive Cincinnati, OH 45240

Abbreviations Used for References		
Primary Resources		Secondary Resources
DOT *North American Emergency Response Guidebook* **NIOSH** *Pocket Guide to Chemical Hazards*	**EAG** **EH**	Association of American Railroads' *Emergency Action Guides* Association of American Railroads' *Emergency Handling of Hazardous Materials in Surface Transportation*
	NFPA	National Fire Protection Association's *Fire Protection Guide to Hazardous Materials*
	CHRIS	U.S. Coast Guard's *Hazardous Chemical Data*
	ACGIH	American Conference of Governmental Industrial Hygienists' *Threshold Limit Values and Biological Exposure Indices*
	FCH	*Farm Chemicals Handbook*
	MSDS	Material safety data sheet

9. Applicable control measures including personal protective equipment; and
10. Emergency and first aid procedures.

Technical resources: A number of technical resources are available, including:

1. *CHEMTREC*, the Chemical Transportation Emergency Center, is part of the Chemical Manufacturers' Association (CMA). CHEMTREC operates around the clock to receive toll-free calls from the United States and Canada through a wide-area telephone service number (800) 424-9300. If in Washington, DC, or Alaska, call (202) 483-7617. CHEMTREC can usually provide immediate hazard information from the CHEMTREC files. CHEMTREC will also contact the shipper; therefore, be prepared to provide CHEMTREC with the following information when available:

   • Caller's name and phone number where caller can be reached;
   • Name of the carrier, shipper, manufacturer, or facility operator;
   • Nature, location, and time of the incident;
   • Name of the material (or other identifying information); and
   • Container type, rail car or truck number, vessel name, or other identifying information such as a rail car's reporting marks (initials) and number.

   CHEMTREC can also provide a teleconferencing bridge, connecting technical specialists with those needing technical information.

2. *National Response Center* (NRC), operated by the U.S. Coast Guard, receives reports required from spillers of hazardous substances. The NRC also acts as the notification, communications, technical assistance, and coordination center for the National Response Team. After receiving a report, the NRC will immediately relay the information to the responsible, predesignated, Federal On-Scene Coordinator and other resources of the Federal Government. The NRC can be contacted by calling (800) 424-8802.

TABLE 10-9F.   *A Cross-Reference Indicating What Information Is Available in the Common Reference Guides*
An "X" in the intersection of the resource and the required information indicates that the information is typically found in that publication.[6]

Required Information	Primary Resources		Secondary Resources						
	DOT	NIOSH	EAG	EH	NFPA	CHRIS	ACGIH	FCH	MSDS
DOT ID No.	X	X	X	X		X			X
STCC No.			X	X					
Synonyms		X	X		X	X		X	X
Hazard class			X	X	X	X		X	X
NFPA 704 marking			X		X	X			
Form		X	X	X	X	X		X	X
Color		X	X	X	X	X		X	X
Odor		X	X	X	X	X		X	X
Chemical formula		X	X		X	X		X	X
Molecular weight		X	X			X			X
Boiling point		X	X		X	X		X	X
Melting point		X	X			X		X	X
Vapor pressure		X	X			X			X
Expansion ratio									
Specific gravity			X	X	X	X			X
Vapor density			X	X	X	X			X
Soluble		X	X	X	X	X		X	X
Degree of solubility		X	X	X	X	X		X	X
Flash point		X	X	X	X	X		X	X
Ignition temperature			X		X	X			X
Flammable range		X	X		X	X			X
Toxic products of combustion			X			X		X	X
Reactive to what	X	X	X	X		X			X
Odor threshold			X			X			X
IDLH		X				X			X
LD$_{50}$						X		X	X
LC$_{50}$						X		X	X
TLV-C		X				X	X		X
TLV-STEL			X			X	X		X
TLV-TWA		X	X				X		X
PEL		X							X
Route of entry	X	X	X	X	X	X		X	X
Toxicity rating					X	X			
Evacuation distances	X		X	X		X			X
First aid	X	X	X	X		X		X	X
Personal protective equip.	X	X	X	X	X	X		X	X
Decontamination	X	X	X					X	X
Extinguishing agents	X		X	X	X	X		X	X
Neutralizing agents			X	X		X			X

3. *Local Poison Control Center.*   For consumer products, comprehensive hazard information is available through Regional Poison Control Centers. The local planning documents should contain the telephone number.

4. *Manufacturer's Technical Medical Staff.*   For bulk industrial chemicals, assistance may be available from the technical medical staff of the manufacturer. CHEMTREC will make the contact.

5. *Agency for Toxic Substances and Disease Registry (ATSDR).*   For chemical mixtures or wastes, when a second opinion is required or when the chemical is unknown, a preferred source of information is the ATSDR, a part of the Center for Disease Control. ATSDR can be reached by calling (404) 639-0615.

6. *Department of Energy.*   For radioactive materials, contact the Department of Energy by calling (202) 586-8100.

7. *Hazardous Material Emergency Response Teams.*   Assistance from these sources ranges from immediate telephone advice to the dispatching of emergency teams to actually assist in field operations. Some common examples include the following:

• Pesticide Safety Team Network

• CHLOREP Teams

• U.S. Coast Guard Strike Teams

• Carrier Response Teams

• U.S. Environmental Protection Agency Regional Response and Technical Assistance Teams

• Association of American Railroads, Bureau of Explosives

On contact with these resources, provide as much information as possible; especially the full name, form, use, and concentration of the chemical, when available. An occupational-health physician may be required to obtain or interpret information from any of these sources.

Assistance in contacting many of these technical resources may be obtained from CHEMTREC or the NRC, if not listed in the local planning documents.

Assessing the Extent of Containment System Damage

Information on the types and extent of damage to a containment system is used in predicting the likely behavior of the containment system and its contents. This task can be initiated simultaneously with collecting information about the materials involved. Provide appropriate personal protective equipment for the response personnel if they could be exposed to released materials. Contact technical personnel with knowledge of the type of containment systems involved for assistance.

The following steps will help determine the extent of damage to a containment system.

1. *Identify containment systems that are damaged.* Indications of damage include abnormal orientation or position of the containment system, damage to other containment systems in the area, or a change in shape of the containment system.
2. *Determine the construction of the damaged containment system.* Before inspecting the damaged containment system, review the containment system's basic design, construction, and secondary containment features.
3. *Determine the type and location of closures on the containment system.* Review the closures on the containment system by name and purpose. Use of standard terminology will help describe the problem to the technical resources.
4. *Determine the type and location of any damage to the container or its closures.* Dents, cracks, gouges, scores, corrosion, and punctures should be noted. Using appropriate personal protective equipment when necessary, the containment system must be evaluated for signs of failure or potential failure. If the containment system has not already failed, it may have been weakened and result in failure.
5. *Determine the extent of the damage to the container or its closures.* Information on the extent of damage (e.g., the radius of curvature for dents or the depth of scores and gouges on pressure tank cars) should be determined. It may be necessary to determine the pressure in a bulk packaging or facility containment system using either a pressure gage or the temperature of the contents.
6. *Estimate the release rate.*

Figure 10-9I provides an example of a form used for determining the extent of damage to tank cars.

Predicting Likely Behavior

After collecting information on the characteristics of the material and the extent of damage to the containment system, the next task is to predict their likely behavior. This prediction, or "mental movie," is used throughout the response activity as a reference point. A mental movie is made for each containment system and hazardous material involved and is based on the worst possible case. As additional information is obtained, the mental movie is modified. The steps for predicting the behavior of the containment system and its contents include the following.

1. *Identify the types of stress or potential stress on a containment system and/or its contents.* Stress is the applied force or system of forces that tend to strain or deform a body (either the containment system or the contents). The types of stress include: thermal, mechanical, chemical, irradiation, and etiologic. Thermal, mechanical, and chemical stresses can typically be identified visually, while irradiation and etiologic are more difficult to identify. Considerations include: type and characteristics of the containment system, type of stress, amount of stress applied, intensity of stress, and duration of stress.
2. *Predict the way in which the containment system will breach.* A breach is an opening (failure) caused by damage to the containment system through which matter and/or energy can or does escape. The types of breach include: disintegration, runaway linear cracking, closures that open up, punctures, and splits or tears. Considerations include: type and duration of the stress being applied, behavior of the containment system under the stresses applied, behavior of the contents, location of the applied stresses, force of opening of the containment system, size of breach, and speed of the breach event. Even if the containment system has already breached, this prediction step must be considered in light of a further breach.
3. *Predict the way in which the contents will be released.* Release is the escape of the contents through a breach in a containment system. The types of release include: *detonation*—the disintegration of the containment system and/or detonation of the contents; *violent rupture*—the runaway cracking of the containment system or the rapid expansion of the contents that burst the containment system abruptly; *rapid relief*—pressurized flow of the contents through closures, splits or tears, and punctures; and *spill or leak*—gradual flow through closures, splits or tears, and punctures. Considerations include: type of breach, location of breach, quantity of contents, flow characteristics of contents, release rate, length of release, and weather conditions.
4. *Predict the dispersion pattern that will create the engulfed area.* Dispersion is the scattering of the matter and/or energy (from the point of release) to form an engulfed area. The engulfed area is the actual or potential area of exposure from a hazardous material or its containment system. The types of dispersion patterns include: hemisphere, cloud, plume, cone, stream, pool, and irregular. Considerations include: type of release, form of the released contents, weather conditions, topography, release rate, solubility of material, secondary reactions, melting point, freezing point, boiling point, vapor density, specific gravity, secondary containment, vapor pressure, and expansion ratio. For fixed facilities, consider tank spacing, impoundment and diking, tank venting and flaring systems, transfer operations, monitoring and detection systems, and fire protection systems.
5. *Predict the length of contact that exposures will have with the released contents.* Contact means to be exposed to something undesirable or injurious. Type of contact is considered in relation to time: short term refers to minutes and hours; medium term is associated with days, weeks, and months; and long term is in terms of years and generations. Considerations include: form of the material upon release, dispersion pattern, weather conditions, topography, distance to exposure, barriers or shielding, speed of dispersing materials, and quantity released.
6. *Predict the hazards that will cause harm.* Harm is the injury or damage caused by being exposed to the hazards of the released contents. Types of hazards include: physical hazards including thermal and mechanical effects; and health hazards including chemical (poisonous, corrosive, and asphyxiation), radiation, and etiologic effects. Considerations include: hazards associated with the released contents, concentrations of released

Tank Car Damage Assessment	Car Initials & Number

Tank Car Characteristics

Type of Car: ☐ Non-pressure ☐ Pressure ☐ Cryogenic
☐ Other _____

Features	Y	N
Jacketed		
Insulated		
Thermal Protection		
Linings		
Claddings		
Heater Coils		

Specification Nº: | **Shell Capacity:**

Year Constructed: | **Tank Test Pressure:**

Underframe: ☐ Continuous ☐ Stub sill

Thermal Protection: ☐ Jacketed ☐ Sprayed-on

Construction Material: Type/Grade _____ Thickness _____

Stress: ☐ Thermal ☐ Mechanical ☐ Chemical ☐ None

Jacket, Tank, and Head Damage

Indicate location and severity of damage (punctures, cracks, scores, gouges, wheel burns, dents, rail burns, underframe, and leaks) on the appropriate diagram(s).

Fitting Damage

Fitting	Damage	Description
☐ Liquid Valve	☐ Yes ☐ Leaking	
☐ Vapor Valve	☐ Yes ☐ Leaking	
☐ Air Valve	☐ Yes ☐ Leaking	
☐ Bottom Outlet Type: _____	☐ Yes ☐ Leaking	
☐ Safety Relief Type: _____	☐ Yes ☐ Leaking	
☐ Vacuum Relief	☐ Yes ☐ Leaking	
☐ Gauging Device Type: _____	☐ Yes ☐ Leaking	
☐ Manway Cover	☐ Yes ☐ Leaking	
☐ Fill Hole	☐ Yes ☐ Leaking	
☐ Sample Line	☐ Yes ☐ Leaking	
☐ Thermometer Well	☐ Yes ☐ Leaking	
☐ Washout	☐ Yes ☐ Leaking	
☐ Sump	☐ Yes ☐ Leaking	
☐ Other Type: _____	☐ Yes ☐ Leaking	

FIG. 10-9I. Tank car damage assessment. This worksheet is used for recording results of damage assessment for rail tank cars. (Courtesy of Union Pacific Railroad)

materials, duration of contact, and how often the exposures will come into contact with the released materials.

To assist in predicting the behavior of the containment system and its contents, a behavior model form is suggested. (See Figure 10-9J.)

If unable to make all the predictions necessary to analyze the incident, seek help from the following resources: manufacturer, shipper and consignee, carrier, and federal, state, and local agencies involved with hazardous materials.

A number of response guides indicate the effects of mixing various chemicals, including NFPA 491M, *Manual of Hazardous Chemical Reactions.*

Estimating Outcomes

This final task in the analysis process results in an indication of the magnitude of the problem in terms of outcomes within the engulfed area. The engulfed area is the actual or potential area of exposure from a hazardous material. The following steps should be taken to estimate outcomes in the engulfed area.

1. *Determine the dimensions of an engulfed area.* Using the Table of Initial Isolation and Protective Action Distances in the *North American Emergency Response Guidebook,* planning documents, or dispersion modeling programs, identify the potential size of the dispersion pattern. (See Table 10-9G.) Considerations include wind speed and direction and topography.
2. *Estimate the number of exposures within the engulfed area.* Exposures include people, property, the environment, and critical systems. Considerations include: direction of wind, time of day, day of the week, setting (rural or urban), occupancies in area (schools, businesses, industry, transportation), bodies of water, sewer systems, and essential services.
3. *Determine the concentrations of a released hazardous material within the engulfed area.* Resources, such as monitoring equipment, computer dispersion modeling systems, toxic chemical release reporting forms, and a programmable calculator with equations, are available for this step. Information about odor may provide an insight on the level of concentration, but not without risk.

In some cases, monitoring equipment may be used to identify or classify unknown materials, verify the identity of hazardous materials, or determine the concentration of the hazardous materials. Monitoring equipment includes, but is not limited to the following:

- Colorimetric tubes
- CO meter
- Combustible-gas meter
- Organic-vapor analyzer
- Oxygen meter
- Passive dosimeter
- Personal air-monitoring equipment
- Photoionization detectors
- pH papers and strips
- Radiation-detection instruments

Where monitoring equipment is available, personnel should be able to select the appropriate monitoring equipment considering its limitations, perform field testing procedures, operate the equipment to collect data, and record and interpret the results. If monitoring equipment is not provided, the potential sources of monitoring equipment and personnel to operate the equipment should be identified as part of the pre-emergency planning activities.

4. *Predict the extent of physical, health, and safety hazards within the engulfed area under the current conditions.* Predicting health hazards requires an understanding of the various exposure limits, including: (1) immediately dangerous to life and health value (IDLH), (2) lethal concentrations (LC_{50}), (3) lethal dose (LD_{50}), (4) permissible exposure limit (PEL), (5) threshold limit value-ceiling (TLV-C), (6) threshold limit value-short-term exposure limit (TLV-STEL), and (7) threshold limit value-time weighted average (TLV-TWA).
5. *Predict the areas of potential harm within the engulfed area.* Considerations address likely concentrations, lethal concentrations, length of contact, frequency, and intensity of harm.
6. *Estimate the outcomes within an engulfed area.* (See Figure 10-9K.) Considerations include number of exposures in the engulfed area, age and condition of the exposures, anticipated reaction of exposures on contact with material, and weather conditions.

The purpose of analyzing the hazardous material problem is to determine the likely magnitude of the problem in terms of outcomes. While time consuming, this process is an essential part of managing a hazardous material incident. As the incident conditions change, analysis information must be reviewed and updated. When decisions regarding protective actions and control measures are made before completing the analysis process, the risk to the responder is greater.

PLANNING THE RESPONSE

Overview

Based on the magnitude of the problem, appropriate preventive measures and control actions are identified. Control actions include containment, confinement, and extinguishment. The potential impact (benefits and losses) to people, property, equipment, and the environment of each decision is considered.

The result of the planning process is the direction that the response effort will take to influence the sequence of events in the emergency and favorably change the outcome. Response planning is a process for identifying and evaluating response objectives (strategy) and response options (tactics). This process considers the available resources (e.g., personnel, personal protective equipment, and other tools and equipment). The planned response should be consistent with the local emergency response plan and the organization's standard operating procedures.

The planning process begins as part of the pre-emergency response planning efforts prior to the incident and continues on the scene of the incident, again from a safe location. Federal, state, and local agencies, industry, and carrier personnel may be called upon to help.

The process of response planning is based on the following tasks.

1. Determine the response objectives.
2. Determine the available response options that could favorably change the outcomes.
3. Identify the personal protective equipment for the response options.
4. Identify an appropriate decontamination process for each response option.
5. Select the response options within the response community's capabilities that will most favorably change the outcomes.
6. Develop a plan of action including safety considerations.

Behavior Model	Containment System ID.

Material: _____   Quantity: _____

Form of Released Material: ☐ Gas   ☐ Liquid   ☐ Solid

Containment System Type: _____

Weather Conditions: Ambient Temperature: _____ ° Wind Speed: _____ m.p.h.   Rain ☐   Snow ☐

Terrain: _____

Event Sequence

Stress	Breach	Release	Engulf	Contact	Harm

Prediction Steps

Identify the Type of Stress	Predict the Type of Breach	Predict the Type of Release	Predict the Dispersion Pattern	Predict the Length of Exposure	Predict the Hazard Causing the Harm

Behavior Options

Thermal	Disintegration		Hemisphere		Physical
	Runaway linear cracking	Detonation	Cloud		*Thermal*
			Plume		*Mechanical*
Mechanical	Closures open up	Violent rupture	Cone	Short term	
		Rapid relief	Stream	Medium term	Health
Chemical	Punctures	Spill or leak	Pool	Long term	*Poisonous*
					Corrosive
	Splits or tears		Irregular		*Asphyxiation*
					Radiation
					Etiologic

(Adapted from Ludwig Benner's *Hazardous Materials Emergencies*, 1978.)

FIG. 10-9J.   *Behavior model. This worksheet is used for predicting the behavior of a containment system and its contents in a hazardous material incident.*[7]

TABLE 10-9G. *Sample of Table of Initial Isolation and Protective Action Distances from the North American 1996 Emergency Response Guidebook*[8]

ID No.	Name of Material	Small Spills (From a small package or small leak from a large package.)						Large Spills (From a large package or from many small packages.)					
		First ISOLATE in All Directions		Then, PROTECT persons Downwind during—				First, ISOLATE in all Directions		Then, PROTECT persons Downwind during—			
				Day		Night				Day		Night	
		m	(ft)	Km	(mi)	Km	(mi)	m	(ft)	Km	(mi)	Km	(mi)
1005 1005	Ammonia, anhydrous Ammonia, anhydrous, liquefied	30 m	(100 ft)	0.2 km	(0.1 m)	0.3 km	(0.2 mi)	95 m	(300 ft)	0.3 km	(0.2 m)	0.8 km	(0.5 mi)
1005	Ammonia solution with more than 50% ammonia	30 m	(100 ft)	0.2 km	(0.1 m)	0.2 km	(0.1 mi)	60 m	(200 ft)	0.2 km	(0.1 m)	0.3 km	(0.2 mi)
1005 1005	Anhydrous ammonia Ammonia, anhydrous, liquefied	30 m	(100 ft)	0.2 km	(0.1 m)	0.3 km	(0.2 mi)	95 m	(300 ft)	0.3 km	(0.2 m)	0.8 km	(0.5 mi)
1008 1008	Boron trifluoride Boron trifluoridecom- pressed	60 m	(200 ft)	0.2 km	(0.1 m)	0.36 km	(0.4 mi)	185 m	(600 ft)	0.6 km 0.8	(0.4 m)	2.4 km	(1.5 mi)
1016 1016	Carbon monoxide Carbon monoxide, compressed	30 m	(100 ft)	0.2 km	(0.1 m)	0.2 km	(0.1 mi)	95 m	(300 ft)	0.2 km	(0.1 m)	0.6 km	(0.4 mi)
1017	Chlorine	60 m	(200 ft)	0.3 km	(0.1 m)	0.8 km	(0.5 mi)	185 m	(600 ft)	0.8 km	(0.5 m)	3.1 km	(1.9 m)
1023 1023	Coal gas Coal gas, compressed	30 m	(100 ft)	0.2 km	(0.1 m)	0.2 km	(0.1 mi)	30 m	(100 ft)	0.3 km	(0.2 m)	0.8 km	(0.5 mi)
1026 1026 1026	Cyanogen Cyanogen, liquefied Cyanogen gas	60 m	(200 ft)	0.3 km	(0.1 m)	1.0 km	(0.6 mi)	215 m	(700 ft)	0.8 km	(0.5 m)	3.5 km	(2.2 m)
1040 1040	Ethylene oxide Ethylene oxide with nitrogen	60 m	(200 ft)	0.2 km	(0.1 m)	0.3 km	(0.2 mi)	125 m	(400 ft)	0.3 km	(0.2 m)	1.0 km	(0.6 mi)
1045 1045	Fluorine Fluorine, compressed	60 m	(200 ft)	0.2 km	(0.1 m)	0.8 km	(0.5 mi)	185 m	(600 ft)	0.6 km	(0.4 m)	2.7 km	(1.7 m)
1048	Hydrogen bromide, anhydrous	60 m	(200 ft)	0.2 km	(0.1 m)	0.3 km	(0.2 mi)	125 m	(400 ft)	0.3 km	(0.2 m)	1.1 km	(0.7 mi)
1050	Hydrogen chloride, anhydrous	60 m	(200 ft)	0.2 km	(0.1 m)	0.5 km	(0.3 mi)	155 m	(500 ft)	0.5 km	(0.3 m)	1.8 km	(1.1 mi)

For SI units: 1 ft = 0.305 m; 1 mi = 1.609 Km.

Determining the Response Objectives (Strategy)

The first task in planning the response is to determine the response objectives (strategy) based on the estimated outcomes. The response objectives, based on the stage of the incident, are the strategic goals for stopping the event now occurring or keeping future events from occurring. Two basic principles apply to these decisions.

1. One cannot influence events that have already happened or change the outcomes of those events; and

2. The earlier that the event sequence can be interrupted, the more acceptable the loss. (See Figure 10-9L.)

The following steps should be taken when determining response objectives.

1. *Estimate the exposures that could be saved.* The level of response and the acceptable risk associated with a response is based on the exposures that can be saved. The number of exposures that could be saved is based on the estimated outcomes minus the exposures already lost. (See Figure 10-9K.)

Exposures	Estimated Exposed	Estimated Type of Harm	Estimated Outcomes	Amount Already Lost	Amount That Could Be Saved
People	#	Deaths	#		
		Injuries	#		
Property	$	Damage	$		
Environment	$	Damage	$		

FIG. 10-9K. *Worksheet for loss estimate and determining the amount that could be saved. This worksheet is used for estimating both the outcomes in an emergency and the amount that could be saved.*

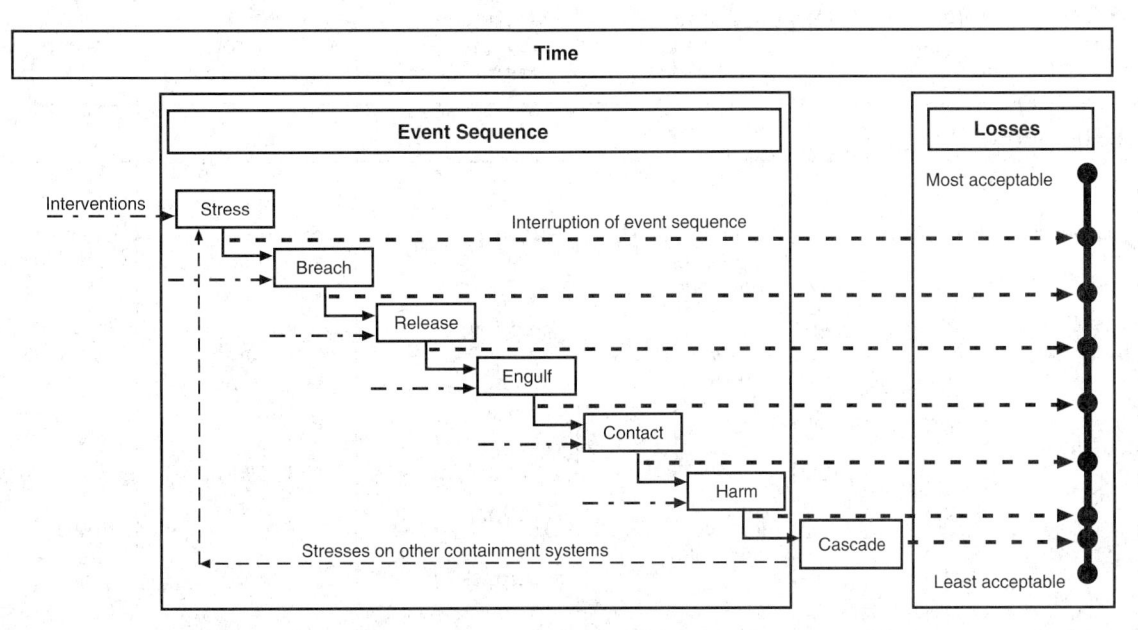

FIG. 10-9L. *A chart depicting the relationship of the event sequence and the losses.*

2. *Determine the response objectives.* The response objectives, based on the stage of the incident, are the strategic goals for stopping the event now occurring or keeping future events from occurring. Decisions should focus on changing the actions of the stressors, the containment system, and the hazardous material.

Determining Response Option (Tactics)

Two types of response options are available. Corrective actions are taken to resolve the immediate problem, while preventive actions are taken to prevent the immediate problem from escalating. These actions are intended to keep losses to a minimum.

1. *Determine the potential response options available by response objective.* Response options are the tactical activities for stopping the event now occurring or keeping a predicted event from occurring. Response options are associated with the specific events in the emergency. (See Figure 10-9M.) An understanding of the procedures, equipment, and safety precautions for each technique is critical to this decision. Response options include one of the following types: defensive, offensive, and nonintervention. Control options involve either containment or confinement options. The responder must determine which type is

appropriate for the incident based on the analysis of the problem.

2. *Estimate how each response option will affect the outcomes.* Before a response option is selected, the effect of that response option or combination of response options on the sequence of events and ultimately the outcomes should be reviewed. (See Figure 10-9N.)

The response options must be based on their expected effect on the outcomes.

Identifying Appropriate Personal Protective Equipment (PPE)

The purpose of identifying appropriate personal protective equipment for each response option is to determine the equipment required to implement the various response options identified. Again, pre-emergency planning cannot be overemphasized in the selection of personal protective equipment.

The following PPE-related definitions are used for the purpose of this chapter.

Personal protective equipment is protective clothing and respiratory protection used to shield or isolate a person from the chemical

Response Objective Analysis Form	Containment
	Material

Event Sequence

Stress	Breach	Release	Engulf	Contact	Harm

Response Objectives

Change Applied Stresses	Change Breach Size	Change Quantity Release	Change Size of Danger Zone	Change Exposures Contacted	Change Severity of Harm

Sample Response Options

Move stressor Move stressed system Shield stressed system	Chill contents Limit stress levels Activate venting devices Mechanical repair	Change container position Minimize pressure differential Cap off breach Remove contents	Barriers Dikes and dams Adsorbents Absorbents Diluents Reactants Overpack	Provide sheltering Begin evacuation Personal protective equipment	Rinse off contaminant Increase distance from source Provide shielding Provide prompt medical attention

(Adapted from Ludwig Benner's *Hazardous Materials Emergencies,* 1978.)

FIG. 10-9M. *Response objective analysis worksheet. This worksheet is used for identifying response options in a hazardous material incident by response objective.*[7]

and biological, physical, or thermal hazards that may be encountered at a hazardous material incident. No single combination of personal protective equipment is capable of protecting against all hazards. Personal protective equipment should be used in conjunction with other protective methods, such as medical monitoring and environmental surveillance. For any given situation, personal protective equipment should be selected to provide an adequate level of protection. Both over-protection and under-protection can be hazardous and should be avoided.

Protective clothing includes those pieces of apparel that are designed to protect the wearer from dermal (skin and eye) contact with the hazardous material. Protective clothing is divided into the following types:

Worksheet for estimating the effect of each response option

Exposures/Harm	Amount that could be saved	Amount that could be saved with option 1	Amount that could be saved with option 2	Amount that could be saved with option 3	Amount that could be saved with option 4	Amount that could be saved with option 5	Amount that could be saved with option 6
People Death	#	#	#	#	#	#	#
People Injuries	#	#	#	#	#	#	#
Property Damage	$	$	$	$	$	$	$
Environ. Damage	$	$	$	$	$	$	$

FIG. 10-9N.   Worksheet for estimating the effect of each response option. This worksheet is used for estimating the effect of each response option on the outcomes of the incident.

1. Structural fire-fighting clothing (NFPA 1971)
2. Vapor protective clothing (NFPA 1991)
3. Liquid splash protective clothing (NFPA 1992)
4. High-temperature clothing

(See bibliography for NFPA titles.)

Respiratory protection is the equipment designed to protect the wearer from inhaling the hazardous material. Respiratory protection is divided into three types.

1. Air-purifying respirators
2. Supplied air respirators (air-line respirators)
3. Positive-pressure self-contained breathing apparatus

Breakthrough time is the time it takes for the hazardous material to move through the protective material.

Penetration is the movement of a chemical through zippers, stitched seams, or imperfections, such as pinholes, in a protective equipment material.

Permeation is the process by which the chemical dissolves in and moves through a protective clothing material on a molecular level.

Degradation is the physical decomposition of the material due to exposure to chemicals, use, or ambient conditions, such as sunlight.

The identification of appropriate personal protective equipment requires completion of the following steps:

1. *Determine the name, concentration, and hazards of each of the materials involved.* This information should have been developed during the process of analyzing the hazardous material problem.
2. *Predict the types of exposure associated with each response option.* The types of exposure include:

   • Immersion (e.g., standing in or handling the chemical),

   • Splash (e.g., splash of a liquid from a drum or decontaminating people or equipment coming in direct contact), and

   • Airborne (e.g., exposure to airborne chemicals as a result of sampling).

3. *Determine the level of personal protective equipment required.* This step provides a list of the required pieces of personal protective equipment, including suits, gloves, and boots. If the chief threat is through liquid contact, the protective clothing should resist chemical degradation and penetration. For

a threat from vapor contact, the protective clothing may have to be airtight and impermeable. The minimum requirement for respiratory protection, set by OSHA in 29 CFR 1910.120, for emergency conditions at hazardous material incidents (until concentrations have been determined) is positive-pressure self-contained breathing apparatus.

The U.S. Environmental Protection Agency identifies four levels of protection:

Level A is the highest level of protection and should be worn when respiratory, skin, eye, and mucous membrane protection is needed.

Level B is the second highest level of protection and should be worn when a lessor level of skin and eye protection is needed. It is the minimum level recommended for initial site entries until the hazards have been determined by monitoring, sampling, or other reliable methods of analysis.

Level C is the third highest level of protection and may be worn only when the type of airborne substance is known, concentration measured, criteria for using air purifying respirators met, and skin and eye exposure is unlikely. Periodic monitoring of the air must be performed while working in level C.

Level D is the lowest level of protection and is primarily a work uniform. It should not be worn on any site where respiratory or skin hazards exist.

Responders must understand the advantages, limitations, and use of each of the levels of protection at hazardous material incidents. If the appropriate level of protection is not available, arrange to obtain it from other resources.

4. *Identify chemically compatible materials for the chemicals involved.* Look for breakthrough times of at least 2 hr. A breakthrough time of 2 hr represents the normal physical limits of a responder's capacity for effective work in protective equipment. Resources for identifying chemically compatible materials include *Guidelines for the Selection of Chemical Protective Clothing*[9] or the *Quick Selection Guide to Chemical Protective Clothing,*[10] various emergency response guides, and chemical compatability charts provided by the manufacturers of chemical protective clothing.

Figure 10-9O provides a worksheet for recording chemically compatible materials identified.

5. *Match the material requirements with personal protective equipment provided or available.* Using chemical-compatibility

Chemical	Chemical concentration	Chemically compatible clothing material required		
		>8 Hr	>2 Hr	$> \frac{1}{2}$ Hr

FIG. 10-9O. *Chemically compatible materials identification worksheet. This worksheet is for recording the chemically compatible protective clothing materials by hazardous material.*

charts provided with the protective equipment available, identify the available equipment that provides the required protection. Multiple materials can be layered to provide the appropriate level of protection.

6. *Identify the appropriate personal protective equipment needed for each response option.* Keep in mind the requirement for the buddy system and backup personnel. Considerations include availability of sufficient equipment in the appropriate sizes and the potential of multiple layers providing required protection. The use of personal protective equipment itself can create significant responder hazards, such as heat stress, physical and psychological stress, as well as impaired vision, mobility, and communication.

Identifying Appropriate Decontamination Procedures

With the potential that personnel, personal protective equipment, apparatus, and tools and equipment may become contaminated in a hazardous material incident, a decontamination (contamination reduction) process is established.

Decontamination (contamination reduction): This is the act of removing or neutralizing contaminants from equipment or personnel to preclude the occurrence of foreseeable adverse health effects outside the engulfed area. Possible decontamination options include absorption, adsorption, chemical degradation, dilution, neutralization, or solidification.

Decontamination procedures should be tailored to the specific hazards of the incident. They may vary in complexity and number of steps, depending on the level of hazard and the level of exposure to the hazard. The steps to be taken to select an appropriate decontamination (contamination reduction) process include the following.

1. *Identify what needs to be decontaminated.* This includes injured persons, response personnel, personal protective equipment, other equipment, and emergency-care facilities.

2. *Determine the contaminant and the type and amount of contamination.* Considerations include whether the material is organic or inorganic, in the solid, liquid, or gas form, a dermal or inhalation hazard, a local or systemic material, and whether the contaminant is life threatening, injurious, or irritating. Another consideration is whether physical trauma is involved.

3. *Seek appropriate advice.* Immediate decontamination guidelines are found in the *North American Emergency Response Guidebook*[8] (under first aid on the numbered guide pages), the NIOSH *Pocket Guide to Chemical Hazards,*[11] or local planning documents. Followup guidance can be obtained from local poison control centers, the shipper or manufacturer through CHEMTREC, or the Agency for Toxic Substances and Disease Registry.

4. *Identify methods for isolating and securing and then disposing of clothing and equipment that cannot be decontaminated.* The local emergency response plan or the organization's standard operating procedures provide guidelines on decontamination (contamination reduction) procedures. Other resources should be identified in pre-emergency planning activities. Followup resources include the Agency for Toxic Substances and Disease Registry, poison control centers, and the shipper or manufacturer through CHEMTREC.

Selecting Response Options

The selection of response options to change the outcomes favorably must be consistent with the local emergency response plan and the organization's standard operating procedures and also within the capability of available personnel, personal protective equipment, and control equipment. The steps to be taken in selecting the response options include the following.

1. *Determine the required resources to implement each response option.* Required resources include the time to implement the option, availability of sufficient personnel with the appropriate training, and appropriate personal protective equipment, materi-

als for decontamination (contamination reduction) activities, and control equipment.

2. *Inventory the resources available and identify those resources available.* These resources should be readily accessible.

3. *Determine how to obtain the needed resources.* A review of the planning documents should provide these answers.

4. *Select the response option(s) consistent with the available resources.* Evaluate the potential reaction of the hazardous material to the control action and the ability to contain the side effects of the response option. Surrounding conditions, including topography and weather conditions, and response times can also affect the decision to select the response options.

Developing a Plan of Action

After selecting the response option for a hazardous material incident, a plan of action including safety and health considerations should be developed. This plan of action describes the response objectives and options and the personnel and equipment required to accomplish the objectives. The plan provides a permanent record of the decisions made at the incident. An organization's standard operating procedures provide the basis of this plan of action. Input from all segments of the response community is considered in developing the plan. Based on the specific incident conditions, the standard operating procedures are modified without having to write an entire plan for each incident.

A plan of action also outlines the safety and health procedures to protect responders and the public from the potential hazards at an incident. These procedures should address incident management, communications protocol (both internal and external), control zones for incident security, personal protective equipment use, decontamination procedures, and documentation. They also include designation of a safety sector and a safety officer, emergency medical care procedures, environmental monitoring, emergency procedures, and personnel monitoring.

Components for a typical plan of action include the following:

1. Site description
2. Entry objectives
3. On-scene organization and coordination
4. On-scene control
5. Hazard evaluation
6. Personal protective equipment
7. On-scene work assignments
8. Communication procedures
9. Decontamination procedures
10. On-scene safety and health considerations including designation of the safety officer, emergency medical care procedures, environmental monitoring, emergency procedures, and personnel monitoring

Response personnel at a hazardous material incident must recognize and understand the potential hazards associated with the incident. They must also be familiar with the procedures contained in the plan of action. Therefore, prior to working on-scene, all personnel must be provided a safety briefing. The safety briefing will describe the assigned tasks and their potential hazards, coordinate activities, identify methods and precautions to prevent injuries, and plan for emergencies. The plan of action can serve as a guide to this safety briefing.

The result of planning a response to a hazardous material incident is a plan of action consistent with the local emergency response plan and the organization's standard operating procedures and within the capability of the resources available. Development of a written plan ensures that all safety aspects of the response are thoroughly examined. Plans should be modified as necessitated by changing conditions on the scene.

IMPLEMENTING THE PLANNED RESPONSE

Overview

Once the plan of action is determined, response personnel must implement the response options in that plan. The following tasks are associated with implementing the planned response to a hazardous material incident:

1. Implement the incident management system (IMS).
2. Implement the selected protective actions.
3. Establish and enforce scene-control procedures.
4. Implement planned control actions.

Initiating the Incident Management System

The incident management system (IMS) is an organized approach to control and manage operations effectively at an emergency through a system of roles, responsibilities, and standard operating procedures. The incident management system should be developed as part of pre-emergency planning efforts, including the local emergency-response plan and an organization's standard operating procedures.

One individual is in charge of managing the incident. This individual may delegate responsibility for performing various tasks to others on the scene. All communications are routed through the command post. Delegation of tasks may be by location or by function. Some of the functions include: medical services, evacuation, water supply, resources, media relations, safety, and site control. The incident management system reduces confusion, improves safety, organizes and coordinates actions, and facilitates effective management of the incident.

The National Inter-Agency Incident Management System provides an example of the elements of an incident management system used to coordinate response activities:

1. Common terminology
2. Modular organizations
3. Integrated communications
4. Unified command structure
5. Consolidated action plans
6. Manageable span of control
7. Predesignated incident facilities
8. Comprehensive resource management

Responders should understand the roles and responsibilities of the various positions in the local incident management system as defined in the local emergency response plan or the organization's standard operating procedures.

The incident management system includes the procedures for notification and use of other than local resources, including private, state, and federal government personnel. The incident commander should be familiar with and comply with the requirements of the following plans:

1. Organization's emergency response plan
2. Organization's standard operating procedures
3. Local emergency response plan
4. State emergency response plan
5. Regional emergency response plan, including the resources available from the regional response team
6. Federal emergency response plans

Response personnel should be aware of the primary local, state, regional, and federal government agencies and understand the scope of their regulatory authority (including the regulations) pertaining to the production, transportation, storage, and use of hazardous materials and the disposal of hazardous wastes. They should also be aware of private-sector resources offering assistance during a hazardous material incident and identify their role and the type of assistance or resources available.

The incident management system should provide a focal point for information transfer to the media and local elected officials. A public-information officer should be designated to provide appropriate information to the media and local, state, and federal officials.

Implementing Selected Protective Actions

Protective measures are those taken to preserve the health and safety of emergency responders and the public during an incident involving the release of hazardous materials. Protective actions include the following.

1. *Isolate the hazard area and deny entry.* Keep all persons away from the area if they are not directly involved in emergency response operations. Unprotected emergency responders should not be allowed within the isolation area. Planning documents should include techniques to isolate the hazard area and deny entry to unauthorized persons at hazardous material incidents.

2. *Evacuate persons in the threatened area.* Evacuation is defined as a process to "move all people from the threatened area to a safer place."[8] To perform an evacuation, there must be enough time for people to be warned, to get ready, and to leave the area. Even after people are evacuated, they should not be allowed to congregate on the perimeter of the control zones. Evacuees should be sent upwind to a definite place, by a specific route, far enough away so that they will not have to be moved again if the conditions change.

3. *Protect persons who cannot be evacuated in-place.* In-place protection is defined as a "means to direct people to quickly go inside a building and remain inside until the danger passes."[8] Certain protective actions must be taken inside the building. In-place protection is used when evacuating the public would cause greater risk than directing them to stay where they are or when an evacuation cannot be performed.

Establishing and Enforcing Scene-Control Procedures

Scene-control procedures must be implemented quickly. Control zones (see Figure 10-9P), decontamination (contamination reduction) activities, and communication must be established and maintained throughout the incident. Personnel should move between control zones through access control points.

Scene control is established through the use of control zones based upon safety and the degree of hazard. The hot zone is the area surrounding a hazardous material incident. It extends far enough to prevent adverse effects from hazardous material releases to personnel outside the zone. The warm zone is the area where personnel and equipment decontamination (contamination reduction) takes place. The cold zone is the area containing the command post and other support functions. Access to various zones is determined by need and level of personal protection provided as specified in the plan of action.

Extreme caution is necessary when providing emergency medical care to victims of hazardous material incidents.

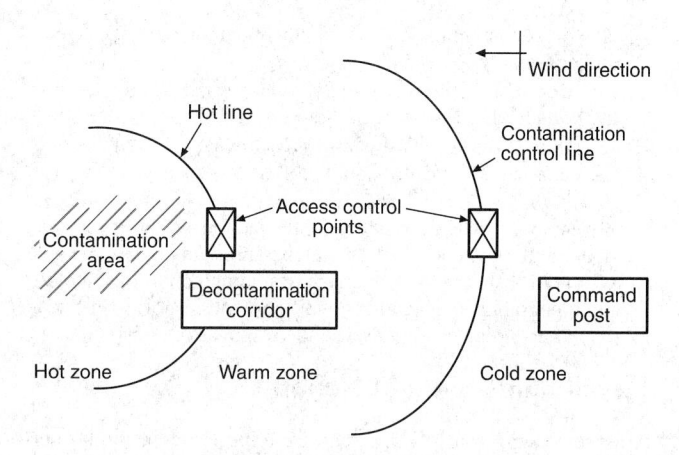

FIG. 10-9P. *Diagram of control zones.*

Performing Control Functions

Responders are expected to perform control functions identified in the plan of action. They will have to select the tools, equipment, and materials for the tasks assigned. They should also understand the precautions for controlling releases from the containment systems involved.

In conjunction with their ability to use personal protective equipment, response personnel should be able to identify the symptoms of heat and cold stress. Response personnel should be aware of the physical and psychological stresses that can affect users of personal protective equipment. Personnel should be able to identify the signs and symptoms of exposure to any hazardous material involved. Response personnel should be able to record the use, repair, and testing of chemical protective clothing.

EVALUATING PROGRESS AND ADJUSTING ACCORDINGLY

The fourth and final duty in responding to hazardous material incidents is to evaluate the progress of the planned response—whether it stabilizes, intensifies, or otherwise changes. This duty evaluates the effectiveness of the following:

1. Personnel being used
2. Personal protective equipment
3. Established control zones
4. Decontamination process
5. Selected action options

A comparison of the actual behavior to that predicted indicates the effectiveness of the operations. Negative feedback initiates a reevaluation of the situation, including another analysis of the problem, response objectives and options, and a modified plan of action. Positive feedback indicates successful operations; however, modification of the plan may still be necessary.

Termination

Termination of a hazardous material incident is that portion of incident management in which personnel are involved in documenting safety procedures, site operations, hazards faced, and lessons from the incident. Termination is divided into three phases: (1) debriefing the responders, (2) post-incident analysis, and (3) critiquing the incident.[12]

Debriefing the responders is the process of reviewing a hazardous material incident focusing on the following factors:

1. Informing responders of any potential exposure to hazardous materials and the signs and symptoms associated with that exposure,
2. Identifying equipment damage and unsafe conditions,
3. Assigning information-gathering responsibilities for a critique, and
4. Summarizing the activities performed.

Post-incident analysis is the reconstruction of the incident to establish a clear picture of the events that took place during the incident. The post-incident analysis focuses on four key topics: (1) command and control, (2) tactical operations, (3) resources, and (4) support services.

Critiquing the incident—emphasizing successful and unsuccessful operations—should be conducted with all response community members. The procedures for conducting a critique of a hazardous material incident should be in the local emergency response plan or the organization's standard operating procedures including written reports to management that suggest ways to improve future operations.

At the termination of the emergency phase, authority may be transferred to those responsible for the recovery activity in the incident. The local emergency response plan or the organization's standard operating procedures outline the steps for terminating the emergency phase of the incident.

Providing Reports and Subsequent Documentation

Responders to hazardous material incidents must comply with the reporting requirements of federal, state, and local agencies. All reports and other incident documents should be consistent with the local, state, and federal requirements.

Training records, exposure records, incident reports, and critique reports should become part of the file for the incident. An activity log and detailed exposure records should be maintained on each hazardous material incident. Typically, requirements for filing documents and maintaining records are found in the local emergency response plan and the organization's standard operating procedures.

SUMMARY

As with any emergency, favorably changing the outcomes requires a logical sequential process involving four duties: (1) analysis, (2) planning, (3) implementation, and (4) evaluation. Each of these duties is made up of tasks that need to be accomplished to complete that duty. Each task requires that certain steps be performed to complete the task. The topics in this chapter describe the duties, tasks, and steps associated with managing the response to hazardous material incidents.

BIBLIOGRAPHY

References

1. Benner, L., Jr., "D.E.C.I.D.E. in Hazardous Materials Emergencies," *Fire Journal*, Vol. 69, No. 4, July 1975, p. 13.
2. Department of Transportation, Research and Special Programs Administration, "Performance-Oriented Packaging Standards; Changes to Classification, Hazard Communication, Packaging and Handling Requirements Based on UN Standards and Agency Initiative; Final Rule," *Federal Register*, Vol. 55, No. 246, Dec. 21, 1990, pp. 52402–52729.
3. *Acute Hazardous Events*, Data Base, Industrial Economics Inc., Cambridge, MA, Dec. 1985.
4. NFPA, *Fire Protection Guide to Hazardous Materials*, 11th ed., National Fire Protection Association, Quincy, MA, 1991.
5. Chemical Manufacturers Association and Association of American Railroads, *Technical Bulletin on Packaging for Hazardous and Nonhazardous Materials*, Chemical Manufacturers Association, Washington, DC, 1989.
6. Ericksen, N. A., Keffer, W. J., and Wright, C. J., *Introduction to Hazardous Materials Incident Response*, U.S. Environmental Protection Agency and Union Pacific Railroad, Omaha, NE, 1989.
7. Benner, L., Jr., *Hazardous Materials Emergencies*, 2nd ed., Lufred Industries, Inc., Oakton, VA, 1978.
8. *North American Emergency Response Guidebook*, 2nd printing, U.S. Department of Transportation, Research and Special Programs Administration, Washington, DC, 1996.
9. *Guidelines for the Selection of Chemical Protective Clothing*, prepared by A. D. Little, Inc., Cambridge, MA, 1987.
10. Forsberg, K., and Mansdorf, S. Z., *Quick Selection Guide to Chemical Protective Clothing*, 2nd ed., Van Nostrand Reinhold, New York, 1995.
11. NIOSH, *Pocket Guide to Chemical Hazards*, U.S. Department of Health and Human Services, Government Printing Office, Washington, DC, 1994.
12. Noll, G. G., Hildebrand, M. S., and Yvorra, J. G., "Terminating the Incident," *Hazardous Materials: Managing the Incident*, 2nd ed., Oklahoma State University, Stillwater, OK, 1995.

NFPA Codes, Standards, and Recommended Practices

Reference to the following NFPA codes, standards, and recommended practices will provide further information on managing response to hazardous material incidents discussed in this chapter. (See the latest version of *The NFPA Catalog* for availability of current editions of the following documents.)

NFPA 49, *Hazardous Chemicals Data*
NFPA 325, *Guide to Fire Hazard and Properties of Flammable Liquids, Gases, and Volatile Solids*
NFPA 471, *Recommended Practice for Responding to Hazardous Materials Incidents*
NFPA 472, *Standard for Professional Competence of Responders to Hazardous Materials Incidents*
NFPA 491M, *Manual of Hazardous Chemical Reactions*
NFPA 704, *Standard System for the Identification of the Hazards of Materials for Emergency Response*
NFPA 1971, *Standard on Protective Ensemble for Structural Fire Fighting*
NFPA 1981, *Standard for Open-Circuit Self-Contained Breathing Apparatus for Fire Fighters*
NFPA 1991, *Standard on Vapor-Protective Suits for Hazardous Chemical Emergencies*
NFPA 1992, *Standard on Liquid Splash-Protective Suits for Hazardous Chemical Emergencies*
NFPA 1993, *Standard on Support Function Protective Clothing for Hazardous Chemical Operations*

U.S. Government Publications. U.S. Government Printing Office, Superintendent of Documents, Washington, DC 20402

Handbook of Chemical Hazard Analysis Procedures (FEMA, DOT, EPA).
Hazardous Materials Emergency Planning Guide (NRT-1), National Response Team, 1987.
Occupational Safety and Health Guidance Manual for Hazardous Waste Site Activities, NIOSH/OSHA/USCG/EPA, U.S. Department of Health and Human Services, NIOSH, 1985.
Occupational Safety and Health Standards, Title 29, Code of Federal Regulations, Part 1910.120, "Hazardous Waste Operations and Emergency Response Final Rule," 1989.
Occupational Safety and Health Standards, Title 29, Code of Federal Regulations, Part 1910.1200, "Hazard Communication Standard," 1987.
Standard Operating Safety Guides, Environmental Response Branch, Office of Emergency and Remedial Response, U.S. Environmental Protection Agency, 1988.
Technical Guidance for Hazard Analysis (EPA, FEMA, DOT).
Title 40 CFR Part 261.33 (EPA).

Title 40 CFR Part 302 (EPA).

Title 40 CFR Part 355 (EPA).

Title 49 CFR Parts 170–179 (DOT).

Additional Readings

Bills, R. W., "Cargo Protection in a Toxic Environment," Technical Report, Army, Dungway Proving Ground, UT, DPB/JOD-91/008, Feb. 1992.

Carroll, T. R. and Schwope, A. D., "Non-Destructive Testing and Field Evaluation of Chemical Protective Clothing," Final Report, Federal Emergency Management Agency, Washington, DC, FA-106, Dec. 1990.

Donahue, M. L., "CHEMTREC: A Vital Link in Chemical Emergencies," *NFPA Journal*, Vol. 87, No. 3, May/June 1993, pp. 86–88, 90.

"Engineering Special Chemical Risks," *Record*, Vol. 70, No. 3, May/June 1993, pp. 3–8.

Eversole, J. M., "Planning, Preparation and Response to Explosions and Fires at Chemical Plants," International Fire Conference and Exhibition in Tokyo for Firesafety Frontier '94, Creating a Safe Tomorrow, 1994, pp. 63–67.

Fischer, S., et. al., "Vadautslapp av brandfarliga och giftiga gaser och vatskor. [Toxic and Inflammable/Explosive Chemicals: A Swedish Manual for Risk Assessment.]," National Defence Research Establishment, Stockholm, Sweden, FOA-D-95-00099-4.9-SE, March 1995.

Isman, W. E., "Chemical Properties Influence Decisions," *NFPA Journal*, Vol. 85, No. 6, Nov./Dec. 1991, pp. 94–95.

Kearney, J., "Bulk Chemical Incidents: Learning the Lessons From Past Emergencies," *Fire*, Vol. 86, No. 1057, July 1993, pp. 41–42.

Klunk, D., and Blaine, B., "Santa Fe Springs' Model Haz Mat Plan. How One City Protects Its Environment," *American Fire Journal*, Vol. 45, No. 2, Feb. 1993, pp. 28–31.

Noll, G. G., CSP, "Handling Haz Mats: Subsurface Spills and Release into a Sewer System," *Fire Engineering*, Vol. 148, No. 4, April 1995, pp. 54–60.

"Oil Pollution Act of 1990," Title VII—Oil Pollution Research and Development Program, Conference Report, 101st Congress, 2nd Session, House of Representatives, Washington, DC, Report 101-653, H. R. 1465, Aug. 1, 1990.

Schwope, A. D., and Renard, E. P., "Estimation of the Cost of Using Chemical Protective Clothing," Performance of Protective Clothing: Challenges for Developing Protective Clothing for the 1990s, Fourth (4th) Volume, ASTM STP1133, ASTM, Philadelphia, PA, 1991, pp. 972–981.

Stull, J. O., "New Comprehensive Performance Standards for Chemical Protective Gloves, Boots, and Other Types of Protective Clothing," Performance of Protective Clothing: Challenges for Developing Protective Clothing for the 1990s, Fourth (4th) Volume, ASTM STP1133, ASTM, Philadelphia, PA, 1991, pp. 322–338.

Stull, J. O., et. al., "Developing and Selecting Test Methods for Measuring Protective Clothing Performance in Chemical Flashover Situations," Performance of Protective Clothing: Challenges for Developing Protective Clothing for the 1990s, Fourth (4th) Volume, ASTM STP1133, ASTM, Philadelphia, PA, 1991, pp. 908–923.

"Threats to the Environment," *Fire Prevention*, No. 231, July/Aug. 1990, pp. 15–16.

Tokle, G., *Hazardous Materials Response Handbook*, 2/e, National Fire Protection Association, Quincy, MA, 1993.

Wackerlig, H., "Risk Assessment of Chemical Warehouses," *Fire Protection*, Vol. 17, No. 4, Dec. 1990, pp. 4–6.

Yates, J., "Phillips Petroleum Chemical Plant Explosion and Fire, Pasadena, Texas, October 23, 1989," USFA Fire Investigation Technical Report Series, Federal Emergency Management Agency, Washington, DC, Report 035, 1990.

RESCUE OPERATIONS

Christopher J. Naum

Rescue operations are a major component of the multifaceted fire and emergency service system. These service professions have traditionally made the protection of life and property the primary focus of their mission in the community, and have consistently exhibited their unique ability to overcome the challenges of a rapidly changing society and working and living environments therein.

The complexity of rescue responses can vary greatly with the incident being handled. Typical training levels and the expertise and innovation of operating companies and personnel determine the effectiveness and efficiency of these responses.

SPECIALIZED TECHNICAL RESCUE

The rescue emergency services system involves more than basic structural fire rescue and extrication capabilities; it extends into other areas that impact incident response capabilities. Specialized technical rescue response teams have evolved as traditional and typical functional areas are transcended.

Specialized technical rescue response capabilities are recognized as a tactical component in the emergency response delivery system. In many instances, the need for specialized technical rescue capabilities has slowly developed due to identified community risk factors in incident responses or through specific incident occurrences that have shown the deficiencies that were present within the response system.

Operational Deployment

Specialized technical rescue encompasses numerous areas of operational deployment, and includes, but is not limited to:

- High- and low-angle rope rescue
- Motor vehicle extrication operations
- Industrial extrication and entrapment rescue
- Confined-space entry and rescue operations
- Trench and excavation rescue
- Below-grade rescue

- Structural collapse rescue
- Surface and underwater rescue
- Swift-water rescue
- Ice rescue
- Wilderness search and rescue
- Urban search and rescue
- Agricultural and farm rescue
- Hazardous materials (HAZMAT) rescue operations
- High-rise rescue
- Helicopter rescue
- Heavy rescue applicable to: air, rail, and maritime
- Large-scale disaster response rescue

The need for specialized rescue teams is typically based on the historical perspective of a community or jurisdiction, local and regional conditions and trends, and capabilities and deficiencies encountered in incident operations, coupled with the degree of risk potential that exists.

The acceptable level of risk is determined by local conditions, and must be identified and balanced with the capabilities of the emergency service delivery system.

Rescue deployment and operations ability is based upon the degree of awareness, knowledge, training, resources, and planning that are present, coupled with the potential, frequency, magnitude, and impact that these factors may have on the specialized agency's ability to carry out the necessary functions dictated by the incident response.

The development of a specialized technical rescue component within any jurisdiction requires a thorough understanding of the inherent hazard potential present at specific technical rescue responses; the degree of commitment required in the way of financial support, training, and skill enhancement; resource allocation and equipment requirements; and logistical planning and legal ramifications that will impact the conceptualization, implementation, and ultimate operation of a specialized rescue team. In many situations, a regional approach to organizing, staffing, and funding a technical rescue team will be the most practical answer.

The development of specialized technical rescue teams will continue to be a major influencing factor in the national emergency service community. The need for specialization *vs.* the generalist

Christopher J. Naum, a 21-year fire service veteran, is president of L.A. Emergency Management & Training Consultants, Syracuse, NY. He is a member of the NFPA 1670 Committee on Technical Rescue and the Structural Collapse Rescue and Confined Space Rescue Task Groups, and chairperson of the Trench Rescue Task Group.

approach toward operations will continue to drive the development of this unique delivery system. The emergency service community must maintain both proficiency in the delivery of their services and the safety of the personnel who perform those various responsibilities. Identification of, and planning for, specialized rescue situations enhances incident response capability. Knowledge, skills, and understanding of these special conditions are limited by unique operational parameters and safety considerations. In summary, successful outcome and termination of special rescue incidents relies on skills, training, preplanning, and deployment of adequately prepared and outfitted resources and personnel.

COMMUNITY RESOURCE PLANNING (CRP)

The typical demands associated with a site-specific technical rescue incident or a regionwide disaster can quickly overwhelm even the largest of responding emergency service agencies. Even with an effective in-place incident management system, the planning and logistical demands placed upon incident managers, coupled with the requirements for meeting tactical deployment requests, can impede or impair the most efficient of organizations or departments.

Due to the potential complexities associated with rescue situations and the variety of conditions and parameters that may be present at site-specific or large-scale disaster incidents, external resource allocation and deployment may become necessary in order to enhance and support operational incident requirements.

The need for comprehensive community risk assessment and planning for complex collapse or disaster responses forms the foundation for subsequent incident operations. One component of community risk assessment planning focuses upon the development of a data base of external resources that could be deployed and utilized in order to facilitate strategic and tactical objectives.

Community resource planning (CRP) provides the means by which an agency or jurisdiction can enhance identified internal deficiencies and provide a system for external resource allocation that will provide on-scene incident commanders with the necessary resources to augment the agency's ability to respond and operate at rescue incidents.

As with any incident response, the successful intervention, operation, and termination of the incident is directly related to the degree of resources available or deployed, coupled with the effectiveness and efficiency of their respective tactical assignments, objectives, and time management parameters.

Effective deployment of a community resource plan provides for a pre-determined and coordinated means of supporting the planning and logistical demands associated with rescue incidents, and allows for the substantial reduction in the time allotments required to identify, locate, and subsequently contact and call-out resources for anticipated or required deployment needs.

Community resource planning is not an entirely new concept within the fire and emergency service delivery system. The identification and utilization of mutual-aid agencies and companies to augment daily fire, rescue, and emergency medical service (EMS) alarm assignments is commonplace in many jurisdictions. Supplementation of local jurisdictional manpower and equipment needs with mutual-aid agencies allows the receiving agency to: (1) deploy adequate resources at a given incident and (2) increase the ability to mitigate the incident conditions.

The size, organizational structure, internal capabilities and resources, operating environment, and technical rescue capabilities of the department, coupled with the identified community risk hazards, will determine to what extent the community resource plan will be developed. The magnitude, severity, and frequency of related incidents and responses will also dictate the methodology of the planning process.

The identification and formulation of select resource criteria and subsequent development of the community resource directory (CRD) is unique to each department or agency's community resource plan. The CRP and CRD augment and enhance the limitations of the jurisdiction's capabilities during rescue operations, and provide the framework and operating structure to build effective, safe, and timely rescue operations.

The basis of the community resource plan is to establish contacts; identify procedures; and initiate agreements with local, jurisdictional, regional, and/or state and national entities for the procurement of identified resource needs. Through careful and thorough research and preplanning, a useful data base can be developed that includes established procurement procedures and memoranda of agreement.

DEVELOPMENT OF A CRP DATABASE

There are four basic functional areas that can be generally associated with the categories of resource planning for rescue response. These include, but are not limited to: (1) equipment, (2) supplies, (3) services, and (4) technical support.

Since the basis of the community resource plan and resource directory are established and defined by the identification of internal agency/department limitations or deficiencies, they are also directly related to the degree of risk potential to which the community and jurisdiction are susceptible. The relationship of current operational capabilities, coupled with the influence of existing mutual-aid agreements and assignments, will provide a measurement by which the parameters of the community risk profile and analysis can be factored in order to determine the existing limitations and deficiencies. This provides the foundation for the research and development of an external resources data base.

The following lists provide an overview of items that could comprise the data base. These lists are not all-inclusive.

Equipment

- Aerial lifts
- Air tools
- Backhoes
- Bulldozers
- Cellular phone companies
- Compressors
- Computer equipment supply and rental companies
- Concrete construction form suppliers
- Concrete cutting, breaking, and sawing firms
- Construction equipment
- Conveyors and material handlers
- Cranes, rigging/erectors
- Excavators
- Fire equipment and supply firms
- Flood lighting and searchlights
- Forklifts
- Generators
- Ham radio operators groups

- Hydraulic, pneumatic, mechanical
- Hydraulic tool distributors
- Light and heavy equipment firms
- Loaders, trenchers, skid-steer loaders
- Portable pumps
- Public works agencies
- Radio communications and equipment suppliers
- Sawing and cutting equipment
- Scaffolding
- Shoring
- Television and communications companies
- Tent suppliers and distributors
- Welding supply companies

Supplies

- Bag, burlap, and canvas manufacturers
- Battery distributors
- Building materials supply firms
- Chemical supply and distributor companies
- Compressed gases supply distributors
- Fire equipment and supply firms (local/regional)
- Fuel distributors
- Grocers and food suppliers
- Hardware supply firms
- Ice
- Lumberyards
- Medical supply companies
- Portable shelter and tent firms
- Power tool suppliers
- Rental service and supply companies
- Restaurants
- Utility companies

Services

- Aviation/helicopter services
- Awning and canopy manufacturers
- Baking companies
- Banquet and party supply companies
- Beverage supply and distributors
- Boat supply companies
- Bottled water distributors
- Buses, public, private; school districts
- Catering companies
- Fencing companies
- Glass/windshield repair companies
- Ice distributors
- Motorcycle, all-terrain/ATV firms
- Portable restroom supply companies

- Rubbish/dumpster suppliers and haulers
- Tire distributors/repair forms
- Towing services
- Transportation service
- Tree- and debris-removal firms
- Trucking and leasing companies

Technical Support

- American Concrete Institute
- American Institute of Architects
- American Institute of Steel Construction
- American Public Works Association
- American Society of Consulting Engineers
- Association of Building Contractors American Society of Safety Engineers
- Association of Engineering Geologists
- Association of General Contractors
- Concrete Reinforcing Steel Institute
- Construction Specification Institute
- International Association of Bridge, Structural, and Ornamental Iron Workers
- Local and regional professional societies, associations, and trade groups
- National Society of Professional Engineers
- Physicians and Surgeons Professional Associations

Contractors

- Demolition
- Drilling and boring
- Excavation
- General contractors
- Heavy-equipment operators
- Steel erectors and Fabricators
- Utility

Consulting Engineers and Architects

- Architects
- Chemical
- Civil engineers
- Construction engineers
- Environmental
- Marine
- Mechanical
- Safety, health, and hygiene
- Sanitary
- Soil
- Structural
- Surveying

Other

- Electric equipment and supply distribution
- Environmental and remedial contractors
- Hazardous waste abatement companies
- Highway and transportation agencies
- Industrial rescue teams
- Land surveyors
- Search-and-rescue dog handlers
- Wilderness search groups/associations

After the appropriate CRP data base items are identified, the data collection process must identify local, regional, statewide, or other appropriate contacts, e.g., companies, firms, agencies, or individuals that will ultimately provide the identified resource needs.

CRP AGREEMENTS

The next phase in the community resource plan requires the initiation, preparation, and adoption of appropriate types of agreements that clearly define the scope and conditions for the provision of services, equipment, and/or supplies established by the CRP. These agreements can take the form of the following:

- Memoranda of agreement (MOA)
- Mutual-aid agreements (MAA)
- Service agreements (SA)
- Service or retainer contracts (SR/SC)

These agreements should be reviewed and approved by the legal council having jurisdiction within the organization structure of the department to ensure that the agreement and its parameters are equitable, concise, and that all liability issues have been addressed.

INFORMATION MAINTENANCE

The documents and data contained within the CRD "yellow pages" for complex incident management provide for the timely identification, call out, and deployment of required resources, based upon specific incident parameters.

As with any resource or reference document, the community resource directory and data base should be updated and revalidated on an annual basis, so to ensure that individual contact names, phone numbers, and conditions are valid and current. In addition, it is important to identify and document any changes in the resource listing that may affect its usefulness within the data base. It is not uncommon for products, services, or materials to change or be withdrawn, especially when dealing with equipment, supply, or service companies. This, obviously, can compromise the dependability and/or function of the CRP.

Financial constraints, budgets curtailments, staffing limitations and reductions, and limited resource acquisitions can prevent the initiation of programs or delay expansion of existing services. All proposals must be thoroughly assessed, properly planned, and adequately developed in order to effectively implement program goals. Technical rescue team development relies heavily on a thorough understanding of the issues surrounding the needs associated with such a team, coupled with an appropriate planning process that will ultimately lead an agency toward its established goal of enhanced rescue capabilities or deployment of specially trained and adequately outfitted personnel for technical rescue responses.

CONCEPTUALIZATION AND PLANNING PROCESS

The conceptualization and planning process is defined by the following ten parameters. Each must be thoroughly identified and addressed.

1. What type of specialized rescue team(s) is required?
2. What risk factors are present or potential within the jurisdiction?
3. What level of expertise currently exists within the organization?
4. How can system enhancement be best achieved?
   - Short term
   - Intermediate
   - Long term
5. What level or degree of technical capabilities can the plan achieve?
6. How can the services be best achieved?
   - Interdepartmental
   - Local
   - Regional
   - Countywide
   - Intercounty
   - Cooperative agreements with industrial or private teams
   - Outside agencies
7. What is the current projected level of interest of staff and personnel?
   - Interest
   - Motivation
   - Availability for staffing requirements
   - Previous levels of special/specific training
   - Employment/trade experience
   - Fire/rescue/EMS response experience
   - Educational Background
   - Time commitment requirements
   - Personal risk factors
   - Dedication
8. What are the projected financial considerations and impacts?
   - Short term
   - Intermediate
   - Long term
9. What influencing regulations/standards must be complied with, or basis achieved?
   - NFPA
   - OSHA
   - ANSI
   - NIOSH
   - DOT
   - DOL
   - API
   - NASAR
   - AAUS
   - Other

10. What is the time commitment for organization, development, and implementation of the program initiative?

Efforts must be made early in the conceptualization and planning process to ensure that the commitments and objectives being developed can be achieved within the organization. Initial interest may be high with any new program, but may quickly wane as the demand of increased training and time commitments take their toll.

State/Federal Regulations

Regulatory impact of associated state/federal regulations, along with the influence of recognized national standards, will also influence and help direct the planning, organization, and training process. NFPA has standards under development that will cover technical rescue operations and the professional qualifications of rescue technicians. These regulations and standards may assist the planning group or committee in identifying relevant issues that may impact the development of objectives, goals, and organizational scope for the subsequent teams deployment.

Research and Data Gathering

Efforts should be taken by respective departments and agencies to research and obtain information from other departments and teams around the country, in order to gain the wealth of information that can be analyzed and disseminated, with applicable information applied and augmented in order to meet with local conditions, factors, and team objectives. This research and data-gathering network can save a tremendous amount of time and effort, by avoiding redundancy in program planning and technical rescue team applications. Those departments that have developed technical response teams over the years, and the process by which they developed the methods utilized for operations and the equipment utilized, coupled with the revisions that have been adopted through field-proven situations, can only help to provide time-proven data and information.

TECHNICAL RESCUE TEAMS

It is imperative that departments and agencies considering developing specialized technical rescue teams understand the long-term commitments and the level of dedication required to fulfill the operability and usefulness of team activation. A department or agency must identify the appropriate types and levels of technical service(s) necessary as the needs or risks within the jurisdiction or community change. In the 1980s, many jurisdictions quickly implemented hazardous materials teams as the wave of this concept and need swept throughout the emergency services only to find that the demands of time, financial support, and commitment levels necessitated a re-evaluation of their level of commitment and degree of involvement.

Rescue Specialties

The following list is designed to assist any agency or organization in identifying the specialties necessary to safely and effectively undertake technical rescue operations.

1. Rope/high-angle rescue operations
2. Confined space rescue operations
3. Trench rescue operations
4. Structural collapse rescue operations
5. Water/ice rescue operations
6. Motor vehicle rescues
7. Industrial/agricultural rescue operations

Technical Rescue Specialties Program Examples

The following examples are typical technical rescue program elements that can facilitate the development of rescue operation capabilities for technical rescue specialties.[1]

Rope rescue: Rope techniques are a basic underlying skill for most other types of rescue. Most fire fighters will be familiar with basic rope techniques and knot tying as part of the basic fire fighter curriculum.

An awareness of rope skills can be taught to rescuers in only a day. It could include topics such as rope characteristics, strengths, basic knots, hardware, hazards to be aware of when using rope, and dangerous techniques to avoid.

An operation-level could cover rope rescue techniques. Rescuers could be taught basic techniques of rappelling, rigging, belaying, safety, anchoring, and simple mechanical advantage systems. Additional operational techniques could include patient packaging, low-angle evacuations, and simple pickoff maneuvers. This could be taught in 2 days.

A detailed technician-level program could be conducted in approximately one week, covering basic and advanced rigging techniques, anchor systems, belays, simple and complex mechanical advantage systems, and advanced patient extrication techniques and stokes basket operations. Low- and high-angle rescue techniques, including telpher and Tyrolean systems, could be included.

Some specialty topics could include advanced techniques for helicopter operations, ladder operations and bridging techniques, and other topics. Practical and teaching experience should be required. Urban rope techniques could be incorporated for areas where high-angle rescues may be adapted to an urban environment.

Confined space rescue: Confined spaces are defined as any area not designed for human occupancy, with limited entrance and egress. OSHA has established one of the few standards that is applicable to technical rescue, 29 CFR 1910.146, which requires confined space rescue personnel who enter these spaces to be trained (although it provides little training specifics).

An awareness of confined space rescue can be taught in a few hours, and could include background on OSHA regulations, recognition of permit-required spaces, confined space hazard recognition, how to secure the scene, available resources for confined space rescue, and what conditions preclude rescue entry into a space.

Operation-level personnel could be taught safe entry and rescue technique, atmospheric monitoring techniques, and how to assess the hazards and risks. An operations level could be achieved with several days of training.

Technician-level personnel could be trained for a wide range of skills and hazards assessment. Skills may include patient evacuation, special retrieval systems, use of communications and command at confined space incidents, familiarity with various types of confined space, atmospheric monitoring, hazard assessment, and ventilation techniques. At least 40–60 hr are necessary to train personnel to the technician level.

Trench rescue: By definition, a trench is deeper than it is wide. Rescuers have been injured or killed after entering and unshored trench that suffered a secondary collapse. Awareness of the dangers of trench incidents can be taught in about 4 hr, covering the basics of hazard recognition, scene security, rescuer safety, types of trench collapses, additional resources, and initial actions.

An operations level of training can be taught in several days, with students gaining knowledge of rescue equipment, different types of shoring, means of securing the site according to departmental standard operation procedures (SOPs), how to perform a safe entry, and other support operations.

Technician-level personnel should become familiar with various rescue techniques, shoring techniques, victim retrieval systems, EMS and patient care skills for trench collapse, control of utilities, and long-term operations skills. The technician level could be taught in about 10–14 days.

Trench rescue shares equipment, rescue techniques, and skills with both confined space rescue and collapse rescue. A course could be designed to include aspects of each discipline.

Structural collapse: Structural collapse shares many techniques with trench and confined space rescue. An awareness of the dangers of structural collapse could cover types of construction and associated hazards, types of collapses, how to secure the scene, and when to call for help. This could be taught in approximately 8–12 hr.

An operations-level of training could also include patterns for conducting a surface debris search for victims, basic stabilization, utility control, and atmospheric monitoring. It could be taught in 5 to 8 days.

A technician-level course covering shoring and building stabilization, rescue equipment, search equipment and operations, tunneling and excavation techniques, hazard stabilization and mitigation, components of urban search and rescue techniques, and patient care could be taught in approximately 5–7 days.

Water Rescue

One of the most dangerous types of special rescue is water rescue. There are several different specialties within the field of water rescue. Rescuers may face incidents involving calm water, swift water, ice, or even surf conditions. Dive rescue is a specialty within itself. Courses in each training level could be designed to address all types of water rescue or individual types (e.g., swift water rescue only).

A basic awareness of water hazards, safety, and shore-based rescue techniques can be taught in a few hours. Different types of water rescue may share similar techniques, but pose different dangers.

Operations-level training could cover techniques for in-water or ice rescue. Rescuers should become familiar with different types of water rescue techniques, ice and current hazards, hypothermia and EMS considerations, ice rescue equipment, and shore-based swift-water rescue techniques. This course could be taught in about one week, and requires personnel to be able to swim.

The technician level could require knowledge in all facets of water rescue and how to perform special rescue techniques, such as victim retrieval using boats or a helicopter. This course could be taught in about one week.

SUMMARY

The daily demands imposed upon fire and emergency service agencies can, at times, overwhelm the best prepared and trained departments and their personnel. Changes in society and the built environment influence fire and emergency incidents. Specialized rescue operations, and the skills necessary for their safe and effective deployment and subsequent operations, require that all depart-

ments and agencies prepare appropriately to meet the challenges and demands associated with the risk potential and hazards present or projected for the jurisdiction and community.

Specialized technical rescue operations is an emergency integrated system. Through proactive planning, community-based risk assessment and evaluation, preparedness training, resource development, and contingency planning, fire and emergency service agencies can effectively prepared and anticipate the manner in which they can handle incident responses, while maintaining the highest degree and consideration for the safety and well-being of their personnel and emergency responders. Technical rescue operations requires a fine balance between training, knowledge and skills, risk management, and resources.

BIBLIOGRAPHY

Reference Cited

1. *Federal Emergency Management Agency Technical Rescue Program Development Manual,* FA-159, Feb. 1996, pp. 7.5–7.9. United States Fire Administration, Washington, DC.

NFPA Codes, Standards, and Recommended Practices

Reference to the following NFPA codes, standards, and recommended practices will provide further information on rescue operations discussed in this chapter. (See the latest version of *The NFPA Catalog* for availability of current editions of the following documents.)

NFPA 220, *Types of Building Construction*
NFPA 1001, *Standard for Fire Fighters Professional Qualifications*
NFPA 1201, *Developing Fire Protection Services for the Public*
NFPA 1470, *Standard on Search and Rescue Training for Structural Collapse Incidents*
NFPA 1500, *Standard on Fire Department Occupational Safety and Health Program*
NFPA 1521, Fire Department Safewty Officer
NFPA 1561, Fire Department Incident Management System
NFPA 1600, Disaster Management
NFPA 1983, Fire Service Life Safety Rope & System Components

Additional Reading

Collins, Larry, "An Innovative Approach to Swiftwater Rescue," *Firehouse,* May 1994, p. 83–111.
Downey, Ray, *The Rescue Company,* Fire Engineering Publications, Tulsa, OK, 1992.
Federal Emergency Management Agency, Earthquake Reduction Series 41, "Rapid Visual Screening for Potential Deismic Hazards, Washington, DC.
Federal Emergency Management Agency, "Earthquake Damaged Buildings; An Overview of Heavy Debris and Victim Extrication," FEMA 158, Sept. 1988, Earthquake Reduction Series 43, Washington, DC.
Final Report; Alfred P. Murrah Federal Building Bombing, Fire Publications, Oklahoma State University, Stillwater, OK. 1996.
Foley, Steven N., "Picking up the Pieces," *NFPA Journal,* May/June 1995. P-52ff
Gallagher, Tim and Steve Storment, "Confined Space Rescue," *Rescue,* May/June 1994, p. 51–69.
IFSTA, *Fire Service Rescue Practices,* Fire protection Publications, Oklahoma State University, Stillwater, OK, 1996.
Mellot, Kevin, D., "Building a Specialized Rescue Capability on a Budget." *Fire Chief,* November 1992, p. 36–39.
Naum, C. J. "Basic Training for Structural Collapse Operations," *On-Call,* Vol. 1., No. 5, Nov./Dec.
Naum, C. J., "Search and Rescue for Structural Collapse Operations," *The Voice,* May 1996; June 1996.
Naum, C. J., "Specialty of the houseTechnical Rescue Teams," *Firehouse,* Vol. 18, No. 4., April 1993, pp. 86, 88, 90.

Naum C. J., "Developing a Technical Rescue Team", *Firehouse*, Vol. 18, No. 5., May 1993, pp. 3841, 4445.

Naum, C. J., "Trench Collapse Response, Rescue & EMS Considerations," *Rescue—EMS*, Vol. 11, No. 3, May/June 1993, pp. 36-40.

Naum, C. J., "Community Resource Planning for Collapse Rescue," *Firehouse*, Vol. 18, No. 8., Aug. 1993, pp. 32–34.

Naum, C. J., "Search and Rescue Marking Systems for Collapse Rescue," *Firehouse*, Vol. 19, No. 11, Dec. 1994, pp. 32–37.

Naum, C. J., "Trench Collapse Rescue Operations," *The Voice*", Vol. 23, No. 4; April/May 1994, pp. 17–19.

Naum, C. J., 'FEMA Urban Search and Rescue Task Force System," *Firefighter News*, June/July 1993, pp. 42–43.

Naum, C. J., "The Essence of Rescue in the 1990's,"*Firefighter News*, June/July 1993, pp. 74–77.

NFPA, "KOBE-NFPA Fire Investigation Report," National Fire Protection Association, Quincy, MA, 1997.

Practical Rescue Manual, Montgomery County Fire and Rescue, 1988.

Superintendent of Documents Publications, Superintendent of Documents, U.S. Government Printing Office, Washington, DC.

Title 29, Code of Federal Regulations, 29 CFR 1910.120 (OSHA), "Hazardous Waste Sites and Emergency Response," March 6, 1989.

Title 29, Code of Federal Regulations, 29 CFR 1910.146 (OSHA), "Confined Space Operations," April 15, 1993.

Title 29, Code of Federal Regulations, 29 CFR 1926.650 (OSHA), "Excavation and Trenching Operations," Oct. 31, 1989.

U.S. Fire Administration, Technical Rescue Technology Assessment, FA-153, Washington, DC, Jan. 1995.

U.S. Fire Administration, "Protective Clothing and Equipment Needs of Emergency Responders for Urban Search and Rescue Missions," FA-136, Washington, DC, Sept. 1993.

U.S. Fire Administration, *Technical Rescue Program Development Manual*, FA-159, U.S. Fire Administration, Washington, D.C., Aug. 1995.

U.S. Fire Administration, *New Technologies in Vehicle Extraction*, FA-152, U.S. Fire Administration, Washington, D.C., Sept. 1994.

Wolf, Alisa, "Oklahoma City Bombing," *NFPA Journal,* Jan./Feb. 1996, p-50ff.

BUILDING CONSTRUCTION CONCERNS OF FIRE DEPARTMENTS

Francis "Frank" L. Brannigan

The building is the fire fighter's enemy. A fire that would be inconsequential in an open field can be a major disaster when it occurs within a building. For example, a few hundred pounds of burning cork wall tile caused a very serious five-alarm fire in New York's Empire State Building.[3]

Fire department personnel have a very strong interest in building construction. They must fight fires in buildings, from those under construction to those built hundreds of years ago. Some building methods encountered are long forgotten, but the hazards still exist. For example, a reference to tile arch floors would be difficult to find in architectural texts; yet there are thousands of them in existence and they do present a specific hazard.

Other fire department concerns include:

1. Reviewing and approving plans for fire protection features in new construction, in cooperation with the building department. In some cases, the fire department has absorbed the building department.

2. Making regular inspections that, among other functions, check on the reliability of code-specified fire protection features—both static, such as fire walls, and dynamic, such as the fire doors that protect openings in the fire walls. A most important function is ensuring that code-required life safety features, e.g., exit facilities, are fully usable as intended.

3. Making prefire surveys to assemble and record information vital to the incident commander at a fire scene. It is most important that the information be retrievable instantly at the fire scene.

4. Acquiring knowledge of building contruction. This is important in investigations of the origin and extent of serious fires.

The western division of the International Association of Fire Chiefs (IAFC) is involved in the International Fire Code Institute (IFCI), International Conference of Building Officials (ICBO), and Building Officials and Code Administrators (BOCA) International model building codes. The southeastern and southwestern divisions are associated with the Southern Building Code Congress (SBCC) and the *Standard Fire Code*.

Francis L. Brannigan, SFPE and NFPA life member, is the author of NFPA's *Building Construction for the Fire Service*, now in its third edition.

FIG. 10-11A. *Some architectural writers, unaware of the entire story, still praise the geodesic dome of the U.S. exhibit at Expo '67 in Montreal. It accommodated a huge number of visitors. After it was turned over to the City of Montreal, a workman accidentally touched a torch to the plastic covering. It totally burned in 10 min. The framework stands today as a monument to a tragedy that, by great good fortune, didn't happen. (Photo by D. Lion)*

Other Relevant Material

This handbook is a formidable document. Its size can be intimidating. But it also serves as a valuable source of information on a wide variety of fire protection topics. There are a number of sections and chapters that deal with many important aspects of building construction. Fire officers should be aware of their content and be familiar with material that pertains to their own local situation. Relevant material should be used for in-service training programs.

There are a number of other relevant chapters. These include all the other chapters in Section 10; as well as Section 1, Chapter 2, "Fundamentals of Firesafe Building Design"; Section 7, Chapter 8, "Structural Fire Safety in One- and Two-Family Dwellings"; and Section 7, Chapter 12, "Fire Hazards of Construction, Alterations, and Building Demolition."

Space in this handbook is necessarily limited. The subject is treated much more fully in *Building Construction for the Fire Service*, third edition, by Francis L. Brannigan, published by NFPA.

FIG. 10-11B.   Heavy fire loads may be found in conference rooms in fire-resistive buildings. This one is decorated with a virtual lumber yard of planks on the overhead. It is possible that planks are treated to be flame resistant. In any case, fire involving the desks and chairs might cause failure of the hangers, usually set in lead anchors, allowing the planks to fall.

FIG. 10-11C.   Sometimes when sprinklers are retrofitted, costly work is done to conceal them. This cost is aesthetic and not required for fire protection. It should not be allowed to be included in sprinkler costs, when the owner argues they are too expensive.

Understanding the Principles of Construction

The fire officer should be easily and accurately conversant with the principles of construction and the proper terms for various building elements. A knowledgeable listener who is presented with gross errors in terminology or principles will probably dismiss even sound fire protection recommendations because of the fire officer's apparent ignorance of building construction. Further, court testimony may be fatally discredited on cross-examination by a competent and knowledgeable attorney.

Every building must endure the force of gravity. By a construction system, the builder defeats, for a time, the force of gravity. A fire of sufficient intensity will accelerate the process and cause the building to fail in whole or in part. This is critical information to the fire service.

Some types of construction, and some specific structural elements, are more likely to fail in a fire than others. If fire fighters are in, on, or adjacent to the structure when it fails, injury or death may result. Difficult and dangerous rescue efforts may be required. Literally only seconds may elapse between the time it is recognized that fire fighters should evacuate a structure and failure of the structure.

This author has recommended that fire department terminology distinguish between "building" and "structural" fires. Units might be dispatched to a "building" fire, as distinguished from a grass or automobile fire. When the incident commander determines that the *structure* is involved in or affected by (e.g., unprotected steel) the fire, the announcement would be made, "This is now a *structural* fire." All fire fighters, therefore, would then be alert for any evidence of potential failure.

Fire departments should have a readily recognized signal that means immediate evacuation. Multiple, short, air horn blasts are used by some departments. Evacuation drills should be carried out regularly to overcome the reluctance of fire fighters to abandon equipment and withdraw. The proudest ships in the navy conduct regular "abandon ship" drills.

Failure on the part of command officers to understand the principles of construction and its repercussions has caused injuries and deaths. Take, for example, a cantilever beam which can be com-

FIG. 10-11D.   These tendons will be post tensioned in several increments as the building grows taller. They remain exposed until the building is finished. They will fail at 800°F (426°C) and thus collapse the structure, possibly causing fire fighter deaths. Wooden falsework and other combustibles provide sufficient fuel.

pared to a seesaw. It is supported in position by connections in the interior of a building. If these connections fail because of fire, the extended portion of the beam, and whatever it supports, will fall. In lieu of a cantilever, a suspended beam may be used. To the building designer there is little difference in whether the reaction goes downward through a column, or upward through a tension member to the wall. However, if the connection fails, the suspended beam becomes an unsupported cantilever and will fall.

Other Hazards

The building being attacked by fire presents other serious hazards to fire fighters. Fire can hide in void spaces with little or no evidence and burst out unexpectedly and increase in intensity when a nearby breach admits oxygen. Buildings of otherwise noncombustible or fire-resistive construction may have extensive, hazardous interior finish of floors, walls, and ceilings.

Fire fighters may be inadequately trained in these or other hazards. When live fire training is conducted in most typical training

FIG. 10-11E. *Skiers attention!! The cold drawn steel cable which supports the tram car, loses its tensile strength at 800°F (426°C). Many of the stations are of wood, sufficient fuel for a destructive fire.*

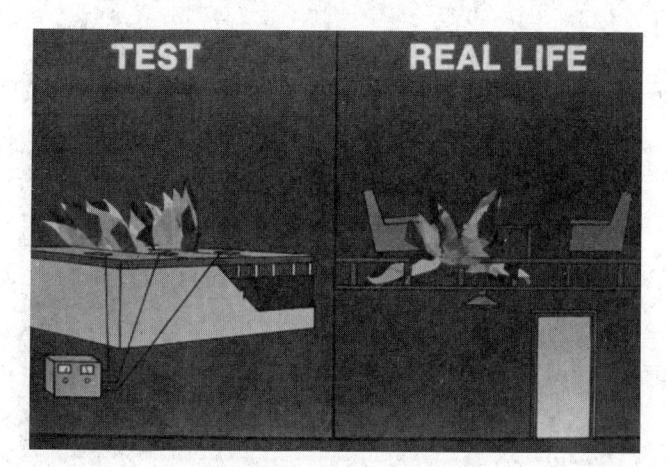

FIG. 10-11F. *In ASTM E119, fire attacks the floor only from below. Real fires are different.*

school structures, the hazards of collapse, hidden fire, and highly combustible interior finish are absent. There have been serious accidents during live fire training in abandoned buildings when these hazards were present. One alternative is adequate classroom instruction, followed by preplanning for the hazards of individual major buildings and general building construction.

Contents

The problems of structure fires cannot be isolated from contents. Most building fires originate in contents. High heat-release fires in contents can destroy noncombustible structures and damage fire-resistive structures. The hazard of today's fast-burning, high rate-of-heat-release contents is a major problem.

In 1975, the Society of Fire Protection Engineers (SFPE) warned, "A high rate of fire development can create a condition that may tax or overpower traditional fire defenses. Defenses of the past, both passive and active (evacuation, alarm, ventilation, and manual fire control), have not been designed by engineer or code to anticipate this hazard. This is primarily a furnishings problem. Many occupancies (residences, office buildings, theaters, hospitals) are being affected."[1]

Fire Resistance

All structures have some resistance to fire-caused collapse. This can be described informally as inherent fire resistance, as distinguished from legal or code-specified. The inherent fire resistance of one structure may be absent in another, similar structure. Direct experience with this fire resistance, whether personal or organizational, is not transferable from one structure to another. To put this simply, "What you know, may not be so."

By experience the fire service has developed some understanding of and criteria for inherent fire resistance. In fact, there are many estimates as to how long fire fighters can safely stay inside a structure on fire. Unfortunately, this is imprecise. Many times fire fighters risk life and limb without knowing the degree of risk involved. Generally accepted indications of imminent collapse—softening, water flowing between bricks, smoke pushing from mortar joints, strange noises—are grossly inadequate, even for older buildings. If they are relied on solely to protect fire fighters from collapse in today's different, lighter buildings, disasters will certainly result.

Building Codes and Building Construction Classification

Building codes classify buildings according to their fire resistance. Here the subject is discussed following the traditional five fundamental construction types:

1. Wood frame,
2. Heavy timber,
3. Ordinary,
4. Noncombustible, and
5. Fire-resistive.

These classifications are very general and by no means precise and should not be relied on for any specific building. Many buildings include several types of construction, some or all of which may not conform in detail to the five standard categories. Some typical variations are given here.

1. Wood frame buildings with truss floors may be structurally considered as platform frame, but because of floor voids interconnected by plumbing voids should be considered balloon frame from the fire point of view.
2. Brick buildings may be partially brick or brick/block masonry on some walls and brick veneer over wood bearing walls on others. In some cases, the lower floor(s) are of masonry and the upper floors are brick veneer. Brick veneer may be applied over steel or concrete.
3. Heavy timber buildings may have dangerous departures from the ideal "mill construction," which is designed in all aspects to resist collapse from fire.
4. Noncombustible buildings may have many combustible features, such as roofs, balconies, overhangs, and combustible metal deck roofs. Noncombustibility does not indicate any designed resistance to collapse.
5. Fire-resistive-rated elements may vary greatly in quality. The test used, ASTM E119, *Standard Test Methods for Fire Tests of Building Construction and Materials,* may give equal time ratings to combustible structures of wood and gypsum. However, Dr. S. H. Ingberg, who developed the concept of applying ASTM E119 to wood structures and is considered the father of the fire resistance testing methods of the National Bureau of Standards (now the National Institute for Standards and Technology), warned:

We must particularly distinguish between the fire resistance of a combustible and a noncombustible structure. While they may be equivalent in time, there is a world of difference in the protection afforded as it concerns the individual building, the hazard to nearby structures, and the hazard to life.[2]

Wood Frame Structures

The most prevalent construction material in the United States and Canada is wood. Wood can carry structural loads such as columns and beams. In older buildings there are huge quantities of nonstructural wood used as lath to support plaster. Wood is also used for decoration and interior trim. A wooden interior in a restaurant in a high-rise building provided such a fierce fire that the fire department was unable to reach three trapped civilians, who finally jumped to their deaths.[3]

For over 150 yr, attempts have been made to render wood fireproof or noncombustible. None have worked. Wood can be surface-treated to render it ignition-resistant or reduce its surface flame spread. One method is to paint wood with an intumescent coating, which swells up like a marshmallow when heated and retards ignition. One problem in using this method to reduce the hazard of installed wood paneling is its untreated reverse side.

Wood can be rendered fire retardant by drawing out the sap and moisture and replacing it with chemicals. Some codes will accept such wood as noncombustible, although this is a misnomer. Fire-retardant-treated (FRT) wood is just that—fire retardant. It is still combustible.

In recent years, the excess wood in wooden structural members has been reduced. This excess wood had provided some inherent fire resistance due to its additional mass. Using sawn or laminated two-by-four lumber or flakeboard in wooden I-beams has reduced the mass of these wooden beams. These lighter beams can be expected to fail earlier in a fire, because as soon as the web starts to burn, the strength of the beam is lost.

In the construction of wood-frame row-type dwellings, no fire protection provision is resisted more than parapeting fire walls to separate the units. Parapeted fire walls present a special problem to the builder because they require skilled craftsmen to flash the wall properly to prevent leaks. It is much easier to place shingles on a flat unbroken roof.

Some years ago the practice of laying a 4-ft (1.2-m) sheet of ignition-resistant plywood on each side of a fire wall which did not penetrate the roof was adopted in lieu of parapetting. The treated plywood is supported on untreated lightweight, wood trusses, which are subject to early failure. It is now known that plywood, laid over the top of masonry firewalls, will delaminate and/or raise up and pass fire over the top of the wall. Furthermore, plywood over a rough-cut, concrete block fire wall is difficult to seal against the passage of fire, smoke, and hot gases.

In an attempt to prevent this extension, some codes specified a 4-ft (1.2-m) roof section of fire-retardant-treated plywood to be laid lengthwise along each side of the fire wall. Some of the chemicals used in the treatment reacted from heat and humidity and destroyed the plywood. Thousands of roofs required replacement. In at least one jurisdiction, the underside of the replaced roofs is covered with a 4-ft (1.2-m) wide layer of gypsumboard along the firewall line. In the 19th century, plastering was used in an attempt to provide fire resistance to wood columns. Rot occurred under the plaster.

Another method of protecting wood from fire is found in floor and ceiling assemblies and wall sections that combine wood framing with gypsumboard membranes. Floor and ceiling assemblies may receive fire resistance ratings of 1 hr or more after being subjected to the NFPA 251/ASTM E119 test, *Standard Methods of*

FIG. 10-11G. Firestopping is at the mercy of any worker with a pipe or cable to run.

Tests of Fire Endurance of Building Construction and Materials (hereinafter referred to as NFPA 251/ASTM E119). One problem is that this fire resistance test was developed for floors of homogeneous, noncombustible materials, which were essentially the same from top down and bottom up. The one-way test in which the fire is applied from the bottom does not adequately determine fire resistance in today's fires where severe temperatures develop at floor level quite early in the fire.

The difficulty with fire fighters accepting the 1-hr resistance rating as indicating a degree of safety is even more serious in the case of lightweight gusset-plate trusses. When solid wood floor joists are used, each joist can act as a fire stop, thus slowing the progress of the fire through the floor construction. In the case of a truss, the entire floor void, along with its air supply, is available to the fire as soon as fire enters the void. The lightweight members of the truss are of small kindling-like dimensions. The voids can extend over sizable areas.

Firestopping is difficult to install and may be ineffective in preventing the extension of fire to adjacent truss voids. Early collapse of a structure in which fire has possession of the truss voids is inevitable. A three-story trussed floor apartment house in San Antonio, TX, collapsed moments after fire fighters evacuated the building. The fire started when a flammable liquid accidentally spilled on a floor, leaked down into the truss void, and ignited.[3]

Trusses can extend out past the exterior wall of the building to form the platforms of the stairway structure. A piece of gypsumboard, plywood, or flakeboard may be used as firestopping to separate the floor void from the balcony void. There are no data indicating that such firestopping is adequate, particularly when it is penetrated for utilities. In such a structure, a fire in the floor void could easily extend to the stair platform void, leading to early collapse of the stairway which is both the exit for the occupants and a principal avenue for attack by the fire fighter.[4]

In one type of truss, the gusset plates and diagonal tension members were stamped from one piece of steel. When a fire caused a floor to collapse, it fell on the floor below, and it too collapsed, because the tension members just buckled out of the wood.[3]

Fire burning down into the floor void may come up on the other side of a rated wall, thus making dubious the concept that compartmentation is provided when a "rated wall" meets a "rated floor."

A fire that started in the truss void was caused by a ceiling light fixture. The fire extended unnoticed from the truss void into the void

FIG. 10-11H. *Many churches are of plank on laminated wood beam or arch construction, often a fast flame spread potential.*

FIG. 10-11I. *Many old buildings are being adapted to other uses. Dropped ceilings provide huge voids through which hidden fire travels. Nailing up "fire-rated" gypsum board does not solve the fire extension problem.*

between the plywood sheathing and the brick veneer, then back into the floor voids above. A collapse resulted.[3]

In addition to collapse and fire spread, the truss floor provides a void in which toxic and combustible CO can accumulate.

A more complete discussion of the hazards of trusses can be found in references 3 and 4.

Heavy Timber

Heavy timbers are slower to reach sustained burning than lighter wood members. This has lead to the term *slow burning* being applied to buildings of masonry exterior walls and timber interiors. Such a building is expected to retain its structural stability for a long time when subjected to a fire. Unfortunately, "a long time" by building standards may not be long in the context of fire department operations, so fire fighters should not assume such a building can be safely occupied indefinitely.

The original concept of heavy timber construction, or mill construction, employed timber columns. Where the column intersected the floor girder, it should have been replaced by a solid cast-iron pintle, which passes through a hole in the girder. (A girder is any beam that supports other beams.) The small cross-section of the pintle would cause it to be protected by the mass of wood girder. However, this procedure was not used universally. Some heavy timber buildings have cast-iron columns that may fail early in a fire. Where these columns pass through the floor girder, there is no reduction in size of the column. As a result, two "half moons" of the timber rest on a shelf protruding from the column. Failure of the shelf, or the loss of a small amount of wood around this column, could cause the floor to fall.

Rehabilitated buildings often are fitted with unprotected steel columns, which may fail early in a fire. Sometimes, Lally columns (steel columns filled with concrete) are used. In the 19th century these Lally columns were often cast iron and were drilled to release the steam from the concrete. Current columns often are not drilled and can explode when heated, because of the steam generated when the concrete core is heated and releases water.

Heavy timber, slow burning, or mill construction buildings should be carefully examined for these potentially disastrous weaknesses, which make invalid the assumption that the building will retain its stability for any appreciable time.

Plank and beam construction is sometimes called *western framing*. The typical 16-in (406-mm) spacing of floor beams may be considered too boxlike and unattractive. Except in the roughest construction, a ceiling is installed to hide floor or roof beams. One take-off from heavy timber construction provides residences, churches, and other public buildings with an aesthetically pleasing interior by eliminating the ceiling, and exposing the wood structure. Heavy beams, widely spaced, are spanned with heavy tongue-and-groove planks instead of floor boards. The wood surface is planed smooth. Often a clear lacquer is used as a finish. This has the excellent advantage of eliminating void spaces. However, fire may still spread rapidly over these wood surfaces.

The hazard to the plank-on-beam roof of a building can be reduced if there are no flammable subsidiary structures, such as an office, to provide kindling to ignite the structure. Such spaces should be fire-resistive enclosures, or completely sprinklered.

Long heavy timbers are often in short supply or are not available. It is common practice to splice timbers to attain the proper length. Laminated arches are commonly shipped in sections to be field spliced. A properly designed spliced timber will carry its load adequately in normal use, but may fail disastrously in a fire. The arches of a Florida jai alai fronton were of spliced laminated timbers. Though the timbers were burned only moderately, the arches fell apart because the heated metal of the splice connections simply burned away the wood to which they were attached.

For a more complete discussion of the hazards of heavy timber buildings and a number of case studies and references, see reference 3.

Ordinary Construction

Ordinary construction is the term applied to buildings of usually combustible interior with masonry exterior walls. Originally, such buildings had brick or stone exterior walls and wood floors. The brick or stone walls were often required by law to limit fire spread from building to building in cities. In most building codes, the area of the city required to have construction that limits fire spread is defined by the "fire limits."

In more recent times, in addition to brick and stone construction, exterior masonry construction may use brick and clay tile composite, brick and block composite, cast-in-place concrete, concrete block, hollow clay tile, and adobe.

FIG. 10-11J.   A peaked truss roof was erected to change the roof line of this flat roofed building. The old combustible built-up roof covering was left in place. The exhaust duct was not extended through the new roof. Ventilating the new roof would be hazardous and useless. (Source: D. Harlow)

Brick is attractive, but expensive. Many buildings that appear to be of brick construction are actually a single layer of brick veneer on the face of a wood building. Brick veneer can be applied to any surface. Brick or stone veneer is often used to rehabilitate an old building. Imitation stone veneer is made of concrete cast in molds.

Brick veneer is totally dependent on the basic structure of the building for stability. If the supporting structure is damaged, the veneer may fall down. A veneer building is not a masonry building. It should be described by its basic construction type, e.g., wood frame, steel, or concrete.

In the past it was possible to recognize brick masonry from veneer, because in solid masonry some brick courses were turned to show the ends (headers), while the others were laid lengthwise (stretchers). The development of the masonry wire truss has made it possible to lay all the bricks as stretchers, so the difference is no longer obvious.

Some causes of masonry wall collapse are:

- Many brick walls are not totally brick. It was common practice to insert wood lintels over windows and doors. In many buildings, wood beams were laid in the wall to provide a level surface for floor beams. When this wood burns out, a collapse is likely. This can be determined only by competent prefire analysis.

- Water-soluble sand-lime mortar was universally used until 1880, and partially used thereafter for many years. It is still used today in the rehabilitation of older brickwork. It can be washed out by hose streams. The loss of mortar to water leaks can cause collapse when there is no fire.

- While lightweight steel bar joists can fail early in a fire, first they will elongate. Elongating steel can push down a masonry wall.

- Collapse of wood floors was early recognized as a hazard to the masonry in which the floor beams were imbedded. Collapse of a floor can precipitate wall collapse. The beams may act as a lever pulling down the brick wall. In response to this danger, builders may corbel out a ledge, then set the beams on the ledge, so that if they fell, the wall would be undisturbed. In other cases, a fire cut might be required on wood beams. In these, a 45-degree cut is made on the end of the beam, so that the top

is shorter than the bottom. Thus, the beam could fall out of the wall without disturbing the brickwork. Since wood is removed, the fire-cut beam will likely collapse earlier than a beam solidly set into the wall. Another method uses beam boxes, which are cast-iron boxes laid in the brick wall to accommodate a girder. When the girder fails, the box remains in place and the wall is not affected.

- The impact load of debris from a wall collapse can collapse adjacent buildings, with further risk to fire fighters.

- Unreinforced masonry buildings are liable to collapse in earthquakes.

- Old masonry buildings may be renovated for other uses, such as conversion from residences into commercial buildings. The floor load from files, safes and merchandise may be much greater than that which previously existed. This can place additional stress on the floor construction as well as the supporting wall construction.

- Some older masonry buildings were built with interior walls of balloon frame construction. Fire in such a wall can spread freely from floor to floor and into the floor voids. The wall is structural and its damage can precipitate an interior collapse.

- More substantial buildings were built with interior masonry structural walls. Since the practical span of wood beams was only about 25 ft (7.6 m) there were many small interior rooms. In many cases, walls have been removed to enlarge the interior space. The heavy masonry above was supported on a steel beam that, in turn, was supported by cast-iron columns, usually unconnected to the beam. The connection is vulnerable to any lateral thrust. Such an arrangement provided the larger space needed for a coffee shop, in the century-old Vendome Hotel in Boston. Nine fire fighters died in 1972 when the beam-column arrangement failed.[5]

Beyond these indications, there are other dangers associated with building construction that the fire service needs to be concerned with. There are some widely accepted beliefs regarding construction-related issues that are open to question. Some of these beliefs may prove fatal to fire fighters or others at the fire scene.[3]

One rule of thumb attempts to specify how far out a wall would fall depending on its height. This rule is of doubtful validity. Even if true for most of the wall, bricks and timbers can land much farther out than anticipated.

Collapse of masonry walls may be attributed to contraction caused by cold water hitting hot bricks. This phenomenon is unproved. The greatest amount of water thrown in a defensive attack is directed through windows, not onto a wall. Considering the low heat transmission rate of brickwork, it is likely that the brick is at best only slightly cooled by the water.

Many older buildings have an outer layer of finished brick, which may be a veneer. A hose stream hitting such a wall can get behind the finished brick, tear the bricks off the surface, and convert bricks into missiles.

The interior of ordinary construction buildings is almost completely of combustible construction. Where steel is used it may be left unprotected or enclosed in combustible voids that provide no protection from heat. The fire department may be completely unaware of the hazard of the hidden steel. Another major hazard of the voids which has been noted is the potential for hidden fires.

Alterations of the interior may leave hazardous floor conditions. When a stairway is relocated, the former opening may not be restored to a strength equal to the original floor. This presents a weak floor through which fire fighters may fall.

FIG. 10-11K. *The roof of this pre-cast tilt slab concrete building was fire damaged. The roof is vital to the stability of the building. The contractor's first move was to replace the tormentors which braced the walls before the roof was in place.*

FIG. 10-11L. *These unprotected steel struts could be loaded to twice the limit for permanent structural steel, thus lowering the failure temperature. A fire in combustible materials, often present in excavations, could cause the steel to fail. The adjacent structures could then collapse into the excavation.*

For more detailed discussions of buildings that have killed fire fighters, see references 3 and 5.

Noncombustible Construction

Noncombustible construction is a building code term that unfortunately may not be literally true. Fire departments should be aware of what the local code permits under this heading. It is possible that substantial combustible components, such as a wooden balcony or roof, or a combustible metal deck roof, are included in a noncombustible building. The fact that a material is noncombustible is not necessarily significant in reducing the fire loss, or in reducing hazards to fire fighters. The two common materials of noncombustible construction are: (1) steel and (2) concrete.

Steel is an extremely strong material. Heavy loads can be carried on very light structural steel elements. Because of the small mass of the steel, all but very heavy members are readily heated to the elongation temperature. Steel beams will elongate 9 in. (229 mm) in 100 ft (31 m) at about 1000°F (538°C). This can be sufficient to push out wall panels and cause toppling. Elongating steel has pushed down many walls in this fashion. At slightly higher temperatures steel reaches its collapse range. Collapse of steel roofs that carry sprinkler piping can cause early sprinkler system failure, resulting in massive destruction in warehouse fires, when the sprinkler system becomes inadequate for the hazard involved.

As steel trusses are heated, connecting ties can transmit undesigned torsional load to other trusses. This may result in failure of trusses away from the fire area.

Height of the steel above the floor is no guarantee of stability during a fire, if there is a significant fire load. Tests at Underwriters Laboratories Inc. (UL), after the devastating 1967 McCormack Place Exhibition Hall Fire in Chicago, showed that failure temperatures were reached in steel 30 ft (9.2 m) above the floor a few minutes after the fire started.[3]

Steel being heated by the fire should be cooled by hose streams if a safe operating position is available. If the steel is elongating, it will retract to its original length when cooled. If it has started to fail,

it will be frozen in the shape it had reached. Contrary to a once-prevalent myth, the water does not cause the distortion of the steel.

For a full explanation of the origin of the "don't put water on hot steel" myth, see reference 6.

Extreme caution should be the rule at any fire where unprotected steel is being heated. The potential consequences are not always evident. Preplanning should be thorough, and limits should be set on fire-fighting procedures to ensure safety of personnel.

Four fire fighters died in Pennsylvania in the collapse of a "concrete" building in which the floor was supported by unprotected steel columns.[3]

Concrete is strong in compression, but weak in tension. Plain concrete, usually found in the form of concrete blocks, is used for compressive loads. Reinforced concrete is a composite material. Steel is incorporated into the concrete to provide the tensile strength. By definition, a composite material is one in which both elements must act together to carry the load. A failure of the bond between the steel and concrete destroys the composite nature and strength of the structural element.

Unlike steel, which is protected from fire by an insulating covering, concrete can be inherently fire resistive. The resistance to fire is accomplished by the nature and thickness of the concrete itself. However, it is impossible to distinguish fire-resistive concrete from that which is merely noncombustible by simply looking at the concrete member.

Non-fire-resistive concrete has no specific fire resistance requirement and can fail early in a fire, even when exposed to only modest heat. No valid assumptions can be made as to when fire-exposed, non-fire-resistive concrete will fail. If the building is of monolithic construction, however, the load may be redistributed throughout the building, and a general collapse is unlikely. However, local collapse may cause injuries from falling concrete or from fire fighters falling through holes.

Precast concrete tilt slab buildings are now commonly used for warehouses and factories. Many of them are several stories high to accommodate high-rack storage. The integrity of the roof is vital to the tilt slab building.

The roof may be carried on several types of beams. Most common are wood, steel bar joists, and concrete pretensioned TEE beams.

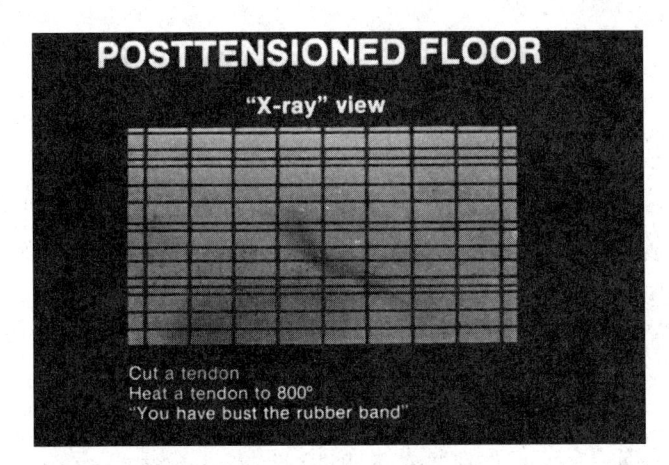

FIG. 10-11M.   *The post tensioned concrete building is a different building forever.*

FIG. 10-11N.   *This collapse was caused by the premature removal of falsework. A falsework fire could have the same effect, possibly burying fire fighters under tons of concrete.*

Concrete TEE beams would not generally be rated fire resistive unless the building was required to be fire resistive. If the concrete below the bottom tendon spalls off, the cold drawn tendon, which typically fails at 800°F (427°C), is exposed to the fire, and the beam is subject to failure.

A Metropolitan Dade County, FL, fire lieutenant lost his life when prestressed tendons in TEE beams in a carpet warehouse roof were heated to failure and the beam(s) collapsed.[3]

Massive reinforced concrete, fire-resistive multistory buildings have a good fire record. However, there have been cases of serious destruction, usually during long-duration, high-intensity fires. A recent fire in a concrete building loaded with tires caused heavy damage.[3] A long-burning fire in the top floor of the Military Records Center near St. Louis, MO, caused considerable elongation of the roof, which resulted in columns being torn off either the roof or the floor. After the fire, the top floor was removed.[3]

Massive concrete is a substantial heat sink. The heat that is stored in the concrete is not available to extend the fire. However, the retained heat in the concrete may cause serious health hazards to fire fighters during operations and overhauling. Personnel should be monitored and regular reliefs provided. Fire fighters required regular relief in the Los Angeles Library fire due to the retained heat.[3]

Finished, post-tensioned concrete buildings cannot be distinguished by appearance from conventional reinforced concrete buildings. They are very different with respect to fire operations. It is not safe to drill through a post-tensioned floor to attack a stubborn fire on a floor below. The drill might cut a tendon and cause collapse. A hole alongside a tendon might permit the tendon to heat to failure. To cut large holes with jackhammers, as was required in the library fire, would be extremely hazardous in a post-tensioned building.[3]

Concrete buildings under construction: The extraordinary rescue and body recovery operations required at the Oklahoma City, OK, bombing in 1995[7] could be repeated at a fire in a concrete building under construction, if conventional fire-fighting tactics are followed.

When concrete buildings are under construction, massive loads are supported on temporary forms that are either combustible, or, if made of steel or aluminum, subject to fire-caused failure. A massive failure of forms that drops one floor on another is almost certain to cause progressive collapse of several floors, possibly all the way to the ground. Fire in such forms is a serious concern and might well justify keeping fire fighters entirely out of the potential collapse zone.

In such cases, the command decision process should be the opposite of the traditional method. Traditionally, fire fighters rush into the structure, and withdraw only when it is determined that the structure is unsafe. In the case of fire in concrete forms, which support thousands of lb (kg) of concrete, it may be advisable to stay out until a decision can be made as to whether it is safe to enter the structure, or even be near it.

In Fairfax County, VA, in 1973, workers were prematurely ordered to remove shores supporting the 23rd floor. The floor collapsed, causing the progressive collapse of the floors below, all the way to the ground. Had the shoring burned, the result would have been the same.[3]

The situation in the case of post-tensioned concrete construction is even more serious than for conventional reinforced concrete. Until the tendons are tensioned, and the cables are stretched, the entire weight of the concrete, even hardened and dry to the touch, is supported on the concrete forms. If the forms burn, the entire floor or floors dependent on it will fall, possibly precipitating a total collapse. A 14-story post-tensioned concrete structure under construction in Cleveland, OH, in 1974 totally collapsed when fire attacked the forms. The fire chief, aware of the hazard, had kept the fire fighters out of the collapse area.[3]

For further information on concrete construction, see reference 3.

Fire-Resistive Construction

The major components of fire-resistive buildings are required to be similar to building components that have been subjected to the NFPA 251/ASTM E119 test. Test results are stated in the time to the

FIG. 10-110. Concrete filled steel tube columns like these, have exploded due to contained moisture. 19th century Lally® columns were drilled to vent the steam.

last full hour (or half hour) the assembly withstood the assault of the test fire. Thus, a column might be rated as having "1-hr fire resistance," or a floor-ceiling assembly as having "2-hr fire resistance."

There may be no specific relationship between the time given in the rating and the time to failure of the structure in a real fire, however. There are many variables that affect the time a structural element can withstand fire exposure. The limit of confidence that one can have is more nearly that a 4-hr fire-resistive building is more fire-resistive than a 2-hr fire-resistive building. Or, that a 1-hr fire-resistive building is less fire resistive than a 2-hr fire-resistive building.

For a more complete discussion of this often misunderstood subject, see reference 3.

Fire-resistive buildings generally fall into three classes, corresponding roughly to the time they were built.

1. During the period 1880 to 1920, emphasis was on the building itself. Floors were of tile arch with or without protection of the steel beams. Vertical opening protection was not even considered. Fire resistance consisted of whatever the architect thought proper, because there were few requirements for fire resistance in the building codes. There are many such buildings still in use.
2. From 1920 to 1945, many building components were rated fire resistive in accordance with the ASTM E119 test. Steel frames were fireproofed with concrete. Floors were cast-in-place concrete. Stairways and elevators were generally enclosed. Fire and occupant loads were low because of limited floor areas due to need for natural light and air. Many exit stairs were designed to provide an atmospheric break between the floor and the stairway, thus providing exits that did not become smoke polluted. Sprinklers were generally not provided, except for such buildings as the headquarters of the National Board of Fire Underwriters, and high-rise factory buildings in New York City as a result of laws passed in the wake of the Triangle Shirtwaist fire. Only a few cities had high-rise buildings of this type. The Empire State building represents the zenith of this type of fire-resistive construction.
3. From 1945 to present, many changes have occurred in building construction techniques. Air conditioning and fluorescent lighting have made extensive floor areas possible. The increase in height and floor area has permitted a huge increase in occupants. Timely evacuation is difficult. Air conditioning often requires nonoperable windows. Building height can increase the stack effect, which can become a serious factor in smoke movement.

AREAS OF CONCERN

The following areas of concern for the fire service were drawn from reports of serious fires in fire-resistive buildings. There are no "minor" deficiencies in buildings being attacked by fire. Any single deficiency may contribute to loss of life, or the destruction of the building.

Alarms and Staff Training

Building staff should be properly trained. A potentially fatal error is for a security guard or maintenance personnel to go to the floor from which the fire alarm has been received—thus walking right into the fire. The lone fatality in the spectacular 1988 First Interstate Bank fire in Los Angeles was the maintenance person who went to the fire floor to investigate.[3] Further, a guard went to the alarm floor of the One Meridian Plaza fire in Philadelphia, PA, and barely escaped death.[8]

Sometimes personnel are temporary employees who have not had adequate instruction in their duties and responsibilities. Fire department personnel should inspect buildings during the night shift and question employees as to the actions they would take in the event of a fire alarm, or for reported smoke.

Building staff have reset alarms to prevent the alarms from being transmitted to the fire department. This was a factor in a 12-fatality fire in the Westchase Hilton Hotel in Houston, TX, in 1982.[3] Delayed alarms may be due to a mistaken management perception that the arrival of the fire department is bad publicity. A policy of charging for false alarms should be carefully monitored to determine if it results in nontransmission of alarms.

Preplanning

A thorough study of a target building should be made before the action preplan is developed. This fire department preplanning should study each building as an individual problem. Problems and/or deficiencies in areas noted below may be present in buildings and can be factors in serious fires. Some problems or deficiencies not listed here will no doubt be listed in the next edition, after being found significant at a serious fire or disaster.

Automatic sprinklers: Full automatic sprinkler protection, properly maintained, is a proven fire protection measure that can provide reasonable assurance of occupant and building fire safety. However, far too often sprinklers are inoperable. The fire department should carefully monitor the maintenance and shutdown of the sprinkler system. The history of serious fires in sprinklered buildings is replete with examples of disasters that occurred when the sprinklers were cut off. Special precautions and watch service should be required during periods of sprinkler impairment.

In many cases, building codes permit alternatives from code provisions if sprinklers are installed. If the sprinklers are turned off when the building is occupied, it is probably in serious violation of the code provisions.

Water supply: The water supply system should be carefully checked for possible failures. "What happens if?" questions should be developed, asked, and answered.

Examples of water supply problems include water supply pumping system disconnections. Also, falling glass may cut lines supplying standpipes. Improperly marked standpipe connections also may be factors in fires.

Compartmentation and structural integrity: A fire-resistive building is generally designed to provide for compartmentation and

FIG. 10-11P.   Due to unprotected openings in floors, fire-resistive buildings often resemble a sieve to smoke and gases. The handful of fiberglass stuffed into the opening will not stop smoke and gases.

FIG. 10-11Q.   A very serious hazard is created when old combustible tile is left in place when a new "code-approved" suspended ceiling is installed below it. (See Reference 10.)

structural integrity in a fire. Unfortunately, the design may be flawed, or the construction faulty. Changes after the building was occupied may compromise the value of the compartmentation or structural integrity.

- Early fire-resistive design did not include compartmentation. It was believed to be sufficient that the building was "fireproof." Several disastrous fires in fire-resistive buildings proved the concept faulty, and compartmentation became required. Many early fire-resistive buildings are still undivided and allow the passage of smoke and toxic gases throughout the structure.

- Floors are penetrated for wiring and other services. Openings may not be properly protected to restrict the passage of heat and smoke.

- The reliability of horizontal compartmentation separating floors into fire areas or smoke compartments must be questioned. The wall may remain intact but there may be utility openings through the walls which are not properly sealed.

- Fire barrier doors into stairways and between fire areas are required to have door closers and latches. These may be removed or doors blocked open for convenience. This problem occurred at a high-rise apartment house that had both interior stairway doors propped open. A fast-spreading fire erupted which filled the stairs with heat and smoke. Four occupants died. It took almost 200 fire fighters to subdue the fire and to evacuate the tenants.[9]

Door closers with fusible links can be purchased so that the door could remain open, but when the link fuses due to heat from the fire the door will close. Such fusible links are not sensitive to smoke and toxic gases. Another method is to release or close doors upon operation of smoke detectors.

Structural integrity: The structural integrity of a building may depend on how the fire resistance was achieved.

- Buildings with tile arch floors may not provide protection for the steel beams. In later buildings, skewback tiles were provided to protect the steel.

- Steel floor and ceiling assemblies using suspended ceilings are common. Unfortunately, ceiling tiles are easily removed and integrity is hard to maintain. The ceiling voids can pass smoke and fire.

- Permitting steel columns to be unprotected in the plenum space is particularly dangerous. A single failure of the ceiling membrane would subject the column to fire temperatures. The heating of only a short length of a column to failure temperature could have costly, or even catastrophic, effects on the building.[2]

Spray-on fire proofing used asbestos fibers for many years. The health hazard of asbestos fibers now requires its removal. A most important question is: what, if anything, is done to fire-protect the steel when the asbestos fireproofing is removed.

Flame spread: Fire departments should be aware of the presence of combustible interior finish, such as plywood and combustible acoustical tile, in buildings. Older acoustical tile was made of compressed wood fibers and will burn furiously. It is usually installed on wood stringers, which enables burning on both sides of the tile. If it is applied with glue, the adhesive can add significantly to the rate of burning. This was discovered in the investigation of the Hartford Hospital Fire in 1961, in which 16 people died. The combustible tile was glued to a suspended gypsumboard ceiling, surface-treated with a flame retardant. When tested, it was learned that the adhesive caused a very rapid flame spread.[3] When new tile is installed, it will probably be required to meet flame spread requirements. However, there is often no requirement that the old tile be removed. Fire may develop on the old combustible tile in the concealed space above the new tile ceiling.

Flame spread was a significant factor in the John Sevier Nursing Home, Johnson City, TN, fire, in which 16 occupants died.[11] Of additional significance is the Indianapolis, IN, Athletic Club fire, in which two fire fighters died.[11]

Construction in occupied buildings: Buildings are often partially occupied before construction is finished. Many major remodeling projects are carried on in occupied buildings. The fire department should be aware of any such construction. Special permits should be required for any construction after the "certificate of occupancy" is issued.

A typical construction fire in such a building is a serious threat to the safety of all the occupants. At least two very serious fires have occurred.

Lightweight construction: The almost complete change from sawn lumber to gusset plate wood trusses and wood I-beams has profoundly changed the fire suppression environment. Tactics learned, and experience gained, on sawn joist and rafter buildings is not transferrable to truss buildings. The truss floor introduces a new hazard, which creates a floor void in which deadly explosive carbon monoxide (CO) gas can accumulate.

The strength of a wood beam can be defined by picturing an I (as in a steel I-beam) within a vertical cross section of the beam. The wood outside the I can be considered as "fat." The burning of this "fat" has little effect on the strength of the beam. The wood I-beam eliminates this "fat." When a fire attacks a wood I-beam, the expendable "fat" is absent and consequently the "muscle" fails rapidly.

Vacant buildings: Baltimore, Philadelphia, Lynn (MA), Minneapolis, Indianapolis, and Montreal are some of the cities that have suffered recent massive downtown fires in old, combustible, interior masonry buildings that were being remodeled. The buildings were often undivided, and sprinkler systems had been removed. In these cases, the fire was in possession of the building on fire department arrival. Multimillion dollar damages were sustained in nearby structures. Many valuable assets and plans for downtown rehabilitation went up in smoke. In the Minneapolis fire, floors six through sixteen of an adjacent fire-resistive high-rise office building were gutted by an exposure fire in a department store being demolished. The loss was almost $100,000,000.[3]

The identical problem is presented by similar huge structures not planned for rehabilitation but simply vacant because of economic changes. Sprinklers are often cut off to prevent freezing, and the structures are an open invitation to arson.

Added roofs: New roofs are often constructed over the original roof to control leaks (rain roof) or for aesthetic reasons. These present serious problems in fire fighting. Reference 12 provides detailed information.

Stadiums: Two recent serious stadium fires in Texas and Georgia have focused attention on the severe difficulties faced in fighting fires in the luxury "sky boxes." Fortunately, in both cases, there was no opportunity for loss of life or panic. The potential for flaming plastic windows falling onto occupied combustible seats in the arena below is a serious risk. Fire departments should perform realistic risk assessments and, where indicated, require fire protection improvements.[13]

SUMMARY

This chapter presents only the highlights of the problems that buildings present to fire suppression forces. Further detailed information can be found in many other materials in the references cited. Needless to say, knowledge and experience with building construction methods and materials can go a long way in improving fire department safety and efficiency.

BIBLIOGRAPHY

References

1. *Hazard Alert Bulletin,* Society of Fire Protection Engineers, 75-2, Apr. 1975, Boston, MA.
2. *Fire Protection Through Modern Building Codes,* American Iron and Steel Institute, 1961, p. 123.
3. Brannigan, F. L., *Building Construction for the Fire Service,* 3rd ed., National Fire Protection Association, Quincy, MA, 1992.
4. Malanga, R., "Fire Endurance of Lightweight Wood Trusses in Building Construction," *Fire Technology,* Vol. 31, No. 1, 1995.
5. Dunn, V., "Collapse of Burning Buildings," *Fire Engineering,* Saddle Brook, NJ, 1988.
6. Brannigan, F. L., "Fire Fictions II," *Fire Chief,* May 1995.
7. Comeau, E. R., "NFPA Report on the Oklahoma City Bombing," National Fire Protection Association, Quincy, MA, 1996.
8. Klem, T., *NFPA Investigative Report: One Meridian Plaza Fire, Philadelphia, PA, 1991,* National Fire Protection Association, Quincy, MA.
9. Isner, M., "Investigation Report, Apartment High-Rise Fire, Manhattan, New York, January 11, 1988," National Fire Protection Association, Quincy, MA. See also DeVita, A., and Dunn, V., "Fatal Fire on E. 50 St.," *With New York Fire Fighters,* First Issue, 1988, New York Fire Department, Fire Academy, Randalls Island, NY.
10. Isner, M., "Fire Investigation Report: Elderly Housing Fire, Johnson City, TN," Dec. 24, 1989, National Fire Protection Association, Quincy, MA.
11. Isner, M., "Fire Investigation Report: Indianapolis Private Club," NFPA, National Fire Protection Association, Quincy, MA.
12. Harlow, T. D., "New Roofs over Old," *Fire Engineering,* Nov. 1987, pp. 51–56.
13. Isner, M., "Summary Fire Investigation Report, Two Stadium Fires," 1993, National Fire Protection Association, Quincy, MA.

NFPA Codes, Standards, and Recommended Practices

Reference to the following NFPA codes, standards, and recommended practices will provide further information on building construction concerns of fire departments discussed in this chapter. (See the latest version of *The NFPA Catalog* **for availability of current editions of the following documents.)**

NFPA 13, *Standard for the Installation of Sprinkler Systems*
NFPA 220, *Standard on Types of Building Construction*
NFPA 241, *Standard for Safeguarding Construction, Alteration and Demolition Operations*
NFPA 251, *Standard Methods of Tests of Fire Endurance of Building Construction and Materials*
NFPA 252, *Standard Methods of Fire Tests of Door Assemblies*
NFPA 255, *Standard Method of Test of Surface Burning Characteristics of Building Materials*

Additional Reading

ASTM E119, *Standard Test Methods for Fire Tests of Building Construction and Materials,* American Society for Testing and Materials, Philadelphia, PA, 1988.
"Balanced Design Approach to Firesafety for Low-Rise Multifamily Construction. Fire Protection Planning Report No. 18 Of a Series," Concrete and Masonry Industry Firesafety Committee, Skokie, IL, Fire Protection Planning 18, 1990.
Barnett, J. R., "New Design Approach for Steel Structures Exposed to Fires," *Journal of Fire Protection Engineers,* Vol. 3, No. 1, 1991, pp. 1–8.
Becker, W., "Role of Intumescents for Construction Products With Improved Fire Safety," *Fire and Materials,* Vol. 15, No. 4, Oct./Dec. 1991, pp. 169–173.
Beilicke, G., "Basic Thoughts on Structural Fire Protection of Protected Historic Buildings," *Fire Science and Technology,* Vol. 11, No. 1–2, 1991, pp. 57–59.
Brannigan, F. L., "Concrete, Fire Resistive or Non-Combustible," *Fire Fighting in Canada,* Vol. 35, No. 3, April 1991, p. 17.
Brannigan, F. L., "Safety at Building Structural Fires," *Fire Engineering,* Vol. 145, No. 2, Feb. 1992, pp. 14, 16–17.
Burgess, I. W., Olawale, A. O. and Plank, R. J., "Failure of Steel Columns in Fire," *Fire Safety Journal,* Vol. 18, No. 2, 1992, pp. 183–201.
Chandrasekaran, V. and Suresh, R., "Analysis of Steel columns Exposed to Fire," Third International Conference and Exhibition on Fire and Materials, Oct. 27–28, 1994, Arlington, VA, 1994, pp. 41–50.
Chandrasekaran, V., Suresh, R. and Grubits, S., "Performance Mechanisms of Fire Walls at High Temperatures," Third International Conference and Exhibition on Fire and Materials, Oct. 27–28, 1994, Arlington, VA, 1994, pp. 51–62.
Clifton, C., "HERA: Fire Resistance of Structural Steel," *BUILD,* Feb. 1991, pp. 28–29.
Collier, P., "Design of Light Timber Framed Walls and Floors for Fire Resistance. Technical Recommendation," Building Research Association

of New Zealand, Judgeford, BRANZ Technical Recomm. 9, Aug. 1991.

Collier, P. C. R., "Design of Loadbearing Light Timber Frame Walls for Fire Resistance. Part 1. Study Report," Building Research Association of New Zealand, Judgeford, BRANZ Study Report SR36, December 1991.

Cooke, G. M. E., "Can Integrity in Fire Resistance Tests Be Predicted?," Sixth International Fire Conference on Fire Safety, Interflam '93, March 30–April 1, 1993, Oxford, England, Interscience Communications Ltd., London, England, 1993, pp. 321–330.

Cooke, G. M. E., Virdi, K. S. and Jeyarupalingam, "Thermal Bowing of Brick Walls Exposed to Fire On One Side," Fire Research Station, Garston, England, City University, London, England, ISBN 0-9516320-9-4; *Proceedings* of the Seventh International Interflam Conference, Interflam '96, March 26–28, 1996, Cambridge, England, Interscience Communications Ltd., London, England, 1996, pp. 915–919.

Corlett, R. C. and Tuttle, M. E., "Testing Structural Composite Materials for Fire Performance," *Proceedings* of the Eighteenth International Conference on Fire Safety, Volume 18, Jan. 11–15, 1993, Millbrae, CA, Product Safety Corp., Sunnyvale, CA, 1993, pp. 194–197.

Coyle, J., "Innovation in Building Design and Safety," VHS Video Tape, Fire Administration, Washington, DC, Video; Federal Fire Forum, Interaction of Fire Safety With Building Design, Conducted in Conjunction With "Safety/Survivability 2000." May 1, 1991, Washington, DC, 1991.

"Design of Loadbearing Light Steel Frame Walls for Fire Resistance," *Build*, Nov. 1995, pp. 28, 30.

Dunn, V., "Building Construction and Fire Spread," *Firehouse*, Vol. 21, No. 6, June 1996, pp. 16, 18-21.

Dunn, V., "Common Construction Types," *Firehouse*, Vol. 19, No. 2, Feb. 1994, pp. 20–23

Eisner, H., "Evaluating Structural Stability: Specialists Observe, Confer to Maintain Safety," *Firehouse*, Vol. 20, No. 10, Oct. 1995, pp. 67–69.

Falke, J., "Comparison of Simple Calculation Methods for the Fire Design of Composite Columns and Beams," Composite Construction in Steel and Concrete II, *Proceedings* of an Engineering Foundation Conference, 1992, Potosi, MO, ASCE, New York, NY, 1992, pp. 226–241.

Favro, P. C., "Fire-Retardant Treatments for Wood Roofs: A Permanent Solution," Fire Safety '91 Conference, Volume 4, Nov. 4–7, 1991, St. Petersburg, FL, Am. Technologies Intl., Pinellas Park, FL, 1991, pp. 136–145.

"Fire Safe Building Design for Architects and Designers," AmerInd, Alexandria, VA, CD-ROM.

"Fire Safe Building Design for Architects and Designers: National Fire Academy Produces First-Class CD-ROM," *Sprinkler Age*, Vol. 75, No. 4, April 1996, pp. 10–12.

Fisher, G. L., "Detection, Suppression and Construction: The Balanced Approach to Fire Protection," Sprinklers in Residential and Commercial Buildings, Aug. 19–20, 1990, Charleston, SC, Natl. Conf. States on Bldg. Codes and Standards, 1990, pp. 173–182.

Fitzgerald, R. W., "Engineering Method for Translating Fire Science Into Building Design," Worcester Polytechnic Inst., MA, University of Ulster and Fire Research Station, CIB W14: Fire Safety Engineering, International Symposium and Workshops Engineering Fire Safety in the Process of Design: Demonstrating Equivalency, Part 2, Symposium: Engineering Fire Safety for People With Mixed Abilities, Sept. 13–16, 1993, Newtownabbey, Northern Ireland, 1993, pp. 39–70.

Fitzgerald, R. W., "Selection of Fire Endurance Requirements for Building Design. [Integration of Fire Science Into Building Requirements.]," *Fire Safety Journal*, Vol. 17, No. 2, 1991, pp. 159–163.

Franssen, J. M. and Dotreppe, J. C., "Fire Resistance of Columns in Steel Frames," *Fire Safety Journal*, Vol. 19, No. 2–3, 1992, pp. 159–175.

Fredlund, B., "Calculation of the Fire Resistance of Wood Based Boards and Wall Constructions," Lund Institute of Science and Technology, Sweden, LUTVDG/TVBB-3053, 1990.

Gerlich, J. T., "Design of Loadbearing Light Steel Frame Walls for Fire Resistance. Fire Engineering Research Report," University of Canterbury, Christchurch, New Zealand, Fire Engineering Research 95/3, Aug. 1995.

Gerlich, J. T., Collier, P. C. R. and Buchanan, A. H., "Design of Light Steel-Framed Walls for Fire Resistance," *Fire and Materials*, Vol. 20, No. 2, March/April 1996, pp. 79–96.

Harada, K. and Terai, T., "Fire Resistance of Concrete Walls," First Asian Conference on Fire Science and Technology (ACFST), Oct. 9–13,

1992, Hefei, China, International Academic Publishers, China, 1992, pp. 215–220.

Harmathy, T. Z., "Design of Buildings Against Fire Spread (A Review)," *Journal of Applied Fire Science*, Vol. 1, No. 1, 1990-1991, pp. 65–81.

Hasemi, Y. and Yoshida, M., "Wall Fire Modeling and Its Relevance to Building Fire Safety Design," Building Research Inst., Ibaraki, Japan, Science University of Tokyo, Japan, '93 Asian Fire Seminar, October 7–9, 1993, Tokyo, Japan, 1993, pp. 117–127.

Holmberg, S. and Anderberg, Y., "Computer Simulations and Design Method for Fire Exposed Concrete Structures," Swedish Fire Research Board, Stockholm, Sweden, Document 92050-1, FSD Project 92-50, Sept. 27, 1993.

Honorowski, C. J., "Fire Protection Design Methodology for Buildings," Kling Lindquist Partnership, Worcester Polytechnic Institute Conference on Firesafety Design in the 21st Century, Part 1, Conference Report, Part 2, Conference Papers, May 8–10, 1991, Worcester, MA, 1991, pp. 212–223.

Hosser, D., Dorn, T. and Richter, E., "Evaluation of Simplified Calculation Methods for Structural Fire Design," Fire Safety Journal, Vol. 22, No. 3, 1994, pp. 249–304

"Importance of Structural Protection," Fire, Vol. 88, No. 1082, Aug. 1995, p. 20.

Inha, T., "Terasrakentamisen palotekniset suojaustavat. [Fire Prevention in Steel Structures.]," Palontorjuntatekniikka, January 1993, pp. 14–17.

Isa, D., "Fire Safety Engineering in the Structural Design in Buildings," Malaysia Fire Dept., Kuala, Lumpur, CIB W14/95/1 (J); Tokyo Fire Department, Firesafety Frontier '94, International Fire Conference and Exhibition in Tokyo, Creating a Safe Tomorrow, Oct. 18–22, 1994, Tokyo, Japan, 1994, pp. 39–45.

Jones, G. H., "Firesafety Design in the 21st Century: A Building Officials Perspective," Director of Codes Admin., Kansas City, MO, Worcester Polytechnic Institute Conference on Firesafety Design in the 21st Century, Part 1, Conference Report, Part 2, Conference Papers, May 8–10, 1991, Worcester, MA, 1991, pp. 224–227.

Karydas, D. M. and Delichatsios, M. A., "Risk Assessment Methodology for Fire Safety Factors in Performance-Based Design of Buildings," Factory Mutual Research Corp., Norwood, MA, NISTIR 5499, Sept. 1994, National Institute of Standards and Technology, Annual Conference on Fire Research: Book of Abstracts, Oct. 17–20, 1994, Gaithersburg, MD, 1994, pp. 43–44.

Klingenhofer, H. G., "Europaische Normung fur den baulichen Brandschutz. [European Standard for Constructional Fire Protection—Fire Behavior of Building Elements.]," VFDB, No. 2, May 1991, pp. 55–58.

Kruppa, J., *et. al.*, "Fire-Endurance Testing and Development in Fireproofing Technology," Chapter 12, Council on Tall Buildings and Urban Habitat, *Fire Safety in Tall Buildings. Tall Building Criteria and Loading*, Committee 8A, McGraw-Hill, Inc., Blue Ridge Summit, PA, 1992, pp. 201–227.

Law, M., "Fire Safety Design Practices in the United Kingdom: New Building Regulations," Ove Arup Partnership, London, England, Worcester Polytechnic Institute Conference on Firesafety Design in the 21st Century, Part 1, Conference Report, Part 2, Conference Papers, May 8–10, 1991, Worcester, MA, 1991, pp. 228–235.

Lawson, M., "How Light Steel Used for Structural Purposes Can Be Made Fire Resistant," *Fire*, Vol. 86, No. 1064, February 1994, pp. 27–28, 30.

Lawson, R. M., "Behavior of Steel Beam-to-Column Connections in Fire," *Structural Engineer*, Vol. 68, No. 14, July 17, 1990, pp. 263–271.

Malanga, R., "Fire Endurance of Lightweight Wood Trusses in Building Construction," Technical Note, *Fire Technology*, Vol. 31, No. 1, First Quarter 1995, pp. 44–61.

Malhotra, H. L., "Specifying the Performance of Elements of Construction," FIREX, First International Conference on Fire Safety in Buildings The Influence of the EEC Construction Products Directive, Conference Papers, June 12-14, 1990, Birmingham, England, 1990, pp. 1–9.

"Methodologies and Available Tools for the Design/Analysis of Steel Components at Elevated Temperatures," Work Package No. FR2, Steel Construction Inst., UK, FR2, July 1991.

Messersmith, J. J., "Fire Resistance Rating Requirements for Exterior Bearing Walls," *Building Official and Code Administrator*, Vol. 24, No. 3, May/June 1990, pp. 34–40.

O'Hagen, J., "High-Rise Fire and Life Safety," *Fire Engineering*, New York, 1989.

O'Meagher, A. J. and Bennetts, I. D., "Modelling of Concrete Walls in Fire," *Fire Safety Journal*, Vol. 17, No. 4, 1991, pp. 315–335.

Quiter, J., "How Today's Fire Codes Impact Building Construction," *Consulting/Specifying Engineer*, Vol. 9, No. 5, April 1991, pp. 46–48, 50, 52.

Richardson, L. R. and Onysko, D. M., "Fire-Endurance Testing of Building Constructions Containing Wood Members," *Fire and Materials*, Vol. 19, No. 1, Jan./Feb. 1995, pp. 29–33.

Sakumoto, Y., *et. al.*, "Fire Resistance of Fire-Resistant Steel Columns," *Journal of Structural Engineering*, Vol. 120, No. 4, April 1994, pp. 1103–1121.

Scawthorn, C., "Structural Damage," *Fire Engineering*, Vol. 147, No. 8, Aug. 1994, pp. 68, 71–72, 74, 76, 78, 80–82.

Schaffer, E. L., "Fire-Resistive Structural Design," *Proceedings* of the International Seminar on Wood Engineering, April 3, 6, 8, 10, 1992, Boston, MA, 1992, pp. 47–60.

Smith, A., "Quick Guide to UBC Building Types,"*American Fire Journal*, Vol. 42, No. 5, May 1990, pp. 22–26.

Smith, C. I., "Building Design and Smoke Control," *Fire Engineers Journal*, Vol. 52, No. 164, March 1992, pp. 13–17.

Szitanyine-Siklosi, M., "Observations Regarding Resistance to Fire of Building Construction Pieces," Third International Symposium on Fire Protection of Buildings, May 10-12, 1990, Eger, Hungary, 1990, pp. 89-99.

Talamona, D., *et. al.*, "Factors Influencing the Behavior of Steel Columns Exposed to Fire," *Journal of Fire Protection Engineering*, Vol. 8, No. 1, 1996, pp. 31–43.

Tanaka, T., "Concept of a Performance Based Design Method for Building Fire Safety," Building Research Institute, Ibaraki-Ken, Japan, NISTIR 4449, 1990, 9 pages; Eleventh Joint Panel Meeting of the U. S./Japan Government Cooperative Program on Natural Resources (UJNR) on Fire Research and Safety, Oct. 19-24, 1989, Berkeley, CA, 1990, pp. 23–31.

Thomas, I., "Progress in Australia Towards Engineering Design for Fire Safety in Buildings," CFR Video Seminar, BHP Melborune Research Labs., Victoria, Australia, Video, March 20, 1990.

White, R. H., "Fire Resistance of Wood-Frame Wall Sections," Forest Service, Madison, WI, Sept. 1990.

Willse, P. J. G. and Smith, A. C., "Fire Walls: The Last Line of Defense," *Chemical Engineering*, Vol. 100, No. 7, July 1993, pp. 110–113.

Wimpey Offshore, "Use of Alternative Materials in the Design and Construction of Blast and Fire Resistant Structures," Work Package No. G5., Wimpey Offshore, UK, June 1991.

Woolley, D., "Fire Safety and the Construction Products Directive," FIREX, First International Conference on Fire Safety in Buildings—The Influence of the EEC Construction Products Directive, Conference Papers, June 12–14, 1990, Birmingham, England, 1990, pp. 1–11.

FIRE LOSS PREVENTION AND EMERGENCY ORGANIZATIONS

Revised by Thomas F. Barry and Richard W. Asa

Fire loss consequences, which can involve property damage, business interruption, life safety hazards, environmental issues, damage to corporate image, and lack of future profitability, can present a major threat to corporate goals and survival. Recognition of these consequences has prompted the development of a formal management structure for fire loss prevention and control in many corporations.

FIRE RISK MANAGEMENT

Fire risk management can be defined as the systematic application of management policies, procedures, and practices to the tasks of analyzing, assessing, and controlling fire risk in order to protect employees, the public, the environment, and company assets while avoiding business interruption.

This section describes the general framework of the fire risk management process, represented in Figure 10-12A, which includes hazard identification, quantified risk assessment, and risk management. Resources, such as *The SFPE Handbook of Fire Protection Engineering*'s section on fire risk analysis;[1] and the Center for Chemical Process Safety's *Guidelines for Chemical Process Quantitative Risk Analysis*[2] provide additional information. Supplementary reading resources that provide detailed discussions of fire risk and decision analysis topics have been added to the bibliography.

Hazard Identification

A fire or explosion hazard can be defined as a characteristic of a system, plant, or process that represents a potential for an unplanned event leading to undesirable losses. The key words in this definition are "undesirable losses." Hazard identification is the process of recognizing hazards that can pose significant undesirable losses. Hazard identification should be a continuous activity in the evaluation of new materials, plant additions, and production modifications, as well as the continuing inspection and evaluation of existing facilities.

Resources, such as NFPA's *Industrial Fire Hazards Handbook*,[3] various chapters within this handbook, and many NFPA standards, describe fire hazards in major industries and special process hazards based on current technology and past loss experience.

Thomas F. Barry, P.E., is a senior engineer, specializing in fire and explosion risk assessment, with HSB Professional Loss Control, Kingston, TN. Richard W. Asa, P.E., is a fire protection engineering consultant in Naperville, IL.

Risk Assessment

The risk assessment process consists of both identifying risks and estimating the probability and magnitude of potential fire losses. In recent years, there has been an increase in the use of formalized risk assessment methods to support decisions on fire safety issues. Predictive methods that integrate statistical data, deterministic models, and expert opinion increasingly are being used.

The primary steps in fire risk assessment include the following:

1. Identification of the fire events that could lead to significant loss.
2. Quantification of the fire risk—that is, probability of fire event occurrences and loss consequences—for the generated fire loss scenarios.
3. Development and evaluation of alternative fire prevention and/or fire protection strategies (recommendations) to reduce the fire risk.
4. Measurement of the estimated change in fire risk (difference in probability and/or consequences) associated with the alternatives.

One of the most difficult tasks in the risk assessment process is addressing probabilities. Fire risk reflects both the probability of fire occurrence and the loss consequences of the fire. In both factors there is an element of uncertainty that must be recognized. Approaches for determining the probabilities of values for fire event occurrence are as follows.

Objective estimation: If valid data on loss event frequencies are available and applicable, then probabilities can be extracted from this source. Often objective data sources are scarce. To evaluate the application of historical data, it is necessary to look at the following, both for the source of the data and for the situation under consideration:

- Design standards and quality
- Inspection methods and quality
- Maintenance philosophy and quality
- Operational philosophy and quality
- Safety standards and enforcement

Subjective estimation: If historical data sources are not available for the loss event(s) of interest, probabilities can be evaluated using inferential judgment based on available loss-trending information, such as equipment failures, human error, ignition sources,

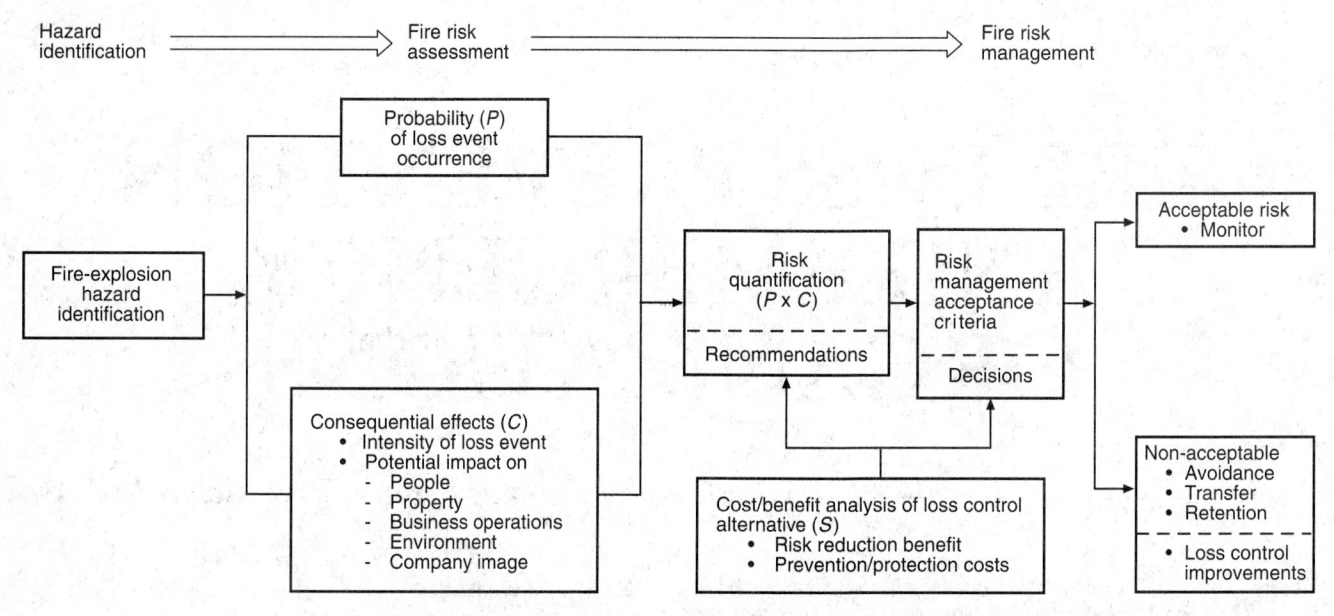

FIG. 10-12A.   General framework for industrial fire risk management process.

loss control elements, and damageability factors. A good source of information is the NFPA Fire Analysis and Research Division.

Subjective estimation processes should be documented and should involve the best judgment a fire protection expert can make based on knowledge of contributing factors from past fire losses, site-specific evidence of loss control deficiencies, and exposed values.

A very good reference on equipment failure data sources and application precautions is the AIChE's Center for Chemical Process Safety book entitled *Guidelines for Process Equipment Reliability Data, with Data Tables.*[4]

The physical intensity of fire-explosion consequences can be quantified in terms of the expected energy released (heat exposure, smoke and/or corrosive gas contamination, explosion blast over-pressures, etc.), the area involved, and the duration of the fire. Prob-abilistic engineering assessment—and, in some situations, deterministic fire-explosion modeling tools—are utilized to aid the evaluation process. *The SFPE Handbook of Fire Protection Engineering*[1] provides information on various analytical tech-niques that can be used in this evaluation.

Once the intensity and duration of the potential fires have been quantified, then the impact on both direct and indirect loss potential must be assessed. Direct losses include damage to buildings, equip-ment, and contents. Indirect losses include business interruption, li-ability for injury or death, environmental contamination, and damage to company image. For most quantification studies, loss po-tential is expressed in equivalent monetary terms.

The type of risk assessment conducted depends on the signifi-cance of the decision, the complexity of the problem, and the time and cost limitations. Routine code-compliance problems can often be handled by simple choices. Complex problems involving new technologies or high-hazard operations require application of more detailed risk assessment and decision analysis methods.

Risk Management

Risk management addresses the value judgments involved in estab-lishing acceptable levels of risk and methods of handling identified risks. The acceptable risk decision-making process is based on spe-

cific organizational goals and generally includes (in any order) the following.

- Profit (competitive market position)
- Protection of company assets (major loss exposures)
- Continued company operation (business interruption)
- Continued growth (expansion)
- Humanitarian concerns (employee and public safety)
- Community goodwill (potential embarrassment)
- Legal requirements (liability, building codes, etc.)
- Insurance company requirements
- Environmental concerns

If the risk is acceptable, no immediate action may be neces-sary, but monitoring for changes that could increase the risk must be done. If the risk is unacceptable, then decisions must be made about how to deal with the risk.

General options available to the risk management decision maker for handling fire risk exposure include the following.

1. Avoiding the risk by nonparticipation in risky operations.
2. Risk transfer by purchasing insurance to cover potential losses.
3. Financing alternate risk transfer arrangements such as captive or corporate funded reserves (self-insurance).
4. Providing loss control improvements.
5. Developing a risk management program that includes a combi-nation of the above.

The fifth option generally provides a cost-effective approach for reducing fire loss potential and reducing the risk funding re-quirements for options 2 and 3 above.

Once a risk management decision has been made that includes loss control improvements, it may be necessary to conduct a cost/benefit analysis, either formally or informally.

Evaluating the cost of fire prevention and protection alterna-tives, which includes design, installation, system maintenance, and training expenses, is normally a straightforward process. A good discussion on cost comparison is provided in Section 11, Chapter 8, "Fire Risk Analysis."

Evaluating the benefit or the amount of risk reduction presents a more difficult task. Estimating the degree of risk reduction benefit involves the assessment of reduced probability of fire occurrence and/or reduction of direct and indirect losses.

Current methods for estimating risk reduction for alternative fire prevention and/or protection strategies are based on available loss experience and fire protection engineering analysis, but also tend to require considerable judgment. Valuable information may be obtained from industry and insurance reports, NFPA's Fire Incident Data Organization (FIDO), the National Fire Incident Reporting System (NFIRS), and internal plant reporting systems. Future development of computer models that integrate deterministic fire hazard modeling, probabilistic models, and risk profile (loss data) information will provide additional tools for risk assessment studies.

An important part of fire risk reduction is the development, implementation, and monitoring of comprehensive loss prevention and control programs. Written documentation should be provided for programs and updated periodically. Management programs should define specific objectives regarding personnel safety (both plant personnel and the general public), property conservation, environmental impact, and minimizing interruption to plant production.

Because of the numerous interfaces between the various engineering disciplines, as well as maintenance and operational programs, it is essential to place the responsibility for the fire protection program with a single individual or small group with direct access to top management. This individual must have sufficient authority to delegate responsibilities affecting plant fire protection across engineering discipline and departmental lines, and even across separate divisions such as engineering, plant services, utilities, safety and loss control, and emergency response.

LOSS PREVENTION AND CONTROL PROGRAM MANAGEMENT

Today, few companies, in the face of severe competition and the high cost of money, could survive a major loss. Over the years, it has become apparent that people can substantially aid in the minimization of losses. It has been statistically demonstrated, by major property insurance companies, that the factor with the largest effect on the size of a loss by fire/explosion is what people do or fail to do.

Management loss control programs are policies and procedures utilized for the protection of assets from loss due to fire or other causes. This section describes the general responsibilities, organization, and content of effective loss prevention and control program management.

Management Echelons

The implementation and enforcement of an effective loss control program is dependent on management involvement and support.

Management can be divided into three echelons, each of which makes important contributions to loss prevention and control: (1) executive management—the board of directors and company officers; (2) middle management—facility managers, department heads, and related staff managers; and (3) line management—the supervisors and section chiefs who direct each phase of the operation.

In a relatively small organization, the owner may perform the functions of all three management levels. Although the owner's loss control problems may be fewer, or less complex, than those of the management group in a larger organization, some of the same judgments must be exercised. The owner who tries to be loss control

manager may be at a disadvantage, however, because of the lack of specialized training in loss control, and the absence of managers with whom to share expertise and creative ideas regarding a sound loss control program. This need not pose a major obstacle. Expertise is available from professional loss control consultants, insurance companies, and fire departments.

Executive management: Members of executive management establish company policy, make major financial decisions, determine the product or service to be provided, and establish production levels. Executive management's primary goal is to guide the organization to produce a profit.

Regardless of insurance coverage, losses to plant property by fire or other causes can disrupt profitability. On the surface, assets diverted to property protection would appear to be a non-productive expenditure. However, after a serious loss, the need for an effective loss prevention/control program becomes apparent. For loss control programs to be effective, they must have complete support from executive management. In addition, once a program has been developed and accepted, executive management must insist on regular updates to monitor progress.

Middle management: Middle management is responsible for the development, implementation, and monitoring of cost-effective loss control programs.

The person with primary responsibility for loss control generally has the title of loss control manager. Depending on the organization, titles assigned to individuals vary, and in some cases, the individual may have several other titles when loss control is not a full-time function. To be effective, the assigned individual must have access to decision makers at the executive management level. Direct access would involve reporting to the officer level (e.g., vice presidents of engineering, finance, production, etc.). Indirect access would involve reporting to the director of risk management or other managers with direct access to executive management.

Major responsibilities of the loss control manager include the following.

Overall supervision.   Defining loss control program functions, clarifying responsibilities, and ensuring that all procedures are properly carried out.

Planning.   Anticipating hazards, and formulating plans and procedures for fire prevention and overall plant safety.

Education and training.   Providing employee training in fire prevention. Making sure that all production processes and operations are carried out safely, and that contractors follow safe procedures according to company regulations. (See Figure 10-12B.)

Assignment.   Delegating specific maintenance and emergency duties to employees, including guards, caretakers, patrol officers, and night workers.

Coordination.   Maintaining liaison with the following corporate departments and outside organizations.

- With production and engineering on new processes or materials that can introduce additional fire hazards within the facility.

- With plant engineering and maintenance for fire protection installations, impaired protection, maintenance of fire protection equipment, and housekeeping.

- With the public fire department and insurance company on plant layout, location and type of hazardous processes, coordination of forces, location and type of fire protection equipment, fire doors, fire walls, defense plans, and main defense lines.

FIG. 10-12B.   *Proper training is an important function of many loss control departments.*

Review.   Checking, periodically, the efficiency of the entire loss control program. Reviewing details of the plan with line management. Conducting drills and classes to maintain employee preparedness.

In smaller organizations, the facility manager or an appointee must assume these responsibilities.

Line management: Implementation of a loss control program extends beyond the middle management level. Unless the members of line management adequately perform their function, the program cannot produce satisfactory results.

Line managers, those with daily personnel contact, should aid in the information collection that middle managers incorporate into loss control programs. Once programs are implemented, line managers are responsible for regular enforcement of company policy.

With regard to employees, and work areas under the line manager's supervision, it is the line manager who is responsible for the following.

1. Trains personnel to carry out individual duties in a loss control program.
2. Works with the loss control manager and staff in selecting employees for the private fire brigade or other program assignments.
3. Works with the loss control manager to establish priorities for inspection, testing, and maintenance of fire protection equipment.
4. Inspires employees to develop a fire-conscious attitude and respect for the loss control program.
5. Insists that access to all protection equipment, as well as all means of egress, be kept open.
6. Works closely with the loss control manager and staff to ensure that new equipment or processes are adequately protected before startup.
7. Reports all occupancy changes that may affect loss control.
8. Works with the loss control manager and staff to ensure that fire exit drills are held regularly.
9. Accounts for each employee, following facility evacuation.
10. Supervises emergency action when a fire or other property-related incident occurs in the absence of members of an established emergency organization.

LOSS PREVENTION PROGRAM MOTIVATION AND DEVELOPMENT

Human error is the leading cause of fire and explosions. Various statistical studies of loss incidents in a variety of industries have indicated the following general breakdown:

- Human factors (70–85 percent) (e.g., management program deficiencies, human operator errors)
- Equipment failures (10–20 percent) (e.g., design, maintenance deficiencies)
- External factors (10–20 percent) (e.g., earthquake, floods)

Table 10-12A identifies some human error problem areas in the Chemical Processing Industry (CPI).[5]

To reduce human error leading to fire incidents, many companies are establishing and expanding policies and practices that prioritize the implementation and auditing of loss prevention programs. Integrated into many of these programs are insurance company guidelines, industry recommended practices, regulatory requirements, and NFPA standards.

Insurance Company Guidelines

Industrial property insurance companies focus on recommending management programs that assist in the prevention and minimization of property damage and business interruption losses. Some of the primary management programs include:[6]

- Impairments to fire protection systems
- Smoking regulations
- Maintenance
- Employee training
- New construction
- Pre-emergency planning
- Hazardous materials evaluation
- Cutting, welding, and other hotwork
- Loss prevention inspection
- Fire protection and security surveillance
- Fire protection equipment inspection
- Hazard identification and evaluation
- Proper housekeeping

Federal Regulations

In 1992, the Occupational Safety and Health Administration (OSHA) mandated a new standard within Subpart H, *Hazardous Materials*, to deal with the risks involved in the storage, handling, and processing of highly hazardous materials. This regulation, CFR 1910.119, *Process Safety Management of Highly Hazardous Chemicals*, focuses on the application of performance-based management programs, rather than specific engineering guidelines for the prevention of toxic releases, fires, and explosions on the plant site.

Process safety management (PSM) is the application of management systems to the identification, understanding, and control of process hazards to prevent hazardous and flammable material accidents and injuries. Regulation CFR 1910.119 applies to:

1. Flammable liquids or gases in quantities greater than 10,000 lb (4,536 kg) that are stored, handled, or transported under pressure on the plant site; and

TABLE 10-12A.   *Studies of Human Error in the Chemical Processing Industry: Magnitude of Human Error Problem*

STUDY	RESULTS	
Garrison (1989)	Human error accounted for $563 million of major chemical accidents up to 1984.	
Joshchek (1981)	80–90% of all accidents in the CPI are due to human error.	
Rasmussen (1989)	Study of 190 accidents in a CPI facility, identified the top 4 causes:	
	• Insufficient knowledge	34%
	• Design errors	32%
	• Procedure errors	24%
	• Personnel errors	16%
Butikofer (1986)	Accidents in petrochemical and refinery units:	
	• Equipment and design failures	41%
	• Personnel and maintenance failures	41%
	• Inadequate procedures	11%
	• Inadequate inspection	5%
	• Other	2%
Uehara and Hoosegow (1986)	Human error accounted for 58% of the fire accidents in refineries:	
	• Improper management	12%
	• Improper design	12%
	• Improper materials	10%
	• Misoperation	11%
	• Improper inspection	19%
	• Improper repair	9%
	• Other errors	27%
Oil Insurance Association Report on Boiler Safety (1971)	Human error accounted for 73% and 67% of total damage for boiler startup and on-line explosions, respectively.	

Source: *Guidelines for Preventing Human Error in Process Safety,* Center for Chemical Process Safety, American Institute of Chemical Engineers.[5]

2. Identified toxic chemicals (approximately 140) that exceed listed threshold quantities.[7]

Many of the highly hazardous chemicals listed in this OSHA regulation that are highly reactive, flammable, or explosive have been drawn from chemicals rated 3 or 4 in NFPA 49, *Hazardous Chemicals Data* (hereinafter referred to as NFPA 49).

The major program segments in CFR 1910.119 are:

- Employee participation
- Process safety information
- Process hazard analysis
- Operating procedures
- Training
- Contractors
- Pre-startup safety reviews
- Mechanical integrity
- Hot work permits
- Management of change
- Incident investigations
- Emergency planning and response
- Compliance safety audits

In addition to OSHA's *Process Safety Management* regulations, the Environmental Protection Agency (EPA) has developed a regulation, 40 CFR Part 68, *Risk Management Program for Chemical Accident Prevention*, that focuses on the exposure of toxic chemical releases and fire and explosions to the public.

EPA's risk management program (RMP) elements include:

- Hazard assessments
- A prevention program
- An emergency response program

EPA's RMP regulation will require development and implementation of management programs at facilities that manufacture, process, use, store, or transport regulated hazardous chemicals in quantities that exceed specified threshold limits as listed in this proposed regulation.

EPA's regulation rule adopts the OSHA CFR 1910.119 PSM regulation as the RMP's prevention program segment.

A Management Loss Prevention and Control Program Framework

Program elements and complexity will vary depending on the type of plant facilities, nature of hazardous operations, complexities of processes, and exposures to plant employees and the public.

Primary management program elements can be categorized as follows:

1. *Process Technology*
   - Process safety information
   - Process hazard analysis
   - Pre-startup safety reviews

2. *Mechanical Integrity*
   - Process equipment
   - Emergency safety devices
   - Protection systems

3. *Operations*
   - Operating procedures
   - Safe work practices
   - Hotwork permits
   - Fire system

- Impairments
- Flammable liquid handling and ignition control

4. *Management*
- Employee training
- Contractors
- Management of change
- New construction

5. *Incidents*
- Incident investigations
- Emergency planning

Process Safety Information

For process, storage, and transportation operations involving flammable and hazardous materials, it is important to develop, document, and update process safety information, including:

1. Information pertaining to hazards of the chemicals,
2. Information pertaining to the process technology, and
3. Information pertaining to the equipment.

Material hazards documentation provides pertinent hazard data for each chemical involved in a process or operation, including physical and chemical properties such as flammability and thermal stability data, reaction data, and any other property relevant to a particular hazard. Toxicity data are an important part of the package. In many cases, material safety data sheets (MSDSs) will satisfy this requirement, provided they contain the necessary details.

Chemical and flammability data can be extracted from sources such as NFPA 49, NFPA 491M, *Manual of Hazardous Chemical Reactions*, NFPA 325, *Guide to Fire Hazard Properties of Flammable Liquids, Gases, and Volatile Solids*, and the NFPA *Fire Protection Handbook*. Much of the data available in these publications are at atmospheric temperature and pressure. Experimental data appropriate to process conditions will sometimes be needed.

Dust explosion data for explosion-venting calculations are presented in NFPA 68, *Guide for Venting of Deflagrations*, which includes additional references. A considerable amount of dust explosion data can be obtained from various U.S. Bureau of Mines and NFPA publications.

Process Hazard Analysis

Hazard evaluation activities can be accomplished by trained plant personnel working with outside specialists where required. It is the hazard evaluation team's responsibility to study the process operation thoroughly, identifying all the potential fire and explosion hazards, and recommending changes or further reviews to eliminate or control the identified hazards.

A written report of the evaluation must be developed and should include:

- List of participants
- Methodology(s) applied
- List of documents, such as part and identification (P&ID) numbers and issue
- Summary of the findings
- Recommendations

All hazard evaluation reports, recommendation status reports, and follow-up documentation should become part of the process documentation file.

An organized approach is essential to conducting effective and efficient hazard evaluations. Hazard evaluation techniques include:

- Safety review
- Checklist analysis
- Relative ranking indexes
- Preliminary hazard analysis
- What-if analysis
- What-if/checklist analysis
- Hazard and operability (HAZOP) analysis
- Failure modes and effects analysis (FMEA)
- Fault-tree analysis
- Event-tree analysis
- Cause-consequence analysis
- Human reliability analysis

Each technique has specific application benefits, limitations, resource needs (manpower, time, budget), and documentation requirements. It would be impossible in this chapter to describe each technique. An excellent reference that addresses those techniques in detail and provides illustrative examples is the Center for Chemical Process Safety's (CCPS) *Guidelines for Hazard Evaluation Procedures with Examples*.[8]

The synonym for hazard evaluation presently in wide use is process hazard analysis (PHA), as this term is used in the OSHA *Process Safety Management* regulation, CFR 1910.119.[7]

In the OSHA regulation, employers are required to perform an analysis identifying and evaluating hazards involved in a process. This must include:

1. Using one or more of the following to perform a hazard analysis:
- Checklists
- What-if checklists
- Hazard and operability study (HAZOP)
- Failure modes and effects analysis (FMEA)
- Or an appropriate equivalent methodology
2. The hazard analysis must address:
- Process hazards
- Engineering and administrative controls
- Consequences of failure of controls
- Consequence analysis of effects on employees
3. The hazard analysis should be performed by a team with engineering and operations expertise, including at least one person with knowledge specific to the process.
4. The employer is required to establish a system for documenting findings and actions taken, then communicating them to employees.
5. The hazard analysis must be reviewed and updated at least once every 5 years.

It should be noted that OSHA's definition of process includes the storage, handling, processing, and transportation of hazardous chemicals and flammable and explosive materials. In addition, provisions of OSHA CFR 1910.119 include:

1. Fire and explosion exposures must be considered, in addition to toxic releases. The process hazard analysis should include fire prevention, fire protection systems, employee fire safety training, and emergency response to fire and explosion incidents.

2. Prior process hazard analyses must be updated and revalidated. Analyses that focused on toxic releases will now have to include fires and explosions if the process materials are flammable.

Pre-Startup Safety Reviews

A pre-startup safety review should be performed for new facilities and for modified facilities when there may be changes in process safety information. The pre-startup safety review should confirm:

1. Construction is in accordance with design specifications.
2. Safety, operating, maintenance, and emergency procedures are in place and adequate.
3. Mitigation actions necessary for startup are in place.
4. Operating procedures are in place and employees are trained.
5. Fire prevention and protection systems are in service.

Mechanical Integrity Programs

A primary fire prevention objective is to prevent the release of flammable and hazardous materials from containment. A quality mechanical integrity program (MIP) is a key element in meeting this objective.

The purpose of a mechanical integrity program is to ensure the safe operation of process and safety equipment through design, inspection, testing, preventive maintenance, and quality assurance monitoring.

The mechanical integrity program policies and practices at facilities handling flammable and hazardous materials will have a major influence on the level of failures and accidents experienced. Establishing comprehensive programs is a key ingredient in successful loss prevention and control management, and should include:

1. Identification of all critical equipment requiring inspection, testing, and preventive maintenance
2. Procedures and schedules for inspecting process vessels, piping, and associated process equipment, and safety and protection systems
3. Procedures and schedules for testing critical protective devices, such as pressure relief valves, detection and alarm interlocks, emergency generators, and fire protection systems
4. Quality assurance procedures to ensure corrective maintenance functions, employee qualifications, replacement part specifications, etc.

Mechanical integrity, addressed in OSHA CFR 1910.119, requires employers to ensure that process equipment is designed, installed, and operating properly, and includes the following provisions:

1. It applies to pressure vessels and storage tanks, piping, valves, relief and vent systems and devices, emergency shutdown systems, controls, monitoring devices and sensors, alarms, interlocks, and pumps.
2. Written procedures must be in place for maintenance.
3. Employees must be properly trained for maintenance.
4. Inspection and testing criteria should include:

   • Following accepted codes, standards, and engineering practices

   • Meeting or exceeding applicable codes and standards intervals of testing frequency

   • Documenting all inspections and tests

5. Equipment deficiencies must be corrected before further use.
6. Quality assurance criteria should include:

   • Ensure equipment meets design specifications

   • Checks and inspections should be performed to ensure proper installation

   • Ensure maintenance materials, spare parts, and equipment meet design specifications

Systems that protect facilities and personnel from hazardous situations (toxic releases, fire, explosion) must operate effectively and reliably. Emergency safety devices and protection systems generally remain static for long periods of time and, therefore, need scheduled inspection, maintenance, and testing to ensure that these systems function correctly during an emergency. Testing and maintenance of emergency safety and protection devices is very important and must be covered by a formal system with full documentation. Some examples of equipment and systems include:

• Pressure relief systems, vent systems, and devices
• Process monitoring devices and sensors
• Explosion vents
• Electrical grounding and bonding systems
• Dikes and drainage systems
• Gas and flame detectors
• Emergency alarm and communication systems
• Emergency shutdown devices and system interlocks
• Emergency generators
• Sprinkler and water spray systems
• Fire pumps
• Special suppression systems (e.g., dry chemical, foam)

Testing and inspection policies and procedures should follow applicable codes and standards, such as those published by the American Society of Mechanical Engineers (ASME), the American Petroleum Institute (API), the American Institute of Chemical Engineers (AIChE), the American National Standards Institute (ANSI), the American Society for Testing and Materials (ASTM), and the National Fire Protection Association (NFPA), where they exist; or recognized and generally accepted engineering practices.

An excellent reference on mechanical integrity programs is the *Mechanical Integrity Supplement to the Maintenance Excellence Guide*, published by the Chemical Manufacturers Association (CMA).[9]

Fire protection systems: An important mechanical integrity program component is the testing and preventive maintenance of fire protection equipment. The evaluation categories necessary to assess the quality of fire protection equipment testing and maintenance programs include code compliance reviews, corrective maintenance tracking, maintenance team qualifications, program monitoring and updating, equipment replacement programs, and compliance with appropriate standards.

1. *Code compliance.*

   • Are documented procedures in place that meet current NFPA standards for the maintenance and testing of fire protection systems?

   • Are procedures being properly applied at the required intervals, and are documented/signed-off testing and maintenance logs being maintained and readily available for review?

2. *Corrective maintenance tracking.*

   • Are documented corrective maintenance and testing procedures in place, and is a tracking method showing completion-date status with sign-offs to verify completion readily available for review?

3. *Maintenance team qualifications.*

- Are the people doing testing, maintenance checks, and corrective actions properly qualified and continually trained in conducting proper and efficient maintenance, and in recognizing and correcting problem areas?

4. *Program monitoring and updating.*

- Do procedures comply with the manufacturer's suggested maintenance for unique and specific equipment functions, and are procedures updated based on the manufacturer's experience, plant specific experience, and/or equipment modifications?

5. *Equipment replacement program.*

- Is equipment age, operating environment, and availability of spare parts considered in the updating and modification of maintenance and testing efforts, frequencies, and replacement of systems and system components?

6. *Compliance with NFPA standards.*

NFPA 11C, *Standard for Mobile Foam Apparatus*

NFPA 12, *Standard on Carbon Dioxide Extinguishing Systems*

NFPA 12A, *Standard on Halon 1301 Fire Extinguishing Systems*

NFPA 17, *Standard for Dry Chemical Extinguishing Systems*

NFPA 25, *Standard for the Inspection, Testing, and Maintenance of Water-Based Fire Protection Systems*

NFPA 69, *Standard on Explosion Prevention Systems*

NFPA 70B, *Recommended Practice for Electrical Equipment Maintenance*

NFPA 72, *National Fire Alarm Code*

NFPA 80, *Standard for Fire Doors and Fire Windows*

NFPA 110, *Standard for Emergency and Standby Power Systems*

NFPA 780, *Standard for the Installation of Lightning Protection Systems*

NFPA 25 was first issued in 1992. This standard addresses the inspection, testing, and maintenance activities for:

- Sprinkler systems
- Standpipe and hose systems
- Private fire service mains
- Fire pumps
- Water storage tanks
- Water-spray fixed systems
- Foam-water systems
- Fire protection valves.

Operating Procedures

It is important to develop and implement written operating procedures that provide clear instructions for safely conducting activities for each process that represents potential hazardous releases, fire, or explosion potential. The written operating procedures should address:

1. *Steps for each operating phase.*

- Initial startup
- Normal operation

- Temporary operations, as needs arise
- Emergency operations (including shutdown)
- Normal shutdown
- Startups following outages

2. *Operating limits.*

- Consequences of deviations
- Steps to correct or avoid deviations
- Safety systems and their functions

3. *Safety and health considerations.*

- Properties and hazards of process chemicals
- Exposure precautions, including controls (engineering, administrative, emergency response)
- Safety procedures for operating equipment
- Control measures in event of exposure
- Any special or unique hazards

Operating procedures should be readily accessible to employees at all times and must be maintained current.

Hotwork permits: A program should be developed and implemented for a permit system for all hotwork (including permanent locations) conducted on or near a covered process. The permit should document fire protection requirements, indicate the authorized date(s), and identify the object to be worked on. The permit must be kept on file until completion of the hotwork operation.

Fire hazards may occur in the use of both gas and electric welding, and in flame cutting. Both cutting and welding operations produce dangerous sparks. Sparks from cutting are more hazardous, as they are more numerous, and are carried greater distances. Most fires start from drops, or globules, of hot slag. Smoldering fires, not apparent when the work is completed, may later burst into flame when no one is present. Sparks, in the presence of flammable vapors, may start fires immediately.

A written cutting and welding permit should be issued by the plant fire chief, or other responsible individual, for all cutting and welding operations. Permits should include the following information:

1. Date
2. Work to be performed
3. Location of job
4. Type of immediate fire fighting equipment available, to include noncombustible covers
5. Fire watch assigned
6. Inspection of area before commencing work
7. Authorized signature of individual in charge of the permit system, or approved alternate
8. Signature of the individual who authorized the job to indicate the area was inspected upon completion of work

Precautions for cutting and welding should include:

1. Welding or cutting should be prohibited in areas where sprinklers are out of service.
2. Welding or cutting should not be done in a flammable liquids room.
3. Sparks or molten metal should not be permitted to pass through doorways, or cracks or holes in walls and floors.
4. All exposed combustibles should be moved a minimum of 50 ft (15.3 m) from cutting and welding operations. Noncombustible curtains must be used between the operation and combustible materials when adequate separation cannot be maintained.

5. Floors should be swept clean. Wood floors should be wet down or covered with a protective shield, preferably sheet metal.

6. Extinguishers should always be available at all cutting and welding operations. Any extinguishers that are used should be replaced promptly.

7. The immediate area, and areas on floors above and below, should be inspected for a minimum of 30 min after completion of an operation.

8. When oxyacetylene equipment is not in use, the gas should be turned off at the cylinder valve.

Work done by an outside contractor that requires cutting or welding should be closely supervised to comply with the cutting and welding permit system. Approximately one of every three cutting and welding fires reported occurs while outside contractors are engaged in cutting and welding operations.

Fire system impairment procedures: Many large fire losses have been directly attributed to impaired fire protection systems or equipment. Major property insurance companies, during routine facility inspections, continue to find fire protection systems of all types out of service. The need for an effective impairment notification program is evident to eliminate the possibility of undetected shutdown of fire protection systems.

A protection system may become impaired for a number of reasons (e.g., maintenance, renovation, construction, equipment failure, or failure to re-activate the system or device).

Facility managers and loss control managers should be notified *immediately* of any impairments during normal working hours. During off hours, depending on the seriousness of the impairment, these same individuals may still want to be notified. This procedure should be addressed in detail in the company policy.

Any impairment program should require:

1. Assigning responsibility and authority to control the impairment to one individual. This is normally a plant engineer or fire protection safety supervisor. In an emergency, a shift supervisor or fire brigade chief may have the authority to impair a system, but the overall responsibility for the impairment remains with the loss control manager.

2. Educating plant personnel in basic precautions to be taken when a protection system or equipment is impaired. Precautions include:

   • Limiting the number, scope, and duration of impairment

   • Supplementing manual fire protection with extra extinguishers

   • Avoiding cutting and welding operations

   • Shutting down any hazardous processes

   • Completing impairment work in a timely manner

   • Restoring protection system upon completion of all work

   • Verifying, by testing, that the protection system is operational

The three forms of impairment are: (1) concealed, (2) emergency, and (3) planned.

A concealed impairment is an unknown impairment. It occurs when a fire protection system is left out of service or removed from service by an unauthorized person. Concealed impairments are typically discovered during a plant's self-inspection or by a security service.

Emergency impairments are caused by unexpected events impairing the normal function of the protection system (e.g., a section of sprinkler piping freezing and bursting).

Emergency situations are normally associated with confusion and a sense of urgency. To eliminate much of the confusion, a written procedure should be in a location available to all personnel.

When handling emergency impairments, the following steps need to be performed:

1. Isolate the area where the situation, or condition, is causing the impairment. If possible, the remaining protection system should be kept in service. This may require temporary connections, or bypasses.

2. Notify the shift supervisor and fire brigade chief that an impairment has occurred.

3. Secure any hazardous production operation in the area where the protection system is impaired.

4. Properly "tag out" the impaired system or equipment.

5. Start repairs on the impaired system once, the area is secured. Work on the impairment should continue until it has been restored to service. Welding or cutting, required for the repair, should be performed in a protected area.

6. Place additional portable extinguishers and/or charged fire hoses in the impaired area, at accessible locations.

7. Notify the public fire department that emergency impairment has occurred, and the extent to which the protection system is out of service.

8. Notify the alarm company of the impairment, indicating whether any alarms have been affected.

A planned impairment is a scheduled impairment, typically involving improvement or modification to the present system. Such impairment might involve adding a new section of sprinkler piping or changing old sprinkler heads.

When work has been completed, it is important to ensure that fire protection is properly restored. Each step, of the several steps required, should be verified by the individual who authorized the impairment. These steps include:

1. Opening all valves that were secured during the impairment and conducting a drain test. (Note: If during the test the pressure drops below normal, the system may have a restricted, or partially closed, valve.)

2. Placing all alarms and/or detection devices back in service.

3. Restoring any fire protection equipment to "automatic" that was either secured or placed in "manual."

4. Verifying that portable extinguishers are in place and fully charged, and hose lines are returned to racks or reels.

5. Notifying plant supervisors (shift supervisors and fire brigade chief) that the fire protection system has been restored.

6. Notifying the alarm service, or central stations, that the fire protection has been restored and that alarms will be verified.

7. Notifying the public fire department that the fire protection system has been restored to service and all alarms activated.

Self-inspection: Engineered fire protection systems are a vital factor in the reduction of large fire losses at industrial facilities. Statistical studies of suppression by properly designed and maintained fire protection equipment have proved these systems to be effective. Unfortunately, reports of large fire losses stemming from fire protection equipment malfunction continue. Some of the recurring problems include:

• Undetected closure of sprinkler system control valves

• Inoperative fire pumps

• Empty water supply tanks

• Malfunctioning special suppression equipment

- Inoperative fire doors

- Inoperative fire detection systems

Management must take a positive approach in establishing programs for periodic inspection, testing, and maintenance of fire protection equipment. Equally important is management's need for a documented self-inspection program as part of ongoing firesafe plant operation.

Once an adequate program has been established and implemented, inspections must be conducted by individuals knowledgeable in the purpose, operation, and testing of the equipment.

To simplify facility self-inspections, forms appropriate for the occupancy and equipment should be used. Forms should be specific enough to be usable, and comprehensive enough so that no element of prevention or protection is overlooked.

As part of a thorough inspection, regardless of frequency, the occupancy should be inspected for:

- Poor housekeeping, waste, and combustible materials accumulations

- Careless smoking

- Improperly stored flammable/combustible liquids

- Blocked electrical switch gear, and storage of combustible material in electrical rooms

- Improper maintenance of electrical conduits and junction boxes

- Abnormal amounts of combustible in-process or finished products in areas where fire protection systems have not been adequately designed for this overload

Upon the completion of inspections and testing, reports should be forwarded to the facility manager. For those deficiencies that are severe and could be of imminent danger, the facility manager should be notified immediately and should take immediate action.

Flammable liquid handling and ignition control: The accidental release of flammable liquids or gases from containment, and potential ignition, presents a major loss exposure. An effective management program should address personnel training in the handling of flammable liquids and control of ignition sources. NFPA 30, *Flammable and Combustible Liquids Code* (hereinafter referred to as NFPA 30) provides information on this subject. The following highlights some of the management issues, extracted from NFPA's *Industrial Fire Hazards Handbook.*[3]

1. Training of personnel should include:

- Thorough indoctrination of all employees, including supervisors, in the hazards associated with the storage, transfer, and use of flammable and combustible liquids

- Indoctrination of employees in the importance of keeping flammable liquids and vapors confined to closed equipment and containers

- Indoctrination of employees in the importance of limiting the quantities of liquids in the work area to that amount needed for efficient operations

- Training of employees in the need for constant attendance during all flammable liquid transfer operations

- Training of employees in the proper procedure for control and cleanup of leaks and spills

2. Control of ignition sources should include:

- Electrical equipment and wiring in the area should be suitable for the hazard. NFPA 30 and NFPA 70, *National Electrical Code®*, specify the requirements for electrical equipment in hazardous areas.

- Do not locate any equipment having open flames, hot surfaces, or radiant heat sources in areas where flammable or combustible liquids are used, handled, or stored.

- Prohibit the use of friction- and spark-producing equipment in areas where flammable liquids are used.

- Provide adequate grounding and bonding of all equipment handling or using flammable liquids, to minimize the accumulation of hazardous static charges. NFPA 77, *Recommended Practice on Static Electricity*, provides the guidelines for grounding and bonding of equipment.

- Prohibit smoking, open-flame torches, and cutting and welding in hazardous areas.

- Establish a program to ensure that all vessels, tanks, piping, and process equipment are properly drained and purged of flammable and combustible liquids *prior* to performing maintenance operations.

Management

Employee training: Employees should be educated in the hazards involved in the job and in the functions of emergency safety control and fire protection equipment.

Employee training should be well documented. This includes initial, refresher, and supplemental training and certification. Initial training requires that all workers involved in a process, as well as newly assigned workers, be trained in the basics of the process, operating procedures, and associated hazards. Refresher and supplemental training should be given at least annually. After training, the employer should certify that workers have received and successfully completed the specified training.

Contractors: Prior to the commencement of work, contractors working on or near a given process should be informed of the potential hazards of fire, explosion, or toxic releases. Contractors must be trained in the plant's safe work practices and be instructed in the provisions of the plant's emergency response plan.

Management of change: Making design or operational changes that may affect the process hazard level without formal review and authorization can lead to serious fire or explosion incidents. To reduce this potential, a management of change (MOC) program should be implemented. This program should establish written procedures to manage any changes to facilities or to process chemicals, technologies, or equipment, including fire protection equipment. Procedures should address:

1. Change of elements

- Technical basis for the change

- Impact on safety and health

- Modifications to operating procedures

- Necessary time period for the change

- Authorization requirements for the change

2. Employee training prior to implementation
3. Revision of process safety analysis information
4. Revision of operating procedures

New construction: Loss control managers should be directly involved in new construction project planning. Responsibilities should include:

1. *New construction site evaluation.* The loss control manager should determine if the following conditions exist at the site under consideration.

   • Adequate water supply to meet fire protection requirements

   • Exposure to flood

   • Exposure conditions involving adjacent facilities

   • Acceptable support forces, such as public fire departments, disaster relief, and medical teams

   • Impediments to quick response by support forces (e.g., lift bridges, railroad crossings, or heavily traveled highways)

2. *Facility and protection systems planning.* Loss control factors to be considered during the planning stage are:

   • Using fire-resistant materials

   • Limiting the size of fire areas by the use of fire walls, fire doors, etc.

   • Segregating unusual concentrations of values (e.g., sensitive equipment and highly hazardous operations)

   • Ensuring emergency exits are adequate in number, size, and location

   • Providing physical protection (e.g., automatic sprinklers, hydrants, and standpipes)

   • Providing access for fire protection vehicles

   • Considering design for earthquake, wind, and heavy snow or rain

3. *Production operations involving new equipment or modifications.* Loss control management should participate in process planning, which may require:

   • Segregation to limit exposure from other operations

   • Special protection/suppression systems

   • Limited access or improved security

   • Special exit facilities

4. *Fire protection during facility construction.* Loss control management personnel should participate in:

   • Advising contractor personnel of loss control management policies (facility personnel must have authority to enforce compliance)

   • Acceptance testing of all fire protection equipment

   • Monitoring construction for compliance with specifications

Incidents

Incident investigations: It is important to investigate every incident that results in, or could have reasonably resulted in, a major release, fire, or explosion accident. Incident investigation program factors include:

1. Incident investigations should be initiated promptly, within 48 hr.
2. Investigation teams should have plant and process knowledge, as well as other specialties, such as fire loss investigation skills.
3. Incident reports should include:

   • Date of incident and date the investigation begins.

   • Description of incident, contributing factors, and resulting recommendations.

4. Reports should be reviewed with appropriate personnel within the facility and retained in loss incident files.
5. Recommendations should be addressed and resolved.

Emergency planning: An emergency planning program should be established and implemented, requiring at minimum training employees in the following procedures:

1. Emergency escape procedures and emergency escape route assignments
2. Procedures for employees who remain to operate critical plant operations before they evacuate
3. Procedures to account for all employees after emergency evacuation has been completed
4. Rescue and medical duties for those employees who are to perform them
5. Preferred means of reporting fires and other emergencies

As part of emergency planning, emergency organizations (EO) should be formed and trained at each facility to provide immediate effective response to emergency situations.

The EO should be effective during all occupied hours, regardless of staffing, and duties should be detailed in writing and posted in a conspicuous location in all buildings.

The EO plan should address fires, explosions, earthquakes, floods, windstorm (tornado and/or hurricane), water/liquid damage, flammable/combustible liquid spills, salvage operations, and cleanup. The primary function of the EO is to respond effectively to all emergency situations, including fires.

A detailed discussion of emergency organization is given in NFPA's *Industrial Fire Hazards Handbook.*[3]

The EO should develop pre-fire plans for all areas of the plant. The plans should be designed to limit large fires, minimize the threat to employees, and reduce the damage resulting from fires. While the pre-fire plans should be detailed, they must be flexible enough to allow for the diversity of fire problems that will be encountered. It is important that plant fire brigade members participate in pre-fire plan development.

A pre-fire emergency plan should be developed for each area and should be organized in a form that is useful to the fire brigade. It should contain, at minimum, the following information:

1. A review of the facility layout and best access and evacuation routes for each area
2. A discussion of facility construction features and their probable reaction during a fire incident
3. An overview of facility operations, equipment, and associated fire hazards
4. A review of available water supplies, fire pumps, sprinkler control valves, fire department connections, detection, alarms, and special suppression equipment
5. A description of available communications and pre-established plans with the responding public fire department
6. The requirements for manual fire-fighting efforts, including the most appropriate tactics, equipment, and approach to be used by the brigade
7. Information on ventilation systems that may cause the spread of smoke or may be utilized to prevent smoke spread
8. An evaluation of chemical hazards and methods to handle spills
9. The location of flammable liquid and gas system isolation valves and electrical power disconnects

QUALITY GRADING OF MANAGEMENT LOSS CONTROL PROGRAMS

It is beneficial to develop a quality rating system for loss prevention and control programs to provide a framework for budgeting and monitoring program investments *vs.* program quality.

TABLE 10-12B. *Example of Management Loss Prevention and Control Program Quality Grading*

PMS Program Safety	Planning (S1)	Accountability (S2)	Implementation (S3)	Monitoring (S4)	Program Score S1 + S2 + S3 + S4
1. Process safety information					
2. Process hazard analysis					
3. Pre-startup safety					
4. MIP—process equipment					
5. MIP—emergency safety systems					
6. MIP—protection systems					
7. Operating procedures					
8. Hot work permits					
9. Fire system impairments					
10. Facility self-inspection					
11. Flammable liquid handling and ignition control					
12. Employee training					
13. Contractors					
14. Management of change					
15. New construction					
16. Incident investigation					
17. Emergency planning					

Note: MIP = Mechanical Integrity Programs.

Table 10-12B provides an example format for grading some primary loss prevention and control programs. This is a general framework that can be expanded and customized, as needed.

The example quality scoring is based on four fundamental quality elements. (Note: points are allocated for example only.)

A description and discussion of the following program factors is presented in *Plant Guidelines for Technical Management of Chemical Process Safety.*[10]

Planning (+40 maximum points)

- Explicit goals and objectives (+5)
- Well-defined scope (+5)
- Clear-cut desired outputs (+5)
- Consideration of alternative achievement (+5)

Mechanisms

- Well-defined inputs and resource (+10)

Requirements

- Identification of needed tools and training (+10)

Accountability (+40 maximum points)

- Strong sponsorship (+5)
- Clear lines of authority (+5)
- Explicit assignments of roles, and responsibilities
- Formal procedures (+10)
- Internal coordination and communication (+10)

Implementation (at the facility level) (+40 maximum points)

- Detailed work plans (+10)
- Specific milestones for accomplishments (+10)
- Initiating mechanisms (+10)
- Execution (+10)

Monitoring (program effectiveness) (+40 maximum points)

- Performance standards (+5)
- Performance measurement and reporting (+5)
- Internal reviews (+10)
- Audit mechanisms (+10)
- Corrective action mechanisms (+10)

Example Program Quality Grading

(Note: for general example only)

- 140–160   Excellent quality program score (i.e., exceeds minimum requirements)
- 120–140   Good quality program score (i.e., meets expected program requirements)
- 100–120   Fair quality program score (i.e., marginal, minor improvements needed)
- Under 100   Poor quality program score (i.e., major improvements needed)

Program Monitoring

Two important aspects of program monitoring are: (1) auditing and (2) recommendation tracking.

Auditing provides a means to measure the performance of loss prevention and control programs. Audits document program progress and success and also demonstrate the interest of plant management in maintaining a high program performance quality.

Effective auditing provides valuable feedback on the program's strength and pinpoints deficiencies in program areas that may need improvement.

A system for tracking recommended corrective actions resulting from the audit should be established. The system should assign responsibilities and provide a follow-up reporting mechanism that tracks progress and completion of action items.

Conducting Community Relations Programs

The loss prevention and control manager should assist the facility manager and/or the public relations manager in publicizing the facility's loss control policies, programs, and practices to the community. This type of publicity can help build a favorable company image. The community has an interest in a facility's fire safety planning, as it relates to employee safety, to the economic impact of loss of wages due to facility shutdown, and to exposure of other property.

BIBLIOGRAPHY

References

1. *The SFPE Handbook of Fire Protection Engineering*, 2nd ed., DiNenno, P. J., ed., National Fire Protection Association, Quincy, MA, 1995.
2. Center for Chemical Process Safety, *Guidelines for Chemical Process Quantitative Risk Analysis*, American Institute of Chemical Engineers, New York, 1989.
3. National Fire Protection Association, *Industrial Fire Hazards Handbook*, 3rd ed., Cote, A. E., ed., NFPA, Quincy, MA.
4. Center for Chemical Process Safety, *Guidelines for Process Equipment Reliability Data, with Data Tables*, AIChE, New York, 1989.
5. *Guidelines for Preventing Human Error in Process Safety*, published by the Center for Chemical Process Safety, American Institute of Chemical Engineers, New York, 1994.
6. *Guidelines for Loss Prevention and Control, Management Programs, Overview*, Industrial Risk Insurers, Hartford, CT, 1990.
7. 29 CFR 1910.119, *Process Safety Management of Highly Hazardous Chemicals*, Occupational Safety and Health Administration, Washington, DC, 1992.
8. Center for Chemical Process Safety, *Guidelines for Hazard Evaluation Procedures with Examples,* 2nd ed., American Institute of Chemical Engineers, New York, 1992.
9. *Mechanical Integrity Supplement to the Maintenance Excellence Guide*, Chemical Manufacturers Association, Washington, DC, 1994.
10. *Plant Guidelines for Technical Management of Chemical Process Safety*, Center for Chemical Process Safety, published by the American Institute of Chemical Engineers, New York, 1992.

NFPA Codes, Standards, Manuals, and Recommended Practices

Reference to the following NFPA codes, standards, manuals, and recommended practices will provide further information on fire loss prevention and control management discussed in this chapter. (See the latest version of the *The NFPA Catalog* for availability of current editions of the following documents.)

NFPA 1, *Fire Prevention Code*
NFPA 10, *Standard for Portable Fire Extinguishers*
NFPA 11, *Standard for Low-Expansion Foam*
NFPA 11A, *Standard for Medium- and High-Expansion Foam Systems*
NFPA 11C, *Standard for Mobile Foam Apparatus*
NFPA 12, *Standard on Carbon Dioxide Extinguishing Systems*
NFPA 12A, *Standard on Halon 1301 Fire Extinguishing Systems*
NFPA 13, *Standard for the Installation of Sprinkler Systems*
NFPA 14, *Standard for the Installation of Standpipe and Hose Systems*
NFPA 15, *Standard for Water Spray Fixed Systems for Fire Protection*
NFPA 16, *Standard for the Installation of Deluge Foam-Water Sprinkler and Foam-Water Spray Systems*
NFPA 16A, *Standard for the Installation of Closed-Head Foam-Water Sprinkler Systems*
NFPA 17, *Standard for Dry Chemical Extinguishing Systems*
NFPA 20, *Standard for the Installation of Centrifugal Fire Pumps*
NFPA 22, *Standard for Water Tanks for Private Fire Protection*
NFPA 24, *Standard for the Installation of Private Fire Service Mains and Their Appurtenances*
NFPA 25, *Standard for the Inspection, Testing, and Maintenance of Water-Based Fire Protection Systems*
NFPA 30, *Flammable and Combustible Liquids Code*
NFPA 49, *Hazardous Chemicals Data*
NFPA 51B, *Standard for Fire Prevention in Use of Cutting and Welding Processes*
NFPA 68, *Guide for Venting of Deflagrations*
NFPA 69, *Standard on Explosion Prevention Systems*
NFPA 70, *National Electrical Code®*
NFPA 70B, *Recommended Practice for Electrical Equipment Maintenance*
NFPA 72, *National Fire Alarm Code*
NFPA 77, *Recommended Practice on Static Electricity*
NFPA 80, *Standard for Fire Doors and Fire Windows*
NFPA 110, *Standard for Emergency and Standby Power Systems*
NFPA 325, *Guide to Fire Hazard Properties of Flammable Liquids, Gases, and Volatile Solids*
NFPA 471, *Recommended Practice for Responding to Hazardous Materials Incidents*
NFPA 472, *Standard for Professional Competence of Responders to Hazardous Materials Incidents*
NFPA 491M, *Manual of Hazardous Chemical Reactions*
NFPA 600, *Standard on Industrial Fire Brigades*
NFPA 601, *Standard on Guard Service in Fire Loss Prevention*
NFPA 780, *Standard for the Installation of Lightning Protection Systems*
NFPA 1403, *Standard on Live Fire Training Evolutions*
NFPA 1500, *Standard on Fire Department Occupational Safety and Health Program*
NFPA 1901, *Standard for Automotive Fire Apparatus*
NFPA 1962, *Standard for the Care, Use, and Service Testing of Fire Hose Including Couplings and Nozzles*
NFPA 1971, *Standard on Protective Ensemble for Structural Fire Fighting*
NFPA 1975, *Standard on Station/Work Uniforms for Fire Fighters*
NFPA 1981, *Standard on Open-Circuit Self-Contained Breathing Apparatus for Fire Fighters*
NFPA 1982, *Standard on Personal Alert Safety Systems (PASS) for Fire Fighters*

Additional Reading

American Institute of Chemical Engineers, *Chemical Engineering Progress, Software Directory of Hazard and Risk Computer Models*, published annually, New York.
American Petroleum Institute, *Recommended Practice 750, Management of Process Hazards*, 1st ed., Washington, DC, 1990.
Brannigan, F. L., "Fire Loss Management," *Fire Engineering*, Vol. 143, 1989.
Brannigan, F. L., "Fire Loss Management. Part 18: Fire Protection Organizations," *Fire Engineering*, Vol. 143, No. 9, Sept. 1990, pp. 43–44.
Bretherick, L., *Handbook of Reactive Chemical Hazards*, 4th ed., Butterworths, London, 1990.
Bryan, J. L., "Damageability of Buildings, Contents, and Personnel from Exposure to Fire," *Fire Safety Journal*, Vol. 1, Nos. 1 and 2, July–Sept. 1986, pp. 15–31.
Chemical Manufacturers Association, *Evaluating Process Safety in the Chemical Industry—A Manager's Guide to Quantitative Risk Assessment*, Washington, DC, June 1989.
Dittman, C. R., "Selling Management the Best Fire Protection Available," *Professional Safety*, Vol. 33, No. 10, 1988, pp. 21–26.
Dow Chemical Company, *Fire and Explosion Index Hazard Classification Guide*, 7th ed., AIChE, New York, NY, 1989.
Factory Mutual Engineering Corp., "Guide to Planning Your Emergency Organization," Factory Mutual Research Corp., Norwood, MA, P8116, 1994.
Federal Emergency Management Agency, *Handbook of Chemical Hazard Analysis Procedures* (ARCHIE Manual), Washington, DC, 1989.
Fire Protection Association, "Fire Safety Structure in the United Kingdom. Fire Safety Data Organization of Fire Safety OR01," Fire Protection Assoc., London, England, OR01, May 1990.
Fire Protection Association, "Fire Test Facilities. Fire Safety Data Organization of Fire Safety OR06," Fire Protection Assoc., London, England, OR06, July 1991.
Grayson, S. J., and Smith, D. A., eds., *New Technology to Reduce Fire Losses and Costs*, Elsevier, New York, 1986.
Greenberg, H. R., and Cramer, J. J., eds., *Risk Assessment and Risk Management for the Chemical Process Industry*, Van Nostrand Reinhold, New York, 1991.

Hall, J. R., *U.S. Experience with Sprinklers*, Fire Analysis and Research Division Publication, National Fire Protection Association, Quincy, MA, 1990.

Hawksley, J. L., "Process Safety Management: A. V. K. Approach," *Plant/Operation Progress*, Vol. 7, No. 4, Oct. 1988, pp. 265–269.

Khishigbaatar, D. and Sultaninkarim, K., "Works Carried Out by Fire-Fighting Organization and Actual Problems," '93 Asian Fire Seminar, Oct. 7–9, 1993, Tokyo, Japan, 1993, pp. 111–116.

Lauriski, D. D., and Guymon, R. M., "Safety Management: What It Means to Us," *Mining Engineering*, Vol. 41, No. 10, Oct. 1989, pp. 1032–1035.

Lecari, Frederick A., *RMPP: Risk Management and Prevention Program Guidance*, California Office of Emergency Services, Hazardous Material Division, Preliminary Advance Copy, June 1989.

Lewis, D. J., *Mond Fire, Explosion and Toxicity Index: A Development of the Dow Index*, 13th Annual Loss Prevention Symposium, Houston, American Institute of Chemical Engineers, New York, 1979.

Lorenzo, D. K., *A Manager's Guide to Reducing Human Errors—Improving Human Performance in the Chemical Industry*, Chemical Manufacturers Association, Washington, DC, July 1990.

Marshall, V. C., *Major Chemical Hazards*, John Wiley & Sons, New York, 1985.

Meister, D., "Human Factors in Reliability," *Reliability Handbook*, W. G. Ireson, ed., McGraw-Hill, New York, 1996, pp. 12.2–12.37.

Moulton, G. A., ed., *NFPA Inspection Manual*, National Fire Protection Association, Quincy, MA, 1989.

"National Fire Service Organizations: Review and Preview '95–'96," *Fire Engineering*, Vol. 148, No. 12, Dec. 1995, pp. 43–48.

Ozog, H., and Bendixen, L. M., "Hazard Identification and Quantification," *Chemical Engineering Progress*, Vol. 83, No. 4, Apr. 1987, pp. 55–64.

Pessemier, R., *Up in Smoke: A Business Guide to Fire and Life Safety*, Phoenix, Redmond, WA, 1987.

Prugh, R. W., "Evaluation of Unconfined Vapor Cloud Explosion Hazards," *International Conference on Vapor Cloud Modeling*, Cambridge, MA, American Institute of Chemical Engineers, New York, 1987.

Ramachandran, G., "Value of Human Life," The *SFPE Handbook of Fire Protection Engineering*, 2nd ed., DiNenno, P. J., ed., National Fire Protection Association, Quincy, MA, 1995.

Robertson, J. C., *Introduction to Fire Prevention*, Macmillan, New York, 1989.

Schiller, C. G., "Standard Operational Procedure Guidance Manual for a Volunteer Fire Organization: Regulations, Standards, and Form," Safety Training and Consultation, Trenton, NJ, 1st Printing, May 1990.

Shortreed, J. H., and Stewart, A. "Risk Assessment and Legislation," *Journal of Hazardous Materials*, Vol. 20, 1988, pp. 315–334.

Slater, D. H., Pitblado, R. M., and Williams, J. C, "Quantitative Assessment of Process Safety Programs," *Plant/Operator Progress*, Vol. 9, No. 3, New York, July 1990.

State of California, Office of Emergency Services, *Guidance for the Preparation of a Risk Management and Prevention Program*, Sacramento, CA, 1989.

Stephenson, J., *System Safety 2000—A Practical Guide for Planning, Managing, and Conducting System Safety Programs*, Van Nostrand Reinhold, New York, 1991.

Suokas, J., and Kakko, R., "On the Problems and Future of Safety and Risk Analysis," *Journal of Hazardous Materials*, 21, 1989, 105–124, Elsevier Science, Publishers B.V., Amsterdam (printed in the Netherlands).

Technica, Ltd., *Techniques for Assessing Industrial Hazards*, World Bank Technical Paper No. 55, World Bank, Washington, DC, 1988.

Tokyo Fire Department, "Organization and Operations," Tokyo Fire Dept., Japan, 1990.

U.S. Department of Agriculture, "Findings From the Wildland Firefighters Human Factors Workshop. Improving Wildland Firefighter Performance Under Stressful, Risky Conditions: Toward Better Decisions on the Fireline and More Resilient Organizations. June 12-16, 1995. Missoula, Montana," Department of Agriculture, Washington, DC, 5100-F&AM, 9551-2855-MTDC, Nov. 1995.

VanderMolen, W., "Entwicklungen hinsichtlich Organisation und Arbeitsverhaltnisse im niederlandischen Feuerwehrwesen. [Development of Organization and Working Conditions of the Fire Service in the Netherlands.]," *VFDB*, Feb. 1994, pp. 52–56.

Watts, J. M., "Fire Risk Assessment Schedules," *SFPE Handbook of Fire Protection Engineering*, 1st ed., Sec. 4, Chap. 11, National Fire Protection Association, Quincy, MA, 1988.

Withers, J., *Major Industrial Hazards: Their Appraisal and Control*, Halsted Press, New York, 1988.

Yoshida, K., "International Maritime Organization Activity on Fire Safety," Third International Conference and Exhibition on Fire and Materials, Oct. 27–28, 1994, Arlington, VA, 1994, pp. 159–165.

EMERGENCY MEDICAL SERVICES

James O. Page

Some fire departments in the United States began providing first aid and medical rescue services to the public as early as the 1920s. In the late 1960s, pioneering advanced life support (ALS) programs were developed in several cities, and local fire fighters were trained to provide sophisticated medical procedures. These developments were widely publicized and began to change internal and external expectations about the role of the fire service in medical emergencies.

Presently, the vast majority of fire departments provide emergency medical services (EMS) on a routine basis. The levels of service range from first aid or basic life support (BLS) engine company first response to advanced life support (ALS) ambulance service staffed by multirole fire fighters cross-trained as paramedics.

The factors that contributed to this trend are numerous and vary somewhat from community to community. Today, in many departments, calls for EMS account for three-fourths or more of all alarms received. Through the evolutionary adoption of its EMS role, the fire service has become a major provider of health care services throughout the United States.

An EMS system traditionally has been defined as a coordinated arrangement of public safety and medical resources to provide the victim of a sudden and unexpected illness or injury with the greatest potential for survival and recovery.

The definition of an EMS system, and the regulations and guidelines that seek to control and govern the system, generally has been created and enforced by one or more government agencies, such as a state public health agency. The system's resources, however, are most often provided by public safety agencies (such as fire departments), volunteer or private ambulance companies, and hospitals.

Many different system configurations have developed. In fact, a national study commissioned by the U.S. Fire Administration in 1981 showed that more than two dozen different approaches to delivery of EMS with fire service resources were in operation. A 1985 study in California was consistent with the earlier findings and revealed 26 different profiles of fire department EMS programs operating in that state.[1]

Often, local history has been a major factor in system design. For example, if local fire chiefs were inspired by EMS develop-

ments in Miami, Columbus, Seattle, or Los Angeles in 1970, they probably worked to implement an EMS program in their fire departments. On the other hand, if local private ambulance operators were politically astute during that era of rapid change, they would have staked out the territory, so to speak, for their companies, to the exclusion of the fire departments.

Within the fire service, the topic of EMS has long been controversial. However, a number of factors combined in the 1990s to make EMS a more respected and sought-after part of the fire service mission.

ROLES AND RESPONSIBILITIES

Depending on state and local regulations there may be several levels of training and certification (or licensure) for fire fighters who are cross-trained for EMS.

First-Responder

This individual generally is a member or employee of a public safety agency and has completed an advanced first aid and CPR (cardiopulmonary resuscitation) course, or been trained as a BLS EMT (emergency medical technician) or an ALS paramedic. First-responder programs have a very positive effect on several types of life-threatening illness or injury. Since first-responders usually serve on fire companies, and since those fire companies are located to provide relatively quick response times, the rapid delivery of first-response services is the key benefit.

EMT (Emergency Medical Technician)

This individual has been trained and certified (or licensed) in one or more categories of prehospital life support and patient transportation. Among the 50 states there are many official types of EMT training and qualification. The generic term, "EMT-Basic," refers to a basic life support technician, trained in a program consisting of at least 110 hr of instruction and observation. Additional modules can be added to the course of instruction to include some limited advanced skills (such as endotracheal intubation).

In some states, the term "EMT-Intermediate" generally refers to an EMT-Basic who has received additional training and has been qualified to provide certain limited advanced skills. This category also includes "EMT-D" (EMT-Defibrillation), an individual who has been trained to recognize certain cardiac arrhythmias and attempts to correct them with electric defibrillation. Since this proce-

James O. Page is publisher, *JEMS* (*Journal of Emergency Medical Services*), Carlsbad, CA. He is also an attorney and retired chief of the Monterey Park, CA, Fire Department.

dure must be performed within a very few minutes of the onset of an arryhthmia, fire department EMT-D responders are ideally located for this level of EMS.

The "EMT-P" (EMT-Paramedic) usually is the most thoroughly trained and highly qualified category of prehospital EMS personnel. Advanced life support (paramedic) services are obliged to operate within the policies and procedures established by their medical director, who may or may not be employed by the fire department.

"Expanded Scope" Training

Residents of some remote and rural areas have to travel long distances for health care or to do without it. A proposed solution is to expand the scope of training and authority of paramedics in those areas. Those expanded-scope paramedics would provide a wider range of services, including minor suturing; treatment of fevers, flu, and earaches; and administration of certain drugs and medications.

Proponents of this type of program have suggested that it has potential application in urban areas. These proposed "mobile medical services" could become an additional service provided by private ambulance companies, or specialized units of fire departments. However, as ever-greater numbers of people become enrolled in managed care plans, it remains to be seen whether those plans will pay for mobile medical services as an alternative to neighborhood clinics that are staffed with allied health care personnel.

ECONOMIC IMPLICATIONS

Initially, the fire service's increased interest in EMS was motivated, in part, by economic factors. The fire protection environment in the United States has changed substantially since 1972, but the need for strong and reliable fire protection services did not disappear. National and international economics have reduced local government tax revenues and created competition between public agencies for their share of a reduced supply of tax dollars.

Much like a business that adds a product line, there are inherent economies in the fire service's use of multirole personnel who are cross-trained for both fire protection and EMS. Some fire departments have reassigned staff from fire companies to EMS duties, as the number of fire calls has decreased.

In recent years, in numerous cities and counties, there has been a lot of controversy over various EMS roles—including emergency ambulance service and whether they should be provided by the fire service or private ambulance companies. In these situations, the fire service generally has some competitive advantages. In addition to the residual goodwill and trust of the community toward its fire department, the department's actual costs for providing the service with multirole personnel will be limited to its incremental costs and equipment. Fire department personnel can also provide prevention services for the public, such as blood pressure and cholesterol checks. A private ambulance company, by contrast, since it uses its personnel in a single role, must generate revenues sufficient to cover the entirety of its costs.

While many fire departments would rather not be involved in ambulance transportation, long-standing policies of the federal government and major health insurers allow for reimbursement only to the agency or company that provides transportation. In other words, even though first-response services may have the greatest impact on outcome in life-death medical emergencies, the first-responder agency cannot be paid or reimbursed for those services *unless* it also provides ambulance transportation. At a time when all avenues of revenue are increasingly important to fire departments, this fact has motivated both fire service administrators and labor organizations to seek the opportunity to provide ambulance service.

THE IMPACT OF EMS ON THE FIRE SERVICE

Today's fire service is very different than it was in 1970. Despite two decades of controversy over whether EMS belonged in the fire service and how it should fit, many of today's most influential fire chiefs actually worked as a cross-trained fire fighter/paramedic. The sheer volume of responses to medical calls *vs.* the number of working fires has affected the way people view the fire service, both internally and externally.

Improved control over the fire problem through prevention and education, coupled with increasingly frequent involvement in EMS incidents, may tend to distract from the vital mission of fire suppression. To prevent decay of their effectiveness, and to ensure safety, fire departments with busy EMS programs should afford sufficient time for training in fire suppression knowledge, skills, and abilities on an ongoing basis.

Despite the mid-1990s Congressional rejection of a comprehensive federal program of health care reform, it is happening nonetheless. Health maintenance organizations (HMOs) and managed care organizations (MCOs) are consolidating the health care industry in an effort to achieve greater efficiency and to save money through economies of scale. These organizations may eventually wield more economic power than federal and state governments, with regard to health care costs and reimbursement.

Among the inefficiencies to be challenged by HMOs and MCOs are the conflicting boundaries of their plans and those of local government agencies (cities or counties). With regard to medical emergencies, it is the goal of most HMOs and MCOs to get their ill or injured members to the "right bed" in their facility on the first trip. The policies of some fire departments stand in the way of this goal; for example, a rule that a fire department's ambulance must take a patient to the nearest hospital.

The private ambulance industry recognizes the reluctance of many fire departments to operate across their boundary lines. Therefore, private companies increasingly will seek contracts with HMOs and MCOs for exclusive rights to transport their members. HMOs and MCOs will establish alternatives to the 911 emergency number for their members (to prevent a public agency from "scrambling" the system by transporting to nonparticipating hospitals). Inevitably, without a more flexible or cooperative approach to meeting the needs of the HMOs and MCOs, many fire department ambulance services may be left with a mostly uninsured or indigent population of patients.

The solution probably lies in development of cooperatives that can represent and coordinate all the EMS resources within an entire medical trade area (which generally will exceed the size of a single city or county). By dispatching the nearest unit (whether public or private) on all EMS calls, and accommodating the destination policies of the HMOs and MCOs, and coordinating the reimbursement for actual services rendered, a cooperative may be able to meet the needs of all parties for EMS in a reformed health care system.

FUNDING AND LEGISLATION

Federal legislation and funding for EMS was an important factor during the 1970s. However, except for development of new training curricula and development of public education materials, the federal government has had little involvement in EMS since 1981.

One exception is the EMS management training course, which is conducted at the National Fire Academy on an ongoing basis. This program is revised and updated periodically and offers to fire service administrators an opportunity to develop a working knowledge of current and future issues that affect the delivery of EMS through a fire department.

NETWORKING AND TECHNICAL ASSISTANCE

In recent years, national fire service organizations have recognized the importance of EMS. The International Association of Fire Chiefs (IAFC) and the International Association of Fire Fighters (IAFF), have created staff positions to assist members with EMS-related questions or problems. The IAFC has an EMS section that meets several times a year, and the IAFF has an EMS committee that meets periodically. The IAFF also conducts a biannual EMS conference that has become one of the major events of its kind. NFPA has assigned EMS-planning issues to its Organization, Deployment, and Evaluation of Fire and EMS Services Committee.

In addition, at least one network has been created by graduates of the National Fire Academy's EMS administration course. It publishes a periodic newsletter to keep members up to date on each other and current EMS issues and challenges.

BIBLIOGRAPHY

Reference Cited

1. U.S. Fire Administration, *EMS PIER Manual: Public Information, Education, and Relations in Emergency Medical Services,* USFA Publications, Washington, DC, 1994.

Additional Reading

Adams, R., "EMS Operations at Crime Scenes," Firehouse, Vol. 19, No. 9, Sept. 1994, p. 14.

Adams, R., "Pediatric Emergencies for Rural EMS Providers," Firehouse, Vol. 21, No. 3, March 1996, p. 10.

Alfred, R., "Statement of Ray Alfred, Fire Chief, District of Columbia, Fire and Emergency Medical Services Department, Washington, D. C. Federal Fire Prevention and Control Act. Hearing Before the Subcommittee on Science of the Committee on Science, Space, and Technology," 102nd Congress, 1st Session, No. 8, District of Columbia Fire Dept., Public Law 93-498, March 21, 1991.

American Society for Testing and Materials, *ASTM Standards on Emergency Medical Services*, ASTM, Philadelphia, PA, 1994.

Brame, K., "What is an EMS System?," Fire Chief, Vol. 37, No. 5, May 1993, pp. 38-40.

Briese, G. L. and McDowell, A. E., "Developing Programs for Citizen Involvement in Emergency Medical Services (EMS)," International Association of Fire Chiefs, Washington, DC, CIB W14/95/1 (J); Tokyo Fire Department, Firesafety Frontier '94, International Fire Conference and Exhibition in Tokyo, Creating a Safe Tomorrow, October 18–22, 1107-112, Tokyo, Japan, 1994, pp. 359–365.

Bruno, H., "Debate on Fire and EMS Issues," Firehouse, Vol. 19, No. 10, October 1994, p. 10.

Burge, C. D., "Estimation, Optimization, and Control in Rural Emergency Medical Service (EMS) Systems," Southern Mississippi Univ., Hattiesburg, Institute of Electrical and Electronics Engineers, Inc. (IEEE), International Emergency Management and Engineering Conference, 10th Anniversary, Research and Applications, 1993, pp. 213–218.

Cowen, A. R., "EMS Operations," Fire Engineering, Vol. 147, No. 8, August 1994, pp. 156–159.

Damioli, C., "EMS Protective Clothing Continues to Unfold," Fire Chief, Vol. 39, No. 4, April 1995, pp. 53–54, 56–57.

Davis, G., "Victims by the Hundreds: EMS Response and Command," Fire Engineering, Vol. 148, No. 10, Oct. 1995, pp. 98–104.

Delbridge, T. R., "Future of EMS: An Achievement," Fire Chief, Vol. 40, No. 4, 44–47, April 1996, pp. 44–47.

Federal Emergency Management Agency, "Guide to Funding Alternatives for Fire and Emergency Medical Services Departments," Federal Emergency Management Agency, Washington, DC, FA-141, Dec. 1993.

Federal Emergency Management Agency, "Resources on Fire and Emergency Medical Services (EMS): A Publications Catalogue of the United States Fire Administration," Federal Emergency Management Agency, Washington, DC, FA-102, Sept. 1993.

Federal Emergency Management Agency; National Highway Traffic Safety Administration, "Make the Right Call: Emergency Medical Services (EMS)," Federal Emergency Management Agency, Washington, DC, National Highway Traffic Safety Admin., Washington, DC, K-80-S, Nov. 1992.

Feehan, W. M., "New York Commissioner Takes Aim at Column on EMS and the Fire Service," Journal of Applied Fire Science, Vol. 2, No. 4, 1992-1993, pp. 333–334.

"First Responders: Standards for Those Providing Emergency Medical Care," ASTM Standardization News, Vol. 22, No. 2, March 1994, pp. 42–45.

Forbuss, R. I., "Understanding the EMS Competitive Forces," Mercy Medical Services, Las Vegas, NV, International Association of Fire Chiefs (IAFC), Fire-Rescue International Conference Proceedings, 1993 Annual Conference, August 28-Sept. 1, 1993, Dallas, TX, 1993, pp. 207.

Goldfarb, Z. and Kuhr, S., "EMS Response to the Explosion," Fire Engineering, Vol. 146, No. 12, Dec. 1993, pp. 74–83.

Hall, R., "Interview With District Chief Gary Davis, EMS Coordinator, Oklahoma City Fire Department," Speaking of Fire, Vol. 2, No. 4, Winter 1995, pp. 17–18, 28.

Hanchar, J. M. and Dolan, F. M., "FDNY Enters World of EMS: New Program Aims to Increase Cardiac Arrest Survival Rate," Firehouse, Vol. 20, No. 7, July 1995, pp. 92–94.

International Association of Fire Chiefs, "Trends in the Ambulance Industry" (a series of reports), IAFC, Fairfax, VA.

International Association of Fire Fighters, "Analysis of Fire and Emergency Medical Service Delivery: Scottsdale, Arizona," International Association of Fire Fighters, Washington, DC, March 6, 1990.

International Association of Fire Fighters, *Effectiveness of Fire-Based EMS*, IAFF, Washington, DC, 1995.

International Association of Fire Fighters, *Emergency Medical Services—A Guidebook for Fire-Based Systems*, IAFF, Washington, DC, 1995.

Kuehl, A. E., *Prehospital Systems and Medical Oversight*, 2nd ed., Mosby Lifeline, St. Louis, MO, 1994.

Lipowitz, S., "Taking the 'E' Out of EMS," Fire Chief, Vol. 39, No. 4, April 1995, pp. 37–38, 40, 42, 44.

Ludwig, G. G., "Matter of Necessity: EMS Risk Management," Fire Chief, Vol. 38, No. 5, May 1994, pp. 54–56, 58–59.

Painter, B., "Firefighter/EMT Interface in the Volunteer Fire Department," Speaking of Fire, Spring 1996, pp. 14–15.

Roush, W., et al., *Principles of EMS Systems*, 2nd ed., American College of Emergency Physicians, Irving, TX, 1994.

Stull, J. O., "NFPA 1999: A Standard of Protective Clothing for EMS Operations," Fire Engineering, Vol. 147, No. 3, March 1994, pp. 79–80, 82, 84, 86.

U.S. Fire Administration, *EMS PIER Manual: Public Information, Education, and Relations in Emergency Medical Services*, USFA Publications, Washington, DC, 1994.

U.S. Fire Administration, *EMS Safety: Techniques and Applications*, USFA Publications, Washington, DC, 1994.

U.S. Fire Administration, "Emergency Medical Services: Management Resource Directory," Fire Admin., Washington, DC, FA-119, Nov. 1992.

U.S. Fire Administration, "Report of the National Forum on Emergency Medical Services Management. February 10-12, 1992. Arlington, Virginia," Fire Admin., Washington, DC, FA-126, Dec. 1992.

FIRE PREVENTION AND CODE ENFORCEMENT

Revised by John F. Bender

Fire prevention includes all fire-service activity that decreases the incidence of uncontrolled fire. Usually, fire prevention methods used by the fire service focus on inspection, which includes engineering and code enforcement, public fire safety education, and fire investigation.

Inspection, including enforcement, is the legal means of discovering and correcting deficiencies that pose a threat to life and property from fire. Enforcement is implemented when other methods fail. Education informs and instructs the general public about the dangers of fire and about fire-safe behavior. Fire investigation aids fire prevention efforts by indicating problem areas that may require corrective educational efforts, inspection emphasis, or legislation. Good engineering practices, including plans review—another fire prevention method—can provide built-in safeguards that help prevent fires from starting and limit the spread of fire should it occur.

In the United States, the full value of fire prevention was not realized until fire departments and agencies began to compile meaningful information concerning the causes and circumstances of fires. Such information highlighted problem areas and led progressive departments to initiate more effective fire prevention efforts. The results of fire prevention efforts are borne out statistically, every year confirming the effectiveness of fire prevention programs. Fire prevention, as the most important priority in cutting fire losses, received an endorsement in 1973, when the National Commission on Fire Prevention and Control published the results of its in-depth study on the fire problem in America. In its report *America Burning*, the commission emphasized the necessity of increased fire prevention activities to reduce fire. It was the commission's position that increasing emphasis on fire prevention would measurably affect the U.S. fire loss picture. Statistics validate this position for areas in which fire prevention activities have been implemented and evaluated.

FIRE PREVENTION PERSONNEL

Effective fire prevention depends on a personnel network technically capable and motivated to enforce fire codes, educate the general public, and investigate fire causes. This network of people is usually organized at the state, county, and/or local level.

John F. Bender, P.E., is chief fire protection engineer for the Maryland State Fire Marshal. He is a member of NFPA's Technical Committee on Fire Prevention Codes.

State and Provincial Fire Marshal/Commissioner

Most states have offices to oversee certain phases of fire protection. (See Table 10-14A.) The state's chief fire protection administrator usually is called the state fire marshal. The organization of state fire marshal offices differs from state to state. (See Table 10-14B.) Most fire marshals receive their authority from the state legislature and are answerable to the governor, a high state officer, or a fire commission. In some states, the fire marshal's office may be a division of the state insurance department, state police, department of public safety, commerce department, or other state agency. Some offices are organized as independent state agencies.

State fire marshals' offices normally function in areas beyond the scope of municipal, county, or fire-district organizations. Depending upon the authority outlined in enabling legislation or administrative regulations, state fire marshal offices may have combination responsibilities in the following areas: code enforcement, fire prevention inspections, plans review, product approval, fire and arson investigation, fire data collection and analysis, fire service training, public fire education, fire legislation development, explosives and other hazardous materials regulation, manufactured-housing regulations, mobile-home regulations, electrical inspections, and state agency and public fire protection consulting.

State fire marshals may share responsibility with local fire officials, or each may have specifically defined areas of concern. Local fire protection officials sometimes are granted the authority to act as agents for the state in stipulated areas of inspection, enforcement, and investigation.

At the national level in Canada, the Dominion Fire Commissioner is responsible for fire safety enforcement in all nonmilitary government properties and for gathering national fire statistics. The Canadian Forces Fire Marshal oversees properties of the armed forces, where fire protection facilities include a number of operating fire departments.

In Canada, each of the provinces has a provincial fire commissioner or fire marshal, while the two territories have territorial fire marshals. In most cases, the provincial/territorial fire officer has a wide variety of responsibilities, including code enforcement, fire-service training, operation of fire-reporting systems, and fire investigation, as well as support of local fire departments. The Association of Canadian Fire Marshals and Fire Commissioners consists of

TABLE 10-14A.   Functions of the State Fire Marshal's Office

STATE	Code Enforcement	Fire & Arson Investigation	Plans Review	Fire Prevention Inspections	Fire Data Collection	Fire Data Analysis	Fire Service Training	Public Fire Education	Fire Legislation Development	Manufactured Housing Regulations	Mobile Home Regulations	Electrical Inspections	Boiler Inspections	Distribution of Funds to Fire Dept.
Alabama	X	X	X	X	X	X	X	X				X		
Alaska	X	X	X	X	X	X	X	X	X					
Arizona	X	X	X	X	X		X	X	X					
Arkansas	X	X	X	X			X	X	X					
California	X	X	X	X	X	X	X	X	X					
Colorado										X				
Connecticut	X	X	X	X	X	X	X	X	X				X	
Delaware	X	X	X	X	X	X			X					
Florida	X	X	X	X	X	X	X	X	X	X	X	X	X	
Georgia	X	X	X	X				X			X	X		
Hawaii*		NONE												
Idaho	X	X		X	X	X		X	X	X				
Illinois	X	X	X	X	X	X	X	X	X				X	
Indiana	X	X	X	X	X	X	X	X	X					
Iowa	X	X	X	X	X	X		X	X	X	X			
Kansas	X	X	X	X	X	X		X	X					
Kentucky	X	X	X	X	X	X	X	X	X	X	X	X	X	X
Louisiana	X	X	X	X	X	X		X	X	X	X	X	X	
Maine	X	X	X	X	X	X		X	X					
Maryland	X	X	X	X	X	X		X	X			X		
Massachusetts	X	X	X	X	X	X		X	X					
Michigan	X	X	X	X	X	X	X	X	X					
Minnesota	X	X	X	X	X	X	X	X	X					
Mississippi	X	X		X	X	X	X			X	X			
Missouri		X	X	X	X	X	X		X				X	
Montana	X	X	X	X	X	X		X	X					
Nebraska	X	X	X	X	X	X		X	X					
Nevada	X	X	X	X			X	X	X			X		
New Hampshire†	X	X	X	X	X	X		X	X			X		
New Jersey	X			X	X	X	X	X	X					X
New Mexico	X	X		X	X	X	X	X	X					X
New York	X	X	X	X	X	X	X	X	X					X
North Carolina	X	X			X	X	X	X	X	X	X	X		
North Dakota	X	X	X	X	X	X	X	X	X					
Ohio	X	X	X	X	X	X	X	X	X					
Oklahoma	X	X	X	X	X	X		X	X					
Oregon	X	X	X	X	X	X	X	X	X					
Pennsylvania	X	X	X		X									
Rhode Island	X	X	X	X	X	X	X	X	X					
South Carolina	X	X	X	X		X	X	X	X			X		X
South Dakota	X	X	X	X	X	X	X	X	X				X	X
Tennessee	X	X	X	X	X	X		X	X	X	X	X		
Texas	X	X	X	X	X	X		X						
Utah	X	X	X	X	X	X	X	X	X					
Vermont‡	X	X	X	X	X	X		X	X			X	X	X
Virginia	X		X	X										
Washington	X	X	X		X	X		X	X					
West Virginia	X	X	X	X	X	X	X	X	X			X	X	X
Wisconsin§		X												
Wyoming	X	X	X	X	X	X	X	X	X			X		

* Hawaii: State fire marshal's office disbanded April 1979; all functions transferred to county fire marshal's offices.
† New Hampshire: Division of fire service created July 1, 1989.
‡ Vermont: Department of Labor and Industry—fire prevention division.
§ Wisconsin: Fire and arson investigation is a function of fire marshal, Department of Justice; other functions Chief of Fire Protection, Department of Industry, Labor, and Human Relations.

TABLE 10-14B. *State Fire Marshal Organizational Patterns*

Under Department of Insurance
Alabama
Florida
Idaho
Mississippi
New Mexico
North Carolina
Tennessee
Texas
Washington

Under Department of Public Safety
Alaska
Colorado
Connecticut
Iowa
Louisiana
Maine
Maryland
Massachusetts
Minnesota
Missouri
New Hampshire
Utah

Under a Separate Government Department
California
Illinois
Indiana
Montana (Justice—Attorney General)
Kansas
Nebraska
Nevada
Oklahoma (Separate Agency)
Oregon
South Carolina
Wisconsin—Fire Investigation (Justice—Attorney General)
Wyoming

Under a Regulatory Agency
Arizona
Kentucky
New Jersey
South Dakota
Vermont
Virginia

Under State Police
Arkansas
Michigan
Pennsylvania

Under Cabinet-Level Official
Georgia (Controller General)
New York (Secretary of State)
North Dakota (Attorney General)
Ohio (Director of Commerce)
Rhode Island (Governor's Office)

Under State Fire Commission
Delaware
West Virginia

the previously mentioned dominion and armed-forces fire officials, plus each provincial and territorial fire commissioner or fire marshal. The association exerts a major influence on fire safety policy in Canada.

Chief of Fire Prevention or Local Fire Marshal

Because of the importance and diversity of fire prevention activities, certain members of the department should specialize in fire prevention functions. Staff size of a fire prevention division, bureau, or fire marshal's office will depend upon the community's needs and the department size. This office may be divided into specialized subsections addressing administration; fire prevention, code inspections, and enforcement; engineering; public education; and investigation. (See Table 10-14C.) These subsections are headed by subordinate officers.

The fire prevention division personnel should consist of those department members best qualified for this work. In addition, qualified technical specialists should be available when possible—the department's overall effectiveness depends upon the technical skills the fire prevention personnel possess. It is also important that such personnel be able to properly communicate technical requirements to building owners, architects, engineers, and other professionals involved in the construction industry.

The laws of the state, county, municipality, or fire district often delegate responsibility and authority for fire prevention to the fire chief or fire department head, who may then delegate this authority to an individual or division, depending upon the size of the department. In some states, the local fire marshal's authority and area of enforcement responsibility is derived from state laws, independent of or in addition to the powers and duties of the fire department chief.

The fire chief may directly supervise fire prevention activities in smaller departments. In larger departments, the division head should be a high-ranking chief officer, functioning as a staff officer to the fire chief. The chief officer for the fire prevention activities may be described as fire marshal, chief of the fire prevention division, or chief fire inspector.

Fire Inspector or Fire Prevention Officer

Fire inspectors or fire prevention officers should be selected for these tasks based on their technical training, expertise, and their ability to motivate people. Individuals in this position are responsible for conducting fire inspections and also may be responsible for division duties, including fire investigations, public education, or plans review, among others.

An inspector conducting an inspection should be able to convince property owners or occupants to maintain or improve fire safe conditions. The inspector who relies only on police power cannot accomplish as much as the inspector who relies on salesmanship. Strict enforcement of the fire law of the jurisdiction is necessary to achieve compliance and to assure fire safe conditions in the community.

Plans Review Specialist

The review of plans and specifications is a code-enforcement process intended to ensure compliance with the fire protection and life-safety provisions of the building code, as well as the fire prevention code. Individuals in plans review must be technically proficient in both the intent and the letter of the applicable codes. The fire department's plans-review specialist examines site plans, building and fire protection system plans, and specifications and may prepare recommendations or consult with individuals involved in building construction, remodeling, or renovations. The plans-review specialist must regularly interact with other public safety officials and agencies at the local, state, and federal level, including building public works, zoning, health, and others, to ensure the construction of fire safe, code-conforming buildings.

TABLE 10-14C. *Sample Organization of Local Fire Prevention Functions in the Fire Marshal*

	Administraton	Fire Prevention, Code Inspections, and Enforcement	Construction Plans and Specification Review	Public Fire Education	Fire Investigation	Research, Statistics, and Evaluation
Personnel:	Fire marshal Secretarial staff Clerical staff	Fire inspector Fire officer Fire company personnel	Fire protection officer Fire protection engineer Plans review specialist	Public fire education officer Fire officer Fire inspector Fire fighter	Fire records specialist Programmers Statisticians	

Functions and Activities:	Overall supervision
	Policy determination
	Authority Having Jurisdiction
	Coordination
	Office planning
	Agency liaison
	Budget
	Support
		Inspections
		Permits
		Violations
		Summons
		Court cases
		Complaints
		Night inspections
		Red tags
		Hazardous materials
			Plans review
			Specification review
			Consultations
			On-site inspections
			Certificate of occupancy sign-off
			Systems testing
			Condemnations
			Use permits
			Liaison with:
			  Public works
			  Building
			  Electrical
			  Mechanical
			  Zoning
			  Housing
				Fire prevention behavior education
				Fire safety
				Behavior education
				Special groups
				Seasonal programs
				Juvenile fire setters
				Special programs
				  Sprinkler
				  Detection
				  EDITH
				  Clean-up
				  Etc.
					Fire cause determination
					Fire investigation
					Arrest
					Case preparation
					Courtroom testimony
					New products evaluation
					New methods, approaches, and alternatives
					Liaison with:
					  Prosecutor
					  Arson task force
					  Local, state, federal investigative/police agencies
					  insurance interests
						Property files
						Violation status reports:
						  monthly,
						  quarterly,
						  annually by
						  special
						  fire experience
						  statistics

Fire Protection Engineer

Due to the complexity and magnitude of fire protection problems, the services of fire protection engineers (FPEs) can be beneficial. Fire protection engineers may be employed on a consulting basis or as full-time staff engineers. The staff fire protection engineer contributes a high level of technical ability to the plans review process, interpretations, consultations, and the preparation of recommendations for a broad range of fire protection problems. The fire protection engineer's credentials, based on professional training, licensure and (registration) as a professional engineer (P.E.), are recognized by architects, builders, and other professionals involved in the building process.

Fire Prevention Education Officer

Due to increasing interest in performance-based building and fire codes, the fire protection engineer is often in the best position to evaluate such proposals and assess the equivalent level of life safety.

The public fire protection education officer creates public awareness of fire as a personal, family, business, and community concern, and then strives to motivate the general public to do something about fire risks, based on proper fire safe behavior. The personnel selected for this function should be knowledgeable in a wide area of fire technology and be able to develop and present effective fire prevention and fire safety programs. This job requires someone with the imagination, creativity, motivation, communication skills, and adaptability needed to develop and convey fire safety programs to all segments of the community.

Depending upon the size of the department and the type of fire prevention and fire safety program it undertakes, all levels of fire department personnel can and should participate in presenting the fire safety and prevention message to the community. Nonuniformed personnel, senior citizens, and citizens with education backgrounds also are being used successfully by the fire department in this area of fire prevention.

Fire Investigators

Fire cause determination and subsequent investigation can be the responsibility of the fire department shared, or of outside agencies. When a fire department is responsible, fire cause detection and investigation may be assigned to the fire chief, fire marshal, battalion chiefs, company officers, fire inspectors, or other personnel specially trained as fire investigators.

Investigative personnel should be perceptive, inquisitive, and thorough in determining facts and conducting an investigation. Investigation experience should be supported by training in areas such as chemistry, criminal law, forensics, and criminal investigation. An increasing number of progressive departments have fire investigation personnel trained in appropriate fields of criminal justice and provided with police powers.

National Professional Qualifications

Professional qualifications standards for fire service career positions are the responsibility of the Professional Qualifications Correlating Committee. Three of the ten standards are of particular interest to the fire prevention community. An interest in the field should include a review of the latest editions of NFPA 1031, *Standard for Professional Qualifications for Fire Inspector;* NFPA 1033, *Standard for Professional Qualifications for Fire Investigator;* and NFPA 1035, *Standard for Professional Qualifications for Public Fire and Life Safety Educator.*

Certification to the career levels defined in these standards is accomplished by agencies based on third-party accreditation. Accreditation for the fire service is currently available through the National Professional Qualifications Board of the National Board on Fire Service Professional Qualifications (NBFSPQ) and the Fire Service Accreditation Congress of Oklahoma State University.

Programs such as the NFPA Fire Inspector Certification Program for fire inspector certification to levels I and II are available on a national level with accreditation from the Professional Qualifications Board on Fire Service.

FIRE PREVENTION INSPECTIONS

Fire prevention inspections are conducted by state or local fire department personnel in compliance with laws and ordinances which usually require specific fire inspections. Occupancies normally inspected include places of public assembly, educational, institutional, residential (except the interiors of dwellings), mercantile, business, industrial, manufacturing, storage, and special-hazards structures. In addition to such mandatory inspections, the fire department also may conduct voluntary fire inspections, such as home fire safety surveys.

The organization and operation of fire prevention inspection programs varies. Inspections required under the fire prevention code traditionally have been conducted by personnel from the fire prevention division. However, this responsibility is being shared increasingly with fire company personnel. The department's training program must train all its members in fire prevention.

Fire Prevention Inspection Objectives

Inspections conducted as part of code enforcement help to ensure reasonable life safety conditions within a structure. The condition of and usability of means of egress, interior finish, emergency lighting, exit signs, and all fire doors should be inspected. Inspection of means of egress should include inspection of the exit discharge area.

Inspections, which are intended to prevent fires from occurring, are effective because the inspector identifies fire hazards that could cause a fire, allow a fire to develop, or allow a fire to spread. In addition to locating and correcting potential fire causes, the fire inspector should check any accumulation of combustible trash and debris, storage practices, maintenance procedures, and safe operation of building utilities.

Inspections determine the proper installation, operation, and maintenance of fire protection features, systems, and appliances within the building. The inspection process should ascertain whether each fire protection system is tested regularly, and records are kept for inspection by the fire department or others.

Fire detection equipment, alarms, annunciation and notification systems, sprinkler-valve operation, supervisory switches, and fire pumps all should be tested regularly as part of the overall inspection process. Other fire protection features, including standpipes and fire escapes, should be tested or closely examined to detect possible malfunctions due to deterioration from weather and corrosion. Portable fire extinguishers should be checked as to proper type, placement, maintenance, testing, and distribution in the structure.

Particular attention should be given to ensure that personnel expected to use portable fire extinguishers are adequately trained in their use; otherwise, valuable time may be wasted before other building occupants and the fire department are alerted.

Technical information on a building and its processes should be recorded during the inspection. When used in pre-fire planning, such information can be valuable to the fire department in case of a fire at the property. The type of construction, vertical openings, util-ity type and placement, fire protection systems, fire department access, tenancy, hazardous materials, or special life-hazard conditions are the kinds of information that should also be noted during inspections and used to develop fire-fighting plans.

In addition, the fire inspector should review written fire safety emergency procedures to ensure they are adequate and up-to-date with respect to the type of occupancy involved and the various types of fire protection features available. In many cases, it may be necessary to witness an actual drill or planned exercise of the emergency plan to properly evaluate its effectiveness. This should be coordinated closely with the facility manager.

Inspections provide an opportunity to educate the owners or occupants of a building about fire safe behavior and the need for adequate fire and life safety conditions in the areas under their control. "Selling" fire prevention is the key to success in obtaining code compliance, and how fire prevention is "sold" should be an important consideration in training programs for inspection personnel. When inspection programs are properly designed and put into practice, inspectors may achieve more through public education and persuasion than through exercising their enforcement authority. The persuasive effect of the inspector's presence, coupled with the inspector's ability to spot and directly see that hazards are corrected, enhances the effectiveness of the inspection program.

Fire Company Inspections

In many communities, in-service fire suppression personnel conduct most or all regular inspections. Company fire inspection procedures may include conducting building surveys, correcting common problems concerning life safety conditions, locating and correcting fire hazards, and testing fire protection systems. Inspections also may include checking use and storage of hazardous materials, which may require issuing a hazardous-use permit. The inspection process helps familiarize fire company personnel with individual buildings and locations in their jurisdiction at which they may have to fight fires or perform other emergency duties. This information can be used in formulating pre-fire plans.

Company inspections usually are conducted in the fire company's first-due area. Inspections and reinspections may be scheduled by the company officer, by the battalion or district officer, or by the fire prevention bureau, according to requirements of the fire prevention code or the department. Before fire fighters perform inspections, they should receive proper training and be qualified and authorized to conduct inspections as fire prevention officers.

The inspection may uncover some situations that are easy to correct or others that require more technical skill. In either case, follow-up responsibilities should be assigned to the fire companies, fire company officers, and district chiefs according to the department's resources. The fire prevention division may help to obtain compliance by building owners after violations are located by fire companies, so it is essential for fire prevention personnel to coordinate their activities closely with those of fire company personnel.

Fire Prevention Division Inspections

In some departments, inspection personnel from the fire prevention division are solely responsible for conducting fire prevention inspections. The objectives of their inspections, like those of the general inspections performed by fire companies, are to ensure compliance with applicable fire codes and to locate conditions that may cause a fire or endanger life and property. Each inspection must be thorough. The owner or occupants of the property being inspected must be notified when unsafe conditions are found. Inspections usually are required by law in order to locate any violations and to see that they are corrected. Legal direction is available in such situations if necessary. To perform such work competently, the inspector should have appropriate technical education, specialized training, and experience.

While fire companies perform some or all regular inspections, fire prevention division personnel may be concerned with the initial inspection, follow-up inspections, or enforcement actions necessary to correct fire code violations. Fire prevention division personnel may perform inspections associated with the issuance of permits as required by the fire prevention code and may be concerned with buildings and premises that require a high level of code application. The inspector may need to confer with the property management or with fire department officers to prepare comprehensive reports, recommendations, or orders.

Both fire company and fire prevention division inspections may reveal conditions hazardous to the health and safety of owners, occupants, and the general public. Even when these hazardous situations are not specifically covered in the fire laws, such conditions must be brought to the attention of the owners or occupants; other inspection authorities should be so notified for the record. This mutual assistance in the inspection of buildings or premises can greatly enhance the level of life safety for the general public. Close cooperation among responsible inspection authorities should be encouraged and, where possible, recognized in municipal ordinances or interdepartmental administrative agreements.

Night inspections of public assembly and business occupancies should be part of routine inspection procedures to assure compliance with occupancy load requirements and to demonstrate that the means of egress are in good operating order.

Inspections based on citizen complaints should be conducted by the fire prevention division in a timely manner. An informed and interested public can be an important asset in uncovering unsafe conditions that might develop after routine inspections are conducted.

Home Inspection Process

Approximately 80 percent of lives lost to fire occur in one- and two-family dwellings and apartments. Home inspections traditionally have not been mandatory because of unfounded concerns associated with the rights of citizens to ensure the sanctity of their homes. For years, however, many fire departments have inspected homes on a voluntary, by-invitation, or planned basis and many have been successful in reducing home fire-loss experience. Inspections of this nature, whether conducted by career or volunteer fire department personnel, should be supported by a strong publicity program. Adequate training must be provided for all personnel involved. Voluntary inspections can be conducted either as a separate part of the fire inspection program or as part of fire prevention programs that include occupant fire escape planning and such practices as Operation EDITH (exit drills in the home), smoke detector or residential sprinkler drives, spring clean-up, burn prevention programs, heating fire safety, and so on. Where the program is not comprehensive, it should target high-fire-rate neighborhoods.

CODE ENFORCEMENT

Fire Prevention Codes

The fire laws of the state, county, fire district, or community delegate general responsibility and authority to the fire officials involved in fire prevention activities. A fire prevention code adopted into law that outlines specific fire prevention requirements and enforcement procedures is essential for an effective fire prevention program.

State law or local ordinances may outline how a fire prevention code is written or adopted. The fire prevention official may have authority to write a local fire prevention code or to adopt a nationally developed code by reference. Using a model fire prevention code is preferable to using one that has been developed locally. Nationally developed consensus codes are based on a broad spectrum of accepted fire prevention experience and may prevent a fire prevention official from being accused of developing a fire code that is too stringent or is biased.

There are additional advantages to adopting a model fire prevention code. A model code provides a document that may be recognized as authoritative by architects, engineers, and builders who already are familiar with code requirements; provides an interpretation process by which the local official may determine the intent of the code-developing body should a question arise; and allows for periodic revision to reflect new technology and current thinking.

Several organizations publish model fire prevention codes that jurisdictions may adopt by reference for enforcement use. NFPA publishes NFPA 1, *Fire Prevention Code* which incorporates many NFPA standards by reference. Each of the three major model building code organizations also has developed its own model fire prevention code. Building Officials and Code Administrators (BOCA) International publishes the *BOCA National Fire Prevention Code;* the International Conference of Building Officials (ICBO), in conjunction with the Western Fire Chiefs Association, publishes the *Uniform Fire Code;* and the Southern Building Code Congress International (SBCCI), with assistance from divisions of the International Association of Fire Chiefs, publishes the *Standard Fire Prevention Code.* Each code references codes, standards, and other NFPA documents. These documents are found in the volumes of the NFPA *National Fire Codes*, published yearly and considered to be the most authoritative set of fire safety regulations in the United States.

Code Administration

Most fire prevention codes specify similar enforcement procedures and contain similar requirements. A typical fire prevention code contains an administrative section that establishes the legal framework and organization for the fire prevention program. This section usually outlines the code's applicability to occupancy types and specifies whether it applies to new or existing buildings, or both.

The code may define the enforcement authority of the fire prevention official and assistants, including lawful right of entry for inspections, the right to issue orders to correct hazardous conditions within a building, and the right to order evacuation of an unsafe building or premises. The code may further assign duties that require the investigation of fires and the maintenance of fire records. Permit requirements, conditions, and procedures to issue notices of violation may be included.

The administrative section should be written in such a way as to provide the fire prevention official with as broad a scope of authority as possible. This facilitates enforcement of the fire prevention code, other county or city ordinances, and state laws that pertain to the fire safety of buildings and premises, as well as to the control of hazardous materials within the jurisdiction. This section also may outline the owner's or occupant's responsibility for properly maintaining a structure and its fire protection features.

Subsequent code sections may define important code-related terminology, outline general fire precautions, and establish the proper installation, operation, testing, and maintenance procedures for fire protection systems and appliances within a building. Final sections of the code may outline the use and maintenance of specific types of equipment, processes, and occupancies, or they may describe handling requirements for various types of hazardous or explosive materials. Appendices may reference supplemental standards or other helpful data in the administration or application of the fire prevention code.

Under the fire prevention code or other ordinances of the jurisdiction, a Fire Prevention Code Appeals Board or similar body may be empowered to hear appeals of orders made by the fire official under the fire prevention code. This board should consist of technically knowledgeable and unbiased members of the community.

Enforcement Procedures

The fire prevention process includes a number of enforcement procedures used to establish compliance with the fire code. They are described in the following paragraphs.

Permits: The permit is an official document issued by the fire prevention division to authorize the performance of a specific activity. Permits are issued in the name of the fire official for the use, handling, storage, manufacturing, occupancy, or control of specific hazardous operations and conditions. The permit should be issued only if the condition meets code requirements.

The permit process provides the fire prevention official with information on what, where, how, or when specific hazards are being installed, stored, or used within the jurisdiction. The process allows cross-checking with building, zoning, public health, or other departments' requirements for the use outlined on the permit. Further, it allows the fire official to review and approve devices, safeguards, and procedures that may be needed to assure the safe use of hazardous materials or operations. The lack of a required permit or failure to meet permit requirements constitutes a misdemeanor in legal terms and is grounds for stopping an operation in, or use of, a structure.

The permit is the property of the issuing agency, not the permit holder. A license or permit authorizes, by law, the right of entry for inspection purposes to ensure compliance with the permit requirements. If an inspector operating under a fire code permit is refused entrance to perform a regulatory inspection, this refusal constitutes grounds to halt operation in, or use of, the structure involved.

Permits usually are issued for the place, process, and operator and are nontransferable so that the notification function of the permit process is maintained.

Certificates: A certificate, which is a written document issued by the authority of the fire official to any person or business, grants permission to conduct or engage in any operation or act for which certification is required. Depending upon the needs of a jurisdiction, certificates of approval may be required before smoke or heat detectors, fire extinguishers, fireproofing materials, or other fire protection devices are offered for sale to the general public. This procedure is designed to ensure that only those fire protection appliances that will function in a satisfactory manner are offered for sale to the public.

Certificates of fitness are issued to individuals or businesses with demonstrated proficiency in skills, training, and testing in areas that affect fire safety. This includes pyrotechnics and persons who handle explosives; persons who install fire protection equipment and systems, including sprinklers; and persons who maintain fire extinguishers and fire detection systems. Certification requirements may include a financial bond or liability insurance where conduct of the regulated activity is inherently hazardous.

Licenses: A license is a permission granted by a competent authority to individuals about to engage in a business, occupation, or other lawful activity. Licenses are issued to provide knowledge of specific business locations, ensure compliance with particular standards, and add a source of revenue to the community. The fire pre-

vention division may issue some licenses directly and may be involved in the check-off process for licenses issued by other regulatory authorities.

Enforcement Notices

When violations of the fire laws or ordinances are discovered during regular or spot inspections, the violations must be called to the attention of the owner or occupants. When the situation is not corrected during the inspection or is a recurring uncorrected violation, several types of enforcement procedures may be available to the fire inspection officer.

Warnings or notices of violation: The inspection official may issue these notices to the owner or occupant, stating that a specific violation of fire regulations and/or a fire prevention code has been identified during the inspection. Using discretion and judgment, the fire official may designate an appropriate period of time for correction of the violation cited. Recommendations for correction of the violation also may be noted on the form. The owner or occupant must sign the form and keep a copy of the document. Reinspection after the allotted time is essential to ensure that the correction has been made. If it has not, further legal action should be taken.

Red tag or condemnation notices: These notices usually are attached to appliances, systems, or equipment that would be unsafe and dangerous if allowed to remain in operation. The red-tag process requires that a competent service technician repair or replace the item before the tag can be removed and the equipment returned to operation. This process also may require that the fire official inspect the completed work before granting approval to resume the operation of the equipment in question.

Citations or summonses: Violation of the fire prevention code requirements or other fire regulations usually are considered misdemeanors. An authorized fire official may issue a citation or summons to an individual who is in violation of the law. These documents constitute a notice for the violator to appear before the appropriate court.

Warrants: A warrant is an order issued by a magistrate or agent of the court that directs a law officer to arrest a violator of the law and bring the violator before the court to respond to the charges specified in the document. Arrest warrants usually are issued in felony cases. The person requesting the warrant must provide factual information to the magistrate or court officer issuing the warrant, showing the existence of sufficient "probable cause" of a violation of a law.

When authorized by state law or local ordinances, fire marshals, fire investigators, and other fire officials may be empowered to issue summonses, serve warrants, and make arrests. Police or peace officer training, in accordance with the standards of the state, is required by most states before a fire official may serve warrants and make arrests for felony violations of state laws.

Where violations represent a clear hazard to life or property, it may be necessary for the fire official to take immediate action to correct unsafe conditions. Entrance into private property, shutdown of an operation, evacuation of a building, or withdrawal of permits are actions that the fire official may find necessary to ensure the public safety. These powers may be explicitly stated or implied in discretionary powers and "duty to act" requirements of appointing laws or ordinances. In all cases, actions must be based on clearly demonstrable threats to public safety, showing that delay would provide an unreasonable danger to residents, occupants, guests, or the public and that this judgment is based on accepted standards or concepts of safety.

RECORD-KEEPING

Keeping records and files on all actions taken by the fire prevention division or the fire marshal's office is an essential part of code enforcement and administration. All instruments and records of code enforcement, all documents relating to inspections, violation notices, summonses, plans review comments and approvals, fire reports, investigations, permits issued, certificates issued, and so on are to be handled as legal documents. Well-organized and well-maintained inspection files and building records are essential foundations for enforcement actions. Complete and accurate records also are needed to measure fire department effectiveness in accomplishing fire prevention goals and to provide department management with information for budgetary and administrative purposes.

For each inspected property, a file should be maintained that summarizes information about the inspected premises and contains copies of all inspection reports. Records and reports of fire prevention activities should be clear, concise, and complete. Every time an inspector or fire prevention officer inspects a location, information about that location should be included in a record or a report. The occupancy file of each building visited should include a complete history of the building site; building plans and specifications, where possible; fire protection system plans; information on and permits issued for the use, storage, and handling of hazardous materials; correspondence; inspection reports; and records of fires at that location.

Files are needed for properties where a certificate of occupancy, a license, or a permit has been issued. Files also should be maintained on properties with automatic sprinklers, standpipe systems, private hydrants, or other private means of fire protection. A special notation or key should identify any system or equipment that is legally required. With a well-maintained system of files, established data need not be recompiled for each inspection. Inspection work planning is expedited, and time is used more effectively in the field for the inspection itself.

Computer Applications

In recent years, computer technology has helped fire prevention officials handle record-keeping and manage fire prevention programs. Many communities have found their records are more accurate and complete when computers are used for processing, organizing, or managing a great deal of the information associated with the overall fire prevention activity. Fire prevention reporting systems should be designed to collect, store, and process data and to schedule periodic activities according to local policy. For example, all data, including occupancy description, activity, and violations, are collected on a standard form during an inspection. After the inspection, forms are completed and processed according to local needs. These data, to be useful, must be collected and maintained in a uniform and consistent manner. NFPA 901, *Standard Classification for Incident Reporting and Fire Protection Data,* provides a basis for systematic information gathering.

Computers can produce management reports related to inspection activity and to fire prevention personnel resource allocation. In addition, separate master file listings can be provided of occupancies currently under inspection and all occupancies to be inspected. Inspection forms or schedule sheets can be printed to show when periodic inspections are required and indicate violations to be checked for correction during follow-up inspections. These management reports should document:

1. The occupancy inspected, location the time of day and the date, and the name of the inspector;
2. The code violations that have been corrected, filed by type, number, and kind of occupancy; and

3. How the fire department used its resources to accomplish fire prevention program objectives.

These systems save the inspector time and provide data in a form useful for decision making by operational management. Because they contain a current inventory of occupancies, the systems can provide information for fire protection fiscal and long-range master planning.

PLANS REVIEW PRACTICES

Traditionally, the building department has been active in the design, construction, and final occupancy inspection of the building. The fire department's role traditionally has begun when the building has been occupied and has concerned the maintenance of life safety conditions and fire protection systems, as well as fire safe storage and handling of contents.

Today, a progressive department's role in the building construction process is changing. An increasingly important and practical fire prevention function involves the participation of the fire marshal or a designated representative in the review of building plans and specifications and in the construction process. In most cases, plans review is conducted in cooperation with the local building, zoning, and public works departments or with state agencies that may have this review authority.

The review of building plans and specifications provides the fire service with its best opportunity to see that fire protection standards are met before construction is completed and the building is occupied. The type and depth of the review process depend upon the community's needs and the functions of other local departments or state agencies with review authority.

Whenever possible, fire officials should participate in pre-construction conferences; address questions relating to fire protection features in the planned building, to the building code, or to fire prevention code requirements; and comment during the plans review process. If questions and issues concerning the effect of construction on fire safety are discussed at this time with the architect, engineers, contractors, and other code officials, misunderstandings and conflicts that could arise during the construction or final finish phase can be prevented. The fire official can emphasize fire safety code requirements and coordinate responsibilities with other code enforcement officials. Design professionals and contractors benefit from this procedure as well, because problems that otherwise would cost them time and money are eliminated before construction begins.

Site plan reviews: The site plan review provides the fire official with the first look at new construction and at additions to existing buildings. The site plan provides an overview of the intended construction in relation to existing conditions and includes information such as building placement, exposures, size, type of construction, occupancy, water supply from both public and private mains, hydrant placement, and access. The site plan also may provide information about existing conditions that must be modified, such as abandoned flammable liquid tanks or pipelines in the area. The contour of the land may be significant. The current and projected uses of adjoining properties, including zoning, also should be reviewed at this time.

Preliminary building plans: The review of preliminary building plans gives the fire official an opportunity to comment on those features of the building that significantly affect life safety and protection of the building from fire. The depth and scope of the review and comments will depend upon local conditions.

The provision for required fire protection systems is a primary concern in plans review. Because most of the building code may relate directly to provision for life safety in case of a fire within the building, the review may include, among other items, the type of occupancy, allowable areas and heights, fire separations, fire resistance of construction, interior finish, occupancy loading, number and location of exitways, protection of vertical openings, fire protection system and special hazards.

Final building plans and specifications: When the final building plans are submitted, they should include modifications required by the review official and agreed to by the design professional submitting the plans. Plans may be approved if they agree with the applicable code requirements. A building permit can then be issued, and construction may begin.

Plans review and approval must be followed by on-site inspections to ensure that the fire protection features and systems are constructed and installed as planned and approved. All deviations from the approved plans should be documented. All agreements reached on-site or by telephone should be documented for all parties in follow-up correspondence or file memos. This correspondence and a copy of the approved plans should be retained in an organized filing system or on microfilm or microfiche or disk so a permanent record of building construction is available for future reference.

As practical, building construction information should be provided to the fire companies responsible for fire suppression and fire inspection of the building. Information provided in all of the plans discussed in this section can aid in pre-fire planning for fire operations.

Certificates of occupancy: If the fire marshal's office is not the primary issuing agency for occupancy certificates, then either the fire marshal's office or the fire prevention division should be involved in the final inspection process. Each agency should certify the building before the certificate of occupancy is issued. This certificate indicates that all requirements under the building and other applicable codes have been met and that the building is safe and habitable. All fire protection systems should be tested and placed in service before occupancy is allowed. The building should not be occupied until inspection agencies have approved it. This final review helps to ensure the life safety of the occupants and to verify that any required corrections have been made before occupancy.

CONSULTATION

The public looks to the fire service for answers and advice concerning fire problems. Due to its unique ability to provide this information, the fire department should offer consulting services to the community. Fire prevention officers must be able to explain fire codes, fire-related sections of building codes, and the application of specific standards to design professionals, contractors, and members of the building trades. Consultation also should be available directly to property owners, managers, occupants, and members of the general public, who may not be as familiar with fire problems and their solutions.

To maximize this service, the fire department should inform and instruct citizens about the best ways to deal with fire hazards. When the fire prevention division receives frequent calls about the same problem, it may wish to develop standard recommendations for that problem. The fire department should maintain a library of up-to-date reference materials. In addition, the chief and other officers should be acquainted with individuals outside the fire department who have specialized knowledge or experience and who could act as resource persons. These and other consulting services usually are based in the fire-prevention division or the fire marshal's office.

FIRE INVESTIGATION

The thorough investigation of fires is an integral part of the fire department's commitment to public safety. Fire investigation includes two areas: (1) fire cause determination and (2) investigation of criminal actions that may have contributed to a fire.

Fire cause determination is of major importance to a fire prevention program. Analysis of the causes of fires in a community is the basis for establishing fire prevention-program priorities and providing fire safety information to the public. Maximum utilization of resources must be based on firm information about the local fire problem. As information is accumulated from fire investigations, data becomes available on which to base corrective programs. Over time, these data will indicate trends in the area's fire problem, and a fire prevention program designed to tackle priority areas can be implemented.

If it is determined that a fire is caused by arson or other unlawful burning of property, a full criminal investigation should be conducted. Other fires for which a cause cannot be readily determined, including those classified as suspicious or undetermined, should also be investigated. Assertive fire investigations by skilled, trained specialists can deter arsonists.

In communities and states where fire service investigators have police powers, personnel may conduct an investigation to the point of arrest and incarceration of perpetrators. Where fire department personnel do not have police powers, fire cause determination must be made by the fire department, with the follow-up investigation conducted by the police, sheriff, state police, or investigators from the state fire marshal's office.

PUBLIC FIRE EDUCATION

Fire safety education is an increasingly valuable area of public fire protection. This function has undergone rapid change and growth, focusing increasingly on the importance of prevention in public fire protection management.

Public fire safety education has two facets: (1) fire prevention education and (2) fire reaction education. Both facets are necessary in order to change fundamentally the way the general public views fire and to encourage people to act in a fire-safe manner. The public needs to be motivated and instructed in actions that minimize the dangerous effects of a fire, should one occur. The basic method of instruction is changing from "preaching" to "teaching." Information is best presented in short, interesting, even humorous messages that require thought and that place responsibility for action on the recipients. Recipients relate most effectively to the fire safety message when it is presented in a dramatic context. Cultural and language factors should be considered when fire safety messages are prepared for communication throughout the community.

National fire safety campaigns can be matched to local efforts. NFPA's *Learn Not to Burn*® effort is one such national campaign, with messages designed to supplement, but not supplant, local fire prevention education programs. NFPA has many public fire safety materials and programs designed for general public use and for specific groups and fire safety concerns. Community programs can be developed using the *Learn Not to Burn* theme and applying it to priority local fire problems.

Community awareness and participation at the grassroots level are the keys to encouraging people to adopt fire safe behaviors. Civic and service clubs, youth and fraternal organizations, neighborhood action groups, the business community, schools, and other groups are contributing to the growth of fire safety education. Highly effective programs in such areas as escape planning, smoke detector installation, fire hazard inspections, burn prevention, juvenile firesetting problems, fire safety for babysitters, and fire safety for the elderly have been conducted by community groups in conjunction with local fire prevention personnel.

Thousands of teachers are giving classroom instruction on fire safe behaviors. NFPA developed the *Learn Not to Burn* Curriculum® to assist in this effort. The curriculum, designed to teach twenty-two basic fire safety and fire prevention behaviors, presents comprehensive teaching strategies for teachers and fire departments to convey key life safety messages to children from kindergarten through the eighth grade. Because the curriculum is designed to integrate easily into existing school subjects, it does not represent an additional course for instructors.

The Public Education Office of the U.S. Fire Administration (USFA) has helped communities and states develop comprehensive public fire education programs by providing funding, technical support, and regional and national workshops to facilitate the exchange of information. State exchange and resource centers for public fire safety education material also have been established through the efforts of the Public Education Office. These efforts have greatly increased the amount of information exchanged among fire educators and encouraged the growth of their educational programs.

Many fire departments and fire prevention divisions have appointed full-time fire safety education officers. As more people have assumed these positions, the level of talent and professionalism of fire safety educators has increased. Many fire departments no longer require that fire safety education officers be trained fire fighters; rather, these departments recognize the value of hiring educators with teaching experience and skills to develop and present educational programs geared to all ages and groups.

Groups with Special Information Needs

Groups with information needs similar to those of the general public, but different enough to require specialized programs include, among others, educational, industrial, institutional, high-rise, civic, service, elderly, professional, handicapped, and commercial groups. Many innovative ways are used to reach such groups with fire safety information.

Seasonal Activities

Informative educational programs of interest to the general public often are built around the four seasons of the year. Public education programs include National Fire Prevention Week, which is always the week that includes the date of October 9, and Operation EDITH. Many topical and effective materials are available from NFPA for use in public education programs, including some featuring Sparky® the fire dog, symbol of personal and property fire safety.

BIBLIOGRAPHY

Reference Cited

1. National Commission on Fire Prevention and Control, *America Burning, The Report of the National Commission on Fire Prevention and Control,* U.S. Government Printing Office, Washington, DC, 1973.

NFPA Codes, Standards, and Recommended Practices

Reference to the following NFPA codes, standards, and recommended practices will provide further information about fire prevention and code enforcement. (See the latest version of *The NFPA Catalog* for availability of current editions of the following documents.)

NFPA 1, *Fire Prevention Code*
NFPA 901, *Standard Classification for Incident Reporting and Fire Protection Data*

NFPA 1031, *Standard for Professional Qualifications for Fire Inspector*
NFPA 1033, *Standard for Professional Qualifications for Fire Investigators*
NFPA 1035, *Standard for Professional Qualifications for Public Fire and Life Safety Educator*

Additional Reading

"1991 Updated: Legislation and Codes Affecting the Fire Sprinkler Industry," Sprinkler Age, Vol. 10 No. 11, Nov. 1991, pp. 12–15, 17.
Albinson, B., "Information Support for a Fire Officer," Fire Engineers Journal, Vol. 53, No. 171, Dec. 1993, pp. 26–30.
Bathurst, D. G. ,and Stroup, D. W., "Federal Fire Safety Act: The Impact on Fire Protection Engineering," Proceedings of the Society of Fire Protection Engineers Engineering Seminars on Performance-Based Fire Safety Engineering, Nov. 15–17, 1993, Phoenix, AZ, SFPE, Boston, MA, 1993, pp. 39–43.
Beever, B., "Regulations, Codes and Standards: The Fire Engineering Approach," Proceedings of the First International Conference on Fire Science and Engineering, ASIAFLAM '95, March 15–16, 1995, Kowloon, Hong Kong, 1995, pp. 1–7.
Buchanan, A. H., "Fire Engineering for a Performance Based Code," Fire Safety Journal, Vol. 23, No. 1, 1994, pp. 1–16.
Bugbee, P., and Cote, A., *Principles of Fire Protection*, 3rd ed., National Fire Protection Association, Quincy, MA, 1993.
Building Officials and Code Administrators International, Inc., *Legal Aspects of Code Administration*, Country Club Hills, IL, 1984.
Bukowski, R. W., and Babrauskas, V., "Developing Rational, Performance-based Fire Safety Requirements in Model Building Codes," Fire and Materials: An International Journal, Vol. 18, No. 3, May—June 1994, pp. 173–192.
Callahan, T., "Acceptance of Emerging Fire Protection Engineering Technology Dilemma of the Fire Protection Engineering Profession," Fire Protection Consultants, Inc., Walnut Creek, CA, Worcester Polytechnic Institute, Conference on Firesafety Design in the 21st Century, Part 1, Conference Report, Part 2, Conference Papers, May 8–10, 1991, Worcester, MA, 1991, pp. 71–83.
Churchward, D. L., "OSHA and the Fire Investigator," Fire and Arson Investigator, Vol. 46, No. 1, Sept. 1995, pp. 11–13.
Coffee, C. W., "Consolidated Fire and Building Department Enforcement—A Success Story," *Building Standards*, Vol. 55, No. 5, Sept.–Oct. 1986, pp. 24–27.
Coggan, R. S., "The Standards Crunch: A Fire Services Dilemma," *Fire Engineering*, Vol. 142, No. 5, 1989, pp. 38–39.
Coleman, R. J., and Granito, J. A., eds., *Managing Fire Services*, 2nd ed., International City Management Association, Washington, DC, 1988.
Computers in Code Administration, Building Officials and Code Administrators International, Country Club Hills, IL, 1987.
Connolly, R. J., "Fire Safety Engineering and the Structural Eurocodes," Fire Engineers Journal, Vol. 56, No. 180, Jan. 1996, pp. 27–30.
Corcoran, D., "Fire Prevention and Building Restoration Activities," Fire Engineering, Vol. 146, No. 12, Dec. 1993, pp. 94–98, 100.
Day, T., "Fire Test Data and the Fire Safety Engineer," Fire Surveyor, Vol. 21, No. 1, Feb. 1992, pp. 14–18.
Deakin, A. G., "Fire Safety in Buildings: Standards for 1992's Europe," Fire International, No. 121, Feb./March 1990, pp. 15–16.
Directory of State Building Codes and Regulations, 4th ed., National Conference of States on Building Codes and Standards, Herndon, VA, 1987.
Factory Mutual Engineering, "Loss Prevention and Equipment Inspection Checklist," Factory Mutual Engineering, Norwood, MA, P9116, December 1993; Factory Mutual Engineering, Business Under Fire! A Manager's Guide to Fire Prevention Programs, 1993, pp. 1–5.
"Fire Protection Engineers and Fire Protection Engineering Technicians: Relationships and Functions," SFPE Bulletin, Nov./Dec. 1991, pp. 7–14.
"Fire Safety Legislation and Enforcement," Fire Safety Engineering, Vol. 1, No. 4, Aug. 1994, pp. 30–32.
Flynn, W. B., "Legal Commentary: The Increasing Importance of OSHA Inspections," Journal of Applied Fire Science, Vol. 2, No. 3, 1992/1993, pp. 239–241.
Franchuk, D. M. ,and Waggoner, W. W., "Engineers, Sprinklers, and the Authority Having Jurisdiction: Working Together to Benefit Fire Fighters and the Public," NFPA Journal, Vol. 89, No. 2, March/April 1995, pp. 85–89.

Friedman, R., "Fire Protection Engineering—Science or Art?," Journal of Fire Protection Engineer, Vol. 2, No. 1, 1990, pp. 25–29.

Gross, J. G., "Harmonization of Standards and Regulations: Problems and Opportunities for the United States," Building Standards, National Institute of Standards and Technologies, Gathersburg, MD, March/April 1990, pp. 32–35.

"Halonisammuttimien tarkastus, huolto ja kayttotarkoituksen muuttaminen. [Inspection, Maintenance and Alteration of Halon Extinguishers.]," Palontorjuntatekniikka, Feb. 1993, p. 16.

Harvey, C. S., "Flexible Approach to Fire-Code Compliance," *Architectural Record*, No. 10, Oct. 1988, pp. 130–135.

Harvey, C. S., "Inspection Overload? Not in Boulder," NFPA Journal, Vol. 89, No. 1, Jan./Feb. 1995, pp. 57–60.

Hayden, W., "Code Enforcement Critical to Fire Prevention," Fire Engineering, Vol. 146, No. 6, June 1993, pp. 49, 51, 53–55.

Holmstedt, G., et. al., "University Training of Senior Fire Brigade Officers: The New Approach in Sweden. Short Communication," Journal of Fire Science, Vol. 6, No. 6, Nov./Dec. 1988, pp. 383-389; Fire Safety Journal, Vol. 15, No. 1, 1989. pp. 95-101; SFPE Bulletin, November/December 1989, pp. 10–11,16; Fifth International Fire Conference on Fire Safety, Interflam '90, Sept. 3–6, 1990, Cantebury, England, Interscience Communications Ltd., London, England, 1990, pp. 355-360.

Johnson, M. A., "Bar-Code Based Inspection Systems Improve Fire Safety," Journal of Applied Science, Vol. 1, No. 4, 1990-1991, pp. 315–319.

Johnson, P. F., "International Implications of Performance Based Fire Engineering Design Codes," Journal of Fire Protection Engineering, Vol. 5, No. 4, 1993, pp. 141–146.

Kaminski, E. J., "Practical Aspects of Wet and Dry Chemical Extinguishing System Inspection and Acceptance," Building Official and Code Administrator, Vol. 29, No. 4, July/Aug. 1995, pp. 34–37.

Kaufman, S., "1990 National Electric Code—Its Impact on the Communication Industry," 38th International Wire and Cable Symposium, U.S. Army Communication Electronics Command, Atlanta, GA, 1989, pp. 301305.

Kerrigan, J. J., "Aggressive Code Enforcement Pays Off," Fire Engineering, Vol. 149, No. 6, June 1996, pp. 74-76, 78.

LaMay, W. L., "Get Firm for Code Enforcement," Fire Command, Vol. 57, No. 9, Sept. 1990, pp. 24–26.

Lathrop, J. K., "Life Safety Code Key to Industrial Fire Safety," NFPA Journal, Vol. 88, No. 4, July/Aug. 1994, pp. 36–46.

Law, M., "What Is a Fire Engineer?," Journal of Applied Fire Science, Vol. 1, No. 1, 1990–1991, pp. 3–6.

"Legal Aspects of Code Enforcement: A Report on the 1993 Annual Conference Education Program," Building Standards, Vol. 63, No. 1, Jan./Feb. 1994, pp. 27–30.

Louderback, J., "Atlantic City Pioneers Hand-Held Computer Inspection System," Firehouse, Vol. 19, No. 7, July 1994, pp. 30, 32, 139.

Lucht, D. A., Kime, C. H., and Traw, J. S., "International Developments in Building Code Concepts," Journal of Fire Protection Engineering, Vol. 5, No. 4, 1993, pp. 125–133.

Magnusson, S. E., "University Degree Education in Fire Safety Engineering," Fire Engineers Journal, Vol. 50, No. 159, Dec. 1990, pp. 7–9.

Magnusson, S. E., et. al., "Sweden's View on University Training of Fire Officers," Fire Engineers Journal, Vol. 50, No. 159, Dec. 1990, pp. 9–11.

"Major Changes to the 1995 Codes," Consensus, Spring 1995, p. 25.

Malhotra, B., "Engineering Fire Safety," Fire Prevention, No. 238, April 1991, pp. 29–32.

Malhotra, H. L., "Role of a Fire Safety Engineer," Fire Safety Journal, Vol. 17, No. 2, 1991, pp. 129–134.

Mawhinney, J. R., "Development of Regulations in the 1990 National Fire Code of Canada on Storage of Dangerous Goods," Fire Technology, Vol. 26, No. 3, Aug. 1990, pp. 266–280.

Meacham, B. J., "Performance-Based Codes and Fire Safety Engineering Methods: Perspectives and Projects of the Society of Fire Protection Engineers," Society of Fire Protection Engineers, Boston, MA, ISBN 0-9516320-9-4; Proceedings of the Seventh International Interflam Conference, Interflam '96, March 26–28, 1996, Cambridge, England, Interscience Communications Ltd., London, England, 1996, pp. 545–553.

Meacham, B. J. and Custer, R. L. P., "Performance-Based Fire Safety Engineering: An Introduction of Basic Concepts," Journal of Fire Protection Engineering, Vol. 7, No. 2, 1995, pp. 35–54.

Moulton, G. A., ed., *NFPA Inspection Manual*, 7th ed., National Fire Protection Association, Inc., Quincy, MA, 1994.

Moss, D., "Fire Safety and Compliance of 1992," FM Journal, July/August 1995, pp. 15–19.

Murphy, J. J., Jr., "Basics of Fire Inspections," Fire Engineering, Vol. 149, No. 6, June 1996, pp. 91–92, 94.

Murphy, J. J., Jr., "Fire Safety Operations and Code Compliance Concerns. Part 2," Fire Engineering, Vol. 148, No. 1, Jan. 1995, pp. 78–82, 84, 86.

Neale, R. A., "When Code Equivalencies Don't Work," American Fire Journal, Vol. 48, No. 1, Jan. 1996, pp. 20–23.

Neuenschwander, H., "Problems Relating to the Inspection of Fire Detection Systems From the Point of View of a Certified Bureau for Inspections," Proceedings of the Tenth International Conference on Automatic Fire Detection "AUBE '95", [Internationale Konferenz uber Automatischen Brandentdeckung.] April 4–6, 1995, Duisburg, Germany, 1995, pp. 64–71.

Oey, K. H., and Passchier, E., "Complying with Practice Codes," *Batiment International Building Research and Practice, Vol. 21, No. 1, 1988.*

Price, R., "Portable Inspection Computers Put Pen to Pad," American Fire Journal, Vol. 46, No. 2, Feb. 1995, pp. 16–19.

Richardson, L. R., "Determining Degrees of Combustibility of Building Materials—National Building Code of Canada," Fire and Materials: An International Journal, Vol. 18, No. 2, March–April 1994, pp. 99–106.

Robbins, G. C., "What's Customer Service Got to Do With Code Enforcement," American Fire Journal, Vol. 46, No. 7, July 1994, pp. 12–13.

Robertson, J. C., "Development and Enactment of Fire Safety Codes," *Introduction to Fire Prevention*, 3rd ed., Macmillan, New York, 1989, pp. 112–132.

Robertson, J. C., "Enforcement of Fire Safety Codes," *Introduction to Fire Prevention*, 3rd ed., Macmillan, New York, 1989, pp. 133–153.

Robertson, J. C., "Fire Prevention Responsibilities of the Public Sector," *Introduction to Fire Prevention*, 3rd ed., Macmillan, New York, 1989, pp. 39–75.

Robertson, J. C., *Introduction to Fire Prevention*, 3rd ed., Macmillan, New York, 1989.

Santavuori, A., "Automaattisten sammutuslaitteistojen tarkastustoiminta. [Inspection of Sprinkler Systems.]," Palontorjuntatekniika, March 1992, pp. 16–17.

"Social Clubs—New York's Happy Land Fire Drew Attention to Code Enforcement Efforts in Social Clubs," Building Official and Code Administrator, Vol. 24, No. 3, May/June 1990, pp. 28–33.

Stollard, P., "Fire assessment and Fire Engineers," Fire Surveyor, Vol. 22, No. 3, June 1993, pp. 21–23.

Strength, R. S., "Status Report Model Building Codes 1992, NEC-93 and IEC-89," Fire Retardant Chemicals Association Fall Conference: Industry Speaks Out on Flame Retardancy: Coatings; Polymers and Compounding; Test Method Development; New Products, Technomic Publishing Co., Lancaster, PA, 1992, pp. 41–46.

Stroup, D. W., "Understanding and Complying With the Federal Fire Safety Act of 1992," FM Journal, July/Aug. 1995, pp. 4–5, 8–9.

Teague, P. E. and Jones, J., "Taking Fire Inspection to a Higher Level," NFPA Journal, Vol. 85, No. 3, May/June 1991, pp. 57, 59, 61.

"World Trade Center Bombing May Bring Code Reviews," Consulting—Specifying Engineer, Vol. 13, No. 5, April 1993, p. 13.

Young, K. R., "Excuses, Excuses: Fire Investigations," *Fire Engineering*, Vol. 141, No. 6, 1988, pp. 34–51.

Young, K. R., "Excuses, Excuses: What the Fire Marshal Didn't Tell You But You Need to Know About Fire Investigations," Fire and Arson Investigator, Vol. 42, No. 1, Sept. 1991, pp. 34–38.

Organization/Agency Information and Program Sources

Building Officials and Code Administrators International, Inc. (BOCA), Country Club Hills, IL.

Conference of States on Building Codes and Standards, Herndon, VA.

International Conference of Building Officials (ICBO), Whittier, CA.

International Fire Chiefs Association (IAFC), Washington, DC.

International Fire Service Training Association (IFSTA), Oklahoma State University, Stillwater, OK.

National Fire Academy, Emmitsburg, MD.

National Institute of Standards and Technology, Center for Fire Research, Washington, DC.

Southern Building Code Congress International, Inc. (SBCCI), Birmingham, AL.

United States Fire Administration, Emmitsburg, MD.

FIRE DEPARTMENT FACILITIES AND FIRE TRAINING FACILITIES

Nicholas J. Cricenti

Fire station or fire-training facility design can range from the very simple to the very complex (see Figure 10-15A). For fire stations, an adequate facility can be a single bay with a desk at the back that offers fire protection in a rural area where the majority of responses are of the wildland type. Or the design can be a multiple-company station that houses engines, ladders, ambulances, and a complete administrative function. In a similar manner, a training facility may simply be a small tower for practicing ladder work, or it could be a campus-like complex that offers both classroom and practical education to hundreds of students at one time. Because of this variation, there is no right or wrong way to build either a fire station or training facility. This chapter will assist those responsible for the design of such facilities to develop proper design criteria to successfully complete the facility.

This chapter by no means lists all of the elements of any kind of facility, but rather includes the major components of these facilities. This is left to the engineer or architect and the users to consider during the planning of the project. For example, a kitchen space in a fire station may be developed to only serve meals, whereas in an-

FIG. 10-15A.   Fire stations exist in virtually every city and town throughout the world. This station is located in Kingston, MA, where it protects a large shopping mall.

Nicholas J. Cricenti, P.E., is a principal of PSC Engineering Partnership Engineering Association, Inc., Concord, NH. He is a member of NFPA's Technical Committee on Fire Service Training.

other fire station the kitchen may also serve as a training room and thus may be arranged slightly different. The objective of this chapter would be to get the designer and fire department personnel to evaluate the use of the kitchen and plan accordingly. This chapter discusses building layout only in broad terms. The designer is responsible for an efficient building layout, including the necessary elements requested or required by the fire personnel.

This chapter first discusses fire stations and the major areas that should be addressed during the planning phase and possibly incorporated into the final design. Then the chapter takes a similar approach to the development of a training facility design. It lists major items to be considered and elements to be incorporated. By no means does this provide a complete list of props found at a training facility, but rather touches upon the major elements of the design and opens possibilities for in-depth research.

FIRE STATION DESIGN

Fire stations exist in virtually every city and town in the world, and there are a variety of functions that the fire stations have to serve. A fire station may be simply a garage to store a truck; while others, in addition to housing trucks, provide housing for a complex hierarchy of offices for managing delivery of safety services. There are, however, many elements of fire stations that need to be present without regard to what is housed in the building. The design professional must be aware of these elements.

One of the most important elements of a fire station is whether or not it represents a safe and efficient place for the people to use. This requires a careful analysis of the proposed building and the functions it is expected to serve. There are obvious program elements consistent for all stations, such as security, fire protection, weather proofing, and structural stability. The building where the fire equipment is to be housed must be designed to withstand the forces of man and nature. Other necessary design elements are lighting, heating or cooling, storage facilities, and writing surfaces.

It is very important when designing a fire station that a clear understanding between the design professional and the fire department be established in the conceptual design phase of the project. It is also necessary that the design professional have a clear understanding of the development costs of program elements to keep the total costs within the budget.

The fire station design must address the needs of its users. Even if it is a single-bay station without any other space required, consideration must still be given to the people who are going to use the station. If it is a complex that is planned for housing a mix of civilians and fire rescue personnel, the building must be functional for both the civilian and the fire rescue personnel. If the design professional is not familiar with the necessary elements of a fire station, then the project will be difficult to bring to a successful conclusion. The design professional cannot and should not depend on the fire chief to be aware of all of the elements necessary for a fire station, such as architectural elements, that need to be considered in the design of the building.

Site Selection

The site for the building must be selected before the structures can be selected. This is usually determined by the analysis of many factors, and sometimes it is determined by a single factor. In any case, the selection of a building site is a task that requires some knowledge of fire department operations and goals. To properly site a fire station, an analysis of the department's response goals must be undertaken. If it is the goal of the department, for example, to respond to any incident within 4 min, the site location will be different than if the goal is a 15-min response time. The analysis of response time must include a traffic study, addressing potential intersection delay, travel times, and access to travel routes. Response time for volunteers to the station should be part of the analysis, taking into consideration where they live.

Often, locating a fire station is done in reaction to past growth. Nevertheless, the site should be preemptive instead of reactive. Future growth, including location and type of uses projected, must be considered when siting a fire station. This means that the designer must look ahead up to 50 years to attempt to determine the growth patterns of an area.

Often fire stations are located on major travel routes, based upon the thinking that the response time is the most efficient; this approach has some validity. But this also removes what may be very expensive real estate from the tax roles. An alternative approach is to locate the station on a secondary route, just off from the major route, and utilize a signalized intersection to allow safe access to the major route. With the recently developed technology to allow intersection signal control from fire apparatus, this method of secondary route location is becoming more popular, and the property on the secondary route often is not taxed at the same high rate as the property on the major route.

Once the site has been tentatively selected, it is necessary to view it, considering improvements that must be made to construct the fire station. It is also necessary to consider such items as parking for both fire fighters and apparatus, keeping in mind that the apparatus must have room to back in and sufficient turning radii, as well as visitors parking; drainage and runoff from storms; and security. The capacity of the soil to carry the loads imposed by the building must be considered. Utilities, must also be reviewed especially for operations that are often ancillary to the fire station use, such as dispatching, which will require multiple phone lines, or an air compressor, which may require three-phase power for efficient operation.

Planning for Building Functions

The designer and the fire department personnel must work closely together on the development of the building plan and the relationship of the spaces to each other. The building must function properly for each type of use. If the building contains one or two bays for fire apparatus, it is not difficult to design the building efficiently. If the building will also function as an office, residence, and community center, then careful analysis of the relationship of the different spaces is necessary. One method of evaluating the spaces is to have the fire department personnel fill out a spread sheet on each proposed room and their relevance to each other in both importance and travel distance.

There are also other items to consider, such as the effect of the apparatus starting, leaving, and returning, upon the other operations in the building. The noise and vibration caused by apparatus movement can be a distraction for other building uses. Fortunately, several methods of removing exhaust gases have been perfected and, with the proper installation of one of these systems, the effect of exhaust can be eliminated. However, the method selected for exhaust gas removal must be carefully considered. Some designers have simply opted to totally change the air in the apparatus bay when the apparatus is started or is returning. This method may not be desirable in areas where the air is conditioned either by heating or cooling, and where a total change of air will result in the necessity of conditioning entirely new air in a relatively large space. The newer systems either filter the existing air or connect to the exhaust pipe of the vehicle and direct the exhaust completely outside the building.

The proximity of personnel to fire apparatus when a response is required must be considered when laying out a fire station. Because the key time for fire suppression and emergency medical operations is the time spent getting equipment out of the door, the building design should minimized the travel time of the personnel to the vehicles. Several spaces need to be in close proximity to the apparatus floor. They can include staff sleeping area, medical supply rooms, and equipment storage for fire trucks such as hose and spare self-contained breathing apparatus (SCBA) bottles. Locating these spaces next to the apparatus floor limits the travel necessary to replace supplies on the apparatus and increases efficiency.

It is construction practice today to attempt to construct a single-story building within the allowable limits of the site. Because of concern over injuries, sliding poles in fire stations are a rapidly diminishing tradition. It is far safer to design and construct a stairway for personnel to travel. When a single-story building is not reasonable, then care must be taken to properly design the building to limit the amount of travel from the first floor to other levels. One approach to this is to locate administrative functions on the second level, and maintain as much as possible response functions, including sleeping accommodations, on the first floor.

When the program for the building has been completed and a conceptual layout for the building has been created, it is then necessary to begin to develop designs for structural stability, sewer, water, heating, cooling, ventilation, electrical, and fire protection. Each of these tasks is very important to the functionality of the building. The structure of the building must be such that it supports all the operations of the building.

The structure must withstand the forces of nature, including snow, wind, rain, and earthquake. A fire station cannot withstand even the slightest failure of a structural component and, thus, must be framed stronger than most buildings. The structural engineer must ensure that the building will not fail. This means extra bracing and stronger components than is normal for a comparable building size. All fire stations, regardless of size, should have a backup power supply in case of emergency. Adequate response time and sufficient power is critical.

Typical Fire Station Space

The following are some of the typical spaces that are programmed into fire stations. Not all of these spaces must be in every building, but, rather, this list is a guide to assist in the development of a program for a fire station.

Apparatus bay: This is the area where the fire trucks, rescue trucks, ambulances, and other support vehicles are housed (see Figure 10-15B). Design considerations must address the fact that

FIG. 10-15B. *Apparatus bays comprise the largest percentage of floor space in a typical fire station.*

heavily loaded trucks may sit for day after day on the floor, and that the trucks are loaded when they enter and leave the building. The trucks will be the full height and width allowed by law, and the station doors should be oversized to reflect this fact. Station bay width for the trucks should be wider than in a normal garage, because the trucks should and do receive some maintenance in place, and there needs to be space to open doors and tilt hoods without forcing the personnel to crawl over or under the trucks. The bay should also have enough space to store minor repair items, perhaps on some shelves. The bay should be equipped with a system to vent exhaust gases from the trucks, be well lighted, and heated. In times past, most floors had drains installed to collect drippings from the trucks; today's conventional wisdom is to forego the floor drains because of the concern of pollutants from the trucks. If floor drains are absolutely necessary, then an oil/water separator must be used to limit pollution.

Sleeping accommodations: In fire stations where personnel are on duty at all times, an area commonly called the "bunk room" was provided for sleeping. Traditionally, bunk beds (two beds, one on top of another) were provided to save space. For a more "home-like" feel, many new sleeping areas are now being designed like college dormitory rooms, with two single beds in one room.

A growing trend in the volunteer fire and rescue service is to require or encourage volunteers to perform a "duty night" where they staff the stations at night. In many cases, this can reduce response time to emergencies by four or more minutes. To support these efforts, many amenities are being added such volunteers.

Access from this area to the emergency equipment room should be as direct as possible. The best case is to have this room on the same level of the station as the equipment room, because injuries have occurred when personnel traveled from the sleeping area to the equipment room using poles or stairs. Although stairs are safer than poles, both have caused serious injuries and should be avoided.

Some of these sleeping spaces have walls of soundproof materials of partial height, and some are full height. With the recent addition of women in the fire and rescue service, many new questions have developed:

- Should separate shower and bathroom facilities be provided? Many of the existing fire stations were designed to accommodate only men. Therefore, they had one large "bunk room" where the entire company slept, and one shower/bathroom fa-

cility. To avoid potential problems, men and women need separate showers and bathrooms. As a temporary fix, a system using prominent signs and inside locks has worked in some situations.

- Can women and men share the same sleeping facilities? In a large "bunk room" area, men and women can and do share sleeping facilities with few problems. This type of traditional arrangement helps to ensure that each member feels part of the company. Fire fighting is a team activity and every effort should be made to support the team concept. One problem that needs to be specifically dealt with is the clothing that is acceptable in these situations. Acceptable sleepwear could be an issued item from the fire department.

- Some departments have provided separate sleeping facilities for women. This can have some negative consequences. Women may feel alienated by feeling that they are not part of the company. Also, from a practical point of view, you simply do not know how many men or women will be sleeping at the station on any given night. You may have a whole crew of women (as reported in one large West Coast fire department) or you may have none on duty. This leads to designs that can double the amount of space and furniture needed for sleeping accommodations.

- Having men and women sleeping in two person "dorm-type" rooms should not be allowed. Enforce a rule that stipulates either all men or all women in each room. Also, using $1/2$ or $3/4$ height separation walls will provide privacy without sacrificing the company feeling that is so important to a well-functioning team of fire fighters.

Another trend occurred after fire departments started providing ambulance service. Separate sleeping rooms have been provided so that the EMS crew can be alerted for a call without disturbing other fire crews. This has many of the disadvantages noted in our previous discussion of women in the fire service. If this is a problem, some other type of alerting should be considered. Designating certain beds for the EMS crew and proving a separate telephone has been successful in some cases.

Adequate clothing storage should be provided for personnel to keep several sets of clean uniforms and other clothing. Many departments provide clothing storage areas for each person assigned to the station. This can mean up to 3 or 4 storage areas for each on duty person. Volunteer departments that commit to having "on duty" companies for immediate response may also need several clothing storage areas for each person "on duty." Several designs use a separate room for clothing storage with metal lockers and wooden benches similar to a dressing area of a sports facility.

Watch room: The watch room, while it has a variety of names, is essentially the office for the apparatus floor. This space, usually about 120 sq ft (11 m^2), generally contains a desk, a chair or two, a phone, a map, and usually a radio. The typical watch room serves as the place where paperwork for the apparatus, the log-in of calls, and station to apparatus communications are handled. The space is generally located adjacent to the apparatus floor to facilitate operations. Often this is the place where visitors to the fire station first enter without going directly onto the apparatus floor.

Kitchen: For a staffed station, the kitchen quickly becomes an essential part of the building. The kitchen is where the fire fighters take their meals and take their breaks. The eating area of the kitchen can also be used for training. The cooking area should be of commercial quality, due to the almost constant use of the kitchen. In a station with two or three full companies and a chief, the kitchen can be called upon to serve upwards of sixty meals per day, every day. If the kitchen is used in conjunction with a community room or other similar space that is, or may become, accessible to the public, then even more meals can be expected to be served in an emergency situation. Part of the kitchen must be reserved for storage of food, and each shift expected to use the station should have a separate space for food. The kitchen should also be equipped with a dishwasher. This will ensure that, even if the station has a busy day, there will be clean dishes for meals. The size of the kitchen should be designed similar to a commercial kitchen, and the dining area also should be designed based upon the number of expected diners.

Lounge or living room area: These areas resemble the typical family room in many homes, except larger. A TV and videotape recorder will be found that can be used during training sessions or for relaxation between alarms. Several couches and chairs to accommodate the normal number of personnel in the station will be provided. The furniture is normally commercial grade to hold up under the constant use. The area should be to provide a "home-like" environment for the comfort of the fire and rescue personnel.

Decontamination room or space: Regardless as to whether or not the personnel are medical or hazardous materials responders, it is reasonable to assume that they will at some time encounter a contamination problem, either a biohazard or other type, that will require complete clothing removal. A space must be available where personnel can be decontaminated, beginning with the removal of outer layer of clothes, and cleansing their skin without contaminating the rest of the station or its personnel. This means a room, or more likely rooms, that contains equipment to allow contaminated personnel to enter without touching anything to remove and properly dispose of clothing, clean themselves, and don new noncontaminated clothing with some degree of dignity and privacy. The equipment involved includes foot-activated hampers, motion- or foot-operated wash basins and soap dispensers, automatic showers, automatic doors, and linen rooms for spare clothes and towels.

Emergency medical supply room: As emergency medical services by the fire service become more common, the appropriate supplies for the service become more important. Such supplies include narcotics, which require a large degree of security. Other medical supplies have "shelf lives" that require frequent rotation. The diversity of the medical equipment needed to provide effective emergency medical services is increasing at a rapid rate. These items, and the need to keep them clean and secure, establish the need to have a room dedicated to the storage of medical supplies. This medical supply room should be located near the emergency medical vehicle(s) and must have a lock. It should be of such size that the equipment can be stored in a neat and orderly fashion. The room should be equipped with shelves and small cabinets, as well as a locked cabinet for narcotics and other regulated equipment. A double sink is also handy in this space. A room 10 by 12 ft (3.1 by 3.7 m) generally will serve this purpose.

Laundry room: As with decontamination of personnel, protective clothing also needs to be cleaned. Laundry rooms are becoming a popular item in fire stations. Turnout gear and other personnel clothing worn on the job need to be cleaned. It is not reasonable to expect personnel to take this equipment home to wash, and few commercial laundries will wash turnout gear. A commercial washer and dryer, as well as adequate venting, are necessary for this space. This space may best be incorporated into an area where the fire personnel hang their personal gear during off-duty time.

Bathing and toilets: A fire station layout must provide space for personnel to wash. Separate facilities for men and women should include sinks, toilets, showers, and lockers, as well as closets for linen. These spaces should be of tile or some other type of construction to withstand constant use with very little care. The equipment in the space should be of such quality as to give an adequate time of use before renovation. The space allocated for these rooms can be based upon the expected number of users. In addition to the space for personnel, additional space should be allocated in the building design for toilet facilities for other personnel working, or visitors to the station.

Training room: If a separate space for training is desired, then it should be set up as a classroom. The room should be spacious, provide desks for the personnel to use, contain blackboards, allow space for slide and overhead projectors, and provide adequate space for the display of training materials. The training room should also contain a lecture table for the instructor.

Offices: There are several types of offices that are found in fire stations. Each subset of administration requires a different type of office. The minimum size of each office type does not vary much, but the equipment found in the offices is vastly different. A chief's office is similar to the CEO of a corporation. The office requires space for the chief to work, as well as a space for meeting with other people. A fire prevention officer's office is essentially an engineering office and requires space for plans review as well as other office functions. The plans review space may be a drafting table or an oversized desk, but should accommodate oversized plans sets. A secretarial office must contain space for the secretary to work, and a file cabinet, fax machine, copy machine, and other equipment are often found in a modern office. A training office, which is basically a teacher's office, should be located next to a training space or seminar room so that training materials do not need to be carried great distances. There are other types of tasks that may be required by fire administration such as receivables and payables, and these office types should be structured the same as any other business.

If the station serves an administrative function, then there should be space in the building for people to wait. This may be a small separate room or it may be a space near the offices, but it should not be a few chairs set up in a corridor. A conference room is also necessary. This room can serve as a place where administrative personnel can meet with individuals or small groups, e.g., the fire prevention officer meeting with a group of prospective developers or builders in a relatively private setting.

FIRE TRAINING FACILITY DESIGN

Design and construction of a fire department training facility can be one of the most complex operations in the construction field. Training facilities incorporate several different types of uses and typically many types of construction. A full-sized training facility contains classrooms, seminar rooms, eating facilities, dormitories, administrative facilities, live fire training spaces, ladder training spaces, and could contain a multitude of other facilities limited only by the imagination of the fire department and the design team. (see Figure 10-15C).

It is very important that a member of the design team be familiar with modern fire-fighting techniques. This is necessary because most architects and engineers have no detailed familiarity with the types of fire-fighting evolutions that are performed at the training grounds and the amount of space each evolution occupies.

The program of the training facility must be very carefully defined in order to create the most efficient use of space possible. While not all training facilities will contain all possible training aids, it is important to define which training aids are to be located at the facility and to allow room for expansion or alteration should the focus of training change. The program developed should also be reflective of the fire department's makeup and response needs. In addition, the safety and welfare of the trainees and instructors is a very

FIG. 10-15C.   *A modern fire training facility.*

important design consideration. Protection of the environment is also one of the prime considerations in developing a program for a training facility.

The selection of a design team for the facility is one of the first tasks involved in the development of a training facility. The team should consist of an architect, civil engineer, electrical engineer, mechanical engineer, structural engineer, and a fire service consultant to aid in the programming and space planning. The fire chief or a representative should have input into the decision of who the design team will be. In fact, a representative of the fire department should be a key player during the entire design and construction process.

Working with the design team, the fire department representative should establish which training aids will be required or desired to be constructed at the facility. These training aids could range from a simple pad for practicing vehicle extrication to a full-sized facility with classrooms, dormitory space, and many training aids. When compiling the list of training aids and facilities, the design team should maintain a concept of cost as well as space requirements for the training aids. This program development is then the basis for all other decisions that are made in relationship to the training facility. That is not to say that some adjustments cannot be made in the program, but the development of the program and then holding to the principles of the program are very important in the development of a training facility.

Site Selection

The search for a site can be a tedious process. There are many factors that are included in a proper site search and, depending on the size and type of facility, the search could encompass a wide geographic area. When searching for a site, the design team must have a clear concept of all of the elements that are to be included in the fire training facility as well as the space required for each element. This will require that each element be considered, including space necessary for parking, safety features, trainee operating area, and separation from the next training prop. This exercise is often completed using large circles (i.e., bubble diagram) or other geometric shapes to represent all of the training features and their necessary spaces. Design development must also allow for what would be a natural expansion of the facility as training elements change to reflect different approaches to fire suppression and prevention. For example, when the Civil Defense organization erected fire training facilities in the 1940s and 50s little thought was given to either

classroom training on site or rehabilitation areas at the site. The modern training facility should include both elements in the program.

Another factor that must be considered is access to the site. If the facility is to serve one municipality then travel time for companies must be considered. Whether or not the companies are to remain available during training must also be considered. Relocating an in-service company far away from their district and expecting them to be available for prompt response to that district is not reasonable. Rather, it may require relocating some companies to cover for the training companies to resolve a travel problem. If the facility is a regional facility, then the travel for all of the participants in the region should be balanced. This may require locating the facility in a central location or balancing the travel time of all of the participants. In some cases, the area covered by the training facility is so vast and the travel times are so long that it may require finding or developing sleeping facilities for the personnel attending the facility. Another minor element of site selection is the ease with which the site can be reached or found by major travel routes without facing major traffic problems.

As stated previously, a major element of site selection is environmental issues. The public is very sensitive to wetlands and other environmental issues, and the perception of training facilities right or wrong is that they are dangerous environmental hazards. Locating a training facility next to a sensitive environmental area may require major design features to protect the area. The protection of the environment is perhaps one of the most important elements of site selection. If it is not possible to select a site away from a sensitive environmental area, then the addition of an environmental consultant to advise on protection of the area must be considered. If the facility is near an environmentally sensitive area, then the training props may have to be redesigned to ensure that no pollution occurs. Another area of environmental concern is the smoke generated by training props. If the facility produces smoke and that smoke affects the neighboring property, then the design is a failure. Locating the facility away from areas that are sensitive to smoke is important. Noise and the absorption or deflection of it is important when searching for a site. Since it is reasonable to assume that at most training facilities night operations will be necessary or desirable, then the site should have physical features that limit the noise produced by fire training operations. Even if no night operations are planned, the noise of the engines of the fire trucks or the noise produced by power tools will carry for a long distance across open ground. Runoff from operations is another environmental concern that must be addressed; there should be space available to allow for the collection and reuse of the water used for training without contaminating the surrounding open waters or groundwater.

Another factor that needs to be considered in site selection is the ability of the ground to carry the building loads imposed by the facility. The necessary size of the building may impose building loads that some soil types are not capable of carrying without special foundation designs, which are very expensive. Also land-use regulations of the area are important. Even if exempt from land-use regulations, the training facility should attempt to comply with the regulations. Locating a training facility, which is essentially industrial in nature, in a residential neighborhood is not good planning practice and may lead to complaints from the very people that the fire service is looking to for support. The facility should fit in with the local surroundings, including land use. Finally, the estimated water and sewerage disposal quantities should be identified and the methods of sewerage disposal should also be identified.

Using the site development criteria and space needs, the available areas of land that meet the requirements need to be identified. This task can be completed in a number of ways, but some materials or methods that can be useful for available land identification are tax maps, local or state planners, other fire departments, and real estate companies. U.S. Geological Survey (USGS) quad sheets and other regional maps are useful in identifying physical features of the land that may be favorable for the development of training facilities. Ideally, several parcels of land should be identified that may meet, within reason, the criteria for a site. Visit the sites and evaluate the land for its suitability for the uses necessary for the training facility. This evaluation should be done on a rating basis, with points given to various features of site suitability depending upon the importance given to them by the design team. It is often helpful to develop a spreadsheet to assist in this process. Even though it might appear that some of the potential sites can be eliminated without performing the rating, it is important to perform the rating in order to validate the process; on occasion, those sites seemingly eliminated by initial inspection can rate fairly well in the point process, and if the negotiations for a prime site fall through then it may be necessary to fall back to a secondary site. The weight given to each element in the rating process may vary from place to place because the importance of each element locally. Each site in the analysis should use the same weight for each element during the process.

Design Considerations

Once the site is selected and acquired, then the actual design process can begin. The site must be surveyed to develop a map of the physical features as they exist, then the design team can begin to lay out the training features of the site and develop a site plan for the facility. The location of each element of the facility should be developed as a team process and should consider such items as smoke drift from training props, noise, visibility, public access, runoff, and security.

It may be desirable to have the classroom element of the facility located so that the public can easily access it without traveling through or past ongoing training where safety may be compromised. It is also not desirable to allow the unescorted public to wander around the site where injury may occur.

As mentioned above, the runoff from operations must also be considered. It is not desirable to have runoff from master streams flow through a class that is performing below-grade rescue. The civil engineer designing the site must be told of the flows and other unusual features of the site so that they can be incorporated into the design early.

It is often better to recycle the water used for training than to use new water for training. It seldom makes economical sense to use fresh water from a municipal system that has been treated with chlorine and other expensive chemicals when there is no intent to drink the water. Establishing a drainage system that will allow the collection and reuse of the water is a better use of the resource. The collection and storage of water from training operations is also desirable should some fire-fighting chemicals be mixed with the water. It is possible to design a pump and hydrant system that will allow the training yard to act exactly as a municipal hydrant system. If designed properly, the pump system itself can be used as a training aid.

It may be possible to utilize other physical features of the site to assist in construction of other training props. For example, an elevation differential of 10 to 13 ft (3.1 to 4 m) will allow a building to be constructed so that it appears to be one story from the high side and two stories from the low side, thus increasing the flexibility of the facility.

General Building Considerations

Prior to laying out the building's exact locations, preliminary test boarings should be done to determine the soil bearing capacity of the site. Should the facility require on-site sewerage disposal, this is the time for final selection of the site for that. These two concerns have a priority for land use, since there are specific soil requirements. The supply for potable water should be determined. If the source is from a municipal system, then estimated flows and pressures and pipe sizes should be developed. If potable water is to come from a source such as a well, then the well location and protective radius should be identified. If potable water is to come from an on-site source, then a determination as to how to treat the water for consumption and the sort of delivery system to be used is necessary.

The area around the individual training props must also be carefully developed. There must be sufficient space to allow for the staging of personnel, safety officials, instructors, and the operation itself. For example, around a training tower there should be sufficient space for the maneuvering of multiple pieces of apparatus in a combined operations scenario. The space required for each training prop can be shared with other spaces; however, if that occurs then a decision of which prop to use at any given time will have to be made.

A training facility with classroom space for one hundred students and outside training for another one hundred students will require a site of about fifteen acres to handle all of the necessary facilities.

Building system selection is the architect's responsibility; however, the architect should be made aware of the types of building uses. If there are to be live hose streams in the building, then a construction type that will withstand being wet all of the time and the power of the hose streams will be necessary. If the building is to be used as an office space, then the type of construction should reflect that use.

It is important that the architect understand the potential uses of the building and the training environment essentials. The architect must be guided by the fire department and a consultant familiar with the design of these facilities. The development of a training facility is similar to the development of a campus for a multiple-building facility. Use of similar fixtures and other replaceable items throughout the project will be helpful in limiting costs.

Main Structures

All possible elements are not included in all training facilities. The following major items and their requirements provide guidance. This is by no means an all-inclusive list, but instead covers most of the major props and other support structures.

Classroom building: The classroom building, which often contains the administrative offices of the facility, provides the shelter and tools necessary for the instruction of the trainees. The building should be equipped similar to a school with desks, blackboards, overhead projectors, and other educational aids (see Figure 10-15D). The classrooms should be flexible enough in design so that multiple configurations can be achieved. Often this flexibility is achieved with moveable partitions that can subdivide a larger space into several smaller classrooms. In addition, larger facilities should contain an auditorium that will seat large groups of people for lectures and other programs suitable to a large audience. Small meeting rooms can also enhance the classroom facility. The small rooms provide a breakout area for small group activities, as well as a meeting place for instructors to develop programs for the trainees. These rooms also provide a space for meeting with a limited number of people without tying up a larger classroom. Often these rooms can be rented to other fire-related groups for small meetings or seminars. Additional space is required for instructor offices, printing, storage, and administration.

FIG. 10-15D.   Large double classroom. (Source: Mississippi Fire Academy.)

FIG. 10-15E.   Drill tower at the University of Kansas.

Some facilities have included a classroom that is accessible from the outside drill area so that groups that have completed an outside fire-fighting evolution can immediately come in to the classroom in their protective clothing and the instructor can review the evolution and the class can then return to the drill yard. This space needs to be constructed in a manner that will withstand the hot, wet, dirty trainees and be relatively easy to clean. The furniture in this space also needs to be somewhat oversized to accommodate the trainees with their protective ensemble. For a space such as this to be effective, it must be in close proximity to the outside training props.

It is often desirable to include a lunchroom or cafeteria in the classroom building. While it may not be necessary to have kitchen facilities, the trainees need a place where they can have their meals. The trainee should not be expected to take all meals offsite. Certainly some of the trainees will travel to a restaurant, and some of the trainees will probably bring a bag lunch. However, a space with tables and chairs as well as vending machines for drinks, etc., will serve the facility well. This space does not have to be too large, but it should be able to accommodate a good percentage of the trainees and staff.

Parking allowance for the classroom building and the entire facility should be based upon the largest reasonable expected occupancy. Parking can be estimated using the ratio of one acre for every one hundred cars. Excess parking area could also double as a driver training area. For some classroom buildings, it may not be necessary to construct all of the parking necessary, as parking may be shared with a facility nearby if the anticipated parking requirement should be exceeded on a weekend when nearby facilities are closed. However, having trainees parking randomly throughout a neighborhood is a quick way to create hard feelings with abutters, and, therefore, on-site parking must be provided.

Training tower: Perhaps the single most important element that is included in a fire training facility is a training tower. These towers range in height from two to ten stories and vary about as much in perimeter (see Figures 10-15E through G). The towers allow the trainees to practice several different evolutions, but are primarily targeted to ladder training. The construction of the tower is typically either wood or concrete, although steel towers have been used.

Design of the tower should reflect conditions found in the area that the training facility serves, and there should be consideration for the use of aerials in the training process. While there is no right or wrong way to develop the design for the towers, there are several items that should be considered. First is the load that will be applied to the tower by the training process, as well as by environmental elements. For example, winter training exercises could create a layer of ice on the structure; this load may cause failure to a marginally designed structure. The structural engineer should be familiar with the unusual loads that may be applied to the tower.

Towers should have a method to be secured. These structures can be attractive to people that may find them irresistible toys and could be hurt by climbing and falling from the tower.

The towers are often constructed from wood or precast prestressed concrete because of the ability of these materials to be erected in a modular fashion. This modular construction lends itself to tower construction, since most towers repeat the same floor plan for level after level. Because of the height, the towers need to be well founded in the soil, and the often slender tall building needs to be designed to withstand sway loads from seismic activity.

The layout of the tower should present the trainee with as many of the situations as possible that the trainee may encounter in real fire situations. There should be interior rooms, balconies, stairways of various types, fire escapes, various roof pitches and flat roof areas. The roof section should be equipped with 4- by 8-ft (1.2- by 2.4-m) replaceable sections that can be used to practice ventilation. There should be a variety of window types, with sills made of heavy pine with a thickness of 1 in. (25.4 mm) or more. These sills can absorb the impact of the ladders and should be designed to be replaced as they become worn. All of the exterior areas need to be protected by substantial railings. If a rappelling area is to be incorporated into the tower, then there must be places to secure ropes and allow the trainees to go over the edge. Many towers have standpipes installed and some even have ductwork with which smoke can be directed onto different levels of the tower. Note: Live fire should only be used in buildings specifically designed for live fire training. (See the following subsection.)

The area around the tower should be kept open for vehicle maneuvering and ground ladder handling by trainees, and kept open for an area of 100 to 200 ft (30.5 to 61 m) in as many directions as possible. If the area on all sides of the tower can be kept open, then it will be better suited for training. Ground cover around the outside of the tower is the subject of some debate. While having the perimeter constructed of a solid maintenance-free material is desirable, what that material actually is should be carefully thought out by the designers. Typically these areas are made of cast-in-place concrete; but, when using ground ladders, it is difficult to set the ladder spurs reliably on concrete. Bituminous concrete (asphalt) is another material that is used at the base of ladder towers. Asphalt, however, commonly softens in hot weather and then breaks up from heavy traffic. In addition, the ladder spurs penetrate the asphalt and eventually tear the asphalt apart. Grass will not stand up to the heavy vehicular and foot traffic. Often the material that works best is a granular material, recovered from stone crushing operations; this packs hard enough to withstand traffic and does not create a dust hazard, but is granular enough to allow ground ladders to be set firmly. The material is also easy to repair should damage occur, and it can be plowed on to allow snow removal.

Live fire training building: A building dedicated to live fire training is a very desirable element of a fire training facility. These buildings are constructed of noncombustible materials and are used to present the trainees a scenario of fire fighting that involves live fire. The building must be designed to withstand the 2,000+ °F (1,093+ °C) that these training fires may impart to the building. The building also must be able to withstand the change in temperature created when water from hose streams strikes the walls, ceilings, and floors of the building after it has been heated to high temperatures. There are many materials that have been tested to create a building that will withstand the high-heat temperature and the change in temperature, e.g., cast-in-place reinforced concrete, precast prestressed concrete, masonry, and steel. All of these materials have met with varying success and all eventually break down under the stress that is applied to them.

Most buildings constructed now are a combination of reinforced concrete, precast prestress concrete, and masonry. Often panels made from calcium aluminate concrete with lightweight aggregate of high-carbon content, refractory blocks, or steel liner is then offset from the walls of live fire rooms, are used to protect the concrete walls. This method of construction has been successful, although the liners need periodic replacement. Steel also has to be installed with an allowance for the great amount of expansion and contraction that will occur when the steel is heated by the fire and cooled by the hose streams.

It appears that the key to longevity of live fire training buildings is control over the heat produced by the fires. It is not necessary

FIG. 10-15F.   Drill tower at the New York Fire Department Academy.

FIG. 10-15G.   Ladder training tower at the Dover Township Fire Academy.

to have fires hot enough to "bake" the trainees in their protective clothing to provide successful live fire training. The reinforcing steel in the concrete should have additional coverage to limit spalling by the expansion and contraction of the steel.

Live fire training obviously requires a fire in the building. This can be accomplished by burning Class A materials or using a gas (either natural gas or propane) fire. Care must be taken with any fire so that an atmosphere dangerous to the users of the building is not developed. When using any fire, only enough heat and smoke should be generated to create an atmosphere necessary for training. Class A materials have been used successfully in training buildings for many years. Buildings have burn rooms or a specific area designed for the kindling of the fires.

A new generation of live fire buildings has recently been developed. These buildings use gas, either natural gas or propane, as the fuel for training fires. These buildings were first developed for the military, but are now being constructed for civilian fire training facilities. They are often operated in conjunction with a computer that controls the scenario in a prescribed manner. There are several of these computer-activated buildings in use; however, there are also several manually controlled gas-fired buildings currently in use. In either type of gas-fired building, several safety items must be addressed. These safety items include ignition verification, gas concentration sensing, safety shutoffs, and burner failure sensing. Because the gas fires in live training buildings do not generate much smoke, these buildings often have an additional smoke source to develop a realistic fire scenario. This smoke source may be a smoke generation machine or simply smoke from a Class A material smoke source, such as wet hay. Most of these newer buildings have been developed with a fan-driven ventilation system for quick smoke and heat evacuation in case of an emergency. The buildings that have computer-controlled systems have sensors that will shut down the gas and activate the emergency ventilation systems at a predefined temperature. In addition, there are emergency shutoff buttons located strategically throughout the building to activate the emergency systems manually.

The room layout inside the live fire training building should be as realistic as possible. While it may not be financially possible to simulate every room that the trainee may encounter, there should be a variety of simulated rooms, e.g., a bedroom, a kitchen, a stockroom, or a living room. The burners can be mounted in or near the simulated furniture for realism.

The exterior of the live fire training building needs to be accessible, so that attacks can be made from fire apparatus. In addition, there should be sufficient room for the placement of apparatus for training multiple-company operations. The surface material outside of the building should be a hard durable material, such as concrete or asphalt. Where ground ladders are used, it is possible to construct boxes in the concrete that are filled with compacted crusher dust; these set the ladders in so the ladders will not slip.

The building should have access from multiple areas, such as bulkhead, doors at grade, doors above grade and as many kinds of windows as possible. The windows in the building should be limited to those that are absolutely necessary; and those necessary windows should have heavy shutters to limit light transmission to simulate fires at night and low visibility in heavy smoke. The window sills should be made of thick pine to absorb the impact of the ladders, and be designed so that they can be changed as they wear out. The building should also be designed with multiple roof pitches, so that the trainees can practice on the various roof types that they may encounter. These roof pitches could also have a panel that the trainees could remove for ventilation.

Gas leak/fire simulators: Perhaps the easiest of the props to construct are gas leak/fire simulators. In its crudest form it merely consists of a gas source connected through a valve to a pipe with several holes. The gas is turned on and the vapor coming from the pipe is ignited and then trainees practice advancing hose lines and directing the flame away from the direction of the hose crews. A safety shutoff must be manned by a qualified safety person should an accident occur. These types of simulators can contain simulated broken flanges, cracked valves, broken pipes, railroad cars, or defective gas cooking appliances. Another popular configuration is the residential gas barbecue. They do not occupy a lot of space and can easily be constructed as a mobile training aid. The valving arrangement should be such that the valve has to be held open in order for gas to flow; thus, in an emergency, all the designated safety person has to do is release the handle and the gas is shut off. The props can be ignited by a variety of means, including a torch on the end of a long pole, a piezometric igniter, or an oil burner igniter. These props do not take a lot of space, just enough for the hose crews and the back-up crews to maneuver their hoses. The ground cover around the props can be crushed stone, which gives a firm stable base without the expense of asphalt.

Flammable liquids training: There is a tremendous amount of disagreement in the fire service as to the best way to conduct flammable liquids training without causing an environmental problem. For small training facilities that have intermittent use, the cost of installing environmental safeguards will far outweigh the benefits of the training. For these facilities it is more cost effective to install a flammable liquids simulator or to send the trainees to a larger facility for this type of training. A simulator can be developed using natural gas or propane in a multiple burner at grade installation. This simulator can have sensors and be controlled by a computer to simulate a flammable liquids scenario. These types of simulators can even be used with Class B foams.

For larger facilities that will have more continuous use, the cost of the environmental controls becomes more feasible. It is important that all of the materials be recovered and not be allowed to pollute the area. This recovery includes all of the water used for the training as well as all of the flammable liquids. This recovery will include the design and construction of extensive drainage facilities capable of handling the flows created by the training. In addition, an oil/water separator will be required to separate the oil from the water so that both can be reused. Only clean oils of known character should be used in these training props. The use of unknown liquids may lead to unexpected pollution concerns. Props that use flammable liquids produce large quantities of thick black smoke. This may be undesirable in some areas, and facilities utilizing these props require careful siting of the facility. The area required for the operation of a facility that uses flammable liquids is comparatively large. There needs to be drainage to collect runoff, and storage for both contaminated flammable liquids and clean flammable liquids. There also needs to be room for the storage of runoff that has been separated from the flammable liquids. There also needs to be storage space for flammable liquids that are awaiting use, and that storage will require the design of a spill control plan and the storage of any equipment necessary for the control of a spill as prescribed by the plan.

Automotive training: There are two types of training necessary for automotive training. Trainees typically receive training in extinguishing automotive fires and also receive training in auto extrication. For automotive fires, a gas simulator can be easily constructed of steel and served by one or more gas burners similar to the way a gas training prop would be created. By constructing a prop of steel it will last longer than using a vehicle adapted for fires. It is recommended that the vehicle be placed upon a reinforced concrete slab that will help prevent erosion from the power of the hose streams. If the choice is made to use a vehicle adapted to training, then certain precautions must be taken to prevent injury. All of the fluid tanks should be drained and removed, and any closed container must be removed, such as bumpers, door or window pistons, some steering columns, and shock absorbers. In addition, in some areas burning of the foam seats will be in violation of air pollution standards, and the seats and dashboards will have to be removed. In short, the vehicle will have to be stripped virtually to bare metal prior to being used for training.

For auto extrication vehicles it is possible to use vehicles from the local auto yard. However, care should be taken that fluids are not leaked from the vehicle during the operations. To ensure that the tools and personnel are not in the dirt, a slab for auto extrication should be constructed. This slab of reinforced concrete should be about 30 ft (9.2 m) long and 12 to 15 ft (3.7 to 4.6 m) wide. In addition, a slab off to the side approximately 8 by 8 ft (2.4 by 2.4 m) for the storage of tools should be constructed.

Air crash fire rescue: An airplane simulator and related equipment is perhaps the most real estate consumptive prop on the training grounds. This prop requires a mockup of an airplane the size of which is governed by the FAA and the size of the airport from which the trainees are from. The FAA criteria also describe the number and types of training aid required, based upon the airport served. There may be a requirement for a wheel and brake fire, a galley fire, and a storage compartment fire, in addition to the engine fire and fuselage mockup. The area around the props also requires maneuvering space for the crash rescue vehicles. A complete fire crash rescue training area will require 3 to 5 acres of land. However, it is possible to simulate several portions of aircraft fires with smaller props. A simple engine fire mockup fired by gas will occupy approximately 100 by 100 ft (30.5 by 30.5 m). This mockup is controlled by a gas valve that will allow the prop to simulate an engine fire. In fact, with the exception of a 727 mockup, it is possible to use small simulators for most of the aircraft fires. The storage and galley fire can also be incorporated into a live fire training building.

Drafting and pump testing area: When instructing trainees in the basics of pumping it is necessary to develop an area where the pumper can flow water and not wet down the entire area. In addition, it is necessary to instruct the trainees in drafting techniques. It is possible to perform these tasks with either a small pond or an underground tank that has access for suction hose from the pumper. The pump operator can then establish a water supply and flow through a master stream device into a horn-type arrangement that will return the water to the drafting area. This will allow the maximum use of the water without just throwing it away. This area can be constructed in a space of approximately 50 by 50 ft (15.3 by 15.3 m) including the tank. By using a master stream device it is possible to measure the flow from the pump to demonstrate to the trainees proper flow and pressure techniques.

Driver training area: A 2-acre paved area would be ideal to instruct driver training and allow maneuvering room for vehicles; however, with careful planning, it is possible to perform the same training in an area much smaller. By utilizing overflow parking areas, existing loading docks at the classroom buildings, and other features incorporated into the site the actual segregated driving area can be quite small.

Storage area: It is necessary to have a space on the site to store equipment, both large and small. A garage to house vehicles with storage space attached out near the training area would be ideal; however, there also needs to be some place where the equipment is stored. The size of this space can be determined only by the actual equipment required for the site.

Rehabilitation area: In order to allow rest for the participants in the training, it is often necessary to construct a space at the training site that will allow participants to go to rest and recover or to retreat from bad weather. This space needs to be equipped with water, emergency showers, first-aid equipment, chairs, and tables. The space can also serve as a critique space for field operations.

Rescue operations area: If there is an intent to teach rescue operations, such as confined space rescue or building collapse, then props for these items must be constructed. For building collapse, the prop can be as simple as two parallel reinforced concrete walls between which building debris is piled in various configurations for practice in tunneling and debris removal. This set of walls also can serve as a support for the development of confined space rescue techniques. This is accomplished by using the two walls and installing a confined entrance at the end of the two walls to represent a confined area.

There are several smaller props that can be constructed on the site without taking up a lot of space. For example, a pitched roof section can be constructed close to the ground to allow the trainees to practice with axes and saws prior to working at heights. This section can be constructed of heavy timber and have a replaceable panel to permit the actual cutting of the roof. For SCBA training, a mobile home can be converted for use. This obviously cannot have live fire, but can be blacked out to allow confidence to be built with air masks prior to entering heat and flame.

The previous list of equipment is not by any means all inclusive, but is intended rather to provide the basis upon which to begin development and a discussion of the major props and equipment that are found in large training facilities.

BIBLIOGRAPHY

NFPA Codes, Standards, and Recommended Practices

Reference to the following codes, standards, and recommended practices will provide further information about fire department facilities and fire training facilities. (See the latest version of *The NFPA Catalog* for availability of the current edition of the following document.)

NFPA 1402, *Building Fire Service Training Centers*

Additional Reading

Adams, B. and Carlson, G., "Unique Way to Develop Technical Training Materials," *Speaking of Fire*, Vol. 2, No. 3, Fall 1995, pp. 11, 22.
"Apparatus for the Long Haul," *Fire Command*, Vol. 57, No. 1, Jan. 1990, p. 36.
"Apparatus Overview," *Fire Command*, Vol. 56, No. 2, Feb. 1989, pp. 33–40.
"A Stitch in Time . . . Selecting the Right Protective Clothing," *Fire Command*, Vol. 56, No. 4, April 1989, pp. 29–38.
"A Survey of Hoses and Nozzles," *Fire Command*, Vol. 56, No. 10, Oct. 1989, pp. 25–30.
Blackett, D., "Essex Links With Sweden in Realistic Flashover Training," *Fire*, Vol. 88, No. 1091, May 1996, pp. 30–31.
Brace, T., *et. al.*, "Training, Publications and Certification," 1992 National Arson Forum, Sept. 9–10, 1992, Arlington, VA, 1992, pp. 1–6.
Bradley, M., "Heavy Foam Cover," *Fire Engineering*, Vol. 140, No. 6, 1987, p. 52.
Bradley, T., "Virtual Training Academy," *Fire Chief*, Vol. 39, No. 1, Jan. 1995, pp. 82–85.
Brewer, R. E. and Marolla, T., "Firefighter Training Facilities," *Proceedings of the 1993 Halon Alternatives Technical Working Conference*, May 11–13, 1993, Albuquerque, NM, 1993, pp. 621–634.
Callan, M. J., "Do Our Training Programs Work?," *NFPA Journal*, Vol. 89, No. 4, July/Aug. 1995, p. 27.
Campbell, C., "Cleaning Up the Fire Training Environment," *Fire International*, Vol. 136, Aug./Sept. 1992, pp. 20, 22.

Campbell, C., "Question One at the Fire Academy—Where Do We Find the Money?," *Fire International*, No. 123, June/July 1990, pp. 31, 33, 36.

Carter, H. R., "Fire Service Training: A View for the Future," *Firehouse*, Vol. 21, No. 2, Feb. 1996, pp. 38–39.

Chubb, M., "Firefighters: Train to Prevent As Well As Protect," *Fire Engineering*, Vol. 148, No. 8, Aug. 1995, pp. 117–118, 120, 122, 124.

Coneley, C. and Reed, S., "Training the Industrial Firefighter of Tomorrow," *Industrial Fire World*, Vol. 9, No. 5, Sept./Oct. 1994, pp. 6–72.

Corbett, G. P., "Connected: Fire Department Connection . . . ," *Fire Engineering*, Vol. 141, No. 1, 1988, p. 40.

Diment, E., "Portland Starts Fire Safety Training Early," *NFPA Fire Journal*, Vol. 88, No. 1, Jan./Feb. 1994, pp. 67–69.

Edwards, S. T. and Schappert, R. J., III, "Training Academy: Maryland Fire and Rescue Institute," *Firehouse*, Vol. 18, No. 3, March 1993, p. 22.

Egsegian, R., *et. al.*, "Practical Applications of Virtual Reality to Firefighter Training," Capitol Safety Systems, Raleigh, NC, North Carolina State Univ., Raleigh, Institute of Electrical and Electronics Engineers, Inc. (IEEE), International Emergency Management and Engineering Conference, 10th Anniversary, Research and Applications, 1993, pp. 155–160.

Ellam, L. D., *et. al.*, "Initial Training as a Stimulul [sic] for Optimal Physical Fitness in Firemen," *Ergonomics*, Vol. 37, No. 5, 1994, pp. 933–941.

Evans, C. W., "Review of Fire Resistant Hose Requirements," *Proceedings of the International Conference on Fire Resistant Hose, Cables and Belting*, Plastics and Rubber Institute, London, England, 1986.

"Fire Safety Engineering Science Program--College of Engineering, University of California at Berkeley. Current Research," *Fire Technology*, Vol. 27, No. 3, Aug. 1991, pp. 273–279.

Forney, G. P. and Davis, W. D., "Analyzing Strategies for Eliminating Flame Blow-down Occurring in the Navy's 19F4 Fire Fighting Trainer," National Institute of Standards and Technology, Gaithersburg, MD, NISTIR 4825, May 1992.

"From Burns To Bytes: Computers Promise to Enhance Firefighter Training," *NFPA Journal*, Supplement 1995, pp. 31–32.

Gamble, C. C., Jr., "Training the Water Supply Officer," *Fire Chief*, Vol. 37, No. 10, Oct. 1993, pp. 43–45.

Garcia, C. S., "Training for Shipboard Fire Fighting," *Firehouse*, Vol. 15, No. 10, Oct. 1990, pp. 52–54.

Garner, M., "Aviation Fire Training School to Have a New Boeing 747 Training Module," *Fire*, Vol. 85, No. 1053, March 1993, p. 26.

Garza, J., MacDonnell, J. and Bradford, G., "Going Live With Wildland Fire Training," *Fire Chief*, Vol. 39, No. 7, July 1995, pp. 53–55, 57.

Hansen, J., "Fire Fighter Trainers," VHS Video Tape, Symtron Corp, National Institute of Standards and Technology, Federal Fire Forum on Technology Update, Nov. 1, 1990, Gaithersburg, MD, 1990.

Hasegawa, Y., "Tokyo's Fire Academy," *Fire International*, No. 149, Winter 1995, p. 27.

Hoffman, S., "Training Should Include Protective Clothing," *Fire*, Vol. 87, No. 1075, Jan. 1995, pp. 21–22.

Holmstedt, G., *et. al.*, "Swedish Brigade Officers Receive University Training," *Fire Chief*, Vol. 34, No. 12, December 1990, pp. 34, 36–37.

Holmstedt, G., *et. al.*, "University Training of Senior Fire Brigade Officers: The New Approach in Sweden. Short Communication," *Journal of Fire Science*, Vol. 6, No. 6, Nov./Dec. 1988, pp. 383-389; *Fire Safety Journal*, Vol. 15, No. 1, 1989, pp. 95-101; *SFPE Bulletin*, Nov./Dec. 1989, pp. 10–11,16; Fifth International Fire Conference on Fire Safety, Interflam '90, Sept. 3-6, 1990, Cantebury, England, Interscience Communications Ltd., London, England, 1990, pp. 355–360.

Keith, G. and Garner, D., "Fighting Fire in Sprinklered Buildings–Training the Fire Service to Use Sprinkler Systems More Effectively," *Sprinkler Age*, Vol. 9, No. 8, Aug. 1990, p. 10, 12.

Kennett, S., Lambrineas, P. and Suendermann, B, "Fire Management Training Aid for the Navy," *Fire International*, Vol. 12, April/May 1991, pp. 15–18.

Kieffer, K. E., "Training Academy: Minnesota Fire Academy," *Firehouse*, Vol. 18, No. 7, July 1993, p. 24.

Klein, L. J. and Lane, S. C., "Emergency Vehicle Driver Training," Federal Emergency Management Agency, Emmitsburg, MD, FA-110, September 1991.

Knapp, J. and Delisio, C., "Survival Training in the Flashover Simulator," *Fire Engineering*, Vol. 148, No. 6, June 1995, pp. 83–84, 87–88, 90–93.

Laing, C., "Training Today's Firefighters," *Fire Prevention*, No. 290, June 1996, pp. 19–22.

Larson, R. D., "21st-Century Training Model at Work," *Fire Chief*, Vol. 40, No. 2, Feb. 1996, pp. 65–66, 68.

Larson, R. D., "Bay Area Wildfire Training '93," *American Fire Journal*, Vol. 45, No. 9, Sept. 1993, pp. 28–31.

Levine, R. S., "Navy Fire Fighter Trainers," BFRL Video Seminar, National Institute of Standards and Technology, Gaithersburg, MD, Video, July 16, 1991.

Levine, R. S., "Safety and Environmental Aspects of Navy Firefighter Trainers," Twelfth Joint Panel Meeting of the U. S./Japan Government Cooperative Program on Natural Resources (UJNR) on Fire Research and Safety, Oct. 27-Nov. 2, 1992, Tsukuba, Japan, Building Research Inst., Ibaraki, Japan, 1992, pp. 334–342.

Levine, R. S. and Greenaugh, K., "Exhaust Gas Analysis for Harmful Species: 191F1A Fire Fighting Trainer at Mayport, Florida," National Institute of Standards and Technology, Gaithersburg, MD, Naval Training Systems Center, Orlando, FL, NISTIR 4318, May 1990.

Lewis, R. J., "Fire Station Location Studies Using a Microcomputer," TR 86-1, Society of Fire Protection Engineers, Boston, MA, 1986.

Loikkanen, P.; Mangs, J., " [Investigation on the Operation and Safe Use of a Flashover Simulator for the Training of Fire Departments.]," Technical Research Centre of Finland, Espoo, Research Report PAL1166/91, June 18, 1991.

Mack, B. J., Jr., *et. al.*, "Fire Prevention Outreach Programs. Zeroing Fire Deaths in the Next Millennium. Virginia Beach's Monster Fire Truck. Getting the Fire Prevention Message Out Year-Round With Theater Advertising. Boy Scouts of America Fire Academy," *Fire Engineering*, Vol. 149, No. 6, June 1996, pp. 107–111.

Magnusson, S. E., "University Degree Education in Fire Safety Engineering," *Fire Engineers Journal*, Vol. 50, No. 159, Dec. 1990, pp. 7–9.

Magnusson, S. E., *et. al.*, "Sweden's View on University Training of Fire Officers," *Fire Engineers Journal*, Vol. 50, No. 159, Dec. 1990, pp. 9–11.

Mahoney, G., "Introduction to Fire Department Apparatus and Equipment," *Fire Engineering*, New York, 1987.

Malinin (Professor), "Fire Academy of Saint-Petersburg," *Fire Engineers Journal*, Vol. 52, No. 167, Dec. 1992, pp. 23–24.

Mangs, J. and Kruse, H., "Fire-Fighter Training in a Room Fire Simulator," *Fire International*, Vol. 16, No. 132, Dec. 1991/Jan. 1992, pp. 32, 34, 38.

Marlatt, F. P., "Regional Training Centers: Marriage Made in Maryland," *Fire Chief*, Vol. 40, No. 2, Feb. 1996, pp. 80, 82, 85–86.

McCoy, L. C., "Simulation in Training and Exercises for Emergency Response," National Institute for Urban Search and Rescue, Santa Barbara, CA, Institute of Electrical and Electronics Engineers, Inc. (IEEE), International Emergency Management and Engineering Conference, 10th Anniversary, Research and Applications, 1993, pp. 15–19.

Meyer, S., "New Training Opportunities for Volunteer Fire Officers," *Firehouse*, Vol. 15, No. 7, July 1990, pp. 58, 60–61.

Miura, K., "Fire Prevention and Disaster Preparedness Training at Business Establishments," Sendai City Fire Bureau, CIB W14/95/1 (J); Tokyo Fire Department, Firesafety Frontier '94, International Fire Conference and Exhibition in Tokyo, Creating a Safe Tomorrow, October 18–22, 1107-112, Tokyo, Japan, 1994, pp. 319–322.

Monigold, G. E., "State Fire Training: What the Numbers Tell Us," *NFPA Journal*, Vol. 89, No. 1, Jan./Feb. 1995, pp. 61–67.

Morrison, N., "Realistic Fire Training for Offshore Industry," *Fire*, Vol. 89, No. 1092, June 1996, pp. 23–24.

Nielsen, M. T., "Realistic Approach to Fire Investigation Training Within the British Fire Service," *Fire Engineers Journal*, Vol. 55, No. 178, 18–21, Sept. 1995, pp. 18–21.

Nussbickel, P., "The Fire Hydrant," *Fire Engineering*, Vol. 142, No. 1, 1989, pp. 41–46.

Perkins, J. and Dittmer, K., "Fire Suppression Training for Latin America: The Tie That Binds," Wildfire–International Wildland Fire Conference *Proceedings*, Meeting Global Wildland Fire Challenges: The People, The Land, The Resources, July 23–26, 1989, Boston, MA, U. S. Forest Service, Washington, DC, 1990, pp. 41.

"Protective Clothing: Options for the 90's," *Fire Command*, Vol. 57, No. 4, April 1990, p. 31.

Rosenhan, A. K., "Smoke Diver Training Tests Firefighter Stamina, Skills," *Fire Chief*, Vol. 35, No. 4, April 1991, pp. 42–45.

Rowe, G. D., "Computer Training Technology: Where Does the Fire Service Fit In?," *Firehouse*, Vol. 18, No. 7, July 1993, pp. 54, 56–57.

Rucky, F. P., "Remotely Controlled Robotic Vehicle for Fire Fighting," *Proceedings* of the IEEE International Symposium on Intelligent Control, IEEE, New York, 1987, pp. 239–245.

Sakamoto, A., "Current Approaches and Ideals Regarding Education in Fire Protection and Safety at the Fire Academy," Tokyo Fire Dept., Japan, CIB W14/95/1 (J); Tokyo Fire Department, Firesafety Frontier '94, International Fire Conference and Exhibition in Tokyo, Creating a Safe Tomorrow, October 18–22, 1107-112, Tokyo, Japan, 1994, pp. 305–312.

Salka, J., Jr., "Training Your Firefighters to Get Out Alive," *Firehouse*, Vol. 21, No. 2, Feb. 1996, pp. 124, 126–128, 130–131, 134.

Sanders, R. E., "Training Is Key to Future Success," *NFPA Journal*, Vol. 87, No. 3, May/June 1993, p. 26.

Schappert, R. J., III, "University of Maryland: Maryland Fire and Rescue Institute. A Showcase in Teaching Excellence," *Firehouse*, Vol. 18, No. 7, July 1993, pp. 94–97.

Schwope, A. D., *et al.*, *Guidelines for the Selection of Chemical Protective Clothing*,. Volume 1: Field Guide.

Shields, J., "Fire Safety Education and Training in Transition," *Fire Engineers Journal*, Vol. 50, No. 159, Dec. 1990, pp. 12–15.

Shuman, W. R., "Basic Training: Electrical Fires," *Fire and Arson Investigator*, Vol. 43, No. 2, Dec. 1992, pp. 52–58.

Sibley, C. W., "Dallas-Fort Worth Airport Aircraft Rescue and Fire Fighting Trainer Center," Department of Public Safety-Fire Service, DFW Airport, TX, International Association of Fire Chiefs (IAFC), Fire-Rescue, International Conference *Proceedings*, 1993 Annual Conference, Aug. 28-Sept. 1, 1993, Dallas, TX, 1993, pp. 191–196.

Smith, J. P., "Benefits of Training," *Firehouse*, Vol. 21, No. 4, April 1996, pp. 16, 18.

Solomon, S. S., "Fire Apparatus Color: Tradition vs. Safety," *Industrial Fire World*, Vol. 3, No. 5, Oct. 1988, pp. 8–11.

Szmanski, D. J,. and Stacy, J. W., "Lesson Plan: Computer-Aided Dispatch System," *Fire Engineering*, Vol. 141, No. 7, 1988, pp. 32–35.

Standing, J., "Airport Fire Training Goes Green," *Fire Prevention*, No. 270, June 1994, pp. 19–22.

Straseski, J. B., "Hands-On Training: A 4-Step Approach to Teaching Firefighting Skills," *Firehouse*, Vol. 21, No. 4, April 1996, pp. 42, 44, 46.

Taylor, J., "Flashover Training," *Fire Engineers Journal*, Vol. 53, No. 169, June 1993, pp. 15–23.

Thomas, M., "Recent Research on Fire Extinction and Fire-Fighting Equipment," *Fire Engineers Journal*, Vol. 47, No. 144, 1987, pp. 6–7, 10.

Thorn, P., "Training Center Has Backdraft Simulator," *Fire*, Vol. 88, No. 1091, May 1996, p. 29.

Thrasher, C. L., "SCBA Cylinder Failure," *Fire Engineering*, Vol. 141, No. 12, 1988, pp. 52–54, 56–60.

"Tools of the Rescue Trade," *Fire Command*, Vol. 56, No. 5, May 1989, pp. 39–44.

"Training Centers," *Fire Prevention*, No. 290, June 1996, pp. 24–25.

Tucker, D. J., "SCBA Management," *Fire Engineering*, Vol. 141, No. 12, 1988, pp. 46–47, 50–51.

Veghte, J. H., *Design Criteria for Fire Fighter's Protective Clothing*, 2nd ed., Janesville Apparel, Dayton, OH, 1988.

Winston, R. M., "Training the Structural Firefighter for SWI Firefighting," *Firehouse*, Vol. 21, No. 2, February 1996, pp. 153-155.

"Women's Training Program Upgrades Firefighting Skills," *Fire Chief*, Vol. 35, No. 2, February 1991, pp. 60–61.

Yasukawa, M. K., and Gagliardi, F. M., "Estimated Apparatus Speeds by Street Type in Chicago, Illinois," *Fire Technology*, Vol. 24, No. 4, 1988, pp. 291–298.

FIRE DEPARTMENT COMMUNICATION SYSTEMS

Revised by Evan E. Stauffer

This chapter is intended to provide the Handbook user with an overview of the total emergency communications responsibilities of the fire service, together with some of the principal management and operational systems and/or techniques through which the respective responsibilities may be achieved. The fire service relies upon four major elements for success: (1) intelligence, (2) personnel, (3) supplies, and (4) good communications.

Intelligent and well-trained personnel are of no value if there are no reliable means to direct them to the right place at the right time. The fire department's communications divisions are entrusted to provide this direction.

For additional information, consult the NFPA documents referenced in this chapter and review the experiences of other similar organizations. One should then make an individual needs assessment of the community and fire department, and from that, determine the adequacy of the emergency communications services provided. In all cases, the requirements of NFPA 1221, *Standard for the Installation, Maintenance, and Use of Public Fire Service Communication Systems* (hereinafter referred to as NFPA 1221), should be followed.

COMMUNICATION CENTERS

In creating any communications network, one must first establish a center point (communication center) through which all related process and control functions will be directed. (See Figure 10-16A.) A communication center may serve a single agency or area, multiple agencies under the control of a single political subdivision, several similar agencies from adjacent political subdivisions, an entire county, or any other large political jurisdiction. Combining fire and police communications within a specific political jurisdiction (and fire communications for several adjacent jurisdictions) has become an increasingly common management concept. If properly planned, such consolidations can provide positive results.

The communication center is where all requests for emergency assistance under the jurisdiction of a fire department are received and translated into an appropriate response. Further, the communication center should be structured to provide logistical support to the responding units until the situation is under control. Accordingly, as an operating entity, the communication center must ensure

---

Evan E. Stauffer, P.E., is chairman of the NFPA Committee on Public Fire Service Communications. He is a fire protection engineer employed by the U.S. Navy.

that the equipment, personnel, and established procedures are adequate to satisfactorily perform in emergencies.

Even the smallest fire department must have some means of receiving notice of a fire. In small communities, the communication center may be located in a private residence or business establishment where someone is always available and where there is a switch to sound a siren or other device to alert fire fighters. Quite frequently, the police station serves as the communication center. As fire departments grow, a room should be provided to house the fire alarm operator and related facilities in a fire station or other suitable structure.

Site Selection and Configuration Considerations

The following is a list of items that should be considered in the development of a communication center:

1. Location
2. Seismic stability
3. Security
4. Emergency electrical power
5. Wiring access (computer floor)
6. Lighting
7. Air conditioning (computers/dispatch)
8. Air-conditioning backup
9. Console layout [per Occupational Safety and Health Administration (OSHA) and Americans with Disabilities Act (ADA) regulations]
10. Console arrangement
11. Acoustics (background noise suppression)
12. Restroom facilities
13. Kitchen facilities
14. Decoration
15. Rest areas (lounges)
16. Dormitories (sleeping quarters)
17. Emergency rations
18. Alternate locations

The communication center should be designed and constructed so the probability of interruption to operations due to fire or other causes will be minimized. A building in a park or other open space is most desirable and can often be designed in harmony with the surroundings. When a separate building is not feasible and the communication center must be housed with other operations, such as city hall offices or in a fire station, the center should be

FIG. 10-16A. A communication center equipped with an auto-mated status board (top left). The status board is controlled by a fully computerized, multiple-operator dispatching network. (Source: Phoenix, AZ, Fire Department)

properly separated from the remainder of the building by vertical and horizontal separations with a fire-resistance rating of at least 2 hr, and preferably with entrance only from the outside. Protection should be provided against fire exposures and unauthorized entry.

Alternate Communication Facility

Each jurisdiction should have an alternate communication facility that, when fully staffed, is capable of performing the emergency functions normally performed at the communication center. For example, a bomb threat or fire in the center requires an alternate facility and plan to evacuate the communication center. This alternate location should be totally removed from the primary location, and should have all the necessary requirements for long-term usage. In the event of fire in the center, provisions should be made to continue operations long enough to transfer operations to the alternate facility.

Communication Center Protection

Because of the types of activities conducted and the records maintained in any modern communication center, building and systems protection must be provided. Security begins at the entrance to the communication center, and can be as simple as a dead-bolt lock or as complex as a computerized entry control system. Only authorized personnel should be allowed access to the center. The center should be protected against damage due to vandalism, sabotage, and civil disturbances. The lowest floor should be above the 100-year flood plain. If the center is located on the first floor (grade level) of the building, windows should not be permitted. Entrances into the center directly from the exterior, if provided, should be protected by two doors and a vestibule. The communication center should be located in a fully sprinklered, fire-resistive, or noncombustible building. Alarms and supervisory signals from the building sprinkler system(s) should be monitored in the operations room. The center and the spaces adjoining the center should be provided with an automatic fire detection system that is monitored in the operations room. Evacuation signals should be designed to avoid interference with communications operations. In any case, hazardous areas and

the center itself should be protected by supervised automatic fire detection and fire suppression system(s). Portable fire extinguishers should be provided throughout. Heating, ventilating, and air-conditioning (HVAC) systems serving the center should be independent of the main building HVAC systems to avoid the possibility of smoke contamination from a fire elsewhere in the building. Fresh-air intakes should be located to minimize the possibility of drawing in smoke from a fire inside or outside the building. Consideration should be given to providing operator-controlled smoke-control modes for the center HVAC system. Interior finishes in the center should be noncombustible. Emergency lighting of sufficient intensity to permit operators to perform their duties during a total failure of normal lighting should be provided. Records must be properly safeguarded. Access to records should be limited to authorized personnel only. Off-premises storage of backup records, software, etc., is essential. This backup information should be updated on a regular basis.

Emergency Power

The communication center must be able to operate in the event of a primary power failure. If the primary power is provided from a utility distribution system, the secondary power may be either an engine-driven generator and a standby storage battery having a 4-hr capacity, a circuit from a separate utility distribution system arranged so that failure of the primary power circuit will not affect the secondary circuit, or two standby generators. Standby generators should be tested weekly for at least 1 hr and should be able to power the system at full load. Fuel storage should be adequate for 24 hr of operation. Where the potential exists for longer interruptions of primary power caused by natural disasters, such as hurricanes and earthquakes, fuel storage should be increased accordingly.

Uninterruptable Power Source

A computer-aided dispatch (CAD) system cannot tolerate the power surges and interruptions normally associated with commercial power systems. A momentary power surge or outage could blank the computer's memory or damage its many sensitive circuits. To prevent this, an uninterruptable power supply (UPS) is used as an isolator element between the commercial ac power lines and the computer system. This UPS consists of a set of batteries, a battery charger, and a dc to ac inverter. The inverter is phased to the commercial power frequency and produces the same power. Thus, any fluctuations or interruptions in commercial power would not affect the computer.

FIRE DEPARTMENT RADIO COMMUNICATIONS SYSTEMS

Radio System Selection

Considerable thought should be given to selecting the proper radio system, and professional personnel should be used to ensure the proper functioning of each system.

Radio Frequency Selection

Several bands of radio frequencies are available to the fire radio service. These bands are VHF low band, VHF high band, UHF 450 MHz, and UHF 800 MHz. Each band of frequencies has its advantages and disadvantages, and the selection of a particular band will depend upon factors such as frequency availability, area to be covered, type of terrain, number of radio units required, frequencies used by bordering fire districts, mutual-aid agreements, type of operation, and use of emergency medical radios.

Due to the lack of available frequencies, it may be necessary for fire departments or districts to share radio frequencies. For small fire districts, sharing radio channels with neighboring departments could be advantageous.

Type of Radio System Operation

The several types of radio systems available are as follows.

Simplex: This system utilizes a single radio frequency for both transmitting and receiving in all radios for each channel. With a simplex system, only one radio can transmit at any time and all others must receive.

Two-frequency half duplex: This system utilizes separate frequencies for transmitting and receiving. As in the simplex system above, half duplex allows only one radio to transmit at any one time.

Two-frequency full duplex: The two-frequency full duplex system utilizes separate transmit and receive frequencies and permits simultaneous conversations in two directions.

Two-frequency repeater system: With this system, a high-powered base station "repeater" is centrally located at a favorable site overlooking the area to be covered, preferably on a mountaintop, a tall building, or a tower. The repeater receives a transmission from any radio in the system on one radio frequency and instantly retransmits or "repeats" the message on a second frequency that is received by all other radios in the system. A repeater system greatly increases the range of all of the radios in the system. It is especially useful in hilly or mountainous terrain where it might be impossible to talk directly from unit to unit. Repeater systems are two-frequency half duplex systems as described above.

Trunking systems: The trunking system uses a group of radio frequencies. The system is controlled by a computer at the communication center or at the base station transmitter site. When a transmitter is keyed, it transmits its unique identity code to the computer. The computer instantaneously selects an available communications frequency and automatically directs the transmitting radio to use that frequency for the transmission. The advantage of this type of system is that it uses a group of available frequencies more efficiently by utilizing the capabilities of the computer to manage the radio traffic on each frequency. Thus, more radio traffic can be handled on the available frequencies, without causing congestion and interference. The computer can prioritize traffic from competing transmitters, based on their individual identity codes. For example, a request for a clear frequency transmitted by a fire department radio can receive a higher priority than a request from a nonfire agency's radio. Provisions should be included to keep the system operational in the event of failure of the computer controlling the system.

Tone-coded squelch: In a tone-coded squelch system, each radio transmits a subaudible tone along with the voice signal. This audio tone is used to "turn on" the receiver so the conversation will be heard. Only the proper tone will activate the receiver, which eliminates the reception of radio interference.

Radio paging systems: Contacting fire department personnel who do not normally listen to the radio system can be accomplished easily by a radio paging system. This system consists of a paging "encoder," which is located in the communication center, and individual personnel pagers. Pagers are also available that will respond to both an individual and group call.

Pagers are available for tone-only or tone-and-voice operation. When utilizing tone-only pagers, each individual is given a telephone number to call or location to respond to when the tone activates. With tone-and-voice pagers, the tone alerts the individual and turns on the receiver audio. At that time, the dispatcher can relay a verbal message to the individual.

Commercial paging service is available in most areas. While suitable for fire service administrative paging, its use for emergency dispatching should be avoided if possible. In some regions, the volume of paging traffic may cause significant delays in the transmission of a message. If necessary to use a commercial paging service for dispatching, the fire department and the provider should establish a protocol to assign fire service emergency messages top priority over all other paging traffic. Standby power and redundant communication paths between the fire dispatcher and the transmitter site should also be provided. (Refer to the discussion of dispatch circuits in this chapter.)

Use of Multiple Radio Channels

The size and complexity of each fire department will determine the number of radio channels necessary for adequate communications. Very small departments may require only one radio channel for their exclusive use, while departments in large metropolitan areas may require several. A separate frequency exclusively for fireground communications is required for each department (or multiple departments sharing the same dispatch channel) receiving a total of 2,500 or more alarms per year. In addition, all departments adjacent to one another should share a common mutual-aid radio channel. This is provided for by the FCC and is usually managed on a statewide basis.

Base Station or Repeater Location

Care should be taken in selecting the main base station or repeater location for the radio system. Each radio coverage area is unique and must be dealt with on an individual basis. A high, central location, such as a mountaintop or large hill overlooking the area to be covered, a tall building, or similar structure, is generally the ideal site for the main base station location. (This depends upon coverage requirements, frequency availability, and requirements for channel sharing.) Where none of the above is present, a high antenna tower can be utilized. Towers are required, even on mountaintops, to ensure proper antenna system operation.

The availability of commercial power and telephone access is also desirable when locating the main base station site. Standby power supplied by a generator or batteries (or both) must be provided to ensure the site remains "on the air" in the event of a commercial power outage.

In some locations, where commercial power is unavailable, solar power (solar panels), batteries, or generators may be used as the primary source of power for the site. If the base station or repeater location is remote and unattended, provisions should be made to secure the area against unauthorized entry and vandalism. Intrusion alarms should be monitored at the communication center. A chain-link fence should surround the area, and all radio equipment should be located inside a proper building, out of sight, and protected from the weather. Security, fire protection, standby power, and circuit protection considerations are similar to those of the communication center.

Standby base station or repeater location: A separate location should be provided for a standby base station or repeater. This standby location should have the same complement of radios as the main station. A good high central location is also desirable for this site. These standby radios will be used if one or all of the radios at the main site fail. A separate commercial power source should be provided, if possible, and standby power should also be available.

The communication center should have the capability of selecting either the main or standby radios. The standby radios should be used periodically to ensure they are operable and ready for use in case of an emergency.

Radio Site Antenna Configuration

When more than one radio is installed at a single site, an antenna system may be required. It may be impractical for each radio to have its own separate antenna, because considerable interference could be generated using separate antennas. When several compatible radios are used, it is customary to provide one or more transmit antenna systems and a separate receive antenna system. A base station repeater using a single antenna requires a duplexer in order to transmit and receive simultaneously. Several transmitters can be connected to a single antenna by use of an antenna combiner; several receivers can be connected to a single antenna through the use of a multicoupler. Professional assistance should be utilized to design the antenna system for all radio sites.

The antenna design should provide maximum radio frequency (RF) isolation between transmitting and receiving antennas. This is usually accomplished by mounting them one above the other on the antenna structure. Coaxial cables should be used for connections between radios and antennas. These cables should be high quality, low loss, with length as short as practical. Good grounding of the antenna system is required to prevent damage due to lightning.

Access from Communication Center to Radio Site

If base station radios are located at the communication center, individual wires or "house cables" can be used to connect the operators' consoles to the proper base station. If the base station radios are located remotely from the communication center, radios can be connected by leased telephone lines, a microwave radio system, or a point-to-point radio. Redundant links should be provided so that failure of a single link will not prevent operation of the remote transmitter site from the communication center. Adequate circuit protection should be provided at the communication center and at the remote site.

Mobile Radios

Mobile radios are ideal for use in fire apparatus or staff vehicles. They have high RF power output (30 to 100 Watts) and, therefore, considerable range. Their use, however, is restricted to vehicles, and they cannot be carried into buildings or structures. All fire department vehicles should have two-way radios. Mobile radios should be water resistant. Radios should be equipped with carrier-control timers to disable a transmitter that remains on the air due to a malfunction. If scanning devices are used, they should have an automatic priority feature whereby the radio will revert to its primary channel automatically when the channel is being used.

Portable Radios

Portable radios are much lower in RF power than mobile radios (1 to 7 Watts), and, therefore, have a considerably shorter operating range. Their main advantage is portability. They can be hand-carried and allow communications outside the vehicle, giving the user considerable flexibility. The reduced RF power, however, can cause problems in communications reliability inside structures in some areas. Portable radios should have carrier-control timers to disable a transmitter that remains on the air due to a malfunction.

Diversity receiver-voter system: Since portable radios have a very low RF power output, this low power creates problems for communications over large areas. Portables can normally receive the high-power base station, but they do not always have sufficient power to transmit back to the main radio site. This problem can be eliminated by a diversity receiver-voter system in which radio receivers are installed in a uniform manner at various locations throughout the coverage area. This provides a receiver for each sub-area in the system. In this manner, a portable radio should not be more than a mile or two from one of the receivers in its sub-area.

The signal from each of these receivers is fed to the communication center. The voter is capable of switching from receiver to receiver in midsyllable, so, as the RF signal rises and fades at each receiver, the dispatcher hears only the best signal received. A diversity receiver-voter system can also be incorporated into a repeater system to greatly exceed portable-to-portable communications. A block diagram of a simple diversity (sub-area) receiver-voter system is shown in Figure 10-16B.

Mobile data terminals: Mobile data terminals are small computer terminals that may be connected to a radio system, allowing that unit to access another mobile terminal, base station terminal, or computer. These units generally consist of a microcomputer that converts digital information into a signal that a radio can transmit or receive. These units may have preprogrammed display capabilities where by pressing one key, a "canned" message, such as unit status, condition, operator name, or even the time of day, can be sent. These units are so reduced in size they can be handheld (including the radio).

Screen formats vary from simple light indicators to complex visual images including text and graphic characters. Some systems even have city maps or building drawings showing specific items relative to access, hydrant location, standpipes, and hazardous materials storage. One major consideration, however, is that these units may require dedicated, conventional radio channels.

Station dispatch notification: There are three items that should be included as part of a fire department dispatch/alerting system: (1) a voice announce or public address system to alert the company(s) to an emergency response requirement; (2) a remote control, either wire or radio controlled, for station lights/enunciators; (3) failsafe communications to ensure the proper sending of the dispatch message. For departments receiving more than 600 alarms per year, at least two means of alerting stations is required. Vehicles may even be included in this alerting process by the use of digital decoders or other suitable audible alarms.

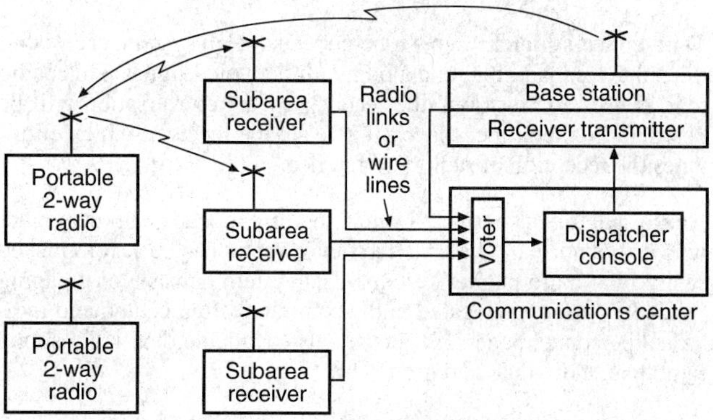

FIG. 10-16B. Block diagram of a simple diversity (sub-area) receiver-voter system.

Mobile Communications

There are several factors to consider when establishing a mobile communications system. This chapter cannot specify the exact number of channels or frequencies needed; a fire department should be aware of the various requirements so a system can be selected more knowledgeably.

An analysis of service area and system configuration will help determine the wattage rating needed and the number of channels or frequencies required to support current usage and provide for future expansion. Small departments may be able to operate with one or two frequencies. However, if a department is considering the possible need of a mutual- or automatic-aid frequency, a police or interagency frequency, a disaster communications frequency, and the required individual departmental fireground frequencies, it is quickly obvious why careful planning is needed.

To ensure quality communications, headsets should be considered for the driver, company officer, and fire fighters. To further support the suppression personnel as much as possible, each apparatus should be equipped with at least two radio units, one of which should be a portable radio and charger unit for the company officer. A portable unit will allow the officer to leave the vehicle while maintaining communications with the apparatus, other units, and the communication center, without returning repeatedly to the vehicle. Many departments provide a portable radio to each fire fighter on the apparatus.

Emergency Medical Service Communications

In many communities, the fire service is involved in providing emergency medical services (EMS), both basic and advanced life support, to the citizens they serve. (See Section 10, Chapter 13, "Emergency Medical Services.") Each state has an EMS communications plan on file with the FCC, and these plans identify how the EMS communications system will be implemented in that state. The EMS system differs from the normal fire communications system in the approach the FCC has taken to frequency allocation. The FCC has set aside UHF frequency pairs for communications between field EMS personnel and base hospital personnel. These channels are shared and used in all fifty states. Therefore, to ensure quality, noninterfering communications on these channels during critical life-threatening incidents, it is imperative that local coordination prevail.

Hazardous Materials Team Communications

The development of hazardous materials (HAZMAT) response teams in the fire service has resulted in new communications requirements for safety. These new requirements are in two major areas: (1) hazardous materials information and (2) HAZMAT team communications from within encapsulating suits.

A major element in the successful outcome of hazardous materials incidents is the ability to rapidly obtain information about the material(s) involved. This generally involves contacting outside agencies for a variety of information. Mobile radio telephone patch(es) may be considered, but this usually ties up the only available radio channel. Cellular telephones should be installed in the hazardous materials team apparatus to facilitate obtaining needed information. This approach also allows a printer to be installed along with the telephone so that one of the hazardous materials computerized information services can be used for on-scene, rapid-printed information.

Cellular telephones also allow the hazardous materials team members direct access to HAZMAT information services, reducing the potential for errors. The medical component of the HAZMAT team should be equipped with a UHF MED radio for medical control or to obtain toxicological information from base hospitals.

Many hazardous materials team members must work inside encapsulating suits. They require special communications systems designed to meet their requirements for operational safety. Two-way wireless communications, preferably hands-free, should be considered. A variety of radio systems that use bone, ear, or throat microphones for voice transmission are available to meet these needs. These radio systems come in simplex, half duplex, or full duplex. They use push-to-talk or voice-actuated circuits to transmit, and mobile repeaters may be integrated into them. The fire department should thoroughly examine its needs in this area prior to purchase.

PERSONNEL

No matter how well designed, installed, and maintained, no system is better than the people who operate it. Communications systems should be supervised by responsible and competent people whose duties include directing, testing, and inspection programs. Fire alarm operators should be temperamentally suited to the position, and be able to efficiently handle assigned duties. They should have the ability to remain calm, take decisive action during emergencies, remain alert during periods of inactivity and while carrying out normal repetitive operations, and work harmoniously with other people. They should also be familiar with fire department operations.

Work schedules should be structured to provide for maximum efficiency. Stress and high activity lead to operator fatigue. Therefore, operators should be provided with a place to take work breaks in the communication center, but outside the operations room.

Communications personnel should not become isolated from the fire department. Their structured training should be directed toward maintaining and improving their call taking and dispatching skills, but it must also include a full understanding of all department operations. Standard operating procedures should be provided in writing to ensure standard performance.

An adequate number of qualified communications operators should be on duty at all times to handle the expected alarm load. For jurisdictions receiving fewer than 600 alarms per year, there should be at least one operator on duty at all times. This operator may be a designated employee of the telephone company; a member of another government agency; or the house watch at a fire station, provided an alternate means is provided to handle alarms if the house watch responds to emergencies. For jurisdictions receiving 600 or more alarms per year, at least one full-time operator should be on duty in the communications center at all times. As the number of alarms per year increases, the number of operators on duty should be increased to ensure the prompt receipt and processing of alarms and other requests for fire department services. NFPA 1221 should be consulted for specific requirements for additional operators.

Provisions should be made for recalling off-duty operators in event of an unexpectedly high alarm load. Extra operators should be placed on duty in advance to deal with anticipated natural disasters, such as blizzards and hurricanes. Arrangements should be made to provide meals and rest periods for operators during natural disasters, since operators may be unable to leave the communication center for extended periods.

CONSOLIDATION OF COMMUNICATION CENTERS

Many small-to-medium-sized fire jurisdictions have found that, by consolidating communications functions with other agencies, they can provide a higher level of service to their citizens at a lower overall cost. One of the major benefits of consolidation for these agencies is the ability to provide highly trained, alert operators who are on duty 24 hr a day. A larger communication center will also have adequate backup to handle multiple, simultaneous incoming alarms, while the smaller center may not have the personnel to cope with them. The major benefit of consolidation is to provide an increased level of service at reduced costs.

Methods of consolidating fire communication centers vary, but the two most popular are forming: (1) a joint powers authority (JPA) to provide this service as an independent agency or (2) one or more departments contracting with another fire department to provide this service. Either system can work well and provide the same benefits to the participants.

Joint Powers Authority

A joint powers authority is an independent government entity, established to provide specific services. There are several benefits of a JPA arrangement for combined fire communications *vs.* other arrangements, but the major benefit is shared administration. With a JPA operating with an independent board of directors, no one agency that might set policy unacceptable to another agency is "in control." In this way, each participant has equal representation and equal responsibility to make the operation a success.

REPORTING A FIRE OR EMERGENCY

Reports of fires and other emergencies under the jurisdiction of the fire service originate from three principal sources: (1) the general public, (2) the business community (industrial, institutional, commercial, and mercantile), and (3) other public service/safety agencies. The reporting process may utilize any of the following four means, individually or in combination: (1) public switched telephone network, (2) public emergency reporting/fire alarm systems, (3) privately operated alarm systems, and (4) two-way radio communications systems.

Telephones

The most common method, and no doubt the least costly to the community involved, is the conventional public switched telephone network.

Universal emergency number 9-1-1 service, where adopted, provides a universal, easily remembered telephone number for use by the public to summon emergency services. "Enhanced 9-1-1" service provides both the calling number and the location of the related telephone instrument directly to the dispatcher. This indication is presented visually on a cathode-ray tube (CRT), and can be automatically recorded if the communication center includes CAD or computerized logging support systems. Use of a cellular telephone negates some of the benefits of enhanced 9-1-1 service, since the actual location of the caller cannot be displayed.

Although there are some disadvantages to total dependency upon a utility-owned, -operated, and -maintained communications network, specifically with regard to emergency reporting during natural disasters, telephones are often the only means of communication available in urban centers, as well as in suburban and rural areas. The installation of outdoor telephone booths increases the availability of telephones for emergency reporting. In general, the public telephone system, widely used for reporting fires and other emergencies, provides a very viable function.

Public Fire Alarm Reporting System

From the perspective of the operator receiving an alarm, use of a public fire alarm reporting station (alarm box) eliminates the diffi-culty of determining the location from which the alarm is being transmitted. When actuated, an alarm box transmits a distinct code to the communication center, which enables the operator to pinpoint the exact location of the box.

Removal of public fire alarm reporting systems began to take place in many major metropolitan centers in the mid-1970s. This trend has since spread to smaller communities. System maintenance costs, vandalism, malicious false alarms, and the nearly universal availability of telephones for alarm reporting all contributed to the elimination of public fire alarm reporting systems in many communities.

Although not an alarm system, the expanding implementation of universal emergency number 9-1-1 service has further reduced the need for public fire alarm reporting systems. This is especially true where enhanced 9-1-1 service provides the alarm operator with the exact location of the caller.

Recently, however, some jurisdictions have shown an increased interest in the use of public emergency reporting systems. In poverty-stricken areas, many households may not have working telephones, and vandalism and other criminal activity have caused the removal of many public "pay phones." Also, public emergency reporting systems have been shown to provide reliable alarm reporting during natural disasters, such as hurricanes, earthquakes, tornadoes, and blizzards, when the public switched telephone network becomes jammed with traffic or is physically damaged.

A typical public fire alarm reporting system consists of a mix of publicly accessible manual fire alarm boxes ("street boxes") and "master boxes" that can be actuated both manually and automatically by a connection from a protected premises fire alarm system. "Auxiliary boxes" that are activated by the protected premises alarm system but lack the ability for manual activation by the public are also sometimes used. In many cases, the operating jurisdictions reduce their legal liability by accepting auxiliarized connections from government buildings only, and referring all others to private supervising station alarm companies. Requirements for installation, maintenance, and use of public fire alarm reporting systems are contained in Chapter 4 of NFPA 72, *National Fire Alarm Code* (hereinafter referred to as NFPA 72).

A public emergency reporting system may be used for the transmission of other emergency signals or calls, provided such transmission does not interfere with the proper handling of fire alarms. For example, systems employing voice communications between the street alarm box and communication center can be used to transmit alarms of nonfire emergency nature. Fire alarm boxes in a radio-type system can be provided with pushbuttons for calling the police department, the ambulance service, or other emergency services, and signals can be transmitted directly to the service called. With parallel-type telephone systems (each box served by a separate circuit), the operator can crossconnect to the proper emergency service, such as the police department, or systems can be arranged so authorized callers, such as law enforcement personnel, can be connected directly to their own departments.

Coded Systems

The coded category is divided into two classifications: (1) the wired telegraphic system and (2) the radio (wireless) system.

The wired telegraph-type box is actuated by depressing a lever that is accessible through a small door. (See Figure 10-16C.) This starts a spring-wound clockwork mechanism and transmits a code number by the rotation of a code wheel that opens and closes the circuit. The transmitted code number of each box is different, so the location of the box transmitting the alarm can be determined. If a

FIG. 10-16C. (Left) A coded telegraphic-type box. (Right) An interior view of the telegraphic alarm box, showing the actuating lever (arrow) that normally protrudes through the hole shown on the inner door at right. When depressed, the handle releases the spring-wound mechanism and sends a coded signal from the code wheel shown at center. (Source: Gamewell Corp.)

circuit wire is broken, the telegraph box system usually is designed to transmit the coded signal through a ground connection.

Radio boxes are similar to wired telegraphic boxes. (See Figure 10-16D.) When activated, they transmit a coded signal, representative of their geographical location; however, in addition to the basic numerical identification code, each box must be equipped to send at least three specific messages identifying the actuation cause. A "test" message must be transmitted at least once in each 24-hr period, assuring the monitoring point (communication center) that the box is working. If a box is struck or physically battered, a specific "tamper" message must be transmitted. If the box is used for public reporting only, the third required message would be "fire." Radio boxes are available that allow transmission of up to 50 messages actuated either publicly (fire, police, medical aid, etc.), privately (through auxiliarized connections), or a combination of both.

Voice Systems

Voice systems are wired telephone networks. The fire alarm box is essentially a telephone handset installed in a specially designed housing. (See Figure 10-16E.) The location from which the alarm is transmitted is definitely established, without dependence upon voice transmission and without interference from other boxes. Removing the handset from its cradle in the box lights a lamp on the communication center switchboard, where the number of the box and the time of message can be recorded. Voice systems technology is divided into parallel and series classifications.

Some jurisdictions have installed emergency voice call boxes (typically along limited-access highways) that utilize cellular telephones arranged to automatically dial a telephone number at the communication center. While these call boxes do not constitute a public fire alarm reporting system as defined by NFPA 72, they are useful for reporting disabled vehicles, traffic accidents, medical emergencies, and similar problems. These call boxes are generally battery operated, with solar-powered chargers to maintain battery charge. They rely upon the commercial cellular telephone network, and, therefore, may be subject to loss of service during major disasters when nearby cell sites are out of service, or if the number of calls from all sources exceeds the cellular network capacity. Having the cellular service provider assign a priority access code to these telephones, if possible, improves the reliability of the service.

Coded-voice combination: Boxes that also contain a telegraph mechanism are called combination telephone-telegraph type, and both voice and coded signals can be transmitted over the same circuit to the communication center. (See Figure 10-16F.) The series telephone box may be intermixed with telegraph street boxes and master boxes.

Two-Way Radio Communications

A commonly overlooked alarm/emergency notification source is the two-way radio networks of public safety agencies. Notification of incidents in watershed areas are typically reported by mobile patrols. Similarly, requests for supplemental dispatches to multiple-alarm fires usually originate from units in the field.

Communications between agencies in adjacent jurisdictions, as well as between agencies of a single jurisdiction, may often be expedited by using two-way radios.

PROCESSING COMMUNICATIONS WITHIN THE DEPARTMENT

Regardless of the method by which an alarm is received, the location of the emergency is the minimum information required to respond

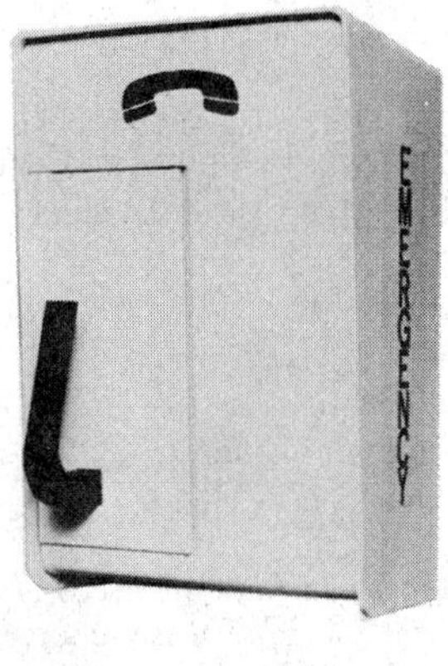

FIG. 10-16D. (Left and center) Typical battery/dc-powered coded radio-type alarm boxes equipped for multiple messages (fire, police, ambulance, etc.). (Source: King Fisher Co.) (Right) A user-power-coded radio box. This type of device must incorporate an automatic test feature to meet standard requirements. (Source: Signal Communications Corp.)

FIG. 10-16E. A voice-telephone alarm box, series type, for handset operation only. (Source: Gamewell Corp.)

FIG. 10-16F. A combination telephone-and-coded telegraphic-type alarm box. (Source: Gamewell Corp.)

to an emergency. The nature of the incident being reported is also obviously essential. However, one must bear in mind that thousands of responses to nonvoice notifications are made each year.

Data Collection Systems

To efficiently operate a communication center, reliable data must be available. To be useful, these data must be both accurate and readily available. The data may take a wide variety of forms, from card indexes to sophisticated computer files, and may include such categories as the status of vehicles, the availability of resources, geographic information, and personnel information. The following categories outline the various types and uses of data that can be collected and maintained by a communication center.

Incident-related data: It is important (and often a legal requirement) to keep accurate records of fire and emergency medical incidents to which the department responds. The types of data worth collecting include:

1. Time of dispatch
2. Address
3. Identifications and times of responding units
4. Arrival times at the scene
5. Departure times from the scene
6. Radio traffic tape logs
7. Telephone tape logs
8. Times of significant events at the scene

Operational-related data: These data consist of information that must be rapidly available at all times to maintain efficient operation of the communication center. The data include:

1. Geographic data
2. Equipment data
3. Response order data
4. Occupancy/hazard information
5. Additional resource data

Reports: Many types of reports are necessary for overall fire department management. Some of these reports will be required at regular intervals, while others will be required on demand to fill a specific departmental information need. A list of probable reports includes:

1. Response time reports
2. Emergency activity reports
3. Occupancy activity reports
4. Specific incident reports
5. Reports for outside agencies
6. Fire loss/injury reports
7. Equipment-related reports

These reports can be used to project future needs of the department, such as location of new fire stations, hiring of additional personnel, and relocation of existing equipment.

Receipt of Alarms

Most alarms are received by telephone over the public switched telephone network. Automatic recording devices must be interconnected to all types of voice-reporting circuits, including radio systems. However, recording devices cannot ensure the credibility or accuracy of the information received.

Correct telephone numbers are the only means of positively identifying the location of an informant. Enhanced 9-1-1 service is designed to provide the answering operator with an immediate reference to the number of the telephone in use and its specific location. The data are usually displayed on a CRT screen. The data input is formulated for electronic data processing (EDP) usage and can be fed concurrently to a departmental CAD system.

All public emergency reporting systems (municipal fire alarm systems) provide an individual code for each box installed, and the proper code is transmitted when the box is activated. All incoming coded signals are permanently recorded on printers integral to the specific systems and, in some cases, are also displayed on direct readout (LED, CRT, etc.) devices.

The translation of the numerically coded data into a usable address is performed with the aid of a running-card file or CAD system, which is necessary for processing of alarms.

Voice Recording and Reproducing Systems

The voice recording and reproducing system should consist of a multichannel recording device that provides a permanent, time-referable, unalterable, and court-admissible recording of all emergency-related telephonic/radio communications within a fire department.

The communication center supervisor(s) and at least one of the operators on duty should be totally familiar with the operation and maintenance of the multichannel recording device. Familiarizing other department personnel is advisable, but responsibility should lie with no more than two individuals per watch or shift.

A multichannel recording system can provide:

1. Verification of emergency messages
2. A log of all emergency communications within the department
3. Protection of the department against civilian or other department complaints
4. Substantiation or negation of claims against the department
5. Added protection to the civilian population
6. A training tool for dispatch personnel
7. Validation of response time

The importance of these various benefits will differ from department to department.

Running-Card Files

A running-card file contains a dispatch plan for every area of the jurisdiction served. The number of running cards in a given file is dependent upon the size of the jurisdiction and the manner in which it has been indexed for responses. The most common method of indexing is to assign numeric designations to all intersections, or to grids superimposed on a map of the jurisdiction. There is a specific running card to each intersection or grid. Public emergency reporting stations (alarm boxes) are normally assigned an identity code corresponding to the intersection or grid closest to their physical location.

When an alarm is received, the operator utilizes the addresses and/or intersection data obtained from the informant to find the correct running card.

Status-Keeping Systems

The ability to maintain an accurate record of the current status of all emergency units in the system is essential to a successful dispatch system and will aid in: (1) minimizing response times, (2) ensuring appropriate response, and (3) collecting nonemergency unit activity information.

Status-keeping systems may involve the tracking of various pieces of information, such as:

1. Availability of the unit for emergency response
2. Service capabilities of the unit (paramedic service, extrication equipment, all-terrain capability, etc.)
3. Location of the unit
4. Means of contacting the unit (radio, phone, etc.)
5. Miscellaneous items based on the particular department's operating philosophy

The mechanism of a status-keeping system will vary, depending upon the number of units on which status must be kept and the actual dispatch process utilized by a department. A small department might utilize a manual system with a small "clipboard," while a large department may require a computer.

Computer-Aided Dispatch (CAD)

Computers are used in all areas of the fire service. These areas include personnel management, budgeting, fire prevention, training, inventory, etc. One major use is the dispatching of emergency units. The degree to which a CAD system can be implemented depends upon the size of the department. A small fire department may not require a complex CAD system with mobile digital terminals, station terminals, and printers, unless it is part of a consolidated dispatch system. Dispatch software packages are available for personal computers, and small departments can use them to rapidly recover the information needed to expeditiously dispatch units. This information includes preplans, hazardous materials information, lists of other responsible public safety agencies, names, addresses, and phone numbers of responsible parties, etc.

NFPA 1221 categorizes CAD systems as Class 1, 2, or 3. The Class 1 system is the most complex, utilizing computers to automatically track the status of all fire apparatus, and to select and recommend the appropriate units for dispatch. Upon approval of the operator, signals are then transmitted to the appropriate stations and field units to dispatch the appropriate personnel and equipment. A Class 1 CAD system requires complete redundancy and automatic switchover to an on-line backup computer, in event of failure of computer hardware or software. A Class 2 CAD system uses computers to support dispatching operations. Redundancy, but not automatic switchover, is required. A Class 3 CAD system requires manual entry from the operator, and is used primarily to keep track of unit status and related information. In a Class 3 system, computer redundancy is not required. In all classes of CAD systems, manual backup facilities should be provided and be readily available for use in the event of problems with, or failure of, the CAD system. There also should be an up-to-date hard copy file of stored information located on the premises to support the dispatch function should the CAD system fail. For more detail on CAD systems, the reader should refer to NFPA 1221.

With the advances in microcomputer (personal computer) hardware and software, many applications previously requiring a main-frame computer can now be handled by a personal computer. Caution must be used in both the selection and use of personal computers. Computer professionals should match the application to the proper computer. The dispatch computer should not be used for administrative applications. It should be kept free of time-consuming processes to allow for rapid retrieval of dispatch information. CAD systems for medium-to-large departments are more complex and must be designed for the needs of an individual department.

Dispatch Circuits and Equipment

A dispatch circuit is the means by which an alarm operator notifies units to respond to an alarm. In general, two separate dispatch circuits must be provided, but only one is required when fewer than 600 alarms per year are received. One dispatch circuit must consist of one of the following:

1. A supervised wired circuit
2. A radio channel with duplicate base stations, transmitters, receivers, microphones, and antenna networks
3. A supervised microwave carrier
4. A polling or self-interrogating radio or microwave radio system with duplicate transmitters and complete redundancy
5. A properly arranged, supervised telephone circuit

A circuit terminating at a telephone instrument only is not considered part of either required circuit.

The second dispatch circuit also should consist of one of the above, but need not be supervised or have duplicate equipment. If voice transmission is used to dispatch units, a distinctive audible alerting signal should precede the voice message to differentiate it from routine messages. Alarms should be automatically recorded as they are transmitted.

It should be noted that CAD is not a circuit. It is a method used to access stored information to expedite dispatching. CAD systems identify the units normally assigned and the units currently available for dispatch. The actual dispatch is made over one of the above mentioned circuits.

Radio circuits are commonly used for dispatching. They have the advantage of transmitting alarms simultaneously to all units, regardless of whether they are in quarters or out in the field. Also, they are not dependent upon wired circuits, unless the transmitter is remote from the communication center.

Wired circuits can be used for data and/or voice transmission. They are limited to transmission to fixed locations. Units in the field must still be contacted by radio.

A telegraph circuit is one example of a wired circuit. Once more commonly used, telegraph circuits are still in use in some jurisdictions. Telegraph systems transmit coded signals to fire stations, where they are received on a variety of instruments, including gongs, and punch or printing registers that register the coded signal with holes in, or marks on, a paper tape. The location of the alarm is then decoded manually or electronically in a manner similar to that previously discussed. Other uses of wired circuits include voice, teletype, and computer data transmission.

Other Alerting Facilities

Many volunteer fire departments have no one on duty at a fire station to receive an alarm. In these instances, bells, air horns, whistles, or sirens are commonly used to alert the volunteers when an alarm has been received. Radio receivers in volunteer fire fighter residences or places of business and radio pagers (i.e., portable receivers carried on the person) are frequently used for receiving alarms.

Supervision and Testing Facilities

The communication center is where trouble signals from public emergency reporting systems are received and testing facilities are generally provided. In a leased telephone system, some test facilities are located in telephone buildings. In the event of trouble, both an audible and a visual signal should be provided. An audible signal may be shut off, but it should be designed to operate again in the event of trouble in any of the other circuits connected to the same supervisory device.

BIBLIOGRAPHY

NFPA Codes, Standards, and Recommended Practices

Reference to the following NFPA codes, standards, and recommended practices will provide further information on fire department emergency communications systems discussed in this chapter. (See the latest version of *The NFPA Catalog* for availability of current editions of the following documents.)

NFPA 70, *National Electrical Code®*
NFPA 72, *National Fire Alarm Code*
NFPA 297, *Guide on Principles and Practices for Communications Systems*
NFPA 1221, *Standard for the Installation, Maintenance, and Use of Public Fire Service Communication Systems*

Additional Reading

Baird, D., "Computer Programs Aid in Fire Station Location," Fire Fighting in Canada, Vol. 34, No. 9, Dec. 1990, pp. 4–5.
"Communications: Scanning the Equipment," *Fire Command*, Vol. 56, No. 9, Sept. 1989, p. 29.
Foot, N., "Radio Interference to Alarm Systems," *Fire Surveyor*, Vol. 17, No. 6, Dec. 1988, pp. 36–39.
"How One City Is Dealing with the Nuisance Alarm Problem," *Fire Journal*, Vol. 82, No. 1, 1988, p. 57.
Maillet, M., "Microprocessors and New Communications Systems in Improved Detection Systems," *Proceedings* of the Conference on New Technology to Reduce Fire Losses and Costs, Elsevier, 1986, pp. 243–247.
"Malicious Alarms or Nuisance Alarms: Which Is the Larger Problem," *Fire Journal*, Vol. 83, No. 1, 1989, p. 5.
Morita, M., and Yamauchi, Y., "Computer Geometry Applied to Locus Approach to Fire Stations," Proceedings of the fourth International Symposium on Fire Safety Science, July 13–17, 1994, Ottawa, Ontario, Canada, Intl. Assoc. for Fire Safety Science, Boston, MA, 1994, pp. 1303–1311.
Nesse, T., "Digital Alarm Communications: Where They Come From, Where They're Going," *Fire Journal*, Vol. 82, No. 4, 1988, p. 47.
Ruda, S. J., "Busiest Station in the Nation: Station 11, Los Angeles City Fire Department—20, 225 Runs in 1990," Firehouse, Vol. 16, No. 11, Nov. 1991, pp. 61–62.
Spahn, E., "Increasing Your Fire Alarm Literacy," *Fire Engineering*, Vol. 143, No. 6, 1990, p. 75.
Sybesma, P., "Equation for Station Location," Fire Chief, Vol. 39, No. 10, Oct. 1995, pp. 55–56, 62–65.
Szymanski, D. J., and Stacey, J. W., "Lesson Plan: Computer-Aided Dispatch System," *Fire Engineering*, Vol. 141, No. 7, 1988, pp. 32–35.
Todd, C. S., "Remote Monitoring of Alarm Signals, Part 2: Monitoring by Central Stations," *Fire Surveyor*, Vol. 17, No. 1, Feb. 1989, pp. 18–31.
Todd, C. S., "Remote Monitoring of Alarm Signals, Part 4: Central Station Operations and Communications with Fire Controls," *Fire Surveyor*, Vol. 17, No. 2, Apr. 1988, pp. 32–37.
Todd, C. S., "Remote Monitoring of Alarm Signals, Part 5: The Future," *Fire Surveyor*, Vol. 17, No. 3, June 1988, pp. 16–21.
vanBowen, J., Jr., "Use of In-House Data for Fire Station Location," Journal of Applied Fire Science, Vol. 1, No. 1, 1990–1991, pp. 39–44.
Walsh, C. P., "Radio Communications: Coming in Loud and Clear," *Fire Engineering*, Vol. 143, No. 11, 1989, pp. 32–36.
Walsh, M. B., "Evanston's New Station Blends Economy and Utility," Fire Chief, Vol. 34, No. 6, June 1990, pp. 36–38.
Walsh, M. B., and Wojcik, K., "What's New in Fire Stations," Fire Chief, Vol. 34, No. 6, June 1990, pp. 32–35.

FIRE DEPARTMENT APPARATUS AND EQUIPMENT

Revised by Ralph Craven

The tools of the fire department for providing fire suppression and emergency services are apparatus and the equipment they carry. Without the proper tools and the knowledge of how to use those tools correctly, a fire department cannot do its job effectively. This section contains information on the standards used and procurement policies that should be followed in specifying and acquiring apparatus and equipment.

APPARATUS

The basic fire apparatus in North America is a diesel- or gasoline-engine-driven vehicle that carries a rather extensive assortment of tools and equipment for fighting fires. Such equipment may include a pump, hose, water tank, aerial ladder or elevated platform, ground ladders, and various portable tools and appliances. The amount and capacities of the various fire-fighting components carried vary in accordance with the intended service of the particular vehicle.

For pumping operations and fire attack, the size of the pump and the amount of water, hose, and equipment carried will vary with the type of service the fire department provides and the nature of the community it protects, be it predominantly urban, suburban, or rural. The fire apparatus standards set 250 gpm (946 L/min) as the smallest fire pump, but pumps of 1,000, 1,250, or 1,500 gpm (3,785, 4,732, or 5,678 L/min, respectively) capacity are popular for general municipal service.

Many fire departments supplement units equipped primarily for pumping and fire attack with other vehicles that provide a large variety of tools and equipment for support functions. These tools include ground ladders, forcible-entry tools, generators, lights, and rescue equipment. Such units are variously termed rescue, squad, or utility vehicles.

A sizeable fraction of the work at an average fire involves the use of tools and equipment in duties classed as rescue or ladder-company work. Apparatus suitably equipped for such service should be available at all structural fires and at major emergencies, such as significant highway accidents. While pumping engines may carry considerable equipment, the larger aerial-ladder and elevating-platform apparatus commonly transport additional equipment for forcible entry, cutting, and extrication. This type of apparatus provides vehicle-supported, power-operated equipment for access to areas above the normal reach of ground ladders, and effective elevated stream service.

In rural districts and in the outlying districts of cities where hydrant distribution is not complete, supplemental water-transport apparatus, defined as "mobile water supply apparatus," is common. Such apparatus has large water tanks with quick-fill and -dump connections, and they may have a permanently installed pump.

A wide combination of fire-fighting equipment is often provided on a single fire apparatus. Besides combinations of water pump, hose, and water tank, commonly termed a "triple combination," it is not uncommon to provide a booster pump and small-stream equipment on aerial-ladder or elevating-platform apparatus or a water tower on a pumping engine. When fire apparatus is equipped with long ground ladders and other usual ladder-company equipment, in addition to the usual pumping engine equipment, the apparatus is referred to as a "quadruple combination," or "quad." If a power-operated aerial ladder or elevating platform is added, such apparatus is termed a "quint." However, operating efficiency and capability tends to decrease if too many functions are to be performed by one piece of apparatus. And, if one feature of the apparatus requires repair, the entire apparatus must be taken out of service.

Increasingly, apparatus is being designed to improve its functional performance. Initial fire-attack capability for pumpers not only includes a variety of pre-connected hose lines but frequently also includes pre-connected master stream capability or elevated-stream equipment. Pumpers are often built with considerable compartment space for emergency service and medical equipment and may serve as combination rescue pumpers.

NFPA has promulgated two standards for the design of fire apparatus. They are NFPA 1901, *Standard for Automotive Fire Apparatus,* and NFPA 1906, *Standard for Wildland Fire Apparatus.*

NFPA 1901 deals with the design, performance, functions, and components of fire apparatus. In previous years there were four standards that dealt with fire apparatus. In 1996 they were combined into one document for ease of use by the fire service. The current NFPA 1901 standard is broken down into chapters that were previously stand-alone standards.

Pumpers are defined as vehicles designed for sustained pumping operations during structural fire fighting and for supporting associated fire department operations. NFPA 1901 requires a pump of at least 750 gpm (2,839 L/min) capacity, a water tank of 500 gal (1,893 L), storage for both supply hose and preconnected fire-fighting hose lines, and miscellaneous equipment. (See Figure 10-17A.)

Ralph Craven is president of the National Institute of Emergency Vehicle Safety, Inc., Reno, NV. He is a member of the NFPA Technical Committee on Fire Department Apparatus.

FIG. 10-17A. A typical pumper. (Source: Saulsbury Fire Equipment Corp.)

Initial-attack fire apparatus are defined as vehicles designed for making the initial fire suppression attack on structural, vehicular, or vegetation fires and for supporting associated fire department operations.

Mobile water-supply vehicles are defined as vehicles designed to transport water to the scene of an emergency to be applied there by other vehicles or pumping equipment. NFPA 1901 requires a water tank of at least 1,000 gal (3,785 L) with quick-fill and -dump capability, as well as limited hose and equipment to support the pickup and dumping of the water.

Aerial-ladder or elevating platform fire apparatus are defined as vehicles designed to provide elevated fire fighting and rescue capability. NFPA 1901 requires a power-operated, self-supporting aerial device capable of attaining an elevation of at least 50 ft (15.3 m). Aerial ladders are required to support at least 250 lb (113.4 kg) at the tip in any position in which they can be placed. The elevated platform must be able to support at least 750 lb (340 kg) in any position in which it can be placed. NFPA 1901 provides requirements if a pump is installed.

NFPA 1906 deals with the design, performance, functions, and components of a wildland fire-fighting vehicle. These are defined as vehicles equipped with a pump, water tank, limited hose, and equipment. The vehicle must be capable of supporting "pump and roll" operations. Generally, these are smaller vehicles, frequently equipped with all-wheel drive, a small water tank, forestry hose, a small-volume high-pressure pump, a number of portable water extinguishers, and various hand tools for fighting wildland fires. Larger vehicles may have a 500-gpm (1,893-L/min) front-mounted or power takeoff (PTO) pump and a 500-gal (1,893-L) or larger water tank to support initial fire attack on all types of fires, whether wildland, vehicular, or structural.

NFPA 1906 deals with various auxiliary systems that are often installed on fire apparatus. These may include auxiliary pumps, a 120/240-V electrical system, a foam system, a booster reel, or a water tower.

Both NFPA 1901 and NFPA 1906 define performance requirements for the apparatus and define tests to measure this performance. In the appendix of each document, there are explanations of the requirements and details, as well as work sheets, to assist in the purchase of the apparatus.

Persons buying the apparatus must specify the options that make it suitable for local needs. For example, requirements for

fighting large brush fires in southern California are quite different from those for fighting fires in below-zero weather in northern Minnesota. NFPA 1901 and NFPA 1906 provide for standardization of many items for which uniform performance, measurable by tests, is desirable but it allows the necessary flexibility to meet local needs that could not be supplied by one assembly-line production. Such differences include rated pump capacity, water-tank capacity, length of aerial ladders, height of elevating platforms, and many other important design features.

Besides the functional characteristics of a piece of fire apparatus, considerable thought should be given to personnel safety and operations. The design of the vehicle's driving compartment and riding spaces, and the location of various components should be studied carefully. Only fully enclosed driving and crew compartments with sufficient seating for all fire fighters who are expected to ride on the apparatus should be specified. Fire fighters should never be allowed to ride standing up or in exposed positions on the rear or sides of the apparatus. The use of seat belts must be mandatory in all cases. All of the equipment carried in the crew compartment area should be restrained in mechanical holding devices or placed in compartments. This minimizes the possibility of injury to fire fighters by flying objects, in the event of an accident.

The pump operator's panel may be positioned to the side, to the rear, or in a top midships position. The position of valve controls and other control equipment will affect the overall efficient and safe operation of the apparatus. The better the operator's view, the more effective he or she can be. Care should also be taken to design the operator's position so that all controls can be safely operated without having to straddle discharge hose lines and so that close working proximity to large suction hoses can be avoided.

Engines in Fire Apparatus

Engines may be either diesel or gasoline types, but there are few gasoline engines capable of supplying sufficient power for modern apparatus. The engine size and horsepower must be chosen to correspond with the conditions of service and apparatus design. Apparatus must have enough power to move it over the road network on which it is expected to operate. Road tests are conducted with apparatus fully loaded with hose, water, ground ladders, the applicable personnel weight, and an equipment allowance weight. From a standing start, through the gears, the vehicle must attain a true speed of 35 mph (56 km/h) within 25 sec. The vehicle must also accelerate from 15 to 35 mph (24 to 56 km/h) in 30 sec without the operator moving the gear selector. This demonstrates that the vehicle has the necessary power to operate safely and efficiently in traffic. The vehicle must attain a top speed of at least 50 mph (80 km/h) and must be able to maintain a speed of at least 20 mph (32 km/h) on any grade up to and including 6 percent. While these tests demonstrate the power needed to negotiate the grades found in most communities, any special requirements needed to climb very steep grades should be specified, and the finished apparatus should be tested to those requirements.

A braking performance test is also required. The vehicle must come to a complete stop within 35 ft (10.7 m) after attaining a true speed of 20 mph (32 km/h). This test is required in NFPA 1901 and NFPA 1906 and the Federal Motor Vehicle standards. This test is only required to be performed one time, which is not realistic for emergency vehicles. Many agencies require secondary braking systems to be installed on their apparatus to enhance braking performance. NFPA 1901 requires that secondary braking systems be installed on vehicles weighing over 36,000 lb (16,330 kg) and recommends a system if the vehicle weighs over 31,000 lb (14,062 kg). There are four types of secondary braking systems: (1) hydraulic retarder, (2) exhaust brake, (3) engine brake, and (4) drive line. Agencies should match their driving conditions to the type of secondary brake desired.

The set or parking brake provided is required to hold the vehicle on grades up to 20 percent. This may not be adequate for every fire department, so the agency needs to specify their particular requirements. Some states require parking brakes to hold on any grade.

Carrying capacity is one of the most important, least understood, features of a vehicle. The gross axle weight rating (GAWR) and gross vehicle weight rating (GVWR) of the apparatus must be adequate to carry the weight of the fully equipped and staffed vehicle. This includes 200 lb (76 kg) per person personnel weight, a full water tank, the specified hose load, ground ladders, and a miscellaneous equipment allowance.

One of the critical factors is the size of the water tank. Water weighs about $8^1/_3$ lb per U.S. gallon (1 kg/L). A value of 10 lb (4.54 kg) per gallon is often used when estimating the weight of a full water tank, making a full 500-gal (1,893-L) tank $2^1/_2$ tons (2.27 metric tons). Larger tanks mean heavier vehicles, which may have limited mobility on rural roads and bridges. The efficiency of mobile water-supply apparatus, in transporting water to a fire, depends upon the vehicle's over-the-road mobility. Any apparatus that weights too much for local operating conditions severely limits the capabilities of the fire department.

Too many fire departments seriously overload apparatus by adding more equipment than the vehicle was designed to carry. NFPA 1901 specifies minimum equipment allowances of 2,000 lb (909 kg) for pumpers, up to 2,000 lb (907 kg) for initial attack vehicles depending upon their size, 1,000 lb (454 kg) for mobile water-supply apparatus, and 2,500 lb (1,136 kg) for aerial ladder and elevating-platform fire apparatus. Part of the purchase process should be to ensure that sufficient additional allowance is provided when needed.

GAWR is the value specified by the vehicle manufacturer as the loaded weight on a single-axle system, and GCWR (gross combination weight rating) is the value specified by the manufacturer as the loaded weight of a combination vehicle, such as a tractor-trailer unit. GVWR is the value specified by the manufacturer as the loaded weight of a single vehicle, and is the sum of the weights of the chassis, body, cab, equipment, water, fuel, crew, and all other loads. A number of components affect the GVWR, such as the springs or suspension system, the rated axle capacity, the rated tire loading, and the weight distribution between the front and rear axles. All chassis are designed with a "rated GVWR," or maximum total weight that should not be exceeded by the apparatus manufacturer or by the fire department after the vehicle has been placed in service.

The improper distribution of weight between the front and rear wheels can make a vehicle difficult to control and may require tires of different sizes to carry the load. Overloading of the vehicle not only affects the handling characteristics but will result in increased maintenance problems with transmissions, clutches, the drive-train, and brakes, as well as suspension systems and tires. Many states exempt fire apparatus from meeting weight limitations; however, it is not a good practice to overload the vehicle. In the event the vehicle is involved in an accident, the overloading issue may be used against the department in any litigation.

Compliance with Federal Standards

The federal government has promulgated motor vehicle safety standards that are applicable to all vehicles, including fire apparatus. This legislation is enforced by the U.S. Department of Transportation (DOT). It is unlawful to sell a vehicle that is not in compliance with the current federal standards. Manufacturers should not build fire apparatus that does not conform to these standards. Because fire apparatus is complex and often requires considerable lead time between the signing of a contract and delivery, the federal regulations provide that the standards in effect at the time the contract is signed must be those complied with, provided the apparatus is delivered within two years.

Additional requirements are placed on the apparatus and engine manufacturers by the Clean Air Act, which is enforced by the Environmental Protection Agency (EPA). Engines cannot be modified once approved by the EPA. These standards have resulted in some downgrading in engine performance, often resulting in the need to use larger engines than previously employed to obtain the same performance. Likewise, more frequent maintenance checks are often required due to the pollution-control equipment now required.

Electrical Power for Apparatus

Sufficient electrical capacity for the vehicle and its components is a major requirement for fire apparatus. Electricity is essential to start the apparatus reliably under all temperature and weather conditions. Emergency warning lights and sirens impose a heavy electrical demand, with vehicle lighting and communication equipment as an added load. Electric rewind hose reels, pump controls, scene lighting, and electrically operated tools can all impose additional loads. Much of the equipment must perform when the apparatus is idling at a fire or other emergency, and the engine is not producing maximum output from the alternator.

Batteries, which are an integral part of the electrical system, must be a minimum of 1,000 cold-cranking amps (CCA) and of the high-cycle type. However, the batteries should not be expected to carry the electrical load when the engine is idling, as many of the electronic components on modern apparatus will not function properly if the voltage drops by 10 percent or more. It is also critical to provide a means for keeping the batteries charged while the apparatus is in the station. This can be done with an onboard battery conditioner or charger or with a polarized plug for connection to an external charger.

A department, when specifying apparatus, should require an electrical audit. The audit would be performed by the department or manufacturer. The audit determines the amperage required to power all electrical accessories; then a determination can be made if the alternator supplied is large enough for the load. With more engine and transmission combinations becoming computer controlled, it is imperative that batteries be maintained properly. NFPA 1901 requires a load management system for the electrical components on the vehicle. The load management system will shut off electrical loads if the alternator and batteries cannot keep up with the demand.

Pumps for Fire Apparatus

A fire pump used on pumpers is defined as a pump of at least 750 gpm (2,839 L/min) at 150 psi (1,034 kPa) net pump pressure. It must be capable of pumping 70 percent of its capacity at 200 psi (1,379 kPa) net pump pressure and 50 percent of its capacity at 250 psi (1,724 kPa) net pump pressure. Pumps larger than 750 gpm (2,839 L/min) are rated in 250-gpm (946-L/min) increments—for example, at 1,000, 1,250, 1,500 gpm, and so on (3,785, 4,732, 5,678 L/min, respectively). When new, fire pumps are certified by tests witnessed by an independent third party. The certification test includes, among other things, a test at draft during which the pump must deliver its rated capacity for 2 hr at 150 psi (1034 kPa) net pump pressure, followed by two half-hr periods during which 70 percent of rated capacity is delivered at 200 psi (1,379 kPa) net pump pressure and 50 percent of rated capacity at 250 psi (1,724 kPa) net pump pressure. The apparatus must also perform a 10-min overload test, discharging rated pump capacity at 165-psi (1,138-kPa) net pump pressure to demonstrate reserve engine power.

An attack pump is defined as a pump of at least 250 gpm (946 L/min) but not more than 700 gpm (4,827 L/min) at 150-psi (1,034-kPa) net pump pressure. Like fire pumps, they must be capable of pumping 70 percent of their capacity at 200-psi (1,379-kPa) net pump pressure and 50 percent of their capacity at 250-psi (1,724-kPa) net pump pressure. Typically, attack pumps are not designed for long-duration pumping, although they must meet a 50-min pump test and can be designed for longer operation. Attack pumps are typically certified by the apparatus manufacturer, and the tests are not witnessed by an independent third party.

A third type of pump, designated as a transfer pump, is used to assist in filling or dumping water tanks on fire apparatus. These pumps have a minimum rated capacity of 250 gpm (946 L/min) at 50 psi (345 kPa) and are typically used with mobile water-supply apparatus.

The design requirements for fire and attack pumps include the ability to develop 22 in. of mercury vacuum at 2,000 ft (610 m) of elevation and the ability to deliver rated capacity at up to 10 ft (3.1 m) of lift through 20 ft (6.1 m) of suction hose. Where a pump on a piece of apparatus is expected to perform at an altitude above 2,000 ft (610 m), the purchaser must advise the apparatus manufacturer of the altitude and advise whether a 10-ft (3.1-m) lift-suction capability is required, so that the proper engine and pump can be provided.

The pump and the attached piping and valves are designed for 500 psi (3,448 kPa) hydrostatic pressure. Depending upon the size of the pump, one or two large suction intakes and at least one auxiliary gated suction intake is required. This auxiliary intake is normally a 2$^1/_2$-in. (64-mm) intake and is provided so an additional water-supply line can be taken into the pump without shutting down. It is also a means of connecting a supply line in mutual-aid operations if hose connections for the large intake are not compatible. As the large suction intake provides the least restriction, normal supply to the pump should be through that intake, whether directly or through a suction siamese.

Discharge piping from the pump feeds a variety of outlets and, in some cases, may feed a master stream device directly. Fire pumps are required to have two 2$^1/_2$-in. (64-mm) discharge outlets and enough additional outlets to allow the capacity of the pump to be discharged. If the apparatus is equipped with an aerial device with a pre-piped waterway, the pump must be piped directly to the waterway.

To control the development of water hammer in the pump and hose lines, any 3-in. (76-mm) or larger valve should be designed to restrict changing the position of the flow-regulating element from full close to full open, or vice versa, in less than 3 sec. The pump should be further protected by a permanently installed intake pressure-relief system that is 2$^1/_2$ in. (64 mm) or larger and discharges directly to the atmosphere. The surplus-water-discharge location should be away from the pump operator's position and terminate in a male fitting so hose can be attached to carry any water away from the apparatus.

The pressure on the discharge side of the pump must also be controlled, either through an automatic relief valve or a pressure control device that controls the excessive discharge pressure. The device is required to limit the pressure rise upon activation to a maximum of 30 psi (207 kPa) and be capable of operating over a range of 90- to 300-psig (620- to 2070-kPa) discharge pressure.

Generally, a midship fire pump is powered by the vehicle engine through a pump transmission that redirects power from the rear wheels to the pump itself. Power is transferred with the vehicle in a stationary position and only after a parking brake has been set. An interlock should be provided to ensure that the engine speed cannot be advanced until the parking brake is set, the transfer is complete, and the chassis transmission is either in neutral or in the correct gear for pumping. Otherwise, the apparatus may move when the engine speed is advanced. New automatic transmission designs allow the use of a PTO to transmit power to a large pump. They should be limited to no more than 1,250 gpm (4,732 L/min), due to horsepower requirements through the PTO.

On smaller pumps, the power is often transferred to the pump through a PTO device. In some designs, the pump is mounted at the front of the chassis where it is driven through a clutch arrangement from the front of the engine.

Hose-Carrying Capability

The amount of hose carried should be sufficient to support the function of the apparatus. Operations involving pumpers typically need the most hose, and pumpers should carry a minimum of 1,200 ft (366 m) of 2$^1/_2$-in. (64-mm) or larger hose and 400 ft (122 m) of 1$^1/_2$-in. (38-mm), 1$^3/_4$-in. (44-mm), or 2-in. (51-mm) hose. However, many fire departments carry a minimum of 2,000 ft (610 m) of 2$^1/_2$-in. (64-mm) or larger hose, and 800 to 1,000 ft (244- to 305-m) of 1$^1/_2$-in (38-mm), 1$^3/_4$-in. (44-mm), or 2-in. (51-mm) hose on their pumpers to permit more versatile use of the apparatus.

Increasingly, fire departments are using hose of diameters larger than the nominal 2$^1/_2$-in. (64-mm) size. While larger diameter hose costs more, it may actually be more economical, since multiple smaller hose lines would be required to meet the increased carrying capacity through one large-diameter line. A standard hose body for pumpers has at least 55 cu ft (1.56 m^3) of space and is designed to provide a split hose load if desired. By splitting a hose load into two bed sections, either one long lay or a shorter double hose line can be made quickly, saving considerably on the time and personnel needed to establish flows approaching the capacity of the pumper. Where larger diameter hose is to be used, a larger hose storage area may be needed.

With larger capacity pumps now commonplace, additional hose capacity is required to move larger volumes of water. A convenient way to relate needed hose capacity to pumper discharge capability is based on the relative water-carrying capacity of various sizes of hose at normal operating pressures, as follows: 250 gpm (946 L/min) for 2$^1/_2$-in. (64-mm) hose; 350 gpm (1,325 L/min) for 3-in. (76-mm) hose; 500 gpm (1,893 L/min) for 3$^1/_2$-in. (89-mm) hose; 750 gpm (2,839 L/min) for 4-in. (100-mm) hose; and 1,300 gpm (4,921 L/min) for 5-in. (125-mm) hose. Thus, moving 500 gpm (1,893 L/min) 1,000 ft (305 m) requires 2,000 ft (610 m) of 2$^1/_2$-in. (64-mm) hose or 1,000 ft (305 m) of 3$^1/_2$-in. (89-mm) hose. In short lines, some increases in flow may be obtained by higher discharge pressures, but in longer hose lines, increased flow will be minimal.

Auxiliary Pumps

Some fire apparatus will have a separate pump for small-volume/high-pressure operations. Other designs use a third stage in the main pump to develop the higher pressures. If an auxiliary pump is provided and is interconnected with a fire or attack pump, suitable valving and controls should be provided to preclude the introduction of higher pressures into piping and fire hoses not designed for the pressure.

Auxiliary pumps used predominantly for fighting grass fires and other small fires usually have a capacity of up to 100 gpm (379 L/min) at pressures not exceeding 250 psi (1,724 kPa). Often, pumps used on grass fires are designed to operate when the apparatus is in motion. Some vehicles will have separate engines to drive the pump, so "pump and roll" operations may be accomplished without mechanical complexity. Likewise, the use of skid-mounted, self-contained fire-fighting packages, complete with pump and engine, water tank, hose, and equipment, allows any vehicle to transport the unit.

Water Tanks

The majority of fire apparatus equipped with pumps are also equipped with water tanks. The tank supplies water to the pump for initial hose streams before water from hydrants or suction sources is available. Pumpers carry a minimum of 500 gal (1,893 L) of water, initial attack vehicles carry a minimum of 200 gal (760 L), and mobile water-supply vehicles carry a minimum of 1,000 gal (3,785 L). Water tanks supplied on aerial apparatus should hold a minimum of 150 gal (568 L). And wildland apparatus carry a minimum of 125 gal (473 L).

All water tanks should be constructed of noncorrosive material or materials that are protected against corrosion and deterioration. Tanks being used today are constructed of plastic material. They are increasing in popularity because of the noncorrosive qualities. They should also be equipped with a water-level indicator visible at the pump operator's position. One or more sumps with a 3-in. (76-mm) or larger removable pipe plug should be provided so that debris in the tank can be cleaned out.

All tanks must have baffles to minimize water surging when the vehicle is in motion. The dimension of any spaces in the tank, either transverse or longitudinal, should not exceed 48 in. (1,220 mm) nor be less than 23 in. (584 mm). The baffle arrangement should allow an adequate flow rate from the tank to the pump, which is typically tested to 80 percent of the tank capacity.

The tank fill opening should be at least 20 sq in. (12,900 mm^2) and vents/overflows at least 12 sq in. (7,742 mm^2). The venting must be matched to the maximum fill and discharge rate so as not to overpressure the tank. The overflow outlet should direct any water behind the rear axle so as not to interfere with rear-tire traction.

Tanks that hold more than 1,000 gal (3,785 L) should have a single outlet capable of allowing water to be dumped from the tank at an average rate of 1,000 gpm (3,785 L/min). Likewise, a single external fill connection should be provided that permits a minimum filling rate of 1,000 gpm (3,785 L/min) from sources outside to the unit.

Aerial Ladders

Aerial ladders have been used by fire departments for more than a century to gain access to the upper floors and roofs of buildings. The first aerials were manually operated before spring-assist and air hoists were introduced. Current aerial ladders are steel or aluminum truss construction with hydraulic hoists and ladder controls.

Common sizes of aerial ladders are 75, 85, 100, 110, and 135 ft (22.8, 25.9, 30.5, 33.5, and 41.1 m, respectively). The rated height of an aerial is measured by a plumb line from the top ladder rung to the ground with the ladder fully extended at its maximum elevation. These ladders can be mounted on a single-chassis vehicle or a tractor-drawn vehicle. The rear-mount aerial ladder has become common. This arrangement places the turntable over the rear wheels, with the ladder extending forward when bedded, which results in shorter overall vehicle lengths. In some cases, a shorter aerial ladder mounted on a pumper chassis provides a useful combination of devices for fire attack.

Aerial ladders must have a minimum width of 18 in. (457 mm) at the narrowest point of the ladder. Rungs must be spaced 14 in. (356 mm) on center, and the top rails along the sides must be at least 12 in. (305 mm) high. The ladder must support 250 lb (114 kg) at

the tip of the fly section with the ladder fully extended in a horizontal position without water in any permanently piped water system. It must support the same weight at the tip of the fly section with the ladder at 45 degrees with water flowing.

Aerial ladders may be equipped with a fixed waterway to the tip of the fly section or some other section. Where such a waterway is provided, it should be capable of flowing 1,000 gpm (3,785 L/min) at 100 psi (690 kPa) nozzle pressure through a 45-degree, side-to-side range of motions and through an arc of 135 degrees from a line parallel to the ladder and downward. Often, the monitor and nozzle on these systems are controlled remotely by electric wires or radio control. The friction loss in the water system between the base of the swivel and the monitor outlet should not exceed 100 psi (690 kPa) with 1,000 gpm (3,785 L/min) flowing and the water system at full extension.

When the aerial ladder is not equipped with a pre-piped water way, a detachable ladder pipe provides elevated fire-stream service. When supplied by 3-in. (76-mm) hose up the ladder, 600 gpm (2,271 L/min) is the practical maximum that can be supplied in normal fire-ground operations. The ladder pipe is provided with 1$^1/_4$-, 1$^3/_8$-, and 1$^1/_2$-in. (32-, 35-, and 38-mm, respectively) tips and with a 500-gpm (1,893-L/min) spray nozzle.

The aerial ladder must be stable when a load one-and-a-half times the rated capacity is suspended from its tip while the aerial ladder is in its least stable position and the vehicle on which the ladder is mounted is on a firm, level surface with the stabilizers extended to a firm footing. The aerial ladder must also be stable when a load one-and-a-third times the rated capacity is suspended from its tip when the vehicle on which it is mounted is placed on a firm surface that slopes downward at 5 degrees in the direction most likely to cause it to overturn. Again, the stabilizers are extended to a firm footing. The stability of the apparatus is checked during the certification tests, and the vehicle cannot show any signs of instability during these tests.

Elevating Platforms

Elevating platforms are of three principal designs. In one, the platform is mounted on an articulated boom that travels in the arc desired by the operator. In the second, the platform is mounted on an extendible or telescopic boom, in much the same fashion as an aerial ladder. In the third design, the platform is mounted on booms that are both articulated and telescopic.

Platforms are available with booms designed for elevations from approximately 75 ft to more than 200 ft (23 m to 61 m). The rated height of an elevating platform is measured by a plumb line from the top rail of the platform to the ground, with the platform raised to its maximum elevation. In selecting an elevating platform, it is important to know its reach and elevation capabilities under operating conditions. The articulated-boom design provides its maximum horizontal reach at approximately half of its maximum elevation. The extendible telescopic boom of the same nominal length generally permits somewhat greater horizontal reach at higher and lower elevations. The telescopic boom often has an extending ladder attached to it, which provides access to and from the platform while it is elevated.

All designs provide stable platforms with controls located both on the platform and at ground level. The ground-level controls must be able to override the platform controls so an operator on the ground can move the platform away from danger if the platform operator is in trouble.

An elevating platform should be able to carry a load of at least 750 lb (340 kg) in all operating positions without the water system charged. Units of 110 ft (34 m) or less should be able to reach their maximum elevation and extension and to rotate 90 degrees within 150 sec. An advantage of an elevating platform in operation is that the platform can be moved quickly from window to window for fire fighting or rescue purposes.

Platforms are designed to provide elevated fire streams. The platform turret or turrets must be able to rotate through at least 45 degrees up, down, and to each side. The mobility of the platform permits the operator to change the turret location quickly, as desired. The apparatus must be designed so that, regardless of the position of the platform and the direction of the stream, the equipment can be operated safely while discharging 1,000 gpm (3,785 L/min) at 100 psi (690 kPa) nozzle pressure. The friction loss in the water system between the base of the swivel and the monitor outlet should not exceed 100 psi (690 kPa) with 1,000 gpm (3,785 L/min) flowing and the water system at full extension.

The stability requirements described above for aerial ladders also apply to elevating platforms. In addition, there are requirements in NFPA 1901 for the hydraulic system, for the structure, and for quality control during manufacture which apply to both aerial ladders and elevating platforms.

Water Towers

The success of elevating platforms on the fireground during the 1960s was a prelude to the use of hydraulically operated water towers designed to apply large flows on a fire from effective heights and positions. Such water towers have proved very useful and have been widely accepted in both large and small communities as an important addition to the attack capabilities of a pumper. Water towers are designed to discharge 1,000 gpm (3,785 L/min) or more at 100-psi (690-kPa) nozzle pressure in either a straight stream or a spray pattern, and to move the boom and the nozzle both horizontally and vertically under the control of the operator for the best application of water on the fire.

Like elevating platforms, water towers may be of either articulated or telescopic boom design. Heights typically range from 50 to 75 ft (15.2 to 22.9 m). The telescopic type may be equipped with a ladder extending the full length of the boom. In those cases, the ladder should meet the same requirements as an aerial ladder.

Because water towers are typically installed on pumpers, the requirements are covered in NFPA 1901. The tower is mounted on a turntable or pedestal to permit it to rotate 360 degrees in either direction and rotate with the nozzle operating at rated capacity. All controls are typically grouped at the pump operator's position so the operator has full control of the pressure, volume, and stream pattern and position.

The stability requirements described for aerial ladders also apply to water towers. NFPA 1901 also provides requirements for the hydraulic system, the structure, and quality control during manufacture.

Foam-Proportioning Equipment

Where the potential for encountering serious flammable liquid fires is present, many fire departments provide on the apparatus various types of foam-making equipment or other agents that are more effective than plain water. Built-in proportioners and foam tanks are often specified for new pumpers. In other cases, foam-making nozzles, portable foam-making equipment, and foam supplies are carried on the apparatus. Foams are available not only to seal off the surface of flammable liquid fires and smother the fire, but foams for Class-A fires are particularly popular for wildland-fire suppression application. Special foams are also available for certain types of polar solvent/ alcohol and unusual liquid fires.

Portable Pumps

In selecting portable pumps, fire departments should be careful to choose models that will give the needed flow characteristics at a safe, continuous engine speed. Portable pumps for fire department service are covered in NFPA 1921, *Standard for Fire Department Portable Pumping Units.*

Portable pumps for fire department service are of the centrifugal type. They are grouped in categories based upon the pressure-volume characteristics that make them suitable for various classes of work. Small streams at high pressure are intended mainly for grass and brush fire operations. Pumps delivering relatively large volumes at low pressures can supply pumps on fire apparatus where the water supply is beyond the reach of suction hose or refilling water tanks. These pumps also may be used as de-watering pumps.

Communications Equipment

Effective communications is essential for any fire department operation. Every fire department vehicle should be equipped with a two-way radio that operates on all frequencies used for dispatch and fireground communication. It is highly desirable to provide a radio speaker and microphone at the apparatus-operator's position so messages can be heard above the usual noise on the fireground. In addition, each vehicle, as well as chief officers' vehicles, should carry portable radios for effective fireground communications.

Vehicles also may be equipped to send and receive hard-copy messages or to receive visual displays of information, including maps and fire-hazard data that may be transmitted from the communications center. Cellular telephones are also extremely popular in chiefs' vehicles to provide communication directly with parties outside the fire department radio network. Cellular phones, when installed on fire apparatus, can be used on the fireground or hazardous materials incident scene.

Specialized Apparatus

There is a wide diversity of fire apparatus designed for special types of service. These units should be designed with the rigors of fire duty in mind, and should conform to the general requirements outlined in the fire apparatus standards.

Many fire departments operate a vehicle to support rescue services carrying equipment that deals with everything from emergency medical service to vehicle extrication to building collapse. The larger of these vehicles are typically equipped with a large 120/240-V electric generator, hydraulic rescue tools, and a wide variety of hand tools, power tools, and devices for jacking, shoring, cutting, and breaching. This emergency equipment typically supplements that carried on pumpers and ladder trucks. Many rescue vehicles have fully enclosed bodies to provide places for personnel to ride, to provide shelter for emergency medical treatment, and to provide an environment for incident scene rehabilitation during inclement weather. (See Figure 10-17B.)

Fire departments are equipping vehicles to support hazardous materials operations. The size and amount of equipment on these vehicles will vary with the role that the fire department sees for itself in such operations. Typically, these vehicles carry a library of materials to assist in identifying a hazardous material, tools and equipment for shutting valves or stopping leaks, materials for diking or absorbing spills, and special drums for recovering leaking drums. Specialized clothing to protect the fire fighters while they are working in the contaminated area is also carried.

A self-contained breathing apparatus (SCBA) service unit is designed to field-service SCBAs and recharge depleted air cylinders on the fireground. These vehicles may carry numerous extra SCBA

FIG. 10-17B. Rescue vehicle with lighting towers extended.

cylinders for merely exchanging cylinders, several large air cylinders arranged to allow for refilling of SCBA cylinders in the field, and both large cylinders and a high-pressure air compressor for filling SCBA cylinders in the field. Small fire departments may not need a separate vehicle, but every fire department should be able to provide full SCBA tanks in the field.

Some fire departments find it advantageous to have separate vehicles with large generators to provide 120/240-V electricity. These units may have a generator of 15,000 W or larger. The generator may be driven by its own engine; hydraulically, with the hydraulic pump driven by the vehicle engine; or directly from a PTO on the vehicle's transmission. These units typically carry floodlights, electric cord and adapters, and various power tools.

It is important that any electrical wiring or equipment on an emergency vehicle be properly installed and grounded. This equipment is typically used under conditions of extreme wetness, where failure to properly install and maintain the electrical integrity of the wiring and equipment will result in injury or death to persons using the equipment.

Chief Officers' Vehicles

Many fire departments find it essential to provide transportation for staff and command officers. In the larger fire departments, sizable fleets of automobiles are maintained. Automobile transportation is necessary for chief officers, fire prevention officers and inspectors, training officers, communications officers, fire investigators, and others. Some fire departments pay a mileage allowance for the use of private cars. This may not be a good arrangement, especially where response to emergencies may be involved, as private vehicles are, generally, not properly equipped with warning lights and sirens or otherwise properly marked. They also may not be properly insured for that type of service.

All fire department vehicles, including automobiles, that respond to fires and emergencies must be equipped with warning lights, sirens, and radio communication facilities with the emergency networks, including mutual-aid frequencies, on which the department operates. The vehicles should be clearly identified as fire department vehicles. They should carry protective clothing and equipment for the officers and their aides, as well as portable fire extinguishers, first-aid equipment, and hand lights. If the vehicle is being used by command officers, it should also carry directories containing pre-fire plans on properties in the area served, water distribution plans, and reference books on hazardous materials.

Many fire departments provide their command officers with station wagons or similar style vehicles in which the necessary equipment or materials is mounted or provided for ready access.

Fireboats

Fireboats are available in various sizes and types, in accordance with local needs, and vary from large tugs to fast, jet-propelled fire and rescue craft. They may be operated by fire departments or port authorities. U.S. Navy and Coast Guard vessels also are equipped to provide fire-fighting services.

Fireboats are principally used to protect vessels in the harbor; protect piers, pier sheds, and cargoes along the waterfront; protect yachts and houseboats in basins and marinas; assist in marine rescue from all types of water accidents; and serve as pumping stations to provide large flows at fires within reach of hose lines supplied by the boats. Fireboats also may provide a valuable emergency source of water for fire protection, should an earthquake or other accident interrupt the normal supply. Some fire departments operate hose trucks or "fireboat tenders" that carry large-diameter hose to allow fireboat pumping capacity to be used at shore fires along the waterfront.

The size, type, pumping capacity, and equipment carried by fireboats will depend upon the type of service expected of the vessels. Fireboats must carry much of the same equipment as pumpers, ladder trucks, and rescue vehicles on land. They need foam-making equipment and may carry quantities of special extinguishing agents. Fireboats need radar, as well as radio communication, for safe operation under all weather conditions and, at times, in heavy smoke. Self-contained breathing apparatus is essential for fireboat crews.

Pumping capacity for individual fireboats varies from 500 gpm (1,893 L/min) for very small craft to 10,000 gpm (37,854 L/min) or more for larger vessels. Pumping capacity is rated at 150 psi (1,034 kPa) discharge pressure, as with pumpers. Some small jet-powered fireboats rely upon their jet pumps to supply water for fire fighting. These pumps may develop substantial volumes at pressures adequate to supply turret nozzles attached to the pump, but it may be difficult to provide the necessary pressure to supply the hose streams to fight ship fires or fires ashore, unless additional pumps are provided to meet standard pressure requirements.

Airport Crash Trucks

Specialized apparatus is required for aircraft rescue and fire-fighting service at airports. In many cities, such equipment is part of the municipal fire department. In others, it is maintained and operated by an airport authority. NFPA 414, *Standard for Aircraft Rescue and Fire Fighting Vehicles,* covers the design and performance requirements for aircraft rescue and fire-fighting vehicles equipped for rescuing occupants and combating fires in aircraft at or near an airport.

NFPA 414 covers three types of vehicles: (1) major fire-fighting vehicles, (2) rapid intervention vehicles (RIV), and (3) combined-agent vehicles. RIVs are intended to reach the emergency site quickly so rescue operations can be started before the major vehicles arrive. Airport crash apparatus should include a small rescue vehicle, which can also serve as a command car, and a "nurse" tanker carrying additional water and extinguishing agents where needed. Crash trucks can be modified to fight fires in hangars and other airport structures, but this should not detract from the primary function of these vehicles.

Because of off-highway performance needs, the vehicle weight of crash trucks should be distributed equally over all wheels. These vehicles need greater axle and chassis clearances than standard fire apparatus, as well as high acceleration characteristics. Positive drive to each wheel is required to negotiate soft ground, snow, and ice. The positive drive can be provided by torque proportioning, or no-spin differentials, or by other automatic devices that ensure that each wheel, rather than the axle, is driven independently.

EQUIPMENT CARRIED ON APPARATUS

Apparatus must be equipped with the tools necessary to accomplish fireground operations. The NFPA fire apparatus standards include listings of equipment and appliances needed with various types of fire apparatus.

When developing specifications for fire apparatus, the fire department must evaluate its operation and determine how the apparatus will be used in its operation and what equipment will be necessary to support its operation.

Most fire apparatus carry some hose. Pumpers typically carry the most hose, as the operational purpose of a pumper is to establish a continuous water supply and apply water on the fire. Use of larger-diameter supply hose [(4 in. (100 mm) and 5 in. (125 mm)] for pumping apparatus is becoming more common, and the length to be carried will depend on such things as hydrant spacing and distances to water supply points. Attack hose in $1^1/_2$-, $1^3/_4$-, or 2-in. (38-, 44-, or 51-mm, respectively) diameters is common for hand-held hose lines for interior attack, and $2^1/_2$-in. (64-mm) diameter hose lines are common when larger flows are required during exterior attack or to protect exposures. A variety of nozzles that will produce both spray and straight streams should be carried for the various sizes of hose.

Fire department ground ladders are an important tool for rescuing persons and gaining access to buildings for fire-fighting purposes. Ladders may be constructed of wood, aluminum, or fiberglass. All ground ladders used in the fire service should be designed and built to conform to NFPA 1931, *Standard on Design of and Design Verification Tests for Fire Department Ground Ladders.*

Loads that fire ladders are designed to support vary according to the type of ladder. Extension ladders, single ladders, and roof ladders built since 1984 are designed for 750 lb (340 kg) maximum load, while folding and pompier ladders are designed for 300 lb (136 kg) maximum load. Load limitations are based on the use of the ladder supported at the top end against a building, and with the ladder set at a $75^1/_2$-degree angle. This angle is measured from the horizontal, with the ladder in the raised position. At angles smaller than this, the ladder may be unable to support the design load.

While the NFPA apparatus standards list the minimum requirements for ground ladders on each type of vehicle, careful consideration should be given to the ladder needs of the area served. Manually raised extension ladders more than 40 ft (12 m) long are generally not required, because aerial ladders and elevating platforms provide adequate emergency access to upper floors and roofs. If congested conditions prohibit the placement of aerial apparatus to provide effective aerial service, however, additional or longer ground ladders should be provided. It must be recognized that additional personnel are required to raise longer ladders, and that the space available for ladders on some vehicles is limited.

Besides hose, nozzles, and ground ladders, fire fighters need a wide variety of other equipment to perform their jobs. This includes cutting tools for forcing entry and ventilating smoke from a building, smoke ejectors for ventilation, salvage covers for protecting unburned contents, and a variety of fittings for use with hose lines. The complement of equipment carried will vary from department to department, based on the needs and operations of the companies. It is critical that sufficient space and carrying weight be allotted for the equipment when the apparatus is designed so the equipment does not cause the apparatus to become overweight.

APPARATUS PROCUREMENT POLICIES

The responsibility for procurement of fire apparatus and equipment should rest with the agency operating the fire department, whether a municipality, a fire district, or private industry. In some cases, procurement is a cooperative measure in which a larger unit of government contributes to the cost of providing fire apparatus for fire departments serving its territory. In a few cases, state governments contribute to local fire apparatus costs. In some areas, several fire departments have banded together to generate standard apparatus specifications to buy apparatus in quantity and realize significant cost savings. Various fire department groups also cooperate in volume purchases of hose and other equipment.

When the fire department is an agency of municipal government, it is the responsibility of the latter to provide for all public fire-fighting equipment. Providing the necessary funds is the responsibility of the appropriate fiscal authorities, but the actual selection of the equipment should be the responsibility of the fire department management. The chief administrative officer of the fire department, aided by a staff of technical specialists, including a master mechanic, should keep the municipal administrator informed of the age and condition of department equipment and should prepare specifications consistent with applicable national standards for items that need to be replaced. In autonomous volunteer fire departments, there may be a purchasing committee appointed to procure apparatus. In municipalities, the actual purchase is usually handled by a municipal purchasing department. The purchasing department should not try to tell the fire department what type of fire apparatus it should use, but should see that required procurement procedures, such as open competitive bidding, are followed.

The appendices of NFPA 1901 and NFPA 1906 contain helpful suggestions covering the purchase of fire apparatus, including developing specifications, reviewing proposals, dealing with manufacturers, and acceptance of the completed apparatus.

In general, the purchase and replacement costs of fire apparatus should be a regular item of the fire department capital budget. In most cases, except for accidents, the requirements can be planned and funded on a long-range basis. Systematic apparatus replacement provides the fire department with reliable apparatus at all times. Improvements in fire apparatus design can be introduced, maintenance costs become more favorable, operating efficiency increases, and equipment remains reliable.

Lease or lease/purchase of fire apparatus can be an alternative to outright purchase. Income tax benefits and depreciation accrue to the actual, commercial owner of the apparatus, while the fire department, which is tax exempt, realizes savings in overall life cycle cost. In addition, the department does not have to expend a large purchase price at one time, but can spread the cost of apparatus over several years. Further, there is no capital tied up in the event the department wishes to trade, exchange, or dispose of the apparatus. Closed and open-ended leases are available, and the finance department of the municipality should evaluate these alternatives carefully. In a smaller or volunteer department, a local banker can advise the purchasers of the advantages of a lease *vs.* a purchase arrangement.

Any lease or lease/purchase contract should clearly spell out who is responsible for maintenance, repairs, and liability. Likewise, the conditions under which the arrangement may be terminated, by either party, should be stated clearly.

The normal life expectancy for first-line fire apparatus will vary from city to city, depending upon the amount of use the equipment receives and the adequacy of the maintenance program. In general, a 10- to 15-year life expectancy is considered normal for first-line pumping engines. First-line ladder trucks should have a normal life expectancy of at least 15 years. In fire departments where ladder trucks make substantially fewer responses to alarms than engines, a planned first-line service of 20 years may be warranted for ladder trucks. Some smaller fire departments that have infrequent alarms operate pumping engines up to 20 years with reasonable efficiency, although obsolescence will make older apparatus less desirable, even if it is mechanically functional. In some types of service, including areas of high-fire frequency, a limit of 10 years may be reasonable for first-line service. The older apparatus may be maintained as part of the reserve fleet, as long as it is in good condition, but in almost no case should the fire department rely on any apparatus more than 25 years old.

Acceptance Tests

Acceptance tests are designed to demonstrate that apparatus will perform as specified in the purchase contract. Tests should be performed within 10 days of delivery and before the apparatus is accepted. The tests should be conducted by the manufacturer's representative in the presence of such persons the purchaser may have designated in the delivery requirements. Normally, the fire chief or a designated representative is the acceptance authority, exercising this authority following satisfactory completion of tests and inspections to ensure compliance with the purchase specifications. Third-party acceptance is gaining in popularity and should be considered by the agency. Costs for third-party testing are generally associated with the purchase price of the vehicle.

The acceptance-test requirements for fire apparatus and its various components should be detailed in the purchase specifications, and should be based on the manufacturer's certification tests and any additional tests the fire department desires to ensure that the apparatus performs to specification. The acceptance tests should always be conducted in the community, so that any problems that developed during the delivery of the apparatus will be detected.

If the apparatus is equipped with a pump, a test plate is required at the pump-operator's position that gives the rated discharge and pressures, together with engine speed determined by the manufacturer's certification test for the unit. The no-load governed speed of the engine, as stated on a certified brake horsepower curve, is also given. Tests are conducted to see that the water tank has the minimum capacity specified and can flow the specified rate to the pump.

Aerial devices are inspected and tested in accordance with the requirements in NFPA 1914, *Standard for Testing Fire Department Aerial Devices,* as part of the certification tests.

In addition, the stability of the aerial device is tested with the aerial device on a level surface, and then on a firm surface sloping downward at 5 degrees in the direction most likely to cause it to overturn. The vehicle cannot show any signs of instability during these tests.

If the aerial device is equipped with a pre-piped waterway, the system is flow-tested to determine whether the water system can flow 1,000 gpm (3,785 L/min) at 100-psi (690-kPa) nozzle pressure with the aerial device fully elevated and extended. If there is a fire pump on the vehicle, the test is conducted using that fire pump, and the intake pressure to the fire pump cannot exceed 20 psi (138 kPa).

MAINTENANCE OF APPARATUS AND EQUIPMENT

Equally important to the procurement of proper fire apparatus and equipment is its maintenance. A regular maintenance program should be set up to ensure that all apparatus and equipment is serviced and tested in accordance with the manufacturer's recommen-dations. The program should define the roles and responsibilities for the maintenance.

Service Tests of Pumps

All pumps should undergo service tests at least annually and after any major repairs. These tests demonstrate that the pump/engine combination is capable of meeting the performance requirements of the original certification or acceptance tests. Records of service tests are important evidence of proper apparatus maintenance.

NFPA 1911, *Standard for Service Tests of Pumps on Fire Department Apparatus,* outlines the procedures for the service test. The test should cover the following items:

1. A test pumping from draft for 20 min at 100 percent of rated capacity, at 150 psi (1039 kPa) net pump pressure; 10 min at 70 percent of rated capacity at 200 psi (1379 kPa) net pump pressure; and 10 min at 50 percent of rated capacity at 250 psi (1724 kPa) net pump pressure
2. An engine speed test to determine if the engine is capable of reaching its no-load governed speed
3. A vacuum test to ensure that the pump and the attached piping are still tight
4. A pressure-control test to ensure that the pressure control device can control the discharge within the specified limits
5. A check of the accuracy of the gages and flow meters.

The purpose of the test is to ensure that the pump is generally in good condition, that the pump casing and various fittings are tight, and that the transfer valve is operating properly, if the pump is a series parallel type. If rated capacity cannot be obtained at 150 psi (1,034 kPa) and the pump is not cavitating, or it appears that the pump, engine, pump accessories, or other parts of the power train and pumping equipment are not in good condition, the apparatus manufacturer, an authorized representative, or a competent mechanic should be contacted for advice so the condition can be corrected.

Service Tests of Aerial Devices

Aerial devices also need to be service-tested periodically. These devices develop problems not only from use at fires but also from responding to and returning from alarms. NFPA 1914 requires a yearly inspection of the aerial device and of the pre-piped water system, if so equipped, and a test of their operation. NFPA 1914 requires additional nondestructive tests, at intervals not exceeding 5 years, that check structural components for hardness, welds for cracks or discontinuities, and bolts and pins for wear or internal flaws. The full testing should be done any time the aerial device has had major repairs or if the department has reason to believe that they have exceeded the manufacturer's design criteria to ensure that it can be returned to service.

The annual inspection and tests can normally be conducted by a qualified fire department mechanic. The more complex nondestructive tests require an expertise that may be beyond the abilities of fire department personnel. The qualifications of the inspection personnel and the test protocols they must use are outlined in NFPA 1914, and only personnel meeting those qualifications and having the proper equipment should attempt to do the nondestructive testing.

Ground Ladder Service Testing

NFPA 1932, *Standard on Use, Maintenance, and Service Testing of Fire Department Ground Ladders,* calls for service testing all ladders at least annually and at any time the ladder is suspected of being unsafe. The tests are designed to show that the ladder is safe for continued use under the conditions for which it was originally designed. The service-test procedures must be carefully observed to ensure that the ladder is not damaged. Ladders also should be inspected visually, as outlined in NFPA 1932, to detect loose rungs, loose belts and rivets, weld defects, cracks, splintering, breaks, discoloration, and other signs of possible weakness that might warrant testing to determine whether the ladder is safe to use or needs repair.

HOSE, COUPLINGS, AND NOZZLES

Fire hose is the vital link between the water supply and the nozzles used to project streams on a fire. Hose must be rugged, dependable, and capable of carrying water under substantial pressures, yet flexible and sufficiently easy to handle. Fire department management should be concerned with selecting the proper grades and types of hose and with maintenance to ensure maximum useful life. These two factors are covered in detail in NFPA 1961, *Standard for Fire Hose,* and NFPA 1962, *Standard for the Care, Use, and Service Testing of Fire Hose Including Couplings and Nozzles.* Aside from unavoidable mechanical injury at fires, a hose's dependability and length of life rest on three factors: (1) the quality and suitability of the hose purchased, (2) the care with which it is handled at fires, and (3) the maintenance and care the hose receives.

Types of Fire Hose

Fire hose is manufactured for at least four uses. These include: (1) fire attack by fire department personnel (attack hose), (2) relaying water through large-diameter hose from a source to the fire (large-diameter supply hose), (3) occupant use in buildings from standpipe systems (rack-and-reel hose), and (4) fire department use in wildland fires (forestry hose). Each of these types has its own minimum requirements for strength.

Attack hose is designed for a minimum service-test pressure of 300 psi (2,069 kPa) to meet a normal highest operating pressure of 275 psi (1,896 kPa). If a higher working or operating pressure is needed, the purchaser needs to specify a higher service-test pressure.

Large-diameter supply hose has a minimum size of 3 1/2 in. (89 mm) and a minimum design service-test pressure of 200 psi (1,380 kPa) to meet a 185-psi (1,276-kPa) normal highest operating pressure.

Rack-and-reel hose has a minimum size of 1 1/2 in. (38 mm) and a minimum design service-test pressure of 150 psi (1,034 kPa) to meet a 135-psi (931-kPa) normal highest operating pressure.

Forestry hose is only available in 1-in. (25-mm) and 1 1/2-in. (38-mm) sizes and has a minimum design service-test pressure of 300 psi (2,069 kPa) to meet a 250-psi (1,724-kPa) normal highest operating pressure.

All hose must meet a proof-test pressure—formerly an acceptance-test pressure—of twice its service-test pressure, a burst-test pressure of three times its service-test pressure, and a kink-test pressure of one-and-a-half times the service-test pressure.

NFPA 1961 provides requirements for the hose jacket, the lining and the lining adhesion, any covering and its adhesion, the tensile strength of the hose, the ultimate elongation of the hose, and cold resistance. Tests are outlined to verify these requirements. Another section in NFPA 1961 defines how the hose must be marked.

Care, Maintenance, and Testing of Fire Hose

To be reliable, fire hose should always be cared for properly. It should not be used for anything but training or fire-fighting service, except in emergencies. Burst hose at a fire may seriously injure fire fighters and other persons, and may mean a loss of time in bringing a fire under control. NFPA 1962 and the manufacturer's instructions should form the basis of a hose care and maintenance program.

Hose carried on fire apparatus should be loaded to reduce edge wear, and loaded so that air can circulate around it. The hose compartments should be protected by waterproof covers to keep the hose dry and to protect the hose from the direct rays of the sun when fire apparatus is out of quarters during inspection and training periods.

Hose installed at yard hydrants at industrial plants should be kept in well-ventilated hose houses in such a way that air can circulate and excessive heat can be avoided.

To prevent damage and permanent set to the rubber tube, hose should be removed from the apparatus or hose house at least quarterly and reloaded so that the folds are in a different position. Whether hose jackets are of cotton or synthetic construction, the hose should be cleaned and dried after use to remove possible abrasive or contaminating material and to protect the hose compartment of the apparatus against rust or water damage.

When used for fighting fires, fire hose is subject to severe strains, pressure surges, and mechanical injury. Care should be taken to lay hose so that injury will not result from contact with sharp or rough objects. Vehicles should not be driven over hose lines. Where it is necessary for fire department vehicles to cross, hose bridges should be used where possible. When it is necessary to hoist hose lines, hose rollers can make the task easier and prevent mechanical injury. Hose lines extended up ladders should be supported by hose-rope tools placed to take the strain off couplings.

Pressure surges are a principal cause of damage to fire hose. Shutoff nozzles should be opened and closed slowly because sudden closure of nozzles can cause severe pressure surges, or water hammer, that can be extremely damaging to both hose and pumping apparatus. Pressure-relief devices on pumping engines should always be used to control sudden increases in pressure.

In cold climates, care should be taken to prevent water from freezing in the hose. Once water has been turned on, some should be left running through the hose because moving water does not freeze readily. Sharp bends should be avoided in any frozen hose. Care must be used in chopping hose free from ice after a fire so as not to cut the hose. The ice should then be allowed to melt off the hose, and any hose that has been frozen should be service tested before being placed back in service.

Care must also be taken to avoid burning the hose at forest and grass fires. As fire is knocked down and the nozzle is advanced, the hose may come in contact with hot spots or embers.

When hose has been returned to quarters it should be laid out where it can be swept and washed as needed to remove grime and contaminants. Hose should be thoroughly dried after cleaning. Hanging in hose towers or laying on racks are common methods for drying hose. However, drying cabinets, which circulate heated air around the hose, are also used.

Good hose records are necessary to keep accurate data on hose performance. Records for hose on racks, on reels, or in hose houses or similar enclosures may be kept at the hose location or at a central location on the premises where the hose is located. Fire department records should include a complete record of all hose. Upon delivery and acceptance, each length of hose should be given an identification number. This is used to record its history throughout its service life, including all testing and repairs, and ultimately the reason it was condemned and removed from service.

All fire hose, except rack-and-reel hose, should be tested at least annually and after repairs, or at any time it has had hard usage and its condition is suspect. NFPA 1962 includes required service-test procedures and a typical annual hose-test record form. The standard also specifies service-test pressures for hose manufactured before July 1987. Hose manufactured after that date has the service-test pressure stenciled on it.

Hose stored on a rack or reel should be examined carefully at least once a year. This annual physical inspection consists of a visual examination to see that the hose has not been vandalized, that the couplings and nozzle are attached and free of debris, and that there is no evidence of mildew, rot, chemicals, vermin, or abrasion. Racked hose should be reracked with the folds in a different location. Five years after the date of purchase and every 3 years thereafter, the hose should be service-tested, as it should be any time the condition of the hose is questionable.

Fire Hose Connections

NFPA 1963, *Standard for Fire Hose Connections*, covers ten sizes of threaded connections, from $3/4$ to 6 in. (19 to 152 mm), used in fire protection. The standard gives the dimensions for screw-thread connections, gages, gaskets, and gasket seats, as well as the size of the threaded connections. These standards apply to fire-hose couplings, suction-hose couplings, relay-supply-hose couplings, fire pump suction, discharge valves, fire hydrants, nozzles, adapters, reducers, caps, plugs, wyes, siamese connections, standpipe connections, sprinkler connections, and all other hose fittings, connections, and appliances that connect to fire pumps, hose, and hydrants.

These threaded connections are defined as the American National Fire Hose Connection Screw Thread, abbreviated throughout NFPA 1963 as NH and also known as NST and NS. Each of the ten standard sizes is designated by specifying in sequence the nominal size of the connection, the number of threads per inch, and the thread symbol as follows:

0.75-8 NH	3.5-6 NH
1-8 NH	4.0-4 NH
5-9 NH	4.5-4 NH
5-7.5 NH	5.0-4 NH
3-6 NH	6.0-4 NH

The need for standard fire service connections is most apparent during mutual aid operations. When hose couplings, nozzles, and other equipment do not conform to standards, it becomes necessary to use adapters, thus complicating fire-fighting operations.

Nozzles

Nozzles used in the fire service are of either the spray or solid-stream design. A spray nozzle typically can produce a straight stream, although that stream does not have the same characteristics as a stream from a solid-bore nozzle. Nozzles for 1-, $1^1/_2$-, $1^3/_4$-, and 2-in. (25- 38- 44- and 51-mm) hose are typically spray nozzles. Nozzles for $2^1/_2$-in. (65-mm) hand lines and for master stream devices may be of the spray or solid-stream design. All nozzles should be equipped with shutoff valves, either as an integral part of the nozzle or as an attached item.

Fire department nozzles for $1^1/_2$-in. (38-mm) hose normally are designed to discharge approximately 95 to 125 gpm (360 to 473 L/min). Nozzles for $1^3/_4$- and 2-in. (44- and 51-mm) hose normally are designed to discharge approximately 125 to 200 gpm (473 to 757 L/min). Nozzles for 1-in. (25-mm) hose generally discharge in the 20- to 30-gpm (76- to 114-L/min) range. A standard hose stream from a $2^1/_2$-in. (65-mm) hose with a $1^1/_8$-in. (29-mm) smooth-bore tip is 250 gpm (946 L/min) at 50-psi (345-kPa) nozzle pressure.

Solid-bore nozzles have a straight open tip that generally is attached to some type of stream shaper. These nozzles are designed to hold the solid stream together as water is thrown a considerable distance. The design of the nozzles provides a gentle taper from the diameter of the inside of the hose to the diameter of the tip to reduce the turbulence that causes a stream to breakup.

Requirements for spray nozzles are covered in NFPA 1964, *Standard for Spray Nozzles (Shutoff and Tip)*. There are essentially three general types of spray nozzles.

The basic spray nozzle is an adjustable-pattern spray nozzle in which the rated flow is delivered at a designated nozzle pressure and nozzle setting. Due to its basic design, the flow [gpm (L/min)] will vary as the pattern is changed from a straight stream through a wide spray pattern. This is caused by the orifice size changing to affect pattern adjustment.

The constant-gallonage spray nozzle is an adjustable-pattern spray nozzle in which the flow is delivered at a designated nozzle pressure. At the rated pressure, the nozzle will deliver a constant gallonage from straight-stream through a wide-spray pattern. This is done by maintaining a constant orifice size during flow-pattern adjustment. These nozzles are popular because the friction-loss characteristics in the hose and the nozzle pressures remain constant regardless of the selected orifice setting.

A constant-pressure (automatic) spray nozzle is an adjustable-pattern nozzle in which the pressure remains constant through a range of flows. The constant pressure provides the velocity for an effective stream-reach at various flow rates, which is accomplished by means of a pressure-activated, self-adjusting orifice baffle. These nozzles can produce streams whose discharge varies considerably but whose visual characteristics and stream reach remain the same. Do not to be mislead by a good-looking stream that lacks the adequate flow for proper fire attack. Remember the volume of water, not pressure alone, extinguishes fire.

There are other nozzles available for special purposes. These include nozzles used in the aspiration and delivery of foaming agents, very-high-pressure spray nozzles that are designed to deliver small flows at pressures in the 400- to 800-psi (2756- to 5516-kPa) range, and spray nozzles used on standpipe hose for first-aid fire protection.

PROTECTIVE EQUIPMENT FOR FIRE FIGHTERS

Each year, fire fighters are injured and some die because they do not have protective clothing and equipment or do not use it properly. No equipment can guarantee a fire fighter's safety, but full protective clothing and self-contained breathing apparatus (SCBA) offer fire fighters a better chance to carry out their duties effectively without suffering injury or death. The safety odds are further improved if that equipment is carefully selected, properly maintained, and used at all appropriate times.

Although the initial cost of good protective clothing may seem high, it is small compared to the expense of hospitalization or the cost of replacing an injured fire fighter. Recruitment and training expenses, medical expenses, disability payments, and early retirements are costly. The loss of a volunteer fire fighter disrupts a community's fire service effort, to say nothing of its economic impact on his or her family.

The protective equipment that fire fighters wear while combating structural fires must be viewed as a system. That system must include a helmet, a protective coat and protective trousers, gloves, footwear, SCBA, and PASS (personal alert safety systems). Together, these can protect fire fighters from toxic fumes and gases, heat, moisture, puncture, impact, and other adverse environmental conditions.

Special situations may call for specialized protective clothing. Approach and proximity garments protect against radiant heat while near a fire and are often used by airport fire fighters. The U.S. Forest Service has special light-weight clothing for wear by wildland fire fighters which provides limited protection to the wearer.

Various types of special protective clothing are available for hazardous-materials incidents, but these must be chosen for protection from a specific exposure. Specific decontamination or disposal procedures must be followed for the chemical to which the garment was exposed. These garments are not designed as protective clothing for fire fighting applications.

Protective clothing should be sized properly for the individual wearer, allowing the person to perform his or her duties without unnecessary hindrance. Most important, its use should be mandated and supported by the fire department and all of its members. NFPA 1500, *Standard on Fire Department Occupational Safety and Health Program*, outlines the type of equipment that must be used for personal protection during fire-fighting operations. A complete ensemble of protective clothing and equipment must be provided for each fire fighter working in a hazardous area.

Helmets

A fire fighter's helmet should meet the minimum performance requirements of NFPA 1971, *Standard on Protective Ensemble for Structural Fire Fighting*, for exposure to convected and radiated heat; resistance to flame; resistance to penetration; resistance to top, front, side, and back impact; and insulation from electrical shock. A fire fighter's helmet should have flame-resistive flaps that protect the ears and the neck; the flaps are intended to be worn in the full down position during all fireground operations. A retention system must provide a secure fit. The helmet should be waterproof and divert falling liquids and debris, but it should not hinder the use of SCBA. It should also have an attached protective face shield. The helmet shell, suspension system, and an energy-absorbing liner are designed to protect the head from impact and penetration.

An important interface item between the helmet and the coat is a lightweight fabric protective hood such as those worn by race-car drivers. These hoods are made from a flame-resistive fabric and should meet the applicable requirements of NFPA 1971, *Standard on Protective Ensemble for Structural Fire Fighting*. The hood is worn over the head and neck and under the coat and helmet, protecting the head, forehead, neck, and throat.

Face shields provide partial eye and face protection from heat, sparks, liquids, and flying debris. The SCBA face piece will also provide protection from these hazards. The face shield does not replace primary eye protection when such is required. Although the polycarbonate-type face shields are susceptible to scratching, replacement face shields are inexpensive and should be used when the scratching interferes with vision.

Coats and Trousers

NFPA 1971 covers minimum requirements for fire fighters' protective coats and trousers and outlines a layered composite protection system. The outer layer of the composite is the shell. The shell of both coat and trousers must be made of a flame-resistive fabric that will not be destroyed through charring, separating, or melting when exposed to 500°F (260°C) in a forced-air laboratory oven for a 5-min period. The outer shell may be of a light color for better visibility. The coat must have a minimum of 325-sq in. (2097-cm²) fluorescent/retroflective area, and the trousers must have a minimum of 80-sq in. (520-cm²) fluorescent/retroflective area.

The second layer is the moisture barrier, which prevents moisture from penetrating through to the wearer. The third layer is a thermal barrier that provides insulation from radiant, conducted, and convective heat. In cold weather, an additional flame-resistive, snap-in winter liner may help to provide warmth.

The moisture barrier and thermal barrier are attached at the neck or waist. The shell of the coat and trousers must never be worn without the thermal barrier on the moisture barrier because the garment would offer little, if any, protection against conducted, convected, and radiated heat. Since fire fighters normally work in a high-temperature environment, protection from heat transfer is critical.

The collar of the protective coat should have a throat enclosure that will protect the neck. The collar should be worn up and closed during all fireground operations. The coat should have a flap closure in front to protect against heat, steam, and water.

The length of the turnout coat is particularly important. It must overlap protective trousers enough to protect the wearer adequately in any working position such as kneeling or crawling. Coats worn without protective trousers do not provide adequate torso and leg protection. Protective coats and trousers must be worn as an ensemble for a more complete system of protection.

Station/work uniforms should be made of flame-resistive fabrics that meet the requirements of NFPA 1975, *Standard on Station/ Work Uniforms for Fire Fighters*. Station uniforms made of synthetic fabrics that melt and drip, such as many "wash and wear" fabrics, should be avoided. Volunteers should pay attention to the street clothes they wear under their protective clothing or don flame-resistive jump suits or coveralls before they put on their coats and trousers.

Footwear

NFPA 1971 defines the requirements for footwear used in structural fire-fighting operations. This footwear must resist heat, puncture, impact, and water. Good fit is important because it will lessen foot and leg fatigue and can prevent blisters. It should have ladder shanks that spread weight across the sole to provide comfort while working from ladders. Safety toe-caps can protect toes from heavy impact even when the outer shell of the footwear is damaged. Shin padding will help to prevent lower leg injury. Thermal insulation and pull loops should be standard in any fire fighting boot.

Gloves

Hand protection is essential for fire fighters. NFPA 1971 covers hand protection by gloves and provides performance criteria for heat and flame resistance, cut and puncture resistance, water penetration, dexterity, and sizing. Gloves come in many fabrics and styles.

Self-Contained Breathing Apparatus

NFPA 1981, *Standard on Open-Circuit Self-Contained Breathing Apparatus for Fire Fighters*, sets forth the minimum requirements for the design, performance, and testing of open-circuit self-contained breathing apparatus used in rescue, fire fighting, and other hazardous duties. SCBA must first be certified by the National Institute for Occupational Safety and Health and the Mine Safety and Health Administration (NIOSH/MSHA) as "positive pressure" before it can be tested to the requirements of NFPA 1981. NFPA standards and OSHA regulations specify that only positive-pressure SCBA with a rated service life of at least 30 min can be used by the fire service.

Open-circuit positive-pressure apparatus is designed so the pressure inside the facepiece, relative to the immediate environment, is positive during both inhalation and exhalation. The older-style, demand-type apparatus in which the pressure inside the facepiece, relative to the immediate environment, is positive during exhalation and negative during inhalation should not be used in structural fire fighting, as inward leakage could occur because of improper facepiece fit.

In closed-circuit breathing apparatus, the wearer rebreathes the exhalation in after the carbon dioxide has been effectively removed and a suitable oxygen concentration restored. Closed-circuit breathing apparatus must be NIOSH/MSHA certified with a rated service life of at least 30 min and operate only in a positive-pressure mode.

Self-contained breathing apparatus (SCBA) for fire service use must have a minimum rated service life of 30 min. Certification from NIOSH/MSHA for the duration of use is based on tests conducted by NIOSH at 40 L/min. In NFPA 1981, open-circuit SCBA is tested with a breathing machine at a use rate of 103 ± 3 L/min and must be able to function in the positive-pressure mode without going negative. (For U.S. unit: 1 gpm = 3.785 L/min.) Closed-circuit SCBA is tested for service life by subjects who wear the apparatus and perform prescribed activity, because the service life of such apparatus is based on metabolic use of oxygen, and no mechanical test is currently available that can simulate this type of use.

Because work performed by the user may be more or less strenuous than the work level of the test procedure, the actual service life of the SCBA is affected. During heavy exertion, the service life may be reduced as much as 50 percent.

The service duration of each unit will depend on such factors as:

1. The degree of physical activity by the user
2. The physical condition of the user
3. The emotional condition of the user, such as stress or excitement, which may increase the user's breathing rate
4. The degree of training or experience the user has had with such equipment
5. Whether the cylinder is fully charged at the beginning of use
6. The possible presence of carbon dioxide (CO_2) in the compressed air supply at levels greater than the 0.4 percent found in normal air
7. Atmospheric pressure. For example, was it used in a pressurized tunnel or caisson? At two atmospheres, the duration will be half as long as the rated service life; at three atmospheres, the duration will be one-third as long as the rated service life.
8. Condition of the breathing apparatus

Fire departments must not use SCBA with less than 30-min-rated service life for entry into hazardous atmospheres because fire fighters can get into dangerous positions without sufficient time to conduct operations or escape safely. SCBA with a rated service life of less than 30 min is not approved for entry protection for fire fighters.

The closed-circuit SCBA used by fire departments is an oxygen-rebreathing type. This SCBA has a cylinder that supplies oxygen to the wearer through a pressure-reducing valve, or regulator, and a container of chemicals to remove carbon dioxide from the exhaled air. In this apparatus, the cylinder valve opens to supply oxygen to a breathing chamber. When the user inhales, air flows from the breathing chamber through an inhalation hose to the facepiece. Exhaled air passes through an exhalation hose to a carbon dioxide absorber, where CO_2 is removed, and the air then enters the breathing chamber, where it is enriched by fresh oxygen and then rebreathed.

Breathing-Air Quality

Air supplies for SCBA should be at least Class D breathing air, as specified by the Compressed Gas Association, Inc. (CGA) in ANSI/CGA G-7.1, *Commodity Specification for Air.* Class D breathing air has less than twenty parts per million (ppm) of carbon monoxide. The water-vapor level of the air supply should be 25 ppm or less.

Fire departments should check the quality of breathing air at least every three months, whether or not they fill their own cylinders. Air samples can be tested by a public-health laboratory equipped for such tests or by a private testing laboratory.

Cylinders for SCBA may be made of steel, of aluminum, or of an aluminum shell with fiberglass wrapping. NIOSH/MSHA requires the maximum weight of the SCBA not exceed 35 lb (15.9 kg), so light-weight cylinders help reduce overall weight. Air tanks used with SCBA should be inspected periodically for external damage and must be recertified periodically through an independent pressure test.

Other Protective Equipment

Fire fighters engaged in ladder work should use life-safety harnesses that meet the requirements of Class I harnesses as defined in NFPA 1983, *Standard on Fire Service Life Safety Rope and System Components.* These harnesses are worn over the protective coat and provide a sturdy, quick-connect clamp to prevent fire fighters from falling.

Personal alert safety systems (PASS) sound an audible alarm for help if the wearer is unable to do so. PASS monitor fire fighter motion and signal an audible alarm when motion is undetected for more than 30 sec. The fire fighter can also actuate the audible alarm if he or she needs assistance. NFPA 1982, *Standard on Personal Alert Safety Systems (PASS) for Fire Fighters,* is the standard covering the design and testing of PASS devices. PASS should be worn by fire fighters whenever they operate in any hazardous area.

Rope used to raise or lower persons is termed life-safety rope and is covered by NFPA 1983. The standard provides design, construction, and testing criteria for the rope and for the associated harnesses and hardware used for rescue operations. Rope is classified as one-person or two-person.

Significant strides have been made in the development of protective clothing for use in the fire service. All members of the fire service who engage in emergency operations should be equipped with and use complete protective clothing and SCBA that meet current minimum standards at all appropriate times on the fireground or emergency scene.

BIBLIOGRAPHY

NFPA Codes, Standards, and Recommended Practices

Reference to the following NFPA codes, standards, and recommended practices will provide further information about fire department apparatus, equipment, and protective clothing. (See the latest version of *The NFPA Catalog* for availability of current editions of the following documents.)

NFPA 414, *Standard for Aircraft Rescue and Fire Fighting Vehicles*
NFPA 1500, *Standard on Fire Department Occupational Safety and Health Program*
NFPA 1901, *Standard for Automotive Fire Apparatus*
NFPA 1906, *Standard for Wildland Fire Apparatus*
NFPA 1911, *Standard for Service Tests of Pumps on Fire Department Apparatus*
NFPA 1914, *Standard for Testing Fire Department Aerial Devices*
NFPA 1921, *Standard for Fire Department Portable Pumping Units*

NFPA 1931, *Standard on Design of and Design Verification Tests for Fire Department Ground Ladders*

NFPA 1932, *Standard on Use, Maintenance, and Service Testing of Fire Department Ground Ladders*

NFPA 1961, *Standard for Fire Hose*

NFPA 1962, *Standard for the Care, Use, and Service Testing of Fire Hose Including Couplings and Nozzles*

NFPA 1963, *Standard for Fire Hose Connections*

NFPA 1964, *Standard for Spray Nozzles (Shutoff and Tip)*

NFPA 1971, *Standard on Protective Ensemble for Structural Fire Fighting*

NFPA 1975, *Standard on Station/Work Uniforms for Fire Fighters*

NFPA 1981, *Standard on Open-Circuit Self-Contained Breathing Apparatus for Fire Fighters*

NFPA 1982, *Standard on Personal Alert Safety Systems (PASS) for Fire Fighters*

NFPA 1983, *Standard on Fire Service Life Safety Rope and System Components*

Additional Reading

Arthur D. Little, Inc., *Guidelines for the Selection of Chemical Protective Clothing*, prepared for U. S. Environmental Protection Agency, Los Alamos National Laboratory, Los Alamos, NM, March 1983.

Bailey, M., "Insulated Boots-Field and Laboratory Evaluation," *Fire Technology*, Vol. 15, No. 4, Nov. 1979, pp. 283–290.

Barnett, P., "Voice Alarm Systems: Loudspeakers—Fire, Flame or Fool Proof?," Fire Surveyor, Vol. 21, No. 6, Dec. 1992, pp. 7–9.

Barnhart, A., and Swing, A., "Linking Fire Communities Through Gateway Communications," Proceedings of the First International Fire Information Conference, Cosponsored by International Fire Information Conference (IFIC) and international network for Fire Information and Reference Exchange (inFIRE), April 27–May 1, 1992, Moreton-in-Marsh, England, Society of Fire Protection Engineers, Boston, MA, 1993, pp. 101–132.

Brummer, D., "Bundeseinheitliche Brandberichterstattung–wurde das gesteckte Ziel erreicht? [Use of Modern Communication Techniques by Fire Brigades and the Hereby Arising Problems.]," VFDB, Vol. 4, Nov. 1990, pp. 158–164.

Brunacini, A. V., *Fire Command*, National Fire Protection Association, Quincy, MA, 1985.

Choppen, B., "Voice of Alarm in Public Areas," Fire Prevention, No. 266, Jan./Feb. 1994, pp. 28,30.

Coombs, P., "Planning and Good Communications are Vital for Rescues From Tunnels," Fire, Vol. 87, No. 1070, Aug. 1994, pp. 19–20.

Creasey, C. R., "Emergency Communications," Fire Engineering, Vol. 147, No. 8, Aug. 1994, pp. 40–42.

Crouch, K. G., "Fire Fighter's SCBA: Belt Radiant Heat Resistance," *Fire Technology*, Vol. 17, No. 1, Feb. 1981, pp. 39–53.

Davis, L., *Rural Firefighting Operations*, 2 Vols., International Society of Fire Service Instructors, Ashland, MA, 1985–1986.

Fire Department Aerial Apparatus, International Fire Service Training Association, Oklahoma State University, Stillwater, OK, 1991.

Fire Department Pumping Apparatus, International Fire Service Training Association, Oklahoma State University, Stillwater, OK, 1989.

Fire Service Ground Ladder Practices, International Fire Service Training Association, Oklahoma State University, Stillwater, OK, 1984.

Foltz, J. L., "Disposable Respirators for Forest Fire Fighters," *Fire Technology*, Vol. 17, No. 3, Aug. 1981, pp. 174–176.

Fritzen, B., "Stand der Informations- und Kommunikationstechnik in Leitstellen am Beispiel der Feuerwehr Koln. [State of Information and Communication Techniques in Control Centers Given by the Example of Fire Service Cologne.]," VFDB, Nov. 1993, pp. 154–158.

Grenados, M. R., Sr., "Communications and Computers: Changes Ahead Will Impact Public Safety Providers," Firehouse, Vol. 18, No. 2, Feb. 1993, pp. 32–33, 60.

Hose Practices, International Fire Service Training Association, Oklahoma State University, Stillwater, OK, 1988.

Jones, R. and Audix, T., "Voice of Warning," Fire Prevention, No. 235, Dec. 1990, p. 18.

Martin, B., "Protection for Communications Systems," Fire International, No. 134, April/May 1992, p. 38.

Mission Research Corporation, *Guide for Preparing Fire Pumper Apparatus Specifications; Part 1: Executive Summary*, FA-29-1, prepared for the Federal Emergency Management Agency, U.S. Fire Administration, Emmitsburg, MD.

Perkins, R. M., "Insulative Values of Single-Layer Fabrics for Thermal Protective Clothing," *Textile Research Journal*, Vol. 49, No. 4, 1979, pp. 202–212.

Saunders, P. B., and Ku, R., "Sequencing the Purchase and Retirement of Fire Engines," *NBS Special Pub. 411*, U.S. National Bureau of Standards, Gaithersburg, MD, Nov. 1974, pp. 201–214.

Sawyer, M. S., "Supplemental Tone Alert for Use With Crash Phone," Industrial Fire Safety, Vol. 2, No. 5, Sept./Oct. 1993, pp. 35–38.

SCBA: A Fire Service Guide to the Selection, Use and Care of Self-Contained Breathing Apparatus, National Fire Protection Association, Quincy, MA, 1981.

Self-Contained Breathing Apparatus, International Fire Service Training Association, Oklahoma State University, Stillwater, OK, 1982.

Simon, C. C., "New Mobilizing and Communication System of Hong Kong Fire Services," Fire Engineers Journal, Vol. 54, No. 172, March 1994, pp. 24–27.

Sittleburg, P. C., "Opening the Lines of Communication," NFPA Fire Journal, Vol. 88, No. 1, Jan./Feb. 1994, p. 20.

Terrett, D., "Voice Alarm Loudspeakers," Fire Surveyor, Vol. 22, No. 5, Oct. 1993, pp. 12–13.

Thompson, B., "Voice Alarm Systems: The Benefits of Integrated Voice Alarm Systems," Fire Surveyor, Vol. 21, No. 6, Dec. 1992, pp. 4–6.

Todd, C., "Voice Alarms and Evacuation," Fire Prevention, No. 274, Nov. 1994, pp. 21–25.

Veghte, J. H., *Design Criteria for Fire Fighter's Protective Clothing*, Janesville Apparel, Dayton, OH, 1981.

Vink, M., VanBrunt, M., and Mastroianni, A., "Preparedness Is Key to Surviving an Asbestos Fire: Emergency Planning and Communication Play Strong Roles," Firehouse, Vol. 18, No. 2, Dec. 1993, pp. 34–36.

Worsham, J. P., *Rural Fire Protection: A Selected Bibliography*, Vance Bibliographies, Monticello, IL, Aug. 1978.

FIRE SERVICE PROTECTIVE CLOTHING AND PROTECTIVE EQUIPMENT

Bruce W. Teele and Stephen N. Foley

Protecting fire fighters from hostile operating environments is a complex issue. Key facets in minimizing fire-fighting injuries and deaths are: adequate training of officers and fire fighters, considerable medical and physical requirements and health maintenance, a detailed and well-functioning incident management system, clear and concise standard emergency operational procedures, and state-of-the-art protective clothing and equipment that is consistently and properly used.

Much has changed in these areas during the past several decades, and more must be done in the future. Fire departments stay well informed of developments in all of these areas and provide the information and training to their personnel to keep them as current and informed as possible. Fire fighters need to realize their human limitations and the dangers they face during operations in hostile environments. They must also understand that the inherent safety of participating as a member of a team as part of well-planned operations under a safety-driven incident management system does far more for their safety than just relying on the protective clothing and equipment they use.

NFPA INVOLVEMENT

In the early 1970s, the National Fire Protection Association (NFPA) became directly involved with fire service safety issues through their technical committees and new standards. The effort began with the Technical Committee on Fire Department Equipment. This committee established a Sectional Committee on Protective Equipment for Fire Fighters to specifically address these issues, and their documents were directed toward equipment standards.

The Sectional Committee developed a new document on respiratory protective equipment, and the association membership adopted NFPA 19B, *Standard on Respiratory Protective Equipment for Fire Fighters*, on May 17, 1971. This new standard prohibited the use of filter-type canister masks for protection of fire fighters from the products of combustion, and only allowed self-contained breathing apparatus (SCBA) for fire fighter respiratory protection. This action was the direct result of fire fighters dying while wearing filter-type masks during fire-fighting operations. At least one court decision found that filter-type canister masks, and other respirators that only filtered the atmosphere and did not provide a source of noncontaminated breathing air, were inappropriate for use by fire fighters, as they did not provide protection from the known and expected respiratory hazards that fire fighters encounter during fire-fighting operations.

The second document from the Sectional Committee, NFPA 19A-T, was a tentative document that was the first to address protective clothing for structural fire fighting. Although NFPA 19A-T never became a standard, it was the basis for a future document on the same subject, NFPA 1971, *Standard on Protective Ensemble for Structural Fire Fighting*.

In 1974, NFPA restructured the Sectional Committee as a separate and full technical committee for protective clothing and equipment. This new committee, the Technical Committee on Protective Equipment for Fire Fighters, was no longer under the direction of the Fire Department Equipment Committee. It assumed the responsibility for NFPA 19B and developed a new document, NFPA 1971, which was adopted on November 18, 1975. This has been followed by a series of documents on protective clothing and protective equipment items for structural fire fighting, proximity fire fighting, wildland fire fighting, hazardous chemicals emergencies, rope rescue operations, and emergency medical operations. The committee was renamed in 1988 as the Technical Committee on Fire Service Protective Clothing and Equipment.

Because of the continuing increase of the types of fire-fighting clothing and equipment being addressed by the committee, and areas other than fire fighting that the committee was involved in, NFPA created a Project on Fire and Emergency Services Protective Clothing and Equipment in 1994. The project is structured with a Technical Correlating Committee managing the project and seven technical committees now responsible for emergency medical services, hazardous chemicals, respiratory protection and personal alarm devices, special operations, specialized fire-fighting applications, structural fire fighting, and wildland fire-fighting protective clothing and equipment. Within this project, there are currently 14 active documents and several more under development or in the planning stages.

In 1973, NFPA created the Technical Committee on Fire Service Occupational Safety and Health. This was the first committee to focus on the issues of safety and health of fire fighters instead of just the clothing and equipment. In its early years, the committee considered many issues, and then wrote a document, NFPA 1501, *Standard for Fire Department Safety Officer*, that was adopted on May 18, 1977.

The committee produced NFPA 1500, *Standard on Fire Department Occupational Safety and Health Program*, which became

Bruce W. Teele is a senior fire service safety specialist at NFPA. Stephen N. Foley is a senior fire service specialist at NFPA.

effective on August 7, 1987. NFPA 1500 was developed as an overall umbrella document, encompassing an entire fire department safety and health program.

NFPA 1500 became the first nationally recognized consensus standard to address fire service occupational safety and health. It was the first national document on safety and health to be written by the fire service and fire service interest groups for the fire service. Prior to NFPA 1500, any safety and health measures had to be based on general industry standards and regulations that either did not adequately address the fire service problems, or were difficult to impose on a nonindustry agency.

The second edition of NFPA 1500 became effective on August 14, 1992. NFPA 1500 is the "user's" requirements as it has the criteria for fire departments and fire fighters, rather than the product requirements that manufacturers must meet. NFPA 1500 is the document that makes it happen, as it requires fire departments to have, and fire fighters to use, protective clothing and equipment.

NFPA 1500 requires protective clothing and protective equipment appropriate for the intended application. The basis of the requirements is to provide for the protection of fire fighters who are responsible for various fire-fighting and other emergency operations. The fire department is required to provide the appropriate protective clothing and protective equipment for the task and operation.

The fire department is also required to ensure that the protective clothing and equipment is properly utilized, inspected, and maintained. Accordingly, each fire fighter is to be provided training in the use and limitations of protective clothing and equipment.

OSHA INVOLVEMENT

In addition to NFPA 1500, in the United States there are federal and state regulations, generally under the federal Occupational Safety and Health Administration (OSHA) or the state equivalent, that include protective clothing and equipment as well as other safety and health issues.

OSHA first addressed fire service occupational safety and health in 1980 when they issued 29 CFR 1910.156. Although the federal OSHA regulations did not apply to state and local government employees, states that have a federally approved state OSHA plan were, and still are, required to cover state and local government employees.

However, when U.S. OSHA issued 29 CFR 1910.120 on hazardous materials emergency response, the U.S. Congress directed the Environmental Protection Agency (EPA) to issue identical regulations that would apply to state and local government employees as well as volunteers in all 50 states. 29 CFR 1910.120 includes a mandatory reference requiring fire departments to be compliant with 29 CFR 1910.156.

Therefore, where fire departments also respond to hazardous materials incidents, the OSHA/EPA regulations apply to all fire departments in the United States.

THIRD PARTY CERTIFICATION

All NFPA standards on fire and emergency services protective clothing and equipment require that the item be certified by an independent third-party certification organization. All items of fire and emergency services protective clothing and equipment that are claimed to be compliant with the appropriate NFPA standard *must* carry the label, symbol, or other identifying mark of that certification organization.

Any item that does not bear the mark of an independent third-party certification organization is not compliant with the appropri-

ate NFPA standard, even if the garment or product label states that the item is compliant.

Third-party certification is an important means of ensuring the quality of fire and emergency services protective clothing and equipment. If purchasers, or would-be purchasers, have any questions concerning whether an item is properly certified, labeled, and listed, they should be sure to require the appropriate evidence of certification from the manufacturer *before* the purchase is made. Purchasers should also contact the third-party certification organization to obtain their "listing," the certification organization's document identifying which products have been tested and found to be in compliance with which standards, to verify that the protective clothing or equipment item has been "listed."

After purchase, questions regarding the certification of any protective clothing or equipment item can be addressed to the indicated certification organization.

Details of the certification and the labeling of protective clothing or equipment can be found in Chapter 2, or in some documents in Chapter 2 and 3, in the applicable NFPA standard.

STRUCTURAL FIRE-FIGHTING PROTECTIVE CLOTHING AND EQUIPMENT

The use of "full protective clothing and equipment," compliant with the respective NFPA standard, for structural fire fighting is required for all structural fire-fighting operations. This includes the protective coat, protective trousers, and protective hood, helmet, gloves, protective footwear, self-contained breathing apparatus (SCBA), and personal alert safety system (PASS) devices.

NFPA 1971, first issued in 1975, was developed to improve the protection offered to fire fighters through their protective clothing when worn in the course of performing fire-fighting and rescue operations in buildings, enclosed structures, vehicles, vessels, and other similar properties.

NFPA 1971 was followed by three other standards on other elements of structural fire-fighting protective clothing and equipment:

NFPA 1972, *Standard on Helmets for Structural Fire Fighting* (first issued in 1979)

NFPA 1973, *Standard on Gloves for Structural Fire Fighting* (first issued in 1983)

NFPA 1974, *Standard on Protective Footwear for Structural Fire Fighting* (first issued in 1987)

In February of 1997, these four standards were combined into a document covering all elements, other than SCBA, of a structural fire-fighting protective ensemble. This document is still numbered NFPA 1971, but with the new title of *Standard on Protective Ensemble for Structural Fire Fighting*.

PROTECTIVE GARMENT ELEMENTS

All protective garments for structural fire-fighting operations must meet the garment element requirements of NFPA 1971.

Protective garments are the coat, trouser, or coverall elements of the protective ensemble that are designed to provide minimum protection to the upper and lower torso, arms, and legs, excluding the head, hands, and feet.

Selection of protective garments should ensure that fire fighters are protected from the hazards of the work environment to which they may be exposed. The protective garments should be suitable for tasks the fire fighter is required to perform, be selected to enhance the mobility of the fire fighter, and be viewed as one element in an ensemble, since each element must interface successfully with the other elements. It must be remembered that structural fire fighter protective garments are designed to protect fire fighters against the hazards associated with structural fire fighting. The performance requirements for protective garments are not intended to establish the limitations of the fire fighter's working environment, but are intended to establish material performance requirements based on empirical field data and laboratory testing. While performing structural fire fighting, users must be aware that the feeling of a continual increase in heat is an indication that the protective garments may be nearing their maximal capability and injury may be imminent.

Protective trousers are considered an integral part of the protective system. Protective trousers provide greater protection than any combination of clothing that includes only a protective coat and boots. Three-quarter boots, even when fully "pulled-up," provide significantly less thermal protection for the fire fighter's lower torso. For this reason, NFPA 1500 requires that protective trousers must be worn by fire fighters when they are engaged in or exposed to the hazards of structural fire fighting.

Proper fit of protective garments has a direct relationship to the safety of the fire fighter. Prior to manufacturing, the protective garment manufacturer should measure the individuals to ensure a proper fit of the protective garments. This is important since an overlap of at least 2 in. (51 mm) of all the required layers of the coat and trouser must exist to avoid any gapping of total thermal protection. In order to ensure proper overlap protection, each fire fighter should be measured in two specific positions while wearing the protective coat and protective trousers without any belt, waist strap, or SCBA. The first position is the wearer standing erect with hands together and reaching overhead as far as possible. The second position, while standing with hands together reaching as far overhead as possible, the wearer bends forward, to each side, and to the back as much as possible. The 2-in. (51-mm) overlap must be maintained in each of these positions.

Particular attention needs to be paid to fire fighters with longer- or shorter-than-average upper or lower torsos, and to fire fighters with longer- or shorter-than-average arm lengths. Proper sizing for these persons is very important to ensure that they will be properly protected. The first two sizing methods discussed below probably will not allow for the proper fitting of these persons, as a good chest fit does not provide for other than the "average" arm or torso length. Waist size does not provide for other than the "average" lower torso length. Custom sizing is the most desirable method.

Sizing of protective garments can be done through several methods. One is the system utilized in standard clothing, where numerical sizes are provided based on increments of chest measurements, such as 36, 38, 40. Numerical sizing is the most popular sizing system, because it can give effective fit in many situations and has a moderate cost factor. Another method uses S/M/L/XL/XXL notation where each size combines a larger increment of chest measurements. For example, a size "medium" is designed to fit 42 through 45. The S/M/L/XL/XXL system is the most economical, but least accurate sizing method. The third method is known as custom sizing, where the individual fire fighter is measured by the manufacturer's representative or according to the manufacturer's instructions and a specific garment is produced to these measurements. The custom sizing obviously allows for superior fit, especially with fire fighters who are shorter or taller than average or who have shorter or longer than the average arms or torsos.

Performance and Testing

NFPA 1971 performance requirements are designed to ensure that minimum levels of flame and heat resistance, thermal insulation, durability, water and viral penetration resistance, barrier tear strength, washing shrinkage, and water absorption are met. Design requirements for construction features are also included.

The most important of these requirements specifies performance for the protective qualities of the fabrics utilized in the construction of the outer shell, moisture barrier, and thermal barrier, including flame resistance, heat resistance, thermal shrinkage, thermal insulation qualities of composites of fabrics used in the various layers, and the viral penetration resistance.

Flame resistance of protective garment fabrics is measured by Method 5903.1 of Federal Test Method Standard 191A, *Textile Test Methods*, which exposes a vertical strip of fabric to a gas flame. Afterglow, or the time that the fabric continues to glow after the flame is extinguished, must be less than an average of 2.0 sec for the fabric to pass the test. Afterflame, or the time that the fabric continues to burn after the flame source is extinguished, must be less than 2.0 sec for the fabric to pass the test. If any of the burning fabric melts and drips during the test, the fabric fails. The charred fabric is measured to establish a char length, which must be less than 4 in. (102 mm).

Heat resistance of protective garment materials is measured by exposing the fabrics to 500°F (260°C) in a forced-air oven for a period of 5 min. The fabrics must not melt, separate, or ignite.

Thermal shrinkage is measured by exposing the fabric to 500°F (260°C) in a forced-air oven for a period of 5 min. In this test, the fabric must not shrink more than 10 percent in all directions.

Thermal protective performance (TPP) provides a laboratory test method to quantitatively evaluate materials and composites of fire- and heat-protective garments. The thermal insulation evaluation of structural fire-fighting protective garments had traditionally been based on subjective field evaluations by individual fire departments, antidotal information, or data provided by manufacturers and fabric suppliers.

This TPP test method measures the amount of heat transfer through a fabric or composite layers of a protective garment that is challenged by a significant thermal exposure. The amount of heat transfer that results in second-degree burn damage to the wearer's skin is related to the time that adequate protection is afforded. Materials with poor protective qualities transmit heat rapidly, while materials that are resistant to heat and are good insulators provide protection for a longer period.

As a result of this test methodology, NFPA 1971 substituted TPP for the thickness requirement. A minimum TPP value of 35 is utilized, based on laboratory examination of garments that met the minimum thickness requirements for thermal insulation and provided effective protection during emergency conditions actually experienced by fire fighters in the field. The TPP performance was then substantiated by laboratory testing.

The TPP test apparatus consists of two gas burners to provide a convective/radiant heat load, and a bank of nine infrared quartz lamps to provide a radiant source. A calibrated heat flux of 2.0 cal/cm^2/sec, similar to that produced during exposure to severe conditions of a flashover, is generated. The rise in heat transfer through the fabric(s) is measured by a copper calorimeter and recorded. When the temperature reached is equivalent to a second-degree burn, the time of exposure is recorded. This time, in seconds, when multiplied by the heat flux, determines the TPP value. For example, a garment ensemble that provides protection for 17.5 sec., multiplied by the heat flux of 2.0 cal/cm^2/sec, yields a TPP value of 35.

Protective qualities of garments and the resultant TPP values measured are related to several properties:

1. Resistance to degradation from heat of all fabric layers
2. Resistance to hot gas penetration through garment layers

3. Weight of the fabric composite
4. Thickness of each of the layers
5. Amount of air trapped within and between the fabric layers

If all other factors are equal, increasing any of the above will yield a higher TPP value. For example, aramid is more heat resistant than cotton, but the heavier weight of cotton outer shells yields similar TPP ratings for both.

Increasing the weight and hot gas penetration resistance of the outer shell by coating it with neoprene will increase the TPP value. A heavier neoprene coating on the moisture barrier will increase weight and TPP value. Thicker thermal barriers will yield a higher TPP than thinner barriers of the same weight, while barriers that are both thicker and heavier will provide even higher TPP values. In addition, a three-layer system will provide a somewhat higher TPP than a two-layer system that is equivalent in all other factors.

However, TPP data can be misleading. For example, TPP testing of materials that do not meet the flame resistance and afterflame requirements of NFPA 1971 will not be representative of true protective performance. Assemblies that ignite and burn can give a TPP value that appears to be good; but since the fabric will continue to burn, the TPP value recorded is meaningless.

A high TPP value does not necessarily equate to a better garment. A garment with an extremely high TPP value could be detrimental to the safety of the fire fighter, due to other factors, e.g., greater heat stress, less freedom of movement, and poorer performance characteristics. A hot, heavy garment must be considered a factor in exacerbating stress, the number one killer of fire fighters.

Another important factor in reviewing and comparing TPP results is the warning time. This is the time period between when pain is felt and the burn occurs. The longer the warning time, the more opportunity there is to reduce the exposure to the heat or to escape before being burned. Independent laboratory investigations of TPP performance found protective garments with nonbreathable moisture barriers had similar or higher TPP ratings than breathable systems (depending on the wet/dry conditions of the outer shell and thermal barrier), but provided less warning time; systems without moisture barriers provide significantly lower warning times. A garment with a lower TPP may allow more sensitivity to heat penetration, providing earlier warning for the wearer, but a warning time period of less duration. Excessive protective envelopes can also lead to a false sense of security for fire fighters and could cause them to over extend themselves, possibly leading to severe injuries.

In summary, the substitution of TPP for a thickness requirement in NFPA 1971 has been extremely beneficial in the development of lighter-weight fabrics. However, TPP is simply one of many test requirements contained within NFPA 1971, and must not be taken out of context with generalized interpretations, such as the higher the TPP the better the garment. Table 10-18A demonstrates the conditions found in structural fire fighting.

Viral penetration resistance testing evaluates the garment's resistance to the penetration of body fluids and bloodborne pathogens that the fire fighter can be exposed to during rescue functions. The garment composite materials, including seams, are exposed to a bacteriophage suspension of a liquid inoculated with *E. coli* C bacteria for 1 hr. Materials are tested in both directions and cannot allow any penetration of the bacteriophage.

Heat Stress

Test methods suitable for evaluating materials used in fire fighter protective garments relative to their contribution to heat stress have been investigated by the committee. A test methodology was developed and standardized based on the heated sweating flat-plate apparatus. A calibration procedure was established to assess comparability of fabric measurements of insulation and permeability from different laboratories. Four independent laboratory groups (Comfort Technology; W. L. Gore & Associates, Inc.; Kansas State University; and North Carolina State University) have used the test method and its calibration procedure in evaluating the insulation and evaporative heat transfer of four, three-layer composites representing a range of available structural fire-fighting protective garment ensembles.

A simple criterion—the Watts/m^2 of heat transferred through the composite by the combined dry and evaporative heat exchanges from a 95°F (35°C), fully sweating test plate surface in a 77°F (25°C), 65 percent RH environment—provides a single number for comparing each fabric ensemble.

The actual differences in garments made up from these fabric ensembles will be smaller, because the insulation of the garments will be greater due to the air layers in garments that are not accounted for on the flat test plate.

Garment Layers

Protective coats and protective trousers must consist of a composite of three functional layers: (1) outer shell, (2) moisture barrier, and (3) thermal barrier. The composite may be configured as a single layer or as multiple layers. Each has a specific function that must meet minimum requirements if adequate protection is to be afforded to the fire fighter. Among the most important choices to be considered in providing proper protective garments are the types of fabrics to be used for the outer shell, moisture barrier, thermal barrier, and winter liner (if utilized). Only fabrics that meet the minimum requirements of NFPA 1971 have been included in this discussion.

The outer shell is the exterior layer of fabric on a protective garment that forms the outermost barrier between the fire fighter and the environment. The outer shell is a flame- and heat-resistant barrier that provides a first line of defense against the external environment and protects the moisture barrier and thermal barrier from physical damage. In addition to being flame and heat resistant, the outer shell must also be resistant to abrasion, tearing, and puncture. Water repellency is also a desirable characteristic for outer shell fabric.

TABLE 10-18A. *Thermal Classification—Structural Fires*

Class I condition occurs in a room or enclosed area that has been involved after the fire has been "knocked down." Environmental temperatures up to 104°F (40°C) and thermal radiation up to 0.05 Watts/cm^2 are encountered for up to 30 min.

Class II condition occurs when a small fire is burning in a room or enclosed area. In this case, environmental temperatures from 105° to 203°F (41° to 95°C) and thermal radiation from 0.05 to 0.100 Watts/cm^2 are encountered for up to 15 min.

Class III condition occurs when a fire in a room or enclosed area is more completely involving the contents. Environmental temperatures from 204° to 482°F (96° to 250°C) and thermal radiation from 0.100 to 0.175 Watts/cm^2 are encountered for up to 5 min.

Class IV condition occurs during ignition of fire gases, a backdraft, or flashover in a room or enclosed area; where environmental temperatures from 483° to 1500°F (251° to 815°C) and thermal radiation from 0.175 to 4.2 Watts/cm^2 are encountered for about 10 sec.

The moisture barrier is that garment layer used to prevent the transfer of water from the outside environment to the fire fighter's body. The outer shell cannot be utilized as the moisture barrier. Garments constructed with separate moisture barriers and thermal barriers provide more effective protection and better durability than combination moisture barrier/thermal barriers. Moisture barriers are currently constructed of fabrics coated with or laminated to waterproof materials. All types of moisture barrier materials require seam sealing to provide a continuous water barrier.

The thermal barrier is that layer of the protective garment that separates the moisture barrier from the body and provides insulation from heat. An incidental effect of this insulation is to provide a measure of insulation from cold. A thermal barrier should not be confused with a optional winter liner, which is a removable liner that can be fastened into the protective garment for added insulation from cold. The thermal barrier must be utilized at all times, even in hot weather, to provide adequate thermal protection to the fire fighter from fire conditions.

Winter liners can be extremely useful in areas where winter temperatures are commonly below freezing. The most commonly utilized configuration consists of a waist-length sleeveless vest that is snapped to the thermal barrier. Winter liners must meet the same thermal and flame-resistance requirements as the thermal liner.

Garment Components and Materials

Fabrics and other components utilized by the garment manufacturers are generally obtained from the same sources. Finished garments differ due to the type of components chosen as well as other factors such as fit, which is directly related to the quality of the manufacturer's design and patterns, and care in construction techniques (i.e., cutting, sewing, and seam finishing).

Patterns utilized in protective garments are one of the most critical elements in producing effective garments. In general, each manufacturer has several basic patterns for each type of garment. Variations in such areas as sleeve design, number of pieces utilized in each garment component, and seam allowances vary among manufacturers. Basic pattern design should always be specified to ensure that manufacturers are preparing bids based on similar construction techniques. For example, a coat shell with a three- or four-piece body provides a superior fit, but will usually be more expensive to produce than a one-piece coat body.

The collar is the component that provides protection to the neck and throat. The shape and size of the collar is critical to the amount of coverage that occurs as well as the degree of comfort to the wearer. In general, a collar should be a minimum of 4.0 in. (102 mm) in height and cut to allow for ease of movement. Taller collars can interfere with head movement when used with modern helmets, which are equipped with integral ear protectors.

The design of the collar closure is also critical. The closure strap should provide effective protection to the neck and throat and prevent debris and water from entering the top of the storm flap when closed, as well as allow easy and effective adjustment with the gloved hand. It should interface effectively with the SCBA utilized by the fire department. Hook-and-pile fastening tape is considered to be the most effective system, as it is very easy to fasten and adjust with the gloved hand.

Protective coat cuffs are an area of high abrasion and tend to wear out far sooner than the rest of the garment. For this reason, reinforcing material is often added to the cuff area. Protective coat wristlets should be designed to provide protection to the wrist interface area. Several configurations are utilized. NFPA 1500 requires that protective coats have wristlets meeting the NFPA 1971 interface component requirement. They may be constructed as a resilient wristlet secured through a thumb opening (over-the-palm) and allowed to be used with a glove of the wristlet or gauntlet type, or a nonsecured wristlet that can only be used with a wristlet glove. Knit wristlets alone provide little protection from water. For this reason, a combination of an inner knit wristlet and outer-coated wristlet with an elastic strip to firmly grip the wrist provides better protection.

The protective coat closure should be designed to provide secure protection from steam and water when the coat is worn, and must allow freedom for leg movement. In general, this is accomplished through the use of a storm flap constructed with a waterproof and insulated layer to ensure that water and heat do not penetrate to the body. The storm flap can be fastened by use of hooks and dee rings, snaps, zipper, hook and pile tape, or combinations thereof. The traditional closure system for protective coats is an outer storm flap fastened with hooks and dee rings. Hooks and dees are durable and reliable; quick and easy to align, fasten, and unfasten; and resist soiling and freezing. Hooks must be of the reverse inward-facing type to resist hooking onto fences, bed springs, wires, and other objects encountered during structural fire fighting.

The use of hook and dees alone can allow the passage of water under some conditions; therefore, additional backup closures are often used to provide increased protection. This is usually accomplished by placing hook-and-pile tape, snaps, or a zipper on the inside of the storm flap.

Hook-and-pile tape is quick to fasten and unfasten and provides a water-resistant closure. Soiling can compromise the security of the fastening tape, thus care should be taken to clean the tape regularly in accordance with the manufacturer's instructions. Snaps provide a less water-resistant backup closure and take longer to align and fasten, but can also be easily and quickly released. Zippers can also be used as a backup to hooks and dees. However, zipper closures are difficult to operate on coats longer than waist-length.

The protective trouser closure should be designed so as to remain fastened during all forms of vigorous physical movement. Combinations of snaps, buttons, zippers, straps and buckles, hook-and-pile tape, and hooks and dee rings are utilized in trouser closures. The use of a hook and dee assembly ensures ease of alignment and security, while the hook-and-pile tape provides a water-resistant seal. A strap and buckle can be added to the hook and dee assembly to allow a range of size adjustment. Proper design and complete overlap of the fly closure of the moisture barrier/thermal barrier ensemble is essential to ensure a water-resistant and protective barrier for the crotch area.

Pocket selection is one that varies from department to department and depends on such factors as type of work done and type of tools carried. Although special pockets for items, such as SCBA facepiece, radio, tools, etc., may be added to the coat, the increase in weight carried and the resulting increase in heat stress to the wearer should be considered. The minimum amount of ancillary tools and equipment should be carried and careful consideration be given to the actual need *vs.* the "image," *vs.* the added stress to the fire fighter.

Stitching for all seams is specified by NFPA 1971 to ensure that adequate seams and stitching are utilized in the specified garments. Points of stress and strain in a protective garment should be reinforced with bar tacks, such as the ends of coat pocket flaps and upper pocket corners, ends of storm flaps, fly flaps, and attachment points of snap tabs at trouser cuffs. Thread must be compatible with the material on which it is used.

The color of the protective coat and trouser has little effect on the protective qualities of the garment; however, there are several factors that should be considered in selecting an outer shell color, including visibility, radiant heat absorption, cleaning, and identification. Yellow, white, and natural provide better visibility. The least visible colors are black and olive drab. Even though fluorescent and retroreflective trim is required on protective garments, the selection of a light-colored garment can provide an extra margin of safety, especially where a fire fighter is not directly illuminated by a light beam. Light-colored shells, if clean, will be cooler than dark shells when worn in direct sunlight. Lighter colors also allow easier inspection of the garment for soiling, such as oil and soot (carbon), which can reduce the fire-resistant and water-resistant properties of the outer shell.

Fluorescent and retroreflective trim is used on protective garments for better visibility. Retroreflective properties mean that, when visible light rays strike the tape, they are reflected back in the direction of the incoming ray, causing the tape to appear to glow in the dark. Fluorescent properties provide more effective daytime visibility because, when the dyes in the trim are struck by sunlight, the ultraviolet (UV) light rays are absorbed and re-emitted in the visible spectrum, causing a brilliant and highly visible color response.

Use of vertical trim on garments has been prohibited, since it is capable of causing severe burns to the wearer, due to the heat that is generated and transmitted through the garment while the trim is burning vertically. This can also be detrimental to the performance of SCBA.

Cleaning and Maintaining Protective Garments

Before any cleaning of fire-fighting garments, the manufacturer's instructions must be followed. The manufacturer provides specific cleaning instructions and these must be consulted and followed, since failure to do so may reduce the protective qualities of the garments or negate warranties.

The cleaning of structural fire-fighting protective garments should be conducted periodically to remove foreign and flammable contaminants. Periodic cleaning should prolong the life of the protective garments and help maintain the performance. NFPA 1500 requires that clothing be cleaned at least every 6 months by the fire department, by either utilizing a cleaning service or a fire department facility equipped to handle contaminated clothing.

Cleaning must recognize that the garment is composed of a combination of materials. Each of these materials has its own unique characteristics, capabilities, and weaknesses. Even fabrics inherently flame resistant can have this characteristic negated by improper care and use. The amount of soil and the cleaning procedures and chemicals utilized can adversely affect a fabric's performance. Thus, procedures utilized to clean such materials must protect the weakest of the materials. Historically, fabric suppliers' care recommendations have failed to consider the other materials that comprise the garment. Many recommendations, while appropriate for the supplier's own material, would destroy the garment because of their adverse impact on the other components.

Clean protective garments reduce safety and health risks, and it is recommended that garments be cleaned frequently to reduce the level of, and bodily contact with, contaminants. User agencies should establish guidelines for frequency and situations for garment cleaning. For gross contamination with products of combustion, fire debris, or body fluids, removal of contaminants by flushing with water as soon as practical is necessary, followed by appropriate cleaning.

Decontamination may not be possible when protective garments are contaminated with chemical, radiological, or biological agents. When decontamination is not possible, garments should be discarded in accordance with local, state, and federal regulations.

Chlorine bleach must *not* be utilized for protective garment cleaning and/or disinfecting. Some components of these garments lose their physical integrity on exposure to chlorine bleach. Other components will actually lose their flame-resistant properties and thermal insulation, on exposure to chlorine bleach. In either case, the protection provided by the garment will be compromised.

There are industrial cleaning products and facilities available for protective garments that the user may wish to investigate. The protective garment manufacturer should be contacted directly for additional information. The original protective garment manufacturer should always be contacted for specific instructions for cleaning various items of protective garments, as the manufacturer knows what materials and components went into the construction of the garment. Not consulting the original manufacturer or following the manufacturer's cleaning instructions could result in damage to the protective garment, or loss of some or all of the protective properties.

Only industrial front-loading washing machines should be used to provide satisfactory cleaning of the garments and to reduce the possibility of cross-contamination. Industrial machines provide for several rinse (extraction) cycles that are more efficient in removal of the contaminants and flushing them from the wash tub so they are not retained on the fabric. Protective clothing should be washed separately from other garments. All hooks and dees should be fastened and the garment turned inside out. A stainless-steel tub should be utilized, if available.

The fire department must maintain protective clothing only in accordance with manufacturer's instructions. Maintenance should include regular inspection, proper repair, and retirement when appropriate. Storage in direct sunlight will substantially reduce the useful life of the garment and reduce protective qualities that might not be visually apparent. Color fading or color change often indicates that the garment has experienced UV degradation. All protective clothing that is retired should be destroyed, rather than using it as spare sets or giving it away to another organization, as the protective qualities have been degraded or eliminated.

The trim should also be inspected to ensure effective performance under daytime and nighttime conditions. It is important that high visibility be maintained at all possible orientations to the light source. A flashlight in a dark room can be utilized to inspect retroreflectivity, comparing trim on the garment to a new specimen of the utilized trim.

PROTECTIVE HELMET ELEMENTS

All protective helmets for structural fire-fighting operations must meet the helmet element requirements of NFPA 1971. Helmets that meet all requirements of this standard must be certified, labeled, and listed by an independent third-party certification organization.

The original function of the helmet was to provide protection to the head and to prevent hot water and embers from reaching the ears and necks. This was accomplished by a design featuring a reinforced crown and an extended back brim.

Older helmet designs were traditionally limited in providing impact resistance. A suspension-type harness, typically made of cloth and synthetic straps, was designed to distribute the force from a blow over the head as evenly as possible for blows near the top of the helmet, but side blows could cause serious impact. Borrowing from transportation and sports equipment, the use of an energy-absorbing liner was developed to overcome the limitations of the web suspension system. This liner is designed to absorb the impact energy and distribute the force over the entire surface of the liner, regardless of where the impact occurs on the helmet.

Structural fire-fighting helmets must protect the head from impact and penetration of flying objects, as well as from heat, flames, electrical shock, and burns. Helmets are constructed of high-temperature resins incorporating materials such as Kevlar®, PBI®, or glass fiber; a variety of plastics; and leather.

There are factors that should be considered in evaluating and purchasing helmets. The helmet should divert liquids and debris from the face and neck of the fire fighter, and should not interfere with the fire fighter's ability to hear. Additionally, the neck must be afforded sufficient mobility so as to allow performance of fire-fighting tasks.

The weight of the helmet is also extremely important for comfort, mobility, and physiological stress. There is a dramatic physiological penalty demonstrated for additional helmet weight. Since the head remains the primary region for metabolic heat dissipation, consideration should be given to helmet designs and suspension systems that increase the evaporative heat loss potential to increase heat dissipation from the head. State-of-the-art composites and helmet designs offer improved head protection plus significantly lighter-weight helmets for fire fighters. Such designs play a significant role in reducing fire fighter's heat stress.

NFPA 1971 contains eight major performance requirements. The standard requires that a helmet element must consist of a helmet shell, energy absorption system, retention system, fluorescent retroreflective markings, ear protectors, and a faceshield or goggles or both; and must provide minimum protection to the head from impact, electric shock, penetration, heat, and flame.

Performance and Testing

Top impact force testing measures the impact-absorbing ability of the helmet's suspension system. The helmet is placed on a model head that is mounted over a load cell that measures transmitted force. An 8-lb (3.6-kg) steel mass is dropped onto the top of the helmet from a height of 5 ft (1.5 m). No helmet must transmit a force of more than 850 lb (3,780 N). This is equal to about 20 percent of the applied test force.

Impact acceleration testing measures the amount of acceleration that is transmitted through the helmet to the head. For this test, a drop assembly, consisting of a model head, guidance assembly, and accelerometer mounted on a test stand, is used. The helmet is mounted on the model head and dropped onto a flat, test anvil mounted on a steel plate. The accelerometer on top of the assembly measures acceleration-time duration values, peak acceleration, and impact velocity. The maximum acceleration permitted is 150 Gs on top, and 300 Gs on the front, sides, and back (a "G" is the force of gravity or 32.2 ft per sec).

Penetration resistance measures the helmet's ability to resist puncture from falling objects. The helmet is placed on a model head and a 2.2-lb (1-kg), sharp plumb bob striker is dropped on the helmet. The point of the striker and the top of the model head are both electrically connected to a contact indicator. If electrical contact is made between the striker and the headform when the striker penetrates the helmet, the helmet fails the test.

Heat resistance testing ensures that the helmet will not distort when exposed to heat. Sample helmets are placed in a circulating-air oven at approximately 500°F (260°C) for 5 min. The shell distortion is then measured. No part of the shell can touch the model head. The distortion in the back of the helmet cannot exceed 1.6 in. (41 mm) below the original position of the helmet. The distortion of the helmet on the front and sides cannot exceed 1.2 in. below the original position of the helmet. There can be no separation, melting, or dripping of the retention system, energy-absorbing system, or ear covers. There can be no ignition of any part of the helmet assembly. No part of the faceshield that was not below the brim line before the test can sag below the brim line.

Flame testing measures the resistance to ignition and sustained burning of the helmet materials. There are three flame-resistance tests. In the first test, a Bunsen burner flame is applied for 15 sec to the outer edge of the helmet shell and to the front, sides, and rear. The helmet cannot show any afterflame or afterglow 5 sec after the removal of the flame. In the second test, the area of the helmet to be tested is exposed to a radiant heat load and the top of the inner cone of a Bunsen burner flame for 15 sec. The helmet cannot show any visible afterflame or afterglow 5 sec after removal of the flame. The third test is for the faceshield, and it cannot show afterflame or afterglow 5 sec after removal of the flame.

Electrical insulation testing measures the helmet's ability not to pass electricity from the helmet to the wearer. An inverted helmet is filled with tap water and then submerged in water to the same level. A 60-Hz ac voltage is applied and increased to 2,200 V. This voltage is maintained for 1 min. Leakage through the helmet shell cannot exceed 3.0 milliamperes. There is a second electrical insulation test where the helmet is in the "as worn" position, mounted on a test headform with the same electrical current and performance as in the first test.

Retention system testing measures the ability of the helmet to remain on the wearer's head. There are two tests. In the first, the chinstrap is placed on a stretching test machine and the force is slowly increased to 100 lb (445 N). The 100-lb (445-N) force is maintained for 60 sec. The chinstrap cannot break or stretch more than 0.8 inches (4.1 mm) when tested. In the second, a 10-lb (44.5-N) pull force is exerted on each attachment point of the crown straps. The strap system must not separate from the helmet.

Ear cover testing measures the flame resistance of ear covers. Textile materials are exposed to an open flame in accordance with Method 5903.1 of Federal Test Method Standard 191A, *Textile Test Methods.* Material must have an average char length of not greater than 4.0 in. (102 mm) with an afterflame of not greater than 2.0 sec.

Visibility is addressed by requirements for fluorescent retroreflective trim on each of the four sides of the helmet.

When specific danger to the eyes is present, such as while operating saws or other tools or pulling ceilings, primary eye protection is required. Primary eye protectors must meet the U.S. OSHA regulation 29 CFR 1910.133 and ANSI Z87.1, *Practice for Occupational and Educational Eye and Face Protection.* Primary eye protection may be provided by safety goggles or SCBA facepieces. The faceshield or goggles affixed to the helmet are not intended to be used as primary eye protection, but are required by NFPA 1971 because they provide partial face and eye protection when specific eye hazards are not anticipated, but unexpected exposures could occur given the nature of emergency scenes.

Cleaning and Maintenance of Helmets

Before any cleaning of the helmet shell or components, the manufacturer's instructions must be followed. The manufacturer provides specific cleaning instructions, and these must be consulted and followed since failure to do so may reduce the protective qualities of the helmet or negate warranties.

Cleaning of the helmet shell is dependent on the shell material. Generally, the cleaning of resin shells utilizing glass fiber, PBI®, or Kevlar® is accomplished by using mild solvents, such as ethyl alcohol, mild detergent and water, mild abrasives, industrial cleaners (such as tar remover), acetone, or paint thinner.

Polycarbonate shells can be cleaned using mild soap and water, or rubbing alcohol. *Do not use solvents or cleaning agents.* Removal of light scratching and/or smoke-staining can be accomplished using a fine, nonchlorinated scouring compound. Use of polishes should be avoided, unless approved by the manufacturer.

Faceshields and goggles are considered to be replaceable items and should be replaced periodically or when their lenses become scratched or damaged so that vision is impaired. Faceshield or goggle lenses should be removed prior to shell cleaning. To clean the faceshield or goggles, use mild cleaning agents, such as ethyl alcohol or a mild detergent and water, and a soft sponge or cloth. *Do not use abrasives, solvents, paint removers, acetone paint or lacquer thinner, or any other chlorinated organic solvents.* Removal of smoke stains can be achieved through the use of scratch removers and polishes available from the manufacturer.

Cosmetic shell repair or shell repainting should not be done without explicit permission from the manufacturer.

The helmet should be inspected utilizing the following procedures. The manufacturer should be contacted for specific inspection techniques for their particular helmet.

1. Inspect helmet visually each time it is worn. Note frayed straps, loose faceshield pivots, and other obvious conditions that require correction. After any major exposure to fire, intense heat, impact, or chemicals, and at least every three months, a complete inspection of the helmet should be conducted. Check all mechanical hardware for proper adjustment regularly. All screws and nuts must be tight. The faceshield clutch mechanism should be lubricated after every fire and kept in proper tension. Chinstrap snaps and buckles should be checked for corrosion.
2. Examine the outside of the shell for blistering or cracking. Cracking is most likely to appear in the vicinity of the faceshield bracket mounting holds. Any cracks that appear both inside and outside the shell may require shell replacement. Voids or "dead spots," bubbling, or roughing of the helmet surface indicates heat damage and the need for shell replacement. Areas that have lost their gloss may have been subject to chemical attack. If extensive, consult the manufacturer or replace the shell. Check the bead trim around the brim; replace if it is blistered, distorted, cut, or frayed. The bead trim performs the function of protecting the edge of the brim from chipping and protects against cuts by the sharp edge it covers. Inspect fluorescent retroreflective trims visually and replace if they are abraded, missing, or no longer reflective.
3. Blistering, wrinkling, and/or swelling of the inside or outside surface are indications of heat damage of helmet impact liners. Often swelling of a damaged impact cap will be visible even without removing it from the shell. Dents or cracks on the surface of the impact cap indicate impact damage. In either case, replace impact cap immediately.
4. Carefully inspect straps and other suspension system parts for signs of fraying or damage. Check bar tacks for ripped stitches. Heat damage will appear nearest the brim. If any damage is noted, replace strap set.
5. If the helmet has sustained a forcible blow (impact), it should be replaced, even if there is no visible damage.

The following precautions must be adhered to for protection of the fire fighter:

1. Do not alter or modify the helmet's design or construction without explicit written permission and instructions from the manufacturer.
2. Do not modify or replace any component of the helmet, such as the shell, energy-absorbing system, retention system, fluorescent retroreflective markings, ear covers, or faceshield, with components or accessories other than those certified for use with the specific helmet. Any such alterations, modifications, or replacements with noncertified components would cause it to be noncompliant with the helmet requirements of NFPA 1971, and may void any manufacturer warranties. The following should be observed:

a. Do not paint the helmet without explicit permission and instructions from the manufacturer.
b. Do not store in direct sunlight.
c. Do not neglect the interior fittings, especially the impact cap and suspension (including chinstrap).
d. Inspect the impact cap and interior fittings regularly. Replace worn head linings. Clean ear covers on a periodic basis.
e. Replace impact caps that have damaged foam or overhead strap assemblies.

GLOVE ELEMENT

All protective gloves for structural fire-fighting operations must meet the glove element requirements of NFPA 1971.

Gloves are designed to protect the hands and wrists from mechanical injuries (cuts and punctures) and thermal injuries (burns, scalds, and frostbite). Injury data identify cuts and lacerations as frequently sustained fire scene injuries or while performing other job-related duties. Wounds, cuts, bleeding, or bruises accounted for 21 percent of 1989–1991 reported U.S. fire fighter injuries, and 42 percent of those injuries were to the hand or arm. Fire fighters frequently encounter hazards that can cause cuts or lacerations. These include broken glass and objects with sharp metal edges, such as heating and air-conditioning ducts, furniture, and light fixtures. Some structural components, such as steel and aluminum studding in dry-wall construction, also pose a hazard. A multitude of fire-scene hazards has been identified as potential causes of puncture injuries. These include nails, screws, and spikes, splintered glass, and fencing; splinters from lumber, wood furniture, fixtures and ladders; and pointed tools, such as screwdrivers and ceiling hooks. These hazards are often obscured by heavy smoke and poor visibility.

Injury data also identify burn injuries as frequently sustained fire-scene injuries. Burns accounted for 12 percent of 1989–1991 reported U.S. fire fighter injuries, and 29 percent of those were to the hand or arm.[2] Thermal hazards encountered by fire fighters include exposure to heat, falling materials and debris (e.g., dripping and burning tar, grease, paint and plastic; burning embers, melting insulation; and hot runoff water and steam); exposure to hot objects (e.g. mattresses and bed springs, ladders, ropes, wires, chains, doors, locks, stairway railings, glass, and automobile components); and exposure to high-heat conditions, such as ignition of fire gases, flame rollover, backdraft, and flashover.

Cold injuries are also sustained at the fire scene. Frostbite, the most common cold injury sustained by fire fighters, develops when the skin is exposed to air temperatures below the freezing point. Exposure times can be a matter of minutes or hours. Superficial forms of frostbite result in loss of sensation and minor tissue damage that is similar to reversible, first-degree burn, but can have aftereffects, such as sensitivity to cold and recurring pain if pressure is applied to the affected area. More severe, deeper frostbite results in extensive tissue death and, in extreme cases, gangrene. Fighting fires in winter can involve exposure to snow, rain, and freezing temperatures. Discomfort, pain, and injury from cold are all intensified by fire fighters' gloves that are soaked through with water.

Test requirements include heat and flame resistance of materials, conductive heat resistance, thermal protective performance (TPP), cut and puncture resistance, liquids and viral penetration resistance, and dexterity and grip testing.

Liquids penetration testing is a method for measuring resistance to several kinds of wet-heat exposures; it partly satisfies the necessity for a glove to resist cold/wet exposures, for it to be dexterous during cold/wet exposures, and for it to be resistant to excessive absorption of, and deterioration by, water. Gloves that do not meet the liquids penetration requirement do not have protection from: (1) hot/cold water (which can produce scald and frostbite, respectively), (2) wetting of an already heated glove, (3) steam jets, (4) saturated water-vapor atmospheres, or (5) cold/wet exposures. This is a safety issue that, in and of itself, is an important performance criteria, as it is also the most appropriate test available to measure water and other liquids penetration resistance in a glove.

In addition to the effects of water penetration, gloves that do not meet the liquid penetration requirements would not have resistance to other liquids commonly encountered during structural fire-fighting activities. The common liquids are hexane, 37 percent sulfuric acid (battery acid), 3 percent concentration of aqueous film-forming foam (AFFF), and hydraulic fluid; fire-resistant, phosphate-ester base; and swimming pool chlorinating chemical containing at least 65 percent free chlorine.

The viral penetration requirement provides protection from exposures to blood or other body fluids. This is important for the structural fire-fighting glove, as it is the glove that will be in use during victim rescue from fires where these exposures are likely to occur. It is also the glove that will be used during an emergency when the puncture and cut protection is needed, such as during vehicle extrication.

Gloves must be available in not less than five separate and distinct sizes (XS, S, M, L, XL). NFPA 1971 specifies a method for taking hand dimensions for selection of properly fitting gloves.

When the design of the coat element does not provide protection for the wrists with a fixed wristlet that fits over the hand with openings for the thumb and for the fingers, protective gloves must have wristlets of at least 4.0 in. (102 mm) in length when the arms are extended upward and outward from the body. Materials used for wristlets must meet or exceed the requirements of the flame-resistance test, which includes tests for a 4-in. (102-mm) maximum char length and maximum of 2-sec afterflame.

Performance and Testing

Heat resistance testing of the glove materials is measured by placing gloves in a forced-air circulation oven heated to 500°F (260°C) for 5 min. Materials may char or discolor, but may not separate, melt, or drip. The materials must not shrink more than 5 percent in length or width.

Flame testing measures the glove material resistance to ignition and sustained heating. The glove is exposed to a flame in accordance with Method 5903.1 of Federal Test Method Standard 191A, *Textile Test Methods.* Sample specimens tested must not melt or drip and must not exceed the maximum allowable afterflame of 2.0 sec, and the maximum char length of 4.0 in (102 mm). Also, consumed materials must not exceed 5 percent of the specimen by weight.

Conductive heat testing measures resistance to conductive heat when the hand is in contact with hot surfaces or objects. Gloves are tested in accordance with ASTM F1060, *Standard Test Method for Thermal Protective Performance of Materials for Protective Clothing for Hot Surface Contact.* Gloves are placed on a hotplate heated to 536°F (280°C). A pressure of 0.5 psi (3.4 kPa) is applied to the glove. A thermocouple inside the glove records the temperature that would be conducted to the wearer. The time in seconds to pain and to second-degree burn (blister), as predicted by the Stoll Human Tissue Burn Tolerance Criteria, is reported. Second-degree burn time must be not less than 10 sec and time to pain not less than 6 sec.

TPP testing measures the thermal insulation of the glove in a severe heat exposure. Environmental temperatures during the ignition of fire gases, flame rollover, backdraft, or flashover range from 500 to 1500°F (260 to 816°C), and thermal radiation ranges from 0.175 to 4.2 Watts/cm². Gloves are tested in accordance with a mod-

ified ASTM D4108, *Standard Test Method for Thermal Protective Performance of Materials for Clothing by Open-Flame Method.* The test involves exposing the gloves to a thermal flux of 2.0 cal/cm²/sec. This heat flux is achieved by utilizing convective and radiant thermal flux sources. The convective heat is developed by two Meker or Fisher burners. The radiant heat is developed by nine quartz infrared tubes affixed beneath and centered between the burners. A thermocouple inside the glove measures the heat that would be transmitted to the wearer. The gloves must provide protection for not less than 17.5 sec before second-degree burn (blister) occurs, as predicted by the Stoll Human Tissue Burn Tolerance Criteria. This corresponds to a TPP value of 35.0 (17.5 sec × 2.0 cal/cm²/ sec).

Viral penetration resistance testing evaluates the glove's resistance to the penetration of body fluids and bloodborne pathogens that the fire fighter can be exposed to during rescue functions. The glove body composite materials, including seams, are exposed to a bacteriophage suspension of a liquid inoculated with *E. coli* C bacteria for 1 hr. Materials are tested in both directions and cannot allow any penetration of the bacteriophage.

Liquids penetration testing measures the resistance of the glove to a selection of common liquids frequently encountered during structural fire-fighting operations. The outer surface of the glove is exposed to the liquid under pressure and the inside of the glove is examined for leakage. The glove must withstand a liquid pressure of 4 psi (27.6 kPa) for 1 min., without detectable leakage or seepage to the inner surface of the glove normally in contact with the hand.

Cut resistance testing is designed to evaluate the cut resistance of glove materials. The static cut test apparatus consists of an L-shaped frame and pivoted arm that lowers a sharp-edged blade onto the sample specimen. The materials used for the outer surface of the glove must not cut completely through under an average applied force of 18 lb (8.2 kg).

Puncture resistance testing is designed to evaluate the puncture resistance of glove materials from sharp objects, such as glass and nails. The test is conducted on materials from the palm and palm side of the fingers. A pointed penetrometer, moving at a uniform speed under load conditions, is driven through the sample specimen. The penetrometer is stopped when it has driven through the specimen and the force is recorded. Materials must not be punctured under an average applied force of 13.2 lb (6 kg). Gloves are tested both wet and dry.

Dexterity testing demonstrates that gloves do not limit a fire fighter's ability to use tools and protective equipment, such as SCBA. The dexterity test used is a standardized procedure, known as the Bennett Hand-Tool Dexterity Test. Test subjects complete a baseline test without gloves. The follow-up test with wet gloves must not exceed 140 percent of the time required to complete the baseline test.

Grip testing demonstrates that gloves do not reduce the ability of a fire fighter to grip and pull things, such as hose and ladder halyards. The test uses a $^3/_8$-in. (9.5-mm) 3-strand pre-stretched polyester halyard attached to a spring scale. Bare-handed weights lift capability is the baseline weight. The test is conducted using both wet and dry gloves. The weight-pulling capacity must not be less than 80 percent of the baseline weight.

Cleaning and Maintenance of Gloves

Before any cleaning of fire-fighting gloves, the manufacturer's instructions must be followed. The manufacturer provides specific cleaning instructions and these must be consulted and followed, since failure to do so may reduce the protective qualities of the glove or negate warranties.

As with all protective clothing and equipment, gloves need to be kept clean. Life expectancy and dexterity are diminished, and protection from flame and thermal insulation are comprised, if abrasives, flammable dirt, and contaminants are not removed.

The use of liquid chemicals or oils to treat or soften the leather is not recommended. Do not use chlorine, chlorine bleach, or other products containing chlorine, as this will degrade the fabrics, materials, and thread used in the glove. It is not recommend that gloves be machine washed or dry cleaned without the explicit instructions from the manufacturer.

Squeeze out (do not twist) excess water from the glove by hand to aid in drying. A clothes dryer using low heat or the delicate/fluff cycle is a quick way to dry the glove. Allowing the gloves to dry through slow evaporation may harden the shells of leather gloves. After the glove has been dried, a crust or stiffness may remain. This stiffness can be removed by putting the glove on and buffing it with a scrub brush until softness returns. Do not lay gloves on radiators or dry them in hot-air dryers, as this severely shortens their service life and causes them to become stiff. Also, do not store or dry gloves in direct sunlight, as exposure may cause degradation of the glove materials and fabrics.

Gloves should be inspected after each use to determine their serviceability. Gloves with cuts, tears, rips, open seams, holes, or frayed backs or cuffs should be removed from service, as their protective qualities will be diminished. Areas of damage to look for include: cuts or worn areas in the shell or lining material, torn seams or cut threads that may allow seams to separate, torn or raveled knit wristlets; and chemical contamination that can not be removed in the outer shell material or liner.

Any repairs must follow the manufacturer's instructions. However, there are few repairs that can be made successfully without reducing the protective qualities of the gloves. Damaged gloves should be removed from service, tagged with the damaged area noted, and replaced. Do not alter or repair gloves without explicit instructions from the manufacturer.

PROTECTIVE FOOTWEAR ELEMENT

All protective footwear for structural fire-fighting operations must meet the footwear element requirements of NFPA 1971. Manufacturers currently produce both rubber and leather boots that meet this standard. Footwear that meet all requirements of this standard must be certified, labeled, and listed by an independent third-party certification organization.

Prior to 1987, the only requirement for fire fighter protective footwear was U.S. OSHA regulation 29 CFR 1910.156, *Fire Brigade Regulations*, which cited ANSI 41.1, *American National Standard for Men's Safety-Toe Footwear*, and covered the areas of impact and compression. In 1987, NFPA 1974, *Standard on Protective Footwear for Structural Fire Fighting*, was adopted. This standard represented the first effort to develop minimum requirements for protective footwear specifically for the fire service. It addressed many critical areas, such as fit, sizing, water penetration, and thermal protection requirements, specifically related to the structural fire-fighting environment.

The footwear element criteria of NFPA 1971 requires that footwear be at least 8 in. (203 mm) in height and be available in a range of full and half sizes and in widths narrow and wide to ensure that fire fighters have properly fitting boots. The standard also has test requirements designed to ensure that footwear can provide adequate protection during structural fire-fighting operations. There are tests for heat resistance, corrosion, puncture resistance of the sole, electrical resistance of the sole, compression and impact testing of the toe area, upper cut and puncture resistance, flame resistance, sole

abrasion resistance, conductive and radiant heat resistance, waterproofness, and ladder shank bend resistance.

An important consideration when selecting footwear is weight. One study examined the physiological and biomechanical changes in fire fighters due to footwear design modifications. Fire fighters performed a series of tests utilizing three types of footwear: (1) sneakers, (2) standard 14-in. (356-mm) rubber fire-fighter boots, and (3) 14-in. (356-mm) leather fire-fighter boots. The tests included stair climbing, cybex leg extension before and after stair climbing to measure thigh muscle fatigue, and an agility run. The rubber boots in the study weighed an average of 8.26 lb (3.75 kg), and the leather boots were significantly lighter at an average of 5.61 lb (2.55 kg). This study reported:

1. There was a 47 percent decrease in the amount of work that fire fighters had to perform in lifting their feet during stair climbing when they were wearing leather boots as opposed to the rubber boots. Thus, there was a significantly less energy expenditure during stair climbing work while wearing leather boots as compared with the conventional heavier rubber boots.

2. There was a 40 percent improvement in leg strength retention when the fire fighters were wearing leather boots in comparison to rubber boots. This would play an important part in postponing fire fighter leg fatigue.

3. There was a 91 percent decrease in the amount of time required to run a 30-ft (9.2-m) agility course when fire fighters were wearing leather boots in comparison to the rubber boots. This could be significant in an emergency situation requiring speed of movement.

4. The amount of energy required by fire fighters to climb stairs in leather boots was 48 percent less than what was required when climbing in the rubber boots. This is significant because it keeps the fire fighter in an aerobic metabolism mode longer, keeps respiratory frequency lower, and delays development of anaerobic lactic acid that would shorten their emergency work time. When the oxygen requirement of the body is greater, the respiratory frequency and the volume of air exchanged will also be greater. The volume of air inhaled and exhaled will also be greater. For fire fighters, the volume of air inhaled during each breath affects the volume of air contained in SCBA cylinders. Higher respiratory frequencies yield shorter SCBA cylinder duration times. The reverse of this is that a lower respiratory frequency, due to a lower oxygen need by the body's muscles, will result in a slower depletion of the breathing air supply.

These conclusions indicate that reduced weight of fire-fighting footwear is a prime consideration and must be addressed to decrease injuries.

Other factors for consideration include:

1. Proper sizing and fitting of fire-fighting footwear is essential. Footwear should be available in both men's and women's sizes and the practice of manufacturers sizing women's feet using men's shoe lasts should be changed. Footwear for female fire fighters should be manufactured utilizing women's shoe lasts.

2. Fire-fighting footwear is intended for use with protective trousers. This eliminates problems with donning and doffing encountered when trying to use footwear for both station use and structural fire-fighting emergency response. This would not prohibit the use of boots for station wear if they were found to be comfortable and would not inhibit donning and doffing of protective trousers.

3. Footwear manufacturers should engineer better ankle fit and support into footwear. The use of heel counters and other methods should be utilized.

4. Footwear should not utilize a calender sole. Elimination of calender soles will increase ground contact and provide better support.
5. The need for oblique toe caps, since their use causes footwear to be aesthetically unappealing, is questioned. However, the continued protection of the toe area is very important and supports further research for the use of light-weight material and shape design.
6. Footwear should offer better cut resistance for the uppers than currently available rubber boots.

Performance and Testing

Heat resistance testing of the footwear materials is measured by placing footwear filled with bone-dry vermiculite in a forced-air circulation oven heated to 500°F (260°C) for 5 min. The footwear is examined on the outside and inside for melting, and all accessories are evaluated to see that they still function.

Corrosion testing evaluates the metal parts of protective footwear, including but not limited to the toe cap, ladder shank, puncture-resistant device, and accessories. These items can show no signs of corrosion and must remain functional after exposure to a 5 percent saline solution.

Puncture resistance testing measures the footwear's resistance to puncture from nails and other sharp objects. Testing for puncture resistance is performed on a testing machine with a moveable platform. The platform is fitted with an 8D common nail. Footwear must not puncture through the sole with heel when tested to an average force of 272 lb (123.5 kg).

Electrical resistance testing evaluates the footwear's ability to provide protection against electrical current if a fire fighter steps on a live wire. A 14,000-volt current at 60 Hz is applied to the sole of the shoe. A metal foil electrode is placed inside the shoe to detect leakage. The footwear must have no leakage in excess of 5.0 milliamperes.

Compression/impact testing measures the impact and compression experienced in the toe area when objects fall on the boot. Compression and impact testing are conducted in accordance with Section 1.4 of ANSI Z41, *Personal Protection—Protective.* For compression testing, a test load is applied to the highest point of the toe box. The toe box must withstand a compressive load of 2,500 lb (11,120 N) without reducing the clearance between the toe box and the insole to a minimum specified clearance. Impact testing consists of dropping a steel weight onto the toe box. The impact requirement is 75 ft-lb without reducing the clearance between the insole and the toe box to a minimum clearance.

Upper cut resistance testing measures the resistance to cutting by sharp objects, such as glass. A weighted sharp blade placed tangential to a upper material specimen is lowered at a rate of 20 in. (508 mm) per min. The material specimen is inspected to determine if the lining was cut through.

Upper puncture resistance testing measures the resistance to puncture of the materials used in the uppers of boots. Material specimens are challenged by a pointed penetrometer operated at a uniform velocity of 20 in. (508 mm) per minute until the penetrometer has driven through the specimen. The upper must not puncture under an average applied force of 13.2 lb (6 kg).

Flame testing measures the footwear material resistance to ignition and sustained burning. The footwear is exposed to a flame in accordance with Method 5903.1, *Flame Resistance of Cloth: Vertical,* of Federal Test Method Standard 191A, *Textile Test Methods,* except that the footwear is suspended so that the burner flame impinges on the surface to be tested at a 90-degree angle to the flame. Testing is done on the sole at the ball of the foot and on the upper at the toe, vamp, quarter, gusset, if present, and shaft. Footwear must have a maximum afterflame of not more than 2.0 sec and must not melt or drip.

Abrasion resistance testing evaluates the sole and heel for abrasion resistance. This test is conducted in accordance with Footwear Industries of America Standard 301, *NBS Abrasion.*

Conductive heat testing measures the amount of heat conducted through the footwear to the foot. The footwear is placed on a 932°F (500°C) hotplate for 30 sec and a thermocouple is affixed to the surface of the insole next to the foot, directly above the ball of the foot. The protective footwear insole surface in contact with the foot must not exceed 110°F (43°C).

Radiant heat resistance testing evaluates the protection against radiant heat encountered during structural fire fighting. The toe, vamp, quarter, gusset, if present, and shaft are exposed to a radiant panel that produces a radiant heat load of 1 Watt/cm^2 over a 3-in. (76-mm) diameter circle. A thermocouple is affixed to the inside surface of the lining next to the foot. The temperature of the upper lining surface in contact with the skin must not exceed 111°F (44°C).

Viral penetration resistance testing evaluates the footwear's resistance to the penetration of body fluids and bloodborne pathogens that the fire fighter can be exposed to during rescue functions. The footwear composite materials, including seams, are exposed to a bacteriophage suspension of a liquid inoculated with *E. coli* C bacteria for 1 hr. Materials are tested in both directions and cannot allow any penetration of the bacteriophage.

Liquids penetration testing measures the resistance of the footwear to a selection of common liquids frequently encountered during structural fire fighting operations. The outer surface of the footwear is exposed to the liquid under pressure and the inside of the footwear is examined for leakage. The footwear must withstand a liquid pressure of 4 psi (27.6 kPa) for 1 min without detectable leakage or seepage to the inner surface normally in contact with the foot. In addition, the overall liquid integrity is tested by flexing the footwear 100,000 times while partially immersed in water. After flexing, the footwear is blotted dry and filled with tap water. The exterior of the footwear must show no signs of water penetration after 2.0 hr.

Ladder shank bend testing measures the deflection of the shank and is designed to ensure that footwear provides a stable base of support when climbing ladders. The ladder shank is placed on mounting blocks as it would be oriented toward the ladder when affixed into protective footwear. It is then subjected to a force on its center of 400.0 lb (1779 N) to determine a calibration value. A 400.0-lb (1779-N) force is then applied at a rate of 2.0 in. (51 mm) per min. Deflection is recorded and the shank must show resistance to bending that does not exceed 0.25 in. (6.4 mm) of the calibration value.

Cleaning and Maintenance of Footwear

Before any cleaning of fire-fighting footwear, the manufacturer's instructions must be followed. The manufacturer provides specific cleaning instructions and these must be consulted and followed since failure to do so may reduce the protective qualities of the footwear or negate warranties.

As with all other elements of the structural fire-fighting ensemble, footwear must be kept clean. Life expectancy and dexterity are diminished and protection from flame and resistance and thermal insulation can decrease, if abrasives, flammable dirt, and contaminants are not removed.

The external surfaces of leather boots should be cleaned periodically with saddle soap or a mild soap, followed by a thorough coating and buffing with a nonflammable shoe cream to soften the top-grain leathers.

Rubber boots should be cleaned after each use, making sure that any oils, dirt, dust, chemicals, etc., are wiped clean with soap and water. Rubber boots should be stored upright, to minimize folding and creasing. Keep linings dry to prevent mildew.

Wet linings should be dried by first draining out any excess liquid, then either stuffing the boot with absorbent toweling or using warm forced-air drying. Do not attempt to dry the boots by placing them in a dryer, on a radiator, or in a convection or microwave oven. Do not dry or store in direct sunlight, and store boots away from electric motors and other ozone-producing elements, such as bright neon lights and fan-blown air.

To combat ozone, which leads to cracks and leaks, some manufacturers add a chemical additive to the rubber. These chemicals may appear on the surface of the rubber in the form of a coating that looks like a frost or powder. This is a sign that the chemicals are reacting to combat ozone.

All footwear should be inspected regularly to ensure that material and manufacturing integrity are maintained. The manufacturer should be contacted if there is any question about the condition of footwear. Do not alter or repair footwear without explicit instructions from the manufacturer. There are few repairs that can be made successfully without reducing the protective qualities of the footwear. Damaged boots should be removed from service, tagged with the damaged area noted, and replaced.

SELF-CONTAINED BREATHING APPARATUS

SCBA Types

The self-contained breathing apparatus (SCBA) that were first used in the U.S. fire service were originally designed for use in the mining industry. Prior to the end of World War II, closed-circuit breathing apparatus was the principal self-contained respirator. Respiratory protection for many fire fighting operations depended upon filter-type canister respirators, and their wide use continued until the late 1960s. Following World War II, open-circuit respirators, which were originally designed for military aviators, began to be used for fire fighting. The principles of operation of these open-circuit SCBA have changed little since their introduction in 1945, with the exceptions of subtle design changes, the use of synthetic materials, and the addition of larger, lighter-weight, high-pressure storage cylinders. Open-circuit SCBA has remained the principal form of respiratory protection used by the U.S. fire service for more than 35 years.

Positive-Pressure and Demand SCBA

All SCBA are classified as either positive pressure or demand (positive pressure is also called pressure demand). U.S. OSHA regulations require use of *only positive-pressure* SCBA for structural fire-fighting operations, and do *not* permit use of *demand* SCBA. The widespread use of positive-pressure SCBA in the U.S. fire service only began in the early to middle 1980s. During normal use, a positive pressure within the facepiece in relation to the ambient atmospheric pressure reduces the potential for inward leakage of a contaminated atmosphere around the perimeter of the facepiece. However, the positive pressure feature cannot be construed to take the place of proper facepiece fitting and fit testing for each user. The positive-pressure feature cannot and will not compensate for any interference with the facepiece-to-face seal, such as from beards or

other facial hair, or from eyeglass temple bows or straps. A positive-pressure SCBA will use little, if any, more air than a demand SCBA, if properly fitted. It should be noted that a poorly fitting or improperly adjusted facepiece can seriously reduce the rated service life by allowing outward leakage, and endanger the user by allowing inward leakage of the contaminated atmosphere.

Demand SCBA permits airflow to the user only during inhalation. U.S. OSHA regulations do *not* permit use of *demand* SCBA for structural fire-fighting operations. Demand SCBA operates on the principle of requiring negative pressure to initiate and maintain breathing airflow during inhalation. The danger of the atmosphere surrounding the facepiece (the contaminated atmosphere) being drawn into the SCBA facepiece is significant.

Tests conducted by the Los Alamos National Laboratory revealed substantial inward leakage of toxic materials within face-pieces of SCBA used in the demand mode. Quantitative tests revealed levels of respiratory protection provided by these SCBA to be far below what was considered safe for emergency operations in toxic and radioactive atmospheres. Positive-pressure SCBA provided adequate levels of protection for such environments. The adoption of positive-pressure SCBA has been one of the most important actions taken by fire departments in providing increased respiratory protection.

Open-Circuit SCBA

Open-circuit SCBA is a respirator from which exhaled breath is vented to the atmosphere and is not rebreathed. It uses an expendable source of breathing air that is stored under pressure in a cylinder. The breathing air is made available to the user through a regulator that automatically functions to reduce high air-storage pressures to lower usable pressures and to supply this breathing air to the user.

Open-circuit SCBA utilizing compressed air cylinders are available in conventional pressure systems [2,216 psi (15,279 kPa)], intermediate pressure systems (3000 psi), and high-pressure systems [4,500 psi (31,027 kPa)]. For SCBA with a 30-min-rated service life, the advantage of the high-pressure cylinder is a slightly lower weight achieved by an aluminum/fiberglass-wrapped cylinder, and the smaller cylinder allows somewhat less bulk than a conventional pressure cylinder. Higher operating pressures and the use of lightweight aluminum/fiberglass-wrapped cylinders also permits use of larger cylinders that are interchangeable with the 30-min rated service life cylinders, providing a rated service life of 60 min.

Open-circuit SCBA that are certified by National Institute for Occupational Safety and Health (NIOSH) include a rated service life based on laboratory tests required by NIOSH. The SCBA is tested using a specified breathing machine with a breathing rate of 40 L/min. NIOSH uses this 40 L/min rate because it represents a moderate work rate that an average user can sustain for a period of time. To attain a rated service life of 30 min, during this 40 L/min test, the typical SCBA cylinder must contain 1,200 L or more of compressed breathing air. A 45-cu ft cylinder has a capacity of 1,273.5 L, based on 28.3 L per cu ft. Because actual work performed by a fire fighter often results in a ventilation rate that exceeds 40 L/min, fire fighters will frequently not attain the rated service life of 30 min. During extreme exertion, for example, actual service life can be reduced by 50 percent or more.

Open-circuit SCBA has the advantages of lower cost than its closed-circuit counterpart on a unit basis, cooler breathing air, familiarity to the fire service, general compatibility, and faster individual recharging capabilities.

Closed-Circuit SCBA

Closed-circuit SCBA is a respirator in which exhaled air is rebreathed by the wearer. This SCBA recirculates the user's exhaled gas. The exhaled breath passes through an exhalation hose and through a carbon dioxide absorber where the carbon dioxide is removed. It then enters a breathing chamber, where it is mixed with additional oxygen from an oxygen source that is stored under pressure in a cylinder, and is available for inhalation.

The primary interest in closed-circuit SCBA is the longer service life duration over open-circuit SCBA. The advantage of the long-duration, closed-circuit SCBA is the additional time available that increases the margin of safety for entering, exiting, operating, and decontaminating at situations, such as hazardous materials incidents, and operations in subways, subcellars, high-rise buildings, and complex structures.

Some disadvantages of closed-circuit SCBA are very warm breathing air as compared to open-circuit SCBA, longer time to perform field recharging, more difficult recharging procedure, and a higher unit cost. There have also been reservations concerning the use of longer-duration closed-circuit SCBA in routine structural fire fighting, because of the tendency to overextend oneself in interior operations.

U.S. OSHA and NIOSH Regulations

Respiratory protection standards in the United States have recognized that unknown types and quantities of toxic elements of the products of combustion and oxygen deficiency encountered during fire fighting do not allow the selection of different respiratory protective equipment for different exposures, as often is the case in the industrial environment. These atmospheres are classified as IDLH (immediately dangerous to life and health) and require SCBA for respiratory protection.

OSHA has utilized NIOSH and other OSHA regulations to establish minimum requirements for fire-fighter respiratory protection. The U.S. OSHA *Fire Brigade Regulations*, 29 CFR 1910.156, requires the use of only positive-pressure SCBA that is certified by NIOSH/MSHA under 30 CFR 11 (now only by NIOSH under 420 CFR 84), and having a minimum-rated service life of at least 30 min. OSHA requires that only positive-pressure SCBA be used during all structural fire-fighting operations; at all hazardous materials incidents; in oxygen-deficient atmospheres; in areas such as sewers, confined spaces, and silos; and wherever a hazardous or oxygen-deficient atmosphere may exist.

OSHA mandates that fire departments have a respiratory protection program meeting the requirements of the general OSHA standard for respiratory protection, 29 CFR 1910.134. This OSHA respiratory program includes requirements for quality of breathing air, maintenance, disinfecting, and special wearing conditions.

SCBA must be given a complete inspection not less than monthly and an operational check before and after each use. Most career fire departments in the U.S. require that this check be done at the start of each shift and after each incident where SCBA is utilized. Written records must be maintained of these procedures. Repairs of SCBA must be done by a factory-certified technician in accordance with the manufacturer's instructions.

SCBA facepieces must also be cleaned and disinfected after each use in accordance with the manufacturer's instructions. The fire department should provide assurance that the method of disinfecting utilized is capable of killing organisms that cause diseases, such as hepatitis, and herpes.

Nothing can be allowed to be in or pass through the area where the facepiece-to-face seal is made. Therefore, beards, drooping mustaches, long or wide side burns, or other facial hair at any point where the face seal with the respirator facepiece is to occur, or wearing of eyeglasses with temple bows or straps, are prohibited for persons who are required to use SCBA. OSHA has ruled (February 15, 1990) these conditions have been shown to prevent a good facepiece-to-face seal, so that use of an SCBA by a person with any of the noted conditions (i.e., beards, drooping mustaches, long or wide sideburns, and other facial hair, temple bows/straps) is in violation of OSHA regulation 29 CFR 1910.134 (e)(5)(i). *This is so, regardless of what fit test measurement can be obtained.* In addition, persons with conditions such as severe skin pitting, loss of teeth, facial deformities, skull caps or head gear, or any other conditions or devices that prevent a proper facepiece fit, are prohibited from using SCBA. SCBA manufacturers offer special prescription lens/frame assemblies that can be affixed inside the facepiece and do not affect the facepiece-to-face seal.

U.S. DOT Regulations

Compressed gas cylinders for breathing air are regulated in the United States by the Department of Transportation (DOT). U.S. federal regulations for the construction, testing, and maintenance of compressed gas containers are contained in 49 CFR 178, Subpart C. These DOT specifications apply to steel SCBA and large storage cylinders only. Cylinders constructed of manganese/steel and conforming to the preceding DOT specifications are designated by a DOT3A marking on the cylinder. Steel cylinders constructed of chrome alloy steel and meeting DOT specifications are designated by a DOT3AA marking on the cylinder, followed by the marked cylinder service pressure. There are currently no federal specifications for aluminum and aluminum/fiberglass-wrapped cylinders similar to those for steel; however, DOT allows their use under a special exemption. The Compressed Gas Association (CGA) has produced a number of pamphlets that are incorporated as part of the federal regulations or stand as nationally recognized good practice. These include:

- C-1 *Methods for Hydrostatic Testing of Compressed Gas Cylinders*

- C-5 *Cylinder Service Life—Seamless Steel, High-Pressure Cylinders*

- C-6 *Standards for Visual Inspection of Steel Compressed Gas Cylinders*

- C-6.1 *Standards for Visual Inspection of High-Pressure Aluminum Compressed Gas Cylinders*

- C-6.2 *Guidelines for Visual Inspection and Requalification of Fiber-Reinforced High-Pressure Cylinders*

NFPA Standards

NFPA 1981, *Standard on Open-Circuit Self-Contained Breathing Apparatus for Fire Fighters*, was the first standard to set performance requirements for respiratory protection equipment worn by fire fighters. This standard specifies minimum acceptable design, certification, and performance criteria for SCBA, as well as defines the test methodology for determining whether a SCBA meets those performance criteria. The compliance testing procedures were designed to relate to many of the environmental extremes and physical demands to which fire fighters are exposed. As a result, NFPA 1981 has become an important benchmark in the evolution of SCBA.

Basic design of the SCBA must be certified by NIOSH and certified as positive pressure, with a rated service life of at least 30 min, and weighing no more than 35 lb (16 kg).

Labeling of the SCBA must include the NIOSH approval plate and a certification organization's label denoting compliance with NFPA 1981.

Manufacturer's instructions must be provided with each SCBA and include instructions and information, supplied by the manufacturer, on maintenance, cleaning, disinfecting, storage, inspection, use, operation, and limitations of the unit.

Performance and testing: Airflow performance testing measures the airflow that the SCBA can supply. A minimum acceptable amount of airflow provided to a fire fighter under exertion of 100 L/min volume is required, surpassing the NIOSH certification requirement of 40 L/min volume.

Thermal resistance testing evaluates the proper function of the SCBA, when exposed to various temperature extremes and temperature cycles, through a series of four tests (cold-cold, hot-hot, cold-hot, and hot-cold). These simulate conditions that might be encountered during actual fire-fighting operations; the SCBA must pass this series of four tests.

Vibration/shock resistance testing evaluates the resistance of the assembled SCBA to vibration and shock experienced when SCBA is transported by fire-fighting vehicles.

Communications testing is designed to evaluate normal voice communications through the SCBA facepiece.

Corrosion resistance testing evaluates the resistance to corrosion of metal components of the SCBA that could interfere with the function and performance of the apparatus.

Particulate resistance testing evaluates the SCBA's ability to function properly after exposure to a specified concentration of particulate, such as the dust conditions commonly present during fire fighting and overhaul operations.

Fabric flame resistance testing measures resistance to ignition and sustained burning of the fabric components and thread utilized in the harness assembly, and tests the fabric components utilized in the harness assembly. This is to ensure that the respirator backframe remains attached to the wearer's body when exposed to flame.

Fabric heat resistance testing measures resistance to heat of the fabric components and thread utilized in the harness assembly. This is to ensure that the respirator backframe remains attached to the wearer's body when exposed to heat.

Facepiece lens testing evaluates the resistance of the facepiece lens to scratching, so that a fire fighter's vision will not be reduced during normal fire-fighting operations.

Heat and flame resistance testing is a total heat and flame test for the entire apparatus. The SCBA, while attached to a breathing machine and operating at a prescribed rate, is conditioned for 15 min in a 203°F (95°C) oven and then exposed to an engulfing direct flame contact at a temperature of between 1500°F to 2102°F (816°C to 1150°C), with an average mean of peak temperature at 1742°F (950°C), for 10 sec. After this exposure, the SCBA is subjected to a drop test. Through the entire testing, the SCBA must operate as prescribed, securing harnesses must not fail, the facepiece or goggle lens must not obscure vision greater than 20/100, and components must not show any afterflame of beyond than 2.2 sec.

NFPA 1404

The Technical Committee on Fire Service Training began work on NFPA 1404, *Standard on Fire Department Self-Contained Breathing Apparatus Program*, in the early 1990s. While OSHA, DOT, and NIOSH address specifics of utilization, training, inspection,

maintenance, etc., these rules and regulations are not applicable outside the United States.

NFPA 1404 has expanded upon those areas addressed in other regulations, as well as other NFPA standards. The standard addresses areas of SCBA provision, including the need for an annual inventory, and how many and where the SCBA are allocated. The provisions go into the specifics of NIOSH certification and an acceptance test upon delivery of the SCBA.

The standard also outlines the use of SCBA in hazardous atmospheres at emergency incidents. This is consistent with OSHA regulations, and is further emphasized in NFPA 1500, Chapters 5 and 6.

The need for training begins at the recruit level. Training is to be conducted by fire service instructors who meet the requirements of NFPA 1041, *Standard for Fire Service Instructor Professional Qualifications*. The performance requirements are included as part of Fire Fighter Level I and II in NFPA 1001, *Standard for Fire Fighter Professional Qualifications*.

The training program requires annual member certification, a qualitative fit testing program, record-keeping process for facepiece fit testing, SCBA safety, how to act properly in case of an emergency, and requirements for progression of training. It further outlines an evaluation of the SCBA training program that is required as a standard operating procedure or policy, recognizing hazards that could be encountered, knowing and understanding the component features of the SCBA, the safety features and limitations of the SCBA, donning and doffing, and practical applications of training programs. Maintenance and testing of SCBA, and storage of reserve SCBA and reserve cylinders are also covered.

The SCBA in-service inspection requirements begin with the weekly and daily service checks. Members must ensure that SCBA is working properly, the cylinder is full, the SCBA is clean, and that it is inspected after each use.

Safety is a common thread that is emphasized in each chapter of this document. SCBA users should be confident that when they are donning the SCBA that it has been properly maintained, and that the air quality has been tested at a minimum of every 3 months.

In order to make a program successful, there must be an ongoing evaluation process. This process is formalized, and reviews every aspect of program. This ongoing process is documented and has a formal review life of its own. There should be input on the process from the agency health and safety personnel/committee, and from the end-users.

Statistical data compiled by NFPA's Fire Analysis and Research tell us that in the years from 1989 to 1993, 10 percent of fire fighter fire ground injuries that were classified as minor or moderate were due to exposure to smoke/toxic products. Of injuries classified as severe/life threatening, 27 percent were due to exposure to smoke/toxic products.[1] Having performance based standards for users, an ongoing training program, and enforcement in the use of SCBA will go a long way in reducing the injuries to fire fighters.

Cleaning and Maintenance of SCBA

Maintenance of open-circuit SCBA must be addressed as a fundamental component of any SCBA program. SCBA, by its very nature, is subject to a variety of different conditions that can prove to be detrimental to its successful operation. For example, SCBA is normally carried on motorized fire apparatus, subjecting it to extremes in temperatures, environments, and to the constant vibration of vehicular movement. When in use, the SCBA is subject to physical abuse and water damage associated with structural fire-fighting operations. As a result, preventive maintenance of SCBA is imperative. The specific requirements for the servicing and testing of

SCBA must be written and cover such policy areas as daily, weekly, and monthly service checks specifically designed to determine the serviceability of the SCBA.

SCBA manufacturers are required to supply the purchasers with periodic inspection and maintenance procedures. The manufacturer's instructions for each specific model SCBA must be carefully followed. In many instances, the periodic maintenance should be performed at more frequent rates than that stated by the manufacturer, based on the department's SCBA work load.

There are various levels of skills required to properly maintain SCBA. User maintenance, the basic level, should be confined to those functions that the individual fire fighter can contribute to the maintenance of the SCBA without the use of tools or special training. Testing and servicing by fire fighters is extremely important in maintaining SCBA in a state of readiness for emergency use.

Another level of maintenance involves the removal and replacement of defective or damaged parts of the facepiece, harness, and regulator. This intermediate level of maintenance requires completion of training provided by the SCBA manufacturer, simple tools, and a complement of spare parts. Departments may choose to have trained personnel within or outside their own organization responsible for the replacement of damaged or defective components. Most manufacturers do not allow intermediate-level maintenance personnel to work with, or replace, items in the high-pressure areas of the SCBA regulator system.

Advanced-level training allows personnel to completely disassemble, rebuild, assemble, and accurately test all portions of the SCBA. This level requires more training and test equipment, making it impractical for every department to provide this level of service by its own personnel.

"Before each use" and "after each use" inspection instructions are a part of the manufacturer's information supplied with each SCBA, and are a critical part of the procedure to ensure safe use of the SCBA. Emphasis must be placed upon before-use checks prior to entry into hazardous environments. Before-use checks of the SCBA are particularly important for the safety of the individual fire fighter. Before entering a hazardous atmosphere, fire fighters must check the facepiece-to-face seal, the operation of the exhalation valve, the low-air alarm, and the bypass valve. Injury or death related to the use of SCBA can be caused by failure to perform these before-use checks to discover a malfunctioning SCBA prior to entry.

Closed-circuit SCBA require the same degree of inspection and servicing as do open-circuit SCBA. The normal inspections are nearly identical to those utilized in open-circuit SCBA.

Air-cylinder maintenance consists of external inspections for damage, hydrostatic testing at specified mandatory intervals, and periodic inspection to detect internal corrosion or contaminants. With a wide variety of cylinders of varying service pressures and construction, the management of cylinder maintenance has taken on new dimensions in its importance in the overall SCBA maintenance program.

As in many areas of SCBA maintenance, the user is in a primary position to observe SCBA cylinders through periodic inspection and recharging of cylinders. Inspection procedures should be developed for fire fighters that indicate what types of damage are of concern to both cylinder exteriors and cylinder valves.

The external inspection of SCBA cylinders is an integral part of the hydrostatic testing and evaluation process. It is also a continual process by which users and personnel responsible for SCBA maintenance can ensure reasonable safety for those involved in SCBA use under emergency conditions. Cylinder necks should be examined for serious crack, folds, and flaws. Neck cracks are normally detected by soap testing the neck during charging operations.

Cylinder neck threads should be examined whenever the valve is removed from the cylinder, particularly at the time of hydrostatic test. The need for external inspection of SCBA cylinders has been even more significant with the advent of new lightweight cylinders.

The hydrostatic testing of SCBA cylinders is regulated in the United States by DOT. The frequency of hydrostatic tests is based upon the type of construction of the cylinder. Damage to the cylinder, or substantial cleaning of the interior or exterior of the cylinder, should require hydrostatic testing of the cylinder. Hydrostatic tests must be conducted at a minimum of 5-year intervals for steel or aluminum cylinders, and 3 years for composite aluminum wrapped cylinders.

Although an internal inspection is required at the time of hydrostatic testing (every 3 to 5 years, depending on the type of cylinder), periodic internal inspections of SCBA cylinders should be done at least annually by departments that utilize SCBA on an infrequent or average usage, and at least semiannually where SCBA is more frequently used. The purpose of the internal inspection is to determine any conditions that may contribute to the deterioration of the cylinder or possibly cause malfunction of the SCBA and injury to personnel. The object of the internal inspection is to check for rust, corrosion, moisture, and damage that may have occurred since previous inspections, and traces or evidence of oil and/or hydrocarbon contamination. Initial corrosion may take the form of a light powder or flaking around cylinder walls, and can be easily rectified. Extensive corrosion can cause pitting and eventual weakening of cylinder walls.

Oxygen cylinders must be treated similarly to air cylinders, with added precautions due to the nature of the contents. To avoid potential explosions, oxygen cylinders must be kept away from any source of hydrocarbon contamination and open flames. The care and maintenance of oxygen cylinders, including such items as valve gaskets, is extremely important. The quality of oxygen for closed-circuit SCBA must be carefully monitored.

Breathing air quality for open-circuit SCBA is of the utmost importance to the health and welfare of fire fighters using SCBA. The purest available breathing air supply should be drawn into the system. Breathing air should be tested from a compressor, *only after* it has been operating for at least 10 min, in order to observe the effects of heating of the motor on the air quality. Air quality testing should also be conducted on any breathing air storage cylinders, such as a cascade system, so that the breathing air quality from the last production point/storage point is verified. An analysis of breathing air should be done at least monthly by those producing it. Twice-monthly testing might be necessary to ensure a more reasonable test of representative breathing air. Test results should be compared with ANSI CGA G-7.1, *Commodity Specification for Air.* NFPA 1500 requires that breathing air quality must meet or exceed requirements for "Grade D" or better breathing air with a dew point of −65°F (−54°C) or dryer (24 ppm water vapor or less). The maximum particulate level cannot exceed 5 mg/m^3 air.

A low-moisture content of the breathing air will help to prevent freezing of regulators (especially in high-pressure systems), waterlogging of the desiccant and catalyst, and rusting and corrosion of cylinders.

PERSONAL ALERT SAFETY SYSTEMS (PASS)

Personal alert safety systems (PASS) encompass devices that are known by several names, including personal distress alarm device, personal distress locator, personal alarm locator, personal alarm monitor, and personal alarm device.

The concept of PASS involves small, lightweight devices with several distinct features. The alarm signal must be able to be activated manually by the user. The device must also contain a motion detector that senses movement or the lack of movement. The alarm signal must be capable of being activated automatically through interaction with the motion detector under a specified series of events. PASS must also have a pre-alert signal to prevent false alarms by alerting the user that the motion detector has activated the automatic alarm signal sequence.

Fire fighters trapped in smoke-filled buildings can become disoriented as they attempt to escape. This scenario can lead to fire fighters' deaths caused by smoke inhalation. Even with the use of SCBA, the fire fighter faces similar dangers brought about by the fact that their breathing supply is limited.

A study that investigated line-of-duty fire fighter deaths, published by the International Association of Fire Fighters (IAFF), found that the capability of fire fighters to let others on the fireground know that they are in trouble and need assistance may have possibly averted approximately one-third of the 41 deaths reported while fighting fires.

A well-publicized example was the deaths of three fire fighters in a Lubbock, TX, restaurant fire. While a number of causal factors have been identified and examined as potentially responsible for the fatalities, including the failure of the breathing apparatus, there was one common element linking the three fire fighters. All three men had become disoriented in a large smoke-filled dining room.

Likewise, four Syracuse, NY, fire fighters were killed while fighting a fire in a three-story wood-frame Syracuse University dormitory. After being overcome by smoke, the fire fighters became disoriented in the narrow hallways on the top floor of the dormitory.

A City of Los Angeles, CA, fire fighter was killed after becoming disoriented in an industrial warehouse. Cries for help from the fire fighter were heard and a search was begun, but dense smoke cut visibility to inches in the 100- by 200-ft (30.5- by 61-m) structure. By accident, a fire fighter who had not heard the cry for help stumbled upon the fire fighter. He provided the unconscious fire fighter with air from his SCBA until he exhausted his own air supply. Finally, the "rescuer" was overcome with smoke himself and had to abandon the unconscious fire fighter. Based on erroneous reports that the fire fighter who cried for help had been found, the search was stopped. When the report could not be confirmed, the search began again. Forty-five min after the shout for help was heard, the fire fighter was finally located but it was too late. This incident eventually prompted the California Occupational Safety and Health Administration (CAL/OSHA) to adopt a standard mandating the use of PASS for all fire fighters in the State of California engaged in fire-fighting activities requiring the use of SCBA. In a report, CAL/OSHA stated, "Fire fighters working in a hazardous location where self-contained breathing apparatus is required may become lost or disoriented, injured, entrapped, and cannot find an exit and, due to location and/or poor visibility, other persons in the area may not know of the emergency, and, if they did, may not be able to locate the fire fighter in time to rescue him. At the present time, no personal rescue alarm device or rescue strobe light is provided fire fighters. The evidence indicated that all fire fighters required to engage in structural fire-fighting operations should be provided personal rescue alarms."

It is evident that the fire fighter needs the capability to contact other workers at the emergency scene when in need of assistance. Due to the expense of equipping each fire fighter with a portable two-way radio, and the fact that an "automatic" notification would still not be available with a portable radio, another device that could summon help was needed.

The recognition of the need for PASS led the Technical Committee on Fire Service Protective Clothing and Equipment to develop a standard for such devices. In 1983, NFPA adopted NFPA 1982, *Standard on Personal Alert Safety Systems (PASS) for Fire Fighters*. The current standard is the 1993 edition.

The purpose of NFPA 1982 is to specify minimum performance criteria, functioning, and test methods for devices that can be worn by fire fighters that emit an audible alarm signal in order to summon assistance in the event that the fire fighter becomes incapacitated or needs assistance.

Several manufacturers are already integrating PASS with SCBA, so that the PASS will be automatically activated when the air supply is turned on. NFPA 1982 already allows for such integration of SCBA/PASS systems.

PASS cannot substitute for an individual's common sense, as it cannot substitute for an active incident management system, communications, and a functional and adequate personnel accountability system to ensure that the assignment and location of each fire fighter is known at all times. While the addition of a PASS does not guarantee safety, it is an important component to the fire fighter's personal protective ensemble.

Function and Design

PASS controls must allow for operation in three modes: (1) off, (2) manual, and (3) automatic. Switches must be clearly labeled for each of the three operations and must be protected against accidental deactivation. Such controls must be able to be operated while wearing a glove. Switches must have a service life of at least 50,000 cycles. To avoid accidental deactivation of the PASS, two separate and distinct manual actions must be required to switch the device from automatic to off. PASS must have a visual or audible signal indicating the mode selection status.

PASS incorporates *motion sensing* (or sensing of lack of motion) and must alarm if left motionless for about 30 sec, independent of where or how the device is worn. PASS must be equipped with an alarm signal that is activated within 25 to 35 sec to warn of failure of the motion detector.

PASS must be equipped with an audible *operational signal* that is emitted within 1 sec after switching to the automatic mode to ensure the wearer that the unit is functioning properly.

A *pre-alert signal* must sound between 7 and 15 sec prior to the activation of the alarm signal. This pre-alert signal must be between 70 and 85 dBA, measured at a distance of 3.3 ft (1 m), and be distinct and different from the alarm signal. Cancellation of the pre-alert signal and return to the automatic mode must be motion-induced without requiring the use of the user's hands.

The *alarm signal* must be audible in a variable or noncontinuous tone, with a minimum of three primary frequencies to ensure an omni-directional signal to enhance locating the victim. The PASS must emit the alarm signal with a minimum sound pressure level of 95 dBA, measured at a distance of 9.9 ft (3 m) for an uninterrupted duration of not less than 1 hr, using a battery charged to 80 percent of full capacity. Each primary frequency must not be less than 1,000 Hz nor more than 4,000 Hz and may be sounded sequentially or simultaneously.

PASS must be equipped with a *low battery indicator* that emits a distinct and different audible signal between 70 and 85 dBA, measured at 3.3 ft (1 m), when the battery can no longer maintain a sound level of 95 dBA for at least 1 hr.

PASS devices must not *weigh* more than 16 oz (454 grams), including batteries.

Performance and Testing

Intrinsic safety testing evaluates if the PASS will cause a spark that ignites. The PASS must be intrinsically safe for use in flammable or explosive atmospheres.

Corrosion testing evaluates the PASS device ability to withstand damage to its operating components caused by battery leakage.

Case integrity testing measures the PASS device's ability to support a 442-lb (200 kg) weight on each surface of the case, without affecting case integrity or causing visible damage.

High-temperature testing evaluates the effect on the operation of the PASS after exposure to the type of elevated temperatures that might be encountered during service life, either in storage or under fire-fighting conditions. All PASS features must operate successfully after being exposed to temperatures gradually being raised to a high of 160°F (71°C) during three 12-hr cycles.

Low-temperature testing evaluates the effect on the operation of the PASS after being exposed to low temperatures during storage. All PASS features must operate successfully after being exposed to a temperature of –40°F (–40°C) for a period of 24 hr.

Temperature shock testing evaluates the effect on the operation of the PASS after it has been exposed to a sudden change in temperature. The PASS must operate successfully after being transferred from temperatures of 160°F (71°C) to –40°F (–40°C).

Salt fog testing evaluates the effect on the operation of the PASS after it has been exposed to a corrosive environment. The PASS must operate successfully when exposed to a salt fog solution at a temperature of 95°F (35°C) for a period of 48 hr.

Dust testing evaluates the effect on the operation of the PASS after it has been exposed to a dry dust-laden environment. The PASS must operate after being exposed to an adequate circulation of dust-laden air at set temperatures.

Immersion/leakage testing evaluates the effect on the operation of the PASS after immersing it in water.

Radiant heat testing evaluates the effect on the PASS when exposed to conditions encountered in Class IV conditions (high temperature). (See Table 10-18 A.) After being preconditioned at 160°F (71°C) for 4 hr, the PASS must operate after the face of the PASS has been exposed to 1.5 cal/cm^2 for 10 sec.

Flame resistance testing evaluates the effects of fire on the PASS. The PASS must not have any visible flame or afterglow 5 sec after removal from the test flame.

Impact testing evaluates the ability of the PASS to perform after impact. The PASS must remain operable in the automatic mode after each of eight drops from a height of 9.9 ft (3 m) onto a concrete surface, where the impact is on each face and on one corner and one edge of the PASS, within 10 min following exposure to a temperature of 77°F (25°C) for at least 4 hr. PASS must also remain operable in the automatic mode after each of eight drops from a height of 9.9 ft (3 m) onto a concrete surface, where the impact is on each face and on one corner and one edge of the device, within 10 min following exposure to a temperature of –40°F (–40°C) for at least 4 hr. PASS must also remain operable in the automatic mode after each of eight drops from a height of 9.9 ft (3 m) onto a concrete surface, where the impact is on each face and on one corner and one edge of the device, within 10 min. following exposure to temperature of 160°F (71°C) for at least 4 hr.

Retention system testing evaluates the ability of the PASS to remain attached to the wearer. After the retention system has been fastened and unfastened 500 times, the retention system must be able to withstand a pulling force of 100 lb (445 N) and remained fastened.

Drainage test testing evaluates that the PASS will not hold water. The alarm signal must operate properly after the PASS is exposed to a rainfall rate of 4 in. (102 mm) per hr and a wind velocity of 20 miles (32.2 km) per hr in two positions, for 30 min each, while in the off mode.

Care and Maintenance of PASS

Before any cleaning of PASS devices, the manufacturer's instructions must be followed. The manufacturer provides specific cleaning instructions, and these must be consulted and followed since failure to do so may reduce the protective qualities of PASS or negate warranties.

Damaged PASS devices should be removed from service, tagged with the damaged area noted, and replaced. Do not paint, alter, or repair PASS devices without explicit instructions from the manufacturer. Most PASS devices do not require adjustment or complicated maintenance. Manufacturers suggest replacing the battery every 6 months, or as required, following the test procedure as outlined in their instructions.

PROXIMITY FIRE-FIGHTING PROTECTIVE CLOTHING AND EQUIPMENT

Garments for protection for high levels of radiant heat, as well as conductive and convective heat encountered during proximity fire-fighting operations, must be certified to NFPA 1976, *Standard on Protective Clothing for Proximity Fire Fighting.*

Proximity fire-fighting operations include aircraft rescue and fire fighting, bulk flammable liquids fire fighting, flammable gas fire fighting, and similar situations releasing high levels of radiant heat. Proximity protective garments are designed to protect fire fighters against the hazards associated with this special operation. They are not designed to provide protection from hazardous materials emergencies, involving biological, chemical, or radioactive agents, or act as an entry-protective garment.

The performance requirements for proximity protective garments were not intended to establish the limitations of the fire fighter's working environment, but are intended to establish material performance requirements, based on empirical field data and laboratory testing. While performing proximity fire fighting, users must be aware that the feeling of continual increase in heat is an indication that the protective garments may be nearing their maximal capability, and injury may be imminent.

Specific definitions have been developed to differentiate between the specialized fire-fighting operations for which structural fire-fighting protective clothing and equipment is not suitable and where special protective clothing and equipment *must be used.* These are defined as follows.

1. *Approach fire fighting.* Limited specialized exterior fire-fighting operations at incidents involving fires producing very high levels of conductive, convective, and radiant heat, such as bulk flammable gas and bulk flammable liquid fires. Specialized thermal protection from exposure to high levels of radiant heat is necessary for the persons involved in such operations, due to the limited scope of these operations and the greater distance from the fire that these operations are conducted. Approach fire fighting is not entry, proximity, or structural fire fighting.
2. *Entry fire fighting.* Extraordinarily specialized fire-fighting operations that can include the activities of rescue, fire suppression, and property conservation at incidents involving fires producing very high levels of conductive, convective, and radiant heat; these include aircraft fires, bulk flammable gas fires, and

bulk flammable liquid fires. Highly specialized thermal protection from exposure to extreme levels of conductive, convective, and radiant heat is necessary for persons involved in such extraordinarily specialized operations, due to the scope of these operations, and when direct entry into flames is made. Usually, these operations are exterior operations. Entry fire fighting is not structural fire fighting.

3. *Proximity fire fighting.* Specialized fire-fighting operations that can include the activities of rescue, fire suppression, and property conservation at incidents involving fires producing very high levels of conductive, convective, and radiant heat; these include aircraft fires, bulk flammable gas fires, and bulk flammable liquid fires. Specialized thermal protection from exposure to high levels of radiant heat, as well as thermal protection from conductive and convective heat, is necessary for persons involved in such operations, due to the scope of these operations and the close distance to the fire that these operations are conducted, although direct entry into flame is not made. Usually, these operations are exterior operations but might be combined with interior operations. Proximity fire fighting is not structural fire fighting, but might be combined with structural fire-fighting operations. Proximity fire fighting is not entry fire fighting.

The proper fit and sizing of proximity protective garments has a direct relationship to the safety of the fire fighter, and follows the same guidelines as stated for structural fire-fighting garments.

Particular attention needs to be paid to fire fighters with longer or shorter than the average upper or lower torsos, and to fire fighters with longer or shorter than average arm lengths. Proper sizing for these persons is very important to ensure that they will be properly protected. A good chest fit does not provide for other than the "average" arm or torso length. Waist size does not provide for other than the "average" lower torso length. Custom sizing is the most desirable method.

Outer Shell Reflectivity

Items that are not 100 percent radiant reflective, such as leather or other reinforcements, lettering, patches, logos, name or number stencils, emblems, and paint or other markings, must not be utilized on the outershell. Fluorescent and retroreflective trim is specifically prohibited from being attached to the outer shell of a proximity protective garment. The use of these items on the outer shell of proximity protective clothing can eliminate or substantially reduce the ability of the garment to reflect radiant heat. Such items act as a heat sink, which accelerates heat transfer through the garment and increases the risk of burn injury to the fire fighter.

Performance and Testing

NFPA 1976 specifies performance requirements for the protective qualities of the fabrics utilized in the construction of the outer shell, moisture barrier, and thermal barrier, including radiant heat resistance, flame and heat resistance, thermal insulation, durability, water penetration, barrier tear strength, washing shrinkage, water absorption, and other functions and construction features.

The most significant difference between proximity garments and those utilized for structural fire fighting is the requirement for radiant heat performance. The outer shell must have a 100 percent radiant reflective minimum value of 20.

Flame resistance of protective clothing fabrics is measured by Method 5903.1 of Federal Test Method Standard 191A, *Textile Test Methods*, which exposes a vertical strip of fabric to a gas flame. Afterglow, or the time that the fabric continues to glow after the flame is extinguished, must be less than an average of 2.0 sec for the fabric

to pass the test. Afterflame, or the time that the fabric continues to burn after the flame source is extinguished, must be less than 2.0 sec for the fabric to pass the test. If any of the burning fabric melts and drips during the test, the fabric fails. The charred fabric is measured to establish a char length, which must be less than 4 in. (102 mm).

Heat resistance of protective clothing materials is measured by exposing the fabrics to 500°F (260°C) in a forced-air oven for a period of 5 min. The fabrics must not melt, separate, or ignite.

Thermal shrinkage is again measured by exposing the fabric to 500°F (260°C) in a forced-air oven for a period of 5 min. In this test, the fabric must not shrink more than 10 percent in all directions.

Thermal protective performance (TPP) provides a laboratory test method to quantitatively evaluate materials and ensembles of fire and heat protective clothing.

Cleaning and Maintaining Proximity Protective Clothing

Before any cleaning of proximity fire-fighting garments, the manufacturer's instructions must be followed. The manufacturer provides specific cleaning instructions and these must be consulted and followed since failure to do so may reduce the protective qualities of the garments or negate warranties.

The cleaning of proximity fire-fighting protective garments should be conducted periodically to remove foreign and flammable contaminants. Periodic cleaning should prolong the life of the proximity protective garments and help maintain the performance.

Cleaning must recognize that the proximity garment is composed of a combination of materials. Each of these materials has its own unique characteristics, capabilities, and weaknesses. Even fabrics inherently flame resistant can have this characteristic negated by improper care and use. The amount of soil and the cleaning procedures and chemicals utilized can adversely affect a fabric's performance. Thus, procedures utilized to clean such materials must protect the weakest of the materials. Historically, fabric suppliers' care recommendations have failed to consider the other materials that comprise the garment. Many recommendations, while appropriate for the supplier's own material, would destroy the garment because of their adverse impact on the other components.

Cleaning protective garments reduces safety and health risks, and it is recommended that garments be cleaned frequently to reduce the level of and bodily contact with contaminants. User agencies should establish guidelines for frequency and situations for garment cleaning. For gross contamination with products of combustion, fire debris, or body fluids, removal of contaminants by flushing with water as soon as practical is necessary, followed by appropriate cleaning.

Decontamination may not be possible when proximity protective garments are contaminated with chemical, radiological, or biological agents. When decontamination is not possible, garments should be discarded in accordance with local, state, and federal regulations.

Chlorine bleach must *not* be used for proximity protective garment cleaning and/or disinfecting. Some components of these garments lose their physical integrity on exposure to chlorine bleach. Other components will actually lose their flame-resistant properties and thermal insulation on exposure to chlorine bleach. In either case, the protection provided by the garment will be compromised.

There are industrial cleaning products and facilities available for protective garments that the user may wish to investigate. The proximity protective garment manufacturer should be contacted directly for additional information. The original proximity protective garment manufacturer should always be contacted for specific instructions for cleaning various items of proximity garments, as the manufacturer knows what materials and components went into the construction of the garment. Not consulting the original manufacturer or following the manufacturer's cleaning instructions could result in damage to the protective garment, or loss of some or all of the protective properties.

Cleaning of outer shells: The outer shells of this ensemble contain a highly reflective surface, in order that it may reflect high levels of radiant heat. *It is extremely important to keep this surface clean so that it may perform at peak efficiency.*

One common reflective surface is an aluminized film laminated to a base fabric. *This material cannot be machine washed without losing radiant reflective qualities of the garment.* Cleaning of this type of garment can be accomplished in the following manner:

1. Clean by gently rubbing surface with a cloth or sponge using mild soap, 1,1,1-trichloroethylene, or isopropanol. When cleaning with 1,1,1-trichloroethylene or isopropanol, ensure that work is performed in an area with adequate ventilation and that proper gloves and eye protection are utilized.
2. Rinse the garment with clear water to ensure that all cleaning compounds have been removed from the garment.
3. Dry the garment by hanging it in a shaded area that receives good cross-ventilation, or hang the garment on a line and use a fan to circulate the air.
4. Do not store the garment until it has been completely dried.

Cleaning compounds that contain ammonia must *not* be used at any time, since the ammonia will react with the aluminum surface and diminish its reflectivity.

Aluminized surfaces can be damaged by contact with the dry powder fire extinguishing agent, potassium bicarbonate powder (PKP). Before storing a garment with PKP residue present, the following is recommended:

1. Dry powder should be removed with a dry brush or vacuum.
2. Wet powder should be removed by thoroughly flushing with water and wiping surface down with a clean soft cloth.
3. Damp or wet garments should be hung to dry in a shaded area that receives good cross ventilation or hang the garment on a line and use a fan to circulate the air.
4. Do not store the garment until it has been completely dried.

Cleaning of liners: *Only industrial front-loading washing machines should be used* to provide satisfactory cleaning of the garment liners and to significantly reduce the possibility of cross-contamination. Industrial machines provide for several rinse (extraction) cycles that are more efficient in removal of the contaminants and flushing them from the washtub so they are not retained on the fabric. The liners should be washed separately from other garments. A stainless-steel tub should be utilized if available.

The fire department must maintain protective clothing only in accordance with manufacturer's instructions. Maintenance should include regular inspection, proper repair, and retirement when appropriate. Storage in direct sunlight will substantially reduce the useful life of the garment and reduce protective qualities that might not be visually apparent. Color fading or color change often indicates that the garment has experienced UV degradation. All protective clothing that is retired should be destroyed, rather than using it as spare sets or giving it away to another organization, as the protective qualities have been degraded or eliminated.

The trim should also be inspected to ensure effective performance under daytime and nighttime conditions. It is important that high visibility be maintained at all possible orientations to the light source. A flashlight in a dark room can be utilized to inspect retroreflectivity, comparing trim on the garment to a new specimen of the utilized trim.

Other Proximity Protective Elements

Proximity protective elements should be selected to enhance the mobility of the fire fighter and must be viewed as one element in an ensemble since each component must interface successfully with the other components. NFPA 1500 requires that other proximity protective elements, i.e., helmets, gloves, and footwear, be certified to the appropriate requirements for the respective items of NFPA 1971 *and* have additional radiant reflective criteria that is considered suitable for the expected exposures where the item will be used. (These requirements were formerly in NFPA 1972 for helmets, NFPA 1973 for gloves, and NFPA 1974 for footwear.) NFPA 1500 also requires that SCBA for use with proximity protective elements be certified to NFPA 1981 *and* have additional radiant reflective criteria that is considered suitable for the expected exposures where the SCBA will be used. The additional radiant reflective criteria for helmets, gloves, footwear, and SCBA should at least meet the performance requirements of NFPA 1976 for radiant reflection.

Some proximity head protection utilizes a hard hat within the hood for impact head protection. Users are cautioned that hard hats may not provide adequate head protection for all situations where proximity protective clothing is needed.

The use of SCBA is required whenever fire fighters are exposed, or may be exposed, to the products of combustion, or to oxygen-deficient, or toxic atmospheres. Since this would cover most, if not all, proximity fire-fighting situations, SCBA will be used with proximity protective clothing. SCBA that is worn over or outside of the proximity garments will be unprotected from the radiant heat exposure of proximity fire-fighting operations and will be exposed to the high radiant levels and, therefore, is required to have proximity protective covers. Specialized proximity protective covers for SCBA will protect the SCBA cylinder, regulator, hoses, and straps from the radiant heat. Specialized faceshields that will mitigate the effects of high levels of radiant energy to the face and the SCBA facepiece, such as gold-coated reflective faceshields, should be utilized during proximity operations. SCBA could be worn under the proximity protective coat if it is so designed to allow access to the apparatus and the controls, and sufficient room for movement without binding or hindering the wearer.

EMERGENCY MEDICAL SERVICE (EMS) PROTECTIVE CLOTHING

NFPA 1500 requires that fire departments actively attempt to identify and limit or prevent the exposure of members to infectious and contagious diseases in the performance of their assigned duties. Fire departments must operate an infection control program that meets the requirements of NFPA 1581, *Standard on Fire Department Infection Control Program.*

When fire department members respond to emergency medical incidents, the fire department should consult with medical professionals and agencies on measures to limit the exposure of members to infectious and contagious diseases. This should include the provision and maintenance of equipment to avoid or limit direct physical contact with patients or contaminated fluids, when feasible.

For members who perform emergency medical care or otherwise may be exposed to blood or other body fluids, NFPA 1500 requires those persons to be provided with emergency medical garments, emergency medical face protection devices, and emergency medical gloves that are certified to the applicable requirements of NFPA 1999, *Standard on Protective Clothing for Emergency Medical Operations.*

NFPA 1581 addresses the provision of minimum requirements for infection control practices within a fire department. This is a comprehensive standard on infection control and requires that the fire department provide the following protective clothing and equipment for each of its members during emergency medical operations:

- Single-use medical gloves
- Fluid-resistant clothing
- Pocket masks
- Splash-resistant eyewear
- Respiratory-assist devices
- Approved sharps containers
- Leakproof bags

Each fire department should provide standard operating procedures for the selection and use of appropriate protective clothing and equipment during emergency medical work. Protective clothing, including garments, face protection, and medical gloves that are certified to NFPA 1999, are required to be provided to, and used by, personnel that may be exposed to blood or other body fluids. Fire fighters should use gloves certified to the glove requirements of NFPA 1971 (formerly NFPA 1973 criteria), including the requirements for water penetration, during situations where sharp or rough surfaces are likely to be encountered (e.g., vehicle extrication).

The Centers for Disease Control (CDC) *Guidelines for Public Safety Workers* discusses the selection and use of appropriate clothing and equipment during medical emergencies. CDC recommends that disposable gloves be worn by all personnel prior *to initiating any* emergency patient care tasks involving exposure to blood or body fluids. Extra pairs of gloves should always be available. For multiple victims, gloves should be changed between patient contacts.

While wearing gloves, members should avoid handling personal items, such as combs and pens, that could become contaminated. Gloves contaminated with blood or other body fluids to which *universal precautions* apply should be removed as soon as possible, taking care to avoid skin contact with the exterior surface.

The use of gloves does not eliminate the need to wash hands after emergency medical incidents. *Handwashing is one of the most important elements of infection control.*

Masks, eyewear, and gowns should be present on all emergency vehicles that respond, or might respond, to medical emergencies or victim rescues. These items should be donned by all personnel prior to any situation where splashes of blood or other body fluids to which *universal precautions* apply is likely to occur. Finally, contaminated gloves should be placed and transported in bags that prevent leakage, and disposed of properly; or, in the case of reusable gloves, cleaned, disinfected, and stored properly.

Because of the risk of salivary transmission of communicable viruses and bacteria (e.g., herpes simplex and Neisseria meningitides, respectively) and the theoretical risk of HIV and hepatitis B transmission during artificial ventilation of victims, CDC recommends the use of mechanical respiratory-assist devices, such as bag-valve masks and oxygen-demand valve resuscitators, or pocket mouth-to-mouth resuscitation masks (i.e., double lumen systems), designed to isolate emergency-response personnel from victims' blood and blood-contaminated saliva, respiratory secretions, and vomitus.

The CDC *Guidelines* include a table with examples of emergency medical tasks and activities, and the recommended personal protective equipment that would be appropriate for these tasks.

Design and Performance

NFPA 1999 was developed to provide biological protective clothing, gloves, and facewear for fire fighters, emergency medical technicians (EMTs), and paramedics to be utilized during emergency medical operations.

NFPA 1999 is the only standard, both from within and outside the fire service, that includes performance tests to ensure that each type of clothing resists penetration of bloodborne pathogens. The standard requires viral penetration resistance that provides resistance to the penetration of body fluids and bloodborne pathogens that the fire fighter can be exposed to during EMS functions. The materials, including seams, are exposed to a bacteriophage suspension of a liquid inoculated with *E. coli* C bacteria for 1 hr. Materials are tested in both directions and can not allow any penetration of the bacteriophage.

Garments also are required to meet performance requirements for liquidtight integrity in those areas of the garment that are designed to provide protection, material strength, and physical hazard resistance, seam strength, and closure strength.

Gloves are tested for tensile strength and elongation both before and after heat aging and isopropyl alcohol immersion. Additionally, gloves must meet requirements for dexterity, puncture resistance, liquidtight integrity, and minimum sizing. ASTM also has a standard for medical gloves, ASTM D 3578, *Standard Specification for Rubber Examination Glove.* This standard includes requirements for sampling to ensure quality control, such as watertightness testing for detecting holes in the gloves, physical dimension testing to ensure proper fit of gloves, and physical testing (tensile strength and ultimate elongation) to ensure that the gloves do not tear easily. This standard was incorporated into the requirements of NFPA 1999.

Like the garments, facewear must meet requirements for liquidtight integrity in those areas of the facewear that are designed to provide biological protection.

EMS protective clothing that is certified to NFPA 1999 must be certified by a third-party certification organization and be labeled and listed.

User instructions and information for all emergency medical garments, gloves, and face protection devices must include the following:

- Donning procedures
- Doffing procedures
- Safety considerations
- Optimum storage conditions
- Recommended storage life
- Decontamination recommendations and considerations
- Retirement considerations
- Disposal considerations

In addition, emergency medical garments must provide the following instruction and information:

- Cleaning instructions
- Marking and storage instructions
- Frequency and details of inspections
- Maintenance criteria

- How to use test equipment, where applicable
- Method of repair, if recommended by manufacturer
- Warranty information

Disinfecting

Each fire department must have procedures for the decontamination of specific items of clothing and equipment. Cleaning, disinfecting, and disposal criteria are included in NFPA 1581, and are required to be utilized by NFPA 1500. The CDC *Public Safety Worker Guidelines* also has recommendations for these procedures.

The CDC recommends, and U.S. OSHA requires, that laundry facilities or services be routinely made available by the employer. Protective clothing and station/work uniforms should be washed and dried according to the manufacturer's recommendations. Also, boots and leather goods can be scrubbed with soap and hot water to remove contamination. The OSHA bloodborne pathogen regulation, 29 CFR 1910.1030, also has specific laundering requirements.

Disinfectants used for decontaminating equipment should be EPA-registered hospital disinfectant chemical germicides that have been documented as effective against mycobacterium tuberculosis.

Care also must be taken in the use of disinfectants. Members should be aware of the flammability and reactivity of disinfectants and should follow manufacturer's instructions for use (e.g., contact time and temperature). Disinfectants should only be used with adequate ventilation and while wearing appropriate garments and equipment for cleaning and disinfecting, including eye protection, gloves, and aprons. It also is important, when disinfecting equipment, to check with the manufacturer of the germicide to determine compatibility of the protective clothing with the disinfectant.

Information on specific label claims of commercial germicides can be obtained by contacting the Disinfectants Branch, Office of Pesticides, EPA, 401 M St. SW, Washington, DC 20460.

CHEMICAL PROTECTIVE CLOTHING

Anyone involved in a hazardous materials incident must be protected against potential hazards. The purpose of personal protective clothing and equipment is to shield or isolate individuals from the chemical, physical, and biological hazards that may be encountered during hazardous materials responses. Adequate chemical protective clothing must be carefully selected and used to protect the respiratory system, skin, eyes, face, hands, feet, head, body, and hearing.

Structural fire-fighting protective clothing and equipment should not be used for hazardous materials incidents. Even when certified to the appropriate NFPA standards for structural fire fighting, these clothing and equipment elements provide little or no protection against hazardous chemicals. Use of structural protective elements for hazardous materials emergency response may result in serious injury or death, for the following reasons:

1. Structural fire-fighting protective clothing materials are easily permeated or penetrated by most hazardous chemicals. Some parts of structural fire-fighting clothing elements may actually absorb chemical liquids or vapors, increasing the likelihood of serious exposure.
2. Many hardware items will fail or lose function when contacted by chemicals (e.g., etching of visors, deterioration of straps, corrosion of hooks, or other metal items).
3. Contamination of structural fire-fighting protective clothing elements may not be effectively removed by laundering. Re-use of contaminated clothing may cause chronic exposure and accelerate physiological effects produced by contact with the chemical.

No single combination of protective clothing and equipment is capable of protecting against all hazards. Therefore, chemical protective clothing should be used in conjunction with other protective methods. The use of such clothing can itself create significant wearer hazards, such as heat stress, physical and psychological stress, impaired vision, impaired mobility, and impaired communications. In general, the greater the level of chemical clothing protection, the greater are the associated risks. For any given situation, protective clothing and equipment should be selected that provide an adequate level of protection. Overprotection, as well as underprotection, can be hazardous and should be avoided.

The approach to selecting chemical protective clothing and equipment must encompass an "ensemble" of clothing and equipment items that are easily integrated to provide both an appropriate level of protection and the ability to carry out emergency response activities. The following is a list of components that may form the chemical protective ensemble:

- Protective clothing (suit, coveralls, hoods, gloves, boots)
- Respiratory equipment (SCBA, combination SCBA/SAR)
- Cooling system (ice vest, air circulation, water circulation)
- Communications device
- Head protection
- Ear protection
- Inner garments
- Outer protection (overgloves, overboots, flashcovers)

Levels of Protection

The U.S. EPA has outlined four levels of protection: A, B, C, and D. The EPA defined these levels of protection primarily for workers at hazardous waste sites, where emergency conditions do not usually exist. These levels of protection are commonly, and often inappropriately, utilized by fire and emergency services. They are inadequate and do not correctly define the chemical protective clothing, with respect to the intended use based on the hazard and the required performance the selected clothing or equipment must offer.

EPA levels of protection can be used as the starting point for ensemble creation; however, each ensemble must be tailored to the specific situation in order to provide the most appropriate level of protection. For example, if the emergency-response activity involves a highly contaminated area or the potential of contamination is high, it may be advisable to wear a disposable covering such, as Tyvek® coveralls or PVC splash suits over the protective ensemble.

It is important to realize that selecting items by their design or configuration alone is not sufficient to ensure adequate protection. In other words, just having the right components to form an ensemble is not enough. Again, the EPA levels of protection do not define what performance the selected clothing or equipment must offer.

For emergency response, the only acceptable types of protective clothing are fully encapsulating suits, or totally encapsulating suits, and non-encapsulating or "splash" suits combined with accessory clothing items, such as chemically resistant gloves and boots. These descriptions apply to how the clothing is designed, and do not apply to performance. NFPA has classified chemical protective suits by their performance in three standards:

1. NFPA 1991, *Standard on Vapor-Protective Suits for Hazardous Chemical Emergencies*
2. NFPA 1992, *Standard on Liquid Splash-Protective Suits for Hazardous Chemical Emergencies*
3. NFPA 1993, *Standard on Support Function Protective Clothing for Hazardous Chemical Operation.*

Chemical protective clothing should completely cover both the wearer and the wearer's respiratory protection. Wearing SCBA or other respiratory equipment outside the suit subjects this equipment to the chemically contaminated environment. SCBA used in hazardous materials emergency response is generally the same as that for structural fire fighting. In general, respiratory protective equipment is not designed to resist chemical contamination, and should be protected from these environments. NFPA 1991 requires that SCBA be worn on the inside of vapor-protective suits. NFPA 1992, liquid splash-protective suits may be configured either with the SCBA on the inside or outside. However, *it is strongly recommended that respiratory equipment be worn inside the ensemble to prevent its failure and to reduce decontamination problems.*

The chemical resistance data provided by the manufacturer of the garment should be studied. Technical data packages are required to be provided by the manufacturers of protective suits that are certified to NFPA 1991 or NFPA 1992. Manufacturers of vapor-protective suits must provide permeation resistance data for their products, while penetration resistance data must accompany liquid splash-protective garments. Data must be provided for every primary material in the suit, including the garment, visor, gloves, and boots.

Permeation data should include the following:

- Chemical name
- Breakthrough time (shows how soon the chemical permeates)
- Permeation rate (shows the rate that the chemical comes through)
- System sensitivity (allows comparison of test results from different laboratories)

Chemical protective suit or garment materials that show no breakthrough or no penetration to a large number of chemicals are likely to have a broad range of chemical resistance. (Breakthrough times greater than 1 hr are usually considered to be an indication of acceptable performance.) If there are specific chemicals within the response jurisdiction that have not been tested, the manufacturer should be asked to supply test data on these chemicals.

The manufacturer's instruction manual should document all the features of the suit and describe what materials are used in its construction. It should cite specific limitations for the suit and what restrictions apply to its use. Procedures and instructions should be supplied for at least donning and doffing, inspection, maintenance, storage, decontamination, and use.

A number of situations can arise when no information is available to judge whether the chemical protective clothing will provide adequate protection. These situations include:

1. Chemicals that have not been tested with the garment materials
2. Mixtures of two or more different chemicals
3. Chemicals that cannot be readily identified
4. Extreme environmental conditions (e.g., hot temperatures)
5. Lack of data in all suit components (e.g., gloves, visors)

Testing material specimens using newly developed field test kits may offer one means for making on-site clothing selection. A portable test kit has been developed by the EPA, using a simple weight-loss method that allows field qualification of protective clothing materials within 1 hr. Use of this kit may overcome the absence of data and provide additional criteria for clothing selection.

Vapor-Protective Suits

NFPA 1991 covers vapor-protective suits, which are designed to provide "gastight" integrity and are intended for response situations where no chemical contact is permissible or where the exposure is unknown. This type of suit is equivalent to the clothing required in EPA's Level A. To determine adequate suit component performance in hazardous chemical environments, the following performance and testing requirements must be followed.

Performance and Testing

Permeation resistance testing specifies a battery of 21 chemicals and gases that were selected because they are *representative of the classes of chemicals* encountered during hazardous materials emergencies. Vapor-protective suits must resist permeation by the chemicals present during a response. Permeation occurs when chemical molecules "diffuse" through the material, often without any evidence of chemical attack. Permeation resistance is measured in terms of breakthrough time. An acceptable material is one where the breakthrough time exceeds the expected period of garment use. A chemical permeation resistance time of at least 1 hr or more must be met for each chemical specified in ASTM F 1001, *Standard Guide for Selection of Chemicals to Evaluate Protective Clothing Materials*, and is required for primary suit materials (garment, visor, gloves, and boots). Any additional chemicals or specific chemical mixtures for which the manufacturer is certifying the suit must meet the same permeation performance requirements.

Gastight integrity testing evaluates the suit for air leakage using a pressurization test to check *each* protective suit. The suit must maintain at least 1.6 in. water gage pressure.

Liquidtight integrity testing evaluates the suit for water leakage, using an overall suit liquid penetration test. No liquid penetration of the suit can occur.

Overall suit function and integrity testing evaluates the entire suit for ability to function and retain gastight integrity while a test subject performs exercises while wearing the suit. The suit must maintain at least 1.6 in. water gage pressure.

Suit ventilation rate testing evaluates the suit's ability to vent internal overpressure, without compromising the suit or the gastight criteria.

Physical durability testing for the materials used for vapor-protective suits evaluates the materials ability to perform and resist chemical breakthrough for at least 1 hr after being subjected to abrasion and flexing, and also tests for burst strength, puncture propagation, flammability resistance, and cold-temperature performance.

Liquid-Splash Protective Suits

NFPA 1992 covers liquid splash-protective suits, which are designed to protect emergency responders against liquid chemicals in the form of splashes, but not against continuous liquid contact or chemical vapors and gases. Liquid splash-protective suits may be acceptable with some chemicals that do not present vapor hazards. Essentially, this type of clothing meets EPA Level B needs. It is important to note, however, that, *by wearing liquid splash-protective clothing, the wearer accepts exposure to chemical vapors and gases.* Materials that prevent liquid penetration can still allow chemical permeation, since this clothing does not offer gastight performance, even if duct tape is used to seal clothing interfaces.

NFPA 1992 specifies a battery of seven liquid chemicals that are representative of the classes of chemicals likely to be encountered during hazardous materials emergencies. Any additional chemicals or specific chemical mixtures for which the manufacturer is certifying the suit must meet the same penetration performance requirements.

Liquid splash-protective suits are for use only for protection from liquids with low vapor pressures and with no known skin absorption toxicity. Liquid splash-protective suits cannot be used for known or suspected carcinogens.

Performance and testing requirements are listed, as follows:

Penetration resistance testing specifies a battery of seven chemicals that were selected because they are *representative* of the classes of chemicals that are encountered during hazardous materials emergencies. Penetration resistance is measured in terms of penetration resistance time. An acceptable material is one where the resistance to liquid transfer time exceeds the expected period of garment use. A chemical penetration resistance time of at least 1 hr or more must be met for each of the seven chemicals specified, and is required for primary suit materials (garment, visor, gloves, and boots). Any additional chemicals or specific chemical mixtures for which the manufacturer is certifying the suit must meet the same penetration performance requirements.

Overall suit function and integrity testing evaluates the entire suit for ability to function and retain liquidtight integrity while a test subject performs exercises while wearing the suit. The suit must not allow any liquid penetration.

Liquidtight integrity testing evaluates the suit for water leakage, using an overall suit liquid penetration test. No liquid penetration of the suit can occur.

Physical durability testing for the materials used for liquid-protective suits evaluates the materials' ability to perform and resist chemical penetration for at least 1 hr after being subjected to abrasion and flexing, and also tests for burst strength, puncture propagation, flammability resistance, and cold temperature performance.

Support Function Protective Clothing

NFPA 1993 covers support function protective clothing that provides liquid splash protection, but offers limited physical protection. These garments can be constructed without the rigid construction requirements for garments that are intended for re-use. They can be designed by the manufacturer for a single-use or a limited-use expectancy.

These garments may comprise several separate protective clothing components (i.e., coveralls, hoods, gloves, and boots). They are intended for use in nonemergency, nonflammable, controlled situations where the chemical hazards have been completely characterized. Examples of support functions include decontamination, hazardous waste cleanup, and training.

Support function protective garments must not be used during emergency response outside of support functions.

Performance and testing requirements are listed, as follows:

Penetration resistance testing specifies a battery of seven chemicals that were selected because they are *representative* of the classes of chemicals that are encountered during hazardous materials emergencies. Penetration resistance is measured in terms of penetration resistance time. An acceptable material is one where the resistance to liquid transfer time exceeds the expected period of garment use. A chemical penetration resistance time of at least 1 hr or more must be met for each seven chemicals specified, and is required for support function materials, including visor, gloves, and boots. Any additional chemicals or specific chemical mixtures for which the manufacturer is certifying the garment must meet the same penetration performance requirements.

Liquidtight integrity testing evaluates the garment for water leakage, using an overall suit liquid penetration test. No liquid penetration of the garment can occur.

Physical durability testing evaluates burst strength and puncture propagation.

BIBLIOGRAPHY

References Cited

1. Karter, M. J., "Patterns of Fire Fighter Fireground Injuries, 1989–1993," NFPA Fire Analysis & Research Division, Mar. 1996.
2. Karter, M. J., "Patterns of Fire Fighter Injuries, 1989–1991," NFPA Fire Analysis & Research Division, Quincy, MA, Dec. 1993.

NFPA Codes, Standards, and Recommended Practices

Reference to the following NFPA codes, standards, and recommended practices will provide further information on fire service protective clothing and equipment discussed in this chapter. (See the latest version of *The NFPA Catalog* for availability of current editions of the following documents.)

NFPA 1001, *Standard for Fire Fighter Professional Qualifications*

NFPA 1041, *Standard for Fire Service Instruction Professional Qualifications*

NFPA 1404, *Standard on Fire Department Self-Contained Breathing Apparatus Program*

NFPA 1500, *Standard on Fire Department Occupational Safety and Health Program*

NFPA 1581, *Standard on Fire Department Infection Control Program*

NFPA 1971, *Standard on Protective Ensemble for Structural Fire Fighting*

NFPA 1976, *Standard on Protective Clothing for Proximity Fire Fighting*

NFPA 1981, *Standard on Open-Circuit Self-Contained Breathing Apparatus for Fire Fighters*

NFPA 1982, *Standard on Personal Alert Safety Systems (PASS) for Fire Fighters*

NFPA 1991, *Standard on Vapor-Protective Suits for Hazardous Chemical Emergencies*

NFPA 1992, *Standard on Liquid Splash-Protective Suits for Hazardous Chemical Emergencies*

NFPA 1993, *Standard on Support Function Protective Clothing for Hazardous Chemical Operations*

NFPA 1999, *Standard on Protective Clothing for Emergency Medical Operations*

Additional Reading

Beason, D. G., Johnson, J. S., and Weaver, W. A., "Summary Report-California Department of Forestry and Fire Protection Evaluation of Full-Face Air-Purifying Respirators for Wildland Fire Fighting Use," UCRL-CR-122559, prepared for Calif. Department of Forestry and Fire Protection, WN-02-19-05-0, Univ. of Calif., Livermore, CA, Feb. 1996.

Behnke, W. P., Geshury, A. J., and Barker, R. L., "Thermo-Man and Thermo-Leg: Large Scale Test Methods for Evaluating Thermal Protective Performance," North Carolina State Univ., Raleigh, ASTM STP 1133; American Society for Testing and Materials (ASTM), Performance of Protective Clothing: Challenges for Developing Protective Clothing for the 1990s, Fourth (4th) Volume, ASTM STP 1133, June 18–20, 1991, Montreal, Canada, ASTM, Philadelphia, PA, 1991, pp. 266–280.

Blirup, N. O., "Protective Clothing: Heavy Shielding. A Safety Guarantee or a Trap?," *Fire Engineers Journal*, Vol. 56, No. 181, March 1996, pp. 30–31.

Dale, J. D. *et. al.*, "Instrumented Mannequin Evaluation of Thermal Protective Clothing," University of Alberta, Edmonton, Canada, ASTM STP 1133; American Society for Testing and Materials (ASTM), Performance of Protective Clothing: Challenges for Developing Protective Clothing for the 1990s, 4th Vol., ASTM STP 1133, June 18–20, 1991, Montreal, Canada, ASTM, Philadelphia, PA, 1991, pp. 717–733.

Damioli, C., "EMS Protective Clothing Continues to Unfold," *Fire Chief*, Vol. 39, No. 4, April 1995, pp. 53–54, 56–57.

Dukes-Dobos, F. N., *et. al.*, "Assessment of Ventilation of Firefighter Protective Clothing," Polytechnic of Wales, UK, University of South Florida, Tampa, ASTM STP 1133; American Society for Testing and Materials (ASTM), Performance of Protective Clothing: Challenges for Developing Protective Clothing for the 1990s, 4th Vol., ASTM STP 1133, June 18–20, 1991, Montreal, Canada, ASTM, Philadelphia, PA, 1991, pp. 629–633.

Federal Emergency Management Agency, "Minimum Standards on Structural Fire Fighting Protective Clothing and Equipment: A Guide for Fire Service Education and Procurement," Federal Emergency Management Agency, Washington, DC, FA-137, Sept. 1993.

Foote, K. L., "Determination of Toxic Material Penetrations for Wildland Respirator Filters," MS thesis, UCRL-LR-122560, Univ. of Calif., Livermore, CA, May 1994.

"Helmets," Fire International, No. 148, Autumn 1995, pp. 22–24.

Hoffman, S., "Training Should Include Protective Clothing," Fire, Vol. 87, No. 1075, Jan. 1995, pp. 21–22.

Lawson, J. R., "Fire Fighter's Protective Clothing and Thermal Environments of Structural Fire Fighting," NISTIR 5804, National Institute of Standards and Technology, Gaithersburg, MD, Aug. 1996.

Lion, D. R., "Protective Clothing Standards," Fire Fighting in Canada, Vol. 39, No. 7, Sept. 1995, p. 14.

McDowell, T., "Firefighter Protective Clothing: Where It Is, Where It's Going," Fire Chief, Vol. 37, No. 9, Sept. 1993, pp. 30–34.

Stanhope, M. T., "Thermal Testing of Fabrics for Protective Clothing," 1992 Fall Conference, Industry Speaks Out on Flame Retardancy: Coatings; Polymers and Compounding; Test Method Development; New Products, Oct. 27–30, 1992, Charleston, SC, Technomic Publishing Co., Lancaster, PA, 1992, pp. 179–209.

Stull, J. O., "NFPA 1999: A Standard of Protective Clothing for EMS Operations," Fire Engineering, Vol. 147, No. 3, March 1994, pp. 79–80, 82, 84, 86.

Stull, J. O., et. al., "Developing and Selecting Test Methods for Measuring Protective Clothing Performance in Chemical Flashover Situations," TRI Environmental, Austin, TX, Biotherm, Inc., Beaver Creek, OH, Kappler Safety Group, Guntersville, AL, Phoenix Fire Dept., AZ, ASTM STP 1133; American Society for Testing and Materials (ASTM), Performance of Protective Clothing: Challenges for Developing Protective Clothing for the 1990s, 4th Vol., ASTM STP 1133, June 18–20, 1991, Montreal, Canada, ASTM, Philadelphia, PA, 1991, pp. 908–923.

Thomas, J., "NFPA vs. ISO: The Protective Clothing Standards Debate," Fire International, No. 146, Spring 1995, pp. 22–23.

U.S. Fire Administration, "Protective Clothing and Equipment Needs of Emergency Responders for Urban Search and Research Missions," Fire Administration, Washington, DC, FA 136, Sept. 1993.

Villa, K. M., "Textile Test Methods for Protective Clothing Standards," Fourth Annual Conference on Protective Clothing: a New Review of Applications, Materials Used, Needs, Concerns, and Trends, April 24–26, 1990, Charlotte, NC, 1990, pp. X/1–5.

FIRE STREAMS

Revised by Robert G. Purington

Water has always been and remains the primary fire extinguishing agent; therefore, the movement of water from its source to the fire is a significant operation in most kinds of fire fighting. A typical fire department delivery system consists of suction hose, a pumper to increase the water pressure to overcome friction loss and provide nozzle pressure, a supply line to move the water from the pump to the fire scene, and "working lines" and nozzles for the final delivery. Water is used in a number of forms and with various additives, and there are numerous combinations in which the water can be delivered to the fire scene.

This chapter includes a description of the most important details of good fire streams, including fire apparatus pumps, nozzles, fire hose friction loss, pump pressure calculations, friction-reducing additives, and reaction forces.

FIRE DEPARTMENT PUMPS

The primary purpose of fire department pumpers is to move water with adequate pressure from a source to the fire. Sources include hydrants, lakes, streams, rivers, or storage tanks in fixed locations or in vehicles. Pumpers serve other functions, including transporting fire fighters, water and other extinguishing agents, and equipment. Pumps are also installed on specialized apparatus, such as tank trucks, aerial ladders, and elevating platform apparatus, aircraft crash vehicles, wildland fire apparatus, and fire boats.

Pumpers, or "engines," are generally equipped with a centrifugal pump and the necessary fittings for pressure regulation and priming operations. Piston, rotary gear, and rotary vane pumps are used for special purposes, such as high-pressure booster pumps and priming. Requirements for standard pumpers are given in NFPA 1901, *Standard for Automotive Fire Apparatus.*

Two important characteristics of fire pumps are: (1) the discharge rate and (2) the discharge pressure. They comprise the final output of the pump. The discharge rate, also known as the quantity, flow rate, or volume, is the amount of water pumped per unit time, such as gallons per minute (gpm) or liters per minute (L/min). The discharge pressure, indicated by gages, is the pressure at which the pump is pumping water. The discharge pressure is in pounds per square inch (psi) or kilopascals (kPa). Pumps are rated according to their flow rate at 150 psi (1034 kPa) discharge and generally range from 750 gpm (2,839 L/min) to 2,000 gpm (7,571 L/min). These

pumps are also required to provide 70 percent of their rated capacity at 200 psi (1,379 kPa) and 50 percent of their rated capacity at 250 psi (1,724 kPa). These ratings are summarized in Table 10-19A.

TABLE 10-19A. *Standard Pumps*

Rated Flow Rate		70 Percent of Rated Flow Rate		50 Percent of Rated Flow Rate	
gpm @ 150 psi	L/min @ 1034 kPa	gpm @ 200 psi	L/min @ 1379 kPa	gpm @ 250 psi	L/min @ 1724 kPa
500	1893	350	1325	250	946
750	2839	525	1987	375	1419
1000	3785	700	2650	500	1893
1250	4731	875	3312	625	2366
1500	5678	1050	3974	750	2839
1750	6624	1225	4637	875	3312
2000	7571	1400	5300	1000	3785
2250	8516	1575	5961	1125	4258
2500	9463	1750	6624	1250	4731

The centrifugal pump, a nonpositive displacement pump, consists essentially of a pair of rotating discs called shrouds, separated by curved partitions, which are called vanes. This assembly makes up the impeller. Typically, fire apparatus pumps are multistage, usually two, thus permitting discharge at either higher flow rates with lower pressures or lower flow rates with higher pressures. The purpose of the multistage pump is to obtain the most effective operation at the lowest possible engine speed.

Pumping operations start with water from either a pressure source, such as a hydrant, or from a static water source, such as a pond. Drafting operations are required in the latter case. The factors that affect the height (lift) limitations to which water can be drafted and the flow rate are interrelated. These factors are: (1) atmospheric pressure, (2) water temperature, (3) friction loss in the suction hose and strainer, and (4) velocity energy.

The lift at a given flow rate can be determined by: Lift equals atmospheric pressure minus water vapor pressure minus strainer and suction hose friction loss minus velocity energy.

Atmospheric pressure varies from day to day according to the weather conditions and the elevation above sea level. Since drafting is essentially the creation of a vacuum or partial vacuum in the pump and suction hose, the total energy available for moving the

Robert G. Purington is a retired fire chief and a consulting engineer based in Arnold, CA.

water from the source to the pump is the atmospheric pressure. The nominal atmospheric pressure, 14.696 psia (101.333 kPa) at sea level, usually rounded to 14.7 psia (100 kPa), can theoretically provide a lift of almost 34 ft (10.5 m). Under actual conditions, a pump in excellent condition and pumping its rated capacity at sea level should be able to draft water at least 25 ft (7.6 m). However, remember atmospheric pressure drops about 0.5 psi (3.4 kPa) for each 1,000 ft (305 m) above sea level.

Often atmospheric pressure is given in inches of mercury. If this is the case, it must be converted to feet of water using:

$$h_w = 1.13 h'_{Hg}$$

where:

$$h_w = \text{feet of water}$$

$$h'_{Hg} = \text{inches of mercury}$$

The vapor pressure reduces the amount of vacuum that is possible and consequently reduces the total lift. Vapor pressure depends on the water temperature and is measured in terms of feet (meters) of water. Table 10-19B lists common values of vapor pressure as a function of water temperature.

Another factor affecting the total lift is the friction loss in the suction hose and the strainer. The friction loss in the suction hose depends on the hose diameter, hose length, and the flow rate of the water. Equations for the friction loss in various sizes of suction hose are shown in Table 10-19C. The friction loss through the strainer is a function of the flow rate and the strainer diameter. The friction equations for the various strainer sizes are shown in Table 10-19D.

The velocity head also affects the total lift of the pump. The velocity head is given by $V^2/2g$, where V is the water velocity in feet per second (meters per second) and g is the acceleration due to gravity and is equal to 32.2 ft/sec^2 (9.81 m/sec^2). The velocity can be found by using the continuity equation, $Q = VA$, where Q is the flow rate in cubic feet per second (cubic meters per second) and A is the cross-section of the suction hose in square feet (square meters). However, the flow rate is usually given in gallons (liters) per minute and the suction hose diameter is in inches (millimeters). Consequently, Table 10-19E can be used to determine the velocity head in the suction hose without tedious calculations.

TABLE 10-19B. *Vapor Pressure*

Water Temperature		Vapor Pressure	
°F	°C	ft of water	kPa
60	16	0.59	1.7
80	27	1.20	3.5
100	38	2.20	6.6
120	49	3.90	11.6

TABLE 10-19C. *Suction Hose Friction Losses per 10 ft (3 m) Length*

Suction Hose Diameter		Friction Loss			
in.	mm	psi	kPa	ft	m
4	100	$0.022q^2$*	$0.0109q^2$†	$0.0519q^2$	$0.00978q^2$
4½	113	$0.013q^2$	$0.0061q^2$	$0.0291q^2$	$0.00054q^2$
5	125	$0.007q^2$	$0.0034q^2$	$0.0162q^2$	$0.00033q^2$
6	150	$0.003q^2$	$0.0014q^2$	$0.0066q^2$	$0.00014q^2$

* q = gpm × 100
† q = L/min × 100.

TABLE 10-19D. *Suction Strainer Friction Losses*

Suction Hose Diameter		Friction Loss			
in.	mm	psi	kPa	ft	m
4	100	$0.0418q^2$*	$0.0202q^2$†	$0.0966q^2$	$0.00229q^2$
4½	113	$0.0367q^2$	$0.0178q^2$	$0.0848q^2$	$0.00142q^2$
5	125	$0.0224q^2$	$0.0108q^2$	$0.0517q^2$	$0.00093q^2$
6	150	$0.009q^2$	$0.0043q^2$	$0.0207q^2$	$0.00054q^2$

* q = gpm × 100
† q = L/min × 100.

TABLE 10-19E. *Velocity Head (V) for Suction Hose, ft (m) Water*

Flow, Q		Suction Hose Diameter, (D) in in. and mm											
		3.0 in.	76 mm	3.5 in.	89 mm	4.0 in.	100 mm	4.5 in.	113 mm	5.0 in.	125 mm	6.0 in.	150 mm
gpm	L/min	ft	m	ft	m	ft	m	ft	m	ft	m	ft	m
100	378	0.32	0.09	0.17	0.05	0.10	0.03	0.06	0.02	0.04	0.01	0.02	0.006
200	757	1.28	0.39	0.69	0.21	0.40	0.12	0.25	0.05	0.17	0.05	0.08	0.02
300	1136	2.88	0.88	1.55	0.47	0.91	0.28	0.57	0.17	0.37	0.11	0.18	0.05
400	1514	5.11	1.56	2.76	0.84	1.62	0.49	1.01	0.31	0.66	0.20	0.32	0.09
500	1892	7.99	2.44	4.31	1.31	2.53	0.78	1.58	0.48	1.04	0.32	0.50	00.15
600	2271	11.51	3.51	6.21	1.90	3.64	1.11	2.57	0.78	1.49	0.45	0.72	0.22
700	2650			8.45	2.58	4.96	1.51	3.09	0.94	2.03	0.62	0.98	0.30
800	3028			11.04	3.36	6.47	1.97	4.04	1.23	2.65	0.81	1.28	0.39
900	3407					8.19	2.50	5.11	1.56	3.36	1.02	1.62	0.49
1000	3785					10.11	3.08	6.31	1.92	4.14	1.26	2.00	0.61
1100	4163					12.24	3.73	7.64	2.33	5.01	1.53	2.42	0.74
1200	4542							9.09	2.78	5.97	1.82	2.88	0.88
1300	4921							10.67	3.25	7.00	2.13	3.38	1.03
1400	5300							12.37	3.78	8.12	2.47	3.92	1.19
1500	5677									9.32	2.84	4.49	1.37

To find the friction loss in parallel suction hoses, divide the friction loss by the square of the number of lines $(\#lines)^2$. For example, with two parallel suction hose lines, of the same size and length, divide the calculated suction hose friction loss by 4.

In solving the lift equation, all units must be compatible, i.e., converted to the equivalent feet (m) of water or kilopascals. Atmospheric pressure is usually given in terms of inches of mercury (or kilopascals). Table 10-19F can be used to convert the atmospheric pressure to the equivalent pressure in feet (m) of water.

TABLE 10-19F. *Inches of Mercury Vacuum Compared with Height of Water and Corresponding Pressure*

Inches of Mercury	Vacuum Pressure psig	psia	Height of Water (ft)
0.00	0.000	14.696	.000
1.00	−0.491	14.205	1.132
2.00	−0.982	13.714	2.263
3.00	−1.473	13.223	3.395
4.00	−1.965	12.731	4.527
5.00	−2.456	12.240	5.658
6.00	−2.947	11.749	6.790
7.00	−3.438	11.258	7.922
8.00	−3.929	10.767	9.054
9.00	−4.420	10.276	10.185
10.00	−4.912	9.784	11.317
11.00	−5.403	9.293	12.449
12.00	−5.894	8.802	13.580
13.00	−5.385	8.311	14.712
14.00	−6.876	7.820	15.844
15.00	−7.367	7.329	16.975
16.00	−7.858	6.838	18.107
17.00	−8.350	6.346	19.239
18.00	−8.841	5.855	20.370
19.00	−9.332	5.364	21.502
20.00	−9.823	4.873	22.634
21.00	−10.314	4.382	23.766
22.00	−10.805	3.891	24.897
23.00	−11.297	3.399	26.029
24.00	−11.788	2.908	27.161
25.00	−12.279	2.417	28.292
26.00	−12.770	1.926	29.424
27.00	−13.261	1.435	30.556
28.00	−13.752	0.944	31.687
29.00	−14.244	0.453	32.819
29.92	−14.696	0.000	32.862

For SI units: 1 in. of mercury = 3.385 kPa; 1 psi = 6.895 kPa; 1 ft = 0.305 m.

EXAMPLE: Find the lift in feet of water (or kilopascals, which can be converted to meters of water by multiplying by 9.84).

Atmospheric pressure = 27.00 in. of mercury (91 kPa)

Water temperature = 60°F (16°C)

Flow rate = 1100 gpm (4163 L/min)

Suction hose diameter (strainer diameter) = 5 in. (125 mm)

Suction hose = Two lengths of 10 ft (3 m) each

SOLUTION: From Table 10-19F, atmospheric pressure of 27.00 in. of mercury (91 kPa) equals 30.556 ft of water. From Table 10-19B, vapor pressure [at 60°F (16°C)] equals 0.59 ft of water (1.7 kPa). Using Table 10-19C, the friction loss for 10 ft (3 m) of 5-in. (125-mm) suction hose is $0.0162q^2$ in feet of water and $0.0034q^2$ in kPa.

$Q = 1100$ gpm (4163 L/min). However, the equation for q is in 100s of gpm or L/min, and $Q/100 = q$. Therefore, use $q = 11(41.63)$. Thus:

$$FL = (0.0162)(11)^2 = 1.96 \text{ ft for 10 ft}$$
$$\text{or } (2)(1.96) = 3.92 \text{ ft for 20 ft}$$

In SI units:

$$FL = (0.0034)(41.63)^2 = 5.89 \text{ kPa for 3 m}$$
$$\text{or } (2)(5.89) = 11.78 \text{ kPa for 6 m}$$

From the suction strainer equation listed in Table 10-19D:

$$FL = 0.0517q^2$$

where $q = 11$ as in the suction equation

$$FL = (0.0517)(11)^2 = 6.26 \text{ ft}$$

In SI units:

$$FL = 0.0108q^2,$$

where

$$q = 41.63$$
$$FL = (0.0108)(41.63)^2 = 18.71 \text{ kPa}$$

The velocity head (from Table 10-19E) is 5 ft (14.9 kPa).

$$\text{Lift} = 30.56 - 0.59 - 4 - 6.26 - 5 = 14.71 \text{ ft}$$

In SI units:

$$\text{Lift} = 91 - 1.7 - 11.78 - 18.71 - 14.9 = 43.91 \text{ kPa}$$

(converting to meters, 43.91/9.84 = 4.46 m)

NOZZLES

The purpose of a nozzle is to shape the stream and convert pressure energy to velocity (kinetic) energy. In this way, the water can be applied to the fire in the appropriate quantity and from an adequate distance. There are many specialized nozzles. However, they all can be classified in two ways: (1) solid stream and (2) fog (spray) nozzles. (See Figure 10-19A.) They both have advantages and disadvantages. These characteristics will be discussed.

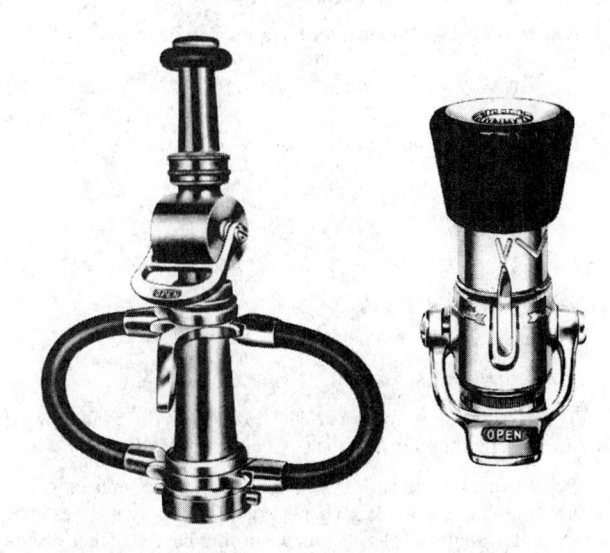

FIG. 10-19A. Straight stream nozzle (left) and fog nozzle (right). (Source: W. S. Darley and Co.)

The governing hydraulic characteristics of nozzles are the flow rate, diameter, and the nozzle pressure. The flow rate is the amount of water flowing out of the nozzle per unit time and is measured in gallons per minute or liters per minute.

Nozzle pressure is the flowing or velocity pressure of the water coming from the nozzle tip, or the pressure at the base of the nozzle with water flowing. For solid stream nozzles, the nozzle pressure can be measured using a pitot tube and gauge, while the nozzle pressure for either a solid stream or spray nozzle can be measured with a ring gauge at the base of the nozzle. Practically speaking, there is usually little numerical difference in the two measurements. The unit of nozzle pressure is pounds per square inch (psi) or kilopascals (kPa).

Solid-Stream Nozzles

Solid-stream nozzles are classified according to the nozzle diameter. The diameters range from $1/4$ in. to $2^1/_2$ in. (6 mm to 64 mm) or larger. The nozzle sizes are generally available in $1/8$ in. (3 mm) graduations. Nozzles up to $1^1/_8$ in. (29 mm) or perhaps $1^1/_4$ in. (32 mm) are generally considered to be for "handlines," i.e., those manually held. Nozzles larger than about $1^1/_4$ in. (32 mm) have such large reaction forces that they must be mechanically restrained. These are usually called master streams, monitors, deluge, or deck guns. The $1^1/_8$-in. (29-mm) nozzle produces the so-called standard stream of 250 gpm (946 L/min) at about 45 psi (310 kPa) nozzle pressure.

Solid streams are useful where extreme range is desired, where it is necessary to penetrate soft material, such as peat, or when thermal degradation of spray streams prevents proper penetration.

Solid-stream nozzles are also useful in measuring water flows, since both their diameter and the Pitot pressure can be measured. From this information, the flow rate can be computed. It is related to the diameter and nozzle pressure by the following equations:

$$Q = 29.7d^2\sqrt{P}$$

where:

Q = flow rate, gpm

d = nozzle diameter, in.

P = nozzle pressure (Pitot), psi

For most field hydraulics, the equation can be rounded to:

$$Q = 30d^2\sqrt{P}$$

For SI units, the flow equation is:

$$Q = 0.066d^2\sqrt{P_m}$$

where:

Q = flow rate (L/min)

d_m = inside nozzle diameter (mm)

P_m = nozzle pressure (kPa)

Theoretical flows from various nozzles at different nozzle pressures can be found in Section 6, Chapter 6, "Hydraulics."

Solid streams, with a lower surface-area-to-volume ratio, do not have as good heat-transfer characteristics as spray nozzles and, consequently, are not as effective in absorbing heat. Another disadvantage of a solid stream is that it is a better conductor of electricity than a spray.

Spray Nozzles

Spray nozzles, or fog nozzles as they are often called, produce varying degrees of water spray. They may have a fixed-spray angle, or may be adjustable from almost a straight stream to a very wide angle spray. The spray angle is customarily measured at the angle between the outer limits of the spray core. Many spray nozzles have predetermined pattern settings, usually at straight-stream, 30-, 60-, and 90-degree spray angles; however, most of them can also be set at intermediate angles.

Spray nozzles are generally classified according to the size of their hose coupling and their flow rate in gpm (L/min). Table 10-19G shows typical flow rates from spray nozzles.

TABLE 10-19G. *Flows from Typical Fog Nozzles*

Nozzle Type	Flow Rate	
	gpm	L/min
$3/4$- to 1-in. [19- to 25-mm] booster spray nozzle	10–40	38–151
$1^1/_2$-in. (38-mm) spray nozzle	70–150	265–568
$2^1/_2$-in. (65-mm) spray nozzle	200–300	757–1136
Master stream nozzles	500	1893
	750	2839
	1000	3785
	1250	4732
	1500	5678
	2000	7570

The designated flow from a spray nozzle is usually rated at 100 psi (690 kPa) nozzle pressure. Some spray nozzles have different flows at various angles. Generally, the flow increases as the angle widens. Other spray nozzles are "constant gallonage," i.e., they have the same flow at all spray angles. In addition, some models have a flow rate setting; they can be set for several different flow rates.

Flow rates are determined experimentally, either by a flow meter or by flowing water into a tank and measuring the time for a given quantity. Flow rate information is obtained from the manufacturer, and is usually reported as the flow rate at 100 psi (690 kPa) nozzle pressure for the various spray angles.

Flows for other pressures may be obtained by using the flow equation:

$$Q = 29.7d^2\sqrt{P}$$

and finding the equivalent nozzle diameter, d_e, using

$$d_e = \sqrt{Q/29.7\sqrt{P}}$$

where:

d_e = equivalent nozzle diameter (in.)

For SI units, the flow formula is:

$$Q = 0.066d_m^2\sqrt{P_m}$$

where:

Q = flow rate (L/min)

d_m = inside nozzle diameter (mm)

P_m = nozzle pressure (kPa)

HOSE LINE FIRE STREAM CALCULATIONS

The pump pressure (commonly called "engine pressure") is usually the end result of hose stream calculations. The methods for obtaining pump pressure are described below. However, algebraic manipulation of the equations can yield any variable, providing the other quantities are known.

Special techniques, known as "field hydraulics" or "rule-of-thumb," are used by pump operators for approximate results. The methods shown below give more precise results, but are generally too involved for field applications. Details on rule-of-thumb fire stream calculations can be found in the fire-fighting hydraulic texts listed in the bibliography of this chapter.

The calculation of the pump pressure depends on a number of variables: discharge pressure, hose friction loss, and back pressure due to elevation changes.

As shown previously, the flow rate is a function of the nozzle pressure and nozzle diameter. The hose friction loss is a function of the hose diameter and the hose length, in addition to losses in various appliances, such as monitors and ladder pipes.

Back pressure is the loss or gain of potential energy due to elevation changes between the pump and the nozzle. This is determined by the equations:

$$BP = 0.434h \quad (\text{SI units: } BP_m = 9.81\, h_m)$$

where:

BP = back pressure, psi (kPa)

h = elevation change or head, ft (m) of water

The basis of these calculations revolves around three fundamental hydraulic principles.

Bernoulli's Equation

A special form of conservation of energy. For a pumper hose line system this becomes:

$$EP = NP + FL + BP$$

where:

EP = pump pressure, psi (kPa)

NP = nozzle pressure, psi (kPa)

FL = friction loss, psi (kPa)

BP = back pressure, psi (kPa)

The Darcy-Weibach Equation

$$FL = \left(fV^2 L\right)/(2gD)$$

where:

g = acceleration due to gravity, ft/sec² (m/s²)

f = friction factor, a dimensionless number

L = hose length, ft (m)

D = hose diameter, in. (mm)

The Continuity Equation

This is a special form of conservation of mass.

$$Q = VA$$

where:

Q = flow rate, ft³/sec (m³/s)

V = velocity, ft/sec (m/s)

A = area, ft² (m²)

These equations can be manipulated to result in friction loss equations and a modern version of the Underwriters equation.[1]

Friction loss is the loss of energy resulting from the internal friction between the moving particles of water and the friction between the water and the lining of the hose. The friction loss depends on water viscosity, flow rate, hose diameter, hose length, and the roughness characteristics of the hose lining. While the viscosity is partly a function of the water temperature, the change is insignificant in the ranges of the temperature of water commonly used for fire fighting. The flow rate, hose diameter, and length are variables that are determined by the fire-fighting requirements. The roughness of the hose lining cannot be easily measured and is generally determined experimentally and reported as the friction factor. Consequently, all variables, except the flow rate, hose diameter, and hose length, are embodied in the friction factor, f, of the Darcy-Weibach equation and in the continuity equation. Putting all constants into one constant factor results in $FL = cq^2L$ for a given hose diameter, where L is traditionally given in hundreds of feet (meters) of hose and q is hundreds of gallons per minute or hundreds of liters per minute ($q = Q/100$).

For a standard 2½-in. (65-mm) double-jacket, lined hose, the currently recommended constant is $c = 2$, or $FL = 2q^2L$. (For SI units: $FL = 3.17q^2L$.) Suggested constants for other hose sizes and friction loss equations are shown in Table 10-19H.

The friction factor for modern fire hose varies from manufacturer to manufaturer. It is prudent to determine a necessary friction factor by flow tests.

TABLE 10-19H. *Recommended Friction Loss Equations*

Hose Diameter and Type in. (mm)	Friction Loss Equation	
	psi	kPa
³/₄ (19) booster	$FL = 1100q^2L$*	$FL = 1741q^2L$ †
1 (25) booster	$FL = 150q^2L$	$FL = 238q^2L$
1¼ (32) booster	$FL = 80q^2L$	$FL = 127q^2L$
1¼ (32) linen	$FL = 127q^2L$	$FL = 201q^2L$
1½ (38) rubber-lined	$FL = 24q^2L$	$FL = 38q^2L$
1½ (38) linen	$FL = 51.2q^2L$	$FL = 81q^2L$
1¾ (44) with 1½ (38) couplings	$FL = 15.5q^2L$	$FL = 24.6q^2L$
2 (51) rubber-lined	$FL = 8q^2L$	$FL = 12.7q^2L$
2 (51) linen	$FL = 12.5q^2L$	$FL = 19.8q^2L$
2½ (65) linen	$FL = 4.26q^2L$	$FL = 6.75q^2L$
2½ (65) rubber-lined	$FL = 2q^2L$ ‡	$FL = 3.17q^2L$
2¾ (70) with 3 (76) couplings	$FL = 1.5q^2L$	$FL = 2.36q^2L$
3 (76) with 2½ (64) couplings	$FL = 0.80q^2L$	$FL = 1.27q^2L$
3 (76) with 3 (76) couplings	$FL = 0.677q^2L$	$FL = 1.06q^2L$
3½ (89)	$FL = 0.34q^2L$	$FL = 0.53q^2L$
4 (100)	$FL = 0.2q^2L$	$FL = 0.305q^2L$
4½ (113)	$FL = 0.1q^2L$	$FL = 0.167q^2L$
5 (125)	$FL = 0.08q^2L$	$FL = 0.138q^2L$
6 (150)	$FL = 0.05q^2L$	$FL = 0.083q^2L$

* L = Hose length in hundreds of ft (m).

† q = gpm × 100.

‡ The equation replaces the old, less accurate, and more complex equation $Fl = (2q^2 + q)L$. The old equation is still used. Those interested in it are referred to the bibliography.

An approximate constant can be determined from $FL = (fQ^2L)/(2gD^5)$ by assuming the same friction characteristics for both hose sizes, so:

$$k_h = (D_1)^5 / (D_2)^5$$

If D_1 is $2\frac{1}{2}$ in., then:

$$k_h = (2.5)^5 / (D_2)^5 = 97.9 / (D_2)^5$$

k_h can then be used to convert a given hose diameter (and length) to $2\frac{1}{2}$ in. equivalent length:

$$L' = k_h L$$

where L' is the equivalent $2\frac{1}{2}$ in. length, k_h is the hose conversion factor, and L is the given hose length. For SI units: $k_h = [(64)^5/(D_2)]^5$, where D_2 is the given hose diameter required to be converted to the 64-mm equivalent length.

The hose conversion factor, k_h, and the friction loss equation constant, c, are functions of both the hose diameter and the friction characteristics of the hose. These two constants are determined experimentally. The classical hose experimentation was done by John R. Freeman.[2] Subsequent tests were made by Underwriters Laboratories Inc.,[3] James Gaskill,[4] and many fire departments. The values in Table 10-19I are based on the best current information and should be used until more current data becomes available.

The pump pressure can be found using the appropriate equations shown below:

Given NP, L, D, and d,

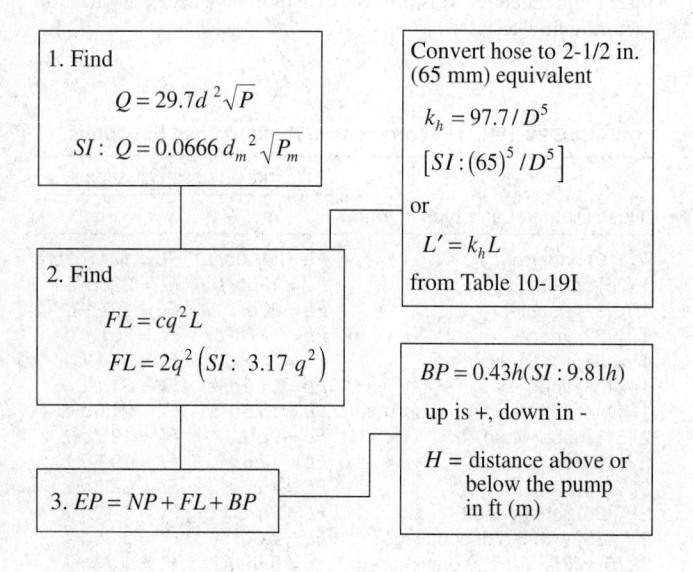

EXAMPLE: What pump pressure is required to establish a theoretical nozzle pressure of 10 psi (69 kPa) at the end of 400 ft (122 m) of 4-in. (102-mm) hose if it rises up through an elevation of 10 ft (3.05 m) from engine to nozzle?

SOLUTION: U.S. custom units:

$$Q = 29.7(4.0)^2 \sqrt{10}$$

$$Q = 1502 \text{ gpm,}$$

i.e.,

$$q = \frac{1502}{100} = 15.02$$

$FL = 0.2\, q^2 L$ (for 4-in. hose)

$$FL = 0.2(1502)^2 \left(\frac{400}{100}\right)$$

$FL = 180$ psi

$BP = 0.434h$

$BP = 0.434 \times 10$

$BP = 4.34$ psi

$EP = NP + F + BP$

$EP = 10 + 180 + 4.3$

$EP = 194.3$ psi

In SI units:

$$Q = 0.0666\,(102)^2 \sqrt{69}$$

$Q = 5756$ (L/min)

$Q = \frac{5756}{100} = 57.6$

$FL = 0.305\, q^2 L$ (for 100 mm hose)

$FL = 0.305(57.6)^2 \left(\frac{122}{100}\right)$

$FL = 1234$ kPa

$BP = 9.81$ L

$BP = 9.81 \times 3.05$

$BP = 29.9$ kPa

$EP = NP + FL + BP$

$EP = 69 + 1234 + 29.9$

$EP = 1333$ kPa (this converts to 193 psi)

This scheme works well for simple straight lines, particularly if the flow rate is given. However, for more complex setups, the following is recommended:[1]

For $2\frac{1}{2}$-in. (65-mm) rubber-lined hose:

$$EP = NP(1 + 1.76k_nL) + BP$$

where:

EP = engine pressure (psi)

NP = nozzle pressure (psi)

k_n = nozzle constant (see Table 10-19K)

L = hose length (in 100s of ft)

BP = back pressure due to elevation change (psi)

In SI units, for the equivalent 65-mm, rubber-lined hose:

$$EP = NP(1 + 2.99k_{nm}L) + BP$$

where:

EP = engine pressure (kPa)

NP = nozzle pressure (kPa)

k_{nm} = nozzle constant (see Table 10-19K)

L = hose length (in 100s of m)

BP = back pressure due to elevation change (kPa)

TABLE 10-19I. ***Factors for Converting Hose to $2^1/_2$-in. (65-mm) Equivalent (kh)***

Hose Diameter in. (mm)	References*						
	IFSTA 1980	Gaskill 1966	Erwin 1972	Holwege 1961	Other	$97.7/D^5$	Recommended Factors
$^3/_4$ (19)	344	550	590	345		412	500
1 (25)	91	75	78.5	90.9		97.7	75
$1^1/_4$ (32)	40					32	40
$1^1/_4$ (32) linen	63.6					32	63.6
$1^1/_2$ (38) rubber-lined	13.5	12	12.25	13.5	9.81	12.86	12
$1^1/_2$ (38)	25.6					12.86	25.6
$1^3/_4$ (44) with $1^1/_2$ (38)					7.76	5.95	7.76
2 (51) rubber-lined	2.94				4.0	3.05	4
2 (51) linen	6.25					3.05	6.25
$2^1/_2$ (65) linen	2.13					1.0	2.13
$2^3/_4$ (70) with 3 (76)					0.75	0.62	0.75
3 (76) with $2^1/_2$ (65)	0.4	0.4	0.4	0.4	0.4	0.4	
3 (76) with 3 (76)	0.385	0.333		0.385		0.4	0.333
$3^1/_2$ (89)	0.17			0.172		0.186	0.17
4 (100)	0.09			0.09	0.1	0.095	0.1
$4^1/_2$ (115)	0.051			0.0513	0.05	0.053	0.05
5 (125)	0.031			0.031	0.03	0.033	0.04
					0.0446		
					0.333		
6 (150)						0.0125	0.0125

*See references 4 through 7 in bibliography.

For other sizes, use either $2^1/_2$-in. (65-mm) equivalent length or a special equation for the particular hose size (shown in Table 10-19J).

The nozzle constant, k_n, is numerically equal to $k_n = d^4/10$ (in SI units: $k_{nm} = d^4/10^6$). Table 10-19K shows values of k_n for solid-stream nozzles and the equivalent k_n for several spray nozzles.

For Siamese or parallel lines, a single $2^1/_2$-in. (65-mm) equivalent length must be determined:

$$L' = k_s L$$

where $L' = 2^1/_2$-in. (65-mm) equivalent length,

and

k_s = the Siamese constant, which is generally taken as the number of lines, assuming that all lines are the same diameter.[8] (However, some authorities use a slightly modified Siamese constant to account for friction in fittings and appliances. Suggested values for k_s are given in Table 10-19L.)

Where Siamesed lines are of unequal length and/or diameter, find the average length and the $2^1/_2$-in. (64-mm) equivalent lengths, respectively. For two lines, all of the above can be combined in one equation:

$$L'' = k_s\left(k_{h1}L_1 + k_{h2}L_2\right)/2$$

TABLE 10-19J. ***Engine Pressure Formulas for Various Hose Sizes***

Hose Size in. (mm)	Formulas	
	U.S. Customary Units	SI Units
$^3/_4$ (19) booster	$EP = NP(1 + 968k_nL)$	$EP = NP(1 + 782k_{nm}L)$
1 (25) booster	$EP = NP(1 + 132k_nL)$	$EP = NP(1 + 106k_{nm}L)$
$1^1/_4$ (32) booster	$EP = NP(1 + 70.4k_nL)$	$EP = NP(1 + 54.5k_{nm}L)$
$1^1/_4$ (32) linen	$EP = NP(1 + 112k_nL)$	$EP = NP(1 + 86.3k_{nm}L)$
$1^1/_2$ (38) rubber-lined	$EP = NP(1 + 21.1k_nL)$	$EP = NP(1 + 16.8k_{nm}L)$
$1^1/_2$ (38) linen	$EP = NP(1 + 4k_nL)$	$EP = NP(1 + 35.9k_{nm}L)$
$1^3/_4$ (44) with $1^1/_2$ (38) couplings	$EP = NP(1 + 13.6k_nL)$	$EP = NP(1 + 11.0k_{nm}L)$
2 (51) rubber-lined	$EP = NP(1 + 7.04k_nL)$	$EP = NP(1 + 5.63k_{nm}L)$
2 (51) linen	$EP = NP(1 + 11k_nL)$	$EP = NP(1 + 8.78k_{nm}L)$
$2^1/_2$ (65) rubber-lined	$EP = NP(1 + 1.76k_nL)$	$EP = NP(1 + 2.99k_{nm}L)$
$2^1/_2$ (65) linen	$EP = NP(1 + 3.75k_nL)$	$EP = NP(1 + 1.40k_{nm}L)$
$2^3/_4$ (70) with 3 (76) couplings	$EP = NP(1 + 1.32k_nL)$	$EP = NP(1 + 1.05k_{nm}L)$
3 (76) with $2^1/_2$ (65) couplings	$EP = NP(1 + 0.704k_nL)$	$EP = NP(1 + 0.56k_{nm}L)$
3 (76) with 3 (76) couplings	$EP = NP(1 + 0.586k_nL)$	$EP = NP(1 + 0.47k_{nm}L)$
$3^1/_2$ (89)	$EP = NP(1 + 0.3k_nL)$	$EP = NP(1 + 0.235k_{nm}L)$
4 (100)	$EP = NP(1 + 0.17k_nL)$	$EP = NP(1 + 0.135k_{nm}L)$
$4^1/_2$ (115)	$EP = NP(1 + 0.88k_nL)$	$EP = NP(1 + 0.074k_{nm}L)$
5 (125)	$EP = NP(1 + 0.07k_nL)$	$EP = NP(1 + 0.061k_{nm}L)$
6 (150)	$EP = NP(1 + 0.022k_nL)$	$EP = NP(1 + 0.037k_{nm}L)$

TABLE 10-19K.   *Nozzle Factors*

Nozzle Diameter, in.	U.S. Custom Units, k_n*	Nozzle Diameter, mm	SI Units, k_{nm}*
$1/16$	1.53×10^{-6}	2	1.6×10^{-5}
$1/8$	2.44×10^{-6}	3	8.1×10^{-5}
$3/16$	1.24×10^{-4}	5	6.25×10^{-4}
$1/4$	3.91×10^{-4}	6	1.30×10^{-3}
$5/16$	9.54×10^{-4}	8	4.10×10^{-3}
$3/8$	1.98×10^{-3}	10	0.010
$7/16$	3.66×10^{-3}	11	0.015
$1/2$	6.3×10^{-3}	13	0.029
$9/16$	0.01	14	0.038
$5/8$	0.0153	16	0.066
$11/16$	0.0223	17	0.084
$3/4$	0.0316	19	0.130
$13/16$	0.0436	21	0.194
$7/8$	0.0586	22	0.234
1	0.1	25	0.416
$1 1/8$	0.16	29	0.707
$1 1/4$	0.244	32	1.05
$1 3/8$	0.357	35	1.50
$1 1/2$	0.506	38	2.11
$1 5/8$	0.697	41	2.83
$1 3/4$	0.938	45	3.90
$1 7/8$	1.24	48	5.31
2	1.6	51	6.66
$2 1/2$	3.9	64	16.3
3	8.1	76	33.7
$3 1/2$	15.0	89	62.4
4	25.6	102	107.0
$4 1/2$	41.0	114	171.0
5	62.5	127	260.0
6	130.0	152	539.0

$$*K_n = \frac{d^4}{10}; \text{ for SI units}: \ K_{nm} = \frac{d^4}{10^6}.$$

L'' is another way of expressing equivalent length. For more than two lines, the equation can be expanded. If the lines are the same diameter, a precise equivalent length can be found from:

$$L'' = \frac{L_1}{\left[1 + \left(\sqrt{\dfrac{L_1}{L_2}}\right)\right]^2}$$

For more than two lines, the general equation is:

$$L'' = \frac{L_1}{\left[1 + \sqrt{\dfrac{L_1}{L_3}} + \sqrt{\dfrac{L_1}{L_3}} + \sqrt{\dfrac{L_1}{L_4}} + \cdots \sqrt{\dfrac{L_1}{L_n}}\right]^2}$$

Figure 10-19B is a schematic illustrating the solution of Siamese line problems, and Figure 10-19C includes hose sizes other than $2^1/_2$ in. (65 mm).

A common hose lay is the wyed or branch lines. This set-up involves a $2^1/_2$-in. (65-mm) or larger supply or lead line branched into

TABLE 10-19L.   *Siamese Conversion Factors (k_s) to Single $2^1/_2$-in. (65-mm) Equivalent Length*

Number of Hose Lines	Siamese Conversion Factor to Single $2^1/_2$-in. Equivalent Length
Two $2^1/_2$ in.	0.25
Three $2^1/_2$ in.	0.11
Four $2^1/_2$ in.	0.0625
Five $2^1/_2$ in.	0.04
Two 3 in. with $2^1/_2$-in. couplings	0.08
Three 3 in. with $2^1/_2$-in. couplings	0.0307

For SI units: 1 in. = 25.4 mm.

two or more lines. The legs or branches are typically smaller diameter hoses than the supply line. A typical set-up involves a $2^1/_2$-in. (65-mm) supply line and two $1^1/_2$-in. (38-mm) branches, each 100 or 150 ft (30 or 45 m) in length.

The solution to engine pressure problems depends on whether or not the branches or legs are symmetrical, i.e., hose length, diameters, and nozzles equal. (See Figure 10-19D for outline of the solution.) If the legs are symmetrical, the solution can be easily determined by finding an equivalent nozzle diameter and equivalent nozzle constant from:

$$L'' = \frac{L_1}{\left[1 + \left(\sqrt{\dfrac{L_1}{L_2}}\right)\right]^2}$$

If the lines are not symmetrical, a more involved calculation is required for an exact solution. (The method described above for symmetrical lines can be used for an approximate solution.)

The technique involves finding the flow through one of the legs and determining the friction loss in that leg and the pressure at the wye. Using the pressure at the wye as engine pressure, calculate the nozzle pressure on the other line and the flow from the second line and then sum up the flow from all the lines. With this flow known, the friction loss in the supply line can be found. The engine pressure

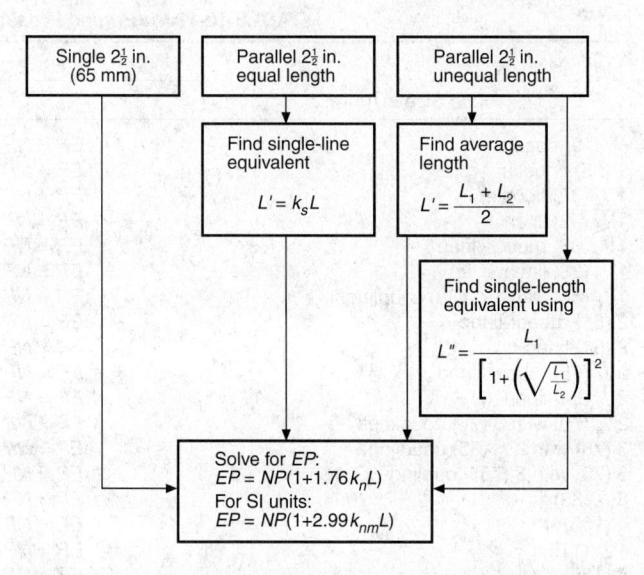

FIG. 10-19B.   *Solving for engine pressure, $2^1/_2$-in. (64-mm) hose.*

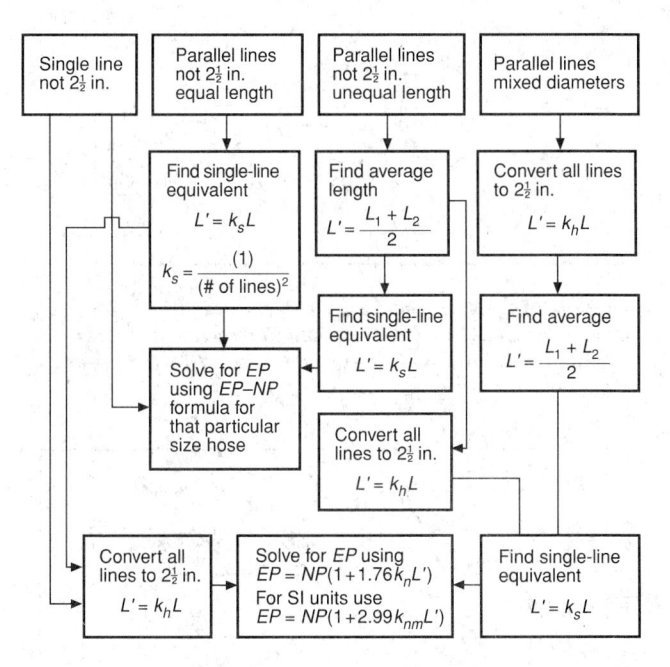

FIG. 10-19C. *Solving for engine pressure, all hose sizes. (For SI units: 1 in. = 25.4 mm.)*

is the supply line friction loss plus the pressure at the wye. This technique is also shown in Figure 10-19D.

LARGE-DIAMETER HOSE

Many fire departments are now using large-diameter hose. The most popular sizes in use include:

- 3-in. (76-mm) hose with $2^1/_2$-in. (64-mm) couplings
- 3-in. (76-mm) hose with 3-in. (76-mm) couplings
- 4-in. (100-mm) hose with 4-in. (100-mm) couplings
- 5-in. (125-mm) hose with 5-in. (125-mm) couplings
- 6-in. (150-mm) hose with 6-in. (150-mm) couplings

The main advantage of large-diameter hose is its lower friction loss. It is especially useful as a supply line with large flows. Because of the low friction loss, some fire departments use large-diameter hose for supply lines into a portable hydrant, from which smaller working lines [$2^1/_2$ in. (65 mm), $1^1/_2$ in. (38 mm), etc.] are then taken off.

Large-diameter hose is also useful for relaying water, i.e., one pumper supplying a second.

There have been some complaints that the large-diameter hose is difficult to handle and, in some cases, wears out rapidly. Nevertheless, even if these disadvantages are not overcome, they appear to be overshadowed by the advantages, at least for certain applications.

FRICTION-REDUCING ADDITIVES

Certain high-molecular-weight polymers injected into a pump or hoseline have the capacity of reducing friction. These polymers include polyethylene oxide, polyacrylamide, polysulfonates, and hydroxethyl cellulose, the first being the most prevalent in current use. Several trade names are used to describe friction-reducing additives. Polymer additives reduce friction loss by as much as 50 percent. This permits flowing greater amounts of water in smaller diameter hose ("rapid water").

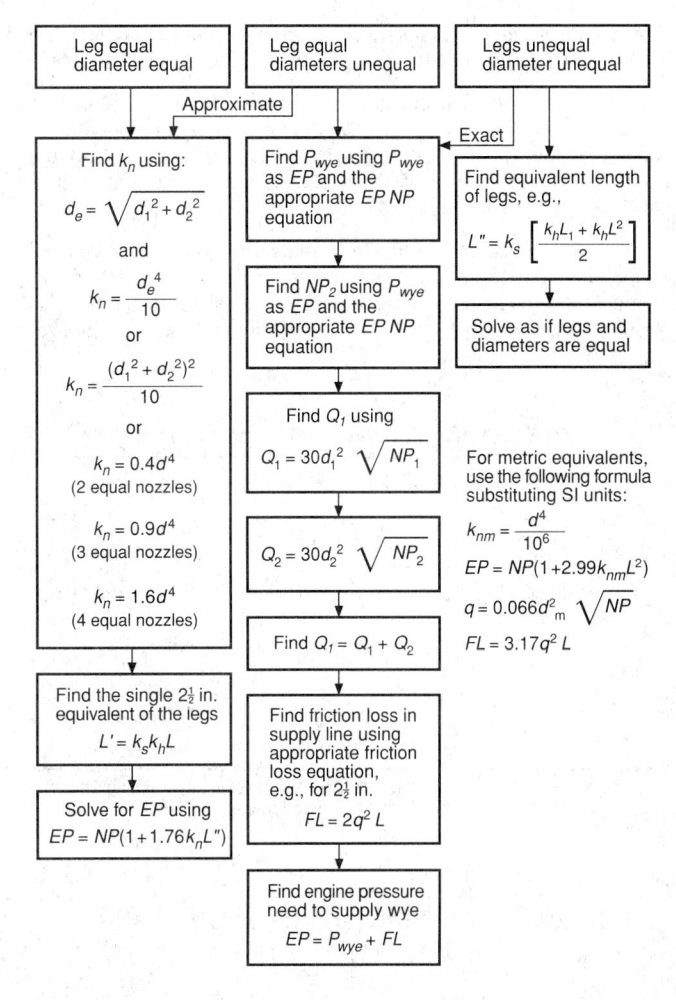

FIG. 10-19D. *Solving for engine pressure for wyed lines.*

For example, 200 gpm (757 L/min) of plain water flowing in $1^1/_2$-in. (38-mm) hose has a friction loss of about 96 psi (662 kPa). With such high friction loss, and assuming a nozzle (spray type) pressure of 100 psi (690 kPa), the maximum distance 200 gpm (757 L/min) could be pumped through $1^1/_2$-in. (38-mm) hose without exceeding the hose test pressure of 250 psi (1,724 kPa) is about 150 ft (45.7 m). On the other hand, with polyethylene oxide additive the friction loss of 200 gpm (757 L/min) through 100 ft (30.5 m) of $1^1/_2$-in. (38-m) hose is about 25 psi (172 kPa). Adding the friction-reducing polymer permits $1^1/_2$-in. (38-mm) lines to provide 200-gpm (757-L/min) flows at 100-psi (690-kPa) nozzle pressures up to 600 ft (183 m) without exceeding hose test pressures of 250 psi (1,724 kPa). In other words, $1^1/_2$-in. (38-mm) hose with the friction-reducing additive can have almost the same "punch" as $2^1/_2$-in. (65-mm) hose lines, for at least distances up to 600 ft (183 m).

The use of the smaller-diameter hose can be a significant advantage, since water in $2^1/_2$-in. (65-mm) hose weighs about 2 lb/ft (3 kg/m) as compared to $^3/_4$ lb/ft (1.1 kg/m) for $1^1/_2$-in. (38-mm) hose. For a 50-ft (15-m) length, this is 100 lb (45 kg) *vs.* about 35 lb (16.8 kg) (neglecting the weight of the hose). This weight reduction is crucial to the fire fighter who is dragging a hose line into position.

The City of New York Fire Department has considerable experience using $1^3/_4$-in. (44-mm) hose [with $1^1/_2$-in. (38-mm) couplings] in conjunction with a friction-reducing agent. With this arrangement, they combined the fire extinguishing capability of a 2$^1/_2$-in. (65-mm) line with the mobility of a 1$^1/_2$-in. (38-mm) line.

There are some technical difficulties with some of the additives that are apparently being overcome. For example, the polymer is nontoxic, but it is difficult to inject into the hose stream. Also, the polymer is subject to deterioration at temperatures exceeding 120°F (48.9°C), or less than 32°F (0°C).

Some equations are shown in Table 10-19M that can be used to establish the friction loss in hose with rapid water. These equations should be used with discretion, however, since the friction loss is a function of polyethylene oxide concentration, which is not reflected in these equations.

TABLE 10-19M.   *Friction Loss, Using "Rapid Water"*

Hose Size in. (mm)	psi	kPa
1$^1/_2$-in. (38) rubber-lined hose*	$FL = 7.5q^2L$	$FL = 11.8q^2L$
1$^3/_4$ in. (45) with 1$^1/_2$-in. (38) couplings	$FL = 6.2q^2L$	$FL = 9.8q^2L$
2$^1/_2$-in. (65) rubber-lined hose*	$FL = 0.7q^2L$	$FL = 1.1q^2L$
	q = 100s of gpm L = 100s of ft	q = 100s of L/min L = 100s of m

*Derived from D. B. Brown, "Reducing Friction Loss in Fire Fighting Streams Using Poly (Ethylene Oxide) Slurry," presented at the 1974 NFPA Meeting, Miami Beach, FL.

REACTION FORCES IN HOSE LINES AND NOZZLES

When water is discharged from a hose line, there are two reaction forces acting on the line: (1) hose line reaction and (2) nozzle reaction. Both contribute in varying degrees to the total reaction that is present in any hose nozzle flow configuration. For purposes of simplicity, each is treated separately in this chapter.

Hose Line Reaction

Reaction forces occur in hose lines when the direction of waterflow is changed by a bend in the hose. Between the hydrant and nozzle, the velocity is constrained in the hose and the acceleration is zero as shown by the equation $Q = AV$, where Q is the flow rate, A is the cross-section area, and V is the average velocity.

Figure 10-19E illustrates a typical hand-line situation where enough push must be applied at the bend to resist the reaction force F_3. There is no problem where the hose bends off the ground.

Figures 10-19F(a) and 10-19F(b) are charts by which values of the reaction F_3, in pounds force, can be directly determined. To obtain reactions in other sizes of hose, Figures 10-19F(a) and 10-19F(b) each include a table of conversion factors.

EXAMPLE: What is the reaction at a 55-degree bend in 1$^1/_2$-in. (38-mm) hose if the pressure is 100 psi (690 kPa)?

SOLUTION: On the deflection angle side, move horizontally from the 55-degree bend to intersection with the 100 psi (690 kPa) curve. From there go down vertically to the reaction scale and read 450 lb (204 kg). From Figures 10-19F(a) and (b), note that the factor for 1$^1/_2$-in. (38-mm) hose is 0.36. Multiplying 450 (204 kg) by 0.36, the answer is 162 lb (73 kg).

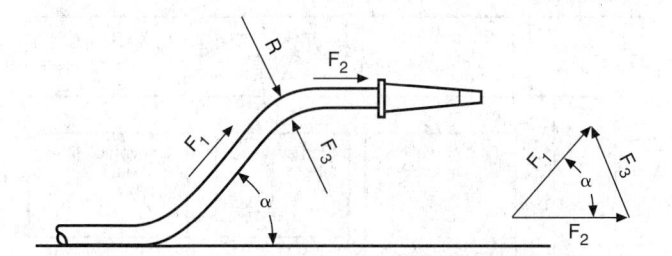

FIG. 10-19E.   *Reaction in a hose line. α = deflection angle at bend. F_3 = resultant of forces F_1, and F_2 (reaction in lb). R = force needed to hold hose against F_3. (For SI units: 1 lb = 0.454 kg.)*

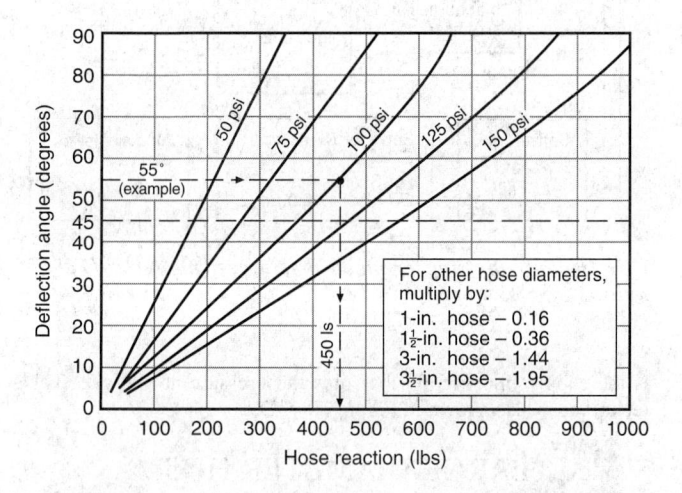

FIG. 10-19F(a).   *Reaction forces in 2$^1/_2$-in. hose. (See table within the graph for conversion factors to use in calculating reaction forces, expressed as pounds force, in hose of other diameters.)*

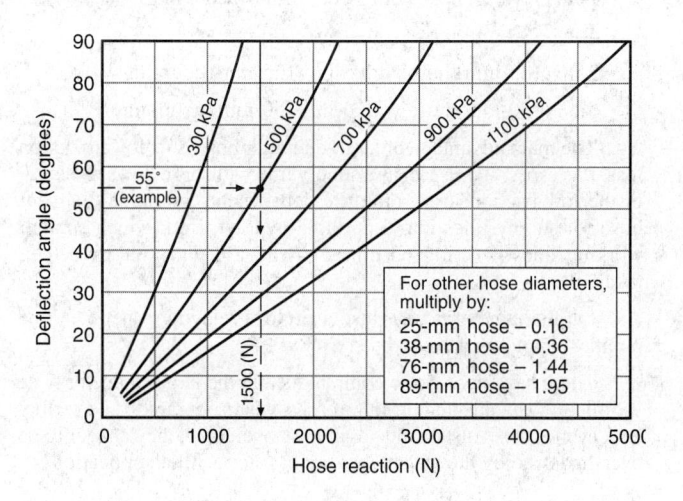

FIG. 10-19F(b).   *Reaction forces in 65-mm hose. (See table within the graph for conversion factors to use in calculating reaction forces, expressed in newtons, in hose of other diameters.)*

Nozzle Reaction

Water discharging from a nozzle produces a reaction that is opposite the flow of water and known as nozzle reaction. As flow and pressure are increased, the nozzle reaction increases. Nozzle reaction of solid-stream nozzles may be calculated by means of the following formula:

$$NR = 1.5d^2 NP$$

where:

NR = nozzle reaction (lb)

d = nozzle diameter (in.)

NP = nozzle pressure (psi)

Another formula that may be used for both solid-stream and spray nozzles is:

$$NR = 0.0505\, Q\sqrt{NP}$$

where Q = flow rate (gpm)

For SI units:

$$NR = \left(1.6 \times 10^{-3}\right)d^2 NP$$

where:

NR = nozzle reaction [N (Newton)]

d = nozzle diameter (mm)

NP = nozzle pressure (kPa)

Reaction Forces in Ladder and Tower Streams

The reaction of heavy streams, either solid or spray, discharged from ladder pipes, platform booms, and towers can easily cause serious problems, such as leverage effects of large magnitude and displacement of centers of gravity. Thus, the stability of the whole structure could be threatened. Likewise, a sudden loss of pressure when a hose line bursts or a pump stops may cause a whipping effect.

A ladder pipe or other large stream from a ladder [including $2^1/_2$-in. (65-mm) hand lines] should always be operated in line with the main beams or trusses. Ladders have little resistance to torsional effects. NFPA 1901 specifies that detachable ladder pipes be incapable of horizontal travel exceeding the ladder manufacturer's recommendations. Rotation of the ladder turntable is the correct way to rotate the stream from ladder pipes. Also, care should be used elevating and lowering the stream, because this changes the direction of the thrust on the ladder mechanism. The fly ladder should not be raised or lowered while a ladder pipe is discharging, and under no circumstances should the vehicle be moved with the ladder pipe discharging. All such streams should be gated to permit movement of the ladder and vehicle without shutting down the pump.

Because there are many variable forces acting on mobile ladders or towers in service, it is quite impractical to devise tables of specific reactions. The problem is one of design and careful operating procedures.

BIBLIOGRAPHY

References Cited

1. Purington, R. G., *Fire-Fighting Hydraulics*, McGraw-Hill, New York, 1974.

2. Freeman, J. R., "Experiments Relating to the Hydraulics of Fire Streams," *Transactions*, American Society of Civil Engineers, Vol. 21, Nov. 1889.

3. UL, "Hydraulic Friction Losses in 1 1/2-in. and 2 1/2-in. Cotton Rubber-Lined Fire Hose," 1939, Underwriters Laboratories Inc., New York.

4. Gaskill, J. R., "Hydraulic Studies of Fire Hose," *Fire Technology*, Vol. 2, No. 1, 1966.

5. IFSTA, *Fire Stream Practices*, 7th ed., International Fire Service Training Association, Stillwater, OK, 1980.

6. Erwin, L. W., *Fire Hydraulics*, Glencoe Press, Beverly Hills, CA, 1972.

7. Holwege, J., "What Size Fire Hose," *Washington Comm. and School*, 1961.

8. Vennard, J. K., *Elementary Fluid Mechanics*, 4th ed., John Wiley & Sons, New York, 1961.

NFPA Codes, Standards, and Recommended Practices

Reference to the following NFPA codes, standards, and recommended practices will provide further information on fire streams discussed in this chapter. (See the latest version of *The NFPA Catalog* for availability of current editions of the following documents.)

NFPA 1901, *Standard for Automotive Fire Apparatus*

NFPA 1911, *Standard for Service Tests of Pumps on Fire Department Apparatus*

NFPA 1961, *Standard for Fire Hose*

NFPA 1962, *Standard for the Care, Use, and Service Testing of Fire Hose Including Couplings and Nozzles*

NFPA 1964, *Standard for Spray Nozzles (Shutoff and Tip)*

Additional Reading

Atreya, A., Crompton, T., and Suh, J., "Experimental and Theoretical Study of Mechanisms of Fire Suppression by Water," Michigan Univ., Ann Arbor, NISTIR 5499, Sept. 1994; National Institute of Standards and Technology, Annual Conference on Fire Research: Book of Abstracts, Oct. 17–20, 1994, Gaithersburg, MD, 1994, pp. 67–68.

Dunn, V., "Master Stream Safety," *Fire Engineering*, Vol. 144, No. 6, June 1991, pp. 32–34, 36, 39, 40.

Dyer, B. D., "Apparatus Placement for Aerial Master Stream Operations," *Firehouse*, Vol. 20, No. 6, June 1995, pp. 50, 52–54.

Eckman, W. F., *The Fire Department Water Supply Handbook*, Fire Engineering Books, Saddle Brook, NJ, 1994.

Fredericks, A. A., "Return of the Solid Stream," *Fire Engineering*, Vol. 148, No. 9, Sept. 1995, 44–46, 48, 50, 52, 55–56.

Johnson, B. P., "Additives for Hose Reel Systems: Trials of Foam on Tire Fires," Home Office, London, England Publication 5/91, SC/88 42/1087/1, 1992.

Loeb, D., "Elevated Master Streams Deliver Straight Out Extinguishing Power," *Fire Chief*, Vol. 35, No. 5, May 1991, pp. 67–69.

Scheffey, J. L., and Williams, F. W., "Extinguishment of Fires Using Low Flow Water Hose Streams. Part 1," *Fire Technology*, Vol. 27, No. 2, May 1991, pp. 128–144.

Scheffey, J. L., and Williams, F. W., "Extinguishment of Fires Using Low Flow Water Hose Streams. Part 2," *Fire Technology*, Vol. 27, No. 4, Nov. 1991, pp. 291–320.

Shapiro, P., "Mastering the Ultimate Stream," *American Fire Journal*, Vol. 47, No. 7, July 1995, pp. 16–20.

PLANNING FIRE STATION LOCATIONS

Robert C. Barr and Anthony P. Caputo

One of the primary responsibilities of a fire department is the delivery of fire and rescue services. The delivery of these services normally originates from fire stations that are placed throughout the area to be protected. To provide effective service, crews must respond in a minimum amount of time after the incident has been reported and with sufficient resources to initiate fire, rescue, or emergency medical activities.

This chapter discusses the basic elements and means by which fire station locations can be identified for the timely and efficient delivery of fire and rescue services.

TIME CONSIDERATIONS

Nothing is more important than the element of time when an emergency is reported. Fire growth can expand at a rate of many times its volume per minute. Time is the critical factor for the rescue of occupants and the application of extinguishing agent.

The time segment between fire ignition and the start of fire suppression activities is critical and has a direct relationship to fire loss. The delivery of emergency medical services is also time critical. Survival rates for some types of medical emergencies are dependent upon the rapid intervention by trained emergency medical personnel. In most cases, the sooner that trained fire or emergency medical rescue personnel arrive, the greater the chance for survival and the conservation of property.

There are a number of factors that determine when flashover may occur. (See Figure 10-20A.) These include the type of fuel, the arrangement of the fuels in the room, room size, etc. Because these factors vary, the time to flashover cannot be predicted.

There are a number of time frames that the fire department can manage, as well as some it cannot manage, in the sequence of events outlined in Figure 10-20A. The time from ignition to discovery and reporting of a fire is indirectly manageable. This time period can be managed through requiring the use of automatic detection and/or suppression systems and automatic reporting to the public safety answering point. In a perfect world, all structures would be equipped with automatic detection and/or suppression systems. One item that has helped to manage this segment of time is the widespread use of automatic smoke detectors in residential occupancies.

What is lacking, however, is the automatic reporting to the public safety answering point.

Fire Department Response Time

Figure 10-20B shows the steps in the fire department response time sequence. Each of the five steps, after receipt of alarm, is defined below.

Dispatch time: Amount of time that it takes to receive and process an emergency call. This includes: (1) receiving the call, (2) determining what the emergency is, (3) verifying where the emergency is located, (4) determining what resources are required to handle the call, and (5) notifying the units that are to respond.

Turn-out time: Amount of time that it takes a crew to: (1) react after receiving dispatch information and (2) prepare to leave the station.

Travel time: Amount of time that it takes for a piece of apparatus to travel from the fire station to the incident scene. (The amount of time from wheel start to wheel stop.)

Access time: Amount of time required for the crew to move from where the apparatus stops to where the emergency exists. This can include moving to the interior or upper stories of a large building and dealing with any barriers in the access to that area.

Set-up time: The amount of time required for fire department units to set up, connect hose lines, position ladders, etc., and prepare to extinguish the fire.

FIRE DEPARTMENT RESPONSE MANAGEMENT

The time required to receive the alarm and dispatch appropriate units is manageable by the way that alarms are received and the way that dispatch activities are handled. Enhanced 9-1-1 (E911) and computer-aided dispatch systems will minimize the time required to receive and handle alarms. The time segment "A" or "respond to the scene " includes turn-out time, travel time, and access time. Turn-out time may be managed to some degree by improving the method of communications between the dispatch center and the fire station, so the time required for crews to receive alarm information is reduced.

Travel time is one of the most manageable segments of time in the entire sequence. This is the amount of time that it takes for a

Robert C. Barr is president of Firescope, Inc., a fire protection consulting firm located in Hingham, MA. Anthony P. Caputo, P.E., is president of Pyrotech Consultants, Inc., a fire protection consulting firm in Sandwich, MA.

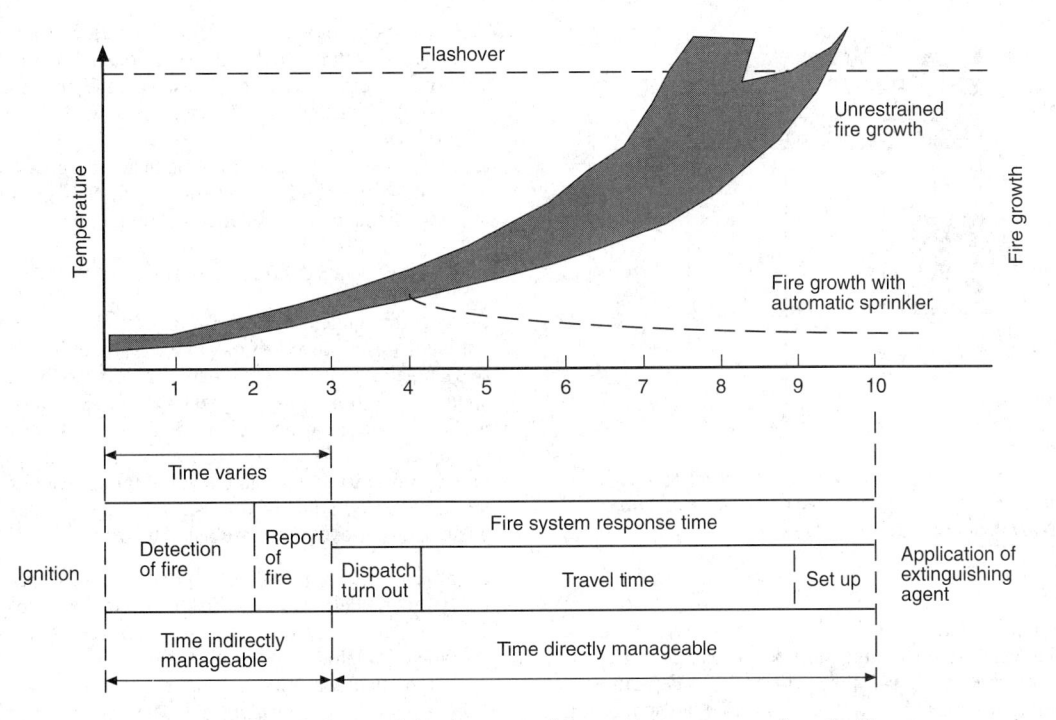

FIG. 10-20A.   *Fire growth over time and the sequence of events that may occur from ignition to suppression.*

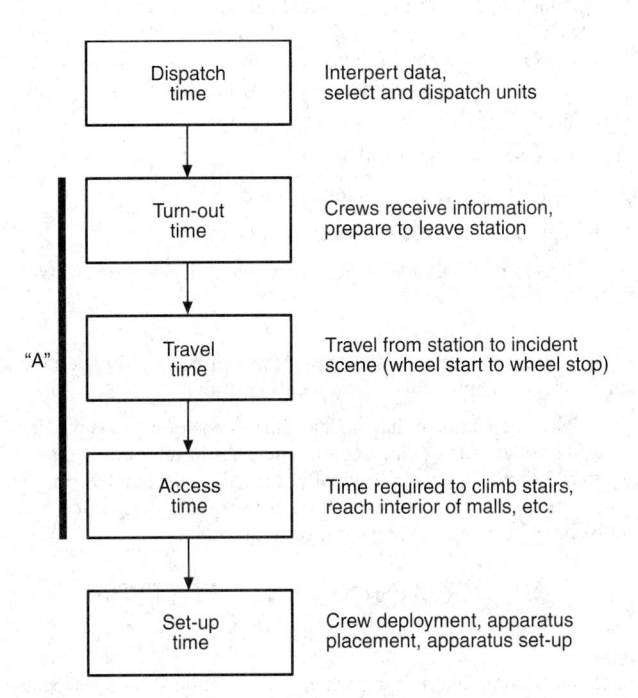

FIG. 10-20B.   *Steps in the fire department response time sequence.*

piece of fire apparatus or an ambulance to travel from a fire station to an incident scene (wheel start to wheel stop). Travel time can be managed by selecting fire station locations based on the amount of time that it takes to travel from the fire station to the incident scene. The maximum amount of time required to travel from a fire station to an incident scene is usually determined by the community, since there are no national standards for travel time. Average travel times in a station's response area may range from 2 min on up.

Access time can be managed through a good prefire planning process that familiarizes the fire fighters with access points and travel routes through the building. The use of key boxes can facilitate getting doors unlocked; and working with security forces at larger facilities can facilitate access, thus reducing access time.

Emergency Medical Services

The delivery of emergency medical services by first responders is also time critical for many types of injuries and events. If a person has a heart attack and cardiopulmonary resuscitation (CPR) is started within 4 min, the victim's chances of leaving the hospital alive are almost four times greater than if the victim did not receive CPR until after 4 min. Figure 10-20C shows the survival rate for heart attack victims when CPR is available.[1]

Response Time Influences

The basis for the placement of fire stations should be the amount of time that it takes to deliver adequate emergency resources to the point of demand from each fire station or combinations of fire stations.

Currently there are no national standards for either response time or travel time. Essentially each community must decide the appropriate response and travel times for their community. There are a number of factors that may influence the selection of a specific response/travel time. All applicable factors must be considered when making a decision on a specific response/travel time for a community. Factors that should be considered are:

1. *What types of services are delivered by the fire department?* Does the department deliver both fire and emergency medical services, or fire service only? The delivery of emergency medical services is an important factor in selecting a response/travel time because of the need to provide initial service as rapidly as possible.

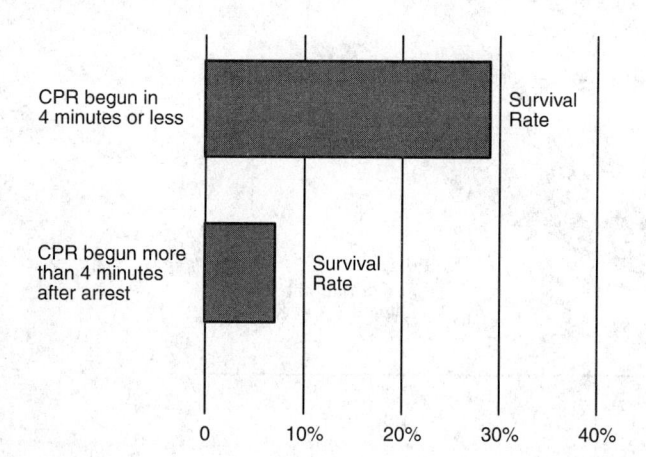

FIG. 10-20C.   Survival rate of heart attack victims when CPR is available. Speed is important. (Source: Heartsaver Manual, American Heart Association)

2. *What is a reasonable travel time for the community?*   The selection of a response/travel time must be reasonable. A short response/travel time, such as 2 min, will enable a department to deliver services within a short period of time, but will require more stations. The time selected must provide a balance between good service and the financial ability of the community to provide the necessary stations and resources.

3. *What is the size of the area being served and the type and amount of resources that are available?*   The community must consider the size of the area being served and the type and amount of available resources. Large rural areas normally have longer response/travel times and fewer resources than urban or suburban areas. For example a countywide fire response and Emergency Medical Service (EMS) that is operated by volunteer personnel will most likely have longer response/travel times than an urban or suburban area that is staffed by career personnel.

   The demographics of the area being served, as well as past fire and EMS activity, must be considered.

4. *What level of risk is the community willing to accept by using longer travel times?*   When a community selects a longer travel time, it is willing to accept a larger level of risk. Accepting a larger level of risk may be based on past experience; e.g., areas that take longer to reach have a significantly lower number of incidents.

   The selection of a response/travel time for a community should be made after all factors are examined. In many cases, this decision should be made after the analysis has started and the decision makers are allowed to review the results of the analysis, using the present configuration of stations with existing travel times, and other configurations of stations using various travel times.

   There are several ways that a community can establish a response/travel time standard. Some of these are: (1) the use of historical fire and EMS response data, (2) demand for service, (3) the level of care that the community wants to provide, and (4) the level of care that the community is able to provide. In some cases, the analysis will assist in establishing the standard after a number of scenarios are examined.

Examples of Response/Travel Time Standards

The following three examples should specify the type of unit necessary and the number of personnel required.

1. After the receipt of a call for assistance, the fire department will respond with the first unit to that location within 3 min.
2. After receipt of a call for assistance, the fire department will respond with a unit to that location, within 4 min, to 90 percent of area served.
3. After receipt of a call for a medical emergency, the fire department will respond with an engine company to that location within 4 min, and an ambulance within 6 min.

Risks *vs.* Response Time Standards

Some communities may elect to adopt several response time standards for various levels of risk in the community, or they may adopt one single response time standard for all risks. Providing several response/travel times based to the level of risk means that some areas will be reached in a shorter period of time than other areas. A single response time standard for the community or area served will provide approximately the same level of initial response to all areas of the community. Some of the risks that may be considered for various response/travel times are listed below.

1. *Sprinklered vs. nonsprinklered:*   The fully sprinklered property theoretically is at less risk than the nonsprinklered property. The community may elect to place fire stations closer to the nonsprinklered property because of the ability of sprinklers to automatically put water on a fire.
2. *Commercial vs. residential:*   The community must decide, in this case, where the greater risk is located. Is there a greater risk with the commercial property? At what time of day is the risk greater? Is the risk greater in residential occupancies at night?
3. *Multifamily vs. single-family residential:*   Is the risk greater in multifamily residential areas than in single-family residential areas? Fire station location should be influenced by population factors.

The risk questions should address:

- High life hazard or high dollar value
- High incident areas *vs.* low incident areas
- High rate of incidents that require a high level of resources *vs.* low rate

   The community that chooses to provide various levels of service to different areas, based on the level of risk, will undertake a much more complex fire station location study.

   It should be noted that most of the information that is provided here addresses career fire departments that have stations that are staffed around the clock. Typically, for call and volunteer fire departments, the turn-out time will be greater. This is significant and must be considered in the overall response time.

DETERMINING FIRE STATION LOCATIONS

The first step in determining where a fire station should be located is the adoption of a response time standard, or standards, for the community, based on the items discussed previously.

   Once the response time standard has been developed, it needs to be broken down into specific time intervals for each of the components that make up "response time," i.e., dispatch time, turn-out time, travel time, access time, and set-up time.

   For most incidents, the dispatch time and turn-out time can be established by using historical data. Set-up time is dependent upon the type and complexity of the incident. A single-family dwelling fire will generally require less time than a major fire in a high-rise building or industrial complex. A thorough analysis will examine set-up times and incident frequencies for several types of incidents. For larger incidents, the standard may be stated in terms of response time for the last of several units required for the attack. Relevant times by type of incident can be established through training and/or historical experience.

Response time—minus the cumulative time for dispatch, turn out, access, and set up—yields the allotment for travel time.

Once the response time standard has been established and a travel time standard selected, the process of determining or planning fire station locations can begin. Even though response time and travel time standards are selected prior to the beginning of the study, this does not mean that other times cannot be examined and used in the analysis. The final response/travel time may be selected after a number of scenarios are examined.

FIRE STATION PLACEMENT

With the advent of modern computer technology, the selection of fire station locations can best be determined with a greater degree of accuracy using geo-based information systems. Unlike previous methods that employed grid and concentric circle analysis, the computer programs simulate the real road network of the area being analyzed. A high degree of accuracy is ensured by using actual travel distances and vehicle speeds, factoring time delays for roadway conditions (e.g., congestion, turning radius, weather, hills, etc.), accounting for one-way or unusable roadways, and implementing user-defined risk factors.

The street network that is used as a basis for the study is typically taken from maps and data that are produced by the U.S. Bureau of Census or other sources. These are actual scale maps that have data attached to them. Each street segment is assigned a travel speed.

There are several geo-based computer programs designed for use on desktop computers. The program output displays the actual street network, including the travel speeds of fire apparatus for each street segment. The computer program is an integrated program that accurately models an emergency vehicle's response over a street/highway network, using actual speeds and distances.

The basic street map (see Figure 10-20D) can also be overlaid with a map of census tracts; aerial photographs; or other types of information, such as the location of past fire and EMS incidents, landmarks, or special boundaries.

The results of a geo-based computer study are a series of color maps of the street/roadway network that are easy to comprehend. These maps can show a variety of information, such as the street segments that can be reached from all stations for a specified travel time, the overlap between stations, the response area for each station, and other information, depending upon the capability of the software program being used.

These maps can effectively illustrate problem areas, as well as proposed solutions. They can show various data for simulated responses to specific locations, including apparatus and personnel on a time-arrival sequence, multiple alarms, and travel time to each street segment in the system.

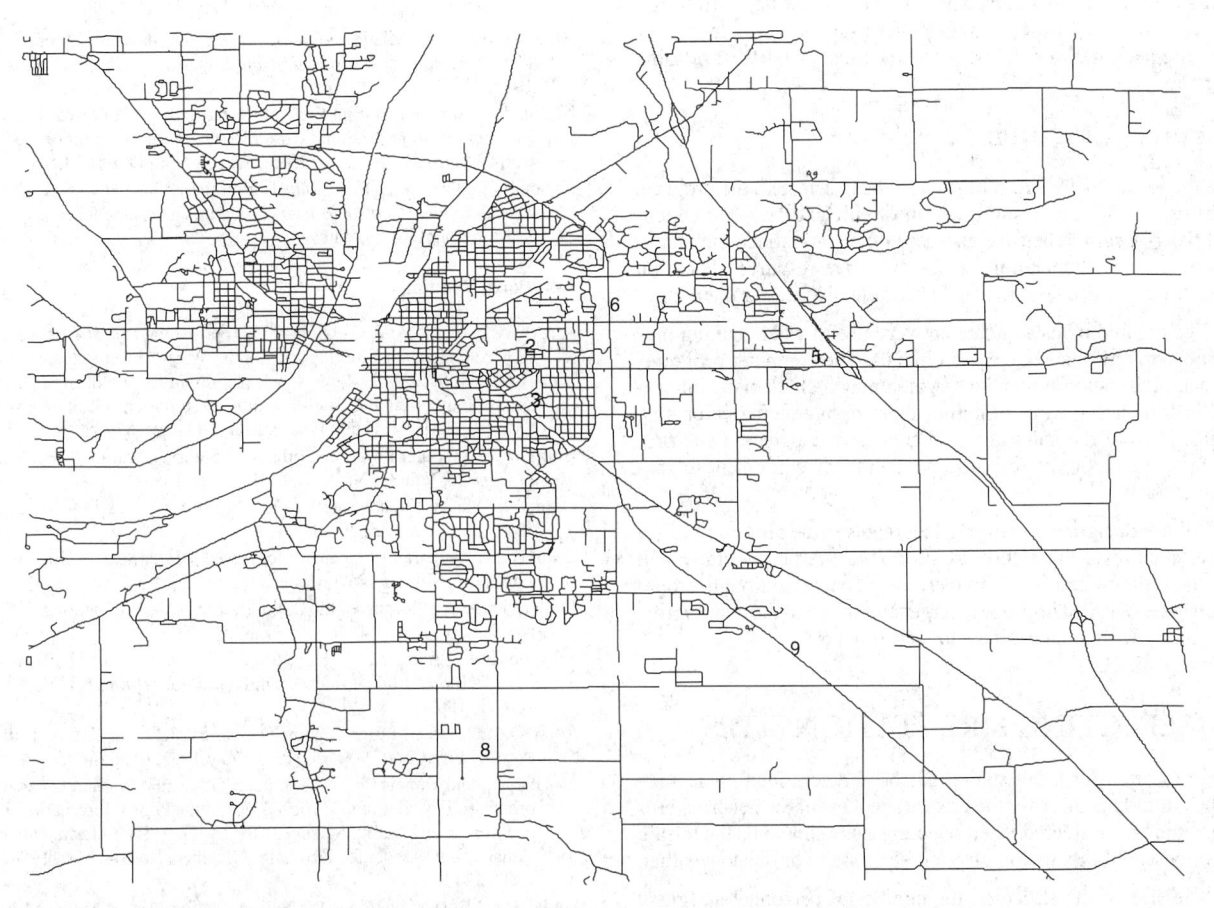

FIG. 10-20D. *Typical computer-generated street map. Overlays can display a variety of information, such as transit time, response areas, and overlap coverage.*

The response area for each fire station, based on the travel time specified, can be calculated and displayed. Each station's response area can be shown as a different color on the map. This is often complemented with a list of the streets and/or street segments for each station's response area.

Some of the computer programs are able to produce color maps that show the overlap between stations for a specified travel time. The color maps show the areas where there are two or more stations whose first-due response areas overlap, with the overlap colors different from the station response area colors. These maps can be used to determine multiple-company response to an area or to show where stations are placed too close together. Areas that receive no coverage or deficient coverage can also be shown.

The response from mutual-aid stations can also be assessed as part of an overall study to show the arrival of additional resources. This is particularly useful to show the response of special apparatus and teams, such as hazardous materials teams, rescue teams, or other special teams. Given limited resources, an entire region can be analyzed using the entire resources of the region.

Alternative and future station locations can be examined through the use of "what if" scenarios using various fire station locations and travel times.

The information from these programs can be integrated with the data from local and regional planning groups to show where new stations and/or roadways may be needed to best serve existing and growing communities.

A comprehensive geo-based fire station location study can provide an essential part of a master plan for fire station locations. This plan can show both the efficiencies and deficiencies of current fire station coverage for a specified travel time, and provide a model for future fire station coverage using the specified or other travel time standard.

Resource Allocation

Once the locations for fire stations have been selected, the program can be used to allocate resources to most efficiently cover the identified risks. A comprehensive geo-based fire station location analysis provides the foundation for cost-effective resource allocation that is critical to fire department and community master planning.

The colored maps and data are very useful when making presentations to citizens and elected officials. The computer programs take complex response scenarios, which may include many interrelated factors, and greatly simplifies them by producing color map images. Most of the maps and data are easily understood and provide factual data on the response of emergency equipment to specific locations.

All fire department planning that involves the allocation of resources requires a fire station location study. A proper study will show how the effective deployment of resources, through proper station locations, can improve emergency service to the community, while minimizing both capital and recurring costs.

SELECTING FIRE STATION SITES

Once a location for a fire station has been determined, there may still be multiple choices for the precise site. One of the general items that should be considered when selecting a site should be the terrain and the susceptibility to flooding or other conditions due to weather.

The size of the structure, the number of personnel assigned, and the provision of a training area will dictate the size of the site that is required.

Another important consideration is the relationship to the adjacent streets and highways. An analysis of traffic patterns in the surrounding area will aid in determining whether the site is suitable for a fire station. The station must be located so that responding apparatus can enter streets and highways as safely as possible, and return to the station without disrupting traffic and placing personnel at risk. It may be more desirable to locate a station on a side street rather than on a heavily traveled street.

Traffic pattern requirements for a specific site will differ for volunteer and career personnel. The major reason for this is that volunteers must respond to the station before responding to an emergency. Therefore, the site must allow easy access for volunteers that respond to the station to answer calls. There may be situations where apparatus is leaving the station and, at the same time, volunteers are still arriving. This is more complex than an all-career station, where all personnel are at the station and respond immediately to a call.

BIBLIOGRAPHY

NFPA Codes, Standards, and Recommended Practices

Reference to the following NFPA codes, standards, and recommended practices will provide further information on planning fire station locations discussed in this chapter. (See the latest version of _The NFPA Catalog_ for availability of current editions of the following documents.)

NFPA 1201, _Standard for Developing Fire Protection Services for the Public_

References

American Heart Association, _Heartsaver Manual: A Student Handbook for Cardiopulmonary Resuscitation and First Aid for Choking_, Dallas, TX, 1987.

National Commission on Fire Prevention and Control, _America Burning_, The Report of the National Commission on Fire Prevention and Control, Government Printing Office, Washington, DC, 1973.

National Fire Prevention and Control Administration, _Urban Guide for Fire Prevention and Control Master Planning_, Government Printing Office, Washington, DC, (no date).

Additional Reading

Atreya, A., Crompton, T., and Suh, J., "Experimental and Theoretical Study of Mechanisms of Fire Suppression by Water," Michigan Univ., Ann Arbor, NISTIR 5499, Sept. 1994; National Institute of Standards and Technology, Annual Conference on Fire Research: Book of Abstracts, Oct. 17–20, 1994, Gaithersburg, MD, 1994, pp. 67–68.

Baird, D., "Computer Programs Aid in Fire Station Location," _Fire Fighting in Canada_, Vol. 34, No. 9, Dec. 1990, pp. 4–5.

Dunn, V., "Master Stream Safety," Fire Engineering, Vol. 144, No. 6, June 1991, pp. 32–34, 36, 39, 40.

Dyer, B. D., "Apparatus Placement for Aerial Master Stream Operations," Firehouse, Vol. 20, No. 6, June 1995, pp. 50, 52–54.

Fredericks, A. A., "Return of the Solid Stream," Fire Engineering, Vol. 148, No. 9, Sept. 1995, 44–46, 48, 50, 52, 55–56.

Johnson, B. P., "Additives for Hose Reel Systems: Trials of Foam on Tire Fires," Home Office, London, England, Publication 5/91, SC/88 42/1087/1, 1992.

Loeb, D., "Elevated Master Streams Deliver Straight Out Extinguishing Power," Fire Chief, Vol. 35, No. 5, May 1991, pp. 67–69.

Morita, M., and Yamauchi, Y., "Computer Geometry Applied to Locus Approach to Fire Stations," _Proceedings_ of the Fourth International Symposium on Fire Safety Science, July 13–17, 1994, Ottawa, Ontario, Canada, Intl. Assoc. for Fire Safety Science, Boston, MA, 1994, pp. 1303–1311.

Ruda, S. J., "Busiest Station in the Nation: Station 11, Los Angeles City Fire Department—20,225 Runs in 1990," _Firehouse_, Vol. 16, No. 11, Nov. 1991, pp. 61–62.

Scheffey, J. L. and Williams, F. W., "Extinguishment of Fires Using Low Flow Water Hose Streams. Part 1," Fire Technology, Vol. 27, No. 2, May 1991, pp. 128–144.

Scheffey, J. L., and Williams, F. W., "Extinguishment of Fires Using Low Flow Water Hose Streams. Part 2," Fire Technology, Vol. 27, No. 4, November 1991, pp. 291–320.

Shapiro, P., "Mastering the Ultimate Stream," American Fire Journal, Vol. 47, No. 7, July 1995, pp. 16–20.

Sybesma, P., "Equation for Station Location," *Fire Chief*, Vol. 39, No. 10, Oct. 1995, pp. 55–56, 62–65.

vanBowen, J., Jr., "Use of In-House Data for Fire Station Location," *Journal of Applied Fire Science*, Vol. 1, No. 1, 1990–1991, pp. 39–44.

Walsh, M. B., "Evanston's New Station Blends Economy and Utility," *Fire Chief*, Vol. 34, No. 6, June 1990, pp. 36–38.

Walsh, M. B., and Wojcik, K., "What's New in Fire Stations," *Fire Chief*, Vol. 34, No. 6, June 1990, pp. 32–35.

INFORMATION AND ANALYSIS FOR FIRE PROTECTION

From a conceptual standpoint, fire protection is achieved through a series of coordinated decisions, which is to say, actions based on processed information. There are decisions in design, construction or manufacture, purchase, installation, maintenance, education, response to emergency, and so forth. Every decision is as good as the information it is based on, and so every aspect of fire protection stands to benefit from better sources of information and better methods of analyzing and interpreting the available information. One of the most exciting developments of recent years has been the rapid evolution of analytical tools to provide better information to support decisions and to coordinate them in a systematic approach to fire safety. Section 11 covers these tools.

Chapters 1 and 2 discuss information-gathering in connection with individual fires, either in-depth investigation, in Chapter 1, or routine incident reporting, in Chapter 2. Chapter 3 then discusses the techniques used to analyze this material—to turn raw data into useful statistics, facts into information, observations into knowledge.

Chapter 4 provides an introduction to the various types of mathematical modeling. Modeling relies on data—fire incident data produced in Chapters 1-3, fire-test-response data from laboratory tests, and other data on parameters and variables ranging from equipment specifications to room and building geometries to patterns of product usage to reliability statistics for fire protection systems. Chapter 5 provides more details on deterministic fire models, while Chapter 6 provides more details on probabilistic fire models.

Chapters 7 and 8 lay out the essential elements of fire hazard assessment and fire risk analysis, two very general methodologies for combining diverse technical measurements into summary characterizations of the overall fire danger posed by some object of analysis, whether they be products, buildings, or something else. These two analytical methods are at the top end of the hierarchy of fire-related models in terms of scope. The fire models described in Chapters 5 and 6 form the core of the broader, more ambitious calculations in fire hazard models, which in turn form the core of fire risk models. Much of the work of fire science and fire protection engineering is now explicitly designed to fill gaps or improve accuracy or flexibility in some comprehensive fire hazard or fire risk model.

With a full array of models to choose from, Chapter 9 describes the use of models in performance-based fire codes and standards. Design to achieve performance produces greater design flexibility and greater clarity on how much safety is being achieved, all while providing latitude for cost savings.

Chapters 10 and 11 provide simplified modeling and analysis techniques. Performance-based codes and standards and related design and assessment activities involve large numbers of calculations. Short cuts can make analysis more affordable and manageable, and these chapters provide a road map to those methods.

Chapters 12 and 13 show how modeling can be applied to fire-safety design, fire investigations, and other fire protection engineering problems. Finally, Chapter 14 examines the special topics of inspecting, surveying and mapping.

Also look for these: Most of the subjects in this section also are examined in much more detail in *The SFPE Handbook of Fire Protection Engineering*, an indispensable reference for anyone wishing to go further in the use of advanced calculation methods as applied to fire protection. This part of the field is changing so fast that the serious practitioner should also keep abreast of new developments in the technical literature through such publications as the *SFPE Journal of Fire Protection Engineering* and NFPA's *Fire Technology*. By the time material of this kind reaches book form, there often are exciting new developments already in use.

—John R. Hall, Jr.
December 1996

FIRE AND ARSON INVESTIGATION

Richard L. P. Custer

This chapter discusses fire loss investigation by examining organizations involved in fire loss investigations and the reconstruction and failure analysis process. Two other pertinent chapters in this section are Chapter 2, "Fire Data Collection and Data Bases," and Chapter 3, "Use of Fire Incident Data and Statistics." A discussion of a fault tree for managing fire risk can be found in Section 1, Chapter 3, "Systems Concepts for Building Fire Safety," and fire-resistive construction information can be found in Section 7, Chapter 2, "Building Construction."

Information about the dynamics of fire spread and growth can be found in Section 1, Chapter 7, "Dynamics of Compartment Fires"; Section 4, Chapter 2, "Combustion Products and Their Effects on Life Safety"; Section 4, Chapter 17, "Upholstered Furniture and Mattresses"; Section 11, Chapter 10, "Simplified Fire Growth Calculations"; and Section 11, Chapter 12, "Applying Models to Fire Protection Engineering Problems and Fire Investigations."

INTRODUCTION

The data gathered during a fire loss investigation have several applications, ranging from filling out the fire department incident report to creating a detailed engineering reconstruction and failure analysis of the incident. Understanding the fire cause and origin can lead to targeted inspection, public education programs, or, perhaps, proposed code changes. A complete failure analysis can reveal factors concerning the fuel, building, and fire suppression that resulted in the ultimate extent of flame and smoke movement. The codes and design standards involved would be included in this analysis, as would construction, operation, and maintenance practices. The results of a failure analysis provide input to insurance loss adjustment and underwriting, improved design practice, and civil and criminal litigation. They can also result in changes to codes and standards.

In most fires cause and origin are expressed in terms of the heat of ignition, the initial fuel source involved, the behavior or malfunction that brought heat and fuel together, and the area of origin. (See NFPA 921, *Guide for Fire and Explosion Investigations*.) In the past, only the large life- or property-loss fires received the attention of a complete investigation. In these cases, the reconstruction and failure analysis was often conducted months or years after the fire,

when only fragmentary evidence, if any, remained. In these situations, the depth and accuracy of the fire department report can be critical. Recently, insurance claims, subrogation activities, and greatly increased litigation have been concerned with smaller and smaller losses. Thus, more detailed examination is needed of all fires that result in a loss, almost regardless of the extent of the loss.

ORGANIZATIONS INVOLVED IN FIRE LOSS INVESTIGATION

The organizations involved in fire investigation can be separated into two types: (1) public agencies, or those required by law to investigate fires at the local, state, and federal levels; and (2) private-sector organizations.

Local and State

Of the public agencies, the local fire department generally has the primary responsibility to document a fire and to provide an initial determination of the cause and origin. If the department uses NFPA 901, *Standard Classifications for Incident Reporting and Fire Protection Data*, as the basis of a reporting system such as the National Fire Incident Reporting System, factors that affected the ultimate course of the fire, such as fire and smoke spread and the performance of fire protection features, may be included.

When a fire department investigation determines that the fire may have been of incendiary or suspicious origin, other organizations may be assigned the legal responsibility to make the final determination of fire cause and to continue the investigation pursuant to possible criminal action. In much of the United States and in all Canadian provinces, the investigation responsibility for incendiary fires is vested in the state or provincial fire marshal's office. In states where this is not the case, the local fire marshal or the local police or fire department may have investigative responsibility, either legally or by delegation from the state.

Federal

Federal agencies are also involved in investigating fires. These agencies are generally charged with investigating incidents to determine compliance with federal regulations and to provide input for the regulatory process to reduce human and property losses.

For example, the National Transportation Safety Board (NTSB) conducts complete reconstruction and failure analysis of accidents, including fire-related incidents, involving air, rail, and

Richard L. P. Custer, P.E., is adjunct associate professor of fire protection engineering at the Center for Firesafety Studies, Worcester Polytechnic Institute, Worcester, MA, and principal consultant in the fire safety consulting firm of Custer Powell, Inc., Wrentham, MA.

highway transportation. Investigations of pipeline incidents and maritime fires are under the authority of the Department of Transportation (DOT). Other federal agencies involved in fire investigations include the Occupational Safety and Health Administration (OSHA) of the Department of Labor; the Bureau of Mines (BoM) of the Department of the Interior; the Federal Bureau of Investigation (FBI); the Bureau of Alcohol, Tobacco, and Firearms (ATF) of the Department of Justice; the Consumer Product Safety Commission (CPSC); the Nuclear Regulatory Commission (NRC); and the National Institute of Standards and Technology (NIST) of the Department of Commerce.

The Federal Emergency Management Agency (FEMA) conducts detailed investigations of fire and hazardous materials incidents and prepares reports that can be obtained through the U.S. Fire Administration. Other agency reports are available from the National Technical Information Service (NTIS) in Atlanta, Georgia.

In addition to investigating fires under their jurisdiction, these federal agencies may serve as resources for lab analysis and technical information for state and local fire officials.

Committees of the U.S. House of Representatives and the Senate are often very interested in the results of these investigations, particularly when the incidents under investigation deal with timely public policy topics, such as standards for elderly care facilities and prison fire safety.

Private Sector Organizations

Most of the complete reconstruction and failure analyses of fires other than those that fall under government jurisdiction are conducted by private-sector organizations for education, improved design, insurance, and litigation purposes.

The National Fire Protection Association's (NFPA) fire investigations program is a long-standing data collection and engineering analysis activity. The purpose of the program is to collect, analyze, and report detailed fire experience data through on-site investigations. This activity assists NFPA in analyzing and documenting facts about fires for technical or educational value; distributing the information to the fire protection community to help prevent future similar fire losses; publishing of investigation reports as individual reports, as well as in *NFPA Journal* and on the NFPA home page; providing technical assistance to state and local officials as an integral service of the NFPA investigation process; determining important "lessons learned" for input to NFPA technical committees and technical programs; and contributing to an increasing fire incident data base for analysis purposes.

Several NFPA fire investigations have included or been supplemented by analysis of human behavior patterns, where the discovery of the fire and subsequent evacuation occurred in sufficient time to permit analysis. The human behavior studies have been conducted through on-site interviews with survivors or mailed surveys and have generally been carried out by NFPA staff in combination with leading authorities on human behavior in fire.

NFPA frequently works with several of the national code organizations during the on-scene investigation. The results of most of these investigations are published as special studies or as reports in various fire protection periodicals.

The insurance industry often reconstructs a fire to evaluate the merits of claims submitted or to identify parties for subrogation action. The reconstruction may be conducted by insurance company personnel, by an adjustment bureau, or by a consultant.

Individual consultants or consulting firms perform reconstruction and failure analysis for a variety of clients, including the insurance industry. Consultants are also retained for litigation purposes by law firms to evaluate fire cause and origin, code compliance, conformance with design standards, and other technical aspects of fire loss.

When major financial consequences are linked to investigators' conclusions, it is not uncommon for two or more parties to retain investigators and for two or more theories of fire origin and development to be presented. The need to resolve such disputes is a major reason for the increasing interest in applying state-of-the-art scientific models, procedures, and test results to fire investigations.

THE RECONSTRUCTION AND FAILURE ANALYSIS PROCESS

The complete reconstruction and failure analysis process involves developing a time history for the incident and establishing the factors that led to ignition and that increased or lessened the severity of the loss. The reconstruction includes the prefire history, as well as the "transfire" events, or those that occur from the point of established burning to extinguishment. Failure analysis is concerned with the design, construction, and performance aspects of the incident. It is also important in the reconstruction to document the successes. For example, did fire protection features inhibit the spread of fire within a portion of the building?

Time is the framework for reconstruction and failure analysis, and it can be viewed in two scales. These are the macro scale, where time is measured in increments of days, weeks, and years, and the micro scale, where the increments are in seconds, minutes, and hours. Macro time is prefire, while micro time may apply to pre- or transfire events.

The best way to organize the reconstruction and failure analysis information is through the use of timelines or event sequences. In constructing a timeline, it is important to establish benchmark events. A benchmark event can be described as an event that is well-documented in time and space and that serves as a point of reference because it can be assumed to have created a set of specific conditions or initiated a sequence of actions that had a significant impact on the fire. The issuance of a building permit or the time of arrival of the fire department apparatus are good examples of benchmark events. Using time of arrival as a reference event, other events for which the time of occurrence is unknown can be placed approximately in time either before or after arrival of the fire department apparatus. Other examples of typical benchmark events are explosions and the collapse of walls or roofs or calling for multiple alarm response.

A single timeline can be adequate for fires in which only a few events occurred. For more complex situations, separate timelines on the same scale can be used for the history of the building, the evolution of the "model codes," the installation and maintenance history of a suppression system, human activity, or any other time-related aspect of the incident. STEP (sequentially timed events plotting) procedures have been developed to deal with complex accident situations such as major fires.[1]

Prefire timeline information can be obtained from building permits, plans and specifications, inspection reports, fire and building codes, and fire research and engineering literature. Sources of information for use in the development of transfire timelines include the fire department incident report, recordings of fire department radio transmissions during the incident, photographs and video footage shot during the event by news media, private citizens or fire department personnel, and interviews with fire fighters and other witnesses.

It should be pointed out that, in many cases, only parts of the reconstruction and failure analysis process are carried out. Many different organizations may be involved in the investigation, often at different times. Since people may have to rely upon information gathered by others, an understanding of the whole process will make it easier to obtain the most complete and accurate data at each step.

The basic steps of overall reconstruction and failure analysis are examination of the scene, collection of background data, testing, incident reconstruction, and specific failure analysis. Many of these steps are outlined in NFPA 921.

Examination of the Scene

Examining the fire scene is critical to the entire reconstruction and analysis process. It is here that the physical and photographic evidence is gathered and the fire damage is documented.

As many photographs as possible should be taken at the scene, and each should be documented on a diagram with a description of the direction of view and contents of the photograph. The techniques of fire-scene photography have been well-reported and can be applied to reconstruction and failure analysis.[2–4] Color photography is recommended.

In addition, maps and diagrams should be prepared that include dimensions and interior layout of the building, the location and degree of fire damage, the location and types of fuel materials, and the locations of victims. The location of fire protection features, such as fire walls, detectors, and suppression systems, should also be noted.

Often, buildings will be modified considerably over their lifetimes, and an individual reviewing the original plans could receive a false impression of the building at the time of the fire. Thus, it is important to record the architectural, construction, and fire protection features of a structure, both as built *and* at the time of the fire.

When fire suppression systems are present, the positions of valves, the locations of discharge nozzles, and the locations and types of detectors used to operate the system should be noted. It may also be necessary to evaluate the water supply in order to determine if it was adequate. With special systems, such as those that use halon, CO_2, or dry chemicals, agent storage tanks should be checked to determine whether the system discharged as designed.

It is important to interview fire department responders to determine what actions they took on the scene. Some of the damage may have occurred during fire fighting and overhaul and not as a direct result of the fire.

A careful study of the damage patterns, combined with on-site interviews and a knowledge of the ignition sources, fuel materials present immediately before the fire, and fire dynamics, should be used by an investigator to determine the area of origin. If possible, the source and form of heat of ignition and the type and form of the materials first ignited should be determined. Burn patterns, such as a "V," low burns, and multiple "points of origin," can often be misleading and should only be used to establish cause and origin when they are unambiguous or are supported by other physical evidence.

The investigator should keep in mind that the burn patterns seen after a fire represent the total burning history of the event, including damage that may have occurred during extinguishment and overhaul. As a rule, patterns observed in fires that are extinguished early in their development are more meaningful and useful than those remaining after the near-total destruction of a building or a portion of a building.

Indicators, such as depth of char, should be used with care. The burning rate of wood traditionally has been calculated as approximately 1.4 in. in 60 min (0.6 mm/min). This figure has been used to estimate the time of burning and to detect the presence of flammable liquids. Research has shown that the traditional calculation varies substantially, depending on the radiant flux and ventilation. The rate of burning can be as high as 10 in. per hr (4.3 mm/min).[5]

Investigators should also be careful not to assume that certain burn patterns can be used as sole proof of the presence of flammable liquids. Factors such as ventilation, room flashover, floor clutter, room contents, fire duration, fire-fighting activities, and overhaul can affect the patterns seen by investigators. Corroborative evidence, such as pertinent chemical traces or the detected absence of an appropriate prefire fuel load, is needed to establish the presence of flammable liquids. The color of observed smoke is not necessarily evidence of the presence of flammable liquids.

In visualizing conditions in the area of origin, it is often helpful to reconstruct the scene physically by replacing the carpet, furniture, and other items in their original locations. While interviews are conducted at many different times in the reconstruction and failure analysis process, those taken at or close to the time of the incident also can be very useful, particularly in describing the course of the fire. Whenever possible, events described in interviews should be related to benchmark events.

Private sector investigators are often not called in to evaluate a fire scene until well after the incident. Frequently, the scene has been disturbed, or, in some cases, the building has been restored or removed. In these situations, the investigator should review physical evidence collected and photographic evidence. Examination of photographs using 7- to 10-power binocular microscopy can often reveal details, particularly when low-speed film has been used. If possible, copies of all photos and videotapes taken of the scene should be obtained. Grouping them for study, chronologically and by area depicted, is useful in documenting the amount of scene disturbance over time. Fire, police, and autopsy reports, as well as witness statements, should be reviewed.

Although an actual on-scene investigation of a fire is most desirable, careful analysis of a well-documented fire can lead to reliable conclusions regarding the origin and cause of many fires, as well as provide information for failure analysis. The effectiveness of a delayed investigation and analysis will depend upon the extent of the fire, as well as the degree of documentation. In some cases where there has been total or near-total destruction, neither an on-scene, well-documented investigation nor subsequent delayed analysis can establish origin or cause.

Collection of Background Data

While some of the background data, such as accounts of activities immediately before the fire, can be gathered at the scene very soon after the fire, most of the information needed for the investigation will be gathered later.

Building plans, permits, and the codes in effect at the time of approval should be obtained, along with the specifications and drawings for the building fire protection features. If any major modifications were made to the building or its fire protection systems or if there was a change of occupancy, the codes, plans, and permits should be obtained again. These materials should be reviewed as part of the reconstruction process, since the changes may have influenced the course of the fire.

Other background data include instruction or training manuals, maintenance records, data on construction and interior finish materials, the previous fire history of the building or equipment involved, the amount of any previous fire damage, the location of contents, and the physical and burning characteristics of the contents. This step also involves obtaining detailed statements from the building occupants to determine the circumstances leading up to ignition and established burning.

Testing

Three types of tests are involved in reconstruction and failure analysis: (1) standard tests of flammability properties, (2) standard analysis of unknown materials, and (3) special tests designed to establish burning behavior under conditions related to a specific fire scenario. The volume and pressure of the water supply could also be tested.

Properties of flammability that may be determined by test include flame spread of materials, fabric flammability, flash point, and rate of heat release, among others.[6–13]

Analysis of unknowns is performed by analytical laboratories using techniques, such as gas chromatography and mass spectroscopy, to identify unknown materials, particularly possible accelerants. Such analysis should be conducted in accordance with the appropriate ASTM standards. These are: ASTM E1385, *Standard Practice for Separation and Concentration of Ignitable Liquid Residues from Debris Samples by Steam Distillation;* ASTM E1386, *Standard Practice for Separation and Concentration of Ignitable Liquid Residues from Fire Debris Samples by Solvent Extraction;* ASTM E1387, *Standard Test Method for Flammable or Combustible Liquid Residues in Extracts from Samples of Fire Debris by Gas Chromatography;* ASTM E1388, *Standard Practice for Sampling of Headspace Vapors from Fire Debris Samples;* and ASTM E1389, *Standard Practice for Cleanup of Fire Debris Sample Extracts by Acid Stripping.* Other useful analysis techniques include metallurgical or X-ray analysis.

In collecting samples for testing, it is important to know the sizes and/or amounts of the sample needed for each test, as well as the conditions under which the samples should be transported. This information can be obtained from a testing laboratory or by reviewing a copy of the test method.

Although special tests, sometimes called "demonstrations," can often be extremely revealing in understanding the course of a particular fire, results of these demonstrations can be misleading because the test may not be truly representative of the actual event. Factors such as ventilation, arrangement of the fuel, or even the point of ignition can have a significant effect on the outcome of the test. In other words, a great deal must be known about the specific conditions at the time of the fire, and these conditions must be reasonably reproduced in the demonstration. Special tests or demonstrations should be conducted under laboratory conditions to permit control of such variables as ambient temperature, relative humidity, and wind effects by personnel familiar with full-scale fire testing.

Reconstruction and Analysis

The fire reconstruction is an organized, step-by-step portrayal of the most likely course of the fire under study, from the ignition sequence through established burning to the limits of flame and smoke movement at the time of extinguishment. It is based upon the facts developed through on-scene investigation, background studies, and testing. The reconstruction should explain the speed and direction of fire growth and smoke spread within the compartment or area of origin and beyond. It also should explain the effects of barriers and automatic suppression systems on fire spread and growth and the effects of manual extinguishment. The reconstruction should also evaluate the performance of the building's life safety features relative to injury or loss of life.

In sorting through the facts to develop the reconstruction, it is often useful to apply the system safety techniques of failure modes and effects analysis (FMEA) and fault tree analysis.[14] These methods help identify alternative sequences of events that could have led to the observed fire situation, such as the development of room flashover or fire breaching a fire wall. The alternatives can then be evaluated against the evidence to determine the most likely scenario. These techniques are particularly valuable where complex or interrelated machinery, buildings, or human activities are involved.

Computer modeling techniques can also be used to predict the outcome of a series of possible fire causes and growth scenarios. The model results can be compared to actual events to narrow down the likely scenarios for a given event or to evaluate the possible effects that fire protection equipment, such as detectors or sprinklers, might have had on the outcome of a fire. The analyses of the DuPont Plaza and the First Interstate Bank fires are examples of model applications.[15,16]

Analysis of the ignition sequence is concerned with three specific items: (1) the size, nature, and source of the ignition energy; (2) the physical and chemical fire properties of the materials ignited; and (3) the circumstances, either human or mechanical, that brought the energy in contact with the fuel. In selecting the most likely ignition source, the energy available from a candidate source must be compared with the ignition energy requirements of the fuel in its given form. For example, given the same material and heat source, ignition is more likely if the material is present in finely divided form than if it is in a solid mass. Energy available at the source, such as temperature and thermal radiation; distance from the fuel to the energy source; and time of exposure to the source of energy are also factors that must be considered in ignition sequence analysis.

The circumstances which result in ignition generally involve human factors or equipment failure. Careless welding, poor housekeeping, and improper selection or misuse of materials are examples of human factors. Valve rupture and combustion control failure could be considered equipment failures. Ultimately, most equipment failures can be traced to design, manufacturing, installation, or maintenance problems or to misuse.

If established burning is achieved after ignition, it must be determined why the fire continued to grow to full room involvement—that is, flashover, or, in the absence of compartmentation such as that in a warehouse or industrial building, why it reached a certain extent. In providing answers, a number of factors must be evaluated, including the rates of heat release of the fuels; continuity of the fuels—that is, the proximity of fuel packages to each other—and location of fuels relative to walls; compartment ceiling height; ventilation; and interior finish.

Heat release rate is one of the more important factors in analysis. In many instances, a fire reported to have grown or spread "unusually fast" is characterized as suspicious or incendiary when, in fact, the growth rate may have been due to a combination of the burning properties of the materials and the nature of the compartment in which the fire burned. A number of studies have been carried out to characterize the burning rates of materials and furnishings. Data can be found in NIST reports and in Appendix B-2.2 of NFPA 72, *National Fire Alarm Code* (hereinafter referred to as NFPA 72).[12–20] The roles of heat release rate, fire location, and ceiling height in fire development also have been reported.[21] Alpert's work includes methods to estimate ceiling temperatures and ceiling gas velocities. Guidance on estimating the likelihood that a given fire size and room ventilation combination can develop flashover also is provided in NIST reports, and *The SFPE Handbook of Fire Protection Engineering.*[22–24]

The performance of automatic suppression systems in controlling the growth and spread of the fire also must be evaluated in the reconstruction process. Here, the factors to be reviewed include the appropriateness of the agent being used; the system response time, as related to the detection devices used, relative to ceiling height and fire growth rate; location of the agent discharge nozzles relative to the source of the fire; agent availability, including rate of discharge, pressure, and adequacy of supply; and conformance of the system design and installation with NFPA standards, industry design, and installation manuals. For example, a sprinkler system may have failed to control a fire due to a very high ceiling that delayed the response time, combined with low water pressure that could not produce the needed water flow rate.

Another component of the reconstruction is barrier performance. Barrier performance evaluation focuses on the nature of the barrier, whether rated or not rated; barrier construction; the thermal stresses to which it was exposed; and the manner in which the barrier failed. Barriers can fail either by passage of flame or hot gases through a small opening, by a hot spot failure, by movement of flame and hot gases through open doors, or by barrier collapse, resulting in a massive failure. Failure of opening protection, poor construction practices, modifications to the barrier after construction, and fire stresses that exceed those assumed in the design can be significant transporters of products of combustion beyond design limits.

Manual suppression activities, whether by a municipal fire department or a plant fire brigade, can have a significant effect on the outcome of a fire. In the absence of fire suppression systems, the response time of the fire department for a given fire will influence the size of the fire at the time of agent application. For ease of analysis, the response time can be broken down into the following segments: detection time, alarm transmission time, dispatch time, travel time, and time from the arrival of fire suppression personnel to agent application. Problems resulting in delays in any of the above time segments will extend the limits of flame and smoke travel. Examples of factors that could cause delays are the absence of automatic detectors, the absence of people in the building, or an improperly designed detection system. Equipment failure or out-of-service fire apparatus can also delay response time. Weather, terrain, traffic, and the condition of apparatus can extend travel time to the incident. Frozen hydrants, blocked access, and shortage of personnel are also examples of factors that could delay application of fire suppression agents.

The reconstruction process also must consider the performance of the building life safety system. This includes the detection and alarm system, the building egress system, and the smoke control system. In addition to the factors discussed above that affect detector response, the audibility of alarms plays a role in life safety. Guidance on this topic can be found in work by Nober,[25] in NFPA 72, and in the *The SFPE Handbook of Fire Protection Engineering*.[26] Building egress factors include the capabilities of the occupants and the number, capacity, and protection of the egress routes. The movement of smoke in buildings is related to the temperature of the fire gases; differences between inside and outside temperatures, or the stack effect; effects of the building's heating and air-conditioning systems; and effects of wind.

Failure Analysis

In a fire safety context, failure can be defined as a fire that results in personal injury, death, or property or monetary losses, or in conditions that prevent buildings or mechanisms from functioning as designed. In fire failure analysis, the objective is to use on-site and background data, testing results, and reconstruction to identify the primary and contributing causes of the failures and their sources. The sources of failures include basic design, material or equipment selection, material or equipment defects, construction or assembly, testing, post-construction modifications, service conditions, and unanticipated conditions or abuse.

The analysis is made by comparing the causes of failures with the codes, standards, design practices, and the state-of-the-art technology present at the time of the fire. In the analysis, it is important to identify noncompliance with the standard practices, but it is even more important to note where compliance with the standards failed to provide the expected or needed degree of fire safety. In either event, the failure analysis should provide recommendations for how the failures could have been prevented or how the knowledge gained can be applied to preventing similar failures in the future.

ORGANIZATION FOR FIRE INVESTIGATION

It is a good idea to be prepared in advance for the investigation of a fire. The preparation may range from identifying an individual or company to hire as needed, to the organization of a permanent fire investigation unit complete with full-time staffing, vehicles, and a laboratory. The level of organization selected is dependent upon the size of the organization, the frequency of need, the resources available, and the legal or statutory requirements. One other factor to be considered is whether or not the organization is in the public sector—the fire department, police department, state and federal government, and so on—or in the private sector—insurance company, product manufacturer, major hotel chain and so on. Frequently, the public-sector needs will include more than just the ability to conduct a fire scene investigation. Should it be determined that the possibility of arson exists, the likelihood of criminal or civil action often requires the involvement of the police or persons trained in the legal aspects of arson and fraud investigation.

Public Sector

In the public sector, most initial investigations are carried out by the fire department personnel responding to the scene. If the cause is easily determined and does not appear to be of a potentially criminal nature, the investigation will end at that level. When the cause cannot be determined, a fatality is involved, or arson is suspected, additional help is generally dispatched. In a small fire department, the local police or the county or state fire marshal's office may be contacted. In some jurisdictions, the assistance may come from the country or state police.

In large cities, a fire investigation unit is usually formed. The unit may be made up of fire department personnel and be wholly within the fire department, or the unit may be within the police department. Regardless of where the unit is placed organizationally, it is generally recommended that both fire investigation and police functions be represented in the background and training of personnel. This objective is often accomplished by selecting individuals from both departments to serve in the fire investigation unit.

In some jurisdictions, an investigations unit is established in the local criminal prosecutor's office. Such a unit will usually have arson detection and prosecution as its major focus where the case load is high. A special prosecutor is often assigned specifically for arson, and that individual frequently is trained in fire investigation techniques.

In many instances, the fire investigation units in both the fire and police departments are called arson investigation units. The determination of arson can usually come only after a thorough fire investigation in which other possible causes are ruled out and physical evidence of an incendiary fire is established. This provision should be stressed to investigative personnel so they do not come to believe that their role is to find arson. The term "fire" investigation is more appropriate to the work that is done than arson investigation. If arson is a major problem in a community, special arson task forces can be established to identify arson-for-profit groups, associating real estate buying and selling and insurance patterns with arson and arson prevention activities. This work is less concerned with actual fire scene investigation, and deals more with the establishment of motivation, insurance manipulation, and patterns of arson occurrence.

Some federal agencies, such as the General Services Administration or the military services, have personnel who investigate fires involving their facilities. These personnel are generally assigned fire investigation in addition to other functions such as facility safety or fire protection design. In the event of a major fire, an agency may conduct its own investigation or use an outside consultant to assist its own personnel.

Private Sector

In the private sector, large insurance companies may have investigative units that conduct fire scene investigations and perform arson and fraud detection and control functions. Some companies will employ adjustment bureaus or private investigators to make the initial cause determination, and follow up with their own personnel if it appears that arson or fraud is involved or if there is a particularly large loss. Small companies may rely exclusively upon outside investigation services.

A number of large industrial companies also maintain a fire investigative capability, often located in their risk management or loss prevention groups. These groups or individuals may investigate fires in company facilities or fires involving products or systems designed, manufactured, supplied, or installed by the company. These investigations may be for the purpose of future loss prevention or plant safety programs, or they may be carried out in response to actual or potential litigation. An important function of this type of investigation is the identification of the causative factors in the fire that could lead to improved design, manufacturing, installation, or operating practices.

Organizations such as NFPA and the U.S. Fire Administration also investigate fires in order to determine "lessons learned."

Small companies may not be able to afford their own investigation staffs. In such cases, an appropriate outside capability should be identified in advance for use if the need arises. The individual or firm selected should either have or be given some knowledge of the specific products, facilities, or operations that are likely to be involved.

EDUCATION AND TRAINING OF FIRE INVESTIGATION PERSONNEL

The nature and extent of education, training, and experience needed by investigators obviously varies greatly with the specific goals and objectives for fire investigations set by the organization where they are employed and by budget resources. Some basic performance requirements for fire investigators are provided in NFPA 1033, *Standard for Professional Qualifications for Fire Investigation*. The level of knowledge needed by a fireground officer faced with a residential fire extinguished early in its development will be less than that needed by an investigator from the department's fire investigation unit. This investigator will have to deal with fires extinguished in more advanced stages, where the evidence of cause is often much less clear. In addition, this investigator will have to deal with fires in the full range of occupancies found in the unit's jurisdiction. This could include residences, office structures, warehousing, manufacturing, and heavy industry.

The private-sector investigator employed by a petrochemical company may need a degree in chemistry or chemical or fire protection engineering in addition to the basic fire scene investigation skills. The specialist in arson investigation will need education and experience in the legal aspects of civil or criminal investigations, or possibly both.

Education

In the past decade, a great deal of data regarding the nature of the ignition process and the forces that control fire spread and growth have emerged. Research by university, government, and insurance laboratories has developed information that can explain much of the evidence remaining after a fire has been extinguished, and assist in determining origin and cause.

To determine adequately the cause and origin of a fire, the modern investigator needs a basic knowledge of physics and chemistry, preferably at the post-secondary-school level. He or she should be familiar with the concepts of heat release rate, fire plume and ceiling layer development, and compartment flashover.

All fire investigators should also have some knowledge of the legal aspects of investigations, such as how to gain access to a scene, how to take and handle evidence, and how to question witnesses.

The fire science courses offered at many post-secondary schools today can provide much of the needed background. Where such facilities are not available, basic science courses supplemented by study using the additional reading at the end of this chapter can be employed. In some areas of the country, the U.S. Fire Administration's Open Learning Program is also available through selected local colleges and universities.

In fire investigation units to which both police and fire personnel are assigned, personnel should be cross-trained in basic fire science and law enforcement. Departments that become familiar with each other's technology and terminology benefit from enhanced communication, more complete and thorough investigations, and more successful prosecutions.

Advanced education might include college-level courses in such areas as forensic science, industrial safety, insurance, accident investigation, and fire protection engineering.

Training

All investigators should be trained in the basics of fire scene investigation, which include adequate documentation of the scene, origin determination, and cause determination. Specific emphasis should be placed on techniques for debris removal and scene reconstruction. Although much of this knowledge can be gained by studying text materials, there is no substitute for field experience. Even limited hands-on work can be of great value in relating the text materials to "the real world."

BIBLIOGRAPHY

References Cited

1. Hendrick, K., and Benner, L., *Investigating Accidents with STEP*, Marcel Deckker, New York, 1987.
2. Lyons, P. R., *Techniques of Fire Photography*, National Fire Protection Association, Quincy, MA, 1978.
3. Kodak, *Using Photography to Preserve Evidence*, Pamphlet M-2, Eastman Kodak Company, Rochester, NY, 1976.
4. Kodak, *Fire and Arson Photography*, Pamphlet M-67, Eastman Kodak Company, Rochester, NY, 1977.
5. Drysdale, D., *Introduction to Fire Dynamics*, John Wiley & Sons, New York, 1985.
6. ASTM E162, *Standard Test Method for Surface Flammability of Materials Using a Radiant Heat Energy Source*, American Society for Testing and Materials, W. Conshohocken, PA, 1994.
7. ASTM E84, *Standard Test Method for Surface Burning Characteristics of Building Materials*, American Society for Testing and Materials, W. Conshohocken, PA, 1995.
8. *Standard for the Flammability of Clothing Textiles (CS-191-53)*, 16CFR1610, Code of Federal Regulations, 1953.

9. *Standard for the Flammability of Children's Sleepwear: Sizes 0 Through 6X (FF3-71) 16CFR1615,* Code of Federal Regulations, 1981.

10. ASTM D92, *Standard Test Method for Flash and Fire Points by Cleveland Open Cup,* American Society for Testing and Materials, W. Conshohocken, PA, 1990.

11. ASTM D56, *Standard Test Method for Flash Point by Tag Closed Tester,* American Society for Testing and Materials, W. Conshohocken, PA, 1993.

12. ASTM E906, *Standard Test Method for Heat and Visible Smoke Release Rate for Materials and Products,* American Society for Testing and Materials, W. Conshohocken, PA, 1983.

13. Babrauskas, V., *et al.,* "Upholstered Furniture Heat Release Rates Measured with a Furniture Calorimeter," NBSIR 82-2604, National Bureau of Standards, Gaithersburg, MD, 1982.

14. Henley, E. J., and Kumamoto, H., *Reliability Engineering and Risk Assessment,* Prentice Hall, Englewood Cliffs, NJ, 1981.

15. Nelson, H. E., "Engineering Analysis of the Early Stages Fire Development—The Fire at the DuPont Plaza Hotel and Casino—December 31, 1986," NBSIR 87-3560, National Institute of Standards and Technology, Gaithersburg, MD, 1987.

16. Nelson, H. E., "Engineering View of the Fire of May 4, 1988, in the First Interstate Bank Building, Los Angeles, California," NISTR 89-4061, National Institute of Standards and Technology, Gaithersburg, MD, 1989.

17. Babrauskas, V., "Combustion of Mattresses Exposed to Flaming Ignition Sources Part I: Full-Scale Tests and Hazard Analysis," NBSIR 77-1290, National Bureau of Standards, Gaithersburg, MD.

18. Babrauskas, V., "Full-Scale Burning Behavior of Upholstered Chairs," *NBSTN 1003,* National Bureau of Standards, Gaithersburg, MD, 1979.

19. Lee, B. T., "Heat Release Rate Characteristics of Some Combustible Fuel Sources in Nuclear Power Plants," NBSIR 85-3195, National Bureau of Standards, Gaithersburg, MD, 1985.

20. Walton, W. D., and Twilley, W. H., "Heat Release and Mass Loss Rate for Selected Materials," NBSIR 84-2960, National Bureau of Standards, Gaithersburg, MD, 1984.

21. Alpert, R. L., and Ward, E. J., "Evaluation of Unsprinklered Fire Hazards," *Technology Report 83-2,* Society of Fire Protection Engineers, Boston, 1983.

22. Babrauskas, V., "Estimating Room Flashover Potential," *Fire Technology,* Vol. 16, No. 2, 1980, p. 94.

23. Babrauskas, V., "Will the Second Item Ignite?" *Fire Safety Journal,* Vol. 4, 1981–1982, p. 281.

24. Walton, W. D., and Thomas, P. H., "Estimating Temperatures in Compartment Fires," *Fire Protection Engineering,* 2nd ed., DiNenno, P. J., ed., National Fire Protection Agency, Quincy, MA, 1995, pp. 3–134 to 3–147.

25. Nober, E. H., *et al.,* "Waking Effectiveness of Household Smoke and Fire Detection Devices," NBSGCR-80-284, National Bureau of Standards, Gaithersburg, MD, 1984.

26. Schifiliti, R. P., Meccham, B. J., and Custer, R. L. P., "Design of Detection Systems," *The SFPE Handbook of Fire Protection Engineering,* DiNenno, P. J., ed., pp. 4–21 to 4–29.

NFPA Codes, Standards, and Recommended Practices

Reference to the following NFPA codes, standards, and recommended practices will provide further information about fire and arson investigation. (See the latest version of *The NFPA Catalog* for availability of current editions of the following documents.)

NFPA 72, *National Fire Alarm Code*

NFPA 422, *Guide for Aircraft Accident Response*

NFPA 901, *Standard Classifications for Incident Reporting and Fire Protection Data*

NFPA 902, *Fire Reporting Field Incident Guide*

NFPA 903, *Fire Reporting Property Survey Guide*

NFPA 904, *Incident Follow-Up Report Guide*

NFPA 921, *Guide for Fire and Explosion Investigations*

NFPA 1033, *Standard for Professional Qualifications for Fire Investigator*

Additional Reading

"Action to Stem Arson Attacks in Europe," *Fire International,* No. 143, May 1994, p. 19.

"Approaching Arson Investigations Armed With Facts, Science and a Computer," *Firehouse,* Vol. 19, No. 2, Feb. 1994, pp. 50–52.

"Arson Fire Kills 87 in Worst Mass Slaying in U. S. History," *Fire Control Digest,* Vol. 16, No. 4, April 1990, pp. 1–4.

"Arson Fires Leave Few Clues," *Fire and Flammability Bulletin,* Nov. 1993, pp. 5–6.

"Arsonist on Board!," *Fire Prevention,* No. 260, June 1993, pp. 18–20.

"Arsonists: And What Happened to Them," *Fire Prevention,* No. 255, Dec. 1992, pp. 20–22.

"Arsonists: And What Happened to Them," *Fire Prevention,* No. 258, April 1993, pp. 15–16.

Andersson, H., "Anlagda branders omfattning: motiv och paverkande faktorer. [Arson in Sweden: Its Extnet, Motives and Contributory Factors.]," Stockholm Univ., Sweden, ISBN 91-7153-322-3, 1995.

Andersson, H., "Swedish Investigation of Arson," *Fire Technology,* Vol. 29, No. 4, Nov. 1993, pp. 350–373.

Babrauskas, V., "Burning Rates," *The SFPE Handbook of Fire Protection Engineering,* 2nd ed., DiNenno, P. J., ed., National Fire Protection Association, Quincy, MA, 1995.

Barber, K., *et al.,* "Social Issues and Juvenile Arson," Insurance Committee for Arson Control, 1992 National Arson Forum, Sept. 9–10, 1992, Arlington, VA, 1992, pp. 1–2.

Beale, R., "Arson: The Problem of Proof," *Fire Prevention,* No. 244, Nov. 1991, pp. 30–31.

Beatty, M. W., "Turning the Corner on Arson," *Fire Chief,* Vol. 36, No. 6, June 1992, pp. 42–44.

Beering, P. S., "Secrets to a Successful Arson Prosecution," *Firehouse,* Vol. 14, No. 9, Sept. 1989, pp. 69–72, 92; *Fire and Arson Investigator,* Vol. 41, No. 2, Dec. 1990, pp. 52–54.

Beering, P. S., "Top 10 Reasons Why Prosecutors Reject Arson Cases," *Firehouse,* Vol. 20, No. 8, Aug. 1995, pp. 44–45.

Bernard, D., "Sniffing Out the Work of Arsonists," *Fire Fighting in Canada,* Vol. 35, No. 3, April 1991, pp. 22, 24.

Brannigan, V., "Arson and the Fourth Amendment," *Fire Chief,* Vol. 37, No. 4, April 1993, pp. 22–24.

Burnett, G. E., Jr., "Checklist for Investigating the Arson-for-Profit Scheme," *Fire and Arson Investigator,* Vol. 42, No. 4, June 1992, pp. 14–16.

Cabe, K., "Firefighter Arson: Local Alarm," *Fire Management Notes,* Vol. 56, No. 1, 1996, pp. 7–9.

Carlsson, G., and Wistedt, I., "Analysis of Arson Accelerant Residues by ATD 400," SKL-National Laboratory of Forensic Science, Linkoping, Sweden, Report 26, March 15, 1993.

Carper, K. L., ed., *Forensic Engineering,* Elsevier, New York, 1989.

Catchpole, L., "Counselling Your Arsonists," *Fire Prevention,* No. 273, Oct. 1994, pp. 25–27.

Chubb, M., "High Temperature Accelerant (HTA) Arson Fires," USFA Fire Investigation Technical Report Series, TriData Corp., Arlington, VA, Federal Emergency Management Agency, Washington, DC, Report 065, 1992.

Corder, D., and Beale, R., "Arson: The Team Approach," *Fire Prevention,* No. 258, April 1993, pp. 21–23.

Corry, R. A., and Golembeski, J. K., "Solving an 'Impossible' Arson Case," *Fire and Arson Investigator,* Vol. 41, No. 1, Sept. 1990, pp. 36–37.

Cotreau, M. R., "Legal Aspects of Arson Investigation," *Fire Engineering,* Vol. 143, No. 1, Jan. 1990, pp. 61–63.

Cronk, P., "Task Force: An Innovative Approach to the Arson Problem. Fighting Fire With Fire," *Fire and Arson Investigator,* Vol. 43, No. 3, March 1993, pp. 26–28.

Cullom, K., "Santa Ana Condition and Arsonist Equal Wildfire," *American Fire Journal,* Vol. 43, No. 1, Jan. 1991, pp. 22–23.

Davies, G., ed., *Forensic Science,* 2nd ed., American Chemical Society, Washington, DC, 1986.

deHaan, J. D., *Kirk's Fire Investigation,* 3rd ed., Prentice-Hall, Englewood Cliffs, NJ, 1990.

Dietz, W. R., "Physical Evidence of Arson: Its Recognition, Collection and Packaging," *Fire and Arson Investigator,* Vol. 41, No. 4, June 1991, pp. 33–39.

Discovery Channel, "Arson: Clues in the Ashes. Broadcast Premiere (52:00). Monday, June 17, 1996 (9 p.m.)," Discovery Channel, Silver Spring, MD, Video, (COPYRIGHT).

Donahue, M. L, "Effective Anti-Arson Programs," *Fire Engineering*, Vol. 145, No. 6, June 1992, pp. 115–116, 119, 121–123.

Donahue, M. L., "Using Case Solvability Factors for Arson Investigation," *Fire Chief*, Vol. 35, No. 41, Sept. 1991, pp. 75–77.

EEI, "Fire and Arson Investigators' Field Index Directory," EEI Alexandria, VA, Federal Emergency Management Agency, Emmitsburg, MD, FA-91, April 1993.

Eisner, H., "Arson Wave Strikes Lawrence, Massachusetts (MA)," *Firehouse*, Vol. 17, No. 8, Aug. 1992, pp. 36–37.

Eisner, H., "Lawrence, Massachusetts: Arson Hits City as Cutbacks Decrease Fire Protection," *Firehouse*, Vol. 19, No. 6, June 1994, pp. 32–35, 116.

Factory Mutual Engineering Corporation, "Pocket Guide to Arson and Fire Investigation," Factory Mutual Engineering Corp., Norwood, MA, 3rd Edition, P7923 (9/94), 1992.

Factory Mutual Engineering, "Prevention of Arson Fires," Factory Mutual Engineering, Norwood, MA, P9213, March 1995; Factory Mutual Engineering, *Business Under Fire! A Manager's Guide to Fire Prevention Programs*, 1995, pp. 1–4.

Farrenkopf, D., "Measures for the Prevention of Arson Fires," Hamburg Fire Dept., Germany, CIB W14/95/1 (J); Tokyo Fire Department, Firesafety Frontier '94, International Fire Conference and Exhibition in Tokyo, Creating a Safe Tomorrow, Oct. 18–22, 1107-112, Tokyo, Japan, 1994, pp. 261–264.

Federal Emergency Management Agency, "Arson Forum Report and Analysis and Recommendations for Federal Arson Policy and Priorities," Federal Emergency Management Agency, Washington, DC, FA-134, June 1993.

Federal Emergency Management Agency, "Arson in America: A Profile of 1989 NFIRS," USFA Fire Investigation Technical Report Series, Special Report, TriData Corp., Arlington, VA, Federal Emergency Management Agency, Washington, DC, Special Report, 1991.

Federal Emergency Management Agency, "Arson Resource Directory," Federal Emergency Management Agency, Washington, DC, FA-74, April 1993.

Federal Emergency Management Agency, "Fire/Arson Investigation Training Resource Catalog," Federal Emergency Management Agency, Emmitsburg, MD, FA-131, March 1993.

Ferry, T. S., *Modern Accident Investigation and Analysis,* 2nd ed., John Wiley & Sons, New York, 1990.

Few, E. W., "Tracking Arsonists in America's National Forests," *Firehouse*, Vol. 15, No. 8, Aug. 1990, pp. 34–35, 83.

Fire Protection Association, "Investigation of Vehicle Fires. Fire Safety Data Arson Dossier AR5," Fire Protection Assoc., London, England, AR5, 1991.

Fisher, M., "Can Automobile Arson Be Stopped? One Case Study," *Firehouse*, Vol. 16, No. 5, May 1991, pp. 70–71, 87.

Forestal, R., "Use of Ultraviolet Light in Fire/Arson, Bomb and Environmental Investigation," *Firehouse*, Vol. 19, No. 9, September 1994, pp. 48–50.

Friedman, R., *Fire Protection Chemistry,* 2nd ed., National Fire Protection Association, Quincy, MA, 1989.

"From Canines to Counseling: Combating Arson," *Record*, Vol. 67, No. 3, May/June 1990, pp. 11–13.

Gilbertson, L. H., "Arson From the Insurer's Perspective," *Fire and Arson Investigator*, Vol. 43, No. 2, Dec. 1992, pp. 43–51.

Gilman, R., "National Arson Issues Paper," Insurance Committee for Arson Control, 1992 National Arson Forum, Sept. 9–10, 1992, Arlington, VA, 1992, pp. 1–2.

Governors Special Arson Task Force, California State Fire Marshals Office and California District Attorneys Association, "Arson Investigation...A Team Approach," Governors Special Arson Task Force, CA, State Fire Marshals Office, CA, District Attorneys Assoc., CA, Team Approach, 1994.

Hall, J. R., Jr., "U.S. Arson Trends and Patterns, 1990," National Fire Protection Assoc., Quincy, MA, Oct. 1991.

Hall, J. R., Jr., "U.S. Arson Trends and Patterns, 1991," National Fire Protection Assoc., Quincy, MA, Nov. 1992.

Hall, J. R., Jr., "U.S. Arson Trends and Patterns, 1992," National Fire Protection Assoc., Quincy, MA, Oct. 1993.

Hall, J. R., Jr., "U.S. Arson Trends and Patterns, 1993," National Fire Protection Assoc., Quincy, MA, Dec. 1994.

Hall, J. R., *U.S. Arson Trends and Patterns,* National Fire Protection Association, Quincy, MA, 1986.

Hankins, R. K., "Arson and the Police Officer," *Fire and Arson Investigator*, Vol. 42, No. 1, Sept. 1991, pp. 22–23.

Hartigan, N. F., "Illinois Attorney General Arson Training and Prosecution Manual," Illinois Attorney General, Chicago.

Higgins, S. E., "Complex Arson: A Specialty of ATF Labs," *Firehouse*, Vol. 18, No. 9, Sept. 1993, pp. 50–51.

Higgins, S. E., "Portrait of a Serial Arsonist. The Newest Tool in the War Against Arson," *Firehouse*, Vol. 15, No. 8, Aug. 1990, pp. 54–57.

Holt, F. X., "Arsonist's Profile," *Fire Engineering*, Vol. 147, No. 6, June 1994, pp. 127–128.

Icove, D. J., and Horbert, P. R., "Serial Arsonists: An Introduction," *Police Chief*, December 1990, pp. 46–47.

Icove, D., "Arson Awareness," Federal Fire Forum on Engineering Aspects of Fire Investigations, April 5, 1990, Gaithersburg, MD, 1990.

"Incendiary Devices: Information and Guidance. Fire Safety Data Arson Dossier AR4," Fire Protection Assoc., London, England, AR41, 1990.

"Intruders and Arsonists Can Sue, Insurers Warn Owners of Empty Buildings," *Fire Prevention*, No. 283, Oct. 1995, pp. 14–15.

"Investigating the Fireground," *Fire Engineering,* New York, 1986.

Jerome, I., "Update on Arson Fires, 1990," *Fire Prevention*, No. 242, Sept. 1991, p. 16.

Jones, N., "Arson-For-Profit Investigations, Success or Failure? Recovering Water Damaged Business Records," *Fire and Arson Investigator*, Vol. 40, No. 3, March 1990, pp. 50–52.

Karchmer, C. L., "Arson 1990," *Firehouse*, Vol. 15, No. 8, Aug. 1990, p. 30.

Karchmer, C. L., "Arson 1992: The Upward Trend Continues," *Firehouse*, Vol. 17, No. 8, Aug. 1992, p. 28.

Karchmer, C. L., "Arson In America: 1995 Overview," *Firehouse*, Vol. 20, No. 8, Aug. 1995, pp. 38–39.

Karchmer, C. L., "Arson Update," *Firehouse*, Vol. 16, No. 8, Aug. 1991, p. 29.

Karchmer, C. L., "Violent Yough [sic] Gangs and Drugs...And Arson," *Firehouse*, Vol. 18, No. 9, Sept. 1993, pp. 48–49.

Keltner, N., *et al.*, "Investigation of High Temperature Accelerant Arsons," *Proceedings* of the Extended Abstracts of the Society of Fire Protection Engineers Engineering Seminars on Large Fires: Causes and Consequences, Nov. 16–18, 1992, Dallas, TX, Society of Fire Protection Engineers, Boston, MA, 1992, pp. 20–22; *Proceedings* of the Sixth International Fire Conference on Fire Safety, Interflam '93, March 30–April 1, 1993, Oxford, England, Interscience Communications Ltd., London, England, 1993, pp. 607–620; *Proceedings* of the Eighteenth International Conference on Fire Safety, Volume 18, Jan. 11–15, 1993, Millbrae, CA, Product Safety Corp., Sunnyvale, CA, 1993, pp. 393–399.

Kennedy, P. M., "The Crisis in Fire Investigation Education," *Fire Journal*, Vol. 83, No. 6, Nov. 1989, p. 13.

Kinard, W. D., and Midfiff, C. R., Jr., "Arson Evidence Container Evaluation. Part 2. " New Generation" Kapak Bags," *Fire and Arson Investigator*, Vol. 43, No. 1, Sept. 1992, pp. 22–26.

Klem, T. J., "Los Angeles High-Rise Bank Fire," *Fire Journal*, Vol. 83, No. 3, May 1989, p. 72.

Klem, T. J., *Investigation Report on the Dupont Plaza Hotel Fire,* National Fire Protection Association, Quincy, MA, 1987.

Kletz, T., *Learning from Accidents in Industry,* Butterworth Co. Ltd., Stoneham, MA, 1988.

Kopf, J., "Arsonists Beware: Task Force at Full Strength," *Arizona Arson Investigator*, Vol. 4, No. 2, Aug. 1993, p. 25.

Krzeszowski, F. E., "What Sets Off an Arsonist," *Security Management*, Vol. 37, No. 1, Jan. 1993, pp. 42–47.

Law, D. K., "Arson Law and Investigation Information," *American Fire Journal*, Vol. 44, No. 6, June 1992, pp. 28–31.

Law, D. K., "Arson Law and Investigation Information," *American Fire Journal*, Vol. 44, No. 6, June 1992, p. 35.

Legault, R., "Four Legs and a Powerful Nose Can Detect Arson," *Fire and Arson Investigator*, Vol. 44, No. 1, Sept. 1993, p. 7.

Lincoln, S., "Case in Review: Charcoal Lighter Fluid Used as an Arson Accelerant," *Fire and Arson Investigator*, Vol. 42, No. 1, Sept. 1991, pp. 46–47.

Lowe, D., "Fire Investigation: Science or Science Fiction?" *Fire and Arson Investigator,* Sept. 1987.

Macmillan, R. J., "Wholesale Xmas Arson," *American Fire Journal*, Vol. 43, No. 3, March 1991, pp. 14–16.

Matthews, D. B., "Identification of Flammable Liquids in Arson Evidence," *Proceedings* of the Eighteenth International Conference on Fire Safety, Volume 18, Jan. 11–15, 1993, Millbrae, CA, Product Safety Corp., Sunnyvale, CA, 1993, pp. 133–139.

May, W., and McDonald, B., "Arson Investigation: It Belongs in the Fire Department," *Fire Chief*, Vol. 32, No. 11, Nov. 1988, pp. 56–59.

Midkiff, C. R., Jr., "Spalling of Concrete as an Indicator of Arson—Let's Look at the Concrete Before We Decide," *Fire and Arson Investigator*, Vol. 41, No. 2, Dec. 1990, pp. 42–44.

Mills, J. S., "Arson During Recessionary Times," *Fire Chief*, Vol. 37, No. 11, Nov. 1993, pp. 50–53.

"Mississippi Story Shows Determined State Officials Can Make Inroads Against Arson," *Fire and Arson Investigator*, Vol. 42, No. 2, December 1991, pp. 54–55.

Myers, K., "Outsmarting the Arsonist," *Fire Prevention*, No. 242, Sept. 1991, p. 17.

National Volunteer Fire Council, "National Volunteer Fire Council Meets With FBI on Firefighter Arson," National Volunteer Fire Council, Washington, DC, Firefighter Arson Special Rpt., Aug. 1994.

Nelson, H. E., "Science in Action: An Engineering View of the Fire at the First Interstate Bank Building," *Fire Journal*, Vol. 83, No. 4, July/Aug. 1989, p. 28.

"New York—Bronx Social Club Arson Claims 87 Lives. On the Job," *Firehouse*, Vol. 15, No. 6, June 1990, pp. 46–48.

Onieal, D. G., "Arson: A Sign of the Times," *Firehouse*, Vol. 16, No. 7, July 1991, pp. 66, 68–70, 88.

Patton, A. J., *Fire Litigation Sourcebook: Technical, Medical and Legal Aspects,* Garland Law Pub., New York, 1986.

Payne, B., "Arson Attacks on Places of Worship," *Fire Prevention*, No. 284, November 1995, pp. 26–27.

Pengra, A. G., "Partial System Not Enough: Arson Fire in Partially Sprinklered Motel Kills Four, Injures Others," *Sprinkler Age*, Vol. 15, No. 3, March 1996, pp. 20–21.

Perroni, C., "Investigating Fires the Scientific Way," *Fire Journal*, Vol. 83, No. 4, July/Aug. 1989, p. 24.

"Perth Disco Blaze Was Arson," *Fire*, Vol. 85, No. 1052, February 1993, p. 18.

Rautaheimo, J., "Making of an Arsonist," *Fire Prevention*, No. 223, Oct. 1989, pp. 30–35.

Rogozinski, R. P., and Knowlton, M. E., "Arson Investigators: The New Breed," *Fire Engineering*, Vol. 143, No. 6, June 1990, pp. 46–48, 54, 56.

Rose, S. E., "Rocket Fuel: A New Weapon for Arsonists?," *Fire Chief*, Vol. 34, No. 12, Dec. 1990, pp. 25–27.

Russo, W. P., "Northwest Battles High-Tech Arson," *NFPA Journal*, Vol. 86, No. 6, Nov/Dec. 1992, pp. 67–69, 72–73.

Rust, P., "Arson for Profit: Follow-Up Investigations," *Arizona Arson Investigator*, Vol. 4, No. 2, Aug. 1993, pp. 22–23.

Sapp, A. D., *et al.*, "Report of Essential Findings From a Study of Serial Arsonists," Central Missouri State Univ., Warrensburg, Federal Bureau of Investigation, Quantico, VA, Bureau of Alcohol, Tobacco and Firearms, Washington, DC, 1994.

Scoones, K., "Serious Arson Fires During 1991," *Fire Prevention*, No. 258, April 1993, pp. 12–14.

Scoones, K., "Serious Arson Fires During 1992," *Fire Prevention*, No. 273, Oct. 1994, pp. 10–12.

Scoones, K., "Serious Arson Fires During 1993," *Fire Prevention*, No. 282, Sept. 1995, pp. 47–49.

Seidel, G. E., "Arson for Profit," *Fire Engineering*, Vol. 144, No. 1, Jan. 1991, pp. 51–52.

Seidel, G. E., "Arson Task Forces," *Fire Engineering*, Vol. 144, No. 1, Jan. 1991, pp. 44–46.

Shriver, K., "Financial Records Investigation Assist in Conviction of Arsonist: One Case Study," *Fire and Arson Investigator*, Vol. 45, No. 4, June 1995, pp. 19–20.

Skelley, M. J., "An Experimental Investigation of Glass Breakage in Compartment Fires," NIST-JCR-90-578, National Institute of Standards and Technology, Gaithersburg, MD, 1990.

Stambaugh, H., "Arson in the United Kingdom," *Fire and Arson Investigator*, Vol. 42, No. 2, Dec. 1991, pp. 43–47.

Stambaugh, H., "Grems Case: How an Arson Case Was Solved and Prosecuted in Colorado," USFA Fire Investigation Technical Report Series, TriData Corp., Arlington, VA, Federal Emergency Management Agency, Washington, DC, Report 047, 1991.

Strube, D., "Fire Investigations Involving Motor Loads and Conductor Overcurrent Protection," *Fire Journal*, Vol. 83, No. 5, Sept. 1989, p. 83.

Sykes, G., "TEAM Approach: Tactical Effort for Arson Management," *Fire Chief*, Vol. 36, No. 11, Nov. 1992, pp. 49–50.

Tobin, W. A., "What Collapsed Springs Really Tell Arson Investigators," *Fire Journal*, Vol. 84, No. 2, March/April 1990, pp. 24–27.

Tobin, W. A., and Monson, K. L., "Collapsed Springs in Arson Investigations: A Critical Metallurgical Evaluation," *Fire Technology*, Vol. 25, No. 4, Nov. 1989.

Trimpe, M. A., "What the Arson Investigator Should Know About Turpentine," *Fire and Arson Investigator*, Vol. 44, No. 1, Sept. 1993, pp. 53–55.

U. S. Senate and U.S. House of Representatives, "Arson Prevention Act of 1994. To Establish a Program of Grants to States for Arson Research, Prevention, and Control, and for Other Purposes. An Act. Public Law 103-254. 103d Congress. H.R. 1727. 108 STAT. 679. May 19, 1994," Senate, Washington, DC, House of Representatives, Washington, DC, Public law 103-254, H. R. 1727, 103d Congress, May 19, 1994.

U. S. Senate; U. S. House of Representatives, "Arson Prevention Act of 1993. To Amend the Federal Fire Prevention and Control Act of 1974 to Establish a Program of Grants to States for Arson Research, Prevention, and Control, and for Other Purposes. A Bill. S. 798. 103d Congress. 1st Session. April 20, 1993," Senate, Washington, DC, House of Representatives, Washington, DC, S. 798, 103d Congress, 1st Session, April 20, 1993.

U. S. Senate; U.S. House of Representatives, "Arson Prevention Act of 1993. To Establish a Program of Grants to States for Arson Research, Prevention, and Control, and for Other Purposes. An Act. H. R. 1727. 103d Congress. 1st Session. July 27, 1993," Senate, Washington, DC, House of Representatives, Washington, DC, H. R. 1727, 103d Congress, 1st Session, July 27, 1993.

Ulley, L., "Arson: Historic Church Lost," *Fire Fighting in Canada*, Vol. 39, No. 7, Sept. 1995, pp. 2–3.

Weldon, C., "How Will Congress Deal With Arson?," *Firehouse*, Vol. 17, No. 8, Aug. 1992, p. 24.

White, E. E., "Profiling Arsonists and Their Motives: An Update," *Fire Engineering*, Vol. 149, No. 3, March 1996, pp. 80–82, 84–86.

Wilby, M. A., "Arsonists—And What Happened to Them," *Fire Prevention*, No. 279, May 1995, pp. 28–29.

Wilby, M. A., "Arsonists—And What Happened to Them," *Fire Prevention*, No. 262, Sept.1993, pp. 12–13.

Wilby, M. A., "Arsonists—And What Happened to Them," *Fire Prevention*, No. 268, April 1994, pp. 16–17.

Wilby, M. A., "Arsonists—And What Happened to Them," *Fire Prevention*, No. 276, Jan./Feb. 1995, pp. 21–22.

Wilby, M. A., "Arsonists—And What Happened to Them," *Fire Prevention*, No. 286, Jan./Feb. 1996, pp. 40–41.

Williams, H. E., Pavllsin, M. J., and Kightlinger, S. M., "Understanding and Effectively Utilizing the Arson Reporting Immunity Laws," *Fire and Arson Investigator*, Vol. 42, No. 2, Dec. 1991, pp. 23–31.

Woodward, C. D., "Motives for Arson," *Fire Prevention*, No. 273, Oct. 1994, pp. 13–16.

Woodward, D., "Are We Only Catching the Incompetent Arsonist?," *Fire Engineers Journal*, Vol. 52, No. 167, Dec. 1992, pp. 17–18.

Yereance, R. A., *Electrical Fire Analysis,* Charles C. Thomas, Springfield, IL, 1987.

Yokota, E., "Japanese Views on Arson Have Paid Dividends," *Fire International*, Vol. 12, April/May 1991, p. 38.

FIRE DATA COLLECTION AND DATA BASES

Revised by Philip S. Schaenman

Data on individual fires provide a valuable tool for both the fire service and the broader fire protection community in understanding the fires, their growth, and the effect that the community's fire protection efforts had on the fires.

Aggregated data on the fire experience of a community, state, or nation are even more useful. They permit assessment of the effectiveness of fire prevention and fire suppression methods, allow comparison of the fire problem among communities and regions, and suggest areas for improvement of resource deployment or redirection of programs.

Fire data also can be used to generate media interest, raise citizen awareness, and justify budgets.

To support this activity, a fire department must produce fire reports that work both as self-standing descriptions of individual incidents and as components in the creation of broad-based fire incident data bases.

The following sections discuss fire reporting and methods for collecting data from fire reports to form broad-based data systems. Use of fire data is discussed in Chapter 3 of this section, "Use of Fire Incident Data and Statistics."

THE FIRE REPORT

A fire report is the written documentation that a fire occurred. It may be as brief as a basic fact statement or as lengthy as an extensive discussion of the fire, supported by photographs, witness statements, laboratory test results, and physical evidence. The length and complexity of the report will depend upon the size and nature of the fire, the local fire service manager's need for specific data, and the resources available for obtaining information and completing reports. They also depend on the training and motivation of the person filling out the report.

The fire report should include, at some level of detail, a time-staged description of the circumstances related to the initiation, discovery, growth, and termination of the fire, along with a description of the casualties or the damage resulting from the incident. This report should be in the words of the fire officer and must be complete, so persons who were not at the fire scene can understand what happened.

Phililp S. Schaenman is president of TriData Corp. of Arlington, VA. He is a member of NFPA's Technical Committee on Fire Reporting.

Purpose of the Fire Report

There are three basic purposes to a fire report at the local level. First, it is the legal record of the fact that the fire occurred and provides official notification to those who may be required legally to know of the incident, such as the state or local fire marshal. It reports facts about the particular property affected, why the fire occurred, how building components and fire protection devices performed, casualties or damage that resulted, and fire department action. Second, it provides information to senior officers and fire department managers so that they are kept informed about what is happening within their areas of responsibility. This allows them to evaluate the performance of their units at the incident and to talk intelligently about the incident to the media and others. Good information about a fire can motivate change in fire protection approaches in a community or even the nation (e.g., the MGM Grand Hotel fire in Las Vegas, NV, in 1980 helped trigger a wave of improvements in fire safety). And finally, the report provides data on the fire problem to fire service management so they can track trends, gauge the effectiveness of fire prevention and fire suppression measures presently in practice, evaluate the impact of new methods, and indicate those areas that may require further attention.

The first two purposes can be served by any report that is an accurate description of the incident. The third purpose, however, requires that information be collected in a consistent format that will permit a meaningful aggregation of the data from reports on many incidents.

It is also important that a single report serve the basic data needs of what may be several types of potential users. The data needed at the state and national levels must be provided from what is collected locally. At the same time, it is important that the locally collected data also have a visible, significant use at the local fire service level. If the data are collected only for the benefit of those outside the local area, the motivation and commitment to quality and completeness may diminish, with a resulting reduction in the usefulness of the data.

It is difficult to routinely collect all the data items that are likely to be needed by all types of potential data users in the future. Compromises are needed between the ease of filling out an incident report and all the potential uses of it. Ease of use also increases reliability, and reliability of the data increases its usefulness. Again, a balance must be struck.

Uniformity in Fire Reporting

To maintain uniformity in fire reporting, the NFPA Technical Committee on Fire Reporting has developed NFPA 901, *Standard Classifications for Incident Reporting and Fire Protection Data*. This standard establishes basic definitions and terminology for use in fire reporting and serves as a means of classifying data so that the information can be aggregated either manually or automatically.

NFPA 901 provides the common language used by nearly all large-scale—that is, state and national—data bases in the United States, and many others around the world. A fire department that uses NFPA 901 will find that it is in a position to contribute data most efficiently to larger data bases and, in turn, to use data from these larger data bases in its management of the local fire problem. It also will find that the quality, uniformity, and usefulness of information within the fire department will improve. NFPA 901 contains data elements that provide a classification of property, whether fixed or mobile; a description of a specific structure prior to an incident; a description of the ignition sequence, including the area of origin of the fire; conditions found upon arrival; what action was taken; and why the fire grew or stayed as small as it did. There are also data elements for describing injuries or fatalities to both civilians and fire fighters, the extent of damage, and the investment in personnel and resources to control the incident. NFPA 901 also contains data elements for reporting the handling of hazardous materials spills and for providing emergency medical service data if an agency is involved in providing those services. Useful analysis can be adapted to the particular data elements adopted by the fire department, no matter how few or how many.

FIRE REPORTING SYSTEMS

There is a major difference between a fire report and a fire reporting system. A fire report is only part of a fire reporting system. A complete fire reporting system must address: fact finding and recording, quality control, and creation of the data base.

Fact Finding and Recording

The traditional functions of a fire report can be satisfied with a minimal written narrative of the basic facts of the incident. To serve as input to a fire reporting system, however, an incident report must be clearly structured and must use uniform definitions and terminology. The collection of such information requires not only the form or other media upon which to record the information desired, but also a set of instructions and related training, so that the information is provided in a uniform manner and a procedure is established for forwarding the information to a central point.

The fire report may include information gathered from several other reports, such as the fire investigator's report, police report, and original draft incident report. The fire report may have to be updated as additional information is collected or corrected.

Quality Control

The persons responsible for reporting the data should be trained and capable of investigating a fire to determine the origin, cause, and circumstances of the fire. Also, the persons need to be motivated to complete all of the relevant data elements for the fire and the associated casualties. This must be done for every fire reported to the fire department. Incomplete forms and incomplete reporting of fires cause major problems in analyzing the data, and learning the life and death lessons from fires.

Once data have been recorded, they must be checked for accuracy, clarity, consistency, and completeness. A procedure of quality-control screening and follow-up corrections is needed for this purpose. For best results, the reports should be screened manually, even when computerized edit checks are employed. Most fire departments use computerized data bases today. Data not collected directly into a computer must be keyed in, and then added to the data base of all incidents.

Once an incident file has been created, whether as a paper file, a computerized data base, or some combination of the two, it has many potential uses.

Small fire departments have special problems in gathering statistically meaningful data because of their low fire incidence. They may be able to analyze the fire problems only by pooling data with several other small departments in their region. Therefore, it is especially important that small fire departments use uniform terminology and uniform coding in collecting information so that data from different fire services can be merged for analysis.

Benefits of a Fire Reporting System

At the local level, a fire department can derive many benefits from a good fire incident reporting system, particularly if it is based on NFPA 901. Some of the following uses involve no more than totaling data from the system. Others require more extensive analysis. Many of these benefits can be derived at the state and national levels when a data base is used that combines the fire experience of many local fire departments.

Describing a community's fire problem: It is possible to pinpoint where fires are occurring, what factors are most responsible for ignitions, and what casualties and damage are occurring as a result of fires. With the problem placed in proper perspective, the most serious and solvable aspects of the fire problem can be tackled first.

Supporting budget requests: In this era of increasing concern about taxes, municipal officials are quick to cut budgets and slow to add new programs. Frequently, fire department managers do not have the statistics to support their requests for additional funds. Good statistics will put the fire problem in perspective with other municipal concerns and help community officials realize the consequences of budget cuts or the value of new programs for the fire department. Such new programs may involve the delivery of other emergency services such as emergency medical services and hazardous materials spill mitigation.

Supporting code refinements: A good data base permits fire departments to identify and describe fires that might have developed differently or might not have occurred at all if certain code changes had been in place. Loss statistics from other areas with more stringent codes also can be used for comparison. Estimating the likely impact of a code change can involve complex analysis, however, and no incident data base can address all the subtleties of code impact.

Evaluating code enforcement programs: It is not sufficient to have codes on the books if they are not properly enforced. In evaluating loss experience, it may be possible to see whether certain losses are occurring in occupancies not up to code, or without desired features, such as sprinkler systems.

Evaluating public fire education programs: Not all problems can be solved by establishing and enforcing codes. There are certain aspects of the fire problem that can best be controlled by public education programs that inform people of the dangers of fire and tell them how to reduce fires, and how to react when hazardous situations arise. It is important to know the exact problem that has to be addressed. Appropriate evaluation criteria must also be in place to measure whether an educational program is in fact helping to solve that aspect of the problem.

Planning future fire protection needs: Many communities and fire departments are becoming very active in planning and are developing master plans. It is essential that the fire service be involved in such planning. A good data base will allow a fire department to compute fire rates relative to population and building inventory, as well as monitor response times. These, with other characteristics of the community fire problem and planning, will support better fire protection in the future based on changing demography and planned community growth. It will also provide input to decisions about the type and level of fire protection a community will provide so that requirements can be established for developers who construct properties that exceed fire department capabilities.

Improving allocation of resources: Proper analysis of fire incident data may show where a redeployment of existing resources can provide the same level of protection or even improve the level of protection within a community.

Scheduling nonemergency activities: Training sessions, in-service inspections, and other activities are important aspects of a fire department's function. A fire department that tracks the times that fires occur and their severity can schedule these activities when they are least likely to be interrupted by emergency calls or when the normal delay caused by such activities will have the least impact on emergencies.

Regulating product safety: Particularly at the national and state levels, a fire reporting system can be useful in measuring the size and severity of problems associated with various types of consumer products. By identifying the most commonly involved products and the ways these products become involved in fire, this reporting system can help manufacturers redesign their products to make them safer, and it can prompt changes in standards and regulations to require safer products. The reported information also can be incorporated into public fire education programs to warn consumers of the dangers of using certain products.

Support for fire engineering models: Engineering models to design or evaluate fire protection depend upon the output of fire reporting systems to guide and calibrate the models.

Support for fire engineering analysis: Analysis of fire data can indicate those methods of fire defense that work best.

DESIGNING A FIRE REPORTING SYSTEM

When a new fire reporting system is to be designed or an existing system redesigned, the first step should be an analysis of the need for the data that will be collected. Some of the questions that should be answered include:

- What data are legally required to be collected?

- What information is this system expected to produce and for what purpose?

- What data elements are necessary to provide the information needed?

- How quickly after the incident is the information needed, by whom, and how long after the incident should the information be kept available?

- Who will be expected to collect the data, assemble them, and screen them for errors?

- What data will be provided from automated systems (e.g., the fire department CAD) and what from manual input?

- Who will be expected to summarize and analyze the data?

- What savings or efficiencies could be realized if the data were available?

- Is any of the needed data already being collected for another use?

- How long does the data have to be retained, and in what form?

- What backup provisions are needed to safeguard the data (e.g., storage of a duplicate copy on computer discs or tapes)?

The next step is to develop a mechanism and written procedures for collecting the data. This step should include a careful definition of the quality control process and the procedures for filing, storing, and disposing of all related material.

Tools for Data Collection

Data usually are recorded either on a paper form or directly onto some electronic media, typically a computer terminal supported by a personal computer or central computer system.

When a paper form is used, data may be keyed from the form to an electronic data base for storage and analysis purposes, but the form is retained as an official record of the incident. When data are collected directly on electronic media, there is no written record in many cases, so more information needs to be captured on the electronic data base, and a backup storage is needed for security.

Paper Forms

If a paper form is to be used, the form should be designed to apply to local needs but also capture the standard data elements needed for state and national reporting systems. Existing forms should be used where appropriate, because forms designed from scratch may have problems that have already been identified and solved by others. Use of existing forms also promotes consistency over time and between departments.

The NFPA Fire Reporting Committee and the National Fire Information Council (NFIC), which directs the National Fire Incident Reporting System (NFIRS), periodically revise the national model forms, and each fire department needs to review those and its own to ensure compatibility over time and the appropriateness of the data being collected.

Enough space to write the answer should be included on the form. Too little space results in illegible handwriting. Never squeeze the space for answers; another page should be added to the form or some of the data elements deleted. If a typewriter or computer word processor is to be used to complete the form, correct spacing between lines is essential.

The form should be of standard size. Forms with odd sizes require special file cabinets, which are more expensive, and oversized forms are more difficult to use in the field. If more than one copy of a paper report will be needed, consider using NCR (no carbon required) paper. Carbon paper is messy and inconvenient. Different colors for different forms help people distinguish the forms at a distance and can facilitate filing.

If electronic data entry is to be done from the form, the form design should be coordinated with the people responsible for the data.

Data should be recorded in the order in which it is most likely to be collected, and information on related topics should be kept together. If the whole form need not be completed in all cases, it can be subdivided. Provide brief instructions on the form as to when each section should be used.

Consider providing space at the top or right margin for case identifiers, such as address, file number, date, or whatever other key is used for filing, so that forms can be readily identified by users thumbing through the file. The form should be kept single-sided, if possible.

The written procedures for completing the report should describe the intended use of the data. Procedures should cover, item by item, the kind of information sought, the format in which it should be recorded, and, if it is to be classified, the appropriate classifications. Definitions should be provided for all data elements, either as part of the instructions or in a separate manual. The type of information and extent of detail expected in the remarks section also should be defined. Samples of properly completed reports can improve the understanding and use of the reports.

Direct Electronic Data Capture

If the only record is on the electronic media, each data entry screen must be designed to capture groups of related data, based on the type of incident. Multiple screens should be used so that a screen does not look jammed.

The system should present screens for completion based on previously entered data. For example, a screen collecting civilian injury data would be presented if the number of civilian injuries was greater than zero.

For many data elements the data base needs to collect both a narrative description and a classification number, since the narrative description will often be more specific. For example, the narrative may say, "portable kerosene space heater," whereas the classification number is that for the general category of portable space heaters. The system must also be able to capture and store comments or remarks by the fire officer in charge of the incident, as well as the full narrative description of the incident.

Help screens are useful in selecting appropriate classification numbers for the data elements rather than having to refer to printed tables.

Data editing should be done as the data are entered or immediately afterward, to ensure completeness and accuracy and allow immediate correction of careless errors. The system must be able to print a hard-copy report, including the narrative descriptions and remarks, on demand for administrative purposes.

If the data are to be routinely summarized into management reports, the written procedures should define how this is to be done, how often, and which persons should receive copies of the summaries.

Training

Actual use of a new form or a new reporting system must be preceded by training for those individuals who will be responsible for collecting and recording the data. The training should be the responsibility of the training division in the department, which should work closely with whomever developed the reporting system to develop the training materials. If state or national data forms are used, in whole or in large part, their associated training manual can be used, with additions made locally for any added or deleted data elements. Written instructions should be the basis of the training. Questions raised during the training may point to a need for additional written instructions.

Once the initial training has been completed, periodic retraining should be scheduled as part of the normal retraining processes. All persons promoted or assigned to positions where they will be responsible for collecting and recording incident data should receive this special training.

Data training must be taken seriously. Incorrect information can kill, by providing misleading facts on which public education and regulations, and resource deployment, are based.

Organizing to Compile Reports

In many fire departments, the data on various aspects of a fire come from different sources. The communications office will be able to provide data on alarm receipt and dispatch times. The officer in charge will be responsible for the incident report. The company officers will be in the best position to record the role of their companies and the resources used in an incident. If there were casualties, data on the injuries or fatalities may best be provided by still other personnel. As the fire is investigated, the fire marshal or fire investigation unit may develop details that the officer in charge could not determine at the time of the fire.

All of this data are often recorded separately by the individuals with direct knowledge of a certain part of the fire incident. It is extremely important that all these reports be compiled into a complete fire report and that the data on the various reports be checked for completeness and consistency.

One division within the department, such as the fire marshal or management information section, should be given the responsibility for compiling all fire reports. Clerical personnel should be properly trained in recording fire data and processing the paper forms or electronically stored data. Some of the responsibilities associated with compiling reports include:

- Ensuring that all required data are recorded and submitted in a timely fashion

- Ensuring that all handwritten forms are properly completed and are legible

- Screening the reports for obvious errors, omissions, or conflicting data

- Following up with the appropriate personnel when necessary to obtain complete or corrected data

- Ensuring that any data entry that is to be automated is timely and correct

- Filing the complete report, if paper forms are used

- Updating the report and the files as new or additional information is received

The updating of incident reports as better information becomes available is often not done and is a major problem in the quality of information available. The lack of updates is especially common for fatal fires and large-loss fires, where the initial incident report is left incomplete pending an investigation, but does not get updated after the investigation is complete. *Updating incident reports is a crucial element in a fire data system.*

Detailed written procedures should be developed for the clerical personnel, outlining how they should perform each step in report compilation. These procedures should address such issues as how the reports are to be edited, processed, and filed, and within what time frame, and how they are to be corrected or updated.

It is important that everyone responsible for completing any portion of the report understand the procedure that the clerical staff uses and the schedule they are expected to follow in completing, correcting, and forwarding the data. Delays in submitting data will result in delays in assembling the complete report and in making it available for use.

Quality Control

An important aspect of any record-keeping function is ensuring the quality of the data collected. This starts with proper training, as discussed above.

All data should be edited as part of the process of compiling it into a master record. This editing can be done by a trained clerical person. At least periodically, a supervisor should spot-check a sample of the reports to ensure that they are completed properly. All reports with errors or omissions should be flagged and returned to the originator for correction or completion as appropriate. This procedure reinforces training because the individual will see where the mistakes are and thus may be able to avoid them in the future. If a person continues to make the same mistakes, special training or other action may be necessary to improve the quality of the reports submitted. Incomplete or poor reports waste time and defeat the whole purpose of collecting data.

It is important that supervisory personnel be involved in the quality control process on a regular basis. A computer can do certain editing tasks and force completion of a report, but it cannot check to see that the incident scene was interpreted correctly or that the report accurately reflects the incident. Nor can it determine if files are being updated properly as new information becomes available. Both of these tasks are as important to ensuring a quality data base as any other part of the process.

Record Retention

An important aspect of any record-keeping system is determining how long to keep the records. The answer varies with the purpose of the record and, to some extent, with local and state legal requirements for record retention.

Obviously, paper records require storage space, and storage space costs money. Likewise, the more voluminous the files become, the more likely material will be lost or misfiled.

There should be a written policy that details the length of time a report is to be retained in the active files, the length of time it should be kept in inactive storage, and when it can be discarded. The city attorney's office can help determine legal requirements for retaining records; that office should review the final written policy before its implementation. There may be different retention requirements for computerized records that require less space and are useful for multiyear analysis.

Setting Standards within the Department

The fire department should have certain standards that apply to all its record-keeping operations. Such standards will help make the data more uniform from report to report and application to application. Standardization aids all users of the data, makes analysis more accurate, and is essential if data is to be automated. Some areas where department standards should be developed include:

- Methods of entering the names of persons—first name first or last name first—on records

- Recording addresses of buildings with multiple or ambiguous addresses, as well as nonstructure locations, such as those on highways or at street intersections

- Designating apparatus. Should it be done by company assigned? Serial number? Shop number?

- Designating employees. Should it be done by name? Badge number? Social security number?

- The common abbreviations that are acceptable to use

A local department should try to maintain consistency with its state data system. NFPA 901 can be used as a basis of terminology, definition, and classification data.

Data Analysis

In many fire reporting systems, a great deal of effort is devoted to data collection, and then very little effort is given to making good use of the data.

Data analyses should be of two kinds: (1) routine, periodic reports that summarize the fire problem and that provide management data; and (2) special reports that answer queries about particular problems, or evaluate programs, support budget requests, etc.

Data analysis can start with simply adding up what has been reported and showing the counts of items reported for each data element.

Data analysis should also consider the interrelationship between various data elements or various data bases. For example, relating fire data to socioeconomic data by census tract show the population groups who are most affected. Comparing fire losses over the years to inflation rates and increasing property values may show that fire protection is currently more effective, even though in raw numbers the total loss is greater.

Data analysis is a special skill that should be sought out among qualified fire department personnel, or provided by someone hired to do analysis, or perhaps done by local universities. See Section 11, Chapter 3, "Use of Fire Incident Data and Statistics," for more detailed guidance on data analysis.

Collecting Data for Special Studies

No one can anticipate all the data that may be needed to understand the fire problem. Once a problem has been identified through an ongoing fire reporting system, a special study may be needed to gather the data required to further define and help solve the problem. For example, the data system may show that misuse of portable space heaters is a problem, but not provide adequate insight into the circumstances that cause the problem, or the specific types of heaters most often involved. The steps in conducting a special study are:

1. Identify the type and sources of data needed to define the problem further. Assess the feasibility and costs of the study, and proceed in the most efficient way. For example, sampling may be appropriate. Previous work to solve this or similar problems should be researched and the results studied.
2. Develop a mechanism or procedure for collecting the needed data. This can be a special form completed by the officer in charge of the incident, additional questions asked on a computer monitor when the data are captured, or a request to include certain data in the remarks section of a report whenever a certain condition is encountered. Alternatively, the procedure may stipulate merely that the officer in charge notify a certain person or agency whenever he or she encounters the particular problem or condition.
3. Collect and analyze the data. Develop a methodology to reduce or solve the problem, based on the results of data analysis and any additional documentation.
4. Terminate the special study. When the methodology is implemented, monitor the progress of the problem-solving methodology using the normal reporting system. Return to Steps 1, 2, or 3 as appropriate if the problem is not being reduced or solved.

Periodic Review

Over time, information needs within a fire department change, and as they do, it is not uncommon for new reporting requirements to be added. New forms or requirements are introduced or existing systems are expanded. Much less common, however, is a reduction in the amount or type of information collected.

INCIDENT REPORT

NFIRS 1

1 ☐ DELETE
2 ☐ CHANGE

FILL IN THIS REPORT
IN YOUR OWN WORDS

_____ FIRE DEPARTMENT

A | FDID | INCIDENT NO. | EXP. NO. | MO | DAY | YEAR | DAY OF WEEK | ALARM TIME | ARRIVAL TIME | TIME IN

B | TYPE OF SITUATION FOUND | TYPE OF ACTION TAKEN | MUTUAL AID 1 ☐ RECEIVED 2 ☐ GIVEN

C | FIXED PROPERTY USE | IGNITION FACTOR

D | CORRECT ADDRESS | ZIP CODE | CENSUS TRACT

E | OCCUPANT NAME (LAST, FIRST, MI) | TELEPHONE | ROOM OR APARTMENT

F | OWNER NAME (LAST, FIRST, MI) | ADDRESS | TELEPHONE

G | METHOD OF ALARM FROM PUBLIC | DISTRICT | SHIFT | NO. ALARMS

H | **NUMBER** FIRE SERVICE PERSONNEL RESPONDED | **NUMBER** ENGINES RESPONDED | **NUMBER** AERIAL APPARATUS RESPONDED | **NUMBER** OTHER VEHICLES RESPONDED

COMPLETE FOR ALL INCIDENTS

I | NUMBER OF INJURIES — FIRE SERVICE | OTHER | NUMBER OF FATALITIES — FIRE SERVICE | OTHER

COMPLETE IF CASUALTY

J | COMPLEX | MOBILE PROPERTY TYPE

K | AREA OF FIRE ORIGIN | EQUIPMENT INVOLVED IN IGNITION

L | FORM OF HEAT OF IGNITION | TYPE OF MATERIAL IGNITED | FORM OF MATERIAL IGNITED

M | METHOD OF EXTINGUISHMENT | LEVEL OF FIRE ORIGIN | ESTIMATED LOSS (DOLLARS ONLY)

COMPLETE FOR ALL FIRES

N | NUMBER OF STORIES | CONSTRUCTION TYPE

O | EXTENT OF FLAME DAMAGE | EXTENT OF SMOKE DAMAGE

P | DETECTOR PERFORMANCE | SPRINKLER PERFORMANCE

Q | IF SMOKE SPREAD BEYOND ROOM OF ORIGIN | TYPE OF MATERIAL GENERATING MOST SMOKE | AVENUE OF SMOKE TRAVEL

R | FORM OF MATERIAL GENERATING MOST SMOKE

COMPLETE IF STRUCTURE FIRE

S | IF MOBILE PROPERTY | YEAR | MAKE | MODEL | SERIAL NO. | LICENSE NO.

T | IF EQUIPMENT INVOLVED IN IGNITION | YEAR | MAKE | MODEL | SERIAL NO.

☐ CHECK IF COMMENTS ON REVERSE SIDE

U | OFFICER IN CHARGE (NAME, POSITION, ASSIGNMENT) | DATE

MEMBER MAKING REPORT (IF DIFFERENT FROM ABOVE) | DATE

FIG. 11-2A.   NFIRS Incident Report.

CIVILIAN CASUALTY REPORT

_____ FIRE DEPARTMENT

FILL IN THIS REPORT
IN YOUR OWN WORDS

NFIRS - 2

A

FDID	INCIDENT NO.	EXP. NO.	MO	DAY	YEAR	DAY OF WEEK	ALARM TIME

CASUALTY NUMBER | 1 ☐ DELETE
2 ☐ CHANGE

GA CASUALTY NAME (LAST, FIRST, MI)

DOB MO. YR. | AGE | TIME OF INJURY

GB HOME ADDRESS TELEPHONE

GC
SEX
1 ☐ MALE
2 ☐ FEMALE

CASUALTY TYPE
1 ☐ FIRE CASUALTY
2 ☐ ACTION CASUALTY
3 ☐ EMS CASUALTY

SEVERITY
1 ☐ INJURY
2 ☐ DEATH

AFFILIATION
2 ☐ OTHER EMERGENCY PERSONNEL
3 ☐ CIVILIAN

GD FAMILIARITY WITH STRUCTURE | LOCATION AT IGNITION | CONDITION BEFORE INJURY

GE CONDITION PREVENTING ESCAPE | ACTIVITY AT TIME OF INJURY | CAUSE OF INJURY

GF NATURE OF INJURY | PART OF BODY INJURED | DISPOSITION

☐ SEE REMARKS ON BACK ☐ SEE ADDITIONAL REPORT

CASUALTY NUMBER | 1 ☐ DELETE
2 ☐ CHANGE

GA CASUALTY NAME (LAST, FIRST, MI)

DOB MO. YR. | AGE | TIME OF INJURY

GB HOME ADDRESS TELEPHONE

GC
SEX
1 ☐ MALE
2 ☐ FEMALE

CASUALTY TYPE
1 ☐ FIRE CASUALTY
2 ☐ ACTION CASUALTY
3 ☐ EMS CASUALTY

SEVERITY
1 ☐ INJURY
2 ☐ DEATH

AFFILIATION
2 ☐ OTHER EMERGENCY PERSONNEL
3 ☐ CIVILIAN

GD FAMILIARITY WITH STRUCTURE | LOCATION AT IGNITION | CONDITION BEFORE INJURY

GE CONDITION PREVENTING ESCAPE | ACTIVITY AT TIME OF INJURY | CAUSE OF INJURY

GF NATURE OF INJURY | PART OF BODY INJURED | DISPOSITION

☐ SEE REMARKS ON BACK ☐ SEE ADDITIONAL REPORT

CASUALTY NUMBER | 1 ☐ DELETE
2 ☐ CHANGE

GA CASUALTY NAME (LAST, FIRST, MI)

DOB MO. YR. | AGE | TIME OF INJURY

GB HOME ADDRESS TELEPHONE

GC
SEX
1 ☐ MALE
2 ☐ FEMALE

CASUALTY TYPE
1 ☐ FIRE CASUALTY
2 ☐ ACTION CASUALTY
3 ☐ EMS CASUALTY

SEVERITY
1 ☐ INJURY
2 ☐ DEATH

AFFILIATION
2 ☐ OTHER EMERGENCY PERSONNEL
3 ☐ CIVILIAN

GD FAMILIARITY WITH STRUCTURE | LOCATION AT IGNITION | CONDITION BEFORE INJURY

GE CONDITION PREVENTING ESCAPE | ACTIVITY AT TIME OF INJURY | CAUSE OF INJURY

GF NATURE OF INJURY | PART OF BODY INJURED | DISPOSITION

☐ SEE REMARKS ON BACK ☐ SEE ADDITIONAL REPORT

U

OFFICER IN CHARGE (NAME, POSITION, ASSIGNMENT) | DATE

GG

MEMBER MAKING REPORT (IF DIFFERENT FROM ABOVE) | DATE

FIG. 11-2B. NFIRS Civilian Casualty Form.

FIRE SERVICE CASUALTY REPORT

NFIRS - 3

_____ FIRE DEPARTMENT

1 ☐ DELETE
2 ☐ CHANGE

FILL IN THIS REPORT
IN YOUR OWN WORDS

FA

FDID	INCIDENT NO.	EXPOSURE NO.	CASUALTY NO.	INJURY OCCURRED	MO	DAY	YEAR	TIME OF INJURY

FB

CASUALTY NAME (LAST, FIRST, MI)	TYPE OF CASUALTY

FC

AGE	SEX	CASE SEVERITY	PRIMARY APPARENT SYMPTOM

FD

PRIMARY PART OF BODY	PATIENT TAKEN TO

FE

ASSIGNMENT	NO. RESPONSES PRIOR TO INJURY	PHYSICAL CONDITION	STATUS BEFORE ALARM

FF

FIRE FIGHTER ACTIVITY	WHERE INJURY OCCURRED

FG

CAUSE OF FIRE FIGHTER INJURY	MEDICAL CARE PROVIDED

PROTECTIVE EQUIPMENT

FH

PROTECTIVE COAT WORN	STATUS	TYPE PROBLEM

FI

PROTECTIVE TROUSERS WORN	STATUS	TYPE PROBLEM

FJ

BOOTS/SHOES WORN	STATUS	TYPE PROBLEM

FK

HELMET WORN	STATUS	TYPE PROBLEM

FL

FACE PROTECTION WORN	TYPE PROBLEM

FM

BREATHING APPARATUS WORN	STATUS	TYPE PROBLEM

FN

GLOVES WORN	TYPE PROBLEM

FO

SPECIAL EQUIPMENT WORN	STATUS	TYPE PROBLEM

FP

MEMBER MAKING REPORT	DATE	OFFICER IN CHARGE (NAME, POSITION, ASSIGNMENT)	DATE

REMARKS

☐ REMARKS CONTINUED ON REVERSE SIDE

FIG. 11-2C.   NFIRS fire service casualty report.

At least yearly and certainly before any form is reprinted, a careful review should be made to determine if the original purpose for collecting the data is still valid and if all the information is still needed in its present format. Do not perpetuate data gathering if the data are not serving a useful purpose. It is easy to fall into the trap of justifying data collection on the basis that "it may be needed someday." This approach generally results in poor quality data, if and when used, because lack of data use results in lack of accuracy and quality control in collecting the data.

There also are changes being made periodically in state and national fire data systems. Local departments need to keep abreast of them, especially when considering any changes to the department's data forms or data system.

STATUS OF UNIFORM FIRE REPORTING

Fire reporting systems based on NFPA 901 continue to be developed. The NFPA Fire Reporting Committee also has established a basic system, published in NFPA 902, *Fire Reporting Field Incident Guide,* that provides the necessary instructions for completing an incident report, supplementary reports on civilian or fire fighter casualties suffered at incidents, and EMS reports that can be used by those fire departments providing emergency medical services. Worksheets are provided for summarizing incident data. This system can be used manually or adapted for electronic data processing.

The U.S. Fire Administration (USFA), within the Federal Emergency Management Agency (FEMA), has developed the National Fire Incident Reporting System (NFIRS), which is an automated system based in part on the work of the NFPA Fire Reporting Committee, as published in NFPA 901 through 1995. The system, which started in 1976, is now operational in 40 states and annually collects data on approximately one million fires from more than 13,400 fire departments. The system has evolved, and some significant changes are planned in data elements and code categories for introduction in the late 1990s.

NFIRS is an incident-based system. Data are reported on an incident-by-incident basis. Computer programs are available to capture and edit the data, maintain a data base, and generate a variety of summary reports. The data collected are similar to the data collected by the basic system published in NFPA 902. Figure 11-2A is the incident form and Figures 11-2B and 11-2C are the two casualty forms used in NFIRS.

At the state level, NFIRS provides for the collection of written reports on incidents to which local communities responded. Communities with automated data systems based on NFPA 901 can participate by providing the state with their data on computer tape or other electronic media. Summary reports are then generated on both individual community and statewide fire experience.

At the national level, NFIRS merges data bases from individual states to form a national data base. FEMA/USFA analyzes this data base and publishes the results. The data are used by NFPA and many others for a wide range of purposes, from public education to changing standards and regulations, to identifying emerging fire problems, to evaluating programs and past regulatory efforts. The data base is available to governments, businesses, or individuals who wish to perform their own analyses.

BIBLIOGRAPHY

NFPA Codes, Standards, and Recommended Practices.

Reference to the following NFPA codes, standards, and recommended practices will provide further information on fire data collection and data bases discussed in this chapter. (See the latest version of *The NFPA Catalog* for availability of current editions of the following documents.)

NFPA 901, *Standard Classifications for Incident Reporting and Fire Protection Data*

NFPA 902, *Fire Reporting Field Incident Guide*

NFPA 903, *Fire Reporting Property Survey Guide*

NFPA 904, *Incident Follow-Up Report Guide*

Additional Reading

Babrauskas, V., "Handling Fire Data: The FDMS," Chapter 18, *Heat Release in Fires*, Elsevier Applied Science, NY, 1992, pp. 591–607.

Fire Protection Association, "Fire Safety Structure in the United Kingdom. Fire Safety Data Organization of Fire Safety OR01," Fire Protection Assoc., London, England, OR01, May 1990.

Fire Protection Association, "Investigation of Vehicle Fires. Fire Safety Data Arson Dossier AR5," Fire Protection Assoc., London, England, AR5, 1991.

Griffith, S. J., and Munday, J. W., "Legal Implications of Real Fire Data Collection and Computer Modeling," Metropolitan Police Forensic Science Laboratory, London, England, ISBN 0-9516320-9-4; *Proceedings* of the Seventh International Interflam Conference, Interflam '96, March 26–28, 1996, Cambridge, England, Interscience Communications Ltd., London, England, 1996, pp. 1027–1031.

Holmes, S., "Extending Information Data Bases for Effective Incident Control," *Fire*, Vol. 86, No. 1067, May 1994, pp. 41–42.

Jenaway, W. F., *Managing Support Systems,* International Association of Fire Service Instructors, Ashland, MA, 1988.

King, M., "Case Study: The Evolution of the Australian Fire Incident Database," Commonwealth Scientific and Industrial Research Organization, Australia, National Fire Protection Association and International Association of Fire Chiefs, *Proceedings* of the inFIRE (international network for Fire Information and Reference Exchange.) Annual Conference, April 28-30, 1993, Norwood, MA, 1993, pp. 42–66.

McGarry, F. A., "Toward a Safer Built Environment: New York State's Building Materials and Finishes Fire Gas Toxicity Data File," Special Conference on Recent Advances in Flame Retardancy of Polymeric Materials—Materials, Applications, Industry Developments, Markets, May 15–17, 1990, Stamford, CT, 1990, pp. 1–4.

Meldrum, G., "Data on the Move," *Fire Engineers Journal*, Vol. 53, No. 171, Dec. 1993, p. 32.

Mowrer, F. W., "Development of the Fire Data Management System," Maryland Univ., College Park, National Institute of Standards and Technology, Gaithersburg, MD, NIST-GCR-94-639, June 1994.

Narayanan, P., "New Zealand Fire Risk Data," Building Research Association of New Zealand, Judgeford, Study Report 64, Dec. 1995.

National Fire Protection Association, "NFPA's One-Stop Data Shop. Fire Data and Statistics Reports and Services From the NFPA Fire Analysis and Research Division," National Fire Protection Association, Quincy, MA, Feb. 1991.

Oehlberg, R. N., *et al.*, "EPRI Fire Events Database," *PSAM-II Proceedings*, An International Conference Devoted to the Advancement of System-Based Methods for the Design and Operation of Technological Systems and Processes, Volume 3, Sessions 73-108, March 20–25, 1994, San Diego, CA, 1994, pp. 082/17-27.

Portier, R. W., "Fire Data Management System, FDMS 2.0," National Institute of Standards and Technology, Gaithersburg, MD, NISTIR 5499, Sept. 1994; National Institute of Standards and Technology, Annual Conference on Fire Research: Book of Abstracts, Oct. 17–20, 1994, Gaithersburg, MD, 1994, pp. 39–40.

Portier, R. W., "Fire Data Management System, FDMS 2.0, Technical Documentation," National Institute of Standards and Technology, Gaithersburg, MD, NIST TN 1407, Feb. 1994.

Schaenman, P. S., "Management Information Systems and Data Analysis," *Managing Fire Services,* Coleman, R., and Granito, J. A., eds., International City Management Association, Washington, DC, 1988, pp. 129–166.

Shelton, W. G., "DOD Fire Incident Reporting System," Fort Belvoir Fire Dept., VA, International Association of Fire Chiefs (IAFC), Fire-Rescue International Conference *Proceedings*, 1993 Annual Conference, Aug. 28-Sept. 1, 1993, Dallas, TX, Swing, 1993, pp. 379–395.

vanBowen, J., Jr., "Use of In-House Data for Fire Station Location," *Journal of Applied Fire Science*, Vol. 1, No. 1, 1990-1991, pp. 39–44.

USE OF FIRE INCIDENT DATA AND STATISTICS

John R. Hall, Jr.

Section 11, Chapter 2, "Fire Data Collection and Data Bases," examined techniques for collecting data on fires. This chapter addresses uses of that collected data in fire protection decision making and planning.

For those familiar with the state of fire experience data prior to the late 1970s, it is still remarkable to see how much the quantity and quality of fire data available for analysis have grown since then. In the early 1970s, even the best fire departments typically had fire incident records only in narrative form. Neither the method of storage (in file cabinets, by incident number) nor the method of description (phrases selected by fire officers, without standardized coded descriptions or even a standard list of characteristics to be documented) facilitated analysis of the community's leading fire problems. At the state level, very few states had any fire experience data base. At the national level, the National Fire Protection Association (NFPA) prepared estimates of U.S. incidents, deaths, injuries, and losses from its annual survey, and provided analyses of fire causes and other detailed factors from its Fire Incident Data Organization (FIDO) data base. The Fire Reporting Committee of NFPA had developed NFPA 901, *Uniform Coding for Fire Protection* (now *Standard Classifications for Incident Reporting and Fire Protection Data*),[1] a standard fire incident reporting format, but few fire departments were using it.

In 1977, NFPA modified its survey techniques to take account of modern statistical design principles and, in the process, greatly improved estimates of the size of the national fire problem. At about the same time, the new National Fire Prevention and Control Administration (now the U.S. Fire Administration, a branch of FEMA, the Federal Emergency Management Agency) launched the National Fire Incident Reporting System (NFIRS). Based on NFPA 901, the NFIRS system provided the framework for a large-scale, truly national fire experience data base. Since 1980, NFIRS has had the participation, in whole or in part, of the majority of states and now receives standardized reports on nearly one-half of all fire incidents. (As sources of the principal fire data bases, NFPA and FEMA also have helped each other over the years, with FEMA providing financial support and technical suggestions on NFPA data bases and NFPA providing developmental work and analysis services on NFIRS for FEMA.) Other countries, such as Australia, have adopted their own versions of NFIRS, making possible much more valid international comparisons.

John R. Hall, Jr., Ph.D., is assistant vice president, Fire Analysis and Research Division, of NFPA.

A parallel development has enlarged the capacity for analysis of the newly generated fire data. The advent of powerful microcomputers and the increasing affordability and flexibility of computer power in general have made a significant impact upon the fire service and other fire data users. Most readers of this handbook now have access to valid national and local data bases in forms suitable for analysis, and to the equipment and skills required to analyze that data.

This chapter will first address the techniques of analyzing fire data, including examples demonstrating the amount of information already obtained as a result of the fire community's new data collection and retrieval capabilities. The major data bases now available to analysts then will be reviewed. See Table 11-3A for a sample of the many potential data users and uses.

USING DATA TO CHARACTERIZE THE FIRE PROBLEM

Table 11-3B lists ten major findings on the nature of the fire problem that have emerged from analyses of the past decade. Many of these analyses could not have been conducted without the new and refined data bases that emerged in the late 1970s. This list is not arranged in any order of priority, nor is it necessarily a list of what everyone would consider the ten most important data-based findings. The list does, however, illustrate the ability of data and analysis to identify emerging problems, describe those problems in ways that can help the design of corresponding programs, and track the impact of those programs.

Top-Down and Topic-Driven Analysis

Most characterizations of the fire problem begin with some measures of the problem's size. Why? Implicit in these measurements is the idea that fire is a problem big enough to justify concern and attention. Each year fire kills thousands of persons, injures tens of thousands more, and destroys billions of dollars of property. Many problems causing far less loss of life and property, such as floods and tornadoes, receive much more attention and have much greater resources devoted to the mitigation of their effects. When the fire community wishes to pursue a program that requires the cooperation of others, that community must first attract their attention. The big numbers on fire—incidents, deaths, injuries, and property damage—accomplish this.

Top-down analysis is an approach that begins with the big numbers on fire, then subdivides the totals into their major parts.

TABLE 11-3A. *Typical Users and Uses of Fire Experience Results*

User	Use
1. Fire protection engineers	Supporting data needs of performance-based codes and standards Defining and quantifying fire scenarios Estimating probabilities of fire incidence and fire protection performance Evaluating fire protection alternatives
2. Technical fire safety standards committees	Supporting data needs of performance-based codes and standards Identifying needs for standards Keeping standards current Monitoring the performance of standards Analyzing the benefits and costs of standards
3. Local and state fire service	Showing value of proposed fire safety and fire protection programs and legislation Targeting fire prevention and suppression programs Backing up budget requests Enacting and enforcing fire codes Developing community fire safety education program Monitoring progress
4. Fire protection planners at national and state levels	Showing value of proposed fire safety and fire protection programs and legislation Comparing fire to other problems competing for resources Planning the most effective deployment of resources to make the greatest impact on the fire problem Monitoring fire protection progress
5. Communicators/educators	Reporting of fire problems Educating for fire safety
6. Committees developing fire fighter safety standards, individuals, fire departments working to improve fire fighter safety	Identifying reasons for fire fighter deaths and injuries Developing safety and training programs Designing protective equipment
7. Researchers	Establishing fire safety research priorities Designing research programs
8. Insurers	Loss prevention Risk selection and underwriting
9. Industry	Worker safety Loss prevention
10. Product developers	Identifying needed products and markets Modifying products to improve fire safety
11. Building designers, architects, managers	Ensuring a firesafe design
12. Legislators, regulators, enforcers, courts	Effective and equitable fire safety laws and regulations

The top-down approach provides the broadest perspective on the overall fire problem. It differs from *topic-driven analysis*, which begins with interest in certain issues involving only certain types of fires. For example, a top-down analysis of fires in residences might identify smoking materials as the leading factor in civilian fatalities in homes, accounting for roughly one-fourth of the fires for which the cause was reported. On the other hand, a topic-driven analysis focused on smoking-material fires—for example, a study of the likely impact of a self-extinguishing cigarette or of a public education program aimed at smokers—would involve more than just homes. Although both topic-driven and top-down analyses might be interested in the role of smoking-related fires in home fire deaths, the topic-driven analysis also might be interested in the involvement of smoking materials in nonhome fire fatalities, or in property damage in home fires. Even though smoking-related fires do not account for the largest shares of the latter two problems (incendiary and suspicious causes lead in both cases), they are still part of the overall smoking-fire problem; any change that affects the smoking-fire problem may affect all smoking-related fires.

In a top-down analysis, the basic rule is to look for the biggest parts of the problem. Why bother subdividing the problem? Why not simply attack the whole problem? The answer is that resources usually are too limited to permit an attack on the whole problem and choices must be made. The use of data to identify the largest manageable parts of the problem is the first step in making those choices. While totals are used to show the overall size of the problem, percentages are used to indicate the largest parts.

Use of percentages to identify the largest parts of the fire problem is not as easy as it sounds. Judgment is involved in selecting the categories for which percentages will be calculated. For example, suppose an analysis showed the biggest subgroup of a particular occupancy group's fire problem consisted of fires involving electrical short circuits. This finding does not fit well into most strategies for attacking the fire problem. If one is going to examine every piece of equipment that could be subject to short circuits, one might as well also check for other electrical faults in that equipment. The available strategy may dictate a broader category, covering fires involving all electrical equipment faults. Or maybe it is easier to design strategies around particular types of electrical equipment—for example, fixed wiring, heating systems, ranges, and ovens—and address all fires involving such equipment, whether the equipment or the user were at fault.

TABLE 11-3B. *Ten Major U.S. Data-Based Findings of Recent Years*

1. *While the U.S. and Canadian fire death rates continue to be among the highest in the developed world, the gap is closing dramatically.* In the late 1970s, U.S. fire death rates were more than double the rates in Japan and Sweden. In the mid-1980s, U.S. fire death rates were roughly 50 percent higher than the rates in Japan and Sweden. In the early 1990s, U.S. fire death rates were only one-fourth to one-third higher than the rates in Japan and Sweden. Fire death rates in all these countries are falling, but the rates in the United States and Canada have been falling fastest.

2. *Within the United States, the states of the Old South still tend to have the highest fire death rates, but the gap is closing, and in some years, the South is not the highest region.* In the early 1980s, fire death rates in the South were typically around 50 percent higher than those in the Northeast and North Central regions and double the rates in the West. In 1989 and 1991, the South fell to second highest region, and on average during 1989–1993, the South's fire death rates have been 12 to 17 percent higher than the rates in the North Central and Northeast regions, respectively. The West continues to do substantially better than any other region. A major statewide fire safety campaign in South Carolina cut that state's death rate in half in two years, and other southern states, like Mississippi, are pursuing similar efforts, demonstrating that the improved fire safety record in the South is no accident.

3. *Large-loss fires of historic size have occurred since 1987, after decades with no such fires.* NFPA maintains a list of the 20 costliest fires and explosions of all time, based on loss values adjusted for inflation. The list is headed by the 1906 San Francisco earthquake fire, with involved losses estimated at $350 million at the time, worth more than $5.7 billion today. Nine of these 20 fires occurred in the brief period from 1987 to 1993, including the Oakland fire storm, third costliest of all time, and four other California fires, three of them wildfires. Only two fires on the list of 20 occurred in the nearly four decades prior to 1987, between 1948 and 1986, and neither of those two fires ranks higher than 15.

4. *Rural areas of the United States have the highest fire death rates, and that gap may be widening.* In 1984–1988, fire death rates in rural communities (less than 2,500 population) were more than double the rates in small cities (25,000 to 50,000 population) and more than 50 percent higher than rates in large cities (500,000 or more in population). In 1989–1993, fire death rates in rural communities were nearly triple the rates in small cities and nearly double the rates in large cities. A caution is needed here because even with 5 years of data, fire death rates for rural areas tend to be volatile and subject to substantial estimation uncertainty and year-to-year variation. Nevertheless, there are signs that the steady, significant declines in fire death rates taking place in communities of all larger sizes have not been taking place as consistently in rural communities.

5. *Usage of automatic sprinkler systems continues to grow, and their fire safety value remains dramatic.* The last 10 years of fire experience statistics indicate that sprinklers reduce both the rate of deaths in fires and the average property loss per fire by one-half to two-thirds. Enough data have now accumulated to extend this statement to American homes, as sprinklers are increasingly proving themselves in the places where most people die in fires. In 1980, only one hotel in nine having a reported fire was found to have sprinklers; by 1992, that had risen to one in three, and the overall rate of sprinkler usage in hotels, without regard to whether they had fires, was over half. In 1980, fewer than half the facilities caring for the sick that reported a fire were found to have had sprinklers; by 1992, two-thirds had sprinklers.

6. *Home smoke detectors are nearly universal, but many aspects of the national use of home smoke detectors still need attention.* According to annual surveys sponsored by *Prevention* magazine, by 1995, 93 percent of U.S. homes had at least one smoke detector, a remarkable change from 20 years earlier, when almost no one had detectors. However, the now tiny group without detectors was experiencing nearly half the reported fires and most of the fire deaths. Also, one-fifth of detectors were found to be nonoperational, a figure that rises to one-third for homes that have fires. Most households did not have escape plans that they had rehearsed, so the extra warning time provided by detectors may often be wasted in real fires.

7. *Children playing with fire, principally matches and lighters, have risen dramatically in relative importance among causes of fire deaths.* From 1980 to 1992, the number of people dying in home fires due to children playing with fire barely decreased, while deaths due to every other significant, well-defined fire cause were falling substantially. As a result, child-playing went from a typical 6 percent share of home fire deaths in the early 1980s to a typical 10 percent share in the early 1990s. Most victims of these fires are preschool children, and child-playing now accounts for a third of all fire deaths in that age range. The U.S. Consumer Product Safety Commission introduced a child-resistance requirement for lighters, which had grown to be the larger part of the problem over the previous decade, but statistics current enough to show the effects of that new rule were not yet available at press time.

8. *Juvenile firesetters became even more dominant as the leading share of the arson problem.* In the early 1980s, juveniles (those under age 18) typically accounted for around 40 percent of those arrested for arson, according to data from the Federal Bureau of Investigation's Uniform Crime Reports. In the mid 1990s, this fraction rose to more than 50 percent. This coincided with two other developments: (1) a leveling off of what had been a steep downward trend in the annual number of incendiary and suspicious fires (normally used in analysis as the most complete definition of the arson problem), and (2) a growing recognition that many of the child-playing fires involved psychological or environmental problems similar to those common among juvenile firesetters classified as arsonists. Taken in combination, these findings supported greatly expanded programmatic attention to the juvenile firesetter problem, which was reflected in fire safety initiatives in the early 1990s.

9. *Steep declines in the number of heating equipment fires dropped heating out of first place among causes of fire incidents.* NFPA statistical analysis had tracked the sharp rise in home heating fires, driven by growth in the usage of portable and space heaters, which account for most home heating fires and associated losses. After peaking in the early 1980s, this heightened home heating fire problem began declining steeply, as the usage of portable and space heaters leveled off and as widespread public education and public awareness campaigns, coupled with the use of applicable codes and standards, took hold. By the early 1990s, heating had fallen to second place, behind cooking, among causes of home fires, and had fallen to third place among causes of structure fires.

10. *Fire fighter on-duty fatalities continue to decline sharply.* From an annual death toll over 150 in the late 1970s, fire fighter fatalities occurring due to injuries or illnesses suffered while on duty fell below 100 on average in the early 1990s. Progress was made across the board, from deaths at the fireground to deaths occurring during response or return.

Take another example. Which of the following findings is more useful: (1) 29 percent of all home fires begin in the kitchen; or (2) 22 percent of home fires involve cooking equipment? The latter finding identifies a smaller part of the problem and uses a completely different categorization (equipment involved in ignition *vs.* area of origin), but it locates the problem much more specifically. In a kitchen, there could be dozens of fire hazards to check out; with cooking equipment, most homes have only a few.

Finally, suppose a categorization includes equipment involvement, material ignited, and the human elements that brought them together. Such a categorization will produce detailed scenarios, but the "leading" category might account for only 2 to 4 percent of all fires. It would be necessary to develop strategies for several dozen scenarios in order to make any noticeable impact upon the total fire problem. If that is so, it may make more sense to select scenarios that fit together logically and can be addressed by variations of the same strategy—that is, use of less specific categories—than to select several dozen distinct scenarios and tackle each one separately. To put it another way, the scenario technique of analyzing several fire characteristics simultaneously is most useful as a second-stage analysis, performed after one characteristic has been used to establish basic clustering.

Therefore, in selecting categories for calculating percentages, it is important to match the structure of the categories to the structure of the strategies available to attack the fire problem. Remember that the size of the problem being attacked is no more or less important than the anticipated leverage on the problem; for example, a 7 percent reduction in a problem that causes 500 deaths a year is worth just as much as a 70 percent reduction in a problem that causes 50 deaths a year. (This is discussed at greater length later in this chapter under the heading "Using Data in Program and Strategy Analysis.") So be sure when you pick out the biggest parts of the problem that they collectively account for a sizable share of the total, but also remember that some parts of the problem may be much more preventable than others.

Figures 11-3A and 11-3B illustrate the effective use of percentages (some implicitly presented through the display of numbers) to describe patterns in the 1993 fire fighter fatality problem. (These figures also illustrate the effective use of graphics to bring potentially dull numbers to life.) In Figure 11-3A, which shows the percentage of fire fighter fatalities by type of duty, the fact that fire-ground activity is the leading type of duty is not as interesting as the fact that its share is less than half. Response/return from alarm, which should involve only controllable dangers and relatively little risk, accounts for roughly one-fourth of all fire fighter deaths. Figure 11-3B implicitly demonstrates that, as fire fighter age increases, the leading share of deaths dramatically shifts to heart attacks. Such patterns have direct implications for targeting of physical fitness programs.

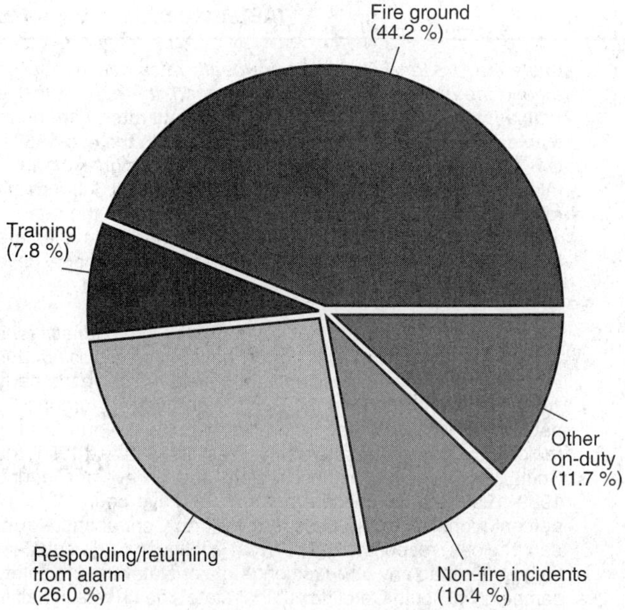

Analysis by Cause or Property Type

The dimensions most frequently chosen for top-down and topic-driven analyses are fire cause and property type, because most fire prevention strategies are structured along those lines. "Analysis by cause" actually refers to a variety of analyses addressed to different aspects of the ignition factors. The question of how to categorize, already discussed in general terms, becomes especially difficult in analyses of cause-related information.

Even the question of how to refer to cause-based descriptions can be a problem. For example, suppose one observes that roughly one-fourth of all civilian fire fatalities in homes with fire cause reported were smoking-related. Does this mean that cigarettes (the principal smoking material involved, by far) were to blame for all those deaths? Not necessarily. If the smokers in question had been

FIG. 11-3A. *Fire fighter fatalities in 1993, by type of duty.*

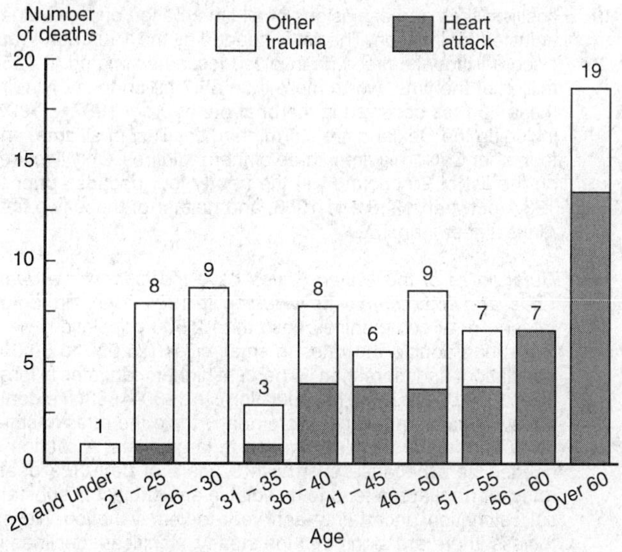

FIG. 11-3B. *Fire fighter fatalities in 1993, by age and cause of death.*

careful not to smoke in bed and always to place cigarettes in sturdy ashtrays, few of the deaths would have occurred. Or if the bedding, mattresses, pillows, and upholstered furniture that are the first items ignited in most fatal smoking-related fires had been more ignition-resistant, many or most of the deaths would have been avoided. It is also true that if the cigarettes used had been designed to self-extinguish more quickly, many or most of the deaths would not have occurred. A decision on which dimension to use in categorizing fire causes—the source of heat, items ignited, or the behaviors that brought them together—should be made in a way that supports analyses of workable prevention strategies. Assessments of blame are tasks for lawyers or philosophers; they do not help in the search for ways to reduce fire problems.

Fire incident data coded according to NFPA 901 format, which is used in most major fire data bases, contains five fire-cause-related elements of information: (1) the form of heat-producing ignition; (2) equipment involved in ignition, if any; (3) type of material that ignited (e.g., wood, plastic, fabric); (4) the form in which that material appeared (e.g., chair, floor covering, structural beam); and (5) ignition factors—the human or equipment failures that brought together the ignition heat and ignited material. Together, these five dimensions can define billions of distinct fire-cause categories, a clearly unworkable partitioning of the problem. In the face of all this detail, several approaches have been developed to extract major patterns. See Table 11-3C for examples of how different approaches can affect the description of "leading cause."

One approach is to divide fire causes into incendiary causes, suspicious causes, all known causes that are accidental in nature, and unknown causes. For many property types, such a partition sufficiently identifies the major fire causes to justify concentrating on arson problems. If not, a different approach is needed.

A second approach has been to use one of the NFPA 901 cause dimensions to sort the fires. One can use the first digit alone to create ten categories, or both digits to create nearly one hundred categories for each dimension. This approach often proves frustrating, however, because the structure of the coding elements often does not coincide with the issues to be analyzed. For example, suppose the ignition-factor dimension is used. Incendiary and suspicious fires will be easy to identify (codes 10–29). Fires caused by children playing, however, involve two code values (36 and 48) that do not have the same first digit and therefore do not fall into a natural common grouping. An analysis using ignition factors also will produce a large number of entries under "abandoned, discarded material." A novice analyst might not recognize that nearly all of these are smoking-related fires and can best be characterized in this manner.

If the ignition-factor dimension has problems when used by itself, what about the other dimensions? Problems arise here because incendiary and suspicious fires are not isolated. Under the category of "heat of ignition," arson fires set with matches (the most common type) are indistinguishable from accidental fires involving matches.

A third approach has been to use scenarios based on two or three of the NFPA 901 cause dimensions. This avoids the tendency, in the one-dimensional sort of scenario, for major problems to be mixed together (e.g., arson and accidental match fires) or disguised (e.g., smoking-related fires listed as discarded or abandoned material). The use of even two dimensions of NFPA 901, however—or the use of three dimensions with the first digit only—creates thousands of cause categories, and again one is faced with the likelihood that even the largest parts of the total will not represent a very large share of the total.

The fourth approach has been to use a small number of major, easily recognizable categories based on NFPA 901 data elements, but not necessarily grouped along the lines laid out in the NFPA 901 coding structure. The most widely used scheme based on this approach is a set of thirteen categories developed at the U.S. Fire Administration (USFA). This scheme uses a hierarchical process to sort fires into categories. For example, one will start with ignition factor and remove all the incendiary, suspicious, and children-playing fires. Then the analyst may switch to heat of ignition and use it to separate fires involving smoking materials.

Criticisms of this or any other hierarchical sorting approach focus on the reasonableness of the sorting priorities and the usefulness of the categories produced. If a fire is caused by a child playing with cigarettes, should that be categorized as a children-playing fire or a smoking-related fire? Is it useful to have a category called "open flame" that combines matches, bonfires, undoused embers, and cutting and welding torches? For most of the categories, the number of

fires falling into each category is not greatly dependent upon the sorting order. The most striking exception is match-related fires, which are scattered among the incendiary/suspicious, children-playing, and open-flame categories. As for usefulness, the categories that seem to be most mixed are, for the most part, categories that account for relatively little of the fire problem in residential properties, the setting for which the categories were designed to be most helpful. For nonresidential properties, particularly manufacturing and industry, the usefulness of the categories may be more questionable. Note the category called "Other Equipment" in Table 11-3C. In nonresidential properties, this includes all kinds of specialized industrial equipment, which may be a leading part of the fire problem. In residential properties, however, these fires tend to be fires reported as involving equipment, but with no details on the type of equipment.

The last approach, which can be used as a refinement of any of the others, is nesting of cause categories. Start with an initial sorting of fires based on either a single dimension of NFPA 901, a scenario structure, or a hierarchically based structure. Then, for each category of the initial sort that has a sufficient number of fires to be worthy of further analysis, use another dimension of NFPA 901 to provide more detail. Table 11-3D shows this nesting approach applied to 1988–1992 store and office fires. Nesting represents the top-down approach at its best—it provides usable levels of detail without sacrificing an overview of the largest parts of the total.

Another matter to consider when analyzing fire data by cause is how to handle incidents for which the cause was unknown or unreported. A sizable share of fires are reported without cause information, particularly severe fires involving deaths or significant property loss. Note that several points in this section have been illustrated with the observation that roughly one-fourth of all civilian fire fatalities in homes with known cause involved smoking materials. Just over one-sixth of all civilian fire fatalities in homes were coded as involving smoking materials. A calculation of percentages based only on cases where cause was reported implicitly assumes (in the absence of contrary evidence) that the cause profile of the fires reported without known cause would look the same (if those causes were known) as the cause profile of the fires reported with known causes. (The two types of percentages are sometimes referred to as causes based on "allocating unknowns over known causes" and causes based on "unallocated unknowns.")

Why is it desirable to allocate unknowns? Suppose two neighboring states have relative cause profiles that look the same, except that fire deaths due to unknown causes appear as 10 percent in one state and 40 percent in the other. Table 11-3E shows the results. State B seems to be doing better than State A with respect to every known cause, but if unknowns are allocated, both states have the same problems to the same degree. If it is necessary or desirable to compare two groups, differences in rates of unknowns need to be taken into consideration.

Why is it difficult to allocate unknowns? Many knowledgeable individuals believe that assessments of probable cause are subject to persistent biases. In fact, no credible evidence, consistent among a variety of fire departments, exists to support the notion that fire officers consistently label fires as suspicious, smoking-related, electrical in origin (all very popular suspicions in certain quarters), or anything else when they really are not sure of the cause.

A more serious argument goes as follows: Probable cause is most difficult to assess in the largest fires because more of the evidence of fire origin is destroyed, and the known-cause fires indicate that cause profiles are different for larger *vs.* smaller fires. For example, incendiary/suspicious and electrical-distribution-system fires tend to have greater loss per fire, and incendiary/suspicious and smoking-related fires are more likely than other fires to result in

TABLE 11-3C. *How "Leading Cause" of Fires Can Depend on How Data Are Sorted**

Emphasis on Accidental vs. Intentional

Accidental	72%
Suspicious	8%
Incendiary	8%
Unknown	13%

Emphasis on Equipment

No equipment	36%
Cooking equipment	17%
Heating equipment	14%
Electrical distribution equipment	8%
Appliance or tool	5%
Other known equipment	6%
Unknown equipment	14%

Emphasis on Behavior

Mechanical failure or malfunction (e.g., short circuit)	24%
Misuse of heat of ignition (e.g., abandoned material)	15%
Operational deficiency (e.g., something left unattended)	15%
Incendiary or suspicious causes	15%
Misuse of material ignited (e.g., combustible placed too close to heat source)	8%
Design, construction, or installation deficiency	5%
Other known ignition factor	5%
Unknown ignition factor	13%

Emphasis on Heat Source

Electrically powered equipment	27%
Fueled equipment	19%
Open flame source involved (e.g., match)	15%
Lighted tobacco product involved (e.g., cigarette)	5%
Exposure (to other hostile fire)	4%
Other known heat source	9%
Unknown heat source	20%

Emphasis on Item Ignited

Cooking materials (e.g., spilled food or grease)	13%
Trash	10%
Structural element	7%
Electrical wire or cable insulation	7%
Mattress, pillow, or bedding	5%
Unclassified item	4%
Exterior sidewall covering	4%
Interior wall covering	3%
Fuel	3%
Multiple items	3%
Clothing	3%
Upholstered furniture	3%
Other known item	23%
Unknown form of item	14%

Hierarchical Sorting Approach

Cooking equipment	16%
Incendiary or suspicious causes	15%
Heating equipment	14%
Other equipment	9%
Electrical distribution system	8%
Appliance, tool, or air-conditioning	6%
Open flame source	5%
Smoking materials	5%
Exposure (to other hostile fire)	4%
Child playing with fire	4%
Other known cause	4%
Unknown cause	11%

*Figures in some groupings may not sum to 100 percent because of rounding errors.
Source: 1989–1993 national estimates of structure fires in apartments, using NFIRS and NFPA survey.

TABLE 11-3D. *Store and Office Fires, 1988–1992 Annual Average*

Cause†	Fires*	
Incendiary or suspicious causes	6,600	(22.1%)
Electrical distribution	5,800	(19.6%)
Lighting fixtures, ballasts, or signs	2,300	(7.6%)
Fixed wiring	1,300	(4.4%)
Cords or plugs	500	(1.6%)
Overcurrent protection devices	400	(1.4%)
Other equipment	3,300	(11.2%)
Appliances, tools, or air conditioning	3,000	(10.0%)
Dryers	1,100	(3.8%)
Central air-conditioning units	400	(1.3%)
Heating equipment	2,400	(8.2%)
Central heating units	800	(2.8%)
Fixed-area heaters	600	(2.0%)
Water heaters	300	(1.1%)
Portable heaters	300	(1.0%)
Open flame, embers, or torches	2,200	(7.5%)
Torches	1,100	(3.6%)
Rekindles or reignitions	400	(1.4%)
Smoking materials	1,900	(6.2%)
Exposure (to other hostile fire)	1,800	(6.0%)
Cooking equipment	1,500	(5.0%)
Stoves	400	(1.4%)
Natural causes	600	(2.0%)
Spontaneous ignitions	300	(1.2%)
Other heat source	400	(1.4%)
Child playing	200	(0.6%)
Total	29,700	(100.0%)

Source: NFIRS, NFPA survey.

*Fires are expressed to the nearest hundred. Percentages are calculated on the actual estimates, so two figures with the same rounded-off estimates may have different percentages. The 12 major cause categories are based on a hierarchy developed by the U.S. Fire Administration. Sums may not equal totals because of rounding errors.

†Unknown-cause fires have been allocated proportionally.

TABLE 11-3E. *The Problem of Comparisons with Unallocated Unknown Cause Fires*

Cause Category	Percentage of Civilian Fire Fatalities in Residences	
	State A	State B
Smoking	36%	24%
Heating	18%	12%
Incendiary/suspicious	18%	12%
Other known	18%	12%
Unknown	10%	40%

deaths. Therefore, if unknowns are to be allocated, the allocation should take into account the cause profiles for fires of comparable size. This is a fire data analysis issue around which no consensus has yet formed. While the ideal solution would be to find ways to reduce the rate of unknowns, there also may be better ways to arrive at estimates in the face of the unknowns that do exist.

Analyses by property use are much more straightforward. Most analyses are concerned with distinctions covered by the "fixed or specific property use" or "mobile property type" categories of NFPA 901. About the only major property class requiring use of

both scales is manufactured homes, and there is little difficulty with unknown or unreported property types. However, unknowns are a data problem for many other fire characteristics, not just cause and property type. Many questions analysts would like to have answered on all reported fires are not asked at all, because fire officers cannot be expected to be able to answer them.

A problem can arise in the size of the data base available for analyses aimed at particular property uses. In 1993, according to NFPA, all nonresidential structures combined accounted for 151,500 fires, 155 civilian deaths, 3,950 civilian injuries, and $2.563 billion in property damage. Typically, roughly half of all fires are represented in the NFIRS, which has individual incident reports and can be used to analyze patterns by specific property use. There are, however, several hundred separate categories of nonresidential properties. A topic-driven analysis focused on one of them might be able to draw upon only a few dozen incidents per year, which is not enough to generate statistical confidence in the results. An analyst who discovers that there is a very small data base available on the property use of interest should do two things. One is to keep the analysis of these fires fairly limited; if each fire represents 2 to 3 percent of the total number of fires being analyzed, there may be no point in conducting fire structure analysis. The other is to reconsider the appropriateness of focusing on this particular property use. If it does not have a large data base, it is not suffering many fires. Even the most successful strategy applied to so small a problem can have only minor impact. There may be more appropriate classes of property in which to attack the fire problem.

Rates and Measures of Fire Risk

Having addressed the usefulness of numbers and percentages as measures of the fire problem, attention can now be turned to the third and last type of measure, which is rates. A rate is a ratio consisting of a measure of the size of a fire problem divided by a measure of the size of the group affected by that problem. In 1993, the U.S. suffered 18 civilian fire deaths per million population. That is a rate consisting of the number of fire deaths divided by the number of people who live in the United States (the so-called resident population). In 1993, the average fire-related property loss in rural communities was $66 per person per year. That is a ratio consisting of the total property damage in rural communities divided by the number of people living in those communities.

Rates provide measures of relative fire risk. They can be used, therefore, in any analysis where the size of the group affected by a problem may change. For example, as has been said, the United States suffered 18 civilian fire deaths in the home per million persons in 1993. Because the total population growth adds about 2 million people to the population each year, the number of civilian fire fatalities will be about 36 deaths higher each year, unless increased fire safety lowers the rate. Along similar lines, the National Safety Council's estimates of fire deaths show total deaths declining by 38 percent from 1973 to 1993, while the fire death rate per million population declined 48 percent in the same period. Increased fire safety is best measured by the decline in the fire death rate; population growth produced a smaller decline in the death toll.

Risk measures do the best job of bringing the fire problem down to a personal level. Rural communities do not account for a majority of the country's fire deaths, but they have by far the highest fire death rates compared to communities of larger size. Person for person, their citizens are in the most danger from fire. Occupants of manufactured homes suffer a substantially higher rate of fire fatalities per million population than do occupants of conventional one- and two-family dwellings. Because there are comparatively few manufactured homes, deaths there do not constitute a large share of the total fire fatality problem, but the individuals living in older manufactured homes are more at risk than their counterparts else-where and should be concerned. An individual wants to know his or her own risk, not the risk to an average American. Barring that, he or she wants to know the risk for people as much like him or her as possible. Risk measured for specifically defined groups provides that information.

Risk measured by rates is of particular interest if there is the potential for large-scale shifts from one group to another of substantially higher or lower risk. In the early 1980s, analysts determined that solid-fueled heating devices, principally wood-burning stoves, accounted for an unusually high risk; that is, there was a high ratio of fires involving those devices divided by homes using those devices. At the time, it appeared that the generation-long decline in use of these devices was about to be reversed, under the pressure of higher prices for oil and natural gas. This was an ideal opportunity to use data to guide action; at a time when decisions were being made, most people did not know the unusual dangers posed by their choices. As an example of risk moving in the other direction, analysts using fire incident data from the early growth years of smoke-detector usage were able to demonstrate that the rate of deaths per home fire was just over half as great for homes with detectors as for homes without them. Here again, decisions with substantial impact on risk were being made regarding equipment, so it was possible to get people's attention and encourage the move toward detectors.

For some analyses, it may be difficult or impossible to obtain the information needed to construct the appropriate rate. One example is an analysis comparing the dangers of fire fighting with risks of other professions. The ratio should be number of on-duty fire service deaths to number of hours of exposure by fire fighters. The latter number cannot be easily calculated, however, because no one has national statistics on how many hours volunteer fire fighters serve. Also, differences in work week need to be considered for career fire service personnel.

Another example is a comparison of fire fatality risks in homes *vs.* hotels. Numbers of persons cannot be used for both, because length of stay is obviously much shorter for hotels than for homes. Numbers of units might work better if one could obtain that data and if occupancy rates could be addressed by a single parameter estimate. The problems of selecting the right denominator and obtaining data for it usually can be solved to some degree of acceptability, but the process is often complex and challenging.

Another problem in constructing rates can occur if the group of interest is defined too narrowly. A household, for example, might like to be able to calculate a risk index that reflects its smoking and drinking habits, its number of children, the ages of its members, its types of equipment and how they are fueled or powered, and many other characteristics. Today this would pose an insoluble problem. There is not enough data to calculate the numbers of households with all those characteristics (so there can be no denominator), and the list of descriptors defines a class so narrow that the number of fires experienced by households of that precise type would be very small and subject to considerable statistical uncertainty. One could separately measure the effects of some of these characteristics, and then combine the results, but such a calculation would involve modeling assumptions that are already known to be false. Considerable statistical modeling and research will be needed to make progress on questions of this type.

USING DATA TO IDENTIFY TRENDS

While the analyses described above provide a "snapshot" of the fire problem, some questions need a "moving picture." Those questions demand trend analysis. Is a part of the fire problem getting better or worse? Is the character of the fire problem changing? If changes are

occurring, do they track with corresponding changes in product use, property use, fire service practices, fire prevention activities, codes and regulations, or other elements of the environment? (For this last question, one may need a full-fledged strategy or program analysis, as described later in this chapter, under "Using Data in Program and Strategy Analysis.")

All trend analyses must address the question of whether or not the past is a consistent guide to the present, let alone the future. It would be nice to be able to say that a trend indicates whether fire risk is going up or coming down. However, a simple trend calculation may not provide this information if the base of comparison has been shifting during the years covered. For example, start with a calculation of the number of fire deaths per year in manufactured homes. If the number dropped over time, that might mean that manufactured homes were becoming safer or it might mean that fewer people were choosing to live in manufactured homes. A trend calculation based on the number of fire deaths per million people living in manufactured homes would factor out the latter possibility and do a better job of showing whether manufactured homes were becoming safer.

Another problem in trend calculations might be a shift in the sampling technique used to create the data base. Suppose that while analyzing fire deaths per million population, as in the previous paragraph, the state of Alaska was added to the data base in the middle of the trend period. The increase in the number of persons would not by itself bias the results because, all other things being equal, numbers of deaths and persons would rise together. But in this case, all other things are not equal. Alaska has an unusually large percentage of households living in manufactured homes, which exerts a sizable influence on the results, and Alaska presents some additional problems—including a severe climate and above-average rates of alcoholism—that differ from those of other states. There would be no way to know how much of any change was due to changes in manufactured home fire safety and how much was due to the addition of Alaska to the data base. In the discussion of "Available Major Data Bases on Fires," more will be said as to which data bases are better designed for trend analysis.

Trends in Fire Fatalities and the Protection of Life

Figure 11-3C shows the trend in NFPA's estimates of civilian fire deaths, starting with 1977 when the NFPA survey was upgraded. The National Safety Council (NSC) has been using a consistent estimating procedure based on state health department records of death certificates for a longer period, although their definition includes nonfire burn deaths and excludes some fire deaths, most notably those involving vehicle accident. Their longer trend line shows a steady decline in the rate of fire deaths and in fire deaths per million population. The cumulative drop in the death rate since the methods changed in 1968 is more than half, and other NSC figures, using slightly different estimating methods, suggest that since 1913 the rates may have dropped by more than four-fifths. The NSC figures, like NFPA's, indicate that the steady decline in the fire death toll, and to a lesser degree the fire death rate, was interrupted by a plateau that lasted from the early 1980s to the late 1980s, during which fire death rates showed no consistent trend up or down.

The most complete and consistent long-term trends of life loss due to fire are those that focus on fires involving large numbers of fatalities per incident. Figure 11-3D traces the rate of deaths per million population per year in fires killing at least 50 persons each, at five-year intervals from 1900 through 1994. A cutoff of 50 fatalities is used because even the poor reporting conditions of the early years are likely to have produced a complete listing of incidents of this magnitude. Figure 11-3E shows the same rates but is limited to fires that began as building fires (or explosions), which excludes mining fires, ship fires, forest fires, post-earthquake fires, and missile silo fires, all of which are represented in Figure 11-3D.

With a longer timeline and a different measure of loss, Figure 11-3D shows an even more dramatic trend than do the National Safety Council figures on total and burn deaths. Instead of a decline by a factor of five or six, this shows a decline by a factor of more than 100, from 1900–1904 to 1990–1994.

This format, focusing on fires that kill at least 50 people each, was suggested as a way of coping with reporting inadequacies in the early years, but it also provides a good match with certain kinds of programs that may be considered worthy of tracking. Specifically, codes and standards, and fire protection engineering generally, normally aim to prevent major incidents, not necessarily all fatal incidents. By focusing on larger incidents, one can see that progress has indeed been most pronounced on the deaths associated with these fires.

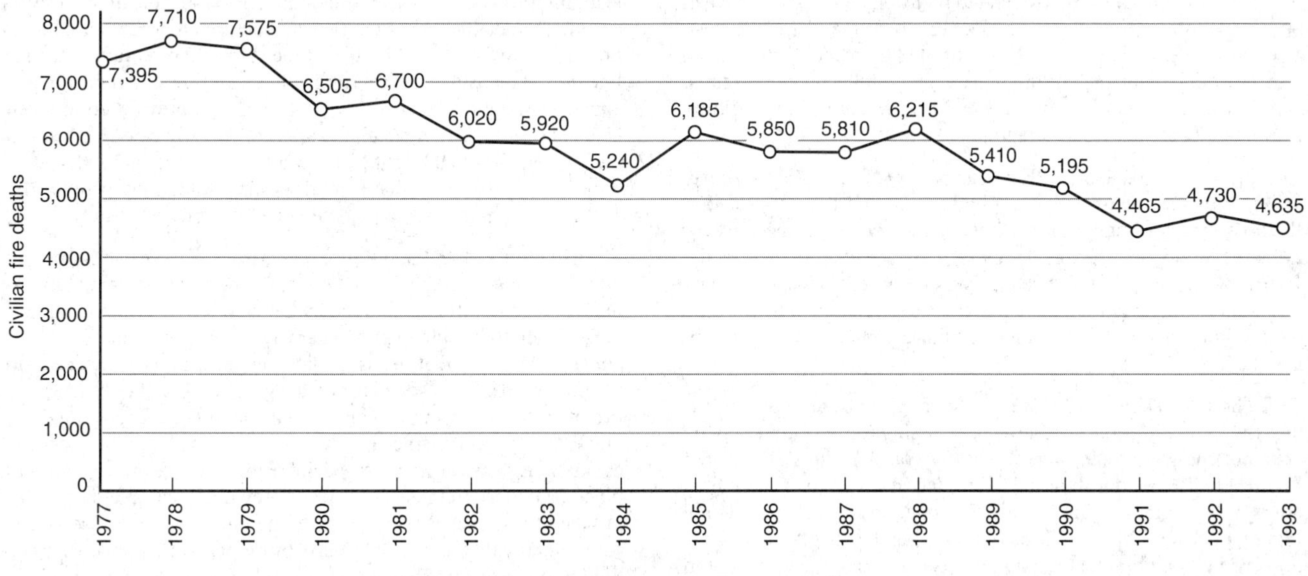

FIG. 11-3C.   *Civilian fire deaths, 1977–1993.*

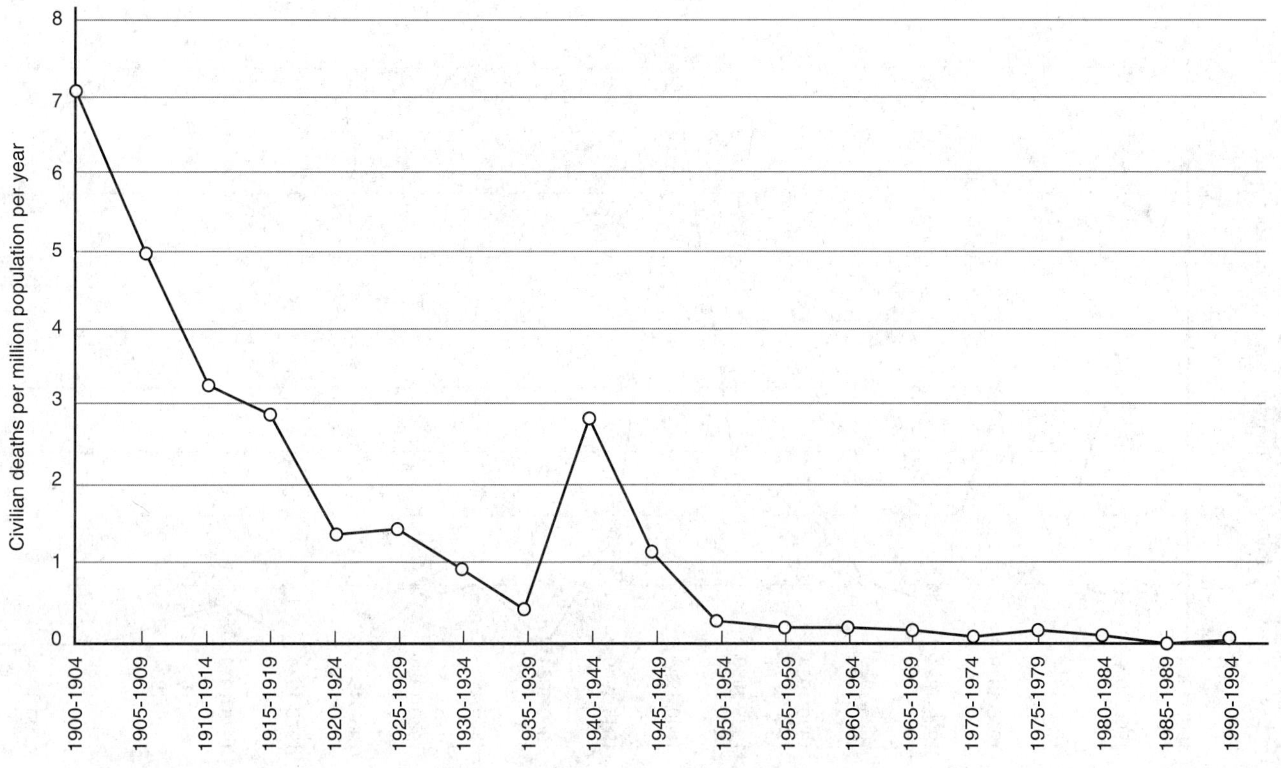

FIG. 11-3D.   *Major loss-of-life fires, taking at least 50 lives, in the United States since 1900.*

Figure 11-3D reflects deaths in just 99 fires, and three of every five of those fires involved mining, usually coal mining. Nearly all of these fires occurred in the first quarter century, so most of the progress seen in the first quarter of the graph reflects progress in making mining safer, as well as the decline in mining activity, as other fuels took on a greater role in supporting America's needs for power. Of the 99 fires, only four occurred after 1969: (1) the Beverly Hills Supper Club fire (165 killed); (2) the Kellogg, ID, silver mine incident (91 killed); (3) the Happy Land social club fire (87 killed); and (4) the MGM Grand Hotel fire (85 killed).

Codes and standards also tend to focus on buildings more than on mines, forests, or vehicles, so Figure 11-3E may be of interest, providing a version of Figure 11-3D limited to incidents that began as fires or explosions in buildings. This graph reflects only 29 fires, and only three of those occurred after 1949, the same three building fires listed in the previous paragraph.

Figures 11-3D and 11-3E paint a picture of dramatic and fairly steady progress, interrupted only by the period around World War II. Fires directly related to wartime hazards were not the only, or even the dominant, factors in the statistical spike shown in the 1940s, but there may have been a diversion of resources and priorities from fire code enforcement and fire safety generally to wartime production objectives, producing an indirect effect not easily isolated.

The rates shown in Figures 11-3D and 11-3E are now so low that they usually reflect the effects of a single fire. Figure 11-3D does not show a 5-year interval with two or more fires after 1969. Figure 11-3E does not show a 5-year interval with two or more fires after 1964. Figure 11-3D has hit zero once in recent years, while Figure 11-3E has hit zero three times. In short, this graph has now devolved to the point that it only reflects the size of the worst fire in each half-decade. It no longer operates effectively to help track progress on the fire problem, because it does not focus on the kinds of fatal fires that are being experienced at present.

Table 11-3F shows another approach to tracking by way of the deadliest of the big fires. Not being a graph, it lacks the visual impact of Figures 11-3D and 11-3E. Instead, the sense of progress is stated more subtly, as the table shows how much less severe the worst fires of the decade have become.

By focusing on specific fires, Table 11-3F can also serve as an introduction to the kind of extended discussion of turning-point fires, i.e., incidents that triggered changes in our approach to fire safety that significantly and permanently altered our loss experience, provided in Section 1, Chapter 1, "America's Fire Problem and Fire Protection." From a statistical standpoint, it is striking that in every decade the deadliest fires tend to include fires that are not building fires, ranging from the ship fires at the start of the century to the airline fires at the end of the century, with forest fires and mine fires occurring throughout the century. Note, too, the prevalence in recent years of deadly fires in which fire was not the only cause of deaths in the incident. The fatal airline fires often had additional deaths due to trauma from the crash that started the fire. And many people were shot to death in the Waco incident where others died in fire.

Remember that the purpose of statistics is to answer important questions that will help indicate what is needed for fire safety. An interest in major fires is justified by the view that fire codes and standards ought to be able to prevent any really large incidents from occurring. This is what leads to the phenomenon of one bad fire leading by itself to a set of code changes, because it only takes one fire of sufficient severity to indicate that an objective as stringent as preventing all very large fires has not been met. This sequence of events is especially likely to occur if the one bad fire occurs in a place or under circumstances never previously associated with really bad fires; because, in that case, the one bad fire will serve as a sign to many that a particular class of properties, equipment, or activities is not as safe as everyone thought it was.

Because this kind of logic is so common and seems so plausible, it is useful to introduce some cautions. First, fire protection en-

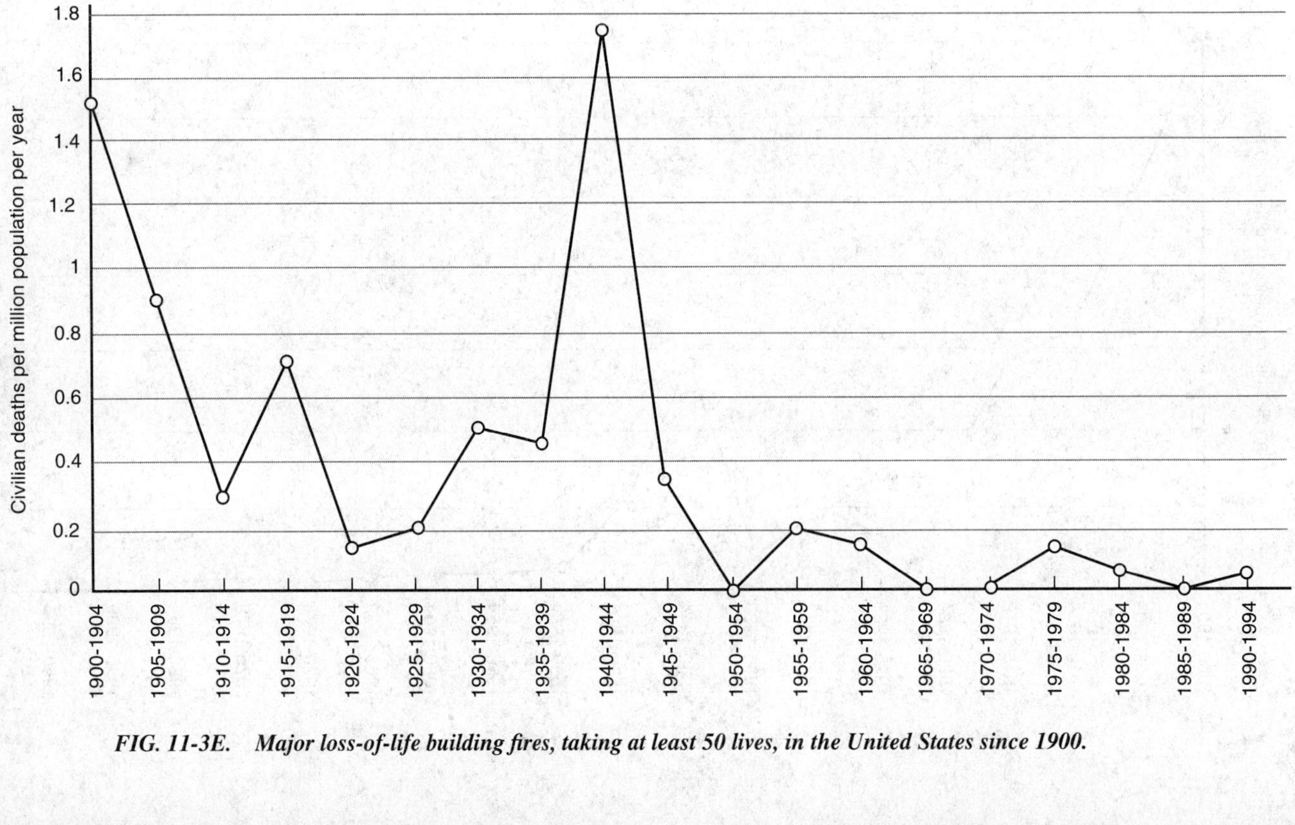

FIG. 11-3E.   Major loss-of-life building fires, taking at least 50 lives, in the United States since 1900.

TABLE 11-3F. *Deadliest U.S. Fires and Explosions of the Past Nine and a Half Decades, 1900–1994*

	Number Killed
1990–1994	
1. Happy Land social club, New York, NY (Mar. 25, 1990)	87
2. Branch Davidian compound, Waco, TX (Apr. 19, 1993)	47
3. Wildland/urban interface, Oakland, CA (Oct. 20, 1991)	26
1980–1989	
1. MGM Grand Hotel, Las Vegas, NV (Nov. 21, 1980)	85
2. Galaxy Airlines #203, Reno, NV (Jan. 21, 1985)	42
3. United Airlines #232, Sioux City, IA (July 19, 1989)	38
1970–1979	
1. Beverly Hills Supper Club, Southgate, KY (May 28, 1977)	165
2. Silver mine, Kellogg, ID (May 2, 1972)	91
3. Capital International chartered airliner, Anchorage, AK (Nov. 28, 1970)	47
1960–1969	
1. Coal mine, Farmington, WV (Nov. 20, 1968)	78
2. Indiana State Fairgrounds Coliseum, Indianapolis, IN (Oct. 10, 1963)	75
3. Golden Age Nursing Home, Fitchville Township, OH (Nov. 23, 1963)	63
1950–1959	
1. Coal mine, West Frankfort, IL (Dec. 21, 1951)	119
2. *U.S.S. Bennington* aircraft carrier, off RI coast (May 26, 1954)	103
3. Our Lady of the Angels Grade School, Chicago, IL (Dec. 1, 1958)	95
1940–1949	
1. Cocoanut Grove Night Club, Boston, MA (Nov. 28, 1942)	492
2. *S.S. Grandcamp* and Monsanto Chemical Plant, Texas City, TX (Apr. 16, 1947)	468
3. Munitions ships and depot, Port Chicago, CA (July 17, 1944)	322
Plus 5 other incidents that each killed at least 100 people.	
1930–1939	
1. Ohio State Penitentiary, Columbus, OH (Apr. 21, 1930)	320
2. Consolidated School, New London, TX (Mar. 18, 1937)	294
3. *S.S. Morro Castle,* off NJ coast (Sept. 8, 1934)	135
1920–1929	
1. Coal mine, Mather, PA (May 19,1928)	273
2. Coal mine, Castle Gate, UT (Mar. 8, 1924)	171
3. Cleveland Clinic, Cleveland, OH (May 15, 1929)	125
Plus 2 other incidents that each killed at least 100 people.	
1910–1919	
1. Forest fire, northern MN (Oct. 12, 1918)	559
2. Coal mine, Dawson, NM (Oct. 22, 1913)	263
3. Aetna Chemical Company, Oakdale, PA (May 18, 1918)	193
Plus 7 other incidents that each killed at least 100 people.	
1900–1909	
1. *S.S. General Slocum* steamship, New York, NY (June 15, 1904)	1,030
2. Iroquois Theater, Chicago, IL (Dec. 30,1903)	602
3. Coal mine, Mononnga, WV (Dec. 6,1907)	361
Plus 13 other incidents that each killed at least 100 people.	

gineering may be thought of as deterministic solutions for a probabilistic world. To put it another way, risk can never be reduced to zero, and it is important that realistic fire safety designs recognize that even the most determined enforcement of the most stringent requirements will not eliminate all possibility of a serious fire.

Second, most fire deaths and other fire losses do not occur in big fires or even in the kinds of places and situations where big fires occur. Big multiple-death fires occur in high-occupancy places like hotels or dormitories or night clubs, while most fire deaths occur in low-occupancy places, such as dwellings or individual apartment units. Big multiple-death fires involve high-occupancy vehicles like commercial airliners or large ships, while most fire deaths, excluding those occurring in homes, occur in small private vehicles like cars and trucks.

Finally, it should be obvious from a casual glance at the deadliest fires of recent years that the problem often is not that the codes aren't tough enough but rather that the adoption isn't universal or the enforcement isn't effective.

Event-driven changes in the fire codes primarily involve the fire community. However, getting those changes adopted into law and backing the law with strong enforcement programs require the cooperation of elected officials and the public—who are concerned about many subjects in addition to fire. Recognizing this, some fire chiefs have prepared legal packages targeted at particular problems in advance, holding them in reserve until a major fire provides the visibility and sense of urgency required for adoption.

In view of this strategy, it is worth noting that "newsworthiness" is not always synonymous with incident severity. Two examples will suffice. The third most deadly fire of the decade 1960–1969 was the Golden Age Nursing Home fire, which killed 63 persons; it occurred, however, on the day after President John F. Kennedy was assassinated. This coincidence of timing sharply reduced the visibility of the fire. By contrast, the 1937 *Hindenburg* zeppelin fire killed "only" 36 persons, a death toll barely a fourth the size of even the third-largest fire of that decade and only one-eighth the size of the New London, TX, school fire in the same year. The zeppelin fire, however, was broadcast "live" and occurred in full view of newsreel cameras, so it was probably the most widely heard and seen fatal U.S. fire until the *Challenger* space shuttle explosion in 1986. It meant the end of the rigid-airship industry.

These examples also have a lesson for analysis of trends in fire deaths, injuries, or incidents. The more severe the type of incident being tracked, the more random the year-to-year fluctuation. This makes it important to look at trends over an appropriate length of time. For fires as severe as those discussed in the last few paragraphs, only decade-by-decade comparisons suffice to separate real trends from random ups and downs. For total U.S. civilian fire deaths, year-to-year tracking is meaningful. For a single large city, however, year-to-year tracking of fire deaths will be subject to considerable statistical noise; and for town and rural areas, there is no good way to track fatality trends except as part of a larger aggregation, such as the state or the rural portion of the entire region.

Trends in Property Damage Due to Fire

The first fire protection standards promulgated by NFPA were directed principally at property protection, reflecting the association's founding by men in the business of insuring property against loss due to fire. Assessments of the success of strategies for property protection are more difficult than for life protection, however, because data on fire damage is subject to several gaps and quirks that do not affect data on fire deaths. While all indicators of fire—incidents, deaths, injuries, and property damage—are clouded by the absence of information on fires not reported to local fire departments, property damage is the only indicator of severity that is left unreported in a significant minority of reported fires. Estimates of property damage involve more guesswork than estimates of people killed or injured. Some very-large-loss fires are never reported to local fire departments, because the affected firms are able to handle them with on-site resources. No comparable omissions occur among multiple-death fires.

Trend analysis must take into account the effect of inflation. Table 11-3G shows NFPA's estimates for total national property damage due to fire in the years 1983–1993, both as published and as recalculated in terms of 1983 dollars, using the U.S. Bureau of Labor Statistics' Consumer Price Index to remove the effects of inflation. When inflation is not removed, the loss totals show a 30 percent *increase* in property damage from 1983 to 1993. When inflation is removed, the loss totals show an 11 percent *decrease* in property damage. Also significant is the effect of changes in the number of fires, because total damage figures involve the total number of fires and the average dollar loss per fire. From 1983 to 1993 the average loss per fire, as published, rose 54 percent, but the average loss per fire adjusted for inflation rose only 6 percent.

TABLE 11-3G. *U.S. Trends in Property Damage Due to Fire, 1983–1993*

Year	Total Estimated Damage	Adjusted to 1983 Dollars
1983	$6.598 billion	$6.598 billion
1984	$6.707 billion	$6.426 billion
1985	$7.324 billion	$6.778 billion
1986	$6.709 billion	$6.107 billion
1987	$7.159 billion	$6.281 billion
1988	$8.352 billion	$7.045 billion
1989	$8.655 billion	$6.964 billion
1990	$7.818 billion	$5.971 billion
1991	$9.467 billion	$6.928 billion
1992	$8.295 billion	$5.897 billion
1993	$8.546 billion	$5.896 billion

To put this even more dramatically, suppose an energetic, creative, intelligent, farsighted fire chief had begun a massive 10-year program in 1972 to reduce by half the property damage due to fire in his city. Then suppose the program had succeeded. Because of the effects of inflation, the actual reported loss would have shown a 9 percent *increase* in property damage over that decade.

This same effect makes it difficult to compare large-loss experience over several years. In the early 1930s, the NFPA defined a large-loss fire as one involving at least $250,000 in direct property damage. In 1978, a large-loss fire was redefined as one involving at least $500,000, and in the early 1980s the amount became at least $1,000,000 in direct property damage. Each minimum represented the cutoff point for the largest several hundred fires of its time. In the late 1980s, it was further decided to shift the large-loss study to a more detailed treatment of a smaller number of fires, so the threshold was raised to $5,000,000.

Table 11-3H lists the 20 largest-loss fires of 1985–1994, first on the basis of their published losses, then on the basis of losses adjusted for inflation. Note that Table 11-3G adjusts backward for inflation, putting all losses in terms of dollars for the earliest year or the period (1983 within 1983–1993), while Table 11-3H adjusts forward, putting all losses in terms of dollars for the latest year in the period (1994 within 1985–1994). Either approach is valid, and, in fact, the year chosen need not even be part of the range of events. (All losses could be converted to 1900 dollars, for example.) Backward-adjusting works well with graphs, because it shows two trend lines emerging from a common starting point, whereas forward-adjusting would show them converging to a common end-point, which looks less natural. In a table, however, as in any display covering a very long timeline, forward-adjusting comes closest to showing what things would cost today. It also yields the largest numbers, which makes the most dramatic impression.

Note how the adjustment for inflation changes the list, as older fires rise and more recent fires fall. For example, No. 11 on the first list is No. 7 on the second list. Each list has one fire the other does not have: No. 19B on the first list, and No. 19 on the second list.

From an interpretation standpoint, note that California accounts for half the fires on the second list, most of them wildfires, for a total in 1994 dollars of nearly $4 billion.

Difficulties in comparison become more pronounced when points even further back in history are considered. The 1947 chemical plant fire in Texas City, TX, had a reported loss of $67 million. Using the Consumer Price Index, that translates into $445 million in 1994. The 1906 earthquake and fire in San Francisco had a reported loss of $350 million. That means more than $5.7 billion today—or more than half the loss attributed to all reported fires in the United States in a typical year. At this point, comparisons can become very sensitive to assumptions. Was the property lost in the Texas City fire sufficiently similar to the items covered by the Consumer Price Index? Or should a more industry-specific price index be used? Were the estimates of loss for the San Francisco fire made to the nearest $10 million or even the nearest $50 million? Then the 1980s version of that loss might be accurate to only the nearest $165 million or $820 million, which would mean its range of uncertainty would be higher than the total loss of almost any recent fire.

The lesson in all this is that trend analysis of loss figures can be very tricky and needs to be done with an eye toward the effects of inflation and population growth (which also affects trends in numbers of fires, deaths, and injuries). Keep in mind that loss trends have a tendency to look bad even if real progress is being made.

USING DATA IN PROGRAM AND STRATEGY ANALYSIS

Earlier in this chapter, techniques were described for using data to analyze the present (size and characteristics of the fire problem) and the past (trends in the size and nature of the fire problem). Program and strategy analysis is the most decision-relevant use of data because it tries to project the future, and, in particular, the ways in which the future will be different if a particular program or strategy is or is not adopted.

The most fundamental question to be answered in using data for program or strategy analysis is whether past fire experience is a good guide to future fire problems. Sometimes it is; sometimes it is not. However, intelligent use of data with other sources of information is an essential part of the best possible projection of the future. After all, what will we use if not data? Models are tied to reality by validation, which is based on experience of the past. Expert judgment is formed, at least in part, by the experience of the past. The issue is not whether to use the past as a guide; it is how best to characterize the important changes that will occur in the future.

Often the most important aspect of the future will be a different mix of the elements already in place in the present. For example, home smoke-detector usage has grown at a phenomenal rate in the past 20 years, with most of the growth occurring in or after 1976. Fire incident data from the early years of growth in detector usage showed that smoke detectors reduce by nearly half the risk (measured by the number of deaths per 1,000 fires) that a person will die if he or she has a fire. At that point, there were uncertainties about the applicability of this finding to the future. Would the life-saving

TABLE 11-3H. *Largest Loss U.S. Fires Reported to NFPA, 1985–1994*

Reported Dollars

1.	Wildland firestorm, Oakland, CA	Oct. 20, 1991	$1,500 million
2.	Polyolefin plant, Pasadena, TX	Oct. 23, 1989	$750 million
3.	Civil disturbance, Los Angeles, CA	Apr. 29, 1992	$567.375 million
4.	"Laguna" wildfire, Orange, CA	Oct. 27, 1993	$527.983 million
5.	Petroleum refinery, Norco, LA	May 5, 1988	$330 million
6.	High-rise office building, Philadelphia, PA	Feb. 23, 1991	$325 million
7.	"Cleveland" wildfire, Placerville, CA	Oct. 1, 1992	$245.525 million
8.	"Paint fire/Goleta" wildfire, Santa Barbara, CA	June 27, 1990	$237.046 million
9.	High-rise office building bombing, New York, NY	Feb. 26, 1993	$230 million
10.	"Malibu/Old Topanga" wildfire, Los Angeles CA	Nov. 2, 1993	$219 million
11.	Chemical manufacturing plant, Pampa, TX	Nov. 14, 1987	$215.3 million
12.	Multitenant warehouse complex, Elizabeth, NJ	Feb. 21, 1985	$123 million
13.	"Fountain" wildfire, Redding, CA	Aug. 20, 1992	$110.471 million
14.	Chemical plant explosion, Monroe, LA	May 1, 1991	$105 million
15.	Chemical manufacturing plant, Henderson, NV	May 4, 1988	$103 million
16.	Cold storage warehouse, Marshall, MO	Mar. 16, 1992	$100 million
17.	Petroleum refinery, Richmond, CA	Apr. 10, 1989	$95 million
18.	Telephone switching station, Hinsdale, IL	May 8, 1988	$90 million
19A.	General warehouse, Richmond, CA	July 11, 1988	$80 million
19B.	Chemical manufacturing plant, Seadrift, TX	Mar. 12, 1991	$80 million

Adjusted to 1994 Dollars

1.	Wildland firestorm, Oakland, CA	Oct. 20, 1991	$1,631.11 million
2.	Polyolefin plant, Pasadena, TX	Oct. 23, 1989	$896.67 million
3.	Civil disturbance, Los Angeles, CA	Apr. 29, 1992	$599.32 million
4.	"Laguna" wildfire, Orange, CA	Oct. 27, 1993	$541.28 million
5.	Petroleum refinery, Norco, LA	May 5, 1988	$413.60 million
6.	High-rise office building, Philadelphia, PA	Feb. 23, 1991	$353.41 million
7.	Chemical manufacturing plant, Pampa, TX	Nov. 14, 1987	$280.69 million
8.	"Paint fire/Goleta" wildfire, Santa Barbara, CA	June 27, 1990	$269.00 million
9.	"Cleveland" wildfire, Placerville, CA	Oct. 1, 1992	$259.35 million
10.	High-rise office building bombing, New York, NY	Feb. 26, 1993	$235.79 million
11.	"Malibu/Old Topanga" wildfire, Los Angeles, CA	Nov. 2, 1993	$224.52 million
12.	Multi-tenant warehouse complex, Elizabeth, NJ	Feb. 21, 1985	$169.10 million
13.	Chemical manufacturing plant, Henderson, NV	May 4, 1988	$129.09 million
14.	"Fountain" wildfire, Redding. CA	Aug. 20, 1992	$116.69 million
15.	Chemical plant explosion, Monroe, LA	May 1, 1991	$114.18 million
16.	Petroleum refinery, Richmond, CA	Apr. 10, 1989	$113.58 million
17.	Telephone switching station, Hinsdale, IL	May 8, 1988	$112.80 million
18.	Cold storage warehouse, Marshall, MO	Mar. 16, 1992	$105.63 million
19.	"Ojai-Wheeler Gorge" wildfire, Camarillo, CA	July 1, 1985	$103.67 million
20.	General warehouse, Richmond, CA	July 11, 1988	$100.27 million

Source: NFPA's Fire Incident Data Organization (FIDO).

impact of detectors remain the same when detectors were installed in poorer households with less-educated occupants?

The answer turned out to be yes, although there was some erosion in the estimated benefit from detectors that coincided with adverse trends in the estimated fraction of detectors that were operational. But the form of the question shows how data can be used intelligently to try to project the future. Start with the future predicted by a simple projection of recent trends. Using judgment, creativity, and brainstorming, try to identify what aspects of the present environment might be changing in ways that would modify the simple projection. Try to obtain data on the speed of those changes. Try to analyze the fire data in hand to determine the sensitivity of your conclusions to those changes. Combine these results to produce a revised projection.

Most analyses to date have found that environmental changes relevant to fire occur slowly and so produce relatively modest changes in what are called baseline projections, i.e., projections of the size and character of the fire problem if no new strategies are adopted. For example, the average age of the U.S. population is rising. This means a drop in the percentage of the population in the high-death-rate years of infancy, a rise in the percentage in the high-death-rate elderly years, and other shifts along the age spectrum. By the end of the century the cumulative effect will be large, but on a year-to-year basis it produces only a small predicted change in the national fire death rate.

An analysis of the projected impact of a strategy or program involves the following five steps:

1. Identify the part of the fire problem that the strategy can affect and measure the size of that problem. For example, a change in the flame resistance of upholstered-furniture-covering materials would affect only fires involving those materials. Keep in mind the need to address both fires that begin with upholstered furniture and fires that begin elsewhere and become severe primarily because of spread to upholstered furniture. A change to self-extinguishing cigarettes would affect only fires whose form of heat of ignition was cigarettes, and so on.

2. Estimate the likely percentage reduction in this target fire problem if the strategy or program were adopted, and prepare estimates for each of the measures of fire severity (deaths, injuries, property loss) because the impact probably will be different for each measure. If the strategy or program is already in use (as was true for smoke detectors), then fire incident data analysis may produce these estimates. If the strategy or program is not in use or is in very limited use, as is true for home sprinklers, then some combination of modeling, laboratory tests, and expert judgment will be needed to produce the estimates. In either case, it makes sense to develop a range of estimates, from optimistic to pessimistic.

3. Estimate how much of the target population will adopt the program or strategy, and how quickly. This typically is a marketing question, and the results can be surprising. Most analyses of the growth in smoke detector coverage were too pessimistic on both how much and how fast. At the same time, most analyses were too optimistic about the installation of detectors in homes most likely to have fires. While 92 percent of U.S. households had detectors in 1993, just over half the home fires that year were reported to occur in homes with detectors. This discrepancy is not unusual for strategies that work through the marketplace. The first purchasers tend to be more affluent and better educated; these people traditionally have lower fire rates than the rest of the population. Also, speed of adoption will reflect the normal life of a product. Changes in cigarettes can be implemented for all cigarettes in months, while changes in the upholstered furniture ignited by cigarettes may take decades to work their way through the whole population.

4. Estimate how often the strategy will be defeated in practice. For example, what proportion of detector-equipped homes will have their detectors out of service because worn-out batteries were not replaced? Note that if the method in step (2) involved the use of actual fire incident data, your estimate already may include the effects of defeating or attenuation (the latter is a term that does not have the connotation of deliberate sabotage that defeating may have for some people).

5. Combine the measures of fire problem size from step (1) and the percentages from steps (2) through (4) to produce estimates of the net percentage reduction in the fire problem and of the new size of the fire problem. With these results, decide how valuable this strategy or program would be and whether to press for its adoption. In some forums, this decision will require a parallel calculation of the cost of the strategy or program, followed by some kind of comparison of the costs and loss reductions.

Analysis can be used not only to make yes or no decisions on adopting a strategy, or to select the strategy with the biggest project impact, but also to fine-tune the design of the strategies themselves. Two examples will illustrate how this can be done.

One city fire department had a small-scale home inspection program, in which fire companies used slack periods to offer home inspections in their first-due areas. Since no one was home at about half the houses visited, this program offered a good opportunity to compare the fire experience of the homes inspected with the fire experience of the similar homes where no one had been home when inspections were offered. Results of the analysis showed no inspection impact, but, more importantly, they showed that the homes chosen for inspection were in areas with some of the lowest fire incident rates in the city. In other words, the program was somehow being targeted at the neighborhoods needing it least. Why? Once the data analysis identified a hitherto unsuspected pattern, the explanation was fairly simple: inspection efforts would be greater in areas with more company slack time, which in turn would tend to be areas requiring fewer company runs, including fewer fires. The analysis pointed to a need to redesign the home inspection program so it would target fire-prone neighborhoods.

Unexpected results of analysis are not the only indicators of the need for design improvements. While designing a public education campaign aimed at heating safety, the Seattle, Washington, Fire Department conducted a day-by-day analysis of fire incidence. By comparing that analysis with historical data on Seattle weather, the analysts were able to identify a significant pattern of cold snaps followed three days later by a sharp rise in heating system fires. Vulnerable heating systems, it seemed, would develop fires after sustained heavy use. This analytical discovery had a direct bearing upon the design of the public education campaign, because it suggested that fire safety messages should be released as soon as a cold snap started. People would already be thinking about their heating systems when the messages appeared, so attention would be greater; also, the need to listen and act at once could be underlined by reference to the established pattern. What is more, the three-day lag meant there was time to get out the messages after the cold snap started but before the fires would begin.

The common thread of these two examples is the use of analysis to narrow the target of a program or strategy in order to maximize impact and conserve scarce program resources. In the first case, the geographical targeting was affected; in the second case, the timing benefited from analysis.

Another consideration in analysis is the need to identify clearly which of many important objectives are to be given priority. For example, is it more important to reduce the fire loss actually being suffered, thereby targeting properties where most loss now occurs? Or is it more important to reduce the potential for catastrophic fire loss, thereby targeting properties with sizable numbers of lives or property value at risk, even if their actual loss has been relatively small? Also, is it most important to reduce numbers of fires, numbers of deaths and injuries, or numbers of dollars in direct property loss? In apartment fires, for example, a focus on fires might mean targeting cooking-related fires, a focus on deaths might mean targeting smoking-related fires, and a focus on property loss might mean targeting incendiary and suspicious fires.

Finally, remember that tracking analysis continues to be useful even after a decision has been made and a strategy or program has been implemented. As noted earlier, most of the estimates used in an analysis involve uncertainty; therefore, it may be useful to see whether the future unfolds as projected. If not, some further refinements in the design of a strategy or program may be warranted. Even early termination of a program might be indicated.

AVAILABLE MAJOR DATA BASES ON FIRES

Three major data bases are available to analyze patterns in U.S. fire experience: (1) annual NFPA survey of fire departments; (2) FEMA/USFA National Fire Incident Reporting System (NFIRS); and (3) NFPA Fire Incident Data Organization (FIDO). Together, these three data bases can provide valid, detailed information on national and regional fire problems, overall or by specific property type and cause. The characteristics of the three data bases and the best ways of using each are the subject of this section.

Annual NFPA Survey of Fire Departments

The NFPA survey is based on a stratified random sample of roughly 3,000 U.S. fire departments (or roughly one of every ten fire departments in the country). The survey collects the following information: (1) the total number of fire incidents, civilian deaths, and civilian injuries and the total estimated property damage (in dollars) for each of the major property-use classes defined by the NFPA 901 standard for fire incident reporting; (2) similar tallies specifically for incendiary and suspicious fires, separated only into structure versus vehicle; (3) the number of on-duty fire fighter injuries, by type of duty and nature of injury or illness; (4) information on the type of community protected (e.g., county *vs.* township *vs.* city) and the size of the population protected, which is used in the statistical formula for projecting national estimates from sample results; and (5) leads on multiple-death and large-loss fires and fire fighter fatalities, which if cited are then captured under FIDO.

The totals in (1) and the special incendiary and suspicious fire results in (2) are analyzed and reported in NFPA's annual study, "Fire Loss in the United States," which traditionally appears in the September/October issue of *NFPA Journal*. The fire fighter injury information in (3) is analyzed and reported in NFPA's annual report, "U.S. Fire Fighter Injuries," now published in the November/December issue of *NFPA Journal*.

The NFPA survey begins with the NFPA Fire Service Inventory, a computerized file of nearly 30,000 U.S. fire departments, which is the most complete and thoroughly validated such listing in existence. The survey is stratified by size of population protected to reduce the uncertainty of the final estimate. Small, rural communities protect fewer people per department and are less likely to respond to the survey, so a larger number must be surveyed to obtain an adequate sample of those departments. (NFPA also makes follow-up calls to a sample of the smaller fire departments that do not respond, to confirm that those that did respond are truly representative of fire departments their size.) On the other hand, large city departments are so few in number and protect such a large proportion of the population that it makes sense to survey all of them. Most respond, resulting in excellent precision for their part of the final estimate.

These methods have been used in the NFPA survey since 1977 and represent a state-of-the-art approach to sample surveying. Because of the attention paid to representativeness and appropriate weighting formulas for projecting national estimates, the NFPA survey provides a valid basis for measuring national trends in fire incidents, civilian deaths and injuries, and direct property loss, as well as for determining patterns and trends by community size and major region.

FEMA/USFA National Fire Incident Reporting System (NFIRS)

FEMA/USFA NFIRS provides annual computerized data bases of fire incidents, with data classified according to a standard format based on NFPA 901. Roughly three-fourths of all states have NFIRS coordinators, who receive fire incident data from participating fire departments and combine the data into a state data base. This data is then transmitted to FEMA/USFA. Participation by the states, and by local fire departments within participating states, is voluntary. NFIRS captures roughly one-half of all U.S. fires each year. Not all U.S. fire departments listed as participants in NFIRS provide data every year.

One of the strengths of NFIRS is that it provides the most detailed incident information of any national data base not limited to large fires. NFIRS is the only data base capable of addressing national patterns for fires of all sizes by specific property use and specific fire cause. (The NFPA survey separates fewer than twenty of the hundreds of property use categories defined by NFPA 901 and provides no cause-related information except for incendiary and suspicious fires.) NFIRS also captures information on the construction type of the involved building, avenues and extent of flame spread and smoke spread, building height, and performance of detectors and sprinklers.

One weakness of NFIRS is that its voluntary character produces annual samples of shifting composition. Despite the fact that NFIRS draws on five times as many fire departments as the NFPA survey, the NFPA survey is more suitable as a basis for projecting national estimates because its sample is truly random and is systematically stratified to be representative.

Analyses based on NFIRS have been widely calculated only since 1982, the year of publication of the second edition of FEMA/USFA *Fire in the United States*—the first study based primarily on NFIRS. Because consensus on how best to address the weaknesses of NFIRS does not yet exist, the next few years may see further revisions in the formulas used to calculate statistics from this system. In the meantime, most analysts use NFIRS to calculate percentages (e.g., the percentage of residential fires that occur in apartments, or the percentage of apartment fire deaths that involve discarded cigarettes). Some analysts combine NFIRS-based percentages with NFPA-survey-based totals to produce estimates of numbers of fires, deaths, injuries, and dollar loss for subparts of the fire problem. This is the simplest approach now available to compensate in the area where NFIRS is weak. It has been documented as an analysis method in a 1989 article in *Fire Technology*.[2]

NFPA Fire Incident Data Organization (FIDO) System

The NFPA FIDO system is a computerized data base that provides the most detailed incident information available, short of a full-scale fire investigation. The fires covered are those deemed to be of high technical interest. The system that identifies fires for inclusion in FIDO is believed to provide virtually complete coverage of incidents reported to fire departments involving three or more civilian deaths, one or more fire fighter deaths, or large dollar loss (redefined periodically to reflect the effects of inflation, and defined since the late 1980s as $5,000,000 or more in direct property damage).

FIDO also captures a selection of smaller incidents as technical interests dictate. These are useful primarily because of the type of property involved (e.g., high-rise buildings), the presence of hazardous materials, or the performance of detectors or sprinklers.

The FIDO system covers these fires from 1971 to date, contains information on more than 75,000 fires, and adds more than 1,000 fires per year. NFPA learns of fires that may be candidates for FIDO through a newspaper-clipping service, insurers' reports, state fire marshals, NFIRS, respondents to the NFPA annual survey, and other sources. Once notified of a candidate fire, NFPA seeks standardized incident information from the responsible fire department and solicits copies of other reports prepared by concerned parties, such as the fire department's own incident report and results of any investigations.

The strength of FIDO is its depth of detail on individual incidents. Information captured by FIDO, but not by NFIRS, includes types and performance of all built-in systems for detection, suppression, and control of smoke and flame; detailed information on factors contributing to flame and smoke spread; estimates of time between major events in fire development (e.g., ignition to detection, detection to alarm); reasons for any unusual delay at various points; indirect loss and detailed breakdowns of direct loss; and escapes, rescues, and numbers of occupants. Additional uncoded information often is available in the hard-copy FIDO files, which are indexed for use in research and analysis.

One weakness of FIDO is that it mostly covers larger incidents. Many questions can best be answered by comparing the characteristics of large and small fires that involved similar types of properties and similar causes of ignition. FIDO does not permit such comparisons.

FIDO supports three annual *NFPA Journal* reports: "U.S. Fire Fighter Deaths," "Catastrophic Multiple-Death Fires in the United

States," and "Large-Loss Fires in the United States." FIDO also supports the anecdotal summaries published in the "Fire Watch" column of *NFPA Journal* and in the annual study cited earlier on U.S. fire fighter injuries.

Comparing Estimates Using Different Data Bases or Analytic Approaches

Occasionally a situation arises for which two different numbers exist, derived from different sources or different analytical approaches. For nonanalysts who may not know the source of either number, such a situation can prove frustrating and can encourage a cynical attitude about the arbitrariness of all fire statistics. Even analysts familiar with the sources of both numbers may have difficulty pinning down the precise reasons for any discrepancies and deciding how important the differences are and which number is best. It is very important, therefore, for all data users—and others who expect to confront arguments based on fire statistics—to understand how and why estimates can differ. For the most part, variations involve the ways different data sources and estimates deal with the inevitable gaps in coverage that affect all sources of fire incident information.

No fire data base can possibly capture all instances of unwanted fires. Few data bases—and none of the three discussed in this section—cover fires that are not reported to fire departments. (Special studies for some property types do provide one-time estimates of the size and composition of the unreported fire problem.) No fire data base even captures all the fires reported to fire departments, but some data bases, notably the NFPA survey, are built on samples designed to be representative of all fire departments.

By their nature, fire data bases are biased in favor of "failures" rather than "successes." The fire that is controlled so quickly it does not need to be reported to a fire department is not captured by the data bases that cover reported fires. Analyses of the impact of devices and procedures that provide early detection or suppression also need to allow for the phenomenon of missing "success" stories. Moreover, data bases like FIDO that provide the most detail on building features and their performance are limited to the largest of the reported fires.

There is also the issue of quality control for a data base. For data bases with limited depth of detail (like the NFPA survey) or limited breadth of coverage (like FIDO, which is confined to large fires), it is possible to invest considerable effort in ensuring that each report is as complete and accurate as possible. Follow-up calls can be used to fill gaps and check possible odd answers. For a data base with the depth and breadth of NFIRS, however, the same level of quality control effort has not been possible. Consequently, NFIRS is missing more entries and has more that are dubious. The trade-off between data quality and data quantity is never easy; an analyst needs to be aware of the strength and dependability of the sources before conducting an analysis.

Two examples indicate how differences in fire data bases and assumptions can produce different results. The U.S. Department of Justice estimates the size of the nation's arson problem through its Uniform Crime Reports (UCR) based on reports from law-enforcement agencies. NFPA, through its annual survey of fire departments, estimates the size of the nation's fire problem due to incendiary or suspicious causes. For 1983, the UCR estimate of the arson problem in structures was 297 fires per million population. For the same year, NFPA estimates were 524 incendiary or suspicious structure fires per million population and 311 structure fires per million population for incendiary fires alone. The UCR arson estimate and the NFPA incendiary fire estimate differ by about 5 percent, which is within the range of statistical uncertainty for an estimate based on a survey the size of NFPA's.

Several points are illustrated by this example. First, the UCR estimate is close to the NFPA incendiary-only estimate, because the UCR definition of arson approximates NFPA's definition of incendiary. NFPA and other fire organizations, however, traditionally regard the combination of incendiary and suspicious fires as the best indicator of the nation's problem with intentionally set fires. The UCR and NFPA systems produce roughly the same estimates where they are attempting to measure the same thing. However, because they differ on how to handle the more ambiguous fires, the "arson" numbers they release may appear significantly different to the casual reader or listener.

Second, it makes sense that the UCR estimate would be the lower of the two because fire officers see every reported fire, while police departments see only those fires that might involve crimes. A fire department may conclude that a fire is incendiary, while the corresponding police department either disagrees or is not notified; but the reverse is much less likely to occur. Also, definitions of arson vary from state to state.

Third, if there are differences in the ways that fire departments and police departments define structures, any such differences appear to have negligible impact on these calculations.

The second example of differing estimates involves the 1981 estimates made by FEMA and NFPA of total U.S. civilian fire fatalities. The NFPA estimate, based on the survey, was 6,700 deaths. The FEMA estimate, based on a multipart procedure that started with death certificate information reported to the National Center for Health Statistics, was 7,600 deaths. At least five significant differences in methods help to account for this discrepancy.

First, NFPA estimates of annual death totals are subject to some uncertainty because they are based upon sample surveys. Using accepted statistical principles, NFPA specified a range for the true value of 6,200 to 7,200. In other words, if someone else had run the same kind of survey and had produced an estimate that was higher or lower by up to 500 deaths, that would be within the range of precision achievable by a survey of that size. The difference would be the result of random variations in the representativeness of the fire departments selected for the sample.

Second, all fire fatality estimating procedures have difficulty with deaths that could have resulted from multiple causes. The principal examples of this problem are vehicle accidents followed by fire. It must be either determined or estimated whether traumatic injuries in the crash or subsequent fire-related injuries caused death. When FEMA began making estimates, one of its principal objections to NFPA's figures was that they tended to include all crash/fire fatalities. As a result, the two organizations pursued different courses to deal with this problem. NFPA began making follow-up inquiries on vehicle fire deaths, permitting the fatalities of suspect cause to be screened out prior to calculation of national projections. The initiation of this screening reduced estimated total fire deaths by nearly one-third from their previous level. FEMA chose to address this problem by use of a correction formula that assumed vehicle fire deaths would continue to be a constant percentage of structure fire deaths and/or total vehicle deaths. This approach was less satisfactory because it could not detect changes in the relative size of the problem. (FEMA no longer uses this formula.)

Third, the NFPA procedure cannot capture fatalities suffered in fires not reported to fire departments. These deaths principally involve clothing ignitions in which the fire is smothered by a civilian, but not before it has caused fatal injuries. The death certificate approach captures these deaths, which now number at most a couple hundred per year. NFPA analyses of clothing-ignition fire deaths suggest that most of these are now reported. It may be that the recent expansion of the fire service into emergency medical service has sharply reduced the number of fire deaths that go unreported. In any event, the principal strength of the death certificate approach is that it collects deaths in unreported fires, which no system based on fire department records can do.

The death certificate data base also does a better job of capturing delayed deaths, in which days or even months pass between fire injury and death. However, such deaths are often recorded in terms of the complications that caused death, so the fire connection may be missed to some degree in both data bases.

Fourth, an approach based on death certificates experienced difficulty in 1981 concerning arson deaths, then classified simply as murder with no indication of the role of fire. Subsequently, the death certificate coding was expanded to increase the number of factors that can be included, so an arson incident coded as murder now can also be coded as involving fire. As of 1981, however, arson deaths were still difficult to identify on death certificates and FEMA had to use other sources to make these estimates. (Suicides by fire could have posed a similar problem, but they are rare in the United States. In Japan, however, where immolation by fire accounts for more than a third of all fire deaths, the decision to include or exclude these victims makes a large difference in the totals.)

The fifth and final difference has to do with timeliness and the methods used to achieve it. As with NFIRS, the death certificate records are assembled on a state-by-state basis with wide variations in speed of delivery. In order to issue an estimated national total in a timely fashion, FEMA was forced to use sources other than death certificates for about one state in four, including some of the largest states. The need to work around late states was a major reason that FEMA abandoned this approach after 1981. Death certificate tracking still has advantages as an after-the-fact check on the accuracy of survey-based estimates, but by itself it cannot provide timely estimates.

SPECIAL DATA BASES ON FIRE

Many special data bases exist. Some concern aspects of the fire problem not well covered by the existing major data bases. Others contain information that may be useful in analysis. The examples cited below are not exhaustive.

Occupational

In addition to NFPA's data bases on fire fighter fatalities and injuries, the International Association of Fire Fighters (IAFF) annually releases totals of on-duty fatalities and injuries among career members of the fire service.

Also, fire injuries suffered on the job in some industries (e.g., the grain storage industry) have been among the occupational injury subjects studied by the National Institute for Occupational Safety and Health (NIOSH). Published reports on these studies are available.

Civilian Injuries

The U.S. Consumer Product Safety Commission (CPSC) maintains a computerized data base drawn from a sample of hospital emergency room cases. The National Electronic Injury Surveillance System (NEISS), dating back to 1972, focuses on product-related injuries in the home. Reports on fire casualties related to consumer products (about five percent of the NEISS total) are published annually. Information collected is similar to that collected for casualties under NFPA 901 but is much more detailed regarding the type of product involved. NEISS is particularly useful for analysis of serious injuries due to electrical shock or burns not caused by fire.

Burn injuries are captured in the National Health Inventory survey done by the U.S. Department of Health and Human Services

(HHS), but results on burns are rarely tabulated. HHS tracking of hospital discharges also can be used to gain a "snapshot" of more serious burns. The small fraction of severely burned patients who pass through burn care units are the subjects of more extensive tracking by the American Burn Association and others.

Transportation

Accidents, including fires, are tabulated by several government agencies. The National Transportation Safety Board (NTSB) publishes accident reports on aircraft and railway accidents and on highway accidents involving hazardous materials. The information collected tends to emphasize the circumstances of the accident with little discussion of the ensuing fire. Some of these reports, however, have unpublished supplements addressing issues (e.g., human factors in escape) and often contain much more fire-related information. The computerized portions of the records have the least information on fires. Human factors studies became routine in the mid-1960s; computerized records date from 1964.

The National Highway Traffic Safety Administration (NHTSA) established a computerized file on fatal highway accidents, beginning in 1975. Statistical summary reports are issued annually.

The U.S. Coast Guard collects reports on accidents involving recreational boats and commercial vessels. As with the other special data bases on vehicle accidents, there is little coded, standardized information on the cause and development of these fires.

International data bases also exist for accidents involving aircraft. Narrative summaries ranging from one sentence to a few paragraphs appear in the *World Airline Accident Summary*, published by the Air Registry Board in England. Standardized narrative reports, usually abstracted from original accident reports prepared by national air safety organizations, are published by the International Civil Aviation Organization, headquartered in Canada, in its *Airline Accident Digest* series.

Forests and Wildlands

The U.S. Forest Service issues annual reports titled *Wildfire Statistics* and *National Forest Fire Report*, and the U.S. Department of the Interior (DOI) Bureau of Land Management issues an annual report of *Public Lands Statistics*. The U.S. Forest Service reports cover fires occurring in national, state, and private forests, including numbers, estimates of the extent of damage (acres burned), and profiles of causes. The DOI reports address fires on the lands it owns and administers. Information addressed includes cause, extent of damage, rate of spread, and method of suppression. These data bases add records on thousands of fires each year.

Military

As with transportation and forest fires, fires on military installations tend to fall outside the jurisdiction of the local fire departments covered by the nation's major data bases (NFIRS and the NFPA survey). The various branches of the military historically have compiled their own fire incident data bases and have produced a number of analyses. Beginning in 1985, the Naval Safety Center at Norfolk, VA, has been the receiving point for fire reporting for all the U.S. military services. Incident records from the center also are submitted to NFIRS.

Electrical

The International Association of Electrical Inspectors (IAEI) and Underwriters Laboratories (UL) maintain data bases generated from clipping files, covering not only electrical fires but also the closely related subject of electrical shocks. The IAEI data is published annually in *IAEI News*. Both organizations discourage the use of this data for statistical analysis because clipping-file records are not representative of incidents of all sizes. By concentrating on large incidents, however, these data bases can be valuable for certain uses.

International

The World Health Organization's *Statistical Annual* includes information on fire death rates by country. It can be difficult, however, to obtain data for many countries for the same year. The World Fire Statistics Centre in England also prepares annual studies with international fire statistic comparisons, covering not only fire fatalities but also property damage and estimates of national expenditures on the various elements of fire protection (e.g., fire departments, insurance, built-in fire protection).

Special Studies

Some special studies have produced one-time data bases or statistics that provide continuing value for fire analysts. In 1985 the CPSC completed its survey of unreported home fires and their characteristics; the resulting data base will be of use for many years. CPSC also has conducted a number of special projects, including in-depth investigations of samples of home fires involving equipment such as electrical systems or alternative heating systems. Reports have been published on all these studies. A similar in-depth study of a sample of manufactured home fires was conducted in the late 1970s by the Department of Housing and Urban Development (HUD).

As of 1989, the NFPA fire investigation program, supported by FEMA and the National Bureau of Standards, had generated more than 300 in-depth incident reports. No one property use accounted for enough incidents to form a statistically significant data base. However, some issues, such as those involving patterns in major fires in buildings with large life exposures, might be able to draw on this data base for statistical significance. These incident reports contain unequaled depth of detail on fire development, smoke spread, human factors in escape, and the performance of fire protection systems and features.

BIBLIOGRAPHY

References Cited

1. NFPA 901, *Uniform Coding for Fire Protection*, 1976 ed. (used in most fire data bases) and 1991 ed.
2. Hall, J. R., Jr., and Harwood, B., "The National Estimates Approach to U.S. Fire Statistics," *Fire Technology*, Vol. 25, No. 2, May 1989, pp. 99–113.

Additional Reading

NFPA's "One-Stop Data Shop" program in the Fire Analysis and Research Division produces dozens of reports on fire experience each year, most of them distributed directly through the division and not published in books or periodicals. A listing of reports that may be ordered is available free. The other major source of analyses is the U.S. Fire Administration's *Fire in the United States* series. For additional, detailed material on statistical methods, see *The SFPE Handbook of Fire Protection Engineering*, 2nd ed., 1995.

INTRODUCTION TO FIRE MODELING

Craig Beyler and Philip J. DiNenno

Modeling is an inherent part of research in science and engineering, and its application to fire is as old as scientific research into fire behavior itself. The first book devoted solely to modeling in fire research was published in 1960.[1] Modern fire research, coupled with the ever-increasing power of computers, has given rise to highly complex models that can only be implemented as computer fire models. While computer fire models have popularized fire modeling, these computer-based models do not differ substantially from earlier fire models; they just are more complex and possess greater capabilities.

Models can be classified into two broad classes: (1) physical models and (2) mathematical models. Physical models attempt to reproduce fire phenomena in a simplified physical situation. Scale models are a very widespread form of modeling, as full-scale experiments are expensive, difficult, and sometimes wholly infeasible. Insights can often be gained by studying fire phenomena at reduced physical scale. Mathematical models are sets of equations that describe the behavior of a physical system. Very often, the goal of physical models is to uncover laws governing the behavior of physical/chemical systems. The resulting mathematical model can then be used to predict the behavior of real physical systems. Thus, physical and mathematical models are interrelated and complementary.

PHYSICAL FIRE MODELS

Physical modeling does not simply mean conducting experiments at reduced physical scale. That is, reducing the linear dimensions of a physical situation and conducting experiments with the reduced-scale model is not sufficient. In addition to geometric scaling, it is necessary to maintain mechanical, thermal, and chemical similarity in the reduced-scale model. The scaling laws required to maintain these similarities can be determined from dimensional analysis or from the fundamental equations describing the physical/chemical phenomena.

The most widespread physical scaling laws in fire are known as "Froude modeling," which is applicable to buoyant flows associated with fires. Froude modeling requires that the ratio of buoyant forces to inertia forces be maintained. The Froude number, Fr, can be expressed as

$$\text{Fr} \sim Q^{2/5}/D \sim V\sqrt{D}$$

where Q is the fire heat release rate, D is the physical scale of the experiment, and V is a characteristic velocity. Thus, if a half-scale experiment is to be performed, the heat release must be reduced to 18 percent of the full-scale heat release to maintain the same Fr number and thus ensure that the flow velocities scale in geometrically similar locations in the half-scale and full-scale models. The velocities themselves scale as the square root of the scale at constant Fr:

$$V \sim \sqrt{D}$$

Froude modeling has been used successfully to understand plume flows, ceiling jet flows, and flame heights. Flame-height correlation models based on Froude modeling have been successful over a wide range of physical scales. An example is shown in Figure 11-4A, which was developed by McCaffrey.[2] The flame height to fire source diameter, L/D, is shown as a function of the $\sqrt{\text{Fr}}$. The correlation is successful over twelve decades of Fr, an amazingly wide range of applicability for any correlation or model.

Because different fire phenomena scale differently, it is not generally possible to study complex fire situations in small scale. This limits the use of scale modeling principles. It is notably difficult to scale convective flows and radiation at the same time. Thus, Froude modeling cannot be applied readily to fire problems where radiation is important, for instance.

Scale modeling using Froude scaling laws has been most successful in addressing smoke movement issues. Two techniques have found important applications: (1) reduced-scale gas-phase modeling and (2) saltwater modeling. The principles of these modeling techniques have been reviewed by Quintiere.[3] Saltwater modeling uses the discharge of high-density salt water into fresh water as an inverse simulation of buoyant flows. This technique has been used successfully in both compartments [4] and in multiple-compartment.[5]

Reduced-scale modeling has been used in a wide range of flow situations.[6] An excellent example of the use of reduced-scale modeling is given by Quintiere and Dillon,[7] and describes smoke movement in an atrium fire. The use of scale modeling in the design of smoke management systems is embraced by NFPA 92B, *Guide for Smoke Management Systems in Malls, Atria, and Large Areas.*

Craig Beyler, Ph.D., is technical director of Hughes Associates, Inc., editor of the *Journal of Fire Protection Engineering,* section editor of *The SFPE Handbook of Fire Protection Engineering,* and adjunct professor of Fire Protection Engineering at Worcester Polytechnic Institute. Philip J. DiNenno, P.E., is vice president, Hughes Associates, Inc., and editor-in-chief of *The SFPE Handbook of Fire Protection Engineering.*

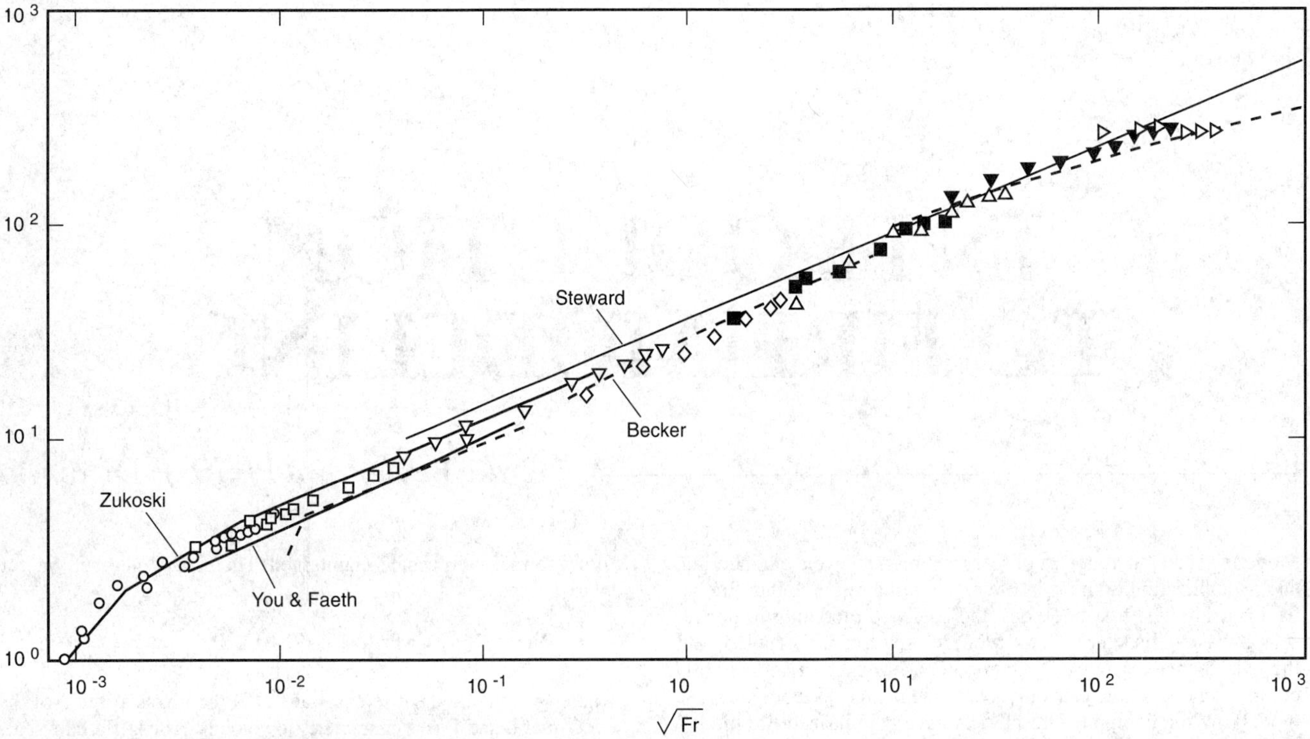

FIG. 11-4A. *Flame-height correlation model based on Froude modeling.*

Deterministic Fire Models

Deterministic fire models can range from simple one-line correlations of data to highly complex models requiring hours of computing time using mainframe computers. The unifying aspect of these models is that the course of a fire is fixed by the variables that estab-

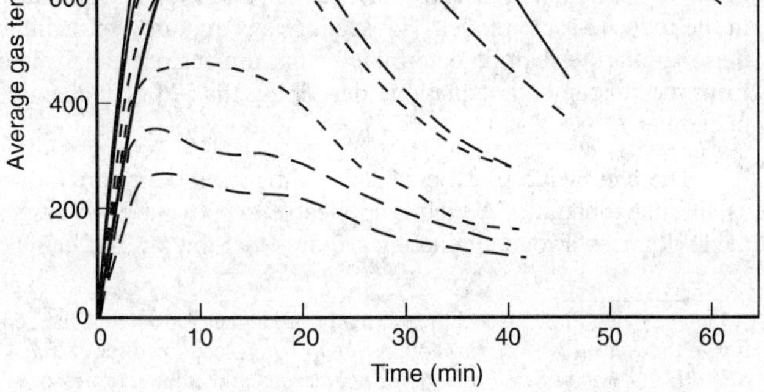

Physical modeling does not always involve major reductions in physical scale. Physical models may simply seek to simplify complex phenomena into a manageable and understandable problem. Examples of this type of physical modeling can be found as far back as the ASTM E119 fire-resistance test. The ASTM E119 time-temperature curve shown in Figure 11-4B was selected as a representative time-temperature curve for testing structural assemblies. The E119 fire-resistance test is a form of physical modeling. The furnace environment is intended to model the fire environment experienced by structural members in a fire. The portions of the building structure exposed in an E119 furnace are intended to model the fire behavior of the building structure in a fire. No single time-temperature exposure can represent all fires, and testing portions of building structures does not fully represent the behavior of the structure, but the results of the simplified E119 test are useful nonetheless. Over the decades the value and the limitations of this widespread physical model have been learned.

All standard fire tests are physical models of fire behavior. The adequacy of these physical models varies widely. There is a trend in the development of modern fire test methods to consider explicitly the adequacy of the test method as a physical model of actual fire behavior and response to fire.

MATHEMATICAL FIRE MODELS

Mathematical fire models can be classified as either probabilistic or deterministic models. Probabilistic models attempt to deal with the random nature of fire behavior, while deterministic models presume that, given a well-defined physical situation, the fire growth and behavior are entirely determined. Both approaches are valuable in understanding fire.

FIG. 11-4B. *Representative time-temperature curve. (For U.S. customary units: °F = °C × 9/5 + 32.)*

lish the environment in which it occurs. The physical conditions that determine the progress and outcome of the fire are called the fire scenario. It includes the fuels involved, the arrangement of the fuels, the characteristics of the building and its fire protection systems, the location of the ignition, the location and capabilities of occupants, and any other variables that affect the fire's outcome. Thus, for all deterministic fire models the formulation of the fire scenario is of critical importance.

An example of a very simple mathematical model is an equation describing the relationship shown in Figure 11-4A:

$$L/D = 0.2\left(Q^{2/5}/D\right)$$

where length units are in meters and heat release rate is in kW. (For U.S. custom units, 1 m = 39.37 in; 1 kW = 1.655 Btu/sec.) This equation is the result of applying Froude modeling principles. Other equations are available that are applicable at even lower values of L/D.

At the other end of the scale of complexity are models that describe the behavior of fire in one or more rooms. These models may be complex because they include many physical/chemical processes or because a few processes are modeled in great detail. These complex models can be broadly classified as zone models and field models.

Perhaps the single most important attribute of computer fire models is their ability to predict accurately and realistically the relevant fire behavior within their stated limitations. The predictive capability of a model depends on both the underlying scientific understanding of the processes being modeled and the translation of that understanding into some calculation scheme. There remain substantial shortcomings in the scientific understanding of fire and related processes. This knowledge shortfall, however, will not necessarily slow the development of calculations with enhanced predictability. It is not necessary to understand a phenomenon fully in the pure scientific context in order to exploit a lower level of understanding for design and practice purposes. In fact, current models take great advantage of "imperfect" knowledge to yield acceptable results. More general discussions of the state of related fire science can be found in the research of Emmons, Pagni, and Friedman.[8–10]

Zone Fire Models

The fire environment in a room is quite complex. Major insights into fire behavior have been achieved by a simple conceptual construct called zone modeling. In essence, a zone model assumes that the compartment may be idealized as consisting of two regions: (1) an upper region, filled with hot combustion gases, and (2) a lower region, filled with essentially cool air. Each region or zone is idealized to have uniform temperatures and gas concentrations. The plane dividing the two zones is the hot layer interface that may move vertically during a fire. Figure 11-4C shows a comparison of the temperature profile in a room fire with the idealized zone model equivalent. In one case, the zone model concept matches reality quite well; in the other, differences are clearly seen. Gas concentrations are idealized in the same way.

The zone model concept simplifies the room fire thermal environment to two temperatures and an interface height, rather than a three-dimensional temperature field. Major simplifications are realized both mathematically and computationally. These simplifications have made many fire problems tractable and have allowed significant progress to be made.

Of course, additional components are needed to specify fully the fire environment. Vent flows, plume entrainment, heat transfer, and combustion models are needed. Even if each of the component sub-

models is fairly simple, the result when coupled together is a complex model requiring computers for efficient, practical calculations.

Zone models by definition will always be approximate. The key is whether the predictions are "close enough" to yield significant insight for the situation under study. Zone modeling yields useful insight into many fire problems. The real question is under what conditions these models yield acceptably accurate results. This question has been largely unanswered. There is a reasonable amount of validation test data available for comparison to models with single or with up to six compartments with length scales of 10 ft (3.05 m) connected via a corridor and a modest (less than 1000 kW) 1,055 Btu/sec, well-characterized fire.[11–19] The extent to which zone models can be effectively applied in large open areas or tall structures is uncertain. The two-zone paradigm does not preclude their use in large or tall structures per se, but rather stretches the assumption of uniform properties within a zone. An excellent summary of the underlying assumptions of zone modeling is given by Quintiere.[20] Further discussions of the status and limitations of zone modeling are given by Emmons, and Kawagoe.[21,22] While the number of zone fire models is very large, Friedman has reviewed the most available model,[23] and Beard has reviewed a few of the most widely used models.[24]

The status of zone modeling as a predictive tool can be evaluated with acceptable confidence as follows. For a relatively small room [approximately 100 sq ft (9.3 m²)] with a well-described, well-ventilated fire source, the temperature, layer position, and gas concentrations can be estimated with acceptable accuracy. Within the limitations of zone modeling, smoke filling a corridor and then adjacent rooms can also be adequately described. This case does not imply that, for conditions outside this narrow description, modeling is inaccurate or unacceptable but rather that the results must be carefully evaluated in the context of what level of prediction is required as well as independent experimental evidence.

The following discussions attempt to briefly describe the current status and near-term possibilities for modeling some important phenomena. No one model is assumed; rather, the discussion summarizes the integrated capability of the HARVARD/FIRST and C-FAST zone models.[25,26]

1. *Ignition.*   Both piloted and unpiloted ignition due to radiative heating can be approximated and if the ignition temperature is known, alternative small-scale ignition test data can be used (ASTM E1354).

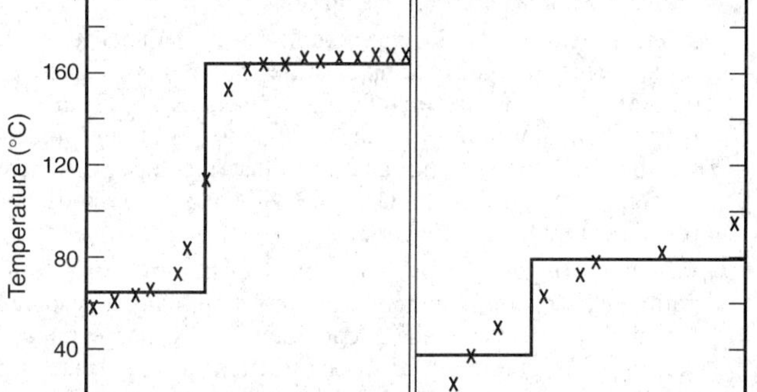

FIG. 11-4C.   *Comparison of the temperature profile in a room fire with the idealized zone model equivalent. (For U.S. customary units:* °F = °C × 9/5 + 32; 1 ft = 3.3 m.)

2. *Burning behavior.* The burning behavior of a single fuel item cannot currently be predicted. The treatment of this obviously important subphenomenon in models is done through empirical correlations and/or large-scale test data. Approximations using small-scale tests are possible.[27] A more fundamental treatment is unlikely in the near term. Empirical methods, small-scale and large-scale data, and "design basis" fire assumptions will continue to form the basis of zone models.

3. *Flame spread.* This area is currently under active development. In conjunction with LIFT apparatus, opposed-flow flame spread is a candidate for near-term modeling. Upward- or concurrent-flow flame spread predictions using a similar LIFT apparatus are quite promising and could be integrated into models readily in the near term.[28,29,30]

4. *Compartment effects:*

   a. *Energy balance.* Zone-type approximations may be improved with the inclusion of plumes or ceiling jets as subzones. Heating of a ceiling in the vicinity of a flame or plume can be calculated. Enhanced radiation levels can be estimated but it is not possible to predict the general effect on burning behavior.

   b. *Walls and corners.* Current methods used to approximate the impact of walls and corners on plumes, flame heights, and ceiling jets are suspect. Some modest development work is needed before these submodels can be appreciably improved.

   c. *Oxygen starvation.* Currently treated by use of heuristics in conjunction with limited-oxygen index for energy release rate. Smoke and gas yields in underventilated conditions are more challenging but can be grossly approximated.[31]

5. *Multiple room fire and smoke spread.* There are several models that treat smoke spread through multiple compartments using the standard two-zone smoke "filling" approximation. Depending on the problem under analysis, this approach may yield acceptable results. Limited empirical studies on smoke flow in corridors could be integrated near term with uncertain results. Flows in shafts, stairs, and through horizontal vents are more problematic. Some experimental work is required.[32,33] It is expected that useful approximations will be available in the near term.

   Fire spread is currently limited to remote ignition of remote objects by heating from flames and the hot layer. Fire spread to multiple compartments is in principle predictable. Unknown effects of multiple compartments burning along a single corridor in close proximity may be important. Fire spread by ignition of a hot layer is currently unresolved.

6. *Forced ventilation.* Several models have attempted to treat forced ventilation under simple vent configurations.[34] There is current active development of a more general model. There appear to be major unresolved technical issues in the full understanding of forced vents and fires that precludes effective modeling for many cases. This area will undergo dramatic improvement in the next decade.

7. *Post-flashover predictions.* The onset of flashover in a compartment poses a difficult prediction problem. Major unresolved technical problems include: prediction of a radiation-enhanced burning rate; prediction of CO production in underventilated conditions; ignition of hot layers; burning at vents; and flame spread down corridors.[35]

   There is one area where post-flashover fires are very well characterized. The prediction of temperature as a function of time for purposes of calculating the fire resistance of a structure is very well developed. Design approaches using post-flashover models are accepted alternatives to prescriptive fire resistance requirements in several countries.

8. *Detection system activation.* Procedures exist for estimating the response of ceiling-mounted heat detectors via heating by a plume or ceiling jet. These calculations have been relatively well established for flat, smooth ceilings, with more limited validation for compartments with high ceiling heights or non-smooth, nonflat ceilings.

   Methods also exist for estimating smoke-detector response. These have not been generally used due to the unavailability of the necessary detector-specific parameters. Calculation methods exist for detector activation where a layer has formed, although the procedures have not been adequately verified. No procedures currently exist for estimating the activation of detectors in situations where the detector responds before a well-developed plume or ceiling jet forms or where mechanical ventilation appreciably perturbs the flow. Incremental progress is expected in this area in the near and moderate term.

9. *Suppression modeling.* Fire suppression with a total-flooding inert gas system has been successfully modeled. While there is research effort in modeling the suppression of fires by water application,[36] no generally applied design or analysis procedures are expected in the next decade.

10. *Fire effects modeling:*

    a. *Heat damage.* If the temperature threshold of a piece of equipment or material is known, standard heat transfer calculations can be used in conjunction with fire models to estimate the time dependence or degree of damage.

    b. *Smoke damage.* Estimates of smoke damage that reflect the sensitivity of the equipment exposed and type of smoke or fire involved are not generally available. Such calculations involve uncertainties associated with cold smoke transport and deposition, prediction of smoke production and properties, the wide variability of smoke properties, and the sensitivity of equipment to this wide variation of smoke types.

11. *Human interaction:*

    a. *Visibility.* Limited experimental data on visibility combined with small-scale constant-yield smoke concentration and optical property measurements enable rough estimates of human visibility in smoke. Using the results of computer models for smoke, layer depth, and smoke properties, approximate visibility estimates can be made.

    b. *Temperature/radiation.* The effects of heat and radiation can be estimated using empirical human tenability data.

    c. *Toxicity.* Toxic effects of CO, CO_2, HCN, and oxygen depletion can be reasonably predicted using N-gas modeling approaches. The effects of gases such as HCl, HF, or HBr can be estimated, but with a lower level of confidence. Bioassay methods can be used to screen for so-called papertoxicant fire products, where their existence is suspected.

The legitimate application of models to problems outside of the relatively narrow ranges discussed above is possible and occurs frequently. If the results of the calculations do not require great accuracy (e.g., ± 50 percent), the important physics of the problem is reflected in the model, and/or the results are used for approximate comparative purposes, these tools can be approximately used in a much broader range of problems.

Field Models

Field models avoid the simplifications inherent in zone models. The temperature, velocity, and gas concentrations are calculated as three-dimensional fields. The compartment is descretized into thousands of computational cells, and the temperatures, velocities, and concentrations are found for each of the thousands of cells throughout the room. The model is a complex fluid mechanical model of turbulent flow. While empirical modeling is required even in these very complex models, the empiricisms are made at a more fundamental level.

While simple field model calculations can be made on the fastest PCs available, larger computers are desirable from a practical standpoint and are often essential.

To date, field models have not progressed to the point where they can include the wide range of physical/chemical processes phenomena included in zone models. Far more detailed knowledge is required for field models, but it is only a matter of time until field models can accommodate a wide range of phenomena.[38,39] Despite the limited scope of field models, they are making important contributions to fire safety today.

The application of field modeling to fire problems has been dramatically increasing. The ready availability of commercial CFD (computational fluid dynamics) packages with increasing sophistication enables more widespread applications. The vast majority of uses of these programs have nothing to do with fire problems but are concerned with fluid flow in pipes, convective flow in enclosures, cooling of electronic packages, etc.[40] In general, there have not been critical evaluations of these methods relative to their use in fire problems.

Since the basic mass-energy and momentum-conservation equations are identical for all problems, these codes have potentially dramatic revolutionary impact on fire modeling. The difficulties in field modeling of fires lie in two areas. First, no direct simulation of turbulent diffusion flames—that is, a fire—is currently possible; hence, the fire source itself must be approximated. Other major phenomena that can only be approximated include turbulence, particularly large eddies associated with strong plumes and flames, and thermal radiation interchange between soot, gases, and solid surfaces. Second, the capability of these codes to handle specific phenomena important in fire modeling varies widely.

The programs require substantial computational effort although most are available in PC versions. More importantly, they require sophisticated users due to the complexity of defining initial and boundary conditions. Field modeling programs have direct and immediate potential application problems driven largely by fluid mechanics and convective heat transfer. Such problems include far-field smoke flow, details of flow and heat transfer to complex geometries (e.g., sprinkler links), and impact of fixed ventilation flows on bouyancy-driven flows. Applications of field models to fire problems have included aircraft terminals and atria, air-supported structures, electronic generating stations, aircraft cabins, tunnels, hospital wards, and shopping malls.[41–49]

Most CFD models use empirical models of turbulence. An alternate approach, known as large eddy models, is being developed and applied and uses no turbulence models.[50] Large-scale turbulence is directly modeled, and very small-scale turbulence is left unmodeled. While computationally intensive, the method has shown great promise.

While validation of models as detailed as field models is an arduous task, papers have begun to appear in the literature in which experimental data and field model predictions are computed.[51–53]

Efforts are ongoing to include the interaction of sprinkler sprays and fire-induced flow fields.[54–56] These models focus solely on the fluid mechanics aspects of the interaction and do not deal with the effect of water on combustion.

Probabilistic Models

Probabilistic fire models can be categorized into three classes: (1) network, (2) statistical, and (3) simulation. Each of these deals with the uncertainties associated with fire growth processes.

Network models are fire growth models in which the transition from one fire stage to another and the effectiveness of fire suppression systems, manual fire fighting, passive fire protection, and so on are governed by user-assigned probabilities that are based on historical data, or engineering evaluations, or both. In some cases, these probabilities are single values, and in other models, the probabilities are time dependent.

Statistical models represent the probability of an occurrence as it is determined from historical data. A classic example of a statistical model is the occurrence of fire alarms. Fire alarms are random events that are, within certain constraints, uniform in nature. That is, a fire or fire alarm might occur at any time with equal probability. Clearly, one of the constraints on this behavior is that the time of day must be controlled. It is well known that fires are a function of the time of day, so one must restrict attention to a single time of day. The probability of k fire alarms occurring over the period s is given by

$$P(k) = \frac{(\lambda s)^k}{k!} e^{-\lambda s}$$

where λ is the mean frequency of fire alarms. This statistical model describes the probability of a given number of alarms, k, occurring during a specified period of time, s. Such a statistical model can be very useful in fire protection resource planning.

Simulation methods try out different sets of conditions to see how they affect the outcome. Probabilities are often used to weigh outcomes for each set of conditions and produce and overall expected outcome. Simulation models may predict outcomes for a given set of conditions using other physical, probabilistic, or deterministic models. In the later case, simulation models regard fires as deterministic once the fire is fully defined. However, the inputs to the models are assumed to follow probabilistic models. Thus, the inputs to the deterministic model are treated as random variables. Such variables might include which furniture item is ignited or how far open the door is when the fire starts. A large series of runs is performed by selecting inputs from the probabilistic models, and the range and frequency of different outcomes are examined. Such modeling combines elements of both probabilistic and deterministic models.[57]

Probabilistic models are taking on increased importance in the fire community in the assessment of risk. Various risk-based methods are under development in Australia, Canada, and the United States. These have been applied to building classes, individual buildings, and ships.

TRENDS

Having briefly discussed the current and near-term capabilities of models to predict fire effects, we will now focus on observable trends and the possibilities for the future of modeling.

Predictive Capability

The predictive capability of computer models is largely uncertain. It is not known with confidence to what limits these models can be applied. It is equally clear that ongoing research and development efforts aimed at resolving some of these uncertainties will result in improved predictive capability. The use of heuristics to aid in simplifying particularly difficult problems will continue. Confidence in the capabilities of models will increase with improved validation and careful use.

Applications

Applications of modeling have included fire litigation and reconstruction, hazard analysis of building materials and contents, and the evaluation of building code exceptions or equivalencies. Computer models and other analytical tools have not, however, been integrated into fire protection engineering design practice.

Several case studies of "reconstructed" fire incidents using computer models have been published.[58–62] They generally demonstrate how a computer model used as part of a more general engineering analysis is a powerful tool. One lesson of these case studies is that the state of the art is not such that one sets up a reasonable input file and the computer describes how the fire evolved. Rather, the models are more appropriately used to evaluate in approximate terms the time sequence of the event or to establish in more detail the reasonableness of one or another ignition growth and spread scenarios. These programs are also used to pose "what if" questions related to the contribution of a particular material, furnishing item, etc. *vs.* one of greater or lesser combustibility, energy release rate, and so on. Most modeling applications occur today, as well as in the recent past, in fire litigation circumstances. Most of these applications are never published.

The use of modeling litigation and fire investigation will expand. The problem of fire reconstruction lends itself readily to modeling tools.

Applications outside of litigation are often associated with the evaluation of the fire hazard posed by a product or material.[30,63–67] Modeling received a major boost in the context of predicting the "toxic hazard" posed by materials. The combination of deterministic models with risk assessment methods into integrated hazard assessment schemes has begun.[68,69] The method enables one to integrate ignition, flammability, and toxicity properties with the end use of the material under varying scenarios to estimate the outcome. As always, the quality of these applications is driven by the methods, data, approximations, and assumptions used. Additional applications that reflect the increasing use of models for a range of applications can be found in work by Bukowski, Bengston and Hagglund, and Hagglund and Wickstrom.

While current applications are somewhat limited in terms of "routine" use, the trend has been for a relatively rapid increase in the application of models across all aspects of fire protection. Clearly, the use of models in fire investigations and litigation will continue, and their application to special facilities and problems not directly dealt with in codes and standards will also increase. Applications of field modeling to unique facilities and problems should dramatically increase.

Computer models are relatively widely used in the fire safety design of buildings and facilities. These applications often arise in the context of "code equivalency," or the so-called "performance-based" approach to fire safety. There are relatively few fire safety problems for which direct comparison of equivalency, based on the prediction of fire growth or smoke spread alone, is possible. There are almost always probabilistic, risk, economic, and political factors involved. Computer models are still helpful, however, given the almost nascent state of development, in that they provide the ability to approximately compare expected fire conditions across a range of scenarios. The potential impact of various intervention strategies can also be evaluated in a relatively crude fashion.

A major question exists on the application of these models in fire safety design relative to building codes and fire safety standards. The primary determinants in this area will likely be the degree to which relatively prescriptive codes and standards can be harmonized with more performance-based design and engineering methods. This is, of course, more complicated than it sounds, given the legal, social and political realities under which buildings are approved. An additional prerequisite for the general use of modeling in building design is the need to deal with the inherently probabilistic nature of fire events and the "reliability" of various construction features and fire safety systems. Several generic approaches have been suggested to integrate deterministic modeling with probabilistic problems.[72,73] This challenge also poses the greatest opportunity. The tremendous potential of this significant technological advance for the purpose of improved fire safety will be recognized and exploited.

Hardware

It has become relatively safe to forecast that more powerful, easy-to-use computer hardware will continue to emerge. The past decade has seen a revolution in computing capacity available to individuals. This has eliminated most computational speed problems with respect to zone models. The computational requirements of field models will continue to be a challenge.

Developments in hardware should also appreciably extend the range and utility of these calculation methods, perhaps into the "real time" world of the fire incident commander. The use of these models in conjunction with interactive video training and simulation will likely result in "fire simulators" becoming widely available to fire departments.[74]

Longer-term hardware developments in the area of intelligent sensors integrated with software derived from fire modeling may revolutionize fire detection. The "smart" building will include fire protection.

Acceptance

The trend toward increased acceptance of predicted results in resolving fire safety questions will continue and likely accelerate. A decade of continued progress in modeling, with the increase in the number of published application studies in *NFPA Journal* and other widely read fire literature, has led to a more open approach to the subject by building and fire officials. Progress related to integration of these methods into the design of facilities is more problematic.

Validation

Numerous validation studies of computer fire models have been undertaken. Most have been in relatively small [approximately 100 sq ft (9.3 m²)] compartments with well-defined fire sources (cribs, pool fires, or gas burners). In general the best agreement that can be expected is on the order of ± 20 percent in terms of smoke layer temperature and depth.

It is unlikely given funding and prioritization realities that any significant systematic and complete validation will ever be performed on any model. It is arguable that complete validation is an impractical objective. Given this, models will be evaluated and the results accepted in a relatively slowly evolving time frame. The obligation that this lack of independent validation places on users and approving authorities is obvious. Particular attention must be paid to using these programs within their stated limits and, perhaps more importantly, to searching out experimental data for cases similar to the problem being evaluated. The dearth of validation studies should not be taken as rationale for rejecting the use or results of a model in any particular case. The validation picture could improve dramatically if it became clear that the lack thereof was a significant barrier to more widespread use.

Material Flammability Testing

The development of small-scale fire test methods, which in principle enables the extrapolation of the results to "real" fires, was essential to the development of modeling. Some of these test methods have been standardized by ASTM and NFPA. The input data required by models, which describe the burning behavior, energy release rate, smoke properties, and combustion gas yields of materials, come largely from these small-scale test methods. It is expected that the availability of these data for a wider range of products and materials will increase. The availability will accelerate dramatically if these relevant test methods are referenced in buildings codes and fire standards in lieu of or in conjunction with more traditional methods.

Since models form the basis for extrapolating this flammability data to full-scale performance, increased use of models in this context is expected. In many ways, this particular class of applications may be ideal for the direct integration of modeling into building design standards.

CONCLUSIONS

Fire modeling is an old and well-developed field, though much remains to be done. Physical models and mathematical models have made and are making important contributions to fire safety and fire protection.

These fire models are an important and growing part of fire protection engineering. Computer fire models are simply the most recent and complex engineering models of fire. Their use requires a knowledge of the underlying fire science and fire protection engineering principles.

It is no longer a question of whether modeling will be used in the solution of fire safety problems but rather to what extent. The degree to which scientifically based engineering models can impact fire safety will be driven largely by their application to design practice. This requires some integration with building codes and fire safety standards. Without such integration, incremental progress in predictive capability and applications can be expected. If fire safety engineering methods and codes are successfully integrated, the forecast will be on of rapid, revolutionary change in all aspects of modeling as well as building design.

BIBLIOGRAPHY

References Cited

1. National Academy of Sciences, *The Use of Models in Fire Research,* Conference Proceeding of November 8–10, Washington, DC, 1959.
2. McCaffrey, B., "Some Measurements of the Radiative Power Output of Diffusion Flames," Paper WSS/CI 81-15, Western States Section Meeting of the Combustion Institute, 1981.
3. Quintiere, J. G., "Scaling Applications in Fire Research," *Fire Safety Journal,* 15, 1989.
4. Fleischmann, C. M., Pagni, P. J., and Williamson, R. B., "Salt Water Modeling of Fire Compartment Gravity Currents," Fire Safety Science—*Proceedings* of the Fourth International Symposium, 1994, pp. 253–264.
5. Steckler, K. D., Baum, H. R., Quintiere, J. G., "Salt Water Modeling of Fire-Induced Flows in Multicompartmental Enclosures," 21st *Symposium (Int'l) on Combustion,* The Combustion Institute, Pittsburgh, PA, 1986, pp. 143–149.
6. Jolly, S., and Saito, K., "Scale Modeling of Fires with Emphasis on Room Flashover Phenomenon," *Fire Safety Journal,* 18(2), 1992.
7. Quintiere, J. G., and Dillon, M. E., "Scale Model Reconstruction of Fire in an Atrium," *Proceedings,* International Conference on Fire Research and Engineering, Sept. 10–15, 1995, Society of Fire Protection Engineers, Boston, MA.

8. Emmons, H. W., "The Needed Fire Science," Fire Safety Science—*Proceedings* of the First International Symposium, Hemisphere Publishing Corp., Washington DC, 1986, pp. 33–53.
9. Pagni, P. J., "Fire Physics—Promises, Problems and Progress," Fire Safety Science—*Proceedings* of the Second International Symposium, Hemisphere Publishing Corp., New York, 1989.
10. Friedman, R., "Fire Protection Engineering—Science of Art?" *Journal of Fire Protection Engineering,* Vol. 2, No. 1, 1990, pp. 25–29.
11. Duong, D., "The Accuracy of Computer Fire Models: Some Comparisons with Experimental Data from Australia," *Fire Safety Journal,* Vol. 16, 1990, pp. 415–431.
12. Jones, W. W., and Peacock, R. D., "Refinement and Experimental Verification of a Model for Fire Growth and Smoke Transport," Fire Safety Science—*Proceedings* of the Second International Symposium, Hemisphere Publishing Corp., New York, 1989, pp. 897–906.
13. Peacock, R. D., Davis, S., and Lee, B. T., "An Experimental Set for Verification of Room Fire Models," NBS 88-3752, Apr. 1988, National Bureau of Standards, Gaithersburg, MD.
14. Rockett, J. A., "Modeling of NBS Mattress Tests with Harvard Mark V Fire Simulations," *Fire and Materials,* Vol. 6, No. 2, 1982, pp. 80–95.
15. Rockett, J. A., *et al.,* "Comparisons of NBS [National Bureau of Standards]/Harvard III Simulations and Data from All Runs of a Full-Scale Multiroom Fire Test Program," *Fire Safety Journal,* Vol. 15, 1989, pp. 115–116.
16. Rockett, J. A., Marta, M., and Cooper, L. Y., "Comparisons of NBS/Harvard VI Simulations and Full-Scale Multiroom Fire Test Data," NBSIR 87-3567, 1987, National Bureau of Standards, Gaithersburg, MD.
17. Nakamura, K., and Tanaka, T., "Predicting Capability of a Multiroom Fire Model," Fire Safety Science—*Proceedings* of the Second International Symposium, Hemisphere Publishing Corp., New York, 1989.
18. Nelson, H. E., and Deal, S., "Comparing Compartment Fires with Compartment Fire Models," Fire Safety Science—*Proceedings* of the Third International Symposium, 1991, pp. 719–728.
19. Peacock, R. D., Jones, W. W., and Bukowski, R. W., "Verification of a Model of Fire and Smoke Transport," *Fire Safety Journal,* 21(2), 1993.
20. Quintiere, J. Q., "Fundamentals of Enclosure Fire Zone Models," *Journal of Fire Protection Engineering,* Vol. 1, No. 3, 1989, pp. 99–119.
21. Emmons, H. W., "Experiments with a Fire Math Model," Fire Safety Science—*Proceedings* of the Second International Symposium, Hemisphere Publishing Corp., New York, 1989.
22. Kawagoe, K., "Real Fire and Fire Modeling," Fire Safety Science—*Proceedings* of the Second International Symposium, Hemisphere Publishing Corp., New York, 1989.
23. Friedman, R., "An International Survey of Computer Models for Fire and Smoke," *Journal of Fire Protection Engineering,* 4(3), 1992.
24. Beard, A., "Evaluation of Deterministic Fire Models: Part I—Introduction," *Fire Safety Journal,* 19(4), 1992.
25. Mitler, H. E., and Rockett, J. A., *User's Guide to FIRST, Comprehensive Single-Room Fire Model,* NBSIR 87-3595, 1987, National Bureau of Standards, Gaithersburg, MD.
26. Jones, W. W., and Peacock, R. D., *Technical Reference Guide for C-FAST Version 18,* NIST Technical Note 1262, 1989, National Institute of Standards and Technology, Gaithersburg, MD.
27. Babrauskas, V., and Wickstrom, U., "The Rational Development of Bench-Scale Fire Tests for Full-Scale Fire Protection," Fire Safety Science—*Proceedings* of the Second International Symposium, Hemisphere Publishing Corp., New York, 1989.
28. Harkleroad, M., Quintiere, J., and Walton, W., "Radiative Ignition and Opposed-Flow Flame Spread Measurements on Materials," DOT/FAA/CT-83/28 Aug. 1983, U.S. Department of Transportation, Federal Aviation Administration, Technical Center, Atlantic City Airport, NJ.
29. Quintiere, J. Q., "Status of Fire Research and Safety," Fire Safety Science—*Proceedings* of the Second International Symposium, Hemisphere Publishing Corp., New York, 1989.
30. DiNenno, P. J., and Beyler, C. L., "Fire Hazard Assessment of Composite Materials, the Use and Limitation of Current Hazard Analysis Methodology," *Proceedings* of the ASTM Symposium on Fire Hazards and Fire Risk Assessment, Dec. 3, 1990, San Antonio, TX, 1990.
31. Babrauskas, V., *et al.,* "The Role of Bench-Scale Test Data in Assessing Real-Scale Fire Toxicity," NIST Technical Note 1284, Jan. 1991, National Institute of Standards and Technology, Gaithersburg, MD.
32. Emmons, H. W., "The Ceiling Jet in Fires," NIST-GCR-90-582, Oct. 1990, National Institute of Standards and Technology, Gaithersburg, MD.

33. Steckler, K., "Fire-Induced Flows in Corridors—A Review of Efforts to Model Key Features," NISTIR 89-4050, Feb. 1989, National Institute of Standards and Technology, Gaithersburg, MD.

34. Steele, S., and Rockett, J. A., "Application of the Harvard Multiroom Fire Simulation Where Forced Ventilation Is Important," Fire Safety Science—Proceedings of the Second International Symposium, Hemisphere Publishing Corp., New York, 1989.

35. Gottuk, D. T., and Roby, R. J., "Effect of Combustion Conditions on Species Production," The SFPE Handbook of Fire Protection Engineering, 2nd ed., DiNenno, P. J., ed., National Fire Protection Association, Quincy, MA, 1995.

36. Madrzkowski, D., and Vettori, R. L., "A Sprinkler Fire Suppression Algorithm," Journal of Fire Protection Engineering, 4(4), 1992.

37. Purser, D. A., "Toxicity Assessment of Combustion Products," The SFPE Handbook of Fire Protection Engineering, 2nd ed., DiNenno, P. J., ed., National Fire Protection Association, Quincy, MA, 1995.

38. Cox, G., "The Challenge of Fire Modeling," Fire Safety Journal, 23(2), 1994.

39. Bilger, R. W., "Computational Field Models in Fire Research and Engineering," Fire Safety Science—Proceedings of the Fourth International Symposium, Hemisphere Publishing Corp., New York, 1994, pp. 95–110.

40. Wolfe, A., "CFD Software: Pushing Analysis to the Limit," Mechanical Engineering, Jan. 1991, pp. 48–54.

41. Waters, R. A., "Stansted Terminal Building and Early Atrium Studies," Journal of Fire Protection Engineering, Vol. 1, No. 2, 1989, pp. 63–76.

42. Pericleous, K. A., "The Field Modeling of Fire in an Air-Supported Structure," Fire Safety Science—Proceedings of the Second International Symposium, Hemisphere Publishing Corp., New York, 1989.

43. Huhaten, R., "Numerical Fire Modeling of a Turbine Hall," Fire Safety Science—Proceedings of the Second International Symposium, Hemisphere Publishing Corp., New York, 1989.

44. Galea, E. R., and Markatos, N. C., "Modeling of Aircraft Cabin Fires," Fire Safety Science—Proceedings of the Second International Symposium, Hemisphere Publishing Corp., New York, 1989.

45. Kumar, S., and Cox, G., "Mathematical Modeling of Fires in Road Tunnels," Proceedings of the Fifth International Conference on Aerodynamics and Ventilation of Vehicle Tunnels, Lille, France, 1985, p. 61.

46. Kumar, S., Hoffman, N., and Cox, G., "Some Validation of Jasmine for Fires in Hospital Wards," Numerical Simulation of Fluid Flow and Heat/Mass Transfer Processes, Springer-Verlag, Berlin, 1986, p. 159.

47. Galea, E., "On-the-Field Modeling Approach to the Simulation of Enclosure Fires," Journal of Fire Protection Engineering, Vol. 1, No. 1, 1989, pp. 11–22.

48. Hamer, A., and Beyler, C. L., "Modeling the Activation of Dry-Pipe Sprinklers in High-Rack Storage," Proceedings International Conference on Fire Research and Engineering, 10–15 Sept. 1995, Society of Fire Protection Engineers, Boston, MA, 1995.

49. Simcox, S., Wilkes, N. S., and Jones, I. P., "Computer Simulation of the Flows of Hot Gases from the Fire at King's Cross Underground Station," Fire Safety Journal, 18, 1992, pp. 49–73.

50. Baum, H. R., McGrattan, K. B., and Rehm, R. G., "Mathematical Modeling and Computer Simulation of Fire Phenomena," Fire Safety Science—Proceedings of the Fourth International Symposium, Hemisphere Publishing Corp., New York, 1994, pp. 185–193.

51. Hadjisophocleous, G. V., and Cacambouras, M., "Computer Modeling of Compartment Fires," Journal of Fire Protection Engineering, 5(2), 1993, pp. 39–52.

52. Luo, M., and Beck, V., "The Fire Environment in a Multi-room Building—Comparison of Predicted and Experimental Results," Fire Safety Journal, 23(4), 1994.

53. Kerrison, L., Mawhinney, N., Galea, E. R., Hoffmann, N., and Patel, M. K., "A Comparison of Two Fire Field Models with Experimental Room Fire Data," Fire Safety Science—Proceedings of the Fourth International Symposium, Hemisphere Publishing Corp., New York, 1994, pp. 161–172.

54. Forney, G. P., and McGratten, K. B., "Computing the Effect of Sprinkler Sprays on Fire-Induced Gas Flow," Proceedings International Conference on Fire Research and Engineering," 10–15 Sept. 1995, Society of Fire Protection Engineers, Boston, MA, 1995.

55. Hadjisophocleous, G., Kim, A., and Knill, K., "Modeling of a Fine Waterspray Nozzle and Liquid Pool Fire Suppression," Proceedings International Conference on Fire Research and Engineering," 10–15 Sept. 1995, Society of Fire Protection Engineers, Boston, MA, 1995.

56. Hoffmann, N. A., and Galea, E. R., "On the Eulerian-Eulerian Approach to Fire Sprinkler Modeling," Journal of Fire Protection Engineering, 3(4), 1991.

57. Blasdell, W., Monte Carlo Simulation of Fire Growth, Center for Fire Safety Studies, Worcester Polytechnic Institute, Worcester, MA, 1987.

58. Bukowski, R., "Fire Models: The Future Is Now," National Fire Protection Association Journal, March/April 1991, p. 60.

59. Nelson, H. E., "An Engineering Analysis of Fire Development in the Hospice of Southern Michigan, December 15, 1985," Fire Safety Science—Proceedings of the Second International Symposium, Hemisphere Publishing Corp., New York, 1989.

60. Levine, R. S., and Nelson, H. E., Full-Scale Simulation of a Fatal Fire and Comparison of Results with Two Multiroom Models, NISTIR 90-4268, 1990, National Institute of Standards and Technology, Gaithersburg, MD.

61. Alvares, N., "Defining Fire and Smoke Spread Dynamics in the DuPont Plaza Fire of 31 December 1986," Proceedings International Conference on Fire Research and Engineering," 10–15 Sept. 1995, Society of Fire Protection Engineers, Boston, MA, 1995.

62. Bukowski, R. W., and Spetzler, R. C., "Analysis of the Happyland Social Club Fire with HAZARD I," Journal of Fire Protection Engineering, 4(4), 1992.

63. Clarke, F., and DiNenno, P. J., "Computer-Based Analysis of the Fire Hazard of Furniture Materials," Proceedings of SPI 28th Annual Technical Conference, San Antonio, TX, 1984.

64. Clarke, F., and DiNenno, P. J., "Fire Safety of Wire and Cable Materials, Part III," Proceedings of International Wire and Cable Symposium, Reno, NV, Nov. 1984.

65. DiNenno, P. J., "Mathematical Fire Modeling—Toward the Rational Integration of Fire Safety Features," BVD/SPI Conference on Fire Protection Concepts, Mar. 1984.

66. Bukowski, R., "Toxic Hazard Evaluations of Plenum Cables," Fire Technology, Vol. 21, No. 4, 1985, pp. 252–266.

67. Peacock, R. D., and Bukowski, R. W., "A Prototype Methodology for Fire Hazard Analysis," Fire Technology, Vol. 26, No. 1, Feb. 1990, p. 15.

68. Clarke, F. B., Bukowski, R. W., Stiefel, S. W., Hall J. R., and Steele, S. A., "A Method to Predict Fire Risk: The Report of the National Fire Protection Research Foundation Risk Project," Report to the National Fire Protection Research Foundation (NFPRF), Quincy, MA, 1990.

69. Hall, J. R., and Sekizawa, A., "Fire Risk Analysis: General Conceptual Framework for Describing Models," Fire Technology, Vol. 27, No. 1, 1991, pp. 33–53.

70. Bengston, S., and Hagglund, B., "The Use of Zone Model in Fire Engineering Applications," Fire Safety Science—Proceedings of the First International Symposium, Hemisphere Publishing Corp., Washington, DC, 1986, pp. 667–675.

71. Hagglund, B., and Wickstrom, U., "Smoke Control in Hospitals—A Numerical Study," Fire Safety Journal, Vol. 16, 1990, pp. 53–63.

72. Ling, W. C., and Williamson, R. B., "Using Fire Tests for Quantitative Risk Analysis," Fire Risk Assessment, ASTM STP 762, 1982, American Society for Testing and Materials, W. Conshohocken, PA.

73. Wakamatsu, T., "Development of Design System for Building Fire Safety," Fire Safety Science—Proceedings of the Second International Symposium, Hemisphere Publishing Corp., New York, 1989.

74. "From Burns to Bytes," NFPA 100 Years, Fire Service Overview, supplement to the NFPA Journal, 1996, pp. 31–32.

NFPA Codes, Standards, and Recommended Practices

References to the following NFPA codes, standards, and recommended practices will provide further information on fire modeling discussed in this chapter. (See the latest version of The NFPA Catalog for availability of current editions of the following documents.)

NFPA 92B, Guide for Smoke Management Systems in Malls, Atria, and Large Areas

Additional Reading

Allan, H. A., "CSIRO Fire Safety Engineering Method Using the Computer Model FIRECALC," Commonwealth Scientific and Industrial Research Organization, Australia, University of Ulster and Fire Research Station, CIB W14: Fire Safety Engineering, International Symposium and Workshops Engineering Fire Safety in the Process of Design: Demonstrating Equivalency, Part 2, Symposium: Engineering Fire Safety for People With Mixed Abilities, September 13–16, 1993, Newtownabbey, Northern Ireland, 1993, pp. 111–144.

Allan, H., Grubits, S., and Quaglia, C., "Predicting Smoke Movement in a Multi-Storey Shopping Center With Interconnecting Voids Using a Zone Model," *Proceedings* of the Society of Fire Protection Engineers (SFPE) Honors Lecture Series, May 20, 1996, Engineering Seminars: Fire Protection Design for High Challenge or Special Hazard Applications, May 20–22, 1996, Boston, MA, 1996, pp. 43–48.

Allen, J. E., and Reynolds, J. R., Jr., "FTI Field Modeling Analysis and Data Animation Dupont Plaza Hotel Fire," *Proceedings* of the International Conference on Fire Research and Engineering, Sept. 10–15, 1995, Orlando, FL, SFPE, Boston, MA, 1995, pp. 588–589.

Andersson, P., and Holmstedt, G., "CFD-Modelling Applied to Fire Detection: Validation Studies and Influence of Background Heating," *Proceedings* of the Tenth International Conference on Automatic Fire Detection "AUBE '95", [Internationale Konferenz uber Automatischen Brandentdeckung.] April 4–6, 1995, Duisburg, Germany, 1995, pp. 429–438.

ASTM E119, *Standard Test Methods for Fire Tests of Building Construction and Materials,* American Society for Testing and Materials, W. Conshohocken, PA, 1995.

ASTM E1354, *Standard Test Method for Heat and Visible Smoke Release Rates for Materials and Products Using an Oxygen Consumption Calorimeter,* American Society for Testing and Materials, W. Conshohocken, PA, 1994.

Babrauskas, V., "Designing Products for Fire Performance: The State of the Art of Test Methods and Fire Models. Commentary," *Fire Safety Journal,* Vol. 24, No. 3, 1995, pp. 299–312.

Babrauskas, V., "Introduction to Mathematical Fire Modelling. [Book Review of An Introduction to Mathematical Fire Modeling by David M. Birk.]," *Fire and Flammability Bulletin,* October 1991, pp. 9–10.

Bailey, J. L., and Tatem, P. A., "Validation of Fire/Smoke Spread Model (CFAST) Using Ex-USS SHADWELL Internal Ship Conflagration Control (ISCC) Fire Tests," Naval Research Laboratory, Washington, DC, Office of Naval Research, Arlington, VA, NRL/MR/6180-95-7781, Sept. 28, 1995.

Baum, H. R., McGrattan, K. B., and Rehm, R. G., "Mathematical Modeling and Computer Simulation of Fire Phenomena," *Proceedings* of the Fourth International Symposium on Fire Safety Science, Intl. Assoc. for Fire Safety Science, Boston, MA, 1994, pp. 185–193.

Beard, A., "Evaluation of Fire Models: Report 9. FIRST: Qualitative Assessment," Edinburgh Univ., Scotland, Report 9, Jan. 1990.

Beard, A., "Evaluation of Deterministic Fire Models. Part 1. Introduction," *Fire Safety Journal,* Vol. 19, No. 4, 1992, pp. 295–306.

Beard, A., "Evaluation of Fire Models: Overview," Edinburgh Univ., Scotland, Report 11, Oct. 1990.

Beard, A., "Evaluation of Fire Models: Report 1. ASET: Qualitative Assessment," Edinburgh Univ., Scotland, Report 1, January 1990, 15 pages.

Beard, A., "Evaluation of Fire Models: Report 10. JASMINE: Qualitative Assessment," Edinburgh Univ., Scotland, Report 10, May 1990.

Beard, A., "Evaluation of Fire Models: Report 2. ASKFRS: Qualitative Assessment," Edinburgh Univ., Scotland, Report 2, Jan. 1990.

Beard, A., "Evaluation of Fire Models: Report 3. HAZARD I: Qualitative Assessment," Edinburgh Univ., Scotland, Report 3, Jan. 1990.

Beard, A., "Evaluation of Fire Models: Report 4. FIRST: Qualitative Assessment," Edinburgh Univ., Scotland, Report 4, Jan. 1990.

Beard, A, "Evaluation of Fire Models: Report 5. JASMINE: Qualitative Assessment," Edinburgh Univ., Scotland, Report 5, April 1990.

Beard, A., "Evaluation of Fire Models: Report 6. ASET: Qualitative Assessment," Edinburgh Univ., Scotland, Report 6, Jan 1990.

Beard, A, "Evaluation of Fire Models: Report 8. HAZARD I: Qualitative Assessment," Edinburgh Univ., Scotland, Report 8, April 1990.

Beard, A., "Limitations of Computer Modelling," Fire Safety Modelling and Building Design, *Proceedings* of the One-Day Conference to Review the Potential and Limitations of Fire Safety Models for Building Design, March 29, 1994, Salford, UK, 1994, pp. 10–26.

Beard, A., "Limitations of Computer Models," *Fire Safety Journal,* Vol. 18, No. 4, 1992, pp. 375–391.

Beard, A. N., "Reliability and Computer Models," *Journal of Applied Fire Science,* Vol. 3, No. 3, 1993/1994, pp. 273–279.

Beard, A. N., *et al.,* "Flashover A1: A Model for Predicting the Conditions for Flashover," University of Edinburgh, Scotland, South Bank University, London, England, University of College, London, England, ISBN 0-9527398-3-6; Institution of Fire Engineers, University of Sunderland, Fire Research Station, CIB W14, Tyne and Wear Metropolitan Fire Brigade, Fire Safety by Design, Conference *Proceedings,* Volume 3, Research Papers, July 10–12, 1995, UK, 1995, pp. 169–179.

Beard, A. N., Drysdale, D. D., and Bishop, S. R., "Non-Linear Model of Major Fire Spread in a Tunnel," *Fire Safety Journal,* Vol. 24, No. 4, 1995, pp. 333–357.

Beck, V., *et al.,* "Experimental Validation of a Fire Growth Model," Victoria University of Technology, Australia, National Research Council of Canada, Ottawa, Ontario, ISBN 0-9516320-9-4; *Proceedings* of the Seventh International Interflam Conference, Interflam '96, March 26–28, 1996, Cambridge, England, Interscience Communications Ltd., London, England, 1996, pp. 653–662.

Beller, D., "Validating Fire Models: FPETOOL, CFAST, WPI/FIRE," Azure Dragon Enterprises, Ltd., USA, ISBN 0-9516320-9-4; *Proceedings* of the Seventh International Interflam Conference, Interflam '96, March 26–28, 1996, Cambridge, England, Interscience Communications Ltd., London, England, 1996, pp. 959–963.

Beller, D., and Till, R, "Computer Fire Model Validation: FPETOOL, CFAST, WPI/FIRE," *Proceedings* of the International Conference on Fire Research and Engineering, Sept. 10–15, 1995, Orlando, FL, SFPE, Boston, MA, 1995, pp. 341–345.

Belles, D., "Practical Application of Computer Fire Models," National Institute of Standards and Technology, Federal Fire Forum on Computer Programs for Fire Protection, November 7, 1991, Gaithersburg, MD, 1991.

Bilger, R. W., "Computational Field Models in Fire Research and Engineering," *Proceedings* of the Fourth International Symposium on Fire Safety Science, Intl. Assoc. for Fire Safety Science, Boston, MA, 1994, pp. 95–110.

Blount, K., "Regulatory Implications of Fire Safety Modelling," Fire Safety Modelling and Building Design, *Proceedings* of the One-Day Conference to Review the Potential and Limitations of Fire Safety Models for Building Design, March 29, 1994, Salford, UK, 1994, pp. 27–35.

Brani, D. M. and Black, W. Z., "Two-Zone Model for a Single-Room Fire," *Fire Safety Journal,* Vol. 19, No. 2–3, 1992, pp. 189–216.

Brooks, W. N., "Comparison of the NFPA 92B Calculation Methods to Several Fire Models," *Proceedings* of Computer Applications in Fire Protection, June 28–29, 1993, Worcester, MA, 1993, pp. 43–50.

Bukowski, R. W., "Fire Models: The Future is Now!," *NFPA Journal,* Vol. 85, No. 6, March/April 1991, pp. 60–62, 64, 66–69.

Bukowski, R. W, "How to Evaluate Alternative Designs Based on Fire Modeling," *NFPA Journal,* Vol. 89, No. 2, March/April 1995, pp. 68–70, 72–74.

Chandrasekaran, V., Grubits, S. J., and Suresh, R., "Zone Models in Fire Safety Engineering. Session 4—The Tools," CSIRO Division of Building, Construction and Engineering, Australia, CSIRO, International Fire Safety Engineering Conference on Concept and the Tools, Presented for FORUM for International Cooperation on Fire Research, Supported by SFPE, NFPA, AFPA, AAFA, Oct. 18–20, 1992, Sydney, Australia, 1992, pp. 1–18.

Chang, X., Laage, L. W., and Greuer, R. E., "User's Manual for MFIRE: A Program Computer Simulation Program for Mine Ventilation and Fire Modeling," Xian Mining Institute, The Peoples Republic of China, Bureau of Mines, Minneapolis, MN, Mining Technological Univ., Houghton, MI, IC 9245, 1990.

Charters, D. A., Gray, W. A., and McIntosh, A. C., "Computer Model to Assess Fire Hazards in Tunnels (FASIT)," *Fire Technology,* Vol. 30, No. 1, First Quarter 1994, pp. 134–154.

Charters, D., and McIntosh, A., "FASIT: A Computer Model for Predicting Fires in Tunnels," *Fire,* Vol. 88, No. 1082, Aug. 1995, pp. 13–14, 16.

Chow, W. K., "Experimental Evaluation of the Zone Models CFAST, FAST and CCFM.VENTS," *Journal of Applied Fire Science,* Vol. 2, No. 4, 1992–1993, pp. 307–332.

Chow, W. K., "Multi-Cell Concept for Simulating Fires in Big Enclosures Using a Zone Model," *Journal of Fire Sciences,* Vol. 14, No. 3, May/June 1996, pp. 186–198.

Chow, W. K., "Short Note on the Simulation of the Atrium Smoke Filling Process Using Fire Zone Models," *Journal of Fire Sciences,* Vol. 12, No. 6, Nov./Dec. 1994, pp. 506–515.

Chow, W. K., "Simulation of Car Park Fires Using Zone Models," *Journal of Fire Protection Engineering,* Vol. 7, No. 3, 1995, pp. 65–74.

Chow, W. K., "Use of the ARGOS Fire Risk Analysis Model for Studying Chinese Restaurant Fires," *Fire and Materials*, Vol. 19, No. 4, July/August 1995, pp. 171–178.

Chow, W. K., "Use of Zone Models on Simulating Compartmental Fires With Forced Ventilation," *Fire and Materials*, Vol. 19, No. 3, 1995, pp. 101–108.

Chow, W. K., "Zone Model Simulation of Fires in Chinese Restaurants in Hong Kong," *Journal of Fire Sciences*, Vol. 13, No. 3, May/June 1995, pp. 235–253.

Chow, W. K., and Cheung, K. C., "Industrial Fire Simulation With Zone Models and Review on the Current Regulations," Hong Kong Polytechnic Univ., Hong Kong, ISBN 0-9516320-9-4; *Proceedings* of the Seventh International Interflam Conference, Interflam '96, March 26–28, 1996, Cambridge, England, Interscience Communications Ltd., London, England, 1996, pp. 601–620.

Chow, W. K., and Cheung, Y. L., "Simulation of Sprinkler-Hot Layer Interaction Using a Field Model," *Fire and Materials*, Vol. 18, No. 6, 1994, pp. 359–379.

Chow, W. K., and Lo, A. C. W., "Scale Modelling Studies on Atrium Smoke Movement and the Smoke Filling Process," *Journal of Fire Protection Engineering*, Vol. 7, No. 3, 1995, pp. 55–64.

Chow, W. K., and Wong, W. K., "Application of the Zone Model FIRST on the Development of Smoke Layer and Evaluation of Smoke Extraction Design for Atria in Hong Kong," *Journal of Fire Sciences*, Vol. 11, No. 4, July/Aug. 1993, pp. 329–349.

Cooper, L. Y., "Overview of a Theory for Simulating Smoke Movement Through Long Vertical Shafts in Zone-Type Fire Models," National Institute of Standards and Technology, Gaithersburg, MD, NISTIR 5499, Sept. 1994; National Institute of Standards and Technology, Annual Conference on Fire Research: Book of Abstracts, Oct. 17–20, 1994, Gaithersburg, MD, 1994, pp. 93–94.

Cooper, L. Y., "VENTCF2: An Algorithm and Associated FORTRAN 77 Subroutine for Calculating Flow Through a Horizontal Ceiling/Floor Vent in a Zone-Type Compartment Fire Model," National Institute of Standards and Technology, Gaithersburg, MD, NISTIR 5470, Aug. 1994.

Cox, G., "Challenge of Fire Modelling," *Fire Safety Journal*, Vol. 23, 1994, pp. 123–132.

Cox, G., "Challenge of Fire Modelling," *Proceedings* of the Technical Conference on Fire Safety by Design: A Framework for the Future, Featuring the Combined 1993 Fire Research Lecture and IFS Inaugural Lecture, November 10, 1993, Borehamwood, England, 1993, pp. 1–14.

Cox, G., "Compartment Fire Modelling," Fire Research Station, Borehamwood, England, ISBN 0-12-194230-9; *Combustion Fundamentals of Fire*, Chapter 6, Academic Press, New York, NY, 1995, pp. 329–404.

Cox, G., "Some Recent Progress in the Field Modelling of Fire," First Asian Conference on Fire Science and Technology (ACFST), Oct. 9–13, 1992, Hefei, China, International Academic Publishers, China, 1992, pp. 50–59.

Cox, G., "State of the Art Modelling," Fire Safety Modelling and Building Design, *Proceedings* of the One-Day Conference to Review the Potential and Limitations of Fire Safety Models for Building Design, March 29, 1994, Salford, UK, 1994, pp. 6–9.

Cox, G., and Rubini, P. A., "Development of a New CFD Fire Simulation Model," Fire Research Station, Borehamwood, England, Cranfield Institute of Technology, UK, VTT-Technical Research Center of Finland and Forum for International Cooperation in Fire Research, Nordic Fire Safety Engineering Symposium, Development and Verification of Tools for Performance Codes, Aug. 30–Sept. 1, 1993, Espoo, Finland, 1993, pp. 1–11.

Curtat, M. R., "Survey of Fire Modelling in France," *Proceedings* of the Third International Symposium on Fire Safety Science, Elsevier Applied Science, New York, 1991, pp. 97–118.

Custer, R. L. P., "Computer Modeling in Fire Reconstruction and Failure Analysis," *Proceedings* of Computer Applications in Fire Protection, June 28–29, 1993, Worcester, MA, 1993, pp. 69–74.

Custer, R. L. P., "Progress on Computer Modeling in Performance-Based Detection System Design in the United States," Worcester Polytechnic Inst., MA, European Society for Automatic Alarm Systems (EUSAS), European Platform for Promoting Security and Fire Protection Engineering, Newsletter, Number 5, June 1994, pp. 17–26.

Custer, R. L. P., "Selecting and Using Computer Fire Models," Worcester Polytechnic Inst., MA, Society of Fire Protection Engineers and WPI Center for Firesafety Studies, *Proceedings* of the Technical Sympo-sium on Computer Applications in Fire Protection Engineering, Final Program, June 20–21, 1996, Worcester, MA, 1996, pp. 7–11.

Daikoku, M., *et al.*, "Modelling of Transport Phenomena in Compartment Fires," *Proceedings* of the International Conference on Fire Research and Engineering, Sept. 10–15, 1995, Orlando, FL, SFPE, Boston, MA, 1995, pp. 15–20.

Davis, W., "Overview of Harwell-Flow 30—A Field Model," Montgomery College, Rockville, MD, National Institute of Standards and Technology, Federal Fire Forum on Computer Programs for Fire Protection, Nov. 7, 1991, Gaithersburg, MD, 1991.

Davis, W. D., Forney, G. P., and Bukowski, R. W., "Field Modeling: Simulating the Effect of Sloped Beamed Ceilings on Detector and Sprinkler Response. International Fire Detection Research Project. Technical Report. Year 2," National Institute of Standards and Technology, Gaithersburg, MD, National Fire Protection Research Foundation, Quincy, MA, Technical Report, Year 2, Oct 1994.

Davis, W. D., Forney, G. P., and Klote, J. H., "Field Modeling of Room Fires," National Institute of Standards and Technology, Gaithersburg, MD, NISTIR 4673, Nov. 1991.

Davis, W. D., Notarianni, K. A., and Tapper, P. Z., "Modelling of Smoke Movement and Detector Performance in High Bay Spaces," *Proceedings* of the International Conference on Fire Research and Engineering, Sept. 10–15, 1995, Orlando, FL, SFPE, Boston, MA, 1995, pp. 307–311.

Deakin, A. G., "Fire Modelling: Where Are We and What Is Needed for Practical Application," Warrington Fire Research Center, UK, TNO Building and Construction Research, Third CIB/W14 Workshop, *Proceedings* of the Fire Engineering Workshop on Modelling, Jan. 25–26, 1993, Delft, The Netherlands, 1993, pp. 15–18.

Deal, S., "Comparison of Various Fire Model Results," National Institute of Standards and Technology, Federal Fire Forum on Computer Programs for Fire Protection, November 7, 1991, Gaithersburg, MD, 1991.

Dembsey, N. A., Pagni, P. J., and Williamson, R. B., "Compartment Fire Experimental Data: Comparison to Models," *Proceedings* of the International Conference on Fire Research and Engineering, Sept. 10–15, 1995, Orlando, FL, SFPE, Boston, MA, 1995, pp. 350–355.

Deubler, J. F., "How to Review a Computer Modeling Report," Schirmer Engineering Corp., Arlington, VA, Society of Fire Protection Engineers and WPI Center for Firesafety Studies, *Proceedings* of the Technical Symposium on Computer Applications in Fire Protection Engineering, Final Program, June 20–21, 1996, Worcester, MA, 1996, pp. 67–72.

Dolph, B. L., "Modeling Fires on Ships," Coast Guard Research and Development Center, Groton, CT, Society of Fire Protection Engineers and WPI Center for Firesafety Studies, *Proceedings* of the Technical Symposium on Computer Applications in Fire Protection Engineering, Final Program, June 20–21, 1996, Worcester, MA, 1996, pp. 1–4.

Duong, D. Q., "Accuracy of Computer Fire Models: Some Comparisons With Experimental Data From Australia," *Fire Safety Journal*, Vol. 16, No. 6, 1990, pp. 415–431.

El-Rimawi, J. A., Burgess, I. W., and Plank, R. J., "Modelling the Behavior of Steel Frames and Subframes With Semi-Rigid Connections in Fire," University of Sheffield, UK, TNO Building and Construction Research, Third CIB/W14 Workshop, *Proceedings* of the Fire Engineering Workshop on Modelling, Jan. 25–26, 1993, Delft, The Netherlands, 1993, pp. 152–168.

Ellington, J. M., "Some Perspectives on Fire Modeling by a Fire Investigator," *Fire and Arson Investigator*, Vol. 45, No. 4, June 1995, pp. 36–38.

Emmons, H. W., "Fire Modeling in the 21st Century," Harvard Univ., Cambridge, MA, Society of Fire Protection Engineers and WPI Center for Firesafety Studies, *Proceedings* of the Technical Symposium on Computer Applications in Fire Protection Engineering, Final Program, June 20–21, 1996, Worcester, MA, 1996, pp. 1–5.

Evans, D. D., "Large Fire Experiments for Fire Model Evaluation," National Institute of Standards and Technology, Gaithersburg, MD, ISBN 0-9516320-9-4; *Proceedings* of the Seventh International Interflam Conference, Interflam '96, March 26–28, 1996, Cambridge, England, Interscience Communications Ltd., London, England, 1996, pp. 329–334.

Fahy, R. F., "EXIT89: An Evacuation Model for High-Rise Buildings," National Fire Protection Assoc., Quincy, MA, NISTIR 4449; Eleventh Joint Panel Meeting of the U. S./Japan Government Cooperative Program on Natural Resources (UJNR) on Fire Research and Safety, Oct. 19–24, 1989, Berkeley, CA, 1990, pp. 306–311.

Fahy, R. F., "EXIT89: An Evacuation Model for High-Rise Buildings. Model Description and Example Applications," *Proceedings* of the Fourth International Symposium on Fire Safety Science, Intl. Assoc. for Fire Safety Science, Boston, MA, 1994, pp. 657–668.

Fahy, R. F., "EXIT89: An Evacuation Model for High-Rise Buildings—Recent Enhancements and Example Applications," *Proceedings* of the International Conference on Fire Research and Engineering, Sept. 10–15, 1995, Orlando, FL, SFPE, Boston, MA, 1995, pp. 332–337.

Fahy, R. F., "EXIT89: High-Rise Evacuation Model–Recent Enhancements and Example Applications," National Fire Protection Association, Quincy, MA, National Institute of Standards and Technology, Gaithersburg, MD, ISBN 0-9516320-9-4; *Proceedings* of the Seventh International Interflam Conference, Interflam '96, March 26–28, 1996, Cambridge, England, Interscience Communications Ltd., London, England, 1996, pp. 1001–1005.

Fangrat, J., "Different Approach to Some Fire Modeling Input Data," Building Research Inst., Warsaw, Poland, CIB W14/93/2 (J); *First Proceedings* of the Japan Symposium on Heat Release and Fire Hazard, Session 2, Fire Modeling as a Tool for Reaction to Fire Evaluation, May 10–11, 1993, Tsukuba, Japan, 1993, pp. II/1–6.

"Fire Models In Use Today," *IRI Sentinel*, Vol. 51, No. 3, Third Quarter, 1994, pp. 10–11.

"Fire Models: A Tale of Four Uses," *IRI Sentinel*, Vol. 51, No. 3, Third Quarter, 1994, pp. 8–9.

"Fire Models: The Pros and Cons," *IRI Sentinel*, Vol. 51, No. 3, Third Quarter, 1994, p. 3.

"Fire Models: Valuable Tools When Used Properly," *IRI Sentinel*, Vol. 51, No. 3, Third Quarter, 1994, pp. 4–7.

"Fire Spread Modeling," *Record*, Vol. 70, No. 3, May/June 1993, pp. 15–16.

Forney, G. P., "Computing Radiative Heat Transfer Occurring in a Zone Fire Model," *Fire Science and Technology*, Vol. 14, No. 1/2, pp. 31–74.

Forney, G. P., and Moss, W. F., "Analyzing and Exploiting Numerical Characteristics of Zone Fire Models," *Fire Science and Technology*, Vol. 14, No. 1/2, 1994, pp. 49–60.

Fransson, C., and Vilselius, H., "Application of a General-Purpose CFD Code to Field Modelling of Tunnel Fires," National Defence Research Establishment, Sundbyberg, Sweden, FOA Report B 20131-2.7, June 1994.

Friedman, R., "Survey of Computer Models for Fire and Smoke. Technical Report," Factory Mutual Research Corp., Norwood, MA, Technical Report, March 1990,.

Galea, E. R., and Galparsoro, J. M. P., "Computer-Based Simulation Model for the Prediction of Evacuation From Mass-Transport Vehicles," *Fire Safety Journal*, Vol. 22, No. 4, 1994, pp. 341–366.

Galea, E. R., and Ierotheou, C. S., "Fire-Field Modelling on Parallel Computers," *Fire Safety Journal*, Vol. 19, No. 4, 1992, pp. 251–266.

Galea, E. R., Owen, M., and Lawrence, P. J., "Emergency Egress From Large Buildings Under Fire Conditions Simulated Using the EXODUS Evacuation Model," University of Greenwich, London, England, ISBN 0-9516320-9-4; *Proceedings* of the Seventh International Interflam Conference, Interflam '96, March 26–28, 1996, Cambridge, England, Interscience Communications Ltd., London, England, 1996, pp. 711–720.

Gallina, G., *et al.*, "Fire Modelling in Building With a CAD-Based Graphical User Interface," CNR-ICITE, Sesto Ulteriano (MI), Italy, TRI srl, Scanzorosicate (BG), Italy, ISBN 0-9527398-3-6; Institution of Fire Engineers, University of Sunderland, Fire Research Station, CIB W14, Tyne and Wear Metropolitan Fire Brigade, Fire Safety by Design, Conference *Proceedings*, Volume 3, Research Papers, July 10–12, 1995, UK, 1995, pp. 239–253.

Golton, C. J., Golton, B. J., and Hinks, A. J., "Human Behavior in Fire: A Background Review for Modelling," Fire Safety Modelling and Building Design, *Proceedings* of the One-Day Conference to Review the Potential and Limitations of Fire Safety Models for Building Design, March 29, 1994, Salford, UK, 1994, pp. 48–62.

Griffith, S. J., and Munday, J. W, "Legal Implications of Real Fire Data Collection and Computer Modelling," Metropolitan Police Forensic Science Laboratory, London, England, ISBN 0-9516320-9-4; *Proceedings* of the Seventh International Interflam Conference, Interflam '96, March 26–28, 1996, Cambridge, England, Interscience Communications Ltd., London, England, 1996, pp. 1027–1031.

Grubits, S. J., and Chandrasekaran, V., "Issues in Fire Modelling," Commonwealth Scientific and Industrial Research Organization, North Ryde, Australia, July 1991.

Gupta, A. K., "CALFIRE: An Interactive Model for Fire Calculations," *Fire Technology*, Vol. 30, No. 3, Third Quarter 1994, pp. 304–325.

Hadjisophocleous, G. V., and Cacambouras, M., "Computer Modeling of Compartment Fires," *Journal of Fire Protection Engineering*, Vol. 5, No. 2, April/May/June 1993, pp. 39–52.

Hadjisophocleous, G. V., and Knill, K., "CFD Modelling of Liquid Pool Fire Suppression Using Fine Watersprays," National Research Council of Canada, Ottawa, Ontario, Advanced Scientific Computing, Ltd., Ontario, Canada, NISTIR 5499, September 1994; National Institute of Standards and Technology, Annual Conference on Fire Research: Book of Abstracts, Oct. 17–20, 1994, Gaithersburg, MD, 1994, pp. 71–72.

Hagglund, B., "Comparing Fire Models With Experimental Data," National Defence Research Establishment, Sundbyberg, Sweden, FOA Report C20864-2.4, Feb. 1992.

Hagglund, B., Fransson, C., and Bengtson, S., "Use of Zone and Field Model to Recommend Fire Protection Measures in New Stores of the Royal Library in Stockholm," National Defence Research Establishment, Sundbyberg, Sweden, FOA Report C20981-2.4, June 1994.

Ham, S., "Fire Modelling From an Architectural Perspective," Fire Safety Modelling and Building Design, *Proceedings* of the One-Day Conference to Review the Potential and Limitations of Fire Safety Models for Building Design, March 29, 1994, Salford, UK, 1994, pp. 1–5.

Hasemi, Y., and Yoshida, M., "Wall Fire Modeling and Its Relevance to Building Fire Safety Design," '93 Asian Fire Seminar, Oct. 7–9, 1993, Tokyo, Japan, 1993, pp. 117–127.

Heins, T., and Dobbernack, R., "FIGARO: A Zone Model for Fire Simulation in Complex Geometries," Technischer Uberwachungs-Verein Hannover/Sachsen-Anhalt e.V., Germany, Universitat Braunschweig, Germany, TNO Building and Construction Research, Third CIB/W14 Workshop, *Proceedings* of the Fire Engineering Workshop on Modelling, Jan. 25–26, 1993, Delft, The Netherlands, 1993, pp. 49–58.

Hinks, A. J., "Modelling and Managing Escape From Fires in Complex Buildings," Fire Safety Modelling and Building Design, *Proceedings* of the One-Day Conference to Review the Potential and Limitations of Fire Safety Models for Building Design, March 29, 1994, Salford, UK, 1994, pp. 73–89.

Hinks, A. J., "Systemic Modelling of Life Safety in Building Fires: LISA," Fire Safety Modelling and Building Design, *Proceedings* of the One-Day Conference to Review the Potential and Limitations of Fire Safety Models for Building Design, March 29, 1994, Salford, UK, 1994, pp. 63–72.

Holen, J., and Magnussen, B. F., "Present Status of Field Models for Fire Growth Developed at NTH/SINTEF," NTH/SINTEF, Trondheim, Norway, VTT-Technical Research Center of Finland and Forum for International Cooperation in Fire Research, Nordic Fire Safety Engineering Symposium, Development and Verification of Tools for Performance Codes, Aug. 30–Sept. 1, 1993, Espoo, Finland, 1993, pp. 1–7.

Hume, B., "Guidance on Fire Models," *Fire Research News*, No. 18, Winter 1994, p. 11.

Janssens, M., "Mathematical Modeling of Compartment Fires," Abstracts of the First U. S. Symposium on Heat Release and Fire Hazard, December 1991, San Diego, CA, 1991, pp. 45–47.

Janssens, M., "Room Fire Models. Part A. General," Chapter 6, *Heat Release in Fires*, Elsevier Applied Science, NY, 1992, pp. 113–157.

Janssens, M. L., "Computer Program Roomfire: A Compartment Fire Model Shell," American Forest and Paper Association, Washington, DC, Chapter 27, ACS Symposium Series 599, American Chemical Society, Fire and Polymers II: Materials and Tests for Hazard Prevention, 208th National Meeting, ACS Symposium Series 599, Aug. 21–26, 1994, Washington, DC, American Chemical Society, Washington, DC, 1995, pp. 409–421.

Johnson, P. F., "Comment on 'The Accuracy of Computer Fire Models: Some Comparisons With Experimental Data From Australia'. Letter to the Editor," *Fire Safety Journal*, Vol. 17, No. 5, 1991, pp. 415–416.

Jones, S., *et al.*, "Zone Model for Smoke Transport Aboard Human-Crewed Spacecraft," *PSAM-II Proceedings*, An International Conference Devoted to the Advancement of System-Based Methods for the Design and Operation of Technological Systems and Processes, Volume 3, Sessions 73-108, March 20–25, 1994, San Diego, CA, 1994, pp. 085/1–4.

Jones, W. W., "Modeling Fires—The Next Generation of Tools," Worcester Polytechnic Inst., MA, Society of Fire Protection Engineers and WPI Center for Firesafety Studies, *Proceedings* of the Technical Symposium on Computer Applications in Fire Protection Engineering, Final Program, June 20–21, 1996, Worcester, MA, 1996, pp. 13–18.

Jones, W. W., "Modeling Smoke Movement Through Compartmented Structures," *Journal of Fire Sciences*, Vol. 11, No. 2, March/April 1993, pp. 172–183; Thirty-ninth Sagamore Army Materials Research Conference, Sept. 16–17, 1992, Plymouth, MA, 1992; Twelfth Joint Panel Meeting of the U.S./Japan Government Cooperative Program on Natural Resources (UJNR) on Fire Research and Safety, Oct. 27–Nov. 2, 1992, Tsukuba, Japan, Building Research Inst., Ibaraki, Japan, Fire Research Inst., Tokyo, Japan, 1992, pp. 34–41.

Jones, W. W., "Progress Report on Fire Modeling and Validation," National Institute of Standards and Technology, Gaithersburg, MD, NISTIR 5835, May 1996..

Jones, W. W., and Baum, H. R., "Modeling Fire Growth and Smoke Transport in the United States," Twelfth Joint Panel Meeting of the U. S./Japan Government Cooperative Program on Natural Resources (UJNR) on Fire Research and Safety, Oct. 27–Nov. 2, 1992, Tsukuba, Japan, Building Research Inst., Ibaraki, Japan, 1992, pp. 4–7.

Jones, W. W., and Forney, G. P., "Programmer's Reference Manual for CFAST, the Unified Model of Fire Growth and Smoke Transport," National Institute of Standards and Technology, Gaithersburg, MD, NIST TN 1283, Nov. 1990.

Kalinich, D. A. and Bailey, R. T., "Parametric Study of Smoke Propagation Using the CFAST Computer Code," Westinghouse Savannah River Co., Aiken, SC, Department of Energy, Washington, DC, WSRC-RP-93-1315, September 1993.

Karlsson, B., "Models for Calculating Flame Spread on Wall Lining Materials and the Resulting Heat Release Rate is a Room," *Fire Safety Journal*, Vol. 23, No. 4, 1994, pp. 365-386.

Karlsson, B., and Magnusson, S. E., "Room Fire Models. Part B. An Example Room Fire Model," Chapter 6, *Heat Release in Fires*, Elsevier Applied Science, NY, 1992, pp. 159–171.

Kasahara, I. ,and Hara, T., "Life Safety Design of an Atrium and an Air Dome With an Applied Zone Model," Taisei Corp., Tokyo, Japan, CIB W14/96/1 (J); Mini-Symposium on Fire Safety Design of Buildings and Fire Safety Engineering, June 12, 1995, Tsukuba, Japan, 1995, pp. VIa/1–3.

Kerrison, L., *et al.*, "Comparison of Two Fire Field Models With Experimental Room Fire Data," *Proceedings* of the Fourth International Symposium on Fire Safety Science, Intl. Assoc. for Fire Safety Science, Boston, MA, 1994, pp. 161–172.

Kerrison, L., *et al.*, "Comparison of a FLOW3D Based Fire Field Model With Experimental Room Fire Data," *Fire Safety Journal*, Vol. 23, No. 4, 1994, pp. 387–411.

Kostreva, M. M., "Sensitivity Analysis for Mathematical Modeling of Fires in Residential Buildings," Clemson Univ., SC, National Institute of Standards and Technology, Gaithersburg, MD, NIST-GCR-95-683, Feb. 1966.

Kumar, S., "Field Model Simulations of Vehicle Fires in a Channel Tunnel Shuttle Wagon," *Proceedings* of the Fourth International Symposium on Fire Safety Science, Intl. Assoc. for Fire Safety Science, Boston, MA, 1994, pp. 995-1006.

Kumar, S., and Yehia, M., "Application of JASMINE to the Modelling of Vehicle Fires in a Channel Tunnel Shuttle Wagon," *Proceedings* of the Second International Conference on Safety in Road and Rail Tunnels, April 3–6, 1995, Granada, Spain, 1995, pp. 321–328.

Li, H. J., and Cheng, X. F., "Design of Building Fire Simulation Models," *Fire Safety Science*, Vol. 3, No. 2, September 1994, pp. 21–26.

Lie, T. T., and Chabot, M., "Fire Resistance Evaluation Using Mathematical Models," Fifth International Fire Conference on Fire Safety, Interflam '90, Sept. 3–6, 1990, Canterbury, England, Interscience Communications Ltd., London, England, 1990, pp. 93–100.

Lie, T. T., Gosselin, G. C. ,and Kim, A. K., "Applicability of Temperature Predicting Models to Building Elements Exposed to Room Fires," *Journal of Applied Fire Science*, Vol. 4, No. 2, 1994/1995, pp. 83–93.

Lucht, D. A., "Changing the Way We Do Business: The Coming of Fire Modeling," *IFCI Fire Code Journal*, Vol. 2, No. 4, Spring 1994, pp. 1, 2, 4–6.

Man, W. K., "Fire Modelling: An Example of Application. Part 1," *Southeast Asia Fire and Security*, May 1993, pp. 18–20.

Marchant, E. W., "Futures of Fire Safety Modelling," Fire Safety Modelling and Building Design, *Proceedings* of the One-Day Conference to Review the Potential and Limitations of Fire Safety Models for Building Design, March 29, 1994, Salford, UK, 1994, pp. 36–47.

Martin, S. B., "Perspective on Fire Modeling," Sixteenth International Conference on Fire Safety, Volume 16, Jan. 14–18, 1991, Millbrae, CA, Product Safety Corp., Sunnyvale, CA, 1991, pp. 370–376.

Mawhinney, R. N., *et al.*, "Critical Comparison of a Phoenics Based Fire Field Model With Experimental Compartment Fire Data," *Journal of Fire Protection Engineering*, Vol. 6, No. 4, 1994, pp. 137–152.

Mitler, H. E., "Mathematical Modeling of Enclosure Fires," National Institute of Standards and Technology, Gaithersburg, MD, NISTIR 90-4294, May 1991.

Mitler, H. E., and Steckler, K. D., "SPREAD: A Model of Flame Spread on Vertical Surfaces," National Institute of Standards and Technology, Gaithersburg, MD, NISTIR 5619, April 1995.

Moss, W. F., "Numerical Analysis Support for Compartment Fire Modeling and Incorporation of Heat Conduction Into a Zone Fire Model. Final Report. Aug. 15, 1988–March 31, 1991," Clemson Univ., SC, National Institute of Standards and Technology, Gaithersburg, MD, NIST-GCR-92-605, March 1992.

Moss, W. F., and Forney, G. P., "Implicitly Coupling Heat Conduction Into a Zone Fire Model," Clemson Univ., SC, National Institute of Standards and Technology, Gaithersburg, MD, NISTIR 4886, July 1992.

Nelson, H. E., "Application of Computer Models for Fire Incident Reconstruction," *Proceedings* of the International Conference on Fire Research and Engineering, Sept. 10–15, 1995, Orlando, FL, SFPE, Boston, MA, 1995, pp. 359–364.

Nelson, H. E., "Computer Models and Fire Reconstruction," *Proceedings* of Computer Applications in Fire Protection, June 28–29, 1993, Worcester, MA, 1993, pp. 75–79.

Nelson, H. E., "Introduction to Computer Fire Modeling," National Institute of Standards and Technology, Gaithersburg, MD, Federal Fire Forum on Computer Programs for Fire Protection. Nov. 7, 1991, Gaithersburg, MD, 1991.

Notarianni, K. A., and Davis, W. D., "Use of Computer Models to Predict the Response of Sprinklers and Detectors in Large Spaces," *Proceedings* of Computer Applications in Fire Protection, June 28–29, 1993, Worcester, MA, 1993, pp. 27–33.

Oar, D. L., "Attachments for Fire Modeling for Building 221-T. T-Plant Canyon Deck and Railroad Tunnel," Westinghouse Hanford Co., Richland, WA, WHC-SD-CP-ANAL-010, Jan. 19, 1995.

Ohnstad, S. J., "Marine Regulatory Applications of the Computer Fire Model HAZARD I," Worcester Polytechnic Inst., MA, Thesis, March 1990.

Opstad, K., "Fire Modelling Using Cone Calorimeter Results," SINTEF NBL, Trondheim, Norway, EUREFIC (European Reaction to Fire Classification), Performance Based Reaction to Fire Classification, International Seminar, Sept. 11–12, 1991, Copenhagen, Denmark, Interscience Communications Ltd., London, England, 1991, pp. 65–71.

Opstad, K. and Magnussen, B. F., "Algorithm for Thermal Flame Spread on Solid Surfaces by the Use of Cone Calorimeter Fire Test Data in a CFD-Model," University of Trondheim, Norway, Norwegian Institute of Technology, Norway, Thesis, May 1995; *Modelling of Thermal Flame Spread on Solid Surfaces in Large-Scale Fires*, 1995, pp. 1–15.

Owen, M., Galea, E. R., and Lawrence, P., "EXODUS Evacuation Model Applied to Building Evacuation Scenarios," University of Greenwich, London, England, ISBN 0-9527398-3-6; Institution of Fire Engineers, University of Sunderland, Fire Research Station, CIB W14, Tyne and Wear Metropolitan Fire Brigade, Fire Safety by Design, Conference *Proceedings*, Volume 3, Research Papers, July 10–12, 1995, UK, 1995, pp. 81–90.

Patnaik, G., Kailasanath, K., and Sinkovits, R., "FLAME3D: A New Time-Dependent, Three-Dimensional Flame Model With Detailed Chemistry," *Proceedings* of the 1995 Fall Technical Meeting of the Combustion Institute/Eastern States Section on Chemical and Physical Processes in Combustion, Oct. 16–18, 1995, Worcester, MA, 1995, pp. 163–166.

Peacock, R. D., "Validation of Computer Models," VHS Video Tape, National Institute of Standards and Technology, Gaithersburg, MD, Federal Fire Forum, Current Technical Fire Issues, Nov. 1, 1993, Gaithersburg, MD, 1993.

Peacock, R. D., Davis, S., and Babrauskas, V., "Data for Room Fire Model Comparisons," *Journal of Research of the National Institute of Standards and Technology*, Vol. 96, No. 4, August 1991, pp. 411–462

Peacock, R. D., *et al.*, "CFAST, The Consolidated Model of Fire Growth and Smoke Transport," National Institute of Standards and Technology, Gaithersburg, MD, NIST TN 1299, Feb. 1993.

Peacock, R. D., Jones, W. W., and Bukowski, R. W., "Verification of a Model of Fire and Smoke Transport," *Fire Safety Journal*, Vol. 21, No. 2, 1993, pp. 89–129.

Pedersen, K., and Magnussen, B. F., "Field Models: their Present and Future Application in Fire Safety Engineering. Session 4—The Tools," SINTEF NBL-Norwegian Fire Research Lab., Trondheim, Norway, NTH/SINTEF, Trondheim, Norway, CSIRO, International Fire Safety Engineering Conference on Concept and the Tools, Presented for FORUM for International Cooperation on Fire Research, Supported by SFPE, NFPA, AFPA, AAFA, Oct. 18–20, 1992, Sydney, Australia, 1992, pp. 1–11.

Pehrson, R., "Computer Modeling of Intumescent Penetration Seals," Worcester Polytechnic Inst., MA, Thesis, May 10, 1993.

Platt, D. G., Elms, D. G., and Buchanan, A. H., "Probabilistic Model of Fire Spread With Timber Effects," *Fire Safety Journal*, Vol. 22, No. 2, 1994, pp. 367–398.

Pollock, A. J., Taylor, I. R., and Donegan, H. A., "Framework for Interfacing Computer Aided Design With Knowledge Based Systems to Support Fire Safety and Other Evaluations," Fire Safety Modelling and Building Design, *Proceedings* of the One-Day Conference to Review the Potential and Limitations of Fire Safety Models for Building Design, March 29, 1994, Salford, UK, 1994, pp. 98–108.

Poon, L. S., "EvacSim: a Simulation Model of Occupants With Behavioral Attributes in Emergency Evacuation of High-Rise Building Fires," *Proceedings* of the Fourth International Symposium on Fire Safety Science, Intl. Assoc. for Fire Safety Science, Boston, MA, 1994, pp. 681–692.

Pucher, K., and Tiefenbock, M., "Three-Dimensional Computer Simulation of a Fire in a Rail Tunnel," *Proceedings* of the Second International Conference on Safety in Road and Rail Tunnels, April 3–6, 1995, Granada, Spain, 1995, pp. 245–252.

Quintiere, J.G., and Dillon, M. E., "Scale Model Reconstruction of Fire in an Atrium," *Proceedings* of the International Conference on Fire Research and Engineering, Sept. 10–15, 1995, Orlando, FL, SFPE, Boston, MA, 1995, pp. 397–402.

Ramachandran, G., "Non-Deterministic Modelling of Fire Spread," *Journal of Fire Protection Engineering*, Vol. 3, No. 2, 1991, pp. 37–48.

Rehm, R. G. and Forney, G. P., "Pressure Equations in Zone-Fire Modeling," *Fire Science and Technology*, Vol. 14, No. 1/2, 1994, pp. 61–73.

Rockett, J. A., "Experience in the Use of Zone Type Building Fire Models," *Fire Science and Technology*, Vol. 13, No. 1/2, 1993, pp. 61–70.

Rockett, J. A., "Zone Model Plume Algorithm Performance," *Fire Science and Technology*, Vol. 15, No. 1/2, 1995, pp. 1–16.

Rockett, J. A., Howe, J. M., and Hanbury, W. L., "Modeling Fire Safety in Multi-Use, Domed Stadia," *Journal of Fire Protection Engineering*, Vol. 6, No. 1, 1994, pp. 11–22.

Satterfield, D. B., "WPI/Harvard Version 2 Computer Fire Model," CFR Video Seminar, Department of the Navy, Washington, DC, Video, Oct. 23, 1990.

Schifiliti, R. P., and Pucci, W. E., "Fire Detection Modeling: State of the Art," R.P. Schifiliti Associates, Inc., Reading MA, Performance Consultants, Holland, MA, Fire Detection Institute, Bloomfield, CT, State of the Art Report, May 6, 1996; Society of Fire Protection Engineers and WPI Center for Firesafety Studies, *Proceedings* of the Technical Symposium on Computer Applications in Fire Protection Engineering, Final Program, June 20–21, 1996, Worcester, MA, 1996, pp. 19–27.

Schneider, V., and Hofmann, J., "Feldmodell-Simulation von Kohlenwasserstoff-Raumbranden und Spruhnebel-Loschversuchen. [Field-Model Simulation of Hydrocarbon Room Fires and Fire Fighting Tests With Water Spray.]," *VFDB*, February 1993, pp. 67–73.

Schneider, V., and Konnecke, R., "Anwendung des Feldmodells KOBRA-3D zur Simulation von komplexen Brandszenarien auf Fragestellungen der automatischen Brandentdeckung. [Application of the Field Model KOBRA-3D to the Simulation of Complex Fire Scenarios for the Purpose of Automatic Fire Detection.]," *Proceedings* of the Tenth International Conference on Automatic Fire Detection "AUBE '95", [Internationale Konferenz uber Automatischen Brandentdeckung.] April 4–6, 1995, Duisburg, Germany, 1995, pp. 404–418.

Shapiro, J. M., "Are the Codes Ready for Fire Models?," *IFCI Fire Code Journal*, Vol. 2, No. 4, Spring 1994, pp. 3, 10.

Sinai, Y. L., and Owens, M. P., "Validation of CFD Modelling of Unconfied Pool Fires With Cross-Wind: Flame Geometry," *Fire Safety Journal*, Vol. 24, No. 1, 1995, pp. 1–34.

Stroup, D. W., "Analyzing Fire Safety Through Computer Modeling and Product Testing," Second International Conference on Fire and Materials, Sept. 23–24, 1993, Arlington, VA, 1993, pp. 7–12.

Stroup, D. W., "FPETOOL for the Personal Computer," General Services Administration, Washington, DC, National Institute of Standards and Technology, Federal Fire Forum on Computer Programs for Fire Protection, Nov. 7, 1991, Gaithersburg, MD, 1991.

Stroup, D. W., "Using Computer Fire Models to Evaluate Equivalent Levels of Fire and Life Safety," *Proceedings* of Computer Applications in Fire Protection, June 28–29, 1993, Worcester, MA, 1993, pp. 21–26.

Stroup, D. W., and Madrzykowski, D., "Modeling Smoke Flow in Corridors," *Proceedings* of the International Conference on Fire Research and Engineering, Sept. 10–15, 1995, Orlando, FL, SFPE, Boston, MA, 1995, pp. 377–382.

Sullivan, P. D., "Recent Applications of Computers in Performance Design of Buildings," Robert W. Sullivan, Inc., Boston, MA, Society of Fire Protection Engineers and WPI Center for Firesafety Studies, *Proceedings* of the Technical Symposium on Computer Applications in Fire Protection Engineering, Final Program, June 20–21, 1996, Worcester, MA, 1996, pp. 87–92.

Taylor, S., *et al.*, "SMARTFIRE: An Intelligent Fire Field Model," University of Greenwich, London, England, ISBN 0-9516320-9-4; *Proceedings* of the Seventh International Interflam Conference, Interflam '96, March 26–28, 1996, Cambridge, England, Interscience Communications Ltd., London, England, 1996, pp. 671–680.

Thomas, P. H., "Fire Modelling: A Mature Technology," *Fire Safety Journal*, Vol. 19, No. 2–3, 1992, pp. 125–140.

Thompson, P. A., and Marchant, E. W., "SIMULEX: Developing New Computer Modelling Techniques for Evaluation," *Proceedings* of the Fourth International Symposium on Fire Safety Science, Intl. Assoc. for Fire Safety Science, Boston, MA, 1994, pp. 613–624.

Thompson, P. A., and Marchant, E. W., "Testing and Application of the Computer Model 'SIMULEX'," *Fire Safety Journal*, Vol. 24, No. 2, 1995, pp. 149–166.

Tubbs, J. S., and Barnett, J. R., "Modeling the NIST High Bay Fire Experiment With JASMINE," *Proceedings* of the International Conference on Fire Research and Engineering, Sept. 10–15, 1995, Orlando, FL, SFPE, Boston, MA, 1995, pp. 383–388.

Urbas, J., "Using ICAL to Obtain Input Data for Fire Models," [Abstract Only], *Proceedings* of the Twentieth International Conference on Fire Safety, 20th, Volume 20, Jan. 9–13, 1995, Millbrae, CA, Product Safety Corp., Sunnyvale, CA, 1995, p. 178.

Walton, W. D., "Introduction to Mathematical Fire Modeling. [Book/Software Review of An Introduction to Mathematical Fire Modeling by David M. Birk.]," *Fire Technology*, Vol. 28, No. 1, Feb. 1992, pp. 87–91.

Walton, W. D., "Zone Computer Fire Models for Enclosures," National Institute of Standards and Technology, Gaithersburg, MD, NFPA SFPE 95, LC Card Number 95-68247, ISBN 0-87765-354-2; *The SFPE Handbook of Fire Protection Engineering*, 2nd Edition, Section 3, Chapter 7, National Fire Protection Assoc., Quincy, MA, 1995, pp. 3/148–151.

Watts, J. M., Jr., "Fire Risk Evaluation Model. Software Review," *Fire Technology*, Vol. 31, No. 4, 4th Quarter, Nov. 1995, pp. 369–371.

Watts, J. M., Jr., "Future of Fire Modeling. Editorial," *Fire Technology*, Vol. 27, No. 2, May 1991, pp. 97–98.

Wilson, J. L., and Bryan, P. Y., "3-D Modeling as a Tool to Improve Integrated Design and Construction," Lehigh Univ., Bethlehem, PA, Source Document 104, Nov. 1994.

Yung, D., "FIRECAM: A Computer Fire Risk Evaluation and Cost Assessment Model," National Research Council of Canada, Ottawa, Ontario, IRC-ORAL-142; 1996 *Proceedings* of the Institution of Fire Engineers (Ontario Branch), Ontario, Canada, 1996, pp. 1–18.

DETERMINISTIC COMPUTER FIRE MODELS

William D. Walton and Edward K. Budnick

Computer models are simply computer programs that model or simulate a process or phenomena. Computer models have been used for some time in the design and analysis of fire protection hardware. The use of computer models, commonly known as design programs, has become the industry's standard method for designing water supply and automatic sprinkler systems. These programs perform large numbers of tedious and lengthy calculations and provide the user with accurate, cost-optimized designs in a fraction of the time required for manual procedures.

In addition to the design of fire protection hardware, computer models may also be used to evaluate the effects of fire on people and property. These computer fire models can provide a faster and more accurate estimate of the impact of a fire and the measures used to prevent or control the fire than many of the methods previously used. While manual calculation methods provide good estimates of specific fire effects (e.g., prediction of time to flashover), they are not well suited for comprehensive analyses involving the time-dependent interactions of multiple physical and chemical processes present in developing fires.

In recent years, increasing attention has been given to the development and use of computer fire models. They have been used by engineers and architects for building design, building officials for plan review, the fire service for pre-fire planning, investigators for post-fire analysis, groups writing fire codes, materials manufacturers, fire researchers, and educators. While these models are not a replacement for building and fire codes, they still can be valuable tools for the fire professional.

The state of the art in computer fire modeling is changing rapidly. Understanding of the processes involved in fire growth is improving, and thus the technical basis for the models is improving. The capabilities, documentation, and support for a given model can change dramatically over a short period of time. In addition, computer technology itself (both software and hardware) is advancing rapidly. A few years ago, a large mainframe computer was required to use most of the computer fire models. Today, almost all of the models can be run on personal computers. Therefore, rather than

provide an exhaustive discussion of rapidly changing state-of-the-art available computer models, the following discussion will focus on a representative selection. The reader should refer to the bibliography at the end of this chapter for in-depth reviews of particular models.

In general, computer models for fire hazard prediction can be grouped into two categories: (1) enclosure fire models and (2) special-purpose fire models.

ENCLOSURE FIRE MODELS

Major advancements have occurred in the development of computational models structured to predict the interaction of multiple fire processes involving heat transfer, fluid mechanics, and combustion chemistry occurring at the same time in an enclosure. These models provide estimates of particular elements of hazard development such as fire growth, temperature rise, and smoke generation and transport. Some models are able to address multiple rooms. Others are confined to the room of fire origin. Generally, the large number of mathematical expressions to be solved simultaneously in any of these models necessitates the use of a computer.

There are two general classes of computer models for analyzing enclosure fire development: (1) probabilistic and (2) deterministic.

Probabilistic Models

Probabilistic models treat fire growth as a series of sequential events or states. These models are sometimes referred to as state transition models. Mathematical rules are established to govern the transfer from one event to another (e.g., from ignition to established burning). Probabilities are assigned to each transfer point based on analysis of relevant experimental data and historical fire incident data. These models do not normally make direct use of the physical and chemical equations describing the fire processes.

Deterministic Models

In contrast, deterministic models represent the processes encountered in a compartment fire by interrelated mathematical expressions based on physics and chemistry. These models may also be referred to as room fire models, computer fire models, or mathematical fire models. Ideally, such models represent the ultimate capability, which means that discrete changes in any physical parameter

William D. Walton, P.E., is a research fire protection engineer in the Building and Fire Research Laboratory of the U.S. National Institute of Standards and Technology, Gaithersburg, MD. Edward K. Budnick, P.E., is a vice president and senior engineer with Hughes Associates, Inc., based in Baltimore, MD.

could be evaluated in terms of the effect on fire hazard. While the state of the art in understanding fire processes will not yet support the ultimate model, a number of computer models are available that provide reasonable estimates of selected fire effects.[1,2]

Zone models: The most common type of physically based fire model in North America is the zone or control volume model, which solves the conservation equations for distinct regions (control volumes). A number of zone models exist, varying to some degree in the detailed treatment of fire phenomena. The dominant characteristic of this class of model is that it divides the room(s) into a hot upper layer and a lower cooler layer. The model calculations provide estimates of key conditions for each of the layers as a function of time.[3] Zone models have proved to be a practical method for providing first-order estimates of fire processes in enclosures.

Field models: The other general type of deterministic model is the field model or computational fluid dynamics (CFD) model. This type of model solves the fundamental equations of mass, momentum, and energy at each element in a compartment space that has been divided into a grid of small elements. Imagine an enclosure filled with a 3-D grid of tiny cubes; a field model will calculate the physical conditions in each cube as a function of time. The calculation will account for physical changes generated within the cube and changes in the cube originating from surrounding cubes. This model will permit the user to determine the conditions at any point in the compartment.

Figure 11-5A shows an example of a field model prediction for the gas velocity in a compartment fire. The figure shows the cross-section of a compartment with the door to the right of center and a fire near the floor to the left. The arrows represent the predicted gas velocity with the arrows pointing in the direction of gas flow, and the speed of the flow indicated by the length of the arrows. The prediction shows air entering the lower level of the compartment to the right, rising in the fire, and flowing under the door lintel as it leaves the compartment.

Generally, field models cannot be used with the current generation of personal computers. The computational demands and memory requirements necessitate the use of powerful desktop workstations, minicomputers, or mainframe/super computers to perform efficiently. (For example, an attempt to model an industrial rack storage fire with a field model on a current generation workstation using a computational domain of around 514,000 cells and a time step of 0.2 sec required 18 days of computer time.)[4] Minimum start-up costs associated with both the software programs and computer hardware to operate typical field models can conservatively range from $50,000 to $100,000.

In spite of the expense, field modeling is growing in popularity as the demand for more detailed fire hazard analyses has evolved. Field models have recently been successfully used to evaluate a wide range of fire problems including the well-known effort to simulate the King's Cross fire.[5] Several efforts to validate specific field models against limited experimental data have been performed.[6–8] An attempt to validate the accuracy of the commonly used "k-e" turbulence model to predict velocity and temperature fields in buoyant flows demonstrated the need to modify significantly the default values of physical constants, such as the turbulent viscosity constant of proportionality and the turbulent Prandtl number, in order to obtain reasonable agreement with experimental measurements.[9] Therefore, caution should be employed in adapting any of the currently available field models while the validation process continues.

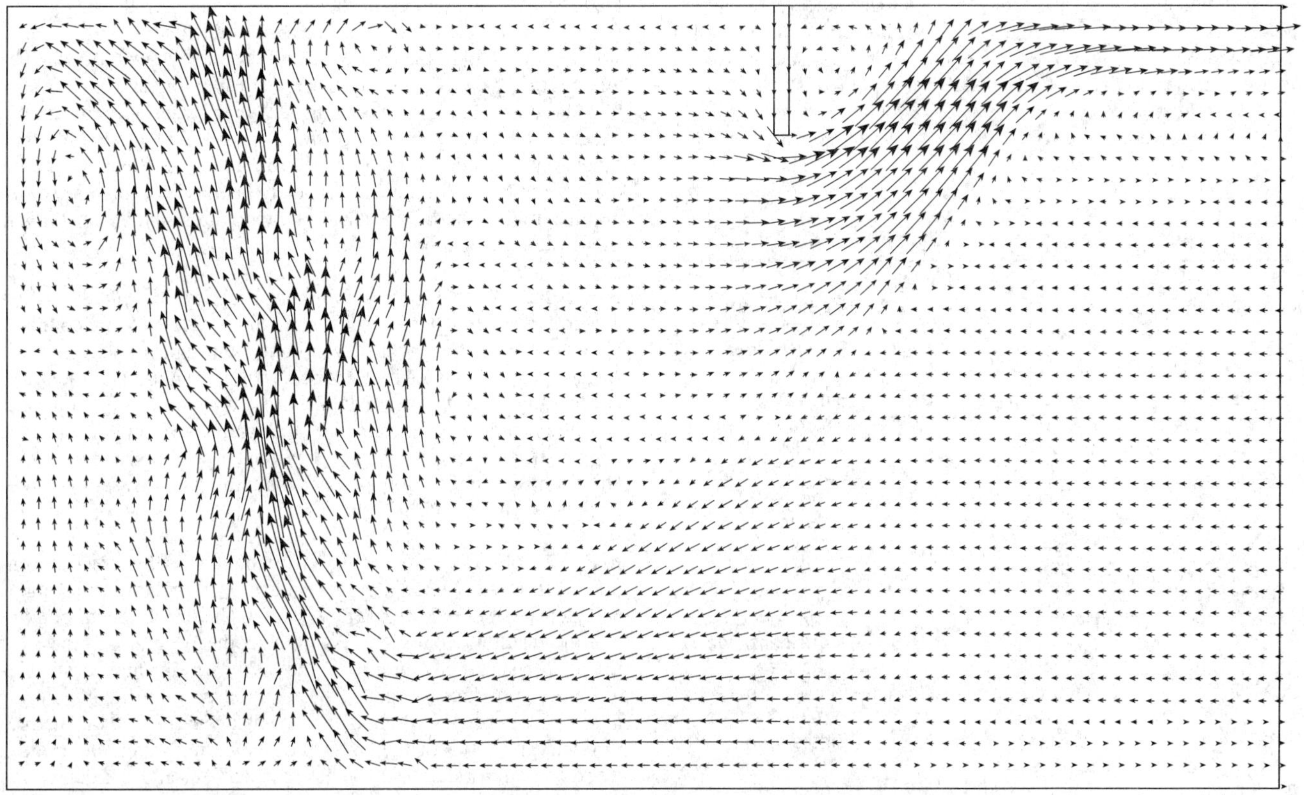

FIG. 11-5A.   *Field model prediction of gas velocity in a compartment fire.*

OVERVIEW OF SELECTED ZONE MODELS

Table 11-5A lists a number of zone-type enclosure fire models that are widely used. A more detailed description of each model is also given.

ASET: ASET (Available Safe Egress Time) is a program for calculating the temperature and position of the hot smoke layer in a single room with closed doors and windows.[10] ASET can be used to determine the time to the onset of hazardous conditions for both people and property. The required program inputs are the heat loss fractions, the height of the fuel above the floor, criteria for hazard and detection, the room ceiling height, the room floor area, a heat release rate, and a species generation rate of the fire (optional). The program outputs are the temperature, thickness, and (optional) species concentration of the hot smoke layer as a function of time, and the time to hazard and detection. ASET can examine multiple cases in a single run.

ASET-B: ASET-B is a program for calculating the temperature and position of the hot smoke layer in a single room with closed doors and windows.[11] ASET-B is a compact version of ASET designed to run on personal computers. The required program inputs are a heat loss fraction, the height of the fire, the room ceiling height, the room floor area, the maximum time for the simulation, and the rate of heat release of the fire. The program outputs are the temperature and thickness of the hot smoke layer as a function of time. Species concentrations and time to hazard and detection calculated by ASET are not calculated in the compact ASET-B version.

CCFM (version VENTS): CCFM (Consolidated Compartment Fire Model) (version VENTS) is a two-layer zone-type compartment fire model computer code.[12–15] It simulates conditions caused by user-specified fires in a multiroom, multilevel facility. The required inputs are a description of room geometry and vent characteristics (up to nine rooms and twenty vents), initial state of the inside and outside environments, fire energy release rates as functions of time (up to twenty fires), and a user-specified heat loss fraction. If simulation of concentrations of products of combustion is desired, then product release rates must also be specified (up to three products). Vents can be simple openings between adjacent spaces (natural vents) or fan/duct-forced ventilation systems between arbitrary pairs of spaces (forced vents). For forced vents, flow rates and direction can be user-specified or included in the simulation by accounting for user-specified fan and duct characteristics. Wind and stack effects can be taken into account. The program outputs for each room are pressure at the floor, layer interface height, upper/lower layer temperature, and (optionally) product concentrations. CCFM (version VENTS) is supported by four-part documentation.

CFAST: The CFAST (Consolidated model of Fire growth And Smoke Transport) program is an upgrade of the FAST (Fire And Smoke Transport) program, and incorporates new and faster numerical solution techniques originally implemented in CCFM.[16,17] CFAST is a multiroom fire model that predicts the conditions within a structure resulting from a user-specified fire. CFAST version 2.0.1 can accommodate up to 15 compartments with multiple openings between the rooms and to the outside. The required program inputs are the geometrical data describing the rooms and connections; the thermophysical properties of the ceiling, walls, and floors; the fire as a rate of mass loss; and the generation rates of the products of combustion. The program outputs are the temperature and thickness of, and species concentrations in, the hot upper layer and the cooler lower layer in each compartment. Also given are surface temperatures and heat transfer and mass flow rates. CFAST also includes mechanical ventilation (up to 30 ducts and 5 fans), new heat transfer algorithms, a ceiling jet algorithm, capability of multiple fires (up to 16), and more accurate solutions due to new pressure equation formulations.

COMPBRN III: COMPBRN III has been generally used in conjunction with probabilistic analysis for the assessment of risk in the nuclear power industry.[18] The model is based on the assumption of a relatively small fire in a large space or fire involving large fuel loads early during the pre-flashover fire growth period. The model's emphasis is on the thermal response of elements within the enclosure to a fire within the enclosure and on modeling simplicity. The temperature profile within each element is computed and an element is considered ignited or damaged when its surface temperature exceeds the user-specified ignition or damage temperature. The model outputs include the total heat release rate of the fire, the temperature and depth of the hot gas layer, the mass burning rate for individual fuel elements, the surface temperatures, and the heat flux at user-specified locations.

COMPF2: COMPF2 is a computer program for calculating the characteristics of a post-flashover fire in a single building compartment, based on fire-induced ventilation through a single door or window.[19] It is intended both for performing design calculations and for the analysis of experimental burn data. Wood, thermoplastic, and liquid fuels can be treated. A comprehensive output format is provided that gives gas temperatures, heat flow terms, and flow variables. The documentation includes input instructions, sample problems, and a listing of the program.

FIRST: FIRST (FIRe Simulation Technique) is the direct descendant of the HARVARD V program developed by Howard Emmons and Henri Mitler.[20] The program predicts the development of a fire and the resulting conditions within a room, given a user-specified fire or user-specified ignition. It predicts the heating and possible ignition of up to three targets. The required program inputs are the geometrical data describing the room and openings, and the thermophysical properties of the ceiling, walls, burning fuel, and targets. The generation rate of soot must be specified, and the generation rates of other species may be specified. The fire may be entered either as a user-specified time-dependent mass loss rate or in terms of

TABLE 11-5A. *Enclosure Zone Fire Models*

Model Name	Author(s)	Maintaining Organization	Special Features
ASET	L. Y. Cooper D. W. Stroup	NIST	Single room
ASET-B	W. D. Walton	NIST	Single room
CCFM-VENTS	L. Y. Cooper G. P. Forney	NIST	Multiroom, multilevel
CFAST	W. W. Jones R. D. Peacock G. P. Forney P. Reneke R. Portier	NIST	Up to 15 compartments, 30 ducts, and 5 fans
COMPBRN III	N. Siu V. Ho	UCLA	Single room, developed for nuclear power facilities
COMPF2	V. Babrauskas	NIST	Single room, post-flashover
FIRST	H. W. Emmons H. E. Mitler	NIST	Single room, burning item
WPI/FIRE	D. B. Satterfield J. R. Barnett	WPI	Single room, ceiling vents

fundamental properties of the fuel. Among the program outputs are the temperature and thickness of, and species concentrations in, the hot upper layer and also in the cooler, lower layer in the compartment. Also given are surface temperatures, and heat transfer and mass flow rates.

WPI/FIRE: WPI/FIRE is a direct descendant of the HARVARD V and FIRST programs.[21] It includes all of the features of the HARVARD program version 5.3 and many of the features of the FIRST program. WPI/FIRE also includes the following features: improved input routine, momentum-driven flow through ceiling vents, two different ceiling jet models for use in detector activation, forced ventilation for ceiling and floor vents, and an interface to a finite difference computer model for the calculation of boundary surface isotherms and hot spots.

OVERVIEW OF SELECTED FIELD MODELS

Table 11-5B provides a list of available field models. Selected models from these lists are described below.

BF3D: BF3D is a computational model of three-dimensional buoyant flow in a single enclosure due to a fire source.[22] No turbulence model or other empirical parameters are introduced. The current algorithms have been verified by comparisons with exact solutions to the equations in simple cases, and predictions of the overall model have been compared with experimental results. The use of Lagrangian particle tracking allows visualization of the three-

dimensional flow patterns. The model is used mainly for research and is generally not available in the public domain.

FISCO-3L: FISCO-3L is a three-dimensional single-room field model for unsteady and compressible buoyant heat flow.[23] Fire can be simulated under both natural and forced ventilation. An option is available for simulating fire suppression by water sprinklers. The program is menu-driven with a graphical user interface and a real-time system for graphical display of results. The algorithms used to calculate turbulence, flame region effects, and combustion processes are simplified in order to reduce computation demands. The model is copyright restricted.

FLOW-3D: FLOW-3D is a general-purpose computational fluid dynamics (CFD) code. The code includes a CFDS Environment for mesh generation and post-processing, and is capable of time-dependent or steady-state heat and mass transfer, and two- or three-dimensional coordinate systems. Model features include body-fitted coordinates, moving and adaptive grids, heat transfer in solid regions, porous media approximation, turbulence modeling, compressibility, scalar transport equations, and discrete particle transport. An optional feature is multifluid modeling. Additional information can be obtained from Harwell Laboratory (U.K.) or Computational Fluid Dynamics Services (U.S.).

JASMINE: JASMINE uses PHOENICS, a CFD code for computation of fluid motion, and provides full three-dimensional solutions for heat and mass transfer.[24] It solves the full partial differential equations describing the conservation of mass, momentum, energy, and species, using a two equation model for turbulence together with simple radiation models. Primary input requirements include a

TABLE 11-5B. *Field Fire Models*

Model Name	Author(s)	Maintaining Organization	Applications and Special Features	Required Hardware
BF3D	H. R. Baum R. G. Rehm D. W. Lozier D. M. Corley	NIST (U.S.)	Single room treatment of buoyant heat-driven flows	Large memory needed for adequate resolution; mainframe, minisuper, or supercomputer
FISCO-3L	V. Scheider J. Hoffmann	INTELLEX, FR (Germany) SINTEF (Norway)	Single room treatment of fire development	80386 chip-based PC computer, MS-DOS operating system with math co-processor, EGA or VGA graphical display and 640 kB memory
FLOW-3D	Harwell Laboratory	Harwell Laboratory (U.K.)	General-purpose computational fluid dynamics (CFD) code	Supercomputer, mainframe, mini, or workstation
JASMINE	G. Cox S. Kumar	Fire Research Station (U.K.)	Analysis of smoke movement	Mini, super mini, VAX preferred
KAMELEON FIRE E-3D	B. F. Magnussen	NTH/SINTEF (Norway)	Single room fire growth model for pool fires	Post-processor needs MS-DOS and VGA or UNIX X-WINDOWS; 640 kB
KAMELEON II	B. F. Magnussen	NTH/SINTEF (Norway)	Multiroom fire and smoke spread	Post- and pre-processor are MS-DOS and VGA display or UNIX X-WINDOWS; 640 kB
KOBRA-3D	INTELLEX	INTELLEX (Germany)	3-D field model for determining hydrodynamical flow in a single fire compartment	IBM-compatible PC, MS-DOS 3.0, EGA graphics, co-processor recommended
PHOENICS	D. B. Spalding	CHAM, LTD. (U.K.)	A general-purpose 3-D transient fluid dynamics code	Supercomputer, mainframe, mini, or workstation
RMFIRE	G. Hadjisophocleous	National Research Council of Canada (Canada)	A 2-D field model for transient calculations of smoke movement in a room fire	Workstation
SPLASH	A. J. Gardiner	South Bank Polytechnic (U.K.)	A quasi-field model describing the interaction of sprinkler sprays with fire gases	VAX
STAR*CD	D. Gossman R. Issa	Computational Dynamics (U.K.)	General-purpose computational fluid dynamics (CFD) code	UNIX workstations and supercomputer

description of the fire source, the thermal properties of the structure, structure geometry, and ventilation conditions. Use of the code is through the Fire Research Station (U.K.).

KAMELEON FIRE E-3D: KAMELEON FIRE E-3D is a three-dimensional field model for transient calculation of pool fires in a single enclosure.[25] The model can be applied to problems involving multiple natural or forced vents. The fire source is characterized by either a predetermined leakage rate or a pool fire (liquid hydrocarbons only). Turbulence is modeled by k-e, and combustion and soot by eddy dissipation. Radiation is included. SINTEF (Norway) provides modeling services using this model, but the model is not directly available commercially.

KAMELEON II: KAMELEON II is an enhanced field model that is optimized for vector performance.[25] It has a graphical pre-processor that allows the user to generate input simply by making drawings. The post-processor provides color graphics of any cross-section and any variable in the calculation domain. The model predicts the spread of smoke and exhaust gases in complex multienclosure geometries and open configurations. Turbulence is modeled by k-e, and combustion and soot by eddy dissipation. Radiation is also modeled. A key input is the burning rate of the fire to be modeled. SINTEF (Norway) will perform customer calculations, but the code is not available commercially.

KOBRA-3D: KOBRA-3D is a three-dimensional field model for calculating the unsteady and compressible heat flow in a natural- or forced-ventilated compartment. The development of a fast converging iteration procedure for solving the hydrodynamic equations allows for effective use on a high-performance personal computer. The fire source can be defined by semi-empirical relations or by defining constant or time-dependent heat release rates. Turbulence modeling is not included. The model is embedded in a menu-driven user surface combined with data bases for fuel and room geometry input data. On-line graphic displays of temperature and velocity contours is possible. Additional information can be obtained from the developer, INTELLEX (Germany).

PHOENICS: PHOENICS is a computational fluid dynamics (CFD) code that includes calculation routines for turbulence effects, heat transfer, chemical reaction, multiphase behavior of fluids, and complex geometries.[26] Output displays include perspective views, contour mapping, vector diagrams, and gradients. Several field models developed for fire hazard applications rely on this code to perform the fluid dynamics processes. It is commercially available.

RMFIRE: RMFIRE is a two-dimensional field model developed for analysis of unsteady smoke movement and heat transfer in a fire compartment.[7] The governing equations are solved in boundary-fitted coordinate systems that allow compartments with complex geometries to be evaluated. Inputs include boundary conditions, initial conditions, and the fire heat release rate history. Output includes temperature contours, as well as velocity and pressure values within the compartment. This model will become available at a later date. Currently NRCC (Canada) will perform the calculations.

SPLASH: SPLASH is a quasi-field model describing the interaction of sprinkler sprays with fire gases in corridors. Inputs include detailed sprinkler parameters, corridor geometry, and smoke layer characteristics.[27] The model provides information on the effects of the spray on the smoke layer conditions, variation in heat transfer and drag to buoyancy ratio in the spray volume, the thermal and physical histories of individual droplets in the spray, and the water delivery pattern on the floor.

STAR*CD: STAR*CD is a general-purpose computational fluid dynamics (CFD) code.[28] STAR*CD is designed to solve a wide range of flow phenomena, including steady and transient, laminar and turbulent (from a choice of turbulence models), incompressible and compressible, heat transfer (convection, conduction, and radiation), mass transfer and chemical reaction (including combustion), porous media, and multiple fluid streams and multiphase flows.

SPECIAL-PURPOSE MODELS

Special-purpose deterministic computer models include computer models designed for special-purpose analyses such as: structural fire resistance, prediction of response time of heat detectors and automatic sprinklers, the design of sprinkler systems, and performance of smoke-control or ventilation systems. These models may require or permit coupling with other special-purpose models or with a more general enclosure fire development model.

Table 11-5C lists a number of special-purpose deterministic fire models that are widely used. A more detailed description of each model follows.

TABLE 11-5C. *Special-Purpose Models*

Model Name	Author(s)	Maintaining Organization	Model Type	Special Features
ASCOS	J. H. Klote	NIST	Smoke control	Steady-state network flow model for smoke-control evaluation, no fire condition
BREAK1	A. A. Joshi P. J. Pagni	U. C. Berkeley	Glass breakage	Calculates glass breakage for window exposed to a compartment fire
DETACT-T2	D. W. Stroup	NIST	Thermal device activation	Calculates actuation time for heat detectors and sprinklers, time-squared fires
DETACT-QS	D. D. Evans	NIST	Thermal device activation	Calculates actuation time for heat detectors and sprinklers, user-defined fires
FIRES-T3	R. H. Idling B. Bresler Z. Nizamuddin	U. C. Berkeley	Structural heat transfer	Three-dimensional heat transfer through structural assemblies
LAVENT	W. D. Davis L. Y. Cooper	NIST	Thermal device activation	Calculates actuation time for sprinklers and link-actuated vents with draft curtains

ASCOS: ASCOS (Analysis of Smoke COntrol Systems) is a program for steady air-flow analysis of smoke-control systems.[29] This program can analyze any smoke-control system that produces pressure differences with the intent of limiting smoke movement in building fire situations. The input consists of the outside and building temperatures, a description of the building flow network, and the flows produced by the ventilation or smoke-control system. The output consists of the steady-state pressures and flows throughout the building.

BREAK1: BREAK1 (Berkeley algorithm for BREAKing window glass in a compartment fire) is a program that calculates the temperature history of a glass window exposed to user-described fire conditions.[30] The calculations are stopped when the glass breaks. The inputs required are the glass thermal conductivity, thermal diffusivity, absorption length, breaking stress, Young's modulus, thermal coefficient of linear expansion, thickness, emissivity, shading thickness, half-width of window, the ambient temperature, numerical parameters and the time histories of flame radiation from the fire, hot layer temperature and emissivity, and heat transfer coefficients. The outputs are temperature history of the glass normal to the glass surface, and the window breakage time.

DETACT-T2: DETACT-T2 (DETector ACTuation-Time squared) is a program for calculating the actuation time of thermal devices below unconfined ceilings.[31] It can be used to predict the actuation time of fixed-temperature and rate-of-rise heat detectors and of sprinkler heads subject to a user-specified fire that grows as the square of time. DETACT-T2 assumes that the thermal device is located in a relatively large area; that is, only the fire ceiling flow heats the device, and there is no heating from the accumulated hot gases in the room. The required program inputs are the ambient temperature, the response time index (RTI) for the device, the activation and rate-of-rise temperatures of the device, the height of the ceiling above the fuel, the device spacing, and the fire growth rate. The program outputs are the time to device activation and the heat release rate at activation.

DETACT-QS: DETACT-QS is a program for calculating the actuation time of thermal devices below unconfined ceilings.[32] It can be used to predict the actuation time of fixed-temperature heat detectors and sprinkler heads subject to a user-specified fire. DETACT-QS assumes that the thermal device is located in a relatively large area; that is, there is no accumulated hot gas layer in the room so only the fire ceiling flow heats the detection device. The required program inputs are the height of the ceiling above the fuel, the distance of the thermal device from the axis of the fire, the actuation temperature of the thermal device, the response time index (RTI) for the device, and the rate of heat release of the fire. The program outputs are the ceiling gas temperature at the device location and the device temperature both as a function of time and the time required for device actuation.

FIRES-T3: FIRES-T3 (FIre REsponse of Structures—Thermal 3-dimensional version) is a finite-element computer model designed to analyze heat transfer through structural assemblies.[33] A wide variety of structural assemblies can be examined with FIRES-T3, including columns, walls, beams, floor/ceiling assemblies, and roof/ceiling assemblies. The structural assembly may be solid or include air cavities. The input requirements include a description of the structural assembly and the fire exposure. The information necessary to describe the column assembly includes geometric factors (dimensions, shape of assembly) and material property values (thermal conductivity, specific heat, and density) as a function of temperature. The fire exposure is characterized in terms of the temperature of the surrounding media and appropriate heat-transfer coefficients. The output is a tabulation of the temperatures within the structural assembly as a function of time.

LAVENT: LAVENT (Link-Actuated VENT) is a program developed to simulate the environment and the response of sprinkler links in compartment fires with draft curtains and fusible-link-actuated ceiling vents.[34] The model used to calculate the heating of the fusible links includes the effects of the ceiling jet and the upper layer of hot gases beneath the ceiling. The required program inputs are the geometrical data describing the compartment, the thermophysical properties of the ceiling, the fire elevation, the time-dependent energy release rate of the fire, the fire diameter or energy release rate per area of the fire, the ceiling vent area, the fusible-link response time index (RTI) and fuse temperature, the fusible-link positions along the ceiling, the link assignment to each ceiling vent, and the ambient temperature. A maximum of five ceiling vents and ten fusible links are permitted in the compartment. The program outputs are the temperature, mass and height of the hot upper layer, the temperature of each link, the ceiling jet temperature and velocity at each link, the radial temperature distribution along the interior surface of the ceiling, the radial distribution of the heat flux to the interior and exterior surfaces of the ceiling, the fuse time of each link, and the vent area that has been opened.

COMBINED MODELS

A new category of deterministic fire models has emerged that can be called combined models or model suites. The first of these were FIREFORM[35] and HAZARD I.[36–38] The combined models contain several models under the control of a single program. The individual models may include zone fire models, egress models, and engineering calculations. The combined models may include fire models that are available in stand-alone form or fire models that are only available as part of the combined model. Table 11-5D lists a number of combined deterministic fire models that are widely used. A more detailed description of each model follows.

ASKFRS: ASKFRS is a collection of fire safety engineering routines.[39] The routines include fire heat release rate, flame height, fire plume calculations, plume rise, compartment hot gas layer temperature, flashover, smoke filling time, roof venting calculations, egress, and toxic threat.

ASMET: ASMET (Atria Smoke Management Engineering Tools) consists of a set of equations and a zone fire model for analysis of smoke management systems for large spaces, such as atria,

TABLE 11-5D. *Combined Models*

Model Name	Author(s)	Maintaining Organization	Special Features
ASKFRS	R. Chitty G. Cox	FRS	Collection of fire safety engineering routines
ASMET	J. H. Klote	NIST	Collection of routines for smoke management in large spaces
FIRECALC	V. O. Shestopal S. J. Grubits	CSIRO	Collection of fire safety engineering routines
FPETOOL	H. E. Nelson S. Deal	NIST	Collection of fire safety engineering routines
HAZARD I	R. D. Peacock W. W. Jones R. W. Bukowski C. L. Forney	NIST	Predicts hazards to building occupants from fire (includes CFAST, EXITT, and DETACT-QS)

shopping malls, arcades, sports arenas, exhibition halls, and airplane hangars.[40] Routines calculate the height of the smoke layer during atrium filling from a steady or an unsteady fire, and the atrium filling time for a steady or an unsteady fire. Routines also calculate the plume mass flow rate, centerline temperature, and average temperature with or without a virtual origin correction. ASMET also contains a C language version of the ASET-B program.

FIRECALC: FIRECALC was developed from the original FPE-TOOL and is a collection of fire safety engineering routines.[41,42] About half of the routines in FIRECALC are the same as routines found in FPETOOL. FIRECALC also contains routines for steel beam load-bearing capacities, atrium smoke temperature, plume flow and temperatures, smoldering fires, smoke control with a common plenum, fire resistance time, and thermal radiation. FIRE-CALC has a one- and two-room hot layer model and a one-room natural ventilation model.

FPETOOL: FPETOOL is the descendant of the FIREFORM program.[43] It contains a computerized selection of relatively simple engineering equations and models useful in estimating the potential fire hazard in buildings. The calculations in FPETOOL are based on established engineering relationships. The FPETOOL package addresses problems related to fire development in buildings and the resulting conditions and response of fire protection systems. The subjects covered include smoke filling in a room, sprinkler/detector activation, smoke flow through (small) openings, temperatures and pressures developed by fires, flashover and fire severity predictions, fire propagation (in special cases), and simple egress estimation. The largest element in FPETOOL is a zone fire model called FIRE SIMULATOR. FIRE SIMULATOR is designed to estimate conditions in both pre- and post-flashover enclosure fires. The inputs include the geometry and material of the enclosure, a description of the initiating fire, and the parameters for sprinklers and detectors being tracked. The outputs include the temperature and volume of the hot smoke layer; the flow of smoke from openings; the response of heat-actuated detection devices, sprinklers and smoke detectors; oxygen, carbon monoxide, and carbon dioxide concentrations in the smoke; and the effects of available oxygen on combustion. FPE-TOOL also contains a routine to predict the characteristics of a moving smoke wave in a corridor, and a routine that predicts smoke conditions developing in a room and the subsequent reduction in human viability resulting from exposure to the conditions.

HAZARD I: HAZARD I is a method for predicting the hazards to the occupants of a building from a fire therein.[36–38] Within prescribed limits, HAZARD I allows one to predict the outcome of a fire in a building populated by a representative set of occupants in terms of which persons successfully escape and which are killed, including the time, location, and likely cause of death for each. HAZARD I is a set of procedures combining expert judgment and calculations to estimate the consequences of a specified fire. These procedures involve four steps: (1) defining the context, (2) defining the scenario, (3) calculating the hazard, and (4) evaluating the consequences. Steps 1, 2, and 4 are largely judgmental and depend on the expertise of the user. Step 3, which involves use of the extensive HAZARD I software, requires considerable expertise in fire safety practice. The heart of HAZARD I is a sequence of procedures implemented in computer software to calculate the development of hazardous conditions over time, calculate the time needed by building occupants to escape under those conditions, and estimate the resulting loss of life based on assumed occupant behavior and tenability criteria. These calculations are performed for a specified building and set of fire scenarios of concern. HAZARD I consists of a three-volume report and a set of computer disks containing the software necessary to conduct hazard analyses of products used in residential occupancies. The HAZARD I software is used to make

detailed predictions of the fire-generated environment within a building; the evacuation process of occupants as they interact with the building, the fire, and each other; and the fate of the occupants as they either successfully escape or are killed. The underlying science—including a detailed presentation of the equations, constants, and assumptions contained in each of the programs—and a set of worked example cases are contained in the *Technical Reference Guide*. The *Software User's Guide* includes detailed instructions on use of the software and an extensively illustrated learning section that "walks" the user through a worked example. Specific applications depend on the user, but some include material/product performance evaluation, fire reconstruction and litigation, evaluation of code changes or variances, fire department pre-planning, and extrapolation of fire test data to additional physical configurations.

EGRESS MODELS

A growing number of deterministic fire-elated models deal with the topic of egress.[44] Although egress models are not fire models, they have been developed in response to the need to evaluate impact of fires on the occupants of a building. Egress models are often used with fire models in performing a hazard analysis. Most egress models describe the building as a network of paths along which the occupants travel. The occupant travel rates are usually derived from studies on people movement and vary with the age and ability of the occupant, crowding, and the type of travel path. Table 11-5E lists a number of prominent fire egress models. A more detailed description of each model is given below.

TABLE 11-5E. *Egress Models*

Model Name	Author(s)	Maintaining Organization	Special Features
ELVAC	J. H. Klote	NIST	Evacuation using elevators
EVACNET+	T. M. Kisko R. L. Francis	U. of Florida	Optimum evacuation routes and times
EXITT	B. M. Levin	NIST	Occupant decisions and actions during fire
EXIT89	R. F. Fahy	NFPA	Large building evacuation

ELVAC: ELVAC (ELevator eVACuation) is an interactive computer program that estimates the time required to evacuate people from a building with the use of elevators and stairs.[45] It is cautioned that elevators are generally not intended as a means of fire evacuation, and they should not be used during fires. However, it is possible to design elevator systems for fire emergencies, and ELVAC can be used to evaluate the potential performance of such systems. ELVAC calculates the evacuation time for one group of elevators. If the building has more than one group of elevators, ELVAC can be run on each group separately. Input consists of floor-to-floor heights, number of people on floors, number of elevators in the group, elevator speed, elevator acceleration, elevator capacity, elevator door type and width, and various inefficiency factors. The output is a table of elevator travel time, round-trip time, people moved, number of round trips for each floor, and the total evacuation time.

EVACNET+: EVACNET+ (EVACuation NETwork computer model) is an interactive computer program that models emergency building evacuation.[46] It consists of a network of nodes and arcs to represent building locations and connecting paths. Input for this model includes a detailed geometry of the building and contents (multiple rooms and floors can be addressed), and information regarding the initial location of the occupants. Output can be selected from a menu and generally takes the form of optimum evacuation times and identification of impediments to smooth, orderly evacuation.

EXITT: EXITT is a discrete-event simulation of occupant decisions and actions in a simulated fire.[47] Before the simulation starts, the characteristics of a residence, a fire in that residence, and the occupants of the residence are entered into the computer. Based on a large set of decision rules, the occupants "make" decisions that are a function of the smoke conditions in the building, the characteristics and status of the occupants (including their capabilities), and the available travel routes. The occupants investigate the fire, alert and assist others, and evacuate the building. The simulation ends when all the occupants either are out of the building or are trapped by the smoke.

EXIT89: EXIT89 is an evacuation model for large buildings.[48] EXIT89 requires as input a network description of the building, geometrical data for each room and for openings between rooms, the number of occupants located at each node throughout the building, and smoke data if the effect of smoke blockage is to be considered. The user can select to have the occupants follow the shortest path out of the building or use familiar routes, whether smoke data is used and if it comes from a fire model or is input as blockages, whether there are delays within specified times, and if any occupants are disabled with a specified reduction in travel speed. The model either calculates the shortest route from each building location to a location of safety or follows user defined routes through the building. It moves people along routes until they are blocked by smoke, then recalculates the routes until everyone who can escape reaches a safe location that is usually outside the building. All occupants can begin evacuation at the same time or be assigned delays, and additional delays can be specified or randomly assigned to occupants. The program can output the location of each occupant with time, floor clearing times, stairwell clearing times, exit clearing times, and how many occupants used each exit.

WILDLAND FIRE MODELS

Wildland fire models used in the United States began as a series of charts, tables, and formulas that were computerized to speed their use.[49] Initially these were implemented as custom chips for programmable calculators and later as computer programs. These deterministic models are used for pre-fire planning like other fire models, but unlike other models they are also used real-time during the fire to aid in fire management. Table 11-5F lists two wildland fire models, the first of which is widely used, and the second of which represents the next generation of programs. A more detailed description of each model is also given.

TABLE 11-5F. *Wildland Fire Models*

Model Name	Author(s)	Maintaining Organization	Special Features
BEHAVE	P. L. Andrews C. H. Chase	U.S. Forest Service	System of modules for a wide range of wildland fire predictions
FARSITE	M. A. Finney	U.S. Forest Service	GIS Interface

BEHAVE: BEHAVE is a collection of wildland fire behavior modules in an interactive program.[50,51] In some cases, output from one of the modules may be used as input to another module. Typical module inputs include fuel description parameters, moisture content, slope, wind speed, temperature, and mode of attack. The modules and their principal output features include CONTAIN—containment time and final fire size; DIRECT—rate of spread, flame height, fireline intensity, and direction of maximum spread; DISPATCH—automatic linking of DIRECT, SIZE, and CONTAIN for containment time and fire size; IGNITE—probability of ignition; MAP—map spread and spotting distance; MOISTURE—fuel level, temperature, wind speed, and relative humidity; MORTALITY—fuel mortality level and crown volume torch; RH—relative humidity; SCORCH—crown scorch height; SITE—rate of spread, flame height, fireline intensity, and direction of maximum spread; SIZE—area, perimeter, width, forward spread distance, and backing spread distance; SLOPE—slope and horizontal distance; and SPOT—maximum spot fire distance.

FARSITE: FARSITE (Fire Area Simulator) is a model for simulating the spread and behavior of wildland fires under conditions of heterogeneous terrain, fuels, and weather.[52] The model requires the support of a geographic information system (GIS) to generate, manage, and provide data for fuels, elevation, slope, topographic aspect, and canopy cover. Outputs include fire area, fire perimeter, fireline intensity, flame length, rate of spread, heat release per unit area, and time of arrival. FARSITE also includes spotting and crown fire routines. FARSITE simulates fire growth as a spreading elliptical wave. The fire is propagated over a finite time step using many small ellipses at points that define the flame front. The boundary formed by the small ellipses, calculated on the original flame front, becomes the new flame front. FARSITE supports graphical display of the input data and the model calculations.

RELATED ACTIVITIES

The development of deterministic fire models is a very active area. The above review is intended to offer the reader a perspective on what is currently available. One should also recognize that the information in this chapter is not exhaustive. Papers listed in the reference section provide additional information on the topic.

Before using any fire model, the user should be aware of the assumptions and limitations for that model. The brief descriptions provided in this chapter should not be used as the basis for the selection of a model for a particular application. Further, the range of validity for an individual computer fire model is difficult to determine and is the topic of a significant amount of continuing research. Before using any model, the user should study the comparisons of model predictions with experimental data to aid in determining if the use of the model is appropriate.

BIBLIOGRAPHY

References Cited

1. Friedman, R., *An International Survey of Computer Models for Fire and Smoke*, 2nd ed., Factory Mutual Research Corporation, Norwood, MA.
2. Friedman, R., "An International Survey of Computer Models for Fire and Smoke," *Journal of Fire Protection Engineering*, Vol. 4, No. 3, 1992, pp. 83–92.
3. Quintiere, J. G., "Compartment Fire Modeling," *The SFPE Handbook of Fire Protection Engineering*, 2nd ed., DiNenno, P. J., ed., National Fire Protection Association, Quincy, MA, 1995, pp. 3-125–3-133.
4. Hamer, A., and Beyler, C. L., "Modeling the Activation of Dry-Pipe Sprinklers in High-Rack Storage," *Proceedings:* International Conference on Fire Research and Engineering, Society of Fire Protection Engineers, Boston, MA, 1995, pp. 65–70.

5. Simcox, S., Wilkes, N. S., and Jones, I. P., "Computer Simulation of the Flows of Hot Gases from the Fire at King's Cross Underground Station," *Fire Safety Journal*, Vol. 18, No. 1, 1991, pp. 49–73.

6. Stroup, D. W., "Using Field Modeling to Simulate Enclosure Fires," *The SFPE Handbook of Fire Protection Engineering*, 2nd ed., DiNenno, P. J., ed., National Fire Protection Association, Quincy, MA, 1995, pp. 3-152-3-159.

7. Hadjisophocleous, G. V., and Cacambouras, M., "Computer Modeling of Compartment Fires," *Journal of Fire Protection Engineering*, Vol. 5, No. 2, 1993, pp. 39–52.

8. Mawhinney, R. N., Galea, E. R., Hoffmann, N., and Patel, M. K., "A Critical Comparison of a Phoenics-Based Fire Field Model with Experimental Compartment Fire Data," *Journal of Fire Protection Engineering*, Vol. 4, No. 3, 1992, pp. 81–92.

9. Nam, S., and Bill, R. G., "Numerical Simulation of Thermal Plumes," *Fire Safety Journal*, Vol. 21, 1993, pp. 231–256.

10. Cooper, L. Y., and Stroup, D. W., "ASET—A Computer Program for Calculating Available Safe Egress Time," *Fire Safety Journal*, Vol. 9, 1985, pp. 29–45.

11. Walton, W. D., "ASET-B: A Room Fire Program for Personal Computers," NBSIR 85-3144, National Bureau of Standards, Gaithersburg, MD, 1985.

12. Cooper, L. Y., and Forney, G. P., "The Consolidated Compartment Fire Model (CCFM) Computer Code Application CCFM.VENTS-Part I: Physical Basis," NISTIR 4342, National Institute of Standards and Technology, Gaithersburg, MD, July 1990.

13. Cooper, L. Y., and Forney, G. P., "The Consolidated Compartment Fire Model (CCFM) Computer Code Application CCFM.VENTS-Part II: Software Reference Guide," NISTIR 4343, National Institute of Standards and Technology, Gaithersburg, MD, July 1990.

14. Cooper, L. Y., and Forney, G. P., "The Consolidated Compartment Fire Model (CCFM) Computer Code Application CCFM.VENTS-Part III: Catalog of Algorithms and Subroutines," NISTIR 4344, National Institute of Standards and Technology, Gaithersburg, MD, July 1990.

15. Cooper, L. Y., Forney, G. P., and Moss, W. F., "The Consolidated Compartment Fire Model (CCFM) Computer Code Application CCFM.VENTS-Part IV: User Reference Guide," NISTIR 4345, National Institute of Standards and Technology, Gaithersburg, MD, July 1990.

16. Peacock, R. D., Forney, G. P., Reneke, P., Portier, R., and Jones, W. W., "CFAST, the Consolidated Model of Fire Growth and Smoke Transport," NIST Tech. Note 1299, National Institute of Standards and Technology, Gaithersburg, MD, Feb. 1993.

17. Portier, R., Reneke, P., Jones, W. W., and Peacock, R. D., "A User's Guide for CFAST Version 1.6," NISTIR 4985, National Institute of Standards and Technology, Gaithersburg, MD, Dec. 1992.

18. Ho, V., Siu, N., Apostolakis, G., "COMPBRN III—A Fire Hazard Model for Risk Analysis," *Fire Safety Journal*, Vol 13, Nos. 2 and 3, pp. 137–154, 1988.

19. Babrauskas, V., "COMPF2—A Program for Calculating Post-Flashover Fire Temperatures," NBS TN 991, National Bureau of Standards, Gaithersburg, MD, 1979.

20. Mitler, H. E., and Rockett, J. A., "User's Guide to FIRST, A Comprehensive Single-Room Fire Model," CIB W14/88/22, National Bureau of Standards, Gaithersburg, MD, 1987.

21. Satterfield, D. B., and Barnett, J. R., "User's Guide to WPI/FIRE Version 2 (WPI-2)—A Compartment Fire Model," Worcester Polytechnic Institute, Center for Fire Safety Studies, Worcester, MA, 1990.

22. Baum, H. R., and Rehm, R. G., "Calculations of Three-Dimensional Buoyant Plumes in Enclosures," *Combustion Science and Technology*, Vol. 40, 1984, pp. 55–77.

23. Scheider, V., and Hoffmann, J., "Modelluntersuchung von Offshore-Kohlenwasserstaff-Branden Feldmodellierung von Loschversuchen," *Intellex*, Frankfurt am Main, Germany.

24. Cox, G., and Kumar, S., "Field Modeling of Fire in Forced-Ventilation Enclosures," *Combustion Science and Technology*, Vol. 52, 1987, pp. 7–23.

25. Magnussen, B. F., NTH/SINTEF, Trondheim, Norway, 1990.

26. Markatos, N. C., Malin, M. R., and Cox, G., "Mathematical Modeling of Buoyancy-Induced Smoke Flow in Enclosures," *International Journal of Heat and Mass Transfer*, Vol. 25, 1982, pp. 63–75.

27. Gardiner, A. J., "The Mathematical Modeling of the Interaction Between Sprinkler Sprays and the Thermally Buoyant Layers of Gases from Fires," Ph.D. Thesis, South Bank Polytechnic, London, 1988.

28. Computational Dynamics, "STAR*CD Version 2.2 Methodology—Part 1, Mathematical Modeling," *Computational Dynamics*, London, UK, 1993.

29. Klote, J. H., "Computer Program for Analysis of Smoke Control Systems," NBSIR 82-2512, National Bureau of Standards, Gaithersburg, MD, 1982.

30. Joshi, A. A., and Pagni, P. J., "User's Guide to BREAK1, The Berkeley Algorithm for Breaking Window Glass in a Compartment Fire," NIST GCR 91-596, National Institute of Standards and Technology, Gaithersburg, MD, Oct. 1991.

31. Evans, D. D., Stroup, D. W., and Martin, P., "Evaluating Thermal Fire Detection Systems (SI Units)," NBSSP 713, National Bureau of Standards, Gaithersburg, MD, Apr. 1986, p. 549.

32. Evans, D. D., and Stroup, D. W., "Methods to Calculate the Response Time of Heat and Smoke Detectors Installed Below Large Unobstructed Ceilings," *Fire Technology*, Vol. 22, No. 1, 1986, pp. 54–65.

33. Iding, R. H., Nizamuddin, Z., and Bresler, B., "FIRES T3—A Computer Program for the Fire Response of Structures—Thermal Three-Dimensional Version," UCB FRG 77-15, University of California, Berkeley, CA, Oct. 1977.

34. Cooper, L. Y., "Estimating the Environment and the Response of Sprinkler Links in Compartment Fires with Draft Curtains and Fusible-Link-Actuated Ceiling Vents Theory," *Fire Safety Journal*, Vol. 16, 1990, pp. 137–163.

35. Nelson, H. E., "FIREFORM—A Computerized Collection of Convenient Fire Safety Computations," NBSIR 86-3308, National Bureau of Standards, Gaithersburg, MD, 1986.

36. Peacock, R. D., Jones, W. W., Bukowski, R. W., and Forney, C. L., "Software User's Guide for the HAZARD I Fire Hazard Assessment Method, Version 1.1, Volume II," NIST HB-146/II, National Institute of Standards and Technology, Gaithersburg, MD, June 1991.

37. Peacock, R. D., Jones, W. W., Bukowski, R. W., and Forney, C. L., "Technical Reference Guide for the HAZARD I Fire Hazard Assessment Method, Version 1.1, Volume I," NIST HB-146/I, National Institute of Standards and Technology, Gaithersburg, MD, June 1991.

38. Peacock, R. D., Jones, W. W., Forney, G. P., Portier, R., Reneke, P., Bukowski, R. W., and, Klote, J. H., "An Update Guide for HAZARD I, Version 1.2," NISTIR 5410, National Institute of Standards and Technology, Gaithersburg, MD, May 1994.

39. Chitty, R., and Cox, G., "ASKFRS: An Interactive Computer Program for Conducting Fire Engineering Estimations," Fire Research Station, Borehamwood, UK, 1988.

40. Klote, J. H., "Method of Predicting Smoke Movement in Atria with Application to Smoke Management," NISTIR 5516, National Institute of Standards and Technology, Gaithersburg, MD, Nov. 1994.

41. *FIRECALC Version 2.3 User's Manual*, CSIRO (Commonwealth Scientific and Industrial Research Organization), Division of Building Construction and Engineering, Sydney, Australia, 1993.

42. Shestopal, V. O., and Grubits, S. J., "Computer Program for an Uninhibited Smoke Plume and Associated Computer Software," *Fire Technology*, Vol. 29, No. 3, 1993

43. Deal, S., "Technical Reference Guide for FPETOOL Version 3.2," NISTIR 5486, National Institute of Standards and Technology, Gaithersburg, MD, Aug. 1994.

44. Thompson, P. A., and Marchant, E. W., "Modeling and Measurement of Escape Movement," *Proceedings*, CIB W14 International Symposium and Workshops, Engineering Fire Safety in the Process of Design: Demonstrating Equivalency, University of Ulster, Ulster, UK, Sept. 1993.

45. Klote, J. H., Alvord, D. M., Levin, B. M., and Groner, N. E., "Feasibility and Design Considerations of Emergency Evacuation by Elevators," NISTIR 4870, National Institute of Standards and Technology, Gaithersburg, MD, Sept. 1992.

46. Kisko, T. M., and Francis, R. L., "EVACNET+: A Computer Program to Determine Optimal Building Evacuation Plans," *Fire Safety Journal*, Vol. 9, 1985, pp. 211–220.

47. Levin, B. M., "EXITT—A Simulation Model of Occupant Decisions and Actions in Residential Fires," NBSIR 88-3753, National Bureau of Standards, Gaithersburg, MD, 1988.

48. Fahy, R. F., "EXIT89: An Evacuation Model for High-Rise Buildings—Recent Enhancements and Example Applications," International Conference of Fire Research and Engineering, Society of Fire Protection Engineers, Sept. 1995, pp. 332–337.

49. Rothermel, R. C., "How to Predict the Spread and Intensity of Forest and Range Fires," INT 143, U.S. Forest Service, U.S. Department of Agriculture, Forest Service, June 1983.

50. Andrews, P. L., "BEHAVE: Fire Behavior Prediction and Fuel Modeling System—BURN Subsystem, Part 1," General Technical Report INT-194, 1986, Ogden, UT, U.S. Department of Agriculture, Forest Service, Intermoutain Research Station, p. 130.

51. Andrews, P. L., and Chase, C.H., "BEHAVE: Fire Behavior Prediction and Fuel Modeling System—BURN Subsystem, Part 2," General Technical Report INT-260, Ogden, UT, U.S. Department of Agriculture, Forest Service, Intermoutain Research Station, 1989, p. 93.

52. Finney, M. A., "FARSITE Users' Guide and Technical Documentation," Systems for Environmental Management, Missoula, MT, 1995.

Additional Reading

Alpert, R. L., "Improvement of the FSG Fire Spread Computer Model Through Use of Data from Large-Scale Experiments in Japan and Canada," U.S./Japan Government Cooperative Program on Natural Resources (UJNR), Fire Research and Safety, 12th Joint Panel Meeting, Oct. 27–Nov. 2, 1992, Tsukuba, Japan, Building Research Inst., Ibaraki, Japan, 1994, pp. 14–15.

Beard, A., "Evaluation of Fire Models—Report 1, ASET: Qualitative Assessment; Report 2, ASKFRS: Qualitative Assessment; Report 3, HAZARD I: Qualitative Assessment; Report 4, FIRST: Qualitative Assessment; Report 5, JASMINE: Qualitative Assessment; Report 6, ASET: Quantitative Assessment; Report 8, HAZARD I: Quantitative Assessment; Report 9, FIRST: Quantitative Assessment; Report 6, JASMINE: Quantitative Assessment; Report 11, Overview; Unit of Fire Safety Engineering, University of Edinburgh, Scotland, 1990.

Beard, A., "Limitations of Computer Models," *Fire Safety Journal*, Vol. 18, No. 4, 1992, pp. 375–391.

Birk, D. M., *Introduction to Mathematical Fire Modeling*, Technomic Publishing Co., Inc., Lancaster, PA, 1991, p. 267.

Brani, D. M., and Black, W. Z., "Two-Zone Model for a Single-Room Fire," *Fire Safety Journal*, Vol. 19, Nos. 2–3, 1992, pp. 189–216.

Bukowski, R. W., "Fire Models: The Future is Now!" *NFPA Journal*, Vol. 85, No. 6, March/April 1991, pp. 60–62, 64, 66–69.

Bukowski, R. W., "Modeling a Backdraft: The Fire at 62 Watts Street," *Fire Journal*, Vol. 89, No. 6, Nov.–Dec. 1995, pp. 85–89.

Bukowski, R. W., "Review of International Fire Risk Prediction Methods," U. S./Japan Government Cooperative Program on Natural Resources (UJNR), Fire Research and Safety, 12th Joint Panel Meeting, Oct. 27–Nov. 2, 1992, pp. 232–240.

Charters, D. A., Gray, W. A., and McIntosh, A. C., "Computer Model to Assess Fire Hazards in Tunnels (FASIT)," *Fire Technology*, Vol. 30, No. 1, first quarter 1994, pp. 134–154.

Custer, R. L. P., "Computer Modeling in Fire Reconstruction and Failure Analysis," Society of Fire Protection Engineers and Worcester Polytechnic Institute, Computer Applications in Fire Protection *Proceedings*, June 28–29, 1993, Worcester, MA, 1993, pp. 69–74.

Duong, D. Q., "The Accuracy of Computer Fire Models: Some Comparisons with Experimental Data from Australia," *Fire Safety Journal*, Vol. 16, No. 6, 1990, pp. 415–431.

Franchuk, D. M., "It's Time We Thought about Using Fire Models to Fight Fires," *Fire Journal*, Vol. 82, No. 6, Nov.–Dec. 1988, p. 13.

Hadjisophocleous, G. V., and Cacambouras, M., "Computer Modeling of Compartment Fires," *Journal of Fire Protection Engineering*, Vol. 5, No. 2, April/May/June 1993, pp. 39–52.

Hokugo, A., Yung, D., and Hadjisophocleous, G. V., "Experiments to Validate the NRCC Smoke Movement Model for Fire Risk-Cost Assessment," International Association for Fire Safety Science, *Proceedings*, 4th International Symposium, July 13–17, 1994, Ottawa, Ontario, Canada, pp. 805–816.

Jackson, L., "Modeling the Future by Computer," *Fire Prevention*, No. 215, Dec. 1988, pp. 34–35.

Janssens, M. L., "Critical Analysis of the OSU Room Fire Model for Simulating Corner Fires," American Society for Testing and Materials, *Fire and Flammability of Furnishings and Contents of Buildings*, ASTM STP 1233, ASTM, Philadelphia, PA, 1992, pp. 169–185.

Jones, W. W., and Peacock, R. D., "Guide for FAST, Version 18," NIST Technical Note 1262, National Institute of Standards and Technology, Gaithersburg, MD, May 1989.

Levine, R. S., and Nelson, H. E., "Full-Scale Simulation of a Fatal Fire and Comparison of Results with Two Multiroom Models," NBSIR 90-4268, National Bureau of Standards, Gaithersburg, MD, 1990.

Mitler, H. E., "Comparison of Several Compartment Fire Models: An Interim Report," NBSIR 85-3233, Oct. 1985, National Bureau of Standards, Gaithersburg, MD.

Nelson, H. E., "Computer Models and Fire Reconstruction," Society of Fire Protection Engineers and Worcester Polytechnic Institute, Computer Applications in Fire Protection, *Proceedings*, June 28–29, 1993, Worcester, MA, 1993, pp. 75–79.

Nelson, H. E., "FPETOOL: Fire Protection Engineering Tools for Hazard Estimation," NISTIR 4380, National Institute of Standards and Technology, Gaithersburg, MD, 1990.

Notarianni, K. A., and Davis, W. D., "Use of Computer Models to Predict the Response of Sprinklers and Detectors in Large Spaces," Society of Fire Protection Engineers and Worcester Polytechnic Institute, Computer Applications in Fire Protection, *Proceedings*, June 28–29, 1993, Worcester, MA, 1993, pp. 27–33.

Peacock, R. D., Jones, W. W., and Bukowski, R. W., "Verification of a Model of Fire and Smoke Transport," *Fire Safety Journal*, Vol. 21, No. 2, 1993, pp. 89–129.

Pietrzak, L. M., and Dale, J. J., "User's Guide for the Fire Demand Model: A Physically Based Computer Simulation of the Suppression of Post-Flashover Compartment Fires," NIST-GCR-92-612, National Bureau of Standards, Gaithersburg, MD, 1992.

Poon, L. S., and Beck, V. R., "Numerical Modelling of Human Behavior during Egress in Multi-Storey Office Building Fires Using EvacSim—Some Validation Studies," Interscience Communications Limited, ASIAFLAM '95, International Conference on Fire Science and Engineering, 1st, *Proceedings*, March 15–16, 1995, Kowloon, Hong Kong, pp. 163–174.

Rocket, J., Morita, M., and Cooper, L. Y., "Comparisons of NBS/HARVARD VI Simulations and Full-Scale, Multi-Room Fire Test Data," NBSIR 87-3567, National Bureau of Standards, Gaithersburg, MD, July 1987.

Stroup, D. W., "Using Computer Fire Models to Evaluate Equivalent Levels of Fire and Life Safety," Society of Fire Protection Engineers and Worcester Polytechnic Institute, Computer Applications in Fire Protection, *Proceedings*, June 28–29, 1993, Worcester, MA, 1993, pp. 21–26.

Todd, N. W., and Ryan, J. D., "Improving Codes by Predicting Product Performance in Real Fires," *Fire Journal*, Vol. 84, No. 2, Mar. 1990, p. 64.

Tubbs, J. S., and Barnett, J. R., "Modeling the NIST High-Bay Fire Experiment with JASMINE," National Institute of Standards and Technology, and Society of Fire Protection Engineers, International Conference on Fire Research and Engineering, *Proceedings*, Sept. 10–15, 1995, Orlando, FL, pp. 383–388.

Walton, W. D., and Notarianni, K. A., "Comparison of Ceiling Jet Temperatures Measured in an Aircraft Hangar Test Fire with Temperatures Predicted by the DETACT-QS and LAVENT Computer Models," NISTIR 4947; Jan. 1993.

Webb, W. A., "Using FPETOOL to Evaluate Fire Safety of a Four-Level Shopping Mall," National Institute of Standards and Technology and Society of Fire Protection Engineers, International Conference on Fire Research and Engineering, *Proceedings*, Sept. 10–15, 1995, Orlando, FL, pp. 365–370.

Ziesler, P. S., and Gunnerson, F. S., "Live Fire Comparisons of the CFAST Code," National Institute of Standards and Technology and Society of Fire Protection Engineers, International Conference on Fire Research and Engineering, *Proceedings*, Sept. 10–15, 1995, Orlando, FL, pp. Ziesler, P. S., and Gunnerson, F. S., "Live Fire Comparisons of the CFAST Code," National Institute of Standards and Technology and Society of Fire Protection Engineers, International Conference on Fire Research and Engineering, *Proceedings*, Sept. 10–15, 1995, Orlando, FL, pp. 346–349.

PROBABILISTIC FIRE MODELS

John M. Watts, Jr.

As indicated in Section 11, Chapter 5, "Deterministic Computer Fire Models," several excellent deterministic models have been developed to describe fire behavior. So an accurate picture of fire growth seems possible if the needed data are available and if the initial and boundary conditions are given. However, it is unlikely the described result is the most probable fire, which is really required for fire hazard assessments.

This chapter presents information on probabilistic models of fire behavior and other aspects of fire protection. It begins with a brief introduction to probability theory and modeling concepts. It then discusses three basic forms of probabilistic models: (1) networks, (2) statistical models, and (3) simulation.

Additional information relating to the material discussed in this chapter can be found in Section 1, Chapter 3, "Systems Concepts for Building Fire Safety"; Section 11, Chapter 4, "Introduction to Fire Modeling"; and Section 11, Chapter 8, "Fire Risk Analysis."

PROBABILITY

Because it plays such a big part in daily life, most people have some idea of what probability is about. They know it involves weighing the chances, or likelihood, that something or other will take place. They often hear, or make statements, such as, "It is likely a fire will start here"; "We have a better chance of controlling the fire if we can ventilate"; "Sprinklers have a very good probability of success." Such statements refer to situations in which the outcome is not certain, but in which one has some degree of confidence that the statement will be correct. The theory of probability provides a mathematical framework for such statements.

The occurrence of fires, aspects of fighting fires, and sprinkler performance are repetitive operations, repeated often under similar circumstances. The percentage of times a certain outcome occurs, under the same conditions, will always remain approximately unchanged, only rarely deviating significantly from some average figure. It can, therefore, be said that this average figure is a characteristic measure of the given repetitive operation, under prescribed conditions. Knowledge of such measures is very important as it enables

one not only to describe the outcome of repetitive operations that have already occurred, but also to predict the outcome in the future.

Probability Defined

A repetitive operation is the recurrence of a large number of similar individual operations, such as fires. Of specific interest is a well-defined result of individual operations, such as fire control, and the number of such results in some repetitive operation, such as how many fires are controlled. The percentage, or fractional part, of "successful" outcomes of a repetitive operation is called the probability of this result. It is important to realize that the probability of an event, or result, has meaning only under the precisely defined conditions in which the repetitive operation takes place. Each significant variation of these conditions can cause a change in the probability of the event under consideration.

If the repetitive operation is such that event A, a success, is observed on the average a times in n individual operations, then the probability of event A under the given conditions is a/n. It can then be said that the *probability* of a successful result for any individual operation is the *ratio of the number of "successful" results observed* to the *total number of individual operations* constituting the prescribed repetitive operation.

$$\frac{\text{Probability}}{\text{of success}} = \frac{\text{Number of outcomes with "successful" result}}{\text{Total number of equally likely outcomes}}$$

In reference to fires, it might be sometimes more appropriate to talk about "unsuccessful" results. In the theory of probability, however, it is conventional to refer to the event of interest as a success.

Evaluating Probabilities

The probability of an event is always a positive number, ranging from zero to unity. The number cannot be greater than one because in the fraction, by which it is defined, the numerator cannot be greater than the denominator. The number of successful operations cannot be greater than the number of operations undertaken. Neither the numerator nor the denominator can be negative, so probability cannot be less than zero. Thus, when probability is measured, a "scale" that runs from zero to one is used.

If the probability of event A is denoted by $P(A)$, then there is the requirement:

$$0 \le P(A) \le 1$$

John M. Watts, Jr., Ph.D., is director of the Fire Safety Institute, a not-for-profit information, research, and educational corporation located in Middlebury, VT. He also serves as editor of NFPA's quarterly technical journal, *Fire Technology*.

The larger $P(A)$ is, the more often event A occurs—for example, the greater the probability that a sprinkler system controls the fire, the more fires that are successfully suppressed by sprinklers, relative to the number of fires that occur. If the probability of an event is very small, then it occurs rarely. If $P(A) = 0$, then the event either never occurs, or it occurs so rarely that, in practice, it is considered to be impossible—for example, the probability that the item first ignited is a reinforced concrete wall.

In contrast, if $P(A)$ is close to one, then the fraction by which the probability is expressed has a numerator close to the denominator, and the overwhelming majority of operations are successful. If $P(A) = 1$, then event A occurs always or almost always so that, in practice, it can be assumed that its occurrence is certain—for example, in a room containing a functional smoke detector and an alert person, it might be assumed the probability of discovering a fire is one.

If $P(A) = \frac{1}{2}$, then event A occurs in approximately half of all cases, or successful operations are observed approximately as often as unsuccessful ones. The classic example is a coin toss. If $P(A) > \frac{1}{2}$, then event A occurs more frequently than it does not occur; for $P(A) < \frac{1}{2}$, the reverse is true.

How small the probability of an event must be before it can be assumed to be impossible depends on how important the event is. If 5 percent of smoke detectors manufactured are not sold due to quality control procedures, 0.05 might appear to be a small number. A manufacturer may be able to accept this cost. If 5 percent of smoke detectors sold are defective, however, 0.05 is a very large number. A manufacturer may find it hard to accept this cost knowing that one out of twenty new, installed smoke detectors will not work. In every individual problem, it must be established in advance, based on practical considerations, how small the probability of an event ought to be, so it can be considered to be impossible and of insignificant consequence.

Assigning Probabilities

In applications of probability to fire problems, the assignment of measures and the analysis of logical possibilities may, at times, depend upon masses of data that are best obtained, analyzed, and interpreted by professional statisticians. At other times, the task depends more on qualitative knowledge and judgment of facts with which the fire protection professional is most familiar, and is, therefore, the person best qualified to assign measures and analyze logical possibilities. Very often, a combination of objective fact and subjective judgment is the best source of information in assigning probabilities.

Having a measure for probability is important to be able to deal with situations involving chance or uncertainty. There are many cases when it is useful to find, knowing the probabilities of certain events, the probabilities of more complicated ones. It would be inconvenient to search for a particular method of solution for each new problem of this sort that is encountered. Science tries to form general rules, the knowledge of which would readily permit one to solve mechanically, or almost mechanically, individual problems that are similar to one another. In the area of repetitive phenomena, the science that takes upon itself the formulation of such general rules is called the theory of probability. Probability is a branch of mathematics, and, therefore, involves precise reasoning, and uses formulas, tables, diagrams, and so on as its tools.

PROBABILITY MODELS AND MODELING

The word "model" has several different meanings. Even when descriptors are added, such as in the phrase "fire model," it is not always clear what is meant. As the word model appears more frequently in

fire protection literature, it becomes more important that its meanings are appreciated.

Analytical Models

Several classifications of models exist that place the models into various groups in an effort to avoid confusion over the term. An important distinction exists between exemplary models and analytical models. An exemplary model is a pattern or design, or otherwise serves as an example for imitation, such as a model code. An analytical model is a tool used to study some aspect of the world so as to make a decision about it. Not all models are easily classified. A scale model may be exemplary, or analytical, or neither.

An analytical model is a description of some part of the environment formulated in a way to make it convenient for analysis. It is an efficient means of viewing a problem. The model may not describe every characteristic of a situation, but only those features that are deemed important. This may, at times, make it difficult to recognize the "real" world in a model description. For example, the Arabic numerals 1, 2, 3, and so on, are more abstract, or unlike what they represent, than their Roman counterparts I, II, III. However, Arabic numerals are much more efficient to use in mathematics.

Mathematical Models

Another way to classify models is by whether they are physical or symbolic. Model trains are physical models, as they look like the thing being modeled. Most analytical models are symbolic, using symbols to represent the real world. Thus, a pie chart is a symbolic model of how something is divided up. Flow charts and fault trees are also models that use pictures, or graphic symbols. Mathematical models are also symbolic. They use numbers and other mathematical symbols to represent various parameters, or conditions, that influence a situation. Einstein's theory of relativity, $E = mc^2$, is a mathematical model of the relationship between energy and mass in the universe.

Dealing with Uncertainty

Depending upon how mathematical models deal with uncertainty, they may be either deterministic or probabilistic. If, for a specific set of input values, there is a uniquely determined output, the model is deterministic. Such models have fixed values for the input variables, and, therefore, have fixed outputs. On the other hand, a model dealing explicitly with the uncertainty inherent in fire is a probabilistic model.

Fire protection is a discipline in which there is significant uncertainty associated with most decisions. The concept of safety itself is one of uncertainty. As there is no such thing as absolute safety, human activity will always, and unavoidably, involve risks. The concept of fire is also uncertain. Unwanted combustion is the least predictable common physical phenomenon. Reliability of manufactured, or fabricated, systems for suppression and confinement is another source of uncertainty.

With a deterministic model, to get a sense of what might happen under different scenarios, it must be decided which scenarios are important, and then treat each separately. This is not difficult when there are only a few variables and, therefore, scenarios. However, real world systems are complex, and there is no practical way to evaluate enough scenarios with a deterministic model.

Probabilistic models recognize variables as having a degree of randomness. This is usually more accurate than simple deterministic models, and is especially useful when evaluating fire risk, since risk, by definition, includes the uncertainty of loss. Probabilistic models can yield a numeric measure of risk.

There are many forms of probabilistic models. Complex models may involve both deterministic and probabilistic components. The remaining sections of this chapter introduce three types of probabilistic model that can be used independently or as part of a more comprehensive fire risk analysis. They are: (1) networks, (2) statistical models, and (3) simulation.

NETWORKS

A network model is a graphic representation of paths, or routes, by which objects, energy, information, or logic may flow, or move, from one point to another. Network models are useful for minimizing the time or distance of travel from point to point and have many applications in solving complex problems.

The connected points in a network are called nodes. The connections themselves are called links, arcs, or branches. The simplest network is two nodes connected by a single link. (See Figure 11-6A).

A path is a sequence of links connecting two nodes and usually passing through other nodes. In driving from Los Angeles to New York, for instance, one might go through Denver and St. Louis. The cities could be considered nodes, and the roads between them links. Then one path from Los Angeles to New York would consist of the links connecting the intermediate cities.

A significant advantage of a graphic model, like a network, is the qualitative representation of the structure of the problem, or system. Such a diagram functions as a visual aid to help convey an intuitive understanding of the process for obtaining probabilities.

Event Trees

The term "tree" is used to describe a special type of network in which there is only one path connecting any two nodes. The simplest probability model, and one of the most powerful, is the event tree.

An event tree is a model of the sequence of possible states of a system and of corresponding events that lead to those states. Nodes represent events, and the links between them show how one event leads to another. Figure 11-6B shows a simple event tree.

By assigning probabilities to the outcome of events, an event tree can be used to calculate the probability of the consequences. Figure 11-6C is an event tree that represents the likelihood of dis-

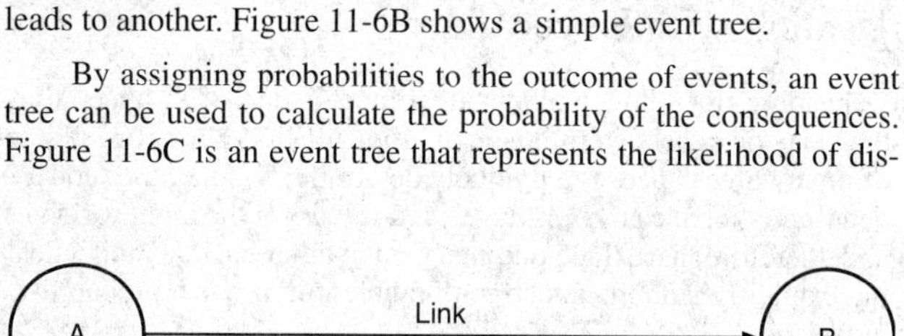

FIG. 11-6A. *Network components.*

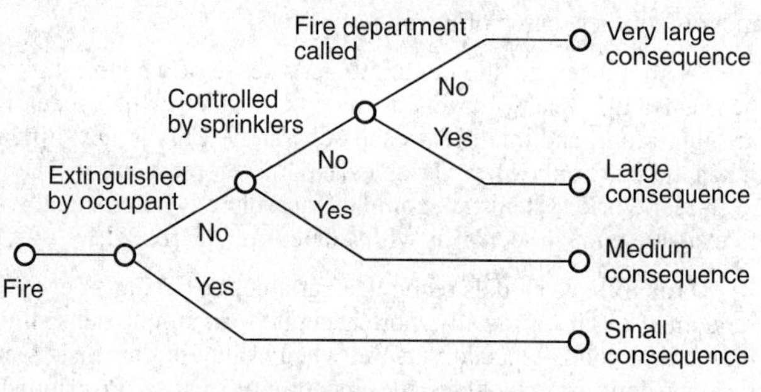

FIG. 11-6B. *Event tree.*

covery of a fire caused by a cigarette in an upholstered chair. Three intermediate events that affect discovery include: (1) someone home, (2) a responsible person awake, and (3) a functional smoke detector. Probabilities of these events are shown on the tree. For each path, a probability of final outcome can be calculated. Assuming the events are independent, the probabilities along each path are multiplied. Thus, looking at the uppermost path, the probability of this path is:

$$(0.95)(0.80)(0.38)(1.00) = 0.038$$

Subsequently, the probabilities for each similar outcome can be summed and results obtained that indicate that the fire will be discovered 82 percent of the time—and that it will not be discovered 18 percent of the time. Using this model, it is relatively easy to see the effect of a smoke detector on the discovery of such a fire.

Fault Trees

According to Recht,[1] fault tree analysis was developed in 1962 by H. A. Watson of Bell Telephone Laboratories. The technique was popularized by the Boeing Company in its application to the Minuteman Ballistic Missile Program[2] and was subsequently made famous by the Rasmussen report on safety of nuclear power plants.[3]

Fault tree analysis (FTA) uses an event tree-like diagram to describe and analyze the undesired or "top" event, so-called because it is at the top of the diagram, with "branches" of the tree extending downward from it. These branches connect the events or conditions that cause the top event to happen. Relationships of these causative events are shown by the connecting lines going through one of two basic "logic gates": (1) the AND gate and (2) the OR gate. The AND gate shows that a "fault" will occur when all the causative events happen simultaneously. If any single one of a group of events can produce the fault, then they are inputs to an OR gate. Figure 11-6D shows a fault tree for fire accidents involving elevator cars stopping at fire-involved floors.

The plus and dot symbols used for OR gates and AND gates are standard symbols for these logic operations used in electronic circuit diagrams and Boolean algebra. They are derived from the algebra of probabilities. For example, consider the flip of a coin. If one wants to calculate the probability of a head OR a tail, then add the probability of a head ($^1/_2$) to the probability of a tail ($^1/_2$), i.e., $^1/_2 + ^1/_2 = 1$. Thus, the symbol for an OR operation is a plus sign. Now suppose one flips the coin twice and wants to calculate the probability of getting a head AND then a tail. This is found by multiplying probabilities, i.e., $^1/_2 \cdot ^1/_2 = ^1/_4$, provided the two events are independent. Thus the symbol for an AND gate is a dot signifying multiplication. Some additional examples of these logic gates and their corresponding Venn diagrams are shown in Figures 11-6E and 11-6F.

Fault trees are based upon setting down a specific failure and examining the systems in a logical, well-organized way to learn what can go wrong to produce the failure. Alternatively, one can consider a desirable top event. A success tree is based upon analysis of requirements and alternatives to achieve a specified goal or objective. The NFPA fire safety concepts tree is a success-type tree.

Use of fault tree analysis requires close knowledge of the systems being analyzed. It is often time consuming but, if thorough, will be revealing. A fault tree can lead to discovery of combinations of factors that otherwise might not have been recognized as causative of the event being analyzed. The tree also becomes a record of the thought process of the analyst and serves as an excellent visual aid for communication with designers and management. More detailed descriptions of fault trees and analysis can be found in cited references 4 and 5. There are many software packages for performing fault tree analysis on personal computers.

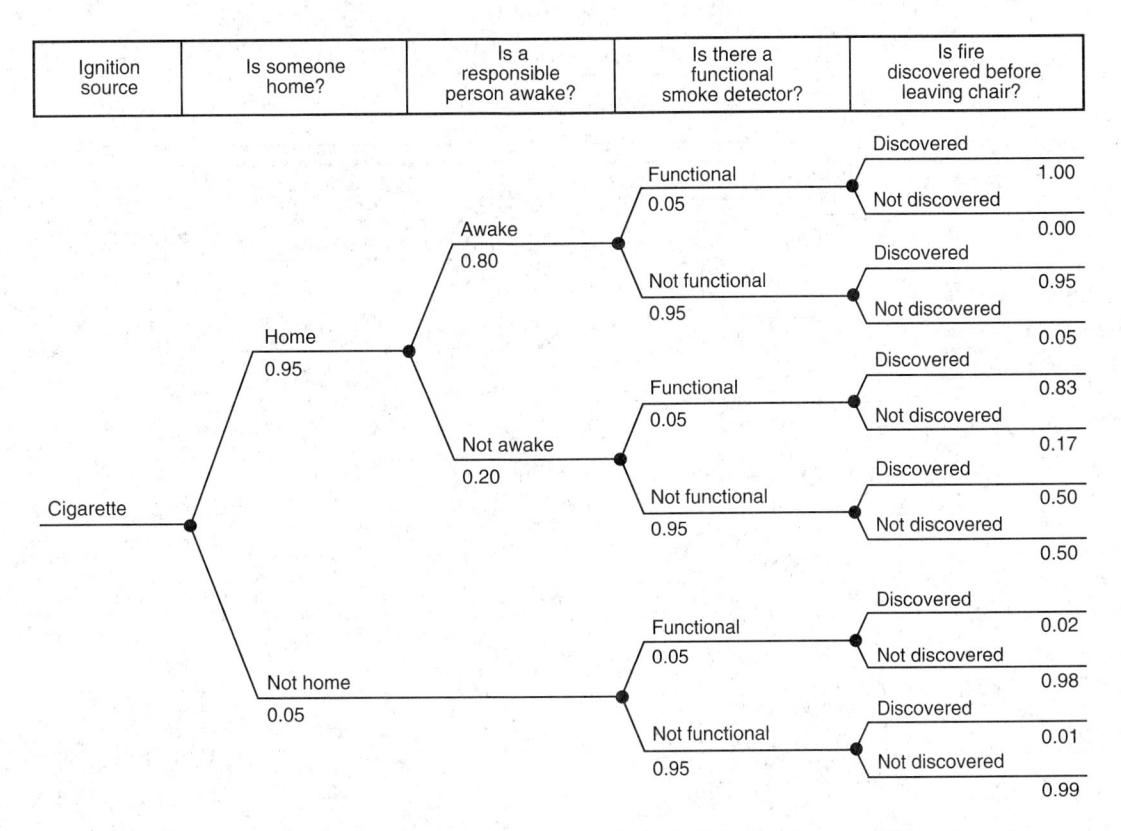

Ignition source	Is someone home?	Is a responsible person awake?	Is there a functional smoke detector?	Is fire discovered before leaving chair?

Cigarette

Home 0.95

Not home 0.05

Awake 0.80

Not awake 0.20

Functional 0.05 — Discovered 1.00 / Not discovered 0.00

Not functional 0.95 — Discovered 0.95 / Not discovered 0.05

Functional 0.05 — Discovered 0.83 / Not discovered 0.17

Not functional 0.95 — Discovered 0.50 / Not discovered 0.50

Functional 0.05 — Discovered 0.02 / Not discovered 0.98

Not functional 0.95 — Discovered 0.01 / Not discovered 0.99

FIG. 11-6C. Event tree for cigarette ignition of upholstered chair (numbers are illustrative only).

STATISTICAL MODELS

The fields of probability and statistics are closely linked. Most statistical concepts, such as sampling and hypothesis testing, have their basis in probability theory. Statistical modeling involves the description of random phenomena by an appropriate probability distribution.

Probability Distributions

A probability distribution may be thought of as a mathematical function that defines the probability of an event. For example, the probability distribution that describes the roll of a die is:

$$p(x) = \tfrac{1}{6}$$

where $x = 1, 2, 3, 4, 5,$ or 6.

All the possible outcomes, one through six, have an equal probability of $\tfrac{1}{6}$. This is called the uniform distribution.

Another example of a uniform distribution is the outcome of a spin of a roulette wheel, which can be modeled mathematically as $P(n) = 0.0263158$. This equation indicates that the probability of any possible outcome, n (the numbers 1 through 36 plus 0 and 00), is $\tfrac{1}{38}$, or 0.0263158.

All models have underlying assumptions. In the roulette model, a perfectly fair wheel was assumed. If one chooses not to accept this assumption, then it becomes necessary to collect a lot of statistical information about the roulette wheel to calculate the individual probability of each outcome.

Poisson Distribution

There are many different probability distributions used to describe many different types of random events. An example is shown in the

data on fire calls in Table 11-6A. The first column shows the number of calls that might be received in any one day, and the second column shows the number of days with that many calls. The data are from a four-month period consisting of a total of 443 calls in 123 days.

Charting the data, as shown in Figure 11-6G, reveals a trend. The frequency starts out low, increases rapidly, decreases rapidly, and then tails off toward higher numbers of calls. If more data were collected, this trend would be "smoother." The shape of this trend can be mathematically described by a Poisson distribution.

The Poisson distribution is widely used to describe random events having a large opportunity to occur, but a small likelihood

TABLE 11-6A. *Number of Days Having a Specified Number of Calls for Service*[6]

Number of Calls for Service per Day (n)	Number of Days Having (n) Public Service Calls
0	6
1	8
2	24
3	22
4	31
5	14
6	7
7	8
8	0
9	0
10	1
11	0

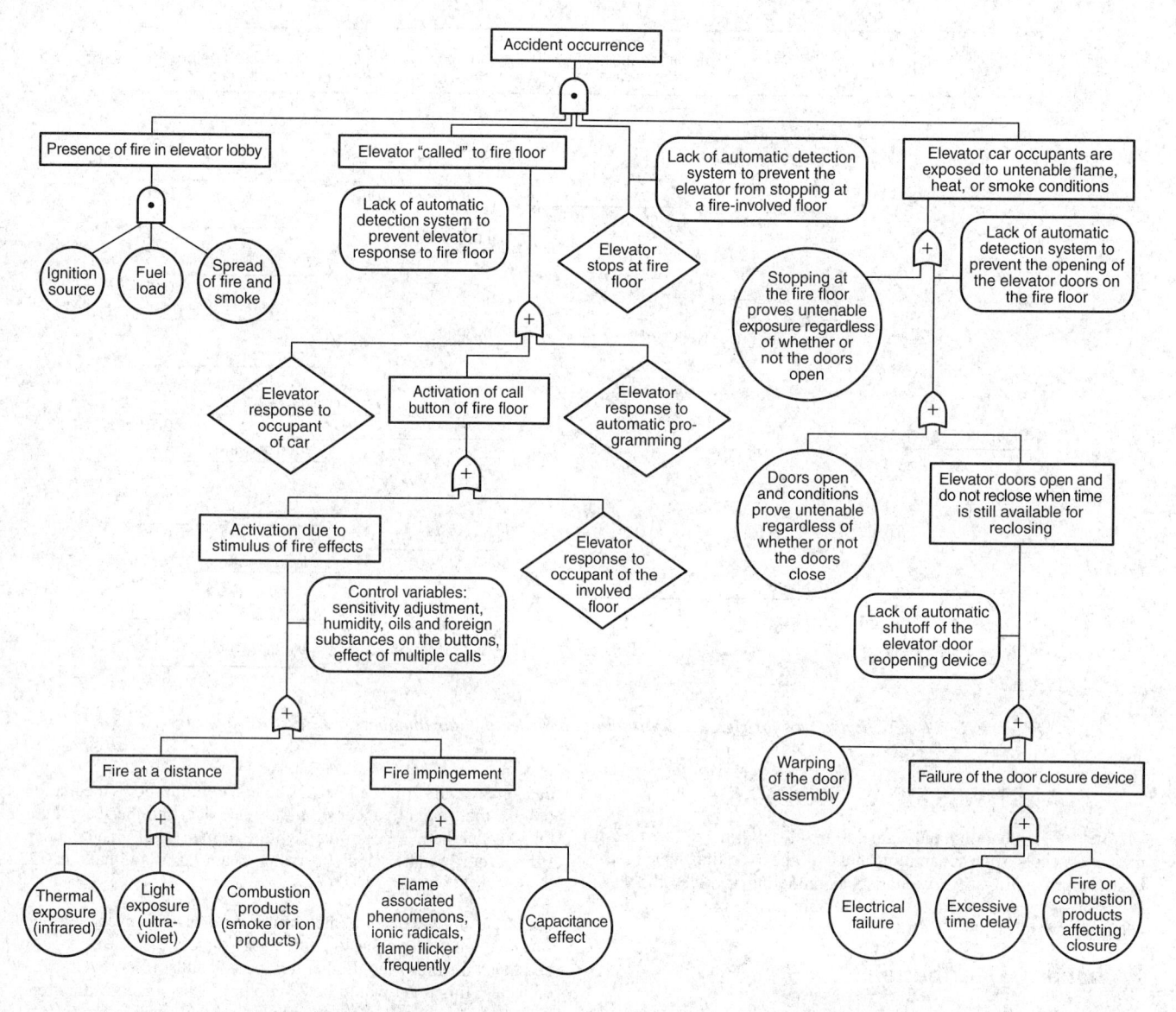

FIG. 11-6D. *Fire accidents involving elevator cars stopping at fire-involved floors.*

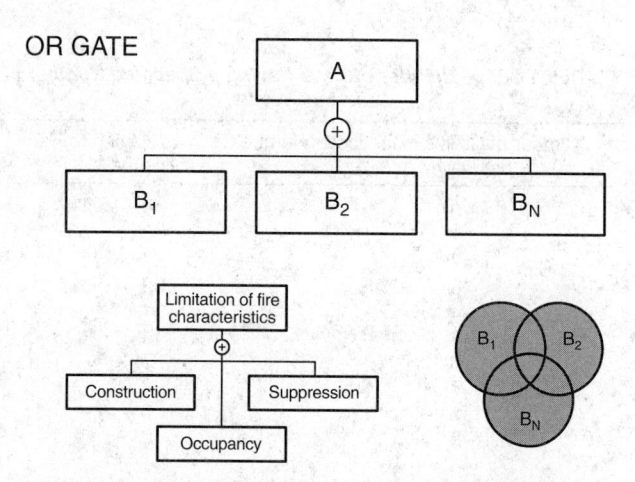

FIG. 11-6E. *Example of OR logic gate, with Venn diagram.*

that any one of these opportunities will actually result in an occurrence. This very accurately fits the description of fire incidents.

Other Distributions

Other probability distributions used to model aspects of fire protection include the familiar normal distribution and the less familiar lognormal distribution.

The normal distribution (see Figure 11-6H) finds wide application in describing biological phenomena, such as human height and IQ. It is also applicable to characteristics, such as shoulder width and walking speed to exits.

The lognormal distribution is skewed like the Poisson distribution (see Figure 11-6I). It has been shown to describe fire load in offices.[7]

The exponential distribution has been suggested as a model of fire growth,[8] and the extreme value distribution is used to characterize fire loss.[9]

AND GATE

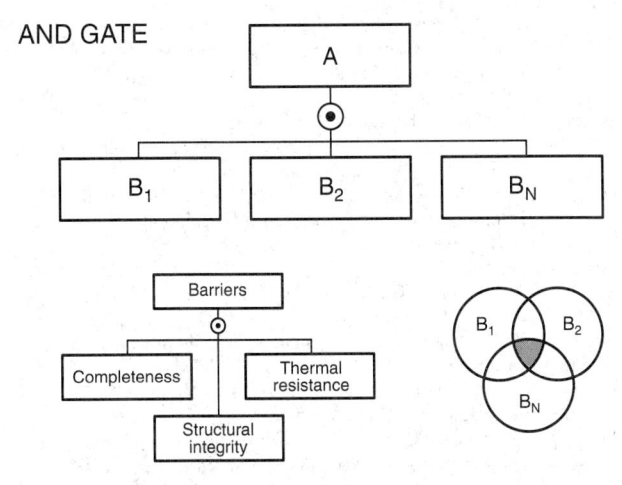

FIG. 11-6F.　Example of AND logic gate, with Venn diagram.

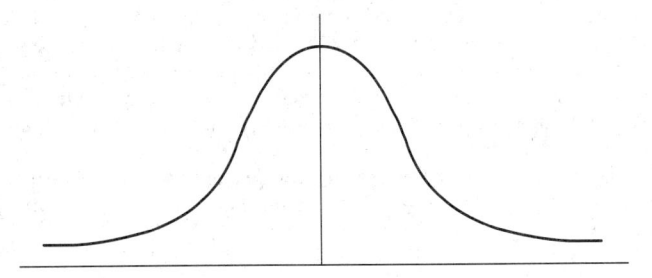

FIG. 11-6G.　Distribution of public service calls.[6]

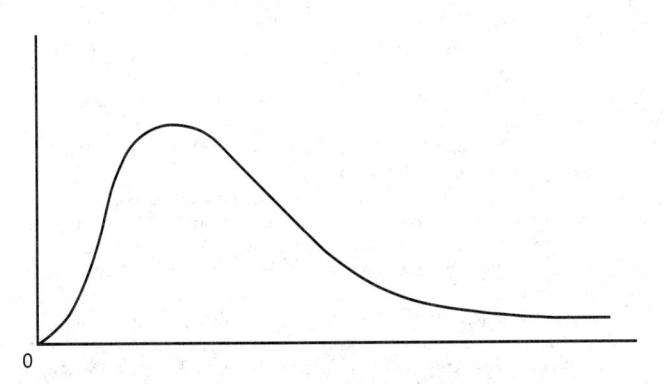

FIG. 11-6H.　Normal distribution.

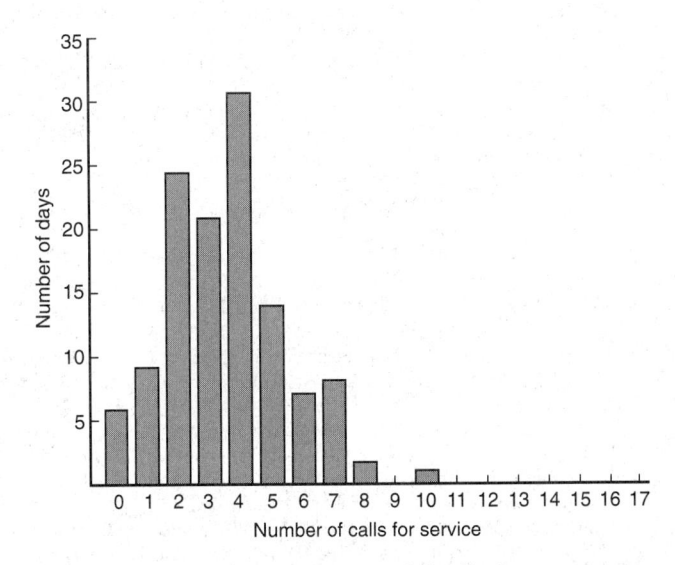

FIG. 11-6I.　Lognormal distribution.

More sophisticated probabilistic models use principles of probability theory to combine probability distributions of two or more random variables.[10,11]

SIMULATION

Often, the relevant variables in a decision problem are too numerous, or the interactions too complex, to be handled mathematically. One way to handle such problems is by computer simulation. Simulation models try out different sets of conditions to see how they affect the outcome. This is what happens in a Link Trainer, a physical simulation model, when an airplane pilot tries various maneuvers to see which work and which don't without risking life or the loss of an airplane. The operation of an airplane is "simulated" by the electro-mechanical training device.

Monte Carlo Simulation

A computer simulation uses mathematical models to predict the range of outcomes from probabilistic situations. For the roulette wheel described in the previous section, letting the computer select a random number from 1 to 38 would simulate a spin of the wheel. Using the computer, the wheel can be "spun" thousands of times in 1 sec.

A roulette wheel analogy can be used to describe a computer simulation of fire scenarios. First, the wheel is divided like a pie chart so that a likely ignition source, such as smoking materials, has a big slice covering several numbers and a less likely ignition source, such as spontaneous combustion, has fewer or only one number. Then, spinning the wheel simulates the first part of an ignition sequence, the heat source. Next, a second wheel is divided according to the most common materials first ignited. Spinning the second wheel will complete the ignition simulation. A refinement would be to provide a different second wheel for each heat source, thereby modeling the observed situation that some materials are more likely to be ignited by some heat sources than by others.

Subsequent stages of fire growth can also be included in the simulation. Fire development could be represented by a state transition model, a type of model that describes a progression from one defined state to another. In a fire, these "transitions" are determined by parameters, or variables of the fire area, such as flame spread, fire load, ventilation, and so on. Now, more roulette wheels should be added to represent these varying state transitions and to allow the computer to "spin" all these wheels hundreds of times. This is referred to as a Monte Carlo simulation model.

Applications

Because of the flexibility of simulation modeling, and the complex nature of fire problems, there is a wide range of applications. One area where simulation is used extensively is in the modeling of building evacuation.[12] Computer simulations of emergency egress range from simple travel time models to detailed representations of human decision making. Other areas of application include fire growth and development,[13,14] and emergency response.[15,16]

An important application of simulation is to confirm that a simpler model may be used safely or even to suggest the form of a simpler model.[17] A simulation model of a complex system is useful, but it also can be time consuming and costly. Whenever possible, it is preferable to use a simple model relating inputs to outputs. A simulation model can provide insight into what form a simpler model might take. It also can be used to provide a level of confidence that a simpler model is a safe substitute for the more accurate simulation.

BIBLIOGRAPHY

References Cited

1. Recht, J. L., "Systems Safety Analysis: The Fault Tree," *National Safety News,* Vol. 93, No. 4, 1966, p. 39.
2. Rogers, W. P., *Introduction to System Safety Engineering,* John Wiley, New York, 1971.
3. *Reactor Safety Study: An Assessment of Accident Risks in U.S. Commercial Nuclear Power Plants,* WASH-1400 (NUREG-75/014), 1975, U.S. Nuclear Regulatory Commission, Washington, DC.
4. Vesely, W. E., Goldber, F. F., Roberts, N. H., and Hasl, D. F., *Fault Tree Handbook* (NUREG-0492), U.S. Nuclear Regulatory Commission, Washington, DC, 1981.
5. Crosetti, P. A., *Reliability and Fault Tree Analysis Guide,* DOE 76-45/22, U.S. Department of Energy, 1982.
6. Nilsson, E. K., and Swartz, J. A., Jr., "Application of Systems Analysis to the Alexandria, Virginia, Fire Department," NBS Report 10-454, Feb. 1972, National Bureau of Standards, Washington, DC.
7. Watts, J. M., Jr., "A Theoretical Rationalization of a Goal-Oriented Systems Approach to Building Fire Safety," NBS-GCR-79-163, Feb. 1979, National Bureau of Standards, Gaithersburg, MD, pp. 70–92.
8. Ramachandran, G., "Stochastic Models of Fire Growth," *The SFPE Handbook of Fire Protection Engineering,* 2nd ed., DiNenno, P. J., ed., National Fire Protection Association, Quincy, MA, 1995.
9. Ramachandran, G., "Extreme Value Theory," *The SFPE Handbook of Fire Protection Engineering,* 2nd ed., DiNenno, P. J., ed., National Fire Protection Association, Quincy, MA, 1995.
10. Watts, J. M., Jr., "A Probability Model for Fire Safety Tree Elements," *Hazard Prevention, Journal of the Systems Safety Society,* Vol. 19, No. 6, 1983, pp. 14–15.
11. Gross, D., "Aspects of Stochastic Modeling for Structural Fire Safety," *Fire Technology,* Vol. 19, No. 2, 1993, pp. 103–114.
12. Watts, J. M., Jr., "Computer Models for Evacuation Analysis," *Fire Safety Journal,* Vol. 12, 1987, pp. 237–245.
13. Berlin, G. N., "A Simulation Model for Assessing Building Fire Safety," *Fire Technology,* Vol. 18, No. 1, 1982, pp. 66–76.
14. Fahy, R. F., "Building Fire Simulation Model," *Fire Journal,* Vol. 77, No. 4, July 1983, pp. 93–95, 102–104.
15. Ignall, E. J., "The Fire Operations Simulation Model," Chapter 13, *Rand Fire Project, Fire Department Deployment Analysis,* North Holland, NY, 1979.
16. *Simulation* (special issue on emergency planning), Society for Computer Simulation, San Diego, Sept. 1989.
17. Ignall, E. J., Kolesar, P., and Walker, W., "Using Simulation to Develop and Validate Analytical Models: Some Case Studies," *Operations Research,* Vol. 26, No. 2, 1978, pp. 237–253.

Additional Reading

Arata, M. J., Jr., "Risk Management and the Fire and Safety Professional," Industrial Fire Safety, Vol. 3, No. 1, Jan./Feb. 1994, pp. 12–13.
Babrauskas, V., "Designing Products for Fire Performance: The State of the Art of Test Methods and Fire Models. Commentary," *Fire Safety Journal,* Vol. 24, No. 3, 1995, pp. 299–312.
Beard, A., "Evaluation of Fire Models: Overview," Edinburgh Univ., Scotland, Report 11, Oct. 1990.
Beck, V. R., and Yung, D., "Cost-Effective Risk-Assessment Model for Evaluating Fire Safety and Protection in Canadian Apartment Buildings," *Journal of Fire Protection Engineering,* Vol. 2, No. 3, 1990, pp. 65–74.

Beck, V. R., "Fire Safety System Design Using Risk Assessment Models: Developments in Australia," Proceedings of the Third International Symposium on Fire Safety Science, Elsevier Applied Science, New York, 1991, pp. 45–59.
Brandyberry, M. D., and Apostolakis, G. E., "Fire Risk in Buildings: Frequency of Exposure and Physical Model," *Fire Safety Journal,* Vol. 17, No. 5, 1991, pp. 339–361.
Brandyberry, M. D., and Apostolakis, G. E., "Fire Risk in Buildings: Scenario Definition and Ignition Frequency Calculations," *Fire Safety Journal,* Vol. 17, No. 5, 1991, pp. 363–386.
Brannigan, V., and Meeks, C., "Computerized Fire Risk Assessment Models: A Regulatory Effectiveness Analysis," *Journal of Fire Sciences,* Vol. 13, No. 3, May/June 1995, pp. 177–196.
Bratley, P., Fox, B., and Schrage, L., *A Guide to Simulation,* 2nd ed., Springer-Verlag, New York, 1987.
Chisman, J., *Introduction to Simulation Modeling Using GPSS/PC,* Prentice-Hall, Englewood Cliffs, NJ, 1992.
Deakin, A. G., "Fire Modelling: Where Are We and What Is Needed for Practical Application," Warrington Fire Research Center, UK, TNO Building and Construction Research, Third CIB/W14 Workshop, *Proceedings* of the Fire Engineering Workshop on Modelling, Jan. 25–26, 1993, Delft, The Netherlands, 1993, pp. 15–18.
"Fire Models In Use Today," *IRI Sentinel,* Vol. 51, No. 3, Third Quarter, 1994, pp. 10–11.
"Fire Models: A Tale of Four Uses," *IRI Sentinel,* Vol. 51, No. 3, Third Quarter, 1994, pp. 8–9.
"Fire Models: The Pros and Cons," *IRI Sentinel,* Vol. 51, No. 3, Third Quarter, 1994, p. 3.
"Fire Models: Valuable Tools When Used Properly," *IRI Sentinel,* Vol. 51, No. 3, Third Quarter, 1994, pp. 4–7.
Forney, C. L., and Jones, W. W., "Fire Risk Assessment Method: Guide to the Risk Methodology Software. Internal Report. Final Report," National Institute of Standards and Technology, Gaithersburg, MD, National Fire Protection Assoc., Quincy, MA, NISTIR 4401, Sept. 1990.
Golton, C. J., Golton, B. J., and Hinks, A. J., "Human Behavior in Fire: A Background Review for Modelling," Fire Safety Modelling and Building Design, *Proceedings* of the One-Day Conference to Review the Potential and Limitations of Fire Safety Models for Building Design, March 29, 1994, Salford, UK, 1994, pp. 48–62.
Grubits, S. J., and Chandrasekaran, V., "Issues in Fire Modelling," Commonwealth Scientific and Industrial Research Organization, North Ryde, Australia, July 1991.
Hadjisophocleous, G. V., and Yung, D., "Fire Risk and Protection Cost Assessment Model for Highrise Apartment Buildings," National Research Council of Canada, Ottawa, Ontario, ASTM STP 1150; American Society for Testing and Materials, Fire Hazard and Fire Risk Assessment, Sponsored by ASTM Committee E-5 on Fire Standards, ASTM STP 1150, Dec. 3, 1990, San Antonio, TX, ASTM, Philadelphia, PA, 1990, pp. 224–233.
Hall, J. R., Jr., "Fire Risk Analysis Model for Assessing Options for Flammable and Combustible Liquid Products in Storage and Retail Occupancies," *Fire Technology,* Vol. 31, No. 4, 4th Quarter, Nov. 1995, pp. 290–308.
Hall, J. R., Jr., "How to Tell Whether What You Have is a Fire Risk Analysis Model," National Fire Protection Association, Quincy, MA, ASTM STP 1150; American Society for Testing and Materials, Fire Hazard and Fire Risk Assessment, Sponsored by ASTM Committee E-5 on Fire Standards, ASTM STP 1150, Dec. 3, 1990, San Antonio, TX, ASTM, Philadelphia, PA, 1990, pp. 131–135.
Hall, J. R., Jr., "Probability Concepts," *The SFPE Handbook of Fire Protection Engineering,* 2nd ed., DiNenno, P. J., ed., National Fire Protection Association, Quincy, MA, 1995.
Hall, J. R., Jr., "Statistics," *The SFPE Handbook of Fire Protection Engineering,* 2nd ed., DiNenno, P. J., ed., National Fire Protection Association, Quincy, MA, 1995.
Ham, S., "Fire Modelling From an Architectural Perspective," Fire Safety Modelling and Building Design, *Proceedings* of the One-Day Conference to Review the Potential and Limitations of Fire Safety Models for Building Design, March 29, 1994, Salford, UK, 1994, pp. 1–5.
Hoover, S. V., and Perry, R. F., *Simulation, A Problem-Solving Approach,* Addison-Wesley, New York, 1989.
Hume, B., "Guidance on Fire Models," *Fire Research News,* No. 18, Winter 1994, p. 11.

Law, A. M., and Kelton, W. D., *Simulation Modeling and Analysis,* 2nd ed., McGraw-Hill, New York, 1991.

Ling, W. C. T., and Williamson, R. B., "The Use of Probabilistic Networks for Analysis of Smoke Spread and the Egress of People in Buildings," Fire Safety Science: *Proceedings* of the First International Symposium, Hemisphere, NY, 1986, pp. 953–962.

Noonan, F., and Fitzgerald, R., "On the Role of Subjective Probabilities in Fire Risk Management Studies," *Proceedings* of the Third International Symposium on Fire Safety Science, Elsevier Applied Science, New York, 1991, pp. 495–504.

Pettitt, G. N., Harvey, B., and Worthington, D. R. E., "Use of Quantitative Risk Assessment as a Decision Making Tool," PSAM-II *Proceedings,* An International Conference Devoted to the Advancement of System-Based Methods for the Design and Operation of Technological Systems and Processes, Volume 1, Sessions 1-36, March 20–25, 1994, San Diego, CA, 1994, pp. 007/1-6.

Phillips, W. G. B., "Computer Simulation for Fire Protection Engineering," *The SFPE Handbook of Fire Protection Engineering,* 2nd ed., DiNenno, P. J., ed., National Fire Protection Association, Quincy, MA, 1995.

Phillips, W. G. B., "Development of a Fire Risk Assessment Model," Fire Research Station, Borehamwood, England, CIB W14/91/01 (UK), BRE IP 08/92, May 1992; Probabilistic Safety Assessment and Management, Elsevier Science Publishing Co., Inc., 1991, pp. 299–306.

Phillips, W. G. B., "Simulation Models for Fire Risk Assessment," *Fire Safety Journal,* Vol. 23, 1994, pp. 159–169.

Phillips, W. G. B., "Simulation Models for Fire Risk Assessment," *Fire Safety Journal,* Vol. 23, No. 2, 1994, pp. 159–169.

Platt, D. G., Elms, D. G., and Buchanan, A. H., "Probabilistic Model of Fire Spread With Timber Effects," *Fire Safety Journal,* Vol. 22, No. 2, 1994, pp. 367–398.

Ramachandran, G., "Non-Deterministic Modelling of Fire Spread," *Journal of Fire Protection Engineering,* Vol. 3, No. 2, 1991, pp. 37–48.

Ramachandran, G., "Probabilistic Approach to Fire Risk Evaluation," *Fire Technology,* Vol. 24, No. 3, 1988, pp. 204–226.

Ramachandran, G., "Probability-Based Building Design for Fire Safety. Part 1," *Fire Technology,* Vol. 31, No. 3, 3rd Quarter, Aug. 1995, pp. 265–275.

Ramachandran, G., "Probability-Based Building Design for Fire Safety. Part 2," *Fire Technology,* Vol. 31, No. 4, 4th Quarter, Nov. 1995, pp. 355–368.

Ramachandran, G., "Probability-Based Fire Safety Code," *Journal of Fire Protection Engineering,* Vol. 2, No. 3, 1990, pp. 75–91.

Ramachandran, G., "Probability-Based Fire Safety Code," *Journal of Fire Protection Engineering,* Vol. 2, No. 3, 1990, pp. 75–91.

Ross, S. M., *A Course in Simulation,* Macmillan, New York, 1990.

Takeda, H., and Yung, D., "Simplified Fire Growth Models for Risk-Cost Assessment in Apartment Buildings," *Journal of Fire Protection Engineering,* Vol. 4, No. 2, 1992, pp. 53–66.

Thomas, P. M., and Singh, J. S., "Fire Risk Quantification Using a Discrete Scenario Model," Third International Conference on Management and Engineering of Fire Safety and Loss Prevention: Onshore and Offshore, Feb. 18–20, 1991, Aberdeen, Scotland, Elsevier Applied Science, New York, 1991, pp. 65–75.

Watts, J. M., Jr., "Criteria for Fire Risk Ranking," *Proceedings* of the Third International Symposium on Fire Safety Science, Elsevier Applied Science, New York, 1991, pp. 457–466.

Watts, J. M., Jr., "Dealing with Uncertainty: Some Applications in Fire Protection Engineering," *Fire Safety Journal,* Vol. 11, 1986, pp. 127–134.

Watts, J. M., Jr., "Fire Risk Evaluation Model. Software Review," *Fire Technology,* Vol. 31, No. 4, 4th Quarter, Nov. 1995, pp. 369–371.

Watts, J. M., Jr., "Fire Risk Evaluation Model. Software Review," *Fire Technology,* Vol. 31, No. 4, 4th Quarter, Nov. 1995, pp. 369–371.

Watts, J. M., Jr., "Fire Risk Ranking. Editorial," *Fire Technology,* Vol. 27, No. 3, Aug. 1991, pp. 193–194.

Watts, J. M., Jr., "Fire Risk Rating Schedules," Fire Safety Institute, Middlebury, VT, ASTM STP 1150; American Society for Testing and Materials, Fire Hazard and Fire Risk Assessment, Sponsored by ASTM Committee E-5 on Fire Standards, ASTM STP 1150, Dec. 3, 1990, San Antonio, TX, ASTM, Philadelphia, PA, 1990, pp. 24–34.

Watts, J. M., Jr., "Future of Fire Modeling. Editorial," *Fire Technology,* Vol. 27, No. 2, May 1991, pp. 97–98.

Weinroth, J., "An Adaptable Microcomputer Model for Evacuation Management," *Fire Technology,* Vol. 25, No. 4, 1990, pp. 291–307.

Woolhead, F., "Fire Risk Management: Which Risk?," *Fire Surveyor,* Vol. 22, No. 5, Oct. 1993, pp. 10–11.

Yung, D., "Fire Risk-Cost Assessment Models," IRC *Fire Research News,* No. 58, April 1990, pp. 1–2.

Yung, D., and Beck, V. R., "Application of a Risk-Cost Assessment Model for the Evaluation of Fire Safety in Buildings," *Proceedings* of the First International Conference on Fire Science and Engineering, ASIAFLAM '95, March 15–16, 1995, Kowloon, Hong Kong, 1995, pp. 51–62.

FIRE HAZARD ANALYSIS

Revised by Richard W. Bukowski

Historically, most fire safety regulation has been on the basis of fire hazard analysis, where such assessments were based on the judgment of "experts." Today formal, scientifically based fire hazard analysis (FHA) is common and increasingly being required as a means to avert certain outcomes, regardless of their likelihood. This chapter will discuss the differences between hazard and risk analysis, the process of performing an FHA, and resources available to assist in this process.

HAZARD *VS.* RISK

The goal of an FHA is to determine the expected outcome of a specific set of conditions called a *scenario*. The scenario includes details of the room dimensions, contents, and materials of construction; arrangement of rooms in the building; sources of combustion air; position of doors; numbers, locations, and characteristics of occupants; and any other details that will have an effect on the outcome of interest. This outcome determination can be made by expert judgment; by probabilistic methods using data on past incidents; or by deterministic means, such as fire models. The trend today is to use models wherever possible, supplemented where necessary by expert judgment. While probabilistic methods are widely used in risk analysis, they find little direct application in modern hazard analyses.

Hazard analysis can be thought of as a component of risk analysis. That is, a risk analysis is a set of hazard analyses that have been weighted by their likelihood of occurrence. The total risk is then the sum of all of the weighted hazard values. In the insurance and industrial sectors, risk assessments generally target monetary losses, since these dictate insurance rates or provide the incentive for expenditures on protection. In the nuclear power industry, probabilistic risk assessment has been the basis for safety regulation. Here they most often examine the risk of a release of radioactive material to the environment, from anything ranging from a leak of contaminated water to a core meltdown.

FHA performed in support of regulatory actions generally look at hazards to life, although other outcomes can be examined as long as the condition can be quantified. For example, in a museum or historical structure, the purpose of an FHA might be to avoid damage to valuable or irreplaceable objects or to the structure itself. It would then be necessary to determine the maximum exposure to heat and combustion products that can be tolerated by these items before unacceptable damage occurs.

PERFORMING AN FHA

Performing an FHA is a fairly straightforward, engineering analysis. The steps include:

1. Selecting a target outcome
2. Determining the scenario(s) of concern that could result in that outcome
3. Selecting design fire(s)
4. Selecting an appropriate method(s) for prediction
5. Performing an evacuation calculation
6. Analyzing the impact of exposure
7. Accounting for uncertainty

Selecting a Target Outcome

The target outcome most often specified is to avoid fatalities of occupants of a building. Another might be to ensure that fire fighters are provided with protected areas from which to fight fires in highrise buildings. The U.S. Department of Energy (DOE) requires that FHAs be performed for all DOE facilities.[1] Their objectives for such FHAs, as stated in DOE 5480.7A, include:

1. Minimizing the potential for the occurrence of fire
2. No release of radiological or other hazardous material to threaten health, safety, or the environment
3. An acceptable degree of life safety to be provided for DOE and contractor personnel, and no undue hazards to the public from fire
4. Critical process control or safety systems that are not damaged by fire
5. Vital programs that are not delayed by fire (mission continuity) and
6. Property damage that does not exceed acceptable levels ($150 million per incident)

In Boston, MA, the Office of the Fire Marshal[2] has established a set of objectives for FHAs performed in support of requests for waivers of the prescriptive requirements of the applicable code. These include:

1. Limit the probability of fatalities or major injuries to only those occupants intimate with the fire ignition.
2. Limit the probability of minor injuries to only those in the dwelling unit of origin.

Richard W. Bukowski, P.E., is senior research engineer, NIST Building and Fire Research Laboratory, Gaithersburg, MD.

3. No occupant outside of the dwelling unit of origin should be exposed to the products of combustion in a manner that causes any injury.

4. Limit the probability of flame damage to the dwelling unit of fire origin (this includes taking into account the possibility of flame extension up the exterior of the building).

5. Limit the probability of reaching hazardous levels of smoke and toxic gases in the dwelling unit of fire origin before safe egress time is allowed. At no time during the incident should the smoke conditions in any compartment, including the compartment of origin, endanger persons in those compartments or prevent egress through those compartments.

6. Limit the incident to one manageable by the Boston Fire Department without major commitment of resources or excessive danger to fire fighters during all phases of fire department operation; i.e., search and rescue, evacuation, and extinguishment.

An insurance company might want to limit the maximum probable loss (MPL) to that which is the basis for the insurance rate paid by the customer; a manufacturer wants to avoid failures to meet orders resulting in erosion of its customer base; and some businesses must guard their public image of providing safe and comfortable accommodations. Any combination of these outcomes may be selected as appropriate for FHAs, depending on the purposes for which they are being performed.

Determining the Scenario(s) of Concern

Once the outcomes to be avoided are established, the task is to identify any scenarios that may result in these undesirable outcomes. Here, the best guide is experience. Records of past fires, either for the specific building or for similar buildings or class of occupancy, can be of substantial help in identifying conditions leading to the outcome(s) to be avoided. Statistical data from the National Fire Incident Reporting System (NFIRS) on ignition sources, first items ignited, rooms of origin, etc., can provide valuable insight into the important factors contributing to fires in the occupancy of interest. (See also Chapter 3, "Use of Fire Incident Data and Statistics," in this section.) Anecdotal accounts of individual incidents are interesting, but may not represent the major part of the problem to be analyzed.

Murphy's Law (anything that can go wrong, will) is applicable to major fire disasters; i.e., all significant fires seem to involve a series of failures that set the stage for the event. Thus, it is important to examine the consequences of things not going according to plan. In DOE-required FHAs, one part of the analysis is to assume both that automatic systems fail and that the fire department does not respond. This is used to determine a worst-case loss and to establish the real value of protective systems. If nothing else, such assumptions can help to identify the factors that mean the difference between an incidental fire and a major disaster so that appropriate backups can be arranged.

Scenarios must be translated into design fires for fire growth analysis and occupant assumptions for evacuation calculation.

Selecting Design Fire(s)

Choosing a relevant set of design fires with which to challenge the design is crucial to conducting a valid analysis. The purpose of the design fire is similar to the assumed loading in a structural analysis; i.e., to answer the question of whether the design will perform as intended under the assumed challenge. Keeping in mind that the greatest challenge is not necessarily the largest fire (especially in a sprinklered building), it is helpful to think of the design fires in terms of their growth phase, steady-burning phase, and decay phase. (See Figure 11-7A.)

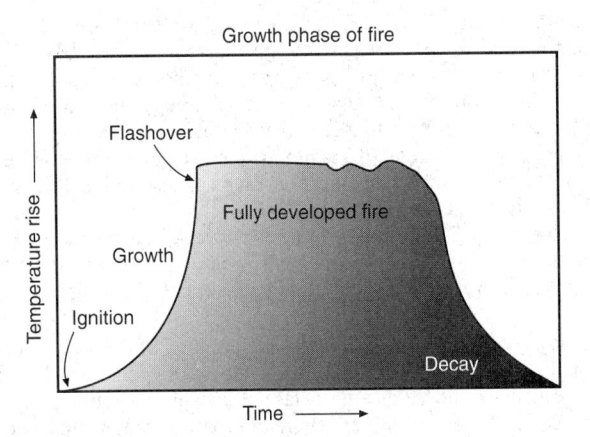

FIG. 11-7A.   *Design fire structure.*

Growth: The primary importance of the appropriate selection of the design fire's growth is in obtaining a realistic prediction of detector and sprinkler activation, time to start of evacuation, and time to initial exposure of occupants.

In 1972, Heskestad first proposed that, for these early times, the assumption that fires grow according to a power law relation works well and is supported by experimental data.[9] He suggested fires of the form:

$$Q = \alpha t^n$$

where

Q = rate of heat release (kW)
α = fire intensity coefficient (kW/s^n)
t = time (sec)
n = 1, 2, 3

Later, it was shown that, for most flaming fires (except flammable liquids and some others), $n = 2$, the so-called t-squared growth rate.[10] A set of specific t-squared fires labeled slow, medium, and fast, with fire intensity coefficients (α) such that the fires reached 1,055 kW (1,000 Btu/sec) in 600, 300, and 150 sec, respectively, were proposed for design of fire detection systems.[11] Later, these specific growth curves and a fourth called "ultra-fast,"[12] which reaches 1,055 kW in 75 sec, gained favor in general fire protection applications.

This set of t-squared growth curves is shown in Figure 11-7B. The slow curve is appropriate for fires involving thick, solid objects (e.g., solid wood table, bedroom dresser, or cabinet). The medium growth curve is typical of solid fuels of lower density (e.g., upholstered furniture and mattresses). Fast fires are thin, combustible items (e.g., paper, cardboard boxes, draperies). Ultra-fast fires are some flammable liquids, some older types of upholstered furniture and mattresses, or other highly volatile fuels.

In a highly mixed collection of fuels, selecting the medium curve is appropriate as long as there is no especially flammable item present. It should also be noted that these t-squared curves represent fire growth starting with a reasonably large, flaming ignition source. With small sources, there is an incubation period before established flaming which can influence the response of smoke detectors (resulting in an underestimate of time to detection). This can be simulated by adding a slow, linear growth period until the rate of heat release reaches 25 kW.

This specific set of fire growth curves has been incorporated into several design methods, such as for the design of fire detection systems in NFPA 72, *National Fire Alarm Code.*[13] They are also referenced as appropriate design fires in several international methods for performing alternative design analyses in Australia and Japan, and in a product fire risk analysis method published in this country.[14] While in the Australian methodology the selection of growth curve is related to the fuel load (mass of combustible material per unit floor area), this is not justified since the growth rate is related to the form, arrangement, and type of material and not simply its quantity. Consider 10 kg (22 lb) of wood: arranged in a solid cube, sticks arranged in a crib, and as a layer of sawdust. (See Figure 11-7C.) These three arrangements would have significantly different growth rates while representing identical fuel loads.

Steady burning: Once all of the surface area of the fuel is burning, the heat release rate goes into a steady burning phase. This may be at a sub-flashover or a post-flashover level; the former will be fuel controlled and the latter ventilation controlled. It should be obvious from the model output (for oxygen concentration or upper layer temperature) in which condition the fire is burning.

Most fires of interest will be ventilation controlled; and this is a distinct advantage, since it is easier to specify sources of air than details of the fuel items. This makes the prediction relatively insensitive to both fuel characteristics and quantity, since adding or reducing fuel simply makes the outside flame larger or smaller. Thus, for ventilation-controlled situations: (1) the heat release rate can be specified at a level that results in a flame out the door, and (2) the heat released inside the room will be controlled to the appropriate level by the model's calculation of available oxygen. If the door flame is outside, it has no effect on conditions in the building; if in another room it will affect that and subsequent rooms. For the much smaller number of fuel-controlled scenarios, values of heat release rate per unit area at a given radiant exposure (from the ASTM E1354, *Standard Test Method for Heat and Visible Smoke Release Rates for Materials and Products Using an Oxygen Consumption Calorimeter*[15]) can be found in handbooks, and used with an estimate of the total fuel area.

Decay: Burning rate declines as the fuel is exhausted. In the absence of experimental data, an engineering approximation specifies this decline as the inverse of the growth curve; this means that fast-growth fuels decay fast and slow decay slow. It is often assumed that the time at which decay begins is when 20 percent of the original fuel is left. While these are assumptions, they are technically reasonable.

This decay will proceed even if a sprinkler system is present and activated. A simple assumption is that the fire immediately goes out; but this is not conservative. A recent National Institute of Science and Technology (NIST) study documents a (conservative) exponential diminution in burning rate under the application of water from a sprinkler. (See Figure 11-7D.)[16] Since the combustion efficiency is affected by the application of water, the use of values of soot and gas yields appropriate for post-flashover burning would represent the conservative approach in the absence of experimental data.

Selecting an Appropriate Method(s) for Prediction

Fire models: A recent survey[3] documented 62 models and calculation methods that could be applied to FHA. Thus, the need is to determine which ones are appropriate to a given situation and which

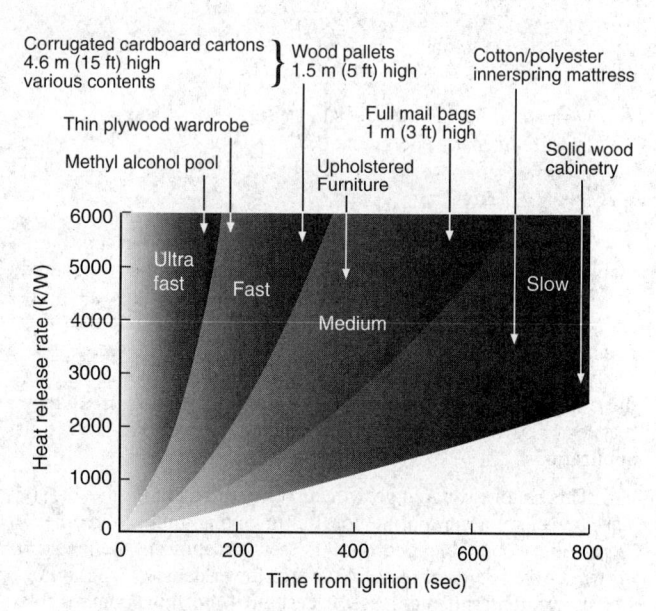

FIG. 11-7B. Set of t-squared growth curves.

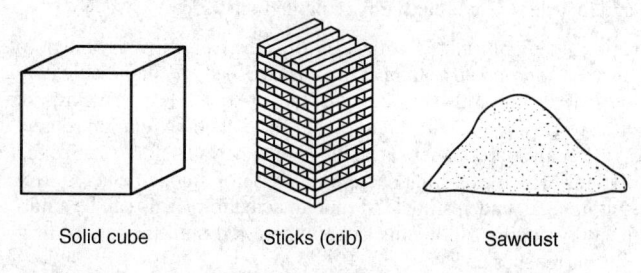

FIG. 11-7C. Fire growth depends on fuel form and arrangements. 10 kg (22 lb) of wood represent identical fuel loads but produce vastly different rates of heat release in a room.

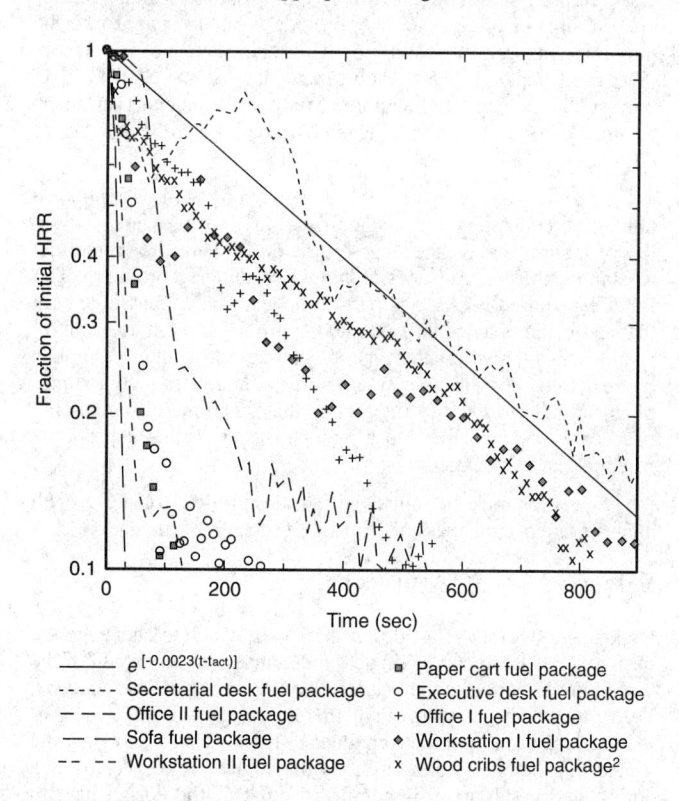

FIG. 11-7D. Decay rates for various fuels under fire sprinkler spray.

are not. The key to this decision is a thorough understanding of the assumptions and limitations of the individual model or calculation and how these relate to the situation being analyzed.

Fire is a dynamic process of interacting physics and chemistry; so predicting what is likely to happen under a given set of circumstances is daunting. The simplest of predictive methods are the (algebraic) equations. Often developed wholly or in part from correlations to experimental data, they represent, at best, estimates with significant uncertainty. Yet, under the right circumstances, they have been demonstrated to provide useful results, especially where used to assist in setting up a more complex model. For example, Thomas' flashover correlation[4] and the McCaffrey/Quintiere/Harkleroad (MQH) upper layer temperature correlation[5] are generally held to provide useful engineering estimates of whether flashover occurs and peak compartment temperatures.

Where public safety is at stake, it is inappropriate to rely solely on such estimation techniques for the fire development/smoke filling calculation. Here, only fire models (or appropriate testing) should be used. Single-room models are appropriate where the conditions of interest are limited to a single, enclosed space. Where the area of interest involves more than one space, and especially where the area of interest extends beyond a single floor, multiple-compartment models should be used. This is because the interconnected spaces interact to influence the fire development and flows.

Many single-compartment models assume that the lower layer remains at ambient conditions (e.g., ASET[6]). Since there is little mixing between layers in a room (unless there are mechanical systems), these models are appropriate. However, significant mixing can occur in doorways, so multiple-compartment models should allow the lower layer to be contaminated by energy and mass. (See Figure 11-7E.)

The model should include the limitation of burning by available oxygen. This is straightforward to implement (based on the oxygen consumption principle) and is crucial to obtaining an accurate prediction for ventilation-controlled burning. For multiple-compartment models, it is equally important for the model to track unburned fuel and allow it to burn when it encounters sufficient oxygen and temperature. Without these features, the model concentrates the combustion in the room of origin, overpredicting conditions there and underpredicting conditions in other spaces.

Heat transfer calculations take up a lot of computer time, so many models take a shortcut. The most common is the use of a con-

stant "heat loss fraction," which is user-selectable (e.g., ASET or CCFM[7]). The problem is that heat losses vary significantly during the course of the fire. Thus, in smaller rooms or spaces with larger surface-to-volume ratios where heat loss variations are significant, this simplification is a major source of error. In large, open spaces with no walls or walls made of highly insulating materials, the constant heat loss fraction may produce acceptable results; but, in most cases, the best approach is to use a model that does proper heat transfer.

Another problem can occur in tall spaces, e.g., atria. The major source of gas expansion and energy and mass dilution is entrainment of ambient air into the fire plume. It can be argued that, in a very tall plume, this entrainment is constrained; but most models do not include this. This can lead to an underestimate of the temperature and smoke density and an overestimate of the layer volume and filling rate—the combination of which may give predictions of egress times available that are either greater or less than the correct value. In the model CFAST[8] this constraint is implemented by stopping entrainment when the plume temperature drops to within one degree (Kelvin) of the temperature just outside the plume, where buoyancy ceases.

Documentation: Only models that are rigorously documented should be allowed in any application involving legal considerations, such as in code enforcement or litigation. It is simply not appropriate to rely on the model developer's word that the physics is proper. This means that the model should be supplied with a technical reference guide that includes a detailed description of the included physics and chemistry, with proper literature references; a listing of all assumptions and limitations of the model; and estimates of the accuracy of the resulting predictions, based on comparisons to experimental data. Public exposure and review of the exact basis for a model's calculations, internal constants, and assumptions are necessary for it to have credibility in a regulatory application.

While it may not be necessary for the full source code to be available, the method of implementing key calculations in the code and details of the numerical solver utilized should be included. This documentation should be freely available to any user of the model, and a copy should be supplied with the analysis as an important supporting document.

Input data: Even if the model is correct, the results can be seriously in error if the data input to the model does not represent the condition being analyzed. Proper specification of the fire is the most critical, and was addressed in detail in the preceding subsection on selecting the design fire(s).

Next in importance is specifying sources of air supply to the fire, i.e., not only open doors or windows, but also cracks behind trim or around closed doors. Most (large) fires of interest quickly become ventilation controlled, making these sources of air crucial to a correct prediction. The most frequent source of errors by novice users of these models is to underestimate the combustion air and underpredict the burning rate.

Two other important items of data are: (1) ignition characteristics of secondary fuel items and (2) the heat transfer parameters for ceiling and wall materials. In each case, the FHA should include a listing of all data values used, their source (i.e., what apparatus or test method was employed and what organization ran the test and published the data), and some discussion of the uncertainty of the data and its result on the conclusions. (See subsection "Accounting for Uncertainty" later in this chapter.)

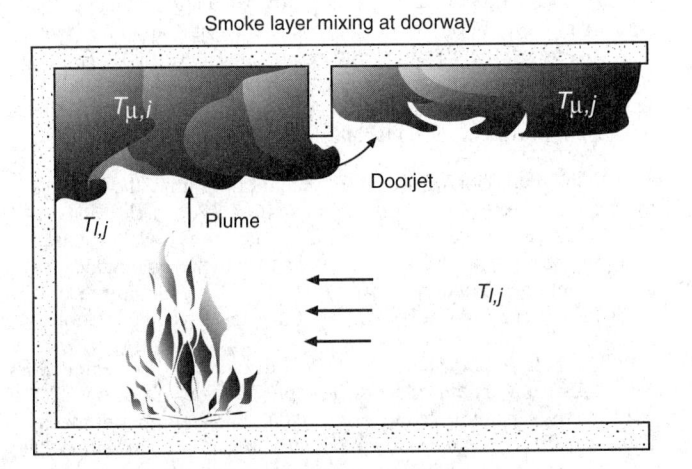

Smoke layer mixing at doorway

FIG. 11-7E.   *Zone models assume that fire gases collect in layers that are internally uniform.*

Performing an Evacuation Calculation

The prediction of the time needed by the building occupants to evacuate to a safe area is performed next, and compared to the time available from the previous steps.

Whether the evacuation calculation is done by model or hand calculation, it must account for several crucial factors. First, unless the occupants see the actual fire, there is time required for detection and notification *before* the evacuation process can begin. Next, unless the information is compelling (again, they see the actual fire), it takes time for people to decide to take action. Finally, the movement begins. All of these factors require time, and that is the critical factor. No matter how the calculation is done, *all* of the factors must be included in the analysis to obtain a complete picture. Excellent discussion of this topic is found in Pauls'[17] and Bryan's[18] chapters in the *The SFPE Handbook of Fire Protection Engineering*.

Models: The process of emergency evacuation of people follows the general concepts of traffic flow. There are a number of models that perform such calculations that may be appropriate for use in certain occupancies. Most of these models do not account for behavior and the interaction of people (providing assistance) during the event. This is appropriate in most public occupancies where people do not know each other. In residential occupancies, family members will interact strongly; and in office occupancies, people who work together on a daily basis would be expected to interact similarly. The literature reports incidents of providing assistance to disabled persons, again especially in office settings.[19] If such behavior is expected, it should be included, as it can result in significant delays in evacuating a building.

Another situation where models (e.g., Fahy's EXIT89[20]) are preferred to hand calculations is with large populations where congestion in stairways and doorways can cause the flow to back up. However, this can be accounted for in hand calculations, as well. Crowded conditions, as well as smoke density, can result in reduced walking speeds.[21] (See Figure 11-7F.) Care should be exercised in using models relative to how they select the path (usually the *shortest* path) over which the person travels. Some models are *optimization* calculations that give the best possible performance. These are inappropriate for a code equivalency determination, unless a suitable safety factor was used.

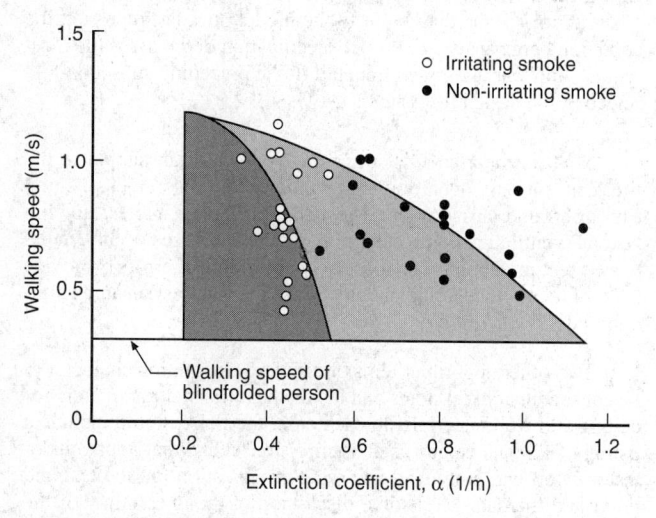

FIG. 11-7F. A person's walking speed decreases in dense smoke until he or she moves as if blindfolded.

Hand calculations: Evacuation calculations are sometimes simple enough to be done by hand. The most thorough presentation on this subject (and the one most often used in alternate design analysis) is that of Nelson and MacLennen.[22] Their procedure explicitly includes all of the factors discussed previously, along with suggestions on how to account for each. They also deal with congestion, movement through doors and on stairs, and other related considerations.

Analyzing the Impact of Exposure

In most cases, the exposure will be to people, and the methods used to assess the impacts of exposure of people to heat and combustion gases involves the application of combustion toxicology models. The HAZARD I software package contains the only toxicological computer model, called TENAB,[23] which is based on research at NIST on lethality to rats[24] and by Purser[25] on incapacitation of monkeys. These methods can also be applied in hand calculations, utilizing the material by Purser[25] and the equations found in reference 22. TENAB accounts for the variation in exposure to combustion products as people move through a building, by reading the concentrations from the fire model in the occupied space during the time the person is in that space. If the person moves into a space with a lower concentration of carbon monoxide, the accumulated dose actually decreases. Details such as these ensure that the results are reasonable. It is important that these details be observed in hand calculations, as well.

Assessing the impact of exposure to sensitive equipment is more difficult, since little data exists in the literature on the effects of smoke exposure on such equipment. Of particular importance here is the existence of acid gases in smoke, which are known to be corrosive and especially harmful to electronics. Fuels containing chlorine (e.g., polyvinyl chlorides) have been studied. However, unless the equipment is close to the fire, acid gases, and especially HCl, deposit on the walls and lower the concentration to which the equipment may be exposed. CFAST in the HAZARD I package contains a routine that models this process and the associated diminution of HCl concentration.

Accounting for Uncertainty

Uncertainty accountability refers to dealing with the uncertainty that is inherent in any prediction. In the calculations, this uncertainty is derived from assumptions in the models and from the representativeness of the input data. In evacuation calculations, there is the added variability of any population of real people. In building design and codes, the classic method of treating uncertainty is with safety factors. A sufficient safety factor is applied such that, if all of the uncertainty resulted in error in the same direction, the result would still provide an acceptable solution.

In the prediction of fire development/filling time, the intent is to select design fires that provide a *worst likely* scenario. Thus, a safety factor is not needed here, unless assumptions or data are used to which the predicted result is very sensitive. In present practice for the evacuation calculation, a safety factor of 2 is generally recommended to account for unknown variability in a given population.

The FHA report should include a discussion of uncertainty. This discussion should address the representativeness of the data used and the sensitivity of the results to data and assumptions made. If the sensitivity is not readily apparent, a sensitivity analysis (i.e., vary the data to the limits and see whether the conclusions change) should be performed. This is also a good time to justify the appropriateness of the model or calculation method.

Final Review

If a model or calculation produces a result that seems counterintuitive, there is probably something wrong. Cases have been seen where the model clearly produced a wrong answer (e.g., the temperature predicted approached the surface temperature of the sun), and those where it initially looked wrong but was not (e.g., a *dropping* temperature in a space adjacent to a room with a growing fire was caused by cold air from outdoors being drawn in an open door). Conversely, if the result is consistent with logic, sense, and experience, it is probably correct.

This is also a good time to consider whether the analysis addressed all of the important scenarios and likely events. Were all the assumptions justified and uncertainties addressed sufficiently to provide a comfort level similar to that obtained when the plan review shows that all code requirements have been met?

CONCLUSION

Quantitative fire hazard analysis is becoming the fundamental tool of modern fire safety engineering practice, and is the enabling technology for the transition to performance-based codes and standards. (For more information on performance-based codes, see Section 11, Chapter 9, of this handbook.) The tools and techniques described in this chapter provide an introduction to this topic, and the motivation for fire protection engineers to learn more about the proper application of this technology.

BIBLIOGRAPHY

References Cited

1. "Department of Energy General Memorandum," Nov. 7, 1991, U.S. Department of Energy, Germantown, MD.
2. Fleming, J. F., Fire Marshal, Boston Fire Department, private communication.
3. Friedman, R., *Survey of Computer Models for Fire and Smoke,* 2nd ed., Factory Mutual Research Corporation, Norwood, MA, 1991.
4. Thomas, P. H., "Testing Products and Materials for Their Contribution to Flashover in Rooms," *Fire and Materials,* 5, 1981, pp. 103–111.
5. McCaffrey, B. J., Quintiere, J. G., and Harkeleroad, M. F., "Estimating Room Temperatures and the Likelihood of Flashover Using Fire Testing Data Correlations," *Fire Technology,* 17, 1981, pp. 98–119.
6. Cooper, L. Y., and Stroup, D. W., "Calculating Available Safe Egress Time (ASET)—A Computer User's Guide," NBSIR 82-2578, National Bureau of Standards, Gaithersburg, MD, 1982.
7. Cooper, L. Y., and Forney, G. P., "The Consolidated Compartment Fire Model (CCFM) Computer Code Application CCFM.VENTS—Part 1: Physical Basis," NISTIR 4342, National Institute of Standards and Technology, Gaithersburg, MD, 1990.
8. Peacock, R. D., Forney, G. P., Reneke, P., and Jones, W. W., "CFAST, the Consolidated Model of Fire Growth and Smoke Transport," NIST Technical Note 1299, National Institute of Standards and Technology, Gaithersburg, MD, 1993.
9. Heskestad, G., "Similarity Relations for the Initial Convective Flow Generated by Fire," FM Report 72-WA/HT-17, 1972, Factory Mutual Research Corporation, Norwood, MA.
10. Shifiliti, R. P., Meacham, B. J., and Custer, R. L. P., "Design of Detection Systems," *The SFPE Handbook of Fire Protection Engineering,* 2nd ed., DiNenno, P. J., *et al.,* eds., National Fire Protection Association, Quincy, MA, 1995.
11. Heskestad, G., and Delichatsios, M. A., "Environments of Fire Detectors—Phase 1: Effect of Fire Size, Ceiling Height, and Material," Vol. 2, Analysis, NBS-GCR-77-95, National Bureau of Standards, Gaithersburg, MD, 1977, p. 100.
12. Stroup, D. W., and Evans, D. D., "Use of Computer Models for Analyzing Thermal Detector Spacing," *Fire Safety Journal,* 14, 1988, pp. 33–45.
13. NFPA 72, *National Fire Alarm Code,* National Fire Protection Association, Quincy, MA, 1993.
14. Bukowski, R. W., "A Review of International Fire Risk Prediction Methods, Interflam '93," Fire Safety 6th International Fire Conference, March 30–April 1, 1993, Oxford England, Interscience Communications, Ltd., London, UK, Franks, C. A., ed., 1993, pp. 437–466.
15. ASTM E1354, *Standard Test Method for Heat and Visible Smoke Release Rates for Materials and Products Using an Oxygen Consumption Calorimeter,* American Society for Testing and Materials, W. Conshohocken, PA, 1994.
16. Madrzykowski, D., and Vettori, R. L., "Sprinkler Fire Suppression Algorithm for the GSA Engineering Fire Assessment System," NISTIR 4833, National Institute of Standards and Technology, Gaithersburg, MD, 1992.
17. Pauls, J., "Movement of People," *The SFPE Handbook of Fire Protection Engineering,* 2nd ed., DiNenno, P. J., *et al.,* eds., National Fire Protection Association, Quincy, MA, 1995.
18. Bryan, J. L., "Behavioral Response to Fire and Smoke," *The SFPE Handbook of Fire Protection Engineering,* 2nd ed., DiNenno, P. J., *et al.,* eds., National Fire Protection Association, Quincy, MA, 1995.
19. Julliet, E., "Evacuating People with Disabilities," *Fire Engineering,* 146, 12, 1993.
20. Fahy, R., "An Evacuation Model for High-Rise Buildings," Interflam '93, Fire Safety 6th International Fire Conference March 30–April 1, 1993, Oxford England, Interscience Communications, Ltd., London, UK, Franks, C. A., ed., 1993, pp. 519–523.
21. Jin, T., "Visibility Through Fire Smoke," Report of Fire Research Institute of Japan, 1971, 2, 33, pp. 12–18.
22. Nelson, H. E., and MacLennen, H. A., "Emergency Movement," *The SFPE Handbook of Fire Protection Engineering,* 2nd ed., DiNenno, P. J., *et al.,* eds., National Fire Protection Association, Quincy, MA, 1995.
23. Peacock, R. D., Jones, W. W., Bukowski, R. W., and Forney, C. L., "Technical Reference Guide for HAZARD I Fire Hazard Assessment Method," Version 1.1, NIST Hb 146, Vol. 2, National Institute of Standards and Technology, 1991, pp. 167–174.
24. Levin, B. C., *et al.,* "Effects of Exposure to Single or Multiple Combinations of Predominant Toxic Gases and Low Oxygen Atmospheres Produced in Fires," *Fundamental and Applied Toxicology,* 9, 2, 1987, pp. 236–250.
25. Purser, D., "Toxicity Assessment of Combustion Products," *The SFPE Handbook of Fire Protection Engineering,* 2nd ed., DiNenno, P. J., *et al.,* eds., National Fire Protection Association, Quincy, MA, 1995.

Additional Readings

Anderson, C. P., "Methodology for the Use of HAZARD I in Residential Fire Investigation," Worcester Polytechnic Institute, MA, Thesis, Dec. 14, 1990.
Aronstein, J., "Fire Hazard Due to High Resistance Connections," Consultant, Poughkeepsie, New York, 1990.
Babrauskas, V., "Effective Measurement Techniques for Heat, Smoke, and Toxic Fire Gases," *Fire Safety Journal,* Vol. 17, No. 1, 1991, pp. 13–26.
Babrauskas, V., and Peacock, R. D., "Heat Release Rate: The Single Most Important Variable in Fire Hazard," *Fire Safety Journal,* Vol. 18, No. 3, 255–272, 1992.
Babrauskas, V., "Toxic Fire Hazards: Control by Limiting Toxic Potency or Control by Limiting Burning Rate?," Flame Retardants '94, Sixth International Conference, Jan. 26–27, 1994, London, England, Interscience Communications Ltd., London, England, 1994, pp. 239–250.
Babrauskas, V., "Toxicity, Fire Hazard and Upholstered Furniture," Third European Conference on Furniture Flammability (EUCOFF'92), November 1992, Brussels, Interscience Communications Ltd., London, England, 1992, pp. 125–133, 1992.
Becker, W. H. K., "Assessment of Fire Hazards Related to Exterior Walls and Facades," First International Conference on Fire and Materials, Sept. 24–25, 1992, Arlington, VA, 1992, pp. 13–19.
Boyt, A., "Saunas: Is There a Fire Hazard?," *Fire Prevention,* No. 261, July/Aug. 1993, pp. 13–15.
Braun, E., *et al.,* "Performance of School Bus Seats: a Fire Hazard Assessment," Fire Safety Developments and Testing: Toxicity—Heat Release—Product Development—Combustion Corrosivity, Oct. 21–24, 1990, Ponte Vedra Beach, FL, Fire Retardant Chemicals Assoc., Lancaster, PA, 1990, pp. 105–115.
Brown, J. R., Fawell, P. D., and Mathys, Z., "Fire-Hazard Assessment of Extended-Chain Polyethylene and Aramid Composites by Cone Calorimetry," *Fire and Materials,* Vol. 18, No. 3, May/June 1994, pp. 167–172.

Bukowski, R. W., "Toxic Hazard Evaluating Plenum Cables," *Fire Technology*, Vol. 21. No. 4, 1985, pp. 252–266.

Bukowski, R. W., "Fire Hazard Prediction: HAZARD I and Its Role in Fire Codes and Standards," *ASTM Standardization News*, Vol. 18, No. 1, Jan. 1990, pp. 40–43.

Bukowski, R. W., "On the Central Role of Fire Calorimetry in Modern Fire Hazard Assessment," National Institute of Standards and Technology, Gaithersburg, MD, DOT/FAA/CT-95/46, AAR-423; National Institute of Standards and Technology, Fire Calorimetry, *Proceedings*, July 27–28, 1995, Gaithersburg, MD, 1995.

Bukowski, R. W., *et al.*, "Technical Reference Guide for the HAZARD I Fire Hazard Assessment Method. Version 1.1. Volume 2," National Institute of Standards and Technology, Gaithersburg, MD, NIST Handbook 146/II, June 1991.

Bukowski, R. W., "Modeling a Backdraft Incident: The 62 Watts St Fire," *NFPA Journal*, Vol. 89, No. 6, 1995, pp. 85–89.

Buszard, D. L., and Bentley, R. L., "Reducing Smoke and Toxic Gases from Burning Polyurethane Foam," Plastics and Rubber Institute and British Plastics Federation Fourth International Conference, Flame Retardants 1990, Jan. 17–18, 1990, London, England, Elsevier Applied Science, New York, 1990, pp. 222–223.

Caldwell, D. J., and Alarie, Y. C., "Method to Determine the Potential Toxicity of Smoke From Burning Polymers. Part 1. Experiments with Douglas Fir," *Journal of Fire Sciences*, Vol. 8, No. 1, Jan./Feb. 1990, pp. 23–62.

Caldwell, D. J., and Alarie, Y. C., "Method to Determine the Potential Toxicity of Smoke from Burning Polymers. Part 2. The Toxicity of Smoke from Douglas Fir," *Journal of Fire Sciences*, Vol. 8, No. 4, July/Aug. 1990, pp. 275–309.

Charters, D. A., Gray, W. A., and McIntosh, A. C., "Computer Model to Assess Fire Hazards in Tunnels (FASIT)," *Fire Technology*, Vol. 30, No. 1, First Quarter 1994, pp. 134–154.

Clarke, F. B., III, vanKuijk, H., and Steele, S., "Predicting the Toxic Hazard of Cable Fires," International Symposium on Characterization and Toxicity of Smoke, ASTM Committee E-5 on Fire Standards, ASTM ASTM STP 1082, Dec. 5, 1988, Phoenix, AZ, ASTM, Philadelphia, PA, 1990, pp. 46–56.

Cleary, T. G., Ohlemiller, T. J., and Villa, K. M., "Influence of Ignition Source on the Flaming Fire Hazard of Upholstered Furniture," *Fire Safety Journal*, Vol. 23, 1994, pp. 79–102.

Coaker, A. W., and Hirschler, M. M., "New Low Fire Hazard Vinyl Wire and Cable Compounds," *Fire Safety Journal*, Vol. 16, No. 3, 1990, pp. 171–196.

Dalzell, G., "Fire Hazard Management," *Fire Prevention*, No. 285, Dec. 1995, pp. 27–30.

DiNenno, P. J., and Beyler, C. L., "Fire Hazard Assessment of Composite Materials: The Use and Limitations of Current Hazard Analysis Methodology," Hughes Associates, Inc., Columbia, MD ASTM STP 1150; American Society for Testing and Materials, Fire Hazard and Fire Risk Assessment, Sponsored by ASTM Committee E-5 on Fire Standards, ASTM STP 1150, Dec. 3, 1990, San Antonio, TX, ASTM, Philadelphia, PA, 1990, pp. 87–99.

Evans, C. B., "Integration of Fire Hazards Analysis and Safety Analysis Report Requirements," Westinghouse Hanford Co., Richland, WA, WHC-SC-GN-FHA-30001, Sept. 9, 1994.

Fechtmann, G. J., "Underwriters Laboratories Fire Hazard Testing," Special Conference on Recent Advances in Flame Retardancy of Polymeric Materials—Materials, Applications, Industry Developments, Markets, May 15–17, 1990, Stamford, CT, 1990, pp. 1–3.

Finucane, M., "Adoption of Performance Standards in Offshore Fire and Explosion Hazard Management," *Fire Safety Journal*, Vol. 23, No. 2, 1994, pp. 171–184.

"Fire Hazard Assessment," *Fire Prevention Design Guide*, Fire Protection Association, London, England, FPDG 7, Aug. 1990.

Fowell, A. J., "Developments Needed to Expand the Role of Fire Modeling in Material Fire Hazard Assessment," National Institute of Standards and Technology, Gaithersburg, MD, DOT/FAA/CT-93/3; Federal Aviation Administration (FAA), International Conference for the Promotion of Advanced Fire Resistant Aircraft Interior Materials, Feb. 9–11, 1993, Atlantic City, NJ, 1993, pp. 255–262.

Frame, S. R., Carakostas, M. C., and Warheit, D. B., "Inhalation Toxicity of an Isomeric Mixture of Hydrochlorofluorocarbon (HCFC) 225 in Male Rats," *Fundamental and Applied Toxicology*, Vol. 18, 1992, pp. 590–595.

Gann, R. G., *et al.*, "Fire Conditions for Smoke Toxicity Measurement," *Fire and Materials*, Vol. 18, No. 3, May/June 1994, pp. 193–199.

Goransson, U., "Model for Predicting the Hazard of Fire Growth," Statens Provingsanstalt, Boras, Sweden, EUREFIC (European Reaction to Fire Classification), Performance Based Reaction to Fire Classification, International Seminar, Sept. 11–12, 1991, Copenhagen, Denmark, Interscience Communications Ltd., London, England, 1991, pp. 15–22.

Goss, J., "Forensic Scientist's Experiments Have Revealed Fire Hazard of Brake Fluids," *Fire*, Vol. 85, No. 1044, June 1992, p. 8.

Hall, J. R., Jr., "Burns, Toxic Gases, and Other Hazards Associated with Fires: Deaths and Injuries in Fire and Non-Fire Situations. Burns VS. Smoke Inhalation, Burn Injuries, Carbon Monoxide, Electrical Current, Explosions," National Fire Protection Assoc., Quincy, MA, April 1996.

Hall, J. R., Jr., "Progress Report on U. S. Research in Fire Risk, Hazard, and Evacuation," National Fire Protection Assoc., Quincy, MA, NISTIR 4449; Eleventh Joint Panel Meeting of the U. S./Japan Government Cooperative Program on Natural Resources (UJNR) on Fire Research and Safety, October 19–24, 1989, Berkeley, CA, 1990, pp. 2–4.

Heikkila, L., "HAZARD I: Estimation Method for Fire Risks in Single-Family Homes," *Palontorjuntatekniikka*, Jan. 1991, pp. 14–16.

Hinderer, R. K., and Hirschler, M. M., "Toxicity of Hydrogen Chloride and of the Smoke Generated by Poly (Vinyl Chloride), Including Effects on Various Animal Species, and the Implications for Fire Safety," International Symposium on Characterization and Toxicity of Smoke, ASTM Committee E-5 on Fire Standards, ASTM ASTM STP 1082, Dec. 5, 1988, Phoenix, AZ, ASTM, Philadelphia, PA, 1990, pp. 1–22.

Hirschler, M. M., "Electrical Cable Fire Hazard Assessment with the Cone Calorimeter," Goodrich (B. F.) Co., Avon Lake, OH, ASTM STP 1150; American Society for Testing and Materials, Fire Hazard and Fire Risk Assessment, Sponsored by ASTM Committee E-5 on Fire Standards, ASTM STP 1150, Dec. 3, 1990, San Antonio, TX, ASTM, Philadelphia, PA, 1990, pp. 44–65.

Hirschler, M. M., "Fire Retardance, Smoke Toxicity and Fire Hazard," Flame Retardants '94, Sixth International Conference, Jan. 26–27, 1994, London, England, Interscience Communications Ltd., London, England, 1994, pp. 225–237.

Hirschler, M. M., "How to Measure Smoke Obscuration in a Manner Relevant to Fire Hazard Assessment: Use of Heat Release Calorimetry Test Equipment," *Journal of Fire Sciences*, Vol. 9, No. 3, May/June 1991, pp. 183–222.

Hirschler, M. M., "Smoke and Heat Release and Ignitability as Measures of Fire Hazard From Burning of Carpet Tiles," *Fire Safety Journal*, Vol. 18, No. 4, 1992, pp. 305–324.

Hirschler, M. M., "Smoke Toxicity Measurements Made So That the Results Can Be Used for Improved Fire Safety," *Journal of Fire Sciences*, Vol. 9, No. 4, July/Aug. 1991, pp. 330–347.

Hirschler, M. M., "Measurement of Smoke in Rate of Heat Release Equipment in a Manner Related to Fire Hazard," *Fire Safety Journal*, Vol. 17, No. 3, 1991, pp. 239–258.

Hoffmann, D., "Analysis of Toxic Smoke Constituents," American Health Foundation, Valhalla, NY, Consumer Product Safety Commission, Washington, DC, Vol. 5, Aug. 1993.

Hoover, J. R., "Practical Fire Hazard Assessment New Approaches," *Proceedings* of the Seventeenth International Conference on Fire Safety, Vol. 17, Jan. 13–17, 1992, Millbrae, CA, Product Safety Corp., Sunnyvale, CA, 1992, pp. 378–380.

International Atomic Energy Agency, "Fire Hazard Analysis for WWER Nuclear Power Plants. Report of the IAEA Extrabudgetary Program on the Safety of WWER Nuclear Power Plants," International Atomic Energy Agency, Vienna, Austria, IAEA-TECDOC-778, 1993.

Jones, G. R. N., and Saito, K., "Cyanide and Fire Victims: a Hazard in Focus," *Fire Technology*, Vol. 26, No. 2, May 1990, pp. 141–148.

Jones, W. W., "Fire Hazard Assessment Methodology," National Institute of Standards and Technology, Gaithersburg, MD, NISTIR 5836, May 1996.

Kaplan, H. L., and Hinderer, R. K., "Extrapolating Animal Smoke Toxicity Results to Human Health Effects: the Need to Consider Animal Species Differences," Recent Advances in Flame Retardancy of Polymeric Materials: Materials, Applications, Industry Developments, Markets, Vol. 2, May 14–16, 1991, Stamford, CT, Business Communications Co., Inc., Norwalk, CT, 1991, pp. 308–318.

Keltner, N. R., Acton, R. U., and Gill, W., "Evaluating the Hazards of Large Petrochemical Fires," Sandia National Labs., Albuquerque, NM,

ASTM STP 1150; American Society for Testing and Materials, Fire Hazard and Fire Risk Assessment, Sponsored by ASTM Committee E-5 on Fire Standards, ASTM STP 1150, Dec. 3, 1990, San Antonio, TX, ASTM, Philadelphia, PA, 1990, pp. 37–43.

Klamerus, E. W., and Ross, S. B., "Fire Hazards Analysis for the Center for National Security and Arms Control (CNSAC) Facility," Sandia National Labs., Albuquerque, NM, Science and Engineering Associates, Inc., Albuquerque, NM, SAND-92-2932, July 1993.

Legg, J., "Cable Flammability Testing—Classifying the Hazard," *Fire Surveyor*, Vol. 20, No. 4, Aug. 1991, pp. 20–24.

LeVan, S., *et al.*, "Assessing the Fire Hazard of Structures in the Wildland-Urban Interface," Forest Products Lab., Madison, WI, Aug. 1990.

Levin, B. C., *et al.*, "Toxicity of Complex Mixtures of Fire Gases," *Toxicologist*, Vol. 10, No. 1, Feb. 1990, p. 84.

Levin, B. C., Paabo, M., and Navarro, M., "Toxic Interactions of Binary Mixtures of NO2 and CO, NO2, and HCN, NO2 and Reduced O2, and NO2 and CO2," National Institute of Standards and Technology, Gaithersburg, MD, Abstract 825, Thirtieth Annual Conference of the Society of Toxicology, Abstracts, Vol. 11, No. 1, Feb. 25–March 1, 1991, Dallas, TX, 1991, p. 222.

Marchant, R., "Furniture Fire Hazards Coming Under Control," *Fire Prevention*, No. 226, Jan./Feb. 1990, pp. 28–30.

Morris, J., "Toxicity of Fire Gases," *Fire Protection*, Vol. 21, No. 4, Dec. 1994, pp. 8–11.

Nelson, G. L., "Carbon Monoxide and Fatalities: a Case Study of Toxicity in Man," Chapter 6, *Carbon Monoxide and Human Lethality: Fire and Non-Fire Studies,* Elsevier Applied Science, New York, 1993, pp. 179–196.

Nelson, H. E., "FPETOOL: Fire Protection Engineering Tools for Hazard Estimation," National Institute of Standards and Technology, Gaithersburg, MD, General Services Administration, Washington, DC, NISTIR 4380, Oct. 1990.

Nelson, H. E., "FPETOOL: Fire Protection Tools for Hazard Estimation: an Overview of Features," Fifth International Fire Conference on Fire Safety, Interflam '90, September 3–6, 1990, Canterbury, England, Interscience Communications Ltd., London, England, 1990, pp. 85–92.

Nelson, H. E., and Forssell, E. W., "Use of Small-Scale Test Data in Hazard Analysis," *Proceedings* of the Fourth International Symposium on Fire Safety Science, Intl. Assoc. for Fire Safety Science, Boston, MA, 1994, pp. 971–982.

"New Hazard Uncovered—Rolled Nonwoven Fabrics," *Record*, Vol. 68, No. 6, Nov./Dec. 1991, pp. 3–8.

Parris, G. E., "Assessing the Risks of Toxic Fumes Produced During Fires," *Fire and Arson Investigator*, Vol. 40, No. 3, March 1990, pp. 26–27.

Pauluhn, J., "Retrospective Analysis of Predicted and Observed Smoke Lethal Toxic Potency Values," *Journal of Fire Sciences*, Vol. 11, No. 2, March/April 1993, pp. 109–130.

Peacock, R. D., and Bukowski, R. W., "Prototype Methodology for Fire Hazard Analysis," *Fire Technology*, Vol. 26, No. 1, Feb. 1990, pp. 15–40; American Society of Safety Engineers (ASSE) Twenty-ninth Professional Development Conference and Exposition, Building Our Professional Heritage–Protecting Our Nation's Resources, June 24–27, 1990, Washington, DC, Am. Soc. of Safety Engineers, Des Plaines, IL, 1990 pp. 20–37.

Peacock, R. D., *et al.*, "Software User's Guide for the HAZARD I Fire Hazard Assessment Method. Version 1.1. Vol. 1," National Institute of Standards and Technology, Gaithersburg, MD, *NIST Handbook 146/I,* June 1991.

Prager, F., and Cabos, H. P., "Fire-Gas Hazards in Rail Traffic," *Fire and Materials*, Vol. 18, No. 3, May/June 1994, pp. 131–149.

Prugh, R. W., "Quantitative Evaluation of "BLEVE" Hazards," *Journal of Fire Protection Engineering*, Vol. 3, No. 1, 1991, pp. 9–24.

Purser, D., "Smoke Toxicity," *Proceedings* on Improved Fire- and Smoke-Resistant Materials for Commercial Aircraft Interiors, Nov. 8–10, 1994, Washington, DC, National Academy Press, Washington, DC, 1994, pp. 175–195.

Purser, D. A., "Development of Toxic Hazard in Fires From Polyurethane Foams and the Effects of Fire Retardants," Plastics and Rubber Institute and British Plastics Federation Fourth International Conference, Flame Retardants 1990, Jan. 17–18, 1990, London, England, Elsevier Applied Science, New York, 1990, pp. 206–221.

Purser, D. A., "Toxicity, Toxic Hazard and Furniture in Fires," Paper 8, Second European Conference on Furniture Flammability, June 27–28, 1990, Brussels, Belgium, 1990, pp. 8/1–17.

Robinson, P., "Duty of Care? Fire Hazard of Multi Layer Paint Surfaces," *Fire Engineers Journal*, Vol. 55, No. 177, June 1995, pp. 29–30.

Sanders, D. C., Endecott, B. R., and Chaturvedi, A. K., "Inhalation Toxicology. Part 12. Comparison of Toxicity Rankings of Six Polymers by Lethality and by Incapacitation in Rats," Federal Aviation Administration, Oklahoma City, OK, Federal Aviation Administration, Washington, DC, DOT/FAA/AM-91/17, Task AM-B-90-TOX-58, Dec. 1991.

Sharma, T. P., *et al.*, "Insight to Museums From Fire Hazards Assessment and Fire Protection Point of View," *Fire Science and Technology,* Vol. 11, No. 1–2, 1991, pp. 61–68.

Snell, J. E., "Fire Hazard and Risk: Evaluating Alternative Technologies," Sprinklers in Residential and Commercial Buildings, Aug. 19–20, 1990, Charleston, SC, Natl. Conf. States on Bldg. Codes and Standards, 1990, pp. 1–23.

Spearpoint, M. J., and Shipp, M. P., "Measurement of Heat Release From Burning Automobiles for Channel Tunnel Hazard Assessment," Fire Research Station, Borehamwood, England, Federal Aviation Administration, Fire Calorimetry, *Proceedings*, July 27–28, 1995, Gaithersburg, MD, 1995, pp. 135–147.

Sundin, D., "Methods of Reducing the Hazards of Fire in Fluid-Filled Indoor Transformers," Cooper Power Systems, American Technologies International. Fire Safety '91 Conference, Volume 4, Nov. 4–7, 1991, St. Petersburg, FL, Am. Technologies Intl., Pinellas Park, FL, 1991, pp. 300–310.

Tewarson, A., "Relationship Between Generation of CO and CO2 and Toxicity of the Environments Created by Materials in Flaming and Non-Flaming Fires and Effect of Fire Ventilation," International Symposium on Characterization and Toxicity of Smoke, ASTM Committee E-5 on Fire Standards, ASTM ASTM STP 1082, Dec. 5, 1988, Phoenix, AZ, ASTM, Philadelphia, PA, 1990, pp. 23–33.

Tewarson, A., "Fire Properties of Materials for Model-Based Assessments for Hazards and Protection," Chapter 30, ACS Symposium Series 599, 208th National Meeting of the American Chemical Society for Fire and Polymers II: Materials and Tests for Hazard Prevention, Aug. 21–26, 1994, Washington, DC, American Chemical Society, Washington, DC, 1995, pp. 450–497.

Tsukagoshi, I., "Progress Report on Risk, Hazard and Evacuation in Japan," Building Research Institute, Ibaraki-ken, Japan, NISTIR 4449; Eleventh Joint Panel Meeting of the U. S./Japan Government Cooperative Program on Natural Resources (UJNR) on Fire Research and Safety, Oct. 19–24, 1989, Berkeley, CA, 1990, pp. 5–7.

Vogman, L. P., and Degtyarev, A. G., "Fire Hazard of Vegetable Raw Materials. Part 1," *Fire Science and Technology*, Vol. 15, No. 1/2, 1995, pp. 47–51.

Zhang, J., Shields, T. J., and Silcock, G. W. H., "Fire Hazard Assessment of Polypropylene Wall Linings Subjected to Small Ignition Sources," *Journal of Fire Sciences*, Vol. 14, No. 1, Jan./Feb. 1996, pp. 67–84.

FIRE RISK ANALYSIS

John R. Hall, Jr.

Fire risk analysis is the most comprehensive form of analysis that can be applied to any fire safety choice. It looks at all the types of fires that may be affected by the choice, e.g., of product, of building design, of anything potentially affecting fire safety, not just one fire or a few selected fires. It provides a flexible framework for estimating the impact of any type of fire safety program or strategy in terms of actual reductions in losses—deaths, injuries, property damage—and in terms that can be compared to the costs of those programs and strategies. There is no other method as well suited for the analysis of strategic options affecting large numbers of properties and their occupants. Product development, related research and marketing, and regulatory decisions can benefit from fire risk analysis, as can any program to manage and oversee fire safety from a financial point of view.*

Fire risk analysis is unusual because the overall framework it uses is taken not from the hard sciences of physics, chemistry, biochemistry, and engineering, which underlie the rest of fire safety research, but from statistical decision theory, which is based upon the fields of economics and operations research. Because fire risk analysis comes from a setting with which much of the fire community may be unfamiliar, this chapter will focus on a discussion of basic concepts, definitions, and approaches.

What Is Fire Risk Analysis?

A measure of fire risk always has two parts: (1) a measure of *severity* and (2) a *probability distribution*. The severity measure may also be separated, for the sake of clarity, into two parts. The first part is a specification of the scale that measures severity, such as deaths, injuries, dollars of property damage, area reached by flames, days of business interruption, and so forth. The second part is a specification of the rules for calculating the specific severity measure to be used for a particular fire.

For example, suppose the basic scale was dollars of damage. One rule would be to set the severity measure equal to the number of dollars of damage. This would set the stage for a risk analysis on expected loss or average loss, the most common approach. Alternatively, a second rule would be to set the severity measure equal to 1 if damage exceeded $100,000,000 and 0 otherwise. This would set the stage for a risk analysis focusing on large losses, which might be more useful for an insurance company that worries less about ordinary claims than about a single loss so large the company cannot cover it.

The probability distribution gives a probability for each value the severity measure can take. Most often, a probability distribution gives a probability for every type of fire, and then rules for calculating the value of the severity measure for each type of fire provide the translation to a probability distribution for the measure of severity.

In general, the fire risk analyst will specify a *scenario structure*, which is a set of classification rules for dividing the range of possible fires into a manageable number of relatively homogeneous groups. Then probabilities can be calculated or estimated for each scenario (a group of like fires), and a severity measure can be calculated or estimated for the average fire in each scenario, usually by detailed examination of a typical or representative fire from each scenario.

The abstract nature of this explanation so far may lead people to ask a simple question: What good is fire risk analysis? Perhaps the best response is that the practical value of fire risk analysis emerges as these general-purpose measures are calculated under specified real or potential conditions. In any fire risk analysis, there will be a certain set of assumptions about key conditions in the environment of interest: some or all of the buildings have sprinklers or detectors; some or all of the mattresses were built to a certain ignition-resistance standard; some or all of the occupants have been regularly trained in fire safe behavior; and so forth.

Fire risk analysis refers to a systematic examination of how measures of fire risk change as the assumptions change. It may be done in conjunction with a parallel examination of changes in costs. On the one hand, if no detector is present, a household can expect to experience fires with a certain probability, and these fires will have a certain expected severity in terms of deaths, injuries, and property damage. On the other hand, if a detector is present, the household will pay purchase and maintenance costs not paid by the household without detectors, but the detector household can expect on average to achieve lower severities of losses—and in particular, fewer deaths per fire. Risk analysis is used to quantify the expected size of the reductions in deaths, injuries, and damage. One can then determine whether these risk reductions are great enough to justify the cost. This is an analysis of cost *vs.* risk-reduction benefits in which the risk portion is measured explicitly.

---

*Much of the material in this chapter was first developed for an in-house concept paper at the National Bureau of Standards, Center for Fire Research (NBS/CFR). The author expresses his appreciation for the valuable support accorded this work by NBS/CFR.

John R. Hall, Jr., Ph.D., is NFPA's assistant vice president for fire analysis and research.

The term *risk analysis*, and similar terms such as *hazard analysis*, may be used to describe this kind of analysis, but other kinds of analysis are also labeled with some of these terms. (See Section 11, Chapter 7, "Fire Hazard Analysis.") Three key elements of the fire risk analysis approach are: (1) the explicit estimation of all relevant probabilities, (2) the use of well-defined measures of severity, and (3) the explicit consideration of the uncertainty attached to all estimates developed in the analysis. Some modeling approaches that are called risk analysis omit one or more of these key elements.

In particular, the explicit consideration of uncertainty is easily the most difficult step, and the one most often given abbreviated treatment in even the better fire risk analyses, but it is essential to sound interpretation of data and separation of meaningful, significant differences from artifacts of statistical "noise."

For example, the term *fire risk analysis* has sometimes been used to refer to an exercise of listing locations under the jurisdiction of the analyst that present the greatest potential for fire damage or for demand on fire suppression resources. Such an exercise lacks most or all of the key elements of a true fire risk analysis. There is no consideration of probabilities or uncertainty at all, and even the severity measures implicit in the definition of potential damage or demand may not be explicitly defined, let alone measured.

Other commonly used analysis approaches present subtler problems. Fire protection engineers often focus on alternative means for dealing with fires, after fire has begun. It may seem natural to construct a fire risk analysis that takes established burning as given but measures the probability of performance for all building features and systems that operate after the fire. Assuming it makes sense to write off fire prevention, there still will be a need to address explicitly different types of initial fires. A fast, accelerant-driven fire or an initial explosion will challenge the built-in fire protection in different ways than an ordinary flaming or smoldering fire. A concealed-space fire will present different problems than a fire in a room, and both will differ from fires started in a means of egress. If each of these fires is analyzed separately, if the analysis leans on physics and chemistry, and if the focus is on preventing major fire extent as an either-or proposition, then the subject is more properly termed a *hazard analysis*. While the terminology is not yet standardized, hazard analysis is best used to describe analyses that do not treat most probabilities explicitly, particularly probabilities of ignition and probabilities of occupant characteristics and behaviors. *Risk analysis* should be reserved for analyses that address all relevant probabilities.

Also, risk analysis usually measures deaths, injuries, and property damage rather than stopping with measures of how many rooms or square feet (m^2) reached a particular temperature or had smoke or gas levels over a particular threshold. In most situations in which risk analysis is used, it is important to address whether people and property are actually damaged, which means implicit or explicit attention to the locations, decisions, movements, and vulnerabilities of occupants—actual, not just potential—and the damageability of property. Risk analysis also requires estimation of the factors that contribute to reliability of systems, such as the probability of features and systems being rendered inoperative or ineffective due to human error or negligence; hazard analysis does not always address field reliability in all its aspects.

Hazard analysis addresses the core concerns of fire development, smoke spread, and sometimes toxicity; it often does so in far greater detail than risk analysis. Risk analysis includes victim decisions, locations, and characteristics; ignition factors; and reliability of systems and building features. In order to be this broad, fire risk analysis cannot always make use of the full power and sensitivity to detail of the hard-science models of hazard analysis; the only models that can be used are those that can be fitted into the complete risk

analysis framework. Thus, hazard analysis has the advantage of being able to examine details of design that fire risk analysis cannot now handle. In the long term, however, work is proceeding on model integration techniques capable of combining the breadth of fire risk analysis with the depth of hazard analysis.

Because real fires reflect all the factors that affect ignition probability and fire severity, fire risk analysis usually begins with calculations from data bases on actual fires. (Nearly two million fires are reported to the fire service each year.) For most analyses of classes of properties, then, one can identify initial data bases of historical fires that will give statistically meaningful probabilities and severities of fires.

Many situations, however, may dictate going beyond incident data bases in performing a fire risk analysis. First, in analyzing new strategies, products, or systems, there will be no data base of fires reflecting the influence of these innovations. Second, there may be interest in so many product or building characteristics that the data base does not have enough fires to cover all possible combinations. Third, more detail may be needed than is available in the typical data base.

If available fire experience data is insufficient, a fire risk analysis will begin to look more like a hazard analysis, with tree diagrams to capture alternative sequences of events, probabilities for the various possible starting conditions, and probabilities for transitions from one stage to the next in a fire. What is measured and in how much detail may differ from a hazard analysis, in the ways noted earlier.

Already you may have begun to sense how difficult it can be to construct a fire risk analysis that is valid, useful, and of manageable size. Concern over the computational burden is one reason why analysts sometimes conduct fire risk analyses using a very small number of scenarios (e.g., only three alternatives for where and how fire begins). There may also be concern over the shakiness of the available data and a desire to minimize the number of different terms supported by expert judgments.

In a situation like that, it is essential to understand what is being implicitly assumed by the use of a stripped-down scenario structure. One way or another, the scenarios that are modeled explicitly are being treated as representative of the far larger number of scenarios that are not modeled explicitly. If only a few scenarios are modeled explicitly, then each one is implicitly required to be representative of a much larger and more varied collection of other scenarios. There may be no good evidence to support this. Is it better, for example, to examine the different types of flaming fires individually or to assume that they all behave like a burning wood crib ignited by a lighter? The variability and uncertainty are there either way; the only choice is how much to examine explicitly and how much to handle by assumption.

Risk Estimation and Risk Evaluation

Fire risk analysis can sometimes be separated into: (1) *risk estimation*, the estimation and analysis of the measures of severity and probability and their associated uncertainties, and (2) *risk evaluation*, the additional steps required to decide on the importance of a particular value of risk or a change in risk. A fire risk analysis that includes risk evaluation may be called a *fire risk assessment* to underline the fact that the analysis will support value judgments.

Risk evaluation should be familiar to anyone who has had to make business decisions, because it essentially involves using analysis to determine whether you will get what you pay for. The most common approach to risk evaluation is *cost-benefit analysis*, a technique in which all benefits of risk reduction are translated into monetary equivalents. This technique permits a proposed new product or fire suppression system to be assessed in terms of its net profit or loss, total cost plus loss, or ratio of profit to loss. In such a context, "profit" means saved lives, avoided injuries, and reduced property damage, all combined in one monetary scale, and "loss" means the cost—both initial and ongoing—of the new product or system. A variation is *cost-effectiveness analysis*, in which the benefits of risk reduction are translated into a single nonmonetary scale. For example, it is possible to derive the cost per life saved for a new system or product.

Cost-benefit and cost-effectiveness analyses require the explicit estimation or derivation of some controversial parameters. Examples include the value of reduced risk of death or injury and the discount rate by which future consequences are compared to present consequences (both discussed in more detail later). Because these parameters are controversial, risk evaluation sometimes uses methods that downplay the role of these parameters.

Acceptable risk is a term used when the method of risk evaluation involves treating risk as a constraint. This method may seem attractive because it refuses to consider costs until or unless a sufficient degree of fire safety has been provided. In an acceptable risk approach, a certain level of risk is defined as acceptable; then all alternatives meeting that level are evaluated strictly on the basis of cost.

This approach can produce unsatisfactory results. If risk is greater than the acceptable level by even a small fraction, no cost is too great to reach acceptable risk. If risk already is acceptably low, not a nickel more should be spent, no matter how much more fire safety could be purchased for very little money. This means that the selected level of acceptable risk is often set with an eye toward affordability and may be reset if technology changes. In effect, this makes the acceptable risk approach a kind of backdoor cost-benefit analysis and runs counter to most approaches to decision making in business.

When acceptable risk is not defined in terms of affordable risk, it is often defined in terms of: (1) historically acceptable risk (i.e., anything in use for a long time is all right), which may be overturned if public understanding of the magnitude of the risk changes dramatically, or (2) unavoidable risk, such as the use of background radiation levels as a guide for acceptable exposure to medical X-rays. In fire protection, acceptable risk has sometimes been inferred from provisions of NFPA codes and standards. The most extreme version of an acceptable risk approach is a minimum risk approach, in which cost is not considered unless all feasible safety improvements have been made.

A logical complement to the acceptable risk approach would be an acceptable cost approach, in which the greatest risk reduction available within the fixed cost budget (but no more) would be sought. Although this approach is rarely mentioned in the literature, it almost certainly describes the way some decisions are made.

Note that the acceptable risk and acceptable cost approaches both can be very sensitive to the initial selection of a single key parameter. If these approaches are used, it is necessary to conduct a sensitivity analysis to see how conclusions would be changed if the reference level of acceptable risk or acceptable cost were different. Sensitivity analysis of key parameters is essential to all other forms of risk analysis as well, and is one way in which the uncertainty surrounding estimates can be addressed systematically.

Overview of a Risk Analysis Conceptual Framework

Figure 11-8A provides an overview of the kind of conceptual framework used to identify necessary models and necessary data in risk analysis. The major aspects to be modeled can be grouped into six models, as shown in the ovals: (1) decision model, (2) ignition-initiation model, (3) post-ignition model, (4) loss evaluation model, (5) cost model, and (6) cost-benefit comparison model. The rectangles indicate numbers, either inputs or outputs, derived from the models. These numbers may in turn be supplied as inputs to other models later in the modeling sequence. The model refers to a "proposed change," which is a general term for anything that might be modeled—a new sprinkler system, increased compartmentation requirements, no-smoking areas, mandatory self-extinguishing cigarettes, a training program for staff, or any other change that could make fires more or less likely or more or less severe.

A decision model is used to describe what the building and its occupants would look like if a proposed change were or were not made. An ignition-initiation model is used to estimate the probabilities of occurrence per year for each fire scenario, while a post-ignition model is used in parallel to estimate the expected losses per fire, for each fire scenario and each type of loss (i.e., deaths, injuries, property damage). The fire scenarios are defined by the requirements of the two models. For example, if the post-ignition model uses different parameters for smoldering and flaming fires, then that distinction must be reflected in the separation of fires into scenarios for the ignition-initiation model. When the outputs of these two models are combined, they produce estimates of expected losses per year by type of loss and for all fire scenarios. Then a loss evaluation model is used to convert all types of loss to a common scale and produce year-by-year projections of overall expected loss, with and without the proposed change.

Meanwhile, on the right side of the framework, the purchase, installation, maintenance, inspection, operating, replacement, and other costs of the new system are combined by a cost model into year-by-year figures of the cost impact of the proposed change.

Finally, a cost-benefit comparison model produces a measure of attractiveness for the proposed change. This is simply a comparison of annualized costs and benefits, with the benefits consisting of changes in risk.

The discussion that follows will address the six component models in more detail. With respect to two terms introduced earlier, *risk estimation* involves the decision model, the ignition-initiation model, and the post-ignition model, while *risk evaluation* covers the other three models.

Much more detailed guidance on the specification of fire risk analysis models can be found in Section 5, Chapters 10, "Product Fire Risk"; 11, "Building Fire Safety Risk Analysis"; and 12, "An Introduction to Quantitative Risk Assessment in Chemical Process Industries," in the second edition of *The SFPE Handbook of Fire Protection Engineering*. That handbook also has guidance on the basics of the mathematical methods used in fire risk analysis, particularly probability theory and statistical analysis.

General Characteristics and Fire Types

Before the models shown in Figure 11-8A come into play, there must be an initial structure to the problem that describes the type of building, the characteristics of its occupants, and the types of fires to be studied. Probability distributions may be needed for any of these factors. Even if only a single building is being analyzed, that building's behavior with respect to fire may vary as a function of randomly varying conditions such as the positions of doors and windows (open or closed) or of forced air heating and cooling systems (on or off). Occupant locations and conditions also may vary in random or patterned ways. It is also useful in setting up a problem to be aware of any factors that cannot be explicitly addressed. This helps in interpreting the results.

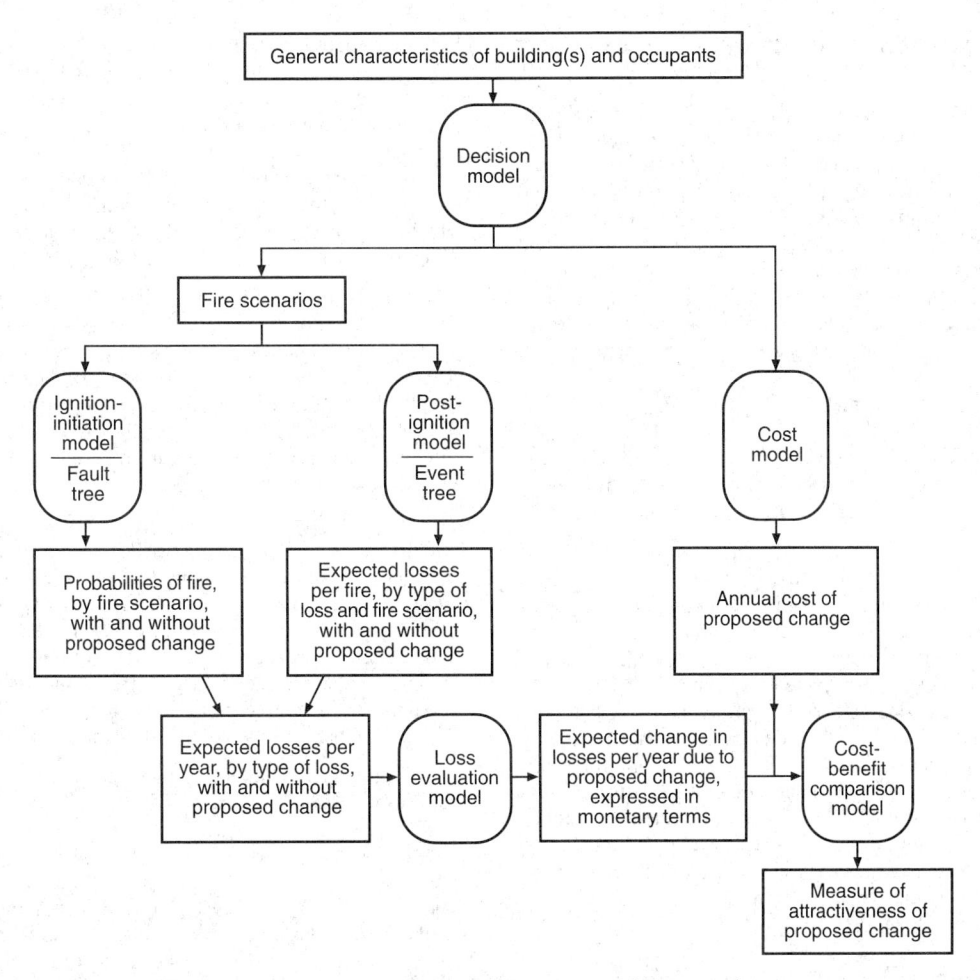

FIG. 11-8A.   A risk analysis conceptual framework.

The most important point to remember in defining types of fires is that all possible fire ignitions must be covered. Fires may have to be grouped into classes that are not entirely homogeneous, but it is not sound practice to exclude certain categories of fires. In the likely event that the post-ignition model requires each fire type to be specified in great detail (e.g., location and heat release rate of first material ignited), one will be left to choose between an unmanageably large computational burden from a large number of fire scenarios or some uncomfortable groupings of very different types of fires into a small number of fire scenarios.

The classification scheme should distinguish among fire scenarios that are affected differently by the systems or strategies being studied. Suppose fast-response sprinklers are being evaluated for possible home use. Then the key properties of fires affecting their response to sprinklers might be identified by answering these questions: Which areas of origin would not be accessible to sprinklers (e.g., the outside of the house, concealed spaces, unsprinklered living areas such as bathrooms)? Could the fire ignition energy best be characterized as smoldering, flaming, or fast-flaming/high energy? The distinction among smoldering, flaming, and fast-flaming fires might be made to recognize the fact that fire growth (which activates the sprinkler) and smoke spread (which is the principal cause of death) do not develop in the same way for all fires.

Decision Model

In addition to defining types of fires and specifying general building and occupant characteristics, it is necessary to specify all the differences that would occur if the proposed change were made. If the change involves new built-in smoke detection systems, for example, specifications for the systems would be needed. If the systems were available in two or more versions (e.g., ionization and photoelectric smoke detectors), the probability of each version being used should be estimated (e.g., from data on usage patterns or projected shares of market). Probabilities also would be needed for the variations in status of a system (e.g., fully operational, operational but blocked in one room, turned off, or removed). These specifications dictate precisely what fire protection and fire-related features would be in place to influence the course of fire if one occurs.

If the proposed change involves training staff or educating occupants, similar questions would be asked: What are the specifics of the program(s)? What is the likelihood that each version of the program would be used? If fire occurs, what would be the status of the program (e.g., everyone was trained and acted accordingly, some missed the training, some forgot, some panicked, etc.)?

It is sometimes useful to separate the parts of the decision model into those characteristics that were selected by a building owner or manager (e.g., a sprinkler was installed) and those that came into play after the selection (e.g., the sprinkler was installed incorrectly). The latter parts are sometimes collectively referred to as the implementation model. The implementation model isolates those factors that are hardest to control, either because they require continued alertness (e.g., without scheduled testing and maintenance, a sprinkler system's reliability will decrease) or because they are not directly under the control of the building owner or manager (e.g., a manager can order a sprinkler system, but has to work through others to make sure it is installed properly).

Neither the implementation model nor the larger decision model needs every bit of information on the proposed changes. The models need only those details that will affect the likelihood of having a fire, the development of that fire, or the reactions of people and property to the fire. For example, some factors will incapacitate the system so that the fire will develop as if no system were present, and that is all the information needed. Some factors will degrade system performance but still permit some impact on fire development; it may be possible to model such reduced impact as simple modifications to the impact expected if the system were fully functional.

Ignition-Initiation Model

The ignition-initiation model is needed to produce estimates of the probability of ignition per year, by type of fire, given a structure of fire scenarios and fire-relevant characteristics of the building and its occupants.

The ignition-initiation model is a device for combining known probabilities in order to infer unknown probabilities. A simple example will illustrate the kinds of calculations that are possible and desirable. Suppose there are n brands of cigarettes. Define p_i as the share of total product usage contributed by product i; in this case p_i is the share of all cigarettes smoked that are brand i. Let r_i be the proportion of brand i cigarettes discarded each year in such a way as to make a fire possible. Let q_i be the probability of ignition given such an exposure for brand i; q_i is a measure of the relative self-extinguishing tendencies of brand i, but it can be affected by changes in the ignitability of materials that are typically ignited. Let F be the number of cigarette-related fires per year and let N be the number of cigarettes smoked per year. Then

$$F = \sum_{i=1}^{n} N_i\, r_i\, q_i$$

Note that there are three ways to cut down on fires from brand i cigarettes: (1) see that almost no one buys them (reduce p_i), (2) see that the people who buy them almost always dispose of used cigarettes properly (reduce r_i), and (3) see that discarded cigarettes usually self-extinguish (reduce q_i).

This model can be used to examine several different types of strategies. A ban on all cigarettes that do not meet a self-extinguishment standard would change the p_i values, because the prohibited brands would have their p_i values cut to zero, and the p_i values of all the other brands would rise to fill the gap in the market. A campaign to educate the public not to discard cigarettes would lower the r_i values; or, a requirement to reduce the flammability of couches would lower all the q_i values, because more cigarettes would self-extinguish if couches were harder to ignite.

To perform the analysis, one must set initial values for the parameters. The purpose of this example is to show how to derive values for the parameters through a combination of direct measurement, reasonable assumptions, and mathematical inference. The values of p_i might be measured through sales records or a survey of product users. The values of q_i might be measured in laboratory tests. The most elusive will be the r_i values, the measures of how often cigarettes are being discarded, whether or not fires result.

If it is assumed or can be proven that all types of cigarettes are equally likely to be discarded, then there is only one common r_i value, and it can be solved for. Or if data can be obtained on numbers of cigarette fires per year by brand of cigarette, then individual r_i values can be solved for.

In a more complex case, suppose the analyst had at least as many years of fire data as there were brands of cigarettes. Also suppose that it can be assumed or shown that year-to-year variations in fire experience were due solely to changes in market share. (This is equivalent to assuming that neither behavior in discarding cigarettes nor the propensity of each brand to start fires if discarded has changed over the years.) Then if p_i values (market shares) could be obtained for each year, it would be possible to solve for the r_i values using the years of fire data as a set of n equations in n unknowns.

These diverse approaches illustrate a general point. To estimate unknown probabilities, it is necessary to: (1) find direct data sources for the unknowns (e.g., survey of couch usage), (2) develop reasonable assumptions as to the values of the unknowns (e.g., why the r_i might all be equal), (3) find valid formulas relating the unknowns to other variables on which there is sufficient data to make inferences back to the unknown probabilities, or (4) fill the gaps with expert estimates.

A full ignition-initiation model needs to combine far more factors than were used in the simple illustration and to accommodate a multitude of logical and probabilistic interdependencies among the variables. There may be many different factors, none necessary for fire to occur and none sufficient to cause fire except in combination with certain other factors. The standard format for constituting such models is the fault tree or success tree, which is discussed in more detail under the more general name of "fire safety concepts tree" in Section 1, Chapter 3, "Systems Concepts for Building Fire Safety."

A more generic flow chart of the steps needed to estimate probabilities is shown in Figure 11-8B.

Post-Ignition Model

The post-ignition model is used for estimating the severity of specified types of fires, given specified building and occupant characteristics and the status of all systems, features, products, and other changes being analyzed. The methodology for tracking the development and final consequences of a specified fire under specified building and occupant conditions may be deterministic, probabilistic, or, more likely, a mix of the two.

Fire protection engineers are aware of the rapid growth in computer-based deterministic models for analyzing the development of a fire and its effects under specified conditions. There are two special requirements for the post-ignition model component of a fire risk analysis that constrain the use of existing deterministic models: (1) it cannot accommodate a more finely divided typology of fires than can be handled by the ignition-initiation model (i.e., for every distinct type of fire whose development and final impact is modeled, it must also be possible to estimate the likelihood of that type of fire); and (2) it should produce not only physical descriptions of flame, gas, and smoke extent but also measures that are useful as end-measures of impact, such as deaths, injuries, and dollar-value property damage.

The post-ignition model—a model of fire development and outcomes—consists of a set of linked time functions of fire characteristics and their effects. For example, the model might need to specify, at each time, the extent of flame, smoke, and gases by type, the status of fire protection systems and other building characteristics, the locations of occupants, the tenability of various locations, etc. No existing model can do all this.

To develop a practical version of the model, one must ask: (1) how many of these characteristics need be known to develop reasonably good estimates of deaths, injuries, and damage from a given fire? and (2) at which moments in time will the development of a particular type of fire change, as a result of the different input values to be modeled (e.g., different initial statuses of fire protection sys-

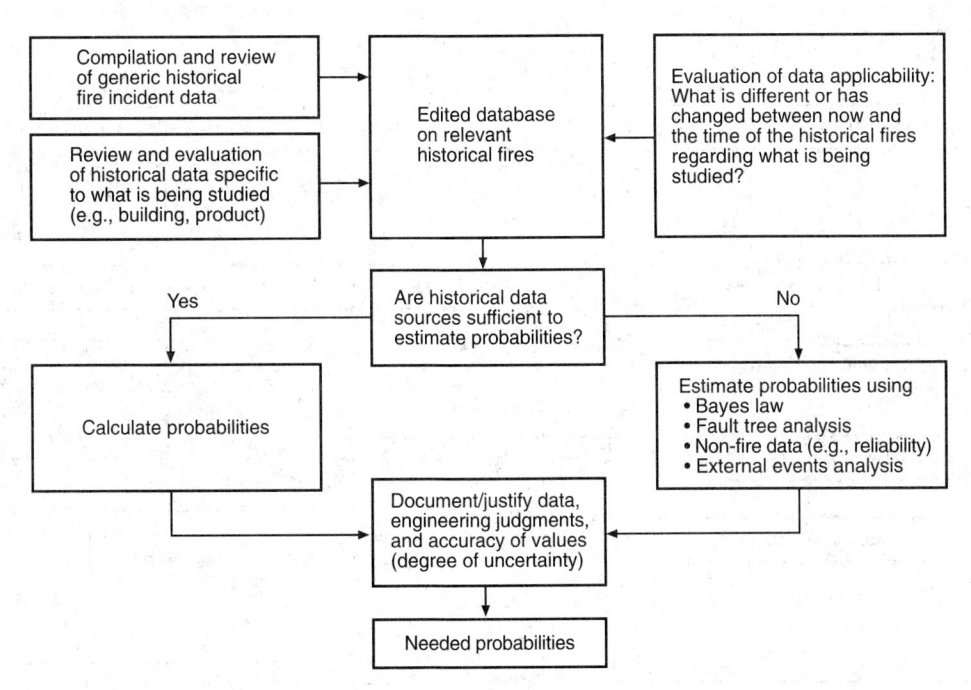

FIG. 11-8B. *General procedure for estimating needed probabilities.*

tems)? In other words, one must identify a limited number of fire descriptors to track and a limited number of points in time at which to check them.

One type of model to do this is called an "event tree," the standard modeling format of statistical decision theory. In an event tree, each point in time is characterized by a set of events. These events indicate all the states in which the fire might be at given points in time (with the states defined by the limited number of fire descriptors previously cited). Then, for each event, one must specify the probabilities that—starting from that event—the fire will develop into each of the possible states (or events) that could characterize the fire at the next point in time. Event trees work well if all the factors being analyzed operate at relatively well-defined moments in time or stages of the fire. Experience in using these models indicates that the performance of most fire protection systems and features can be tied to events, which are themselves defined by characteristics of the fire. Figure 11-8C is a simple but still practical and useful example.

Figure 11-8D shows a fire risk analysis that includes a more integrated system of fire and smoke growth models and treats fire development deterministically rather then probabilistically.

An event tree (see Figure 11-8C), like any tree model, consists of nodes and arcs. Each node represents an event or stage in the development of the fire—typically a point where a particular system or feature will activate if it is operational or where the speed of the fire changes dramatically (e.g., flashover). The condition of the fire at extinguishment is captured by the outcome events, which are the various terminal events of the tree. Each outcome event is assigned values of expected losses per fire (i.e., average severity—deaths, injuries, property damage) that are the estimated losses if the fire ends in the way described by the outcome event. However, the only losses covered by these values are those that occur during the final stage of the fire or are otherwise correlated with the ultimate size of the fire. For computational reasons, deaths and some other losses that take place early in the fire may have to be associated with the stage of the fire where they occur. For example, suppose the first event in a fire is ignition and the second is the point where a fast-response sprinkler would activate. Fatal injuries occurring before

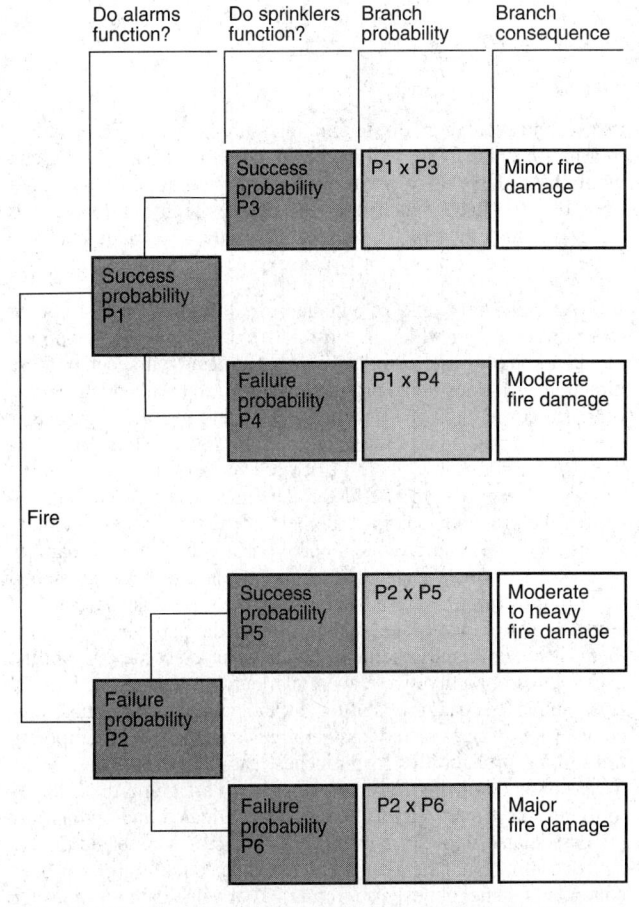

FIG. 11-8C. *Event tree. (Source: Factory Mutual's Record)*

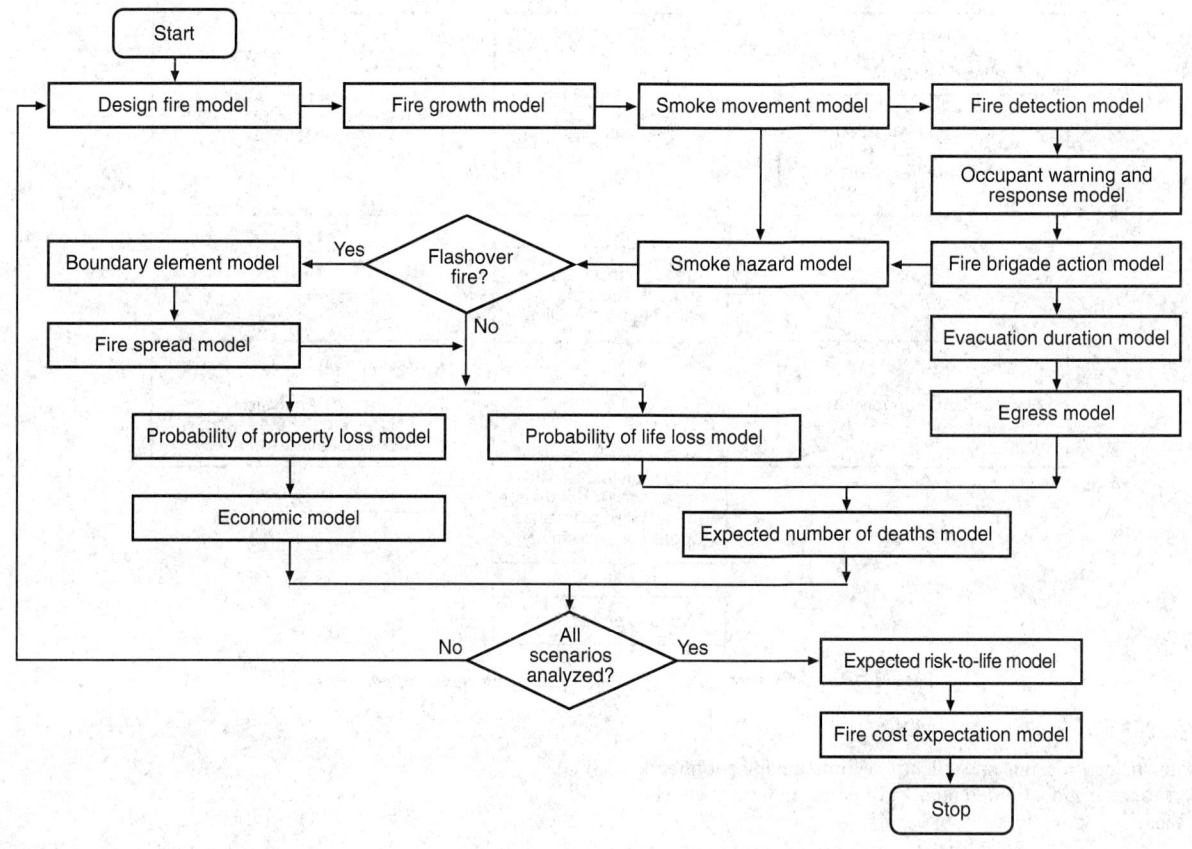

FIG. 11-8D. *Risk-cost assessment model.*

that second event (e.g., due to clothing ignitions) would be captured in a loss-per-fire figure assigned to that second critical event, and would be understood to occur no matter how the fire does or does not grow after that event. These losses are sometimes referred to as "tolls" because any fire that passes through the event acquires those losses, just as a car pays a toll on an express highway.

At each event (except outcome events), there will be two or more directions in which the fire may subsequently develop, and there will be transitional probabilities associated with each of these directions. All transitional probabilities from an event add up to one. They are conditional probabilities that a fire will reach a particular event, given that it has already reached the immediately preceding event. A path is a possible sequence of events for the fire from the first event (ignition) to one of the outcome events. Each path describes the growth and extinguishment of the fire with as much detail as the model can provide. A path probability is the probability of a complete path, calculated as the product of the transitional probabilities for all the events along the path. Because several outcome events often are identical except for the paths used to reach them, it is sometimes useful to compute an outcome probability. This is the probability of a set of similar outcome events and is equal to the sum of the path probabilities for the paths ending in those outcome events. The expected loss per fire is calculated by multiplying transitional probabilities for branches leading to outcomes by the losses associated with those outcomes, then assigning those losses to the events from which the branches originated and adding any tolls associated with those events. This process is repeated (and is called "rolling back the tree") until an expected loss figure has been computed for the first event (ignition). That value then becomes the expected loss for the event tree.

Loss Evaluation Model

The explicit or implicit assignment of monetary values to lives saved and injuries averted is the key element of this model. It is a difficult step that many people find distasteful or even immoral. The first and most important point to make is that individuals are not being asked to name a price for which they would be willing to die or suffer crippling injury. Instead, they are being asked to name a price they would be willing to accept to allow their current low risk of incurring death or injury in fire to increase or what they would pay to make that risk still smaller. With a resident population of about 260 million and an annual fire death toll in the range of 4,000 to 5,000, an average U.S. citizen has less than one chance in 50,000 each year of dying in a fire. Even for the highest-risk groups, the risk is probably less than one chance in 5,000 each year or less than one chance in 65 over an entire lifetime. A person could rationally attach a price to a 10 or 50 percent change in such a risk and still be consistent in believing life (i.e., the certainty of losing it) is beyond price. A rational person would pay much more to reduce the probability of dying from 1.0 to 0.8 than he or she would pay to cut that risk from 0.3 to 0.1.

If that point is made, the next task is identifying what particular figures should be used for the value of life and the value of injury when considering alternatives that change risks in the range characteristic of fire risk. In the 1960s and earlier, the value of life was generally calculated on the basis of discounted foregone future earnings. This approach implicitly assigned no value to the lives of retired people and full-time homemakers, and negligible value to the lives of older workers and young children. Such distinctions were philosophically objectionable. Even for prime wage earners, the methodology did not afford any guarantee that the value obtained would match the price people wanted to pay for risk reduction. In recent years, this approach has been largely abandoned in favor of calculations of willingness to pay to reduce risk of death. Practically speaking, the shift in approach roughly tripled the standard values of life.[1]

For all the philosophical disagreements, the actual values attached to lives saved, however calculated, tend to be concentrated within two orders of magnitude. Most studies estimate the value of life in hundreds of thousands of dollars or millions of dollars. Some of the higher values are taken from jury awards that compensate deaths. Few estimates go as high as tens of millions of dollars or as low as tens of thousands of dollars.

It is difficult to set up fully persuasive methodologies to assess a popular consensus on value of life because people do not like to think about death. If asked about the value of a whole life, they refer to the sanctity of life and say the value is infinite. If asked about the value of a shift in the risk of dying, they find it difficult to relate to such a choice. If presented with forced-choice situations that contain implicit values of life, they give answers that can reflect the way the questions were posed. Nevertheless, a 1988 study of assessments used in evaluating a wide range of proposed federal regulations concluded that "there has recently been some convergence around a figure of $1 to $2 million per statistical life."[2]

Another alternative is to use a value per year of life saved. A large number of regulatory cases were examined and the ranking of alternatives found was not drastically affected by the use of life-year value vs. life value.[1] However, use of life-year value tends to give more credit to saving children (by up to double, since their expected life spans are about double those of the population at large) and less credit to saving older adults (by a factor of four or more). Fire safety in schools would be boosted and fire safety in nursing homes might be scaled back if life-year value calculations were used.

Even after deciding to use willingness to pay as the standard for value of life, some difficult technical problems remain. One is the question of whether to calculate separately the willingness to pay for each individual (or each major group) affected by a proposed change. In an analysis aimed at the individual property owner or manager, such differentiation is unavoidable and should be an explicit, or at least implicit, part of any analysis of the market for a new product, system, or approach.

There also have been several studies of factors that affect willingness to pay. Willingness to pay is lower for the poor, older Americans, the seriously ill or handicapped, and the risk takers. For the poor, of course, ability to pay is lower, too. For older Americans and the seriously ill, the lower value given to life seems to reflect the fact that the quantity (for the older American) or the quality (for the sick) of life remaining is well below the national norm. However, all these groups with lower willingness to pay also tend to have relatively high risks of becoming fire fatalities. They are precisely the groups to target if total lives saved were the criterion of choice. Conversely, the people most willing to pay—affluent, healthy, risk-averse, young heads of families—are the ones least likely to benefit because their current risks of dying in fire are already below average.

Another reason for variations in the willingness to pay involves the nature of the risks rather than the characteristics of those who experience these risks. Risks that are voluntary, nonessential, occupational, or results of product misuse are deemed less serious than risks that are involuntary, essential, public, or results of normal product use. A risk of death to someone who lives near a nuclear reactor is valued more highly than an equal risk of death to someone who works in a coal mine. The difference is based on the assumption that occupational risks are more likely to be voluntary and more likely to be financially compensated. (Both of these assumptions are questionable. Workers in hazardous occupations such as mining may have few realistic occupational alternatives, while residents of hazardous areas, like floodplains, may have many alternative places to live and may have received financial compensation in the form of lower housing costs that at least equal any financial benefits received by the workers.) Similarly, risks of death associated with voluntary nonessential activities, such as smoking and hang gliding, are valued less than equal risks associated with voluntary but essential activities, such as driving a car. In fire risk, this argument appears in the debate over the fairness of imposing flame resistance standards (and accompanying costs) on all mattresses to protect people who choose to smoke in bed.

Deaths occurring in major multifatality incidents are valued differently—and generally more highly—than deaths occurring in smaller incidents. Major incidents are termed dread hazards in the risk analysis literature; it is the factor of dread—the greater fear of death occurring in a major incident—that inflates the value of risk in such cases. The effect of major incidents on families and communities has been used to argue for both higher and lower weighting of such deaths—higher because familial bloodlines may be extinguished, lower because multiple deaths in one family mean fewer survivors to mourn per fatality.[3]

Dread incidents constitute an especially dramatic example of the phenomenon of risk aversion. For example, most people feel that if loss A is ten times as great as loss B but only one-tenth as likely, losses A and B still are not equally onerous. The general public tends to be more concerned about fire scenarios that may kill, say, 100 people once every three years than they are about fire scenarios that kill one person at a time every week, year after year.

Technical adjustments can be made to incorporate some risk aversion into a benefit calculation. Such adjustments will have less effect on dwelling fire risk calculations, where really large incidents are impossible, than on risk calculations for large residential (e.g., hotel), institutional, or public assembly properties.

Values for injuries avoided can be estimated more directly than values for fatalities avoided because direct costs such as medical expense and lost wages seem more appropriate as indicators of value. A survey was used to estimate direct injury-related costs for residential fires.[4] Based on their figures, after adjusting for inflation and for the fact that their cost-per-injury figures are dominated by very small injuries from unreported fires, an estimate of $5,000 might be obtained for actual costs per injury received in a reported fire. Willingness to pay to avoid an injury is greater because of pain and suffering considerations, so a figure of $35,000 was derived in the late 1980s by economists at the U.S. Consumer Product Safety Commission.[5]

This average is based on a highly skewed distribution. The vast majority of injuries can be valued in the low hundreds of dollars or less, but a small number of serious burn injuries each year—considerably fewer than the corresponding number of fire fatalities—can cost hundreds of thousands or even millions of dollars in medical expenses. These few injuries account for most of the overall cost average. This suggests that analyses of expected impacts of new systems or programs on injuries should, if possible, separate serious and nonserious injuries when average values are used, to make sure that the average values are not understated as typical values.

For property damage, the losses probably will already have emerged from the post-ignition model expressed in monetary terms, but that is not necessarily so, and even if it is, some conversions may still be necessary. If fire growth and smoke spread models have been used, property damage may have been calculated in such terms as rooms exposed to fire or smoke for various lengths of time, or areas where structural integrity has been lost. Converting such descriptions to monetary equivalents would require data and models that do not now exist, despite repeated efforts to develop them.[6]

Even if damage is expressed in dollars, one may wish to take account of the fact that the loss faced by the property owner may be mediated by such things as insurance. In that case, it would be necessary to estimate the likely reduction in out-of-pocket, uninsured damage plus insurance premiums, rather than the likely reduction in total direct damage achievable.

Another consideration in the loss evaluation model is whether to include an adjustment for indirect loss—such items as lost wages, costs of a temporary location, and lost revenue for days that a business is closed. These losses can be very large in individual cases, such as the 1980 MGM Grand Hotel fire, but the best studies indicate that in the aggregate they tend to be an order of magnitude smaller than the direct losses.[7]

Cost Model

Costs may be divided into: (1) initial costs of the proposed changes being studied (2) the ongoing costs of these changes once they have been made and (3) the ripple effects on other costs, such as the need to increase the water supply to support a sprinkler system. The last could involve cost increases or cost reductions, including calculation of costs for many years into the future. To make this task manageable, the analysis can be set up in terms of the normal periods of maintaining, repairing, and replacing the items being analyzed. This is called life-cycle costing. An overview of major components of each of these three types of costs is shown below.

1. Initial Costs of Changes Being Studied

a. *Equipment costs.* For new products, it may be necessary to estimate what costs will be when mass production is under way. In many cases, the mass production cost continues to drop as further development occurs. (Smoke detectors have shown this pattern, for example.)

b. *Installation costs.* Estimation of costs of installation may require an analysis of the steps required for installation, because the person-hours and skills required for those steps may be higher or lower than for comparable products already in use. (For example, plastic pipe may be faster to install than iron pipe, and it may require less time-consuming effort to protect carpets and furniture from soiling during installation.)

Labor costs per hour may vary considerably from one place to another, as may overhead rates; these variations argue in favor of a serious effort to collect representative data.

c. *Financing costs.* These will be relevant if the systems are financed through time-payment plans (e.g., as part of what is covered by the building mortgage).

d. *Permit/license costs.* There may be some one-time fees required to install the systems.

e. *Some costs offset in resale.* If the new systems and features add to the resale value of the property, this will partially offset the initial costs.

2. Ongoing Costs of Changes Being Studied

a. *Operating costs.* A new system or product may need labor, power, or some other continuing input to operate. These costs need to be included.

b. *Inspection and testing costs.* Many systems require periodic inspection and testing after installation. These costs should be included. Labor usually will be the main cost element, but some tests (such as sprinkler tests) may involve materials costs, and other tests may require destruction of a sample of system components that would need replacement.

c. *Repair, maintenance, and replacement costs.* Most systems will require repair and maintenance, and—if the study period is long enough—periodic replacement will need to be considered.

d. *Costs of nonfire damage caused by the systems.* An example would be water damage due to accidental discharge of a sprinkler.

e. *Permit/license costs.* An example would be the standby water charge levied in some jurisdictions on buildings equipped with sprinklers.

f. *Salvage revenues for cost offsets.* Equipment that is replaced may be resellable. If so, salvage revenues help reduce net system costs.

3. Ripple Effects on Other Building Costs

a. *Costs of supporting systems.* Many new products may require replacement, modification, or addition of critical supporting systems (e.g., extra water supply for home sprinkler system in a rural area). The equipment and installation costs of these changes in supporting systems need to be identified and included. So do any changes in operating costs, repair and maintenance costs, inspection and testing costs, etc., for the modified supporting systems, and any changes in these ongoing costs for unmodified supporting systems.

b. *Special incentives or credits.* Insurance premium reductions that reflect the expected reduction in direct loss should be counted in the loss evaluation model. Extra reductions offered as inducements to buy systems, as well as incentives or credits in property or income taxes, should be counted here.

c. *Property value and tax impacts.* Changes in property taxes reflecting changed property value assessments should be considered. There may be tax consequences if the features add value to the property.

d. *Changes in land costs or required building features.* Added safety features may permit trade-offs in the form of increased density or reduced requirements for other building features. These need to be accommodated as costs, and any trade-offs in other safety features need to be addressed in the loss evaluation models as well.

e. *Changes in costs of public fire protection.* If buildings in a group receive similar modifications, it may be possible to accept longer response times or reduced sizes of fire suppression teams, resulting in reduced costs of public fire protection.

These lists are not exhaustive, but they indicate the need to estimate the effects of different decisions and assess their cost impacts.

Cost-Benefit Comparison Model

The cost model and loss evaluation model produce time streams of costs and risk-reduction benefits—that is, year-by-year estimates of costs and of reductions in fire deaths, injuries, and property damage, with the latter expressed as total monetized losses. To compare the costs to the benefits, the two time streams need to be combined into a single, manageable, indicator of net benefits.

To compare future and present costs and benefits, it is necessary to decide what the future costs and benefits are worth in the present. This involves the concept of opportunity cost. Suppose $20 was spent now on a fire safety system and $20 received back 10 years later in the form of reduced property damage in a fire. This would not be a break-even proposition because alternative investments could pay interest over that period.

Assumptions about the attractiveness of such investments are reflected in an assumed discount rate—a proportion between 0 and 1 used to reduce the value of future costs and benefits. Most fire risk reducing strategies involve greater costs than benefits in the near years and greater benefits than costs in the later years; this makes the discount rate a critical factor in overall assessment of whether the benefits justify the costs. Also, even if opportunity costs were not involved, there would be a cost associated with delayed consumption. All other things being equal, people usually prefer to have goods and services now rather than later, and a discount rate reflects that fact.

If a cost is incurred 10 years from now, for instance, the discount rate must be applied ten times to translate that cost into a figure comparable with today's costs. This figure is called the present value of a future cost or benefit. It is calculated as the discount rate raised to a power equal to the number of years in the future when the cost or benefit will occur, then multiplied by the value of that cost or benefit.

A reasonable discount rate can be assumed for the purpose of analysis or can be calculated as the discount rate required just to balance benefits and costs. If the latter is done, the derived discount rate is called the internal rate of return. It can be used to compare alternatives in the same way that a benefit-cost ratio can be used.

The two principal objections to discounting of future safety benefits are: (1) the possibility of very large, perhaps even irreversible, effects at a remote point in the future, and (2) the cumulative effects of the short-term biases induced by rigorous application of discounted assessments. The first objection is not a great concern for fire risk problems because fire does not produce irreversible effects on the scale contemplated by this argument. At most, several small towns could be wiped out by a wildfire (ignoring, for the moment, the possibility of wartime firestorms). Nevertheless, as a technical matter, it is worth considering the possibility that discount rates undervalue the real value people assign to events beyond the next decade or so. For example, most people would regard benefits in 105 years as equal to benefits in 100 years; but under constant discounting of say, 10 percent, the former would be only 59 percent of the latter.[8]

As for the cumulative problems of short-term bias, this has been discussed more in the context of business research, development, and innovation in general, than in regard to safety innovations in particular. In business, investments are expected to balance benefits and costs within 3 to 7 years, but many analysts believe such requirements are too demanding and tend, over time, to choke off truly dramatic breakthroughs. The result, in business, can be eventual loss of competitive edge to a competitor willing to take a longer view. One pertinent article[9] was particularly forceful on this point, arguing that the implied opportunity cost model underlying a short payback period requirement assumes a standard reference alternative investment that, contrary to the model's assumptions, is not itself immune to the cumulative effects of a stream of choices driven by short-term considerations. The fallacy, then, is in assuming that there always is an alternative investment that pays back in 3 to 7 years; the short-term-driven decisions may have the cumulative effect of eroding all such alternatives.

The technical approach to addressing this concern is to check the sensitivity of any conclusions to the use of a lower discount rate. Any innovation that year by year, after the initial cost period, produces more benefits than costs, can be made to look attractive through the selection of a sufficiently low discount rate. It is risky, however, to use too low a discount rate, because that will give a misleading picture of what people will be willing to pay.

Other approaches, such as using a higher discount rate for costs than for benefits, can produce perverse results. For example, such an approach could mean that an attractive safety program would seem even more attractive if its implementation were delayed. In this way, a program can be made to seem attractive but may never be implemented because further delay will always make it seem even more attractive.[10]

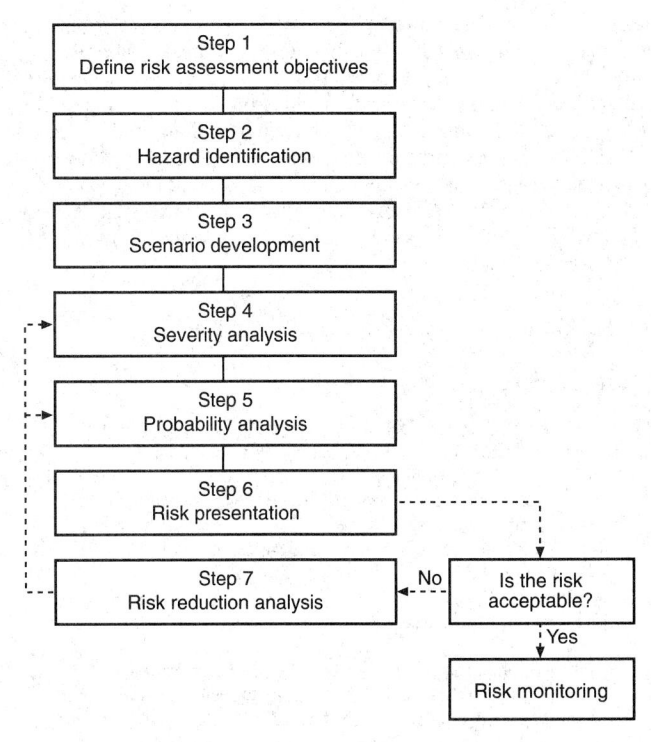

FIG. 11-8E.   Fire risk assessment steps.

Summary of Conclusions

Fire risk analysis is a technique capable of pulling together, reducing, and providing perspectives on large quantities of data from many sources. Recent rapid advances in the size and quality of fire-related data bases and in the power and affordability of computers have helped turn fire risk analysis into a practical tool for individual decision making and especially for large-scale decision making by governments and companies. The accelerating pace of development of computerized models of fire development and occupant behavior will eventually produce integrated deterministic/probabilistic models, providing even more powerful tools for fire risk analysis.

At present there are only a few abbreviated published examples of fire risk analysis,[11–13] but other more elaborate examples, providing more detailed illustrations and more broadly applicable models, are in progress. The largest single source of developmental studies in general-purpose fire risk analysis in the United States has been the National Fire Protection Research Foundation, whose multiyear developmental project led to a method called *FRAMEworks* (for Fire Risk Assessment MEthod), which is designed to address fire risk of burnable products used in buildings.[14] The Nuclear Regulatory Commission (NRC) has sponsored some of the best work in the field, but it is highly specialized to the problems of nuclear power plants and may need more work to translate to other settings. Individual fire risk analyses have appeared in materials prepared as background for regulatory decisions,[15] but these also tend to be highly specialized.

Until better materials are available to demonstrate the specific analytic techniques of fire risk analysis, it is useful to recognize the essence of fire risk analysis: creation of the simplest possible framework for identifying the available choices and estimating their consequences. Whether the estimates consist of nothing more than a series of guesses or are the result of an elaborate network of sophisticated models, laboratory tests, and fire experience, the framework in principle remains the same. Its purpose is to assemble all the information available, however slight or extensive, and focus that information on how and how much things would change as a result of the choices made. (See Figure 11-8E for a simple fire risk assessment flow chart that specifically focuses on the crucial step of presenting all this information to set up decisions and choices.) Identifying choices, predicting consequences, evaluating those consequences, and finally making choices—those are the steps of fire risk analysis. However complicated the methods used, the purpose is always to perform these familiar tasks that are the basics of rational decision making.

BIBLIOGRAPHY

References Cited

1. Graham, J. K., and Vaupel, J. W., "Value of a Life: What Difference Does It Make?" *Risk Analysis*, March 1981, pp. 89–95.
2. Gillette, C. P., and Hopkins, T. D., "Federal Agency Valuations of Human Life," *Report to the Administrative Conference of the United States*, unpublished, April 1988.
3. Starr, C., and Whipple, C., "Risks of Risk Decisions," *Science*, June 6, 1980, p. 1114ff.
4. Munson, M. J., and Ohls, J. C., *Indirect Costs of Residential Fires*, FA-61, Federal Emergency Management Agency, Washington, DC, April 1980.
5. Hall, J. R., Jr., "Expected Changes in Fire Damage from Reducing Cigarette Ignition Propensity," *Final Report to Technical Study Group of Cigarette Safety Act of 1984*, National Fire Protection Association, Quincy, MA, July 16, 1987.
6. Hall, J. R., Jr., "Reduce Fire Loss Guesstimates," *Fire Service Today*, Nov. 1982, pp. 11–13.
7. Hall, J. R., Jr., *The Total Cost of Fire in the United States Through 1988*, NFPA Fire Analysis and Research Division, Quincy, MA, Nov. 1990.
8. Whipple, C., "Energy Production Risks: What Perspective Should We Take?" *Risk Analysis*, March 1981, pp. 29–35.
9. Hayes, R. H., and Garvin, D. A., "Managing as If Tomorrow Mattered," *Harvard Business Review*, May–June 1982, p. 70ff.
10. Keller, E. B., and Cretin, S., "Discounting of Life-Saving and Other Non-Monetary Effects," *Management Science*, 1983, pp. 300–306.
11. Ruegg, R. T., and Fuller, S. K., "A Benefit-Cost Model of Residential Fire Sprinkler Systems," *NBS Technical Note 1203*, National Bureau of Standards, Gaithersburg, MD, Nov. 1984.
12. Helzer, S. G., *et al.*, "Decision Analysis of Strategies for Reducing Upholstered Furniture Fire Losses," *NBS Technical Note 1101*, National Bureau of Standards, Washington, DC, June 1979.
13. Hall, J. R., Jr., "A Fire Risk Analysis Model for Assessing Options for Flammable and Combustible Liquid Products in Storage and Retail Occupancies," *Fire Technology*, Vol. 31, No. 4, 1995.
14. Bukowski, R. W., *et al.*, "Fire Risk Assessment Method: Description of Methodology," National Institute of Standards and Technology, Gaithersburg, MD, National Fire Protection Assoc., Quincy, MA, Benjamin/Clarke Associates, Inc., Kensington, MD, National Fire Protection Research Foundation, Quincy, MA, NISTIR 90-4242, May 1990.
15. Hall, J. R., Jr., and Stiefel, S. W., "Decision Analysis Model for Passenger-Aircraft Fire Safety with Application to Fire-Blocking of Seats," NBSIR 84-2817, National Bureau of Standards, Washington, DC, March 1984.

Additional Reading

Barry, T. F., "An Introduction to Quantitative Risk Assessment in Chemical Process Industries," *The SFPE Handbook of Fire Protection Engineering*, 2nd ed., DiNenno, P. J., ed., National Fire Protection Association, Quincy, MA, 1995.
Barry, T. F., "Fire and Explosion Risk Assessment. Part 1," *Industrial Fire Safety*, Vol. 2, No. 5, Sept./Oct. 1993, pp. 49–53.
Barry, T. F., "Fire and Explosion Risk Assessment. Part 2. Procedures and Major Issues," *Industrial Fire Safety*, Vol. 2, No. 6, Nov./Dec. 1993, pp. 30–37.
Barry, T. F., "Fire and Explosion Risk Assessment. Part 3. Hazard Identification and Scenario Development," *Industrial Fire Safety*, Vol. 3, No. 1, Jan./Feb. 1994, pp. 32–39.
Barry, T. F., "Fire and Explosion Risk Assessment: an Integral Part of Safety Analysis Reports," *Fire and Current Technologies (FACTs)*, Vol. 3, No. 1, Summer 1994, pp. 1–2, 4–7.
Beck, V. R., "Fire Safety System Design Using Risk Assessment Models: Developments in Australia," *Proceedings* of the Third International Symposium on Fire Safety Science, Elsevier Applied Science, New York, 1991, pp. 45–59.
Beck, V. R., and Yung, D., "Cost-Effective Risk-Assessment Model for Evaluating Fire Safety and Protection in Canadian Apartment Buildings," *Journal of Fire Protection Engineering*, Vol. 2, No. 3, 1990, pp. 65–74.
Brandyberry, M. D., and Apostolakis, G. E., "Fire Risk in Buildings: Frequency of Exposure and Physical Model," *Fire Safety Journal*, Vol. 17, No. 5, 1991, pp. 339–361.
Brandyberry, M. D., and Apostolakis, G. E., "Fire Risk in Buildings: Scenario Definition and Ignition Frequency Calculations," *Fire Safety Journal*, Vol. 17, No. 5, 1991, pp. 363–386.
Brannigan, V., and Meeks, C., "Computerized Fire Risk Assessment Models: A Regulatory Effectiveness Analysis," *Journal of Fire Sciences*, Vol. 13, No. 3, May/June 1995, pp. 177–196.
Bukowski, R. W., "Review of International Fire Risk Prediction Methods," Sixth International Fire Conference on Fire Safety, Interflam '93, March 30–April 1, 1993, Oxford, England, Interscience Communications Ltd., London, England, 1993, pp. 437–446.
Hadjisophocleous, G. V., and Yung, D., "Fire Risk and Protection Cost Assessment Model for Highrise Apartment Buildings," National Research Council of Canada, Ottawa, Ontario, ASTM STP 1150; American Society for Testing and Materials, Fire Hazard and Fire Risk Assessment, Sponsored by ASTM Committee E-5 on Fire Standards, ASTM STP 1150, Dec. 3, 1990, San Antonio, TX, ASTM, Philadelphia, PA, 1990, pp. 224–233.
Hall, J. R., Jr., "How to Tell Whether What You Have is a Fire Risk Analysis Model," National Fire Protection Association, Quincy, MA, ASTM STP 1150; American Society for Testing and Materials, Fire Hazard and Fire Risk Assessment, Sponsored by ASTM Committee E-5 on Fire Standards, ASTM STP 1150, Dec. 3, 1990, San Antonio, TX, ASTM, Philadelphia, PA, 1990, pp. 131–135.
Hall, J. R., Jr., "Product Fire Risk," *The SFPE Handbook of Fire Protection Engineering*, 2nd ed., DiNenno, P. J., ed., National Fire Protection Association, Quincy, MA, 1995.
Hall, J. R., Jr., "Key Distinctions in and Essential Elements of Fire Risk Analysis," *Proceedings* of the Third International Symposium on Fire Safety Science, Elsevier Applied Science, New York, 1991, pp. 467–474.
Kazarians, M., Siu, N., and Apostolakis, G., "Risk Management Application of Fire Risk Analysis," *Proceedings* of the First International Symposium on Fire Safety Science, Hemisphere, 1986, pp. 1029–1038.
Pate-Cornell, E., "Managing Fire Risk Onboard Offshore Platforms: Lessons from Piper Alpha and Probabilistic Assessment of Risk Reduction Measures," *Fire Technology*, Vol. 31, No. 2, May 1995, pp. 99–119.
Raiter, S., *Risk Analysis and Risk Management: A Selected Bibliography*, Vance Bibliographies, Monticello, IL, 1987.
Ramachandran, G., "Probabilistic Approach to Fire Risk Evaluation," *Fire Technology*, Vol. 24, No. 3, 1988, p. 204.
Rowe, W. D., *An Anatomy of Risk*, reprint, R. E. Krieger, Malabar, FL, 1988.
Watts, J. M., Jr., "Fire Risk Rankings," *The SFPE Handbook of Fire Protection Engineering*, 2nd ed., DiNenno, P. J., ed., National Fire Protection Association, Quincy, MA, 1995.
Yung, D., and Beck, V. R., "Application of a Risk-Cost Assessment Model for the Evaluation of Fire Safety in Buildings," *Proceedings* of the First International Conference on Fire Science and Engineering, ASIAFLAM '95, March 15–16, 1995, Kowloon, Hong Kong, 1995, pp. 51–62.
Yung, D., and Beck, V. R., "Building Fire Safety Risk Analysis," *The SFPE Handbook of Fire Protection Engineering*, 2nd ed., DiNenno, P. J., ed., National Fire Protection Association, Quincy, MA, 1995.

PERFORMANCE-BASED FIRE CODES AND STANDARDS

Milosh T. Puchovsky

The concept of performance-based design for fire safety has become a topic of considerable discussion in recent times. Performance-based design methods are seen by many as providing a number of advantages over the current prescriptive-based design methods. Proponents claim that they will allow fire safety and building construction to be accomplished more efficiently. In an effort to realize these benefits, a number of initiatives across the globe have been enacted to develop a means that will better support performance-based design.[1] A key area of activity is the development and implementation of performance-based building and fire codes.

PERFORMANCE-BASED *VS.* PRESCRIPTIVE-BASED CODES AND STANDARDS

In general, a true performance-based code or standard explicitly states its fire safety goals and clearly defines the desired levels of safety or risk.[2] Any solution that demonstrates completion of the fire safety goals would be permitted. Fire safety would be designed for a specific use or application, rather than for a generic occupancy. Means to achieve compliance with the code begin with a scientific understanding of fire safety, followed by proven engineering methods and calculations. This process lends itself to the use of computer fire models that serve as tools to better facilitate the performance-based design process.

The majority of fire safety codes and standards used today are considered to be prescriptive based. These documents work differently than those that are described as performance-based. Prescriptive-based documents provide fire safety in a generic fashion by prescribing a combination of specific requirements, such as construction materials, limiting dimensions, or protection systems, but without referring to how these measures achieve a desired level of safety or outcome. In fact, a measurable level of safety is usually neither stated nor defined.

Performance-based and prescriptive-based are best considered as the endpoints of a range of documents. They are not sharply defined exclusive categories. Few, if any, NFPA codes or standards are pure versions of either performance-based or prescriptive-based documents.

Prescriptive codes find their roots in the 19th century when major conflagrations created the need for specific building provi-

sions.[3-5] Revisions to the codes resulted primarily from other significant fires that revealed deficiencies. To prevent recurrence of such disasters, changes were incorporated. However, these changes were instituted without effectively evaluating their adequacy, excessiveness, or conflicts with other requirements. This has created codes based on empiricism and experience, rather than a scientific understanding of fire.

Many advances in fire safety have been made in recent time, but they are not being incorporated into everyday fire safety practice primarily due to prescriptive-based codes and standards. Some believe that this lack of efficient technology transfer has allowed fire safety to fall behind other engineering disciplines such as structural design, which is thought of as relying heavily on scientific- and engineering-based principles. As a result, current prescriptive codes and standards may specify fire safety whose level of protection is unclear or inappropriate. Usually there is only one way of providing the specified fire safety.[6] Excessive redundancy exists and leads to increased costs. It is argued that performance-based codes and standards would promote freedom to develop more appropriate and cost-effective designs and bring fire safety to an optimal level.

THE PERFORMANCE-BASED DESIGN PROCESS

The intent of performance-based design is to provide a fire safety solution that achieves specific fire safety goals for a specific use or application. A proposed solution would be stated in quantifiable terms so that achievement of the fire safety goals can be demonstrated and measured. Developing a solution through a performance-based approach will typically require the completion of several key tasks, such as the ones described below.[7,8]

1. Establish fire safety goals, including performance objectives and performance criteria.
2. Evaluate the pre-fire characteristics of the people and property to be protected.
3. Identify potential hazards and applicable fire scenarios.
4. Select suitable calculation methods (e.g., fire models).
5. Develop the proposed solution.
6. Verify that the proposed solution meets the fire safety goals for the specified fire scenarios.

To better facilitate and promote the performance-based design and approval process, tasks 1 through 4 should be specifically addressed by codes, standards, and manuals of engineering practice. A means of consensus, technical review, and standardization of these

---

Milosh T. Puchovsky, P.E., is a senior fire protection engineer at NFPA.

tasks is currently lacking. Tasks 5 and 6 are specific to finding a solution for the project but are dependent on the completion of the other tasks.

Fire Safety Goals, Objectives, and Criteria

Fire safety goals need to be established and explicitly stated in codes and standards. Fire safety goals are the overall outcomes to be achieved with regard to fire.[9,10] They are nonspecific and are measured on a qualitative basis. They should be stated in terms of conditions that are intrinsically desirable and do not rely on any assumptions. For example, "avoidance of flashover" would not be a goal because it relies on assumptions about what kinds of fires cause harm. Goals should be further stated in terms that are potentially measurable, even if the precise measurement scale is not specified. Thus, they would typically be expressed in terms of impact on people or property, business interruption, or environmental impact.

Examples of a fire safety goal stated in terms of impact on people might be to maintain that the risk of death due to fire is no higher than the current level for similar properties or no higher than x per 100,000 people exposed, or to increase the likelihood that anyone capable of self-evacuation and not intimate with ignition can safely escape. An example of a fire safety goal stated in terms of impact on property might be to limit fire damage to x square feet (m^2) of building space. More specifically, the goal of a sprinkler system installation standard might be to provide for a system that will aid in the detection and control of residential fires, and reduce the likelihood of injury and loss of life due to fire.

In order to assess the degree of achievement of a particular fire safety goal, intermediate measures, such as performance objectives and performance criteria, also need to be defined. Performance objectives can be described as the requirements of the fire, building, or occupants that need to be obtained in order to achieve a fire safety goal.[11] They define a series of actions necessary to make the achievement of a goal much more likely. Performance objectives are stated in more specific terms than goals and are measured on a more quantitative basis.

For example, a standard might require that a sprinkler system be designed, installed, and maintained to prevent instantaneous or cumulative exposure to conditions that exceed a certain survivability criteria, or that the sprinkler system prevent flashover in the room of origin for the period of time necessary for occupant evacuation. Objectives could also include statements that call for prevention of structural damage, no life loss in the room of fire origin, the separation of occupants from fire effects for a specified length of time, or the containment of the fire to the room of origin.

Performance criteria can be described as performance objectives for individual products, systems, assemblies, or areas that are further quantified and stated in engineering terms.[11] Engineering terms can include temperatures, radiant heat flux, or levels of exposure to fire and products of combustion. Performance criteria provide threshold values that are treated as data for calculations used to develop a proposed solution.

For example, a standard might require that the following survivability criteria be maintained at a distance of 5 ft (1.5 m) above the floor in an area where sprinklers are required:

1. Temperature does not exceed 150°F (65.5°C).
2. Instantaneous carbon monoxide concentration does not exceed 10,000 ppm.
3. Cumulative carbon monoxide concentration does not exceed 25 percent carbon monoxide hemoglobin level.
4. Oxygen concentration is maintained at 14 percent or greater.

Criteria with regard to the prevention of flashover in areas where sprinklers are required might be:

1. Upper layer temperature does not exceed 1100°F (or the threshold might be stated as 600°C, which is roughly the same), and
2. Radiant heat flux at the floor does not exceed 20 kW/m^2 (or the threshold might be stated as 1.75 Btu/sec/sq ft, which is roughly the same).

Characteristics of the Occupants, Contents, and Building

It is unlikely that any proposed solution, performance-based or prescriptive-based, will be able to achieve the fire safety goals for everyone and everything under all possible conditions. This is due primarily to limitations and uncertainty with available technology in combination with society's unwillingness to accept the restrictions and costs associated with zero risk. Therefore, a proposed solution needs to include the characteristics or assumptions of who or what is intended to be protected and under what conditions.[11]

An evaluation of the characteristics of the occupants, contents, and building features is needed. All assumptions about who or what is being protected and all other factors that could invalidate the proposed solution need to be identified. For example, a proposed solution may only be valid for occupants who are capable of self-rescue and who have an unobstructed means of egress available to them, or for occupants who are familiar with their egress routes. The effectiveness of the proposed solution for occupants who fall outside the scope of these assumptions would be much less certain.

Fire Hazards and Scenarios

The potential for ignition and development of a fire which presents a threat to somebody or something exists in virtually all locations. The conditions under which these hazards exist and their likelihood of occurring in a given location need to be considered. A risk analysis of some type will identify the hazards that are intended to be guarded against. This is addressed to some degree in prescriptive-based documents through the documents scope or purpose. However, the specific hazards to be guarded against need to be more precisely defined through a set of fire scenarios.

Fire scenarios specify the fire conditions under which a proposed fire safety solution is expected to meet the fire safety goals. A fire scenario can be considered as the "fire load" that an occupant or building would likely be subjected to without interference from a fire safety system. It poses the challenge that the fire safety system must overcome. The fire scenario describes factors critical to the outcome of the fire such as the ignition source, initial fuel package, the location of initial fire development, aspects of the building arrangement, and the available fuel load. Fire scenarios describe the specifications needed for a fire test or for the input for a fire modeling run.

As the number of fire scenarios for any given space can be quite large, fire scenarios should be separated into categories.[11] Those scenarios representing the most severe case of each category would be used. For example, a set of fire scenarios might include those scenarios that are common (i.e., high probability) for the occupancy type in question, and those that are general-purpose high-challenge scenarios (i.e., likely to result in rapid, early onset of severe fire conditions). With regard to a single-family residence, a set of scenarios might include the following. (Obviously certain terms, such as free-burning fire, would need to be more explicitly defined.)

Common scenario #1—ordinary fire in occupied room: This represents a free-burning fire in ordinary combustibles, ignited by a small open-flame source, in one of the principal occupied rooms of a dwelling, such as a living room or bedroom.

Common scenario #2—fire with initial smoldering stage in occupied room: This represents a fire started by cigarette ignition of upholstered furniture, in one of the principal occupied rooms, such as a living room or bedroom.

Common scenario #3—fire originating in a structural area: This represents fires originating outside the dwelling or in structural areas, such as an attic space, and must be specifically designed to be representative of a free-burning fire in ordinary combustibles in an attic space.

High-challenge scenario #1—fire originating in means of egress: This represents a free-burning fire in ordinary combustibles ignited by a small open-flame source, in the entryway.

High-challenge scenario #2—fire originating in small unprotected room: This represents a free-burning fire in ordinary combustibles ignited by a child using a small open-flame source in a bedroom closet.

High-challenge scenario #3—shielded room-corner fire in occupied room: This represents a shielded room-corner fire in ordinary combustibles having a substantial initial smoldering period in one of the principal occupied rooms of a dwelling, such as a living room or bedroom.

A performance-based code for one- and two-family dwellings might require the fire safety goals to be met for all of the six identified fire scenarios. Any proposed solution that demonstrates this would be permitted. A residential sprinkler system designed and installed in accordance with a performance-based sprinkler standard might be presented as the proposed solution. However, the sprinkler system alone might not be sufficient, as residential sprinkler systems are currently not intended to protect against fires in areas that are not sprinklered (high-challenge scenario #2) or fires that enter the dwelling from outdoors (common scenario #3); additional fire safety features, such as a smoke detection system and certain construction materials, might also be needed.

Calculation Methods (e.g., Fire Models)

Calculation methods and computer fire models serve as tools that help assess whether the fire safety goals have been met for each applicable fire scenario.[12–14] Due to the complex nature of the principles involved and the numerous calculations required, models are often packaged as computer software. However, hand calculations are available which can model certain aspects of a fire event.[15,16] Attached to the calculation methods and fire models should be all relevant input data, assumptions, limitations, and safety factors needed to properly implement the model.

The calculation methods do not provide the final solution, but rather help in the evaluation of a performance-based design. This fundamental concept is often misunderstood. The models contain the physics that describe the fire phenomenon, occupant response, fire safety system performance, and their interaction, and quickly perform the necessary calculations, making performance-based design much more likely. The ultimate fire safety decision lies with the designer's and Authority Having Jurisdiction's (AHJ's) interpretation of the model's results.

Fire protection engineers appear eager to apply performance-based methods and fire modeling to their projects, as this increases their design options.[17,18] Currently these approaches are primarily used in special applications, such as fire reconstructions, in the development of unique design arrangements not addressed by current

codes and standards, and to demonstrate that an alternative means of protection meets or exceeds code requirements. Performance-based design requires increased time and effort by fire protection engineers and creates the potential for higher initial costs.[9]

The increased application of performance-based design approaches and fire models warrants a higher level of scrutiny than required for prescriptive-based design. Calculation methods have limitations which may not be identified or known by their developers or users. Misuse of these methods and misinterpretation of their results, like the misuse and misapplication of prescriptive codes and standards, can have a detrimental effect on fire safety. Guidance is needed for users and AHJs so that they either understand or are aware of the fire model's assumptions, limitations, and proper application.[19,20] Further issues exist as many models have not been validated to show under what conditions their results are valid, and much of the input data needed to run the models is not available.

Although the development of calculation methods and fire models continues, more are needed if total performance-based design is to be realized on a routine basis. Additional scientific research and testing are necessary for the development of new models and their input data. Existing models and methods do not provide all that is needed. However, it can be argued that fire models can become more effective once codes and standards explicitly state their fire safety goals, objectives, and criteria.

Proposed Solution

This is the fire safety design arrangement intended to achieve the fire safety goals. It is expressed in terms that make it possible to assess whether the fire safety goals have been achieved. The proposed solution also needs to address how reliably it can achieve the fire safety goals. This will include addressing the reliability of the individual components of the fire safety system as well as the system in its entirety.

In addition to relying on testing and maintenance of the system and its components, safety factors in all likelihood will also be applied. Safety factors can be considered as adjustments to various elements of the design process, such as performance criteria. They reflect conservatism due to the uncertainty in the methods and assumptions employed in measuring performance. To aid in its assessment, the proposed solution should identify all fire models, assumptions, and safety factors employed.

Verification

This is confirmation by the AHJ that a proposed solution meets the established fire safety goals.[11] This will be a function similar to the one employed for current design approval. However, unlike prescriptive-based methods, the provisions presented by a performance-based design cannot be easily confirmed by visual inspection. As solutions need to be developed by the designer and AHJ working together as a team, a "cookbook" mentality does not properly serve performance-based design. A solid understanding and/or confidence level with the fire protection engineering design concepts by both parties is essential.

Verification involves several steps not currently required under current prescriptive-based methods. These include:

1. Verification that the proposed solution shows the building will meet the fire safety goals for the assumptions and fire scenarios established
2. Verification during construction that the building as built fits the design and data used in the proposed solution
3. Verification that the development of the proposed solution was completed by an individual or group qualified to perform such a task

4. Verification that the proposed solution implemented appropriate models, test methods, and other calculation procedures, and that they were used in an appropriate manner

5. Verification that the assessment used appropriate data, and that it was used in an appropriate manner

In summary, the verification process confirms that the building is constructed as proposed to a design that will achieve the intended level of safety, and that the building's ability to achieve the level of safety has been demonstrated by qualified people using the correct methods applied to the correct data. Unfortunately, a means to accomplish all the steps of the verification process has not been established. As a result, the confidence level with performance-based design is not where it needs to be. Additional efforts, especially with gaining acceptance of calculation methods and fire models, establishing qualifications for model users, providing input data and training of both AHJ and designers, need to be undertaken.

PRECEDENTS FOR PERFORMANCE-BASED DESIGN IN EXISTING CODES

Although considered to be a relatively new concept, various versions of performance-based design have been and are currently being performed. Many building and fire safety codes and standards provide an option to develop alternative means of protection, provided the level of safety prescribed by the document is not lowered.[21,22] Primary examples include the development of fire safety for unique architectural or functional features not anticipated by current codes and standards, or the development of alternative means of protection to existing code requirements. Performance-based approaches are also employed in the preparation of fire reconstructions.

Unfortunately, all the pieces needed to properly support this process are not in place. Performance-based approaches today require a major undertaking by both the designer and AHJ. Current codes and standards do not serve performance-based design well. Any performance-based design attempted today must demonstrate a level of safety equivalent to the existing prescriptive standards. However, existing codes and standards typically do not explicitly state their fire safety goals, nor do they identify the assumptions and fire scenarios under which the prescribed fire safety solution is to be effective. The task of determining what the code is intending to achieve, for whom, and under what conditions can prove to be difficult as each concerned individual typically may have a differing opinion of what is intended.

To further complicate matters, current codes and standards do not reference computer models or calculation methods that have been demonstrated to be effective for the design and assessment of a proposed solution. Efforts to validate models, qualify their input data, and train or certify their users need to be increased. In addition, the current level of technology cannot produce the calculation methods and models that will properly address all of the relevant concerns. As a result, the necessary confidence level with these methods does not yet exist.

Once performance-based approaches and their supporting pieces have undergone an acceptable level of standardization, e.g., product listings and installation standards, as other fire safety technologies have, they will enjoy greater acceptance overall. What now exists are general provisions allowing something vague called "equivalency," and a collection of experts, models, data bases, and other resources that may lead to better solutions if used properly. What is needed are better definitions and procedures that allow the concerned parties to agree more explicitly on the goals against which equivalence is defined and how to demonstrate equivalence.

FUTURE CODES AND STANDARDS

Although the application of performance-based methods will continue to increase, it is unlikely that performance-based approaches will entirely replace prescriptive-based methods. Prescriptive-based methods work, although not as effectively or efficiently as they could in certain situations. Additionally, prescriptive-based methods are more straightforward, require less of an initial effort, and are more easily enforced. It is likely that future codes and standards will be revised to include both performance-based and prescriptive-based requirements.[11] This provides the option for either method, as one may be better suited than the other for a certain project. It also supports the philosophy that performance-based options are the next level of sophistication for codes and standards.

Maintaining a single document relates the two design methods and provides a more formalized procedure for performance-based approaches. Future codes and standards will serve as a means of standardizing the pieces necessary to employ performance-based design. They will better support the overall performance-based design and verification process by explicitly identifying their fire safety goals, including objectives, and by providing needed guidance with regard to the selection and use of specific fire safety criteria, assumptions, fire scenarios, and calculation methods. This should promote the overall greater application and acceptance of performance-based design methods by both design engineers and AHJs.

Additionally, codes and standards that contain performance-based design options are intended to reflect the current levels of science and technology. They will provide a better direction for research efforts, greatly improve the transfer of technology, and encourage the prescriptive-based requirements to be based more on science and technology.

PURSUING PERFORMANCE-BASED DESIGN

Major initiatives to increase the greater acceptance of performance-based methods are underway around the world.[23,24] Since the 1980s, governments in a number of nations have been urging the reform of their entire building construction industries and encouraging the use of performance-based design methods. This has prompted the development and implementation of new performance-based building codes in the United Kingdom and New Zealand, as well as development in several other countries that have not yet completed implementation, notably Australia. In addition, regulations have been presented to introduce competition into the plans approval process.

The primary incentives for pursuing a new performance-based building code for the United Kingdom and Australia were within the context of more global regulatory reform rather than an attempt to address specific fire safety concerns.[25] Many government officials believed that the level of protection prescribed by the old building codes could be achieved by more cost-effective solutions, resulting in increased economic growth and international competitiveness. Overall, the governments believed that market forces should rule as much as possible and that regulatory interference should be minimized. New regulations were written to increase technology transfer, and to promote innovation, flexibility, and deregulation with regard to building design.

New regulations in the United Kingdom and New Zealand call for simplified building codes identifying only major fire and life safety goals for an adequate and reasonable level of safety for persons within and in the vicinity of the building, as well as referencing design practices to achieve these goals.[26,27] Designers have been given the freedom to choose any method to demonstrate compliance.

To provide guidance in achieving these goals, engineering design practices and "approved documents" are being presented in New Zealand and the United Kingdom.[28,29] "Approved documents" are essentially user-friendly versions of the old codes and other standards. Currently, compliance with "approved documents" virtually guarantees design acceptance. The use of other practices or methods does not. In the United Kingdom, it was anticipated that the creation of new performance-based building codes will encourage an influx of new engineering-based "approved documents" and other design tools. Unfortunately, the downturn in the economy during the late 1980s and early 1990s has reduced construction projects and tempered the development of new methods.[26]

Competition within the enforcement process has been introduced, as some nations are permitting either the AHJ or other approved inspector to grant design approval. In the United Kingdom, the lack of available liability insurance has limited this activity. In Australia, private certification is currently allowed in two of the eight Australian jurisdictions.[30] However a performance-based code has not yet been introduced.

Overall, the new regulations in the United Kingdom and New Zealand allow the design team and the enforcement authorities to work together to develop acceptable solutions. Building officials are more willing to recognize calculation methods, alternative designs, and trade-offs, as the new rules have removed many of the constraints placed on their professional judgment. Increased training programs have also been a factor.

Although various versions of performance-based codes have been introduced in a number of nations, many designers currently feel that the prescriptive "approved documents" will continue as the solution of choice for most projects.[24,25] Alternative methods require more of an initial financial commitment but with no guarantee of acceptance, as necessary confidence with these methods does not yet exist. Performance-based approaches are used primarily for special projects similar to the way they are used in North America through a code's equivalency provision.

CONCLUSIONS

The primary incentive for employing performance-based design is to attain the required level of fire safety at reduced cost. Performance-based documents offer a means to accomplish this by increasing design freedom and addressing specific fire safety concerns. They remove many of the restrictions associated with prescriptive-based methods. Even so, prescriptive-based methods will likely continue to be the design method of choice for the majority of fire protection projects. However, the pursuit of formalized performance-based design methods will result in greater benefit overall. It will provide a better direction for research efforts and will greatly improve the transfer of technology. This promotes the standardization and acceptance of performance-based approaches and encourages prescriptive-based methods to be based more on science and technology.

The concept of performance-based design is gaining momentum on a global scale. Governments from several countries encourage the use of engineering approaches. Fire protection engineers are incorporating the new methods into more of their designs. An increasing number of AHJs are being asked to make decisions about fire safety designs incorporating performance-based methods. Research organizations and scientific groups continue to improve existing engineering methods and develop new ones. However, a number of difficulties must be overcome before the benefits of performance-based design are realized on a routine basis.

Developing documents that support performance-based design is only a starting point. Implementation of these methods into every-day practice requires more than a rewrite of existing codes and standards. Major efforts to standardize fire models, test methods, and the qualifications of model users are essential, as are efforts with legal issues, building regulations, and education.[31] Success requires cooperation and coordination by all parties concerned.

BIBLIOGRAPHY

References Cited

1. Lucht, D. A., *Strategies for Shaping the Future—A Report on the Conference on Fire Safety Design in the 21st Century*, Worcester Polytechnic Institute, Worcester, MA, 1991.
2. Buchanan, A. H., "Fire Engineering for a Performance-Based Code," *Proceedings*—Interflam '93, Interscience Communications Limited, London, UK, 1993, p. 457.
3. Richardson, J. K., "Moving Toward Performance-Based Fire Codes," *NFPA Journal*, May 1994, p. 70
4. Nelson, H. E., "History of Fire Technology," Conference on Fire Safety Design in the 21st Century—Pre-Conference Papers, Worcester Polytechnic Institute, Worcester, MA, 1991, p. 180.
5. Lucht, D. A., "Changing the Way We Do Business," *Fire Technology*, Vol. 28, No. 3, Aug. 1992.
6. Bukowski, R. W., and Tanaka, T., "Toward the Goal of a Performance Fire Code," *Fire and Materials*, Vol. 15, 1991, p. 175.
7. Bukowski, R. W., "A Review of International Fire Risk Prediction Methods," *Proceedings*—Interflam '93, Interscience Communications Limited, London, England, 1993, p. 437.
8. Hall, J. R., Jr., "The State of the Art in Fire Risk and Hazard Analysis for Use in Performance Standards," white paper prepared for the International Electrotechnical Commission, National Fire Protection Association, Quincy, MA, 1993.
9. Custer, R. L. P., "Selection and Specification of the Design Fire for Performance-Based Fire Protection Design," *Performance-Based Fire Safety Engineering*, Society of Fire Protection Engineers, Boston, MA, 1993, p. 17.
10. Bukowski, R. W., and Babrauskas, V., "Developing Rational, Performance-Based Fire Safety Requirements in Model Building Codes," *Fire and Materials*, Vol. 18, 1994, p. 173.
11. Puchovsky, M. T., Cote, A. E., Cote, R., Fahy, R. F., Grant, C. C., Hall, J. R., and Vondrasek, R. J., *NFPA's Future in Performance-Based Codes and Standards*, National Fire Protection Association, Quincy, MA, July 1995.
12. Friedman, R., "An International Survey of Computer Models for Fire and Smoke," *Journal of Fire Protection Engineering*, Vol. 4, No. 3, 1992, p. 81.
13. Cote, A. E., "The Computer and Fire Protection Engineering—25 Years of Progress," *Fire Technology*, Vol. 26, No. 3, 1990, p. 195.
14. Emmons, H. W., "Fire Safety Science—The Promise of a Better Future," *Fire Technology*, Vol. 26, No. 1, 1990, p. 5.
15. *The SFPE Handbook of Fire Protection Engineering*, 2nd ed., DiNenno, P. J., ed., National Fire Protection Association, Quincy, MA, 1995.
16. Drysdale, D., *An Introduction to Fire Dynamics*, John Wiley and Sons, New York, 1985.
17. Koffel, W. E., private communication, Koffel Associates, Inc., Ellicott City, MD, 1993.
18. Maddox, J. A., private communication, Rolf Jensen & Associates, Inc., Concord, CA, 1993.
19. Lavine, R. S., "User's Needs in Fire Models—Proceedings of the Ad Hoc Mathematical Fire Modeling Working Group," *Fire Technology*, Vol. 24, No. 2, 1988, p. 163.
20. Bukowski, R. W., "Fire Models—The Future Is Now!," *NFPA Journal*, March 1991, p. 61.
21. Mawhinney, J. R., "Evaluating Equivalencies to Building Code Requirements," National Research Council of Canada, Ottawa, Canada, 1991.
22. Bukowski, R. W., "How to Evaluate Alternative Designs Based on Fire Modeling," *NFPA Journal*, March 1995, p. 68
23. Meacham, B. J., "International Development and Use of Performance-Based Building Codes and Fire Safety Design Methods," *SFPE Bulletin*, March 1995, p. 7.

24. Snell, J. E., Babrauskas, V., and Fowell, A. J., "Elements of a Framework for Fire Safety Engineering," *Proceedings*—Interflam '93, Interscience Communications Limited, London, UK, 1993, p. 447.

25. Lucht, D. A., Kime, C. H., and Traw, J. S., "International Developments in Building Code Concepts," *Journal of Fire Protection Engineering*, Vol. 5, No. 4, 1993, p. 141.

26. Law, M., "Fire Safety Design Practices in the United Kingdom—New Building Regulations," Ove Arup Partnership, London, UK, 1991.

27. Hunt, J., "A Performance-Based Code at Work—The New Zealand Experience," 1995 Annual Meeting, National Fire Protection Association, Quincy, MA, May 1995.

28. Buchanan, A. H., *Fire Engineering Design Guide*, Center for Advanced Engineering, University of Canterbury, Christchurch, New Zealand, 1994.

29. Warrington Fire Research Consultants, *Draft British Standard Code of Practice for the Application of Fire Safety Engineering Principles to Fire Safety in Buildings*, Warrington, UK, 1993.

30. Lovegrove, K., "Private Certification in Australia," *Proceedings*—Pacific Rim Conference of Building Officials, Darwin, Australia, May 1995.

31. Richardson, J. K., "Regulatory System Changes Needed to Gain Acceptance of Fire Safety Engineering Methods," *Issues in International Fire Protection Engineering Practice*, Society of Fire Protection Engineers, Boston, MA, 1993, p. 37.

Additional Reading

Anderson, R. C., "New Requirements in Building Codes," Special Conference on Recent Advances in Flame Retardancy of Polymeric Materials—Materials, Applications, Industry Developments, Markets, May 15–17, 1990, Stamford, CT, 1990, pp. 1–7.

Babrauskas, V., "Designing Products for Fire Performance: the State of the Art of Test Methods and Fire Models," *Fire Safety Journal*, Vol. 24, No. 3, 1995, pp. 299–312.

Barnett, C. B., and Simpson, M. R., "Fire Code Review: New Zealand's Performance Based Fire Code," *Proceedings* of the First International Conference on Fire Science and Engineering, ASIAFLAM '95, March 15–16, 1995, Kowloon, Hong Kong, 1995, pp. 27–40.

Beck, V. R., "Performance Based Fire Safety Design: Recent Developments in Australia," *Fire Safety Journal*, Vol. 23, No. 2, 1994, pp. 133–158.

Brannigan, V., "Performance-Based Codes: No Panacea," *Fire Chief*, Vol. 40, No. 4, April 1996, pp. 31–32, 34.

Brannigan, V. M., and Lehner, P., "Performance Based Fire Safety Codes: A Regulatory Effectiveness Analysis," *Proceedings* of the International Conference on Fire Research and Engineering, Sept. 10–15, 1995, Orlando, FL, SFPE, Boston, MA, 1995, pp. 181–186.

Brannigan, V., Smidts, C., and Kilpatrick, A., "Regulatory Requirements for Performance Based Codes Using Mathematical Risk Assessment," Maryland Univ., College Park, Caledonian Univ., Glasgow, Scotland, National Institute of Standards and Technology, Gaithersburg, MD, ISBN 0-9516320-9-4; *Proceedings* of the Seventh International Interflam Conference, Interflam '96, March 26–28, 1996, Cambridge, England, Interscience Communications Ltd., London, England, 1996, pp. 621–630.

Buchanan, A. H., "Fire Engineering for a Performance Based Code," *Fire Safety Journal*, Vol. 23, No. 1, 1994, pp. 1–16.

Bukowski, R. W., "International Activities for Developing Performance-Based Fire Codes," National Institute of Standards and Technology, Gaithersburg, MD, CIB W14/96/1 (J) ; Mini-Symposium on Fire Safety Design of Buildings and Fire Safety Engineering, June 12, 1995, Tsukuba, Japan, 1995, pp. IV/1–3.

Bukowski, R. W., "Setting Performance Code Objectives: How Do We Decide What Performance the Codes Intend?," National Institute of Standards and Technology, Gaithersburg, MD, ISBN 0-9516320-9-4; *Proceedings* of the Seventh International Interflam Conference, Interflam '96, March 26–28, 1996, Cambridge, England, Interscience Communications Ltd., London, England, 1996, pp. 555–561.

Bukowski, R. W., and Tanaka, T., "Toward the Goal of a Performance Fire Code," *Fire and Materials*, Vol. 15, No. 4, Oct./Dec. 1991, pp. 175–180.

Bukowski, R.W., and Babrauskas, V., "Developing Rational, Performance-based Fire Safety Requirements in Model Building Codes," *Fire and Materials: An International Journal*, Vol. 18, No. 3, May—June 1994, pp. 173–192.

Butcher, E. G., and Parnell, A. C., "Fire Safety Engineering: What Performance Standard is Acceptable and What Does It Mean?," *Fire Safety*, Vol. 3, No. 3, June 1996, pp. 29–30.

Cohn, B. M., "Characterization and Use of Design Basis Fires in Performance Codes," Bert Cohn Associates, Inc., ISBN 0-9516320-9-4; *Proceedings* of the Seventh International Interflam Conference, Interflam '96, March 26–28, 1996, Cambridge, England, Interscience Communications Ltd., London, England, 1996, pp. 581–588.

Cohn, B. M., "Synthesis of a Goal-Oriented Building Code," *Journal of Applied Science*, Vol. 1, No. 4, 1990–1991, pp. 301–314.

Crowley, M. A., "Performance Based Smoke Exhaust Systems for High Piled Storage: A Case Study," *Proceedings* of the Society of Fire Protection Engineers (SFPE) Honors Lecture Series, Engineering Seminars: Fire Protection Design for High Challenge or Special Hazard Applications, May 20–22, 1996, Boston, MA, 1996, pp. 55–58.

Custer, R. L. P., "Progress on Computer Modeling in Performance-Based Detection System Design in the United States," Worcester Polytechnic Inst., MA, European Society for Automatic Alarm Systems (EUSAS), European Platform for Promoting Security and Fire Protection Engineering, Newsletter, Number 5, June 1994, pp. 17–26.

Custer, R. L. P., "Selection and Specification of the 'Design Fire' for Performance-Based Fire Protection Design," *Proceedings* of the Society of Fire Protection Engineers Engineering Seminars on Performance-Based Fire Safety Engineering, Nov. 15–17, 1993, Phoenix, AZ, SFPE, Boston, MA, 1993, pp. 17–22.

Custer, R. L. P., Meacham, B. J., and Wood, C. B., "Performance-Based Design Techniques for Detection and Special Suppression Applications," *Proceedings* of the Technical Symposium on Halon Alternatives, June 27–28, 1994, Knoxville, TN, 1994, pp. 17–32.

Dolph, B. L., "Ship Fire Safety Engineering Methodology: a Performance Based Fire Safety Analysis," Third International Conference and Exhibition on Fire and Materials, Oct. 27–28, 1994, Arlington, VA, 1994, pp. 199–206.

Emmons, H. W., "Strategies for Performance Codes in the U. S.," Harvard Univ., Cambridge, MA, Worcester Polytechnic Institute, Conference on Firesafety Design in the 21st Century, Part 1, Conference Report, Part 2, Conference Papers, May 8–10, 1991, Worcester, MA, 1991, pp. 166–179.

Feld, J. M., "Developing a Performance Based Sprinkler System Standard," *Proceedings* of the Society of Fire Protection Engineers Engineering Seminars on Performance-Based Fire Safety Engineering, Nov. 15–17, 1993, Phoenix, AZ, SFPE, Boston, MA, 1993, pp. 31–38.

Finucane, M., "Adoption of Performance Standards in Offshore Fire and Explosion Hazard Management," *Fire Safety Journal*, Vol. 23, No. 2, 1994, pp. 171–184.

Frantzich, H., "En Modell for Dimensionering av Forbindelser for Utrymning Utifran Funktionsbaserade Krav. [Model to Design Escape Routes in a Building Based on Performance Based Requirements.]," Lund Univ., Sweden, LUTVDG/TVBB-1011-SE, Dec. 1994.

Fraser, R. J., "Performance-Based Codes: the New Zealand Experience," Proceedings of the Society of Fire Protection Engineers Engineering Seminars on Performance-Based Fire Safety Engineering, Nov. 15–17, 1993, Phoenix, AZ, SFPE, Boston, MA, 1993, pp. 1–7.

Fryer, R., "Building Regulations Are Barrier to Performance Testing of Fire Doors," Fire International, No. 138, Feb. 1993, p. 28.

Groner, N. E., and Chubb, M. D., "Latent Human Error Is the Principal Threat to the Adoption and Use of Performance-Based Fire Safety Requirements," Proceedings of the International Conference on Fire Research and Engineering, Sept. 10–15, 1995, Orlando, FL, SFPE, Boston, MA, 1995, pp. 203–208.

Gross, J. G., "Codes, Standards, and Institutions—Pressures for Change," *Journal of Professional Issues in Engineering Education and Practice*, Vol. 117, No. 2, April 1991, pp. 75–87.

Hotta, H., "Example of Performance-Based Design for Fire Protection System: New Sprinkler System," Bosai Consultants Corp., Tokyo, Japan, CIB W14/96/1 (J) ; Mini-Symposium on Fire Safety Design of Buildings and Fire Safety Engineering, June 12, 1995, Tsukuba, Japan, 1995, pp. VII/1–20.

Johnson, P. F., and Timms, G. R., "Performance Based Design of Shopping Center Fire Safety," *Proceedings* of the First International Conference on Fire Science and Engineering, ASIAFLAM '95, March 15–16, 1995, Kowloon, Hong Kong, 1995, pp. 41–49.

Johnson, P. F., Beck, V. R., and Horasan, M., "Use of Egress Modelling in Performance-Based Fire Engineering Design: a Fire Safety Study at the National Gallery of Victoria," Proceedings of the Fourth International Symposium on Fire Safety Science, Intl. Assoc. for Fire Safety Science, Boston, MA, 1994, pp. 669–680.

Johnson, P. F., "International Implications of Performance Based Fire Engineering Design Codes," *Journal of Fire Protection Engineering*, Vol. 5, No. 4, 1993, pp. 141–146.

Karlsson, B., and Kokkala, M. A., "New Developments in Performance Based Test Methods for Fire Safety Assessment of Products," Lund Univ., Sweden, VTT-Technical Research Center of Finland, Espoo, CIB W60 Workshop on New Developments in Performance Test Methods, April 10–11, 1995, Horsholm, Denmark, 1995, pp. 1–16.

Karydas, D. M., and Delichatsios, M. A., "Risk Assessment Methodology for Fire Safety Factors in Performance-Based Design of Buildings," Factory Mutual Research Corp., Norwood, MA, NISTIR 5499, Sept. 1994; National Institute of Standards and Technology, Annual Conference on Fire Research: Book of Abstracts, Oct. 17–20, 1994, Gaithersburg, MD, 1994, pp. 43–44.

Kilpatrick, A., "Performance Based Fire Safety Codes: The UK Model?," *Proceedings* of the International Conference on Fire Research and Engineering, Sept. 10–15, 1995, Orlando, FL, SFPE, Boston, MA, 1995, pp. 193–194.

Koffel, W., "Model Codes—New Directions," VHS Video Tape, Federal Fire Forum, Interaction of Fire Safety with Building Design, Conducted in Conjunction with "Safety/Survivability 2000," May 1, 1991, Washington, DC, 1991.

Koffel, W. E., "Are We Ready for Performance Codes?," *NFPA Journal*, Vol. 84, No. 3, May/June 1992, p. 20.

Koffel, W. E., "Ready or Not: Designing for Performance Codes," *NFPA Journal*, Vol. 89, No. 5, Sept./Oct. 1995, p. 20.

Koffel, W. E., "Who Sets Performance-Based Codes?," *NFPA Journal*, Vol. 88, No. 6, Nov./Dec. 1994, pp. 16, 102.

Larsen, G. P., "Nordic Model: Performance Building Code," Nordic Fire Safety Engineering Symposium, Development and Verification of Tools for Performance Codes, Aug. 30–Sept. 1, 1993, Espoo, Finland, 1993, pp. 1–6.

Lathrop, J. K., and Birk, D., "Building Life Safety with Codes," *NFPA Journal*, Vol. 84, No. 3, May/June 1992, pp. 42–46, 48, 50, 52.

Lathrop, J. K., "Life Safety Code Key to Industrial Fire Safety," *NFPA Journal*, Vol. 88, No. 4, July/Aug. 1994, pp. 36–46.

Lippiatt, B., "Cost-Effective Compliance with Performance-Based Life Safety Codes," VHS Video Tape, National Institute of Standards and Technology, Gaithersburg, MD, Federal Fire Forum, Providing Fire Safety for Equipment Use in Buildings. [This Forum Will Also be a Part of the BFRL Building Technology Symposia Series.], June 6, 1994, Gaithersburg, MD, 1994.

Lucht, D., "What is Meant by Performance-Based Design," Worcester Polytechnic Inst., MA, Video; Federal Fire Forum, Performance-Based Design Issues of Concern to the A/E Community, Nov. 6, 1995, Gaithersburg, MD, 1995.

Lucht, D. A., Kime, C. H., and Traw, J. S., "International Developments in Building Code Concepts," *Journal of Fire Protection Engineering*, Vol. 5, No. 4, 1993, pp. 125–133.

Magnusson, S. E., "Performance Based Codes," Sixth International Fire Conference on Fire Safety, Interflam '93, March 30–April 1, 1993, Oxford, England, Interscience Communications Ltd., London, England, 1993, pp. 413–425.

Magnusson, S. E., *et al.*, "Determination of Safety Factors in Design Based on Performance," *Proceedings* of the Fourth International Symposium on Fire Safety Science, Intl. Assoc. for Fire Safety Science, Boston, MA, 1994, pp. 937–948.

McCormick, J. W., "Building Code Applications Using the Emerging Technology: a Challenge Requiring a Strategy," Journal of Applied Science, Vol. 1, No. 4, 1990–1991, pp. 321–330.

Meacham, B. J., "International Development and Use of Performance-Based Building Codes and Fire Safety Design Methods," *SFPE Bulletin*, March/April 1995, pp. 7–16; Selected Readings in Performance-Based Fire Safety Engineering, 1995, Boston, MA, Society of Fire Protection Engineers, Boston, MA, 1995, pp. 64–84.

Meacham, B. J., "Performance-Based Codes and Fire Safety Engineering Methods: Perspectives and Projects of the Society of Fire Protection Engineers," Society of Fire Protection Engineers, Boston, MA, ISBN 0-9516320-9-4; *Proceedings* of the Seventh International Interflam Conference, Interflam '96, March 26–28, 1996, Cambridge, England, Interscience Communications Ltd., London, England, 1996, pp. 545–553.

Meacham, B. J., and Lucht, D. A., "Performance-Based Fire Safety Design in the United States: an Update from the Society of Fire Protection Engineers," *SFPE Bulletin*, Fall 1995, pp. 6–9.

Meacham, B. J., and Lucht, D. A., "Performance-Based Fire Safety Design in the United States: An Update From the Society of Fire Protection Engineers," Society of Fire Protection Engineers, Boston, MA, Center for Fire Safety Studies, Worcester, MA, ISBN 0-9527398-3-6; Institution of Fire Engineers, University of Sunderland, Fire Research Station, CIB W14, Tyne and Wear Metropolitan Fire Brigade, Fire Safety by Design, Conference *Proceedings*, Volume 3, Research Papers, July 10–12, 1995, UK, 1995, pp. 9–14.

Meacham, B. J., and Custer, R. L. P., "Performance-Based Fire Safety Engineering: An Introduction of Basic Concepts," *Journal of Fire Protection Engineering*, Vol. 7, No. 3, 1995, pp. 35–54.

Meeks, C. B., and Brannigan, V., "Performance Based Codes: Economic Efficiency and Distribution Equity," University of Georgia, Athens, Maryland Univ., College Park, National Institute of Standards and Technology, Gaithersburg, MD, ISBN 0-9516320-9-4; *Proceedings* of the Seventh International Interflam Conference, Interflam '96, March 26–28, 1996, Cambridge, England, Interscience Communications Ltd., London, England, 1996, pp. 573–580.

Milke, J. A., "Performance-Based Design of fire Resistant Assemblies," *Proceedings* of the Society of Fire Protection Engineers (SFPE) Honors Lecture Series, Engineering Seminars: Fire Protection Design for High Challenge or Special Hazard Applications, May 20–22, 1996, Boston, MA, 1996, pp. 13–18.

Milke, J. A., "Providing a Performance-Based Design Method for Smoke Management Systems in Atria and Covered Malls," *SFPE Bulletin*, May/June 1992, pp. 6–8.

Nakaya, I., "Our Activities Toward Performance Based Fire Regulation in Japan," Nordic Fire Safety Engineering Symposium, Development and Verification of Tools for Performance Codes, Aug. 30–Sept. 1, 1993, Espoo, Finland, 1993, pp. 1–5.

National Fire Protection Association, "Making of Codes and Standards," National Fire Protection Assoc., Quincy, MA, Video, 1991.

Neale, R. A., "When Code Equivalencies Don't Work," *American Fire Journal*, Vol. 48, No. 1, Jan. 1996, pp. 20–23.

Nelson, H. E., "Performance-Based Smoke Movement Design," Hughes Associates, Inc., Baltimore, MD, Video; Federal Fire Forum, Performance-Based Design Issues of Concern to the A/E Community, Nov. 6, 1995, Gaithersburg, MD, 1995.

Oleszkiewicz, I., "Transition to a Performance-Based NBC," *NBC/NFC News*, No. 136, Fall 1993, pp. 2–3.

Pauls, J., "Egress Time and Safety Performance Related to Requirements in Codes and Standards," Hughes Associates, Inc., Columbia, MD, Building Officials and Code Administration International, Inc. (BOCA) and OBOA, Workshop Landout, June 1990, Ontario, Canada, 1990, pp. 1–10.

Puchovsky, M. T., "Developing Performance-Based Documents One Step at a Time," *NFPA Journal*, Vol. 90, No. 1, Jan./Feb. 1996, pp. 46–49.

Puchovsky, M. T., "Performance Based Design in the United States," *Proceedings* of the International Conference on Fire Research and Engineering, Sept. 10–15, 1995, Orlando, FL, SFPE, Boston, MA, 1995, pp. 187–192.

Puchovsky, M. T., "Supplementing NFPA's Codes and Standards with Performance-Based Provisions," National Fire Protection Association, Quincy, MA, ISBN 0-9516320-9-4; *Proceedings* of the Seventh International Interflam Conference, Interflam '96, March 26–28, 1996, Cambridge, England, Interscience Communications Ltd., London, England, 1996, pp. 945–949.

Quiter, J., "How Today's Fire Codes Impact Building Construction," *Consulting/Specifying Engineer*, Vol. 9, No. 5, April 1991, pp. 46–48, 50, 52.

Ramachandran, G., "Probability-Based Building Design for Fire Safety. Part 1," *Fire Technology*, Vol. 31, No. 3, 3rd Quarter, Aug. 1995, pp. 265–275.

Ramachandran, G., "Probability-Based Building Design for Fire Safety. Part 2," *Fire Technology*, Vol. 31, No. 4, 4th Quarter, Nov. 1995, pp. 355–368.

Ramachandran, G., "Probability-Based Fire Safety Code," *Journal of Fire Protection Engineering*, Vol. 2, No. 3, 1990, pp. 75–91.

Richardson, J. K., "Moving Toward Performance-Based Codes," *NFPA Journal*, Vol. 88, No. 3, May/June 1994, pp. 70–72, 77–78.

Richardson, J. K., "Performance Codes and the Fire Marshal," National Research Council of Canada, Ottawa, Ontario, IRC-ORAL-88; National Fire Protection Association, Fire Marshal's Forum, Oct. 28, 1944 [sic], 1994, pp. 1–22.

Richardson, J. K., "Practice of Fire Protection Engineering in a Performance-Based Regulatory Environment," Selected Readings in Performance-Based Fire Safety Engineering, 1995, Boston, MA, Society of Fire Protection Engineers, Boston, MA, 1995, pp. iii–iv.

Richardson, J. K., Oleszkiewicz, I., and Yung, D., "Toward a Performance-Based Code in Canada," Nordic Fire Safety Engineering Symposium, Development and Verification of Tools for Performance Codes, Aug. 30–Sept. 1, 1993, Espoo, Finland, 1993, pp. 1–5.

Rosenbaum, E. R., "Using Performance Based Analysis to Assess Fire Hazard and Develop Governing Criteria for a Combustible Fuel Package in Facilities," *Proceedings* of the Society of Fire Protection Engineers Engineering Seminars on Performance-Based Fire Safety Engineering, Nov. 15–17, 1993, Phoenix, AZ, SFPE, Boston, MA, 1993, pp. 57–61.

Scherfig, S., "Development of a Performance Fire Code and a Design System for Fire Safety in Buildings," Nordic Fire Safety Engineering Symposium, Development and Verification of Tools for Performance Codes, Aug. 30–Sept. 1, 1993, Espoo, Finland, 1993, pp. 1–10.

Scholten, N. P. M., "Performance Based Building Regulations: Modelling of Works in Relation to Fire and Smoke Compartments and Means of Escape," TNO Building Construction Research Delft, The Netherlands, TNO Building and Construction Research, Third CIB/W14 Workshop, *Proceedings* of the Fire Engineering Workshop on Modelling, Jan. 25–26, 1993, Delft, The Netherlands, 1993, pp. 232–234.

Snell, J. E., "Status of Performance Fire Codes in the USA," Nordic Fire Safety Engineering Symposium, Development and Verification of Tools for Performance Codes, Aug. 30–Sept. 1, 1993, Espoo, Finland, 1993, pp. 1–9.

Sober, G., "Acceptance of Performance-Based Design by the Authority Having Jurisdiction," Video; Federal Fire Forum, Performance-Based Design Issues of Concern to the A/E Community, Nov. 6, 1995, Gaithersburg, MD, 1995.

Strength, R. S., "Status Report Model Building Codes 1992, NEC-93 and IEC-89," Fire Retardant Chemicals Association Fall Conference: Industry Speaks Out on Flame Retardancy: Coatings; Polymers and Compounding; Test Method Development; New Products, Technomic Publishing Co., Lancaster, PA, 1992, pp. 41–46.

Sullivan, P. D., "Recent Applications of Computers in Performance Design of Buildings," Robert W. Sullivan, Inc., Boston, MA, Society of Fire Protection Engineers and WPI Center for Firesafety Studies, Proceedings of the Technical Symposium on Computer Applications in Fire Protection Engineering, Final Program, June 20–21, 1996, Worcester, MA, 1996, pp. 87–92.

Sullivan, P. D., World's First Performance-Based Building Code "The New Zealand Experience," Worcester Polytechnic Inst., MA, Thesis, Dec. 14, 1993.

Talbert, J. H., "Application of Performance-Based Techniques for the Design of Fire Safety in Buildings," *Proceedings* of the International Conference on Fire Research and Engineering, Sept. 10–15, 1995, Orlando, FL, SFPE, Boston, MA, 1995, pp. 475–480.

Tanaka, T., "Concept and Framework of a Performance Based Fire Safety Design System for Buildings," *Journal of Applied Fire Science*, Vol. 3, No. 4, 1993–1994, pp. 335–358.

Tanaka, T., "Concept of a Performance Based Design Method for Building Fire Safety," Building Research Institute, Ibaraki-Ken, Japan, NISTIR 4449, 1990; Eleventh Joint Panel Meeting of the U. S./Japan Government Cooperative Program on Natural Resources (UJNR) on Fire Research and Safety, Oct. 19–24, 1989, Berkeley, CA, 1990, pp. 23–31.

Tanaka, T., "Performance Based Design Method for Fire Safety of Buildings," Building Research Institute, Ibaraki-ken, Japan, NISTIR 4449;Eleventh Joint Panel Meeting of the U. S./Japan Government Cooperative Program on Natural Resources (UJNR) on Fire Research and Safety, Oct. 19–24, 1989, Berkeley, CA, 1990, pp. 269–291.

Tanaka, T., "State of Art: Development of Performance-Based Fire Safety Design Methods of Buildings in Japan," Building Research Inst., Ibaraki-ken, Japan, CIB W14/96/1 (J); Mini-Symposium on Fire Safety Design of Buildings and Fire Safety Engineering, June 12, 1995, Tsukuba, Japan, 1995, pp. II/1–7.

Taylor, P., Timms, G., and Johnson, P., "Performance Based Approach to Fire Safety Design of Cardboard Box Manufacturing Plant," *Proceedings* of the Society of Fire Protection Engineers (SFPE) Honors Lecture Series, Engineering Seminars: Fire Protection Design for High Challenge or Special Hazard Applications, May 20–22, 1996, Boston, MA, 1996, pp. 49–54.

Todd, N. W., and Ryan, J. D., "Improving Codes by Predicting Product Performance in Real Fires," *Fire Journal*, Vol. 84, No. 2, March/April 1990, pp. 64–68, 70–72, 93–94.

VanRickley, C. W., "Survey of Code Officials on Performance-Based Codes and Risk-Based Assessment," *Codes Forum*, Jan./Feb. 1996, pp. 42–43, 45–46.

Watts, J. M. Jr., "Performance-Based Codes. Editorial," *Fire Technology*, Vol. 30, No. 4, 4th Quarter, pp. 385–386.

White, D. A., Beyler, C. L., and DiNenno, P. J., "Performanced-Based Engineering of Industrial Facilities," Hughes Associates, Inc., Columbia, MD, Society of Fire Protection Engineers and Worcester Polytechnic Institute, *Proceedings* of Computer Applications in Fire Protection, June 28–29, 1993, Worcester, MA, 1993, pp. 51–55.

Woolley, D., "Developments in the UK to Support Performance Based Codes," Nordic Fire Safety Engineering Symposium, Development and Verification of Tools for Performance Codes, Aug. 30–Sept. 1, 1993, Espoo, Finland, 1993, pp. 1–12.

Yung, D., "Development of a Tool to Support the Introduction of Performance-Based Codes in Canada," National Research Council of Canada, Ottawa, Ontario, IRC-Oral-93; National Fire Protection Association, Fall Meeting, Nov. 13–16, 1994, Toronto, Canada, 1994, pp. 1–25.

SIMPLIFIED FIRE GROWTH CALCULATIONS

Edward K. Budnick, David D. Evans, and Harold E. Nelson

As part of a fire protection analysis, it is often desirable to estimate the burning characteristics of selected fuels and their effects in enclosures. Also important for many analyses is the estimation of when fire protection devices such as heat detectors or automatic sprinklers will activate for specific fire conditions. Equations are available, based principally on experimental correlations, which permit the user to estimate these effects.

In this chapter, a brief introduction to enclosure fire effects is presented, along with equations that can be evaluated using hand calculators to provide estimates of particular effects. Generally, the equations presented are well documented and are widely used for such estimates. However, the user is cautioned that most of the equations were developed based on data from experiments that were conducted for very specific, and sometimes idealized, conditions. Therefore, some judgment must be exercised when applying these equations to complex conditions occurring in enclosure fires of general interest.

The equations in this chapter are primarily intended to be used in evaluating fire conditions in enclosures during the pre-flashover fire growth period. Most of the methods presented do not apply to fully developed room fires, such as post-flashover conditions. In addition, these shorthand calculations apply only to the room of fire origin and to a single burning fuel package such as a contiguous grouping of combustibles like an upholstered chair and ottoman or a bookcase full of books. More complicated methods are available for multiroom analysis, but they are beyond the scope of this chapter. See Section 11, Chapter 5, "Deterministic Computer Fire Models." Methods to address multiple fuel package involvement are under development but are not yet available.

For some of the effects, more than one equation is presented. In these cases, one equation may be preferred over the other based on the best match of the experimental basis for the equation to the specific case of interest.

Material properties, such as heat of combustion (Δh_c) and stoichiometric air/fuel mass ratio (r_s), are listed in this handbook in Appendix A, "Tables and Charts." For additional information on material properties or the topics introduced in this chapter, re-

fer to Section 3 of *The SFPE Handbook of Fire Protection Engineering*.[1]

All calculations in this chapter are presented in SI units. For U.S. customary units, see Table 11-10A.

TABLE 11-10A. *Conversion Factors*

To Convert from SI Units	To U.S. Customary Units	Multiply by
Kilograms	lb (avdp)	2.2046226
Kilojoules	Btu	0.948608
Kilowatts	Btu/hr	3414.99
Meters	ft	3.2808399

Also, if T_k is a temperature in degrees Kelvin, then the same temperature in degrees Celsius or Centigrade (T_c) is given by $T_c = T_k - 273.15$. The same temperature in degrees Fahrenheit (T_F) is given by $T_F = 1.8\,T_c + 32$.

ENERGY RELEASE RATE

Calculation procedures for fire effects in enclosures require knowledge of the *energy release rate* of the burning fuel. The term energy release rate is frequently used interchangeably with *heat release rate*, and it is usually expressed in units of kilowatts (kW) and symbolized by $\dot{Q}$.

Currently, no broadly accepted methods exist for prediction of energy release rates soley on basic measurements of material properties. Recent efforts in this area show promise.[2] However, it is expected that generalized methods will not be available for some time. In addition, in any enclosure fire, the actual rate of heat release is dependent not just on the burning fuel, but also on the fire environment, the manner in which the fuel is volatilized, the efficiency of the vapor combustion, and other physical and chemical effects. Therefore, for the immediate future one must rely on available laboratory test data for the specific or similar fuels. In addition, a knowledge of the complete energy release rate history may be required for many situations. This is particularly desirable where the fuel package exhibits unsteady burning. (See Figure 11-10A.) For those cases where only limiting conditions or worst-case analysis is required, it may be reasonable to assume that the fuel is burning at a constant rate, which simplifies the calculation considerably.

Edward K. Budnick, P.E., is a senior engineer and vice president at Hughes Associates, Inc., Baltimore, MD. David D. Evans, Ph.D., P.E., is chief of the Fire Safety Engineering Division of the Building and Fire Research Laboratory, U.S. National Institute of Standard and Technology, Gaithersburg, MD. Harold E. Nelson, P.E., is a senior research engineer at Hughes Associates, Inc., Baltimore, MD.

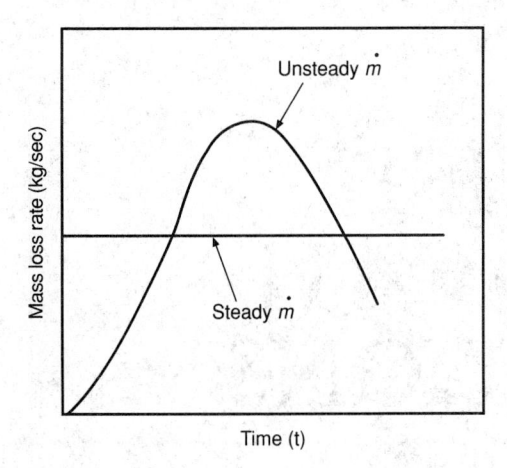

FIG. 11-10A.   An illustration of steady and unsteady burning rates.

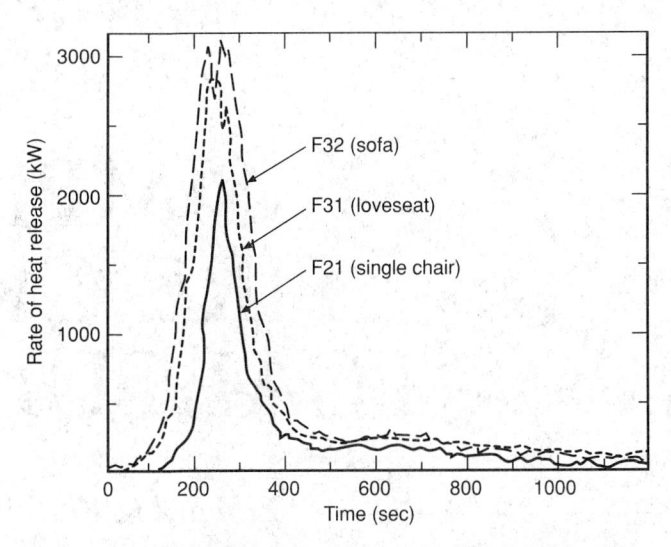

FIG. 11-10B.   Free burn heat release rates for selected furniture items.[7]

For the equations presented here, the more simplified condition of constant energy release rate is generally assumed. However, techniques are available that represent a growing fire by a series of constant energy release rate fires. This approach can require a great deal of calculation time, depending upon the desired accuracy. Such analysis is generally more suited to computer simulation.

For the complete combustion of a fuel, energy release rate and mass loss rate are related by the equation:

$$\dot{Q} = \Delta h_c \cdot \dot{m} \qquad (1)$$

where

$\dot{Q}$ = energy release rate (kJ/s) or (kW)

Δh_c = heat of combustion (kJ/kg)

$\dot{m}$ = mass loss rate (kg/s)

(The heat of combustion is a material property and is tabulated for selected materials in Appendix A, "Tables and Charts." The mass loss rate is typically found experimentally.)

It should be recognized that most enclosure fires of interest do not exhibit constant energy release rates. Rather, as illustrated in Figure 11-10B for selected items of furniture, the mass loss rate, and therefore the energy release rate, varies over time. Depending upon the detail required, one might select a constant mass loss rate, such as a peak value or an average value as the basis for analysis. Data on mass loss rates for selected fuel packages are available in several publications.[1,3–6]

Most information available on fuel package burning rates is reported for "free burn" conditions—that is, the data are collected for items burning in the open rather than in an enclosure. While enclosure effects are of little importance in evaluating early fire growth, they are important in fully developed room fires. The effects of most importance are those related to radiation feedback to the fuel from the hot smoke and enclosure linings and those related to the ability of the fire to obtain sufficient air for combustion. When fire conditions reach a stage where the smoke and heated room linings approach 932°F (500°C), the radiant feedback normally increases the burning rate above that observed in a free burn situation. The difference between the free burn rate and the radiation-enhanced burning rate increases as the room temperature and resulting radiant impact on the fuel package increase. Once flashover conditions are reached, rates greater than double the free burn rate are not unusual.

The second enclosure effect is the availability of oxygen for combustion. If the air in the space, plus that drawn in through openings, plus that provided to the space by HVAC systems or other means is insufficient to burn all the combustible products driven from the fuel package, only that amount of combustion supportable by the oxygen available in the air will burn within the room or other space involved. This situation is referred to as ventilation-limited burning. When ventilation-limited burning occurs, the combustible products driven from the fuel package and not burned in the room often burn when they combine with air outside the room and appear as flame extensions from the room.

The following equation for stoichiometric fuel pyrolysis can be used to estimate the mass loss rate at which these effects begin to dominate:

$$\dot{m}_{st} = \frac{1}{r_s} \cdot 0.5\, A_v \sqrt{h_v} \qquad (2)$$

where

$\dot{m}_{st}$ = stoichiometric mass loss rate (kg/s)

r_s = stoichiometric air/fuel mass ratio

A_v = area of ventilation opening (m²)

h_v = height of ventilation opening (m)

For wood fuel, $r_s = 5.7$. Values of r_s for other materials can be found in Appendix A, "Tables and Charts."

An estimate of the maximum burning rate possible for an enclosure with a particular opening can be determined from Equation 2. If the mass loss rate for a particular fuel package is less than this value, the condition is referred to as fuel-controlled, and results from Equation 1 provide a reasonable estimate of the energy release rate. If the free burn mass loss rate is higher than the stoichiometric rate from Equation 2, then the rate determined for stoichiometric conditions should be used for combustion within the room.

A more rigorous treatment of energy release rates is available for selected material types such as wood cribs, wood and plastic slabs, and liquid pool fires where experimental correlations have been established. Section 3, Chapter 1, in *The SFPE Handbook of Fire Protection Engineering*[1] provides a detailed discussion of the prediction of burning rates for liquid pool fires. Detailed discussions of energy release rates for specific fuels are available elsewhere.[7,8]

FLAME HEIGHTS

Axisymmetric Flames

Estimates of flame height L can be important in determining exposure hazards associated with a burning fuel. (See Figure 11-10C.) Experimentally determined "mean" flame heights have been correlated by several researchers. A simple correlation for flame heights for pool or horizontal burning fuels has been developed by Heskested:[9]

$$\frac{L}{D} = -1.02 + 15.6 N^{1/5} \tag{3}$$

where

L = mean flame height (m)
D = diameter of fire source (m)
N = nondimensional parameter

where

$$N = \left[\frac{C_p T_\infty}{g\rho_\infty^2 \left(\frac{\Delta h_c}{r_s}\right)^3}\right] \frac{\dot{Q}^2}{D^5} \tag{4}$$

C_p = specific heat of air at constant pressure [(kJ/kg)K]
T_∞ = ambient temperature (K)
g = acceleration of gravity (9.81 m/s²)
ρ_∞ = ambient air density (kg/m³)
Δh_c = heat of combustion (kJ/kg)
r_s = stoichiometric air/fuel mass ratio
$\dot{Q}$ = total heat release rate (kJ/s) or (kW)

For noncircular fuel packages, an effective D can be estimated by

$$D = 2\left(\frac{A_f}{\pi}\right)^{1/2} \tag{5}$$

where

D = effective diameter (m)
A_f = area of fire (m²)

For a broad range of experimental conditions, $\Delta h_c / r_s$ is nearly constant, representing the heat liberated per unit mass of air entering the combustion reaction. Assuming $\Delta h_c / r_s = 3100$ kJ/kg and atmospheric conditions such as $T_\infty = 293$ K, and $\rho = 760$ mm H$_g$, Equation 3 can be simplified to

$$L = -1.02 D + 0.23 Q^{2/5} \tag{6}$$

Since flames are unstable, the mean flame height L is generally taken to be the height above the fire source where the flame tip is observed to be at or above this point 50 percent of the time. The above correlation is considered suitable for pool fires or for horizontal surface burning. In addition, the correlation will produce negative values for L at small heat release rates. The available experimental data indicate that the most reliable region of application is where $\dot{Q}^{2/5}/D$ is greater than 16.5. For more detail on flame height calculations, the

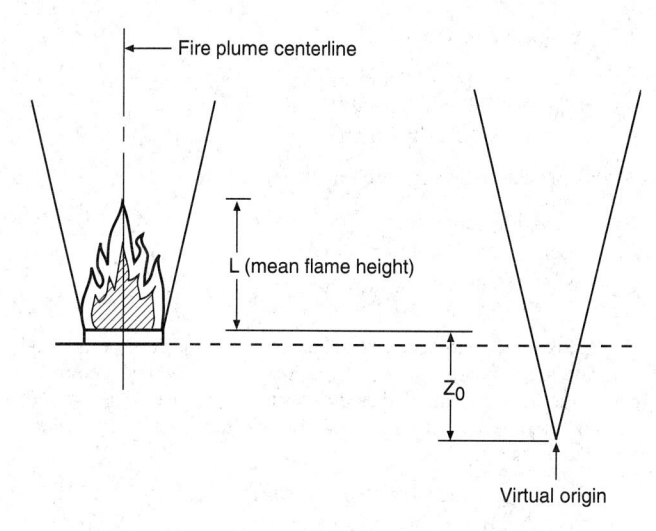

FIG. 11-10C. *Flame and fire plume characteristics.*

reader is referred to Beyler[10] and *The SFPE Handbook of Fire Protection Engineering.*[1]

Wall and Line Fires

Equations have also been developed for elongated fires that are either: (1) against a wall so that air is entrained from one side only (i.e., wall fires), or (2) sufficiently in the open so that air is entrained along both of the longitudinal sides (i.e., line fires).[11] In these equations, the flame height is based on the rate of heat release per meter of length of the fire source.

Wall fire flame height:

$$L = 0.034 \, \dot{Q}'^{2/3} \tag{7}$$

Line fire flame height:

$$L = 0.017 \, \dot{Q}'^{2/3} \tag{8}$$

where

$\dot{Q}'$ = rate of heat release per meter length of the fire source

PLUME CENTERLINE TEMPERATURE AND VELOCITY

The plume centerline excess temperature and velocity at elevations above the mean flame height can be estimated from the following equations.[12] (See Figure 11-10C.)

$$\Delta T_o = 9.1 \left(\frac{T_\infty}{g C_p^2 \rho_\infty^2}\right)^{1/3} \dot{Q}_c^{2/3} (Z - Z_o)^{-5/3} \tag{9}$$

$$U_o = 3.4 \left(\frac{g}{C_p \rho_\infty T_\infty}\right)^{1/3} \dot{Q}_c^{1/3} (Z - Z_o)^{-1/3} \tag{10}$$

where

ΔT_o = excess centerline mean temperature $(T_g - T_\infty)$(K)
T_g = gas temperature (K)

T_∞ = ambient temperature (K)

g = acceleration of gravity (9.81 m/s²)

C_p = specific heat of air at constant pressure [(kJ/kg)/K]

ρ_∞ = ambient air density (kg/m³)

$\dot{Q}_c$ = convective heat release rate (kJ/s) or (kW)

Z = elevation above burning fuel fire source (m)

Z_o = location of virtual fire source (m)

U_o = centerline mean velocity (m/s)

Equations 9 and 10 above are based on extensive experimental data and are known as *strong plume* correlations, which accommodate large density deficiencies present in fire plumes. These equations do not apply to fire plumes with small temperature rises, such as $\Delta T_o / T_\infty \ll 1$.

For normal atmospheric conditions, for example:

T_∞ = 293 K

g = 9.81 m/s²

C_p = 1.00 [(kJ/kg)/K]

ρ_∞ = 1.2 kg/m³

Equations 9 and 10 can be simplified to:

$$\Delta T_o = A \dot{Q}_c^{2/3} \left(Z - Z_o\right)^{-5/3} \qquad (11)$$

$$U_o = B \dot{Q}_c^{1/3} \left(Z - Z_o\right)^{-1/3} \qquad (12)$$

where

A = 25.0 K m^{5/3} kW^{-2/3}

B = 1.03 m^{4/3} s^{-1} kW^{-1/3}

While methods exist to calculate excess temperature and velocities at locations other than along the plume centerline, the highest confidence is placed on centerline estimates. The reader is referred to DiNenno, Beyler, and Heskestad for a detailed discussion of noncenterline temperature and velocity estimates.[1,10,12]

The use of centerline excess temperature and velocity for evaluating exposure conditions is conservative since the centerline values are the highest values at any elevation. The estimates are sensitive to values for the convective heat release rate— $\dot{Q}_c$ — which can vary from 40 to 80 percent of the total heat release rate, depending on the type and arrangement of the burning fuel.

CALCULATING THE HYPOTHETICAL VIRTUAL ORIGIN (Z₀)

In order to estimate the plume centerline mean temperature and velocity, one must first determine the virtual origin. The virtual origin is the hypothetical location or elevation associated with a substitution of a point source fire for the fire in question. (See Figure 11–10C.) The consideration of virtual origin is most important for evaluating centerline conditions near the fire. As the distance above the fire increases, the impact of the discrepancy that results from neglecting the virtual origin decreases. It is common practice to ignore virtual source considerations for calculations where the distance above the fire is many times the diameter of the fire. For centerline elevations near the fire, however, a more accurate estimate of the position of the virtual source is necessary. The following expression, limited to pool fires and horizontal burning, provides an estimate of the location of the virtual source:[9]

$$Z_o = -1.02 D + 0.083 \dot{Q}^{2/5} \qquad (13)$$

where

Z_o = location of virtual fire source (m)

D = diameter of burning fuel surface (m)

$\dot{Q}$ = total heat release rate (kJ/s) or (kW)

The virtual source can be at, above, or below the base of the burning fuel.

RADIANT HEAT FLUX TO A TARGET

For many enclosure fires, it is of interest to estimate the radiation transmitted from a burning fuel array to a target fuel positioned some distance from the fire to determine if secondary ignitions are likely. Figure 11-10D depicts the configuration used in developing the expression:

$$\dot{q}_0'' \approx \frac{P}{4\pi R_o^2} \approx \frac{\chi_r \dot{Q}}{4\pi R_o^2} \qquad (14)$$

where

$\dot{q}_0''$ = incident radiation on the target (kW/m²)

R_o = distance to target fuel (m)

P = total radiative power of the flame (kW)

χ_r = radiative fraction

$\dot{Q}$ = total heat release rate (kJ/s) or (kW)

Usually, χ_r ranges from 20 to 60 percent, depending upon the fuel type. Refer to Appendix A, "Tables and Charts," for values for χ_r for selected liquid pools. Experimental measurements indicate that Equation 14 has good accuracy for:

$$\frac{R_o}{R} > 4$$

where R (in meters) is the radius of the base of the fire. For radiation at:

$$\frac{1}{2} < \frac{R_o}{R} < 4$$

refer to DiNenno for a more exact analysis.[1]

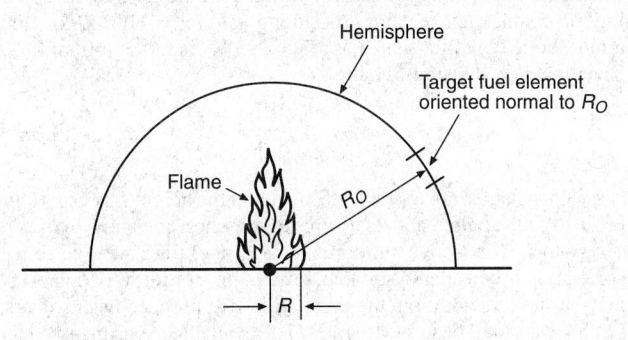

FIG. 11-10D. *Illustration of radiant heat transfer to a target fuel.*[8]

PRE-FLASHOVER TEMPERATURE ESTIMATES

Several researchers have developed correlations for predicting temperature rise in developing enclosure fires before. McCaffrey et al. suggest the following expression for naturally ventilated fires, based on correlation of an extensive number of enclosure experiments:[13]

$$\frac{\Delta T}{T_\infty} = 1.63 \left(\frac{\dot{Q}}{C_p \rho_\infty T_\infty A_v \sqrt{g h_v}} \right)^{2/3} \left(\frac{h_k A_s}{C_p \rho_\infty A_v \sqrt{g h_v}} \right)^{-1/3} \quad (15)$$

where

ΔT = temperature rise of upper gas $(Tg - T_\infty)$ (K)

T_g = gas temperature (K)

T_∞ = ambient air temperature (K)

$\dot{Q}$ = total heat release rate (kJ/sec) or (kW)

g = acceleration of gravity (m/s)

C_p = specific heat of air at constant pressure [(kJ/kg)/K]

ρ_∞ = density of air (kg/m³)

A_s = total surface area of enclosure interior excluding vent area (m²)

A_v = vent area (m²)

h_v = vent height (m)

h_k = effective enclosure conductance [(kW/m)/K]

The terms $h_k A_s$ and $A_v \sqrt{h_v}$ should be summed in Equation 15 for multiple structural materials and openings, respectively. In addition, while it is recognized that the enclosure gas temperature varies within the compartment, this equation is based on the assumption that an average upper-layer temperature and an average lower-layer, or ambient, temperature reasonably approximate temperature conditions in the enclosure. By substituting values for ambient conditions for key variables in Equation 15:

C_p = 1.0 (kJ/kg)/K, at one atmosphere

ρ_∞ = 1.18 kg/m³, density of ambient air

T_∞ = 295 K, ambient air temperature

g = 9.81 m/s² gravitational constant

A simplified expression can be provided of the form:

$$\Delta T = 6.85 \left(\frac{\dot{Q}^2}{h_k A_s A_v \sqrt{h_v}} \right)^{1/3} \quad (16)$$

where

$$h_k = \left(\frac{k \rho c_p}{t} \right)^{1/2} \quad \text{for } t \le t_p$$

and

$$h_k = \frac{k}{\delta} \quad \text{for } t > t_p$$

where

k = thermal conductivity of enclosure surface material [(kW/m)/K]

ρ = density of enclosure materials (kg/m³)

c_p = specific heat of enclosure material [(kJ/kg)/K]

δ = enclosure material thickness (m)

t = time (s)

$$t_p = \frac{\rho c_p}{k} \left(\frac{\delta}{2} \right)^2, \text{ thermal penetration time (sec)}$$

For an enclosure lined with gypsum board, key variable values are

c_p = 1.1 (kJ/kg)/K

ρ = 960 kg/m³

k = 0.00017 (kW/m)/K

δ = 0.016 m

Thus,

$$h_k = (0.18/t)^{1/2} \text{ for } t \le t_p, (0.00017/\delta \text{ for } t > t_p)$$

where

t_p = 400 sec for 16-mm (⁵⁄₈-in.) thickness of gypsum board

For the case of completely forced ventilation conditions, a correlation for enclosure temperature rise has been developed by Foote et al.[14]

$$\frac{\Delta T}{T_\infty} = 0.63 \left(\frac{\dot{Q}}{\dot{m}_v C_p T_\infty} \right)^{0.72} \left(\frac{h_k A_s}{\dot{m}_v C_p} \right)^{-0.36} \quad (17)$$

where

$\dot{m}_v$ = compartment forced mass ventilation rate (kg/sec)

For detailed discussions of methods that address multiple vents as well as forced ventilation, the reader is referred to DiNenno, Foote et al. and Deal and Beyler.[1,14,15]

PREDICTION OF FLASHOVER

A critical point in room fire growth is an event often referred to as "flashover." While a universal definition does not exist, this event is generally associated with rapid transition in fire behavior from localized burning of fuel to involvement of all the combustibles in the enclosure. Experimental work indicates that this transition can occur when upper room temperatures are between 750 and 1112°F (400 and 600°C).[16] Using a value of 932°F (500°C), Equation 15 may be solved for the heat release rate necessary to achieve flashover in a naturally ventilated enclosure.[8] The resulting equation is:

$$\dot{Q}_{fo} = 610 \left(h_k A_s A_v \sqrt{h_v} \right)^{1/2} \quad (18)$$

where

$\dot{Q}_{fo}$ = heat release rate at flashover (kJ/s) or (kW)

h_k = enclosure conductance [(kW/m²)/K]

A_s = total enclosure area (m²), excluding vent area

A_v = area of vent opening (m²)

h_v = height of vent opening (m)

Assuming that an enclosure has been heated thoroughly before flashover, i.e., $t > t_p$, Equation 18 can be simplified to:*

$$\dot{Q}_{fo(min)} = 610\left[(k/\delta)A_s A_\upsilon \sqrt{h_\upsilon}\right]^{1/2} \qquad (19)$$

where

k = thermal conductivity of enclosure material [(kW/m)/K]
δ_s = enclosure material thickness (m)

POST-FLASHOVER TEMPERATURE ESTIMATES

An alternative approach to estimating peak enclosure temperature was originally developed by Thomas[17] and extended by Law[18] to include both natural and forced ventilation. The correlations were based initially on post-flashover enclosure fire data, but they were extended by Law[18] through the evaluation of extensive pre-flashover room fire data. The results indicate that the predictions reasonably, but not exactly, predict the temperatures reported in the test fires. The equation does not consider variations in the thermophysical properties of room linings.

Most of the tests used to justify the equation involved rooms lined with gypsum board or concrete block. Caution should be exercised in applying these equations to rooms lined with highly insulating materials, such as fiberglass or foamed materials, or rooms in which major portions of the lining are of thermally thin materials such as steel or glass.

Forced ventilation is considered to occur when significant air is supplied by a ventilation system. The natural ventilation equation assumes that, at the time of peak temperature, all of the air for combustion will be drawn into the room through the vent openings and that these same openings will vent the product gases. This results in a natural sharing of the opening by the incoming air and outflowing gases that obey the laws of conservation of mass. The forced ventilation equation assumes that enough air is supplied to ensure that the fire will be free burning. It should be used only when the rate of air supply is sufficient to ensure such burning. Where forced ventilation is not enough to ensure free burning but is sufficient to cause concern that the assumed conservation in the natural venting equation has not been preserved, the peak temperature is expected to lie between the predictions of the two approaches.

The expression for natural ventilation conditions is:

$$\Delta T_{max} = 6000\left[\frac{\left(1 - e^{-0.036\eta}\right)}{(\eta)^{1/2}}\right]\left(1 - e^{-0.05\psi}\right) \qquad (20)$$

where

ΔT_{max} = peak temperature rise (K)
η = $[A_s/A_v(h)^{1/2}]$
ψ = $L_f/(A_v A_s)^{1/2}$
A_s = total surface area of enclosure interior excluding vent area (m²)
A_υ = vent area (m²)
h = height of compartment (m)
L_f = total enclosure fire load (equivalent weight of wood) (kg)

---

*For 13-mm (½-in.) thickness of gypsum board enclosure liner, $k/\delta = 0.014$. For 16-mm (⅝-in.) thick gypsum, $k/\delta = 0.011$.

The expression for forced ventilation conditions is:

$$\Delta T_{max} = 1200\left(1 - \theta^{-0.04\psi}\right) \qquad (21)$$

EQUIVALENT FIRE DURATION

Equivalent fire duration or fire severity is an approximation of the potential destructive impact of the burnout of all the available fuel in a room or space with at least one opening. The correlation presented here was developed by Law.[19] The results predict the potential impact of a post-flashover fire in terms of equivalent exposure in a fire-endurance furnace fired to follow the European equivalent exposure of the ASTM E119, *Standard Test Methods for Fire Tests of Building Construction and Materials* (NFPA 251, *Standard Methods of Tests of Fire Endurance of Building Construction and Materials*) standard time-temperature curve. Law based her correlation on data developed through an international research program carried out under the auspices of the Conseil International du Batiment (CIB). The results of this CIB effort are reported by Thomas and Heselden.[20] All of the tests were conducted with wood crib fuel sources. Law reports about 20 percent variation, depending on the porosity of the fuel. In wood cribs, porosity is based on the ratio of openings between the sticks of the crib and the space filled by those sticks, the greater fire severity being experienced with the more loosely packed cribs.

This correlation is not appropriate for rooms that do not have openings for ventilation. While no precise minimum can be stated, it is suggested that this equation not be used unless the area of the opening is at least greater than that of a typical residential window. The equation also assumes that virtually all of the potential energy in the fuel is released in the involved room. This holds true for the wood cribs used in the CIB tests. This assumption may not hold true where there are large surface areas, such as in rooms having combustible linings or in rooms that contain extensive materials with low thermal inertia, such as foam plastics. In these cases, the generation of pyrolized fuel may significantly exceed the combustion ability of the air drawn through the ventilation openings into the fire room. When this occurs, some of the fuel leaves the room in the vented gases. This often burns in the expelled gases causing the extension of flame from the room. In such cases, Equation 22 should be expected to overpredict the equivalent fire duration by an amount approximately proportional to the portion of the fuel that does not burn in the room being evaluated.

$$t = 60\left(\frac{L_f}{\sqrt{A_s A_\upsilon}}\right) \qquad (22)$$

where

t = fire severity (s)
A_s = surface area of enclosure interior surfaces, excluding vent area (m²)
A_υ = vent area (m²)
L_f = wood fuel mass (kg)

SMOKE PRODUCTION RATE

The rate of smoke-filled gas produced by a fire is nearly equal to the rate of air entrained into the rising fire plume, so the mass production rate of smoke-filled gas can be estimated as equal to the mass flow rate of gas in the fire plume. This mass flow rate into a plume above the visible flame height may be estimated using an expression developed by Zukoski:[21]

$$\dot{m}_s = 0.18\dot{Q}^{1/3}\rho_\infty^{2/3}C_p^{-1/3}T_\infty^{-1/3}g^{1/3}Y^{5/3} \qquad (23)$$

where

$\dot{m}_s$ = rate of smoke-filled gas production (kg/s)

$\dot{Q}$ = total heat release rate (kJ/s or kW)

ρ_∞ = density of air (kg/m^3)
C_p = specific heat of air at constant pressure [(kJ/kg)/K]
T_∞ = ambient gas temperature (K)
g = acceleration of gravity (m/s^2)
Y = distance from the virtual point source for the fire to bottom of smoke layer (m)

Assuming an ambient air temperature of 293 K (20°C), Equation 23 can be reduced to

$$\dot{m}_s = 0.065\,\dot{Q}^{1/3}\,Y^{5/3} \qquad (24)$$

When the flame height exceeds Y in Equation 24, Equation 24 tends to overpredict gas production.

Equation 24 does not apply to elevations in the flame region. However, McCaffrey[22] has investigated gas temperatures and velocity distributions within the flame and intermittent flame regions for fires up to 250 kW. Under these conditions, the mass flow rate of combustion products was found to correspond to the expression:

$$\dot{m}_s = 0.055\,\dot{Q}^{1/2}\,Y \qquad (25)$$

Equations 24 and 25 are based on the assumption that the fire can be reasonably approximated as a circular pool fire. Experiments have shown that reasonable results will be obtained with fires that are not circular, provided that the aspect ratio of length to width is relatively small. The equations are not suitable for conditions where entrainment is restricted (e.g., the fire is against a wall) or if the fire is long and narrow (e.g., a line fire).

An alternative approach to predicting smoke production rates in enclosures has been developed by Butcher and Parnell,[23] based on the size of the fire perimeter and vertical distance to the smoke layer. This approach assumes a constant heat release rate. Based on this approach, smoke production rate can be expressed as:

$$\dot{m}_s = 0.096\,P\rho_o\,y^{3/2}\left(g\,\frac{T_o}{T_{fl}}\right)^{1/2} \qquad (26)$$

where

$\dot{m}_s$ = rate of smoke production (kg/s)

P = perimeter of fire (m)
y = distance from floor to bottom of smoke layer (m)
T_o = ambient temperature (K)
T_{fl} = flame temperature (K)
ρ_o = density of ambient air (kg/m^3)
g = gravitational acceleration (9.81 m/s^2)

Since the expression assumes a constant or steady burning rate, its application has limitations. Yet it will provide a reasonable estimate of smoke generation rate for many enclosure configurations of practical interest.

This expression can be further simplified, based on value assignments for selected parameters. That is, for:

ρ_o = 1.22 kg/m^3 at 17°C
T_o = 290 K
T_{fl} = 1100 K
g = 9.81 m/s^2

Equation 26 is reduced to:

$$\dot{m} = 0.188\,Py^{3/2} \qquad (27)$$

Figure 11-10E provides graphical results based on the calculation of the smoke-filled gas mass production rate in Equation 26 for selected values of P and y. The mass rate of smoke-filled gas production can be changed to a volume rate by dividing by the density of air at the appropriate gas temperature.

ENCLOSURE SMOKE FILLING

Smoke from a fire begins to fill an enclosure as it accumulates below the ceiling. The rate of smoke filling depends on the amount of smoke produced and the size and location of vents. The mass rate of smoke flow at any distance above a fire of known heat release rate can be calculated using Equation 23. The rate at which a smoke-filled layer descends toward the floor depends on the plan area of the enclosure, the distance of the lower edge of the smoke layer above the fire, and the temperature of the layer.

For an enclosure vented in the lower layer, the upper layer descends with a velocity given by:

$$U_t = \frac{\dot{m}_s}{\rho_l A_p} \qquad (28)$$

where

U_t = rate of layer descent (m/s)

$\dot{m}_s$ = mass rate of smoke production (kg/s)

ρ_l = density of smoke layer (kg/m^3)
A_p = enclosure floor area (m^2)

The lower limit for the velocity of descent is obtained by using Equation 28 and the ambient density ρ_∞.

Fires in enclosures in which the upper layer is vented can stabilize at a constant-depth smoke layer. Venting can occur naturally through openings, such as doors and windows, or it can be forced by mechanical smoke control systems.

For the case of a known vent flow rate, the height of the bottom of the smoke layer stabilizes above the fire at the position where smoke mass inflow from the fire plume equals vent outflow. This is calculated using a known vent mass flow rate, $\dot{m}_{sv}$, and solving Equation 24 for position Y_v as:

$$Y_v = 1.9\,\dot{Q}^{-1/5}\,\dot{m}_{sv}^{3/5} \qquad (29)$$

This shows that the height at which a smoke layer may be stabilized by venting depends mostly on the vent capacity and is relatively insensitive to changes in the fire heat release rate.

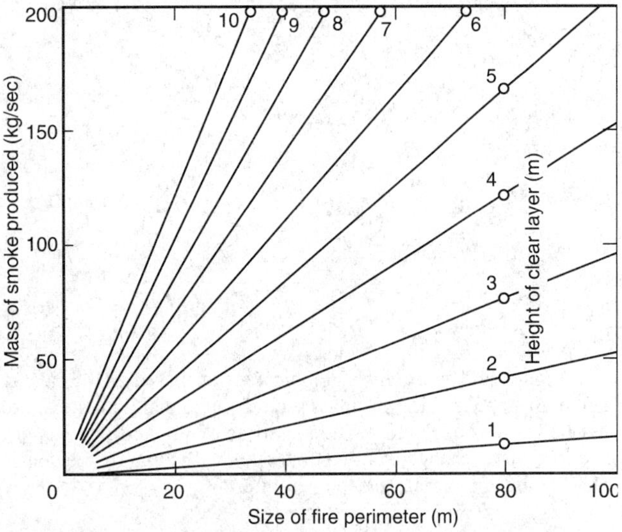

FIG. 11-10E.   Smoke production rate for steady fires at various distances (m) from virtual origin to bottom of smoke layer.[23]

BUOYANT GAS HEAD

During a fire, a pressure differential develops between the fire-heated areas and other spaces. Essentially, expressions for pressure differential are derived from the basic hydrostatic equation. The equations below resulted from rearrangement of terms and simplification based on assumed values for selected variables. These expressions are useful in evaluating a range of enclosure effects caused by pressure differentials. Examples include potential for smoke flow to overcome normal air flow, the pressure loading on a door due to the fire, and the uplift pressure on ceiling tiles.

The pressure calculated by these equations results from the difference between the density of a heated gas column from the fire and the density of the surrounding environment. A pressure surge can also result from the expansion of heated gases in those situations where the rate of fire development is fast and the space is not vented sufficiently to relieve the resultant increase in gas volume. The equations given below, however, address only the pressure difference caused by the heated gas column condition.

The equations are universal to any heated gas column and can apply to buoyant gas heads caused by building stack effects or other temperature differential effects, including, but not limited to, those caused by fire. The equations, as presented, assume that the entire column of heated gas is at the same temperature. This is a reasonable approximation in many fires but is not exact and would be inappropriate for a fire where the condition consisted of a plume freely entraining cooling air over an extensive portion of its length or any other condition where a significant temperature gradient existed in the heated gas column being appraised.

The general equation can be expressed as:

$$\Delta P = (\rho_o - \rho_c)gh \tag{30}$$

where

ΔP = pressure difference (Pa)
ρ = gas (air) density outside the heated gas column (kg/m³)
ρ_c = gas (smoke or flame) density of the heated gas column (kg/m³)
g = gravitational constant (m/s²)

h = distance above point where gas column density is same as density outside the heated column. [In a fire, this is normally the base of the hot gas column (m).]

If it is assumed that the outside atmosphere and the gas column are predominantly air at standard atmospheric pressure, the equation can be expressed as:

$$\Delta P = 3460\left(\frac{1}{T_\infty} - \frac{1}{T_c}\right)h \tag{31}$$

where

ΔP = pressure difference (Pa)
T_∞ = absolute temperature of air outside the heated gas column (K)
T_c = temperature of the heated gas column (K)
h = height of the portion of interest of the hot gas column (m)

For further discussion, see Klote and Milke.[24]

THERMAL FIRE DETECTOR RESPONSE

Computer programs have been developed to calculate the response time of heat detectors and sprinklers installed below ceilings in large rooms.[25,26] These programs can determine the time to operation for a user-specified fire energy release rate history. They are convenient to use because the tedious repetitive calculations needed to analyze a growing fire can be avoided. However, the same calculations can be performed easily with a scientific hand calculator for steady fires that have a constant energy release rate. In cases where a more detailed analysis of a fire that has important changes in energy release rate over time is required, the fire may be represented as a series of steady fires occurring immediately after one another.

A useful calculation directly related to thermal detection is to find the plume temperature at positions directly above the flame produced by burning materials. This can be done using the centerline plume temperature correlation or the simplified correlation:[27]

$$T_{m(\text{plume})} = 16.9\frac{\dot{Q}^{2/3}}{h^{5/3}} + T_\infty \tag{32}$$

where

h = distance above fuel surface (m)
$\dot{Q}$ = fire energy release rate [(kJ/s) or (kW)]
$T_{m(\text{plume})}$ = plume gas temperature above fire (K)
T_∞ = ambient room temperature (K)

This equation was developed from analysis of experiments with large-scale fires having energy release rates from 670 kW to 100 MW.[27]

As an example, using Equation 32, the plume gas temperature [$T_{m(\text{plume})}$] 5 m (h) above the fuel surface of a 500 kW ($\dot{Q}$) fire is found to be 366 K (93°C) for an ambient room temperature (T_∞) of 293 K (20°C).

For the case of fixed-temperature detectors, the minimum fire energy release rate, $\dot{Q}$, needed to operate a fixed-temperature detection or suppression device located directly above the fire can be estimated using Equation 32, solving for $\dot{Q}$ with $T_{m(\text{plume})}$ set equal to the activation temperature of the thermal device. In this form, Equation 32 becomes:

$$\dot{Q} = 0.0144 \left(T_{m(\text{plume})} - T_\infty \right)^{3/2} \times h^{5/2} \qquad (33)$$

Based on cases where the hot gases have begun to spread under a ceiling located above the fire, Equation 33 also applies for a small radial distance, r, from the impingement point. (See Figure 11-10F.) Over this distance, up to $r/h = 0.18$, where the gas is turning to flow out under the ceiling, the highest temperature in the flow remains equal to the value at the impingement point directly over the fire, calculated using Equation 32.

At radial distances greater than $r/h = 0.18$, the maximum temperature in the ceiling jet flow depends upon the distance from the impingement point, according to:

$$T_{m(\text{jet})} = 5.38 \frac{\left(\dot{Q}/r \right)^{2/3}}{h} + T_\infty \qquad (34)$$

where

h = distance above fuel surface (m)

$\dot{Q}$ = fire energy release rate [(kJ/s) or (KW)]

r = radial distance from plume centerline to device (m),

$T_{m(\text{jet})}$ = temperature of ceiling jet (K)

T_∞ = ambient room temperature (K)

Correlations are also available for maximum velocities in the ceiling jet flow, U_m, under a ceiling. As with the temperature correlations, there are two regions: (1) one close to the impingement point where velocities are nearly constant and (2) the other farther away where velocities vary with radial position. The two correlations are:

$$U_m = 0.96 \left(\frac{\dot{Q}}{h} \right)^{1/3} \quad \text{for } r/h < 0.15 \qquad (35)$$

and

$$U_m = 0.195 \left(\frac{\dot{Q}^{1/3} h^{1/2}}{r^{5/6}} \right) \quad \text{for } r/h > 0.15 \qquad (36)$$

where

r = radial distance from plume centerline to device (m)

U_m = gas velocity (m/s)

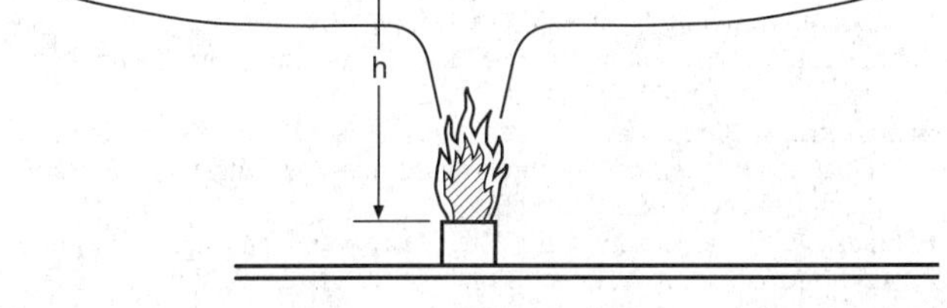

FIG. 11-10F.   Parameters h and r both related to calculation of sprinkler or heat-detector actuation time.

The previous equations can be used to determine whether the temperature of the fire-driven gas flow past a detection device is high enough to operate the device. However, more information is needed to calculate the amount of time needed to heat the detector or sprinkler sensing element to the operating temperature. Often, these elements are made of metal, such as the ordinary solder-type fusible link used in the link and lever sprinklers, or they are liquid-filled glass vials, such as those used in bulb-type sprinklers. Both of these sensing elements require some time to absorb the heat transferred from the hot gas flowing around the device.

For steady fires, the time required to heat the sensing element of a thermal detection or suppression device from room temperature to operation temperature is given by:

$$t_{\text{operation}} = \frac{\text{RTI}}{\sqrt{U_m}} \log_e \left(\frac{T_m - T_\infty}{T_m - T_{\text{operation}}} \right) \qquad (37)$$

where RTI, the response-time index, is a measure of the ease of heating thermal elements in heat detectors and sprinklers. The larger the RTI value, the slower the response of the sensing element. RTI values for sprinklers have been measured[28] in the range of 15 $\text{m}^{1/2}\text{s}^{1/2}$ to 400 $\text{m}^{1/2}\text{s}^{1/2}$.

In the previous example, it was found using an earlier equation that a 500 kW $\left(\dot{Q} \right)$ fire would produce a gas temperature of 366°K [$T_{m(\text{plume})}$] at 5 m (h) above the fuel surface in a room with 293°K (T_∞) ambient temperature. From Equation 35, the gas velocity at this position would be 4.4 m/s (U_m). For a sprinkler with an RTI of 200 $\text{m}^{1/2}\,\text{s}^{1/2}$ and operation temperature of 347°K ($T_{\text{operation}}$), the time to operation in response to the steady fire can be calculated from Equation 37 as:

$$t_{\text{operation}} = \frac{200}{\sqrt{4.4}} \log_e \left(\frac{366 - 293}{366 - 347} \right) = 128 \text{ sec.}$$

The measurement of thermal lag for sprinkler is a topic of current research investigations and verification testing.[29,30] It has been found that, for cases when the gas temperature does not substantially exceed the activation temperature for the sprinkler, significant error can occur in the prediction for activation time. In these cases it is possible for small changes in predicted gas temperatures to result in large changes in predicted activation time.[29] In the case of constant or slowly varying gas temperatures this effect may be important where :

$$\frac{T_m - T_{\text{operation}}}{T_m - T_\infty} < \frac{1}{4}$$

Another factor that contributes to the inaccuracy of predicted results, using Equation 37 under low gas temperatures and gas velocities, is that no means is included to account for loss of heat from the sensing element, either link or glass bulb, to the sprinkler frame and piping by conduction. Heskestad and Bill[31] and Ingason[32] have studied means to account for the effects of conduction loss, and measured values for it for sprinkler hardware. Following Heskestad and Bill,[31] Equation 37 can be modified to account for the conduction losses to the frame and piping which are assumed to be at constant temperature equal to the orginal ambient temperature (T_∞) as:

$$t_{\text{operation}} = \frac{\text{RTI}}{\sqrt{U_m} \left(1 + \frac{C}{\sqrt{U_m}} \right)} \log_e \left[\frac{T_m - T_\infty}{T_m - T_{\text{operation}} - \frac{C}{\sqrt{U_m}} \left(T_{\text{operation}} - T_\infty \right)} \right] \qquad (38)$$

Where C is a conduction loss parameter with units $(m/s)^{1/2}$ obtained by measurement.[31,32] Measured values[31] for the conduction loss parameter range from 0.5 to 1.6 $(m/s)^{1/2}$. Using a value of $C = 0.5(m/s)^{1/2}$ the expected operation time for the sprinkler in the example would increase to 190 sec as predicted using Equation 38, which is over 1 min slower than the prediction of 128 sec (using Equation 37 or Equation 38 with $C = 0$). At a value of $C > 0.74$ $(m/s)^{1/2}$, the heat loss would be great enough to prevent sprinkler operation even though the gas temperature was sustained above the indicated operating temperature of the sprinkler.

All calculations in use today for determining times to operation only consider the convective heating of sensing elements by the hot fire gases. They do not account explicity for any direct heating by radiation from the flames. Research is continuing to evaluate and improve calculations of operation for heat-activated devices, such as sprinklers.

BIBLIOGRAPHY

References Cited

1. DiNenno, P. J., *et al.* ed., *The SFPE Handbook of Fire Protection Engineering*, 2nd. ed., National Fire Protection Association, Quincy, MA, 1995.
2. Nelson, H. E., and Forssel, E. W., "Use of Small-Scale Test Data in Hazard Analysis," *Proceedings* of the Fourth International Symposium, International Association for Fire Safety Science, 1994, pp. 971–982.
3. Babrauskas, V., and Krasny, J. F., "Fire Behavior of Upholstered Furniture," NBS Monograph, National Bureau of Standards, Gaithersburg, MD, 1985.
4. Lawson, J. R., *et al.*, "Fire Performance of Furnishings as Measured in the NBS Furniture Calorimeter, Part 1," NBSIR 83-2787, National Bureau of Standards, Gaithersburg, MD, Jan. 1984.
5. Alpert, R. L., and Ward, E. J., "Evaluating Unsprinklered Fire Hazards," SFPE TR 83-2, Society of Fire Protection Engineers, Boston, MA, 1983.
6. Babrauskas, V., *et al.*, "Upholstered Furniture Heat Release Rates Measured with a Furniture Calorimeter," NBSIR 82-2604, National Bureau of Standards, Gaithersburg, MD, 1982.
7. Babrauskas, V., "Free-Burning Fires," *Proceedings*, SFPE Symposium: Quantitative Methods for Fire Hazard Analysis, University of Maryland, College Park, MD, 1985.
8. Lawson, J. R., and Quintiere, J. G., "Slide-Rule Estimates of Fire Growth," NBSIR 85-3196, National Bureau of Standards, Gaithersburg, MD, 1985.
9. Heskestad, G., "Virtual Origins of Fire Plumes," *Fire Safety Journal*, Vol. 5, No. 109, 1983.
10. Beyler, C. L., "Fire Plumes and Ceiling Jets," *Fire Safety Journal*, Vol. 11, No. 53, 1986.
11. Delischatsios, M., "Flame Heights in Turbulent Wall Fire with Significant Flame Radiation," *Combustion Science and Technology*, Vol. 39, 1984, p. 195.
12. Heskestad, G., "Luminous Heights of Turbulent-Diffusion Flames," *Fire Safety Journal*, Vol. 7, No. 25, 1984.
13. McCaffrey, B. J., Quintiere, J. G., and Harkleroad, M. F., "Estimating Room Temperatures and the Likelihood of Flashover Using Fire Test Data Correlations," *Fire Technology*, Vol. 17, No. 2, 1981, p. 98.
14. Foote, K. L., Pagni, P. J., and Alvares, N. J., "Temperature Correlations for Forced-Ventilated Compartment Fires," *Proceedings* of the First International Symposium, International Association for Fire Safety Science, Hemisphere Publishing, 1986, pp. 139–148.
15. Deal, S. D., and Beyler, C. L., "Correlating Preflashover Room Fire Temperatures," *Journal of Fire Protection Engineering*, Vol. 2, No. 2, 1990, pp. 33–48.
16. Thomas, P. H., "Testing Products and Materials for Their Contribution to Flashover in Rooms," *Fire and Materials*, Vol. 5, No. 3, 1981, pp. 103–111.
17. Thomas, P. H., *Fire in Model Rooms*, CIB Research Program, Building Research Establishment, Borehamwood, Hertfordshire, England, 1974.

18. Law, M., "Fire Safety of External Building Elements—The Design Approach," *AISC Engineering Journal*, Second Quarter, American Institute of Steel Construction, 1978.
19. Law, M., "Prediction of Fire Resistance," *Proceedings*, Symposium No. 5: Fire-Resistance Requirements for Buildings—A New Approach, Joint Fire Research Organization, H. M. Stationery Office, London, 1973.
20. Thomas, P. H., and Heselden, A. J. M., *Fully Developed Fires in Single Compartments*, A Co-operative Research Programme of the Conseil International du Batiment, Joint Fire Research Organization Research Note No. 923, H. M. Stationery Office, London, 1972.
21. Zukoski, E. E., "Development of a Stratified Ceiling Layer in the Early Stages of a Closed-Room Fire," *Fire and Materials*, Vol. 2, No. 2, 1978.
22. McCaffrey, B. J., "Purely Buoyant Diffusion Flames: Some Experimental Results," NBSIR 79-1910, National Bureau of Standards, Gaithersburg, MD, 1979.
23. Butcher, E. G., and Parnell, A. C., *Smoke Controlling Fire Safety Design*, E. and F. N. Spon, Ltd., London, 1979.
24. Klote, J. H., and Milke, J. A., *Design of Smoke Management Systems*, American Society of Heating, Refrigerating, and Air-Conditioning Engineers, Atlanta, GA, 1992.
25. Evans, D. D., and Stroup, D. W., "Methods to Calculate the Response Time of Heat and Smoke Detectors Installed Below Large Unobstructed Ceilings," *Fire Technology*, Vol. 22, No. 1, 1985, pp. 54–65.
26. Walton, W. D., and Notarianni, K. A., "Comparison of Ceiling Jet Temperatures Measured in an Aircraft Hanger Test Fire with Temperatures Predicted by the DETACT-QS and LAVENT Computer Models," NISTIR 4947, National Institute of Standards and Technology, Gaithersburg, MD, 1993.
27. Alpert, R. L., "Calculation of Response Time of Ceiling-Mounted Fire Detectors," *Fire Technology*, Vol. 8, 1972, pp. 181–195.
28. Heskestad, G., and Smith, H., "Investigation of a New Sprinkler Sensitivity Approval Test: The Plunge Test," FMRC Serial No. 22485, Factory Mutual Research Corporation, Norwood, MA, 1976.
29. Madrzykowski, D., "Evaluation of Sprinkler Actuation Prediction Methods," *Proceedings* of the First International Asiaflam Conference 1995, Mar. 15–16, 1995, Kowloon, Hong Kong, InterScience Communications Limited, London, pp. 211–218.
30. Notarianni, K. A., and Davis, W. D., "The Use of Computer Models to Predict Temperature and Smoke Movement in High Bay Spaces," NISTIR 5304, National Institute of Standards and Technology, Gaithersburg, MD, 1993.
31. Heskestad, G., and Bill, R. G., "Quantification of Thermal Responsiveness of Automatic Sprinklers Including Conduction Effects," *Fire Safety Journal*, Vol. 14, Nos. 1–2, July 1988, pp. 113–125.
32. Ingason, H., "Thermal Response Models for Glass Bulb Sprinklers, An Experiment and Theoretical Analysis," SP Report 1992: 12, 1992, Swedish National Testing and Research Institute, Boras, Sweden.

Additional Reading

ASTM E906, *Standard Test Method for Heat and Visible Smoke Release Rates for Materials and Products*, American Society for Testing and Materials, W. Conshohocken, PA, 1993.

Babrauskas V., "A Closed-Form Approximation for Post-Flashover Compartment Fire Termperatures," *Fire Safety Journal*, Vol. 4, 1981, pp. 63–73.

Babrauskas, V., "Applications of Predictive Smoke Measurements," *Journal of Fire and Flammability*, Vol. 12, 1981, pp. 51.

Babrauskas, V., "Burning Rates," *The SFPE Handbook of Fire Protection Engineering,* 2nd Edition, Section 3/Chapter 1, National Fire Protection Association, Quincy, MA, 1995, pp. 3/1–3/15.

Babrauskas, V., "Estimating Room Flashover Potential," Fire Technology, Vol. 16, May 1980, pp. 94–103, 112.

Babrauskas, V., "Free Burning Fires," *Proceedings*, SFPE Symposium: Quantitative Methods for Fire Hazard Analysis, University of Maryland, College Park, MD, 1985.

Babrauskas, V., and Krasny, J. F., "Fire Behavior of Upholstered Furniture," NBS Monograph, National Bureau of Standards, Washington, DC, 1985.

Babrauskas, V., and Peacock, R. D., "Heat Release Rate: the Single Most Important Variable in Fire Hazard," *Fire Safety Journal*, Vol. 18, No. 3, 255–272, 1992.

Curtat, M. R., and Bodart, X. E., "Simple and Not So Simple Models for Compartment Fires," Proceedings of the First International Symposium of Fire Safety Science, Hemisphere, 1986, pp. 637–646.

Delichatsios, M. A., "Fire Growth Rates in Wood Cribs," *Combustion and Flame,* Vol., 27, 1976, pp. 267–278.

Drysdale, D., *An Introduction to Fire Dynamics*, John Wiley & Sons, New York, 1985.

Evans, D. D., "Calculation of a Fire Plume in a Two-Layer Environment," *Fire Technology*, Vol. 20, No. 3, 1984, p. 39.

Gross, D., "Data Sources for Parameters Used in Predictive Modeling of Fire Growth and Smoke Spread," NBSIR 85-3223, National Bureau of Standards, 1985, Gaithersburg, MD.

Heskestad, G., "Modeling of Enclosure Fires," 14th International Symposium on Combustion, The Combustion Institute, Pittsburgh, PA, 1973, pp. 1021.

Huggett, C., "Estimation of Rate of Heat Release by Means of Oxygen Consumption Measurements," *Fire and Materials*, Vol. 4, 1980, pp. 61–65.

Kawagoe, K., and Sekine, T., *Estimation of Fire Temperature-Time Curve for Rooms*, Japanese Ministry of Construction Building Research Institute, June 1963.

Krasner, L. M., *Burning Characteristics of Wooden Pallets as a Test Fuel*, Factory Mutual Research Corp., Norwood, MA, 1968.

Law, M., "A Relationship Between Fire Grading and Building Design and Contents," Fire Research Note 877, British Fire Research Station, 1971.

Law, M., "Notes on the External Fire Exposure Measured at Lehrte," Fire Safety Journal, Vol. 4, 1981–82, pp. 243–246.

Modak, A. T., "Thermal Radiation from Pool Fires," *Combustion and Flame*, Vol. 29, 1977, pp. 177–192.

Modak, A. T., and Croce, P. A., "Plastic Pool Fires," *Combustion and Flame*, Vol. 30, 1977, pp. 251–265.

Peacock, R. D., Davis, S., and Lee, B. T., "An Experiment Data Set for the Accuracy of Room Fire Models," NBSIR 88-3752, National Bureau of Standards, Gaithersburg, MD, April 1988.

Quintiere, J. G., "A Simple Correlation for Predicting Temperature in a Room Fire," NBSIR 83-2712, National Bureau of Standards, Washington, DC, 1983.

Quintiere, J. G., "Smoke Measurements: An Assessment of Correlations between Laboratory and Full-Scale Experiments," *Fire and Materials*, Vol. 6, Nos. 3 and 4, 1982.

Quintiere, J. G., and Rhodes, B., "Fire Growth Models for Materials. Final Report. June 1992–Dec. 1993," Maryland Univ., College Park, National Institute of Standards and Technology, Gaithersburg, MD, NIST-GCR-94-647, June 1994.

Quintiere, J., "A Perspective on Compartment Growth," *Combustion Science and Technology*, Vol. 39, 1984, pp. 11–54.

Ramachandran, G., "Exponential Model of Fire Growth," *Proceedings* of the First International Symposium on Fire Safety Science, Hemisphere, 1986, pp. 657–666.

Williams, F. A., "Mechanisms of Fire Spread," 16th International Symposium on Combustion, The Combustion Institute, 1976, pp. 1281–1294.

Yamashika, S., and Kurimoto, H., "Burning Rate of Wood Cribs," Report of the Fire Research Institute of Japan, Tokyo, No. 41 (Mar. 1976), pp. 8–15.

Zdanowski, M., Teadorczyk, A., and Wojcicki, S., "A Simple Mathematical Model of Flashover in Compartment Fires," *Fire and Materials*, Vol. 10, 1986, p. 145.

SIMPLIFIED FIRE HAZARD AND RISK CALCULATIONS

Thomas F. Barry and John M. Watts, Jr.

This chapter provides an overview of fire hazard and risk estimation used in making hazard and risk reduction decisions. The techniques described were developed for use on facilities that manufacture, process, or store hazardous chemicals but may be adapted for use in other settings, particularly industrial facilities. This assessment can help to identify the uncertainty that exists concerning the probability of accidental events, the magnitude of their consequences, and the determination of cost-effective risk reduction measures. The steps described in this chapter are normally used as part of a comprehensive risk management decision making process. Usefulness and credibility of this information for management decision-makers will depend largely on the methodology, procedures, documentation, and quality assurance controls integrated into the application.

The fire hazard and risk assessment is presented here in a structured five-step approach, as shown in Figure 11-11A.

These steps are based on the quantitative risk assessment (QRA) steps developed by the Center for Chemical Process Safety (CCPS) and the American Institute of Chemical Engineers (AIChE), and described in AIChE's *Guidelines for Chemical Process Quantitative Risk Analysis.*[1] This approach has been generally accepted and adopted by U.S. chemical, oil, and gas industries undertaking QRA projects. *The SFPE Handbook of Fire Protection Engineering*, Section 5, "Fire Risk Analysis," provides additional information on this subject.[2]

STEP 1: HAZARD IDENTIFICATION AND EVALUATION

Hazard is defined as a chemical or physical condition that has the potential for causing damage to people, property, or the environment. To assess the risk associated with fire hazards it is first necessary to identify potential hazards and screen them for those that appear to merit additional analysis.

The hazard identification and evaluation process should include:

1. Identification of the physical and chemical properties of materials processed, stored, or transported on site that can harm employees, the public, property, environment, or other selected risk targets.
2. Identification of weaknesses in the design, operation, and protection of facilities that could lead to fires or explosions.
3. Evaluation of the significance of potential hazardous events associated with a process or activity to allow categorization of the consequences and ranking the risk.

The general tasks associated with hazard identification and evaluation consist of:

1. Gathering technical information
   (a) Plant and process data
   (b) Material properties
2. Gathering historical accident data
   (a) Facility tour and process review
   (b) Plant personnel interviews
   (c) Process safety management documentation evaluation
3. Documenting the hazard evaluation(s)
   (a) Use of one or more selected hazard evaluation methods
   (b) Review of results for completeness and accuracy

Chemical and flammability data can be extracted from sources, such as NFPA 49, *Hazardous Chemicals Data;* NFPA 325, *Guide to Fire Hazard Properties of Flammable Liquids, Gases, and Volatile Solids;* NFPA 491M, *Manual of Hazardous Chemical Reactions;* and other chapters in this handbook. Much of the data available in these publications are at atmospheric temperature and pressure. Experimental data appropriate to process conditions will sometimes be needed.

The SFPE Handbook of Fire Protection Engineering[2] provides information on mass burning and heat release rates, and toxic combustion products. Dust explosion data for explosion-venting calculations are presented in NFPA 68, *Guide for Venting of Deflagrations*, which includes additional references.

Methodologies used for hazard identification and evaluation include the following:

- Engineering checklists
- Hazard indices
- Hazard and operability study (HAZOP)
- Preliminary hazard analysis (PHA)
- Failure mode and effect analysis (FMEA)

Thomas F. Barry, P.E., is a senior engineer specializing in fire and explosion risk assessment with HSB Professional Loss Control, Kingston, TN.

John M. Watts, Jr., Ph.D., is director of the Fire Safety Institute, a not-for-profit information, research, and educational corporation located in Middlebury, VT. He also serves as editor of NFPA's quarterly technical journal, *Fire Technology*.

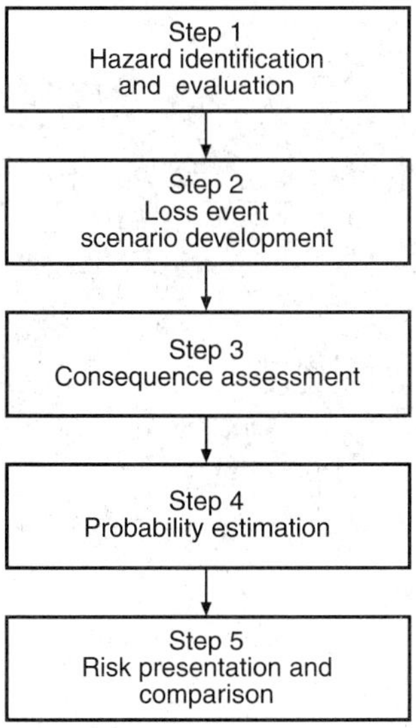

Step 1
Hazard identification and evaluation

Step 2
Loss event scenario development

Step 3
Consequence assessment

Step 4
Probability estimation

Step 5
Risk presentation and comparison

FIG. 11-11A.   Fire hazard and risk assessment.

- Fault-tree analysis (FTA)
- Cause-consequence diagrams
- Reliability analysis
- Fire hazard analysis
- Loss expectancy analysis

Details of these methodologies are beyond the scope of this chapter. Descriptions can be found in *Guidelines for Hazard Evaluation Procedures with Examples*[3] and *Systems Safety Analysis Handbook.*[4]

Integrating the evaluation of existing risk reduction measures and historical loss incidents (or test data, as needed) into the hazard method provides a preliminary risk categorization and screening approach as illustrated in Figure 11-11B.

STEP 2: LOSS EVENT SCENARIO DEVELOPMENT

Hazards (i.e., potential unsafe conditions) screened and ranked as unacceptable or requiring further analysis (in step 1) generally require additional loss event scenario development to identify and evaluate risk contributors in more detail. Structuring credible fire and explosion loss scenarios is an important aspect of the risk estimation process. The primary components that must be evaluated in fire and explosion loss scenario development include initiating failure events, intermediate events, and incident outcome(s).

Initiating Failure Events

Failure events generally are initiated by one or a combination of the following factors: containment failures, human error, ignition, and external exposures.

Containment failures: Accidental releases of flammable liquid or gas are the initiating events in most fire and explosion risk estimations. Review of past fire and explosion incidents associated with containment failures suggests the following to be major causes:

- Rupture of flexible hose
- Overfilling of tanks

- Release from drainage and sampling valves
- Leaks from gaskets and pump seals
- Leaks from flanged joints and small pipe connections
- Piping failures
- Tank rupture
- Internal tank overpressures

Containment failures can be created by equipment malfunctions (mechanical or electrical breakdown, instrumentation failures, etc.), human errors, loss of utilities (electricity, cooling water, etc.), and external events (floods, earthquakes, vandalism, etc.)

Human error: Human error assessment methods and details are related to the complexity of the process or equipment; the operating environment; human-equipment interface factors; hardware design interfaces for emergency actions; and procedures for operations, maintenance, testing, and training.

Human error analysis or human reliability analysis (HRA) generally involves the following evaluations:

1. Describing the characteristics of the personnel, the work environment, and the tasks performed;
2. Evaluating the human-machine interfaces;
3. Performing a task analysis of intended operator functions; and
4. Performing a human error analysis of intended operator functions to assess the likelihood of human error.

Ignition: Primary ignition sources include electrical (e.g., wiring or motors), smoking, incendiarism, overheated materials, hot surfaces, open flames, cutting and welding, friction, spontaneous ignition, combustion sparks and embers, exposure from a nearby fire, chemical reaction (exothermic), mechanical sparks, static sparks, molten substance, and lightning. NFPA and industrial insurance companies are good reference sources for statistical ignition and fire cause data.

External exposures: External exposures include flooding, earthquakes, tornadoes, hurricanes, vehicle impacts, and train derailments.

Intermediate Event(s)

Intermediate events affect the growth or propagation and mitigation of the initiating event source. These events are a function of time, which is very important to recognize when addressing the conditional probabilities. Both propagating and mitigating intermediate factors need to be identified.

Propagating factors: Among the growth or propagation factors within the fire and explosion development scenario are the following:

1. Process parameter deviations: pressure, temperature, flow rate, concentration, phase or state change;
2. Material release type: combustibles, explosive materials, toxic materials, reactive materials;
3. Ignition: immediate or delayed ignition;
4. Energy release rates: explosion or flash fire;
5. Ventilating effects;
6. Meteorological conditions;
7. Liquid spill containment; and
8. Operator emergency response errors.

Mitigating factors: The following mitigating factors generally are included in the fire and explosion scenario:

1. Safety system responses: relief valves, backup utilities, emergency shutdown systems, backup systems;

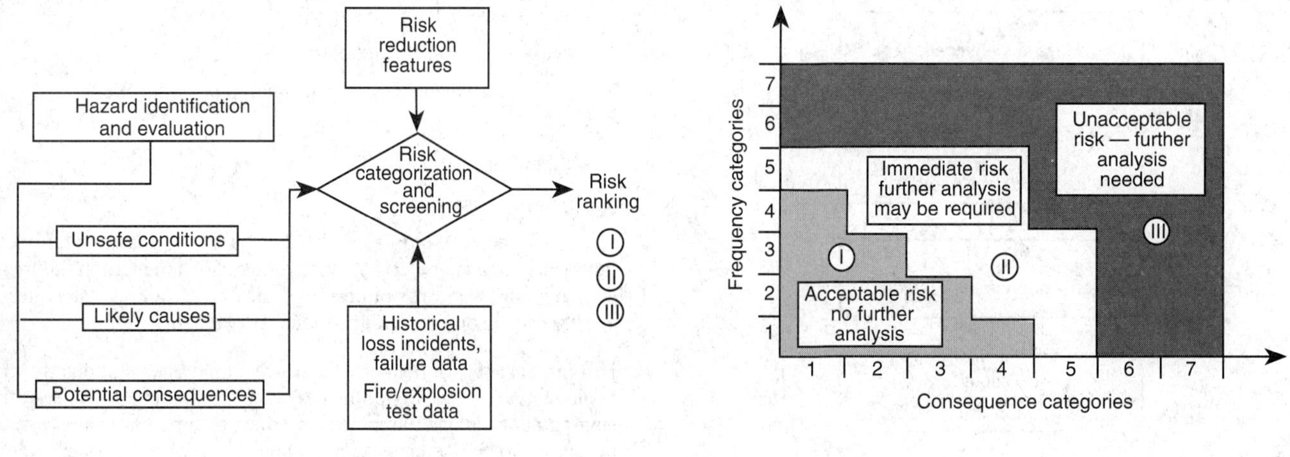

FIG. 11-11B. A preliminary risk screening and ranking approach.

2. Barrier effectiveness with time, explosion relief systems;
3. Mitigation system responses: vents, dikes, detection, alarms, shutdown, fire protection, suppression;
4. Control responses and operator responses: planned or emergency;
5. Contingency operations: emergency procedures, personnel safety equipment, evacuations, security; and
6. Fire detection and protection systems.

Incident Outcomes

Potential incidents of primary interest to the fire risk analyst include the following:

1. Heat from a fire: pool fire, torch fire, flash fire, boiling-liquid, expanding-vapor explosion (BLEVE);
2. Explosion overpressures: unconfined vapor cloud explosions (UVCE), tank or equipment rupture or fragmentation; and
3. Corrosive smoke or fire products concentration: toxic gas concentrations.

Fire exposure is very time dependent. The most widely used technique in the structuring of fire scenarios has involved the use of event tree logic to convey the initiation, propagation, and consequences of potential fire events for probability *vs.* time assessments. Event tree analysis (ETA) provides an inductive, forward logic framework that identifies a failure process, such as fire propagation. It has a major advantage of being able to incorporate time and sequential conditionality into a scenario. Sometimes, it may be weak in identifying specific details contributing to the consequences. In these cases, fault tree analysis (FTA) is used to supplement the event tree structure and provide detail to the top events leading to the final consequence. Cited references 3 through 5 provide information on structuring of and mathematical calculations for ETA and FTA.

Figures 11-11C and 11-11D are event trees for accidental flammable gas and liquid releases.[6]

STEP 3: CONSEQUENCE ASSESSMENT

Consequence modeling is the quantitative evaluation of expected consequences (undesirable results of an accidental fire or explosion) independent of frequency or probability. Quantitative consequence estimation can be used as a screening tool to identify and rank consequence categories as discussed in step 1.

The general consequence modeling concept involves assessment of source, pathway, and target, as illustrated in Figure 11-11E.

Consequence assessment involves evaluating two offsetting issues: (1) the susceptibility or vulnerability of people, or damage to targets (structures, equipment, etc.) being evaluated in the risk assessment project; and (2) the rate of development of a hazardous environment (intensity, distance, time) within the boundaries of the predicted hazardous event (gas dispersion, fire, explosion.) When these issues are evaluated quantitatively, they provide estimates of the loss vulnerability to selected targets *vs.* the predicted exposure from the fire or explosion scenario.

Target Vulnerability

Two general methods are used to assess the vulnerability of selected targets. The first method involves using various tables and references to establish generic threshold damage criteria for estimating exposure to people and property from fire and explosion intensities, radiant heat, and explosion overpressures. This "first-order" level of vulnerability evaluation provides a good approximation; however, it is weak in specific time relationships. Since time is usually a very uncertain variable, many risk estimation projects have employed generic threshold damage levels.[6]

The second method applies sophisticated heat transfer or structural impact models to support vulnerability assessments where time or specific conditions (design, layout, emergency response) warrant this level of detail, usually termed "second-order vulnerability assessment." *The SFPE Handbook of Fire Protection Engineering*[2] provides additional information on this subject.

Various tables have been developed to estimate exposure to people and property from fire. Sometimes they are expressed as radiant heat intensity and sometimes as a power dosage. The effect on people is expressed as the probability of death and different degrees of injury for different levels of heat radiation. (See Table 11-11A.)

Vapor cloud explosion exposures can be generically correlated with the energy of the explosion. This correlation can relate distances to various levels of building and equipment damage and exposure to people. (See Table 11-11B.)

The general method for assessing damage and human exposure for flash fires and torch fires is to estimate a level of structure and equipment damage and to assume fatalities would occur within the flame envelope of the incident.

Source Pathway Modeling

A fire or explosion consequence estimation model is a mathematical representation. Using computers helps the "number crunching"

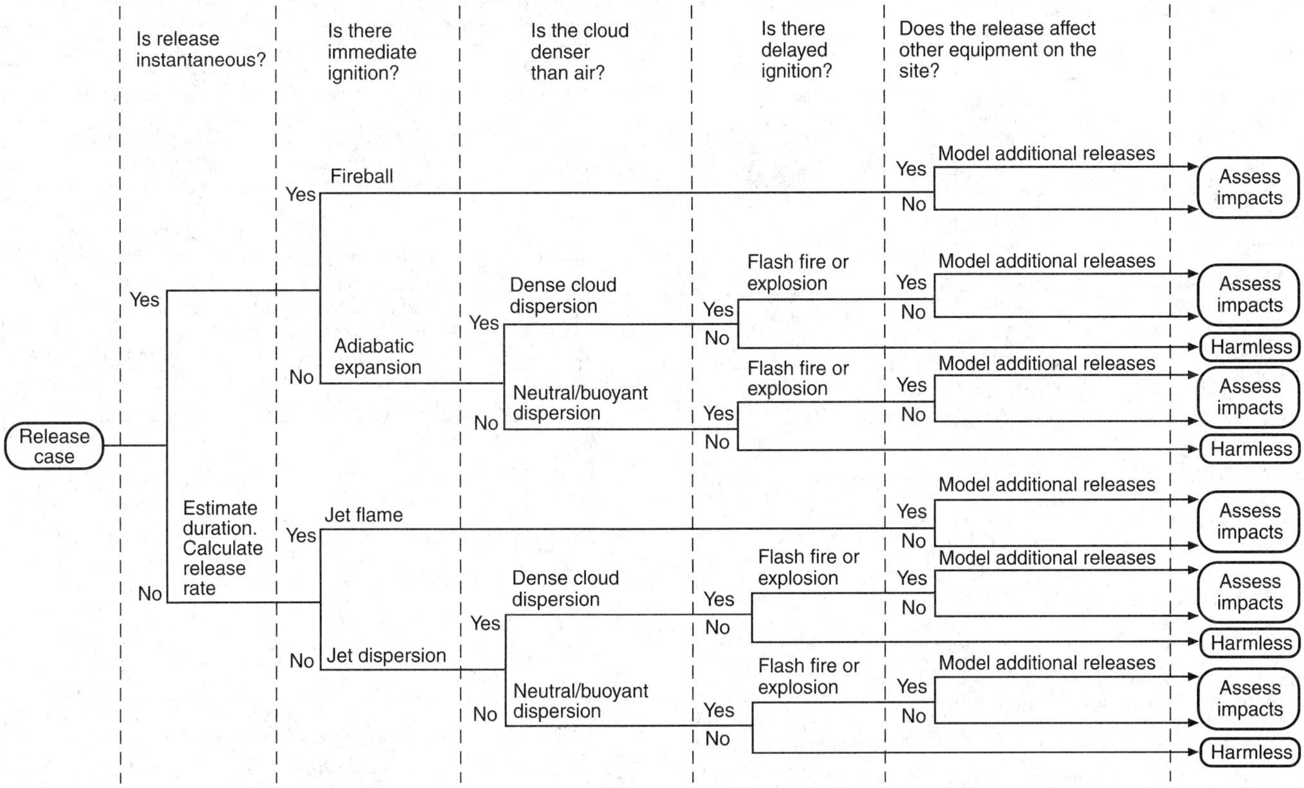

FIG. 11-11C.   *Flammable gas event tree.*[6]

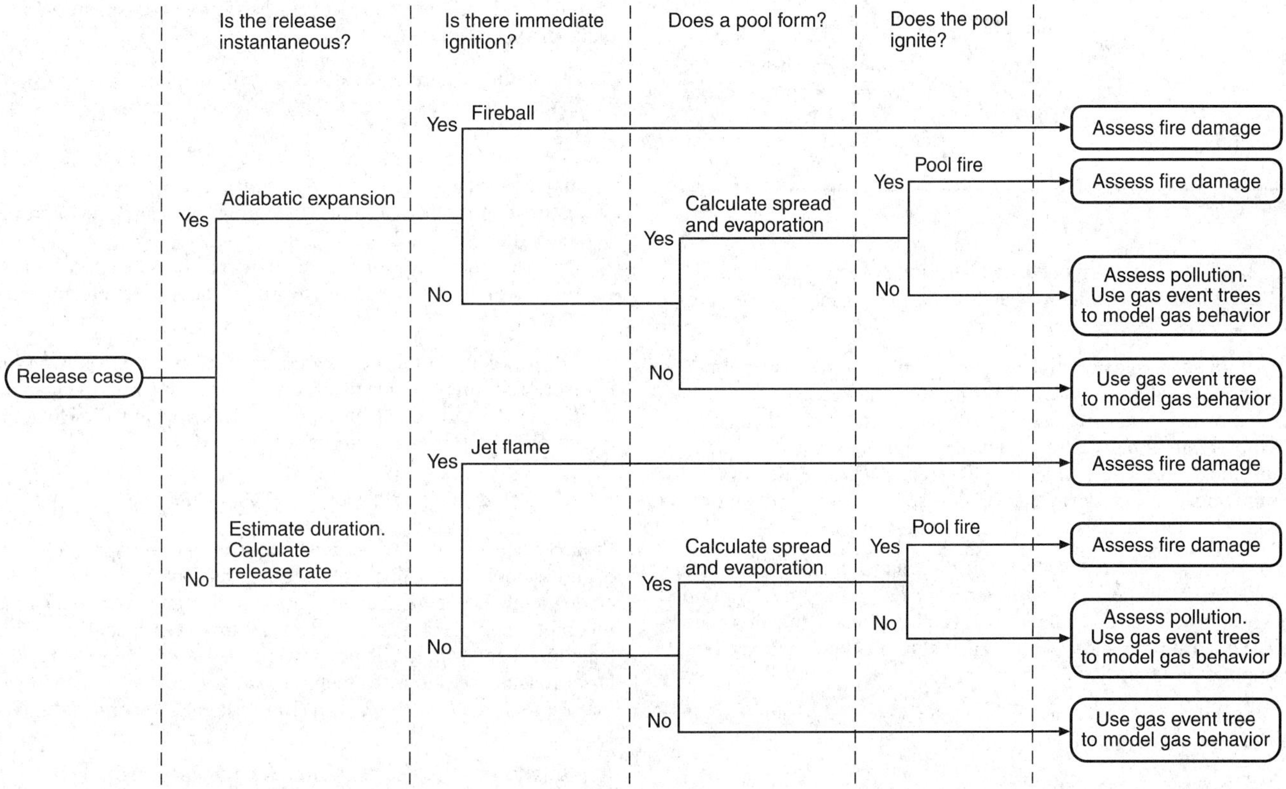

FIG. 11-11D.   *Flammable liquid event tree.*[6]

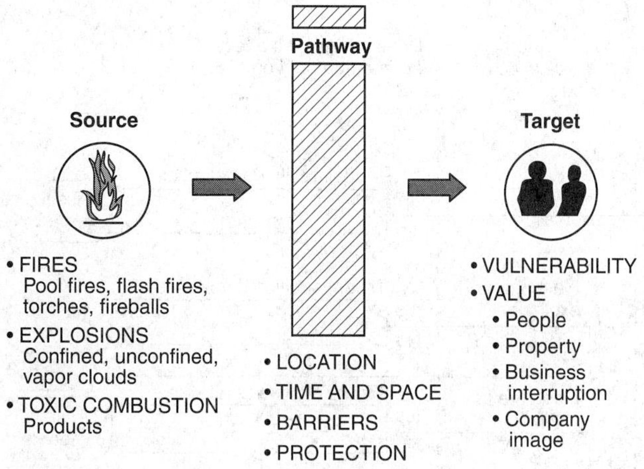

FIG. 11-11E. *Source—pathway–target consequence modeling concepts.*

TABLE 11-11A. *Damage Caused at Different Incident Levels of Thermal Radiation* [6]

Incident Flux (kW/m²)	Type of Damage Caused	
	Damage to Equipment	Exposure to People
37.5	Damage to process equipment	100% lethality in 1 min; 1% lethality in 10 sec
25.0	Minimum energy to ignite wood at indefinitely long exposure without a flame	100% lethality in 1 min; significant injury in 10 sec
12.5	Minimum energy to ignite wood with a flame; melts plastic tubing	1% lethality in 1 min; first-degree burns in 10 sec
4.0		Causes pain if duration is longer than 20 sec, but blistering is unlikely
1.6		Causes no discomfort for long exposure

TABLE 11-11B. *Explosion Overpressure Damage Estimates* [6]

Over-pressure (psig)	Characteristic Damage	
	To Equipment	To People
2.5–5	Heavy damage to buildings and to process equipment	1% death from lung damage > 50% eardrum rupture > 50% serious wounds from flying objects
1–2.4	Repairable damage to buildings and damage to the facades of dwellings	1% eardrum rupture; 1% serious wounds from flying objects
0.5–1	Glass damage	Injury from flying glass
0.15–0.30	Glass damage to about 10% of panes	Slight injury from flying glass

For SI unit: 1 psig = 6.895 kPa.

involved. The user must understand the fire and explosion dynamics involved and the assumptions and limitations of the models being used to assess specific consequences.

Fire and explosion modeling tools can be used to supplement engineering judgment when analyzing potential loss events for time to damage and potential areas of damage. This will enhance the credibility of property damage, business interruption loss estimates, and human life safety exposure. This can open avenues of "risk communication" and provide decision support in prioritizing and obtaining approval for risk reduction improvements.

Deterministic models have been developed to support fire and explosion intensities, and more sophisticated models are continuously being refined and validated. Several chapters within this handbook describe fire and explosion modeling methods and available computer programs.

Available consequence computer models[7–9] generally include the capabilities to estimate or evaluate:

• Discharge rates of liquid or gas

• Area of liquid pools

• Vaporization rate of liquid pools

• Toxic vapor dispersion hazards

• Pool fire radiation hazards

• Fireball radiation hazards

• Flame jet hazards

• Vapor cloud or plume fire hazards

• Vapor cloud explosion hazards

• Tank overpressurization rupture and fragmentation hazard

• Solid or liquid explosion overpressure hazards

In addition, various analytical assessment methods described in this handbook and *The SFPE Handbook of Fire Protection Engineering*[2] can be applied to consequence assessment and include methods for:

1. Estimating heat release rates and flame heights;
2. Estimating exposure temperatures from uncontrolled fires, using fire plume modeling equations;
3. Estimating the time to critical damage thresholds for specified targets, using heat transfer approaches;
4. Estimating the response of existing fire protection or improvement strategies; and
5. Applying the results of fire modeling efforts to support loss estimates and the benefits of recommended fire protection improvements.

Effort required for a consequence assessment depends on the number of different accident scenarios being analyzed, the number of effects the accident sequence produces, and detail with which the effects on the targets of interest are estimated.

STEP 4: PROBABILITY ESTIMATION

Probability estimations must be made using a team approach. The team assembled to support the fire risk analyst generally will include members experienced in process design and instrumentation, reliability engineering, safety and environmental concerns, and team members from the plant knowledgeable in the specific facility hazards and operations. Sometimes, the services of a statistician may be required to analyze data from insurance reports and injury logs.

In selecting fire and explosion event frequencies and probabilities, an experienced risk estimation team will initially use failure frequencies and probabilities based on generic industrywide data.

The team will need to provide supporting rationale for cases in which frequencies in other ranges are used. These variations or adjustments from generic industrywide data must often be made to reflect specific experience at a facility or the quality of the facility's loss prevention and safety management programs.

Two basic approaches are commonly employed in fire and explosion risk assessments to estimate initiating event frequencies and conditional probabilities:

1. Use of relevant historical data; and
2. Synthesis of event frequencies and probabilities using techniques, such as fault trees, human reliability analysis, and expert engineering judgment. (See references 1 and 3 for further information on these methods.)

Often the two approaches are used complementarily to provide an independent check on one another and to increase the validity and confidence in the results.

The first approach examines: (1) relevant historical data to assess the frequency with which these events have occurred in the past, and (2) the likelihood of their occurrence in the future. Where sufficient relevant past data is available, historically based frequencies may be adequate in making a reasonable assessment of fire and explosion risks. However, frequencies derived in this way represent only average values and are most applicable to simple systems where few variables (i.e., propagation and mitigation factors) can significantly change the results.

In actual application, assessments of potential fire and explosion risks at a specific facility will usually require adjustments to historical data to reflect the particular facility management and protection deviations. When evaluating the relevance of historical data for use in a specific facility or process risk assessment, the following factors should be taken into account:

1. For many types of hazardous industrial operations, historical data to make confident predictions about consequence likelihoods is limited or not available.
2. The general distribution between the primary causes leading to fire explosions shows that human error is the leading cause:
   (a) Human factors (e.g., management controls, human operator errors): 70 to 85 percent,
   (b) Equipment failures: 10 to 20 percent, and
   (c) External factors (e.g., earthquakes, floods): 10 to 20 percent.

As most causes are associated with human factors, it would be expected that most risk research would be in this area; however, the quantification of the human element is very difficult and much work still needs to be done.

3. Fire prevention and protection technology is continually being expanded and updated, based on new codes, standards, and industry practices developed after loss occurrences, and continued fire and explosion research efforts.

Considering these factors, the fire risk analyst must take a systematic approach to first determine the best approach for estimating frequencies and probabilities, based on available data and specific plant conditions.

Historical Accident Data

An important part of fire and explosion risk estimation is a review of the history of loss incidents similar to the hazard being analyzed. A review of the available information on loss incidents or the available loss trending data provides:

1. A relative breakdown of consequential effects, as to type of fire or explosion or as to resulting damage (can generally be used for estimating conditional event probabilities);
2. Identification of representative or dominant failure modes [equipment related, human error, system(s) related] that have led to fire or explosion accidents;
3. Identification of ignition sources and fire propagation factors;
4. Information concerning the duration of the fire and the general effect of loss mitigation factors; and
5. Information to support the generation of credible fire and explosion incident loss scenarios and the structuring of event tree analysis.

Data bases are the information foundation of hazard analysis and risk estimation. One of the most effective ways to learn if a system has a fire or explosion potential is to review past loss incident records. Accident data from specific plant operations (if available) are usually the best source and probably more accurate for specific equipment and operations, since the data reflect the operating and maintenance practices of the specific facility. Fire records of NFPA and American Petroleum Institute (API) provide fire loss incident data for many processes, plants, and types of equipment. Federal and state agencies also collect a variety of data related to safety and loss prevention issues. A few loss incident data sources are listed in Table 11-11C.

Equipment Failure Data Sources

An excellent source of generic equipment failure data is CCPS *Guidelines for Process Equipment Reliability Data, with Data Tables.*[10] The failure data can also be obtained as a computerized data base on a diskette. This book provides a listing and general description of all the data sources used to construct the CCPS generic failure rate data base. It is essential for the fire risk analyst to understand the sources of this data, the limitations and applicability to the particular environment, and management controls for the specific facility being assessed. This reference provides a good summary of these concerns.

Human Error Probability

Human failure rate data is very limited; however, some data have been developed on the reliability of human judgment.[1,11,12] The data show that the reliability of human judgment is greatly influenced by the stress level that accompanies the decision-making process. At very low stress levels, the probability of human failure is high because the operator may be bored and inattentive. At very high stress levels, the probability of failure is also high because fear and anxiety interfere with the operator's judgment. Human failure rates are lowest when the task is sufficiently interesting to keep the operator alert and when the operator does not feel endangered. Table 11-11D provides a generalized range of human error potential, based on some human reliability data.

The general procedure in applying human error probabilities includes:

1. Select the human error probability range from relevant data sources (see Table 11-11D); and
2. Select and justify a probability within that range, based on plant-specific knowledge and engineering evaluation of the facility and the human operation factors[1] that can include:

- Types of tasks
- Environmental conditions
- System elements and characteristics
- Type of systems

TABLE 11-11C. *Some Sources of Information for Use in Fire Risk Assessments*

Source	Nature of Information
NFPA (National Fire Protection Association) Fire Analysis and Research Division 1 Batterymarch Park, P.O. Box 9101 Quincy, MA 02269-9101 National Transportation Safety Board (NTSB) U.S. DOT Washington, DC	Fire incident data FIDO (Fire Incident Data Organization) NFIRS (National Fire Incident Reporting System-based analysis) Various occupancy reports, research, statistics Accident reports A detailed report is produced for transportation accidents involving hazardous materials Hazardous materials accident spill maps These give a map showing the location of the spill, any airborne plume, site of fatalities, and injured people, at one or more times after the start of the incident
American Petroleum Institute (API) Publications and Distribution 1220 L Street, NW Washington, DC 20005	Annual summaries of petroleum industry loss incidents Loss incidents
Association of American Railroads/Federal Railroad Administration (AAR/FRA) U.S. Environmental Protection Agency (EPA) 401 M Street, SW Washington, DC 20460	Railroad facts (annual editions) Accident/incident bulletins (annual) Reports on various aspects of hazardous material release incidents
U.S. Department of Transportation (DOT) Research and Special Programs Administration Office of Pipleline Safety Washington, DC	Pipeline leak reports for onshore gas transmission and gathering lines, and liquid lines
Gas Research Institute (GRI) 8600 West Bryn Mawr Avenue Chicago, IL 60631 American Institute of Chemical Engineers (AIChE) 345 East 47th Street New York, NY 10017	Failure rate data for 21 system categories, including cryogenic valves, heat exchangers, fire protection primarily associated with LNG facilities *Guidelines for Process Equipment Reliability Data with Data Tables*, and other excellent books on risk assessment
National Technical Information Service (NTIS) U.S. Department of Commerce P.O. Box 100955 Atlanta, GA 30384	Excellent source for out-of-print documents, government documents, and foreign articles and documents

TABLE 11-11D. *Some Typical Human Operator Error Rates*

Activity	Error Rate
Error of omission/item embedded in procedure*	3×10^{-3}
Simple arithmetic error with self-checking*	3×10^{-2}
Inspector error of operator oversight*	10^{-1}
General rate/high stress/dangerous activity*	0.2–0.3
Checkoff provision improperly used†	0.1–0.9 (0.5 avg.)
Error of omission/10-item checkoff list†	0.0001–0.005 (0.001 avg.)
Carryout plan policy/no check on operator†	0.005–0.05 (0.01 avg.)
Select wrong control/group of identical, labeled, controls†	0.001–0.01 (0.003 avg.)

*WASH-1400 (NUREG-75/014), "Reactor Safety Study—An Assessment of Accident Risks in U.S. Commercial Nuclear Power Plants," 1975.[13]
† See cited reference 12.

- Quality of human engineering of controls and displays

- Motivation

- Level of perceived psychological stress

- Skill and training

- Presence and quality of written instructions and methods

Fire Suppression Failure Probability

Fire suppression system assessment involves evaluation of three factors: (1) availability, (2) reliability, and (3) effectiveness. Failure to consider these aspects will lead to lack of confidence or incorrect decisions. Typical problems in many fire risk estimations result from two major mistakes:

1. System suppression success is only associated with the system being available. This approach assumes perfect operational reliability and perfect suppression effectiveness. Neither of these conditions are realistic.

2. Time is not properly integrated into fire suppression success probability. This includes the time at which damage starts *vs.* time at which system automatically operates, and time delays involving manually actuated suppression systems or manual application of an extinguishing agent.

Figure 11-11F presents an example of a partial fault tree structure for a water-based fire suppression system. Cited references 1 and 3 through 5 provide information on the structuring of fault trees and the Boolean algebra used to calculate failure probabilities.

A distinct advantage of applying an analytical fault tree approach is that it can be made highly facility or process specific. Another advantage is that it can suggest the relative likelihood of fire protection options as they affect the consequential outcomes and, thus, identify a framework for risk reduction by identifying the most beneficial improvement for optimal fire protection strategies. Documenting the probability selection basis and uncertainty to support

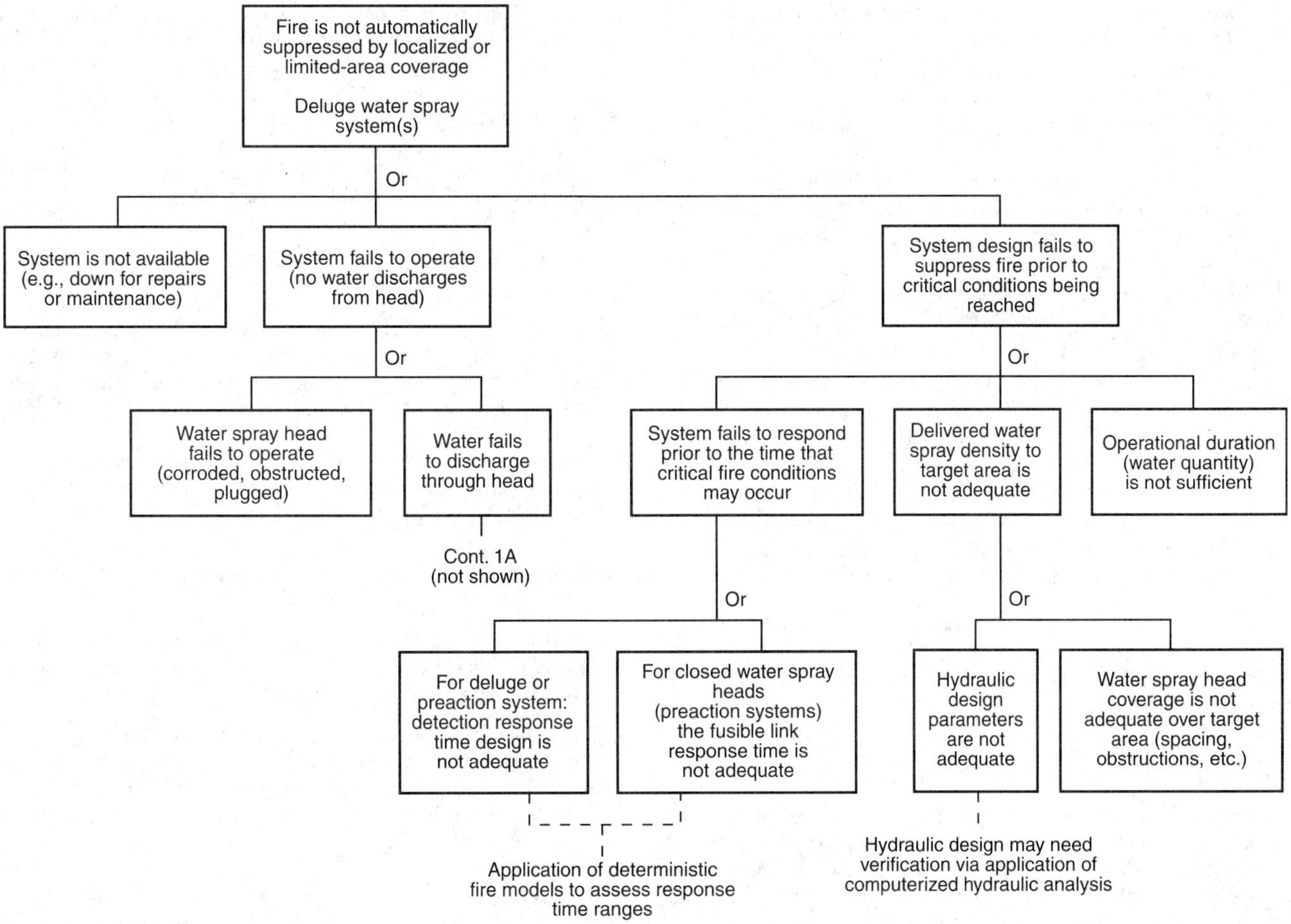

FIG. 11-11F.   *Example of a partial fault tree structure.*

the credibility of the risk assessment process is essential. Developing probability data for protection system fault trees requires a large degree of engineering judgment.

Integration of Deterministic Fire Models

As identified at the bottom of Figure 11-11F, often deterministic fire models can be used to support conditional probabilities on the success or failure of existing fire systems or alternative design strategies. Modeling times to fire detection and suppression activation can support and validate engineering judgments concerning conditional probabilities of whether a fire system will operate and respond before a critical threshold damage is reached at the target. (See Figure 11-11G.)

Fire dynamics and modeling are discussed in several other chapters in this handbook, and will not be elaborated on further in this chapter. Also, *The SFPE Handbook of Fire Protection Engineering*[2] provides additional information on this subject.

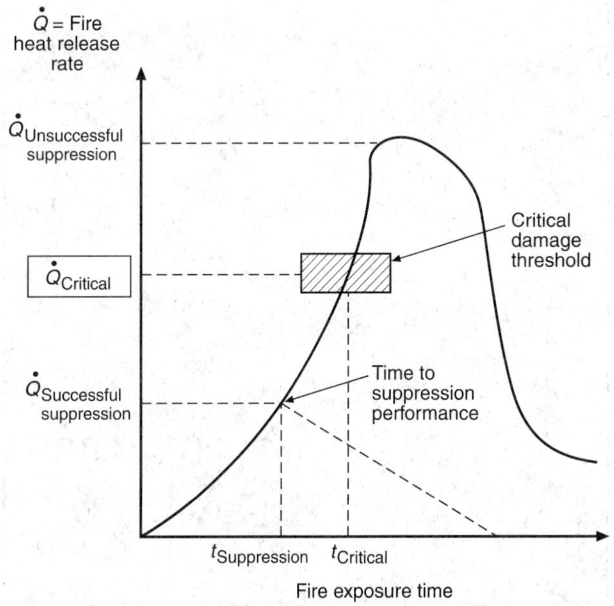

STEP 5: RISK PRESENTATION AND COMPARISON

It is very important to present fire and explosion incident risks into a presentation format that will be easy to interpret and use by fire risk management decision makers. The following factors should be considered in deciding which risk presentation forms are chosen.[1]

FIG. 11-11G.   *Example of modeling fire suppression response effectiveness.*

1. *User requirements.* The user may have a specific need to see risk estimate in a certain format.

2. *User knowledge.* Where the user is unfamiliar with presentation formats, sample formats should be presented to and approved by the user before any effort is made to secure approval for the scope of work.

3. *Effectiveness of communicating results.* It is vital that the presentation adequately express the results. The presentation should be as simple as necessary to ensure comprehension, but not so simple that resolution is lost.

4. *Need for comparative presentations.* Presenting comparisons of the results of a study with other risk estimations or risk data may be desirable. This type of presentation might involve:

   (a) A comparison of alternate design, operation, or protection options.

   (b) A comparison of the current risk estimates with risk estimates of other similar systems studied previously, to highlight areas for risk reduction or further study.

   (c) A comparison of risk estimates with other voluntary and involuntary risks, to rank the current risk estimate among these reference values.

The purpose of risk presentation is to communicate risk to the decision-makers effectively. Further information is available on various risk presentation methods (e.g., individual risk graphs, risk contours, societal risk F-N curves, etc.).[1,3,14,15]

Risk Contours

Individual risk exposure is presented in many risk estimation studies as a risk contour plot. (See Figure 11-11H.) Developing a risk contour (sometimes called a risk isopleth) consists of structuring a closed-line graphical depiction connecting lines of constant risk. Points within the contour represent a risk greater than or equal to the risk of the contour edge. Risk contour plots provide a good way of illustrating individual risk *vs.* distance from defined fire or explosion incidents.

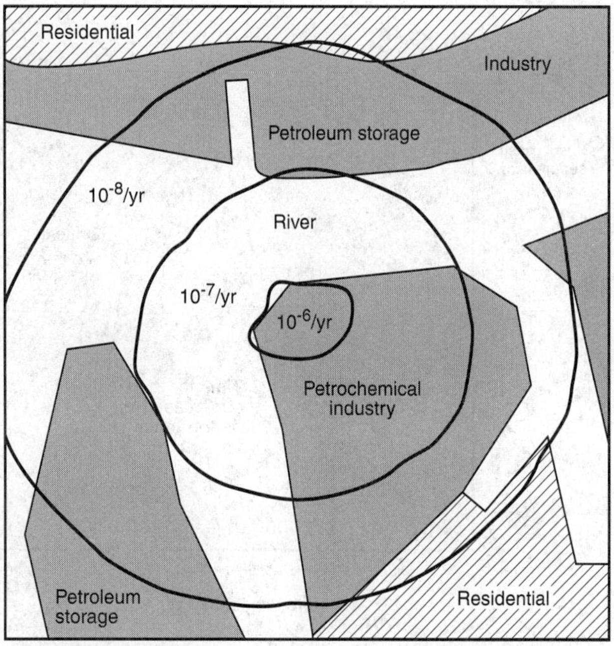

FIG. 11-11H.   Example of an individual risk contour plot.

Individual risk contours show the geographical distribution of individual risk. The risk contours show the expected frequency of fire or explosion loss events capable of causing a specified level of harm to an individual at a specified location. This is regardless of whether or not anyone is present at that location to suffer that harm.

Risk contours can also be used to graphically depict property damage risk from fire and explosion overpressures.

Risk Communication Profiles

Graphical risk profiles can be used to present the "risk difference" between alternative designs or various risk reduction options. Corporate risk managers and insurance companies are using risk profiles to communicate the "risk quality" of facilities and plant sites.

Many organizations use the notion of loss expectancy analysis to specify normal, most probably, and maximum loss potentials. Although definitions may vary, in general terms:

1. *Normal Loss Expectancy (NLE):*   Loss expected with all existing protection systems in service and working.

2. *Probable Maximum Loss (PML):*   Loss expected with the primary automatic protection system out of service (typically this is the automatic sprinkler system).

3. *Maximum Foreseeable Loss (MFL):*   Loss expected with automatic protection out of service and a delay in manual fire-fighting efforts (i.e., delayed or no fire department response). Loss limitation is primarily by passive means only (i.e., fire barrier walls, etc.).

The focus of many of these studies is more toward the severity of the expected loss. As an example, two plants are evaluated and result in the loss expectancy levels as shown in Table 11-11E.

From a severity viewpoint, the two plants seem similar in terms of potential property damage exposure from fire or explosion. However, from a "risk-based" evaluation that differentiates the likelihood of the three loss expectancy levels, Plant No. 2 represents a much lower risk. Figure 11-11I, is a simplified risk management communication matrix that shows this result.

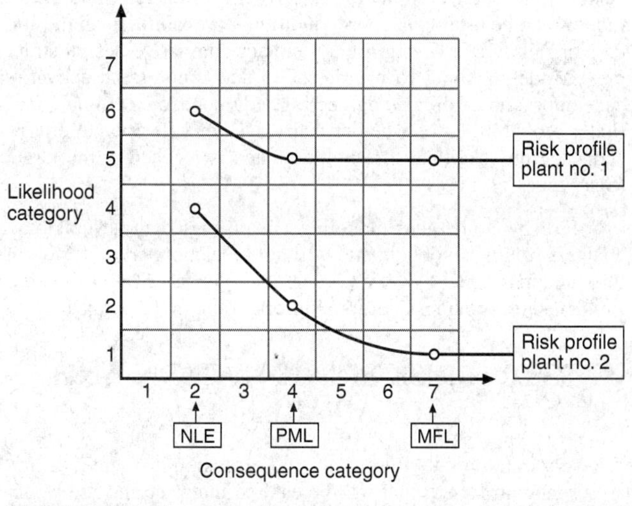

FIG. 11-11I.   Example of loss expectancy risk communication profile.

TABLE 11-11E. *Example Loss Expectancy Levels*

Loss Expectancy	Consequence Category	Likelihood Category	
		Plant No. 1	Plant No. 2
NLE $500,000	2	6	4
PML $2,000,000	4	5	2
MFL $25,000,000	7	5	1

BIBLIOGRAPHY

References Cited

1. Center for Chemical Process Safety, *Guidelines for Chemical Process Quantitative Risk Analysis,* American Institute of Chemical Engineers, New York, 1989.
2. DiNenno, P. J., ed., "Fire Risk Analysis," *The SFPE Handbook of Fire Protection Engineering*, 2nd ed., National Fire Protection Agency, Quincy, MA, 1995.
3. Center for Chemical Process Safety, *Guidelines for Hazard Evaluation Procedures with Examples,* 2nd ed., American Institute of Chemical Engineers, New York, 1992.
4. *Systems Safety Analysis Handbook,* The System Safety Society, Albuquerque, NM, 1993.
5. Vesely, W. E., Goldber, F. F., Roberts, N. H., and Hasl, D. F., *Fault Tree Handbook* (NUREG-0492), U.S. Nuclear Regulatory Commission, Washington, DC, 1981.
6. Technica Ltd., *Techniques for Assessing Industrial Hazards,* World Bank Technical Paper No. 55, 1988, The World Bank, Washington, DC.
7. *Handbook of Chemical Hazard Analysis Procedures* (ARCHIE Manual), Federal Emergency Management Agency, Washington, DC, 1989.
8. *Chemical Engineering Progress*, Software Directory of Hazard and Risk Computer Models, American Institute of Chemical Engineers, New York, published annually.
9. *Computer Systems for Chemical Emergency Planning*, Chemical Emergency Preparedness and Prevention Technical Assistance Bulletin #5, U.S. Environmental Protection Agency, Washington, DC, 1989.
10. Center for Chemical Process Safety, *Guidelines for Process Equipment Reliability Data, with Data Tables*, American Institute of Chemical Engineers, New York, 1989.
11. Meister, D., "Human Factors in Reliability," *Reliability Handbook*, W. G. Ireson, ed., McGraw-Hill, New York, 1966, pp. 12.2–12.37.
12. Swain, A. D., Guttmann, H. E., *Handbook of Human Reliability Analysis with Emphasis on Nuclear Power Plant Applications*, Report NUREG/CR-1278 (draft), 1983, United States Nuclear Regulatory Commission, Washington, DC.
13. Wash-1400 (NUREG-75/014), "Reactor Safety Study—An Assessment of Accident Risks in U.S. Commercial Nuclear Power Plants," 1975.
14. Whithers, J., *Major Industrial Hazards: Their Appraisal and Control*, Halsted Press, New York, 1988.
15. Covello, V. T., Sandman, P. M., and Slovic, P., *Risk Communication, Risk Statistics, and Risk Comparisons: A Manual for Plant Managers*, Chemical Manufacturers Association, Washington, DC, 1988.

NFPA Codes, Standards, and Recommended Practices

Reference to the following NFPA codes, standards, and recommended practices will provide further information on simplified fire hazard and risk estimation discussed in this chapter. (See the latest version of *The NFPA Catalog* for availability of current editions of the following documents.)

NFPA 49, *Hazardous Chemicals Data*
NFPA 68, *Guide for Venting of Deflagrations*
NFPA 325, *Guide to Fire Hazard Properties of Flammable Liquids, Gases, and Volatile Solids*
NFPA 491M, *Manual of Hazardous Chemical Reactions*

Additional Reading

American Petroleum Institute, *Recommended Practice 750*, Management of Process Hazards, 1st ed., Washington, DC, 1990.
Anderson, C. P., "Methodology for the Use of HAZARD I in Residential Fire Investigation," Worcester Polytechnic Institute, MA, Thesis, Dec. 14, 1990.
Babrauskas, V., and Peacock, R. D., "Heat Release Rate: The Single Most Important Variable in Fire Hazard," *Fire Safety Journal*, Vol. 18, No. 3, 255–272, 1992.
Beck, V. R., "Fire Safety System Design Using Risk Assessment Models: Developments in Australia," *Proceedings* of the Third International Symposium on Fire Safety Science, Elsevier Applied Science, New York, 1991, pp. 45–59.
Beck, V. R., and Yung, D., "Cost-Effective Risk-Assessment Model for Evaluating Fire Safety and Protection in Canadian Apartment Buildings," *Journal of Fire Protection Engineering*, Vol. 2, No. 3, 1990, pp. 65–74.
Brandyberry, M. D., and Apostolakis, G. E., "Fire Risk in Buildings: Frequency of Exposure and Physical Model," *Fire Safety Journal*, Vol. 17, No. 5, 1991, pp. 339–361.
Brandyberry, M. D., and Apostolakis, G. E., "Fire Risk in Buildings: Scenario Definition and Ignition Frequency Calculations," *Fire Safety Journal*, Vol. 17, No. 5, 1991, pp. 363–386.
Brannigan, V., and Meeks, C., "Computerized Fire Risk Assessment Models: A Regulatory Effectiveness Analysis," *Journal of Fire Sciences*, Vol. 13, No. 3, May/June 1995, pp. 177–196.
Bretherick, L., *Handbook of Reactive Chemical Hazards*, 4th ed., Butterworths, London, 1990.
Budnick, E., "Systematic Approach for Fire Risk Assessment," VHS Video Tape, Hughes Associates, Inc., Columbia, MD, Federal Fire Forum, Current Technical Fire Issues, Nov. 1, 1993, Gaithersburg, MD, 1993.
Cappuccio, J. A., "Development of Analytical Techniques for Risk Management Training," Worcester Polytechnic Inst., MA, Thesis, Jan. 8, 1992.
Carter, H. R., "Fire Risk Analysis," *Firehouse*, Vol. 21, No. 7, July 1996, pp. 128–129.
Casada, M. L., *et al.*, "Facility Risk Review as an Approach to Prioritizing Loss Prevention Efforts," *Chemical Engineering Progress*, 1990.
Center for Chemical Process Safety, *Plant Guidelines for Technical Management of Chemical Process Safety*, American Institute of Chemical Engineers, New York, 1992.
Chemical Manufacturers Association, *Evaluating Process Safety in the Chemical Industry—A Manager's Guide to Quantitative Risk Assessment*, CMA, Washington, DC, June 1989.
Chow, W. K., "Use of the ARGOS Fire Risk Analysis Model for Studying Chinese Restaurant Fires," *Fire and Materials*, Vol. 19, No. 4, July/Aug. 1995, pp. 171–178.
Dalzell, G., "Fire Hazard Management," *Fire Prevention*, No. 285, December 1995, pp. 27–30.
Dow Chemical Company, Fire and Explosion Index Hazard Classification Guide, 7th ed., AIChE, New York, 1989.
Forney, C. L., and Jones, W. W., "Fire Risk Assessment Method: Guide to the Risk Methodology Software. Internal Report. Final Report," National Institute of Standards and Technology, Gaithersburg, MD, National Fire Protection Assoc., Quincy, MA, NISTIR 4401, Sept. 1990.
Fowell, A. J., "Developments Needed to Expand the Role of Fire Modeling in Material Fire Hazard Assessment," National Institute of Standards and Technology, Gaithersburg, MD, DOT/FAA/CT-93/3; Federal Aviation Administration (FAA), International Conference for the Promotion of Advanced Fire Resistant Aircraft Interior Materials, Feb. 9–11, 1993, Atlantic City, NJ, 1993, pp. 255–262.
Greenberg, H. R., and Cramer, J. J., eds., *Risk Assessment and Risk Management for the Chemical Process Industry*, Van Nostrand Reinhold, New York, 1991.
Hall, J. R., Jr., "Progress Report on U. S. Research in Fire Risk, Hazard, and Evacuation," National Fire Protection Assoc., Quincy, MA, NISTIR 4449; Eleventh Joint Panel Meeting of the U. S./Japan Government Cooperative Program on Natural Resources (UJNR) on Fire Research and Safety, Oct. 19–24, 1989, Berkeley, CA, 1990, pp. 2–4.
Hall, J. R., *U.S. Experience with Sprinklers*, Fire Analysis Research Division Publication, National Fire Protection Agency, Quincy, MA, 1990.
Hall, J, R., Jr., "Key Distinctions in and Essential Elements of Fire Risk Analysis," *Proceedings* of the Third International Symposium on Fire

Safety Science, Elsevier Applied Science, New York, 1991, pp. 467–474.

Heikkila, L., "ARGOS, tulipaloriskien arviointimenetelma. [ARGOS, An Estimation Method for Fire Risks.]," *Palontorjuntatekniika*, March 1992, pp. 28–29.

Heikkila, L., "HAZARD I: Estimation Method for Fire Risks in Single-Family Homes," *Palontorjuntatekniikka*, January 1991, pp. 14–16.

Lewis, D. J., *Mond Fire, Explosion and Toxicity Index: A Development of the Dow Index*, 13th Annual Loss Prevention Symposium, Houston, American Institute of Chemical Engineers, New York, 1990.

Lorenzo, D. K., *A Manager's Guide to Reducing Human Errors—Improving Human Performance in the Chemical Industry*, Chemical Manufacturers Association, Washington, DC, July 1990.

Mundan, K., and Gustafson, R., "Ignition Potential Distribution for Heavy Gas Plumes," International Conference on Vapor Cloud Modeling, AIChE, New York, 1987.

National Fire Protection Association, *Industrial Fire Hazards Handbook*, 3rd ed., Cote, A. E., ed., NFPA, Quincy, MA, 1990.

Nelson, H. E., "FPETOOL: Fire Protection Engineering Tools for Hazard Estimation," National Institute of Standards and Technology, Gaithersburg, MD, General Services Administration, Washington, DC, NISTIR 4380, Oct. 1990, 120.

Nelson, H. E., "FPETOOL: Fire Protection Tools for Hazard Estimation: An Overview of Features," Fifth International Fire Conference on Fire Safety, Interflam '90, Sept. 3–6, 1990, Canterbury, England, Interscience Communications Ltd., London, England, 1990, pp. 85–92.

Noonan, F., and Fitzgerald, R., "On the Role of Subjective Probabilities in Fire Risk Management Studies," *Proceedings* of the Third International Symposium on Fire Safety Science, Elsevier Applied Science, New York, 1991, pp. 495–504.

Occupational Safety and Health Administration, 29 CFR 1910.119, *Process Safety Management of Highly Hazardous Chemicals*, OSHA, Washington, DC, 1992.

Pettitt, G. N., Harvey, B., and Worthington, D. R. E., "Use of Quantitative Risk Assessment as a Decision Making Tool," *PSAM-II Proceedings*, An International Conference Devoted to the Advancement of System-Based Methods for the Design and Operation of Technological Systems and Processes, Volume 1, Sessions 1–36, March 20–25, 1994, San Diego, CA, 1994, pp. 007/1–6.

Phillips, W. G. B., "Development of a Fire Risk Assessment Model," Fire Research Station, Borehamwood, England, CIB W14/91/01 (UK), BRE IP 08/92, May 1992; *Probabilistic Safety Assessment and Management*, Elsevier Science Publishing Co., Inc., 1991, pp. 299–306.

Prugh, R. W., "Evaluation of Unconfined Vapor Cloud Explosion Hazards," International Conference on Vapor Cloud Modeling, Cambridge, MA, American Institute of Chemical Engineers, New York, 1987.

Shortreed, J. H., and Stewart, A., "Risk Assessment and Legislation," *Journal of Hazardous Materials*, 20, 1988, pp. 315–334.

Slater, D. H., Pitblado, R. M., and Williams, J. C., "Quantitative Assessment of Process Safety Programs," *Plant/Operator Progress*, Vol. 9, No. 3, July 1990.

Snell, J. E., "Fire Hazard and Risk: Evaluating Alternative Technologies," Sprinklers in Residential and Commercial Buildings, Aug. 19–20, 1990, Charleston, SC, Natl. Conf. States on Bldg. Codes and Standards, 1990, pp. 1–23.

Soukas, J., and Kakko, R., "On the Problems and Future of Safety and Risk Analysis," *Journal of Hazardous Materials*, Elsevier Science, Publishers B. V., Amsterdam, 21, 1989, pp. 105–124.

Special Report—Consequence Modeling, Risk Management Program (RMP) Publications, Thompson Publishing Co., Washington, DC, 1995.

Special Report—Consequence Modeling, Risk Management Program (RPM) Publications, Thompson Publishing Co., Washington, DC, 1995.

State of California, Office of Emergency Services, *Guidance of the Preparation of a Risk Management and Prevention Program*, Sacramento, CA, 1989.

Stephenson, J., *System Safety 2000—A Practical Guide for Planning, Managing, and Conducting System Safety Programs*, Van Nostrand Reinhold, New York, 1991.

Stollard, P., "Use of Mathematical Fire Risk and Fire Safety Audits in Health Care Facilities," Fifth International Fire Conference on Fire Safety, Interflam '90, Sept. 3–6, 1990, Canterbury, England, Interscience Communications Ltd., London, England, 1990, pp. 323–329.

Takeda, H., and Yung, D., "Simplified Fire Growth Models for Risk-Cost Assessment in Apartment Buildings," *Journal of Fire Protection Engineering*, Vol. 4, No. 2, 1992, pp. 53–66.

Watts, J. M., Jr., "Criteria for Fire Risk Ranking," *Proceedings* of the Third International Symposium on Fire Safety Science, Elsevier Applied Science, New York, 1991, pp. 457–466.

Watts, J. M., Jr., "Fire Risk Evaluation Model. Software Review," *Fire Technology*, Vol. 31, No. 4, 4th Quarter, Nov. 1995, pp. 369–371.

Watts, J. M., Jr., "Fire Risk Ranking. Editorial," *Fire Technology*, Vol. 27, No. 3, Aug. 1991, pp. 193–194.

Watts, J. M., Jr., "Fire Risk Rating Schedules," Fire Safety Institute, Middlebury, VT, ASTM STP 1150; American Society for Testing and Materials, Fire Hazard and Fire Risk Assessment, Sponsored by ASTM Committee E-5 on Fire Standards, ASTM STP 1150, Dec. 3, 1990, San Antonio, TX, ASTM, Philadelphia, PA, 1990, pp. 24–34.

Yung, D., "Fire Risk-Cost Assessment Models," *IRC Fire Research News*, No. 58, April 1990, pp. 1–2.

Yung, D., and Beck, V. R., "Application of a Risk-Cost Assessment Model for the Evaluation of Fire Safety in Buildings," *Proceedings* of the First International Conference on Fire Science and Engineering, ASIAFLAM '95, March 15–16, 1995, Kowloon, Hong Kong, 1995, pp. 51–62.

APPLYING MODELS TO FIRE PROTECTION ENGINEERING PROBLEMS AND FIRE INVESTIGATIONS

Revised by Harold E. "Bud" Nelson and Richard L. P. Custer

Starting in the 1980s, computerized fire growth models were developed and applied to fire safety problems. In general, these models estimate changes to a building or space within a building and its environment as the result of a fire in that structure. For the most part, these models do not model fire but rather the impact of fire. A more extensive discussion of computer fire models and their elements is covered in other chapters in this section. (See Section 11, Chapter 4, "Introduction to Fire Modeling," and Section 11, Chapter 5, "Deterministic Computer Fire Models.") The purpose of this chapter is briefly to advise the user on some of the types of applications where modeling can be of assistance and provide some guidance for the user to establish his or her own criteria for selection of the model chosen.

As an outgrowth of the major advances in understanding the dynamics of fire and its effects on the operation of fire protection devices, workers in the field have developed a number of predictive models. These models range from prediction of upper layer gas temperature in a compartment fire[1,2] and the time of sprinkler head operation[3,4] to human response to the heat and toxic species produced.[5] Many of the models have, in turn, been assembled into larger consolidated models and hazard assessment tools[6-11] that provide a wide capability for the calculation and assessment of the effects of fire using today's state-of-the-art personal computers. (See Section 11, Chapter 7, "Fire Hazard Analysis"; Section 11, Chapter 10, "Simplified Fire Growth Calculations"; and Section 11, Chapter 11, "Simplified Fire Hazard and Risk Calculations.")

The reality of accidental fire is more varied and complex than can be exactly described by any existing collection of equations and data. Users should not expect that the model will produce an exact duplication of reality. Rather, the results of models should be considered as engineering estimates, valid to the degree that the physical relationships being calculated are those that actually dominate the situation and that the data entered into the model is accurate. Figures 11-12A through 11-12F demonstrate the type of approximation that should be expected. In Figure 11-12A, four different models and one simple correlation were used in an attempt to reproduce the results from a well-documented test of a fire in a room.* In Figure 11-12A, the curve using the letter *x* to indicate data points plots the average upper layer temperature computed from thermocouple readings during the test. The other curves graph the predictions by the various models. The complexity of the models varied widely in this exercise. The simplest model was a single correlation equation, while the most complex involved the coordination of almost a hundred variables. In the case of this test, the most sophisticated model was the most accurate, but other much less sophisticated models grouped in the same range. Figure 11-12B is a reproduction of the actual data obtained in the burn test. Figure 11-12B reports the actual measured temperatures detected by a thermocouple tree near the center of the room but away from the fire plume. Every third thermocouple is plotted. Each of the lines depicts the temperature recorded by one of the thermocouples. The thermocouples were mounted with one near the ceiling and the others at 6-in. (0.15-m) intervals, with the bottom thermocouple located about 6 in. (0.15 m) above the floor. Figure 11-12C plots the same information as Figure 11-12B but rearranges the data to show the temperature profile from floor to ceiling as 1-min intervals through the first 6 min of the test.

All of the models used to produce the results in Figure 11-12A are classed as zone models; i.e., they predicted a single "upper layer" (smoke) temperature. As shown in Figures 11-12B and 11-12C, an upper layer or "zone" of hot fire gases with an approximately equal temperature did exist. While the type of approximation produced by these models is often more accurate than previous methods of estimation, the results must still be recognized as an engineering approximation. Figure 11-12D plots the smoke layer interface calculated by the models with that observed by the investigators. Figure 11-12E compares the oxygen concentration measured in the smoke in the test to concentrations calculated by the models. Figure 11-12F compares the measured mass flow through the vent opening with the same flow as calculated by the models. These four measures (i.e., temperature, smoke layer interface, oxygen concentration, and mass flow through the vent opening) were chosen to represent the full range of model outputs.

---

*The fire test used for this comparison is test PRC17W4-$^1/_4$. This was one of a series of compartment fire tests conducted under the sponsorship of the Product Research Committee. Test PRC17W4-$^1/_4$ was conducted in a room 2.18 m[2] and 2.41 m high. The room had a single vent opening 0.19 m wide and 2 m high. The fuel for this test consisted of four wood cribs having a total weight of approximately 14 kg located near the center of the room.[12] (For conversion: 1 ft = 0.305 m; 1 sq ft = 0.0929 m[2]; and 1 lb = 0.4536 kg.)

Harold E. "Bud" Nelson is a senior research engineer at Hughes Associates, Inc., Baltimore, MD. Richard L. P. Custer is president of Custer Powell, Inc., Wrentham, MA.

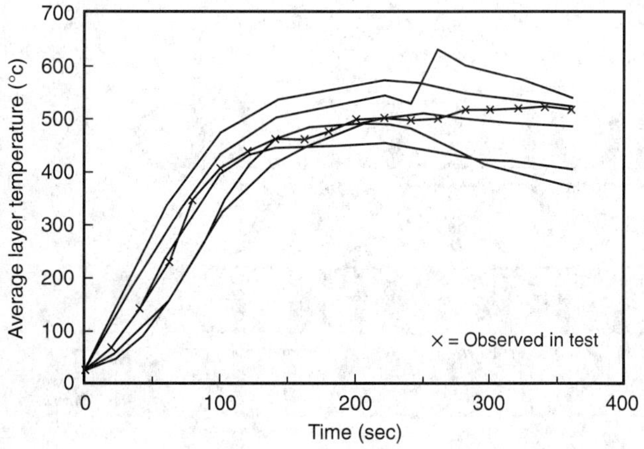

FIG. 11-12A. *Predicted layer temperature by zone models vs. test measurements.[2] (For conversion: °F = °C × 9/5 + 32.)*

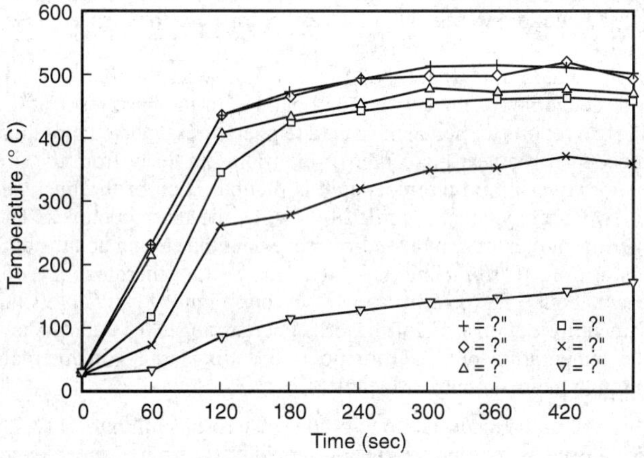

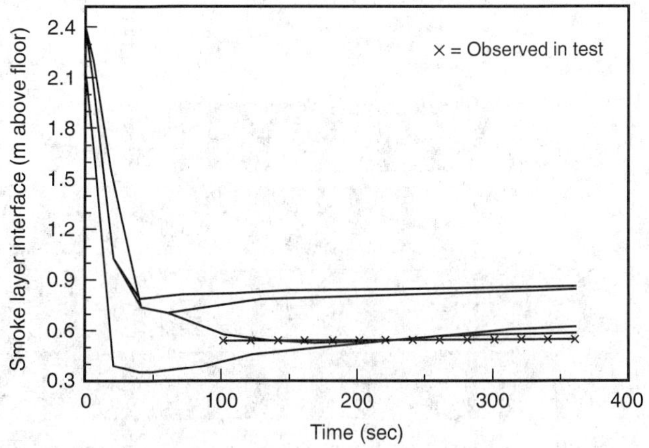

FIG. 11-12D. *Prediction of smoke layer by zone models vs. test measurements.[4] (For conversion: 1 ft = 0.305 m.)*

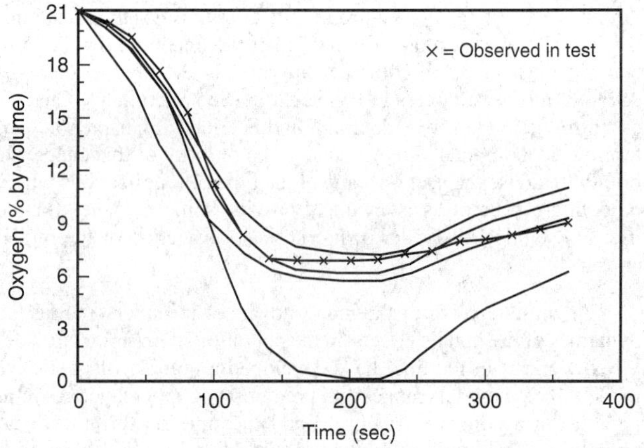

FIG. 11-12B. *Actual temperature plots at various distances below ceiling.[3] (For conversion: °F = °C × 9/5 + 32.)*

FIG. 11-12E. *Prediction of oxygen concentration by zone models vs. test measurements.[5]*

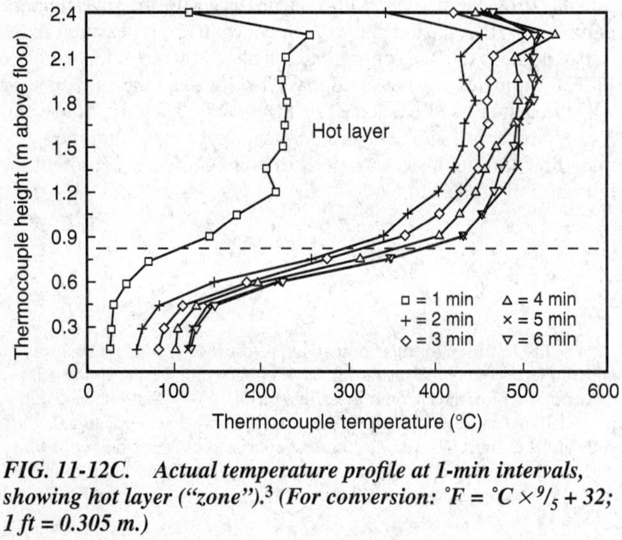

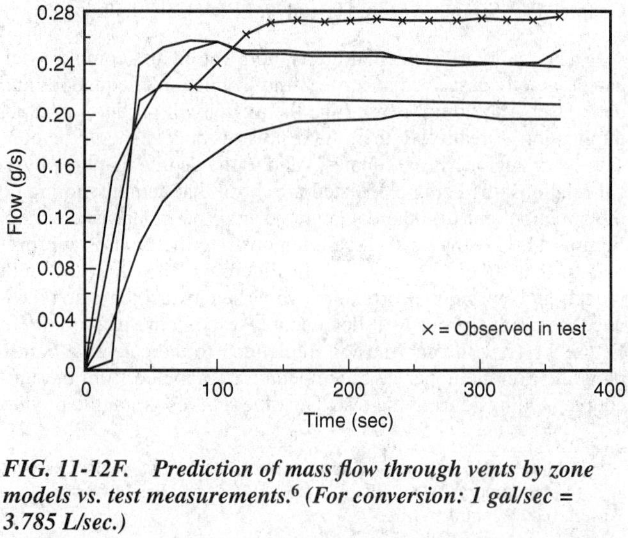

FIG. 11-12C. *Actual temperature profile at 1-min intervals, showing hot layer ("zone").[3] (For conversion: °F = °C × 9/5 + 32; 1 ft = 0.305 m.)*

FIG. 11-12F. *Prediction of mass flow through vents by zone models vs. test measurements.[6] (For conversion: 1 gal/sec = 3.785 L/sec.)*

DEFINITION OF FIRE GROWTH MODELS

The term *model* can be applied to many physical and mathematical procedures designed to simulate reality. For the purpose of this chapter, *fire growth models* are defined as mathematical procedures developed to estimate the change in the environment of a space or building caused by the existence of a fire in that space that varies in intensity and/or area of involvement with time. All of these models predict the temperature of the smoke layer produced by that fire. Most fire growth models also predict the depth of the smoke layer. In addition, many models predict other variables such as the oxygen, carbon monoxide, carbon dioxide, and other gas species concentrations in the smoke; the rate of flow of smoke, gases, and unburned fuel from one space to another; and the temperatures of the walls, ceiling, floor, and air not yet drawn into the fire. Some models also predict the time of smoke and heat detector activation (including sprinkler heads) and the effects of opening or closing doors, breakage of windows, or other physical events that take place during an accidental fire. Some models are limited to the room of fire involvement, others track the impact of a fire in that room through a series of rooms, and a few follow the spread of fire outside the room of origin.

Typically, the models involve the simultaneous solution of two or more differential equations. The more sophisticated models involve the simultaneous solution of such equations for many variables and use sophisticated techniques to obtain the results. The range of complexity and sophistication is wide, but all models follow the basic conservation principles of mass, energy, and momentum.

TYPES OF MODELS

As discussed more thoroughly in Section 11, Chapters 4 and 5, prediction models can generally be classed into the two categories of zone models that solve the basic conservation equation for distinct regions (control volumes), and computational fluid dynamic (CFD) or field models that solve the fundamental equations of mass, momentum, and energy in a space that has been divided into a grid of small-volume cells.

Zone Models

Most of the fire growth models that can be run on modern personal computers are *zone models*. These usually divide each room into two spaces or zones: (1) an upper zone that contains the hot gases produced by the fire and (2) a lower zone that contains a space beneath the upper zone that is the source of the air for combustion. Zone sizes change during the course of the fire. The upper zone can expand to occupy virtually all the space in the room.

Computational Fluid Dynamics (CFD) Models

Conversely, *CFD models* usually require large-capacity computer work stations or mainframe computers. By dividing the space into many small cells (frequently hundreds of thousands), CFD models can examine much greater detail than zone models. Where such detail is needed, it is often necessary to use the sophistication of a CFD model. In general, however, CFD models are much more expensive to use, require more time to set up and run, and often require a high level of expertise to make the decisions required in setting up the problem and interpreting the output produced by the model. The use of CFD models in fire protection problems, however, is increasing. CFD models are particularly well-suited for situations where the space or fuel configuration is irregular, turbulence is a critical element, or very fine detail is sought.

Hybrid Models

A *hybrid model* is defined as one that uses elements of both zone models and CFD models. It is possible that some hybrid models will appear while this handbook is still in use. However, at the time of this writing, none are available, and the level of sophistication that will be necessary to operate one properly is unknown. It is visualized that hybrid models will be able to use the simplicity of zone models for those spaces or portions of time in the fire where the general average calculations typical of a zone model are sufficient but switch to some type of smaller zones or CFD model cells when needed.

MODEL APPLICATIONS

In the broadest sense, a fire growth model can be applied as a tool in the solution of any fire protection problem where the fire-produced conditions inside a building or space or the response of fire protection devices in that space are needed as input to the decision. The following are some typical examples of application classes.

1. *Predicting a hazard.* In this case, *hazard* is defined as the expected fire-produced conditions. To start, a set of circumstances that represents either a specific case of interest or a design case representing the maximum expected fire is postulated. All of the conditions are fixed and the model is run to make a prediction of the rise in smoke (upper layer) temperature, the volume or depth of that smoke, the response of devices, the onset of serious conditions such as flashover, smoke gases or toxicity, or other items of concern. Often, this type of analysis searches for the most serious case and is used to answer questions such as, "How serious are the fire-produced conditions that can occur in this space?" Frequently, a number of case variations will have to be run to establish the most serious case. For example, consider an analysis of the hazard of a hotel room that contains a type of bed that fire tests indicate will grow, if ignited, at a moderate rate, eventually reaching a fire size of 2 megawatts (MW); a chair having a very fast fire-growth rate if ignited but a maximum rate of energy of less than 1 MW; and a number of pieces of furniture that would not be expected to be points of ignition but could join a fire at a later state. Discovery of the worst case might require a series of computer runs in which the ignition source was placed first on the bed and then on the chair, runs where the distance between the bed and the chair were varied, runs with the door open and the door closed, and other possible variations.

2. *Reconstructing a fire.* Fire models can be useful in reconstructing incidents that occur during a fire or testing suggested hypotheses or scenarios. In the perfect case, there is enough physical evidence, data on conditions at the time of the fire, and reliable witness statements to produce all of the information necessary to run the model to describe the course of conditions and the phenomena that did occur. In actual practice, only part of the necessary information is usually available, and running the model may in itself involve some rational hypotheses as to what the conditions were most likely to have been. Often, running the model will help demonstrate which hypotheses are plausible and which are not. In the case of the example of the hotel room, the comparison of the known data to the various scenarios may demonstrate that some would not have run the course known to have occurred. In some cases, a single dominant fire scenario will emerge. The use of computer modeling in fire reconstruction and failure analysis is of particular interest to many fire protection practitioners, and papers have begun to appear in the literature.[13,14] This topic is discussed in more detail next.

3. *Interpolating between or extrapolating beyond test results.* In many cases, fire tests that simulate a real-world condition have

been conducted to obtain data on an arrangement, protection system, product, or other material. In translating that data to real-world applications, the desired use frequently does not match the test array. In many cases, models can be used to appraise the impact of the difference between the test and reality. In such cases, the usual procedure is first to run the model in exactly the arrangement of the test to demonstrate that the model can reasonably approximate the results obtained in the test. In making the comparison between the model and the test result, the user should recognize that there can be a degree of inaccuracy or uncertainty in test results as well as in model predictions. Perfect matches between models and tests seldom occur. Once a reasonable reproduction of the test has been achieved, the model is rerun, changing those variables or parameters that differ between the test and reality. The less change, of course, the more dependable the results.

4. *Evaluating a parametric variation.* In all of the above examples, it is frequently necessary to answer the question, "What if a change were made?" Models provide excellent tools for such parametric studies. Most frequently, the changes would involve the change of a material (construction, finish, furnishings), a change of an arrangement (door opened or closed, air conditioner on or off, spacing variations between fuel packages), or the presence or absence of a protection system such as smoke detectors or sprinklers. Using the example of the hotel room, typical questions might include the change in hazard or risk that would occur should the mattress be changed from a moderate-burning mattress to one that is fast-burning or one that is treated to burn at a very slow rate. A similar question might relate to the use of a combustible wall material. In the case of a fire analysis, questions might relate to the difference in the situation had there been a sprinkler in the space, had the door been equipped with a smoke-operated door closer, or had some particular material been different. A first-order approximation of the contribution of an item of furnishing or a material can be made by exercising the model with and without that material present and noting the differences.

COMPUTER MODELING IN FIRE RECONSTRUCTION AND FAILURE ANALYSIS

Among the applications that have evolved for computer modeling has been the analysis of actual fire incidents after the fact to determine the most likely fire scenario or group of scenarios that could have produced the end result—a process called *fire reconstruction.* (See Section 11, Chapter 1, "Fire and Arson Investigation.") Often reconstruction includes an analysis of the role played by building fire protection features and how the fire might have progressed had additional fire protection measures been present, particularly when none existed at the time of the fire. *Failure analysis*[13] involves identification of the primary and contributing factors that caused the loss. Examples of factors that may result in failure include design, material or equipment selection, maintenance, and post-construction modifications. Traditionally, these activities have been carried out by field investigators based on physical evidence, witness statements, and the investigator's knowledge and experience, but with limited analytical tools.

Application to Reconstruction and Failure Analysis

One of the early attempts at application of the emerging analytical tools was carried out by Fleming[15] focusing on the role that sprinklers might have played had they been present in the Haunted Castle

of the Great Adventure Six Flags Amusement Park, Jackson Township, New Jersey. Fleming employed a combination of modeling and results of a variety of full-scale tests of manufactured home and compartment fires in reaching his conclusions. Since that time a number of analyses have been carried out of major fires using progressively more sophisticated tools and approaches[16–21] in fire reconstruction and failure analysis modeling.

Since each fire incident will be different, there will be no "standard" set of models and techniques to apply. The purpose of this section is to outline the process of collecting the necessary data and setting up the modeling "experiments." The reader is encouraged to study the referenced reports as examples for specific fires.

Before a modeling effort can begin, a great deal of information needs to be obtained about the incident, beginning with an accurate layout of the building at the time of the fire. One also needs to know about the location and position of doors and windows, the type of heating, ventilating, and air-conditioning (HVAC) system, and the size and location of stairways and other vertical openings between floors. Although it may not be possible, a visit to the scene is valuable. Photos of the building during and after the fire should be studied if available. Often, news videos can be obtained.

Information needs to be gathered about the materials of construction, including the interior finishes and the furnishings and their distribution. Sources for this information include building occupants or their relatives, building owners or managers, contractors, interior decorators, or others who may have been involved in the construction, renovation, or operation of the building.

Figure 11-12G is a form that may prove helpful in gathering the type of data needed to use compartment models or similar analytical methods.

Where there are fire victims, important information, such as location and severity of burn injuries or blood chemistry, can be obtained from hospital or autopsy reports.

Since fire growth and the effect of fire on the building and its systems and occupants is largely time dependent, it is essential to obtain as much information as possible regarding the timing of events leading up to ignition (pre-fire events), during the fire (trans-fire events), and after the fire (post-fire events).[13]

The best way to organize this information is through the use of timelines or event sequences. Benchmark events, those for which the time is known and not merely relative to a known event, are important to establish. If there are only a few events, a single timeline may suffice, otherwise multiple timelines or event sequences[22] on the same scale may be useful. The timelines are then used for comparison with the results of the modeling experiments.

Once the data have been gathered and organized, a modeling plan should be developed based on the objectives of the study. The objective may be to assess the validity of a given hypothesis for the origin and cause of a fire or to develop new alternative hypotheses. In this case, the questions to be answered are:

• Would the fuels, fire location, ventilation conditions, and building and occupant characteristics hypothesized results in the actual fire conditions?

• Which scenarios, given the range of available fuels, possible ventilation conditions, and building characteristics, could result in the actual fire conditions?

Another objective might be to determine whether or not someone might have been injured had a detector been in use. In this case, the question to be answered could be:

• For the set of scenarios that could result in the actual fire conditions, what conditions would the potential victim be exposed to, given the range of initial locations of the victim relative to

Data for Compartment Fire Modeling

Room Number _____   Use _____

Size (use diagrams if possible)   Wall/Floor/Ceiling Construction

    Length _____   _____

    Width _____   _____

    Height _____   _____

Lining Materials (that represent over 10 percent of room lining)

(Include thickness, density, and other material characteristics, if known)

    Wall Material   % of walls or area involved

_____   _____

_____   _____

_____   _____

Ceiling Material

_____   _____

_____   _____

Floor or Floor Covering Material

_____   _____

_____   _____

Doors, windows and other openings [Enter all heights as distance above floor. If door sill is at floor enter zero (0).]

Opening	To	Top	Sill	Width	Changed During Fire (How?)[1]
_____	_____	_____	_____	_____	_____
_____	_____	_____	_____	_____	_____
_____	_____	_____	_____	_____	_____
_____	_____	_____	_____	_____	_____

Heating ventilating and air conditioning (HVAC) Include air flows from HVAC systems. Give rates and positions of supply and return or exhaust in this room. Also sizes and types of ducts/diffusers.

_____

_____

_____

Tightness of walls, closed windows, door fits, etc. (Unless fit is very loose, classify fit as tight, average, or loose. If fit is very loose try to get size, number & location of cracks, etc.)

    Doors _____

    Windows _____

    Inside walls _____

    Exterior walls _____

[1]For example: "window broke at 10:33" or "door was closed until opened by escaping occupant, then left open—Exact time 10:30."

FIG. 11-12G.   A sample compartment fire modeling data form.

the fire, the conditions at the time of operation of the detector, and the actions and route taken during an escape attempt?

Using the questions as a guide, a list of scenarios to be modeled and the associated variable can be prepared and perhaps arrayed as a "test matrix." For the detector case above, the matrix could be multidimensional and very large. The matrix can be pruned by eliminating highly unlikely scenarios and by starting with the most conservative scenarios, i.e., those that place the occupant at greatest risk. For example:

- Fire remote from the detector
- Fire between occupant and detector
- Occupant impaired
- Fuel has highest heat-release rate

Once the matrix has been developed, it will probably be noted that there are not enough facts to provide all the needed model inputs. If the initial fuel package is furniture, and the construction materials and design characteristics are not known, a data base for several items that bound the actual item might be used. If the characteristics are known, an exemplar might be tested. If nothing is known about door position, this becomes a variable, as would the starting position of victims.

Most models are driven by a heat-release curve input by the user. The user may apply one of several "standard" t2 fires (slow, medium, fast, or ultra fast[23]) or develop a curve more specific to the fire under investigation. The closer the rate of the heat-release curve is to the actual occurrence, the more accurate the results. If "standard" t2 fires are used, it is incumbent on the user to justify their appropriateness. Developing a case-specific heat-release curve is one of the more difficult tasks in a reconstruction. One begins with the first item ignited, if this is known. If the first item is not known, part of the reconstruction will be to experiment with several possible "first items." Next, the ignition of nearby items is determined and the heat-release curve(s) added to that of the first item, making adjustments for the radiation effects from the burning first item, as necessary. It has been shown[24] that the developing hot layer in a compartment fire increases the burning rate of items before the onset of flashover. Adjustments must be made to reflect this phenomenon.[19] Considerable data and reference material on heat-release rates for a variety of items and materials are available to assist in the development of case-specific heat-release curves.[25,26]

Once a fire curve has been developed, one way to prune the matrix is to plan a set of experiments to test the sensitivity of the results to the unknowns. If, for example, the HVAC has little or no effect on the time to flashover, it may be ignored; but, it may have a major effect on tenability in an area remote from the fire, in which case HVAC would need to be included.

Keeping detailed records of the modeling experiments is important. It is very easy to "check a few things out" at the keyboard and not remember what the results were. Keep a "lab" notebook. It is suggested that a file-naming convention be used that tracks the scenario number as well as the key variables.

It is clear that many, many runs will be necessary for even the most straightforward analysis. It is not the intent of this chapter to thoroughly examine all aspects of modeling for fire construction and failure analysis. There are, however, a number of guidelines that should be observed, as follows.

1. *Know the limitations of the models used.* ASET-B[2] for example, assumes no vent and the data is questionable after flashover. ASET-B[2] also requires that the user assign an arbitrary heat loss factor to the space being evaluated. Most models do not consider the effect of radiation from the upper layer on the heat release of the burning items.[17] Heat detector models may not include

the transit time for heated gas to reach the sensor. This can be a consideration for low-heat output fires (weak plumes) or high ceilings. The importance of transport time may be evaluated using a technique proposed by Mowrer.[27]

2. *Use more than one model.* For example, different models use different correlations for plumes[28] and for pre-flashover layer temperature.[29]

3. *Where possible, check the model against actual experiments*[8] or test data that was collected under similar conditions.

4. *Use ranges of values for variables* to which the model is sensitive, unless the value is known to be correct. For example, the heat of combustion of materials with generic names, such as "polyurethane," can vary from product to product. Modeling with both upper and lower bounds will give a range of results that brackets the problem.

5. *Consider the ambient conditions.* For example, the hot layer formed under a metal deck roof on a blistering hot day will delay the transport of smoke to a detector. If the model being used for detector operation relies on temperature rise above ambient, the "ambient" used should be that of the hot layer, not the temperature lower in the compartment or outside the building. When there are open windows, the effects of wind should be considered in modeling the effects of low-energy fires (i.e., smoldering or small flaming fires).

Cautions on the Use of Computer Models in Fire Reconstruction and Failure Analysis

One should always keep in mind that there are some situations where a real-world fire situation is so complicated that the use of models at the present level of sophistication to analyze the entire event may not be valid. However, parts of the event, particularly the early stage of growth within the compartment or on the floor of origin, may be studied.

The use of fire models for forensic applications is expanding, and some are being used as "black boxes." In order for fire models to gain credibility in the forensic arena, they must be applied with care, keeping in mind their limitations.

SOME ADDITIONAL CONSIDERATIONS

Fire models are engineering tools. Like engineering handbooks, they are available to all persons, regardless of competence. Unlike engineering handbooks, however, many models have been made user-friendly and can be executed by persons who may not understand their capabilities, limitations, or the impact of the choices made in setting up the problems. When decisions produced by a model impact on the safety of life and property, one of the two following cases must hold.

1. Users must understand what the model is doing and how it is achieving the results, thereby treating it as an engineering tool. In such cases, the users are as responsible for the results as they would be had they gone to the engineering handbooks and extracted the equations themselves. Engineers are encouraged to use models, understand them, and become fluent with them. They will become an excellent tool under their control. Or,

2. Users must ascertain the limits to which the model has been demonstrated as rational and restrain their use within these bounds. If the occasion necessitates exceeding these bounds, the user should obtain the advice and/or service of individuals competent to examine the model and determine its appropriateness for the use desired.

Many models require that the user provide inputs beyond a simple description of the space and its contents. Some of these inputs can greatly affect the result. The user may or may not have the data available or an understanding of the impact of his or her decision. Some models provide information to assist users; others do not. It is important that users exercise the model, find the variables required, and ensure that they can properly assign these before using the model. An important thing to remember is that many models include default values for the variables. A default value is one that has been included in the model and will be used by the model unless changed by the user. Usually, the values are ones felt to be reasonable by the model developer. However, they may or may not be appropriate for the case involved. The user must examine and responsibly accept or alter the default values as is appropriate for the case at hand.

The sophistication or detail of a model may or may not be related to its accuracy for the case involved. A general principle is to use the simplest model that meets the user's needs for the type and quality of information desired and the information required to run the model. Many of the more sophisticated models provide masses of detail that are not available in the simpler version. Conversely, some of the very simple models omit important fire phenomena. The balance between ensuring that the necessary phenomena are modeled and maintaining the sophistication of the model is best achieved by frequent use and examination of the models of interest.

Finally, many computer fire models undergo regular review and possible adjustments. Users should try to stay abreast of updates, revisions, or other changes to models they use. They should be wary of models that are not being maintained and adjusted as needed by a competent organization.

BIBLIOGRAPHY

References Cited

1. Cooper, L. Y., and Stroup, D. W., "Calculating Safe Egress Time from Fires," NBSIR 82-2587, National Bureau of Standards, Gaithersburg, MD, 1982.
2. Walton, W. D., "ASET-B, A Room Fire Program for Personal Computers," NBSIR 85-3144, National Bureau of Standards, Gaithersburg, MD, 1985.
3. Alpert, R. L., "Calculations of Response Time of Ceiling-Mounted Fire Detectors," *Fire Technology*, Vol. 8, No. 3, 1972.
4. Evans, D. D., and Stoup, D. W., "Methods to Calculated the Response Time of Heat and Smoke Detectors Installed Below Large Unobstructed Ceilings," NBSIR 85-3167, National Institute of Standards and Technology, Gaithersburg, MD, 1985.
5. Purser, D. A., "Toxicity Assessment of Combustion Products," *The SFPE Handbook of Fire Protection Engineering*, 2nd ed., DiNenno, P. J., *et al.*, eds., National Fire Protection Association, Quincy, MA, 1995.
6. Mittler, H. E., "Guide to FIRST, A Comprehensive Single-Room Fire Model," NBSIR 87-3595, National Institute of Standards and Technology. Gaithersburg, MD, 1987.
7. Jones, W. W., and Peacock, R. D., "Technical Reference Guide for FAST Version 18," NIST Technical Note 1262, National Institute of Standards and Technology, Gaithersburg, MD, 1989.
8. Nelson, H. E., FPETOOL: "Fire Protection Engineering Tools for Hazard Estimation," NISTIR 4380, National Institute of Standards and Technology, Gaithersburg, MD, 1990.
9. Satterfield, D. B., and Barnett, J. R., "User's Guide for WPI/Fire Version 2, Compartment Fire Model," Worcester Polytechnic Institute, Worcester, MA, 1990.
10. Bukowski, R. W., Peacock, R. D., Jones, W. W., and Forney, C. L., "Technical Reference Guide for the HAZARD I Fire Hazard Assessment Method, Handbook 146, Volume II," National Institute of Standards and Technology, Gaithersburg, MD, 1991.
11. Peacock, R. D., Jones, W. W., Forney, G. P., Renke, P., and Portier, R., "CFAST, the Consolidated Model of Fire and Smoke Transport," NIST Technical Note 1299, National Institute of Standards and Technology, Gaithersburg, MD, 1992.
12. Nelson, H. E., and Deal, S., "Comparing Compartment Fire Tests with Compartment Fire Models," Fire Safety Science—*Proceedings* of the Third International Symposium, International Association for Fire Safety Science, pp. 719–728.
13. Custer, R. L. P., "Computer Modeling in Fire Reconstruction and Failure Analysis," *Proceedings,* SFPE Symposium on Computer Applications in Fire Protection Engineering, June, 28–29 1993, Society of Fire Protection Engineers and Worcester Polytechnic Institute.
14. Nelson, H. E., "Computer Models in Fire Reconstruction," *Proceedings*, SFPE Symposium on Computer Applications in Fire Protection Engineering, June 28–29, 1993, Society of Fire Protection Engineers and Worcester Polytechnic Institute.
15. Fleming, R. P., "Analysis of Potential Fire Sprinkler Performance in the Great Adventure Fire, May 11, 1984," National Fire Sprinkler Association, Patterson, NY, 1985.
16. Nelson, H. E., "An Engineering Analysis of the Early Stages of Fire Development—The Fire at the DuPont Plaza Hotel and Casino—December 31, 1986," NBSIR 87-3560, National Institute of Standards and Technology, Gaithersburg, MD, 1987.
17. Nelson, H. E., "Engineering Analysis of Fire Development in the Hospice of Southern Michigan," *Proceedings*, International Association for Fire Safety Science, and International Symposium, June 13–17, 1988, Tokyo, Japan, Hemisphere Publ. Corp., New York, 1989, pp. 927–938.
18. Nelson, H. E., "Engineering View of the Fire of May 4, 1988, in the First Interstate Bank Building, Los Angeles, California," NISTIR 89-4061, National Institute of Standards and Technology, Gaithersburg, MD, 1985.
19. Nelson, H. E., "Engineering Analysis of the Fire Development in the Hillhaven Nursing Home Fire, October 5, 1989," NISTIR 4665, National Institute of Standards and Technology, Gaithersburg, MD, 1991.
20. Moodie, K., and Jagger, S. F., "The Technical Investigation of the Fire at London's King's Cross Station," *Journal of Fire Protection Engineers*, Vol. 3, No. 2, 1991.
21. Bukowski, R. W., and Spetzer, R. C., "Analysis of the Happyland Social Club Fire with HAZARD I," *Journal of Fire Protection Engineers*, Vol. 4, No. 4, 1992.
22. Hendrick, K., and Benner, L., *Investigating Accidents with STEP*, Marcel Deckker, New York, 1987.
23. Evans, D. D., "Ceiling Jet Flows," *The SFPE Handbook of Fire Protection Engineering*, 2nd ed., DiNenno, P. J., *et al.*, eds., National Fire Protection Association, Quincy, MA, 1995.
24. Parker, W., Tu, K., Nurbakhsh, S., and Damant, G., "Furniture Flammability: An Investigation of the California Bulletin 133 Test. Part III: Full-Scale Chair Burns," NISTIR 4375, National Institute of Standards and Technology, Gaithersburg, MD, 1990.
25. Babrauskas, V., "Burning Rates," *The SFPE Handbook of Fire Protection Engineering*, 2nd ed., DiNenno, P. J., *et al.*, eds. National Fire Protection Association, Quincy, MA, 1995.
26. Babrausks, V., and Grayson, S. J., eds. *Heat Release in Fires*, Elsevier Applied Science, New York, 1992.
27. Mowrer, F., "Lag Timer Associated with Fire Detection and Suppression," *Fire Technology*, Vol. 26, No. 3, 1990.
28. Beyler, C. L., "Fire Plumes and Ceiling Jets," *Fire Safety Journal*, Vol. 11, pp. 53–75, 1986.
29. Deal S., and Beyer, C. L., "Correlating Preflashover Room Fire Temperatures," *Journal of Fire Protection Engineers*, Vol. 2, No. 2, 1990.

Additional Reading

Allan, H. A., "CSIRO Fire Safety Engineering Method Using the Computer Model FIRECALC," Commonwealth Scientific and Industrial Research Organization, Australia, University of Ulster and Fire Research Station, CIB W14: Fire Safety Engineering, International Symposium and Workshops Engineering Fire Safety in the Process of Design: Demonstrating Equivalency, Part 2, Symposium: Engineering Fire Safety for People With Mixed Abilities, Sept. 13–16, 1993, Newtownabbey, Northern Ireland, 1993, pp. 111–144.
Allan, H., Grubits, S., and Quaglia, C., "Predicting Smoke Movement in a Multi-Storey Shopping Center With Interconnecting Voids Using a Zone Model," *Proceedings* of the Society of Fire Protection Engineers (SFPE) Honors Lecture Series, May 20, 1996, Engineering Seminars: Fire Protection Design for High Challenge or Special Hazard Applications, May 20–22, 1996, Boston, MA, 1996, pp. 43–48.
Allen, J. E., and Reynolds, J. R., Jr., "FTI Field Modeling Analysis and Data Animation Dupont Plaza Hotel Fire," *Proceedings* of the International

Conference on Fire Research and Engineering, Sept. 10–15, 1995, Orlando, FL, SFPE, Boston, MA, 1995, pp. 588–589.

Andersson, P., and Holmstedt, G., "CFD-Modelling Applied to Fire Detection: Validation Studies and Influence of Background Heating," *Proceedings* of the Tenth International Conference on Automatic Fire Detection "AUBE '95", [Internationale Konferenz uber Automatischen Brandentdeckung.] April 4–6, 1995, Duisburg, Germany, 1995, pp. 429–438.

Babrauskas, V., "Designing Products for Fire Performance: The State of the Art of Test Methods and Fire Models. Commentary," *Fire Safety Journal*, Vol. 24, No. 3, 1995, pp. 299–312.

Babrauskas, V., "Introduction to Mathematical Fire Modelling. [Book Review of An Introduction to Mathematical Fire Modeling by David M. Birk.]," *Fire and Flammability Bulletin*, Oct. 1991, pp. 9–10.

Bailey, J. L., and Tatem, P. A., "Validation of Fire/Smoke Spread Model (CFAST) Using Ex-USS SHADWELL Internal Ship Conflagration Control (ISCC) Fire Tests," Naval Research Laboratory, Washington, DC, Office of Naval Research, Arlington, VA, NRL/MR/6180-95-7781, Sept. 28, 1995.

Baum, H. R., McGrattan, K. B., and Rehm, R. G., "Mathematical Modeling and Computer Simulation of Fire Phenomena," *Proceedings* of the Fourth International Symposium on Fire Safety Science, Intl. Assoc. for Fire Safety Science, Boston, MA, 1994, pp. 185–193.

Beard, A., "Evaluation of Fire Models: Report 9. FIRST: Qualitative Assessment," Edinburgh Univ., Scotland, Report 9, Jan. 1990.

Beard, A., "Evaluation of Deterministic Fire Models. Part 1. Introduction," *Fire Safety Journal*, Vol. 19, No. 4, 1992, pp. 295–306.

Beard, A., "Evaluation of Fire Models: Overview," Edinburgh Univ., Scotland, Report 11, Oct. 1990.

Beard, A., "Evaluation of Fire Models: Report 1. ASET: Qualitative Assessment," Edinburgh Univ., Scotland, Report 1, Jan. 1990.

Beard, A., "Evaluation of Fire Models: Report 10. JASMINE: Qualitative Assessment," Edinburgh Univ., Scotland, Report 10, May 1990.

Beard, A., "Evaluation of Fire Models: Report 2. ASKFRS: Qualitative Assessment," Edinburgh Univ., Scotland, Report 2, Jan. 1990.

Beard, A., "Evaluation of Fire Models: Report 3. HAZARD I: Qualitative Assessment," Edinburgh Univ., Scotland, Report 3, Jan. 1990.

Beard, A., "Evaluation of Fire Models: Report 4. FIRST: Qualitative Assessment," Edinburgh Univ., Scotland, Report 4, Jan. 1990.

Beard, A., "Evaluation of Fire Models: Report 5. JASMINE: Qualitative Assessment," Edinburgh Univ., Scotland, Report 5, April 1990.

Beard, A., "Evaluation of Fire Models: Report 6. ASET: Qualitative Assessment," Edinburgh Univ., Scotland, Report 6, Jan. 1990.

Beard, A., "Evaluation of Fire Models: Report 8. HAZARD I: Qualitative Assessment," Edinburgh Univ., Scotland, Report 8, April 1990.

Beard, A., "Limitations of Computer Modelling," Fire Safety Modelling and Building Design, *Proceedings* of the One-Day Conference to Review the Potential and Limitations of Fire Safety Models for Building Design, March 29, 1994, Salford, UK, 1994, pp. 10–26.

Beard, A., "Limitations of Computer Models," *Fire Safety Journal*, Vol. 18, No. 4, 1992, pp. 375–391.

Beard, A. N., "Reliability and Computer Models," *Journal of Applied Fire Science*, Vol. 3, No. 3, 1993/1994, pp. 273–279.

Beard, A. N., *et al.*, "Flashover A1: A Model for Predicting the Conditions for Flashover," University of Edinburgh, Scotland, South Bank University, London, England, University of College, London, England, ISBN 0-9527398-3-6; Institution of Fire Engineers, University of Sunderland, Fire Research Station, CIB W14, Tyne and Wear Metropolitan Fire Brigade, Fire Safety by Design, Conference *Proceedings*, Volume 3, Research Papers, July 10–12, 1995, UK, 1995, pp. 169–179.

Beard, A. N., Drysdale, D. D., and Bishop, S. R., "Non-Linear Model of Major Fire Spread in a Tunnel," *Fire Safety Journal*, Vol. 24, No. 4, 1995, pp. 333–357.

Beck, V., *et al.*, "Experimental Validation of a Fire Growth Model," Victoria University of Technology, Australia, National Research Council of Canada, Ottawa, Ontario, ISBN 0-9516320-9-4; *Proceedings* of the Seventh International Interflam Conference, Interflam '96, March 26–28, 1996, Cambridge, England, Interscience Communications Ltd., London, England, 1996, pp. 653–662.

Beller, D., "Validating Fire Models: FPETOOL, CFAST, WPI/FIRE," Azure Dragon Enterprises, Ltd., USA, ISBN 0-9516320-9-4; *Proceedings* of the Seventh International Interflam Conference, Interflam '96, March 26–28, 1996, Cambridge, England, Interscience Communications Ltd., London, England, 1996, pp. 959–963.

Beller, D., and Till, R., "Computer Fire Model Validation: FPETOOL, CFAST, WPI/FIRE," *Proceedings* of the International Conference on Fire Research and Engineering, Sept. 10–15, 1995, Orlando, FL, SFPE, Boston, MA, 1995, pp. 341–345.

Belles, D., "Practical Application of Computer Fire Models," National Institute of Standards and Technology, Federal Fire Forum on Computer Programs for Fire Protection, Nov. 7, 1991, Gaithersburg, MD, 1991.

Bilger, R. W., "Computational Field Models in Fire Research and Engineering," *Proceedings* of the Fourth International Symposium on Fire Safety Science, Intl. Assoc. for Fire Safety Science, Boston, MA, 1994, pp. 95–110.

Blount, K., "Regulatory Implications of Fire Safety Modelling," Fire Safety Modelling and Building Design, *Proceedings* of the One-Day Conference to Review the Potential and Limitations of Fire Safety Models for Building Design, March 29, 1994, Salford, UK, 1994, pp. 27–35.

Brani, D. M., and Black, W. Z., "Two-Zone Model for a Single-Room Fire," *Fire Safety Journal*, Vol. 19, No. 2–3, 1992, pp. 189–216.

Brooks, W. N., "Comparison of the NFPA 92B Calculation Methods to Several Fire Models," *Proceedings* of Computer Applications in Fire Protection, June 28–29, 1993, Worcester, MA, 1993, pp. 43–50.

Bukowski, R. W., "Fire Models: The Future is Now!," *NFPA Journal*, Vol. 85, No. 6, March/April 1991, pp. 60–62, 64, 66–69.

Bukowski, R. W., "How to Evaluate Alternative Designs Based on Fire Modeling," *NFPA Journal*, Vol. 89, No. 2, March/April 1995, pp. 68–70, 72–74.

Chandrasekaran, V., Grubits, S. J., and Suresh, R., "Zone Models in Fire Safety Engineering. Session 4—The Tools," CSIRO Division of Building, Construction and Engineering, Australia, CSIRO, International Fire Safety Engineering Conference on Concept and the Tools, Presented for FORUM for International Cooperation on Fire Research, Supported by SFPE, NFPA, AFPA, AAFA, Oct. 18–20, 1992, Sydney, Australia, 1992, pp. 1–18.

Chang, X., Laage, L. W., and Greuer, R. E., "User's Manual for MFIRE: A Program Computer Simulation Program for Mine Ventilation and Fire Modeling," Xian Mining Institute, The Peoples Republic of China, Bureau of Mines, Minneapolis, MN, Mining Technological Univ., Houghton, MI, IC 9245, 1990.

Charters, D. A., Gray, W. A., and McIntosh, A. C., "Computer Model to Assess Fire Hazards in Tunnels (FASIT)," *Fire Technology*, Vol. 30, No. 1, First Quarter 1994, pp. 134–154.

Charters, D., and McIntosh, A., "FASIT: A Computer Model for Predicting Fires in Tunnels," *Fire*, Vol. 88, No. 1082, August 1995, pp. 13–14, 16.

Chow, W. K., "Experimental Evaluation of the Zone Models CFAST, FAST and CCFM.VENTS," *Journal of Applied Fire Science*, Vol. 2, No. 4, 1992–1993, pp. 307–332.

Chow, W. K., "Multi-Cell Concept for Simulating Fires in Big Enclosures Using a Zone Model," *Journal of Fire Sciences*, Vol. 14, No. 3, May/June 1996, pp. 186–198.

Chow, W. K., "Short Note on the Simulation of the Atrium Smoke Filling Process Using Fire Zone Models," *Journal of Fire Sciences*, Vol. 12, No. 6, Nov./Dec. 1994, pp. 506–515.

Chow, W. K., "Simulation of Car Park Fires Using Zone Models," *Journal of Fire Protection Engineering*, Vol. 7, No. 3, 1995, pp. 65–74.

Chow, W. K., "Use of the ARGOS Fire Risk Analysis Model for Studying Chinese Restaurant Fires," *Fire and Materials*, Vol. 19, No. 4, July/Aug. 1995, pp. 171–178.

Chow, W. K., "Use of Zone Models on Simulating Compartmental Fires With Forced Ventilation," *Fire and Materials*, Vol. 19, No. 3, 1995, pp. 101–108.

Chow, W. K., "Zone Model Simulation of Fires in Chinese Restaurants in Hong Kong," *Journal of Fire Sciences*, Vol. 13, No. 3, May/June 1995, pp. 235–253.

Chow, W. K., and Cheung, K. C., "Industrial Fire Simulation With Zone Models and Review on the Current Regulations," Hong Kong Polytechnic Univ., Hong Kong, ISBN 0-9516320-9-4; *Proceedings* of the Seventh International Interflam Conference, Interflam '96, March 26–28, 1996, Cambridge, England, Interscience Communications Ltd., London, England, 1996, pp. 601–620.

Chow, W. K., and Cheung, Y. L., "Simulation of Sprinkler-Hot Layer Interaction Using a Field Model," *Fire and Materials*, Vol. 18, No. 6, 1994, pp. 359–379.

Chow, W. K., and Lo, A. C. W., "Scale Modelling Studies on Atrium Smoke Movement and the Smoke Filling Process," *Journal of Fire Protection Engineering*, Vol. 7, No. 3, 1995, pp. 55–64.

Chow, W. K., and Wong, W. K., "Application of the Zone Model FIRST on the Development of Smoke Layer and Evaluation of Smoke Extraction Design for Atria in Hong Kong," *Journal of Fire Sciences*, Vol. 11, No. 4, July/Aug. 1993, pp. 329–349.

Cooper, L. Y., "Overview of a Theory for Simulating Smoke Movement Through Long Vertical Shafts in Zone-Type Fire Models," National Institute of Standards and Technology, Gaithersburg, MD, NISTIR 5499, Sept. 1994; National Institute of Standards and Technology, Annual Conference on Fire Research: Book of Abstracts, Oct. 17–20, 1994, Gaithersburg, MD, 1994, pp. 93–94.

Cooper, L. Y., "VENTCF2: An Algorithm and Associated FORTRAN 77 Subroutine for Calculating Flow Through a Horizontal Ceiling/Floor Vent in a Zone-Type Compartment Fire Model," National Institute of Standards and Technology, Gaithersburg, MD, NISTIR 5470, Aug. 1994.

Cox, G., "Challenge of Fire Modelling," *Fire Safety Journal*, Vol. 23, 1994, pp. 123–132.

Cox, G., "Challenge of Fire Modelling," *Proceedings* of the Technical Conference on Fire Safety by Design: A Framework for the Future, Featuring the Combined 1993 Fire Research Lecture and IFS Inaugural Lecture, Nov. 10, 1993, Borehamwood, England, 1993, pp. 1–14.

Cox, G., "Compartment Fire Modelling," Fire Research Station, Borehamwood, England, ISBN 0-12-194230-9; *Combustion Fundamentals of Fire*, Chapter 6, Academic Press, New York, NY, 1995, pp. 329–404.

Cox, G., "Some Recent Progress in the Field Modelling of Fire," First Asian Conference on Fire Science and Technology (ACFST), October 9-13, 1992, Hefei, China, International Academic Publishers, China, 1992, pp. 50–59.

Cox, G., "State of the Art Modelling," Fire Safety Modelling and Building Design, *Proceedings* of the One-Day Conference to Review the Potential and Limitations of Fire Safety Models for Building Design, March 29, 1994, Salford, UK, 1994, pp. 6–9.

Cox, G., and Rubini, P. A., "Development of a New CFD Fire Simulation Model," Fire Research Station, Borehamwood, England, Cranfield Institute of Technology, UK, VTT-Technical Research Center of Finland and Forum for International Cooperation in Fire Research, Nordic Fire Safety Engineering Symposium, Development and Verification of Tools for Performance Codes, Aug. 30-Sept. 1, 1993, Espoo, Finland, 1993, pp. 1–11.

Curtat, M. R., "Survey of Fire Modelling in France," *Proceedings* of the Third International Symposium on Fire Safety Science, Elsevier Applied Science, New York, 1991, pp. 97–118.

Custer, R. L. P., "Computer Modeling in Fire Reconstruction and Failure Analysis," *Proceedings* of Computer Applications in Fire Protection, June 28–29, 1993, Worcester, MA, 1993, pp. 69–74.

Custer, R. L. P., "Progress on Computer Modeling in Performance-Based Detection System Design in the United States," Worcester Polytechnic Inst., MA, European Society for Automatic Alarm Systems (EUSAS), European Platform for Promoting Security and Fire Protection Engineering, Newsletter, Number 5, June 1994, pp. 17–26.

Custer, R. L. P., "Selecting and Using Computer Fire Models," Worcester Polytechnic Inst., MA, Society of Fire Protection Engineers and WPI Center for Firesafety Studies, *Proceedings* of the Technical Symposium on Computer Applications in Fire Protection Engineering, Final Program, June 20–21, 1996, Worcester, MA, 1996, pp. 7–11.

Daikoku, M., *et al.*, "Modelling of Transport Phenomena in Compartment Fires," *Proceedings* of the International Conference on Fire Research and Engineering, Sept. 10–15, 1995, Orlando, FL, SFPE, Boston, MA, 1995, pp. 15–20.

Davis, W., "Overview of Harwell-Flow 30—A Field Model," Montgomery College, Rockville, MD, National Institute of Standards and Technology, Federal Fire Forum on Computer Programs for Fire Protection, Nov. 7, 1991, Gaithersburg, MD, 1991.

Davis, W. D., Forney, G. P., and Bukowski, R. W., "Field Modeling: Simulating the Effect of Sloped Beamed Ceilings on Detector and Sprinkler Response. International Fire Detection Research Project. Technical Report. Year 2," National Institute of Standards and Technology, Gaithersburg, MD, National Fire Protection Research Foundation, Quincy, MA, Technical Report, Year 2, Oct. 1994.

Davis, W. D., Forney, G. P., and Klote, J. H., "Field Modeling of Room Fires," National Institute of Standards and Technology, Gaithersburg, MD, NISTIR 4673, Nov. 1991.

Davis, W. D., Notarianni, K. A., and Tapper, P. Z., "Modelling of Smoke Movement and Detector Performance in High Bay Spaces," *Proceed-*

ings of the International Conference on Fire Research and Engineering, Sept. 10–15, 1995, Orlando, FL, SFPE, Boston, MA, 1995, pp. 307–311.

Deakin, A. G., "Fire Modelling: Where Are We and What Is Needed for Practical Application," Warrington Fire Research Center, UK, TNO Building and Construction Research, Third CIB/W14 Workshop, *Proceedings* of the Fire Engineering Workshop on Modelling, Jan. 25–26, 1993, Delft, The Netherlands, 1993, pp. 15–18.

Deal, S., "Comparison of Various Fire Model Results," National Institute of Standards and Technology, Federal Fire Forum on Computer Programs for Fire Protection, Nov. 7, 1991, Gaithersburg, MD, 1991.

Dembsey, N. A., Pagni, P. J., and Williamson, R. B., "Compartment Fire Experimental Data: Comparison to Models," *Proceedings* of the International Conference on Fire Research and Engineering, Sept. 10–15, 1995, Orlando, FL, SFPE, Boston, MA, 1995, pp. 350–355.

Deubler, J. F., "How to Review a Computer Modeling Report," Schirmer Engineering Corp., Arlington, VA, Society of Fire Protection Engineers and WPI Center for Firesafety Studies, *Proceedings* of the Technical Symposium on Computer Applications in Fire Protection Engineering, Final Program, June 20–21, 1996, Worcester, MA, 1996, pp. 67–72.

Dolph, B. L., "Modeling Fires on Ships," Coast Guard Research and Development Center, Groton, CT, Society of Fire Protection Engineers and WPI Center for Firesafety Studies, *Proceedings* of the Technical Symposium on Computer Applications in Fire Protection Engineering, Final Program, June 20–21, 1996, Worcester, MA, 1996, pp. 1–4.

Duong, D. Q., "Accuracy of Computer Fire Models: Some Comparisons With Experimental Data From Australia," *Fire Safety Journal*, Vol. 16, No. 6, 1990, pp. 415–431.

El-Rimawi, J. A., Burgess, I. W., and Plank, R. J., "Modelling the Behavior of Steel Frames and Subframes With Semi-Rigid Connections in Fire," University of Sheffield, UK, TNO Building and Construction Research, Third CIB/W14 Workshop, *Proceedings* of the Fire Engineering Workshop on Modelling, Jan. 25–26, 1993, Delft, The Netherlands, 1993, pp. 152–168.

Ellington, J. M., "Some Perspectives on Fire Modeling by a Fire Investigator," *Fire and Arson Investigator*, Vol. 45, No. 4, June 1995, pp. 36–38.

Emmons, H. W., "Fire Modeling in the 21st Century," Harvard Univ., Cambridge, MA, Society of Fire Protection Engineers and WPI Center for Firesafety Studies, *Proceedings* of the Technical Symposium on Computer Applications in Fire Protection Engineering, Final Program, June 20–21, 1996, Worcester, MA, 1996, pp. 1–5.

Evans, D. D., "Large Fire Experiments for Fire Model Evaluation," National Institute of Standards and Technology, Gaithersburg, MD, ISBN 0-9516320-9-4; *Proceedings* of the Seventh International Interflam Conference, Interflam '96, March 26–28, 1996, Cambridge, England, Interscience Communications Ltd., London, England, 1996, pp. 329–334.

Fahy, R. F., "EXIT89: An Evacuation Model for High-Rise Buildings," National Fire Protection Assoc., Quincy, MA, NISTIR 4449; Eleventh Joint Panel Meeting of the U. S./Japan Government Cooperative Program on Natural Resources (UJNR) on Fire Research and Safety, Oct. 19–24, 1989, Berkeley, CA, 1990, pp. 306–311.

Fahy, R. F., "EXIT89: An Evacuation Model for High-Rise Buildings. Model Description and Example Applications," *Proceedings* of the Fourth International Symposium on Fire Safety Science, Intl. Assoc. for Fire Safety Science, Boston, MA, 1994, pp. 657–668.

Fahy, R. F., "EXIT89: An Evacuation Model for High-Rise Buildings—Recent Enhancements and Example Applications," *Proceedings* of the International Conference on Fire Research and Engineering, Sept. 10–15, 1995, Orlando, FL, SFPE, Boston, MA, 1995, pp. 332–337.

Fahy, R. F., "EXIT89: High-Rise Evacuation Model—Recent Enhancements and Example Applications," National Fire Protection Association, Quincy, MA, National Institute of Standards and Technology, Gaithersburg, MD, ISBN 0-9516320-9-4; *Proceedings* of the Seventh International Interflam Conference, Interflam '96, March 26–28, 1996, Cambridge, England, Interscience Communications Ltd., London, England, 1996, pp. 1001–1005.

Fangrat, J., "Different Approach to Some Fire Modeling Input Data," Building Research Inst., Warsaw, Poland, CIB W14/93/2 (J); *First Proceedings* of the Japan Symposium on Heat Release and Fire Hazard, Session 2, Fire Modeling as a Tool for Reaction to Fire Evaluation, May 10–11, 1993, Tsukuba, Japan, 1993, pp. II/1–6.

"Fire Models In Use Today," *IRI Sentinel*, Vol. 51, No. 3, Third Quarter, 1994, pp. 10–11.

"Fire Models: A Tale of Four Uses," *IRI Sentinel*, Vol. 51, No. 3, Third Quarter, 1994, pp. 8–9.

"Fire Models: The Pros and Cons," *IRI Sentinel*, Vol. 51, No. 3, Third Quarter, 1994, p. 3.

"Fire Models: Valuable Tools When Used Properly," *IRI Sentinel*, Vol. 51, No. 3, Third Quarter, 1994, pp. 4–7.

"Fire Spread Modeling," *Record*, Vol. 70, No. 3, May/June 1993, pp. 15–16.

Forney, G. P., "Computing Radiative Heat Transfer Occurring in a Zone Fire Model," *Fire Science and Technology*, Vol. 14, No. 1/2, pp. 31–74.

Forney, G. P., and Moss, W. F., "Analyzing and Exploiting Numerical Characteristics of Zone Fire Models," *Fire Science and Technology*, Vol. 14, No. 1/2, 1994, pp. 49–60.

Fransson, C., and Vilselius, H., "Application of a General-Purpose CFD Code to Field Modelling of Tunnel Fires," National Defence Research Establishment, Sundbyberg, Sweden, FOA Report B 20131-2.7, June 1994.

Friedman, R., "Survey of Computer Models for Fire and Smoke. Technical Report," Factory Mutual Research Corp., Norwood, MA, Technical Report, March 1990.

Galea, E. R., and Galparsoro, J. M. P., "Computer-Based Simulation Model for the Prediction of Evacuation From Mass-Transport Vehicles," *Fire Safety Journal*, Vol. 22, No. 4, 1994, pp. 341–366.

Galea, E. R., and Ierotheou, C. S., "Fire-Field Modelling on Parallel Computers," *Fire Safety Journal*, Vol. 19, No. 4, 1992, pp. 251–266.

Galea, E. R., Owen, M., and Lawrence, P. J., "Emergency Egress From Large Buildings Under Fire Conditions Simulated Using the EXODUS Evacuation Model," University of Greenwich, London, England, ISBN 0-9516320-9-4; *Proceedings* of the Seventh International Interflam Conference, Interflam '96, March 26–28, 1996, Cambridge, England, Interscience Communications Ltd., London, England, 1996, pp. 711–720.

Gallina, G., *et al.*, "Fire Modelling in Building With a CAD-Based Graphical User Interface," CNR-ICITE, Sesto Ulteriano (MI), Italy, TRI srl, Scanzorosicate (BG), Italy, ISBN 0-9527398-3-6; Institution of Fire Engineers, University of Sunderland, Fire Research Station, CIB W14, Tyne and Wear Metropolitan Fire Brigade, Fire Safety by Design, Conference *Proceedings*, Volume 3, Research Papers, July 10–12, 1995, UK, 1995, pp. 239–253.

Golton, C. J., Golton, B. J., and Hinks, A. J., "Human Behavior in Fire: A Background Review for Modelling," Fire Safety Modelling and Building Design, *Proceedings* of the One-Day Conference to Review the Potential and Limitations of Fire Safety Models for Building Design, March 29, 1994, Salford, UK, 1994, pp. 48–62.

Griffith, S. J., and Munday, J. W., "Legal Implications of Real Fire Data Collection and Computer Modelling," Metropolitan Police Forensic Science Laboratory, London, England, ISBN 0-9516320-9-4; *Proceedings* of the Seventh International Interflam Conference, Interflam '96, March 26–28, 1996, Cambridge, England, Interscience Communications Ltd., London, England, 1996, pp. 1027–1031.

Grubits, S. J., and Chandrasekaran, V., "Issues in Fire Modelling," Commonwealth Scientific and Industrial Research Organization, North Ryde, Australia, July 1991.

Gupta, A. K., "CALFIRE: An Interactive Model for Fire Calculations," *Fire Technology*, Vol. 30, No. 3, Third Quarter 1994, pp. 304–325.

Hadjisophocleous, G. V., and Cacambouras, M., "Computer Modeling of Compartment Fires," *Journal of Fire Protection Engineering*, Vol. 5, No. 2, April/May/June 1993, pp. 39–52.

Hadjisophocleous, G. V., and Knill, K., "CFD Modelling of Liquid Pool Fire Suppression Using Fine Watersprays," National Research Council of Canada, Ottawa, Ontario, Advanced Scientific Computing, Ltd., Ontario, Canada, NISTIR 5499, Sept. 1994; National Institute of Standards and Technology, Annual Conference on Fire Research: Book of Abstracts, Oct. 17–20, 1994, Gaithersburg, MD, 1994, pp. 71–72.

Hagglund, B., "Comparing Fire Models With Experimental Data," National Defence Research Establishment, Sundbyberg, Sweden, FOA Report C20864-2.4, Feb. 1992.

Hagglund, B., Fransson, C., and Bengtson, S., "Use of Zone and Field Model to Recommend Fire Protection Measures in New Stores of the Royal Library in Stockholm," National Defence Research Establishment, Sundbyberg, Sweden, FOA Report C20981-2.4, June 1994.

Ham, S., "Fire Modelling From an Architectural Perspective," Fire Safety Modelling and Building Design, *Proceedings* of the One-Day Confer-

ence to Review the Potential and Limitations of Fire Safety Models for Building Design, March 29, 1994, Salford, UK, 1994, pp. 1–5.

Hasemi, Y., and Yoshida, M., "Wall Fire Modeling and Its Relevance to Building Fire Safety Design," '93 Asian Fire Seminar, Oct. 7–9, 1993, Tokyo, Japan, 1993, pp. 117–127.

Heins, T., and Dobbernack, R., "FIGARO: A Zone Model for Fire Simulation in Complex Geometries," Technischer Uberwachungs-Verein Hannover/Sachsen-Anhalt e.V., Germany, Universitat Braunschweig, Germany, TNO Building and Construction Research, Third CIB/W14 Workshop, *Proceedings* of the Fire Engineering Workshop on Modelling, Jan. 25–26, 1993, Delft, The Netherlands, 1993, pp. 49–58.

Hinks, A. J., "Modelling and Managing Escape From Fires in Complex Buildings," Fire Safety Modelling and Building Design, *Proceedings* of the One-Day Conference to Review the Potential and Limitations of Fire Safety Models for Building Design, March 29, 1994, Salford, UK, 1994, pp. 73–89.

Hinks, A. J., "Systemic Modelling of Life Safety in Building Fires: LISA," Fire Safety Modelling and Building Design, *Proceedings* of the One-Day Conference to Review the Potential and Limitations of Fire Safety Models for Building Design, March 29, 1994, Salford, UK, 1994, pp. 63–72.

Holen, J., and Magnussen, B. F., "Present Status of Field Models for Fire Growth Developed at NTH/SINTEF," NTH/SINTEF, Trondheim, Norway, VTT-Technical Research Center of Finland and Forum for International Cooperation in Fire Research, Nordic Fire Safety Engineering Symposium, Development and Verification of Tools for Performance Codes, Aug. 30-Sept. 1, 1993, Espoo, Finland, 1993, pp. 1–7.

Hume, B., "Guidance on Fire Models," *Fire Research News*, No. 18, Winter 1994, p. 11.

Janssens, M., "Mathematical Modeling of Compartment Fires," Abstracts of the First U. S. Symposium on Heat Release and Fire Hazard, Dec. 1991, San Diego, CA, 1991, pp. 45–47.

Janssens, M., "Room Fire Models. Part A. General," Chapter 6, *Heat Release in Fires*, Elsevier Applied Science, NY, 1992, pp. 113–157.

Janssens, M. L., "Computer Program Roomfire: A Compartment Fire Model Shell," American Forest and Paper Association, Washington,DC, Chapter 27, ACS Symposium Series 599, American Chemical Society, Fire and Polymers II: Materials and Tests for Hazard Prevention, 208th National Meeting, ACS Symposium Series 599, Aug. 21–26, 1994, Washington, DC, American Chemical Society, Washington, DC, 1995, pp. 409–421.

Johnson, P. F., "Comment on 'The Accuracy of Computer Fire Models: Some Comparisons With Experimental Data From Australia.' Letter to the Editor," *Fire Safety Journal*, Vol. 17, No. 5, 1991, pp. 415–416.

Jones, S., *et al.*, "Zone Model for Smoke Transport Aboard Human-Crewed Spacecraft," *PSAM-II Proceedings*, An International Conference Devoted to the Advancement of System-Based Methods for the Design and Operation of Technological Systems and Processes, Volume 3, Sessions 73–108, March 20–25, 1994, San Diego, CA, 1994, pp. 085/1–4.

Jones, W. W., "Modeling Fires—The Next Generation of Tools," Worcester Polytechnic Inst., MA, Society of Fire Protection Engineers and WPI Center for Firesafety Studies, *Proceedings* of the Technical Symposium on Computer Applications in Fire Protection Engineering, Final Program, June 20–21, 1996, Worcester, MA, 1996, pp. 13–18.

Jones, W. W., "Modeling Smoke Movement Through Compartmented Structures," *Journal of Fire Sciences*, Vol. 11, No. 2, March/April 1993, pp. 172–183; Thirty-ninth Sagamore Army Materials Research Conference, Sept. 16–17, 1992, Plymouth, MA, 1992; Twelfth Joint Panel Meeting of the U.S./Japan Government Cooperative Program on Natural Resources (UJNR) on Fire Research and Safety, Oct. 27–Nov. 2, 1992, Tsukuba, Japan, Building Research Inst., Ibaraki, Japan, Fire Research Inst., Tokyo, Japan, 1992, pp. 34–41.

Jones, W. W., "Progress Report on Fire Modeling and Validation," National Institute of Standards and Technology, Gaithersburg, MD, NISTIR 5835, May 1996.

Jones, W. W., and Baum, H. R., "Modeling Fire Growth and Smoke Transport in the United States," Twelfth Joint Panel Meeting of the U. S./Japan Government Cooperative Program on Natural Resources (UJNR) on Fire Research and Safety, Oct. 27–Nov. 2, 1992, Tsukuba, Japan, Building Research Inst., Ibaraki, Japan, 1992, pp. 4–7.

Jones, W. W., and Forney, G. P., "Programmer's Reference Manual for CFAST, the Unified Model of Fire Growth and Smoke Transport," National Institute of Standards and Technology, Gaithersburg, MD, NIST TN 1283, Nov. 1990.

Kalinich, D. A., and Bailey, R. T., "Parametric Study of Smoke Propagation Using the CFAST Computer Code," Westinghouse Savannah River Co., Aiken, SC, Department of Energy, Washington, DC, WSRC-RP-93-1315, Sept. 1993.

Karlsson, B., "Models for Calculating Flame Spread on Wall Lining Materials and the Resulting Heat Release Rate is a Room," *Fire Safety Journal*, Vol. 23, No. 4, 1994, pp. 365–386.

Karlsson, B., and Magnusson, S. E., "Room Fire Models. Part B. An Example Room Fire Model," Chapter 6, *Heat Release in Fires*, Elsevier Applied Science, NY, 1992, pp. 159–171.

Kasahara, I., and Hara, T., "Life Safety Design of an Atrium and an Air Dome With an Applied Zone Model," Taisei Corp., Tokyo, Japan, CIB W14/96/1 (J); Mini-Symposium on Fire Safety Design of Buildings and Fire Safety Engineering, June 12, 1995, Tsukuba, Japan, 1995, pp. VIa/1–3.

Kerrison, L., *et al.*, "Comparison of Two Fire Field Models With Experimental Room Fire Data," *Proceedings* of the Fourth International Symposium on Fire Safety Science, Intl. Assoc. for Fire Safety Science, Boston, MA, 1994, pp. 161–172.

Kerrison, L., *et al.*, "Comparison of a FLOW3D Based Fire Field Model With Experimental Room Fire Data," *Fire Safety Journal*, Vol. 23, No. 4, 1994, pp. 387–411.

Kostreva, M. M., "Sensitivity Analysis for Mathematical Modeling of Fires in Residential Buildings," Clemson Univ., SC, National Institute of Standards and Technology, Gaithersburg, MD, NIST-GCR-95-683, Feb. 1966.

Kumar, S., "Field Model Simulations of Vehicle Fires in a Channel Tunnel Shuttle Wagon," *Proceedings* of the Fourth International Symposium on Fire Safety Science, Intl. Assoc. for Fire Safety Science, Boston, MA, 1994, pp. 995–1006.

Kumar, S., and Yehia, M., "Application of JASMINE to the Modelling of Vehicle Fires in a Channel Tunnel Shuttle Wagon," *Proceedings* of the Second International Conference on Safety in Road and Rail Tunnels, April 3–6, 1995, Granada, Spain, 1995, pp. 321–328.

Li, H. J., and Cheng, X. F., "Design of Building Fire Simulation Models," *Fire Safety Science*, Vol. 3, No. 2, Sept. 1994, pp. 21–26.

Lucht, D. A., "Changing the Way We Do Business: The Coming of Fire Modeling," *IFCI Fire Code Journal*, Vol. 2, No. 4, Spring 1994, pp. 1, 2, 4-6.

Man, W. K., "Fire Modelling: An Example of Application. Part 1," *Southeast Asia Fire and Security*, May 1993, pp. 18–20.

Marchant, E. W., "Futures of Fire Safety Modelling," Fire Safety Modelling and Building Design, *Proceedings* of the One-Day Conference to Review the Potential and Limitations of Fire Safety Models for Building Design, March 29, 1994, Salford, UK, 1994, pp. 36–47.

Martin, S. B., "Perspective on Fire Modeling," Sixteenth International Conference on Fire Safety, Volume 16, Jan. 14–18, 1991, Millbrae, CA, Product Safety Corp., Sunnyvale, CA, 1991, pp. 370–376.

Mawhinney, R. N., *et al.*, "Critical Comparison of a Phoenics Based Fire Field Model With Experimental Compartment Fire Data," *Journal of Fire Protection Engineering*, Vol. 6, No. 4, 1994, pp. 137–152.

Mitler, H. E., "Mathematical Modeling of Enclosure Fires," National Institute of Standards and Technology, Gaithersburg, MD, NISTIR 90-4294, May 1991.

Mitler, H. E., and Steckler, K. D., "SPREAD: A Model of Flame Spread on Vertical Surfaces," National Institute of Standards and Technology, Gaithersburg, MD, NISTIR 5619, April 1995.

Moss, W. F., "Numerical Analysis Support for Compartment Fire Modeling and Incorporation of Heat Conduction Into a Zone Fire Model. Final Report. Aug. 15, 1988–March 31, 1991," Clemson Univ., SC, National Institute of Standards and Technology, Gaithersburg, MD, NIST-GCR-92-605, March 1992.

Moss, W. F., and Forney, G. P., "Implicitly Coupling Heat Conduction Into a Zone Fire Model," Clemson Univ., SC, National Institute of Standards and Technology, Gaithersburg, MD, NISTIR 4886, July 1992.

Nelson, H. E., "Application of Computer Models for Fire Incident Reconstruction," *Proceedings* of the International Conference on Fire Research and Engineering, Sept. 10–15, 1995, Orlando, FL, SFPE, Boston, MA, 1995, pp. 359–364.

Nelson, H. E., "Computer Models and Fire Reconstruction," *Proceedings* of Computer Applications in Fire Protection, June 28–29, 1993, Worcester, MA, 1993, pp. 75–79.

Nelson, H. E., "Introduction to Computer Fire Modeling," National Institute of Standards and Technology, Gaithersburg, MD, Federal Fire Forum on Computer Programs for Fire Protection. Nov. 7, 1991, Gaithersburg, MD, 1991.

Notarianni, K. A., and Davis, W. D., "Use of Computer Models to Predict the Response of Sprinklers and Detectors in Large Spaces," *Proceedings* of Computer Applications in Fire Protection, June 28–29, 1993, Worcester, MA, 1993, pp. 27–33.

Oar, D. L., "Attachments for Fire Modeling for Building 221-T. T-Plant Canyon Deck and Railroad Tunnel," Westinghouse Hanford Co., Richland, WA, WHC-SD-CP-ANAL-010, Jan. 19, 1995.

Ohnstad, S. J., "Marine Regulatory Applications of the Computer Fire Model HAZARD I," Worcester Polytechnic Inst., MA, Thesis, March 1990.

Opstad, K., "Fire Modelling Using Cone Calorimeter Results," SINTEF NBL, Trondheim, Norway, EUREFIC (European Reaction to Fire Classification), Performance Based Reaction to Fire Classification, International Seminar, Sept. 11–12, 1991, Copenhagen, Denmark, Interscience Communications Ltd., London, England, 1991, pp. 65–71.

Opstad, K., and Magnussen, B. F., "Algorithm for Thermal Flame Spread on Solid Surfaces by the Use of Cone Calorimeter Fire Test Data in a CFD-Model," University of Trondheim, Norway, Norwegian Institute of Technology, Norway, Thesis, May 1995; *Modelling of Thermal Flame Spread on Solid Surfaces in Large-Scale Fires*, 1995, pp. 1–15.

Peacock, R. D., "Validation of Computer Models," VHS Video Tape, National Institute of Standards and Technology, Gaithersburg, MD, Federal Fire Forum, Current Technical Fire Issues, Nov. 1, 1993, Gaithersburg, MD, 1993.

Peacock, R. D., Davis, S., and Babrauskas, V., "Data for Room Fire Model Comparisons," *Journal of Research of the National Institute of Standards and Technology*, Vol. 96, No. 4, Aug. 1991, pp. 411–462.

Peacock, R. D., *et. al.*, "CFAST, The Consolidated Model of Fire Growth and Smoke Transport," National Institute of Standards and Technology, Gaithersburg, MD, NIST TN 1299, Feb. 1993.

Peacock, R. D., Jones, W. W., and Bukowski, R. W., "Verification of a Model of Fire and Smoke Transport," *Fire Safety Journal*, Vol. 21, No. 2, 1993, pp. 89–129.

Pedersen, K., and Magnussen, B. F., "Field Models: their Present and Future Application in Fire Safety Engineering. Session 4—The Tools," SINTEF NBL-Norwegian Fire Research Lab., Trondheim, Norway, NTH/SINTEF, Trondheim, Norway, CSIRO, International Fire Safety Engineering Conference on Concept and the Tools, Presented for FORUM for International Cooperation on Fire Research, Supported by SFPE, NFPA, AFPA, AAFA, Oct. 18–20, 1992, Sydney, Australia, 1992, pp. 1–11.

Pucher, K., and Tiefenbock, M., "Three-Dimensional Computer Simulation of a Fire in a Rail Tunnel," *Proceedings* of the Second International Conference on Safety in Road and Rail Tunnels, April 3–6, 1995, Granada, Spain, 1995, pp. 245–252.

Quintiere, J.G., and Dillon, M. E., "Scale Model Reconstruction of Fire in an Atrium," *Proceedings* of the International Conference on Fire Research and Engineering, Sept. 10–15, 1995, Orlando, FL, SFPE, Boston, MA, 1995, pp. 397–402.

Rehm, R. G., and Forney, G. P., "Pressure Equations in Zone-Fire Modeling," *Fire Science and Technology*, Vol. 14, No. 1/2, 1994, pp. 61–73.

Rockett, J. A., "Experience in the Use of Zone Type Building Fire Models," *Fire Science and Technology*, Vol. 13, No. 1/2, 1993, pp. 61–70.

Rockett, J. A., "Zone Model Plume Algorithm Performance," *Fire Science and Technology*, Vol. 15, No. 1/2, 1995, pp. 1–16.

Rockett, J. A., Howe, J. M., and Hanbury, W. L., "Modeling Fire Safety in Multi-Use, Domed Stadia," *Journal of Fire Protection Engineering*, Vol. 6, No. 1, 1994, pp. 11–22.

Satterfield, D. B., "WPI/Harvard Version 2 Computer Fire Model," CFR Video Seminar, Department of the Navy, Washington, DC, Video, Oct. 23, 1990.

Schifiliti, R. P., and Pucci, W. E., "Fire Detection Modeling: State of the Art," R.P. Schifiliti Associates, Inc., Reading MA, Performance Consultants, Holland, MA, Fire Detection Institute, Bloomfield, CT, State of the Art Report, May 6, 1996, 57 pages; Society of Fire Protection Engineers and WPI Center for Firesafety Studies, *Proceedings* of the Technical Symposium on Computer Applications in Fire Protection Engineering, Final Program, June 20–21, 1996, Worcester, MA, 1996, pp. 19–27.

Schneider, V., and Hofmann, J., "Feldmodell-Simulation von Kohlenwasserstoff-Raumbranden und Spruhnebel-Loschversuchen. [Field-Model

Simulation of Hydrocarbon Room Fires and Fire Fighting Tests With Water Spray.]," *VFDB*, February 1993, pp. 67–73.

Schneider, V., and Konnecke, R., "Anwendung des Feldmodells KOBRA-3D zur Simulation von komplexen Brandszenarien auf Fragestellungen der automatischen Brandentdeckung. [Application of the Field Model KOBRA-3D to the Simulation of Complex Fire Scenarios for the Purpose of Automatic Fire Detection.]," *Proceedings* of the Tenth International Conference on Automatic Fire Detection "AUBE '95", [Internationale Konferenz uber Automatischen Brandentdeckung.] April 4–6, 1995, Duisburg, Germany, 1995, pp. 404–418.

Shapiro, J. M., "Are the Codes Ready for Fire Models?," *IFCI Fire Code Journal*, Vol. 2, No. 4, Spring 1994, pp. 3, 10.

Sinai, Y. L., and Owens, M. P., "Validation of CFD Modelling of Unconfied Pool Fires With Cross-Wind: Flame Geometry," *Fire Safety Journal*, Vol. 24, No. 1, 1995, pp. 1–34.

Stroup, D. W., "Analyzing Fire Safety Through Computer Modeling and Product Testing," Second International Conference on Fire and Materials, Sept. 23–24, 1993, Arlington, VA, 1993, pp. 7–12.

Stroup, D. W., "FPETOOL for the Personal Computer," General Services Administration, Washington, DC, National Institute of Standards and Technology, Federal Fire Forum on Computer Programs for Fire Protection, Nov. 7, 1991, Gaithersburg, MD, 1991.

Stroup, D. W., "Using Computer Fire Models to Evaluate Equivalent Levels of Fire and Life Safety," *Proceedings* of Computer Applications in Fire Protection, June 28–29, 1993, Worcester, MA, 1993, pp. 21–26.

Stroup, D. W., and Madrzykowski, D., "Modeling Smoke Flow in Corridors," *Proceedings* of the International Conference on Fire Research and Engineering, Sept. 10–15, 1995, Orlando, FL, SFPE, Boston, MA, 1995, pp. 377–382.

Sullivan, P. D., "Recent Applications of Computers in Performance Design of Buildings," Robert W. Sullivan, Inc., Boston, MA, Society of Fire Protection Engineers and WPI Center for Firesafety Studies, *Proceedings* of the Technical Symposium on Computer Applications in Fire Protection Engineering, Final Program, June 20–21, 1996, Worcester, MA, 1996, pp. 87–92.

Taylor, S., *et al.*, "SMARTFIRE: An Intelligent Fire Field Model," University of Greenwich, London, England, ISBN 0-9516320-9-4; *Proceedings* of the Seventh International Interflam Conference, Interflam '96, March 26–28, 1996, Cambridge, England, Interscience Communications Ltd., London, England, 1996, pp. 671–680.

Thomas, P. H., "Fire Modelling: A Mature Technology," *Fire Safety Journal*, Vol. 19, No. 2–3, 1992, pp. 125–140.

Thompson, P. A., and Marchant, E. W., "SIMULEX: Developing New Computer Modelling Techniques for Evaluation," *Proceedings* of the Fourth International Symposium on Fire Safety Science, Intl. Assoc. for Fire Safety Science, Boston, MA, 1994, pp. 613–624.

Thompson, P. A., and Marchant, E. W., "Testing and Application of the Computer Model 'SIMULEX'," *Fire Safety Journal*, Vol. 24, No. 2, 1995, pp. 149–166.

Tubbs, J. S., and Barnett, J. R., "Modeling the NIST High Bay Fire Experiment With JASMINE," *Proceedings* of the International Conference on Fire Research and Engineering, Sept. 10-15, 1995, Orlando, FL, SFPE, Boston, MA, 1995, pp. 383–388.

Walton, W. D., "Introduction to Mathematical Fire Modeling. [Book/Software Review of An Introduction to Mathematical Fire Modeling by David M. Birk.]," *Fire Technology*, Vol. 28, No. 1, Feb. 1992, pp. 87–91.

Walton, W. D., "Zone Computer Fire Models for Enclosures," National Institute of Standards and Technology, Gaithersburg, MD, NFPA SFPE 95, LC Card Number 95-68247, ISBN 0-87765-354-2; *SFPE Handbook of Fire Protection Engineering*, 2nd Edition, Section 3, Chapter 7, National Fire Protection Assoc., Quincy, MA, 1995, pp. 3/148–151.

Watts, J. M., Jr., "Future of Fire Modeling. Editorial," *Fire Technology*, Vol. 27, No. 2, May 1991, pp. 97–98.

Wilson, J. L., and Bryan, P. Y., "3-D Modeling as a Tool to Improve Integrated Design and Construction," Lehigh Univ., Bethlehem, PA, Source Document 104, Nov. 1994.

Yung, D., "FIRECAM: A Computer Fire Risk Evaluation and Cost Assessment Model," National Research Council of Canada, Ottawa, Ontario, IRC-ORAL-142; 1996 *Proceedings* of the Institution of Fire Engineers (Ontario Branch), Ontario, Canada, 1996, pp. 1–18.

INTEGRATION OF FIRE MODELS WITH THE DESIGN PROCESS

Frederick W. Mowrer

The methods of quantitative building fire hazard assessment have been developed and refined during the past decade to the point where they can be used innovatively as part of the building fire safety design process. This chapter addresses a rational design framework for the application of quantitative fire analysis models. Design elements and attributes appropriate for different types of quantitative fire analysis models are developed. Traditional and computer-based design implementations are discussed.

Concurrent with the development of quantitative fire analysis models, the technology of building design has been changing with the development of computers and software suitable for design purposes. This change in design technology, commonly called computer-aided design (CAD), permits the design process to be implemented in terms of meaningful design elements. This new technology permits unprecedented integration of engineering analyses into the building design and documentation process. Features of CAD systems that allow this integration of design and analysis are discussed in this chapter.

The integration of fire models with the more traditional tools of engineering and design is the ultimate challenge to both fire safety scientists and fire safety engineers—the ultimate application of research. If fire models are integrated into design or engineering, the result may be termed "performance-based design." Section 11, Chapter 9, described this exercise as applied to codes and standards where it is called performance-based codes. This chapter describes the same exercise applied by an individual engineer taking advantage of the increasing computerization of the basic design process.

A number of concepts presented here are also discussed in other chapters. The fire safety concepts tree, discussed in Section 1, Chapter 3, "Systems Concepts for Building Fire Safety," provides a logical and fairly comprehensive qualitative framework for the analysis and design of building fire safety. Fire modeling provides a mechanism to address quantitatively the topics addressed qualitatively in the fire safety concepts tree. Section 11, Chapter 5, "Deterministic Computer Fire Models," Section 11, Chapter 10, "Simplified Fire Growth Calculations," and Section 11, Chapter 11, "Simplified Fire Hazard and Risk Calculations," discuss some of the currently available fire modeling techniques. Section 11, Chapter 7, "Fire Hazard Analysis," also addresses some concepts similar to those presented here.

Frederick W. Mowrer, Ph.D., is assistant professor of fire protection engineering at the University of Maryland.

WHAT IS DESIGN?

Design is decision making. It is the selection and orientation of basic design elements, which may be materials, products, and assemblies, to serve functional and aesthetic objectives within the constraints imposed by cost, available technology, and safety. The suitability of a product for a particular application is determined by analysis. Different analyses use different properties or attributes of design elements to measure their performance. Design elements selected for use must demonstrate acceptable performance based on these analyses.

Historically, the analysis and design of building fire safety has occurred largely within the context of building code or other regulatory compliance. Within this context, a building is evaluated in terms of the conformance of individual design elements with code requirements. Such compliance should be considered as a necessary but not sufficient condition of adequate fire safety. The selection of design elements for code compliance is necessary from a legal standpoint, but it is not sufficient to demonstrate that adequate fire safety has been provided because current building regulations do not necessarily address the actual fire hazards imposed by products in end use. These hazards may include additive or synergistic effects among design elements that can only be evaluated in the context of use.

The methods available to analyze the actual hazards of products in building fires and the fire mitigation strategies used to counter these fires continue to expand and improve. These analysis tools provide a mechanism to consider the consequences of building fires in quantitative terms with physical meaning. With this ability, a rational design framework similar to ones used in other engineering disciplines can begin to be formalized.

AN ENGINEERING DESIGN FRAMEWORK

A general design framework can be considered in terms of the processes and relationships depicted in Figure 11-13A. A design project begins with problem definition and continues with an iterative design process, composed of analysis, synthesis, and documentation. During this process, the performance of proposed designs is evaluated in terms of design goals and constraints. Ultimately, this process yields a final design.

The first step of the design framework, problem definition, should include the clear specification of design goals and constraints in terms amenable to analysis. These goals and constraints

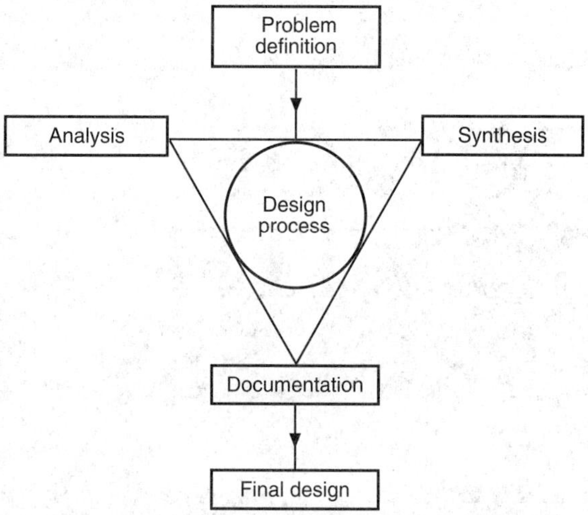

FIG. 11-13A.   General design framework.

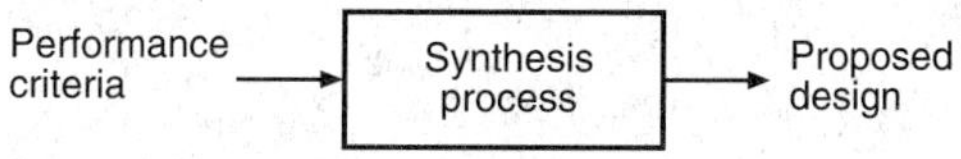

FIG. 11-13C.   The synthesis process.

provide the basis for evaluation of design solutions. The more specifically and operationally a goal or constraint can be defined, the better it can be addressed by quantitative analysis.[1] The topic of problem definition, as applied to fire safety analysis and design, is addressed in more detail below.

Figure 11-13B illustrates the basic concept of analysis in terms of a simple flow diagram. In the analysis segment of the design process, design data are used within an algorithm or model to yield some measurement of performance. The purpose of analysis is to evaluate the suitability of design elements for their anticipated uses. A number of different analyses may be needed to determine this suitability. Foremost, a design element must fulfill its function while maintaining a reasonable level of safety. Other factors, such as cost, will aid in the selection among those design elements meeting the primary criteria.

Models with different levels of complexity and sophistication are now available for the analysis of building fire safety. Computer-based models, such as those discussed in Section 11, Chapter 5, "Deterministic Computer Fire Models," support more comprehensive analyses than simple calculation models, such as those described in Section 11, Chapter 10, "Simplified Fire Growth Calculations." But the use of the more sophisticated models may not result in changes in design decisions because only a discrete number of suitable design alternatives may exist. From a design standpoint, differences in the capabilities of analytical models are important only insofar as they permit distinctions in the selection of design elements.

Figure 11-13C illustrates the concept of solution synthesis in terms of a basic flow diagram. Solution synthesis is the process of evaluating the results of the various analyses in terms of program requirements and performance criteria to develop acceptable design solutions. The final design should be selected from the subset of solutions that fulfill all the design goals and satisfy all constraints.

Solution synthesis is an extremely complicated task, requiring all aspects of design to be evaluated before a decision is made. Some formal synthesis techniques to reduce the number of alternative designs to be analyzed do exist for specialized applications, but

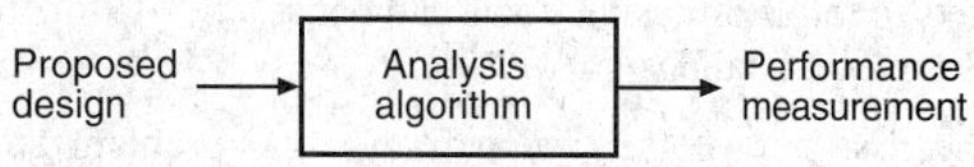

few synthesis algorithms short of exhaustive search procedures have been developed. Normally, it is the experience of a designer that permits the universe of potential designs to be pared to a manageable list of realistic solution alternatives to be evaluated in detail.

Design documentation completes the design process trilogy. Documentation is essential for the recording of design decisions and for the communication of these decisions to clients, consultants, and contractors. This documentation consists of two types of information: geometric data and physical data. Geometric data are most effectively recorded graphically on drawings to depict the dimensions, shapes, locations, and spatial relationships of design elements. Physical attributes of elements, such as models, weights, material properties, or costs, normally are recorded and conveyed by means of specifications, schedules, and calculations.

Ideally, the design process iterates until a clearly superior final design that meets all program requirements and constraints results. Since many issues are likely to be defined in fuzzy terms and because conflicting goals and constraints may exist, the concept of an optimum design may not be easy to characterize or measure. Typically, the final design is the design deemed most acceptable to be developed within the constraints of a limited design period.

This general design framework operates at various levels in the building design process because one designer's constraints may become another designer's goals. Product manufacturers can use this framework to evaluate and communicate the uses and limitations of the design elements they produce. These limitations then become constraints for building designers, who attempt to coordinate a multitude of design elements.

The goals of the various engineers involved in the design of a building's component subsystems include the assurance that cost, technical, and safety constraints imposed by the architect's functional design are adequately considered. Thus, the architect's constraints frequently become part of the engineering designer's goals. Since both goals and constraints must be satisfied, perhaps the major impact of this distinction is in terms of financing; there seems to be a greater inclination to spend money to fulfill goals than to satisfy constraints, particularly if a constraint is ill-defined or its purpose is not clear.

DEFINITION OF FIRE SAFETY GOALS AND CONSTRAINTS

The implementation of a rational fire safety design framework requires the definition of operationally specific goals and constraints. While a general requirement for adequate fire safety is satisfactory as a universal goal, designers need specific performance criteria to evaluate the acceptability of design solutions. Until standardized fire safety criteria that address actual performance are developed, the use of quantitative fire analysis models will be restricted in practice. Nonetheless, such models remain valuable tools that provide better insight for design decision making.

Traditional fire safety goals have been defined in terms of pass-fail criteria specified in building and fire regulations. These criteria frequently are specified in terms of various fire indices (e.g., flame spread ratings), which do not describe or appraise the fire hazard or fire risk of a product's performance under actual fire conditions.

FIG. 11-13B.   The analysis process.

Pass-fail criteria for fire, such as those used by building codes, have difficulty addressing important factors of fire safety, including:

1. The transient characteristics of fires
2. The transitory placements of combustibles in buildings
3. The number of possible ignition and fire scenarios
4. Differences in damage susceptibility or vulnerability
5. The continuous nature of hazard and damage caused by fire

In other words, it is difficult for such criteria to address the specific context of use of a product, but it is just this context that determines the actual hazard represented by a product. Fire modeling provides a means to evaluate the fire performance of a product in a particular application.

COMPONENTS OF FIRE SAFETY DESIGN

A rational framework for fire safety design can be considered in terms of four component parts:

1. Fire hazard development
2. Building response
3. Occupant response
4. Fire protection

The relationships among these components of fire safety design can be considered qualitatively in terms of a "tetrahedron of fire safety," as illustrated in Figure 11-13D. Just as the tetrahedron of fire has been suggested to show the direct linkages between each of the four elements required to sustain flaming combustion, the fire safety tetrahedron shown in Figure 11-13D is intended to show that each component of fire safety design is connected to each other component and that all the components and the interrelationships are time dependent. (See Section 1, Chapter 8, "Theory of Fire Extinguishment.")

Time is the common dimension of fire. Most meaningful fire parameters can be considered as functions of time. The physical hazards of fires are measured in terms such as fire growth rates, flame spread rates, heat release rates, heat fluxes, and burning durations, all functions of time. Similarly, the response of buildings, occupants, and fire protection systems to different fire scenarios can be considered in terms such as fire endurance periods, evacuation times, and fire detection and suppression response times. Consequently, time is an appropriate parameter to use for the quantitative evaluation of alternative fire mitigation strategies.

Fire mitigation strategies for particular scenarios can be evaluated in terms of two opposing times:

1. The time to the onset of unacceptable damage or hazard
2. The time required for fire burnout or control, or for egress

If the first time exceeds the second for all foreseeable fire scenarios, a fire protection strategy can be considered successful. Cooper[2] applied this concept to the evaluation of available safe egress time. Section 11, Chapter 7, "Fire Hazard Analysis," discusses these concepts in more detail.

All unwanted fires cause some damage and present some hazard. To permit quantitative evaluation of fire mitigation strategies, a threshold level of acceptable damage or hazard must be established. The time to the onset of unacceptable damage or hazard for a particular fire scenario depends on two offsetting factors:

1. The susceptibility to damage of an object being evaluated
2. The rate of development of a hazardous environment at the location of the object being evaluated

If these two issues can be assessed quantitatively, they can yield either a time-dependent scale of damage or a time of critical damage onset for a particular fire scenario. The susceptibility to damage of inanimate objects is the subject of the building response component of fire safety design, while the susceptibility of people is part of the occupant response component. The rate of hazard development depends on the fire scenario as well as the fire protection methods employed.

Damage to some objects may be related primarily to the total quantity of a fire signature, such as heat or a smoke component, released during a fire. This is represented as the area under a time-fire signature release rate curve, as illustrated in Figure 11-13E. Other objects may be considered damaged once a critical release rate of a particular fire signature is reached, represented as a single point on a time-release rate curve. As indicated in Figure 11-13E, these threshold damage values yield critical times for particular fire hazard scenarios.

Once threshold damage criteria are established, the first question to be addressed by a hazard analysis is whether critical conditions are ever reached for a particular fire scenario. If so, the time at which critical conditions occur provides the maximum acceptable time for fire burnout or control. Once a time scale of damage is established for a particular scenario, the effectiveness of alternative

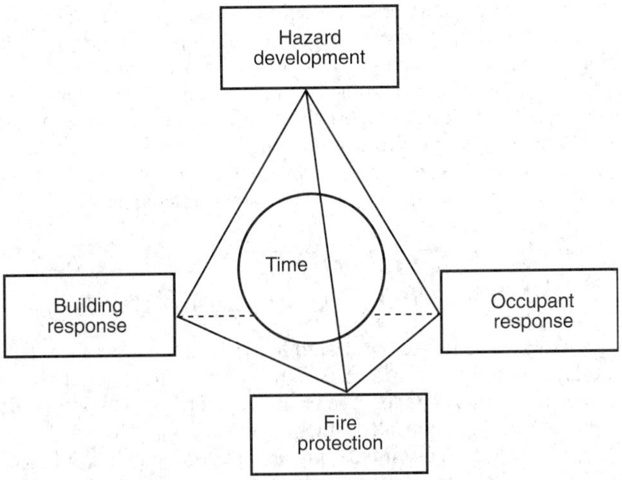

FIG. 11-13D. *Components of fire safety design.*

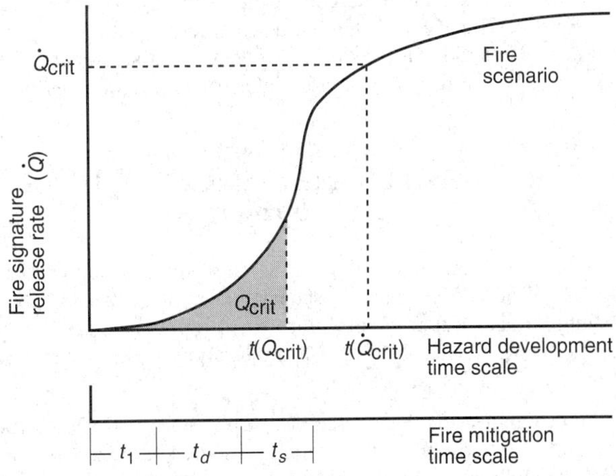

FIG. 11-13E. *Concepts of hazard development and fire-mitigation time scales.*

fire mitigation strategies can be considered in terms of their response characteristics.

The total time for control of fires that do not burn out without intervention can be characterized in terms of three distinct intervals:[3]

1. A transport time lag, t_1
2. A detection time lag, t_d
3. A suppression time lag, t_s

These periods are shown qualitatively on the fire mitigation time scale in Figure 11-13E.

The transport time lag represents the interval between the actual generation of heat or another fire signature and the transport of that signature to a fire detection device. The detection time lag represents the period from the first transport of a fire signature to a sprinkler or other fire detector until the device actuates due to detection of a threshold value of that signature. The final time delay, the suppression time lag, represents the period from fire detection until fire suppression commences. For example, wet-pipe sprinkler systems will have a nil suppression lag period, while dry-pipe systems designed in accordance with NFPA 13, *Standard for the Installation of Sprinkler Systems*, should have not more than a 1-min delay associated with the time needed for the dry-pipe valve to trip and for water to fill the piping. For manual fire fighting, suppression lag periods in excess of 10 min are common because they include the times associated with signal verification, with fire department notification and response, and with preliminary fireground operations.

Within a rational fire safety design framework, the primary goal of fire mitigation strategies is to ensure that the combined period of these three lag times is less than the time to the onset of unacceptable damage. If not, two general fire mitigation alternatives exist: either (1) the total time to fire control can be reduced through the selection of appropriate fire protection systems, or (2) strategies or the time to the onset of critical damage must be increased. This can be accomplished by decreasing the rate of hazard development or by increasing the damage threshold level of target design elements. Within this context, the duties of the fire safety designer can be characterized as:

1. Recognize and characterize potential fire hazard scenarios
2. Establish anticipated damage histories for these scenarios
3. Evaluate alternative fire mitigation strategies
4. Communicate findings and recommendations to the client

The last duty should include the clear enunciation of expected consequences for each of the alternative fire mitigation strategies evaluated. With this information, more educated assessments of fire risk and decisions regarding appropriate protection strategies can be made.

THE TRADITIONAL ENGINEERING DESIGN PROCESS

The traditional implementation of the engineering design process can be considered as depicted schematically in Figure 11-13F. A drawing or sketch is normally prepared to represent a preliminary design iteration. This drawing constitutes a physical representation of the proposed design.

The engineering designer, depicted in Figure 11-13F as a stick figure, must interpret the information presented in the drawing and must manually extract pertinent geometric and physical data from the drawing. Typically, the engineer also manually extracts some physical data needed for an analysis from catalogs, handbooks, manuals, or other similar library references.

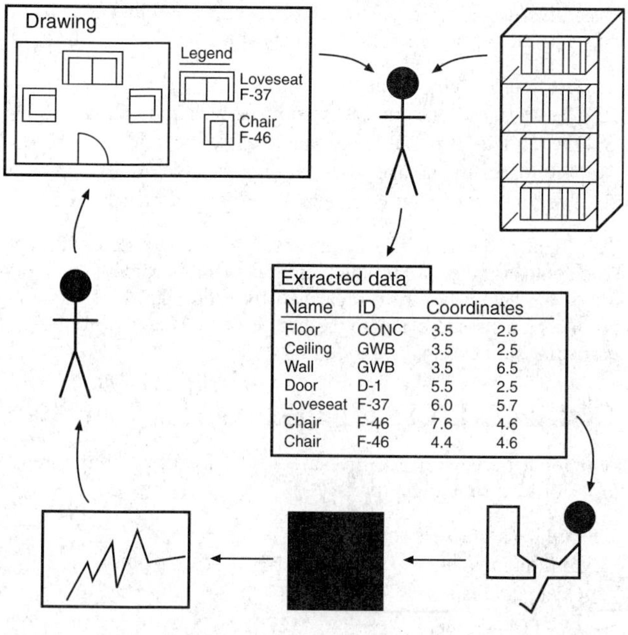

FIG. 11-13F.   *Traditional implementation of the engineering design process.*

Geometric and physical data extracted from design drawings, and reference materials are compiled for input to an analysis model. Analysis might be performed by hand or with the aid of a calculator, particularly for preliminary design iterations, but for the present discussion it is assumed that the analysis model, is computer-based. In this case, the extracted data must be translated into a format appropriate for input to the analysis model and the data must be entered into the computer, typically through a keyboard. This translation process is conceptually simple, but it can be troublesome in practice. For complex analytical models, the data entry process can be tedious and prone to errors in the values of numerical entries and in the format of input data.

Once the data are entered, the analysis model, depicted in Figure 11-13F as a black box to represent its common treatment by designers, is run. Analysis models may be conceptually simple as checks for code compliance or as complex as enclosure fire models. In any case, the model outputs some measure of performance. This output is reviewed and interpreted by the designer, who decides if the performance is acceptable. If not, the engineer may change the solution directly and iterate through the design process until an acceptable solution is found or may report findings and recommendations to the primary building designer for consideration in the next design iteration. This traditional design process has numerous shortcomings. The process is time consuming and thus limits the number of alternative designs that can be evaluated within a limited design period. The process also has much potential for data extraction and input errors because these procedures are handled manually.

A COMPUTER-AIDED DESIGN (CAD) APPROACH TO ENGINEERING

A CAD-based engineering design implementation offers some advantages over the traditional design process. Unlike any previous design medium, computers can be used to represent and process design elements symbolically rather than physically. The capabilities of the computer as a symbolic processor permit the design process to be conceived and implemented in terms of meaningful design elements.[4]

An overview of a computer-based appraisal methodology is illustrated in Figure 11-13G. The CAD system serves as a graphic interface between the designer and a graphics data base maintained in the computer. Design development becomes an act of inserting appropriate design elements into the design-drawing data base, rather than one of drafting on paper physical representations of these elements that require interpretation.

The key distinction of computer-based building descriptions that permits implementation of this design methodology is known as *object orientation*. Physical representations of design, such as manually developed drawings, require human identification and interpretation; symbolic representations permit automatic identification. For example, the seven lines used to represent the chair shown in Figure 11-13G can be defined as a chair within a computer. The symbol representing the chair then may be inserted, deleted, or otherwise manipulated and processed as a distinct unit. It also may be stored in a library of symbols for use in other graphic data bases. The computer recognizes the occurrence of this symbol and thus can be used, for example, to extract the number and locations of chairs in a drawing.

The second key feature of CAD programs is the ability to associate attributes with objects. While most current microcomputer-based CAD programs support object orientation, some do not support attribute association and extraction.[5] This feature permits the association of physical or geometric data with objects. These data then may be extracted along with the object for use in an analysis model.

Graphics data bases generated with most current microcomputer-based CAD systems are best suited for storing geometric data regarding objects, such as their locations and dimensions. This geometric data then may be extracted from the graphics data base for use in an analysis model. In general, separate text data bases are more suitable for storing physical attributes of objects. The only nongeometric data other than the object type that must be assigned and stored in a graphics data base is a key attribute used by the CAD system to identify each object uniquely. The key attribute provides cross-reference to text data bases, where physical properties of objects are maintained.

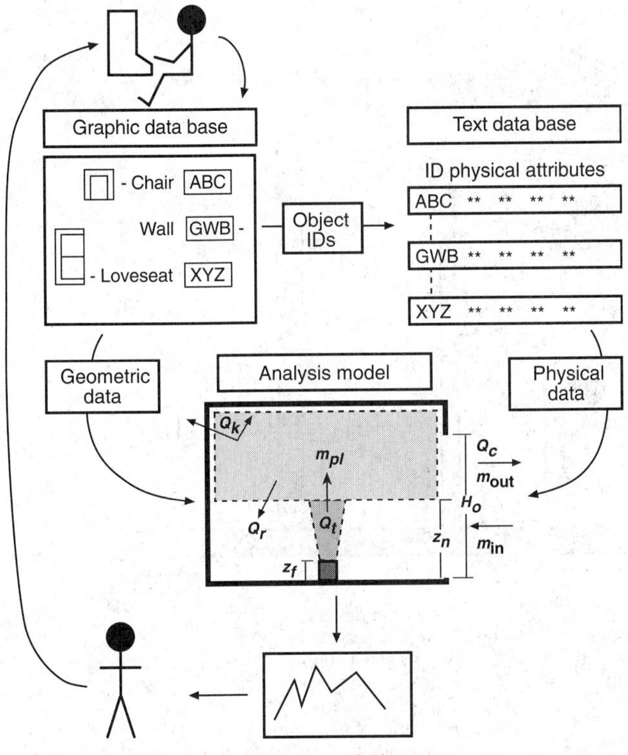

This arrangement minimizes the nongeometric data storage requirements for a graphics data base as well as the data needed at the immediate disposal of the designer. This tends to reduce the potential for user input error, but does not eliminate it entirely because the designer still might enter the wrong ID for an object.

This practice also has the benefit that any changes or additions to the physical attributes or objects only need to be made within the text data base. If these properties were stored with each occurrence of objects in a graphics data base, then each occurrence of an object would have to be edited to revise or append the attributes associated with the object. As new analyses are developed, the necessary attributes can be appended to the existing text data bases and the new analyses performed. In this way, changes to graphics data bases are necessary only to document changes involving design elements, not their attributes.

Geometric attributes of objects are extracted from the graphics data base, and physical attributes are extracted from one or more text data bases to serve as input to the analysis model. Once established, this extraction process is straightforward, and the details can be transparent to the user. The difficulty lies in assuring that the extracted data has the structure needed for input to the analysis model. Since the output formats of current CAD programs are not very flexible, this may require revision of an analysis model's input data structure or the development of separate intermediate programs, known as preprocessors, to translate from CAD output data structures to analysis model input data structures.

Once the data are extracted for use in the analysis model, the model is run to produce some measure of performance. The model output is interpreted by the engineer, who decides if the performance is acceptable or if further design iterations are necessary. This process is conceptually the same as for the traditional methodology at present, although the details of the implementation may differ. In the future, expert systems might be used at this step to evaluate performance and to help synthesize subsequent design alternatives. The capability of the computer as a symbolic processor would permit this cognitive type of task.

CAD systems that support object orientation and attribute association permit the user to define the desired objects and attributes. The definitions of objects and their attributes depend on the analysis to be performed. For different analyses, it is appropriate to utilize descriptions based on different types of elements.[6] One goal of the design community is to establish an integrated building-description data base that can be used for documentation throughout the life cycle of a building.

Design elements and attributes for the different components of fire safety design are considered next. These same elements and attributes apply for traditional analyses; the object orientation of CAD simply highlights the treatment of design as a process of selecting and orienting suitable design elements based on their attributes.

ELEMENTS AND ATTRIBUTES FOR FIRE SAFETY DESIGN

Analysis models have two primary components: (1) data and (2) algorithms. These components cannot be considered independently of one another during model development because the type and structure of the data depends on the algorithm to be used, and the structure of an algorithm depends on the type of data available. A model

FIG. 11-13G.   *Computer-based implementation of the engineering design process.*

that requires unavailable data serves no current practical design purpose, although it can serve to underscore the need for the acquisition of a particular type of data. This discussion focuses on the design element and attribute data needs of some current fire analysis models. It is intended to provide an overview of what the data needs are for quantitative fire analysis, not to be a comprehensive summary.

Quantitative fire analysis models can be grouped into the categories identified above as the components of fire safety design. Hazard development models calculate expected fire conditions in buildings, while models of building response, occupant response, and fire protection calculate the expected reactions of these components to the imposed conditions. In general, the component models cannot be considered independently because of their influences on each other. For example, the operation of an automatic suppression system will alter the hazard development; ventilation may be changed during the course of a fire due to automatic or occupant-induced changes in the natural or mechanical ventilation of spaces. For the present discussion, these coupled influences are ignored, and the models are considered separately to identify the elements and attributes of each component. The general types of design elements for fire safety design are illustrated schematically in Figure 11-13H.

FIRE GROWTH MODELS

Fire growth models, so-called, are really models of the growth and spread of fire effects. They may also be referred to as hazard development models, although a full calculation of "fire hazard," in the sense of Section 11, Chapter 7, "Fire Hazard Analysis," includes factors beyond the scope of fire growth models. These models typically calculate physical conditions resulting from fire, such as fire gas temperatures, velocities, and compositions, as functions of time and location. Models, ranging in complexity from fairly simple data correlations through network and zone models to field models, are described in Section 11, Chapter 5, "Deterministic Computer Fire Models," and Section 11, Chapter 10, "Simplified Fire Growth Calculations." The distinctions among models address differences in the methods and algorithms used to treat the physics more so than differences in the design elements being analyzed. The more sophisticated a model is, the more physical attributes will need to be associated with the design elements. The basic design elements, however, remain the same.

The elements needed to characterize the growth of fire effects within a space can be considered in terms of the elements numbered one through four in Figure 11-13H. A burning fuel element, which may be a furnishing or a finish, releases heat and products of combustion. Some of the heat is transferred to solid boundaries, some of it convects through open boundaries via gas flow, and some of it heats other combustible objects within the enclosure. Smoke signatures generated by the combustion reaction are transported by the buoyant plume from the burning object.

Design elements and attributes needed for quantitative enclosure fire hazard analyses are listed in Table 11-13A. These elements include, in general, the fuels, the openings available for natural ventilation, the mechanical ventilation systems, and the solid boundaries enclosing the space.

Fuels that burn in enclosure fires can be distinguished as either furnishings or finishes. Furnishings are discrete objects of fixed initial dimensions and mass. Finishes have dimensions and masses that depend on the geometry of the application; they include wall, ceiling, and floor linings. In residential buildings, combustible furnishings include items such as chairs, sofas, beds, desks, and wardrobes. In industrial buildings, furnishings would include pieces of machinery and equipment, which may use flammable or combustible liquids, as well as pallets or racks of stored products.

The attributes to be associated with these elements are both geometric and physical, as indicated in Table 11-13A. Geometric attributes typically include the dimensions and locations of design el-

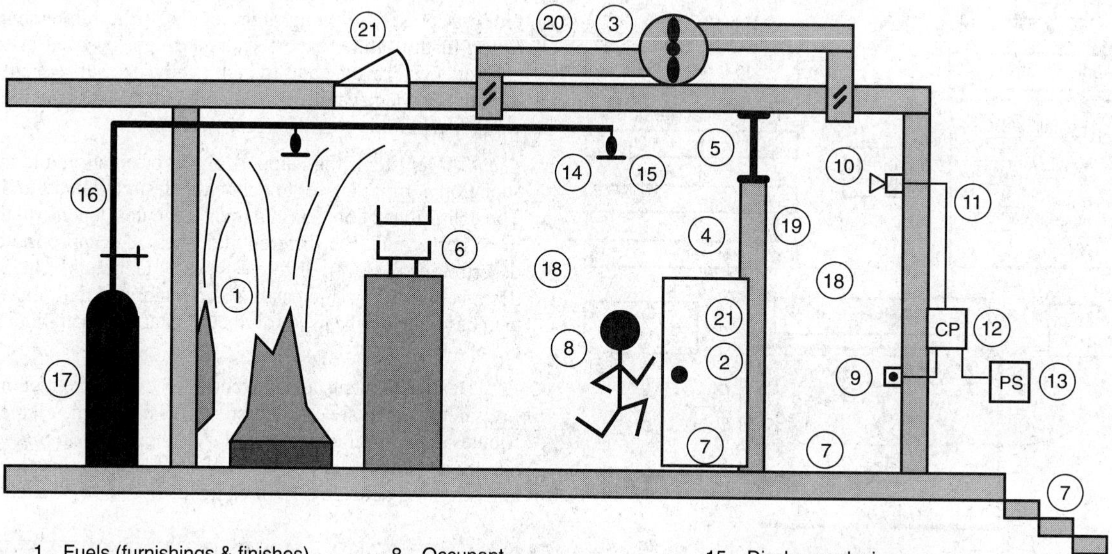

1. Fuels (furnishings & finishes)
2. Natural ventilation openings
3. Mechanical ventilation systems
4. Enclosure boundaries
5. Structural elements
6. Operational elements
7. Egress path components
8. Occupant
9. Alarm initiating devices
10. Alarm indicating devices
11. Alarm system circuits
12. Control panel
13. Power supply
14. Detection devices
15. Discharge devices
16. Distribution system
17. Suppressant supply
18. Smoke zones
19. Smoke barriers
20. Smoke-control systems
21. Smoke vents

FIG. 11-13H. General types of building fire safety design elements.

TABLE 11-13A. *Elements and Attributes of Fire Growth Modeling*

General Type	Design Element	Physical Attributes	Geometric Attributes
Fuels	Furnishing finish	Material quantity release rate	Locations Dimensions
Natural ventilation openings	Door Window Vent	Status Act. parameter	Locations Dimensions
Mechanical ventilation systems	Injection Extraction Recirculation	Flow rate Status Act. parameter	Locations
Boundaries	Floor Wall Ceiling	Material	Locations Dimensions

TABLE 11-13B. *Elements and Attributes of Building Response Analysis*

General Type	Design Element	Physical Attributes	Geometric Attributes
Structural	Beam Column Slab	Material Damage parameter	Locations Dimensions
Operational	Equipment Machinery Commodity	Damage parameter	Locations Dimensions

ements. The specific physical attributes to be associated with objects depend on the algorithms used for a particular analysis.

The primary physical attributes of furnishings and finishes needed for fire growth models include descriptions of heat and smoke release rate histories. These may be expressed as characterizations of large-scale fire test data[7] or in terms of fundamental material properties, depending on the algorithms used by a particular model to consider burning rates. For models that calculate the conditions resulting from a prescribed burning history, the first characterization would be appropriate. For models that attempt to predict such burning histories, fundamental data would be needed.

Normally, open natural ventilation openings have only geometric attributes, unless they are designed to close automatically upon detection of a fire signature. These devices, as well as normally closed devices such as roof vents or windows that may open or break during fires, require the association of one or more activation parameters. For these devices, an activation parameter, such as temperature, and a thermal response parameter, such as a time constant or response-time index, could be associated with the device to indicate the conditions under which it will operate. For mechanical ventilation systems, similar actuation parameters would be necessary for systems that are designed to turn on or off automatically under fire conditions.

Material properties are the physical attributes required to model the response of solid boundaries. These properties can be associated with a single parameter that identifies the boundary material. Only this single material attribute needs to be associated with the boundary in the graphics data base; it can then be used to access the thermal properties from a text data base, as illustrated schematically in Figure 11-13G.

BUILDING RESPONSE MODELS

Design elements and attributes for the analysis of building response to fire are listed in Table 11-13B. The general types of building response design elements are illustrated in Figure 11-13H as elements five and six. Two general types of design element are considered under building response: (1) structural and (2) operational. The design elements for detailed structural response models include slabs, beams, and columns. Operational design elements include equipment, such as electrical cable trays and cabinets, machinery, and stored commodities that may be damaged by fire. If combustible,

these same elements may also constitute furnishings with respect to the analysis of hazard development.

Physical attributes of structural and operational elements include their thermal properties, normally represented by a material attribute, and one or more damage parameters, such as a critical temperature or heat flux. The location is needed to evaluate the incident fire conditions imposed on an element for a given scenario, while the dimensions may be needed to evaluate the fire conditions at an element as well as the response of the element to the imposed conditions.

The response of buildings to fire conditions can be considered at different levels of complexity. Simplified methods to evaluate the thermal response of building elements, such as exposed structural steel members, have been developed.[8] These models use lumped capacitance formulations to consider the transfer of heat from hot fire gases to heat-sensitive design elements. More sophisticated models employ finite element analysis techniques to calculate in greater detail the thermal and structural response of building elements to fires of specified severity.[9,10] Again, these distinctions relate more to the algorithms used and the attributes needed to analyze the response of building elements than to the definition of the basic design elements.

OCCUPANT RESPONSE MODELS

Unlike the other design elements described here, occupants are expected to move during a fire scenario. This movement not only influences the fire conditions to which an occupant is exposed, but may also affect the fire ventilation, the movement of smoke, and the operation of fire protection systems. Nonetheless, with respect to occupant safety, it is possible to list the design elements and attributes normally considered as part of the fire safety design process.

There are two primary aspects of occupant response to fire: (1) tenability and (2) evacuation. For a given fire scenario, tenability criteria are used to evaluate the time to the onset of untenable conditions, which establishes a maximum time for safe egress. Tenability criteria might include individual toxic combustion product concentrations, smoke obscuration, or a combination of factors.

Evacuation times are a function of the number of people in a space or building and their flow rates. The flow rate of people can be calculated in general as a function of travel speed, egress width, and the density of people.[11,12]

The speed of travel depends on the mobility of the occupants, on their density, and on the type of egress component (e.g., stair, ramp, or corridor). The density of people starts at the normal occupant load of the space, but typically increases as people queue at bottlenecks in the egress path, such as doors leading to exit stairs. Once evacuation begins, the situation becomes dynamic, with varying densities and travel speeds. While the complete analysis of evacuation time problems may require complex computer-based solutions, the basic design elements and attributes can be described simply.

Table 11-13C lists the elements and attributes of the occupant response component of fire safety design. The fundamental types of design elements for occupant response include egress path components and people, identified as elements seven and eight in Figure 11-13H. People are represented for design purposes in terms of rooms or other occupied spaces. Therefore, the specific types of design elements include occupied spaces and egress path components. Occupied spaces are distinguished by their uses, which will have the physical attributes of an occupant load factor, an occupant mobility factor, and tenability criteria. Geometric attributes include the floorplan dimensions, from which the area and total occupant load can be determined along with the travel distance within the space.

TABLE 11-13C.   *Elements and Attributes of Occupant Response Analysis*

General Type	Design Element	Physical Attributes	Geometric Attributes
Occupant	Room	Load factor Mobility factor Tenability factor	Locations Dimensions
Egress Path	Door Corridor Ramp stair	Speed factor	Length Width

Egress path design elements include doors, corridors, ramps, and stairs. These elements have just a single physical attribute, an evacuation speed parameter, which, for stairs, is a function of the riser/tread design and whether the egress direction is upward or downward. This speed parameter is also a function of the occupant density. Geometric attributes for egress path elements include an effective width as well as a length, measured along the path of travel.

FIRE PROTECTION SYSTEMS MODELS

Building fire protection includes many types of systems, including fire detection and alarm, fire suppression, and smoke control. The basic design elements and attributes of these systems are discussed here to demonstrate the treatment of the design of fire protection systems as the selection and orientation of design elements.

The primary design elements of fire detection and alarm systems include alarm initiating devices, alarm indicating devices, control panels, and power supplies. Alarm system circuits can also be considered as distinct design elements. These elements are listed in Table 11-13D and illustrated in Figure 11-13H as elements nine through thirteen.

Alarm initiating devices may be manually or automatically actuated. For automatic alarm initiating devices, the physical attributes will include an actuation parameter and a response parameter. The actuation parameter is associated with the threshold concentration of a fire signature required to actuate the device. For heat detectors, this would be the actuation temperature. The response parameter is associated with the inertia of the device. For heat detectors, this would be a time constant or response-time index. Manual initiating devices do not have similar physical attributes; they rely on human intervention for actuation. For sizing power supplies and circuits, the power-demand attributes of the different types of alarm initiating devices are needed.

TABLE 11-13D.   *Elements and Attributes of Fire Alarm Systems*

General Type	Design Element	Physical Attributes	Geometric Attributes
Initiating device	Pull station Fire detector Supervisory	Actuation parameter Response parameter Current demand	Locations
Indicating device	Control panel	Input capacity Output capacity Control logic	Sizes Locations
Control panel	Initiating indicating control power	Current capacity	Locations
Power supply	Primary Secondary	Current capacity duration	Locations Sizes

Indicating devices include audible and visual alarm signaling devices as well as trouble and supervisory signals. For alarm signaling devices, the physical parameters include the intensity and frequency of the device output. With this data, the audibility or visibility of the device can be evaluated for a range of use conditions. The power demand requirements for these devices are also needed to evaluate power supply and circuit size requirements.

The other elements of fire detection and alarm systems are characterized primarily in terms of their electrical attributes. The primary quantitative issues for control panels, circuits, and power supplies are their current capacities. The input and output capacities of control panels establish limitations on their specific applications, both in terms of the number of zones that can be served and the number of devices per circuit. Each alarm initiating device and alarm indication device imposes a current demand on the system. The primary and secondary power supplies must be able to support the power demands imposed by components of the fire alarm signaling system for a specified duration, which is usually established by code.

The control logic of the control panel is an essential component of fire alarm system design. So is system reliability. These components do not fit conveniently into the quantitative design framework presented here. Suffice to say here that control logic and reliability are important design components for fire alarm signaling systems, but they are outside the scope of this discussion.

The general types of design elements for fire suppression systems, the specific design elements, and their physical and geometric attributes are listed in Table 11-13E. Automatic fire suppression systems have four general types of components: (1) detection devices, (2) agent discharge devices, (3) a distribution system, and (4) a supply of extinguishing agent. These are illustrated in Figure 11-13H as elements fourteen through seventeen.

The physical attributes of detection devices required to evaluate their response to a given fire scenario include an actuation parameter and a response parameter. For thermally actuated devices such as sprinklers and heat detectors, the actuation parameter is the temperature rating of the device, while the response characteristic is an appropriate thermal time constant or response-time index.[13] These attributes can be used in conjunction with appropriate models of the environmental fire conditions at the detection device to yield an estimate of the fire detection time for a given fire scenario. These environmental conditions depend on the locations of the detection

TABLE 11-13E. *Elements and Attributes of Fire Suppression Systems*

General Type	Design Element	Physical Attributes	Geometric Attributes
Detection device	Sprinkler Heat detector Smoke detector	Actuation parameter Response parameter	Locations
Discharge device	Sprinkler nozzle	K factor	Locations
Distribution system (hydraulic)	Piping Fittings Auxiliary devices	Friction factors	Lengths Sizes Locations
Suppressant supply	Tank Pump/reservoir	Pressure quantity	Locations Elevations

TABLE 11-13F. *Elements and Attributes of Smoke Control Systems*

General Type	Design Element	Physical Attributes	Geometric Attributes
Smoke zone	Room Corridor shaft	Pressure Temperature	Locations Dimensions
Smoke barrier	Floor Wall/Ceiling Door (closed)	Effective flow area	Locations Dimensions
Mechanical ventilation system	Injection Extraction	Pressure/Flow Status Act. parameter	Locations
Vent	Door (open) Window vent	Flow coeff. status Act. parameter	Locations Dimensions

devices relative to the fire source, so detector locations are the required geometric attributes.

The primary physical attribute to be associated with discharge devices is a flow coefficient that permits calculation of a pressure-flow relationship for a device. The locations of discharge devices are required to calculate head losses and pressure distributions between points. Algorithms used for these friction-loss calculations apply mass and energy conservation principles between nodes to maintain physical consistency.

Another aspect of discharge devices that is currently considered indirectly during normal design practice is the agent distribution characteristics of the discharge device. Currently, this aspect of discharge device attributes is treated in general by the selection of an appropriate type of discharge device for a given application. This selection is based largely on the discharge tests required to list these devices. In the future, it may become possible to evaluate water-spray/fire interactions with greater precision and confidence. At that time, association of one or more characteristic spray attributes with discharge devices would be appropriate.

The primary interests in the distribution system are with its friction-loss characteristics and its support. Elements of the distribution system that influence friction loss include all components of the flow conduit, including pipe, fittings, valves, and any other auxiliary devices. The attributes needed to characterize friction loss include the length, diameter, and friction coefficient of pipe segments and the equivalent lengths of fittings, valves, and auxiliary devices. The types and locations of support components are required to evaluate stresses in the distribution system under static and flow conditions.

The extinguishing agent supply is the last design element for automatic suppression systems. A pressure-flow relationship of the supply is needed along with the total capacity of the supply. For water-based suppression systems, the capacity of the supply may be virtually limitless, but for other types of systems this is not generally the case. For carbon dioxide, halon, and foam systems, the supply capacity normally will be sized for a specific discharge rate and duration.

The general types of design elements for smoke control systems include smoke zones, smoke barriers, natural ventilation openings, and mechanical ventilation systems. These are illustrated in Figure 11-13H. Table 11-13F lists the basic elements and attributes of smoke control systems.

Smoke zones may be individual rooms or floors, or they may be groups of rooms or floors. They are defined by the smoke barriers that enclose them and by the vents and mechanical air-handling systems used to confine or direct the movement of smoke between zones.

The physical attributes of smoke zones include their initial pressures and temperatures. Subsequent pressures and temperatures may be calculated or may be specified as fixed boundary conditions. The physical attribute required for smoke barriers is an effective leakage area. The attributes of natural ventilation openings and mechanical ventilation systems used for smoke control are virtually the same as described above for hazard analyses.

This concludes the discussion on the basic elements of fire safety design and their attributes. Twenty-one different general types of elements have been identified for the various types of analyses discussed. Many of these elements serve multiple purposes, thus reducing the number of discrete elements that form the basis for fire safety design. For example, the boundary elements for hazard analysis will normally be the same as the smoke barrier elements for smoke control analysis and design, but the physical attributes needed for each type of analysis may be different. It is up to the fire safety designer to select a combination of these design elements that will fulfill all fire safety design goals while satisfying applicable constraints.

The reduction of fire safety design to the selection of a few types of design elements should not be misconstrued as a suggestion that the fire safety design process is simple. The relationships among design components are highly coupled and the number and types of design components in a building can be quite large. Rather, the intent of this discussion is to emphasize two primary points: (1) ultimately, fire safety design must be reduced to the selection and orientation of appropriate design elements to fulfill goals and satisfy constraints; and (2) the purpose of fire analysis models is to aid this selection process, regardless of the sophistication or complexity of the model used.

LIMITATIONS

The potential benefits of a rational design framework for building fire safety are substantial, but they are hindered by a number of limitations at present. These limitations include two types of uncertainty:

1. Uncertainty regarding physical and chemical relationships, and
2. Uncertainty regarding the situation of objects in buildings.

The first uncertainty is being reduced through research. Considerable work remains in the prediction of burning and entrainment rates for real objects. The physical interaction between fire and suppression or other mitigation activities is largely undeveloped. Research continues on these topics.

The second uncertainty is likely to remain and must be addressed through appropriate risk-assessment techniques. The importance of this second uncertainty cannot be overemphasized; fire loads in building spaces can change by orders of magnitude simply due to a rearrangement of furniture.

Finally, there is the lack of standardized performance criteria against which a designer can evaluate the acceptability of a design solution. Such criteria should be developed to permit more widespread implementation of a rational fire safety design framework. In the meantime, fire safety designers can establish their own criteria in conjunction with their clients and can apply rational design concepts to the consideration of their current fire safety problems.

SUMMARY

Design is the selection of building elements to achieve functional goals and to satisfy applicable constraints. The selection of design elements is based on the synthesis of appropriate analyses to determine the suitability of a design element to fulfill its anticipated use. Design elements are the basic components chosen by the designer; attributes are the properties of these elements that distinguish their performance and serve as the basis for the selection process.

Appropriate basic design elements and attributes for different types of quantitative fire analysis models have been discussed. Combustible furnishings and finishes are the design elements of major importance for fire hazard development. Their heat and combustion product release rate histories drive the development of hazardous conditions in buildings and consequently the response of building, occupant, and fire protection system design elements.

Computer-aided design programs use the capability of computers as symbolic processors to implement design in terms of relevant design elements. This permits unprecedented integration of engineering analysis with the building design and documentation process. Once established, CAD implementations reduce the potential for user input error and permit more analyses to be performed within a fixed design period. With CAD-based design documentation, new analyses can be applied to existing situations as the analyses become available. Similarly, design changes that occur through the life cycle of a facility can be analyzed more conveniently with a computer-based building description.

Implementation of a rational design framework requires specification of performance-based design criteria. For fire hazard development and fire protection strategies, time is the common parameter for performance evaluation. Successful fire protection strategies mitigate the effects of fire hazards before they reach unacceptable threshold levels.

In view of the limitations discussed above, quantitative fire analyses should be considered at present in terms of their utility to augment and justify, rather than to supersede, many traditional fire protection techniques. Such analyses serve as quantitative tools that can provide insight for the synthesis of rational fire safety design decisions. These tools do not make decisions, however, nor do they determine appropriate situations for their use. These remain duties of the fire safety designer.

BIBLIOGRAPHY

References Cited

1. Dixon, J. R., *Design Engineering: Inventiveness, Analysis, and Decision Making*, McGraw-Hill, New York, 1966, p. 55.
2. Cooper, L. Y., "A Concept for Estimating Available Safe Egress Time in Fires," *Fire Safety Journal*, Vol. 5, 1983, pp. 135–144.
3. Mowrer, F. W., "Lag Times Associated with Fire Detection and Suppression," *Fire Technology*, Vol. 26, No. 3, 1990, pp. 244–265.
4. Radford, A., and Stevens, G., *CAD Made Easy*, McGraw-Hill, New York, 1987, pp. 130–139.
5. Hart, G., "Complex CAD Software for the PC," *PC Magazine*, Vol. 5, No. 5, 1986, pp. 111–148.
6. Mitchell, W. J., *Computer-Aided Architectural Design*, Van Nostrand Reinhold Company, New York, 1977, p. 112.
7. Mowrer, F. W., and Williamson, R. B., "Methods to Characterize Heat Release Rate Data," *Fire Safety Journal*, Vol. 16, 1990 pp. 367–387.
8. Alpert, R. L., and Ward, E. J., "Evaluation of Unsprinklered Fire Hazards," *Fire Safety Journal*, Vol. 7, 1984, pp. 127–143.
9. Bresler, B., "Analytical Prediction of Structural Response to Fire," *Fire Safety Journal*, Vol. 9, 1985, pp. 103–117.
10. Jeanes, D. C., "Application of the Computer in Modeling Fire Endurance of Structural Steel Floor Systems," *Fire Safety Journal*, Vol. 9, 1985, pp. 119–135.
11. Pauls, J., "Movement of People," *The SFPE Handbook of Fire Protection Engineering*, 2nd ed., DiNenno, P. J., ed., National Fire Protection Association, Quincy, MA, 1995, pp. 3–263–3–285.
12. Nelson, H. E., and MacLennan, H. A., "Emergency Movement," *The SFPE Handbook of Fire Protection Engineering*, 2nd ed. DiNenno, P. J., ed., National Fire Protection Association, Quincy, MA, 1995, pp. 3–286–3–295.
13. Heskestad, G., and Smith, H. F., "Investigation of a New Sprinkler Sensitivity Approval Test: The Plunge Test," Technical Report Serial No. 22485, RC 76-T-50, 1976, Factory Mutual Research Corporation, Norwood, MA.

NFPA Codes, Standards, and Recommended Practices

Reference to the following NFPA codes, standards, and recommended practices will provide further information on integration of fire models with the design process discussed in this chapter. (See the latest version of *The NFPA Catalog* for availability of current editions of the following document.)

NFPA 13, *Standard for the Installation of Sprinkler Systems*

Additional Reading

Allan, H. A., "CSIRO Fire Safety Engineering Method Using the Computer Model FIRECALC," Commonwealth Scientific and Industrial Research Organization, Australia, University of Ulster and Fire Research Station, CIB W14: Fire Safety Engineering, International Symposium and Workshops Engineering Fire Safety in the Process of Design: Demonstrating Equivalency, Part 2, Symposium: Engineering Fire Safety for People With Mixed Abilities, Sept. 13–16, 1993, Newtownabbey, Northern Ireland, 1993, pp. 111–144.

Allan, H., Grubits, S., and Quaglia, C., "Predicting Smoke Movement in a Multi-Storey Shopping Center With Interconnecting Voids Using a Zone Model," *Proceedings* of the Society of Fire Protection Engineers (SFPE) Honors Lecture Series, May 20, 1996, Engineering Seminars: Fire Protection Design for High Challenge or Special Hazard Applications, May 20–22, 1996, Boston, MA, 1996, pp. 43–48.

Allen, J. E., and Reynolds, J. R., Jr., "FTI Field Modeling Analysis and Data Animation Dupont Plaza Hotel Fire," *Proceedings* of the International Conference on Fire Research and Engineering, Sept. 10–15, 1995, Orlando, FL, SFPE, Boston, MA, 1995, pp. 588–589.

Andersson, P., and Holmstedt, G., "CFD-Modelling Applied to Fire Detection: Validation Studies and Influence of Background Heating," *Proceedings* of the Tenth International Conference on Automatic Fire Detection "AUBE '95", [Internationale Konferenz uber Automatischen Brandentdeckung.] April 4–6, 1995, Duisburg, Germany, 1995, pp. 429–438.

Babrauskas, V., "Designing Products for Fire Performance: The State of the Art of Test Methods and Fire Models. Commentary," *Fire Safety Journal*, Vol. 24, No. 3, 1995, pp. 299–312.

Babrauskas, V., "Introduction to Mathematical Fire Modelling. [Book Review of An Introduction to Mathematical Fire Modeling by David M. Birk.]," *Fire and Flammability Bulletin*, Oct. 1991, pp. 9–10.

Bailey, J. L., and Tatem, P. A., "Validation of Fire/Smoke Spread Model (CFAST) Using Ex-USS SHADWELL Internal Ship Conflagration Control (ISCC) Fire Tests," Naval Research Laboratory, Washington, DC, Office of Naval Research, Arlington, VA, NRL/MR/6180-95-7781, Sept. 28, 1995.

Baum, H. R., McGrattan, K. B., and Rehm, R. G., "Mathematical Modeling and Computer Simulation of Fire Phenomena," *Proceedings* of the Fourth International Symposium on Fire Safety Science, Boston, MA, 1994, Intl. Assoc. for Fire Safety Science, Boston, MA, 1994, pp. 185–193.

Beard, A., "Evaluation of Fire Models: Report 9. FIRST: Qualitative Assessment," Edinburgh Univ., Scotland, Report 9, Jan. 1990.

Beard, A., "Evaluation of Deterministic Fire Models. Part 1. Introduction," *Fire Safety Journal*, Vol. 19, No. 4, 1992, pp. 295–306.

Beard, A., "Evaluation of Fire Models: Overview," Edinburgh Univ., Scotland, Report 11, Oct. 1990.

Beard, A., "Evaluation of Fire Models: Report 1. ASET: Qualitative Assessment," Edinburgh Univ., Scotland, Report 1, Jan. 1990.

Beard, A., "Evaluation of Fire Models: Report 10. JASMINE: Qualitative Assessment," Edinburgh Univ., Scotland, Report 10, May 1990.

Beard, A., "Evaluation of Fire Models: Report 2. ASKFRS: Qualitative Assessment," Edinburgh Univ., Scotland, Report 2, Jan. 1990, 18 pages.

Beard, A., "Evaluation of Fire Models: Report 3. HAZARD I: Qualitative Assessment," Edinburgh Univ., Scotland, Report 3, Jan. 1990.

Beard, A., "Evaluation of Fire Models: Report 4. FIRST: Qualitative Assessment," Edinburgh Univ., Scotland, Report 4, January 1990.

Beard, A., "Evaluation of Fire Models: Report 5. JASMINE: Qualitative Assessment," Edinburgh Univ., Scotland, Report 5, April 1990.

Beard, A., "Evaluation of Fire Models: Report 6. ASET: Qualitative Assessment," Edinburgh Univ., Scotland, Report 6, Jan. 1990.

Beard, A., "Evaluation of Fire Models: Report 8. HAZARD I: Qualitative Assessment," Edinburgh Univ., Scotland, Report 8, April 1990.

Beard, A., "Limitations of Computer Modelling," Fire Safety Modelling and Building Design, *Proceedings* of the One-Day Conference to Review the Potential and Limitations of Fire Safety Models for Building Design, March 29, 1994, Salford, UK, 1994, pp. 10–26.

Beard, A., "Limitations of Computer Models," *Fire Safety Journal*, Vol. 18, No. 4, 1992, pp. 375–391.

Beard, A. N., "Reliability and Computer Models," *Journal of Applied Fire Science*, Vol. 3, No. 3, 1993/1994, pp. 273–279.

Beard, A. N., *et al.*, "Flashover A1: A Model for Predicting the Conditions for Flashover," University of Edinburgh, Scotland, South Bank University, London, England, University of College, London, England, ISBN 0-9527398-3-6; Institution of Fire Engineers, University of Sunderland, Fire Research Station, CIB W14, Tyne and Wear Metropolitan Fire Brigade, Fire Safety by Design, Conference *Proceedings*, Volume 3, Research Papers, July 10–12, 1995, UK, 1995, pp. 169–179.

Beard, A. N., Drysdale, D. D., and Bishop, S. R., "Non-Linear Model of Major Fire Spread in a Tunnel," *Fire Safety Journal*, Vol. 24, No. 4, 1995, pp. 333–357.

Beck, V., *et al.*, "Experimental Validation of a Fire Growth Model," Victoria University of Technology, Australia, National Research Council of Canada, Ottawa, Ontario, ISBN 0-9516320-9-4; *Proceedings* of the Seventh International Interflam Conference, Interflam '96, March 26–28, 1996, Cambridge, England, Interscience Communications Ltd., London, England, 1996, pp. 653–662.

Beller, D., "Validating Fire Models: FPETOOL, CFAST, WPI/FIRE," Azure Dragon Enterprises, Ltd., USA, ISBN 0-9516320-9-4; *Proceedings* of the Seventh International Interflam Conference, Interflam '96, March 26–28, 1996, Cambridge, England, Interscience Communications Ltd., London, England, 1996, pp. 959–963.

Beller, D., and Till, R., "Computer Fire Model Validation: FPETOOL, CFAST, WPI/FIRE," *Proceedings* of the International Conference on Fire Research and Engineering, September 10–15, 1995, Orlando, FL, SFPE, Boston, MA, 1995, pp. 341–345.

Belles, D., "Practical Application of Computer Fire Models," National Institute of Standards and Technology, Federal Fire Forum on Computer Programs for Fire Protection, Nov. 7, 1991, Gaithersburg, MD, 1991.

Bilger, R. W., "Computational Field Models in Fire Research and Engineering," *Proceedings* of the Fourth International Symposium on Fire Safety Science, Intl. Assoc. for Fire Safety Science, Boston, MA, 1994, pp. 95–110.

Blount, K., "Regulatory Implications of Fire Safety Modelling," Fire Safety Modelling and Building Design, *Proceedings* of the One-Day Conference to Review the Potential and Limitations of Fire Safety Models for Building Design, March 29, 1994, Salford, UK, 1994, pp. 27–35.

Brani, D. M., and Black, W. Z., "Two-Zone Model for a Single-Room Fire," *Fire Safety Journal*, Vol. 19, No. 2–3, 1992, pp. 189–216.

Brooks, W. N., "Comparison of the NFPA 92B Calculation Methods to Several Fire Models," *Proceedings* of Computer Applications in Fire Protection, June 28–29, 1993, Worcester, MA, 1993, pp. 43–50.

Bukowski, R. W., "Fire Models: The Future is Now!," *NFPA Journal*, Vol. 85, No. 6, March/April 1991, pp. 60–62, 64, 66–69.

Bukowski, R. W., "How to Evaluate Alternative Designs Based on Fire Modeling," *NFPA Journal*, Vol. 89, No. 2, March/April 1995, pp. 68–70, 72–74.

Chandrasekaran, V., Grubits, S. J., and Suresh, R., "Zone Models in Fire Safety Engineering. Session 4—The Tools," CSIRO Division of Building, Construction and Engineering, Australia, CSIRO, International Fire Safety Engineering Conference on Concept and the Tools, Presented for FORUM for International Cooperation on Fire Research, Supported by SFPE, NFPA, AFPA, AAFA, Oct. 18–20, 1992, Sydney, Australia, 1992, pp. 1–18.

Chang, X., Laage, L. W., and Greuer, R. E., "User's Manual for MFIRE: A Program Computer Simulation Program for Mine Ventilation and Fire Modeling," Xian Mining Institute, The Peoples Republic of China, Bureau of Mines, Minneapolis, MN, Mining Technological Univ., Houghton, MI, IC 9245, 1990.

Charters, D. A., Gray, W. A., and McIntosh, A. C., "Computer Model to Assess Fire Hazards in Tunnels (FASIT)," *Fire Technology*, Vol. 30, No. 1, First Quarter 1994, pp. 134–154.

Charters, D., and McIntosh, A., "FASIT: A Computer Model for Predicting Fires in Tunnels," *Fire*, Vol. 88, No. 1082, August 1995, pp. 13–14, 16.

Chow, W. K., "Experimental Evaluation of the Zone Models CFAST, FAST and CCFM.VENTS," *Journal of Applied Fire Science*, Vol. 2, No. 4, 1992–1993, pp. 307–332.

Chow, W. K., "Multi-Cell Concept for Simulating Fires in Big Enclosures Using a Zone Model," *Journal of Fire Sciences*, Vol. 14, No. 3, May/June 1996, pp. 186–198.

Chow, W. K., "Short Note on the Simulation of the Atrium Smoke Filling Process Using Fire Zone Models," *Journal of Fire Sciences*, Vol. 12, No. 6, Nov./Dec. 1994, pp. 506–515.

Chow, W. K., "Simulation of Car Park Fires Using Zone Models," *Journal of Fire Protection Engineering*, Vol. 7, No. 3, 1995, pp. 65–74.

Chow, W. K., "Use of the ARGOS Fire Risk Analysis Model for Studying Chinese Restaurant Fires," *Fire and Materials*, Vol. 19, No. 4, July/Aug. 1995, pp. 171–178.

Chow, W. K., "Use of Zone Models on Simulating Compartmental Fires With Forced Ventilation," *Fire and Materials*, Vol. 19, No. 3, 1995, pp. 101–108.

Chow, W. K., "Zone Model Simulation of Fires in Chinese Restaurants in Hong Kong," *Journal of Fire Sciences*, Vol. 13, No. 3, May/June 1995, pp. 235–253.

Chow, W. K., and Cheung, K. C., "Industrial Fire Simulation With Zone Models and Review on the Current Regulations," Hong Kong Polytechnic Univ., Hong Kong, ISBN 0-9516320-9-4; *Proceedings* of the Seventh International Interflam Conference, Interflam '96, March 26–28, 1996, Cambridge, England, Interscience Communications Ltd., London, England, 1996, pp. 601–620.

Chow, W. K., and Cheung, Y. L., "Simulation of Sprinkler-Hot Layer Interaction Using a Field Model," *Fire and Materials*, Vol. 18, No. 6, 1994, pp. 359–379.

Chow, W. K., and Lo, A. C. W., "Scale Modelling Studies on Atrium Smoke Movement and the Smoke Filling Process," *Journal of Fire Protection Engineering*, Vol. 7, No. 3, 1995, pp. 55–64.

Chow, W. K., and Wong, W. K., "Application of the Zone Model FIRST on the Development of Smoke Layer and Evaluation of Smoke Extraction Design for Atria in Hong Kong," *Journal of Fire Sciences*, Vol. 11, No. 4, July/Aug. 1993, pp. 329–349.

Collier, P., "Model for Predicting the Fire-Resisting Performance of Small-Scale Cavity Walls in Realistic Fires," *Fire Technology*, Vol. 32, No. 2, Second Quarter/April/May 1996, pp. 120–136.

Cooke, G. M. E., "Can Integrity in Fire Resistance Tests Be Predicted?," Sixth International Fire Conference on Fire Safety, Interflam '93, March 30–April 1, 1993, Oxford, England, Interscience Communications Ltd., London, England, 1993, pp. 321–330.

Cooper, L. Y., "Overview of a Theory for Simulating Smoke Movement Through Long Vertical Shafts in Zone-Type Fire Models," National Institute of Standards and Technology, Gaithersburg, MD, NISTIR 5499, Sept. 1994; National Institute of Standards and Technology, Annual Conference on Fire Research: Book of Abstracts, Oct. 17–20, 1994, Gaithersburg, MD, 1994, pp. 93–94.

Cooper, L.Y., "VENTCF2: An Algorithm and Associated TRAN 77 Subroutine for Calculating Flow Through a Horizontal Ceiling/Floor Vent in a Zone-Type Compartment Fire Model," National Institute of Standards and Technology, Gaithersburg, MD, NISTIR 5470, Aug. 1994.

Cox, G., "Challenge of Fire Modelling," *Fire Safety Journal*, Vol. 23, 1994, pp. 123–132.

Cox, G., "Challenge of Fire Modelling," *Proceedings* of the Technical Conference on Fire Safety by Design: A Framework for the Future, Featuring the Combined 1993 Fire Research Lecture and IFS Inaugural Lecture, Nov. 10, 1993, Borehamwood, England, 1993, pp. 1–14.

Cox, G., "Compartment Fire Modelling," Fire Research Station, Borehamwood, England, ISBN 0-12-194230-9; *Combustion Fundamentals of Fire*, Chapter 6, Academic Press, New York, NY, 1995, pp. 329–404.

Cox, G., "Some Recent Progress in the Field Modelling of Fire," First Asian Conference on Fire Science and Technology (ACFST), Oct. 9–13, 1992, Hefei, China, International Academic Publishers, China, 1992, pp. 50–59.

Cox, G., "State of the Art Modelling," Fire Safety Modelling and Building Design, *Proceedings* of the One-Day Conference to Review the Potential and Limitations of Fire Safety Models for Building Design, March 29, 1994, Salford, UK, 1994, pp. 6–9.

Cox, G., and Rubini, P. A., "Development of a New CFD Fire Simulation Model," Fire Research Station, Borehamwood, England, Cranfield Institute of Technology, UK, VTT-Technical Research Center of Finland and Forum for International Cooperation in Fire Research, Nordic Fire Safety Engineering Symposium, Development and Verification of Tools for Performance Codes, Aug. 30–Sept. 1, 1993, Espoo, Finland, 1993, pp. 1–11.

Curtat, M. R., "Survey of Fire Modelling in France," *Proceedings* of the Third International Symposium on Fire Safety Science, Elsevier Applied Science, New York, NY, 1991, pp. 97–118.

Custer, R. L. P., "Computer Modeling in Fire Reconstruction and Failure Analysis," *Proceedings* of Computer Applications in Fire Protection, June 28–29, 1993, Worcester, MA, 1993, pp. 69–74.

Custer, R. L. P., "Progress on Computer Modeling in Performance-Based Detection System Design in the United States," Worcester Polytechnic Inst., MA, European Society for Automatic Alarm Systems (EUSAS), European Platform for Promoting Security and Fire Protection Engineering, Newsletter, Number 5, June 1994, pp. 17–26.

Custer, R. L. P., "Selecting and Using Computer Fire Models," Worcester Polytechnic Inst., MA, Society of Fire Protection Engineers and WPI Center for Firesafety Studies, *Proceedings* of the Technical Symposium on Computer Applications in Fire Protection Engineering, Final Program, June 20–21, 1996, Worcester, MA, 1996, pp. 7–11.

Daikoku, M., *et al.*, "Modelling of Transport Phenomena in Compartment Fires," *Proceedings* of the International Conference on Fire Research and Engineering, Sept. 10–15, 1995, Orlando, FL, SFPE, Boston, MA, 1995, pp. 15–20.

Davis, W., "Overview of Harwell-Flow 30—A Field Model," Montgomery College, Rockville, MD, National Institute of Standards and Technology, Federal Fire Forum on Computer Programs for Fire Protection, Nov. 7, 1991, Gaithersburg, MD, 1991.

Davis, W. D., Forney, G. P. ,and Bukowski, R. W., "Field Modeling: Simulating the Effect of Sloped Beamed Ceilings on Detector and Sprinkler Response. International Fire Detection Research Project. Technical Report. Year 2," National Institute of Standards and Technology, Gaithersburg, MD, National Fire Protection Research Foundation, Quincy, MA, Technical Report, Year 2, Oct. 1994.

Davis, W. D., Forney, G. P., and Klote, J. H., "Field Modeling of Room Fires," National Institute of Standards and Technology, Gaithersburg, MD, NISTIR 4673, Nov. 1991.

Davis, W. D., Notarianni, K. A., and Tapper, P. Z., "Modelling of Smoke Movement and Detector Performance in High Bay Spaces," *Proceedings* of the International Conference on Fire Research and Engineering,

Sept. 10–15, 1995, Orlando, FL, SFPE, Boston, MA, 1995, pp. 307–311.

Deakin, A. G., "Fire Modelling: Where Are We and What Is Needed for Practical Application," Warrington Fire Research Center, UK, TNO Building and Construction Research, Third CIB/W14 Workshop, *Proceedings* of the Fire Engineering Workshop on Modelling, Jan. 25–26, 1993, Delft, The Netherlands, 1993, pp. 15–18.

Deal, S., "Comparison of Various Fire Model Results," National Institute of Standards and Technology, Federal Fire Forum on Computer Programs for Fire Protection, Nov. 7, 1991, Gaithersburg, MD, 1991.

Dembsey, N. A., Pagni, P. J., and Williamson, R. B., "Compartment Fire Experimental Data: Comparison to Models," *Proceedings* of the International Conference on Fire Research and Engineering, Sept. 10–15, 1995, Orlando, FL, SFPE, Boston, MA, 1995, pp. 350–355.

Deubler, J. F., "How to Review a Computer Modeling Report," Schirmer Engineering Corp., Arlington, VA, Society of Fire Protection Engineers and WPI Center for Firesafety Studies, *Proceedings* of the Technical Symposium on Computer Applications in Fire Protection Engineering, Final Program, June 20–21, 1996, Worcester, MA, 1996, pp. 67–72.

Dolph, B. L., "Modeling Fires on Ships," Coast Guard Research and Development Center, Groton, CT, Society of Fire Protection Engineers and WPI Center for Firesafety Studies, *Proceedings* of the Technical Symposium on Computer Applications in Fire Protection Engineering, Final Program, June 20–21, 1996, Worcester, MA, 1996, pp. 1–4.

Duong, D. Q., "Accuracy of Computer Fire Models: Some Comparisons With Experimental Data From Australia," *Fire Safety Journal*, Vol. 16, No. 6, 1990, pp. 415–431.

El-Rimawi, J. A., Burgess, I. W., and Plank, R. J., "Modelling the Behavior of Steel Frames and Subframes With Semi-Rigid Connections in Fire," University of Sheffield, UK, TNO Building and Construction Research, Third CIB/W14 Workshop, *Proceedings* of the Fire Engineering Workshop on Modelling, Jan. 25–26, 1993, Delft, The Netherlands, 1993, pp. 152–168.

Ellington, J. M., "Some Perspectives on Fire Modeling by a Fire Investigator," *Fire and Arson Investigator*, Vol. 45, No. 4, June 1995, pp. 36–38.

Emmons, H. W., "Fire Modeling in the 21st Century," Harvard Univ., Cambridge, MA, Society of Fire Protection Engineers and WPI Center for Firesafety Studies, *Proceedings* of the Technical Symposium on Computer Applications in Fire Protection Engineering, Final Program, June 20–21, 1996, Worcester, MA, 1996, pp. 1–5.

Evans, D. D., "Large Fire Experiments for Fire Model Evaluation," National Institute of Standards and Technology, Gaithersburg, MD, ISBN 0-9516320-9-4; *Proceedings* of the Seventh International Interflam Conference, Interflam '96, March 26–28, 1996, Cambridge, England, Interscience Communications Ltd., London, England, 1996, pp. 329–334.

Fahy, R. F., "EXIT89: An Evacuation Model for High-Rise Buildings," National Fire Protection Assoc., Quincy, MA, NISTIR 4449; Eleventh Joint Panel Meeting of the U. S./Japan Government Cooperative Program on Natural Resources (UJNR) on Fire Research and Safety, Oct. 19–24, 1989, Berkeley, CA, 1990, pp. 306–311.

Fahy, R. F., "EXIT89: An Evacuation Model for High-Rise Buildings. Model Description and Example Applications," *Proceedings* of the Fourth International Symposium on Fire Safety Science, Intl. Assoc. for Fire Safety Science, Boston, MA, 1994, pp. 657–668.

Fahy, R. F., "EXIT89: An Evacuation Model for High-Rise Buildings—Recent Enhancements and Example Applications," *Proceedings* of the International Conference on Fire Research and Engineering, Sept. 10–15, 1995, Orlando, FL, SFPE, Boston, MA, 1995, pp. 332–337.

Fahy, R. F., "EXIT89: High-Rise Evacuation Model—Recent Enhancements and Example Applications," National Fire Protection Association, Quincy, MA, National Institute of Standards and Technology, Gaithersburg, MD, ISBN 0-9516320-9-4; *Proceedings* of the Seventh International Interflam Conference, Interflam '96, March 26–28, 1996, Cambridge, England, Interscience Communications Ltd., London, England, 1996, pp. 1001–1005.

Fangrat, J., "Different Approach to Some Fire Modeling Input Data," Building Research Inst., Warsaw, Poland, CIB W14/93/2 (J); *First Proceedings* of the Japan Symposium on Heat Release and Fire Hazard, Session 2, Fire Modeling as a Tool for Reaction to Fire Evaluation, May 10–11, 1993, Tsukuba, Japan, 1993, pp. II/1–6.

"Fire Models In Use Today," *IRI Sentinel*, Vol. 51, No. 3, Third Quarter, 1994, pp. 10–11.

"Fire Models: A Tale of Four Uses," *IRI Sentinel*, Vol. 51, No. 3, Third Quarter, 1994, pp. 8–9.

"Fire Models: The Pros and Cons," *IRI Sentinel*, Vol. 51, No. 3, Third Quarter, 1994, p. 3.

"Fire Models: Valuable Tools When Used Properly," *IRI Sentinel*, Vol. 51, No. 3, Third Quarter, 1994, pp. 4–7.

"Fire Spread Modeling," *Record*, Vol. 70, No. 3, May/June 1993, pp. 15–16.

Forney, G. P., "Computing Radiative Heat Transfer Occurring in a Zone Fire Model," *Fire Science and Technology*, Vol. 14, No. 1/2, pp. 31–74.

Forney, G. P., and Moss, W. F., "Analyzing and Exploiting Numerical Characteristics of Zone Fire Models," *Fire Science and Technology*, Vol. 14, No. 1/2, 1994, pp. 49–60.

Fransson, C., and Vilselius, H., "Application of a General-Purpose CFD Code to Field Modelling of Tunnel Fires," National Defence Research Establishment, Sundbyberg, Sweden, FOA Report B 20131-2.7, June 1994.

Friedman, R., "Survey of Computer Models for Fire and Smoke. Technical Report," Factory Mutual Research Corp., Norwood, MA, Technical Report, March 1990.

Galea, E. R., and Galparsoro, J. M. P., "Computer-Based Simulation Model for the Prediction of Evacuation From Mass-Transport Vehicles," *Fire Safety Journal*, Vol. 22, No. 4, 1994, pp. 341–366.

Galea, E. R., and Ierotheou, C. S., "Fire-Field Modelling on Parallel Computers," *Fire Safety Journal*, Vol. 19, No. 4, 1992, pp. 251–266.

Galea, E. R., Owen, M., and Lawrence, P. J., "Emergency Egress From Large Buildings Under Fire Conditions Simulated Using the EXODUS Evacuation Model," University of Greenwich, London, England, ISBN 0-9516320-9-4; *Proceedings* of the Seventh International Interflam Conference, Interflam '96, March 26–28, 1996, Cambridge, England, Interscience Communications Ltd., London, England, 1996, pp. 711–720.

Gallina, G., *et al.*, "Fire Modelling in Building With a CAD-Based Graphical User Interface," CNR-ICITE, Sesto Ulteriano (MI), Italy, TRI srl, Scanzorosicate (BG), Italy, ISBN 0-9527398-3-6; Institution of Fire Engineers, University of Sunderland, Fire Research Station, CIB W14, Tyne and Wear Metropolitan Fire Brigade, Fire Safety by Design, Conference *Proceedings*, Volume 3, Research Papers, July 10–12, 1995, UK, 1995, pp. 239–253.

Golton, C. J., Golton, B. J., and Hinks, A. J., "Human Behavior in Fire: A Background Review for Modelling," Fire Safety Modelling and Building Design, *Proceedings* of the One-Day Conference to Review the Potential and Limitations of Fire Safety Models for Building Design, March 29, 1994, Salford, UK, 1994, pp. 48–62.

Griffith, S. J., and Munday, J. W., "Legal Implications of Real Fire Data Collection and Computer Modelling," Metropolitan Police Forensic Science Laboratory, London, England, ISBN 0-9516320-9-4; *Proceedings* of the Seventh International Interflam Conference, Interflam '96, March 26–28, 1996, Cambridge, England, Interscience Communications Ltd., London, England, 1996, pp. 1027–1031.

Grubits, S. J., and Chandrasekaran, V., "Issues in Fire Modelling," Commonwealth Scientific and Industrial Research Organization, North Ryde, Australia, July 1991.

Gupta, A. K., "CALFIRE: An Interactive Model for Fire Calculations," *Fire Technology*, Vol. 30, No. 3, Third Quarter 1994, pp. 304–325.

Hadjisophocleous, G. V., and Cacambouras, M., "Computer Modeling of Compartment Fires," *Journal of Fire Protection Engineering*, Vol. 5, No. 2, April/May/June 1993, pp. 39–52.

Hadjisophocleous, G. V., and Knill, K., "CFD Modelling of Liquid Pool Fire Suppression Using Fine Watersprays," National Research Council of Canada, Ottawa, Ontario, Advanced Scientific Computing, Ltd., Ontario, Canada, NISTIR 5499, September 1994; National Institute of Standards and Technology, Annual Conference on Fire Research: Book of Abstracts, Oct. 17–20, 1994, Gaithersburg, MD, 1994, pp. 71–72.

Hagglund, B., "Comparing Fire Models With Experimental Data," National Defence Research Establishment, Sundbyberg, Sweden, FOA Report C20864-2.4, Feb. 1992.

Hagglund, B., Fransson, C. and Bengtson, S., "Use of Zone and Field Model to Recommend Fire Protection Measures in New Stores of the Royal Library in Stockholm," National Defence Research Establishment, Sundbyberg, Sweden, FOA Report C20981-2.4, June 1994.

Ham, S., "Fire Modelling From an Architectural Perspective," Fire Safety Modelling and Building Design, *Proceedings* of the One-Day Conference to Review the Potential and Limitations of Fire Safety Models for Building Design, March 29, 1994, Salford, UK, 1994, pp. 1–5.

Hasemi, Y., and Yoshida, M., "Wall Fire Modeling and Its Relevance to Building Fire Safety Design," '93 Asian Fire Seminar, Oct. 7–9, 1993, Tokyo, Japan, 1993, pp. 117–127.

Heins, T., and Dobbernack, R., "FIGARO: A Zone Model for Fire Simulation in Complex Geometries," Technischer Uberwachungs-Verein Hannover/Sachsen-Anhalt e.V., Germany, Universitat Braunschweig, Germany, TNO Building and Construction Research, Third CIB/W14 Workshop, *Proceedings* of the Fire Engineering Workshop on Modelling, Jan. 25–26, 1993, Delft, The Netherlands, 1993, pp. 49–58.

Hinks, A. J., "Modelling and Managing Escape From Fires in Complex Buildings," Fire Safety Modelling and Building Design, *Proceedings* of the One-Day Conference to Review the Potential and Limitations of Fire Safety Models for Building Design, March 29, 1994, Salford, UK, 1994, pp. 73–89.

Hinks, A. J., "Systemic Modelling of Life Safety in Building Fires: LISA," Fire Safety Modelling and Building Design, *Proceedings* of the One-Day Conference to Review the Potential and Limitations of Fire Safety Models for Building Design, March 29, 1994, Salford, UK, 1994, pp. 63–72.

Holen, J., and Magnussen, B. F., "Present Status of Field Models for Fire Growth Developed at NTH/SINTEF," NTH/SINTEF, Trondheim, Norway, VTT-Technical Research Center of Finland and Forum for International Cooperation in Fire Research, Nordic Fire Safety Engineering Symposium, Development and Verification of Tools for Performance Codes, Aug. 30–Sept. 1, 1993, Espoo, Finland, 1993, pp. 1–7.

Holmberg, S., and Anderberg, Y., "Computer Simulations and Design Method for Fire Exposed Concrete Structures," Swedish Fire Research Board, Stockholm, Sweden, Document 92050-1, FSD Project 92-50, Sept. 27, 1993.

Hume, B., "Guidance on Fire Models," *Fire Research News*, No. 18, Winter 1994, p. 11.

Janssens, M., "Mathematical Modeling of Compartment Fires," Abstracts of the First U. S. Symposium on Heat Release and Fire Hazard, Dec. 1991, San Diego, CA, 1991, pp. 45–47.

Janssens, M., "Room Fire Models. Part A. General," Chapter 6, *Heat Release in Fires*, Elsevier Applied Science, NY, 1992, pp. 113–157.

Janssens, M. L., "Computer Program Roomfire: A Compartment Fire Model Shell," American Forest and Paper Association, Washington,DC, Chapter 27, ACS Symposium Series 599, American Chemical Society, Fire and Polymers II: Materials and Tests for Hazard Prevention, 208th National Meeting, ACS Symposium Series 599, Aug. 21–26, 1994, Washington, DC, American Chemical Society, Washington, DC, 1995, pp. 409–421.

Johnson, P. F., "Comment on 'The Accuracy of Computer Fire Models: Some Comparisons With Experimental Data From Australia'. Letter to the Editor," *Fire Safety Journal*, Vol. 17, No. 5, 1991, pp. 415–416.

Jones, S., *et al.*, "Zone Model for Smoke Transport Aboard Human-Crewed Spacecraft," *PSAM-II Proceedings*, An International Conference Devoted to the Advancement of System-Based Methods for the Design and Operation of Technological Systems and Processes, Volume 3, Sessions 73–108, March 20–25, 1994, San Diego, CA, 1994, pp. 085/1–4.

Jones, W. W., "Modeling Fires—The Next Generation of Tools," Worcester Polytechnic Inst., MA, Society of Fire Protection Engineers and WPI Center for Firesafety Studies, *Proceedings* of the Technical Symposium on Computer Applications in Fire Protection Engineering, Final Program, June 20–21, 1996, Worcester, MA, 1996, pp. 13–18.

Jones, W. W., "Modeling Smoke Movement Through Compartmented Structures," *Journal of Fire Sciences*, Vol. 11, No. 2, March/April 1993, pp. 172–183; Thirty-ninth Sagamore Army Materials Research Conference, Sept. 16–17, 1992, Plymouth, MA, 1992; Twelfth Joint Panel Meeting of the U.S./Japan Government Cooperative Program on Natural Resources (UJNR) on Fire Research and Safety, Oct. 27–Nov. 2, 1992, Tsukuba, Japan, Building Research Inst., Ibaraki, Japan, Fire Research Inst., Tokyo, Japan, 1992, pp. 34–41.

Jones, W. W., "Progress Report on Fire Modeling and Validation," National Institute of Standards and Technology, Gaithersburg, MD, NISTIR 5835, May 1996.

Jones, W. W., and Baum, H. R., "Modeling Fire Growth and Smoke Transport in the United States," Twelfth Joint Panel Meeting of the U. S./Japan Government Cooperative Program on Natural Resources (UJNR) on Fire Research and Safety, Oct. 27–Nov. 2, 1992, Tsukuba, Japan, Building Research Inst., Ibaraki, Japan, 1992, pp. 4–7.

Jones, W. W., and Forney, G. P., "Programmer's Reference Manual for CFAST, the Unified Model of Fire Growth and Smoke Transport," National Institute of Standards and Technology, Gaithersburg, MD, NIST TN 1283, Nov. 1990.

Kalinich, D. A,. and Bailey, R. T., "Parametric Study of Smoke Propagation Using the CFAST Computer Code," Westinghouse Savannah River Co., Aiken, SC, Department of Energy, Washington, DC, WSRC-RP-93-1315, Sept. 1993.

Karlsson, B., "Models for Calculating Flame Spread on Wall Lining Materials and the Resulting Heat Release Rate is a Room," *Fire Safety Journal*, Vol. 23, No. 4, 1994, pp. 365–386.

Karlsson, B., and Magnusson, S. E., "Room Fire Models. Part B. An Example Room Fire Model," Chapter 6, *Heat Release in Fires*, Elsevier Applied Science, NY, 1992, pp. 159–171.

Kasahara, I., and Hara, T., "Life Safety Design of an Atrium and an Air Dome With an Applied Zone Model," Taisei Corp., Tokyo, Japan, CIB W14/96/1 (J); Mini-Symposium on Fire Safety Design of Buildings and Fire Safety Engineering, June 12, 1995, Tsukuba, Japan, 1995, pp. VIa/1–3.

Kerrison, L., et al., "Comparison of Two Fire Field Models With Experimental Room Fire Data," *Proceedings* of the Fourth International Symposium on Fire Safety Science, Intl. Assoc. for Fire Safety Science, Boston, MA, 1994, pp. 161–172.

Kerrison, L., et al., "Comparison of a FLOW3D Based Fire Field Model With Experimental Room Fire Data," *Fire Safety Journal*, Vol. 23, No. 4, 1994, pp. 387–411.

Kostreva, M. M., "Sensitivity Analysis for Mathematical Modeling of Fires in Residential Buildings," Clemson Univ., SC, National Institute of Standards and Technology, Gaithersburg, MD, NIST-GCR-95-683, Feb. 1966.

Kumar, S., "Field Model Simulations of Vehicle Fires in a Channel Tunnel Shuttle Wagon," *Proceedings* of the Fourth International Symposium on Fire Safety Science, Intl. Assoc. for Fire Safety Science, Boston, MA, 1994, pp. 995–1006.

Kumar, S., and Yehia, M., "Application of JASMINE to the Modelling of Vehicle Fires in a Channel Tunnel Shuttle Wagon," *Proceedings* of the Second International Conference on Safety in Road and Rail Tunnels, April 3–6, 1995, Granada, Spain, 1995, pp. 321–328.

Li, H. J., and Cheng, X. F., "Design of Building Fire Simulation Models," *Fire Safety Science*, Vol. 3, No. 2, Sept. 1994, pp. 21–26.

Lie, T. T., and Chabot, M., "Fire Resistance Evaluation Using Mathematical Models," Fifth International Fire Conference on Fire Safety, Interflam '90, Sept. 3–6, 1990, Canterbury, England, Interscience Communications Ltd., London, England, 1990, pp. 93–100.

Lie, T. T., Gosselin, G. C., and Kim, A. K., "Applicability of Temperature Predicting Models to Building Elements Exposed to Room Fires," *Journal of Applied Fire Science*, Vol. 4, No. 2, 1994/1995, pp. 83–93.

Lucht, D. A., "Changing the Way We Do Business: The Coming of Fire Modeling," *IFCI Fire Code Journal*, Vol. 2, No. 4, Spring 1994, pp. 1, 2, 4–6.

Man, W., K., "Fire Modelling: An Example of Application. Part 1," *Southeast Asia Fire and Security*, May 1993, pp. 18–20.

Marchant, E. W., "Futures of Fire Safety Modelling," Fire Safety Modelling and Building Design, *Proceedings* of the One-Day Conference to Review the Potential and Limitations of Fire Safety Models for Building Design, March 29, 1994, Salford, UK, 1994, pp. 36–47.

Martin, S. B., "Perspective on Fire Modeling," Sixteenth International Conference on Fire Safety, Volume 16, January 14–18, 1991, Millbrae, CA, Product Safety Corp., Sunnyvale, CA, 1991, pp. 370–376.

Mawhinney, R. N., et al., "Critical Comparison of a Phoenics Based Fire Field Model With Experimental Compartment Fire Data," *Journal of Fire Protection Engineering*, Vol. 6, No. 4, 1994, pp. 137–152.

Mitler, H. E., "Mathematical Modeling of Enclosure Fires," National Institute of Standards and Technology, Gaithersburg, MD, NISTIR 90-4294, May 1991.

Mitler, H. E. , and Steckler, K. D., "SPREAD: Model of Flame Spread on Vertical Surfaces," National Institute of Standards and Technology, Gaithersburg, MD, NISTIR 5619, April 1995.

Moss, W. F., "Numerical Analysis Support for Compartment Fire Modeling and Incorporation of Heat Conduction Into a Zone Fire Model. Final Report. August 15, 1988–March 31, 1991," Clemson Univ., SC, National Institute of Standards and Technology, Gaithersburg, MD, NIST-GCR-92-605, March 1992.

Moss, W. F., and Forney, G. P., "Implicitly Coupling Heat Conduction Into a Zone Fire Model," Clemson Univ., SC, National Institute of Standards and Technology, Gaithersburg, MD, NISTIR 4886, July 1992.

Nelson, H. E., "Application of Computer Models for Fire Incident Reconstruction," *Proceedings* of the International Conference on Fire Research and Engineering, Sept. 10–15, 1995, Orlando, FL, SFPE, Boston, MA, 1995, pp. 359–364.

Nelson, H. E., "Computer Models and Fire Reconstruction," *Proceedings* of Computer Applications in Fire Protection, June 28–29, 1993, Worcester, MA, 1993, pp. 75–79.

Nelson, H. E., "Introduction to Computer Fire Modeling," National Institute of Standards and Technology, Gaithersburg, MD, Federal Fire Forum on Computer Programs for Fire Protection. Nov. 7, 1991, Gaithersburg, MD, 1991.

Notarianni, K. A., and Davis, W. D., "Use of Computer Models to Predict the Response of Sprinklers and Detectors in Large Spaces," *Proceedings* of Computer Applications in Fire Protection, June 28–29, 1993, Worcester, MA, 1993, pp. 27–33.

Oar, D. L., "Attachments for Fire Modeling for Building 221-T. T-Plant Canyon Deck and Railroad Tunnel," Westinghouse Hanford Co., Richland, WA, WHC-SD-CP-ANAL-010, Jan. 19, 1995.

Ohnstad, S. J., "Marine Regulatory Applications of the Computer Fire Model HAZARD I," Worcester Polytechnic Inst., MA, Thesis, March 1990.

Opstad, K., "Fire Modelling Using Cone Calorimeter Results," SINTEF NBL, Trondheim, Norway, EUREFIC (European Reaction to Fire Classification), Performance Based Reaction to Fire Classification, International Seminar, Sept. 11–12, 1991, Copenhagen, Denmark, Interscience Communications Ltd., London, England, 1991, pp. 65–71.

Opstad, K., and Magnussen, B. F., "Algorithm for Thermal Flame Spread on Solid Surfaces by the Use of Cone Calorimeter Fire Test Data in a CFD-Model," University of Trondheim, Norway, Norwegian Institute of Technology, Norway, Thesis, May 1995; *Modelling of Thermal Flame Spread on Solid Surfaces in Large-Scale Fires*, 1995, pp. 1–15.

Owen, M., Galea, E. R., and Lawrence, P., "EXODUS Evacuation Model Applied to Building Evacuation Scenarios," University of Greenwich, London, England, ISBN 0-9527398-3-6; Institution of Fire Engineers, University of Sunderland, Fire Research Station, CIB W14, Tyne and Wear Metropolitan Fire Brigade, Fire Safety by Design, Conference *Proceedings*, Volume 3, Research Papers, July 10–12, 1995, UK, 1995, pp. 81–90.

Patnaik, G., Kailasanath, K., and Sinkovits, R., "FLAME3D: A New Time-Dependent, Three-Dimensional Flame Model With Detailed Chemistry," *Proceedings* of the 1995 Fall Technical Meeting of the Combustion Institute/Eastern States Section on Chemical and Physical Processes in Combustion, Oct. 16–18, 1995, Worcester, MA, 1995, pp. 163–166.

Peacock, R. D., "Validation of Computer Models," VHS Video Tape, National Institute of Standards and Technology, Gaithersburg, MD, Federal Fire Forum, Current Technical Fire Issues, Nov. 1, 1993, Gaithersburg, MD, 1993.

Peacock, R. D., Davis, S., and Babrauskas, V., "Data for Room Fire Model Comparisons," *Journal of Research of the National Institute of Standards and Technology*, Vol. 96, No. 4, Aug. 1991, pp. 411–462.

Peacock, R. D., et al., "CFAST, The Consolidated Model of Fire Growth and Smoke Transport," National Institute of Standards and Technology, Gaithersburg, MD, NIST TN 1299, Feb. 1993.

Peacock, R. D., Jones, W. W., and Bukowski, R. W., "Verification of a Model of Fire and Smoke Transport," *Fire Safety Journal*, Vol. 21, No. 2, 1993, pp. 89–129.

Pedersen, K., and Magnussen, B. F., "Field Models: Their Present and Future Application in Fire Safety Engineering. Session 4—The Tools," SINTEF NBL-Norwegian Fire Research Lab., Trondheim, Norway, NTH/SINTEF, Trondheim, Norway, CSIRO, International Fire Safety Engineering Conference on Concept and the Tools, Presented to FORUM for International Cooperation on Fire Research, Supported by SFPE, NFPA, AFPA, AAFA, Oct. 18–20, 1992, Sydney, Australia, 1992, pp. 1–11.

Pehrson, R., "Computer Modeling of Intumescent Penetration Seals," Worcester Polytechnic Inst., MA, Thesis, May 10, 1993.

Pettersson, O., "Rational Structural Fire Engineering Design, Based on Simulated Real Fire Exposure," *Proceedings* of the Fourth International Symposium on Fire Safety Science, Intl. Assoc. for Fire Safety Science, Boston, MA, 1994, pp. 3–26.

Platt, D. G., Elms, D. G., and Buchanan, A. H., "Probabilistic Model of Fire Spread With Timber Effects," *Fire Safety Journal*, Vol. 22, No. 2, 1994, pp. 367–398.

Pollock, A. J., Taylor, I. R., and Donegan, H. A., "Framework for Interfacing Computer Aided Design With Knowledge Based Systems to Support Fire Safety and Other Evaluations," Fire Safety Modelling and Building Design, *Proceedings* of the One-Day Conference to Review the Potential and Limitations of Fire Safety Models for Building Design, March 29, 1994, Salford, UK, 1994, pp. 98–108.

Poon, L. S., "EvacSim: a Simulation Model of Occupants With Behavioral Attributes in Emergency Evacuation of High-Rise Building Fires," *Proceedings* of the Fourth International Symposium on Fire Safety Science, Intl. Assoc. for Fire Safety Science, Boston, MA, 1994, pp. 681–692.

Pucher, K., and Tiefenbock, M., "Three-Dimensional Computer Simulation of a Fire in a Rail Tunnel," *Proceedings* of the Second International Conference on Safety in Road and Rail Tunnels, April 3–6, 1995, Granada, Spain, 1995, pp. 245–252.

Quintiere, J.G., and Dillon, M. E., "Scale Model Reconstruction of Fire in an Atrium," *Proceedings* of the International Conference on Fire Research and Engineering, September 10–15, 1995, Orlando, FL, SFPE, Boston, MA, 1995, pp. 397–402.

Ramachandran, G., "Non-Deterministic Modelling of Fire Spread," *Journal of Fire Protection Engineering*, Vol. 3, No. 2, 1991, pp. 37–48.

Rehm, R. G., and Forney, G. P., "Pressure Equations in Zone-Fire Modeling," *Fire Science and Technology*, Vol. 14, No. 1/2, 1994, pp. 61–73.

Rockett, J. A., "Experience in the Use of Zone Type Building Fire Models," *Fire Science and Technology*, Vol. 13, No. 1/2, 1993, pp. 61–70.

Rockett, J. A., "Zone Model Plume Algorithm Performance," *Fire Science and Technology*, Vol. 15, No. 1/2, 1995, pp. 1–16.

Rockett, J. A., Howe, J. M., and Hanbury, W. L., "Modeling Fire Safety in Multi-Use, Domed Stadia," *Journal of Fire Protection Engineering*, Vol. 6, No. 1, 1994, pp. 11–22.

Satterfield, D. B., "WPI/Harvard Version 2 Computer Fire Model," CFR Video Seminar, Department of the Navy, Washington, DC, Video, Oct. 23, 1990.

Schifiliti, R. P., and Pucci, W. E., "Fire Detection Modeling: State of the Art," R.P. Schifiliti Associates, Inc., Reading MA, Performance Consultants, Holland, MA, Fire Detection Institute, Bloomfield, CT, State of the Art Report, May 6, 1996, 57 pages; Society of Fire Protection Engineers and WPI Center for Firesafety Studies, *Proceedings* of the Technical Symposium on Computer Applications in Fire Protection Engineering, Final Program, June 20–21, 1996, Worcester, MA, 1996, pp. 19–27.

Schneider, V., and Hofmann, J., "Feldmodell-Simulation von Kohlenwasserstoff-Raumbranden und Spruhnebel-Loschversuchen. [Field-Model Simulation of Hydrocarbon Room Fires and Fire Fighting Tests With Water Spray.]," *VFDB*, Feb. 1993, pp. 67–73.

Schneider, V., and Konnecke, R., "Anwendung des Feldmodells KOBRA-3D zur Simulation von komplexen Brandszenarien auf Fragestellungen der automatischen Brandentdeckung. [Application of the Field Model KOBRA-3D to the Simulation of Complex Fire Scenarios for the Purpose of Automatic Fire Detection.]," *Proceedings* of the Tenth International Conference on Automatic Fire Detection "AUBE '95", [Internationale Konferenz uber Automatischen Brandentdeckung.] April 4–6, 1995, Duisburg, Germany, 1995, pp. 404–418.

Shapiro, J. M., "Are the Codes Ready for Fire Models?," *IFCI Fire Code Journal*, Vol. 2, No. 4, Spring 1994, pp. 3, 10.

Sinai, Y. L., and Owens, M. P., "Validation of CFD Modelling of Unconfined Pool Fires With Cross-Wind: Flame Geometry," *Fire Safety Journal*, Vol. 24, No. 1, 1995, pp. 1–34.

Stroup, D. W., "Analyzing Fire Safety Through Computer Modeling and Product Testing," Second International Conference on Fire and Materials, Sept. 23–24, 1993, Arlington, VA, 1993, pp. 7–12.

Stroup, D. W., "FPETOOL for the Personal Computer," General Services Administration, Washington, DC, National Institute of Standards and Technology, Federal Fire Forum on Computer Programs for Fire Protection, Nov. 7, 1991, Gaithersburg, MD, 1991.

Stroup, D. W., "Using Computer Fire Models to Evaluate Equivalent Levels of Fire and Life Safety," *Proceedings* of Computer Applications in Fire Protection, June 28–29, 1993, Worcester, MA, 1993, pp. 21–26.

Stroup, D. W., and Madrzykowski, D., "Modeling Smoke Flow in Corridors," *Proceedings* of the International Conference on Fire Research and Engineering, Sept. 10–15, 1995, Orlando, FL, SFPE, Boston, MA, 1995, pp. 377–382.

Sullivan, P. D., "Recent Applications of Computers in Performance Design of Buildings," Robert W. Sullivan, Inc., Boston, MA, Society of Fire Protection Engineers and WPI Center for Firesafety Studies, *Proceedings* of the Technical Symposium on Computer Applications in Fire Protection Engineering, Final Program, June 20–21, 1996, Worcester, MA, 1996, pp. 87–92.

Sullivan, P. J. E., Terro, M. J., and Morris, W. A., "Critical Review of Fire-Dedicated Thermal and Structural Computer Programs," *Journal of Applied Fire Science*, Vol. 3, No. 2, 1993/1994, pp. 113–135.

Taylor, S., *et al.*, "SMARTFIRE: An Intelligent Fire Field Model," University of Greenwich, London, England, ISBN 0-9516320-9-4; *Proceedings* of the Seventh International Interflam Conference, Interflam '96, March 26–28, 1996, Cambridge, England, Interscience Communications Ltd., London, England, 1996, pp. 671–680.

Thomas, P. H., "Fire Modelling: A Mature Technology," *Fire Safety Journal*, Vol. 19, No. 2–3, 1992, pp. 125–140.

Thompson, P. A., and Marchant, E. W., "SIMULEX: Developing New Computer Modelling Techniques for Evaluation," *Proceedings* of the Fourth International Symposium on Fire Safety Science, Intl. Assoc. for Fire Safety Science, Boston, MA, 1994, pp. 613–624.

Thompson, P. A., and Marchant, E. W., "Testing and Application of the Computer Model 'SIMULEX'," *Fire Safety Journal*, Vol. 24, No. 2, 1995, pp. 149–166.

Tubbs, J. S., and Barnett, J. R., "Modeling the NIST High Bay Fire Experiment With JASMINE," *Proceedings* of the International Conference on Fire Research and Engineering, Sept. 10–15, 1995, Orlando, FL, SFPE, Boston, MA, 1995, pp. 383–388.

Urbas, J., "Using ICAL to Obtain Input Data for Fire Models," [Abstract Only], *Proceedings* of the Twentieth International Conference on Fire Safety, 20th, Volume 20, Jan. 9–13, 1995, Millbrae, CA, Product Safety Corp., Sunnyvale, CA, 1995, p. 178.

Uesugi, H., *et al.*, "Computer Modeling of Fire Engineering Design for High Rise Steel Structure," '93 Asian Fire Seminar, Oct. 7–9, 1993, Tokyo, Japan, 1993, pp. 21–32.

Wade, C.A., "Report on a Finite Element Program for Modelling the Structural Response of Steel Beams Exposed to Fire. Study Report," Building Research Association of New Zealand, Judgeford, Public Good Science Fund, New Zealand, Building Research Levy, New Zealand, BRANZ Study Report 55, Feb. 1994.

Walton, W. D., "Introduction to Mathematical Fire Modeling. [Book/Software Review of An Introduction to Mathematical Fire Modeling by David M. Birk.]," *Fire Technology*, Vol. 28, No. 1, Feb. 1992, pp. 87–91.

Walton, W. D., "Zone Computer Fire Models for Enclosures," National Institute of Standards and Technology, Gaithersburg, MD, NFPA SFPE 95, LC Card Number 95-68247, ISBN 0-87765-354-2; *The SFPE Handbook of Fire Protection Engineering*, 2nd Edition, Section 3, Chapter 7, National Fire Protection Assoc., Quincy, MA, 1995, pp. 3/148–151.

Watts, J. M., Jr., "Fire Risk Evaluation Model. Software Review," *Fire Technology*, Vol. 31, No. 4, 4th Quarter, Nov. 1995, pp. 369–371.

Watts, J. M., Jr., "Future of Fire Modeling. Editorial," *Fire Technology*, Vol. 27, No. 2, May 1991, pp. 97–98.

Wilson, J. L., and Bryan, P. Y., "3-D Modeling as a Tool to Improve Integrated Design and Construction," Lehigh Univ., Bethlehem, PA, Source Document 104, Nov. 1994.

Yung, D., "FIRECAM: A Computer Fire Risk Evaluation and Cost Assessment Model," National Research Council of Canada, Ottawa, Ontario, IRC-ORAL-142; 1996 *Proceedings* of the Institution of Fire Engineers (Ontario Branch), Ontario, Canada, 1996, pp. 1–18.

FORMATS FOR FIRE HAZARD INSPECTING, SURVEYING, AND MAPPING

Revised by Lydia A. Butterworth

The reasons for inspecting a property are: (1) to evaluate the danger to life from fire and associated hazards, (2) to evaluate and determine ways for minimizing fire danger to buildings and contents, (3) to prepare pre-fire plans to familiarize responding emergency personnel with the existing fire protection and fire-fighting challenges associated with the property, and (4) to make recommendations to correct deficiencies or unsafe conditions. Inspections are made by fire protection engineers, fire and building officials, insurance representatives, and others similarly qualified.

In order to properly evaluate the existing or needed protection for a property or specific hazard, an inspection or survey of the premises must be made. Three essential results of this evaluation should be:

1. A precise and complete narrative report describing the fire protection features and fire hazards of the property
2. A plan indicating the physical characteristics and layout of the premises
3. Recommendations for improvement, if necessary

Functions performed during an inspection or survey and the compilation of the narrative report are discussed in detail in the *NFPA Inspection Manual.*[1]

This chapter provides: (1) a checklist of the important features to observe when inspecting or surveying a property; (2) guidelines for drawing (mapping) a site or building plan; (3) a chart of standard symbols used on architectural, engineering, and shop drawings and fire risk and loss analysis diagrams; (4) pre-fire planning symbols; and (5) a legend of standard abbreviations. All are key items that contribute to or are part of the narrative report.

INSPECTION OR SURVEYING

There are many different routes to choose from when conducting an inspection or survey. A particular route is selected according to the layout of the property, personal preference of the inspector, and convenience. It is important for the inspector to pass through the building and surrounding site systematically, without leaving any space uncovered. The checklist shown below provides a general outline of

key items that should be addressed in the written report. By formulating similar checklists, the inspector will be assured of being prepared for the survey and of having all the necessary information to prepare a detailed sketch afterwards. Utilizing such checklists to make rough, handdrawn sketches of every individual section, floor, building, or site will reduce the difficulty in preparing a finished plan.

Checklist

I. *Property Identification*

    A. Name and address
    B. Building identification
    C. Date of construction
    D. Date of report
    E. Name of inspector

II. *Property Use*

    A. General property use(s) (e.g., educational, mercantile, industrial, warehouse)
    B. Specific uses; state each principal use and its location; for storage areas, identify the product(s)[2] stored
    C. Names of tenants in a building or site of multiple occupancy, including their location and amount of space occupied

III. *Site Information*

    A. Fire department access
    B. Streets, roadways, parking, and traffic
    C. Natural barriers
    D. Bodies of water
    E. Fences
    F. Exposures: type and separation distances
        1. Other buildings or structures
        2. Outside storage or processes
        3. Natural exposures (e.g., forest, grass, brush)
        4. Railways, highways, and runways

IV. *Construction*

    A. Dimensions
        1. Area
        2. Height and number of stories

Lydia A. Butterworth, P.E., is a fire protection engineer with the Smithsonian Institution and Chairman of NFPA's Technical Committee on Fire Safety Symbols.

3. Ceiling heights

B. Classification of building construction—Types I, II, III, IV, and V. (See NFPA 220, *Standard on Types of Building Construction*). For mixed construction, a table may be shown to report the location of each type.

C. Compartmentation
   1. Locations and ratings of fire areas, fire walls, and fire partitions
   2. Locations and ratings of vertical openings (e.g., atria, stairways, elevators, shafts, chutes)
   3. Protection of openings (e.g., fire doors, fire windows, fire dampers, firestopping)
   4. Concealed spaces and voids

D. Building exterior
   1. Doors
   2. Windows
   3. Loading docks

V. *Life Safety*

A. Exit facilities
   1. Number and arrangement
   2. Capacity
   3. Door hardware, special locking arrangements, and security

B. Exit marking, illumination, and emergency lighting

C. Interior finish
   1. Furnishings
   2. Decorations

D. Evacuation plans and drills

VI. *Water Supply and Distribution*

A. General description including adequacy, deficiencies, and reliability
   1. Fire flow requirements *vs.* existing conditions, as determined by tests
   2. Storage requirements *vs.* existing conditions

B. Storage tanks—gravity, suction, pressure
   1. Capacity [gal (L)]
   2. Construction
   3. Location and dimensions
   4. Pipe arrangement to yard mains or fire pump(s)
   5. Percent of capacity for fire protection

C. Public and private water distribution systems
   1. Type of system—gravity, direct pumping, or a combination of both
   2. Size of water mains and arrangement in relation to the property
   3. Type of pipe (e.g., cement lined, cast iron, etc.)
   4. Connection arrangement to all supply sources available.

D. Other water storage methods (e.g., rivers, ponds, streams, harbors, wells)

E. Hydrants
   1. Locations of public and private hydrants
   2. Types—dry barrel or wet barrel

F. Location and types of check valves and water meters

G. Fire pumps
   1. Location of pump and control valves
   2. Rated capacity [gpm, psi (L/min, kPa)]
   3. Types—split case, horizontal end suction, in-line, or vertical shaft turbine
   4. Name of manufacturer
   5. Type of motor drive—diesel, electric, gas, gasoline, or steam

   6. Suction pressure, head, or lift [ft (m)]
   7. Starting mechanism—automatic, manual, or both

H. Inspections, maintenance, and tests

VII. *Extinguishing Systems and Devices*

A. Automatic sprinkler systems
   1. Types—wet pipe, wet pipe with antifreeze, dry pipe, deluge, preaction
   2. Area covered by type
   3. Coverage, occupancy classification, and spacing of sprinklers
   4. Riser location and size
   5. Temperature rating of sprinklers
   6. Control valves for water supplying sprinklers—location, size, type, and status if normally shut
   7. Fire department connections—location, size, and area covered
   8. Inspection, maintenance, and tests

B. Standpipe and fire hose systems
   1. Classification—Classes, I, II, or III
   2. Types—wet, dry
   3. Outside systems
   4. Inspections, maintenance, and tests

C. Special hazard systems
   1. Types (e.g., carbon dioxide, foam, dry chemical, wet chemical, clean agent)
   2. Hazard protected
   3. Location and capacity of storage containers
   4. Inspection, maintenance, and tests

D. Portable extinguishers
   1. Types of coverage
   2. Inspections and maintenance

VIII. *Fire Alarm and Detection Systems*

A. Location and manufacturer of main control panel

B. Monitoring and fire department notification

C. Audible and visual device types (e.g., bells, horns, speakers, strobes, etc.)

D. Products of combustion detection
   1. Heat detectors (e.g., rate of rise, fixed temperature, rate compensation, pneumatic line type)
   2. Smoke detectors (e.g., photoelectric, ionization, air sampling, etc.)
   3. Gas detectors
   4. Flame detectors (e.g., ultraviolet, infrared, etc.)

E. Water-flow detection on sprinkler risers

F. Trouble and supervisory indications

G. Power supplies

H. Inspection, maintenance, and tests

IX. *Electrical Systems*

A. Condition of conduit, raceways, cables, conductors, and cords

B. Clear space around switchboards and panelboards

C. Grounding

D. Overcurrent protection

E. Ground-fault circuit-interrupters

F. Motors

G. Hazardous (classified) locations

H. Transformers—capacity, protection, liquids, and cutoffs

I. Lightning protection

X. *Heating, Ventilation, and Air-Conditioning (HVAC) Systems*

   A. Location of equipment
   B. Smoke control systems
   C. Inspection, maintenance, and tests

XI. *Fire Prevention*

   A. Common hazards
      1. Heat, light, power, and air-conditioning
      2. Housekeeping, brush, and grass
      3. Ordinary combustibles
      4. Electrical appliances
      5. Smoking
   B. Building inspection program
      1. Frequency
      2. Scope
      3. Records
   C. Employee fire safety training
      1. Adequacy and frequency of training
      2. Reference materials
      3. Records
   D. Fire brigade program
      1. Type of brigade
      2. Training
      3. Special equipment provided
      4. Records

XII. *Special Hazards and Equipment*

   A. Flammable liquids
   B. Flammable gases
   C. Chemicals and hazardous materials
   D. Tanks and cylinders—location, size, contents, and construction
   E. Finishing processes
   F. Industrial processes
   G. Welding and cutting
   H. Cooking
   I. Shops
   J. Materials handling
   K. Waste removal—incinerators, chutes, dumpsters, compactors, recycling
   L. Tunnels
   M. Electronic equipment
   N. Dust collectors
   O. Boilers and furnaces
   P. Water-cooling towers
   Q. Chimneys—location, construction, and height
   R. Silos
   S. Cranes
   T. Conveyers

XIII. *Construction, Demolition, and Modifications*

   A. Fire protection—water supply, fire department access
   B. Fire prevention—control of debris and combustibles, welding and cutting, roofing operations, fire watches, and storage and handling of hazardous materials
   C. Life safety, temporary or alternate means of egress

MAPPING

Once the necessary information has been collected, a site plan can be drawn readily. A practical approach is to split the preparation into three parts:

1. Building sketch

2. Addition of fire protection equipment

3. Detailing occupancy and specific processes or hazards

The initial step is to outline the shape of the building, using an appropriate architect/engineer scale on an adequately sized sheet of paper. Once the shell is complete, the other important construction features referenced in the preceding section on inspection or surveying should be added, and various construction materials labeled.

The inspector should draw sufficient sectional views to show all vertical areas of the facility or building. Figure 11-14A illustrates what a typical, simple "site sketch" looks like when completed.

The second step is to plot the fire protection facilities and equipment on the finished building sketch. Two important points should be noted. First, water supplies—public or private (yard) mains—are continuous from a point of origin to the individual automatic sprinkler risers. Conversely, the automatic sprinkler systems within the buildings are not entirely shown. (Note: Sprinkler occupancy classification and spacing are referred to in the body of the narrative inspection report.) Second, when space limitations prohibit drawing all the equipment on the plan at approximate actual locations (due to size, quantity of symbols, or labeling requirements), a small side sketch (not drawn to scale) can be used effectively.

The last step in completing the plan is to label the internal occupancy or processes of each area of the facility or building, and any yard storage or processes of interest.

The symbols and plans in Figures 11-14A and 11-14B will enable any inspector to develop a building plan in accordance with widespread convention. These symbols are based on NFPA 170, *Standard for Fire Safety Symbols.* However, many organizations use their own, slightly different set of symbols and layouts. When trying to take information from a plan developed from another source, it is essential to refer to the legend of symbols used by that organization to interpret the information accurately.

Occasionally, the use of a symbol alone will not sufficiently describe the situation. In that case, it is also necessary to label that point on the plan. Table 11-14A gives a list of standard abbreviations that can be used. Although brief, the list nevertheless provides all of the abbreviations commonly needed. *Abbreviations for Use on Drawings and in Text*, prepared by the American National Standards Institute,[2] lists additional, less commonly used abbreviations.

At one time, colors were used to describe the type of construction materials used for walls. Today, color-coded plans are being replaced by black and white drawings to facilitate preparation and for economic reasons, so construction type is being noted narratively. Table 11-14B identifies colors that should be used if the plan is to be colored. To avoid confusion on complicated plans, construction materials can be labeled in accordance with the abbreviations in Table 11-14A.

TABLE 11-14A. *Legend of Common Abbreviations**

Above	ABV	Liquid	LIQ
Accelerator	ACC	Liquid Oxygen	LOX
Acetylene	ACET	Manufacture	MFR
Aluminum	AL	Manufacturing	MFG
Asbestos	ASB	Maximum Capacity	MAX CAP
Asphalt-Protected Metal	APM	Mean Sea Level	MSL
Attic	A	Metal	MT
Automatic	AUTO	Mezzanine	MEZZ
Automatic Fire Alarm	AFA	Mill Use	MU
Automatic, Sprinklers	AS	Normally Closed	NC
Avenue	AVE	Normally Open	NO
Basement	B	North	N
Beam	BM	Number	No
Board on Joist	BDOJ	Open Sprinklers	OS
Brick	BR	Outside Screw & Yoke Valve	OS&Y
Building	BLDG	Partition (label composition)	PTN (e.g., WD PTN)
Cast Iron	CI	Plaster	PLAS
Cement	CEM	Plaster Board	PLAS BD
Centrifugal Fire Pump	CFP	Platform	PLATF
Cinder Block	CB	Pound (unit of force)	LB
Composition Roof	COMPR	Pressure	PRESS
Concrete	CONC	Unit of Pressure (psi)	PSI
Construction	CONST	Protected Steel	PROT ST
Corrugated Iron	COR IR	Private	PRIVATE
Corrugated Steel	COR ST	Public	PUB
Diameter	DIA	Railroad	RR
Diesel Engine	D ENG	Reinforced Concrete	RC
Domestic	DOM	Reinforcing Steel	RST
Double Hydrant	DH	Reservoir	RES
Dry-Pipe Valve	DPV	Revolutions per Minute	RPM
East	E	Roof	RF
Electric Motor Driven	EMD	Room	RM
Elevator	ELEV	Slate Shingle Roof	SSR
Engine	ENG	Space	SP
Exhauster	EXH	South	S
Feet	FT	Stainless Steel	SST
Fibre Board	FBR BD	Steam Fire Pump	SFP
Fire Escape	FE	Steel	ST
Fire Department Pumper Connection	FDPC	Steel Deck	ST DK
Fire Detection Units	FDU	Stone	STONE
Products of Combustion	POC	Story	STO
Rate of Heat Rise	RHR	Street	STREET
Fixed Temperature	FTEP	Stucco	STUC
Fire Pump	FP	Suspended Acoustical Plaster Ceiling	SAPL
Floor	FL	Suspended Acoustical Tile Ceiling	SATL
Frame	FR	Suspended Plaster Ceiling	SPC
Fuel Oil (label with grade number)	FO #_____	Suspended Sprayed Acoustical Ceiling	SSAL
Gallon	GAL	Tank (label capacity in gallons)	TK
Gallons per Day	GPD	Tenant	TEN
Gallons per Minute	GPM	Tile Block	TB
Galvanized Iron	GALVI	Timber	TMBR
Galvanized Steel	GALVS	Tin Clad	TIN CL
Gas, Natural	GAS	Triple Hydrant	TH
Gasoline	GASOLINE	Truss	TR
Gasoline-Engine Driven	GED	Under	UND
Generator	GEN	Vault	VLT
Glass	GL	Veneer	VEN
Glass Block	GLB	Volts (indicate number of)	450V
Gypsum	GYM	Wall Board	WLBD
Gypsum Board	GYM BD	Wall Hydrant	WLH
High Voltage	HV	Water Pipe	WP
Hollow Tile	HT	West	W
Hose Connection	HC	Wire Glass	WGL

TABLE 11-14A.   *(Continued)*

Hydrant	HYD	Wire Net	WN
Inch, Inches	IN	Wood	WD
Iron	IR	Wood Frame	WD FR
Iron Clad	IR CL	Yard	YD
Iron Pipe	IP		
Joist, Joisted	J		

*Some words that have a common abbreviation, e.g., "ST" for "street," are spelled out fully to avoid confusion with similar abbreviations used herein for other terms. See reference 1.
For SI units: 1 gal = 3.785 L; 1 psi = 6.895 kPa/1 lb (force) = 4.448 N.

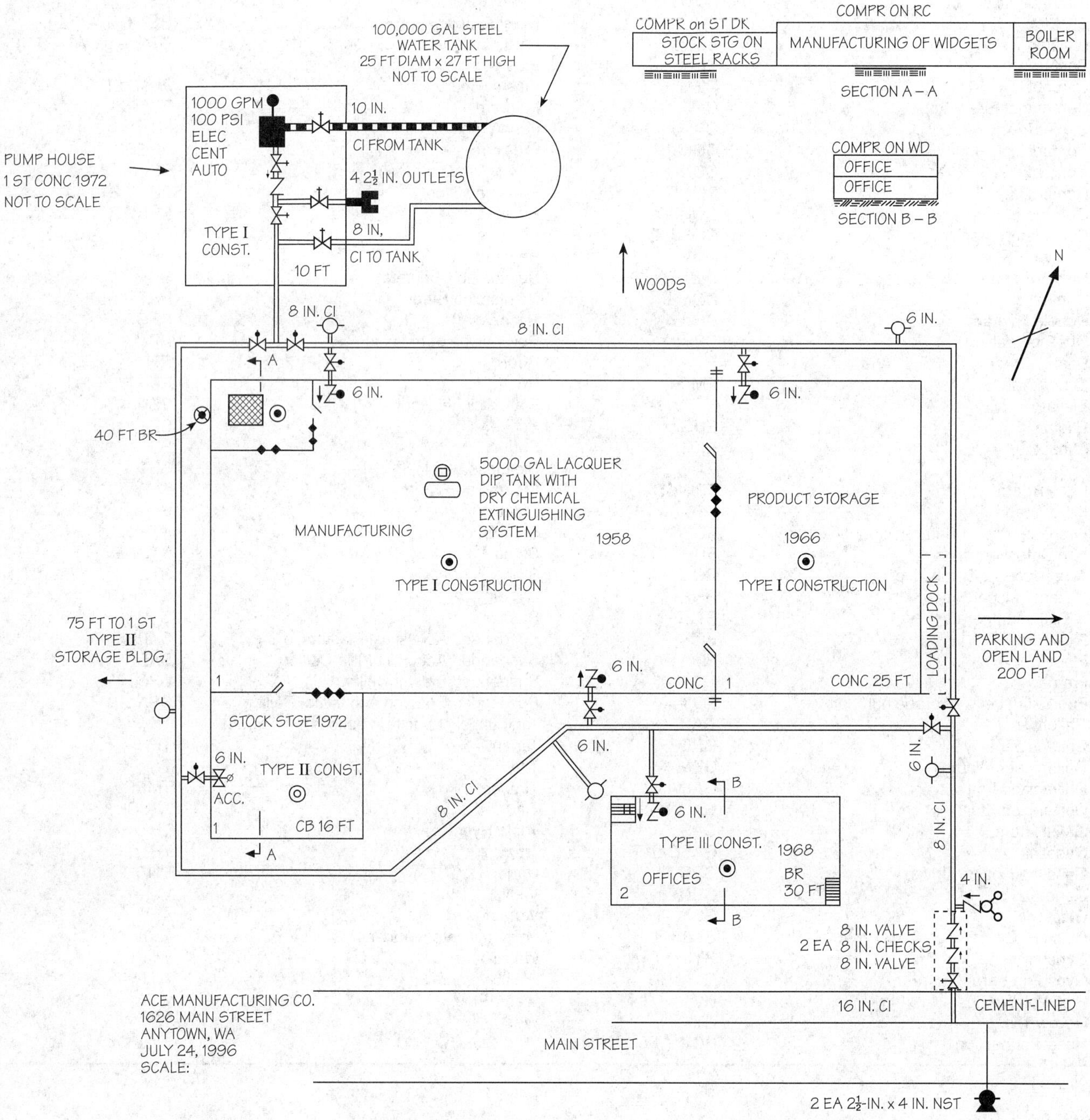

FIG. 11-14A.   *A typical site sketch showing fire protection facilities and equipment.*

SYMBOLS FOR SITE FEATURES

Buildings.

1. The exterior walls of buildings are outlined in single thickness lines if other than fire-rated, and double thickness lines if fire-rated.

2. The perimeter of canopies, loading docks, and other open-walled structures are shown by broken lines.

LOAD-ING

SHED

Railroad tracks. Railroad tracks are shown by a single line with crossed dashes.

Streets. Streets are shown, usually at property lines.

10   12-14

Downing Street

Bodies of water. Rivers, lakes, etc., are outlined.

POND

CREEK

Fences.

Fences are shown by lines with "x's" every in. (25.4 mm).

Gates are shown.

Property lines.

Fire department access.

F.D.

SYMBOLS FOR BUILDING CONSTRUCTION

Types of building construction. Types of construction are shown narratively.

| FIRE RESISTIVE CONST. (TYPE I) | WOOD-FRAME CONST. (TYPE V) |

Height. Indicate number of stories above ground, number of stories below ground, and height from grade to eaves.

A	B	C 1=2 24 ft	F
3 40 ft	1B 20 ft	1 D	13B
	1 E		UNDER-GROUND
			G

A   Three stories, no basement, 40 ft to eaves.
B   One story with basement, 20 ft to eaves.
C   One equals two stories, no basement, 24 ft to eaves.
D   One-story open porch or shed.
E   One-story addition.
F   Thirteen stories with basement.
G   Underground structure.

(Includes copyrighted material of Insurance Services Office with its permission. Copyright, Insurance Services Office 1975.)

Walls. Indicate construction.

Wall

— S —   Smoke barrier
— ► —   1/2-hr fire-rated
— ►S —   1/2-hr fire-rated/ smoke barrier
— ◆ —   3/4-hr fire-rated
— ◆S —   3/4-hr fire-rated/ smoke barrier
— ◆ —   1-hr fire-rated
— ◆S —   1-hr fire-rated/ smoke barrier

◆◆   2-hr fire-rated
◆◆S   2-hr fire-rated/ smoke barrier
◆◆◆   3-hr fire-rated
◆◆◆S   3-hr fire-rated/ smoke barrier
◆◆◆◆   4-hr fire-rated
◆◆◆◆S   4-hr fire-rated/ smoke barrier

Parapets. The symbol for parapets utilizes one cross for each 6 in. (152 mm) that the parapet extends above the roof. The cross is drawn through an extension of the wall line for the parapeted wall (in plan view).

Symbol used to note wall ratings and parapets on life safety plans and risk analysis plans/cross sections.

Floor openings, wall openings, roof openings, and their protection.

Opening in wall. Indicate floors. — —

Rated fire door in wall (less than 3 hr). Indicate floors.

Fire door in wall (3-hr rated). Indicate floors.

Elevator in combustible shaft.   E

Elevator in non-combustible shaft.   E

Opening hoistway   E

Escalator

Stairs in combustible shaft.

Stairs in fire-rated shaft.

Stairs in open shaft.

Skylight   SL

Roof, floor assemblies. These symbols indicate features in cross sections. Descriptive notes are often required.

Fire-resistive floor or roof

Wood joisted floor or roof

Other floors or roofs. Note construction.   (Steel deck on steel joists)

Floor/ceiling or roof/ceiling assembly. Details indicated as necessary.

Floor on ground

Truss roof. Note construction.

Miscellaneous features. For tanks, indicate type, dimensions, construction, capacity, pressurization, and contents.

Boiler

Chimney. Describe height and construction.

Fire escape

Horizontal tank, above ground.

Vertical tank, above ground.

Horizontal tank, below ground.

FIG. 11-14B.   Standard drawing symbols.

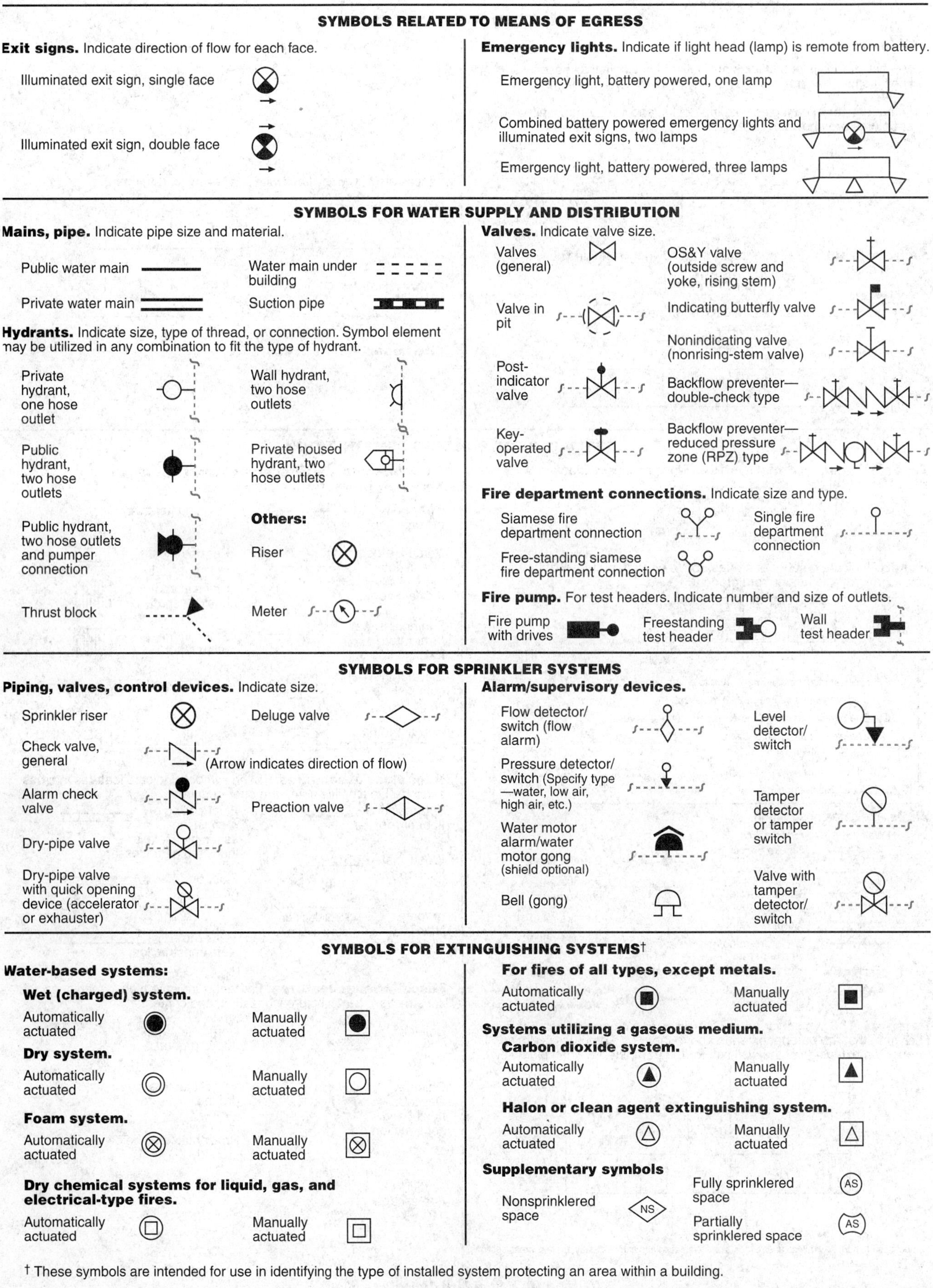

SYMBOLS RELATED TO MEANS OF EGRESS

Exit signs. Indicate direction of flow for each face.

Illuminated exit sign, single face

Illuminated exit sign, double face

Emergency lights. Indicate if light head (lamp) is remote from battery.

Emergency light, battery powered, one lamp

Combined battery powered emergency lights and illuminated exit signs, two lamps

Emergency light, battery powered, three lamps

SYMBOLS FOR WATER SUPPLY AND DISTRIBUTION

Mains, pipe. Indicate pipe size and material.

Public water main

Private water main

Water main under building

Suction pipe

Hydrants. Indicate size, type of thread, or connection. Symbol element may be utilized in any combination to fit the type of hydrant.

Private hydrant, one hose outlet

Public hydrant, two hose outlets

Public hydrant, two hose outlets and pumper connection

Thrust block

Wall hydrant, two hose outlets

Private housed hydrant, two hose outlets

Others:

Riser

Meter

Valves. Indicate valve size.

Valves (general)

Valve in pit

Post-indicator valve

Key-operated valve

OS&Y valve (outside screw and yoke, rising stem)

Indicating butterfly valve

Nonindicating valve (nonrising-stem valve)

Backflow preventer— double-check type

Backflow preventer— reduced pressure zone (RPZ) type

Fire department connections. Indicate size and type.

Siamese fire department connection

Free-standing siamese fire department connection

Single fire department connection

Fire pump. For test headers. Indicate number and size of outlets.

Fire pump with drives

Freestanding test header

Wall test header

SYMBOLS FOR SPRINKLER SYSTEMS

Piping, valves, control devices. Indicate size.

Sprinkler riser

Check valve, general

Alarm check valve

Dry-pipe valve

Dry-pipe valve with quick opening device (accelerator or exhauster)

Deluge valve

(Arrow indicates direction of flow)

Preaction valve

Alarm/supervisory devices.

Flow detector/ switch (flow alarm)

Pressure detector/ switch (Specify type —water, low air, high air, etc.)

Water motor alarm/water motor gong (shield optional)

Bell (gong)

Level detector/ switch

Tamper detector or tamper switch

Valve with tamper detector/ switch

SYMBOLS FOR EXTINGUISHING SYSTEMS†

Water-based systems:

Wet (charged) system.

Automatically actuated

Manually actuated

Dry system.

Automatically actuated

Manually actuated

Foam system.

Automatically actuated

Manually actuated

Dry chemical systems for liquid, gas, and electrical-type fires.

Automatically actuated

Manually actuated

For fires of all types, except metals.

Automatically actuated

Manually actuated

Systems utilizing a gaseous medium. Carbon dioxide system.

Automatically actuated

Manually actuated

Halon or clean agent extinguishing system.

Automatically actuated

Manually actuated

Supplementary symbols

Nonsprinklered space

Fully sprinklered space

Partially sprinklered space

† These symbols are intended for use in identifying the type of installed system protecting an area within a building.

FIG. 11-14B. Continued

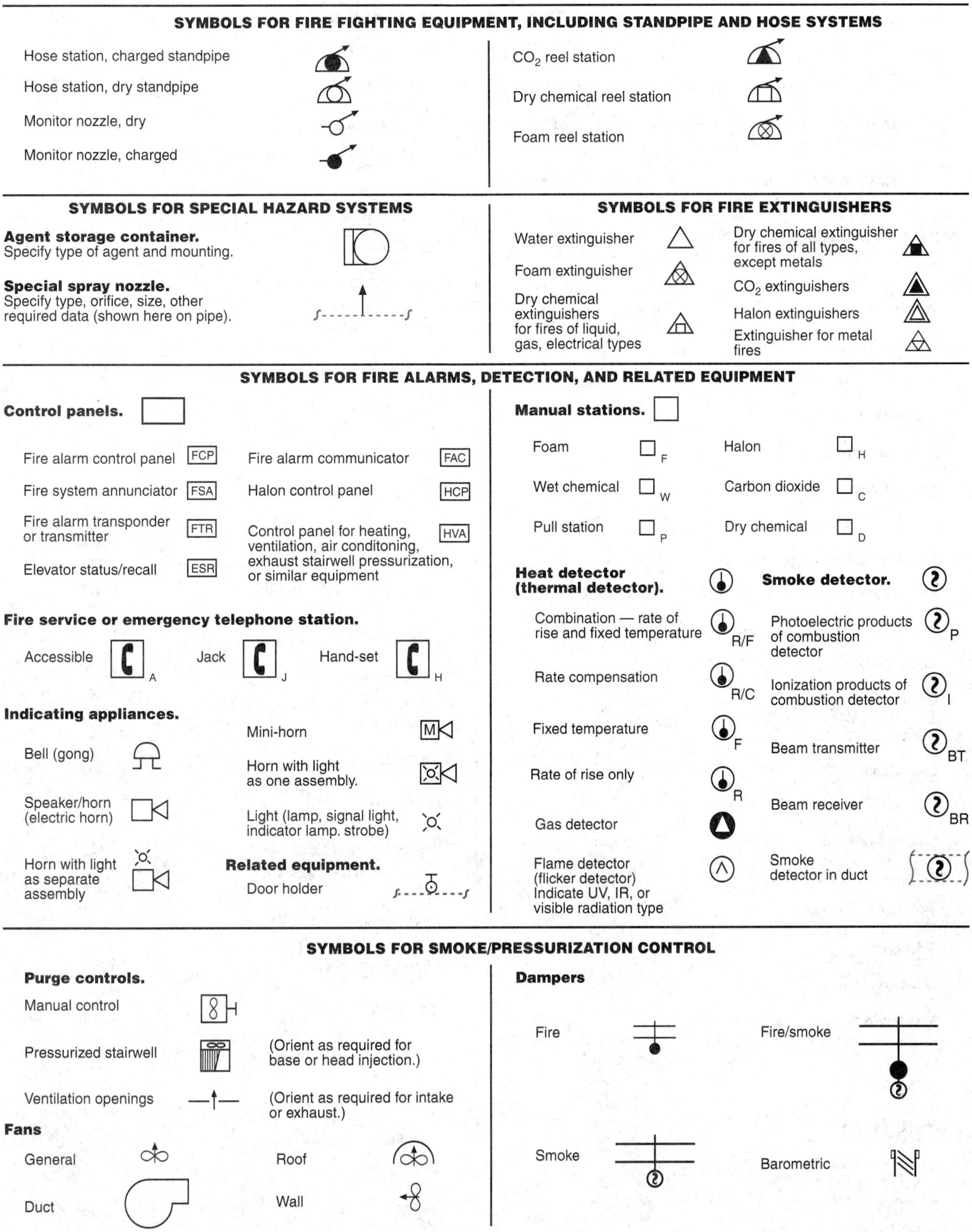

SYMBOLS FOR FIRE FIGHTING EQUIPMENT, INCLUDING STANDPIPE AND HOSE SYSTEMS

Hose station, charged standpipe

Hose station, dry standpipe

Monitor nozzle, dry

Monitor nozzle, charged

CO₂ reel station

Dry chemical reel station

Foam reel station

SYMBOLS FOR SPECIAL HAZARD SYSTEMS

Agent storage container.
Specify type of agent and mounting.

Special spray nozzle.
Specify type, orifice, size, other
required data (shown here on pipe).

SYMBOLS FOR FIRE EXTINGUISHERS

Water extinguisher

Foam extinguisher

Dry chemical
extinguishers
for fires of liquid,
gas, electrical types

Dry chemical extinguisher
for fires of all types,
except metals

CO₂ extinguishers

Halon extinguishers

Extinguisher for metal
fires

SYMBOLS FOR FIRE ALARMS, DETECTION, AND RELATED EQUIPMENT

Control panels.

Fire alarm control panel	FCP
Fire system annunciator	FSA
Fire alarm transponder or transmitter	FTR
Elevator status/recall	ESR

Fire alarm communicator	FAC
Halon control panel	HCP
Control panel for heating, ventilation, air conditoning, exhaust stairwell pressurization, or similar equipment	HVA

Fire service or emergency telephone station.

Accessible   [C]ₐ   Jack   [C]ⱼ   Hand-set   [C]ₕ

Indicating appliances.

Bell (gong)

Speaker/horn
(electric horn)

Horn with light
as separate
assembly

Mini-horn   [M]◁

Horn with light
as one assembly.

Light (lamp, signal light,
indicator lamp. strobe)

Related equipment.

Door holder

Manual stations.

Foam	□F	Halon	□H
Wet chemical	□W	Carbon dioxide	□C
Pull station	□P	Dry chemical	□D

**Heat detector
(thermal detector).**

Combination — rate of
rise and fixed temperature   R/F

Rate compensation   R/C

Fixed temperature   F

Rate of rise only   R

Gas detector

Flame detector
(flicker detector)
Indicate UV, IR, or
visible radiation type

Smoke detector.

Photoelectric products
of combustion
detector   P

Ionization products of
combustion detector   I

Beam transmitter   BT

Beam receiver   BR

Smoke
detector in duct

SYMBOLS FOR SMOKE/PRESSURIZATION CONTROL

Purge controls.

Manual control

Pressurized stairwell   (Orient as required for
base or head injection.)

Ventilation openings   (Orient as required for intake
or exhaust.)

Fans

General   Roof

Duct   Wall

Dampers

Fire

Smoke

Fire/smoke

Barometric

FIG. 11-14B. Continued

SYMBOLS FOR PRE-FIRE PLANNING

Triangle symbols can point at a specific location or direction.

Diamond symbols identify a specific location by touching a wall.

Circle symbols are used for all piping system appendatures, such as valves, since most pipes are round.

Square symbols are used for all room designations, as they represent most rooms having four sides.

Access features, assessment features, ventilation features, and utility shutoffs.

Access features

Fire department access point — FD

Fire department key box — K

Roof access — RA

Assessment features

Fire Alarm annunciator panel — AP

Fire alarm reset panel — RP

Fire alarm voice communication panel — CP

Smoke control and pressurization panel — SP

Sprinkler system water flow bell — WB

Utility shutoffs

Electric shutoff — E

Domestic water shutoff — W

Gas shutoff — G

Specific variations

LP-gas shutoff — LPG

Natural gas shutoff — NG

Ventilation features

Sky light — SL

Smoke vent — SV

Detector/extinguishing equipment.

Duct detector — DD

Heat detector — HD

Smoke detector — SD

Flow switch (water) — FS

Manual pull station — PS

Tamper switch — TS

Halon system — HL

Dry chemical system — DC

CO_2 system — CO2

Wet chemical system — WC

Foam system — FO

Clean agent system — CA

Beam smoke detector — BSD

Water flow control valves and water savers.

Post-indicator valve — PIV

Riser valve — RV

Sprinkler zone valve — ZV

Hose cabinet or connection — HC

Wall hydrant — WH

Test header (fire pump) — TH

Inspector's test connection — TC

Fire hydrant — FH

Fire department connection — FDC

Drafting site — DS

Water tank — WT

Equipment rooms.

Air-conditioning equipment room
Air-handling units (AHUs) — AC

Elevator equipment room — EE

Emergency generator room — EG

Fire pump room — FP

Telephone equipment room — TE

Boiler room — BR

Electrical/transformer room — ET

FIG. 11-14B. Continued

TABLE 11-14B. *Color Code for Denoting Construction Materials for Walls*

Color	Interpretation
Brown	Fire-resistive protected steel
Red	Brick, hollow tile
Yellow	Frame-wood, stucco
Blue	Concrete, stone, or hollow concrete block
Gray	Noncombustible (sheet metal or metal lath and plaster) unprotected steel

BIBLIOGRAPHY

References Cited

1. *NFPA Inspection Manual*, 7th ed., National Fire Protection Association, Quincy, MA, 1994.

2. ANSI, *Abbreviations for Use on Drawings and in Text*, ANSI Y1.1, American National Standards Institute, New York, 1989.

NFPA Codes, Standards, and Recommended Practices

Reference to the following NFPA codes, standards, and recommended practices will provide further information on formats for fire hazard inspection, surveying, and mapping. (See the latest version of *The NFPA Catalog* for availability of current editions of the following documents.)

NFPA 170, *Standard for Fire Safety Symbols*
NFPA 220, *Standard on Types of Building Construction*

Additional Reading

Clyde, J. E., *Construction Inspection*, Wiley, New York, 1983.
Electrical Inspection Illustrated, 2nd ed., National Safety Council, Chicago, IL, 1984.
Pilborough, L., *Inspection of Industrial Plants*, 2nd ed., Gower Technical, Brookfield, VT, 1989.
Robertson, J. C., "Fire Safety Inspection Procedures," *Introduction to Fire Prevention*, 3rd ed., Macmillan, New York, 1989, pp. 154–181.

Appendices

TABLES AND CHARTS

Compiled by Vytenis Babrauskas

TABLE A-1. *Heats of Combustion and Related Properties For Pure, Simple Substances**

Material	Composition	W Molecular Weight	Δh_c^u Gross (MJ/kg)	Δh_c^l Net (MJ/kg)	$\Delta h_c^l/r_o$ (MJ/kg O_2)	r_o Oxygen-fuel Mass ratio	T_b Boiling temp. (°C)	Δh_v Latent Heat of Vaporization (kJ/kg)	C_{pl} Liquid Heat Capacity (kJ/kg-°C)	C_{pv} Vapor Heat Capacity (kJ/kg-°C)
acetaldehyde	C_2H_4O	44.05	27.07	25.07	13.81	1.816	20.8	—	1.94	1.24
acetic acid	$C_2H_4O_2$	60.05	14.56	13.09	12.28	1.066	118.1	395		1.11
acetone	C_3H_6O	58.08	30.83	28.56	12.96	2.204	56.5	501	2.12	1.29
acetylene	C_2H_2	26.04	49.91	48.22	15.70	3.072	−84.0	—	—	1.69
acrolein	C_3H_4O	56.06	29.08	27.51	13.77	1.998	52.5	505	—	1.17
acrylonitrile	C_3H_3N	53.06	33.16	31.92	14.11	2.262	77.3	615	2.10	1.20
(allene) → propadiene										
ammonium perchlorate†	NH_4ClO_4	117.49	2.35	2.16	3.97	0.545	—	—	—	
iso-amyl alcohol	$C_5H_{12}O$	88.15	37.48	34.49	12.67	2.723	132.0	501	2.90	1.50
aniline	C_6H_7N	93.12	36.44	34.79	13.06	2.663	184.4	478	2.08	1.16
benzaldehyde	C_7H_6O	106.12	33.25	32.01	13.27	2.412	179.2	385	1.61	
benzene	C_6H_6	78.11	41.83	40.14	13.06	3.073	80.1	389	1.72	1.05
benzoic acid†	$C_7H_6O_2$	122.12	26.43	25.35	12.90	1.965	250.8	415	—	0.85
benzyl alcohol	C_7H_8O	108.13	34.56	32.93	13.09	2.515	205.7	467	2.00	1.19
bicyclohexyl	$C_{12}H_{22}$	166.30	45.35	42.44	12.61	3.367	236.	263		
1,2-butadiene	C_4H_6	54.09	47.95	45.51	13.99	3.254	10.8	—	—	1.48
1,3-butadiene	C_4H_6	54.09	46.99	44.55	13.69	3.254	−4.4	—	—	1.47
(1,3-butadiyne) → diacetylene										
n-butane	C_4H_{10}	58.12	49.50	45.72	12.77	3.579	−0.5	—	2.30	1.68
iso-butane	C_4H_{10}	58.12	48.95	45.17	12.62	3.579	−11.8	—	—	1.67
1-butene	C_4H_8	56.10	48.44	45.31	13.24	3.422	−6.2	—	—	1.53
n-butylamine	$C_4H_{11}N$	73.14	41.75	38.45	12.84	2.994	77.8	372	2.57	1.62
d-camphor†	$C_{10}H_{16}O$	152.23	38.75	36.44	12.84	2.838	203.4	—	—	0.82
carbon†	C	12.01	32.80	32.80	12.31	2.664	4200.	—	—	0.71
carbon disulfide	CS_2	76.13	6.34	6.34	5.03	1.261	46.5	351	1.00	0.60
carbon monoxide	CO	28.01	10.10	10.10	17.69	0.571	−191.3	—	—	1.04
cellulose†	$C_6H_{10}O_5$	162.14	17.47	16.12	13.61	1.184	—	—	1.16	—
(chloroethylene) → vinyl chloride										
(chloroform) → trichloromethane										
chlorotrifluoroethylene	C_2F_3Cl	116.47	2.00	2.00	3.64	0.549	−28.3	188	1.34	0.72
m-cresol	C_7H_8O	108.14	34.26	32.64	12.98	2.515	202.2	399	2.00	1.13
cumene	C_9H_{12}	120.19	43.40	41.20	12.90	3.195	152.3	312	1.77	1.26
cyanogen	C_2N_2	52.04	21.06	21.06	17.12	1.230	−21.2	—	—	1.12
cyclobutane	C_4H_8	56.10	48.91	45.77	13.38	3.422	12.9	—	—	1.29
cyclohexane	C_6H_{12}	84.16	46.58	43.45	12.70	3.422	80.7	357	1.84	1.26
cyclohexene	C_6H_{10}	82.14	45.67	42.99	12.99	3.311	82.8	371	1.80	1.28
cyclohexylamine	$C_6H_{13}N$	99.18	41.05	38.17	12.79	2.984	134.5			
cyclopentane	C_5H_{10}	70.13	46.93	43.80	12.80	3.422	49.3	389	2.23	1.18

Vytenis Babrauskas, Ph.D., is president of Fire Science and Technology, Inc., Kelso, WA.

TABLE A-1. *Heats of Combustion and Related Properties For Pure, Simple Substances* (continued)*

Material	Composition	W Molec-ular Weight	Δh_c^u Gross (MJ/kg)	Δh_c^l Net (MJ/kg)	$\Delta h_c^l/r_o$ (MJ/kg O_2)	r_o Oxygen-fuel Mass ratio	T_b Boiling temp. (°C)	Δh_v Latent Heat of Vaporization (kJ/kg)	C_{pl} Liquid Heat Capacity (kJ/kg-°C)	C_{pv} Vapor Heat Capacity (kJ/kg-°C)
cyclopropane	C_3H_6	42.08	49.70	46.57	13.61	3.422	−32.9	—	1.92	1.33
(decahydronaphthalene) → *cis*-decalin										
cis-decalin	$C_{10}H_{18}$	138.24	45.49	42.63	12.70	3.356	195.8	309	1.67	1.21
n-decane	$C_{10}H_{22}$	142.28	47.64	44.24	12.69	3.486	174.1	276	2.19	1.65
diacetylene	C_4H_2	50.06	46.60	45.72	15.89	2.877	10.3	—	—	1.47
(diamine) → hydrazine										
diborane	H_6B_2	27.69	79.80	79.80	23.02	3.467	−92.5	—	—	1.75
dichloromethane	CH_2Cl_2	84.94	6.54	6.02	10.65	0.565	39.7	330	1.18	0.60
diethyl cyclohexane	$C_{10}H_{20}$	140.26	46.30	43.17	12.58	3.422	174.		1.87	
diethyl ether	$C_4H_{10}O$	74.12	36.75	33.79	13.04	2.590	34.6	360	2.34	1.52
(2,4 diisocyanotoluene) → toluene diisocyanate										
(diisopropyl ether) → *iso*-propyl ether										
dimethylamine	C_2H_7N	45.08	38.66	35.25	13.24	2.662	6.9	—	—	1.60
(dimethyl aniline) → xylidene										
dimethyldecalin	$C_{12}H_{22}$	166.30	45.70	42.79	13.15	3.254	220.	260		
(dimethyl ether) → methyl ether										
1,1-dimethylhydrazine (UDMH)	$C_2H_8N_2$	60.10	32.95	30.03	14.10	2.130	25.	578	2.73	
dimethyl sulfoxide	C_2H_6SO	78.13	29.88	28.19	15.30	1.843	189.	677	1.89	1.14
1,3 dioxane	$C_4H_8O_2$	88.10	26.57	24.58	9.66	2.543	105.	404		
1,4 dioxane	$C_4H_8O_2$	88.10	26.83	24.84	9.77	2.543	101.1	406	1.74	1.07
ethane	C_2H_6	30.07	51.87	47.49	12.75	3.725	−88.6	—	—	1.75
ethanol	C_2H_6O	46.07	29.67	26.81	12.87	2.084	78.5	837	2.43	1.42
(ethene) → ethylene										
ethyl acetate	$C_4H_8O_2$	88.10	25.41	23.41	12.89	1.816	77.2	367	1.94	1.29
ethyl acrylate	$C_5H_8O_2$	100.12	27.44	25.69	13.39	1.918	100.	290		1.14
ethylamine	C_2H_7N	45.08	38.63	35.22	13.23	2.662	16.5	—	2.89	1.61
ethyl benzene	C_8H_{10}	106.16	43.00	40.93	12.93	3.165	136.1	339	1.75	1.21
ethylene	C_2H_4	28.05	50.30	47.17	13.78	3.422	−103.9	—	2.38	1.56
ethylene glycol	$C_2H_6O_2$	62.07	19.17	17.05	13.22	1.289	197.5	800	2.43	1.56
ethylene oxide	C_2H_4O	44.05	29.65	27.65	15.23	1.816	10.7	—	1.97	1.10
(ethylene trichloride) → trichloroethylene										
(ethyl ether) → diethyl ether										
formaldehyde	CH_2O	30.03	18.76	17.30	16.23	1.066	−19.3	—	—	1.18
formic acid	CH_2O_2	46.03	5.53	4.58	13.15	0.348	100.5	476	2.15	0.98
furan	C_4H_4O	68.07	30.61	29.32	13.86	2.115	31.4	398	1.69	0.96
α-D-glucose†	$C_6H_{12}O_6$	180.16	15.55	14.08	13.21	1.066	—	—	—	—
(glycerine) → glycerol										
glycerol	$C_3H_8O_3$	92.10	17.95	16.04	13.19	1.216	290.0	800	2.42	1.25
(glycerol trinitrate) → nitroglycerin										
n-heptane	C_7H_{16}	100.20	48.07	44.56	12.68	3.513	98.4	316	2.20	1.66
n-heptene	C_7H_{14}	98.18	47.44	44.31	12.95	3.422	93.6	317	2.17	1.58
hexadecane	$C_{16}H_{34}$	226.43	47.25	43.95	12.70	3.462	286.7	226	2.22	1.64
hexamethyldisiloxane	$C_6H_{18}Si_2O$	162.38	38.30	35.80	15.16	2.364	100.1	192	2.01	—
(hexamethylenetetramine) → methenamine										
n-hexane	C_6H_{14}	86.17	48.31	44.74	12.68	3.528	68.7	335	2.24	1.66
n-hexene	C_6H_{12}	84.16	47.57	44.44	12.99	3.422	63.5	333	2.18	1.57
hydrazine	H_4N_2	32.05	52.08	49.34	49.40	0.998	113.5	1180	3.08	1.65
hydrazoic acid	HN_3	43.02	15.28	14.77	79.40	0.186	35.7	690	—	1.02
hydrogen	H_2	2.00	141.79	130.80	16.35	8.000	−252.7	—	—	14.42
(hydrogen azide) → hydrazoic acid										
hydrogen cyanide	HCN	27.03	13.86	13.05	8.82	1.480	25.7	933	2.61	1.33
hydrogen sulfide	H_2S	34.08	48.54	47.25	16.77	2.817	−60.3	548	—	1.00
maleic anhydride†	$C_4H_2O_3$	74.04	18.77	18.17	14.01	1.297	202.0	—	—	—
melamine†	$C_3H_6N_6$	126.13	15.58	14.54	12.73	1.142	—	—	—	—
methane	CH_4	16.04	55.50	50.03	12.51	4.000	−161.5	—	—	2.23
methanol	CH_4O	32.04	22.68	19.94	13.29	1.500	64.8	1101	2.37	1.37
methenamine†	$C_6H_{12}N_4$	140.19	29.97	28.08	13.67	2.054	—	—	—	—
2-methoxyethanol	$C_3H_8O_2$	76.09	24.23	21.92	13.03	1.682	124.4	583	2.23	—
methylamine	CH_5N	31.06	34.16	30.62	13.21	2.318	−6.3	—	—	1.61
(2-methyl 1-butanol) → *iso*-amyl alcohol										
(methyl chloride) → dichloromethane										
methyl ether	C_2H_6O	46.07	31.70	28.84	13.84	2.084	−24.9	—	—	1.43
methyl ethyl ketone	C_4H_8O	72.10	33.90	31.46	12.89	2.441	79.6	434	2.30	1.43
1-methylnaphthalene	$C_{11}H_{10}$	142.19	40.88	39.33	12.95	3.038	244.7	323	1.58	1.12

TABLE A-1. *Heats of Combustion and Related Properties For Pure, Simple Substances* (continued)*

Material	Composition	W Molecular Weight	Δh_c^u Gross (MJ/kg)	Δh_c^l Net (MJ/kg)	$\Delta h_c^l/r_o$ (MJ/kg O_2)	r_o Oxygen-fuel Mass ratio	T_b Boiling temp. (°C)	Δh_v Latent Heat of Vaporization (kJ/kg)	C_{pl} Liquid Heat Capacity (kJ/kg-°C)	C_{pv} Vapor Heat Capacity (kJ/kg-°C)
methyl methacrylate	$C_5H_8O_2$	100.11	27.37	25.61	12.33	2.078	101.0	360	1.91	—
methyl nitrate	CH_3NO_3	77.04	8.67	7.81	75.10	0.104	64.6	409	2.04	0.99
(2-methyl propane) → *iso*-butane										
naphthalene†	$C_{10}H_8$	128.16	40.21	38.84	12.96	2.996	217.9	—	1.18	1.03
nitrobenzene	$C_6H_5NO_2$	123.11	25.11	24.22	14.90	1.625	210.7	330	1.52	—
nitroglycerin	$C_3H_5N_3O_9$	227.09	6.82	6.34	—	—	unstable	462	1.49	—
nitromethane	CH_3NO_2	61.04	11.62	10.54	15.08	0.699	101.1	567	1.74	0.94
n-nonane	C_9H_{20}	128.25	47.76	44.33	12.69	3.493	150.6	295	2.10	1.65
octamethyl-cyclotetrasiloxane	$C_8H_{24}Si_4O_4$	296.62	26.90	25.10	14.56	1.725	175.0	127	1.88	—
n-octane	C_8H_{18}	114.22	47.90	44.44	12.69	3.502	125.6	301	2.20	1.65
iso-octane	C_8H_{18}	114.22	47.77	44.31	12.65	3.502	117.7	272	2.15	1.65
1-octene	C_8H_{10}	112.21	47.33	44.20	12.92	3.422	121.3	301	2.19	1.59
(1-octylene) → 1-octene										
1,2-pentadiene	C_5H_8	68.11	47.31	44.71	13.60	3.288	44.9	405	2.21	1.55
n-pentane	C_5H_{12}	72.15	48.64	44.98	12.68	3.548	36.0	357	2.33	1.67
1-pentene	C_5H_{10}	70.13	47.77	44.64	13.04	3.422	30.0	359	2.16	1.56
phenol†	C_6H_6O	94.11	32.45	31.05	13.05	2.380	181.8	433	1.43	1.10
phosgene	$COCl_2$	98.92	1.74	1.74	10.74	0.162	8.3	247	1.02	0.58
propadiene	C_3H_4	40.06	48.54	46.35	14.51	3.195	−34.6	—	—	1.44
propane	C_3H_8	44.09	50.35	46.36	12.78	3.629	−42.2	—	2.23	1.67
n-propanol	C_3H_8O	60.09	33.61	30.68	12.81	2.396	97.2	686	2.50	1.45
iso-propanol	C_3H_8O	60.09	33.38	30.45	12.71	2.396	80.3	663	2.42	1.48
propene	C_3H_6	42.08	48.92	45.79	13.38	3.422	−47.7	—	—	1.52
(*iso*-propylbenzene) → cumene										
(propylene) → propene										
iso-propyl ether	$C_6H_{14}O$	102.17	39.26	36.25	12.86	2.819	67.8	286	2.14	1.55
propyne	C_3H_4	40.06	48.36	46.17	14.45	3.195	−23.3	—	—	1.51
styrene	C_8H_8	104.14	42.21	40.52	13.19	3.073	145.2	356	1.76	1.17
sucrose†	$C_{12}H_{22}O_{11}$	342.30	16.49	15.08	13.44	1.122	—	—	1.24	—
(1,2,3,4-tetrahydronaphthalene) → tetralin										
tetralin	$C_{10}H_{12}$	132.20	42.60	40.60	12.90	3.147	207.0	425	1.64	1.19
tetranitromethane	CN_4O_8	196.04	2.20	2.20	—	—	125.7	196—	—	—
toluene	C_7H_8	92.13	42.43	40.52	12.97	3.126	110.4	360	1.67	1.12
toluene diisocyanate	$C_9H_6N_2O_2$	174.16	24.32	23.56	13.50	1.746	120.0	—	1.65	—
triethanolamine	$C_6H_{15}NO_3$	149.19	29.29	27.08	15.30	1.770	360.0	—	—	—
triethylamine	$C_6H_{15}N$	101.19	43.19	39.93	12.95	3.083	89.5	303	2.22	1.59
1,1,2-trichloroethane	$C_2H_3Cl_3$	133.42	7.77	7.28	11.02	0.660	114.0	260	1.11	0.67
trichloroethylene	C_2HCl_3	131.40	6.77	6.60	12.05	0.548	86.9	245	1.07	0.61
trichloromethane	$CHCl_3$	119.39	3.39	3.21	9.60	0.335	61.7	249	0.97	0.55
trinitromethane	CHN_3O_6	151.04	3.41	3.25	—	—	unstable	—	—	—
trinitrotoluene†	$C_7H_5N_3O_6$	227.13	15.12	14.64	19.80	0.740	240.0	322	1.40	—
trioxane	$C_3H_6O_3$	90.08	16.57	15.11	14.17	1.066	114.5	450	—	—
urea†	CH_4ON_2	60.06	10.52	9.06	11.34	0.799	—	—	—	1.55
vinyl acetate	$C_4H_6O_2$	86.09	24.18	22.65	13.54	1.673	72.5	167	2.00	1.05
vinyl acetylene	C_4H_4	52.07	47.05	45.36	14.76	3.073	5.1	—	—	1.41
vinyl bromide	C_2H_3Br	106.96	12.10	11.48	13.95	0.823	15.6	—	2.42	0.53
vinyl chloride	C_2H_3Cl	62.50	20.02	16.86	11.97	1.408	−13.8	—	—	0.86
(vinyl trichloride) → 1,1,2-trichlorolthane										
xylenes	C_8H_{10}	106.16	42.89	40.82	12.90	3.165	138–144	343	1.72	1.21
xylidene	$C_8H_{11}N$	121.22	38.28	36.29	12.79	2.838	192.7	366	1.77	—

† Denotes substance in crystalline solid form; otherwise, liquid if $T_b > 25°C$, gaseous if $T_b < 25°C$.
* Sources: 1, 2, 3, 4, 5, 6, 7, 8, 9, 10, 11, 12, 13, 14, 15.

TABLE A-2. *Heats of Combustion and Related Properties for Plastics**

Material	Unit Composition	W Molecular Weight	Δ_c^u Gross (MJ/kg)	Δ_c^l Net (MJ/kg)	Δ_c^l (MJ/kg O_2)	r_o Oxygen-fuel Mass ratio	C_{ps} Heat Capacity Solid (kJ/kg-°C)
acrylonitrile-butadiene styrene copolymer	—	—	35.25	33.75			1.41–1.59
bisphenol A epoxy	$C_{11.85}H_{20.37}O_{2.83}N_{0.3}$	212.10	33.53	31.42	13.41	2.343	
butadiene-acrylonitrile 37% copolymer	—	—	39.94				
butadiene/styrene 8.58% copolymer	$C_{4.18}H_{6.09}$	56.30	44.84	42.49	13.11	3.241	1.94
butadiene/styrene 25.5% copolymer	$C_{4.60}H_{6.29}$	61.55	44.19	41.95	13.07	3.209	1.82
cellulose acetate (triacetate)	$C_{12}H_{16}O_8$	288.14	18.88	17.66	13.25	1.333	1.34
cellulose acetate-butyrate	$C_{12}H_{18}O_7$	274.27	23.70	22.3	14.67	1.517	1.70
epoxy, unhardened	$C_{31}H_{36}O_{5.5}$	496.63	32.92	31.32	13.05	2.400	
epoxy, hardened	$C_{39}H_{40}O_{8.5}$	644.74	30.27	28.90	13.01	2.221	
melamine formaldehyde (Formica)	$C_6H_6N_6$	162.08	19.33	18.52	12.51	1.481	1.46
nylon 6	$C_6H_{11}NO$	113.08	30.1–31.7	28.0–29.6	12.30	2.335	1.52
nylon 6,6	$C_{12}H_{22}N_2O_2$	226.16	31.6–31.7	29.5–29.6	12.30	2.405	1.70
nylon 11 (Rilsan)	$C_{11}H_{21}NO$	183.14	36.99	34.47	12.33	2.796	1.70–2.30
phenol formaldehyde -foam	$C_{15}H_{12}O_2$	224.17	27.9–31.6 / 21.6–27.4	26.7–30.4 / 20.2–26.2	11.80	2.427	1.70
polyacenaphthalene	$C_{12}H_8$	152.14	39.23	38.14	12.95	2.945	
polyacrylonitrile	C_3H_3N	53.04	32.22	30.98	13.70	2.262	1.50
polyallylphthalate	$C_{14}H_{14}O$	198.17	27.74	26.19	9.54	2.745	
(polyamides) → nylon							
poly-1,4-butadiene	C_4H_6	54.05	45.19	42.75	13.13	3.256	
poly-1-butene	C_4H_8	56.05	46.48	43.35	12.65	3.426	1.88
polycarbonate	$C_{16}H_{14}O_3$	254.19	30.99	29.78	13.14	2.266	1.26
polycarbon suboxide	C_3O_2	68.03	13.78	13.78	14.64	0.941	
polychlorotrifluoroethylene	C_2F_3Cl	116.47	1.12	1.12	2.04	0.549	0.92
polydiphenylbutadiene	$C_{16}H_{10}$	202.18	39.30	38.2	13.05	2.928	
polyester, unsaturated	$C_{5.77}H_{6.25}O_{1.63}$	101.60	21.6–29.8	20.3–28.5	11.90	2.053	1.20–2.30
polyether, chlorinated	$C_5H_8OCl_2$	154.97	17.84	16.71	12.45	1.342	
polyethylene	C_2H_4	28.03	46.2–46.5	43.1–43.4	12.63	3.425	1.83–2.30
polyethylene oxide	C_2H_4O	44.02	26.65	24.66	13.57	1.817	
polyethylene terephthalate	$C_{10}H_8O_4$	192.11	22.18	21.27	12.77	1.666	1.00
polyformaldehyde	CH_2O	30.01	16.93	15.86	14.88	1.066	1.46
poly-1-hexene sulfone	$C_6H_{12}SO_2$	148.13	29.78	28.00	14.40	1.944	
polyhydrocyanic acid	HCN	27.02	23.26	22.45	15.17	1.480	
(polyisobutylene) → poly-1-butene							
polyisocyanurate foam	—		23–26.3	22.2–26.2			
polyisoprene	C_5H_8	68.06	44.90	42.30	12.90	3.291	
poly-3-methyl-1-butene	C_5H_{10}	70.06	46.55	43.42	12.67	3.426	
polymethyl methacrylate	$C_5H_8O_2$	100.06	26.64	24.88	12.97	1.919	1.44
poly-4-methyl-1-pentene	C_6H_{12}	84.08	46.52	43.39	12.67	3.425	2.18
poly-α-methylstyrene	C_9H_{10}	118.11	42.31	40.45	13.00	3.116	
polynitroethylene	$C_2H_3O_2N$	73.03	15.96	15.06	19.64	0.767	
polyoxymethylene	CH_2O	30.01	16.93	15.65	14.68	1.066	
polyoxytrimethylene	C_3H_6O	58.04	31.52	29.25	13.27	2.205	
poly-1-pentene	C_5H_{10}	70.06	45.58	42.45	12.39	3.426	
polyphenylacetylene	C_8H_6	102.09	40.00	38.70	13.00	2.978	
polyphenylene oxide	C_8H_8O	120.09	34.59	33.13	13.09	2.531	1.34
polypropene sulfone	$C_3H_6SO_2$	106.10	23.82	22.58	16.64	1.357	
poly-β-propiolactone	$C_3H_4O_2$	72.14	19.35	18.13	13.62	1.331	
polypropylene	C_3H_6	42.04	46.37	43.23	12.62	3.824	2.10
polypropylene oxide	C_3H_6O	58.04	31.17	28.90	13.11	2.205	
polystyrene	C_8H_8	104.10	41.4–42.5	39.7–39.8	12.93	3.074	1.40
polystyrene-foam	—		39.7	35.6–40.8			
polystyrene-foam, FR	—		41.2–42.9				
polysulfones, butene	$C_4H_8SO_2$	120.11	24.04–26.47	22.25–25.01	14.79	1.598	1.30
polysulfur	S	32.06	9.72	9.72	9.74	0.998	
polytetrafluoroethylene	C_2F_4	100.02	5.00	5.00	7.81	0.640	1.02
polytetrahydrofuran	C_4H_8O	72.05	34.39	31.85	13.04	2.443	
polyurea	$C_{15}H_{18}O_4N_4$	318.20	24.91	23.67	13.45	1.760	

TABLE A-2.   *(continued)*

Material	Unit Composition	W Molecular Weight	Δ_c^u Gross (MJ/kg)	Δ_c^l Net (MJ/kg)	Δ_c^l (MJ/kg O_2)	r_o Oxygen-fuel Mass ratio	C_{ps} Heat Capacity Solid (kJ/kg-°C)
polyurethane	$C_{6.3}H_{7.1}NO_{2.1}$	130.30	23.90	22.70	13.16	1.725	1.75–1.84
polyurethane-foam	—		26.1–31.6	23.2–28.0			
polyurethane-foram, FR	—		24.0–25.0				
polyvinyl acetate	$C_4H_6O_2$	86.05	23.04	21.51	12.86	1.673	
polyvinyl alcohol	C_2H_4O	44.03	25.00	23.01	12.66	1.817	1.70
polyvinyl butyral	$C_8H_{14}O_2$	142.10	32.90	30.70	13.00	2.365	
polyvinyl chloride	C_2H_3Cl	62.48	17.95	16.90	12.00	1.408	0.90–1.20
polyvinyl chloride foam	—		22.83				1.30–2.10
polyvinyl fluoride	C_2H_3F	46.02	21.70	20.27	10.60	1.912	
polyvinylidene chloride	$C_2H_2Cl_2$	96.93	10.52	10.07	12.21	0.825	1.34
polyvinylidene fluoride	$C_2H_2F_2$	64.02	14.77	14.08	11.26	1.250	1.38
urea formaldehyde	$C_3H_6O_2N_2$	102.05	15.90	14.61	13.31	1.098	1.60–2.10
urea formaldehyde-foam	—	—	14.80				

* Sources: 16, 17, 18, 19, 20, 21, 22, 23, 24.

Notes to Tables A-1, A-2, A-3, and A-4

Heats of Combustion: The heat of combustion is, by definition, the enthalpy of reaction when fuel and oxidant at standard conditions are reacted and form products at standard conditions. A unique value for the heat of combustion is possible only if these conditions are fully specified.[2,34] In normal combustion work the standard conditions are taken as:

1. Fuel and oxidant enter at 1 atmosphere pressure and 25°C (298 K) temperature. An amount of heat, which is equal to the heat of combustion, is extracted, so that the products are also at 25°C and 1 atmosphere.
2. The oxidant is gaseous oxygen.
3. The main products consist of liquid H_2O, gaseous CO_2, and gaseous N_2. There is no CO formed.
4. For fuels containing sulfur, the standard products include liquid $H_2SO_4.115H_2O$. For chlorine-containing fuels reference states consisting of either liquid HCl in water solution or gaseous Cl_2 have been used.

5. In the combustion of silicones, the silicon goes to amorphous silica, SiO_2.

The state of the fuel—gaseous, liquid or solid—is not standardized and must be specified. The heat of combustion as defined above is termed the gross or upper value and is customarily determined in an oxygen bomb calorimeter.[35] For common materials the value is a negative number; however, customarily a minus sign is included in the definition to make heat of combustion a positive value.[36] Heat of combustion, enthalopy of combustion, calorific value and heating value are synonyms, the latter two being used more commonly in the heating industry.

In many cases the products are not cooled down to 25°C. For modest temperature differences the change in the effective value could be taken as $C_p(T - 25°)$, where C_p is the constant pressure heat capacity of the products. For this relationship to hold, the enthalpy curve has to be smooth and cannot include condensation. Thus it often becomes convenient to define a lower, or net, heat of combustion where the products are as before except that H_2O stays in gaseous

TABLE A-3.   *Heat of Combustion of Miscellaneous Substances**

Material	Δ_c^u Gross (MJ/kg)	Δ_c^l Net (MJ/kg)
acetate (see cellulose acetate)		
acrylic fiber	30.6–30.8	
blasting powder	2.1–2.4	
butter	38.5	
celluloid (cellulose nitrate and camphor)	17.5–20.6	16.4–19.2
cellulose acetate fiber, $C_8H_{12}O_6$	17.8–18.4	16.4–17.0
cellulose diacetate fiber, $C_{10}H_{14}O_7$	18.7	
cellulose nitrate, $C_6H_9N_1O_7/C_6H_8N_2O_9/C_6H_7N_3O_{11}$	9.11–13.48	
cellulose triacetate fiber, $C_{12}H_{16}O_8$	18.8	17.6
charcoal	33.7–34.7	33.2–34.2
coal-anthracite	30.9–34.6	30.5–34.2
-bituminous	24.7–36.3	23.6–35.2
coke	28.0–31.0	28.0–31.0
cork	26.1	
cotton	16.5–20.4	
dynamite	5.4	
epoxy, $C_{11.9}H_{20.4}O_{2.8}N_{0.3}C_{6.064}H_{7.550}O_{1.222}$	32.8–33.5	31.1–31.4

TABLE A-3. *Heat of Combustion of Miscellaneous Substances* *(continued)*

Material	Δ_c^u Gross (MJ/kg)	Δ_c^l Net (MJ/kg)
fat, animal	39.8	
flint powder	3.0–3.1	
fuel oil-No. 1	46.1	
-No. 6	42.5	
gasketing-chlorosulfonated polyethylene (Hypalon)	28.5	
-vinylidene fluoride/hexafluoropropylene (Fluorel, Viton A)	14.0–15.1	
gasoline	46.8	43.7
jet fuel-JP1		43.0
-JP3		43.5
-JP4	46.6	43.5
-JP5	45.9	43.0
kerosene (jet fuel A)	46.4	43.3
lanolin (wool fat)	40.8	
lard	40.1	
leather	18.2–19.8	
lignin, $C^{2.6}H_3O$	24.7–26.4	23.4–25.1
lignite	22.4–33.3	
modacrylic fiber	24.7	
naphtha	43.0–47.1	40.9–43.9
neoprene, C_5H_5Cl-gum	24.3	
-foam	9.7–26.8	
Nomex (polymethaphenylene isophthalamide)		
fiber, $C_{14}H_{10}O_2N_2$	27.0–28.7	
oil-castor	37.1	
-linseed	39.2–39.4	
-mineral	45.8–46.0	
-olive	39.6	
-solar	41.8	
paper-brown	16.3–17.9	
-magazine	12.7	
-newsprint	19.7	
-wax	21.5	
paraffin wax	46.2	43.1
peat	16.7–21.6	
petroleum jelly ($C_{7.118}H_{12.957}O_{0.091}$)	45.9	
rayon fiber	13.6–19.5	
rubber-buna N	34.7–35.6	
-butyl	45.8	
-isoprene (natural) C_5H_8	44.9	42.3
-latex foam	33.9–40.6	
-GRS	44.2	
-tire, auto	32.6	
silicone rubber (SiC_2H_6O)	15.5–16.8	
-foam	14.0–19.5	
sisal	15.9	
spandex fiber	31.4	
starch	17.6	16.2
straw	15.6	
sulfur-rhombic		9.28
-monoclinic		9.29
tobacco	15.8	
wheat	15.0	
wood-beech	20.0	18.7
-birch	20.0	18.7
-douglas fir	21.0	19.6
-maple	19.1	17.8
-red oak	20.2	18.7
-spruce	21.8	20.4
-white pine	19.2	17.8
-hardboard	19.9	
woodflour	19.8	
wool	20.7–26.6	

* Sources: 7, 9, 16, 25, 26, 27, 28, 29, 30.

TABLE A-4.   *Heats of Combustion for Metals**

Material	Δh_c (MJ/kg)
Pure elements	
aluminum	31.04
beryllium	66.43
copper	2.45
iron	7.39
magnesium	24.72
manganese	7.01
molybdenum	6.13
nickel	4.10
tantalum	5.66
tin	3.73
titanium	19.71
zinc	5.37
zirconium	12.07
Copper alloys	
bronze (88 Cu/10 Sb/2 Zn)	2.64
red brass (85 Cu/15 Zn)	2.89
cartridge brass (70 Cu/30 Zn)	3.33
yellow brass (60 Cu/40 Zn)	3.62
Iron alloys	
carbon steels	7.4–7.5
stainless steels	7.7–8.4
Nickel alloys	
Inconel 600	5.40
Monel 400	3.60

* Sources: 31, 32, 33.

form (and sulfur, if any, goes into gaseous SO_2). The relationship between the values for simple fuels can be expressed as:

$$\Delta h^l_c = \Delta h^u_c - 0.2196\ [\%H]$$

where:

Δh^l_c = lower heat of combustion (MJ/kg)

Δh^u_c = upper heat of combustion (MJ/kg)

[%H] = percent, by mass, of hydrogen in fuel

In practice, combustion is rarely complete. For example, some carbon often goes to CO instead of CO_2. Reactions with fuels containing sulfur, or chlorine or other halogens can go into a variety of products, depending on exact reaction conditions. Incomplete combustion acts to lower the heat of combustion. There are no simple general rules to aid in determining the degree of incomplete combustion. Also explosive and propellant materials rarely exhibit reactions where H_2O and CO_2 are the main products; thus, their heats of explosion or reaction will tend to be substantially different from the heat of combustion.

Even though the state of reactants is not described in the definition, most commonly it is taken as the natural state at ambient pressure and temperature. In some references, however, the gaseous state of fuels such as benzene or methanol is assumed, even though they would be normally considered as liquid fuels. The heat of combustion of fuels from the gaseous state is higher than from the liquid state by an amount equal to the fuel's heat of vaporization at 25°C. For combustion of plastic fuels, the most convenient reference state is the solid phase. In such cases, to compute the heat liberated in a given application, it is not necessary to subtract the heat of gasification from the heat of combustion.

Heats of combustion (per kg of fuel) are listed in Tables A-1 through A-3. Also listed in Tables A-1 and A-2 is r_o, the stoichiometric oxygen-fuel mass ratio. The heats can be seen to vary quite a bit for different fuels. Recently, however, increasing engineering use is

TABLE A-5.   *Oxygen Index Values**

Material	Oxygen Index Value†
methanol	11–12
benzene	13–16
sulfur	13.6
polyacetal	14.9
polyoxymethylene	15.7
acetone	16.0
kitchen candle	16.0
cotton	16.0–18.5
mineral oil	16.1
polyurethane foam, typical	16.5
cellulose acetate	16.8
styrene-butadiene-rubber foam	16.9
polybutadiene foam	17.1
natural rubber foam	17.2
polymethylmethacrylate	17.3
polyethylene	17.4
polypropylene	17.4
polystyrene	17.6–18.3
polyacrylonitrile	18.0
paper, cellulose filter	18.2
polybutadiene	18.3
ABS	18.3–18.8
cellulose triacetate	18.4
polyisoprene rubber	18.5
rayon	18.7–18.9
polymethylsiloxane (silicone fluid)	18.9
cellulose	19.0
cellulose acetate-butyrate	19.6
epoxy resin	19.8
polyethylene terephthalate	20.0
nylon	20.1–26.0
birch wood	20.5
polyester fibers	20.6
phenolic resin	21.0
phenolic-paper laminate	21.7
polyvinyl alcohol	22.5
polycarbonate	22.5–28.0
polyvinyl fluoride	22.6
red oak	23.0
wool	23.8
neoprene rubber	26.3–40.0
Nomex fibers	26.7–28.5
modacrylic fibers	26.8
polyphenylene oxide	29.0
silicone rubber	30.0
polysulfone	30–32
leather	34.8
carbon powder	35.0
phenol-formaldehyde resin	35.0
polyimide	36.5
polyvinyl chloride	37.1–49
Kynol fibers	38.0
polybenzimidazole	40.6–41.5
polyvinylidene fluoride	43.7
polyvinyl chloride, chlorinated	45–60
carbon black	56–63
polyvinylidene chloride	60
polychlorotrifluoroethylene	95
polytetrafluoroethylene	95

† The values in the table above were determined using the ASTM Oxygen Test (D2863), and represent typical values for materials without added fire retardants. Adding fire retardants to polymers can increase substantially the measured oxygen index value.

* Sources: 41, 42, 43.

TABLE A-6. *Miscellaneous Properties or Plastics and Other Combustibles**

Material	Density (kg/m³)	Min. Radiant Flux for Ignition (kW/m²)	Min. Mass Loss Rate for Ignition (g/m²-s)	Energy Required for Ignition (kJ/m²)	Ignition Temperature (K)	Effective Heat of Gasification (MJ/kg)	Smoke Specific Extinction Area (m²/kg)	Soot Yield (kg soot/ kg specimen)
ABS						3.2		
benzene, liquid							830–900	
cellulose						3.5	10–15	
Douglas Fir				650–770		1.8		
fiberglass, FR	—	7	—	3950	740	1.7–2.2	250–370	
nylon 6/6						2.3	410	0.14
paper, corrugated						2.2		
phenolic, foams, rigid	67	45	5.5	390	944	3.7		
polycarbonate						2.1		
polyethylene	920	19	2.5	1500–5100	761	1.5–2.7	290–510	
polyethylene, foams	36–143	19–22	2.5–2.6	1430–1780	761–807	1.5–2.7	360–810	
polyethylene, foam, chlorinated						2.1–3.1	1010–1360	
polyisocyanurate, foams, ridgid	33–36	23	5.5–6.8	850–930	798	4.5		
polymethylmethacrylate	1170	18	4.4	1300–4200	751	1.6	110–260	0.012
polyoxymethylene	1400	17	4.5	2100–6000	740	2.4	0–10	
polypropylene	905	20	2.7	1000–3500	771	1.4–2.0	380–610	0.07–0.09
polystyrene	1050	29	4.0	1300–6400	846	1.7	730–1430	
polystyrene, foams, rigid	16–34	18–27	4.9–6.3	1200–3200	751–831	1.3–3.1	940–1300	
polytetrafluoroethylene		43		9000	933			0.02
polyurethane, foams, flexible	29–47	16–30	5.3–7.2	150–770	729–852	1.2–2.7	100–500	
polyurethane, foams, rigid	36–321	20–28	6.9–8.4	150–1300	771–838	2.0–4.5	100–1000	
polyvinyl chloride	1305	21	—	3320	780	1.7–2.5	1100–1400	0.23
red oak	740	11	2.5	420–920	730	1.7–5.5	80–130	.0002–0.018
styrene, liquid						0.64	780–920	

* Sources: 18, 45, 46, 47, 48, 49, 50.

made of the observation that the heat of combustion per kg of oxygen consumed is nearly constant for most organic fuels. It can be shown that a value of $Dh^1_c/r_o = 13.1$ MJ/kg O_2 is near-constant.[37] An assumption of constant heat of combustion per kg of oxygen is useful in heat release rate measurements and for air-limited combustion problems.

For certain fuels it is more convenient to express heats of combustion by use of equations or charts, rather than tables. Two such instances are woods and certain petroleum products. The gross heat of combustion for dry woods is typically in the vicinity of 18-21 MJ/kg. This value varies, however, according to the species and according to the part of the plant. Furthermore, when wood normally burns, pyrolysates are first released and burned, which are lower in calorific value, followed by burning of the remaining char, which has a higher value. The following empirical correlation can be derived, applicable to all species of dry woods, provided the percent carbon, by weight is known.[38]

$$\Delta h^u{}_c = 0.433 \, [\%C] - 1.527$$

where $Dh^u{}_c$ is the gross heat of combustion for dry wood (MJ/kg) and [C] is the percent of carbon, by weight. This correlation is equally applicable to charred wood and to wood pyrolysates, provided their ultimate analysis is known. For wet fuels, an additional correction should be made,

$$\Delta h^u{}_{c,eff} = Dh_c \, (1 - M/100) - 0.0257 \, M$$

where M is the percent water in the fuel. Lower heats of combustion, if needed, can then be determined as noted above.

For petroleum products a series of empirical rules is available for estimating thermochemical properties, based on specific gravity or on API degrees[39] (See Table A-9.) A determination of the exact heat of combustion by calorimetry is usually not required for typical petroleum products, which include crude oils, fuel oils, kerosenes, and volatile products. These empirical rules are:

Heat of combustion

$$\Delta h^u{}_c = 52.12 - 8.79 \, d^2 - 0.14 \, d \text{ (MJ/kg)}$$

Percent hydrogen

$$[\%H] = 26 - 15 \, d$$

Heat of vaporization, at 25°C

$$\Delta h_v = \frac{240}{d} \text{ (kJ/kg)}$$

Specific heat at 25°C, liquid

$$C_{pl} = \frac{1770}{\sqrt{d}} \text{ (J/kg°C)}$$

Specific heat at 25°C, vapor

$$C_{pv} = C_{pl} - \frac{380}{d} \text{ (J/kg°C)}$$

In all the above relationships, d = specific gravity at 15.6°C (60°F). For conversion from degrees API, see Table A-9. The relationship for [%H] can be used to obtain net heat of combustion, according to the general equation given earlier. A more detailed method, requiring additionally the determination of percent aromatics, volatility, and sulfur content, is available as an ASTM method.[40]

TABLE A-7. *Specific Gravity Equivalents for Degrees Baumé*
(Liquids Lighter than Water)

Baumé	Specific Gravity	Lbs per Gallon*	Baumé	Specific Gravity	Lbs per Gallon*	Baumé	Specific Gravity	Lbs per Gallon*
10	1.0000	8.33	45	0.8000	6.66	80	0.6667	5.55
11	0.9929	8.27	46	0.7955	6.63	81	0.6635	5.52
12	0.9859	8.21	47	0.7910	6.59	82	0.6604	5.50
13	0.9790	8.16	48	0.7865	6.55	83	0.6573	5.48
14	0.9722	8.10	49	0.7821	6.52	84	0.6542	5.45
15	0.9655	8.04	50	0.7778	6.48	85	0.6512	5.42
16	0.9589	7.99	51	0.7735	6.44	86	0.6481	5.40
17	0.9524	7.93	52	0.7692	6.41	87	0.6452	5.38
18	0.9459	7.88	53	0.7650	6.37	88	0.6422	5.36
19	0.9396	7.83	54	0.7609	6.34	89	0.6393	5.33
20	0.9333	7.78	55	0.7568	6.30	90	0.6364	5.30
21	0.9272	7.72	56	0.7527	6.27	91	0.6335	5.28
22	0.9211	7.67	57	0.7487	6.24	92	0.6306	5.25
23	0.9150	7.62	58	0.7447	6.20	93	0.6278	5.23
24	0.9091	7.57	59	0.7407	6.17	94	0.6250	5.21
25	0.9032	7.53	60	0.7368	6.14	95	0.6222	5.18
26	0.8974	7.48	61	0.7330	6.11	96	0.6195	5.16
27	0.8917	7.43	62	0.7292	6.07	97	0.6167	5.14
28	0.8861	7.38	63	0.7254	6.04	98	0.6140	5.11
29	0.8805	7.34	64	0.7216	6.01	99	0.6114	5.09
30	0.8750	7.29	65	0.7179	5.98	100	0.6087	5.07
31	0.8696	7.24	66	0.7143	5.95	101	0.6061	5.05
32	0.8642	7.20	67	0.7107	5.92	102	0.6034	5.03
33	0.8589	7.15	68	0.7071	5.89	103	0.6009	5.00
34	0.8537	7.11	69	0.7035	5.86	104	0.5983	4.98
35	0.8485	7.07	70	0.7000	5.83	105	0.5957	4.96
36	0.8434	7.03	71	0.6965	5.80	106	0.5932	4.94
37	0.8383	6.98	72	0.6931	5.78	107	0.5907	4.92
38	0.8333	6.94	73	0.6897	5.75	108	0.5882	4.90
39	0.8284	6.90	74	0.6863	5.72	109	0.5858	4.88
40	0.8235	6.86	75	0.6829	5.69	110	0.5833	4.86
41	0.8187	6.82	76	0.6796	5.66			
42	0.8140	6.78	77	0.6763	5.63			
43	0.8092	6.74	78	0.6731	5.60			
44	0.8046	6.70	79	0.6699	5.58			

* one lb per gallon equals 119.8 kg/m³

In some cases the combustion properties, including heat of combustion, are desired for burning metals. The specified combustion product in each case is a stable metal oxide in crystalline form. Table A-4 lists the heats of combustion for some common metals. For further discussion consult the references.[31,41]

Also listed in the tables, where available, are T_b, the boiling point; the heat capacities C_{pl}, C_{pv}, and C_{ps} of the liquid, vapor, and solid phases respectively, taken at constant pressure at T_b; and Dh_v, the latent heat of vaporization at T_b. For purposes where the total heat of vaporization (i.e., heat to bring a liquid at 25°C up to a gas at T_b) is desired, an estimate can be made by adding to Dh_v the quantity $C_{pl} \cdot (T_b - 25)$. More precise values can be obtained by using enthalpy tables.[5,32]

In some common applications, especially for building materials, a laboratory technique for assessing the possible effect of incomplete combustion is desired. The potential heat test procedure used to be determined by NFPA 259, *Standard Test Method for Potential Heat of Building Materials*. The potential heat test is now

rarely specified. Values of "effective heat of combustion," are more commonly determined using ASTM E-1354, *Standard Test for Heat and Visible Smoke Release Rates for Materials and Production Using an Oxygen Consumption Calorimeter*.[51]

Conversion Units: To convert kcal/kg to megajoules/kg

multiply by 4.184×10^{-3}

to convert kcal/mol to MJ/kg multiply by $4.184/W$

where W = molecular weight (g/mol)

to convert kJ/mol to MJ/kg multiply by $1/W$

to convert Btu/lb to MJ/kg multiply by 2.322×10^{-3}

Notes to Table A-6

In recent years measurement data have started to be tabulated on the engineering performance of plastics. These data are usually measured in heat release rate calorimeters on a bench scale (specimens

TABLE A-8. *Specific Gravity Equivalents for Degrees Baumé*
(Liquids Heavier than Water)

Baumé	Specific Gravity	Lbs per Gallon*	Baumé	Specific Gravity	Lbs per Gallon*	Baumé	Specific Gravity	Lbs per Gallon*
0	1.0000	8.33	25	1.2083	10.07	50	1.5263	12.72
1	1.0069	8.38	26	1.2185	10.16	51	1.5426	12.85
2	1.0140	8.46	27	1.2288	10.24	52	1.5591	12.99
3	1.0211	8.51	28	1.2393	10.32	53	1.5761	13.13
4	1.0284	8.56	29	1.2500	10.41	54	1.5934	13.27
5	1.0357	8.63	30	1.2609	10.51	55	1.6111	13.42
6	1.0432	8.69	31	1.2719	10.59	56	1.6292	13.57
7	1.0507	8.75	32	1.2832	10.69	57	1.6477	13.72
8	1.0584	8.81	33	1.2946	10.78	58	1.6667	13.87
9	1.0662	8.88	34	1.3063	10.84	59	1.6860	14.04
10	1.0741	8.94	35	1.3182	10.98	60	1.7059	14.21
11	1.0821	9.01	36	1.3303	11.09	61	1.7262	14.38
12	1.0902	9.09	37	1.3426	11.18	62	1.7470	14.55
13	1.0985	9.15	38	1.3551	11.29	63	1.7683	14.72
14	1.1069	9.21	39	1.3679	11.39	64	1.7901	14.91
15	1.1154	9.29	40	1.3810	11.51	65	1.8125	15.10
16	1.1240	9.36	41	1.3942	11.61	66	1.8354	15.29
17	1.1328	9.43	42	1.4078	11.72	67	1.8590	15.48
18	1.1417	9.51	43	1.4216	11.84	68	1.8831	15.68
19	1.1508	9.59	44	1.4356	11.96	69	1.9079	15.89
20	1.1600	9.67	45	1.4500	12.08	70	1.9333	16.10
21	1.1694	9.74	46	1.4646	12.21	71	1.9595	16.32
22	1.1789	9.81	47	1.4796	12.33	72	1.9864	16.55
23	1.1885	9.90	48	1.4948	12.46	73	2.0139	16.78
24	1.1983	9.99	49	1.5104	12.58	74	2.0423	17.01
						75	2.0714	17.25

This table and Table 5-11G have been adopted by the U.S. National Bureau of Standards from the formula:

Liquids heavier than water, degrees Baumé $= 145 - \dfrac{145}{R}$

Liquids lighter than water, degrees Baumé $= \dfrac{140}{R} - 130$

Where R is the ratio of the density of a liquid to that of distilled water, both densities being taken at 60°F (15.5°C).

NOTE: Degrees API, used to measure the density of petroleum oils, differ slightly from the Baumé scale for liquids lighter than water. API specific gravity is determined by a similar formula, using the figures 141.5 and 131.5 instead of 140 and 130.

* One lb per gallon equals 119.8 kg/m^3.

typically 0.1 m dimension). A number of these data are summarized in Table A-6. In many cases wide variations for a property are shown. These can be attributed to (a) specimen size differences, (b) specimen orientation differences, (c) air flow rate differences, (d) heating flux (irradiance) differences, and (e) specimen composition differences. The data for the plastics listed do not represent pure polymers, but rather engineering plastics, which vary in additives, degree of polymerization, etc. When only a single number is indicated, this should be taken as reflecting a paucity of data, rather than a high precision of measurement.

The minimum flux needed for ignition has sometimes been considered to be a constant. A-6 shows that it varies among materials, but only over a range of about 2:1. Ignition temperatures are seen to vary little for common plastics. The total absorbed energy (kJ/m^2) required for ignition has more recently been considered a measure of ignitability.

The effective heat of gasification is a parameter intended to represent the ease of achieving a given heat release rate. The tabulated values, however, do not show the wide range of heat release rate behaviors seen with commercial products. This is partly because for most materials the heat release progress is highly transient, whereas the definition of the effective heat of gasification (average energy required to gasify a unit mass of material) requires assuming steady-state conditions.

Smoke properties can be measured in two ways: (1) Optically, by determining the obscuration along a light beam. This can be expressed as smoke specific extinction area per unit specimen mass (m^2/kg). The unit system implied here is natural-log (Base-e). (2) Gravimetrically, by determining the soot mass evolved, per unit specimen mass.

TABLE A-9. *Specific Gravity Equivalents, Degrees API*

The specific gravity of gasoline and other petroleum products is commonly given in Degrees API (American Petroleum Institute). The API degree is based on the following formula:

$$\text{Degrees API} = \frac{141.5}{\text{specific gravity}} - 131.5.$$

The standard method of test for gravity of petroleum and petroleum products by means of a hydrometer specifies 60°F as the test temperature, and the formula is based on the specific gravity of the liquid tested at 60°F as compared with water at 60°F.

Degrees API	Specific Gravity at 60/60°F	Pounds per Gallon at 60°F	Degrees API	Specific Gravity at 60/60°F	Pounds per Gallon* at 60°F	Degrees API	Specific Gravity at 60/60°F	Pounds per Gallon* at 60°F
0	1.076	8.962	35	0.8498	7.076	70	0.7022	5.845
1	1.066	8.895	36	0.8448	7.034	71	0.6988	5.817
2	1.060	8.828	37	0.8398	6.993	72	0.6953	5.788
3	1.052	8.762	38	0.8348	6.951	73	0.6919	5.759
4	1.044	8.698	39	0.8299	6.910	74	0.6886	5.731
5	1.037	8.634	40	0.8251	6.870	75	0.6852	5.703
6	1.029	8.571	41	0.8203	6.830	76	0.6819	5.676
7	1.022	8.509	42	0.8155	6.790	77	0.6787	5.649
8	1.014	8.448	43	0.8109	6.752	78	0.6754	5.622
9	1.007	8.388	44	0.8063	6.713	79	0.6722	5.595
10	1.0000	8.328	45	0.8017	6.675	80	0.6690	5.568
11	0.9930	8.270	46	0.7972	6.637	81	0.6659	5.542
12	0.9861	8.212	47	0.7927	6.600	82	0.6628	5.516
13	0.9792	8.155	48	0.7883	6.563	83	0.6597	5.491
14	0.9725	8.099	49	0.7839	6.526	84	0.6566	5.465
15	0.9659	8.044	50	0.7796	6.490	85	0.6536	5.440
16	0.9593	7.989	51	0.7753	6.455	86	0.6506	5.415
17	0.9529	7.935	52	0.7711	6.420	87	0.6476	5.390
18	0.9465	7.882	53	0.7669	6.385	88	0.6446	5.365
19	0.9402	7.830	54	0.7628	6.350	89	0.6417	5.341
20	0.9340	7.778	55	0.7587	6.316	90	0.6388	5.316
21	0.9279	7.727	56	0.7547	6.283	91	0.6360	5.293
22	0.9218	7.676	57	0.7507	6.249	92	0.6331	5.269
23	0.9159	7.627	58	0.7467	6.216	93	0.6303	5.246
24	0.9100	7.578	59	0.7428	6.184	94	0.6275	5.222
25	0.9042	7.529	60	0.7389	6.151	95	0.6247	5.199
26	0.8984	7.481	61	0.7351	6.119	96	0.6220	5.176
27	0.8927	7.434	62	0.7313	6.087	97	0.6193	5.154
28	0.8871	7.387	63	0.7275	6.056	98	0.6166	5.131
29	0.8816	7.341	64	0.7238	6.025	99	0.6139	5.109
30	0.8762	7.296	65	0.7201	5.994	100	0.6112	5.086
31	0.8708	7.251	66	0.7165	5.964			
32	0.8654	7.206	67	0.7128	5.934			
33	0.8602	7.163	68	0.7093	5.904			
34	0.8550	7.119	69	0.7057	5.874			

* "Apparent weight," or weight when weighed in air. True weight corrected for weight of air, differs by less than 1/10 of 1 percent. One lb per gallon equals 119.8 kg/m³. 60°F equals 15.55°C.

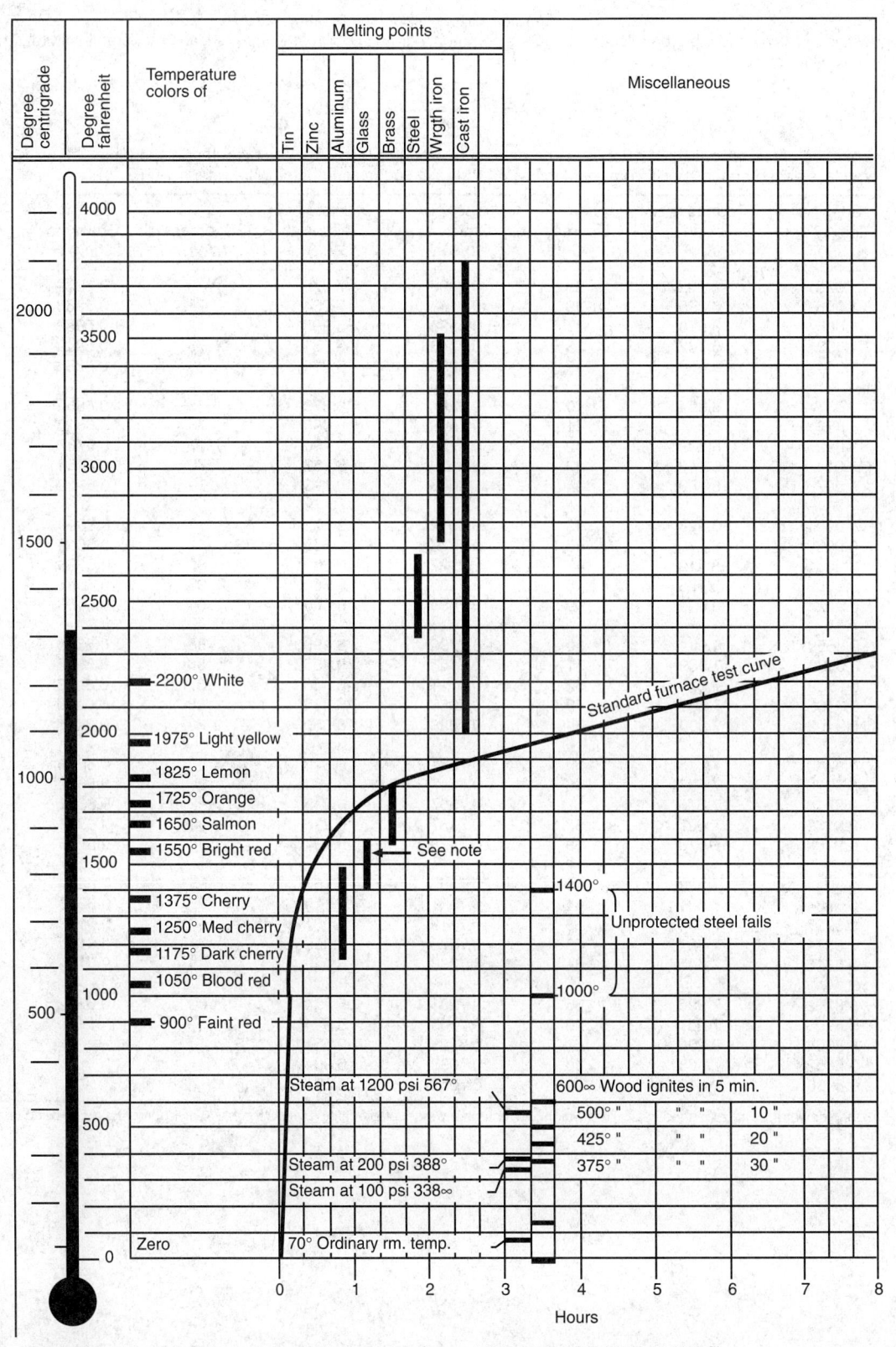

NOTE: Glass has no definite melting point but softens between 1400 and 1600°F (750 and 850°C)

FIG. A-1A. *Temperature colors, melting points of various substances, and data on the behavior of some materials at elevated temperatures and pressures.*

TABLE A-10. *Materials Subject to Spontaneous Heating*

(Originally prepared by the NFPA Committee on Spontaneous Heating and Ignition which has been discontinued. Omission of any material does not necessarily indicate that it is not subject to spontaneous heating.)

Name	Tendency to Spontaneous Heating	Usual Shipping Container or Storage Method	Precautions Against Spontaneous Heating	Remarks
Alfalfa Meal	High	Bags, Bulk	Avoid moisture extremes. Tight cars for transportation are essential.	Many fires attributed to spontaneous heating probably caused by sparks, burning embers, or particles of hot metal picked up by the meal during processing. Test fires caused in this manner have smoldered for 72 hours before becoming noticeable.
Burlap Bags "Used"	Possible	Bales	Keep cool and dry.	Tendency to heat dependent on previous use of bags. If oily would be dangerous.
Castor Oil	Very slight	Metal Barrels, Metal Cans in Wooden Boxes	Avoid contact of leakage from containers with rags, cotton, or other fibrous combustible materials.	Possible heating of saturated fabrics in badly ventilated piles.
Charcoal	High	Bulk, Bags	Keep dry. Supply ventilation.	Hardwood charcoal must be carefully prepared and aged. Avoid wetting and subsequent drying.
Coal, Bituminous	Moderate	Bulk	Store in small piles. Avoid high temperatures.	Tendency to heat depends upon origin and nature of coals. High volatile coals are particularly liable to heat.
Cocoa Bean Shell Tankage	Moderate	Burlap Bags, Bulk	Extreme caution must be observed to maintain safe moisture limits.	This material is very hygroscopic and is liable to heating if moisture content is excessive. Precaution should be observed to maintain dry storage, etc.
Cocoanut Oil	Very slight	Drums, Cans, Glass	Avoid contact of leakage from containers with rags, cotton, or other fibrous combustible materials.	Only dangerous if fabrics, etc., are impregnated.
Cod Liver Oil	High	Drums, Cans, Glass	Avoid contact of leakage from containers with rags, cotton, or other fibrous combustible materials.	Impregnated organic materials are extremely dangerous.
Colors in Oil	High	Drums, Cans, Glass	Avoid contact of leakage from containers with rags, cotton, or other fibrous combustible materials.	May be very dangerous if fabrics, etc., are impregnated.
Copra	Slight	Bulk	Keep cool and dry.	Heating possible if wet and hot.
Corn-Meal Feeds	High	Burlap Bags, Paper Bags, Bulk	Material should be processed carefully to maintain safe moisture content and to cure before storage.	Usually contains an appreciable quantity of oil which has rather severe tendency to heat.
Corn Oil	Moderate	Barrels, Tank Cars	Avoid contact of leakage from containers with rags, cotton, or other fibrous combustible materials.	Dangerous heating of meals, etc., unlikely unless stored in large piles while hot.
Cottonseed	Low	Bags, Bulk	Keep cool and dry.	Heating possible if piled wet and hot.
Cottonseed Oil	Moderate	Barrels, Tank Cars	Avoid contact of leakage from containers with rags, cotton, or other fibrous combustible materials.	May cause heating of saturated material in badly ventilated piles.
Distillers' Dried Grains with oil content (Brewers' grains)	Moderate	Bulk	Maintain moisture 7 percent to 10 percent. Cool below 100°F (38°C) before storage.	Very dangerous if moisture content is 5 percent or lower.
No oil content	Moderate	Bulk	Maintain moisture 7 percent to 10 percent. Cool below 100°F (38°C) before storage.	Very dangerous if moisture content is 5 percent or lower.
Feeds, various	Moderate	Bulk, Bags	Avoid extremely low or high moisture content.	Ground feeds must be carefully processed. Avoid loading or storing unless cooled.
Fertilizers Organic, Inorganic, Combination of both	Moderate	Bulk, Bags	Avoid extremely low or high moisture content.	Organic fertilizers containing nitrates must be carefully prepared to avoid combinations that might initiate heating.
Mixed, Synthetic, containing nitrates and organic matter	Moderate	Bulk, Bags	Avoid free acid in preparation.	Insure ventilation in curing process by small piles or artificial drafts. If stored or loaded in bags, provide ventilation space between bags.

TABLE A-10. *Materials Subject to Spontaneous Heating (Continued)*

Name	Tendency to Spontaneous Heating	Usual Shipping Container or Storage Method	Precautions Against Spontaneous Heating	Remarks
Fish meal	High	Bags, Bulk	Keep moisture 6 percent to 12 percent. Avoid exposure to heat.	Dangerous if overdried or packaged over 100°F (38°C).
Fish Oil	High	Barrels, Drums, Tank Cars	Avoid contact of leakage from containers with rags, cotton, or other fibrous combustible materials.	Impregnated porous or fibrous materials are extremely dangerous. Tendency of various fish oils to heat varies with origin.
Fish Scrap	High	Bulk, Bags	Avoid moisture extremes.	Scrap loaded or stored before cooling is extremely liable to heat.
Foam Rubber in Consumer Products	Moderate		Where possible remove foam rubber pads, etc., from garments to be dried in dryers or over heaters. If garments containing foam rubber parts have been artificially dried, they should be thoroughly cooled before being piled, bundled, or put away. Keep heating pads, hair dryers, other heat sources from contact with foam rubber pillows, etc.	Foam rubber may continue to heat spontaneously after being subjected to forced drying as in home or commercial dryers, and after contact with heating pads and other heat sources. Natural drying does not cause spontaneous heating.
Grain (various kinds)	Very slight	Bulk, Bags	Avoid moisture extremes.	Ground grains may heat if wet and warm.
Hay	Moderate	Bulk, Bales	Keep dry and cool.	Wet or improperly cured hay is almost certain to heat in hot weather. Baled hay seldom heats dangerously.
Hides	Very slight	Bales	Keep dry and cool.	Bacteria in untreated hides may initiate heating.
Iron Pyrites	Moderate	Bulk	Avoid large piles. Keep dry and cool.	Moisture accelerates oxidation of finely divided pyrites.
Istle	Very slight	Bulk, Bales	Keep cool and dry.	Heating possible in wet material. Unlikely under ordinary conditions. Partially burned or charred fiber is dangerous.
Jute	Very slight	Bulk	Keep cool and dry.	Avoid storing or loading in hot wet piles. Partially burned or charred material is dangerous.
Lamp Black	Very slight	Wooden Cases	Keep cool and dry.	Fires most likely to result from sparks or included embers, etc., rather than spontaneous heating.
Lanolin	Negligible	Glass, Cans, Metal Drums, Barrels	Avoid contact of leakage from containers with rags, cotton, or other fibrous combustible materials.	Heating possible on contaminated fibrous matter.
Lard Oil	Slight	Wooden Barrels	Avoid contact of leakage from containers with rags, cotton, or other fibrous combustible materials.	Dangerous on fibrous combustible substances.
Lime, unslaked (Calcium Oxide, Pebble Lime, Quicklime)	Moderate	Paper Bags, Wooden Barrels, Bulk	Keep dry. Avoid hot loading.	Wetted lime may heat sufficiently to ignite wood containers, etc.
Linseed	Very slight	Bulk	Keep cool and dry.	Tendency to heat dependent on moisture and oil content.
Linseed Oil	High	Tank Cars, Drums, Cans, Glass	Avoid contact of leakage from containers with rags, cotton, or other fibrous combustible materials.	Rags or fabrics impregnated with this oil are extremely dangerous. Avoid piles, etc. Store in closed containers, preferably metal.
Manure	Moderate	Bulk	Avoid extremes of low or high moisture contents. Ventilate the piles.	Avoid storing or loading uncooled manures.
Menhaden Oil	Moderate to high	Barrels, Drums, Tank Cars	Avoid contact of leakage from containers with rags, cotton, or other fibrous combustible materials.	Dangerous on fibrous product.
Metal Powders*	Moderate	Drums, etc.	Keep in closed containers.	Moisture accelerates oxidation of most metal powders.
Metal Turnings*	Practically none	Bulk	Not likely to heat spontaneously.	Avoid exposure to sparks.
Mineral Wool	None	Pasteboard Boxes, Paper Bags	Noncombustible. If loaded hot may ignite containers and other combustible surroundings.	This material is mentioned in this table only because of general impression that it heats spontaneously.

TABLE A-10. *Materials Subject to Spontaneous Heating (Continued)*

Name	Tendency to Spontaneous Heating	Usual Shipping Container or Storage Method	Precautions Against Spontaneous Heating	Remarks
Mustard Oil, Black	Low	Barrels	Avoid contact of leakage with rags, cotton or other fibrous combustible materials.	Avoid contamination of fibrous combustible materials.
Oiled Clothing	High	Fiber Boxes	Dry thoroughly before packaging.	Dangerous if wet material is stored in piles without ventilation.
Oiled Fabrics	High	Rolls	Keep ventilated. Dry thoroughly before packing.	Improperly dried fabrics extremely dangerous. Tight rolls are comparatively safe.
Oiled Rags	High	Bales	Avoid storing in bulk in open.	Dangerous if wet with drying oil.
Oiled Silk	High	Fiber Boxes, Rolls	Supply sufficient ventilation.	Improperly dried material is dangerous in form of piece goods. Rolls relatively safe.
Oleic Acid	Very slight	Glass Bottles, Wooden Barrels	Avoid contact of leakage from containers with rags, cotton, or other fibrous combustible materials.	Impregnated fibrous materials may heat unless ventilated.
Oleo Oil	Very slight	Wooden Barrels	Avoid contact of leakage from containers with rags, cotton, or other fibrous combustible materials.	May heat on impregnated fibrous combustible matter.
Olive Oil	Moderate to Low	Tank Cars, Drums, Cans, Glass	Avoid contact of leakage from containers with rags, cotton, or other fibrous combustible materials.	Impregnated fibrous materials may heat unless ventilated. Tendency varies with origin of oil.
Paint containing drying oil	Moderate	Drums, Cans, Glass	Avoid contact of leakage from containers with rags, cotton, or other fibrous combustible materials.	Fabrics, rags, etc., impregnated with paints that contain drying oils and driers are extremely dangerous. Store in closed containers, preferably metal.
Paint Scrapings	Moderate	Barrels, Drums	Avoid large unventilated piles.	Tendency to heat depends on state of dryness of the scrapings.
Palm Oil	Low	Wooden Barrels	Avoid contact of leakage from containers with rags, cotton, or other fibrous combustible materials.	Impregnated fibrous materials may heat unless ventilated. Tendency varies with origin of oil.
Peanut Oil	Low	Wooden Barrels, Tin Cans	Avoid contact of leakage from containers with rags, cotton, or other fibrous combustible materials.	Impregnated fibrous materials may heat unless ventilated. Tendency varies with origin of oil.
Peanuts, "Red Skin"	High	Paper Bags, Cans, Fiber Board Boxes, Burlap Bags	Avoid badly ventilated storage.	This is the part of peanut between outer shell and peanut itself. Provide well ventilated storage.
Peanuts, shelled	Very slight or Negligible	Paper Bags, Cans, Fiber Board Boxes, Burlap Bags	Keep cool and dry.	Avoid contamination of rags, etc., with oil.
Perilla Oil	Moderate to High	Tin Cans, Barrels	Avoid contact of leakage from containers with rags, cotton, or other fibrous combustible materials.	Impregnated fibrous materials may heat unless ventilated. Tendency varies with origin of oil.
Pine Oil	Moderate	Glass, Drums	Avoid contact of leakage from containers with rags, cotton, or other fibrous combustible materials.	Impregnated fibrous materials may heat unless ventilated. Tendency varies with origin of oil.
Powdered Eggs	Very slight	Wooden Barrels	Avoid conditions that promote bacterial growth. Inhibit against decay. Keep cool.	Possible heating of decaying powder in storage.
Powdered Milk	Very slight	Wooden and Fiber Boxes, Metal Cans	Avoid conditions that promote bacterial growth. Inhibit against decay. Keep cool.	Possible heating by decay or fermentation.
Rags	Variable	Bales	Avoid contamination with drying oils. Avoid charring. Keep cool and dry.	Tendency depends on previous use of rags. Partially burned or charred rags are dangerous.
Red Oil	Moderate	Glass Bottles, Wooden Barrels	Avoid contact of leakage from containers with rags, cotton, or other fibrous combustible materials.	Impregnated porous or fibrous materials are extremely dangerous. Tendency varies with origin of oil.

TABLE A-10. *Materials Subject to Spontaneous Heating (Continued)*

Name	Tendency to Spontaneous Heating	Usual Shipping Container or Storage Method	Precautions Against Spontaneous Heating	Remarks
Roofing Felts and Papers	Moderate	Rolls, Bales, Crates	Avoid over-drying the material. Supply ventilation.	Felts, etc., should have controlled moisture content. Packaging or rolling uncooled felts is dangerous.
Sawdust	Possible	Bulk	Avoid contact with drying oils. Avoid hot, humid storage.	Partially burned or charred sawdust may be dangerous.
Scrap Film (Nitrate)	Very slight	Drums and Lined Boxes	Film must be properly stabilized against decomposition.	Nitrocellulose film ignites at low temperature. External ignition more likely than spontaneous heating. Avoid exposure to sparks, etc.
Scrap Leather	Very slight	Bales, Bulk	Avoid contamination with drying oils.	Oil-treated leather scraps may heat.
Scrap Rubber or Buffings	Moderate	Bulk, Drums	Buffings of high rubber content should be shipped and stored in tight containers.	Sheets, slabs, etc., are comparatively safe unless loaded or stored before cooling thoroughly.
Sisal	Very slight	Bulk, Bales	Keep cool and dry.	Partially burned or charred material is particularly liable to ignite spontaneously.
Soybean Oil	Moderate	Tin Cans, Barrels, Tank Cars	Avoid contact with rags, cotton, or fibrous materials.	Impregnated fibrous materials may heat unless well ventilated.
Sperm Oil—See Whale Oil				
Tankage	Variable	Bulk	Avoid extremes of moisture contents. Avoid loading or storing while hot.	Very dry or moist tankages often heat. Tendency more pronounced if loaded or stored before cooling.
Tung Nut Meals	High	Paper Bags, Bulk	Material must be very carefully processed and cooled thoroughly before storage.	These meals contain residual oil which has high tendency to heat. Material also susceptible to heating if over-dried.
Tung Oil	Moderate	Tin Cans, Barrels, Tank Cars	Avoid contact of leakage from containers with rags, cotton, or other fibrous combustible materials.	Impregnated fibrous materials may heat unless ventilated. Tendency varies with origin of oil.
Turpentine	Low	Tin, Glass, Barrels	Avoid contact of leakage from containers with rags, cotton, or other fibrous combustible materials.	Has some tendency to heat but less so than the drying oils. Chemically active with chlorine compounds and may cause fire.
Varnished Fabrics	High	Boxes	Process carefully. Keep cool and ventilated.	Thoroughly dried varnished fabrics are comparatively safe.
Wallboard	Slight	Wrapped Bundles, Pasteboard Boxes	Maintain safe moisture content. Cool thoroughly before storage.	This material is entirely safe from spontaneous heating if properly processed.
Waste Paper	Moderate	Bales	Keep dry and ventilated.	Wet paper occasionally heats in storage in warm locations.
Whale Oil	Moderate	Barrels and Tank Cars	Avoid contact of leakage from containers with rags, cotton, or other fibrous combustible materials.	Impregnated fibrous materials may heat unless ventilated. Tendency varies with origin of oil.
Wool Wastes	Moderate	Bulk, Bales, etc.	Keep cool and ventilated or store in closed containers. Avoid high moisture.	Most wool wastes contain oil, etc., from the weaving and spinning and are liable to heat in storage. Wet wool wastes are very liable to spontaneous heating and possible ignition.

*Refers to iron, steel, brass, aluminum, and other common metals, for information on magnesium, sodium, zirconium, etc.

BIBLIOGRAPHY

References Cited

1. Domalski, E. S. "Selected Values of Heats of Combustion and Heats of Formation of Organic Compounds Containing The Elements C, H, N, O, P, and S," *Journal of Physical and Chemical Reference Data.* Vol 1. pp 221–227, 1972.

2. Gray, D. E., ed. *American Institute of Physics Handbook.* Section 41. McGraw-Hill. New York, 1972.

3. Rossini, F. D. *Selected Values of Physical and Thermodynamic Properties of Hydrocarbons and Related Compounds.* American Petroleum Institute/Carnegie Press. Pittsburgh, PA, 1953.

4. Tinnermans, J. *Physio-Chemical Constants of Pure Organic Compounds.* 2 Vols. Elsevier. New York, 1950–65.

5. Thermodynamics Research Center. *Selected Values of Properties of Chemical Compounds; and Selected Values of Properties of Hydrocarbons and Related Compounds.* Texas A & M University, College Station, TX, undated.

6. Stull, D. R., Westrom, Jr., E. F. and Sinke, G. C. *Thermodynamics of Organic Compounds.* Wiley. New York, 1969.

7. *Landolt-Bornstein Physikalisch-Chemische Tabellen.* Fifth series: main work (vol 2, 1923); first supplement (vol 2, 1927); second supplement (vol 2, 1931); third supplement (vol 3, 1936). Sixth series: Part 2 (vol 4, 1961). Springer-Verlag. Berlin, Republic of Germany, 1927–61.

8. Reid, R. C., Prausnitz, J. M., and Sherwood, T. K. *The Properties of Gases and Liquids.* McGraw-Hill. New York, 1977.

9. NACA. *Basic Considerations in the Combustion of Hydrocarbon Fuels in Air* (NACA Report 1300). National Advisory Committee on Aeronautics, 1957.

10. Belenyessy, L. I. *et al. Heats of Combustion of Complex Saturated Hydrocarbons, J. Chem. and Engrg. Data.* Vol 7, 1962, pp 66–68.

11. Rand Corp. *Physical Properties and Thermodynamic Functions of Fuels, Oxidizers, and Products of Combustion, I. Fuels.* NTIS No AD 605 967. Battelle Memorial Institute, Columbus, OH, 1949.

12. Dreisbach, R. R. *Physical Properties of Chemical Compounds.* v. 1 (1955), v. 2 (1959), v. 3 (1961). American Chemical Society. Washington, DC, 1955–61.

13. Cox, J. D., and Pilcher, G. *Thermochemistry of Organic and Organometallic Compounds.* Academic Press. London, England, 1970.

14. Kharasch, M. S., *Heats of Combustion of Organic Compounds.* Bur. of Standards J. of Research. Vol 2, 1929, pp 359–430.

15. Lipowitz, J., *Flammability of Poly(dimethylsiloxanes).* J. Fire Flamm. Vol 7, 1976, pp 482–503 and 504–529.

16. Throne, J. L., and Griskey, R. G. *Heating Values and Thermochemical Properties of Plastics, Modern Plastics.* Vol 49. pp 96–200. November 1972.

17. Hognon, B. Contribution possible des principaux isolants synthetiques a l'aggravation des dangers en cas d'incendie. *Cahiers du Centre Scientifique et Technique du Batiment.* p 1392 Sept. 1976.

18. NBS. Unpublished tests. U.S. National Bureau of Standards, undated.

19. Roff, W. J., and Scott, J. R. *Handbook of Common Polymers.* Butterworth, London, England, 1971.

20. Joshi, R. M. *A New Generalized Bond Energy/Group Contribution Scheme for Calculating the Standard Heat of Formation of Monomers and Polymers.* Part V. Oxygen Compounds. J. Macromol. Sci. Chem. A9, pp 1309–1383, 1975.

21. Van Krevelen, D. W. *Properties of Polymers.* Elsevier. Amsterdam, Holland, 1976.

22. Berlin, A. A. *et al. Intermolecular Interaction in Polyconjugated Systems.* Acad. of Sciences (USSR) Bulletin. Div. of Chemical Science. pp 1392–1395. July 1969.

23. Franz, J., Mische, W., and Kuzay, P. Zur Bestimmung Von Verbrennungswarmen Von Plasten. *Plaste Und Kautschuk.* Vol 14, pp 472–476, 1967.

24. Janssens, M., "Some Experience with the Cone Calorimeter, and Instrument to Measure the Rate of Smoke and Heat Release," in *Fundamental Aspects of Polymer Flammability,* G. Cox and G Stevens, eds. Institute of Physics, London, pp 111–125, 1987.

25. Moore, L. D. Full-Scale Burning Behavior of Curtains and Draperies (NBSIR 78-1448). National Institute of Standards and Technology, 1978.

26. Domalski, E. S., Evans, W. H., and Jobe, T. L., Jr. Thermodynamic Data for Waste Incineration (NBSIR 78-1479). National Institute of Standards and Technology, 1978.

27. Bostic, J. E., Jr., Yeh, K-N., and Barker, R. H. Pyrolysis and Combustion of Polyester. *Journal of Applied Polymer Science.* Vol 17, pp 471–482, 1973.

28. Lobanov, G. A., and Martynovskaya, L. I. Estimation of the Enthalpies of Combustion of Resin and Rubber Type Compounds. *Izv. Vyssh. Ucheb. Zaved., Khim. Tekhnol.* Vol 15, pp 1747–1748, 1972.30.

29. Ohe, H., Mastsuura, K., and Sakai, N. The Heat of Combustion and Oxygen Index of Fibrous Materials. *Textile Research J.* Vol 47, 1977, pp 212–216.

30. Lowrie, R. Heat of Combustion and Oxygen Compatibility. *Flammability and Sensitivity of Materials in Oxygen-Enriched Atmospheres* (ASTM STP 812). pp 84–96. Amer. Soc. for Testing and Materials. W. Conshohocken, PA, 1983.

31. Hust, J. G., and Clark, A. F. "A Survey of Compatibility of Materials with High Pressure Oxygen Service." Cryogenics. Vol 13, pp 325–336, 1973.

32. Stull, D. R., and Prophet, H. *JANAF Thermochemical Tables* (NSRDS-NBS 37). U.S. National Bureau of Standards, 1971.

33. Brandes, E. A. *Smithells Metals Reference Book.* Butterworths, London, England, 1983.

34. Rossini, F. D., ed. *Experimental Thermochemistry.* Vol 1. Interscience, New York, 1956.

35. ASTM, *Standard Method of Test for Gross Calorific Value of Solid Fuel by the Isothermal-Jacket Bomb Calorimeter* (D3286). American Society for Testing and Materials, undated.

36. ASTM, Standards Definitions of Terms Relating to Coal and Coke (ANSI/ASTM D 121). American Society for Testing and Materials, W. Conshohocken, PA, undated.

37. Huggett, C. "Estimation of Rate of Heat Release by Means of Oxygen Consumption Measurements." *Fire and Materials,* Vol 4, 1980.

38. Susott, R. A., DeGroot, W. F., and Shafizadeth, F. Heat Content of Natural Fuels. *J. of Fire and Flammability.* Vol 6, 1975, pp 311–325.

39. ASTM, *Standard Method for Estimation of Heat of Combustion of Aviation Fuels* (ANSI/ASTM D3338). American Society for Testing and Materials, W. Conshohocken, PA, undated.

40. Cragoe, C. S. *Thermal Properties of Petroleum Products* (Miscellaneous Publication 97). National Institute of Standards and Technology, 1929.

41. Skinner, H. A., ed. *Experimental Thermochemistry,* Vol 2. Interscience, New York, 1962.

42. Cullis, C.F. and Hirschler, M.M. *The Combustion of Organic Polymers.* Clarendon Press. Oxford, England, 1981.

43. Factory Mutual Research Corp. unpublished data, Factory Mutual Research Corp., Norwood, MA.

44. Hilado, C.J. *Flammability Handbook for Plastics.* Technomic. Westport, CT, 1982.

45. Tewarson, A. *Flame-Retardant Polymeric Materials,* Chapter 3, Plenum. New York, NY, 1982.

46. Tewarson, A. *Particulate Formation in Fires.* Paper presented at Conference on Large Scale Fire Phenomenology. Washington, DC. 1984.

47. Tewarson, A. *Physico-Chemical and Combustion/Pyrolysis Properties of Polymeric Materials* (NBS-GCR-80-295). National Institute of Standards and Technology, 1980.

48. Tewarson, A., Khan, M. M., and Steciak, J. *A Study of Combustibility of Electrical Wires and Cables Used in Rail Rapid Transit Systems.* Transportation Systems Center. U.S. Dept. of Transportation, 1982.

49. Tewarson, A., Lee, J. L., and Pion, R. F. *Categorization of Cable Flammability.* Part I (EPRI NP-1200, Part 1). Electric Power Research Inst. Palo Alto, CA. 1979.

50. Tewarson, A., Lee, J. L., and Pion, R. F. *Fuel Parameters for Evaluation of the Fire Hazard of Red Oak* (FMRC J.I.OC6N2.RC). Factory Mutual Research Corp. Norwood, MA.1981.

51. *Standard Test Method for Heat and Visible Smoke Release Rates for Materials and Products Using an Oxygen Consumption Calorimeter,* (ASTM E 1354), American Society for Testing and Materials, W. Conshohocken, PA, 1990.

SI UNITS AND CONVERSION TABLES

Robert P. Benedetti

The International System of Units, defined in *Le Système International d' Unités*, known by the abbreviation SI, and colloquially known as the "modernized metric system," is the dominant system of measurement used in the world, the United States of America being the major exception. In fact, it can be said that the rest of the world is now officially metric, while the United States is still in transition from the "foot/pound" system that has its roots in colonial ancestry and with which we are accustomed. Since the NFPA *Fire Protection Handbook* is major fire protection reference and is used internationally, it is appropriate that it be "metricized" to the greatest extent possible, while still retaining those U.S. customary units with which most users of *Handbook* are familiar.

The extent to which SI units have been incorporated in the *Handbook* varies. In general, a quantity expressed in U.S. customary units is followed by an approximation of that quantity in the appropriate SI unit, e.g., 14.7 psi (101 kPa). Formulas, however, are unit-dependent; they only give the correct answer when consistent units are used. Where possible, formulas have been supplemented with equivalent SI-based formulas and examples illustrating their use. Likewise, some tables have been expanded with columns for SI units, and some graphs have been duplicated in SI form. Where this has not been done, appropriate conversions have been added in captions or as footnotes. In general, the use of SI units in the *Handbook* follows the current edition of ASTM E 380, *Standard Practice for Use of the International System of Units* (SI) *(the Modernized Metric System)*.[1]

HISTORY OF SI

A decimal system of units, one based on powers of 10, was conceived as early as the 16th century. A familiar example of such a system is represented by the U.S. system of currency, i.e., dollars and cents. International efforts at developing standardized units of measurement began in 1870, with a meeting of representatives of fifteen countries in Paris, France. This led to the first International Metric Convention in 1875 and the establishment of the International Bureau of Weights and Measures, publisher of Le Système International d' Unités (SI), in Sèvres, France. The General Conference on Weights and Measures was established to address international issues. The General Conference meets at least once every six years, and has control of the International Bureau. Today's metric system is the result of the 11th

General Conference, held in 1960. The U.S. is represented in these activities by the National Institute of Standards and Technology (NIST).

As stated above, the United States is still in transition to SI. In December 1975, the 94th Congress passed Public Law 94-168, the *Metric Conversion Act* of 1975. This law committed the U.S. to converting to SI. The *Omnibus Trade and Competitiveness Act* of 1988 amended the 1975 Act, in order to encourage the use of metric units in government procurement and grants. In 1991, then-President George Bush signed Executive Order 12770, requiring that federal agencies develop timetables for transition to SI. Most recently, in May 1996, the U.S. Department of Commerce published guidelines for acquisition of certain construction products in SI.

BASE UNITS, SUPPLEMENTARY UNITS, AND DERIVED UNITS

SI consists of three classes of units: (1) base, (2) supplementary, and (3) derived. The base units are those on which the SI system is founded and consist of seven dimensionally independent units that measure seven fundamental physical quantities. They are given in the following table:

Base Unit

Quantity	Unit	Symbol
Length	meter	m
Mass	kilogram	kg
Time	second	s
Electric current	ampere	A
Thermodynamic temperature	kelvin	K
Amount of substance	mole	mol
Luminous intensity	candela	cd

Supplementary units consist of the radian, a measure of plane angle, and the steradian, a measure of solid angle, as given in the following table:

Supplementary Units

Quantity	Unit	Symbol	Expressed in Terms of Other Units
Plane angle	radian	rad	$m/m=1$
Solid angle	steradian	sr	m^2/m^2

Mr. Benedetti is NFPA's Senior Flammable Liquids Engineer and is responsible for NFPA's technical and code development activities in the areas of storage handling, use, and transportation of flammable and combustible liquids.

All other units are "derived units" that are formed by combining base units, supplementary units, and other derived units according to specific algebraic relations. Some are named according to the units from which they are derived, such as the unit for velocity, "meter per second." Others are provided with their own names, such as the unit for force, the "newton." The following table lists those quantities that are provided with specially named SI units.

Derived Units (specific)

Quantity	Unit	Symbol	Expressed in Terms of Other Units
Absorbed dose	gray	Gy	J/kg
Celsius temperature	degree Celsius	°C	K
Dose equivalent	sievert	Sv	J/kg
Electrical capacitance	farad	F	C/V
Electrical charge, quantity of	coulomb	C	A•s
Electrical conductance	siemens	S	A/V
Electrical inductance	henry	H	Wb/A
Electrical potential difference, electromotive force	volt	V	W/A
Electrical resistance	ohm	Ω	V/A
Energy, work, quantity of heat	joule	J	N•m
Force	newton	N	kg•m/s^2
Frequency	hertz	Hz	1/s
Illuminance	lux	lx	lm/m^2
Luminous flux	lumen	lm	cd•sr
Magnetic flux	weber	Wb	V•s
Magnetic flux density	tesla	T	Wb/m^2
Power, radiant flux	watt	W	J/s
Pressure, stress	pascal	Pa	N/m^2
Radioactivity	becquerel	Bq	1/s

Many commonly used SI units that have no special names are given as:

Derived Units (nonspecific)

Quantity	Name	Symbol
Acceleration	meter per second squared	m/s^2
Area	square meter	m^2
Concentration	mole per cubic meter	mol/m^3
Density, current	ampere per square meter	A/m^2
Density, mass	kilogram per cubic meter	kg/m^3
Density, electrical charge	coulomb per cubic meter	C/m^3
Electric field strength	volt per meter	V/m
Electric flux density	coulomb per square meter	C/m^2
Energy Density	joule per cubic meter	J/m^3
Entropy	Joule per kelvin	J/K
Heat capacity	joule per kelvin	J/K
Heat flux density	watt per square meter	W/m^2
Luminance	candela per square meter	cd/m^2
Moment of force	newton meter	N•m
Permittivity	farad per meter	F/m
Power density	watt per square meter	W/m^2
Radiance	watt per square meter steradian	W/(m^2•sr)
Radiant intensity	watt per steradian	W/sr
Specific energy	joule per kilogram	J/kg
Specific entropy	joule per kilogram kelvin	J/(kg•K)
Specific heat capacity	joule per kilogram kelvin	J/(kg•K)
Specific volume	cubic meter per kilogram	m^3/kg
Thermal conductivity	watt per meter kelvin	W/(m•K)
Velocity	meter per second	m/s
Volume	cubic meter	m^3
Wave number	1 per meter	1/m

TEMPERATURE SCALES

The kelvin (k) is the SI unit for thermodynamic or absolute temperature and it is properly used throughout the *Handbook* for expressing thermodynamic temperature and temperature intervals and quantities related to heat transfer. For example, heat capacity is expressed as joule per kelvin (J/K). Degree Celsius (°C) is also used extensively. Degree Celsius is a derived SI unit widely used for expressing a measured temperature or temperature difference. Flash points of flammable liquids are commonly given in degrees Celsius, as are the operating temperatures of sprinkler heads.

The relationship between °C and K is: K = °C + 273.15. [Note that a similar relationship exists between the measured temperature in degrees Fahrenheit (°F) and the thermodynamic temperature in degrees Rankine (°R) in U.S. customary units: R = °F + 459.7.]

Formulas for converting between °C and °F are:

$$°F = (°C \times 1.8) + 32$$

$$°C = \frac{(°F - 32)}{1.8}$$

Prefixes

Standard prefixes are used to express any SI unit as a multiple of submultiple of 10. The usual practice is to select a prefix that limits the numerical value of a quantity to not more than four digits. For example, 28000 meters is better written as 2.8 kilometers. Likewise, 0.0017 grams is better written as 1.7 milligrams. Commonly used prefixes are shown in the following table:

Standard Prefixes

Multiplication Factors		Prefix	Symbol
10^{12}	= 1 000 000 000 000	tera	T
10^{9}	= 1 000 000 000	giga	G
10^{6}	= 1 000 000	mega	M
10^{3}	= 1000	kilo	k
10^{2}	= 100	hecto	h
10^{1}	= 10	deka	da
10^{0}	= 1		
10^{-1}	= 0.1	deci	d
10^{-2}	= 0.01	centi	c
10^{-3}	= 0.001	milli	m
10^{-6}	= 0.000 000 1	micro	μ
10^{-9}	= 0.000 000 000 1	nano	n
10^{-12}	= 0.000 000 000 000 1	pico	p

SOME COMMENTS

Meter vs. Metre and Liter vs. Litre

One departure from strict SI usage has been to use the spellings "meter" and "liter" throughout the *Handbook*, instead of the internationally accepted "metre" and "litre." The former terms are more commonly used and identified in the U.S.

Minute vs. Second

A second difference is the use of liter per minute (L/min) and cubic meter per minute (m³/min) instead of the internationally accepted liter per second (L/s) and cubic meter per second (m³/s), particularly for measurements in hydraulics and water supply analysis. The minute was chosen over the second because L/min and m³/min more nearly equate to gallons per minute, an easier quantity to use in fire protection engineering calculations, and to maintain consistency with the usage of these units in other NFPA publications.

Force and Mass

One area of confusion for those not familiar with SI units is the conversion of force and mass. In the U.S. customary system, a body with a mass of 20 pounds (a unit of mass) experiences a gravitational force of very nearly 20 pound-force. Because the numerical values and unit names are so close, the distinction between the two units is not often appreciated. Although the unit "pound" appears in both terms, the unit "pound force" symbol, "lb_f," is more accurate for the latter.

In SI, two different units are used: kilogram for mass and newton for force. It is important to distinguish whether one is converting mass of material of, e.g., 14 lb to the equivalent number of kilograms, or a force of 14 lb_f to the equivalent value in newtons.

Mass: 1 lb = 0.453 25 kg
Force: 1 lb_f = 4.448 N

The newton, and not the kilogram, appears in force-related terms, such as pressure (N/m²), energy (N • m = J), and power (N • m/s = J/s = W).

Pressure

The U.S. customary system expresses pressure either in gauge or absolute units, depending on whether the measurement is made relative to standard atmospheric pressure or to an absolute vacuum. This is identified by the familiar abbreviations "psig" and "psia" for "pounds per square inch, gage" and "pounds per square inch, absolute," respectively. (Properly, pressure in U.S. customary units should be given as "*x* pound-force per square inch.") No such conventions are allowed in the SI system. For expressing a pressure differential, use of the unit "kPa" is sufficient. But, where necessary to specify that a pressure measurement is relative to a standard atmosphere or to absolute vacuum, then the measurement is qualified, respectively, as follows:

"... at a gauge pressure of 17.7 kPa," or

"... at an absolute pressure of 1.4 kPa."

ACCURACY OF CONVERSIONS AND CONVERSION FACTORS[1,3]

The approach taken in converting from U.S. customary units to SI units has been to retain the precision of the original measurement. For example, "It is about 8 miles," properly translates to "It is about 13 kilometers." It would be inappropriate to convert the 8-mile measurement to 12.875 kilometers, as an exact conversion would give. The degree of precision implied by such an exact conversion is simply not implied by the original measurement. Therefore, conversions have been chosen to properly reflect the precision of the original measurement. An example: a 12-foot run of sprinkler piping converts to 3.7 meters, if the piping is measured to the nearest inch. If measured to the nearest foot, then the appropriate conversion is 4 meters.

The SI units into which quantities have been converted have, for the most part, followed the SI practice of expressing a quantity so that its numerical value falls between 0.1 and 1000. An example is the conversion of 14.7 psi to 101 kPa, not 101 000 Pa. Some deviations from this practice have been allowed, such as where several values of the same quantity are being compared or presented in tabular form. For example, if four samples have masses of 500 kg, 800 kg, 900 kg, and 1200 kg, it would be inappropriate to express the last entry as 1.2 Mg.

Tables B-1 and B-2 give two sets of conversion factors used in this *Handbook*. These tables are taken from *The SFPE Handbook of Fire Protection Engineering*, 2nd ed.,[3] and include a list of units arranged alphabetically and a list arranged by physical quantity, i.e., area, length, etc., respectively. In the alphabetical list (Table B-1), the first two digits of each conversion factor represent the power of 10 by which the conversion factor must be multiplied. An asterisk indicates that the conversion factor is exact. All other conversion factors are either approximate or the result of physical measurements. The physical quantity list (Table B-2) includes only those conversion factors that are frequently used. However, conversion factors for many specialized units can be found in *Lange's Handbook of Chemistry*, 14th ed.[4]

BIBLIOGRAPHY

References Cited

1. ASTM E 380-93, Standard Practice for Use of the International System of Units (SI) (the Modernized Metric System), American Society for Testing and Materials, W. Conshohocken, PA.

2. Wildi, T., Metric Units and Conversion Charts, 2nd ed., IEEE Press, New York, 1995.

3. SFPE Handbook of Fire Protection Engineering, 2nd ed., DiNenno, P. J., et al., eds., National Fire Protection Association, Quincy, MA, 1995.

4. Lange's Handbook of Chemistry, 14th ed., McGraw-Hill Companies, New York.

Additional Reading

American National Standard for Metric Practice, ANSI/IEEE Standard 268-1992, Institute of Electrical and Electronics Engineers, Inc., Piscataway, NJ.

Canadian Metric Practice Guide, CAN/CSA-Z 234.1-89, Canadian Standards Association, Toronto, Ontario, Canada.

Guide for the Use of the International System of Units—The Modernized Metric System, Special Publication 811, National Institute of Science and Technology, Washington, DC.

Interpretation of the SI for the United States and Metric Conversion Policy for Federal Agencies, Special Publication 814, National Institute of Science and Technology, Washington, DC.

Le Syst 6e

"SI (Metric) Handbook," KSC-DM-3673, National Aeronautics and Space Administration, John F. Kennedy Space Center, FL.

"The International System of Units (SI)," Special Publication 330, National Institute of Science and Technology, Washington, DC, 1991.

Note: An extensive bibliography can be found in the May, June, and December, 1994, issues of *ASTM Standardization News,* published by the American Society for Testing and Materials, W. Conshohocken, PA.

TABLE B-1.   *Alphabetic List of Conversion Factors*

To Convert from	to	Multiply by
abampere	ampere	+01 1.00*
abcoulomb	coulomb	+01 1.00*
abfarad	farad	+09 1.00*
abhenry	henry	−09 1.00*
abmho	siemens	+09 1.00*
abohm	ohm	−09 1.00*
abvolt	volt	−08 1.00*
acre	meter^2	+03 4.046 856 422 4*
angstrom	meter	−10 1.00*
are	meter^2	+02 1.00*
astronomical unit (IAU)	meter	+11 1.496 00
astronomical unit (radio)	meter	+11 1.495 978 9
atmosphere	newton/meter^2	+05 1.013 25*
bar	newton/meter^2	+05 1.00*
barn	meter^2	−28 1.00*
barrel (petroleum, 42 gallons)	meter^3	−01 1.589 873
barye	newton/meter^2	−01 1.00*
board foot (1′ × 1′ × 1″)	meter^3	−03 2.359 737 216*
British thermal unit:		
(IST before 1956)	joule	+03 1.055 04
(IST after 1956)	joule	+03 1.055 056
British thermal unit (mean)	joule	+03 1.055 87
British thermal unit (thermochemical)	joule	+03 1.054 350
British thermal unit (39°F)	joule	+03 1.059 67
British thermal unit (60°F)	joule	+03 1.054 68
bushel (U.S.)	meter^3	−02 3.523 907 016 688*
cable	meter	+02 2.194 56*
caliber	meter	−04 2.54*
calorie (International Steam Table)	joule	+00 4.1868
calorie (mean)	joule	+00 4.190 02
calorie (thermochemical)	joule	+00 4.184*
calorie (15°C)	joule	+00 4.185 80
calorie (20°C)	joule	+00 4.181 90
calorie (kilogram, International Steam Table)	joule	+03 4.1868
calorie (kilogram, mean)	joule	+03 4.190 02
calorie (kilogram, thermochemical)	joule	+03 4.184*
carat (metric)	kilogram	−04 2.00*
Celsius (temperature)	kelvin	$t_K = t_c + 273.15$
centimeter of mercury (0°C)	newton/meter^2	+03 1.333 22
centimeter of water (4°C)	newton/meter^2	+01 9.806 38
chain (engineer or ramden)	meter	+01 3.048*
chain (surveyor or gunter)	meter	+01 2.011 68*
circular mil	meter^2	−10 5.067 074 8
cord	meter^3	+00 3.624 556 3
cubit	meter	−01 4.572*
cup	meter^3	−04 2.365 882 365*
curie	disintegration/second	+10 3.70*

*Source: E.A. Mechtly, "The International System of Units, Physical Constants and Conversion Factors," 2nd revision, National Aeronautics and Space Administration, Washington, DC (1973). (No copyright.)

TABLE B-1.   *Continued*

To Convert from	to	Multiply by
day (mean solar)	second (mean solar)	+04 8.64*
day (sidereal)	second (mean solar)	+04 8.616 409 0
degree (angle)	radian	−02 1.745 329 251 994 3
denier (international)	kilogram/meter	−07 1.00*
dram (avoirdupois)	kilogram	−03 1.771 845 195 312 5*
dram (troy or apothecary)	kilogram	−03 3.887 934 6*
dram (U.S. fluid)	meter^3	−06 3.696 691 195 312 5*
dyne	newton	−05 1.00*
electron volt	joule	−19 1.602 191 7
erg	joule	−07 1.00*
Fahrenheit (temperature)	kelvin	$t_K = (5/9)(t_F + 459.67)$
Fahrenheit (temperature)	Celsius	$t_C = (5/9)(t_F − 32)$
faraday (based on carbon 12)	coulomb	+04 9.68 70
faraday (chemical)	coulomb	+04 9.649 57
faraday (physical)	coulomb	+04 9.652 19
fathom	meter	+00 1.828 8*
fermi (femtometer)	meter	+15 1.00*
fluid ounce (U.S.)	meter^3	−05 2.957 352 956 25*
foot	meter	−01 3.048*
foot (U.S. survey)	meter	+00 1200/3937*
foot (U.S. survey)	meter	−01 3.048 006 096
foot of water (39.2°F)	newton/meter^2	+03 2.988 98
footcandle	lumen/meter^2	+01 1.076 391 0
footlambert	candela/meter^2	+00 3.426 259
free fall, standard	meter/second^2	+00 9.806 65*
furlong	meter	+02 2.011 68*
gal (galileo)	meter/second^2	−02 1.00*
gallon (U.K. liquid)	meter^3	−03 4.546 087
gallon (U.S. dry)	meter^3	−03 4.404 883 770 86*
gallon (U.S. liquid)	meter^3	−03 3.785 411 784*
gamma	tesla	−09 1.00*
gauss	tesla	−04 1.00*
gilbert	ampere turn	−01 7.957 747 2
gill (U.K.)	meter^3	−04 1.420 652
gill (U.S.)	meter^3	−04 1.182 941 2
grad	degree (angular)	−01 9.00*
grad	radian	−02 1.570 796 3
grain	kilogram	−05 6.479 891*
gram	kilogram	−03 1.00*
hand	meter	−01 1.016*
hectare	meter^3	+04 1.00*
hogshead (U.S.)	meter^3	−01 2.384 809 423 92*
horsepower (550 foot lbf/second)	watt	+02 7.456 998 7
horsepower (boiler)	watt	+03 9.809 50
horsepower (electric)	watt	+02 7.46*
horsepower (metric)	watt	+02 7.354 99
horsepower (U.K.)	watt	+02 7.457
horsepower (water)	watt	+02 7.460 43
hour (mean solar)	second (mean solar)	+03 3.60*
hour (sidereal)	second (mean solar)	+03 3.590 170 4
hundredweight (long)	kilogram	+01 5.080 234 544*
hundredweight (short)	kilogram	+01 4.535 923 7*
inch	meter	−02 2.54*
inch of mercury (32°F)	newton/meter^2	+03 3.386 389
inch of mercury (60°F)	newton/meter^2	+03 3.376 85
inch of water (39.2°F)	newton/meter^2	+02 2.490 82
inch of water (60°F)	newton/meter^2	+02 2.4884
kayser	1/meter	+02 1.00*
kilocalorie (International Steam Table)	joule	+03 4.186 8
kilocalorie (mean)	joule	+03 4.190 02
kilocalorie (thermochemical)	joule	+03 4.184*
kilogram mass	kilogram	+00 1.00*
kilogram force (kgf)	newton	+00 9.806 65*
kilopound force	newton	+00 9.806 65*

TABLE B-1. *Continued*

To Convert from	to	Multiply by
kip	newton	+03 4.448 221 615 260 5*
knot (international)	meter/second	−01 5.144 444 444
lambert	candela/meter²	+04 1/π*
lambert	candela/meter²	+03 3.183 098 8
langley	joule/meter²	+04 4.184*
lbf (pound force, avoirdupois)	newton	+00 4.448 221 615 260 5*
lbm (pound mass, avoirdupois)	kilogram	−01 4.535 923 7*
league (U.K. nautical)	meter	+03 5.559 552*
league (international nautical)	meter	+03 5.556*
league (statute)	meter	+03 4.828 032*
light year	meter	+15 9.460 55
link (engineer or ramden)	meter	−01 3.048*
link (surveyor or gunter)	meter	−01 2.011 68*
liter	meter³	−03 1.00*
lux	lumen/meter²	+00 1.00*
maxwell	weber	−08 1.00*
meter	wavelengths Kr 86	+06 1.650 763 73*
micron	meter	−06 1.00*
mil	meter	−05 2.54*
mile (U.S. statute)	meter	+03 1.609 344*
mile (U.K. nautical)	meter	+03 1.853 184*
mile (international nautical)	meter	+03 1.852*
mile (U.S. nautical)	meter	+03 1.852*
millibar	newton/meter²	+02 1.00*
millimeter of mercury (0°C)	newton/meter²	+02 1.333 224
minute (angle)	radian	−04 2.908 882 086 66
minute (mean solar)	second (mean solar)	+01 6.00*
minute (sidereal)	second (mean solar)	+01 5.983 617 4
month (mean calendar)	second (mean solar)	+06 2.628*
nautical mile (international)	meter	+03 1.852*
nautical mile (U.S.)	meter	+03 1.852*
nautical mile (U.K.)	meter	+03 1.853 184*
oersted	ampere/meter	+01 7.957 747 2
ounce force (avoirdupois)	newton	−01 2.780 138 5
ounce mass (avoirdupois)	kilogram	−02 2.834 952 312 5*
ounce mass (troy or apothecary)	kilogram	−02 3.110 347 68*
ounce (U.S. fluid)	meter³	−05 2.957 352 956 25*
pace	meter	−01 7.62*
parsec (IAU)	meter	+16 3.085 7
pascal	newton/meter²	+00 1.00*
peck (U.S.)	meter³	−03 8.809 767 541 72*
pennyweight	kilogram	−03 1.555 173 84*
perch	meter	+00 5.0292*
phot	lumen/meter²	+04 1.00
pica (printers)	meter	−03 4.217 517 6*
pint (U.S. dry)	meter³	−04 5.506 104 713 575*
pint (U.S. liquid)	meter³	−04 4.731 764 73*
point (printers)	meter	−04 3.514 598*
poise	newton second/meter²	−01 1.00*
pole	meter	+00 5.0292*
pound force (lbf avoirdupois)	newton	+00 4.448 221 615 260 5*
pound mass (lbm avoirdupois)	kilogram	−01 4.535 923 7*
pound mass (troy or apothecary)	kilogram	−01 3.732 417 216*
poundal	newton	−01 1.382 549 543 76*
quart (U.S. dry)	meter³	−03 1.101 220 942 715*
quart (U.S. liquid)	meter³	−04 9.463 592 5
rad (radiation dose absorbed)	joule/kilogram	−02 1.00*
Rankine (temperature)	kelvin	$t_K = (5/9)t_R$
rayleigh (rate of photon emission)	1/second meter²	+10 1.00*
rhe	meter²/newton second	+01 1.00*
rod	meter	+00 5.0292*
roentgen	coulomb/kilogram	−04 2.579 76*
rutherford	disintegration/second	+06 1.00*

TABLE B-1. *Continued*

To Convert from	to	Multiply by
second (angle)	radian	− 06 4.848 136 811
second (ephemeris)	second	+ 00 1.000 000 000
second (mean solar)	second (ephemeris)	Consult American Ephemeris and Nautical Almanac
second (sidereal)	second (mean solar)	− 01 9.972 695 7
section	meter²	+ 06 2.589 988 110 336*
scruple (apothecary)	kilogram	− 03 1.295 978 2*
shake	second	− 08 1.00
skein	meter	+ 02 1.097 28*
slug	kilogram	+ 01 1.459 390 29
span	meter	− 01 2.286*
statampere	ampere	− 10 3.335 640
statcoulomb	coulomb	− 10 3.335 640
statfarad	farad	− 12 1.112 650
stathenry	henry	+ 11 8.987 554
statohm	ohm	+ 11 8.987 554
statute mile (U.S.)	meter	+ 03 1.609 344*
statvolt	volt	+ 02 2.997 925
stere	meter³	+ 00 1.00*
stilb	candela/meter²	+ 04 1.00
stoke	meter²/second	− 04 1.00*
tablespoon	meter³	− 05 1.478 676 478 125*
teaspoon	meter³	− 06 4.928 921 593 75*
ton (assay)	kilogram	− 02 2.196 666 6
ton (long)	kilogram	+ 03 1.016 046 908 8*
ton (metric)	kilogram	+ 03 1.00*
ton (nuclear equivalent of TNT)	joule	+ 09 4.20
ton (register)	meter³	+ 00 2.831 684 659 2*
ton (short, 2000 pound)	kilogram	+ 02 9.071 847 4*
tonne	kilogram	+ 03 1.00*
torr (0°C)	newton/meter²	+ 02 1.333 22
township	meter²	+ 07 9.323 957 2
unit pole	weber	− 07 1.256 637
yard	meter	− 01 9.144*
year (calendar)	second (mean solar)	+ 07 3.1536*
year (sidereal)	second (mean solar)	+ 07 3.155 815 0
year (tropical)	second (mean solar)	+ 07 3.155 692 6
year 1900, tropical, Jan., day 0, hour 12	second (ephemeris)	+ 07 3.155 692 597 47*
year 1900, tropical, Jan., day 0, hour 12	second	+ 07 3.155 692 597 47

TABLE B-2. *Conversion Factors—by Physical Quantity*

To Convert from	to	Multiply by
ACCELERATION		
foot/second²	meter/second²	− 01 3.048*
free fall, standard	meter/second²	+ 00 9.806 65*
gal (galileo)	meter/second²	− 02 1.00*
inch/second²	meter/second²	− 02 2.54*
AREA		
acre	meter²	+ 03 4.046 856 422 4*
are	meter²	+ 02 1.00*
barn	meter²	− 28 1.00*
circular mil	meter²	− 10 5.067 074 8
foot²	meter²	− 02 9.290 304*
hectare	meter²	+ 04 1.00*
inch²	meter²	− 04 6.4516*
mile² (U.S. statute)	meter²	+ 06 2.589 988 110 336*

TABLE B-2.   *Continued*

To Convert from	to	Multiply by
AREA (continued)		
section	meter²	+ 06 2.589 988 110 336*
township	meter²	+ 07 9.323 957 2
yard²	meter²	− 01 8.361 273 6*
DENSITY		
gram/centimeter³	kilogram/meter³	− 03 1.00*
lbm/inch³	kilogram/meter³	+ 04 2.767 990 5
lbm/foot³	kilogram/meter³	+ 01 1.601 846 3
slug/foot³	kilogram/meter³	+ 02 5.153 79
ENERGY		
British thermal unit:		
(IST before 1956)	joule	+ 03 1.055 04
(IST after 1956)	joule	+ 03 1.055 056
British thermal unit (mean)	joule	+ 03 1.055 87
British thermal unit (thermochemical)	joule	+ 03 1.054 350
British thermal unit (39°F)	joule	+ 03 1.059 67
British thermal unit (60°F)	joule	+ 03 1.054 68
calorie (International Steam Table)	joule	+ 00 4.1868
calorie (mean)	joule	+ 00 4.190 02
calorie (thermochemical)	joule	+ 00 4.184*
calorie (15°C)	joule	+ 00 4.185 80
calorie (20°C)	joule	+ 00 4.181 90
calorie (kilogram, International Steam Table)	joule	+ 03 4.1868
calorie (kilogram, mean)	joule	+ 03 4.190 02
calorie (kilogram, thermochemical)	joule	+ 03 4.184*
electron volt	joule	− 19 1.602 191 7
erg	joule	− 07 1.00*
foot lbf	joule	+ 03 1.355 817 9
foot poundal	joule	− 02 4.214 011 0
joule (international of 1948)	joule	+ 00 1.000 165
kilocalorie (International Steam Table)	joule	+ 03 4.1868
kilocalorie (mean)	joule	+ 03 4.190 02
kilocalorie (thermochemical)	joule	+ 03 4.184*
kilowatt hour	joule	+ 06 3.60*
kilowatt hour (international of 1948)	joule	+ 06 3.600 59
ton (nuclear equivalent of TNT)	joule	+ 09 4.20
watt hour	joule	+ 03 3.60*
ENERGY/AREA TIME		
Btu (thermochemical)/foot² second	watt/meter²	+ 04 1.134 893 1
Btu (thermochemical)/foot² minute	watt/meter²	+ 02 1.891 488 5
Btu (thermochemical)/foot² hour	watt/meter²	+ 00 3.152 480 8
Btu (thermochemical)/inch² second	watt/meter²	+ 06 1.634 246 2
calorie (thermochemical)/cm² minute	watt/meter²	+ 02 6.973 333 3
erg/centimeter² second	watt/meter²	− 03 1.00*
watt/centimeter²	watt/meter²	+ 04 1.00*
FORCE		
dyne	newton	− 05 1.00*
kilogram force (kgf)	newton	+ 00 9.806 65*
kilopound force	newton	+ 00 9.806 65*
kip	newton	+ 03 4.448 221 615 260 5*
lbf (pound force, avoirdupois)	newton	+ 00 4.448 221 615 260 5*
ounce force (avoirdupois)	newton	+ 01 2.780 138 5
pound force, lbf (avoirdupois)	newton	+ 00 4.448 221 615 260 5*
poundal	newton	− 01 1.382 549 543 76*
LENGTH		
angstrom	meter	− 10 1.00*
astronomical unit (IAU)	meter	+ 11 1.496 00
astronomical unit (radio)	meter	+ 11 1.495 978 9
cable	meter	+ 02 2.194 56*
caliber	meter	− 04 2.54*

TABLE B-2.   *Continued*

To Convert from	to	Multiply by
LENGTH (continued)		
chain (surveyor or gunter)	meter	+01 2.011 68*
chain (engineer or ramden)	meter	+01 3.048*
cubit	meter	−01 4.572*
fathom	meter	+00 1.8288*
fermi (femtometer)	meter	−15 1.00*
foot	meter	−01 3.048*
foot (U.S. survey)	meter	+00 1200/3937*
foot (U.S. survey)	meter	−01 3.048 006 096
furlong	meter	+02 2.011 68*
hand	meter	−01 1.016*
inch	meter	−02 2.54*
league (U.K. nautical)	meter	+03 5.559 552*
league (international nautical)	meter	+03 5.556*
league (statute)	meter	+03 4.828 032*
light year	meter	+15 9.460 55
link (engineer or ramden)	meter	−01 3.048*
link (surveyor or gunter)	meter	−01 2.011 68*
meter	wavelengths Kr 86	+06 1.650 763 73*
micron	meter	−06 1.00*
mil	meter	−05 2.54*
mile (U.S. statute)	meter	+03 1.609 344*
mile (U.K. nautical)	meter	+03 1.853 184*
mile (international nautical)	meter	+03 1.852*
mile (U.S. nautical)	meter	+03 1.852*
nautical mile (U.K.)	meter	+03 1.853 184*
nautical mile (international)	meter	+03 1.852*
nautical mile (U.S.)	meter	+03 1.852*
pace	meter	−01 7.62*
parsec (IAU)	meter	+16 3.085 7
perch	meter	+00 5.0292*
pica (printers)	meter	−03 4.217 517 6*
point (printers)	meter	−04 3.514 598*
pole	meter	+00 5.0292*
rod	meter	+00 5.0292*
skein	meter	+02 1.097 28*
span	meter	−01 2.286*
statute mile (U.S.)	meter	+03 1.609 344*
yard	meter	−01 9.144*
MASS		
carat (metric)	kilogram	−04 2.00*
gram (avoirdupois)	kilogram	−03 1.771 845 195 312 5*
gram (troy or apothecary)	kilogram	−03 3.887 934 6*
grain	kilogram	−05 6.479 891*
gram	kilogram	−03 1.00*
hundredweight (long)	kilogram	+01 5.080 234 544*
hundredweight (short)	kilogram	+01 4.535 923 7*
kgf second^2 meter (mass)	kilogram	+00 9.806 65*
kilogram mass	kilogram	+00 1.00*
lbm (pound mass, avoirdupois)	kilogram	−01 4.535 923 7*
ounce mass (avoirdupois)	kilogram	−02 2.834 952 312 5*
ounce mass (troy or apothecary)	kilogram	−02 3.110 347 68*
pennyweight	kilogram	−03 1.555 173 84*
pound mass, lbm (avoirdupois)	kilogram	−01 4.535 923 7*
pound mass (troy or apothecary)	kilogram	−01 3.732 417 216*
scruple (apothecary)	kilogram	−03 1.295 978 2*
slug	kilogram	+01 1.459 390 29
ton (assay)	kilogram	−02 2.196 666 6
ton (long)	kilogram	+03 1.016 046 908 8*
ton (metric)	kilogram	+03 1.00*
ton (short, 2000 pound)	kilogram	+02 9.071 847 4*
tonne	kilogram	+03 1.00*
POWER		
Btu (thermochemical)/second	watt	+03 1.054 350 264 488
Btu (thermochemical)/minute	watt	+01 1.757 250 4

TABLE B-2. *Continued*

To Convert from	to	Multiply by
POWER (continued)		
calorie (thermochemical)/second	watt	+00 4.184*
calorie (thermochemical)/minute	watt	−02 6.973 333 3
foot lbf/hour	watt	−04 3.766 161 0
foot lbf/minute	watt	−02 2.259 696 6
foot lbf/second	watt	+00 1.355 817 9
horsepower (550 foot lbf/second)	watt	+02 7.456 998 7
horsepower (boiler)	watt	+03 9.809 50
horsepower (electric)	watt	+02 7.46*
horsepower (metric)	watt	+02 7.354 99
horsepower (U.K.)	watt	+02 7.457
horsepower (water)	watt	+02 7.460 43
kilocalorie (thermochemical)/minute	watt	+01 6.973 333 3
kilocalorie (thermochemical)/second	watt	+03 4.184*
watt (international of 1948)	watt	+00 1.000 165
PRESSURE		
atmosphere	newton/meter^2	+05 1.013 25*
bar	newton/meter^2	+05 1.00*
barye	newton/meter^2	−01 1.00*
centimeter of mercury (0°C)	newton/meter^2	+03 1.333 22
centimeter of water (4°C)	newton/meter^2	+01 9.806 38
dyne/centimeter^2	newton/meter^2	−01 1.00*
foot of water (39.2°F)	newton/meter^2	+03 2.988 98
inch of mercury (32°F)	newton/meter^2	+03 3.386 389
inch of mercury (60°F)	newton/meter^2	+03 3.376 85
inch of water (39.2°F)	newton/meter^2	+02 2.480 82
inch of water (60°F)	newton/meter^2	+02 2.4884
kgf/centimeter^2	newton/meter^2	+04 9.806 65*
kgf/meter^2	newton/meter^2	+00 9.806 65*
lbf/foot^2	newton/meter^2	+01 4.788 025 8
lbf/inch^2 (psi)	newton/meter^2	+03 6.894 757 2
millibar	newton/meter^2	+02 1.00*
millimeter of mercury (0°C)	newton/meter^2	+02 1.333 224
pascal	newton/meter^2	+00 1.00*
psi (lbf/inch^2)	newton/meter^2	+03 6.894 757 2
torr (0°C)	newton/meter^2	+02 1.333 22
SPEED		
foot/hour	meter/second	−05 8.466 666 6
foot/minute	meter/second	−03 5.08*
foot/second	meter/second	−01 3.048*
inch/second	meter/second	−02 2.54*
kilometer/hour	meter/second	−01 2.777 777 8
knot (international)	meter/second	−01 5.144 444 444
mile/hour (U.S. statute)	meter/second	−01 4.4704*
mile/minute (U.S. statute)	meter/second	+01 2.682 24*
mile/second (U.S. statute)	meter/second	+03 1.609 344*
TEMPERATURE		
Celsius	kelvin	$t_K = t_C + 273.15$
Fahrenheit	kelvin	$t_K = (5/9)(t_F + 459.67)$
Fahrenheit	Celsius	$t_C = (5/9)(t_F - 32)$
Rankine	kelvin	$t_K = (5/9)t_R$
TIME		
day (mean solar)	second (mean solar)	+04 8.64*
day (sidereal)	second (mean solar)	+04 8.616 409 0
hour (mean solar)	second (mean solar)	+03 3.60*
hour (sidereal)	second (mean solar)	+03 3.590 170 4
minute (mean solar)	second (mean solar)	+01 6.00*
minute (sidereal)	second (mean solar)	+01 5.983 617 4
month (mean calendar)	second (mean solar)	+06 2.628*
second (ephemeris)	second	+00 1.000 000 000
second (mean solar)	second (ephemeris)	Consult American Ephemeris and Nautical Almanac

TABLE B-1. *Continued*

To Convert from	to	Multiply by
TIME (continued)		
second (sidereal)	second (mean solar)	− 01 9.972 695 7
year (calendar)	second (mean solar)	+ 07 3.1536*
year (sidereal)	second (mean solar)	+ 07 3.155 815 0
year (tropical)	second (mean solar)	+ 07 3.155 692 6
year 1900, tropical, Jan., day 0, hour 12	second (ephemeris)	+ 07 3.155 692 597 47*
year 1900, tropical, Jan., day 0, hour 12	second	+ 07 3.155 692 597 47
VISCOSITY		
centistoke	meter²/second	− 06 1.00*
stoke	meter²/second	− 04 1.00*
foot²/second	meter²/second	− 02 9.290 304*
centipoise	newton second/meter²	− 03 1.00*
lbm/foot second	newton second/meter²	+ 00 1.488 163 9
lbf second/foot²	newton second/meter²	+ 01 4.788 025 8
poise	newton second/meter²	− 01 1.00*
poundal second/foot²	newton second/meter²	+ 00 1.488 163 9
slug/foot second	newton second/meter²	+ 01 4.788 025 8
rhe	meter²/newton second	+ 01 1.00*
VOLUME		
acre foot	meter³	+ 03 1.233 481 837 547 52*
barrel (petroleum, 42 gallons)	meter³	− 01 1.589 873
board foot	meter³	− 03 2.359 737 216*
bushel (U.S.)	meter³	− 02 3.523 907 016 688*
cord	meter³	+ 00 3.624 556 3
cup	meter³	− 04 2.365 882 365*
dram (U.S. fluid)	meter³	− 06 3.696 691 195 312 5*
fluid ounce (U.S.)	meter³	− 05 2.957 352 956 25*
foot³	meter³	− 02 2.831 684 659 2*
gallon (U.K. liquid)	meter³	− 03 4.546 087
gallon (U.S. dry)	meter³	− 03 4.404 883 770 86*
gallon (U.S. liquid)	meter³	− 03 3.785 411 784*
gill (U.K.)	meter³	− 04 1.420 652
gill (U.S.)	meter³	− 04 1.182 941 2
hogshead (U.S.)	meter³	− 01 2.384 809 423 92*
inch³	meter³	− 05 1.638 706 4*
liter	meter³	− 03 1.00*
ounce (U.S. fluid)	meter³	− 05 2.957 352 956 25*
peck (U.S.)	meter³	− 03 8.809 767 541 72*
pint (U.S. dry)	meter³	− 04 5.506 104 713 575*
pint (U.S. liquid)	meter³	− 04 4.731 764 73*
quart (U.S. dry)	meter³	− 03 1.101 220 942 715*
quart (U.S. liquid)	meter³	− 04 9.463 592 5
stere	meter³	+ 00 1.00*
tablespoon	meter³	− 05 1.478 676 478 125*
teaspoon	meter³	− 06 4.928 921 593 75*
ton (register)	meter³	+ 00 2.831 684 659 2*
yard³	meter³	− 01 7.645 548 579 84*

ORGANIZATIONS WITH FIRE PROTECTION INTERESTS

Leonard Belliveau, Jr.

Since fire is a universal threat, it is natural that a large number of organizations take an active interest in fire prevention and control. The degree of interest and activity varies widely. This appendix provides information on some representative organizations that have demonstrated interest in fire prevention and control. It is not a complete listing but may serve to indicate the breadth of the interest and the availability of resources. This appendix is presented as follows:

A. National Fire Protection Association
B. Insurance Organizations in the United States with Fire Protection Interests
C. National Fire Service Organizations in the United States
D. Fire Research Laboratories
E. Fire Testing Laboratories
F. Society of Fire Protection Engineers
G. Federal Involvement in Fire Protection
H. Additional Organizations with Fire Protection Interests in the United States
I. International Organizations with Fire Protection Interests

A. NATIONAL FIRE PROTECTION ASSOCIATION

The mission of the National Fire Protection Association (NFPA), which was organized in 1896, is to reduce the burden of fire on the quality of life by advocating scientifically-based consensus codes and standards and research and education for fire and related safety issues. The Association was incorporated in 1930 under laws of the Commonwealth of Massachusetts.

In the spring of 1981, the Association moved from downtown Boston, where it rented office space for some 80 years, to Quincy, about 10 miles to the south. There, at Batterymarch Park, NFPA occupies its first permanent headquarters, situated on a 52-acre hilltop site.

NFPA is an independent, voluntary membership, nonprofit (tax-exempt) organization. A 26-member Board of Directors has general charge of the affairs of the Association, which has a staff of 156 professional men and women, plus 144 support personnel.

NFPA is financed principally by revenues from publications, membership dues, and seminars. NFPA has an annual expense budget of $50 million and annual revenues of about $55 million.

In 1996, NFPA celebrated its one hundredth anniversary. Many noteworthy events and activities marked this important milestone, including the production of a brand new series of broadcast public service announcements designed to honor NFPA's centennial by teaching life-saving fire safety messages to television viewers across North America.

Membership: Membership in NFPA totals more than 67,000 individuals and over 100 national trade and professional organizations. The vast majority of the members are residents of the United States, but because of its international status, there are many members from Canada and 70 other nations. Members are drawn from fire departments (24%); health care facilities (11%); business and industry (20%); insurance (6%); federal, state, and local government (7%); architects and engineers (8%); fire equipment manufacturers and distributors (6%); trade and professional associations (2%); and other fields (16%).

There are six categories of voting members: regular, life, senior, organization, sustaining, and honorary. There are two categories of nonvoting members: affiliate and student.

There are 14 sections within the voting membership. The Fire Marshals Association of North America is the oldest. It was founded in 1906 and reorganized in 1927 as the Fire Marshals Section of NFPA. The other sections, in order of their founding, are Electrical (1948); Rail Transportation Systems (1963); Industrial Fire Protection (1963); Metropolitan Fire Chiefs Committee was organized in 1965 and established as a section in October 1996; Fire Service (1973); Health Care (1976); Fire Science and Technology Educators (1976); Architects, Engineers and Building Code Officials (1979); Aviation (1980); Education (1981); Research (1987); Wildland Fire Management (1988); Lodging Industry (1988); and Building Fire Safety Systems (1993).

Member Publications: The *NFPA Journal*, the official journal of the NFPA, reaches all of the Association's various fire safety professionals. Covering major topics in fire protection and suppression, the *Journal* carries investigation reports; special NFPA statistical studies on large-loss fires, multiple death fires, fire fighter deaths and injuries, and other annual reports; articles on fire protection advances and public education; and information of interest to NFPA members. Timely articles on fire-fighting techniques and fire department management are covered.

Fire Technology, a quarterly peer-reviewed technical journal, reaches some 3,000 international subscribers interested in receiving information about the scientific and technical aspects of fire, including reports on fire modeling test results from laboratory experiments.

Fire News is the bimonthly member newsletter. Issues go to every NFPA member, providing standards information and Association reports on events of interest to the entire membership.

The *NFPA Buyers' Guide* is an annual NFPA member publication that provides a guide to the names, offices, and products or services of all leading fire protection and fire service manufacturers. *Noticiero Tecnico sobre Incendios* is the NFPA's Spanish-language bulletin containing reports, articles, and meeting notices of interest to the international Spanish-speaking community. In addition to these publications, many NFPA sections publish their own informational newsletters.

The *NFPA Journal* and the *NFPA Buyers' Guide* are supported, in part, by advertising. The Association solicits paid ads through traditional avenues, such as direct selling by representatives and attendance at trade shows. The Association has formulated an advertising standards policy that is widely known throughout the industries served by the publications. The Association also maintains a standing Advertising Review Board that checks every ad submitted against the published standards policy.

Codes and standards: Activities of NFPA generally fall into two broad, interrelated areas: technical and educational.

The basic technical activity involves development, publication, and dissemination of timely consensus codes and standards intended to minimize the possibility and effects of fire in all aspects of contemporary activity.

In addition, efforts continue to educate people of all ages from all regions in preventing the loss of life and property from fire. Key to this effort are the teaching of codes and standards and the importance of fire safety as a way of life.

Codes and standards are developed by more than 211 NFPA committees, each of which represents a balance of affected interests. More than 5,519 individuals serve voluntarily on the Association's committees on an unpaid basis. Committees operate according to the detailed official *Regulations Governing Committee Projects* and are administered by the Standards Council, which reports to the Association's Board of Directors.

Built into the codes and standards development and adoption process is the publication of calls for proposals to amend existing documents or on the proposed draft of new documents. These public proposals and the committee action on each proposal, as well as committee-generated proposals, are published in the *Technical Committee Report on Proposals* (ROP) for public review and comment. Public comments and the committee action on each comment are then published in the *Technical Committee Report on Comments* (ROC). Only after this public review and comment cycle has been completed is the final committee report brought before the membership for action. This democratic legislative procedure allows proponents and opponents to be freely heard. Once adopted by the NFPA membership at either an annual meeting or a fall meeting, and issued by the Standards Council, codes and standards are published and made available for voluntary adoption.

Tentative Interim Amendments (TIAs) and Formal Interpretations (FIs) are issued from time to time. Notices of the issuance of TIAs and FIs are published in *Fire News* and other media.

Each official NFPA document is published in a pamphlet edition. All documents appear in the multivolume set of *National Fire Codes.*® Among the most widely used documents are the *National Electrical Code*® (NFPA 70); the *Life Safety Code*® (NFPA 101®); the *Flammable and Combustible Liquids Code* (NFPA 30); the *Standard for the Installation of Automatic Sprinkler Systems* (NFPA 13); the *Standard for the Storage and Handling of Liquefied Petroleum Gases* (NFPA 58); and the *Standard for Health Care Facilities* (NFPA 99).

NFPA codes and standards, which currently number more than 285, have great influence because they are widely used as the basis of legislation and regulation at all levels of government, from local to national. Many are referenced by agencies of the federal government. In the U.S., this includes the regulations of the Occupational Safety and Health Administration (OSHA), National Institute of Health (NIH), and Department of Energy (DOE). The documents also are used by insurance authorities for risk evaluation and premium rating.

A National Fire Codes Subscription Service is offered, through which over 19,000 subscribers automatically receive the Fire Codes and all other standards-related information, such as TIAs, FIs, ROPs, and ROCs, in loose-leaf format.

Engineering: The Engineering Division is responsible for providing staff liaison support to NFPA Technical Committees in the areas of fire protection systems and equipment, special hazards and occupancies, chemical and hazardous material storage and handling, building fire protection and life safety, and electrical installations. Technical advisory services regarding questions concerning NFPA codes and standards are also provided. Four advisory committees provide guidance for NFPA activity in specialized project areas. These projects include Electrical, Flammable Liquids, Gases, and Marine. Through the Marine Chemist Qualification Board, the Association administers the NFPA Certificated Marine Chemist Program. Engineering staff perform the executive secretarial function for seven NFPA membership sections: Architects, Engineers, and Building Officials; Aviation; Building Fire Safety Systems; Electrical; Health Care; Industrial Fire Protection; and Rail Transportation Systems. Engineering Division technical staff also serve as instructors, lecturers, and panelists in a variety of Association programs and programs of other organizations throughout the world.

The Engineering Division's fire investigators conduct investigations of significant fire incidents of technical interest. Important lessons learned from fires provide input to NFPA technical committees and technical programs and are distributed to the fire community. The fire investigators regularly coordinate with state fire marshals, metropolitan fire chiefs, building code groups, and federal agencies on their fire investigation activities.

Technical publications: The Association's technical and general-interest publications program ranges across the fire protection spectrum. In addition to codes and standards, it includes handbooks, technical/reference books, fire service publications, textbooks, and training manuals. More than 290 codes and standards are available as single copies, as part of a 13-volume softbound or loose-leaf set, or as part of a subscription service. The *Fire Protection Handbook*, now in its 18th edition, is the "bible" in its field. The *SFPE Handbook of Fire Protection Engineering* is the standard technical reference for fire protection engineers and researchers. The *National Electrical Code* is equally important to electricians, architects, building trades people, inspectors, and other municipal officials. NFPA's many handbooks include *Automatic Sprinkler Systems, National Electrical Code, Life Safety Code, Flammable and Combustible Liquids Code,* NFPA 1500, *LP-Gases Code, National Fuel Gas Code, National Fire Alarm Code, Health Care Facilities,* and the *Hazardous Materials Response.* Other titles include *Industrial Fire Hazards, Fire Alarm Signaling Systems, Fire Litigation Handbook, NFPA Inspection Manual, Fire Command, Engine Company and Truck Company Fireground Operations, Fire Protection Systems: Test, Inspection and Maintenance Manual, Automatic Sprinkler and Standpipe Systems, Management in the Fire Service, Fire*

Service Administration, Principles of Fire Protection Chemistry, and *Building Construction for the Fire Service.*

Public Education: *Learn Not to Burn®* (LNTB) is the theme and focus of NFPA's comprehensive public education initiatives. Based on NFPA's long held belief that fire safety information should be presented in a positive, nonthreatening manner, *Learn Not to Burn* teaches people of all ages how to make responsible choices regarding health and safety, without using scare tactics. The program stresses how to prevent fires and, in the event of a fire, how to protect oneself. One of NFPA's most important commitments is to developing future generations of fire safe adults. Children in preschool through eighth grade continue to receive critical life safety skills through *Learn Not to Burn* fire safety education materials. The *Learn Not to Burn* curriculum, first released in 1979 and now in its third edition, teaches 22 key fire safety behaviors and is organized into three learning levels.

The *Learn Not to Burn Resource Books* were released in 1991 as a grade-based alternative to the *Learn Not to Burn* curriculum. This four-volume program teaches children in kindergarten through grade three 14 basic fire safety behaviors over a 4-year period. The program provides lesson guidelines and reproducible activity sheets to help the teacher present fire safety lessons teaching core subject areas such as math, language arts, art, and physical education.

In 1994, NFPA launched a $1 million strategic fire safety initiative which made the Association's educational materials more accessible to diverse communities across North America. As part of this unprecedented effort, the Association reduced the price of *Learn Not to Burn* program materials by 50 percent and created the *Learn Not to Burn* Champion Award Program to provide free *Learn Not to Burn* materials and training for 60 communities in 1994.

The response from the field was so positive that NFPA decided to extend the Champion Program by selecting up to 60 communities each year to receive awards. In addition, in 1995 NFPA launched *"Learn Not to Burn: Safe Cities,"* a program designed to help meet the special fire and burn prevention education needs of major urban centers. Each year ten "safe cities" are selected to receive *Learn Not to Burn* materials and technical assistance for a three-member team representing the fire department, school department, and a local organization willing to commit ongoing support as the *Learn Not to Burn* program expands in that city.

Learn Not to Burn really works! More than 350 people have been saved from fire injury or death because of lessons learned through *Learn Not to Burn* fire safety programs. A regional team of fire safety education representatives located throughout the U.S. and Canada helps fire departments and communities to implement or better understand *Learn Not to Burn,* to promote awareness of the fire/burn problem and solutions through education, and to communicate with those in the fire prevention community.

Through its Education Section, NFPA supports an expanding international network of individuals committed to applying innovative educational methods to reduce the loss of life and property from fire.

NFPA Center for High-Risk Outreach: The new NFPA Center for High-Risk Outreach was created to address the needs of people at greatest risk from fire and burns: young children, older adults, and people in low-income communities. One priority of the Center is to reduce deaths and injuries among preschool children in low-income communities.

Preschool children: After researching the location of low-income preschool children, the Center implements the *Learn Not to Burn® Preschool Program* in day care programs. It also works to reach parents, family members, grandparents and other caregivers through public service announcements, parent videos and other means. The Center also looks for creative and effective ways of reaching children who are not in organized day care programs.

Older adults: The Center is developing a comprehensive fire safety education program for older adults and will test that program in low-income communities.

The Center looks for emerging issues that relate to factors that make people at higher risk for deaths and injuries, such as people being trapped in fires because of bars and grates and other security devices, and then provides engineering, enforcement and educational solutions.

A presidential advisory committee appointed by NFPA President George Miller provides expert advice to the Center on technical and educational issues of fire and burn safety related to high-risk audiences.

Public Fire Protection: The Public Fire Protection Division advises the public, and fire and emergency service personnel in response to inquires related to NFPA fire service standards and other documents. They clarify and provide insight into provisions that provide for the safety and effectiveness of fire service and other emergency personnel. Staff liaison service is provided to over 80 fire and emergency service standards that are created and updated by NFPA committees. Division staff, all with extensive fire service experience, also serve as executive secretary's of four NFPA membership sections: Fire Marshal's Association of North America, Fire Service Section, Fire Science and Technology Educators Section, and Wildland Fire Management Section. In addition, division staff help develop NFPA fire service audio-visual educational products; investigate fire and emergency incidents of national consequence; and present lectures at seminars and conferences throughout North America and abroad.

Fire Analysis and Research: The Association maintains one of the world's most extensive fire experience databases. NFPA's annual survey of fire departments and its Fire Incident Data Organization (FIDO) file on large fires and other fires of major technical interest are used along with special studies and other major databases, most notably the Federal Emergency Management Agency's (FEMA) National Fire Incident Reporting System (NFIRS). The Fire Analysis and Research Division produces two dozen annual reports and many special studies on aspects of the nation's fire problem; responds to thousands of individual data requests; supports the information needs of NFPA's committees, membership sections, and technical programs; and provides analysis and data services for government agencies and private sector organizations. The Division also provides the Executive Secretary for the NFPA Research Section and maintains liaisons and activities in several areas of research, notably fire risk assessment, and evacuation modeling.

Reports and packages on a wide range of topics are available through the Division's One-Stop Data Shop program. Special analyses and summaries of sample incidents relating to specific interests can be obtained, with great flexibility in meeting the needs of requesters.

Public Affairs: NFPA's public information program is carried out by the Public Affairs Division. Through public awareness campaigns, public service announcements, news releases, feature articles, and media interviews, the staff communicate NFPA's mission, technical activities, and fire-safety messages to diverse audiences. The Association is recognized by the electronic and print media as the leading authoritative source of technical background, data, and consumer advice on the fire problem and fire protection and prevention. NFPA regularly works with international, national, and local media outlets to provide timely advice and information. The Public

Affairs division also facilitates creation and distribution of public service announcements featuring NFPA's mascot, Sparky the Fire Dog®, and develops a community awareness campaign kit every year that is distributed at no cost to all U.S. fire departments and thousands of other fire safety advocates to support local Fire Prevention Week activities.

Charles S. Morgan Technical Library: The Charles S. Morgan Technical Library maintains one of the largest collections of information relevant to the fire problem facing us today. Founded in 1945, it holds vital data on the topics of fire prevention, fire safety, fire research, fire service, and the world's only complete holdings of past and present NFPA codes and standards.

The Library's comprehensive collection of materials, from both national and international sources, includes approximately 3,000 books, 6,500 technical reports, 200 periodicals, films, videocassettes, and NFPA published archives dating from the Association's founding in 1896. The Library also prepares information that is uploaded to NFPA's Home Page on the World Wide Web and is the representative to inFIRE, an international organization of fire information professions with substantial collections of fire literature.

Regional Offices: NFPA has established regional representation by creating offices in various parts of the country to better serve members in these areas. An office in Tallahassee, FL serves the southern states, an office in Smyrna, DE serves the mid-Atlantic states, another in Louisville, KY serves the central states, and a representative in Ontario, CA serves the western states. The northeastern part of the U.S. is covered by the Regional Director operating out of NFPA headquarters in Quincy, MA and another representative in Quincy covers Canada.

Washington Office: The Association works with the U.S. Congress and Federal Agencies to promote the adoption and use of NFPA Codes and Standards and to promote a uniform national approach to combating the fire problem in America. Consequently, NFPA maintains an office in Washington, DC. In addition, the NFPA Washington office has the responsibility of working with the Association's Washington-based members and allied organizations concerned with fire and related safety issues.

Annual and Fall Meetings: The four-day NFPA Annual Meeting and Fire Safety Exhibit takes place each May, with an average attendance of 6,000 persons and the largest fire-safety exhibit of its kind in North America. The Fall Meeting, which lasts three or four days and attracts some 1,200 persons, occurs each November. To reach a larger segment of members, meetings are held in various locations throughout the United States. Some take place in Canada, and one, thus far, has been in Europe. Technical presentations on new developments and concepts, discussions, voting on technical committee standards actions, NFPA section meetings, conduct of other Association business, and special events make up the meeting programs. Registration is open to all who are interested, although only voting members may vote.

Member Advisory Council and Regional Meetings: The Member Advisory Council was established in 1975 to facilitate communication between Association members and management. Its members are geographically well diversified, and they represent the varied interests in the Association membership. The Council is managed on a local level by eight regional vice Chairs, each of whom are Council members. The Council voluntarily serves to communicate to NFPA management the thoughts and suggestions of members of the Association. It meets once a year in November during the Fall Meeting.

Since 1982, regional membership meetings have taken place in various cities throughout the U.S. and Canada each year. They have now extended to Mexico as well. These one-day events enable NFPA members in a region to meet each other; to meet and hear from the NFPA Board of Directors, administration, and staff members; to discuss openly any problems of concern; and to learn firsthand about new developments in fire safety.

Marketing and Sales: The Marketing and Sales Division serves the broad needs of members and customers, providing NFPA products and membership services on an efficient and timely basis. To make the nationwide ordering of products easier, an 800 number (1-800-344-3555) operates from 8:30 am to 8 pm eastern time. Division staff market products through direct response marketing methods. The director of sales markets products such as the *National Electrical Code* through a small group of distributors.

Film and Video Products: The Association is also a major distributor of video programs designed to serve the training and information needs of the fire service, business, industry, education, and health care.

NFPA films consistently receive top honors in major film festivals throughout the country. Recent awards include the Gold Camera from the United States Industrial Film Festival, the Gold Medal from the International Film and Television Festival of New York, the Blue Ribbon from the American Film and Video Association, the Golden Slate from the International Television Association, and the Gold Cindy from the Association of Visual Communicators.

Seminars: In other educational activities, NFPA disseminates knowledge and methodology on fire-related problems and solutions through NFPA-sponsored and contracted seminars and workshops. These programs involve discussion and application of codes and standards, plus application of state-of-the-art fire protection technology. Current programs support the various NFPA codes and standards, such as the *Life Safety Code*® (NFPA 101®), the *National Electrical Code*® (NFPA 70), the *Fire Alarm and Signaling Systems* (NFPA 72), and *Automatic Sprinklers* (NFPA 13 and NFPA 25). All of the Division's programs are available for on-site, contracted training.

Fire Prevention Week: The History of National Fire Prevention Week has its roots in the Great Chicago Fire, which occurred on October 9, 1871. This tragic conflagration killed more than 250 people, left 100,000 homeless, and destroyed more than 17,000 structures. The origin of the fire has generated speculation since its occurrence, with fact and fiction becoming blurred over the years. One popular legend has it that Mrs. Catherine O'Leary was milking her cow when the animal kicked over a lamp, setting the O'Leary barn on fire and starting the spectacular blaze. How ever the massive fire began, it swiftly took its toll, burning more than 2,000 acres in 27 hours. The City of Chicago quickly rebuilt, however, and within a couple of years residents began celebrating their successful restoration by memorializing the anniversary of the fire with festivities.

Intending to observe the fire's anniversary with a more serious commemoration, the Fire Marshals Association of North America (FMANA), the oldest membership section of the National Fire Protection Association (NFPA), decided that the 40th anniversary of the Great Chicago Fire should be observed not with festivities, but in a way that would keep the public informed about the importance of fire prevention. So on October 9, 1911, FMANA sponsored the first National Fire Prevention Day.

In 1920, President Woodrow Wilson issued the first national Fire Prevention Day proclamation. For more that 70 years, the nonprofit NFPA has officially sponsored and selected the theme for the national commemoration of Fire Prevention Week, honoring the anniversary of the Great Chicago Fire and using the event to increase awareness of the dangers of fire. And every year since 1925, the President of the United States has signed a proclamation pronouncing the Sunday-through-Saturday period in which October 9 falls a national observance.

When President Calvin Coolidge proclaimed the first National Fire Prevention Week, October 4–10, 1925, he noted that in the previous year some 15,000 lives were lost to fire in the United States. Calling the loss "startling," President Coolidge's proclamation stated, "This waste results from conditions which justify a sense of shame and horror; for the greater part of it could and ought to be prevented... It is highly desirable that every effort be made to reform the conditions which have made possible so vast a destruction of the national wealth."

In 1996, NFPA released the first in a series of broadcast public service announcements, reinforcing the '96 Fire Prevention Week theme, "Let's Hear it for Fire Safety, Test Your Detectors!" The public service announcements, which combine live action with animation, are known as Sparky's Safety Spots and feature the voice of veteran actor Dick Van Dyke as Sparky the Fire Dog.

The Association produces posters, booklets, and flyers covering a broad range of fire safety topics, such as match and lighter fire safety, school fire drills, cooking fire safety, and fire safety in health care facilities, high rises, and hotels and motels. *Sparky's Coloring Book, Little Folks' Fire Safety Fun,* and *Sparky's Inspection Checklist* are some of the educational and fun products designed to teach children important fire safety lessons.

Research Foundation: The National Fire Protection Research Foundation is the world's only charitable fire research institution. Established in 1982, it provides a catalyst for conducting independent research on fire risk and new technologies and strategies. Broad-based technical advisory committees guide each project, and the foundation collaborates with laboratories throughout the world.

International Department: NFPA has been "international" since its beginning in 1896. Today many NFPA works are de facto global standards, in healthcare, high rise construction, aviation, industrial storage, hazardous material identification and handling, fire service operations, and many more applications. The NEC® was translated into Spanish as early as 1943, and is presently used by 160 million people, or about half the Spanish speaking world. Many other codes and reference texts are now available in Spanish. In 1995 The Korea Fire Protection Association signed an accord with NFPA, and translated the nearly 300 NFPA codes into Korean. NFPA's proven public education program, *Learn Not to Burn*®, has been translated into French, Spanish and British English.

B. INSURANCE ORGANIZATIONS IN THE UNITED STATES WITH FIRE PROTECTION INTERESTS

American Insurance Services Group, Inc.
85 John St.
New York, NY 10038

Incorporated in 1984 as a not-for-profit organization, the American Insurance Services Group, Inc. (AISG) provides specialized services to the property-casualty insurance industry. AISG is affiliated with the American Insurance Association, the nation's oldest and largest insurance trade association. The American Insurance Association was created in 1964 by merging the National Board of Fire Underwriters (organized 1866), the Association of Casualty and Surety Companies (organized 1926), and the former American Insurance Association (founded 1953).

Among the important services provided by AISG is its Engineering and Safety Service, a direct successor to the National Board of Fire Underwriters. As of 1996, the Engineering and Safety Service (E & S) had over 350 subscriber insurance companies, representing nearly 90 percent of the property-casualty insurance business. E & S is dedicated to providing information, education, and consultation to the technical and loss control personnel of its subscribing companies through a series of technical reports and bulletins, training programs, and conferences that deal with a variety of technical categories. E & S supports a total loss control concept, including boilers and machinery, building technology, commercial vehicles, construction hazards, crime prevention, environmental science, fire protection, industrial hygiene, occupational safety and health, pollution control, product safety, and special and chemical hazards. The Service also publishes a wide variety of accident prevention publications, fire prevention education materials, a specialized publication on fire resistance of construction assemblies, and numerous other related items, many of which are available to the public and risk managers.

Alliance of American Insurers
1501 Woodfield Road, Suite 400W
Schaumburg, IL 60173-4980

The Alliance, organized in 1922, is a national organization of 260 property loss control and casualty insurance companies providing engineering, legal, legislative, educational, and public relations services for its member companies.

The Alliance offers technical seminars and publishes training materials. Technical seminars have addressed such topics as fire protection, property safety, machine safeguarding, fleet safety, and industrial hygiene. Firesafety materials include the series "Judging the Fire Risk" and "Simplified Water Supply Testing." The Alliance, in conjunction with its members, monitors and advocates public policy safety on affecting the insurance industry.

Factory Mutual System
1151 Boston-Providence Turnpike
P.O. Box 9102
Norwood, MA 02062

Based in Norwood, Mass., Factory Mutual is a world leader in property loss prevention engineering, research and training. With more than 1,800 employees including hundreds of scientists, engineers and researchers, the organization provides information on property loss prevention to companies worldwide.

Three major industrial and commercial insurance companies—Allendale Insurance, Arkwright, and Protection Mutual Insurance—own or direct the Factory Mutual organizations.

Factory Mutual comprises three entities:

Factory Mutual Engineering Association (FMEA) provides field inspection and loss consulting services, appraisal services, adjusts losses, prepares building and property plans, reviews contractors' plans, and helps insureds manage loss prevention.

Factory Mutual Engineering Corporation (FMEC) comprises the business operations of the organization: Legal, Treasury, Human Resources Departments, and Information Services Division. It also conducts loss prevention seminars, and provides print and audio visual materials for insureds at its Training Resource Center in Norwood.

Factory Mutual Research Corporation (FMRC) is an independent, nonprofit organization that conducts research and development activities, including fire research at the FMRC Test Center in West Glocester, RI. Recognized worldwide as a leader in standards-setting, FMRC provides loss prevention engineering guidelines that are published in the *Loss Prevention Data Books.*

Through its Approvals Division, FMRC also provides third-party evaluation of fire prevention products and services, which are listed in its annual *Approval Guide.*

Industrial Risk Insurers
85 Woodland St.
P.O. Box 5010
Hartford CT 06102

Industrial Risk Insurers (IRI), association of leading insurance organizations, provides specialized insurance and loss prevention for a wide variety of high-valued, well-protected risks around the world. A Canadian affiliate, Canadian Industrial Risks Insurers (CIRI), formed in 1973, underwrites Canadian accounts under terms and conditions similar to the IRI in the United States.

IRI provides insurance to the following industries: aeronautics, auto, electronics, light metal workers, hospitals and institutions, pharmaceuticals, electric generation, heavy metal workers, light hazard chemical and non-ferrous metals. Also, rubber, service risks, cloth working, grain leather goods, pulp and paper, textiles, basic steel, building materials, and woodworking, among others.

IRI's home office is in Hartford, Connecticut. Other locations are in Atlanta, Boston, Charlotte, Chicago, Cleveland, Dallas, Los Angles, New York, Philadelphia and San Fransico. Outside the U.S., offices are located in Frankfurt, Germany; London, England; Toronto, Canada; San Juan, Puerto Rico; Sydney, Australia; and Tokyo, Japan.

IRI offers field surveys and loss prevention activities. These include:

1. Loss prevention review of new plant layouts and the design of protection facilities, including sprinklers, alarms and fire walls.
2. Field Surveys as appropriate.
3. Investigation of losses.

A loss prevention training center is located in Hartford. A variety of courses, including IRI's *Industrial Fire Control Concepts,* are offered to interested parties on-site, as well as at locations of their choosing. Those attending the Hartford sessions receive the added benefit of working with the center's extensive collection of fire protection hardware, one of the largest in the world.

Insurance Services Office Inc.
7 World Trade Center.
New York, NY 10048-1199

The Insurance Services Office (ISO) is a national corporation providing a wide range of advisory services to property-casualty insurance companies. ISO functions, as provided by law, as an insurance rating organization, as an insurance actuarial service or advisory organization, and as a statistical agent. In 1996, it served approximately 1,300 affiliated insurance companies, and had a countrywide staff of 2,000.

It was formed in 1971 through the consolidation of several national insurance industry service organizations, including the Fire Insurance Research and Actuarial Association, the Inland Marine Insurance Bureau, the Insurance Rating Board, the Multi-Line Insurance Rating Bureau, and the National Insurance Actuarial and Statistical Association. Most of the former state and regional fire rating organizations in the United States are now a part of ISO.

ISO serves many kinds of personal and commercial lines of insurance, including commercial general liability, homeowners, dwelling fire, and commercial and personal allied lines. The other lines are: boiler and machinery, commercial auto, commercial multiple line, commercial crime, commercial glass, commercial inland marine, nuclear energy liability and property, private passenger auto, personal inland marine, professional liability, farm, and personal insurance coverages (PIC).

The municipal grading and classification function for the larger cities was transferred to ISO from the American Insurance Association (formerly the National Board of Fire Underwriters) in 1971. ISO has responsibility for the evaluation and classification of cities, both large and small, in 45 states, Puerto Rico and Washington, DC. "Specific Commercial Property Evaluation Schedule" (SCOPES) and the "Fire Suppression Rating Schedule" (FSRS) are important publications available from ISO.

Other Insurance Organizations

There are many other important groups performing varied fire protection and inspection services on behalf of the insurance industry and their insureds. Among these are:

American Hull Insurance Syndicate, 14 Wall St., New York, NY 10005: An insurance syndicate created to insure vessels of high value.

American Institute of Marine Underwriters (AIMU), 14 Wall Street, New York, NY 10005: Organized in 1898 to promote, protect and advance the interests of ocean marine insurers domiciled in the United States.

American Nuclear Insurers (ANI), Town Ctr., 29 S. Main St., Suite 300-S, West Hartford, CT 06107-2445: A voluntary, unincorporated association of insurance companies formed for the purpose of providing property and liability insurance protection to the nuclear energy industry.

Formerly known as NEL-PIA, it was formed in 1974 through the merger of two organizations—the Nuclear Energy Liability Insurance Association (NELIA) and the Nuclear Energy Property Insurance Association (NEPIA)—both of which began writing nuclear insurance in 1957. Its new name, American Nuclear Insurers, was adopted early in 1978.

Assurex International, 445 Hutchinson Ave., Suite 820, Columbus, OH 43235: A corporation owned by independent agents and brokers throughout the North American continent to provide education, marketing, engineering, and selling support services for its agency owners.

Improved Risk Managers (IRM), 4401 Barclay Downs Drive, Charlotte, NC 28209-4604: Improved Risk Managers is the management arm of IRM Insurance, which is an association of general insurance companies, their affiliates and subsidiaries, providing property insurance for large commercial, industrial and institutional risks since 1921. Improved Risk Managers (formerly Improved Risk Mutuals) provides loss control training and inspections for insureds, member company personnel, agents and brokers, and offers these services to non-members for a fee.

Independent Insurance Agents of America, 127 S. Peyton St., Alexandria, VA 22201: Independent Insurance Agents of America operates as a federation of state associations to identify, assess and meet the needs of its members by coordinating association resources to achieve favorable legislative, regulatory and legal environments for independent agents and consumers.

Insurance Committee for Arson Control, 3601 Vincennes Road, Indianapolis, IN 46268: The Insurance Committee for Arson Control (ICAC) serves as a national resource, education and communications organization working to increase public awareness of the arson problem and how the industry is responding. Through its publications and training seminars, ICAC provides information to help insurers avoid arson-prone risks, resist fraud-motivated arson claims and otherwise help bring arson under control. ICAC also sponsors and coordinates the activities of the National Arson Forum.

Insurance Information Institute, 110 William St., New York, NY 10038: An educational, fact-finding, and communications organization for the property and liability insurance business. It has more than 200 company members and also provides public relations services to other organizations in the fields of property, liability, fidelity, surety, and marine insurance.

Insurance Institute for Highway Safety (IIHS) 1005 N. Glebe Rd., Suite 800, Arlington, VA 22201: The IIHS is an independent, nonprofit, scientific, and educational organization dedicated to reducing losses/deaths, injuries, and property damage resulting from crashes on the nation's highways. The Institute is supported by the American Insurance Highway Safety Association, the American Insurers Highway Safety Alliance, the National Association of Independent Insurers Safety Association, and several individual insurance companies.

MAERP (Mutual Atomic Energy Reinsurance Pool) Association, 330 N. Wabash, Suite 2600 IBM Plaza, Chicago, IL 60611: A pool that has the capacity to insure in excess of $55 million per occurrence on any one nuclear reactor installation or any other operation involving radiation hazards.

Mutual Reinsurance Bureau, 1550 Pearl St., P. O. Box 188, Belvidere, IL 61008: An organization to provide property and casualty reinsurance for a large segment of the mutual insurance industry, plus statistical services to member companies.

National Association of Insurance Brokers, Inc., 1511 K St., N.W., Suite 316, Washington, D.C. 20005: An association to keep track of proposed and current state and national legislation supporting what is in the best interests of brokers and the insurance-buying public.

National Association of Insurance Commissioners, 633 W. Wisconsin Ave., Suite 1015, Milwaukee, WI 53203: A voluntary association of the chief insurance regulatory officials of the 50 states, the District of Columbia, Guam, Puerto Rico, and the Virgin Islands to promote legislative uniformity and provide a means for protecting the insurance-consuming public.

National Association of Mutual Insurance Companies, 3601 Vincennes Road, Indianapolis, IN 46268: A national trade association representing the interests of nearly 1,200 property and casualty insurance companies. It is the largest property/casualty trade association, both in number of member companies and amount of business underwritten. NAMIC's membership is diverse, ranging from the largest to the smallest property/casualty companies of North America. Farm mutuals, single-state companies, multiple lines carriers, regional and national insurers are all a part of NAMIC. Together these companies underwrite approximately 30 percent of the property/casualty business.

Educational opportunities and services for insurance companies are offered through NAMIC-affiliated entities, which include: NAMIC Insurance Co. (NAMICO), NAMIC Insurance Agency (NIA), Conference of Casualty Insurance Companies (CCIC), Crop Insurance Research Bureau Inc. (CIRB), Insurance Loss Control Association (ILCA), Professional Insurance Communicators of America (PICA) and Urban Insurance Partners Foundation.

National Association of Professional Insurance Agents (PIA), 400 N. Washington St., Alexandria, VA 22314: This association represents 80,000 independent insurance agents who primarily sell and service property and casualty insurance.

National Insurance Crime Bureau (NICB), 10330 South Roberts Road, Palos Hills, IL 60465: The National Insurance Crime Bureau (formerly the National Automobile Theft Bureau) is a not-for-profit organization supported by approximately 1,000 property-casualty insurers and several self-insured companies dedicated to combating vehicle theft and insurance fraud.

National Cargo Bureau, Inc. (NCB), 30 Vesey Street, New York, NY 10007-2914: NCB is a not-for-profit membership corporation, organized in 1952 to promote the safety of life and property at sea by formulating and recommending procedures, practices, rules and regulations for the safe loading and unloading, stowing, and securing of cargo, including hazardous materials, on vessels and to provide related inspection and certification services. Certificates of loading issued by NCB are recognized by the U.S. Coast Guard as evidence of compliance with applicable U.S. and I.M.O. regulations.

American Association of Insurance Services (AAIS), 1035 South York Road, Bensenville, IL 60106: AAIS (formerly the Transportation Insurance Rating Bureau) is a non-profit advisory organization that creates policy forms and manuals and statistical plans for property/casualty insurers. Licensed in all states, the District of Columbia, and Puerto Rico, AAIS supplements its core services with product development, actuarial, and filing services to let companies develop customized programs.

Volunteer Firemen's Insurance Services, Inc. (VFIS), Inc., P.O. Box 2726, York, PA 17405: VFIS, Inc., was founded in 1969 to fill a serious need for insurance coverage to American and Canadian volunteer firefighters. VFIS, today, is the largest insurance agency specializing in this field, with over 12,000 volunteer and career emergency service organizations.

C. NATIONAL FIRE SERVICE ORGANIZATIONS IN THE UNITED STATES

The following national and international associations are the most prominent groups serving the needs of the fire service in the United States.

Fire Department Safety Officers Association
P.O. Box 149
Ashland, MA 01721-0149

The Fire Department Safety Officers Association (FDSOA) is a nonprofit association dedicated to promoting safety standards and practices in the fire and emergency services. Its purpose is to provide safety officers, with the tools necessary to perform their duties through education. Through these educational programs FDSOA has taken steps toward decreasing the injury and death statistics associated with fire and emergency services.

Fire Marshals Association of North America
One Batterymarch Park
Quincy, MA 02269

Organized in 1906, the Fire Marshals Association of North America was reorganized in 1927 as the Fire Marshals' Section of the National Fire Protection Association.

Constitution and By-Laws adopted March 31, 1947 (revised May 1982, May 1987 and May 1992) list the following objectives:

a. to unite for mutual benefit those officials engaged primarily in the prevention of fire, the control of arson, and/or public fire safety education.
b. to provide educational and professional development opportunities through technological, certification and prevention programs.
c. to provide a resource service to the members of the Fire Marshals Association of North America.
d. to actively market and promote a positive, dynamic and proactive profile for the Fire Marshals Association of North America.
e. to actively participate in the codes and standard-making process at the national, state and local level.
f. to expand the Fire Marshals Association of North America membership, create new chapters, and enhance membership participation at all levels.
g. to monitor and support research and development of solutions to fire protection and fire prevention problems.

Membership in the Fire Marshals Association of North America is contingent upon membership in the NFPA and is open to the fire marshal, fire prevention officer, or such other official of each state, province, county, municipality, or fire protection district charged with the legal responsibility for fire prevention or investigation.

Member: The State, Provincial, Municipal, or Local fire official, who has been lawfully appointed and authorized by the authority having jurisdiction and charged with the statutory responsibilities and duties for fire prevention accomplished through enforcement of fire laws and regulations, property inspections, public fire safety education, or investigation of the cause and origin of fire.

Associate Member: Any individual who was previously an Association Member but cannot now qualify because of change in assignment or retirement.

Membership applications are available upon request.

Fire Service Section of the NFPA
One Batterymarch Park
Quincy, MA 02269

Organized in 1973, this organization is open to all members of the National Fire Protection Association who are:

a. an active or retired member of a fire department providing public fire prevention, fire suppression services, and/or EMS service to a state, county, municipality or organized fire district; or
b. an active or retired member of a fire department providing fire prevention, fire suppression services, and/or EMS service to airfields and military bases; or

c. principally engaged in the training and/or education of fire department members, whether or not an active or retired member of a particular department.

Its objectives are:

1. to unite for mutual professional benefit those members of NFPA who are members of the fire service;
2. to act as a vehicle for the exchange of information among its members;
3. to advance the interests of profession in the fields of fire protection, public life safety education, emergency medical services, code enforcement and fire suppression;
4. to stimulate awareness of the need for continually improving programs in management, training and education;
5. to promote occupational health and safety issues by conducting meetings, conferences, seminars, and such other forums as may be practicable for the exchange of information and the encouragement of professionalism;
6. to encourage participation of its members on the Technical Committees of the NFPA;
7. to advance and encourage the development of improved fire suppression equipment, apparatus, and personal protective clothing ensembles;
8. to encourage public authorities to specify and purchase fire protection equipment on the basis of performance standards;
9. to bring to the attention of its members such matters of legislation and regulations as would be of interest;
10. to promote cooperation within the fire service and between the fire service and other fire protection practitioners; and
11. to provide for the establishment, within the Section, of a fire service professional society.

International Association of Arson Investigators
300 S. Broadway, Suite 100
St. Louis, MO 63102-2808

The International Association of Arson Investigators (IAAI) was formed at Purdue University, West Lafayette, IN, in 1949, when insurance industry, fire service, and law enforcement personnel from the United States and Canada met to discuss the growing arson problem, and the need for training and education in fire investigation. The Association has the following objectives:

a. to unite for mutual benefit those public officials and private persons engaged in the control of arson and kindred crime;
b. to provide for exchange of technical information and developments;
c. to cooperate with other law enforcement agencies and associations to further fire prevention and the suppression of crime;
d. to encourage high professional standards of conduct among arson investigators; and
e. to continually strive to eliminate all factors that interfere with administration of crime suppression.

Active membership in the IAAI is open to any representative (21 years of age or over) of government or of a governmental agency, and any representative of a business or industrial concern who is actively engaged in some phase of the suppression of arson and whose qualifications meet the requirements of the Membership Committee of the Association. Associate Membership is open to persons not qualified for Active Membership after determination of their qualification by the Membership Committee. The IAAI publishes a quarterly magazine, *The Fire and Arson Investigator*, and conducts an annual meeting in conjunction with a conference on fire investigation and with a chapter of the IAAI. Various Committees are organized to assist the Association in its attack on the arson problem, such as the Engineering Forensic Science Committee, the Fraud Fire, the Insurance Advisory Committee, the Juvenile Firesetter Committee, the Training and Education Committee, and the Wildland Arson Committee.

International Association of Black Professional Fire Fighters
8700 Central Avenue, Suite 206
Landover, MD 20785

The IABPFF was organized in 1970 to: (1) create a liaison between black fire fighters across the nation; (2) compile information concerning injustices that exist in the working conditions in the fire service, and to implement action to correct them; (3) collect and evaluate data on all deleterious conditions where minorities exist; (4) see that competent blacks are recruited and employed as fire fighters where they reside; (5) promote interracial progress throughout the fire service; and (6) aid in motivating African Americans to seek advancement to elevated ranks. The organization is a Life Member of the National Association for the Advancement of Colored People.

International Association of Fire Chiefs
4025 Fair Ridge Drive
Fairfax, VA 22033-2868

The International Association of Fire Chiefs (IAFC) is a professional organization open to all fire and emergency services administrators and managers, career and volunteer, interested in improving the delivery of fire and emergency medical services.

The IAFC has over 11,000 members from all over the world, from rural volunteer departments to career metropolitan services. The Association's mission is to provide vision, information, education, services and representation to enhance the professionalism and capabilities of its members.

The IAFC is comprised of eight geographical divisions as well as technical committees and sections that enable members to interact a wide range of emergency services issues. The Association currently has six sections including Apparatus Maintenance, Industrial, EMS, Federal Military, Metropolitan Chiefs, and Volunteer Chief Officers Section. The IAFC is also served by over 14 committees.

Established in 1873, the IAFC has both a rich history and a strong commitment to the future. *IAFC On Scene* is the twicemonthly publication of the IAFC. Each year the IAFC holds a fire and emergency services conference and exhibition in addition to several regional workshops. Its headquarters are located in Fairfax, VA and has a staff of 27.

International Association of Fire Fighters
1750 New York Ave., NW
Washington, DC 20006

Organized in 1918, the IAFF has approximately 225,000 members in the United States and Canada. The IAFF is affiliated with the American Federation of Labor and Congress of Industrial Organizations in the United States and the Canadian Labour Congress (AFL-CIO/CLC). Any person, who is engaged as a permanent and paid employee of a fire department, is eligible for active membership through the chartered locals, state or provincial associations, and joint councils. Conventions of the Association are biennial. The IAFF headquarters provides a variety of services to the membership including: technical assistance in the occupational safety and health area; labor relations expertise through in-house personnel and field representatives; an extensive educational seminar program; legislative representation at the national level; and public relations guidance. The Association's official publication, *The International Fire Fighter,* is published monthly under the direction of the International president. Among other publications are the annual Death and Injury Survey, a Public Relations Manual, and Salary and Working Conditions Surveys.

The International Fire Service Accreditation Congress
1700 W. Tyler
Stillwater, OK 74078

The International Fire Service Accreditation Congress (IFSAC) is a non-profit peer driven association which accredits programs that issue certificates for emergency services training and/or academic degrees in fire-related fields. IFSAC was founded in 1990 in order to ensure the quality of training received by individuals in the field of fire protection. IFSAC does this through accrediting organizations which comply with applicable codes or standards. This accreditation is made at the state/provincial/territorial level for firefighter certification courses, at the college level for degree granting programs, and at the national level for federal government entities involved in fire service training.

An international organization, IFSAC is represented on 5 continents, in 5 countries and in 50 states and provinces. Separate assemblies specialize in the issues peculiar to degree granting institutions and certificate granting institutions. Both of these are represented in a Congress of the whole which deals with the general business of IFSAC and is overseen by a regularly elected Council of Governors. Administrative offices located in Oklahoma State University handle the daily operations of the organization.

International Fire Service Training Association
Fire Protection Publications
Oklahoma State University
930 N. Willis
Stillwater, OK 74078-8045

The International Fire Service Training Association was established in 1934 as a nonprofit educational association of fire fighting personnel who are dedicated to upgrading fire fighting techniques and safety through training. IFSTA's purpose is to validate training materials for publication, develop training materials for publication, check proposed rough drafts for errors, add new techniques and developments, and delete obsolete and outmoded methods. To carry out the mission of IFSTA, Fire Protection Publications was established as a entity of Oklahoma State University. Fire Protection Publications' primary function is to publish and disseminate training texts as proposed and validated by IFSTA. IFSTA manuals are the official teaching texts of most states and provinces in North America. Additionally, numerous US and Canadian government agencies as well as other English-speaking countries have officially accepted the IFSTA manuals.

International Municipal Signal Association
P. O. Box 539
Newark, NY 14513

IMSA was organized in 1896 and currently has 5,000 members. It is an educational, nonprofit organization dedicated to conveying knowledge, technical information, and guidance to its membership, which consists of municipal signal and communication department heads and their first assistants. The range of communications covered includes traffic control, fire alarm, and police alarm. There are nineteen Sections of IMSA (based on geographical areas in the United States and Canada) and a Sustaining Section. The former serve the regional needs of its members. The *IMSA Journal* is the Association's bimonthly publication.

International Society of Fire Service Instructors
P.O. Box 2320
Stafford, VA 22555

The International Society of Fire Service Instructors was organized in 1960. The Society is comprised of those persons who are responsible for the training of fire officers, fire fighters, and rescue and emergency medical personnel. The goal is to assist in the development of fire service instructors through better training and educational opportunities, to provide the means for continuous upgrading of instructors through in-service training, and to actively promote the role of the fire service instructor in the total fire service organization in industry and the public sector. It has membership in all fifty states and fifteen foreign countries.

National Board on Fire Service Professional Qualifications
P. O. Box 492
Quincy, MA 02269

The primary mission of the NBFSPQ is to accredit fire service training agencies that use National Fire Protection Association's fire service professional qualification standards through a national system of fire service professional qualifications certification and accreditation. The NBFSPQ supports the development of an NFPA 1000 on certification and accreditation, using the standard, a National Registry of Fire Service Professional Qualifications Certification, and the NFPA Fire Service Professional Qualification Standards.

Metropolitan Fire Chiefs Section of the National Fire Protection Association and the International Association of Fire Chiefs

The Executive Secretary of the NFPA Section can be reached at 3257 Beals Branch Road, Louisville, KY 40206. The IAFC Section can be reached at 4025 Fair Ridge Drive, Fairfax, VA 22033-2868.

Membership in the Metropolitan Fire Chiefs Section is limited to members of the NFPA and the IAFC who are fire chiefs of cities or jurisdictions having a population more than 200,000 or a minimum authorized strength of 400 fully-paid, uniform personnel. Membership classifications include Regular (Active) Members; Past (Retired) Members; Honorary Members; and, Affiliate Members.

The purpose of the Section is to bring together fire chiefs from large, metropolitan fire departments in order to share information and to affect policy changes. The Metro Chiefs was originally organized in 1965.

National Volunteer Fire Council
1050 17th Street, NW, Suite 1212
Washington, DC 20036

The National Volunteer Fire Council is a non-profit membership association representing the interests of the volunteer fire, EMS and rescue services. Organized in 1976, the NVFC serves as the information source for the emergency services. Its membership includes

state-level organizations that represent volunteer firefighters and EMS personnel in 49 states, individual firefighters, fire departments and corporate members.

Its purpose is to represent the volunteer fire service in the national policy arena and on numerous national and international committees and organizations. The NVFC also serves on the Board of Visitors of the National Fire Academy, the Congressional Fire Services Institute National Advisory Board, the Fallen Firefighters Foundation and the Federation of World Volunteer Firefighters Association.

The NVFC National Office also serves as a clearinghouse for informational publications and videos. The Council has recently entered the information superhighway by including a home page as one of its membership services (http://www.nvfc,org).

Women in the Fire Service
P.O. Box 5446
Madison, WI 53705

Women in the Fire Service is a non-profit organization providing networking, advocacy and peer support for fire service women, and information resources on women's issues to the fire service at large. Established in 1982, WFS offers publications (both brochures and periodicals), workshops, consulting, and national conferences on topics relating to the gender integration of the fire service.

D. FIRE RESEARCH LABORATORIES

There are many laboratories in the United States capable of performing, to varying degrees, fire-related research. Broadly speaking, these laboratories can be classified into three general categories: (1) private and industrial laboratories, (2) university laboratories, and (3) government laboratories.

Representative of laboratories in the private sector are Underwriters Laboratories, Inc., Factory Mutual Research Corporation, and Southwest Research Institute; in the academic sector—Ohio State University; and in the government sector—the National Institute of Standards and Technology. Fire research is by no means limited to the representative laboratories mentioned above. In fact, there has been a marked increase in research activity in recent years as a result of a heightened interest nationally in firesafety and health related issues.

A good source of information on the extent of fire research and the laboratories involved in it is the Directory of Fire Research in the United States which is published periodically by the Committee on Fire Research of the Division of Engineering, National Research Council, National Academy of Sciences, 2101 Constitution Avenue, Washington, DC 20418.

The following descriptions of the activities of the representative laboratories mentioned earlier in this text are given as an indication of the type of activities fire research oriented laboratories are engaged in. Inclusion in the listing is in no way an endorsement of the laboratories by the NFPA.

Underwriters Laboratories Inc.

Underwriters Laboratories Inc. (UL), with headquarters at 333 Pfingsten Road, Northbrook, IL 60062, is a not-for-profit corporation having as its sole objective the promotion of public safety through conduct of "scientific investigation, study, experiments, and tests, to determine the relation of various materials, devices, products, equipment, constructions, methods, and systems to hazards appurtenant thereto or to the use thereof affecting life and property and to ascertain, define, and publish standards, classifications, and specifications for materials, devices, products, equipment, constructions, methods, and systems affecting such hazards, and other information tending to reduce or prevent bodily injury, loss of life, and property damage from such hazards."

The organization was founded in 1894 by the fire insurance industry under whose sponsorship it operated until 1968 when it became an independent, public service-type corporation. It has no capital stock, nor shareholders, and exists solely for the service it renders in the fields of fire, crime, and casualty prevention.

Factory Mutual Research Corporation

Factory Mutual Research Corporation (FMRC), with headquarters at 1151 Boston-Providence Turnpike, Norwood, MA 02062, conducts research and development in the field of property loss control and operates a third party certification program (called Approvals) for materials, products and services beneficial to prevention and control of property loss.

FMRC Approved products and services are made and distributed around the world. They include fire detection and suppression equipment, building materials, explosion protection equipment, flammable liquid handling equipment, industrial heating and combustion control equipment, electrical equipment for use in explosion hazard locations and materials handling equipment. ISO 9000 Quality System Registrations, product follow-up audits and reliability certification of control systems are also offered.

The research and development activity serves the needs of the Factory Mutual System and is available to other entries, such as government agencies, trade associations and individual businesses, through contracts. The scope of the research ranges from basic research into the nature of fire to the development of standards for direct use by industry in establishing practices designed to minimize property loss. Although such standards are intended for use by Factory Mutual System members, they are recognized and widely used by others. Between the extremes of basic research and standards, major effort is made in the area of applied research. This activity includes: studies, surveys, quantitative risk assessment, experimentation, laboratory-scale testing, and full scale testing to evaluate hazards and protection of industrial and commercial property.

Southwest Research Institute

Located at 6220 Culebra Road, San Antonio, TX 78284, Southwest Research Institute (SwRI) is a nonprofit independent organization dedicated to research and development for industry and government both nationally and abroad. The Department of Fire Technology is one of four departments within the Applied Chemistry and Chemical Engineering Division of SwRI. With professionals in the fields of mechanical, civil, structural, aerospace, fire protection, and chemical engineering, computer science, mathematics, physics, biology and chemistry, the Department of Fire Technology offers services in three major areas: fire testing, engineering and consulting, and third party quality assurance.

The Fire Testing Services Group evaluates materials, products, and assemblies in accordance with standard and nonstandard, small, intermediate, and large scale tests to measure parameters such as flame propagation, smoke generation, ignition temperatures, rate of heat release, toxicity, fire endurance, fire containment, heat transfer, and time to failure.

The Engineering Services Group designs and develops protocols to evaluate the fire performance of new materials, products, and assemblies. Services include computer modeling, code consultation and custom design testing.

The Listing, Labeling, and Follow-Up Inspection Services Group provides third party quality assurance services for materials, products, and assemblies, to ensure that the product that reaches the consumer is manufactured similarly to the product evaluated and tested. A Quality Control Manual is developed, and unannounced follow-up inspections, random sampling, and testing are conducted to verify compliance.

Building and Fire Research Laboratory, National Institute of Standards and Technology

Formerly the National Bureau of Standards' Center for Fire Research, the Building and Fire Research Laboratory (BFRL) of the National Institute of Standards and Technology (NIST) is located in the NIST complex at Gaithersburg, MD. The laboratory conducts research designed to improve codes and standards. This work is done by developing improved or new test methods and by using large-scale tests to validate the test method developments. Work is also conducted on developing knowledge of the properties of materials and their performance in fire conditions. The laboratories also develop mathematical models to predict performance of materials under fire conditions.

E. FIRE TESTING LABORATORIES

Many fire research laboratories are also testing laboratories that act as third-party certifiers of products and materials. Both Underwriters Laboratories and Factory Mutual Research Corporation function as testing laboratories, as well as research laboratories.

For many years, the phrase "a nationally recognized testing laboratory" was used somewhat indiscriminately to call attention to the sources for testing of products and assemblies to determine if they meet the performance (and sometimes specification) requirements of fire protection standards. Thus, a standard user would be directed to use only products and assemblies that had been "listed" or "approved" by a nationally recognized testing laboratory if compliance with certain provisions of the standard in question was the goal. Unfortunately, there was always a doubt about what exactly was a "nationally recognized testing laboratory," thus, the phrase has been dropped from NFPA published documents.

An effort to bring order to the question of identifying technically competent testing laboratories in a consistent and uniform fashion has been established within the framework of the U.S. Department of Commerce (DOC). The National Voluntary Laboratory Accreditation Program (NVLAP), established in 1976, has as its goal "to serve on a timely basis the needs of industry, consumers, the government, and others by accrediting the nation's testing laboratories." To date, however, the NVLAP program only covers a few of the many products, materials, and devices that are tested and "listed" or "approved" by testing laboratories.

F. SOCIETY OF FIRE PROTECTION ENGINEERS

Organized in 1950, the Society of Fire Protection Engineers (SFPE), with headquarters at One Liberty Square, Boston, MA 02109-4825, is the professional society for engineers involved in the multi-faceted field of fire protection engineering. The purposes of the Society are to advance the art and science of fire protection engineering and its allied fields, to maintain a high ethical standing among its members, and to foster fire protection engineering education. Its worldwide members include engineers in private practice, in industry, and in local, regional, and national government, as well as technical members of the insurance industry. Forty-six chapters of the Society are located in the United States, Canada, Europe, New Zealand, and Australia.

Membership in the Society is open to those possessing engineering or physical science qualifications, coupled with experience in the field, and to those in associated professional fields.

The Society serves as an international clearinghouse for fire protection engineering state-of-the-art advances and information. It publishes the *SFPE Bulletin*, a newsletter with regular features, the *Journal of Fire Protection Engineering*, *The Handbook of Fire Protection Engineering* plus monographs, proceedings and other technical books.

G. FEDERAL INVOLVEMENT IN FIRE PROTECTION

The Federal Government has been involved in fire protection for many decades. Prior to the 1970's the federal role was minimal and decisively non-regulatory so far as the general public was concerned; it provided for the protection of federal property, and employees and the public while on federal property. Developments in various fields of technology and in communications, a growing awareness of public safety in general, and an increasingly responsive social conscience manifested by Congress intensified federal activities in fire protection beginning in the 1970's. As a result, the related federal concerns became multi-faceted, complex, and regulatory in nature. These concerns are administered within the jurisdiction of 12 Executive Branch departments and 10 independent agencies. These departments and agencies are headquartered in the Washington, DC area, and many oversee regional and field offices in other parts of the country. Their functions encompass fire research, fire training, actual fire prevention, detection and suppression, technical and financial assistance, intra-agency self-regulatory programs, the promulgation and enforcement of both prescriptive and conditioning rules and regulations, fire investigations, criminal investigation and enforcement, and the development and administration of safety policy. The NFPA enjoys a cooperative relationship with these agencies. A number of agencies rely upon NFPA standards and participate in the NFPA standards-making process.

Lead Agencies

In 1974, Congress established the first major federal program to enhance the fire safety of the general public. While recognizing that state and local governments have primary concern for fire safety, the Federal Fire Prevention and Control Act of 1974 (15 U.S.C. 2201) provides training, assistance and coordination in the field of fire protection. Pursuant to the findings and recommendations of the underlying report "America Burning," the Act authorized the establishment of an Administration to oversee these activities, and assigned specific responsibilities regarding fire protection to the National Institute of Standards and Technology (NIST) within the U.S. Department of Commerce (DOC).

United States Fire Administration (USFA)

Headquartered in Emmitsburg, MD, the USFA administers an extensive fire data and analysis program, provides overall fire policy and coordination, and acts as the lead agency in federal fire prevention and arson control programs in coordination with state and local fire service and law enforcement agencies. In conjunction with the National Institute of Occupational Safety and Health, the USFA administers a program concerned with fire fighter health and safety. The USFA Administrator reports to the Director of the Federal

Emergency Management Agency (FEMA). As originally established the agency functioned under the name National Fire Prevention and Control Administration (NFPCA) within the DOC and was transferred in 1978 to the then recently created FEMA.

National Fire Academy (NFA)

The NFA is part of FEMA's Office of Training, and its National Emergency Training Center (NETC), located in Emmitsburg, MD. The NFA provides training programs ranging from fire service management to the hazard mitigation of various materials, and from arson investigation techniques to fire code application and enforcement by architects and local officials. Some fire-related training is also provided by the NETC's Emergency Management Institute (EMI). The EMI's main focus, however, is on the training of civil defense forces which include the fire services. The NFA and the EMI are headed by superintendents reporting to the FEMA Associate Director for Training.

PROGRAMS AFFECTING PUBLIC FIRE SAFETY

Department of Agriculture (DOA)

The Department's fire protection programs are aimed at fire prevention and education in rural areas, and are carried out by the U.S. Forest Service and the Farmers' Home Administration.

U.S. Forest Service (USFS)

Known to millions through its familiar symbol Smokey the Bear, the USFS provides fire protection to more than 200 million acres of forests, grasslands, and nearby private lands (National Forest System), conducts research and develops improved methods in forest fire management (Forest Fire and Atmospheric Research), and provides technical and financial assistance to state forestry organizations to improve fire protection efficiency on non-federal wildlands (Cooperative Fire Protection). Research is also conducted on wood and wood-based products.

The Farmers' Home Administration (FHA) makes loans to public bodies and nonprofit corporations in rural areas for the construction of fire stations, the provision of water supply, and for the purchase of fire suppression apparatus and equipment.

Department of Housing and Urban Development (HUD)

The Manufactured Housing and Construction Standards Division promulgates and enforces rules regarding the safety and durability of manufactured housing, including firesafety standards for consumer protection to determine eligibility for HUD loans and mortgage insurance policies.

Department of Interior (DOI)

The Bureau of Land Management (BLM) provides protection against wildfires on 545 million acres of public land, natural resources, and other values, and services and supports the Interagency Fire Center in Boise, ID. The Center provides logistic support for the DOA's Forest Service, the DOC's National Oceanic and Atmospheric Administration (NOAA), and the DOI's BLM, Bureau of Indian Affairs, National Park Service (NPS), and Fish and Wildlife Service. The NPS also provides presuppression and suppression services, and administers a fire safety program to protect visitors, national park employees, resources and facilities.

Department of Labor (DOL)

Congress, in 1970, established the Occupational Safety and Health Administration (OSHA) within the DOL to develop and promulgate mandatory occupational safety and health standards—rules and regulations applicable at the workplace. As part of the enforcement of its rules, OSHA conducts investigations and inspections, and cites and penalizes workplaces for violations.

The Mine Safety and Health Administration was established in 1977 with a functional scope similar to that of OSHA, but with a focus on the mining industry.

Department of Transportation (DOT)

Ten major operational units within the DOT investigate, research, analyze, and provide for the safety of vehicles, avenues, environment, passengers, and cargoes in all modes of transportation. All of these functions heavily involve fire safety. The organizational units are: five divisions of the U.S. Coast Guard (ports, marine environment, merchant marine, marine firesafety, safety programs), three divisions within the Federal Aviation Administration (airports, special programs, aircraft safety and airport technology), the Federal Highway Administration (interstate highways), the Federal Railroad Administration, the National Highway Traffic Safety Administration (on-road vehicle safety hazards, accidents), the Maritime Administration (merchant marine), the Urban Mass Transportation Administration (buses, subways), the Materials Transportation Bureau (hazardous materials), the Transportation Safety Institute (safety and security management, training), and the Transportation Systems Center (applied research). Related to, but independent from the DOT, the National Transportation Safety Board investigates and makes recommendations with respect to traffic accidents, including vehicular fire incidents.

Department of Treasury

Through its network of agents throughout the United States, the Bureau of Alcohol, Tobacco and Firearms (BATF) conducts a vigorous arson investigation program, including training and technical assistance to state and local law enforcement and fire authorities. The BATF's Firearms and Explosives programs protects interstate and foreign commerce from interference and disruption by reducing hazards to persons and property stemming from the insecure storage of explosives.

Consumer Product Safety Commission (CPSC)

The CPSC's Fire and Thermal Burn Program encompasses the investigation of injury patterns, data collection, research, and the promulgation and enforcement of mandatory standards, with respect to consumer products.

OTHER FEDERAL AGENCIES

Within NIST, the Building and Fire Research Laboratory conducts laboratory, field, and analytical research, and develops technologies to predict, measure, and test the performance—including under conditions of fire—of building materials, components, systems, and practices. In the Department of Defense, the Office of Safety Policy at the Secretary's level provides guidelines for all military departments. Each branch of the Armed Forces deals with its fire safety concerns individually. The Directorate of Engineering and Services in the Air Force, the Corps of Engineers in the Army, the Naval Material Command, and the Marine Installations and Logistics staff address fire prevention and protection concerns.

The Department of Energy's (DOE) Safety and Engineering Analysis Division, located in Germantown, MD, provides for the fire protection of all DOE facilities, including those that manufacture the Nation's nuclear weapons stockpile. The DOE's authority—and thus, its fire protection concerns—is extremely broad, ranging from nuclear research reactors to petroleum reserves—assets valued at more than $75 billion. The Department of Health and Human Services (HHS), and in particular the National Institute of Health (NIH), administers its own fire protection program; and the Department of Justice (DOJ) Bureau of Prisons provides fire protection to all federal correctional institutions. The U.S. Department of State also has an extensive and unique fire protection program, one that involves the safeguarding of U.S. Government property abroad.

Among the independent agencies, the General Services Administration (GSA) acts as the "housekeeper" within the government, providing supplies, services, and facilities for government agencies. Larger than many blue-chip corporations, the GSA manages some 325 million square feet of existing space in about 8,000 buildings, and oversees major construction projects. The GSA's Public Building Services (PBS) provides for the fire protection of these buildings. As part of the GSA, the National Archives and Records Service presents a special risk because of the priceless historic documents and government records stored by that service.

Beside its large-scale fire protection program, the National Aeronautics and Space Administration (NASA) has an active technology transfer program by which concepts, materials, and products used in the nation's space program are being applied in earth-bound, civilian programs. A number of fire fighter safety devices and materials have their origins at NASA.

The Environmental Protection Agency's (EPA) Solid Waste Management underground storage tank, and hazardous materials storage programs have fire safety implications, and the Nuclear Regulatory Commission's function by which it inspects, investigates, and certifies privately owned and operated nuclear reactors contains a rather significant and unique fire protection program. In addition to the above, the Smithsonian Institution, the U.S. Postal Service, and the Veterans Administration have their own, unique fire safety and fire protection programs.

By virtue of the Occupational Safety and Health Act of 1970, administered by the DOL, each federal department and agency, and frequently some subordinate units of these departments and agencies, has its own OSHA program.

Finally, a number of federal programs affect or are of interest to the fire services. Federal Fire Fighters' Retirement funds are administered by the Office of Personnel Management (OPM). Public Safety Officers' Death Benefits are provided by the DOL. Overtime exemptions for federal fire fighters are administered by the Wage and Hour Division of the DOL, and Surplus Military Equipment is made available to state and local fire fighting agencies by the DOD, Bidder Control Office in Battle Creek, MI. The provision of Public Safety Awards was authorized by the Federal Fire Prevention and Control Act of 1974, and so were Reimbursements for Fire Fighting on Federal Property. Fire Suppression Disaster Assistance was authorized by the Disaster Relief Act of 1974. These programs are administered by FEMA.

H. ADDITIONAL ORGANIZATIONS WITH FIRE PROTECTION INTERESTS IN THE UNITED STATES

American Fire Sprinkler Association, Inc.
12959 Jupiter Road, Suite 142
Dallas, TX 75238

The American Fire Sprinkler Association was founded in 1981 as a non-profit, international association to provide training for open-shop fire sprinkler contractors. As the association has grown, the mission has changed to provide complete trade association services, including a full range of industry resources. In recent years, AFSA has greatly expanded efforts to provide training to the fire service on topics relating to the design, installation and inspection of automatic fire sprinkler systems. AFSA has developed information and materials to assist local organizations and fire officials in developing sprinkler ordinances. AFSA is headquartered in Dallas, Texas and includes in its membership sprinkler contractors, manufacturers, suppliers, and fire service officials dedicated to the educational and professional advancement of the fire sprinkler industry.

Building Officials and Code Administrators, International, Inc.
4051 W. Flossmoor Road
Country Club Hills, IL 60478-5795.

Building Officials and Code Administrators, International, Inc. (BOCA) is the nation's original professional organization of code enforcement officials and publisher of the *BOCA National Code* series of model construction codes. Building and fire code officials participate actively in the ongoing development of BOCA'S progressive, performance-oriented code series. BOCA is one of the three founders and statutory members of the International Code Council (ICC) which is working to develop the *International Code*, a single set of model construction codes intended for adoption and use throughout the nation. The ICC has already published the *International Plumbing Code* and the *International Mechanical Code* and anticipates issuing the *International Building Code* and the *International Fire Code* in the year 2000.

Certified Fire Protection Specialist Board
P.O. Box 1658
Hockessin, DE 19707

The CFPS board is a non-profit, worldwide organization whose purpose is to recognize and document those professionals who, through education, experience and written exam, have demonstrated a high level of expertise and professionalism within any of the fire protection fields.

The goal of the CFPS program is to: Utilize the resources of the organization to assist fire protection professionals in job related issues, provide a communications and information exchange forum, support the objectives of other fire protection/prevention organizations and promote the advancement of fire protection and prevention. CFPS is the only program specifically addressing the fire protection engineering technologist which does not require an engineering degree. The credential is a balance of education and experience and provides opportunity for the hands-on fire protection specialist to achieve professional recognition. Certification provides the candidate with both a period of self-assessment and self-improvement during the process of preparing for the examination and also recognition from peers and within the industry once certified.

Chemical Manufacturers Association, CHEMTREC® Center
1300 Wilson Boulevard
Arlington, VA 22209

The Chemical Manufacturers Association (CMA) operates CHEMTREC® the Chemical Transportation Emergency Center. This round-the-clock emergency communications center provides immediate technical assistance to emergency responders in the event of hazardous materials emergencies. CHEMTREC® links on-scene personnel with the manufacturer, shipper or other expert to assist in dealing with the released product information from CHEMTREC®'S Vast Library of *Material Safety Data Sheets* (MS-DSs) and other resources which are also available for emergency fax transmission to the incident scene or other locations. Emergency medical service assistance is also available through MEDTREC, a program that provides emergency response information for incidents involving chemical exposure.

CMA/CHEMTREC® also provides non-emergency services including its lending library of hazardous materials training packages, chemical information referrals, MSDS information and other assistance. The Center also participates in drills.

Council of American Building Officials
5203 Leesburg Pike
Falls Church, VA 22041

The Council of American Building Officials (CABO) is a national not-for-profit organization established by three nationally recognized model code organizations: Building Officials and Code Administrators International, International Conference of Building Officials, and Southern Building Code Congress. The Council's sole purpose is to enhance the efforts of the nation's building officials at the national level in their efforts to improve public health, safety, and general welfare in the built environment. It does so through the development and use of model codes produced by its constituent organizations, and facilitating acceptance of innovative building products and systems.

Fire and Emergency Manufacturers & Services Association
808 17th Street, N.W., Suite 200
Washington, DC 20006

The Fire and Emergency Manufacturers and Services Association (FEMSA) represents manufacturers of fire apparatus, protective clothing and equipment, hoses, nozzles, rescue tools, and other products and services used in the fire service. FEMSA is dedicated to the advancement of the fire service through development and improved fire fighting equipment, techniques and procedures.

In representing the manufacturers and service providers in the US fire service industry, FEMSA promotes the development of new fire fighting technologies to enhance fire fighter health and safety. The association assists in supporting legislative efforts which benefit the US fire service. FEMSA represents 176 companies in the fire and emergency services industry.

International City/County Management Association
777 North Capitol Street, N.E., Suite 500
Washington, DC 20002

The mission of this association is to enhance the quality of local government and to support and assist professional and local government administrators internationally. The specific goals that support this mission are to support and actively promote council, manager, government, and professional management in all forms of local government; to provide training and development programs and publications for local government professionals that improve their skills, increase their knowledge of local government, and strengthen their commitment to the ethics, values and ideals of the profession; to support members in their efforts to meet professional, partnership, and personal needs; to serve as a clearinghouse for the collection, analysis, and dissemination of local government information and data to enhance current practices and serve as a resource to public interest groups in the formulation of public policy, and; to provide a strong association capable of accomplishing these goals.

International Fire Buff Associates, Inc.
7509 Chesapeake Avenue
Baltimore, Maryland 21219

The International Fire Buff Associates serves as a common ground for fire buff groups and individuals active in promoting the general welfare of fire departments and allied emergency services. Many of their clubs furnish, maintain and/or operate a canteen service at fires or other disasters and other clubs are interested in fire prevention, fire fighting techniques and preservation of antique fire apparatus.

National Association of State Foresters
Hall of the States, Suite 526
444 North Capitol Street, NW
Washington, DC 20001

The National Association of State Foresters (NASF) represents the directors of the state forestry agencies from the 50 States and three U.S. Territories (Guam, Puerto Rico and the U.S. Virgin Islands), as well as the District of Columbia. Its members provide a variety of management and protection services on state and private forest lands emphasizing fire prevention and control. Through interstate compacts, cooperative agreements, and other efforts, state forestry agencies work with one another, with rural fire departments, and federal land management agencies to provide coordinated and rapid response to wildland fires. In addition to engaging in and coordinating direct suppression activities, NASF works with a number of partners to help prevent fires, and is a major partner with the USDA Forest Service and the Ad Council in the Cooperative Forest Prevention Campaign. Smokey Bear is this program's highly visible spokesman and ambassador.

National Fire Sprinkler Association
Robin Hill Corporate Park, Route 22
P.O. Box 1000
Patterson, NY 12563

Established in 1914, the National Fire Sprinkler Association is a trade association comprised of installers and manufacturers of fire sprinklers and related equipment and services. Professional, subscriber, and international memberships are available. NFSA provides publications, seminars, representation in codes and standards making, market development, labor relations, and other services to its membership. Headquartered in Patterson, New York, NFSA has regional offices throughout the country.

National Truck Equipment Association
37400 Hills Tech Dr.
Farmington Hills, MI 48331-3414

The National Truck Equipment Association represents over 1,400 distributors and manufacturers of specialized commercial and vocational trucks, truck bodies, and equipment including manufacturers of fire and rescue vehicles. The Association advocates high standards of service in the industry by conducting research and promoting legislation that will encourage safety on public highways and in the vehicles thereon.

The Tobacco Institute
1875 I Street, N.W.
Washington, DC 20006

The Tobacco Institute is the trade association representing U.S. cigarette manufacturers. The Institute's Fire Safety Education Program works with the fire community at national, state and local levels to develop and distribute public education and prevention materials aimed at reducing the number and severity of accidental fires.

I. INTERNATIONAL ORGANIZATIONS WITH FIRE PROTECTION INTERESTS

Appendix C includes a representative listing of international organizations outside the United States that are involved in fire protection and prevention. Many of these organizations have extensive libraries and data bases on fire protection matters. This list was prepared from a variety of sources including an extensive search of the Internet. However, the editors recognize that it undoubtedly fails to include many worthwhile fire protection organizations. NFPA welcomes suggestions for additions to future editions.

ADPC (Asian Disaster Preparedness Center), Asia Institute of Technology, PO Box 4, Klong Luang, Pathumthdn, Thailand. Tel: 66-2-524-5391, Fax: 66-2-524-5360, E-mail: adpc@ait.ac.th: The role of the ADPC is to assist countries in Asia and the Pacific in formulating their policies and developing their capabilities in all aspects of disaster management.

AFAC (Australasian Fire Authorities Council), 1st Floor, 312 Stephensons Road, Mount Waverley, Victoria 3149, Australia. Tel: 61-3-988-1644, Fax: 61-3-988-1557, E-mail: afac@ozemail.com.au: The AFAC was established in 1993 to coordinate the diverse range of activities of the fire protection organizations in Australia and New Zealand. Its mission is to help develop a safer environment through leadership and innovation.

AFPA (Australian Fire Protection Association, Ltd.), 689 Burke Road, First Floor, Hawthorn East, Victoria 3123, Australia or PO Box 456, Camberwell, Victoria 3124. Tel: 61-3-988-2800, Fax: 61-3-982-4748, E-mail: afpa@werple.net.au: The AFPA is a nonprofit technical and educational organization. Its mission is to safeguard life and property from fire. AFPA distributes NFPA materials.

AMHS (Asociación Mexicana de Higiene y Seguridad, A.C.), Lirio No. 7, Col. Santa Maria la Ribera, CP 06400, Mexico, D.F. Tel: 52-5-547-8782, Fax: 52-5-541-1566: This association's specialized group, the "Comité Técnico de Prevención y Combate de Incendios" was founded in August of 1979 and participates in standardization and training. Together with Texas A & M University, its Spanish fire school has trained more than 15,000 firefighters from Mexico, Central, South America and Spain.

In 1996, 615 students were trained in Mexico in cooperation with the Mexican Fire Chiefs Association; it runs 2 courses each year for industrial and municipal fire fighters.

ANPI-NVBB (Association Nationale Pour la Protection Contre L'incendie et L'intrusion—Nationale Vereniging Voor Beviliging Tegen Brand en Binnendringing), Parc Scientifique, 1348 Louvain-la-Neuve, Belgium. Tel: 32-10-475-231, Fax: 32-10-475-270: E-mail: info@anpi_nvbb.be: ANPI-NVBB is the Belgian national fire and intrusion protection association. ANPI is a member of the Confederation of Fire Protection Associations Europe (CFPA-Europe).

ANPI Laboratories is certified as meeting ISO 9002 requirements by the Bureau Veritas Quality International and CEBEC and can test and certify as wide range of fire protection equipment.

In 1994, ANPI Laboratories was certified as meeting ISO 9002 requirements by the Bureau Veritas Quality International and CEBEC. The labs can test and certify a wide range of fire protection equipment.

APAC (Asia Pacific Centre for Fire Safety), No. 58, Jalan 19/3, 46300 Petaling Jaya, Selangor, Malaysia. Tel: 6-03-756-2177, Fax: 6-03-757-8769: APAC is a key distributor of NFPA materials in Asia. APAC has participated in seminars, trade shows, and conferences with NFPA in the Pacific region.

CAFC (Canadian Association of Fire Chiefs), 2425 Don Reid Drive, Unit 1, Ottawa, Ontario K1H 1A4, Canada. Tel: (613) 736-0567: CAFC is a national association located in Ottawa, Ontario, Canada representing many segments of society sharing a common interest in fire protection, fire safety education, and fire prevention. The membership includes professionals working in the fire service, health care, education, architecture, engineering, insurance, industry and government.

The CAFC represents the membership by identifying the public agenda issues affecting fire chiefs and by contributing to the decision-making process. CAFC addresses issues that affect the fire service community and is committed to the promotion of the fire chief profession.

CEPREVEN (Centro Nacional de Prevención de Daños y Perdidas), c/Sagasta 18, 28004 Madrid, Spain. Tel: 34-1-455-7381, Fax: 34-1-455-7136: CEPREVEN is the fire protection association of Spain. It has translated several NFPA materials into Spanish and has organized several fire protection conferences and forums in Spain. CEPREVEN is active with the CFPA in Europe.

CFPA (China Fire Protection Association), No. 14 Dong Chang An Street, Beijing, China 100741. Tel: 86-1-660-6-8788, Fax: 86-1-660-6-9640: CFPA is an independent membership organization dedicated to fire protection. The CFPA has a membership of about 28,000, and its purpose is to promote and develop fire science and technology.

CNPP (Centre National de Prévention et de Protection), Pôle Européen de Sécurité, B.P. 2265, 27950—Saint-Marcel, France. Tel: 4-94-91-1104, Fax: 4-94-91-1090: Founded in 1956 by members of the French Insurance profession, the National Prévention and Protection (insurance) Center ("Centre National de Prevention et de Protection"—CNPP) is dedicated to general risk prevention for companies, group organizations, and individuals through its study, research, training information, and advisory operations.

Its primary areas of specialization are risk management, general safety, fire, environmental and vandalism protection, security, and occupational health.

COPANT (Comisión Panamericana de Normas Téchnicas), Chile 1192, 1098 Buenos Aires, Argentina. Tel: 54-1-383-3751, Fax: 54-1-383-8463: COPANT promotes the development of technical standardization and related activities in its member countries with the aim of promoting their industrial, scientific, and technological development. Almost all countries in the Americas and the Caribbean are members.

While its legal seat is in Buenos Aires, Argentina, its executive secretariat has been located in Caracas, Venezuela since April 1989. The COPANT document center is located in Buenos Aires. In 1994 the leadership of COPANT signed an agreement with NFPA to seek the translation, adoption, and recognition of NFPA codes and standards where there are no comparable ISO Standards.

CSN (Comhairle Sábháilteacht Náisiúnta, National Safety Council), 4 Northbrook Road, Ranelagh, Dublin 6, Ireland. Tel: 01-496-3422, Fax: 01-496-3306: The National Safety Council was established in December 1987 under the Local Government Services (Corporate Bodies) Act, 1971. The function of the Council is to promote Road Safety, Water Safety, and Fire Safety through education, training programs, and publicity campaigns.

The Council replaced the Fire Prevention Council, the National Road Safety Association, and the Irish Water Safety Association, which were dis-established on the formation of the new Council.

CTIF (International Technical Committee for the Prevention and Extinction of Fire) Secretariat General, B.P. 351, 9 rue Bruat F-68006, Colmar, Cedex, France. Tel: 33-8-922-6886, Fax: 33-8-941-2819: The CTIF is an international technical organization created to promote, facilitate, and develop international technical and scientific cooperation in the fields of fire prevention, fire fighting, and life saving as well as technical assistance in the event of fire or natural disasters, excluding those aspects covered by civil defense.

The five official languages of the CTIF are German, French, English, Russian and Italian. The International Technical Committee for the Prevention and Extinction of Fire (CTIF) has its registered office in Paris, and the secretary-general's office is in Colmar, France.

DIFT (Dansk Brandteknisk Institut, Danish Institute of Fire Technology) Datavej 48, DK 3460 Birkerod, Denmark. Tel: 45-82-0099, Fax: 45 45-82-2499: The Danish Institute of Fire Technology (DIFT) is an independent nonprofit institution approved by the Minister of Industry as a technological service institute. The principal object of DIFT is to improve fire safety in society both in Denmark and internationaly.

Editorial MAPFRE, S.A., Sor Angela de la Cruz, 28020 Madrid, Spain. Tel: 34-1-581-5360, Fax: 34-1-581-1883: MAPFRE is the publishing arm of a leading Spanish mutual insurance company, which has produced the Spanish version of the FPH since the 14th edition. MAPFRE has also produced NFPA 921, *Guide for Fire and Explosion Investigations*. The MAPFRE Foundation also conducts seminars on safety subjects including fire protection, and publishes a quarterly magazine. MAPFRE is a multinational company with branches throughout the Americas, and also provides scholarships for post-graduate study in safety areas.

EFPA (Egyptian Fire Protection Association), PO Box 11, Cairo Airport, 11776 Cairo, Egypt. Tel: 20-2-345-5611, Fax: 20-2-347-4658: The EFPA is the national fire protection organization of Egypt. It campaigns to promote fire safety and offers fire safety related technical and educational services.

Fire Prevention Canada, 2425 Don Reid Drive, Unit 2, Ottawa, Ontario K1H 1A4, Canada. Tel: (613) 736-0576, Fax: (613) 736-0684: Fire Prevention Canada is the education arm of the fire service in Canada. It is the primary agency for the development of public resource material for fire safety in Canada. Fire Prevention Canada developed the concept of "Partners in Fire Prevention" to provide a vehicle for professional organizations and groups to collaborate in the promotion of fire safety through education at national, provincial and local levels.

Fire Prevention Canada's mission is to achieve a fire safe environment for Canadians through education and to provide Canada with dynamic leadership and a national focus in the field of fire prevention.

FPA (Fire Protection Association), Melrose Avenue, Borehamwood, Hertfordshire, WD6 2BJ, England. Tel: 44-181-207-2345, Fax: 44-181-236-9701, E-mail: fpa@lpc.co.uk: The Fire Protection Association is the U.K.'s national fire safety organization. The FPA campaigns to promote fire safety management and offers a range of services for business and industry including fire safety management training and fire investigation. The services offered range from inspections of specific features, such as fire protection systems, to complete fire safety audits of company premises.

FPASA (Fire Protection Association of South Africa), PO Box 15467, Impala Park, 1472, South Africa. Tel: 27-12-428-7911, Fax: 27-12-344-1568: The FPASA is the national fire protection organization of South Africa. It campaigns to promote fire safety and offers fire safety related technical and educational services.

IEC (International Electrotechnical Commission), 3 rue de Varembè, PO Box 131, 1211 Geneva 20, Switzerland. Tel: 41-22-919-0211, Fax: 41-22-919-0300: The International Electrotechnical Commission (IEC) promotes through its members, international cooperation on all questions of standardization and related matters, such as the assessment of conformity to standards, in the fields of electricity, electronics, and related technologies. It therefore provides a forum for the preparation and implementation of consensus-based voluntary international standards, facilitating international trade in its field.

The IEC publishes international standards and technical reports; the international standards serve as a basis for national standardization and as references when drafting international tenders and contracts. IEC publications are bilingual in English and French, while the Russian Federation National Committee issues Russian-language editions.

IFE (Institution of Fire Engineers), 148 New Walk, Leicester LE1 7QB, England. Tel: 44-116-255-3654, Fax: 44-116-247-1231: The Institution of Fire Engineers promotes, encourages, and works to improve the science and practices of fire extinction, fire prevention, and fire engineering. The IFE has chapters in several British Commonwealth nations.

IRAM (Instituto Argentino de Normalización), Chile 1192, 1098 Buenos Aires, Argentina. Tel: 54-1-383-3751, Fax: 54-1-383-8463: IRAM is the national standards organization for Argentina,

and also holds the Fire Protection Code Secretariat for COPANT. IRAM is licensed to publish certain NFPA codes and standards in Spanish.

ISS (Swiss Institute for the Promotion of Safety and Security, Institute of Safety and Security), Sicherheitsinstitut, Nüscheler-strasse 45, CH-8001 Zürich, Switzerland. Tel: 41-1-217-4333, Fax: 41-1-211-7030: The Swiss Institute for the Promotion of Safety and Security (Institute of Safety and Security) is a private nonprofit organization established in 1945 to promote integral safety in industry and trade including protection against fire and explosion, intrusion, environmental protection, and labor safety. The Institute also educates specialists in safety and security and qualifies safety experts in the testing, inspection, and certification of systems, products, experts and specialized firms according to Swiss and/or European standards.

A quarterly technical journal *Securité-Sicherheit-Sicurezza* is published.

JFPA (Japan Fire Protection Association), Nippon Shobo Kaikan, 2-9-16 Toranomon, Minatoku, Tokyo 105, Japan. Tel: 81-3-3503-1481: The JFPA is the national fire protection organization of Japan. It campaigns to promote fire safety and offers fire safety related technical and educational services.

KFPA (Korean Fire Protection Association), PO Box 27 Yeo-euido, Seoul 150, Korea. Tel: 82-2-780-8111, Fax: 82-2-783-4094: The Korean Fire Protection Association is a professional risk management organization. Its objectives are to scientifically inspect the fire protection and anti-disaster systems of buildings and present proper countermeasures. The KFPA is devoted to the technical development of industries related to risk management and the propagation of advanced technology for risk management. KFPA has translated the complete set of approximately 300 NFPA standards into Korean.

LPA (Loss Prevention Association of India, Ltd.), Warden House, 4th Floor, Sir Pherozeshah Mehta Road, Bombay 400 001, India. Tel: 91-22-287-4523, Fax: 91-22-287-4512: The LPA deals with general safety issues, including fire prevention. It has branch offices in Calcutta, Hyderabad, Kochi, Madras, and New Delhi.

LPC (Loss Prevention Council), Melrose Avenue, Borehamwood, Hertfordshire, WD6 2BJ, England. Tel: 44 181 207 2345, Fax: 44-181-207-6305, E-mail: info@lpc.co.uk: The Loss Prevention Council is a leading international authority in the field of loss prevention and control. The aim of the LPC is to improve loss prevention practice by addressing the scientific, technical and environmental factors which contribute to risk. The LPC carries out research, testing, approvals, certification, drafts standards and codes of practice and supplies information in the fields of fire, security, health and safety, construction and the environment. Supported by the Association of British Insurers and Lloyd's, the Loss Prevention Council also works with a large number of organizations in both the public and private sectors.

NZFPA (New Zealand Fire Protection Association), PO Box 11129, Ellerslie, Auckland, New Zealand. Tel: 64-3-579-2300, Fax: 64-3-579-3106: The NZFPA is the national fire protection organization of New Zealand. It campaigns to promote fire safety and offers fire safety related technical and educational services.

OPCI (Organización Iberoamericana de Protección Contra Incendios), Avenida 75, No. 19-09, Santafe de Bogota, D.C., Colombia. Tel: 57-1-217-5001, Fax: 57-1-217-5001: OPCI is a not-for-profit entity founded in the early 1980s as NFPA's representative for Latin America. OPCI has produced dozens of translated NFPA documents and a Spanish edition of *NFPA Journal* since 1982. It has organized and participated in fire protection conferences in many Latin American cities, and organizes training sessions based on NFPA-related topics.

SASO (Saudi Arabian Standards Organization), PO Box 3437, Riyadh 11471, Saudi Arabia. Tel: 966-1-479-3332, Fax: 966-1-479-3063: SASO promulgates standards, technical rules, and technical certification in Saudi Arabia — including fire safety standards.

SFPA (Swedish Fire Protection Association), Svenska Brandförsvarsföreningen, Tegeluddsvägen 100, S 115 87 Stockholm, Sweden. Tel: 8-783 7470, Fax: 8-661 2284: The Swedish Fire Protection Association (SFPA) is a private, nonprofit organization was founded in 1919 by fire fighting brigades, insurance companies, and Swedish industry.

SFPA offers a wide range of consulting services, training programs, and information to private business, industry, municipal governments, organizations, and the public. It specializes in the technical aspects of fire protection and risk analysis with regard to fire and rescue hazards. SFPA distributes information through books, lectures, movies, training courses, informational campaigns, and its magazine *Brand & Räddning* (Fire & Rescue). Regional fire protection associations cover Sweden's 24 counties. These associations are partially financed by provincial insurance companies and work with training courses, advisory service, and constancy.

SPEK (Suomen Pelastusalan Keskusjärjestö), The Finnish Central Organization I Finland, Ratamestarinkatu 11, 00520 Helsinki, Finland. Tel: 358-9-47-6112, Fax: 358-9-4761-1400: Finland's first fire protection association was established in 1906 and its first civil defense organization in 1927. They worked independently up until 1993 when they were merged under the name of Suomen Pelastusalan Keskusjärjestö (SPEK).

SPEK's main activities include fire prevention, fire protection, training, inspection of fire alarm equipment, and providing fire prevention and safety materials, information, and education.

SZPV (Slovenian Fire Protection Association), Sr. Gameljne 41, SI-1211 Ljubljana-Smartno, Slovenia. Tel: 386-64-710-593, Fax: 386-64-710-593: SZPV is a nonprofit organization that assists government organizations preparing national fire regulations and promotes the community's interest in fire safety by introducing fire safety education programs on all levels, promotes national fire research and development projects. SZPV, together with the Slovenian Fire Fighters Association and the Slovenian Professional Fire Officers Association publishes a journal *Pozar* and other fire safety publications. SZPV also organizes and conducts conferences and seminars on fire safety in conjunction with government organizations.

WFC (World Fire Statistics Centre), 32 Westmoreland Road, London SW13 9RY, England. Tel: 1-81-748-1988: Since 1981, the World Fire Statistics Centre has encouraged national governments to solve their fire problem in the most cost-effective manner by collecting, collating, and publishing a set of internationally comparable fire cost statistics, that are analyzed under seven main headings: direct losses, indirect losses, human losses, fire brigade costs, fire insurance, building protection, and fire research, training, and publicity. Current participants in the Centre's annual statistical inquiry include the USA, Canada, Japan, Australia, New Zealand, and most West European countries. The Centre is sponsored by the International Association for the Study of Insurance Economics (commonly known as "the Geneva Association"), reports annually to the United Nations and thereafter issues a statistical bulletin, copies of which are available on request.

International Libraries

Listed below are a few of the libraries located outside the United States which have extensive collections of materials related to fire protection and prevention.

BRE, Fire Information Services, BRE Library, Garston, Watford, Herts. WD2 7JR, United Kingdom. Tel: 923-664071, Fax: 923-664605, E-mail: FIS@bre.co.uk: Arson, building codes, building fires, building materials, building research, chemical engineering, civil defense, combustion, computer sciences, disaster planning, emergency management, environmental pollution, failure analysis, fire alarms and detection, fire codes, fire education, fire fighting, fire investigations, fire models, fire protection engineering, fire research, fire safety, fire sciences, fire services management, fire tests, hazardous cargo, hazardous materials, risk analysis, smoke control, statistics, suppression/ extinguishment, toxicity.

CSIRON, Building Information Resource Centre, CSIRO Division of Building, Construction and Engineering, PO Box 310, Riverside Corporate Park, North Ryde, New South Wales 2113, Australia. Tel: 61-2-9934-3418, Fax: 61-2-9934-9335: Building fires, building materials, building research, combustion, combustion toxicology, computer sciences, fire alarms and detection, fire education, fire litigation, fire models, fire protection engineering, fire research, fire sciences, fire services management, fire tests, hazardous materials, suppression/extinguishment.

JIBC, Library, Justice Institute of British Columbia, 715 McBride Blvd., New Westminster, B.C., Canada, V3L 5T4. Tel: (604) 528-5594, Fax: (604) 528-5593, E-mail: AHaddad@jibc.org: Arson, disaster planning, emergency management, emergency medical services, fire education, fire fighting, fire investigations, fire litigation, fire safety, fire sciences, fire services management, hazardous materials, suppression/extinguishment.

NRCC, IRC Information Service, Institute for Research in Construction, National Research Council Canada, Ottawa, K1A OR6 Ontario, Canada. Tel.: (613) 993-9710, Fax: (613) 952-7671, E-mail: mellon@IRC.LAN.NRC.CA: Building codes, building fires, building materials, building research, combustion, failure analysis, fire models, fire protection engineering, fire research, fire sciences, fire tests, smoke control, toxicity.

UCNZ, Engineering Library, University of Canterbury, Private Bag 4800, Christchurch, New Zealand. Tel: 64-3-364-2155, Fax: 64-3-364-2755, E-mail: library-eng@canterbury.ac.nz: Building codes, building fires, building materials, building research, chemical engineering, combustion, computer sciences, consulting, disaster planning, environmental pollution, failure analysis, fire alarms and detection, fire codes, fire protection engineering, fire research, fire safety, fire sciences, fire tests, hazardous materials, risk analysis, smoke control.

International Testing Laboratories

Listed below are the principle testing laboratories located outside the United States that can test and certify a wide range of fire protection equipment.

BSI (British Standards Institute), PO Box 375, Milton Keyney MK14 6ll, United Kingdom. Tel: 44-908-220908, Fax: 44-908-220617

NVBB (Nationale Vereniging Voor Beviliging Tegen Brand en Binnendringing), Parc Scientifique, 1348 Louvain-la-Neuve, Belgium. Tel.: 32-10-47-5231, Fax: 32-10-47-5270, E-mail: info@anpi_nvbb.be

SISIR (Singapore Institute of Standards), 1 Science Park Drive, Singapore 0511, Singapore. Tel: 65-778-7777, Fax: 65-778-0086

UL (Underwriters Laboratories Inc.), 333 Pfingsten Road, Northbrook, IL 60062. Tel: (847) 272-8800, Fax: (847) 272-8129

UL operates testing laboratories and provides conformity assessment services at several locations in the United States, and through subsidiaries at locations throughout the world.

DEMKO A/S, PO Box 514, Lyskaer 8, DK-2730 Herlev, Denmark.
Tel: 45-44-85-6565, Fax: 45-44-85-6500

UL International Ltd., Veristrong Industrial Centre, Block B, 14th Floor,
34 Au Pui Wan Street, Fo Tan, Dha Tin, New Territories, Hong Kong.
Tel: 852-2695-9599, Fax: 852-2695-8196

UL Japan Co., Ltd., L.Kakuei Sasazuka Bldg., 2-18-3 Sasazuka Shibuya-ku,
8th Floor, Tokyo 151, Japan.
Tel: 81-3-5351-1971,
Fax: 81-3-5351-1974

UL Korea, Ltd., Manhattan Bldg.
801-806, #36-2 Youido-Dong, Young Dungpo-Gu, Seoul 150-010, Korea.
Tel: 82-2-784-4346, Fax: 82-2-784-4446

UL International Services Ltd., Taiwan Branch, 3rd Floor, 35 Chung Yang South Road, Sec. 2, Pei Tou, Taipei 112, Taiwan. Tel: 886-2-896-7790,
Fax: 886-2-891-7644

UL of Canada, 7 Crouse Road, Scarborough, Ontario M1R 3A9, Canada. Tel: (416) 757-3611, Fax: (416) 757-1781

VDE (Pruf-und Zertifizierungsinstitut), Merianstrabe 28, D-6050 Offenbach (Main), Germany. Tel: 49-69-8306-656, Fax: 49-69-8306-581

ZAG (Fire Research and Testing Laboratory), Sr. Gameljne 41, SI-1211 Ljubljana-Smartno, Slovenia. Tel: 386-64-710-593, Fax: 386-64-710-593: ZAG, Fire Research and Testing Laboratory is Slovenia's national laboratory for fire testing building materials, construction elements, and different types of equipment for fire protection. It is an independent nonprofit organization. The laboratory provides technical assistance to other governmental organizations on fire research projects and collaborates with other European fire testing laboratories.

COMPLETE TITLES OF ALL OFFICIAL NFPA DOCUMENTS AS OF NOVEMBER 27, 1996

Complied by Leona Attenasio Nisbet

NFPA 1	*Fire Prevention Code*
NFPA 10	*Standard for Portable Fire Extinguishers*
NFPA 10R	*Recommended Practice for Portable Fire Extinguishing Equipment in Family Dwellings and Living Units*
NFPA 11	*Standard for Low-Expansion Foam*
NFPA 11A	*Standard for Medium- and High-Expansion Foam Systems*
NFPA 11C	*Standard for Mobile Foam Apparatus*
NFPA 12	*Standard on Carbon Dioxide Extinguishing Systems*
NFPA 12A	*Standard on Halon 1301 Fire Extinguishing Systems*
NFPA 13	*Standard for the Installation of Sprinkler Systems*
NFPA 13D	*Standard for the Installation of Sprinkler Systems in One- and Two-Family Dwellings and Manufactured Homes*
NFPA 13E	*Guide for Fire Department Operations in Properties Protected by Sprinkler and Standpipe Systems*
NFPA 13R	*Standard for the Installation of Sprinkler Systems in Residential Occupancies up to and Including Four Stories in Height*
NFPA 14	*Standard for the Installation of Standpipe and Hose Systems*
NFPA 15	*Standard for Water Spray Fixed Systems for Fire Protection*
NFPA 16	*Standard for the Installation of Deluge Foam-Water Sprinkler and Foam-Water Spray Systems*
NFPA 16A	*Standard for the Installation of Closed-Head Foam-Water Sprinkler Systems*
NFPA 17	*Standard for Dry Chemical Extinguishing Systems*
NFPA 17A	*Standard for Wet Chemical Extinguishing Systems*
NFPA 18	*Standard on Wetting Agents*
NFPA 20	*Standard for the Installation of Centrifugal Fire Pumps*
NFPA 22	*Standard for Water Tanks for Private Fire Protection*
NFPA 24	*Standard for the Installation of Private Fire Service Mains and Their Appurtenances*
NFPA 25	*Standard for the Inspection, Testing, and Maintenance of Water-Based Fire Protection Systems*
NFPA 30	*Flammable and Combustible Liquids Code*
NFPA 30A	*Automotive and Marine Service Station Code*
NFPA 30B	*Code for the Manufacture and Storage of Aerosol Products*
NFPA 31	*Standard for the Installation of Oil-Burning Equipment*
NFPA 32	*Standard for Drycleaning Plants*
NFPA 33	*Standard for Spray Application Using Flammable or Combustible Materials*
NFPA 34	*Standard for Dipping and Coating Processes Using Flammable or Combustible Liquids*
NFPA 35	*Standard for the Manufacture of Organic Coatings*
NFPA 36	*Standard for Solvent Extraction Plants*
NFPA 37	*Standard for the Installation and Use of Stationary Combustion Engines and Gas Turbines*

Ms Nisbet is Director of NFPA's Standards Administration Department and Recording Secretary of the Standards Council

NFPA 40	*Standard for the Storage and Handling of Cellulose Nitrate Motion Picture Film*
NFPA 40E	*Code for the Storage of Pyroxylin Plastic*
NFPA 43B	*Code for the Storage of Organic Peroxide Formulations*
NFPA 43D	*Code for the Storage of Pesticides*
NFPA 45	*Standard on Fire Protection for Laboratories Using Chemicals*
NFPA 46	*Recommended Safe Practice for Storage of Forest Products*
NFPA 49	*Hazardous Chemicals Data*
NFPA 50	*Standard for Bulk Oxygen Systems at Consumer Sites*
NFPA 50A	*Standard for Gaseous Hydrogen Systems at Consumer Sites*
NFPA 50B	*Standard for Liquefied Hydrogen Systems at Consumer Sites*
NFPA 51	*Standard for the Design and Installation of Oxygen-Fuel Gas Systems for Welding, Cutting, and Allied Processes*
NFPA 51A	*Standard for Acetylene Cylinder Charging Plants*
NFPA 51B	*Standard for Fire Prevention in Use of Cutting and Welding Processes*
NFPA 52	*Standard for Compressed Natural Gas (CNG) Vehicular Fuel Systems*
NFPA 53	*Guide on Fire Hazards in Oxygen-Enriched Atmospheres*
NFPA 54	*National Fuel Gas Code*
NFPA 55	*Standard for the Storage, Use, and Handling of Compressed and Liquefied Gases in Portable Cylinders*
NFPA 57	*Standard for Liquefied Natural Gas (LNG) Vehicular Fuel Systems*
NFPA 58	*Standard for the Storage and Handling of Liquefied Petroleum Gases*
NFPA 59	*Standard for the Storage and Handling of Liquefied Petroleum Gases at Utility Gas Plants*
NFPA 59A	*Standard for the Production, Storage, and Handling of Liquefied Natural Gas (LNG)*
NFPA 61	*Standard for the Prevention of Fires and Dust Explosions in Agricultural and Food Products Facilities*
NFPA 65	*Standard for the Processing and Finishing of Aluminum*
NFPA 68	*Guide for Venting of Deflagrations*
NFPA 69	*Standard on Explosion Prevention Systems*
NFPA 70	*National Electrical Code®*
NFPA 70A	*Electrical Code for One- and Two-Family Dwellings and Mobile Homes*
NFPA 70B	*Recommended Practice for Electrical Equipment Maintenance*
NFPA 70E	*Standard for Electrical Safety Requirements for Employee Workplaces*
NFPA 72	*National Fire Alarm Code®*
NFPA 73	*Residential Electrical Maintenance Code for One- and Two-Family Dwellings*
NFPA 75	*Standard for the Protection of Electronic Computer/Data Processing Equipment*
NFPA 77	*Recommended Practice on Static Electricity*
NFPA 79	*Electrical Standard for Industrial Machinery*
NFPA 80	*Standard for Fire Doors and Fire Windows*
NFPA 80A	*Recommended Practice for Protection of Buildings from Exterior Fire Exposures*
NFPA 82	*Standard on Incinerators and Waste and Linen Handling Systems and Equipment*
NFPA 86	*Standard for Ovens and Furnaces*
NFPA 86C	*Standard for Industrial Furnaces Using a Special Processing Atmosphere*
NFPA 86D	*Standard for Industrial Furnaces Using Vacuum as an Atmosphere*
NFPA 88A	*Standard for Parking Structures*
NFPA 88B	*Standard for Repair Garages*
NFPA 90A	*Standard for the Installation of Air Conditioning and Ventilating Systems*
NFPA 90B	*Standard for the Installation of Warm Air Heating and Air Conditioning Systems*
NFPA 91	*Standard for Exhaust Systems for Air Conveying of Materials*
NFPA 92A	*Recommended Practice for Smoke-Control Systems*
NFPA 92B	*Guide for Smoke Management Systems in Malls, Atria, and Large Areas*
NFPA 96	*Standard for Ventilation Control and Fire Protection of Commercial Cooking Operations*
NFPA 97	*Standard Glossary of Terms Relating to Chimneys, Vents, and Heat-Producing Appliances*
NFPA 99	*Standard for Health Care Facilities*
NFPA 99B	*Standard for Hypobaric Facilities*
NFPA 99C	*Standard on Gas and Vacuum Systems*
NFPA 101®	*Code for Safety to Life from Fire in Buildings and Structures*
NFPA 101A	*Guide on Alternative Approaches to Life Safety*
NFPA 102	*Standard for Grandstands, Folding and Telescopic Seating, Tents, and Membrane Structures*
NFPA 105	*Recommended Practice for the Installation of Smoke-Control Door Assemblies*
NFPA 110	*Standard for Emergency and Standby Power Systems*
NFPA 111	*Standard on Stored Electrical Energy Emergency and Standby Power Systems*
NFPA 115	*Recommended Practice on Laser Fire Protection*
NFPA 120	*Standard for Coal Preparation Plants*

NFPA 121	*Standard on Fire Protection for Self-Propelled and Mobile Surface Mining Equipment*
NFPA 122	*Standard for Fire Prevention and Control in Underground Metal and Nonmetal Mines*
NFPA 123	*Standard for Fire Prevention and Control in Underground Bituminous Coal Mines*
NFPA 130	*Standard for Fixed Guideway Transit Systems*
NFPA 150	*Standard on Fire Safety in Racetrack Stables*
NFPA 170	*Standard for Fire Safety Symbols*
NFPA 203	*Guide on Roof Coverings and Roof Deck Constructions*
NFPA 204M	*Guide for Smoke and Heat Venting*
NFPA 211	*Standard for Chimneys, Fireplaces, Vents, and Solid Fuel-Burning Appliances*
NFPA 214	*Standard on Water-Cooling Towers*
NFPA 220	*Standard on Types of Building Construction*
NFPA 221	*Standard for Fire Walls and Fire Barrier Walls*
NFPA 231	*Standard for General Storage*
NFPA 231C	*Standard for Rack Storage of Materials*
NFPA 231D	*Standard for Storage of Rubber Tires*
NFPA 231E	*Recommended Practice for the Storage of Baled Cotton*
NFPA 231F	*Standard for the Storage of Roll Paper*
NFPA 232	*Standard for the Protection of Records*
NFPA 232A	*Guide for Fire Protection for Archives and Records Centers*
NFPA 241	*Standard for Safeguarding Construction, Alteration, and Demolition Operations*
NFPA 251	*Standard Methods of Tests of Fire Endurance of Building Construction and Materials*
NFPA 252	*Standard Methods of Fire Tests of Door Assemblies*
NFPA 253	*Standard Method of Test for Critical Radiant Flux of Floor Covering Systems Using a Radiant Heat Energy Source*
NFPA 255	*Standard Method of Test of Surface Burning Characteristics of Building Materials*
NFPA 256	*Standard Methods of Fire Tests of Roof Coverings*
NFPA 257	*Standard for Fire Test for Window and Glass Block Assemblies*
NFPA 258	*Standard Research Test Method for Determining Smoke Generation of Solid Materials*
NFPA 259	*Standard Test Method for Potential Heat of Building Materials*
NFPA 260	*Standard Methods of Tests and Classification System for Cigarette Ignition Resistance of Components of Upholstered Furniture*
NFPA 261	*Standard Method of Test for Determining Resistance of Mock-Up Upholstered Furniture Material Assemblies to Ignition by Smoldering Cigarettes*
NFPA 262	*Standard Method of Test for Fire and Smoke Characteristics of Wires and Cables*
NFPA 263	*Standard Method of Test for Heat and Visible Smoke Release Rates for Materials and Products*
NFPA 264	*Standard Method of Test for Heat and Visible Smoke Release Rates for Materials and Products Using an Oxygen Consumption Calorimeter*
NFPA 264A	*Standard Method of Test for Heat Release Rates for Upholstered Furniture Components or Composites and Mattresses Using an Oxygen Consumption Calorimeter*
NFPA 265	*Standard Methods of Fire Tests for Evaluating Room Fire Growth Contribution of Textile Wall Coverings*
NFPA 266	*Standard Method of Test for Fire Characteristics of Upholstered Furniture Exposed to Flaming Ignition Source*
NFPA 267	*Standard Method of Test for Fire Characteristics of Mattresses and Bedding Assemblies Exposed to Flaming Ignition Source*
NFPA 268	*Standard Test Method for Determining Ignitibility of Exterior Wall Assemblies Using a Radiant Heat Energy Source*
NFPA 269	*Standard Test Method for Developing Toxic Potency Data for Use in Fire Hazard Modeling*
NFPA 291	*Recommended Practice for Fire Flow Testing and Marking of Hydrants*
NFPA 295	*Standard for Wildfire Control*
NFPA 297	*Guide on Principles and Practices for Communications Systems*
NFPA 298	*Standard on Fire Fighting Foam Chemicals for Class A Fuels in Rural, Suburban, and Vegetated Areas*
NFPA 299	*Standard for Protection of Life and Property from Wildfire*
NFPA 302	*Fire Protection Standard for Pleasure and Commercial Motor Craft*
NFPA 303	*Fire Protection Standard for Marinas and Boatyards*
NFPA 306	*Standard for the Control of Gas Hazards on Vessels*
NFPA 307	*Standard for the Construction and Fire Protection of Marine Terminals, Piers, and Wharves*
NFPA 312	*Standard for Fire Protection of Vessels during Construction, Repair, and Lay-Up*
NFPA 318	*Standard for the Protection of Cleanrooms*
NFPA 325	*Guide to Fire Hazard Properties of Flammable Liquids, Gases, and Volatile Solids*
NFPA 326	*Standard Procedures for the Safe Entry of Underground Storage Tanks*
NFPA 327	*Standard Procedures for Cleaning or Safeguarding Small Tanks and Containers Without Entry*
NFPA 328	*Recommended Practice for the Control of Flammable and Combustible Liquids and Gases in Manholes, Sewers, and Similar Underground Structures*
NFPA 329	*Recommended Practice for Handling Underground Releases of Flammable and Combustible Liquids*

NFPA 385	*Standard for Tank Vehicles for Flammable and Combustible Liquids*
NFPA 386	*Standard for Portable Shipping Tanks for Flammable and Combustible Liquids*
NFPA 395	*Standard for the Storage of Flammable and Combustible Liquids at Farms and Isolated Sites*
NFPA 402	*Guide for Aircraft Rescue and Fire Fighting Operations*
NFPA 403	*Standard for Aircraft Rescue and Fire Fighting Services at Airports*
NFPA 407	*Standard for Aircraft Fuel Servicing*
NFPA 408	*Standard for Aircraft Hand Portable Fire Extinguishers*
NFPA 409	*Standard on Aircraft Hangars*
NFPA 410	*Standard on Aircraft Maintenance*
NFPA 412	*Standard for Evaluating Aircraft Rescue and Fire Fighting Foam Equipment*
NFPA 414	*Standard for Aircraft Rescue and Fire Fighting Vehicles*
NFPA 415	*Standard on Airport Terminal Buildings, Fueling Ramp Drainage, and Loading Walkways*
NFPA 418	*Standard for Heliports*
NFPA 422	*Guide for Aircraft Accident Response*
NFPA 423	*Standard for Construction and Protection of Aircraft Engine Test Facilities*
NFPA 424	*Guide for Airport/Community Emergency Planning*
NFPA 430	*Code for the Storage of Liquid and Solid Oxidizers*
NFPA 471	*Recommended Practice for Responding to Hazardous Materials Incidents*
NFPA 472	*Standard for Professional Competence of Responders to Hazardous Materials Incidents*
NFPA 473	*Standard for Competencies for EMS Personnel Responding to Hazardous Materials Incidents*
NFPA 480	*Standard for the Storage, Handling and Processing of Magnesium Solids and Powders*
NFPA 481	*Standard for the Production, Processing, Handling and Storage of Titanium*
NFPA 482	*Standard for the Production, Processing, Handling and Storage of Zirconium*
NFPA 485	*Standard for the Storage, Handling, Processing, and Use of Lithium Metal*
NFPA 490	*Code for the Storage of Ammonium Nitrate*
NFPA 491M	*Manual of Hazardous Chemical Reactions*
NFPA 495	*Explosive Materials Code*
NFPA 496	*Standard for Purged and Pressurized Enclosures for Electrical Equipment*
NFPA 497A	*Recommended Practice for Classification of Class I Hazardous (Classified) Locations for Electrical Installations in Chemical Process Areas*
NFPA 497B	*Recommended Practice for the Classification of Class II Hazardous (Classified) Locations for Electrical Installations in Chemical Process Areas*
NFPA 497M	*Manual for Classification of Gases, Vapors, and Dusts for Electrical Equipment in Hazardous (Classified) Locations*
NFPA 498	*Standard for Safe Havens and Interchange Lots for Vehicles Transporting Explosives*
NFPA 501A	*Standard for Fire Safety Criteria for Manufactured Home Installations, Sites, and Communities*
NFPA 501C	*Standard on Recreational Vehicles*
NFPA 501D	*Standard for Recreational Vehicle Parks and Campgrounds*
NFPA 502	*Recommended Practice on Fire Protection for Limited Access Highways, Tunnels, Bridges, Elevated Roadways, and Air Right Structures*
NFPA 505	*Fire Safety Standard for Powered Industrial Trucks Including Type Designations, Areas of Use, Conversions, Maintenance, and Operation*
NFPA 512	*Standard for Truck Fire Protection*
NFPA 513	*Standard for Motor Freight Terminals*
NFPA 550	*Guide to the Fire Safety Concepts Tree*
NFPA 555	*Guide on Methods for Evaluating Potential for Room Flashover*
NFPA 560	*Standard for the Storage, Handling, and Use of Ethylene Oxide for Sterilization and Fumigation*
NFPA 600	*Standard on Industrial Fire Brigades*
NFPA 601	*Standard for Security Services in Fire Loss Prevention*
NFPA 650	*Standard for Pneumatic Conveying Systems for Handling Combustible Materials*
NFPA 651	*Standard for the Manufacture of Aluminum Powder*
NFPA 654	*Standard for the Prevention of Fire and Dust Explosions in the Chemical, Dye, Pharmaceutical, and Plastics Industries*
NFPA 655	*Standard for Prevention of Sulfur Fires and Explosions*
NFPA 664	*Standard for the Prevention of Fires and Explosions in Wood Processing and Woodworking Facilities*
NFPA 701	*Standard Methods of Fire Tests for Flame-Resistant Textiles and Films*
NFPA 703	*Standard for Fire Retardant Impregnated Wood and Fire Retardant Coatings for Building Materials*
NFPA 704	*Standard System for the Identification of the Fire Hazards of Materials*
NFPA 705	*Recommended Practice for a Field Flame Test for Textiles and Films*
NFPA 750	*Standard on Water Mist Fire Protection Systems*
NFPA 780	*Standard for the Installation of Lightning Protection Systems*
NFPA 801	*Standard for Facilities Handling Radioactive Materials*
NFPA 802	*Recommended Practice for Fire Protection for Nuclear Research and Production Reactors*

NFPA 803	*Standard for Fire Protection for Light Water Nuclear Power Plants*
NFPA 804	*Standard for Fire Protection for Advanced Light Water Reactor Electric Generating Plants*
NFPA 820	*Standard for Fire Protection in Wastewater Treatment and Collection Facilities*
NFPA 850	*Recommended Practice for Fire Protection for Electric Generating Plants and High Voltage Direct Current Converter Stations*
NFPA 851	*Recommended Practice for Fire Protection for Hydroelectric Generating Plants*
NFPA 901	*Standard Classifications for Incident Reporting and Fire Protection Data*
NFPA 902	*Fire Reporting Field Incident Guide*
NFPA 903	*Fire Reporting Property Survey Guide*
NFPA 904	*Incident Follow-up Report Guide*
NFPA 906	*Guide for Fire Incident Field Notes*
NFPA 910	*Recommended Practice for the Protection of Libraries and Library Collections*
NFPA 911	*Recommended Practice for the Protection of Museums and Museum Collections*
NFPA 912	*Recommended Practice for Fire Protection in Places of Worship*
NFPA 914	*Recommended Practice for Fire Protection in Historic Structures*
NFPA 921	*Guide for Fire and Explosion Investigations*
NFPA 1000	*Standard for Fire Service Professional Qualifications Accreditation and Certification Systems*
NFPA 1001	*Standard for Fire Fighter Professional Qualifications*
NFPA 1002	*Standard for Fire Department Vehicle Driver/Operator Professional Qualifications*
NFPA 1003	*Standard for Airport Fire Fighter Professional Qualifications*
NFPA 1021	*Standard for Fire Officer Professional Qualifications*
NFPA 1031	*Standard for Professional Qualifications for Fire Inspector*
NFPA 1033	*Standard for Professional Qualifications for Fire Investigator*
NFPA 1035	*Standard for Professional Qualifications for Public Fire and Life Safety Educator*
NFPA 1041	*Standard for Fire Service Instructor Professional Qualifications*
NFPA 1051	*Standard on Wildland Fire Fighter Professional Qualifications*
NFPA 1061	*Standard for Professional Qualifications for Public Safety Telecommunicator*
NFPA 1122	*Code for Model Rocketry*
NFPA 1123	*Code for Fireworks Display*
NFPA 1124	*Code for the Manufacture, Transportation, and Storage of Fireworks*
NFPA 1125	*Code for the Manufacture of Model Rocket and High Power Rocket Motors*
NFPA 1126	*Standard for the Use of Pyrotechnics before a Proximate Audience*
NFPA 1127	*Code for High Power Rocketry*
NFPA 1141	*Standard for Fire Protection in Planned Building Groups*
NFPA 1201	*Standard for Developing Fire Protection Services for the Public*
NFPA 1221	*Standard for the Installation, Maintenance, and Use of Public Fire Service Communication Systems*
NFPA 1231	*Standard on Water Supplies for Suburban and Rural Fire Fighting*
NFPA 1401	*Recommended Practice for Fire Service Training Reports and Records*
NFPA 1402	*Guide to Building Fire Service Training Centers*
NFPA 1403	*Standard on Live Fire Training Evolutions*
NFPA 1404	*Standard for a Fire Department Self-Contained Breathing Apparatus Program*
NFPA 1405	*Guide for Land-Based Fire Fighters Who Respond to Marine Vessel Fires*
NFPA 1410	*Standard on Training for Initial Fire Attack*
NFPA 1420	*Recommended Practice for Pre-Incident Planning for Warehouse Occupancies*
NFPA 1451	*Standard for a Fire Service Vehicle Operations Training Program*
NFPA 1452	*Guide for Training Fire Service Personnel to Make Dwelling Fire Safety Surveys*
NFPA 1470	*Standard on Search and Rescue Training for Structural Collapse Incidents*
NFPA 1500	*Standard on Fire Department Occupational Safety and Health Program*
NFPA 1521	*Standard for Fire Department Safety Officer*
NFPA 1561	*Standard on Fire Department Incident Management System*
NFPA 1581	*Standard on Fire Department Infection Control Program*
NFPA 1582	*Standard on Medical Requirements for Fire Fighters*
NFPA 1600	*Recommended Practice for Disaster Management*
NFPA 1901	*Standard for Automotive Fire Apparatus*
NFPA 1906	*Standard for Wildland Fire Apparatus*
NFPA 1911	*Standard for Service Tests of Pumps on Fire Department Apparatus*
NFPA 1914	*Standard for Testing Fire Department Aerial Devices*
NFPA 1921	*Standard for Fire Department Portable Pumping Units*
NFPA 1922	*Standard for Fire Service Self-Contained Pumping Units*
NFPA 1931	*Standard on Design of and Design Verification Tests for Fire Department Ground Ladders*

NFPA 1932	*Standard on Use, Maintenance and Service Testing of Fire Department Ground Ladders*
NFPA 1961	*Standard for Fire Hose*
NFPA 1962	*Standard for the Care, Use and Service Testing of Fire Hose, Including Couplings and Nozzles*
NFPA 1963	*Standard for Fire Hose Connections*
NFPA 1964	*Standard for Spray Nozzles (Shutoff and Tip)*
NFPA 1971	*Standard on Protective Ensemble For Structural Fire Fighting*
NFPA 1975	*Standard on Station/Work Uniforms for Fire Fighters*
NFPA 1976	*Standard on Protective Clothing for Proximity Fire Fighting*
NFPA 1977	*Standard on Protective Clothing and Equipment for Wildland Fire Fighting*
NFPA 1981	*Standard on Open-Circuit Self-Contained Breathing Apparatus for Fire Fighters*
NFPA 1982	*Standard on Personal Alert Safety Systems (PASS) for Fire Fighters*
NFPA 1983	*Standard on Fire Service Life Safety Rope and System Components*
NFPA 1991	*Standard on Vapor-Protective Suits for Hazardous Chemical Emergencies*
NFPA 1992	*Standard on Liquid Splash-Protective Suits for Hazardous Chemical Emergencies*
NFPA 1993	*Standard on Support Function Protective Clothing for Hazardous Chemical Operations*
NFPA 1999	*Standard on Protective Clothing for Emergency Medical Operations*
NFPA 2001	*Standard on Clean Agent Fire Extinguishing Systems*
NFPA 8501	*Standard for Single Burner Boiler Operation*
NFPA 8502	*Standard for Prevention of Furnace Explosions/Implosions in Multiple Burner Boilers*
NFPA 8503	*Standard for Pulverized Fuel Systems*
NFPA 8504	*Standard on Atmospheric Fluidized-Bed Boiler Operation*
NFPA 8505	*Recommended Practice for Stoker Operation*
NFPA 8506	*Standard on Heat Recovery Steam Generators*

Index